TWELFTH EDITION

Modern Nutrition

in Health and Disease

Editors:

Katherine L. Tucker, PhD

Christopher P. Duggan, MD, MPH

Gordon L. Jensen, MD, PhD

Karen E. Peterson, ScD

A. Catharine Ross, PhD

Riva Touger-Decker, PhD, RD

JONES & BARTLETT
LEARNING

World Headquarters
Jones & Bartlett Learning
25 Mall Road
Burlington, MA 01803
978-443-5000
info@jblearning.com
www.jblearning.com

Jones & Bartlett Learning books and products are available through most bookstores and online booksellers. To contact Jones & Bartlett Learning directly, call 800-832-0034, fax 978-443-8000, or visit our website, www.jblearning.com.

Substantial discounts on bulk quantities of Jones & Bartlett Learning publications are available to corporations, professional associations, and other qualified organizations. For details and specific discount information, contact the special sales department at Jones & Bartlett Learning via the above contact information or send an email to specialsales@jblearning.com.

26447-0

Production Credits
Vice President, Product Management: Marisa R. Urbano
Vice President, Content Strategy and Implementation: Christine Emerton
Director, Product Management: Matthew Kane
Product Manager: Whitney Fekete
Director, Content Management: Donna Gridley
Manager, Content Strategy: Orsolya Gall
Outsource Manager, Content Strategy and Design: Carol Brewer Guerrero
Content Coordinator: Samantha Gillespie
Director, Project Management and Content Services: Karen Scott
Manager, Intellectual Properties and Content Production: Kristen Rogers
Manager, Project Management: Jackie Reynen
Content Product Managers: Roberta Sherman and Karrie Larsson
Senior Digital Project Specialist: Carolyn Downer
Senior Product Marketing Manager: Susanne Walker
Content Services Manager: Colleen Lamy
Senior Director of Supply Chain: Ed Schneider
Procurement Manager: Wendy Kilborn
Composition: Exela Technologies
Project Management: Exela Technologies
Cover Design: Scott Moden
Text Design: Scott Moden
Media Development Editor: Faith Brosnan
Rights & Permissions Manager: John Rusk
Rights Specialist: Robin Silverman
Cover Image (Title Page, Part Opener, Chapter Opener): © 2013 ESB Professional/Shutterstock
Printing and Binding: Lakeside Book Company

Library of Congress Cataloging-in-Publication Data
Names: Tucker, Katherine, editor.
Title: Modern nutrition in health and disease / [edited by] Katherine L. Tucker, PhD, Christopher P. Duggan, MD, MPH, Gordon L. Jensen, MD, PhD, Karen E. Peterson, ScD, A. Catharine Ross, PhD, Riva Touger-Decker, PhD, RD.
Description: Twelfth edition. | Burlington, MA : Jones & Bartlett Learning, [2026] | Includes bibliographical references and index.
Identifiers: LCCN 2023051295 | ISBN 9781284264432
Subjects: LCSH: Nutrition. | Diet therapy.
Classification: LCC QP141 .M64 2026 | DDC 613.2–dc23/eng/20240813
LC record available at https://lccn.loc.gov/2023051295

6048

Printed in the United States of America
28 27 26 25 24 10 9 8 7 6 5 4 3 2 1

Brief Contents

Online Chapters and Resources

The following chapters and resources are available in the Navigate eBook.

Contents

CHAPTER 80 Nutrition in Inflammatory Bowel Diseases . . . 1197

CHAPTER 81 Celiac Disease 1207

CHAPTER 86 Parenteral Nutrition 1259

E. Skeletal and Joint Disorders 1277

CHAPTER 87 Bone Biology in Health and Disease 1277

Online Chapters and Resources

The following chapters and resources are available in the Navigate eBook.

CHAPTER 112 The Nutrition Transition: Global Trends, Causes, and Implications 1643

CHAPTER 113 Food-Based Dietary Guidelines For Healthier Populations: International Considerations................1653

CHAPTER 114 Global Food Systems......................1667

About the *Twelfth* Edition

Welcome to the twelfth edition of *Modern Nutrition in Health and Disease*, a seminal text in the field of Nutrition. Originally published in 1950, this revised print and digital product is both a higher education learning resource and an authoritative reference for nutrition practitioners.

Authored and edited by distinguished experts worldwide, *Modern Nutrition in Health and Disease, Twelfth Edition* is grounded by physiologic nutrition principles, with new chapters on interprofessional practice, global food systems, precision nutrition, and much more.

Fully referenced and representing the latest scientific research, the new *Twelfth Edition* demonstrates major scientific progress relating to nutrition's role in disease prevention.

Modern Nutrition in Health and Disease, Twelfth Edition is appropriate for adoption into the following advanced undergraduate and graduate Nutrition courses:

- Medical Nutrition Therapy
- Nutrition Macronutrients and Micronutrients
- Human Nutrition and Metabolism
- Clinical Nutrition
- Nutrition and Disease
- Nutrition Assessment
- Nutritional Biochemistry and Pathophysiology
- Vitamins and Minerals

Full-Color Art

For the first time, *Modern Nutrition in Health and Disease, Twelfth Edition* is presented in full color! More captivating figures and tables are presented in color to foster maximum comprehension.

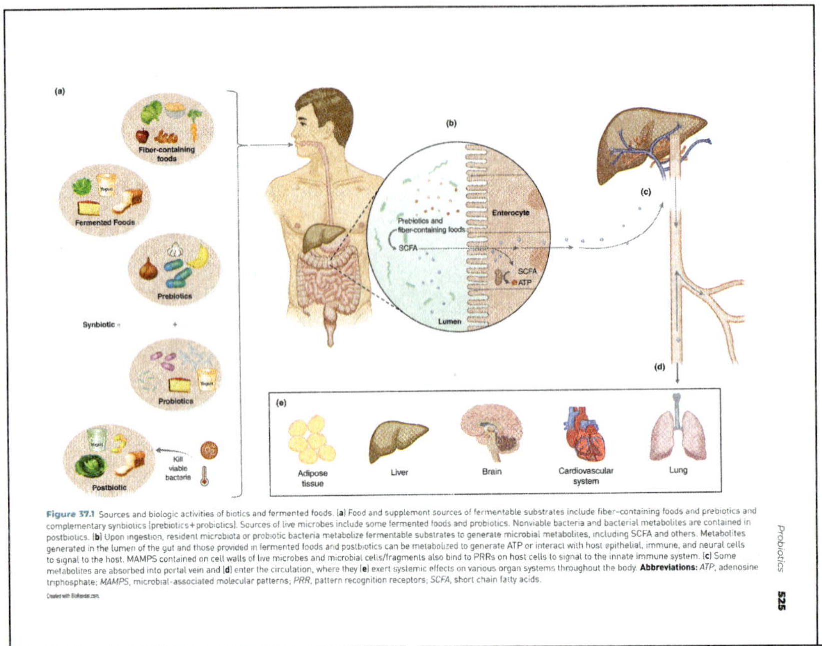

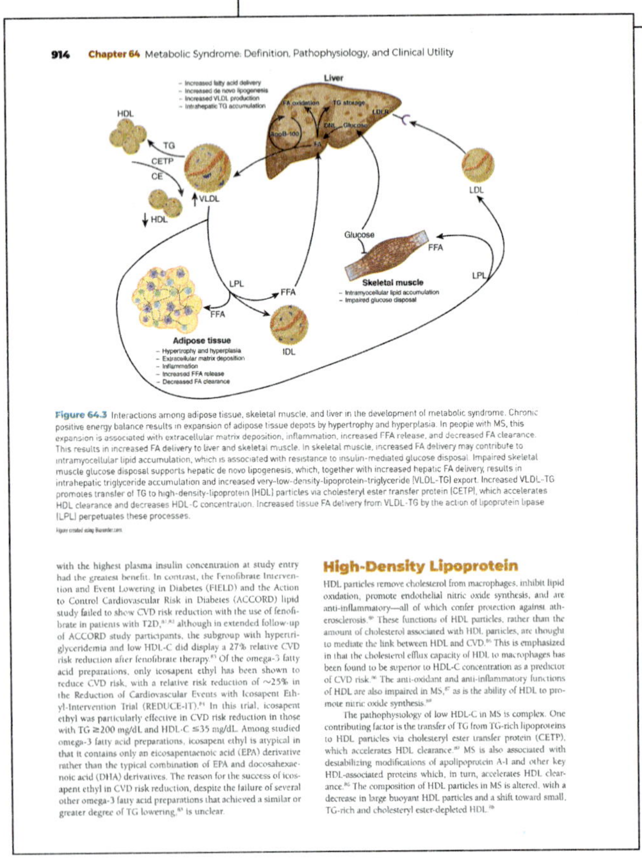

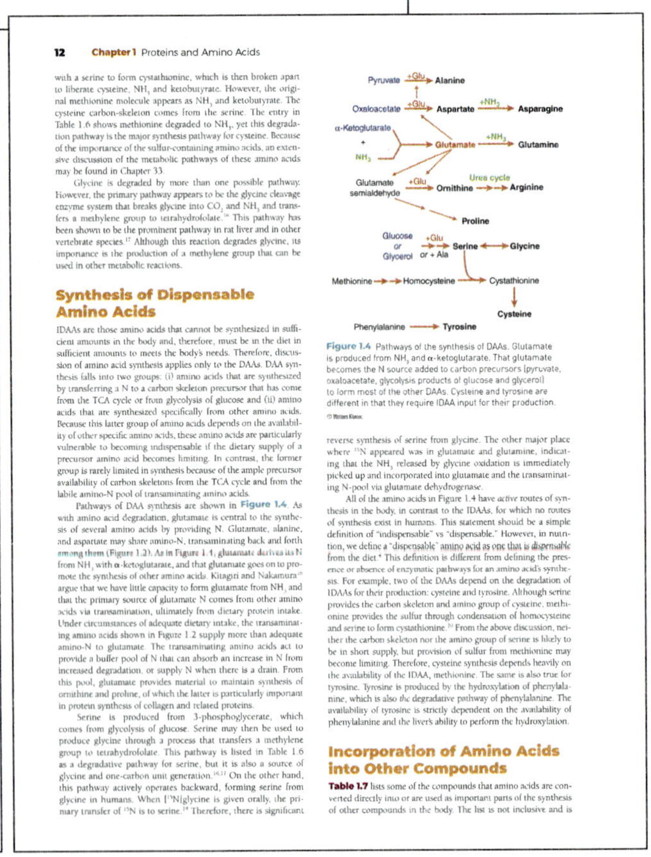

New Chapters

Twelve new chapters are included in this edition, as follows:

- Chapter 49: Dietary Assessment Methods
- Chapter 58: Nutrition Assessment
- Chapter 59: Interprofessional Teams and Nutrition
- Chapter 90: Malnutrition in Disease and Inflammatory States
- Chapter 94: Nutrition and Intellectual and Development Disabilities
- Chapter 100: Nutrition and Age-Related Eye Diseases
- Chapter 114: Global Food Systems
- Chapter 119: Nutrition in the Military
- Chapter 120: Nutrition and Spaceflight
- Chapter 121: Metabolomics and Proteomics in Nutrition Research
- Chapter 122: Use of Artificial Intelligence in Nutrition Research
- Chapter 123: Precision Nutrition

Navigate eBook

Read the digital text online or offline, enhance learning with interactive tools, and print pages easily via the Navigate eBook. This online resource includes:

- All chapters in the printed book (Chapters 1–100)
- Chapters 101 to 123 (available only in the eBook)
- All Appendices
- Weblinks to key resources such as Food Composition Tables, Growth Charts, and Dietary Guidelines
- A comprehensive Index

Each new print purchase of the textbook includes a bound-in eBook access code for 365-day access. Additional eBook access codes are available at www.jblearning.com.

Instructor Resources

The following instructor resources are available to qualified instructors from Jones & Bartlett Learning:

- Slides in PowerPoint format, containing every image and table from the textbook

About the Editors

Katherine L. Tucker, PhD, is Professor of Nutritional Epidemiology and Director of the Center for Population Health at the University of Massachusetts Lowell. From 1989–2009, she was a research scientist at the USDA Human Nutrition Research Center on Aging at Tufts University. She earned her PhD from Cornell University and her BSc from the University of Connecticut, both in nutritional sciences. Her research focuses on dietary intake, metabolism, and chronic disease (osteoporosis, diabetes, heart disease, and cognitive decline) in diverse populations, including as principal investigator of the Boston Puerto Rican Health Study, a longitudinal cohort. She is a past Chair of the Nutritional Sciences Council, past board member of the American Society for Nutrition (ASN), and served two terms on the Food and Nutrition Board of the Health and Medicine Division of the National Academies. She recently completed a 10-year term as Editor-in Chief of *Advances in Nutrition*, the ASN international review journal; prior to that she was an associate editor for *The Journal of Nutrition*. She is a Fellow of the ASN, the American Society for Bone and Mineral Research, and the Gerontological Society of America.

Christopher P. Duggan, MD, MPH, earned his BA from Dartmouth College, his MD from the Johns Hopkins University School of Medicine, and his MPH from the Harvard TH Chan School of Public Health. He is a pediatric gastroenterologist and nutrition physician at Boston Children's Hospital, where he directs the Center for Nutrition and is Medical Director of the Center for Advanced Intestinal Rehabilitation. His research interests are in maternal and child health in low-resource settings as well as diarrheal disease and undernutrition. He is the Samuel Meltzer Professor of Pediatrics and Director of the Division of Nutrition at Harvard Medical School, and a Professor in the Departments of Nutrition and Global Health and Population at the Harvard T.H. Chan School of Public Health. He is the course cofounder of the Advanced Integrated Science Course at Harvard Medical School "Nutrition, Metabolism and Lifestyle Medicine." Since 2019, he has been Editor-in-Chief of The *American Journal of Clinical Nutrition*.

Gordon L. Jensen, MD, PhD, became Senior Associate Dean for Research and Professor of Medicine and Nutrition at the Larner College of Medicine, University of Vermont in 2016. He retired with Emeritus recognition in 2022. From 2007–2015, he was Professor and Head of Nutritional Sciences at Penn State University and Professor of Medicine at the Penn State College of Medicine. From 1998–2007, he was Director of the Center for Human Nutrition and Professor of Medicine at Vanderbilt University Medical Center. He earned his medical degree from Cornell University Medical College and his PhD in nutritional biochemistry from Cornell University. He completed residency training in internal medicine and fellowship training in clinical nutrition at New England Deaconess Hospital, Harvard Medical School. He is a past-president of the American Society for Nutrition (ASN), apast-president of the American Society for Parenteral and Enteral Nutrition (ASPEN), and a past-chair of the Association of Nutrition Programs and Departments. He served two terms on the Food and Nutrition Board of the Health and Medicine Division of the National Academies. His research interests have focused on the impact of nutritional status on health and functional outcomes in older persons. His contributions have been recognized with the 2014 Jonathan Rhoads Lecture, ASPEN's most prestigious award, his election to honorary membership in the Academy and Nutrition and Dietetics, and his selection as a fellow in both ASPEN and ASN.

Karen E. Peterson, ScD, is the Stanley M. Garn Collegiate Professor and inaugural Chair of the Department of Nutritional Sciences at the University of Michigan School of Public Health, where she also holds joint appointments as Professor of Environmental Health Sciences and Professor of Global Public Health. She earned her ScD in Nutrition at the Harvard T.H. Chan School of Public Health and BA in Foods and Nutrition from the University of Utah. Her research focuses on interactions of toxicants and diet on growth, maturation and cardiometabolic risk across the life course and the evaluation of interventions addressing lifestyle behaviors related to obesity and chronic conditions. Since 2010, she has served as Associate Director of the Michigan Nutrition and Obesity Research Center. She is an Associate Editor of *Pediatric Obesity*. She teaches courses on physical growth and maturation and on the evaluation of public health nutrition programs.

A. Catharine Ross, PhD, received her undergraduate education at the University of California at Davis (zoology), and masters (MNS) in nutrition and PhD in biochemistry from Cornell niversity, Ithaca, NY. She was a postdoctoral fellow in the Department of Medicine, Columbia University. She began her academic career at the Medical College of Pennsylvania and moved to Pennsylvania State University in 1994. In 2022, she retired as Professor Emerita and occupant emerita of the Dorothy Foehr Huck Chair in the Department of Nutritional Sciences, Penn State. In 2023, she joined Texas A&M University, College Station, as Professor of Nutrition. Dr. Ross has conducted research on vitamin A nutrition and metabolism and on nutrition and immune function for more than four decades. She served as Editor-in-Chief of the *Journal of Nutrition* from 2004–2013, and was the senior Editor for the eleventh edition of *Modern Nutrition in Health and Disease*. She is a Fellow of the American Association for the Advancement of Science and the American Society for Nutrition, member of the National Academy of Science (USA), and has served on the Food Advisory Committee of the U.S. Food and Drug Administration, Food and Nutrition Board of the Health and Medicine Division of the National Academies, and Board of Scientific Counselors for the National Institute of Diabetes and Digestive and Kidney Diseases. She was awarded a Fulbright fellowship for teaching and research in Japan, 2017–2018.

Riva Touger-Decker, PhD, RD, CDN, FADA, is Professor and Associate Dean of Global Affairs at the Rutgers School of Health Professions and director of the nutrition division in the School of Dental Medicine. She chaired the Department of Clinical and Preventive Nutrition Sciences from 2007–2021 at Rutgers where she developed and directed the first Doctor of Clinical Nutrition Program in the United States. She earned her BS at the State University of New York College at Buffalo and her Master's and PhD at New York University. She is a registered dietitian, internationally recognized for her scholarship and leadership in nutrition and oral health, nutrition-focused physical examination, and dietetics education. Her research has explored diet/nutrition and orofacial pain, tooth loss in older adults, head and neck cancer, nutrition-focused physical exam practices of dietitians, and hybrid approaches to training health professionals globally. Dr. Touger-Decker was the lead editor of the first and second editions of *Nutrition and Oral Medicine*. Awards include the American Dietetic Association Medallion, Excellence in Dietetic Education Award, and the American Society for Clinical Nutrition Dannon Institute Award for Excellence in Medical/Dental Nutrition Education. She is a Fellow of the New York Academy of Medicine, and is a past chair of the Commission on Dietetic Registration. She is the co-lead for the Rutgers IIE—International Academic Partnership Program-Greece initiative.

Preface

Modern Nutrition in Health and Disease, Twelfth Edition continues its long tenure as an authoritative textbook and reference for the study of human nutrition, nutritional disease, and beyond. The textbook's contents span nutritional sciences to applications for improving clinical outcomes. It is appropriate for adoption into undergraduate and graduate courses like Medical Nutrition Therapy; Nutrition Macronutrients and Micronutrients; Human Nutrition and Metabolism; Clinical Nutrition; Nutrition and Disease; Nutrition Assessment; Nutritional Biochemistry and Pathophysiology; and Vitamins and Minerals.

The goal of the *Twelfth Edition*, like that of its predecessors, is to provide a comprehensive and masterful text authored by experts in their fields. More than 260 authors have contributed to the *Twelfth Edition*, and nearly 70% are new authors. All authors' scholarship is reflected in this new edition.

The history of *Modern Nutrition in Health and Disease* now spans 7 decades, as outlined in the following table, which highlights the editions, publication year, editors, and publisher of this text.

The title and objectives of this work evolved from a book originally designated *Dietotherapy*, initiated by Michael G. Wohl, MD, and Robert S. Goodhart, MD, as Coeditors, in 1950. The second edition was the first to use the title *Modern Nutrition in Health and Disease*, subtitled *Dietotherapy*. Drs. Wohl and Goodhart continued to edit the first four editions. Beginning with the fifth edition in 1973, Maurice E. Shils, MD, ScD, joined in editing the book with Dr. Goodhart. The sixth edition is the same as the fifth edition. Dr. Shils became the Senior Editor beginning with the seventh edition in 1988 and continued in this role through the tenth edition. The tenth edition celebrated the 50th anniversary of *Modern Nutrition in Health and Disease*. A. Catharine Ross was the senior editor of the eleventh edition.

Edition Number	Publication Year[a]	Editors	Publisher
1	1950	Drs. Michael Wohl, Robert Goodhart	Lea & Febiger
2	1955	Drs. Michael Wohl, Robert Goodhart	Lea & Febiger
3	1964	Drs. Michael Wohl, Robert Goodhart	Lea & Febiger
4	1968	Drs. Michael Wohl, Robert Goodhart	Lea & Febiger
5	1973	Drs. Robert Goodhart, Maurice E. Shils	Lea & Febiger
6	1980	Drs. Robert Goodhart, Maurice E. Shils	Lea & Febiger
7	1988	Drs. Maurice E. Shils, Vernon Young	Lea & Febiger
8	1994	Drs. Maurice E. Shils, James A. Olson, Moshe Shike	Lea & Febiger[b]
9	1998	Drs. Maurice E. Shils, James A. Olson, Moshe Shike, A. Catharine Ross	Lippincott Williams & Wilkins
10	2005	Drs. Maurice E. Shils, Moshe Shike, A. Catharine Ross, Benjamin Caballero, Robert J. Cousins	Lippincott Williams & Wilkins
11	2012	Drs. A. Catharine Ross, Benjamin Caballero, Robert J. Cousins, Katherine L. Tucker, Thomas R. Ziegler	Lippincott Williams & Wilkins
12	2025	Drs. Katharine L. Tucker, Christopher P. Duggan, Gordon L. Jensen, Karen E. Peterson, A. Catharine Ross, and Riva Touger-Decker	Jones & Bartlett Learning

[a]Year of the first printing.

[b]Lea & Febiger, Philadelphia, was purchased shortly before the publication of the eighth edition by the Waverly Company, owner of Williams & Wilkins Publishers, Baltimore. Shortly before the publication of the ninth edition, the Waverly Company was purchased by Wolters Kluwer Publishers and merged with Lippincott, the medical publisher in Philadelphia.

The Editors of the *Twelfth Edition*, both those continuing and those who are new to this work, wish to extend our most sincere gratitude to the editors of previous versions for the guidance they provided and for their dedication to this book and rigorous approach to overseeing the tasks that are involved. We hope that, as the book makes its new transition, it will continue to remain highly respected as the authoritative reference for the study and practice of Nutrition.

Modern Nutrition in Health and Disease, Twelfth Edition encompasses state-of-the-art nutrition topics, some of which have been consolidated to keep modern in approach. Beginning with this twelfth edition, select introductory and niche chapters, along with the appendices, are offered online only, allowing a similar size and scope to comparable textbooks. These online materials are freely available when the hard copy of the book is purchased.

The *Twelfth Edition* presents content on the most current concepts and practical concerns in nutrition and nutritional management of disease. New chapters, as indicated above, include Dietary Assessment Methods; Nutritional Assessment in Clinical Practice; Nutrition in Interprofessional Practice; Malnutrition in Disease and Inflammatory States; Nutrition and Intellectual and Development Disabilities; Nutrition and Eye Diseases; Global Food Systems; Nutrition and the Military; Nutrition and Space; Omics in Nutrition Research: Metabolomics, Proteomics; Machine Learning in Nutrition Research; and Precision Nutrition.

The Editors wish to acknowledge outstanding support in the preparation, editing, and production of this extensive work. The authors who have contributed their expertise are listed alphabetically on the following pages. The Editors have worked personally with some of the staff from our publishing company, whereas other members have been involved "behind the scenes" at the editorial, publication, distribution, and marketing stages. We wish to thank all of them for their support. Whitney Fekete, Product Manager, helped move the *Twelfth Edition* forward. The project owes an enormous thank you to Jones & Bartlett Learning content, production, and media staff whose efforts enhanced the quality of the work.

THE EDITORS
Katherine L. Tucker, PhD
Christopher P. Duggan, MD, MPH
Gordon L. Jensen, MD, PhD
Karen E. Peterson, ScD
A. Catharine Ross, PhD
Riva Touger-Decker, PhD, RD

Contributors

The editors acknowledge the work of all involved with the previous edition of this text. Although some chapters are new and some completely rewritten, many chapters have been updated and contain anywhere from a few paragraphs to large sections of the previous version. In these cases, the authors of the previous version are noted in the acknowledgments and their important contributions are appreciated.

Shabina Roohi Ahmed, MD
Assistant Professor
Johns Hopkins University School of Medicine
Medical Director
Johns Hopkins Community Physicians Endocrinology at Downtown Bethesda
Bethesda, Maryland

Amanda S. Alexander, MD
Assistant Professor
Department of Medicine
Division of Clinical Immunology & Rheumatology
University of Alabama at Birmingham Marnix E. Heersink School of Medicine
Birmingham, Alabama

Johane P. Allard, MD, FRCP
Professor
Department of Medicine
Division of Gastroenterology
Toronto General Hospital
University of Toronto
Toronto, Ontario, Canada

Olivia S. Anderson, PhD, MPH, RD
Associate Clinical Professor
Department of Nutritional Sciences
School of Public Health
University of Michigan
Ann Arbor, Michigan

Gabriela Andries, MD, FACC
Cardiologist
AtlantiCare Regional Medical Center
Atlantic City, New Jersey

Aśok C. Antony, MD, MACP
Chancellor's Professor
Indiana University School of Medicine
Indianapolis, Indiana

Gene R. Arnold III, MS, RD, CDCES
Program Director
Baltimore Metropolitan Diabetes Regional Partnership
Johns Hopkins University School of Medicine
Baltimore, Maryland

Bregje S.M. van Asperen, MSc
Program Associate
Standing Together for Nutrition
Micronutrient Forum
Washington District of Columbia

David A. August, MD
Professor Emeritus
Robert Wood Johnson Medical School
Rutgers Cancer Institute of New Jersey
New Brunswick, New Jersey

Joseph E. Baggott, PhD
Assistant Professor
School of Health Professions
University of Alabama at Birmingham
Birmingham, Alabama

Regan Bailey, PhD, MPH, RD
Associate Director
Institute for Advancing Health through Agriculture
Texas A&M University
College Station, Texas

Connie Watkins Bales, PhD, RD
Professor
Duke University School of Medicine
Associate Director
Geriatric Research Education and Clinical Center
Durham VA Medical Center
Durham, North Carolina

Linda G. Bandini, PhD, RD
Professor
Eunice Kennedy Shriver Center
University of Massachusetts Chan Medical School
Worcester, Massachusetts

Vickie Baracos, PhD
Professor
Faculty of Medicine & Dentistry
University of Alberta
Edmonton, Canada

Dalia Barayan, PhD
Post Doctoral Research Associate
Centre for Burn Research
Hamilton Health Sciences
McMaster University
Hamilton, Ontario, Canada

Joseph L. Baumert, PhD
Professor
Department of Food Science & Technology
Director
Food Allergy Research and Resource Program
University of Nebraska-Lincoln
Lincoln, Nebraska

Chantal Bémeur, RD, PhD
Professor
Faculté de Médecine
Université de Montréal
Montréal, Quebec, Canada

Gerard T. Berry, MD
Professor
Harvard Medical School
Division of Genetics & Genomics
Boston Children's Hospital
Boston, Massachusetts

Shilpa N. Bhupathiraju, PhD
Assistant Professor
Harvard Medical School
Channing Division of Network Medicine
Brigham and Women's Hospital
Boston, Massachusetts

Zulfiqar Ahmed Bhutta, MD, PhD
Co-Director
Centre for Global Child Health
Hospital for Sick Children
Toronto, Ontario, Canada
Founding Director
Centre of Excellence in Women and Child Health
Aga Khan University
Karachi, Pakistan

David G. Binion, MD
Professor
University of Pittsburgh School of Medicine
Co-Director
Inflammatory Bowel Disease (IBD) Center
Director
Translational IBD Research and Nutrition Support Service
Division of Gastroenterology, Hepatology and Nutrition
Medical Director
Intestinal Rehabilitation and Transplant Center
UPMC Presbyterian Hospital
Pittsburgh, Pennsylvania

Sarah L. Booth, PhD
Center Director and Senior Scientist
Jean Mayer USDA Human Nutrition Research Center on Aging
Tufts University
Boston, Massachusetts

Megan W. Bourassa, PhD
Program Lead
Micronutrient Forum
New York, New York

Rebecca A. Brody, PhD, RD
Professor
Department of Clinical and Preventive Nutrition Sciences
School of Health Professions
Rutgers University
Newark, New Jersey

Jason L. Burkhead, PhD
Professor
Department of Biological Sciences
University of Alaska Anchorage
Anchorage, Alaska

David L. Burns, MD
Associate Professor
Tufts University School of Medicine
Boston, Massachusetts
Director
Nutrition Services
Senior Staff Gastroenterologist
Lahey Hospital and Medical Center
Burlington, Massachusetts

Douglas G. Burrin, PhD
Professor
Baylor College of Medicine
Senior Research Scientist
USDA Children's Nutrition Research Center
Houston, Texas

Britt Burton-Freeman, PhD, MS
Professor
Department of Food Science and Nutrition
Illinois Institute of Technology
Chicago, Illinois

Roger F. Butterworth, PhD
Professor
Faculté de Medicine
Université deMontréal
Montréal, Quebec, Canada

Philip C. Calder, PhD, DPhil
Professor
School of Human Development & Health
University of Southampton
Southampton, United Kingdom

Mona S. Calvo, PhD, MA
Adjunct Professor
Icahn School of Medicine at Mount Sinai
New York, New York

Stephen J. Carter, MS, PhD
Assistant Professor
Department of Kinesiology
Indiana University Bloomington
Bloomington, Indiana

Lingtak-Neander Chan, PharmD, BCNSP
Professor
School of Pharmacy
University of Washington
Seattle, Washington

Uday Kumar Chekkilla, PhD
Post-doctoral Fellow
Wilmer Eye Institute
Johns Hopkins University School of Medicine
Baltimore, Maryland

Lawrence J. Cheskin, MD, FACP, FTOS
Professor and Chair
Nutrition and Food Studies Department
George Mason University College of Public Health
Fairfax, Virginia

Alexandre D.P. Chiavegatto Filho, PhD
Professor
School of Public Health
University of São Paulo
São Paulo, Brazil

Steven K. Clinton, MD, PhD
Professor and Medical Oncologist
Department of Internal Medicine
Comprehensive Cancer Center
The Ohio State University
Columbus, Ohio

Paul M. Coates, PhD
Adjunct Professor
Indiana University School of Public Health
Bloomington, Indiana

Marc Cohen, MD, FACC
Chair
Department of Medicine
Formerly, Director
Division of Cardiology
Newark Beth Israel Medical Center
Professor of Medicine
Rutgers New Jersey Medical School
Rutgers University
Newark, New Jersey

Suzanne M. Cole, PhD
Senior Lecturer
Department of Nutritional Sciences
University of Michigan School of Public Health
Ann Arbor, Michigan

Sarah C. Collica, MD
Assistant Professor
Department of Psychiatry and Behavioral Sciences
Johns Hopkins University School of Medicine
Baltimore, Maryland

James F. Collins, PhD
Professor
Food Science and Human Nutrition Department
University of Florida
Gainesville, Florida

Beth F. Coplan, RDN, CDCES, CCTD
Clinical Transplant Dietitian and Bariatric Dietitian
Robert Wood Johnson Barnabas Health Care System
Newark Beth Israel Medical Center
Newark, New Jersey

Rebecca B. Costello, PhD
Senior Research Fellow
Center for Magnesium Education & Research
Pahoa, Hawaii

Mélina Côté, DtP, MSc
Graduate Research Assistant
Centre Nutrition, Santé et Société
Université Laval
Montréal, Québec, Canada

Ruy J Cruz, MD
Associate Professor
Department of Surgery
University of Pittsburgh School of Medicine
Pittsburgh, Pennsylvania

Carol Curtin, PhD, MSW
Professor
Department of Family Medicine and Community Health
Morningside Graduate School of Biomedical Sciences
Eunice Kennedy Shriver Center
University of Massachusetts Chan Medical School
Worcester, Massachusetts

Sai Krupa Das, PhD
Professor and Senior Scientist
Jean Mayer USDA Human Nutrition Research Center on Aging
Tufts University
Boston, Massachusetts

Catherine Hambleton Davis, MD, MPH
Clinical Assistant Professor
Division of Surgical Oncology
Texas A&M University School of Medicine
Baylor University Medical Center
Dallas, Texas

Teresa A. Davis, PhD, MS
Professor
USDA Children's Nutrition Research Center
Baylor College of Medicine
Houston, Texas

Bess Dawson-Hughes, MD
Professor and Senior Scientist
Jean Mayer USDA Human Nutrition Research Center on Aging
Tufts University
Boston, Massachusetts

Lilian de Jonge, PhD
Professor
Department of Nutrition and Food Studies
George Mason University
Fairfax, Virginia

Augusto César de Moraes, PhD, MS
Assistant Professor
Department of Epidemiology
UT HealthHouston School of Public Health in Austin
Michael & Susan Dell Center for Healthy Living
University of Texas Health Science Center at Houston
Houston, Texas

Russell Debose-Boyd, PhD
Professor
Department of Molecular Genetics
University of Texas Southwestern Medical Center
Dallas, Texas

John K. DiBaise, MD, FACP, FASPEN
Professor
Department of Internal Medicine
Division of Gastroenterology and Hepatology
Mayo Clinic in Arizona
Scottsdale, Arizona

Michelle DiBaise, DHSc, PA-C, DFAAPA
Professor
Health Sciences Campus
Creighton University
Scottsdale, Arizona

Dana C. Dolinoy, PhD, MSc
Professor and Chair
Environmental Health Sciences
University of Michigan
Ann Arbor, Michigan

Sharon M. Donovan, PhD, RD
Professor and Melissa M. Noel Endowed Chair
Division of Nutritional Sciences
Director
Personalized Nutrition Initiative
University of Illinois at Urbana-Champaign
Urbana, Illinois

Valerie B. Duffy, PhD, RD
Professor
Department of Allied Health Science College of Agriculture, Health, and Natural Resources
University of Connecticut Storrs, Connecticut

Ifechukwude Ebenuwa, MD, MHSc
Assistant Clinical Investigator
Molecular and Clinical Nutrition Section
National Institute of Diabetes and Digestive and Kidney Diseases
Bethesda, Maryland

Curtis D. Eckhert, PhD, MS
Professor Emeritus
Department of Environmental Health and Molecular Toxicology
Fielding School of Public Health
University of California
Los Angeles, California

Megan Edelman, MS, RDN
Graduate Research Assistant
College of Food, Agricultural, and Natural Resource Sciences
University of Minnesota—Twin Cities
St. Paul, Minnesota

Keith M. Erikson, PhD, MS
Professor
School of Health and Human Sciences
University of North Carolina Greensboro
Greensboro, North Carolina

Abby G. Ershow, ScD, RD
Senior Nutrition Scientist
Office of Dietary Supplements
National Institutes of Health
Bethesda, Maryland

Emily K. Farina, PhD, RD
Research Scientist
Military Nutrition Division
U.S. Army Research Institute of Environmental Medicine
Natick, Massachusetts

Edward A. Fisher, MD, PhD
Leon H. Charney
Professor
Department of Medicine
Director
Center for the Prevention of Cardiovascular Disease
New York University School of Medicine
New York, New York

Luigi Fontana, MD, PhD, FRACP
Professor and Ullmann Chair
Scientific Director
Charles Perkins Centre Royal Prince Alfred Clinic
University of Sydney
Sydney, NSW, Australia

Steve C. Fordahl, PhD, MS
Associate Professor
Department of Nutrition
University of North Carolina Greensboro
Greensboro, North Carolina

Anna Gaddy, MD, FASN
Assistant Professor
Nephrology Center for Advanced Care
Medical College of Wisconsin
Milwaukee, Wisconsin

Xiang Gao, MD, PhD
Dean and Distinguished Professor
Institute of Nutrition
Fudan University
Shanghai, China

Christine D. Garner, PhD, MS, RD, CLC
Assistant Professor
Department of Pediatrics
Texas Tech Health Sciences Center
Amarillo, Texas

Zachary P. Gersten, PhD, MPH
Research Nutritionist
USDA Western Human Nutrition Research Center
Davis, California

Edward L. Giovannucci, DSc, MD
Professor
Departments of Medicine and Nutrition
Harvard TH Chan School of Public Health
Boston, Massachusetts

Jaclyn M. Goodrich, PhD
Research Associate Professor
Department of Environmental Health Sciences
University of Michigan
Ann Arbor, Michigan

Aria A. Grabowski, MPH, RD
Doctoral Candidate
Nutritional Sciences Department
School of Public Health
University of Michigan
Ann Arbor, Michigan

Elizabeth M. Grainger, PhD, RD
Clinical Research Nutritionist
Comprehensive Cancer Center
The Ohio State University
Columbus, Ohio

Leah Gramlich, MD, FRCP(C)
Professor of Medicine
Scientific Director
Diabetes, Obesity & Nutrition Strategic Clinical Network
University of Alberta
Edmonton, Alberta, Canada

Cara L. Green, PhD
Postdoctoral Researcher
Department of Medicine
University of Wisconsin-Madison
Madison, Wisconsin

Ralph Green, MD, PhD
Distinguished Professor
Department of Pathology and Laboratory Medicine
School of Medicine
University of California, Davis
Davis, California

Jesse F. Gregory III
Professor Emeritus
Department of Food Science and Human Nutrition
University of Florida
Gainesville, Florida

Angela S. Guarda, MD
Stephen and Jean Robinson Professor
Department of Psychiatry and Behavioral Sciences
Johns Hopkins University School of Medicine
Director
Johns Hopkins Eating Disorders Program
Johns Hopkins Hospital
Baltimore, Maryland

Craig Gundersen, PhD
Snee Family Endowed Professor
Department of Economics
Baylor University
Waco, Texas

David Hakimian, MD
Senior Gastroenterologist
Department of Gastroenterology
Hadassah Medical Center
Hebrew University of Jerusalem
Jerusalem, Israel

Andrew G. Hall, PhD
Associate Project Scientist
Department of Nutritional Sciences & Toxicology
University of California, Berkeley
Berkeley, California

Peter J. Havel, DVM, PhD
Professor
Department of Molecular Biosciences
School of Veterinary Medicine
University of California, Davis
Davis, California

Robert A. Hegele, MD, FRCPC
Jacob J. Wolfe Distinguished Medical Research Chair
Edith Schulich Vinet Canada Research Chair
Martha G. Blackburn Chair
Departments of Medicine and Biochemistry
The University of Western Ontario
London, Ontario, Canada

Stephen R. Hennigar, PhD
Assistant Professor
Pennington Biomedical Research Center
Louisiana State University
Baton Rouge, Louisiana

Steve Hertzler, PhD, RD, LD
Senior Scientist
Clinical Research
Abbott Nutrition
Columbus, Ohio

Steven B. Heymsfield, MD
Professor
Pennington Biomedical Research Center
Louisiana State University
Baton Rouge, Louisiana

Deanna M. Hoelscher, PhD, RD, LD, CNS
John P McGovern Professor and Associate Dean for Research
Department of Health Promotion and Behavioral Sciences
Director
Michael & Susan Dell Center for Healthy Living
School of Public Health
The University of Texas Health Science Center at Houston
Houston, Texas

L. John Hoffer, MD, PhD
Professor
Faculty of Medicine
McGill University
Montréal, Quebec, Canada

Hannah D. Holscher, PhD, RD
Associate Professor
Department of Food Science & Human Nutrition
University of Illinois, Urbana-Champaign
Urbana, Illinois

Bridget M. Hron, MD
Assistant Professor
Department of Pediatrics
Harvard Medical School
Associate Director
Center for Nutrition
Boston Children's Hospital
Boston, Massachusetts

Maureen B. Huhmann, DCN, RD
Global Research & Development Lead – Acute Care
Nestlé Health Science
Bridgewater, New Jersey

Gary R. Hunter, PhD
Distinguished Professor Emeritus
School of Education and Human Sciences
University of Alabama at Birmingham
Birmingham, Alabama

Susan L. Hyman, MD
Professor and Chief
Division of Neurodevelopmental and Behavioral Pediatrics
University of Rochester Medical Center
Rochester, New York

Aamer Imdad, MBBS, MPH
Assistant Professor
Department of Pediatrics
University of Iowa
Iowa City, Iowa

Khursheed N. Jeejeebhoy, MBBS, PhD, FRCP, FRCPC
Emeritus Professor
Department of Nutritional Sciences
University of Toronto
Toronto, Ontario, Canada

Gordon L. Jensen, MD, PhD
Professor Emeritus
Larner College of Medicine
University of Vermont
Burlington, Vermont

Marc Jeschke, MD, PhD
Professor
Department of Surgery
Faculty of Health Sciences
McMaster University
Director
Ross Tilley Burn Centre at Sunnybrook Health Sciences Centre
Hamilton Health Sciences
Hamilton, Ontario, Canada

Margaret M. Johnson, MD
Pulmonologist
Department of Pulmonary Medicine
Mayo Clinic
Jacksonville, Florida

Mary Ann Johnson, PhD
Professor Emerita
Department of Nutrition and Health Sciences
University of Nebraska-Lincoln
Lincoln, Nebraska

Andrew D. Jones, PhD
Associate Professor
Department of Nutritional Sciences School of Public Health
University of Michigan
Ann Arbor, Michigan

Glenville Jones, PhD
Professor
Department of Biomedical and Molecular Sciences
Queen's University
Kingston, Ontario, Canada

Loretta L. Jophlin, MD, PhD
Assistant Professor
School of Medicine
University of Louisville
Louisville, Kentucky

J. Philip Karl, PhD, RD
Nutrition Physiologist
Military Nutrition Division
U.S. Army Research Institute of Environmental Medicine
Natick, Massachusetts

Sindhu Kashyap, MSc
Graduate Research Assistant
Division of Nutrition
St. John's Research Institute
Bangalore, Karnataka, India

Richard M. Katz, MD, MBA
Associate Professor
Johns Hopkins University School of Medicine
Chief Medical Officer
Mt. Washington Pediatric Hospital
Baltimore, Maryland

Claudine J. Kavanaugh, PhD, MPH, RD
Director
Office of Nutrition and Food Labeling
Center for Food Safety and Applied Nutrition
Food and Drug Administration
College Park, Maryland

Nancy L. Keim, PhD, RDN
Research Nutritionist
USDA Western Human Nutrition Research Center
Davis, California

Kathleen L. Keller, PhD
Professor
Department of Nutritional Sciences
The Pennsylvania State University
State College, Pennsylvania

Erin Kenny, DCN, RD, CSO
Clinical Dietitian
Memorial Sloan Kettering Cancer Center
New York, New York

Yeonsoo Kim, PhD, RDN
Professor
Nutrition & Dietetics Program
Central Michigan University
Mount Pleasant, Michigan

James B. Kirkland, PhD
Retired Professor
Department of Human Health & Nutritional Sciences
University of Guelph
Guelph, Ontario, Canada

Sharon I. Kirkpatrick, PhD, RD
Associate Professor and University Research Chair
School of Public Health Sciences
University of Waterloo
Waterloo, Ontario, Canada

Irina A. Kirpich, PhD, MPH
Professor
Department of Microbiology & Immunology
University of Louisville
Louisville, Kentucky

Joel D. Kopple, MD
Professor Emeritus
Division of Nephrology and Hypertension
David Geffen School of Medicine at UCLA
Los Angeles, California
The Lundquist Institute at Harbor-UCLA Medical Center
Torrance, California

Rebecca Kuang, MD
Internal Medicine Resident
University of Pittsburgh Medical Center
Pittsburgh, Pennsylvania

Catherine Kubrak, PhD, RN
Research Associate
Department of Oncology
Faculty of Medicine & Dentistry
University of Alberta
Edmonton, Alberta, Canada

Toshinobu Kuroishi, PhD
Senior Assistant Professor
Graduate School of Dentistry
Tohoku University
Sendai, Miyagi, Japan

Anura V. Kurpad, MD, PhD, FRCP
Professor
Department of Physiology and Nutrition
St. John's Medical College
Bangalore, Karnataka, India

Benoît Lamarche, PhD, FAHA
Professor
École de Nutrition
Université Laval
Québec City, Québec, Canada

Nathalie J. Lambrecht, PhD
Postdoctoral Research Associate
Institute of Public Health
Charité - Universitätsmedizin Berlin
Berlin, Germany

Dudley W. Lamming, PhD
Associate Professor
Department of Medicine
University of Wisconsin-Madison
Madison, Wisconsin

Sun Y. Lee, MD, MSc
Assistant Professor
Chobanian & Avedisian School of Medicine
Boston University
Boston, Masaschusetts

Xin Gen Lei, PhD, MS, BS
Professor and Associate Dean
College of Agricultural and Life Sciences
Cornell University
Ithaca, New York

Edgar V. Lerma, MD, FACP, FASN
Clinical Professor
Section of Nephrology
University of Illinois at Chicago
Chicago, Illinois

Mark Levine, MD
Section Chief
Molecular and Clinical Nutrition Section, Digestive Disease Branch
National Institute of Diabetes and Digestive and Kidney Diseases
Bethesda, Maryland

Alice H. Lichtenstein, DSc
Gershoff Professor and Senior Scientist
Jean Mayer USDA Human Nutrition Research Center on Aging
Tufts University
Boston, Massachusetts

Berkeley Limketkai, MD, PhD
Associate Clinical Professor
Director of Clinical Research
UCLA Center for Inflammatory Bowel Diseases
University of California, Los Angeles School of Medicine
Los Angeles, California

Nicola M. Lowe, BSc, PhD, RNutr, FAfN, FHEA
Professor and Director
UCLan Research Centre for Global Development
University of Central Lancashire
Preston, United Kingdom

Thomas H. Magnuson, MD
Associate Professor
Johns Hopkins University School of Medicine
Chair
Department of Surgery
Johns Hopkins Bayview Medical Center
Baltimore, Maryland

Molly G. Mailes, MSN, RN, MEDSURG-BC
Registered Nurse
Mayo Clinic
Jacksonville, Florida

Sarah Malik, MD
Norton College of Medicine
SUNY Upstate Medical University
Syracuse, New York

Monica Mars, PhD
Associate Professor
Division of Human Nutrition and Health
Wageningen University
Wageningen, The Netherlands

Teresa A. Marshall, PhD, RDN/LDN
Professor
College of Dentistry
University of Iowa
Iowa City, Iowa

Laura E. Matarese, PhD, RDN, LDN, FADA, FASPEN, FAND
Professor
Department of Internal Medicine
Adjunct Professor
Department of Surgery
Brody School of Medicine
East Carolina University
Greenville, North Carolina

Dwight E. Matthews, PhD
Pomeroy Professor Emeritus
Department of Chemistry
University of Vermont
Burlington, Vermont

Craig J. McClain, MD, FACN, AGAF, FACG, FAASLD
Professor and Distinguished University Scholar
Departments of Medicine and Pharmacology & Toxicology
Associate Vice President for Translational Research and Health Affairs Research
Division of Gastroenterology, Hepatology and Nutrition
University of Louisville School of Medicine
Louisville, Kentucky

James P. McClung, PhD
Research Scientist
Military Nutrition Division
U.S. Army Research Institute of Environmental Medicine
Natick, Massachusetts

Michelle K. McGuire, PhD
Professor and Director
School of Family and Consumer Sciences
University of Idaho
Moscow, Idaho

Robin A. McKinnon, PhD, MPA
Senior Advisor
Center for Food Safety and Applied Nutrition
US Food & Drug Administration
College Park, Maryland

Sarah A. McNaughton, PhD
Professor
Institute for Physical Activity and Nutrition
Deakin University
Geelong, Victoria, Australia

Nilesh M. Mehta, MD
Professor
Harvard Medical School
Chair
Critical Care Nutrition & Metabolism
Boston Children's Hospital
Boston, Massachusetts

Ralph G. Meyer, PhD, MS
Professor
College of Veterinary Medicine
Utah State University
Logan, Utah

Mirella L. Meyer-Ficca, PhD, MS
Associate Professor
College of Agriculture and Applied Sciences
Utah State University
Logan, Utah

Joshua W. Miller, PhD
Professor and Chair
Department of Nutritional Sciences
Rutgers, The State University of New Jersey
New Brunswick, New Jersey

Sarah L. Morgan, MD, MS, RD, FADA, FACP, CCD
Emeritus Professor
Division of Clinical Immunology and Rheumatology
Heersink School of Medicine
University of Alabama at Birmingham
Birmingham, Alabama

Gerard E. Mullin MD
Associate Professor
Johns Hopkins University School of Medicine
Baltimore, Maryland

Jillian Murphy, PhD
Licensed Psychologist
Mt. Washington Pediatric Hospital
Baltimore, Maryland

Aviva Must, PhD
Professor
Public Health and Community Medicine
Tufts University School of Medicine
Boston, Massachusetts

Marcus V. Nascimento-Ferreira, PhD
Assistant Professor
Food and Nutrition Program
Universidade Federal do Tocantins
Palmas, Tocantins, Brazil

Lynnette M. Neufeld, PhD
Director
Food and Nutrition Division
Food and Agriculture Organization of the United Nations
Rome, Italy

Sara O'Brien, MS, RD-AP, LDN, CNSC
Lecturer
Sargent College of Health & Rehabilitation Sciences
Boston University
Boston, Massachusetts

Deborah L. O'Connor, PhD
Earle W. McHenry Professor and Chair
Department of Nutritional Sciences
University of Toronto
Toronto, Ontario, Canada

Susan Oh, PhD, MPH
Director
Research Nutrition Core
Institute for Clinical and Translational Research
Johns Hopkins University School of Medicine
Baltimore, Maryland

Stephen J.D. O'Keefe, MD, MSc, FRCP
Professor
Department of Medicine
Division of Gastroenterology and Nutrition
University of Pittsburgh
Pittsburgh, Pennsylvania

Endashaw M. Omer, MD, MPH
Associate Professor
School of Medicine
University of Louisville
Director of Endoscopy
University of Louisville Hospital
Louisville, Kentucky

Jose M. Ordovas, PhD
Professor and Senior Scientist
Jean Mayer USDA Human Nutrition Research Center on Aging
Tufts University
Boston, Massachusetts
Nutritional Genomics and Epigenomics Group
Instituto Madrileño de Estudios Avanzados (IMDEA) Food, CEI UAM + CSIC
Madrid, Spain

Saskia J.M. Osendarp, PhD
Executive Director
Micronutrient Forum
Washington District of Columbia

Alyssa M. Parian, MD
Associate Professor
Gastroenterology
Johns Hopkins University School of Medicine
Baltimore, Maryland

Elizabeth N. Pearce, MD, MSc
Professor
Chobanian & Avedisian School of Medicine
Boston University
Boston, Massachusetts

Rafael Pérez-Escamilla, PhD
Professor
School of Public Health
Yale University
New Haven, Connecticut

Max C. Petersen, MD, PhD
Instructor
Division of Endocrinology, Metabolism & Lipid Research
Milliken Department of Medicine
Washington University School of Medicine in St Louis
Saint Louis, Missouri

Karen E. Peterson, ScD
Professor and Chair
Nutritional Sciences Department
School of Public Health
University of Michigan
Ann Arbor, Michigan

Kavita H. Poddar, PhD, RD, CD
Senior Registered Dietitian and Clinical Nutritionist
University of Wisconsin Hospitals and Clinics
Madison, Wisconsin

Daniel J. Pomerantz, MD
Post-Doctoral Fellow
Division of Genetics and Genomics
Boston Children's Hospital
Boston, Massachusetts

Andrea Mario Pompeo Romani, MD, PhD
Associate Professor
Department of Physiology and Biophysics
Case Western Reserve University
Cleveland, Ohio

Lu Qi, MD, PhD
HCA Regents Distinguished Chair and Professor
Department of Epidemiology
Director
Obesity Research Center
Tulane University School of Public Health & Tropical Medicine
New Orleans, Louisiana

Monica Ramakrishnan, PhD
Research Associate
Department of Nutrition Science
Purdue University
West Lafayette, Indiana

Vikram Rangan, MD
Assistant Professor
Division of Gastroenterology and Hepatology
Harvard Medical School
Beth Israel Deaconess Medical Center
Boston, Massachusetts

Uzma Rani, MD, MPH
Assistant Professor
Division of General Pediatrics
SUNY Upstate Medical University
Syracuse, New York

Shristi Rawal, PhD
Assistant Professor
Department of Clinical and Preventive Nutrition Sciences
Rutgers, The State University of New Jersey
Newark, New Jersey

Margaret P. Rayman, PhD
Professor
Faculty of Health and Medical Sciences
University of Surrey
Surrey, United Kingdom

Casey M. Rebholz, PhD, MS, MNSP, MPH
Associate Professor
Department of Epidemiology
Johns Hopkins Bloomberg School of Public Health
Baltimore, Maryland

Dominic N. Reeds, MD
Professor
Goldfarb School of Nursing
Washington University School of Medicine in St. Louis
St. Louis, Missouri

Deborah L. Renaud, MD
Professor
Division of Neurology and Pediatrics
Mayo Clinic College of Science and Medicine
Rochester, Minnesota

Connie Rhee, MD, MSc
Professor
David Geffen School of Medicine at UCLA
Chief of Nephrology
Veterans Affairs Greater Los Angeles Healthcare System
Los Angeles, California

Jon Marc Rhoads, MD
Professor
Department of Pediatrics
Division of Gastroenterology
McGovern Medical School
The University of Texas Health Science Center at Houston
Houston, Texas

Charis Ripley-Hager, MD
Assistant Professor
Lewis Katz School of Medicine
Temple University
Philadelphia, Pennsylvania

Susan B. Roberts, PhD
Professor and Senior Associate Dean
Geisel School of Medicine
Dartmouth College
Hanover, New Hampshire

Andrea Rosanoff, PhD
Director
Center for Magnesium Education & Research
Pahoa, Hawaii

A. Catharine Ross, PhD
Professor
College of Agriculture & Life Sciences
Texas A&M University
College Station, Texas

Edward A. Ruiz-Narváez, ScD, MS
Associate Professor
Department of Nutritional Sciences
University of Michigan, School of Public Health
Ann Arbor, Michigan

Alexa Neomi Sasson, MD, MPH, FRCPC
Gastroenterologist
University Health Network, Toronto Western Hospital, University of Toronto
Toronto, Ontario
Canada

Dennis Savaiano, PhD
Meredith Professor of Nutrition Science
Dean Emeritus
Purdue University
West Lafayette, Indiana

Ernst J. Schaefer, MD
Distinguished University Professor
Tufts University School of Medicine and Friedman School of Nutrition Science & Policy
Boston, Massachusetts
Founder, Chief Medical Officer & Laboratory Director
Boston Heart Diagnostics
Framingham, Massachusetts

Michael Schweitzer, MD
Associate Professor
Department of Surgery
Johns Hopkins University School of Medicine
Baltimore, Maryland

Carol E. Semrad, MD
Professor
Section of Gastroenterology
The University of Chicago Department of Medicine
Chicago, Illinois

Young Ah Seo, PhD
Associate Professor
Department of Nutritional Sciences
University of Michigan School of Public Health
Ann Arbor, Michigan

Anuja Shah, MD
Associate Professor
Fellowship Training Program
David Geffen School of Medicine
University of California, Los Angeles
Investigator
The Lindquist Institute
Los Angeles, California

Raanan Shamir
Director
Institute of Gastroenterology, Nutrition and Liver Diseases
Schneider Children's Medical Center of Israel
Petah Tikva, Israel
Professor and Chair
Eduarda and Dr. Moshe Ishay Institute for the Study of the Effects of Natural Food on Human Health
Faculty of Medicine
Tel Aviv University
Tel Aviv, Israel

Martin J. Shearer, PhD, FRCPath
Consultant Clinical Scientist Emeritus
Guy's and St Thomas' NHS Foundation Trust
London, United Kingdom

Barbara Shukitt-Hale, PhD
Research Psychologist
Jean Mayer USDA Human Nutrition Research Center on Aging
Tufts University
Boston, Massachusetts

Joanne L. Slavin, PhD, RDN
Professor
Department of Food Science and Nutrition
College of Food, Agricultural, and Natural Resource Sciences
University of Minnesota—Twin Cities
St. Paul, Minnesota

Melissa J. Slotnick, PhD, MPH, MESc
Research Associate
Department of Nutritional Sciences
University of Michigan School of Public Health
Ann Arbor, Michigan

Paul A.M. Smeets, PhD
Senior Scientist
Division of Human Nutrition and Health
Wageningen University
Wageningen, The Netherlands

Ellen Smit, PhD, MS, RDN
Professor
School of Public Health & Nutrition
Oregon State University
Corvallis, Oregon

Kimberly R. Smith, PhD
Assistant Professor
Department of Psychiatry and Behavioral Sciences
Johns Hopkins University School of Medicine
Baltimore, Maryland

Scott M. Smith, PhD
Nutritionist and Manager
Nutritional Biochemistry Laboratory
Biomedical Research and Environmental Sciences Division
Human Health and Performance Directorate
NASA Johnson Space Center
Houston, Texas

Ryan G. Snodgrass, PhD
Research Molecular Biologist
Immunity and Disease Prevention Research Unit
USDA Western Human Nutrition Research Center
Davis, California

Abigail A. Sommer, PhD, MS
Graduate Research Fellow
Aeronautical and Astronautical Research Laboratories
The Ohio State University
Columbus, Ohio

Meir J. Stampfer, MD, DrPH
Research Professor
Harvard T.H. Chan School of Public Health
Professor
Harvard Medical School
Brigham and Women's Hospital
Boston, Massachusetts

Kimber L. Stanhope, PhD, MS, RD
Research Nutritional Biologist
School of Veterinary Medicine
University of California, Davis
Davis, California

Charles B. Stephensen, PhD
Research Leader and Center Director
Immunity and Disease Prevention Research Unit
USDA Western Human Nutrition Research Center
Davis, California

Patrick J. Stover, PhD
Professor and Director
Institute for Advancing Health Through Agriculture
Texas A&M University
College Station, Texas

Liang Sun, MD, PhD
Professor
Department of Nutrition and Food Hygiene
School of Public Health
Institute of Nutrition
Fudan University
Shanghai, China

Roger A. Sunde, PhD
Professor Emeritus
Department of Nutritional Sciences
University of Wisconsin-Madison
Madison, Wisconsin

Fred K. Tabung, PhD, MSPH
Assistant Professor
Department of Internal Medicine
College of Medicine and Comprehensive Cancer Center
The Ohio State University
Columbus, Ohio

Alice M. Tang, PhD, ScM
Professor
Department of Public Health and Community Medicine
Tufts University School of Medicine
Boston, Massachusetts

Sherry Tarleton, RD, CNSC
Clinical Dietitian
Mayo Clinic
Scottsdale, Arizona

Allen Taylor, PhD
Professor Emeritus
Friedman School of Nutrition Science & Policy
Tufts University
Boston Massachusetts

Beth Taylor, DCN, RD-AP, FASPEN, FCCM
Senior Clinical Research Coordinator
John T Milliken Department of Medicine
Washington University School of Medicine in St Louis
St. Louis, Missouri

Steve L. Taylor, PhD
Professor Emeritus and Founding Director
Food Allergy Research & Resource Program
Department of Food Science & Technology
University of Nebraska
Lincoln, Nebraska

Jennifer L. Temple, PhD
Professor
Department of Exercise and Nutrition Sciences
School of Public Health and Health Professions
University at Buffalo
Buffalo, New York

Diana Thomas, PhD, FTOS
Professor
Department of Mathematical Sciences
United States Military Academy
West Point, New York

Paul R. Thomas, EdD, RDN
Scientific Consultant
Office of Dietary Supplements
National Institutes of Health
Bethesda, Maryland

Madeline Timm, MS, RD
Graduate Research Assistant
College of Food, Agricultural, and Natural Resource Sciences
University of Minnesota—Twin Cities
St. Paul, Minnesota

Francisco Leonardo Torres-Leal, PhD, MS
Associate Professor
Department of Biophysics and Physiology
Federal University of Piaui
Teresina City, Brazil

Riva Touger-Decker PhD, RD, CDN, FADA
Associate Dean of Global Affairs
School of Health Professions
Professor and Director
Division of Nutrition
School of Dental Medicine
Rutgers University
Newark, New Jersey

Maret G. Traber, PhD
Professor Emeritus
Linus Pauling Institute
Oregon State University
Corvallis, Oregon

Isis Trujillo-Gonzalez, PhD
Assistant Professor
UNC Chapel Hill
Nutrition Research Institute
Chapel Hill, North Carolina

Paula R. Trumbo, PhD
Adjunct Professor
Liberty University
Lynchburg, Virginia

Katherine L. Tucker, PhD
Professor
Department of Biomedical & Nutritional Sciences
University of Massachusetts Lowell
Lowell, Massachusetts

Andrew Ukleja, MD
Director
Center for Human Nutrition
Department of Gastroenterology, Hepatology and Nutrition
Cleveland Clinic
Weston, Florida

Jaime Uribarri, MD
Professor
Department of Medicine
Icahn School of Medicine at Mount Sinai
New York, New York

Ayman A. Usmani, MD
Gastroenterology Fellow
Lahey Hospital and Medical Center
Burlington, Massachusetts

Pierre-Christian Violet, PhD
Staff Scientist
National Institute of Diabetes and Digestive and Kidney Diseases, NIH
Bethesda, Maryland

Jerry Vockley, MD, PhD
Professor and Cleveland Family Endowed Chair
Division of Genetic and Genomic Medicine
University of Pittsburgh School of Medicine
Director
Center for Rare Disease Therapy
UPMC Children's Hospital of Pittsburgh
Pittsburgh, Pennsylvania

Yael Vodovotz, PhD
Professor
Department of Food Science and Technology
The Ohio State University
Columbus, Ohio

Helen Walls, PhD, MPH
Associate Professor
Faculty of Public Health and Policy
London School of Hygiene and Tropical Medicine
London, United Kingdom

Xiang-Dong Wang, MD, PhD
Professor and Senior Scientist
Jean Mayer USDA Human Nutrition Research Center on Aging
Tufts University
Boston, Massachusetts

Kate A. Ward, PhD, BSc
Professor
MRC Lifecourse Epidemiology Unit
University of Southampton
Southampton, Hampshire, United Kingdom

Connie M. Weaver, PhD
Distinguished Research Professor
School of Exercise & Nutritional Sciences
San Diego State University
San Diego, California

Maxine Z. Webb, MCR, RD, LD, CCRP
Clinical Research Specialist
Comprehensive Cancer Center
The Ohio State University
Columbus, Ohio

Jasper Weinberg, PhD
Graduate Research Assistant
USDA Jean Mayer Human Nutrition Research Center on Aging
Tufts University
Boston, Massachusetts

Elizabeth A. Whitcomb, PhD
Scientist III
Jean Mayer USDA Human Nutrition Research Center on Aging
Tufts University
Boston Massachusetts

Walter Willett, MD, DrPH
Professor
Department of Nutrition
Harvard T. H. Chan School of Public Health
Boston, Massachusetts

Ellen Wingert, OTR/L
Manager
Feeding Day Program
Mt. Washington Pediatric Hospital
Baltimore, Maryland

Kate Wingrove, BNutrDiet, PhD, APD, RPHNutr
Research Fellow
School of Exercise and Nutrition Sciences
Deakin University
Geelong, Victoria, Australia

Lynne A. Wolfe, MS, CRNP
Senior Nurse Practitioner
National Human Genome Research Institute
National Institutes of Health
Bethesda, Maryland

Guoyao Wu, PhD
University Distinguished Professor
Department of Animal Science
Texas A&M University
College Station, Texas

Ann L. Yaktine, PhD, MS, RD
Director
Food and Nutrition Board
The National Academies of Sciences, Engineering, and Medicine
Washington, District of Columbia

Alejandra Yu Lee-Mateus, MD
Research Fellow
Division of Pulmonary, Allergy, and Sleep Medicine
Mayo Clinic
Jacksonville, Florida

Steven H. Zeisel, MD, PhD
Professor Emeritus
Nutrition Research Institute
The University of North Carolina at Chapel Hill
Chapel Hill, North Carolina

Janos Zempleni, PhD
Willa Cather Professor
College of Education and Human Sciences
Director
Nebraska Center for the Prevention of Obesity Diseases
University of Nebraska-Lincoln
Lincoln, Nebraska

Vivian M. Zhao, PharmD, BCNSP, FASPEN
Clinical Pharmacy Specialist II—Nutrition Support
Emory University Hospital
Atlanta, Georgia

Thomas R. Ziegler, MD
Professor
Division of Endocrinology, Metabolism, and Lipids
Director
Center for Clinical and Molecular Nutrition
Emory University School of Medicine
Atlanta, Georgia

Sara R. Zwart, PhD
Senior Scientist
University of Texas Medical Branch
Biomedical Research and Environmental Sciences Division
NASA Johnson Space Center
Houston, Texas

Reviewers

Amir Alakaam, PhD, RDN
Associate Professor
The University of Tennessee at Chattanooga
Chattanooga, Tennessee

Guoxun Chen, PhD
Associate Professor
Department of Nutrition
The University of Tennessee, Knoxville
Knoxville, Tennessee

Rawia Khasawneh, PhD, RDN
Assistant Professor
Nutrition and Dietetics Program
School of Rehabilitation and Medical Sciences
Herbert H. & Grace A. Dow College of Health Professions
Central Michigan University
Mount Pleasant, Michigan

Suzanne Kitts, PhD
Department Chair
Applied Exercise Science
Gannon University
Erie, Pennsylvania

David Lightsey, MS
Food and Nutrition Science Advisor to Quackwatch.org
Adjunct Professor
Bakersfield College
Bakersfield, California

Mary Katherine Lockwood, PhD, MS
Clinical Associate Professor
University of New Hampshire
Durham, New Hampshire

Libo Tan, PhD
Assistant Professor
University of Alabama
Tuscaloosa, Alabama

Annie Wetter, PhD
Professor
School of Health Sciences & Wellness
University of Wisconsin—Stevens Point
Stevens Point, Wisconsin

PART I

Specific Dietary Components

A. MAJOR DIETARY CONSTITUENTS

CHAPTER 1

Proteins and Amino Acids

Dwight E. Matthews

CHAPTER OUTLINE

Abbreviations

AV, arteriovenous; **AAS**, amino acid score; **BCAA**, branched-chain amino acid; **CO_2**, carbon dioxide; **CoA**, coenzyme A; **DAA**, dispensable amino acid; **DAAO**, direct amino acid oxidation; **EAR**, estimated average requirement; **FAO/WHO/UNU**, Food and Agriculture Organization/World Health Organization/United Nations University; **IAA**, indicator amino acid; **IAAO**, indicator amino acid oxidation; **IDAA**, indispensable amino acid; **KIC**, α-ketoisocaproate; **N**, nitrogen; **NH_3**, ammonia; **PDCAAS**, protein digestibility corrected amino acid score; **PER**, protein efficiency ratio; **RDA**, recommended dietary allowance; **SLC**, solute-linked carrier protein; **TCA**, tricarboxylic acid; **TML**, trimethyllysine; **VLDL**, very low-density lipoprotein

Introduction

Proteins are associated with all forms of life, and much of the effort to determine how life began has centered on how proteins were first produced. Amino acids joined together in long strings by peptide bonds form proteins that twist and fold in 3-dimensional space, producing centers to facilitate the biochemical reactions of life that either would run out of control or not run at all without them. Life could not have begun without enzymes, for which there are thousands of different types in the body. Proteins are prepared and secreted to act as cell-cell signals in the form of hormones and cytokines. Plasma proteins produced and secreted by the liver stabilize the blood by forming a solution of the appropriate viscosity and osmolarity. These secreted proteins also transport a variety of compounds through the blood.

The largest source of protein in higher animals resides in muscle. Through complex interactions, entire sheets of proteins slide back and forth to form the basis of muscle contraction and all aspects of our mobility. Muscle contraction provides for pumping oxygen and nutrients throughout the body, for inhalation and exhalation of our lungs, and for movement. Many of the underlying causes of noninfectious diseases are due to derangements in the sequence of proteins. Advances in molecular biology have provided tremendous information about DNA and RNA and introduced the field of genomics. This research is not driven to understand DNA per se, but to understand the purpose and function of the proteins that are translated from the genetic code. Now, we have the emerging field of proteomics that studies the expression, modification and regulation of proteins.

Three major classes of substrates are used for energy: carbohydrates, fat, and protein. Protein differs from the other two primary

Table 1.1 Body Composition of a 70-kg Man in Terms of Energy Components

Component	Mass (kg)	Energy (kcal)	Availability[a] (days)
Body water and minerals	49	0	0
Protein	6	24,000	13
Glycogen	0.2	800	0.4
Fat	15	140,000	78
Total:	70	164,800	91.4

[a]Availability is the duration for which the energy supply would last, assuming an 1800 kcal/day resting energy consumption.

Data from Cahill GF Jr. Starvation in man. N Engl J Med. 1970;282(12):668-675. doi:10.1056/NEJM197003192821209

sources of dietary energy by inclusion of nitrogen (N). Protein, on average, is 16% N by weight. The amino acids of proteins contain one N in the form of an amino group and additional N, depending on the amino acid. When amino acids are oxidized to CO_2 and water to produce energy, N is also produced as a waste product that must be eliminated via incorporation into urea. Conversely, N must be available when the body synthesizes amino acids *de novo*. Other N-containing compounds in the body (e.g., nucleic acids for DNA and RNA synthesis) get their N during synthesis from a donation of N from amino acids. Therefore, when we discuss amino acid metabolism in the body, we really mean N metabolism.

Protein and amino acids are also important to energy metabolism of the body. As Cahill has pointed out,[1] protein is the second largest store of energy in the body after adipose tissue fat stores (**Table 1.1**). Carbohydrate is stored as glycogen and, while important to short-term energy needs, is of very limited capacity to provide energy beyond a few hours. Amino acids from protein are converted to glucose by the process called gluconeogenesis, to provide a continuing supply of glucose after glycogen is consumed during fasting. On the other hand, protein stores must be conserved for the numerous critical roles in which protein functions in the body. Loss of more than about 30% of body protein results in reductions in muscle strength for breathing, immune function, and organ function, and can result in death. Hence, the body must adapt to fasting by conserving protein, as is seen by a dramatic decrease in N excretion within the first week of onset of starvation.

Body protein is made up of 20 different amino acids, each with different metabolic fates in the body, with different activities in different metabolic pathways in different organs, and with differing compositions in different proteins. When amino acids are liberated after absorption of dietary protein, the body makes a complex series of decisions concerning the fate of those amino acids: to oxidize them for energy, to incorporate them into proteins in the body, or to use them in the formation of other N-containing compounds. The purpose of this chapter is to elucidate the complex pathways and roles amino acids play in the body with a focus on nutrition.

Amino Acids

Basic Definitions

The amino acids that we are familiar with and all of those incorporated into mammalian protein are "alpha" amino acids. By definition, these amino acids have a carboxyl-carbon group and an amino N group attached to a central α-carbon (**Figure 1.1**). Amino acids differ in structure by the substitution of one of the two hydrogens on the α-carbon with another functional group. Amino acids can be characterized by their functional groups, which are often organized at neutral pH into the classes of (a) nonpolar, (b) uncharged but polar, (c) acidic (negatively charged), and (d) basic (positively charged) groups. Within any class, there are considerable differences in shape and physical properties. Thus, amino acids are often grouped into other functional subgroups. For example, amino acids with an aromatic group, phenylalanine, tyrosine, tryptophan, and histidine are often associated together, although tyrosine is clearly polar and histidine is also basic. Other common groupings are the aliphatic or neutral amino acids (glycine, alanine, isoleucine, leucine, valine, serine, threonine, and proline). Proline is different in that it forms a five-membered ring that includes the amino group. Due to the ring, proline is really an *imino* acid, not an amino acid. Serine and threonine contain hydroxyl groups. There is another important subgroup: the branched-chain amino acids (BCAAs: isoleucine, leucine and valine), which share common enzymes for the first two steps of their degradation. The acidic amino acids, aspartic and glutamic acids, are often referred to as their ionized, salt forms: *aspartate* and *glutamate*. These amino acids become *asparagine* and *glutamine* when an amine is added in the form of an amide group to their carboxyl tails.

The sulfur-containing amino acids are methionine and cysteine. Cysteine is often found in the body as an amino acid dimer called *cystine* where the thiol groups (the two sulfur atoms) are connected to form a di-sulfide bond. Particular attention should be paid when reading the literature to note the distinction between *cysteine* and *cystine*, because the former is a single amino acid, and the latter is a dimer with different properties. Other amino acids that contain sulfur, such as homocysteine, are not incorporated into protein.

All amino acids exist as charged particles in solution: in water, the carboxyl group rapidly loses a hydrogen to form a carboxyl anion (negatively charged), while the amino group gains a hydrogen to become positively charged. The amino acids, therefore, become "bipolar" (often called a *zwitterion*) in solution, but without a net charge (the positive and negative charges cancel each other). However, the attached functional group may distort that balance. The acidic amino acids will lose the hydrogen on the second carboxyl group and become negatively charged in solution. In contrast, the basic amino acids will, in part, accept a hydrogen on the second N and form a molecule with a net positive charge. Although the other amino acids do not specifically accept or donate additional hydrogens in neutral solution, their functional groups do influence the relative polarity and acid-base nature of the bipolar portion of the amino acids, giving each amino acid different properties in solution.

The functional groups of amino acids also vary by size. Molecular weights of the amino acids are shown in **Table 1.2**. Amino acids range from the smallest, glycine, to the large and bulky molecule, tryptophan. Most amino acids crystallize as uncharged molecules when purified and dried. The molecular weights shown in Table 1.2 are reflective of their molecular weight as crystalline amino acids. However, the basic and acidic amino acids tend to form more stable crystals as salts, rather than as free amino acids. Glutamic acid can be obtained as a free amino acid with a molecular weight of 147, and as its sodium salt, monosodium glutamate (MSG), which has a crystalline weight of 169. Lysine is typically found as a hydrogen chloride-containing salt. Therefore, when amino acids are represented by weight, it is important to know whether the weight is based on the free amino acid or on its salt.

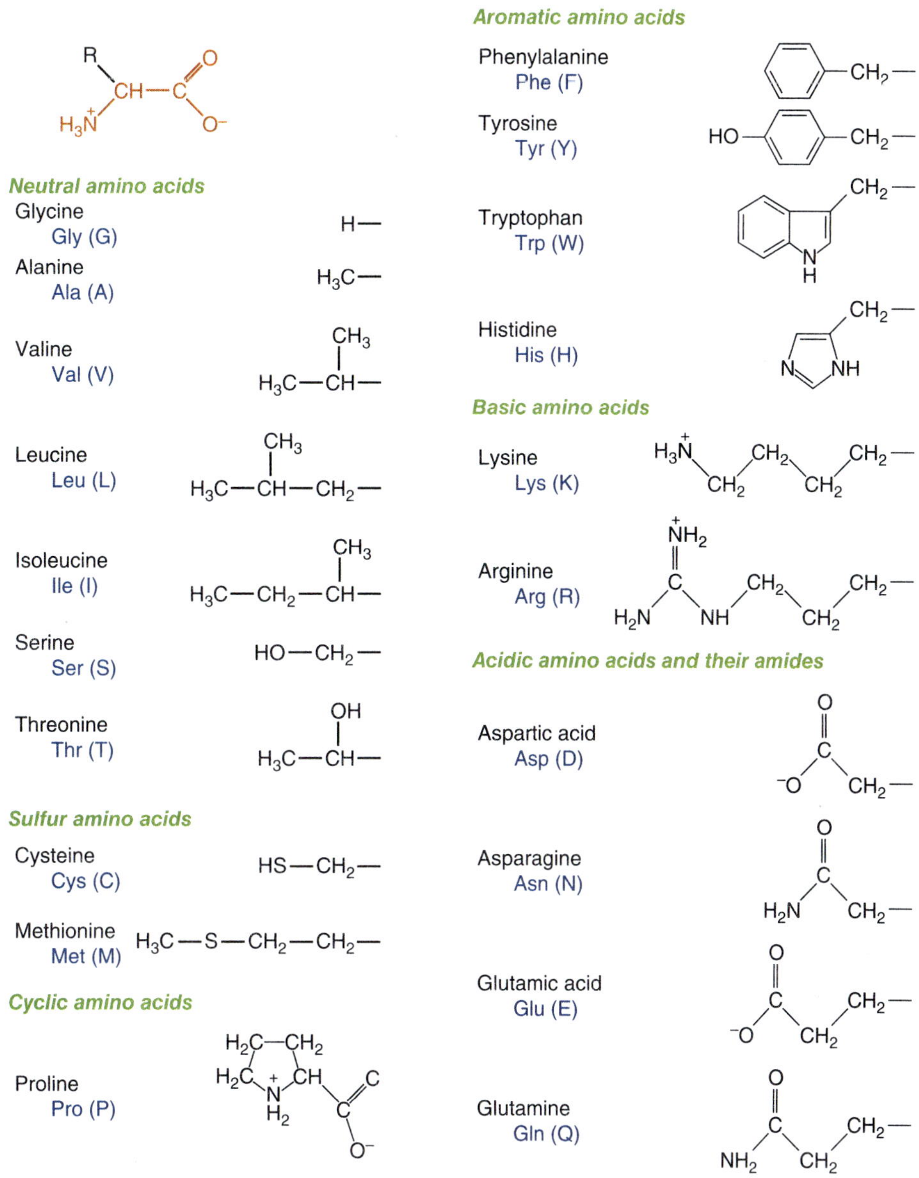

Figure 1.1 Structural formulae of the 21 common α-amino acids. The α-amino acids all have (i) a carboxyl group, (ii) an amino group, and (iii) a differentiating functional group attached to the α-carbon. The generic structure of amino acids is shown in the upper left corner with the differentiating functional marked by an "R." The functional group for each amino acid is shown below. Amino acids have been grouped by functional class. Proline is actually an imino acid due to its cyclic structure involving its N.

Another important property of amino acids is optical activity. Except for glycine, all amino acids have at least one chiral center: the α-carbon. The term "chiral" comes from Greek for *hand* in that these molecules have a left ("*levo*" or "L") or right ("*dextro*" or "D") handedness to them around the α-carbon atom. Because of the tetrahedral structure of the carbon bonds, there are two possible arrangements of a carbon center, with the same four different groups bonded to it that are not superimposable. The two configurations, called *stereoisomers*, are mirror images of each other. The body recognizes only the "L" form of amino acids for most reactions in the body, although some enzymatic reactions will operate with a lower efficiency when given the "D" form. Because we do encounter some "D" form amino acids in the foods that we eat, the body has mechanisms for clearing these amino acids through renal filtration.

Any number of molecules could be designed that complete the basic definition of an amino acid: a molecule with a central carbon to which an amino group, a carboxyl group, and a functional group are attached. However, there are relatively few appearing in nature that are incorporated directly into mammalian protein. Amino acids are selected for protein synthesis when coupled to transfer RNA (tRNA). To synthesize protein, strands of DNA are transcribed into messenger RNA (mRNA). tRNA binds to mRNA in three-base groups. Different combinations of three consecutive RNA molecules in the mRNA code for different tRNA molecules. However, the three-base combinations of mRNA are recognized by only 20 different tRNA molecules, and 20 different amino acids are incorporated into protein during protein synthesis. There is one exception, selenocysteine, the serine equivalent where the serine hydroxyl has been replaced with a selenide. Selenocysteine is incorporated into several specific proteins and has been termed the "21st amino acid" (see Chapter 14 for a discussion of selenium and selenocysteine).

Table 1.2 Common Amino Acids in the Body

	Standard Abbreviation		
	3-Letter	1-Letter	Molecular Weight[a]
Indispensable (essential) amino acids			
Isoleucine	Ile	I	131.2
Leucine	Leu	L	131.2
Lysine	Lys	K	146.2
Methionine	Met	M	149.2
Phenylalanine	Phe	F	165.2
Threonine	Thr	T	119.1
Tryptophan	Trp	W	204.2
Valine	Val	V	117.2
Histidine[b]	His	H	155.2
Dispensable (nonessential) amino acids			
Alanine	Ala	A	89.1
Arginine	Arg	R	174.2
Aspartic acid	Asp	D	133.2
Asparagine	Asn	N	132.2
Glutamic acid	Glu	E	147.2
Glutamine	Gln	Q	146.2
Glycine	Gly	G	75.1
Proline	Pro	P	115.1
Serine	Ser	S	105.1
Conditionally dispensable amino acids			
Cysteine	Cys	C	121.2
Tyrosine	Tyr	Y	181.2
Some special amino acids			
Ornithine	Orn		132.2
Citrulline	Cit		175.2
Homocysteine	Hcy		135.2
Hydroxylysine	Hyl		162.2
Hydroxyproline	Hyp		131.2
3-Methylhistidine			169.2
Selenocysteine	Sec	U	168.1

[a]Molecular weight (Daltons) is rounded to the nearest tenth and represents the number of grams per mole of amino acid. Because glutamine is degraded to glutamate when proteins are hydrolyzed, the sum of the glutamine and glutamate together is often abbreviated *Glx*. The same is true also for the sum of asparagine and aspartate: *Asx*. The 1-letter abbreviations are often used to indicate protein sequences.

[b]The essentiality for histidine has only been shown for infants, but probably small amounts are needed for adults as well. To date the indispensability of histidine has not been documented in healthy adults.[2]

Of the 20 common amino acids in proteins, some are synthesized *de novo* in the body, either from other amino acids or simpler precursors. These *nonessential* amino acids may be deleted from our diet without impairing health or blocking growth; they are also termed *dispensable amino acids (DAAs)* from the diet. However, no pathways exist for the synthesis of several other amino acids in humans, and hence, these amino acids are *essential* or *indispensable amino acids (IDAAs)*. The classification of amino acids as nondispensable/dispensable or essential/nonessential for humans is shown in Table 1.2. Both the standard three-letter abbreviation and the one-letter abbreviation used in representing amino acid sequences in proteins are also presented in Table 1.2 for each amino acid. Some DAAs may become *conditionally indispensable* under conditions when synthesis becomes limited or when adequate amounts of precursors are unavailable to meet the needs of the body.[3-5] The history and rationale of the classification of amino acids in Table 1.2 will be discussed in greater detail later in this chapter.

Several other amino acids appear in blood and in cells besides the amino acids that are recognized by tRNA for incorporation into protein. These amino acids have important metabolic functions. Examples are ornithine and citrulline, which are linked to arginine through the urea cycle. Other amino acids appear as modifications of amino acids after they have been incorporated into proteins. Examples are hydroxyproline and hydroxylysine, which are produced when proline and lysine residues in collagen protein are hydroxylated, and 3-methylhistidine, which is produced by post-translational methylation of select histidine residues of actin and myosin proteins. Because there is no tRNA to code for these amino acids, they cannot be reused when a protein containing them is broken down (hydrolyzed) to its individual amino acids.

Amino Acid Pools and Distribution

The distribution of amino acids is complex. Not only are different amino acids incorporated into a variety of proteins in a variety of different organs in the body, but amino acids are consumed in the diet from a variety of protein sources. In addition, each amino acid is maintained in part as a free amino acid in solution in blood and inside cells. Overall, there is a wide range of concentrations among amino acids across the various protein and free pools that exist. We consume protein in food that is enzymatically hydrolyzed in the alimentary tract, releasing free individual amino acids that are then absorbed by the gut lumen and transported into portal blood. Amino acids then pass into the systemic circulation and are extracted by different tissues. Although the concentrations of individual amino acids vary among different free pools, such as plasma and intracellular muscle, the abundance of individual amino acids is relatively constant in a variety of proteins throughout the body and nature. **Table 1.3** shows the composition of amino acids in hen egg protein, mammalian muscle and liver proteins[6] and human milk.[2] The data are expressed as *moles* of amino acid. The historical expression of amino acids is on a weight basis (e.g., *grams* of amino acid). Comparing amino acids by weight skews the comparison toward the heaviest amino acids, making them appear more abundant than they are. For example, tryptophan (molecular weight 204) appears almost three times as abundant to glycine (molecular weight 75) when quoted in terms of weight.

An even distribution of all 20 amino acids would be 5% per amino acid per protein, and the median amino acid content centers around this value for the proteins shown in Table 1.3. Tryptophan is the least common amino acid in many proteins.

Table 1.3 Amino Acid Composition of Several Different Protein Sources

Amino Acid	Composition (μmol/g protein)			
	Hen	Mammalian		
	Hen Egg	Muscle	Liver	Human Milk
Alanine	810	730	750	426
Arginine	360	380	328	132
Aspartate + Asparagine	530	600	600	679
Cysteine	190	120	140	182
Glutamate + Glutamine	810	990	800	1206
Glycine	450	670	610	306
Histidine	150	180	170	148
Isoleucine	490	360	380	434
Leucine	650	610	690	770
Lysine	425	580	510	472
Methionine	200	170	170	107
Phenylalanine	340	270	310	242
Proline	350	430	430	695
Serine	770	480	510	476
Threonine	410	390	390	395
Tryptophan	80	55	80	88
Tyrosine	220	170	200	259
Valine	600	470	520	538

Data from Block RJ, and Weiss KW. (1956). Amino Acid Handbook: Methods and Results of Analysis, Charles C. Thomas, Springfield, Ill.; FAO/WHO/UNU. (2007). Protein and Amino Acid Requirements in Human Nutrition Vol. technical report series #935, World Health Organization, Geneva, Switzerland;

Table 1.4 Typical Concentrations of Free Amino Acids in the Body

Amino Acid[a]		Concentration (mM)		Gradient Intracellular/ Plasma
		Plasma	Intracellular Muscle	
Aspartic acid	D	0.02		
Phenylalanine	I	0.05	0.07	1.4
Tyrosine	CI	0.05	0.10	2.0
Methionine	I	0.02	0.11	5.5
Isoleucine	I	0.06	0.11	1.8
Leucine	I	0.12	0.15	1.3
Cysteine	CI	0.11	0.18	1.6
Valine	I	0.22	0.26	1.2
Ornithine		0.06	0.30	5.0
Histidine	I	0.08	0.37	4.6
Asparagine	D	0.05	0.47	9.4
Arginine	D	0.08	0.51	6.4
Proline	D	0.17	0.83	4.9
Serine	D	0.12	0.98	8.2
Threonine	I	0.15	1.03	6.9
Lysine	I	0.18	1.15	6.4
Glycine	D	0.21	1.33	6.3
Alanine	D	0.33	2.34	7.1
Glutamic acid	D	0.06	4.38	73.0
Glutamine	D	0.57	19.45	34.1
Taurine[b]		0.07	15.44	221.0

[a] I = indispensable; D = dispensable; CI = conditionally indispensable.
[b] Taurine is not an amino acid per se but is highly concentrated in free form in muscle.

Data from Bergström J, Fürst P, Noree LO, et al. (1974). Intracellular free amino acid concentration in human muscle tissue. J Appl Physiol 1974;36:693-7

Considering the effect that tryptophan's large size has upon the conformation of proteins, it is not surprising to find less tryptophan in protein. Other amino acids of modest size and limited polarity, such as alanine, leucine, serine, and valine, are relatively abundant in protein (8%-10% per amino acid). While the abundance of the IDAAs is similar across the protein sources in Table 1.3, there are a variety of vegetable proteins that are deficient or low in some IDAAs. In the body, there are also a variety of proteins that are particularly rich in specific amino acids to produce specific functions in the protein. For example, collagen is a fibrous protein abundant in connective tissues in tendons, bone, and muscle. Collagen fibrils are arranged in different ways, depending on the functional type of collagen. Glycine comprises about one-third of collagen, and there is also considerable proline and hydroxyproline (proline converted after it has been incorporated into collagen). The glycine and proline residues allow the collagen protein chain to turn tightly and intertwine, and the hydroxyproline residues provide for hydrogen bond cross-linking. Generally, the alterations in amino acid concentrations do not vary dramatically among proteins as they do in collagen, but such examples demonstrate the diversity and functionality of the different amino acids in proteins.

The abundance of amino acids varies among amino acids over a far wider range in the free pools of extracellular and intracellular compartments. Typical values of free amino acid concentrations in plasma and in intracellular muscle are shown in **Table 1.4**. The primary points of Table 1.4 are: (1) amino acid concentrations vary widely, and (2) free amino acids are generally concentrated inside cells. Although there is a significant correlation between plasma and muscle-free intracellular amino acids, the relationship is not linear.[7] Concentrations of plasma amino acids range from as low as approximately 20 μM for aspartic acid and methionine, to a high of approximately 500 μM for glutamine. The median concentration for plasma amino acids is 100 μM. There is no defined relationship between amino acid concentration and type of amino acid (e.g., the three similar BCAAs have plasma concentrations ranging from 50-250 μM). One notable point is that the concentrations of the acidic amino acids, aspartate, and glutamate, are very low outside cells (e.g., plasma). In contrast, the concentration of glutamate is among the highest inside cells (e.g., muscle).

It is important to note the differences in relative amounts of N contained in extracellular and intracellular amino acid pools and in protein itself. For a normal person, there are ≈55 mg of amino acid N per liter in extracellular space, and an ≈800 mg of amino acid N/liter inside cells, which means that free amino acids are ≈15-fold more abundant inside than outside cells.[7] The second point is that the total pool of free amino acid N is small compared with protein-bound amino acids. Multiplying the free pools by estimates of extracellular water (0.2 l/kg) and intracellular water (0.4 l/kg) provides a measure of the total amount of N present in free amino acids: 0.33 g N/kg body weight. In contrast, body composition studies have shown that the N content of the body is 24 g N/kg body weight.[8,9] Therefore, free amino acids make up about 1% of the total amino N pool vs 99+% of the amino acids that reside in proteins.

Amino Acid Transport

The gradient of amino acids within and outside cells is maintained by active transport. Table 1.4 suggests that different transport mechanisms must exist for different amino acids to produce the range of concentration gradients observed. Different types and groups of amino acids have different transporters.[10-13] Amino acid transport is one of the more difficult areas of amino acid metabolism to quantify and characterize. The affinities of the transporters and their mechanisms of transport define the intracellular concentrations of the amino acids. In general, the IDAAs have lower intracellular/extracellular gradients than the DAAs (Table 1.4). Amino acid transporters are membrane-bound proteins that recognize different amino acid shapes and chemical properties (e.g., neutral, basic, or anionic). Transport occurs both into and out of cells, and may be thought of as a process that sets the intracellular/extracellular gradient; or the transporters may be thought of as processes that set the rates of amino acid influx into and efflux from cells that then define the intracellular/extracellular gradients. It is the gradient (e.g., intracellular muscle amino acid levels) that is readily measured; the inward/outward fluxes of amino acids are much more difficult to define.

Most information concerning transport has been accrued through kinetic studies of membranes using amino acids and competitive inhibitors or amino acid analogues to define and characterize individual systems. However, these methods are difficult to apply because a single transporter will transport several different amino acids with different efficiencies, and a single amino acid may be transported by several different transporters. These various transporters were given abbreviation identifiers based on the types of amino acid they transport (e.g., neutral, anionic, basic). This classical view changed with the advent of genomics and assignment of protein function based on sequence similarities with proteins of known function. Thus, transporters were reclassified into solute-linked carrier (SLC) families of membrane transport proteins.[14] Individual membrane transport proteins were then numbered after their family assignment. Most of these proteins already had gene name assignments, and an individual transporter may be seen under several different names. For example, the "excitatory amino acid transporters" of anionic amino acids go by the gene name EEAT and fall within the SLC1A family. The transporters originally classified as for transport of glutamate and aspartate are EAAT1 assigned as SLC1A3 and EAAT3 assigned as SLC1A1.

Table 1.5 highlights a broad range of important amino acid transporters and the amino acids they transport, but this list is not inclusive of all transporters. Transporters fall into two classes: sodium-dependent and sodium-independent carriers. The sodium-dependent carriers co-transport a sodium atom into the cell along with the amino acid. The high extracellular/intracellular sodium gradient (140 mEq/L outside and 10 mEq/L

Table 1.5 Examples of Important Amino Acid Transporters

SLC Family	Gene Identifier	Older Identifier	Amino Acids Transported
Sodium dependent			
1A1-3 & 1A6,7	EAAT1-5	X_{AG^-}	Anionic (Glu, Asp)
1A4,5	ASCT1,2	ASC	Small neutrals (Ala, Ser, Cys)
6A5,9	GLYT1,2	Gly	Gly, sarcosine
6A15,19	B^0AT1,2	B	Most neutrals
38A2,4	SNAT2,4	A	Small neutrals (Ala, Ser)
38A3,5	SNAT3,5	N	Gln, Asn, His
Sodium independent			
1A4,5	ASCT1,2	asc	Small neutrals (Ala, Ser, Cys, Thr)
7A1-3	CAT1-3	y^+	Cationic (Arg, Lys, Orn)
16A10	TAT1	T	Aromatics (Trp, Phe, Tyr)
7A5,8 & 43A1,2	LAT1-4	L	Large neutrals (Leu, Ile, Val, Met, Phe, Tyr, Trp, His)

Data from Poncet N, and Taylor PM. (2013). Curr. Opin. Clin. Nutr. Metab. Care 16, 57-65; Gauthier-Coles G, Vennitti J, Zhang Z, et al. (2021) Nat. Commun. 12, 5282; Broer S, and Palacin M. (2011). Biochem. J. 436, 193–211.

inside) facilitates the inward transport of amino acids by sodium-dependent carriers. These transporters generally produce larger gradients and accumulations of amino acids inside cells. The sodium entering the cell can be transported out via the sodium-potassium pump that transports a potassium ion into the cell and removal of a sodium ion from the cell.

Some overall generalizations can be made in terms of the type of amino acid transported by a given carrier in Table 1.5, but the amino acid transport system is not readily simplified. One generalization is that amino acids highly concentrated within cells are transported via sodium-dependent transporters. For example, anionic transporter (X_{AG}^-, EAAT1-5) produces a steep intracellular/extracellular gradient for the dicarboxylic amino acids, glutamate, and aspartate. In contrast, bulky, neutral, amino acids (the BCAAs, phenylalanine, methionine, and histidine) are transported by system L (LAT1-4) that is sodium independent, operates with a high rate of exchange, and produces small intracellular/extracellular gradients. Amino acid transporters are sensitive to the charge of the amino acid, with dedicated transporters for basic (CAT1-3) and anionic (EAAT1-5) amino acids.

Pathways of Amino Acid Synthesis and Degradation

Several amino acids have their metabolic pathways linked to the metabolism of other amino acids. These co-dependencies that link the pathways of amino acids become important when nutrient intake is limited or when metabolic requirements are increased. There are two aspects of metabolism to be reviewed here: (i) synthesis of amino acids and (ii) amino acid degradation. Degradation serves two useful purposes: (i) production of energy from the oxidation of individual amino acids (≈4 kcal/g protein, almost the same energy production as for carbohydrate), and (ii) conversion of amino acids into other products. The latter is also related to amino acid synthesis: the degradation pathway of one amino acid may be the synthetic pathway of another amino acid. The other important aspect of amino acid degradation is production of other non-amino acid N-containing compounds in the body. The need for synthesis of these compounds may also drain the pools of their amino acid precursors, increasing the need for these amino acids in the diet. When amino acids are degraded for energy rather than converted to other compounds, the ultimate products become CO_2, water, and urea. CO_2 and water are produced through classical pathways of intermediary metabolism involving the tricarboxylic acid (TCA) cycle. Urea is produced because other forms of waste N, such as NH_3, are toxic if their concentrations rise in the blood and inside cells. Urea production is a means of removal of waste N from the oxidation of amino acids in the form of a non-toxic, water-soluble compound.

Detailed descriptions of amino acid pathways can be found in standard textbooks of biochemistry, but keep in mind when consulting such texts that pathways for nonmammalian systems (e.g., *E. coli* and yeast) are often presented, and these pathways may have little importance to human biochemistry. Discussion here is relevant to human biochemistry, including routes of degradation of each amino acid when the pathway is directed toward oxidation of the amino acid for energy, followed by pathways of amino acid synthesis and, finally, use of amino acids for other important compounds in the body.

Amino Acid Degradation Pathways

Complete amino acid degradation ends up with the production of N, which is removed by incorporation into urea. Carbon skeletons are eventually oxidized as CO_2 via the TCA cycle (also known as the Krebs or citric acid cycle). Inputs to the cycle are acetyl CoA and oxaloacetate forming citrate, which is degraded to α-ketoglutarate and then to oxaloacetate. The carbon skeletons from amino acid may enter the Krebs cycle via acetate as acetyl CoA or via oxaloacetate/α-ketoglutarate. These latter two precursors are direct metabolites of the amino acids aspartate and glutamate. Alternatively to the complete oxidation of the carbon skeletons to CO_2 is the use of these carbon skeletons for the formation of fat and carbohydrate. Fat is formed from elongation of acetyl units, and amino acids whose carbon skeletons degrade to acetyl CoA and ketones may be used for synthesis of fatty acids. Glucose is split in glycolysis to pyruvate, and pyruvate is the immediate product of alanine. Pyruvate may be converted back to glucose by elongation to oxaloacetate. Amino acids whose degradation pathways go toward formation of pyruvate, oxaloacetate, or α-ketoglutarate may be used for synthesis of glucose. Therefore, the degradation pathways of many amino acids can be partitioned into two groups with respect to the disposal of their carbon: amino acids whose carbon skeleton may be used for synthesis of glucose (gluconeogenic amino acids) or those whose carbon skeletons degrade for potential use for fatty acid synthesis (ketogenic amino acids).

It is of particular interest to note that the amino acids that degrade directly to the primary gluconeogenic and TCA cycle precursors, pyruvate, oxaloacetate, and α-ketoglutarate, do so by rapid and reversible transamination reactions. For example, by the enzyme aspartate aminotransferase:

$$\text{L-aspartate} + \alpha\text{-ketoglutarate} \leftrightarrows \text{oxaloacetate} + \text{L-glutamate}$$

and by the enzyme alanine aminotransferase:

$$\text{L-alanine} + \alpha\text{-ketoglutarate} \leftrightarrows \text{pyruvate} + \text{L-glutamate}$$

The amino-N of these three amino acids rapidly exchange and each amino acid is rapidly converted to/from a primary compound of gluconeogenesis and the TCA cycle. As discussed below, compartmentation among different organ pools is the only limiting factor for complete and rapid exchange of the N of these amino acids.

The BCAAs, leucine, isoleucine, and valine, share the first two steps in their degradation pathway. First, there is reversible transamination with α-ketoglutarate to form glutamate and a branched-chain keto acid:

$$\left.\begin{array}{l}\text{Leucine}\\\text{Isoleucine}\\\text{Valine}\end{array}\right\} + \alpha\text{-ketoglutarate} \leftrightarrows \text{glutamate} + \left\{\begin{array}{l}\alpha\text{-ketoisocaproate}\\\alpha\text{-keto-}\beta\text{-methylvalerate}\\\alpha\text{-ketovalerate}\end{array}\right.$$

The reversible transamination to keto acids is then followed by irreversible decarboxylation of the carboxyl group to liberate CO_2. The BCAAs are the only IDAAs that undergo transamination and, therefore, are unique among IDAAs.

Together, the BCAAs, alanine, aspartate, and glutamate make up the pool of amino-N that can move among amino acids via reversible transamination. Glutamic acid is central to the transamination process (**Figure 1.2**). In addition, N can leave the transaminating pool via removal of the glutamate N

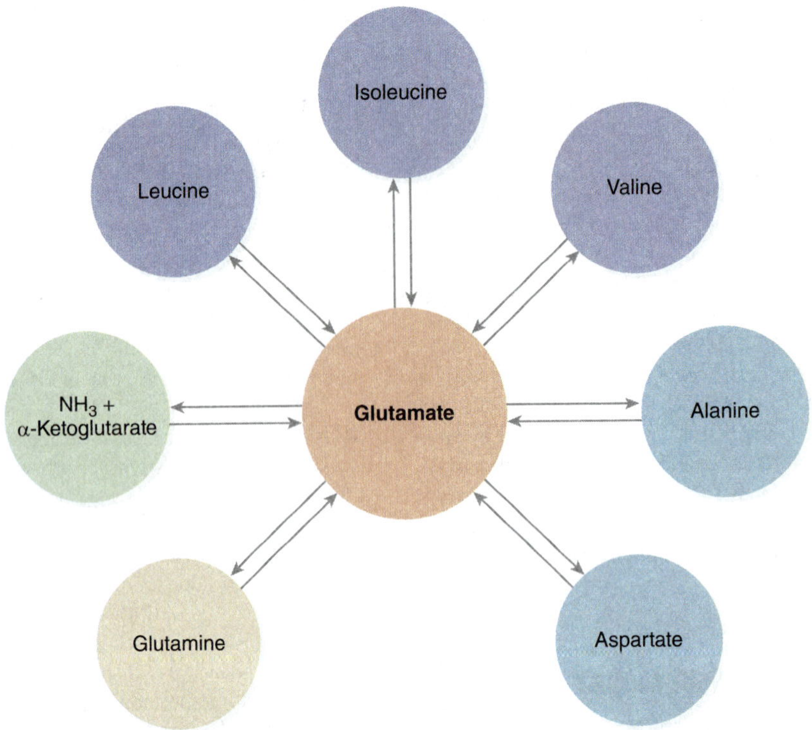

Figure 1.2 Movement of amino-N around glutamic acid. Glutamate undergoes reversible transamination with several amino acids. Nitrogen is also removed from glutamate by glutamate dehydrogenase producing an α-ketoglutarate and an NH_3. In contrast, the enzyme glutamine synthetase will add an NH_3 to glutamate to produce glutamine. Glutamine is degraded back to glutamate by liberation of the amide-N to release NH_3 by a different enzymatic pathway (glutaminase).

via glutamate dehydrogenase or enter by the reverse process. The amino acid glutamine is intimately tied to glutamate as well: all glutamine is made from amidation of glutamate and glutamine is degraded by removal of the amide-N to form NH_3 and glutamate. A similar process occurs for formation and degradation of asparagine from aspartate.

Glutamate's central role in N metabolism becomes even clearer when we look at how urea is synthesized in the liver. The inputs into the urea cycle are CO_2, ATP, and NH_3 to form carbamoyl phosphate, which condenses with ornithine to form citrulline (**Figure 1.3**). The second N enters via aspartate to form arginosuccinate, which is then cleaved into arginine and

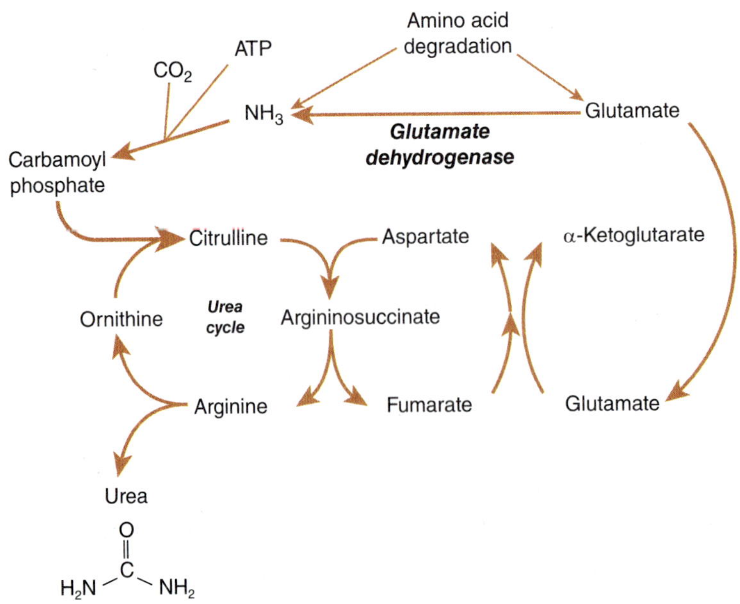

Figure 1.3 Urea cycle disposal of amino acid N. Urea synthesis incorporates one N from NH_3 and another from aspartate. Ornithine, citrulline, and arginine sit in the middle of the cycle. Glutamate is the primary source for the aspartate N; glutamate is also an important source of the NH_3 into the cycle.

fumarate. Arginine is hydrolyzed by arginase to ornithine, liberating urea. The resulting ornithine can reenter the urea cycle. Some amino acids may liberate NH_3 directly (e.g., glutamine, asparagine, and glycine), but most transfer N through glutamate first, which is then degraded to α-ketoglutarate and NH_3. The pool of aspartate is small in the body, and aspartate cannot be the primary transporter of the second N into urea synthesis. Rather, aspartate must act like arginine and ornithine: as a vehicle for the introduction of the second N. If so, the second N is delivered by transamination via glutamate, again placing glutamate at another integral point in the degradative disposal of amino acid N.

The major pathways for degradation, including the primary end-products, are presented in **Table 1.6**. The individual steps may be found in textbooks of biochemistry. Because of the importance of transamination, most of the N from amino acid degradation appears via N transfer to α-ketoglutarate to form glutamate. In some cases, the aminotransferase catalyzes the transamination reaction with glutamate bidirectionally, as indicated in Figure 1.2, and these enzymes are distributed in multiple tissues. In other cases, the transamination reactions are liver specific, compartmentalized, and act to degrade, not reversibly exchange N. For example, when leucine labeled with the stable isotope tracer ^{15}N was infused into dogs for 9 hours, considerable amounts of the ^{15}N tracer were found in circulating glutamine + glutamate, alanine, the other two BCAAs, but not in tyrosine,[15] indicating that the transamination of tyrosine did not proceed backward.

The primary pathways of the metabolism of some amino acids are blurry for humans. For example, two pathways for cysteine are shown. Both are active, but how much cysteine is metabolized by which pathway is not as clear. Methionine is metabolized by conversion to homocysteine. Homocysteine is not directly converted to cysteine, rather homocysteine condenses

Table 1.6 Pathways of Amino Acid Degradation

Metabolic Pathway	Important Enzymes	Nitrogen End-Products	Carbon End-Products
Amino acids converted to other amino acids			
Asparagine	*Asparaginase*	Aspartate + NH_3	
Glutamine	*Glutaminase*	Glutamate + NH_3	
Arginine	*Arginase*	Ornithine + Urea	
Phenylalanine	*Phenylalanine hydroxylase*	Tyrosine	
Proline		Glutamate	
Serine	*Serine hydroxymethyltransferase*	Glycine	
Cysteine		Taurine	
Amino acids transaminating to form glutamate			
Alanine		Glutamate	Pyruvate
Aspartate		Glutamate	Oxaloacetate
Cysteine		Glutamate	Pyruvate + SO_4^{-2}
Isoleucine		Glutamate	Succinate
Leucine		Glutamate	Ketones
Ornithine		Glutamate	α-Ketoglutarate
Serine		Glutamate	3-Phosphoglycerate
Valine		Glutamate	Succinate
Tyrosine		Glutamate	Ketone + fumarate
Other pathways			
Glycine		NH_3	CO_2
Histidine		NH_3	Urocanate
Methionine		NH_3	Ketobutryate
Serine	*Serine dehydratase*	NH_3	Pyruvate
Threonine	*Serine dehydratase*	NH_3	Ketobutyrate
Tryptophan		NH_3	Kynurenine
Lysine		2 Glutamates	Ketones

with a serine to form cystathionine, which is then broken apart to liberate cysteine, NH_3 and ketobutyrate. However, the original methionine molecule appears as NH_3 and ketobutyrate. The cysteine carbon-skeleton comes from the serine. The entry in Table 1.6 shows methionine degraded to NH_3, yet this degradation pathway is the major synthesis pathway for cysteine. Because of the importance of the sulfur-containing amino acids, an extensive discussion of the metabolic pathways of these amino acids may be found in Chapter 33.

Glycine is degraded by more than one possible pathway. However, the primary pathway appears to be the glycine cleavage enzyme system that breaks glycine into CO_2 and NH_3 and transfers a methylene group to tetrahydrofolate.[16] This pathway has been shown to be the prominent pathway in rat liver and in other vertebrate species.[17] Although this reaction degrades glycine, its importance is the production of a methylene group that can be used in other metabolic reactions.

Synthesis of Dispensable Amino Acids

IDAAs are those amino acids that cannot be synthesized in sufficient amounts in the body and, therefore, must be in the diet in sufficient amounts to meets the body's needs. Therefore, discussion of amino acid synthesis applies only to the DAAs. DAA synthesis falls into two groups: (i) amino acids that are synthesized by transferring a N to a carbon skeleton precursor that has come from the TCA cycle or from glycolysis of glucose and (ii) amino acids that are synthesized specifically from other amino acids. Because this latter group of amino acids depends on the availability of other specific amino acids, these amino acids are particularly vulnerable to becoming indispensable if the dietary supply of a precursor amino acid becomes limiting. In contrast, the former group is rarely limited in synthesis because of the ample precursor availability of carbon skeletons from the TCA cycle and from the labile amino-N pool of transaminating amino acids.

Pathways of DAA synthesis are shown in **Figure 1.4**. As with amino acid degradation, glutamate is central to the synthesis of several amino acids by providing N. Glutamate, alanine, and aspartate may share amino-N, transaminating back and forth among them (Figure 1.2). As in Figure 1.4, glutamate derives its N from NH_3 with α-ketoglutarate, and that glutamate goes on to promote the synthesis of other amino acids. Kitagiri and Nakamura[18] argue that we have little capacity to form glutamate from NH_3 and that the primary source of glutamate N comes from other amino acids via transamination, ultimately from dietary protein intake. Under circumstances of adequate dietary intake, the transaminating amino acids shown in Figure 1.2 supply more than adequate amino-N to glutamate. The transaminating amino acids act to provide a buffer pool of N that can absorb an increase in N from increased degradation, or supply N when there is a drain. From this pool, glutamate provides material to maintain synthesis of ornithine and proline, of which the latter is particularly important in protein synthesis of collagen and related proteins.

Serine is produced from 3-phosphoglycerate, which comes from glycolysis of glucose. Serine may then be used to produce glycine through a process that transfers a methylene group to tetrahydrofolate. This pathway is listed in Table 1.6 as a degradative pathway for serine, but it is also a source of glycine and one-carbon unit generation.[16,17] On the other hand, this pathway actively operates backward, forming serine from glycine in humans. When [^{15}N]glycine is given orally, the primary transfer of ^{15}N is to serine.[19] Therefore, there is significant

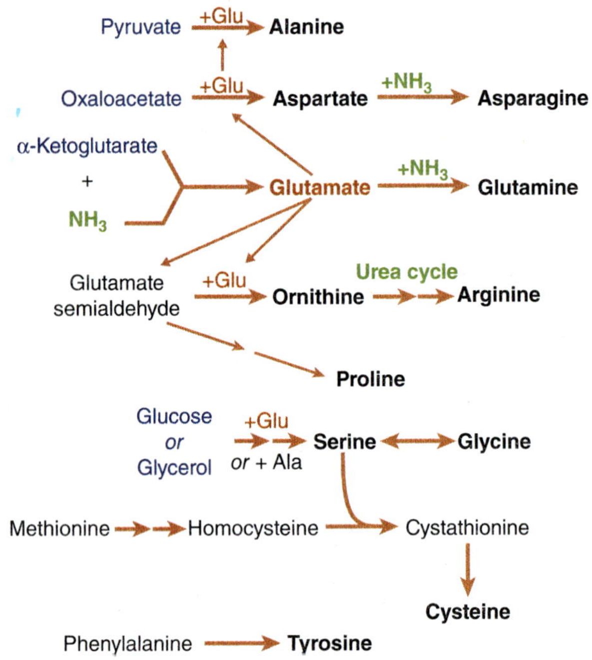

Figure 1.4 Pathways of the synthesis of DAAs. Glutamate is produced from NH_3 and α-ketoglutarate. That glutamate becomes the N source added to carbon precursors (pyruvate, oxaloacetate, glycolysis products of glucose and glycerol) to form most of the other DAAs. Cysteine and tyrosine are different in that they require IDAA input for their production.

reverse synthesis of serine from glycine. The other major place where ^{15}N appeared was in glutamate and glutamine, indicating that the NH_3 released by glycine oxidation is immediately picked up and incorporated into glutamate and the transaminating N-pool via glutamate dehydrogenase.

All of the amino acids in Figure 1.4 have *active* routes of synthesis in the body, in contrast to the IDAAs, for which no routes of synthesis exist in humans. This statement should be a simple definition of "indispensable" vs "dispensable." However, in nutrition, we define a "dispensable" amino acid as one that is *dispensable* from the diet.[4] This definition is different from defining the presence or absence of enzymatic pathways for an amino acid's synthesis. For example, two of the DAAs depend on the degradation of IDAAs for their production: cysteine and tyrosine. Although serine provides the carbon skeleton and amino group of cysteine, methionine provides the sulfur through condensation of homocysteine and serine to form cystathionine.[20] From the above discussion, neither the carbon skeleton nor the amino group of serine is likely to be in short supply, but provision of sulfur from methionine may become limiting. Therefore, cysteine synthesis depends heavily on the availability of the IDAA, methionine. The same is also true for tyrosine. Tyrosine is produced by the hydroxylation of phenylalanine, which is also *the* degradative pathway of phenylalanine. The availability of tyrosine is strictly dependent on the availability of phenylalanine and the liver's ability to perform the hydroxylation.

Incorporation of Amino Acids into Other Compounds

Table 1.7 lists some of the compounds that amino acids are converted directly into or are used as important parts of the synthesis of other compounds in the body. The list is not inclusive and is

Table 1.7 Important Products Synthesized from Amino Acids

Amino Acid	Incorporated Into
Arginine	Creatine
	Nitric oxide
Aspartate	Purines and pyrimidines
Cysteine	Glutathione
	Taurine
Glutamate	Glutathione
	Neurotransmitters
Glutamine	Purines and pyrimidines
Glycine	Creatine
	Glutathione
	Porphyrins (hemoglobin and cytochromes)
	Purines
Histidine	Histamine
Lysine	Carnitine
Methionine	One-carbon methylation/transfer reactions
	Creatine
	Choline
Serine	One-carbon methylation/transfer reactions
	Ethanolamine and choline
Tyrosine	Catecholamines
	Thyroid hormone
Tryptophan	Serotonin
	Nicotinic acid

meant to highlight important compounds in the body that depend on amino acids for their synthesis. Other important uses of amino acids are for synthesis of taurine,[20,21] which is the "amino acid like" 2-aminoethanesulfonate, found in far higher concentration inside skeletal muscle than any amino acid.[7] Another important, sulfur-containing compound is glutathione,[20,22,23] a tripeptide composed of glycine, cysteine, and glutamate.

Carnitine[24] is important in the transport of long-chain fatty acids across the mitochondrial membrane before fatty acids can be oxidized and is synthesized from ε-N,N,N-trimethyllysine (TML). TML synthesis occurs from posttranslational methylation of specific lysine residues in specific proteins. TML is liberated when proteins containing TML are broken down. TML can also arise from hydrolysis of ingested meats. In contrast to 3-methylhistidine, TML can be found in proteins of both muscle and other organs such as the liver. In rat muscle, TML is about one-eighth as abundant as 3-methylhistidine.

Amino acids are the precursors for a variety of neurotransmitters that contain N. Glutamate may be an exception in that it serves both as a precursor for neurotransmitter production and is a primary neurotransmitter.[25] Glutamate appears important in a variety of neurodegenerative diseases, from amyotrophic lateral sclerosis to Alzheimer's disease.[26] Tyrosine is the precursor for catecholamine synthesis. Tryptophan is the precursor for serotonin synthesis. A variety of studies have reported the importance of plasma concentrations of these and other amino acids upon the synthesis of their neurotransmitter products. The most common putative relationship cited is the administration of tryptophan to increase brain serotonin.

Creatine and Creatinine

Most of the creatine in the body is found in muscle, where it exists primarily as creatine phosphate.[27,28] When muscular work is performed, creatine phosphate provides energy through hydrolysis of its "high-energy" phosphate bond, forming creatine with transfer of the phosphate to form an ATP. The reaction is reversible and mediated by the enzyme ATP-creatine transphosphorylase (also known as creatine phosphokinase).

The original pathway of creatine synthesis from amino acid precursors was defined by Bloch and Schoenheimer in an elegant series of experiments using ^{15}N labeled compounds.[29] Creatine is synthesized outside muscle in a two-step process (**Figure 1.5**). The first step occurs in the kidney and involves the transfer of a guanidino group of arginine onto the amino group of glycine to form ornithine and guanidinoacetate. Methylation of the guanidinoacetate occurs in the liver via S-adenosylmethionine to create creatine. Although glycine donates a nitrogen and carbon backbone to creatine, arginine must be available to provide the guanidino group, as well as methionine for donation of the methyl group. Creatine is then transferred to muscle where creatine is phosphorylated. When creatine phosphate is hydrolyzed in muscle to form creatine, most of the creatine is recycled back to the phosphate form. However, a non-enzymatic process forming creatinine continually dehydrates some of the muscle creatine pool. Creatinine is not retained by muscle but is released into body water, removed by the kidney from blood, and excreted into urine.

The daily rate of creatinine formation is remarkably constant (≈1.7% of the total creatine pool per day) and dependent on the mass of the creatine/creatine-phosphate pool, which is proportional to muscle mass. Thus, daily urinary output of creatinine has been used as a measure of total muscle mass in the body.[30] Urinary creatinine excretion increases within days after a creatine load has been added to the diet, and several more days are required after removal of creatine from the diet before urinary creatinine excretion returns to baseline, indicating that creatine in the diet per se affects creatinine production. Therefore, consumption of creatine and creatinine in meat-containing foods will increase urinary creatinine measurements. With adequate control of food composition and intake, creatinine excretion measured from 24-hour urine collection is a useful index of body muscle mass.[31] More recently, a stable isotopic tracer method, where subjects receive an oral dose of [$^2H_3$]creatine, has been developed to determine skeletal muscle mass using a single urine sample taken between 48–96 hours after the oral dose of [$^2H_3$]creatine has been given.[32,33]

Purine and Pyrimidine Biosynthesis

The purines (adenine and guanine) and the pyrimidines (uracil, cytosine, and thymine) form the building blocks of DNA and RNA. Purines are heterocyclic double-ring compounds that require

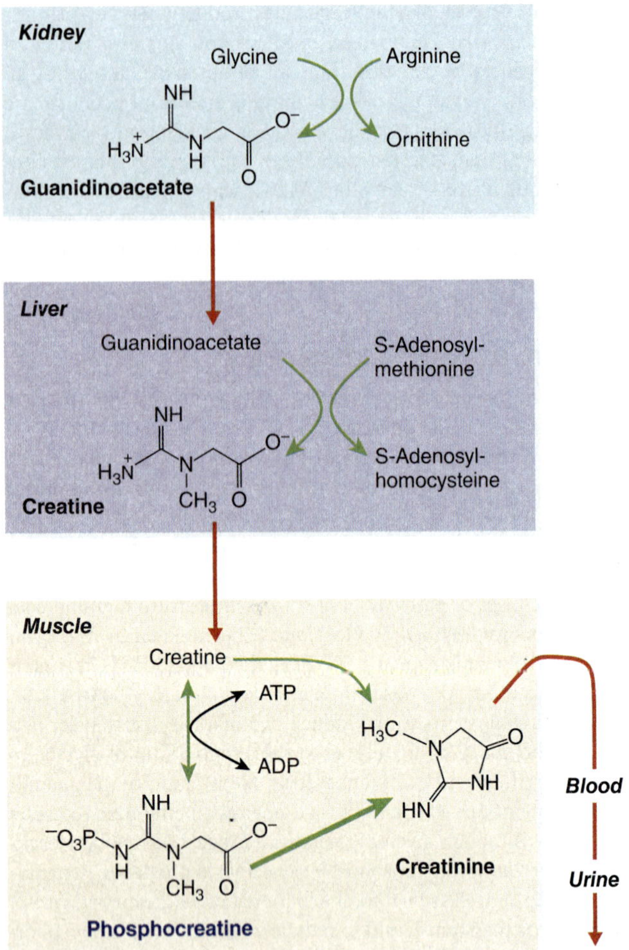

Figure 1.5 Synthesis of creatine and creatinine. Creatine is synthesized in the liver from guanidinoacetic acid that is synthesized in the kidney. Creatine taken up by muscle is primarily converted to phosphocreatine. Although there is some limited direct dehydration of creatine directly to creatinine, the majority of the creatinine comes from dehydration of phosphocreatine. Creatinine is rapidly filtered by the kidney into urine.

incorporation of two glutamine molecules (donation of the amide N), a glycine molecule, a methylene group from tetrahydrofolate, and the amino N of aspartic acid for their synthesis as inosine monophosphate. Adenine and guanine are formed from inosine monophosphate by the addition of another glutamine amide-N or aspartate amino-N.

Pyrimidines are synthesized after an amide-N of glutamine is condensed with a CO_2 to form carbamoyl phosphate, which is further condensed with aspartic acid to make orotic acid–the pyrimidine's heterocyclic six-member ring. The enzyme that forms this carbamoyl phosphate is present in many tissues for pyrimidine synthesis but is not the enzyme found in the liver that makes urea (Figure 1.3). However, a block in the urea cycle causing a lack of adequate amounts of arginine to prime urea synthesis cycle in the liver will result in diversion of unused carbamoyl phosphate to orotic acid and pyrimidine synthesis.[34] Uracil is synthesized from orotic acid, and cytosine is synthesized from uracil by adding an amide group from glutamine to uridine triphosphate to form cytidine triphosphate.

Turnover of Proteins in the Body

Proteins in the body are not static molecules. Just as every protein is synthesized, each is also degraded. The concept that proteins are continually made and degraded at different rates was first described by Schoenheimer and Rittenberg, who first applied isotopically labeled tracers of amino acids to the study of amino acid metabolism and protein turnover in the 1930s.[35] We now know that the rate of protein turnover in the body spans a broad range and that the rate of turnover of individual proteins tends to follow their function in the body, i.e., those proteins whose concentrations need to be regulated (e.g., enzymes) or that act as signals (e.g., peptide hormones) have relatively high rates of synthesis and degradation as a means of regulating concentrations. On the other hand, structural proteins, such as collagen and myofibrillar proteins or secreted plasma proteins, have relatively long lives. However, there must be an overall balance between synthesis and breakdown of proteins. In healthy adults who are neither gaining or losing weight, the amount of N consumed as protein in the diet will match the amount of N lost in urine, feces, and other routes.

A general balance of the processes occurring with protein is shown in **Figure 1.6**. For an average adult, about 250 g of protein will turn over each day.[36] That is 250 g of new protein synthesized and 250 g of old protein broken down to component amino acids. Muscle protein turnover accounts for about 75 g/d of this 250 g/d. The proportion of skeletal muscle mass in the body is consistent with skeletal muscle's contribution to whole body protein turnover: skeletal muscle comprises about one third of the protein in the body[8] and accounts for about one-fourth of the turnover. Viscera and other organ protein turnover accounts for an additional 127 g/d. White and red blood cell synthesis accounts for about 28 g/d of protein, and proteins are synthesized and secreted by liver into plasma (≈20 g/d).

Protein is also added directly into the gut lumen in the form of secreted proteins, and the small intestine is continually being remodeled, as cells formed in the crypts migrate toward the villus tips where cells are sloughed off the tips. A reasonable estimate is that 20 g/d of secreted protein and 50 g/d protein from sloughed cells contribute 70 g/d of protein into the gut, which is then very efficiently reabsorbed.

As shown in Figure 1.6, if 90 g of protein are eaten in a day and 10 g are lost in feces, the net absorption will be 80 g. At the same time, considerably more protein is synthesized and degraded in the body. The total turnover of protein, including both dietary intake and endogenous metabolism, is $90 + 250 = 340$ g/d, of which oxidation of dietary protein accounts for $(75 + 5)/340 = 24\%$ of the turnover of protein in the body per day. When dietary protein intake is restricted, adaptation occurs where the body reduces N losses, and protein intake/oxidation becomes a much smaller proportion of total protein turnover.

Although there is no definable entity such as "whole body protein," the term is useful for understanding the amount of energy and resources spent by the body in producing and breaking down protein. Several methods using isotopically labeled tracers have been defined to quantitate the whole-body turnover of protein. An important point of Figure 1.6 is that overall protein turnover is severalfold greater than the input of dietary amino acids and amino acid oxidation. Methods to assess turnover of protein in the body range from simple and noninvasive to expensive and complicated.

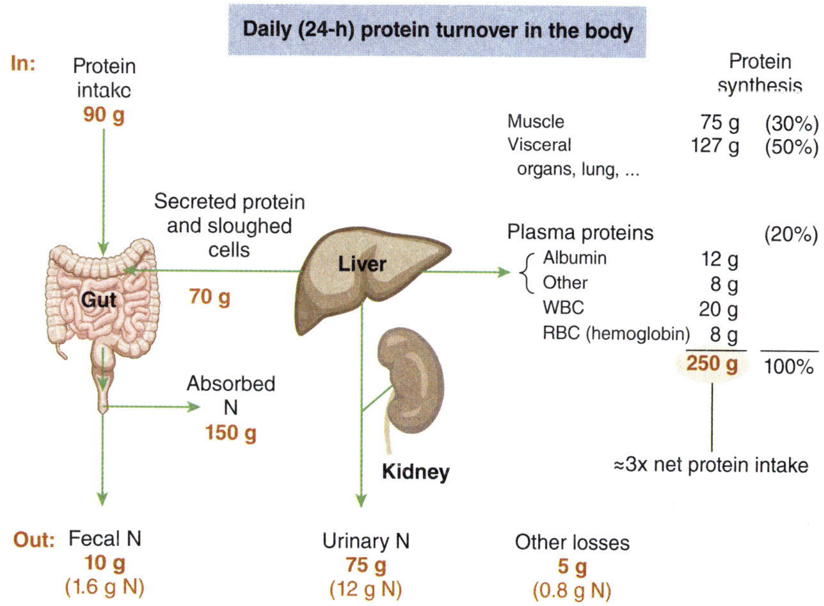

Figure 1.6 Relative rates of protein turnover and intake in a healthy 70-kg human. Under normal circumstances, dietary intake (IN = 90 g) will match N losses (OUT = 90 g). Protein breakdown will then match synthesis. Protein intake is only 90/(90 + 250) ≈ 25% of total turnover of N in the body per day.

Redrawn with permission from Hellerstein MK, and Munro HN. (1988). Interaction of liver and muscle in the regulation of metabolism in response to nutritional and other factors. In The Liver: Biology and Pathobiology (Arias IM, Jakoby WB, Popper H, Schachter D, and Shafritz DA, eds) pp. 965-83, Raven Press, New York.

Methods of Measuring Protein Turnover and Amino Acid Kinetics

Nitrogen Balance

The oldest (and most widely used) method to define changes in the body N pool is the N balance method. Because of its simplicity, the N balance technique has been the gold standard to define minimum levels of dietary protein and IDAA intakes in humans of all ages. Subjects are placed for several days on a specific amount of amino acid and/or protein intake and their urine and feces are collected over several 24-hour periods to measure N excretion. A week or more may be required before collection will reflect adaptation to a dietary change. An example of adaptation is the placement of healthy subjects on a diet containing a minimal amount of protein. As shown in **Figure 1.7**, urinary N excretion drops dramatically in response to a protein-deficient diet over the first three days and stabilizes at a new lower level of N excretion by day 8.[37]

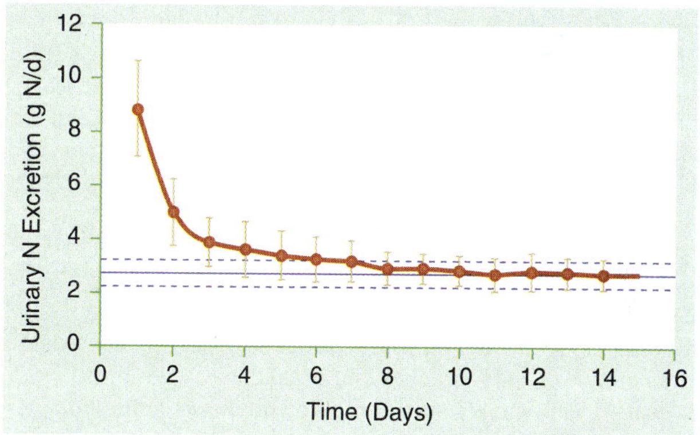

Figure 1.7 Time required for urinary N excretion to stabilize after changing from an adequate to a deficient protein intake in young men. Horizontal solid and broken lines are mean ±1 standard deviation for N excretion at the end of the measurement period.

Data from Scrimshaw NS, Hussein MA, Murray E et al. Protein requirements of man: variations in obligatory urinary and fecal nitrogen losses in young men. J Nutr 1972;102:1595-604.

The N end-products excreted in the urine are not only end products of amino acid oxidation (urea and NH_3) but also other N-containing species, such as uric acid from nucleotide degradation and creatinine. However, these nonurea, nonammonia N species are a relatively small proportion of the total N in the urine and are excreted at a relatively constant rate over a variety of dietary situations. Although urea is the predominant N species excreted, ammonia N excretion is variable and needs to be measured.

Nitrogen appears in the feces because the gut does not completely absorb all dietary protein and reabsorb all N secreted into the gastrointestinal tract (Figure 1.6). N is also lost from skin via sweat and from shedding of dead skin cells. There are also losses through hair, menstrual fluid, nasal secretions, and so forth. As N excretion in the urine decreases among subjects on a minimal-protein diet (Figure 1.7), it becomes increasingly important to account for N losses through nonurine, nonfecal routes. Most of the losses that are not readily measurable are minimal (<10% of total N loss) and can be discounted by use of a simple offset factor for nonurine, nonfecal N losses. Where the assessment of losses comes into play is in the finer definition of where zero balance occurs as a function of dietary protein intake when determining amino acid and protein requirements. Small changes in N balance corrections make significant changes in the assessment of protein requirements using N balance.

Although the N balance technique is useful and simple, it provides no information about the inner workings of the system.

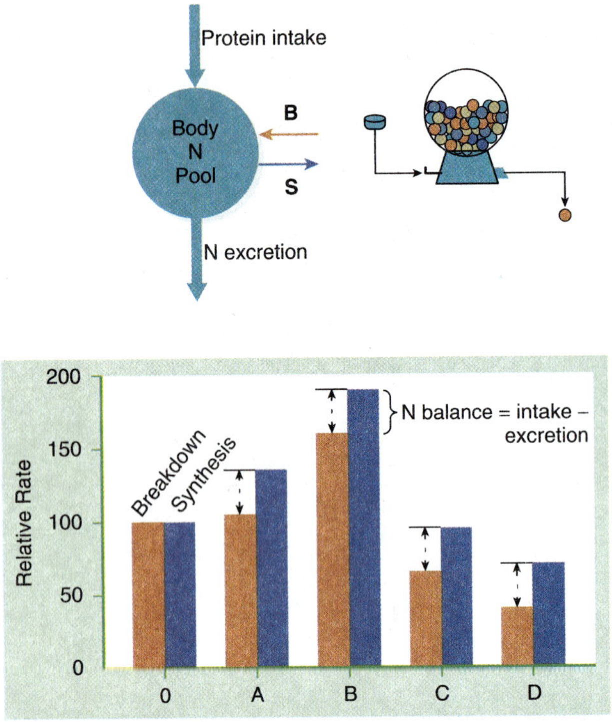

Figure 1.8 Illustration of the N balance technique. Nitrogen balance is simply the difference between input and output that is similar to the introduction of a coin into a gumball machine resulting in a gumball being released. The perception of only the "in" and "out" observations is that the machine changed the coin directly into a gumball or that the dietary intake becomes directly the N excreted without consideration of amino acid entry from protein breakdown (B) or uptake for protein synthesis (S). This point is further illustrated with four different hypothetical responses to a change from a zero N balance (case 0) to a positive N balance (cases A-D). A positive N balance can be obtained by increasing protein synthesis (A), by increasing synthesis more than breakdown (B), by decreasing breakdown (C), or by decreasing breakdown more than synthesis (D). The N balance method does not distinguish among the four possibilities.

An interesting analogy for the N balance technique is illustrated in **Figure 1.8** where the simple model of N balance is represented by a gumball machine. Balance is taken between "coins in" and "gumballs out." However, we should not conclude that the machine turns coins into gum, although that conclusion is easy to reach with the N balance method. What the N balance technique fails to provide is information about what occurs *within* the system (i.e., inside the gumball machine). Inside the system is where the changes in whole body protein synthesis and breakdown occur (shown as the smaller arrows into and out of the *Body N Pool* in Figure 1.8). A further illustration of this point is made at the bottom of Figure 1.8 where a positive increase in N balance has been observed going from zero (case 0) to positive balance (cases A-D). A positive N balance could be obtained with identical increases in N balance by any of four different alterations in protein synthesis and breakdown: a simple increase in protein synthesis (case A), a decrease in protein breakdown (case C), an increase in both protein synthesis and breakdown (case B), or a decrease in both (case D). The effect is the same positive N balance for all four cases, but the energy implications are considerably different. Because protein synthesis costs energy, cases A and B are more expensive, while cases C and D require less energy than the starting case 0. To resolve these four cases, we must look directly at rates of protein turnover (breakdown and synthesis) using a labeled tracer.

Tracer Methods to Measure Amino Acid Kinetics

To follow flows of *endogenous* metabolites in the body, isotopically labeled tracers are used that are identical to the endogenous metabolites in terms of chemical structure but have substitution of one or more atoms with isotopes different from those usually present. The substitution of isotopes is done to make the tracers distinguishable (measurable) from normal metabolites. We often think first of radioactive isotopes (e.g., ^{3}H for hydrogen and ^{14}C for carbon) as tracers, but there are also nonradioactive, stable isotopes that are used extensively. Because isotopes only differ in the number of neutrons that are contained, they can be distinguished in a compound by mass spectrometry. Most of the lighter elements have one *very abundant* stable isotope and one or two isotopes of higher mass of *minor abundance*. The major and minor isotopes, respectively, are protium (^{1}H) and deuterium (^{2}H) for hydrogen, ^{12}C and ^{13}C for carbon, ^{14}N and ^{15}N for nitrogen, and ^{16}O, ^{17}O, and ^{18}O for oxygen. Except for some effects that can be significant for hydrogen isotopes, a compound that is isotopically labeled will behave identically to the corresponding unlabeled endogenous compound in the body.

Because they do not exist in nature and so little of the radioactive material is administered, radioisotopes are considered "weightless" tracers that do not add material to the system. Radioactive tracer data are expressed as counts or disintegrations per minute per unit of compound. Because the stable isotopes are naturally occurring (e.g., ≈1% of all carbon in the body is ^{13}C), stable isotope tracers are administered and measured as "the excess above the naturally occurring abundance" of the isotope in the body the *mole ratio* of the amount of tracer isotope divided by the amount of unlabeled material (the *tracer-to-tracee ratio* or TTR).[38]

The basis of most tracer measurements to determine amino acid kinetics is the simple concept of tracer dilution. This concept is illustrated in **Figure 1.9** for the determination of the flow of water in a stream. If you infuse a dye of known concentration (i.e., isotopic enrichment) into the stream, go downstream after the dye has mixed well with the stream water, and take a sample

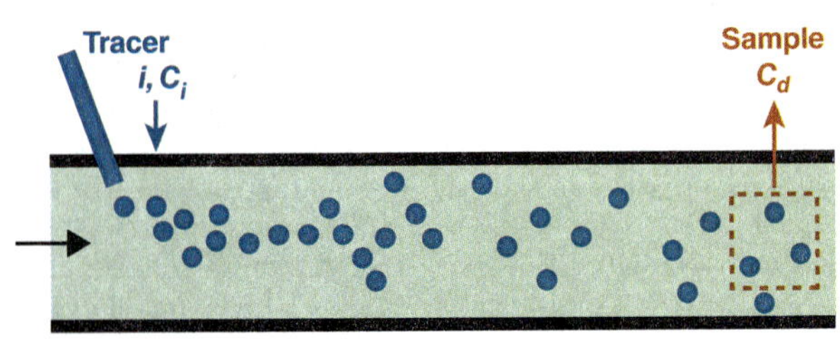

$$\text{Production rate (Flux, } Q\text{)} = \text{Tracer infusion rate, } i \left(\frac{\text{Initial tracer concentration, } C_i}{\text{Tracer concentration diluted, } C_d} \right)$$

Figure 1.9 Basic principle of the "dye-dilution" method of determining tracer kinetics.

of the dye, then you can calculate from the measured dilution of the dye the rate at which water must be flowing in the stream to make that dilution. The necessary information required is infusion rate of dye (tracer infusion rate) and measured concentration of the dye (enrichment or specific activity of the tracer). The calculated value is the flow of water through the stream (flux of unlabeled metabolite) causing the dilution. This simple dye-dilution analogy is the basis for most kinetic calculations in a wide range of formats for a wide range of applications.

Models for Whole Body Amino Acid and Protein Metabolism

Limitations to using tracers to define amino acid and protein metabolism are largely driven by how the tracer is administered and where it is sampled. The simplest method of tracer administration is an oral dose, but intravenous administration is preferred to deliver the tracer *systemically* (to the whole body) into the free pool of amino acids. The simplest site of sampling of the tracer dilution is also from the free pool of amino acids via blood. Therefore, most approaches to measuring amino acid and protein kinetics in the whole body using amino acid tracers have been to assume a single, free pool of amino N, as shown in **Figure 1.10**. Amino acids enter the free pool from dietary amino acid intake (I, enteral or parenteral) and from amino acids released from protein breakdown (B). Amino acids leave the free pool by amino acid oxidation (C) to end-products (CO_2, urea and NH_3) and via amino acid uptake for protein synthesis (S). The free amino acid pool can be viewed from the standpoint of all of the amino acids together (as discussed for the end-product method) or from the viewpoint of a single amino acid, and its metabolism per se. The model in Figure 1.10 is called a "single pool model" because protein is not viewed as a pool per se, but rather as a source of entry of unlabeled amino acids into the free pool, on the one hand, and as route of amino acid removal for protein synthesis, on the other. Only a small portion of the body protein is assumed to turn over during the time course of the experiment. Obviously, these assumptions are not true: many proteins in the body are turning over rapidly (e.g., most enzymes). Those proteins that do turn over during the experiment will become labeled and appear as part of the free amino acid pool.[39] However, these make up only a fraction of the total protein; the remainder turns over slowly (e.g., muscle protein). The bulk of amino acids entering via protein breakdown and leaving for new protein synthesis are coming from slowly turning over proteins. These flows are the "B" and "S" arrows of the traditional single-pool model of whole-body protein metabolism (Figure 1.10).

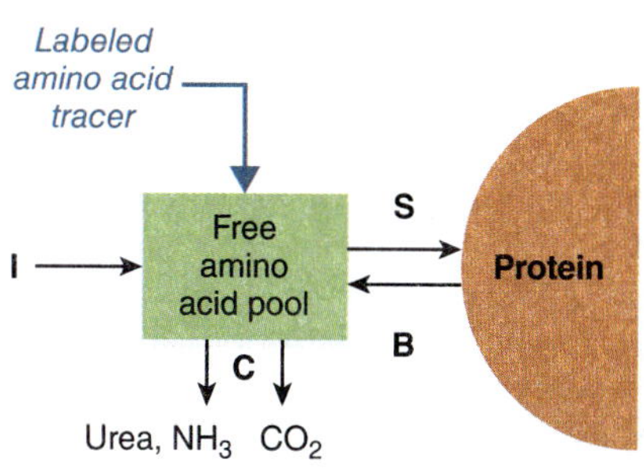

Figure 1.10 Single-pool model of whole-body protein metabolism measured using a labeled amino acid tracer. Amino acid enters the free pool from dietary intake (I) and amino acid released from protein breakdown (B) and leaves the free pool via amino acid oxidation (C) to urea, NH_3 and CO_2 and via uptake for protein synthesis (S).

End-Product Approach

The earliest model of whole body protein metabolism in humans was applied by San Pietro and Rittenberg in 1953 using [^{15}N]glycine.[40] Glycine was used as the first tracer because it is the only amino acid without an optically active α-carbon center and, therefore, easy to synthesize with a ^{15}N label. At that time, measurement of the glycine tracer in plasma was very difficult. Thus, San Pietro and Rittenberg proposed a model based on something that could be readily measured, urinary urea, and NH_3. The assumption was that the urinary N end-products reflected the average enrichment of ^{15}N of all free amino acids being oxidized. Although glycine ^{15}N was the tracer, the tracee was assumed to be *all* free amino acids (assumed to be a single pool). However, it became obvious that the system was more complicated and that a more complex model and solution were required. The method languished until 1969 when Picou and Taylor-Roberts[41] proposed a simpler method that also followed the glycine ^{15}N tracer into urinary N. Their method dealt only with the effect of the dilution of the ^{15}N tracer in the whole free amino acid pool, rather than invoking solution of tracer-specific equations from a specific model. Their assumptions were like the earlier Rittenberg approach in that they assumed that the ^{15}N tracer mixes (scatters) among the free amino acids in some distribution that is not required to be known, but is representative of amino acid metabolism per se. The [^{15}N]glycine tracer is administered (usually orally), and urine samples are obtained to measure the ^{15}N dilution in the free amino acid pool.[42] The ^{15}N in the free amino acid pool is diluted with unlabeled amino acid entering from protein breakdown and from dietary intake. The turnover of the free pool (Q, typically expressed as mg N/kg/d) is calculated from the measured the dilution of ^{15}N in the end products via the same approach, illustrated in Figure 1.9:

$$Q = i/E_{UN}$$

where i is the rate of [^{15}N]glycine infusion (mg ^{15}N/kg/d) and E_{UN} is the ^{15}N enrichment in urinary N (either urea and/or NH_3). The free pool is assumed to be in steady state and; therefore, the turnover of amino acid will be equal to the rate of amino acids entering via whole body protein breakdown (B) and dietary intake (I) and also equal to the rate of amino acids leaving via uptake for protein synthesis (S) and via amino acid oxidation to the end-products urea and NH_3 (C):

$$Q = I + B = C + S$$

Because dietary intake should be known and urinary N excretion is measured, the rate of whole-body protein breakdown can be determined: $B = Q - I$, as well as the rate of protein synthesis: $S = Q - C$. In these calculations, the standard value of 6.25 g protein = 1 g N is used to interconvert protein and urinary N. Attention to the units (g of protein vs. g of N) is important, as both units are often used concurrently in the same equation.

The end-product method is not without its problems. When the [^{15}N]glycine tracer is given orally at short intervals (e.g., every 3 h) the time required to reach a plateau in urinary urea ^{15}N is about 60 h regardless of whether adults,[43] children, or infants[44] are studied. The delay in attaining a plateau is due to the time required for the ^{15}N tracer to equilibrate within the free glycine, serine, and urea pools.[19,42] Waterlow and colleagues[45] suggested measuring the ^{15}N in NH_3 after a *single dose* of [^{15}N]glycine. The advantage is that the ^{15}N tracer passes through the body NH_3 pool within 24 hours. The caveat is the dependence of the single-dose, end-product method upon NH_3 metabolism. Urinary $^{15}NH_3$ enrichment is also usually different from the urinary urea ^{15}N enrichment[46] because the amino-^{15}N precursor for NH_3 synthesis is of renal origin, while the amino-^{15}N precursor for urea synthesis is of hepatic origin. The ^{15}N-glycine end-product method is rarely applied today. However, from the end-product method, the method of measuring direct turnover of specific amino acids arose.

Measurement of the Kinetics of Individual Amino Acids

Whole-body kinetics of individual amino acids can also be determined by infusing (intravenously) an amino acid tracer into the free amino-acid pool and measuring the dilution of the tracer in that amino acid's free pool (via blood samples). The simplest models consider only IDAAs without de novo synthesis components. The kinetics of IDAAs mimic the kinetics of protein turnover as shown in Figure 1.10. The same type of model can be constructed, but cast specifically in terms of a single IDAA, and the same steady-state balance equation can be defined:

$$Q_{aa} = I_{aa} + B_{aa} = C_{aa} + S_{aa}$$

where Q_{aa} is the turnover rate (or flux) of the IDAA, I_{aa} is the rate that the amino acid is entering the free pool from dietary intake, B_{aa} is the rate of amino acid entry from protein breakdown, C_{aa} is the rate of amino acid oxidation, and S_{aa} is the rate of amino acid uptake for protein synthesis. The most common method for defining amino acid kinetics has been a primed infusion of an amino acid tracer to quickly reach isotopic steady state (constant dilution of tracer) in blood. The flux for the amino acid is measured from the dilution of the tracer in the free pool. Knowing the tracer enrichment and infusion rate and measuring the tracer dilution in blood samples taken at plateau, the rate of unlabeled metabolite appearance is determined[38,47]:

$$Q_{aa} = i_{aa}\left[{E_i}/{E_p} - 1\right]$$

where i_{aa} is the infusion rate of tracer with enrichment E_i (mole % excess) and E_p is the blood amino acid enrichment.

For a carbon-labeled tracer, the amino acid oxidation rate can be measured from the rate of $^{13}CO_2$ or $^{14}CO_2$ excretion.[38,47] The choice of a carbon label that is quantitatively oxidized is critical. For example, the ^{13}C of a L-[1-^{13}C]leucine tracer is quantitatively released at the first irreversible step of leucine catabolism.[47] In contrast a ^{13}C-label in the leucine tail will end up in acetoacetate or acetyl CoA, which may or may not be quantitatively oxidized.[48] Although several other amino acids, such as lysine and phenylalanine, have more nebulous pathways of oxidation, their oxidation is often reported using ^{13}C tracers.[49,50]

Before the oxidized carbon-label is recovered in exhaled air, it must pass through the body bicarbonate pool. Therefore, we must know what fraction of bicarbonate pool turnover goes to release as CO_2 into exhaled air, vs retention for alternative fates in the body. In general, about 80% of the bicarbonate produced is released immediately as expired CO_2, as determined from infusion of a labeled bicarbonate and measurement of the fraction infused that is recovered in exhaled CO_2.[51] The other 20% is retained in bone and metabolic pathways that "fix" carbon. The amount of bicarbonate retained is variable (ranging from 0%–40% of its production) and needs to be determined when different metabolic situations are investigated. In cases where the retention of bicarbonate in the body may change with metabolic perturbation, parallel studies measuring the recovery of an administered dose of ^{13}C- or ^{14}C-labeled bicarbonate are indispensable in interpretation of the oxidation results.[52]

The rates of amino acid release from protein breakdown and amino acid uptake for protein synthesis are calculated by subtracting dietary intake and oxidation from the flux of an IDAA, just as is done with the end-product method. The primary distinction is that the measurements are specific to the kinetics of a single amino acid (micromole of amino acid per unit of time), rather than directly in terms of N per se. Flux components are then extrapolated to whole body protein kinetics by dividing the amino acid rate by the assumed concentration of the amino acid in body protein (as shown in Table 1.3).

The principal advantages of measuring the kinetics of an individual amino acid are (a) that results are specific to *that* amino acid, improving the confidence of the measurement, and (b) that measurements can be performed quickly because turnover time of the free pool is rapid (usually less than 4 hours using a priming dose). Drawbacks to the measurement of the kinetics of an individual amino acid are (a) that an appropriately labeled tracer may not be available to follow the pathways of the amino acid being studied, especially for amino acid oxidation, and (b) that metabolism of amino acids occurs within cells, but tracers are typically administered into and sampled from the blood outside of cells.

Amino acids do not freely pass through cell membranes; they are transported. For the neutral amino acids (leucine, isoleucine, valine, phenylalanine, and tyrosine), transport in and out of cells may be rapid and bidirectional (Table 1.5). These amino acids have a small concentration gradient between plasma and intracellular milieus (Table 1.4). However, that small gradient will limit exchange of intracellular and extracellular amino acids. For leucine, this phenomenon can be defined using α-ketoisocaproate (KIC), which is formed from leucine inside cells by transamination. Some of the KIC formed is then decarboxylated, but most is reaminated to reform leucine[53] or released from cells into plasma. Thus, plasma KIC enrichment can be used as a marker of intracellular leucine enrichment from which it came.[54]

Generally, plasma KIC enrichment is about 25% lower than plasma leucine enrichment.[34] If plasma KIC enrichment is substituted for the plasma leucine tracer enrichment into the calculation of leucine kinetics, the measured leucine flux and oxidation rate (and estimates of protein breakdown and synthesis) are increased by ≈25%. However, when protein metabolism is studied under two different conditions and the resulting leucine kinetics are compared, the same relative response will be obtained regardless of whether leucine or KIC enrichment is used for the calculation of kinetics.[54]

Most amino acids do not have an intracellular metabolite that can be readily measured in plasma to define the gradient of enrichment differences between intracellular and extracellular milieus, and there is no correction factor such as using KIC measurements when leucine is a tracer. The turnover rate of many amino acids, both indispensable and dispensable, have been measured in humans in several studies. As depicted in Figure 1.10, the turnover (flux) of every individual IDAA's should be reflective of each IDAA's release from protein breakdown when studied in the postabsorptive state (i.e., there no intake, I, so turnover= breakdown, Q=B.)[55] If the model is correct, each individual IDAA turnover rate should be proportional to the content of that amino acid in body protein: amino acids more, vs less, abundant in protein will have higher rates of release from protein breakdown, and a linear relationship of amino acid flux and amino acid abundance in body protein should exist. That relationship is shown in **Figure 1.11** for data gleaned from studies in humans measured in the postabsorptive state (without dietary intake during the infusion studies) previously consuming diets of adequate N and energy intake.[55] A correlation of amino acid flux and amino acid composition in protein exists across IDAA tracers and studies. This correlation suggests that even with problems defining intracellular/extracellular concentration gradients of tracers to assess true intracellular events, changes in flux measured for the various IDAAs are still reflective of changes in whole body protein breakdown.

Because DAAs are synthesized in the body, their fluxes are expected to be higher than their expected flux based on the regression line in Figure 1.11 by the amount of de novo synthesis that occurs. Because de novo synthesis and disposal of the DAAs is expected to be based on the metabolic pathways of individual amino acids, the degree to which individual DAAs lie above the line should also be variable. For example, tyrosine is a DAA because it is made from the hydroxylation of phenylalanine, which is also the pathway of phenylalanine disposal. The rate of tyrosine de novo synthesis *is* the rate of phenylalanine disposal. In the postabsorptive state, 10%–20% of IDAA turnover goes to oxidative disposal. For phenylalanine, with a flux of ≈40 µmol/kg/h, disposal produces ≈6 µmol/kg/h

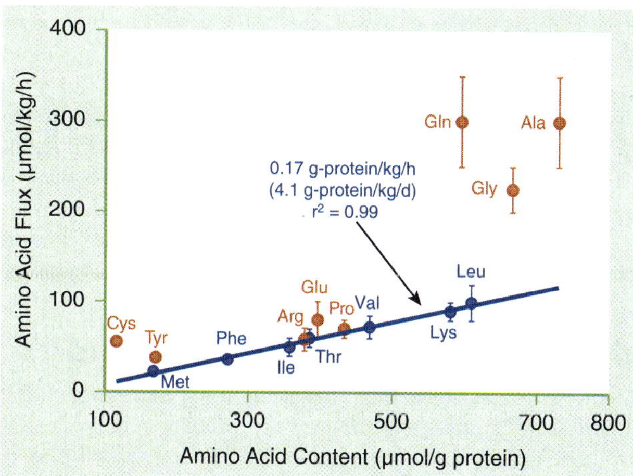

Figure 1.11 Fluxes of individual amino acids measured in postabsorptive humans plotted against amino acid concentration in protein. The closed circles are for DAAs, and the open circles represent IDAAs. The regression line is for the flux of the IDAAs vs their content in protein. Error bars represent the range of reported values that were taken from various reports in the literature of studies of amino acid kinetics in healthy humans eating adequate diets of N and energy intake studied in the postabsorptive state. The amino acid content of protein data are taken for muscle values from Table 1.3. The regression line slope of 4.1 g-protein/kg/d is similar to other estimates of whole-body protein turnover. Figure redrawn from reference with additional data added.

Data from Bier DM. Intrinsically difficult problems: the kinetics of body proteins and amino acids in man. Diabetes Metab Rev 1989;5:111–32, with additional data.

of tyrosine. We would predict from the tyrosine content of body protein that tyrosine release from protein breakdown would be 21 µmol/kg/h, and that the flux of tyrosine (tyrosine release from protein breakdown plus tyrosine production from phenylalanine) would be 21 + 6 = 27 µmol/kg/h. The measured tyrosine flux approximates this prediction (Figure 1.11).[56]

Other DAAs have very large de novo synthesis components because of the metabolic pathways in which they are involved. For example, arginine is at the center of the urea cycle (Figure 1.3). Normal synthesis for urea is 8–12 g of N per day. That amount of urea production translates into an arginine de novo synthesis of ≈250 µmol/kg/h, which is four times the expected rate of arginine released from protein breakdown. As seen in Figure 1.11, the *measured* arginine flux approximates the arginine release from protein breakdown.[57] The expected large de novo synthesis component does not appear in the measured flux. The explanation for this low flux is that the arginine involved in urea synthesis is highly compartmentalized in liver cells, and this arginine does not exchange with the tracer arginine infused intravenously.

Similar disparities are seen between the measured fluxes of glutamine and glutamate determined with intravenously infused tracers and their expected fluxes. The predicted flux for glutamate should include transamination with the BCAAs, alanine, and aspartate, as well as glutamate's contribution to the production and degradation of glutamine. However, the glutamate flux is barely above expected rate of glutamate release from protein breakdown (Figure 1.11). The size of the free glutamate pool has also been determined from tracer dilution and was found to be small, approximately the size of glutamate in extracellular water.[58] The larger intracellular pool that exists in muscle (Table 1.4) was not seen with the intravenously administered tracer. The flux measured for glutamine is considerably larger than that for glutamate, reflecting a large de novo synthesis component (Figure 1.11). However, the pool size determined with the [^{15}N]glutamine tracer was also small, not much larger than glutamine in extracellular water.[58] The large intracellular muscle free pool of glutamine was not found. Thus, the large intracellular pools of glutamate and glutamine (especially those in muscle) are tightly compartmentalized and do not readily mix with extracellular glutamine and glutamate. Glutamine and glutamate tracers, administered intravenously, define pools of glutamine and glutamate that reflect primarily extracellular free glutamine and glutamate.[59] The glutamate tracer does not detect intracellular events such as glutamate transamination. However, the prominent role of glutamine in the body *is interorgan transport*, i.e., production by muscle and release for use by other tissues,[60] and that event is measured by the glutamine tracer (as is obvious from Figure 1.11where the tracer determined glutamine flux shows the highest flux measured of any amino acid).

The model in Figure 1.10 is designed for intravenous administration of tracers. When a tracer is administered enterally, the tracer first is absorbed in the small intestine, passes through the portal vein and liver before entering the systemic circulation (from which the tracer is sampled). Both the gut and the liver can remove tracer prior to its entry into systemic circulation. Any metabolism of these amino acids by gut or liver on the first pass during absorption will not be "seen" by an intravenously infused tracer in terms of systemic kinetics. Therefore, another pool, with a second arrow showing the first-pass removal by gut and liver, should precede the input arrow for "*I*" (**Figure 1.12**) to indicate the role of the splanchnic bed. A fraction "*f*" of the dietary intake

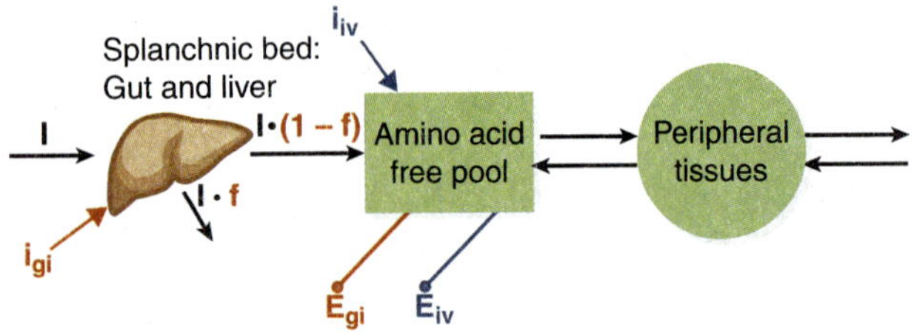

Figure 1.12 Model of whole-body protein metabolism for the fed state where the first-pass uptake of dietary intake is considered. A labeled amino acid tracer is administered by the gastrointestinal route (i_{gi}) to follow dietary amino acid intake (I). The fraction of dietary amino acid sequestered on the first-pass by the splanchnic bed (f) can be determined by administering the tracer by both the gastrointestinal and the intravenous route (i_{iv}) and comparing the enrichments in blood for the two tracers (E_{gi} and E_{iv}, respectively).

($I \cdot f$) is sequestered on the first pass, and only $I \cdot (1 - f)$ enters systemic circulation.

There are two approaches to addressing this problem. The first does not evaluate the fraction sequestered explicitly but builds the tracer administration scheme into the first pass losses. Simply add the amino acid tracer to the dietary intake so that the tracer administration is the oral route (I_{gi}) and enrichments in blood (E_{gi}) come after first pass metabolism by the splanchnic bed.[61,62] This approach is especially useful for studying the effect of varying levels of amino acid intake but does not evaluate the amount of material sequestered by the splanchnic bed per se.

The second approach applies the tracer both by the intravenous route and by the enteral route. IV tracer infusion (I_{iv}) and plasma enrichment (E_{iv}) are used to determine systemic kinetics, and the enteral tracer infusion and its plasma enrichment determines systemic kinetics plus the effect of the first pass. By difference, the fraction, f, is readily calculated.[63] This approach can be applied even in the postabsorptive state to determine basal uptake of amino acid tracers by the splanchnic bed. Several IDAAs and DAAs have been studied, and first pass fractional uptake values for these different amino acids have been determined. In general, the splanchnic bed extracts between 20%–50% of the IDAAs leucine,[63] phenylalanine,[63,64] and lysine.[65,66] More than half of the DAAs are extracted by the splanchnic bed on the first pass, including alanine,[67] arginine,[68] and glutamine,[64,69] but splanchnic bed extracts almost all of enteral glutamate.[69,70]

Synthesis of Specific Proteins

The above methods deal with measures at the whole-body level, but do not address specific proteins and their individual rates of synthesis and degradation. To do so requires obtaining measured samples of the proteins and tracer enrichment that has been incorporated. Some proteins are readily sampled, e.g., lipoproteins, albumin, fibrinogen, and other proteins secreted into blood. Other proteins require tissue sampling (e.g., obtaining muscle biopsies). If a protein or group of proteins can be sampled and their tracer enrichment measured, their synthetic rates can be determined directly from the rate of tracer incorporation into the proteins. Proteins that turn over slowly (e.g., muscle protein or albumin) incorporate only a small amount of tracer during a tracer infusion. Because the incorporation rate of tracer is approximately linear during this time, protein synthesis can be measured by obtaining only two samples. This technique has been especially useful for evaluating protein synthesis of myofibrillar protein with a limited number of muscle biopsies. Once a tissue biopsy has been obtained, the sample may be fractionated into cellular components and the synthetic rate of proteins in organelles (e.g., mitochondria) or specific proteins of muscle (e.g., actin and myosin) can be determined.[71,72] This approach has been extended through a laborious effort to separate proteins by 2-dimensional gel electrophoresis, excise individual protein spots, hydrolyze each protein spot, and measure its amino acid enrichment to determine rates of synthesis of individual proteins[73] and even individual modifications of the same protein.[74]

The determination of the protein fractional synthetic rate is a "precursor-product" method that requires both knowledge of the rate of tracer incorporation into the protein being synthesized and knowledge of the enrichment of the amino acid precursor used for synthesis. For muscle, L-[1-^{13}C]leucine has often been used as the tracer, and plasma KIC ^{13}C enrichment is used to approximate the intracellular muscle leucine enrichment. Alternatively, the intracellular amino acid enrichment of free amino acids have been measured, and some researchers has measured the tracer enrichment in tRNA, the direct site of the precursor enrichment for protein synthesis.[75] For proteins that turn over at a faster rate, the tracer concentration rises exponentially during the course of the tracer infusion toward a plateau value of enrichment that matches that of the precursor amino acids used for its synthesis, i.e., the intracellular amino acid enrichment. The types of protein that have been measured under these conditions include lipoproteins, especially apolipoprotein B in very low-density lipoprotein particles.[76]

Another approach to defining precursor enrichment for slow turnover proteins is to use deuterated water (^{2}H-water) as the tracer.[77] Deuterated water is inexpensive and can be given as multiple oral doses over multiple days to label body water at a constant low enrichment (≈0.5%).[71,77,78] Several compounds have hydrogens that exchange with body water. The hydrogens on the second carbon of alanine will exchange during rapid reversible transamination[79] and quickly label free alanine with deuterium. The ^{2}H-alanine will then be incorporated into protein to label slow turnover proteins, and protein synthesis of individual proteins can be determined from measurement of ^{2}H-alanine enrichment.[39] The key point is that the precursor enrichment for the ^{2}H-alanine in protein is that of ^{2}H-water, easily measured in blood or urine.

Degradation of Specific Proteins

Methods to measure protein degradation are more limited.[39,72] To measure protein degradation, the protein must be prelabeled. Three methods have been used: (a) removal of plasma proteins, followed by iodination with radioactive iodine, and reinjection back into the body to follow the disappearance of the labeled protein, (b) use of posttranslational amino acids such as 3-methylhistidine, and (c) administration of a tracer amino acid to label proteins via incorporation of the tracer (via protein synthesis) followed by measurement of release of the labeled amino acid degradation of the protein. The first two methods are limited in terms of what can be measured. Only the method of pre-labeling proteins with a tracer amino acid has universal options for measuring rates of protein breakdown.

Slow turnover proteins may first be labeled for degradation measurement by long infusions of amino acid tracer or by chronic administration of deuterated water to label amino acids, such as alanine. This approach has been used to measure rates of protein synthesis but can also be used to label proteins for measurement of protein breakdown.[39,72] After the tracer infusion is stopped, the tracer enrichment will disappear quickly from plasma. At that point, serial sampling of proteins and measurement of the decrease in tracer enrichment with time will give protein degradation rates. However, 80% or more of the amino acids released from protein breakdown are reused for synthesis of new proteins, causing recycling of amino acid tracer from protein degradation back into new proteins. Because there is generally low tracer enrichment in the proteins being measured, the recycling of tracers of low enrichment becomes a complicating problem to the interpretation of the labeled protein data.[39]

Several specific amino acids in a select group of specific proteins are modified by enzymes after the protein has been synthesized. Posttranslational modifications are generally modest, occur to specific amino acids in specific proteins, and are often the addition of a hydroxyl group (e.g., conversion of proline to hydroxyproline in collagen) or methylation of N-moieties of amino acid residues such as histidine or lysine. Because t-RNAs do not code for these hydroxylated or methylated amino acids, these modified amino acids are not reused for protein synthesis once the protein containing them is degraded, and their release and collection in urine can be used as a measure of the degradation rate of the proteins that contained them.

Because of the quantitative importance of muscle to whole body protein metabolism, measurement of the release of 3-methylhistidine is an important tool for following myosin and actin breakdown, both primary proteins in skeletal muscle and the primary proteins containing 3-methylhistidine, and the approach is noninvasive using 24-h urine collection to measure urinary 3-methylhistidine.[80] There are caveats; dietary meat intake can contribute as much as 5% of the 3-methylhistidine obtained in urine.[81] And, 3-methylhistidine is not specific to skeletal muscle myofibrillar protein.[82] Although skin and gut only contain a small pool of myofibrillar protein (compared with the large mass of myofibrillar protein in skeletal muscle), skin and gut protein turn over rapidly and, therefore, can also contribute a significant amount of 3-methylhistidine to the urine.[83] A way around these issues is to measure 3-methylhistidine release directly from muscle by AV blood measurement across the leg or arm.[84] This AV difference can be combined with the AV difference measurement of an IDAA that is not metabolized in muscle, such as tyrosine. The AV difference of tyrosine across an arm or leg defines net protein balance, i.e., the difference between protein breakdown and synthesis. If we subtract this protein balance measurement from the myofibrillar protein breakdown measurement from the AV difference of 3-methylhistidine, an estimate of muscle protein synthesis is obtained.[85] A final twist is to include co-infusion of an isotopically labeled 3-methylhistidine tracer and measure the AV difference of the tracer in conjunction with the concentration of 3-methylhistidine to provide an additional measurement of myofibrillar protein breakdown.[86]

Contribution of Specific Organs to Protein Metabolism

Whole Body Metabolism of Protein

The preceding discussion defines turnover of protein in various parts of the body but does not integrate flows of material or highlight the relationship of amino acids to metabolites that are used for energy, such as glucose and fatty acids. Clearly, there must be interorgan cooperation to maintain protein homeostasis because tissues such as muscle have large amino acid reservoirs, yet all tissues have amino acid needs. A regular feeding schedule means that part of the day is a fasting period where endogenous protein is used for energy and gluconeogenesis. The fed period then supplies amino acids from dietary protein to replenish these losses and provide additional amino acids that can be used for energy during the feeding portion as well. This diurnal feeding and fasting pattern causes movement of amino acids among organs and takes on particular importance in situations of trauma and stress where adaptation, or lack of adaptation, of amino acid metabolism to physiologic insults or pathophysiologic states occur.

As Cahill has emphasized,[1,87] the primary consideration of the body is to maintain and distribute energy supplies (oxygen and oxidative substrates). The brain makes up only ≈2% of body weight yet has 20% of the energy needs.[88,89] The brain also lacks the ability to store energy (e.g., glycogen depots) and depends on constant delivery of energy substrates via the blood from other organs. As shown in **Figure 1.13A** the primary source of fuel for the brain is glucose derived from the liver. In infancy and early childhood, where the brain makes up a significantly greater proportion of body mass, glucose production and utilization rates are proportionately higher.[90] The pioneering studies of Cahill, Felig, Owen, and Wahren have provided us with a wealth of data concerning flows of amino acids and glucose from organ balance studies in humans studied over a range of nutritional states.[91] Some basic concepts may be defined from these studies.

In the postabsorptive state, the liver has glycogen stores available to provide most of the glucose exported through glycogenolysis and secondarily from glucose synthesis (gluconeogenesis) from amino acids, derived primarily by uptake from the blood. The source of amino acids is primarily from breakdown of muscle protein. Other substrates, e.g., glycerol released from triglyceride lipolysis, may also be used for gluconeogenesis, but amino acids provide the bulk of the gluconeogenic substrate. As discussed above, many amino acids provide gluconeogenic precursors. The remaining amino acids released from muscle protein breakdown not used for gluconeogenesis may be oxidized. The amino acid N released by this process is removed from the body by incorporation into urea via synthesis in the liver and excretion into urine via the kidney. Gluconeogenesis also occurs in the kidney, but the effect and magnitude are masked from AV measurements because the kidney is also a glucose consumer.[92,93]

Role of Skeletal Muscle in Whole Body Amino Acid Metabolism

An interesting observation from the early AV balance studies across the human leg and arm was that more than 50% of the amino acids released from skeletal muscle were in the form of alanine and glutamine,[91] yet alanine and glutamine comprise less than 20% of amino acids in protein (Table 1.3). Several possible reasons exist as to why alanine and glutamine are released from muscle in such large amounts. First, skeletal muscle oxidizes DAAs and BCAAs in situ for energy. Because amino acid oxidation produces waste N and because NH_3 is neurotoxic, muscle release of waste N as NH_3 must be avoided. Because both alanine and glutamine are readily synthesized from intermediates derived from glucose (alanine from transamination of pyruvate from glycolysis and glutamine from α-ketoglutarate), they are excellent vehicles

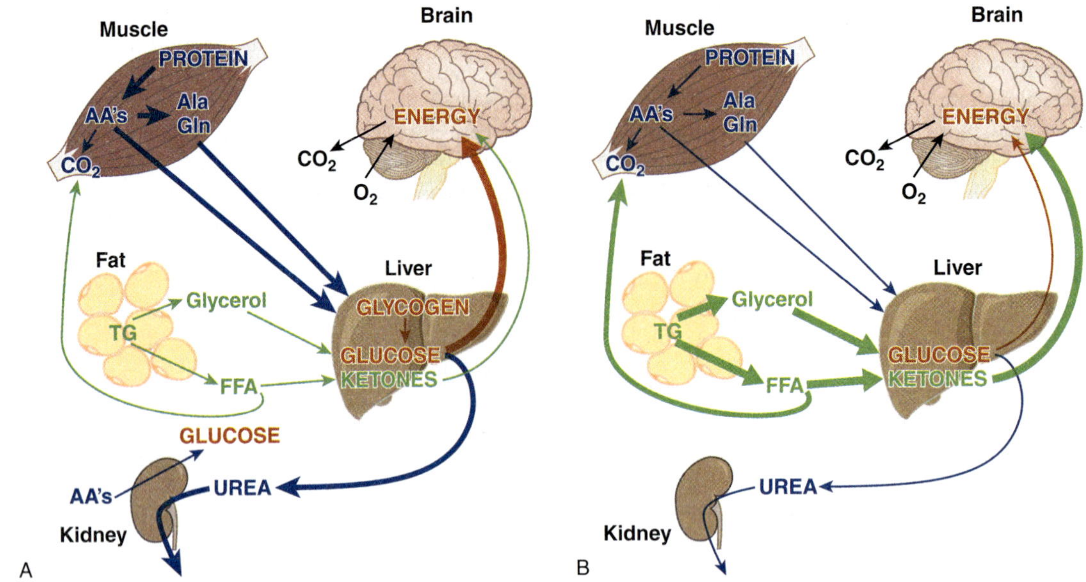

Figure 1.13 Interorgan flow of substrates in the body to maintain energy balance in the postabsorptive state (panel **A**) and after adaptation to starvation (panel **B**). The schematic diagrams are patterned after the work of Cahill. In all states, energy needs of the brain must be satisfied. In the postabsorptive state, glucose from liver glycogenolysis provides the majority of the glucose needed by the brain. After liver glycogen stores have been depleted (fasting state), gluconeogenesis from amino acids from muscle protein breakdown predominates as the glucose source. Eventually, the body adapts to starvation by production of ketone bodies by the liver from free fatty acids that then are used as a fuel by the brain instead of glucose, thereby sparing amino acid loss for gluconeogenesis. "AAs" means "amino acids; "Ala" and "Gln" are alanine and glutamine, respectively; "TG" and "FFA" are triglycerides and free fatty acids, respectively.

Modified from Cahill GF, Jr., and Aoki TT. (1980). Partial and total starvation. In Assessment of Energy Metabolism in Health and Disease, Report of the First Ross Conference on Medical Research (Kinney JM, ed) pp. 129-34, Ross Laboratories, Columbus, Ohio.

to remove waste N while avoiding NH_3 release. Alanine removes one and glutamine removes two Ns per amino acid. These observations led to the proposal of a glucose-alanine cycle where glucose made by the liver is taken up by muscle where glycolysis liberates pyruvate. The pyruvate is then transaminated to alanine and released from muscle. That alanine is extracted by the liver and transaminated to pyruvate, which is then used for glucose synthesis.[91] This scheme has been expanded to explain the use of BCAAs by muscle for energy and disposal through alanine of their amino-N groups. Such a scheme resolves a problem related to the BCAAs. In contrast, other IDAAs that cannot be metabolized by muscle are released to be metabolized in liver.

Metabolic Adaptation to Fasting and Starvation

As indicated in Figure 1.13A, lipolysis (breakdown of adipose triglyceride to free fatty acids and glycerol) plays a minor role in postabsorptive energy supply to the brain. However, liver glycogen stores are limited and will become depleted in less than 24 hours. That point in time when liver glycogen stores are exhausted is the beginning of the *fasting* state. In fasting, the glucose supply to the brain must come completely from gluconeogenesis, which means sacrificing amino acids from protein. Because protein is critical to body function, from enzyme activity to muscle function related to breathing and circulation, unrestrained use of amino acids for glucose production would rapidly deplete protein stores, causing death in a matter of days. However, people may survive weeks without food. As shown in **Figure 1.13B**, adaptation occurs with prolonged fasting by the brain switching from a glucose-based to a ketone-based fuel supply. Free fatty acids now released in larger quantities from lipolysis are converted into ketone bodies in the liver that can then be used by the brain and other tissues for energy. That conversion begins early in the fasting state and is complete under long-term periods of fasting. In starvation, tissues such as muscle may use free fatty acids directly for energy while the brain uses ketone bodies. The body's dependence on glucose as a fuel has been greatly reduced, thereby conserving protein. This adaptation process is complete within a week of onset of starvation (87) and is evidenced even for protein-restricted diets in terms of greatly reduced urea N production (Figure 1.7).

Although the body can accommodate to starvation, it is not a normal occurrence. The adaptations that normally occur are the daily shift between postabsorptive and fed periods. We spend our nights in the postabsorptive period, drawing from the nutrient stores of glycogen and protein for glucose production (Figure 1.13A). During the fed portion of the 24-hour day, dietary amino acids absorbed from digested protein are used to replete protein degraded to amino acids and used during the postabsorptive period via new protein synthesis. When dietary protein intake exceeds what is needed to replete nighttime losses, the excess amino acids are either oxidized or converted to glycogen and fat. Although muscle contains the bulk of body protein, all organs are expected to lose protein during the postabsorptive period and, therefore, need repletion during the fed period. What is poorly understood is how individual amino acids that enter through the diet are distributed among the various tissues in the correct amounts needed for each tissue. Just as each amino acid has its own separate metabolic pathway, the rates of absorption and use are expected to be different among amino acids. Thus, dietary protein requirements cannot be discussed without also considering the requirements of individual amino acids.

Digestion and Absorption of Protein

All dietary protein must be hydrolyzed to component amino acids to be used in the body. Digestion of protein begins with pepsin secretion in gastric juice in the stomach and with inactive proteolytic enzymes secreted from pancreas and mucosa of the small intestine.[94] These inactive enzymes or zymogens are activated in the small intestine after proteolytic cleavage by enteropeptidases secreted by intestinal mucosa. For example, the pancreas secretes the proenzyme (zymogen) trypsinogen that is then cleaved to the active proteolytic enzyme trypsin. The presence of dietary protein in the gut signals the secretion of the enzymes. As trypsin becomes activated, it binds to proteins at lysine and/or arginine residues breaking the peptide bond on the C-terminus side of these amino acids to form peptides of 2 to 20 or more amino acids. Some plants, such as soybeans, contain inhibitors of proteolytic enzymes. To use these plant proteins, the proteins need to be denatured by heating (i.e., cooking). For example, feeding uncooked soybeans to rats results in hypertrophy of the pancreas, presumably due to hypersecretion of trypsin that binds to these proteins but does not cleave them.[95]

The events of protein digestion and absorption are well established.[94,96] Proteins are successively broken down into smaller peptides based on the amino acid residues targeted by the proteolytic enzymes. For example, trypsin specifically targets lysine and arginine residues while pepsin has a relatively low specificity for neutral amino acids such as leucine and phenylalanine. In addition, exopeptidases attack the free ends of the peptide chains such as pancreatic-produced carboxypeptidases that cleave at the carboxyl terminus and aminopeptidases secreted in intestinal juice that cleave at the amino terminus of proteins.

Free amino acids are absorbed from the gut lumen by active transport into the mucosa by transporters specific to different types of amino acids.[96] At the same time, di- and tripeptides are also absorbed by the luminal side intact and are then hydrolyzed by peptide hydrolases present in the brush border and cytosol of the mucosal cells. There are specific transport systems for peptide uptake into mucosal cells separate from the transporters for amino acids. It is thought that a quarter of dietary protein is absorbed as di- and tripeptides.[97,98] For example, patients with the rare genetic Hartnup disease have a defect in renal and gut transport of selected amino acids and cannot transport tryptophan into mucosal cells but do absorb tryptophan when it is administered as a dipeptide.[99]

To a limited but important extent, some proteins and large peptides pass from the gut intact into the basolateral blood. Absorption of intact proteins or partially cleaved portions of proteins is a tenable physiologic explanation for diseases involving food allergies. The gut is generally viewed as an impermeable barrier where nutrients cross by active transport or where a break in the barrier occurs through cell injury. Small amounts of some proteins may pass this barrier by several possible mechanisms, such as through "leaks" between epithelial cell junctions or possibly by transport through uptake into vesicles from the lumen to the submucosal side of the epithelial cells.[100] Although the amount of protein passing through the gut intact may be minor, such absorption may be important in situations of immune response to the proteins or in delivery of some peptide drugs.

Protein and Amino Acid Requirements

A fundamental question in nutrition is *what amount of protein is required in the diets of humans to maintain health?* This question has several subparts. First, we must evaluate the intake of both protein and the amounts of the individual amino acids in that protein. Second, we need to evaluate requirements across (a) the complete range of life and development, (b) in sickness and in health, and (c) under different conditions of work and environment. For these reasons, protein requirement has been defined as "the lowest level of dietary protein intake that will balance the losses of N from the body, and thus maintain the body protein mass, in persons at energy balance with modest levels of physical activity plus, in children or in pregnant or lactating women, the needs associated with the deposition of tissues or the secretion of milk at rates consistent with good health".[6]

When we discuss the amino acid composition of a specific protein source, we generally focus on IDAA content because this is the *indispensable* portion of amino acid intake. Which amino acids are dispensable and which are indispensable were originally determined by testing whether a diet deficient in a particular amino acid would support growth in a rat. However, there are important species differences between rats and humans that limit the comparison.

A technique for studying amino acid requirements in humans is the N balance method. A diet that is adequate in total N, but deficient in a specific IDAA, cannot produce a positive N balance because protein can only be synthesized if every amino acid is present in adequate amounts. The body is then faced with a dietary excess of the other *nonlimiting* amino acids that it cannot put into protein, and these excess amino acids must be oxidized to urea, producing a negative N balance.

The classic studies of Rose and colleagues measured N balance in humans fed diets deficient in individual amino acids. They determined that eight amino acids produced a negative N balance when they were deficient in the diet of adult humans.[101,102] Although the IDAAs cannot be synthesized de novo, several IDAAs have a catabolic pathway in which the first step in their catabolism is reversible transamination to form a keto acid. For example, BCAAs transaminate to form branched-chain keto acids, but this process is equally reversible.[103] Growth can be supported in rats by replacing the dietary IDAAs that transaminate with their keto acid analogues, and various formulations have been proposed for supplying the carbon skeleton of IDAAs without adding N, which is detrimental in disease states such as renal disease.[104]

Do DAAs ever become indispensable? If a DAA is used in the body faster than it can be synthesized, it becomes indispensable for that *condition*.[2] For example, tyrosine and cysteine are synthesized from phenylalanine and methionine, respectively. If phenylalanine or methionine is consumed in insufficient amounts, tyrosine and cysteine will become deficient and *indispensable*. This question must be evaluated across the range of life from infancy to the elderly as well as in sickness and health. For example, enzymes for amino acid metabolism mature at different rates in the growing fetus and newborn infant. Histidine is indispensable in infants but not necessarily in healthy children or adults.[6,105] Therefore, the classification of "indispensable" or "dispensable" is dependent upon (a) species, (b) life stage (e.g., infant, growing child, or adult), (c) diet, (d) nutritional status and (e) pathophysiologic condition. There is also the question of whether a particular amino acid *in excess of requirement* has properties that may ameliorate or improve a clinical condition. These considerations must be evaluated for each population group where they are important.

Protein Requirements

Determination of protein requirement must consider both the amount of amino acid N as well as its quality, that is, the ability to digest and absorb protein and its IDAA content (6). The simplest approach to the measurement of the nutritional quality of a protein is to measure the ability of that protein to promote growth in young, growing animals. Their growth is dependent upon synthesis of new protein that is dependent on IDAA intake. Because alterations in rat-growth can be measured in only a few days, the growing rat has been often used as the model to compare differences in quality (composition) of protein/amino acid diets. Because this approach cannot ethically be applied to humans, other approaches for assessing *human* requirements are used.

Factorial Method

When a person is placed on a protein-free diet, amino acid oxidation and resultant urea production will decrease over a several-day period as the body tries to conserve its resources, but amino acid oxidation and urea production do not drop to zero (Figure 1.7). There will always be some *obligatory* oxidation of amino acids producing some urea formation and miscellaneous losses of N. The factorial method assesses all routes of losses possible for adult humans on a N-free diet. The minimum daily requirement of protein is assumed to be that amount that matches the sum of the various obligatory N losses.

A variety of studies have been performed to assess these losses, and the results have been tabulated and used as a basis for determining protein requirements in the 1985 FAO/WHO report.[106] A total obligatory daily loss of endogenous N of 54 mg/kg/d was assumed for men in a temperate climate, which corresponds to a protein intake of 0.34 g/kg/d (where 1 g N = 6.25 g protein). However, there are additional obligatory N losses for people living in a warmer climate that must be included in this group. These values were then adjusted upward to account for inefficiency of use of dietary protein and for the quality (amino acid composition and digestibility) of the source of protein consumed. For children and pregnant or lactating women, an additional (theoretically determined) amount of protein is added to this recommendation to account for growth and milk formation.

The factorial approach is based on extrapolation of N losses from protein-starvation conditions and may reflect an adaptation to N deprivation that may not be reflective of normal metabolism and N requirements of healthy humans near the actual requirement level. Rand and Young[107] have also shown that the relationship between protein intake and N retention is curvilinear, making it troublesome to extrapolate obligatory N loss to a protein requirement. Hence, the 2002 report put less weight on the factorial method in assessing protein requirements and more weight on the balance method.[6,108]

Balance Method

In the balance method, subjects are fed varying amounts of protein or amino acids, and the *balance* of a particular parameter, usually *N*, is measured. An adequate amount of dietary protein is that level of intake that will maintain a neutral or slightly positive N balance. The balance method can be used to titrate N intake in infants, children, and women during pregnancy where the end point is a balance positive enough to allow for appropriate accretion of new tissue. The balance method is also useful for testing the validity of factorial method estimates. In general, N-balance studies in which dietary protein intake is titrated give higher measures of protein requirements than predicted by the factorial method.

The N balance method has important errors.[6,107,109,110] Urine collections tend to underestimate N losses, and intake tends to be overestimated. Miscellaneous losses are "best guesses" and may have small but substantial errors. Although these factors affect both methods, the balance method "titrates" dietary intake to determine zero balance, and that response is nonlinear when protein intake is increased from grossly deficient toward an adequate status.[107,110] Rand et al[110] systematically reviewed N-balance studies determining protein requirements in a meta-analysis. Through a very careful analysis of all factors, the 2002 Food and Nutrition Board report adopted the results of this study and set the median estimated average requirement (EAR) at 0.66 g/kg/d for protein for adult men and women ages 19 and older.[108] This recommendation assumes that most studies of N balance have been performed at adequate levels of energy intake, because N balance is affected by energy intake. Decreasing energy intake causes the N balance measurement to go from zero to negative when the protein intake is near the requirement. The recommendation also considers protein quality and digestibility of the protein being consumed. Generally, it is assumed that protein of lesser quality and digestibility than egg white will be consumed, and a correction factor is added.

Recommended Dietary Allowances for Protein

In 1989, the Food and Nutrition Board subcommittee of the U.S. National Research Council updated their recommended dietary allowances (RDA) for protein and amino acids,[111] which were previously based on the 1985 FAO/WHO/UNU committee report.[106] The Food and Nutrition Board prepared a new report in 2002 on dietary reference intakes for a range of macronutrients, including protein and amino acids.[108] The RDA values for protein shown in **Table 1.8** are based on this report and reflect N balance data (rather than factorial method data) from studies where a high-quality, highly digestible source of protein was used. Data are presented for the EAR for protein. The EAR represents protein intake that produced zero N balance in half of the population. That value was then increased by two standard deviations to encompass 97.5% of the population to get the RDA for the reference protein. For example, from studies of young adult men, the EAR value of 0.66 g/kg/d was increased to 0.80 g/kg/d of protein for the RDA.[108]

Special cases must be considered where growth and accretion of tissue occur: during pregnancy, during lactation, in infants, and in children. In pregnancy, total protein deposited was estimated to be 925 g, based on maternal weight gain and average birth weight at term. Rates of protein accretion were then divided by trimesters with adjustment for variation in birth weight (+15%) and an assumed efficiency of conversion of dietary protein to fetal, placental and maternal tissues (+70%) to produce increments in reference protein intake of +1.0, +6.3, and +10.6 g protein/day for the 1st, 2nd, and 3rd trimesters, respectively.[108] The amount of additional protein needed to be added to the diet during the last two trimesters to compensate for uncertainties about rates of tissue deposition and maintenance of those increases is estimated to be an EAR of +21 g

Table 1.8 Recommended Intakes of High-Quality Protein for Normal Humans

Age (years)	EAR[a] (g protein/kg/d)	RDA[b] (g protein/kg/d)
0–0.5		1.52[c]
0.5–1	1.1	1.5
1–3	0.88	1.10
4–8	0.76	0.95
9–13	0.76	0.95
14–18	0.72	0.85
>18	0.66	0.80
Pregnancy	0.88	1.10
Lactation	1.05	1.10

[a]EAR–estimated average requirement: the intake that meets the estimated nutrient needs of half of the individuals in a group.

[b]RDA–recommended dietary allowance: the intake that meets the nutrient need of almost all (97.5%) of individuals in a group.

[c]Value for infants in the first half-year of life is the adequate intake estimate determined in that population that appears to sustain a defined nutritional status, including growth rate, normal circulating nutrient values and other functional indicators of health. This value is not equivalent to an RDA.

Data from Food and Nutrition Board - Institute of Medicine (U.S.). (2002). Proteins and Amino Acids. In Dietary Reference Intakes for Energy, Carbohydrate, Fiber, Fat, Fatty Acids, Cholesterol, Protein, and Amino Acids pp. 589–768, National Academy Press, Washington, DC

protein/d or an RDA of +25 g/d additional protein over prepregnancy needs.[108]

Women who are lactating also require additional protein intake. Using the factorial method and data for the protein content of human milk, volume of milk produced, and adding an adjustment for the estimated 50% conversion efficiency of dietary protein into newly synthesized milk protein, an increase of +23.4 g protein/d was added to the EAR of women in their first month of lactation. The EAR drops to +22 g/d in month 2 and to +18.3 g/d for months 4–6 of lactation.[108] To compensate for variance among women, the EAR value is increased to an RDA of +25 g /d additional protein for women in their first month of lactation.

Amino Acid Requirements

Recommendations for intake of individual amino acids were largely based on the pioneering work of W. C. Rose and his colleagues in the 1950s.[101] Irwin and Hegsted reviewed these and other studies of amino acid requirements published before 1971.[112] Rose's studies were N-balance measurements of young males placed on diets where N intake consisted of a mixture of crystalline amino acids. The intake of a single amino acid could be altered, and the N balance measured. Because of the expense of crystalline amino acid diets and the difficulty in performing serial N-balance studies at different intakes, Rose and colleagues only studied a handful of subjects per amino acid. Problems with interpreting the N-balance data for a limited number of subjects cloud the extrapolation of these data to populations[113,114]; yet, the data of Rose have been the primary basis for amino acid recommendations in adults.

Direct Amino Acid Oxidation (DAAO) Method

An alternative approach was taken by Young and colleagues, based on the method of Harper and others to assess amino acid requirements in growing animals using amino acid oxidation as an index of dietary sufficiency. Animals fed an insufficient amount of a specific individual test amino acid reduce their oxidation of that amino acid to obligatory levels. The oxidation of the dietary-deficient amino acid will remain at obligatory oxidation levels as intake is increased until the requirement level is met. As dietary intake of the test amino acid rises above requirement, the excess amino acid intake will be oxidized. Therefore, a two-line curve of amino acid oxidation should appear when plotted against the test amino acid intake: a flat line below requirement (indicating obligatory oxidation) and a rising curve above the requirement (indicating oxidation of excess amino acid intake). The requirement level for the test amino acid should be the intersection of the two curves, i.e., where oxidation of excess amino acid begins.

The DAAO method uses the breakpoint in amino acid oxidation as a function of the intake of the test amino acid to determine requirement. Amino acid oxidation is determined by administering a ^{13}C- or ^{14}C-labeled amino acid tracer of the test amino acid. The tracer amino acid is administered at the end of each diet period. The DAAO method was used by Young and coworkers to estimate amino acid requirements for isoleucine, leucine, lysine, phenylalanine/tyrosine, and valine in healthy adults.[115,116]

Indicator Amino Acid Oxidation (IAAO) Method

Zello and colleagues[117] took a different approach to the measurement of amino acid requirements. Rather than administer and measure the oxidation of a tracer of the same amino acid manipulated in the diet, they use the oxidation of a tracer of another IDAA as an *indicator* of N balance. Nitrogen balance becomes negative when an IDAA is deficient in the diet because protein synthesis is limited, and the excess dietary intake of all other amino acids are oxidized. The excess oxidation is measured by a ^{13}C tracer of the indicator amino acid (IAA). As the dietary intake of the test amino acid is decreased below requirement level, oxidation of the IAA will increase.[118] An example is shown in **Figure 1.14** where [1-^{13}C]phenylalanine is infused as the IAA in young adult men consuming different levels of dietary threonine intake.[119] The intakes of all other amino acids are held constant (including that of the IAA, phenylalanine). At threonine intake above requirement, phenylalanine oxidation is constant, but phenylalanine oxidation progressively increases as threonine intake is decreased below the threonine requirement. The breakpoint between the two curves in Figure 1.14 indicates the mean EAR for threonine intake. The RDA for threonine intake is set at two standard deviations above the EAR.

The key to this method is the availability of an IAA tracer whose oxidation can be accurately and precisely measured, which is different from the test amino acid being manipulated in the diet. Using this approach and [1-^{13}C]phenylalanine as the IAA, the Pencharz group redetermined requirement levels for several different amino acids.[118,120,121] These estimates are largely in agreement with estimates by the DAAO method. However, a concern about the IAAO method has been the

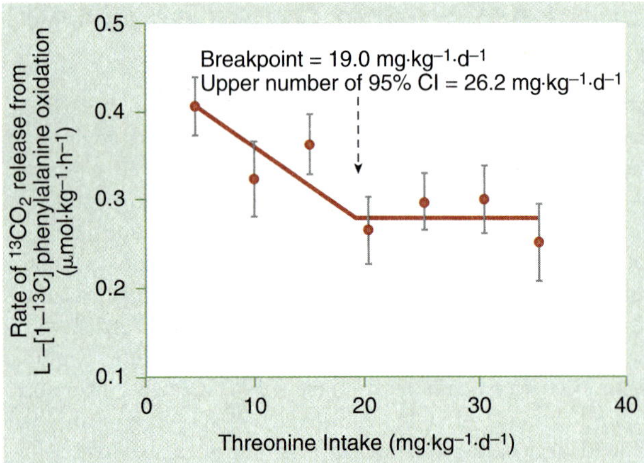

Figure 1.14 Oxidation of the indicator amino acid tracer [1-^{13}C]phenylalanine to $^{13}CO_2$ in young men fed different dietary intakes of threonine. Phenylalanine oxidation is constant above the dietary requirement threonine but progressively increases as threonine intake drops below requirement because limitation of threonine intake limits the body's ability to synthesize protein, causing the excess amino acids to be oxidized, including the indicator amino acid, phenylalanine. Thus, the breakpoint between the two lines indicates the threonine requirement in these subjects.
CI = confidence interval.

Reprinted with permission from Wilson DC, Rafii M, Ball RO et al. Threonine requirement of young men determined by indicator amino acid oxidation with use of L-[1-(13)C] phenylalanine. Am J Clin Nutr 2000;71:757–64. Copyright American Society for Clinical Nutrition.

relatively short (e.g., 3 day) adaptation periods used for the different dietary intakes tested. Classical N balance studies require 7–10 days for equilibration to occur in urinary N output, but this constraint is not required using the IAA to measure oxidation directly. Although Elango et al. demonstrated that short adaptation periods can be used with the IAAO method,[122] the impact of these short adaptation periods remains controversial.

Twenty-Four Hour Tracer Balance Method

The DAAO and IAAO methods use very short ^{13}C tracer administration periods and do not account for the fact that we oxidize amino acids 24 hours a day, not just during daytime periods of feeding. El-Khoury et al. infused a [1-^{13}C]leucine tracer for 24 hours into subjects receiving different intakes of leucine to determine leucine requirements by the DAAO method.[123,124] Borgonha et al. infused [1-^{13}C]leucine for 24 hours as an IAAO tracer into subjects receiving different intakes of threonine to determine threonine requirements.[125] Similar studies have been performed to redefine amino acid requirements in humans and have been used to estimate recommendations for amino acid intakes[126] that are considerably greater for several IDAAs than have been determined previously by the N balance method. Overall, there is consistency in these increased requirements among the three approaches used: DAAO, IAAO, and the 24-hour tracer balance methods.

Both the Food and Nutrition Board/Institute of Medicine and the FAO/WHO/UNU considered the myriad of new ^{13}C tracer data in making their current recommendations.[6,108] The first four columns shown in **Table 1.9** for infants, children, and adults come from the 2002 Food and Nutrition Board/Institute of Medicine report.[108] The RDAs for infants decreased with this report for most amino acids. The RDAs for children were reduced primarily for the BCAAs, but the RDAs for amino acids

Table 1.9 Estimates of Recommended Dietary Amino Acid Allowances (mg/kg/d) by Age Group

	Infants[a]	Children[a]		Adults (>18 y)	
Amino Acid	**7–12 mo**	**1–3 y**	**4–13 y**	**FNB[a,b] 2002**	**FAO/WHO[c] 2007**
Histidine[d]	32	21	16	14	10
Isoleucine	43	28	22	19	20
Leucine	96	62	48	42	39
Lysine	89	58	45	38	30
Methionine + cysteine	43	28	22	19	15
Phenylalanine + tyrosine	84	54	41	33	25
Threonine	49	32	24	20	15
Tryptophan	13	8	6	5	4
Valine	58	37	28	24	26

[a]Data for the first four columns obtained from Food and Nutrition Board - Institute of Medicine (U.S.). (2002).[108]
[b]FNB is the Food and Nutrition Board
[c]Data in the last column is from FAO/WHO/UNU. (2007).[6]
[d]Although no requirement for histidine has been quantified beyond infancy, histidine has been recommended for children and adults based on the histidine content of the RDA for protein of each of these age groups.

Data from FAO/WHO/UNU. (2007). Protein and Amino Acid Requirements in Human Nutrition Vol. technical report series #935, World Health Organization, Geneva, Switzerland; Food and Nutrition Board - Institute of Medicine (U.S.). (2002). Proteins and Amino Acids. In Dietary Reference Intakes for Energy, Carbohydrate, Fiber, Fat, Fatty Acids, Cholesterol, Protein, and Amino Acids pp. 589-768, National Academy Press, Washington, DC.

where ^{13}C tracer data were available increased significantly from the older N balanced-based 1985 recommendations.[106] The last column in Table 1.9 reports recommendations from the 2007 FAO/WHO/UNU report for adults.[6] Only small adjustments were made going from 2002 to 2007. However, requirements increased from the 1985 recommendations to almost double for those amino acids where ^{13}C tracer data were available. For example, the EAR for leucine increased from 14 to 39 mg/kg/d and that of lysine increased from 12 to 30 mg/kg/d.

Histidine

Although histidine has been shown to be indispensable to the diet of the rat, it has been hard to define as indispensable to the diet of adult humans.[112] Older studies of adults suggest that the requirement for histidine may be less than 2 mg/kg/d,[127] but a requirement for histidine has not been clearly documented.[102] Dietary histidine restriction was studied in adults over a 48-day period using both the IAAO method and other measurements of protein metabolism to assess metabolic changes. The results showed adaptation to a histidine-free diet but did not provide evidence of histidine indispensability.[105] Nonetheless, histidine is listed as an IDAA and assigned an ERA of 10 mg/kg/d (Table 1.9) based on histidine content of protein and the RDA for protein.

Why is it so difficult to define whether histidine is indispensable in adults when there is little evidence that a metabolic pathway for histidine synthesis exists in humans? The difficulties occur because the requirement for histidine is small and the stores of histidine in the body are large.[4,102] Histidine is particularly abundant in hemoglobin and carnosine (the dipeptide β-alanylhistidine, which is present in large quantities in muscle). Furthermore, gut flora synthesize an unknown amount of histidine, which may be absorbed and used. Histidine must be removed from the diet for more than a month to observe effects, and those effects are indirect measures of histidine insufficiency (a fall in hemoglobin and rise in serum iron) rather than alteration in conventional indices (N balance).

Use of the IAAO Method to Assess Protein Requirements

The use of the IAAO method has been extended beyond IDAA requirements in humans. It has also been used as an alternative to the N-balance method for assessing protein requirements.[121,128] Again, because the IAAO method is applied over very short dietary adaptation periods, it has been applied to states of dynamic change in metabolism and to assess protein requirements of children[129] and in pregnancy[130] without requiring additional factors. However, these estimates of EAR for protein in children and in pregnancy are about 40% higher than that of the current recommendations (Table 1.8). The same authors found similar +40% increases in their estimates for protein in both young and elderly adults.[131] These higher protein requirements determined that using the IAAO method would have significant impact on nutritional and health care policies if the recommendations were accepted. However, concerns about the IAAO method and how the studies are conducted remain.[132]

Assessment of Protein Quality

The "quality" of a protein is defined by its ability to support growth in animals. High-quality protein produces a faster growth rate. Growth rate measurements evaluate the factors important for a protein: (a) pattern and abundance of IDAAs, (b) relative amounts of DAAs ves IDAAs, (c) digestibility when consumed, and (d) presence of toxic materials, trypsin inhibitors or allergenic stimuli. Methods to define the quality of a formula or protein source have fallen generally into two categories: empirical biological assays and scoring systems.

Biological Assays

It is assumed that the "highest-quality protein" is protein that supports maximal growth in a young animal. Because rats grow quickly, have limited protein stores, and a high metabolic rate, it is easy to detect deficiencies and imbalances in amino acid patterns in young growing rats in a short period of time. The *protein efficiency ratio (PER)* has been defined as the weight gained (in grams) divided by the amount of test protein consumed (in grams) by a young growing rat over a several-day period. Obviously, duration of diet, age, starting body weight, and species of rat employed are important variables. Typically, 21-day old male rats fed 9%–10% protein (by weight) for 10 days to 4 weeks have been used. In one series of tests, casein produced a PER of 2.8, soy protein 2.4, and wheat gluten 0.4, demonstrating that gluten is a poor-quality protein. Such an approach has been useful in defining the relative efficacy of clinical formulas used in enteral and parenteral nutrition.[133] The formula that provides the optimal mixture of indispensable and DAAs should induce the most rapid growth. However, this method's results will be skewed in application to humans, depending on the extent that human requirements for individual amino acids mimic those of the rat. Nonetheless, the method has been useful in comparing a new protein source against reference proteins, such as egg protein, and allows evaluation of other factors, such as relative digestibility.

Scoring Systems

As an alternative to animal growth as an indicator of protein quality, scoring methods have been developed to assign a quantitative value to the pattern of amino acids in a specific protein source or to a diet containing multiple, different proteins. Assignment of quality is based on the amounts of individual IDAAs in the protein source referenced against the IDAA content of an ideal protein source. The scoring system is easy to apply because no animal studies or clinical studies are required to compare different nutritional formulations.

The ideal content of individual IDAAs in the reference protein is not based on a specific protein's IDAA content, but rather on protein requirements in Table 1.8 and amino acid requirements found in Table 1.9.[134] For example, the content of leucine in the reference protein (mg leucine/g protein) would be

$$\text{Reference leucine content} = \frac{\text{leucine requirement}}{\text{protein requirement}} = \frac{39\ \dfrac{\text{mg Leu}}{\text{kg}\cdot\text{d}}}{0.66\ \dfrac{\text{g protein}}{\text{kg}\cdot\text{d}}} = 59\ \frac{\text{mg Leu}}{\text{g protein}}$$

The *chemical score* of a protein is calculated in two steps. First an amino acid score (AAS) is calculated for each IDAA in the protein against the reference protein:

$$\text{AAS} = \frac{\text{content of the IDAA in the } \textit{test} \text{ protein/misture}}{\text{content of the IDAA in the } \textit{reference} \text{ protein/mixture}} \cdot 100$$

Because the IDAA with the lowest score is the *limiting amino acid* in the protein, that lowest score is assigned to the protein. Protein digestibility is also considered, so that the final value assigned is the protein digestibility corrected amino acid score (PDCAAS).

A commonly limiting amino acid is lysine, which is particularly low in cereal grains and rice.[49] The next limiting amino acids are the sulfur-containing amino acids, threonine and/or tryptophan. The BCAAs and phenylalanine/tyrosine are not usually limiting. The scoring method is particularly useful for balancing intakes of multiple proteins that have individual different deficiencies. For example, combining protein from beans (AAS for lysine = 1.5) with that of corn (AAS for lysine = 0.6) provides a protein mixture with a PDCAAS >1.0.

Protein and Amino Acid Requirements in Disease and Special Conditions

Although the effect of disease and special conditions on amino acid and protein requirements is beyond the limits of this introductory chapter, a few general points can be made. The first is that energy and protein needs are tied together, as illustrated in Figure 1.13. When metabolic rate rises, body protein is mobilized for use as fuel (amino acid oxidation) and for supply of carbon for gluconeogenesis. Several disease states produce an increase in metabolic rate. The first is infection, where the onset of fever is a hallmark of increased metabolic rate. The second is injury, be it trauma, burn injury, or surgery. Along with onset of a hypermetabolic state comes a characteristic increase in the loss of protein measured by increased urea production. In 1930, Sir David Cuthbertson observed that a simple bone fracture causes significant loss of N in the urine.[135] Since then, numerous studies of the hypermetabolic state of injury and infection have been performed. For most people, injuries are minimal and self-limiting, i.e., the fever goes away in a couple of days or the injury heals. In normal, healthy people, the impact of the injury on overall protein metabolism has a minimal impact similar to that produced by a period of fasting. However, in critical or chronic, long-term illness, or in patients otherwise weakened due to age or other factors, the onset of a hypermetabolic state may produce a significant and dangerous loss of body N. Extensive information about hypercatabolic states and nutritional support in surgery, trauma, sepsis, and burn injury can be found in Chapters 94 and 95.

Ratio of Indispensable to Dispensable Amino Acids in Protein

Protein requirements are greatest in infancy (Table 1.8) because of the high rates of accretion of new protein with growth. Protein requirements then decrease with maturity. When the change in requirements for IDAAs in Table 1.9 are compared with the total protein requirements in Table 1.8, a greater drop with age is seen in IDAA requirements than in protein requirements. IDAAs make up >30% of protein requirements in infancy and early childhood, drop to 20% in later childhood, and to 11% in adulthood. As IDAAs become a less important part of the amino acid requirements with age, DAA intake could increase and become an increasingly greater proportion of our intake. Tessari addressed the question of DAA usage for protein replenishment through a series of parallel calculations and estimations.[136] Based on these and other calculations, it appears that when we consume high-quality protein at or above the RDA, we consume an excess of individual IDAAs beyond requirements. In general, it is not hard for adults to meet the minimum for IDAA intake, as recommended[6,108] when protein is consumed at or above requirement.

Do Some DAAs Become Indispensable?

Are there specific amino acids that provide therapeutic benefit when administered in a pathophysiologic state such as injury, trauma, infection, or cancer? Glutamine and arginine have probably been the most thoroughly studied and tested for therapeutic benefit in a range of pathophysiologic states.[137-140] Glutamine is the most highly concentrated amino acid inside muscle cells and in plasma.[141] Glutamine is an important nutrient to many cells, especially the gut and white blood cells, where glutamine may be used as a source of energy and for critical processes, such as the synthesis of nucleotides. Glutamine is an essential nutrient for cell culture media. Because a hallmark of injury is a drop in the intracellular level of muscle glutamine, presumably due to increased use by other tissues, glutamine has been proposed as a nutrient that becomes conditionally indispensable in trauma and infection.[112,113] Arginine is another DAA with important properties for promoting immune system function. Arginine is the precursor for nitric oxide synthesis[144] and has been proposed as a nutrient for altering immune function and improving wound healing.[145,146] Although supplementation of specific amino acids or cofactors may produce beneficial responses, there may be occasions where supplementation may produce undesirable effects on the disease state. Thus, the benefit of specific amino acid supplementation in specific pathophysiologic states remains difficult to assess and requires a multi-factorial approach in clinical trials. Detailed information about glutamine, arginine, and other amino acids can be found in Chapter 34.

Are there normal physiologic states that have increased needs for specific DAAs? As shown in Table 1.7, glycine is a DAA used for synthesis of creatine, glutathione, heme, nucleic, and uric acids. Wang et al estimated that use of glycine for the synthesis of these various compounds may be greater than the sum of glycine de novo synthesis plus dietary glycine intake.[147] Under such conditions, glycine would become conditionally indispensable and would need to be supplemented in the diet. Pregnancy is a physiologic condition that increases protein requirements due to accretion of new tissue. During this period of rapid accretion of new tissue, could glycine become conditionally indispensable? Rasmussen et al applied the IAAO method along with feeding of crystalline amino acids in adequate amounts, but with variable amounts of glycine, to determine possible glycine requirements in various stages of pregnancy. Because of the short adaptation periods to the various diets used with the IAAO method, this study could be conducted in pregnant women. They also added other measurements of glycine synthetic capacity (e.g., 5-oxoproline formation indicative of limited glutathione synthesis) to the study. No requirement for glycine was found except in late gestational stages of pregnancy, where a breakpoint (requirement) of 37 mg glycine/kg/d was noted.[148] A pregnant woman is recommended to consume 1.1 g protein/kg/d (RDA Table 1.8), Consumption of this amount of high-quality protein contains at least 37 mg glycine/kg/d, suggesting that the RDA provides for adequate glycine intake. However, this novel work provides a new method at addressing some of these questions about DAA synthesis and requirements under various physiologic conditions.

Of the three macronutrients, protein may be considered the most important. Proteins, composed of a diverse group of amino acids, play critical roles ranging from enzymes that regulate and control metabolism to structural proteins that provide muscle for movement and breathing. Thus, we have a complex set of mechanisms to keep the balance of amino acids and protein stable as we daily go from overnight fasting to repletion by meal feeding and to restrict use of amino acids as fuel during starvation. Because we cannot synthesize de novo several amino acids, dietary intake of appropriate amounts of these indispensable amino acids is critical. For all these reasons, much research has gone into defining required amounts of proteins and individual amino acid intakes in diets of humans of all ages. Even in the 21st century, our understanding of protein and amino requirements continues to evolve and expand as new research methods are developed. As will be discussed in later chapters, our understanding of protein and amino acid requirements is particularly limited in many disease states. There is still much research yet to be done to understand alterations of protein and amino acid metabolism in states of pathophysiology.

References

1. Cahill GF. N Engl J Med 1970;282:668–75.
2. FAO/WHO/UNU. *Protein and Amino Acid Requirements in Human Nutrition* Vol. technical report series #935, World Health Organization, Geneva, Switzerland, 2007.
3. Chipponi JX, Bleier JC, Santi MT, et al. Am J Clin Nutr 1982; **35**:1112–6.
4. Harper AE. Dispensable and indispensable amino acid interrelationships. In *Amino Acids: Metabolism and Medical Applications* (Blackburn GL, Grant JP, and Young VR, eds) pp. 105–21, John Wright, Boston, 1983.
5. Laidlaw SA, Kopple JD. Am J Clin Nutr 1987;46:593–605.
6. Block RJ, Weiss KW. *Amino Acid Handbook: Methods and Results of Analysis*, Charles C. Thomas, Springfield, Ill, 1956.
7. Bergström J, Fürst P, Noree LO, et al. J Appl Physiol 1974;36:693–7.
8. Cohn SH, Vartsky D, Yasumura S, et al. Am J Physiol Endocrinol Metab 1980;239:E524–E30.
9. Heymsfield SB, Waki M, Kehayias J, et al. A. J Physiol Endocrinol Metab 1991;261:E190–E8.
10. Poncet N, Taylor PM. Curr Opin Clin Nutr Metab Care 2013;16:57–65.
11. Gauthier-Coles G, Vennitti J, Zhang Z, et al. Nat Commun 2021;12:5282.
12. Bröer S, Gauthier-Coles G. J Nutr 2021;152:16–28.
13. Broer S, Palacin M. Biochem J 2011;436:193–211.
14. Hundal HS, Taylor PM. Am J Physiol Endocrinol Metab 2009;296:E603–E13.
15. Ben Galim E, Hruska K, Bier DM, et al. J Clin Invest 1980;66: 1295–304.
16. Yoshida T, Kikuchi G. Arch Biochem Biophys 1970;139:380–92.
17. Yoshida T, Kikuchi G. J Biochem (Tokyo) 1972;72:1503–16.
18. Katagiri M, Nakamura M. Biochem Biophys Res Commun 2003;312:205–8.
19. Matthews DE, Conway JM, Young VR, et al. Metabolism 1981;30:886–93.
20. Stipanuk MH, Dominy JE, Jr., Lee JI, et al. J Nutr 2006;136: 1652s–9s.
21. Hayes KC. Nutr Rev 1985;43:65–-70.
22. Franco R, Schoneveld OJ, Pappa A, et al. Arch Physiol Biochem 2007;113:234–58.
23. Wu G, Fang YZ, Yang S, et al. J Nutr 2004;134:489–92.
24. Rebouche CJ, Seim H. Annu Rev Nutr 1998;18:39–61.
25. Meldrum BS. J Nutr 2000;130:1007S–15S.
26. Rothstein JD, Martin LJ, Kuncl RW. N Engl J Med 1992;326:1464–8.
27. Wyss M, Kaddurah-Daouk R. Physiol Rev 2000;80:1107–213.
28. Brosnan JT, da Silva RP, Brosnan ME. Amino Acids 2011;40: 1325–31.
29. Bloch K, Schoenheimer R. J Biol Chem 1941;138:167–94.
30. Heymsfield SB, Arteaga C, McManus C, et al. Am J Clin Nutr 1983;37:478–94.
31. Wang Z, Gallagher D, Nelson M, et al. *Am J Clin Nutr* 1996;63: 863–9.
32. Evans WJ, Hellerstein M, Orwoll E, et al. J Cachexia Sarcopenia Muscle 2019;10:14–21.
33. Clark RV, Walker AC, O'Connor-Semmes RL, et al. J Appl Physiol 2014;116:1605–13.
34. Milner JA, Visek WJ. Nature 1973;245:211–2.
35. Schoenheimer R. *The Dynamic State of Body Constituents*, Harvard University Press, Cambridge, Mass, 1942.
36. Hellerstein MK, Munro HN. Interaction of liver and muscle in the regulation of metabolism in response to nutritional and other factors. In *The Liver: Biology and Pathobiology* (Arias IM, Jakoby WB, Popper H, Schachter D, and Shafritz DA, eds) pp. 96–83, Raven Press, New York, 1988.
37. Scrimshaw NS, Hussein MA, Murray E, et al. J Nutr 1972;102: 1595–604.
38. Wolfe RR, Chinkes DL. *Isotope Tracers in Metabolic Research: Principles and Practice of Kinetic Analysis*, Wiley-Liss, Hoboken, NJ, 2004.
39. Holm L, O'Rourke B, Ebenstein D, et al. (2013) Am J Physiol Endocrinol Metab 2004;308:E895–E907.
40. San Pietro A, Rittenberg D. J Biol Chem 1953;201:457–73.
41. Picou D, Taylor-Roberts T. Clin Sci 1969;36:283–96.
42. Bier DM, Matthews DE. Fed Proc 1982;41:2679–85.
43. Steffee WP, Goldsmith RS, Pencharz PB, et al. Metabolism 1976;25: 281–97.
44. Yudkoff M, Nissim I, McNellis W, et al. Pediatr Res 1987;21:49–53.
45. Waterlow JC, Golden MHN, Garlick PJ. Am J Physiol Endocrinol Metab 1978;235:E165–E74.
46. Fern EB, Garlick PJ, McNurlan MA, et al. Clin Sci 1981;61:217–28.
47. Matthews DE, Motil KJ, Rohrbaugh DK, et al. Am J Physiol Endocrinol Metab 238:E473–E9.
48. Toth MJ, MacCoss MJ, Poehlman ET, et al. Am J Physiol Endocrinol Metab 2001;281:E233–E41.
49. Matthews DE. J Nutr 2020;150:2548S–55S.
50. Matthews DE. J Nutr 2007;137:1549S–55S.
51. Allsop JR, Wolfe RR, Burke JF. J Appl Physiol 1978;45:137–9.
52. El-Khoury AE, Sánchez M, Fukagawa NK, et al. J Nutr 1994;124:1615–27.
53. Matthews DE, Bier DM, Rennie MJ, et al. Science 1981;214:1129–31.
54. Matthews DE, Schwarz HP, Yang RD, et al. Metabolism 1982;31:1105–12.
55. Bier DM. Diabetes Metab Rev 1989;5:111–32.
56. Tessari P, Barazzoni R, Zanetti M, et al. Am J Physiol Endocrinol Metab 1989;271:E733–E41.
57. Castillo L, Sánchez M, Vogt J, et al. Am J Physiol Endocrinol Metab 1995;268:E360–E7.
58. Darmaun D, Matthews DE, Bier DM. Am J Physiol Endocrinol Metab 1986;251:E117–E26.

59. van Acker BAC, Hulsewé KWE, Wagenmakers AJM, et al. Clin Sci 1998;95:339–46.
60. Souba WW. Annu Rev Nutr 1991;11:285–308.
61. Cortiella J, Matthews DE, Hoerr RA, et al. Am J Clin Nutr 1988;48:988–1009.
62. Tessari P, Pehling G, Nissen SL, et al. Diabetes 1988;37:512–9.
63. Matthews DE, Marano MA, Campbell RG. Am J Physiol Endocrinol Metab 1993;264:E109–E18.
64. Haisch M, Fukagawa NK, Matthews DE. Am J Physiol Endocrinol Metab 2000;278:E593–E602.
65. Hoerr RA, Matthews DE, Bier DM, et al. Am J Physiol Endocrinol Metab 1993;264:E567–E75.
66. Metges CC, El Khoury AE, Henneman L, et al. Am J Physiol Endocrinol Metab 1999;277:E597–E607.
67. Battezzati A, Haisch M, Brillon DJ, et al. Metabolism 1999;48: 915–21.
68. Castillo L, Chapman TE, Yu YM, et al. Am J Physiol Endocrinol Metab 1993;265:E532–E9.
69. Matthews DE, Marano MA, Campbell RG. Am J Physiol Endocrinol Metab 1993;264:E848–E54.
70. Battezzati A, Brillon DJ, Matthews DE. Am J Physiol Endocrinol Metab 1995;269:E269–E76.
71. Wilkinson DJ, Brook MS, Smith K. Clin Nutr Open Sci 2021;36:111–25.
72. Millward DJ, Smith K. J Physiol 2019;597:1235–49.
73. Jaleel A, Short KR, Asmann YW, et al. Am J Physiol Endocrinol Metab 2008;295:E1255–E68.
74. Jaleel A, Henderson GC, Madden BJ, et al. Diabetes 2010;59: 2366–74.
75. Toffolo G, Albright R, Joyner M, et al. Am J Physiol Endocrinol Metab 2003;285:E1142–E9.
76. Reeds PJ, Hachey DL, Patterson BW, et al. J Nutr 1992;122:457–66.
77. Busch R, Kim YK, Neese RA, et al. Biochim Biophys Acta 2006;1760:730–44.
78. Hellerstein M, Evans W. Curr Opin Clin Nutr Metab Care 2017;**20.**
79. Yang RD, Matthews DE, Bier DM, et al. Am J Physiol Endocrinol Metab 1984;247:E634–E8.
80. Wang Z, Deurenberg P, Matthews DE, et al. JPEN J Parenter Enteral Nutr 1998;22:82–6.
81. Elia M, Carter A, Bacon S, et al. Clin Sci 1980;59:509–11.
82. Rennie MJ, Millward DJ. Clin Sci 1983;65:217–25.
83. Rathmacher JA, Flakoll PJ, and Nissen SL. Am J Physiol Endocrinol Metab 1995;269:E193–E8.
84. Lundholm K, Bennegård K, Edén E, et al. Cancer Res 1982:42:4807–11.
85. Möller-Loswick A-C, Zachrisson H, Hyltander A, et al. Am J Physiol Endocrinol Metab 1994;266:E645–E52.
86. Vesali RF, Klaude M, Thunblad L, et al. Metabolism 2004;53: 1076–80.
87. Cahill GF, Jr., Aoki TT. In *Assessment of Energy Metabolism in Health and Disease, Report of the First Ross Conference on Medical Research* (Kinney JM, ed) pp. 12–34, Ross Laboratories, Columbus, Ohio, 1980.
88. Elia M. Organ and tissue contribution to metabolic rate. In *Energy Metabolism: Determinants and Cellular Corollaries* (Kinney JM, and Tucker HN, eds) pp. 61–79, Raven Press, New York, 1992.
89. Wang Z, Ying Z, Bosy-Westphal A, et al. Am J Clin Nutr 2010;92: 1369–77.
90. Bier DM, Leake RD, Haymond MW, et al. Diabetes 1977;26:1016–23.
91. Felig P. Annu Rev Biochem 1975;44:933–55.
92. Stumvoll M, Chintalapudi U, Perriello G, et al. J Clin Invest 1995;96:2528–33.
93. Cersosimo E, Judd RL, Miles JM. J Clin Invest 1994;93:2584–9.
94. Alpers DH. Digestion and absorption of carbohydrates and proteins. In *Physiology of the Gastrointestinal Tract* (Johnson LR, Alpers DH, Christensen J, Jacobson ED, and Walsh JH, eds) Vol. 2 pp. 1723–49, Raven Press, New York, 1994.
95. Green GM, Olds BA, Matthews G, et al. Proc Soc Exp Biol Med 1973;142:1162–7.
96. Matthews DM. *Protein Absorption: Development and Present State of the Subject*, Wiley-Liss, New York, 1991.
97. Alpers DH. *Fed. Proc.* 1986;45:2261–7.
98. Broer S, Fairweather SJ. Compre Physiol 2018;9:343–73.
99. Asatoor AM, Cheng B, Edwards KDG, et al. Gut 1970;11:380–7.
100. Gardner ML. Absorption of intact proteins and peptides. In *Physiology of the Gastrointestinal Tract* (Johnson LR, Alpers DH, Christensen J, Jacobson ED, and Walsh JH, eds) Vol. 2 pp. 1795–820, Raven Press, New York, 1994.
101. Rose WC. Nutr Abstr Rev 1957;27:631–47.
102. Visek WJ. Annu Rev Nutr 1984;4:137–55.
103. Matthews DE, Harkin R, Battezzati A, et al. Metabolism 1999;48:1555–63.
104. Walser M. Clin Sci 1984;66:1–15.
105. Kriengsinyos W, Rafii M, Wykes LJ, et al. J Nutr 2002;132: 3340–8.
106. FAO/WHO/UNU. *Energy and Protein Requirements* Vol. technical report series #724, World Health Organization, Geneva, 1985.
107. Rand WM, Young VR. J Nutr 1999;129:1920–6.
108. Food and Nutrition Board – Institute of Medicine (U.S.). Proteins and Amino Acids. In *Dietary Reference Intakes for Energy, Carbohydrate, Fiber, Fat, Fatty Acids, Cholesterol, Protein, and Amino Acids* pp. 589-768, National Academy Press, Washington, DC, 2002.
109. Munro HN. Amino acid requirements and metabolism and their relevance to parenteral nutrition. In *Parenteral Nutrition* (Wilkinson AW, ed) pp. 34–67, Churchill-Livingstone, London, 1972.
110. Rand WM, Pellett PL, Young VR. Am J Clin Nutr 2003;77:109–27.
111. Protein and Amino Acids. In *Recommended Dietary Allowances* ((U.S.) FaNB-NRC, ed) pp. 52-77, National Academy Press, Washington, DC, 1989.
112. Irwin MI, and Hegsted DM. J Nutr 1971;101:539–66.
113. Millward DJ, Price GM, Pacy PJH, et al. Proc Nutr Soc 1990;49:473–87.
114. Young VR, Bier DM, Pellett PL. Am J Clin Nutr 1989;50:80–92.
115. Young VR. J Nutr 1994;124 Suppl.:1517S–23S.
116. Young VR, El-Khoury AE. Proc Natl Acad Sci U S A 1995;92: 300–4.
117. Zello GA, Wykes LJ, Ball RO, et al. J Nutr 1995;125:2907–15.
118. Elango R, Ball RO, Pencharz PB. J Nutr 2008;138:243–6.
119. Wilson DC, Rafii M, Ball RO, et al. Am J Clin Nutr 2000;71:757–64.
120. Elango R, Ball RO, Pencharz PB. Amino Acids 2009;37:19–27.
121. Elango R, Ball RO, Pencharz PB. Br J Nutr 2012;108:S22–S30.
122. Elango R, Humayun MA, Ball RO, et al. J Nutr 2009;139: 1082–7.
123. El-Khoury AE, Fukagawa NK, Sánchez M, et al. Am J Clin Nutr 1994;59:1000–11.
124. El-Khoury AE, Fukagawa NK, Sánchez M, et al. Am J Clin Nutr 1994;59:1012–20.
125. Borgonha S, Regan MM, Oh SH, et al. Am J Clin Nutr 2002;75:698–704.
126. Young VR, Borgonha S. J Nutr 2000;130:1841S–9S.
127. Kopple JD, Swendseid ME. J Nutr 1981;111:931–42.
128. Humayun MA, Elango R, Ball RO, et al. Am J Clin Nutr 2007;86:995–1002.
129. Elango R, Humayun MA, Ball RO, et al. Am J Clin Nutr 2011;94:1545–52.
130. Elango R, Ball RO. Adv Nutr 2016;7:839s–44s.
131. Courtney-Martin G, Ball RO, Pencharz PB, et al. Nutrients 2016;8.
132. Millward DJ, Jackson AA. Am J Clin Nutr 2012;95:1498–501; author reply 501–2.
133. Bjelton L, Sandberg G, Wennberg A, et al. Assessment of biological quality of amino acid solutions for intravenous nutrition. In *Perspectives in Clinical Nutrition* (Kinney JM, and

Borum PR, eds) pp. 31–41, Urban & Schwarzenberg, Baltimore, DC, 1989.
134. Millward DJ. Br J Nutr 2012;108:S31–S43.
135. Cuthbertson DP. Injury 1980;11:175–89.
136. Tessari P. Am J Clin Nutr 2019;110:255–64.
137. Mundi MS, Shah M, Hurt RT. Nutr Clin Pract 2016;31:445–50.
138. Apostolopoulou A, Haidich A-B, Kofina K, et al. Nutrition 2020;78:110960.
139. Severo JS, da Silva Barros VJ, Alves da Silva AC, et al. Clin Nutr 2021;42:53–60.
140. Arribas-López E, Zand N, Ojo O, et al. Nutrients 2021;13:2498.
141. Souba WW, Herskowitz K, Austgen TR, et al. *JPEN* 1990;14: 237S–43S.
142. Wilmore DW. N Engl J Med 1991;325:695–702.
143. Labow BI, Souba WW. World J Surg 2000;24:1503–13.
144. Griffith OW, Stuehr DJ. Annu Rev Physiol 1995;57:707–36.
145. Brittenden J, Heys SD, Ross J, et al. Clin Sci 1994;86:123–32.
146. Ziegler TR, Gatzen C, Wilmore DW. Annu Rev Med 1994;45:459–80.
147. Wang W, Wu Z, Dai Z, et al. Amino Acids 2013;45:463–77.
148. Rasmussen BF, Ennis MA, Dyer RA, et al. J Nutr 2021;151:361–9.

A. MAJOR DIETARY CONSTITUENTS

CHAPTER 2

Carbohydrates

Kimber L. Stanhope, Nancy L. Keim, and Peter J. Havel

CHAPTER OUTLINE

Abbreviations

AI, adequate intake; **ATP**, adenosine triphosphate; **Ca^{2+}** calcium; **GLUT**, glucose transporter; **K^+**, potassium; **K_m**, Michaelis Menten constant; **Na^+**, sodium; **RDA**, recommended dietary allowance; **RS**, resistant starch; **SGLT**, sodium-linked glucose transporter

Historical Highlights

Modern man began consuming domesticated grains approximately 10,000 years ago, as agrarian societies emerged. Prior to this time, humans were hunter-gathers, and their diets consisted mainly of meats and wild plants. Relative to the history of *homo sapiens*, consumption of diets high in grains is a fairly recent event in human evolution. Rice grown in the near East is the oldest domesticated grain, and the cultivation of oats in Europe occurred about 3,000 years ago. The origin of sugar cane is thought to be Papua New Guinea, where it was likely cultivated from wild plants, also at the time of the global Neolithic agricultural revolution. The slow diffusion of migrants brought sugar cane to India, Southeast Asia, and China. After the Arabs defeated the Romans, they brought the sugar cane from Persia to Europe and the Mediterranean, where it failed to thrive, apart from the Moroccan coast. The returning Crusaders brought sugar to the European courts, where it became an important and desirable luxury dietary constituent. Sugar cane was introduced to the Caribbean by Christopher Columbus on his second voyage, in 1493. These plants thrived and were dispersed to Central and South America and throughout the Caribbean. The boiling process to extract the raw sugar from the cane occurred at or near the plantations. By the early seventeenth century, the raw sugar was being shipped to and handled by refineries in England and France.

The chemistry of carbohydrates was launched in 1812 when Kirchoff, a Russian chemist, reported that starch, when boiled with dilute acid, produced a free sugar known to be contained in grapes (glucose). Schmidt, in 1844, designated carbohydrates as compounds that contained carbon, hydrogen, and oxygen and showed that sugar was found in the blood. Glycogen, the animal storage form of carbohydrate in liver and muscle, was discovered by the renowned French physiologist Claude Bernard in 1856.

Today, sugar is produced and consumed worldwide, along with eight major cereal grains: wheat, rye, barley, oats, corn, rice, sorghum, and millet. Technological advances in harvesting techniques and plant breeding to produce disease-resistant plants in the past 50 years have yielded plants that are very different genetically from their ancient counterparts. Furthermore, the refinement of grains to produce palatable, inexpensive foods coincided with a 48% increase in grain consumption (wheat and corn) from the 1970s to the new millennium. As awareness around the inverse relationship between whole grain consumption and chronic disease grows, there has been a resurgence of efforts to increase consumption of whole grains. Better access to products and information through technology has spurred increased public demand for whole grains such as high fiber wheat products, gluten-free products (quinoa, rice, amaranth), and other diverse grains (bulgar, kamut, rye).

Definition

What are carbohydrates? The formal definition is a class of compounds having the formula $C_n(H_2O)_n$; that is, the molar ratio of carbon to hydrogen to oxygen is 1:2:1. Simple carbohydrates include the hexose monosaccharides (e.g., glucose, galactose, and fructose) and the disaccharides: maltose (glucose-glucose), sucrose (glucose-fructose), and lactose (glucose-galactose). Complex carbohydrates include the oligosaccharides that yield three to 10 monosaccharides on hydrolysis; examples include trioses (glycerose, $C_3H_6O_3$), tetroses (erythrose, $C_4H_8O_4$), and pentoses (ribose, $C_5H_{10}O_5$). Pentoses are important constituents of nucleic acids. Polysaccharides are larger complex carbohydrates containing more than 10 monosaccharide units. Common polysaccharides include starch, glycogen, pectins, cellulose, and gums. The polysaccharides serve both energy storage and structural functions. Chitin is a modified polysaccharide containing nitrogen as n-acetylglucosamine that forms the exoskeleton of arthropods such as insects and crustaceans. Starch is the storage form carbohydrate of plants, whereas animals store carbohydrate as glycogen (liver contains up to 6% and muscle about 1% glycogen by weight). Many different types of starch exist, depending on the plant source. Inulin, for example, is a starch found in the tubers and roots of dahlias, artichokes, and dandelions and, when hydrolyzed, yields only fructose; hence it is a fructosan. Cellulose consists of glucose units linked by β (1–4) bonds to form long, straight chains strengthened by hydrogen bonding. It is the chief structural framework of plants and cannot be digested by humans because we do not produce an intestinal carbohydrase that hydrolyzes the β (1–4) linkage. Thus, cellulose is considered to be a dietary fiber that provides bulk to plant-based foods. Bacterial enzymes, however, can break down cellulose. A small amount of fiber/cellulose is hydrolyzed by this process in the human colon; however, microbial digestion of cellulose provides only negligible amounts of energy for humans.

Dietary Carbohydrates

As discussed above, carbohydrates represent a large family of naturally occurring compounds and derivatives of these compounds (**Figure 2.1**). However, only a relatively small number of carbohydrates are produced commercially and used in the food industry or are of significant metabolic importance. Dietary carbohydrate is a major macronutrient for both humans and omnivorous animals. Human adults in Western countries obtain approximately half of their daily energy requirements from dietary carbohydrates. In other countries, it has traditionally contributed an even higher proportion of energy, at least until the more recent introduction of Westernized diets. Of ingested carbohydrate, approximately 60% is in the form of polysaccharides, mainly starch, while the disaccharides sucrose and lactose contribute 30% and 10%, respectively (**Table 2.1**). Monosaccharides (glucose and fructose) are naturally present in fruits, but other foods contain small amounts. A recent compilation of monosaccharides in food can be found in the Davis Food Glycopedia.[1] Other sources of monosaccharides are manufactured foods and drinks, primarily in the form of high-fructose corn syrup. Some oligosaccharides, such as raffinose and stachyose, are found in small amounts in legumes. They cannot be digested by pancreatic and intestinal enzymes (**Table 2.2**) but are digested by bacterial enzymes, especially in the colon.

Digestible polysaccharides need to be broken down into their constituent monosaccharides before they can be absorbed and metabolized. This breakdown is initiated during mastication and gastric passage by the carbohydrase α-amylase secreted by the salivary glands, continued by pancreatic amylase in the duodenum, and completed by disaccharidases located in the brush-border membrane of the enterocytes in the small intestine (see Table 2.2 for the major intestinal glycosidases).[2]

Starch

Starch, the predominant dietary polysaccharide, consists only of glucose units and is, thus, a homopolysaccharide, designated a glucosan or glucan. It is composed of two such homopolymers (**Figure 2.2**): amylose, which has linear (1–4) linked α-D-glucose, and amylopectin, a highly branched form containing both (1–4) and (1–6) linkages at the branch points. Plants contain both forms as insoluble, semicrystalline granules and differing ratios of amylopectin and amylose, depending on the plant (**Table 2.3**). Salivary and pancreatic amylases act on the interior (1–4) linkages but cannot break the outer glucose-glucose links. Thus, the final breakdown products formed by the amylases are α-(1–4)–linked disaccharides (maltose) and trisaccharides (maltotriose).

Starch Breakdown

The breakdown of starch begins in the mouth, with salivary amylase. It is often assumed that as this enzyme is swallowed into an acid stomach, the enzymic carbohydrate breakdown is stopped (although acid hydrolysis may still occur) because salivary amylase is inhibited by a pH lower than 4. However, starch and its end products, and the proteins and amino acids present in a mixed meal, all buffer the acid of the stomach and allow some amount of hydrolysis to continue. Thus, the quantitative involvement of salivary amylase in the breakdown of starch may be underestimated. Pancreatic α-amylase added to the emptying gastric contents (chyme) in the duodenum cannot hydrolyze the (1–6) branching links and has little specificity for the (1–4) links adjacent to the branching points. Amylase action produces large oligosaccharides (α-limit dextrins) containing, on average, about eight glucose units with one or more (1–6) links. These α-limit dextrins are split by the enzymatic action of glucoamylase (α-limit dextrinase), which sequentially removes a single glucose unit from the nonreducing end of a linear α-(1–4)-glucosyl oligosaccharide. Maltose and maltotriose are then broken down by secreted and brush-border

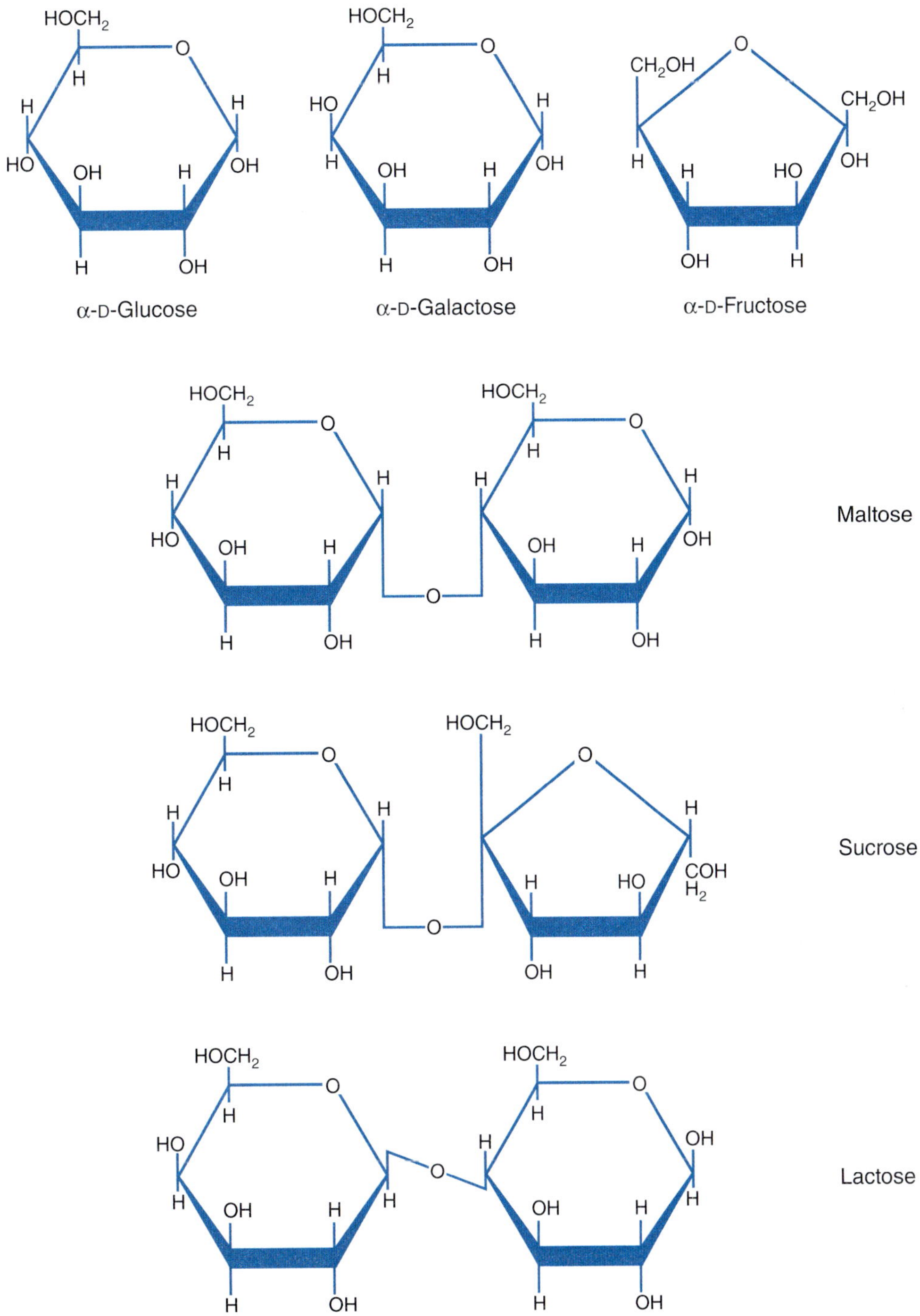

Figure 2.1 Structures of the common dietary monosaccharides and disaccharides in perspective. Haworth representation.

disaccharidases, especially sucrase-isomaltase, into free glucose, which is then transported into and across the enterocytes by hexose transporters (**Table 2.4**).

The initial breakdown of starch into α-limit dextrins, the intraluminal or cavital digestion phase, occurs mainly in the bulk fluid phase of the intestinal contents. In humans, there appears to be little contact or membrane digestion, in which adsorption of amylase onto the brush-border surface of enterocytes facilitates its enzymatic activity.[3]

Normally, α-amylase is not a limiting factor in the assimilation of starches in humans, but newborn babies, and especially premature ones, cannot assimilate starch because the pancreas secretes insufficient α-amylase to digest it. However, within a month, the secretion of α-amylase is usually sufficient for full digestion.[4]

Dietary Fiber

Dietary fiber was originally defined as "the remnants of plant cell walls not hydrolyzed by the alimentary enzymes of man," but the definition was subsequently modified to include "all plant polysaccharides and lignins, which are resistant to hydrolysis by the digestive enzymes of man".[5]

Table 2.1 Principal Dietary Carbohydrates

Food Source	Grains	Starchy Vegetables	Legumes	Fruits	Sugars and Sweeteners	Milk
	Rice	Yam	Soybeans	Apple	Cane sugar	
	Wheat	Potato	Dried peas	Orange	Beet sugar	
	Oats	Sweet corn	Lima beans	Grapes	Sorghum	
	Barley	Cassava		Peach	Honey	
	Rye			Pineapple	Corn syrup	
	Maize			Banana		
Polysaccharide Oligosaccharide	Starch	Starch	Starch Raffinose, stachyose			
Disaccharide Monosaccharide	Maltose			Sucrose Fructose Glucose	Sucrose Fructose Glucose	Lactose

Table 2.2 Major Glycosidases of the Mammalian Enterocyte Brush Border

Glycosidase	Enzyme Complex	Enzyme Activity
Maltase-sucrase	Sucrase-isomaltase	80% of maltase; some α-limit dextrinase; all of sucrase; most of isomaltase
Maltase-isomaltase		
Maltase-glucomylase (2)	Glucoamylase	All glucoamylase; most of α-limit dextrinase; 20% maltase; small % isomaltase
Trehalase		All trehalase
Lactase	β-glycosidase	All neutral lactase and cellobiose
Glycosyl-ceramidase (phlorizin hydrolase)		Most of aryl-β-glycosidase

Modified from Semenza, Giorgio. Disaccharidases of Small-Intestinal Mucosa. Journal of Pediatric Gastroenterology and Nutrition 4(6):p 857-864, December 1985.

Resistant Starch

Starch is most frequently eaten after cooking. The heat of cooking gelatinizes the starch granules, thus increasing their susceptibility to enzymatic (α-amylase) digestion. However, a proportion of the starch, known as resistant starch (RS), is indigestible even after prolonged incubation with amylase. In cereals, RS represents 0.4% to 2% of the dry matter; in potatoes, 1% to 3.5%; and in legumes, 3.5% to 5.7%. RS has been categorized as the sum of the starch and degradation products not absorbed in the small intestine of a healthy person.[6] Three main categories are recognized: RS1, physically enclosed starch (partially milled grains and seeds), RS2, ungelatinized crystalline granules of the B-type x-ray pattern (as found in bananas and potatoes); and RS3, retrograded amylose

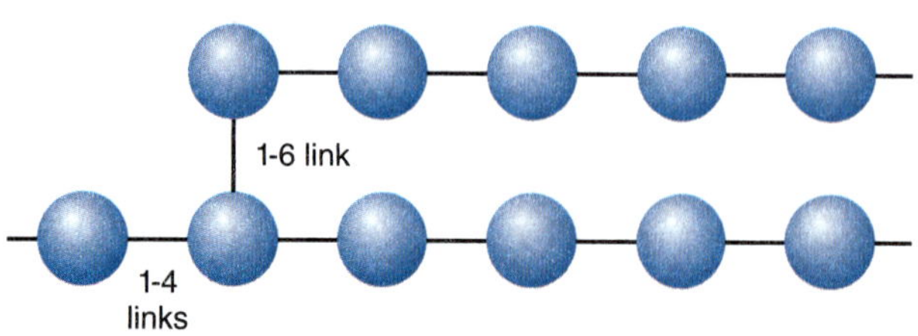

Figure 2.2 Starch is composed of amylose (15%–20%) and amylopectin (80%–85%). Amylose is a nonbranching helical chain structure of glucose residues, whereas amylopectin (a portion shown above) has branched chains of 24 to 30 glucose residues *(solid black)* joined by (1→4) glucosidic linkages with (1–6) linkages creating the branching points.

Table 2.3 Amylose and Amylopectin Content of Various Plant Starches

Plant	Amylose (%)	Amylopectin (%)
Maize (standard)	24	76
Potato	20	80
Rice	18.5	81.5
Tapioca	16.7	83.3
Wheat	25	75

Table 2.4 Human Facilitated-Diffusion Glucose Transporter Family (GLUT1–5)

Type	Amino Acids (N)	Chromosome Location	K_m (MMOL/L) for Hexose Uptake[a]	Major Expression Sites
GLUT1 (red cell)	492	1	1–2 (red cells)	Placenta, brain, kidney, colon
GLUT 2 (liver)	524	3	15–20 (hepatocytes)	Liver, β-cell, kidney, small intestine
GLUT 3 (brain)	496	12	10 (*Xenopus* oocytes)	Brain, testis
GLUT 4 (muscle/fat)	509	17	5 (adipocytes)	Skeletal and heart muscle, brown and white fat
GLUT 5 (small intestine)	501	1	6–11*	Small intestine, sperm

K_m, Michaelis-Menten constant; N, nitrogen.

[a]The approximate K_m values refer to the uptake of glucose (fructose in the case of GLUT 5) in the designated tissue or cells in parentheses and are shown to give an approximate index of the affinity of the transporter for glucose.

*(fructose) (Xenopus oocytes).

(formed during the cooling of starch gelatinized by moist heating). The resistant starches escape digestion in the small intestine, but then enter the colon, where they can be fermented by the local resident bacteria. In this respect, RS is similar to dietary fiber. Estimates of the RS and unabsorbed starch represent approximately 2% to 5% of the total starch ingested in the average Western diet, approximately 10 g/day.[7] The end products of the fermentation of the RS in the colon are short-chain fatty acids (e.g., acetate, butyrate, propionate), carbon dioxide, hydrogen, and methane (released as flatus).

Resistant starches stimulate bacterial growth in the colon. Short-chain fatty acids stimulate crypt cell mitosis in animals and humans.[8] However, if the human colon is bypassed surgically, colonocytes lose their absorptive function, and ionic absorption is reduced. Luminal short-chain fatty acids from bacterial fermentation are used by colonocytes as metabolic substrates and appear to be required for normal colonic function.[9] Volatile fatty acids (such as butyrate and propionate), produced by the microbial digestion of resistant starches and oligosaccharides (such as inulin and oligofructose) and dietary fiber, can stimulate the expression and production of hormones produced by the distal gastrointestinal tract, including glucagon-like peptide-1 (GLP-1) and peptide-YY (PYY). GLP-1 and PYY. These can contribute to satiety, in part by inhibiting gastric emptying. GLP-1, in particular, has beneficial effects on insulin secretion and carbohydrate and lipid metabolism.[10,11]

Soluble and Insoluble Fibers

Soluble dietary fiber includes pectin and hydrocolloids, and insoluble fiber includes cellulose and hemicellulose. Grains, cereals, legumes, fruits, and vegetables are heterogeneous with respect to the mix of the various fiber types they contain, but the concentration of dietary fiber within a food group is generally similar. Pectins are found in greatest proportions in fruits and vegetables, hemicelluloses are the major fiber type in grains, and legumes are a significant source of both soluble and insoluble fibers.[12] Soluble and insoluble fibers are fermented by the luminal bacteria of the colon. High-fiber diets maintained for the long term reduce the incidence of colon cancer, but the mechanisms involved in the effects of fiber to prevent colon cancer are not well understood. Investigators have suggested that the bulk action of fiber speeds colonic transit and reduces the absorption of luminal chemicals, or that fiber absorbs the carcinogenic agents.[8] Chapter 3 in this textbook reviews Fiber in detail.

Sugars: Functions and Properties

Sugars, unlike starch, have an obvious impact on human taste because they are sweet. Sweet is one of five distinct tastes linked to specific receptors, with all other taste sensations considered mixtures of these. The prevailing thought is that sweetness is not a unitary quality, and individual variation exists in the ability to "taste" different sweetness qualities of different sweeteners. Human neonates recognize and like sweetness, a finding that is not surprising because the lactose in their major food, human milk, gives it a sweet taste. Estimates of relative sweetness of various carbohydrates by humans are usually made against the standard, sucrose (100%). On this scale, glucose is less sweet (sweetness rating = 61 to 70, whereas fructose is sweeter with a fruity taste (sweetness rating = 130 to 180). The sweetness of maltose is 43 to 50 and that of lactose is between 15 to 40. The sweetness ratings of the high-fructose corn syrup (HFCS) sweeteners are 128 for HFCS-55 (55% fructose) and 116 for HFCS-42 (42% fructose). Investigators have speculated that during human evolution, the quest for foods containing maximal energy resulted in primitive humans acquiring the ability to recognize that sweetness indicated safety and energy.

Today, sugars (primarily sucrose, glucose, and fructose) are used extensively in foods to provide sweetness, energy, texture, and bulk, and also for appearance, preservation (by raising the osmotic pressure), and fermentation (in bread, alcoholic beverages). The palatability, appearance, and shelf life of a huge variety of foods and drinks are enhanced by adding sucrose; examples include: breads, cakes, and biscuits; preserves and jellies; confectionery; dairy products; cured, dried, and preserved meats; breakfast cereals; and frozen and canned vegetables. As a result of the addition of sugars to so many food products, sugar consumption has increased by 20% overall since the 1970s and corn-based sweetener use has increased by 277%.[13] Soft drinks, "juice" drinks, and other beverages, sweetened with sucrose or high-fructose corn syrup, are major sources of dietary sugar worldwide. Incorporation of sweeteners into beverages and many other commonly consumed foods makes accurate assessment of dietary sugar intake difficult.

Getting Glucose into Cells: the Transporters

A major source of metabolic energy for most, if not all, mammalian cells is the oxidation of D-glucose. The lipid-rich membranes of such cells, however, are relatively impermeable to hydrophilic polar molecules such as glucose. Specific transport processes have evolved to allow the cellular entry and exit of glucose. Carrier proteins located in the plasma membranes of cells can bind glucose and allow it to traverse the lipid membrane barrier, thus releasing the hexose into the cellular cytoplasm or body fluids.

Two distinct classes have been described: (a) a family of facilitative glucose transporters (see Table 2.4) and (b) sodium (Na^+)-glucose cotransporters (symporters). The former class consists of membrane integral proteins found on the surface of all cells. They transport D-glucose down its concentration gradient (from high to low), a process described as *facilitative diffusion*. The energy for the transfer is derived from the concentration gradient of glucose across plasma membranes. Glucose transporters allow glucose to enter cells readily, but they can also allow it to exit from cells, according to the prevailing concentration gradient. In contrast, the Na^+-glucose co-transporters participate in the "uphill" movement of D-glucose against its concentration gradient; that is, they perform *active transport*. They are especially expressed in the specialized brush borders of the enterocytes of the small intestine and the epithelial cells of the kidney (proximal) tubule. They occur at lower levels in the epithelial cells lining the lung, and in the liver.[14] Cooperation between the two classes of glucose transporters, together with the hormones involved in carbohydrate metabolism, allows a fine control of glucose concentration in the plasma, thus maintaining a continuous supply of the body's main source of cellular energy.

Human Facilitative-Diffusion Glucose Transporter Family

Several major hexose transporters have been identified and since the characterization of the first, glucose transporter 1 (GLUT1), more have been identified by molecular cloning.[15] The classical glucose transporters, GLUT1 to GLUT4, are well-characterized proteins with similar molecular structures, containing between 492 and 524 amino acid residues. Mueckler and colleagues,[15] using hydropathic and secondary structure predictions, proposed a two-dimensional orientation model of GLUT1 in the plasma membrane (**Figure 2.3**). The molecule has three major domains: (a) 12 α-helical segments spanning the membrane with the N and C termini of the protein on the cytoplasmic side of the cell membrane, (b) an intracellular domain of 65 hydrophilic amino acids (between membrane [M] regions 6 and M7 of Figure 2.3), and (c) an extracellular 33-amino acid segment (between Ml and M2), containing the site for an asparagine-linked oligosaccharide at asparagine 45.

It was thought that the polypeptide backbone of the molecule traverses or spans the plasma membrane 12 times. Both the

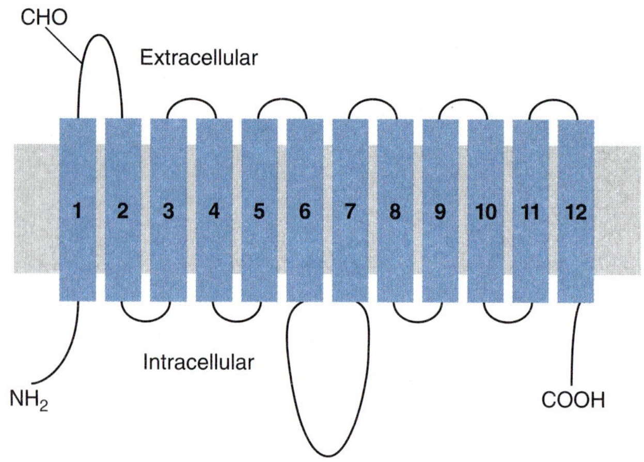

Figure 2.3 Highly schematic diagram illustrating the predicted secondary structure model of the glucose transporter molecule (GLUT1) in the cell membrane *(shaded)*. The putative membrane-spanning α-helices are shown as rectangles numbered *1* to *12* connected by chains *(lines)* of linked amino acids.

Modified from Mueckler M, Caruso C, Baldwin SA et al. Sequence and structure of a human glucose transporter. Science 1985;229:941-5.

amino- and carboxy-terminal ends of the molecule are on the cytoplasmic side of the membrane, whereas an N-glycosylation site is present on the first extracytoplasmic loop (Ml and M2). These basic topologic features have been confirmed by studies using proteolytic digestion and sequence-specific antibodies. GLUT1, purified from human red cells and reconstituted in liposomes, appears to be predominantly in the α-helical form, and the transmembrane segments form α-helices at right angles to the plane of the lipid membrane.[16] The molecular structure of GLUT1, shown in Figure 2.3, is a two-dimensional model. Studies using radiation inactivation of the carrier in intact red cells have indicated that GLUT1 probably exists as a homotetramer.[17] The structures, properties, expression sites, and roles of each of the five facilitative glucose transporter isoforms are briefly described here and are summarized in Table 2.4. Because of the recognized importance of these transporters in health and disease, numerous reviews have been published,[18,19] and they should be consulted for greater detail.

GLUT1 (Erythroid-Brain Carrier)

GLUT1 is the glucose transporter in the human red cell. The first to be characterized by molecular cloning,[15] it consists of 492 amino acid residues (see Table 2.4). The gene for its expression is located on chromosome 1. GLUT1 is widely distributed in many other tissues, including heart, kidney, adipose cells, fibroblasts, placenta, retina, and brain, but little is expressed in muscle or liver. There is particularly high expression in the endothelial cells of the microvessels of the brain, where GLUT1 forms part of the blood-brain barrier.[20] The transport process for D-glucose in the red cell is asymmetric, because the affinity (Michaelis-Menten constant [K_m]) for D-glucose uptake is approximately 1 to 2 mmol/L, whereas the K_m for the exit of glucose is 20 to 30 mmol/L. This asymmetry appears to be allosterically regulated by the binding of intracellular metabolites and inhibited by adenosine triphosphate (ATP).[21] The asymmetry allows the transporter to be effective when the extracellular glucose is low and the intracellular demand is high.

GLUT 2 (Hepatic Glucose Transporter)

Many biochemical studies indicated that the glucose transporter in liver cells was distinct from that of the red cells. Moreover, adult liver cells had very low levels of GLUT1 mRNA. Cloning of the second glucose carrier, GLUT2, was accomplished by screening rat and human cDNA libraries with a cDNA probe for GLUT1. GLUT2 has 55% identity in amino acid sequence with GLUT1, and it displays the same topologic organization in the cell membrane as predicted for GLUT1. Human GLUT2 contains 524 amino acids (see Table 2.4), compared with rat GLUT1 of 522 residues, and they show 82% identity in amino acid sequences, an excellent example of conservation of structure among species. GLUT2 is preferentially expressed in liver (sinusoidal membranes), kidney (tubule cells), small intestine (enterocytes), and the insulin-secreting β-cells of the pancreas.

In the liver cell, GLUT2 has a low affinity for glucose (K_m = 17 mmol/L) and shows symmetric transport, that is, a similar K_m for influx and efflux. This high-capacity, low-affinity transporter is useful for rapid glucose efflux following gluconeogenesis. GLUT2 can also transport galactose, mannose, and fructose.[22]

GLUT3 (Brain Glucose Transporter)

GLUT3 was originally cloned from a human fetal muscle cDNA library.[23] It contains 496 amino acid residues (see Table 2.4) and shows 64% identity with GLUT1 and 52% identity with GLUT2. Its amino acid sequence again suggests that its membrane topology is similar to that of GLUT1 (see Figure 2.3). GLUT3 mRNA appears to be present in all tissues, but its highest expression is in adult brain, kidney, and placenta. Adult muscle, however, shows very low levels. In the brain, it is mainly expressed in neurons. GLUT3 mRNA is found in fibroblasts and in smooth muscle. Because both of these cell types are found in practically all tissues, the ubiquitous expression of GLUT3 is understandable. Its affinity for glucose transport is relatively low ($K_m \approx$ 10 mmol/L) but significantly higher than that of GLUT2 (17 mmol/L). GLUT3 is also found in spermatozoa. Such cells undertake glycolysis in the male genital tract and take up glucose from epididymal fluid.

GLUT4 (Insulin-Responsive Glucose Transporter)

Glucose is transported across the cell membranes of adipocytes (fat cells), and its rate of transport can be speeded up 20- to 30-fold within 2 or 3 minutes by addition of insulin, without evidence of protein synthesis. Studies showed that this stimulation of glucose transport resulted in part from translocation of GLUT1 from an intracellular pool into the membrane. Careful quantitative measurements showed; however, that this could account for only a 12- to 15-fold increase in glucose transport. It became obvious that another transporter would have to be involved to account for the much larger insulin-stimulated transport. This new transporter, GLUT4, was first identified in rat adipocytes by use of a monoclonal antibody. Subsequently, it has been cloned from rat, mouse, and human DNA.[22] It is a protein with 509 amino acid residues (see Table 2.4), with 65% identity with GLUT1, 54% identity with GLUT2, and 58% identity with GLUT3. Rat and mouse GLUT4s have 95% and 96% identity, respectively, with human GLUT4. As with the previous GLUT transporters, the two-dimensional orientation of the structure in the cell membrane is similar to that proposed for GLUT1 (see Figure 2.3).

GLUT4 is the major glucose transporter of the insulin-sensitive tissues, brown and white fat, and skeletal and cardiac muscle. It occurs primarily in intracellular vesicles in the cells of these tissues. Insulin stimulation causes a rapid increase in the number of glucose transporters on the membranes of these cells because the vesicles are translocated toward the membrane and then fuse with it, releasing the molecule. This process ensures a high density of glucose transporters and enhances the ability to move glucose from the surrounding cellular fluid into the interior of the cell, that is, increased maximal velocity for glucose uptake. Because of this mechanism, the position of GLUT4 and its regulation are important components of glucose homeostasis, and its role in diabetes has been and continues to be the subject of intense investigation.

GLUT5—the Fructose Transporter

GLUT5 was isolated from human,[24] rat, and rabbit enterocyte cDNA libraries. It consists of 501 amino acid residues (see Table 2.4), and it has only 42%, 40%, 39%, and 42% identity with GLUTs 1, 2, 3, and 4, respectively. It is said to be primarily expressed in the jejunum (both in the brush border and basolateral membrane), but its mRNA has been detected, albeit at low levels, in human kidney, skeletal muscle, and adipocytes, microglial cells, and the blood-brain barrier. GLUT5 appears to transport glucose poorly and is really the transporter for fructose. It is found in high concentration in mature human spermatozoa,[25] which are known to use fructose as an energy source (human seminal fluid contains high concentration of fructose, which is manufactured by the seminal vesicles). In oocytes injected with GLUT5 mRNA to effect GLUT5 expression, the K_m for fructose uptake was 6 to 11 mmol/L. With importance to the regulation of energy homeostasis, the expression of GLUT5 in pancreatic β-cells is very low[26] and consequently, fructose has little if any effect on stimulation of insulin secretion.[27]

Other Transporters

There are currently a total of 14 isoforms in the family of sugar transport proteins including GLUT1 to GLUT14.[28] Thus, the list of both Na^+-dependent and facilitative carbohydrate transport proteins identified continues to grow. Because these transporters exhibit a wide range of properties, with variable combinations distributed across different cell and tissue types,[29,30] the capacity exists for far greater complexity in sugar transport, storage, and metabolism than originally considered at the time when the first transporters were initially identified.

Study of Glucose Transporters by Use of Transgenic and Knockout Mice

Although many metabolic inhibitors are available for use in examining metabolic pathways, the specificity of these agents is often questionable. With molecular techniques; however, metabolic pathways can be altered in specific ways, even in intact animals. A protein (e.g., enzyme/carrier) can be overexpressed, expressed in a tissue that normally does not contain it, or eliminated in a particular cell type. Site-directed mutations allow a molecule to be dissected and particular component groups removed or altered so their role in the molecule's functioning can be studied (see Chapter 38). Application of these techniques to the investigation of metabolic pathways is yielding interesting insights into the biologic roles of glucose transport proteins.

Transgenic mice were established that expressed high concentration of human GLUT1, properly located in the muscle sarcolemma. The increase in the GLUT1 expression resulted in a three- to fourfold increase in glucose transport into specific muscles, a finding confirming that GLUT1 plays a major role in controlling glucose entry into resting muscle. Strangely, insulin did not increase the entry of glucose into the transgenic mice muscles even though GLUT4 concentration in these mice was the same as in control mice. Possibly, the GLUT1 was elevated in the transgenic animals to a degree that glucose transport was not limited by transporter activity. Muscle glucose concentration was four- to fivefold and glycogen tenfold higher in the transgenic mice, even though they showed 18% (fed) to 30% (fasted) decreases in plasma glucose concentration. Oral glucose loads did not increase plasma glucose as much as in normal mice and glucose disposal was enhanced. Thus, increasing the number of GLUT1 transporters affected not only muscle metabolism, but also whole-body glucose homeostasis.

Selective overexpression of GLUT4 in muscle or adipose tissue protects against the development of diabetes in several rodent models.[31] Mice with genetic ablation of GLUT4 in all tissues exhibit impaired glucose tolerance, despite postprandial hyperinsulinemia, and decreased glucose lowering after insulin injection, indicating that they are insulin resistant. However, these animals do not develop overt diabetes.[32] Investigators had shown that glucose disposal in rodents, and mice in particular, has a large insulin-independent component that may protect against diabetes induced by ablation of GLUT4. In addition, the background strain of the animals used for genetic manipulations can have major influence on the phenotypic outcome. For example, overt diabetes and glucose toxicity were observed in mice with a muscle-specific inactivation of GLUT4.[33] Finally, adipose-specific GLUT4 ablation not only markedly impairs glucose uptake in isolated adipocytes from these animals but also induces insulin resistance in skeletal muscle and liver, despite preservation of GLUT4 expression in these tissues.[34] These results suggest that some factors regulated by adipose glucose transport are involved in the control of insulin action in extra-adipose tissue and, thus, whole-body insulin sensitivity.

Because increased expression of the insulin-responsive transporter GLUT4 occurs in human and rodent adipocytes, and is associated with obesity, the question arises whether increasing GLUT4 in adipocytes plays a role in obesity. Transgenic mice were produced that expressed human GLUT4 in their adipocytes. Basal-level glucose transport into adipocytes was increased approximately 20-fold compared with wild-type animals, but insulin only stimulated glucose uptake by a factor of 2.5, rather than the 15-fold increase of controls. Again, the possible explanation is that glucose transport is already so high in the adipocytes of transgenic animals that the number of transporters activated by insulin has a reduced relative contribution to the overall transport. Although the fat cell size was unchanged in the transgenic mice, the number of fat cells was more than doubled, and total body lipid nearly tripled, reflecting the increase in cell number. These results suggest that a specific increase in GLUT4 in adipocytes can contribute to obesity.

One major limitation of using transgenic and "knockout" mice is that the induced genetic alterations will often occur very early in the developing animal. Thus, observed phenotypic outcomes in these animals may result from the presence or absence of the transgene at the time of the laboratory measurements, or, alternatively, the genetic alteration could initiate a plethora of other events contributing to the observed phenotype. The development of conditional transgenic or knockout models in which transgenic manipulations allow temporal expression or inactivation of specific genes will help to overcome this limitation[35] (see Chapter 38).

Sodium-Glucose Cotransporters—Transepithelial Hexose Transport: Intestine and Kidney

The intestine and the kidney are two major organs that have epithelia with the specific function of transferring hexoses across their cells into the bloodstream. In the intestine, the transporters of the mature enterocytes capture the hexoses from the lumen after breakdown of dietary polysaccharides into simple hexoses, D-glucose, D-galactose, and D-fructose. In the kidney, the cells of the proximal tubule capture the glucose from the glomerular filtrate to return it to the blood. These glucose transporters, localized in the brush-border membranes of the epithelial cells, differ from the GLUT1 to GLUT5 types and share no sequence homology. They are, thus, members of a different protein family.

Moreover, they transport glucose across the cell membrane by having both hexose and Na^+ binding sites, hence the name Na^+-glucose cotransporters. They couple cellular glucose transfer to the inwardly directed electrochemical gradient of Na^+. The low intracellular concentration of Na^+ ions, maintained by Na^+-K^+/ATPase or the Na^+ pump at the basolateral borders of the cells, powers the uphill transfer of glucose through the agency of the cotransporter. The affinity of the sugar molecule for its cotransporter binding site is greater when the Na^+ ions are attached to the transporter than when they are removed. Thus, the external binding of Na^+ and its subsequent intracellular dissociation (because of the lower intracellular Na^+ ion concentration) causes the binding, then release, of glucose, allowing it to be transported uphill against its concentration gradient. It is then transported across the basolateral membranes of the cells of the small intestine and kidney, usually by GLUT2, but in the S3 segments of straight kidney tubules GLUT1 is found. In this part of the kidney, GLUT1 is probably involved both in the transepithelial transfer of glucose and in its uptake from the blood to provide energy for cellular glycolysis.

The low concentration of the cotransporters in cell membranes (0.05%–0.7%), their hydrophobic nature, and their sensitivity to proteolysis and denaturation made them nearly impossible to prepare by normal biochemical extraction and purification techniques. The first to be cloned and sequenced was sodium-linked glucose transporter-1 (SGLT-1), the form found in the rabbit small intestine.[36] Amphibian eggs of the toad *Xenopus laevis* can express almost any RNA that is injected into them and make the corresponding protein. Poly(A)$^+$ mRNA isolated from the rabbit small intestinal mucosa and microinjected into *Xenopus* oocytes stimulated Na^+-dependent uptake of the hexose analogue α-methyl glucoside that could be blocked by the plant glycoside phlorizin, a high-affinity competitor for the sugar site of the transporter.[18] Phlorizin has no effect on the GLUT1–5 transporters; they are inhibited by the mold metabolite phloretin, which is the aglycone of phlorizin. Phloretin has no effect on the Na^+-glucose transporter but blocks the GLUT1–5 transporters. The predicted topologic organization of SGLT-1 in the cell membranes was surmised from its amino acids and, like the glucose transporter family, it is a large polypeptide with 12 putative membrane-spanning α-helices. The polypeptide is glycosylated at one site, but this has little effect on its function.[37] Radiation inactivation analysis of SGLT-1 suggests that the functional form in the membrane is a tetramer. It is composed of 664 amino acids. The gene for human SGLT-1 is located on chromosome 22 at qll.2 $\rightarrow$ qter.

More recently, it has been shown that three different isoforms of the SGLT cotransporters exist, designated SGLT-1, SGLT-2 (672 amino acids), and SGLT-3.[38] The cotransporters SGLT-1 and SGLT-2 have different glucose-Na^+ coupling ratios. SGLT-1 is a high-affinity cotransporter (Km $\approx$ 0.8 mmol glucose/L), primarily expressed in the small intestine, that transports each glucose molecule with two Na^+ ions, whereas the SGLT-2 is a lower-affinity (Km $\approx$ 1.6 mmol glucose/L) cotransporter, expressed in the kidney tubules, that transports glucose with one Na^+. Of interest is that SGLT1 is also considered to be a water channel and for each 2 Na^+ and 1 sugar molecule, 264 molecules of water are co-transported by human SGLT1. This cotransport of water with glucose and sodium helps explains why glucose/NaCl administration in rehydration fluids is so effective in treatment of the dehydration in diarrheal diseases of any etiology. SGLT-3, isolated from pig intestine, is a low-affinity cotransporter. It has approximately 60% homology in amino acid sequence to SGLT-2.[39] Inhibitors of SGLT-2 that prevent renal tubular glucose reabsorption are currently being explored as a method for markedly increasing urine glucose spillover to reduce hyperglycemia in diabetes.[40]

Glucose-Galactose Malabsorption

The importance of human SGLT-1 for intestinal glucose absorption is exemplified in glucose-galactose malabsorption, a rare inborn error of glucose transport. This condition gives rise to severe watery diarrhea in neonates that is lethal unless glucose- and galactose-containing foods are removed from the diet. The diarrhea occurs because the unabsorbed hexoses enter the colon and are fermented into compounds that promote diarrhea. The lack of hexose absorption in two sisters identified with this condition appeared to result from a single base change at nucleotide position 92, where a guanine is replaced by adenine. The mutation changed amino acid 28 of SGLT-1 from aspartate to asparagine, rendering the SGLT-1 cotransporter inactive. Thus, single amino acid alteration in the 664 that comprise the molecule results in an inability to function as a cotransporter[41] and humans who lack a functional SGLT-1 cannot absorb glucose and galactose. Experimental studies measuring the absorption of glucose in human jejunum in vivo showed that more than 95% of glucose absorption occurs via a carrier-mediated process, a finding that agrees with the described pathophysiology of glucose-galactose malabsorption.[42,43]

Electrogenic Glucose-Linked Sodium Transfer

Because SGLT cotransporters transport both glucose and Na^+ ions across cell membranes without counter-charged ions, movement of the charged Na^+ ions creates an electrical potential difference across the cell membrane and, subsequently, across the epithelium. The transfer of glucose (or galactose) across the intestine or kidney tubule is called electrogenic (potential generating) or rheogenic (current generating). This electrical activity has been of inestimable value in the assessment of the kinetics of active hexose transport in native tissues and injected *Xenopus* eggs. This linking of the electrogenic Na^+ ion transfer with the hexose also enhances the net absorption of fluid across the small intestine. It is so effective that it overcomes the terrible excessive fluid secretory consequences of cholera toxin action in the small bowel. The application of this principle—oral rehydration therapy—is a highly effective and inexpensive treatment to keep patients hydrated and alive. A simple solution of NaCl and glucose has saved countless lives.[44]

Blood Glucose: Metabolic, Hormonal, and Transcriptional Regulation

Metabolic Regulation of Carbohydrate Metabolism

Glucose is among the most highly regulated circulating substrates. The concentration of glucose in the blood after an overnight fast has a normal range of 3.9 to 5.8 mmol/L (70–105 mg/dL). When a carbohydrate-containing meal is ingested, this may temporarily rise to 6.5 to 7.4 mmol/L, and during prolonged fasting, it can fall to 3.3 to 3.9 mmol/L. One of the major reasons that blood glucose concentration is so tightly regulated is that the brain normally depends on a continuous supply of glucose for its energy needs, although it can adapt to use ketone bodies from fat breakdown if the adaptation occurs slowly during prolong fasting or starvation.[45] The adaptation to use ketones becomes essential during starvation (see Chapter on Metabolic Consequences of Starvation), because the adult human brain uses approximately 140 g/day of glucose,[46] and only about 130 g/day of glucose can be obtained from noncarbohydrate sources. The importance of maintaining blood glucose concentration within the physiologic range is underscored by the fact that an acute reduction in blood glucose resulting from an overdose of insulin can rapidly lead to convulsions, coma, and even death, if not promptly treated.

Glucose enters the circulating pool from both exogenous (diet) and endogenous (hepatic production via glycogenolysis and gluconeogenesis) sources. In physiologically normal postabsorptive humans, the plasma appearance rate is 8 to 10 g/hour, with the circulating pool being replaced every 2 hours. At normal blood glucose concentration, the liver is a net producer of glucose. After digestion of dietary carbohydrates, the absorbed monosaccharides are transported via the portal vein to the liver. The rising concentrations of glucose and insulin, as well as the indirect effect of insulin to inhibit lipolysis from adipose tissue to reduce the systemic and portal delivery of free fatty acids to the liver, lead to a decrease of hepatic glucose production. As glucose arrives at the liver and peripheral tissues, the first metabolic step is phosphorylation by hexokinase. Hexokinase has tissue-specific isoforms that catalyze the same reaction, but with different kinetics and regulatory mechanisms. Hexokinase I, found in skeletal muscle, has a low Km and is coordinated with the low Km of GLUT4. Together hexokinase I and GLUT4 work to balance glucose uptake and phosphorylation. Hexokinase I is subject to feedback inhibition by its product, glucose-6-phosphate. The liver enzyme glucokinase (hexokinase IV) is not inhibited by glucose-6-phosphate and has a lower affinity for glucose. Glucokinase activity is coordinated with the high Km GLUT2, such that both are active when portal glucose delivery is elevated. Thus, glucokinase can increase its activity over the relatively larger range of glucose concentration in the portal blood. The hexokinases, coupled with the effects of increases of circulating insulin and translocation of GLUT4, possess the characteristics required for the efficient uptake of the substantial amounts of glucose arriving at the liver and peripheral tissues following consumption of a carbohydrate-containing meals.

Gluconeogenesis and the Cori Cycle

As discussed above, in the postabsorptive state, the normal plasma appearance rate of glucose in humans is 8 to 10 g/hour, and at normal blood glucose concentration, the liver is a net producer of glucose with the circulating glucose pool being replaced about every 2 hours. In addition to the liver, gluconeogenesis also occurs in the kidney and, to a lesser extent, the small intestine. Metabolic precursors for the formation of glucose include the glucogenic amino acids (especially alanine during starvation), glycerol, and propionate. Both muscle and red blood cells metabolize glucose to form lactate which, on entering the liver, can be resynthesized into glucose. This newly formed glucose is then available for recirculation back to the tissue, a process known as the Cori, or lactic acid, cycle. During exercise, the Cori cycle may account for approximately 40% of the normal plasma glucose turnover.

Hormonal Regulation of Carbohydrate Metabolism

The concentration of glucose in the blood is regulated by both hormonal and metabolic mechanisms. The major hormones involved are insulin, glucagon, and epinephrine (adrenaline), but other hormones, including thyroid hormone, glucocorticoids, growth hormone, leptin, and adiponectin also play a role.

Insulin and Diabetes (Type 1 and Type 2)

Insulin has a central role in regulating glucose metabolism. It is secreted by the β-cells of the islets of Langerhans in the pancreas. The daily output of insulin by the human pancreas is approximately 40 to 50 U, or 15% to 20% of pancreatic insulin stores. Circulating glucose concentration is the major signal controlling insulin secretion. Increasing blood glucose (hyperglycemia) stimulates the secretion of insulin, whereas low glucose (hypoglycemia) results in reduced insulin secretion. When the pancreas is unable to secrete insulin or to secrete sufficient insulin to compensate for insulin resistance, the medical condition is known as diabetes (see the chapter on Nutritional Management of Diabetes). This disease, the third most prevalent in industrialized countries, is normally classified as type 1, or insulin-dependent diabetes or type 2, non–insulin-dependent diabetes. Type 2 diabetes is strongly associated with obesity and accounts for more than 90% of all cases of diabetes. Type 1 diabetes is caused by autoimmune-mediated destruction of pancreatic β-cells, resulting in an absolute inability to produce insulin and the need for daily injections of the hormone to prevent severe hyperglycemia that can progress to diabetic ketoacidosis and death. The onset of insulin-dependent diabetes occurs predominantly in children and younger adults and becomes manifest when greater than 80% to 90% of the β-cells are destroyed. Most cases of type 2 diabetes occur in adults, although the number of cases reported in adolescents and children has increased dramatically since the mid-1990s in association with the increasing prevalence of childhood obesity. A form of type 2 diabetes occurs in pregnancy and is termed gestational diabetes. In type 2 diabetes, patients have reduced secretion of insulin accompanied by reduced metabolic responses to insulin in certain key insulin-sensitive tissues, including liver and skeletal muscle (i.e., peripheral insulin resistance). The precise molecular mechanisms underlying insulin resistance are yet to be fully identified; however, ectopic fat (triglyceride) deposition in liver and muscle is strongly associated with insulin resistance. Findings about the role of GLUT4 in insulin insensitivity are conflicting. Some researchers reported no changes in GLUT4 mRNA expression or protein levels,[47] but others have found a small (18%) decrease.[48] Thus, a defect in translocation of GLUT4 to the muscle membrane could contribute to insulin resistance.[22] In addition, defects at numerous control points in insulin receptor function, including impaired phosphorylation of tyrosine and increased phosphorylation of serine residues in the insulin receptor and post-receptor signal transduction cascade, such as insulin receptor substrate-1 (IRS-1), have been implicated in the etiology of insulin resistance. Inflammation and oxidative stress are now considered likely to have causal roles in the etiology of insulin resistance (For an in-depth discussion of insulin action and insulin resistance, see Chapters on Nutritional Management of Diabetes and on Metabolic Syndrome.)

Mechanism of Insulin Secretion

The mechanism of the regulation of insulin secretion by the external glucose concentration has been studied using patch clamp techniques to control the ionic channels in the β-cell membrane. The resting membrane potential of the β-cells is maintained by the Na^+-K^+/ATPase and ATP-sensitive K^+ channels (K_{ATP} channels). Normally, these channels are open, but they close in response to events triggered by the metabolism of glucose, when there is a concomitant increase in the ratio of ATP to adenosine diphosphate.[49] This depolarizes the cell membrane and opens voltage-gated Ca^{2+} channels. The resulting increase in the intracellular free Ca^{2+} concentration activates secretion of insulin through exocytosis—the fusion of insulin-containing granules with the plasma membrane and the release of their contents.[50] Certain drugs (sulfonylureas), such as tolbutamide and glibenclamide, induce insulin secretion by inhibiting the K_{ATP} channels of the β-cells and are used medically in the treatment of type 2 diabetes. K_{ATP} channels are a heteromultimeric complex of the inward rectifier K^+ channel (K_{IR} 6.2) and a receptor for sulfonylureas, SUR1, a member of the ATP-binding cassette or traffic family of plasma membrane proteins.[51]

In addition to glucose, other nutrients, including certain amino acids and fatty acids, can contribute to increased insulin secretion. Gastrointestinal hormones, including glucagon, secretin, and the known incretin hormones, glucagon-like peptide-1 (GLP-1), and glucose-dependent insulinotropic polypeptide (GIP), augment meal-induced insulin secretion.[52] The pancreatic islet is well innervated by the autonomic nervous system and the classic neurotransmitters, including acetylcholine and norepinephrine, as well as neuropeptides including galanin, vasoactive intestinal polypeptide, and pituitary adenylate cyclase–activating polypeptide, modulate insulin secretion.[53] Neural regulation by the parasympathetic nervous system augments insulin responses to meal ingestion and improves postprandial glucose tolerance,[54,55] whereas the sympathetic nervous system inhibits insulin secretion during periods of stress to increase glucose availability for the central nervous system.[56] In pregnancy, the hormones placental lactogen, estrogens, and progestin, all increase insulin secretion. Detailed overviews of the biology of insulin secretion have been published.[57,58]

Insulin lowers blood glucose concentration by facilitating its entrance into insulin-sensitive tissues and uptake by the liver. It does this by increasing GLUT4 translocation in tissues, such as muscle and adipose tissue. In the liver; however, insulin stimulates the storage of glucose as glycogen or enhances its metabolism via the glycolytic pathway. Glucose entry into liver cells is not mediated by changes in glucose transporter function, even though these transporters are present in hepatocyte membranes.[59] There is a functional specialization in the liver for disposition of GLUT1 and GLUT2. GLUT2 has higher expression in the periportal hepatocytes than in the perivenous hepatocytes. In the perivenous region; however, GLUT1 is present in the sinusoidal membranes of the hepatocytes, which form rows around the terminal hepatic venules. Periportal hepatocytes are more gluconeogenic than the more glycolytic perivenous cells.[60] Why hepatocytes have GLUT2 in their membranes is an enigma, because it is certainly not necessary for the entrance or release of glucose. Investigators have suggested that GLUT2 may be involved in transporting fructose, because GLUT5, the fructose transporter, is not well expressed in the liver. However, GLUT1 expression correlates well with the glycolytic activity of cells; in general, the higher the activity, the greater the concentration of GLUT1. Thus, the presence of GLUT1 in the perivenous liver cells may aid the efficient functioning of glucose entry into the glycolytic pathway.

Although insulin has a primary influence on glucose homeostasis, it also influences many other cellular functions (**Table 2.5**). Glucose has a profound effect on the secretion of insulin, and insulin strongly affects the normal storage of ingested fuels, including fatty acid and amino acid/protein metabolism, as well as cellular growth and differentiation (as exemplified in Table 2.5). Thus, indirectly, glucose also influences these cellular events, a finding that underscores the crucial role of glucose in influencing metabolism and catabolism, both directly and indirectly.

Table 2.5 Influence of Glucose via Insulin

Positive Effects	Negative Effects
Glucose uptake	Pyruvate→glucose
Amino acid uptake	Apoptosis
Acetyl-coenzyme A→fatty acid	Gene expression
Glucose→glycogen	
Protein synthesis	
DNA synthesis	
Sodium-potassium pump	
Gene expression	

Glucagon

Glucagon is secreted by the α-cells of the pancreatic islets of Langerhans. A major stimulus for its secretion is hypoglycemia (low blood glucose concentration). Glucagon acts on the liver to activate glycogenolysis, the breakdown of glycogen, by activating the enzyme phosphorylase. It also enhances gluconeogenesis (formation of glucose) from amino acids and lactate. Thus, the major actions of glucagon oppose those of insulin. Pancreatic α- and β-cells in the islets have a close anatomic and functional relationship with intra-islet regulation of glucagon by insulin and of insulin by glucagon.[61] The suppression of glucagon secretion by glucose during hyperglycemia is mediated, in part, by the actions of increased intra-islet insulin release and by the islet hormone, somatostatin.[62]

Glucagon binds to its specific receptor in the plasma membrane to activate cellular responses. This glucagon receptor is a member of a superfamily of G-protein–coupled receptors, and is also a member of a smaller subfamily of homologous receptors for glucagon-like peptide-1 peptides. This subfamily is a product of the proglucagon gene produced by endocrine L-cells in the distal small intestine, as well as other peptides, including glucose-dependent insulinotropic polypeptide, vasoactive intestinal peptide, secretin, growth hormone–releasing factor, and pituitary adenylate cyclase–activating polypeptide. Using specific glucagon receptor mRNA expression in rat tissues, glucagon receptor mRNA was observed to be abundant in liver, adipose tissue, and pancreatic islets, as expected, but also in heart, kidney, spleen, thymus, and stomach. Low concentrations were found in adrenal glands, small intestine, thyroid, and skeletal muscle. No expression was observed in testes, lung, large intestine, or brain.[63] Importantly, increased glucagon secretion is the first line of defense against low blood glucose (hypoglycemia). Activation of the autonomic nervous system (both parasympathetic and sympathoadrenal divisions) is an important mediator of increased glucagon secretion during hypoglycemia, and this mechanism becomes impaired in people with diabetes, leaving these patients at increased risk of hypoglycemia during insulin treatment.[64]

Other Counterregulatory Hormones

Epinephrine. Epinephrine is secreted by the chromaffin cells of the adrenal medulla. It is often called the "fight-or-flight" hormone because adrenal epinephrine release is triggered in response to many types of stress, such as fear, excitement, hypoglycemia, hypoxia, and blood loss (hypotension). Epinephrine acts in the liver, along with norepinephrine released from hepatic sympathetic nerves, to increase glycogenolysis directly by activating glycogen phosphorylase, and indirectly by stimulating glucagon secretion and inhibiting insulin secretion, thereby releasing glucose for use by muscle and the central nervous system.

Thyroid Hormone. In humans, fasting blood glucose concentration is elevated in hyperthyroid patients and is lower than normal in hypothyroid patients. Thyroid hormones enhance the action of epinephrine in increasing glycolysis and gluconeogenesis and can potentiate the actions of insulin on glycogen synthesis and glucose use. Thyroid hormones have a biphasic action in animals; enhancing glycogen synthesis in the presence of insulin at low doses but increasing glycogenolysis at higher doses.

Glucocorticoids. Glucocorticoids (cortisol and corticosterone) are secreted by the adrenal cortex in response to adrenocorticotrophic hormone (ACTH) released from the anterior pituitary. Glucocorticoids increase gluconeogenesis and inhibit the use of glucose in the extrahepatic tissues and, thus, are antagonistic to insulin's actions. The increased gluconeogenesis stimulated by the glucocorticoids is enhanced by increased protein catabolism, leading to greater availability of glucogenic amino acids to the liver and increased activity of transaminases and other enzymes involved with hepatic gluconeogenesis.

Growth Hormone. Growth hormone is secreted by the anterior pituitary. Its secretion is enhanced by hypoglycemia. It has direct and indirect effects on decreasing glucose uptake in specific tissues, such as muscle. Part of this effect may be the result of the liberation of fatty acids from adipose tissue, which then inhibits glucose metabolism. If growth hormone is administered on a long-term basis or is released from a pituitary tumor, it results in a persistent modest elevation of circulating glucose. However, if the pancreatic β-cells' capacity to secrete insulin becomes exhausted, overt diabetes will ensue.

Transcriptional Regulation of Carbohydrate Metabolism

Excess blood sugar can have pathologic consequences. Thus, when liver and muscle glycogen stores are replete, surplus glucose is converted into and stored as fat. The conversion of carbon from carbohydrate (mainly glucose or fructose) into fatty acids is known as *de novo* lipogenesis. Homeostatic control of blood glucose is essential to prevent pathologies associated with diabetes. This control is achieved through hormonal and intermediate regulation of constitutive enzymes (as discussed above), and also by inducing transcription of enzymes either via direct actions of glucose itself or indirectly via hormonal mechanisms (e.g., insulin).

Integration of Carbohydrate and Lipid Metabolism

Excess dietary carbohydrate is lipogenic. In the liver, surplus pyruvate derived from carbohydrates is converted to triglycerides. Unlike synthesis of other lipids, such as cholesterol, which is highly controlled, hepatic triglyceride synthesis is primarily driven by the presence of excess carbon produced from the glycolytic metabolism of dietary carbohydrates.[65] Numerous nuclear and membrane bound factors involved in transcriptional regulation of substrate metabolism have been identified.[66] There are several involved with lipid, and a few with carbohydrate metabolism. Specific transcription factors that play a key role in lipogenesis following excess carbohydrate consumption are SREBP1-c and ChREBP.

Sterol Regulatory Element Binding Proteins (SREBP) are a family of membrane bound protein factors that bind to Sterol Regulatory Element (SRE), present in the transcription domain of key genes that regulate lipid metabolism, and induce transcription of target proteins.[67] Three isoforms of SREBP's– 1a, 1c, and 2 have been identified. SREBP-1a and 1c are primarily involved in transcription of proteins involved in glucose and lipid metabolic homeostasis, while SREBP-2 is involved in cholesterol synthesis.[65] Hepatic and adipose tissues have been known to express a threefold and ninefold higher abundance of the SREBP1-c gene, as opposed to SREBP-2.[67]

SREBP1-c is bound to the nuclear membrane and homodimerizes upon activation; it is cleaved from the membrane, and binds to SRE-1 (5'-ATC-ACCCCAC-3)[68] to activate transcription of target proteins. These include enzymes involved in fatty acid synthesis, such as acetyl CoA carboxylase (ACC), fatty acid synthase and glycerol-3-phosphate acyl transferase.[65] Insulin-mediated activation of SREBP1-c occurs through insulin receptor substrate 1 (IRS1). Activation of IRS1 has been shown to induce activation of SREBP1-c at several steps. Primary among these involves IRS1 activation of Akt, which is directly involved in mobilizing nascent SREBP from endoplasmic reticulum to Golgi bodies, where it undergoes final transcriptional activation.[69-71]

When a high carbohydrate diet is consumed, hepatic glucose uptake via GLUT2 occurs concomitant to activation of SREBP1-c via IRS1, thereby promoting lipogenesis, particularly when dietary carbohydrate is consumed beyond energy requirements. Excess glucose is converted to triglycerides and either stored or oxidized; thereby normal blood glucose concentration is maintained. This insulin-mediated lipogenic response acts in conjunction with glucose-induced lipogenesis.[72,73] Glucose regulates transcription via a carbohydrate responsive element binding protein (ChREBP)-mediated pathway. ChREBP is a nuclear transcription factor that is activated by intracellular glucose. Once activated, ChREBP translocates into the nucleus to bind to and activate carbohydrate response elements (ChoRE) on target genes.[74,75] ChoRE have been found in genes responsible for liver pyruvate kinase, glucose-6-phosphate dehydrogenase and, importantly, fatty acid synthase and ACC.[73] ChREBP activation results in induction of both glycolysis and lipogenesis. ChREBP knock-out mice exhibit reductions in both pathways.[76] Hence, the combined activation of ChREBP and SREBP1-c mediated by glucose and insulin, results in *de novo* lipogenesis, particularly following excess carbohydrate consumption.

Galactose Metabolism and Transport

Galactose is a hexose monosaccharide whose dietary intake is usually in the form of the disaccharide lactose (milk sugar). Lactose is hydrolyzed by the digestive enzyme lactase into its hexose moities, glucose and galactose. Galactose shares the same transport mechanisms as glucose in enterocytes, specifically, apical SGLT cotransporters and the basolateral GLUT2. It enters the portal blood and is cleared in its passage through the liver, so little or no galactose above 1 mmol/L is seen in the systemic blood, even after ingesting as much as 100 g of lactose. Ingestion of galactose without glucose, however, induces higher plasma concentration. Alcohol has been shown to decrease galactose uptake and metabolism by the liver, leading to increased circulating galactose (galactosemia). In the liver cells, galactose is converted by the enzyme galactokinase into galactose-1-phosphate. This, in turn, is converted by a two-stage enzymic transformation into glucose-1-phosphate, which can be converted into glycogen. Although, in theory, glucose-1-phosphate can enter the glycolytic pathway, it does not normally do so. Most tissues have enzymes that can metabolize galactose. However, even in the complete absence of dietary galactose, glucose can be converted into galactose and supply cellular needs for galactose when required. Many structural elements of cells and tissues (glycoproteins and mucopolysaccharides) contain galactose, and in mammals endogenously synthesized, galactose is produced and secreted into breast milk.

Cataracts and Inborn Errors of Galactose Metabolism. Galactose concentration in peripheral blood normally does not exceed 1 mmol/L. When higher than this (as in galactosemia), several tissues can remove galactose from the blood and convert it into galactitol (dulcitol) via actions of the enzyme aldehyde reductase. Because it is not metabolized, galactitol can accumulate in tissues, resulting in pathologic changes due to increased osmotic pressure. In the lens of the eye, this results in cataract formation.[77] Cataracts are also observed in two inborn errors of galactose metabolism, resulting from deficiencies of the enzymes galactose-1-phosphate uridyltransferase and galactokinase. The former enzyme deficiency produces classic galactosemia. Unless it is treated promptly in the neonate by withdrawing galactose from the diet (from the lactose component of milk), either death or severe mental retardation can occur. Cataracts can also occur as a complication of diabetes, when high blood glucose results in excess glucose transport into the lens, where it is metabolized to sorbitol, thus causing the lens to swell and become opaque.

Fructose Absorption and Metabolism

Fructose, a monosaccharide ketohexose, is present as the free hexose naturally occurring in honey and fruit, or produced by isomerization of glucose from corn and added to soft drinks and many other sweetened beverages and foods as high-fructose corn syrup. Fructose is also produced from hydrolysis of the dietary disaccharide sucrose (yielding glucose and fructose). Fruits contain various combinations of free fructose, free glucose, and sucrose, usually yielding between 45% and 70% fructose. Although fructose is absorbed across the enterocytes of the small intestine, it is not a substrate for the SGLT cotransporters. The evidence for this is threefold: (a) fructose absorption is normal in those with glucose-galactose malabsorption, who have defective SGLT-1 cotransporters; (b) fructose absorption is not reduced by phlorizin, the classic inhibitor of SGLT 1 cotransporters; and (c) fructose absorption is neither Na^+-sensitive nor electrogenic like that of glucose or galactose. Studies on the expression of human GLUT5 transporter in *Xenopus* oocytes showed that the transporter exhibited selectivity for high-affinity fructose transport that was not blocked by cytochalasin B, a potent inhibitor of facilitative glucose transport by glucose transporters.[25] Because GLUT5 is also expressed in the brush border of enterocytes in the small intestine,[78] this isoform is likely to be the major fructose transporter of the small intestine. Indirect evidence for the likelihood of fructose transport by GLUT5 is the finding that it is expressed in high concentration in human spermatids and spermatozoa,[78] gamete cells known to metabolize fructose. GLUT2 is localized to the basolateral membrane of enterocytes and, although having a much lower affinity for fructose transport than GLUT5, probably mediates the exit of the absorbed fructose from the enterocytes into the portal circulation. Investigators have reported that GLUT5 is also localized on the basolateral membrane in the human jejunum,[79] so fructose could also exit from the enterocytes via this transporter. In humans, absorption of fructose from sucrose ingestion is more rapid than that from equimolar amounts of fructose ingestion. Explanations for this phenomenon include differences in gastric emptying, the close association of sucrase activity with proximity to the intestinal brush border membrane, enhanced fluid absorption initiated by the glucose entraining fructose, and cotransport of fructose and glucose by a disaccharidase-related transport system.[80,81]

Fructose Metabolism. Approximately 85% of the fructose transported into the portal circulation is cleared in a single passage through the liver.[82] Thus, the concentration of fructose in the systemic circulation after consumption of substantial amounts of fructose with meals is low. After a large oral dose of fructose (25% of energy requirement), the plasma concentration increases to 0.4 mmol/L in 30 minutes and then slowly decreases to < 0.1 mmol/L during the next 90 minutes.[83] In the liver, fructose is phosphorylated by the abundant enzyme fructokinase into fructose-1-phosphate, which is cleaved by hepatic aldolase into the trioses, glyceraldehyde and dihydroxyacetone phosphate. Dihydroxyacetone phosphate is an intermediary metabolite in both the glycolytic and gluconeogenic pathways. Glyceraldehyde, although not an intermediary in either pathway, can be converted into glycolytic intermediary metabolites. The fate of either trios can be glucose/glycogen, lactate or energy and, to a lesser extent, triglyceride. However, it is likely that the proportion converted to triglyceride increases as the dose of fructose increases, and as the energy state of the liver increases.

When large amounts of fructose are ingested, such as occurs when a large serving of beverage sweetened with sucrose (50% fructose) or high-fructose corn syrup (55% fructose) is rapidly consumed, the glycolytic pathway becomes saturated with intermediates. These intermediates can be used for the glycerol moiety of triglyceride synthesis or can enter the pathway of de novo lipogenesis to form fatty acids that are then esterified to triglycerides, packaged with apolipoprotein-B and exported as very low density lipoproteins (VLDL). The preferential increase of lipogenic precursors after fructose ingestion occurs in large part because, unlike glucose metabolism via phosphofructokinase, fructokinase is not subject to allosteric negative feedback inhibition by ATP and citrate[84] (**Figure 2.4**). Thus, although only a small percentage (1%–3%) of ingested glucose-containing carbohydrate enters de novo lipogenesis and is incorporated into triglyceride in normal-weight, insulin-sensitive individuals, a proportionally greater amount of carbon from ingested fructose is metabolized to form triglycerides. The unregulated metabolism of fructose mediated by fructokinase also increases uric acid concentration. Specifically, the conversion of fructose to fructose-1-phosphate results in conversion of ATP to AMP and a depletion of inorganic phosphate, leading to uric acid production via the purine degradation pathway.[85] The key role of fructokinase in mediating the metabolic dysregulation induced by fructose consumption was demonstrated by the first phase II clinical trial testing the effects of a fructokinase inhibitor.[86] The intervention group had decreased liver fat and circulating uric acid, compared with the placebo group.

Small "catalytic" amounts of fructose appear to enhance hepatic glucose uptake, perhaps by activation of glucokinase,[87] and this has led to the idea that including limited amounts of dietary fructose may be beneficial in the management of postprandial blood glucose excursions in patients with diabetes.[88] However, caution should be exercised in recommending fructose in the dietary management of diabetes because larger amounts could contribute to weight gain and visceral adipose deposition and could exacerbate hyperlipidemia or insulin resistance (see later) or induce protein fructosylation and/or oxidative damage[89,90] involved in the pathogenesis of diabetic complications.

Inborn Errors of Fructose Metabolism

Six genetically determined abnormalities in the metabolism of fructose have been described in humans.[91] These abnormalities are caused by deficiencies in fructokinase, aldolase A and B,

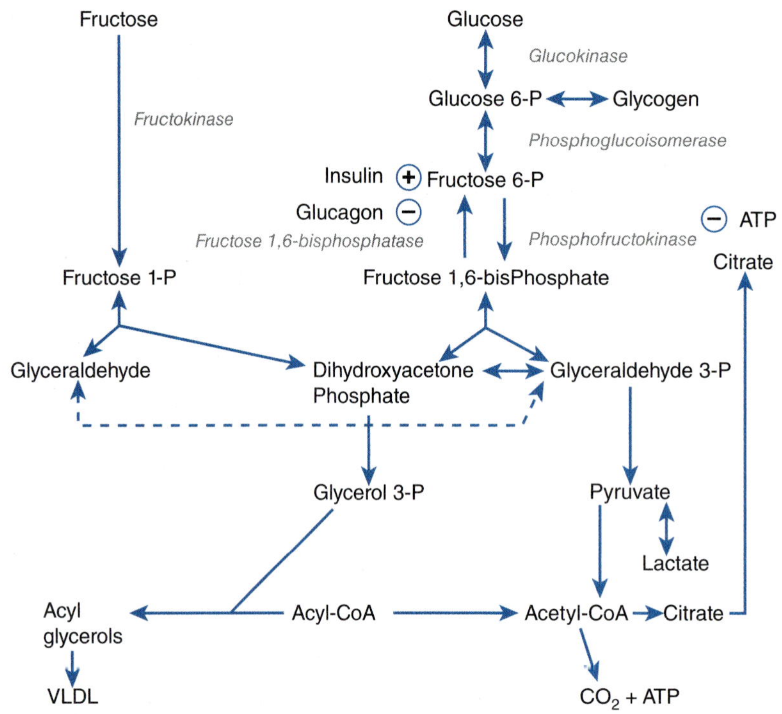

Figure 2.4 Fructose and glucose use in the liver. Hepatic fructose metabolism begins with phosphorylation by fructokinase. Fructose carbon enters the glycolytic pathway at the triose phosphate level (dihydroxyacetone phosphate to and glyceraldehyde-3-phosphate [P]). Thus, fructose bypasses the major control point by which glucose carbon enters glycolysis (phosphofructokinase) where glucose metabolism is limited by feedback inhibition by citrate and adenosine triphosphate (ATP). This allows fructose to serve as an unregulated source of both glycerol-3-phosphate and acetyl-coenzyme A (CoA) for hepatic lipogenesis. VLDL, very low density lipoprotein.

Reproduced from Havel PJ. Dietary fructose: implication for dysregulation of energy homeostasis and lipid/carbohydrate metabolism. Nutr Rev 2005;63:133-7.

fructose-1,6-diphosphatase, and glycerate kinase, and by fructose malabsorption. Limiting dietary fructose produces a favorable result in each of these conditions except aldolase A deficiency. Fructokinase deficiency, manifest in the liver, causes fructosemia (elevated blood fructose) and fructosuria (fructose excretion in urine). In contrast to the low concentration of fructose observed in the blood of physiologically normal persons after ingestion of 1 g of free fructose/kg, the concentration in the fructokinase-deficient individuals approaches 3 mmol/L and is sustained for many hours. Despite the sustained high concentration of fructose in the blood, cataracts do not develop, in sharp contrast to the cases of galactokinase deficiency and diabetes.

The three aldolases, A, B, and C, catalyze the reversible conversion of fructose-1,6-diphosphate into glyceraldehyde-3-phosphate and dihydroxyacetone phosphate. Each aldolase is coded for by a different gene: A is on chromosome 16, B on 9, and C on 17. Expression of the enzymes is regulated during development, so A is produced in embryonic tissues and adult muscle; B in adult liver, kidney, and intestine; and C in adult nervous tissue. Deficiency of aldolase A produces a syndrome of mental retardation, short stature, hemolytic anemia, and abnormal facial appearance. Deficiency of aldolase A probably results in these defects because it is involved in fetal glycolysis. No treatment exists for the condition. Deficiency of aldolase B (hereditary fructose intolerance), the most frequent of the three deficiencies, was first described in 1956.[92] When fructose is ingested, vomiting, failure to thrive, and liver dysfunction occur. The treatment is to avoid dietary fructose.

Deficiency of fructose-1,6-diphosphatase was first described in 1970. Patients exhibit hypoglycemia, acidosis, ketonuria, and hyperventilation. Urinalysis shows many changes in organic acids, but excretion of glycerol is diagnostic. The treatment is to avoid dietary fructose. D-glyceric aciduria is rare and is caused by D-glycerate kinase deficiency. The presentation of the disease is variable, from no clinical symptoms to severe metabolic acidosis and psychomotor retardation, findings that suggest that among the 10 described cases, other enzyme deficiencies are also present.

In fructose malabsorption, ingestion of moderate to large quantities of fructose produces abdominal bloating, flatulence, and diarrhea. No assessments of intestinal GLUT5 or its controlling gene have yet been made in any of these patients. If either glucose or galactose is ingested with fructose, fructose absorption is enhanced, and often symptoms of malabsorption do not occur.[81,91]

Other Disorders of Carbohydrate Digestion, Absorption, or Metabolism

Carbohydrate Intolerance

There are several clinical disorders in which sugar digestion or absorption is disturbed and gives rise to carbohydrate intolerance. The symptoms are a result of undigested or unabsorbed sugar and water entering the intestine, activating peristalsis and inducing passage of frequent fluid stools. The undigested carbohydrate can also enter the colon and undergo fermentation by colonic microflora into products that promote diarrhea. These disorders are classified as either congenital or secondary to some other disease, to impaired digestion of disaccharides, or to impaired absorption of the monosaccharides. The congenital deficiencies, although rare, can be life-threatening: examples are sucrase-maltase deficiency (watery diarrhea after ingesting sucrose-containing foods), alactasia (absence of lactase, diarrhea from ingestion of milk), glucose-galactose malabsorption (diarrhea from ingestion of glucose, galactose, or lactose), and the very rare trehalase deficiency (intolerance to trehalose contained in mushrooms). Sugar intolerance secondary to underlying gastrointestinal disease is more common, especially in pediatrics (see Chapter 51). Infections of the gastrointestinal tract, for example, often result in a temporary intolerance to lactose.

Lactose Intolerance

Adult mammals and most human groups after weaning retain only a fraction of the intestinal lactase activity of neonates (who need it to digest the lactose from breast milk). The persistence of lactase activity in Europeans has been regarded as the exception to the rule, because most human populations are hypolactasic and, therefore, malabsorb lactose.[93] However, small amounts of dietary lactose, up to 250 mL of milk, can be tolerated by most adults who do not fully digest lactose. The decrease in lactase in adults is a developmentally programmed event, and feeding high-lactose diets does not prevent the decrease. The mechanisms of the decline in activity have been studied in rats. As the animal matures, more mRNA message for lactase is needed to maintain the decreasing lactase activity in the enterocytes, suggesting that translational or posttranslational events are of greater importance than decreased lactase gene expression[94] (see also Chapter 74).

Diagnostic Tests to Evaluate Carbohydrate Digestion, Absorption, or Metabolism

Breath Hydrogen Tests

Carbohydrates that have not been digested or absorbed reach the colon and undergo fermentation by the resident bacteria. Hydrogen gas is produced and some of this is absorbed by the colon, enters the bloodstream, and is excreted in the breath as it reaches the lungs. Measuring breath hydrogen, thus, provides an estimate of sugar or carbohydrate malabsorption. This test was first used to detect lactose intolerance and has since been used in numerous studies on carbohydrate intolerance.[95] It has certain weaknesses; for example, it gives no indication of the amount of carbohydrate absorbed before the sugar reached the colon, some patients don't excrete hydrogen, and the patient needs to be kept off antibiotics and nil per os before testing.

Sugar Tolerance Tests

Clinical quantitative assessment of the efficiency of the digestion and absorption of carbohydrates in humans rests mainly on relatively simple tests in which carbohydrate loads (≥50 g) are ingested, and blood samples collected to estimate the sugar concentrations attained at various time intervals after ingestion. The concentrations are then compared with those obtained in physiologically normal individuals. The most commonly used test is the oral glucose tolerance test. Typically, nonpregnant adults consume 75 g of glucose as a liquid solution over 5 minutes, and glucose is measured in serum at 0, 30, 60, 90, and 120 minutes. For assessment of glucose intolerance and gestational diabetes in pregnancy, woman consume 75–100 g of glucose and samples for glucose measurement are obtained. In children, the test consists of 1.75 g/kg of glucose up to the maximum of 75 g.[96] Values greater than normal indicate impaired glucose tolerance/insulin sensitivity or possibly diabetes. Often, the criterion for diabetes is glucose concentration greater than 200 mg/dL 2 hours after glucose ingestion. The reproducibility of the oral glucose tolerance test has been considered to be poor.[97] An oral tolerance test is sometimes also performed with galactose. Because the liver is a major site of galactose metabolism, the test has been used to assess liver function. Similar oral tolerance tests exist for fructose and the disaccharides lactose (lactase deficiency) and sucrose (sucrase deficiency).

The Glycemic Index

The glycemic index was formulated to assess the glycemic potential of different carbohydrate-containing foods with the intention of assisting patients with diabetes to better control their blood glucose. For each food item under evaluation, a measured quantity containing 50 grams of available carbohydrate is ingested, and blood glucose concentration is measured over a period of 2 hours. The glycemic response, typically the incremental area under the curve over 2 hours, is compared with that obtained by ingesting a reference food, with a 50-gram glucose load considered the gold standard. Alternately, a portion of white bread containing 50 grams of carbohydrate has been used as the reference food. This normalized value, expressed as a percentage of the value obtained with the reference food, is designated the *glycemic index* of the food.[98] Several factors are known to affect the glycemic index of a given food including the nature of the starch structure, the particle size, pH, content of fiber, fat, and protein in the food matrix, as well as the cooking methods and time. Recent studies have demonstrated that these factors contribute large within (intraindividual CV 20%) and between variability (interindividual CV 25%, ranging between 2% to 77%) in measuring glycemic index of white bread[99] or meals.[100] These studies suggest that the glycemic index it is not reliable for use by healthy consumers who are purchasing foods based on mean GI values. The average glycemic index of a meal can be calculated by summing the products of the glycemic index for each food multiplied by the amount of carbohydrate in the food portion and dividing by the total amount of carbohydrate in the meal, although there are reports published that contest this approach. Another concept, the *glycemic load*, combines the glycemic index with the total amount of carbohydrate consumed to characterize the full glycemic potential of a mixed meal or diet plan. The glycemic load is determined by calculating the sum of the products of the glycemic index for each constituent food multiplied by the amount of carbohydrate in each food. These classifications are useful for the management of diabetes and hypoglycemia (see also Chapter 61), but other applications to the healthy public are not currently supported by research.

Epidemiologic evidence has linked the glycemic index and the glycemic load with the risk of developing chronic diseases such as type 2 diabetes,[101] cardiovascular disease,[102] and diet-related cancers of the colon and breast,[103-105] thus raising the issue of whether restricting foods with a high glycemic index and the overall glycemic load may be potentially useful in disease prevention. There has been considerable interest in applying the concept of glycemic index to the management of body weight.

Some of the current evidence suggests that, under free-living conditions, when food intake is not controlled, diets with lower glycemic loads are associated with weight reduction whereas higher glycemic loads are associated with weight gain. There are data to suggest this is particularly true for persons with insulin resistance or glucose intolerance.[106] However, it is not clear if this relationship is driven by the glycemic index, per se, or other differences between high and low glycemic index diets, particularly dietary fiber, which lowers the glycemic index of foods.[107] The debate between competing models of obesity (carbohydrate insulin based on GI vs energy balance) continues.[108-111]

Dietary fructose is another potential contributing factor to the inconsistent effects of dietary glycemic index/load. In a controlled dietary intervention study in overweight to obese men and women, when a high complex carbohydrate diet (GI = 64) was replaced with a lower glycemic index/load diet containing fructose-sweetened beverages (GI = 38) for 10 weeks, there was an increase in visceral adipose tissue mass and a number of adverse changes in the lipid/lipoprotein profiles, including increases of LDL cholesterol and apolipoprotein-B, and decreased insulin sensitivity, compared with the baseline diet.[112,113] In contrast, in the individuals consuming a higher glycemic index diet with glucose sweetened beverages (GI = 83), these adverse effects on visceral adiposity, plasma lipids/lipoproteins and insulin sensitivity were not observed. Another study in which diets with high vs low glycemic indices were compared in overweight men and women for 11 weeks reported no differences in fasting glucose or insulin, lipids or several inflammatory markers between the diets.[114] These and other results have led to the suggestion that a dietary fructose index may be more relevant than the glycemic index.[115]

Before public health recommendations can be made, well-controlled, long-term clinical trials, in which the effects of the diets on postprandial circulating glucose and insulin concentrations are assessed over the course of the study, are needed to determine whether the glycemic index/load of a diet has a role in body weight regulation or directly influences risk factors for chronic diseases such as type 2 diabetes and cardiovascular disease. The value of the glycemic index remains controversial; there are arguments in support of[116] and opposed to[117] the use of the glycemic index in health and disease. In addition, foods are not consumed in isolation (as when they are tested for their glycemic indices) but generally in the form of meals containing a mixture of macronutrients and carbohydrate types, including fiber. Thus, the glycemic effects of any particular food in the context of a mixed meal may be different from when a food is tested as the sole ingested food item. However, the glycemic index does illustrate that carbohydrate foods can differ widely in their effects on blood glucose and hormonal responses after a meal (see later).

Dietary Reference Intakes for Carbohydrate

The recommended dietary allowance (RDA) for carbohydrate is set at 130 g/day for adults and children ages 1 through 18 years.[118] This value is based on the amount of available carbohydrate that can provide an adequate supply of glucose for brain and central nervous system cells without the need for glucose production from ingested proteins or triglycerides. It also assumes that energy intake is sufficient, and that the central nervous system is not relying on a partial replacement of glucose fuel by ketones.

For infants, an RDA value has not been established, but the adequate intake (AI) is set at 60 g/day for those 0 to 6 months of age. This value is equal to the amount of carbohydrate consumed in human milk and is considered optimal for growth and development during the first 6 months of life. For infants 7 to 12 months of age, the AI is set at 95 g/day. This value is based on the amount of carbohydrate consumed from human milk and complementary foods in the diets of infants of this age group. No gender differences exist for the RDA or AI values for recommended amounts of carbohydrate.

The amount of dietary carbohydrate that supports optimal health is not known, but an acceptable macronutrient distribution range has been set, with carbohydrate contributing between 45% to 65% of energy intake. The potential for adverse effects of overconsumption of carbohydrate was considered. Specifically, the potential effects of glycemic index, total sugar intake, and added sugar intake on increasing the risk for coronary heart disease, cancer, diabetes, or obesity were examined. Currently, the evidence available is insufficient to support an upper limit of carbohydrate intake based on the glycemic index of the diet.

The World Health Organization and the US Dietary Guideline for Americans have recommended that added sugar intake should not exceed 10% of total energy intake.[119] Recommendations from the American Heart Association set an upper daily limit of energy intake from added sugar at 100 kcal for women and 150 kcal for men.[120] Data from the US Department of Agriculture (USDA) show that the US per capita availability of caloric sweeteners increased by more than 40 pounds per year from 1960 to 1999 but has declined by about 19% over the last 2 decades.[121] Nevertheless, the latest estimate for loss-adjusted per capita intake of caloric sweeteners is 344 kcal/day,[121] which suggests that the most Americans are consuming considerably more added sugar than the recommended 10% of energy intake. The data in the USDA Sugar and Sweeteners Yearbook Tables show that worldwide total use of sugar has increased by approximately 37% since 2000.[122]

Carbohydrates and Athletic Performance

Carbohydrate, present in limited amounts in muscle (300 g glycogen), liver (90 g glycogen), and body fluids (30 g glucose), is a major fuel for physical performance. The ATP stored in muscle cells can only provide energy for high-power output for a few seconds. It can be resynthesized anaerobically for a further few seconds (5–8 seconds) by using the phosphate from creatinine phosphate. These short, intense bursts of muscular activity occur in sprints (100 m), track and field events, and sports such as tennis, hockey, football, gymnastics, and weightlifting. If the maximum effort lasts for 30 seconds or longer, muscle glycogenolysis can supply the energy for physical activity, with concurrent buildup of lactic acid in muscle. Most physical activity, however, requires an energy source that can power muscles for longer periods. Both duration and intensity of exercise determine the mix of fuel used. At light to moderate activity levels, and as duration of exercise lengthens, the contribution of fat to energy used for physical activity increases. In contrast, as the intensity of activity increases from rest to light to moderate to intense, the contribution of carbohydrate to energy production increases. The change to using carbohydrate is not a linear response but accelerates with the intensity of the work. At higher exercise intensities, the versatility of carbohydrate as a fuel source is demonstrated, because it can produce energy under conditions of limited oxygen supply. Endurance athletes use more fat and so conserve the carbohydrate stored in muscle and liver and maintain blood glucose concentrations for longer periods. Ultimately, the availability of carbohydrates sets the limits for continued endurance performance. The store of carbohydrate usually suffices for just 1 to 3 hours of physical exertion, depending on the intensity of effort. It has been suggested that avoiding hypoglycemia during exercise plays a key role in prolonging the time to fatigue. Several homeostatic mechanisms exist to keep blood glucose in the normal range during exercise and have been reviewed by Noakes.[123]

Dietary Manipulation of Glycogen Stores: Carbohydrate Loading

Dietary manipulation can be used to increase the stores of glycogen in muscle and liver. Glycogen increases when more carbohydrate is consumed. The practice is known as carbohydrate loading. The traditional protocol called for 3 days of exhausting physical exercise on a low-carbohydrate diet followed by 3 days of rest on a high-carbohydrate diet. In general, athletes dislike both phases; in the first, they often feel both mentally and physically exhausted and are at increased risk of injury; and in the second, they may feel bloated because the glycogen is stored with water. For these reasons, the traditional protocol has been modified. One protocol calls for the elimination of the initial carbohydrate depletion phase and relies solely on the tapering of exercise with a high-carbohydrate diet several days before the event to augment glycogen stores. Another protocol shortens the process of glycogen depletion and loading into a one-day time frame by having the athletes perform a short-duration (~3 minutes), high-intensity (supramaximal) workout and then consume a high carbohydrate diet during the next 24 hours. For athletes in general, it makes sense to eat plenty of carbohydrate to maximize glycogen storage, because the usual training periods of several hours per day can deplete it. Little doubt exists that a high-carbohydrate diet increases glycogen storage and can improve athletic performance. What to advise athletes to ingest just before an event is still debated and constantly evolving. A meal or snack taken 3 to 4 hours before exercise should include about 200 to 300 g of carbohydrate, or about 1 hour before exercise, about 13 to 60 g of carbohydrate can be taken to maximize maintenance of blood glucose. Consumption of solid food is not advisable immediately before strenuous exercise. During endurance events, beverages containing simple carbohydrates (solutions of glucose, fructose, or sweetened fruit juices) can be provided but should not exceed the maximal absorption rate, which is 60 g per hour for glucose. Along with glucose, fructose ingestion can avoid gastrointestinal upset, as a different transporter (GLUT5) is used, and this effectively increases carbohydrate absorption to about 90 g per hour and promotes carbohydrate oxidation rate.[124] After glycogen-depleting exercise, optimal carbohydrate intake of 1.2 grams per kilogram per hour spaced over 4 to 24 hours after exercise will aid in the restoration of liver and muscle glycogen.[125]

Carbohydrates and Chronic Disease

Sugar and Dental Caries

Dental caries is a disease created by bacterial plaque on the enamel of teeth. Gradual and progressive demineralization of the enamel, dentin, and cementum occurs. The literature consistently shows that excessive sugar consumption is the main cause of dental caries.[126,127] The relationship between sugar and caries is complex because the formation of caries involves interactions such as nutrients and food components of diet, plaque bacteria, salivary flow and composition, minerals and fluoride status, genetics, age, and even race (see also Chapter on dental health). The most common organism in dental plaque associated with caries is *Streptococcus mutans*, but other bacteria also appear to contribute. Most studies have focused on the acids (lactic and acetic) generated from sugars (sucrose) by the bacteria, but the complex formation and accumulation of plaque from the insoluble dextran made from sucrose may also be important.[128,129] A role for acids such as phosphoric acid added to many soft drinks in the demineralization in dental caries is possible as well.

Health Impacts of Excess Consumption of Fructose and Fructose-Containing Added Sugars

Fructose has been proposed to contribute to metabolic diseases including hyperlipidemia, insulin resistance, and obesity.[130] The idea that fructose has these adverse metabolic effects was originally based on a substantial number of studies reporting that feeding high-fructose and high-sucrose diets to experimental animals produced adverse metabolic consequences, including weight gain, dyslipidemia, insulin resistance, and hypertension (see Review.[131] In fact, many of the high-fat diets used to induce obesity and metabolic dysfunction in rodent models are also high in sucrose. High-fructose diets have been employed as a model of hypertension in rats.[132] Inclusion of a moderate amount of fructose (20% of energy) in the diet accelerates the onset of diabetes in a rat model of type-2 diabetes.[133] High fructose and high fructose corn syrup sweetened beverages have been used to induce weight gain, hyperlipidemia, insulin resistance and systemic inflammation as a diet-induced model of metabolic dysfunction in adult rhesus macaques.[134-136] High doses of fish oil prevent the effects of the high fructose diet to increase plasma triglycerides and to induce insulin resistance.[137] The diet-induced rhesus model has also been used to investigate the metabolic effects of oxytocin administration on energy balance[138] and the roles of apo-lipoprotein-C3 and angiopoietin-like protein-3 in sugar-induced hyperlipidemia.[139,140]

Because of the marked differences in the hepatic metabolism of fructose and glucose discussed earlier, fructose is more lipogenic than glucose and is, therefore, more readily converted in the liver to triglyceride, which can be exported as VLDL containing apolipoprotein-B and stored in adipose tissue. In addition, several studies have demonstrated that fructose increases circulating triglyceride concentrations in the postprandial period,[141-143] and evidence indicates that this effect is more pronounced in individuals with existing hyperlipidemia or insulin resistance.[83,144,145] Thus, long-term consumption of diets high in fructose may increase the risk of atherosclerosis or other cardiovascular disease. In addition, recent data indicate that compared with glucose, consuming fructose with meals, which does not stimulate insulin secretion, results in a reduction of circulating leptin and an attenuated postprandial suppression of ghrelin, a hormone produced by the stomach that stimulates hunger and increases food intake[143] (see Chapters 46 and 63). Thus, with respect to the hormones insulin, leptin, and ghrelin that are involved in the long-term endocrine regulation of food intake, energy balance, and body adiposity,[146,147] dietary fructose behaves more like dietary fat[148] than do other types of

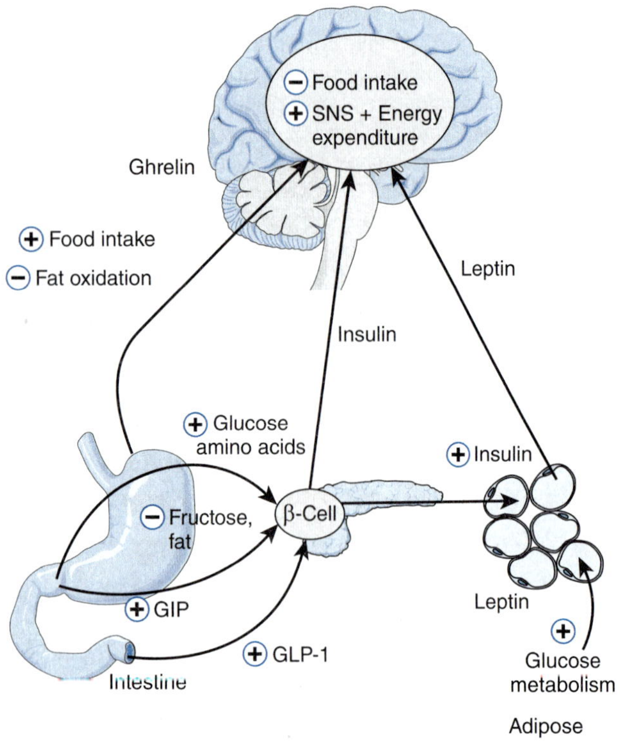

Figure 2.5 Long-term signals regulating food intake and energy homeostasis. Insulin and leptin are important long-term regulators of food intake and energy balance. Both insulin and leptin act in the central nervous system to inhibit food intake and to increase energy expenditure, most likely by activating the sympathetic nervous system (SNS). Insulin is secreted from the β-cells in the endocrine pancreas in response to circulating nutrients (glucose and amino acids) and to the incretin hormones, glucose-dependent insulinotropic polypeptide (GIP) and glucagon-like peptide-1 (GLP-1), which are released during meal ingestion and absorption. Insulin can also act indirectly by stimulating leptin production from adipose tissue via increased glucose metabolism. In contrast, dietary fat and fructose do not stimulate insulin secretion and, therefore, do not increase leptin production. Ghrelin, a hormone produced by endocrine cells in the stomach, increases food intake and decreases fat oxidation and appears to have an anabolic role in long-term regulation of energy balance. Ghrelin secretion is normally suppressed after meals, but it is not suppressed by fat or fructose consumption. The long-term signals interact with the short-term signals in the regulation of energy homeostasis and appear to set sensitivity to the satiety-producing effects of short-term signals such as cholecystokinin.

Modified from Havel PJ. Peripheral signals conveying metabolic information to the brain: short-term and long-term regulation of food intake and energy homeostasis. Exp Biol Med (Maywood) 2001;226:963-77.

carbohydrate (**Figure 2.5**). The lack of effect of fructose on these endocrine signals suggests that chronic consumption of a diet high in fructose could contribute, along with dietary fat and inactivity, to increased energy intake, weight gain, and obesity. However, this suggestion is not supported by short-term clinical trials in which fructose consumption induced lower 24-h leptin concentration, but not higher gains in body weight.[149,150] Longer studies in free-living participants are warranted, but will be difficult to conduct, due to the sample size that would be required to achieve detectable differences in body weight between diet groups, the confounding effects of other dietary components (e.g., protein and fiber) and physical activity, and the difficulty of ensuring dietary compliance in long-term nutrition studies.

In a carefully controlled diet intervention study designed to address and compare the metabolic effects of fructose and glucose consumption on body composition, lipid and carbohydrate metabolism, glucose or fructose-sweetened beverages were consumed for 10 weeks at 25% of energy requirements by older (age 40–72 years) overweight to obese men and women.[113] During the first 8 weeks of the intervention, when the individuals consumed the sweetened beverages along with their usual diets *ad libitum*, both groups gained about 1.5 kg in body weight. However, in the individuals consuming fructose-sweetened beverages, there was a significant increase in intra-abdominal (visceral) fat that was not observed in the group consuming glucose-sweetened beverages, in which the increase of fat in the abdominal area was primarily in the subcutaneous compartment. In addition, in individuals consuming fructose-sweetened beverages, there were increases of isotopically determined *de novo* lipogenesis, 24-hour postprandial triglyceride acid profiles, LDL cholesterol, apolipoprotein-B, small-dense LDL cholesterol, oxidized LDL and remnant lipoproteins, and a 20% decrease in insulin sensitivity that did not occur in the individuals consuming glucose-sweetened beverages.

A substantial number of subsequent clinical dietary intervention studies[151-155] have corroborated many of these results and have also demonstrated that consumption of fructose decreases fat oxidation[154-156] and hepatic insulin sensitivity,[154,157] and increases hepatic lipid content[154,155] and circulating uric acid[152,158]. Clinical dietary intervention studies have demonstrated that these risk factors are adversely affected when sucrose and HFCS are consumed.[152,159-163] Several more recent studies have shown that these metabolic risk factors are beneficially affected when research participants consumed reduced amounts of added sugars.[164-168] Previous[113,141,169,170] and more recent studies[162,163] suggest that consumption of fructose or fructose-containing sugars leads to greater increases in several metabolic risk factors in men than in women, including visceral adiposity, liver fat content, circulating lipids/lipoproteins and uric acid. A question that has yet to be fully explored is whether sugars naturally contained in fruit juices such as apple juice have the same adverse metabolic impact as beverages sweetened with added sugars.[171] Several mechanisms by which dietary fructose and fructose-containing sugars may affect adiposity and adipose distribution[172-175] and adversely alter lipid profiles and insulin sensitivity have been proposed, and are the subject of a number of recent reviews.[176-180]

Acknowledgments

The authors would like to acknowledge Roy J Levin for his contributions to the previous version of this chapter.

Dr. Havel acknowledges research support from the National Institutes of Health Grants HL-075675, HL-091333, HL-121324, DK-087307, DK-095960 and OD-0111107 and the American Diabetes Association.

Dr. Keim acknowledges research funding from the US Department of Agriculture CRIS 2032-51530-025-00D.

Dr. Stanhope acknowledges research support from National Institutes of Health Grants HD-051958 and HL-137716.

References

1. Castillo JJ, Couture G, Bacalzo, Jr. NP, et al. Nutrients 2022;14.
2. Dahlquist A, Semenza, G. J Pediatr Gastroenterol Nutr 1985;4(6): 857–65.
3. Ugolev AM, De Laey P. Biochim Biophys Acta 1973;300:105–28.
4. Gray GM. J Nutr 1992;122:172–7.
5. Trowell H. Suthgate DA, Wolever TM, Leeds AR, Gassull MA, Jenkins DJ. Lancet 1976;1:967.
6. Asp NG. Am J Clin Nutr 1994;59(3 Suppl):679S–81S.
7. Wursch P. World Rev Nutr Diet 1989;60:199–256.
8. Wursch P, Gracey M, Kretchmer N, Rossi E, eds. Raven Press: New York, 1991.
9. Roediger WE. Dis Colon Rectum 1990;33(10):858–62.
10. Cani PD, Delzenne NM. Curr Pharm Des 2009;15:1546–58.
11. Delzenne NM, Cani PD, Neyrinck AM. J Nutr 2007; 137:2547S–51S.
12. Marlett JA. J Am Diet Assoc 1992;92: 175–86.
13. Drewnowski A. Epidemiol Rev 2007;29:160–71.
14. Lee WS, Kanai Y, Wills RG, Hediger MA. J Biol Chem 1994;269:12032–9.
15. Mueckler M, Caruso C, Baldwin SA, et al. Science 1985;229: 941–5.
16. Alvarez J, Lee DC, Baldwin SA, Chapman D. J Biol Chem 1987;262: 3502–9.
17. Cuppoletti J, Jung CY, Green FA. J Biol Chem 1981;256:1305–6.
18. Hediger MA, Rhoads DB. 1994;74:993–1026.
19. Silverman M. Annu Rev Biochem 1991;60757–94.
20. Maher F, Vannucci SJ, Simpson IA. FASEB J 1994;8:1003–11.
21. Diamond DL, Carruthers A. J Biol Chem 1993;268:6437–44.
22. Gould GW, Holman GD. Biochem J1993;295:329–41.
23. Kayano T, Fukumoto H, Eddy RL, et al.. J Biol Chem 1988; 263:15245–8.
24. Kayano T, Burant CF, Fukumoto H, et al. J Biol Chem 1990; 265:13276–82.
25. Burant CF, Takeda J, Brot-Laroche E, Bell GI, Davidson NO. J Biol Chem 1992;267:14523–6.
26. Sato Y, Ito T, Udaka N, et al. Tissue Cell 1996;28:637–43.
27. Curry DL. Pancreas 1989;4:2–9.
28. Chen, L-Q, Cheung LS, Feng L, Tanner W, Frommer WB. Annu Rev Biochem 2015;84:865–94.
29. Uldry M, Thorens B. Pflugers Arch 2004;447:480–9.
30. Wood IS, Trayhurn P. Br J Nutr 2003;89:3–9.
31. Wallberg-Henriksson H, Zierath JR. Mol Membr Biol 2001;18:205–11.
32. Katz EB, Stenbit AE, Hatton K, DePinhot R, Charron MJ. Nature 1995;377:151–5.
33. Kim JK, Zisman A, Fillmore JJ, et al. J Clin Invest 2001;108: 153–60.
34. Abel ED, Peroni O, Kim JK, et al. Nature 2001;409:729–33.
35. Misra RP, Duncan SA.. Endocrine 2002;19: 229–38.
36. Hediger MA, Coady MJ, Ikeda TS, Wright EM. 1987330:379–81.
37. Wright EM, Turk E, Zabel B, Mundlos S, Dyer J. J Clin Invest 1991;88:1435–40.
38. Hediger MA, Kanai Y, You G, Nussberger S. J Physiol 1995482:7S–17S.
39. Mackenzie, B., Panayotova-Heiermann M, Loo DD, Lever JE, Wright EM. J Biol Chem 1994269: 22488–91.
40. Bays H. Curr Med Res Opin 2009;25:671–81.
41. Turk E., Zabel B, Mundlos S, Dyer J, Wright EM. Nature 1991;350:354–6.
42. Fine KD, Santa Ana CA, Porter JL, Fordtran JS. Gastroenterology 1993;105:1117–25.
43. Levin R.J. Am J Clin Nutr 1994;59:690S–8S.
44. Black, R., Fontaine O, Lamberti L, et al. J Glob Health 2019;9:020801.
45. Owen, O.E., Morgan AP, Kemp HG, Sullivan JM, Herrera MG, Cahill Jr., GF. J Clin Invest 1967. 46(10): p. 1589–95.
46. Cahill GF, Jr., Owen OE, Felig P. Physiologist 1968;11:97–102.
47. Pedersen, O., Bak JF, Andersen PH, et al. Diabetes 1990;39: 865–70.
48. Dohm GL, Elton CW, Pilch PF, et al., Am J Physiol 1991;260: E459–63.
49. Ashcroft FM, Harrison DE, Ashcroft SJ. Nature 1984;312:446–8.
50. Dunne MJ, Petersen OH. Biochim Biophys Acta 1991;1071: 67–82.
51. Inagaki N, Gonoi T, Clement JP, et al., Science 1995;270:1166–70.
52. Vahl T, D'Alessio D. Curr Opin Clin Nutr Metab Care 2003;6: 461–8.
53. Ahren B. Diabetologia 2000;43:393–410.
54. Ahren B, Holst JJ. Diabetes 2001;50:1030–8.
55. D'Alessio DA., Kieffer TJ, Taborsky GJ, Havel Jr., PJ. J Clin Endocrinol Metab 2001;86:1253–9.
56. Havel PJ, Taborsky Jr., Jr. *Stress-induced activation of the neuroendocrine system and its effects on carbohydrate metabolism.* 6th ed. Ellenberg and Rifkin's Diabetes Mellitus, ed. D. Porte, Jr., R. Sherwin, and A.D. Baron. New York: McGraw-Hill, 2003.
57. Rorsman P, Ashcroft FM. Physiol Rev 2018;98:117–214.
58. Taborsky GJ, Jr., Ahren B. *Beta-cell function and insulin secretion.* 6th ed. Ellenberg and Rifkin's Diabetes Mellitus, ed. D. Porte, Jr., R. Sherwin, and A.D. Baron. New York: McGraw-Hill, 2003.
59. Thorens, B., Cheng Z-Q, Brown C-D, Losish HF. Am J Physiol 1990;259:C279–85.
60. Jungermann KKatz N. Physiol Rev 1989;69:708–64.
61. Pipeleers D. Diabetologia 1987;30:277–91.
62. Greenbaum CJ., Havel PJ, Taborsky Jr., GJ, Klaff LJ. J Clin Invest 1991;88:767–73.
63. Hansen LH, Abrahamsen N, Nishimura E. Peptides 1995; 16:1163–6.
64. Taborsky GJ, Jr., Ahren B, Havel PJ. Diabetes 199847:995–1005.
65. Shimano H. Prog Lipid Res 200140:439–52.
66. Chawla, A., Repa JJ, Evans RM, Mangelsdorf DJ. Science 2001;294:1866–70.
67. Osborne TF. J Biol Chem 2000;275:32379–82.
68. Yokoyama C, Wang X, Briggs MR, et al. Cell1993;75:187–97.
69. Matsumoto M, Ogawa W, Teshigawara K, et al. Diabetes 2002;51:1672–80.
70. Osborne TF, Espenshade PJ. Genes Dev 2009;23:2578–91.
71. Yellaturu CR, Deng X, Cagen LM, et al. J Biol Chem 2009284: 7518–32.
72. Dentin R, Girard J, Postic C. Biochimie 2005:87:81–6.
73. Iizuka K, Horikawa Y. Endocr J 2008;55:617–24.
74. Fukasawa M, Ge Q, Wynn RM, Ishii S, Uyeda K. Biochem Biophys Res Commun 2010391:1166–9.
75. Kawaguchi T, Takenoshita M, Kabashima T, Uyeda K. Proc Natl Acad Sci U S A 200198:13710–5.
76. Iizuka K., Bruick RK, Liang G, Hoton JD, Uyeda K. Proc Natl Acad Sci U S A 2004101:7281–6.
77. Van Heyningen R. Exp Eye Res 197111:415–28.
78. Davidson NO, Hausman AM, Ifkovits CA, et al. Am J Physiol 1992;262:C795–800.
79. Blakemore SJ, Aledo JC, James J, Campbell FC, Lucocq JM, Hundal HS. Biochem J 1995;309:7–12.
80. Fujisawa T, Riby J, Kretchmer N. Gastroenterology 1991;101: 360–7.
81. Riby JE, Fujisawa T, Kretchmer N. Am J Clin Nutr 1993. 58:748S–753S.
82. Francey C, Cros J, Rosset R, et al. Clin Nutr ESPEN 2019;29: 125–132.

83. Teff KL, Grudziak J, Townsend RR, et al. J Clin Endocrinol Metab 2009;94:1562–9.
84. Havel PJ. Nutr Rev 2005;63:133–57.
85. Mayes PA. Am J Clin Nutr 1993;58:754S–765S.
86. Kazierad DJ, Chidsey K, Somayaji VR, Bergman AJ, Birnbaum MJ, Calle RA. Med (N Y) 20212:800–813 e3.
87. Petersen KF, Laurent D, Yu C, Cline GW, Shulman GI. Diabetes 2001;50:1263–8.
88. Moore MC, Davis SN, Mann SL, Cherrington AD. Diabet Care. 2001.;24:1882–7.
89. Bell, R.C., Carlson JC, Storr KC, Herbert K, Sivak J. Br J Nutr 2000;84:575–82.
90. Levi B, Werman MJ. J Nutr 1998;128:1442–9.
91. Hommes FA. Am J Clin Nutr 199358:788S–795S.
92. Cori GT. Harvey Lect 1954;48:148–71.
93. Kretchmer N. Gastroenterology 1971;61:805–13.
94. Nudell DM, Santiago NA, Zhu JS, Cohen ML, Majuk Z, Gray GM. Am J Physiol 1993;265:G1108–15.
95. Levitt MD, Donaldson RM. J Lab Clin Med 197075:937–45.
96. Potparic O, Gibson J. Parthenon Publishing: Lancashire, UK, 1993.
97. McDonald GW, Fisher GF, Burnham C. Diabetes 1965;14:473–80.
98. Jenkins DJ, Wolever TM, Taylor RH, et al. Am J Clin Nutr 1981;34:362–6.
99. Matthan, NR, Ausman LM, Meng H, Tighiouart, Lichtenstein AH. Am J Clin Nutr 2016;104:1004–1013.
100. Meng H, Matthan NR, Ausman LM, Lichtenstein AH. Am J Clin Nutr 2017;105:842–853.
101. Salmerón J, Ascherio A, Rimm EB, et al. Diabet Care 1997;20:545–50.
102. Liu S, et al., Am J Clin Nutr 2000;71:1455–61.
103. Augustin LS, Dal Maso L, La Vecchia C, et al.,. Ann Oncol 2001;12:1533–8.
104. Franceschi S, Dal Maso L, Augustin L, et al. Ann Oncol 2001;12:173–8.
105. Hu FB, Manson JE, Liu S, et al. J Natl Cancer Inst 1999;91: 542–7.
106. Stanhope KL, Goran MI, Bosy-West[ja; A, et al. Obes Rev 2018;19:1205–1235.
107. Livesey, G., Taylor R, Hulshof T, Howlett J. Am J Clin Nutr 2008;87:258S–268S.
108. Ludwig DS, Aronne LJ, Astrup A, et al., Am J Clin Nutr 2022;115:595–597.
109. Drewnowski A, et al. Am J Clin Nutr 2022;115:591–592.
110. Ludwig DS, Aronne LJ, Astrup A, et al. Am J Clin Nutr 2021;114:1873–1885.
111. Hall KD, Farooqi S, Friedman JM, et al. Am J Clin Nutr 2022;115:1243–1254.
112. Stanhope KL, Havel PJ. Ann NY Acad Sci, 2010;1190:15–24.
113. Stanhope KL, Schwarz JM, Keim NL, et al. J Clin Invest 2009;119:1322–34.
114. Vrolix R, Mensink RP. Am J Clin Nutr 2010;92:366–74.
115. Johnson RJ, et al. Am J Clin Nutr 2007;86:899–906.
116. Jenkins DJ, Kendall CW, Marchie A, et al. Am J Clin Nutr 2002;76:266S–73S.
117. Pi-Sunyer FX. Am J Clin Nutr 2002;76:290S–8S.
118. United States Department of Agriculture and Department of Human Health and Services. December 2020; 9th Edition:[Available from: DietaryGuidelines.gov, 2020.
119. Nishida C, Uauy R, Kumanyika S, Shetty P. Public Health Nutr 2004;7:245–50.
120. Johnson RK. Appel LJ, Brands M, et al. Circulation 2009;120: p.1011–20.
121. United States Department of Agriculture, E.R.S. *Loss-Adjusted Food Availability Documentation* Jan 20, 2022]; Available from: www.ers.usda.gov/data-products/food-availability-per-capita-data-system/loss-adjusted-food-availability-documentation/.
122. United States Department of Agriculture, E.R.S. *Sugar and Sweeteners Yearbook Tables* Jan 20, 2022]; Available from: www.ers.usda.gov/data-products/sugar-and-sweeteners-yearbook-tables.aspx.
123. Noakes TD. Nutrients 2022;14.
124. Mata F, Valenzuela PL, Gimenez J, et al. Nutrients 2019;11:1084.
125. Burke LM, van Loon LJ, Hawley JA, et al. J Appl Physiol; 2017;122:1055–1067.
126. Moynihan PJ, Kelly SA. J Dent Res 2014;93:8–18.
127. Sheiham A, James WP. J Dent Res 2015;94:1341–7.
128. Bowen W., *Simple carbohydrates as microbiological substrates.* Biological Functions of Carbohydrates, ed. D. Conning.: International symposium, 1993.
129. Navia JM. Am J Clin Nutr 1994;59:719S–727S.
130. Elliott SS, et al., *Fructose, weight gain, and the insulin resistance syndrome.* Am J Clin Nutr, 2002. 76(5): p. 911–22.
131. Kleinert M, et al. Nat Rev Endocrinol 2018;14:140–162.
132. Chan AM, Ng AM, Yunus MH, et al. Nutrients 2021;13.
133. Cummings BP, Stanhope KL, Graham JL, et al. Am J Physiol Regul Integr Comp Physiol 2010;298:R1343–50.
134. Bremer AA, Stanhope KL, Graham JL, et al., Clin Transl Sci 2011;4:243–52.
135. Ghoshal S, Stevens JR, Billon C, et al. Mol Metab 2018;8:51–64.
136. Havel PJ, Kievit P, Comuzzie AG, Bremer AA. ILAR J 2017; 58:251–268.
137. Bremer AA, et al. J Nutr 2014;144:5–11.
138. Blevins JE, Graham JL, Morton GJ, et al., Am J Physiol Regul Integr Comp Physiol 2015;308:R431–8.
139. Butler AA, Graham JL, Stanhope KL, et al. J Lipid Res 2020; 61:376–386.
140. Butler AA, Price CA, Graham JL, et al. J Lipid Res 2019;60: 805–818.
141. Bantle JP, Raatz SK, Thomas W, Georgopoulos A. Am J Clin Nutr 2000;72:1128–34.
142. Stanhope KL, Havel PJ. Am J Clin Nutr 2008;88:1733S–1737S.
143. Teff KL, Elliott SS, Tschöp M, et al. J Clin Endocrinol Metab 2004;89:2963–72.
144. Abraha A, Humphreys SM, Clark ML, Matthews DR. Br J Nutr 1998;80:169–75.
145. Jeppesen J, Chen YI, Zhou MY, Schaaf P, Coulston A, Reaven GM. Am J Clin Nutr 1995;61:787–91.
146. Havel PJ. Exp Biol Med (Maywood) 2001;226:963–77.
147. Havel PJ, Diabetes 2004;53:S143–51.
148. Havel PJ, Townsend R, Chaump L, Teff K. Diabetes 1999;48:334–41.
149. Rezvani R, Cianflone K, McGahan JP, et al. Obesity (Silver Spring) 2013;21:2471–80.
150. Sigala DM, Widaman AM, Hieronimus B, et al. Nutrients 2020;12(12).
151. Aeberli, I., Gerber PA, Hochuli M, et al. Am J Clin Nutr 2011;94:479–85.
152. Hieronimus B, Medici V, Bremer AA, et al. Metabolism 2020;112:154356.
153. Johnston RD, Stephenson MC, Crossland H, et al. Gastroenterology 2013;145:1016–1025 e2.
154. Schwarz J-M, Noworolski SM, Wen MJ, et al., J Clin Endocrinol Metab 2015;100:2434–42.
155. Taskinen MR, Söderlund S, Bogl LH, et al. J Intern Med 2017;282:187–201.
156. Cox C.L, Stanhope KL, Schwarz JM, et al. Eur J Clin Nutr 2012;66:201–8.
157. Aeberli I., Hochuli M, Gerber PA, et al. Diabetes Care 2013. 36:150–6.
158. Cox CL, Stanhope KL, Schwarz JM, et al. Nutr Metab (Lond) 2012;9:68.
159. Bruun JM, Maersk M, Belza A, Astrup A, Richelsen B. Eur J Clin Nutr 2015;69:949–53.

160. Maersk M, Belza A, Stødkilde-Jørgensen H, et al. Am J Clin Nutr 2012;95:283–9.
161. Sigala DM, Hieronimus B, Medici V, et al. Nutrients 2022;14:1648..
162. Sigala DM, Hieronimus B, Medici V, et al. J Clin Endocrinol Metab 2021;106:3248–3264.
163. Stanhope KL, Medici V, Bremer AA, et al. Am J Clin Nutr 2015;101:1144–54.
164. Campos V, Despland C, Brandejsky V, et al. Obesity (Silver Spring) 2015;23:2335–9.
165. Ibarra-Reynoso LDR, Løpez-Lemus HL, Garay-Sevilla ME, Malacara JM. Obes Facts 2017;10:332–340.
166. Schwarz J-M, Noworolski SM, Erkin-Cakmak A, et al. Gastroenterology 2017;153:743–752.
167. Schwimmer JB, Ugalde-Nicalo P, Welsh JA, et al. JAMA 2019;321:256–265.
168. Simons N, Pandichelvam V, Simons PI, et al. Am J Clin Nutr 2021;113:391–400.
169. Stanhope KL, Griffen SC, Bair BR, et al. Am J Clin Nutr 2008;87:1194–1203.
170. Tran C, Jacot-Descombes, Lecoultre V, et al., Br J Nutr 2010;104: 1–9.
171. Price CA, Medici V, Nunez MV, et al.. Nutrients 2021;13:760.
172. Sigala DM, Stanhope KL. Curr Obes Rep 2021;10:39–52.
173. Payant MA, Chee MJ. Neurosci Biobehav Rev 2021;128: 346–357.
174. Helsley RN, Moreau F, Gupta MK, Radulescu A, DeBosch B, Softic S. Curr Diab Rep 2020;20: 64.
175. DiNicolantonio JJ, Mehta V, Onkaramurthy N, O'Keefe JH. Prog Cardiovasc Dis 2018;61:3–9.
176. Herman MA, Birnbaum MJ. Cell Metab 2021;33:2329–2354.
177. Jensen, T., Abelmalek MF, Sullivan S, et al. J Hepatol 2018;68:1063–1075.
178. Softic S, Kahn CR. Crit Rev Clin Lab Sci 2020;57:1–15.
179. Stanhope KL. Crit Rev Clin Lab Sci 2016;53:52–67.
180. Taskinen MR, Packard CJ, Boren J. Nutrients 2019;11(9).

A. MAJOR DIETARY CONSTITUENTS

CHAPTER 3

Dietary Fiber

Madeline Timm, Megan Edelman, and Joanne Slavin

CHAPTER OUTLINE

Abbreviations

AI, adequate intake; **CAC**, Codex Alimentarius Commission; **DP**, degree of polymerization; **ECP**, extracellular polysaccharides; **FOS**, fructo oligosaccharides; **GI**, gastrointestinal; **GOS**, galacto-oligosaccharides; **IOM**, Institute of Medicine; **LDC**, low-digestible carbohydrates; **MMPI**, Minnesota multiphasic personality inventory; **NDF**, neutral detergent fiber; **PHGG**, partially hydrolyzed guar gum; **RS**, resistant starch; **SCFA**, short chain fatty acid; UL, upper limit

What Is the Definition of Dietary Fiber?

Defining dietary fiber and measuring dietary fiber have always been challenging. From its start as crude fiber to its evolution to dietary fiber, chemical methods have been developed and accepted as the standards for labeling and research purposes. Fiber has historically been described as any nondigestible portion of a plant cell wall.[1] In 2002, the Institute of Medicine (now, National Academy of Medicine) stated that *total fiber* is the sum of *dietary fiber* plus *functional fiber*.[2] Dietary fiber consists of nondigestible carbohydrates and lignin that are intrinsic and intact in plants, whereas functional fiber consists of isolated, nondigestible carbohydrates that have presumed beneficial physiologic effects in humans. Regulatory definitions for dietary fiber have been described by governments and organizations worldwide (**Table 3.1**). Despite nuanced differences, all definitions agree that fiber is mostly carbohydrate, that is not completely digested or absorbed in the small intestine, but that may be fermented in the large intestine.

Dietary fiber is defined by the FDA as nondigestible soluble and insoluble carbohydrates with >3 monomeric units and lignin that are either intrinsic and intact in plants or isolated and synthetic and demonstrate a physiologic health benefit in humans. FDA reviewed the scientific literature and determined that 15 isolated and synthetic nondigestible carbohydrates had >1 beneficial physiologic effect. Included were cellulose, pectin, guar gum, locust bean gum, hydroxypropylmethlcellulose, beta-glucan, psyllium husk, mixed plant cell waters, arabinoxylan, alginate, inulin, and inulin-type fructans (fructo-oligosaccharides, high amylose starch/soluble corn fiber (resistant starch 2,), galactooligosaccharides, polydextrose, and resistant maltodextrin/dextrin. Other isolated and synthetic NDCs are being evaluated for accepted physiologic effects. It should be noted that the FDA has accepted the following physiologic benefits: 1. Attenuation of blood glucose or insulin; 2. Lower fasting cholesterol concentration; 3. Improved laxation; 4. Increased gastrointestinal (GI) mineral absorption; and 5. Reduced energy intake and increased satiety. Production of fermentative end products and change in the GI microbiota are not accepted physiologic endpoints, although they are the endpoints of interest in prebiotic research.

Table 3.1 Regulatory Global Definitions of Dietary Fiber

Regulatory Body	Definition of Fiber
WHO/FAO	Dietary fiber denotes carbohydrate polymers with 10 or more monomeric units that are not hydrolyzed by the endogenous enzymes found in the small intestine of humans, belonging to the categories below: 1. Edible carbohydrate polymers naturally occurring in consumed foods. 2. Carbohydrate polymers that have been obtained from food raw material by physical, enzymatic, or chemical means and which have been shown to have physiologic benefit to health, as demonstrated by generally accepted scientific evidence to competent authorities. 3. Synthetic carbohydrate polymers that have been shown to have a physiologic benefit to health, as demonstrated by generally accepted scientific evidence to competent authorities. (Noting, this also includes lignin and other compounds quantified by AOAC 991.43 and that the decision to include carbohydrates with 3–9 monomeric units should be left to the discretion of national authorities).
FDA	1. Nondigestible soluble and insoluble carbohydrates (with >3 monomeric units) and lignin that are intrinsic and intact in plants. 2. Isolated and synthetic nondigestible carbohydrates (with >3 monomeric units) that the FDA has granted to be included in the definition of dietary fiber, in response to a petition submitted to the FDA demonstrating that such carbohydrates have a physiologic effect that is beneficial to human health. 3. Isolated and synthetic nondigestible carbohydrates (with >3 monomeric units) that are the subject of an authorized health claim.

Reproduced from Carlson, J. L., Erickson, J. M., Lloyd, B. B., & Slavin, J. L. (2018). Health Effects and Sources of Prebiotic Dietary Fiber. Current developments in nutrition, 2(3), nzy005. https://doi.org/10.1093/cdn/nzy005

What Are the Characteristics of Fiber?

No matter which definition one chooses to accept, many different types of fiber exist, and each is unique. In the United States, fiber must be included on the Nutrition Facts panel on food packaging, but soluble and insoluble fiber can also be specifically identified. These fiber values are measured by methods accepted by the Association of Official Analytical Chemists. Fiber is undeniably a complex substance so, to characterize a fiber by solubility alone would be remiss. In fact, in 2001, the Institute of Medicine (IOM) Fiber Panel recommended that this practice be abandoned because the solubility of fiber was not a predictor of physiologic effects (2). Characteristics such as viscosity and fermentability may be more important in predicting the health benefits of fiber in humans.[3]

Viscosity is similar to solubility and is often (but not always) associated with the fiber's water-holding properties.[4] Determining the viscosity of a liquid product is relatively simple, but the methods used to determine the viscosity of fiber as part of a food or diet are difficult, and the results are inconsistent among methods. For example, a particular fiber may be extremely viscous in water, but when it is baked into bread with other ingredients, the same fiber may behave quite differently. It is unreasonable to extrapolate the results of viscosity at a set point in the digestive process because viscosity likely changes at different points in the digestive tract and at different times throughout the digestive process.

The fermentability of fiber is also important, yet difficult to assess. Because fiber is not digested in the small intestine, it arrives in the large intestine intact and is available for fermentation by the resident microflora.[5] The fermentation process yields short-chain fatty acids (SCFAs), which are available for uptake by colonocytes. Fiber fermentation is believed to play a key role in colonic health. Neither in vitro nor in vivo assessments clarify the way in which a specific fiber would be fermented in a specific individual, however. In vitro methods attempt to determine fermentability by inoculating various fibers with human fecal samples, but this closed, static system does not represent the dynamic and changing environment of the human colon. In vivo measurements of fiber fermentation are impossible to extrapolate to the in vivo situation because the large intestine of every individual is colonized with different types and amounts of microflora.

The dilemma is that viscosity and fermentability are two important characteristics of fiber, but no "gold standards" exist for measuring either property. Ultimately, this limitation makes discussions of fiber challenging, and should be considered when one interprets research on fiber and health.

Which Foods Contain Fiber, and How Much?

Most foods consumed in the United States are low in dietary fiber (**Table 3.2**). Standard servings of fruits or vegetables contain 1 to 3 g of fiber, while higher fiber contents (10 g or higher) are found in high fiber cereals that contain wheat bran, oat bran, or other added fiber. Other fiber sources include over-the-counter laxatives containing fiber, fiber supplements, and fiber-fortified foods and beverages. Recently, soluble fibers have been added to snack foods, beverages, and waters to enhance the fiber content of the diet.

What is Dietary Fiber?

Recent definitions of dietary fiber agree that dietary fiber includes diverse carbohydrate polymers and oligomers (and lignin) that escape digestion in the small intestine and pass into the large bowel, where they are partially, or completely, fermented by the gut microbiota. Differences among definitions for fiber are found in three main areas: 1. Degree of polymerization; 2. Relationship to food and food processing; and 3. Physiologic effects.[6] Disagreements on these issues have led to confusion in fiber definition and labeling. For example, in the United States, fibers currently include oligosaccharides with a degree of polymerization (DP) of 3–9, whereas other bodies such as the Codex Alimentarius

Table 3.2 Total Dietary Fiber in Common Foods

Food	Quantity	Fiber (g)
White bread	1 slice	0.6
Whole wheat bread	1 slice	1.9
Brown rice	½ cup	1.7
White rice	½ cup	0.3
All Bran Original	½ cup	8.8
Product 19	1 cup	1.0
Raisin Bran	1 cup	7.3
Wheat Chex	1 cup	3.3
Rice Chex	1 cup	0.2
Oatmeal, cooked	1 cup	4.0
Apple, with skin	1 medium	3.3
Orange	1 medium	3.1
Prunes, dried	5	3.0
Raspberries	½ cup	4.0
Broccoli, raw	½ cup	1.1
Cauliflower, raw	½ cup	1.2
Sweet corn	½ cup	2.1
Iceberg lettuce, raw	½ cup	0.35
Kidney beans	½ cup	6.6
Peas	½ cup	4.4
Pinto beans	½ cup	7.7
Baked potato	1 small	2.3
Yellow squash, cooked	½ cup	1.25

Data from US Department of Agriculture, Agricultural Research Service, Nutrient Database for Standard Reference, Release 22. Washington, DC: US Department of Agriculture, 2009.

Commission (CAC) support that the national authorities should decide if carbohydrate polymers with a DP 3–9 should be included. This creates challenges for measuring dietary fiber and databases of dietary fiber, as total dietary fiber values will depend on whether the lower chain DP fibers are included.

The North American definition includes all resistant carbohydrates >3 DP but requires that isolated fibers must show a physiologic benefit to be included as dietary fiber sources on the Nutrition Facts label. Physiologic benefits long accepted include improved laxation, serum lipid lowering, and glucose control.[7] Newer benefits, such as improved mineral absorption or blood pressure control, can also be accepted as physiologic benefits of dietary fiber.

Definitions recognize that dietary fibers can be extracted from edible material (intrinsic) or modified and added back into a food (extrinsic). Because recommendations for dietary fiber are based on protective effects of intrinsic fiber on risk of cardiovascular disease, regulatory bodies have discussed the need to divide fiber into dietary fiber (intrinsic) and functional or added fiber (extrinsic). Earlier discussions linking fiber to physiologic effects divided fiber into soluble and insoluble fractions.[8] Soluble fibers dissolve in water and may form a gel. They have been linked to blood lipid lowering and glucose control.[7] Soluble fibers survive transit through the small intestine and may be fermented in the gut to form short-chain fatty acids (SCFAs) such as butyrate, acetate, and propionate. Insoluble fibers do not dissolve in water and are the fibers we traditionally think of as roughage, fibers like cellulose and hemicellulose. These fibers are more closely linked to bowel function. Because of these early associations between the solubility of fibers and physiologic effects, many regulatory bodies allow foods to list total dietary fiber, soluble fiber, and insoluble fiber on labels.

Classification of Dietary Fiber

Classifying fiber has challenged analytical chemists through the ages. Animal feeds originally measured crude fiber as the nondigestible carbohydrate that might be used by ruminants, but it was not well tolerated in monogastric animals. Crude fiber was referred to as the residue of plant-based food left after extraction with solvent, dilute acid, and dilute alkali.[9] In 1953, Hipsley defined dietary fiber as remnants of edible plant cells, polysaccharides, lignin, and associated substances resistant to digestion by the alimentary enzymes of humans, or "unavailable carbohydrate." Dietary fiber includes cellulose, hemicellulose, lignin, gums, mucilage, oligosaccharides, pectin, and other associated substances, including waxes, cutin, and suberin. By contrast, crude fiber includes only insoluble fibers such as some hemicelluloses, cellulose, and lignin.

Dietary fiber chemistry is critical to classifying dietary fiber. Dietary fiber can be divided into noncellulosic polysaccharides, cellulose, and lignin. Human foodstuffs contain mainly noncellulosic polysaccharides, some cellulose, and little lignin.[7] Generally, fruits and vegetables tend to be higher in cellulose than cereals. Lignin is highest in fruits with edible seeds or in mature vegetables such as carrots or other root vegetables. Dietary fiber content of foods depends on plant species, maturity, and components (i.e., leaf, root, or stem).

The solubility of dietary fiber was thought to help explain physiologic effects, but more recent work suggests that solubility does not always predict physiologic effects. Viscous, soluble fibers, such as oat bran or psyllium, are known to lower serum cholesterol and have health claims for their ability to lower blood lipids. Most other soluble fibers, for example, fructo-oligosaccharides, have no effect on blood lipids. Generally, only viscous, soluble fibers play a role in blood glucose control. Even the 2001 Dietary Reference Intakes concluded that separating soluble and insoluble fiber to predict physiologic effects was problematic.[2] Yet, nutrition labels had already included this designator and consumers were told that insoluble fibers would aid in laxation and soluble fibers would lower blood lipids and improve glucose control.

Similar problems are seen with the belief that insoluble fibers aid bowel function and soluble fibers do not. The increase in stool weight is caused by the presence of fiber in the colon, by the water that the fiber holds, and by partial fermentation of the fiber, which increases the number of bacteria in stool.[10] Fiber in mixed diets, including insoluble cellulose and hemicellulose, are particularly effective promoters of bowel function, as are cereal brans, psyllium seed husk, and methylcellulose in the form of supplements. If fiber is rapidly fermented in the large bowel, as are many oligosaccharides, there is little increase in stool weight. But this rapid fermentation of easily fermented fiber can cause intestinal gas and discomfort in some individuals.

Studying dietary fiber in humans is difficult, as balance studies require collection of all food consumed and feces produced, and measurement of the fiber consumed and the fiber left over in feces. Balance studies must be done on controlled food intake and 5-day fecal composites collected as indicators of fiber surviving gut transit. Few contemporary studies exist to include in this review. Stephen et al.[11] summarized the current state of knowledge on fiber definitions, sources, recommendation, intakes, and relationships to health in a comprehensive review paper of European fiber work. Disagreements on definitions of dietary fiber and methods of analysis continue, but there is more agreement that standardized definitions are first needed before there is agreement on analytical methods. Nutrient databases for dietary fiber, soluble fiber, and insoluble fiber continue to be lacking.

Cumming et al.[12] fed 20 g/day of concentrated dietary fiber from carrot, cabbage, apple, bran, and guar gum to the controlled basal diet of nineteen healthy volunteers. Fiber from four commonly eaten food sources and guar gum produced a very different response in colonic function. Fecal weight increased by 127% on bran and 20% on guar gum with carrot, cabbage, and apple producing intermediate changes in fecal weight. Differences were related to the intake of pentose-containing polysaccharides in the fiber, generally hemicelluloses. Individual variation in response to a given fiber was significant with stool weights ranging from 65 to 194 g/day on a controlled diet with tightly controlled fiber intake.

Mechanisms of action of dietary fiber in the human colon, assessed with a novel modified method for isolating bacteria, found that the main component of human feces is bacteria.[13] Cabbage fiber, which is extensively broken down in the gut, provides a readily usable substrate for stimulation of microbial growth, whereas wheat fiber remains largely undigested and retains water in the gut lumen. Additional work by this research group examined water-holding capacity and its relationship to fecal output in human adults. Pectin had the greatest water holding capacity but produced the smallest change in fecal weight, while bran had the lowest water holding capacity and produced the largest increase in fecal weight.[14]

Early work with consumption of purified cellulose, a poorly fermentable fiber, in humans found that when purified cellulose, 16 g/day, was ingested with a semi-purified liquid diet, only 8% of the cellulose was broken down.[15] In contrast, more than half of the fiber in a habitual low-fiber diet containing fruits, vegetables, and refined grains was degraded. This early work supports that the breakdown of fiber in the large intestine varies greatly, depending on the chemical composition of the dietary fiber.

Fiber fermentation on enteral diets with and without fiber supplementation were examined and related to breath gas production.[16] Individuals consumed enteral diets with 0, 30, and 60 g/day of soy polysaccharide, an isolated fiber source. Feces were collected, homogenized, dried, and analyzed for neutral detergent fiber (NDF). NDF was also determined in the diets to measure fiber balance. Fiber breakdown was extensive on all doses of fiber, about 90%. No relationship was found between fiber fermentability and production of breath gases, although there was large individual variability in the breath gas production levels. NDF from the soy polysaccharide fiber in the enteral formula was extensively fermented.

Fiber fermentation in human adults was measured on controlled enteral diets with quick bread with 0 g, 10 g, and 30 g of dietary fiber as wheat bran and mixed vegetable fiber.[17] Each treatment was consumed for 23 days. Women and men seemed to respond differently to wheat bran and vegetable fiber regarding NDF excretion and breakdown. Fecal fiber excretion was greater in men than women; correspondingly, women tended to digest more fiber than men. This study found large interindividual responses to dietary fiber and identified factors beyond fiber, such as sex, that alter the gut response to dietary fiber consumption.

Few metabolic studies on rigidly controlled diets have examined the effect of different fiber sources on fiber breakdown and physiologic outcomes. Tucker et al[18] published a study on dietary fiber and personality factors such as social introversion as measured by the MMPI (Minnesota Multiphasic Personality Inventory) as determinants of stool output. They collected extensive fecal samples while subjects consumed rigidly defined diets at the USDA laboratory in Grand Forks, ND. Even when food and fiber intake was tightly controlled, they found substantial individual differences in stool output. Personality measures including how outgoing participants were predicted stool weight and fecal frequency better than dietary fiber intake. They concluded that personality factors such as introversion predispose humans to low stool output. This early work supports current thinking on the relationship between the gut health and brain health.

Grabitske and Slavin[19] reviewed the gastrointestinal effects of low-digestible carbohydrates (LDCs) including definitions, classifications, and mechanisms. LDCs are carbohydrates that are incompletely or not absorbed in the small intestine but are at least partly fermented by bacteria in the large intestine. Fiber, resistant starch, and sugar alcohols are types of LDCs. LDCs are increasingly added to processed foods because of potential health benefits, including a reduced caloric content, reduced or no effect on blood glucose levels, and not linked to tooth decay as they are not fermented by bacteria in the mouth. The benefits of LDCs are related to the inability of human digestive enzymes to break down completely the carbohydrates into absorbable saccharides and the subsequent fermentation of unabsorbed carbohydrates in the colon.

Thus, LDCs may cause gastrointestinal effects, including abdominal discomfort, flatus, and diarrhea, especially at higher or excessive intakes. Current recommendations for fiber intake do not consider total LDC consumption nor recommend an upper limit for LDC intake based on potential gastrointestinal effects. In fact, the IOM did not set an Upper Limit (UL) for dietary fiber intake when fiber recommendations were finalized as it is accepted that humans are omnivores and can adapt to high intakes of dietary fiber. A vegan individual may consume 75 g of fiber per day or more. Of course, this fiber comes from foods and is not isolated and potentially fed as a large bolus.

The term "laxation" refers to the elimination of feces.[20] Normal laxation varies widely among individuals. It is clinically defined as having a range of three bowel movements per week to three bowel movements per day. Stools are generally about 75% water, although this is highly variable and water content increases greatly with diarrhea. Of the remaining stool, about one-third is dead bacteria; one-third is undigested carbohydrate; and one-third is protein, fat mucus, dead cells, and inorganic material.

The frequency, consistency, composition, and weight of bowel movements are all indicators of bowel function and digestive health, but unlike blood, stool samples are generally not collected in large cohorts of healthy subjects. Stool weights are significantly related to fiber intake and studies find a wide range of stool weights, from 72 g/day to 470 g/day.[21] Different fibers have different effects on stool weight. Average increase in fecal weight per gram of fiber fed ranged from 5.4 g for wheat bran to 1.2 g for pectin.[22] This information supports why fiber sources high in insoluble fiber (wheat bran) were linked to increased stool weight and fiber sources high in soluble fiber (pectin) were linked to low stool weight.

Many factors affect laxation and digestive health. Lifestyle factors, diet, drugs, physical activity, and stress are all known to alter digestive function.[20] Personal factors, including age, sex, genetics, the microflora, personality, food intolerances, and gastrointestinal pathology, such as celiac disease or lactose intolerance, will also alter digestive health. Bowel habits are also affected by changes in routine, such as travel or new work schedules. Typically, regular bowel habits depend on eating breakfast, a morning cup of coffee, or other regular behaviors. Drugs to alter bowel function and motility range from fiber supplements to irritants or other mechanisms. Antibiotics that decrease the gut microbiota will alter fiber fermentation in the gut and must be considered in any study of dietary fiber fermentation.

Limitations to human studies of fiber fermentation and metabolism suggest the need for more support for animal studies. Pigs are considered the monogastric animal with the closest digestive tract to the human, so findings from pig studies may provide insight to human physiology. The classification of potentially fermentable carbohydrates into soluble and insoluble is no longer enough for the information required to elucidate mechanisms by which dietary fiber has beneficial effects on monogastric health.[23] Characteristics including fermentability (both kinetics of fermentation and end-products) are needed to help us understand how dietary fiber affects health in monogastric animals. Unfortunately, these in vivo studies are expensive, so results for in vitro studies must currently inform our understanding of dietary fiber fermentation.

What Is Resistant Starch?

Describing and measuring such a diverse mixture of nondigestible carbohydrates and associated substances is a major undertaking. Should nondigestible starch or resistant starch be included in dietary fiber? According to Codex, if resistant starch is naturally present in food, it could be classified as dietary fiber. However, if it is derived from an artificial synthesis, such as physical, enzymatic, or chemical synthesis, it should provide desirable physiologic benefits to be considered dietary fiber.

Starch has been divided into three fractions based on its digestive rate, including rapidly digesting starch, slowly digesting starch, and resistant starch. Resistant starch is a broad and diverse term and was originally divided into four types, i.e., physical inaccessible starch (RS1), ungelatinized starch granules (RS2), retrograded starch (RS3), and chemically modified starch (RS4). More recently, RS5 has been introduced as the amylose-lipid complex, which has a similar resistant mechanism toward starch digestive enzymes as RS3.[24]

In the 1980s, RS was proposed and categorized as a kind of insoluble fiber, as it could not be digested in the small intestine. Interest in RS grew as clinical data supported that resistant starch increased bacterial mass in feces, was a good substrate for growth of colonic microbiota, but preferentially increased production of the short chain fatty acid butyrate.[25] Butyrate is the most important energy source for the colonocyte, increasing interest in resistant starch fermentation beyond its dietary fiber content.[26]

Potatoes, cereals, whole grains, beans, legumes, rice, and bananas are considered top sources of resistant starch in human diets.[27] Barley, potatoes, and rice with higher amylose concentration have more resistant starch than lower amylose varieties, and foods cooked and then chilled have higher resistant starch than foods cooked without chilling. Resistant starch can also be produced from isolated starches and then added back to foods and beverages as dietary fiber, if allowed by local regulations on fiber definitions and labeling.

Resistant starch intervention studies have been conducted on aging, insulin resistance, metabolic syndrome, kidney disease, and schizophrenia and it is thought to be particularly relevant for diseases characterized by dysregulated epithelial integrity and immune function, like inflammatory bowel disease.[28] The variable effects of RS on the gut microbiome are striking, with one individual showing increased butyrate production and another showing less butyrate production. Measuring SCFA production in humans is challenging, as SCFAs are quickly absorbed in the colon and blood concentrations are not useful biomarkers of SCFA production.

Fermentation of Dietary Fiber as a Biomarker

Dietary fiber is not only classified based on chemical analysis and solubility but also fermentability. As noted above, in vivo studies to measure fiber fermentation in humans are difficult to conduct and few recent studies have been published to add to the body of evidence on this topic. Thus, in vitro studies give us our best idea of differences in fermentation of dietary fibers of interest to human nutrition.

Although databases exist for the total dietary fiber, soluble fiber, and insoluble fiber content of many foods and fiber supplements, no accepted databases exist for fermentability of isolated fibers. Although fermentability is just one attribute of dietary fiber, the fermentation of fiber to yield "good bacteria" or associated metabolites (SCFAs) are critical data points. The physiologic importance of SCFAs from nondigestible carbohydrate fermentation has been described and it is proposed that the degree of fermentability of an NDC, rather than the endpoints of clinical trials, may be appropriate for classifying it as a dietary fiber.[29]

Thus, in vitro studies that compare the fermentation aspects of isolated fibers are of mechanistic interest but are not accepted as a measurement of prebiotic potential or fiber fermentation dynamics. Standard in vitro methods for fiber fermentability do not exist but, generally, human donor fecal samples are used from healthy volunteers and then fiber samples incubated with the fecal slurries. Endpoints include fiber fermentation, gas production, pH changes, SCFA production, and changes in the microbiota. Gas production with different resistant starches varied greatly, but all were fermentable and promoted formation of beneficial SCFAs.[30] Examining the in vitro fermentation of commercial fiber supplements promoted as prebiotics found differences in the production of *Bifidobacteria* and *Collinsella*.[31] All fibers were extensively fermented and produced beneficial SCFAs.

Shorter chain fructo-oligosaccharides exhibited more rapid fermentation than long-chain inulin in an in vitro fermentation.[32] Particle size and fraction of wheat bran also influenced SCFAs production in vitro.[33] Five samples were compared: large-particle bran, small-particle bran, aleurone, coarse by-product, and fine-by product. Small/fine particle size increased SCFA production compared with large/coarse particle size. The molar percentage of butyrate at 24 hours was significantly greater for large-particle bran than the other sample. Bran characteristics and composition should be considered when manufacturing foods due to the diversity of physiologic effects.

Blends of fibers with different fermentation rates were compared in vitro.[34] Fructo oligosaccharides (FOS) and inulin were fermented most readily but, when included in blends, fermentation was slower and less gas was produced. This study suggested that fiber blends may be more useful in enteral products to combine the desirable solubility profile and fermentation profile for optimum health.

What Is a Prebiotic?

The health effects of dietary fiber have been extensively reviewed and are accepted worldwide.[35] Prebiotics were first defined in 1995 and their definition has continued to evolve over time. Prebiotic dietary fibers are specific, microbiota-shaping compounds that function as a carbon source for growth of beneficial taxa, thus delivering a specific or selective change that confers the host health related to its metabolism of the prebiotic dietary fibers. Oligosaccharides in human milk are the prototypical prebiotic, owing to their ability to engender a *bifidobacter* and *lactobaccilus* predominant colonic flora.

Most of the original prebiotics were also dietary fibers, including inulin, oligofructose, lactulose, and resistant starch. As there is no official analytical method for prebiotics, other carbohydrates, including galacto-oligosaccharides, trangalacto-oligosaccharides, polydextrose, wheat dextrin, acacia gum, psyllium, and partially hydrolyzed guar gum (PHGG), to name a few, have been studied to show that consumption of these fibers increases fecal levels of bifidobacterial or lactobacillus.[36] Often, production of SCFAs in vitro has been used as a biomarker of prebiotic effect, as SCFAs are the byproduct of fiber fermentation.

Reviews have attempted to classify prebiotics based on in vitro studies and short-chain fatty acid production[37]. Isolated dietary fibers, including resistant starch, pectin, hemicellulose, beta-glucan, and fructans, were reviewed and their support for being a prebiotic offered. Regulatory acceptance of the importance of prebiotics beyond their ability to list their fiber content on the label is lacking.

The effect of fructans, prebiotics, and biers on the gut microbiome was reviewed.[38] Molecular studies confirmed the selective bifidogenic effect of fructans and galacto-oligosaccharides (GOS) in humans. Resistant starches, polydextrose, and beta-glucan showed broader effects with more and different types of gut microbial species being enhanced, often including phylotypes of *Ruminococcaceae*. There was significant variation in the magnitude of response and in individual responses to a specific fiber that may be due to numerous factors, including presence and relative abundance of microbial types, diet, genetics of the host, and intervention parameters, including intervention duration and fiber dose. The field could benefit from a more systematic approach that will support defining the impact of prebiotics and fibers on the gut microbiome, identify biomarkers that link gut microbes to health, and address the personalized responses of an individual's microbiota to prebiotics and dietary fibers. Because the definition of prebiotics requires a health benefit, manufacturers will tend to classify these as dietary fibers since definitions and regulatory approval are already realized for fibers.

Another recent review described dietary fibers as prebiotic, prebiotic candidates, or fibers without prebiotic potential.[39] Resistant oligosaccharides, fructans (FOS, oligofructose, and inulin) and galactans are recognized in the literature as prebiotics according to the authors. Galactosides and some nonstarch polysaccharides and resistant starches may act as prebiotic fibers or have prebiotic potential. They suggest that associated substances (noncarbohydrates) are considered dietary fibers but apparently have no prebiotic effect on humans. Intrinsic dietary fiber in plants with associated substances have never been studied for prebiotic potential, so this conclusion can be questioned. Proteins and fats have been described as prebiotics, so the lack of specificity for the term will continue to limit the labeling of products as "prebiotics."[40]

So et al[41] published a systematic review and meta-analysis of dietary fiber intervention on gut microbiota composition in healthy adults. A total of 64 studies involving 2099 participants were included. Dietary fiber intervention resulted in higher abundance of *Bifidobacterium* spp. and *Lactobacillus* spp. as well as fecal butyrate concentration compared with placebo/low-fiber comparators. Subgroup analysis found that fructans and galacto-oligosaccharides led to significantly greater abundance of both *Bifidobacterium* spp. and *Lactobacillus* spp. compared with comparators. No differences in effect were found between fiber intervention and comparators for alpha-diversity, abundances of other prespecified bacteria, or other SCFA concentrations. Dietary fiber interventions, particularly for fructans and GOS, lead to high fecal abundance of Bifidobacterium and Lactobacillus, but do not affect alpha-diversity.

Health Benefits of Fiber, Prebiotics, and Probiotics

Gastrointestinal health and the gut microbiome is a rapidly emerging field and fiber, prebiotics, and probiotics are all dietary components that play a role in maintaining a healthy gut microflora.[27] Generally, we consider prebiotics as fiber-like compounds that generate bacteria in the colon while probiotics are live bacteria that you consume. Synbiotics are often combinations of prebiotics and probiotics. Unfortunately, the marketplace is awash in these products and little regulatory oversight is available for these products. Prebiotics operate in the safe space of dietary fiber, an accepted nutrient in the human diet that can be measured by accepted analytical tools and regulatory approval throughout the world.

What Are Postbiotics?

A consensus statement by the International Scientific Association of Probiotics and Prebiotics (ISAPP) on the definition and scope of postbiotics defined postbiotic as a "preparation of inanimate microorganisms and/or their components that confers a health benefit on the host."[42] The review paper lists past proposed definitions of the term postbiotic, which perhaps explains the confusion in the literature and with consumers on this concept. Wegh et al[43] define postbiotics as functional bioactive compounds, generated in a matrix during fermentation, which may be used to promote health. The term postbiotics can be regarded as an umbrella term for all synonyms and related terms of these microbial fermentation compounds. Therefore, postbiotics can include many different constituents, including metabolites, SCFAs, microbial cell fractions, functional proteins, extracellular polysaccharides (EPS), cell lysates, teichoic acid, peptidoglycan-derived muropeptides, and pili-type structures. The concept suggests that the microbes produced by fermentation of fiber are not the important mechanisms. The byproducts of fiber fermentation, for example, SCFAs are more important.

The need for established clinical markers and microbial endpoints are still essential to determine the causal role of the GI microbiota in protecting health.[44] Few studies collect fecal samples and even when fecal samples are collected, there is little information provided on the dietary exposure and the kinds and amounts of dietary fiber that were consumed. Collection of fecal samples to measure the gut microbiota with the suggestion that we can predict health from such information is premature and potentially misleading.

Links Between Dietary Fiber Intake and Health Outcomes

Intake of intrinsic fiber is linked to lower risk of cardiovascular disease. Additionally, dietary fiber is accepted as an important piece in gut health and a first step in treating idiopathic constipation and some forms of diarrhea, including irritable bowel syndrome. Fibers differ in their chemistry and there are some relationships between fiber chemistry and physiologic effect. Other aspects of fiber, including fermentability, will also predict physiologic effects.

Additional attributes of fiber, for example, viscosity, will determine the possibility that fiber will lower blood lipids or modulate blood glucose. Because of the inconsistency of fiber's effect on physiologic parameters, the FDA now requires isolated fiber to show at least one physiologic benefit before the fiber can list its content on the Nutrition Facts label.

Fiber Intake and Recommendations

We have long appreciated the importance of dietary fiber in the diet and have promoted the need to consume adequate dietary fiber across the lifecycle.[45] Additionally, the importance of dietary fiber in treating diseases such as constipation and high blood lipids is well accepted. Yet, intakes of dietary fiber are inadequate across the globe and despite public health guidance to eat more plant foods and fiber, usual intake is about half of recommended levels. In the past, public health measures to add deficient nutrients to the diet in foods have been promoted to solve dietary deficiencies. As fibers and related dietary components gain acceptance by the FDA as having a physiologic benefit, adding fiber to popular foods or beverages to increase dietary fiber consumption to recommended levels might be proposed as a public health measure. The large individual variability in response to dietary fiber and concerns with intestinal gas and bloating must be considered as those are often reported as the reasons that consumers are not willing to consume more dietary fiber.

Conclusion

Even in 300 BC, Hippocrates said "wholemeal bread makes larger feces than refined bread." The importance of dietary fiber in laxation and bowel health has long been appreciated. More recently, prospective, cohort studies find that higher intakes of dietary fiber are linked to decreased risk of developing cardiovascular disease. Recommendations for intake of dietary fiber in North America are based on these epidemiologic studies that find that 14 g dietary fiber/1,000 kcalories is associated with lower cardiovascular risk. Thus, the Nutrition Facts panel recommends 28 g of dietary fiber for the average consumer of 2,000 kilocalories. Usual intake of dietary fiber is approximately half that amount, which makes dietary fiber a nutrient of concern. Although dietary guidance supports consumption of foods concentrated in dietary fiber, whole grains, legumes, vegetables, nuts, and fruit, many consumers turn to fiber supplements to meet dietary requirements.

Current research interest in the effect of dietary fiber on the gut microbiota has expanded the interest of fiber in disease prevention into fields such as immunology, brain health, and diabetes control. Fiber is active throughout the gastrointestinal tract, but much of the current interest is in the fermentation of dietary fiber in the gut. Although earlier thinking supported that certain microbiota were "healthy," more recent thinking is that the products of fiber's fermentation are important players in fiber's protective role in disease prevention. SCFAs from fiber fermentation help maintain the integrity of the colonic cells and are thought to trigger a cascade of additional health benefits.

We know that intake of dietary fiber is protective, but we also know that fibers cover a range of physiologic benefits, solubility, fermentability, and viscosity. Thus, public health guidance recommends that consumers choose a variety of fiber-rich foods to obtain the Adequate Intake (AI) of dietary fiber daily. Ideally, fiber should be provided by whole foods, but because dietary fiber intake continues to lag significantly behind dietary recommendations, isolated fibers with health benefits supported by research trials are also part of the solution to increasing dietary fiber intakes to recommended levels.

Dietary fiber continues to be a nutrient of concern in the Dietary Guidelines for Americans,[46] which noted that more than 90% of women and 97% of men do not meet recommended intakes. Recommendations for plant foods, including grains, vegetables, pulses, and fruits are important public health messages, but fortification or supplementation with dietary fiber may be necessary to meet fiber intakes associated with health benefits and chronic disease prevention.

Acknowledgments

The authors would like to acknowledge Holly Willis for her contributions to the previous version of this chapter.

References

1. Hipsley EH. Br Med J 1953;2:420–2.
2. Food and Nutrition Board, Institute of Medicine. Dietary Reference Intakes. Dietary, Functional, and Total Fiber. Washington, DC: National Academy Press, 2002.
3. Jones JR, Lineback DM, Levine MJ. Nutr Rev 2006;64:31–8.
4. Dikeman CL, Fahey GC. Crit Rev Food Sci Nutr 2006;46:649–63.
5. Barry JL, Hoebler C, Macfarlane GT, et al. Br J Nutr 1995;74: 303–22.
6. Fuller S, Beck E, Salman H, et al. Plant Foods Human Nutr 2016;71:1–12.

7. Slavin JL. J Am Diet Assoc 1987;87:1164–1171.
8. Slavin JL, Savarino V, et al. J Int Med Res 2009;37:1–17.
9. Dai F, Chau C. J Food Drug Anal 2017;25:37–42.
10. Slavin JL, Greenberg NA. Nutrition 2003;19:549–552.
11. Stephen AM, Champ MM, Cloran SJ, et al. Nutr Res Rev 2017;30:149–190.
12. Cummings JH, Southgate DA, Branch W, et al. Lancet 1978.
13. Stephen AM, Cummings JH. Nature 1980;284:283–284.
14. Stephen AM, Cummings JH. Gut 1979;20:722–729.
15. Slavin JL, Brauer PM, Marlett JA. J Nutr 1981;111:287–297.
16. McNamara EA, Levitt MD, Slavin JL. Am J Clin Nutr 1986;43: 898–902.
17. Lampe JW, Fredstrom SB, Slavin JL, et al. Gut 1993;34:531–536.
18. Tucker DM, Sandstead HH, Logan GM, et al. Gastroenterology 1981;81:879–883.
19. Grabitske HA, Slavin JL. Crit Rev Food Sci Nutr 1009;49: 327–360.
20. Grabitske HA, Slavin JL. Nutr Today 2008;43:193–198.
21. Cummings JH, Bingham SA, Heaton KW, et al. Gastroenterology 1992;103:1783–1789.
22. Cummings JH. In: Spiller, G.A., (ed.) *CRC Handbook of Dietary Fiber in Human Nutrition*, pp 263-333, Boca Raton, FL: CRC Press, 1993.
23. Williams BA, Mikkelsen D, Flanagan BM, et al. J An Sci Biotech 2019;10:45.
24. Li C, Hu Y. Cr Rev Food Sci Nutr 2022. https://doi.org/10.1080/10408398.2022.2031101
25. Bird AR, Conlon MA, Christophersen CT, et al. Benefic Micr 2010;1:423–431.
26. Carlson JL, Slavin JL. Quality Assur Safety Crops Foods 2016;8:539–554.
27. Patterson MA, Maiya M, Stewart ML. J Acad Nutr Diet 2020;120:230–244.
28. Dobranowski PA, Stintzi A. Gut Microbes 2021;13:e1926842.
29. Alexander C, Swanson KS, Fahey GC, et al. Adv Nutr 2019;10:576–589.
30. Erickson JM, Carlson JL, Stewart ML, et al. Foods 2018;7:18.
31. Carlson JL, Erickson JM, Hess JM, et al. Nutrients 2017;9:1361.
32. Stewart M, Timm DA, Slavin JL. Nutr Res 2008;28:329–334.
33. Stewart ML, Slavin JL. Br J Nutr 2009;102:1404–1407.
34. Koecher KJ, Noack JA,Timm DA, et al. J Agric Food Chem 2014;62:1332–1337.
35. Carlson JL, Erickson JM, Lloyd, BB, et al. Curr Develop Nutr 2018;2:NZY005.
36. Slavin JL. Nutrients 2013;5:1417–1435.
37. Ashaolu TJ, Ashaolu JO, Adeyeye SAO. J Applied Microbiol 2020;130:677–687.
38. Swanson KS, de Vos W, Martens EC, et al. Benef Micr 2020;11: 101–129.
39. Rezende ESV, Lima GC, Naves MMV. Nutrition 2021;89:111217.
40. Sanders ME, Merenstein DJ, Reid G, et al. Nature Rev Gastroenter Hepatol 2019;16:605–616.
41. So D, Whelan K, Rossi M, et al. Am J Clin Nutr 2018;107: 965–983.
42. Salminen S, Collado MC, Endo A, et al. Nature Rev Gastroenterol Hepatol 2021;18:649–667.
43. Wegh CAM, Geerlings SY, Knol J, et al. Int J Molecular Sci 2019;20:4673.
44. Deehan EC, Duar RM, Armet AM, et al. Microbiol Spectrum 2017;5(5):BAD-0019.
45. Korczak R, Slavin JL. Nutr Rev 2020;78:6–12.
46. U.S. Department of Agriculture and U.S. Department of Health and Human Services. Dietary Guidelines for Americans, 2020-2025. 9th Edition. December 2020. Available at DietaryGuidelines.gov. Accessed at: www.dietaryguidelines.gov/sites/default/files/2020-12/Dietary_Guidelines_for_Americans_2020-2025.pdf

A. MAJOR DIETARY CONSTITUENTS

CHAPTER 4

Lipids and Lipid Metabolites

Alice H. Lichtenstein

CHAPTER OUTLINE

Abbreviations

AA, arachidonic acid; **ACAT**, acyl CoA cholesterol acyltransferase; **AEA**, anandamide; **ALA**, α-linolenic acid; **CETP,** cholesterol ester transfer protein; **COX** cyclo-oxygenase; **DHA** docosahexaenoic acid; **DHEA**, docosahexaenoyl ethanolamine; **DPA,** docosapentaenoic acid; **EPA,** eicosapentaenoic acid; **EPEA,** eicosapentaenoyl ethanolamine; **GRAS**, Generally Recognized as Safe; **HDL**, high-density lipoproteins; **HET,** hydoxyeicosatetranoic acid; **HMGCoA**, 3-hydroxy 3-methylglutaryl CoA; **IDL**, intermediate density lipoproteins; **LA,** linoleic acid; **LCAT,** lecithin cholesterol acyl protein transferase; **LDL**, low density lipoproteins; **LPL**, lipoprotein lipase; **LT,** leukotrienes; **Mar1,** maresins; **PD1**, DHA-derived RvD1-6, protectins; **PG**, prostaglandins; **PLTR**, phospholipid transfer protein; **RvE,** E-series resolvins; **SPM**, specialized proresolving lipid mediator; **TX**, thromboxanes; **VLDL,** very low-density lipoproteins; **2-AG**, 2-arachidonyglycerol

Introduction

The essentiality of dietary lipids was first documented in 1913. The observation was made that an ether extract of butter or eggs was essential for sustaining growth in rats, whereas an ether extract of lard or olive oil did not confer the same benefits.[1] This led to the speculation that there were certain 'accessory articles' in the organic extract of some fats that were essential for growth. These substances were eventually unified under the term "vitamins" and have since been labeled fat soluble vitamins A, D, E and K.[2] Subsequently, in 1929, the essentiality of polyunsaturated fatty acids was elucidated, as well as the identification of two positional fatty acid isomers, omega-3 and omega-6.[3]

More recently, issues related to dietary lipids have gone beyond essentiality to center on the amounts of total dietary fat, fatty acid subclasses, some individual fatty acids, and dietary cholesterol, which are optimum to promote health and minimize development of diet-related chronic diseases, particularly atherosclerotic cardiovascular diseases. Of note, lipid is the general term for organic compounds that are insoluble in water (polar) and soluble in nonpolar solvents. The major lipid categories in terms of food and nutrition include glycolipids (e.g., triglyceride, phospholipid) and sterols (e.g., cholesterol). They are often referred to collectively as dietary fats, with reference to their occurrence in food. The other source is endogenous synthesis of nonessential lipids and cholesterol.

Although frequently cast in a negative light, particularly in terms of diets or excess accumulation in the body (adiposity), the general class of lipids perform critical roles in all living forms. In humans, lipids serve as structural components of cell membranes and are integral to maintaining membrane fluidity. In plasma, through their capacity to form lipoprotein particles, they provide a mechanism for the transport of nonpolar compounds through the aqueous milieu of blood to their targets. Lipids, particularly triglyceride in adipose tissue, represent an energy-dense reserve of metabolic energy, particularly important in times of depravation. Metabolically, lipids provide substrates for the synthesis of bioactive compounds such as eicosanoids, vitamin D, and bile acids. Subcutaneous adipose tissue serves as an insulator, minimizing wide fluctuations in body temperature. Visceral adipose tissue surrounds internal organs, providing a cushion, reducing the risk of physical damage. Lipids in food are carriers of fat-soluble vitamins, essential fatty acids, and nonpolar phytonutrients, and facilitate their absorption. When consumed in adequate quantities, dietary fat minimizes the development of carbohydrate-induced hypertriglyceridemia.

Major Types of Lipids

Fatty Acids

Fatty acids are the basic components of complex lipids, such as triglycerides, phospholipids, and cholesteryl esters. Fatty acids are composed of an acyl (hydrocarbon) chain bracketed by methyl and carboxyl groups (**Figure 4.1**). Chain lengths for common fatty acids typically range from 12 to 22 carbons, although shorter- and longer-chain fatty acids occur in nature (**Table 4.1**). The majority of fatty acids have an even number of carbons and linear chains. Trace amounts of odd numbered and branched chain fatty acids occur. In addition to chain length, fatty acids differ among each other in number, conformation, and location of double bonds. Some polyunsaturated fatty acids are "essential," defined as those necessary to support growth, development, and maintenance but cannot be synthesized by humans. Hence, they must be obtained from the diet.

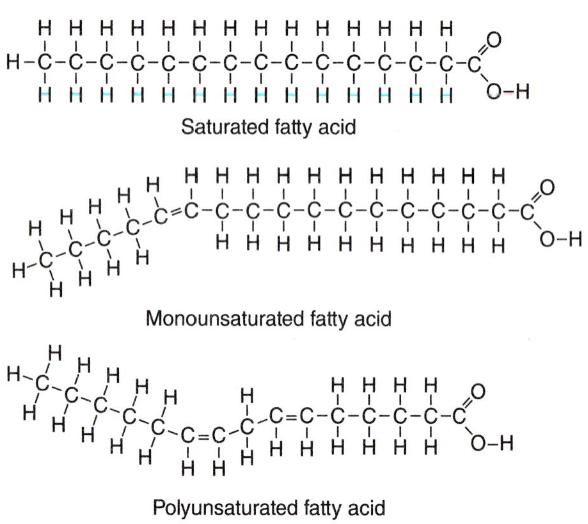

Figure 4.1 Structures of saturated, monounsaturated, and polyunsaturated fatty acids.

Fatty acids containing no double bonds are referred to as saturated, one double bond as monounsaturated, and two or more double bonds as polyunsaturated (Figure 4.1). The majority of double bonds are in the *cis* rather than the *trans* configuration (**Figure 4.2**). The more double bonds in the acyl chain, the less linear the physical conformation. Lipids with a high proportion of saturated fatty acids tend to be solid at ambient temperature, because the straight acyl chains can align with each other (e.g., butter and lard) (**Figure 4.3**). In contrast, unsaturated fatty acids have a "kink" at each double bond, thereby creating a nonlinear configuration, which interferes with acyl chain alignment. Unsaturated fatty acids with a single double bond (monounsaturated) tend to be liquid at ambient temperature and appear cloudy when cooled (e.g., olive and peanut oils). Unsaturated fatty acids with multiple double bonds (polyunsaturated) tend to be liquid at ambient temperature and when cooled (e.g., soybean and corn oils).

The nomenclature used to refer to fatty acids takes into consideration acyl chain length, and number, location, and conformation (stereochemistry) of the double bonds. There are two systems in common usage. For the beta system, so named according to the designation of the bond broken during fatty acid oxidation, the location of the double bond is designated by the Greek letter delta and a superscript referring to the number of the first carbon of the double bond after the carboxylic group. For example, the double bond in oleic acid is designated Δ^9, the double bonds in linoleic acid are $\Delta^{9,12}$ (Table 4.1). For the omega system, carbon atoms are numbered from the methyl end of the acyl chain, and location of double bonds is demarcated by the number of the first carbon of each double bond. For example, the double bonds in linoleic acid are represented by ω-6,9 or n-6,9. Some fatty acids have the same number of carbon atoms and double bonds but occur as geometric or positional isomers. Geometric fatty acid isomers differ in the conformation of one or more of the double bonds, *cis* or *trans* (Figure 4.2). Examples include oleic acid (C18:1, n-9 *cis*) and elaidic acid (C18:1, n-9 *trans*). Positional fatty acid isomers differ in the location of one or more of the double bonds in the acyl chain. Examples include alpha-linolenic acid (18:3, n-3) and gamma-linolenic acid (18:3, n-6). The metabolic fate of fatty acid isomers is determined by their structure.

Triglycerides (Triacylglycerides)

Triglyceride is the predominant form of lipid in plants and animals. It is composed of three constituent fatty acids esterified to a glycerol (3-carbon) molecule, in positions designated as sn-1, sn-2, and sn-3. The mass of a triglyceride molecule is contributed primarily by the constituent fatty acids. Triglyceride molecules containing three identical fatty acids are termed simple triglycerides. They are rare in nature. Triglyceride molecules containing two or three different fatty acids are termed mixed triglycerides and represent the majority that occur naturally. The physical properties of a triglyceride molecule, including melting point and susceptibility to oxidation, are determined by the

Table 4.1 Common Fatty Acids

Common Name	International Union of Pure and Applied Chemistry Name	Abbreviation
Saturated Fatty Acids		
Butyric acid	Butanoic acid	C4:0
Caproic acid	Hexanoic acid	C6:0
Caprylic acid	Octanoic acid	C8:0
Capric acid	Decanoic acid	C10:0
Lauric acid	Dodecanoic acid	C12:0
Myristic acid	Tetradecanoic acid	C14:0
Palmitic acid	Hexadeconoic acid	C16:0
Stearic acid	Octadecanoic acid	C18:0
Arachidic acid	Eicosanoic acid	C20:0
Behenic acid	Docosanoic acid	C22:0
Lignoceric acid	Tetracosanoic acid	C24:0
Cerotic acid	Hexacosanoic acid	C26:0
Monounsaturated Fatty Acids		
Palmitoleic acid	Δ^{9}-Hexadecaenoic acid	C16:1, n-7 *cis*
Oleic acid	Δ^{9}-Octadecaenoic acid	C18:1, n-9 *cis*
Elaidic acid	Δ^{9}-Octadecaenoic acid	C18:1, n-9 *trans*
Vaccenic acid [(11E)-11-octadecenoic acid]	Δ^{11}-Octadecenoic acid	C18:1, n-11 *cis*
Vaccenic acid [(11Z)-11-octadecenoic acid]	Δ^{11}-Octadecenoic acid	C18:1, n-11 *trans*
Eicosenoic acid	Δ^{11}-Eicosenoic acid	C20:1, n-9 *cis*
Erucic acid	Δ^{13}-Docosaenoic acid	C22:1, n-9 *cis*
Brassidic acid	Δ^{13}-Docosaenoic acid	C22:1, n-9 *trans*
Nervonic acid	Δ^{15}-Tetracosaenoic acid	C24:1, n-9 *cis*
Polyunsaturated Fatty Acids		
Linoleic acid	$\Delta^{9,12}$-Octadecadienoic acid	C18:2, n-6,9 all *cis*
α-Linolenic acid	$\Delta^{9,12,15}$-Octadecatrienoic acid	C18:3, n-3,6,9 all *cis*
γ-Linolenic acid	$\Delta^{6,9,12}$-Octadecatrienoic acid	C18:3, n-6,9,12 all *cis*
Columbinic acid	$\Delta^{5,9,12}$-Octatrienoic acid	C18:3, n-6 *cis*, 9 *cis*, 13 *trans*
Mead acid	$\Delta^{5,8,11}$-Eicosatrienoic acid	C20:3, n-9,12,15 all *cis*
Dihomo-γ-linolenic acid	$\Delta^{8,11,14}$-Eicosatetraenoic acid	C20:3, n-6,9,12 all *cis*
Arachidonic acid	$\Delta^{5,8,11,14}$-Eicosatetraenoic acid	C20:4, n-6,9,12,15 all *cis*
Timnodonic acid	$\Delta^{5,8,11,14,17}$-Eicosapentaenoic acid	C20:5, n-3,6,9,12,15 all *cis*
Clupanodonic acid	$\Delta^{7,10,13,16,19}$ -Docosapentaenoic acid	C22:5, n-3,6,9,12,15 all *cis*
Docosahexaenoic acid	$\Delta^{4,7,10,13,16,19}$ -Docosahexaenoic acid	C22:6, n-3,6,9,12,15,18 all *cis*

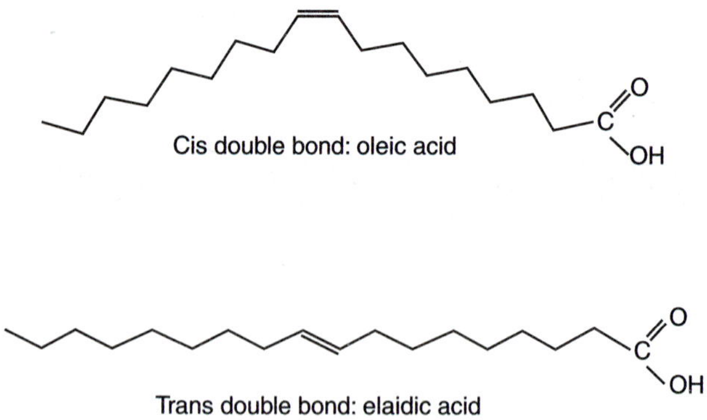

Figure 4.2 *Cis* and *trans* double bonds.

Reproduced from R P Mensink, M B Katan. Effect of dietary trans fatty acids on high-density and low-density lipoprotein cholesterol levels in healthy subjects. *N Engl J Med.* 1990 Aug 16;323(7):439-45. PMID: 2374566.

fatty acid position on the glycerol molecule (sn-1, sn-2, or sn-3) and characteristics (chain length and number, location, and conformation of the double bonds). In animals and humans, adipose tissue triglyceride profiles reflect, in part, dietary fatty acid composition. This property has been used in humans to estimate long-term fat intake, and in animal husbandry, to manipulate the physical characteristics of fat stores (e.g., hay or corn-fed pigs). In contrast to triglycerides, mono- and diglycerides contain one and two fatty acids, respectively, with hydroxyl groups in place of the other fatty acids. They occur in trace amounts in plants and animals, and are primarily intermediate products of triglyceride digestion, clearance from the bloodstream, and intracellular metabolism. In food products, they are used as emulsifiers to improve stability and shelf life.

Phospholipids (Phosphoglycerides)

Phospholipids are composed of two fatty acids esterified to a glycerol molecule in positions sn-1 and sn-2. A phosphate group, sometimes referred to as a polar head group, is esterified to the third carbon, sn-3. Due to the amphipathic nature conferred by the fatty acids and phosphate group, phospholipids are the main constituents of cell membranes (bilayer) and surface components of lipoproteins (monolayer). In both cases, the polar (phosphate) hydrophilic head groups orient toward the aqueous phase (cytoplasm, plasma) and hydrophobic acyl chains (fatty acids) orient toward the lipid phase. Phospholipid fatty acids in the sn-2 position tend to be unsaturated.

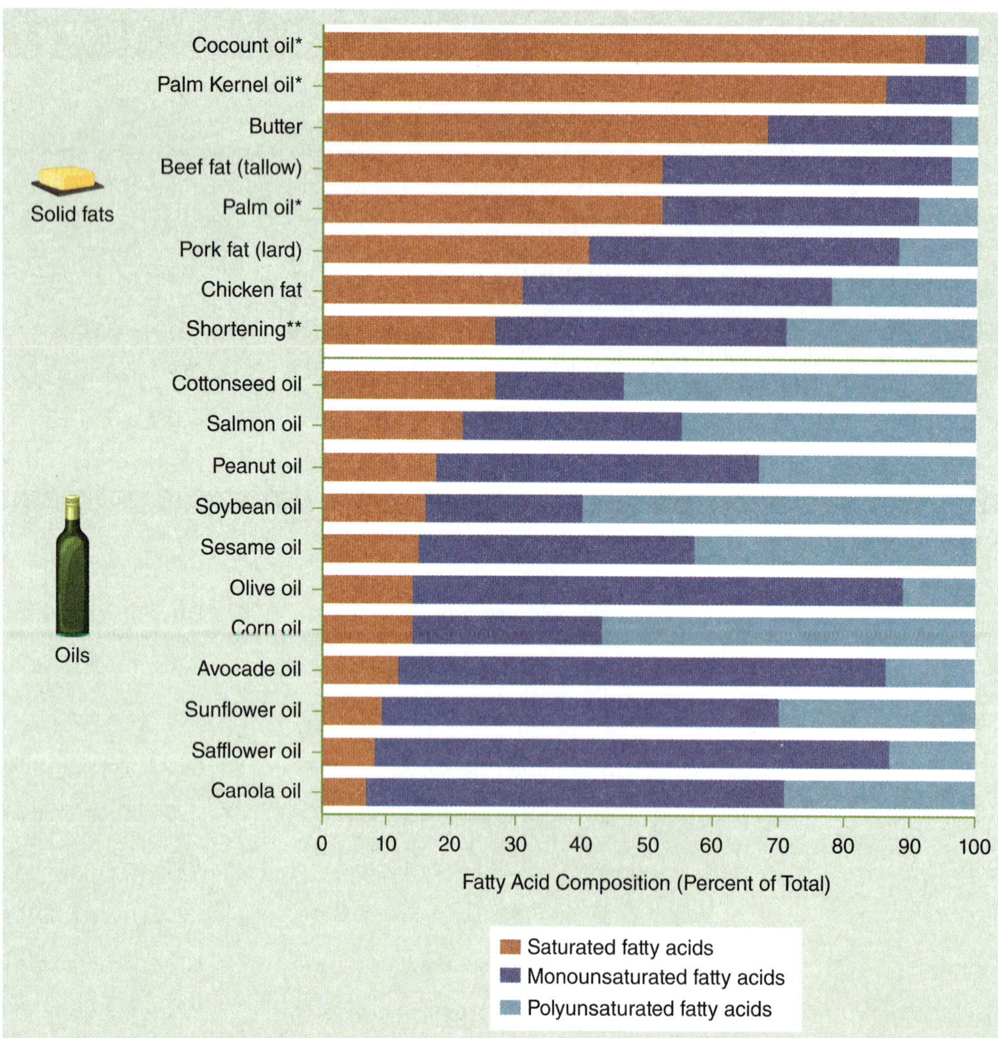

Figure 4.3 Saturated, monounsaturated, and polyunsaturated fatty acid profiles of different types of dietary fat.

Reproduced from U.S. Department of Health and Human Services and U.S. Department of Agriculture. *2015 – 2020 Dietary Guidelines for Americans.* 8th Edition. December 2015. Available at https://health.gov/our-work/food-nutrition/previous-dietary-guidelines/2015.

In cell membranes, these fatty acids influence membrane fluidity and serve as a substrate reservoir for subsequent metabolism (e.g., synthesis of ceramides, eicosanoids and steroid hormones). The predominant phospholipid phosphate groups are choline, serine, inositol, and ethanolamine, forming phosphatidylcholine (lecithin), phosphatidylserine, phosphatidylinositol, and phosphatidylethanolamine, respectively.

Steroids/Sterols

Steroids are compound that contain a core four ring structure, referred to as A, B, C, and D. Differences among steroids are due to the presence of methyl and functional groups, oxidation state of the rings, and structure of the hydrocarbon tails. Sterols are a subgroup of steroids, characterized by having a hydroxyl group at the 3-carbon of the A-ring. Sterols are naturally present in plants (phytosterols) and animals (zoosterols). In plants, common phytosterols include β-sitosterol, campesterol and stigmasterol. The major difference among these sterols is their side chain. They regulate growth and development. In animals, the most common sterol is cholesterol, which occurs in nonesterified and esterified forms. Nonesterified cholesterol serves as a precursor for the synthesis of vitamin D, steroid hormones (e.g., estrogen, testosterone), adrenal steroids (e.g., hydrocortisone, aldosterone), and bile acids. In cell membranes, as needed, it decreases fluidity by intercalating between highly unsaturated phospholipid fatty acids, restricting mobility. Esterified cholesterol, termed cholesteryl ester, occurs primarily in the core of circulating lipoprotein particles and intercellular lipid storage droplets, along with triglyceride. High concentrations of cholesterol are found in the brain, resulting primarily from *de novo* synthesis.[4] Stanol is the term for the saturated form of sterols, which contain no double bonds.

Fat-Soluble Nutrients

Fat-soluble vitamins include vitamin A (retinol and retinyl ester) and its provitamins (β-carotene, alpha-carotene, gamma-carotene, and xanthophyll beta-cryptoxanthin), vitamin D (cholecalciferol and ergocalciferol), vitamin E (tocopherols and tocotrienols), and vitamin K (phylloquinone [vitamin K1] and menaquinone [vitamin K2]). They are discussed in Chapters 17, 18, 19 and 20. Phytochemicals are produced by plants and are thought to be involved in resisting fungi, bacteria, and viral infections. Many of the colors, flavors, and odors in plants are attributed to their photochemical content. The health effects in humans have yet to be fully elucidated. An exception is carotenoids.[5] (see Chapter 31).

Digestion

The digestion of dietary lipids presents unique challenges, compared with carbohydrates and proteins, due to their hydrophobic nature. As such, an elegant system has evolved to accommodate the digestion of lipids.

Mouth and Esophagus

The process of lipid digestion starts in the mouth, where mastication, salivation, and lingual lipase, an enzyme secreted by the von Ebner (serous) glands of the tongue, cause the hydrolysis of some short chain fatty acids from the sn-3 position of triglycerides. The major products are nonesterified fatty acids and diglycerides. From a teleologic perspective, this initial step in fat digestion best accommodates milk fat, which contains a relatively high proportion of short chain fatty acids, critical to the survival of human infants. Triglyceride hydrolysis continues as the material (bolus) travels from the mouth through the esophagus to the stomach. Once the mass enters the stomach, a decrease in pH inhibits the activity of lingual lipase.

Stomach

Upon entry of the bolus into the stomach, gastric lipase is secreted by chief cells located in the gastric mucosa. Consistent with its environment, gastric lipase functions optimally at a relatively low pH. The primary substrates for gastric lipase are short and medium chain fatty acids. Like lingual lipase, the major products of hydrolysis are nonesterified fatty acids and diglycerides. It has been estimated that 10%–30% of fatty acids are released from triglycerides as a result of lingual and gastric lipases in adults and up to 50% in breastfed infants.[6]

Intestine

Most lipid digestion occurs in the small intestine. Key components include bile acids and pancreatic lipase. In response to masticated food (chyme) entering the upper part of the small intestine (duodenum), the hormone cholecystokinin is secreted by mucosa cells. This hormone stimulates the pancreas to release pancreatic lipase and the gallbladder to release bile (containing bile acids). Bile acids are synthesized from cholesterol *de novo* (rate limiting enzyme 7α-hydroxylase) and recycled by resorption after reaching the lower part of the small intestine (ilium). The amphipathic nature of bile acids allows them to emulsify chyme to form small particles, termed micelles. The formation of micelles increases the surface area of the mixture, maximizing interaction with pancreatic lipase and other hydrolytic enzymes. Colipase, also released in bile, facilitates the interaction of pancreatic lipase with the surface of micelles, enhancing triglyceride hydrolysis. The products of triglyceride digestion include nonesterified fatty acids, glycerol, and monoglycerides. Phospholipase A_2 hydrolyzes micellar phospholipid. The products are nonesterified fatty acids and lysolecithin. The fatty acids on cholesteryl ester are hydrolyzed by pancreatic cholesterol esterase.

Absorption

The products of lipid digestion are absorbed from the last and longest part of the small intestine, the ilium. For absorption to take place, micelles must penetrate an unstirred water layer that separates the intestinal contents from epithelial cells that line the intestinal wall, termed the brush border. The surface of intestinal epithelial cells is covered with villi and the villi with microvilli. These finger-like projections maximize the surface area, facilitating absorption. As the micelles interact with microvilli on the epithelial cells, the constituent lipids traverse the cell membrane down the concentration gradient by passive diffusion. Once inside the epithelial cells, non-esterified fatty acids are re-esterified to monoglycerides, lysolecithin and cholesterol to re-form triglycerides, phospholipids, and cholesteryl esters, respectively. This occurs on the outer surface of the endoplasmic reticulum.

These components, along with other fat-soluble molecules, traverse the endoplasmic reticulum and, with the addition of a protein—apoprotein (apo) B-48, form triglyceride rich lipoprotein particles termed chylomicrons. The triglyceride-rich nascent chylomicron particles move from the endoplasmic reticulum to the Golgi apparatus and are released into the lymphatic system to the thoracic duct, which feeds into peripheral circulation via the subclavian vein. The exception is short- and medium-chain nonesterified fatty acids (≤10 carbons). They are absorbed directly into the hepatic portal vein bound to albumin.

Transport and Metabolism

The majority of lipids are transported in blood as components of lipoprotein particles. Lipoprotein particles are spherical in shape and are composed of a nonpolar lipid core, primarily containing triglycerides and cholesteryl esters, and an amphipathic surface, primary made up of a phospholipid monolayer, as well as nonesterified fatty acids, cholesterol, and apoprotein(s). There are four major classes of lipoprotein particles, distinguished by the relative ratio of core to surface (particle size), lipid composition and constituent apoprotein(s): chylomicrons, very low-density lipoproteins (VLDL) (including intermediate density lipoproteins ([IDL]), low density lipoproteins (LDL) and high-density lipoproteins (HDL) (**Figure 4.4**, **Table 4.2**). The different types of lipoprotein particles provide a mechanism for delivering lipids to cells for subsequent use, storage, metabolism, or excretion.

Chylomicrons

Chylomicrons are intestinally derived lipoprotein particles. Their major lipid constituent is triglycerides. The primary function of chylomicrons is to ferry dietary fat-soluble compounds (e.g., triglycerides, fat soluble vitamins) from the intestine to peripheral tissues and the liver. Due to the core to surface ratio, chylomicrons are the least dense and largest lipoprotein particles (Figure 4.4). The distinguishing feature of chylomicron particles is the presence of the protein apo B-48. Additional apoproteins identified in isolated chylomicron particles, acquired after secretion from the intestine, include apo C, apo E, apo A-I, apo A-II. As chylomicrons circulate, they interact with an enzyme attached to heparan sulfated peptidoglycans on the surface of capillaries, lipoprotein lipase (LPL). LPL hydrolyzes the core triglycerides, releasing nonesterified fatty acids. These nonesterified fatty acids enter the underlying cells, and can be oxidized, stored, or metabolized into bioactive compounds. Hepatic lipase serves a similar role in hepatic capillaries, having the capacity to hydrolyze fatty acids from both triglyceride and phospholipid. As chylomicron particles are delipidated of triglycerides, they are referred to as chylomicron remnants and taken up by the liver via a receptor-dependent mechanism. The uptake is mediated by apo B-48 and apo E.

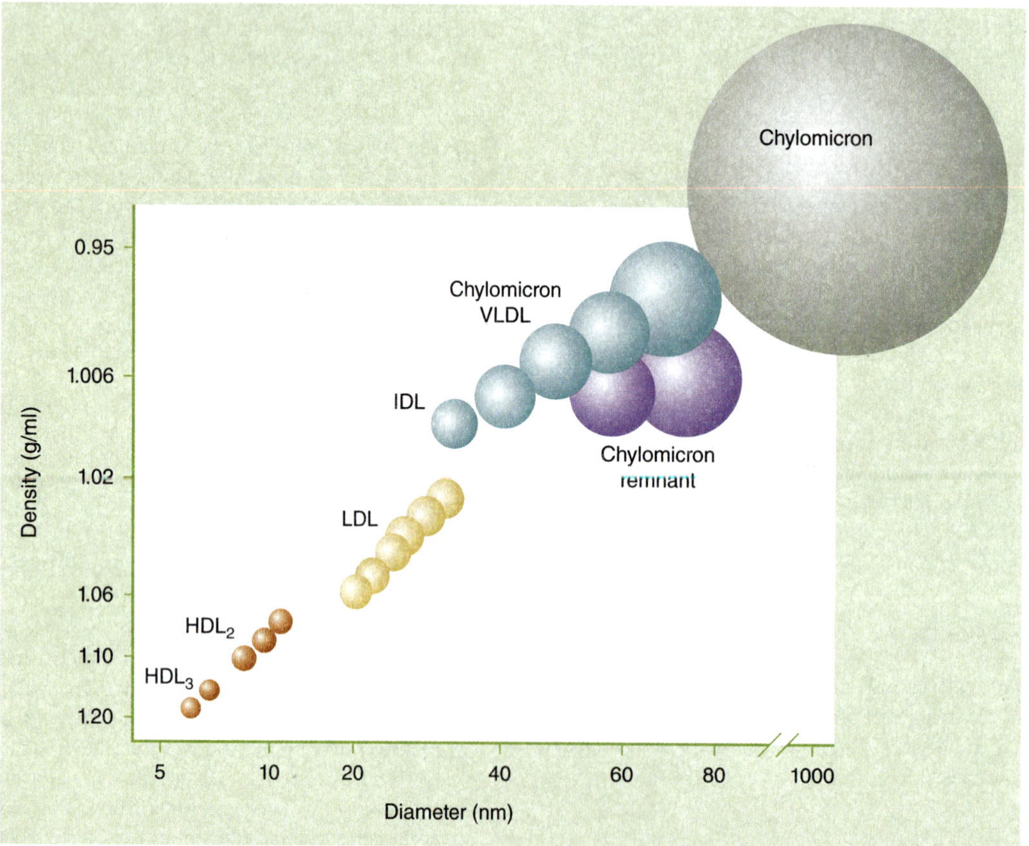

Figure 4.4 Lipoprotein particles.

Table 4.2 Lipoprotein Classes

Class	Origin	Major Function	Apo-Proteins	Lipid Composition			
				TG	CE	FC	PL
Chylomicron	Intestine	Transport dietary triglyceride and fat-soluble compounds from the small intestine to peripheral tissue for metabolism and storage	Apo B-48, apo C, apo E, apo A-I, apo A-II	81-89	2-4	1-3	7-9
VLDL	Liver	Transport triglyceride and fat-soluble compounds out of the liver for transport to peripheral tissue	Apo B-100, apo C, apo E	50-58	15-23	4-9	19-21
IDL	Lipolysis VLDL	Intermediate between VDL and LDL caused by dilapidation of triglyceride in core of VLDL particle	Apo B-100, apo C, apo E	25-27	28-31	6-7	13-14
LDL	Lipolysis IDL	Deliver cholesterol to peripheral tissue	Apo B-100	7-11	47-51	10-12	28-30
HDL	Intestine and liver	Shift cholesterol from peripheral tissue to liver for metabolism or excretion (reverse cholesterol transport)	Apo A-I, apo A-II, apo C, apo E	6-7	24-45	6-8	42-51

VLDL, very low-density lipoprotein; IDL, intermediate-density lipoprotein; LDL, low-density lipoprotein; HDL, high-density lipoprotein; apo, apoprotein.

Very Low-Density Lipoproteins (VLDL) and Intermediate-Density Lipoproteins (IDL)

VLDL particles are synthesized and secreted by the liver. Elevated concentrations have been associated with higher cardiovascular disease risk.[7] Their primary function is to transport fatty acids, in the form of triglycerides, cholesterol, and fat-soluble compounds, out of the liver to peripheral tissues. The triglyceride in VLDL particles is derived from chylomicron remnants and other lipoproteins taken up by the liver, and *de novo* lipogenesis. The latter is fueled by glucose that enters the circulation after the digestion of carbohydrate in quantities that exceed the capacity of peripheral tissue to clear. Whereas intestinally derived lipoprotein particles are distinguished by the presence of apo B-48, hepatically derived lipoprotein particles are distinguished by the presence of apo B-100. Either prior to or immediately after secretion from the liver, VLDL acquires additional apoproteins, including apo Es and apo Cs. Although smaller than chylomicrons, the major lipid component of VLDL is triglyceride (Table 4.2). In circulation, the triglyceride is hydrolyzed by LPL. As the core triglyceride of the VLDL particle is depleted, it becomes smaller and denser and is reclassified as an IDL particle (or VLDL remnant). These particles are either taken up by a receptor-mediated mechanism involving apo B-100 or apo E, or remain in circulation where further delipidation occurs, as does loss of apoproteins, except for apo B-100.

Low-Density Lipoprotein (LDL)

LDL is the metabolic product of VLDL. Commensurate with a dramatic depletion of triglyceride content, the resulting lipoprotein particle is cholesterol rich and contains only apo B-100. The primary function of LDL is to deliver cholesterol to peripheral tissues. Most LDL particles are cleared from plasma by the LDL receptor, for which apo B-100 (and apo E) are ligands. Both the particle and receptor are internalized. Once internalized, the LDL receptor disassociates from the particle and can be recycled for subsequent use. The LDL particle fuses with a lysosome and the constituent components are degraded. Release of the nonesterified cholesterol into the cytoplasm is critical, as it influences whole body cholesterol homeostasis. High intracellular nonesterified cholesterol is cytotoxic. There are several mechanisms that prevent high levels from accumulating. Intracellular nonesterified cholesterol (*i*) inhibits the synthesis of new LDL receptors, thereby downregulating the uptake of additional cholesterol through negative feedback inhibition; (*ii*) inhibits the activity of 3-hydroxy 3-methylglutaryl CoA (HMGCoA) reductase, the rate-limiting enzyme in *de novo* (endogenous) cholesterol biosynthesis; and (*iii*) stimulates the activity of acyl CoA cholesterol acyltransferase (ACAT), an enzyme that esterifies nonesterified cholesterol, forcing it out of the cytoplasm and into intercellular lipid droplets. Genetic abnormalities in the receptor-mediated LDL uptake from plasma result in elevated plasma LDL cholesterol concentration, due to decreased clearance and subsequent lack of inhibition of endogenous cholesterol biosynthesis and is associated with elevated atherosclerotic cardiovascular disease.[8]

LDL can also be taken up by macrophage scavenger receptors. Unlike receptor-mediated uptake, the scavenger receptors are not feedback regulated. Rather, the presence of high plasma LDL cholesterol concentration results in excessive uptake of cholesterol. In the arterial wall, resident macrophages become engorged with cholesterol, leading to their transformation into what is termed "foam" cells, so named due to their appearance. The unregulated accumulation of cholesterol in foam cells ultimately results in cell death and promotes the formation of atherosclerotic lesions (plaque). As the plaque enlarges, it causes a narrowing of the blood vessel lumen, thereby impeding blood flow. In coronary arteries, this can result in angina and coronary events, in carotid arteries it can result in stroke, and in extremities it can result in phlebitis. Because of the strong relation of elevated LDL cholesterol concentration and cardiovascular risk, it is frequently referred to as the "bad" cholesterol. Reducing LDL cholesterol concentration, by diet or drugs, results in a decrease in atherosclerotic cardiovascular events.[9]

Lipoprotein (a) (Lp[a])

Lp(a) is an LDL particle that contains an additional apoprotein, apo(a), bound covalently to apo B-100.[10] The distribution of Lp(a) concentration in a population is skewed to the left, that is, most individuals have relatively low concentration. The few with high Lp(a) concentration are at elevated risk of developing atherosclerotic cardiovascular disease.[11] Lp(a) concentration is, for the most part, genetically determined.

High-Density Lipoprotein (HDL)

HDL particles originate from the liver, intestine, and excess surface components after the delipidation of triglyceride-rich lipoprotein remnant particles (chylomicrons and VLDL). There is a broad range of HDL particles, distinguished by their apoprotein content and size. Apo A-I is the constituent apoprotein of HDL. Other apoproteins found on isolated particles include apo A-II, apo C, and apo E. The main function of HDL is to ferry cholesterol from peripheral tissue to the liver. This cholesterol is used for the synthesis of bile acids, for the formation and secretion of VLDL and HDL, or is excreted directly into the bile. The process is referred to as "reverse cholesterol transport." High HDL cholesterol concentration is associated with low risk of developing cardiovascular disease.[12-14] For this reason, HDL is frequently referred to as the "good" cholesterol.

Several HDL-related mechanisms promote "reverse cholesterol transport." HDL has the capacity to acquire nonesterified cholesterol from peripheral tissue. Once this occurs, an enzyme that circulates on HDL, lecithin cholesterol acyl protein transferase (LCAT), esterifies the nonesterified cholesterol to form cholesteryl ester.[15] The apoprotein on HDL, apo AI, is the cofactor for LCAT. The now nonpolar form of cholesterol migrates to the core of the HDL particle, freeing up the surface to acquire additional nonesterified cholesterol from peripheral tissues. The protein phospholipid transfer protein (PLTP) enhances the ability of HDL to promote reverse cholesterol transport by increasing the particle's phospholipid content, enhancing its capacity to take up nonesterified cholesterol from peripheral tissues.[16,17] The membrane-bound protein, ABCA1, promotes the net flux of nonesterified cholesterol from peripheral tissues to HDL, enhancing reverse cholesterol transport.[18] The hepatic HDL receptor, scavenger receptor (SR)-B1, facilitates the transfer of cholesteryl ester from HDL to the liver. This protein selectively translocates cholesteryl ester, allowing HDL particles to continue participating in reverse cholesterol transport. HDL particles are also involved in whole body cholesterol homeostasis through the activity of cholesterol ester transfer protein (CETP).[19] This protein facilitates the exchange of cholesteryl ester for triglyceride from HDL to chylomicrons and VLDL. The latter particles are cleared by the LDL receptors.

Essential Fatty Acids

Humans cannot synthesize linoleic acid (LA, 18:2 n-6) and α-linolenic acid (ALA, 18:3 n-3) because they lack the Δ12 and Δ15 desaturases necessary to insert a double bond at positions omega-6 and omega-3 of the fatty acyl chain. Therefore, LA and ALA are classified as essential fatty acids and must be obtained through the diet. *In vivo*, LA can be elongated and desaturated to form arachidonic acid (AA [n-6]), whereas ALA can be elongated and desaturated to form eicosapentaenoic acid (EPA [n-3]), docosapentaenoic acid (DPA [n-3]) and docosahexaenoic acid (DHA [n-3]). Because LA and ALA are both substrates for the Δ12 and Δ15 desaturases, competition between them impacts the rate of conversion to their final products.[20] Hence, the relative intakes of LA and ALA are determinants of the endogenous synthesis of their subsequent metabolites, and availability for incorporation into plasma membrane phospholipids and use as precursors of other lipid metabolites.[21] In humans, the conversion of ALA to EPA and DHA is relatively inefficient, estimated to range from 3% to 5%.[22] Thus, dietary intake of the latter fatty acids is recommended. Essential fatty acid deficiency is rare but has been documented in very young infants. Deficiency is associated with dry skin, thrombocytopenia and, in children, intellectual disability. According to the Dietary Reference Intakes, the Adequate Intake value for LA is 12 g for adult women and 17 g for adult men, and for ALA is 1.1 g for adult women and 1.6 g for adult men.[23] (**Table 4.3**) Soybean and corn oils are rich sources of LA. Soybean and canola oils, as well as some nuts and seeds, are rich sources of ALA. Seafood is the primary dietary source of EPA and DHA.

Lipid Metabolites

The hydrolysis of AA, EPA, and DHA from the cell membrane phospholipid bilayer provides substrate for the synthesis of a wide range of metabolically active compounds. The reaction is catalyzed by phospholipase A_2. The specific classes of lipid

Table 4.3 Adequate Intake Values for Essential Fatty Acids

Age	6–11 m	1–3 y	4–8 y	9–13 y	14–18 y	19–30 y	31–50 y	50+ y
Linoleic acid, AI								
Males	4.6	7	10	12	16	17	17	14
Females	4.6	7	10	10	11	12	12	11
α-linolenic acid, AI								
Males	0.5	0.7	0.9	1.2	1.6	1.6	1.6	1.1
Females	0.5	0.7	0.9	1.0	1.1	1.1	1.1	1.1

AI, Adequate Intake.

metabolites are determined by the substrate fatty acid; AA, EPA, or DHA. These include eicosanoids, docosanoids, and specialized proresolving lipid mediators (SPMs).[24] The first report of a molecule with biological function derived from dietary fat goes back to the 1930s, when von Euler observed that a lipid-soluble extract from sheep vesicular glands lowered blood pressure and stimulated uterine smooth muscle contraction. The putative fatty acid-derived compound was termed prostaglandin.[25] Cyclo-oxygenase (COX) 1 and 2 convert AA to the 2-series prostaglandins (PG), molecules with proinflammatory activity, to thromboxanes (TX), molecules with proprothrombotic activities.[24] 5-Lipoxygenase converts AA to the 4-series leukotrienes (LT). EPA is also a substrate in these enzymes, generating 3-series PG and TX and 5-series LT, which have less, anti-inflammatory effects compared with those of AA-derived eicosanoids.[24,26] As a class, eicosanoids, once secreted, have autocrine and paracrine functions dictated by their fatty acid precursor, and relatively short half-life (seconds to minutes).

In the 2000s, several SPMs derived from EPA, DPA, DHA, and AA were identified. These compounds have potent anti-inflammatory and pro-resolution effects on inflammation. SPMs are synthesized in large part by immune cells via COX, lipoxygenases, and cytochrome P450 enzymes, including resolvins, protectins, maresins, and lipoxins. Both the intermediate and final fatty acid metabolites in these cascades have biological functions. The AA metabolites 12-hydoxyeicosatetranoic acid (12-HETE) and 15-HETE are prothrombotic and proinflammatory and, hence, are proatherogenic.[27,28] The EPA-derived E-series resolvins 1-4 (RvE1-4) and DHA-derived RvD1-6, protectins (PD1) and maresins (Mar1) have potent anti-inflammatory and inflammation resolution actions. SPMs orchestrate the resolution of inflammation by inhibiting the chemotactic recruitment and migration of proinflammatory neutrophils and other immune cells into the inflammatory site and promote macrophage-mediated efferocytosis of apoptotic cells. These effects restore homeostasis.[29] SPMs modulate innate and adaptive response by binding to cell membrane G-protein coupled receptors, including Chem23, GPR18, GPR32, GPR37, BLT1, and LGR6.[30]

AA, EPA, and DHA serve as precursors for the synthesis of endocannabinoids. Endocannabinoids have signaling functions in the brain that modulate neurotransmitter release, and have been demonstrated to influence appetite, energy metabolism, pain perception, memory and learning.[31] AA is a precursor for anandamide (AEA) and 2-arachidonyglycerol (2-AG), while EPA and DHA are precursors for eicosapentaenoyl ethanolamine (EPEA) and docosahexaenoyl ethanolamine (DHEA), respectively. In the brain, the endocannabinoid AEA is thought to have interpretation of stimulus properties, whereas EPA and DHEA have anti-inflammatory properties that are greater than their precursor fatty acids.[32]

Dietary Fat Guidance for Cardiovascular Disease Risk Reduction

Total Dietary Fat

Current evidence suggests that the type of dietary fat, rather than amount, has the largest impact on prevention and treatment of cardiometabolic disorders. Over the past few decades, fads have shifted between high- and low-fat diets (and, by extension, low- and high-carbohydrate diets). The two major issues to consider are the impact of total dietary fat on lipoprotein profiles and body weight. Low-fat diets tend to be high in carbohydrate, particularly refined carbohydrate, which can lead to carbohydrate-induced hypertriglyceridemia (high triglyceride and low HDL cholesterol concentrations).[33] Unless implemented thoughtfully, high-fat diets tend to be high in animal fats, resulting in lower unsaturated to saturated fat ratios, which can lead to an increase in LDL-cholesterol concentration.[34] In both cases, the result, dyslipidemia, is associated with elevated cardiovascular disease risk.[35] It was once reasoned that because fat has a higher energy density than protein or carbohydrate (9 kcal/g vs 4 kcal/g, respectively), limiting dietary fat would result in a lower caloric density, automatically leading to lower energy intake and weight loss. Neither observational nor interventional data support this hypothesis. Two long-term intervention studies (1-2 y) demonstrated that low fat compared with high-fat diets promotes similar weight loss.[36,37] For the most part, current dietary guidance is silent regarding the amount of dietary fat and instead focuses on the type of dietary fat.[38,39]

Dietary Fat Type

Saturated, Monounsaturated, And Polyunsaturated Fatty Acids

Data from observational studies indicate that dietary patterns lower in saturated fat and higher in unsaturated fat (monounsaturated and polyunsaturated), particularly polyunsaturated fat, are associated with lower risk of cardiovascular diseases.[35] When assessing the totality of the literature, it is important to distinguish between studies that compared types of fat, for example, saturated fat relative to polyunsaturated fat, rather than other energy sources, e.g., carbohydrate.[40] Major sources of dietary saturated fat include animal fat—meat and dairy, and tropical oils—coconut and palm (Figure 4.3). Major sources of monounsaturated fat include olive and canola oils, as well as some nuts, such as almonds, cashews, hazelnuts, and pecans. Major sources of dietary monounsaturated fat depend on cultural cuisines. Traditionally, in Mediterranean regions, the main source of monounsaturated fat is olive oil. In regions following what is commonly referred to as Western-type dietary patterns, a significant contributor of monounsaturated fat is meat. The latter source also comes with saturated fat. Hence, cross-cultural comparisons of diet composition should be approached with caution. Major sources of dietary polyunsaturated fat include soybean, corn, and unmodified safflower and sunflower oils, as well as some nuts such as Brazil nuts, pine nuts, and walnuts.

Omega-3 Polyunsaturated Fatty Acids

As noted previously, fatty acids can occur as positional isomers. A common example is omega-3 and omega-6 polyunsaturated fatty acids. Although these isomers share common metabolic enzymes, their products are different and have different biological properties. The dietary plant sources of the most abundant omega-3 fatty acid, alpha-linolenic acid (18:3n-3), include soybean and canola oils and walnuts. The major dietary source of longer chain omega-3 polyunsaturated fatty acids, EPA (20:5n-3), and DHA (22:6n-3), is seafood. EPA and DHA are the biologically active omega-3 polyunsaturated fatty acids. Observational data have suggested benefits of dietary patterns high in seafood in terms of cardiovascular disease.[41] Potential mechanisms include lower ventricular fibrillation, triglyceride concentration, platelet aggregation and blood pressure.[42] Whether the benefit of dietary patterns rich in seafood is due to the presence of omega-3 fatty acids or replacement of foods high in animal (saturated) fat, or a combination of the two, has yet to be elucidated.[39] Initial reports suggested a benefit of fish oil supplements in terms of cardiovascular disease prevention and treatment.[43] However, more recent studies have not confirmed these findings.[44-46] Differences in study design and chemical form of the supplements preclude direct comparisons.[47] Hence, the current recommendation is to regularly consume seafood, at least twice a week.[39]

Trans Fatty Acids

Trans fatty acids contain at least one double bond in the *trans* configuration. They occur in both monounsaturated and polyunsaturated fatty acids. *Trans* double bonds have a wider bond angle than the more common *cis* double bonds (geometric isomers) (Figure 4.2). This results in a straighter acyl chain, more similar in conformation to saturated than unsaturated fatty acids. Dietary *trans* fatty acids occur naturally in meat and dairy fat, resulting from anaerobic bacterial fermentation in ruminant animal stomachs.[48] They are absorbed and deposited in milk and muscle. *Trans* fatty acids are also introduced when foods are prepared with partially hydrogenated fats. The process of partial hydrogenation had been favored by food manufacturers because it increases the viscosity of the fat (conversion of a liquid to semi-solid or solid) and decreases the susceptibility to oxidation (development of off flavors and odors—rancidity). Changing the textural characteristics of the fat facilitates its use as a replacement for animal fats (e.g., lard, butter) and the reduced susceptibility to oxidization increases shelf-life. However, during the partial hydrogenation process, in addition to saturation of double bonds, others are cleaved and some randomly reform into the *trans* configuration. In the early 1990s, data emerged that *trans* fatty acids had a similar effect as saturated fatty acids on LDL cholesterol concentration.[49] These data were subsequently confirmed.[50] Eventually, in the United States and other countries, this finding triggered restrictions on the use of partially hydrogenated fats for commercial food preparation, mandatory inclusion of *trans* fatty acid content on Nutrient Facts labels, and eventual removal of partially hydrogenated fats from the Generally Recognized as Safe (GRAS) list.[51] It has been suggested that there is a difference in response to *trans* fatty acid isomers predominant in ruminant fat vs partially hydrogenated fat. If any, the differences are small. Current recommendations are to limit dietary *trans* fatty acid intake.[39]

Dietary Cholesterol and Plant Sterols

Historically, dietary cholesterol has been associated with elevated plasma total and LDL-cholesterol concentrations, and cardiovascular disease risk. However, within the range currently consumed in the United States, the evidence suggests that dietary fat type, rather than amount of dietary cholesterol, has a greater effect.[52] This has resulted in the lack of target dietary cholesterol recommendations, although increasing current levels of intake is discouraged.

Plant oils contain phytosterols (β-sitosterol, campesterol, and stigmasterol). These compounds are similar in structure to cholesterol. In humans, phytosterols are poorly absorbed and, in the gastrointestinal tract, compete with cholesterol for incorporation into micelles. This results in a decrease in cholesterol bioavailability. At the concentrations naturally present in plant oils, phytosterols have little effect. However, when consumed in amounts up to 2 g per day, they lower LDL-cholesterol concentration 5%–10%, in addition to that achieved with statin therapy (cholesterol synthesis inhibitors).[53] Phytosterols have been added to a wide variety of foods (e.g., orange juice, chocolate candy), using either the sterol (unsaturated) or stanol (saturated) forms. When provided as such, the foods containing them are commonly referred to as functional foods, because they are enriched in biologically active components.

Heart Healthy Dietary Pattern

Current evidence-based dietary guidance to promote cardiometabolic health includes adjusting energy intake and expenditure to achieve and maintain a healthy body weight; eating plenty and a variety of fruits and vegetables; choosing whole grain foods and products; choosing healthy sources of protein (mostly plants, regular intake of fish and seafood, low-fat or fat-free dairy products and, if meat or poultry is desired, lean cuts and unprocessed forms); using liquid plant oils rather than tropical oils and partially hydrogenated fats; choosing minimally processed foods instead of ultra-processed foods; minimizing the intake of beverages and foods with added sugars; choosing and preparing foods with little or no salt; if not drinking alcohol, do not start; if choosing to drink alcohol, limit intake; and adhering to a healthy dietary pattern, regardless of where food is prepared or consumed (**Figure 4.5**).[39]

Summary

Although frequently cast in a negative light, particularly in terms of foods or excess accumulation in the body, *in vivo*, the general class of lipids performs critical roles essential for survival and optimal functioning. In cells, lipids serve as structural components of membranes; in plasma, they are an integral part of lipoprotein particles, which provide a mechanism for the transport of nonpolar compounds in the aqueous milieu of blood; in adipose tissue, they are a dense reserve of metabolic energy; in food, they carry fat soluble vitamins, essential fatty acids, and nonpolar phytonutrients, as well as facilitating their absorption; *in vivo*, they provide substrates for the synthesis of bioactive compounds; in subcutaneous adipose tissue, they serve as an insulator to minimize wide fluctuations in body temperature; in visceral adipose tissue, they offer cushioning to protect internal organs from trauma; and, when consumed in adequate quantities relative to protein and carbohydrate, they minimize the development of carbohydrate-induced hypertriglyceridemia. Fatty acids are the basic components of complex lipids, including triglycerides and phospholipids, and substrates for the synthesis of steroids and sterols. The metabolites of lipids have a wide range of biological effects that have been related to both health and disease. Current evidence suggests that the type of dietary fat, rather than amount, has the most significant impact on prevention and treatment of cardiometabolic disorders. Data from observational studies indicate that dietary patterns lower in saturated fat and higher in unsaturated fat (monounsaturated and polyunsaturated),

Figure 4.5 Heart Healthy Dietary Pattern.

Reproduced from Lichtenstein AH, Appel LJ, Vadiveloo M, Hu FB, Kris-Etherton PM, Rebholz CM, Sacks FM, Thorndike AN, Van Horn L, Wylie-Rosett J. 2021 Dietary Guidance to Improve Cardiovascular Health: A Scientific Statement From the American Heart Association. Circulation. 2021 Dec 7;144(23):e472-e487. doi: 10.1161/CIR.0000000000001031. Epub 2021 Nov 2. PMID: 34724806.

particularly polyunsaturated fat, are associated with lower risk of cardiovascular diseases. Current evidence-based dietary guidance to promote cardiometabolic health includes adjusting energy intake and expenditure to achieve and maintain a healthy body weight; including a variety of fruits and vegetables; choosing whole grain foods and products; choosing healthy sources of protein (plants, fish and seafood, low-fat or fat-free dairy products, and lean cuts and unprocessed forms of meat); using liquid plant oils; choosing minimally processed foods instead of ultra-processed foods; minimizing the intake of beverages and foods with added sugars; choosing and preparing foods with little or no salt; if not consuming alcohol, not to start; if consuming alcohol, limit intake; and adhering to a healthy dietary pattern regardless of where food is prepared or consumed.

Acknowlegments

The author acknowledges the authors of the previous version of this chapter, Peter Jones and Todd Rideout, for their contributions.

References

1. McCollum EV, Davis M. J Biol Chem 1913;15:167–75.
2. Blaner WS. J Lipid Res 2013;54:1716–8.
3. Burr GO, Burr MM. Nutr Rev 1973;31:248–9.
4. Orth M, Bellosta S. Cholesterol 2012;2012:292598.
5. Liu RH. Adv Nutr 2013;4:384S–92S.
6. Mu H, Hoy CE. Prog Lipid Res 2004;43:105–33.
7. Huang JK, Lee HC. Int J Mol Sci 2022;23.
8. Shah NP, Ahmed HM, Wilson Tang WH. Cleve Clin J Med 2020;87:109–20.
9. Arnett DK, Blumenthal RS, Albert MA, et al. Circulation 2019;140:e596–e646.
10. Koschinsky ML, Kronenberg F. Atherosclerosis 2022;349:1–6.
11. Coassin S, Kronenberg F. Atherosclerosis 2022;349:17–35.
12. Cheng W, Rosolowski M, Boettner J, et al. Lipids Health Dis 2022;21:47.
13. Kjeldsen EW, Thomassen JQ, Frikke-Schmidt R. Biochim Biophys Acta Mol Cell Biol Lipids 2022;1867:159063.
14. Tall AR. Clin Chem 2021;67:19–23.
15. Yang K, Wang J, Xiang H, et al. Biomed Pharmacother 2022; 147:112677.
16. Chowaniec Z, Skoczynska A. Adv Clin Exp Med 2018;27: 429–36.
17. Yazdanyar A, Yeang C, Jiang XC. Curr Atheroscler Rep 2011;13: 242–8.
18. Phillips MC. J Lipid Res 2018;59:749–63.
19. Schmidt AF, Hunt NB, Gordillo-Maranon M, et al. Nat Commun 2021;12:5640.
20. Kinsella JE, Broughton KS, Whelan JW. J Nutr Biochem 1990; 1:123–41.
21. Sun Q, Ma J, Campos H, et al. Am J Clin Nutr 2007;86:74–81.
22. Burns-Whitmore B, Froyen E, Heskey C, et al. Nutrients 2019;11.
23. Institute of Medicine. *Dietary Reference Intakes for Energy, Carbohydrate, Fiber, Fat, Fatty Acids, Cholesterol, Protein, and Amino Acids.*, The National Academies Press. https://doi.org/10.17226/10490, Washington, DC, 2005.
24. Calder PC. Essays Biochem 2020;64:423–41.
25. von Euler US. J Physiol 1936;88:213–34.
26. Funk CD. Science 2001;294:1871–5.
27. Lundqvist A, Sandstedt M, Sandstedt J, et al. PLoS One 2016; 11:e0161629.
28. Pascale JV, Lucchesi PA, Garcia V. J Cardiovasc Pharmacol 2021; 77:707–17.
29. Dalli J, Serhan CN. Front Immunol 2017;8:1400.
30. Chiang N, Serhan CN. Essays Biochem 2020;64:443–62.
31. Cascio MG, Marini P. Handb Exp Pharmacol 2015;231:39–58.
32. de Bus I, Witkamp R, Zuilhof H, et al. Prostaglandins Other Lipid Mediat 2019;144:106351.

33. Lichtenstein AH. J Lipid Res 2006;47:1661–7.
34. Gardner CD, Landry MJ, Perelman D, et al. Am J Clin Nutr 2022.
35. Sacks FM, Lichtenstein AH, Wu JHY, et al. Circulation 2017;136:e1–e23.
36. Gardner CD, Trepanowski JF, Del Gobbo LC, et al. JAMA 2018;319:667–79.
37. Sacks FM, Bray GA, Carey VJ, et al.) N Engl J Med 2009;360: 859–73.
38. U.S. Department of Agriculture and U.S. Department of Health and Human Services. Dietary Guidelines for Americans, 2020-2025. 9th Edition. Available at DietaryGuidelines.gov, 2020.
39. Lichtenstein AH, Appel LJ, Vadiveloo M, et al. Circulation 2021;144:e472–e87
40. Jakobsen MU, O'Reilly EJ, Heitmann BL, et al. Am J Clin Nutr 2009;89:1425–32.
41. Rimm EB, Appel LJ, Chiuve SE, et al. Circulation 2018;138: e35–e47.
42. Drenjancevic I, Pitha J. Int J Mol Sci 2022;23.
43. Bhatt DL, Miller M, Brinton EA, et al. Circulation 2020;141: 367–75.
44. Costenbader KH, MacFarlane LA, Lee IM, et al. Clin Chem 2019;65:1508–21.
45. Kalstad AA, Myhre PL, Laake K, et al. Circulation 2021;143: 528–39.
46. Nicholls SJ, Lincoff AM, Garcia M, et al. JAMA 2020;324: 2268–80.
47. Ridker PM, Rifai N, MacFadyen J, et al. Circulation 2022;146: 372–9.
48. Lichtenstein AH. Curr Atheroscler Rep 2014;16:433.
49. Mensink RP, Katan MB. N Engl J Med 1990;323:439–45.
50. Lichtenstein AH, Ausman LM, Jalbert SM, et al. N Engl J Med 1999;340:1933–40.
51. Hyseni L, Bromley H, Kypridemos C, et al. Bull World Health Organ 2017;95:821–30.
52. Carson JAS, Lichtenstein AH, Anderson CAM, et al. Circulation 2020;141:e39–e53.
53. Gylling H, Strandberg TE, Kovanen PT, et al. Nutrients 2020;12.

A. MAJOR DIETARY CONSTITUENTS

CHAPTER 5

Energy and Energy Metabolism

Sai Krupa Das and Susan B. Roberts

CHAPTER OUTLINE

Abbreviations

ATP, adenosine triphosphate; **BAT**, brown adipose tissue; **BEE**, basal energy expenditure; **BMI**, body mass index; **BMR**, basal metabolic rate; **CO_2**, carbon dioxide; **DIT**, diet-induced thermogenesis; **DLW**, doubly labeled water; **DRI**, Dietary Reference Intake; **EE**, energy expenditure; **EER**, estimated energy requirement; **EI**, energy intake; **FDA**, US Food and Drug Administration; **FFA**, free fatty acid; **FFM**, fat-free mass; **FM**, fat mass; **FQ**, food quotient; **GE**, gross energy; **HR**, heart rate; **IOM**, Institute of Medicine; **ME**, metabolizable energy; **NPRQ**, nonprotein respiratory quotient; **PAL**, physical activity level; **P:O ratio**, phosphorylation-to-oxidation ratio; **REE**, resting energy expenditure; **RMR**, resting metabolic rate; **RQ**, respiratory quotient; **SMR**, sleeping metabolic rate; **TEE**, total energy expenditure; **TEF**, thermic effect of food; **UCP1**, uncoupling protein-1; **VCO_2**, carbon dioxide production; **$\dot{V}O_2$**, oxygen consumption; **$\dot{V}O_{2max}$**, maximal oxygen consumption

Energy Metabolism

To sustain life, humans must obtain energy from food. Complex biochemical, structural, and physiologic mechanisms underlie the transformation of chemical free energy from food into energy forms that can be used by the body to maintain its structural and biochemical integrity. Conversion of food energy into high-energy biochemical compounds includes energy that is lost as heat. The energy of food that is not initially lost as heat is converted and stored primarily as the molecule adenosine triphosphate (ATP), which is eventually lost from the body in the form of additional heat or used to perform work (e.g., circulation, respiration, muscle contraction).[1-3] The heat byproducts of energy conversion are critical to maintaining body temperature.

Energy Intake

Energy Sources

Energy is provided in the diet by protein, carbohydrate, fat, and alcohol. Humans can survive on foods with varying proportions of carbohydrates, fats, and proteins.[4-8] The total energy in food is called its "gross energy" (GE), which is larger than the amount of energy available to the body (called "metabolizable energy", [ME]), due primarily to obligatory losses during digestion, in feces, and energy-containing nitrogenous compounds in urine.[9] The metabolizable energy available from different nutrients and foodstuffs was measured more than 100 years ago by Atwater, in a small number of individuals who participated in meticulous energy balance techniques. Atwater's original study included measurement of all consumed food and energy losses in feces and urine. These values for the ME content of nutrients and types of foods, and modern-day methodologies for estimating ME, are further discussed below, in **Energy of Food.**

Conversion of Food Energy to Physiological Energy

The conversion of food energy to forms of energy used by the body is a biochemical process of oxidation of the substrates fat, carbohydrate, protein, and alcohol into ATP, with heat as a primary byproduct. Substrate oxidation rates are a function of dietary macronutrient intake and level of energy turnover.[10] Protein oxidation is largely determined by protein intake, whereas the relative contributions of glucose and free fatty acids (FFAs) to the fuel mix are more variable. Glucose oxidation is adjusted to carbohydrate intake to maintain stable glycogen stores within physiological limits. Fat intake does not promote its own oxidation and, under conditions of positive energy balance (metabolizable energy intake [EI] in excess of needs; see below for further discussion), most dietary fat will be deposited. Most cells can use the metabolic intermediates of carbohydrates, fats, and proteins interchangeably to regenerate ATP, with a few exceptions. The brain preferentially uses glucose and is able to use ketone bodies after adaptation to starvation, but it does not use FFAs.[11] Red blood cells also depend on glucose. At rest, the brain (20%), internal organs (25%–30%), and skeletal muscle (20%) account for most of the energy needs or consumption. During vigorous activity, skeletal muscle dominates energy use over other tissues. In the postabsorptive state, FFAs are mainly oxidized by muscle. During exertion, the glycogen reserve in muscles is used initially, with a subsequent shift toward use of FFAs mobilized from muscle fat stores and adipose tissue. The ability to shift from carbohydrate to fat as the main source of energy, coupled with substantial reserves of body fat, makes it possible for humans to accommodate large fluctuations in EI, balance of macronutrients, and energy expenditure (EE).

When alcohol (predominantly ethanol in the human diet) is consumed, it promptly appears in the circulation and is oxidized at a rate determined largely by its concentration and by the activity of the liver enzyme alcohol dehydrogenase. The oxidation of alcohol rapidly reduces the oxidation of the other substrates used for ATP regeneration. Alcohol oxidation proceeds in large part through conversion to acetate and oxidative phosphorylation. Approximately 80% of the energy liberated by alcohol oxidation is used to drive ATP regeneration, and approximately 20% is released as heat.[12] The effects of alcohol intake on energy balance are unclear and may vary based on the level of consumption and energy status of the individual.[13]

Flatt and Tremblay (10) computed the ATP yield from oxidation of macronutrients based on the phosphorylation-to-oxidation (P:O) ratio (defined as the number of ATP molecules produced by oxidative phosphorylation for each pair of electrons passing from a substrate through the respiratory chain to O_2), and the ATP required to initiate degradation, transport, activation, and handling of the metabolic fuels (**Figure 5.1**). Assuming a P:O ratio of 3:1 for the reoxidation of mitochondrial reduced nicotinamide adenine dinucleotide, the oxidation of 1 mol of glucose produces 38 mol ATP, but 2 mol ATP is used for activation; therefore, the net ATP yield is 95%. After using energy to recycle through the Cori cycle, the glucose-alanine cycle, and gluconeogenesis, the

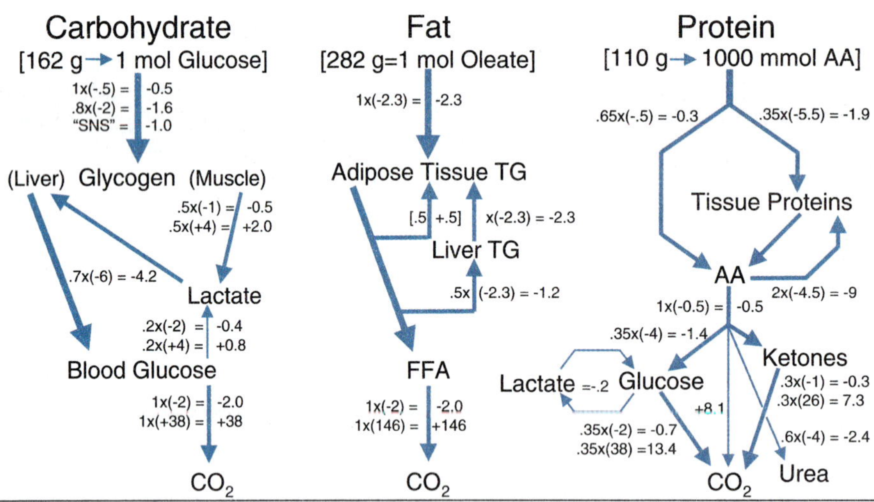

	Carbohydrate	Fat	Protein
kcal/ATP made	18	18. 3	18
Post-absorptive ATP yield	**82%**	**96%**	**65%**
kcal/ATP replaced	22	19	28
Overall ATP yield	**75%**	**90%**	**55%**
kcal/ATP replaced	24	20.3	33

Figure 5.1 Adenosine triphosphate (ATP) yields from oxidation of carbohydrate, fat, and protein. Moles of substrate flowing through the metabolic pathways are in *brackets*, and the moles of ATP produced and expended per mole of substrate metabolized are in *parentheses*, assuming a phosphorylation-to-oxidation (P:O) ratio of 3 for the reoxidation of mitochondrial-reduced nicotinamide adenine dinucleotide. For example, 38 ATPs are produced during the oxidation of 1 mol glucose, but because of substrate handling, storage, and recycling costs, the postabsorptive ATP yield is approximately 82%, and the overall yield is 75%. *AA*, amino acid; *FFA*, free fatty acid; *SNS*, sympathetic nervous system; *TG*, triglyceride.

net postabsorptive ATP yield is approximately 82%. Accounting for the postprandial phase of digestion, absorption, and transport, the final net ATP yield from dietary carbohydrate is 75%, such that the oxidation of 24 kcal of dietary carbohydrate is required to replace 1 mol of ATP.

To calculate the ATP yield from dietary fat, the fatty acid oleate was used as the example substrate. The oxidation of 1 mol of oleate yields 146 mol ATP but expends 5.5 mol ATP in lipolysis/re-esterification and activation to oleyl-coenzyme A; therefore, the ATP yield for fat oxidation is approximately 96%. Accounting for the postprandial phase, the net ATP yield from dietary fat is approximately 90%.

In the case of proteins, the oxidation of 1 mol of amino acids generates approximately 28.8 mol of ATP or 18 kcal/mol ATP. The costs of gluconeogenesis, ureagenesis, and protein resynthesis reduce the net postabsorptive ATP yield to 65%. Accounting for the postprandial phase, the net ATP yield is 55%. Based on these estimates, transport, storage, recycling, and activation dissipate approximately 10%, 25%, and 45% of the ATP produced in the oxidation of dietary fat, carbohydrate, and protein, respectively. Therefore, the corresponding net ATP yields are estimated to be 90%, 75%, and 55%, respectively.

Lipogenesis from the conversion of dietary carbohydrate to fat is an inefficient process, estimated at 25% (i.e., 25% of energy in fat is converted to energy in carbohydrate, with the remaining energy dissipated as heat). This pathway is not primarily triggered in humans because large amounts of dietary carbohydrate first expand glycogen reserves and increase oxidation rates to match intake.[14,15] Lipogenesis, therefore, does not account for the higher dissipation of dietary energy by carbohydrate compared with fat. However, when dietary carbohydrate intake far exceeds weight maintenance energy needs, or TEE (total energy expenditure), then *de novo lipogenesis* is triggered and may be a significant contributor to fat deposition.[14,15] Similarly, futile cycles (cycles in which the only net change is energy dissipation) or substrate cycles that dissipate ATP with no net change in the organism also appear to make only a minor contribution to overall energy economy and, therefore, account for only a small percentage of TEE.[16] However, brown adipose tissue (BAT) contains uncoupling proteins that can induce thermogenesis through futile cycles and does exist in humans, especially in those who are younger and without obesity, raising the potential for futile cycles to be significant contributors to energy physiology under some circumstances, as yet to be fully elucidated.[17]

Energy Expenditure

Human TEE is composed of basal EE (BEE), the thermic effect of food (TEF), and physical activity, as well as other thermogenic factors.

Basal Energy Expenditure

BEE represents the energy needed to sustain the metabolic activities of cells and tissues plus the energy to maintain blood circulation and respiration in the awake state. BEE is defined as the rate of EE in the postabsorptive state after a 12-hour overnight fast. BEE is measured while an individual is supine, awake, and motionless in a thermoneutral environment. The terms basal metabolic rate (BMR) or BEE (i.e., BMR extrapolated to 24 hours) and resting metabolic rate (RMR) or resting energy expenditure (REE) (RMR extrapolated to 24 hours) are often used interchangeably and refer to the same component of EE. BEE is used throughout this chapter to refer to BMR (except when exact text from a publication is used, such as in an equation). The sleeping metabolic rate (SMR) is approximately 5%–10% lower than BEE.[18] BEE is affected by age, biological sex, body composition, and nutritional and health status, as well as other individual-level factors.

The contribution of fat-free mass (FFM) and fat mass (FM) to the variability in BEE was examined in a meta-analysis of seven published studies.[19] FFM was the single best predictor of BEE, accounting for 73% of the variability; FM accounted for only an additional 2%. In another meta-analysis, the relationship between BEE and FFM was found to be nonlinear across a wide range of ages, from infants to adults.[20] BEE/kg of weight or BEE/kg FFM falls as mass increases because the relative contributions made by the most metabolically active tissues (brain, liver, and heart) decline as body size increases.

FFM includes not only muscle but also organs, which are the most metabolically active compartments of the body[21] and, therefore, are the major predictors of BEE. The contributions of organ and tissue masses to the variability in BEE has been investigated using magnetic resonance imaging and echocardiography methods to measure liver, kidney, spleen, heart, and brain masses.[22-26] BEEs of skeletal muscle (14.5 $kcal \cdot kg^{-1} \cdot d^{-1}$) and adipose tissue (4.5 $kcal \cdot kg^{-1} \cdot d^{-1}$) are low, relative to the metabolic rates of brain (240 $kcal \cdot kg^{-1} \cdot d^{-1}$), liver (200 $kcal \cdot kg^{-1} \cdot d^{-1}$), heart, and kidneys (440 $kcal \cdot kg^{-1} \cdot d^{-1}$). Together, the brain, liver, heart, and kidneys account for approximately 60%–70% of BEE in adults, but they represent less than 6% of body weight. Skeletal muscle accounts for only 20%–30% of BEE and comprises 40%–50% of body weight. Furthermore, accounting for the mass of the organs reduced the variability in BEE attributed to age, biological sex, and race.[25] The decline in BEE in growing children is attributed to both a decrease in the proportion of some metabolically active organs and tissues, and changes in the metabolic rate of specific organs and tissues.[23]

BEE declines with age. The pattern of decline has, for many years, been reported as a rate of approximately 1%–2% per decade in weight-constant persons.[27] This decline was attributable to loss of FFM and gain of less metabolically active fat associated with aging. However, more recent data from an international database[28] has instead reported essentially constant BEE in adult life until approximately 47 years of age, in parallel with a similar pattern of change in TEE (constant until 63 years), both with and without adjustment for FFM and FM. These two very different trajectories of how EE changes over the adult lifespan cannot be reconciled at this time, as differences are likely due to differences in data sets used for analysis. This discrepancy high lights the shortage of high-quality, within-person longitudinal data to address this topic. Biological sex differences in BEE are also evident, but there is no known differential effect of biological sex on changes in BEE with aging.[28] Lower BEE in women is largely attributed to differences in FFM and FM, although hormonal differences may also play a role. BEE varies throughout the menstrual cycle[29] and is approximately 6%–15% lower in the preovulatory (follicular) phase than in the premenstrual (luteal) phase of the cycle.[29,30] Even when BEE data are adjusted for biological sex differences in FFM and FM, differences in BEE still exist, possibly because of variations in relative contributions of organs and tissues to FFM. A recent publication shows greater variability in objectively measured EE in males compared with females throughout the adult life course.[31]

Race may also affect BEE. Numerous studies have reported lower BEE in Black adults and children compared with White adults[32-35] and children[36-39]. BEE, expressed per kilogram of body weight or per kilogram of FFM, is on the order of 5%–10% lower in Black vs. White subjects. Differences in relative contributions of organs and tissues to FFM may explain the differences in BEE among racial groups. Lower BEE in Black, compared with White, women has been attributed to the greater proportion of low-metabolic-rate skeletal muscle and bone relative to organs in the former group.[26]

BEE also changes acutely in response to weight flux.[40,41] However, whether EE is permanently shifted after equilibration at a different weight remains unclear. Some studies reporting metabolic adaptation to weight loss show lower EE than predicted after equilibration at a lower weight,[42] while other studies of longer duration show no significant changes in EE.[43] One small study (n = 14), of competitors from the television show "Biggest Loser," following massive weight loss at 6 years, suggested the potential for such weight loss to trigger a profound, permanent decrease in BEE, with a reported lack of increase in BEE even after 41 kg weight regain.[44] However, this finding has not yet been replicated, and studies of massive weight loss in gastric bypass patients show no measurable metabolic adaptation to mean weight loss of 53 kg after accounting for changes in body FFM and FM.[45] Overall, these studies demonstrate the potential for BEE to adapt acutely to changes in energy balance; however, in the long term, the predominance of data to date suggest minimal or no accommodations beyond those explained by changes in FFM and FM.

Thermic Effect of Food

TEF refers to the increase in EE elicited by food consumption (1). TEF is sometimes referred to as diet-induced thermogenesis (DIT). A component of TEF is considered obligatory thermogenesis, that is, the energy cost of digesting, absorbing, and processing or storing nutrients. TEF varies by the macronutrient content of the food consumed, ranging from 5%–10% of energy content for carbohydrate, 0%–5% for fat, and 20%–30% for protein. A mixed meal elicits an increase in EE equivalent to approximately 10% of the energy consumed.[46] Beyond macronutrient content, obligatory thermogenesis of TEF is also affected by exercise and meal size and frequency.[47-49]

TEF also has a facultative thermogenesis component.[14,50] Facultative thermogenesis, broadly, represents the additional EE not accounted for by the known energy costs of obligatory thermogenesis. The facultative thermogenesis component of TEF is thought to be modulated primarily by the sympathetic nervous system and thyroid, with active sites in BAT, skeletal muscle, and potentially, other organs, such as the liver (see below discussion of **Additional Factors Inducing Thermogenesis**).[51] A blunted ability to oxidize fat in old age can be attributed to impaired facultative thermogenesis, and might be one factor contributing to the increase in body fatness and obesity observed in aging.[52]

Physical Activity

EE for physical activity represents the most variable component of TEE. Physical activity level (PAL) is defined as the ratio of daily TEE to BEE, and is strongly associated with body composition (higher PAL with higher FFM relative to FM).[53] This ratio is commonly used to describe typical activity levels. For sedentary individuals, PAL varied from 1.3 to 1.5, with an average value of 1.35 among nine studies.[54] In whole room (indirect) calorimetry studies, the TEE/BEE ratio averaged 1.32 in groups with no exercise, 1.42 in those who did 30–75 minutes/day of exercise, and 1.60 in those who did 100–180 minutes/day.[55] The value of 1.4 × BEE represents maintenance energy requirements and covers BEE, TEF, and minimal activity. In more active groups, the PAL ranges from 1.4 to 1.7, and from 2.0 to 2.8 in very active groups.

The energy costs of discrete physical activities (e.g., walking, running) have been measured using indirect calorimetry. Ainsworth et al[56] provided comprehensive tables to estimate the energy expended in discrete physical activities for adults, and Butte et al. provide comparable tables for children.[57]

The energetic efficiency for the conversion of dietary energy into physical work is remarkably constant in humans for non–weight-bearing activities.[1,3,58-60] The metabolic cost of performing specific physical activities is highly reproducible under standardized test conditions. Under optimal conditions, the network efficiency of the body is approximately 25%–27%, but under typical circumstances, the mechanical efficiency of the body is considerably less. This finding does not imply that the energy cost of activities is constant among individuals, however. Energy cost of activities among individuals varies because of differences in body weight, skill, and adaptation. For weight-bearing physical activities, the cost is roughly proportional to body weight.

Excess post-exercise oxygen consumption ($\dot{V}O_2$) refers to the small increase in EE, which occurs for some time after the exercise has been completed. Excess postexercise $\dot{V}O_2$ is estimated to be approximately 14% of the increment in expenditure that occurs during the exercise itself.[61] A sustained increase in post-exercise BEE occurs only after intense and prolonged exercise (70%–75% maximal $\dot{V}O_2$ [$\dot{V}O_2$max] for 80–90 minutes or longer of activity), and even this increase is small, relative to the energy expended in exercise. Moderate levels of exercise do not appear to increase subsequent EE markedly. Substrate use during exercise depends mainly on relative intensity although other factors, such as exercise duration, age, biological sex, training status, environmental conditions, chronic disease, and recent dietary intake, influence the relative contribution of carbohydrate, fat, and protein use.[62,63] Whole body fuel use is predominantly from FFA at low exercise intensities.[63-65] As exercise intensity increases, a shift from the predominant use of fat to carbohydrate occurs. The peak rate of fat oxidation is achieved at approximately 45% of $\dot{V}O_2$max, and for exercises at greater than 50% of $\dot{V}O_2$max, the oxidation of FFAs declines in muscle, both as a percentage of total energy and on an absolute basis. The main carbohydrate energy source is muscle glycogen, supplemented by blood glucose and lactate. If exercise persists beyond 60–90 minutes, fat oxidation will rise as carbohydrate fuel sources become depleted. In this case, the intensity of exercise is impacted by the depletion of muscle glycogen, decreased blood glucose, and fatigue.[58]

Additional Factors Inducing Thermogenesis

As noted, facultative thermogenesis augments BEE in response to stimuli unassociated with muscular activity or other obligatory thermogenesis and can be rapidly turned on or suppressed by the nervous system. Stimuli include exercise and cold and heat exposure. Cold- and heat-induced thermogenesis refer to the increase in EE that is induced at temperatures lower or higher than the zone of thermoneutrality. Studies consistently suggest that low-normal temperatures of 20–22°C and high temperatures of 28–30°C are associated with a small increase in sedentary EE of 2%–5%, compared with temperatures of 24–27°C. Thus, facultative thermogenesis plays a role in thermal regulation (i.e., regulation of body temperature), via mechanisms such as shivering in muscle and nonshivering in BAT, which balance heat loss. Because people usually adjust their clothing and environment to maintain comfort, the additional energy cost of thermoregulation makes a minimal contribution to TEE.

Facultative thermogenesis also plays a small role in energy balance, in which thermogenesis induced, for example, by exercising in muscle, or by diet in BAT, balances control of EI. While the maintenance of body temperature is more stringently controlled than energy balance, the two processes are not unrelated. In laboratory animals with obesity, there is evidence of reduced EE and deficient (relative to normal-weight controls) DIT in BAT, that may also be associated with hyperphagia. Similarly, reduced cold-induced thermogenesis in BAT, as well as in other organs, is indicative of altered thermal balance. This, in turn, results in altered energy balance and obesity, potentially exacerbated by hyperphagia. In some hyperphagic animals with obesity, it is possible that the exaggerated obligatory TEF is sufficient to alter thermal balance such that BAT thermogenesis is suppressed. In all animals with obesity, there is evidence of deficient hypothalamic control of both facultative thermogenesis and EI.[66,67]

BAT has long been recognized to play a unique role in facultative thermogenesis in rodents. Uncoupling protein-1 (UCP1) in the inner mitochondrial membrane of BAT is responsible for this adaptive thermogenic process. UCP1 allows brown adipocytes to dissipate the mitochondrial proton electrochemical gradient that normally drives ATP synthesis.[68] Evidence of the presence of BAT in humans has prompted a reappraisal of its role in human physiology. Fluorodeoxyglucose positron emission tomography used to trace tumor metastasis revealed symmetric areas of increased tracer uptake in the upper parts of the body corresponding to BAT.[69] Human depots of BAT were found in the supraclavicular and the neck regions with some additional paravertebral, mediastinal, para-aortic, and suprarenal localizations. The presence of BAT-unique UCP1 in samples of adipose tissue from the neck of 35 individuals confirmed the presence of BAT.[70] BAT activity is induced acutely by cold exposure and is stimulated by the sympathetic nervous system.[71] BAT has the potential to be of metabolic significance in normal human physiology.

Other substances, such as caffeine, can increase BEE by 10%–30% for 1–3 hours.[72] On a daily basis, normal caffeine consumption may cause a modest 3% increase in TEE.[73] Drugs such as amphetamines, ephedrine, and some antidepressants stimulate the sympathetic nervous system and, in turn, increase metabolism, whereas propranolol, reserpine, or bethanidine may depress it. The effect of smoking on the BEE is unclear,[74,75] but one study showed a 10% increase in 24-hour EE in a room calorimeter associated with smoking 24 cigarettes.[76]

Energy Balance

Energy balance is the difference between EI and the sum of energy losses, which are primarily as heat resulting from EE and energy losses in feces, urine, and from secretions and combustible gasses.[77] Energy balance may be expressed as follows:

$$\text{E Balance} = \text{GE}_{\text{intake}} - \text{E}_{\text{losses}}\,(\text{E}_{\text{expenditure}} + \text{E}_{\text{feces}} + \text{E}_{\text{urine}} + \text{E}_{\text{secretion}} + \text{E}_{\text{combustible gas}})$$

and a streamlined equation using ME rather than GE is:

$$\text{E Balance} = \text{ME}_{\text{intake}} - (\text{E}_{\text{expenditure}} + \text{E}_{\text{secretions}})$$

This second equation accounts for energy losses from feces and urine within ME and assumes negligible energy losses in combustible gasses.

Implicit in both energy balance equations is the concept that energy is conserved. EI and EE are not, however, independent of each other. Changes in EI or EE trigger physiologic and downstream behavioral responses that attenuate changes in energy balance, regulated by complex physiologic systems, including neuroendocrine feedback. For example, an increase in EE (e.g., beginning an exercise program) can result in increased hunger or a reduction in otherwise typical physical activity across the rest of the day. In response to overeating without intentional change in physical activity, EE increases slightly, but 85%–90% of excess kcal are deposited.[78] In response to a reduction in EI by intentional dieting, greater compensation is observed to attenuate changes in energy balance with acute reductions in EE for BEE and physical activity that reduce the loss of body energy by more than one-third of the reduction in EI.[79,80]

Measurement of Energy of Food, Energy Intake, and Energy Expenditure

Energy of Food

The energy in foods is expressed as a unit of heat, such as the kilocalorie (kcal). A kcal is defined as the amount of heat required to raise the temperature of 1000 g of water by 1°C (e.g., from 14.5°C to 15.5°C). The scientific international unit of energy is the joule (J), defined as the energy expended when 1 kg is moved 1 m by a force of 1 Newton in the direction of the force applied. In 1956, an international committee standardized the equivalency of these units as 1 calorie = 4.1868 J, but the value of 4.184 is more commonly used in nutrition studies. For practicality, a kcal, which is 1000 times the energy of 1 calorie, is commonly used in nutrition and dietetics. Hence, 1 kcal = 4.184 kJ, and 1 kJ = 0.0239 kcal.

The GE content of food, defined as the total heat that can be liberated from food, is determined experimentally by measuring the heat produced inside a bomb calorimeter when foodstuffs are completely combusted to carbon dioxide (CO_2) and water.[77] The bomb calorimetry method is the gold standard method for quantifying GE and uses dried, homogenized samples of food. The actual amount of heat evolved per gram of foodstuff varies according to its chemical composition. Subtracting obligatory energy losses in stool and urine from GE, as measured across the period of food ingestion, results in values of ME.

As noted previously, in the early 1900s, Atwater developed values of ME still used today. Average values are 4.1 kcal/g of carbohydrate, 9.3 kcal/g of fat, and 5.4 kcal/g of protein. However, the body cannot oxidize nitrogen; therefore, energy resulting from the oxidation of the nitrogenous component of protein is unavailable. Given the significant presence of nitrogen in protein, only 4.2 kcal/g protein is potentially available to the body. Therefore, the simplified physiologic fuel values, known as the Atwater factors, are 4 kcal/g for carbohydrate, 9 kcal/g for fat, and 4 kcal/g for protein. The physiologic fuel value for alcohol is 7 kcal/g (**Table 5.1**).

Several different methods are used to apply the Atwater factors to estimating the energy content of foods. The U.S. Food and Drug Administration (FDA) of the U.S. government defines

Table 5.1 Heats of Combustion, Physiologic Energy Values, Heat Equivalents, and Corresponding Volumes of Oxygen and Carbon Dioxide for Carbohydrate, Protein, Fat, and Ethanol Oxidation

	ENERGY (kcal/g)			Heat Equivalents			Volume	
Food	Heat of Combustion	Human Oxidation	Physiologic Value	$\dot{V}O_2$ (kcal/L)	VCO_2 (kcal/L)	RQ	OXYGEN (L/g)	CO_2 (L/g)
Carbohydrate	4.1	4.1	4	5.05	5.05	1.00	0.81	0.81
Protein	5.4	4.2	4	4.46	5.57	0.80	0.94	0.75
Fat	9.3	9.3	9	4.74	6.67	0.71	1.96	1.39
Ethanol	7.1	7.1	7	4.86	7.25	0.67	1.46	0.98

RQ, respiratory quotient; VCO_2, carbon dioxide production; $\dot{V}O_2$, oxygen consumption.

Data from Brown AC. Energy metabolism. In: Ruch TC, Patton HD, eds. Physiology and Biophysics III: Digestion, Metabolism, Endocrine Function and Reproduction. Philadelphia: WB Saunders, 1973:85–104, with permission.

five alternative methods, most of which are loosely derived from Atwater factors, to legally describe the energy content of a food. These methods are summarized in **Figure 5.2** and elsewhere.[81] The simplest method is to determine the macronutrient content of the food and apply the general 4, 4, and 9 kcal/g Atwater factors for the dietary energy contents of carbohydrate, protein, and fat, respectively, which accounts for estimated energy losses in stool and urine, based on data from a small number of human participants studied over 100 years ago. Of note, there is remarkable variability in calculated dietary energy across the different FDA methods. A high-fiber bread (illustrated in Figure 5.2) has an 18% difference in calculated dietary energy among methods, demonstrating that dietary energy is not the simple and accurate number it is widely assumed to be. The Food and Agriculture Organization of the United Nations has its own methods for calculating dietary energy[82] but, for the most part, these are based on the same Atwater factors used by the FDA.

Energy Intake

Several methods are used to assess dietary intake, including weighed or observed diet records, dietary recalls and diaries, image-assisted records, and food frequency questionnaires. It is now generally recognized that self-reported EI tends to underestimate actual EI.[83,84] Evidence of underreporting has been substantiated from measurements of TEE by the doubly labeled water (DLW) method, further discussed below.[54,85] Implausibly low EIs have been revealed when TEE was substantially greater than reported usual EI in weight-stable individuals. Underreporting of food intake is pervasive, ranging from 10%–45%, depending on the age, biological sex, and body composition of study participants[86] but there is still an important place for well-conducted dietary assessments in nutrition research that minimize error.[87] The 24-hour recall method has less variation and likely less underreporting than other methods,[88] making it the method of choice for energy intake assessment. Mobile phone-based apps are widespread and include features like food imaging and portion guides. A recent methodological advance includes using analysis of natural spoken language to automatically code food records without requirement for individuals to identify a suitable food code for each consumed food.[89] However, these newer assessment methods of EI require further refinement to improve accuracy and decrease user burden.[90]

Energy Expenditure

Methods for measuring EE include direct calorimetry, indirect calorimetry, and noncalorimetric methods.[50] Direct calorimetry is the measurement of the heat emitted from the body over a given period.[1,91] A direct calorimeter chamber measures heat loss by radiation, convection, conduction, and latent heat arising from vaporization of water.

Indirect calorimetry estimates heat production by measuring the respiratory gas exchange as an indirect measure of heat production. Indirect calorimetry measures oxygen consumption (VO_2) and carbon dioxide production (VCO_2), to derive the respiratory quotient (RQ), which is equal to the ratio of the VCO_2 to VO_2.[92] RQ reflects substrate use (i.e., the metabolic conversion and use of consumed energy as an oxidative process) in the fasted or post-prandial states. Substrate use can be determined from rates of $\dot{V}O_2$, VCO_2, and urinary nitrogen.[50,92] First, gas exchange must be corrected for the incomplete oxidation of protein. One gram of urinary nitrogen represents the combustion of an amount of protein that would require 5.92 L of oxygen and produce 4.75 L of CO_2. The $\dot{V}O_2$ and VCO_2 associated with the oxidized protein are subtracted from the total and are used to compute a nonprotein RQ (NPRQ). The amount of protein oxidized may be calculated directly from urinary nitrogen, assuming that 1 g of nitrogen represents 6.25 g of protein. The NPRQ is then used to calculate the proportions of carbohydrate and fat oxidized when the NPRQ is less than 1.0. When the NPRQ is greater than 1.0, net fat synthesis occurs and carbohydrate is used both for storage of energy and for oxidation.[92]

The complete oxidation of glucose results in an RQ equal to 1.0. The value of 1.0 reflects complete reliance on carbohydrate oxidation for metabolic needs. A RQ value equal to 0.7 indicates a state of pure fat oxidation. The complete oxidation of fat and protein results in an RQ averaging about 0.71 and 0.84, respectively, depending on the chemical structure of the foodstuff. Specific RQs for FFAs range from 0.69 to 0.81. RQs of amino acids range from 0.56 to 1.00, with conventional food proteins ranging from 0.81 to 0.87. In mixed diets, the RQ is approximately 0.85. Macronutrient balance is achieved when the RQ is equal to the food quotient (FQ), which represents the relative proportion of macronutrients in the diet. When energy balance prevails, the average RQ may be estimated from the composition of the diet by using the FQ.[93] Modifying FQ (e.g., exchanging dietary carbohydrate with fat while maintaining the same EI) will result in

FDA Methods for Calculation of Dietary Energy

How you calculate dietary energy using FDA-approved methods can result in very different calorie counts on the nutrition label.

Method 1: Specific Atwater Factors
Dietary energy (kcal/g) =
Carbohydrate g × A + Protein g × B + Fat g × C
A, B and C are the specific Atwater factors for metabolizable energy of macronutrients in different foods

Amount per serving
Calories 76

Method 2: General Atwater Factors
Dietary energy (kcal/g) =
Carbohydrate g × 4 + Protein g × 4 + Fat g × 9
4,4, and 9 are general Atwater factors for metabolizable energy in carbohydrate, protein and fat

Amount per serving
Calories 90

Method 3: General Atwater Factors plus Correction for Fiber
Dietary energy (kcal/g) =
(Carbohydrate g – non-digestible carbohydrate g) × 4 + Protein g × 4 + Fat g × 9 + Soluble non-digestible fiber g × 2 + Sugar alcohol g × D
Energy from fermentable fiber is 2 kcal/g, D is energy content of sugar alcohols

Amount per serving
Calories 80

Method 4: Modified Gold Standard Method
Dietary energy (kcal/g) = Gross food energy (kcal/g) – (protein g × 1.25)
1.25 is energy from protein lost in urine and stool; assumes complete digestibility of fat and carbohydrate

Method 5: Individualized Method
Dietary energy determined using specific food factors approved by the FDA

Nutrition Facts
25 slices

Serving size	**1 slice**
Amount per serving **Calories**	**?**
	% Daily Value*
Total Fat 8g	**10%**
Saturated Fat 1g	**5%**
Trans Fat 0g	
Cholesterol 0mg	**0%**
Sodium 160mg	**7%**
Total Carbohydrates 37g	**13%**
Dietary Fiber 4g	**14%**
Total Sugars 12g	
Includes 10g Added Sugars	**20%**
Protein 3g	
Vitamin D 2mg	10%
Calcium 260mg	20%
Iron 8g	45%
Potassium 240mg	6%

The % Daily Value (DV) tells you how much a nutrient in a serving of food contributes to a daily diet. 2,000 calories a day is used for general nutrition advice.

Figure 5.2 The US FDA allows dietary energy in Nutrition Fact panels on commercial foods to be calculated by any one of five different methods. As illustrated for a single food example (a whole-grain bread), these methods can give substantially different dietary energy values for the same micronutrient content. *FDA*, Food and Drug Administration.

a state of imbalance until the body's fuel selection adapts to the new FQ.[40]

Whole body (indirect) calorimeters, which are small rooms in which a study participant may reside comfortably for days to weeks, provide a controlled experimental environment to measure TEE and its components and substrate oxidation. In these rooms, the concentrations of O_2 and CO_2 and airflow through the system are monitored continuously.

Portable indirect calorimetric systems also have been devised to measure EE in field, clinical, and laboratory settings.[94-96] The Douglas bag method has been used historically for many measurements of BMR and RMR. With this method, all expired air is collected into a nonpermeable bag with a capacity up to 150 L. After a known period, the volume of air expired into the bag dries at standard temperature and pressure, and the concentrations of O_2 and CO_2 are measured, from which $\dot{V}O_2$, VCO_2, and RQ are calculated. Commercial metabolic carts in laboratory and clinical settings have largely replaced the Douglas bag method. For field measurements, several portable respirometers have been designed with oxygen analyzers as well as gas flow meters and electronics to process and store data. These systems require airtight masks, breathing valves, and nose clips to measure respiratory gas exchange quantitatively. Hood devices and canopies have been designed for more comfort, but they restrict movement.

Noncalorimetric methods to assess EE applicable to field conditions include DLW and heart rate (HR) monitoring. DLW is a stable (nonradioactive) isotope method that provides an estimate of TEE in free-living individuals. The DLW method was originally developed by Lifson et al. for use in small animals,[97,98] and it was later adapted for humans.[99,100] Two stable isotopic forms of water ($H_2{}^{18}O$ and ^2H_2O) are administered to the individual, and their ^{18}O and ^2H disappearance rates from the body are monitored for 7–21 days, equivalent to one to three half-lives for these isotopes. The disappearance rate of ^2H_2O reflects water flux, whereas the disappearance rate of $H_2{}^{18}O$ reflects water flux plus VCO_2, because of the rapid equilibration of body water and bicarbonate pools by carbonic anhydrase. The difference between the two disappearance rates is used to calculate VCO_2.

In weight-stable individuals, the DLW method may be used to assess energy requirements. If substantial gains or losses of body constituents are known to occur during the period of measurement (e.g., FFM loss or gain), appropriate adjustments must be made in estimating the RQ. Under field conditions, DLW is accurate within 5% or better. The advantage of this technique is the noninvasive way that it measures TEE. The disadvantages of this method are the high cost of ^{18}O, expensive, sophisticated mass spectrometric equipment, and the expertise required to measure ^{18}O and ^2H.

The HR monitoring method is based on the linear relationship between HR and EE.[101] Because of variations resulting from age, biological sex, body size, fitness, and nutritional status, the relationship must be calibrated on an individual basis. Simultaneous measurements of EE and HR are performed in individuals across a range of activities to calibrate estimates to the individual. Other confounding factors such as ambient conditions, time of day, emotional state, hydration status, food and caffeine intake, and smoking may influence the EE:HR relationship. As a result, HR data on individuals are subject to error and may yield unreliable estimates of EE. When applied to groups of individuals, the HR monitoring method provides an acceptable average estimate of TEE.

Energy Expenditure Estimating Equations

Weir developed the most widely used equation for estimating EE based on respiratory exchange as measured by indirect calorimetry[102]:

$$\text{EE (kcal)} = 3.941 \times \dot{V}O_2\,(L) + 1.106\, VCO_2\,(L) - (2.17 \times \text{UrN [g]}) \text{ or EE (kcal)} = 3.941 \times \dot{V}O_2\,(L) + 1\, VCO_2\,(L)/(1 + 0.082\,p)$$

where UrN is urinary nitrogen and p is the fraction of kcal resulting from protein. Assuming approximately 12.5% of total kcal will arise from protein; the prior equation can be reduced to the following:

$$\text{EE (kcal)} = 3.9 \times \dot{V}O_2\,(L) + 1.1\, VCO_2\,(L)$$

Because the calorimetric methods described above require physiologic measurements as well as technical skill, a variety of equations have been developed to estimate EE or its components (e.g., BEE) that rely on noninvasive and more easily measurable attributes, notably those of body size, biological sex, and age.

In 1919, Harris and Benedict[103] published prediction equations for BEE based on biological sex, height, age, and weight:

$$BMR_{females}\text{ (kcal/day)} = 665 + (9.6 \times \text{Weight [kg]}) + (1.8 \times \text{Height [cm]}) - (4.7 \times \text{Age [y]})$$

$$BMR_{males}\text{ (kcal/day)} = 66 + (13.7 \times \text{Weight [kg]}) + (5 \times \text{Height [cm]}) - (6.8 \times \text{Age [y]})$$

Subsequently, Schofield et al[104] derived widely used equations for predicting BEE using data on 7,549 persons. In the last few decades, equations for BEE from Mifflin-St. Jeor et al. and for TEE from the Institute of Medicine (now the National Academy of Medicine, NAM) have become more prevalent, displacing the use of several previously developed equations. The Mifflin-St. Jeor equation was developed in 1990 using BEE measured by indirect calorimetry in 498 healthy adults with normal weight or obesity, age 19–78 years.[105] The simplified version of the equation for females and males is:

$$REE_{females} = (10 \times \text{Weight [kg]}) + (6.25 \times \text{Height [cm]}) - (5 \times \text{Age [y]}) - 161$$

$$REE_{males} = (10 \times \text{Weight [kg]}) + (6.25 \times \text{Height [cm]}) - (5 \times \text{Age [y]}) + 5$$

The predictive value of these equations was not significantly affected by body-weight distributions, and they outperformed the Harris and Benedict equations, which were found to overestimate measured BEE by 5%. The authors also observed that FFM was the best single predictor of BEE.

Predictive equations continue to be developed (e.g., those from the World Health Organization) and assessed against prior iterations. The Mifflin-St. Jeor equation continues to be among the most accurate, using easily assessed measures against indirect calorimetry, although it and most others tend to lose accuracy among individuals with obesity.[106] Thus, in addition, equations for unique populations (e.g., obese, hospitalized, elderly) continue to be evaluated and refined.[106-110]

Finally, considerable effort by Hall and others has recognized that perturbations in weight result in, at least temporary and sometimes longer-term, changes in BEE that are differential from BEE in static weight conditions or those predicted for a given achieved weight following gain or loss.[40,41]

The IOM equation for TEE is based on the DLW method.[111,112] In the most recent NAM equations, a DLW database with 8600 values was compiled on TEE in children and adults.[112] Four PALs were defined to reflect inactive, low active, active, and very active levels of EE. Linear models were developed to predict TEE from age, biological sex, weight, height, and PAL category. The equations are given in **Table 5.2**. (See also, section below on **Dietary Reference Intakes: Estimated Energy Requirements** for details regarding the equations for estimating energy requirements).

Energy Requirements and Dietary Reference Intakes by Life Stage and Physiological Status

The estimated energy requirement (EER) is defined as the average dietary energy intake that is predicted to maintain energy balance in an adult of a defined age, biological sex, weight, height, and level of physical activity.[113] In children and pregnant and lactating women, the EER includes the needs associated with the deposition of tissues or the secretion of milk at rates consistent with health. Recommendations for the nutrient intakes of individuals

Table 5.2 Equations for Total Energy Expenditure for Biological Sex and Age Groupings and Different Physical Activity Levels

Biological Sex and Age Group	Physical Activity Level	Equation for Total Energy Expenditure
Males, ≥19 y	Inactive	753.07 – (10.83 × Age [y]) + (6.50 × Height [cm]) + (14.10 × Weight [kg])
	Low active	581.47 – (10.83 × Age [y]) + (8.30 × Height [cm]) + (14.94 × Weight [kg])
	Active	1,004.82 – (10.83 × Age [y]) + (6.52 × Height [cm]) + (15.91 × Weight [kg])
	Very active	−517.88 – (10.83 × Age [y]) + (15.61 × Height [cm]) + (19.11 × Weight [kg])
Females, ≥19 y	Inactive	584.90 – (7.01 × Age [y]) + (5.72 × Height [cm]) + (11.71 × Weight [kg])
	Low active	575.77 – (7.01 × Age [y]) + (6.60 × Height [cm]) + (12.14 × Weight [kg])
	Active	710.25 – (7.01 × Age [y]) + (6.54 × Height [cm]) + (12.34 × Weight [kg])
	Very active	511.83 – (7.01 × Age [y]) + (9.07 × Height [cm]) + (12.56 × Weight [kg])
Males, 3–18 y	Inactive	−447.51 + (3.68 × Age [y]) + (13.01 × Height [cm]) + (13.15 × Weight [kg])
	Low active	19.12 + (3.68 × Age [y]) + (8.62 × Height [cm]) + (20.28 × Weight [kg])
	Active	−388.19 + (3.68 × Age [y]) + (12.66 × Height [cm]) + (20.46 × Weight [kg])
	Very active	−671.75 + (3.68 × Age [y]) + (15.38 × Height [cm]) + (23.25 × Weight [kg])
Females, 3–18 y	Inactive	55.59 – (22.25 × Age [y]) + (8.43 × Height [cm]) + (17.07 × Weight [kg])
	Low active	−297.54 – (22.25 × Age [y]) + (12.77 × Height [cm]) + (14.73 × Weight [kg])
	Active	−189.55 – (22.25 × Age [y]) + (11.74 × Height [cm]) + (18.34 × Weight [kg])
	Very active	−709.59 – (22.25 × Age [y]) + (18.22 × Height [cm]) + (14.25 × Weight [kg])
Males, 0–2 y	—	−716.45 – (1.00 × Age [y]) + (17.82 × Height [cm]) + (15.06 × Weight [kg])
Females, 0–2 y	—	−69.15 + (80.0 × Age [y]) + (2.65 × Height [cm]) + (54.15 × Weight [kg])
Pregnant females in their second and third trimester of pregnancy	Inactive	1,131.20 – (2.04 × Age [y]) + (0.34 × Height [cm]) + (12.15 × Weight [kg]) + (9.16 × Gestation [wk])
	Low active	693.35 – (2.04 × Age [y]) + (5.73 × Height [cm]) + (10.20 × Weight [kg]) + (9.16 × Gestation [wk])
	Active	−223.84 – (2.04 × Age [y]) + (13.23 × Height [cm]) + (8.15 × Weight [kg]) + (9.16 × Gestation [wk])
	Very active	−779.72 – (2.04 × Age [y]) + (18.45 × Height [cm]) + (8.73 × Weight [kg]) + (9.16 × Gestation [wk])

Adapted with permission from National Academies of Sciences, Engineering, and Medicine. Dietary Reference Intakes for Energy. Washington, DC: The National Academies Press, 2023 (123).

are generally set to provide enough to meet or exceed the requirements of 95% of healthy persons in each biological sex–age group, and enough to allow reasonably fast recovery of losses that may occur. This approach is reasonable for nutrients for which modest excess intakes present no health risks. However, excess EI is eventually deposited in the form of body fat, which provides a means to maintain metabolism during periods of limited food intake, but can result in obesity.

This definition implies that desirable EI should support healthy body weights and composition, as well as adequate PALs. Although it is possible to maintain energy balance and avoid excess weight gain by reducing dietary EI only, optimizing both EI *and* EE has important advantages. First, some evidence suggests that the ability to control food intake may be reduced at very low PALs.[114] Second, marked reductions in food intake may make it difficult to fulfill the requirements for essential nutrients such as vitamins and minerals. Implicit in this statement is that recommended EI for persons with obesity is less than their EE, under the assumption that lowering body weight is associated with better health status. For underweight persons, conversely, recommended EI are greater than EE to permit weight gain and maintenance of a higher body weight. Unlike other nutrients, body weight can be used to monitor the adequacy or inadequacy of habitual EI. Body weight provides a readily monitored and noninvasive indicator of the adequacy or inadequacy of habitual EI. Chronic energy deficiency or energy excess will eventually manifest as wasting or obesity. Weight-for-height indices and body mass index (BMI) are used to assess the weight status of individual persons as well as population groups.[115,116]

The factorial method has been used historically to assess energy requirements.[117,118] In this approach, TEE is estimated from BEE and activity EE derived from the time devoted to different activities and the energy costs of each activity. Limitations of this approach include the accuracy of the BEE predictions, data availability of energy costs of all activities, and the difficulty in estimating random, spontaneous movement. In addition, compared with the currently used DLW method and its expansive database of TEE measurements, the factorial method has been found to give significantly higher estimates of TEE.[119,120]

Dietary Reference Intakes: Estimated Energy Requirements

Dietary reference intakes (DRIs) are published by the Food and Nutrition Board of the National Academies of Sciences, Engineering and Medicine and are intended for persons in the United States and Canada. The 2005 edition of the DRIs developed the methodology of predicting energy requirements using DLW data[111] and the approach was reapplied with a larger dataset in the new 2023 DRIs.[112] Their equations are based on the principle that energy requirements are equal to TEE (derived from the DLW method for individuals leading their normal daily life), plus an allowance for growth in the case of children, energy deposition in pregnancy and energy excretion in lactation,[111] as well as varying PALs. After TEE estimation, the EER equations were derived for specific biological sex-age groups. The 2023 DRI equations for estimating energy requirements[112] are presented in **Table 5.3**, and the following sections discuss research for different physiologic groups.

Table 5.3 Equations for Estimated Energy Requirements from Biological Sex and Age Groupings and Different Physical Activity Levels, Together with Energy Deposition for Relevant Groups

Biological Sex	Age Group	Physical Activity Level	Equation For Estimated Energy Requirement (Kcal/D)
Male	0–2.99 mo	—	−716.45 – (1.00 × Age [y]) + (17.82 × Height [cm]) + (15.06 × Weight [kg]) + 200
Female	0–2.99 mo	—	−69.15 + (80.0 × Age [y]) + (2.65 × Height [cm]) + (54.15 × Weight [kg]) + 180
Male	3–5.99 mo	—	−716.45 – (1.00 × Age [y]) + (17.82 × Height [cm]) + (15.06 × Weight [kg]) + 50
Female	3–5.99 mo	—	−69.15 + (80.0 × Age [y]) + (2.65 × Height [cm]) + (54.15 × Weight [kg]) + 60
Male	6 mo–2.99 y	—	−716.45 – (1.00 × Age [y]) + (17.82 × Height [cm]) + (15.06 × Weight [kg]) + 20
Female	6–11.99 mo	—	−69.15 + (80.0 × Age [y]) + (2.65 × Height [cm]) + (54.15 × Weight [kg]) + 20
Female	12–35.99 mo	—	−69.15 + (80.0 × Age [y]) + (2.65 × Height [cm]) + (54.15 × Weight [kg]) + 15
Male	3 y	Inactive	−447.51 + (3.68 × Age [y]) + (13.01 × Height [cm]) + (13.15 × Weight [kg]) + 20
		Low active	19.12 + (3.68 × Age [y]) + (8.62 × Height [cm]) + (20.28 × Weight [kg]) + 20
		Active	−388.19 + (3.68 × Age [y]) + (12.66 × Height [cm]) + (20.46 × Weight [kg]) + 20
		Very active	−671.75 + (3.68 × Age [y]) + (15.38 × Height [cm]) + (23.25 × Weight [kg]) + 20
Male	4–8 y	Inactive	−447.51 + (3.68 × Age [y]) + (13.01 × Height [cm]) + (13.15 × Weight [kg]) + 15
		Low active	19.12 + (3.68 × Age [y]) + (8.62 × Height [cm]) + (20.28 × Weight [kg]) + 15
		Active	−388.19 + (3.68 × Age [y]) + (12.66 × Height [cm]) + (20.46 × Weight [kg]) + 15
		Very active	−671.75 + (3.68 × Age [y]) + (15.38 × Height [cm]) + (23.25 × Weight [kg]) + 15

Biological Sex	Age Group	Physical Activity Level	Equation For Estimated Energy Requirement (Kcal/D)
Male	9–13 y	Inactive	−447.51 + (3.68 × Age [y]) + (13.01 × Height [cm]) + (13.15 × Weight [kg]) + 25
		Low active	19.12 + (3.68 × Age [y]) + (8.62 × Height [cm]) + (20.28 × Weight [kg]) + 25
		Active	−388.19 + (3.68 × Age [y]) + (12.66 × Height [cm]) + (20.46 × Weight [kg]) + 25
		Very active	−671.75 + (3.68 × Age [y]) + (15.38 × Height [cm]) + (23.25 × Weight [kg]) + 25
Female	3–8 y	Inactive	55.59 – (22.25 × Age [y]) + (8.43 × Height [cm]) + (17.07 × Weight [kg]) + 15
		Low active	−297.54 – (22.25 × Age [y]) + (12.77 × Height [cm]) + (14.73 × Weight [kg]) + 15
		Active	−189.55 – (22.25 × Age [y]) + (11.74 × Height [cm]) + (18.34 × Weight [kg]) + 15
		Very active	−709.59 – (22.25 × Age [y]) + (18.22 × Height [cm]) + (14.25 × Weight [kg]) + 15
Female	9–13 y	Inactive	55.59 – (22.25 × Age [y]) + (8.43 × Height [cm]) + (17.07 × Weight [kg]) + 30
		Low active	−297.54 – (22.25 × Age [y]) + (12.77 × Height [cm]) + (14.73 × Weight [kg]) + 30
		Active	−189.55 – (22.25 × Age [y]) + (11.74 × Height [cm]) + (18.34 × Weight [kg]) + 30
		Very active	−709.59 – (22.25 × Age [y]) + (18.22 × Height [cm]) + (14.25 × Weight [kg]) + 30
Male	14–18.99 y	Inactive	−447.51 + (3.68 × Age [y]) + (13.01 × Height [cm]) + (13.15 × Weight [kg]) + 20
		Low active	19.12 + (3.68 × Age [y]) + (8.62 × Height [cm]) + (20.28 × Weight [kg]) + 20
		Active	−388.19 + (3.68 × Age [y]) + (12.66 × Height [cm]) + (20.46 × Weight [kg]) + 20
		Very active	−671.75 + (3.68 × Age [y]) + (15.38 × Height [cm]) + (23.25 × Weight [kg]) + 20
Female	14–18.99 y	Inactive	55.59 – (22.25 × Age [y]) + (8.43 × Height [cm]) + (17.07 × Weight [kg]) + 20
		Low active	−297.54 – (22.25 × Age [y]) + (12.77 × Height [cm]) + (14.73 × Weight [kg]) + 20
		Active	−189.55 – (22.25 × Age [y]) + (11.74 × Height [cm]) + (18.34 × Weight [kg]) + 20
		Very active	−709.59 – (22.25 × Age [y]) + (18.22 × Height [cm]) + (14.25 × Weight [kg]) + 20
Male	≥19 y	Inactive	753.07 – (10.83 × Age [y]) + (6.50 × Height [cm]) + (14.10 × Weight [kg])
		Low active	581.47 – (10.83 × Age [y]) + (8.30 × Height [cm]) + (14.94 × Weight [kg])
		Active	1,004.82 – (10.83 × Age [y]) + (6.52 × Height [cm]) + (15.91 × Weight [kg])
		Very active	−517.88 – (10.83 × Age [y]) + (15.61 × Height [cm]) + (19.11 × Weight [kg])
Female	≥19 y	Inactive	584.90 – (7.01 × Age [y]) + (5.72 × Height [cm]) + (11.71 × Weight [kg])
		Low active	575.77 – (7.01 × Age [y]) + (6.60 × Height [cm]) + (12.14 × Weight [kg])
		Active	710.25 – (7.01 × Age [y]) + (6.54 × Height [cm]) + (12.34 × Weight [kg])
		Very active	511.83 – (7.01 × Age [y]) + (9.07 × Height [cm]) + (12.56 × Weight [kg])
Female	2nd and 3rd trimester of pregnancy	Inactive	1,131.20 – (2.04 × Age [y]) + 0.34 × Height [cm]) + (12.15 × Weight [kg]) + (9.16 × Gestation [wk]) + Energy deposition [kcal/d]
		Low active	693.35 – (2.04 × Age [y]) + (5.73 × Height [cm]) + (10.20 × Weight [kg]) + (9.16 × Gestation [wk]) + Energy deposition [kcal/d]
		Active	−223.84 – (2.04 × Age [y]) + (13.23 × Height [cm]) + (8.15 × Weight [kg]) + (9.16 × Gestation [wk]) + Energy deposition [kcal/d]
		Very active	−779.72 – (2.04 × Age [y]) + (18.45 × Height [cm]) + (8.73 × Weight [kg]) + (9.16 × Gestation [wk]) + Energy deposition [kcal/d]

Energy deposition (kcal/d) in pregnant females is estimated for underweight (BMI <18.5 kg/m^2) females as +300 kcal/d; for normal weight (BMI 18.5–24.9 kg/m^2) females as +200 kcal/d; for overweight (BMI 25–29.9 kg/m^2) females as +150 kcal/d; and for obese (BMI ≥30 kg/m^2) females as −50 kcal/d.

Adapted with permission from National Academies of Sciences, Engineering, and Medicine. Dietary Reference Intakes for Energy. Washington, DC: The National Academies Press, 2023.[113]

Growth (Infants and Children)

In infants and children, energy requirement includes the energy required for the deposition of tissues. The energy requirement for growth relative to maintenance is low, except for the first months of life. As a percentage of total energy requirements, the energy cost of growth decreases from 35% at 1 month to 3% at 12 months of age, and it remains low until puberty, at which time it increases to 4%.[60] During childhood, girls grow slightly more slowly than boys, and girls have slightly more body fat. During adolescence, the biological sex differences in body composition are accentuated.[121-124] Adolescence in boys is characterized by rapid acquisition of FFM, a modest increase in FM in early puberty, followed by a decline. Adolescence in girls is characterized by a modest increase in FFM and continual FM accumulation.

The energy requirements of infants and young children should balance EE at PALs conducive to normal development and should allow for deposition of tissues at rates consistent with health. Because of the dominant contribution of the brain to energy needs (60% to 70%), BEE is highest during the first years of life.[125] The BEE of term infants ranges from 43 to 60 kcal/kg/day, or two to three times more than in adults.[104] BEE and TEE are influenced by age (older greater than younger), biological sex (males greater than females), and feeding mode (breastfed less than formula-fed infants).[60]

Energy requirements of older children and adolescents are defined to promote normal growth and maturation and to support a desirable PAL consistent with health. Energy requirements of children and adolescents are highly variable because of differences in growth rate and physical activity. Mean PALs, estimated by DLW, HR monitoring, time-motion/diary, and time allocation records, range from 1.3 to 1.5 for children less than 5 years of age and from 1.5 to 1.9 for children 6–18 years who are living in urban, industrialized settings.[126] Although absolute EE increases with age, weight-specific EE decreases across adolescence, primarily because of the decrease in BEE.

Haschke[127] estimated changes in body composition during adolescence from literature values of total body water, potassium, and calcium. FFM increases in boys, with peak deposition coinciding with peak rates of height gains. The percentage of FM increases during this period in girls, and it declines in boys.

The energy cost of growth is more accurately estimated from the individual costs of protein and fat deposition because the composition of weight gain varies with age. The energy cost of growth ranges from 2.4–6.0 kcal/g (10–25 kJ/g), depending on the composition of the tissues deposited.[128,129] For the DRI, the energy cost of growth is estimated to be 175 kcal/day for the age interval 0–3 months, 60 kcal/day for 4–6 months, and 20 kcal/day for 7–35 months. Although the composition of newly synthesized tissues varies in childhood and adolescence, these variations have a minor impact on total energy requirements, because only approximately 20–25 kcal/day are estimated to be required for growth.

Adults

In weight-stable adults, energy requirements are equal to TEE. As noted, the DLW database was used to derive separate TEE predictive equations for men and women, based on age, height, weight, and PAL category. The age-related decline in TEE was found to amount to approximately 11 and 7 kcal/year for men and women, respectively.[112] Although a separate analysis of compilations of DLW data suggested that TEE does not decrease progressively with age, but instead remains relatively constant to age 63 after which a decline is noted,[28] that report has been replaced by the new DRIs,[112] which were derived from a larger and more representative database. These different results highlight the importance of continuing to prioritize collection of TEE data from normative populations to update future dietary energy requirements. In all these analyses, however, marked variation is apparent in PALs, which depend on the occupational and recreational lifestyles of adults.

Pregnancy and Lactation

The additional energy requirements of pregnancy include increased BEE and the energy costs of physical activity and energy deposition in maternal and fetal tissues. The BEE increases as a result of the metabolic contribution of the uterus and fetus and the increased internal work of the maternal heart and lungs.[130] In the latest 2023 *Dietary Reference Intakes for Energy*, the total energy requirement considers activity level, age, height and weight, and during the second and third trimesters, there is also an increment for additional energy requirements for pregnancy beyond those associated with maternal body weight, which are estimated at 9.16 kcal/day × weeks gestation. This new approach reflects a progressive increase in energy requirements starting in the second trimester, due to needs for the energy deposition in the fetus, placenta, amniotic fluid, uterus, breasts, blood, extracellular fluid, and in maternal adipose tissue stored in preparation for lactation.

Recommended weight gain across pregnancy is now based on maternal pregravid weight. For women with BMI in the normal range (18.5–24.9 kg/m^2) prior to pregnancy, the recommended weight gain is 11.5–16 kg; for those with underweight (BMI <18.5 kg/m^2), 12.5–18 kg; for those with overweight (BMI 25.0 ≤ 30.0 kg/m^2), 7–11 kg, and for those with obesity (BMI ≥ 30 kg/m^2), 5–9 kg.[131]

The energy cost of lactation is estimated from milk production rates and the energy density of human milk.[112,132-135] Milk production rates averaged 0.73 L/day from 0–2 months postpartum, increasing to 0.83 L/day at 5 months and decreasing to 0.57 L/day at 7–12 months. Considering the energy content of breast milk, and energy mobilized from maternal tissue stores that subsidizes the energy cost of lactation, the net additional energy requirement during lactation is estimated to be 353 kcal/day during 0–2 months, 385 kcal/day during 3–4 months, 416 kcal/day at 5 months, 462 kcal/day at 6 months, and 381 kcal/day at 7–12 months. Gradual weight loss during lactation is, therefore, anticipated.[131]

Older Adults

Aging is characterized by the steady loss of FFM and FM, as well as functional and physiologic deterioration of almost all body compartments. Shifts in appetite and other hormonal and metabolic control mechanisms further affect both EI and EE, with malnutrition a common byproduct of the aging process.

Recent research using DLW, conducted in 6,421 participants (64% female, age 8 days to 95 years) from 29 countries, suggested a distinct metabolic inflection point at approximately age 60 years.[28] At age 60 years, notable declines in both EE and BEE were reported that are not solely accounted for by FFM or FM; regression estimates indicate that TEE declines by 0.7% per year beginning at 60 years, and BEE declines similarly. However, the 2023 DRIs used a larger database of 8600 DLW data points and found a progressive decrease in energy requirements over the adult lifespan (approximately 11 kcal/year for males and 7 kcal/year for females) even if physical activity category remains constant. Thus, even allowing for maintenance of activity, by age 90, men are expected to have a typical decrease in energy requirements of nearly 800 kcal/day compared with age 20, and women a decrease of 500 kcal/day. Decreases in both BEE and energy expenditure for physical activity are expected for older adults, but the exact contributions of these components of EE to decreasing energy requirements is not well characterized.

Therefore, two specific goals of ongoing research and policy include improving the accuracy of EER equations in this age group and assuring that EI is comparable to requirements while also maintaining adequate nutrient density, which becomes increasingly important in older age. To meet the same or increased recommended intakes of macro- and micronutrients in a diet with declining levels of energy needs, the proportion of nutrient-dense foods in the diet must continue to increase as adults age, with a concomitant decrease in the quantity of foods consumed of lower nutrient density. The diet must become increasingly healthier in older age. In fact, it is estimated that the dietary density of some micronutrients must increase in older adults by 50%, and up to nearly 100% for nutrients that are required in greater absolute amounts.[136]

Acknowledgments

This authors acknowledge the authors of the previous version of this chapter, Nancy Butte and Benjamin Caballero, for their contributions.

References

1. Kleiber M. Rev. ed. Huntington, N.Y: R. E. Krieger Pub. Co; 1975.
2. Kinney J. In: Kinney J, Jeejeebhoy K, Hill G, Owen O, eds. Nutrition and Metabolism in Patient Care. Philadelphia: WB Saunders; 1988:3–34.
3. Brown A. In: Ruch T, Patton H, eds. Physiology and Biophysics III: Digestion, Metabolism, Endocrine Function and Reproduction. Philadelphia: WB Saunders; 1973:85–104.
4. Flatt J. In: Garrow J, Halliday D, eds. Substrate and Energy Metabolism in Man. London: John Libbey; 1985:58–69.
5. Flatt J. Recent Adv Obes Res 1978;2:211–28.
6. Flatt JP. Diabetes Metab Rev 1988;4:571–81.
7. Flatt J. Am J Clin Nutr 1995;62:820–36.
8. Flatt J. Am J Clin Nutr 1987;45:296–306.
9. Consolazio C, Johnson R, Pecora L. Physiological Measurements of Metabolic Functions in Man. New York: McGraw-Hill; 1963:313–25.
10. Flatt J, Tremblay A. In: Bray G, Bouchard C, James W, eds. Handbook of Obesity. New York: Marcel Dekker; 1998.
11. Elia M. In: Garrow J, James W, Ralph A, eds. Human Nutrition and Dietetics. Edinburgh: Churchill Livingstone; 2000:37–59.
12. Siler SQ, Neese RA, Hellerstein MK. Am J Clin Nutr 1999;70: 928–36.
13. Sayon-Orea C, Martinez-Gonzalez MA, Bes-Rastrollo M. Nutr Rev 2011;69:419–31.
14. Acheson KJ, Schutz Y, Bessard T, et al. 1984;246:E62–70.
15. Hellerstein MK. Eur J Clin Nutr 1999;53 Suppl 1:S53–65.
16. Wolfe R. In: Muller M, Danforth E, Burger A, eds. Hormones and Nutrition in Obesity and Cachexia. New York: Springer; 1990.
17. Cypess AM, Lehman S, Williams G, et al. N Engl J Med 2009;360:1509–17.
18. Garby L, Kurzer MS, Lammert O, et al. Hum Nutr Clin Nutr 1987;41:225–33.
19. Nelson KM, Weinsier RL, Long CL, et al. Am J Clin Nutr 1992;56:848–56.
20. Weinsier RL, Schutz Y, Bracco D. Am J Clin Nutr 1992;55:790–94.
21. Elia M. In: Kinney JM, Tucker HN, eds. Energy Metabolism: Tissue Determinants and Cellular Corollaries. New York: Raven Press; 1992:61–79.
22. Gallagher D, Belmonte D, Deurenberg P, et al. Am J Physiol-Endocrinol Metab 1998;275:E249–58.
23. Hsu A, Heshka S, Janumala I, et al. Am J Clin Nutr 2003;77: 1506–11.
24. Javed F, He Q, Davidson LE, et al. Am J Clin Nutr 2010;91: 907–12.
25. Wang Z, Heshka S, Heymsfield SB, et al. Am J Clin Nutr 2005;81:799–806.
26. Jones A, Shen W, St-Onge M-P, et al. Am J Clin Nutr 2004;79: 780–86.
27. Keys A, Simonson E, Sturgeon Skinner A, et al. Minneapolis: University of Minnesota Press; 1950.
28. Pontzer H, Yamada Y, Sagayama H, et al. Science 2021;373: 808–12.
29. Bisdee JT, James WPT, Shaw MA. Br J Nutr 1989;61:187–99.
30. Solomon SJ, Kurzer MS, Calloway DH. Am J Clin Nutr 1982;36:611–16.
31. Halsey LG, Careau V, Pontzer H, et al. J Hum Evol 2022;171:103229.
32. Albu J, Shur M, Curi M, et al. Am J Clin Nutr 1997;66:531–38.
33. Carpenter WH, Fonong T, Toth MJ, et al. Am J Physiol-Endocrinol Metab 1998;274:E96–101.
34. Foster GD, Wadden TA, Vogt RA. Obes Res 1997;5:1–8.
35. Jakicic JM, Wing RR. Int J Obes Relat Metab Disord J Int Assoc Study Obes 1998;22:236–42.
36. Zemel MB, Sun X, Sobhani T, et al. Am J Clin Nutr 2010;91: 16–22.
37. Treuth MS, Butte NF, Wong WW. Am J Clin Nutr 2000;71: 893–900.
38. Wong WW, Butte NF, Ellis KJ, et al. J Clin Endocrinol Metab 1999;84:906–11.
39. Yanovski SZ, Reynolds JC, Boyle AJ, et al. Obes Res 1997;5: 321 25.
40. Hall KD. Am J Physiol-Endocrinol Metab 2010;298:E449–66.
41. Hall KD, Sacks G, Chandramohan D, et al. Lancet Lond Engl 2011;378:826–37.
42. Redman LM, Smith SR, Burton JH, et al. Cell Metab 2018;27: 805–15.e4.
43. Ravussin E, Redman LM, Rochon J, et al. J Gerontol A Biol Sci Med Sci 2015;70:1097–104.
44. Fothergill E, Guo J, Howard L, et al. Obes Silver Spring Md 2016;24:1612–19.
45. Das SK, Roberts SB, McCrory MA, et al. Am J Clin Nutr 2003;78:22–30.
46. Westerterp KR. Nutr Metab 2004;1:5.
47. de Jonge L, Bray GA. Obes Res 1997;5:622–31.
48. Tai MM, Castillo P, Pi-Sunyer FX. Am J Clin Nutr 1991;54: 783–87.
49. Thearle MS, Pannacciulli N, Bonfiglio S, et al. J Clin Endocrinol Metab 2013;98:2791–99.

50. Jequier E, Acheson K, Schutz Y. Annu Rev Nutr 1987;7:187–208.
51. van Marken Lichtenbelt WD, Schrauwen P. Am J Physiol-Regul Integr Comp Physiol 2011;301:R285–96.
52. Blaak EE, van Baak MA, Saris WH. J Clin Endocrinol Metab 1999;84:3764–69.
53. Westerterp KR, Yamada Y, Sagayama H, et al. Am J Clin Nutr 2021;114:1583–89.
54. Goldberg GR, Black AE, Jebb SA, et al. Eur J Clin Nutr 1991;45:569–81.
55. Warwick PM. In: Truswell AS, Dreosti IE, English RM, eds. Recommended Nutrient Intakes: Australian Papers. Sydney, Australia: Australian Professional Publications; 1990:295–320.
56. Ainsworth BE, Haskell WL, Leon AS, et al. Med Sci Sports Exerc 1993;25:71–80.
57. Butte NF, Watson KB, Ridley K, et al. Med Sci Sports Exerc 2018;50:246–56.
58. Pahud P, Ravussin E, Jequier E. J Appl Physiol 1980;48:770–5.
59. Graham TE, Adamo KB. Can J Appl Physiol Rev Can Physiol Appl 1999;24:393–415.
60. Butte NF, Wong WW, Hopkinson JM, et al. Am J Clin Nutr 2000;72:1558–69.
61. Bahr R, Ingnes I, Vaage O, et al. J Appl Physiol Bethesda Md 1985 1987;62:485–90.
62. Brooks GA, Fahey TD, White TP, Baldwin KM, eds. Exercise physiology: human bioenergetics and its applications. 3rd ed. Mountain View, California: Mayfield Pub; 2000.
63. Hargreaves M, Spriet LL. Cold Spring Harb Perspect Med 2018;8:a029744.
64. Brooks GA, Mercier J. J Appl Physiol Bethesda Md 1985 1994;76:2253–61.
65. van Loon LJ, Greenhaff PL, Constantin-Teodosiu D, et al. J Physiol 2001;536:295–304.
66. Saito M, Matsushita M, Yoneshiro T, et al. Front Endocrinol 2020;11:222.
67. Ho KKY. J Endocrinol 2018;238:R185–91.
68. Ricquier D. Int J Obes 2005 2010;34 Suppl 1:S3-6.
69. Nedergaard J, Bengtsson T, Cannon B. Am J Physiol-Endocrinol Metab 2007;293:E444–52.
70. Zingaretti MC, Crosta F, Vitali A, et al. FASEB J 2009;23: 3113–20.
71. Cannon B, Nedergaard J. Int J Obes 2005 2010;34 Suppl 1: S7–16.
72. Acheson KJ, Zahorska-Markiewicz B, Pittet P, et al. Am J Clin Nutr 1980;33:989–97.
73. Garrow J, Webster J. In: van Es A, ed. Human Energy Metabolism. Wageningen, Netherlands: Agricultural University; 1985.
74. Warwick PM, Chapple RS, Thomson ES. Int J Obes 1987;11: 229–37.
75. Dallosso HM, James WP. Int J Obes 1984;8:365–75.
76. Hofstetter A, Schutz Y, Jéquier E, et al. N Engl J Med 1986;314: 79–82.
77. Blaxter KL. Energy metabolism in animals and man. Cambridge [England]; New York: Cambridge University Press; 1989.
78. Roberts SB, Young VR, Fuss P, et al. Am J Physiol 1990;259: R461–9.
79. Hill JO, Wyatt HR, Peters JC. Eur Endocrinol 2013;9:111–15.
80. Heyman MB, Young VR, Fuss P, et al. Am J Physiol 1992;263: R250–57.
81. Roberts SB, Flaherman V. Adv Nutr 2022:nmac092.
82. MacLean WC, Warwick P. In: Food and Agriculture Organization of the United Nations, eds. Calculation of the Energy Content of Foods - Energy Conversion Factors. Food Energy: Methods of Analysis and Conversion Factors: Report of a Technical Workshop, Rome, 3-6 December 2002. FAO food and nutrition paper. Rome: Food and Agriculture Organization of the United Nations, 2003.
83. Black AE, Prentice AM, Goldberg GR, et al. J Am Diet Assoc 1993;93:572–9.
84. Dhurandhar NV, Schoeller D, Brown AW, et al. Int J Obes 2005 2015;39:1109–13.
85. Schoeller DA. Metabolism 1995;44:18–22.
86. Johnson RK, Soultanakis RP, Matthews DE. J Am Diet Assoc 1998;98:1136–40.
87. Subar AF, Freedman LS, Tooze JA, et al. J Nutr 2015;145: 2639–45.
88. Burrows TL, Ho YY, Rollo ME, et al. Front Endocrinol 2019;10:850.
89. Taylor S, Korpusik M, Das S, et al. J Med Internet Res 2021;23:e26988.
90. Das SK, Miki AJ, Blanchard CM, et al. Adv Nutr 2022;13: 1–15.
91. Holmes FL. Lavoisier and the chemistry of life: an exploration of scientific creativity. Wisconsin publications in the history of science and medicine. Madison, Wis: University of Wisconsin Press; 1985.
92. Livesey G, Elia M. Am J Clin Nutr 1988;47:608–28.
93. Black AE, Prentice AM, Coward WA. Hum Nutr Clin Nutr 1986;40:381–91.
94. Webb P. Human calorimeters. Endocrinology and metabolism series. New York: Praeger; 1985.
95. McLean JA. Animal and human calorimetry. Cambridge [Cambridgeshire] ; New York: Cambridge University Press; 1987.
96. Murgatroyd PR, Shetty PS, Prentice AM. Int J Obes Relat Metab Disord J Int Assoc Study Obes 1993;17:549–68.
97. Lifson N, McClintock R. J Theor Biol 1966;12:46–74.
98. Lifson N, Gordon GB, McClintock R. J Appl Physiol 1955;7: 704–10.
99. Schoeller DA, van Santen E. J Appl Physiol 1982;53:955–9.
100. Schoeller DA, Leitch CA, Brown C. Am J Physiol-Regul Integr Comp Physiol 1986;251:R1137–43.
101. Schutz Y, Weinsier RL, Hunter GR. Obes Res 2001;9:368–79.
102. Weir JB de V. J Physiol 1949;109:1–9.
103. Harris JA, Benedict FG. A biometric study of basal metabolism in man. Washington, DC: Carnegie Institution; 1919.
104. Schofield WN, Schofield C, James WPT. Human nutrition Clinical nutrition. London: Libbey; 1985.
105. Mifflin MD, St Jeor ST, Hill LA, et al. Am J Clin Nutr 1990;51:241–7.
106. Frankenfield DC. Clin Nutr Edinb Scotl 2013;32:976–82.
107. Noreik M, Maurmann M, Meier V, et al. Exp Gerontol 2014;59:47–50.
108. Ocagli H, Lanera C, Azzolina D, et al. Nutrients 2021;13:458.
109. Ndahimana D, Go N-Y, Ishikawa-Takata K, et al. Nutr Res Pract 2019;13:256–62.
110. Vinken AG, Bathalon GP, Sawaya AL, et al. Am J Clin Nutr 1999;69:920–6.
111. Panel on Macronutrients, Panel on the Definition of Dietary Fiber, Subcommittee on Upper Reference Levels of Nutrients, et al. Dietary Reference Intakes for Energy, Carbohydrate, Fiber, Fat, Fatty Acids, Cholesterol, Protein, and Amino Acids. Washington, D.C.: National Academies Press; 2005:10490.
112. Committee on the Dietary Reference Intakes for Energy, Food and Nutrition Board, Health and Medicine Division, National Academies of Sciences, Engineering, and Medicine. Dietary Reference Intakes for Energy. Washington, D.C.: National Academies Press; 2023:26818.
113. Institute of Medicine. Dietary Reference Intakes: The Essential Guide to Nutrient Requirements. Washington, D.C.: National Academies Press; 2006:11537.
114. Stubbs RJ, Hughes DA, Johnstone AM, et al. Am J Clin Nutr 2004;79:62–69.

115. World Health Organization. Obesity: Preventing and Managing the Global Epidemic. Report of a World Health Organization Consultation on Obesity. 1998:1–276.
116. Kuczmarski RJ, Ogden CL, Grummer-Strawn LM, et al. CDC growth charts: United States. Adv Data 2000:1–27.
117. National Research Council (US) committee on the Tenth Edition of the Recommended Dietary Allowances 10. rev. ed., Washington: National Academy Press; 1989.
118. Dewey KG, Finley DA, Lönnerdal B. J Pediatr Gastroenterol Nutr 1984;3:713–20.
119. Haggarty P, McNeill G, Manneh MK, et al. Br J Nutr 1994;72: 799–813.
120. Jones PJ, Martin LJ, Su W, et al. Can J Public Health Rev Can Sante Publique 1997;88:314–9.
121. Ellis KJ. Am J Clin Nutr 1997;66:1323–31.
122. Ellis KJ, Abrams SA, Wong WW. Am J Clin Nutr 1997;65: 724–31.
123. Forbes GB. Human body composition: growth, aging, nutrition, and activity. New York: Springer-Verlag; 1987.
124. Tanner JM. Growth at Adolescence: With a General Consideration of the Effects of Hereditary and Environmental Factors Upon Growth and Maturation from Birth to Maturity. 2nd ed. Oxford: Blackwell Scientific Publications; 1962.
125. Holliday MA, Potter D, Jarrah A, et al. Pediatr Res 1967;1: 185–95.
126. Torun B, Davies PS, Livingstone MB, et al. Eur J Clin Nutr 1996;50 Suppl 1:S37 80; discussion S80–1.
127. Haschke F. In: Klish WJ, Kretchmer N, eds. Body composition measurements in infants and children: report of the 98th Ross Conference on Pediatric Research. Columbus, Ohio: Ross Laboratories; 1989:76–83.
128. Butte NF, Wong WW, Garza C. Proc Nutr Soc 1989;48:303–312.
129. Roberts SB, Young VR. Am J Clin Nutr 1988;48:951–955.
130. Hytten FE, Chamberlain G. Clinical physiology in obstetrics. Oxford: Blackwell Scientific Publications; 1991.
131. Rasmussen KM, Yaktine AL eds. Weight Gain During Pregnancy: Reexamining the Guidelines. Washington, D.C.: National Academies Press; 2009:12584.
132. Butte NF, Wong WW, Hopkinson JM. J Nutr 2001;131:53–58.
133. Forsum E, Kabir N, Sadurskis A, Westerterp K. Am J Clin Nutr 1992;56:334–342.
134. Goldberg GR, Prentice AM, Coward WA, et al. Am J Clin Nutr 1991;54:788–798.
135. Lovelady CA, Meredith CN, McCrory MA, et al. Am J Clin Nutr 1993;57:512–518.
136. Roberts SB, Silver RE, Das SK, et al. Adv Nutr 2021;12:1438–1448.

A. MAJOR DIETARY CONSTITUENTS

CHAPTER 6

Water, Electrolytes, and Acid–Base Metabolism

Anna Gaddy and Edgar Lerma

CHAPTER OUTLINE

Abbreviations

ADH, antidiuretic hormone; **AG**, anion gap; **ATPase**, adenosine triphosphatase; **Ca^{2+}**, calcium; **CKD**, chronic kidney disease; **Cl^-**, chloride; **CO_2**, carbon dioxide; **DASH**, Dietary Approaches to Stop Hypertension; **DI**, diabetes insipidus; **ECG**, electrocardiographic; **ENaC**, epithelial sodium channel; **FDA**, Food and Drug Administration; **FFA**, free fatty acid; **GFR**, glomerular filtration rate; **GI**, gastrointestinal; **HCl**, hydrochloric acid; **HCO_3^-**, bicarbonate; **K^+**, potassium; **Mg^{2+}**, magnesium; **Na^+**, sodium; **NaCl**, sodium chloride; **NAD + (NAD^+)**, nicotinamide adenine dinucleotide; **NADH**, reduced nicotinamide adenine dinucleotide; **NAE**, net acid excretion; **$NaHCO_3$**, sodium bicarbonate; **NH_4^+**, ammonium; **P**, phosphorus; **PO_4^-**, phosphate; **PRAL**, potential renal acid load; **ROMK**, renal outer medulla potassium; **RTA**, renal tubular acidosis; **SIADH**, syndrome of inappropriate antidiuretic hormone secretion; **SPS**, sodium polystyrene sulfonate

Water

Humans can survive only a few days without a source of water. This essential nutrient plays an integral role in the maintenance and regulation of normal cellular and metabolic processes. Drinking liquids accounts for some of our water intake, but humans also consume significant amounts of water from fruits and vegetables. Water is also formed as a byproduct of the metabolism of foods, although the amount made in this way is less than daily losses. Urine losses account for most excretion, but sweat, respiration, and stool losses are important contributors.

Water Content and Distribution

Water constitutes approximately 54% of body weight in hospitalized adults without fluid and electrolyte disorders.[1] The fraction of body weight that is water is highest in infants (up to 90% in premature infants[2]) and children and progressively decreases with aging; it also varies depending on body fat content. Women and obese persons, who have higher body fat content, tend to have less water for any given weight. Therefore, age and body fat content must be taken into account when total body water is calculated.

Water is present in both the intracellular and extracellular fluid compartments of the body as an aqueous solution containing electrolytes. Each cell has its own separate environment, but it also communicates with other cells through the extracellular space and intercellular connection. Cell membranes are permeable to water, so the concentration of ions per liter of solution (i.e., osmolality) is the same throughout both compartments, although it is composed of distinct ions.[3] To maintain normal metabolic functions, optimal ionic composition is critically important, particularly in the intracellular fluid.

Because sodium is the predominant extracellular osmole and water flux is controlled by osmotic forces, the quantity of sodium (Na^+) determines the volume of the extracellular compartment. Estimates of total body water vary from 30% to 53%, depending on whether chloride (Cl^-), inulin, or sulfate is used in the determination.[4] It is greater in older adults and in females.[1,5] Generally, a value of 40% of total body water is considered to represent the extracellular volume. The extracellular volume can be further divided into three fractions: interstitial (space outside of blood vessels and between cells), plasma (space inside blood vessels and inside cells), and transcellular (sequestered) water volume, which constitute 28%, 8%, and 4%, respectively, of the total body water (**Figure 6.1**).[6] Thus, most extracellular fluid is partitioned between the extravascular and intravascular compartments, which are in equilibrium with each other (**Table 6.1**). Transcellular water represents fluids that are sequestered out of osmotic equilibrium, including luminal fluid of the gastrointestinal (GI) tract, the fluids of the central nervous system, and fluid in the eye, as well as the lubricating fluids at serous surfaces.[4,7]

Table 6.1 Estimated Volumes of Body Fluid Compartments in a healthy Adult Male weighing 73 kg and with total body water of 40 L

Intracellular volume: 24.0 L (60%)
Extracellular volume: 16.0 L (40%)
Interstitial volume: 11.2 L (28%)
Plasma volume: 3.2 L (8%)
Transcellular volume: 1.6 L (4%)

Reprinted with permission from Oh MS, Uribarri J. Electrolytes, water, and acid-base balance. In: Shils ME, Shike M, Ross AC et al, eds. Modern Nutrition in Health and Disease. 10th ed. Burlington: Jones & Bartlett Learning, LLC, 2006:149–93.

Composition of Body Fluid

Clinically, we measure electrolyte concentrations in the extracellular compartment only: mean plasma Na^+ is 140 mEq/L, potassium (K^+) is 4 mEq/L, Cl^- is 104 mEq/L, and bicarbonate (HCO_3^-) is 24 mEq/L. Although Na^+, Cl^-, and HCO_3^- are the main solutes in the extracellular fluid, K^+, magnesium (Mg^{2+}), phosphate (PO_4^-), and proteins (with negative charges) are the dominant solutes in the cell (**Table 6.2**). If red blood cells are hemolyzed during a blood draw, plasma potassium can appear spuriously high due to the release of intracellular potassium. The maintenance of this transmembrane electrochemical gradient, called the transmembrane potential, is critically important for conductivity of electrical impulses throughout the body and any aberrancy can lead to neurologic, cardiac, and skeletal muscle dysfunction. The concentrations of individual electrolytes inside the cell cannot be measured but, under most circumstances, the osmolality is exactly the same inside and outside the cell.[8-10]

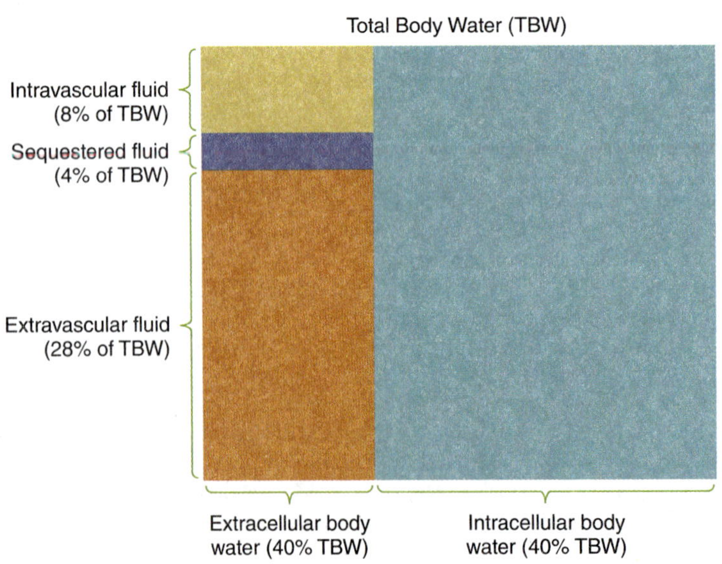

Figure 6.1 Total body water.

Table 6.2 Electrolyte Concentrations in Extracellular and Intracellular Fluids

	Plasma		Interstitial Fluid		Plasma Water		Cell Water (Muscle)	
	(mEq/L)	(mmol/L)	(mEq/L)	(mmol/L)	(mEq/L)	(mmol/L)	(mEq/L)	(mmol/L)
Na^+	140	140	145.3	145.3	149.8	149.8	13	13
K^+	4.5	4.5	4.7	4.7	4.8	4.8	140	140
Ca^{2+}	5.0	2.5	2.8	2.8	5.3	5.3	1×10^{-7}	1×10^{-7}
Mg^{2+}	1.7	0.85	1.0	0.5	1.8	0.9	7.0	3.5
Cl^-	104	104	114.7	114.7	111.4	111.4	3	3
HCO_3^-	24	24	26.5	26.5	25.7	25.7	10	10
SO_4^{2-}	1	0.5	1.2	0.6	1.1	0.44	—	—
PO_4^-	2.1	1.2[a]	2.3	1.3	2.2	1.2[a]	107	57[b]
Protein	15	1	8	0.5	16	1	40	2.5[c]
Organic anions	5	5[d]	5.6	5.6	5.3	5.3[d]	—	—

Ca^{2+}, calcium; Cl^-, chloride; HCO_3^-, bicarbonate; K^+, potassium; Mg^{2+}, magnesium; Na^+, sodium; PO_4^-, phosphate; SO_4^-, sulfate.

[a]The calculation is based on the assumption that the pH of the extracellular fluid is 7.4 and the dissociation constant (pK_a) of dihydrogen phosphate ($H_2PO_4^-$) is 6.8.

[b]The intracellular molal concentration of phosphate is calculated from the assumption the pK_a of organic phosphate is 6.1 and the intracellular pH 7.0.

[c]The calculation is based on the assumption that each millimole of intracellular protein has on the average 15 mEq, but the nature of cell proteins is not clearly known.

[d]The assumption has been that all organic anions are univalent.

Reprinted with permission from Oh MS, Uribarri J. Electrolytes, water, and acid-base balance. In: Shils ME, Shike M, Ross AC et al, eds. Modern Nutrition in Health and Disease. 10th ed. Burlington: Jones & Bartlett Learning, LLC, 2006:149–93.

The Difference Between Serum Sodium Concentration and Total Body Sodium

Because we measure extracellular fluid in which Na^+ is the dominant cation, serum Na^+ concentration is the main determinant of body fluid osmolality.[11] Counterintuitively, dietary intake of Na^+ does not usually change the blood Na^+ concentration. An increase in dietary Na^+ is accompanied by a transient increase in osmolality, which triggers thirst, leading to a proportional increase in water content as the body maintains the serum osmolality, and serum Na^+ concentration remains unchanged. (This process is interrupted in, for example, patients who are fed wholly by gastrostomy feeds or are otherwise unable to increase free water intake.) If dietary Na^+ decreases, the kidney ensures that a proportional amount of water is also lost, and serum osmolality is again maintained. Therefore, total body Na^+ content is most accurately reflected by the extracellular (plasma) volume, which is the main determinant of the vascular volume. When the total body Na^+ content rises, an increase in the vascular and interstitial volume is expected, whereas a drop in the total body Na^+ content predicts a decrease in the vascular and interstitial volume. Blood pressure and physical signs of volume status such as the presence or absence of edema are much better markers of total body Na^+. For this reason, a collection of urine sodium or a corrected single urine sodium concentration does not predict body sodium content but can be an indicator of circulating volume as perceived by chemoreceptors in the kidney.

Dietary Salt Intake, Edema, and Blood Pressure

As the kidney retains Na^+, total body Na^+ content increases. This results in increases in intravascular volume, which usually results in increased cardiac output and arterial blood pressure. At some point, a variety of mechanical and hormonal mechanisms cause the kidney to excrete the excess Na^+, restoring balance toward normal intravascular volume. This process is called "pressure natriuresis".[12] In individuals with hypertension, a long-term increase in cardiac output results in arteriolar constriction, an autoregulatory mechanism that prevents transmission of systemic blood pressures to the capillary beds. With chronic arteriolar constriction, cardiac output gradually returns to baseline; however, peripheral vascular resistance remains elevated,[13] leading to elevated blood pressure. This distinction is why hypertension is a systemic disease process rather than a description of a single elevated vital sign. The role of primary renal Na^+ retention as a cause of hypertension is documented in various renal diseases, primary hyperaldosteronism, and many congenital disorders that are characterized by increased renal Na^+ reabsorption.

In some disease processes, an increase in vascular volume is not accompanied by a rise in cardiac output and arterial blood pressure—this is the case with congestive heart failure. In other disease processes, such as nephrotic syndrome and cirrhosis, increased total body sodium leads to increased interstitial volume rather than intravascular volume, leading to edema or other extravascular fluid expansion such as ascites or pleural effusions.

Although aberrantly high renal Na^+ retention is known to be the primary cause of secondary hypertension, the exact role of Na^+ intake in causing essential hypertension is unknown, and the degree to which dietary Na^+ should be restricted is highly debated. The National Heart, Lung, and Blood Institute of the National Institutes of Health supports the position of the National High Blood Pressure Education Program and recommends that adults should consume no more than 2,400 mg/day Na^+ (6 g salt).[14] This amount is reduced to 1,500 mg/day in individuals with hypertension or with renal disease and increased in certain individuals with high Na^+ losses. These recommendations are in stark contrast to current intake patterns in the United States, where the 2007–2010 National Health and Nutrition Examination Survey estimates that females consume nearly 3,000 mg and males consume 4200 mg of sodium per day.[15]

A reduction in salt intake is not accompanied immediately by a reduction in renal Na^+ excretion, and so salt excretion temporarily exceeds salt intake. This imbalance leads to a reduction in extracellular volume and the effective vascular volume. Eventually, the kidney reduces salt excretion in response to a reduction in the extracellular volume, and a new balance between Na^+ intake and output is achieved. Until then, salt excretion exceeds salt intake. A reduction in the total body Na^+ content is accompanied by a reduction in the extracellular volume and a decrease in blood pressure—this scenario can be provoked by the administration of natriuretics (nonosmotic diuretics). Those persons who have substantial salt loss before a new balance is achieved are likely to have a greater decrease in blood pressure than are those who lose little salt before a new salt balance is achieved.[16]

Chloride

Whereas we consider that Na^+ largely determines extracellular volume, total body Cl^- is usually regulated in exactly the same proportion as Na^+ to maintain electroneutrality, so it is difficult to measure the effect of Cl^- on extracellular fluid volume. Thus, one could use either the Na^+ or the Cl^- concentration to calculate changes in osmolality, as when total body Na^+ rises, total body Cl^- rises as well. The Cl^- concentration does vary in certain acid–base disorders; however, so for practical reasons, the clinical standard is to use the serum or plasma Na^+ concentration for calculation of osmolality. Evidence indicates that dietary Cl^- has effects independent of Na^+: for example, a dose of sodium chloride (NaCl) raises blood pressure to a much greater extent than an equal dose of sodium bicarbonate ($NaHCO_3$).[17] Moreover, administration of Cl^-, but not Na^+, relieves metabolic alkalosis.

Sodium and Chloride Content of Food

Due to the prevalence of sodium chloride as a seasoning and preservative agent, Na^+ and Cl^- are consumed together in most foods. Although Na^+ and Cl^- are major extracellular solutes, the amount of Na^+ contained in food is quite small because the interstitial fluid represents a small fraction of the total fluid content of foods. Moreover, although the intracellular content of Cl^- is somewhat higher than Na^+, intracellular content of both ions is still low.[18] For these reasons, the salt content of food is low before preparation. The high intake of Na^+ and Cl^- results

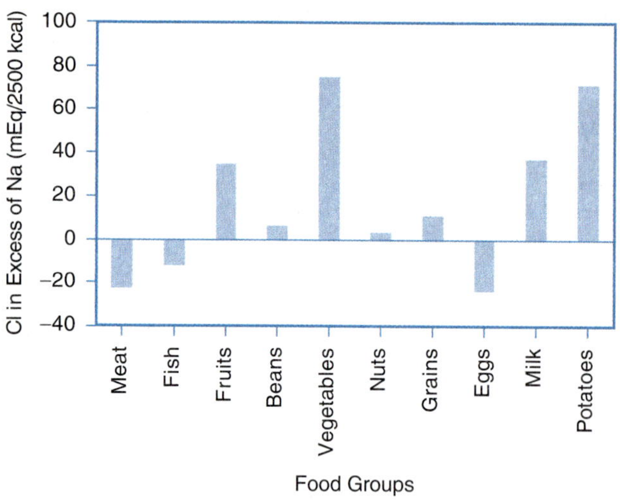

Figure 6.2 Sodium chloride difference of major food groups. Most sodium (Na) and chloride (Cl) in food is now in the form of added salt (1:1 sodium-to-chloride ratio). In natural foods without added salt, chloride content is greater than sodium, except for meat, fish, and eggs, which contain more sodium than chloride.

Reprinted with permission from Oh MS, Uribarri J. Electrolytes, water, and acid-base balance. In: Shils ME, Shike M, Ross AC et al, eds. Modern Nutrition in Health and Disease. 10th ed. Burlington: Jones & Bartlett Learning, LLC, 2006:149–93.

from salt added to food in its preparation or during cooking. While, on average, the Na^+ and Cl^- content in foods before processing tends to be equal; many plant-derived foods such as nuts, vegetables, fruits, and cereals contain more Cl^- than Na^+,[18] whereas meat, fish, and eggs contain more Na^+ than Cl^- (**Figure 6.2**).

Reduced-sodium salt substitutes have long been available and recommended as a strategy to help patients reduce sodium intake, but evidence for their use on clinical outcomes has been varied. A meta-analysis of such trials in 2014 concluded that salt substitution was effective at lowering systolic and diastolic blood pressures.[19] The more recent cluster-randomized controlled Salt Substitute and Stroke Study demonstrated that this substitute also decreased the risk of stroke, albeit in a rural Chinese population, without increasing risk of hyperkalemia.[20]

Pathophysiology of Water and Osmolality

Osmolar Relations and Regulations

Measurement of Plasma Osmolality

Osmolality refers to the number of particles into which a given solute dissociates when in solution and is physiologically important because it is related to the amount of osmotic pressure that solute exerts. Plasma osmolality can be measured with an osmometer or estimated as the sum of the concentration of all the solutes in the plasma. NaCl, glucose, and urea are the major constituents of the plasma that contribute to plasma osmolality. Na^+ always partners with its anion, Cl^-, to preserve electroneutrality, whereas the contribution of glucose and urea to the osmolality depends on fractional molecular weight. The molecular weight of glucose is 180 daltons, and that of urea is 28 daltons. Therefore, plasma osmolality is estimated from the following formula:

$$\text{Plasma osmolality} = \text{Plasma Na}^+ \text{ (mEq/L)} \times 2 + \text{Glucose (mg/dL)}/18 + \text{Urea (mg/dL)}/2.8$$

Unlike NaCl or glucose, which largely remain in the plasma, urea can cross cell membranes and is not restricted to the extracellular fluid. As such, it is considered an ineffective osmole. Although urea can attain substantial concentration in the plasma, its normal concentration is only 5 mOsm/L. Because of urea's small contribution to the total osmolality, the total osmolality is nearly equal to the effective osmolality in normal plasma.

Dangers of Changes in Osmolality

Solutes that are restricted to the extracellular fluid and that contribute to the osmolality are called effective osmoles, whereas solutes that can enter the cell freely are called ineffective osmoles. Examples of effective osmoles include glucose and Na^+; examples of ineffective osmoles are urea and alcohol. When the concentration of effective osmoles increases, osmotic equilibrium is reestablished by water shifting from the cell to the extracellular fluid. Intracellular osmolality then increases to the same level as the extracellular osmolality.[21-23] If ineffective osmoles are added to the extracellular fluid, the osmotic equilibrium is reestablished by entry of those solutes into the cell. Because most of the solutes normally present in the extracellular fluid are effective osmoles, loss of extracellular water, which can occur through insensible losses, will result in an increase in the effective osmolality and will cause a shift of water from the cells into the extracellular fluid. If extracellular osmolality is reduced, either by loss of normal extracellular solutes or by retention of water, a shift of water into the cells will occur to maintain osmolality. When the effective osmolality changes, cellular metabolism is affected, and cell swelling or shrinkage occurs with intracellular volume changes. Some of the most serious manifestations of altered osmolality are related to changes in brain cell volume because the brain is confined to a fixed space. Brain cells have the capacity to regulate their volume with time, and this explains why rapidity of alteration in osmolality is a determinant of severity of symptoms.[24]

Most of the signs and symptoms of a reduced concentration of Na^+ (hyponatremia, representing a low osmolality) are caused by brain swelling and increased intracranial pressure and include nausea or vomiting, headache, papilledema, and mental confusion.[25] With increasing severity, lethargy, weakness, hyperreflexia and hyporeflexia, delirium, coma, psychosis, focal weakness, ataxia, aphasia, generalized rigidity, and seizures occur, due to an increase in cell volume and reduced electrolyte concentration of the brain cells. GI manifestations include abdominal cramps, temporary loss of sense of taste and flavor, decreased appetite, nausea, vomiting, salivation, and paralytic ileus. Cardiovascular effects of hypo-osmolality are usually manifested as hypotension and other signs of low effective vascular volume. Hyponatremia can also be accompanied by muscle cramps, twitching, and rigidity.[26,27]

Increased effective osmolality need not be accompanied by a high serum Na^+ concentration (hypernatremia), but hypernatremia is always accompanied by hyperosmolality. Signs and symptoms of hyperosmolality depend on the rapidity of development as well as its severity. In both humans and animals, acute hyperosmolality from hypernatremia leads to subdural, cortical, and subarachnoid hemorrhages; it causes sudden shrinkage of brain cells and creates negative pressure in the brain.[28] Depression of mental status ranges from lethargy to coma. If the condition is severe, generalized seizure may also be observed but less commonly than in hypo-osmolality. Muscular symptoms of hyperosmolality include muscular rigidity, tremor, myoclonus, hyperreflexia, spasticity, and rhabdomyolysis. In children with chronic hyperosmolality, spasticity, chronic seizure disorder, and cognitive disturbances may occur.[28]

Regulation of Thirst and Antidiuretic Hormone Release

Stimulation and inhibition of osmoreceptor cells affect production by the hypothalamus and ADH secretion by the posterior pituitary. If the effective osmolality of plasma rises, hypothalamic osmoreceptor cells will shrink; this process then stimulates the thirst center in the cerebral cortex and increases antidiuretic hormone (ADH) production in the supraoptic and paraventricular nuclei.[29,30] If the effective osmolality declines, the osmoreceptor cells will swell and ADH production will be inhibited. ADH produced in the hypothalamus is carried through long axons and is secreted from the posterior pituitary.[31-33] ADH secretion is extremely sensitive to changes in effective osmolality. A rise in the effective osmolality by just 2% to 3% stimulates ADH secretion sufficiently to result in maximally concentrated urine,[28] whereas a decline in plasma osmolality of only 2% to 3% produces maximally dilute urine (<100 mOsm/L). ADH release is also regulated by nonosmotic factors such as nausea, pain, and intravascular volume.[25] A low, effective vascular volume (~10% decrease) provokes thirst and ADH release.[34] These effects are mediated through baroreceptors and some humoral factors released in response to reduced blood flow. This response explains the severe thirst despite hyponatremia seen in heart or liver failure. Other factors, including β-catecholamines, angiotensin II, and physical and emotional stress, enhance ADH output. Ethanol and catecholamines inhibit the output of ADH, even in the presence of an appropriate stimulus. Lithium, certain tetracycline antibiotics (demeclocycline), foscarnet, methoxyflurane, amphotericin B, and V–2 receptor antagonists (Vaptans) inhibit the effect of ADH on the kidney[25]—clinically, this inability of the kidney to respond to ADH is called nephrogenic diabetes insipidus and is characterized by large volumes of inappropriately dilute urine.

To understand the effect of ADH, consider that in healthy adults, 180 L of water are filtered through the kidney daily; 120 L are reabsorbed in the proximal tubule, and 35 L are reabsorbed in the descending limb of Henle. All this water absorption is accompanied by salt absorption (in the case of the loop of Henle in the ascending limb); however, no net change in osmolality occurs.[35] The distal convoluted and collecting tubules reabsorb salt without water, so approximately 25 L of dilute urine are delivered to the collecting duct. When ADH is totally absent, approximately 5 L of water are reabsorbed in the inner medullary collecting duct, and 20 L are excreted as the final urine. In the presence of maximal ADH, urine volume can be as low as 0.5 L/day as the urine is concentrated to as high as 1,200 mOsm/L and water is reabsorbed in the cortical and medullary collecting duct (**Figure 6.3**). The reabsorption of water in the collecting duct is regulated by ADH. Water is conserved as the urine is concentrated. The net effect is osmotically concentrated urine.[36-45]

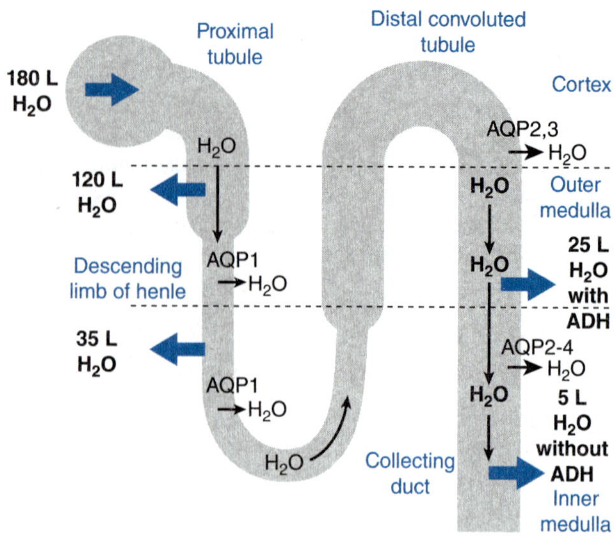

Figure 6.3 Of the 180 L of water filtered through the kidney daily, 120 L are reabsorbed in the proximal tubule and 35 L are reabsorbed in the descending limb of Henle. Most of the remaining 25 L are reabsorbed in the collecting duct in the presence of antidiuretic hormone (ADH). When ADH is totally absent, approximately 5 L are reabsorbed in the collecting duct, and the remaining 20 L are excreted as the final urine. *AQP*, aquaporin.

Nonrenal Control of Water and Electrolyte Balance

Besides urinary losses of water, water is also lost from the skin and through normal respiration. Water is lost from the skin primarily as a means of eliminating heat, and the amount of water lost depends on the amount of heat generated in the body. In the absence of sweat or febrile illness, water loss from the skin is called insensible perspiration. Water loss from the skin depends mainly on the amount of heat generated in the body. Water loss from the skin is 30 mL/100 kcal or approximately (~300 to 1000 mL/24 hours). Sweat contains Na^+ and K^+ at a concentration of approximately 50 mEq/L and 5 mEq/L, respectively, equivalent to 0.45% normal saline. The Na^+ content of sweat varies, depending on the conditioning of the individual. An unconditioned individual placed in a hot environment (e.g., a new recruit in basic training) may have sweat that contains up to 100 mEq/L Na^+, whereas after training, the Na^+ sweat content may be as low as 30 mEq/L. This difference explains why a higher dietary Na^+ intake is required for unconditioned individuals.[46]

Both fats and carbohydrates, the major energy sources for the body, are broken down into carbon dioxide (CO_2) and water. Both are lost through ventilation. Water is lost during normal ventilation because the water content of inspired air is less than that of the expired air. Ventilation is determined by the amount of CO_2 production, which is determined by the caloric expenditure. The amount of water lost during ventilation also depends on caloric expenditure:

$$\text{Respiratory water loss} = 13 \text{ mL/100 kcal at normal partial pressure of } CO_2 \ (P_{CO_2})$$

By this metabolism of macronutrients, water is produced in an amount that is largely lost during normal respiration. In calculating water balance, respiratory water loss may be ignored in the measurement of insensible water loss, provided metabolic water gain is also ignored. In cases of hyperventilation or fever, respiratory water loss increases disproportionately to metabolic water production.[1] Such patients may become hypernatremic if water is not administered.

The net activity of the GI tract to the level of the jejunum is secretion of water and electrolytes. The net activity from jejunum to colon is reabsorption of water and electrolytes. Most of the fluid entering the small intestine is absorbed there, and the remainder is absorbed in the colon, leaving only approximately 100 mL of water to be excreted daily in the feces. The contents of the GI tract are isotonic with plasma, and any fluid that enters the GI tract becomes isotonic. Thus, if water is ingested and vomited, solute is lost from the body. Conversely, depending on the etiology, diarrhea can have a wide range of tonicity.

Water and Volume Depletion

In any discussion of salt and water losses, the terms *dehydration* and *volume depletion* must be used precisely. Dehydration is characterized by water loss alone or an excess loss of water to salt. Volume depletion occurs when both solute and water are lost. Depending on the quantity of solute losses in relation to water losses, mixed forms of volume depletion and dehydration are encountered. In hypotonic dehydration, NaCl losses exceed water losses.

Volume Depletion

NaCl may be lost isotonically (i.e., at the same concentration as in the plasma) through the GI tract, directly from aspiration of the extracellular fluid from pleural effusions or ascites, or from bleeding. With GI fluid loss, NaCl is lost with an equal or larger amount of water loss, and the osmolality of the body fluids is subsequently adjusted to isotonicity by changes in oral intake or urinary excretion of water. Isotonic fluid loss reduces only the extracellular fluid volume and can be treated with isotonic crystalloid solution.

Total Body Water Depletion

The primary aberration in total body water depletion (Formerly referred to as dehydration[47]) is water loss, and hypernatremia results from an increase in the concentration of Na^+ in the extracellular space. This disproportionate excess of NaCl to water can occur if water intake is inadequate or water loss is excessive. Dehydration resulting from excessive water loss usually develops more rapidly than dehydration caused by reduced water intake. A lack of water intake is always caused by one of two mechanisms: (a) a defect in the thirst sensing mechanism or impaired consciousness or access to water.[48,49]

Hypotonic Volume depletion. Hypotonic volume depletion (volume depletion with more Na^+ than water loss) occurs when the patient loses NaCl and replaces the salt with water or with water containing less NaCl than the fluid lost (see the later discussion of hyponatremia). In the presence of normal renal function, NaCl loss in excess of water loss is difficult to achieve because the kidney readily excretes the excess water through the suppression of ADH. This response is blunted or absent in patients with hypotonic dehydration.[50]

Principles of Fluid Therapy

Goals of Salt and Water Replacement

The goal of therapy is to restore the patient to normal plasma volume while concurrently maintaining normal tonicity. Deficits in volume and water must be identified and repleted; basal requirements for electrolytes and water must be supplied daily, and ongoing salt and water losses must be quantified and provided for in the treatment plan.[51,52] It is crucial to note whether a patient has an acute kidney injury or chronic kidney disease, which can impair ability to dilute and concentrate urine. It is also crucial to make note of whether a patient has agency to drink water in response to thirst.

Basal Requirements

The basal requirement for water depends on sensible (urinary) and insensible losses of water.[53] Fever increases both respiratory and skin water losses because of an increase in the basal metabolic rate. To some degree, urinary loss of water declines to compensate for these losses; however, this depends, in part, on the total amount of solute excreted and the degree to which the kidney can concentrate the urine. Solute excretion depends mainly on salt ingestion and protein intake, but severe glycosuria causes osmotic diuresis and increases urinary water losses.

Daily Water Requirements

In the absence of fever or exercise, water loss through the skin is relatively fixed, but urinary water losses vary greatly and depend on the total amount of solute to be excreted and urinary concentrating ability. For example, for a total solute excretion of 600 mOsm/day, the urine volume will be 500 mL if urine is concentrated to 1,200 mOsm/L and 6 L if urine osmolality is 100 mOsm/L. For the former individual who can maximally concentrate the urine, the minimum water requirement would be 1,100 mL (500 mL for urinary water loss plus 600 mL for skin water loss at 2,000 kcal/day). For the latter individual who is unable to concentrate the urine, the maximal allowable water intake would be 6.6 L. In the absence of an abnormality in urinary concentrating or diluting ability, large ranges of water intake are well tolerated, as the kidney adjusts and maintains fluid homeostasis.[3,54] Clinically, in hospitalized patients, it is best not to overestimate water requirements to avoid water intoxication. Impairment in urinary dilution, as occurs in the syndrome of inappropriate ADH secretion (SIADH), is more common than impairment in urine concentration. In a conscious patient, thirst is an effective defense mechanism, whereas patients with severe hyponatremia often lapse into coma without warning.[1]

Polyuria

Polyuria, which is arbitrarily defined as an unintentional urine volume greater than 2.5 L/day in adults, can be caused by either osmotic diuresis or water diuresis.[1] In osmotic diuresis, urine output increases because of an excessive rate of solute excretion. Certain solutes, such as glucose, urea, mannitol, radiopaque media, and NaCl can cause osmotic diuresis in which the solute excretion rate exceeds 60 mOsm/hour or 1,440 mOsm/day in the adult.[1] In water diuresis, urinary osmolality is lower than plasma osmolality because the kidney excretes dilute urine and water is not reabsorbed in the collecting duct. Reduced water reabsorption in the collecting duct can be attributed to a large volume of water intake (typically psychogenic polydipsia), lack of ADH[55-64] (central diabetes insipidus [DI]), or unresponsiveness to ADH (nephrogenic DI).

Lack of ADH can stem from a primary deficiency (central DI), from physiologic suppression of ADH production by a low serum osmolality, or from excessive production of vasopressinase during pregnancy (gestational DI).[65,66] The latter results from the consumption or infusion of large amounts of water and is common among patients with psychosis, particularly among those with schizophrenia.[55-57,67,68] Various gradations of ADH deficiency can be seen. In the setting of a partial ADH deficiency, urine osmolality is close to normal and clinical manifestations are typically only seen when the patient loses the agency to drink water. ADH deficiency can be congenital or acquired.[58-63] Causes of polyuria are listed in **Table 6.3**. Note that urine outputs of more than 2.5 L may be considered desirable in patients with kidney stone formation.

Primary polydipsia is an increase in water intake in the absence of a physiologic stimulus such as hyperosmolality or volume depletion.[55-57,67,68] It is usually psychogenic in origin, hence the term *psychogenic polydipsia*. An increase in urine output is caused by physiologic suppression of ADH secretion, and the serum Na^+ is usually at the low range of normal. Occasionally, the serum Na^+ may be low and indicates that the capacity of the GI tract to absorb water exceeds the normal capacity of the kidney to excrete water. In contrast, secondary polydipsia results from thirst stimulation in response to hyperosmolality. This condition is seen in patients with DI or patients with diabetes with severe glycosuria; the serum Na^+ is usually in the high normal range.

Differential Diagnosis

The first step in the initial workup of polyuria should be measurement of urine osmolality.[69] Osmotic diuresis can be ruled out or diagnosed solely on the rate of osmole excretion: if the excretion of osmoles exceeds a rate greater than 60 mOsm/hour or 1,440 mOsm/day, this suggests osmolar diuresis. In contrast, an excretion of a large volume of maximally dilute urine at 100 mOsm/L constitutes water diuresis.

In determining the cause of water diuresis, the first step is to determine the serum Na^+ concentration. In DI, the serum Na^+ tends to be high normal, whereas in primary polydipsia, it tends to be low normal. These conditions overlap, however, and a water deprivation test[70] is needed to confirm the diagnosis. To perform this test, water is either restricted overnight or until loss of 5% of body weight occurs. If the patient has a failure to concentrate the urine maximally, the diagnosis of DI is confirmed. If the concentrating ability improves significantly on administration of ADH, the DI is central. If the patient fails to concentrate the urine maximally and fails to respond to ADH, the DI is nephrogenic. Primary polydipsia and partial nephrogenic DI cannot be distinguished by

Table 6.3 Causes of Polyuria

Water diuresis
- **A. Lack of ADH**
  1. Central diabetes insipidus
     - Congenital
     - Acquired (destruction of posterior pituitary)
       - Tumors
       - Granulomas
       - Pituitary surgery
       - Trauma
       - Infarction
       - Infection of the pituitary or hypothalamus
  2. Primary (psychogenic) polydipsia
  3. Gestational DI
- **B. Failure of the kidney to respond to ADH**
  1. Congenital nephrogenic diabetes insipidus
     - Defect in the ADH receptor
     - Defect in aquaporin expression
  2. Chronic renal failure
  3. Acquired nephrogenic diabetes insipidus
     - Drugs
       - Lithium
       - Demeclocycline
       - Methoxyflurane
       - Heavy metals
     - Interstitial kidney disease
       - Amyloidosis
       - Sickle cell anemia or trait
       - Sarcoidosis
     - Electrolyte imbalance
       - Hypercalcemia
       - Hypokalemia
       - Obstructive uropathy

Solute diuresis
- **A. Saline loading**
- **B. Postobstructive diuresis**
- **C. Hyperglycemia**
- **D. High-protein tube feedings**
- **E. Salt-wasting nephropathy**

ADH, antidiuretic hormone.

this test. Patients with partial nephrogenic DI are frequently hypernatremic, however, whereas those with primary polydipsia are frequently hyponatremic.

Treatment

Treatment of polydipsia is aimed at the underlying etiology. To treat osmotic diuresis, the cause of the increased solute excretion, such as poor glucose control or excessive protein intake, must be ascertained and controlled. A careful dietary history can be helpful. For central DI, desmopressin (DDAVP), a synthetic analog of ADH, can be administered intranasally, subcutaneously, or intravenously.[71,72] Nephrogenic DI cannot be treated with ADH preparations, but measures to reduce the distal delivery of salt and water (i.e., low-salt diet and thiazide diuretics) are helpful.[73,74] Statin drugs may be useful to increase water channel abundance by reducing their cell surface removal in patients with nephrogenic DI.[75] In contrast, reports have noted the use of clozapine and propranolol to treat primary polydipsia in patients with schizophrenia,[76] but drugs that interfere with urinary dilution, such as thiazide diuretics, will exacerbate the problem and should be avoided.

Disorders of Sodium Metabolism

Hyponatremia

Hyponatremia occurs when the plasma Na^+ concentration falls to less than 135 mEq/L. It is the most common electrolyte disorder and generally causes clinical concern when less than 130 mEq/L. Pseudohyponatremia is a spurious reduction in reported serum Na^+ concentration resulting from a systematic error in measurement. Changes in methodology have largely reduced this problem[77]; however, the presence of a nonosmotic substance (in vitro hemolysis, hyperlipidemia, hyperproteinemia, and mannitol), may still cause pseudohyponatremia.[78]

Causes and Pathogenesis

The mechanisms responsible for a reduction in extracellular Na^+ concentration (hyponatremia) are as follows:

1. Water can shift from the cell in response to an accumulation of extracellular solutes other than Na^+ salts.[79-81] Hyperglycemia causes hyponatremia by this mechanism, and the serum Na^+ decreases by 1.6 mEq/L for every 100 mg/dL rise in the serum glucose. Serum osmolality is unchanged.
2. The kidney can fail to excrete excess water.
3. The body can fail to retain Na^+.[82]
4. Na^+ can shift into cells.

In examples 2, 3, and 4, hypotonicity results, and the appropriate physiologic response is suppression of ADH release, which leads to rapid excretion of excess water and correction of hyponatremia. Therefore, persistence of hyponatremia indicates the failure of this compensatory mechanism. In most instances, hyponatremia is maintained because the kidney fails to produce a water diuresis, but sometimes, ingestion of water above the limits of normal renal compensation is responsible. Reasons for the inability of the kidney to excrete water include reduced glomerular filtration rate (acute or chronic kidney disease), reduced delivery of glomerular filtrate to the distal nephron, inadequate solute intake, and the presence of ADH. After ruling out causes of pseudohyponatremia, an assessment of extracellular fluid volume provides a useful working classification of hyponatremia.[83]

In most cases of hyponatremia, the main reason for the fall of serum Na^+ is abnormal retention of water, which is either ingested or administered in the form of intravenous fluids.[82] Increased ADH leads to excretion of hypertonic urine. The response is considered appropriate when ADH is released in response to hypertonicity of the body fluid or when the effective vascular volume is reduced. Hyponatremia in clinical states such as congestive heart failure and cirrhosis of the liver is associated with reduced effective vascular volume and; therefore, ADH secretion is appropriate (if not physiologically advantageous). Similarly, decreased perfusion may cause ADH secretion despite hyponatremia in hypothyroidism.[84] Cerebral salt-wasting syndrome, which is defined as renal loss of salt caused by humoral substances released in response to cerebral disorders, such as acute subarachnoid hemorrhage, causes volume depletion that results in hyponatremia by the same mechanism of appropriate ADH release.[85]

Table 6.4 Causes of Hyponatremia

Disorders in which renal water excretion is impaired
- **A. Effective circulating volume depletion**
  1. Gastrointestinal losses
     - Vomiting
     - Diarrhea
     - Tube drainage
     - Intestinal obstruction
  2. Renal losses
     - Diuretics
     - Hypoaldosteronism
     - Sodium-wasting nephropathy
  3. Skin losses
     - Marathon runners
     - Burns
     - Cystic fibrosis
  4. Edematous states
     - Heart failure
     - Hepatic cirrhosis
     - Nephrotic syndrome
  5. potassium depletion
- **B. Thiazide diuretics**
- **C. Renal failure**
- **D. Nonhypovolemic states of ADH excess**
  1. Syndrome of inappropriate ADH secretion
  2. Cortisol deficiency
  3. ypothyroidism
- **E. Decreased solute intake**

Disorders in which renal water excretion is normal
- **A. Primary polydipsia**
- **B. Reset osmostat**
  1. Pregnancy
  2. Psychosis
  3. Quadriplegia
  4. Severe malnutrition

ADH, antidiuretic hormone.

The term syndrome of inappropriate ADH (SIADH) is reserved for ADH secretion that occurs despite hyponatremia, with a normal or increased effective vascular volume. Causes of SIADH include nausea, pain, tumors, pulmonary diseases such as tuberculosis and pneumonia, central nervous system diseases, and drugs (**Table 6.4**).[25]

Adrenal insufficiency causes ADH release by a unique mechanism: When cortisol is low, corticotropin-releasing hormone (CTRH) is secreted by the hypothalamus. CTRH acts as an ADH secretagogue, which could be considered inappropriate as it is not accompanied by a low vascular volume. Interestingly, in animal models, CTRH also increases sodium excretion by inhibiting release of atrial natriuretic peptide.[86]

Finally, mild hyponatremia may be caused by resetting of the osmostat to an osmolality lower than usual. In such cases, urine dilution occurs normally when the plasma osmolality is brought down to less than the reset level. Patients with chronic debilitating diseases such as pulmonary tuberculosis often manifest this phenomenon.[87]

Diagnosis

The presence of a low plasma Na^+ concentration and normal osmolality suggests pseudohyponatremia (see Table 6.4). With modern analyzers, high glucose concentration is the most common cause, and serum Na^+ decreases by 1.6 mEq/L for every 100 mg/dL rise in the serum glucose. Hyponatremia caused by glucose is suspected from the history or by simultaneous measurements of plasma Na^+, osmolality, and glucose. Pseudohyponatremia resulting from hyperlipidemia occurs only with certain analyzers and is caused by the accumulation of chylomicrons, which consist mostly of triglycerides. Substantial hyponatremia resulting from hyperlipidemia requires accumulation of more than 5 to 6 g/dL of lipids, and such a degree of hyperlipidemia does not occur with hypercholesterolemia alone.

In evaluating hyponatremia associated with hypo-osmolality, the main concern is distinguishing between SIADH and hyponatremia resulting from other causes, mainly volume depletion and edematous states. The major distinction between SIADH and other causes of hyponatremia lies in the status of effective vascular volume. Effective vascular volume is normal or increased in SIADH and reduced in other disorders that cause hyponatremia. No single diagnostic test measures effective vascular volume with certainty. Physical examination is inaccurate in determining mild-to-moderate volume depletion. A more reliable method is the measurement of certain laboratory parameters, all of which depend on renal responses to changes in effective vascular volume. Urinary Na^+ excretion of greater than 20 mEq/L, serum urea nitrogen less than 10 mg/dL, serum creatinine less than 1 mg/dL, and serum urate less than 4.0 mg/dL all suggest normal or increased effective vascular volume. The fractional excretion of urea is more reliable than the serum urea nitrogen value in determining the status of the effective vascular volume because the serum urea nitrogen also depends on protein intake. A fractional excretion of urea of less than 35% is considered an indicator of low effective vascular volume,[88] whereas a fractional excretion of Na^+ less than 0.5% is thought to represent a low effective vascular volume.[89]

Contrary to common belief, urine osmolality in SIADH need not be greater than plasma osmolality. Furthermore, high urine osmolality does not necessarily support the diagnosis of SIADH because most other causes of hyponatremia are also accompanied by urine osmolality higher than plasma osmolality. The only situation in which urine osmolality may be appropriately low in the presence of hyponatremia is hyponatremia caused by primary polydipsia. In all other causes of hyponatremia, urine osmolality is inappropriately increased (i.e., >100 mOsm/L).[90]

Treatment

The treatment of hyponatremia is directed at the underlying cause and may range from the addition of Na^+, the removal of water, or improvement of the organ dysfunction (cardiac, renal, or hepatic). Salt supplementation should be given only to patients with hyponatremia resulting from salt depletion or, as a last resort, to increase tonicity.[91,92] The speed of correction of hyponatremia depends on the speed of development and on the patient's symptoms. Severe symptomatic hyponatremia is a life-threatening condition and should be treated with hypertonic saline solution,[93,94] but volume overload and central pontine myelinolysis (also known as osmotic demyelinating disease) are associated with the administration of a large quantity of salt containing solutions.[95,96] Central pontine myelinolysis, a demyelinating disease of the central pons and other areas of the brain, is characterized by motor nerve dysfunction. If it is severe enough, quadriplegia occurs. It is seen more often during treatment of chronic than of acute hyponatremia, and is more frequent in malnourished and debilitated patients. Commonly accepted rates of correction of chronic hyponatremia range from 0.5 to 1.0 mEq/L/hour or less, but central pontine myelinolysis has been reported to occur when the rate of correction of hyponatremia was less than 0.5 mEq/L. To prevent overcorrection of hyponatremia, 6 to 8 mmol/L in 24 hours, 12 to 14 mmol/L in 48 hours, and 14 to 16 mmol/L in 72 hours have been recommended.[94] Because the danger of central pontine myelinolysis is limited mainly to patients with asymptomatic chronic hyponatremia, rapid correction (at a rate of 1 to 2 mEq/L/hour) should be restricted to those with acute symptomatic hyponatremia.[24,97] Clinical outcomes are not improved by a rapid correction of the serum Na^+ to a level greater than 120 mEq/L.[77] For patients admitted with volume depletion and chronic asymptomatic hyponatremia, the traditional recommendation has been administration of isotonic crystalloid or other appropriate volume resuscitation (i.e., blood transfusion if hemorrhage is the underlying etiology). With volume expansion, ADH release is suppressed. Water excretion and a rise in serum Na^+ concentration ensue. A 4 to 6 mmol/L increase in serum Na^+ concentration is usually sufficient to improve the most severe symptoms in patients with acute hyponatremia, so a therapeutic goal of 6 mmol/day is reasonable even when the serum Na^+ falls to extremely low levels.[98] For K^+ depletion, the appropriate treatment is with 0.45% NaCl containing 40 mEq/L K^+. Regardless, serum electrolytes should be monitored closely, with blood monitoring every 2 hours and adjustments in infusion rates made to avoid too rapid a correction.

Acute Treatment. For hyponatremia with Na^+ depletion and symptomatic hypo-osmolality (e.g., confusion), intravenous administration of Na^+ as hypertonic saline solution corrects hypoosmolality effectively. The amount of Na^+ necessary to increase the Na^+ to a desired level is calculated as follows[71,99]:

$$Na^+ \text{ requirement (mEq)} = \text{Body weight} \times \Delta Na$$

where and ΔNa^+ is the desired serum Na^+ (120 mEq/L) minus the current serum Na^+. Na^+ is administered as a 3% NaCl solution. In emergency situations, 100 mL or 2 mL/kg bolus infusions of 3% saline can be administered rapidly over minutes in the setting of seizures and can be repeated, if necessary.[94,100] Very close serial monitoring of serum Na^+ is required until blood levels are stable.

When accumulation of excess water is primarily responsible for hyponatremia, as in SIADH, water may be rapidly removed by administration of intravenous osmotic diuretics such as mannitol or urea. Sodium glucose 2- transporter inhibitors (SGLT2i) have been used for their osmotic diuretic effect in patients with SIADH.[101] Loop diuretics, such as furosemide, hinder the kidney's ability to retain both Na^+ and water. When loop diuretics are given in conjunction with hypertonic saline, the net effect is a rise in the serum Na^+ because Na^+ replacement exceeds water replacement. The response to furosemide cannot be accurately predicted, and frequent follow-up measurements of serum Na^+ must be made. Administration of hypertonic saline solution alone usually causes salt and water diuresis, but the addition of a loop diuretic makes correction of hyponatremia more predictable by preventing excretion of a concentrated urine and prevents fluid overload. Vasopressin antagonists (vaptans) have been used to facilitate free water excretion and correct hyponatremia (102).[102] Vasopressin mediates its biologic effects by binding to three receptor subtypes: V1A, V1B, and V2. V1A receptors are located on vascular smooth muscle, platelets, and the liver. Activation of these receptors results in vasoconstriction, platelet aggregation, and gluconeogenesis. V1B receptors are in the anterior pituitary, and their activation stimulates adrenocorticotropic hormone (ACTH) release, whereas V2 receptors are located in the principal cells of the renal collecting duct. Stimulation of these receptors results in water retention; conversely, antagonism of these receptors results in dilute urine.[103]

Tolvaptan, the first oral selective V2 receptor antagonist approved for use in the United States by the FDA, was shown to increase serum Na^+ concentration significantly, compared with placebo, in patients with euvolemic hyponatremia (SIADH) and hypervolemic hyponatremia (from cirrhosis, congestive heart failure), but it did not alter disease progression or mortality.[103]

Long-Term Treatment. Chronic hyponatremia may be treated by water restriction or by an increase in renal water excretion. Reduction of water intake is preferable, but compliance is difficult because of severe thirst. Hard candy and chewing gum may be helpful in keeping the mouth moist, and ice chips are more thirst quenching than an equal volume of water. If water restriction is unsuccessful, increased renal water excretion can be achieved with pharmacologic agents that interfere with urine concentration. Lithium and demeclocycline increase urine output by interfering with the renal effects of ADH. Demeclocycline is more effective and has fewer side effects, but it may cause nephrotoxicity in patients with liver disease.

Administration of a loop diuretic such as furosemide in conjunction with increased salt and K^+ intake is safer than the foregoing methods. The diuretic prevents high medullary interstitial osmolality by limiting the reabsorption of salt in the loop of Henle and, hence, prevents high urine concentration. Increased intakes of salt and K^+ increase water output by increasing the rate of solute excretion. Although vasopressin antagonists are commercially available, their prohibitive cost makes them difficult to use in long-term treatment.[104] Still, these agents have proven useful in the treatment of heart failure, cirrhosis, and SIADH, as well as in patients with psychotic disorders.[105,106]

In patients with poor solute intake, supplementation of solute in the diet (preferably in the form of protein to avoid the edema, which can accompany salt supplementation) can be effective. Urea supplements are a useful tool to treat low-solute intake as well as mild chronic SIADH, and are generally inexpensive and well-tolerated in an oral formulation.[29]

Hypernatremia

Hypernatremia is defined as an increased Na^+ concentration in plasma water. Whereas hyponatremia may not be accompanied by hypo-osmolality, hypernatremia is always associated with an increase in effective plasma osmolality and reduced cell volume. An increase in plasma osmolality should stimulate thirst. Hence, hypernatremia can occur only if the thirst mechanism is blocked, such as in the setting of altered mental status, or when an immobile patient has no access to water. Although the extracellular volume in hypernatremia may be normal, decreased, or increased, hypernatremia almost always occurs in the setting of volume depletion.

Causes and Pathogenesis

Theoretically, hypernatremia is caused by loss of water, reduced water intake, gain of Na^+, or a combination of these (**Table 6.5**). Gain of Na^+ in a person who has normal perception of thirst and the ability to drink water, however, does not result in hypernatremia because a proportional amount of water is retained to

Table 6.5 Causes of Hypernatremia

Water loss
- **A. Insensible loss**
  1. Increased sweating: fever, exercise
  2. Burns
  3. Respiratory infections
- **B. Renal loss**
  1. Central diabetes insipidus
  2. Nephrogenic diabetes insipidus
  3. Osmotic diuresis
     - Glucose
     - Mannitol
- **C. Gastrointestinal loss**
  1. Osmotic diarrhea
     - Lactulose
     - Malabsorption
     - Infectious enteritides
- **D. Hypothalamic disorders**
  1. Primary hypodipsia
  2. Reset osmostat from volume expansion in primary mineralocorticoid excess
- **E. Water loss into cells**
  1. Seizures
  2. Severe exercise
  3. Rhabdomyolysis

Reduced water intake
- **A. Defective thirst**
  1. Altered mental status
  2. Thirst center defect
- **B. Inability to drink water**
- **C. Lack of access to water**

Sodium retention
- **A. Administration of hypertonic sodium chloride or sodium bicarbonate**
- **B. Ingestion of sodium**

maintain normal body fluid osmolality. Because thirst is such an effective and sensitive defensive, it is virtually impossible to increase serum Na^+ by more than a few milliequivalents (mEq) per liter if the mechanism for drinking water is intact. Therefore, hypernatremia is only seen in patients without the agency to take in free water: patients with altered mental status, patients with a defective thirst mechanism, patients with continuous vomiting, patients including children who lack access to water, or patients with mechanical obstruction preventing enteral intake.

The excess gain of Na^+ leading to hypernatremia is rare and usually iatrogenic. This occurs in the setting of hypertonic saline infusion, accidental entry into the maternal circulation during abortion with hypertonic saline solution, or administration of hypertonic $NaHCO_3^-$ (i.e., during cardiopulmonary resuscitation.). Reduced renal Na^+ excretion leading to hypernatremia usually occurs in response to dehydration caused by a primary water deficit. Water depletion resulting from DI, osmotic diuresis, or insufficient water intake leads to secondary Na^+ retention in patients who continue to ingest Na^+ or who are given Na^+.[107]

Diagnosis

Whether hypernatremia is the result of Na^+ retention or water loss can be determined by examination of the patient's volume status. For example, if a patient with a serum Na^+ concentration of 170 mEq/L does not have obvious evidence of dehydration, hypernatremia is not caused entirely by water loss. To increase the serum Na^+ to 170 mEq/L through water loss alone would require a loss of more than 20% of the total body water.

Treatment

Acute Treatment. Hypernatremia is generally treated by the addition of water; when the cause is iatrogenic, Na^+ must be removed. When hypernatremia is associated with volume depletion, isotonic crystalloid may be given initially to stabilize circulatory dynamics, followed by administration of hypotonic solutions to normalize the tonicity. Because rapid reduction of the plasma osmolality may result in cerebral edema, the serum Na^+ should be reduced by 6 to 8 mEq/L in the first 3 to 4 hours for acute symptomatic hypernatremia.[103] As with hyponatremia, chronic hypernatremia usually does not cause central nervous system symptoms and, therefore, does not require rapid correction. Although a slow reduction of the serum Na^+ concentration, by less than 10 mmol/L/day or 0.5 mmol/L/hour, is desirable, little documented evidence exists regarding what constitutes a safe rate of rehydration.

In one study in a children's hospital in China, risk factors for cerebral edema were an initial fluid bolus, the severity of hypernatremia, and the overall hydration rate.[107] Cerebral edema appeared to be mitigated if the overall hydration rate in the first 24 hours was less than 0.5 mmol/L/hour. Therefore, a safe rate of correction is 0.5 mEq/L/hour and one should not exceed a 10% change in the serum Na^+ concentration during the initial 24-hour period.

The amount of water needed to correct hypernatremia can be estimated using the following equation[71]:

$$\text{Water deficit (L)} = \text{TBW} \times (\text{Actual Na}^+ - \text{Desired Na}^+)/\text{Desired Na}^+ = \text{TBW} \times (\Delta\text{Na}^+/\text{Desired Na}^+)$$

where ΔNa^+ is the difference between the desired and actual serum Na^+ concentration.

In the setting of hypernatremia with excess Na^+, a reduction in serum Na^+ with fluids usually initiates natriuresis. If natriuresis does not occur promptly, Na^+ may be removed with diuretics. Furosemide in combination with 5% dextrose solution may be an appropriate regimen for treatment of hypernatremia associated with excess Na^+.[71] If a patient with Na^+ excess has renal failure, salt can be removed by dialysis. The total water requirement must also include insensible water losses (~300 to 500 mL/24 hours) and urinary electrolyte-free water (the volume of urine, which is free of electrolytes if isotonic amounts of electrolytes are removed.) losses. The latter represents the urine water losses above the volume necessary to contain urinary Na^+ in addition to K^+ at the same concentration as serum Na^+. When the urine electrolyte-free water excretion is a positive number, the serum Na^+ rises further; when it is a negative number, serum Na^+ is decreased. Urinary electrolyte-free water excretion is calculated as follows:

$$\text{Electrolyte-free water excretion} = V - (U_{Na^+} + U_{K^+})V/S_{Na^+}$$

where V is urine volume, $U_{(Na^+ + K^+)}$ is the sum of concentrations of urine Na^+ and K^+, and S_{Na^+} is serum Na^+ concentration.

Long-Term Treatment. Hypernatremic disorders that require long-term preventive therapy include DI and primary hypodipsia. Although DI is often listed as a cause of hypernatremia, it does not produce hypernatremia in the absence of a thirst defect. Treatment is directed toward the curtailment of polydipsia and polyuria, which are the patients' main complaints. Patients with primary hypodipsia should be educated to drink water on schedule. In some instances, stimulation of the thirst center with chlorpropamide has been effective.[71]

Potassium Metabolism and its Disorders

Dietary Sources of Potassium and Handling

Because K^+ is the major intracellular cation, it is widely distributed in all foods, but the content varies greatly, depending on the type of food (**Figure 6.4**). The highest K^+ content is in fruits and vegetables; the K^+ content of vegetables is particularly high when it is expressed as the content per kcal.[18,108] Among the starchy foods, the K^+ contents of polished rice and light wheat flour are particularly low, whereas the K^+ contents of potato, soybean, and buckwheat are high (**Figure 6.5**). Although citrus fruits and bananas are often cited by health professionals as rich sources of K^+, many other foods contain K^+ at higher concentration. Tomatoes, apricots, and cantaloupes contain far more K^+ than oranges and bananas when expressed as milliequivalents of K^+ per kcal (**Figure 6.6**).[18] Although protein sources are not as rich in K^+ as fruits or vegetables, meats and fish contain approximately 80 to 100 mEq/oz. Treating foods with salt and then discarding the liquid (brining, boiling) induces Na^+ for K^+ exchange and reduces the K^+ content of the foods.

Dietary K^+ is almost completely absorbed in the normal small intestine; approximately 10% of dietary K^+ is excreted in the stool.[109] K^+ absorption in the small intestine is passive and depends on mechanisms of Na^+ and glucose absorption. Thus, it is not surprising that factors decreasing Na^+ and water absorption also decrease K^+ absorption. Although losses are less than that of Na^+, sweating does result in the loss of approximately 0.2 g (5 mEq)/L of K^+. This value is relatively consistent with acclimation.

Patients with chronic kidney disease and end-stage renal disease develop increased secretion of potassium in the large bowel, which helps maintain serum concentration.[110] This becomes clinically relevant in cases of colonic disruption such as colitis or colectomy. Decreased absorption of enteral potassium is a pharmacologic target in this patient population, both with older binding resins such as sodium polystyrene and newer selective cation exchange polymers such as sodium zirconium cyclosilicate.

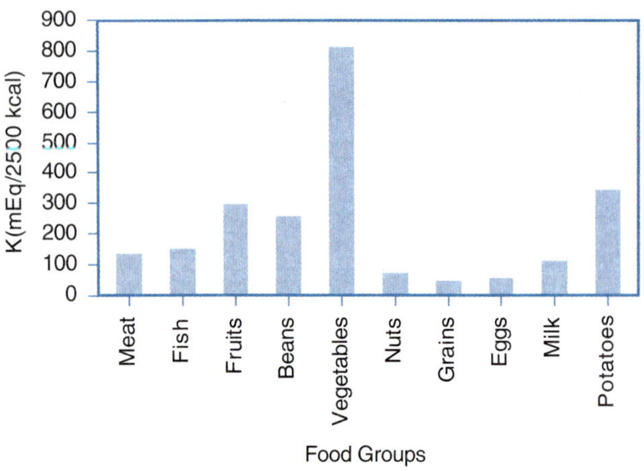

Figure 6.4 Potassium (K) content of major food groups. When expressed by the caloric content of food, vegetables contain the highest amounts of potassium. The potassium content of grains is low.

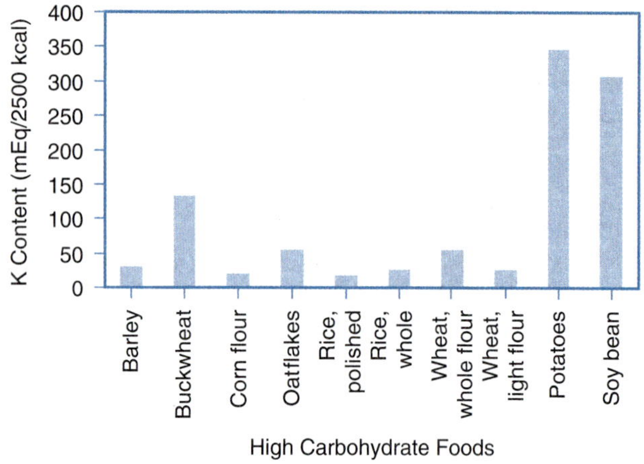

Figure 6.5 Potassium (K) content of major carbohydrate-containing foods. Grains, especially polished rice and light wheat flour, contain little potassium, whereas potatoes and soy beans contain large quantities.

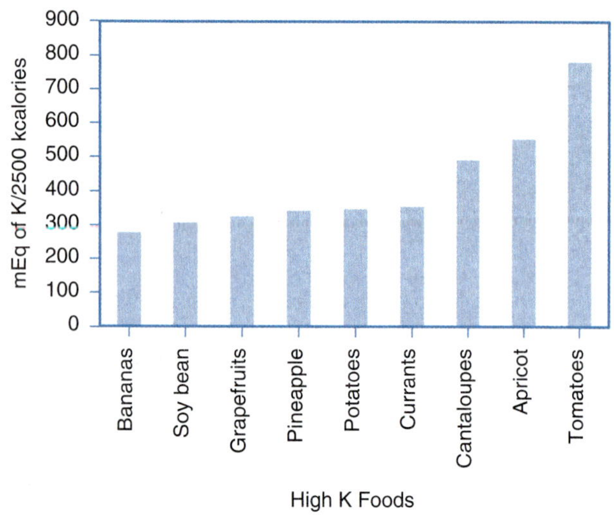

Figure 6.6 Potassium (K) content of major fruits. Although oranges and bananas are most often cited as fruits of high potassium content, many other fruits contain more potassium. For example, the potassium content of apricots is more than twice that of oranges.

Control of Intracellular and Extracellular Potassium

Although total body K^+ is approximately 43 mEq/kg body weight, only approximately 2% of this K^+ is found in the extracellular fluid, and measured serum K^+ reflects only extracellular stores.[111] The extracellular K^+ concentration is responsible for most clinical abnormalities and measures alterations in total body K^+ most sensitively. In patients with a deficit of K^+, both intracellular and extracellular K^+ decrease; but the extracellular K^+ decreases proportionately more than the intracellular concentration.[112] Similarly, when patients have an excess of K^+ in the body, the increase in extracellular K^+ is proportionately greater than that in intracellular K^+.

Factors that affect the movement of K^+ into and out of cells change extracellular K^+ concentration independently of total body stores. The rise in insulin after a meal lowers serum K^+ by increasing its transport into cells, thus blunting the ability of dietary K^+ to cause significantly high serum K^+.[113] Catecholamines (through β_2-adrenergic receptors) have a similar effect.[114-116] Acidemia raises serum K^+, whereas alkalemia lowers it because of shifts of hydrogen ions (H^+) into and out of the cell in exchange for K^+.[117] In general, metabolic acidosis causes greater K^+ efflux than respiratory acidosis.[118] Metabolic acidosis that results from inorganic acids, such as sulfuric acid and hydrochloric acid (HCl), causes greater K^+ efflux than metabolic acidosis resulting from organic acids, such as lactic acid and ketoacids.[117,119] Importantly, meals high in net acid are generally lower in K^+, and this function protects patients from hypokalemia when such meals are consumed. In respiratory alkalosis, K^+ influx is less than that seen in metabolic alkalosis, due to a drop in cellular HCO_3^- concentration.[120]

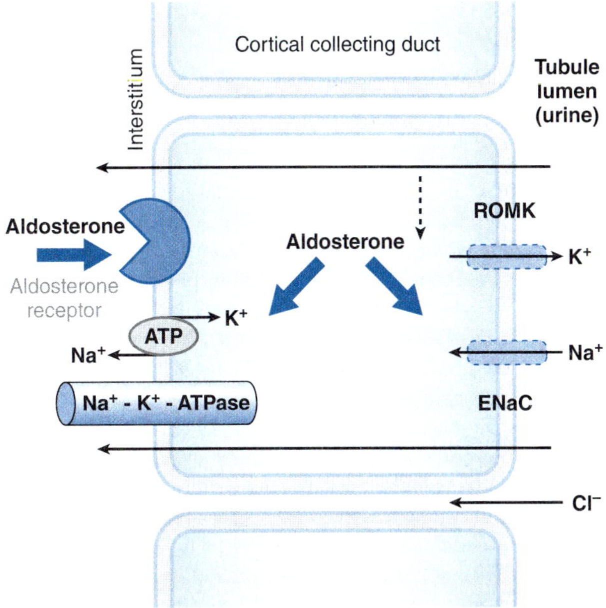

Figure 6.7 Control of potassium (K^+) secretion at the cortical collecting duct. Sodium (Na^+) enters from the luminal fluid into the cell through the epithelial Na^+ channel (ENaC) and is transported out of the cell through Na^+/K^+-adenosine triphosphatase (ATPase) on the basolateral membrane. These processes create the luminal electrical potential that is more negative than the electrical potential of the peritubular fluid. The electrical charge imbalance created by Na^+ reabsorption is partly offset by entry of K^+ into the lumen through the renal outer medulla K^+ channel (ROMK), a K^+ channel. *ATP*, adenosine triphosphate; *Cl^-*, chloride.

Control of Renal Excretion of Potassium

Approximately 90% of the daily K^+ intake (40 to 100 mEq) is excreted in the urine.[109] Because K^+ filtered at the glomerulus is almost completely reabsorbed by the proximal tubule and the ascending limb of the loop of Henle, net urinary K^+ excretion is determined at the cortical collecting duct by mechanisms shown in **Figure 6.7**. Secretion of K^+ occurs through the renal outer medulla K channel (ROMK) as Na^+ enters the cell through the epithelial Na channel (ENaC).[121,122] Thus, despite a dedicated single exchanger, K^+ is excreted in exchange for Na^+. Either high Na^+ in the tubule fluid or high blood K^+ concentration increases urinary K^+. This is why thiazide diuretics cause urinary K^+ loss. Diuretics that affect transporters in the proximal tubule (carbonic anhydrase inhibitors) and the thick ascending limb (loop diuretics) partly cause K^+ depletion by preventing K^+ absorption in these segments, but they also promote K^+ secretion by increasing Na^+ delivery to the cortical collecting duct.[123-125]

When anions other than Cl^- (e.g., HCO_3^-) are excreted, more Na^+ is delivered to the distal tubule, and K^+ secretion is enhanced. HCO_3^- in the tubular fluid also enhances K^+ secretion by directly stimulating ROMK activity.[122] This mechanism is important because many foods with high K^+ content are also high in bicarbonate (see later)—in this way, base excretion helps the kidney excrete a K^+ load. This mechanism is also illustrated by the increased K^+ excretion when patients vomit. The loss of stomach acid leaves base behind in the bloodstream, and serum HCO_3^- rises. When this HCO_3^- is filtered in the kidney, it is not reclaimed in the proximal tubule, and it is delivered with Na^+ to the distal nephron, where it enhances K^+ loss.

In addition to delivery of Na^+ to the distal nephron, plasma aldosterone concentration controls K^+ excretion.[126] High plasma aldosterone increases the activity of ENaC to reabsorb Na^+ and ROMK to increase the excretion of K^+. The retention of salt and water, and the resulting volume expansion, leads to increased distal delivery of Na^+, which can further enhance K^+ excretion.

On a typical Western high-Na^+ diet, the kidney easily excretes K^+ because of high distal Na^+ delivery. Most of the K^+ excreted in the urine is derived from secretion at the cortical collecting duct and is determined by the amount of Na^+ delivered to that portion of the nephron.[127,128,129] Indeed, people may develop hypokalemia on a high-Na^+ low-K^+ diet, because of the obligate loss of approximately 15 mEq/L (0.6 g) of K^+ in the urine. Individuals with normal kidney function have the capacity to excrete more than 400 mEq/day (16 g/day) of K^+ without causing a clinically significant change in the serum K^+ concentration.

In addition, an adaptation increases K^+ excretion in the stool as K^+ intake increases: there is active transport of K+ across the basolateral membrane through Na–K-ATPase coupled with luminal NKCC2. Negative potential in the colonic lumen and high intracellular K+ also facilitates secretion of K+ through the large conductance K+ channel (BK), and absorption occurs in the distal colon via a colonic H–K-ATPase.[110] It is well documented that this distal secretion is increased in patients with end-stage renal disease, independently of serum K+ concentration.[130] Although the mechanism is not completely elucidated, this can become clinically relevant when patients with very low eGFR have changes such as colectomy or colitis.

The diet of humans in preagricultural eras was much higher in K^+, due to higher intake of fruits, vegetables, and meat but lower intake of grain; Na^+ intake was low because of a lack of availability. The daily dietary load of K^+ in those times probably exceeded 300 mEq/day (~12 g/day), whereas dietary Na^+ was likely less than 90 mEq/day (~2 g/day). Based on the mechanisms discussed thus far, one would predict that the amount of Na^+ delivered to the cortical collecting duct was far less in prehistoric times than in modern times. So how do the kidneys adapt to excrete more K^+ on a low-Na^+ diet? In addition to bicarbonate increasing K secretion (even on a low Na^+ diet) by delivering more Na^+ to the distal nephron, additional mechanisms explain K^+ excretion. High-K^+ diets decrease the activity of the thiazide-sensitive NaCl cotransporter in the distal nephron.[131] The thiazide-sensitive transporter is in the segment immediately before the segment expressing ENaC, so this decreased cotransporter activity allows more Na^+ to reach ENaC and be exchanged for K^+. It would cause more Na^+ excretion if ENaC activity did not change, but ENaC is also increased by high dietary K^+. When diets are high in both K^+ and Na^+,[132] the K^+ effect on the thiazide-sensitive transporter increases Na^+ excretion to a greater extent than the effect on ENaC reduces it, thus causing a net increase in Na^+ excretion. As discussed earlier, blood pressure rises with Na^+ intake to increase the glomerular filtration rate (GFR) and Na^+ excretion. Thus, in the setting of high Na^+, high dietary K^+ increases the amount of Na^+ excretion to lower blood pressure effectively.

The converse is also true. Low K^+ intake makes it more difficult to excrete a high Na^+ intake, and blood pressure rises because of Na^+ retention. When low Na+ is sensed by the kidney, increased aldosterone upregulates ENaC, so that net Na^+ excretion is less influenced by K^+. A diet high in K^+ and low in Na^+ results in an increase in Na^+ reabsorption while allowing excretion of K^+. This mechanism fails only when Na^+ intake is so low that blood pressure falls and the kidney itself begins to decrease filtration rate.

Role of Potassium Intake on Sodium Excretion and Blood Pressure

Diets high in fruit and vegetables were shown to be beneficial in reducing blood pressure in the Dietary Approaches to Stop Hypertension (DASH) study.[133,134] Some of the blood pressure–lowering effects of these diets probably resulted from their high K^+ content. Although the combination of low Na^+ intake and the DASH diet provided the lowest blood pressure, the DASH diet lowered the blood pressure the most when Na^+ intake was high.[133,134] The DASH investigators showed that the K^+ content of the diet enhanced Na^+ excretion for any given blood pressure. Several other studies confirmed that an increase in K^+ by 40 mEq/day reduced blood pressure more than reduction in Na^+ intake of 60 to 80 mEq/day.[133-136] A high-K^+ diet likely reduces blood pressure through increased delivery of Na^+ to the distal nephron; Na^+ diuresis ensues. As discussed earlier, dietary K^+ increases Na^+ excretion to a greater extent when Na^+ intake is high. Nevertheless, an effect is still seen at a Na^+ intake of as little as 1.6 g (an amount that is difficult to achieve on a Western diet). These data suggest that when a dietary approach is used to treat hypertension, increased K^+ intake is an effective adjunct to low Na^+ to lower blood pressure.

Hypokalemia

Causes and Pathogenesis

Because intracellular K^+ concentration greatly exceeds extracellular concentration, K^+ shift into the cell can cause severe hypokalemia with little change in its intracellular concentration[137] (**Table 6.6**). Alkalosis, insulin, and β_2-agonists can lower serum potassium by shifting K^+ inside cells[138,139], which is harnessed clinically to treat hyperkalemia but can be pathologic in cases of refeeding. In states of malnutrition, poor Na^+ intake decreases K^+ excretion, and low energy intake causes catabolism and release of K^+ from tissues.[137] The result of these effects is that it is difficult to achieve low serum potassium purely from poor potassium intake. During recovery from starvation; however, insulin release causes K^+ to be transported inside the cells. Because cell mass increases during nutritional recovery, K^+, the main intracellular cation becomes trapped inside cells, thereby leading to a drop in the extracellular level.[140] Thus, K^+ must be watched carefully in patients with refeeding following a period of starvation or prolonged undernutrition (e.g., alcoholic patients, hospital patients with poor

Table 6.6 Causes of Hypokalemia

Decreased net intake
- **A. Low dietary intake**
- **B. Clay ingestion (pica)**

Increased entry into cells that causes transient hypokalemia
- **A. Elevation in intracellular pH**
- **B. Increased availability of insulin**
- **C. Elevated β-adrenergic activity**
  1. Stress
  2. Coronary ischemia
  3. Delirium tremens
  4. Administration of β-adrenergic agonists
- **D. Periodic paralysis, hypokalemic form**
- **E. Treatment of megaloblastic anemia**
- **F. Hypothermia**
- **G. Nutritional recovery state**

Increased gastrointestinal losses
- **A. Diarrhea**
- **B. Vomiting**
- **C. Intestinal drainage**
- **D. Laxative abuse**

Increased urinary losses
- **A. Diuretics**
- **B. Mineralocorticoid excess**
  1. Primary hyperaldosteronism (adrenal adenoma or hyperplasia)
  2. Secondary hyperaldosteronism (malignant hypertension, renal artery stenosis)
- **C. Increased flow to the distal nephron**
  1. Salt-wasting nephropathies
  2. Diuretics
- **D. Sodium reabsorption with a nonabsorbable anion**
  1. Vomiting or nasogastric suction
  2. Metabolic acidosis
  3. Penicillin derivatives
- **E. Amphotericin B**
- **F. Hypomagnesemia**
- **G. Polyuria**

Increased sweat losses

Dialysis

nutritional intake before or after admission). Phosphorus (P) and Mg^{2+} follow K^+ in many of these cases and may also become acutely decreased in the blood with refeeding of at-risk patients.[140]

Vomiting and diarrhea are the most common causes of hypokalemia.[137] Poor intake of K^+ in patients with these conditions contributes to hypokalemia but is not the major cause. Diarrhea causes direct K^+ loss in the stool, but in vomiting, hypokalemia occurs because of direct loss in the vomitus, as well as K^+ loss in the urine. Vomiting causes metabolic alkalosis, and the subsequent renal excretion of HCO_3^- leads to increased delivery of Na^+ to the distal nephron. Because of poor Na^+ intake and increased urinary losses of Na^+, aldosterone rises. As discussed earlier, renal K^+ wasting occurs when increased aldosterone concentration is accompanied by increased distal delivery of Na^+.[141-154]

In primary aldosteronism, overactivity of the adrenal gland occurs without a secondary cause. Distal retention of Na^+ is increased, and blood pressure rises due to volume expansion.[155] In patients with secondary causes of high aldosterone, hypokalemia occurs only in conditions accompanied by increased distal Na^+ delivery. Examples include conditions in which blood pressure rises, such as renal artery stenosis and malignant hypertension. Heart failure does not lead to hypokalemia despite secondary aldosteronism unless renal Na^+ transporters are blocked. This situation occurs in the setting of loop or thiazide diuretic therapy. Defective Na^+ reabsorption, proximal to the aldosterone effective site, results in delivery of Na^+ to the cortical collecting duct and subsequent hypokalemia. Bartter's and Gitelman's syndromes are genetic diseases that lead to decreased activity of the same transporters inhibited by loop or thiazide diuretics, respectively.[121,156] Hypokalemia occurs by the same mechanisms.[148-153] In chronic metabolic acidosis, hypokalemia develops because metabolic acidosis directly stimulates aldosterone secretion, reduces proximal reabsorption of NaCl, and allows increased delivery of NaCl to the distal nephron.[157] With natural licorice intake, renal K^+ wasting results from the sustained mineralocorticoid activity of cortisol because licorice contains a glycyrrhizic acid, which inhibits the enzyme 11-β-hydroxysteroid dehydrogenase and prevents the rapid metabolism of cortisol in the kidney[154] Liddle's syndrome, another genetic cause of hypokalemia, is characterized by increased Na^+ channel activity, resulting increased K^+ secretion.[158]

Clinical Manifestations

Low serum K^+ can be life threatening because of potentially negative changes in cardiac rate, rhythm, and conduction, as well as numerous structural and functional alterations in various organs, especially skeletal muscle.[159] Hypokalemia produces abnormalities of rhythm and of rate of cardiac conduction through alteration in several physiologic states: alteration in ventricular repolarization leads to depression of the ST segment, flattening and inversion of T waves, and appearance of U waves—the most common electrocardiographic (ECG) abnormalities of hypokalemia. Combinations of altered states of polarization and conduction can produce arrhythmias, most commonly supraventricular and ventricular ectopic beats and tachycardia, atrioventricular conduction disturbances, and ventricular fibrillation. Rapidly developing hypokalemia is more likely to produce cardiac arrhythmias than is more slowly developing hypokalemia.[159] Changes in organ function include cardiac and skeletal muscle cell necrosis and acute skeletal muscle rhabdomyolysis. Reduced insulin secretion and reduced intestinal motility are important nutritional effects of hypokalemia. Chronic hypokalemia may be associated with hypertension that results from decreased Na^+ excretion, and kidney stone formation that results from inhibition of citrate excretion.[160,161]

Treatment

Hypokalemia is usually treated either by K^+ administration or by prevention of the renal loss of K^+. In a nonemergency setting, K^+ should be given orally in the diet or pharmacologically as KCl, K^+ phosphate, or the salt of organic acids. Administration of K^+ must take the intracellular and extracellular K^+ distribution into account. Patients who have been depleted of K^+ for prolonged periods have lower intracellular stores and require more K^+ to correct the disorder than do patients with normal K^+ stores. K^+ should be given gradually to allow time for it to shift inside cells and to prevent the extracellular concentration from rising too quickly. Only 40 mEq may be given safely as a single dose to allow time for K^+ to shift into the cell. The safe amount for a mixed meal has not been established,[162,163] but it is significantly greater because insulin promotes a shift of K^+ into cells. The US Institute of Medicine recommends 125 mEq (~5 g) of K^+ daily for people with normal kidney function.[164] Unless ongoing losses (e.g., diarrhea, vomiting, diuretics) occur, the body's K^+ level should rise if intake is increased by as little as 40 mEq/day. In patients with prolonged hypokalemia and depleted intracellular stores, a decline in serum K^+ of 1 mEq/L generally indicates total body losses of 150 to 200 mEq of K^+, whereas a decline of 2 mEq/L indicates losses greater than 500 mEq.

In the critical care setting, K^+ is usually given intravenously, as KCl, at a rate less than 10 mEq/hour. In life-threatening hypokalemia, it may be useful to estimate the number of liters of extracellular fluid as body weight in kilograms multiplied by 0.2.[165] This value multiplied by the desired increment in serum K^+ per liter provides an estimate of the amount of K^+ that can be safely given in 20 to 30 minutes without danger of hyperkalemia. A glucose-containing solution should not be used as a vehicle for KCl administration whenever serum K^+ is to be increased rapidly; glucose stimulates release of insulin which, in turn, drives K^+ into the cells and decreases blood levels. K^+, at concentrations exceeding 40 mEq/L, may lead to sclerosis of smaller vessels. While it is less peripherally toxic to give central venous infusion of K^+ at high concentration, this may have higher risk of arrhythmia. Evidence for either approach, however, is lacking.

Renal loss of K^+ is prevented by treating its cause (e.g., removal of aldosterone-producing adenoma or by discontinuation of diuretics), by lowering distal delivery of Na^+, or by administering K^+-sparing diuretics.[166] Because reduced delivery of Na^+ to the distal nephron reduces K^+ secretion, a low-salt diet helps to reduce renal K^+ loss unless a separate mechanism increases distal delivery of Na^+ (e.g., metabolic alkalosis or loop diuretics). The K^+-sparing diuretics in current use are aldosterone antagonists (e.g., spironolactone and eplerenone) and blockers of the ENaC, triamterene, and amiloride. Aldosterone antagonists are most effective in preventing renal K^+ loss if an increased mineralocorticoid concentration is responsible for hypokalemia; otherwise, EnaC inhibitors are preferred.

Hyperkalemia

Causes and Pathogenesis

Hyperkalemia may be caused by either a shift of K^+ from the cells to the extracellular fluid[1,22,167,168] or an increase in total body K^+ (**Table 6.7**). A shift of K^+ from inside cells can be caused by

Table 6.7 Causes of Hyperkalemia

Causes of Hyperkalemia
Increased intake
A. **Oral (usually with decreased excretion)**
B. **Intravenous**
Movement from cells into extracellular fluid
A. **Pseudohyperkalemia**
1. Mechanical puncture during venipuncture
2. Leukocytosis (white blood cell count >100,000/mm^3)
3. Thrombocytosis (platelet count >400,000/mm^3)
B. **Metabolic acidosis**
C. **Insulin deficiency and hyperglycemia in uncontrolled diabetes mellitus**
D. **Tissue catabolism**
E. **β-Adrenergic blockade**
F. **Severe exercise**
G. **Digitalis overdose**
H. **Periodic paralysis, hyperkalemic form**
I. **Succinylcholine**
J. **Arginine**
Decreased urinary excretion
A. **Renal failure**
B. **Effective circulating volume depletion**
C. **Hypoaldosteronism**
D. **Type I renal tubular acidosis, hyperkalemic form**

inadequate insulin (starvation type 1 diabetes) or, more commonly, by insulin resistance seen in type 2 diabetes, hyperkalemic familial periodic paralysis, administration of muscle paralytic agents,[1,22,167,168] administration of excess cationic amino acids such as arginine and lysine and acute acidosis. All cause hyperkalemia by extracellular K^+ shift. Cell death causing leakage of intracellular contents (i.e., rhabdomyolysis or hemolysis) can be associated with massive K^+ shifts. Although hyperkalemia is not as predictable in organic acidosis as in inorganic acidosis in experimental situations, it is common in diabetic ketoacidosis because of the lack of an insulin effect in shifting K^+ into cells.[169] Hyperkalemia can also occur in severe digitalis intoxication by extracellular shift of K^+ because digitalis inhibits the Na^+/K^+-adenosine triphosphatase (ATPase) pump.[170]

The kidney's ability to excrete dietary K^+ is so great that hyperkalemia rarely occurs solely from increased intake of K^+ in food. Rather, hyperkalemia almost always occurs in the setting of impaired renal excretion. The three major mechanisms of diminished renal K^+ excretion are reduced aldosterone or aldosterone responsiveness, reduced distal delivery of Na^+, and renal failure (acute or chronic). Aldosterone deficiency may be part of a generalized deficiency of adrenal hormones (e.g., Addison's disease), or it may represent a selective process (e.g., hyporeninemic hypoaldosteronism). Hyporeninemic hypoaldosteronism is the most common cause of all aldosterone deficiency states and is the most common cause of chronic hyperkalemia among patients who are not undergoing dialysis.[22,171] Selective hypoaldosteronism can also occur with heparin therapy, which inhibits steroid production in the zona glomerulosa.[172]

In patients with reduced aldosterone secretion, any agent that limits the supply of renin or angiotensin II may provoke hyperkalemia. Examples include angiotensin-converting enzyme inhibitors, nonsteroidal anti-inflammatory agents, and β-blockers. The last category of drugs may compound the tendency to cause hyperkalemia by interfering with K^+ transport into cells. Chronic kidney disease (CKD) stage 3 or greater is often associated with decreased aldosterone, produced by Na^+ retention and high blood pressure, but it may also cause hyperkalemia as a result of renal tubular unresponsiveness to aldosterone. This acquired aldosterone resistance may result from destruction of the distal nephron from urinary obstruction of the kidney or from interstitial nephritis. Stage 4 or greater CKD is associated with an inability to excrete a K^+ load regardless of aldosterone. Genetic pseudohypoaldosteronism may also occur; this disorder may involve only K^+ secretion (pseudohypoaldosteronism type II) or Na^+ reabsorption in addition to K^+ secretion (pseudohypoaldosteronism type I).[173,174] Severe volume depletion may cause hyperkalemia despite secondary hyperaldosteronism because of a marked reduction in delivery of Na^+ to the cortical collecting duct.

Pseudohyperkalemia, defined as a measured K^+ value higher than that present in the blood, is often caused by lysis of blood cells during the process of obtaining blood.[175,176] Repeating the blood draw is required to verify hyperkalemia. A severe increase in platelets and leukocytes can cause pseudohyperkalemia through K^+ release during the process of coagulation (see Table 6.7).

Clinical Manifestations

K^+ helps maintain the polarization of skeletal and cardiac muscle. In severe hyperkalemia, paralysis of the skeletal muscle occurs with rapidly ascending neuromuscular weakness or paralysis. Before cardiac paralysis can develop, abnormalities of cardiac rhythm and of its rate of conduction occur. Characteristic ECG findings are used to assess the degree of cardiac involvement. The earliest, but most nonspecific, sign of hyperkalemia consists of tall, peaked T waves with shortened QT intervals. As hyperkalemia worsens, P waves flatten and QRS complexes widen progressively; then the P waves disappear entirely, and the QRS complexes merge with the T waves to stimulate a sine wave. This last rhythm is associated with a decline in cardiac output that is fatal unless rapidly treated. Other ECG findings include fascicular block and complete heart block (especially in digitalized patients), ventricular tachycardia, flutter and fibrillation, and cardiac arrest without the sine wave pattern.[177,178] Mild ECG changes have been noticed with serum K^+ concentration of as low as 5.5 mEq/L. As in hypokalemia, the rate of development of hyperkalemia is important in the development of cardiac abnormalities. In chronic hyperkalemia, patients with K^+ higher than 7.0 mEq/L have had normal ECG tracings. Hyperkalemia can also cause confusion and paresthesias.[179]

Treatment

Acute hyperkalemia may be treated by removal of K^+ from the body, by shifting extracellular K^+ into the cells, or by antagonizing K^+ action on the membrane of the cardiac conduction system (**Table 6.8**).[180–182] Antagonism of the action of K^+ on the heart with intravenous Ca^{2+} salts has the fastest effect against hyperkalemia and is used in life-threatening cases. $CaCl_2$ and Ca gluconate are equally effective,[183] but Ca gluconate is preferable because tissue damage is less when extravasation of the drug occurs during intravenous infusion. Shift of K^+ into cells can be accomplished with insulin (often given with glucose to prevent hypoglycemia), by β-adrenergic agonists (which have the advantage of the inhalation route if the patient does not have intravenous access), or by increasing the blood pH with $NaHCO_3^-$. HCO_3^- is less effective in the short term than the other approaches, and partly works by diluting K^+ and increasing its

Table 6.8 Treatment of Hyperkalemia

Reduction in body potassium content
A. Reduction of intake
B. Increased fecal secretion; potassium exchange resin and sorbitol
C. Increased renal excretion; mineralocorticoids, increased salt intake, diuretics
D. Peritoneal or hemodialysis
Intracellular shift of potassium
A. Glucose and insulin
B. Administration of alkali
C. β-agonists: salbutamol, albuterol
Antagonism of the membrane effect of hyperkalemia
A. Calcium salts
B. Hypertonic sodium salts

excretion.[182] It is imperative to remember that these redistribution techniques are temporary measures while the underlying cause of hyperkalemia is addressed.

Removal of K^+ may be accomplished by several routes: through the GI tract with K^+ exchange resins or laxatives; through the kidney by diuretics, mineralocorticoids, and increased salt intake or saline infusion; by $NaHCO_3$ administration; or by hemodialysis or peritoneal dialysis (see Table 6.8). A K^+ exchange resin, Na polystyrene sulfonate (SPS [Kayexalate]), is more effective when given with agents such as sorbitol or mannitol that cause osmotic diarrhea, but no clinical trials have compared its effectiveness with other agents. Intestinal necrosis has been noted when sorbitol is given as an enema,[184] so other laxatives may be preferable.[185,186] In response to reports of these complications, the FDA issued an advisory against concomitant administration of SPS with sorbitol in 2009. Because investigators thought that toxicity was related to 70% sorbitol and not 30% sorbitol, a premixed preparation of SPS in 30% sorbitol was allowed to stay on the market. More recently, reports exist showing colon necrosis following the use of premixed preparations of SPS in 30% sorbitol.[187,188] More data are needed on the benefit and harm associated with SPS and sorbitol.[189] As an alternative, saline solution works by dilution and increasing distal delivery of Na^+ in patients with intact renal function. $NaHCO_3$ can also be used. When the kidney is responsive and volume overload is present, diuretics are effective.[190,191] Mineralocorticoids are too slow in onset for acute use. Glycyrrhizic acid, the active ingredient of licorice, which inhibits 11 β-hydroxysteroid dehydrogenase, has been shown to decrease the serum K^+ concentration in patients undergoing dialysis. More data on pharmacokinetics and on toxicity are needed before the drug can be adopted for long-term use.[192]

The emergence of unabsorbable cation exchange resins such as sodium zirconium cyclosilicate and patiromer calcium sorbitex have dramatically changed clinical management of hyperkalemia. These compounds are administered orally and can interfere with oral absorption of other drugs, however they do not have the colonic complications of sodium polystyrene sulfonate.[193]

For chronic hyperkalemia, reduced intake of K^+ is theoretically effective in controlling high K^+ but this approach is not practical. Problems include palatability, insufficient intake of fruit and vegetables (and thus fiber, antioxidants, etc.), and compliance. An unresolved problem is that despite increased cardiac risk in patients with CKD, a low-K^+ diet restricts the very foods associated with reduction in cardiovascular disease risk. Diuretics, if needed for control of Na^+, are effective at wasting K^+ in the long term and may allow a healthier diet in some patients. Mg^{2+} wasting may be a problem when diuretics are combined with a low-K^+ diet. Some authorities advocate diuretics combined with a high-salt diet in patients without volume expansion. Renal salt retention is an important mechanism of hypertension and hypertensive cardiac damage, however, so monitoring Na^+ balance is critical. Mineralocorticoids (most often fludrocortisone) have been used in these patients as a means of wasting K^+,[171] but mineralocorticoids have been implicated in the pathogenesis of cardiac and renal disease. $NaHCO_3$ can mildly increase K^+ excretion in patients with CKD. Because it is beneficial for its effect on acid–base balance, $NaHCO_3$ is a useful adjunct for managing these patients.

Acid-Base Balance and Disorders

Terminology

Various states of nutrition such as feeding, vomiting, diarrhea, and catabolism are affected by and, in turn, affect the pH of the blood. These disorders are complex in etiology and diagnosis. Clinicians use numerous models (e.g., HCO_3^--anion gap [AG], base excess, strong ion difference) to classify acid–base disorders; of these, the HCO_3^--AG model is most suitable for nutrition science because it classifies disorders by the predominant species of acid or base accumulated.[70,194] In this model, acid and base come in two forms: respiratory, derived from dissolved CO_2; and metabolic, derived largely from metabolic or dietary acids and bases.[22] Chemically, a substance is an acid or a base, depending on whether it will donate or accept a net H^+ load after metabolism. CO_2 is an acid because it reacts with water to form carbonic acid. For example, citric acid (in fruit sodas) and phosphoric acid (in colas) are chemically acids, but citric acid becomes a base after metabolism in the liver whereas phosphoric acid is unchanged. Thus, fruit sodas provide base and colas provide acid to the body. The term *acidosis* or *alkalosis* refers to a pathologic process leading to acidic or alkaline pH, whereas *acidemia* and *alkalemia* refer to acidic and alkaline pH.[70] This means that patients could have acidosis but have alkaline pH if, for example, they had combined respiratory acidosis (retaining CO_2) and metabolic alkalosis (excess consumption of base, e.g., calcium carbonate tablets).

Whole Body Acid-Base Balance

Net Acid Production

In healthy individuals, respiratory acids are derived from CO_2 from cellular respiration, whereas metabolic acids are usually derived from the diet.[22,70] On a typical Western diet, the daily production of nonvolatile acid is approximately 90 mEq/day.[195] The main acid is sulfuric acid (~40 mEq/day), which originates from metabolism of the sulfur-containing amino acids methionine and cystine. Additional acid comes from organic acids that are not completely metabolized. The acid load varies greatly with the cystine and methionine content of proteins that are ingested and can be calculated from databases containing amino acid composition.[195,196] In general, when sulfur content is expressed as mEq/100 g of protein, proteins of animal sources (meat, fish, milk, and egg) contain higher amounts of sulfate for a given

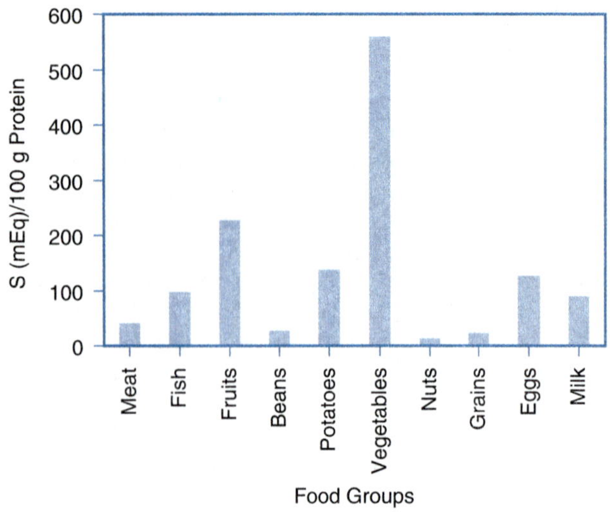

Figure 6.8 Sulfur (S) content of food groups.

Reprinted with permission from Oh MS, Uribarri J. Electrolytes, water, and acid-base balance. In: Shils ME, Shike M, Ross AC et al, eds. Modern Nutrition in Health and Disease. 10th ed. Burlington: Jones & Bartlett Learning, LLC, 2006:149–93.

amount of protein than proteins of plant origin (cereal, beans, and nuts) (**Figure 6.8**). The sulfur content per kcal is greater in fruits, vegetables, and potatoes, but these food groups are not important sources of protein in the amounts usually eaten. Inorganic phosphates as food additives may also increase the acid content of food (e.g., colas).

The total amount of acid load also depends on the alkali content of food, which is present mainly as salts of organic acids.[197] When both factors are considered, fruits and vegetables contain net alkali; milk can be mildly acidic to alkaline, depending on the species; and meat, fish, and grain have a net acid value (**Figure 6.9**). Purified vegetable protein, which is devoid of organic bases, still provides a net acid load. The same is true for isolated amino acids used in parenteral nutrition and

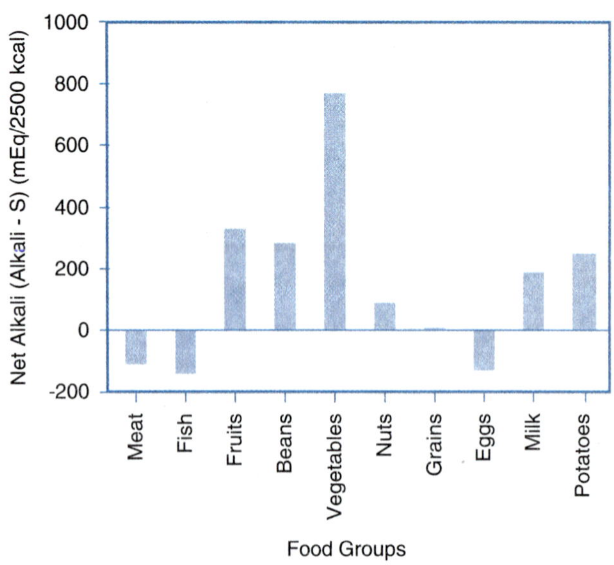

Figure 6.9 Net alkali content of food expressed as the alkali content of food minus sulfur (S) content. Only meat, fish, and eggs have negative alkali content. Vegetables have the highest alkali content when expressed by energy content.

Reprinted with permission from Oh MS, Uribarri J. Electrolytes, water, and acid-base balance. In: Shils ME, Shike M, Ross AC et al, eds. Modern Nutrition in Health and Disease. 10th ed. Burlington: Jones & Bartlett Learning, LLC, 2006:149–93.

in amino acid infusions (e.g., glutamine), which are included in some critical care protocols. Parenteral amino acid infusions are often buffered with additional base to prevent acidosis. On typical US diets, the amount of alkali absorbed from the GI tract is approximately 30 mEq/day.[198] If organic acids are normally metabolized to a base such as citrate and are excreted rather than metabolized, they leave their H^+ behind, thus creating a net acid load. Furthermore, stool loss of HCO_3^- and other organic bases also contribute to the acid load.

Measuring Dietary Acid or Base Load

Precise and accurate measurement of the amount of acid or base provided by the diet are difficult to obtain. As discussed earlier, measuring the acid load of the diet involves determining the number of sulfur-containing amino acids in addition to any added inorganic phosphate. Errors arise in measuring the net alkali content of the diet; however, because these measurements are based on the metabolic fates of the components after metabolism. Ions are often used because the final metabolism of accepting an H^+ (base) involves release of cations and that of releasing an H^+ by release of anions.[195] Thus, the net amount of organic base in food could be estimated from its ionic content (so-called strong ion difference) after metabolism.[194] For example, the total amount of noncombustible cations (Na^+, K^+, Ca^{2+}, and Mg^{2+}) relative to the total amount of noncombustible anions (Cl^- and P) provides a simple estimate of alkali content:

$$\text{Net alkali content} = (Na^+ + K^+ + Ca^{2+} + Mg^{2+}) - (Cl^- + 1.8\ P)$$

All units are expressed as milliequivalents per day, except P, which is expressed as millimoles per day multiplied by 1.8 and reflects the dependency of P valence on pH. Only the foregoing six ions are considered in the equation because other noncombustible ions are present in negligible amounts in the normal food. Sulfate is not included here because it is measured in the determination of the acid load.

Of course, foods induce differences in absorption, as well as stool losses of base.[199] A simple method to measure net GI alkali absorption involves running the analysis on urine electrolytes instead of the diet and stool electrolytes. The method is based on the principle that noncombustible ions absorbed from the GI tract would eventually be excreted in the urine, and the individual amounts of these electrolytes excreted in the urine would equal those absorbed from the GI tract. The formula would be the same as described earlier and uses a 24-hour urine collection performed while individuals eat the diet of interest.[199,200] If the clinician has no need to know the acid or base content of a diet, but only its net acid–base effects, renal net acid excretion (NAE) can be measured directly (see the later discussion of renal acid–base excretion) or estimated from dietary components.[201,202] Estimates of NAE have been derived from food frequency questionnaires and are based on electrolyte and protein content.[201,202] The sulfur content can be estimated from total dietary protein with a correction for dietary P. The potential base load is estimated by subtracting the effect of base using electrolytes, thus creating the so-called potential renal acid load (PRAL)[201]:

$$\text{PRAL (mEq/day)} = 0.49 \times \text{Protein (g/day)} + 0.037 \times \text{P (mg/day)} - 0.21 \times K^+ \text{(mg/day)} - 0.026 \times Mg^{2+} \text{(mg/day)} - 0.13 \times \text{Calcium } (Ca^{2+}) \text{(mg/day)}$$

PRAL correlates with renal NAE, so it is useful in epidemiologic studies, but it does not track the measured NAE in a linear fashion. An adjustment based on the metabolism or urinary loss of organic acids in the diet has been proposed. A correction factor (OA) for these organic acids on acid–base balance was measured to be OA = body surface area × 41/1.73.[203] This formula correlates with estimated intake from food frequency questionnaires and with 24-hour urine measurements.[203,204] Thus, the best estimate of net acid load in a diet is as follows:

$$NAE = PRAL + OA$$

A problem with this approach is that all ions in most foods in the diet must be known, but this is not the case for many prepackaged foods. In particular, the added P content of packaged foods is poorly known. Investigators wanted a simpler equation that used easily measurable components in the diet for use in cohort studies.[202] They chose K^+, the dominant marker of organic base in the diet (and one whose content in packaged food is reported), and they performed a ratio with dietary protein, the source of dietary acid, and fit the data to measured NAE. This formula, which approximates PRAL, is useful when knowledge of the foods is more limited: NAE (mEq/day) = −10.2 + 54.5 (dietary protein [g/day]/K^+ [mEq/day]).[202] One concern with this formula is that it was verified on populations obtaining most of their protein from meat (containing ample protein and K^+). Furthermore, the usefulness of this formula in epidemiologic studies is confounded by strong dependence on components that have independent effects on health outcomes.

Buffering

Because buffering occurs, protons (H^+) added or removed from body fluids do not result in an instant change in pH.[70,205] All body buffers, including the HCO_3^--CO_2 system, are in chemical equilibrium with protons and affect the pH according to the equilibrium equation[206]:

$$pH = \text{dissociation constant } (pK_a) + \log A^-/HA$$

where A is the buffer and A^- is a conjugate base of an acid HA. Because HCO_3^- and CO_2 are the major buffers of the body, pH is typically expressed as a function of their ratio, as in the Henderson–Hasselbalch equation:

$$pH = 6.1 + \log HCO_3^-/(Pco_2 \times 0.03)$$

where 6.1 is the pK_a of the HCO_3^- and CO_2 buffer system, and 0.03 is the solubility coefficient of CO_2.

The equation can be further simplified by combining the two constants, pK_a and the solubility coefficient of CO_2:

$$\begin{aligned} pH &= 6.1 + \log HCO_3^-/(Pco_2 \times 0.03) \\ &= 6.1 + \log 1/0.03 + \log HCO_3^-/Pco_2 \end{aligned}$$

Hence,

$$\begin{aligned} &= 7.62 + \log HCO_3^-/Pco_2 \\ pH &= 7.62 - \log Pco_2/HCO_3^- \\ &= 7.62 + \log HCO_3^- - \log CO_2 \end{aligned}$$

When H^+ is expressed in nanomolars instead of a negative log value (pH), Pco_2 can be related to HCO_3^- in the following equation:

$$H(nM) = 24 \times Pco_2 \text{ (mm Hg)}/HCO_3^- \text{ (mM)}$$

Clinically, the Henderson–Hasselbalch equation is useful because it points out that pH depends on the HCO_3^-/Pco_2 ratio.[76] pH increases when the ratio increases (alkalosis), and pH decreases when the ratio decreases (acidosis). The ratio may be increased by an increase in HCO_3^- (metabolic alkalosis) or by a decrease in Pco_2 (respiratory alkalosis). The ratio may be decreased by a decrease in HCO_3^- (metabolic acidosis) or by an increase in Pco_2 (respiratory acidosis).

Although the CO_2 buffer system is crucial in understanding acid–base disturbances, the complications of acidemia and alkalemia are better explained by bone and protein buffering.[207] The carbonate apatite in the bone matrix acts as a reserve of base during an acid load, which commonly occurs after a high-protein meal. Protons interact with the carbonate to form CO_2 and water, thus liberating Ca^{2+} and P from bone. The direct effect of an acid load is bone dissolution.[208] Proteins have carboxylic acid side groups that can donate or accept protons. This change in electrical charge alters protein folding or protein–protein interactions and causes cellular dysfunction; free Ca^{2+} is altered.[208,209] Acidosis increases free (ionic) Ca^{2+} by increasing the positive charges on protein and decreasing Ca^{2+} binding; alkalosis decreases free (ionic) Ca^{2+} by increasing the negative charges on protein and increasing Ca^{2+} binding.

Shifts of Protons Within the Body

Protons move in a regulated fashion into cells in net exchange for K^+, but this movement is more important for K^+ than for pH regulation.[118] The movement of acids and base into sequestered spaces has a far greater effect on serum pH.[70,209] For example, the movement of protons into the stomach to form stomach acid and the movement of HCO_3^- into the intestine to form bile and pancreatic secretions affect serum pH.[70] The release of stomach acid causes a postprandial "alkaline tide." This effect is caused by the delay between acid secretion into the stomach before and immediately on feeding and base secretion into the intestine that occurs on gastric emptying. The urine pH rises transiently after meals as the kidney excretes some of this base and increases the net acid load associated with the meal. The urine pH then falls as alkaline bile and pancreatic secretions are replenished. This transient rise in urine pH is eliminated by drugs that block gastric acid secretion.

Renal Acid–Base Excretion

The kidney protects the body from rises in pH by simply failing to reabsorb HCO_3^- when the pH is high, but renal excretion of acid is more complex.[70,205] Acid is excreted in the form of titratable acid (free protons and buffered protons on PO_4^- and sulfate anions) and ammonium (NH_4^+). Excretion of titratable acid is usually modest because of the limited amount of buffer that produces titratable acid (i.e., P, creatinine, and urate), but it may be increased markedly in disease states (e.g., β-hydroxybutyric acid in diabetic ketoacidosis). Normally, approximately two-thirds of acid excretion is in the form of NH_4^+, but in acidosis, NH_4^+ excretion may increase as much as 10-fold.

Nutritionally, it is critical to recognize that ammonia is required to excrete an acid load and is created from the destruction of the amino acid glutamine.[210] In response to acidemia, muscle releases branched-chain amino acids that the liver converts into glutamine. The proximal tubule deaminates glutamine to form a proton, ammonia, and HCO_3^-. Thus, the mechanism for renal acid secretion relies on increased glutamine production from muscle and liver and contributes to the catabolic effects of chronic metabolic acidosis (see later).[211] High K^+ blocks ammonia production, but this effect is blunted in the setting of protein malnutrition and the metabolic syndrome.[205,212]

The HCO_3^- formed from glutamine returns to the body, but the ammonia and proton are excreted into the lumen of the proximal tubule[70,205] (**Figure 6.10**). The proton is excreted in exchange for Na^+ through the Na^+ hydrogen antiporter 3. This exchange explains why altering Na^+ reabsorption affects the absorption of base in the proximal tubule, because the proton reacts with filtered HCO_3^- present in the tubule to form CO_2 and water (so-called HCO_3^- reclamation). If more HCO_3^- is filtered than the proximal tubule can reabsorb (because the blood level of HCO_3^- is high or the GFR suddenly rises), the excess is delivered to the distal nephron. Most often, this results in a loss of HCO_3^- from the body and net alkali secretion. Electroneutrality is maintained by coupling the HCO_3^- with Na^+. As discussed earlier, this HCO_3^- drag influences Na^+ and K^+ balance.[119] The distal nephron contains both hydrogen pumps (primarily the vacuolar H ATPase) to add acid to the final urine and Cl^--HCO_3^- exchangers (primarily pendrin) to add base to the final urine.[205] Acidemia or increased activity of the renin-angiotensin-aldosterone system increases acid secretion and decreases base secretion, whereas alkalemia or reduced activity of this physiologic system decreases acid secretion and increases base secretion.

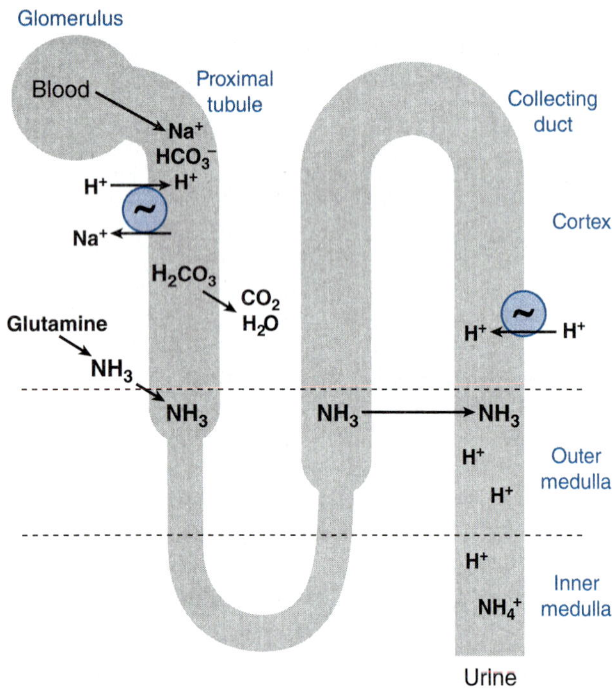

Figure 6.10 Renal acid excretion. Bicarbonate (HCO_3^-) and sodium (Na^+) are filtered from the blood at the glomerulus. In the proximal tubule, hydrogen ions (H^+) are excreted in the urine in exchange for sodium, and the hydrogen and bicarbonate form carbonic acid, which dissolves into carbon dioxide and water. The carbon dioxide and water are reabsorbed, and the net result is that the filtered bicarbonate is not lost in the urine. The proximal tubule makes ammonia (NH_3) from glutamine, which is transported to the collecting duct. Hydrogen ions are pumped into the lumen. Some react with NH_3 to form ammonium (NH_4), which does not lower the urine pH, whereas others remain in solution to lower the urine pH. Thus, acid is excreted as hydrogen ions and ammonium, and the relative amounts of each determine the final pH of the urine.

On a Western diet, net acid secretion in the distal nephron always occurs, and the final urinary pH is academic.[195] Potential bases in the urine such as uric acid buffer the pH.[70,205] The ammonia created in the proximal tubule is also excreted into the lumen and travels to the distal tubule, where it accepts protons being secreted in the distal nephron to form NH_4^+. Thus, NH_4^+ allows acid excretion without lowering the final pH of the urine.

This final urinary pH is relevant to the effects of diet on kidney stone formation.[212,213] Western diets are associated with the formation of kidney stones in an acid urine pH such as calcium oxalate and uric acid stones. Uric acid dissolves at a pH higher than 6, so uric acid stones require a persistently acidemic urine pH. As discussed earlier, protein malnutrition or metabolic syndrome lowers urinary pH during acidosis by suppressing ammonia production, and these conditions stimulate the formation of uric acid kidney stones.[212] Consumption of excess base (e.g., calcium citrate or carbonate dietary supplements) raises the pH of the urine. At a pH higher than 6.8, PO_4^- accepts a second proton. This shift makes $CaPO_4^-$ much less soluble and leads to $CaPO_4^-$ stones, thus explaining the increased risk of kidney stones in patients taking these supplements.[213]

In summary, NAE, which is equivalent to net renal production of alkali, can be determined by subtracting HCO_3^- excretion from acid excretion:

$$\text{NAE} = \text{Acid excretion} - HCO_3^- \text{ excretion} = \text{Urinary } NH_4^+ + \text{Urinary titratable acid}$$

where titratable acid includes urinary pH + phosphates and other titratable acids–urinary HCO_3^-.

Maintenance of acid–base balance requires that net acid production equals NAE. Metabolic acidosis develops when net acid production exceeds NAE, and metabolic alkalosis develops when NAE exceeds net acid production.

Metabolic Acidosis

Metabolic acidosis is defined as an excess of protons and thus a fall in the HCO_3^- content of the body.[22,70] A reduction in HCO_3^- content may be caused by a primary increase in acid production (extrarenal acidosis) or by a primary reduction in NAE (renal acidosis) (**Table 6.9**). In this classification, nonrenal loss of HCO_3^- or an alkali precursor is considered part of increased acid production. In extrarenal acidosis, NAE is markedly increased as the kidney compensates to overcome acidosis. Conversely, NAE may be restored to normal in chronic renal acidosis because acidosis stimulates renal H^+ excretion. Normal NAE in the presence of an acidic pH suggests a defect in renal acid excretion and, therefore, renal acidosis. If the renal acid excretion capacity is normal, NAE should be supernormal in the presence of an acidic pH.

Complications of Acidemia

The body readily tolerates mild acute metabolic acidosis.[208] This condition is seen during strenuous exercise (e.g., sprinting) when the blood pH falls to less than 7.2 without immediate adverse effects. In some ways, acute acidosis helps the body in the setting of exercise or acute illness. Principally, increased oxygen delivery to the tissues occurs through the Bohr effect as oxygen is shifted from hemoglobin to the tissues. Vasodilation and stimulation of

Table 6.9 Causes of Metabolic Acidosis

High anion gap	Anion
A. Lactic acidosis	Lactate
B. Ketoacidosis	β-Hydroxybutyric acid, acetoacetic acid
C. Renal failure	Sulfate, phosphate, urate
D. Ingestions	
1. Salicylate	Salicylate, lactate
2. Methanol	Formate
3. Ethylene glycol	Glycolate, oxalate
4. Paraldehyde	Acetate, chloroacetate
5. Toluene	Hippurate
E. Massive rhabdomyolysis	
Normal anion gap (hyperchloremic acidosis)	
A. Gastrointestinal loss of bicarbonate	
1. Diarrhea	
2. Pancreatic, biliary, or intestinal fistulas	
3. Ureterosigmoidostomy	
B. Renal bicarbonate loss	
1. Type 2 (proximal) renal tubular acidosis	
C. Renal dysfunction	
1. Some cases of renal failure	
2. Hypoaldosteronism	
3. Type 1 (distal) renal tubular acidosis	
D. Ingestions	
1. Ammonium chloride	

the respiratory drive act to aid tissue respiration and minimize lactic acidosis.[214] Severe metabolic acidosis (pH <7.1), however, can lead to ventricular irritability, arrhythmias, resistance to the effects of catecholamines, and lower blood pressure.[207,215] Severe acidemia at a pH less than 6.8 is not well tolerated; shock, coma, respiratory failure, and death may ensue.

On a long-term basis, the body does not tolerate even small degrees of acidemia.[208] Catabolism of both bone and muscle occurs with chronic acid loading even with a normal blood pH. This is because much of the catabolic response to acidemia is caused by chronic activation of the body's own mechanisms that buffer and excrete excess H^+.[208,211] The chronic buffering of acid by bone leads to both a direct loss of bone mineral and an indirect loss related to hormonal changes.[207] Elevated cortisol and parathyroid hormone increase bone turnover. At the same time, acidosis induces a direct inhibition of tubular Ca^{2+} reabsorption that is independent of Ca^{2+} concentration, so the usual feedback loops of high Ca^2 are of no assistance.[207] In addition, acidosis causes a fall in the urinary concentration of citrate, a major urinary inhibitor of Ca^{2+} crystallization, so an increase in urinary Ca^{2+} is associated with increased risk of calcium oxalate kidney stones.[216] Acidosis induces both osteoporosis in adults and reduced bone growth in children, effects that are reversed by correction of the acidosis. The catabolic effects of acidosis on bone may apply to patients who merely eat a diet with a high net acid load. Adding base to the diet of middle-aged women had a positive effect on bone density.[217]

As discussed earlier, muscle protein is broken down to yield amino acids that are converted to the glutamine that forms the substrate for NH_4^+ secretion.[210,211] Acidemia also induces negative nitrogen balance and stimulates renal ammonia excretion. By this mechanism, acid directly stimulates glutamine production by increasing muscle protein degradation and enhancing oxidation of branched-chain amino acids to provide substrate for the liver to synthesize glutamine.[212] Furthermore, acidemia increases catabolic hormones such as cortisol and parathyroid hormone to increase muscle wasting, and induces muscle-specific resistance to the action of anabolic hormones (insulin and insulinlike growth factor-I). This catabolic effect may apply to a net acid diet: middle-aged women demonstrated marked improvement in nitrogen balance when their diets were supplemented with base.[108] Base supplements are an effective ergonomic aid for sports performance,[218] may be useful for muscle rehabilitation in elderly patients,[219,220] and may improve muscle mass in CKD.[221] Thus, correction of acidemia has major nutritional benefits on Ca^{2+} and muscle metabolism.

High-protein diets can increase kidney damage in patients with CKD, and acidemia is one potential mediator of this effect.[211] The adrenal hormone aldosterone is stimulated by acid and augments renal net acid secretion.[22,70] Aldosterone has been implicated in the progression of congestive heart failure and CKD.[211,221,222] Although no cardiac studies have been conducted, reducing acidemia with $NaHCO_3$ has been shown to slow the loss of kidney function in patients with advanced CKD.[221]

Hyperchloremic Acidosis

Metabolic acids exist in two forms: (a) they are associated with the addition of HCl or the removal of equivalent amount of HCO_3^- or (b) they originate from the addition of a nonmetabolizable organic acid.[22,70,205] When HCl is added or HCO_3^- is lost, serum Cl^- concentration rises and the serum HCO_3^- concentration falls, thus causing so-called hyperchloremic metabolic acidosis. This condition results from the loss of HCO_3^- in the stool with diarrhea or from the retention of H^+ in CKD. When a nonmetabolized acid is added, hyperchloremic acidosis results as the organic anion is lost in the urine following release of H^+. As HCO_3^- falls, Cl^- shifts out of cells to maintain electroneutrality, whereas the sulfur anion is lost in the urine in the individual eating a high-protein diet. The extant proton combines with Cl^- and causes hyperchloremic acidosis. A similar effect occurs when parenteral amino acid infusions are not buffered with sufficient base.

Causes of and Therapies for Hyperchloremic Acidosis. Correct diagnosis of the cause of metabolic acidosis is important because diets and therapy vary.[22,70,205] Diarrheal loss of HCO_3^- is the most common cause of hyperchloremic acidosis and is usually suspected from the patient's history. When diarrhea is not obvious, the culprit is often laxatives, lactose intolerance, gluten-sensitive enteropathy, hyperosmolar enteral liquid feeding products, drugs or dietary supplements that increase stool volume, or surgical drainage of HCO_3^--rich biliary or pancreatic fluid. Because urinary excretion of NH_4^+ increases as a part of renal compensatory mechanisms, direct or indirect measurements of urine NH_4^+ can be used to confirm this diagnosis.[211] Given that acidosis delays restoration of protein stores and enhances bone anabolism, the catabolic effects of chronic acidosis play a major role in the nutritional abnormalities resulting from chronic diarrhea. The use of base supplementation as K^+, Ca^{2+}, or Na^+ salts can provide significant benefit.[223] In selecting feeding products, it is important to provide adequate, but not excessive, protein and to consider the osmolar load and ease of digestion to hasten recovery.

Uremic acidosis seen in acute kidney injury or CKD can be readily diagnosed by the measurement of serum creatinine and blood urea nitrogen[22,70,222] Because development of renal acidosis depends on acid production, it varies greatly according to the protein and vegetable content of the diet.[211] Unfortunately, the low-K^+ diet that is sometimes required in advanced CKD is often deficient in organic base, and supplementary HCO_3^- tablets are required. Studies also suggest that acidosis frequently occurs at a GFR of 40 mL/minute (stage 3 CKD) when excessive protein or insufficient base is consumed or when renal tubular acidosis (RTA) is present (see later). With moderate protein restriction, uremic acidosis may develop as GFR falls to less than 20 mL/minute (stage 4 CKD).[211] Dietary protein restriction and $NaHCO_3$ are effective therapies for acidosis in CKD.[211,221,224] Both therapies have been shown to have benefits on muscle and bone anabolism and in slowing the loss of kidney function. Protein restriction must be part of a carefully designed and monitored diet, however (see the chapter on kidney disease).

RTA occurs when the kidney cannot excrete acid and advanced CKD is absent.[22,70,224] Three types of RTA are known. Type I RTA, also called classic RTA or distal RTA, is characterized by an inability to reduce urine pH maximally to less than 5.5. Type I RTA can develop as a primary disorder or secondary to drug toxicity, tubulointerstitial renal diseases, autoimmune disease, or other renal diseases.[224] It is associated with low serum K^+ and kidney stones.[70,224,225] Dietary management hinges on a K^+-rich- and base-rich diet high in fruit and vegetables and low in animal protein (see the DASH diet). K^+ base supplements are effective in treatment.

Type II RTA, also called proximal RTA, causes defective proximal HCO_3^- reabsorption.[22,70,224] Most patients with proximal RTA have evidence of generalized proximal tubular dysfunction (i.e., Fanconi syndrome), which is manifested by bicarbonaturia, aminoaciduria, glycosuria, phosphaturia, and uricosuria. Of these conditions, renal glycosuria (glycosuria in the presence of normal blood glucose) is most useful in diagnosing Fanconi syndrome. Type II RTA may be a primary disorder, secondary to genetic or acquired renal dysfunction, or induced by drugs that inhibit carbonic anhydrase. Hypokalemia is a characteristic finding of both type I and type II RTA, but it tends to be more severe in type I than in type II RTA. Because HCO_3^- losses in the urine cannot be repleted by diet alone, dietary considerations often focus on K^+ and PO_4^- repletion. High-dose therapy using K^+, Na^+, and Ca^{2+} base is required to normalize acid–base status.

Type IV RTA is caused by either aldosterone deficiency or tubular unresponsiveness to aldosterone and results in impaired renal tubular K^+ secretion and hyperkalemia.[171] Although reduced H^+ secretion in the collecting duct plays a role, the major mechanism of acidosis in type IV RTA is hyperkalemia-induced impairment in ammonia production in the proximal tubule. Thus, control of K^+ is the major nutritional consideration. As discussed earlier, a diet low in K^+ is often low in base content, and $NaHCO_3$ is an effective supplement.

Anion Gap Acidosis

Massive overproduction of organic acid occurs in two syndromes (lactic acidosis and ketoacidosis) that result from catastrophic failure in energy metabolism.[22,70,223] In this situation, the kidney is unable to excrete the organic acid as quickly as it is produced, and the anion is retained in the body. As the anion accumulates in the blood as a negative charge, the released H^+ causes a fall in the serum HCO_3^- concentration without any change in the serum Cl^- concentration. Because the anion is not measured on routine clinical chemistry tests, the patient's serum appears to have a missing anion.[22,70,224] This "anion gap" is used clinically to detect acidosis secondary to nonmetabolizable organic acids that are not lost in the urine.

It is important to remember that the term *anion gap* refers to a gap between measures and actual cation and anion differences; the concentration of total cations in the serum must be exactly equal to the concentration of total anions.[22,70] A change in serum Na^+ usually does not alter the AG because the serum Cl^- usually changes in the same direction. Given that normal serum K^+ concentration is quantitatively a minor component of serum electrolytes, the AG is estimated from the concentrations of Na^+, Cl^-, and HCO_3^- as follows:

$$AG = Na^+ - (Cl^- + HCO_3^-)$$

Because unmeasured cations outnumber unmeasured anions, the normal value is approximately 10 mEq/L (8 to 14 mEq/L). Although the total concentration of unmeasured anions (i.e., all anions other than Cl^- and HCO_3^-) is approximately 23 mEq/L, the AG is only 12 mEq/L because of the presence of approximately 11 mEq/L of unmeasured cations (i.e., all cations other than Na^+).[226] An increased AG is most often caused by accumulation of anions of acids, such as sulfate, lactate, or ketones. Because albumin is a negatively charged protein, it is responsible for a large part of the normal AG—a decreased AG is most commonly the result of a reduction in serum albumin concentration.[224,227] The AG also varies with other influences, such as severe changes in pH and hypergammaglobulinemia. Proper interpretation of a serum AG requires the knowledge of the existence of conditions that influence the AG even though they may have no direct effect on metabolic acidosis. For example, if a person with hypoalbuminemia develops lactic acidosis with renal failure, the AG could be normal because the low albumin concentration and the accumulated lactate have opposite effects on the AG. Not knowing the effect of serum albumin on AG, one could overlook the existence of lactic acidosis based on a normal AG.

When the AG is increased, likely causes include organic acidosis (lactic and ketoacidosis, uremic acidosis, and acidosis resulting from certain toxins)[22,70,224] (see Table 6.9). Most cases of acidosis in CKD are accompanied by normal AG, because the kidney still clears the sulfur from proteins adequately. Only in advanced chronic and acute renal failure is the AG increased. When the GFR falls to less than 15 mL/minute, sulfates from dietary amino acids accumulate in the blood and lead to AG acidosis.

Lactic Acidosis. In cellular metabolism, carbohydrates are first reduced to pyruvic acid (pyruvate) through glycolysis. The pyruvic acid can either enter the mitochondria to be metabolized to CO_2 and water in the presence of oxygen in the tricarboxylic acid cycle (Krebs cycle) or be recycled to lactic acid and eventually back to glucose. Thus, availability of oxygen is usually the major regulator of lactate production and explains the clinical use of lactic acidosis as a critical marker of tissue hypoxia.[225] Hypoxia may arise from increased oxygen demand (e.g., exercise, seizures, cancer) or from decreased oxygen delivery (e.g., shock, severe anemia respiratory failure, carbon monoxide poisoning).[224] Lactate dehydrogenase and the cofactor reduced nicotinamide adenine dinucleotide (NADH) are required for both lactate production (in the peripheral tissue, usually muscle) and for lactate conversion back to pyruvic acid (in the liver and, to a lesser extent, in the kidney.) For this reason, hypoxia sometimes affects both production and metabolism of lactic acid. An elevated concentration of pyruvic acid and an increased $NADH/NAD^+$ ratio raise lactic acid production while reducing its metabolism. Decreased metabolism of lactate is largely responsible for lactic acidosis in acute alcoholism (ethanol increases the NADH/NAD^+ ratio), salicylate and metformin poisoning, and severe liver disease.[223,224]

Ketoacidosis. Ketoacids, acetoacetic acid, and β-hydroxybutyric acid are produced in the liver from free fatty acids (FFAs) and are metabolized by the extrahepatic tissues.[22,70,228] Increased production of ketoacids, the main mechanism for ketoacid accumulation, requires a high concentration of FFAs and their conversion to ketoacids in the liver. Insulin deficiency is responsible for increased mobilization of FFAs from the adipose tissue, whereas glucagon excess and insulin deficiency stimulate conversion of FFAs to ketoacids in the liver. The initial step in ketoacid production from FFAs is the entry of FFAs into the mitochondria, a process that requires acylcarnitine transferase. This step is stimulated by glucagon excess. The next step is metabolism of FFAs to acetyl-coenzyme A, and then finally to ketoacids (**Figure 6.11**). Diversion of acetyl-coenzyme A to fatty acid resynthesis requires the enzyme acetyl-coenzyme A carboxylase. Inhibition of this enzyme by insulin deficiency, glucagon excess, or an excess of stress-induced hormones further contributes to increased ketoacid synthesis.

It follows that the absence of an insulin effect, FFA availability, and increased glucagon are required for ketone formation in the liver. Although these conditions are present during prolonged starvation (starvation ketosis), full expression of life-threatening ketoacidosis occurs only in diabetic and alcoholic ketoacidosis. Only those patients with diabetes with very low insulin production (type 1 diabetes) or with the most severe insulin resistance develop severe ketoacidosis.[228,229] In patients with type 1 diabetes, the lack of insulin stimulates glucagon production to a greater extent than starvation alone and mobilizes larger amounts of FFAs. The hyperglycemia leads to Na^+ loss in the urine through osmotic diuresis, whereas nausea and vomiting lead to a decreased intake of Na^+. The lack of Na^+ compromises kidney function, traps the ketones in the body, and blocks acid excretion. Insulin, intravascular volume expansion, and usually additional glucose are required to treat diabetic ketoacidosis. Although ethanol is metabolized to a ketone, acetoacetate, by alcohol dehydrogenase, casual alcohol consumption does not lead to ketosis.[70] Prolonged alcohol consumption in the absence of carbohydrate intake leads to a marked increase in glucagon and insulin suppression. FFA release is stimulated, whereas Na^+ depletion results from ethanol's inhibition of vasopressin secretion.[129] Carbohydrate and NaCl are sufficient to reverse alcoholic ketoacidosis. Because starvation is associated with relatively higher insulin and lower glucagon and FFAs, the amount of ketone formation is lower and does not lead to life-threatening acidemia.

The bases of the ketoacids are rapidly cleared in the urine following loss of their respective proton.[22,70,228] The two major ketoacids, acetoacetic acid and β-hydroxybutyric acid, can be converted from one to another by β-hydroxybutyric acid dehydrogenase and the cofactor NADH. Consequently, the ratio of NADH to NAD^+ is the major determinant of the ratio of acetoacetic acid to β-hydroxybutyric acid. This is important because urinary testing for ketoacids detects acetoacetic acid only. When β-hydroxybutyric acid is the primary ketone, the diagnosis of ketoacidosis can be missed. This situation can occur during shock or during alcoholic ketoacidosis.

Various poisons can lead to AG acidosis. Salicylate (aspirin) overdose can lead to lactic acidosis.[22,70] Like ethanol, other alcohols, including methanol (wood alcohol) and propylene glycol (antifreeze), are metabolized by alcohol and aldehyde dehydrogenase.[230] Instead of the formation of acetone and subsequently acetoacetic acid with ethanol, methanol forms formaldehyde and formic acid, and ethylene glycol forms glycolic acid. Glycolic acid is eventually metabolized to oxalate. Formaldehyde causes the blindness characteristic of methanol poisoning, whereas oxalate crystallizes in the urine with Ca^{2+} at high concentrations to cause acute kidney injury. Finally, gut bacteria can create numerous organic acids that are difficult to measure when they overgrow in the GI tract.

Treatment. Correction of the underlying disorder is the best therapy for AG metabolic acidosis.[231,232] When possible, promoting metabolism of the organic acid is extremely effective. For

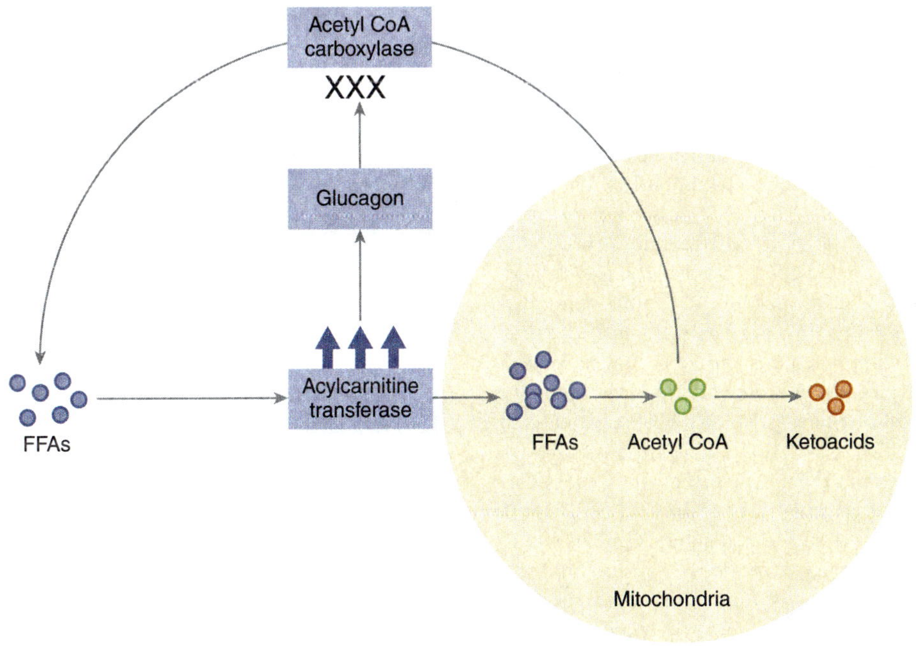

Figure 6.11 Metabolism of free fatty acids to ketoacids.

example, when ketoacidosis is treated with insulin, the acetoacetate and β-hydroxybutyric acid are metabolized in the Krebs cycle–producing base. Thus, exogenous alkali is seldom necessary in ketoacidosis. When lactic acid results from physiologic overproduction, in exercise or during seizures, it is rapidly metabolized. Restoration of tissue oxygen delivery in hypoxia-induced lactic acid is equally effective, but it is often not possible because of the severity of the disease.

Rapid restoration of normal pH with base therapy is usually unnecessary and may be undesirable for several reasons.[232] A sudden increase in extracellular pH can cause paradoxical acidosis of the cerebrospinal fluid. Rapid restoration of a normal serum HCO_3^- concentration in metabolic acidosis would also be undesirable because persistent hyperventilation would produce high blood pH. The initial aim in the treatment of severe metabolic acidosis should be to increase the blood pH to a level at which the adverse cardiovascular effects of severe acidemia can be avoided. Although the risk of acidosis varies with the age and the cardiovascular status of patients, critical care guidelines suggest acutely targeting replacement to a pH of 7.15.[215]

Respiratory Acidosis

Causes and Pathogenesis

Decreasing ventilation of the lungs causes CO_2 accumulation in the body.[22,70,204] This causes respiratory acidosis and diseases of the lung (most common), respiratory muscle, respiratory nerve, thoracic cage, and airways and suppression of the respiratory center by stroke, drugs such as phenobarbital, or severe hypothyroidism (**Table 6.10**). Although studies suggest subtle differences between the complications of respiratory acidosis and those of metabolic acidosis, in clinical practice, they are largely indistinguishable, except for nutritional effects on muscle. Renal compensation for respiratory acidosis involves retaining HCO_3^-, so ammonia production (and thus glutamine use) is not as severely affected.[233]

Table 6.10 Causes of Acute and Chronic Respiratory Acidosis

Inhibition of the medullary respiratory center
- **A. Acute**
  1. Drugs: opiates, anesthetics, sedatives
  2. Central sleep apnea
  3. Cardiac arrest
- **B. Chronic**
  1. Extreme obesity

Disorders of the respiratory muscles and chest wall
- **A. Acute**
  1. Muscle weakness
     - Myasthenia gravis
     - Periodic paralysis
     - Guillain-Barré syndrome
     - Severe hypokalemia
     - Severe hypophosphatemia
- **B. Chronic**
  1. Muscle weakness
     - Poliomyelitis
     - Amyotrophic lateral sclerosis
     - Multiple sclerosis
     - Myxedema
  2. Kyphoscoliosis
  3. Extreme obesity

Upper airway obstruction
- **A. Acute**
  1. Aspiration of foreign body or vomitus
  2. Obstructive sleep apnea
  3. Laryngospasm

Disorders affecting gas exchange across the pulmonary capillary
- **A. Acute**
  1. Exacerbation of underlying lung disease (increased carbon dioxide production with a high-carbohydrate diet)
  2. Adult respiratory distress syndrome
  3. Acute cardiogenic pulmonary edema
  4. Severe asthma or pneumonia
- **B. Chronic**
  1. Chronic obstructive pulmonary disease
     - Bronchitis
     - Emphysema
     - Mechanical ventilation

Treatment

All cases of respiratory acidosis are caused by alveolar hypoventilation, but in severe respiratory acidosis, efforts should be made to normalize ventilation.[204] When restoration of effective ventilation is delayed and the patient is comatose or has cardiac arrhythmias, acidosis can be treated temporarily with alkali. Administration of HCO_3^- is not effective in correcting the brain pH, however, because of the slow penetration of HCO_3^- into the central nervous system. After 24 to 96 hours, the kidneys respond to respiratory acidosis and control the pH with renal compensation.[233] As a result, hypoxia poses a greater problem than the chronic respiratory acidosis. Various drugs have been used to stimulate respiration in acute and chronic respiratory acidosis.[22] The apnea of prematurity has been treated with respiratory stimulants such as doxapram and the methylxanthines, caffeine citrate, and theophylline. Caffeine occurs naturally in coffee and is a common food additive. Theophylline occurs naturally in black tea and related plants. Caffeine citrate is the consensus drug of choice for premature babies with respiratory depression. In chronic obstructive pulmonary disease, theophylline, doxapram, and progesterone derivatives including progesterone, medroxyprogesterone, and chlormadinone have been used.[229]

Metabolic Alkalosis

Causes and Pathogenesis

At a normal serum HCO_3^- concentration on a Western diet, HCO_3^- filtered at the glomerulus is virtually completely reabsorbed.[70,205,209] As serum HCO_3^- concentration rises, HCO_3^- reabsorption is incomplete, and bicarbonaturia begins. A slight increase in serum HCO_3^- to more than 24 mEq/L causes marked bicarbonaturia. Hence, when renal tubular HCO_3^- handling and GFR are normal, it is difficult to maintain high plasma HCO_3^- concentration unless an enormous amount of HCO_3^- is given. Therefore, maintenance of metabolic alkalosis requires two conditions: a mechanism to increase plasma HCO_3^- and a mechanism to maintain the increased concentration. HCO_3^- concentration may be increased by administration of alkali, gastric loss of HCl through vomiting or nasogastric suction, or renal generation

Table 6.11 Causes of Metabolic Alkalosis

Chloride sensitive (low urine chloride)

A. Gastrointestinal loss of acid and chloride
1. Vomiting, nasogastric suction
2. Villous adenoma

B. Renal
1. Diuretics
2. Posthypercapnic status
3. Low chloride intake
4. Hypercalcemia
5. Anion drag (usually penicillin)

C. Base administration
1. Calcium base (calcium carbonate, calcium citrate) Milk alkali syndrome
2. Sodium or potassium base (sodium bicarbonate, potassium citrate)
3. Blood products (citrate)
4. High acetate parenteral nutrition

With low urine chloride but no response to chloride

A. Effective circulating volume depletion
1. Kidney failure
2. Heart failure
3. Cirrhosis of liver

Chloride resistant (normal urine chloride)

A. Increased aldosterone action
1. Hyperaldosteronism
2. Cushing syndrome
3. Natural licorice
4. Pseudohyperaldosteronism
5. Severe hypokalemia

of HCO_3^-. Plasma HCO_3^- concentration can be maintained at a high level if HCO_3^- is not filtered at the glomerulus because of advanced renal failure or if the filtered HCO_3^- is reabsorbed avidly because of an increased renal threshold for HCO_3^-.[22] The two most common causes of increased renal HCO_3^- threshold are volume depletion and K^+ depletion, but excess aldosterone may also be a cause (**Table 6.11**).

When metabolic alkalosis results from volume depletion, urinary excretion of Cl^- is reduced.[22,70,209] Measurement of urinary Na+ is an unreliable index of volume depletion in metabolic alkalosis because excretion of HCO_3^- causes obligatory loss of Na+ despite volume depletion. Metabolic alkalosis accompanied by low urinary Cl^- can be corrected by administration of Cl^--containing food or intravenous solution, such as NaCl or KCl, hence the term Cl^--responsive metabolic alkalosis. Patients with Cl^--responsive metabolic alkalosis are volume depleted, and the response to Cl^- alone suggests that Cl^- helps regulate extracellular volume.[17] When volume depletion is caused by primary renal Na+ loss (e.g., with diuretics), urinary loss of Cl^- is not reduced. In edema-forming conditions (heart failure, cirrhosis), administration of Cl^- may not improve the metabolic alkalosis. Although the pattern of urinary excretion of Cl^- would suggest that Cl^--responsive metabolic alkalosis, fluid administration usually does not restore the effective vascular volume to normal.[209]

Metabolic alkalosis accompanied by normal excretion of Cl^- in urine is called Cl^--resistant metabolic alkalosis (e.g., hypokalemia-induced alkalosis); administration of Cl^- does not correct the alkalosis in such condition.[209] Increased activity of aldosterone is the most common cause of Cl^--resistant alkalosis, but severe hypokalemia can also produce this condition. The renal threshold for HCO_3^- is increased in K^+ depletion because of enhanced tubular reabsorption of HCO_3^- and a decrease in GFR, which could reduce the filtered load of HCO_3^-. This condition responds rapidly to K^+ repletion.

Complications

Chronic mild alkalemia is extremely well tolerated and may be beneficial nutritionally to counteract the negative effects of the acid Western diet.[108,217] Moderate alkalemia, however, may be a sign of volume depletion (Cl^- deficiency) from overdiuresis, chronic vomiting, heart failure, or other states.[209] Ca^{2+} deposition in the kidney (stones or nephrocalcinosis) and perhaps in blood vessels may become a serious problem if the alkalemia is driven by excess base intake, especially in the form of calcium salts.[209,213]

Acutely, the major problems with metabolic alkalosis are decreased tissue oxygen delivery and a fall in Ca^{2+}.[209] Just as acidemia is a vasodilator, alkalemia causes vasoconstriction. In addition, the Bohr effect reduces the release of oxygen from hemoglobin at a high pH. Thus, alkalemia is associated with decreased tissue oxygen delivery. The respiratory compensation for metabolic alkalosis is a decreased respiratory drive, so alkalosis can complicate the management of respiratory failure. Ca^{2+} (a divalent cation) binds the carboxylic acids of proteins, lowering its effective concentration. The free or ionized Ca^{2+} falls with alkalemia as proteins buffer the rise in pH by donating protons. The free negative charges bind Ca^{2+} and lower its bioavailability. Thus, symptoms of acute severe alkalemia are largely those of hypocalcemia. At a pH higher than 7.75, sufficient free Ca^{2+} is not usually present for normal cardiac contractility, and death rapidly ensues.

Treatment

When an increased renal HCO_3^- threshold in metabolic alkalosis is caused by reduced effective vascular volume and hypokalemia, correction of these abnormalities leads to rapid restoration of serum HCO_3^- concentration in most patients. Correction of low effective vascular volume can be accomplished by administration of saline as a crystalloid, which both expands plasma volume and provides Cl^-; however, other crystalloids or expansion products can be used as well. Sometimes, discontinuation of an offending agent (e.g., a diuretic) and restoration of salt intake are sufficient. If volume depletion is to be corrected, Cl^- must be given to replace the excreted HCO_3^-, either as NaCl or KCl.[214] In certain clinical situations, such as edema-forming states, treatment of reduced effective vascular volume with salt solution is not effective. In such situations, acetazolamide (Diamox), a carbonic anhydrase inhibitor, treats metabolic alkalosis as well as edema.[209] Acetazolamide administration usually reduces the renal HCO_3^- threshold to a subnormal level, but the HCO_3^- threshold may remain supernormal despite the drug in patients with severe volume depletion. In renal failure, metabolic alkalosis can be treated by administration of dilute HCl or acidifying salts or by dialysis. Acidifying salts include NH_4^+Cl, arginine Cl, and lysine Cl. Metabolism of these salts results in release of HCl, which then neutralizes HCO_3^-. Direct administration of HCl into large veins has also been used. If continuous acid loss from the stomach is the cause of metabolic alkalosis, an inhibitor of acid secretion such as histamine (H_2) blockers or proton pump inhibitors may be used.

Respiratory Alkalosis

Causes and Pathogenesis

Except for respirator-induced alkalosis and voluntary hyperventilation, respiratory alkalosis is the result of stimulation of the respiratory center[22,48,204] (**Table 6.12**). The two most common causes of respiratory alkalosis are hypoxic stimulation of the respiratory center and stimulation through the pulmonary receptors caused by various lung lesions, such as pneumonia, pulmonary congestion, and pulmonary embolism. Certain drugs, such as salicylate and progesterone, stimulate the respiratory center directly.[234] Respiratory alkalosis is common in Gram-negative sepsis and liver disease through an unknown mechanism. Blood pH tends to be extremely high when respiratory alkalosis is caused by psychogenic stimulation of the respiratory center because the condition is usually superacute, and; therefore, no time exists for compensation. Complications of severe respiratory alkalemia are related to low ionized Ca^{2+} concentration.

Table 6.12 Causes of Respiratory Alkalosis

Hypoxemia
- **A. Pulmonary disease**
  - Pulmonary edema
  - Pneumonia
  - Pulmonary fibrosis
  - Pulmonary emboli
- **B. Congestive heart failure**
- **C. Severe anemia**
- **D. High-altitude residence**

Direct stimulation of the medullary respiratory center
- **A. Psychogenic or voluntary hyperventilation**
- **B. Hepatic failure**
- **C. Gram-negative septicemia**
- **D. Salicylate intoxication**
- **E. Pregnancy**
- **F. Status after correction of metabolic acidosis**
- **G. Neurologic disorders: cerebrovascular accidents, pontine tumors**

Treatment

In chronic respiratory alkalosis, treatment is usually not needed because renal compensation restores the blood pH to nearly normal values.[204] Mild alkalosis may be beneficial nutritionally. In acute respiratory alkalosis resulting from psychogenic hyperventilation, Pco_2 can be increased using a rebreathing bag. If this fails, the patient may need sedation to depress the respiratory center.

Acknowledgments

The authors would like to acknowledge James L. Bailey, Jeff M. Sands, Harold A. Franch, Man S. Oh, and Jaime Uribarri for their contributions to previous versions of this chapter.

References

1. Carroll HJ, Oh MS. Water, electrolyte, and acid-base metabolism. Philadelphia: JB Lippincott, 1989.
2. Young A, Brown LK, Ennis S, et al. Arch Dis Childhood Fetal Neon Ed 2021;106:542–548.
3. Berl T, Robertson GL. Pathophysiology of water metabolism. In: Brenner BM, ed. The Kidney. 6th ed. Philadelphia: WB Saunders, 2000:866–924.
4. Altman PL, Dittmer DS, eds. Blood and Other Body Fluids. Washington, DC: Federation of American Societies for Experimental Biology, 1961.
5. Gamble JL Jr, Robertson JS, Hannigan CA, et al. J Clin Invest 1953;32:483–9.
6. Sterns RH, Palmer BF, eds. Nephrol Self Assess Program 2007;6:21–272.
7. Hendry EF. Clin Chem 1961;155:154 64.
8. Conway EJ. Physiol Rev 1957;37:84–132.
9. Maffly RH, Leaf A. Gen Physiol 1959;42:1257–75.
10. Conway EJ, McCormack JI. J Physiol (Lond) 1953;120:1.
11. Maffly RH, Leaf A. Nature 1958;182:60–1.
12. Selkurt EE, Womack I, Dailey WN. Am J Physiol 1965;209:95–9.
13. Mulvaney MJ. Structural changes in the resistance vessels in human hypertension. In: Laragh JH, Brenner BM, eds. Hypertension: Pathophysiology, Diagnosis, and Management, vol 1. New York: Raven Press, 1995:503–13.
14. Whelton PK, He J, Appel LJ, et al. JAMA 2002;288:1882–8.
15. Centers for Disease Control and Prevention (CDC). National Center for Health Statistics (NCHS). National Health and Nutrition Examination Survey Questionnaire (or Examination Protocol, or Laboratory Protocol). Hyattsville, MD: U.S. Department of Health and Human Services, Centers for Disease Control and Prevention, 2007-2010.275. Peng YG, Li W, Wen XX, et al. Am J Clin Nutr 2014;100:1448–54.
16. Oh MS, Carroll HJ. External balance of electrolytes and acids and alkali. In: Seldin DW, Giebisch G, eds. The Kidney. 3rd ed. Philadelphia: Lippincott Williams & Wilkins, 2000: 33–60.
17. Papadoyannakis NJ, Stefanidis CJ, McGeowan M. Am J Clin Nutr 1984;40:623–7.
18. Lentner C. Geigy Scientific Tables, vol 1: Units of Measurement, Body Fluids, Composition of the Body, Nutrition. Basel: Ciba-Geigy, 1986:243–60.
19. Peng YG, Li W, Wen XX, et al. Am J Clin Nutr 2014;100(6): 1448–54.
20. Neal B, Wu Y, Feng X, et al. N Engl J Med. 2021;16;385(12): 1067–1077.
21. Hill LL. Pediatr Clin North Am 1990;37:241–56.
22. Oh MS, Uribarri J. Electrolytes, water, and acid-base balance. In: Shils ME, Shike M, Ross AC et al, eds. Modern Nutrition in Health and Disease. 10th ed. Burlington: Jones & Bartlett Learning, LLC, 2006:149–93.
23. Weisberg HF. Ann Clin Lab Sci 1978;8:155–64.
24. Wilson CS, Mongin AA. Curr Top Membr 2018;81:385–455.
25. Ellison DH, Berl T. N Engl J Med 2005;356:2064–72.
26. Arieff AI, Llack F, Massry SG. Medicine (Baltimore) 1976; 55: 121–9.
27. Fishman RA, Brain ED. N Engl J Med 1975;293:706–11.
28. Androgue HJ, Madias NE. N Engl J Med 2000;342:1494–9.
29. Decaux G, Gankam Kengne F, Couturier B, Vandergheynst F, Musch W, Soupart A. J Clin Med. 2014;3(3):1043–1049.
30. Overgaard-Steensen C, Stø´dkilde-Jø´rgensen H, Larsson A, et al. Am J Physiol Regul Integr Comp Physiol 2010;299:R521–32.

31. McKinley MJ, Allen AM, Burns P, et al. Clin Exp Pharmacol Physiol Suppl 1998;25:61S–7S.
32. Ibata Y, Okamura H, Tanaka M, et al. Front Neuroendocrinol 1999;20:241–68.
33. Wells T. Mol Cell Endocrinol 1998;136:103–7.
34. Bourque CW, Oliet SH. Annu Rev Physiol 1997;59:601–19.
35. Eaton DC, Pooler J. Vander's Renal Physiology. 7th ed. New York: McGraw-Hill, 2009.
36. Pallone TL, Turner MR, Edwards A et al. Am J Physiol Regul Integr Comp Physiol 2003;284:R1153–75.
37. Hogg RJ, Kokko JP. Urine concentrating and diluting mechanisms in mammalian kidneys. In: Brenner BM, Rector FC, eds. The Kidney. Philadelphia: WB Saunders, 1986:251–79.
38. de Rouffignac C, Jamison RL. Kidney Int 1987;31:501–672.
39. Oh MS, Halperin ML. Nephron 1997;75:84–93.
40. Sands JM, Kokko JP. Kidney Int Suppl 1996;57:93S–9S.
41. Burg MB. Am J Physiol 1995;268:F983–9.
42. Schmidt-Nielson B. Fed Proc 1977;36:2493.
43. Knepper MA. Am J Physiol 1983;245:F634–9.
44. Hogg RJ, Kokko JP. Kidney Int 1978;14:428–36.
45. Gregger R, Schlatter E, Lang F. Pflugers Arch 1983;396: 308–14.
46. Shirreffs SM, Maughan RJ. Am J Physiol 1998;274:F868–75.
47. Bhave G, Neilson EG. Am J Kidney Dis 2011;58(2):302–9.
48. Thornton SN. Physiol Behav 2010;100:15–21.
49. Mavrakis AN, Tritos NA. Am J Kidney Dis 2008;51:851–9.
50. Elkinton JR, Winkler AW, Danowski TS. J Clin Invest 1947; 26:1002–9.
51. Arieff AI. Principles of parenteral therapy. In: Maxwell MH, Kleeman CR, eds. Clinical Disorders of Fluid and Electrolyte Metabolism. 2nd ed. New York: McGraw-Hill, 1972:567–89.
52. Shoemaker WC, Walker WF. Year Book of Surgery. Chicago: Year Book Medical Publishers, 1970.
53. Adolph EF. Physiology of Man in the Desert. New York: Hafner, 1969.
54. Gaskill MB, Reilly M, Robertson GL. Clin Res 1983;31:780a.
55. Rendell M, McGrane D, Cuesta M. JAMA 1978;240:2557–9.
56. Hariprassad MK, Eisinger RP, Nadler IM, et al. Arch Intern Med 1980;140:1639–42.
57. Levine S, McManus BM, Blackbourne BD, et al. Am J Med 1987;82:153–5.
58. Vokes TJ, Gaskill MB, Robertson GL. Ann Intern Med 1988;108:190–5.
59. Arai K, Akimoto H, Inokami T, et al. Nippon Jinzo Gakkai Shi 1999;41:804–12.
60. Leggett DA, Hill PT, Anderson RJ. Australas Radiol 1999;43; 104–7.
61. Siggaard C, Rittig S, Corydon TJ, et al. J Clin Endocrinol Metab 1999;84:2933–4.
62. Rutishauser J, Kopp P, Gaskill MB, et al. Mol Genet Metab 1999;67:89–92.
63. Ito M, Jameson JL, Ito M. J Clin Invest 1997;99:1897–905.
64. Shibata S, Mori K, Teramoto S. No Shinkei Geka 1978;6: 795–801.
65. Nielsen S, Frokiaer J, Marples D, et al. Physiol Rev 2002; 82: 205–44.
66. Aleksandrov N, Audibert F, Bedard MJ, et al. J Obstet Gynaecol Can 2010;32:225–31.
67. Siegel AJ. Harv Rev Psychiatry 2008;16:13–24.
68. Dundas B, Harris M, Narasimhan M. Curr Psychiatry Rep 2007;9:236–41.
69. Oyama H, Kida Y, Tanaka T, et al. Neurol Med Chir (Tokyo) 1995;35:380–4.
70. Rose BD. Clinical Physiology of Acid-Base and Electrolyte Disorders. Chicago: R.R. Donnelley & Sons, 1989:657–60.
71. Oh MS, Carroll HJ. Crit Care Med 1992;20:94–103.
72. Loh JA, Verbalis JG. Endocrinol Metab Clin North Am 2008;37:213–34.
73. Fukuda I, Hizuka N, Takano K. Endocr J 2003;50:437–43.
74. Magaldi AJ. Nephrol Dial Transplant 2000;15:1903–5.
75. Bouley R, Hasler U, Lu HA, et al. Semin Nephrol 2008;28: 266–78.
76. Kirchlechner V, Koller DY, Seidl R, et al. Arch Dis Child 1999;80:548–52.
77. Nguyen MK, Ornekian V, Butch AW, et al. Am J Physiol 2007;292:F1652–6.
78. Oh MS, Dawood M, Carroll HJ. Proceedings of the American Society of Nephrology. Boston: American Society of Nephrology, 1993.
79. Milionis HJ, Liamis GL, Elisaf MS. Can Med Assoc J 2002;166:1056–62.
80. Agraharkar M, Agraharkar A. Am J Kidney Dis 1997;30:717–9.
81. Akan H, Sargin S, Turkseven F, et al. Br J Urol 1996;78:224–7.
82. Agarwal R, Emmett M. Am J Kidney Dis 1994;24:108–11.
83. Berl T, Anderson RJ, McDonald KM. Kidney Int 1976;10: 117–21.
84. Friedmann AS, Memoli VA, North WG. Cancer Lett 1993;75: 79–85.
85. Hill AR, Uribarri J, Mann J, et al. Am J Med 1990;88:357–64.
86. Gutkowska J, Jankowski M, Mukaddam-Daher S, et al. Proc Natl Acad Sci U S A 20004;97(1):483–8.
87. Koide Y, Oda K, Shimizu K, et al. Endocrinol Jpn 1982;29: 363–8.
88. Carvounis CP, Nisar S, Guro-Razuman S. Kidney Int 2002;62:2223–9.
89. Musch W, Thimpont J, Vandervelde D, et al. Am J Med 1995;99:348–55.
90. Oh MS, Carroll HJ. Nephron 1999;82:110–4.
91. Halperin ML, Bichet DG, Oh MS. Clin Nephrol 2001;56: 339–45.
92. Sterns RH. Am J Med 1990;88:557–60.
93. Sterns RH, Hix JK, Silver S. Curr Opin Nephrol Hypertens 2010;19:493–8.
94. Sterns RH, Nigwekar SU, Hix JK. Semin Nephrol 2009;29: 282–99.
95. Sterns RH. Semin Nephrol 1990;10:503–14.
96. Sterns RH. Crit Care Med 1992;20:534–9.
97. Verbalis JG, Goldsmith SR, Greenberg A, et al. Am J Med. 2013;126:S1–42.
98. Sterns RH, Hix JK. Am J Kidney Dis 2010;56:774–9.
99. Sterns RH. Ann Intern Med 1987;107:656–64.
100. Moritz ML, Ayus JC. Metab Brain Dis 2010;25:91–96.
101. Nobbenhuis R, Refardt J, Vogt D, et al Biomarkers. 2021;26(7):647–655.
102. Kumar S, Berl T. Semin Nephrol 2008;28:279–88.
103. Rozen-Zvi B, Yahav D, Gheorghiade M, et al. Am J Kidney Dis 2010;56:325–37.
104. Oh MS, Kim HJ. Nephron 2002;92:S56–9.
105. Nemerovski C, Hutchinson DJ. Clin Ther 2010;32:1015–32.
106. Josiassen RC, Curtis J, Filmyer DM, et al. Expert Opin Pharmacother 2010;11:637–48.
107. Fang C, Mao J, Dai Y, et al. J Paediatr Child Health 2010;46: 301–3.
108. Frassetto L, Morris RC Jr, Sellmeyer DE, et al. Eur J Nutr 2001;40:200–13.
109. Wingo C, Weiner ID. In: Brenner BM, ed. Brenner's and Rector's The Kidney, vol 2. Philadelphia: WB Saunders, 2000:998–1035.
110. Batlle D, Boobés K, Manjee KG. EBioMedicine 2015;2:1562–3.
111. Rastegar A, DeFronzo RA. In: Schrier RW, Gottschalk CW, eds. Diseases of the Kidney. 4th ed. Boston: Little, Brown, 1988:2921–45.
112. Shirreffs SM, Maughan RJ. Am J Physiol 1998;274:F868–75.

113. Meister B, Aperia A. Semin Nephrol 1993;13:41–9.
114. Powell WJ Jr, Skinner NS. Am J Cardiol 1966;18:73–82.
115. de la Lande IS, Manson J, Parks VJ, et al. J Physiol (Lond) 1961;157:177–84.
116. Vick R, Todd E, Luedhe D. J Pharmacol Exp Ther 1972;181: 139–46.
117. Adrogue HJ, Madias NE. Am J Med 1981;71:456–67.
118. Guillerm R, Radziszewski E. Undersea Biomed Res 1979;6: S91–114.
119. Perez GO, Oster JK, Vaamondi CA. Nephron 1981;27:233–43.
120. Krapf R, Caduff P, Wagdi P, et al. Kidney Int 1995;47:217–24.
121. Wang WH, Giebisch G. Pflugers Arch 2009;458:157–68.
122. Welling PA, Ho K. Am J Physiol 2009;297:F849–83.
123. Giebisch GH. Kidney Int 2002;62:1498–512.
124. Halperin ML, Kamel KS. Lancet 1998;352:135–40.
125. Giebisch G. Am J Physiol 1998;274:F817–33.
126. Bock HA, Hermle M, Brunner FP, et al. Kidney Int 1992;41: 275–80.
127. Stokes JB. J Clin Invest 1982;70:219–29.
128. Jamison RL. Kidney Int 1987;31:695–703.
129. Wile D. Ann Clin Biochem 2012;49:419–31.
130. Sandle GI, Gaiger E, Tapster S, et al. Clin Sci (Lond) 1986;71:393–401.
131. Wang MX, Cuevas CA, Su XT, et al. Kidney Int 2018;93(4): 893–902.
132. Van Brummelen P, Schalekamp M, DeGraeff J. Acta Med Scand 1978;204:151–7.
133. Sacks FM, Willett WC, Smith A, et al. Hypertension 1998;31:131.
134. Sacks FM, Svetkey LP, Vollmer WM, et al. N Engl J Med 2001;344:3–10.
135. Geleijnse JM, Kok FJ, Grobbee DE. J Hum Hypertens 2003;17:471–80.
136. Espeland MA, Kumanyika S, Yunis C, et al. Ann Epidemiol 2002;12:587–95.
137. Steen B. Acta Med Scand Suppl 1981;647:61–6.
138. Aronson PS, Giebisch G. J Am Soc Nephrol 2011;22(11): 1981–9.
139. Elliott MJ, Ronksley PE, Clase CM, et al. CMAJ 2010;182(15): 1631–5.
140. Gariballa S. Nutr Clin Pract 2008;24:604–6.
141. Torpy DJ, Gordon RD, Lin JP, et al. J Clin Endocrinol Metab 1998;83:3214–8.
142. Stowasser M, Bachmann AW, Jonsson JR, et al. Clin Exp Pharmacol Physiol 1995;22:444–6.
143. Stowasser M, Bachmann AW, Jonsson JR, et al. J Hypertens 1995;13:1610–3.
144. Abdelhamid S, Lewicka S, Vecsei P, et al. J Clin Endocrinol Metab 1995;80:737–44.
145. Litchfield WR, New MI, Coolidge C, et al. J Clin Endocrinol Metab 1997;82:3570–3.
146. Litchfield WR, Coolidge C, Silva P et al. J Clin Endocrinol Metab 1997;82:1507–10.
147. Vargas-Poussou R, Huang C, Hulin P, et al. J Am Soc Nephrol 2002;13:2259–66.
148. Sakakida M, Araki E. J Clin Endocrinol Metab 2003;88:781–6.
149. Finer G, Shalev H, Birk OS, et al. J Pediatr 2003;142:318–23.
150. Kunchaparty S, Palcso M, Berkman J, et al. Am J Physiol 1999;277:F643–9.
151. Seyberth HW, Rascher W, Schweer H, et al. J Pediatr 1985;107:694–701.
152. Zelikovic I, Szargel R, Hawash A. Kidney Int 2003;63:24–32.
153. Schulthesis PJ, Lorenz JN, Menton P, et al. J Biol Chem 1998;273:29150–5.
154. Omar HR, Komarova I, El-Ghonemi M, et al. Ther Adv Endocrinol Metab 2012;3(4):125–38.
155. Rossi G, Pessera A, Heagarty A. J Hypertens 2008;26:613–21.
156. Hebert SC, Desir G, Giebisch G, et al. Physiol Rev 2005;85: 319–71.
157. Nagami G. Am J Physiol 2008;294:F874–80.
158. Warnock DG. Contrib Nephrol 2001;136:1–10.
159. McKenna M, Bangsbo J, Renaud J. J Appl Physiol 2008;104: 288–95.
160. Agarwal R. Hypertension 2008;52:1012–3.
161. Barri YM, Wingo CS. Am J Med Sci 1997;314:37–40.
162. Savica V, Bellinghieri G, Kopple JD. Annu Rev Nutr 2010;30: 365–401.
163. Kovesdy CP, Shinaberger CS, Kalantar-Zadeh K. Semin Dial 2010;23:353–8.
164. Sterns RH, Palmer BF, eds. Nephrol Self Assess Program 2009;8:84–6.
165. Crop M, Hoorn E, Lindemans J, et al. Nephrol Dial Transplant 2007;22:3471–7.
166. Cohn JN, Kowey PR, Whelton PK, et al. Arch Intern Med 2000;160:2429–36.
167. Wasserman K, Stringer WW, Casaburi R, et al. J Appl Physiol 1997;83:631–43.
168. Perazella MA, Biswas P. Am J Kidney Dis 1999;33:782–5.
169. Sterns RH, Emmett M. Nephrol Self Assess Program 2011;10:129–30.
170. Reza MJ, Kovick RB, Shine KI, et al. N Engl J Med 1974; 291:777–8.
171. Phelps KR, Lieberman RL, Oh MS, et al. Metabolism 1980;29:186–99.
172. Phelps KR, Oh MS, Carroll HJ. Nephron 1980;25:254–8.
173. Segal A. Nat Clin Pract 2008;4:102–8.
174. Wilson FH, Kahle KT, Sabath E, et al. Proc Natl Acad Sci U S A 2003;100:680–4.
175. Belot A, Ranchin B, Fichtner C, et al. Nephrol Dial Transplant 2008;23:1636–41.
176. Don BR, Sebastian A, Cheitlin M, et al. N Engl J Med 1990;322:1290–2.
177. Ong YL, Deore R, El-Agnaf M. Int J Lab Hematol 2010;32: e151–7.
178. Montague B, Ouellette J, Buller G. Clin J Am Soc Nephrol 2008;3:324–30.
179. Krishnan AV, Kiernan MC. Muscle Nerve 2007;35:273–90.
180. Greenberg A. Semin Nephrol 1998;18:46–57.
181. Mandelberg A, Krupnik Z, Houri S, et al. Chest 1999;115:617–22.
182. Wong SL, Maltz HC. Ann Pharmacother 1999;33:103–6.
183. Martin TJ, Kang Y, Robertson KM, et al. Anesthesiology 1990;73:62–5.
184. McGowan CD, Saha S, Chu G, et al. South Med J 2009;102: 493–7.
185. Emmett M, Hootkins RE, Fine KD, et al. Gastroenterology 1995;108:752–60.
186. Gruy-Kapral C, Emmett M, Santa Ana CA, et al. Am Soc Nephrol 1998;9:1924–30.
187. Thomas A, James BR, Landsberg D. Am J Med Sci 2009;337: 305–6.
188. Trottier V, Drolet S, Morcos MW. Can J Gastroenterol 2009;23:689–90.
189. Sterns RH, Rojas M, Bernstein P, et al. J Am Soc Nephrol 2010;21:733–5.
190. Palmer BF. Am J Kidney Dis 2010;56:387–93.
191. Nyirenda MJ, Tang JI, Padfield PL, et al. BMJ 2009;339:b4114.
192. Farese S, Kruse A, Pasch A, et al. Kidney Int 2009;75:877–84.
193. Beccari MV, Meaney CJ. Core Evid 2017;12:11–24.
194. Rastegar A. Clin J Am Soc Nephrol 2009;4:1267–74.
195. Remer T, Manz F. J Am Diet Assoc 1995;95:791–7.
196. US Department of Agriculture. USDA National Nutrient Database for Standard Reference. Release 18. Washington, DC: US Department of Agriculture, 2006.

197. Osuna-Padilla IA, Leal-Escobar G, Garza-García CA, et al. Nefrologia (Engl Ed) 2019;39(4):343–354.
198. Oh MS. Nephron 1991;59:7–10.
199. Oh MS, Carroll HJ. Contrib Nephrol 1992;100:89–104.
200. Lennon EJ, Lemann J Jr, Litzow JR. J Clin Invest 1966;45:1601–7.
201. Remer T, Dimitriou T, Manz F. Am J Clin Nutr 2003;77:1255–60.
202. Frassetto LA, Todd KM, Morris RC Jr, et al. Am J Clin Nutr 1998;68:576–83.
203. Berkemeyer S, Remer T. J Nutr 2006;136:1203–8.
204. Oh YK. Electrolyte Blood Press 2010;8:66–71.
205. Koeppen BM. Adv Physiol Educ 2009;33:275–81.
206. Soupart A, Silver S, Schrooeder B, et al. J Am Soc Nephrol 2002;13:1433–41.
207. Lemann J Jr, Bushinsky DA, Hamm LL. Am J Physiol 2003;285:F811–32.
208. Franch HA, Mitch WE. J Am Soc Nephrol 1998;9:S78–81.
209. Laski ME, Sabatini S. Semin Nephrol 2006;26:404–21.
210. Welbourne TC. Am J Physiol 1987;253:F1069–76.
211. Franch HA, Mitch WE. Annu Rev Nutr 2009;29:341–64.
212. Maalouf NM, Cameron MA, Moe OW, et al. Clin J Am Soc Nephrol 2010;5:1277–81.
213. Patel AM, Goldfarb S. J Am Soc Nephrol 2010;21:1440–3.
214. Hsia CC. N Engl J Med 1998;338:239–47.
215. Dellinger R, Levy M, Carlet J, et al. Intensive Care Med 2008;34:17–60.
216. Taylor EN, Stampfer MJ, Curhan GC. J Am Soc Nephrol 2003;15:3225–32.
217. Sebastian A, Harris ST, Ottaway JH, et al. N Engl J Med 1994;330:1776–81.
218. Webster M. Sodium bicarbonate. In: Bahrke M, Yesalis C, eds. Performance-Enhancing Substances in Sport and Exercise. Champaign, IL: Human Kinetics, 2002.
219. Mitch WE, Price SR, May RC, et al. Am J Kidney Dis 1994;23:224–8.
220. Dawson-Hughes B, Castaneda-Sceppa C, Harris S, et al. Osteoporos Int 2010;21:1171–9.
221. de Brito-Ashurst I, Varagunam M, Raftery MJ, et al. J Am Soc Nephrol 2009;20:2075–84.
222. Kraut JA, Madias NE. Pediatr Nephrol 2011;26:19–28.
223. Morris CG, Low J. Anaesthesia 2008;63:396–411.
224. Kraut JA, Madias NE. Nat Rev Nephrol 2010;6:274–85.
225. Wagner CA. Kidney Int 2008;73:1103–5.
226. Umpierrez GE, Khajavi M, Kitabchi AE. Am J Med Sci 1996;311:225–33.
227. Chawla LS, Shih S, Davison D, et al. BMC Emerg Med 2008;8:18.
228. Rewers A. Adv Pediatr 2010;57:247–67.
229. Oh MS, Carroll HJ. Nephron 2002;91:379–82.
230. Kraut JA, Kurtz I. Clin J Am Soc Nephrol 2008;3:208–25.
231. Oh MS, Phelps KR, Traube M, et al. N Engl J Med 1979;301:249–52.
232. Sabatini S, Kurtzman NA. J Am Soc Nephrol 2009;20:692–5.
233. Madias NE. J Nephrol 2010;23:S85–91.
234. Saaresranta T, Polo-Kantola P, Irjala K, et al. Chest 1999;115:1581–7.

B. MINERALS

CHAPTER 7

Calcium

Connie M. Weaver*

CHAPTER OUTLINE

Abbreviations

$1,25(OH)_2D$, 1,25–dihydroxyvitamin D; **AI**, adequate intake; **ATP**, adenosine triphosphate; **ATPase**, adenosine triphosphatase; **Ca^{2+}**, calcium ion; **CaSR**, calcium-sensing receptor; **CaT_1**, calcium channel; **CT**, calcitonin; **DAG**, diacylglycerol; **ECaC**, calcium channel; **ECF**, extracellular fluid; **IGF-I**, insulinlike growth factor-I; **$InsP_3$**, inositol-1,4,5-triphosphate; **Na^+**, sodium ion; **NHANES**, National Health and Nutrition Examination Survey; **PIP_2**, phosphatidylinositol-4,5-bisphosphate; **PTH**, parathyroid hormone; **RyR**, ryanodine receptor; **VDR**, vitamin D receptor

Biologic Roles of Calcium

In higher mammals, the most obvious role of calcium is structural or mechanical and is expressed in the mass, hardness, and strength of the bones and teeth. Calcium has another fundamental function; however: shaping key biologic proteins to activate their catalytic and mechanical properties. A significant portion of the regulatory apparatus of the body is concerned with the protection of this second function (e.g., all activities and roles of parathyroid hormone [PTH], calcitonin [CT], and a key activity of vitamin D). Calcium is the most tightly regulated ion in the extracellular fluid (ECF). The structural role is discussed in greater detail in the chapter on osteoporosis, whereas the cell metabolic, regulatory, and nutritional aspects of this critical element are discussed in this chapter.

Calcium and the Cell

The calcium ion (Ca^{2+}) has an ionic radius of 0.99 Å and can form coordination bonds with up to 12 oxygen atoms.[1] The combination of these two features makes calcium nearly unique among all cations in its ability to fit neatly into the folds of the peptide chain. Cytoplasmic proteins are extremely flexible to the point of being literally floppy. They typically assume hundreds of different three-dimensional configurations each second. Some of these configurations have the capacity to bind critical ligands or to assume catalytic functions. Without calcium, these

*Robert Heaney, deceased, was a contributor to the first edition of this chapter.

configurations are so short lived as to be of little functional significance. When present in the cytosol in sufficient concentration, calcium binds to aspartate and glutamate side chains on the peptide backbone and, thus, builds intramolecular linkages that bind together different folds of the peptide chain and "freeze" the protein into a functionally active shape. Magnesium and strontium, which are chemically similar to calcium in the test tube, have different ionic radii and do not bond as well with proteins. Lead and cadmium ions, by contrast, substitute well for calcium. In fact, lead binds to various calcium-binding proteins with greater avidity than calcium itself. Fortunately, neither element is present in significant quantity in the milieu in which living organisms thrive. Nevertheless, the ability of lead to bind to calcium-binding proteins is part of the basis for lead toxicity.

The binding of calcium to cell proteins triggers changes in protein shape that govern function.[2] These proteins range from those involved with cell movement and muscle contraction to nerve transmission, glandular secretion, and even cell division. In most of these situations, calcium acts both as a signal transmitter from the outside of the cell to the inside, and as an activator or stabilizer of the functional proteins involved. In fact, ionized calcium is the most common signal transmitter in all of biology. It operates from bacterial cells all the way up to cells of highly specialized tissues in higher mammals.

When a cell is activated (e.g., a muscle fiber receives a nerve stimulus to contract), the first thing that happens is that calcium channels in the plasma membrane open to admit calcium ions into the cytosol. These bind immediately to a wide array of intracellular activator proteins which, in turn, release a flood of calcium from the intracellular storage vesicles (the sarcoplasmic reticulum, in the case of muscle). This second step quickly raises the cytosol calcium concentration and leads to activation of the contraction complex. Two of the many reactions involving calcium-binding proteins are of particular interest here: (a) troponin C, after it has bound calcium, initiates a series of steps that lead to the actual muscle contraction; and (b) calmodulin, a second and widely distributed calcium-binding protein, activates enzymes that break down glycogen to release energy for contraction. In this way, calcium ions both trigger the contraction and fuel the process. When the cell has completed its assigned task, the various pumps quickly lower the cytosol calcium concentration, and the cell returns to a resting state. These processes are described in more detail later in this chapter.

If all of the functional proteins of a cell were fully activated by calcium at the same time, the cell would rapidly self-destruct. For that reason, cells must keep the free calcium ion concentration in the cytosol at extremely low levels, typically on the order of 100 nmol. This is 10,000-fold lower than the concentration of calcium ions in the extracellular water outside of the cell. Cells maintain this concentration gradient by a combination of mechanisms: (a) a cell membrane with limited calcium permeability; (b) ion pumps that move calcium rapidly out of the cytosol, either to the outside of the cell or into storage vesicles within the cell; and (c) a series of specialized proteins in the storage vesicles that have no catalytic function in their own right, but serve only to bind (and hence sequester) large quantities of calcium. Low cytosolic $[Ca^{2+}]$ ensures that the various functional proteins remain dormant until the cell activates some of them. It does this simply by letting $[Ca^{2+}]$ rise in critical cytosolic compartments. In contrast to proteins that are activated by rising cytosolic $[Ca^{2+}]$ are enzymes, such as several proteases and dehydrogenase, which are activated or stabilized by bound calcium independent of changes in $[Ca^{2+}]_i$.

Occurrence and Distribution in Nature

Calcium is the fifth most abundant element in the biosphere (after iron, aluminum, silicon, and oxygen). It is the stuff of limestone and marble, coral and pearls, seashells and egg shells, antlers and bones. Because calcium salts exhibit intermediate solubility, calcium is found both in solid form (rocks) and in solution. It was probably present in abundance in the watery environment in which life first appeared. Today, seawater contains approximately 10 mmol calcium per liter (approximately eight times higher than the calcium concentration in the extracellular water of higher vertebrates); and even fresh waters, if they support abundant biota, typically contain calcium at concentrations of 1 to 2 mmol. In most soils, calcium exists as an exchangeable cation in the soil colloids. It is taken up by plants, whose parts typically contain from 0.1% to as much as 8% calcium. Generally, calcium concentrations are highest in the leaves, lower in the stems and roots, and lowest in the seeds.

In land-living mammals, calcium accounts for 2% to 4% of gross body weight. A 60-kg woman typically contains approximately 1000 to 1200 g (25 to 30 mol) of calcium in her body. More than 99% of that total is in the bones and teeth. Approximately 1 g is in the plasma and ECF bathing the cells, and 6 to 8 g are in the tissues themselves (mostly sequestered in calcium storage vesicles inside of cells, as discussed earlier).

In the circulating blood, calcium concentration is typically 2.25 to 2.5 mmol. Approximately 40% to 45% of this quantity is bound to plasma proteins, approximately 8% to 10% is complexed with ions such as citrate, and 45% to 50% is dissociated as free ions. In the ECF outside of the blood vessels, total calcium is on the order of 1.25 mmol, which differs from plasma concentration because of the absence of most plasma proteins in the ECF. It is the calcium concentration in the ECF that the cells see and that is tightly regulated by the parathyroid, CT, and vitamin D hormonal control systems.

With advancing age, humans commonly accumulate calcium deposits in various damaged tissues, such as atherosclerotic plaques in arteries, healed granulomas, other scars left by disease or injury and, often, in the rib cartilages as well. These deposits are called *dystrophic calcification* and rarely amount to more than a few grams of calcium. These deposits are not caused by dietary calcium but by local injury, coupled with the widespread tendency of proteins to bind calcium. Calcification in tissues other than bones and teeth is generally a sign of tissue damage and cell death. This process is greatly exaggerated in conditions such as end-stage kidney disease when the calcium × phosphorus product of the ECF exceeds 2.5 to 3.0 $mmol^2/L^2$.

Metabolism

Calcium metabolic pathways and factors that perturb these pathways are illustrated in **Figure 7.1**. Calcium balance is the result of calcium intake minus loss regulated at the gut, kidney, and bone.

Part of dietary calcium is absorbed into the bloodstream where it is in intimate exchange with ECF calcium. Part of the absorbed calcium is returned as endogenous secretion to the gut, where it is excreted along with unabsorbed calcium. Part is excreted in the urine through the kidney, and part enters the slower exchange pools of soft tissue and bone. Dietary calcium

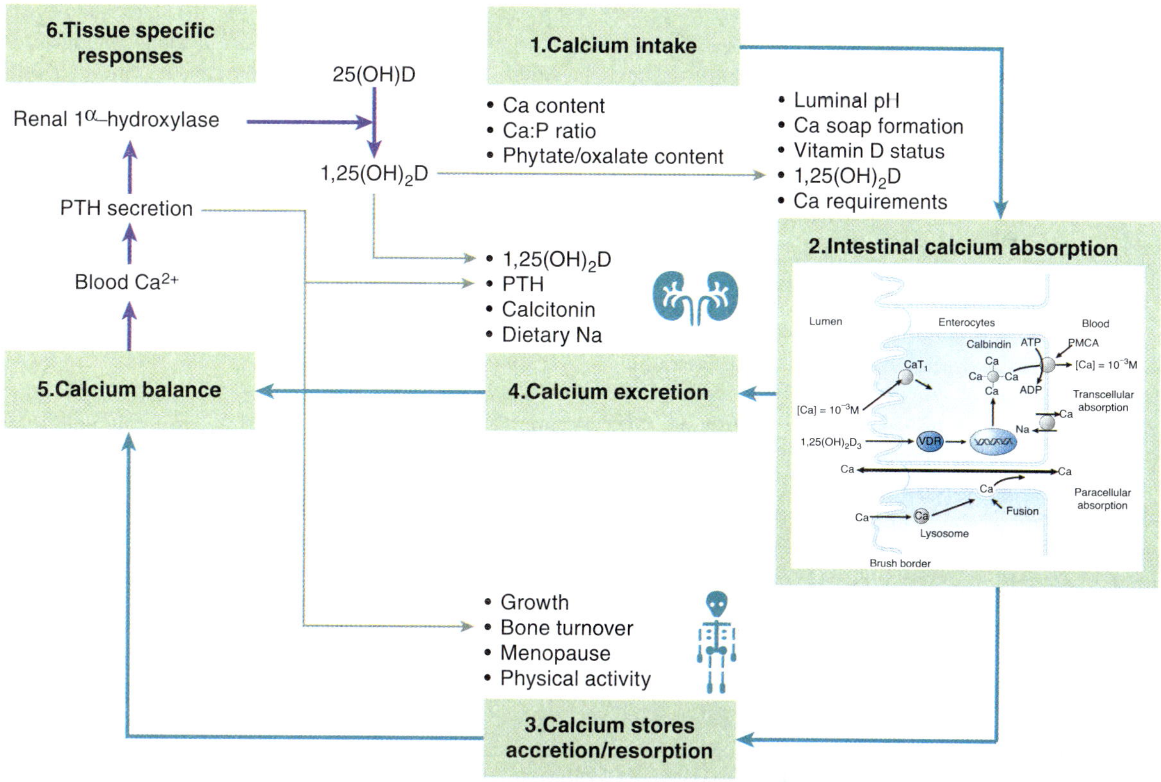

Figure 7.1 Calcium metabolic pathways and potentially perturbing factors.

Reproduced from Ziaul et al. Annals NYAS, 2022; and Weaver CM and Heaney RP chapter 7 Modern Nutrition in Health and Disease, 11th Ed. 2013; pp. 133-149.

influences calcium absorption and, consequently, fecal calcium and, to a lesser extent, urinary calcium excretion. An obligatory loss of calcium occurs through endogenous secretion, urine, and skin. Sex, age, and racial differences in calcium metabolism exist as shown in **Table 7.1**. Adolescents are more efficient at using calcium than young adults, and elderly persons are the least efficient. Boys are more efficient at calcium metabolism than girls, and Black individuals are more efficient than White individuals.

Table 7.1 Calcium (Ca) Metabolism as Influenced by Race and Age

Life Stage (Age [Y])	mg/d[a]							
	Intake	Absorbed	Endogenous Secretion	Fecal	Urine	Bone Formation	Bone Resorption	Bone Balance
White pubertal girls (12–14)	1,330	494 ± 232	112 ± 35	918 ± 253	100 ± 54	1,459 ± 542	1,177 ± 436	282 ± 269
Black pubertal girls (11–14)	1,128	636 ± 188	109 ± 50	680 ± 178	46 ± 38	1,976 ± 540	1,496 ± 528	484 ± 180
Asian pubertal girls (11–15)	1,068	567 ± 27	104 ± 17	604 ± 19	87 ± 6	1,369 ± 86	992 ± 89	378 ± 22
Asian pubertal boys (11–15)	1,211	662 ± 30	154 ± 19	702 ± 20	78 ± 6	2,416 ± 95	1,986 ± 97	430 ± 24
Young White women (19–31)	1,330	283 ± 122	121 ± 39	1,138 ± 143	203 ± 79	501 ± 129	542 ± 212	−41 ± 165
Postmenopausal women (46–67)	1,083	221 ± 58	151 ± 49	1,092 ± 256	121 ± 63	307 ± 138	415 ± 192	−108 ± 110

[a]1 mg Ca = 25 μmol.

Data from Wastney ME, Ng J, Smith D et al. Am J Physiol 1996;271:R208–16; Bryant RJ, Wastney ME, Martin BR et al. J Endocrinol Metab 2003;88:1043–7[31]; Spence LA, Lipscomb ER, Cadogan J et al. Am J Clin Nutr 2005;81:916–22[70]; and Wu L, Martin BR, Braun MM et al. J Bone Miner Res 2010;25:1842–9.

Homeostatic Regulation

Plasma calcium is tightly regulated at approximately 2.5 mM (9 to 10 mg/dL). When plasma calcium is more than 10% away from the population mean, one has reason to suspect disease. The regulation of calcium concentration involves a system of controlling factors and feedback mechanisms.

Plasma calcium concentration is detected by surface calcium-sensing receptors (CaSRs) found in the parathyroid and clear cells of thyroid glands, kidney, intestine, bone marrow, and other tissues. When plasma calcium concentration is elevated, PTH release is inhibited, and CT release is stimulated.

When plasma calcium concentration falls, the parathyroid gland is stimulated to release PTH. PTH increases renal phosphate clearance and renal tubular reabsorption of calcium; it activates bone resorption loci, augments osteoclast activity at existing resorption loci, and activates vitamin D to enhance intestinal calcium absorption. Activation of vitamin D occurs in two steps. An initial hydroxylation is catalyzed by vitamin D-25-hydroxylase (CYP27), a microsomal cytochrome P-450 enzyme system in the liver. The second hydroxylation by 25-OH D-1-α-hydroxylase (CYP27B1) in the proximal convoluted tubule cells of the kidney converts the vitamin to its active potent form, 1,25-dihydroxyvitamin D [1,25(OH)$_2$D] or calcitriol. (See the chapter on vitamin D for additional details.) This latter step is stimulated by PTH and is augmented by a fall in serum phosphate. PTH and 1,25(OH)$_2$D act synergistically to enhance renal tubular reabsorption of calcium and to mobilize calcium stores from bone. PTH acts in a classical negative feedback loop to raise the ECF [Ca^{2+}], thereby closing the loop and reducing PTH release. FGF23 is a potent counter regulatory hormone to 25-OH-D-1-α-hydroxlase, which is blunted when serum calcium concentration falls.[3] Some evidence indicates that the intestine also has CYP27B1 activity that could produce 1,25(OH)$_2$D for local use; this would explain observations of increased calcium absorption with increased 25(OH)D concentration without change in serum 1,25(OH)$_2$D.[4]

Although the sophisticated regulatory mechanism described earlier allows a rapid response that corrects transient hypocalcemia, in the presence of a chronically calcium-deficient diet, it maintains ECF [Ca^{2+}] at the cost of depleting the skeleton. The three tissues supporting serum calcium concentration (i.e., gut, kidney, and bone) operate independently of one another, and altered responsiveness of any of these can increase bone fragility. For example, low fractional calcium absorption capacity was associated with increased hip fracture risk in elderly postmenopausal women.[5]

When plasma calcium concentration rises in response to increased calcium absorption or increased bone resorption, extracellular Ca^{2+} binds to CaSR on the surface of parathyroid cells and, thus, stimulates a conformational change in the receptors leading to an inhibition of PTH secretion from the parathyroid.[6] PTH augments tubular reabsorption of calcium. That reabsorption has a maximum capability (the T_mCa), and when that maximum is exceeded, additional filtered calcium is excreted.

In infants and children, a principal defense against hypercalcemia is release of CT by the C cells of the thyroid gland. CT is a peptide hormone with binding sites in the kidney, bone, and central nervous system. Absorption of calcium from an 8-oz feeding in a 6-month-old infant dumps 150 to 220 mg calcium into the ECF. This is enough, given the small size of the ECF compartment at that age (1.5 to 2 L), to produce fatal hypercalcemia if other adjustments are not made. Instead, CT is released, in part in response to the rise in serum calcium, but even before that, in response to gut hormones signaling coming absorption. This burst of CT slows or halts osteoclastic resorption, thus stopping bony release of calcium. Then, when absorption stops, CT levels also fall, and osteoclastic resorption resumes. By contrast, CT has little significance in adults because absorption is lower to begin with, and the ECF is vastly larger. As a result, absorptive calcemia from a high-calcium diet raises ECF [Ca^{2+}] by only a few percentage points, and the absence of CT (as with thyroid ablation) has little impact on calcium homeostasis.

Absorption

Calcium is usually freed from complexes in the diet during digestion and is released in a soluble and typically ionized form for absorption. However, small-molecular-weight complexes such as calcium oxalate and calcium carbonate can be absorbed intact.[7]

Fractional calcium absorption (absorptive efficiency) generally varies approximately inversely with the logarithm of intake, but the absolute quantity of calcium absorbed increases with intake.[8,9] However, only 20% of the variation in calcium absorption can be accounted for by usual calcium intake.[10] Rather, individuals seem to have preset absorptive efficiencies; approximately 60% of the variance in calcium absorption among individuals can be accounted for by their individual fractional calcium absorption.[11]

Mechanisms of Absorption

Calcium absorption occurs by two pathways (Figure 7.2):

1. Transcellular: This saturable (active) transfer involves a calcium-binding protein, calbindin.
2. Paracellular: This nonsaturable (diffusional) transfer is a linear function of calcium content of the chyme.

Paracellular absorption is bidirectional; transcellular absorption is unidirectional. Ca enters the cytosol down a concentration gradient. Ca enters the cell through TRPV6 and is transported across the enterocyte against an uphill gradient with the aid of vitamin D–induced calbindin, probably at least partially through endosomes and lysosomes. Finally, it is extruded at the basolateral membrane, primarily by the plasma membrane calcium

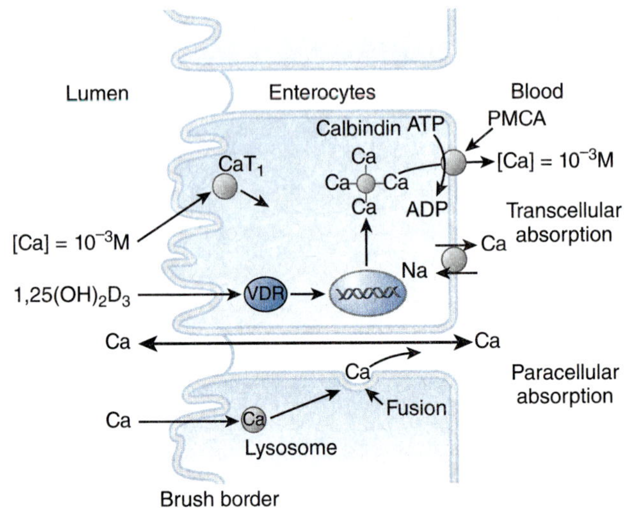

Figure 7.2 Intestinal calcium absorption.

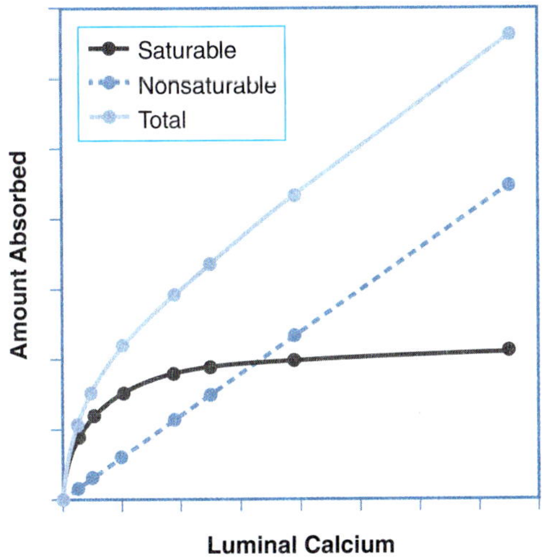

Figure 7.3 Calcium is absorbed by both saturable and nonsaturable pathways. Total calcium transport (the sum of a saturable component [A] defined by the Michaelis-Menten equation and a concentration-dependent, nonsaturable component [B] defined by a linear equation) is described by a curvilinear function.

adenosine triphosphatase (ATPase) pump (PMCA) and, secondarily, by the sodium (Na^+)/Ca^{2+} exchanger or by exocytosis.

The relationship between calcium intake and absorbed calcium is shown in **Figure 7.3**. At lower calcium intakes, the active component contributes most to absorbed calcium. The Michaelis-Menten constant (K_m) for the active component in adults is calculated to be 3.2 to 5.5 mM (equivalent to a calcium load of 230 to 400 mg).[4] As calcium intakes increase and the active component becomes saturated, an increasing proportion of calcium is absorbed by passive diffusion.

Active absorption is most efficient in the duodenum and next in the jejunum, but total calcium absorbed is greatest in the ileum, where residence time is the longest. In one rat study, net calcium absorption was distributed as 62% in the ileum, 23% in the jejunum, and 15% in the duodenum.[12] Absorption from the colon accounts for approximately 5% to 23% (or ~1% of ingested calcium) of the total absorption in normal individuals, but this may be important in patients with small bowel resections and when colonic bacteria break down dietary complexes.

Transcellular Calcium Transport. Calcium entry into the epithelial cells occurs primarily through a calcium channel, TRPV6 (formerly CaT_1),[13] although it is not a rate-limiting step.[14] Calcium transfer occurs down a steep electrochemical gradient and does not require energy. The main regulator of transport across the epithelial cell against the energy gradient is $1,25(OH)_2D$. As illustrated in Figure 7.1, $1,25(OH)_2D$, which is responsive to serum calcium concentration, regulates the synthesis of calbindin by binding with vitamin D receptor (VDR) in the cytoplasm and translocating to the nucleus, where it binds to response elements to initiate transcription of calbindin mRNA. The essentiality of VDR and $1,25(OH)_2D$ in the control of calcium absorption was established with transgenic mice.[15] Intestinal calbindin, a 9-kDa protein in mammals and a 28-kDa protein in birds, is capable of binding 2 Ca^{2+} per molecule. Calbindin operates by binding Ca^{2+} on the surface of the cell and then internalizes the ions through endocytic vesicles that may fuse with lysosomes. After release of the bound calcium in the acidic lysosomal interior, the calbindin returns to the cell surface, and the Ca^{2+} ions exit the cell through the basolateral membrane.[16] Using ion microscopic imaging of injected $^{44}Ca^{2+}$, calcium entry into the villus was observed in vitamin D–deficient chicks, but the rapid transfer of Ca^{2+} through the cytoplasm to the basolateral pole did not occur in the absence of the ability to synthesize calbindin.[17] Thus, calbindin serves both as a Ca^{2+} translocator and a cytosolic Ca^{2+} buffer to resist toxicity in chick intestine.[18] However, its role in mammalian intestinal epithelial cells has been questioned.[4] Much remains to be understood about calcium transport across the intestine because, in a double-calbindin D9k/TRPV6 knockout mouse model, calcium absorption did respond to $1,25(OH)_2D$, although that response was reduced by 60% compared with the wild-type mouse.[19]

Vitamin D–induced calcium transport also involves activation of a Ca^{2+}-dependent adenosine triphosphate (ATP) pump (PMCAlb) to effect extrusion of calcium against an electrochemical gradient into the ECF.[20] Relative Ca^{2+}-binding capacities across the enterocyte are brush border, 1, calbindin, 4, and the ATP-dependent Ca^{2+} pump, 10; this gradient ensures unidirectional transfer of Ca^{2+}.[21] A rapid increase in calcium absorption resulting from transcaltachia, which involves $1,25(OH)_2D$-mediated (but not transcriptional) events, also appears to be at work.[4]

Paracellular Calcium Transport. In the paracellular pathway, calcium transfer occurs between the cells. Theoretically, this transfer can occur in both directions, but normally, the predominant direction is from lumen into blood because much of the transfer is by solute drag, which is predominantly from lumen into ECF. The rate of transfer depends on ingested calcium load and tightness of the junctions. Calcitriol also enhances flux of Ca^{2+} by increasing expression of claudin proteins, the major tight junction components, and by suppressing aquaporin 8, a tight junction channel, to regulate permeability of the endothelium.[22] Water probably carries calcium through the junctions by solvent drag, which is stimulated by $1,25(OH)_2D$ through induction of tight junction proteins.[23]

Physiologic Factors Affecting Absorption

Various host factors affect fractional calcium absorption. Vitamin D status, intestinal transit time, and mucosal mass are the best established.[24] Phosphorus deficiency, as may occur through prolonged use of aluminum-containing antacids, can cause hypophosphatemia, increased circulating $1,25(OH)_2D$, and elevated calcium absorption.

Stage of life also influences calcium absorption. In infancy, absorption is dominated by diffusion. Therefore, the vitamin D status of the mother does not affect fractional calcium absorption of breastfed infants. Both active and passive calcium transport is increased during pregnancy and lactation. Calbindin, plasma $1,25(OH)_2$ and PTH increase during pregnancy. From midlife onward, absorption efficiency declines by approximately 0.2 absorption percentage points per year and, at menopause, an additional 2% decrease occurs.[25] Decreased calcium absorption efficiency with age is related to increased intestinal resistance to $1,25(OH)_2D$, as illustrated by a steeper slope in the relationship between fractional calcium absorption and serum $1,25(OH)_2D_3$ in elderly postmenopausal women than in young premenopausal women.[26] The age-related decrease in calcium absorption from intestinal resistance to 1,25$(OH)_2D_3$ has been associated with lower VDR,[27] as well as with reduced estrogen concentrations.[24]

Decreased stomach acid, as occurs in achlorhydria, reduces the solubility of insoluble calcium salts (e.g., carbonate, phosphate) and thus could, in theory, reduce absorption of calcium unless fed with a meal.[28] Absorption of calcium supplements improves when they are taken with food irrespective of gastric acid status, perhaps by slowing gastric emptying and thereby extending the time in which the calcium-containing chyme is in contact with the absorptive surface.

VDR polymorphisms have been studied for their relationship with calcium absorption efficiency. One study showed a significant association between the VDR Fokl polymorphism and calcium absorption in children.[29] Using recombinant inbred mouse lines, gene clusters that regulate calcium absorption in response to low calcium diets were identified.[30]

Excretion

Loss of calcium from the body occurs in urine, feces, and sweat. Differences in losses between adult women and adolescent girls on equal and adequate calcium intakes were given in Table 7.1, which shows the conservation of calcium at the kidney for building bone during the rapid period of skeletal growth during puberty. Black girls absorb more calcium and excrete less calcium than do White girls, and this characteristic results in greater net bone deposition.[31] Black women average 10% higher bone mineral content than White women.[32]

Turnover of the miscible calcium pool in healthy adults is approximately 16%/day, and the rapidly exchanging component (of which the ECF is a part) is approximately 40%/day. The filtered load of the kidney is determined by the glomerular filtration rate and the plasma concentration of ultrafiltrable calcium (ionized plus that bound to small-molecular-weight anions). In adults, this is approximately 175 to 250 mmol/day (7 to 10 g/day). More than 98% of this calcium is reabsorbed by the renal tubule as the filtrate passes through the nephron, but 2.5 to 5 mmol (100 to 200 mg) are excreted in the urine daily. Endogenous fecal excretory loss is similar to the amount excreted in the urine. Loss in the sweat is typically 0.4 to 0.6 mmol (16 to 24 mg)/day,[33] and additional diurnal losses occur from shed skin, hair, and nails, thus bringing the total to as much as 1.5 mmol (60 mg)/day. Cutaneous losses from children average 1.3 mmol (52 mg)/day.[34] Moderate exercise can increase calcium loss.[35]

Endogenous Fecal Calcium

Fecal calcium includes the calcium that is unabsorbed from the diet, plus calcium that enters the gut from endogenous sources, including shed mucosal cells and digestive secretions. Endogenous fecal calcium losses are approximately 2.5 to 3.0 mmol (100 to 120 mg)/day. These losses are inversely proportional to absorption efficiency and are directly related to gut mass (and hence to food intake). Urinary calcium increases during childhood up to adolescence. Endogenous fecal calcium values in adolescent girls do not differ significantly from those of young women (as shown in Table 7.1).

Urinary Excretion

In the kidney, an increase in ECF calcium ion concentration decreases the glomerular filtration rate, has a diuretic action in the proximal tubule, and inhibits the actions of antidiuretic hormone.[36] Machinery for calcium transport, described earlier for the intestinal epithelial cells, is also present in the nephron. Paracellular transport dominates in the proximal tubule, as reabsorption occurs across a concentration gradient, and it also occurs in the thick ascending limb of the loop of Henle, the distal nephron, and the collecting ducts.

Similar to intestinal calcium absorption, calcium transport in the kidney depends on calcium load, calcium concentration is detected by CaSR, calcium transport is stimulated by PTH and 1,25$(OH)_2D$, and calcium transport uses a microvillar myosin I–calmodulin complex.[37] PTH acts on proximal tubular cells to upregulate CYP1α expression. Calcium enters renal epithelial cells through a calcium channel, ECaC or CaT_2.[38] Active transport occurs in the distal convoluted tubule against a concentration gradient. In the mammalian kidney, vitamin D regulation works through calbindin-D_{28}k, which binds 4 Ca^{2+} per molecule and shares no sequence homology with calbindin-D_9k of the intestine. This calcium-binding protein has been cloned and is regulated by both transcriptional and posttranscriptional mechanisms. Administration of 1,25$(OH)_2D$ to rats induces calbindin-D_{28}k mRNA and VDR mRNA in vitamin D–sufficient animals.[39] However, in the absence of vitamin D, hypercalciuria is not observed, as would be predicted if mechanisms were similar to those in the gut. A fall in filtered load is associated with slight reductions in urine calcium. Even so, renal calcium clearance is reduced in vitamin D deficiency and is increased in PTH deficiency—indicating that the major effect on conservation of calcium is exerted by PTH. Nevertheless, CaSR in the kidney can regulate serum calcium concentration directly, independent of PTH.[40]

During the rapid growth of adolescence, urinary calcium is little influenced by load. Absorbed calcium is diverted to bone growth at calcium intakes typically ingested, except for obligatory losses in urine, skin, and endogenous secretions. Tubular reabsorption decreases in postmenopausal women.

Dietary Considerations

Dietary sources and calcium intakes have changed considerably during human evolution. Early humans derived calcium from roots, tubers, nuts, and beans in quantities believed to exceed 37.5 mmol (1500 g)/day[41] and perhaps up to twice this when they were consuming food to meet the caloric demands of a hunter-gatherer of contemporary body size. After domestication of grains, calcium intakes decreased substantially because the staple foods became grains (fruits), the plant parts that accumulate the least calcium. Pre–Iron Age milling practices were based on limestone and hence added appreciable calcium as calcium carbonate to the otherwise low-calcium flour. Consequently, the modern human, on average, consumes insufficient calcium to optimize bone density. The food group that supplies the bulk of the calcium in the Western diet is now the dairy food group.

Food Sources and Bioavailability

Milk and other dairy products supply more than 70% of the calcium in the U.S. diet.[42] Although corn tortillas processed with lime and dried beans provide the bulk of dietary calcium for some ethnic groups, it is difficult for most individuals to ingest enough calcium from foods available in a cereal-based economy without liberal consumption of dairy products. Thus, food manufacturers have developed calcium-fortified products. Many individuals have turned to dietary supplements to meet their calcium needs. However, it is prudent to remember that calcium is not the only nutrient important to health supplied by dairy products. Milk intake is not only associated with calcium but also with potassium, magnesium, zinc, riboflavin, vitamin A, folate, and vitamin D intakes.[43] Median intake of milk in the United States meets the recommended intake in children aged 1 to 8 years, although 25% of children do not consume the recommended 2 cups daily.[44] In contrast, the median intake for older groups falls well below the recommended 3 cups daily (i.e., 1.9 cup equivalents for girls and 2.4 for boys aged 9 to 13 years, 1.5 cup for girls and 2.3 cups for boys aged 14 to 18 years, and 1.2 cup for women and 1.6 for men).

Aside from gross calcium content, potential calcium sources vary importantly in bioavailability. Fractional calcium absorption from various dairy products is similar, at approximately 30%.[45] The calcium from most supplements is absorbed as well as from milk, as solubility of the salts at neutral pH has little impact on calcium absorption.[46] A few calcium salts, including calcium citrate malate and calcium ascorbate, have superior absorbability. However, adjuvants added to supplements or food matrices can alter bioavailability substantially.

Several plant constituents form indigestible salts with calcium and, thereby, decrease absorption of their calcium. The most potent inhibitor of calcium absorption is oxalic acid, found in high concentration in spinach, rhubarb, and, to a lesser extent, sweet potatoes and dried beans.[47] Calcium absorption from spinach is only 5% compared with 27% from milk ingested at a similar load.[48] When these two foods of dissimilar bioavailability are ingested during the same meal, calcium fractional absorption from milk is depressed by 30% of the difference between milk and spinach fed alone; and calcium fraction absorption from spinach is enhanced by 37% of the difference between milk and spinach alone.[49] The absence of complete exchange and the failure to find equal absorption from the two foods intermediate between the values for the foods fed singly suggest that calcium does not completely form a common dietary pool, as has been reported for iron and zinc.

Phytic acid, the storage form of phosphorus in seeds, is a modest inhibitor of calcium absorption. The phytic acid content of seeds, which depends on the phosphorus content of the soil where the plants are grown, influences calcium absorption.[50] Fermentation, such as occurs during bread making, reduces phytic acid content by virtue of the phytase present in yeast. This process results in increased calcium absorption.[51] Since the early balance studies of McCance and Widdowson, who reported negative calcium balance during consumption of whole wheat products,[52] it has been assumed that fiber negatively affects calcium balance through physical entrapment or through cationic binding with uronic acid residue.[53] However, it is more likely that the phytic acid associated with fiber-rich foods affects the balance, as purified fibers do not negatively affect calcium absorption.[54] Only concentrated sources of phytate such as wheat bran ingested as extruded cereal,[51] or dried beans,[55] have substantially reduced calcium absorption. For other plants rich in calcium (primarily the *Brassica* genus, which includes broccoli, kale, bok choy, cabbage, and mustard and turnip greens), calcium bioavailability is as good as that from milk.[56] The *Brassicas* are an anomaly in the plant kingdom in that they do not accumulate oxalate as a mechanism to detoxify excess calcium to protect against cell death.

A comparison of several foods for calcium content, bioavailability, and number of servings needed to equal the amount of calcium absorbed from one serving of milk is given in **Table 7.2**.

True enhancers of calcium absorption have not been well characterized. Lactose appears to enhance calcium absorption in infants. However, in adults, calcium absorption from various dairy products is equivalent regardless of the lactose content, chemical form of calcium, or presence of flavorings.[57] Nondigestible carbohydrates can increase calcium absorption in the lower bowel, where they are fermented and where the resulting short-chain fatty acids produced lower pH and increase solubility of calcium.[58] Some proteins can enhance calcium absorption acutely, but the effect disappears with long-term feeding, when calcium absorption adapts by upregulation of transport proteins.[59]

Table 7.2 Food Sources of Bioavailable Calcium

Food	Serving Size (G)	Calcium Content (Mg)	Absorption Efficiency[a] (%)	Estimated[b] Absorbable Calcium/Serving (Mg)	Servings Needed To Equal 1 C Milk
Milk (or 1 c yogurt or 1½ oz cheddar cheese)	260	300	32.1	96.3	1.0
Beans, dried	177	50	15.6	7.8	12.3
Broccoli	71	35	61.3	21.5	4.5
Bok choy	85	79	52.7	41.6	2.3
Kale	65	47	58.8	27.6	3.5
Spinach	90	122	5.1	6.2	15.5
Tofu, calcium set	126	258	31.0	80.0	1.2

C, cup.

[a]Adjusted for load; for milk, this is fractional absorption (Fx abs) = 0.889−0.0964 ln load; for low-oxalate vegetables, after adjusting by the ratio of fractional absorption determined for kale relative to milk at the same load, the equation becomes Fx abs = 0.959−0.0964 ln load.

[b]Calcium content (mg) × Fx abs.

Data from Weaver CM, Proulx WR, Heaney RP. Am J Clin Nutr 1999;70(Suppl):543S–8S.[55]

Nutrient–Nutrient Interactions

Several nutrients and food constituents affect aspects of calcium homeostasis by means other than a simple effect on digestibility, as described earlier. Several dietary components influence urinary calcium excretion. Dietary calcium has relatively little influence on urinary calcium loss, especially during growth.[60] In contrast, a major determinant of urinary calcium is urinary sodium, which reflects dietary sodium.[61,62] Sodium and calcium share some of the same transport systems in the proximal tubule. In adults, each 100 mmol (2.3 g) increment of sodium excreted by the kidney pulls out approximately 0.6 to 1.0 mmol (24 to 40 mg) of accompanying calcium.[63] Because urinary calcium losses account for 50% of the variability in calcium retention, dietary sodium has a tremendous potential to influence bone loss at suboptimal calcium intakes in women; each extra gram of sodium per day is projected to produce an additional rate of bone loss of 1% per year if the calcium loss in the urine comes from the skeleton.[64] A longitudinal study of postmenopausal women showed a negative correlation between urinary sodium excretion and bone density of the hip.[61] The investigators concluded, from the range of values available to them, that bone loss could have been prevented by either a daily dietary calcium increase of 891 mg calcium or by halving the daily sodium intake. Racial differences in the effect of dietary sodium on urinary sodium and calcium excretion are observed as early as puberty.[65] White female children excrete more sodium and calcium on high-salt diets, compared with Black female children, a finding that may partially account for lesser vulnerability to hypertension through water retention, but greater vulnerability to osteoporosis with bone loss as they mature.[65,66]

Another dietary component that influences urinary calcium excretion is protein. Each gram of protein metabolized increases urinary calcium by approximately 1 mg; thus, doubling purified dietary proteins or amino acids in the diet increases urinary calcium by approximately 50%.[67] The acid load of the sulfate produced in the metabolism of sulfur-containing amino acids that produces acid ash is mainly responsible for this increase. However, a meta-analysis concluded that little evidence exists for the acid ash effect on calcium balance.[68] Increases in calcium absorption,[69] decreases in endogenous secretion,[70] or the hypocalciuric effect of phosphorus in high-protein foods can offset the hypercalciuric effect of protein. At the other extreme, inadequate protein intake compromises bone health and contributes to osteoporosis in elderly persons.[71] There appear to be dietary protein–calcium interactions such that calcium absorption increases to offset the calciuric effects of high dietary protein more at low than at high calcium intakes.[72] The benefits of calcium supplementation in mitigating bone loss in elderly persons are greater with higher protein intakes, however.[73]

Concerns about high phosphate consumption, especially with the popular trend toward high consumption in soft drinks that contain phosphoric acid, has been raised for bone. A met-analysis of calcium balance studies in response to phosphate intake showed decreased urine calcium and increased calcium retention despite increased endogenous secretion with increasing phosphate intake.[74] Cola beverages have been associated with reduced bone gain in children,[75] but it is possible that at least part of the explanation is the displacement of milk. However, in adult women, one study showed a dose response relationship between cola intake, but not non-cola soft drinks, and lower bone mineral density with no difference in milk intake by cola consumption.[76]

Although caffeine in high amounts acutely increases urinary calcium,[77] 24-hour urinary calcium was not altered in a double-blind, placebo-controlled trial.[78] In a systematic review of the safety of caffeine on health in healthy adults, pregnant women, adolescents, and children, intakes of caffeine up to 400 mg/d were found to be safe, with no adverse effects on calcium metabolism or bone end points.[79]

Fat intake may have a negative impact on calcium balance, but only during steatorrhea. In this condition, calcium forms insoluble soaps with fatty acids in the gut.

Increasing use of calcium supplements and fortified foods raised concern about the possibility of high calcium intake producing relative deficiencies of several minerals. High calcium intake led to relative magnesium deficiency in rats.[80] However, calcium intake does not appear to affect magnesium retention in humans.[81] Similarly, except for a single report in postmenopausal women,[82] decreased zinc retention has not been associated with high calcium intakes. The nature of this interaction is unclear and requires further study. Iron absorption from nonheme sources has been shown to decrease by half from radiolabeled test meals in the presence of calcium intakes up to 300 mg calcium/day, after which no further reduction occurs. Thus, practically speaking, it is prudent to set iron requirements assuming that individuals are going to ingest the amount of calcium in at least one glass of milk with each meal.[83] The inhibition of iron absorption by calcium does not appear to be a gut effect and may involve competition with the transport of iron in the intestinal mucosa,[84] possibly at the level of mobilferrin (see the chapter on iron). Calcium supplementation for up to 12 weeks does not appear to produce changes in iron status,[85] probably because of compensating up-regulation of iron absorption, nor does long-term supplementation reduce total body iron mass accumulation in adolescent girls.[86] Single-meal iron absorption studies may exaggerate inhibitory effects that disappear in the context of the whole diet. Iron deficiency in growing rats has a detrimental effect on bone morphology that is exacerbated by calcium deficiency.[87] The extent to which changing ratios of minerals predispose to chronic diseases is not well understood.

Functions

Intracellular Messenger

Ionized calcium is the most common signal transduction element in cells because of its ability to bind reversibly to proteins. To affect a regulatory change, an internal or external stimulus (physical, electrical, or chemical) causes a change in [Ca^{2+}] at a specific site in the cell by releasing a store of Ca^{2+} from within, or by causing Ca^{2+} to enter the cell from the outside (**Figure 7.4**). [Ca^{2+}] is maintained at approximately 100 nM in the cytosol by many binding and specialized extrusion proteins. This is necessary because Ca^{2+} is not metabolized in the same way as other second-messenger molecules. A released Ca^{2+} ion probably migrates less than 0.1 to 0.5 μm and exists as a free ion for only approximately 50 milliseconds before encountering a binding protein. The endoplasmic reticulum (sarcoplasmic reticulum in muscles) with its Ca^{2+}-ATPase pumps is the major intracellular calcium sink housing Ca^{2+}-binding proteins. Accumulation of Ca^{2+} in the cytosol would lead to cell death because it would precipitate phosphate (vital in energy transfer).

The [Ca^{2+}] is perceived by the body through CaSR. Thus, Ca^{2+} itself is one stimulus represented in Figure 7.4 detected

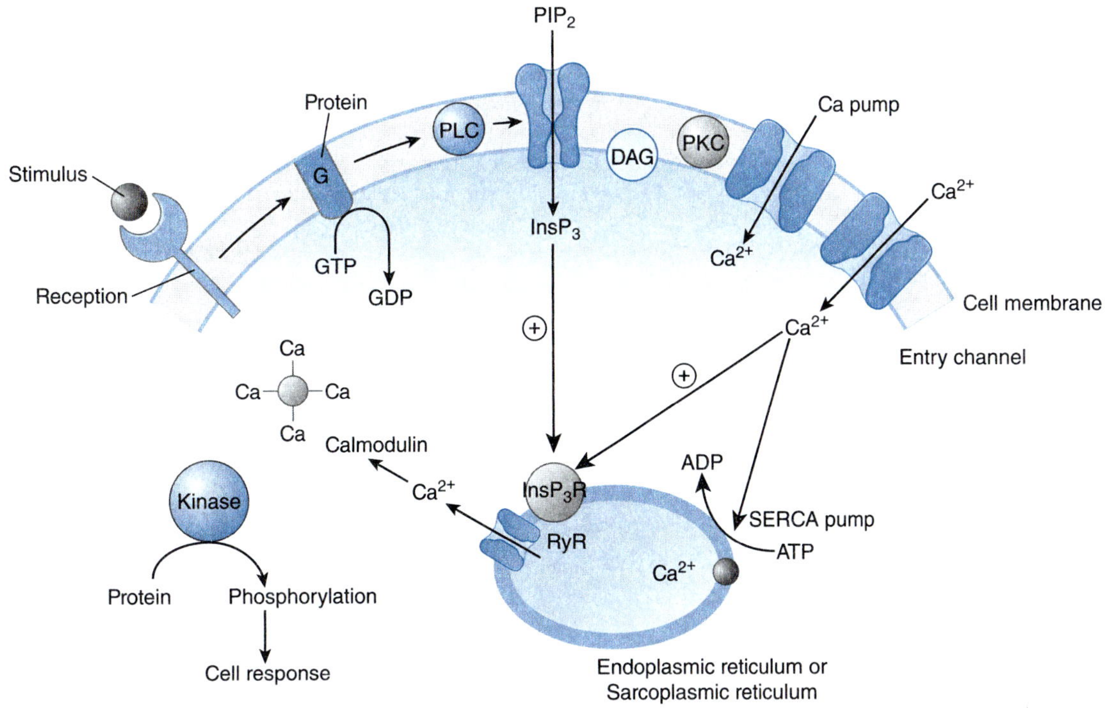

Figure 7.4 Intracellular calcium (Ca^{2+}) signaling. *ADP*, adenosine diphosphate; *ATP*, adenosine triphosphate; *CA PUMP*, plasma membrane calcium adenosine triphosphatase (Ca^{2+} ATPase) pump; *DAG*, diacylglycerol; *GDP*, guanosine diphosphate; *GTP*, guanosine triphosphate; *InsP₃*, inositol-1,4,5-triphosphate; *InsP₃R*, $InsP_3$ receptor; *PIP₂*, phosphatidylinositol-4,5-bisphosphate; *PKC*, protein kinase C; *PLC*, phospholipase C; *RYR*, ryanodine receptor; *SERCA pump*, plasma membrane calcium adenosine triphosphatase (Ca^{2+} ATPase) pump.

Data from Clapham DE. Cell 2007;131:1047–58.

by the G protein–coupled receptor, CaSR. In this way, calcium is an important extracellular "first" messenger, as well as a key intracellular second messenger. The plasma membrane is important in maintaining calcium homeostasis because the resting membrane is only slightly permeable to entering Ca^{2+}, and a Ca^{2+}–magnesium ion (Mg^{2+})–ATPase pumps Ca^{2+} out of the cell cytosol back to the ECF space. The pump is activated by calmodulin, an intracellular Ca^{2+} receptor protein that lowers the K_m of [Ca^{2+}] from a level of 400 to 800 nM to 200 nM and increases the total capacity of the pump. Thus, a momentary increase of cytosolic [Ca^{2+}] caused by an influx of Ca^{2+} is quickly returned to pre-excitation levels. Other less-important pathways of Ca^{2+} flux across the plasma membrane include influx pathways, potential-operated (voltage-gated) channels in excitable cells, receptor-operated channels in postsynaptic membranes, sodium ion (Na^+) channels, and efflux pathway, which is an Na^+/Ca^{2+} exchange pathway maintained by the Na^+ pump. Calcium messenger systems include sustained responses and trigger proteins.[1]

Sustained Responses

When an external or internal stimulus such as a hormone or neurotransmitter binds to a receptor in the plasma membrane, a series of responses occurs. Receptors can be G protein–coupled, as shown in Figure 7.4, or receptor tyrosine kinases. Phospholipase C is activated, and this hydrolyzes phosphatidylinositol-4,5-bisphosphate (PIP_2) bound to the inner layer of the plasma membrane into inositol-1,4,5-triphosphate ($InsP_3$) and diacylglycerol (DAG). Released into the cytosol, $InsP_3$ binds to G protein–coupled receptors in the endoplasmic reticulum (or sarcoplasmic reticulum in muscles) membrane, a process that induces liberation of Ca^{2+} from internal stores. Ca^{2+} can also enter the cytosol through plasma membrane Ca^{2+}-selective voltage-independent channels. Cytosolic Ca^{2+} concentration can change from 100 to 2 mM or 20,000 times higher. The increased cytosolic Ca^{2+} binds to calmodulin which, in turn, activates kinases to phosphorylate-specific proteins. This system accounts for the secretion of aldosterone from adrenal cells in response to angiotensin II, insulin secretion from β cells, contraction of smooth muscles, exocytosis, activation of T cells and B cells, adhesion of cells to the extracellular matrix, apoptosis, and many other processes.

Meanwhile the lipid portion of PIP_2, DAG, remains in the membrane and activates another membrane-attached enzyme, protein kinase C, which stimulates activity of the calcium pump. Thus, waves of Ca^{2+} are initiated by extra Ca^{2+} cycling in and out of cells.[88] As Ca^{2+} concentration returns to resting levels following action of Ca^{2+} pumps, recovery occurs in approximately 1 second and sets the stage for another Ca^{2+} spike. With the cloning of CaSR, signal transduction pathways influenced by CaSR are rapidly being discovered. CaSR activates phospholipases and mitogen-activated protein kinase (MAPK) and inhibits adenylate cyclase.[89]

Trigger Proteins

Calcium receptor protein pathways are almost universal and are present in both excitable and nonexcitable cells. They are important for fast switch processes in which Ca^{2+} acts as an on–off switch. Examples of excitable cells are skeletal muscle, neurons, smooth muscle, and taste cells that respond to salt. Excitable cells contain voltage gated Ca^{2+} channels in the plasma membrane, in addition to the system described earlier for nonexcitable cells that allows dramatic increases in intracellular Ca^{2+}. Entering Ca^{2+} activates ryanodine receptors (RyRs) to release Ca^{2+} from internal stores.

This one diffusible ion can regulate such diverse cellular processes as proliferation, differentiation, neuronal adaptation, and movement because it can be localized at one spot in a cell or spread throughout a cell by adjusting the amount released (amplitude modulation); and it can be released in pulses of different frequencies (frequency modulation).

Cofactor for Extracellular Enzymes and Proteins

Calcium is necessary to stabilize or allow maximal activity for certain proteases and blood clotting enzymes. These functions are not significantly affected by changes in extracellular Ca^{2+} concentration. Those substances that do not seem to be calmodulin activated by the system described earlier include glyceraldehyde phosphate dehydrogenase, pyruvate dehydrogenase, and α-ketoglutarate dehydrogenase.

Bones and Teeth

The role of calcium in bones and teeth is described more fully in the chapter on osteoporosis. Calcium exists primarily as the insoluble hydroxyapatite with the general formula $Ca_{10}(PO_4)_6(OH)_2$. Calcium comprises 39.9% of the weight of bone mineral.

Aside from the obvious structural role, the skeleton is an important reservoir of calcium to maintain plasma calcium concentration. Mobilization of calcium from bone may involve yet unidentified calcium-binding sites.[90]

The bone calcium pool in adults turns over every 8 to 12 years, on average, but turnover does not occur in the teeth. Remodeling of bone continues throughout life. Bone-resorbing osteoclasts begin this process by attaching to a bone surface and then extruding packets of citric and lactic acids (to dissolve bone mineral) and proteolytic enzymes (to digest organic matrix). Later, bone-forming osteoblasts synthesize new bone to replace resorbed bone. Usually, these processes are coupled. Bone formation exceeds resorption during growth. Bone resorption exceeds formation during development of osteoporosis. Osteoblasts have receptors for PTH, $1{,}25(OH)_2$ vitamin D, estrogen, and prostaglandin E_2. Osteoclasts have receptors for CT and various cytokines. Bone resorption is enhanced by PTH and is inhibited by CT.

Assessment of Calcium Status

A team of international calcium experts evaluated potential biomarkers of calcium status and concluded there has been little progress in decades.[91] Assessment of calcium nutrient status presents challenges unique among the nutrients. The skeleton, as noted in the chapter on osteoporosis, functions as a very large calcium reserve, both for the maintenance of ECF calcium concentration and for the critical cellular functions of calcium. This reserve is so large that deficiency of calcium at a cell or tissue level is essentially never encountered, at least for nutritional reasons. However, because the mechanical function of the skeleton is directly proportional to skeletal mass (i.e., to the size of the calcium reserve), it follows that any reduction in the reserve will result in a decrease in bone strength. In this sense, calcium is the only nutrient for which the reserve has a major function of its own. The size of that reserve can be assessed by total body bone mineral estimation using dual-energy x-ray absorptiometry (DXA) (see the chapter on osteoporosis). A problem arises in the interpretation of the results: the reserve can be low not only for nutritional causes but also for other reasons, such as lack of adequate physical activity, weight loss, gonadal hormone deficiency, or various medical diseases and their treatments.

In a research setting, calcium balance (intake minus excretion) can be used to determine whether losses of calcium from the body are being met by the intake of the controlled diet. If an individual is in negative balance, calcium is being lost from bone. Ideally, nonpregnant or lactating adults are in zero balance. The ideal positive during growth is unknown for many populations. Furthermore, the calcium status of a free-living population on self-selected calcium intakes cannot be readily assessed.

The other aspect of calcium metabolism, the concentration of $[Ca^{2+}]$ in blood and ECF, can readily be measured, however. Altered serum $[Ca^{2+}]$ usually means some abnormality of parathyroid function. Although a postabsorptive rise in serum calcium is detectable following large calcium loads, serum $[Ca^{2+}]$ is rarely ever low because of dietary calcium deficiency or high because of high calcium intake. This is basically because (as noted earlier) the skeleton serves as a very large calcium reserve and protects the ECF $[Ca^{2+}]$ essentially without limit. As described elsewhere in this chapter, it is the function of the parathyroid glands to draw down calcium from these reserves for the maintenance of ECF $[Ca^{2+}]$.

Deficiency

Overt, uncomplicated, calcium *metabolic* deficiencies are almost nonexistent given the large skeletal reserves as discussed earlier. Nevertheless, nutritional rickets is the most frequent bone disease worldwide, and is related to deficiency of both vitamin D and calcium or either one alone.[92]

Adequate calcium intake has been established as protective against osteoporosis. A meta-analysis of RCTs of calcium and vitamin D showed a significant 15% reduction in community dwelling individuals and 30% reduction in institutionalized postmenopausal women in total and hip fractures.[93] The primary strategies for reducing risk of osteoporosis are to maximize development of peak bone mass during growth and to reduce bone loss later in life. Achieving optimal calcium intake is a goal for both these aims. Further details on the role of calcium in preventing this debilitating disease may be found in the chapter on osteoporosis.

The intracellular messenger function of calcium, described earlier, is not affected by variation in calcium intake in the range usually encountered in the populations of industrialized nations. It, nevertheless, plays a role in calcium deficiency indirectly. Some of the consequences of low calcium intake involve systems not directly related to the calcium economy. High circulating $1{,}25(OH)_2D$, such as would occur in response to low calcium intake (see earlier), open calcium channels in the membrane of certain cells (e.g., smooth muscle and adipocytes) and thereby elevate cytosolic $[Ca^{2+}]$, with all the consequences described earlier (i.e., activation of various tissue-specific responses, such as contraction in arteriolar smooth muscle, upregulation of fat synthesis, and downregulation of lipolysis in adipocytes). In this way, low calcium intake contributes to the development or severity of disorders such as obesity and hypertension.

Calcium intakes required to prevent bone loss may also improve serum lipid concentrations and protect against hypertension.[94,95] An inverse association exists between calcium intake and the risk of some cancers including colon[96] and breast.[97] Recurrence of colorectal adenomas has been reduced by approximately 20% with calcium supplementation.[98]

Adequate calcium intake also decreases the risk of kidney stones (see the toxicity section later in this chapter).[99] Unabsorbed calcium in the gut forms a highly insoluble oxalate salt, thereby reducing absorption of oxalate from the diet.[100] Calcium supplements are accepted therapy for the kidney stone problem of intestinal hyperoxalosis.

An emerging set of etiologic factors accompanying the increase in the overweight population has a dietary component. Consumption of dairy products (which may be partially or wholly related to calcium consumption) is associated with lower risk of developing insulin resistance syndrome and its components (e.g., obesity, hyperinsulinemia, and insulin resistance).[101] Low-fat dairy products are also part of the Dietary Approaches to Stop Hypertension (DASH) diet recommended for managing hypertension by the Joint National Committee on Prevention, Detection, Evaluation, and Treatment of High Blood Pressure.[102]

Requirements and Recommended Intakes

The *calcium requirement* is the amount of dietary calcium required to replace losses in the urine, feces, and sweat, plus the calcium needed for bone accretion during periods of skeletal growth. Recommendations across the life span for optimal calcium intakes by the Institute of Medicine Food and Nutrition Board are given in **Table 7.3**. Currently, nutrient recommendations are determined by various regions and globally by the World Health Organization. Setting globally harmonized average requirements has been advocated for ease of cross-cultural comparisons and food labeling.[103] One attempt at setting harmonized, average requirements for several nutrients, including calcium, used whichever value was more recently updated between the U.S. Food and Nutrition Board and the European Food Safety Authority.[104] The basis for harmonizing nutrient recommendations is the assumption that biology is the same for any population, allowing that requirements may have to be adjusted for bioavailability of the nutrient from local food supplies.

Bone health is the primary consideration for determining adequate calcium intake because extraskeletal needs are easily met through accessing skeletal calcium when dietary calcium is inadequate. Calcium requirements for bone health throughout life are not uniform because of changes in skeletal growth and age-related changes in absorption and excretion.

Infancy

A term human infant contains approximately 0.65 to 0.75 mol (26 to 30 g) calcium. By the end of the first year of life, total body calcium increases to approximately 2 mol (80 g). The rate of calcium deposition in relation to body size is higher than at any other period during life. Required intakes of calcium are based on adequate intakes (AIs) determined from the mean intake of human milk for infants fed primarily human milk during the first 6 months and milk plus solids during the second 6 months.

Table 7.3 Calcium Recommendations During the Life Span

Group	1997 Adequate Intake (AI) (mg/d)	2010 Recommended Dietary Allowances (mg/d)
Infant		
Birth–6 mo	210	200 (AI)
6 mo–1 y	270	260 (AI)
Children		
1–5 y	500	700
4–8 y	800	1,000
Adolescents		
8–18 y	1,300	1,300
Adults		
19–50 y	1,000	1,000
>50 y	1,200	1,200
Pregnant and Lactating		
14–18 y	1,300	1,300
>19 y	1,000	1,000

Data from Institute of Medicine (2011). Dietary Reference Intakes for Calcium and Vitamin D. Washington, DC, The National Academies Press, with permission.[42]

Childhood and Adolescence

Calcium accretion continues throughout childhood. The rate of growth slows between the ages of 2 and 8 years. Between 9 and 17 years, approximately 45% of the adult skeleton is acquired, but the rate is not uniform. Expected growth in children aged 1 to 4 years is achieved at intakes of approximately 470 mg/day.[105] Maximal accretion occurs during the pubertal growth spurt, which, for most girls, is between the ages of 12 and 14 years and, for boys, between 14 to 16 years.[106] Calcium intake, race, markers of puberty such as serum insulin-like growth factor-I (IGF-I), and premenarcheal age are the largest predictors of skeletal calcium retention.[107] The lowest estimate of the intake required for mean maximal calcium retention in adolescent girls is 1300 mg/day.[60] Calcium intake higher than this, compared with lower, increases bone calcium accretion through increased absorbed calcium and suppressed bone resorption.[108] However, whether calcium intakes lower than recommended in Table 7.3 will lead to suboptimal adult bone mass is uncertain.[109] Calcium intakes estimated to yield observed, but not necessarily optimal, rates of bone accretion in a longitudinal study of White children were 1,000 mg/day for girls and 1,200 mg/day for boys, aged 14 to 18 years.[110] Animal studies showed that calcium deficiency during the pubertal growth period results in partial, but incomplete, catch-up growth.[111,112] Retention of the differential gain in bone mass between calcium supplementation and placebo groups in randomized controlled trials in humans waned after cessation of calcium supplementation.[109,113]

Peak Bone Mass

After adult height is achieved, calcium accretion continues to occur during the phase of bone consolidation. At the end of consolidation, when the maximum amount of bone has been accumulated, the adult is said to have achieved his or her peak bone mass. In girls and women, 90% of total body bone mineral content is achieved by age 16.9 years, 95% by 19.8 years, and 99% by 22.1 years.[114] However, the timing of peak bone mass varies with the skeletal site. The hip achieves peak bone mineral density first, at approximately age 14.2 years for the greater trochanter, 18.5 years for the femoral neck, and 15.8 years for Ward's triangle, whereas the spine can add mass throughout most of the third decade of life in women.[115] The skull accumulates bone throughout life, as does the femur shaft.[116]

Approximately 60% to 80% of peak bone mass is genetically predetermined. Diverse genes control all aspects of calcium use, including those influencing skin pigmentation, which affects vitamin D synthesis and, thereby, influences active calcium absorption efficiency, those controlling renal tubule reabsorption efficiency, and those controlling body size. A position paper from the National Osteoporosis Foundation that evaluated 18 factors affecting development of peak bone mass gave an A grade to calcium intake and physical activity based on the strength of the evidence.[117] During adolescence, urinary calcium is relatively unaffected by calcium intake,[60] indicating an ability to use all the absorbed calcium from the range of intakes studied for bone accumulation. Adequate dietary calcium influences bone size and geometry in addition to bone mass, both of which also contribute to bone strength.[118]

Aside from calcium intake, other lifestyle choices that affect peak bone mass include physical activity, intake of other nutrients that affect calcium balance (covered earlier in this chapter), anorexia, and substance abuse. As may be expected, dietary calcium and exercise positively interact in forming strong skeletons.[119-121] Beyond the timing of peak bone mass, lifestyle choices can affect rate of bone loss, but the window of opportunity to build bone has passed.

Adults

The mature woman has 23 to 25 mol (920 to 1,000 g) body calcium, and the mature man has approximately 30 mol (1200 g) total body calcium. The population coefficient of variation around these means is approximately 15%. Total body bone mass remains relatively constant over the reproductive years, as decreases in the proximal femur and other sites after age 18 years are offset by continued growth of the forearm, total spine, and head. Then, age-related bone loss varies with the individual, but occurs most rapidly during the first 3 years after menopause in women. The average adult loses bone at a rate of approximately 1% per year. Age-related decreases in calcium absorption and increases in urinary calcium contribute to this loss. These physiologic changes are more abrupt at menopause in women. Loss of estrogen and aging are associated with loss of intestinal VDR.[4] Furthermore, explanations for bone loss during aging include declining calcium intakes (discussed later) and physical activity and decreased gonadal hormones. The calcium intake required by older adults to achieve mean maximal retention or minimal loss was determined to be 1,200 mg/day by the Panel on Calcium and Related Nutrients (see Table 7.3).[42]

Pregnancy

Fetal skeletal calcium accretion is not great until the third trimester. During the third trimester, approximately 5 mmol/day (200 mg/day) of calcium are required for fetal growth. The mother's calcium absorption and renal conservation increase beginning by the second trimester to meet fetal demands and to store calcium for the subsequent lactational drain governed by PTH and IGF-I.[12,122] From before pregnancy status to the third trimester, fractional calcium absorption increases 60% to 70%.[123] At low calcium intakes, the mother's skeleton is compromised to meet calcium demands of the fetus, and the fetal skeleton is protected except at exceptionally low calcium intakes.[124] These changes are accompanied by a fall in biologically active PTH, increases in CT during early pregnancy, and increases in prolactin by up to ten- to twenty-fold. Prolactin may coordinate with calcitriol to increase intestinal calcium absorption.[22] Calcium supplementation increased bone density of neonates of malnourished women in India[125] and improved calcium balance and bone formation rates across pregnancy and lactation in women with habitual intakes of less than 500 mg/day.[126] No benefits to bone mineral status of infants occurred with calcium supplementation to pregnant Gambian women whose habitual calcium intakes were 9 mmol (360 mg)/day, however.[127]

Calcium supplementation during pregnancy can safely reduce the risk of maternal hypertensive disorders (a major cause of maternal death worldwide), including preeclampsia, as well as maternal morbidity and preterm birth, according to a calcium task force assembled by the New York Academy of Sciences.[128] The World Health Organization recommends 1–1.5 g calcium supplementation daily for pregnant women from areas with low calcium intake, based on current evidence that this level of supplementation reduces preeclampsia by 50%. Ongoing studies are exploring whether lower levels of calcium supplementation or fortification are sufficient to reduce incidence of maternal hypertensive disorders.

Lactation

Calcium transfer to breast milk varies mainly with changes in volume; calcium concentration is relatively constant at 7 ± 0.65 mmol/L (280 ± 26 mg/L) and is independent of the calcium content of the mother's diet. Wide variability in the amount of calcium transferred to milk daily has not generally been associated with bone mineral growth or status in infancy.[129] However, low dairy consumption by pregnant Black adolescents was associated with decreased fetal femur length.[130] Daily calcium transfer from maternal serum to breast milk increases from 4.2 mmol/day (168 mg/day) at 3 months following parturition to 7 mmol/day (280 mg/day) at 6 months following parturition. The increase in intestinal calcium absorption at the end of pregnancy gradually disappears after childbirth and during the lactation period. To meet the need of milk production, some renal conservation occurs but, more importantly, the maternal skeleton is depleted at a rate of approximately 1% per month; this loss is not prevented with calcium and vitamin D supplementation.[131] Increased bone turnover during lactation may be under the control of PTH-related peptide (PTHrP) produced by the lactating mammary gland.[132] A postlactation anabolic phase allows recovery of bone density to prelactation levels. Whether this recovery is complete in all individuals, such as older lactating women, is not known.

Epidemiologic studies have found no association between pregnancy or lactation and the risk of osteoporotic fractures.

Adequacy of Calcium Intake

Calcium intakes relative to requirements are deficient globally, meeting only two-thirds of the global population's requirement.[133] Mean calcium intakes of various regions of the world are shown in **Figure 7.5**. As dairy products are the most concentrated source of calcium, calcium inadequacy is of particular concern in Asian and African countries, where dairy products are less available and affordable.[90] Individuals at risk for calcium deficiency in Western countries include postmenopausal women, adolescents, vegans, and others who avoid dairy products. In the United States, according to the 2007–2010 National Health and Nutrition Examination Survey (NHANES), 49% of children aged 4–18 years and 39% of all individuals aged 4 and older consume less than the EAR for calcium from foods and supplements.[134] Using national food balance sheets and demographically weighted calcium requirements, inadequacy of the food supply to meet calcium requirements globally was estimated as 69%.[90] In the United States, calcium supplements make up some of the intake gaps from food, but supplements are taken primarily by adults and there is unequal use across race, rural vs urban, and socioeconomic status.[135,136]

Assessment of calcium intakes of populations is important for determining nutritional status and for drawing conclusions about the relationship between diet and health and disease.[90] Assessing usual calcium intake of an individual is fraught with error, however. Calcium intake can be assessed with food frequency questionnaires, diet recalls, diet records, or duplicate plate analysis. Duplicate plate analysis eliminates many of the errors associated with other methods but is not practical for large groups of individuals or for assessing long-term, usual intake. Food frequency questionnaires assess calcium better than they do some other nutrients because dairy foods are the major source of calcium, and individuals recall dairy product consumption reasonably well. Hidden calcium taken as food additives (e.g., anticaking agents), water, fortified foods, and components of pharmaceuticals can be easily overlooked, however. When calcium intakes from fortified foods were considered in assessing diets of Asian, Hispanic, and White 10- to 18-year-old children and adolescents, higher calcium intakes were observed than previously reported in national surveys, but most subgroups still fell below the recommended intakes.[137] The gap between calcium intakes and recommended intakes is greatest for the U.S. Black population.[138] Diet recall and diet records suffer from errors in estimating portion size, from variability in food composition, and from inadequacies of existing food composition tables. Multiple diet records can improve the estimate of an individual's average calcium intake. However, the generally large variability in calcium intake from day to day precludes confidence in estimates of usual calcium intake of an individual.[139,140]

Risks of Excess Dietary Calcium

Nutritional toxicity of calcium means an elevation of blood calcium concentration (hypercalcemia) by reason of overconsumption of calcium or an elevation of urine calcium excretion (hypercalciuria) to the point that either the kidneys calcify or renal stones develop. Hypercalcemia, particularly if severe, results in

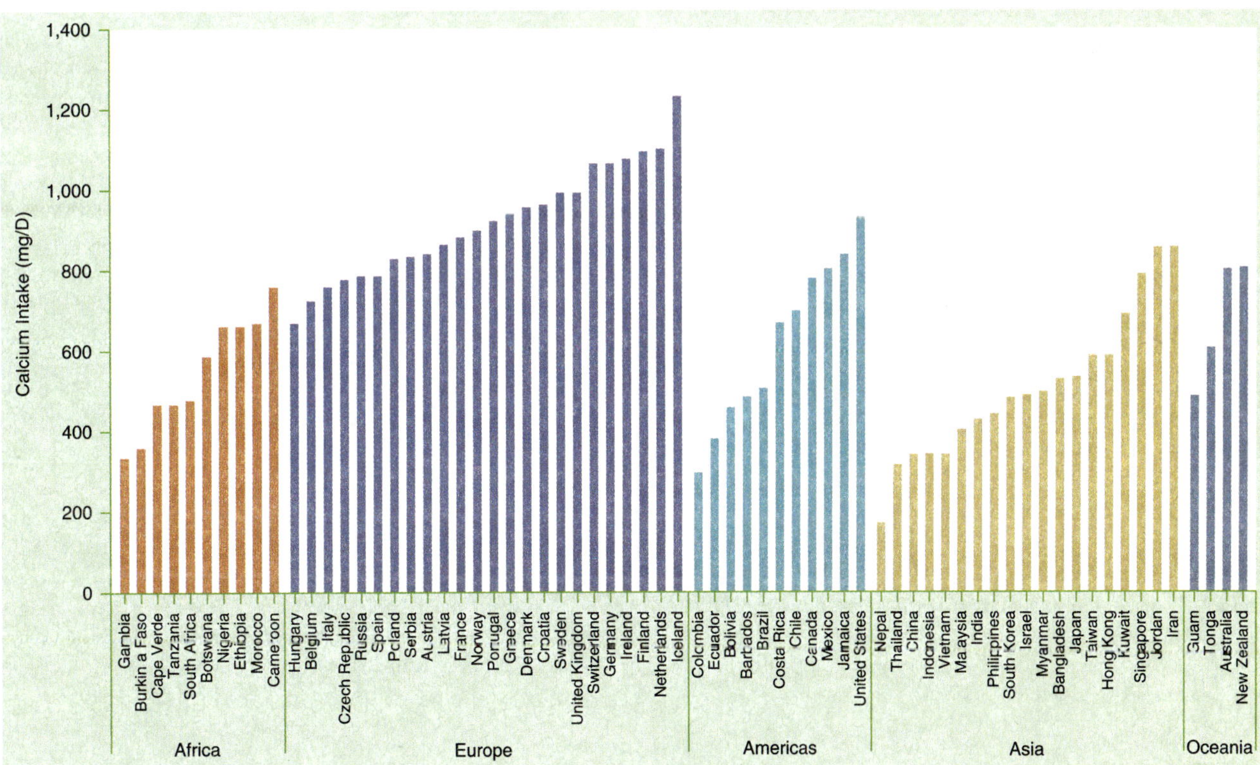

Figure 7.5 Average calcium intake in regions around the world.

Courtesy of Dennis Cladis.

lax muscle tone, constipation, large urine volumes, nausea, and ultimately, confusion, coma, and death. It essentially never occurs from ingestion of natural food sources. A good illustration of the safety of food calcium sources is provided by nomadic, pastoralist peoples, such as the Masai.[141] Because their diet consists mostly of the milk of their herds and flocks, they have calcium intakes higher than 5000 mg/day (and often appreciably higher), roughly 5 to 10 times what people of industrialized nations ingest. Such pastoral peoples are not known to have any unusual incidence of hypercalcemia or kidney stones.

Concern over potential risk of calcium supplementation and cardiovascular risk led to a position paper and clinical guideline from the National Osteoporosis Foundation and the American Society for Preventative Cardiology concluding that calcium supplementation up to the Upper Level is safe from a cardiovascular perspective.[142] They based this position on data analyses from RCTs, prospective cohort studies, a nestled case-controlled study, and a controlled feeding study in a pig model evaluating mechanisms and found no consistent dose-response relationships between calcium intake and CVD risk in generally healthy adults.

Kidney stones are not usually caused by dietary calcium. More often, individuals with kidney stones have high urine calcium because they have a renal leak of calcium. Accordingly, they often have some degree of reduction of their skeletal calcium reserves. Lowering calcium intake in such individuals rarely affects their kidney stone problem, but it always leads to further reduction in bone mass. High calcium intake may contribute to kidney stone formation in certain susceptible individuals. The only documented risk associated with calcium and vitamin D supplementation was in a reanalysis of the 7-year Women's Health Initiative trial, which found a 17% increase in the risk of kidney stones (HR = 1.17, 95% CI: 1.02 – 1.34),[143] but the events labeled "kidney stones" were not medically confirmed. In individuals with recurrent calcium oxalate stones, the stone problem may actually be helped by *increasing* calcium intakes to 30 mmol (1,200 mg)/day together with restricted animal protein and salt, compared with individuals on low-calcium diets of 10 mmol (400 mg) calcium/day.[97] This is because urinary oxalate excretion is an important risk factor for stones, and dietary calcium binds oxalate of dietary origin in the gut, prevents oxalate absorption, and thereby reduces the urinary oxalate load.

Clinical Disorders Involving Calcium

As noted earlier, low calcium intake, coupled with low calcium absorption efficiency and high obligatory calcium losses from the body, deplete skeletal calcium reserves. In other words, low intake leads to subnormal bone mass (and strength). This is one of the contributing causes of the disorder called *osteoporosis*, which is covered in the chapter on osteoporosis. Genetic mutations can alter intracellular calcium signaling. For example, alterations in the RyR protein can lead to hypertrophy and stroke. Familial benign hypercalcemia occurs when CaSR is partially or totally inactivated. Cell injury, damage, or serious dysfunction is always associated with a rise in cytosolic calcium concentration, an association that probably reflects an impaired ability of the cell to maintain the normal 10,000-fold gradient between the interior and exterior of the cell. The rise in cytosolic calcium may worsen cell damage and hasten cell death.[144]

The most common disorders of calcium metabolism (other than osteoporosis, which is multifactorial in origin), involve regulation of ECF [Ca^{2+}]. Usually, these conditions result from disorders of parathyroid gland function and are not nutritional. As noted earlier, the skeletal calcium reserves are so vast, relative to the size of the ECF [Ca^{2+}] compartment, that simple dietary deficiency of calcium essentially never compromises ECF [Ca^{2+}] regulation. A few rare exceptions are worth noting, however, because they illustrate how the system operates.

During growth, when the demands of skeletal mineralization are highest, extremely low-calcium diets may lead to hypocalcemia, despite maximal secretory output of the parathyroid glands. One consequence of the hypersecretion of PTH is lowering of serum phosphate. The combination of low calcium and low phosphorus in the ECF results in both undermineralization of newly deposited bone matrix and in osteoblast dysfunction. The clinical result is rickets. Rickets can be caused by either a severe deficiency of calcium or vitamin D, but they act synergistically. The usual etiology is a deficiency of both and is common in India, Africa, and the Middle East. As calcium intake decreases, more vitamin D is necessary to maintain normal calcium homeostasis.[90]

Another example of nutritional hypocalcemia is in magnesium deficiency, most often noted in severe alcoholism or as a result of intestinal fistula or malabsorption that causes excessive magnesium loss from the body. Magnesium, of course, is an essential cation for many cellular metabolic processes (see the chapter on magnesium). With severe magnesium depletion, many organs and systems function abnormally. The system regulating ECF [Ca^{2+}] is an example. Both PTH release from the parathyroid glands and bone response to PTH depend on magnesium, and both are defective in magnesium deficiency. Evidence that both steps are impaired is provided by the findings that, in magnesium-deficient patients, PTH fails to rise adequately in response to hypocalcemia and exogenous PTH fails to elevate bone remodeling, as it should. Magnesium repletion corrects both problems.

Acknowledgments

The author would like to acknowledge the contributions of Robert Heany to the previous version of this chapter.

This work was supported in part by Public Health Service (PHS) grants HD098589/HD 061908.

Disclosure: CMW is a member of boards for FDA, HHS, the International Life Sciences Institute, the California Prune Board and the California Walnut Board.

References

1. Clapham DE. Cell 2007;131:1047–58.
2. Carafoli E, Penniston JT. Sci Am 1985;253:70–8.
3. Quinn SI, Thomsen ARB, Pang JL et al. Am J Physiol Endocrinol Metab 2013;304:E310–E320.

4. Fleet JC, Schoch RD. In Crit Rev Clin Lab Sci 2010;47:181–195.
5. Ensrud KE, Duong T, Cauley JA et al. Ann Intern Med 2000;132:345–53.
6. Chattopadhyay N, Brown EM. Cell Signal 2000;12:361–6.
7. Hanes D, Weaver CM, Wastney ME. FASEB J 1995;9:A283 (abstract1642).
8. Heaney RP, Saville PD, Recker RR. J Lab Clin Med 1975;85: 881–90.
9. Heaney RP, Weaver CM, Fitzsimmons ML. J Bone Miner Res 1990;5:1135–8.
10. Heaney RP. Am J Clin Nutr 1991;54(Suppl):242S–57S.
11. Heaney RP, Weaver CM, Fitzsimmons ML et al. J Bone Miner Res 1990;5:1139–42.
12. Marcus CS, Lengermann FW. J Nutr 1962;77:155–60.
13. Cui M, Li Q, Johnson R et al. J Bone Miner Res 2012;27(10): 2097–2107.
14. Song Y, Kato S, Fleet JC. J Nutr 2003;133:374–80.
15. Xue Y, Fleet JC. Gastroenterology 2009;136:1317–27.
16. Nemere I, Leathers V, Norman AW. J Biol Chem 1986;261: 16106–14.
17. Fulmer CA. J Nutr 1992;122:644–50.
18. Nemer I. J Nutr 1992;122:657–61.
19. Benn BS, Ajibade D, Porta A et al. Endocrinology 2008;149: 3196–205.
20. Wasserman RH, Fullmer CS. J Nutr 1995;125(Suppl): 1971S–79S.
21. Wassermann RH, Chandler JS, Meyer SA et al. J Nutr 1992;122:662–71.
22. Christakos S, Dhawan P, Porta A et al. Mol Cell Endocrinol 2011;347(1–2):25–29.
23. Fujita H, Sugimoto K, Inatomi S et al. Mol Biol Cell 2008;19: 1912–21.
24. Barger-Lux MJ, Heaney RP, Lanspa SJ et al. J Clin Endocrinol Metab 1995;80:406–11.
25. Heaney RP, Recker RR, Steagman MR et al. J Bone Miner Res 1989;4:469–75.
26. Pattanaungkul S, Riggs BL, Yergey AL et al. J Clin Endocrinol Metab 2000;85:4023–27.
27. Ebeling PR, Sandgren ME, Dimagno EP et al. J Clin Endocrinol Metab 1992;75:176–182.
28. Recker RR. N Engl J Med 1985;43:133–7.
29. Ames SK, Ellis KJ, Gunn SK et al. J Bone Miner Res 1999;14: 740–6.
30. Reyes PC, Replogle RA, Wang L et al. J Bone Miner Res 2016;31:994–1002.
31. Bryant RJ, Wastney ME, Martin BR et al. J Endocrinol Metab 2003;88:1043-7.
32. Looker AC, Wahner HW, Dunn WL et al. Osteoporos Int 1998;8:468–89.
33. Charles P, Jenson FT, Mosekilde L et al. Clin Sci 1983;65:415–22.
34. Palacios C, Wigertz K, Martin B et al. Nutr Res 2003;23: 401–11.
35. Martin BR, Davis S, Campbell WW et al. Med Sci Sports Exerc 2007;39:1986–6.
36. Humes HD, Ichikawa I, Troy JL et al. J Clin Invest 1978;61: 32–40.
37. Coluccio LM. Eur J Cell Biol 1991;56:286–94.
38. Hoenderop JG, van der Kemp AW, Hartog A et al. J Biol Chem 1999;274:8375–78.
39. Christakos S, Gill R, Lee S et al. J Nutr 1992;122:678–82.
40. Loupy A, Ramakrishnan SK, Wootla B et al. J Clin Invest. 2012;122(9):3355–67.
41. Eaton SB, Konner M. N Engl J Med 1985;312:283–9.
42. Institute of Medicine Dietary Reference Intakes for Calcium and Vitamin D. Washington, DC, The National Academies Press, 2011.
43. Ballow C, Kuester S, Gillespie C. Arch Pediatr Adolesc Med 2000;154:1148–2.
44. 2010 Dietary Guidelines Advisory Committee Report. Accessed June 30, 2010. Available at: www.health.gov/dietaryguidelines.
45. Nickel KP, Martin BR, Smith DL et al. J Nutr 1996;126:1406–11.
46. Heaney RP, Recker RR, Weaver CM. Calcif Tissue Int 1990;46: 300–4.
47. Heaney RP, Weaver CM. Am J Clin Nutr 1989;50:830–2.
48. Heaney RP, Weaver CM, Recker RR. Am J Clin Nutr 1988;47: 707–9.
49. Weaver CM, Heaney RP. Calcif Tissue Int 1991;56:436–42.
50. Heaney RP, Weaver CM, Fitzsimmons ML. Am J Clin Nutr 1991;53:745–7.
51. Weaver CM, Heaney RP, Martin BR et al. J Nutr 1991;121: 1769–75.
52. McCance RA, Widdowson EM. J Physiol 1942;101:44–85.
53. James WPT, Branch WJ, Southgate DAT. Lancet 1978;1:638–9.
54. Heaney RP, Weaver CM. J Am Geriatr Soc 1995;43:1–3.
55. Weaver CM, Proulx WR, Heaney RP. Am J Clin Nutr 1999;70(Suppl):543S–8S.
56. Weaver CM, Heaney RP, Connor L et al. J Food Sci 2002;67: 3144–7.
57. Recker RR, Bammi A, Barger-Lux MG et al. Am J Clin Nutr 1988;47:93–5.
58. Cashman KD. Br J Nutr 2002;87(Suppl):169S–77S.
59. Zhao Y, Martin BR, Wastney ME et al. Exp Biol Med 2005;230: 536–42.
60. Jackman LA, Millane SS, Martin BR et al. Am J Clin Nutr 1997;66:327–33.
61. Devine A, Criddle RA, Dick IM et al. Am J Clin Nutr 1995;62: 740–5.
62. Matkovic V, Ilich JZ, Andon WB et al. Am J Clin Nutr 1995;62: 417–25.
63. Itoh R, Suyama Y. Am J Clin Nutr 1996;63:735–40.
64. Shortt C, Madden A, Fllynn A et al. Eur J Clin Nutr 1988;42: 595–603.
65. Wigertz K, Palacios C, Jackman LA et al. Am J Clin Nutr 2005;l81:845–50.
66. Palacios C, Wigertz K, Martin BR et al. J Clin Endocrinol Metab 2004;89:1858–63.
67. Heaney RP. J Am Diet Assoc 1993;93:1259–60.
68. Fenton TR, Lyon AW, Eliasziw M et al. J Bone Miner Res 2009;24:1835–40.
69. Kerstetter JE, O'Brien KO, Caseria DM et al. J Clin Endocrinol Metab 2005;90:26–31.
70. Spence LA, Lipscomb ER, Cadogan J et al. Am J Clin Nutr 2005;81:916–22.
71. Dawson-Hughes B. J Nutr 2003;133:852S–4S.
72. Hunt JR, Johnson LK, Roughead ZKF. Am J Clin Nutr 2009;89:1354–65.
73. Dawson-Hughes B, Harris SS. Am J Clin Nutr 2002;75:773–9.
74. Fenton TR, Lyon AW, Eliasziw M et al. Nutr J 2009;8:41–56.
75. Whiting SJ, Vatanparast H, Baxter-Jones A et al. J Nutr 2004;134(Suppl):696S–700S.
76. Tucker KL, Morita K, Qiao N et al. Am J Clin Nutr 2006;84(4): 936–42.
77. Hasling C, Sondergraad K, Charles P et al. J Nutr 1992;122: 1119–26.
78. Barger-Lux MJ, Heaney RP, Stegman MR. Am J Clin Nutr 1990;52:722–5.
79. Wikoff D, Welsh BT, Henderson R et al. Food Chem Tox 2017;109:585–648.
80. Evans GH, Weaver CM, Harrington DD et al. J Hypertens 1990;8:327–37.
81. Sojka J, Wastney ME, Abrams S et al. Am J Phys 1997;273(42):R710–R715.

82. Wood RJ, Zheng JJ. Am J Clin Nutr 1997;65:1803–9.
83. Gleerup A, Rossander-Hulten L, Gramatkovski E et al. Am J Clin Nutr 1995;61:97–104.
84. Halberg L, Rossander-Hulten L, Brune M et al. Eur J Clin Nutr 1992;46:317–27.
85. Whiting SJ. Nutr Rev 1995;53:77–80.
86. Ilich-Ernst JZ, McKenna AA, Badenhop NE et al. Am J Clin Nutr 1998;68:880–7.
87. Medeiros DM, Plattner A, Jennings D et al. J Nutr 2002;132: 3135–44.
88. Berridge MJ. Nature 1993;361:315–25.
89. Brown EM. Annu Rev Nutr 2000;20:501–33.
90. Bronner F, Stein WD. J Nutr 1995;125(Suppl):1987S–95S.
91. Ziaul R, Bourassa MW, Gomes F et al. Ann NYAS In Press. 2022.
92. Shlisky J, Mandlik R, Askara S et al. Ann NY Acad Sci In press. 2022.
93. Weaver CM, Alexander DD, Boushey CJ et al. Osteoporos Int 2016;27(1):367–376.
94. Reid IR, Mason B, Horne A et al. Am J Med 2002;112:343–47.
95. Van Vierlo LA, Arends LR, Streppel MT et al. J Hum Hypertens 2006;20:571–80.
96. Chia V, Newcomb RA. Nutr Rev 2004;62:115–20.
97. Moorman PG, Terry PD. Am J Clin Nutr 2004;80:5–14.
98. Shaukat A, Scouras N, Schunemann HJ. Am J Gastroenterol 2005;100:390–4.
99. Borghi L, Schianchi T, Meschi T et al. N Engl J Med 2002;346: 77–84.
100. Heaney RP, Weaver CM. Am J Clin Nutr 1989;50:830–2.
101. Weaver CM. Nutr Rev 2022;8:4–15.
102. Chobanian AV, Bakris GL, Black HR et al. JAMA 2003:289: 2560–72.
103. Yaktine AL, King JC, Allen LH. Adv Nutr 2020;11(5):1102–1107.
104. Allen LH, Carriquiry AL, Murphy SP. Adv Nutr 2020;11(3): 469–83.
105. Lynch MF, Griffin IJ, Hawthorne KM et al. Am J Clin Nutr 2007;85:750–4.
106. Bailey DA, McKay HA, Mirald RL et al. J Bone Miner Res 1999;14:1672–9.
107. Hill K, Braun MM, Kern M et al. J Clin Endocrind Metab 2008;93:4743–8.
108. Wastney ME. J Clin Endocrinol Metab 2000;85:4470–5.
109. Atkinson S, McCabe GP, Weaver CM et al. J Nutr 2008;138: 1182–6.
110. Vatanparast H, Bailey DA, Baxter-Jones ADG et al. Br J Nutr 2010;103:575–80.
111. Peterson CA. J Bone Miner Res 1995;10:81–95.
112. Weaver CM, Janle E, Martin B et al. J Bone Miner Res 2009:4:1411–9.
113. Bonjour J-P. Lancet 2001;358:1208–13.
114. Teegarden D, Proulx WR, Martin BR et al. J Bone Miner Res 1995;10:711–5.
115. Lin Y-C, Lyle RM, Weaver CM et al. Bone 2003;35:546–53.
116. Heaney RP, Barger-Lux MJ, Davis KM et al. Osteoporos Int 1997;7:426–30.
117. Weaver CM, Gordon CM, Janz KF, et al. Osteoporos Int 2016;27(4):1281–1386.
118. Cheng S, Lyytikainen A, Kroger H et al. Am J Clin Nutr 2005;82:1115–26.
119. Specker B, Binkley T, Wermers J. J Bone Miner Res 2002;17(Suppl):S398.
120. Stear SJ, Prentice A, Jones SC et al. Am J Clin Nutr 2003;77: 985–92.
121. Bass SL, Naughton G, Saxon L et al. J Bone Miner Res 2007;22:458–64.
122. Heaney RP, Skillman TG. J Clin Endocrinol Metab 1971;331: 661–70.
123. Ritchie LD, Fung EB, Holloran BP et al. Am J Clin Nutr 1998;67:693–701.
124. Zapatas CLV, Donangelo CM, Woodhouse LR et al. Am J Clin Nutr 2004;80:417–22.
125. Wargovich MJ. J Am Coll Nutr 1988;7:295–300.
126. O'Brien KO, Donangelo CM, Zapato CLV et al. Am J Clin Nutr 2006;83:317–23.
127. Jarjou LM, Prentice A, Sawo Y et al. Am J Clin Nutr 2006;83; 657–66.
128. Gomes F, Ashorn P, Askari S et al. Ann N Y Acad Sci. In Press:2021.
129. Prentice A, Laskey A, Jarjou LMA. Nutrition and Bone Development, vol 41. Philadelphia: Lippincott-Raven, 1999:127–145.
130. Chang S-C, O'Brien KO, Nathanson MS et al. Am J Clin Nutr 2003;77:1248–54.
131. Kalkwarf HJ, Specker BC, Henbi JE et al. Am J Clin Nutr 1996;63:526–31.
132. Kalkwarf HJ, Specker BL. Endocrine 2002;17:49–53.
133. Smith NW, Fletcher AJ, Dav LA, et al. J Nutr 2021;151: 3253–3261.
134. Wallace TC, McBurney M, Fulgoni VL, 3rd. J Am Coll Nutr 2014;33(2):94–102.
135. Blumberg JB, Frei BB, Fulgoni VL, et al. Nutrients 2017;9: 1295–1304.
136. Blumberg JB, Frei B, Fulgoni III, VL et al. Nutrients 2018;10: 1–11.
137. Novotny R, Peck L, Auld G et al. J Am Coll Nutr 2003;224: 64–70.
138. Wallace TC, Reider C, Fulgoni VL, III. JACN 2013;32(5): 321–330.
139. Weaver CM, Martin BR, Peacock M. Nutritional Aspects of Osteoporosis, vol 7. New York: Raven Press, 1995:123–8.
140. Barger-Lux MJ, Heaney RP. Nutritional Aspects of Osteoporosis, vol 7. New York: Raven Press 1995:243–51.
141. Jackson RT, Latham MC. Am J Clin Nutr 1979;32:779–82.
142. Kopecky SL, Bauer DC, Gulati M, et al. Ann Intern Med. 2016;165(12):867.
143. Prentice RL, Pettinger MB, Jackson RD et al. Osteoporos Int. 2013;24(2):567–80.
144. Rasmussen H, Palmieri GMA. Calcium in Biological Systems. New York: Plenum Publishing, 1985:551–60.

B. MINERALS

CHAPTER 8

Phosphorus

Jaime Uribarri and Mona S. Calvo

CHAPTER OUTLINE

Abbreviations

ADHR, Autosomal Dominant Hypophosphatemic Rickets; **AI**, Adequate intake; **ARHR**, Autosomal Recessive Hypophosphatemic Rickets; **CKD**, chronic kidney disease; **CVD**, cardiovascular disease; **DRI**, dietary reference intake; **EAR**, Estimated Average Requirement; **ESRD**, end-stage renal disease; **FDA**, Food and Drug Administration; **FGF-23**, fibroblast growth factor-23; **GMP**, good manufacturing practice; **GRAS**, Generally Regarded as Safe; **GWAS**, genome wide association studies; **HHRH**, Hereditary Hypophosphatemic Rickets with Hypercalciuria; **IOM**, Institute of Medicine**; NFL**, Nutrition Facts Label; **NHANES**, National Health and Nutrition Examination Survey, **PTH**, parathyroid hormone; **RDA**, Recommended Dietary Allowance; **SNP**, single nucleotide polymorphism; **TIO**, Tumor-Induced Osteomalacia; **UL**, tolerable Upper intake Level; **USRDS**, United States Renal Data System; **XLH**, X-Linked Hypophosphatemic Rickets (XLH); **DKA**, Diabetic Ketoacidosis

History of Discovery and Current Concerns

The element phosphorus was discovered more than 350 years ago, yet it has only been since late in the last century that significant progress in unraveling the complexities of how deficiency and excess of this critical nutrient impacts all life on earth.[1] Among elements comprising the earth's crust, phosphorus is the 11th most abundant and is not present in nature as elemental phosphorus due to its extremely high reactive properties, mainly occurring bound to oxygen as phosphate.[2] Phosphate rock in the earth's crust slowly releases phosphate by chemical weathering, building up phosphate in soil over time, which enters the food chain when taken up by soil microbes, crop plants, and ultimately livestock.

Geologic deposits rich in phosphate rock are limited across the globe and most are actively mined to produce fertilizers that enable vigorous crop growth when the soil becomes depleted of phosphate.[3] With low geologic abundance and slow weathering release from rock, phosphate is often the key growth-limiting factor for all living things. Phosphate rock deposits are finite in number and content, are not widely dispersed across countries, and are being aggressively mined to produce chemical fertilizers needed to meet the demands of modern agriculture and the growing human population food supply.[2] As human populations urbanize, the ability to recirculate phosphate back into food production dwindles and concern over the need for sustainable sources of phosphate critical to the global food supply now takes center stage.

Chemical Properties, Functions and Analysis

Figure 8.1 illustrates the three forms of phosphate that occur in nature: ionic anion, inorganic salt, and organic compound. Elemental phosphorus is highly reactive in nature mainly occurring bound to oxygen as orthophosphates. In solution, it forms an anion that readily binds to cations forming phosphate salts. Inorganic or mineral phosphate largely comprises the different salts of orthophosphate, as the quantity present as pyrophosphate is relatively small. Within the pH range of the human body, the two main forms of orthophosphates are $H_2PO_4^{-1}$ and HPO_4^{-2}. The organic form of phosphate occurs when bound to a carbon atom of protein, lipids, nucleic acids, and other organic compounds, usually through phosphate ester linkages. Phosphorus in its elemental form is toxic and not found in the human body or in food sources.[4] Phosphorus exists in three isotopic states: ^{31}P, an abundant isotope; ^{33}P, a longer-lived radioisotope (25-d half-life, weak beta radiation emitter); and ^{32}P, a shorter 14-d half-life and strong beta emitter.[4,5] With respect to human medical research, only the ^{33}P isotope can be used safely. Radioisotopic tracers are used to measure intestinal absorption of minerals, as they allow direct measurement of the movement of a tracer from the intestine into the circulation and, in the case of phosphate, the rate of transport between physiologic compartments.[6] Recently ^{33}P radioisotopic tracer methods were used to determine the difference in intestinal absorption of typical dietary phosphorus intake in patients with normal renal function compared with age, sex, and race-matched patients with chronic kidney disease (CKD). This important study found no differences in intestinal absorption with mild renal impairment.[7]

Phosphate is an essential nutrient that functions in critical pathways and cellular components in all life forms on earth, from subcellular virus to complex plants and animals dependent on phosphate for energy, growth, reproduction, structure, and homeostasis facilitated through signal transduction.[4] Phosphates participate in all biological processes, providing energy and its storage in phosphodiester bonds of ATP, the phosphodiester backbone of RNA and DNA, structural integrity of cell membranes as phospholipids, regulation of acid base balance, mineralization of teeth and bones, lipid transport in blood, and signaling pathways. With phosphates critical to so many biological processes, homeostasis must be tightly regulated.[8] If phosphate sources are deficient, limited growth or reproduction will occur; in contrast, if phosphate sources are in excess, overgrowth and toxicity can occur, as with algal eutrophication or toxicity in humans that can manifest as disease.

The measurement of phosphorus in biological fluid and tissues does not measure organic phosphorus, only inorganic phosphate.[5] Widely used clinical autoanalyzer methods for serum inorganic phosphate are based on the color reaction of phosphate ions with ammonium molybdate, forming a phosphomolybdate complex that turns blue after reduction. The molybdenum blue complex is then measured by spectrophotometry at 600–700 nm wave lengths.[9] Soft and hard tissues, as well as food and feces, must first be ashed and then solubilized to produce a neutral pH, colorless solution for analysis using the above spectrophotometric method or directly by atomic absorption spectrophotometry.

Physiology

Serum inorganic phosphate measured by clinical laboratories is normally 3.0–4.5 mg/dL (56% is ionized, 20% bound to protein, and 24% bound to other cations); however, there is a significant amount of organic phosphate in serum (7.5–8.0 mg/dL) not included in this clinical determination.[4] This concentration of serum phosphate follows an inherent diurnal variation pattern, lowest at 9AM and highest at 7PM (**Figure 8.2A** and **B**). Figure 8.2A shows the effect of different dietary phosphate intakes[10]; the intake of foods with different phosphate bioavailability shown in Figure 8.2B follows the same diurnal variation pattern, but with different levels of magnitude.[11] There is also seasonal variation, higher during summer than winter, thought to be due to the effect of sunlight on vitamin D synthesis stimulating phosphate absorption. Serum phosphate is also higher in women

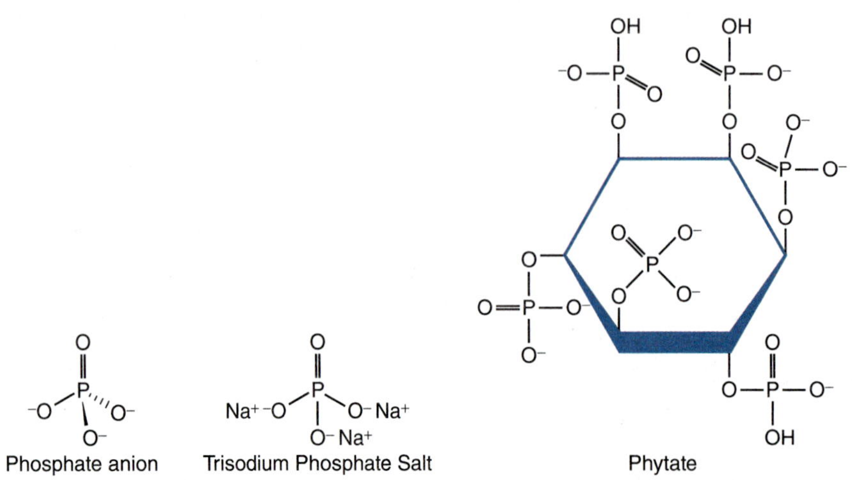

Figure 8.1 Phosphate ions, salts, and organic compounds.

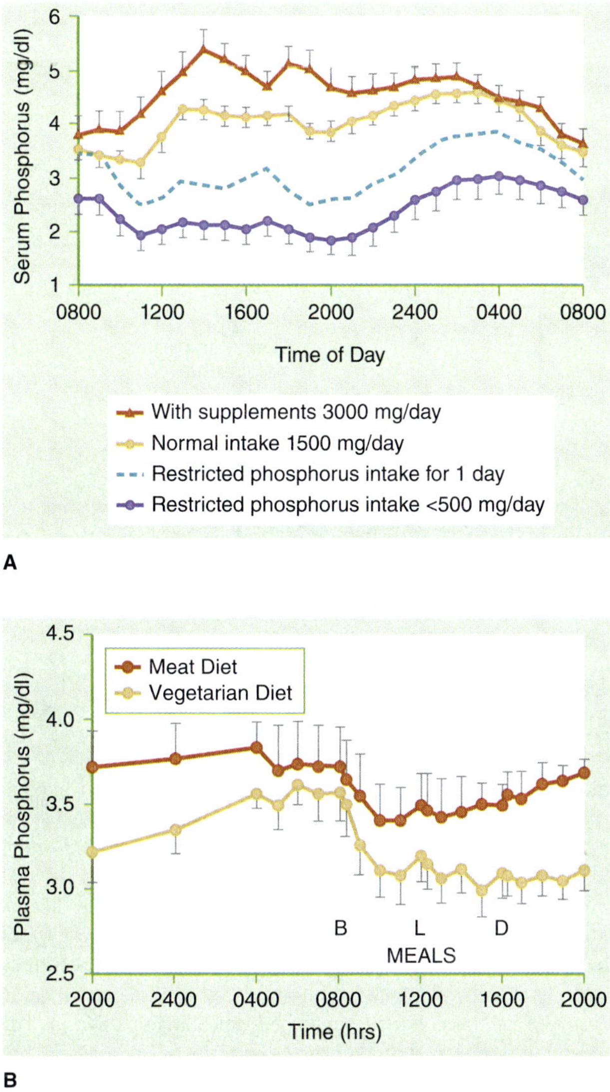

Figure 8.2 Diurnal variation of serum phosphate: (**A**) Inherent diurnal pattern and effect of total phosphate intake (**B**) Effect of dietary phosphate bioavailability from food on magnitude of diurnal variation when Pi content is equal.

(A): Reproduced from Portale AA, Halloran BP, Morris RC, Jr. Dietary intake of phosphorus modulates the circadian rhythm in serum concentration of phosphorus. Implications for the renal production of 1,25-dihydroxyvitamin D. J Clin Invest 1987;80(4): 1147-1154[10]; (B): Reproduced from Moe SM, Zidehsarai MP, Chambers MA, et al. Vegetarian compared with meat dietary protein source and phosphorus homeostasis in chronic kidney disease. Clin J Am Soc Nephrol 2011;6(2):257-264. doi:10.2215/CJN.05040610.[11]

than men (about 0.31 mg/dL) and higher in children than adults (normal range 4.5–5.5 mg/dL). It is important to remember that the terms *phosphorus* and *phosphate* are used interchangeably by clinical laboratories reporting serum concentrations, but they are not the same. To determine actual elemental phosphorus expressed in molar concentration, divide the serum phosphate concentration (reported in mg/dL) by the atomic weight of 31 for elemental phosphorus. A serum phosphate concentration of 4 mg/dL is 1.29 mM/L elemental phosphorus.

Phosphate Homeostasis

Total body phosphate content in a 70 kg man is about 700 g. Approximately 85% of this phosphate is in bone and teeth (as calcium phosphate hydroxyapatite, $Ca_{10}(PO_4)_6(OH)_2$), 14% in soft tissues and less than 1% in the extracellular space.

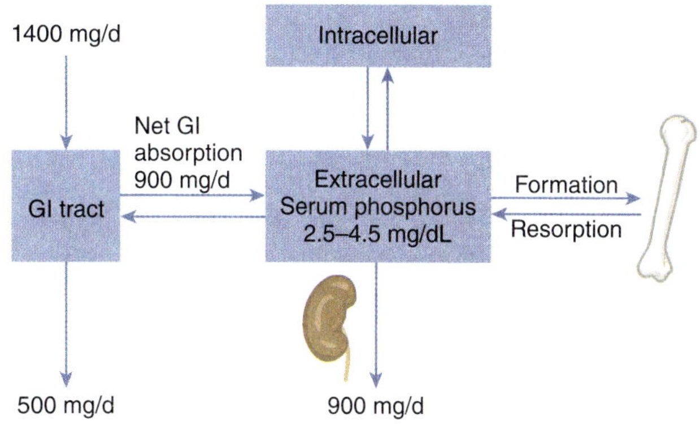

Figure 8.3 Phosphate balance.

Reproduced from Uribarri J, Calvo, MS. Excess Phosphorus Intake: Association with Disruption of Mechanisms Regulating Phosphorus Homeostasis and Disease risk. In: Collins J, ed. Molecular Genetic and Nutritional Aspects of Major and Trace Minerals. San Diego: Elsevier; 2016.[12]

Intracellular phosphate consists largely of organic molecules such as creatine phosphate, ATP, nucleic acids, phospholipids, and phosphoproteins, while the concentration of intracellular inorganic phosphate is only about 3–5 mM. At low intracellular concentration, free inorganic phosphate is unable to participate in sustaining a significant accumulation of inorganic phosphate in the absence of increased cell mass, shifting reliance on bone, teeth, and soft tissue calcification (including vascular calcification) to serve as the main homeostatic storage tissues for excess body phosphate. Body phosphate homeostasis is the result of the interaction of intestinal absorption of dietary phosphate, renal phosphate excretion, and phosphate exchange between extracellular and bone and intracellular phosphate pools (**Figure 8.3**). At present, the only easily available parameter to study total body phosphate in clinical practice is serum phosphate concentration, although the latter represents just a minute portion of total body phosphate content and, at least acutely, may be markedly affected by shifts among compartments.[12,13]

Renal Handling

The kidneys are major players in maintaining phosphate homeostasis, securing a match between urinary excretion (output) and net gastrointestinal absorption (input) of this mineral. In a normal individual, the kidneys reabsorb about 89% of the filtered load of phosphate and the rest is excreted in the urine. Phosphate filtered in the glomerulus is reabsorbed in the proximal tubule (75%) and in the distal tubule (10%) and about 10%–15% is lost in urine. Two main transporters have been identified as responsible for renal phosphate reabsorption: NaPi-2a and NaPi-2c. NaPi-2a is responsible for 70% of reabsorption, with 30% handled by NaPi-2c. A small amount of phosphate is reabsorbed by PiT-2 phosphate transporter (**Figure 8.4**). All these transporters are located on the apical brush border membrane of the renal proximal tubule cells.[12]

Intestinal Phosphate Transport

Gastrointestinal (GI) phosphate absorption in humans has traditionally been measured as the difference between dietary and fecal phosphate content; this net phosphate absorption is a linear function of dietary phosphate intake (**Figure 8.5**).[14] For a dietary phosphate intake range of 4 to 30 mg/kg/day, the percentage of net absorption is approximately 60%–65%.

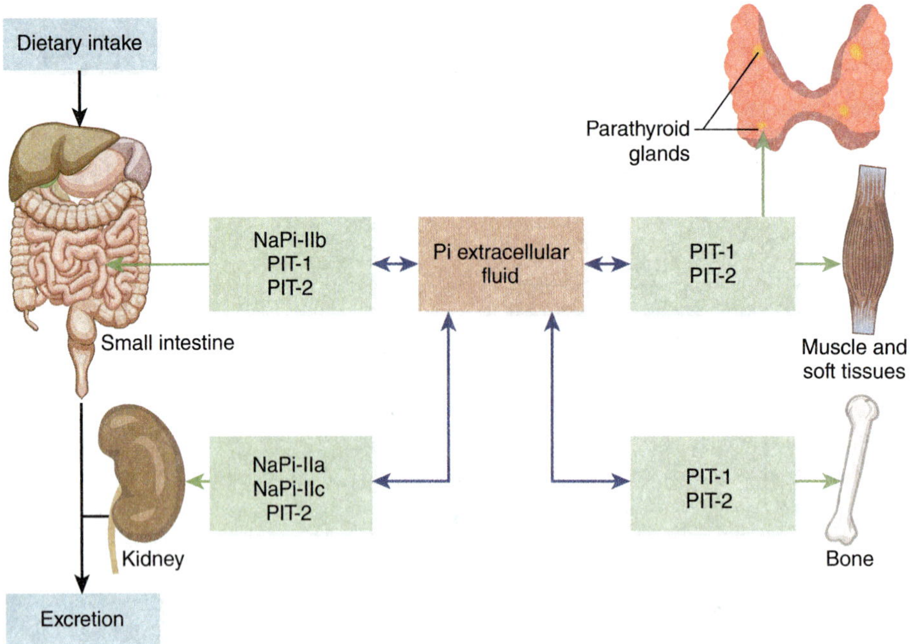

Figure 8.4 Phosphate membrane transporters in different organs.

Reproduced from Uribarri J, Calvo, MS. Excess Phosphorus Intake: Association with Disruption of Mechanisms Regulating Phosphorus Homeostasis and Disease risk. In: Collins J, ed. *Molecular Genetic and Nutritional Aspects of Major and Trace Minerals*. San Diego: Elsevier; 2016.[17]

Similar percentages of fractional intestinal phosphorus absorption have been found recently in a group of CKD patients using a radioisotopic tracer method.[7] There are two main transport systems for intestinal phosphorus absorption, one sodium-dependent, saturable, and transcellular (NaPi-2b) (Figure 8.4) and another that is sodium-independent, nonsaturable, and paracellular.[15,16] Phosphate concentration in the intestinal lumen and transporter abundance at any point will determine which transporter pathway will dominate. NaPi-2b transporter activity is regulated by vitamin D and parathyroid hormone (PTH) (Figure 8.4).[17] The intestinal paracellular phosphorus absorption pathway lacks tight regulation and depends on the phosphate concentration gradient across the epithelium, the electrical gradient (lumen negative), and tight junction permeability, controlled by several claudins.[16,17] Some phosphate transport may also occur through Na-Pi-III cotransporters (PiT-1, PiT-2, members of the SLC20 family).[8] The large intestine exhibits high paracellular phosphate permeability, which may explain the development of hyperphosphatemia following the use of high phosphate-containing enemas and may also be important in phytate phosphate bioavailability.[18] It is currently held that phytate is poorly absorbed in humans because of the lack of phytase. Many colonic bacteria produce phytases (*myo*-inositol hexakisphosphate phosphohydrolases) capable of sequentially hydrolyzing phytate. Liberated inorganic phosphate may then be absorbed in the colon through the paracellular pathways. The overall importance of this remains uncertain, as increasing solidity of distal colonic contents could make soluble phosphate less accessible for absorption. A recent review concluded that at least 50% of phosphate present in phytate is recovered as phosphate in 24-hour urine collections, based on past human studies.[19]

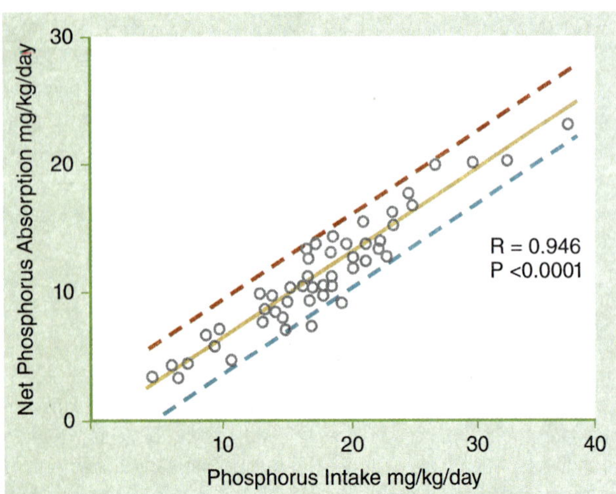

Figure 8.5 Phosphate absorption is linear with phosphate dietary intake.

Hormonal Regulation of Serum Phosphate

The tight regulation of serum phosphate concentration within the normal range involves the interplay of four organs—the kidneys, bone, parathyroid glands, and small intestine—controlled through the action of three hormones—PTH, fibroblast growth factor-23 (FGF23), and calcitriol (1,25-dihydroxy vitamin D), which act on organ-specific phosphate transporters and putative phosphate sensors.[18] As shown in **Figure 8.6**, the major regulatory impact is achieved by affecting intestinal phosphate absorption and/or renal tubular phosphate reabsorption. However, no hormone has been identified with a key role in controlling serum phosphate through action on bone. Deposition and release of phosphate in and out of bone is thought to simply follow that of calcium. FGF23 is a hormone produced in bone in response to excess phosphate functioning to maintain phosphate homeostasis by suppressing PTH secretion, 1,25-dihydroxy vitamin D renal synthesis, and renal phosphate transport. FGF23 action may be dependent or independent of Klotho, a beta-glucuronidase enzyme that occurs as both a transmembrane protein and a secreted renal protein and can function as a co-receptor to FGF23.

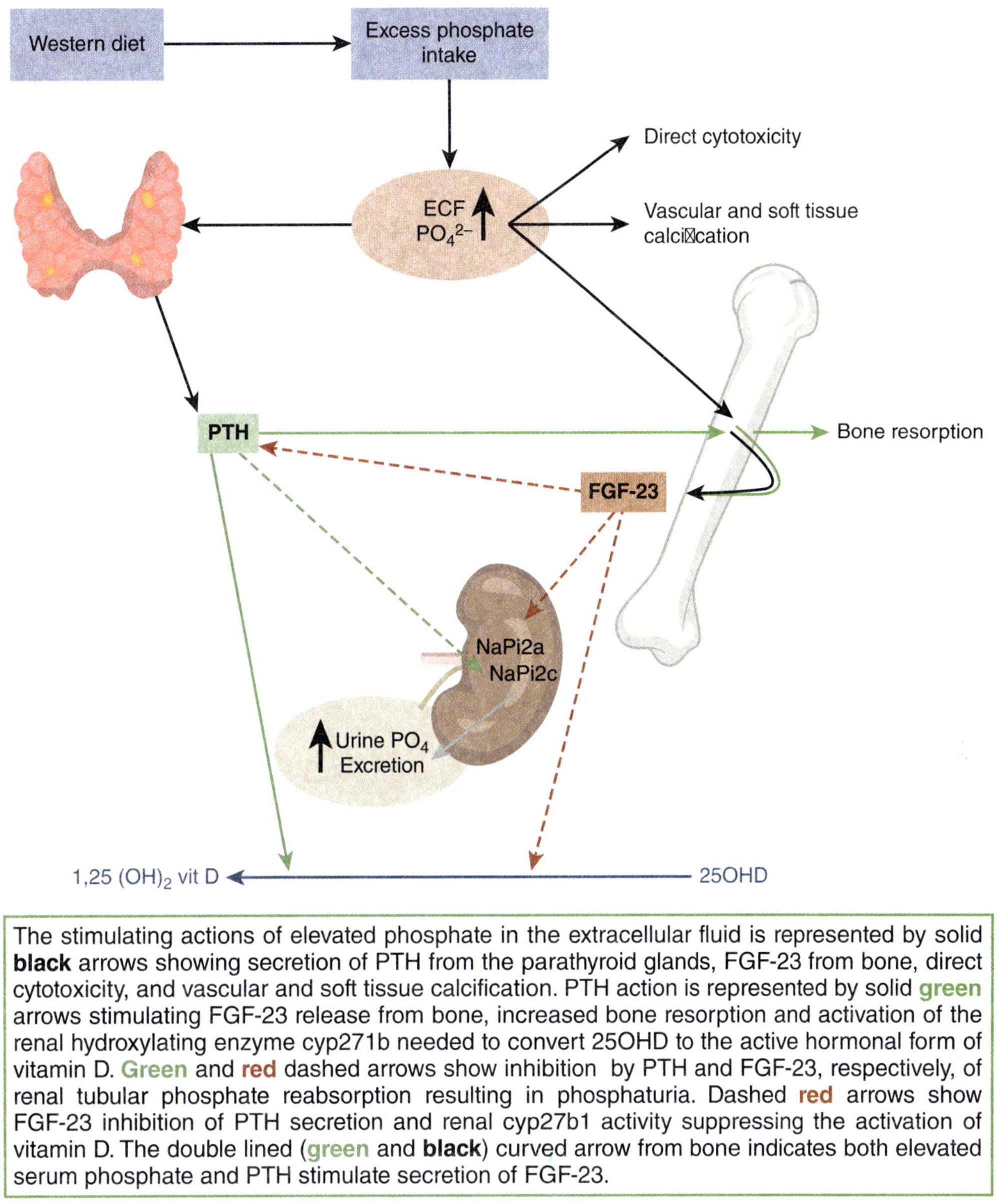

Figure 8.6 Complexity of phosphate endocrine regulation. Black denotes phosphate action; purple shows PTH action; and red shows FGF23 with dashed arrows indicating inhibitory action and solid arrows representing stimulatory action. White arrows shows incomplete calcitriol pathway and action.

Data from Rausch S, Föller M. The regulation of FGF23 under physiological and pathophysiological conditions. Pflugers Arch 2022;474(3): 281-292.[20]

Although many of Klotho's functions remain unclear, it has a proven role in phosphate regulation.[21]

FGF23 induces renal phosphate excretion by decreasing surface expression of NaPi-2a and NaPi-2c, the major Na^+-dependent phosphate transporters of the proximal tubule. FGF23 downregulates renal *cytochrome P450 (Cyp)27b1* expression, the key enzyme for 1,25-dihydroxy vitamin D production, and enhances *Cyp24a1* production catalyzing the inactivation of 1,25-dihydroxy vitamin D. These two effects are Klotho mediated (not shown in Figure 8.6). It is unclear how calcitriol impacts FGF23 action, but FGF23 clearly suppresses calcitriol synthesis and, with time, PTH secretion. Initially, in response to high serum phosphate, PTH is believed to stimulate FGF23 secretion from osteocytes.[20] Calcitriol upregulates Na-Pi co-transporters (an opposite action of FGF23) in both the intestine and proximal renal tubules, leading to increased intestinal absorption of phosphate and increased renal phosphate reabsorption. Calcitriol suppresses PTH secretion and has a negative feedback action on renal 1-alpha hydroxylase, thereby reducing its own production while upregulating 24-hydroxylase, increasing production of 24,25-dihydroxy vitamin D, thus inactivating calcitriol hormonal action (not shown in Figure 8.6).

Whole Body Phosphate Balance

On average, the mixed American diet contains 1,400 mg of phosphate and about 900 mg (60%–65%) will be absorbed into the extracellular fluid each day and excreted in the urine. In an adult person with no net bone growth, the amount of phosphate deposited into and removed from bone is essentially the same; therefore, a 24-hr urinary phosphate excretion should almost equal the net GI absorption of phosphate. This is independent of whatever renal mechanisms or hormonal factors exist to influence phosphate balance. Transient discrepancies between intake and output may be present. For example, as phosphate is the major intracellular anion, its transcellular flux is affected by any condition that increases glycolysis, such as acute alkalosis, insulin, or carbohydrate administration, with increased intracellular shift of phosphate leading to hypophosphatemia; or by acidic pH, which will have the opposite effect. Eventually, balance must be reached for survival, as prolonged marked electrolyte imbalance, including phosphate, is not compatible with life. In the presence of normal kidney function, an excess oral load of phosphate will produce only a transient elevation of serum phosphate followed by rapid elimination of the excess phosphate in the urine, lowering the serum phosphate to baseline in a few hours.

Abnormalities of Serum Phosphate

Clinically, patients may present with either hyperphosphatemia (serum phosphate >4.5 mg/dL) or hypophosphatemia (serum phosphate <2.5 mg/dL), with these conditions characterized by different manifestations and causes that cannot always be attributed to dietary phosphorus intake. For easy reference, differences in clinical manifestation and causes for both hypophosphatemia and hyperphosphatemia are summarized in **Box 8.1** and **Box 8.2**.

Box 8.1 Serum Phosphate Abnormalities: Hypophosphatemia

Clinical Manifestations

Muscular:

- Muscle weakness
- Rhabdomyolysis
- Cardiomyopathy
- Respiratory insufficiency

Bone:

- Osteomalacia and rickets

Hematologic:

- Hemolysis
- Impaired phagocytic activity
- Thrombocytopenia
- Impaired production of 2,3 Diphosphoglycerate resulting in tissue hypoxia

Nervous system dysfunction:

- Irritability, seizures, confusion, and coma
- Paresthesias, numbness

Renal:

- Hypercalciuria
- Proximal Renal Tubular Acidosis and reduced production of ammonia

Metabolic:

- Impaired insulin secretion

Causes

Decreased dietary intake

Decreased GI absorption:

- Malabsorption syndrome
- Phosphate binding antacids

Shift into cells or bone:

- Respiratory alkalosis
- Refeeding syndrome
- Administration of carbohydrates (fructose more than glucose)
- Hyperthermia
- Rapid cell proliferation
- Hungry bone syndrome after parathyroidectomy
- Recovery phase of DKA
- Hyperthermia and neuroleptic malignant syndrome
- Hypokalemic periodic paralysis

Increased urinary losses:

- Primary hyperparathyroidism
- Secondary hyperparathyroidism as in vitamin D deficiency
- Metabolic acidosis
- Volume expansion

- Fanconi's syndrome
- Oncogenic osteomalacia
- Tumor production of PTH-like peptide
- Genetic disorders of phosphate metabolism:
  - X-linked hypophosphatemic rickets
  - Autosomal dominant hypophosphatemic rickets
  - Hereditary hypophosphatemic rickets with hypercalciuria
- Neurofibromatosis
- Acute falciparum malaria
- Medications
  - Gallium nitrate
  - Bisphosphonates (etidronate, pamidronate, zoledronic acid)
  - Glucocorticoids
  - Ifosfamide
  - Cisplatin
  - Foscarnet
  - Barium poisoning
  - Theophylline

Nutrient Essentiality

Biomarkers of Phosphate Nutritional Status

Phosphate balance, a proxy for phosphate nutritional status in clinical practice, is best assessed by serum phosphate concentration, usually done in the fasting state. Nevertheless, single fasting serum phosphate measures do not necessarily reflect body phosphate stores, because acute shifts of phosphate between body compartments may temporarily affect serum phosphate without affecting total body phosphorus content. Acute alkalosis, for example, can produce significant hypophosphatemia by

Box 8.2 Serum Phosphate Abnormalities: Hyperphosphatemia

Clinical Manifestations

- Hypocalcemia with acute hyperphosphatemia
- Metastatic calcifications
- Progression of renal disease and secondary hyperparathyroidism

Causes

Increased oral intake

Exogenous phosphate loads through phosphate-containing enemas

Shift from cells

- Acute metabolic acidosis
- Respiratory acidosis
- Chronic respiratory alkalosis
- Tumor lysis syndrome
- Rhabdomyolysis
- Malignant hyperthermia
- Heat stroke

Reduced urinary excretion

- Advanced renal insufficiency
- Hypoparathyroidism
- Pseudohypoparathyroidism
- Vitamin D intoxication
- Growth hormone excess
- Glucocorticoid deficiency
- Biphosphonates
- Hypophosphatasia
- Tumoral calcinosis
- Fluoride intoxication

intracellular shift of phosphate without affecting total body phosphate. These transcellular phosphate shifts are usually transient in nature. However, when hyperphosphatemia or hypophosphatemia are chronic, there is usually a correlation between serum and total body phosphate. Moreover, there is a significant circadian variation in serum phosphate concentration that is influenced by time of blood draw, as well as gender and age, unrelated to phosphate intake effect on balance. Given these problems, serum phosphate was not considered an appropriate biomarker for establishing dietary phosphorus requirements or nutritional status by the European Food Safety Authority and other countries.[23] In sharp contrast, the Institute of Medicine (IOM) deemed measuring serum inorganic phosphorus acceptable for determining adult phosphate requirements in the United States and Canada.[24,25]

Dietary Recommended Intakes (DRI)

In 1997, the IOM determined adult age-specific Estimated Average Requirement (EAR) values derived from studies using serum phosphate as a biomarker, while the biomarker for children was based on published factorial estimates of accretion of phosphate into bone.[25] The EAR for adults 19–50 years was based on the relationship between serum phosphate and absorbed intake published by Nordin,[4] expanded to estimate intake associated with the normal range of phosphate intake and then extrapolated from absorbed phosphate to ingested phosphate based on the assumed efficiency of absorption of 62.5% from a mixed diet not high in phytate.[25] An intake of 580 mg was estimated to meet the needs of 50% of the population and, therefore, was set as the EAR. The EAR values serve as the basis for determining the Recommended Dietary Allowance (RDA) for phosphorus covering 97.5% of the population's phosphate intake needs. The age-specific EAR and RDA intake levels for males, females, and physiologic states of pregnancy and lactation recommended in 1997 are shown in **Table 8.1** for comparison to the most recent usual mean dietary intakes for these population groups, based on

Table 8.1 Dietary Reference Intake Values Set in 1997 for Phosphorus by the Institute of Medicine[a] and Mean Usual Intakes from NHANES 2015–2018[b]

Age Group	Estimated Average Requirement EAR, mg/day	Recommended Daily Allowance RDA, mg/day	Tolerable Upper Intake Levels UL, mg/day	Mean Usual Phosphorus Intake mg/day (SE)			
				Male		Female	
				N	Mean(SE)	*N*	Mean(SE)
Infants[c]	~	~	~	~	~	~	~
0–6 mo	~	100[c]	~	~	~	~	~
7–12 mo	~	275[c]	~	~	~	~	~
1–3 y	380	460	3000	*527*	1069 (21)	*488*	994 (15)
4–8 y	405	500	3000	*720*	1248 (22)	*731*	1089 (17)
9–13 y	1055	1250	4000	*740*	1343 (44)	*770*	1256 (26)
14–18 y	580	1250	4000	*715*	1482 (47)	*706*	1108 (36)
19–30 y	580	700	4000	*919*	1613 (21)	*904*	1210 (20)
31–50 y	580	700	4000	*1469*	1644 (29)	*1564*	1232 (20)

(continues)

Table 8.1 Dietary Reference Intake Values Set in 1997 for Phosphorus by the Institute of Medicine[a] and Mean Usual Intakes from NHANES 2015–2018[b] *(continued)*

Age Group	Estimated Average Requirement EAR, mg/day	Recommended Daily Allowance RDA, mg/day	Tolerable Upper Intake Levels UL, mg/day	Mean Usual Phosphorus Intake mg/day (SE)			
				Male		Female	
				N	Mean(SE)	*N*	Mean(SE)
51–70 y	580	700	4000	*1700*	1592 (25)	*1745*	1189 (20)
71+ y	580	700	3000	*726*	1435 (28)	*750*	1104 (27)
Pregnancy						Age 20 to 44 years	
≤ 18 y	1055	1250	3500	~		*111*	1297 (50)
≥ 18 y	580	700					
Lactation							
≤ 18 y	1055	1250	4000	~		78	1490 (47)
US FDA Intake Labeling Guidelines for Packaged Food[d]							
Target Group, ≥ 1 y	Daily Value = 1250 mg P/day Represented on the Nutrition Facts label as the P content in mg/serving and/or as % of Daily Value (DV) per serving						

[a]Data Source: Institute of Medicine. 1997. Dietary Reference Intakes for Calcium, Phosphorus, Magnesium, Vitamin D, and Fluoride, Washington, DC. National Academy Press, pp. 454.[24]

[b]Data Source: Moshfegh, A.J., Goldman, J.D., Rhodes, D.G., Clemens, J.C. 2021. Usual Nutrient Intake from Food and Beverages, What We Eat In America, NHANES 2015–2018. Worldwide Web Site: Food Surveys Research Group. Available at: www.ars.usda.gov/northeast-area/beltsville-md-bhnrc/beltsville-human-nutrition-research-center/food-surveys-research-group/docs/wweia-usual-intake-data-s[26]

[c]Infant values are AIs (Adequate Intake), not RDAs and are based on the composition of breast milk.

[d]Labeling guidelines for nutrient intakes that enable determination of adequate intakes are required on the Nutrition Facts label for four required nutrients: calcium, vitamin D, iron, and potassium. However, there is no requirement for phosphorus content to be listed. The intake guideline for phosphorus if manufacturers decide to list it on the label is 1,250 mg/day and is based on the RDA for 9- to 18-year-olds that reflects increased need for growing bones.

Data from Institute of Medicine. 1997. Dietary Reference Intakes for Calcium, Phosphorus, Magnesium, Vitamin D, and Fluoride, Washington, DC. National Academy Press, pp.454; and Moshfegh, A.J., Goldman, J.D., Rhodes, D.G., Clemens, J.C. 2021. Usual Nutrient Intake from Food and Beverages, What We Eat In America, NHANES 2015-2018. Worldwide Web Site: Food Surveys Research Group. Available at: www.ars.usda.gov/northeast-area/beltsville-md-bhnrc/beltsville-human-nutrition-research-center/food-surveys-research-group/docs/wweia-usual-intake-data-tables

nationally representative surveys.[26,27] Usual intakes account for the day-to-day variability in intake, and, for most groups, phosphate intake clearly exceeds the dietary guidelines.

A tolerable upper intake level (UL) for phosphate was also set as part of the DRI in 1997. The UL is considered a safe intake level, but as intake increases above the UL, risk for adverse events increases.[27,28] The UL for adults is 4,000 mg/d. The method used by the IOM panel in arriving at this value, which is nearly six times higher than the RDA, is in question and suggests a need to revisit the UL. The increased use of phosphate-containing food additives in processing led EFSA to reassess the safety of phosphate additive use and cumulative phosphate intake, resulting in a revision of the "group acceptable daily intake" (ADI) of 2.8 g phosphorus/day for an average 70 kg adult, representing a decrease from the 4.2 g/day set earlier.[27,28]

Labeling guidelines for nutrient intakes enable consumers to determine adequate intakes on the Nutrition Facts Label (NFL) required on all packaged foods in the United States and Canada. The FDA requires four nutrients to be listed on the NFL (calcium, vitamin D, iron, and potassium). Although there is no required listing for phosphorus,[29] if manufacturers opt to list phosphorus on the label, the content needs to be expressed in mg and/or as a percent Daily Value, the FDA's label intake guideline of 1250 mg/day. The phosphorus Daily Value is based on the RDA for 9- to 18-year-olds and reflects their increased need for growing bones. Recent usual intakes of phosphate plotted over percentiles of intake for adult men and women 19–50 years (**Figure 8.7**) illustrates how phosphate consumption in the United States greatly exceeds the DRI. Current mean phosphate intakes are two to three times higher than the RDA for all age groups >1 year (except for rapidly growing adolescents). Specific to the phosphate and calcium needs of growing adolescents, the FDA's Daily Value of 1250 mg/day intake guideline may unwittingly promote excessive phosphate intake for the greater proportion of the US population (19–50 y), particularly with the UL almost six times higher than the RDA. This likely creates a false sense of safety with such a large gap between recommended intake and the threshold for safe intakes of phosphate.[27] Consideration should be given to revising the 25-year-old DRIs, notably the UL, given current excessive phosphate intakes and growing evidence of potential disease risk associated with high phosphate intake in the general population.

Disorders of Excess Phosphate Intake With or Without Overt Hyperphosphatemia

High phosphate intake has always been an important concern for patients with chronic kidney disease in view of their impaired ability to excrete phosphate loads and the known association between excess phosphate in the body and bone and cardiovascular

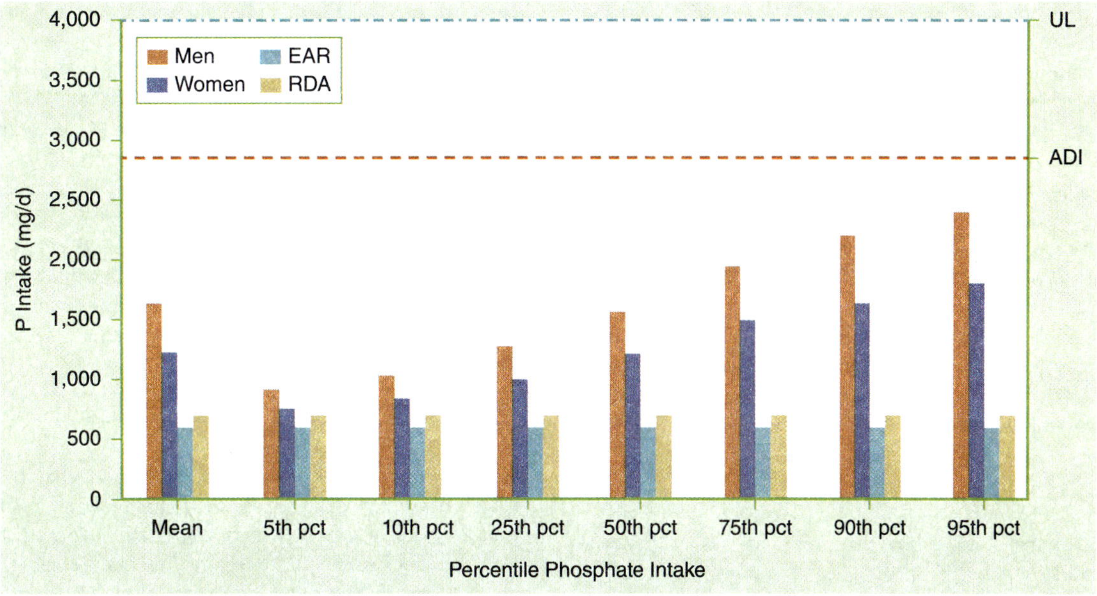

Figure 8.7 Percentile usual phosphate intakes for men and women aged 19–50 years.

Data from USDA Agricultural Research Service, 2019. Usual Nutrient intake from Food and Beverages by Gender and Age, from What We Eat in America, NHANES 2015-20p18. Available: www.ars.usda.gov/nea/bhnrc/fsrg

diseases in this population. Relatively recently, there has been growing recognition that even in the absence of overt hyperphosphatemia, high intake of phosphate may be associated with disruption of mechanisms regulating phosphate homeostasis and poor health outcomes. Recently, dietary intake of phosphate exceeding nutrient needs has been shown to be associated with disease risk in large epidemiologic studies in the general population, including a variety of chronic nontransmissible diseases.[30-33]

Hyperparathyroidism and Bone Health

Adequate dietary phosphate and calcium intake are essential for normal bone mineralization. However, it is not just the absolute amounts but also the relationship between both of them that matters. An imbalance between calcium and phosphorus intake can lead to bone loss.[4] High total phosphate intake over time that out paces dietary calcium intake has been shown in Western cultures; national surveys reveal that approximately 25% of the US population has a calcium to phosphorus intake ratio of less than 0.5, while intake guidelines recommend ratios of 1 to 1.2 (phosphorus to calcium).[34] The increasing dietary phosphate content is, at least in part, related to greater consumption of highly processed foods, which contain phosphate additives. This situation is compounded by the high efficiency of phosphate absorption, especially from inorganic phosphate constituents of additives. Resulting higher postprandial serum phosphate lowers serum ionized calcium concentration, and both act to stimulate PTH secretion. Elevated serum PTH will, in turn, stimulate bone resorption, thus liberating both minerals from bone into circulation. Serum mineral balance is restored through PTH action in the kidneys, which increases urinary phosphate excretion. These responses have been documented acutely using oral phosphate loads in young adults,[35] and more chronically in dietary intervention studies using foods containing phosphate additives,[36,37] and in large cross-sectional population studies.[38,39] This secondary hyperparathyroidism due to oral phosphate loading is clearly recognized in CKD, leading to significant bone disease (renal osteodystrophy).[40]

Cardiovascular Disease Risk and Mortality

Overt sustained hyperphosphatemia occurs almost exclusively in renal patients and, traditionally, most of the health risks attributed to hyperphosphatemia were related to secondary hyperparathyroidism and bone disease. In 1998, Block et al., using the United States Renal Data System (USRDS), data showed that hemodialysis patients with serum phosphate greater than 6.5 mg/dL had higher relative risk of death compared with those with serum phosphate between 2.7–6.5 mg/dL.[41] The higher risk of death was due to cardiovascular disease, adding a new dimension to the potential long-term side effects of hyperphosphatemia. A few years later, the same association between serum phosphate and mortality was confirmed in a large cohort of CKD patients not on dialysis, with moderate kidney impairment. Elevated serum phosphate has been associated with changes in left ventricular mass (measured by cardiac magnetic resonance), a known marker of cardiovascular disease (CVD), in the CKD population.[42] Whether the association between phosphate and CVD is the result of the direct action of phosphate or mediated through elevations of PTH, FGF23 and Klotho remains to be determined. Importantly, several studies have demonstrated that high serum phosphate concentrations were significantly associated with risk factors for cardiovascular disease or mortality in healthy adults without renal impairment.[43,44] A strong association between serum phosphate and mortality has been documented in the National Health and Nutrition Examination Survey (NHANES).[45]

Chronic Kidney Disease

Experimental work in animals has shown that a high phosphate diet will impair kidney function, while dietary phosphate restriction has the opposite effect. More recently, several epidemiologic studies have supported the association between higher serum phosphate and worsening kidney function. Among 985 veterans, those with higher serum phosphate had a higher likelihood of either reaching end-stage renal disease (ESRD) or doubling their serum creatinine.[46] Later epidemiologic findings over the past decade provide similar evidence of high phosphorus intake on the decline of renal function and progression to ESRD.[47-49] Analysis of 94,989 adults followed as part of an integrated health system showed a significant association between serum phosphate and risk of developing ESRD.[50] In a group of 10,672 CKD patients followed in the community, those with higher serum phosphate had higher risk of developing ESRD, although the risk became nonsignificant after adjusting for potential confounders.[51] A recent meta-analysis of 12 cohort studies with 25,546 participants determined that every 1-mg/dL increase in serum phosphate was independently associated with increased risk of developing kidney failure (defined as doubling of serum creatinine level, 50% decline in estimated GFR, or development of ESRD) and mortality.[52]

Cancer Risk

In vitro and animal studies show that phosphate has a significant cancer-modulatory effect. For example, high phosphate concentration in cell cultures has been shown to directly activate specific metabolic pathways, including cell cycle, altered protein translation, oncogene activation—K-ras pathways and ERK1/2-protein kinase C and AKT-ERK signaling that promote cell transformation and tumorigenesis in skin and lungs.[53-58] Such findings in animal models have sparked interest in large human cohort studies focused on cancer, and these have begun to show associations between serum phosphate and cancer risk, particularly prostate cancer. In more than 47,000 men, higher dietary phosphorus was associated with greater risk of total, deadly, and high-grade prostate cancer.[59] These early findings from follow-up studies need to be interpreted cautiously, as they are limited by inaccurate measurement of phosphorus intake attributed to the incomplete data describing phosphorus content of foods in the nutrient composition databases.[34] A different perspective, illustrating the complexities of dietary-cancer relationships is presented in a recent study looking at food consumption and malignancies using multiple global databases.[60] The excessive intake of terrestrial animal food, especially of protein, nitrogen, and phosphate in developed countries, was associated with higher prevalence of cancer, while equivalent intake from vegetables was associated with lower prevalence. Unfortunately, residual confounding factors associated with the lifestyle of countries consuming more animal foods was not evaluated and could be masking these relationships. In a recent study, individuals with single nucleotide polymorphisms (SNP = 125) independently associated with serum phosphate concentration were selected from the UK databank genome wide association studies (GWAS, n = 291,408) for meta-analysis, and observational studies were used to assess associations of dietary phosphate intake and serum phosphate (four studies) with risk of prostate cancer.[61] High dietary phosphate intake was associated with an 8% increase in prostate cancer risk and circulating phosphate with a 7% increase. These findings suggest potential causality for high phosphate intakes and serum concentrations and warrant further study.

Anemia

An association between hyperphosphatemia and anemia has been suggested by several epidemiologic studies. Hyperphosphatemia, defined as serum phosphate ≥4.4 mg/dL, was associated with low hemoglobin in participants of the NHANES (2005–2010) without chronic kidney disease.[62] A Southern California Health System study involving 155,974 adult participants demonstrated a 16% incremental risk of moderate anemia for each 0.5 mg/dL increment of serum phosphate, even in individuals with normal kidney function.[63] Mechanisms involved in the relationship between high dietary phosphate and anemia are currently under study in rodent models. Mice fed a high phosphate diet developed anemia, inflammation, and muscle wasting even in the presence of normal kidney function. These complications were still present after ablation of pathologic FGF23/FGFR4 but could be ameliorated with a low phosphate diet.[63] The authors suggested that high phosphate induces inflammatory signaling and hepcidin expression in hepatocytes, as a potential causative link between hyperphosphatemia, anemia, and skeletal muscle dysfunction.[63]

Senescence-Aging

A role for hyperphosphatemia in senescence has been postulated, especially because of the marked aging phenotype of Klotho-deficient mice, which also have hyperphosphatemia. Mice lacking Klotho present with a premature aging syndrome characterized by shortened life span, growth retardation, hypogonadotropic hypogonadism, skin atrophy, sarcopenia, vascular calcification, osteopenia, cognition impairment, hearing disturbance, and motor neuron degeneration.[20] The same phenotype is also observed in mice lacking FGF23. FGF23 often requires Klotho to exert its function, which explains the striking similarity in phenotypes between FGF23 knockout mice and Klotho-deficient mice.[64] A low phosphate diet improves life span and vascular calcification in both FGF23-deficient mice and Klotho-deficient mice, strongly suggesting a major role for hyperphosphatemia in the short life span of these inherited syndromes.[64] Some authors[20,65] have suggested a striking similarity between these congenital syndromes and advanced CKD in humans, which is also characterized by high serum phosphate, FGF23 and Klotho, as well as senescence of many organs/tissues throughout the body, and increased mortality.[21]

Kidney Stones

Although most idiopathic kidney stones in the Western world are made up of calcium oxalate and only a small fraction contain phosphate, calcium phosphate may be important as a predisposing factor for calcium oxalate kidney stone formation.[66] It is generally accepted that most idiopathic calcium oxalate stones start to form from Randall plaques, which are subepithelial deposits of calcium phosphate in the renal papilla.[67,68] It has been postulated that deposition of calcium phosphate in the renal interstitium, and its growth into a Randall plaque formation, is like pathological biomineralization seen elsewhere in the body. Once this plaque is exposed to urine, which is met as with respect to calcium oxalate, calcium oxalate stone formation begins. The role of dietary phosphate in this theory of kidney stone formation requires experimental confirmation. However, attention to both calcium and phosphate intake is needed for the prevention of calcium oxalate kidney stone formation, not only to avoid urine supersaturation for calcium oxalate but also for reducing calcium phosphate deposition.

Dietary Sources

Phosphate is present in most food sources, usually highest in protein rich foods, but its bioavailability differs between that derived from animal sources compared with plant sources.

Plants store phosphate as phytate (Figure 8.1), which requires enzymatic or physical action to release phosphate, increasing its bioavailability, which can occur with some processing methods of whole grains and other plant-based foods.[19] Animal-derived proteins do not require enzymatic treatment or mechanical action to release phosphate for absorption and, therefore, have greater bioavailability. Excessive intake of phosphorus can arise from more than consumption of animal vs plant-derived food sources with their differences in phosphorus content and bioavailability. It also involves the methods used to process the food, the amount of added phosphorus used in processing, and the habitual pattern of the types of foods consumed (**Table 8.2**). Foods classified according to the NOVA system[69] reflect their degree of processing and include four groups with

Table 8.2 Characteristics and Sources of Phosphate in Top Consumed Foods Classified According to the NOVA system[a]

NOVA Food Classification Definitions	Typical Food Sources in NOVA Category	Main Forms of Phosphate Natural Components and Added Ingredients in NOVA Food Groups	NOVA Food Groups Speculated Phosphate Absorption Rates %
Group 1. Minimally Processed, Unprocessed Foods: Natural plant and animal foods and water processed by drying, crushing, grinding, fracturing, filtering, roasting, boiling non-alcoholic fermentation, pasteurizing, refrigeration, chilling, freezing, placing in container and vacuum packaging for storage	Cooked Oatmeal Lentils Lima beans Soy flour Black beans	Organic Plant-derived Phytate	<40–50%
	Chicken breasts Tuna fillet 2% Milk Beef steak Egg white	Organic Animal-derived Protein	>60–70%
Group 2. Processed Culinary Ingredients:[b] Includes oils, butter, sugar, and salt that allows for seasoning, cooking and preparing unprocessed foods into palatable dishes at home	ND	ND	ND
Group 3. Processed Foods: Foods processed by various preservation or cooking methods or fermentation as in bread and usually contains two to three ingredients. Examples include canned fish and vegetables, fruits in syrup, cheeses, and freshly made bread using salt, oil, sugar or other substances from group 2	Frozen Chicken Patty Aged Cheddar Cheese Canned Tuna* Bakery Sour Dough Bread Canned Green Beans* Canned Corn* Frozen Blue Berries	Organic Animal Protein, Plant Protein and Phytate with modest use of Phosphate additives* (e.g., Firming agent or Sequestrant used in canning)	Mixed diet: >60%–90%
Group 4. Ultraprocessed Foods: Foods industrially formulated from substances derived from foods, additives, and ultraprocessed products not usually used in home cooking such as hydrogenated oils, hydrolyzed proteins, soya protein, high fructose corn syrup and added other sugars and salt	Chicken Nuggets* Hot Dogs & Sausages** Fish Sticks* Frozen Pizza* Processed Cheese* Infant Formulas** R-T-E-Breakfast Cereals** Cake Mixes* Savory Crackers* Industrial Produced Bread**	Organic Animal Protein derived Pi with added Inorganic Pi from additive use and Organic added protein concentrates** Organic rigorously processed Plant protein with lower Phytate but higher Inorganic Pi from additive use and Organic Pi from added protein concentrates**	Ultra-processed Animal foods ~90%–100% Ultra- processed Plant foods >70%–100%

[a]Data from Monteiro C.A. et al. (2017) The UN Decade of Nutrition, the NOVA food classification and the trouble with Ultra-processing. Public Health N.21(1): 5-17.[69]

*Indicates presence of one or more inorganic phosphate additive used in processing in many top selling packaged foods.

**Indicates the frequent use of one or more inorganic phosphate additive and added processed protein isolates or concentrates in processing popular food categories.

[b]No determination made for processed culinary ingredients, which do not allow estimates of the amount of ingredient used in home food preparation. To illustrate major phosphorus food sources, focus is on packaged food components rather than those from home prepared foods.

increasing degrees of processing. A cross-sectional study of the US food supply found that 71% of packaged foods and beverages were classified as ultra-processed and 60% of energy intake came from ultra-processed foods.[70] Examined over 2 decades of NHANES Surveys (2001–2018), the extent of ultra-processed food consumption has increased across all populations in the United States, while minimally processed food consumption has decreased.[71] Importantly, higher ultra-processed food consumption patterns have been shown to be associated with risk of CVD and mortality,[72] CKD,[73] and various other serious chronic diseases.

As the use of food additives, including phosphate-containing additives, are one of the hallmarks defining ultra-processed foods, dietary patterns high in processed and ultra-processed foods are likely higher in bioavailable phosphate, given their higher efficiency of absorption, as they readily dissolve in stomach acid. The use of phosphates in processing markedly improves the shelf life, texture, taste, and perceived quality of many foods,[74] are widely present as excipients in frequently used prescription and OTC drugs,[75] have global widespread use in processing food and beverages,[76,77] and are believed to contribute to the addictive characteristics of highly processed packaged foods[78]—all of which explains the increase in processed food consumption. It has been hypothesized that the increased cumulative use of phosphate additives in food processing will result in dietary phosphorus intake greatly exceeding adult requirements and potentially, with time, disrupting the delicate balance in the tightly regulated serum phosphate pool. This raises the issue of the potential toxicity of phosphate when consumed in excess and urges a closer look at the role of phosphate additive use in food processing.

Ultra-processed foods rarely list the phosphate content on the label, but unquantified use of all ingredients is required to be listed in the NFL ingredients list where they may or may not be recognized as containing phosphate. For most phosphate additives with Generally Regarded as Safe (GRAS) status (the amount that can be added to foods has no clear limitation other than GMP (good manufacturing practice). The FDA requires all food additives to have approved function prior to marketing or use in food production. Table 8.2 shows top-selling grocery store foods classified according to the NOVA system, indicating those containing two or more phosphate additives or added processed protein isolates or concentrates—all contributing to hidden phosphate content of the foods. **Table 8.3** illustrates a unique characteristic of phosphate food additives, showing how many have multiple approved technical functions.

Table 8.3 Phosphate Additives and Protein Isolate Ingredients with Multiple FDA-Approved Functional Uses Contributing to High Cumulative Phosphate Content of Some Processed Foods[a]

FDA-Approved Technical Use	Examples of Phosphate Ingredients with the Specific Technical Use	Number of Ingredients Containing Phosphate Approved for Specific Technical Use
Nutrient supplements	Ammonium phosphate, dibasic; Calcium phosphate, mono- di-, tribasic and pyrophosphate; Dipotassium phosphate; Ferric phosphate; Ferric pyrophosphate; Fish protein concentrate; Lecithin; Magnesium phosphate, di-, and tribasic; Manganese hypophosphate; Micro-particulated protein powder; Potassium phosphate, mono-, and tribasic; Potassium tripolyphosphate; Protein hydrolysate, unspecified; Protein, vege, hydrolyzed; Sodium ferricitropyrophosphate; Sodium ferritripolyphosphate; Sodium phosphate, tribasic; Soy protein concentrate	30
Stabilizer and thickeners	Starch, food, modified: Distarch phosphate; Starch food, modified: Acetylated distarch phosphate; Whey protein isolate; Sodium tripolyphosphate; Protein hydrolysate, unspecified; Potassium phosphate, monobasic; Magnesium phosphate, dibasic	24
Emulsifiers and emulsifier salts	Sodium metaphosphate; Sodium phosphate, monobasic; Sodium phosphate, tribasic, Sodium tripolyphosphate; Soy protein isolate; Starch, Food, modified; Sodium acid pyrophosphate; Lecithin	20
pH control agents	Ammonium phosphate, monobasic, Calcium phosphate, mono-, di- and tribasic; Dipotassium phosphate; Phosphorus oxychloride; Sodium phosphate, mono- and tribasic; Potassium phosphate, mono- and tribasic; Sodium metaphosphate	18
Texturizers	Calcium phosphate, tribasic; Lecithin; Milk powder, whole enzyme modified; Potassium acid, monophosphate; Potassium pyrophosphate; Potassium tripolyphosphate; Protein hydrolysate, unspecified; Sodium Aluminum Phosphate, acid or basic; Sodium metaphosphate; Sodium pyrophosphate and tripolyphosphate; Soy protein concentrate; Whey protein concentrate	16
Flavoring agents and adjuvants	Ammonium phosphate, dibasic; Calcium phosphate, tribasic; N-Gluconyl-ethanolamine phosphate; Milk powder, whole enzyme modified; Phosphoric acid; Protein, animal, hydrolyzed; Protein, milk, hydrolyzed; Sodium hexametaphosphate, Sodium metaphosphate, dibasic and monobasic; Starch, Food, modified; Whey protein concentrate	15

FDA-Approved Technical Use	Examples of Phosphate Ingredients with the Specific Technical Use	Number of Ingredients Containing Phosphate Approved for Specific Technical Use
Anticaking agent and free-flow agent	Calcium phosphate, di- and tribasic; Lecithin; Milk powder, whole enzyme modified; Protein, vege, hydrolyzed; Sodium aluminum phosphate, acid or basic; Sodium phosphate mono-, di-, and tri-basic; Sodium pyrophosphate and tripolyphosphate; Starch, Food, modified	12
Humectants	Calcium phosphate, dibasic; Calcium phosphate, tribasic; Lecithin; Milk powder, whole enzyme modified; Potassium poly-metaphosphate; Protein, vege, hydrolyzed; Sodium aluminum phosphate, acidic or basic; Sodium phosphate, monobasic and tribasic; Sodium tripolyphosphate; Starch, Food, modified	12
Drying agent	Sodium pyrophosphate; Sodium tripolyphosphate; Starch, Food, modified; Sodium phosphate, mono-, di-, tri-basic; Protein, vege, hydrolyzed; Lecithin; Calcium phosphate, di- and tri-basic	12
Flavor enhancer	Ammonium phosphate, dibasic; Milk powder, whole enzyme modified; Phosphoric acid; Protein, animal, hydrolysate; Protein hydrolysate, unspecified; Protein, vege, hydrolyzed; Sodium metaphosphate; Starch, Food, modified; Whey protein concentrate	12
Formulation aids	Sodium pyrophosphate; Soy protein isolate; Whey; Sodium phosphate, tribasic; Sodium acid pyrophosphate; milk powder, whole enzyme modified; Calcium phosphate, tribasic; Magnesium glycerophosphate	9
Sequestrants	Calcium hexametaphosphate; Calcium phosphate, monobasic; Dipotassium phosphate; Phosphoric acid; Sodium acid pyrophosphate; Sodium metaphosphate; Sodium phosphate, tribasic; Sodium pyrophosphate	9
Leavening agent	Calcium phosphate, tribasic; Ammonium phosphate, dibasic; Sodium acid pyrophosphate; Sodium aluminum phosphate, acidic or basic; Sodium phosphate, monobasic; Mono-/Diglycerides, monophosphate derivatives	9
Antioxidant	Lecithin; Calcium hypophosphite; Sodium hypophosphite; Sodium tripolyphosphate	4
Antimicrobial agent	Phosphoric acid; Sodium phosphate, monobasic; Sodium phosphate, dibasic	3
Dough strengtheners	Calcium phosphate, monobasic; Sodium acid pyrophosphate; Calcium phosphate, dibasic	3
Solvents and vehicles	Whey protein concentrate; Whey; Starch, food, modified	3
Surface active agents	Sodium metaphosphate: Lecithin	3
Color and coloring adjuncts	Sodium acid pyrophosphate; Whey	2
Firming agents	Ammonium phosphate, dibasic; Calcium phosphate, monobasic	2
Lubricant and release agents	Mono-, diglycerides, mono phosphate derivatives; Lecithin	2
Flour treating and release agents	Calcium phosphate, monobasic; Sodium acid pyrophosphate	2
Fumigants	Phosphoric acid	1
Oxidizing and reducing agents	Sodium acid pyrophosphate	1
Processing aids	Whey	1
Non-nutritive sweeteners	Phosphoric acid	1

[a]Data from Substances Added to Foods (formerly EAFUS). Database available at: www.cfsanappsexternal.fda.gov/scripts/fdcc/?set=FoodSubstances) last updated March 2022.

Sixty-one commonly used phosphate-containing ingredients conservatively identified from the *FDA's Substances Added to Foods* list are approved for 26 of the 32 FDA-allowed uses (technical functions) in processing food. Multiple functional uses also contribute to the cumulative phosphorus content of these preferred ultra-processed foods.

A few human studies focused on the phosphate additive content of commercially available foods have provided some insight into how the widespread use of phosphate additives may contribute to the adverse health effects of ultra-processed food consumption. Dietary intervention using processed foods with additives have shown significant associations with persistently elevated PTH, decreased calcitriol and other biomarker changes.[36,37] Higher intake of cured meats rich in phosphate additives was associated with carotid intimal thickening,[79] use of an algorithm determining bioavailability of phosphate from food frequency questionnaires showed an association with decreased renal function,[80] and estimates of added phosphate intake were inversely associated with HDL cholesterol.[81] Avoidance of phosphate additive containing foods has been associated with a decrease in serum phosphorus in CKD,[82] and higher consumption of specific foods processed with phosphoric acid (Cola drinks) and Asian noodles made with multiple phosphate additives was associated with lower bone mineral density[83] and higher serum phosphate.[84] To get a better understanding of how phosphate additives may affect our health, more studies are needed to evaluate foods in the marketplace, rather than adding phosphate salts to meals to determine effects of phosphate loading. The addition of phosphate additives during processing may change the food matrix in a way not duplicated with phosphate salt loading. Although data are suggestive, causality between phosphate additive use in ultra-processed food and increased risk of chronic disease or mortality remain to be established.

Inherited and Tumor-Induced Disorders of Phosphate Metabolism

Some disorders of phosphorus metabolism are not related to dietary intake, other than the use of dietary phosphorus to treat genetic disorders. Unlike disorders associated with excess phosphate intake, genetic-related disorders of phosphate metabolism are very rare in most populations.

X-Linked Hypophosphatemic Rickets (XLH)

XLH is the most common genetic disorder of phosphate metabolism. It is transmitted as an X-linked dominant pattern with 100% penetrance but variable severity of symptoms. The disease is due to mutation of the PHEX gene sequence in the chromosome X and subsequent lack of activity of the PHEX protein, which leads to elevated FGF23, in turn causing urinary phosphate wasting. High FGF23 will also interfere with 1,25-dihydroxy vitamin D concentration (low-normal) followed by impaired intestinal absorption of calcium and phosphate. Clinical manifestations are dominated by evidence of rickets in many bones and severe hypophosphatemia. Bone manifestations respond, in part, to the administration of oral phosphate supplements and calcitriol.[85]

Autosomal Dominant Hypophosphatemic Rickets (ADHR)

ADHR results from a missense mutation that prevents cleavage of FGF23, leading to failure of its posttranslational processing and increased circulating concentration. Clinical manifestations are, therefore, like XLH: hypophosphatemia with urinary phosphate wasting, rickets, high FGF23, normal serum calcium, low-normal 1,25-dihydroxy vitamin D, normal PTH. Inheritance is via an autosomal pattern that includes transmission from male to male with incomplete penetrance.[86]

Autosomal Recessive Hypophosphatemic Rickets (ARHR)

ARHR can mimic the biochemical characteristics of XLH. The main phenotype is high FGF23, but the mechanisms operative in three specific genetic alterations of ARHR leading to overproduction of FGF23 remain elusive. All three syndromes are rare (RHR1, ARHR2, ARHR3).[86]

Hereditary Hypophosphatemic Rickets with Hypercalciuria (HHRH)

HHRH is an autosomal recessive genetic phosphate disorder characterized by reduced renal phosphate reabsorption, hypophosphatemia, and rickets. It can be distinguished from other forms of hypophosphatemia by increased serum calcitriol resulting in hypercalciuria. Its gene locus was mapped to the end of the long arm of chromosome 9 in two consanguineous families. The candidate region contained a sodium-phosphate cotransporter gene, SLC34A3, which has been shown to be expressed in renal proximal tubule cells. Sequencing of this gene revealed disease-associated mutations in five families, including two frameshift and one splice-site mutations. Loss of function of the SLC34A3 sodium-phosphate transporter protein presumably results in a primary renal tubular defect and is compatible with the HHRH phenotype. Identification of the gene mutated in a further form of hypophosphatemia adds to the understanding of phosphate homeostasis and may help to elucidate the interaction of the proteins involved in this pathway.[87] Mutations in the sodium-phosphate co-transporter NaPi-2c result in the development of urinary phosphate wasting and hypophosphatemic rickets, bowing, and short stature, as well as appropriately elevated calcitriol concentration, which sets this FGF23-independent disorder apart from the more common X-linked hypophosphatemia. Elevated calcitriol, in turn, results in hypercalciuria due to enhanced intestinal calcium absorption and reduced PTH-dependent calcium-reabsorption in the distal renal tubules, leading to kidney stones and/or nephrocalcinosis in approximately half of the individuals with HHRH.[88]

Tumor-Induced Osteomalacia

Tumor-induced osteomalacia (TIO), also known as oncogenic osteomalacia, is a rare, acquired disorder caused by tumors that secrete FGF23. The biochemical features of TIO resemble those in XLH, ADHR, and ARHR. TIO is characterized by hypophosphatemia resulting from decreased tubular phosphate reabsorption resulting from excessive tumor secreted FGF23, causing bone pain, fractures, and muscle weakness.[89]

Summary

This chapter aims to summarize our current knowledge about phosphate as an essential structural and biochemical functioning component of our bodies and necessary dietary requirement. Most of the body's phosphate is contained within the bones, teeth, and intracellular space, with less than 1% present in the extracellular serum. The small fraction of the body's phosphate in serum is regulated within a narrow physiologic range and is active in endocrine hormone regulation of total body phosphate homeostasis and signal transduction for organ-to-organ communication. Despite transient and inherent diurnal variations, serum phosphate is the most accessible component for clinical laboratory measurement to inform about physiologic and nutritional phosphate status. Under most chronic conditions, serum phosphate concentration correctly reflects the state of total body phosphate balance, where a patient may present as normal, or with hyperphosphatemia (serum phosphate >4.5 mg/dL) or hypophosphatemia (serum phosphate <2.5 mg/dL), with the two latter conditions characterized by different manifestations and causes that may or may not relate to dietary phosphate intake.

Unlike most essential nutrients, phosphate is ubiquitously distributed in foods; however, foods differ in phosphate bioavailability, with animal-derived sources having greater efficiency of absorption than plant-based sources, which require enzymatic cleavage of phosphate from the plant storage form, phytate. The modern food supply can offer even higher total phosphate content and bioavailability in widely consumed ultra-processed foods produced with mechanical processing and use of phosphate additives. These ultra-processed packaged foods now dominate the global marketplace in higher income countries. Middle- to low-income countries with limited supply of processed and ultra-processed foods are more likely to experience poorer phosphate bioavailability and lower total intakes; however, inadequate phosphate intake is not a global concern. On the contrary, higher income countries are increasingly faced with excessive intake of this mineral due to the use of added phosphate to improve many desirable qualities in what has become our top food choices. Large-scale Western population studies are starting to reveal high or excessive intakes of dietary phosphate beyond our nutritional requirements and, even in the absence of overt hyperphosphatemia, associations with disruption of endocrine pathways that may connect phosphate intake with pathology known to contribute to a variety of nontransmissible diseases. Further studies are needed to adequately inform us about the potential toxicity of excess intakes of an essential mineral like phosphate and urge a closer look at the role human activity in the refinement of the foods we eat may play in disease risk.

Acknowledgments

The authors would like to acknowledge Kimberly O'Brien, Jane Kerstetter, and Karl Insogna, the authors of the previous version of the chapter, for their contributions.

References

1. Sharpley A, Jarvie H, Flaten D, et al. J Environ Qual 2018;47: 774–7.
2. Ferro C. The Effects of Changing Global Food Choices on Human Health and a Sustainable Supply of Phosphorus. In: Uribarri J, Calvo, MS, ed. *Dietary Phosphorus: Health, Nutrition and Regulatory Aspects*. Boca Raton: CRC Press; 2018.
3. Elser J, Chan, NI, Corman, JR, Stoltzfus, J. Save the P(ee)! The Challenges of Phosphorus Sustainability and Emerging Solutions. In: Uribarri J, Calvo, MS, ed. *Dietary Phosphorus: Health, Nutrition and Regulatory Aspects*. Boca Raton: CRC Press; 2018.
4. Nordin B. J Food Nutr 1989;45:62–75.
5. Peacock M. Calcif Tissue Int 2021;108:3–15.
6. Farrington K, Mohammed MN, Newman SP, et al. Clin Sci (Lond) 1981;60:55–63.
7. Stremke ER, Wiese GN, Moe SM, et al. J Am Soc Nephrol 2021;32:2057–69.
8. Hernando N, Gagnon K, Lederer E. Physiol Rev 2021;101: 1–35.
9. Endres DB RR. *Minerals and Bone Metabolism*. St Louis: Elsevier Suanders, 2006.
10. Portale AA, Halloran BP, Morris RC, Jr. J Clin Invest 1987;80: 1147–54.
11. Moe SM, Zidehsarai MP, Chambers MA, et al. Clin J Am Soc Nephrol 2011;6(2):257–64.
12. Uribarri J, Calvo, MS. Excess Phosphorus Intake: Association with Disruption of Mechanisms Regulating Phosphorus Homeostasis and Disease risk. In: Collins J, ed. *Molecular Genetic and Nutritional Aspects of Major and Trace Minerals*. San Diego: Elsevier; 2016.
13. Uribarri J, Oh, MS. Phosphorus Intake and Whole-Body Phosphorus Homeostasis. In: Uribarri J, Calvo, MS, ed. *Dietary Phosphorus: Health, Nutrition and Regulatory Aspects*. Boca Rtaon: CRC Press; 2018.
14. Marshall DH, Nordin BE, Speed R. Proc Nutr Soc 1976;35: 163–73.
15. Forster IC, Hernando N, Biber J, et al. Mol Aspects Med 2013;34:386–95.
16. Marks J. Pflugers Arch 2019;471:165–73.
17. Saurette M, Alexander RT. Exp Biol Med (Maywood) 2019;244: 646–54.
18. Figueres L, Beck-Cormier S, Beck L, et al. Int J Mol Sci 2021;22.
19. Calvo MS, Uribarri J. Adv Nutr 2021;12:2056–67.
20. Rausch S, Föller M. Pflugers Arch 2022;474:281–92.
21. Kuro OM. Nat Rev Nephrol 2019;15:27–44.
22. Buchanan S, Combet E, Stenvinkel P, et al. Front Endocrinol (Lausanne) 2020;11:560.
23. EFSA Panel on Dietetic Products, Nutrition and Allergies NaAN. EFSA Journal 2015.
24. Phosphorus. *Institute of Medicine (US) Standing Committee on the Scientific Evaluation of Dietary Reference Intakes*. Washington, DC: National Academic Press; 1997.
25. Calvo M, Whiting, SJ. The Regulatory Aspects of Phosphorus Intake: Dietary Guidelines. In: Uribarri J, Calvo, MS, ed. *Dietary*

Phosphorus: Health, Nutrition and Regulatory Aspects. Boca Raton: CRC Press; 2018.

26. Moshfegh AJ, Goldman, J.D., Rhodes, D.G., et al. 2021. Usual Nutrient Intake from Food and Beverages, What We Eat In America, NHANES 2015–2018.
27. Whiting S, Calvo, MS. Dietary Guidelines for Safe Levels of Phosphorus Intake in North America and Europe. In: Uribarri J, Calvo, MS, ed. *Dietary Phosphorus: Health, Nutrition and Regulatory Aspects*. Boca Raton: CRC Press; 2018.
28. EFSA FAF Panel (EFSA Panel on Food Additives and Flavourings); Younes MA, Castle GL, Engel K-H, et al. Scientific Opinion on the re-evaluation of phosphoric acid– phosphates–di-, tri- and polyphosphates (E 338–341, E 343, E 450–452) as food additives and the safety of proposed extension of use. In: Panel EF, ed. EFSA J. 2019, 17, e056742019.
29. Food Labeling: Revision of the Nutrition and Supplement Facts Labels. Fed Regist 2016;81:33741–999.
30. Uribarri J, Calvo MS. Ann N Y Acad Sci 2013;1301:iii–iv.
31. Mehta S, Uribarri, J. Overview of Health Risks Associated with Phosphorus Excess. In: Uribarri J, Calvo, MS, ed. *Dietary Phosphorus: Health, Nutrition and Regulatory Aspects*. Boca Raton: CRC Press; 2018.
32. Calvo MS, Sherman RA, Uribarri J. Am J Kidney Dis 2019;73: 542–51.
33. Calvo MS, Tucker KL. Ann N Y Acad Sci 2013;1301:29–35.
34. Calvo MS, Moshfegh AJ, Tucker KL. Adv Nutr 2014;5:104–13.
35. Kärkkäinen M, Lamberg-Allardt C. J Bone Miner Res 1996;11(12): 1905–1912.
36. Calvo MS, Kumar R, Heath H. J Clin Endocrinol Metab 1990;70(5):1334–40.
37. Gutiérrez OM, Luzuriaga-McPherson A, Lin Y, et al. J Clin Endocrinol Metab 2015;100:4264–71.
38. Kemi VE, Rita HJ, Kärkkäinen MU, et al. Public Health Nutr 2009;12:1885–92.
39. Pinheiro MM, Schuch NJ, Genaro PS, et al. Nutr J 2009;8:6.
40. Ikizler TA, Burrowes JD, Byham-Gray LD, et al. Am J Kidney Dis 2020;76:S1–S107.
41. Block GA, Hulbert-Shearon TE, Levin NW, et al. Am J Kidney Dis 1998;31:607–17.
42. Chue CD, Edwards NC, Moody WE, et al. Heart 2012;98: 219–24.
43. Dhingra R, Sullivan LM, Fox CS, et al. Arch Intern Med 2007;167(9):879–85.
44. Foley RN, Collins AJ, Herzog CA, et al. J Am Soc Nephrol 2009;20:397–404.
45. Chang AR, Lazo M, Appel LJ, et al. Am J Clin Nutr 2014;99: 320–7.
46. Schwarz S, Trivedi BK, Kalantar-Zadeh K, et al. Clin J Am Soc Nephrol 2006;1:825–31.
47. Zoccali C, Ruggenenti P, Perna A, et al. J Am Soc Nephrol 2011;22:1923–30.
48. Chue CD, Edwards NC, Davis LJ, et al. Nephrol Dial Transplant 2011;26:2576–82.
49. O'Seaghdha CM, Hwang SJ, Muntner P, et al. Nephrol Dial Transplant 2011;26:2885–90.
50. Sim JJ, Bhandari SK, Smith N, et al. Am J Med 2013;126:311–18.
51. Mehrotra R, Kermah DA, Salusky IB, et al. Kidney Int 2009;76:977–83.
52. Da J, Xie X, Wolf M, et al. Am J Kidney Dis 2015;66:258–65.
53. Jin H, Chang SH, Xu CX, et al. Toxicol Sci 2007;100:215–23.
54. Jin H, Xu CX, Lim HT, et al. Am J Respir Crit Care Med 2009;179:59–68.
55. Camalier CE, Young MR, Bobe G, et al. Cancer Prev Res (Phila) 2010;3:359–70.
56. Camalier CE, Yi M, Yu LR, et al. J Cell Physiol 2013;228: 1536–50.
57. Beck G. Preclinical Evidence of the Nonskeletal Adverse Health Effects of High Dietary Phosphorus. In: Uribarri J, Calvo, MS, ed. *Dietary Phosphorus: Health, Nutrition and Regulatory Aspects*. Boca Raton: CRC Press; 2018.
58. Arnst JL, Beck GR, Jr. Biochem Pharmacol 2021;183:114305.
59. Wilson KM, Shui IM, Mucci LA, et al. Am J Clin Nutr 2015;101: 173–83.
60. Penuelas J, Krisztin T, Obersteiner M, et al. Int J Environ Res Public Health 2020;17.
61. Lv L, Ye D, Chen J, et al. Am J Clin Nutr 2022;115:534–43.
62. Wojcicki JM. H BMC Nephrol 2013;14:178.
63. Czaya B, Heitman K, Campos I, et al. Elife 2022;11.
64. Kuro OM. Clin Sci (Lond) 2021;135:1915–27.
65. Yamada S, Giachelli CM. Bone 2017;100:87–93.
66. Tiselius HG. Urol Res 2011;39:231–43.
67. Sethmann I, Wendt-Nordahl G, Knoll T, et al. Urolithiasis 2017;45:235–48.
68. Khan A. Int Urol Nephrol 2018;50:799–806.
69. Monteiro CA, Cannon G, Moubarac JC, et al. Public Health Nutr 2018;21:5–17.
70. Baldridge AS, Huffman MD, Taylor F, et al. Nutrients 2019;11.
71. Juul F, Parekh N, Martinez-Steele E, et al. Am J Clin Nutr 2022;115:211–21.
72. Juul F, Vaidean G, Lin Y, et al. J Am Coll Cardiol 2021;77: 1520–31.
73. Cai Q, Duan MJ, Dekker LH, et al. Am J Clin Nutr 2022;116: 263–73.
74. D'ziezak J. Food Technology 1990;44:80.
75. Calvo MS, Uribarri J. Semin Dial 2013;26:54–61.
76. Chazelas E, Deschasaux M, Srour B, et al. Sci Rep 2020;10: 3980.
77. Fulgoni K, Fulgoni VL, 3rd. Nutrients 2021;13.
78. Tseng M, Grigsby CJ, Austin A, et al. Front Nutr 2021;8:762814.
79. Itkonen ST, Karp HJ, Kemi VE, et al. Nutr J 2013;12:94.
80. Duong CN, Akinlawon OJ, Gung J, et al. Am J Clin Nutr 2022;116:541–550.
81. Fulgoni K, Fulgoni VL, 3rd, Wallace TC. Nutrients 2022;14.
82. Sullivan CM, Leon JB, Sehgal AR. J Ren Nutr 2007;17:350–4.
83. Tucker KL, Morita K, Qiao N, et al. Am J Clin Nutr 2006;84: 936–942.
84. Saito Y, Sakuma M, Narishima Y, et al. J Clin Biochem Nutr 2021;68:78–85.
85. Carpenter TO, Imel EA, Holm IA, et al. J Bone Miner Res 2011;26:1381–8.
86. Athonvarangkul D, Insogna KL. Calcif Tissue Int 2021;108: 143–57.
87. Bergwitz C, Miyamoto KI. Pflugers Arch 2019;471:149–163.
88. Dhir G, Li D, Hakonarson H, et al. Bone 2017;97:15–19.
89. Brandi ML, Clunie GPR, Houillier P, et al. Bone 2021;152: 116064.

B. MINERALS

CHAPTER 9

Magnesium

Rebecca B. Costello, Andrea Mario Pompeo Romani, and Andrea Rosanoff

CHAPTER OUTLINE

Abbreviations

AI, adequate intake; **ATP**, adenosine triphosphate; **Ca^{2+}**, calcium; **cAMP**, cyclic adenosine monophosphate; **Cl^-**, chloride; **CLMD**, chronic latent magnesium deficit; **CRP**, C-reactive protein; **CVD**, cardiovascular disease; **DRI**, dietary reference intake; **EAR**, estimated average requirement; **EGF**, epidermal growth factor; **EGFR**, epidermal growth factor receptor; **ER**, endoplasmic reticulum; **FGF-23**, fibroblast growth factor-23; **K^+**, potassium; **Mg^{2+}**, magnesium; **MRS2**, mitochondrial magnesium channel; **Na^+**, sodium; **PGI_2**, prostacyclin; **PTH**, parathyroid hormone; **RDA**, recommended dietary allowance; **RER**, rough endoplasmic reticulum; **SLC41**, solute carrier 41; **SLC41A1**, solute carrier family 41 member 1; **SR**, sarcoplasmic reticulum; **tRNA**, transfer RNA; **TRPM6/7**, transient receptor potential ion channel melastatin subtypes 6 and 7; **UL**, tolerable upper intake level

Introduction

Magnesium (Mg) plays an essential role in a wide range of fundamental biologic reactions, so it is not surprising that Mg deficiency may lead to serious clinical symptoms. As described previously by Rude,[1] the first systematic observations of Mg deficiency in rats and dogs were made in the 1930s, and the first description of clinical depletion in humans in 1934 involved a small number of patients with various underlying diseases. In the early 1950s, studies documented depletion of this ion in individuals with alcoholism and in patients receiving Mg-free intravenous solutions. Although the diets ordinarily consumed by healthy Americans contain less Mg than the recommended dietary allowance (RDA) set by the US Institute of Medicine,[2] they do not seem to lead to immediately detectable symptomatic Mg deficiency but may lead to greater risk of disease if chronic. This chapter describes

1) Mg biochemistry and physiology, 2) Mg body composition and homeostasis, 3) Mg requirements, 4) Mg supplementation practices, 5) risk factors for and causes of Mg deficiency, and 6) management of Mg depletion. Some clinical disorders associated with a low-Mg diet, Mg-depleting medications, or genetic conditions are also discussed.

Biochemistry and Physiology

Mg is widely distributed in nature and is the eighth most abundant element on earth and the second most abundant cation in seawater.[1] Mg is the fourth most abundant cation in the human body and the second most prevalent intracellular cation after potassium.[3] Hydrated Mg presents with a radius far larger than that of calcium (Ca). Because of its positive charge, Mg binds to negatively charged molecules.[1] Intracellularly, Mg is highly compartmentalized within the nucleus, mitochondria, and rough endoplasmic reticulum (RER; or discreet portions of the sarcoplasmic reticulum [SR]) and is bound to nucleic acids, ribosomes, membranes, and phosphate groups.[3] In the cytoplasm, Mg is largely bound to adenosine triphosphate (ATP) and in part to other phosphonucleotides, organelle membranes, and other macromolecules (**Figure 9.1** and **Table 9.1**).

Cellular Homeostasis

Mg plays an essential role in the function of the glycolytic cycle, citric acid cycle, protein kinases, RNA and DNA polymerases, lipid metabolism, amino acid activation, vitamin D activation, and the cAMP and phospholipase C second-messenger systems, among others. The role of Mg^{2+} as an intracellular second messenger was first shown in 2011.[4]

The relative constancy of the Mg^{2+} in the intracellular milieu is attributed to the limited permeability of the plasma membrane to Mg and to the operation of specific Mg transport proteins, which regulate the rates at which Mg is taken up or extruded from cells.[3,5,6]

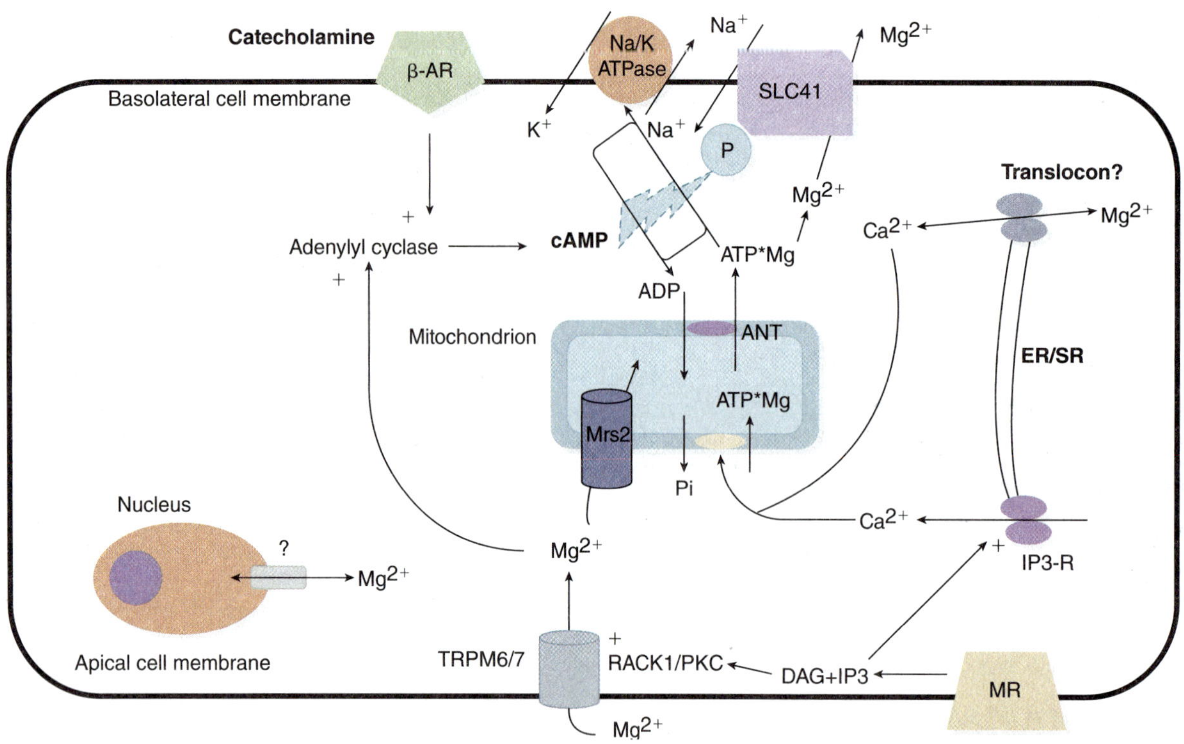

Figure 9.1 Cellular magnesium ion (Mg^{2+}) distributions and transport in mammalian cells. Plasma membrane mechanisms that control Mg^{2+} extrusion (top) and Mg^{2+} entry (bottom) and the main cellular compartments in which Mg^{2+} redistributes are shown. SLC41 (solute carrier 41, or Na^+/Mg^{2+} exchanger) is the main cellular Mg^{2+} extrusion mechanism and is located at the basolateral side of the cell membrane and is regulated via direct phosphorylation by cyclic adenosine monophosphate (cAMP). SLC41 operates as an electrogenic (1:1) Na^+/Mg^{2+} exchanger. TRPM6/7 (transient receptor potential ion channel, melastatin subfamily, isoforms 6/7) is a tetrameric channel that allows Mg^{2+} entry into the cell. Protein kinase C (PKC) activates this channel by binding to receptor for activated C kinase 1 (RACK1), which maintains the channel close under resting conditions. Within the cell, Mg^{2+} is mainly localized within the sarcoplasmic/endoplasmic reticulum (SR/ER), mitochondria, and nucleus. MRS2 is the main mitochondrial channel that allows Mg^{2+} to enter the organelle matrix and plays a major role in mitochondria bioenergetics. In this organelle, Mg^{2+} also plays a key role in the extrusion of ATP via the adenine nucleotide translocase (ANT). The transport mechanisms that allow Mg^{2+} movement across the membrane of SR/ER and nucleus are not characterized. In the cytoplasm, Mg^{2+} is mainly bound to ATP and possibly to guanosine-5′-triphosphate (GTP) and other phosphonucleotides. It is unclear whether Mg^{2+} binds to cytosolic proteins. Question marks indicate an unknown transporter. To note, the TRMP6/7 channel has been placed at the apical side for graphic purposes only. Table 9.1 reports the most commonly measured or estimated total and free Mg^{2+} concentrations within mammalian cells and cellular compartments. ADP, adenosine diphosphate; β-AR, β-adrenergic receptor; Ca, calcium; DAG, diacyl-glycerol; IP_3, inositol 1,4,5, trisphosphate; IP_3-R, inositol 1,4,5, trisphosphate receptor; MR, muscarinic receptor; Pi, inorganic phosphate.

Data from Rude RK. (2012). Magnesium. In *Modern Nutrition in Health and Disease* (Ross AC, ed) pp. 159–75, Jones & Bartlett Learning, LLC, Burlington.

Table 9.1 Total and Free Magnesium Concentrations in Whole Cells and Intracellular Organelles[a]

Cellular Compartment	Total Mg Concentration (mM)	Free Mg Concentration (mM)
Whole cell	18–20	0.5–0.7 in cytosol
Nucleus	18–20	Undetermined
Mitochondria	18–20	0.6–0.7 (matrix)
RER/SR	~18	0.7 (lumen, estimated)
Cytoplasm	4–5 bound to ATP	0.5–0.7
Other organelle	Undetermined	Undetermined

Abbreviations: ATP, adenosine trisphosphate; RER, rough endoplasmic reticulum; SR, sarcoplasmic reticulum.
[a]Data are from Romani A. Arch Biochem Biophys 2007;458:90–102.

The processes that maintain or modify the relationships between total and ionized internal and external Mg are not completely understood. Changes in cytosolic Mg^{2+} regulate some channels, such as transient receptor potential melastatin TRPM6 (kidney, colon) and the ubiquitously expressed TRPM7.[7] Researchers have identified several additional proteins critical to cellular Mg^{2+} homeostasis, including Mg2 transporter 1 (MagT1) and solute carrier family 41 member 1 (SLC41A1). Tissue-specific Mg^{2+} transporters include transient receptor cyclin M2 (CNNM2; kidney) and cyclin M4 (CNNM4; colon).[8] The regulation of Ca^{2+} and Mg^{2+} reabsorption in the kidney is regulated by claudin-14, claudin-16, and claudin-19 channels that, through protein interactions, confer the permeability of these ions through tight junctions.[9]

Mg transport in mammalian cells may be influenced by hormonal, pharmacologic, and growth factors.[3] Insulin and dextrose were found to increase ^{28}Mg uptake by several tissues, including the pancreas and skeletal and cardiac muscle.[5,6] The mechanism of insulin-induced Mg transport is likely the result of an effect on protein kinase C.[5,6] Insulin-induced transport of Mg into cells could explain the decrease in serum Mg concentration observed during diabetes and insulin therapy of diabetic ketoacidosis.[10] Recently, evidence for a dynamic exchange of Mg^{2+} between the ER and mitochondria to modulate cellular energy metabolism has been reported,[11] adding to the regulatory role of Mg^{2+} within the cell.

Enzyme Interactions

Overall, the predominant action of Mg is related to ATP utilization. ATP has a strategic position in "free-energy" currency for virtually all cellular processes, by providing high-energy phosphate. As described previously in Rude,[1] ATP exists in all cells, primarily as $MgATP^{2-}$. Beyond this, Mg is involved in >600 essential metabolic reactions. The Mg ion (Mg^{2+}) forms complexes with a variety of organic molecules. Enzymatically, Mg^{2+} is an essential cofactor and has two general interactions: a) Mg^{2+} binds to the substrate, thereby forming a complex with which the enzyme interacts, as in the reaction of kinases with MgATP; and b) Mg^{2+} binds directly to the enzyme and alters its structure or serves a catalytic role (e.g., exonucleases, topoisomerases, and RNA and DNA polymerases).[6,12,13]

Structural Stability of Nucleic Acids and Membranes

An important role of Mg is its ability to form complexes with nucleic acids. The negatively charged ribose phosphate structure of nucleic acids has a high affinity for Mg^{2+}; the resulting stabilization of ribonucleotides and deoxyribonucleotides induces important physicochemical changes that affect chromatin condensation, DNA maintenance, duplication, and transcription.[6,12-16] In addition, the binding of hydrated Mg^{2+} by transfer RNA (tRNA) and by modified tRNA and its DNA analogs results in structure configurations that cannot be duplicated by the binding of other metals.[6,12-15]

Mg^{2+}, Ca^{2+}, and some other cations react with hydrophilic polyanionic carboxylates and phosphates present in various subcellular membranes to stabilize membrane integrity and consequently membrane fluidity and permeability, with major implications for the proper operation of ion channels, transporters, and signal transduction pathways.[6]

Ion Channel Regulation

Ion channels constitute a class of proteins across the cell membrane, which allow passage of ions into or out of cells when the channels are open. Ion channels are classified according to the ion moiety they carry, such as sodium (Na^+), potassium (K^+), or Ca^{2+}.[17] Mg^{2+} plays an important role in the function of certain ion channels. A deficit of Mg results in cellular K^+ depletion.[17] Mg^{2+} is necessary for the active transport of K^+ into the cells via the Na^+/K^+-ATPase.[3] Another mechanism for K^+ loss is an increased efflux of K^+ from cells via Mg^{2+}-sensitive K^+ channels, as observed in skeletal and heart muscle.[18,19] Therefore, Mg^{2+} deficiency leads to reduced intracellular K^+ content. The arrhythmogenic effect of Mg deficiency, as discussed later, may be related to its effect on intracellular K^+. Furthermore, as cellular Mg^{2+} is extruded across the cell membrane in exchange for external Na^+ via the solute carrier 41 (SLC41) antiporter, the resulting increase in cellular Na^+, and ultimately, Ca^{2+} via the Na^+/Ca^{2+} exchanger can elicit arrhythmogenic events in the heart, including early afterdepolarization events.[20]

Mg has been called nature's physiologic Ca channel blocker. As detailed previously by Rude,[1] intracellular Ca^{2+} rises during Mg depletion. This may result from both increased Ca^{2+} entry from the extracellular compartment and release from intracellular Ca^{2+} stores (endoplasmic reticulum [ER] or SR). Mg^{2+} has been shown to decrease the inward Ca^{2+} flux through slow Ca channels. In addition, Mg^{2+} decreases the transport of Ca^{2+} out of the SR into the cytosol. Inositol triphosphate has an inverse ability to release Ca^{2+} from intracellular stores in response to changes in Mg^{2+} concentrations, further contributing to a rise in intracellular Ca^{2+} during a fall in Mg^{2+} (see Rude[1] and references therein).

Mg also plays a major role in regulating intracellular transporters such as the H^+/K^+ exchanger located in the mitochondrial membrane. By modulating the operation of this exchanger, Mg contributes to the maintenance of proper mitochondrial volume and functioning of the various components of the organelle electron chain complexes to generate ATP. Lastly, mitochondrial ATP is extruded through the activity of the adenine nucleotide translocase, which also requires Mg for correct functioning.[21]

Body Composition and Homeostasis

Composition

The distribution of Mg in various body compartments of apparently healthy adult individuals is summarized in **Table 9.2**. The concentration of Mg in the human body varies between 20 and 28 g.[8,22] Mg is mainly an intracellular cation. Approximately 50%–53% of total body Mg content is in the skeleton, of which two-thirds is within the hydration shell and one-third is on the crystal surface.[23] The latter may serve as a reservoir for maintaining whole-body Mg homeostasis. Skeletal muscles represent 27% and soft tissues 19% of body Mg. The extracellular fluid compartment contains 1%–2% Mg.[24] Serum Mg accounts for only 0.3% of total Mg content, and approximately one-third of this is bound to proteins, whereas the residual portion is ionized and filtered through the kidneys.

Table 9.2 Distribution and Concentrations of Magnesium in a Healthy Adult (Total Body: 833–1170 mmol[a] or 20–28 g)[b]

Site	Total Body Mg (%)	Concentration/Content
Tissues		
Bone	53	(0.5% of bone ash)
Muscle	27	9 mmol/kg wet weight
Soft tissue	19	9 mmol/kg wet weight
Adipose tissue	0.012	0.8 mmol/kg wet weight
Serum and Blood Components		
Serum	0.3	0.88 ± 0.06 mmol/L[c]
Free (ionized) 55%–70%		■ Concentration varies
Complexed 5%–15%		■ Concentration varies
Bound 20%–30%		■ Concentration varies
Erythrocytes	0.5	1.65–2.73 mmol/L[d]
Mononuclear blood cells[e]		2.91 ± 0.6 fmol/cell[f]
		2.79 ± 0.6 fmol/cell
		3.00 ± 0.4 fmol/cell
Platelets		2.26 ± 0.29 mmol/L
$[Mg^{2+}]_i$[g]		0.5–1.0 mmol/L
Other Compartments		
Cerebrospinal fluid		1.25 mmol/L
Free 55%		
Complexed 45%		
Secretions		
Saliva, gastric, bile		0.3–0.7 mmol/L
Sweat		0.3 mmol/L (38°C)
		0.09 mmol/h

[a] 1 mmol = 2 mEq = 24.3 mg.
[b] Data from Rude RK. Magnesium. In *Modern Nutrition in Health and Disease* (Ross AC, ed) pp. 159–75. Jones & Bartlett Learning, LLC, Burlington; 2012.
[c] Similar at various ages.
[d] Mg decreases slowly with aging.
[e] Monocytes and lymphocytes in venous blood.
[f] 1 fmol = 24.3 fg.
[g] Intracellular free Mg concentration.

Body Homeostasis

Mineral homeostasis of the individual depends on the amounts ingested, the efficiency of intestinal and renal absorption and excretion, tissue/blood translocation, and all other factors affecting each of these steps. A schema for human Mg balance is given in **Figure 9.2**, and factors affecting Mg absorption are presented in **Table 9.3**.

Intestinal Absorption

Data on Mg absorption from balance studies using differing diets have been quite variable, ranging from 35%–70%. Molecular mechanisms for Mg homeostasis have been reviewed.[38] In humans, the primary sites of intestinal Mg absorption are the jejunum and ileum, although absorption can occur at other sites, including the colon.[30] Under normal dietary Mg intake, 30%–40% of it is absorbed. Net Mg absorption increases with increasing Mg intake, by up to 80% of dietary intake if serum Mg is low. Following oral ingestion, Mg appears in the blood within 1 hour, stabilizes at a rate of 4%–6% per hour from hours 2 to 8, then decreases rapidly, and ceases at hour 10.[39] Mg absorption has both a passive paracellular mechanism and an active transport process (**Figure 9.3**). The paracellular mechanism depends on a potential difference across the membrane generated by Na^{2+} transport and accounts for approximately 80%–90% of intestinal Mg absorption.[30] The ileum and distal parts of the jejunum are the most permeable because they lack junction-tightening claudins. TRPM6 accounts for the remainder.[40] Absorption of intake is curvilinear, and this pattern reflects the sum of this active saturable process and passive diffusion occurring mostly in the ileum. Mg absorption in the cecum and colon is thought to be transcellular, mediated by TRPM6 and TRPM7.[8]

Regulation of Intestinal Magnesium Absorption

The fractional absorption of ingested Mg by healthy persons is not only influenced by its dietary concentration but also by the presence of dietary components and hormonal modulation either inhibiting or promoting absorption (see Table 9.3). Several hormones may influence the action of TRPM6. Vitamin D and its active metabolites were shown to increase intestinal Mg absorption in several studies.[30] 1,25 dihydroxyvitamin D [$1,25(OH)_2D$] increases intestinal Mg absorption in healthy human subjects and in patients with chronic renal failure.[41] Mg is also a cofactor for the binding of vitamin D to its transport protein, vitamin D binding protein. Moreover, conversion of vitamin D into the active, hormonal form [$1,25(OH)_2D$] is Mg dependent. Both in vitro and in

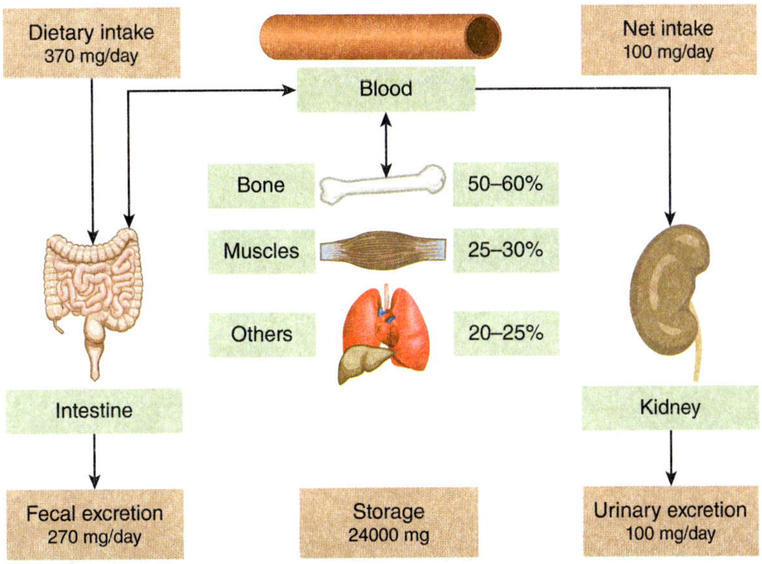

Figure 9.2 Magnesium (Mg) homeostasis in humans. Panels represent the daily amount of Mg^{2+} intake and excretion. On a daily basis, the intestines absorb ~120 mg and secrete 20 mg of Mg^{2+}, resulting in a net absorption of 100 mg. In the kidney daily, ~2,400 mg Mg^{2+} is filtered by the glomerulus, of which 2,300 mg is reabsorbed along the kidney tubule. This results in a net excretion of 100 mg, which matches the intestinal absorption. Bone and muscle provide the most important Mg^{2+} stores, which restore homeostasis when intake, absorption, and/or excretion are disturbed.

Reproduced from de Baaij JH, Hoenderop JG, Bindels RJ. Magnesium in man: implications for health and disease. Physiol Rev 2015;95(1):1-46. doi:10.1152/physrev.00012.2014

vivo studies indicate that Mg deficiency affects enzymes that synthesize and metabolize 25(OH)D and $1,25(OH)_2D$[42-44] and alters bone-specific binding sites for $1,25(OH)_2D_3$ in Mg deficiency.[45]

Recent data from a mouse model suggest a role for fibroblast growth factor-23 and parathyroid hormone (PTH) in suppressing small intestinal absorption of Mg.[33]

Renal Regulation

Renal Filtration. The kidney is the critical organ regulating Mg homeostasis. Mg handling is a process of filtration and reabsorption. The fractional absorption of the filtered load in the various segments of the nephron is summarized in **Figure 9.4**.

Table 9.3 Factors Affecting Magnesium Gastrointestinal Absorption

Factor	Effect on Mg Absorption	Comment
Dietary Factors		
Protein	Decreases	Low protein intakes have been shown to decrease Mg absorption, retention, and balance. High intakes may increase renal Mg excretion[25]
Calcium	No effect or may decrease	Physiologic doses of Ca tend not to decrease Mg absorption.[1,26] High Ca intakes coupled with low Mg intake can exacerbate a low Mg absorption[27]
Phosphate	May decrease or have no effect	Increased Mg intake may decrease phosphate absorption; high phosphate intake did not alter Mg balance due to an associated decrease in urinary excretion of Mg[1]
Zinc	Can decrease Mg absorption and balance significantly	Very high zinc intakes of 53 and 142 mg/day decreased Mg balance in adult men and postmenopausal women, respectively[25,28]
Oxalic acid	Impairs absorption[1]	Affects Mg bioavailability by chelating Mg[29]
Phytates	Decreases Mg; may be dependent on the cereal grain	Binds Mg to hexakisphosphate groups[26]
Fiber (nonfermentable)	Decreases Mg	Cellulose and lignins bind Mg[25,26]
Short-chain fatty acids	Promotes Mg solubility	Produces acidic environment[1] favorable to Mg absorption
Oligosaccharides	Increases	Inulin and a high-fructose diet[25,26]

(continues)

Table 9.3 Factors Affecting Magnesium Gastrointestinal Absorption *(continued)*

Factor	Effect on Mg Absorption	Comment
Hormonal Factors		
Vitamin D and active metabolites	Increases Mg	Pharmacologic doses of vitamin D increase Mg[31] absorption in both vitamin D–deficient and vitamin D–replete animals. A substantial amount of Mg absorption, however, occurs independent of vitamin D[30] Influences TRPM6[1,30]
Estrogens	Stimulate absorption	Regulation of TRPM6/7 (see Romani[3] and references therein)
EGF	Stimulates absorption	TRPM6 is key in mediating Mg influx in intestinal cells and is stimulated by the action of EGF[32]
PTH	Suppresses small intestine Mg absorption	Mechanism of action is unknown, but possibly through TRPM6 channel protein expression and CNNM4 (a Na^+/Mg^{2+} antiporter) expression in the small intestine[33]
Other Factors		
Aging	Decreases Mg	May be reflective of decreased dietary intake of Mg, intestinal absorption, or polypharmacy[34-36]
Intestinal inflammatory diseases	Decreases Mg	Crohn's disease, ulcerative colitis, or short bowel syndrome[26]
Type of formulation in dietary supplements	Varies	Higher absorption with gelatin capsules than enteric-coated tablets[1,26,37]

Abbreviations: CNNM4, cyclin M4; Mg, magnesium; PTH, parathyroid hormone; TRPM, transient receptor potential melastatin channel.

(For details on molecular mechanisms of renal Mg reabsorption, see also Ellison et al.[46])

Approximately 2,400 mg of Mg is filtered by the glomeruli on a daily basis, with the nephron recovering 95%–99% by coordinated transport processes.[8] Only 10%–25% of Mg is reabsorbed via passive diffusion in the proximal tubule, which is associated with Ca, Na, and water transport,[8,47] with hormonal effects on Na reabsorption affecting Mg reabsorption. Proximal tubule Mg reabsorption does not parallel Na reabsorption, which may explain why thiazides do not reduce urinary Mg excretion.[46] The molecular pathways of proximal tubule Mg reabsorption have not been identified, but two members of the claudin family, paracellin-1 (claudin-16) and claudin-19, expressed here at significant levels appear to mediate this transport.[40,46,48]

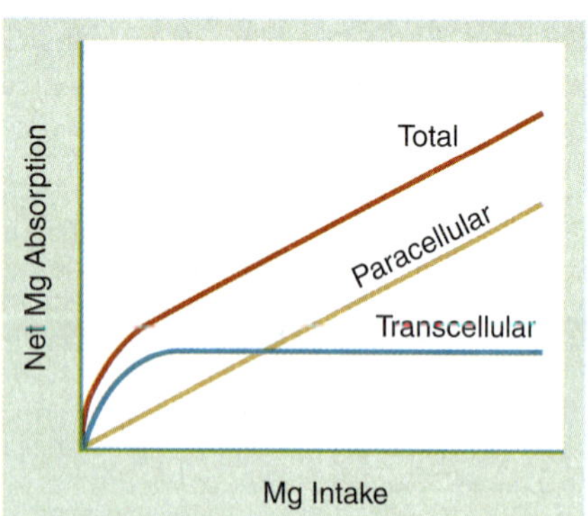

Figure 9.3 Proposed pathways for magnesium (Mg) absorption across the intestinal epithelium. When dietary Mg intake is normal, the active transcellular transport mediates 30% of intestinal magnesium absorption. This fraction increases when dietary Mg intake is lower. When dietary Mg is higher, the majority of intestinal magnesium absorption occurs via the passive paracellular pathway, which is regulated by proteins comprising the tight junction, including claudins, occludin, and zona-occludens 1.

Reprinted with permission from Blaine J, Chonchol M, Levi M. Renal control of calcium, phosphate, and magnesium homeostasis [published correction appears in Clin J Am Soc Nephrol. 2015 Oct 7;10(10):1886-7]. Clin J Am Soc Nephrol 2015;10(7):1257-1272. doi:10.2215/CJN.09750913

Renal Absorption. Unlike most other electrolytes that are absorbed in the proximal convoluted tubule, approximately 50%–70% of Mg reabsorption occurs at the thick ascending limb of the loop of Henle via passive diffusion.[8] Reabsorption along this segment is highly regulated and claudins are critical for paracellular Mg absorption.

The distal convoluted tubule reabsorbs 5%–10% of filtered Mg via an active transcellular pathway in contrast with the passive reabsorption process in the proximal and thick ascending limb segments. No evidence of further reabsorption occurs beyond this segment. Several proteins may be involved, including the sodium chloride (NaCl) cotransporter as well as TRPM6, which is also expressed in the distal tubule and interacts with TRPM7 for maximal reabsorption (see Schlingmann et al.[40] and references therein).

Flow through the distal convoluted tubule determines final urinary Mg concentrations because no absorption takes place beyond this segment. TRPM6 is specifically expressed in this segment and its activity is regulated by intracellular Mg.[8,49] Activation of the EGF receptor (EGFR) has been shown to induce TRPM6 expression and activity in vitro.[46] Mutations of TRPM6 result in lower intestinal Mg absorption and renal Mg wasting.[40,48,50]

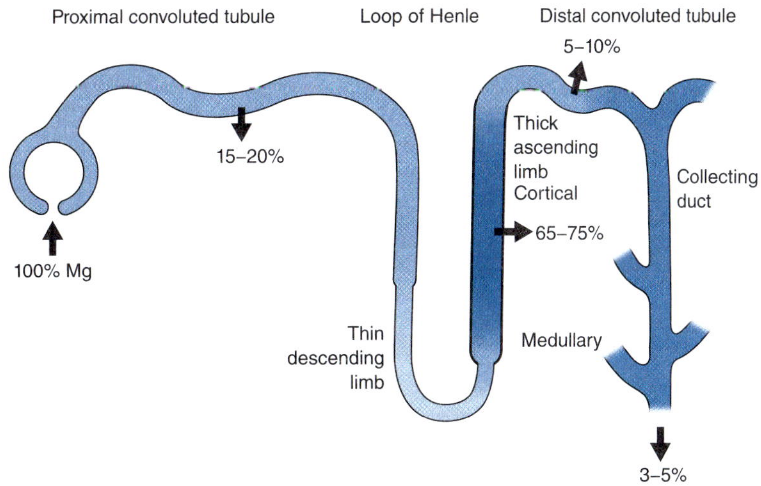

Figure 9.4 Fractional segmental reabsorption of filtered magnesium (Mg^{2+}) in the nephron. The percentage absorption of filtered Mg^{2+} has been determined by micropuncture techniques in various laboratory animals as the Mg^{2+} proceeds through the nephron. Approximately 15%–20% of the Mg^{2+} is reabsorbed in the proximal convoluted tubule. The major site for Mg^{2+} reabsorption is the thick ascending limb of the loop of Henle, primarily in its cortical portion. Here, 65%–75% of Mg^{2+} leaves the lumen. In the distal convoluted tubule, 5%–10% of Mg^{2+} is reabsorbed.

Adapted with permission from Cole DE, Quamme GA. Inherited disorders of renal magnesium handling. J Am Soc Nephrol 2000;11:1937–47.

Renal Excretion. When Mg intakes are excessive, the kidneys excrete high amounts of Mg; during Mg deficiency, the kidneys conserve Mg by excreting <12–24 mg/day.[47] Diets high in Na, Ca, and protein may increase renal Mg excretion, as does caffeine and alcohol consumption.[35,47,51]

Estrogen and PTH are involved in renal Mg excretion. Estrogen stimulates TRPM6 expression.[52] This expression seems to be regulated by serum Mg levels but not by $1,25(OH)_2D$ or PTH.[53] Hormonal effects on Na reabsorption in the proximal tubule also affect Mg reabsorption because sufficient Na transport is required to drive water transport, which is a prerequisite for Mg reabsorption.[8] PTH can have profound effects on Mg homeostasis by enhancing Mg reabsorption in the kidney, absorption in the gut, and release from bone.[54]

Translocation of Magnesium from Body Stores

Intracellular Mg concentrations are tightly regulated, as they are vital to intracellular functions.[3] Extracellular and intracellular Mg and bone Mg all decrease during Mg depletion. Bone may serve as an important reservoir for Mg. Human iliac crest changes indicate a broad range of Mg loss during depletion, with a weighted average of 18% or 1.2 mmol/kg body weight.[55] When Mg intakes are low, intracellular Mg concentrations can be maintained by releasing Mg from phospholipids in cell membranes.[8,56] If dietary Mg intake remains so low that the body's physiologic mechanisms to increase intestinal Mg absorption and decrease renal excretion are exceeded, resorption of surface bone Mg stores in the hydroxyapatite form occurs.

Insulin administration causes Mg to shift from extracellular to intracellular spaces, as indicated by a decline in plasma Mg levels and a rise in red cell Mg levels, with minimal alteration in renal Mg reabsorption.[57] Under diabetic conditions, instead, a systemic loss of Mg from erythrocytes, liver, and skeletal muscles into the bloodstream and ultimately the urine is observed.[3] Stimulation of β-adrenergic receptors by specific agonists at the tissue or systemic level also results in a significant extrusion of Mg^{2+} into the bloodstream and ultimately the urine (see Rude and references therein). Lastly, the amount of Mg lost in sweat is very small compared with the loss of other cations.

Assessing Magnesium Requirements and Status

Magnesium Requirements and Intakes

Half of the US population aged 20 years or older consumes less than the estimated average requirement (EAR) of Mg from foods, and percentages are higher for teens and those aged >70 years.[58] In 2005, the Dietary Guidelines for Americans identified Mg as an underconsumed nutrient; this is still true today.[59] A diet incorporating high-Mg foods includes unrefined grains, legumes or beans, nuts, seeds, and green vegetables. Mg content is generally low in processed foods.

Mg requirements aim to prevent development of Mg deficiency in healthy humans. They are usually set for various age/gender groupings. Published Mg requirements for the United States[2] and the United Kingdom[60] populations are based on Mg balance studies. The World Health Organization/Food and Agriculture Organization of the United Nations[61] requirements are based on Mg balance studies plus usual intakes, whereas the European Food Safety Authority[62] requirements use usual intakes only. All of these recommended intakes for healthy humans are probably underestimates, as none take into account translocation of body Mg stores in the face of rising human body weights[63,64] nor the impact of long-term medications that can gradually deplete body Mg (see Table 9.2). Thus, Mg requirements need re-evaluation if they are to prevent development of a chronic latent Mg deficit (CLMD; discussed in "Chronic Latent Magnesium Deficit") in otherwise healthy people.

Table 9.4 compares the published 1997 Mg dietary reference intakes (DRIs)[2] for adults with a set of revised RDAs

Table 9.4 US Institute of Medicine vs Body Weight–Corrected Dietary Reference Intakes for Magnesium in Adults

Gender and Age	DRIs (mg/day)			
	1997 (US IOM)		Recalculated[a]	
	EAR[b]	RDA[c]	EAR[d]	RDA[e]
Male, y				
19–30	330	400	386	540
>30	350	420	409	573
Female, y				
19–30	233	310	319	447
>30	265	320	332	465

Abbreviations: CV, coefficient of variation; DRI, dietary reference intake; EAR, estimated average requirement; IOM, Institute of Medicine; RDA, recommended dietary allowance.
[a]Data are from Rosanoff A. Adv Nutr 2021;12:298–304; Fryar CD et al. Natl Health Stat Rep 2018;122:1–16.
[b]The 1997 EAR values use standard ref body weights of 166 lb for males and 133 lb for females.
[c]The 1997 RDA uses 10% CV to convert EAR to RDA.
[d]The recalculated EAR uses mean adult US body weights of 195 lb for males and 168 lb for females (from Rosanoff 2021).
[e]Recalculated RDA uses 20% CV to convert EAR to RDA (from Hunt CD and Johnson LK. Am J Clin Nutr 2006;84:843–52; and Nielsen FH. Crop Pasture Sci 2015;66:1259–64).

recalculated with actual mean body weights of the populations and more appropriate coefficients of variation.[65] These re-evaluated requirements for Mg show that most mean Mg intakes among adults in modern cultures are below the EAR and all are below the revised RDA for adults.[65] This means that the average healthy adult in these populations is in chronic negative daily Mg balance, gradually depleting their body Mg stores as they build up a CLMD, putting them at risk for development of cardiovascular and metabolic syndrome risk factors. This has led to a call for a chronic disease risk reduction estimate for Mg intakes,[54] which would be 450–550 mg Mg/day to prevent CLMD in healthy adults and as much as 600 mg Mg/day[66,67] to bring individuals with mild or no symptoms back into Mg replete status.

Using the same assumptions and methodology, the 1997 and body weight–corrected EARs and RDAs for children and adolescents are presented in **Table 9.5**.

The DRIs[2] established tolerable upper intake levels (Uls) for supplemental Mg for adolescents and adults at 350 mg/day (14.6 mmol/day). This is based on the lowest observed adverse (diarrhea) effect level of 360 mg/day (15 mmol/day). Because high Mg intake from foods has not been shown to cause adverse effects, the UL for Mg applies only to Mg from supplemental sources.

Table 9.5 US Institute of Medicine vs Body Weight–Corrected Dietary Reference Intakes for Magnesium in Children and Adolescents

Gender and Age	DRIs (mg/day)				
	1997 (US IOM)			Corrected for Actual Mean BW[a]	
	AI	EAR	RDA	EAR	RDA
Male, y					
0–0.5	30	—[b]	—	—	—
0.5–1.0	75	—	—	—	—
1–3	—	65	80	69	96
4–8	—	110	130	125	174
9–13	—	200	240	229	320
14–18	—	340	410	382	535
Female, y					
0–0.5	30	—	—	—	—
0.5–1.0	75	—	—	—	—
1–3	—	65	80	65	91
4–8	—	110	130	121	170
9–13	—	200	240	228	320
14–18	—	300	360	346	485

Abbreviations: AI, adequate intake; BW, body weight; DRI, dietary reference intake; EAR, estimated average requirement; IOM, Institute of Medicine; RDA, recommended dietary allowance.
[a]Data for mean US body weights are from Fryar CD, Carroll MD, Gu Q, et al. Vital Health Stat 2021;3:1–44; coefficient of variation = 20%.
[b]Dashes indicate not applicable.

Assessment of Magnesium Status

A simple, rapid, and reliable single measurement to determine the presence of Mg deficiency is lacking. Each of the three most commonly used indicators of Mg status (intake, circulating level, and urinary excretion) has shortcomings, as does measurement of Mg body stores via the Mg load retention test.[68] **Table 9.6** summarizes the current methodologies to assess Mg status. A circadian effect on Mg cellular concentration[69] is reflected in Mg blood levels and has been demonstrated for urinary,[70,71] serum,[70] and plasma ionized

Table 9.6 Other Methods That Can Be Used to Determine Magnesium Levels[a]

Magnesium Test	Result Analysis	Collection Method
Serum Mg	Normal level not standardized. Three general levels are deemed normal in the most recent research (see Figure 9.5): ■ 0.6–1.1 mmol/L (1.46–2.68 mg/dL; 1.2–2.2 mEq/L) ■ 0.75–0.95 mmol/L (1.8–2.3 mg/dL; 1.5–1.9 mEq/L) ■ 0.85–0.95 mmol/L (2.06–2.3 mg/dL; 1.7–1.9 mEq/L)	Blood test Measured in NHANES I 1971–1974 for age ≥12 y[b]

Magnesium Test	Result Analysis	Collection Method
24-h urine excretion	Normal urine Mg levels are 3–5 mmol (~100 mg)	Patient urinates in a provided urine collection device over a 24-h period
Mg loading test	Assesses intestinal absorption and indirectly assesses bone Mg status	Patient is started on an infusion of 33 mmol (7.5 g) mg sulfate continuously for 8 h. Urine samples are collected for 24 h and Mg uptake is calculated
	Fairly reproducible clinically	
	Invasive and cumbersome	
	Wide normal range (18–30 mmol/24 h)	
	Results should be presented in percent retention	
RBC concentration	Normal RBC Mg levels are 4.2–6.8 mg/dL	Blood test sent to laboratory for centrifugation
	Not readily available	
Ionized Mg	Can be measured in serum, plasma, and whole blood	Blood test
	Best for critically ill patients	
	Not affected by low albumin	
	Reference range of 0.55–0.75 mmol/L	

Abbreviations: Mg, magnesium; NHANES, National Health and Nutrition Examination Survey; RBC, red blood cell.
[a]Data from Costello RB, Nielsen F. Curr Opin Clin Nutr Metab Care 2017;20:504–11; Elin RJ. Magnes Res 2010;23:S194–8; and Rude RK. Magnesium. In Encyclopedia of Dietary Supplements (Coates PM, Betz JM, Blackman MR, Cragg GM, Levine M, Moss J, et al., eds) pp. 527–37, Informa Healthcare USA, Inc., New York, NY, 2010.
[b]Data from Lowenstein FW, Stanton MF. J Am Coll Nutr 1986;5:399–414.[75]

Mg.[72] Plasma levels peak around 10 a.m. and 8 p.m. and vary by approximately 6% throughout the day. The circadian rhythm is also present after supplementation.[73] This significant within-day variability should be accounted for in clinical and research settings.

Even though serum Mg does not have a single, uniform, standardized normal range, it remains the most used and most available measure of Mg status. Recent literature shows three very different general ranges of serum Mg now recommended for use (see **Figure 9.5**). The differences in diagnostic uniformity of these three ranges are substantial, yet all are currently used. The range of 1.46–2.68 mg Mg^{2+}/dL most recently reported by Ehrenpreis et al.[74] is wider than the range of 1.8–2.3 mg/dL reported by Lowenstein and Stanton,[75] which is based on 1974 measures in healthy US adults. The lower cutoff values—<1.46 from Ehrenpreis et al.[74] and <1.8 mg/dL from Lowenstein and Stanton[75]—show the discrepancy of serum Mg values now used for diagnosis where overt symptoms of hypomagnesemia arise and are mostly used in the hospital setting. Regardless, these highly used "normal" but differing ranges make no allowance for persons with CLMD. The 2.06–2.3 mg/dL range proposed by Rosanoff et al.[76] tries to set a lower boundary defining hypomagnesemia that will include seemingly healthy persons who may have an undetected Mg deficit (i.e., CLMD, see below).

Chronic Latent Magnesium Deficit

As shown in Figure 9.5, several factors are needed to achieve and maintain healthy Mg balance, including adequate intake of Mg, normal absorption of Mg from the gastrointestinal tract, and a level of urinary Mg excretion that does not waste Mg. Persistent or chronic disorder of any one of these organs that remains chronic may lead to Mg translocation from body stores and development of a CLMD that progresses over a lifetime.[77]

CLMD can be expected to gradually develop in an otherwise healthy person when low dietary Mg intake and/or increased urinary Mg excretion, usually from certain medications (see the discussion of renal disorders in "Risk Factors for and Causes of Magnesium Deficiency"), are maintained day after day, year after year. CLMD puts the individual at risk for developing hypertension,[65,78] cardiovascular disease (CVD),[65,79] and diabetes-related metabolic risk factors[65,80] even though their daily Mg intake falls within published recommended requirements and their serum Mg level is within the currently recommended "normal" range.

Supplementation Practices

Mg supplements are available in a variety of salt forms, including Mg oxide, citrate, Cl, malate, glycinate, and others.[81] The Supplement Facts panel on a dietary supplement label declares the amount of elemental Mg in the product, not the weight of the entire Mg-containing compound.

According to pooled data from 30,899 US adults aged ≥20 years who participated in six NHANES cycles (1999–2000 to 2009–2010) and provided information on dietary supplement use, more than half reported using dietary supplements in the past 30 days.[82,83] Of 24,763 dietary supplement users, 33.3% used Mg supplements (mean 146.8 mg/day). Although the number of adults with intakes below the EAR decreased as a result of Mg supplementation, 49.7% of those surveyed still remained below the EAR.

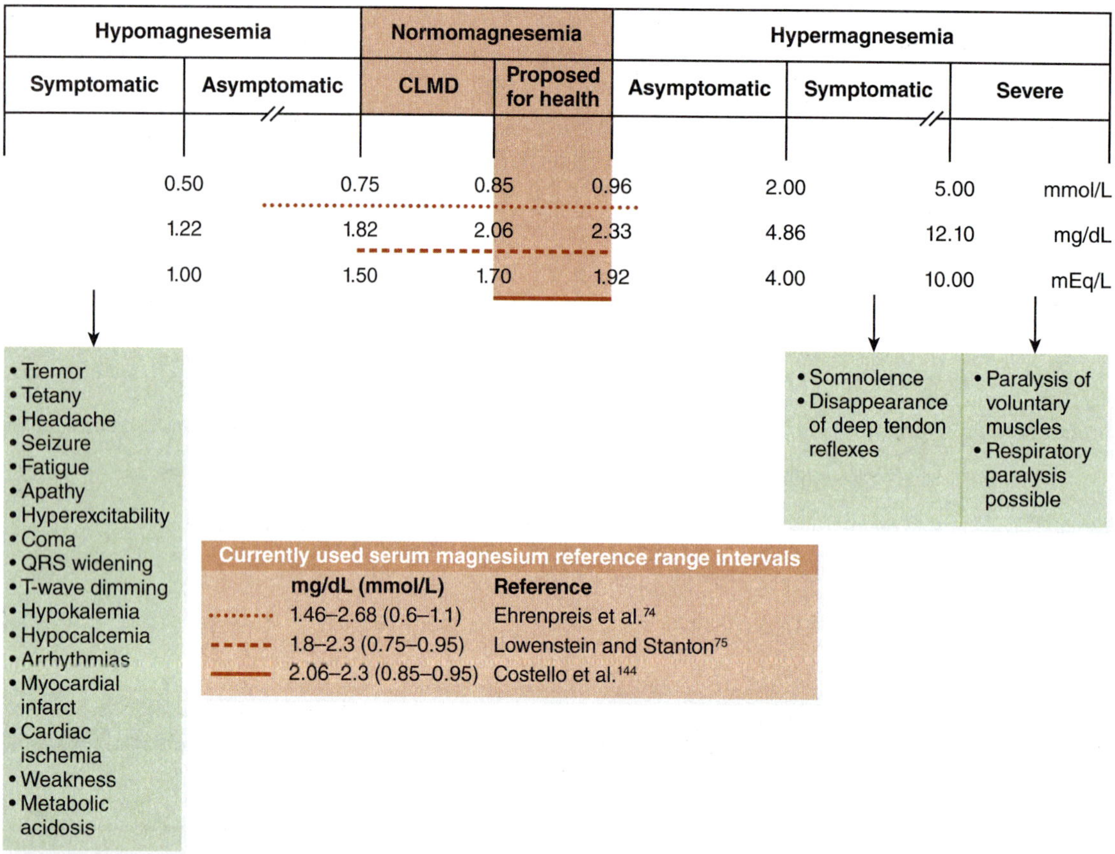

Figure 9.5 Serum magnesium symptoms range with currently used reference range intervals.

Data from Costello RB, Elin RJ, Rosanoff A, et al. Perspective: The Case for an Evidence-Based Reference Interval for Serum Magnesium: The Time Has Come. Adv Nutr. 2016;7(6):977-993. Published 2016 Nov 15. doi:10.3945/an.116.012765

Absorbability of Magnesium Salts

Studies on the bioavailability of different Mg salts consistently demonstrate that organic salts of Mg have higher bioavailability than inorganic salts.[84] But higher absorption does not always mean higher retention. A recent randomized clinical trial demonstrated that both urinary excretion and serum levels of Mg were significantly higher after single-dose administration of Mg citrate compared with Mg oxide.[85] A meta-analysis of the effect of Mg oral supplements on hypertension showed that the inorganic forms were better at lowering a high blood pressure than were the organic Mg supplements.[86]

The fractional absorption of a salt depends on its solubility in intestinal fluids, moderated by pH, and the amounts ingested. In one study, Mg citrate had high solubility even in water, whereas Mg oxide was poorly soluble even in acid solution; better absorption of the citrate salt was demonstrated in humans.[87] Using an in vitro gastrointestinal simulator system along with dissolution tests, investigators tested 15 different Mg formulations and found wide variation in absorption and dissolution, with organic compounds being more easily absorbable than inorganic compounds.[37]

Mg oxide and various salts in large doses act as an osmotic laxative, with resultant diarrhea. Thus, the physician treating a patient with diarrhea of uncertain origin should consider measuring fecal Mg.[88] Higher solubility has been seen with effervescent tablets as opposed to single-dose tablets.[26]

Mg is also a primary ingredient in some laxatives. For example, Phillips' Milk of Magnesia provides 500 mg elemental Mg (as Mg hydroxide) per tablespoon. Mg is also included in some remedies for heartburn and upset stomach due to acid indigestion. Extra-strength Rolaids provides 55 mg elemental Mg (as Mg hydroxide) per tablet.[81]

Safety of Oral Magnesium Supplementation

The safety of oral Mg dietary supplements is difficult to summarize. As described earlier, the US Institute of Medicine[2] established a UL of 350 mg/day for Mg from supplemental intake alone. Individual responses to Mg supplementation and gastrointestinal symptomatology vary greatly, owing to the variety of Mg formulations available, differing durations of use, and differences in underlying diet and health status. For the healthy individual, the efficiency of the kidney in reabsorbing or eliminating Mg is so great that even excessive ingestion rarely leads to significant increases in serum Mg concentration.[89]

Oral Mg supplementation can be used carefully in adequate doses without causing diarrhea. Mg gluconate and MgCl have been preferred for oral replacement because they caused diarrhea less often than other salt forms.[90] Oral MgCl solution has also been successfully used for replacement purposes.[91] Mg carbonate should be avoided as a supplement because it may not be soluble enough to adequately replace Mg levels.[84,91]

Mg-containing antacids are generally considered safe due to the healthy kidney's ability to excrete Mg, but hypermagnesemia may develop when renal function is compromised.[92] Mg-containing enemas may lead to excessive amounts of Mg absorption and should be avoided even in individuals with normal renal function.[93]

Very large doses of Mg-containing laxatives and antacids (typically providing >5 g Mg/day) have been associated with Mg toxicity[94] and there are a few reports of fatal hypermagnesemia due to excessive oral supplementation.

Risk Factors for and Causes of Magnesium Deficiency

As noted earlier, dietary Mg intake for a large proportion of the United States and world population has been shown to be lower than the recommended intake[2,65,95] and is one of the primary causes predisposing individuals to Mg deficiency. However, poor Mg status may be observed in the presence of other conditions that impair Mg balance.

Aging

Aging is often associated with a total body Mg deficit, which can be manifested by insufficient Mg intake, reduced intestinal Mg absorption, increased urinary Mg excretion, and drug therapies.[36]

Generalized Inflammation

Poor Mg status has been associated with increased production of oxygen free radicals, with low-grade systemic inflammation and increased levels of inflammation markers and proinflammatory molecules.[36] Epidemiologic studies have shown increased levels of C-reactive protein (CRP) with favorable lowering of CRP levels in response to Mg supplementation dependent on baseline CRP levels.[96-102]

Gastrointestinal Disorders

Mg depletion is common in acute and chronic diarrhea, regional enteritis, ulcerative colitis, and intestinal and biliary fistulas. Malabsorption syndromes may also result in Mg deficiency. Steatorrhea and resection or bypass of the small bowel, particularly the ileum, often results in intestinal Mg loss or malabsorption.[1] Mg for constipation is a standard use.[103,104]

Proton pump inhibitors have been reported to cause severe hypomagnesemia in some patients[105] (**Table 9.7**).

Renal Disorders

As described in Rude,[1] excessive excretion of Mg into the urine may be the basis of Mg depletion (**Table 9.8**).[115,116] Renal Mg reabsorption is proportional to tubular fluid flow and to Na^{2+} and Ca^{2+} excretion. Hypercalcemia has been shown to decrease renal Mg reabsorption, which is likely mediated by Ca binding to the Ca-sensing receptor in the thick ascending limb of the loop of

Table 9.7 Medications Capable of Altering Magnesium Status

Medication	Indication	System Affected	Action
Antidiabetic Agents			
Insulin	IM and oral hypoglycemic	GI (Liver), Renal[189]	Can inhibit Na^+/Mg^{2+} exchanger SLC41A1 and promote efflux of Mg^{2+} from mitochondria or other organelles to maintain physiological Mg^{2+} concentration in the cytoplasm ■ Caution hypomagnesemia; monitor laboratory results ■ Increase dietary Mg[106,107]
Do Biguanide (e.g., metformin)	Oral hypoglycemic	GI (Liver), Renal[36]	May decrease serum Mg and increase total intra-erythrocyte Mg. May decrease urinary Mg excretion[57]
Pioglitazone (SGLT2; e.g., canagliflozin, dapagliflozin)	Oral hypoglycemic	Renal	Inhibits glucose reabsorption in the proximal tubule. Increases serum Mg; potential for hypermagnesemia, exact mechanism unknown[8,57,106]
Anti-EGFR modulators (e.g., cetuximab)	Antineoplastic	Renal	Increases renal Mg loss due to reduced TRPM6 activity. Significant risk for hypomagnesemia ■ Monitor laboratory results ■ Increase dietary or supplemental Mg[8,107]
Antidepressants, tricyclics (e.g., imipramine)	Depression, anxiety	Renal	Interferes with the Na^+/Mg^{2+} exchanger SLC41A1 and Mg efflux
Aminoglycoside antibiotics (e.g., IM or IV use, gentamicin, tobramycin, amikacin)	Infection	Renal	Possible acute tubular necrosis, inhibiting TRPM6 and impairing renal tubular reabsorption. ■ Caution hypomagnesemia; monitor laboratory results and urinalysis ■ Supplementation may be needed in some patients[106]

(continues)

Table 9.7 Medications Capable of Altering Magnesium Status *(continued)*

Medication	Indication	System Affected	Action
Antimicrobials			
Amphotericin B	Antifungal	Renal	Impairs Mg reabsorption ■ Caution hypomagnesemia (IV use); monitor laboratory results ■ Increase dietary or supplemental Mg[8,106,107]
Pentamidine (e.g., NebuPent)	Pneumonia	Renal	Increases renal Mg loss from the DCT requiring Mg replacement therapy ■ Increase dietary or supplemental Mg[8]
β-adrenergic agonists (e.g., theophylline, albuterol, salbutamol)	Bronchodilation	Renal	May increase renal Mg excretion, metabolic abnormalities (Mg shift into cells) May result in renal Mg wasting in patients after organ transplantation that is caused by down-regulation of TRPM6[108] ■ Monitor for hypomagnesemia[106]
Bisphosphonates (pamidronate IV)	Adjunct to treat hypercalcemia, Paget's disease, multiple myeloma, and bone metastases	GI	Contributes to renal impairment, increases Mg excretion ■ Monitor for hypomagnesemia[106]
Calcineurin inhibitors (e.g., cyclosporine A and tacrolimus)	Immunosuppressant	Renal	*May* reduce TRPM6 expression (DCT), increase renal Mg loss. May decrease serum Mg levels ■ Monitor laboratory results ■ Increase dietary Mg[8,107]
Cardiac glycosides (e.g., digoxin)	Heart failure	Renal	Reduces renal tubular reabsorption, and increases Mg excretion ■ Caution hypomagnesemia; monitor laboratory results ■ Increase dietary Mg[106,107]
Corticosteroids	Anti-inflammatory	Renal	Chronic use increases Mg excretion ■ Some patients may require Mg supplementation[109]
Diuretics			
Loop (e.g., furosemide)	Chronic heart failure	Renal	Reduces paracellular Mg reabsorption (TAL), increases renal Mg loss ■ Monitor for hypomagnesemia ■ Increase dietary Mg and potassium[8,106]
Thiazide (e.g., hydrochlorothiazide)	Antihypertensive	Renal	Blocks NaCl transport and may reduce TRPM6 expression in DCT decreasing reabsorption, increasing renal Mg loss ■ Monitor for hypomagnesemia ■ Increase dietary Mg[8,106,107]
Potassium-sparing (e.g., amiloride, triamterene, spironolactone)	Antihypertensive	Renal	Alters renal Ca^{2+} and Mg^{2+} transport. Decreases renal Mg losses and may increase serum Mg ■ Monitor laboratory results
Estrogen (e.g., Estrace, Premarin, and Prempro)	Hormone replacement therapy	Soft tissues and bone	Enhances Mg uptake by soft tissues and bone. Reduces renal excretion; *may* decrease serum Mg levels ■ Increase dietary Mg
Laxatives (e.g., over-the-counter preparations; sodium phosphates)	Bowel cleansing	GI/renal	Ingestion of excessive amounts or compromised kidney function can lead to significant hypermagnesemia, hospitalization, and possibly death[8]

Medication	Indication	System Affected	Action
Platinum derivatives (e.g., IV cisplatin and carboplatin)	Antineoplastic agent	Renal	Have been shown to cause renal lesions that result in hypermagnesuria and hypomagnesemia[110-113] may reduce TRPM6 expression ■ Monitor for hypomagnesemia ■ Increase dietary Mg[8,106,107]
Proton pump inhibitors (e.g., omeprazole, pantoprazole)	Acid reflux	GI/renal	Reduces intestinal absorption by inhibiting pH-sensitive TRPM6 and TRPM7 transporters by reducing their ionized/un-ionized ratio.[114] Long-term use *may* increase renal Mg loss, especially if taken with thiazide diuretics ■ Monitor for hypomagnesemia ■ Increase dietary Mg[8,106,107]

Abbreviations: DCT, distal convoluted tubule; EGFR, epidermal growth factor receptor; IM, intramuscular; Mg, magnesium; IV, intravenous; SGLT2, sodium-glucose cotransporter2; SLC41A1, solute carrier family 41 member 1; TAL, thick ascending limb; TRPM, transient receptor potential melastatin channel.

Table 9.8 Causes of Magnesium Deficiency[a]

Causes
Gastrointestinal disorders
Nutritional deficiency
Prolonged nasogastric suction/vomiting
Acute and chronic diarrhea
Intestinal and biliary fistulas
Malabsorption syndromes
Extensive bowel resection or bypass
Acute hemorrhagic pancreatitis
Primary intestinal hypomagnesemia (mutation of TRPM6 channel)
Proton pump inhibitors
Renal loss
Long-term parenteral fluid therapy
Osmotic diuresis (glucose, urea, mannitol)
Hypercalcemia
Polyuric phase of acute renal failure, renal transplant, history of renal obstruction
Non-drug-associated tubulointerstitial nephropathy
Alcohol
Diuretics (furosemide, hydrochlorothiazide)
Epidermal growth factor blockers (cetuximab, panitumumab)
Renal tubular nephrotoxins (aminoglycosides, cisplatin, amphotericin B, pentamidine)
Calcineurin inhibitors (cyclosporin, tacrolimus)
Genetic mutations of Mg transport channels
Activating mutation of the calcium-sensing receptor
Endocrine and metabolic disorders
Diabetes (glycosuria, osmotic diuresis)
Phosphate depletion
Primary hyperparathyroidism
Hypoparathyroidism
Primary aldosteronism
Excessive lactation
Cutaneous loss
Sweat (athletics)
Burns
Redistribution of Mg to bone and soft tissues
Hungry bone syndrome
Parenteral nutrition/refeeding syndrome

Abbreviations: Mg, magnesium; TRPM, transient receptor potential melastatin channel.
[a]Data from Rude RK. Magnesium. In *Modern Nutrition in Health and Disease* (Ross AC, ed) pp. 159–75, Jones & Bartlett Learning, LLC, Burlington; 2012.

Henle and decreasing transepithelial voltage.[117] Osmotic diuresis caused by glucosuria results in urinary Mg wasting.[116] Alcohol ingestion may also cause renal Mg wasting and is one factor contributing to the high prevalence of Mg deficiency in patients with chronic alcoholism.[118] Many prescription drugs may cause renal Mg wasting and Mg depletion, as described in Table 9.8.

In a large sample of US middle-aged adults, low total serum Mg was independently associated with incident chronic kidney disease.[119] This finding was recently confirmed.[120]

Disorders of Glucose Metabolism

In epidemiologic studies, inadequate dietary Mg has been associated with an increased risk of glucose intolerance, metabolic syndrome, and type 2 diabetes.[121-126] Diabetes is the most common disorder associated with Mg deficiency[127] and the incidence of type 2 diabetes is significantly greater in people with lower Mg intake[128,129] in a dose-related linear fashion. Investigators generally believe that the mechanism for Mg depletion in diabetes is renal Mg wasting secondary to osmotic diuresis generated by hyperglycosuria.

Cardiovascular Disease

Although the diets ordinarily consumed by healthy Americans contain less Mg than the RDA, they are currently not considered leading to symptomatic Mg depletion. However, some clinical disorders have been associated with a low-Mg diet. Investigators have suggested that milder degrees of Mg deficiency as seen with CLMD may, over time, contribute to disease states such as hypertension, coronary artery disease, and preeclampsia. In addition, low intracellular Mg levels in conjunction with elevated cytosolic free Ca have been shown as a characteristic of several aspects of CVD and/or metabolic syndrome.[130]

A large body of evidence from epidemiologic studies, randomized clinical trials, and meta-analyses supports inverse associations between Mg intake or serum Mg and decreased risk for incident CVD.[78,79,131-136]

Epidemiologic studies have related water hardness (Ca and Mg content) inversely to cardiovascular death rates. Studies have shown that water with high Mg content (10–100 ppm) benefits heart health[137] and has been inversely associated with coronary artery disease mortality[138] and stroke in postmenopausal women.[139]

Mg is believed to be linked to CVD risk through a broad range of physiologic roles. These roles include low circulating levels of Mg associated with impaired glucose homeostasis and insulin action, dyslipidemia, elevated blood pressure, chronic inflammation, impaired vasomotor tone, and increased carotid intima-media thickness, impaired peripheral blood flow, and electrocardiogram abnormalities.[86,140-143] Prospective observational studies in generally healthy populations link low circulating Mg with higher risk of CVD morbidity and mortality.[144,145]

Platelet hyperactivity is a recognized risk factor in the development of CVD. Suppression of intracellular Mg levels in conjunction with elevated cytosolic free Ca has been shown to be characteristic of platelet aggregation and increased thrombosis.[130] Mg has been shown to inhibit the platelet aggregation promoted by specific agents or conditions. Patients with diabetes and Mg depletion have been shown to have high platelet aggregations. Mg therapy in these subjects returned the response toward normal. The antiplatelet effect of Mg may be related to the inhibitory effect of Mg on the synthesis of thromboxane A_2 and 12-hydroxyeicosatetraenoic acid (12-HETE), eicosanoids thought to be involved in platelet aggregation.[146,147] Also, Mg inhibits thrombin-induced Ca influx in platelets, stimulates synthesis of prostacyclin (PGI_2, a potent antiaggregatory eicosanoid), and modulates arachidonic acid metabolism.[3]

Hypertension

Several studies have demonstrated that blood pressure is inversely related to dietary Mg intake, hypomagnesemia, and reduction of intracellular Mg levels.[7,148,149] Patients with essential hypertension were found to have low free Mg^{2+} concentrations in red blood cells. The Mg^{2+} concentrations were inversely related to both systolic and diastolic blood pressure. Intervention studies with Mg therapy in hypertension have generated conflicting results.[66] Several studies showed a positive blood pressure–lowering effect of Mg supplements in hypertensive subjects taking an adequate oral Mg dose, whereas studies on normotensive subjects did not. Based on the findings of 38 intervention studies, the US Food and Drug Administration[150] concluded that there is some credible evidence suggesting a relationship between the intake of elemental Mg from conventional foods and dietary supplements and reduced risk of high blood pressure (hypertension). The mechanism by which Mg deficiency affects blood pressure is not clear but may involve decreased PGI_2 production, increased thromboxane A_2 production, and enhanced vasoconstrictive effects of angiotensin II and norepinephrine. Investigators have also suggested that the vascular TRPM7 Mg channel may be altered in hypertension.[7]

Preeclampsia and Eclampsia

Preeclampsia affects an estimated 4.6% of pregnancies globally. Mg therapy has been used for decades in both preeclampsia and eclampsia and contributes to the very low mortality rate in developed countries.[151] Oral Mg therapy during pregnancy was shown to lead to 36 fewer hospitalizations per 1,000 pregnant women compared with placebo.[152] Although subtle deficits in total body Mg may contribute to hypertension during pregnancy, the role of Mg may relate more to its stabilizing neuronal and vascular effects than to correction of an electrolyte deficit. Mg therapy is clearly indicated for women with preeclampsia to decrease the incidence of eclampsia and likely to decrease overall mortality and newborn complications.[151]

Bone Disorders

The equilibrium between bone formation and resorption is necessary and depends on the action of several local and systemic factors, including hormones, cytokines, chemokines, and biomechanical stimulation. Several direct and indirect mechanisms contribute to the effects of low Mg on bone density. As noted earlier in the discussion on translocation, Mg deficiency rapidly leads to hypomagnesemia, which is in part buffered through the mobilization of surface Mg from the bone. Additionally, the newly formed crystals of apatite are larger and more highly structured in Mg-deficient animals than controls, and this increases bone stiffness.[153] Osteoblastic bone formation has been shown to be reduced while an increase in the number and activity of osteoclasts in the Mg-deficient rat and mouse was reported. Low extracellular Mg inhibits osteoblast growth by increasing the release of nitric oxide through the upregulation of inducible nitric oxide synthase. One study suggested that the TRPM7 Mg channel is critical for osteoblast function and that Mg deficiency may thereby decrease bone formation. Mg deficiency causes both serum PTH and $1,25(OH)_2D$ to decrease, as discussed earlier. Given that both hormones are trophic for bone, impaired secretion or skeletal resistance may result in osteoporosis. Rude[1] and colleagues noted that an increased release of inflammatory cytokines was shown to result in activation of osteoclasts and increased bone resorption in rodents.

Dietary Mg intakes in humans have been shown to be positively associated with bone mineral density in cross-sectional studies.[154-158] However, this may[159] or may not[160-162] translate into lower risk of fracture as demonstrated in several cohort studies.

Circulating Mg concentrations are closely related to bone metabolism because bone surface Mg is continually exchanged with blood Mg. Low serum Mg concentration (<0.74 mmol/L) are independently associated with an increased risk of total and femoral fracture over a median follow-up of 25.6 years.[162] Red blood cell Mg concentration has been shown to be lower in postmenopausal women with osteoporosis vs healthy controls.[163]

Rude[1] noted that low serum and red blood cell Mg concentrations as well as high retention of parenterally administered Mg in patients with osteoporosis, suggested a Mg deficit, but these results were not consistent among studies. Similarly, while some studies reported low skeletal Mg content, others found normal or even high Mg content.[1] Dietary and serum Mg have been shown to be inversely associated with knee osteoarthritis in several large epidemiologic studies.[153,164-167]

Cancer

Mg has metabolic[168] and physiological functions that can influence carcinogenesis; however, the relation between Mg and cancer is complex. Animal models have shown that Mg may exert both anti- and protumor effects, depending on the tumor site and concentration of Mg bathing the tumor cells.[36]

High dietary Mg intake has been associated with reduced risk for several cancers. The association of dietary Mg and colorectal cancer risk was evaluated in 386,792 participants in the EPIC (European Prospective Investigation into Cancer and Nutrition) study, which suggested lower risk at higher intakes of Mg.[169] In addition, Dana et al[170] found that Mg intakes of <300 mg/day significantly reduced lung cancer risk by 19%; however, no significant association of intakes >300 mg/day and lung cancer was observed.

As reviewed in Costello et al,[171] several studies (including one from China and nine from the United States) have a priori evaluated the dietary Ca:Mg ratio in relation to breast, prostate, gastric, and colorectal cancer. The majority of studies showed positive or improved cancer outcomes when the Ca:Mg intake ratio fell within the range of 1.70–2.60.[171] An evaluation of seven cohort studies showed a significant inverse association between dietary Mg intake and risk of cancer mortality but not between supplemental Mg or total Mg intake and cancer deaths.[172] The data also suggest that cancer pathologies may act differently independent of biological sex and/or gender and under different dietary conditions (food vs supplements), making associations more difficult to interpret.[171]

Other Disorders

Mg deficiency has been associated with migraine headache, and Mg therapy has been reported to be effective in the treatment of migraine.[152] Because Mg deficiency results in smooth muscle bronchial reactivity and contractility and facilitates bronchoconstriction in subjects predisposed to asthma, some studies have shown that Mg therapy has been effective in treating asthma.[173] Mg losses may occur from the skin through sweating and in patients with extensive burn injury.[174,175] Recent reviews have gathered evidence of Mg treatment for chronic and acute pain in several conditions, a growing area of research.[176-178]

A number of systematic reviews and meta-analyses for many of the chronic diseases noted above suggest improvement in CVD risk factors with Mg supplementation, including hypertension and lipid profile,[86,101,145,179] diabetes,[99,180] migraine,[152,181,182] and muscle cramps.[183] However, a recent US Preventive Services Task Force report did not identify any Mg supplementation studies examining all-cause mortality, CVD, or cancer.[184]

Management of Magnesium Depletion

Hypomagnesemia

Hypomagnesemia is generally defined as a serum Mg level <1.8 mEq/L (see Figure 9.5). Presently, it is not clear which percentage of the US population falls into this category, as serum Mg levels are not routinely assessed in outpatients. Furthermore, the decreased Mg content in the typical Western diet, which is hypercaloric and high in refined and overprocessed foods, significantly limits daily intake of this mineral in the general population and contributes to the occurrence of modest and mild hypomagnesemia in a large percentage of the population, estimated to be at least 50%, if not more.[185]

Patients with hypomagnesemia and/or abnormal Mg homeostasis typically fall into one of the following three groups:

- Hypomagnesemia (low serum Mg concentration) in the absence of Mg deficiency (i.e., a normal total body Mg content)
- Mg deficiency (low total body Mg content) but no evidence of hypomagnesemia (i.e., normal serum Mg concentration)
- Mg deficiency (low total body Mg content) and a resultant hypomagnesemia (low serum Mg concentration)

Clinical Features of Magnesium Deficiency and/or Hypomagnesemia

Individuals with a mild decrease in serum Mg levels (i.e., <1.8 mEq/L or <0.9 mmol/L) are usually asymptomatic or may present with nonspecific symptoms such as fatigue, occasional headaches or muscle cramps, and a general sense of unwellness.

Severe Mg deficiency is defined as a serum Mg level of <1 mEq/L or 0.5 mmol/L.[1] Patients with such severe Mg depletion present with symptoms including hallucinations, seizures, muscle spasms, swallowing difficulty or inability (spasms of esophageal muscles), and irregular heartbeat. Severe hypomagnesemia can be life-threatening due to recurring seizures or irregular heartbeat that can progress toward arrhythmias, extrasystoles, ventricular tachycardia, torsades de pointes, and ultimately, ventricular fibrillation. There is no *direct* correlation between serum Mg levels and the severity of these symptoms. Furthermore, most of the symptoms are nonspecific, as they include the following:

- Neuromuscular irritability similar to that produced by hypocalcemia, manifesting with extensor plantar reflexes, positive Trousseau and Chvostek signs, and, in severe cases, tetany (hypoparathyroidism should be considered if these signs are present)
- Cardiovascular features such as rapid heartbeat and elevated blood pressure, tachycardia, and/or ventricular arrhythmias
- Central nervous system symptoms of vertigo, ataxia, depression, and seizure activity

Clinically, Mg deficiency should be suspected in patients who present with a relevant chronic disease causing abnormalities in Mg homeostasis, symptoms of Mg deficiency, or persistent associated electrolyte abnormalities such as hypocalcemia or hypokalemia.[186,187] However, often Mg deficiency may only be detected following a blood test.

Cardiac Arrhythmias

Mg depletion produces characteristic electrocardiogram (ECG) changes. Modest depletion causes widening of the QRS complex, with resultant prolongation of the QT interval, and peaking of T waves. As Mg depletion worsens, the QRS complex continues to widen, prolongation of the PR interval occurs, and the T wave presents with decreased amplitude. Ventricular arrhythmias can result, usually in patients with concurrent ischemic heart disease. Arrhythmias include ventricular extrasystoles, ventricular tachycardia (particularly torsades de pointes), and ventricular fibrillation. It is not clear whether these arrhythmias are due to hypomagnesemia, associated hypokalemia, or both conditions. Furthermore, because cellular Mg levels are inversely proportional to cellular Na^+ and Ca^{2+} content, any abnormal increase in the cellular content of these two cations directly affects upstroke in fast and slow action potentials in cardiac myocytes and pacemaker cells as well as cardiac contractility, a direct function of cellular Ca^{2+} content.[185]

Patients who present with ECG changes require oral Mg replacement therapy. The usual therapeutic dose of oral elemental Mg is 500 mg/day for adults and 3–6 mg/kg/day (maximum dose: 400 mg/day) for children, although local dosing recommendations may vary. Patients with arrhythmias require intravenous Mg sulfate. For adults, the usual protocol is 1–2 g administered intravenously as a single dose; local protocols for adult dosing may vary and should be consulted. Specialist advice should be sought for dosing in children. Patients with sustained ventricular tachycardia that does not respond to initial therapy require cardioversion. Defibrillation is required in cases of ventricular fibrillation in accordance with advanced life support algorithms.

Prolonged QT intervals are often observed in severe hypomagnesemia. Prolonged QT intervals due to inherited channelopathies[188] can be complicated or aggravated by underlying hypomagnesemia, especially if severe. Restoration of proper serum Mg levels by oral supplementation (500 mg/day) for mild hypomagnesemia or by 1,500 mg Mg intravenously for severe hypomagnesemia is recommended to attenuate the incidence of prolonged QT intervals and to prevent occurrence of torsades de pointes.

Eclampsia

Eclampsia is characterized by the occurrence of seizures in a patient with preeclampsia in the absence of any other identified cause. Eclampsia is treated with intravenous Mg sulfate. The usual protocol is to give a 4-g intravenous bolus initially, followed by a 2- to 3-g/hour infusion. Local protocols may vary and should be consulted. Any pregnant individual who presents with seizures requires immediate treatment, even if the diagnosis is not yet established. The diagnosis is confirmed by the presence of hypertension and proteinuria and the exclusion of other causes of seizures. Although controversy still exists regarding the beneficial effect of Mg therapy in pregnant women, especially if symptoms are mild, Mg therapy has consistently been shown to be beneficial for newborns because it decreases the occurrence of neurologic damage and cerebral palsy.[185]

Diabetes

Type 1 and type 2 diabetes are among the most common pathologies associated with Mg deficiency.[127] Although Mg depletion due to renal Mg wasting secondary to osmotic diuresis generated by hyperglycosuria is considered a possible mechanism for hypomagnesemia, clinical and experimental evidence suggests a direct effect of insulin in regulating cellular Mg homeostasis and retaining physiologic Mg levels in various tissues, including erythrocytes, skeletal muscle, and hepatocytes, through a close but ill-defined interaction with glucose homeostasis.[189] Mg deficiency has been reported to impair insulin secretion and promote insulin resistance,[128,129] contributing to the hypertensive state observed in patients with diabetes.[190,191] The role of Mg in insulin secretion is not fully defined but has been attributed to abnormal glucose metabolism, since Mg is a cofactor in several enzymes in this cycle.[3] As for the role of Mg in insulin resistance, Mg depletion decreases the tyrosine kinase activity of the insulin receptor and impairs insulin secretion by the β cell.[3] Mg therapy has been shown to improve diabetes control. Two studies reported that the incidence of type 2 diabetes was significantly greater in people with lower Mg intake.[128,129] Genetic variants of TRPM6 and TRPM7 have been reported to increase the risk of type 2 diabetes in women with intakes of <250 mg/day.[192] Mg status should, therefore, be assessed in patients with diabetes because a vicious cycle may occur: diabetes may lead to Mg loss, and the subsequent Mg deficiency may result in impaired insulin secretion and action, with worsening diabetic conditions.

Metabolic Syndrome

Hypomagnesemia has been indicated as a major risk factor for the development of metabolic syndrome and ultimately the onset of type 2 diabetes in these patients,[193] especially if hypomagnesemia is severe, as it occurs for a dietary intake <280 mg/day.[194] The underlying mechanism has been ascribed to the increase in basal inflammation level and the associated increase in serum CRP and tumor necrosis factor-α.[195]

Refractory Hypocalcemia and Refractory Hypokalemia

Refractory hypocalcemia and refractory hypokalemia refer to hypocalcemia and hypokalemia in the presence of hypomagnesemia that do not resolve unless Mg levels are normalized using appropriate Mg supplementation. Normomagnesemic Mg deficiency should always be considered as a cause of refractory hypocalcemia or refractory hypokalemia.[196]

Gastrointestinal Disorders

As discussed earlier, dietary Mg intake decreases to less than the recommended intake in a large proportion of the US population.[2] Nutritional Mg deficiency may be observed in the presence of other conditions that impair Mg balance. Gastrointestinal disorders (Table 9.8) may lead to Mg depletion in various ways.[116] The Mg content of upper intestinal tract fluids is approximately 1 mEq/L. Vomiting and nasogastric suction may, therefore, contribute to Mg depletion. The Mg content of diarrheal fluids and fistulous drainage is much higher (≤15 mEq/L); consequently, Mg depletion is common in acute and chronic diarrhea, regional enteritis, ulcerative colitis, and intestinal and biliary fistulas. Malabsorption syndromes may also result in Mg deficiency. Steatorrhea and resection or bypass of the small bowel, particularly the ileum, often results in intestinal Mg loss or malabsorption. Acute severe pancreatitis is associated with hypomagnesemia, which may result from the clinical problem causing the pancreatitis, such as alcoholism, or from saponification of Mg in necrotic para-pancreatic fat.[197]

Proton pump inhibitors have been reported to cause hypomagnesemia in some patients.[105] The evidence implicates intestinal Mg malabsorption. A primary defect in intestinal Mg absorption, which manifests early in life with hypomagnesemia, hypocalcemia, and seizures, has been described as an autosomal recessive disorder linked to chromosome 9q22. This disorder appears to be caused by mutations in TRPM6, the channel involved with active intestinal Mg transport.[39]

Renal Disorders

As described previously, the kidneys—together with the various segments of the gastrointestinal tract—are responsible for the regulation of systemic Mg homeostasis by regulating the level of Mg retention and Mg intake from the diet. Several renal disorders present with hypomagnesemia and/or hypermagnesuria. The majority of these conditions are on genetic bases and, for the most part, affect either specific proteins like claudins or their signaling regulatory mechanisms, which directly relate to Mg reabsorption in various segments of the nephron. As Table 9.8 shows, excessive urinary excretion of Mg also contributes to Mg depletion.[115,116] Renal Mg reabsorption is proportional to tubular fluid flow and to Na and Ca excretion. Therefore, long-term parenteral fluid therapy, particularly with saline solution, and volume expansion conditions such as primary aldosteronism and hypercalciuria may result in Mg depletion. Hypercalcemia has been shown to decrease renal Mg reabsorption, probably via Ca binding to the Ca-sensing receptor in the thick ascending limb of the loop of Henle and decreased transepithelial voltage.[117] Osmotic diuresis caused by glucosuria also results in urinary Mg wasting.[116]

Hypermagnesuria also occurs during the polyuric phase of recovery from acute renal failure in a native kidney, during recovery from ischemic injury in a transplanted kidney, and in postobstructive diuresis. In all of these conditions, it is likely that residual tubule reabsorptive defects persisting from the primary renal injury, together with polyuria, play as important a role in inducing renal Mg wasting.[112] Renal Mg wasting has occasionally been reported in patients with acute or chronic tubulointerstitial nephritis caused by acute renal allograft rejection or chronic pyelonephritis.[112] Alcohol ingestion also causes renal Mg wasting, thus explaining the high prevalence of Mg deficiency in patients with chronic alcoholism.[118]

Several identified renal Mg wasting disorders now have a genetic basis.[198] For example, the autosomal recessive mutations in the paracellin-1 gene on chromosome 3 (claudin 16) result in low serum Mg, hypercalciuria, and nephrocalcinosis. Another autosomal dominant form of isolated renal Mg wasting and hypomagnesemia has been linked to chromosome 11q23 and identified as a mutation on the Na^+/K^+-ATPase γ-subunit of gene *FXYD2*. A mutation of the Mg channel, TRPM6, also results in Mg wasting. Gitelman syndrome (familial hypokalemia-hypomagnesemia syndrome) is an autosomal recessive disorder caused by a genetic defect of the thiazide-sensitive NaCl cotransporter gene on chromosome 16. Bartter's syndrome also results in Mg wasting. This autosomal recessive disease affects the Na/K/2Cl cotransporter in the thick ascending limb of the loop of Henle and results in polyuria and polydipsia, with dehydration, which triggers hyperreninemia and hyperaldosteronism in the attempt to compensate for Na^+ and water loss. The large increase in urine loss and the reduced reabsorption of Na and K in the nephron contribute to the loss of Mg in the urine. Other undefined genetic defects also exist.

Other Disorders

Hypomagnesemia may accompany several other disorders.[116] Phosphate depletion has been shown experimentally to result in urinary Mg wasting and hypomagnesemia. Hypomagnesemia may also accompany "hungry bone" syndrome, a condition characterized by rapid bone mineral accretion in subjects with hyperparathyroidism or hyperthyroidism following surgical treatment, as well as mineral accretion into soft tissue during the refeeding syndrome.[199,200]

Therapeutic Pearls

- Patients with mild Mg depletion are usually asymptomatic and can be treated with oral Mg replacement. The usual therapeutic dose of oral elemental Mg is 500 mg/day in adults and 3–6 mg/kg/day (maximum dose: 400 mg/day) in children. Local dosing recommendations may vary and should be followed and a diet rich in whole grains, cereals, nuts, beans, and green leafy vegetables should be consumed.
- If serum Mg is ≤1 mEq/L (0.5 mmol/L; 1.22 mg/dL; i.e., severe hypomagnesemia) or if the patient is symptomatic (see Figure 9.5), intravenous Mg replacement is indicated. Local protocols should be consulted and followed. Specialist advice should be sought for dosing in children. The underlying cause should also be treated.
- An example of an intravenous replacement protocol is to give a loading dose of 1,500 mg of Mg intravenously over 15 minutes, followed by continuous infusion, as required. The infusion rate should not exceed 150 mg/minute or 125 mg/kg/hour.[196] An infusion rate of ≤2 g/hour decreases the risk of hypotension and is preferred in most cases. Higher infusion rates may be used in emergencies (e.g., seizures).

Acknowledgments

The authors would like to acknowledge the contributions of Robert Rude to the previous version of this chapter. Special acknowledgment to Christina West, BA, for editorial review, graphic design, and copy editing support.

References

1. Rude RK. Magnesium. In *Modern Nutrition in Health and Disease* (Ross AC, ed) pp. 159–75, Lippincott Williams & Williams; 2012.
2. US Institute of Medicine. *Dietary Reference Intakes for Calcium, Phosphorus, Magnesium, Vitamin D, and Fluoride*, National Academies Press, Washington, DC; 1997.

3. Romani A. Arch Biochem Biophys 2007;458:90–102.
4. Li FY, Chaigne-Delalande B, Kanellopoulou C, et al. Nature 2011;475:471–6.
5. Maguire ME, Cowan JA. Biometals 2002;15:203–10.
6. Wolf FI, Cittadini A. Mol Aspects Med 2003;24:3–9.
7. Touyz RM. Am J Physiol Heart Circ Physiol 2008;294:H1103–18.
8. de Baaij JH, Hoenderop JG, Bindels RJ. Physiol Rev 2015;95: 1–46.
9. Hou J. Curr Opin Nephrol Hypertens 2016;25:308–13.
10. Kumar D, Leonard E, Rude RK. Arch Intern Med 1978;138:660.
11. Daw CC, Ramachandran K, Enslow BT, et al. Cell 2020;183: 474–89.
12. Cowan J. Introduction to the biological chemistry of magnesium. In *The Biological Chemistry of Magnesium* (JA C, ed) pp. 1–24, VCH Publishers, New York; 1995.
13. Black C, Cowan J. Magnesium dependent enzymes in nucleic acid biochemistry; and magnesium dependent enzymes in general metabolism. In *The Biological Chemistry of Magnesium* (Cowan J, ed) pp. 137–58, VCH Publishers, New York; 1995.
14. Cowan JA. Biometals 2002;15:225–35.
15. Smith D. Magnesium as the catalytic center of RNA enzymes. In *The Biological Chemistry of Magnesium* (JA C, ed) pp. 111–36, VCH Publishers, New York; 1995.
16. Ohyama T. Int J Mol Sci 2019;20:4232.
17. Ackerman MJ, Clapham DE. N Engl J Med 1997;336:1575–86.
18. Dorup I. Acta Physiol Scand 1994;150:7–46.
19. White RE, Hartzell HC. Biochem Pharmacol 1989;38:859–67.
20. Chrysant SG, Chrysant GS. Int J Cardiol Hypertens 2019;1: 100005.
21. Pilchova I, Klacanova K, Tatarkova Z, et al. Oxid Med Cell Longev 2017;2017:6797460.
22. Saris NE, Mervaala E, Karppanen H, et al. Clin Chim Acta 2000;294:1–26.
23. Wallach S. Magnes Trace Elem 1990;9:1–14.
24. Endres DB, Rude RK. Disorders of bone. In *Tietz Textbook of Clinical Chemistry* (Burtis C, Ashwood E, and Burns D, eds) pp. 711–34, WB Saunders, Philadelphia; 2008.
25. Nielsen FH. Biol Trace Elem Res 2019;188:52–9.
26. Schuchardt JP, Hahn A. Curr Nutr Food Sci 2017;13:260–78.
27. Seelig MS. Am J Clin Nutr 1964;14:242–90.
28. Spencer H, Norris C, Williams D. J Am Coll Nutr 1994;13: 479–84.
29. Bohn T, Davidsson L, Walczyk T, et al. Br J Nutr 2004;91:601–6.
30. Kerstan D, Quamme GA. Physiology and pathophysiology of intestinal absorption of magnesium. In *Calcium in Internal Medicine* (Massry SG, Morii H, and Nishizawa Y, eds) pp. 171–83, Springer-Verlag, Surrey, UK; 2002.
31. Hardwick LL, Jones MR, Brautbar N, et al. J Nutr 1991;121: 13–23.
32. Pietropaolo G, Pugliese D, Armuzzi A, et al. Nutrients 2020;12.
33. Suksridechacin N, Thongon N. Physiol Rep 2022;10:e15247.
34. Coudray C, Feillet-Coudray C, Rambeau M, et al. J Trace Elem Med Biol 2006;20:73–81.
35. Rosanoff A, Dai Q, Shapses SA. Adv Nutr 2016;7:25–43.
36. Barbagallo M, Veronese N, Dominguez LJ. Nutrients 2021;13:463.
37. Blancquaert L, Vervaet C, Derave W. Nutrients 2019;11:1663.
38. Alexander RT, Hoenderop JG, Bindels RJ. J Am Soc Nephrol 2008;19:1451–8.
39. Quamme GA. Curr Opin Gastroenterol 2008;24230–5.
40. Schlingmann KP, Waldegger S, Konrad M, et al. Biochim Biophys Acta 2007;1772:813–21.
41. Rude RK, Shils ME. Magnesium. In *Modern Nutrition in Health and Disease* (Shils ME, Shike M, Ross AC, Caballero B, and Cousins R, eds) pp. 223–47, Jones & Bartlett Learning, LLC, Burlington; 2006.
42. Risco F, Traba ML. Magnes Res 1992;5:5–14.
43. Risco F, Traba ML. Magnes Res 1994;7:169–78.
44. Risco F, Traba ML, de la Piedra C. Magnes Res 1995;8:27–35.
45. Risco F, Traba ML. J Physiol Biochem 2004;60:199–203.
46. Ellison DH, Maeoka Y, McCormick JA. J Am Soc Nephrol 2021;32:2125–36.
47. Rude RK. J Bone Miner Res 1998;13:749–58.
48. Hou J, Renigunta A, Gomes AS, et al. Proc Natl Acad Sci USA 2009;106:15350–5.
49. Voets T, Nilius B, Hoefs S, et al. J Biol Chem 2004;279:19–25.
50. Günzel D, Yu AS. Pflugers Arch 2009;458:77–88.
51. Palacios C, Wigertz K, Braun M, et al. Am J Clin Nutr 2013;97:1014–9.
52. Groenestege WM, Hoenderop JG, van den Heuvel L, et al. J Am Soc Nephrol 2006;17:1035–43.
53. de Baaij JH, Hoenderop JG, Bindels RJ. Clin Kidney J 2012;5: i15–i24.
54. Cao S, Hodges JK, McCabe LD, et al. Nutr Today 2019;54: 195–206.
55. Wallach S. Magnesium 1988;7:262–70.
56. Rubin H. Magnes Res 2005;18:268–74.
57. Filippatos T, Tzavella E, Rizos C, et al. Expert Opin Drug Saf 2017;16:1121–32.
58. USDA Agricultural Research Service Food Surveys Research Group. Table 1. Nutrient Intakes from Food and Beverages. Mean Amounts Consumed per Individual, by Gender and Age, in the United States, 2015–2016. Vol. 2019; 2016.
59. US Department of Health and Human Services, and US Department of Agriculture. 2015–2020 Dietary Guidelines for Americans, 2015.
60. Panel on Dietary Reference Values of the Committee on Medical Aspects of Food Policy. Rep Health Soc Subj (Lond) 1991;41: 1–210.
61. World Health Organization, and Food and Agriculture Organization of the United Nations. Vitamin and mineral requirements in human nutrition: report of a joint FAO/WHO expert consultation, Bangkok, Thailand, 21–30 September 1998. Second Edition. WHO, Geneva, Switzerland; 2004.
62. EFSA Panel on Dietetic Products N, and Allergies. EFSA J 2015;13:4186.
63. Finucane MM, Stevens GA, Cowan MJ, et al. Lancet 2011;377: 557–67.
64. Nielsen FH. Crop Pasture Sci 2015;66:1259–64.
65. Rosanoff A. Adv Nutr 2021;12:298–304.
66. Rosanoff A, Costello RB, Johnson GH. Nutrients 2021;13:195.
67. Liebscher DH, Liebscher DE. J Am Coll Nutr 2004;23:730S–1S.
68. Witkowski M, Hubert J, Mazur A. Magnes Res 2011;24:163–80.
69. Sun Q, Bertrand KA, Franke AA, et al. Am J Clin Nutr 2017;105:159–68.
70. van Ooijen G, O'Neill JS. Cell Cycle 2016;15:2997–8.
71. Kanabrocki EL, Sothern RB, Ryan MD, et al. Clin Ter 2008;159: 329–46.
72. Siener R, Jahnen A, Hesse A. Urol Res 2011;39:123–7.
73. Newhouse IJ, Johnson KP, Montelpare WJ, et al. BMC Physiol 2002;2:6.
74. Ehrenpreis ED, Jarrouj G, Meader R, et al. Dis Mon 2022;68:101285.
75. Lowenstein FW, Stanton MF. J Am Coll Nutr 1986;5:399–414.
76. Rosanoff A, West C, Elin RJ, et al. Eur J Nutr 2022;61:3697–706.
77. Elin RJ. Magnes Res 2010;23:S194–8.
78. Han H, Fang X, Wei X, et al. Nutr J 2017;16:26.
79. Del Gobbo LC, Imamura F, Wu JH, et al. Am J Clin Nutr 2013;98:160–73.
80. Hruby A, McKeown NM, Song Y, et al. Nutrients 2013;5: 4990–5011.
81. US National Insitutes of Health Office of Dietary Supplements. Magnesium: fact sheet for health professionals. 2022.

82. Kantor ED, Rehm CD, Du M, et al. JAMA 2016;316:1464–74.
83. Chen F, Du M, Blumberg JB, et al. Ann Intern Med 2019;170: 604–613.
84. Ranade VV, Somberg JC. Am J Ther 2001;8:345–57.
85. Kappeler D, Heimbeck I, Herpich C, et al. BMC Nutrition 2017;3:7.
86. Zhang X, Li Y, Del Gobbo LC, et al. 2017 Hypertension 2017;68:324–33.
87. Lindberg JS, Zobitz MM, Poindexter JR, et al. J Am Coll Nutr 1990;9:48–55.
88. Fine KD, Santa Ana CA, Fordtran JS. N Engl J Med 1991;324: 1012–7.
89. Rude RK. Magnesium. In *Encyclopedia of Dietary Supplements* (Coates PM, Betz JM, Blackman MR, Cragg GM, Levine M, Moss J, et al., eds) pp. 527–37, Informa Healthcare USA, Inc., New York, NY, 2010.
90. Anderson PO, Knoben JE. *Handbook of clinical drug data*, McGraw Hill; 1997.
91. Rodríguez-Morán M, Guerrero-Romero F. Diabetes 2003;Care 26:1147–52.
92. Musso CG. Int Urol Nephrol 2009;41:357–62.
93. Gerard SK, Hernandez C, Khayam-Bashi H. Ann Emerg Med 1988;17:728–31.
94. Kutsal E, Aydemir C, Eldes N, et al. Pediatr Emerg Care 23: 570–2.
95. Cowan AE, Jun S, Tooze JA, et al. Nutrients 2019;12:38.
96. King DE, Mainous AG, 3rd, Geesey ME, et al. J Am Coll Nutr 2005;24:166–71.
97. Song Y, Ridker PM, Manson JE, et al. Diabet Care 2005;28: 1438–44.
98. Mazidi M, Rezaie P, Banach M. Arch Med Sci 2018;14:707–16.
99. Veronese N, Pizzol D, Smith L, et al. Nutrients 2022;14:679.
100. Talebi S, Miraghajani M, Hosseini R, et al. Biol Trace Elem Res 2022;200:1538–50.
101. Simental-Mendia LE, Sahebkar A, Rodriguez-Moran M, et al. Curr Pharm Des 2017;23:4678–86.
102. Nielsen FH. Adv Chronic Kidney Dis 2018;25:230–5.
103. Guerrera MP, Volpe SL, Mao JJ. Am Fam Physician 2009;80: 157–62.
104. Hicks MA, Tyagi A. Magnesium sulfate. In *StatPearls*, StatPearls Publishing, Treasure Island (FL), 2022.
105. Cundy T, Dissanayake A. Clin Endocrinol (Oxf) 2008;69: 338–41.
106. Gröber U. Int J Mol Sci 2019;20:2094.
107. Pronsky ZM. *Powers and Moore's Food-Medication Interactions*, Food-Medication Interactions, Birchrunville, PA; 1999.
108. Navaneethan SD, Sankarasubbaiyan S, Gross MD, et al. Transplant Proc 2006;38:1320–2.
109. TRC Healthcare. Natural Medicines Comprehensive Database. 2022.
110. Goldman RD, Koren G. J Pediatr Hematol Oncol 2004;26: 421–6.
111. Lajer H, Kristensen M, Hansen HH, et al. Cancer Chemother Pharmacol 2005;55:231–6.
112. Rude R. Magnesium disorders. In *Fluids and Electrolytes* (Kokko J, and Tannen R, eds) pp. 421–45, WB Saunders, Philadelphia; 1996.
113. Wilkinson R, Lucas GL, Heath DA, et al. Br Med J (Clin Res Ed) 1986;292:818–9.
114. Bai JP, Hausman E, Lionberger R, et al. Mol Pharm 2012;9: 3495–505.
115. Quamme GA, de Rouffignac C. Front Biosci 2000;5:D694–711.
116. Rude RK. Magnesium homeostasis. In *Principles of Bone Biology* (Bilezikian JB, Raisz L, and Rodan G, eds) pp. 487–513, Academic Press, San Diego; 2008.
117. Quamme GA. Kidney Int 1997;52:1180–95.
118. Romani AM. Magnes Res 2008;21:197–204.
119. Tin A, Grams ME, Maruthur NM, et al. Kidney Int 2015;87: 820–7
120. Oost LJ, van der Heijden A, Vermeulen EA, et al. Diabet Care 2021;44:1757–65.
121. Dibaba DT, Xun P, He K. Eur J Clin Nutr 2014;68:510–6.
122. Hruby A, Meigs JB, O'Donnell CJ, et al. Diabetes Care 2014;37:419–27.
123. Papanikolaou Y, Brooks J, Reider C, et al. J Hum Nutr Food Sci 2014;2:1038.
124. Xu T, Chen GC, Zhai L, et al. Biomed Environ Sci 2015;28: 527–34.
125. Moore-Schiltz L, Albert JM, Singer ME, et al. Br J Nutr 2015;114:924–35.
126. Zhao B, Deng H, Li B, et al. Diabetes Metab Res Rev 2020;36:e3243.
127. McNair P, Christensen MS, Christiansen C, et al. Eur J Clin Invest 1982;12:81–5.
128. Lopez-Ridaura R, Willett WC, Rimm EB, et al. Diabet Care 2004;27:134–40.
129. Song Y, Manson JE, Buring JE, et al. Diabet Care 2004;27: 59–65.
130. Resnick L. Prog Cardiovasc Dis 1999;42:1–22.
131. Zhang W, Iso H, Ohira T, et al. Atherosclerosis 2012;221: 587–95.
132. Qu X, Jin F, Hao Y, et al. PLoS One 2013;8:e57720.
133. Wu J, Xun P, Tang Q, et al. Nutr J 2017;16:60.
134. Pickering RT, Bradlee ML, Singer MR, et al. Nutrients 2021;13.
135. Hruby A, O'Donnell CJ, Jacques PF, et al. JACC Cardiovasc Imaging 2014;7:59–69.
136. Zhao L, Hu M, Yang L, et al. J Cardiovasc Pharmacol 2019;74: 516–27.
137. Rosanoff A. Med Hypotheses 2013;81:1063–5.
138. Jiang L, He P, Chen J, et al. Nutrients 2016;8:5.
139. Helte E, Säve-Söderbergh M, Larsson SC, et al. Am J Clin Nutr 2022;2022:nqac186.
140. Seelig M. *Magnesium Deficiency in the Pathogenesis of Disease. Early Roots of Cardiovascular, Skeletal, and Renal Abnormalities*, Plenum, New York, NY; 1980.
141. He K, Song Y, Belin RJ, et al. J Cardiometab Syndr 2006;1: 351–5.
142. Chacko SA, Sul J, Song Y, et al. Am J Clin Nutr 2011;93: 463–73.
143. Kieboom BC, Niemeijer MN, Leening MJ, et al. J Am Heart Assoc 2016;5:e002707.
144. Costello RB, Elin RJ, Rosanoff A, et al. Adv Nutr 2016;7: 977–93.
145. Verma H, Garg R. J Hum Nutr Diet 2017;30:621–33.
146. Hwang DL, Yen CF, Nadler JL. Am J Hypertens 1992;5:700–6.
147. Nadler JL, Malayan S, Luong H, et al. Diabet Care 1992;15: 835–41.
148. Sontia B, Touyz RM. Arch Biochem Biophys 2007;458:33–9.
149. Cheteu Wabo TM, Wu X, Sun C, et al. Nutr Res Pract 2022;16:74–93.
150. US Food and Drug Administration. RE: Petition for a qualified health claim for magnesium and reduced risk of high blood pressure (hypertension) (docket No. FDA-2016-Q-3770); 2022.
151. Belfort MA, Anthony J, Saade GR, et al. N Engl J Med 2003;348:304–11.
152. Veronese N, Demurtas J, Pesolillo G, et al. Eur J Nutr 2020;59:263–72.
153. Castiglioni S, Cazzaniga A, Albisetti W, et al. Nutrients 2013;5:3022–33.
154. Welch AA, Skinner J, Hickson M. Nutrients 2017;9:1189.
155. Tucker KL, Hannan MT, Chen H, et al. Am J Clin Nutr 1999;69:727–36.

156. Angus RM, Sambrook PN, Pocock NA, et al. Bone Miner 1988;4:265–77.
157. Ilich JZ, Brownbill RA, Tamborini L. Eur J Clin Nutr 2003;57: 554–65.
158. Groenendijk I, van Delft M, Versloot P, et al. Bone 2022;154:116233.
159. Veronese N, Stubbs B, Solmi M, et al. Br J Nutr 2017;117: 1570–6.
160. Farsinejad-Marj M, Saneei P, Esmaillzadeh A. Osteoporos Int 2016;27:1389–99.
161. Orchard TS, Larson JC, Alghothani N, et al. Am J Clin Nutr 2014;99:926–33.
162. Kunutsor SK, Whitehouse MR, Blom AW, et al. Eur J Epidemiol 2017;32:593–603.
163. Odabasi E, Turan M, Aydin A, et al. Ann Acad Med Singapore 2008;37:564–7.
164. Zeng C, Li H, Wei J, et al. PLoS One 2015;10:e0127666.
165. Qin B, Shi X, Samai PS, et al. Arthritis Care Res (Hoboken) 2012;64:1306–11.
166. Shmagel A, Onizuka N, Langsetmo L, et al. Osteoarthritis Cartilage 2018;26:651–8.
167. Zeng C, Wei J, Li H, et al. J Rheumatol 2015;42:1231–6.
168. Rubin H. Magnes Res 2014;27:94–102.
169. Papadimitriou N, Bouras E, van den Brandt PA, et al. Clin Gastroenterol Hepatol 2022;20:864–73.e13.
170. Dana N, Karimi R, Mansourian M, et al. Int J Vitam Nutr Res 2021;91:539–46.
171. Costello RB, Rosanoff A, Dai Q, et al. Adv Nutr 2021;12:291–7.
172. Bagheri A, Naghshi S, Sadeghi O, et al. Adv Nutr 2021;12: 1196–210.
173. Mohammed S, Goodacre S. Emerg Med J 2007;24:823–30.
174. Nielsen FH, Lukaski HC. Magnes Res 2006;19:180–9.
175. Berger MM, Rothen C, Cavadini C, et al. Am J Clin Nutr 1997;65:1473–81.
176. Urits I, Jung JW, Amgalan A, et al. Anesth Pain Med 2021;11: e112348.
177. Morel V, Pickering ME, Goubayon J, et al. *Nutrients* 2021;13.
178. Shin HJ, Na HS, Do SH. Nutrients 2020;12.
179. Asbaghi O, Hosseini R, Boozari B, et al. Biol Trace Elem Res 2021;199:413–24.
180. Simental-Mendia LE, Sahebkar A, Rodriguez-Moran M, et al. Pharmacol Res 2016;111:272–82.
181. Chiu HY, Yeh TH, Huang YC, et al. Pain Physician 2016;19: E97–112.
182. Okoli GN, Rabbani R, Kashani HH, et al. Can J Neurol Sci 2019;46:224–33.
183. Liu J, Song G, Zhao G, et al. Taiwan J Obstet Gynecol 2021;60:609–14.
184. Mangione CM, Barry MJ, Nicholson WK, et al. JAMA 2022; 327:2326–33.
185. Romani AM. Met Ions Life Sci 2013;13:49–79.
186. Flink EB. J Am Coll Nutr 1985;4:17–31.
187. Günther T, Vormann J. Magnes Trace Elem 1990;9:279–82.
188. Romani A, Scarpa A. FEBS Lett 1990;269:37–40.
189. Voma C, Romani AM. Role of magnesium in the regulation of hepatic glucose homeostasis. In *Glucose Homeostasis* (Szablewski L, ed) pp. 95–111, InTech, London, UK; 2014.
190. Barbagallo M, Dominguez LJ, Resnick LM. Am J Ther 2007;14: 375–85.
191. Günther T. Magnes Res 2010;23:5–18.
192. Song Y, Hsu YH, Niu T, et al. BMC Med Genet 2009;10:4.
193. Dwipayana P, Saraswati MR, Gotera W, et al. Fam Med Commun Health 2013;1:14.
194. Jiao Y, Li W, Wang L, et al. Nutrients 2022;14:2013.
195. Guerrero-Romero F, Bermudez-Peña C, Rodríguez-Morán M. Magnes Res 24:45–53.
196. Romani A. BMJ Point of Care 2021.
197. Ryzen E, Rude RK. West J Med 1990;152:145–8.
198. Naderi AS, Reilly RF, Jr. Nat Clin Pract Nephrol 2008;4:80–9.
199. Farese S. Ther Umsch 2007;64:277–80.
200. Ziegler TR. N Engl J Med 2009;361:1088–97.

B. MINERALS

CHAPTER 10

Iron

James P. McClung and Stephen R. Hennigar

CHAPTER OUTLINE

Abbreviations

2,5-DHBA, 2,5-dihydroxybenzoic acid; **AI**, anemia of inflammation; **BMP**, bone morphogenetic protein; **CHr**, reticulocyte hemoglobin; **DcytB**, duodenal cytochrome B; **DMT1**, divalent metal transporter-1; **ERFE**, eythroferrone; **HCP1**, heme carrier protein-1; **HJV**, hemojuvelin; **IRE**, iron response element; **IRIDA**, iron-refractory iron deficiency anemia; **IRP**, iron regulatory protein; **MCH**, mean corpuscular hemoglobin; **NTBI**, non-transferrin-bound iron; **PCBP**, poly(rC)-binding protein; **RDA**, recommended dietary allowance; **sTfR**, soluble transferrin receptor; **TBI**, total body iron; **UL**, tolerable upper intakes level; **WHO**, World Health Organization

Historical Perspective

Iron (from the Latin: "ferrum") is required for nearly all organisms. As early as the sixteenth and seventeenth centuries, chlorosis (iron deficiency anemia) was reported as a medical condition that could be treated with iron supplementation, but it was not until the turn of the twentieth century that our understanding of the essential nature of iron for heme synthesis slowly emerged.[1] Since that time, the pace of discovery in the field of iron metabolism has accelerated to the current explosion in molecular information in the 21st century.[2]

Chemistry

Iron has four naturally occurring stable isotopes, ^{54}Fe (5.8%), ^{56}Fe (91.7%), ^{57}Fe (2.1%), and ^{58}Fe (0.3%) with a recommended atomic mass of 55.845 u. Of the many iron radioisotopes, ^{59}Fe has a half-life of 45 days and is the most widely used radioisotope in preclinical research. Iron is a transition metal that exists in one of two oxidation states: the ferrous form (Fe^{2+}) or the ferric form (Fe^{3+}). This chemical property results in iron's catalytic role in a multitude of redox reactions necessary to support basic metabolic functions for life. In fact, iron's central role in oxygen and energy metabolism underscores the biologic significance of this element and helps to explain why it is one of the best-studied metals in nutrition and health. These same catalytic properties of iron also confer its well-known toxicity resulting from Fenton chemistry, a reaction that generates free radicals, including superoxide. Thus, iron is an essential nutrient as well as a powerful toxicant, and it is important to understand how both features are kept in balance.

Biology

Total body iron content is estimated to be 3.8 g in men and 2.3 g in women. Most body iron is found as heme iron. Heme iron is the essential constituent for oxygen transport in hemoglobin, oxygen storage in myoglobin, and electron transport for cytochrome function in aerobic respiration, and it is even necessary for signal transduction as a cofactor for nitric oxide synthase and guanylyl cyclase. The second largest pool of iron is found in its storage form ferritin (also hemosiderin). Ferritin is a large assembly of 24 protein subunits of two types, light chain (19 kDa) and heavy chain (21 kDa), which form a large sphere around a mineralized ferric core of several thousands of iron atoms.[3] In times of demand, iron is liberated from ferritin to fulfill essential functions in oxygen transport and energy metabolism. Damage generated by reactive oxygen species arising from redox-reactive free iron is prevented by its storage in ferritin.

In a similar fashion, newly absorbed iron is bound and transported by transferrin in serum, thereby limiting the redox-reactive and potentially toxic effects of "free" or non–transferrin-bound iron (NTBI). Transferrin-bound iron is destined to be taken up by the transferrin receptor in peripheral tissues for storage or utilization. Transferrin has two binding sites for one iron atom each. Under normal circumstances, 30% to 40% of these iron-binding sites are filled with approximately 3 mg of total body iron. Circulating transferrin-bound iron represents a highly dynamic storage pool that can be drawn from to fulfill immediate demands. Hence the saturation state of serum transferrin plays a key role in the regulation of iron metabolism and is one of the indices used clinically to evaluate iron status. In cases of iron overload and certain pathologic conditions, transferrin may become saturated and result in NTBI.[4,5]

Iron absorption, utilization, and storage are finely tuned to maintain the metal's homeostasis (**Figure 10.1**). Unlike the situation with other essential elements, iron metabolism is subject to a high degree of conservation. Rather than eliminate excess iron that is not immediately required, iron is stored in ferritin, as described earlier, and can be released in times of need by NCOA4-mediated ferritinophagy where NCOA4 binds and traffics ferritin to the lysosome where ferritin is degraded and iron is released for use by the cell.[6,7] The nature of iron homeostasis reflects the metal's key chemical role in oxygen and energy metabolism, processes that are necessary for life and must, therefore, rely on a substantial reservoir of iron to support the ultimate demands of human physiology. Approximately 20 to 25 mg of iron is turned over daily with the erythrophagocytosis of senescent red blood cells, and iron that is released from heme is captured for reutilization in the production of new erythrocytes. Small amounts of iron are lost in feces (~0.6 mg/day), urine (<0.1 mg/day), and sweat (<0.3 mg/day). Menstruating women experience an average blood loss of approximately 40 mL/cycle or 0.4 to 0.5 mg iron/day. Dietary iron contributes ~1–2 mg of absorbed iron each day, which can offset these losses; however, conditions resulting in excessive blood loss or increased physiologic requirements (e.g., infants, children, adolescents, pregnancy) affect iron demands. A key feature of iron homeostasis is that the body's iron status is generally maintained at the level of dietary iron absorption. When iron is depleted from the body, dietary iron absorption increases to meet the demand for iron. Although no known regulated pathway exists for excretion of excess iron, dietary iron absorption is reduced under high-iron conditions to prevent toxic accumulation.

Dietary Sources

Iron is absorbed from the diet as either heme or nonheme iron. Heme iron is typically derived from hemoglobin or myoglobin and is contained in foods such as red meats, fish, and poultry. Various sources of nonheme iron, such as plant foods, are also available. This form of iron is also added to enrich and fortify foods such as cereals. Although nonheme iron is the predominant form in the diet, heme iron is more bioavailable.[8] Approximately 15% to 35% of heme iron will be assimilated, compared

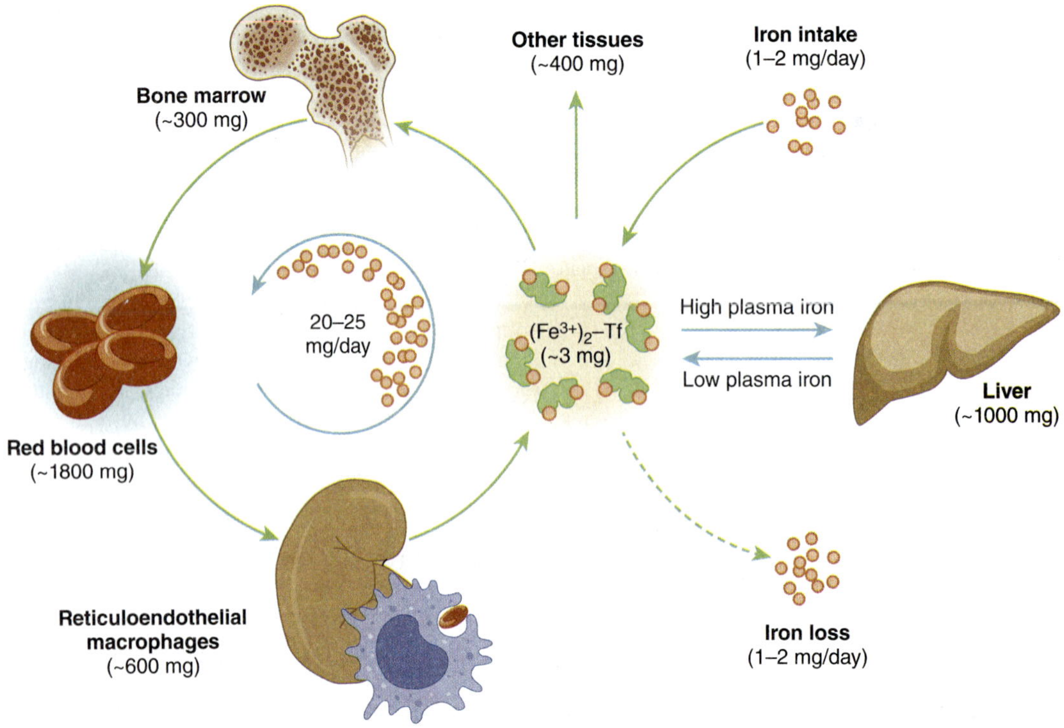

Figure 10.1 Iron dynamics in the human body. Iron is absorbed, transported, and exchanged dynamically to maintain iron homeostasis. Pool sizes and amounts transferred per day in a typical adult are indicated. Tf, transferrin.

with 2% to 20% of nonheme iron. Consumption of a heme iron source (e.g., meat, fish, poultry) stimulates absorption of nonheme iron.[9] Although the mechanism for the so-called "MFP (meat, fish, and poultry) factor" is not completely understood, studies suggest that partially digested peptides from muscle proteins bind iron via their cysteine and histidine residues to form soluble complexes for absorption.[10] As noted earlier, the levels of stored iron in the body influence the extent of absorption. Under low-iron conditions, dietary absorption is promoted, whereas high-iron conditions reduce absorption. Many other dietary and endogenous factors can influence iron uptake (**Table 10.1**). For example, ascorbic acid (vitamin C) can help to reduce ferric iron to the more bioavailable form of ferrous iron.[11,12] Polyphenols and phytates can interfere with nonheme iron uptake.[12] Calcium blocks uptake of both heme and nonheme iron,[13] whereas other metals can inhibit nonheme iron absorption by sharing the same pathway for absorption.[14] In particular, not only is lead a competitive inhibitor for uptake but it can also disrupt steps of iron metabolism required for heme synthesis.[15] Because low iron status enhances metal absorption, lead poisoning is sometimes associated with iron deficiency in children.[16]

Table 10.1 Factors Influencing Iron Absorption

	Nutrients	Endogenous Factors
Enhancers	Ascorbic acid (vitamin C)	Enhanced erythropoiesis due to
	Fructose	Hypoxia
	Citric acid	Hemorrhage
	Dietary protein	Hemolysis
	Lysine	Androgens
	Histidine	Cobalt
	Cysteine	Low iron stores
	Methionine	
Inhibitors	Oxalic acid	Infection/inflammation
	Tannins	Lack of stomach acid
	Phytate	High iron stores
	Polyphenols	
	Carbonate	
	Phosphate	
	Fiber	
	Other metal ions	

Dietary Reference Intakes

The dietary reference intakes (DRIs) set for iron by the Institute of Medicine of the National Academy of Sciences, Engineering and Medicine (NASEM) are listed in **Table 10.2**. Iron is a micronutrient: adult men require 8 mg iron/day, and during their reproductive years, girls and women require 18 mg iron/day. The typical North American diet of 12 to 18 mg iron/day should be adequate to fulfill these needs, but the requirement for iron markedly increases to 27 mg/day during pregnancy, and iron supplements are often needed to match this high demand. Infants are born with a 4- to 6-month iron supply, and a recommended dietary allowance (RDA) has not been established for this early age. An adequate intake of 0.27 mg/day is recommended for infants 0–6 months. Beyond this age, the iron content in milk cannot entirely meet the needs of the developing child, and food sources are required to meet RDAs (7 mg/day at age 1–3 years and 10 mg/day at age 4–8 years). Iron toxicity can also pose a risk for children and may commonly occur from ingestion of excess iron supplements. Death occurs at levels of 200 to 300 mg/kg. The tolerable upper intake levels (UL) for iron intake are listed in Table 10.2.

Table 10.2 Dietary Reference Intakes and Tolerable Upper Intake Levels for Iron

	Recommended Dietary Allowances			Tolerable Upper Intake Levels		
	Age (y)	Males (mg/d)	Females (mg/d)	Age (y)	Males (mg/d)	Females (mg/d)
Infants	0.5–1.0	11	11	0.5–1.0	40	40
Children	1–3 4–8	7 10	7 10	1–13	40	40
Adolescents	9–13 14–18	8 11	8 15	14–18	45	45
Adults	19–50 51+	8 8	18 8	19+	45	45
Pregnancy	14–18 19–50	—	27 27	14–18 19+	—	45 45
Lactation	14–18 19–50	—	10 9	14–18 19+	—	45 45

Data from Food and Nutrition Board, Institute of Medicine. Dietary Reference Intakes for Vitamin A, Vitamin K, Arsenic, Boron, Chromium, Copper, Iodine, Iron, Manganese, Molybdenum, Nickel, Silicon, Vanadium, and Zinc. Washington, DC: National Academy Press, 2001.

Iron Metabolism and its Regulation

The field of iron biology has made rapid advances since 2000. Many of the proteins involved in iron transport and homeostatic regulation have been identified, and their physiologic roles have been uncovered. A breakthrough in the understanding of iron metabolism came with the discovery of the iron regulatory hormone, hepcidin. Features of hepcidin regulation of iron metabolism have been compared to insulin action in glucose metabolism, thus creating a new field of iron endocrinology.[17] How hepcidin regulates systemic iron homeostasis is a major focus of current investigation. At the cellular level, molecular insights into the regulation of iron-binding proteins have yielded information about regulation of iron transport, utilization, and storage that have important clinical considerations. Finally, transcriptional and posttranscriptional networks that may be activated by hepcidin-induced signaling are beginning to emerge and provide clues into the relationships between iron and inflammation.

Intestinal Iron Absorption

Body iron status is precisely tuned by the absorption of dietary iron. Therefore, it is important to understand the mechanisms involved in this process and the multiple layers that regulate the flux of iron into the system (**Figure 10.2**). Because the body does not eliminate excess iron, dysregulation of intestinal iron absorption causing assimilation of too much iron will result in iron overload. Conversely, if sufficient iron is not absorbed to make up for demand, the risk for iron deficiency increases. Iron is absorbed as nonheme or heme forms. Heme iron is absorbed more effectively,[8] and this process does not appear to be subject to the same regulatory mechanisms as nonheme iron uptake.[18]

These findings indicate that heme and nonheme iron are taken up by independent mechanisms. A putative heme transporter called heme carrier protein-1 (HCP1) was initially identified,[19] but was later shown to be a high-affinity folate transporter and renamed proton-coupled folate transporter.[20] HCP1 may possibly be a low-affinity transporter for heme, but its physiologic relevance remains to be better established. A different molecule, heme-responsive gene-1, or *HRG1*, has been identified as a heme transporter in *Caenorhabditis elegans*.[21] HRG1 is found on erythrophagosomal membranes and imports heme into the cytosol of macrophages during iron recycling from red blood cells.[22] Although HRG1 is expressed in the human small intestine,[21] its activity has yet to be defined. The majority of heme iron entering the intestinal absorptive enterocyte is released by heme oxygenase to join the pool of newly absorbed nonheme iron entering the cell.[23] In theory, intact heme may also be released across the basolateral surface. Feline leukemia virus C receptor (FLVCR) has been identified to function as a heme exporter in erythroid cells,[24] and a second possible efflux pathway involving the ABC transporter ABCG2 (also known as breast cancer regulated protein, BCRP) has been suggested,[25] but their possible roles in heme assimilation by the intestine have yet to be fully explored.

Uptake of nonheme iron by enterocytes is better understood. Although less effectively absorbed, nonheme iron is present in a greater range of foods, most typically in the ferric (Fe^{3+}) form. Reduction to Fe^{2+} is the first step in intestinal nonheme iron assimilation and is mediated by brush-border ferrireductase activity. An enzyme, duodenal cytochrome B (DcytB), has been implicated in this process.[26] Although it does not appear to be an essential gene,[27] DcytB is highly regulated in response to iron status in animals and humans,[28-30] and a promoter polymorphism observed in the human population appears to modify serum ferritin concentrations in *HFE*-associated hereditary hemochromatosis.[31]

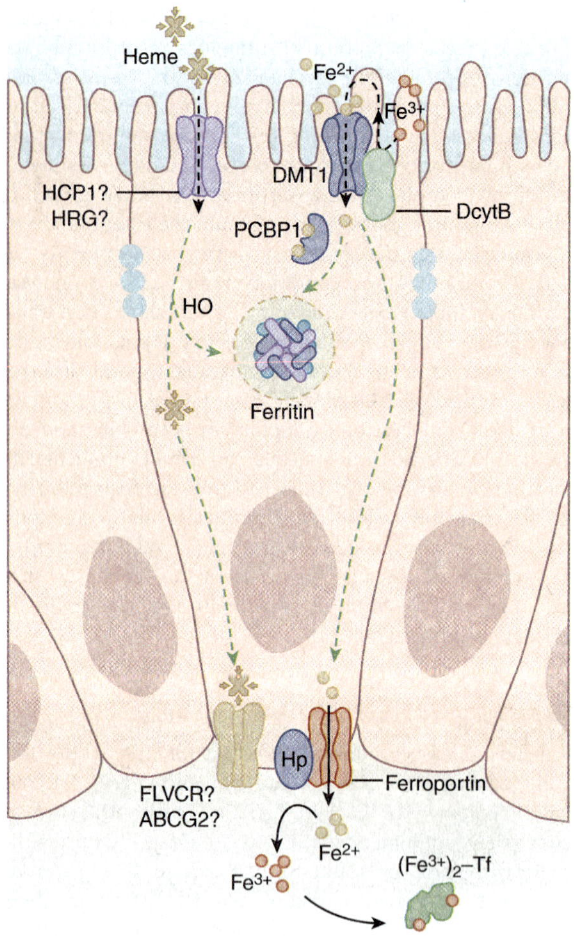

Figure 10.2 Intestinal iron (Fe) absorption. *ABCG2*, ABC transporter; *DcytB*, duodenal cytochrome B; *DMT1*, divalent metal transporter-1; *FLVCR*, feline leukemia virus C receptor; *HCP1*, heme carrier protein-1; *HRG*, heme-responsive gene; *PCBP1*, poly(rC)-binding protein-1; $(Fe^{3+})_2$–Tf, diferric transferrin.

After reduction by DctyB or another brush-border ferrireductase, uptake of the ferrous (Fe^{2+}) form of iron is mediated by divalent metal transporter-1 (DMT1).[32-34] The low pH of the intestinal lumen is important for these initial steps because DMT1 is a proton-coupled transporter, and acidification is, therefore, necessary for its optimal activity.[33] Like DcytB, DMT1 is also highly regulated by iron status. The small intestine expresses four transcripts of DMT1, and mRNA levels are regulated both transcriptionally and posttranscriptionally.[35,36] Different isoforms of the protein appear to have tissue-specific function and subcellular localization.[37] Iron status appears to control not only the protein and mRNA levels for DMT1 but also the distribution of the protein in various enterocyte compartments.[38] Studies have shown that intestinal DMT1 is necessary for iron absorption in mice after birth, but it appears to be dispensable for other tissues—a finding suggesting that redundant activities fulfill that role.[34] Human mutations in DMT1 are associated with microcytic anemia,[39] consistent with its major function in dietary iron absorption. The finding that these patients also load iron is consistent with an important DMT1 function in delivering iron to erythroid cells (see the later description of the transferrin cycle).

Molecular details of the mechanism by which imported iron is transferred across the intestinal mucosal cell for release into circulation remain largely unknown. Investigators have long speculated that a cytosolic iron chaperone directs the fate of newly absorbed intestinal iron. The poly(rC)-binding protein (PCBP) family of metallochaperones have been identified[40,41]; however, their function in the intestine has yet to be fully characterized. When efflux from the enterocyte is impaired, intestinal iron is known to accumulate in the ferritin storage compartment.[42,43] Deletion of intestinal ferritin in mice promotes increased dietary iron absorption and dysregulated systemic iron metabolism.[44] Iron stored within ferritin as a result of excess dietary absorption would probably be lost from the body as enterocytes are shed from the villus tip. PCBP1 (and to a lesser extent, PCBP2) has been shown to deliver iron to ferritin and is ubiquitously expressed.[40] It, therefore, seems likely that PCBP1 may function in the intestine to help load iron onto ferritin. In vitro studies in hepatocyte-derived cells demonstrate that heme oxygenase can transfer the iron released from heme to PCBP2,[45] and that PCBP2 interacts with DMT1 and ferroportin.[46,47] Whether an iron chaperone function is necessary to traffic cytosolic iron across the mucosa for its entry into portal circulation is unclear, however.

An alternate model is that iron crosses the enterocyte through a vesicular trafficking pathway,[48,49] which involves intracellular transfer to a compartment with the membrane iron exporter ferroportin and the ferroxidase hephaestin, along the ultimate target, iron-free apotransferrin. Each of these factors plays an important role in export of iron from the enterocyte, but whether these factors act within the lumen of intracellular vesicles or directly at the basolateral surface is not certain because they are topologically equivalent. Ferroportin is essential for iron efflux from the intestine.[50] It requires calcium as a cofactor[51] and is believed to export iron in concert with hephaestin, a membrane-bound ceruloplasmin homolog that oxidizes ferrous iron to the ferric form.[43] Ceruloplasmin itself can also fulfill this function,[52] and both ferroxidases provide iron to transferrin in the correct oxidation state. Transferrin binds two atoms of ferric iron and circulates in serum to deliver iron to peripheral tissues.

The Transferrin Cycle

Circulating transferrin delivers iron by binding to cell surface receptors. Two receptors that specifically and uniquely recognize transferrin as a ligand are known. Transferrin receptor-1 has long been studied as the functional partner in iron uptake and is ubiquitously expressed.[53,54] Its closely related homolog, transferrin receptor-2, has a more restricted expression pattern and predominates in the liver, where it plays an iron-sensing role in metabolism.[55-57]

Uptake of iron by cells from the transferrin-receptor binding complex begins with its internalization by clathrin-mediated endocytosis (**Figure 10.3**). Clathrin-coated vesicles deliver their cargo to acidic intracellular compartments called

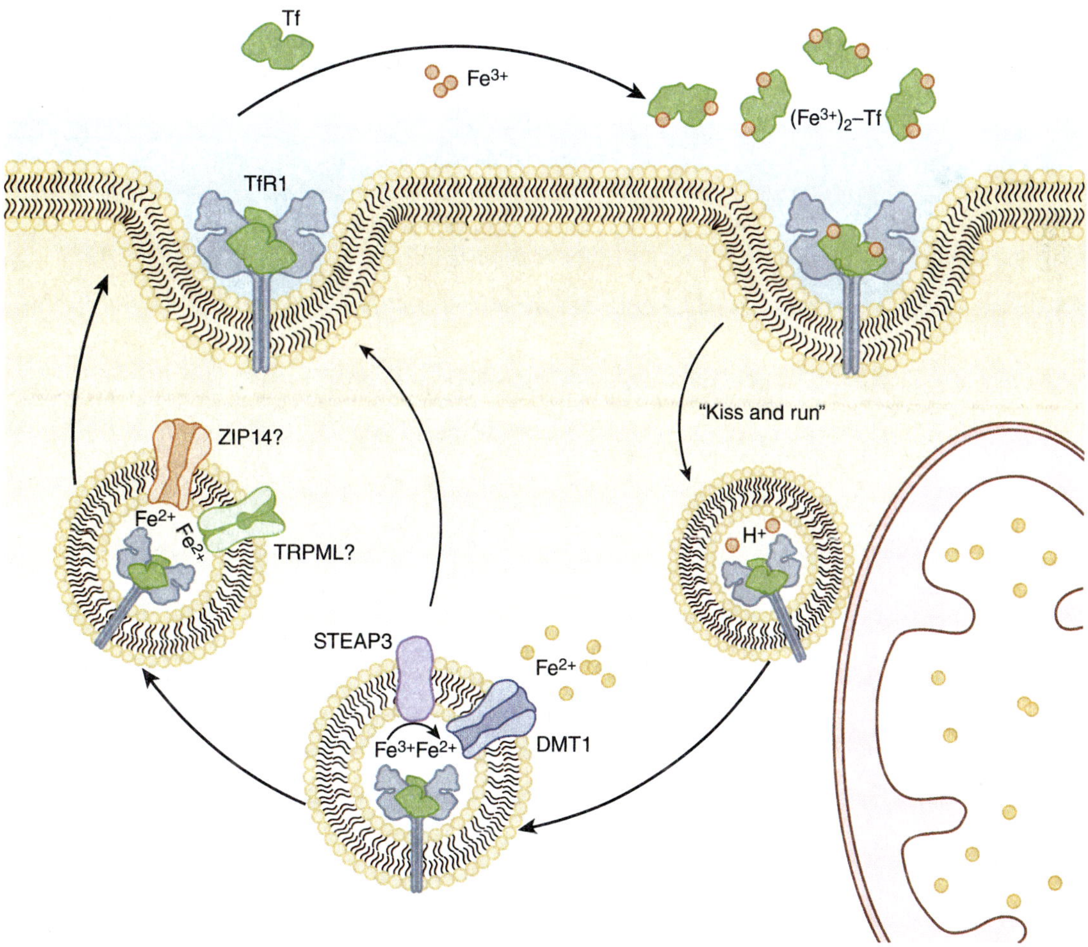

Figure 10.3 Transferrin (Tf) cycle. *DMT1*, divalent metal transporter-1; *Fe*, iron; *STEAP3*, a ferrireductase; *TfR1*, transferrin receptor 1; *TRPML*, transient receptor potential cation channel, mucolipin subfamily (a transporter); *ZIP14*, a zinc transporter.

early endosomes. The low pH of this environment promotes release of iron and stabilizes the binding of apotransferrin to the receptor. Together, they are recycled back to the cell surface, where apotransferrin dissociates from the receptor at neutral pH.[58] Reduction of ferric iron released in the lumen of the endosome is supported by the ferrireductase STEAP3.[59] Subsequent transport of ferrous iron in erythroid cells is mediated by DMT1.[60] In one model, investigators proposed that reticulocyte endosomes bearing iron-bound transferrin directly deliver cargo to mitochondria for heme biosynthesis, the "kiss-and-run hypothesis".[61]

Other transporters may also provide for iron transfer in the endosomal-lysosomal system in peripheral tissues, including ZIP14[62] and TRPML1.[63] On entering the cell cytoplasm, iron is either rapidly metabolized or stored in ferritin. A small labile iron pool exists and is in the micromolar range in most cell types. In excess, this labile iron can produce reactive oxygen species that can cause cellular damage. For example, loss of iron storage as a result of tissue-specific deletion of ferritin heavy chain in mice leads to liver damage.[64] Defects in transferrin's intracellular trafficking are also known to give rise to anemia in the *hbd* hemoglobin-deficit mouse model as a result of mutation of Sec15l1 of the exocyst complex.[65] Emerging evidence indicates that the iron-regulated myotonic dystrophy kinase–related CDC42-binding kinase α (MRCKα) may be involved in modulating transferrin-mediated iron uptake, possibly by regulating the internalization of transferrin receptors.[66,67] Clearly, the transferrin cycle must be regulated tightly and in a coordinated manner to control the distribution of iron sufficient to meet metabolic demands but limited to avoid toxicity.

Regulation of Iron Status at the Cellular Level

A major mechanism regulating iron status at the cellular level involves posttranscriptional regulation of the transferrin receptor, ferritin, and other key metabolic factors. Uptake of iron into cells is proportional to concentration of the transferrin receptor and reflects the needs of the cell; otherwise, excess iron must be stored in ferritin to prevent oxidative damage, as described earlier. Transcripts for the transferrin receptor and ferritin are known to contain iron response elements (IREs), stem-loop RNA structures that control stability and translation of the mRNAs, respectively.[68,69] IREs are bound by iron regulatory proteins (IRP1 and IRP2) to regulate expression of these and many other important factors in iron transport, utilization, and storage in a coordinated way (**Table 10.3**). Under low-iron conditions, IRPs confer translational control by binding to IREs present in the 5′ ends of iron storage or efflux proteins (ferritin and ferroportin) to reduce protein synthesis. At the same time, IRP binding to 3′ ends of mRNAs for iron uptake factors (transferrin receptor-1 and DMT1) increases message stability to enhance transport into cells. Conversely, under high-iron conditions, IRP–IRE binding is lost, and synthesis of proteins involved in storage or efflux is increased, whereas iron uptake is reduced.

Table 10.3 Factors with Iron Response Elements or Iron Response Element–like Elements Present in mRNA Transcripts

	Factors	Metabolic Role
3′ IRE	Transferrin receptor	Iron uptake
	Divalent metal transporter-1 (SLC11A2)	Iron uptake
	Cell cycle phosphatase (CDC14A)	Cell cycle
	Myotonic dystrophy kinase–related CDC42-binding kinase α (MRCKα)	Actin organization; iron uptake (?)
	Hydroxyacid oxidase (Hao1)	Oxidative injury response (?)
5′ IRE	L-ferritin	Iron storage
	H-ferritin	Iron storage
	Amino levulinate synthase (ALAS2)	Heme biosynthesis (erythroid specific)
	Ferroportin (SLC40A1)	Iron exporter
	Aconitase (ACO2)	Citric acid cycle
	Drosophila succinate dehydrogenase (dSDH)	Citric acid cycle
	Endothelial PAS domain protein-1 (EPAS1) Hypoxia-inducible factor-2α (HIF2α)	Oxygen sensor/transcriptional regulator of iron metabolism
	Alzheimer precursor protein (APP)	Iron-induced injury response; iron chelator and/or ferroxidase
	α-Hemoglobin stabilizing protein (AHSP)	Hemoglobin synthesis
	Glycolate oxidase (GOX)	Oxidoreductase with flavin mononucleotide (FMN) binding site

IRE, iron response element.

Cellular iron also exerts transcriptional effects through antioxidant response elements (AREs) found in genes for ferritin heavy and light chains.[70,71] In addition, transcription of the ferritin genes is responsive to heme (iron in protoporphyrin X), exerted through its interactions with BACH1, a DNA-binding transcriptional repressor.[72] Indirectly, ferritin levels also respond to regulation of protein levels of IRP1 and IRP2 by the iron sensing hemerythrin domain of FBXL5.[73,74] This E3 ligase is destabilized when iron (or oxygen) is low; conversely, FBXL5 is stabilized by high iron and oxygen and is enabled to target the IRPs for degradation. To add an additional layer of complexity, IRP1 is a bifunctional protein that becomes protected from iron-dependent degradation when it binds an iron-sulfur cluster that converts its RNA-binding activity to the enzymatic activity of aconitase (**Figure 10.4**). In cells that are iron deficient, IRP1 fails to bind the iron/sulfur (Fe/S) cluster, promoting its binding to IREs. Although Fe/S cluster assembly and disassembly are known to be linked to mitochondrially derived cluster biogenesis, the process is poorly understood.[75,76] Investigators have suggested that IRP1 levels may correspond to mitochondrial iron metabolism, whereas IRP2 is more directly controlled by cytosolic iron metabolism.[77] In sum, the cellular response to iron status is complex and occurs through transcriptional, posttranscriptional, translational, and posttranslational mechanisms controlled by iron deficiency or surplus.

Mitochondrial Iron Transport and Metabolism

Because mitochondria are the central compartment for heme biosynthesis, Fe/S cluster biogenesis, and energy generation, this organelle is paramount to cellular iron homeostasis. Although a comprehensive picture of mitochondrial iron transport and metabolism has yet to emerge, advances have begun to provide clues about how handling of iron by mitochondria interfaces with the rest of cellular storage and utilization.

One important observation is that loss of IRP1 and IRP2 function in mice causes severe mitochondrial iron deficiency,[78] a finding suggesting that cellular regulation of cytosolic iron is intimately linked with properties of mitochondrial iron supply. Under these conditions, mitoferrin-2, which is a homolog of the erythroid mitochondrion iron importer mitoferrin-1,[79,80] is upregulated in the liver of tissue-specific knockout mice. Mitoferrin-1 is known to transport iron across the mitochondrial membrane in a manner that is stabilized by ABCB10, another erythroid mitochondrial membrane protein.[81] ABCB10 is regulated by heme

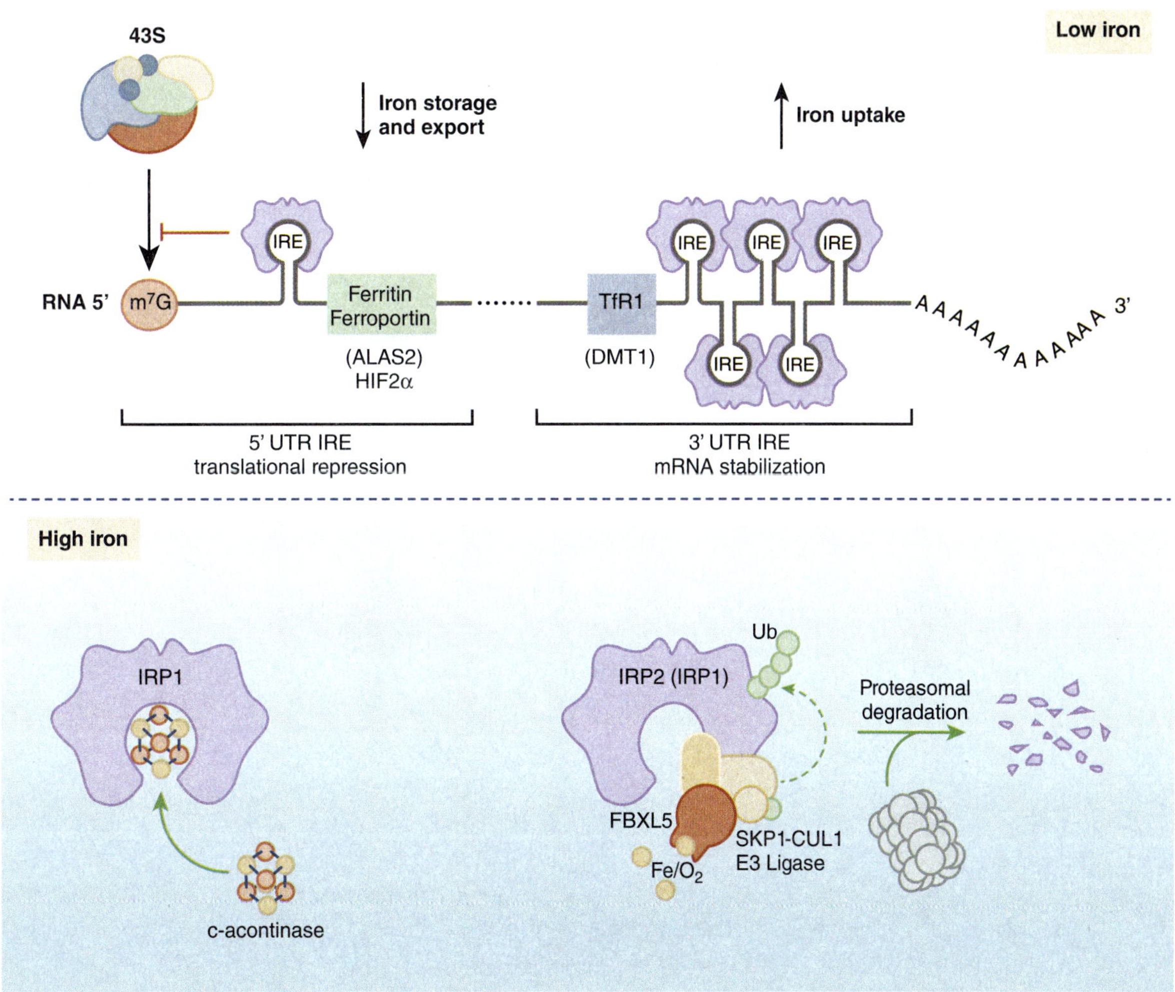

Figure 10.4 Regulation of cellular iron status. *ALAS2*, amino levulinate synthase; *DMT1*, divalent metal transporter-1; *Fe*, iron; *HIF2α*, hypoxia-inducible factor-2α; *IRE*, iron response element; *IRP*, iron regulatory protein; *SKP1-CUL1*, S-phase kinase-associated protein-1–Cullin-1; *TfR1*, transferrin receptor-1; *UTR*, untranslated region.

levels such that when demand for heme synthesis has been met, protein levels decrease to limit mitoferrin-1 activity and adjust the mitochondrial iron supply. Whether a similar regulatory pathway occurs with mitoferrin-2 in nonerythroid cells remains to be elucidated, but this indirect evidence suggests that heme biosynthesis is coordinated with mitochondrial iron levels, and that cytosolic factors are involved in sensing the surplus or deficiency in metabolically active iron.

Another mechanism of cellular regulation of both extramitochondrial and mitochondrial iron supplies is through Fe/S cluster assembly. This pathway also appears to involve close communication between the two compartments, as highlighted by inborn diseases such as Friedreich ataxia, which is associated with loss of the mitochondrial matrix protein frataxin, and sideroblastic anemia, which is associated with genetic defects in the inner mitochondrial membrane protein ABCB7.[82] Deficiency of frataxin disrupts Fe/S cluster assembly and mitochondrial iron handling because of a loss of its iron chaperone activity thought to be involved in regulation of cluster biogenesis. ABCB7, conversely, is thought to be involved in cluster transfer to the cytosol for incorporation into extramitochondrial enzymes.[83] When cluster assembly is disrupted, or if heme biosynthesis is impaired, iron accrues in mitochondria. In contrast, mitochondrial iron insufficiency is a relatively unexplored area. Investigators have shown that 2,5-dihydroxybenzoic acid (2,5-DHBA) binds iron and that depletion of this metabolite leads to high levels of cytosolic iron, increased reactive oxygen species, and iron-deficient mitochondria.[84] 2,5-DHBA associates with a protein known as lipocalin-2, which can deliver iron to cells; conversely, lipocalin-2 that is devoid of iron can liberate cellular iron to reduce the intracellular labile iron pool. Lipocalin-2 and 2,5-DHBA may contribute to mitochondrial iron import, either working together with mitoferrin-1 and mitoferrin-2 or in a parallel pathway.

Systemic Regulation of Iron Homeostasis

The discovery of hepcidin advanced the understanding of how iron metabolism is regulated systemically. The peptide is produced by the liver in response to increased body iron,[85,86] and it binds to the iron exporter ferroportin to induce its internalization and degradation.[87] Decreased ferroportin results in reduced dietary iron absorption and promotes iron retention in macrophages of the reticuloendothelial system. Hepcidin synthesis is also induced during inflammatory responses,[88-90] ultimately producing consequent systemic iron withdrawal and sequestration.[91] The regulation of hepcidin has been an active area of research, particularly with the recent discovery of erythroferrone (ERFE), a hormone produced by erythroblasts, which acts upon hepatocytes to suppress hepcidin production, thereby increasing dietary iron absorption and mobilizing iron from stores.[92,93]

The coordinated regulation of systemic iron metabolism by this hormone signaling mechanism has been largely uncovered by studies of human disease (**Figure 10.5**). Mutations in the

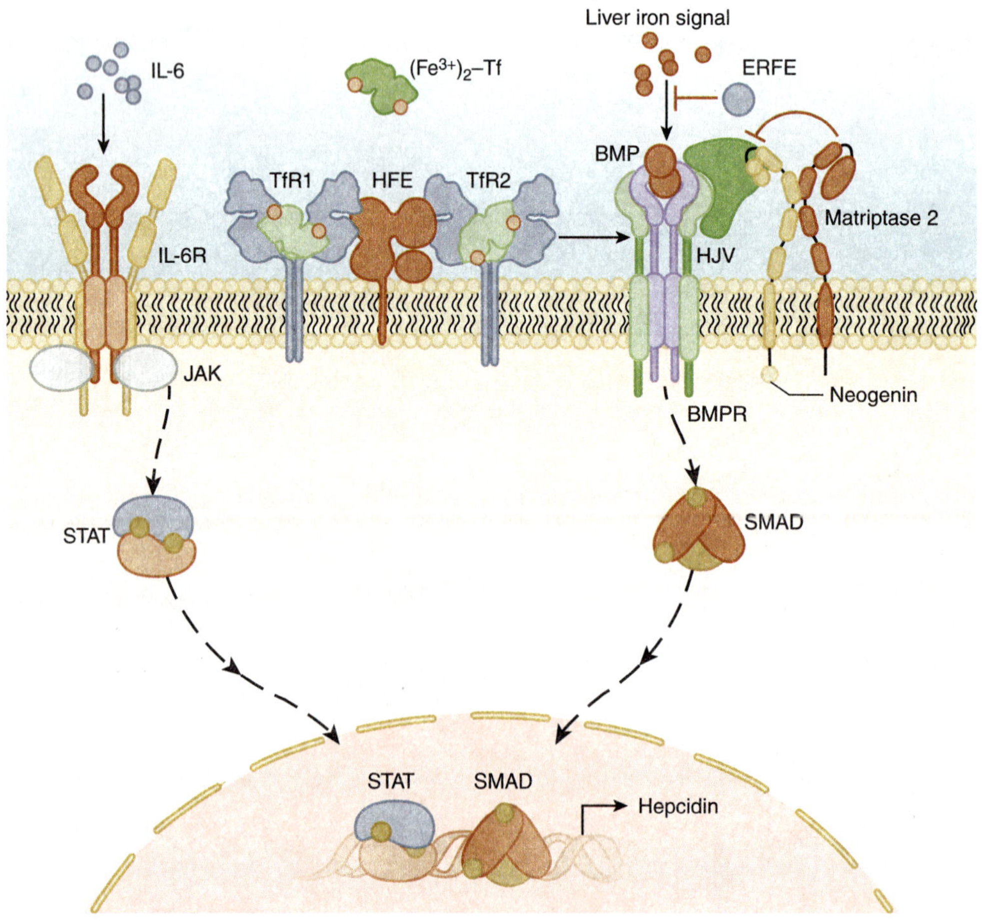

Figure 10.5 Hepcidin regulation. *BMPR*, bone morphogenetic protein receptor complex; *HFE*, the *HFE* gene; *HJV*, hemojuvelin; IL-6R, interleukin-6 receptor; *SMAD*, a signaling pathway; *STAT*, a signaling pathway; *Tf*, transferrin; *TfR*, transferrin receptor.

human hepcidin gene (*HAMP*) itself were identified to promote iron loading, and defects associated with the gene encoding hemojuvelin (HJV) were found to be responsible for upstream regulation of this pathway.[94] HJV associates with bone morphogenetic proteins (BMPs) as a coreceptor to signal through Smad transcription factors that promote gene expression of hepcidin.[95] Recent evidence indicates that the transmembrane protein, neogenin, interacts with HJV on the plasma membrane and may act as a scaffold to induce BMP signaling and hepcidin expression.[96] Another human disease, iron-refractory iron deficiency anemia (IRIDA), is caused by mutations in *TMPRSS6* (matriptase-2),[97,98] a negative regulator of BMP/Smad signaling.[99] Other genes associated with hereditary hemochromatosis, *HFE* and *TFR2*, have also been found to regulate hepcidin levels through an iron-sensing mechanism.[29,100,101] *HFE* associates with transferrin receptor-1, but in the presence of iron-saturated transferrin, it is displaced, associates with transferrin receptor-2, and regulates hepcidin expression by an unknown mechanism.[102,103]

Although these regulatory pathways function to maintain iron balance and protect against overload and deficiency under healthy conditions, the induction of hepcidin during the inflammatory response promotes anemia of inflammation (AI, also referred to as anemia of chronic disease).[91,104,105] Inflammatory control of hepcidin gene expression results from activation of the STAT pathway by interleukin-6 and other cytokines.[88-90] This transcriptional control pathway is the major regulator of the acute phase response; thus, hepcidin can be considered a type II acute phase reactant. Because of its prominent role in hypoferremia associated with infection and chronic inflammation, hepcidin and its regulators represent potential therapeutic targets to improve iron status in disease states.[106]

Whereas increased circulating hepcidin typically reflects iron loading or inflammation to "turn down" intestinal iron absorption, hypoxia is also known to "turn on" iron intake. Hypoxic control of iron absorption appears to operate at the intestinal level because of regulatory control of transporters involved in apical iron uptake, namely DMT1 and DcytB, by the transcription factor hypoxia-inducible factor 2α.[107,108] This same mechanism is also thought to explain the mucosal block that occurs when ingestion of iron reduces subsequent absorption.[109] Thus, although enterocytes receive the systemic signals through hepcidin regulation to downregulate basolateral uptake by ferroportin, local changes in iron status also promote a unique response to environmental conditions that refines entry of the nutrient into the system. Furthermore, during periods of increased erythropoietic demand, or in some conditions of dysfunctional erythropoiesis associated with iron overload, ERFE acts to suppress hepcidin, increasing iron absorption and mobilization, making it available for heme and hemoglobin synthesis.[92]

Assessment of Iron Status

Blood loss, increased physiologic demands, and limited diet promote iron deficiency and iron deficiency anemia. AI is also prevalent in patients with medical illness. Important symptoms include fatigue resulting from the requirement for iron in oxidative metabolism. Clinical tests involve analysis of several indicators in blood samples (**Table 10.4**), with values reflecting iron depletion or iron-restricted red blood cell production.[110,111] Problems with excessive iron absorption and overload are also common, and diagnosis generally relies on the same blood tests.

Reduced serum ferritin is the key indicator of iron depletion because this storage protein circulates in a manner dependent on iron availability. Iron deficiency is typically characterized by ferritin lower than 30 ng/mL. In the presence of inflammation, ferritin concentration may be normal, or even high, because ferritin is an acute phase reactant protein. In this setting, acute phase proteins (e.g., C-reactive protein (CRP) and alpha-1-acid glycoprotein (AGP)) should be assessed.[112] Iron deficiency is also reflected by low transferrin saturation, elevation in a truncated fragment of transferrin receptor 1 present in serum and referred to as soluble transferrin receptor (sTfR), and decreased total body iron (TBI), which can be estimated as the log ratio of sTfR to serum ferritin. In iron deficiency without anemia, hemoglobin may be normal because a large amount of iron is stored in the body. However, as iron stores are depleted, iron deficiency anemia may develop due to iron-restricted red blood cell production. In addition to the markers of iron depletion described above, the distinguishing characteristic of frank iron deficiency anemia is low hemoglobin concentration (male, <13 g/dL; and female, <12 g/dL). Inflammation should also be absent to rule out AI (see later). Red blood cell distribution width (RDW), mean corpuscular hemoglobin (MCH), and mean corpuscular volume (MCV) can also be used to diagnose anemia. Low MCH and MCV are reflective of microcytic anemia and, thus, can be used as indicators of iron deficiency in red blood cells. Vitamin B_{12} or folate deficiency can also promote anemia with or without iron deficiency, and cell size (microcytosis) is, therefore, important in distinguishing iron deficiency anemia.

Patients with chronic or inflammatory disease are identified with high serum CRP, > 1.0 mg/dL. AI can be diagnosed by low hemoglobin concentration (male, <13 g/dL; female, <12 g/dL) and low transferrin saturation (<20%). Serum ferritin concentrations are normal or increased (>100 ng/mL). For patients with AI and iron deficiency, serum ferritin may fall into the normal range of 30 to 100 ng/mL. To determine the diagnosis of AI with iron deficiency, sTfR is useful. sTfR is proportional to the amount of surface receptors and, therefore, the number of progenitor erythroid cells.[113] A calculated TBI greater than 2 suggests AI coexisting with iron deficiency, whereas a TBI less than 1 is typical in AI without iron deficiency.[111] Two other useful markers are CHr and hypochromic red cells, which are less than 28 pg and less than 5% in iron deficiency, respectively. Investigators have argued that these values are the most direct indicators of functional iron deficiency whereas iron stores fail to be mobilized from the reticuloendothelial system to bone marrow in AI.[114]

Hemochromatosis, or iron overload, results from accrual of toxic levels of iron in liver, heart, and endocrine tissues. Because iron is primarily stored in the liver, the determination of iron content in liver biopsies is a definitive means of diagnosis. Transferrin saturation and serum ferritin are perhaps more useful immediate indices of the disease state, however. Both indicators reflect increased iron stores in the absence of inflammation. Transferrin saturation greater than 45% can identify most cases of *HFE*-associated hemochromatosis.[115] A rise in serum ferritin usually occurs subsequent to increased transferrin saturation on iron loading, and ferritin higher than 1,000 ng/mL is used as an indication for liver biopsy.[116] Detection of NTBI may also be a useful tool in determining iron overload.[117]

Table 10.4 Biomarkers of Iron Status and Anemia

Clinical Test	Interpretation	Comments
Serum ferritin	Reflects body iron stores	Increases with infection/inflammation independent of iron status
Serum iron	Transferrin saturation is calculated by dividing serum iron by TIBC (i.e., transferrin), and corresponds to transport iron	Serum iron and transferrin saturation peak in the morning and are lowest in the evening
TIBC		
Transferrin saturation		
sTfR	Reflects cellular iron demands and erythroid mass; increases with iron deficiency and decreases with sufficient iron	Concentrations increase with increased rates of erythropoeisis
Log ratio of sTfR to serum ferritin	Estimate of total body iron; proportional to body iron stores	
Hemoglobin	Protein in red blood cells that carries oxygen; decreases with anemia	
MCH	Average amount of hemoglobin in red blood cells; low values are indicative of microcytic anemia	
MCV	Average size of red blood cells; low values are indicative of microcytic anemia	
RDW	Measures variability in the volume and size of red blood cells; high variability may indicate anemia	
Hypochromic red blood cells	The proportion of hypochromic red blood cells; early indicator of functional iron deficiency	
CHr	Hemoglobin content of reticulocytes; early indicator of functional iron deficiency	
ZPP	Zinc can be substituted for iron in heme when iron is not available; increases are proportional to the severity of functional iron deficiency.	Lead poisoning can increase ZPP independent of iron deficiency

CHr, reticulocyte hemoglobin; MCH, mean corpuscular hemoglobin; MCV, mean corpuscular volume; RDW, red blood cell distribution width; sTfR, soluble transferrin receptor; TIBC, total iron binding capacity; ZPP, erythrocyte zinc protoporphyrin.

Manifestations of Iron Deficiency and Iron Overload

Iron Deficiency

Iron deficiency affects more than 2 billion people. Because the presence of iron in foods is ubiquitous, the amount of iron in the diet is generally related to energy intake.[118] When iron requirements are greater than energy demands, deficiency occurs and anemia evolves. Iron deficiency anemia is particularly problematic in women and children. Estimates are disturbing: 43% of children less than 5 years of age, 29% of nonpregnant women, and 38% of pregnant women suffer from anemia globally, and iron deficiency is present in half of these individuals.[119]

Mild iron deficiency may be compensated by more efficient oxygen delivery from hemoglobin to tissues, redistribution in blood flow to protect brain and heart, and enhanced cardiac output. However, work performance can become significantly impaired by reduced oxidative energy production as iron deficiency anemia evolves,[120] and acidosis can occur with severe iron deficiency.[121] Another characteristic of anemia is lack of body temperature regulation under cold conditions that is associated with decreased thyroid hormone synthesis resulting from reduced activity of thyroid peroxidase, a heme-dependent enzyme.[122] Restless legs syndrome arises from impaired dopaminergic function and is also linked to iron deficiency.[123] Iron deficiency is thought to impair immune function secondary to the important role of iron in generating free radical response to infection (e.g., myeloperoxidase activity), and low iron status is associated with changes in lymphocyte and neutrophil function.[124] Respiratory infections are observed to be more frequent and to last longer in iron-deficient children.[125] During early childhood, low brain concentrations of iron can also create developmental issues that may not be reversed on iron repletion (see later).[126]

Effects of low iron status during pregnancy and in infants and young children are associated with adverse pregnancy outcomes, including increased maternal morbidity and mortality, preterm birth, low birth weight, and impaired growth and intellectual development.[127] The requirements for iron in brain function and development are poorly understood, but problems with auditory performance identified in infants with iron deficiency anemia point to conditions associated with hypomyelination.[128] Iron is known to play a role in oligodendritic function, as well as myelination.[129] Important aspects of motor and behavioral development also appear to be impaired in persons with low iron status.[129,130] Evidence from animal studies indicates that neurotransmitter function is disrupted by iron deficiency, and dopamine metabolism is specifically affected.[131] It has also been suggested that iron deficiency alters the energy metabolism of the brain.[126] Because the profound influence of iron deficiency on brain development and function does not appear to be reversed by later iron repletion, early life represents a particularly vulnerable stage.[126,131,132] Paradoxically, neonatal iron loading in rodent models also produces learning and motor defects.[133,134] Neuronal impairment is seen in early childhood cases of Friedreich ataxia, neurodegeneration with brain iron accumulation (formerly known as Hallervorden-Spatz syndrome), and other inherited conditions associated with iron loading. Thus, both iron deficiency and iron overload present unique challenges to the developing brain.[135]

Iron Overload

The converse problem of iron overload can be caused by inherited or acquired disease. Hereditary hemochromatosis arises from defective *HFE*, *TfR2*, *HJV*, *HAMP*, and *FPN* genes. Other genetic causes include aceruloplasminemia, hypotransferrinemia, iron-loading anemia (e.g., thalassemia intermedia), Friedreich ataxia, and porphyria cutanea tarda. Mutations in DMT1 and ferritin heavy chain genes also promote iron loading. However, more than 80% of patients with hereditary hemochromatosis who are of northern European descent are homozygous for the *C282Y HFE* gene and, although many carriers do not develop conditions related to iron loading, 1 in 200 is homozygous for this allele and may be considered at risk.[136] Causes of acquired iron overload are also common and include transfusion-dependent anemias (hemolytic, sideroblastic, and thalassemic anemia), hemodialysis, dietary or parenteral loading, and chronic liver diseases (alcoholic, viral, or metabolic disease). The toxicity of iron associated with oxidative stress makes the cardiac and endocrine systems particularly sensitive in the early stages of iron loading because these tissues have more mitochondria and less antioxidant capacity. As a result, heart failure and endocrinopathies (e.g., diabetes) often occur.[17] The liver is the primary storage site for iron and, as such, has a large capacity for this function. More severe iron loading results in hepatotoxicity, however, with fibrosis, cirrhosis, and hepatocellular carcinoma. Iron depletion therapies, including phlebotomy and chelation, are commonly employed to improve survival.

Strategies to Prevent or Correct Iron Deficiency

The frequency, severity, and global nature of the adverse effects of nutritional deficiency in iron have led to great efforts in correcting the problem. Iron fortification to prevent nutritional iron deficiency, either through the open market or government-regulated programs, has been successful in high income countries, but it has not proven as effective in low- and middle-income countries where iron deficiency is more prevalent.[137] This may be due to an individual basis, dietary recommendations include increasing the intake of foods with high iron bioavailability (e.g., meat sources), consuming foods with the ability to enhance iron absorption (e.g., vitamin C–rich sources), and avoidance of dietary inhibitors of iron absorption with meals (e.g., coffee and tea). Iron supplements may also be used to satisfy greater needs. Pregnant women, preterm or low birth weight infants, young children, and women in their reproductive years are likely to benefit from iron supplementation. People with kidney failure who are undergoing dialysis and individuals with inflammatory bowel diseases who have difficulty with iron absorption may also require iron supplements. The anemia resulting from hookworm infection and its associated blood loss is also ameliorated by iron supplementation. The use of supplements may be contraindicated in AI, however.[106]

Iron fortification policies have been criticized because they may pose problems for people with iron overload or susceptibility to genetic or acquired iron loading. Similarly, the benefits of routine iron supplementation in pregnant women are controversial, given that high iron concentration can promote oxidative stress and complications such as gestational diabetes.[138,139] As a result, weekly supplementation (as opposed to daily) and strategies focused on prevention of iron deficiency (rather than treatment of anemia) have been proposed.[138,140]

Fortification with multiple micronutrient powders (MNPs) or supplementation with multiple micronutrients may be an effective approach to prevent or treat micronutrient deficiency in areas where deficiencies are common. Folic acid is commonly provided with iron supplements/MNPs in pregnant women and children and is justified not only because of the possibility that anemia stems from folic acid deficiency but also because folate reduces the risk of neural tube defects. Nonetheless, vitamin A deficiency affects iron mobilization and often coexists with iron deficiency such that providing iron alone is not as effective as supplementation with combination of these micronutrients.[139] Riboflavin has similar effects. Similarly, iron deficiency can disrupt vitamin A metabolism,[141] and it can limit the effectiveness of iodine supplementation.[122] Thus, micronutrient supplements must be used judiciously to address metabolic problems correctly.

Future Prospects

Despite significant advances in the understanding of iron biology, billions of people continue to suffer from poor iron status and the resulting physiologic sequelae. Conversely, iron overload disorders, including hemochromatosis, remain a significant health concern. As such, efforts to understand the myriad of factors affecting iron balance, along with the development of safe and effective interventions to prevent and treat both poor iron status and overload should be a focus of future research. Advanced noninvasive and continuous monitoring devices for hemoglobin and other biomarkers of anemia and iron status may be of use in some populations,[142] and serum immunoassays for hepcidin may become routinely applied to differentiate between patients with AI and iron deficiency and patients with AI who do not have iron deficiency.[143,144] Magnetic resonance imaging holds great promise to determine iron content of liver in patients with iron-loading disorders,[145] as do contemporary methods of assessing NTBI.[117] Hepcidin-replacement therapies are under development,[17] and the role of ERFE in iron metabolism is an active area of research. The development of contemporary strategies to prevent and treat poor iron status, including the use of MNPs and alternative forms of iron supplementation, will be critical, as will advanced understanding of the coordinated regulation of systemic iron metabolism—which will form the basis for the discovery of potential treatment targets.

Acknowledgments

This chapter is dedicated to the memory of Dr. Marianne Wessling-Resnick. Much of the content of this chapter is her original work.

Disclaimer: The opinions or assertions contained herein are the private views of the authors and are not to be construed as official or reflecting the views of the Army or the Department of Defense.

References

1. Guggenheim KY. J Nutr 1995;125:1822–5.
2. Andrews NC. Blood 2008;112:219–30.
3. Theil EC. Annu Rev Nutr 2004;24:327–43.
4. Hershko C, Graham G, Bates GW et al. Br J Haematol 1978; 40:255–63.
5. Brissot P, Ropert M, Le Lan C et al. Biochim Biophys Acta 2012;1820:403–10.
6. Dowdle WE, Nyfeler B, Nagel J et al. Nat Cell Biol 2014;16: 1069–79.
7. Mancias JD, Wang X, Gygi SP et al. Nature 2014;509:105–9.
8. Hunt JR, Roughead ZK. Am J Clin Nutr 2000;71:94–102.
9. Hallberg L, Bjørn-Rasmussen E, Howard L et al. Scand J Gastroenterol 1979;14:769–79.
10. Hurrell RF, Reddy MB, Juillerat M et al. J Nutr 2006;136: 2808–12.
11. Hunt JR, Gallagher SK, Johnson LK. Am J Clin Nutr 1994;59: 1381–5.
12. Siegenberg D, Baynes RD, Bothwell TH et al. Am J Clin Nutr 1991;53:537–41.
13. Hallberg L, Brune M, Erlandsson M et al. Am J Clin Nutr 1991;53:112–9.
14. Miret S, Simpson RJ, McKie AT. Annu Rev Nutr 2003;23: 283–301.
15. Goyer RA. Environ Health Perspect 1993;100:177–87.
16. Kwong WT, Friello P, Semba RD. Sci Total Environ 2004;330: 21–37.
17. Pietrangelo A. Gastroenterology 2010;139:393–408, 408.e1-2.
18. Roughead ZK, Hunt JR. Am J Clin Nutr 2000;72:982–9.
19. Shayeghi M, Latunde-Dada GO, Oakhill JS et al. Cell 2005;122:789–801.
20. Qiu A, Jansen M, Sakaris A et al. Cell 2006;127:917–28.
21. Rajagopal A, Rao AU, Amigo J et al. Nature 2008;453:1127–31.
22. White C, Yuan X, Schmidt PJ et al. Cell Metab 2013;17:261–70.
23. Raffin SB, Woo CH, Roost KT et al. J Clin Invest 1974;54:1344–52.
24. Quigley JG, Yang Z, Worthington MT et al. Cell 2004;118: 757–66.
25. Krishnamurthy P, Ross DD, Nakanishi T et al. J Biol Chem 2004;279:24218–25.
26. McKie AT, Barrow D, Latunde-Dada GO et al. Science 2001;291:1755–9.
27. Gunshin H, Starr CN, Direnzo C et al. Blood 2005;106: 2879–83.
28. Latunde-Dada GO, Van der Westhuizen J, Vulpe CD et al. Blood Cells Mol Dis 2002;29:356–60.
29. Muckenthaler M, Roy CN, Custodio AO, et al. Nat Genet 2003;34:102–7.
30. Nelson JE, Mugford VR, Kilcourse et al. Am J Physiol Gastrointest Liver Physiol 2010;298:G57–62.
31. Constantine CC, Anderson GJ, Vulpe CD et al. Br J Haematol 2009;147:140–9.
32. Fleming MD, Trenor CC, Su MA et al. Nat Genet 1997;16: 383–6.
33. Gunshin H, Mackenzie B, Berger UV et al. Nature 1997;388: 482–8.
34. Gunshin H, Fujiwara Y, Custodio AO et al. J Clin Invest 2005;115:1258–66.
35. Hubert N, Hentze MW. Proc Natl Acad Sci U S A 2002;99: 12345–50.
36. Lee PL, Gelbart T, West C et al. Blood Cells Mol Dis 1998;24: 199–215.
37. Lam-Yuk-Tseung S, Gros P. Biochemistry 2006;45:2294–301.
38. Ma Y, Specian RD, Yeh K-Y et al. Am J Physiol Gastrointest Liver Physiol 2002;283:G965-974.
39. Mims MP, Guan Y, Pospisilova D et al. Blood 2005;105:1337–42.
40. Shi H, Bencze KZ, Stemmler TL et al. Science 2008;320:1207–10.
41. Philpott CC, Ryu M-S, Frey A et al. J Biol Chem 2017;292: 12764–71.
42. Edwards JA, Hoke JE, Mattioli M et al. J Lab Clin Med 1977; 90:68–76.
43. Vulpe CD, Kuo YM, Murphy TL et al. Nat Genet 1999;21:195–9.
44. Vanoaica L, Darshan D, Richman L et al. Cell Metab 2010;12: 273–82.
45. Yanatori I, Richardson DR, Toyokuni S et al. J Biol Chem 2017; 292:13205–29.
46. Yanatori I, Yasui Y, Tabuchi M et al. Biochem J; 2014;462: 25–37.
47. Yanatori I, Richardson DR, Imada K et al. J Biol Chem 2016; 291:17303–18.
48. Alvarez-Hernandez X, Smith M, Glass J. Blood 2000;95:721–3.
49. Moriya M, Linder MC. Am J Physiol Gastrointest Liver Physiol 2006;290:G301–309.
50. Donovan A, Lima CA, Pinkus JL et al. Cell Metab 2005;1: 191–200.
51. Deshpande CN, Ruwe TA, Shawki A et al. Nat Commun 2018;9:3075.
52. Cherukuri S, Potla R, Sarkar J et al. Cell Metab 2005;2:309–19.
53. Testa U, Pelosi E, Peschle C. Crit Rev Oncog 1993;4:241–76.
54. Enns CA, Rutledge EA, Williams AM. The transferrin receptor. In: Biomembranes: a multi-volume treatise. p. 255–87, 1996.
55. Kawabata H, Yang R, Hirama T et al. J Biol Chem 1999;274: 20826–32.
56. Robb A, Wessling-Resnick M. Blood 2004;104:4294–9.
57. Johnson MB, Enns CA. Blood 2004;104:4287–93.
58. Richardson DR, Ponka P. Biochim Biophys Acta 1997;1331: 1–40.
59. Ohgami RS, Campagna DR, Greer EL et al. Nat Genet 2005;37:1264–9.
60. Fleming MD, Romano MA, Su MA et al. Proc Natl Acad Sci U S A 1998;95:1148–53.
61. Zhang A-S, Sheftel AD, Ponka P. Blood 2005;105:368–75.
62. Zhao N, Gao J, Enns CA et al. J Biol Chem 2010;285:32141–50.
63. Dong XP, Cheng X, Mills E et al. Nature 2008;455:992–6.
64. Darshan D, Vanoaica L, Richman L et al. Hepatology 2009;50: 852–60.
65. Lim JE, Jin O, Bennett C et al. Nat Genet 2005;37:1270–3.
66. Cmejla R, Ptackova P, Petrak J et al. Biochem Biophys Res Commun 2010;395:163–7.
67. Richard C, Viret S, Cantero Aguilar L et al. Am J Hematol; 2021;96:480–92.
68. Wallander ML, Leibold EA, Eisenstein RS. Biochim Biophys Acta 2006;1763:668–89.
69. Theil EC, Goss DJ. Chem Rev 2009;109:4568–79.

70. Wasserman WW, Fahl WE. Proc Natl Acad Sci U S A 1997;94:5361–6.
71. Torti FM, Torti SV. Blood 2002;99:3505–16.
72. Hintze KJ, Katoh Y, Igarashi K et al. J Biol Chem 2007;282: 34365–71.
73. Vashisht AA, Zumbrennen KB, Huang X et al. Science 2009;326:718–21.
74. Salahudeen AA, Thompson JW, Ruiz JC et al. Science 2009;326:722–6.
75. Ye H, Rouault TA. Biochemistry 2010;49:4945–56.
76. Sheftel A, Stehling O, Lill R. Trends Endocrinol Metab 2010;21:302–14.
77. Hentze MW, Muckenthaler MU, Cell 2010;142:24–38.
78. Galy B, Ferring-Appel D, Sauer SW et al. Cell Metab 2010;12:194–201.
79. Shaw GC, Cope JJ, Li L, et al. Nature 2006;440:96–100.
80. Paradkar PN, Zumbrennen KB, Paw BH et al. Mol Cell Biol 2009;29:1007–16.
81. Chen W, Paradkar PN, Li L et al. Proc Natl Acad Sci U S A 2009;106:16263–8.
82. Richardson DR, Lane DJR, Becker EM et al. Proc Natl Acad Sci U S A 2010;107:10775–82.
83. Sheftel AD, Lill R. Ann Med 2009;41:82–99.
84. Devireddy LR, Hart DO, Goetz DH et al. Cell 2010;141: 1006–17.
85. Nicolas G, Bennoun M, Porteu A et al. Proc Natl Acad Sci U S A 2002;99:4596–601.
86. Nicolas G, Viatte L, Bennoun M et al. Blood Cells Mol Dis 2002;29:327–35.
87. Nemeth E, Tuttle MS, Powelson J et al. Science 2004;306: 2090–3.
88. Verga Falzacappa MV, Vujic Spasic M, Kessler R et al. Blood 2007;109:353–8.
89. Wrighting DM, Andrews NC. Blood 2006;108:3204–9.
90. Lee P, Peng H, Gelbart T et al. Proc Natl Acad Sci U S A 2005;102:1906–10.
91. Nemeth E, Rivera S, Gabayan V et al. J Clin Invest 2004;113: 1271–6.
92. Coffey R, Ganz T. Hemasphere 2018;2:e35.
93. Kautz L, Jung G, Valore EV et al. Nat Genet 2014;46:678–84.
94. Lee PL, Beutler E. Annu Rev Pathol 2009;4:489–515.
95. Babitt JL, Huang FW, Wrighting DM et al. Nat Genet 2006;38:531–9.
96. Enns CA, Jue S, Zhang A-S. Blood 2021;138:486–99.
97. Du X, She E, Gelbart T et al. Science 2008;320:1088–92.
98. Finberg KE, Heeney MM, Campagna DR et al. Nat Genet 2008;40:569–71.
99. Silvestri L, Pagani A, Nai A et al. Cell Metab 2008;8:502–11.
100. van Dijk BAC, Laarakkers CMM, Klaver SM et al. Br J Haematol 2008;142:979–85.
101. Nemeth E, Roetto A, Garozzo G et al. Blood 2005;105:1803–6.
102. Schmidt PJ, Toran PT, Giannetti AM et al. Cell Metab 2008;7:205–14.
103. Gao J, Chen J, Kramer M et al. Cell Metab 2009;9:217–27.
104. Nemeth E, Valore EV, Territo M et al. Blood 2003;101:2461–3.
105. Nicolas G, Chauvet C, Viatte L et al. J Clin Invest 2002;110: 1037–44.
106. Wessling-Resnick M. Annu Rev Nutr 2010;30:105–22.
107. Shah YM, Matsubara T, Ito S et al. Cell Metab 2009;9:152–64.
108. Mastrogiannaki M, Matak P, Keith B et al. J Clin Invest 2009;119:1159–66.
109. Hahn PF, Bale WF, Ross JF et al. J Exp Med 1943;78:169–88.
110. Cook JD. Best Pract Res Clin Haematol 2005;18:319–32.
111. Weiss G, Goodnough LT. N Engl J Med 2005;352:1011–23.
112. Suchdev PS, Williams AM, Mei Z et al. Am J Clin Nutr 2017;106:1626S–1633S.
113. Cook JD, Flowers CH, Skikne BS. Blood 2003;101:3359–64.
114. Punnonen K, Irjala K, Rajamäki A. Blood 1997;89:1052–7.
115. McLaren CE, McLachlan GJ, Halliday JW et al. Gastroenterology 1998;114:543–9.
116. Clark P, Britton LJ, Powell LW. Clin Biochem Rev 2010;31:3–8.
117. Angoro B, Motshakeri M, Hemmaway C et al. Clin Chim Acta; 2022;531:157–67.
118. Zimmermann MB, Hurrell RF. Lancet 2007;370:511–20.
119. Stevens GA, Finucane MM, De-Regil LM et al. Lancet Glob Health 2013;1:e16–25.
120. Gardner GW, Edgerton VR, Barnard RJ et al. Am J Clin Nutr 1975;28:982–8.
121. Finch CA, Gollnick PD, Hlastala MP. J Clin Invest 1979;64: 129–37.
122. Zimmermann MB. Annu Rev Nutr 2006;26:367–89.
123. Salas RE, Gamaldo CE, Allen RP. Curr Opin Neurol 2010;23: 401–6.
124. Bhaskaram P. Br J Nutr 2001;85:S75–80.
125. de Silva A, Atukorala S, Weerasinghe I et al. Am J Clin Nutr 2003;77:234–41.
126. Beard J. J Nutr 2003;133:1468S–72S.
127. Sangkhae V, Fisher AL, Chua KJ et al. Blood 2020;136:2206–16.
128. Roncagliolo M, Garrido M, Walter T et al. Am J Clin Nutr 1998;68:683–90.
129. Lukowski AF, Koss M, Burden MJ et al. Nutr Neurosci 2010;13:54–70.
130. Shafir T, Angulo-Barroso R, Jing Y et al. Early Hum Dev 2008;84:479–85.
131. Beard JL, Connor JR. Annu Rev Nutr 2003;23:41–58.
132. Lozoff B, Beard J, Connor J et al. Nutr Rev 2006;64:S34-43; discussion S72–91.
133. Perez VP, de Lima MNM, da Silva RS et al. Curr Neurovasc Res 2010;7:15–22.
134. Fredriksson A, Archer T. J Neural Transm (Vienna) 2007;114:195–203.
135. Rao R, Georgieff MK. Semin Fetal Neonatal Med 2007;12: 54–63.
136. Allen KJ, Gurrin LC, Constantine CC et al. N Engl J Med 2008;358:221–30.
137. Lynch SR. Best Pract Res Clin Haematol 2005;18:333–46.
138. Peña-Rosas JP, Viteri FE. Cochrane Database Syst Rev 2009;CD004736.
139. Allen LH. J Nutr 2002;132:813S–9S.
140. Scholl TO. Am J Clin Nutr 2005;81:1218S–1222S.
141. Hodges RE, Rucker RB, Gardner RH. Ann N Y Acad Sci 1980;355:58–61.
142. Bamberg R. Clin Lab Sci 2008;21:225–31.
143. Ganz T, Olbina G, Girelli D et al. Blood 2008;112:4292–7.
144. Koliaraki V, Marinou M, Vassilakopoulos TP et al. PLoS One 2009;4:e4581.
145. Tziomalos K, Perifanis V. World J Gastroenterol 2010;16: 1587–97.

B. MINERALS

CHAPTER 11

Zinc

Nicola M. Lowe and Andrew G. Hall

CHAPTER OUTLINE

Abbreviations

AD, Alzheimer's disease; **AE**, acrodermatitis enteropathica; **AI**, adequate intake; **Akt**, protein kinase; **BAMD**, age-related macular degeneration; **Å**, angstrom; **CD4**, cluster of differentiation 4; **CD8α**, cluster of differentiation 8 alpha; **CREBP**, cyclic adenosine monophosphate response element binding protein; **CRY1**, cryptochrome circadian regulator 1; **DNA**, deoxyribonucleic acid; **EAR**, estimated average requirement; **ERK**, extracellular signal-regulated kinase; **FA**, fatty acid; **FAO**, Food and Agriculture Organization of the United Nations; **HIV**, human immunodeficiency virus; **HRG**, histidine-rich glycoprotein; **IOM**, Institute of Medicine; **kDa**, kilodalton; **Lck**, lymphocyte-specific protein tyrosine kinase; **LMICs**, low- and middle-income countries; **LOAEL**, lowest observed adverse effect level; **MAPK**, mitogen activated protein kinase; **MTF1**, metal sensing transcription factor; **NAD^+/NADH**, nicotinamide adenine dinucleotide +/− hydrogen; **NAM**, National Academy of Medicine; **NF-κB**, nuclear factor kappa light chain enhancer of activated B cells; **NFAT**, nuclear factor of activated T cells; **NO**, nitric oxide; **NOAEL**, no observed adverse effects level; **PDB**, protein data bank; **PER2**, period circadian regulator 2; **PET**, positron emission tomography; **PI3K**, phosphoinositide 3-kinase; **PTP**, protein tyrosine phosphatase; **RDA**, recommended dietary allowance; **RE**, reticuloendothelial; **ROS**, reactive oxygen species; **SCD**, sickle cell disease; **SLC**, solute carrier; **SNP**, single nucleotide polymorphism; **STAT**, signal transducer and activator of transcription; **T1D**, type 1 diabetes; **T2D**, type 2 diabetes; **TNF-α**, tumor necrosis factor alpha; **UL**, tolerable upper intake level; **WHO**, World Health Organization; **ZIP**, Zrt, Irt-like protein; **ZnT**, zinc transporter family

Historical and Global Context

The essential role of zinc in the biochemistry of plants was established in 1869, and the essentiality in animals first identified in rodents in 1934.[1] Its biochemical role as a cofactor for carbonic anhydrase was discovered in 1940.[2,3] However, it wasn't until the 1960s that the essentiality of zinc in the human diet was clearly demonstrated by Prasad and coworkers, who were exploring the etiology of a syndrome of iron deficiency anemia, stunting and delayed sexual maturation associated with geophagia (eating clay) in adult males living in a rural community in Iran.[4] As iron deficiency had not previously been linked with stunting, Prasad hypothesized that zinc deficiency was the underlying cause. He tested this hypothesis in a landmark trial conducted in a cohort of 40 malnourished males from rural communities in the Nile delta region of Egypt, who displayed similar clinical characteristics to those previously observed in Iran.[5] Although multiple micronutrient deficiencies were present, indices of growth and development responded dramatically to supplementation with zinc alone, providing conclusive evidence of the essentiality of zinc in the human diet.

It is now more than 60 years since these initial discoveries, and current estimates indicate that inadequate dietary zinc intake results in the death of an estimated half a million children under 5 years of age each year, and impacts the health of 17.3% of the world's population.[6] Zinc deficiency has been identified as one of the three most prevalent nutrient deficiencies in low- and middle-income countries (LMICs) (along with iron and vitamin A) and is a target micronutrient for international intervention programs to address the United Nations Sustainable Development Goal number 2: Zero hunger.[7,8] The biochemical basis for the essentiality of zinc rests within the catalytic, structural, intra- and intercellular signaling roles of this micronutrient, underpinned by its unique chemistry.

Chemistry, Biochemical and Physiological Functions

Zinc is a divalent transition metal with no redox activity. In this chapter, the authors use the notation Zn^{2+} in reference to zinc that is not bound to protein, also referred to as free zinc. Although this notation is traditionally used in chemistry to describe free ions, it should be noted that zinc is not known to exist in a truly free ionic form in any kind of steady state at physiologic conditions.[9] When unbound to protein or other low molecular weight zinc ligands in biological systems, aqueous Zn^{2+} binds six water molecules (zinc hexa-aquo complex), although it alternatively may be bound to four or five water molecules. In contrast to other ions, Zn^{2+} bonds with water are partially covalent in nature. Due to the filled d-shell electron configuration of zinc, its binding geometry is flexible and lacks the stereochemical constraints typical of other transition metals. Zinc can, thus, move between ligand geometries with only minor differences in bond energy. Free zinc is a powerful mediator of biological function and will be discussed more in the section on regulatory roles of zinc.

Zinc binds specifically to more proteins than any other metal ion nutrient, an estimated one-tenth of all human proteins.[10] In addition to these, zinc interacts with low molecular weight zinc binding ligands, including peptides and amino acids, nucleic acids, various metabolites or other organic molecules, and inorganic anions, in a specific manner determining a broad array of essential cellular functions.[11] Zinc binding in biological systems may be within one molecule or may be multimeric in nature where multiple zinc binding molecules or protein structural domains bind the same zinc atom.

The binding of zinc in cells is regulated through a combination of buffering and muffling.[12,13] Zinc buffering is analogous to pH buffering, where the concentration of Zn^{2+} is maintained within narrow limits at a steady state due to the affinities of zinc-binding ligands (i.e., proteins or low molecular weight ligands). Zinc muffling refers to the mechanism where a cell responds to transient changes in cytosolic Zn^{2+}. It includes the movement of Zn^{2+} across vesicular or cellular membrane zinc transporters, promoting the rapid return to a steady state. Buffering and muffling allow cells to maintain the cytosolic concentration of free zinc ions in the picomolar range, i.e., several orders of magnitude lower than the absolute concentration of zinc in cells. This regulation of Zn^{2+} prevents zinc from interfering with the binding sites of other divalent metals, such as iron or copper. At the same time, these processes assist in the control of the transient fluctuations in Zn^{2+} that serve in cell signaling.

Zinc complexes with low molecular weight ligands are needed for the transfer or transport of zinc.[9] Although not well understood, the speciation of these low molecular weight zinc complexes may determine how zinc interacts with proteins. For example, zinc transfer is enhanced in the presence of a low molecular weight zinc binding ligand when its affinity for zinc is intermediate of the donor and acceptor.[14] The ligands that bind zinc in intracellular vesicles, where zinc concentration can be the highest, may also be important for the respective functions of these vesicles. Within zinc-rich vesicles, zinc is bound with sulfur, nitrogen, and oxygen in proportions resembling protein zinc binding sites.[15] Although zinc is known to be co-released with other species, e.g., with glutamate in neuronal signaling, with insulin and amylin from pancreatic β-cells, or with proteases secreted during digestion, the specific vesicular zinc binding ligands are not currently well understood.

Zinc-dependent biochemical mechanisms that determine physiologic functions have received extensive study, but clear relationships to phenotype have not been defined. Zinc has a ubiquitous subcellular distribution, which increases the challenge of isolating changes in zinc-dependent function in response to changes in dietary zinc intake. Consequently, zinc contrasts with iron, which exists in defined cellular components and has defined physiologic roles. Three general functional classes—catalytic, structural, and regulatory—comprise zinc's role in biology.[16]

Catalytic

The zinc dependence of carbonic anhydrase, carboxypeptidase, and alcohol dehydrogenase was discovered due to reversible loss of activity in the absence of zinc.[2,17,18] Zinc dependent enzymes are found in all enzyme classes: oxido-reductases, transferases, hydrolases, lyases, isomerases, ligases, and translocases.[19] Recently, translocases have been added as a seventh enzyme classification and includes zinc transporters and other enzymes that catalyze the movement of ions across membranes.[20] Zinc catalytic domains are present in about three fourths of all zinc enzymes, although structural roles of zinc are also frequent, occurring in more than one-third.[21] Zinc catalytic domains typically coordinate zinc between three amino acids and one water molecule.[22] Due to the presence of a water molecule at the catalytic domain, zinc serves as a Lewis acid for catalysis. In contrast, proteins maintain the neutrality of structural zinc through a combination of steric repulsion and charge reduction by cysteine sulfur.[23]

Hydrolases, such as proteinases, nucleases, and lipases, make up most known zinc enzymes. These enzymes use zinc catalytic sites to cleave chemical bonds by hydrolysis. Examples of hydrolases in humans include carboxypeptidase A, phospholipase C, leukotriene A4 hydrolase, TNF-α converting enzyme, and matrix metalloproteinases such as collagenases.[24] In contrast, lyases use zinc to catalyze the cleavage of bonds without hydrolysis by acidifying the substrate. Examples include carbonic anhydrase, glyoxalase, and 6-pyruvoyl-tetrahydropterin synthase. A similar mechanism is used by oxido-reductases, where zinc catalyzes the oxidation of a substrate hydrogen or electron donor from a redox-active cofactor. Examples include alcohol dehydrogenase, which uses NAD^+/NADH as a cofactor; and copper-zinc superoxide dismutase, which uses Cu^{2+}/Cu^+.[25]

How zinc is donated to apometalloenzymes has not been established. Zinc binding is a posttranslational protein modification, which likely requires a metal donor molecule, perhaps coordinated by events in the endoplasmic reticulum or a vesicular compartment. This process may require zinc transporter activity. An example is that ZnT5/ZnT6 and ZnT7 complexes provide Zn^{2+} to activate tissue-nonspecific alkaline phosphatase.[19,26]

Unequivocal evidence of a direct link between signs of zinc deficiency or toxicity and a specific metalloenzyme has not been found in complex organisms. Most likely, an overt physiologic defect would occur only if the zinc-requiring enzyme was rate limiting in a critical biochemical pathway. Some older literature has examples of zinc/enzyme/disease relationships, e.g., alcohol dehydrogenase/liver disease, and RNA polymerases/growth retardation. Such enzyme changes are no longer considered as representing a critical function for zinc. Reports documenting zinc-responsive control of enzymes for intermediary metabolism, perhaps operating through effects on intracellular zinc concentrations, have also been advanced.[27,28]

Structural

The structural function for zinc was first reported in 1985 with the discovery of repeated zinc coordinating motifs in the *xenopus* TFIIIA transcription factor.[29,30] These motifs ("zinc fingers") use cysteine (Cys) and histidine (His) to form a tetrahedral Zn^{2+} coordination complex. Typical zinc finger domains follow the sequence [-Z-X_2-Cys-$X_{2\text{-}4}$-Cys-X_3-Z-X_5-Z-X_2-His-$X_{3\text{-}5}$-His-], where Z represents the amino acids (e.g., tyrosine, phenylalanine, or leucine) forming the hydrophobic core, and X_n indicates other amino acid residues and the number.[31] Zinc finger motifs allow the formation of domains that recognize specific nucleic acid sequences.[32]

Classic examples of zinc finger transcription factors are the retinoic acid and calcitriol nuclear receptors, and MTF1. An example of typical zinc finger structure is illustrated in **Figure 11.1**. Dissociation constants of zinc fingers vary widely (log K_d = 4.6 – 15.0), indicating a wide range of affinities for zinc.[31] Apparent dissociation constants of selected zinc binding ligands are shown in **Table 11.1**. Both nitrosative and oxidative stress readily disrupt zinc finger motifs, causing loss of function.[33] Because zinc fingers exhibit a spectrum of binding

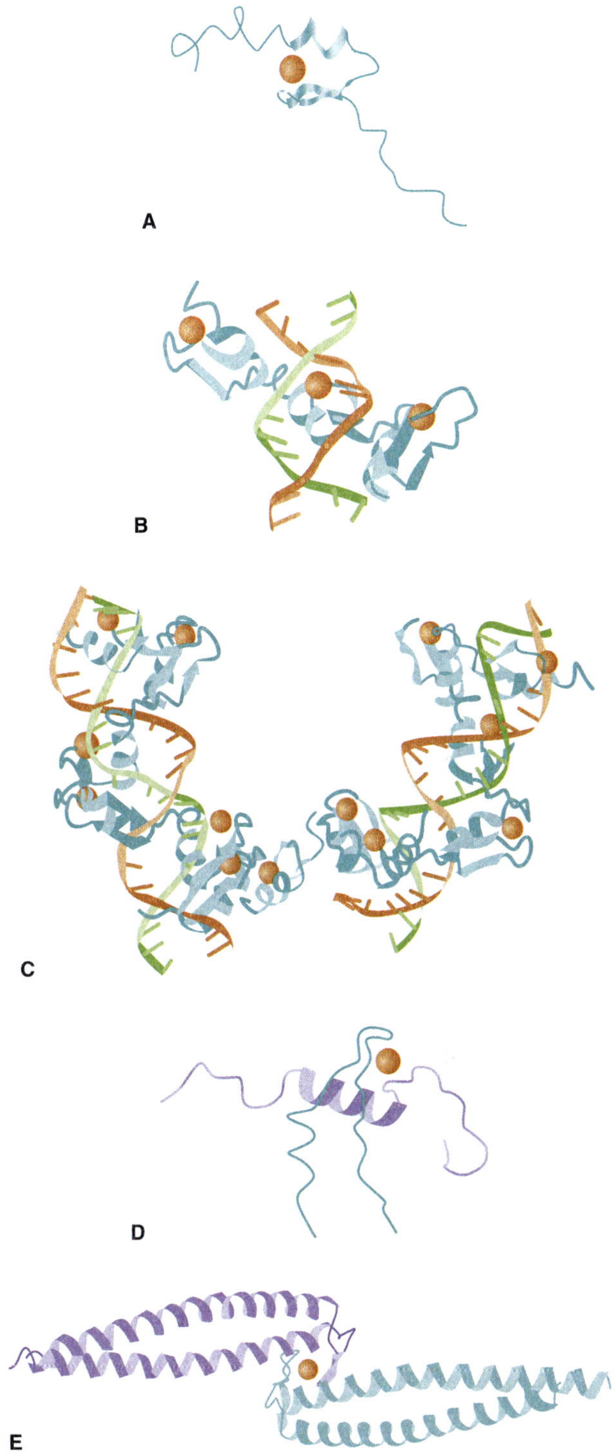

Figure 11.1 Zinc-dependent protein structural domains. Zinc atoms are represented as ●. (**A**) Zinc finger domain of human zinc finger protein 473 (PDB 2YRH), representing a typical His2/Cys2-class zinc finger, the most common DNA-binding domain. (**B**) A complex of three His2/Cys2 zinc fingers from transcription factor Zif268 in complex with DNA (PDB 1AAY).[34] (**C**) DNA sequence recognition by the human meiosis-mediating PR/SET domain 9 protein (PRDM9) requires 13 zinc fingers (PDB 5V3G).[35] (**D**) Zinc clasp structure joins Lck (blue) with T-cell coreceptor CD4 (violet) (PDB 1Q68) or CD8 (not shown).[36] (**E**) Zinc hook motif interlocking the coiled coil regions of two Rad50 proteins (PDB 1L8D).[37]

Data from Wang J, Youkharibache P, Zhang D, et al. (2020) Bioinformatics 36, 131-5 311314.; Madej T, Lanczycki CJ, Zhang D, et al. (2014) Nucleic acids research 42, D297-303.[38,39]

Table 11.1 Apparent Dissociation Constants of Selected Zinc Binding Ligands

Zinc-Binding Ligand	Dissociation Constant(s) (log K_d)	Source
Zinc finger domains	4.6–15.0	31
Metallothionein zinc binding sites	11.4–13.5	76
Albumin primary zinc-binding site	7.5	82
Cysteine (Zn + 2 Cys)	8.5	9
Histidine (Zn + 2 His)	11	9
Picolinic acid (Zn + 2 picolinate)	11.9	9
D-penicillamine (Zn + 2 D-penicillamine)	11.6	9
Phytic acid	10.4	9

affinities, some sites may be particularly facile and potentially influenced by dietary zinc and through zinc transporter activity. Zinc finger proteins have broad biological roles to include transcriptional and translational control, modulation of those processes, and signal transduction.

Two other known zinc structural motifs include the zinc hook and zinc clasp (Figure 11.1). A zinc binding domain on the DNA repair protein, Rad50, was first identified as a half zinc finger by sequence analysis, and later found to form a zinc-hook structure between two Rad50 coiled coils.[37] Zinc coordinates with four sulfurs between the Cys-X-X-Cys sequences of adjacent Rad50 proteins. In the presence of zinc, Rad50 and the DNA double stranded break repair nuclease, Mre11, form a complex allowing them to tether two DNA ends within close proximity (1200 Å) for repair of strand breaks. Two $(Mre11)_2/(Rad50)_2$ heterotetramers further link to form an octameric complex. The Mre11/Rad50 zinc hook complex is highly conserved in nature and variants are found in all life. A third member of the complex, Nbs1, is found in all vertebrates. Although the removal of any of these three proteins is nonviable, alterations of the amino acid sequence in the zinc binding domain of Rad50 merely impair tolerance to radiation.

It is not clear whether functions dependent on the formation of the zinc hook structure are sensitive to dietary zinc; however, there is evidence from clinical studies that DNA repair mechanisms are responsive to changes in dietary zinc intake. For example, zinc supplementation reduced double stranded DNA breaks among Ethiopian women,[40] and, short-term dietary zinc depletion increased double stranded breaks in DNA in apparently healthy adult men, who returned to normal after repletion of dietary zinc.[41]

The zinc clasp structure was discovered after observation of the zinc-dependence of a heterodimeric protein complex between the T-cell CD4 or CD8α coreceptor with the cytosolic tyrosine kinase, $p56^{lck}$ (Lck).[36] Like the zinc hook, the CD4/CD8α Lck zinc clasp forms through coordination of zinc with four cysteine residues: two from a Cys-X-Cys-Pro motif in CD4 or CD8α, and two from a Cys-X-X-Cys motif in Lck. The formation of this complex is necessary for T-cell activation and is highly conserved in mammals.[31] Through its roles in the zinc hook and zinc clasp superstructures, zinc thus has a fundamental role in the formation of multi-protein structures needed in DNA damage repair and immune cell signaling.

Regulatory

Free zinc is a powerful regulator of signaling between and within cells, analogous to the roles of Ca^{2+} and Mg^{2+}, though tightly controlled at a much lower range of concentration, and with a much broader array of potential binding sites.[42] The intracellular and extracellular regulatory targets of Zn^{2+} have not been fully identified and are an active area of research. However, several notable examples exist and are discussed here.

Within the cell, a Zn^{2+} sensing property is conferred to transcription factors that have zinc fingers with lower affinities for Zn^{2+}, allowing the Zn^{2+} concentration-dependent formation of the zinc finger structure.[43] The metal sensing transcription factor, MTF1, has six zinc finger domains, with a range of affinities at the lower end of those generally observed in zinc fingers. Although most known for regulating metallothionein gene transcription, MTF1 can activate the transcription of different sets of genes, depending on the binding of zinc and other metals.[44]

Polymorphisms in the zinc finger domain of the human MTF1 gene[45] suggest the possibility of genetic variation in the response of MTF1-regulated genes to dietary zinc intakes. A homolog transcription factor MTF2, documented to be involved in stem-cell development,[46] may repress genes during normal zinc status and produce activation upon zinc depletion. Reciprocal expression of zinc-responsive genes, including zinc transporters that maintain zinc homeostasis, may be regulated by these transcription factors.

Free zinc is also a potent inhibitor of phosphatases. For example, protein tyrosine phosphatases (PTP) are inhibited by Zn^{2+} in the nanomolar to picomolar range.[47] This inhibition is an important component of insulin signaling. In the absence of Zn^{2+}, PTP-1B dephosphorylates the β subunit of the insulin receptor and, thus, reduces the initiation of the intracellular signaling cascade in response to insulin binding. However, with increased intracellular Zn^{2+}, insulin sensitivity increases[48] (see Diabetes section).

Zn^{2+} further serves as a signal in the activation of ERK kinases necessary for cellular migration and synaptic plasticity of neurons.[49] Such control of phosphorylation/dephosphorylation could explain many of the effects attributed to zinc on activity of phosphorylated transcription factors, cell surface receptor binding of growth factors and cytokines, and activity of key phosphorylated substrates within cells.

Zn^{2+} also acts in parallel to Ca^{2+} in cell signaling. Signaling pathways induced by exposure of monocytes to lipopolysaccharide involving MAPK, ERK, and NF-κB are blocked in the absence of Zn^{2+}, preventing the release of proinflammatory cytokines,[50] a process that is partly regulated by the L-type calcium channel.[51] The profound effects of zinc on the immune system may similarly be traceable to effects of zinc availability to indirectly regulate transcription factors such as the STATs, NFAT, and CREBP, as well as the phosphatase, calcineurin and the cytoplasmic protein-tyrosine kinase, Lck.

Another important regulatory role of zinc is in the control of reactive oxygen species (ROS). Through its catalytic role in copper-zinc superoxide dismutase, zinc mediates the redox activity of copper at physiologic pH in the catalytic disproportionation of superoxide into dioxygen and hydrogen peroxide.[25]

When this catalytic role of zinc to remove ROS is overwhelmed, Zn^{2+} takes on a regulatory role by signaling the presence of oxidative stress, through its displacement from cysteine-rich protein-binding sites. The thiol group of cysteine is readily oxidized by ROS. Where a thiol is bound to zinc, its oxidation causes the release of Zn^{2+}. This mechanism is exemplified by metallothionein which, in the presence of ROS, forms intermolecular and intramolecular disulfide bonds, resulting in its polymerization and release of a wave of free zinc.[52,53]

In cardiomyocytes, Zn^{2+} modulates both the frequency and amplitude of Ca^{2+} waves through its action on the ryanodine receptor-mediated Ca^{2+} release,[54] where it is also implicated in protective effects of nitric oxide via protein kinase G.[55] Circulating histidine-rich glycoprotein (HRG) has a Zn^{2+} sensing region with functional significance in the regulation of the healing response to tissue injury.[56]

Zinc is abundant in the central nervous system. Zinc is concentrated in the presynaptic vesicles of glutamatergic neurons, regulated by the zinc transporter ZnT3.[57,58] Zn^{2+} co-released with glutamate at synapses in the hippocampus activates the ZnR zinc sensing receptor, stimulating zinc-dependent signaling needed for contextual discrimination and special memory.[58,59]

A Zn^{2+} sensing mechanism may also regulate circadian rhythms. Formation of a complex between the circadian clock period protein PER2 and cryptochrome CRY1 is dependent on zinc. The zinc binding affinity is consistent with sensitivity to fluctuations in intracellular Zn^{2+}, while disulfide bond formation in cysteine residues at the zinc binding site may confer a competing zinc and redox regulation of circadian rhythms.[60,61]

Dietary zinc deficiency has been shown to alter brain zinc homeostasis.[62-64] Cerebral ischemia leads to Zn^{2+} release, which participates in the activation of downstream signaling cascades (particularly PI3K/Akt), resulting in oxidative and nitrosative stress, followed by necrotic, apoptotic, and autophagic death of neuronal and glial cells.[57,65]

Some regulatory roles of zinc are closely related to structural roles. For example, a hybrid function between structural and regulatory roles is that movement of large quantities of zinc are associated with acid secretion by parietal cells of the stomach, zinc metallo-digestive enzyme secretion by pancreatic acinar cells, insulin secretion by β-cells, and regulation of fibrin structure. In stomach parietal cells, the Zn^{2+}/H^+ antiport stabilizes the secretion of H^+ ions from parietal cells.[66] In the exocrine pancreas, zinc output correlates with digestive enzyme output,[67] and is responsible for both stabilizing and inactivating carboxypeptidase A prior to its release.[68] In the endocrine pancreas, zinc and calcium are required in the assembly and crystallization of insulin hexamers, thus stabilizing insulin storage.[69] Upon insulin release, Zn^{2+} provides a feedback signal for the downregulation of further insulin and glucagon release.[70,71]

Platelets, which can release Zn^{2+} to increase its local concentration, may also regulate the formation of clots based on the action of secreted Zn^{2+} in determining the morphology of fibrin networks.[72] In the absence of Zn^{2+}, fibrin forms dense networks of thin fibers. As Zn^{2+} increases, the fibers become thicker and form looser networks, consistent with observations that clot stiffness decreases and clot permeability increases with increasing Zn^{2+} concentration.[73,74]

Due to the sensitivity of these regulatory processes to small changes in unbound zinc, Zn^{2+} inside and outside of the cell must be tightly controlled. The next section reviews how this is accomplished through a complex network of zinc carrier and transport proteins.

Zinc Carriers and Transporters

The concentration of free zinc is tightly regulated, and intricate control of zinc movement across membranes is maintained by zinc carrier proteins, including metallothionein in the cell and albumin in blood plasma, and by an array of membrane zinc transporters (**Figure 11.2**). Although most zinc transporters appear to have zinc transport as their primary role, they are regulated by complex networks of proteins and signaling processes in the cellular interactome. In contrast, most zinc carriers have several roles in addition to carrying zinc. Determination of the proteins primarily responsible for regulation of zinc concentration and zinc flux throughout the body is thus a challenging area of study.

Within the cell, metallothioneins are important regulators of cytoplasmic Zn^{2+}. Metallothioneins are compact (7 kDa), cysteine-rich proteins present in all life forms. Although they were discovered due to their association with cadmium in equine kidneys,[75] metallothioneins are highly specific to, and primarily bind, zinc. Mammalian metallothioneins bind up to seven zinc ions in tetrahedral coordination, with four zinc ions coordinated by eleven cysteine residues in the high affinity α-domain, and three zinc ions coordinated by nine cysteines in the lower affinity β-domain.[76] Metallothionein zinc-binding sites have a range of affinities, with log K_d from 11.4 to 13.5. Through its high affinity zinc-binding sites, metallothionein participates in zinc buffering by maintaining a low concentration of Zn^{2+} in the cytoplasm.

In normal steady state physiologic conditions, the divalent metal-binding sites on metallothionein are not saturated. However, in the presence of elevated cytoplasmic Zn^{2+}, the zinc-binding sites of metallothionein may become saturated, and in this form, signal events responsive to high Zn^{2+}. Some of metallothionein β-domain zinc-binding sites have affinities low enough to support zinc exchange with other proteins. Human metallothionein 1A is thus capable of donating zinc to the active sites of enzymes including carbonic anhydrase.[76,77]

In addition to metallothionein, several proteins may have auxiliary zinc binding domains with intermediate affinities for zinc that would potentially allow them to serve as temporary zinc carriers or chaperones for the delivery of zinc to active sites.[11,22] For example, in the endoplasmic reticulum, Zn^{2+} is buffered by calreticulin, which has the capacity to bind 14 zinc ions.[78] Upon binding zinc, conformational changes in calreticulin determine its ability to bind unfolded protein and glycoprotein substrates.[79] Regulation of the expression of such multi-role zinc carriers could provide a mechanism for intricately controlling zinc delivery to catalytic or structural zinc-binding sites without variation in free zinc.

In blood plasma, albumin is the primary carrier of zinc, binding approximately 80% of circulating extracellular zinc in healthy individuals.[80] Albumin is also a carrier for a variety of other metabolites (e.g., heme, fatty acids).[81] Albumin has at least one zinc binding site (log K_d = 7.5), although there is evidence for one or more additional, lower-affinity zinc binding sites.[82] Due to the relatively low affinity of albumin zinc binding, plasma Zn^{2+} readily exchanges with albumin. Albumin also appears to be the major carrier for dietary zinc entering portal circulation after transfer across the basolateral surface of enterocytes.[83]

At a typical plasma concentration of 600 μM, albumin has a molar ratio with zinc of 40:1. This confers a high capacity to buffer Zn^{2+} in plasma, a feature important for the release of insulin. Albumin readily scavenges Zn^{2+} from insulin hexamer crystals, facilitating their rapid disassembly to monomeric insulin upon

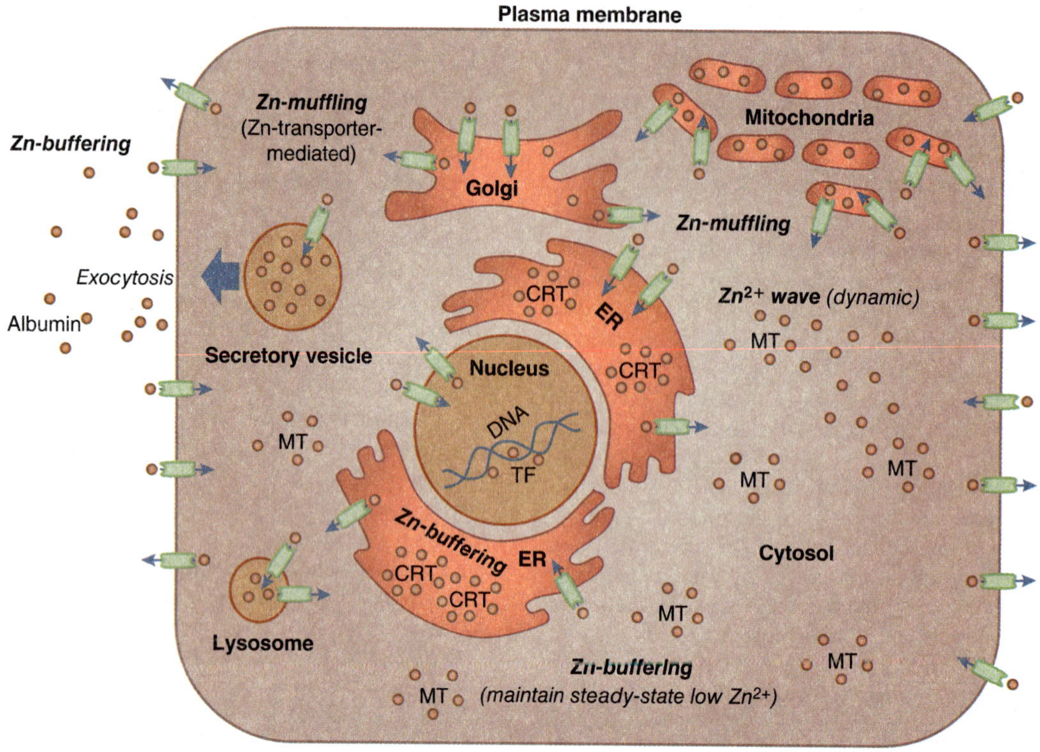

Figure 11.2 Intracellular zinc homeostasis. Zinc is represented as ●. A low steady-state concentration of free zinc (Zn^{2+}) is maintained through a combination of zinc buffering and muffling in response to transient changes in Zn^{2+}, for example, due to regulatory zinc waves, the release of zinc from zinc carriers, or an increased cellular demand for zinc. **Buffering** is accomplished by high affinity zinc-binding sites on zinc carrier proteins, other zinc-binding domains, or low molecular weight zinc-binding ligands. Predominant zinc-binding proteins that contribute to zinc buffering include albumin in blood plasma, metallothionein (MT) in the cytosol, and calreticulin (CRT) in the endoplasmic reticulum (ER). In the nucleus, transcription factor (TF) zinc fingers have a wide range of zinc binding affinities. **Muffling** is accomplished via zinc transport across membranes (plasma membrane, vesicular or organelle membranes, all represented as black lines), facilitating the rapid return to a steady state. Zinc transporters are indicated by ●▭→ (additional detail of the identities, regulation, and functional roles of specific zinc transporters is provided in the section on zinc carriers and transporters).

entering the plasma.[84] During prolonged exposure to elevated blood glucose, albumin becomes glycated, which leads to a conformational change in its zinc binding domain. This may contribute to the aberrations in zinc metabolism observed in patients with diabetes.[85]

In addition to its zinc carrier role, albumin is the primary carrier of circulating free fatty acids and mediates the delivery of fatty acids (FA) to cells following lipolysis. The FA carrier and zinc carrier roles of albumin are interactive. When FA binds albumin, a conformational change in the zinc-binding site causes zinc release.[82] This may be an important feature in the regulation of energy metabolism. Postprandial lipolysis results in FA release from lipoprotein particles, and subsequent FA transfer to cells is mediated by albumin. Zn^{2+} thus released from albumin may also be trafficked into nearby cells. Given the role of cytoplasmic Zn^{2+} waves in PTP-mediated activation of insulin receptor signaling, such localized changes in postprandial extracellular and intracellular Zn^{2+} are of particular interest toward understanding zinc's role in regulating energy metabolism.

Most of the remaining zinc in plasma is a structural component of α-2-macroglobulin, a major plasma protease inhibitor that also carries growth factors and cytokines. NMR studies revealed that α-2-macroglobulin has two high-affinity zinc-binding sites and at least two additional lower-affinity zinc-binding sites.[86] Although early *in vitro* studies did not find that zinc readily exchanged with α-2-macroglobulin, the properties of its lower-affinity zinc-binding sites and potential interaction with its carrier roles have not been studied. Transferrin, an iron carrier protein, also has a significant capacity to bind zinc in plasma, although its contribution to the distribution of zinc in plasma is unclear. Finally, approximately 1% of plasma zinc is bound to low molecular weight ligands, including amino acids.[87]

Zinc transport proteins, driven by transmembrane electrochemical gradients, regulate the uptake and excretion of dietary zinc, the exchange of zinc between tissues and plasma, and transfer of zinc into and out of cells, cellular organelles, and vesicles. Two families of mammalian zinc transporters have been defined, and twenty-four zinc transport proteins identified. They are composed of the Zinc Transporter (ZnT) family (solute carrier; encoded by genes SLC30A1-10) and Zrt, Irt-like Protein (ZIP) family (encoded by genes SLC39A1-14) reviewed in[88,89] ZnT proteins (ZnT1 to ZnT10) direct zinc efflux from the cytoplasm across the cell membrane out of the cell, or into intracellular vesicles. In contrast, the 14 mammalian ZIP transporters (ZIP1 to ZIP14) import zinc into the cytoplasm from extracellular space or from vesicles. Both families exhibit tissue-specific expression. Depending on dietary zinc intake and demands for trafficking cellular zinc, zinc transporters undergo varying rates of synthesis, trafficking, endocytosis, and degradation. The tissue-specific localization of individual zinc transporters, and their interaction with other cellular proteins, is an active area of research.

The zinc transporter primarily responsible for the absorption of dietary zinc into small intestine enterocytes is the apically-localized ZIP4 (SLC39A4). Upon stimulation by zinc, endocytosis moves ZIP4 from the plasma membrane into the cell, and thus reduces the rate of zinc absorption. In higher zinc environments, ubiquitination of human ZIP4 leads to its active proteasomal and lysosomal degradation following internalization.[90,91] In human studies, ileal ZIP4 protein was reduced to undetectable levels following 14 days of a 25 mg zinc supplement taken with food.[92] In murine models, expression of ZIP4 and localization to the plasma membrane may also be increased in response to dietary zinc deficiency.[91,93,94] The ability to readily upregulate zinc absorption in response to a low zinc environment likely differs between mice and humans. Human ZIP4 does not contain the zinc-sensing domain implicated in the high sensitivity of murine ZIP4 to low zinc environments.[95] A mutation of human ZIP4 is responsible for the zinc malabsorption disorder, acrodermatitis enteropathica (AE).[96] AE further leads to zinc-responsive immune dysfunction and cognitive abnormalities.[97]

ZnT1, localized to the basolateral membrane of small intestine enterocytes, regulates zinc efflux to the portal circulation.[98] In contrast to the ZIP family, which has multiple members localized to the plasma membranes of cells, ZnT1 is the only member of its family that is predominantly localized to the plasma membrane. Transcription of ZnT1 is regulated by the same metal sensing transcription factor, MTF1, that regulates metallothionein production.[99] ZnT1 and ZIP4 work in concert to regulate zinc absorption from diet. ZnT1 further works in concert with a complex interactome in cell-specific settings. For example, in cardiomyocytes, ZnT1 inhibits trafficking of the L-type calcium channel to the plasma membrane.[100]

ZnT5 is localized throughout small intestine enterocytes, depending on the splicing variant. Variant A is localized to the Golgi, where it recruits ZnT6 to form a heterodimer needed in the activation of alkaline phosphatase.[26] ZnT5, variant B, is localized throughout the cell as well as on the apical membrane.[101] However, its role in regulating dietary zinc is not clear. ZnT family zinc transporters typically move zinc out of the cytosol, either across the plasma membrane out of the cell, or into intracellular vesicles or organelles. Cell studies demonstrate that apically localized ZnT5 is capable of bidirectional zinc transport. Furthermore, the overexpression of ZnT5 paradoxically increases cellular zinc in cultured enterocytes.[102] Consistent with a role in regulating dietary zinc, total ZnT5 protein is reduced in response to dietary zinc supplementation.[92]

ZIP5 and ZIP14 are both localized to the enterocyte basolateral membrane.[103,104] ZIP5 expression increases in response to high extracellular zinc along the basolateral membrane, suggesting a role in regulating intestinal zinc excretion.[104] ZIP14 is needed for the transport of endogenous zinc from circulation into the enterocytes for the maintenance of tight junctions. Gut permeability increases in the absence of ZIP14.[103]

Other ZnTs are typically localized to vesicular or cellular organelle membranes, where they are responsible for moving cytoplasmic zinc into those compartments. ZnT3 allows the concentration of zinc for intracellular signaling into presynaptic vesicles of neurons, for release into the synaptic cleft.[58] ZnT2 moves zinc into vesicles for secretion into milk by mammary epithelial cells,[105] or into the intestine by Paneth cells.[106] ZnT8 in β-cells concentrates zinc into insulin secretory vesicles and is also necessary for hepatic insulin clearance.[107]

Zinc transporters have been linked to several specific genetic diseases: e.g., ZnT2 (low Zn in breast milk); ZIP4 (AE – Zn malabsorption) and ZIP13 (Ehlers-Danlos syndrome).[96,108-110] Others have undefined associations with human disease e.g., ZnT8 (protection against type 2 diabetes) and ZnT3 (Alzheimer's disease).[111,112] A SNP mutation in ZIP8 is associated with increased risk for Crohn's disease, idiopathic scoliosis, and cardiovascular disease.[113] Several ZIPs are upregulated in cancer cells and are implicated in the progression of pancreatic, colon, breast, and prostate cancers.[114-117] Other diseases caused by zinc transporters may be due to alterations in manganese or iron homeostasis. For example, a mutation in ZIP8 causes a disorder of manganese transport and congenital transferrin deglycosylation.[118] A mutation in ZIP14 causes accumulation of manganese and child-onset parkinsonism.[119]

Bioavailability

The bioavailability of zinc is the fraction of zinc intake that is absorbed, retained, and used for physiologic functions. The fraction of dietary zinc absorbed in the human gastrointestinal tract is typically in the range of 16%–50%[120] and is inversely related to intake:

$$\text{Fractional zinc absorption} = \frac{\text{quantity of zinc absorbed}}{\text{total zinc intake}}$$

The quantity of zinc absorbed depends on the bioavailability of zinc from the diet, and it is important to understand the implications of low and high dietary zinc bioavailability, as this has a significant impact on the total amount of dietary zinc required to meet the needs for normal physiologic processes. For example, dietary zinc recommendations for healthy adult, nonpregnant females, set by the European Food Safety agency (EFSA) range from 7.5–12.7 mg/d, depending on whether the zinc in the diet has a low or high bioavailability.[121] In other words, individuals consuming a diet with low zinc bioavailability need to consume a greater total amount of zinc to meet their physiologic needs because a greater proportion of the zinc consumed will pass through the gastrointestinal tract unabsorbed, compared with those consuming a high zinc bioavailable diet. This will be discussed further in the sections on zinc homeostasis and dietary zinc recommendations.

Factors Affecting Bioavailability

Zinc absorption is most efficient from aqueous sources in the absence of food. With intakes below 5 mg, absorption is close to 100% when consumed in an aqueous solution in the fasting state.[122] More typically, zinc is consumed within a food matrix that impacts its accessibility to intestinal enterocytes for transport into the portal system. A number of dietary components have been implicated as having a positive or negative effect on zinc bioavailability in adults, which are summarized in a systematic review by Bel-Serrat.[123] However it is clear that zinc bioavailability in healthy adults is determined primarily by the quantities of zinc and phytate in the diet.

Phytate (myoinositol hexaphosphate) is present in plant products, especially cereals and legumes. It irreversibly binds zinc in the intestinal lumen, rendering it inaccessible for uptake into enterocytes. Thus, the efficiency of zinc absorption from plant foods is lower than that from animal-source foods.[124] When plant foods are fermented (leavened breads and porridges from fermented cereals), the fermenting organisms produce phytases that break down phytate and release the zinc, making it available for absorption.[125-127] Because phytate is a major inhibitor of zinc absorption, the phytate-zinc molar ratio is used to estimate the likely absorption of zinc from a mixed diet. It can be calculated as follows: (phytate content of food/660)/(zinc content of food/65.4), where 660 and 65.4 represent the molecular or atomic weights of phytate and zinc, respectively.[128] It is generally considered that phytate-zinc molar ratios greater than 15 have relatively poor zinc bioavailability, those between 5 and 15 have medium bioavailability, and those less than 5 have good bioavailability. Most plant foods have phytate-zinc molar ratios above 15 (seeds and nuts, whole grain cereals, beans and lentils, and tubers). Zinc absorption from those plant foods generally is less than 1 mg/100 g of food whereas about 2–2.5 mg zinc is absorbed from 100 g of meat. A mathematical model of zinc absorption, based on human physiology, has been developed to predict the amount of zinc absorbed as a function of dietary zinc and phytate.[129] In LMIC countries, plants are the major source of food, thus biofortification strategies have been employed to increase the zinc content of staple crops. Biofortification is achieved through agricultural techniques (traditional breeding or development of transgenic varieties) and/or agronomic methods (addition of zinc-rich fertilizer to the soil or crop directly).[130] The evidence indicates that it is possible to achieve this without a concurrent increase in the phytate content of the crop, but further research is needed to be able to combine this with low phytate traits without compromising yield and stress tolerance.[123,131]

As calcium has the propensity to form complexes with phytate and zinc that are insoluble, it has been proposed that the phytate/zinc molar ratio should be multiplied by the dietary calcium concentration to improve the prediction of zinc bioavailability.[132] The interaction with calcium is complex; however, and the effect of calcium on the phytate-zinc interaction is not consistent. Thus, the effect of calcium is usually ignored when determining the effect of phytate on zinc absorption. Supplemental or dietary calcium without phytate has little to no effect on zinc absorption at an adequate level of intake.

Much of the research relating to the impact of dietary components on zinc bioavailability has been undertaken in adults. The evidence of a negative effect of phytate in zinc bioavailability in children under 5 years of age is conflicting. A recent systematic review and meta-analysis examined the data from studies conducted in children under 5 years old and found that although the absorption of zinc was reported in some to increase following dephytinization of food products, other studies failed to find a difference in zinc absorption from foods with high vs low phytate contents.[133] Modeling of data from studies conducted in children between 8 and 50 months[134] suggests that dietary phytate has little or no impact on zinc absorption in this age range, which could be explained by more favorable conditions (higher pH) for gastric hydrolysis of phytate in the young children compared with adults.

Nutrient Interactions

Interaction with other divalent cations in the intestinal lumen may also influence zinc bioavailability. The interaction between iron and zinc has been extensively studied in human supplementation trials and it has been hypothesized that there is a competition at a specific transport protein during intestinal absorption. However, the exact mechanism of the interaction is unclear, and the data are conflicting, with the presence or absence of an interaction depending on the relative quantities of iron or zinc consumed and the mode of delivery (in aqueous solution or with food). A review of placebo-controlled, randomized, controlled supplementation trials concluded that zinc supplementation alone did not have a clinically important negative impact on iron status. However, when both zinc and iron are given together, the beneficial effects of iron on hematologic indices of iron status may be attenuated. Iron supplementation did not impact biochemical indices of zinc status in most studies.[135] A systematic review and meta-analysis of the adverse effects of excessive zinc intake in infants and children aged 0–3 years found that zinc interventions with 3 to 20 mg/d had adverse effects on serum/plasma copper, ferritin, hemoglobin, and serum transferrin receptor in children aged 0–3 years. However, the changes observed may not be clinically significant in healthy populations.[136]

The interaction between zinc and copper has been less frequently studied. Shifts in dietary zinc:copper ratios from 2:1 to 15:1 had no effect on copper absorption.[137] However, some studies have shown a negative effect of large-dose zinc supplements (50 mg/d) on indicators of copper status.[138] High levels of tin and cadmium inhibit zinc absorption, but the extent to which lower, physiologic levels affect absorption in humans is unknown.

Food Sources

The quality of dietary zinc sources is determined by the amount and bioavailability of zinc from the foods. Organs and flesh of mammals, fowl, fish, and crustaceans are the richest food sources of zinc and as these foods do not contain phytate, they are particularly good sources of absorbable zinc. Eggs and dairy foods are also free of phytate, but they have lower zinc concentrations than organ or flesh foods. Cereals and legumes contain a modest amount of zinc, but their high phytate content reduces the amount of zinc available. Fruits and vegetables are low in zinc. The zinc content of foods can be enhanced by biofortification (see bioavailability section) or by the addition of zinc (alone or in combination with other micronutrients) postharvest, during food processing. Many countries have mandated food fortification of staple foods, however zinc is not always included, even in countries where zinc deficiency is prevalent.[135] Pooled data from studies designed to evaluate the health benefit of consuming zinc-fortified foods indicate that where populations are at risk of inadequate zinc intake, consuming zinc-fortified products can have a beneficial impact on zinc status and health outcomes.[139]

Metabolism

Body Zinc

Total body zinc in adult humans is between 1.5 g (females) and 2.5 g (males), making it slightly less abundant than iron. The tissue distribution of zinc, however, is ubiquitous, with some tissues, e.g., the prostate, having a particularly high overall concentration. Zinc concentrations in some tissues of an adult male are shown in **Table 11.2**. Eighty-six percent of the total body zinc is in skeletal muscle and bone. Within organs, regional differences in zinc abundance can be striking, e.g., hippocampus of the cerebral hemispheres and β-cells of the pancreas and kidney cortex. These foci of zinc concentration are believed to be functionally relevant. About 95% of body zinc is intracellular.

Table 11.2 Approximate Zinc Content of Major Organs and Tissues in a Normal Adult Man

Tissue	Approximate Zn Concentration		Contribution to Body Zn	
	Wet Weight (μM/g)	(μg/g)	g	(%)
Skeletal muscle	0.78	51	1.53	57 (approx.)
Bone	1.54	100	0.77	29
Skin	0.49	32	0.16	6
Liver	0.89	58	0.13	5
Brain	0.17	11	0.04	1.5
Kidneys	0.85	55	0.02	0.7
Heart	0.35	23	0.01	0.4
Hair	2.30	150	<0.01	0.1 (approx.)
Blood plasma	0.02	1	<0.01	0.1 (approx.)

Data from Mills CF. (1989) *In Zinc in Human Biology* (Mills CF, ed) pp. 371-81, Springer-Verlag, London.[140]

Intestinal Uptake and Absorption

Upon ingestion of food, dietary zinc is freed from the food matrix and hydrolyzed in the acidic gastric environment. When the chyme enters the duodenum, its acidity is reduced from pH 2 to pH 5 by bicarbonate from pancreatic secretions and other sources, reaching pH 6 at the junction of the duodenum and jejunum.[141] In addition to the zinc present in food, endogenous secretions, including those from the pancreas, contribute to luminal zinc abundance.

The nature of absorbed zinc is not clear. As zinc-binding macromolecules of food are liberated and degraded to smaller zinc-binding molecules during digestion, bioavailability is generally improved. Beginning in the duodenum and proximal jejunum, apically localized ZIP4 mediates the uptake of Zn^{2+} into brush border epithelial cells. With increasing pH, some zinc-binding molecules form precipitates and, thus, reduce zinc absorption. A notable example is dietary phytate, which forms an insoluble complex with zinc and calcium (see bioavailability section). However, molecules that bind zinc and remain soluble can also increase zinc absorption.

Soluble protein hydrolysates and animal sources of protein, for example, bind zinc and reduce phytate inhibitory effects in animal models.[142,143] An enhancement of zinc absorption due to dietary protein is also observed in combined data from zinc tracer studies in humans.[144] The addition of milk increases zinc absorption from high phytate meals.[145] Overall, the extent of zinc absorption appears to be a function of the solubility of ingested zinc. Evidence from model systems suggests that zinc is transported as free solute (Zn^{2+}) and/or as a complex that may liberate Zn^{2+} prior to or subsequent to membrane transport. How zinc is liberated from complexes in the neutral pH of the intestinal lumen for transport as a free solute is not known.

ZIP4 has an additional, lower-affinity zinc binding site that increases the efficiency of Zn^{2+} absorption, perhaps facilitating the transfer of zinc to the transport site.[146] Zinc-binding molecules may similarly serve as intermediaries for zinc delivery to ZIP4 or ZnT5, by nature of similar affinities for zinc binding that can facilitate the transfer of Zn^{2+} to the transport protein active site. It is also conceivable that some dietary zinc may be co-absorbed with zinc-binding small molecules, e.g., amino acids or dipeptides via amino acid transporters abundant over the length of the small intestine.

Studies of AE caused by loss of function mutation in the gene encoding ZIP4, provide further insight into the nature of dietary zinc absorption. Clinical signs of zinc deficiency in AE generally appear after weaning, and returning human milk to the diet recovered the absorption of an orally administered zinc tracer.[147] This, and the responsiveness of AE to zinc supplementation, suggest that mechanism(s) for zinc absorption exist that do not depend on functional ZIP4 and may be more active in the presence of a component or components of human milk. Citrate and picolinic acid, both of which are present in milk, were identified as zinc-binding molecules that could be responsible for these observations.[148,149] Zinc deficiency in AE is also prevented by the administration of the zinc-binding antibiotic drugs di-iodohydroxyquinoline and clioquinol, which rescue zinc absorption and synergistically increase the effectiveness of zinc supplements.[150,151]

Zinc is absorbed from the entire intestinal tract.[152] Perfusion studies in humans demonstrate the highest rates of zinc absorption in the proximal jejunum compared with the duodenum.[153] Although 80% of the length of the small intestine was not covered by these measures, a significant zinc absorption rate (~1/3 that of the duodenum, or ~1/5 the proximal jejunum rate) was observed in the most distal region of the ileum, just proximal to the ileocecal valve. If this rate of absorption is typical of the whole ileum, which makes up more than half of the overall length of the small intestine, the ileum would contribute substantially to total zinc absorption. This is consistent with observations that 30% to 60% of total small intestine zinc absorption occurs in the ileum in animal models.[154,155] There is further evidence from rat models that the cecum and colon can contribute minimally to zinc absorption, providing an adaptive mechanism to recover zinc absorption when the small intestine is impaired.[156] However, no human studies have measured zinc absorption in the large bowel.[157]

Zinc absorption is down a concentration gradient from a relatively high luminal concentration (μM range) after a meal. Kinetic measurements also place the K_m (affinity) for the active site of human ZIP4 at 0.9 μM.[158,159] Zinc uptake kinetics by the small intestine show predominantly carrier-mediated saturable transport, although there is also a nonmediated (or passive transport) component.[158,160] The saturable component may represent the sum of zinc transporter activity of enterocytes.

Systemic factors that influence endogenous intestinal zinc secretion or result in poor hydrolysis of luminal constituents (e.g., pancreatic insufficiency; inflammatory bowel diseases) can have substantial impacts on intestinal zinc absorption/retention. For example, zinc deficiency increased more than fivefold in patients who underwent the bariatric procedure, biliopancreatic diversion with duodenal switch.[161]

Zinc Turnover, Transport, and Storage

Plasma zinc comprises only 0.1% of total body zinc and, depending upon species, represents 20%–30% of the zinc in whole blood. Plasma concentration is typically maintained within a range close to 100 μg/dL (15 μM). Serum concentrations are slightly higher, due to release of zinc from platelets during coagulation of blood samples. Severe human zinc depletion causes a fall in plasma zinc within about 2 weeks.[162] With zinc intake between 3 to 10 mg/d, plasma zinc increases slowly until a plateau is reached with intakes between 25–30 mg/d.[163] Plasma zinc may fluctuate by as much as 20% during a day largely due to effects of food ingestion. The highest values of the day occur following an overnight fast.[164]

A total of 70%–80% of circulating zinc is cellular, with the concentration of leukocytes (6 mg Zn/10^6 cells) greater than that of erythrocytes (1 mg Zn/10^6 cells).[165] Erythrocyte zinc is found mostly with carbonic anhydrase (>85%); copper-zinc superoxide dismutase; and various other proteins, including metallothionein.[166] In mice, erythrocyte zinc transporters are regulated in response to dietary zinc deficiency.[167] Some leukocyte genes are sensitive to dietary zinc[168] and may reflect conditions of zinc status of progenitor cells in the bone marrow.

Kinetic data from radioactive and stable zinc isotopic tracers has provided important information on zinc pool turnover in humans. Two metabolic pools (rapid [≈ 12.5 days] and slow [≈ 300 days]) have been identified.[169,170] Kinetically active tissues are liver, pancreas, kidney, and spleen. Slow turnover is found in muscle and red blood cells, followed by bone and the nervous system. Using a kinetic model, cAMP administration to rats appeared to alter the distribution and metabolism of zinc in the thymus, skin, spleen, intestine, and especially, the bone marrow.[171] An exchangeable zinc pool (EZP) has been identified in humans using kinetic analysis and zinc-stable isotopes.[172] The pool represents approximately 10% of whole-body zinc, and fully exchanges with the plasma over a 2-day period. As most zinc in the body is bound to protein, the size of the pool is influenced by the amount of lean-body-mass.[173] A severe dietary zinc restriction causes the size of the pool to decrease by about one-third. This may reflect the redistribution of this more rapidly turning over zinc pool to other tissues.[162] General features of zinc metabolism are illustrated in **Figures 11.3** and **11.4**.

Recycling of zinc through the erythron is analogous to that for iron. Red blood cells contain between 20–40 μg Zn/g hemoglobin[165] and the average circulating hemoglobin content is 750 g in adults. This represents a red blood cell zinc pool of 15–30 mg. The average life span of a red blood cell is 120 days, therefore, turnover for this zinc pool is between 0.12 and 0.25 mg/day (15 to 30 mg per 120 days). A substantial amount of zinc, thus, needs to be available in the body for erythropoiesis.

Zinc does not have a specific "storage site" that is analogous to tissue reserves found with type-I nutrients such as iron or vitamin A. However, free zinc has been observed in vesicles and cellular organelles, including the mitochondria, Golgi apparatus, and endoplasmic reticulum.[174,175] Zinc in cellular organelles or vesicles may serve as a temporary store of cellular zinc, and as a way to protect the cell from the cytotoxicity due to transient increases in cytoplasmic Zn^{2+}.[176] For example, supplemental zinc above the requirement continued to support normal growth in chicks for 8 days after withdrawal.[177] Also, micronutrient supplements containing zinc given to Vietnamese infants either daily or weekly improved growth similarly in both groups.[178]

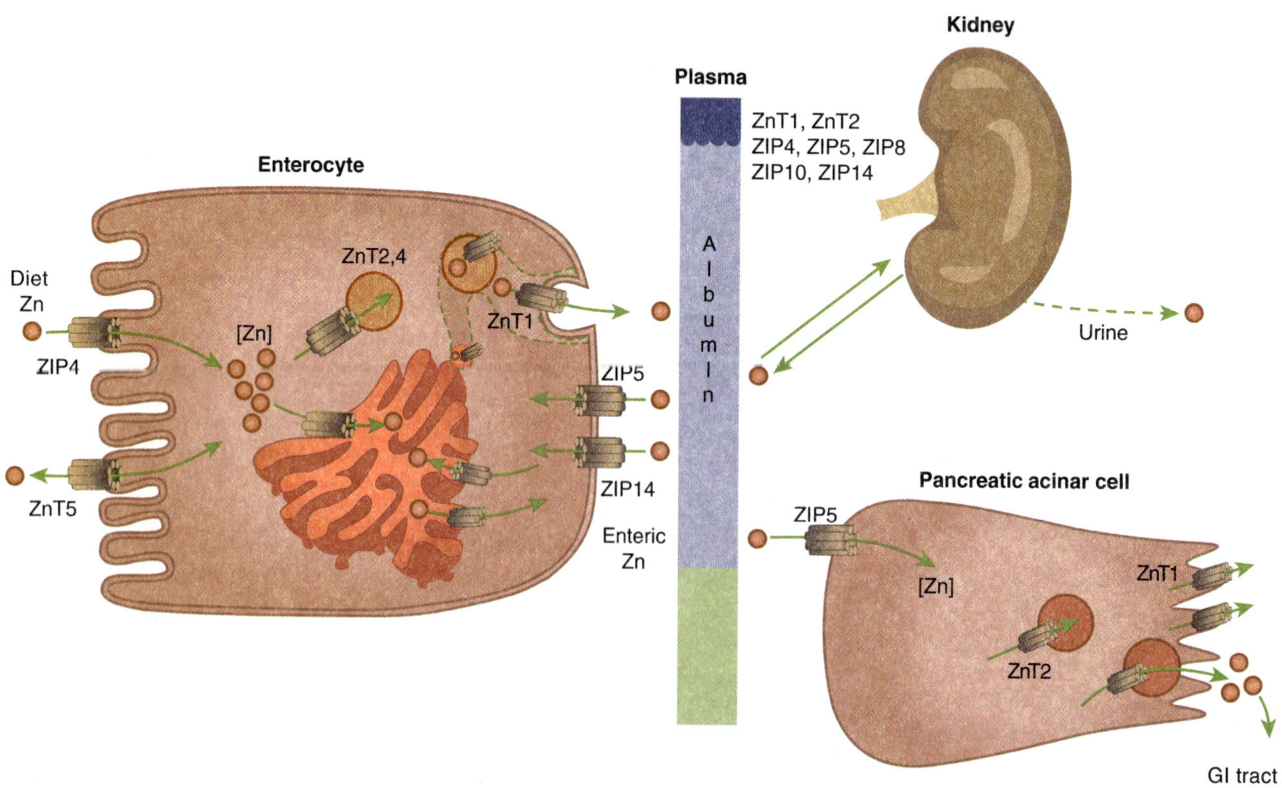

Figure 11.3 Major transporter-mediated zinc pathways for enteric absorption, pancreatic and renal output.

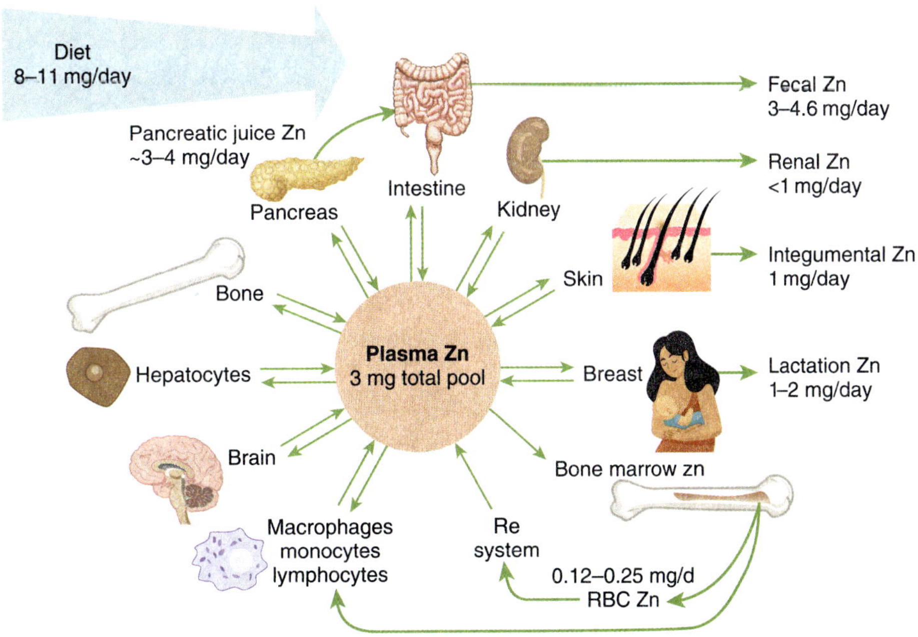

Figure 11.4 Diagrammatic representation of mammalian zinc metabolism. Tissues of high metabolic activity and/or functional significance are shown as deriving zinc from the plasma pool (double arrows). Systems contributing to absorption, recycling, and loss of zinc are shown as unidirectional arrows.

Excretion, Conservation, and Losses

The endogenous secretion of zinc was first proposed following observations that fecal zinc concentration remained high several days following a high zinc meal.[179] Secretion into the gastrointestinal tract is the major route of zinc excretion. This is the combined contribution from pancreatic secretions, sloughing of mucosal cells into the intestinal lumen, and transepithelial flux of intestinal zinc in the serosal to mucosal direction.[140] Digestive zinc follows enterohepatic recirculation,[180] allowing it to be efficiently conserved similarly to the recycling of digestive enzymes. Zinc is also transferred from the peripheral circulation directly to the small intestine,[162] as has been observed in positron emission tomography (PET) imaging following an intravenous zinc tracer.[181]

Pancreatic secretions deliver 6 to 20 g of digestive enzymes and 2.5 liters of fluid per day to the duodenum.[182] Pancreatic juices released in response to secretin administration contain 1.4 mg zinc (1.7 mg zinc, adjusted) per liter,[183] and secretions into the small intestine in response to a breakfast meal contain 2.4 to 4.8 mg of zinc.[184] Zinc concentrated in pancreatic secretions comprises an ill-defined mixture. Zinc is needed for presecretory inhibition of carboxypeptidases.[185] Whether zinc may have a similar role in inhibiting other co-secreted proteases (e.g., trypsin, chymotrypsin, and elastase) is not clear. However, other trypsin or chymotrypsin-like proteases have been found to be inhibited directly by Zn^{2+},[186] or by inhibitory proteins with increased activity in the presence of Zn^{2+}.[187] The pancreas is also sensitive to excessive dietary zinc, which can produce necrosis in chickens.[188] Pancreatic acinar cells produce large amounts of metallothionein, which closely reflects dietary zinc intake and may serve a protective role.[189]

The total amount of zinc secreted into the gastrointestinal tract during the day usually exceeds that consumed in the diet, but much is reabsorbed to maintain zinc balance.[190] Net gastrointestinal zinc excretion is a function of dietary zinc intake. Estimates are as low as <0.5 mg/day in a severe dietary zinc restriction (0.3 mg/day).[162] At typical intakes of 7–15 mg/day, endogenous zinc excretion via the gastrointestinal tract ranges from 3.0–4.6 mg/day in adults,[191] and increases proportionately at higher intakes.

Urinary zinc output is low (<1 mg/day) and is refractory to change over a wide zinc intake range (4–25 mg/day).[191] Starvation, trauma, and other conditions that increase muscle protein catabolism will increase urinary zinc as the load of amino acids filtered by the kidney increases. Some supplements that bind zinc tenaciously (e.g., picolinic acid) may promote zinc loss in the urine.[192] Glucagon has been shown to regulate zinc reabsorption by the renal tubular system.[193] Other losses of zinc include integument (1 mg/day), semen (1 mg/ejaculate), menstruation (0.1–0.5 mg total), and parturition (100 mg/fetus, 100 mg/placenta). Lactation produces losses of 2.2 mg/day at 4 weeks and 0.9 mg/day at 35 weeks.[194]

Dietary and Physiologic Adaptation Mechanisms

Humans experience a circadian variation in plasma zinc[195] and reproducible reduction in plasma zinc concentration postprandially,[196] perhaps related to metal-induced changes in insulin and other hormones. Following a meal, zinc moves rapidly from peripheral circulation into slowly exchanging tissue compartments.[197] Likewise, plasma zinc increases during acute fasting are caused by hormonally-influenced muscle catabolism, with concomitant zinc release.[198] Stress and myocardial infarction reduce plasma zinc in humans.[199] Plasma zinc is transiently reduced following other acute stresses (infection, trauma, surgery),[200] and during recovery from aerobic exercise.[201] The biological significance of hypozincemia associated with stress and the acute phase is not well understood. Underlying mechanisms involve zinc transport into the liver and other organs, perhaps involving cytokine responsive transporters.[202] Plasma free fatty acids are also typically elevated with stress and may displace albumin zinc,[82] triggering its uptake by tissues.

Hypozincemia may be beneficial in decreasing zinc availability to microbial pathogens, providing zinc for protein synthesis and/or maintaining zinc for immune and metabolic signaling pathways. Comparable processes mediated by hepcidin occur in iron metabolism and initiate the anemia of inflammation. Hemodilution, as occurs during pregnancy, oral contraceptive use, and other hormonal treatments, also reduce plasma zinc. Conditions increasing blood cell hemolysis will cause a rise in plasma zinc because intracellular zinc concentrations are higher than plasma zinc.

Homeostatic Regulation

The gastrointestinal tract plays a primary role in maintaining a stable whole body zinc content, or homeostasis. At very low levels of ingested zinc, less than 1 mg/d, nearly all zinc from a low phytate diet is absorbed.[162] Concomitantly, there is a reduction in zinc secretion into the intestinal lumen via pancreatic secretions and from the intestine via transepithelial flux in the serosal to mucosal direction. Mechanisms regulating intestinal zinc secretion are not well understood.

As the amount of ingested zinc increases and more zinc is absorbed, the efficiency of zinc absorption declines, and endogenous losses increase accordingly to maintain zinc balance or homeostasis. In humans, decreases in the fraction of zinc absorbed with increasing zinc intake best fits a saturable response model.[203] These observations are consistent with the molecular mechanism of human ZIP4 regulation which, in the presence of high zinc, undergoes endocytosis and degradation, resulting in reduced cellular uptake.[95]

Between 1 to about 9 mg per day of dietary zinc intake, the efficiency of zinc absorption declines from nearly 100% to ~40%. With intakes above ~9 mg per day (close to the estimated average requirement) the FZA declines further, which helps maintain homeostasis by minimizing absorption of excess zinc. Although zinc absorption responds rapidly to changes in intake, changes in absorption with shifts in zinc status appear to be minor. With low zinc intake, homeostasis is maintained primarily by a reduction in the amount of zinc secreted into the small intestine.[162] This contrasts with murine models, where an upregulation of ZIP4 facilitates a compensatory increase in the efficiency of zinc absorption in response to low zinc intake.[94] Thus, in humans, even after a period of dietary zinc depletion, the fractional absorption of zinc is determined primarily by the zinc content of a meal rather than previous zinc intake.[122] Zinc absorption, however, does increase in response to changes in physiologic requirements during late pregnancy and lactation[204]; it may also decline with aging.[163]

Pregnancy, Lactation, and Growth

The additional zinc required for pregnancy, estimated from the weight of tissues gained and the zinc concentration of those tissues, totals about 100 mg.[205] The additional daily need for zinc increases with the rate of fetal growth; needs in the first half of pregnancy are below 0.25 mg/d but increase to between 0.5–0.75 mg/d in the last half. There is little evidence that pregnant women increase their zinc intake to meet this additional need, although modest adjustments may occur in zinc absorption or endogenous fecal excretion.[204] As the additional daily zinc need is small, changes in zinc homeostasis may not be apparent. Severe zinc deficiency in pregnant experimental animals causes multiple teratogenic anomalies and limits fetal growth.[206] Similar effects of zinc deficiency and fetal development have been observed in women with AE. Randomized controlled trials of zinc supplementation show little benefit to women consuming typical diets marginal to adequate in zinc worldwide.[207]

The zinc requirement for lactation varies with changes in milk volume and zinc concentration throughout the lactation period. The need is highest in the first month of lactation when the concentration is at its peak (about 2.8 mg zinc/mL). Total breast milk zinc varies from 1–2 mg/d and declines by about 75% by the 9th month.[208] Due to this sharp fall in milk zinc concentration after the first 6 months of lactation, human milk alone is an inadequate source of zinc.[194] Some women do not produce milk with normal amounts of zinc. In some individuals, this due to a mutation of ZnT2, which reduces zinc uptake into secretory vesicles of the mammary gland and reduces the normal zinc content of milk.[108] Milk zinc concentration does not appear to vary with maternal zinc intake. The effects of human milk on zinc bioavailability from complementary foods has not been studied.

The additional need for zinc in growing infants and children, estimated from the average zinc concentration of wet tissue weight, 20 mg/g, is used in dietary recommendations by the U.S. National Academy of Medicine (NAM, formerly the Institute of Medicine (IOM)) and the European Food Safety Authority.[209] It is assumed that each gram of new lean and adipose tissue gained during growth requires this amount of zinc, bringing the average additional needs for absorbed zinc to about 840 mg/d in infants between 7 and 12 months and about 750 mg/d in children 1 through 3 years of age.

Zinc Deficiency—Mechanisms and Phenotypic Outcomes

Tissue damage (including peroxidation, decreased cytotoxicity/cytoprotection, necrosis, decreased cell proliferation), stress intolerance, immune deficiency including cytokine imbalances, and developmental changes have been reported in zinc-deficient animals and humans.[177,179,210,211] However, although the biochemical functions of zinc, described in the previous sections, are so fundamental to cellular growth, development, and activity, the mechanisms responsible for these effects of zinc deficiency have yet to be fully defined. Phenotypic outcomes of zinc functions in integrative systems can be placed in four general categories: a) redox, peroxidation, and tissue damage; b) apoptosis regulation; c) cell proliferation and growth; and d) immune regulation. These provide a framework to help understand the complexities of zinc deficiency.

Zinc is not a redox metal, yet tissue peroxidation and oxidative injury have been observed in zinc-deficient animals. Iron accumulation leading to reactive oxygen or nitrogen radical generation could be a factor, as at least one zinc transporter (ZIP14) also transports iron.[212,213] Nitric oxide (NO) causes oxidative zinc release from sulfhydryl binding sites, leading to dysfunction.[33,214] Upregulation of inducible nitric oxide synthetase in zinc deficiency may exacerbate NO-induced zinc release and cell injury.[212,215,216] Animal-based evidence suggests that zinc protects against xenobiotic radical damage. Similarly, zinc may be a factor in regulating apoptosis through influences at various steps in the signaling cascades involved. Cells with characteristically high turnover rates, e.g., immune, and epithelial cells, could be most vulnerable. Consequently, immune dysfunction and skin and intestinal disorders associated with zinc deficiency could be outcomes of altered apoptosis.

Reduced growth and cell proliferation observed in zinc deficiency could also be related to abnormal apoptosis. Furthermore, direct effects on hormones that influence cell division or food intake (e.g., insulin-like growth factor or leptin), or the genes that produce these hormones or their receptor, or alter their signal transduction pathways, could also explain the decreased growth as an outcome of zinc deficiency.[217] Similarly, immune dysfunction and susceptibility to infection in zinc deficiency could involve atypical regulation of cytokine gene function and signaling pathways which, in turn, disrupt the balance of cell-mediated vs. humoral immunity.[218-220] Alternatively, failure of zinc-dependent structural factors needed for antigen presentation or microbial killing could be an outcome that leads to parasitic and microbial infections secondary to zinc deficiency.

Clinical Presentation and Causes of Zinc Deficiency in Humans

The clinical presentation of severe zinc deficiency in humans was first described by Prasad[4] and is characterized by erythematous, pustular skin lesions, reductions in growth, delayed puberty, hypogonadism, host defense defects, including infections of the epithelium, and poor appetite (**Table 11.3**). Skin lesions appear primarily adjacent to the body orifices and at the extremities and hair, may become hypopigmented and acquire a reddish hue. Patchy loss of hair is a common feature. Severe zinc deficiency, such as that observed by Prasad and coworkers, is uncommon and is most likely to occur when zinc absorption is severely impeded such as in the genetic condition, AE. In most cases, chronic low dietary intake of zinc (or habitual diets with low zinc bioavailability) results in marginal zinc deficiency. An analysis of dietary survey data from 20 countries that measured plasma zinc concentration as an indicator of zinc status revealed that at least 17% of the global population is at risk of inadequate zinc intake[221] with the highest burden of deficiency in low- and middle-income countries where zinc deficiency is one of the ten significant factors contributing to the burden of disease.[222]

Table 11.3 Clinical Manifestations of Marginal to Severe Human Zinc Deficiency

Growth retardation
Delayed sexual maturation and impotence
Hypogonadism and hypospermia
Diarrhea and intestinal inflammation
Alopecia
Acro-orificial skin lesions
Other epithelial lesions: glossitis, alopecia, nail dystrophy
Immune deficiencies: lymphopenia, thymic defects, reduced phagocytosis, depressed T-cell function, impaired cytokine production
Behavioral disturbances, including impaired hedonic tone
Impaired taste (hypogeusia) and smell (hyposmia)
Delayed healing of wounds, burns, and decubitus ulcers
Impaired appetite and food intake
Eye lesions, including photophobia and lack of dark adaptation and photic injury

Some signs have been observed in severe deficiency or reversed by zinc supplementation.

In addition to stunted growth and poor immune function, other possible signs of marginal zinc deficiency include dysfunctions in taste (hypogeusia) and smell (hyposmia). Zinc deficiency is thought to affect the number and structure of taste bud cells and reduce nerve sensitivity. Supplementation with zinc has been shown to be efficacious in improving the symptoms of hypogeusia and hyposmia.[223] There is general agreement, from studies done in many parts of the world, that physical growth of children in some population groups benefit from zinc supplementation. In several studies, cognitive performance and other measures of neuropsychological performance have been improved upon zinc supplementation. A systematic review and meta-analysis revealed that zinc supplementation with doses of 10 mg per day or more significantly reduced the risk of all-cause mortality.[224] Most of this relates to a reduction in secretory diarrhea and upper respiratory infections, including pneumonia.[225]

Zinc deficiency can be attributed to five general causes occurring either in isolation or in combination.[226] These include inadequate intake, increased requirements, malabsorption, increased losses, and impaired use. Primary, or diet-induced, zinc deficiency occurs when there is an inadequate intake of absorbable zinc.

Zinc deficiency is characterized as a Type II nutrient deficiency.[140] Type II nutrient deficiencies cause growth to stop, the nutrient is avidly conserved and, if necessary, weight is lost to make the nutrient internally available and thus maintain the concentration of the nutrient in tissues. As such, reduced growth occurs without a concomitant reduction in zinc concentration of tissues, coupled with nonspecific signs, particularly in the most metabolically active tissues. Depletion of zinc is mitigated by adaptive mechanisms that involve homeostatic control of absorption and endogenous losses; therefore, the biochemical and physiologic consequences are not realized until the homeostatic mechanisms are no longer able to compensate for low zinc intake.

Low intakes of absorbable zinc are exacerbated by physiologic or pathological conditions that increase zinc requirement. Increased zinc needs for growth place infants, children, adolescents, and pregnant and lactating women at greater risk for zinc depletion. Pathological conditions such as preterm birth, low birth weight, and diarrheal disorders reduce zinc absorption due to immaturity of the gastrointestinal tract or increase intestinal losses and further increase the risk of zinc deficiency among infants and children.[225]

Malabsorptive Disorders

Malabsorptive syndromes and inflammatory bowel diseases (IBD) that alter the integrity of the enteric mucosa, or interfere with the enteric recirculation of zinc, may reduce zinc absorption and precipitate secondary zinc deficiency states. Zinc deficiency has been observed in Crohn's disease, celiac disease, short bowel syndrome, and following bariatric surgery.

Crohn's disease, or regional enteritis, causes chronic inflammation most frequently in the small intestine and proximal large intestine. Low serum zinc concentration has been reported in adult and pediatric patients with Crohn's disease.[227,228] Zinc depletion in Crohn's disease may result from impaired absorption without appropriate reduction in excretion, hypoalbuminemia, or an internal redistribution of zinc.[229,230] Zinc may further be involved in the etiology of Crohn's disease. Dietary zinc, from foods more so than from zinc supplements, was inversely associated with the incidence of Crohn's disease in the Nurses Health Studies I and II, although not with the incidence of ulcerative colitis.[231] A missense SNP in the gene encoding the zinc transporter Zip8 has also been associated with Crohn's disease.[232]

Low plasma zinc has been reported in patients with celiac disease,[233,234] while treatment with a gluten-free diet increases plasma zinc concentration.[235,236] Low zinc status in celiac patients appears to be caused primarily by increased endogenous losses rather than reduced intestinal absorption. Zinc absorption is not different between treated and untreated celiac,[237] or between those with mild compared with severe disease,[238] although absorption is positively correlated with gut function in celiac patients. Endogenous zinc losses are higher in untreated celiac patients compared with healthy controls and may be reduced to levels similar to healthy controls after treatment with a gluten-free diet.[237]

Low plasma zinc concentration has been reported in patients following bariatric surgery,[239] although the risk of zinc deficiency differs, depending on the type of surgery.[161,240] The greatest postoperative increase in the proportion of patients falling below a plasma cutoff for zinc deficiency was observed after biliopancreatic diversion with duodenal switch. Over the course of 12 months after surgery, rates of low plasma zinc increased to 92%. Zinc deficiency also increased to 35% or 38% after Roux-en-Y gastric bypass, and less severely, increased to 15% or 19% following gastric sleeve.[161,240]

These observations illustrate the importance of the architecture of the digestive tract in maintaining zinc homeostasis. The simple reduction in stomach volume with the gastric sleeve procedure does not substantially change the architecture of the small intestine, where zinc-rich pancreatic secretions mix with ingested foods and are subsequently reabsorbed. The effect of gastric sleeve on plasma zinc may, however, be related to reduced zinc absorption in the proximal small intestine. Absorption of an orally administered zinc tracer in the fasted state (i.e., with no interference from food) was reduced by 72% 1 year following gastric sleeve.[240]

In Roux-en-Y gastric bypass, a small stomach pouch is created and connected directly to the jejunum, bypassing the duodenum. The duodenum, in turn, is connected to the proximal small intestine, forming a "Y" junction, thus allowing pancreatic secretions to mix with ingested foods beyond the "Y." Although zinc absorption is also reduced following Roux-en-Y, this impairment is less severe than with gastric sleeve.[240] The more severe risk for plasma zinc deficiency in Roux-en-Y compared with gastric sleeve may, therefore, be related to impaired reabsorption of secreted zinc. Biliopancreatic diversion with duodenal switch similarly forms a "Y" in the small intestine; however, the junction is aggressively located at the distal ileum, preventing pancreatic juices from mixing with food until they have nearly reached the colon.

Alcoholism

Patients with alcoholic cirrhosis often have hyperzincuria, hypozincemia, and low liver zinc concentrations.[241] Elevated urinary zinc excretion is attributed to a shift in the binding of plasma zinc from albumin to ligands that are easily excreted and that inhibit tubular zinc reabsorption. Long-term alcohol feeding in monkeys, rats, and pigs lowers hepatic zinc concentration. This may be related to dysfunctional tight-junction barrier proteins in the distal small intestine leading to alcoholic endotoxemia and hepatic apoptosis.[242,243] Combined vitamin A and zinc deficiency are common in alcoholics[244] and may result in poor dark adaptation due to a reduction in the activity of the retinal zinc-dependent enzyme, retinol alcohol dehydrogenase.[245] It has been proposed that resulting impairments in zinc homeostasis mediate several adverse effects of prolonged ethanol exposure, including reduced gut and lung barrier functions, inflammation, and neurotoxicity.[246]

Diabetes

Zinc is involved in multiple aspects of glucose homeostasis and its deficiency impacts the synthesis, storage, and secretion of insulin, as well as its transport and binding to other cells. Therefore, it is not surprising that zinc metabolism appears to be altered in both type 1 (insulin dependent) and type 2 (non–insulin dependent) diabetes (T1D, T2D). Rats with genetic or chemically induced diabetes accumulate zinc in the liver and kidney and have hyperzincuria.[247] Both T1D and T2D patients may exhibit hyperzincuria, which tends to increase with severity of the disease.[248]

The mechanisms underlying altered zinc metabolism in diabetes have not been completely identified. Zinc ions have an insulin-like effect[249,250] and zinc is released from β-cells as an insulin complex and may be involved in regulating the phosphorylation site of the insulin receptor.[251] Zinc also plays a role in insulin secretion. The ZnT8 zinc transporter is uniquely expressed in pancreatic β-cells to facilitate the uptake of zinc from the cytoplasm for incorporation into secretory vesicles.[252] Subsequently, ZnT8 was identified in multiple studies as a risk locus for type 2 diabetes and single-nucleotide polymorphisms identified.[253] In addition, ZnT8 is an autoantigen for type 1 diabetes,[254] suggesting that autoantibodies to this transporter may be used as an independent predictive marker for T1DM reviewed in Lichten and Cousins (2009).[88]

In the context of T2DM, β-cells that lack sufficient zinc produce fewer insulin granules and show greater signs of oxidative stress than cells with sufficient zinc.[255] Outside of the pancreas, zinc deficiency can alter insulin sensitivity and resistance through the activation of various cell-signaling cascades.[256] This ultimately leads to increased concentration of circulating glucose, thus increasing risk of T2D or exacerbating existing T2D. Importantly, individuals with existing T2D who exhibited good glycemic control had a higher zinc status than those who exhibited poor glycemic control.[256] A recent meta-analysis of 36 trials found that zinc supplementation significantly improved risk factors for both CVD [plasma triglycerides, very-low-density lipoprotein (VLDL), total cholesterol] and T2D [fasting blood glucose (FBG), 2-hour postprandial glucose, HOMA-IR (homeostatic model for assessment of insulin resistance), and glycated hemoglobin (HbA1c)], compared with placebo.[257] Similarly, a meta-analysis of 16 longitudinal studies found that individuals with the highest dietary zinc intake had 13% lower risk of T2D than individuals in the lowest dietary zinc intake category.[258]

Infections

Zinc is required for the synthesis and function of immune regulatory proteins and maintenance of immune function.[259] Zinc deficiency may, thus, impair resistance to new infections and recovery. Preventive zinc supplementation in low- and middle-income countries reduced the incidence of diarrhea by about 27% among children over 12 months of age and decreased the incidence of acute lower respiratory tract infections by about 15%.[260] Furthermore, zinc supplementation reduced child mortality from infection by about 6%; in children above 12 months of age, the impact was higher—an 18% reduction in mortality.[260]

Beneficial effects of therapeutic zinc supplementation have been observed in the treatment of a variety of infections, including diarrhea, shigellosis, common cold, lower respiratory tract infections, and chronic hepatitis C.[261,262] Therapeutic zinc supplementation reduces the duration and severity of diarrhea, particularly in malnourished children over 6 months age.[263,264] The World Health Organization and the United Nations Children's Fund recommended that therapeutic zinc be included in diarrhea control programs.[265]

Infection with HIV alters zinc nutrition by reducing zinc intake, absorption, and metabolism.[266] Excessive loss of zinc with diarrhea, which often accompanies HIV, increases dietary zinc requirement. Low serum or plasma zinc are common in HIV-infected individuals but may not reflect deficiency, as both HIV and other opportunistic infections can depress circulating zinc in the presence of adequate zinc intake.[266] In view of the important role zinc plays in maintaining immunity, supplemental zinc is expected to improve the management of HIV-infected individuals. However, inconsistent results of zinc supplementation trials of HIV patients have prevented the formulation of a supplemental zinc recommendation.

Other Diseases

Chelators, such as D-penicillamine (for treatment of copper overload in Wilson's disease) and diethylenetriamine pentaacetate (for treatment of iron overload in thalassemia and sickle cell disease, SCD), may chelate zinc as well as the other targeted cations and cause secondary zinc deficiency.[267] The stunting and low plasma zinc concentration seen in prepubertal children with SCD has been attributed to poor zinc status possibly associated with chelation therapy. In a randomized controlled trial, SCD children given 10 mg of supplemental zinc per day were significantly taller than children given placebos after 12 months.[268] Although zinc is frequently given with penicillamine to treat Wilson's disease, the combination therapy did not prevent a decline in bone mineral density 1 year after treatment was initiated.[269]

Zinc is found in high concentration in the retina and is hypothesized to reduce risk of age-related macular degeneration (AMD), the leading cause of visual impairment and blindness in the United States. In a 6-year study of 3640 individuals, aged 55–80 y, a supplement of antioxidants (vitamins C, E, and beta-carotene) plus zinc (80 mg/d) reduced the risk of developing advanced AMD by about 25%.[270] Interestingly, mortality declined by about 27% in those receiving supplemental zinc.[271] The mechanism of zinc action in reducing the risk of AMD has not been defined but may be related to reduction in oxidative injury to the zinc-rich retinal epithelium.[270,271] Supplemental zinc plus antioxidants may also be detrimental in some cases, paradoxically increasing disease progression in individuals with a specific genetic predisposition to AMD.[272]

Alzheimer's disease (AD) is a common neurodegenerative disease that has been associated with altered zinc, copper, and iron metabolism.[273] When categorizing AD, severely low plasma zinc concentration in the range of 50 to 60 μg/dL appears with distinct disease features, including an initial loss of long-term memory maintenance that precedes loss of new memory formation, and typically, no family history of AD.[274] The exact molecular mechanisms are unclear but involve modulation of the expression of key ZnT and ZIP transporters and metallothionein in brain tissue at various stages of the disease.[275] This abnormal cellular zinc homeostasis is thought to contribute to the formation of the extracellular amyloid plaques seen in AD, through the binding of zinc to β-amyloid protein, which alters the secondary structure of the protein, resulting in its aggregation and accumulation in the form of a plaque.[57,276]

Dietary Considerations and Requirements

The World Health Organization/Food and Agriculture Organization/International Atomic Energy Agency (WHO/FAO/IAEA) regularly convenes expert committees to estimate human zinc requirements.[128,277] In addition, expert panels from specific global regions and individual countries may set their own recommendations[278] such as the NAM in the USA and EFSA in Europe.[194,279] Although the approach used by international groups is similar, recommendations differ over time as new data become available in the published literature. In addition, expert panels have differing opinions on the most appropriate way to interpret and use the published research data within a regional or international context. Thus, there is variation in the dietary requirement values used around the world and in the nomenclature used to describe them.[209] This is illustrated in **Table 11.4** where the dietary recommendations from the NAM and EFSA are summarized. For the purposes of simplicity, the remainder of this section will use the nomenclature and values used by the NAM.

Absorption vs. Endogenous Losses

Estimates of dietary zinc requirements are derived for age groups except infants under 1 year of age using the factorial method, which is defined as the amount of zinc that must be absorbed to offset the amount of endogenous zinc lost from both intestinal and nonintestinal sites.[128,194] Nonintestinal losses include urine and "surface losses" (desquamated skin, hair, nails, sweat). Other nonintestinal losses include semen, menstrual losses, growth needs for children and pregnant women, and breast milk

Table 11.4 Recommended Zinc Intakes Derived by NAM and EFSA (mg/d)

	Age, (Years)	Male RDA[a]	Female RDA[a]	Male UL	Female UL	Age, (Years)	Male PRI[c]	Female PR1[c]
Infants	0–0.5	–[b]	–[b]	4	4	0–0.5	–	–
	0.5–1	3	3	5	5	0.5–1	2.9	2.9
Children or Adolescents	1–3	3	3	7	7	1–3	4.3	4.3
	4–8	4	4	12	12	4–6	5.5	5.5
	9–13	8	8	23	23	7–10	7.4	7.4
	14–18	11	9	34	34	11–14	9.4	9.4
						15–17	12.5	10.4
Adults	19–71+	11	8	40	40	≥18 Level of phytate 600 mg/d 900 mg/d 1200 mg/d	 9.4 11.7 14.0 16.3	 7.5 9.3 11.0 12.7
Pregnancy	> 19	–	11	–	40	Pregnancy	–	+1.6
Lactation	> 19	–	12	–	40	Lactation	–	+2.9

[a]From the Institute of Medicine (2001).[194]
[b]Adequate Intake (AL) that reflects the observed mean zinc intakes of infants principally fed human milk. AL is 2 mg/d for both males and females.
[C]From EFSA Panel on Dietetic Products Nutrition and Allergies (2014).[279]
From the Institute of Medicine (2001).[194]

zinc. When estimating zinc requirements, nonintestinal losses are considered constant over the usual range of zinc intakes, whereas intestinal losses vary with the quantity of zinc absorbed. Thus, intestinal losses are the amount lost when total absorbed zinc is just sufficient to meet physiologic needs. To estimate the total need for absorbed zinc to meet physiologic needs, nonintestinal and intestinal losses are summed and regressed against absorbed zinc. The intercept of the line of perfect agreement (i.e., where absorbed zinc equals endogenous zinc) crosses the regression line is the amount of absorbed zinc needed to replace total endogenous losses.

Estimated Average Requirement

The Estimated Average Requirement (EAR) is defined as the nutrient intake estimated to meet the needs of half of the healthy individuals in a specific life-stage group.[194] The relationship between ingested and absorbed zinc is asymptotic (i.e., the efficiency of zinc absorption declines as the quantity of ingested zinc increases). The average amount of dietary zinc required to absorb enough zinc to replace total endogenous losses is derived from the relationship between ingested and absorbed zinc. For example, if adult men need to absorb 3.84 mg/d to replace their total endogenous loss, they need to consume 9.4 mg zinc/day, assuming that, on average, 40% of the dietary zinc is bioavailable. Similar estimates are made for other life-stage and sex groups. The Recommended Dietary Allowance (RDA) is set at two standard deviations above the EAR. Thus, it is the amount of dietary zinc that meets the needs for 97.5% of the population. Because the variance and standard deviation of requirement is not known, it is assumed that the coefficient of variation for zinc requirement is 10%. Thus, 20% of the EAR is added to the mean requirement to establish an RDA.

A tolerable upper intake level (UL), which is defined as "the highest average daily nutrient intake level likely to pose no risk of adverse health effects for nearly all people," is the third component of the Dietary Reference Intakes (DRIs) for zinc (Table 11.4). After considering adverse effects of high zinc intakes on copper status, a Lowest Observed-Adverse-Effect Level (LOAEL) was derived based on the reduced activity of a copper-dependent enzyme, erythrocyte copper-zinc superoxide dismutase, when zinc intakes exceeded 50 mg/day.[194] Assuming dietary zinc intake at about 10 mg/day, the LOAEL was calculated to be 60 mg Zn/day (50 mg + 10 mg). An uncertainty factor for the LOAEL of 1.5 was applied, and the UL was estimated at 40 mg zinc/day (60/1.5) for adults of ≥19 years of age.

The UL for infants is 4 mg/d (Table 11.4), based on a study of 68 infants in which 4 mg Zn was added to formula providing 1.8 mg Zn/d.[194] As no adverse effects were observed, the UL was set at 4 mg/d for infants 0–6 months, increasing to 5 mg/d between 7–12 months of age and to 7 mg/d for toddlers 1–3 years old. A national survey of infants and toddlers in the United States showed that 92% of the children (0–6 mo), 86% (7–12 mo), and 51% (1–3 y) exceeded the UL for zinc.[280] The lack of health problems associated with current zinc intakes by infants and toddlers calls for a reassessment of their UL. This is currently being undertaken by the World Health Organization.[136]

Evaluation of Zinc Status

Plasma Zinc

The efficient regulation of zinc homeostasis at the whole body and cellular levels complicates the assessment of zinc status and hampers identification of a single, sensitive, specific biomarker of zinc status that reflects the entire spectrum from deficiency through adequacy to excess and toxicity. Plasma zinc concentration, which normally ranges from 80–120 μg/dL (12–18 μM), is typically used. In a zinc depletion/repletion study of men, changes in total body zinc estimated from balance data correlated better with changes in plasma zinc ($r^2 = 0.826$, $p<0.001$) than any other indicator measured (plasma zinc flux, alkaline phosphatase, retinol binding protein).[281] Also, a systematic review of 32 potential zinc biomarkers from 46 publications identified plasma zinc concentration as the only biomarker to respond in a dose-dependent manner to dietary manipulation.[282] The sensitivity and specificity of plasma zinc as a biomarker of zinc status is limited by its responsiveness to stimuli unassociated with zinc status. Concentrations change with food intake, acute stress such as infection, and steroidal hormone concentrations such as occurs during pregnancy and oral contraceptive therapy, and zinc concentration increases with muscle catabolism during illnesses or weight loss.[283]

Nevertheless, plasma zinc concentration has important characteristics that make it a good indicator of zinc status for populations: it reflects dietary zinc intakes over the long-term, it responds to zinc supplementation, and reference values are available for most age and sex groups.[283] Cut-off levels, based on NHANES II data that control for time of day or fasting status, sex, and physiologic state, have been established for use in population surveys.[284,285] The suggested cut-off for the 2.5th percentile is 74 μg/dL (11.3 μM) in fasting males and 70 μg/dL (10.7 μM) in fasting females. Based on this, the International Zinc Nutrition Consultative Group (IZiNCG) suggested lower cutoffs for serum zinc concentration by age group, sex, time of day and time since last meal.[286] An adjustment for the fall in zinc concentration due to inflammation may be applied in populations where inflammation is prevalent.[287,288]

Zinc concentrations of other tissues and fluids have been explored as indices of zinc status, including leukocytic zinc, erythrocyte zinc, and salivary zinc; however these are not viewed as good indices of zinc status due to lack of sensitivity in response to changes in zinc intake.[282,289] Hair zinc concentration was considered to have potential by the BOND zinc review group; however, there is currently a lack of availability of reference values.[290] Similarly, nail zinc concentration appears to respond to changes in dietary zinc intake; however, time lag between the formation of the nail in the nail bed (in response to quantity of zinc available in the blood supply at the time) to the tip of the nail where the clipping is measured, must be taken into consideration.[291]

Functional Indicators of Zinc Status

Severe zinc deficiency causes several nonspecific clinical changes, listed in Table 11.3. These symptoms facilitate a definitive diagnosis of zinc deficiency. However, severe zinc deficiency is rare, and functional indicators of mild or marginal zinc depletion are needed. Functional zinc indicators can be divided into two types—cellular or metabolic and whole-body processes.

It is well known that several metabolic changes occur with zinc deficiency. Establishing a direct relationship between zinc status and these changes has proved challenging. Examples include shifts in the activity of zinc-containing enzymes, including plasma alkaline phosphatase, erythrocyte copper-zinc superoxide dismutase, lymphocyte ecto-5-nucleotidase, aminolevulinic acid dehydratase and erythrocyte carbonic anhydrase. Zinc metalloenzyme activity has not proven to be a consistent biomarker of zinc status. Metabolic indicators of increased oxidative stress have also been associated with zinc depletion. One such indicator is the presence of DNA strand breaks that can be detected using a technique referred to as the "comet assay," which involves isolation of peripheral cell DNA and gel electrophoresis to separate the DNA fragments according to their size. DNA fragmentation has been associated with marginal zinc deficiency, which may be related to an increase in cellular oxidative stress, and levels of fragmentation have been shown to fall in response to zinc supplementation.[40,292]

Numerous molecularly based zinc responsive markers have been established in rodents. Metallothionein (MT) is expressed in most tissues and in some, such as the pancreas, is sensitive to zinc status at the mRNA level.[189,219] Plasma MT reflects zinc intake in rats but is sensitive to infection and stress.[293] Specific zinc transporters are also sensitive to zinc status in rodents in PBMCs or RBCs.[167,294] Several studies have explored the response of MT and zinc transporter expression in circulating blood cells to zinc supplementation in humans. A systematic review of studies conducted in healthy adults reported that Zip1, ZnT1, and MT were the most measured proteins. The response of Zip1, ZnT1 to zinc supplementation or depletion was not consistent across studies; however, MT measured in leukocyte subtypes responded in a dose dependent manner and was more reliable and responsive than plasma zinc concentration.[295] The measure of labile zinc using fluorometric analysis[296] has yet to be tested as a status assessment tool. These methods and concepts, which are becoming more widely used in diagnostic medicine, await testing in the rigor of field surveys and intervention studies.

Functional, whole-body indicators of zinc status have been explored, including incidence and prevalence of diarrhea, lower-respiratory tract infections and linear growth in children. A systematic review identified 80 randomized trials that explored the impact of preventative zinc supplementation on morbidity in children aged 6 months to 12 years.[297] Many studies included in the review were conducted in LMICs where the prevalence of dietary zinc insufficiency is high. Meta-analysis revealed that zinc supplementation resulted in reduction in all-cause diarrhea, and a small increase in linear growth. There was no evidence of an effect on lower respiratory tract infections. Some investigators propose that risk of zinc deficiency can be assessed in a population by measuring plasma/serum zinc concentration, dietary zinc intake, and stunting prevalence.[298] The risk of zinc deficiency is considered elevated if 1) prevalence of low serum zinc concentration is >20%, 2) prevalence of low dietary zinc intake is >25% (i.e., less than the EAR for the population), and 3) prevalence of stunting is >20% (i.e., height-for-age is <2 standard deviations of the reference median).[284]

Zinc Upper Limits and Toxicity

The tolerable upper limit (UL) for zinc, defined as the highest level of daily zinc intake that is likely to pose no risk of adverse health effects for almost all individuals, has been set at 45 mg/d for adult men, and 35 mg/d for adult women by WHO/FAO, and 40 mg/d men and women by NAM.[121] The UL for children are more difficult to define due to a lack of data, but the WHO/FAO extrapolated from adult data to set the UL for children at 23–28 mg/d, depending on age, which is higher than the UL of 4–12 mg/d for children set by the NAM using the approach of defining a "no observed adverse effects level" (NOAEL) and adjusting this according to an uncertainty factor, with adjustment for age ranges based on body weight.[209]

Adverse effects of consuming zinc from food alone are not expected, as zinc absorption is tightly regulated over a wide range of intakes and zinc is not stored in body tissues, limiting the potential for toxic levels (see metabolism section). However, toxic effects due to high-level supplement intake are possible, and several case studies have reported severe consequences of exposure to high quantities of zinc. The effects of zinc toxicity have been recently reviewed,[136,194,299] and include gastric distress, dizziness, and nausea. Zinc can have an emetic effect at doses as low as 50 mg. Long-term high supplement use can negatively impact iron, copper, lipoprotein, and cholesterol metabolism.[300] One case study reported death with an accidental infusion of more than 7 g of zinc in a 60-hour period.[301] Two case studies reported data from infants; one infant was given a dose of 14–24 mg/d[302] and another infant received a dose of 45 mg/d.[303] Both resulted in adverse effects on serum copper and iron. Several cases of severe zinc toxicity have been reported in patients with metal pica involving post-1981 pennies, which are made primarily of zinc.[304]

Chronic exposure to zinc by galvanization workers caused an elevation in serum zinc and decreases in serum copper and calcium, but chest x-rays were normal.[305] Other symptoms of chronic toxicity include gastric problems, reduced immune function (lymphocyte stimulation by phytohemagglutinin), and decreased high-density lipoprotein cholesterol. Low serum copper concentration has been seen at intakes as low as 50 mg zinc/d for several weeks.[306] In animals (dogs and chicks), the pancreas is a target of zinc toxicity.[211] This toxicological effect has not been observed in humans. Hypocupremia was observed in zinc-treated (150 mg zinc/day) Wilson's disease patients. This may be due to reduced copper absorption, possibly through zinc induction of intestinal metallothionein, which preferentially binds copper over zinc, rendering copper unavailable for release from enterocytes.[307] Clinically, this zinc-copper interaction is the basis for an FDA-approved treatment for Wilson's disease, a copper accumulation condition. Sideroblastic anemia was observed secondarily to copper deficiency induced by high zinc intake.[308]

The development of neurological disease has been associated with high intakes of zinc over time.[309] Several cases of neuropathy were linked to excessive amounts of denture cream containing up to 17 to 34 mg zinc/g cream.[310] Along with the neurological symptoms of numbness and loss of balance, patients presented with hyperzincinemia and copper deficiency, which resolved on cessation of the use of the dental cream.[309] In one case, copper supplementation improved hematologic abnormalities, but not the neurologic disorders.[311] These clinical observations likely relate to zinc induced metallothionein synthesis with subsequent copper binding or zinc-induced influences on genes associated with copper absorption.

Not directly linked to zinc consumption, but potentially relevant in clinical situations, is the toxic effect of zinc ions on cortical neurons leading to neuronal cell death.[312] After transient forebrain ischemia, zinc accumulates and may contribute to neurodegeneration and neuronal death. Overexpression of ZnT1 reduces this neuronal damage.[313]

In general, zinc is a relatively nontoxic nutrient with moderate levels of supplementation (<50 mg/d). However, there is the possibility that nutrient imbalances with selective supplementation could cause toxicity symptoms.

Future Directions

Zinc malnutrition is a global health problem affecting individuals of all age groups and life stages in both developing and developed countries. One of the key challenges in zinc nutrition is the persisting high prevalence of zinc deficiency, particularly among women and children in LMICs.[8] Zinc is critical for a wide range of metabolic functions that influence growth, immune function, gene expression, DNA synthesis and repair, and energy metabolism. Although zinc is uniquely regulated at the cellular, tissue, and whole-body levels, no unique indicator of individual zinc status is currently available. Consequently, the etiology of zinc deficiencies, the role of zinc in acute and chronic disease, and the efficacy of intervention strategies to correct zinc deficiency, are not well understood.

The effective development of new knowledge in these areas will depend on enhanced techniques to assess and monitor changes in zinc status in response to potential zinc intervention. Progress will depend on the integration of a wide range of expertise, from zinc's molecular nature to dietary and physiologic determinants of its bioavailability, and the movement of zinc from diet and circulating pools to its active roles at the cellular, tissue, and whole-body levels.

Alongside strategies to increase dietary zinc intake is the ability to monitor their impact on individuals, communities, and populations. This requires sensitive and reliable biomarkers to measure change in zinc status, which have so far been elusive. Future developments in this area are essential for providing a robust evidence base for informing policy. New technologies are enabling a better understanding of the role of zinc in disease pathology. Advances in subcellular imaging methodologies have accelerated our understanding of zinc signaling pathways, paving the way for novel therapeutic interventions.[314]

Long term, sustainable solutions to combating zinc deficiency will require a "systems approach," that looks beyond a single micronutrient to the broader context within the food system and beyond. This involves embedding nutrition sensitive interventions within all elements of the food system from sustainable fisheries and livestock farming to food processing, distribution, and waste management. The potential impact of climate change on zinc in the food system cannot be ignored, with evidence of an inverse relationship between atmospheric CO_2 concentration and food crop zinc content and predicted concomitant increased risk of global zinc deficiency[315] (**Figure 11.5**). Thus, development of new climate-resilient crop varieties with enhanced micronutrient densities should be considered, while improving food security to ensure affordable access to a diverse diet.

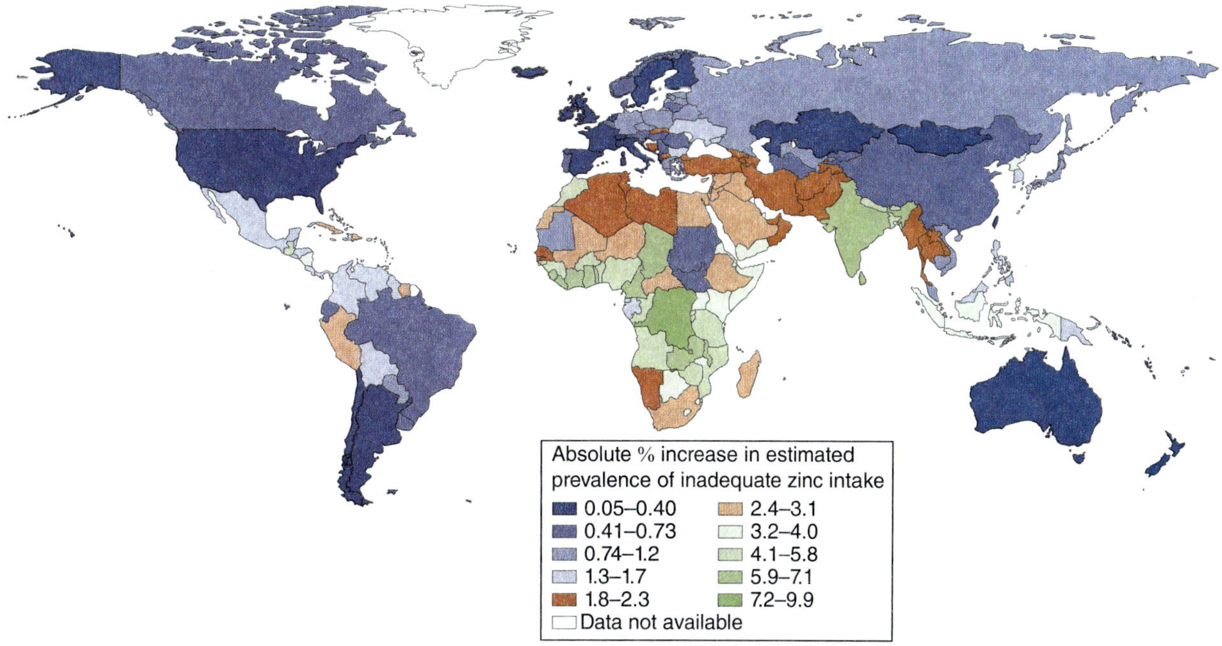

Figure 11.5 Absolute percentage increase in risk of zinc deficiency in response to elevated atmospheric [CO_2].

Reproduced from Myers, S. S., Wessells, K. R., Kloog, I., Zanobetti, A., & Schwartz, J. (2015). Effect of increased concentrations of atmospheric carbon dioxide on the global threat of zinc deficiency: a modelling study. The Lancet. Global health, 3(10), e639–e645. https://doi.org/10.1016/S2214-109X(15)00093-5.[310]

Acknowledgments

The authors would like to acknowledge the previous authors, Janet C. King and Robert J. Cousins for their contributions to this chapter.

References

1. Vallee BL. Physiol Rev 1959;39:443–90.
2. Keilin D, Mann T. Biochem J 1940;34:1163–76.
3. Keilin D Mann T. Nature 1944;153:107–8.
4. Prasad AS, Miale A, Jr., Farid Z, et al. J Lab Clin Med 1963;61:537–49.
5. Sandstead HH, Prasad AS, Schulert AR, et al. Am J Clin Nutr 1967;20:422–42.
6. Wessells KR, Singh GM, Brown KH. PloS One 2012;7:e50565.
7. Black RE, Allen LH, Bhutta ZA, et al. Lancet 2008;371:243–60.
8. Victora CG, Christian P, Vidaletti LP, et al. Lancet 2021;397:1388–99.
9. Krezel A, Maret W. Arch Biochem Biophysics 2016;611:3–19.
10. Andreini C, Banci L, Bertini I, et al. J Proteome Res 2006;5:196–201.
11. Kocyla A, Tran JB, Krezel A. Trends Biochem Sci 2021;46:64–79.
12. Colvin RA, Holmes WR, Fontaine CP, et al. Metallomics Integ Biometal Sci 2010;2:306–17.
13. Krezel A, Maret W. J Biol Inorg Chem 2006;11:1049–62.
14. Jacob C, Maret W, Vallee BL. Proc Nat Acad Sci U S A 1998; 95:3489–94.
15. Wellenreuther G, Cianci M, Tucoulou R, et al. Biochem Biophys Res Commun 2009;380:198–203.
16. Cousins RJ. Zinc. In *Present knowledge in nutrition* (Brown Bowman BA, Russell RM, and International Life Sciences Institute-Nutrition Foundation., eds) 2006;445–57, ILSI Press, International Life Sciences Institute, Washington, D.C.
17. Vallee BL, Neurath H. J Am Chem Soc 1954;76;5006–7.
18. Vallee BL, Hoch FL. Proc Nat Acad Sci U S A 1955;41;327–38.
19. Suzuki T, Ishihara K, Migaki H, et al. J Biol Chem 2005;280: 637–43.
20. McDonald AG, Tipton KF. *FEBS J* 2021;290:2214–2231.
21. Andreini C, Bertini I. J Inorgan Biochem 2012;111:150–6.
22. Andreini C, Bertini I, Cavallaro G. *PloS one* 2011;6:e26325.
23. Lee YM, Lim C. *J Mol Biol* 2008;379:545–53.
24. Auld DS. Biometals 2001;14:271–313.
25. Tainer JA, Getzoff ED, Richardson JS, et al. Nature 1983;306:284–7.
26. Suzuki E, Ogawa N, Takeda TA, et al. J Biol Chem 2020; 295:5669–84.
27. Pedrosa FO, Pontremoli S, Horecker BL. Proc Nat Acad Sci U S A 1977;74:2742–5.
28. Brand IA, Kleineke J. J Biol Chem 1996;271:1941–9.
29. Brown RS, Sander C, Argos P. *FEBS Lett* 1985;186:2712–4.
30. Miller J, McLachlan AD, Klug A. *EMBO J* 1985;4:1609–14.
31. Padjasek M, Kocyla A, Kluska K, et al. J Inorganic Bioch 2020;204:110955.
32. Klug A. Annu Rev Biochem. 79;213–31.
33. Kroncke KD, Klotz LO, Suschek CV, et al. J Biol Chem 2002;277:13294–301.
34. Elrod-Erickson M, Rould MA, Nekludova L, et al. Structure 1996;4:1171–80.
35. Patel A, Zhang X, Blumenthal RM, et al. J Biol Chem 2017;292:15994–6002.
36. Kim PW, Sun ZY, Blacklow SC, et al. Science 2003;301:1725–8.
37. Hopfner KP, Craig L, Moncalian G, et al. Nature 2002;418: 562–6.
38. Wang J, Youkharibache P, Zhang D, et al. Bioinformatics 2020;36:131–5.
39. Madej T, Lanczycki CJ, Zhang D, et al. Nucleic Acids Res 2014;42:D297–303.

40. Joray ML, Yu TW, Ho E, et al. Nutr Res 2015;35:49–55.
41. Zyba SJ, Shenvi SV, Killilea DW, et al. Am J Clinical Nutr. 2017;105:343–51.
42. Maret W. Int J Molec Sci 2017;18:2285.
43. Kocyla A, Adamczyk J, Krezel A. Metallomics 2018;10:120–31.
44. Sims HI, Chirn GW, Marr MT, 2nd. Proc Nat Acad Sci U S A 2012;109:16516–21.
45. Otsuka F, Okugaito I, Ohsawa M, et al. Biophysica Acta 2000;1492:330–40.
46. Walker E, Chang WY, Hunkapiller J, et al. Cell Stem Cell 2010;6:153–66.
47. Bellomo E, Birla Singh K, Massarotti A, et al. Coord Chem Rev 2016;327-328: 70–83.
48. Cruz KJC, de Oliveira ARS, Morais JBS, et al. Biol Trace Element Research 2018;186:407–12.
49. Anson KJ, Corbet GA, Palmer AE. Proc Nat Acad Sci U S A 2021;118.
50. Haase H, Ober-Blobaum JL, Engelhardt G, et al. J Immunol 2008;181:6491–502.
51. Yamasaki S, Hasegawa A, Hojyo S, et al. PloS One 2012;7:e39654.
52. Haase H, Maret W. Electrophoresis 2008;29:4169–76.
53. Maret W, Vallee BL. Proc NatAcad Sci of the U S A 95, 3478–82.
54. Woodier J, Rainbow RD, Stewart AJ, et al. *J Biol Chem* 2015;290:17599–610.
55. Jang Y, Wang H, Xi J, et al. Cardiovasc Res 2007;75:426–33.
56. Priebatsch KM, Kvansakul M, Poon IK, et al. Biomolecules 2017;7.
57. Sensi SL, Paoletti P, Bush AI, et al. Nat Rev Neurosci 2009;10: 780–91.
58. Sindreu C, Palmiter RD, Storm DR. Proc Nat Acad Sci U S A 2011;108:3366–70.
59. Besser L, Chorin E, Sekler I, et al. J Neurosci 2009;29: 2890–901.
60. Schmalen I, Reischl S, Wallach T, et al. Cell 2014;157: 1203–15.
61. Baidanoff FM, Trebucq LL, Plano SA, et al. Biomolecules 2022;12.
62. Takeda A, Minami A, Takefuta S, et al. J Neurosci Res 2001;63:447–52.
63. Chowanadisai W, Kelleher SL, Lonnerdal B. J Nutr 2005;135: 1002–7.
64. Cote A, Chiasson M, Peralta MR, 3rd, et al. J Physiol 2005;566: 821–37.
65. Kwak YD, Wang B, Pan W, et al. *J Biol Chem* 2010;285: 9847–57.
66. Naik HB, Beshire M, Walsh BM, et al. Am J Physiol Cell Physiol 2009;297:C979–89.
67. Dominguez-Munoz JE, Martinez SM, Leodolter A, et al. Pancreatology 2004;4:57–62.
68. Larsen KS, Auld DS. Biochemistry 1991;30:2613–8.
69. Dunn MF. Biometals 2005;18:295–303.
70. Bancila V, Cens T, Monnier D, et al. J Biol Chem 2005;280: 8793–9.
71. Zhou H, Zhang T, Harmon JS, et al. Diabetes 2007;56: 1107–12.
72. Xia J, Cai LH, Wu H, et al. Proc Natl Acad Sci U S A 2021;118.
73. Henderson SJ, Xia J, Wu H, et al. Thromb Haemost 2016;115:533–42.
74. Marx G. Thromb Haemost 1988;59:500–3.
75. Kagi JH, Valee BL. J Biol Chem 1960;235:3460–5.
76. Pinter TB, Stillman MJ. Biochem J 2015;471:347–56.
77. Krezel A, Maret W. J Am Chem Soc 2007;129:10911–21.
78. Baksh S, Spamer C, Heilmann C, et al. FEBS Lett 1995;376: 53–7.
79. Saito Y, Ihara Y, Leach MR, et al. EMBO J 1999;18:6718–29.
80. Giroux EL. Biochem Med 1975;12:258–66.
81. Fasano M, Curry S, Terreno E, et al. IUBMB Life 2005;57: 787–96.
82. Lu J, Stewart AJ, Sleep D, et al. J Am Chem Soc 2012;134: 1454–7.
83. Smith KT, Failla ML, Cousins RJ. Biochem J 1979;184:627–33.
84. Pertusa JAG, Leon-Quinto T, Berna G, et al. PloS One 2017;12:e0187547.
85. Iqbal S, Qais FA, Alam MM, et al. J Biol Inorg Chem 2018;23: 447–58.
86. Gettins P, Cunningham LW. Biochemistry 1986;25:5004–10.
87. Harris WR, Keen C. *J Nutr* 1989;119:1677–82.
88. Lichten LA, Cousins RJ. Ann Rev Nutr 2009;29:153–76.
89. Kambe T, Taylor KM, Fu D. J Biol Chem 2021;296:100320.
90. Mao X, Kim BE, Wang F, et al. J Biol Chem 2007;282: 6992–7000.
91. Weaver BP, Dufner-Beattie J, Kambe T, et al. Biol Chem 2007;388:1301–12.
92. Cragg RA, Phillips SR, Piper JM, et al. Gut 2005;54:469–78.
93. Dufner-Beattie J, Wang F, Kuo YM, et al. J Biol Chem 2003;278:33474–81.
94. Hennigar SR, Olson CI, Kelley AM, et al. J Nutr Biochem 2022;101:108927.
95. Chun H, Korolnek T, Lee CJ, et al. J Biol Chem 2019;294: 2815–26.
96. Wang K, Zhou B, Kuo YM, et al. Am J Human Genet 2002;71: 66–73.
97. Thyresson N. Acta Derm Venereol 1974;54:383–5.
98. McMahon RJ, Cousins RJ. Proc Nat Acad Sci U S A 1998;95: 4841–6.
99. Nishito Y, Kambe T. J Biol Chem 2019;294:15686–97.
100. Levy S, Beharier O, Etzion Y, et al. J Biol Chem 2009;284: 32434–43.
101. Jackson KA, Helston RM, McKay JA, et al. J Biol Chem 2007;282:10423–31.
102. Valentine RA, Jackson KA, Christie GR, et al. J Biol Chem 2007;282:14389–93.
103. Guthrie GJ, Aydemir TB, Troche C, et al. Am J Physiol Gastrointest Liver Physiol 2015;308:G171–8.
104. Geiser J, De Lisle RC, Andrews GK. PloS One 2013;8:e82149.
105. Lee S, Hennigar SR, Alam S, et al. J Biol Chem 2015;290: 13064–78.
106. Podany AB, Wright J, Lamendella R, et al. Cell Mol Gastroenterol Hepatol 2016;2:369–83.
107. Tamaki M, Fujitani Y, Hara A, et al. J Clin Invest 2013;123: 4513–24.
108. Chowanadisai W, Lonnerdal B, Kelleher SL. J Biol Chem 2006;281:39699–707.
109. Giunta C, Elcioglu NH, Albrecht B, et al. Am J Human Gen 2008;82:1290–305.
110. Itsumura N, Inamo Y, Okazaki F, et al. PloS One 2013;8: e64045.
111. Adlard PA, Parncutt JM, Finkelstein DI, et al. J Neurosci 2010;30:1631–6.
112. Flannick J, Mercader JM, Fuchsberger C, et al. Nature 2019;570:71–6.
113. Waterworth DM, Ricketts SL, Song K, et al. Arterioscler Thromb Vasc Biol 2010;30:2264–76.
114. Kagara N, Tanaka N, Noguchi S, et al. Cancer Sci 2007;98: 692–7.
115. Hogstrand C, Kille P, Ackland ML, et al. Biochem J 2013;455: 229–37.
116. Taylor KM, Vichova P, Jordan N, et al. Endocrinology 2008;149:4912–20.
117. Miyai T, Hojyo S, Ikawa T, et al. Proc Natl Acad Sci U S A 2014;111:11780–5.

118. Park JH, Hogrebe M, Gruneberg M, et al. AmJ Human Gen 2015;97:894–903.
119. Tuschl K, Meyer E, Valdivia LE, et al. Nature Comm 2016;7:11601.
120. Maares M, Haase H. Nutrients 2020;12:762.
121. Gibson RS, King JC, Lowe N. Food Nutr Bull 2016;37:443–60.
122. Chung CS, Stookey J, Dare D, et al. Am J Clin Nutr 2008;87:1224–9.
123. Bel-Serrat S, Stammers A-L, Warthon-Medina M, et al. Nutr Rev 2014;72:334–52.
124. Lonnerdal B. J Nutr 2000;130:1378S–83S.
125. Hotz C, Gibson RS, Temple L. Int J Food Sci Nutr 2001;52: 133–42.
126. Troesch B, Jing H, Laillou A, et al. Food Nutr Bull 2013;34: S90–101.
127. Gupta RK, Gangoliya SS, Singh NK. J Food Sci Technol 2015;52:676–84.
128. World Health Organization (WHO), International Atomic Energy Agency (IAEA), and Food and Agriculture Organization of the United Nations (FAO). *Trace Elements in Human Nutrition and Health*, WHO, Geneva, 1996.
129. Miller LV, Krebs NF, Hambidge KM. J Nutr 2007;137:135–41.
130. Gupta S, Brazier AKM, Lowe NM. J Hum Nutr Diet 2020;33: 624–43.
131. Raboy V. Plants 2020;9:140.
132. Fordyce EJ, Forbes RM, Robbins KR, et al. J Food Sci 1987;52:440–4.
133. Marena Ceballos-Rasgado NML, Victoria H Moran, Andrew Clegg, et al. Nutr Rev In press.
134. Miller LV, Hambidge KM, Krebs NF. J Nutrition 2015;145: 1763–9.
135. Walker CF, Kordas K, Stoltzfus RJ, et al. AJ Clin Nutr 2005;82: 5–12.
136. Ceballos-Rasgado M, Lowe NM, Mallard S, et al. Adv Nutr. 2022;6:2488–2519.
137. August D, Janghorbani M, Young VR. *Am J Clin Nutr* 1989; 50:1457–63.
138. Davis CD, Milne DB, Nielsen FH. Am J Clin Nutr 2000;71: 781–8.
139. Tsang BL, Holsted E, McDonald CM, et al. Adv Nutr. 2021;12:1821–37.
140. Mills CF. The Biological Significance of Zinc for Man: Problems and Prospects. In *Zinc in Human Biology* (Mills CF, ed) pp. 371–81, Springer-Verlag, London, 1989.
141. Ovesen L, Bendtsen F, Tage-Jensen U, et al. Gastroenterology 1986;90:958–62.
142. Heth DA, Hoekstra WG. J Nutr 1965;85:367–74.
143. Nielsen FH, Sunde ML, Hoekstra WG. J Nutr 1966;89:24–34.
144. Miller LV, Krebs NF, Hambidge KM. Brit J Nutr 2013;109: 695–700.
145. Talsma EF, Moretti D, Ly SC, et al. J Nutr. 2017;147:1086–93.
146. Zhang T, Kuliyev E, Sui D, et al. *Biochem J* 2019;476:1791–803.
147. Lombeck T, Schnippering HG, Ritzl F, et al. Lancet 1975;1:855.
148. Lonnerdal B, Stanislowski AG, Hurley LS. 1980 J Inorg Biochem 1975;12:71–8.
149. Krieger I, Cash R, Evans GW. J Pediatr Gastroenter Nutr 1984;3:62–8.
150. Geiser J, De Lisle RC, Finkelstein D, et al. PloS One 2013;8:e72543.
151. Jackson MJ. J Clin Pathol 1977;30:284–7.
152. Cousins RJ. Adv Exp Med Biol 1989;249:3–12.
153. Lee HH, Prasad AS, Brewer GJ, et al. Am J Physiol 1989;256: G87–91.
154. Antonson DL, Barak AJ, Vanderhoof JA. J Nutr 1979;109: 142–7.
155. Naveh Y, Bentur L, Diamond E. J Nutr 1988;118:61–4.
156. Hara H, Konishi A, Kasai T. J Nutr 2000;130:83–9.
157. Gopalsamy GL, Alpers DH, Binder HJ, et al. Nutrients 2015; 7:572–83.
158. Hoadley JE, Leinart AS, Cousins RJ. Am J Physiol 1987; 252:G825–31.
159. Lee DY, Prasad AS, Hydrick-Adair C, et al. J Lab Clin Med 1993;122:549–56.
160. Raffaniello RD, Lee SY, Teichberg S, et al. J Cell Physiol 1992;152:356–61.
161. Salle A, Demarsy D, Poirier AL, et al. Obes Surg 2010;20: 1660–70.
162. King JC, Shames DM, Lowe NM, et al. Am J Clin Nutr 2001;74:116–24.
163. Fairweather-Tait SJ, Harvey LJ, Ford D. Exper Gerontol 2008;43:382–8.
164. Wallock LM, King JC, Hambidge KM, et al. Am J Clin Nutr 1993;58:695–701.
165. Milne DB, Ralston NV, Wallwork JC. Clin Chem 1985;31:65–9.
166. Grider A, Bailey LB, Cousins RJ. Proc Nat Acad Sci U S A 1990;87:1259–62.
167. Ryu MS, Lichten LA, Liuzzi JP, et al. J Nutr 2008;138: 2076–83.
168. Cousins RJ, Blanchard RK, Popp MP, et al. Proc Nat Acad Sci U S A 2003;100:6952–7.
169. Foster DM, Aamodt RL, Henkin RI, et al. Am J Physiol 1979;237:R340–9.
170. Wastney ME, Aamodt RL, Rumble WF, et al. Am J Physiol 1986;251:R398–408.
171. Dunn MA, Cousins RJ. Am J Physiol 1989;256:E420–30.
172. Miller LV, Hambidge KM, Naake VL, et al. J Nutr 1994;124: 268–76.
173. Pinna K, Woodhouse LR, Sutherland B, et al. J Nutr 2001; 131:2288–94.
174. Lu Q, Haragopal H, Slepchenko KG, et al. Int J Physiol Pathophysiol Pharmacol 2016;8:35–43.
175. Chabosseau P, Tuncay E, Meur G, et al. ACS Chem Biol 2014;9:2111–20.
176. Haase H, Beyersmann D. Biometals 1999;12:247–54.
177. Emmert JL, Baker DH. Poult Sci 1995;74:1011–21.
178. Thu BD, Schultink W, Dillon D, et al. Am J Clin Nutr 1999;69: 80–6.
179. Drinker KR, Fehnel JW, Marsh M. J Biol Chem 1927;72:375–83.
180. Krebs NF. *J Nutr* 2000;130:1374S–7S.
181. DeGrado TR, Kemp BJ, Pandey MK, et al. Molec Imag 2016;15.
182. Case RM, Ishiguro H, and Steward MC. Pancreatic Bicarbonate Secretion. In *Encyclopedia of Gastroenterology* (Johnson LR, ed) pp. 38–40, Elsevier, New York, 2004.
183. Carrigan PE, Hentz JG, Gordon G, et al. Cancer Epidemiol Biomarkers Prev 2007;16:2656–63.
184. Matseshe JW, Phillips SF, Malagelada JR, et al. Am Journal Clin Nutr 1980;33:1946–53.
185. Larsen KS, Auld DS. Biochemistry 1989;28:9620–5.
186. Debela M, Goettig P, Magdolen V, et al. J Mol Biol 2007;373: 1017–31.
187. Janc JW, Clark JM, Warne RL, et al. Biochemistry 2000;39: 4792–800.
188. Lu J, Combs GF, Jr. J Nutr 1988;118:681–9.
189. Guo L, Lichten LA, Ryu MS, et al. Proc Nat Acad Sci U S A 2010;107:2818–23.
190. Wang F, Kim BE, Petris MJ, et al. J Biol Chem 2004;279: 51433–41.
191. King JC, Shames DM, Woodhouse LR. J Nutr 2000;130: 1360S–6S.
192. Seal CJ, Heaton FW. J Nutr 1985;115:986–93.
193. Victery W, Levenson R, Vander AJ. Am J Physiol 1981;240: F299–305.

194. Institute of Medicine (US) Panel on Micronutrients. Zinc. In *Dietary Reference Intakes for Vitamin A, Vitamin K, Arsenic, Boron, Chromium, Copper, Iodine, Iron, Manganese, Molybdenum, Nickel, Silicon, Vanadium, and Zinc*, National Academies Press, Washington (DC), 2001.
195. Lifschitz MD, Henkin RI. J Appl Physiol 1971;31:88–92.
196. King JC, Hambidge KM, Westcott JL, et al. J Nutr 1994;124: 508–16.
197. Lowe NM, Woodhouse LR, King JC. Brit J Nutr 1998;80: 363–70.
198. Fell GS, Fleck A, Cuthbertson DP, et al. Lancet 1973;1:280–2.
199. Prasad AS. J Am Coll Nutr 1985;4:591–8.
200. Falchuk KH. N Engl J Med 1977;296:1129–34.
201. Chu A, Petocz P, Samman S. Sports Med 2017;47:127–34.
202. Liuzzi JP, Lichten LA, Rivera S, et al. Proc Nat Acad Sci U S A 2005;102:6843–8.
203. Hambidge KM, Miller LV, Westcott JE, et al. Am J Clin Nutr 2010;91:1478S–83S.
204. Hambidge KM, Miller LV, Mazariegos M, et al. J Nutr 2017;147:1079–85.
205. Swanson CA, King JC. A J Clin Nutr 1987;46:763–71.
206. King JC. Am J Clin Nutr 2000;71:1334S–43S.
207. Hess SY, King JC. Food Nutr Bull 2009;30:S60–78.
208. King JC, Turnlund JR. Human zinc requirements. In *Zinc in Human Biology.* (Mills CF, ed) pp. 335–50, Springer-Verlag, London, 1989.
209. Gibson RS, King JC, Lowe N. Food Nutr Bull 2016;37:443–60.
210. Hambidge MH, Casey CE, Krebs NF. Zinc. In *Trace Elements in Human and Animal Nutrition* (Mertz W, ed) pp. 1–137, Academic Press, New York, Ny, 1986.
211. Prasad AS. J Pharmacol 1985;16:344–52.
212. Mackenzie GG, Keen CL, Oteiza PI. Dev Neurosci 2002;24: 125–33.
213. Liuzzi JP, Aydemir F, Nam H, et al. Proc Nat Acad Sci U S A 2006;103:13612–7.
214. Spahl DU, Berendji-Grun D, Suschek CV, et al. Proc Nat Acad Sci U S A 2003;100:13952–7.
215. Cui L, Blanchard RK, Cousins RJ. J Nutr 2003;133:51–6.
216. Liuzzi JP, Guo L, Chang SM, et al. Am J Physiol Gastrointest Liver Physiol 2009;296:G517–23.
217. MacDonald RS. J Nutr 2000;130:1500S–8S.
218. Haase H, Rink L. Ann Rev Nutr 2009;29:133–52.
219. Moore JB, Blanchard RK, Cousins RJ. Proc Nat Acad Sci U S A 2003;100:3883–8.
220. Koski KG, Scott ME. Ann Rev Nutr 2001;21:297–321.
221. Hess SY. Food Nutr Bull 2017;38:3–17.
222. Khalid N, Ahmed A, Bhatti MS, et al. Crit Rev Food Sci Nutr 2014;54:1222–40.
223. Guan G, Mei P. Modern approaches in dentistry and oral health care. 2018;2(5). doi:10.32474/MADOHC.2018.02.000146
224. Rouhani P, Rezaei Kelishadi M, Saneei P. Eur J Nutr 2022;61: 37–54.
225. Fischer Walker C, Black RE. Ann Rev Nutr 2004;24:255–75.
226. Solomons NW, Cousins RJ. Zinc. In *Absorption and Malabsorption of Mineral Nutrients* (Solomons NW, and Rosenberg IH, eds) pp. 125–97, Alan R. Liss, Inc., New York, 1984.
227. McClain CJ. J Am Coll Nutr 1985;4:49–64.
228. Fritz J, Walia C, Elkadri A, et al. Inflamm Bowel Dis 2019;25: 445–59.
229. Matsui T. J Gastroenterol 1998;33:924–5.
230. Griffin IJ, Kim SC, Hicks PD, et al. Pediatr Res 2004;56: 235–9.
231. Ananthakrishnan AN, Khalili H, Song M, et al. Int J Epidemiol 2015;44:1995–2005.
232. Li D, Achkar JP, Haritunians T, et al. Gastroenterology 2016;151:724–32.
233. Solomons NW, Rosenberg IH, Sandstead HH. Am J Clin Nutr 1976;29:371–5.
234. Deora V, Aylward N, Sokoro A, et al. J Pediatr Gastroenterol Nutr 2017;65:185–9.
235. Hogberg L, Danielsson L, Jarleman S, et al. Acta Paediatr 2009;98:343–5.
236. Rawal P, Thapa BR, Prasad R, et al. J Trop Pediatr 2010;56: 391–7.
237. Crofton RW, Aggett PJ, Gvozdanovic S, et al. Am J Clin Nutr 1990;52:379–82.
238. Tran CD, Katsikeros R, Manton N, et al. Am J Clin Nutr 2011;94:1026–32.
239. Mahawar KK, Bhasker AG, Bindal V, et al. Obes Surg 2017;27:522–9.
240. Ruz M, Carrasco F, Rojas P, et al. Am J Clin Nutr 2021;114: 322–9.
241. Halsted CH, Keen CL. Eur J Gastroenterol Hepatol 1990;2: 399–405.
242. Zhong W, McClain CJ, Cave M, et al. Am J Physiol Gastrointest Liver Physiol 2010;298:G625–33.
243. Szuster-Ciesielska A, Plewka K, Daniluk J, et al. Toxicol Appl Pharmacol 2008;229:1–9.
244. Russell RM. Am J Clin Nutr 1980;33:2741–9.
245. McClain CJ, Su LC. Alcohol Clin Exp Res 1983;7:5–10.
246. Skalny AV, Skalnaya MG, Grabeklis AR, et al. Eur J Nutr 2018;57:2313–22.
247. Uriu-Hare JY, Stern JS, Keen CL. Diabetes 1989;38:1282–90.
248. Walter RM, Jr., Uriu-Hare JY, Olin KL, et al. Diabetes Care 1991;14:1050–6.
249. Beletate V, El Dib RP, Atallah AN. Cochrane Data System Rev 2007;CD005525.
250. El Dib R, Gameiro OL, Ogata MS, et al. Cochrane Data System Rev 2015;CD005525.
251. Haase H, Maret W. Biometals 2005;18:333–8.
252. Chimienti F, Devergnas S, Favier A, et al. Diabetes 2004;53: 2330–7.
253. Staiger H, Machicao F, Stefan N, et al. PloS One 2007;2:e832.
254. Wenzlau JM, Juhl K, Yu L, et al. Proc Nat Acad Sci U S A 2007;104:17040–5.
255. Choi S, Liu X, Pan Z. Acta Pharma Sinica 2018;39:1120–32.
256. Sinha S, Sen S. Int J Diabet Dev Countries 2014;34:220–3.
257. Wang X, Wu W, Zheng W, et al. Am J Clin Nutr 2019;110: 76–90.
258. Fernandez-Cao JC, Warthon-Medina M, V HM, et al. Nutrients 2019;11.
259. Wessels I, Maywald M, Rink L. Nutrients 2017;9.
260. Brown KH, Baker SK, Committee IZS. Food Nutr Bull 2009; 30:S179–84.
261. Prasad AS. Curr Opin Clin Nutr Metab Care 2009;12:646–52.
262. Read SA, Obeid S, Ahlenstiel C, et al. Adv Nutr 2019;10: 696–710.
263. Walker CL, Black RE. Int J Epidemiol 2010;39 Suppl 1:i63–9.
264. Lazzerini M, Wanzira H. Cochrane Database Syst Rev 2016; 12:CD005436.
265. World Health Organization (WHO). 2004 Clinical management of acute diarrhoea: WHO/UNICEF joint statement., WHO, Geneva, 2016.
266. Kupka R, Fawzi W. Nutr Rev 2002;60:69–79.
267. Weismann K. Dan Med Bull 1986;33:208–11.
268. Zemel BS, Kawchak DA, Fung EB, et al. Am J Clin Nutr 2002;75:300–7.
269. Selimoglu MA, Ertekin V, Doneray H, et al. J Clin Gastroenterol 2008;42:194–8.
270. Age-Related Eye Disease Study Research Group. Arch Ophthalmol 2001;119:1417–36.

271. Clemons TE, Kurinij N, Sperduto RD, et al. Arch Ophthalmol 2004;122:716–26.
272. Vavvas DG, Small KW, Awh CC, et al. Proc Nat Acad Sci U S A 2018;115:E696–E704.
273. Fasae KD, Abolaji AO, Faloye TR, et al. J Trace Elem Med Biol 2021;67:126779.
274. Bredesen DE. Aging (Albany NY) 2015;7:595–600.
275. Xu Y, Xiao G, Liu L, et al. Mol Brain 2019;12:106.
276. Overk CR, Masliah E. Biochem Pharmacol 2014;88:508–16.
277. World Health Organization (WHO). *Vitamin and mineral requirements in human nutrition*, WHO, Geneva, 2004.
278. Doets EL, de Wit LS, Dhonukshe-Rutten RA, et al. Eur J Nutr 2008;47 Suppl 1:17–40.
279. EFSA Panel on Dietetic Products Nutrition and Allergies. EFSA J 2014;12:3844.
280. Arsenault JE, Brown KH. Am J Clin Nutr 2003;78:1011–7.
281. Lowe NM, Woodhouse LR, Sutherland B, et al. J Nutr 2004;134:2178–81.
282. Lowe NM, Fekete K, Decsi T. Am J Clin Nutr 2009;89:2040S–51S.
283. Hess SY, Peerson JM, King JC, et al. Food Nutr Bull 2007; 28:S403–29.
284. Hotz C, Peerson JM, Brown KH. Am J Clin Nutr 2003;78: 756–64.
285. Wessells KR, King JC, Brown KH. J Nutr 2014;144:1204–10.
286. International Zinc Nutrition Consultative Group (IZiNCG). *Assessing population zinc status with serum zinc concentration* Vol. IZiNCG Technical Brief no. 2, 2nd Ed., University of California, Davis, CA, 2012.
287. McDonald CM, Suchdev PS, Krebs NF, et al. Am J Clin Nutr 2020;111, 927–37.
288. International Zinc Nutrition Consultative Group (IZiNCG). *Adjusting plasma or serum zinc concentrations for inflammation* Vol. IZiNCG Technical Brief no. 10, IZiNCG, Oakland, CA, 2020.
289. Gibson RS, Hess SY, Hotz C, et al. Brit J Nutr 2008;99: S14–S23.
290. King JC, Brown KH, Gibson RS, et al. J Nutr 2015;146(4): 858S–885S. doi:10.3945/jn.115.220079
291. Wessells KR, Brown KH, Arnold CD, et al. Biol Trace Elem Res 2021;199:442–52.
292. Song Y, Chung CS, Bruno RS, et al. Am J Clin Nutr 2009;90: 321–8.
293. Bremner I, Morrison JN, Wood AM, et al. J Nutr 1987;117; 1595–602.
294. Liuzzi JP, Bobo JA, Lichten LA, et al. Proc Nat Acad Sci U S A 2004;101:14355–60.
295. Hennigar SR, Kelley AM, McClung JP. Adv Nutr 2016;7: 735–746.
296. Haase H, Hebel S, Engelhardt G, et al. Anal Biochem 2006;352:222–30.
297. Mayo-Wilson E, Imdad A, Junior J, et al. BMJ Open 2014;4:e004647.
298. Brown KH, Rivera JA, Bhutta Z, et al. Food Nutr Bull 2004;25:S99–203.
299. Wuehler S, Lopez de Romana D, Haile D, et al. Nutrients 2022;14.
300. European Food Safety Authority (EFSA). *Tolerable Upper Intake Levels for Vitamins and Minerals*, EFSA, Parma, Italy, 2006.
301. Fosmire GJ. Am J Clin Nutr 1990;51:225–7.
302. Botash AS, Nasca J, Dubowy R, et al. Am J Dis Child 1992;146:709–11.
303. Sugiura T, Goto K, Ito K, et al. Acta Paediatr 2005;94:1333–5.
304. Pawa S, Khalifa AJ, Ehrinpreis MN, et al. Am J Med Sci 2008;336:430–3.
305. El Safty A, El Mahgoub K, Helal S, et al. Ann New York Acad Sci 2008;1140:256–62.
306. Greger JL. Zinc: Overview from deficiency to toxicity. In *Risk Assessment of Essential Elements* (Mertz W, Abernathy CO, and Olin SS, eds), ILSI Press, Washington, D.C., 1994.
307. Yuzbasiyan-Gurkan V, Grider A, Nostrant T, et al. *J Lab Clin Med* 1992;120:380–6.
308. Fiske DN, McCoy HE, 3rd, Kitchens CS. 1994;46:147–50.
309. Hedera P, Peltier A, Fink JK, et al. NeuroToxicology 2009:30: 996–9.
310. Nations SP, Boyer PJ, Love LA, et al. Neurology 2008;71:639–43.
311. Afrin LB. Am J Med Sci 2010;340:164–8.
312. Choi DW and Koh JY. Annu Rev Neurosci 1998;21:347–75.
313. Kim AH, Sheline CT, Tian M, et al. Brain Res 2000;886:99–107.
314. Carpenter MC, Lo MN, Palmer AE. Arch Biochem Biophys 2016;611:20–9.
315. Myers SS, Wessells KR, Kloog I, et al. Lancet Glob Health 2015;3:e639–45.

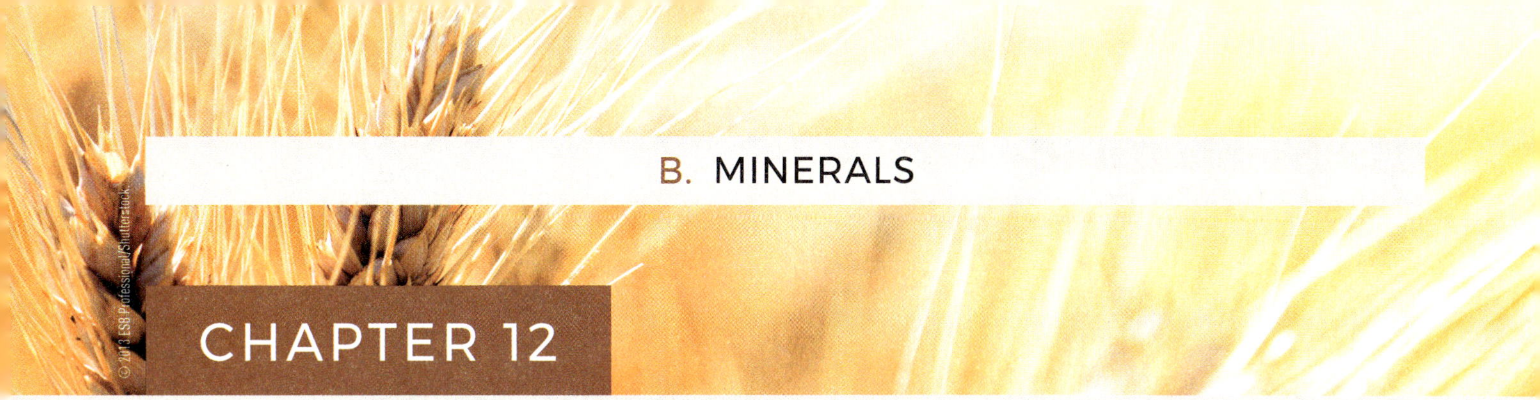

B. MINERALS

CHAPTER 12

Copper

Jason L. Burkhead and James F. Collins

CHAPTER OUTLINE

Abbreviations

AO, amine oxidase; **AOC3**, amino oxidase copper-containing 3; **CNS**, central nervous system; **CP**, ceruloplasmin; **Cu**, copper; **MAO**, monoamine oxidase; **SOD**, superoxide dismutase; **WD**, Wilson Disease

Historical Highlights

It was recognized over two centuries ago that plants and lower marine invertebrates contained copper, but it was not firmly established that mammalian tissues also contained copper until 1921, when Bodansky showed that the human brain contains copper.[1] Also around this time, it was noted that copper and iron were necessary to cure experimental anemia in rats,[2] highlighting a specific physiologic role for copper. Direct evidence of copper being involved in human pathophysiology was provided in the early 1900s, with the first description of Wilson disease (WD)[3]; however, it was not appreciated until several decades later that WD was an inborn error of metabolism.[4] An association between low body copper levels and anemia in humans was suspected in the 1930s, but conclusive experimental proof came decades later. Copper deficiency in humans was first described in patients with Menkes disease in 1962,[5] but the underlying physiologic defect was not discovered until a decade later.[6] It is now widely accepted that copper is an essential nutrient for humans and other mammals. Copper is present in body fluids and tissues in the parts per million (μg/g) to parts per billion (ng/g) range. As copper is an essential enzyme cofactor, yet toxic when in excess, mammals have developed complex physiologic systems for regulating copper absorption, transport, storage, utilization, and excretion.

Chemistry

Copper has an atomic mass of ~63.5 Daltons, with two stable isotopes, ^{63}Cu and ^{65}Cu. Seven radioisotopes of copper also exist. The two with the longest half-lives, ^{67}Cu (~70 hours) and ^{64}Cu (~13 hours), and the two stable isotopes, are most frequently used for experimental analyses of copper metabolism. Copper has two predominant oxidation states in human biology, Cu^{2+} (cupric) and Cu^{+} (cuprous), and it commonly shifts back and forth during enzymatic reactions. Cuprous copper (Cu^{+}) is insoluble in aqueous solutions and is, thus, strongly complexed.[7] Very little free, or unbound, copper exists within biological systems; most is bound to proteins via specific interactions with amino acid side chains, thus mitigating its reactivity.

Biochemical and Physiologic Functions

Copper serves a predominant role in mammalian biology as an enzymatic cofactor for a host of cuproenzymes (**Table 12.1**). Most cuproenzymes are oxidases that facilitate single electron transfer reactions between a substrate and molecular oxygen using either oxidized (Cu^{2+}) or reduced copper (Cu^{+}) atoms. Detailed descriptions of these proteins and their physiochemical properties and functions have been published elsewhere.[8,9] Moreover, a host of copper-binding proteins have been identified, including transmembrane transporters and intracellular chaperones, copper transport proteins in blood and a copper storage protein (**Table 12.2**). Select examples of these proteins will be briefly discussed below.

Catalytic Functions of Selected Cuproenzymes

Amine Oxidases (AO)

AOs function in the oxidative deamination of biogenic, primary amines. In the plasma, amine oxidases act on the physiologically active amines histidine, tyramine, and polyamines. AOs may also play a role in intracellular signaling via production of hydrogen peroxide.[10] Included in this family of proteins is amino oxidase copper-containing 3 (AOC3), which is purported to be involved in leukocyte trafficking[11] and metabolic regulation of adipocyte function.[12]

Monoamine Oxidases (MAO)

Two isoforms of MAO have been identified, MAOA and MAOB, each with distinct tissue localization. These copper-containing enzymes are involved in the catabolism of catecholamines such as serotonin, norepinephrine, tyramine, and dopamine. Abnormal regulation of MAOs has been associated with depression, substance abuse, attention deficit disorder and irregular sexual maturation.[13] Additionally, recent evidence suggests that MAOs possess proinflammatory properties, as inhibitors of these proteins have anti-inflammatory effects.[14]

Diamine Oxidases (DO)

One DO is involved in the catabolism of histamine. In the stomach, acid production is inhibited, while, throughout the body, allergic reactions are attenuated, by the inactivation of histamine via a diamine oxidase. These enzymes also inactivate polyamines and, thus, limit excessive cell growth, potentially having relevance to apoptosis and cancer.[15]

Lysyl Oxidase (LOX)

LOX is a copper-dependent, amine oxidase that initiates crosslinking and stabilization of elastin and collagen fibers. LOX is involved in the formation of connective tissues, including bone, blood vessels, skin, lungs, and teeth. LOX has been implicated in diverse pathophysiologic processes including fibrosis, tumor progression and metastasis, and neurodegenerative and cardiovascular diseases. LOX is, in fact, considered a potential therapeutic target for these pathologic conditions.[16] LOX may also regulate transcription and mediate the epithelial-to-mesenchymal transition in response to hypoxia or TGFβ.[17] Moreover, a family of at least four LOX genes has now been identified (termed LOX-like proteins), with all encoded proteins having similar catalytic domains and predicted copper and cofactor binding sites.[18,19]

Table 12.1 Selected Mammalian Cuproenzymes

Gene Symbol	Biological Function	Outcomes Associated with Copper Deficiency or Genetic Mutations
AOC1	Inactivation of histamine, putrescine, spermidine	Phlyctenulosis; radiation proctitis
AOC3	Cell adhesion; leukocyte trafficking	Diabetes; microvascular complications of diabetes
CP	Iron release from stores; ferroxidase activity	Aceruloplasminemia; hemosiderosis
CCO	Electron transport chain; ATP production	Decreased capacity for oxidative phosphorylation; hypomyelination
DβH	Neurotransmitter (norepinephrine) synthesis	Dopamine β-Hydroxylase deficiency; orthostatic hypotension
HEPH	Intestinal iron absorption; ferroxidase activity	Low serum iron and iron-restricted erythropoiesis; anemia
HEPHL1	Placental iron efflux; alias-Zyklopen	Cytochrome C oxidase deficiency; hair defects
LOX	Crosslinking of elastin and collagen fibers	Aortic aneurysm, familial thoracic aortic aneurysm and dissection
MAOA/B	Metabolism of neuroactive/vasoactive amines	Atypical depressive disorder; Brunner syndrome; Norrie disease
PAM	Activation of biologically-active peptides	Phaeohyphomycosis; Menkes disease
SOD1	Antioxidant defense; cytoplasm, mitochondria	Motor neuron disease; amyotrophic lateral sclerosis
SOD3	Antioxidant defense; extracellular fluids	Myocardial shunting; abestosis
TYR	Melanin biosynthesis; pigmentation	Albinism, oculocutaneous, types, 1a and 1b

Table 12.2 Selected Mammalian Copper-Interacting Proteins

Gene Symbol	Physiologic Function
A2M	Transport of Cu^{2+} in portal and systemic blood
ALB	Transport of Cu^{2+} in portal and systemic blood
APP	Neurite growth; neuronal adhesion; axogenesis
ATOX1	Copper chaperone for ATP7A and ATP7B; copper-dependent transcription factor
ATP7A	Pumps copper into TGN; copper export; Menkes disease gene
ATP7B	Pumps copper into TGN; copper excretion in bile; Wilson disease gene
COMMD1	Biliary Cu excretion; SOD1 dimerization; endosomal trafficking of ATP7A/ATP7B
CCS	Copper chaperone for cytosolic and mitochondrial SOD1
CTR1	Cu^{+} transporter; plasma membrane
COX11	Mitochondrial chaperone to support CCO biosynthesis
COX17	Mitochondrial chaperone to support CCO biosynthesis
MTI/II	Intracellular copper storage/detoxification (also interacts with zinc and cadmium)
PRNP	Function unknown; possible antioxidant defense or copper transport into the CNS
SCO1	Mitochondrial copper chaperone; supports CCO biosynthesis
SCO2	Mitochondrial copper chaperone; supports CCO biosynthesis
SLC25A3	Copper transporter; mitochondrial inner membrane; supports CCO biosynthesis
XIAP	Ubiquitination of COMMD1 (and CCS) for targeting to proteasome for degradation

Peptidylglycine α-Amidating Monooxygenase (PAM)

PAM is a highly conserved, copper- and ascorbate-dependent enzyme that is essential for the activation of many bioactive peptides. A few examples of peptides amidated by PAM are insulin, glucagon, vasopressin, α-melanocyte stimulating hormone, cholecystokinin, gastrin, neuropeptide Y, and substance P.[20,21] Mice lacking PAM die in mid-gestation,[22] whereas those heterozygous for *PAM* exhibit notable deficits, including the inability to maintain body temperature in cold temperatures, increased anxiety-like behavior, and altered inhibitory synaptic neurotransmission.[20] Moreover, polymorphisms in *PAM* were recently suggested to be associated with increased risk for developing type 2 diabetes.[23]

Multi-Copper Ferroxidases (MCFs)

Three MCFs have been identified to date, ceruloplasmin (CP), hephaestin (HEPH), and hephaestin-like 1 (HEPHL1). CP is a liver-derived, circulating glycoprotein that functions in iron release from some tissues by oxidizing ferrous iron to allow subsequent binding of ferric iron to transferrin for distribution in the blood. Mutations in *CP* cause aceruloplasminemia in humans, which is typified by iron-overload, but copper homeostasis is apparently unaffected. A cell-membrane associated, GPI-anchored CP isoform, which is expressed in hepatocytes, brain, and macrophages, was also recently discovered. GPI-CP may be important for iron release from hepatocytes and also hepatic Kupffer cells.[24]

HEPH is a CP-related protein (50% homology), which was originally described as a membrane-anchored, intestinal ferroxidase. It was originally discovered as the mutant gene causing perturbations in iron homeostasis in the sex-linked anemia (*sla*) mouse.[25] Intestinal HEPH expression may respond to body copper concentration in a manner that modulates its activity and, concomitantly, absorption of dietary iron.[26] Recent evidence demonstrates that HEPH is also expressed in the antrum of the stomach, the enteric nervous system, and pancreatic β-cells,[27] but the function of HEPH in these tissues is currently not understood. HEPH-like 1 (HEPHL1) (or zyklopen) was recently discovered in mice, with a proposed function in placental iron efflux.[28] A recent case report, however, described an adolescent male with compound heterozygous *HEPHL1* mutations that caused an abnormal hair phenotype.[29] HEPHL1-mediated ferroxidase activity may thus be essential for normal hair growth and development. A recent review article provides more detail on the mammalian MCFs.[30]

Cytochrome C Oxidase (CCO)

Mitochondrial CCO, the terminal component of the electron transport chain, reduces molecular oxygen to form water, and ultimately allows ATP to be produced by the generation of a proton gradient. CCO is a large complex, consisting of 13 proteins. It contains two heme groups and three copper ions, which are involved in electron transfer. CCO requires a specific mitochondrial copper delivery system for normal function. Mutations affecting function/activity of this copper delivery system are likely lethal.[31]

Dopamine β-Hydroxylase (DβH)

DβH requires copper in each of its four subunits and ascorbate as a co-factor. It catalyzes the conversion of dopamine to norepinephrine, which functions as a neurotransmitter in the central and peripheral nervous systems. Altered noradrenergic signaling has been associated with the development of nervous system pathologies. Circulating DβH has been evaluated as a biomarker of norepinephrine function in several human diseases.[32] Expression of DβH is highest in the adrenal medulla, sympathetic neurons in the peripheral nervous system, and in noradrenergic and adrenergic neurons in the brain.[21] In mice, *DβH* gene inactivation leads to embryonic lethality, exemplifying its key role in nervous system physiology.[33]

Superoxide Dismutases (SOD)

SOD proteins scavenge superoxide free radicals to protect against oxidative damage. Cu/Zn-superoxide dismutase (SOD1) and extracellular SOD (SOD3) require catalytic copper (and structural zinc) for function. Cytosolic SOD1 is a homodimer of ~32 kDa, while extracellular SOD3 is a tetramer of ~135 kDa. SOD1 may also function in gene transcription,[32] stabilization of specific RNA species,[32] and in oxygen sensing and nutrient signaling.[34,35] SOD3, the predominant extracellular dismutase, is present in lymph, synovial fluid, and plasma.[23] SOD3 may also be involved in the development of chronic obstructive pulmonary disease in humans,[36] while more recent work suggests broader inflammatory and immunomodulatory functions.[37]

Tyrosinase (TYR)

TYR is required for melanin biosynthesis and, thus, pigmentation. Loss of activity leads to albinism. TYR catalyzes the conversion of tyrosine to dopamine and the subsequent oxidation of dopamine to dopaquinone, steps along the pathway of melanin synthesis. The copper dependency of this process is best exemplified by the achromotrichia observed in copper-depleted domestic and laboratory animals.[21]

Physiologic Functions of Copper

The requirement of copper for several enzymes discussed above gives us clues as to the outcomes of copper depletion. In many cases, symptoms of copper inadequacy can be directly linked to decreased activity of one or more of these cuproenzymes.

Connective Tissue Formation

The copper-dependent enzyme LOX is required for normal formation of connective and bone tissue, as well as the integrity of connective tissue in the heart and vasculature. Copper deficiency may, thus, result in connective tissue disorders, osteoporosis, and bone defects. Skeletal perturbations have been documented in copper-deficient neonates, mirroring the bone abnormalities of scurvy (vitamin C deficiency).[38] Moreover, recent data demonstrated that long-term copper supplementation may decrease bone loss in adult humans,[39] but contradictory results have also been obtained.[40] Although whether copper supplementation positively impacts bone health is unclear, low copper intake is clearly associated with poor bone health,[41] consistent with the connective tissue weaknesses associated with copper transport defects in Menkes disease and other ATP7A-related disorders.[42]

Iron Metabolism

Copper homeostasis is intimately entwined with that of iron.[43] The most obvious link is the multi-copper ferroxidases, CP and HEPH. During copper depletion, HEPH activity in the intestine is impaired and CP activity is extremely low. The net effect is that iron absorption is reduced, and iron efflux from liver (and other tissues) is impaired, leading to reductions in serum iron.[44] This hypoferremia can result in development of an iron-deficiency-like anemia, probably due to reduced iron delivery to the erythron and an inability of erythroid precursors to properly use iron for hemoglobin synthesis.

Central Nervous System Function

Copper plays well known roles in the physiology of the central nervous system (CNS), including in brain development. Copper is deposited in the brain late in gestational development and during the perinatal period and, as such, copper deprivation of pregnant or lactating females results in pathological outcomes in offspring. Many of the effects of copper deprivation can be ascribed to altered expression/activity of cuproenzymes in the CNS, as their activity is impaired by copper depletion.[21] The essentiality of copper in brain development is perhaps best exemplified by the neuropathological phenotype of infants with the genetic copper deficiency disorder Menkes disease, as well as the milder *ATP7A*-related disorders Occipital Horn Syndrome (OHS) and Distal Motor Neuropathy (DMN).[42] The tremors, ataxia, hypo- or demyelination of nerve fibers, and reductions in some neurotransmitters observed during copper deficiency likely result from decreased production of sphingolipids (as mediated by CCO) and decreased activity of DβH and MAOs.[32,45]

Melanin Pigment Production

Copper is necessary for normal pigmentation given the copper dependency of TYR, which is a key enzyme in melanin synthesis. During copper depletion in humans and animals, depigmentation of skin and hair is commonly observed.

Cardiovascular Function

Several pathologic anomalies in the cardiovascular system are noted in severely copper-deficient, young animals. Copper deficiency may also increase chances for developing cardiovascular disease in humans. For example, some clinical trials demonstrated that cardiac arrhythmias developed after consumption of a low-copper diet.[21] Supporting this observation, several observational studies have shown that lower serum copper concentration was associated with increased incidence of coronary heart disease.[46] Copper deficiency may also perturb normal lipid metabolism and result in hypercholesterolemia and hypertriglyceridemia, thus increasing risk for development of atherosclerotic cardiovascular disease.[47]

Immunity

Copper likely plays an important role in immune system function. For example, many studies have shown that systemic copper deficiency is associated with increased risk of infection,[48] probably because cellular and humoral immune system factors require copper for normal function. Moreover, neutropenia is commonly associated with copper depletion in humans, and macrophage and lymphocyte functions are impaired by even marginal copper deficiency. Studies in adult males demonstrated that in vitro stimulation of T lymphocytes was suppressed by consuming a diet with 0.36 mg copper/day for 42 days (which is well below the RDA of 0.9 mg/d). These individuals further had decreased plasma copper concentration and reduced activity of some copper-dependent enzymes, but their hematologic parameters were normal.[49] Furthermore, at the cellular level, copper may have a direct antimicrobial function, such as in the phagolysosome.[50,51] These, and other recent findings,[52,53] implicate copper in the ability of immune cells to respond to infectious stimuli. Definitive proof, however, is lacking mainly due to the inability to detect marginal copper deficiencies in experimental subjects.

Lipid Metabolism

A reciprocal relationship between copper and lipid metabolism has been defined through experimental animal model and human studies over the last 4 decades. For example, elevated liver cholesterol synthesis and serum triglycerides are noted in copper deficiency, while copper excess has the opposite effect. Copper modulation of lipid metabolism may be tissue specific. For example, in white adipose tissue, copper depletion decreases cAMP and lipolysis, while copper excess increases both.[54] Moreover, a recent meta-analysis of eight cross-sectional studies and 27 case-control studies indicated a positive association between elevated serum copper and obesity in children and adults, compared with healthy weight controls and no difference between overweight and control.[55] Copper-dependent amine oxidase, AOC3, regulates adipocyte size, while obese patients have increased circulating AOC3 and amine oxidase activity.[12,56]

Cell Cycle Regulation

Although the primary role of copper in cells is typically thought of as a catalytic cofactor, recent work suggests a signaling role that impacts cell cycle, metabolism and potentially cancer progression and treatment. For example, one investigation concluded that copper is an allosteric regulator, acting via interaction with MEK1 and MEK2, which influences cell proliferation and growth.[57] Chelator-mediated copper depletion has been proposed and is in clinical trials as an anti-cancer therapy.[58] Recent work to investigate how excess copper may selectively kill cancer cells has led to a description of copper toxicity named "cuproptosis," where high intracellular copper damages lipoylated TCA cycle proteins, resulting in a specific copper-dependent mechanism of cell death.[59]

Bioavailability

The relative amount of copper in the diet seems to be the major predictor of absorption. Several factors, including certain amino acids and proteins, iron, zinc, molybdenum, vitamin C, and carbohydrates have been reported to exert adverse effects on the bioavailability of dietary copper.[60] High doses of zinc induce symptoms of systemic copper deficiency, as recently reported in several patients that used excessive amounts of zinc-containing denture cream.[61] The impact of dietary components on copper absorption may be more pronounced in neonates, as digestive function and homeostatic regulation of biliary copper excretion are not yet fully developed.

Nutrient Interactions

Cu metabolism is known to be affected by iron, zinc, and vitamin C. Variations in dietary copper may also affect the homeostasis of other nutrients, a topic that will not be considered in detail here.

Copper-Iron Interactions

Copper and iron interact in numerous ways.[62] Important iron-copper interactions in the gut include the regulation of HEPH by dietary copper intakes and the regulation of ATPase Copper Transporting Alpha (ATP7A; a copper efflux transporter) expression by iron concentration.[63,64] Furthermore, liver copper varies inversely with iron status, for unexplained reasons.[65] Interestingly, an unknown aspect of erythroblast iron use is copper dependent as, during copper deficiency, hemoglobin production is inefficient despite normal serum iron.[66]

Copper-Zinc Interactions

High dietary zinc intake impairs copper absorption. This may, in part, be explained by the induction of metallothionein, a copper-binding protein, in enterocytes. Moreover, copper depletion has been observed in humans consuming supplements containing 50 mg of zinc daily for extended periods of time; this is, in fact, the rationale for the UL for zinc of 40 mg/d for adults.[67] Reciprocally, studies focused on copper toxicity in Wilson Disease suggest that excess copper inhibits specific zinc-dependent biochemical processes.[68]

Copper-Ascorbate Interactions

Ascorbic acid supplementation may induce copper deficiency in experimental animals and could have a similar effect in humans. In premature infants, plasma vitamin C was negatively correlated with serum CP and antioxidant activity.[69] Other human studies have also suggested that ascorbic acid supplementation may perturb serum ferroxidase activity.

Food Sources

The typical diet of an adult in the U.S. supplies slightly more copper than is recommended by the RDA (0.9 mg/d). The richest sources of dietary copper are shellfish, seeds, nuts, organ meats, wheat bran cereal, whole grain products and chocolate containing foods. Vegan diets contain ample copper, but absorption seems to be lower from plant foods than from animal-based sources.[70] Other sources of copper include vitamin and mineral supplements, although copper is often in the cupric oxide form which has low bioavailability.

Metabolism

Genetic Regulation

Expression of transcripts encoding many proteins involved in copper metabolism, including, for example, copper transporters (e.g., CTR1, ATP7A, ATP7B), does not change in response to variations in dietary copper intake or according to body copper status.[71] Regulation of copper homeostasis, thus, seems to be mostly at a post-transcriptional level, predominantly via protein trafficking mechanisms.[72]

Body Copper Homeostasis

Copper intake from a typical, balanced diet is ~1.2–1.3 mg/d, with around 0.8 mg being absorbed daily in the upper portion of the small intestine (**Figure 12.1**). Excretion occurs predominantly into the bile (~0.4 mg/d), with total fecal losses being ~1 mg/d. In the liver, copper is incorporated into CP and other cuproenzymes. CP is secreted into the blood, along with atomic copper, which binds to serum proteins for copper delivery to cells of the body. Homeostatic control of body copper includes modulation of enteral copper absorption and biliary copper excretion.

Vectorial Intestinal Copper Transport

Dietary and supplemental copper, and endogenous copper from saliva, gastric juices, and pancreatic and biliary secretions, constitute the intestinal copper pool (Figure 12.1). Enteral copper (mainly Cu^{2+}) must be reduced to the Cu^{+} state for absorption (**Figure 12.2**). Three cupric reductases have been identified, cytochrome b [558] ferric/cupric reductase, Steap 2 and Cybrd1, but the relative contribution of each has not been definitively established.[62] Subsequently, copper is transported into enterocytes mainly via copper transporter 1 (CTR1).[73] Within intestinal epithelial cells, copper is bound by chaperones that traffic copper within the cytosol to support SOD1 function (CCS) or deliver it to the *trans*-golgi network (ATOX1) for incorporation into cuproenzymes in the secretory pathway. Copper is also trafficked into mitochondria, but the chaperone involved has not been definitively established. Copper movement within mitochondria involves an additional host of copper chaperones and a copper transporter that moves copper from the inter membrane space into the matrix (Figure 12.2). Excess copper may be bound in cells by metallothionein (MT). Finally, copper can be transported out of enterocytes by ATP7A. Once copper exits enterocytes, the oxidizing environment of the interstitial fluids presumably converts cuprous copper (Cu^{+}) to cupric copper (Cu^{2+}), which binds to albumin or α_2-macroglobulin for delivery in the portal blood to the liver.

Copper Transport and Transfer

After transport to the liver, Cu^{2+} is reduced and then imported into hepatocytes via CTR1 (**Figure 12.3**).[74] Inside liver cells, copper is bound to chaperones and distributed similarly to what occurs in enterocytes. ATP7B pumps copper into the TGN where it is incorporated into CP and other cuproproteins. Excess copper stimulates the translocation of ATP7B from the TGN to the canalicular membrane of the hepatocyte, facilitating excretion of copper into bile.[75]

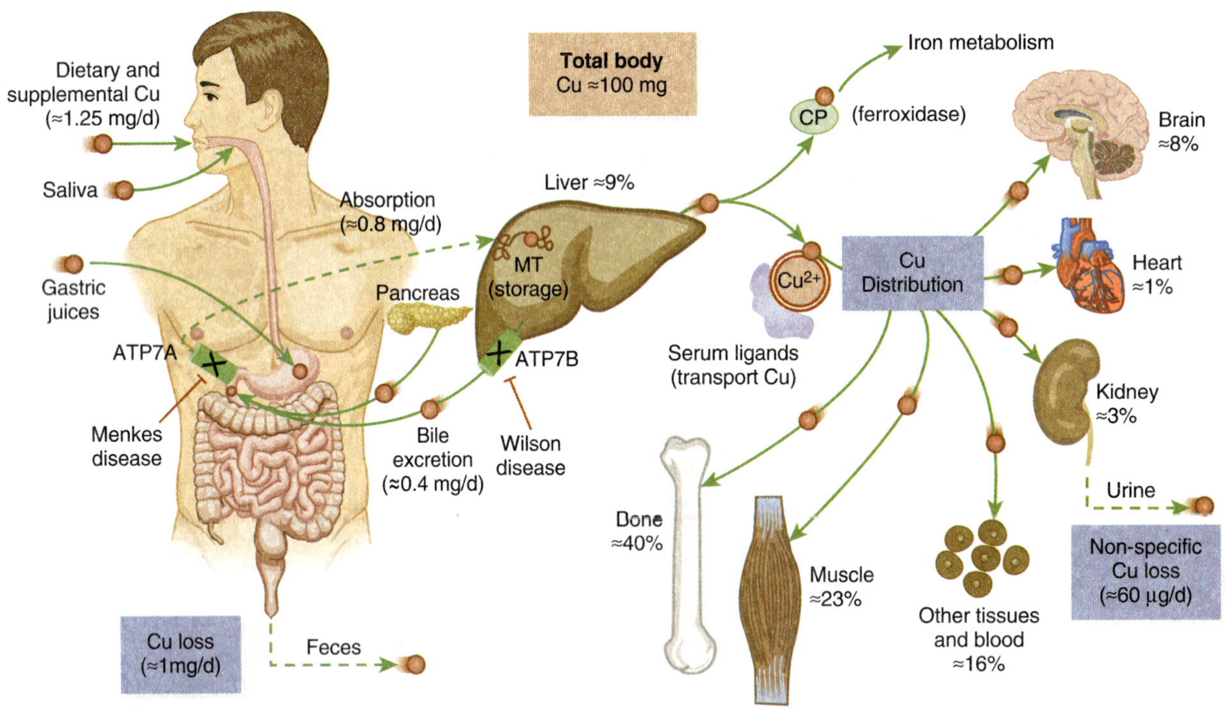

Figure 12.1 Total Body Copper Homeostasis. The major regulatory mechanisms that control body copper levels, including absorption, transport, and excretion are depicted in this diagram. Enteral copper derives from the diet and from endogenous GI tract and pancreatic secretions. The ATP7A copper-transporting ATPase mediates copper efflux from enterocytes. Copper can be bound within MT in the liver, or released and distributed to body cells, tissues, and organs. Numbers under various organs indicate the approximate percentage of body copper that is present in that organ/tissue. Although most copper in the serum is bound to CP, other copper binding proteins must exist as the absence of CP (in aceruloplasminemia) does not cause copper depletion of peripheral tissues. Copper excretion occurs mainly in the bile and is mediated by the ATP7B copper-transporting ATPase. Small amounts of copper are lost in the urine and excess enteral copper is eliminated in the feces. Mutations in *ATP7A* underlie Menkes Disease (MD) and mutations in *ATP7B* cause Wilson Disease (WD).

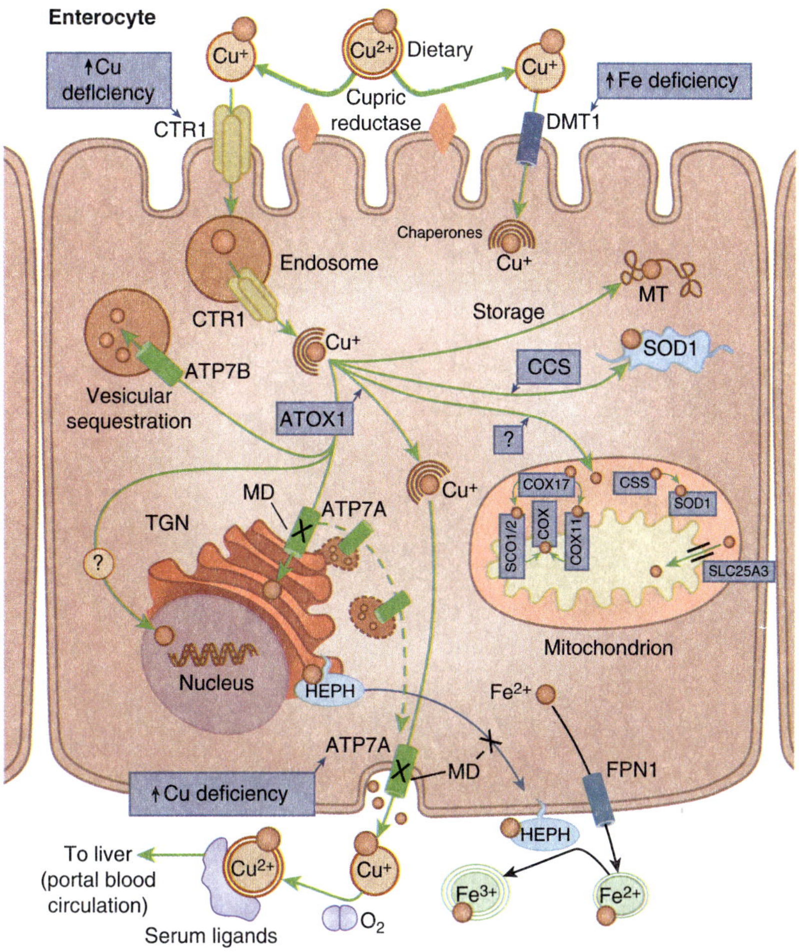

Figure 12.2 Enterocyte Copper Homeostasis. A single enterocyte is depicted in this diagram, which shows the processes that are involved in enteral copper absorption. Dietary Cu^{2+} is reduced to Cu^{+}, probably by three identified cupric reductases, and then transported into enterocytes by CTR1. This process may involve endocytosis followed by copper export from endosomes into the cytosol. DMT1 could mediate copper import as well, but probably mainly during iron deficiency. Once in cells, Cu^{+} interacts with chaperones for intracellular distribution. The CCS chaperone transfers copper to cytosolic SOD1. ATOX1 delivers copper to ATP7A in the TGN and possibly the basolateral membrane, and into the nucleus (where ATOX1 may function as a transcription factor). ATOX1 may also deliver copper to ATP7B contained within intracellular vesicles for storage. Copper transfer to mitochondria is mediated by an unknown chaperone. Within mitochondria, CCS delivers copper to SOD1, while COX17, SCO1/2, and COX11 provide copper to support biosynthesis of CCO. Copper enters the mitochondrial matrix via SLC25A3, also supporting synthesis of CCO. In the TGN, copper is incorporated into cuproenzymes, including HEPH, which functions as a ferroxidase to allow iron release from cells, and subsequent binding to transferrin (which is not shown). When intracellular copper is elevated, ATP7A-containing vesicles traffic to the BLM, thus facilitating copper efflux by vesicular exocytosis. Cu^{+} that exits cells is oxidized by O_2 (to Cu^{2+}) and then bound by ALB and A2M for transport to the liver. CTR1 and ATP7A expression and/or activity may increase during copper depletion. Dysfunctional copper transport by ATP7A underlies MD in humans. Excess intracellular copper can be stored bound to MT.

Copper Excretion

The primary excretory route for endogenous copper is via hepatocytes into bile, as mediated by ATP7B and a host of accessory proteins. Biliary copper and unabsorbed enteral copper are eliminated from the body in the feces. Copper excretion is less effective during fetal and neonatal life, which increases risk for copper toxicosis. Impaired bile secretion, or cholestasis, can also increase liver copper content.

Copper Storage

Total copper content of adult humans ranges from 50–120 mg. In general, copper is not stored in the human body; tissue copper, thus, likely reflects quantities of cuproenzymes.

Copper: Homeostatic Mechanisms

Humans have developed efficient adaptive mechanisms designed to protect against copper deficiency and toxicity. Absorption of dietary copper is regulated, with percent absorption increasing when intakes are low.[75] Under conditions of high copper intake, copper may be sequestered in metallothionein in enterocytes and biliary excretion may also increase. Copper absorption under normal conditions is ~10%, reflecting combined absorption and subsequent excretion of newly absorbed copper.[75] These adaptive mechanisms become inefficient with chronic intake of copper <0.7 mg/d; fortunately, this intake level is well below the estimated average intake in the United States of ~1.2 mg/d.

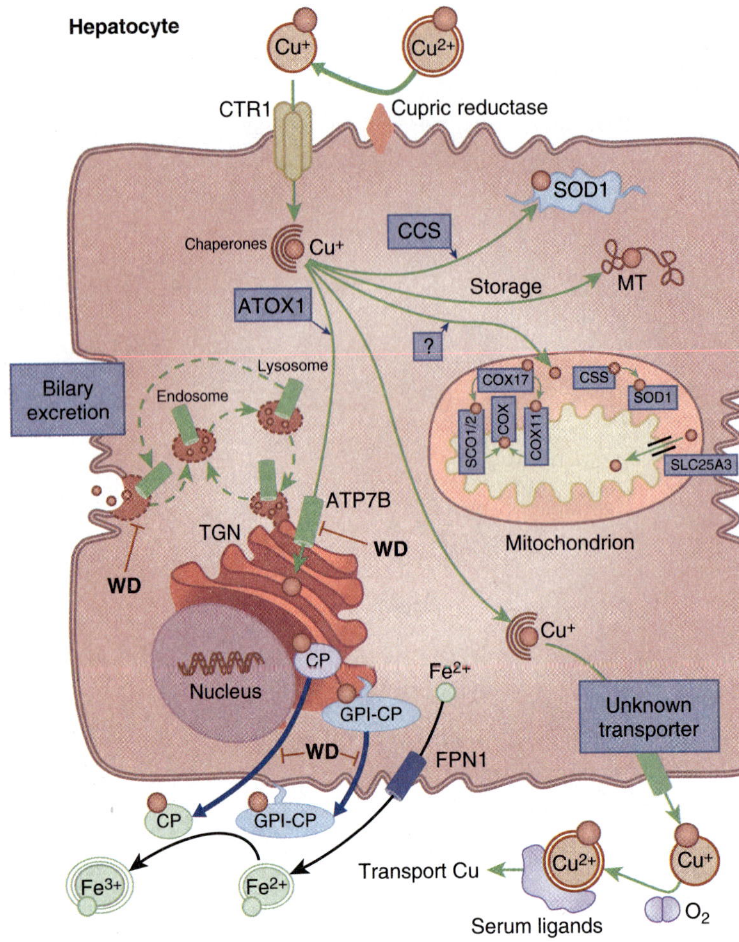

Figure 12.3 Hepatocyte Copper Homeostasis. A single hepatocyte is depicted in this diagram, which shows the major proteins involved in copper metabolism. Cu^{2+} in the portal blood circulation is reduced to Cu^{+} prior to being imported into the cell by CTR1. As in enterocytes, intracellular chaperones distribute copper, and excess copper can be stored in MT. Mitochondrial copper handling is probably the same as that depicted for the enterocyte. The ATP7B copper-transporting ATPase pumps copper into the TGN for incorporation into CP and other cuproenzymes. CP and GPI-CP are ferroxidases that oxidize Fe^{2+} to Fe^{3+} after release by various tissues and cells, including hepatocytes. ATP7B-containing endosomes derived from the TGN fuse with lysosomes, which can then facilitate copper excretion in bile via vesicular exocytosis or cycle back to the TGN. Atomic copper also exits hepatocytes via an unknown transport process, followed by oxidization and binding to various serum ligands (e.g., ALB, A2M) for distribution throughout the body. Dysfunctional ATP7B underlies the copper-loading disorder WD in humans.

Genetic Defects in Copper Metabolism

Copper-related pathologies in humans most frequently arise due to defects in copper exporters encoded by *ATP7A* and *ATP7B*, as detailed below.

Menkes Disease (MD)

MD is an X-linked, recessive disorder of copper metabolism affecting multiple organ systems. As expected, most patients are males. Typical manifestations include progressive neurodegeneration, connective tissue disturbances and unusual "kinky" hair. The disease is usually fatal by the age of 3 years; no cure exists although copper histidine treatment (via subcutaneous injection) of affected individuals early in life has shown promise, particularly in partial amelioration of neurologic symptoms.[76] The prevalence of MD varies in different regions of the world with a lower incidence noted in Japan and Europe (1:300,000–1:360,000), but a much higher frequency in Australia (1:50,000–100,000).[77] The underlying genetic defect is in the *ATP7A* gene, encoding a copper-transporting ATPase required for copper delivery to cuproenzymes in the secretory pathway and for cellular copper export. Less deleterious *ATP7A* mutations cause Occipital Horn Syndrome (OHS) and Distal Motor Neuropathy (DMN), which present with milder phenotypes than Menkes disease.[42]

Defective copper elimination from cells is the basic physiologic disturbance in MD, with most tissues (except for liver and brain) accumulating excess copper. Copper does not, however, accumulate to toxic levels, at least partially due to an intestinal block to absorption of dietary copper. The lack of copper in some peripheral tissues (e.g., liver and brain), despite accumulation in other tissues (e.g., intestinal mucosa, muscle, spleen, and kidney), leads to signs and symptoms of systemic copper deficiency.[78] These include low serum copper and CP activity, and impaired synthesis of SOD and CCO. LOX activity is also impaired leading to defective artery formation in the CNS, and osteoporosis. Progressive nerve degeneration is noted in the brain, leading to the classic neurologic symptoms of MD.[79]

Wilson Disease (WD)

WD is an autosomal recessive disease in which affected individuals exhibit excessive copper storage.[80] The underlying defect is in the *ATP7B* gene, which encodes a copper-transporting ATPase. The worldwide prevalence of WD was reported to be 1:30,000.[81] In WD patients, copper accumulation in liver, brain, and cornea (Kayser-Fleisher rings) results in multiorgan damage, being particularly severe in brain and liver. Neurological damage and cirrhosis may ensue if left untreated. Acute hepatitis, hemolytic crisis and hepatic failure may also occur. Other observations include abnormally high urinary copper excretion and low CP values. Permanent organ damage can be avoided in WD with proper medical treatment, especially if initiated early in life. Typical treatments use decoppering chelation therapy with penicillamine and/or trientine,[82] with life-long compliance required, or high zinc dosing (which interferes with absorption of dietary copper).

Huppke-Brendel Syndrome

Huppke-Brendel Syndrome is caused by defects in the *SLC33A1* gene and is a lethal autosomal recessive disorder characterized by low or undetectable serum copper and CP, severe developmental delay, congenital cataracts, and significant defects in cerebellar structure. *SLC33A1* encodes a highly conserved acetyl-CoA transporter that is required for acetylation of specific gangliosides and glycoproteins, including CP.[83]

MEDNIK Syndrome

MEDNIK syndrome (Mental retardation, Enteropathy, Deafness, Neuropathy, Ichthyosis, Keratodermia) is caused by mutations in the *AP1S1* gene, which encodes the small subunit of the adapter protein 1 complex (σA1).[84] Lack of AP1S1 activity impacts clathrin coat assembly and subcellular trafficking between the *trans*-Golgi, the endosomal system, and the plasma membrane, which is critical for the copper export functions of ATP7A and ATP7B. Biochemical characteristics include features of both Menkes Disease and Wilson Disease, while treatment is by zinc supplementation, as in Wilson Disease.

Copper Deficiency in Humans and Animals

Severe copper deficiency in animals results in abnormalities in the immune, skeletal, and cardiovascular systems. Further consequences of copper deprivation are hypochromic anemia (which does not respond to iron supplementation), hypopigmentation, thrombocytopenia, and neutropenia.[38] Across many species, other hallmarks of copper deficiency in addition to anemia include neutropenia, and osteoporosis. More specific features include: 1) skeletal abnormalities, fractures, and spinal deformities; 2) neonatal ataxia; 3) depigmentation of hair and wool; 4) abnormal keratinization of hair, wool, and fur; 5) reproductive failure; 6) cardiovascular abnormalities; and 7) impaired immune function. It should be noted here that some of these symptoms have only been observed in one or two species and, further, that manifestations of copper deficiency in experimental and other animals is typically more severe than is what observed in humans (as described in the next paragraphs).

Systemic copper deficiency in humans, which is rare, can result from ineffective absorption of dietary copper or excessive copper loss through the endogenous biliary excretion system. Several groups of individuals are susceptible to copper deficiency, including: 1) individuals chronically receiving total parenteral nutrition without proper copper supplementation; 2) premature infants consuming milk-based formulas lacking adequate copper; 3) neonates experiencing chronic diarrhea or malnutrition; 4) hospitalized patients undergoing long term peritoneal dialysis; 5) severe burn patients; 6) renal dialysis patients; and 7) persons consuming large doses of supplemental zinc, antacids or copper chelators.[19] Malabsorption may also result in copper deficiency. Moreover, recent evidence suggests an association between surgical bowel resection in the management of morbid obesity and acquired copper deficiency.[85]

In humans, moderate to severe copper deficiency is accompanied by low serum copper concentration, and reduced serum ferroxidase (e.g., CP) activity. Usual pathophysiologic features include anemia, leukopenia, and neutropenia. During periods of rapid growth, osteoporosis is a common feature. Furthermore, mild, or subclinical, copper deficiency because of long-term low copper intakes is a possibility in humans. This may result in additional manifestations, including increased risk for developing arthritis, arterial disease, depigmentation, myocardial disease, and neurological abnormalities.[75] Possible additional effects of marginal copper deficiency include cardiac arrhythmias, increased serum cholesterol and glucose intolerance.[86] It is also important to note that these observations were not duplicated in all human studies, indicating that future experimental work is necessary.

Dietary Considerations and Requirements

Dietary Requirements and Recommendations

Dietary reference intakes for copper were established almost 2 decades ago[87] and are listed in **Table 12.3**. Based on a lack of experimental data, Adequate Intake (AI) levels for copper have been established for infants 0–6 months-of-age and for those between 7 and 12 months. The RDA increases throughout childhood and adolescence and is increased from adult levels during pregnancy and lactation. Upper tolerable intake levels (ULs) have also been established for copper (also listed in Table 12.3).

Evaluation of Copper Status

Analytical Methods

The most widely used methods to quantify copper are inductively coupled plasma optical emission spectroscopy (ICP-OES) and atomic absorption spectroscopy (AAS).[13] For AAS, samples are atomized with a graphite furnace for electrothermal ionization (GFAAS) or with an air-acetylene flame (flame AAS) for

Table 12.3 Copper Dietary Reference Intakes

Copper DRIs (μg/d)			
AGE (years)	RDA[a] (males/females)	AI[b] (males/females)	UL[c]
0–0.5	—	200	ND[d]
0.5–1.0	—	220	ND
1–3	340		1,000
4–8	440		3,000
9–13	700		5,000
14–18	890		8,000
19–30	900		10,000
31–50	900		10,000
51–70	900		10,000
>70	900		10,000
Pregnancy			
≤18	1,000		8,000
19–50	1,000		10,000
Lactation			
≤18	1,300		8,000
19–50	1,300		10,000

[a]RDA = Recommended Dietary Allowance.
[b]AI = Adequate Intake (mean intake of healthy infants receiving human milk).
[c]UL = Tolerable Upper Intake Level (total intake from food, water, and supplements should not exceed this value).
[d]Not Determinable (the source of intake should be from food only for these age groups).

electrothermal ionization. ICP-OES is often used when more than one mineral is being quantified. Studies to determine the distribution of copper in living animals typically use stable isotopes of copper to track the absorption, utilization, excretion, and turnover of copper in biological systems.[75] The most common method to measure the copper isotope ratios is mass spectrometry (MS), with ICP-MS and thermal ionization mass spectrometry (TIMS) being the most commonly used techniques, including more recent capacity for single-cell analysis and spatial localization of metals in tissue sections.[88] X-ray fluorescence microscopy (XFM) is capable of high-resolution spatial quantitation of metals in tissues and cells.[89]

Assessment of Copper Status

Much effort has recently gone into identifying biomarkers of copper status that are sensitive to even marginal deficiency, and are noninvasive and consistently reliable.[78] The most frequently utilized method in humans has been to quantify copper concentration and the activity of various cuproenzymes in blood.[90] Reductions in plasma copper and CP activity are noted in severely copper-deficient humans; intakes in experimental settings of 0.6 mg/d or less for at least a month-and-a-half are required to consistently see these decreases. Observed reductions in serum Cu and CP activity are, however, complicated by the fact that several (patho)physiologic alterations can increase copper content and CP activity in blood, including the acute phase response to infection and inflammation, pregnancy and other hormonal perturbations, and some carcinogenic phenotypes.[90] Other utilized copper biomarkers, including SOD1 activity in erythrocytes, CCO activity in platelets and mononuclear cells, and copper content of various circulating blood cells, have shown limited usefulness in determining the copper status of humans. Erythrocyte CCS expression, however, was shown to be responsive to dietary copper depletion and repletion.[91] Finally, studies in rats revealing alterations in serum and tissue PAM activity[92] have been correlated with similar observations in copper-deficient humans.[93] Additional experimentation with these promising, new biomarkers, is, however, required.

Copper Toxicity and Upper Limits

Copper toxicity is rare, as humans and other mammals have evolved adequate homeostatic mechanisms to regulate body copper content, since free (unbound) copper is potentially toxic (as it can potentiate the generation of reactive oxygen species). Free copper in cells and in the body is, thus, extremely low; copper almost always exists in biological systems bound to proteins. Ingestion of high copper levels may, however, override the innate checkpoints designed to regulate overall body copper, leading to enhanced intestinal absorption in the absence of a physiological demand for copper. Due to possible adverse consequences of high copper ingestion and potential liver damage, a UL of 10 mg/d has been established for adults.

Copper is often included in complete nutrition and micronutrient supplements without consequence. In a recent study, adults were supplemented with 10 mg of cupric gluconate daily for 12 weeks without evidence of liver damage or gastrointestinal distress.[94] Copper toxicity risks are, however, higher for neonates and infants given an immature biliary excretion system and an enhanced intestinal absorption apparatus. Copper loading is observed clinically today in the setting of Wilson disease and other disorders in which biliary copper excretion is impaired such as biliary cirrhosis and biliary atresia.

Acknowledgments

The writing of this chapter was supported by grants R01 DK074867 from the National Institute of Diabetes and Digestive and Kidney Diseases (NIDDK) & R01 DK109717 from NIDDK and the Office of Dietary Supplements (to JFC), and R15 DK114747 & P20GM103395-7525 (to JLB).

References

1. Bodanksy MJ. J Biol Chem 1921;48:361.
2. Cohn EJ, Fulton JF, Minot GR, et al. J Biol Chem 1927;74: 69–72.
3. Wilson SA. Brain 1912;34:295–509.
4. Mason KE. J Nutr 1979;109:1979–2066.
5. Menkes JH, Alter M, Steigleder GK, et al. Pediatrics 1962;29: 764–79.
6. Danks DM, Campbell PE, Stevens BJ, et al. Pediatrics 1972; 50:188–201.
7. Dyer FF, Leddicotte GW. National Research Council: The Radiochemistry of Copper, The National Academies Press, Washington, DC, 1961.
8. Harris ED. Copper Vol. 2, Marcel Dekker, New York, NY, 1997.
9. Linder MC. Int J Mol Sci 2020;21:4932.
10. McDonald A, Tipton K, O'Sullivan J, et al. J Neural Transm (Vienna) 2007;114:783–6.
11. Dunkel P, Gelain A, Barlocco D, et al. Curr Med Chem 2008;15:1827–39.
12. Yang H, Ralle M, Wolfgang MJ, et al. PLoS Biol 2018;16:e2006519.
13. Turnlund JR. Copper, Lippincott, Williams & Wilkins, Baltimore, MD, 2006.
14. Ostadkarampour M, Putnins EE. Front Pharmacol 2021;12: 676239.
15. Toninello A, Pietrangeli P, De Marchi U, et al. Biochim Biophys Acta 2006;1765:1–13.
16. Rodriguez C, Rodriguez-Sinovas A, Martinez-Gonzalez J. Drug News Perspect 2008;21:218–24.
17. Yang N, Cao DF, Yin XX, et al. Biomed Pharmacother 2020; 131:110791.
18. Grau-Bove X, Ruiz-Trillo I, Rodriguez-Pascual F. Sci Rep 2015;5: 10568.
19. Molnar J, Fong KS, He QP, et al. Biochim Biophys Acta 2003;1647:220–4.
20. Merkler DJ, Hawley AJ, Eipper BA, et al. Br J Pharmacol 2022;179:3306–24.
21. Prohaska JR. Copper, ILSI Press, Washington, DC, 2006.
22. Czyzyk TA, Morgan DJ, Peng B, et al. J Neurosci Res 2003;74: 446–55.
23. Thomsen SK, Raimondo A, Hastoy B, et al. Nat Genet 2018;50:1122–31.
24. Jonczy A, Lipinski P, Ogorek M, et al. Metallomics 2019;11: 1079–92.
25. Vulpe CD, Kuo YM, Murphy TL, et al. Nat Genet 1999;21: 195–9.
26. Chen HJ, Huang G, Su T, et al. J Nutr 2006;136:1236–41.
27. Hudson DM, Curtis SB, Smith VC, et al. Am J Physiol Gastrointest Liver Physiol 2010;298:G125–32.
28. Chen H, Attieh ZK, Syed BA, et al. J Nutr 2010;140:1728–35.
29. Sharma P, Reichert M, Lu Y, et al. PLoS Genet 2019;15:e1008143.
30. Helman SL, Zhou J, Fuqua BK, et al. BioMetals 2023;263–281.
31. Baker ZN, Cobine PA, Leary SC. Metallomics 2017;9:1501–12.
32. Gonzalez-Lopez E, Vrana KE. J Neurochem 2020;152:157–81.
33. Thomas SA, Matsumoto AM, and Palmiter RD. (1995) Nature 374, 643–6.
34. Reddi AR, Culotta VC. Cell 2013;152:224–35.
35. Montllor-Albalate C, Kim H, Thompson AE, et al. Proc Natl Acad Sci U S A 2022;119.
36. Oberley-Deegan RE, Regan EA, Kinnula VL, et al. COPD 2009;6:307–12.
37. Nguyen NH, Tran GB, Nguyen CT. J Mol Med (Berl) 2020;98: 59–69.
38. Uauy R, Olivares M, Gonzalez M. Am J Clin Nutr 1998;67: 952S–9S.
39. Eaton-Evans J, Mcllrath EM, Jackson WE, et al. J Trace Elem Experi Med 1996;9:87–94.
40. Baker A, Harvey L, Majask-Newman G, et al. Eur J Clin Nutr 1999;53:408–12.
41. Nielsen FH, Lukaski HC, Johnson LK, et al. Br J Nutr 2011;106:1872–9.
42. Kaler SG, DiStasio AT. In: GeneReviews(R) (Adam MP, Mirzaa GM, Pagon RA, Wallace SE, Bean LJH, Gripp KW, et al., eds), Seattle, WA, 1993.
43. Fox PL. Biometals 2003;16:9–40.
44. Thackeray EW, Sanderson SO, Fox JC, et al. J Clin Gastroenterol 2011;45:153–8.
45. Lutsenko S, Washington-Hughes C, Ralle M, et al. J Biol Inorg Chem 2019;24:1179–88.
46. Klevay LM. Am J Clin Nutr 2000;71:1213–4.
47. Klevay LM. Nutr Res Rev 2016;29:172–9.
48. Prohaska JR, Failla ML. Copper and Immunity, Plenum Press, New York, NY, 1993.
49. Kelley DS, Daudu PA, Taylor PC, et al. Am J Clin Nutr 1995;62:412–6.
50. Ladomersky E, Khan A, Shanbhag V, et al. Infect Immun 2017;85.
51. Djoko KY, Ong CL, Walker MJ, et al. J Biol Chem 2015;290: 18954–61.
52. White C, Lee J, Kambe T, et al. J Biol Chem 2009;284: 33949–56.
53. Sheldon JR, Skaar EP. Curr Opin Immunol 2019;60:1–9.
54. Blades B, Ayton S, Hung YH, et al. Biochim Biophys Acta Gen Subj 2021;1865:129979.
55. Gu K, Li X, Xiang W, et al. Biol Trace Elem Res 2020;194: 336–47.
56. Yang H, Liu CN, Wolf RM, et al. Metallomics 2019;11:1363–71.
57. Brady DC, Crowe MS, Turski ML, et al. Nature 2014;509:492–6.
58. Lopez J, Ramchandani D, Vahdat L. Met Ions Life Sci 2019;19.
59. Xu S, Liu D, Chang T, et al. Front Genet 2022;13:938259.
60. Lonnerdal B. Am J Clin Nutr 1998;67:1046S–53S.
61. Nations SP, Boyer PJ, Love LA, et al. Neurology 2008;71:639–43.
62. Gulec S, Collins JF. Annu Rev Nutr 2014;34:95–116.
63. Ravia JJ, Stephen RM, Ghishan FK, et al. J Biol Chem 2005; 280:36221–7.
64. Collins JF, Franck CA, Kowdley KV, et al. Am J Physiol Gastrointest Liver Physiol 2005;288:G964–71.
65. Yokoi K, Kimura M, Itokawa Y. Biol Trace Elem Res 1991;29: 257–65.
66. Collins JF, Prohaska JR, Knutson MD. Nutr Rev 2010;68: 133–47.
67. Board FaN. Zinc, National Academy Press, Washington DC, 2002.
68. Barber RG, Grenier ZA, Burkhead JL. Biomedicines 2021;9.
69. Powers HJ, Loban A, Silvers K, et al. Free Radic Res 1995;22: 57–65.
70. Hunt JR, Matthys LA, Johnson LK. Am J Clin Nutr 1998;67: 421–30.
71. Prohaska JR, Gybina AA. J Nutr 2004;134:1003–6.
72. van den Berghe PV, Klomp LW. J Biol Inorg Chem 2010;15: 37–46.
73. Nose Y, Kim BE, Thiele DJ. Cell Metab 2006;4:235–44.
74. Kim H, Son HY, Bailey SM, et al. Am J Physiol Gastrointest Liver Physiol 2009;296:G356–64.
75. Danks DM. Annu Rev Nutr 1988;8:235–57.
76. Kaler SG, Holmes CS, Goldstein DS, et al. N Engl J Med 2008;358:605–14.
77. Tumer Z, Moller LB. Eur J Hum Genet 2010;18:511–8.

78. Harvey LJ, Ashton K, Hooper L, et al. Am J Clin Nutr 2009; 89:2009S–24S.
79. Nishihara E, Furuyama T, Yamashita S, et al. Neuroreport 1998;9:3259–63.
80. Mak CM, Lam CW. Crit Rev Clin Lab Sci 2008;45:263–90.
81. Scheinberg IH, Sternlieb, I. Wilson's disease, Saunders, Philadelphia, PA, 1984.
82. LeWitt PA. Mov Disord 1999;14:555–6.
83. Huppke P, Brendel C, Kalscheuer V, et al. Am J Hum Genet 2012;90:61–8.
84. Martinelli D, Dionisi-Vici C. Ann N Y Acad Sci 2014;1314: 55–63.
85. Griffith DP, Liff DA, Ziegler TR, et al. Obesity (Silver Spring) 2009;17:827–31.
86. Davis GK, Mertz, W. Copper Vol. 1, Academic Press, San Diego, CA, 1987.
87. Trumbo P, Yates AA, Schlicker S, et al. J Am Diet Assoc 2001;101:294–301.
88. Ogra Y, Tanaka YK, Suzuki N. J Clin Biochem Nutr 2022;71: 2–6.
89. Leary SC, Ralle M. Curr Opin Chem Biol 2020;55:19–25.
90. Prohaska JR. Copper, In: Erdman Jr, J.W., Macdonald, I.A., Zeisel, S.H. eds. *Present Knowledge in Nutrition*, 10th ed. Washington, D.C.: ILSI Press, 2012;540–553.
91. Lassi KC, Prohaska JR. Nutr Res 2011;31:698–706.
92. Prohaska JR, Gybina AA. J Neurochem 2005;93:698–705.
93. Prohaska JR. Neurochemical Roles of Copper as Antioxidant or Prooxidant, Springer, Boston, MA, 1997.
94. Pratt WB, Omdahl JL, Sorenson JR. Am J Clin Nutr 1985;42: 681–2.

B. MINERALS

CHAPTER 13

Iodine

Sun Y. Lee, Abby Ershow, and Elizabeth N. Pearce

CHAPTER OUTLINE

Abbreviations

CDC, Centers for Disease Control and Prevention; **DV**, daily value; **FDA**, U.S. Food and Drug Administration; **GFDx**, Global Fortification Data Exchange; H_2O_2, hydrogen peroxide; **ICCIDD**, International Council for the Control of Iodine Deficiency Disorders; **IGN**, Iodine Global Network; **KI**, potassium iodide; **MCT8**, monocarboxylate transporter 8; **NIS**, sodium iodide symporter; T_3, triiodothyronine; T_4, thyroxine; **TPO**, thyroid peroxidase; **TRH**, thyrotropin-releasing hormone; **TSH**, thyroid-stimulating hormone; **UNICEF**, United Nations Children's Fund; **USDA**, U.S. Department of Agriculture; **WHO**, World Health Organization

Overview

The only established role of iodine in humans is to serve as a component of thyroid hormones. Thyroid hormones are essential for development and growth, and severe iodine deficiency may lead to developmental brain damage.[1] Moreover, thyroid hormones participate in regulation of the daily activity of every tissue. Appropriate tissue-specific thyroid hormone concentrations are maintained through multiple complex systems. Iodine exerts strong autoregulatory effects on the thyroid gland to allow for ongoing daily hormone production despite large variations in the iodine supply. However, both inadequate and excess iodine intakes may result in thyroid dysfunction. In addition, over time, activation of iodine autoregulatory mechanisms may lead to thyroid functional abnormalities in many individuals. Thus, the epidemiology of thyroid disease in populations may be associated with iodine intake levels, even if neither chronic, severe iodine deficiency nor excess is present.

Historical Background

Iodine discovery is credited to Bernard Courtois, in 1811, and the first use of iodine for treatment of goiter was published in 1820, by Coindet in Geneva.[2,3] The use of iodine in high doses for medical purposes in the early nineteenth century led to the first reports of clinical thyrotoxicosis after intake of iodine.[4] A landmark trial performed in Ohio between 1917 and 1922 by Marine and Kimball in 4,495 school children showed that iodine supplements dramatically decreased the incidence of goiter.[5] Voluntary iodized salt prophylaxis was introduced in Michigan in 1924 and rapidly spread throughout the United States. During the same period, Hunziker, from iodine-deficient Switzerland, observed that as little as 100 μg iodine per day was effective in preventing goiter, and in 1922, voluntary iodine prophylaxis was introduced in parts of Switzerland.[6]

Despite considerable knowledge regarding the prevention of iodine deficiency, developmental brain damage caused by iodine deficiency occurred in many parts of the world until recent decades. The 1985 formation of the International Council for the Control of Iodine Deficiency Disorders (ICCIDD; now the Iodine Global Network [IGN], ign.org)[7] and its subsequent activities in collaboration with the World Health Organization (WHO) and the United Nations Children's Fund (UNICEF) have led to an estimated 76% reduction in the global prevalence of iodine deficiency disorders,[8] but continuing efforts are necessary.

Dietary Sources

Table 13.1 shows values for the iodine content of selected U.S. foods, drawn from a recently released database.[9] The database was developed by direct sampling of foods, beverages, and water through established U.S. Department of Agriculture (USDA) and Food and Drug Administration (FDA) protocols, followed by assay with inductively coupled plasma mass spectrometry. Rich sources of iodine were found to include iodized salt, seaweed, eggs, fish, and other seafoods, breads (only if made with iodate dough conditioners), and dairy products. Noniodized sea salts contain minimal iodine. Similarly, there is little iodine in plant-based beverages used as dairy milk substitutes. Iodine is present in breast milk and infant formula. Also, labeled amounts of iodine in U.S. dietary supplements are available through the Dietary Supplement Label Database[10] and analytically estimated amounts through the Dietary Supplement Ingredient Database.[11]

Table 13.1 Iodine Content of Selected Foods[9,12]

Food	Micrograms (μg) per serving	Percent DV*
Bread, white, enriched, made with iodate dough conditioner, two slices**	320	213
Bread, whole-wheat, made with iodate dough conditioner, two slices**	309	206
Cod, baked, 3 ounces	158	106
Seaweed, nori, dried, 2 tablespoons, flaked (5 g)	116	77
Oysters, cooked, 3 ounces	93	62
Yogurt, Greek, plain, nonfat, ¾ cup	87	58
Milk, nonfat, 1 cup	85	57
Iodized table salt, ¼ teaspoon	76	51
Fish sticks, cooked, 3 ounces	58	39
Pasta, enriched, boiled in water with iodized salt, 1 cup	38	25
Ice cream, chocolate, ⅔ cup	28	19
Egg, hard boiled, 1 large	26	17
Cheese, cheddar, 1 ounce	15	10
Liver, beef, cooked, 3 ounces	14	9
Shrimp, cooked, 3 ounces	13	9
Tuna, canned in water, drained, 3 ounces	7	5
Fruit cocktail in light syrup, canned, ½ cup	6	4
Fish sauce, 1 tablespoon	4	3
Beef, chuck, roasted, 3 ounces	3	2
Soy beverage, 1 cup	2	1
Chicken breast, roasted, 3 ounces	2	1
Apple juice, 1 cup	1	1
Bread, whole-wheat, made without iodate dough conditioner, two slices**	1	1
Bread, white, enriched, made without iodate dough conditioner, two slices**	1	1
Rice, brown, cooked, ¾ cup	1	1
Sea salt, noniodized, ¼ teaspoon	<1	<1
Corn, canned, ½ cup	0	0
Broccoli, boiled, ½ cup	0	0
Banana, 1 large	0	0

Food	Micrograms (μg) per serving	Percent DV*
Soy sauce, 1 tablespoon	0	0
Lima beans, boiled, ½ cup	0	0
Green peas, frozen, boiled, ½ cup	0	0
Pasta, enriched, boiled in water without iodized salt, 1 cup	0	0

*DV = Daily Value. The FDA developed DVs to help consumers compare the nutrient contents of foods and dietary supplements within the context of a total diet. The DV for iodine is 150 mcg for adults and children aged 4 years and older.[13] The FDA does not require food labels to list iodine content unless iodine has been added to the food. Foods providing 20% or more of the DV are considered good sources of a nutrient, but foods providing lower percentages of the DV also contribute to a healthful diet.

**About 20% of bread products in the United States list iodate dough conditioners on their labels. Products made without these conditioners contain very little iodine.

Data from Roseland, JM, Spungen JH, Patterson KY, et al. (2022). USDA, FDA, and ODS-NIH Database for the Iodine Content of Common Foods, Release 2. Accessed March 2023. www.ars.usda.gov/ARSUSERFILES/80400535/DATA/IODINE/IODINE%20DATABASE_RELEASE_2_DOCUMENTATION.PDF; National Institutes of Health Office of Dietary Supplements. Iodine Fact Sheet for Health Professionals. Accessed March 2023. https://ods.od.nih.gov/factsheets/Iodine-HealthProfessional/

Recommended Intakes

Daily intake needs have been derived from iodine balance studies[14-16] and from radioisotope studies assessing thyroidal iodine turnover.[17] National and international organizations have made similar recommendations for iodine intake. Because approximately 90% of iodine in the diet is excreted by the kidney, and because iodine nutrition is most frequently assessed by urinary iodine measurement, recommendations have also been provided for median population urinary iodine excretion values. Recommendations from the WHO, UNICEF, and IGN are outlined in **Table 13.2**.[1,18]

The iodine intake recommendations of the Food and Nutrition Board of the US Institute of Medicine[20] are shown in **Table 13.3**. These recommendations are for iodine intakes of population groups to minimize the risk of disease and for average intakes of individuals over time. The recommendations are not for evaluation of individuals' daily intake. Because of the adaptive capacities of the thyroid, most individuals can adapt to days of very low or high iodine intake without incident.

The American Thyroid Association and American Academy of Pediatrics recommend that women living in the United States and Canada receive daily dietary supplements that contain 150 μg iodine during preconception, pregnancy, and lactation to cover the increase in need of iodine during these periods.[20,21] Detailed recommendations on iodine intake during pregnancy and lactation have been published by WHO/UNICEF/IGN (see Table 13.3).[22]

Table 13.2 World Health Organization, United Nations Children's Fund, and Iodine Global Network epidemiologic criteria for assessing iodine nutrition based on median urinary iodine concentration

School-age children (≥6 y) and nonpregnant adults	
Median urinary iodine concentration (μg/L)	**Iodine status**
<20	Insufficient; severe iodine deficiency
20–49	Insufficient; moderate iodine deficiency
50–99	Insufficient; mild iodine deficiency
100–299	Adequate iodine nutrition
≥300	Excessive; risk of adverse health consequences (iodine-induced hyperthyroidism, autoimmune thyroid diseases)
Pregnant women*	
Median urinary iodine concentration (μg/L)	**Iodine status**
<150	Insufficient
150–249	Adequate
250–499	Above requirements
≥500	Excessive

*For lactating women and for children <2 years of age, a median urinary iodine concentration of 100 μg/L can be used to define adequate iodine intake, but no other categories of iodine intake are defined. Although lactating women have higher intake requirements than pregnant women, the median urinary iodine is lower because some ingested iodine is excreted in breast milk.[19]

World Health Organization, United Nations Children's Fund, International Council for the Control of Iodine Deficiency Disorders. (2007). Assessment of Iodine Deficiency Disorders and Monitoring Their Elimination: A Guide for Programme Managers (3rd ed.). Geneva: World Health Organization. https://iris.who.int/bitstream/handle/10665/43781/9789241595827_eng.pdf?sequence=1

UNICEF. (2019). Guidance on the monitoring of salt iodization programmes and determination of population iodine status. Retrieved from https://stats4sd.org/download/128/guidance-on-the-monitoring-of-salt-iodization-programmes-and-determination-of-population-iodine-status_2019-05-14_14:02:40/2019-05-14_14:02:40_1928-unicef-full-layout-301117-rev.pdf/1928%20UNICEF%20Full%20layout%20301117%20rev.pdf

Effects of Iodine in the Human Body

The effects of iodine in the human body can be regarded as belonging to one of three groups:

1. Effects of deficient production of thyroid hormone caused by iodine deficiency.
2. Effects of iodine excess on the thyroid gland.
3. Potential extrathyroidal effects of iodine that are poorly understood.

The central role of iodine is to serve as a component of thyroid hormones produced in the thyroid gland. Prevention of iodine deficiency is especially important in pregnancy, to avoid developmental brain damage in the fetus.

Table 13.3 Dietary Reference Intakes for Iodine

	RDA[a] (μg/d)	AI[b] (μg/d)	UL[c] (μg/d)
Infants			
0–6 mo		110	
7–12 mo		130	
Children			
1–3 y	90		200
4–8 y	90		300
9–13 y	120		600
14–18 y	150		900
Women	150		1,100
Men	150		1,100
Pregnancy	220		1,100
Lactation	290		1,100

[a]RDA, recommended dietary allowance: the average daily intake that meets the estimated iodine needs of almost all (97.5%) individuals in the group.
[b]AI, adequate intake: the iodine intake that appears to sustain normal thyroid structure and function in the group. Data are insufficient to establish an RDA.
[c]UL, tolerable upper intake level: the highest level of daily iodine intake that is likely to pose no risk of adverse health effects for almost all individuals in the group.

Data from Food and Nutrition Board, Institute of Medicine. Dietary Reference Intakes: Iodine. Washington, DC: National Academy Press, 2001:258–89, with permission.

Autoregulatory effects of iodine on the thyroid gland tend to compensate for a low iodine supply by increasing the activity of processes involved in utilization of iodine for thyroid hormone production.

After intake of excess iodine, homeostatic processes rapidly and transiently shut down thyroid utilization of iodine. Thus, the thyroid gland normally keeps thyroid hormone production stable, despite large changes in iodine supply. However, this compensatory process may fail in susceptible individuals and excess iodine ingestion may lead to either hypo- or hyperthyroidism. The price for the complex ability to compensate for a fluctuating iodine supply is a tendency to develop thyroid disease. Because the thyroid processes activated by low and high iodine intake are different, the pattern of thyroid disease in a population depends on the level of iodine intake.[23] Rarely, very large amounts of iodine (exceeding many grams) may lead to nausea, vomiting, fever, skin eruptions, and various other toxicities.[20]

Extrathyroidal effects and iodine handling have received less attention. The exception is the processes involved in transport of iodine from the mother to the breastfed child, and from the mother to the fetus through the placenta. This iodine is necessary for thyroid hormone production by the fetal and infant thyroid. Investigators have suggested that iodine may function in the human body (e.g., in breast and gastrointestinal tract) to protect against reactive oxygen species.[24] However, more evidence is necessary before a conclusion can be drawn.

Metabolism

Although iodine in the diet occurs in various forms, most iodine-containing compounds are readily broken down in the gut, and the iodide is rapidly absorbed. The absorbed iodide enters the circulating pool of inorganic iodide together with iodide released by metabolism of thyroid hormones. Most circulating iodide is either excreted by the kidney (clearance of 30 to 50 mL/minute, independent of iodine status) or concentrated in the thyroid gland (clearance from very low to more than 100 mL/minute). Thyroid uptake of iodine is low after excessive iodine intake and high in iodine deficiency, and it depends on the functional status of the thyroid (low in a nonfunctional thyroid, high in the stimulated thyroid). In people ingesting the recommended iodine intake of approximately 150 μg/day, thyroid clearance of iodine is approximately half that of the kidney, and the half-life of iodide in blood is approximately 6 hours. In lactating women, a considerable fraction of the circulating iodide is taken up by the mammary glands and is excreted in breast milk.

The iodide taken up by the thyroid is built into tyrosine groups to create monoiodotyrosines and diiodotyrosines in the 660-kDa glycoprotein thyroglobulin present in the colloid of thyroid follicles. This process is catalyzed by the membrane-bound enzyme thyroid peroxidase (TPO) in the presence of hydrogen peroxide (H_2O_2). TPO is a heme-containing protein, and interaction exists between the effects of iodine and iron deficiency.[25] TPO also catalyzes the coupling of iodinated tyrosine residues in thyroglobulin to form the double-ring iodothyronines L-thyroxine (tetraiodothyronine, T_4) and L-triiodothyronine (T_3).

Thyroid hormone secretion is initiated by uptake into the follicular cells of thyroglobulin through micropinocytosis and macropinocytosis. The thyroid hormones T_4 and T_3 and iodotyrosines are liberated by hydrolysis of thyroglobulin catalyzed by lysosomal enzymes. The iodotyrosines are nearly entirely deiodinated in the follicular cells, thus preserving the iodide in the cells for hormone production. Release of T_4 and T_3 can be immediately and reversibly blocked by certain iodine-containing compounds; this finding is an indication of active transport of the hormones out of the follicular cells.[26] Some thyroid hormone transporters have been identified, but their potential role in thyroid hormone secretion has not been fully elucidated.[27]

The two most important regulators of thyroid activity are thyrotropin (thyroid-stimulating hormone [TSH]) and iodide. TSH is secreted from the anterior lobe of the pituitary, regulated by classical feedback inhibition from T_4 and T_3, and by hypothalamic signals. The most important hypothalamic factor is thyrotropin-releasing hormone (TRH), which modulates the set point of the thyroid system. Vitamin A plays a role in the activity of the pituitary-thyroidal axis, and thus, vitamin A deficiency may exacerbate the thyroidal effects of iodine deficiency.[28] TSH activates all the processes involved in thyroid hormone synthesis by binding to the TSH receptor.

Sodium Iodide Symporter

The sodium iodide symporter (NIS) is central in the transport of iodine in the body.[29] It is an 85-kDa membrane protein that couples the translocations of sodium (Na^+) (from high to low concentration) and iodine (I^-) (from low to high concentration) into the cells. NIS is a member of solute carrier family 5A and is encoded by the *SLC5A5* gene. The most important sites of NIS are the basolateral membrane of the follicular cells of the thyroid and the lactotrophs of the lactating mammary gland, but NIS is also present in salivary glands, in the gut mucosae, in sweat glands, and in the choroid plexus of the cerebral lateral ventricles. Mutations in *SLC5A5*, which lead to defects in thyroidal iodine transport, are a rare cause of congenital hypothyroidism.[29]

Iodide Autoregulation of the Thyroid

Iodide has a strong autoregulatory function in the thyroid independent of thyroidal TSH regulation.[30] The mechanisms involved have been only partially elucidated, but they may occur through organic iodinated compounds, possibly iodolactones.[31] When iodine is scarce, practically all processes involved in iodine trapping and hormone synthesis are enhanced. Conversely, excess iodine exposure is rapidly followed by blocking of iodide organification (the acute Wolff–Chaikoff effect) and of hormone secretion (an effect used therapeutically in patients with thyroid storm) and blood flow (used therapeutically before thyroid surgery to prevent bleeding). Moreover, high iodine concentrations inhibit thyroid growth and promote apoptosis of thyrocytes.[32] Shortly after exposure to a large load of iodine, downregulation of NIS occurs,[33] with a reduction of thyroid iodine uptake and intrathyroidal iodine content. This process tends to re-establish iodide organification (escape from the acute Wolff–Chaikoff effect).

Iodide Transport into Milk

The lactating mammary gland concentrates iodide from blood into milk by way of the NIS.[34] A major difference between iodide transport in the mammary gland and transport in the thyroid is that TSH stimulation has little or no effect on the mammary gland, although there may be some capacity for autoregulation, with iodine preferentially partitioned into breast milk at very low iodine intakes.[19]

Certain chemicals, most importantly thiocyanate and perchlorate, are competitive inhibitors of iodide transport by NIS,[35] both in the thyroid gland and in the lactating mammary gland. Because of the lack of iodide autoregulation in the mammary gland, intake of such compounds affects iodide excretion into milk much more directly than such intake affects thyroid iodide uptake. Environmental exposures to thiocyanate and perchlorate may worsen iodine deficiency and lead to an increase in the risk of goiter and thyroid nodules when iodine intake is low.[36] The leading cause of high thiocyanate in blood is tobacco smoking, because cyanide in the tobacco smoke is detoxified in the liver to thiocyanate. Thus, tobacco smoking during the period of breastfeeding leads to a low iodine concentration in breast milk, and the neonates of smoking mothers have a low urinary iodine excretion because their iodine intake from mother's milk is low.[37] Smokers also have a higher risk of goiter if their iodine intake is low.[38] Thiocyanate is also present in some "goitrogenic" foods.

Actions of Thyroid Hormones

The main actions of thyroid hormones are exerted by binding to nuclear receptors that modulate transcription of many proteins. The active hormone is T_3, which may be directly synthesized in the thyroid or derived by outer ring 5'-deiodination of T_4. In physiologically normal persons, approximately 80% of circulating T_3 is derived from T_4 in peripheral tissues. Outer ring deiodination of T_4 to T_3 is catalyzed by type 1 or type 2 iodothyronine deiodinase, whereas type 3 iodothyronine deiodinase exclusively inactivates thyroid hormones by inner ring deiodination.[39] The three deiodinases are selenoenzymes, and severe selenium deficiency interferes with normal thyroid function. Selenium may also influence thyroid autoimmunity and growth by several mechanisms.[40] The iodothyronine deiodinases, as well as different types of thyroid hormone transporters and receptors, are differently distributed in tissues. The complex roles of the deiodinases and thyroid hormone transporters in health and disease are still an area of research.[39,41]

Thyroid hormone is important for brain development, as discussed below. In adults, the most significant effects of thyroid hormones are their overall regulation of levels of activity in many organs and tissues. Thus, excess thyroid hormone production is associated with the clinical picture of thyrotoxicosis, characterized by psychomotor hyperactivity, high pulse rate, sweating, weight loss, and tremor. Deficiency of thyroid hormone (hypothyroidism) is associated with symptoms and signs including fatigue, cold intolerance, weight gain, irregular menses, and constipation.

Effects of Iodine Deficiency on Human Health

Exposure to mild-to-moderate iodine deficiency may not lead to obvious thyroid dysfunction in adults, because in most individuals, the thyroid gland adapts to maintain euthyroidism. However, chronic severe iodine deficiency with daily intakes below 50–60 μg can lead to hypothyroidism. Chronic iodine deficiency also increases risk for the development of goiter and thyroid nodules, although the mechanism is not entirely understood. Stimulation from increased TSH production in the pituitary gland in response to decreased thyroid hormone levels in the peripheral circulation may be partially responsible. Upregulation of thyroidal processes to compensate for a lack of iodine availability and increase in thyroidal production of H_2O_2 may also play a role.[23] The presence of increased autonomous function in thyroid nodules due to chronic iodine deficiency makes iodine-deficient individuals more susceptible to developing hyperthyroidism when exposed to a sudden increase in iodine intake, such as after implementation of salt iodization or exposure to iodine-containing medications (including amiodarone) or iodinated intravenous contrast used for CT scans and angiography. The prevalence of iodine–deficiency-induced hypothyroidism and goiter varies geographically, suggesting some impact of other environmental factors and possibly genetics.

Iodine deficiency during pregnancy may be particularly harmful, as adequate thyroid hormone is critical for normal fetal development, especially neurodevelopment.[42,43] Pregnant women have higher iodine requirements compared with non-pregnant adults, due to increased thyroid hormone production, increased renal iodine clearance, and transfer of iodine to the developing fetus. Lactating women also have increased iodine requirements as iodine is secreted in breast milk to support infant nutrition. Infants are born with limited intrathyroidal iodine stores and, therefore, are highly susceptible to the effects of dietary iodine deficiency.[15] Chronic severe iodine deficiency in utero can cause cretinism, characterized by intellectual disability as well as stunted growth and delay in motor development.[44] Severe iodine deficiency during gestation has been associated with a 7–10 point decrease in offspring IQ.[45] In addition to adverse effects in fetal neurodevelopment, severe iodine deficiency in pregnancy has been associated with increased risks of miscarriage, stillbirth, and neonatal mortality.[46] The effects of mild-to-moderate iodine deficiency in pregnancy are more controversial, likely due to heterogeneity of observational studies regarding population iodine status and timing of iodine status assessment. However, there is increasing evidence suggesting a link between mild-to-moderate iodine deficiency and poorer neurodevelopment in children.[47] Well-designed, adequately powered, randomized, controlled studies currently are lacking regarding the impact of iodine supplementation on mild-to-moderate iodine deficient populations.[48]

Effects of Iodine Excess on Human Health

Exposure to excess iodine can paradoxically cause hypothyroidism. With a sudden increase in iodine intake, the thyroid gland autoregulates its thyroid hormone production to prevent hyperthyroidism via the acute Wolff-Chaikoff effect described above. After several days of exposure to excess iodine, the thyroid gland will escape from the acute Wolff-Chaikoff effect, with resumption of normal thyroid hormone production. However, individuals with underlying thyroid autoimmunity or previously treated Graves' disease may not be able to escape appropriately, leading to hypothyroidism.[49] A comparative epidemiologic study of elderly people living in Jutland, Denmark, an area of mild-to-moderate iodine deficiency, showed a high prevalence of low serum TSH concentration, while those living in Iceland, an area of high iodine intake, had a high prevalence of high serum TSH concentration among those with abnormal thyroid function tests.[50] Chronic high iodine intake has also been associated with higher TSH reference ranges in areas of historically high iodine intake, in Korea[51] and China.[52] A 2017 systematic review reported an association between chronic high iodine exposure and increased risk of developing hypothyroidism.[53] High iodine exposure has also been implicated in development of thyroid autoimmunity in animal studies.[54] The association between high iodine exposure and thyroid autoimmunity in humans is less clear, with some studies suggesting that there may be initial increases in lymphocytic changes in thyroid gland from acute excess iodine exposure that may decrease over time.[55,56]

While iodine deficiency in pregnant women has long been recognized as a risk factor for adverse pregnancy and child neurodevelopmental outcomes, the effects of maternal iodine excess are still not well understood. In a recent study of 214 Brazilian pregnant women, UIC ≥250 μg/L (more than adequate iodine nutrition per WHO guidelines[22]) was associated with increased risk of gestational diabetes and hypertensive disorders of pregnancy, compared with UIC 150–249 μg/L.[57] In another large study, of 1569 Chinese pregnant women, UIC ≥250 μg/L was associated with macrosomia in newborns, compared with UIC 150–249 μg/L.[58] On the other hand, a 2020 meta-analysis did not show any associations between maternal iodine excess and adverse pregnancy outcomes in euthyroid women, suggesting that the effects of iodine excess may be mediated through thyroid dysfunction.[59] It is important to note that the fetus does not have the ability to escape from the acute Wolff-Chaikoff effect until late in gestation and is susceptible to developing hypothyroidism from excess iodine even if the pregnant woman remains euthyroid.

Given the potential adverse effects of both iodine deficiency and excess in human health, it is important to into take account the underlying iodine status of populations to determine the need for and degree of iodine supplementation or fortification to minimize the risk of thyroid dysfunction.

Assessment of Iodine Nutrition

Iodine intake may be estimated from a combination of diet records and data on the iodine content of foods.[60] The recommended primary tool to evaluate and monitor iodine nutrition of a population is measurement of urinary iodine excretion in representative groups of the population.[1] Spot urine samples from approximately 125 individuals are needed to estimate the iodine concentration in a population with 95% confidence within a precision range of ±10% is approximately 125, and within a precision range of ±5%, it is approximately 500. To estimate an individual's habitual iodine intake with even 20% precision, iodine must be measured in 10–12 repeated spot or 24-hour urine samples.[61] Because the concentration of iodine in a spot urine sample generally reflects iodine intake within the last few hours, urinary concentration in an individual may vary considerably from day to day, even within the same day, and median values from groups of samples are necessary for evaluation.[62] Thus, iodine deficiency in an individual cannot be diagnosed based on a single urine iodine measurement. Studies of urinary iodine excretion in populations require that several other technical details be considered.[63,64]

The clinical consequences of iodine deficiency may be monitored by examination of school-age children for goiter. A simple classification system is used to describe goiter, but the sensitivity and specificity for finding small goiters via palpation are low. The goiter frequency should be less than 5%. Thyroid ultrasonography gives a much more precise estimate of thyroid size. Normative values dependent on age and sex have been developed.[65]

Thyroglobulin is released from the thyroid in amounts that depend on thyroid size and activity. In population studies, the level of serum thyroglobulin is a sensitive indicator of iodine deficiency, but this value is nonspecific in individuals.[66,67]

Serum TSH, measured as part of neonatal screening for congenital hypothyroidism, has been used for evaluation of iodine deficiency during pregnancy. The rationale is that low thyroid hormone production in the fetus or neonate caused by iodine deficiency would lead to a compensatory increase in fetal or neonatal pituitary TSH secretion. Careful control of the TSH measurement method is necessary. The prevalence of neonatal TSH concentration >5.0 mIU/L is typically less than 3% in iodine-sufficient regions.[1]

National Iodine Nutrition Programs

In the United States, agencies of the Federal government have key responsibilities related to iodine nutrition. These include population surveys for monitoring indices of thyroid function and iodine intake performed by the Centers for Disease Control and Prevention[68] and the National Center for Health Statistics[69]; standardization of analytical laboratory quality control and provision of reference materials for measuring iodine in foods, dietary supplements, and biological fluids by the Centers for Disease Control and Prevention (CDC)[70] and National Institute of Standards and Technology[71]; and assessment of the iodine content of foods, beverages, and dietary supplements, as well as development and provision of databases and other educational materials for consumers, researchers, and health professionals by the US Department of Agriculture Agricultural Research Service,[72] the FDA Center for Food Safety and Applied Nutrition[73]; and the National Institutes of Health Office of Dietary Supplements.[74]

Elsewhere, the agencies responsible for such efforts vary by country. Most countries have national programs for iodine fortification of salt or other foods, with universal salt iodization being the mainstay of iodine deficiency prevention in most regions. Information about the iodine fortification of foods is listed in the Global Fortification Data Exchange (GFDx) database.[75] In 2022, 126 countries had legislation for mandatory salt iodization and about 21 countries (including the United States) had legislation allowing voluntary salt iodization.[75]

Global Iodine Nutrition Status

A landmark in the history of iodine nutrition was the report on endemic goiter in the world, by Kelly and Snedden, published by the WHO in 1960.[76] At that time, endemic goiter was present in large areas of all continents and most countries. Although certain environmental agents may lead to endemic goiter because of their interaction with iodine use and thyroid function,[77] the predominant cause of endemic goiter is iodine deficiency. Although global iodine nutrition status has improved greatly over the last three decades, areas of iodine deficiency and excess remain common.[8,78] Continuous monitoring and adjustment of food fortification programs are essential. The recommended approach to achieve optimal iodide nutrition in most countries is the iodization of salt.[1] If all salt used by households and by the food industry is iodized (universal salt iodization), this will distribute iodine evenly to most population groups.[79] Efforts to reduce population sodium intake to reduce cardiovascular disease can be harmonized with iodine deficiency disorder prevention efforts via careful population-based studies of both salt intake and increased salt iodization levels as salt intakes decline.[80]

Iodide as a Thyroid-Blocking Agent in Radiation

The Chernobyl reactor accident in Ukraine in 1986 resulted in massive releases of radioiodines, including iodine-131 (^{131}I). With a latency of approximately 4 years, this accident was followed by a sharp increase in the incidence of thyroid cancer among children and adolescents in Belarus and the Ukraine.[81] Iodine deficiency of the population with high thyroid uptake of ^{131}I may have added to the high frequency of thyroid cancer. In Poland, where fallout from Chernobyl was fortunately only moderate, more than 10 million children received potassium iodide (KI) after the Chernobyl accident to protect against thyroid cancer. Only mild and clinically insignificant side effects of iodine were observed.[82]

The March 2011 Fukushima nuclear power plant disaster in Japan caused by an earthquake and subsequent tsunami renewed focus on the distribution of iodine for the prevention of thyroid cancer. Fortunately, iodine intake with food is high in Japan (from seaweed), and this reduces the risk of significant thyroid uptake of radioactive iodine. Shortly after the disaster, Japanese authorities distributed 230,000 doses of KI to evacuation centers from the area around the power plant as a precautionary measure. To date, there has been no evidence of an increase in thyroid cancer incidence related to the Fukushima accident.[83]

The WHO,[84] the FDA,[85] and the American Thyroid Association[86] have put forward detailed recommendations on the use of KI as a thyroid-blocking agent in case of radiation emergencies.

Acknowledgments

The authors would like to acknowledge the contributions of Carol Haggans, MS, RD, to this chapter, and of Peter Laurberg, author of the previous version of this chapter.

References

1. World Health Organization. Assessment of iodine deficiency disorders and monitoring their elimination: a guide for programme managers, 3rd ed. World Health Organization, 2007. Accessed March 2023. https://apps.who.int/iris/handle/10665/43781
2. Coindet JF. Ann Chim Phys Ser 1820;15:49–59.
3. Zimmermann MB. J Nutr 2008;138:2060–3.
4. Stanbury JB, Ermans AE, Bourdoux P, et al. Thyroid 1998;8:83–100.
5. Carpenter KJ. J Nutr 2005;135:675–80.
6. Bürgi H, Supersaxo Z, Selz B. Acta Endocrinol (Copenh) 1990;123:577–90.
7. Hetzel B.S. Asia Pac J Clin Nutr 2012;21:164–70.

8. Gorstein JL, Bagriansky J, Pearce EN, et al Thyroid 2020;30:1802–9.
9. Roseland JM, Spungen JH, Patterson KY, et al. USDA, FDA, and ODS-NIH Database for the Iodine Content of Common Foods, Release 3, 2023. Accessed March 2023. www.ars.usda.gov/northeast-area/beltsville-md-bhnrc/beltsville-human-nutrition-research-center/methods-and-application-of-food-composition-laboratory/mafcl-site-pages/iodine/
10. National Institutes of Health Office of Dietary Supplements. Dietary Supplement Label Database (DSLD). Accessed March 2023. https://ods.od.nih.gov/Research/Dietary_Supplement_Label_Database.aspx
11. National Institutes of Health Office of Dietary Supplements. Dietary Supplement Ingredient Database. Accessed March 2023. https://dietarysupplementdatabase.usda.nih.gov/
12. National Institutes of Health Office of Dietary Supplements. Iodine Fact Sheet for Health Professionals. Accessed March 2023. https://ods.od.nih.gov/factsheets/Iodine-HealthProfessional/
13. Food and Drug Administration. Food Labeling: Revision of the Nutrition and Supplement Facts Labels, 2016. Fed Reg 05/27/2016. Accessed March 2023. www.federalregister.gov/documents/2016/05/27/2016-11867/food-labeling-revision-of-the-nutrition-and-supplement-facts-labels
14. Harrison MT, Harden RM, Alexander WD, et al. J Clin Endocrinol Metab 1965;25:1077–84.
15. Dold S, Zimmermann MB, Baumgartner J, et al. Am J Clin Nutr 2016;104:620–8.
16. Dworkin HJ, Jacquez JA, Beierwaltes WH J Clin Endocrinol Metab 1966;26:1329–42.
17. Fisher DA, Oddie TH J Clin Endocrinol Metab 1969;29:721–7.
18. UNICEF. Guidance on the monitoring of salt iodization programmes and determination of population iodine status, 2018. Accessed March 2023. www.ign.org/cm_data/2018-guidance-Monitoring-of-Salt-Iodization-English.pdf
19. Dold S, Zimmermann MB, Aboussad A, et al. J Nutr 2017;147:528–37.
20. Institute of Medicine. *Dietary Reference Intakes for Vitamin A, Vitamin K, Arsenic, Boron, Chromium, Copper, Iodine, Iron, Manganese, Molybdenum, Nickel, Silicon, Vanadium, and Zinc*, National Academies Press (US) Washington DC, 2001.
21. Rogan WJ, Paulson JA, Baum C, et al. Pediatrics 2014;133:1163–6.
22. Andersson M, de Benoist B, Delange F, et al. Public Health Nutr 2007;10:1606–11.
23. Laurberg P, Cerqueira C, Ovesen L, et al. Best Pract Res Clin Endocrinol Metab 2010;24:13–27.
24. Venturi S, Venturi M. Nutr Health 2009;20:119–34.
25. Zimmermann MB. Annu Rev Nutr 2006;26:367–89.
26. Laurberg P. Endocrinology 1985;117:1639–44.
27. Heuer H, Visser TJ. Endocrinology 2009;150:1078–83.
28. Zimmermann MB. Int J Vitam Nutr Res 2007;77:236–40.
29. Ravera S, Reyna-Neyra A, Ferrandino G, Annu Rev Physiol 2017;79:261–89.
30. Leung A, Pearce EN, Braverman LE. Expert Rev Endocrinol Metab 2010;5:593–602.
31. Gärtner R, Dugrillon A, Bechtner G. Acta Med Austriaca 1996;23:47–51.
32. Chen W, Man N, Shan Z, et al. Exp Clin Endocrinol Diabetes 2011;119:1–8.
33. Eng PH, Cardona GR, Fang SL, et al. Endocrinology 1999;140:3404–10.
34. Tazebay UH, Wapnir IL, Levy O, et al., Nat Med 2000;6:871–8.
35. Leung AM, Pearce EN, Braverman LE. Curr Opin Endocrinol Diabetes Obes 2014;21:372–6.
36. Steinmaus CM. Curr Environ Health Rep 2016;3:136–43.
37. Laurberg P, Nøhr SB, Pedersen KM, et al. J Clin Endocrinol Metab 2004;18:225–48.
38. Knudsen N, Brix TH. Best Pract Res Clin Endocrinol Metab 2014;28:495–506.
39. Russo SC, Salas-Lucia F, Bianco AC. Endocrinology 2021;162:bqab059.
40. Köhrle J. Curr Opin Endocrinol Diabetes Obes 2015;22:392–401.
41. van Geest FS, Gunhanlar N, Groeneweg S, et al. Front Endocrinol (Lausanne) 2021;12:723750.
42. de Escobar GM, Obregón MJ, del Rey FE. Best Pract Res Clin Endocrinol Metab 2004;18:225–48.
43. Bernal J. Vitam Horm 2005;71:95–122.
44. Syed S. Pediatrics 2015;135:594–6.
45. Bougma K, Aboud FE, Harding KB, et al. Nutrients 2013;5:1384–1416.
46. Toloza FJK, Motahari H, Maraka S. Front Endocrinol (Lausanne) 2020;11:409.
47. Rodriguez-Diaz E, Pearce EN. Best Pract Res Clin Endocrinol Metab 2020;34:101430.
48. Dineva M, Fishpool H, Rayman MP, et al. Am J Clin Nutr 2020;112:389–412.
49. Leung AM, Braverman LE. Nat Rev Endocrinol 2014;10:136–42.
50. Laurberg P, Pedersen KM, Hreidarsson A, et al. J Clin Endocrinol Metab 1998;83:765–9.
51. Jeon MJ, Kim WG, Kwon H, et al. Thyroid 2017;27:967–72.
52. Li Y, Teng D, Ba J, et al. Thyroid 2020;30:568–79.
53. Katagiri R, Yuan X, Kobayashi S, et al. PLoS One 2017;12:e0173722.
54. Braverman LE, Paul T, Reinhardt W, et al. Acta Endocrinol Suppl (Copenh) 1987;281:70–6.
55. Mazziotti G, Premawardhana LD, Parkes AB, et al. Eur J Endocrinol 2003;149:103–10.
56. Zimmermann MB, Moretti D, Chaouki N, et al. Thyroid 2003;13:199–203.
57. Silva de Morais N, Ayres Saraiva D, Corcino C, et al. Thyroid 2020;30:1792–1801.
58. Xiao Y, Sun H, Li C. Biol Trace Elem Res 2018;182:231–7.
59. Nazarpour S, Ramezani Tehrani F, Behboudi-Gandevani S, et al. Biol Trace Elem Res 2020;197:411–20.
60. Ershow AG, Goodman G, Coates PM, et al. Am J Clin Nutr 2016;104:859s–863s.
61. Andersen S, Karmisholt J, Pedersen KM, et al. Br J Nutr 2008;99:813–8.
62. König F, Andersson M, Hotz K, et al. J Nutr 2011;141:2049–54.
63. Pearce EN, Caldwell KL. Am J Clin Nutr 2016;104:898s–901s.
64. Rohner F, Zimmermann MB, Jooste P, et al. J Nutr 2014;144:1322s–42s.
65. Zimmermann MB, Hess SY, Molinari L, et al. Am J Clin Nutr 2004;79:231–7.
66. Zimmermann MB, Moretti D, Chaouki N, et al. Am J Clin Nutr 2003;77:1453–8.
67. Ma ZF, Skeaff SA. Thyroid 2014;24:1195–1209.
68. Centers for Disease Control and Prevention. Micronutrients. Last reviewed 2021. Accessed March 2023. www.cdc.gov/nutrition/micronutrient-malnutrition/index.html
69. Centers for Disease Control and Prevention. National Center for Health Statistics, National Health and Nutrition Examination Survey. Accessed March 2023. www.cdc.gov/nchs/nhanes/index.htm
70. Centers for Disease Control and Prevention. EQUIP: Ensuring the Quality of Urinary Iodine Procedures. Accessed March 2023. www.cdc.gov/labstandards/equip.html

71. NIST Store. Accessed March 2023. https://shop.nist.gov/
72. U.S. Department of Agriculture, The Agricultural Research Service Home. Accessed March 2023. www.ars.usda.gov/
73. Food and Drug Administration. The Center for Food Safety and Applied Nutrition Home. Accessed March 2023. www.fda.gov/about-fda/fda-organization/center-food-safety-and-applied-nutrition-cfsan
74. National Institutes of Health, Office of Dietary Supplements Home. Accessed March 2023. https://ods.od.nih.gov/
75. Global Fortification Data Exchange. Providing actionable food fortification data all in one place. Accessed March 2023. https://fortificationdata.org/
76. Kelly FC, Snedden WW. Geneva: Monogr Ser World Health Organ 1960;44:27–233. Accessed March 2023. www.ign.org/prevalence-and-geographical-distribution-of-endemic-goitre-who-monograph-series-1960.htm?from=0142002801
77. Gaitan E. Annu Rev Nutr 1990;10:21–39.
78. Zimmermann MB, Andersson M. Eur J Endocrinol 2021;185: R13–R21.
79. Dold S, Zimmermann MB, Jukic T, et al. J Nutr 2018;148: 587–98.
80. World Health Organization. Salt reduction and iodine fortification strategies in public health, 2014.
81. Cardis E, Howe G, Ron E, et al. J Radiol Prot 2006;26:127–40.
82. Nauman J, Wolff J. Am J Med 1993;94:524–32.
83. Ory C, Leboulleux S, Salvatore D. Endocrine 2021;71:298–309.
84. World Health Organization. Iodine thyroid blocking: Guidelines for use in planning and responding to radiological and nuclear emergencies, 2017. Accessed March 2023. www.who.int/publications/i/item/9789241550185
85. Food and Drug Administration, USA. Radiological and Nuclear Emergency Preparedness Information from FDA, 2022. Accessed November 2023. www.fda.gov/emergency-preparedness-and-response/mcm-issues/radiological-and-nuclear-emergency-preparedness-information-fda
86. Leung AM, Bauer AJ, Benvenga S, et al. Thyroid 2017;27: 865–877.

B. MINERALS

CHAPTER 14

Selenium

Xin Gen Lei, Margaret P. Rayman, and Roger A. Sunde

CHAPTER OUTLINE

Abbreviations

AD, Alzheimer's disease; **AI**, adequate intake; **ApoER2**, apolipoprotein E receptor; **CH_3-SeGalNac**, methyl selenosugar; **CH_3SeH**, methylselenol; **DIO**, iodothyronine 5′ deiodinase; **DMSe**, dimethylselenide; **DRVs**, dietary reference values; **EFSA**, European Food Safety Authority; **ER**, endoplasmic reticulum; **ERAD**, ER-associated degradation; **GPX**, glutathione peroxidase; **GSH**, glutathione; **GS-SeGalNac**, glutathionyl-selenosugar; **GS-Se-SG**, seleno-diglutathione; **GWAS**, genome wide association study; **i.v.**, intravenous(ly); **INMT**, indolethylamine N methyltransferase; **KBD**, Kashin-Beck disease; **KD**, Keshan disease; **MSRB1**, methionine sulfide reductase; **RDA**, recommended dietary allowance; **RDI**, recommended dietary intake; **RNI**, reference nutrient intake; **S**, sulfur; **Se**, selenium; **SEC (U)**, selenocysteine, **SECIS**, selenocysteine insertion sequence; **SeGalNac**, selenosugar; **SELENO**, selenoprotein (e.g., SELENOP-selenoprotein P); **SeMet**, selenomethionine; **SEPHS2**, selenophosphate synthase 2; **$SePO_3H^=$**, selenophosphate; **SNP**, single nucleotide polymorphism; **TMS**, trimethyl sulfonium; **TMSe**, trimethylselenonium; **TPMT**, thiopurine S methyltransferase; **TXNRD**, thioredoxin reductase; **UGGT**, glucosyltransferase; **UL**, tolerable upper intake level (USA) or upper Level of Intake (Australia, New Zealand); **WHO**, World Health Organization

Introduction

Selenium (Se) was identified as a new element in 1817 by Berzelius in Sweden and attracted biologic interest in the 1930s for poisoning livestock grazing in areas with high-Se soils.[1] In 1957, Schwarz and Foltz[2] found that Se prevented liver necrosis in vitamin E-deficient rats, revealing its nutritional essentiality for the first time. Thereafter, several Se/vitamin E deficiency diseases were reported in domestic animals.[1] In 1973, glutathione peroxidase-1 (GPX1)[3] was discovered as a Se-dependent enzyme, which provided the first evidence for a biochemical function of Se in vivo. Because GPX1 is a major scavenging enzyme for reactive oxygen species (ROS), and supplemental Se can spare vitamin E (a well-known antioxidant) to prevent Se/vitamin E deficiency diseases in animals, Se has been widely regarded as an antioxidant for many years. In the human genome, 25 Se-dependent protein (selenoprotein) genes have been identified with broad functional implications beyond redox controlling.[4]

Most Se in biologic systems is reported to be present in proteins as a constituent of amino acids: usually selenocysteine (SEC or U) and selenomethionine (SeMet). Being the functional Se form in selenoproteins and referred to as the 21st amino acid, SEC contains Se as selenol that is largely ionized at physiologic pH (pK: 5.3) and is a stronger nucleophile than the thiol of cysteine (pK, 8.3).[5] These two features enable Se in selenol to react with ROS faster than sulfur (S) in thiol. Although selenol- and thiol-dependent enzymes are both susceptible to inactivation by oxidation, the resultant oxidized enzymes of the former, but not the latter, can be readily re-reduced and activated, providing perhaps the strongest explanation to the preferred utilization and retention of Se vs S in biological systems.[5,6] Conversely, SeMet contains Se bound covalently to two carbon atoms (Se is shielded) and, thus, is considerably less reactive than SEC. Numerous proteins contain Se as SeMet in nonstoichiometric amounts, because enzymes that attach methionine to the methionine-tRNA for incorporation into protein or that metabolize methionine cannot distinguish SeMet from methionine.

The importance of Se in human nutrition was documented in 1979 with effectiveness of Se supplementation in preventing Keshan disease (KD), a cardiomyopathy in children and young women living in low-Se areas of China,[7] and a clinical response to Se in a Se-depleted patient in New Zealand.[8] The recommended dietary allowance (RDA) for Se was established in 1989[9] and revised in 2000,[10] and the World Health Organization (WHO) issued its dietary recommendations for Se in 1996.[11] During the past two decades, novel tools have been used to enhance our understanding of Se metabolism and selenoprotein functions in animals and humans. Exciting progress has been made in exploring health potentials of Se beyond its nutritional essentiality.

Digestion, Intermediate Metabolism, and Utilization

Dietary Sources and Bioavailability

Sources of Dietary Se in the Food Chain

Plant foods and feeds contain Se mainly in the form of SeMet, along with SEC and Se-methylselenocysteine, and their Se content varies with the Se content of the soils where they grow, and their capacity to uptake Se from soils.[1,12] Se-deficient soils (0.1–0.5 mg/kg)[1] are found in a geographic belt in China where KD occurred and in scattered regions in the United States, New Zealand, and Finland, where Se deficiency in livestock was seen.[13] In contrast, soil Se content in the northern high plains of the United States tends to be high, and could reach 80 mg/kg in seleniferous areas of the United States, and >100 mg/kg in Enshi, Hubei Province, China.[14-16] Organ meats and seafood are rich sources of Se (0.4 to 1.5 μg/g fresh weight), followed by muscle meats (0.1 to 0.4), cereals and grains (<0.1 to >0.8), dairy products (<0.1 to 0.3), and fruits and vegetables (<0.1).[17] Brazil nuts contain 0.4 to 158 μg Se/g.[18] Feed Se supplementation and body Se homeostasis render less variation of Se content in animal- than plant-sourced foods. The US FoodData Central-USDA provides analytic or inferred values for Se contents of various food items.[19] Drinking water, except from highly seleniferous areas, contributes negligibly to total Se intake.[17,20]

Bioavailability of Dietary Se

Across various domestic animal species, bioavailability of dietary Se ranges from 30% to >90%.[17,20] A commonly used procedure to estimate Se availability is to follow increases in hepatic GPX activity after feeding various food sources of Se to rodents previously depleted of Se.[21] On this basis, Se fed as mushrooms, tuna, and wheat was 5%, 57%, and 83% as available to rats, respectively, as sodium selenite. A human trial performed in Finland with men of moderately low Se status showed significant differences in bioavailability among various forms of Se tested (e.g., selenate, wheat, yeast), depending on the criterion used (increase in platelet GPX activity, elevation of plasma or red blood cell Se content, or retention of Se).[21] A 16 week study of US adults supplemented with 200 to 600 μg Se/day as selenite, SeMet, or selenized yeast showed no effect of any of these supplements on plasma GPX or selenoprotein P (SELENOP) in this already Se-adequate population, nor any increase with selenite in plasma Se, but showed rapid increases in plasma Se with high levels of SeMet or selenized yeast.[22]

Absorption and Transport

Uptake

Dietary Se is well digested and readily absorbed from the small intestine by humans and animals, independent of body Se status.[17] Uptakes of selenate and selenite in the intestine are mediated by sulfate transporters and simple diffusion, respectively.[23,24] A proton-coupled monocarboxylate transporter (JEN1P) catalyzes a high-affinity uptake of selenite in yeast, and is responsible for the majority of selenite uptake in *Saccharomyces cerevisiae*.[25] Because carboxymethyl transferases (transporting small molecules such as formate, acetate, pyruvate, and lactate) are ubiquitously present in many organisms, including humans, selenite, at ~1000-fold lower concentration than these negatively-charged molecules, possibly hitch-hikes along with them in humans. SeMet, SEC, and other low Mol-weight Se metabolites are readily transported by respective S-metabolite transporters.[17,20]

Apparent vs True Absorption Rates

Absorption plays little role in homeostatic regulation of body Se.[26,27] This is because absorption of Se, as selenite, selenate, or SeMet, is most likely at levels approaching 90%.[28] However, differences in Se absorption between inorganic and organic forms might be unrealistically stressed in literature. Many reported Se absorption rates were obtained from balance studies in which endogenous Se losses were not corrected.[29] In fact, an intravenous (i.v.) administration of selenite or SeMet led to similar biliary secretion of Se, as seleno-diglutathione (GS-Se-SG).[30] Because the small intestine has SELENOP transcript expression levels second only to the liver,[31] absorbed Se in the intestine may be used for the onsite biosynthesis of SELENOP. Although the fate of this intestinal SELENOP is unknown, it may contribute to increases in plasma SELENOP and (or) endogenous Se losses. Thomson and colleagues performed several studies in rats and humans using orally or i.v. administrated [^{75}Se]-labeled selenite, selenate, or Se-Met at various Se intakes for short and long durations. They found similarly high true absorption rates (>90%) of Se from both inorganic and organic forms.[28,32-34] However, absorption rates of Se, estimated using dietary and fecal Se without measuring endogenous fecal losses, were 75% for SeMet, 46% for selenite, and 66% for Se in mackerel, respectively, when adults in New Zealand consumed 100 μg Se/day as SeMet or selenite for 10–11 weeks, or 65 μg Se/day in mackerel for 4 weeks.[35] Similar high Se absorption obtained in stable isotope studies with North American men also showed only small effects of Se form. Results from metabolic tracer studies in rats using three inorganic and two organic selenocompounds[36] and urinary Se excretion studies in humans[37,38] further reinforced that the form of administered Se had a limited long-term effect. In addition, administering 60 mg vs 290 mg of ascorbate (vitamin C) in humans produced little effect on Se absorption from selenite (68 vs 61%, respectively).[39] As summarized by Thomson,[40] observed differences among various forms of Se, including selenate, wheat-Se, and yeast-Se in absorption or retention from human studies largely depend upon methods used to measure it.

Transport

SELENOP is needed for supplying Se to brain and testes, as a targeted deletion of SELENOP decreases testis Se substantially, even in mice supplemented with adequate Se,[41,42] and results in sperm defects indistinguishable from those elicited by dietary Se deficiency.[43,44] This deletion also causes motor incoordination that can be prevented, but not reversed, by supranutrition of Se[42,45] that largely restores Se concentration in the brain but not testes.[41] A SELENOP-specific receptor, apolipoprotein E receptor (ApoER2), is expressed in the brain and testes, and deletion of ApoER2 in mice leads to marked reduction in brain and testis Se concentrations, and neurologic dysfunction and sperm defects that are identical to those present in *Selenop*-deleted mice or in Se-deficient mice and rats.[46,47] Transgenic restoration of SELENOP production in the liver alone is sufficient to restore Se uptake by the brain and testes and to prevent the neurologic and fragile sperm phenotypes.[44] In addition, the megalin receptor and the low-density lipoprotein receptor-related protein 1 receptor[48] can also mediate SELENOP uptake by the kidney and other tissues.[46] However, the ability of higher levels of dietary Se to restore brain Se[41] or to provide Se for development in fetuses and nursing pups[49] in the Selenop$^{-/-}$ mice indicates that other, perhaps low-molecular-weight, forms of Se may provide additional transport routes. Furthermore, Se may also enter the brain as selenosugars[50] or as selenite, using anion transporters, and can be sequestered by Se binding protein 1 in astrocytes.[51]

Intermediate Metabolism and Excretion

Feature of Metabolites

Selenometabolites are small (<2 kDa) products of Se metabolism that include selenite, selenate, SEC, SeMet, and selenosugars. Once absorbed, both inorganic and organic forms, are readily or rapidly metabolized to the same precursor of selenide (see below).[36] An overall metabolic scheme of Se is shown in **Figure 14.1**.

SeMet

Most plants and bacteria can synthesize methionine, cysteine, and related intermediates from inorganic S. Because of similar properties of S and Se, enzymes in those S-metabolism pathways

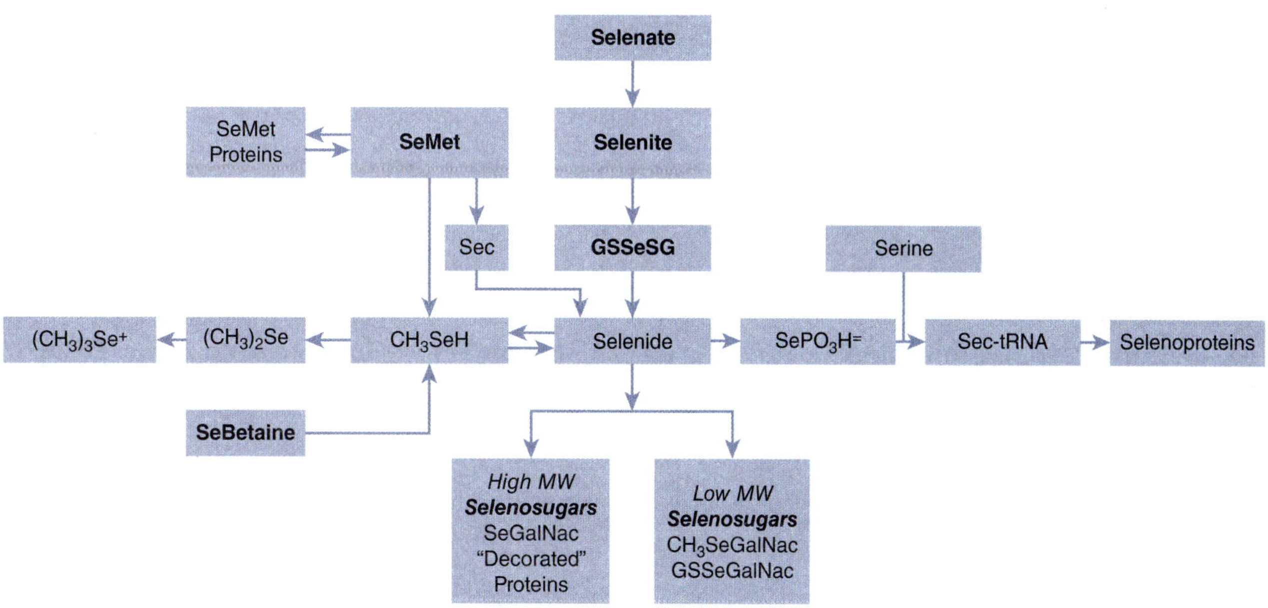

Figure 14.1 Diagram of Se metabolism. Shown are dietary forms of Se (selenite, selenate, SeMet), the three major forms of high molecular-weight Se-containing proteins (true selenoproteins, SeMet-containing proteins, and selenosugar-decorated proteins), and major forms of excreted low molecular-weight selenospecies [CH_3SeH, $(CH_3)_2Se$, $(CH_3)_3Se^+$, $CH_3SeGalNac$]. Key intermediates are shown in brackets. Underlined selenocompounds are very rapidly metabolized to the common selenide-level intermediate used for synthesis and incorporation of Se (^{75}Se) into major selenoproteins in a variety of tissues.[36]

Data from Evenson JK and Sunde RA. FrontiersNutr 2021;8. doi:10.3389/fnut.2021.699652

also readily synthesize SeMet, SEC, and related metabolites from inorganic Se.[52] Simple-stomached animals, however, lack the enzymes or microbes for *de novo* synthesis of methionine and cysteine or selenoamino acids from inorganic Se.[53] Recent metabolomics studies detected no SeMet in the liver of turkeys or rats fed a Se-deficient diet (0.005 μg Se/g) supplemented with up to 5 μg Se/g as selenite.[54,55] Animals as well as plants and bacteria, however, readily incorporate SeMet as a methionine analog via the Met-tRNA into general proteins in place of methionine.[56] Plants do not have UGA-encoded selenoproteins[52] and, thus, contain SeMet as the major form of Se.[57] The incorporation of SeMet into general tissue proteins in place of methionine in animals can markedly increase total tissue Se beyond what is present as bona fide selenoproteins. Se in SeMet, however, is not a highly effective reservoir for maintaining selenoprotein levels.[58,59] Replacement of methionine with SeMet does not seem to affect protein functions.[60] Degradation of SeMet is catalyzed by S metabolic enzymes: either via the transsulfuration pathway to selenocystathionine and then to SEC, or via the transamination pathway to methaneselenol (CH_3SeH) and then to selenide.[61]

SEC

Arising from SeMet metabolism, diet, or selenoprotein degradation, SEC is catabolized by a Se-specific enzyme, selenocysteine lyase (SCLY). This enzyme releases Se in the zero oxidation state (Se^0) as an intermediate, which is then rapidly reduced nonenzymatically to selenide.[62] Knockout of *Scly* diminishes hepatic GPX1 and selenoprotein S (SELENOS) levels and circulating SELENOP concentration in mice.[63]

Selenate and Selenite

Selenate is reduced to selenite via S metabolite pathways.[52] Selenite is further reduced to selenide by thioredoxin reductase or glutathione-S-transferase to Se^0 intermediates, such as seleno-diglutathione (GS-Se-SG), which are then further reduced nonenzymatically.[64]

Selenide

This compound has three major fates in simple-stomached animals: methylation, incorporation into selenoproteins, and formation of selenosugar (SeGalNac, 1 β-methylseleno-N-acetylgalactosamine) compounds. Recent ^{76}Se metabolism studies in rats demonstrated that a diverse set of three inorganic selenocompounds (selenite, selenate, and GS-Se-SG), and two organic selenocompounds (SeMet and selenobetaine), were very rapidly metabolized to the common selenide-level intermediate used for synthesis and incorporation of ^{76}Se-SEC into major selenoproteins in a variety of tissues.[36] Subsequently, there were no major differences in the resultant selenoprotein profiles or tissue Se distribution associated with forms of the initially administered ^{76}Se.

Methylselenol (CH_3SeH), Dimethylselenide (DMSe), and Trimethylselenonium (TMSe)

Methylation is the mechanism used to detoxify and eliminate Se. Largely, CH_3SeH and TMSe are excreted in the urine, and volatile DMSe is lost in breath. Recent research has substantiated that two methyltransferases are key to syntheses of CH_3SeH, DMSe and TMSe.[65] Inorganic Se as selenide is sequentially methylated by thiopurine S methyltransferase (TPMT) to CH_3SeH (methyl selenol) and then to DMSe by TPMT. DMSe metabolism to TMSE is catalyzed by indolethylamine N methyltransferase (INMT), although INMT can also methylate CH_3SeH. *Tpmt* transcripts are expressed in most mouse tissues and preferably expressed in the liver and kidneys, while *Inmt* transcripts are abundantly expressed in the lung and, at lower levels, in the liver and kidneys. Thus, excess Se is converted to CH_3SeH and DMSe in the liver, and DMSe is converted to TMSe in the lung. This scenario would explain why large ingestion of Se may overwhelm the ability of the lung to convert DMSe to the urinary excretion product TMSe, thus releasing DMSe in breath, the source of garlic breath in individuals with high Se intakes.

These methylated selenometabolites, however, cannot be used as biomarkers of body or tissue Se status. Blood Se, urinary Se, and urinary TMSe were determined in Bangladeshi and Andean women. Only 33% of Bangladeshi women and 3.6% of Andean women released TMSe in their urine.[66] In a genome wide association study (GWAS) with Bangladeshi women, a single nucleotide polymorphism (SNP) in the *INMT* gene was strongly associated with percentage of urinary TMSe. In both cohorts, high TMSe producers had the AA genotype, intermediate producers had the AG genotype, and low urinary TMSe producers had the reference GG genotype. The TMSe producers also produced the urinary S analog, trimethyl sulfonium (TMS), at substantially higher concentration, illustrating that Se was metabolized via these S metabolic pathways. There were no significant effects of the SNPs on erythrocyte Se or total urinary Se.[66] It has been estimated that up to 75% of the European population has lost the ability to excrete normal amounts of TMSe in urine,[67] further indicating that methylated selenometabolites cannot be used as biomarkers of Se status.

Selenosugars

In 2002, methyl-SeGalNac was first reported to be the monomethylated Se species in urine of animals with Se-deficient to "low-toxic" Se status.[68] This discovery showed the formation of a new Se–C bond, with Se bound to the 1-carbon of the galactose ring, distinctly different from a Se–S bond. The same group also identified methyl-SeGalNac and GS-SeGalNac in the liver, and inferred a metabolic pathway for selenosugars arising from an inorganic selenide-GSH pool to formation of the precursor, GS-SeGalNac, which was reduced and methylated to produce methyl-SeGalNac (the selenide to methyl-SeGalNac shown in Figure 14.1).[68] Targeted analysis by LC MS/MS has shown the presence of methyl-SeGalNac in the liver and kidney.[69-71] Recent metabolomics studies found that the parent selenosugar (SeGalNac) was also linked *via* Se–S bonds to general proteins in turkey liver.[54] Thus, Se is present in tissues in three types of high molecular-weight proteins (Figure 14.1): 1) as true Sec-containing selenoproteins, 2) as SeMet substituting for methionine in general proteins, and 3) as selenosugars decorating general proteins. At least in Se-adequate turkeys, a far greater portion of liver Se is present as SeGalNac than as SEC in selenoproteins. When high Se supplementation elevated selenosugar formation and further enhanced these high molecular-weight, selenosugar-decorated proteins, it could also increase selenosugar linked to low molecular-weight thiols.[54] This leads logically to the formation of methyl-SeGalNac in the liver, the urinary Se excretion species. The finding on the abundance of Se as selenosugar-decorated protein relative to selenoproteins will revise our knowledge of Se metabolism and may help identify new biomarkers to assess Se status and new mechanisms to explain health benefits of Se supplementation.

Other Metabolites

A novel Se compound, selenoneine (2-Se-N,N,N-trimethyl-L-histidine) with Se attached to an imidazole ring, was identified as the major organic form of Se in the blood of tuna and was also found in chicken and pig tissues.[72] Similarly, selenohomolanthionine was found as a novel, major selenometabolite in Torula yeast,[73] and was also present in selenized Japanese pungent radish[74] and in the Se accumulator Indian mustard.[75]

Se "hitch-hikes" along S Metabolic Pathways

While humans and higher animals downregulate functional expression of selenoproteins as a homeostatic adaptation to low Se status, the opposite does not happen. Because Se and S share metabolic pathways, the "hitchhiking" of Se allows these species to adapt to high Se intakes by using responsible S enzymes and pathways to metabolize excessive Se. Because Se is at $<1/1{,}000$th of the level of S, such an alternation, even at high Se intakes, would have little impact on regulating the S-related enzymes or pathways.[76]

Biosynthesis of Selenoproteins

The novel metabolic fate of selenide is cotranslational incorporation of Se as SEC into selenoproteins [see reviews[77-79]]. As shown in Figure 14.1, synthesis of true selenoproteins starts with selenide[80] and serine.[81] Selenide is metabolized to selenophosphate ($SePO_3H^-$). Serine is esterified to a novel tRNA[Ser]Sec with an anticodon specific for UGA, and selenophosphate is used to form Sec-tRNA (Figure 14.1). Selenoprotein transcripts must possess an in-frame UGA codon at the position of SEC incorporation (**Figure 14.2**). What distinguishes selenoprotein transcripts from transcripts with a premature or nonsense UGA stop codon is a stemloop SECIS (SEC insertion sequence) element in the 3′UTR of eukaryotic selenoprotein transcripts.[82] For SEC incorporation, the SECIS element recruits several novel factors (SBP2 and EFsec) that bind the Sec-tRNA, such that the Sec-tRNA out-competes release factors and inserts the SEC into the growing peptide chain during protein translation.[4,77] This Sec-tRNA-mediated pathway is the sole path for specific, targeted incorporation of Se into true selenoproteins in animals. Deletion of tRNA$^{[ser]sec}$ disallows biosynthesis of any selenoprotein, leading to embryonic lethality.[83]

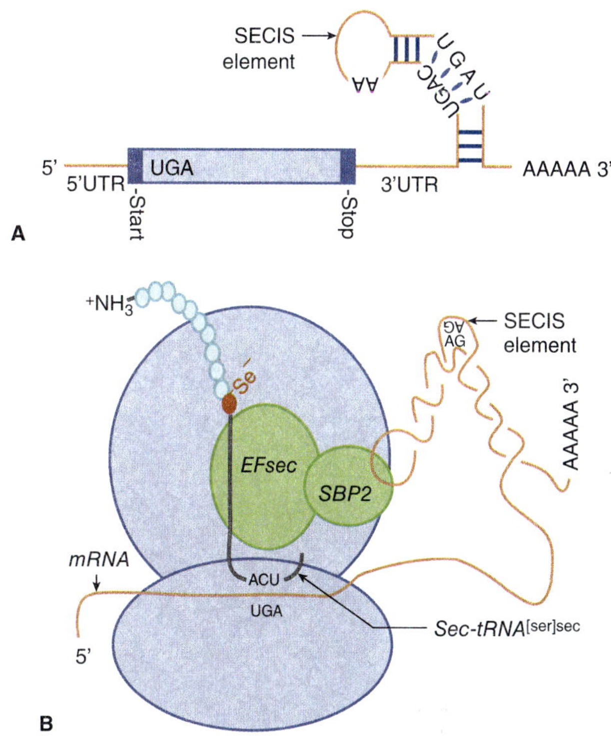

Figure 14.2 Selenoprotein synthesis. **(A)** Hypothetical selenoprotein mRNA showing the necessary UGA codon specifying the position for SEC incorporation during protein synthesis, and the necessary SECIS element in the 3′-UTR. **(B)** Diagram showing the selenoprotein mRNA on the ribosome during protein synthesis, with the 3′-UTR SECIS element recruiting the necessary factors, SBP2, and EFsec, that bind and position the Sec-tRNA at the UGA codon to facilitate SEC incorporations into the growing protein chain.

Functional Genomics of Selenoproteins

Revealing Selenogenomes

Selenoproteins identified in early years using biochemical methods share a common feature in their transcripts: containing at least a UGA codon for SEC in the open reading frame and a SECIS element in the 3′-UTR. Using these two transcript elements as a signature to screen annotated genomes, 25, 24, and 25 selenoprotein genes were found in genomes of humans, rodents, and chickens, respectively (**Table 14.1**).[4,84-86] Unlike in humans, rodent GPX6 contains no SEC in the peptide. Chickens are missing *GPX6* and *SELENOV*, but have two unique genes of *SELENOU* and *SELENOP2*.[87] *SEPHS2* was not found in the initial genome assembly of chickens,[87] but its transcript and mRNA expression were later shown in EST sequences from *Gallus gallus*[88] and in adipose tissue of chickens,[89] respectively. Both mammalian and avian *SELENOP* genes contain two SECIS elements in the 3′ UTR. In birds, 12 of the 13 SEC residues in the peptide of SELENOP are in the C terminus. In contrast, *SELENOP2* in chickens contains only a single SEC and SECIS each.[87] Human *DIO2* seems to have a second UGA codon serving as a stop codon, because there is little actual protein sequence information to confirm the presence of two SECs in the peptide. Notably, five putative candidate selenoproteins were identified using a SEC-specific mass spectrometry-based approach, but no SECIS element was found in their transcripts.[90]

Functions of Selenoproteins

In most selenoproteins, Se is presented as part of a SEC-Cys pair (or SEC-serine or SEC-threonine), allowing these proteins to function as oxidoreductases. The SEC moieties are in the polypeptide backbone of the active site (Table 14.1). In fact, catalytic activity and enzymatic kinetics are characterized in 13 of the 25 human selenoproteins. These include GPXs (GPX1–4), iodothyronine 5′ deiodinases (DIO1–3), thioredoxin reductases (TXNRD1–3), methionine sulfide reductase (MSRB1), selenophosphate synthase 2 (SEPHS2), and phosphatidyl ethanolamine

Table 14.1 Biochemical Features, Functions, and Health Relevance of Selenoproteins of Humans, Rodents, and Avians

Selenoprotein	Abbreviation	Cellular Location & Tissue Distribution	Number of AA and Position of SEC	Function	Health Relevance
Selenoenzymes					
Glutathione peroxidase-1 (cellular glutathione peroxidase)	GPX1	In cytosol & mitochondria, ubiquitously expressed in all tissues, high in the liver, lung, kidney, and erythrocytes	H: 203(49) R: 201(47) A: 195(39)	Reduces peroxides to control cellular signaling and redox status, regulating PDX1, AKT, insulin receptor, REG2/Ca^{2+} influx to affect insulin synthesis, secretion, and sensitivity; regulating metabolism of glucose, lipid, and protein, modulating cell survival and death, and promoting female reproduction function	*Gpx1*$^{-/-}$ mice: viable; overexpression and knockout of *Gpx1* led to T2DM and T1DM-like phenotypes, respectively; polymorphism related to obesity, T2DM, and cancer risk in humans; involved in endothelial NO generation, anti-inflammation, vasoreactivity, neoangiogenesis, vascular remodeling, and cardiovascular protection; related to pre-eclampsia
Glutathione peroxidase-2 (gastrointestinal glutathione peroxidase)	GPX2 (GI-GPX)	In cytosol, high in GI tract and endothelial cells, but also in other tissues including placenta	H: 190(40) R: 190(40) A: 191(41)	Reduces peroxide, inhibits inflammation-induced carcinogenesis in the gut; Implicated in protections of embryos and extraembryonic tissues	*Gpx2*$^{-/-}$ mice: viable; promoting growth of some cancers including bladder cancer
Glutathione peroxidase-3 (plasma glutathione peroxidase)	GPX3 (plasma GPX)	Secreted to extracellular fluids (plasma, milk)	H: 226(73) R: 226(73) A: 219(71)	Serve as a main antioxidant in extracellular fluids/plasma, implicated in the implantation and other reproductive process, affecting bioavailable NO, platelet activation, and vasoreactivity; antithrombotic	*Gpx3*$^{-/-}$ mice: viable; cardiovascular protective, deficiency facilitates platelet aggregation and increases risk of stroke; involved in MetS, IR and T2D complications; acts as a tumor suppressor in many cancers (lung cancer)
Glutathione peroxidase-4 (phospholipid hydroperoxide glutathione peroxidase)	GPX4 (PHGPX)	In cytosol, mitochondria, and nucleus, highly expressed in the testis, spermatozoa and kidney, followed by liver, spleen, pancreas, heart, and brain	H: 197(73) 170(46) R: 197(73) 170(46) A: 202(70) 194(70)	Serves as a membrane antioxidant to reduce phospholipid hydroperoxides and as a structure protein in sperm; a key inhibitor of ferroptosis/cell death; anti-inflammation, anti-atherogenic, anti-thrombotic, and cardioprotective; vital for embryonic development	*Gpx4*$^{-/-}$ mice: lethal, but deficiency tolerant to humans; implicated in male infertility, immune disorders, HIV, clinical outcomes of gestational diabetes mellitus, and several cancers; non-sense mutations cause Sedaghatian disease (skeletal dysplasia, cardiorespiratory failure), loss of white cortical matter, hypogenesis of corpus callosum, and cerebellar hypoplasia
Glutathione peroxidase-6	GPX6	Cellular location is unclear, detected in embryos and adult olfactory epithelium	H: 221(73) R: NA A: NA	Function unknown; homolog to GPX3?	*Gpx6*$^{-/-}$ mice: NA; SEC in only humans and pigs; reduces the motor defects found in Huntington's disease

Selenoprotein	Abbreviation	Cellular Location & Tissue Distribution	Number of AA and Position of SEC	Function	Health Relevance
Selenoenzymes					
Iodothyronine 5´ deiodinase-1	DIO1	Plasma membrane, in peripheral tissues (e.g., liver and kidney)	H: 249(126) R: 257(126) A: 246(124)	Conversion of T_4 to T_3 in peripheral tissues	*Dio1*$^{-/-}$ mice: viable, but deficiency is intolerant to humans; implicated in immune thyroid disease and thyroid dysfunctions
Iodothyronine 5´ deiodinase-2	DIO2	ER membrane, in brain, pituitary and brown fat tissue	H: 265 (133, 266, the 2nd as a stop codon?) R: 262(130) A: 264(130)	Conversion of T_4 to T_3 locally to control intracellular T_3 in brain, pituitary, and brown fat tissue	*Dio2*$^{-/-}$ mice: viable but deficiency is intolerant to humans; implicated in immune thyroid disease and thyroid dysfunctions
Iodothyronine 5´ deiodinase-3	DIO3	Plasma membrane/ER(?)	H: 304(170) R: 304(170) A: 274(138)	Conversion of T_4 to reverse T_3 and T_3 to T_2	*Dio3*$^{-/-}$ mice: viable but deficiency is intolerant to humans; implicated in immune thyroid disease and thyroid dysfunctions
Thioredoxin reductase 1	TXNRD1	In cytosol (nucleus?), in all tissues	H: 649(648) R: 499(498) A: 499(498)	As a oxidoreductase to reduce thioredoxin, Trx1, or glutaredoxin in all tissues; cell signaling and controls of transcription factors; antiapoptotic and pro-inflammatory	*Txnrd1*$^{-/-}$ mice: lethal and the deficiency is also intolerant to humans; roles in embryogenesis & vasoreactivity; overexpressed in several types of cancer; missense mutations cause generalized epilepsy in humans
Thioredoxin reductase 2	TXNRD2	In cytosol and mitochondria	H:524(523) R: 527(526) A: 514(513)	Similar to TXNRD1, but reduces Trx2 and related to mitochondrial function	*Txnrd2*$^{-/-}$ mice: lethal, but deficiency is tolerable to humans, vasoreactivity, cardiac development, cardioprotective, polymorphism associated with T2DM complications; non-sense mutations cause glucocorticoid deficiency in humans
Thioredoxin reductase 3	TXNRD3	In cytosol & mitochondria of testis	H: 643(642) R: 652(651) A: 606(605)	Similar to TXNRD1 and TXNRD2, but is expressed and functions at mitochondria in testis only	*Txnrd3*$^{-/-}$ mice: N/A; role in sperm maturation and embryogenesis
Methionine sulfide reductase	MSRB1	In cytosol & nucleus (?), highly expressed in immune-activated macrophages	H: 116(95) R: 116(95) A: 111(93)	Reduces methionine-R-sulfoxide to methionine in proteins	*Msrb1*$^{-/-}$ mice: viable; Involved in shaping immune responses of cells and organs

(continues)

Table 14.1 Biochemical Features, Functions, and Health Relevance of Selenoproteins of Humans, Rodents, and Avians *(continued)*

Selenoprotein	Abbreviation	Cellular Location & Tissue Distribution	Number of AA and Position of SEC	Function	Health Relevance
Selenoenzymes					
Selenophosphate synthase 2	SEPHS2	Cytosol	H: 448(60) R: 452(63) A: ?(?)	Catalyzing formation of monoselenophosphate from selenide and ATP for synthesis of SEC	*Sephs2*$^{-/-}$ mice: N/A Not found in early genome of chickens, but identified from EST sequences and expression
Selenoprotein I	SELENOI	Plasma membrane & cytosol	H: 397(387) R: 398(388) A: 400(384)	Regulation of phospholipid biosynthesis by catalyses the transfer of phosphoethanolamine from CDP-ethanolamine to diacylglycerol to form phosphatidylethanolamine	*Selenoi*$^{-/-}$ mice: lethal; involved in mouse embryogenesis and the formation and maintenance of vesicular membranes; missense mutations cause complex spastic paraplegia and progressive cerebello-cerebral atrophy
Selenium Transporter					
Selenoprotein P	SELENOP	Secreted into extracellular fluids, cytoplasm & mitochondria	H: 381 (59, 300, 318, 330, 345, 352, 367, 369, 376, 378) R: 380 (59, 264, 277, 318, 330, 352, 366, 368, 375, 377) A: 393 (55, 287, 300, 313, 331,343, 356, 358, 365, 379, 381, 388, 390)	Main plasma Se transporter to brain and testis; antioxidant	*Selenop1*$^{-/-/-}$ mice: viable; 10 SEC in mammals and 13 SEC in avian species; used as an indicator of Se status; implicated in cancer, neurodegenerative diseases, male fertility and maternal-fetal Se transfer, apoptosis, the endocrine system, intracellular signaling, and insulin resistance and T2DM
Cxxu-Motif Containing Selenoproteins					
Selenoprotein F	SELENOF (15kD selenoprotein, Sep15, SelF)	ER	H: 165(96) R: 162(93) A: 160(91)	As a thioredoxin-like oxidoreductase to regulate ER stress, enhancing activity of UGGT, involved in glycoprotein folding quality control by rearranging or reducing the disulphide bonds of UGGT-recognized misfolded proteins, sequence homologous to protein disulphide isomerases (PDI)	*Selenof*$^{-/-}$ mice: viable; may be involved in apoptosis

Selenoprotein	Abbreviation	Cellular Location & Tissue Distribution	Number of AA and Position of SEC	Function	Health Relevance
Cxxu-Motif Containing Selenoproteins					
Selenoprotein H	SELENOH (SelH, C11 or F31)	In nucleus & Golgi apparatus	H: 122(44) R: 116(38) A: 129(51)	Regulation of redox sensing and status, transcription, and GSH synthesis	*Selenoh$^{-/-}$ mice: N/A; implicated in placenta oxidative stress*
Selenoprotein M	SELENOM (SelM)	ER lumen & mitochondria	H: 145(48) R: 145(48) A: 152(37)	Involved in body weight and energy metabolism control; homologous to SELENOF and PDI, likely function as a thiol-disulphide oxidoreductase and involved in protein folding	*Selenom*$^{-/-}$ mice: viable; highly expressed in the brain with neuroprotective properties
Selenoprotein O	SELENOO	Mitochondria	H: 669(667) R: 667(665) A: 670(668)	Possible redox and kinase function	*Selenoo*$^{-/-}$ mice: viable
Selenoprotein T	SELENOT (SelT)	ER membrane	H: 195(49) R: 195(49) A: 199(53)	Oxidoreductase in Golgi, ER, and cytosol, involved in redox regulation, anchorage, ER homeostasis, Ca flux and store, and hormone secretion	*Selenot*$^{-/-}$ mice: lethal, and the deficiency is also intolerant to humans; protection of dopaminergic neurons against oxidative stress in PD; control of glucose homeostasis in pancreatic beta-cells
Selenoprotein V	SELENOV	ER (?), expressed in testis, brain, and adipose tissue	H: 346(273) R: 328(255) A: N/A	Interacting with UDP-*N*-acetylglucosaminyl transferase (OGT) and regulating O-GlcNAcylation, lipogenesis, energy expenditure, body temperature, tissue Se distribution and other selenoprotein expression, and redox status and ER stress	*Selenov*$^{-/-}$ mice: viable expressed in placental animal only (not in birds), high expression in testis, but role the in spermatogenesis is unclear
Selenoprotein W	SELENOW (SelW, SEPW1)	Cytosol, highly expressed in skeletal muscle, heart, and brain	H: 87(13) R: 88(13) A: 85(13)	Putative antioxidant, involved in muscle growth and differentiation, protection of neurons from oxidative stress during development	*Selenow*$^{-/-}$ mice: viable

(continues)

Table 14.1 Biochemical Features, Functions, and Health Relevance of Selenoproteins of Humans, Rodents, and Avians *(continued)*

Selenoprotein	Abbreviation	Cellular Location & Tissue Distribution	Number of AA and Position of SEC	Function	Health Relevance
Other ER Resident Selenoproteins					
Selenoprotein K	SELENOK	ER & plasma membrane	H: 94(92) R: 94(92) A: 95(93)	Involved in oxidation resistance, Ca flux regulation and chaperone-like function in ER-associated degracation of misfolded proteins	*Selenok*$^{-/-}$ mice: viable; expression decreased by glucose
Selenoprotein N	SELENON (SelN, SEPN1)	ER membrane	H: 556(428) R: 557(428) A: 530(402)	Regulating RyR-mediated Ca mobilization for muscle development & differentiation, uncertain such role in other tissues	*Selenon*$^{-/-}$ mice: viable; mutations (deletion, missense, non-sense) cause multiminicore disease and other myopathies in humans
Selenoprotein S	SELENOS (SEPS1, Sel S, Tanis or VIMP- valosin-containing protein-interacting membrane protein)	ER & plasma membrane	H: 189(188) R: 190(189) A: 194(193)	Involved in ER-associated protein degradation, protecting against ER-stress and the induced apoptosis, anti-inflammation	*Selenos*$^{-/-}$ mice: N/A; levels associated with T2DM and T2DM macrovascular complications and polymorphisms associated with the risk of T2DM and macroangiopathy
Avian Only Selenoproteins					
Selenoprotein P2	*SELENOP2*	unknown	H: NA R: NA A: 263(59)	Unknown	*SELENOP2*$^{-/-}$: N/A Avian and fish only
Selenoprotein U	*SELENOU*	unknown	H: NA R: NA A: 224(85)	Unknown	*SELENOU*$^{-/-}$: N/A Avian only

Only the predominant isoform is listed if there are multiple forms for a given selenoprotein except for GPX4.
Abbreviations: AA, amino acid; ER, endoplasmic reticulum; H/R/A, humans/rodents/avian species; NA, not applicable or existing in the given species.

Data from Lei XG, Combs Jr GF, Sunde RA, et al. Annu Rev Nutr 2022;42:337-75[17]; Li S, Gao F, Huang J, et al. Biochem Biophys Acta. Gen subjects 2018;1862:2473-9[86]; Sun LH, Huang JQ, Deng J, et al. Poult Sci 2019;98:4247-54[91]; Barchielli G, Capperucci A, Tanini D. Antioxidents 2022;11:251[92]; Ojeda M, Carreras O, Nogales F. Antioxidents 2022;11:394[93]; Gorini F, Vassalle C. Antioxidents 2022;11:1188[94]; Hogan C, Perkins AV. Nutr 2022;14:628[95]; Weaver K, Skouta R. Biomedicines 2022;10:891[96]; Schweizer U, Fabiano M. Free Radic Biol Med 2022;190:105-15[97].

transferase (SELENOI). A CXXU motif located in a thioredoxin-like fold was found in seven selenoproteins (SELENOF, H, M, O, T, V, and W), which provides a structural potential for them to function as oxidoreductases. Seven selenoproteins (SELENOF, K, M, N, S, and T and DIO2), and perhaps SELENOI, are localized in ER (endoplasmic reticulum)[98,99] and are involved in regulating proteostasis, ER and redox stresses, and Ca^{2+} flux.[100] SELENOP is recognized as a Se transporter to facilitate target transfer of Se to the brain and testes, with peroxidase activity. Global knockouts of 19 of the 24 selenoprotein genes in mice, except for *Selenoh*, *Selenoo*, *Selenos*, *Sephs2*, and *Txnrd3*, were generated.[17] Only 5 homozygous lines (*Gpx4*$^{-/-}$, *Trxnd1*$^{-/-}$, *Trxnd2*$^{-/-}$, *Selenoi*$^{-/-}$, and *Selenot*$^{-/-}$) of these knockouts exhibited embryonic lethality (Table 14.1). Functional expressions of TXNRD1 and SELENOT are essential to both humans and rodents. However, loss of any of the DIO enzymes is well tolerated by rodents, but not by humans, whereas the opposite is true for the tolerance to the loss of GPX4 or TXNRD2.[101] Biochemical properties, main functions, and health relevance of human, rodent, and avian selenoproteins are summarized in Table 14.1. Special considerations and new progress are highlighted as follows.

Glutathione Peroxidases

GPX1 is not only the very first identified, but also the most abundant, member of selenoproteins, representing 60% of the total hepatic Se in mice.[102,103] Comparatively, GPX4 differs from other GPXs: 1) it is the only GPX that catalyzes reduction of phospholipid hydroperoxides; 2) it is a monomer, instead of a homotetramer; 3) it has three alternative cellular locations (cytosolic, mitochondrial, and nucleic); and 4) it displays two-phased functions in spermatozoa: as an active enzyme before maturation and as an oxidized structural protein in the mitochondrial capsule after maturation.[104] Most notably, deletion of *Gpx4*, but not *Gpx1*, *Gpx2*, or *Gpx3*, is embryonically lethal.[105] This lethality is involved in ferroptosis, and can be rescued by feeding dams a high amount of vitamin E (500 IU/kg diet).[106] The central role of GPX4 in ferroptosis was revealed using specific inhibitors of GPX4 and cysteine transport (important for maintaining GSH).[107] Recently, GPX4 was found to bind prothymosin alpha in regulating apoptotic signaling and cell death of pancreatic acinar cells.[108] Despite no effect on survival under nonstressed conditions, global knockout and overexpression of *Gpx1* decreases and enhances resistance of mice to various acute and chronic oxidant insults, respectively.[109-112] Remarkably, these alterations of GPX1 produce contrasting effects on pancreatic beta cell mass, insulin synthesis, secretion, and signaling, and glucose, lipid, and protein metabolism by dysregulating related genes and proteins at transcriptional, translational, epigenetic, and(or) protein functional levels.[113-116] Double knockouts of *Gpx1* and *Gpx2* in mice lead to ileocolitis.[117] GPX6 is present in the olfactory apparatus,[4] but its function is not known.

Iodothyronine Deiodinases

DIO1-3 are transmembrane selenoproteins, and DIO2 is also an ER-resident protein. DIO1 or DIO2 catalyzes 5′-deiodination of prohormone thyroxine (3,3′,5,5′-tetraiodothyronine, T_4) to active 3,3′,5-triiodothyronine (T_3), the principal receptor-binding molecule interacting with the nuclear T_3-receptors.[118] Meanwhile, the 5-deiodinations of T_4 and T_3 by DIO3 lead to inactive reverse T_3 (3,3′,5′-triiodothyronine, rT_3) and 3,3′-T_2, respectively. While DIO1 controls T_3 production in the thyroid and many tissues, DIO2 and DIO3 are fine regulators of local activation and inactivation of thyroid hormones, respectively. DIO2 and DIO3 have distinct expression patterns, and their expression switches in many organs including the brain, leading to amplification of the T_3 signal during developmental transitions.[119] As mentioned above, loss of DIO1-3 is tolerated by rodents, but not by humans.[101] Several thiols can serve as the reducing substrate for these enzymes, but GSH is likely to be the physiologic substrate. Because of thyroid hormone-linked feedback mechanisms; however, thyroid Se status may not reflect Se status of the organism.[118] In addition, DIO2 activity in the brain was shown to be a determinant of well-being and neurocognitive function, and a Thr92Ala polymorphism was associated with ER stress, hypothyroidism[120,121] and Alzheimer disease (AD) development in African Americans.[122] The DIO2 gene was related to bipolar disorder in an Asian population.[123]

Thioredoxin Reductases

TXNRD1–3 are flavin-containing selenoproteins and contain SEC as the penultimate C-terminal amino acid. Using reduced nicotinamide adenine dinucleotide phosphate (NADPH), these enzymes reduce the internal disulfide of cytosolic and(or) mitochondrial thioredoxin and (or) glutaredoxin in all tissues (TXNRD1, 2), or specifically in testis (TXNRD3).[124] Neither mice nor humans tolerate deletion of TXNRD1, but TXNRD2 seems to be essential to only mice (Table 14.1). These reductases provide reducing equivalents to a variety of enzymes, many of which are oxidant defense enzymes, with some function in DNA synthesis and cell signaling.[125,126]

Other Selenoenzymes

Although two selenophosphate synthetases were identified in animals,[4] only SEPHS2 is a Se-dependent enzyme required for biosynthesis of SEC in selenoproteins.[127] It catalyzes the ATP-dependent formation of selenophosphate from selenide in the first step to assemble Sec-$tRNA^{Sec}$ in all vertebrates. Interestingly, SEPHS1 is not a selenoprotein and has no known function. But, SEPHS1 deletion in mice is embryonically lethal.[127] MSRB1 contains zinc and repairs oxidized methionine residues in proteins.[129] SELENOI has phosphatidyltransferase activity that catalyzes formation of phosphatidyl ethanolamine. Global knockout of *Selenoi* is embryonically lethal.[130-132]

Selenoprotein P

This extracellular glycoprotein was identified in plasma in 1977,[47] and four isoforms of the protein were identified in rat plasma. SELENOP accounts for approximately 47% of plasma Se in a typical North American.[17] The Se transport function of SELENOP is attributed to those SEC moieties located in the C terminus and requires the binding of circulating SELENOP to ApoER2 and other receptors in the testes and brain (see Se transport section). With a SEC moiety in the middle of the peptide, SELENOP was reported to possess peroxidase activity in vitro.[47] However, clinical evidence or relevance for such activity is lacking.[133] Plasma concentration of SELENOP declines in Se deficiency, which is used as an indicator of Se status.[47] Elevation of SELENOP was suggested to induce or potentiate diabetes or insulin resistance, but not all data support that notion (see below).

Selenoproteins with a CXXU Motif

The presence of CXXU motif in SELENOF, H, M, O, T, V, and W provides a structural potential for them to act as oxidoreductases. SELENOF is an ER resident selenoprotein and serves as a partner of glucosyltransferase (UGGT), a folding sensor of the ER transferring a glucose residue to the A branch of oligomannoses from misfolded glycoproteins to make it join the quality control process. Knockout of SELENOF in mice induces early onset of cataract, probably due to an increased misfolded protein or to the lack of interaction with a protein (RDH11) involved in vitamin A metabolism.[100] SELENOH may regulate redox sensing and transcription.[134] SELENOM is located in ER and is involved in body weight control and energy metabolism. The protein is highly produced in the brain, but knockout of SELENOM leads to obesity without cognitive impairment.[135] SELENOO is the largest selenoprotein located in mitochondria, with potentials of redox-control and kinase functions.[17]

SELENOT functions as an oxidoreductase in Golgi, ER, and cytosol and is involved in redox regulation, anchorage, ER stress, and cytosolic Ca^{2+} flux.[136,137] As the most evolutionarily conserved selenoprotein, SELENOT, similar to TXNRD1, is one of the selenoproteins with the highest priorities.[90,101,138,139] SELENOT and TXNRD1 are the only two selenoproteins known to be essential to both mice and humans.[101,136] SELENOT deletion results in oxidative injuries of dopaminergic neurons, leading to brain growth retardation and increased vulnerability to neurodegeneration.[136] With its interaction with the oligosaccharyltransferase complex, and its modulation of N-glycosylation, inhibiting SELENOT causes secretion defects and ER stress in endocrine cells.[137]

Deletion of SELENOV affects tissue Se distribution,[140] cellular and tissue defenses against oxidative and ER stresses,[141] and elevated lipogenesis and decreased energy expenditure in male mice, leading to the accumulation of body fat via a cascade of SELENOV–O-GlcNAc transferase–AMP-activated protein kinase.[142] SELENOW was originally identified in muscle and was postulated to play a role in the development of white muscle disease, a classical Se/vitamin E deficiency disease in sheep and cattle.[143] It has since been identified in many other tissues in several forms. One form has GSH bound to it, a finding suggesting that SELENOW undergoes redox changes. Some evidence indicates that it can protect against oxidative injury.[144]

Other Selenoproteins

Highly expressed in the brain and muscle[145] as transmembrane glycoprotein, SELENON senses Ca^{2+} fluctuations in ER by binding it through a luminal EF-hand domain. After activation by low luminal (ER) Ca^{2+}, SELENON turns on its redox activity toward its partner, the sarcoplasmic/endoplasmic reticulum Ca^{2+}-ATPase (SERCA pump), leading to a reduction of the luminal cysteines of this Ca^{2+} pump that are hyperoxidized by ER oxidoreductin 1 (ERO1)-generated peroxides.[146] Loss of SELENON function leads to hypersensibility to ERO1, defects in Ca^{2+} reuptake, and myopathies associated with insulin resistance. There are 65 *SELENON* mutations identified, mainly in exons.[147] Congenital muscular dystrophy resulting from these mutations was initially characterized by development of a rigid spine[148] and early onset of muscle weakness.[149] Mouse fetuses and pups lacking the SELENON gene develop normally, suggesting a role for SELENON in maturation of organs or protection under stress.[150] Patients with *SELENON* variants do not show intellectual disability or central nervous system change,[147] but their muscles contain lower amounts of Ca regulatory proteins and skeletal muscle excitation-contraction coupling.[151] In mice, SELENON deficiency does not affect limb muscles at rest but triggers inappropriate response to ER stress and dysfunction in highly active muscles.[152] SELENON also protects skeletal muscle from saturated fatty acid-induced ER stress and insulin resistance[153] and is thought to be involved in white muscle diseases in animals.[154]

Both SELENOK and SELENOS have a single transmembrane domain, but lack the catalytic CXXU found in a number of selenoproteins.[100] SELENOS is involved in the AD-associated APP-C99 fragment degradation through ER-associated degradation (ERAD),[155] and its expression correlates with neurofibrillary tangles formation in post-mortem brains of AD patients.[156] SELENOS upregulation can lead to insulin resistance.[100] SELENOK interacts with SELENOS and other partners in controlling ER stress.[157-159] It also interacts with a protein (DHHC6) responsible for IP_3 receptor palmitoylation, a necessary step for its stabilization,[160] and a link with Ca^{2+} regulation for microglial migration and Aβ deposit clearance.[161] SELENOK contains a SEC residue in the third position from the C-terminal end, supporting an antioxidant function.[162,163] No function has been identified for the two avian selenoproteins, SELENOU and SELENOP2.

Regulation of Selenoprotein Expression by Dietary Se

Dietary Se is the best studied and perhaps the most predominant regulator of functional expression of selenogenome in various tissues across species. However, there seems to be no universal mechanism for regulation across selenoproteins, species, or tissues. The lack of a global regulatory pattern may be due to biological heterogeneity of selenoproteins and(or) limitations of data from past studies that tested responses of only partial selenogenome to a few concentrations of dietary Se, with relatively large gaps in several tissues.

Transcript

Although dietary Se deficiency, compared with adequate Se intake, often decreases mRNA levels of many selenoproteins, it could also upregulate transcripts of certain selenoproteins (e.g., *Sephs2, SELENOS*, and *SELENOT*).[164-168] Dietary Se excess (e.g., 3 mg/kg) does not further increase transcripts of selenoproteins or even downregulate their transcripts.[164,165] The overall susceptibility or magnitude of such changes varies with selenoproteins and (or) tissues. A comprehensive study measured transcript expression of 24 selenoproteins in the liver, kidney, muscle, and/or blood of rats in responses to 10 graded levels (0–0.8 mg Se/kg) of selenite supplementation in a basal diet (0.007 mg Se/kg).[168] Dietary Se deficiency decreased hepatic transcripts of 11 of the 24 selenoproteins. Full selenoprotein transcript expression occurred at <0.07 mg Se/kg, lower than that (0.1 to 0.15 mg Se/kg) for GPX activities or Se concentrations in the liver and other tissues.[169] Likewise, dietary Se deficiency downregulated hepatic mRNA levels of 13 genes in mice, whereas five genes (*Gpx2*, *Selenon*, *Selenoo*, *Selenov*, and *Sephs2*) showed no or little response.[116] Further, dietary Se deficiency decreased mRNA levels of *GPX1*, *GPX4*, *SELENOM*, *SELENOO*, and *SELENOU* in the liver, muscle, and pancreas of chickens, but exerted relatively weak and (or) tissue-specific effects on expression of other selenoprotein genes.[87,91]

Protein

Much less is known on regulation of selenoprotein production than that of selenotranscript expression, due to lack of usable antibodies against many selenoproteins, and the unavailability or high cost of quantitative mass spectrometry analysis of them. A limited number of studies, conducted in several species,[84,87,170-172] indicate that production of many, but not all, selenoproteins is decreased by Se deprivation and restored by Se supplementation. It remains elusive if these effects are mediated through *de nova* synthesis or protein turnover. Dietary Se excess (3 mg/kg) seemed to increase SELENOP in the thyroid, testis, and muscle (up to 41-fold), GPX3 and SELENOH in the liver, and SELENOS in the thyroid, over those of adequate Se (0.3 mg/kg) in pigs.[115,165] Responses of selenotranscript and protein levels to dietary Se do not always synchronize. In chickens and turkeys, *GPX3* and *GPX4* transcript changes did not consistently lead to respective selenoprotein alteration.[173,174] Two forms of SELENOP are detected in chicken plasma or tissues, and the short form is elevated, while the long-form is decreased, in Se deficiency.[173]

Activity and Function

Total GPX activities in the plasma (GPX3) and (or) liver (mainly GPX1) are readily decreased by dietary Se deficiency, and their responses to dietary Se repletion are widely used to determine nutrient requirements of Se by humans and (or) domestic species. These activities reach plateaus in various species at dietary Se concentration between 0.1 to 0.3 mg/kg, which is the main biochemical basis for the currently recommended nutrient requirements of Se.[17] Supranutrition of Se was widely believed to produce no further increases in plasma or tissue GPX activities over those of adequate Se intake. However, feeding pigs, chickens, and mice 3 mg Se/kg, compared with 0.3 mg Se/kg, resulted in moderate (21% to 57%), but significant increases in GPX activities in the plasma, liver, muscle, and(or) pancreas.[110,115,165] Total activities of TXNRD are decreased by dietary Se deficiency in various tissues and species, and DIO functions are often measured by circulating T_4 and T_3 concentrations. However, responses of activities or functions of other selenoenzymes or proteins at different dietary Se concentrations have not been determined, or are difficult to be assayed. Nevertheless, moderate changes in protein or activity, let alone mRNA levels, of selenoproteins, despite statistical significance, may not necessarily translate into meaningful metabolic impacts. This is because, as described above, 14 of the 19 selenoprotein genes that are deleted in mice are not essential for survival or breeding under nonstressed conditions.

Other Regulation of Selenoproteins and Nutrient Interactions

Vitamin E

Se and vitamin E were indelibly linked by the necessity of their double deficiencies for inducing Se/vitamin E deficiency diseases in domestic animals. Their interaction was reinforced by subsequent discoveries of antioxidant potential of many, if not all, selenoproteins. Dietary vitamin E deficiency induced GPX4 and transcripts of several selenoproteins in the liver and (or) pectoral muscle of chickens,[166,171] but in otherwise healthy rats, it showed no direct effect on protein or activity of GPX1, GPX3, or GPX4 in the plasma, red blood cell, or liver.[175] Feeding $Gpx1^{-/-}$ and wild type mice high amounts of dietary vitamin E (all-rac-α-tocopheryl acetate at 750 or 7500 mg/kg) elevated their hepatic GPX and GPX4 activities by 17% to 48%, but this high vitamin E failed to replace the protection by GPX1 against paraquat-induced lethality.[176] In contrast, $Gpx4^{-/-}$ pups born to dams fed 41 IU of vitamin E/kg (as DL-α-tocopheryl acetate) died shortly after birth but survived when their dams were fed 500 IU of vitamin E/kg.[106] This GPX4 deficiency-induced death, and rescue by high vitamin E, were explained by their common role in protecting against ferroptosis, a regulated cell death entailing cellular iron accumulation and lipid peroxidation.[177]

Lipid

High dietary Se intake promoted lipogenesis and inhibited lipolysis in several species.[115,140,178] There was also a positive relationship between blood Se status and lipid concentrations.[179] Likewise, knockout or overexpression of GPX1 impacted tissue lipid profiles and functional expression of lipid metabolism-related genes and proteins.[116,180,181] Conversely, a high-fat diet altered mRNA levels of most selenoprotein genes in various tissues of pigs.[172,182] Some of the affected genes, such as *DIO2*, *GPX6*, and *SELENOK*, responded only to dietary fat, but not Se. In pigs, 12 genes were affected by both dietary Se and fat concentrations, including five genes encoding ER-resident proteins (DIO2, SELENOF, SELENOK, SELENOM, and SELENOS).[182,183] The high-fat diet induced GPX1 and GPX4 in the liver but decreased GPX1 and SELENOP in the testis of mice.[140] Importantly, the high Se and high-fat diets exerted tissue-specific and GPX1 expression–dependent impacts on lipid metabolism and related gene expression in mice.[116] Intriguingly, the high-Se diet suppressed the expression of essentially all assayed selenoprotein genes in the adipose tissue of WT mice, whereas the high-fat diet enhanced the expression of many selenoprotein genes in the liver of WT mice. The interaction between dietary Se and fat concentrations also affected SREBP1 (sterol regulatory element-binding protein 1) in the adipose tissue of both genotypes and C-JUN in the WT mice.[116]

Vitamin C

With its antioxidant activity, an interaction of ascorbate and Se is easily predicted. Indeed, high dietary vitamin C (1% of the diet) reduced tissue Se concentrations and interfered with the ability of selenite to prevent mammary tumors in rat models.[184] However, high vitamin C intake in humans showed little effect on Se absorption from selenite.[39]

Toxic Elements

As Se protects against toxicities of mercury, cadmium, and silver, a role for Se in counteracting heavy metal pollutants has been proposed.[61] The novel selenometabolite, selenoneine, has been found in chicken and pig tissues,[72] and its presence in zebrafish embryos may protect against methylmercury accumulation and toxicity.[185] The low bioavailability of Se in tuna may result from this form or from complexation with mercury.[186] In the International Polar Year Inuit Health study, in which adults had an average daily intake of 200 μg Se/day and 80 μg Hg/day,[187] high blood Hg was associated with higher blood Se concentration. Adults in the high Se and low Hg quadrant had the lowest prevalence of cardiovascular outcomes, except for stroke, and the prevalence of diabetes was more than doubled in those adults with high blood Se (>280 μg/L) regardless of the low Hg quadrant (10.3%) or high Hg quadrant (9.1%) vs low Se (3.9% and 4.4%) quadrants.

Other Factors

Supplemental dietary serine, administration of lipopolysaccharides, and heat stress were shown to affect mRNA levels and/or activities of several selenoproteins in porcine tissues or cell lines.[172,188] The importance of tRNA[ser]sec and SCLY[63] in biosynthesis and degradation of selenoproteins, roles of SELENOP in transport of Se to the brain and testis for synthesis of other selenoproteins, and effects of SELENOV on tissue Se distribution and other selenoprotein expression,[140] are described above. A unique dependence for GPX1 structure and activity on SOD1 was reported.[189] Knockout of SOD1 in mice did not change liver GPX1 protein but accelerated the conversion of SEC (38% less) to the Se-lacking dehydroalanine (77% more) in the active site of hepatic GPX1, compared with the wild-type. This conversion led to aggravated dissociation of the tetramer structure and approximately 40% decreases in the apparent kcat for peroxide substrates.

Dietary Intake and Body Status

Recommended Levels of Intake and Amounts Believed to Be Safe

Nutrient intake recommendations for a population are known as dietary reference intakes (DRI) or dietary reference values (DRV). Among other parameters, they include the amount of a nutrient that is thought to be an adequate intake for 97.5% of a population: that amount is referred to variously as the RDA[190] or Reference Nutrient Intake (RNI)[191]; alternatively the term Adequate Intake (AI) may be chosen.[192] The recommended values for Se intake vary by authority, as shown in **Table 14.2**, and range from 55 to 75 μg/d in adults, with most countries increasing the recommended intake in pregnant and/or lactating women. By the same token, those same authorities have defined the upper intake level (UL) of Se believed to be safe, although we will query this later.

Dietary Intake of Se Across the World

Dietary intake of Se in various countries from the year 2000 onwards has been published by Winther et al. (**Figure 14.3**).[193] Se intake is high in parts of China, Venezuela, North America, and Japan, and considerably lower in Europe and other parts of China.[194,195] In many European countries, individuals do not achieve recommended levels of intake.[192] As mentioned above, a belt of Se deficiency runs from Northeast to Southwest China.[195] However, currently used dietary assessment methods are imprecise measures of Se intake due to the large variability in the Se content of specific foods.[194]

Biomarkers of Body Se Status

Measurement of plasma or serum concentration of Se is the most common way of determining Se status and reflects Se intake.[194] Values cover a wide range, with both the highest and lowest values occurring in China (e.g., plasma concentrations of Se in the range 22–550 μg/l).[196,197] Data from almost 2,000 individuals from 10 European countries, who were control participants in the EPIC study, showed a mean serum Se concentration of 85.6 μg/l,[198] whereas the mean (SD) serum Se concentration in 6,683 U.S. residents measured in the 2011–2016 National Health and Nutrition Examination Survey (NHANES) was much higher, at 127.5 (17.8) μg/l.[199] Human blood Se concentration reflects, indirectly, the associated soil, forage, and food Se status in regions across the world.[10,200-204] Although seldom used, whole blood Se is a measure of Se status over a longer time period, as Se in erythrocytes is included; erythrocytes have a lifespan of ~120 days.[205] In addition, Se in toenails, often measured by neutron activation analysis, is a marker of Se status over a period of up to 1 year.[206] Both serum and plasma Se, and whole blood Se, to some extent, are decreased in response to inflammation, which can be checked by measuring C-reactive protein concentration; this situation is often relevant to patient samples and the Se measurement needs to be interpreted accordingly.[207] Se in erythrocytes is not affected by inflammatory responses and can be used to assess Se status reliably over a wide range of concentrations.[208] The technique most used to measure Se in serum/plasma, whole blood, or erythrocytes is inductively coupled plasma mass spectrometry (ICP-MS). For quality control, matrix-matched certified reference materials must be used in the analytical procedure, and the values obtained for the reference materials must be reported.[209]

Functional techniques are used to determine activities or concentrations of selenoproteins, such as GPX (plasma, erythrocyte, or platelet) and SELENOP. Measurements of SELENOP are

Table 14.2 Recommended Intake for Se (μg/day) and the Upper Intake Level (UL) in Adults and Pregnant and Lactating Women from Various Authorities

RDA (USA)[10]	RNI (UK)[191]	AI (EFSA)[192]	RDI (Australia, New Zealand)[190]	UL (USA)[10]; (Australia, New Zealand)[190]	UL (UK)[210]	UL (EFSA)[211]
M: 55 F: 55 Pregnant: 60 Lactating: 70	M: 75 F: 60 Pregnant: 60 Lactating: 75	M: 70 F: 70 Pregnant: 70 Lactating: 85	M: 70 F: 60 Pregnant: 65 Lactating: 75	400	450	300 (new UL = 255)

AI: Adequate Intake; EFSA: European Food Safety Authority; F: Female; M: Male; RDA: Recommended Dietary Allowance; RNI: Reference Nutrient Intake; UL: Tolerable upper intake level (USA), Upper Level of Intake (Australia, New Zealand). The reference for the new UL: EFSA Journal, Scientific opinion on the tolerable upper intake level for selenium, Jan 2023, www.efsa.europa.eu/en/efsajournal/pub/7704.[210]

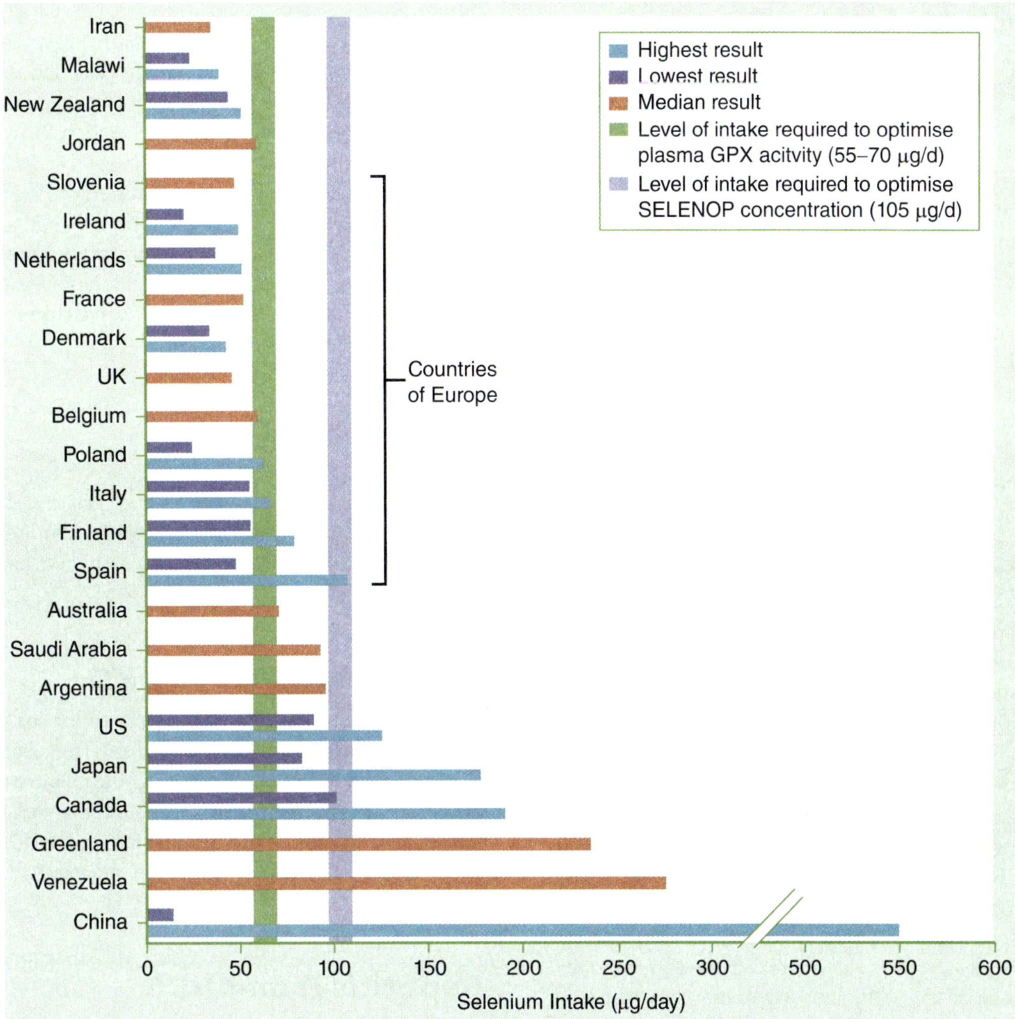

Figure 14.3 Selenium intake in different countries and level of intake required to achieve proposed plasma GPX and SELENOP levels. Se intake in various countries is shown, based on data published since 2000. Where countries had more than one result, the highest values (teal) and the lowest values (dark blue) are shown. Where there was only one result, it is shown in orange. Dietary intake in China ranges from 14 to 550 μg/d. The break in the Selenium Intake (μg/day) axis represents a break in intake of 150 μg/day in order to fit the upper limit of intake (550 μg/day) on the figure. The figure for Venezuela is pre-2000 but is included as the value is notably high at 250–300 μg/d. Dietary intake data are only estimates and are recognized to be of variable quality and accuracy. GPX, glutathione peroxidase; SELENOP, selenoprotein P.

Reproduced from: Winther KH, Rayman MP, Bonnema SJ, et al. Nat Rev Endocrinol 2020;16(3):165-176. doi: 10.1038/s41574-019-0311-6

often made by enzyme-linked immunosorbent assay (ELISA), but the results can vary from one laboratory to another owing to differences in the assays used.[212] This variation is particularly relevant to the measurement of SELENOP where the C-terminal domain might not contain the full complement of nine selenocysteine residues.[213] Measurement of more than one biomarker is proposed to be a better way of assessing Se status.[214]

Health Effects of Se Deficiency and Supplementation

Keshan Disease

Se deficiency is widely accepted as a fundamental cause of KD, an endemic cardiomyopathy first reported in 1935 and found in parts of China that were low in Se, most notably in Keshan county, Heilongjiang Province of China.[215] It particularly affects children and young women.[215,216] Adults living in the affected areas were found to have Se intake of 11 μg (0.14 μmol)/day or less.[10] Seasonal variation of KD suggested an additional causal factor, which we now know to be the Coxsackie virus, which was isolated from frozen blood and tissues of KD patients.[217] Increased oxidative stress in people (or mice) infected with the Coxsackie virus that causes cardiomyopathy is reduced by adequate selenoprotein expression in the host; this is missing in Se deficiency (see also *Viral Disease*).[217] A systematic review and meta-analysis that spanned 50 years found 17 articles (including 41 studies) that looked at the effect of Se supplementation on residents of KD endemic areas; supplementation with Se significantly reduced the risk of KD with an overall effect (risk ratio) of 0.14 (95% CI 0.12 to 0.16).[218] Recent progress on molecular mechanisms of pathogeneses for KD and KBD (below) has been reviewed.[17]

Kashin-Beck Disease

KBD is a degenerative and endemic osteoarthropathy found throughout North Korea, Russia, and China (Tibet) that is associated with poor Se status.[219,220] This is a preadolescent/adolescent condition with necrotic degeneration of the chondrocytes, with dwarfism and joint deformation resulting from the cartilage abnormalities.[220] Many studies have confirmed that Se supplementation has a beneficial effect on KBD prevention and control.[220] In the Chinese Han population, the genotypic and allelic frequency of *GPX1* Pro198Leu was found to be significantly different among KBD patients and controls: the *GPX1* Leu allele had higher KBD risk and lower GPX enzyme activity.[221]

Aside from Se deficiency, numerous other etiologic factors have been suggested to be responsible for this condition (e.g., mycotoxin contamination of barley, organic contaminants in drinking water, and mineral imbalance).[219,220] Analysis of hair from children with and without KBD showed high manganese and iron with low Se and iodine as characteristic features of the elemental profile of KBD.[220] A study of iodine and Se status in 575 children from endemic and nonendemic areas of KBD in Tibet suggested that abnormal thyroid function might be related to the occurrence of KBD.[222]

Mortality

In the US Third NHANES, serum Se was measured at baseline in 13,887 adult participants.[223] Participants were followed for mortality for up to 18 years, i.e., up to end of 2000.[194] Risk of all-cause and cancer mortality showed a U-shaped response, being greater at low and high Se status, but there was no association with cardiovascular (CVD) mortality. The minimum all-cause mortality was at a serum Se of 135 μg/l, after which mortality rose.[195,223] Later NHANES data (1999–2006) with 17,867 individuals also showed a U-shaped association between Se intake and all-cause mortality.[224] The effect of excessive Se status on increasing the risk of mortality in these two studies was also observed in a multiple-dose, randomized, controlled trial in Denmark, where 491 participants, of relatively low baseline plasma Se status (89 μg/l), took Se for 5 years.[225,226] After an additional 10 years of follow-up, it appeared that a dose of 300 μg/day Se (as Se-enriched yeast), compared with placebo, significantly increased all-cause mortality by 11.3% (95% CI, 0.0–22.6%), although there was no effect on CVD.[225] Toxicity from high-dose SeMet, the major component of Se-enriched yeast, can persist for ≥5 years, as it is extensively recycled with a whole-body turn-over time of 363 days.[227]

Immune Function

The immune system needs adequate Se.[228] However, even in apparently Se-replete individuals, Se supplementation enhanced the proliferation of activated T cells, cytotoxic lymphocyte-mediated tumor cytotoxicity, and natural killer cell activity.[194] Interleukin 2 (IL-2) is a necessary growth factor for most T cells,[229] and Se supplementation upregulated the IL-2-receptor enabling T and B lymphocytes to respond better to IL-2.[230,231] Selenoproteins have a vital function in T cells[229]: (i) SELENOP delivers Se to the T-cell; (ii) SELENOK raises the cytosolic Ca^{2+} content to μM levels, activating pathways for optimal T-cell activation; (iii) GPX1, GPX4, TXNRD1, TXNRD2 control redox regulation during T-cell receptor signaling; and (iv) GPX4 prevents the accumulation of lipid-based reactive oxygen species (ROS), which drives the ferroptosis of T-cells. The importance of Se for immune status and the ability to handle viruses was shown in UK men with plasma Se ≤95 μg/L who were supplemented with 50 or 100 μg Se as Na_2SeO_3 or placebo daily for 15 weeks in a double-blind study.[232] All participants received an oral live attenuated poliomyelitis vaccine after 6 weeks. Those given Se had improved immune response through an earlier peak T-cell proliferation and an increase in T-helper cells and showed more rapid clearance than those without extra Se.

Viral Diseases

Se deficiency is linked to incidence, virulence, or disease progression of viral infections. The Se-poor belt in China[233] has been associated with a number of viral-related conditions, including KD, hepatocellular carcinoma (HCC), hemorrhagic fever, and SARS-CoV-2.

Coxsackie Virus and KD

Se deficiency in mice, with its associated low or absent activity of protective selenoproteins, allowed the coxsackie virus to replicate to a higher titer, thus increasing the chances for a mutation to occur.[217,234] This finding may explain the myocarditis-inducing mutations in the coxsackie virus that resulted in KD cardiomyopathy in Chinese individuals, referred to previously.[217] Supplemental Se significantly reduced the risk of KD.[218]

Hepatitis B and HCC

In Qidong county near Shanghai, 11%–15% of adults were infected with hepatitis B and were 200-times more likely to develop HCC.[235,236] Of 2,065 hBsAg-positive men from Qidong, 1,112 received 0.5 mg Se/day (as sodium selenite) and the remaining 953 men received a placebo.[236] After 3 years of treatment and follow-up, 34 and 57 cases of new HCC had developed in the Se vs. placebo group, respectively ($P<0.01$).[236]

Hantavirus and Epidemic Hemorrhagic Fever

Hantaviruses are RNA viruses that are zoonotic pathogens; transmission among rodents and from rodents to humans generally occurs through inhalation of aerosolized excreta. There was a 6-times higher incidence of Hantavirus infection in severe Se-deficient regions than in non-deficient regions in China.[237] An epidemic of hemorrhagic fever caused by hantavirus in Henan Province, China, was successfully treated with oral sodium selenite, with an overall 80% reduction in mortality.[238]

Influenza Virus

Mice infected with influenzaA/Bangkok/1/79 were fed a Se-adequate or a Se-deficient diet.[234] More severe lung inflammation developed on those on the Se-deficient diet than those on the adequate diet and, following infection, there were 29 mutations in one segment of the viral genome.[234]

Polio Viruses

As described under Immune Function, UK participants were inoculated with attenuated polio virus 6 weeks after starting supplementation with 100 or 50 μg Se/day (as sodium selenite) or placebo for 15 weeks.[232] Those given selenite, particularly at the higher dose, cleared the virus more quickly as measured at 7, 14, and 21 days and with fewer mutations in the viral genome.[232]

HIV and AIDS

Humans infected with HIV are under chronic oxidative stress.[239] A dramatic consequence of this oxidative stress is the fatal decrease in the number of CD4 T-cells by apoptosis and, ultimately, failure of the immune system leading to death.[239] Low Se status is associated with fewer CD4 T-cells, faster progression of AIDS and 20% increase in the risk of death.[239] Se deficiency in HIV disease has been associated with disease progression and mortality, independent of antiretroviral therapy (ART).[240] In Botswana, 878 ART-naïve, HIV-infected adults were supplemented daily for 24 months with multivitamins and 200 μg Se.[241] There was a significant reduction in the risk of immune decline and morbidity. In Rwanda, supplementation with 200 μg Se daily for 24 months significantly reduced the rate of $CD4^+$ cell count decline among ART-naïve patients.[242] Se supplementation delayed HIV progression through maintenance of CD4 cell counts.[240]

SARS-CoV-2 and COVID-19

Effect of Se status on the SARS-CoV-2-related COVID-19 was first reported in China in 2020, when the cure rate in cities with up to 50 COVID-19 patients appeared to be different and to relate to the Se status of those cities.[243] Cure rate was highly correlated with background Se status, as represented by hair Se concentration in those cities ($R^2 = 0.72$).[244,245] Hair-Se status corresponding to an intake of 188 μg/day was the highest point on the relationship with the COVID-19 cure rate, suggesting that a factor other than a selenoprotein might be relevant to the mechanism by which Se affected SARS-CoV-2 and COVID-19.[246,247] A later study corroborating these findings showed that the case fatality rate of COVID-19 was linked to Se deficiency in soils and crops at the city level in China.[248]

A German study on COVID-19 patients showed a pronounced deficit in total serum Se and SELENOP concentrations compared with reference data from a cross-sectional analysis of the European EPIC study.[249] Se status was significantly higher in samples from surviving COVID patients than in nonsurvivors. Other studies have assessed Se in COVID-19 patients and found lower Se status in those with COVID than in healthy individuals.[249-252]

SARS-CoV-2 may interfere with the human selenoprotein system, which can scavenge ROS, and control immunity, reduce inflammation, ferroptosis and endoplasmic reticulum (ER) stress.[253] SARS-CoV-2 was found to significantly suppress mRNA expression of *GPX4*, of the ER selenoproteins, *SELENOF*, *SELENOM*, *SELENOK*, and *SELENOS* and downregulate *TXNRD3*.[253] Redox-active Se species may also be involved, given the optimal dose for maximal cure rate seen in Hubei.[244]

While the studies reviewed above concerning viral infections provide some evidence that Se supplementation of low Se status individuals may be protective, there is little consistent evidence that supranutritional supplementation protects against viral infections.

Brain Function and Neurodegeneration

During Se depletion, brain Se is maintained at the expense of other tissues.[47] Most selenoproteins are expressed in all major brain regions, mainly in neurons.[254] Interruption of Se supply to the brain by impairing the SELENOP-apoER2 pathway causes mice to develop severe neurologic disorders.[255,256] SELENOP knock-down increases neurotoxicity in neuronal cells by amyloid-β.[257] Early evidence showed that Se deficiency caused seizures in children that were remedied by Se supplementation.[258] We now know that the survival of a specific type of interneuron depends on SEC-containing GPX4, thereby preventing fatal epileptic seizures.[259] Recent evidence implicates selenoprotein dysfunction in Parkinson's Disease, but results are difficult to summarize.[260] However, evidence from mouse models implicates SELENOT, GPX1, and plausibly SELENOP and GPX4 in counteracting Parkinson's Disease through Se transportation to the brain and reduction in oxidative stress.[260]

In the French Eva cohort of 1,166 adults aged 60–70, an increased risk of cognitive decline was found over a 4-year period in those with low plasma Se concentration at baseline.[261] In the same cohort, cognitive decline was associated with the magnitude of plasma Se decrease over a 9-year period.[262] Lower nail Se was associated with lower cognitive score in rural elderly Chinese.[263] A meta-analysis of 14 studies with 40 observations on Se concentration in AD and control brains showed impaired Se status in AD vs non-AD brains.[264] Se concentrations were decreased in the temporal, hippocampal, and cortex regions in AD. However, in the PREADViSE trial (3,800 volunteers), which began as a double-blind, randomized clinical trial in May of 2002, and was transformed into a cohort study (no further supplementation) from September of 2009 to May of 2015, there was no effect of Se (or vitamin E) on dementia incidence among asymptomatic men.[265] However, this study has been criticized on a number of fronts; notably the low age of the participants at study entry, which meant that most of the cohort was observed in their 60s when dementia incidence is still low.[266] Hence it is very unlikely that the PREADViSE trial could have detected an effect from the interventions.[266]

The most dynamic area of the neurobiology of Se is the importance of GPX4 in compromising cell death from ferroptosis.[254] Recent studies have shown that ferroptosis induced by lipid peroxidation contributed to neuronal death in AD.[267] Significant iron accumulation and lipid peroxidation, combined with reductions in GSH and GPX4, were observed in the hippocampus of AD patients.[267] Furthermore, GPX4 knockout in the forebrain neurons of mice directly caused age-related neurodegenerative changes and obvious neuronal loss, showing that GPX4 was negatively associated with AD.[268] A use for Se has been shown in recovering neurologic function after intracerebral hemorrhage (ICH) by driving antioxidant GPX4 expression.[269] In a hemorrhagic stroke model, a single dose of Se delivered into the brain increased antioxidant GPX4 expression, reduced ferroptopic cell death, protected neurons, and improved behavior.[269]

Fertility and Reproduction

Se is required for testosterone biosynthesis and formation and normal development of spermatozoa.[270] At least two selenoproteins are essential for male fertility, GPX4 and SELENOP.[47,104] Early studies showed that infertile men with defective or low

GPX4 protein content in spermatozoa had poor sperm viability, morphologic integrity, and forward motility.[194] A Turkish study enrolled 59 men, 44 with idiopathic infertility, and 15 with proven fertility.[270] Serum and semen Se were higher in the fertile than in the infertile group, and concentration was positively correlated with motility and morphology of the spermatozoa.[271] In a randomized trial in Glasgow in 1998, Se supplementation (100 μg per day) of subfertile men with low Se intake increased sperm motility and enabled 11% of the men to achieve paternity, compared with none in the placebo group.[272] A 2019 review tabulated eight human studies from 2009–2019, including four randomized trials that involved Se supplementation to improve male fertility, reproductive status, or clinical outcomes.[273] Improved sperm count, motility, and morphology were observed in six of the studies, but there was no change in the two studies conducted in the United States[273,274]; this is almost certainly because American men had sufficient Se status so additional Se conferred no benefit.[274] However, it reminds us of the successful study in Glasgow where men who achieved paternity had a low Se intake before supplementation.[272]

Low serum Se was found in women who miscarried in the first trimester or had recurrent miscarriages.[194] Low Se status also appears to be a risk for pre-eclampsia.[275] A cross-sectional study in the Netherlands on 1,129 pregnant women found that women in the lowest quartile of serum Se at 12-weeks' gestation had twice the risk of a preterm birth as the rest.[194] Similarly, in 72,025 women with singleton live births from the population-based, prospective Norwegian Mother, Father and Child Cohort Study (MoBa), maternal dietary Se intake (but not intake of Se-containing supplements) during the first half of pregnancy was significantly associated with decreased risk of preterm delivery.[276]

Thyroid Function and Autoimmune Thyroid Disease

A large number of selenoproteins, in particular GPX and DIO enzymes, are essential to thyroid action.[277] Hashimoto's thyroiditis (HT) and Graves' disease (GD) are examples of autoimmune thyroiditis (AIT), the most common autoimmune condition.[278] Antibodies to thyroid peroxidase (TPO), the enzyme that catalyzes thyroid-hormone production, are characteristic of HT, while antibodies to the receptor for the thyroid-stimulating hormone (TSH) are characteristic of GD.[278] Genetic susceptibility, immune disorders and environmental factors that include iodine, iron and Se contribute to development of AIT.[278] More than 20 RCTs have shown that supplementation of AIT patients with Se decreased inflammation and TPO autoantibody (TPO-Ab) concentrations.[193] However, little evidence was found of clinical importance.[193] Nonetheless, a SNP in the *SELENOS* gene was linked to the risk of HT.[279] In a large cross-sectional study in two counties of Shaanxi (China) with similar iodine status but different Se status, the county with lower Se status had a higher risk of enlarged thyroid, hypothyroidism, and HT (significantly raised TPO-Abs).[280] To explore longitudinal effects of low Se status, a prospective cohort study was conducted in the same area from 2013 to 2019, where TPO-Ab and disease incidence were compared.[281] Newly diagnosed HT in the high-Se county was less frequent than in the low-Se county, with conversion rates to seropositive TPO-Ab of 5.6% vs. 10.2%, supporting the notion that Se deficiency contributed to HT as a modifiable risk factor. Lower Se status has been linked to higher risk of Graves' disease.[193] Supplementation with Se in mild Graves' orbitopathy resulted in reduced eye involvement, disease progression and improved quality of life.[282] An RCT of 169 Italian pregnant women positive for TPO-Ab supplemented with 200 μg Se/day as SeMet during pregnancy and the postpartum period decreased postpartum AIT and permanent hypothyroidism.[283]

Cardiovascular Disease

Although Se was thought to have a role in CVD owing to its antioxidant properties,[194] evidence from observational studies and RCTs has been inconsistent.[284] However, a 2020 systematic review and meta-analysis of RCTs investigated CVD and all-cause mortality following Se supplementation alone or by antioxidant supplement mixtures.[285] Decreased risk with antioxidant mixtures was seen for CVD mortality and all-cause mortality, but only when Se was part of the mixture.[285] A meta-analysis of 16 prospective observational studies suggested a nonlinear relationship between baseline blood Se concentration and risk of incident CVD, where the beneficial range of Se concentration ranged from 55 to 145 μg/L.[284] The same authors also carried out a meta-analysis of nine RCTs and found no overall effect of oral Se supplements on CVD.[284] It may be that Se is only effective in reducing CVD risk at relatively low concentrations, as shown in the following study published in 2019. Swedish adults without a history of CVD showed a markedly increased risk of CVD morbidity and mortality in the 20% with the lowest SELENOP concentration.[286] Those in the lowest Se quintile had serum SELENOP concentration below 4.3 mg/L, corresponding to serum Se concentration <70μg/l, which is insufficient to raise GPX3 to a maximum level.[287]

Cancer

Probably the most telling study on Se and cancer mortality was a prospective cohort from the US Third NHANES, which measured serum Se at baseline in 13,887 adult participants who were then followed up for mortality for up to 12 years.[223] An inverse association with cancer mortality was seen at low serum Se (<130 μg/L). Minimum mortality was reached at around 150 μg/l serum Se, after which cancer mortality began to rise. The relationship between colorectal, prostate and lung cancer was different, with lung cancer risk rising at a lower Se status whereas colorectal cancer continued to fall.[223]

Results from other observational studies showed a significant beneficial association between Se status (serum/plasma/toenail) and risk or mortality of the following cancers: lung, bladder, colorectal, liver, esophageal, gastric cardia, thyroid, and prostate.[194] A 2016 meta-analysis of Se exposure and cancer risk showed that higher Se status was associated with lower cancer risk and that the risk of cancer fell as Se status rose, i.e., up to 170 μg/l in plasma/serum, and up to 1.15 μg/g in toenails.[288]

Two RCTs with Se were carried out in the United States. In the Nutritional Prevention of Cancer (NPC) trial, 200 μg/d Se, as high-Se yeast or placebo-yeast were randomized to 1,312 participants with a previous history of skin cancer.[280] After a 7.5-year follow-up, significant reductions were seen in cancer mortality, cancer incidence and, most notably, in prostate cancer incidence, i.e., HR 0.48 (95% CI 0.28, 0.80); the effect of supplementation was strongest in the lowest Se tertile (<105.2 μg/l).[287] Following the NPC trial, the Selenium and Vitamin E in Cancer Prevention Trial (SELECT) was set up with Se 200 μg/d (SeMet) and/or vitamin E 400 mg/d and/or placebo in 32,400 men, with a prostate-cancer endpoint.[289] After 5.5 years of follow-up, no effect of Se or Se/vitamin E was found on prostate-cancer risk, neither in the 2009 report[289] nor in the 2011 follow-up.[290] However, in a later analysis, an increased risk of high-grade prostate cancer was found in men supplemented with Se with/without vitamin E in the upper 40th percentile of toenail Se at baseline.[291] Such a result is not surprising, as baseline Se status (median 136 μg/l in SELECT men) was already higher than that in men in the top tertile of NPC (>123 μg/l) in whom no benefit of supplementation was seen.[286] Together, meta-analyses on Se supplements in cancer prevention showed no significant effect of Se supplementation on overall cancer risk.[288,292] In contrast, data from SELECT imply that giving extra Se to men who already have adequate Se intake will *not* reduce risk, but may actually increase risk.[291]

Type-2 Diabetes

Animal experiments have shown diabetogenesis risk with prolonged high Se intakes (0.4 to 3.0 mg/kg),[115,178,293] or overproduction of selenoproteins.[113] Since the NPC data showed an increased risk of type-2 diabetes (T2D) with Se supplementation,[294] it has been suggested that Se may increase the risk of T2D. While a number of observational studies have found an association between Se intake or status and T2D or insulin resistance, no relationship was found with toenail Se status.[48] Furthermore, the interaction between Se status and glucose control is not limited only to hyperglycemia, but apparently extends to hypoglycemia risk in Se deficiency.[295] Importantly, dietary Se supplementation, as shown by the large SELECT trial,[290] does not appear to be a major causal factor for the development of T2D in humans.[296] Elevated Se biomarkers such as SELENOP, which are often measured in T2D patients, are more likely to be a consequence, rather than a cause, of diabetes.[296]

For instance, we know that high glucose concentrations increase hepatic SELENOP biosynthesis and that SELENOP and gluconeogenic enzyme expression are similarly regulated by methylation of the same transcription factors.[296] Furthermore, SELENOP, at artificially high levels (e.g., when injected), increases glucose release and insulin resistance.[296] Ultimately, results from RCTs must carry more weight than those of observational studies when evaluating whether there is a truly causal relationship between Se, T2D and insulin resistance.[297]

Overall Benefit and Risk of Se Supplementation

Se status varies widely in different parts of the world in line with Se intake (see Figure 14.3).[193] SELENOP (and most selenoproteins) are at maximum concentration at 125 μg/L.[245] For those with low Se intake, i.e., plasma/serum concentration <100 μg/l, there may be benefit of Se supplementation at no more than 100 μg/day. The serum concentration associated with minimal all-cause mortality was 135 μg/l, after which mortality rose (see **Figure 14.4**).[194,223] There would normally be no advantage of raising Se status above that point, except for a short period in the case of viral disease[238,244] or sepsis.[298]

Adverse Effects of Excess and Toxicity

Overexposure of livestock to Se from grazing on high-Se forage was linked to diseases such as blind staggers.[297] Overexposure to Se has also been linked to detrimental health effects in humans.[299] The symptoms of selenosis, most notably loss of hair and nails, were first described in 1983 in Enshi, Hubei Province, China.[299]

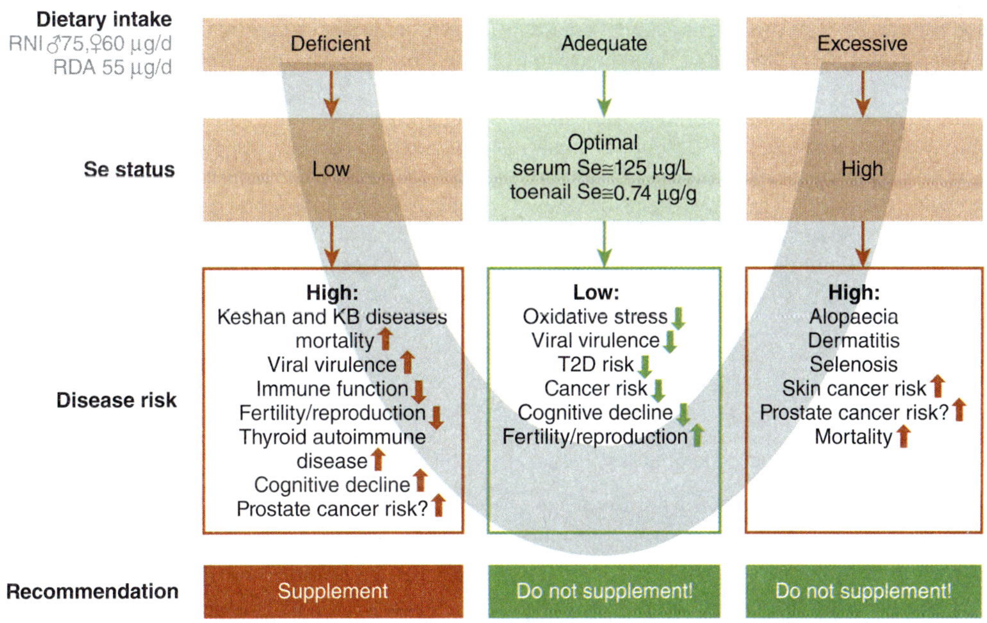

Figure 14.4 U-shaped relationship between Se status and disease risk. The effects of deficient, adequate, and excessive dietary Se intake on Se status are shown. When Se status is low, disease risk will be high, and the recommendation is to increase dietary Se intake. When Se status is optimal (meets the recommendations shown for dietary intake according to the RNI or the RDA), disease risk will be low, and the recommendation is not to supplement. When Se status is high, disease risk will be high, and the recommendation is not to supplement.

Modified from Winther KH, Rayman MP, Bonnema SJ, et al. Nat Rev Endocrinol 2020;16(3):165-176. doi: 10.1038/s41574-019-0311-6

Daily dietary intakes of Se, estimated after the peak prevalence had subsided, averaged 4.99 (range 3.20–6.69) mg, and hair and blood Se concentrations averaged 32.2 and 3.2 mg/l, respectively. The source of Se was a stony coal of very high Se content, which entered the soil by weathering, and was available for uptake by crops because of the traditional use of lime as fertilizer in that region.[299] In a 1994 summary of Chinese findings, safe Se intake, that is, the "no adverse effect level," was stated to be ~800 μg/day, with 400 μg designated as the maximum safe daily dietary Se intake.[300] Maximal tolerable levels of dietary Se range from 3 to 5 mg/kg for various domestic species.[301] Other examples of Se toxicity were noted during Se supplementation trials in North America, where baseline Se status was already high. Supplementation of men in the SELECT trial, with 200 μg Se/day (as Se-Met) for a mean of 5.5 years, statistically significantly increased the risk of alopecia (RR 1.28, 95% CI 1.01–1.62) and dermatitis (RR 1.17, 95% CI 1.00–1.35).[289] Although no adverse effect of Se supplementation on prostate cancer risk was found overall, a follow-up study showed that men in the top 40% of Se concentration in toenails had an increased risk of high-grade prostate cancer (HR 1.62, 95% CI 0.95–2.77).[291] Furthermore, the same Se dose delivered (as Se-enriched yeast) for 4.5 years increased risk of squamous cell carcinoma (HR 1.25, 95% CI 1.03–1.51)[302] and T2D (HR 1.55, 95% CI 1.03–2.33)[294] in participants of the NPC trial. Figure 14.4 shows the U-shape associated with different doses of Se and human health.

The current upper tolerable intake level of Se (UL), which includes Se obtained from foods and supplements, has been set at 255–450 μg/day (Table 14.2).[10,190,192,210] While short-term "acute" intake at those levels is probably harmless and may be beneficial in some cases,[238,244,297] long-term "chronic" intake is of concern, as Se can build up in the body as shown in a long-term RCT in Denmark where 300 μg Se/day significantly increased all-cause mortality.[223] In contrast, Inuit populations in Northern Canada, whose diet includes high levels of selenoneine, exhibit little evidence of selenosis, despite high whole blood Se concentration and dietary intake, often exceeding the UL.[303] Separate ULs should perhaps be developed for different Se species that reflect Se requirements and toxicity.

Currently, there are no good biomarkers for excess and toxic Se status. As discussed above, tissue Se concentrations continue to rise above the nutrient requirement and are further modulated by dietary SeMet intake. Abundance of selenoproteins does not progressively increase above Se-adequate intakes in animals or humans provided graded levels of dietary Se. Furthermore, Se transcriptomics studies have failed to identify any mRNA levels of genes that could serve as effective biomarkers of high Se status to distinguish between successfully coping with super-nutritional Se intake (as suggested to occur in Se anti-cancer studies) and the onset of toxicity.[76] Excretion of methylated Se species (e.g., trimethyl selenonium ion) is dramatically affected by SNPs in the human methyltransferase gene, with varied prevalence in different human populations.[66] Because metabolomic analysis of turkey liver found increased low-molecular weight and high molecular weight selenosugars associated with increasing dietary selenite intakes, selenosugars may have potential as biomarkers of high body Se status.[54]

Emerging Issues

Although a full human selenoproteome is likely revealed, complete biochemical and metabolic functions of many selenoproteins are far from clear. It is recognized today that selenoproteins not only protect against oxidative stress but also play vital roles in a broad spectrum of metabolism, such as cellular signaling, metabolic regulation, Ca^{2+} flux, and protein folding/degradation. Such novel functions are highlighted by recent findings on the role and mechanism of GPX1 overproduction in regulating glucose-induced Ca^{2+} influx and insulin release via transcriptional inhibition of islet-derived regenerating protein 2, and decreasing its interaction with transmembrane Ca channel CaV1.2 in pancreatic islet beta cells.[304,305] It is challenging to determine complete functions of each of these selenoproteins because of their unique expression patterns associated with cell types and subcellular locations, their induced transient redox boundaries, their nonspecific redox interactions, and their functional dependence on developmental stage and stress conditions.[100] Furthermore, the recombinant selenoproteins produced are mainly cysteine variants, as SEC isoforms are difficult to express in mammalian cells. Although plasma GPX3 activity and SELENOP concentration are widely used as biomarkers of Se status to estimate dietary Se requirement by humans, their representation or reflection of Se status in different tissues (e.g., liver vs adipose) remains to be validated. Meanwhile, biomarkers for high Se status are lacking. Because Se deficiency and selenoprotein mutations are either identified or potential etiologic factors of many genetic or metabolic disorders, function-based selenopeptides or mimetics[306] targeting specific cellular processes might be developed for supplementation and treatment. Above all, there remains a major need for human studies to better clarify impacts and mechanism of Se status, metabolism, and function on human health and disease.

References

1. National Research Council. *Selenium in Nutrition: Revised Edition*, National Academies Press, Washington, DC, 1983.
2. Schwarz K, Foltz CM. J Am Chem Soc 1957;79:3292–3.
3. Rotruck JT, Pope AL, Ganther HE, et al. Science (New York, NY) 1973;179:588–90.
4. Kryukov GV, Castellano S, Novoselov SV, et al. Science (New York, NY) 2003;300:1439–43.
5. Reich HJ, Hondal RJ. ACS Chem Biol 2016;11:821–41.
6. Maroney MJ, Hondal RJ. Free Radic Biol Med 2018;127:228–37.
7. Keshan Disease Research Group of the Chinese Academy of Medical Sciences, Antiepidemic Station of Sichuan Province, Antiepidemic Station of Xichang District, et al. Chinese Med J 1979;471–6.
8. van Rij AM, Thomson CD, McKenzie JM, et al. Am J Clin Nutr 1979;32:2076–85.
9. National Research Council. Recommended Dietary Allowances. *National Academy Press, Washington, DC*, 1989.
10. Institute of Medicine (US), Panel on Dietary Antioxidants and Related Compounds. Dietary Reference Intakes for Vitamin C, Vitamin E, Selenium, and Carotenoids., National Academies Press, Washington, DC, 2000.
11. World Health Organization. *Trace Elements in Human Nutrition and Health*. 1996. www.who.int/publications/i/item/9241561734
12. Wu L, Guo X, Banuelos GS. Environ Tox Chem 1997;16:491–7.
13. Yang X, Chen W, Feng Y. Environ Geochem Health 2007;29: 413–28.

14. Dhillon KS, Dhillon SK. Sci Total Environ 2009;407:6150–6.
15. Van Mantgem PJ, Wu L, Banuelos GS. Ecotox Envir Safety 1996;34:228–38.
16. Zhu J, Wang N, Li S, et al. Sci Total Environ 2008;392:252–61.
17. Lei XG, Combs Jr GF, Sunde RA, et al. Annu Rev Nutr 2022;42:337–75.
18. Lemire M, Fillion M, Barbosa Jr F, et al. Sci Total Environ 2010;408:4076–84.
19. USDA, Food Data Central. 2022. https://fdc.nal.usda.gov.
20. Panel on Additives and Products or Substances used in Animal Feed, European Food Safety Authority (EFSA). EFSA J 2013;11:3219.
21. Burk RF LO. *Selenium. In Modern Nutrition in Health and Disease*, Lippincott Williams and Wilkins, Philadelphia, 2005.
22. Burk RF, Norsworthy BK, Hill KE, et al. Canc Epi Biomarkers Prev 2006;15:804–10.
23. Turner JC, Osborn PJ, McVeagh SM. Comp Biochem Physiol A, Comp Physiol. 1990;95:297–301.
24. Van Dael P, Davidsson L, Ziegler EE, et al. Pediatr Res. 2002;51: 71–5.
25. McDermott JR, Rosen BP, Liu Z. Molr Biol Cell 2010;21: 3934–41.
26. National Research Council. *Diet, Nutrition, and Cancer*, National Academies Press, Washington, DC, 1982.
27. Sunde R. Selenium. In: Shils ME, Shike M, Ross CA, Caballero B, Cousins RJ, editors. Modern Nutrition in Health and Disease. Philadelphia: Lippincott Williams and Wilkins, 2012.
28. Thomson CD, Robinson MF. Am J Clin Nutr 1986;44:659–63.
29. Lutwak L. Am J Clin Nutr 1969;22:771–85.
30. Takahashi K, Ogra Y. Metallomics 2020;12:241–8.
31. Fagerberg L, Hallström BM, Oksvold P, et al. Mol Cell Proteomics 2014;13:397–406.
32. Thomson CD, Stewart R. Br J Nutr 1973;30:139–47.
33. Thomson CD, Stewart R. Br J Nutr 1974;32:47–57.
34. Thomson CD, Burton CE, Robinson MF. Br J Nutr 1978;39: 579–87.
35. Robinson MF, Rea HM, Friend GM, et al. Br J Nutr 1978;39: 589–600.
36. Evenson JK, Sunde RA. Front Nutr 2021;8:699652.
37. Jäger T, Drexler H, Göen T. Arch Tox 2016;90:1069–80.
38. Jäger T, Drexler H, Göen T. Arch Tox 2016;90;149–58.
39. Janghorbani M, Young VR. Selenium metabolism in North Americans: Studies based on stable isotope tracers. In: Combs GF, Jr., Spallholz JE, Levander OA, Oldfield JE, editors. Selenium in Biology and Medicine (the 1st ed). New York, NY, AVI; p. 450–71, 1987.
40. Thomson C. Eur Clin Nutr 2004;58:391–402.
41. Hill KE, Zhou J, McMahan WJ, et al. J Biol Chem 2003;278: 13640–6.
42. Schomburg L, Schweizer U, Holtmann B, et al. BioChem J 2003; 370:397.
43. Olson GE, Winfrey VP, NagDas SK, et al. Biol Reprod 2005;73:201–11.
44. Renko K, Werner M, Renner-Müller I, et al. BioChem J 2008; 409:741–9.
45. Hill KE, Zhou J, McMahan WJ, et al. J Nutr 2004;134:157–61.
46. Olson GE, Winfrey VP, Hill KE, et al. J Biol Chem 2008;283: 6854–60.
47. Burk RF, Hill KE. Biochim Biophys Acta-General Subjects 2009;1790:1441–7.
48. Vinceti M, Filippini T, Wise LA, et al. Env Res 2021;197:111210.
49. Schweizer U, Michaelis M, Köhrle J, et al. Bio Chem J 2004; 378:21–6.
50. Burk RF, Hill KE. Annu Rev Nutr 2015;35:109–34.
51. Prohaska J, Ganther H. J Neurochem 1976;27:1379–87.
52. White PJ. Biochim Biophys Acta. General Subjects 2018;1862: 2333–42.
53. Sunde RA Selenium. In: Stipanuk MH, editor. Biochemical, Physiological, and Molecular Aspects of Human Nutrition. 2nd ed. New York, NY: W.B. Sanders, p. 1091–126, 2006.
54. Katarzyna B, Taylor RM, Szpunar J, et al. Metallomics 2020;12: 758–66.
55. Bierla K, Szpunar J, Lobinski R, et al. Metallomics. 2023;15: mfad066.
56. McConnell KP, Hoffman JL. FEBS Letters 1972;24:60–2.
57. Beilstein M, Whanger P. J Nutr 1986;116:1711–9.
58. Waschulewski IH, Sunde RA. J Nutr 1988;118:367–74.
59. Waschulewski IH, Sunde RA. Br J Nutr 1988;60:57–68.
60. Huber R, Criddle R. Biochim Biophys Acta-General Subjects 1967;141:587–99.
61. Sunde RA Selenium. In: Shils ME, Shike M, Ross AC, Caballero B, Cousins RJ, editors, *Modern Nutrition in Health and Disease*, 11th edition, Lippincott Williams and Wilkins, Philadelphia, p 225–37, 2012.
62. Esaki N, Tanaka H, Uemura S, et al. Biochemistry. 1979;18: 407–10.
63. Seale LA, Hashimoto AC, Kurokawa S, et al. Mol Cell Biol 2012;32:4141–54.
64. Hatfield DL, Berry MJ, Gladyshev VN. *Selenium: its molecular biology and role in human health*, Springer Science & Business Media, 2011.
65. Fukumoto Y, Yamada H, Matsuhashi K, et al. Chem Res Tox 2020;33:2467–74.
66. Kuehnelt D, Engström K, Skröder H, et al. Am J Clin Nutr 2015;102:1406–15.
67. Lajin B, Francesconi KA. PLOS One 2016;11:e0167013.
68. Kobayashi Y, Ogra Y, Ishiwata K, et al. Proc Natl Acad Sci USA 2002;99:15932–6.
69. Lu Y, Pergantis SA. Metallomics 2009;1:346–52.
70. González-Iglesias H, Fernández-Sánchez ML, Lu Y, et al. J Anal Atomic Spectrometry 2015;30:267–76.
71. Anan Y, Ohbo A, Tani Y, et al. Anal Bioanal Chem 2014;406: 7959–66.
72. Yamashita Y, Yamashita M. J Biol Chem 2010;285:18134–8.
73. Bierła K, Suzuki N, Ogra Y, et al. Food Chem 2017;37: 1196–201.
74. Anan Y, Mikami T, Tsuji Y, et al. Anal Bioanal Chem 2011;399: 1765–72.
75. Ogra Y, Ogihara Y, Anan Y. Metallomics 2017;9:61–8
76. Sunde RA. J Nutr 2021;151:772–84.
77. Labunskyy VM, Hatfield DL, Gladyshev VN. Physiol Rev 2014;94:739–77.
78. Carlson BA, Xu XM, Kryukov GV, et al. Proc Natl Acad Sci USA 2004;101:12848–53.
79. Schweizer U, Fradejas-Villar N. FASEB J 2016;30:3669–81.
80. Sunde RA, Hoekstra WG. Biochem Biophys Res Commun 1980;93:1181–8.
81. Sunde RA, Evenson JK. J Biol Chem 1987;262:933–7.
82. Berry MJ, Banu L, Larsen PR. Nature 1991;349:438–40.
83. Kumaraswamy E, Carlson BA, Morgan F, et al. Mol Cell Biol 2003;23:1477–88.
84. Sunde RA, Sunde GR, Sunde CM, et al. PLOS One 2015;10: e0129801.
85. Gladyshev VN, Arnér ES, Berry MJ, et al. J Biol Chem 2016;291: 24036–40.
86. Mariotti M, Ridge PG, Zhang Y, et al. PLOS One 2012;7: e33066.
87. Li S, Gao F, Huang J, et al. Biochim Biophys Acta. General Subjects 2018;1862:2473–9.
88. Mariotti M, Santesmasses D, Capella-Gutierrez S, et al. Genome Res 2015;25:1256–67.

89. Liang Y, Lin S-l, Wang C-W, et al. Biolog Trace Element Res 2014;160:41–8.
90. Guo L, Yang W, Huang Q, et al. Cell Chem Biology 2018;25: 1380–8. e4
91. Sun LH, Huang JQ, Deng J, et al. Poult Sci 2019;98:4247–54.
92. Barchielli G, Capperucci A, Tanini D. Antioxidants 2022:11:251.
93. Ojeda M, Carreras O, Nogales F. Antioxidants 2022;11:394.
94. Gorini F, Vassalle C. Antioxidants 2022;11:1188.
95. Hogan C, Perkins AV. Nutrients 2022;14:628.
96. Weaver K, Skouta R. Biomedicines 2022;10:891.
97. Schweizer U, Fabiano M. Free Radic Biol Med 2022;190: 105–15.
98. Shchedrina VA, Zhang Y, Labunskyy VM, et al. Antiox Redox Signal 2010;12:839–49.
99. Varlamova E, Goltyaev M, Novoselov V, et al. Doklady Biochemistry and Biophysics 2017;476:320–22.
100. Jehan C, Cartier D, Bucharles C, et al. Redox Biol 2022;55: 102412.
101. Santesmasses D, Mariotti M, Gladyshev VN. Mol Biol Evol 2020;37:341–54.
102. Cheng WH, Combs Jr GF, Lei XG. Biofactors 1998;7:311–21.
103. Cheng W-H, Ho Y-S, Ross DA, et al. J Nutr 1997;127:675–80.
104. Ursini F, Heim S, Kiess M, et al. Science (New York, NY) 1999;285:1393–6.
105. Yant LJ, Ran Q, Rao L, et al. Free Radic Biol Med 2003;34: 496–502.
106. Carlson BA, Tobe R, Yefremova E, et al. Redox Biol 2016;9: 22–31.
107. Yang WS, SriRamaratnam R, Welsch ME, et al. Cell 2014;156: 317–31.
108. Huang J-Q, Jiang Y-Y, Ren F-Z, et al. Redox Biol 2022;57:102482.
109. Cheng WH, Fu Y, Porres J, et al. FASEB J 1999;13:1467–75.
110. Cheng W-H, Ho Y-S, Valentine BA, et al. J Nutr 1998;128: 1070–6.
111. Fu Y, Cheng W-H, Porres JM, et al. Free Radic Biol Med 1999;27:605–11.
112. Fu Y, Cheng W-H, Ross DA, et al. Proc Soc Exp Biol Med 1999;222:164–9.
113. McClung JP, Roneker CA, Mu W, et al. Proc Natl Acad Sci USA 2004;101:8852–7.
114. Wang XD, Vatamaniuk MZ, Wang SK, et al. Diabetologia 2008;51:1515–24.
115. Zhao Z, Barcus M, Kim J, et al. J Nutr 2016;146:1625–33.
116. Zhao Z, Kim J, Lei XG. J Nutr 2020;150:1738–48.
117. Esworthy RS, Yang L, Frankel PH, et al. J Nutr 2005;135:740–5.
118. Köhrle J, Gärtner R. Best Pract Res Clin Endocr Metab 2009;23:815–27.
119. Hernandez A, Martinez ME, Ng L, et al. Endocrinology 2021;162.
120. Jo S, Fonseca TL, Bocco BM, et al. J Clin Invest 2019;129: 230–45.
121. Joshi G, Wang Y. Bioessays 2015;37:240–7.
122. McAninch EA, Rajan KB, Evans DA, et al. J Clin Endocrinol Metab 2018;103:1818–26.
123. He B, Li J, Wang G, et al. Progr Neuro-Psychopharmacol Biol Psych 2009;33:986–90.
124. Arnér ES. Biochim Biophys Acta-General Subjects 2009;1790: 495–526.
125. Hill KE, McCollum GW, Boeglin ME, et al. Biochem Biophys Res Commun 1997;234:293–5.
126. Gencheva R, Arnér ES. Annu Rev Pharm Tox 2022;62:177–96.
127. Lobanov AV, Hatfield DL, Gladyshev VN. Prot Sci 2008;17: 176–82.
128. Tobe R, Carlson BA, Huh JH, et al. BioChem J 2016;473: 2141–54.
129. Lee BC, Lee S-G, Choo M-K, et al. Scientific Rep 2017;7:1–9.
130. Horibata Y, Elpeleg O, Eran A, et al. J Llipid Res 2018;59: 1015–26.
131. Avery JC, Yamazaki Y, Hoffmann FW, et al. Arch Biochem Biophys 2020;689:108444.
132. Ma C, Martinez-Rodriguez V, Hoffmann PR. Intl J Mol Sci 2021;22:11174.
133. Schomburg L. Free Radic Biol Med 2022;191:150–163.
134. Zachara BA. Adv Clin Chem 2018;86:157–78.
135. Reeves M, Hoffmann P. Cell Mol Life Sci 2009;66:2457–78.
136. Boukhzar L, Hamieh A, Cartier D, et al. Antiox Redox Signal 2016;24:557–74.
137. Prevost G, Arabo A, Jian L, et al. Endocrinology 2013;154: 3796–806.
138. Sunde RA. PLOS One 2018;13:e0191449.
139. Touat-Hamici Z, Legrain Y, Bulteau A-L, et al. J Biol Chem 2014;289:14750–61.
140. Chen LL, Huang JQ, Xiao Y, et al. J Nutr 2020;150:483–91.
141. Zhang X, Xiong W, Chen L-L, et al. Free Radic Biol Med 2020;160:670–9.
142. Chen LL, Huang JQ, Wu YY, et al. Redox Biol 2021;45:102048.
143. Whanger P. Biochim Biophys Acta-General Subjects 2009;1790: 1448–52.
144. Whanger P. Cell Mol Life Scis CMLS 2000;57:1846–52.
145. Zhang Y, Zhou Y, Schweizer U, et al. J Biol Chem 2008;283: 2427–38.
146. Marino M, Stoilova T, Giorgi C, et al. Human Mol Genetics 2015;24:1843–55.
147. Villar-Quiles RN, von Der Hagen M, Métay C, et al. Neurology 2020;95:e1512–e27.
148. Moghadaszadeh B, Petit N, Jaillard C, et al. Nat Genetics 2001;29:17–8.
149. Lescure A, Rederstorff M, Krol A, et al. Biochim Biophys Acta-General Subjects 2009;1790:1569–74.
150. Castets P, Maugenre S, Gartioux C, et al. BMC Devel Biol 2009;9:1–12.
151. Bachmann C, Noreen F, Voermans NC, et al. Human Mutation 2019;40:962–74.
152. Pozzer D, Varone E, Chernorudskiy A, et al. Redox Biol 2019;20:354–66.
153. Varone E, Pozzer D, Di Modica S, et al. Redox Biol 2019;24:101176.
154. Congdon EE, Sigurdsson EM. Nat Rev Neurol 2018;14: 399–415.
155. Jang JK, Park KJ, Lee JH, et al. Biochem Biophys Res Commun 2017;486:444–50.
156. Rueli RH, Torres DJ, Dewing AS, et al. J Alzheimer's Dis 2017;55:749–62.
157. Lee JH, Park KJ, Jang JK, et al. J Biol Chem 2015;290: 29911–52.
158. Liu J, Zhang Z, Rozovsky S. FEBS Lett 2014;588:3311–21.
159. Shchedrina VA, Everley RA, Zhang Y, et al. J Biol Chem 2011;286:42937–48.
160. Fredericks GJ, Hoffmann FW, Rose AH, et al. Proc Natl Acad Sci USA 2014;111:16478–83.
161. Meng X-L, Chen C-L, Liu Y-Y, et al. Neuroscience 2019;406: 38–49.
162. Lu C, Qiu F, Zhou H, et al. FEBS Lett 2006;580:5189–97.
163. Röseler A, Prieto JH, Iozef R, et al. Antiox Redox Signal 2012;17: 534–43.
164. Zhou JC, Zhao H, Li JG, et al. J Nutr 2009;139:1061–6.
165. Liu Y, Zhao H, Zhang Q, et al. J Nutr 2012;142:1410–6.
166. Huang J-Q, Li D-L, Zhao H, et al. J Nutr 2011;141:1605–10.
167. Raines AM, Sunde RA. BMC Genomics 2011;12:26.
168. Barnes KM, Evenson JK, Raines AM, et al. J Nutr 2009;139: 199–206.
169. Sunde RA, Li JL, Taylor RM. Adv Nutr 2016;7:1129–38.

170. Chen XD, Zhao ZP, Zhou JC, et al. Free Radic Biol Med 2018; 127:116–23.
171. Huang JQ, Zhou JC, Wu YY, et al. Free Radic Biol Med 2018; 127:108–15.
172. Zhao H, Zhao Z, Lei XG. Chapter 2: Function and regulation of porcine selenogenome and selenoproteome. In *Selenium in Pg Nutrition and Health, Surai, P. F. ed, Wageningen Academic Publishers, The Netherlands*, pp. 39–59, 2021.
173. Huang JQ, Ren FZ, Jiang YY, et al. Free Radic Biol Med 2015;83:129–38.
174. Sunde RA, Hadley KB. Exp Biol Med (Maywood, N.J.) 2010;235:23–-31.
175. Sunde RA, Thompson KM, Fritsche KL, et al. Biol Trace Ele Res 2017;177:139–47.
176. Cheng W-H, Valentine BA, Lei XG. J Nutr 1999;129:1951–7.
177. Friedmann Angeli JP, Schneider M, Proneth B, et al. Nat Cell Biol 2014;16:1180–91.
178. Zeng MS, Li X, Liu Y, et al. Free Radic Biol Med 2012;52: 1335–42.
179. Rayman MP, Stranges S, Griffin BA, et al. Ann Inter Med 2011;154:656–65.
180. Yan X, Pepper MP, Vatamaniuk MZ, et al. J Nutr 2012;142: 1975–82.
181. Wang X, Vatamaniuk MZ, Roneker CA, et al. Antiox Redox Signaling 2011;14:391–401.
182. Zhao H, Li K, Tang JY, et al. J Nutr 2015;145:1394–401.
183. Liu T, Yang T, Pan T, et al. Biol Trace Ele Res 2018;183:102–13.
184. Ip C. J Natl Canc Inst 1986;77:299–303.
185. Yamashita M, Yamashita Y, Suzuki T, et al. Marine Biotechnol (New York, N.Y.) 2013;15:559–70.
186. Ganther HE, Goudie C, Sunde ML, et al. Science (New York, N.Y.) 1972;175:1122–4.
187. Hu XF, Eccles KM, Chan HM. Env Intl 2017;102:200–6.
188. Long J, Liu Y, Zhou X, et al. Biol Trace Ele Res 2020;199:148–53.
189. Wang SK, Weaver JD, Zhang S, et al. (2011) *Free Radic Biol Med* 51, 197–204.
190. National Health and Medical Research Council (NHMRC), Ministry, Health, et al. Nutrient Reference Values for Australia and New Zealand Including Recommended Dietary Intakes, 2006. www.nhmrc.gov.au/ ISBN 1864962437
191. Department of Health. Dietary Reference Values of the Committee on Medical Aspects of Food Policy (COMA), in Dietary Reference Values for Food, Energy and Nutrients for the United Kingdom, The Stationery Office: London, 1991.
192. Panel on Dietetic Products, Nutrition and Allergies, European Food Safety Authority (EFSA). EFSA J 2014;12:3846.
193. Expert Group on Vitamins and Minerals (2003) Safe Upper Levels for Vitamins and Minerals, by Food Standards Agency, May 2003, ISBN 1-904026-11-7, https://cot.food.gov.uk/sites/default/files/vitmin2003.pdf
194. Winther KH, Rayman, MP, Bonnema SJ, et al. Nat Rev Endocr 2020;16:165–76.
195. Rayman MP. Lancet 2012;379:1256–68.
196. Johnson CC, Fordyce FM, Rayman MP. Proc Nutr Soc 2010;69:119–32.
197. Xia Y, Hill KE, Byrne DW, et al. Am J Clin Nutr 2005;81:829–34.
198. Yang GQ, Xia YM. Biomed Env Sci: BES 1995;8:187–201.
199. Hughes DJ, Fedirko V, Jenab M, et al. Intl J Canc 2015;136: 1149–61.
200. Bastola MM, Locatis C, Maisiak R, et al. BMC Cardiovasc Disord 2020;20:45.
201. Stoffaneller R, Morse NL. Nutrients 2015;7:1494–537.
202. World Health Organization. *Vitamin and Mneral Requirements in Human Nutrition*, World Health Organization, Geneva, 2004.
203. Alfthan G, Aro A, Arvilommi H, et al. Am J Clin Nutr 1991;53:120–5.
204. Duffield AJ, Thomson CD, Hill KE, et al. Am J Clin Nutr 1999;70:896–903.
205. Naderi M, Puar P, Zonouzi-Marand M, et al. Sci Tot Environ 2021;767:144329.
206. Rayman MP, Searle E, Kelly L, et al. Br J Nutr 2014;112:99–111.
207. Rayman MP, Bode P, Redman CWG. Am J Obstet Gynecol 2003;189:1343–9.
208. Duncan A, Talwar D, McMillan DC, et al. Am J Clin Nutr 2012;95:64–71.
209. Stefanowicz FA, Talwar, D, O'Reilly DS, et al. Clin Nutr (Edinburgh, Scotland) 2013;32:837–42.
210. Accreditation E. Selection and Use of Reference Materials In *Accreditation EA-4/14 INF:2003*, 2003.
211. EFSA Journal, Scientific opinion on the tolerable upper intake level for selenium, Jan 2023, www.efsa.europa.eu/en/efsajournal/pub/7704
212. Ashton K, Hooper L, Harvey LJ, et al. Am J Clin Nutr 2009;89:2025S–39S.
213. Mita Y, Nakayama K, Inari S, et al. Nat Commun 2017;8:1658.
214. Demircan K, Bengtsson Y, Sun Q, et al. Redox Biol 2021;47: 102145.
215. Yang J, Wang T, Wu C, et al. Biol Trace Element Res 2010;138: 53–9.
216. Ge K, Yang G. Am J Clin Nutr 1993;57:259s–63s.
217. Beck MA. Nutr Rev 1998;56:S140–6.
218. Zhou H, Wang T, Li Q, et al. Biolog Tr Element Res 2018;186: 98–105.
219. Stone R. Science (New York, N.Y.) 2009;324:1378–81.
220. Zhang F, Wu C, Zhang P, et al. Biol Trace Element Res 2022;200:4278–88.
221. Xiong YM, Mo XY, Zou XZ, et al. Osteoarthritis Cartilage 2010;18:817–24.
222. Moreno-Reyes R, Suetens C, Mathieu F, et al. Int Orthop 2001;25:164–6.
223. Bleys J, Navas-Acien A, Guallar E. Arch Int Med 2008;168: 404–10.
224. Xie B, Wang J, Zhang J, et al. Asia Pac J Clin Nutr 2020;29: 827–38.
225. Rayman MP, Winther KH, Pastor-Barriuso R, et al. Free Radic Biol Med 2018;127:46–54.
226. Rasmussen LB, Schomburg L, Köhrle J, et al. Eur J Endocrinol 2011;164:585–90.
227. Swanson CA, Patterson BH, Levander OA, et al. Am J Clin Nutr 1991;54:917–26.
228. Huang Z, Rose AH, Hoffmann PR. Antiox Redox Signal 2012;16:705–43.
229. Ma C, Hoffmann PR. Semin Cell Dev Biol 2021;115:54–61.
230. Roy M, Kiremidjian-Schumacher L, Wishe HI, et al. Biol Trace Element Res 1994;41:103–14.
231. Kiremidjian-Schumacher L, Roy M, Wishe HI, et al. Biol Trace Element Res 1994;41:115–27.
232. Broome CS, McArdle F, Kyle JA, et al. Am J Clin Nutr 2004;80:154–62.
233. Sun GX, Meharg AA, Li G, et al. Sci Rep 2016;6:20953.
234. Beck MA, Handy J, Levander OA. Trends Microbiol 2004;12: 417–23.
235. Yu SY, Zhu YJ, Li WG. Biol Trace Element Res 1997;56:117–24.
236. Li W, Zhu Y, Yan X, et al. Zhonghua Yu Fang Yi Xue Za Zhi 2000;34:336–8.
237. Fang LQ, Goeijenbier M, Zuo SQ, et al. Viruses 2015;7:333–51.
238. Hou JC. Biol Trace Element Res 1997;56:125–30.
239. Guillin OM, Vindry C, Ohlmann T, et al. Nutrients 2019;11:2101.
240. Martinez SS, Huang Y, Acuna L, et al. Int J Mol Sci 2022;23:280.
241. Baum MK, Campa A, Lai S, et al. JAMA 2013;310:2154–63.
242. Kamwesiga J, Mutabazi V, Kayumba J, et al. Aids 2015;29: 1045–52.

243. Baidu. 2020. Accessed 18-02-2020. https://voice.baidu.com/act/newpneumonia/newpneumonia/?from=osari_pc_3
244. Zhang J TE, Bennett K, Saad R, et al. Am J Clin Nutr 2020;111;1297–9.
245. Li S, Banuelos GS, Wu L, et al. Nutrients 2014;6:1103–14.
246. Zhang J, Saad R, Taylor EW, et al. Redox Biol 2020;37:101715.
247. Hurst R, Armah CN, Dainty JR, et al. Am J Clin Nutr 2010;91: 923–31.
248. Zhang HY, Zhang AR, Lu QB, et al. BMC Infect Dis 2021; 21:452.
249. Moghaddam A, Heller RA, Sun, Q, et al. Nutrients 2020;12: 2098.
250. Du Laing G, Petrovic M, Lachat C, et al. Nutrients 2021;13: 3304.
251. Fakhrolmobasheri M, Mazaheri-Tehrani S, Kieliszek M, et al. Biol Trace Element Res 2022;200:3945–56.
252. Younesian O, Khodabakhshi B, Abdolahi N, et al. Biol Trace Element Res 2022;200:1562–7.
253. Rayman MP, Taylor EW, Zhang J. Proc Nutr Soc 2023;82:1–12.
254. Schweizer U, Bohleber S, Zhao W, et al. Front Neurosci 2021; 15:652099.
255. Schweizer U, Bräuer AU, Köhrle J, et al. Brain Res Brain Res Rev 2004;45:164–78.
256. Valentine WM, Abel TW, Hill KE, et al. J Neuropathol Exp Neurol 2008;67:68–77.
257. Takemoto AS, Berry MJ, Bellinger FP. Ethn Dis 2010;20: S1-92-5.
258. Ramaekers VT, Calomme M, Berghe DV, et al. Neuropediatrics 1994;25:217–23.
259. Ingold I, Berndt C, Schmitt S, et al. Cell 2018;172:409–22e21.
260. Zhang X, Liu R-P, Cheng W-H, et al. Mech Ageing Devel 2019;180:89–96.
261. Berr C, Balansard B, Arnaud J, et al. J Am Geriatr Soc 2000; 48:1285–91.
262. Akbaraly TN, Hininger-Favier I, Carriere I, et al. Epidemiology (Cambridge, Mass.) 2007;18:52–8.
263. Gao S, Jin Y, Hall KS, et al. Am J Epidemiol 2007;165:955–65.
264. Varikasuvu SR, Prasad VS, Kothapalli J, et al. Biol Trace Element Res 2019;189:361–9.
265. Kryscio RJ, Abner EL, Caban-Holt A, et al. JAMA Neurol 2017;74:567–73.
266. DeKosky ST, Schneider LS. JAMA Neurol 2017;74:508–10.
267. Zhang ZH, Song GL. Front Neurosci 2021;15:646518.
268. Hambright WS, Fonseca RS, Chen L, et al. Redox Biol 2017;12: 8–17.
269. Alim I, Caulfield JT, Chen Y, et al. Cell 2019;177:1262–79.e25.
270. Behne D, Weiler H, Kyriakopoulos A. J Reprod Fertil 1996;106: 291–7.
271. Eroglu M, Sahin S, Durukan B, et al. Biolog Tr Element Res 2014;159:46–51.
272. Scott R, MacPherson A, Yates RW, et al. Br J Urol 1998;82: 76–80.
273. Qazi IH, Angel C, Yang H, et al. Antioxidants (Basel) 2019;8:268.
274. Steiner AZ, Hansen KR, Barnhart KT, et al. Fertil Steril 2020;113:552–60.e3.
275. Perry A, Stephanou A, Rayman MP. BMJ Nutr Prev Hlth 2022;5:118–33.
276. Barman M, Brantsaeter AL, Nilsson S, et al. Br J Nutr 2020;123: 209–19.
277. Schomburg L. Nat Rev Endocrinol 2011;8:160–71.
278. Effraimidis G, Wiersinga WM. Eur J Endocrinol 2014;170: R241–52.
279. Santos LR, Duraes C, Mendes A, et al. J Clin Endocrin Metab 2014;99:E719–23.
280. Wu Q, Rayman, MP, Lv H, et al. J Clin Endocrin Metab 2015;100:4037–47.
281. Wu Q, Wang Y, Chen P, et al. J Clin Endocrin Metab 2022;107:e3603–e3611.
282. Marcocci C, Kahaly GJ, Krassas GE, et al. N Engl J Med 2011;364:1920–31.
283. Negro R, Greco G, Mangieri T, et al. J Clin Endocrin Metab 2007;92:1263–8.
284. Zhang X, Liu C, Guo J, et al. Eur J Clin Nutr 2016;70:162–9.
285. Jenkins DJ, Kitts D, Giovannucci EL, et al. Am J Clin Nutr 2020;112:1642–52.
286. Schomburg L, Orho-Melander M, Struck J, et al. Nutrients 2019;11:1852.
287. Cai X, Wang C, Yu W, et al. Sci Rep 2016;6:19213.
288. Duffield-Lillico AJ, Reid ME, Turnbull BW, et al. Canc Epi, Biomarkers Prev 2002;11:630–9.
289. Lippman SM, Klein EA, Goodman PJ, et al. JAMA 2009;301: 39–51.
290. Klein EA, Thompson IM, Jr., Tangen CM, et al. JAMA 2011;306: 1549–56.
291. Kristal AR, Darke AK, Morris JS, et al. J Natl Canc Inst 2014;106:djt456.
292. Vinceti M, Filippini T, Del Giovane C, et al. Cochrane Database Syst Rev 2018;1:CD005195.
293. Zhou J, Huang K, Lei XG. Free Radic Biol Med 2013;65: 1548–56.
294. Stranges S, Marshall JR, Natarajan R, et al. Ann Intern Med. 2007;147:217–23.
295. Wang Y, Rijntjes E, Wu Q, et al. Redox Biol 2020;37:101709.
296. Steinbrenner H, Duntas LH, Rayman MP. Redox Biol 2022;50: 102236.
297. Forceville X. J Trace Elem Med Biol 2007;21:62–5.
298. O'Toole D, Raisbeck M, Case JC, et al. Vet Pathol 1996;33: 109–16.
299. Yang GQ, Wang SZ, Zhou RH, et al. Am J Clin Nutr 1983;37: 872–81.
300. Yang G, Zhou R. J Trace Elem Electrolytes Hlth Dis 1994;8: 159–65.
301. National Research Council. *Mineral Tolerance of Animals: Second Revised Edition*, The National Academies Press, Washington, DC, 2005.
302. Duffield-Lillico AJ, Slate EH, Reid ME, et al. J Natl Canc Inst 2003;95:1477–81.
303. Achouba A, Dumas P, Ouellet N, et al. Environ Intl 2016;96: 8–15.
304. Yan X, Zhao Z, Weaver J, et al. Redox Biol 2022;56:102457.
305. Yun J-W, Zhao Z, Yan X, et al. Free Radic Biol Med 2019;134: 385–93.
306. Rocca C, De Bartolo A, Granieri MC, et al. Antioxidants 2022;11:571.

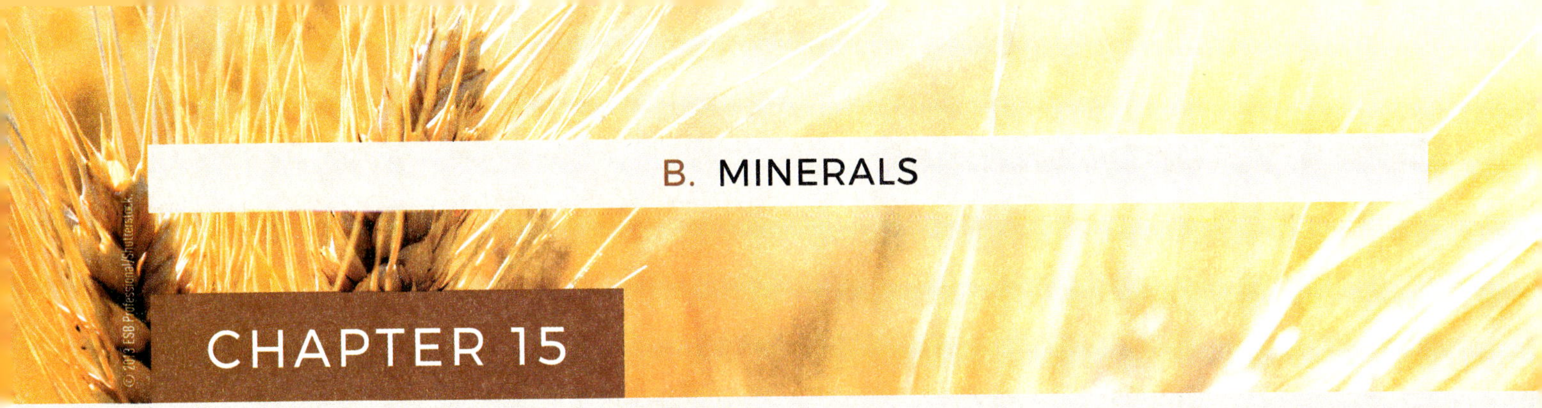

B. MINERALS

CHAPTER 15

Manganese

Steven C. Fordahl and Keith M. Erikson

CHAPTER OUTLINE

Abbreviations

AI, adequate intake; **Ca**, calcium; **Mn**, manganese; **MRI**, magnetic resonance imaging; **PN**, parenteral nutrition; **SOD**, superoxide dismutase; **UL**, tolerable upper intake level

History, Chemistry, and Biochemistry

Manganese (Mn) was first isolated as a free metal in 1774, following the reduction of its dioxide with carbon. It was first found as a constituent of animal tissues in 1913, although a state of deficiency (in animals) was not described until 1931.[1-3] Mn is a hard, brittle metal. Its oxidation state ranges between −3 and +7, although the most stable valence is +2 and the most abundant is 4+. Mn^{2+}, the only form absorbed by humans, is oxidized to Mn^{3+}, the oxidative state, over time in plasma. The physical properties of Mn make it useful in several industrial settings, such as in the manufacturing of steel and batteries, as a constituent of welding rods, and as an antiknock agent added to gasoline. It is these industrial settings where toxic exposures to Mn are a relevant health concern (discussed in the Toxicity section). The human body contains approximately 10 to 20 mg of Mn, with 25% to 40% present in bone and 5 to 8 mg turned over daily. The biologic half-life of Mn ranges from approximately 12 to 40 days.[4]

Associated Enzymes

Mn is essential as a cofactor for the metalloenzymes superoxide dismutase (SOD), xanthine oxidase, arginase, galactosyltransferase, and pyruvate carboxylase.[5] It functions as a constituent of these metalloenzymes or as an enzyme activator. SOD activity is depressed in Mn-deficient animals.[6] SOD protects the cell against antioxidant processes, including injury associated with radiation, chemicals, and ultraviolet light. Mn binding to arginase has significant importance in nitrogen metabolism through the ornithine cycle.[7] It hydrolyzes L-arginine to urea and L-ornithine. Decreased arginase results in increased plasma ammonia in rats.[8] Pyruvate carboxylase is involved in gluconeogenesis, but its activity appears minimally affected by Mn deficiency, except in newborns.[6,9] Mn also activates numerous enzymes, including various decarboxylases, glutamine synthetase, hydrolases, kinases, and transferases, such as glycosyltransferases, the last of which are involved in polysaccharide biosynthesis.[10] Depressed galactosyltransferase activity may account for the connective tissue defects observed in Mn-deficient animals.[11] Activation of these enzymes by Mn probably involves Mn binding to the protein that induces a conformational change or binding to a substrate, such as adenosine triphosphate (ATP). Mn is not essential to most of these enzyme systems, which can also be activated by other metals, except for glycosyltransferases. Therefore, at least in nonprimate animals, Mn deficiency can result in defective cartilage formation.

Dietary Considerations

Manganese is commonly found in the food supply and is higher in plant sources than animal sources. It is particularly abundant in whole grains (oat bran and wheat), fruits (pineapple), and vegetables (spinach). For a comprehensive list of food sources and their Mn concentrations, see the review by Freeland-Graves et al., 2016.[12] A 1982 study of 10,000 French households found that the average daily Mn intake was 2 mg/day, based on purchases of foods designed to total 2000 kcal/day (8360 kJ/day).[13] Other dietary surveys in the United States, Canada, and New Zealand have shown intake to range between 2.0 and 4.7 mg/day, with vegetarians having significantly greater intake.[14] Daily food intake ranges between 2 and 6 mg/day and up to 11 mg/day in vegetarian diets.[14] Dietary supplements and vitamins are another source of Mn, some of which contain up to 20 mg of the element. Manganese is taken as a supplement for a variety of conditions, including osteoarthritis and osteoporosis.[15]

Human milk is generally low in Mn (1.8–27.5 μg/L); however, concentrations in infant formulas can vary dramatically (33–300 μg/L).[16-19] In human (and animal) milk, Mn concentrations vary with the stage of lactation.[3,17,19-21] For example, Stastny and coworkers (1984) reported that mean (±SD) human milk Mn concentrations in the fourth week of lactation were 6.6±4.7 μg/L, significantly higher than those collected during the 12th week of lactation (3.5±1.4 μg/L).[17] Notably, in human milk, Mn is in the trivalent oxidation state where it is bound to lactoferrin, the major iron-binding protein in milk. Receptors for this protein are abundant in the brush border membranes of epithelial cells throughout the length of the small intestine, thus allowing for regulation of the uptake of Mn across the gastrointestinal tract. In formula, Mn is in the divalent oxidation state, and absorption through the GI tract cannot be regulated by lactoferrin receptors. Accordingly, transport of infant formula-derived Mn is likely governed by mechanisms that are different from those from Mn in human breast milk. It should be noted that, while most infant formulas do not contain lactoferrin, in the past few years, formula companies have produced some formulas that add bioactive components, including lactoferrin, based on emerging research.[22] Thus, in these new, lactoferrin-containing formulas, Mn delivery may be more like human milk.

In general, infant formulas contain much higher Mn concentrations compared with those observed in human milk.[3] Given the risk posed by high Mn in infant formulas,[23] their composition has changed significantly over the last 20 years. For example, Enfamil™ (Mead Johnson and Company) contained 1,289 μg Mn/L in 1983, but dropped this to 105 μg/L shortly thereafter.[17] It has been suggested that consumption of soy-based infant formulas is a potential area of concern for human infants, due to low bioavailability.

The dietary reference intakes of the Food and Nutrition Board, Institute of Medicine for Mn are given in **Figure 15.1**. From the criteria list, it is apparent that none of the values are based on quantitative biochemical data. Because insufficient data were available for a Recommended Dietary Allowance to be formulated, the values set for Mn are Adequate Intakes (AI). For infants, the AI reflects the mean intake of Mn from breast milk. For adults, the AI was set based on median intakes reported in the US Food and Drug Administration (FDA) Total Diet Study. Although no documented need exists for dietary Mn supplementation, absorption of Mn supplements is substantially greater in the fasting state.[24] The tolerable upper intake level

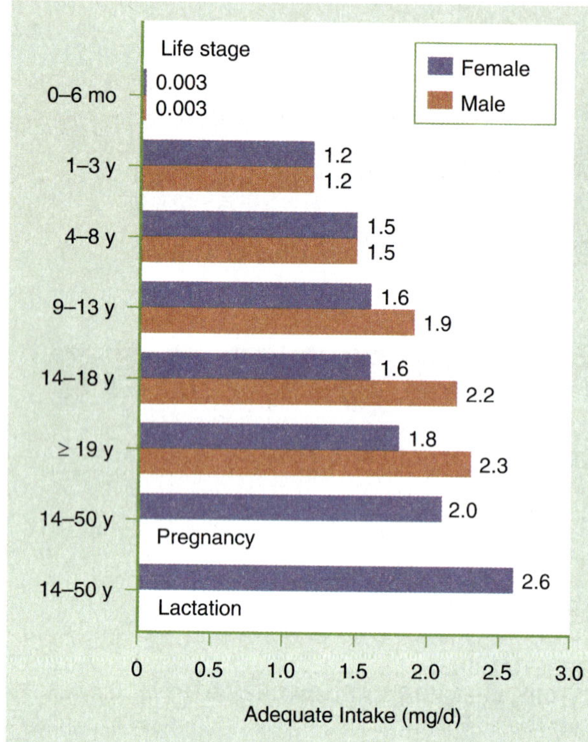

Figure 15.1 Adequate Intake for Mn by life stage group: Adequate Intake (AI). The observed average or experimentally determined intake by a defined population or subgroup that appears to sustain a defined nutritional status, such as growth rate, normal circulating nutrient values, or other functional indicators of health. The AI is used if sufficient scientific evidence is not available to derive an estimated average requirement (EAR). For healthy infants receiving human milk, the AI is the mean intake. The AI is not equivalent to a recommended dietary allowance.

Data from Food and Nutrition Board, Institute of Medicine. Dietary reference intakes for vitamin A, vitamin K, arsenic, boron, chromium, copper, iodine, iron, manganese, molybdenum, nickel, silicon, vanadium, and zinc. Washington, DC: National Academy Press, 2001.

(UL) is described later in this chapter. Toxicity from usual dietary intake is unusual, given that only approximately 5% of dietary Mn is absorbed.[25,26]

Parenteral Nutrition

For patients who require parenteral nutrition (PN), the American Society for Parenteral and Enteral Nutrition recommended 0.06 to 0.10 mg/day in adults and 0.001 to 0.150 mg/kg for children, depending on age.[27] Contamination in PN components is low (<3 to 20 μg/day); therefore, nearly all Mn in PN is derived from the addition of a multi-trace metal complex.[28-33] At the low end of probable requirement; however, such contamination could supply up to one third of the daily need. Data on Mn in patients who received long-term PN indicated that adequate blood concentrations could be maintained at 60 to 120 μg/day (1.5 to 3.0 μg/kg).[34] Human deficiency, even in the absence of Mn supplementation, has not clearly been described in patients receiving PN, however, and supplementation may not be required. Mn supplementation should cease in the presence of biliary obstruction or cholestasis jaundice, because of decreased Mn excretion, with subsequent accumulation in tissues (see the later discussion of toxicity).

Nutrient–Nutrient Interactions

The addition of large doses of Mn (four to eight times the AI) leads to a decrease in iron absorption by approximately one-third.[35] Mn supplementation leads to decreased iron absorption, even in iron-deficient animals, although this effect has not been demonstrated in humans.[36] In the gut, DMT1 expression is influenced primarily by systemic iron status[28] with duodenal biopsy data showing increased protein concentration of DMT1, correlating with decreased serum ferritin.[29] In studies evaluating Mn concentration in males and females following ingestion of a test dose of ^{54}Mn, serum ferritin was associated with prolonged Mn retention in males, but females with lower ferritin had increased Mn absorption.[30] There is an inverse relationship between iron and Mn in absorption and tissue distribution where, generally, iron deficiency (ID) increases Mn absorption[37] and, in cases of Mn overexposure, ID accelerates its tissue accumulation.[31,32] Specifically, under iron-deficient conditions, Mn is reported to accumulate in the brain, most notably in dopamine-rich nuclei like the caudate putamen and globus pallidus.[38] In no other known situations does Mn ingestion have an effect on nutrient, metal, or medication absorption. Adding calcium (Ca) to human milk and increasing dietary phytate have been shown to reduce Mn absorption.[37,39] Because of biliary excretion,[40] balance studies are not useful in the determination of daily requirement. Therefore, most estimates of absorption are based on whole body retention after 10 to 30 days of using ^{54}Mn.

Absorption, Transport, and Excretion

Dietary Mn is absorbed by diffusion and transport mechanisms that are rapidly saturable.[41,42] Approximately 6% to 16% of dietary Mn is absorbed (mean, 9%), with a retention half-life of 8 to 33 days.[28,43,44] Mena[45] found that retention was 15.4% in premature infants at 10 days, but only 8.0% in term newborns, and 1.0% to 3.0% in adults. The better absorption from human milk than from bovine milk or soy-based formula may be related to the lower concentration of Mn in human milk, the increased binding of Mn in human milk to lactoferrin, the increased Ca content of bovine milk and, for soy-based formula, the relatively large amounts of phytic acid.[46,39] No other dietary factors are known to affect Mn absorption, including ascorbic acid.

Homeostatic mechanisms control intestinal absorption; lower amounts are absorbed during periods of significant exposure.[17] The cellular mechanisms that govern and control absorption are unknown, however. Absorption is increased in patients with hemochromatosis,[43] as well as in patients with iron deficiency.[17] Mn absorption is decreased in the presence of a large Ca load.[18] Mn sulfate is the most soluble salt and is, therefore, the form found in most nutritional supplements.[19] After absorption into the portal circulation, Mn may remain either free or bound, preferably to transferrin,[20] but also to α_2-macroglobulin[21,22] and, to a lesser extent, albumin,[23] all three of which are rapidly taken up by the liver. Healthy liver function is important in Mn absorption and utilization, as patients with liver cirrhosis are reported to have a 67% increase in circulating Mn.[47] Impaired liver function can also affect proper biliary excretion of Mn (discussed below).

In serum, Mn appears to be bound primarily to transferrin[22] and an α_2-macroglobulin,[48] although binding to transferrin does not appear to be essential for Mn uptake by extrahepatic tissues.[22] The divalent form of Mn is firmly bound within the erythrocyte. The mechanism by which Mn is transported to and taken up by extrahepatic tissues has not yet been clearly elucidated, but it appears to involve internalization within endosomal vesicles as well as voltage-regulated Ca channels.[19] Mn crosses the blood–brain barrier by carrier-mediated transport, although the specific carrier has not been elucidated.[50,51] Some evidence suggests that the primary carrier protein is divalent metal transport-1 (DMT1).[52] Conflicting data suggest otherwise, however.[53] Crossgrove and Yokel[53] suggested that store-operated Ca channels are largely responsible for the transport of Mn across the blood–brain barrier. Over time, Mn^{2+} is oxidized to Mn^{3+} in plasma[54,55] possibly by ceruloplasmin,[48] which is also bound to transferrin.[48] Transferrin may be the moiety that accumulates in tissues.[55] When Mn is oxidized, it becomes more tightly bound to transferrin.[48] Newer data suggest a possible role of the zinc transporters, ZIP8 (SLC39a8) and ZIP14 (SLC39a14), in manganese uptake[56,57] (see also Chapter 11). These transport processes and the potential competition of Mn ions with other divalent and trivalent metal ions are illustrated schematically in **Figure 15.2**.

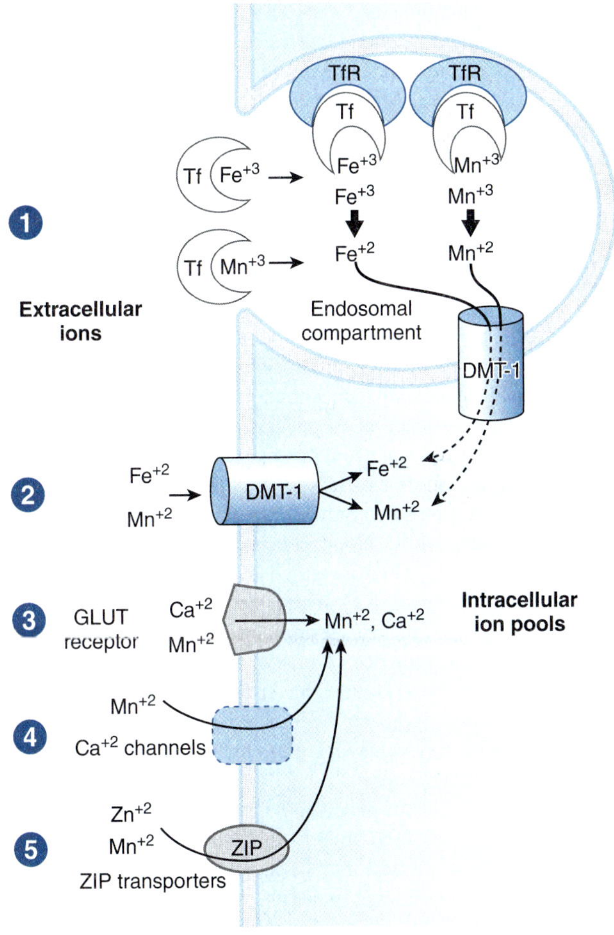

Figure 15.2 Various transport mechanisms responsible for manganese uptake: 1, transferrin (Tf)–transferrin receptor (TfR) mediated; 2, mediated by divalent metal transporter 1 (DMT-1); 3, mediated by glutamic acid ionotropic receptor (GLUT receptor); 4, mediated by channels typically considered calcium ion channels; 5, mediated by divalent metal transporters of the zinc transporter (ZIP/SLC39) family.

Data from Davidsson L, Lonnerdal B, Sandstrom B et al. J Nutr 1989;119:1461–4.[22]; Harris WR, Chan Y. J Inorg Chem 1994;54:1–19.[48]; Ruth JA, Garrick MD. Biochem Pharmacol 2003;66:1–13.[49]; Wu LJ, Leenders AG, Cooperman S et al. Brain Res 2004;1001:108–17.[52]; Crossgrove JS, Yokel RA. Neurotoxicology 2004;25:451–60.[53]; He L, Girijashanker K, Dalton TP et al. Mol Pharmacol 2006;70:171–80.[56]; Girijashanker K, He L, Soleimani M et al. Mol Pharmacol 2008;73:1413–23.[57]; Wasserman GA, Liu X, Parvez F et al. Environ Health Persp 2006;114:124-9.[115]

The trivalent form of Mn is transported and bound to transferrin, albumin, and the β(1)-globulin transmanganin.[58] Mn is taken up by the liver, pancreas, and kidney, although it is not clear that Mn^{3+} is stored in tissues except for bone, where 25% of the estimated 10 to 20 mg of total body Mn can be found.[59] Metabolically active tissues with high numbers of mitochondria as well as pigmented structures appear to have greater Mn concentration.[6] Animal studies indicated that Mn is then secreted into bile against a concentration gradient.[60]

Excretion occurs primarily through bile and nearly all Mn is excreted in the feces.[40] Studies in the rat indicated that approximately 11% of intravenously infused Mn excreted into the biliary system is reabsorbed in the intestine, although some variation may exist among species. As little as 1% of an intravenously administered dose of ^{54}Mn was present in the blood 10 minutes after injection, however.[61] This finding indicates enterohepatic circulation of Mn and shows that all is not eliminated by the biliary tract[62] Minimal Mn is excreted in the urine, and urinary excretion does not correlate with dietary intake.[58] Because biliary excretion is the primary route of Mn removal, healthy liver function is necessary to remove excess Mn from the body. T-1 weighted magnetic resonance (MR) imaging can detect Mn in the brain, and one study reported elevated Mn in the pallidum of 73% of cirrhotic patients.[63] The study also analyzed pallidum brain tissue obtained through necropsy of deceased cirrhotic patients and detected 3-fold the amount of Mn per gram of tissue compared with age-matched, cirrhotic-free controls.[63]

Analytic Methods

Flame atomic absorption spectrophotometry is the accepted method for quantitation of Mn in biologic samples, although in the presence of low Mn concentration (brain, bone), graphite furnace atomic absorption is preferable.[40] Inductively coupled plasma source mass spectrometry may also be used, although this technique is more expensive.[28] Whole blood concentration reflects current exposure, rather than chronic toxicity or deficiency.[64] Normal whole blood concentration is 4 to 15 μg/L (73 to 274 nmol/L) when measured by atomic absorption, and it is increased in cirrhosis.[65] Serum concentration of Mn is not generally useful, given that elevated intracerebral concentration may occur in the presence of normal serum concentration. In addition, serum concentrations are subject to error because the concentration may approach the limits of detection. Finally, even slight sample hemolysis may increase either plasma or serum Mn concentration dramatically.

Red blood cell Mn may also be measured, because erythrocytes account for 60% to 80% of the Mn in whole blood and are a reliable indicator of tissue stores.[66] When blood is obtained for analysis, the potential for contamination from Mn-containing disposable steel syringe needles must be considered; the error value may be as great as 80%.[67] It is, therefore, appropriate to discard the initial sample obtained by syringe before obtaining blood for analysis. EDTA is the preferred anticoagulant because heparin may also be contaminated with Mn. Given that the concentration of Mn in erythrocytes is approximately 25 times greater than in serum, contamination has a less significant effect on the measurement of erythrocyte Mn. Care must also be taken to avoid Mn-contaminated water used for dilutions during analysis.[68] A three-step purification procedure including deionization, double distillation, and re-deionization is advised to achieve true blanks. In addition, because of significant contamination potential from dust, the analytic apparatus should be covered.

Mn can also be detected by magnetic resonance (MR) imaging because the metal's atom has unpaired electrons in the level 3d orbit.[69] Brain Mn content can be estimated using the Pallidal Index, a ratio of the intensity of the T1 signal in the globus pallidus to the frontal white matter signal.[70] This technique may be used to help differentiate between Parkinson's disease and other neurological movement disorders.

Deficiency

Human Deficiency

Human Mn deficiency has not been well documented. One patient studied in a metabolic ward who accidentally received 0.34 mg/day of Mn for 17 weeks developed delayed blood clotting, hypocholesterolemia, weight loss, slowed nail and hair growth, and reddening of his beard.[71] This patient had been placed on an experimental vitamin K–free diet to induce vitamin K deficiency. The Mn in food was determined from food tables created before 1970, when analytic methods were not as accurate as more recent methods, and additional Mn from contamination was not considered. Unfortunately, neither blood Mn nor Mn loss was measured. Prothrombin time failed to respond to a 0.5-mg dose of vitamin K, but it responded incompletely to a 10-mg dose injected intramuscularly. Cholesterol fell, but only by 20 mg/dL. Whether these findings, other than the prothrombin time, improved following resumption of a normal diet was not reported. Finally, only one patient in the study developed these characteristics despite identical experimental diets. Improvement in clotting ability occurred when a standard diet was provided without any additional Mn supplementation.

Another case of purported Mn deficiency described a neonate who was dependent on total PN and who developed irregularities in bone calcification as well as bone demineralization in the presence of very low serum Mn concentration.[72] These abnormalities were reportedly corrected following an unspecified amount of Mn supplementation over a 4-month period.

Finally, Friedman et al.,[73] described clinical and laboratory findings in a group of seven male college students who were fed a purported Mn-deficient diet for 39 days. The serum cholesterol concentration decreased, although the students had been placed on a low-cholesterol diet 3 weeks before baseline evaluation, for unclear reasons. The cholesterol concentration continued to decline, however, following Mn supplementation during the repletion phase of the study. *Miliaria crystallina* dermatitis developed in five of the seven students during the Mn depletion period and resolved following Mn supplementation. Plasma, serum, and whole blood Mn concentration did not change during the study. Based on measured Mn losses and calculated retention (difficult to determine accurately because of the enterohepatic circulation of Mn), the investigators calculated the minimum Mn requirement to be 98.5 to 1037 μg/day (mean, 743 μg/day).

Experimental Animals

In nonprimate animal species, Mn deficiency has been associated with skeletal abnormalities,[10,74] ataxia,[75] decreased fertility,[76] corneal degeneration,[77] and abnormalities in carbohydrate and lipid metabolism, including impaired insulin production[78] and decreased serum high-density lipoprotein concentration.[79-82]

Toxicity

Toxicity affects the central nervous system and was first described in 1837 in Chilean miners who were exposed to Mn-containing dust and developed *locura manganica* (Mn madness).[83] Workers with significant occupational exposure to Mn have been observed to exhibit a manic stage initially, with insomnia, depression, and delusions, followed by anorexia, apathy, arthralgias, asthenia, headaches, irritability, lethargy, and lower-extremity weakness. Eventually, progressive alterations in gait and balance as well as tremor and Parkinson-like symptoms develop (including tremor and rigidity), consistent with Mn deposition in the basal ganglia. Symptoms may improve, but not completely resolve, and they may even continue to progress despite withdrawal from Mn exposure and improvement of MRI findings.[84-86] More recently, the potential for airborne Mn toxicity was suggested in response to release of Mn^{2+} in the exhaust fumes from methylcyclopentadienyl Mn tricarbonyl, a lead-replacement additive in gasoline used to increase octane. This additive is used in parts of Europe and Canada.[87]

Mn deposits primarily in the globus pallidus and subthalamic area and in the internal capsule and white matter. Mn deposition results in an increased signal observed on T1-weighted images of a brain MRI scan. However, these lesions may also be seen in other diseases including cirrhosis and neurofibromatosis type 1, as well as in patients with basal ganglia calcification. Mn appears to alter dopaminergic neurotransmission by some unknown mechanism. Investigators have postulated that Mn binding to dopaminergic receptors leads to auto-oxidation of dopamine with subsequent dopamine depletion and the formation of free radicals,[88] although conflicting data suggested that Mn may function as a potent antioxidant.[89] This may be associated with degeneration of γ-aminobutyric acid (GABA)–ergic neurons within the globus pallidus.[90,91] Mn-induced apoptosis may also contribute to toxicity,[92,93] but it is unlikely to be the primary reason, because inhibition of apoptotic markers does not prevent cytotoxicity.[94]

Several cases of possible Mn toxicity were reported in patients receiving PN at home, although not all patients were symptomatic.[95-98] Whole blood concentration correlates with both MRI intensity of the globus pallidus and T1 value.[99] Brain Mn deposition may occur even with a daily Mn dose of 0.1 mg.[24] In some patients, the increased signal intensity in the globus pallidus dissipates following removal of supplemented Mn (although not Mn present through contamination) from the PN solutions, although the Parkinson-like symptoms often fail to improve in the absence of medical therapy.[95,96] Tremor has improved in other patients following withdrawal of Mn and a decrease in whole blood Mn concentration.[100,101] The homeostatic mechanism that controls Mn absorption becomes irrelevant when Mn is intravenously infused. Despite the lack of well-demonstrated evidence that Mn deficiency occurs in patients who require long-term PN, supplementation of PN solutions has been advised.[102] This supplementation is in addition to that contained in the various PN components as a contaminant.[103] Whether Mn accumulation and brain deposition of Mn are direct results of Mn toxicity, or are caused by a decrease in biliary excretion mediated by PN and PN-associated liver disease, is unknown.[104]

Significantly elevated whole blood Mn concentration has been described in patients with cholestatic jaundice.[96-98] Investigators have postulated that PN-associated liver disease may, in part, be related to Mn toxicity. Given that Mn excretion is decreased in liver disease, especially cholestasis, and that bile flow is decreased during PN,[105] the elevated Mn concentration and brain deposition in these patients are likely the sequelae of the liver disease, rather than its cause.

Toxicity from high dietary intake of Mn has not been well described in healthy humans. No adverse effects have been observed in persons ingesting an estimated 13 to 20 mg/day of Mn,[106-108] although blood concentration may increase significantly in association with lymphocyte Mn-dependent SOD activity.[106] A single case report described an increased whole blood Mn concentration in a patient who received long-term cyclic enteral nutrition with concomitant substantial tea consumption.[109]

Prolonged exposure to elevated Mn intake may be associated with chronic toxicity.[110] In one study, Mn sulfate (15 to 20 mg/kg per week) was administered intravenously as a weekly bolus for a period of approximately 28 weeks to cynomolgus macaque monkeys. Whole blood Mn concentration increased significantly, similarly to humans with toxicity, although the administered dose was greatly in excess of what humans would receive, even in PN. This Mn concentration was accompanied by mild deficits in spatial working memory and more significant deficits in nonspatial working memory, and performance was inversely associated with brain Mn concentration. Task performance improved in vehicle-treated animals. In humans, however, the level of exposure and time required for the development of chronic toxicity are unclear. Toxicity, manifested by neurologic symptoms, may worsen even though toxic exposure ceased years earlier.[111]

Yin et al.[112] reported that, in mice, the cytoplasmic iron exporter ferroportin (Fpn) may function to transport Mn actively from cells in the presence of toxicity. Although the mechanism of Mn toxicity remains unknown, data suggest a role for Mn-induced oxidative stress.[113] Mn is transported by a calcium (Ca^{2+}) uniporter into mitochondria, where Mn may accumulate because of slow clearance and may inhibit oxidative phosphorylation.[114]

While acute exposure manifests neurotoxic symptoms quickly (within months of exposure), chronic exposure to low-dose airborne Mn in the form of Mn-contaminated drinking water may present future health concerns. This is particularly true in vulnerable infant and iron-deficient populations, where chronic low-grade exposure to Mn has been associated with cognitive impairment.[115-117] In Bangladesh, children drinking from wells with high Mn content (793 μg/L, compared to the EPA established safe limit of 50 μg/L) had significantly lower verbal scores and overall intellectual performance compared with children consuming water with Mn concentration under the EPA's established safe limit.[115] The concentration of Mn in drinking water varies by location, ranging between 1 and 100 μg/L (but can exceed 200 μg/L in well water). A recent study found that more than 1 million people who rely on well water in parts of the southeastern United States reside in an area where soil Mn concentrations are exceptionally high; most of the wells tested had unhealthy amounts of Mn.[118]

Few treatment options have been proposed for Mn neurotoxicity. Removal from the Mn toxic environment is the first course of action. Only two clinical treatments have been tested; calcium disodium EDTA ($CaNa_2EDTA$) and para-Aminosalicylic Acid (PAS), each yielding success.[119,120] However, the sample sizes in these studies were small and, to date, no progress has evolved from these putative therapies. Additionally, these treatments are intended to relieve or improve symptoms secondary to Mn toxicity, when neuronal damage may have already occurred. It is imperative that treatment strategies shift to prevent the onset of Mn neurotoxicity.

Tolerable Upper Intake Levels

The Food and Nutrition Board chose the following reasonable no-observed-adverse-effect levels of total Mn from food, water, and supplements as UL: 2, 3, and 6 mg/day for children at ages 1 to 3 years, 4 to 8 years, and 9 to 13 years, respectively; 9 mg/day for ages 14 to 18 years and during pregnancy and lactation for adolescents; and 11 mg/day for adults more than 19 years old and for adults during pregnancy and lactation. Based on the U.S. Food and Drug Administration Total Diet Study, the highest daily dietary intake, at the 95th percentile, was 6.3 mg (by men aged 31 to 50 years).[13]

Acknowledgments

The authors would like to acknowledge the author of the previous version of this chapter, Alan Buchman for contributions.

References

1. Kemmerer AR, Elvehjem CA, Hart EB. J Biol Chem 1931;92:623–30.
2. Orent ER, McCollum EV. Science 1931;73:501–6.
3. Orent ER, McCollum EV. J Biol Chem 1931;92:651–78.
4. Mahoney JP, Small WJ. J Clin Invest 1968;47:643–53.
5. Schroder HA, Balassa JJ, Tipton IH. J Chronic Dis 1966;19:545–71.
6. Keen CL, Ensunsa JL, Clegg MS. Metal Ions Biol Systems 2000;37:89–121.
7. Kuhn NJ, Ward S, Piponski M, et al. Arch Biochem Biophys 1995;320:24–34.
8. Brock AA, Chapman SA, Ulman EA, et al. J Nutr 1994;124:340–4.
9. Baly DL, Keen CL, Hurley LS. J Nutr 1985;115:872.
10. Staley GP, Van der Lugt JJ, Axsel G, et al. J S Afr Vet Assoc 1994;65:73–7.
11. Baly DL, Keen CL, Hurley LS. J Nutr 1985;115:872–9.
12. Freeland-Graves JH, Mousa TY, Kim S. J Trace Elem Med Biol 2016;38:24–32.
13. Food and Nutrition Board, Institute of Medicine. Dietary Reference Intakes for Vitamin A, Vitamin K, Arsenic, Boron, Chromium, Copper, Iodine, Iron, Manganese, Molybdenum, Nickel, Silicon, Vanadium, and Zinc. Washington, DC: National Academy Press, 2001.
14. Couzy F, Aubree E, Magnolia C, et al. J Trace Elem Electrolytes Health Dis 1988;2:79–83.
15. Horning KJ, Caito SW, Tipps KG, et al. Annu Rev Nutr 2015;35:71–108.
16. Davidson L, Almgren A, Juillerat MA, et al. Am J Clin Nutr 1995;62:984–7.
17. Mena I, Horiuchi K, Burke K, et al. Neurology 1969;19:1000–6.
18. Freeland-Graves JH, Lin PH. J Am Coll Nutr 1991;10:38–43.
19. Wong-Valle J, Henry PR, Ammerman CB, et al. J Anim Sci 1989;67:2409–14.
20. Rabin O, Hegedus L, Bourre JM, et al. J Neurochem 1993;61:509–17.
21. Davis CD, Wolf TL, Greger JL. J Nutr 1992;122:1300–8.
22. Davidsson L, Lonnerdal B, Sandstrom B, et al. J Nutr 1989;119:1461–4.
23. Davis CD, Zech L, Greger JL. Proc Soc Exp Biol Med 1993;202:103–8.
24. Sandstrom B, Davidsson L, Eriksson R, et al. J Trace Elem Electrolytes Health Dis 1987;1:33–8.
25. Davidsson L, Cederblad A, Lonnerdal B, et al. Am J Clin Nutr 1989;49:170–9.
26. Johnson PE, Lykken GI, Korynta ED. J Nutr 1991;121:711–7.
27. National Advisory Group on Standards and Practice Guidelines for Parenteral Nutrition. JPEN J Parenter Enteral Nutr 1998;22:49–66.
28. Stobbaerts RFJ, Ieven M, Deelstra H, et al. Z Emahrungswiss 1992;31:138–46.
29. Krachler M, Rossipal E. Ann Nutr Metab 2000;44:68–74.
30. Stastny D, Vogel RS, Picciano MF. Am J Clin Nutr 1984;39:872–8.
31. Lonnerdal B. Physiol Rev 1997;77:643–9.
32. Finley JW, Johnson PE, Johnson LK. Am J Clin Nutr 1994;60:949–55.
33. Bertinet DB, Tinivella M, Balzola FA, et al. J Parenter Enteral Nutr 2000;24:223–7.
34. Shike M, Ritchie ME, Shils ME. Clin Nutr 1986;34:804A.
35. Rossander-Hulten L, Brune M, Sandstrom B, et al. Am J Clin Nutr 1991;54:152–6.
36. Thompson ABR, Olatunbosun P, Valberg LS. J Lab Clin Med 1971;78:642–55.
37. Davidsson L, Cedarblad A, Lonnerdal B, et al. Am J Clin Nutr 1991;54:1065–70.
38. Anderson JG, Fordahl SC, Cooney PT, et al. Neurotoxicol 2008;29:1044–53.
39. Davidsson L, Almgren A, Jullerat MA, et al. Am J Clin Nutr 1995;62:984–7.
40. Ishihara N, Matsushiro T. Arch Environ Health 1986;41:324–30.
41. Garcia-Aranda JA, Wapnir RA, Lifshitz F. J Nutr 1983;113:2601–7.
42. Bell JG, Keen CL, Lonnerdal B. J Toxicol Environ Health Res 1989;26:387–98.
43. MacDonald NS, Figueroa WG. UCLA Rep (US Atomic Energy Com) 1969;June 30:51–33.
44. Davidsson L, Cederblad A, Hagebo E, et al. J Nutr 1988;118:1517–21.
45. Mena I. In: Bronner FL, Coburn JW, eds. Disorders of Mineral Metabolism. New York: Academic Press, 1981:233–70.
46. Ekmekcioglu C. Nahrung 2000;44:390–7.
47. Spar L, Butterworth RL, Fontaine S, et al., Hepatology 1996;24(5):1116–20.
48. Harris WR, Chan Y. J Inorg Chem 1994;54:1–19.
49. Ruth JA, Garrick MD. Biochem Pharmacol 2003;66:1–13.
50. Ascher M, Ascher JL. Neurosci Biobehav Rev 1991;15:333–40.
51. Rabin O, Hegedus L, Bourren JM, et al. J Neurochem 1993;61:509–17.
52. Wu LJ, Leenders AG, Cooperman S, et al. Brain Res 2004;1001:108–17.
53. Crossgrove JS, Yokel RA. Neurotoxicology 2004;25:451–60.
54. Aisen P, Aesa R, Redfield AG. J Biol Chem 1969;244:4628–33.
55. Gibbons RA, Dixon SN, Hallis K, et al. Biochem Biophys Acta 1976;444:1–10.
56. He L, Girijashanker K, Dalton TP, et al. Mol Pharmacol 2006;70:171–80.
57. Girijashanker K, He L, Soleimani M, et al. Mol Pharmacol 2008;73:1413–23.
58. Foradori AC, Bertinchamps A, Gulibon JM, et al. J Gen Physiol 1967;50:2255.

59. Sumino K, Hayakawa K, Shibata T, et al. Arch Environ Health 1975;30:487–94.
60. Klassen C. Toxicol Appl Pharmacol 1974;29:458–68.
61. Cotzias GC, Horiuchi K, Fuenzalida S, et al. Neurology 1968;18: 376–82.
62. Cikrt M. Arch Toxikol 1973;31:51–9.
63. Pomier-Layrargues G, Spahr L, Butterworth R. Lancet 1995; 345(8951):735.
64. Tsalev DL, Langmyhr FJ, Gunderson N. Bull Environ Contam Toxicol 1977;17:660–6.
65. Hauser RA, Zesiewica TA, Martinez C, et al. Can J Neurol Sci 1996;23:95–8.
66. Milne DB, Sims RL, Ralston NVC. Clin Chem 1990;36:450–2.
67. Versieck J. Crit Rev Clin Lab Sci 1985;22:97–184.
68. Neve J, Leclercq N. Clin Chem 1991;37:723–8.
69. Aschner M, Erikson KM, Hernandez EH, et al. Neuromol Med 2009;11:252–66.
70. Krieger D, Krieger S, Jansen O, et al. Lancet 1995;346:270–4.
71. Doisy EA Jr. Trace Sub Environ Health 1972;6:193–9.
72. Norose N, Terai M, Norose K. J Trace Elem Exp Med 1992;5: 100–1.
73. Friedman BJ, Freeland-Graves JH, Bales CW, et al. J Nutr 1987;117:133–43.
74. Smart ME. Vet Clin North Am Food Anim Pract 1985;1:13–23.
75. Erway L, Hurley LS, Fraser AS. J Nutr 1970;100:643–54.
76. Hidiroglou M. J Diary Sci 1979;62:1195–206.
77. Gong H, Amemiya T. Cornea 1999;18:472–82.
78. Baly DL, Curry DL, Keen CL, et al. J Nutr 1984;114:1438–46.
79. Leach RM Jr, Lilburn MS. World Rev Nutr Diet 1978;32: 123–34.
80. Everson GJ, Shrader RE. J Nutr 1968;94:89–94.
81. Amdur MO, Norris LC, Heuser GF. J Biol Chem 1946;164: 783–4.
82. Kawano J, Ney DM, Keen CL, et al. J Nutr 1987;117:902–6.
83. Couper J. Br Ann Med Pharm Vital Statis Gen Sci 1837;1:41–2.
84. Rodier J. Br J Ind Med 1955;12:21–35.
85. Huang CC, Lu CS, Chu NS, et al. Neurology 1993;43:1479–83.
86. Huang CC, Chu NS, Lu CS, et al. Neurology 1998;50:698–700.
87. Kaiser J. Science 2003;300:926–8.
88. Mergler D. Can J Neurol Sci 1996;23:93–4.
89. Sziraki I, Rauhala P, Koh KK, et al. Neurotoxicology 1999;20: 455–6.
90. Olanow CW, Good PF, Shinotoh H, et al. Neurology 1996;46: 492–8.
91. Pal KP, Samii A, Caline DB. Neurotoxicology 1999;20:227–38.
92. Desole MS, Sciola L, Delogu MR, et al. Neurochem Int 1997;31: 169–76.
93. Latchoumycandane C, Anantharam V, Kitazawa A, et al. J Pharmacol Exp Ther 2005;313:46–55.
94. Roth JA, Walowitz J, Browne RW. J Neurosci Res 2000;61: 162–71.
95. Mirowitz SA, Westrich TJ. Radiology 1992;18:535–6.
96. Ejima A, Imamura T, Nakamura S, et al. Lancet 1992;339:426.
97. Taylor S, Manara AR. Anaesthesia 1994;49:1013.
98. Azaz A, Thomas A, Miller V, et al. Arch Dis Child 1995;73:89.
99. Takagi Y, Okada A, Sando K, et al. Am J Clin Nutr 2002;75: 112–8.
100. Komaki H, Maisawa SI, Sugai K, et al. Brain Dev 1999;21:122–4.
101. Nagatomo S, Umehara F, Hanada K, et al. J Neurol Sci 1999;162:102–5.
102. American Society for Parenteral and Enteral Nutrition Board of Directors. J Parenter Enteral Nutr 1998;22:49–66.
103. Buchman AL, Neely M, Grossie VB Jr., et al. Nutrition 2001;17:600–6.
104. Alves G, Thiebot J, Tracqui A, et al. J Parenter Enteral Nutr 1997;21:41–5.
105. Messing B, Bories C, Kunstlinger F, et al. Gastroenterology 1983;84:1012–9.
106. David CD, Greger JL. Am J Clin Nutr 1992;55:747–52.
107. Greger JL. Neurotoxicology 1999;20:205–12.
108. Schroeder HA, Balassa JJ, Tipton IH. J Chronic Dis 1966;19: 545–71.
109. Ross C, O'Reilly DS, McKee R. Ann Clin Biochem 2006;43: 226–8.
110. Schneider JS, Decamp E, Clark K, et al. Brain Res 2009;1258: 86–95.
111. Rosenstock HA, Simons DG, Meyer JS. JAMA 1971;217:1354–8.
112. Yin Z, Jiang H, Lee ES, et al. J Neurochem 2010;112:1190–8.
113. Stredrick DL, Stokes AH, Worst TJ, et al. Neurotoxicology 2004;25:543–53.
114. Gavin CE, Gunter KK, Gunter TE. Toxicol Appl Pharmacol 1992;115:1–5.
115. Wasserman GA, Liu X, Parvez F, et al. Environ Health Persp 2006;114:124–9.
116. Sahni V, Léger Y, Panaro L, et al. Environ Health Perspect 2007;115:1776–9.
117. Chen P, Totten M, Zhang, Z, et al. Exp Rev of Neurotherap 2019; 19:243–60.
118. Gillispie EC, Austin RE, Rivera NA, et al. Environ Sci Tech 2016;50:9963–71.
119. Herrero Hernandez E, Discalzi G, Valentini C, et al. Neurotoxicology 2006;27:333–9.
120. Jiang YM, Mo XA, Du FQ, et al. J Occup Environ Med 2006;48:644–9.

B. MINERALS

CHAPTER 16

Trace Elements

Curtis D. Eckhert

CHAPTER OUTLINE

Abbreviations

AI, adequate intake; **ARE**, antioxidant response element; **As**, arsenic; **AsO_2^-**, arsenite; **AFT4**, activating transcription factor 4; **ATF6**, activating transcription factor 6; **ATP**, adenosine triphosphate; **B**, boron; **Ca^{2+}**, calcium; **BOR**, boric acid/borate transporter; **cADPR**, cyclic adenosine diphosphate ribose; **CD38**, CD38 molecule; **CHOP**, DNA damage inducible transcript 3; **CO_2**, carbon dioxide; **Cr**, chromium; **DARP**, dissimilatory arsenate-reducing prokaryote; **DRI**, dietary reference intake; **EDEM**, ER degradation enhancing alpha-mannosidase like protein 1; **EPA**, Environmental Protection Agency; **FAD**, flavin adenine dinucleotide; **FDA**, U.S. Food and Drug Administration; **GADD34**, protein phosphatase 1 regulatory subunit 15A; **GCLC**, glutamate-cysteine ligase catalytic subunit; **GRP78/BiP**, glucose regulated protein 78; **GRP94**, heat shock protein Hsp90 family protein; **GTF**, glucose tolerance factor; **IRAC**, Interagency for Research on Cancer; **Mo**, molybdenum; **mARC**, mitochondrial amidoxime reducing component; **MOT**, molydate transporter; **NAD**, nicotinamide adenine dinucleotide; **Ni**, nickel; NIDDM, non–insulin-dependent diabetes mellitus; **NiP**, boric acid/borate channel; **NQO1**, NAD(P)H quinone dehydrogenase 1; **PERK**, eukaryotic translation initiation factor 2; **PO_3^-**, phosphite; **Si**, silicon; **TPN**, total parenteral nutrition; **UL**, tolerable upper intake level; **V**, vanadium

Introduction

Elements ingested in milligrams or less per day are referred to as trace elements. Chemists originally used the term *trace* to indicate that a concentration was below the detectable limit of the analytic procedure. Statistical analysis cannot use words, so "trace" was replaced with an estimated number, often the midpoint been the lowest detectable limit and zero.

Trace elements enter the food chain from soil, water, and atmospheric particles derived from weathered geologic formations and volcanic eruptions. Two concepts are important when considering the essentiality of a trace elements.[1] First, evolving biologic systems were based on geochemistry and aquatic chemistry that used metals and metalloids to perform catalytic, structural, and signaling functions. In metal-poor environments such as the ocean, organisms survive because many functions can be maintained using different metals with similar ionic radii and electronic structures. The second concept is chemical reactivity. In living systems, reactive metal atoms are not "free," but rather are stabilized by coordinate bonds to functional groups of amino acids in energetically strained (entatic) catalytic sites of proteins or are bound to ligands such as nucleotides and tetrapyrroles. In metalloenzymes, the strained state of metal coordination geometry stores most energy required to reach the critical high-energy transition state of the enzyme-substrate complex, so only small geometric changes are required to produce the activation energy needed to initiate enzymatic catalysis. One of the coordination sites on the metal atom is open for substrate binding and, in the relaxed state, binds an easily replaced ligand such as water (H_2O). The reactivity of a metal confined in this manner performs an essential biologic function, but at concentrations that exceed the capacity to coordinate or bind metal atoms, the same reactivity can damage neighboring molecules. Thus, at low levels of intake, the benefit of a trace element metal may be relative to the availability of other elements, but at high intakes, the probability of toxicity is great.

This chapter addresses several of the trace elements: arsenic (As), boron (B), chromium (Cr), molybdenum (Mo), nickel (Ni), silicon (Si), and vanadium (V). These elements are present in tissues at concentrations in the micrograms per kilogram range and have been reported to alter some biologic process in some species. The elements Cr, Mo, Ni, and V are metals, whereas As, B, and Si are metalloids with properties of both metals and nonmetals. The term *essential* has two parts: essentiality of the biologic function and the requirement of a specific element to achieve that function. B and Mo are the only trace elements covered in this chapter that are essential micronutrients for plants with transporters that regulate internal concentrations and specific molecular functions. B is unique in that it is the only trace element formed during the Big Bang and has been proposed to be essential in the origin of life.[2]

Arsenic

Historical Overview

Arsenic (As) has been used as a poison for thousands of years. The ancient Syrians used inorganic salts of As as agricultural pesticides and, today, organic arsenics such as roxarsone (4 hydroxy 3-nitrophenyl arsenic acid) are used to prevent coccidiosis in swine and improve the growth of poultry. As trioxide was such an effective human poison during the Middle Ages; it was referred to as "inheritance powder." One of the curious events of European history involved the metabolism of As. William Morris (1834 to 1896) was a member of the family that owned the largest As mine in Europe. The mine polluted the ecology of Devon, England, and caused pocked skin and lung disease in the local population. To divorce himself from the mine, William sold his share and used the proceeds to manufacture expensive wallpaper colored with the dye Scheele's green, which contained copper arsenite. Fungi in the wallpaper paste methylated the As salt into toxic trimethyl arsine. This volatile substance rose to high concentrations in the poorly ventilated rooms of the time and is presumed to be the source of the high As levels in the hair of exiled Napoleon Bonaparte.[3]

Terminology, Chemistry, Metabolic Roles, Interactions with Other Compounds, and Basic Importance in Normal Functions

As is widely distributed in nature in association with ores of metals such as copper, lead, and gold.[4] It is a metalloid existing in four oxidation states: As(−III), As(0), As(III), and As(V). The predominant form of inorganic As in aqueous and aerobic environments is arsenate (As[V] as $H_2AsO_4^-$ and $HAsO_4^{2-}$), whereas in anoxic environments, arsenite (As[III] as $H_3AsO_3^0$ and $H_2AsO_3^-$) predominates. Adsorption of arsenate on the surface of minerals such as ferrihydrite and alumina constrains its hydrologic mobility. The most common form of arsenite (AsO_2^-) is less strongly adsorbed to minerals, so its oxyanion is more mobile in environmental water. As has not been shown to be required for any physiologic functions in animals or humans, but deficiencies have been reported to result in myocardial damage. Its principal importance in the diet is as a toxin that can induce damage to the nervous and cardiovascular systems and increase the risk of cancer of the skin, lung, and bladder.

Dietary Sources

The total dietary intake of As from food is approximately 50 μg of As/day, of which 10 μg is inorganic.[5,6] Less than 4 μg/day is derived from drinking water. Salt-water fish contain the highest concentration of As (1662 ng/g) as arsenobetaine, a nontoxic organic form. Cereals and bakery products provide approximately 23.5 ng/g, and fats and oils contain 19 ng/g. The major contributors of inorganic As are rice, flour, spinach, and grape juice. As intakes in North America range from 0.5 to 0.81 μg/kg/day, with median intakes of 2.0 to 2.9 and 1.7 to 2.1 μg/day for men and women, respectively. The concentration in human milk ranges from 0.2 to 6 μg/kg wet weight. Drinking water is the primary source of inorganic As(III) and As(V).

Recommended Dietary Allowances

The Food and Nutrition Board of the Institute of Medicine has not established a dietary reference intake (DRI) or a tolerable upper intake level (UL) for dietary As[5] (**Table 16.1**).

Table 16.1 Recommended Dietary Allowances and Upper Limits

Element	AI	RDA	UL
Arsenic	ND		
Boron	ND		
1–3 y			3 mg/d
4–8 y			6 mg/d
9–13 y			11 mg/d
14–18 y			17 mg/d
Adults			20 mg/d
Chromium			
0–6 mo	0.2 μg/d		
7–12 mo	5.5 μg/d		
1–3 y	11 μg/d		
4–8 y	15 μg/d		
9–13 y	25 μg/d		
Adult male	35 μg/d		
Adult female	25 μg/d		
Molybdenum			
0–6 mo		2 μg/d	
7–12 mo		3 μg/d	
1–3 y		17 μg/d	300 μg/d
4–8 y		22 μg/d	600 μg/d
9–13 y		34 μg/d	1,100 μg/d
14–18 y		43 μg/d	1,700 μg/d
Adults		45 μg/d	2,000 μg/d
Pregnancy and Lactation		50 μg/d	
Nickel	ND		
1–3 y			0.2 mg/d
4–8 y			0.3 mg/d
9–13 y			0.6 mg/d
Adults			1.0 mg/d
Silicon	ND		ND
Vanadium	ND		ND

AI, adequate intake; ND, no data; RDA, recommended dietary allowance; UL, tolerable upper intake level.

Sites of Intestinal Absorption, Blood Transport, and Intracellular Forms

Soluble forms of ingested As are readily absorbed from water (90%) and food (60% to 70%) by the human gastrointestinal tract.[7] Cooking does not alter bioavailability but changes the species from As(V) and dimethylarsinic acid to As(III).[8] Less soluble arsenosugars occur in plant products such as seaweed and are poorly absorbed. The proportion of inhaled As absorbed ranges from 30% to 34%. As bound to the skin is slowly released into the circulation.[9] As is cleared from the blood in humans, although some remains, bound to the cysteine residues of hemoglobin.[10]

It is methylated in the liver using *S*-adenosylmethionine as the methyl donor to methylarsonic acid and dimethylarsinic acid.[11] Arsenics containing As(III) are the preferred substrates for enzymatically catalyzed methylation. Inorganic and organic As(V) are first reduced to As(III) by glutathione or other thiols and then methylated and cycled between the As(V) and As(III) oxidation states to form dimethylated products,

$$As^{V}O_4^{3-} + 2e \rightarrow As^{III}O_3^{3-} + CH_3^+ \rightarrow CH_3As^{V}O_3^{2-}$$
$$+ 2e \rightarrow CH_3As^{III}O_2^{2-} + CH_3^+(CH_3)_2As^{V}O_2^- + 2e$$
$$\rightarrow (CH_3)_2As^{III}O^- + CH_3^+$$

The mammalian enzyme responsible for catalyzing the transfer of the methyl group from *S*-adenosyl-L-methionine to trivalent and dimethylated arsenics is *S*-adenosyl-L-methionine: As(III) methyltransferase. Methylation efficiency in humans decreases when concentrations are high and when the liver's methylating capacity is exceeded, inorganic As accumulates in soft tissues. Pretreatment of cells with small amounts of As over prolonged periods increases the methylating efficiency and, thereby, decreases the risk of toxicity.[12]

Tissue accumulation is influenced by methyltransferases, and polymorphisms of these enzymes may explain the variability in individual risk for As toxicity.[12] As accumulates in the liver, kidney, muscles, heart, spleen, pancreas, lungs, and brain.[11] Following low-level exposure to inorganic As, its methylated metabolites are rapidly excreted in urine with small amounts of inorganic As eliminated in feces, sweat, skin desquamation, hair, and nails. The relative proportions of urinary As metabolites are 40% to 60% dimethylarsinic acid, 20% to 25% inorganic As, and 15% to 25% methylarsinic acid. One study, in which a single intravenous injection of radiolabeled trivalent inorganic As(III) was administered to human volunteers, showed that most of the As(III) was removed by urinary excretion within 2 days, and a small amount of excretion continued during the subsequent 2 weeks. The biologic half-life of As from fish is estimated to be less than 20 hours, with total clearance occurring over 48 hours. Blood concentrations may appear normal while amounts in the urine remain elevated.

Functions in Metabolism and Biology

A group of bacteria called dissimilatory arsenate-reducing prokaryotes (DARPs) uses As(V) as a nutrient.[13] DARPs occur in anoxic environments, in the gastrointestinal tracts of animals, and in the subsurface aquifer sediments of Bangladesh. DARPs use As in respiration by linking the oxidation of lactate to the reduction of As(V) to As(III). No unique biologic functions have been discovered for As in vertebrates.

Assessment of Nutrient Status

Methods are not available to assess nutritional status, but blood concentrations of As are affected by the status of folate. Methylation of ingested inorganic As to monomethylarsonic and dimethylarsinic acids requires folate-dependent one-carbon metabolism that facilitates urinary As elimination. Gamble et al.[14] used folic acid supplementation (400 μg/day) to reduce total blood As concentrations from a preintervention mean ± SE of 9.86 ± 0.62 μg/L to 8.20 ± 0.50 μg/L ($P < 0.0001$). A nonsignificant decline from 9.59 ± 0.63 μg/L to 9.14 ± 0.61 μg/L occurred in individuals receiving a placebo ($P = 0.10$).

Specific Causes and Manifestations of Deficiency and Excess

The toxicity of As is based on the ability of As(III) to react with the sulfhydryl groups in proteins, thus leading to inactivation of enzymes.[12] Mitochondria are the primary cellular targets of As(III), and that is where it accumulates, uncouples oxidative phosphorylation, and reduces the synthesis of adenosine triphosphate (ATP). As is also a cocarcinogen with ultraviolet radiation. The underlying mechanism may be inhibition of DNA repair by AsO_2^- following ultraviolet damage.[15] Methylation of As competes for *S*-adenosylmethionine and leads to hypomethylation of DNA and potential damage.[16] Acute As poisoning causes an acute paralytic syndrome characterized by cardiovascular collapse and loss of brain function resulting from necrosis of white and gray matter secondary to vasodilation.[11] The symptoms of As toxicity are dose dependent and include encephalopathy, gastrointestinal symptoms, skin pigmentation and dermatitis, peripheral vascular disease and neuropathy, genotoxicity, and cancer. Acute ingestion of 1 mg/kg/day of inorganic As causes anemia and hepatotoxicity.[17] Ingestion of 10 mg/kg/day or more can result in encephalopathy and gastrointestinal disturbances. Long-term ingestion of 10 μg/kg/day in drinking water can produce arsenicism. This is an occlusive peripheral vascular disease, commonly referred to as black foot disease, in which, in extreme cases, feet turn black and develop gangrene. The U.S. Environmental Protection Agency's maximum contaminant level (MCL) for drinking water As is 10 μg/L.[18]

Long-term ingestion of low levels of inorganic As occurs over an extensive region of Southern Asia and, to a lesser extent, in China, increasing the risk of more than 140 million people to cancers of the skin, bladder, and lung.[19-21] The problem is a consequence of trying to meet the water demand for the large populations of Bangladesh and West Bengal, India. In the 1970s, The United Nations Children's Fund (UNICEF) and other relief international agencies drilled 6 to 10 million shallow drinking water wells to bypass sewage-tainted surface waters contaminated with cholera. The sediments in the area contained As adsorbed on the surface of iron oxides. By 1998, 61% of the shallow wells were found to be contaminated with As, thus exposing millions of people to high levels, with 200,000 reported cases of arsenicosis. The release of As from the minerals may have been initiated by the reduction of iron oxide coating on sand grains in the sediment by DARPs, inorganic carbon from peat, and methane. Rice grown in this region is often prepared as gruel that contains appreciable levels of As (1.7 mg/kg). Parboiling or washing gruel in nonarsenic containing water is effective in reducing As (0.4 mg/kg).[8]

Boron

Historical Overview

In 1857, Wittstein and Apoiger detected B in plants (cited in)[22] and, in 1912, Bertrand and Agulmon detected it in 27 animal species.[23] Wiley, who was responsible for the *Pure Food and Drug Act of 1906*, gave 5-gram doses of boric acid to a group of human adults and determined that it was rapidly absorbed and excreted.[24,25] To determine if B was essential for growth and reproduction in animals, Hart's laboratory compared rats fed a synthetic diet that provided 0.6 μg B/day with another group provided 75 μg B/day.[26] No differences were observed between the groups. They concluded that 0.6 μg-B/day was too high to identify a deficiency, but sufficient for growth. The question remained, what did boric acid do between the time it was absorbed and excreted. Finding the answer to this question faced three challenges. First, B-deficient conditions had to be achieved to expose a deficiency symptom that could be corrected by B repletion. The widespread occurrence of B in diet ingredients made this impossible to achieve. Second, analytical techniques were needed to accurately quantify B in biological matrices. Currently, B can be reliably measured at the mg-B/kg level, but tissue concentrations in the low μg-B/kg range are subject to contamination from reagents used to solubilize biological matrices and from spectral interferences.[27] The third challenge was to identify biological functional molecular targets in the absence of a stable radioactive isotope for tracing.

The inability to formulate a B-deficient diet forced researchers to turn to another approach. Rats fed basal diets deficient in another nutrient were compared with those fed the same diet supplemented with B.[28] Hunt[29] determined that B supplementation partially alleviated the effects of Vitamin D deficiency in chicks. Nielsen et al.[30] were not able to identify a critical deficiency that could be linked to B in humans provided a low-B, low-magnesium diet (0.25 mg-B/2000 kcal) for 119 days followed by a supplement of 3 mg-B/day for 48 days. B supplementation reduced plasma calcium, urinary calcium, and magnesium loss and elevated the serum concentrations of 17 beta-estradiol and testosterone. Penland[31] determined that the major change in humans subjected to 62 days of low-B diet (0.23 mg-B/2000 kcal) and then supplemented with 3 mg-B/day for 49 days was a decrease in brain executive function. A review of the effects of B supplementation in growing farm animals found that the major impact was on bone, presumably because B accumulates in bone, but not in soft tissue.[32] Nielsen and Meacham reviewed the evidence from laboratory animals and humans in 2011 and concluded that boron had beneficial effects on inflammation and arthritis, the immune system, oxidative stress, bone growth, prevention of cancer, and the central nervous system. They postulated that the known formation of boroesters with ribose in S-adenosylmethionine and nicotinamide adenine dinucleotide (NAD^+) might be the reason for boron activity.[33]

In 1998, Eckhert showed that the diet problem could be bypassed using fish as a model. Trout eggs (Oncorhynchus mykiss) fertilized in water pretreated with an ion exchange resin to remove B were transferred to graded levels of boric acid. Embryonic growth was dose dependent up to a concentration of 10 μM B, demonstrating that 10 μM B was adequate for growth.[34] Essentiality was convincingly demonstrated in zebrafish maintained in water with a concentration of 0.1 μM-B or in water supplemented with ultrapure boric acid to 45 μM-B. Both groups were fed brine shrimp grown in a B deficient salt solution.[35] B deficient fertilized eggs failed to cleave normally from the zygote stage to the two- and four-cell stage. Forty-six percent of the B deficient fertilized eggs did not complete the blastula stage and 92% died within 10 days. Repletion with B during the first hour after fertilization rescued them from death. This observation was reinforced by studies that showed a low B diet impaired the morphogenesis of frog (Xenopus) embryos[36] and trophoblasts of mice that failed to implant under low boron conditions.[37] The major symptom of B deficiency in adult zebrafish was retinal degeneration, an observation that reinforced Penland's observation that B was important for the nervous system.[31,38]

At this point, it was known that B intake resulted in rapid absorption and distribution of boric acid to all cells and that B was excreted without a change in structure. The effect of B at the one- to four-cell stage in zebrafish demonstrated essentiality. The next step was to identify a physical interaction between boric acid and a target molecule that had a function that could be measured in a cellular model. Biological relevance of the interaction could then be assessed by the ability of postprandial concentrations of boric acid to modulate the target's function in an appropriate timeframe.

Terminology, Chemistry, Metabolic Roles, Interactions with Other Compounds, and Basic Importance in Normal Functions

B was formed, along with hydrogen, carbon, nitrogen, and oxygen, during the nucleosynthesis of low-weight elements following the Big Bang.[2] Borates from meteorites could stabilize glyceraldehyde, thus allowing it to combine with enediolate and stabilize ribose in the interstellar environment.[39] This postulated role, in the transition between interstellar chemistry and the RNA world, has placed B center stage at the origin of life.

B, atomic number 5, has an atomic weight of 10.81 and exists as a mixture of stable isotopes ^{10}B and ^{11}B with respective abundances of 19.8% and 80.2% in the natural environment. Boric acid evaporates easily in the steam of geysers, and this is believed to be the origin of borax deposits in Tuscany, California, and Turkey.[40] The principal geologic forms of borate include the following: tincal (borax), $Na_2B_4O_7.10H_2O$; kernite (borax pentahydrate), $Na_2[B_4O_5(OH)_4].2H_2O$; colemanite, $Ca[B_3O_4(OH)_3].H_2O$; and ulexite, $NaCa[B_5O_6(OH)_6].5H_2O$.[41]

B is a metalloid with an electronic structure of $1s^2 2s^2 2p^1$ and oxidation state of +3. The chemistry of B in nature is dominated by its affinity for oxygen. Three (trigonal) covalent bonds with oxygen form boric acid and four (tetrahedral) borates. Soluble forms of B include boric acid $B(OH)_3$ and the monovalent anion $B(OH)_4^-$, with the predominate form dependent on the pH of the solvent. Boric acid is a weak Lewis acid with a negative logarithm of the constant for the ionization equilibrium (pKa) of 9.2. Boric acid is the major form of B in physiologic fluids, at concentrations with reported ranges from 2 to 100 μM B, but usually between 2 and 10 μM B. Boric acid and borate form complexes with cis-diol groups on the five-carbon sugars apiose in RG-II in plants and ribose in animals.[41]

The observation in zebrafish that B was involved in cell replication led the Eckhert laboratory to search for B-responsive mammalian cells that could serve as a model. Epidemiology was used as a tool to screen for B-responsive health effects in cancer, a cell replication disease.[42] Estimated B intake from food composition tables were entered into the NHANES III cancer database and uncovered an inverse association between B intake and the risk of prostate cancer. This inverse association held in a regression of regional cancer rates on Texas groundwater B concentrations.[43] Mahabir subsequently reported that B intake was associated with a dose-dependent decrease in the incidence of lung cancer in postmenopausal women.[44]

The development of the boron-specific ion exchange resin Amberlite IRA 743[45] provided the means to deplete cell culture media for B deficiency/repletion experiments using human prostate cancer cells. B repletion inhibited prostate cancer cell growth and caused cells to develop stress granules and become senescent in a dose-dependent manner.[46] Stress granules enable cells to remain in a quiescent state without degrading the transcription apparatus.

Smith had reported that borate inhibited alcohol dehydrogenase derived from yeast but not liver. This species difference emphasized the need to assess the function of B molecular targets for biological relevance. In yeast, alcohol dehydrogenase inhibition was based on borate competing for nicotinamide adenine dinucleotide (NAD^+).[47] In the absence of a tracer radioactive isotope to identify potential molecular targets, the Eckhert laboratory focused on the presumed target NAD^+ using mass spectrometry to characterize boric acid/borate binding to the ribose moiety.[48,49] Extracellular NAD^+ derived from active cellular secretion or necrosis binds to CD38 on the plasma membrane of neighboring cells. CD38 is a multifunctional enzyme that converts NAD^+ to cADPR, and this was inhibited by boric acid.[50] cADPR is an intracellular messenger which, when released into the cytoplasm, binds to FKBP12, an accessary protein of the ryanodine receptor complex. The ryanodine receptor is a Ca^{2+} channel that controls the release of endoplasmic reticulum Ca^{2+} stores into the cytoplasm.[51] Mass spectrometry showed a physical interaction between boric acid and cADPR.[52] Confocal Ca^{2+} imaging using the boron responsive DU 145 prostate cell model identified boric acid as a dose dependent reversible competitive inhibitor of cADPR function.[53] The ability to modulate cADPR-stimulated Ca^{2+} release occurred within the dietary blood range of healthy humans. The effect occurred within seconds and led to a 30% drop in endoplasmic reticular Ca^{2+}. Depletion of ER Ca^{2+} activates transducers of the unfolded protein response (UPR). Treatment with 10 μM boric acid activated transducers PERK and ATF6, but not IRE1. Immunoblots showed treatment with 10 μM boric acid increased eIF2α protein 30 min and ATF4 and ATF6 proteins at 1 h and 30 min, respectively. Genes downstream of ATF4 that were activated included growth arrest and DNA damage-induced protein GADD34 and homocysteine-induced ER protein (Herp). Pro-apoptotic gene CHOP decreased. Genes downstream of ATF6 that were increased included GRP78/BiP, calreticulin, Grp94, and EDEM. The boric acid interaction with cADPR lead to activation of two UPR transducers ATF6 and PERK and Nrf2 and the downstream antioxidant response element (ARE).

$$\text{BA} + \text{cADPR} \rightarrow \text{BA-cADPR} \rightarrow [[\text{ER}]\text{i } Ca^{2+}\downarrow] \rightarrow \text{ATF6}\uparrow$$

$$\text{BA} + \text{cADPR} \rightarrow \text{BA-cADPR} \rightarrow [\text{ER}]\text{i } Ca^{2+}\downarrow] \rightarrow \text{PERK}\uparrow \rightarrow$$

$$\text{2 pathways}$$

$$\text{PERK} + \text{eif2}\alpha \rightarrow \text{eif2}\alpha\text{P} \rightarrow \text{ATF4} \rightarrow \text{GADD34 and ARE}$$

$$\text{PERK} + \text{eif2}\alpha \rightarrow \text{eif2}\alpha\text{P} \rightarrow \text{ATF4}\uparrow$$

$$\text{PERK} + \text{Nrf2} \rightarrow \text{Nrf2P} \rightarrow \text{ARE}$$

Mild UPR induces changes in transcription and translation to adapt cells to a change in the local environment for survival. Numerous factors are involved in whether the UPR activates pathways for survival or death. Survival was the case with dietary B, which makes evolutionary sense. B is ingested primarily as a component of plants. Plants synthesize chemicals for self-protection that Yang has postulated drove the evolution of detoxification in animals.[54] The rapid absorption of B and activation of mild UPR institutes a quiescent period to delay cell replication until DNA damage is repaired and toxic dietary chemicals are detoxified. Inhibiting the CD38/cADPR pathway protected rats against sepsis-associated damage to the brain, heart, liver, kidney, and lung.[55] B supplementation of rats has also been reported to ameliorate arsenic-induced DNA damage, proinflammatory cytokine gene expression, oxidant and bisphenol-A induce oxidative stress and inflammatory gene expression.[56,57]

Gene functions that were activated within 120 minutes of 10 μM B treatment included[58]: **eiF2α** (induces a pause in protein synthesis, a strategy that protects against viral infection, accumulation of misfolded proteins, cell differentiation, and replication), **Gadd34** (arrests growth and induces DNA repair), **GRP78/BiP** (promotes osteogenesis and bone remodeling, aggregation of misfolded opsins), **ATF4** (bone remodeling, osteogenesis, activation of the antioxidant response element ARE), **HERP** (promotes survival by stabilizing ER Ca^{2+}), **ATF6** (essential for human cone development, activates genes in response to ER stress), **Grp94** (anti-apoptotic, ER chaperone required for properly folded insulin growth factor, activates calreticulin required for p53 function, folding of histocompatibility complex, mineralization of teeth), **EDEM1** (prevents aggregation of misfolded opsins), and **Nrf2/ARE** (counteracts intracellular oxidative stress and inflammation). The B-activated transcription downstream of Nrf2 included **GCLC** (rate-limiting step of glutathione synthesis), and **NQO1** (encodes a two-electron reductase that reduces quinones to hydroquinones and is activated by dioxin. Mutations in NQO1 occur in many tumors and are associated with Alzheimer disease).

Dietary Sources

B content of foods reflects local soil and water conditions in which crops are grown.[59] Plants and their byproducts contain B as an essential structural component of the pectin network of cell walls. Seeds, nuts, and vegetables contain a higher concentration than fruits and grains. Milk and eggs reflect the intake of their producers. The major contributors of B to the diet include those associated with the Mediterranean diet, such as apples, avocados, legumes, dates, prunes, nuts, wine, whole grain breads, tomato sauce, and potatoes.[60]

Recommended Dietary Allowances

The Food and Nutrition Board has not established a DRI for B. The ULs for different age groups are as follows: 3 mg B/day, 1 to 3 years of age; 6 mg B/day, 4 to 8 years; 11 mg B/day, 9 to 13 years, 17 mg B/day, 14 to 18 years; and 20 mg/day for pregnant and lactating women more than 19 years old and all adults[5] (Table 16.1). An adequate level of B intake has not been determined. When the amount required for women's bone health is selected, the report by Kim et al., in Korean women suggests that 0.9 mg/day would be adequate for women.[61] When prostate health is chosen as the endpoint for men, the report by Muezzinoglu et al.,[62] suggests that 6 mg/day would be an adequate intake.

Sites of Intestinal Absorption, Blood Transport, and Intracellular Forms

All forms of B have been reported to be absorbed from the gastrointestinal tract as boric acid, but the site of absorption has not been identified.[63] Borates are not absorbed through skin, but small amounts can be absorbed by inhalation of dust from occupational and consumer product exposures. The major form in blood and other body fluids is boric acid, which is distributed to all tissues and excreted unchanged in structure. Toxicokinetic and toxicodynamic studies demonstrated that >90% of ingested B is eliminated as boric acid in the urine of humans and rats following first-order kinetics. The half-life of renal clearance is approximately 21 hours with renal reabsorption when the ratio of B to creatinine is less than 1. Ku et al.,[64] evaluated the tissue concentrations of male rats fed a diet containing 1,575 mg B/kg for 7 days. After bone, the seminal vesicles accumulated the highest concentration. Healthy men living in low-B regions have semen B concentrations that are four times higher than in blood. The ratio is more equivalent in men living in higher-B regions.[65]

Functions in Metabolism and Biology

B is required for vascular plant growth, root elongation, and fertility. B-deficient soils are a major cause of crop failure worldwide. Two types of transporters regulate the internal B concentrations of plants.[59] NIP (nodulin-26-like intrinsic proteins) channels function in B uptake from soil and BOR efflux-type borate transporters function in loading B into xylem under low B conditions and in mitigating toxic levels. Growth requires borate diesters to cross-link rhamnogalacturonan II (RG-II) moieties between polysaccharide chains to form a three-dimensional pectic network in the primary cell wall. This pectic network strengthens cell adhesion, enabling cells to withstand the high turgor pressure required for the elongation of cells to hundreds of times their length, and the formation of pollen tubes. Evolution has also found use for a single B atom in an antibiotic synthesized by fungi and a quorum-sensing molecule (AI-2) in bacteria. Gram positive and negative bacteria release AI-2 to coordinate gene expression in response to cell population density.[66]

Positive health effects of B intake have been reported in human studies and include an increase in brain executive function and prevention of cancer. Humans depleted of B for up to 65 days exhibited deficits in executive brain function, with minor alterations in steroid metabolism and hematologic indicators that were reversible by B supplementation.[31] The risk of prostate cancer decreased with increasing B intake from either food[42] or the water supply.[43] The biologic plausibility of this epidemiologic association was shown in studies using human cell lines[46,67] mice,[68] and humans.[69] Dietary boric acid inhibited the growth of xenograph human prostate cell tumors and the amount of serum antigen (prostate-specific antigen) in a murine model of cancer.[68] Enlarged prostates are a major risk factor for prostate cancer. Turkish urologists used ultrasound to measure prostate volume in men who were 59 years old, on average, in two villages with different concentrations of B in the water supply. Men living in the high-B village had intakes of 6.2 mg B/day and significantly smaller prostates ($P < 0.0001$) than men living in the low-B village with intakes of 0.6 to 0.8 mg/day.[62] One epidemiologic study in the United States did not observe a protective effect of B on the risk of prostate cancer. One of the problems with the study was that participants were recruited from both low-B and high-B watersheds of Northwest Washington State.[69] The authors did not state this and were probably not aware of it. The recruited individuals were pooled and assigned to either high- or low-boron intakes, based on a food composition database. Foods not in the database were estimated.[69] Both problems could be solved in future epidemiologic studies by using a biochemical marker of boron status.

The protective effect of B is not limited to prostate cancer. The risk of lung cancer in women was also reported to be reduced by B in a dose-dependent manner[44] and cervical dysplasia was lower in high-B regions compared with low-B regions.[70] In vitro studies have also shown that boric acid reduces proliferation of breast cancer[71] and melanoma cells.[72]

B was shown in numerous studies to alter bone in chicks, pigs, rats, and rabbits.[73,74] In pigs, supplementation levels of 5 mg B/kg diet increased bone-bending moment in males, but not females.[75] In male rats, B supplementation did not change tibia or femur resistance to bending, but dietary levels of 200 mg/kg diet increased vertebral resistance to crush force.[76] In contrast, B supplementation of ovariectomized female rats did not provide protection against osteopenia or improve vertebrae strength. An epidemiologic study of Korean women who were 41 years old, on average, did not observe a significant relationship between B intake (0.9 mg/day) and bone density.[76]

Assessment of Nutrient Status

The most sensitive method of quantitative analysis is inductively coupled plasma mass spectrometry (ICP-MS). It has a detection limit of 0.01 mg-B/L and requires a sample volume of 5 ml.[27] Measuring B excretion in a 24-hour urine collection is the most objective approach to measuring B intake.[63]

Specific Causes and Manifestations of Deficiency and Excess

Excessive B is a reproductive toxin and has been shown to damage seminiferous tubules in male rats.[77] No reports exist of B toxicity induced by diet in humans. Toxicity in animals requires a blood concentration greater than 1,000 μM, a level 20 times higher than reported in B mine workers in the People's Republic of China who had very high intakes of 41 mg/day.[65] In rats fed a diet containing 9,000 mg/kg diet B concentrations in plasma and all tissues except adipose rapidly increased (2 to 20-fold) reaching steady state (range 12–30 micrograms/g) in 3 to 4 days. This included the epididymis, testis, epididymis, accessory sex organs, and brain. Bone boron concentration continued to increase up to 7 days when the study was ended.[64]

Boron as a Therapeutic Agent, Food Preservative, and Supplement

Boric acid suppositories are used to treat vaginal yeast infections; boric acid is used to shorten the duration of cold sores; and solutions of sodium borate-boric acid-sodium chloride are used for eye wash.[78] Boric acid was used as a food preservative from the 1870s until the 1950s. Currently, the European Union authorizes boric acid and sodium borax as preservative for sturgeon eggs (caviar) with a maximum concentration of 4 g/kg.[79]

Dietary supplements contain boron in more than 10 different forms. Boric acid is rapidly absorbed and there are no data showing that any other form has higher bioavailability or circulates in a form other than boric acid. In the pH of the stomach, most forms will be converted into boric acid before absorption. Clinically, boronated phenylalanine (BPA) is infused intravenously for boron neutron capture therapy (BNCT) where irradiation of ^{10}B nuclei form lithium-7 (^{7}Li) ions and highly energetic helium (^{4}He) nuclei (alpha particles) that destroy cells in the range of 5–9 μm. Urine analyzed by ^{10}B-NMR 15 h following a 1.5 h infusion found significant breakdown of BPA, with 11% of total ^{10}B present as borate.[80]

Chromium

Historical Overview

Rats fed a necrogenic diet develop liver degeneration and glucose intolerance. Schwarz and Mertz[81] discovered that dietary selenium protected against the liver degeneration, and glucose intolerance could be corrected by intubating crude fractions of homogenized pig kidney or brewer's yeast. They named the unknown factor in the crude fractions glucose tolerance factor (GTF). The fraction's activity was lost during storage, but it could be restored with wet ash of pig kidney. The researchers concluded that the ash contained a trace element that functioned as a cofactor. They screened 43 different salts for their ability to improve the rate of glucose removal when taken orally and reported the three most active salt preparations: a combination of Cr(II) and Cr(VI), vanadyl sulfate and Cr(VI), and Cr(III) chloride alone. From this they concluded that Cr(III) was required for GTF activity.[81] In 1998, Mertz[82] discussed the four major criticisms others had made about their conclusion and against the use of Cr(III) to treat glucose intolerance. These criticisms were as follows: (a) Cr(VI) is a carcinogen; (b) the bioavailability of Cr(III) is very low; (c) Cr(III) does not ameliorate glucose intolerance in all organisms; and (d) despite great effort, GTF had not been characterized, and different laboratories did not agree on its composition or structure. These criticisms remain valid today.

Terminology, Chemistry, Metabolic Roles, Interactions with Other Compounds, and Basic Importance in Normal Functions

Cr (atomic number 24, molecular weight 59) occurs in each oxidative state from −2 to +6, with Cr(III) and Cr(VI) the most important in human health. Vincent[83] isolated a low-molecular-weight Cr-binding substance from adipocytes called chromodulin. He proposed the following mechanism: activation of the insulin receptor allows Cr to enter the cell, which binds to apo-chromodulin, converting it to holochromodulin, the active form. Holochromodulin binds to the insulin receptor, thus activating its receptor kinase activity and physiologic function. Wang et al.,[84] reported that several different Cr(III) compounds were effective at enhancing insulin receptor phosphorylation in intact cells but they were not effective in activating a recombinant insulin receptor kinase. Clinical trials in humans and farm animals have failed to demonstrate clinically significant effects of Cr(III) on insulin resistance or body composition.[85]

Dietary Sources

Plants bioaccumulate heavy metals and the levels of Cr in dietary plants is determined by local soil and water conditions. Plants do not require Cr and, in unpolluted soils, they contain less than 0.2 mg/kg. Cr levels are high in plants grown in soil polluted from Cr-emitting industries or when sewage sludge is used as a fertilizer.[86] Whole grains, cereals, and unrefined sugars contain the highest concentrations, with lesser amounts in fruits and vegetables. Most servings contain less than 1 to 2 μg.[87] Cr also enters foods from stainless steel cooking pots and pans during cooking, and in the processing of meats.[88]

Recommended Dietary Allowances

The Institute of Medicine set an adequate intake (AI) for Cr (5) (see Table 16.1). The infant AI was calculated from Cr in milk. A rounded adult diet is estimated to contain 13.4 μg Cr/1000 kcal. The AI for men and women has been set at 35 and 25 μg/day, respectively. A UL value for Cr has not been established. In the EU, the opinion of the Panel on Dietetic Products, Nutrition and Allergies (NDA) was that chromium is not an essential element.[89] Diets that are low in Cr do not produce abnormalities that can be reversed by Cr supplementation. There is no evidence of beneficial effects associated with Cr intake in healthy individuals. The Panel concluded that setting of an AI for chromium was not appropriate.[89]

Sites of Intestinal Absorption, Blood Transport, and Intracellular Forms

Ingested Cr(III) is poorly absorbed (0.4% to 2.5%), with the remainder excreted in feces. In rats, 80% of absorbed dietary $^{51}CrCl_3$ was reported to bind to transferrin.[90] Many different synthetic Cr(III) chelates have been synthesized in an attempt to increase its bioavailability, including amino acids, vitamins, and picolinic acid.

Functions in Metabolism and Biology

It has been proposed that Cr(III) binds to apo-chromodulin and that this complex interacts with the insulin receptor. Clinical trials, meta-analyses and systematic reviews have failed to demonstrate clinically significant effects from Cr(III) supplementation on body composition or symptoms of insulin resistance and related conditions in humans or farm animals.[91]

Assessment of Nutrient Status

No indicator of Cr status is known at this time. Urinary Cr is related to recent Cr intake and is not a good indicator of statu.[5,91]

Specific Causes and Manifestations of Deficiency and Excess

No cases of Cr deficiency in a healthy population have been reported. Three patients receiving total parenteral nutrition (TPN), including one patient for more than 3 years, exhibited symptoms of impaired glucose removal, elevated free fatty acids, peripheral neuropathy, and unexplained weight loss. The addition of 250 μg Cr to the TPN solution corrected the glucose resistance within 2 weeks.[5]

Cr(VI) is well established as a teratogen, genotoxin, and carcinogen. The toxicity of Cr(III) was thought for a long time to be low because of its poor bioavailability and reactivity. Cr(III) toxicity stems from its ability to bind to nucleic acids in DNA and to the sulfhydryl groups in proteins. Cr(III) is a hapten and binds to proteins triggering immune responses that result in allergic reactions.[92] Vincent[83] reviewed the toxicity of nutrient supplements containing Cr(III). Two clinical case reports of chronic interstitial nephritis attributed to the ingestion of Cr picolinate have been published.[5] Cr(III), given as chromic chloride, is mildly teratogenic to the nervous system of mice.[92] Cr(III) has not been considered genotoxic based on tests for mutations, DNA fragmentation, unscheduled DNA synthesis, and sister chromatid exchange assays. A test for the loss of genes (gene deletion) showed that pharmacologic concentrations of Cr(III) were more potent than Cr(VI) in causing gene deletions in both mice and yeast.[93]

Molybdenum

Historical Overview

Mo serves an essential role in the nitrogen cycle through its role in the molybdopterin cofactor of molybdoenzymes that are involved in nitrogen fixation and in nitrate reductase, an enzyme required for the conversion of nitrate to ammonia. Mo was recognized as essential for human xanthine oxidase activity in 1953 and for sulfite oxidase activity in 1971.[94,95] Human essentiality is based on observations of genetic defects that cause a deficiency in the Mo cofactor, which results in seizures and death of newborns within days of birth.[96]

Terminology, Chemistry, Metabolic Roles, Interactions with Other Compounds, and Basic Importance in Normal Functions

The earth's crust contains approximately 1 mg/kg Mo. Physiologically relevant oxidations states for Mo are between +IV and +VI with redox potentials of −0.3 V.[97] At these oxidation states, Mo has an affinity toward negatively charged O and S ligands such as oxide, sulfide, thiolates, or hydroxide and nitrogen ligands. Mo is an essential element for most plants and animals. The stable hexavalent form, molybdate(VI), MoO_4^{2-}, is very soluble at pH 7 and resembles the sulfur-transporting ion, SO_4^{2-}. Molybdate aggregates into clusters at oxidation states below +VI, but this action is suppressed in biologic systems by coordination of the Mo atom to dithiolene sulfurs on molybdopterin to form the molybdopterin cofactor. The molybdopterin cofactor is identical in all species and serves as the coenzyme in molybdoenzymes.

Mo binds to molybdopterin to form molybdopterium cofactor, which is required for the function of four enzymes: aldehyde oxidase, sulfite oxidase, xanthine oxidase, and mitochondrial amidoxime reducing component (mARC). These enzymes metabolize sulfur-containing amino acids and heterocyclic compounds including purines and pyrimidines. Aldehyde oxidase, xanthine oxidase, and mARC are also involved in the detoxification of N-hydroxylated compounds.[98]

Mo accumulates as the molybdopterin cofactor in the liver, kidney, adrenal gland, and bone at concentrations that range from 0.1 to 1 mg/g wet weight.[99] A pool of enzyme free metal-pterin complex, the Mo cofactor, is present in the mitochondrial outer membrane. Molybdoenzyme sulfite oxidase is in the mitochondrial intermembrane space, and xanthine dehydrogenase and aldehyde oxidase are cytosolic enzymes.[100]

Dietary Sources

Mo occurs in dry plant matter over a concentration range from less than 0.5 to more than 100 mg/kg.[101] The Mo concentration in foods reflects the level of the soil and irrigation water in which they were grown. Rich sources include legumes, grains, and nuts. Low amounts are found in animals, fruits, and vegetables.[102,103]

Recommended Dietary Allowances

Data from the Total Diet Study indicated that mean Mo intakes in the United States are 76 μg/day for women and 109 μg/day for men.[102] The recommended AI for term infants is 0.3 μg/(kg/day),[5] but concern has been expressed that the value should be between 4 and 6 μg/(kg/day) for premature infants.[104] The Institute of Medicine used a lowest observed adverse effect level of 0.9 mg/kg/day and uncertainty factor of 30 to determine the UL. The ULs for different age groups are as follows: 300 μg Mo/day, 1 to 3 years; 600 μg Mo/day, 4 to 8 years; 1,100 μg Mo/day, 9 to 13 years, 1,700 μg Mo/day, 14 to 18 years, and 2,000 μg Mo/day for pregnant and lactating women more than 19 years old.[5]

Sites of Intestinal Absorption, Blood Transport, and Intracellular Forms

Two molybdate [MoO_4^{2-}] specific transporters are present in nature (*MOT1* and *MOT2*). Humans only express *HsMOT2*, which is also present in algae and plants.[105] The cellular location of the MOT2 transporter has not been characterized for humans.

In its most stable hexavalent form as molybdate (VI) (MoO_4^{2-}), Mo is water soluble and absorbed over a wide range of intakes. The extrinsic addition of stable isotopes of Mo to diets was used to measure absorption, retention, and excretion.[106,107] Mo is rapidly absorbed and excreted from the kidney, with retention regulated primarily by urinary excretion. Absorption averaged 89% when daily Mo intake ranged from 25 to 122 μg/day, and 93% when average intakes ranged from 466 to 1488 μg/day. Mo absorption is inhibited by high intakes of sulfate, possibly because sulfate anions compete for the same transport proteins.[108,109] Excess intake of Mo produced a copper deficiency in ruminants and nonruminants grazing on a pasture contaminated with high concentrations of Mo from industrial and mining waste.[110]

Intrinsically labeled food was produced by incorporating ^{97}Mo into soybeans and kale grown in hydroponic systems.[107,111] Purees of extrinsically and intrinsically labeled soybeans and kale were then fed to 12 women. The mean 8-day absorption of Mo was 87% from extrinsic Mo, 86.1% from kale, and 56.7% from soybeans. The mean urinary excretion was 60.8% of absorbed dose for extrinsic Mo, and it was 56.6% from kale and 63.9% from soybeans.

Functions in Metabolism and Biology

The most important Mo function in mammalian systems is the transfer of oxygen to a two-electron substrate using one-electron transferring compounds, such as flavin adenine dinucleotide (FAD). Coupling electron transfer and oxide exchange transfer an oxygen atom from the metal center to the substrate.

$$LMo^{VI}O_2 + X \leftrightarrow LMo^{IV}O + XO$$

where L is a ligand.

Three mammalian hydroxylases are molybdoenzymes. They include the mitochondrial enzyme sulfite oxidase, which catalyzes the oxidation of sulfite to sulfate in the metabolism of sulfur from methionine and cysteine; and two enzymes that hydroxylate heterocyclic substrates including purines and pyridines, xanthine oxidase, and aldehyde oxidase. Xanthine oxidase catalyzes the conversion of xanthine and its derivatives such as caffeine to uric acid and uric acid derivatives. Aldehyde oxidase is a metalloflavoprotein composed of FAD, Mo, and iron in a 1:1:4 ratio. It is involved in the formation of cotinine, a major metabolite of nicotine that occurs in the urine of cigarette smokers.

Assessment of Nutrient Status

Mo blood concentrations range widely in the literature.[112] Isotope dilution studies reported plasma values of 5 nmol/L in individuals with an intake of 22 μg/day, 20 μmol/L at an intake of 467 μg/day, and 44 μmol/L at an intake of 1,490 μg/day.[113]

Specific Causes and Manifestations of Deficiency and Excess

Tungsten, the element immediately below Mo in Group 6B in the Periodic Table, has a similar ionic radius and electronic structure and forms a complex with molybdopterin that activates molybdoenzymes. Tungsten can induce Mo deficiency as measured by a decrease in molybdoenzyme activity, but it is not considered to be significant to livestock or humans because it is rarely found in the environment.

Evidence for Mo essentiality rests on clinical observations.[114] The first was a case of sulfite oxidase deficiency in a child with an inborn error of metabolism. The symptoms included seizures, mental retardation, and dislocated ocular lenses, with the appearance of the unusual amino acid *S*-sulfocysteine in the plasma and urine, as well as high urinary levels of sulfite, thiosulfate, and taurine. A postmortem examination confirmed sulfite oxidase deficiency. Since then, nearly 50 additional cases of sulfite oxidase deficiency have been identified. The second observation was loss of sulfite oxidase activity in a patient with Crohn disease who was supported on TPN for 18 months.[115] Symptoms developed after 1 year and included tachycardia, tachypnea, night blindness, and coma. Biochemical evaluation showed an elevation in plasma methionine, low serum uric acid, and reduced urinary sulfate, thiosulfate, and uric acid. All symptoms were eliminated by the addition of 300 μg/day of ammonium molybdate to the TPN solution.

Mo deficiency has not been observed in human populations, and no beneficial effects of taking supplements of the element have been documented.[5] Observations of Mo deficiency have been limited to genetic defects that interfere with the Mo cofactor's ability to activate molybdoenzymes. Mo toxicity was induced in rats and caused renal insufficiency at levels of 80 mg/kg/day, but not at 40 mg/kg/day.[97] In rabbits, 5 mg/kg/day induced weight loss and histopathologic changes in the kidney and liver.[116]

Nickel

Historical Overview

Ni was first shown to promote the growth of bacteria in 1965.[117] The biologic importance of Ni was subsequently shown in both bacteria and plants to be based on its catalytic role in four enzymes: urease, hydrogenases, CD dehydrogenase, and methyl-coenzyme M reductase. Between 1975 and 1978, diets were developed to induce deficiency symptoms in rats, chicks, pigs, and goats.[118-121] The nutritional importance of Ni in humans is unknown and remains largely unstudied.

Terminology, Chemistry, Metabolic Roles, Interactions with Other Compounds, and Basic Importance in Normal Functions

Ni is in the first transition series of the Periodic Table with oxidation states of −I, 0, II, III, and IV. State II is the most important in biologic systems. Ni coordinates with the amino acids histidine, glutamic acid, and aspartate in metal centers of proteins,[122] and it binds to histidine and cysteine on albumin and a macroglobulin called nickeloplasmin. Tissue concentrations decrease with age and are lower in 90-year-old adults than in 1-year-old children.

Dietary Sources

The 1984 Total Diet Study by the U.S. Food and Drug Administration (FDA) showed that the mean Ni consumption in infants and children was 69 to 90 μg/day; and the median for adolescents, adults, and the elderly was 71 to 97 μg/day, 74 to 100 μg/day, and 80 to 97 μg/day, respectively.[102] The major contributors vary in different countries. The major contributors in the United States diet are mixed dishes and soups (19% to 30%), grains and grain products (12% to 30%), vegetables (10% to 24%), legumes (3% to 16%), and desserts (4% to 18%). The major contributors in Canada are meat and poultry.[5] Foods with the highest concentration of Ni include nuts, legumes, and chocolate.[102]

Recommended Dietary Allowances

No AI or recommended dietary allowance (RDA) has been determined for Ni.[5] For the UL the Institute of Medicine used a no observed adverse effect level of 5 mg/kg/day, based on decreased weight gain in rats. An uncertainty factor of 300 was derived by multiplying uncertainties of 10 each for extrapolation from rat to human, human variation, and three for potential toxic reproductive effects. The ULs for children 1 to 3, 4 to 8, and 9 to 13 years old were 0.2, 0.3, and 0.6 mg/day, respectively, of soluble Ni salts. The UL for adolescents and all adults was 1.0 mg/day.[5]

Sites of Intestinal Absorption, Blood Transport, and Intracellular Forms

Dietary Ni is poorly absorbed, with reported values from 1% to 5%.[123,124] A stable isotope study using ^{62}Ni reported that 29% to 40% of the metal was absorbed.[125] The highest concentration in the blood occurred 2 to 3 hours after oral intake of Ni sulfate or oxide.[126,127] The amount retained varied from 0% to 11%.

Functions in Metabolism and Biology

Ni has not been shown to be essential for any biochemical processes in humans.[5,123] Ni is an essential component of ureases in jack beans, ruminal bacteria, and several other plants, algae, and fungi. Ureases catalyze the degradation of urea to carbon dioxide (CO_2) and ammonia. Ni is also essential for hydrogenases in methanogenic bacteria that catalyze the conversion of hydrogen (H_2) and CO_2 to methane (CH_4).[122]

Assessment of Nutrient Status

No methods are available to assess Ni nutritional status, and urine concentrations have been reported to be higher in smokers.[128]

Specific Causes and Manifestations of Deficiency and Excess

Consuming dietary Ni has no known beneficial health effects in humans. Nielsen[123] reviewed the health effects of dietary Ni restriction in animals. The major observations in Ni-deprived pigs and rats were delayed sexual maturity, perinatal mortality, rough coat, and disorganization of the rough endoplasmic reticulum of the liver.

Silicon

Historical Overview

Abnormalities in bone, joints, skin, feathers, and hair in Si-deficient chickens and rats were first reported in 1972 by Carlisle[128] and by Schwarz and Milne.[129] The results quickly became controversial when several other laboratories failed to observe similar changes.[130] In studies in which Si supplementation of Si-deficient animals did have an effect, that effect was on the extracellular matrix and at the active site of bone formation during growth. A discussion of the controversy can be found in a review by Sripanyakorn et al.[130] An epidemiologic study in 2004 showed a positive association between Si intake and bone mineral density in the hip of men and in premenopausal women, but not in postmenopausal women.[131] The best evidence of biologic essentiality is in plants and diatoms. A gene family of Si transport proteins was identified in diatoms in 1997.[132]

Terminology, Chemistry, Metabolic Roles, Interactions with Other Compounds, and Basic Importance in Normal Functions

Si is the second most abundant element in the Earth's crust. The chemistry of Si in the natural world is dominated by its affinity for oxygen, to which it is tetrahedrally coordinated in minerals.[133] In aqueous solution at neutral pH, Si takes the form of monosilicic acid, $Si(OH)_4$.[134] It forms stable complexes with mannitol and other polyhydroxy aliphatic hexose sugars containing two hydroxyls in the *theo* position. This results in the formation of stable polyolate complexes containing five and six coordinated Si atoms. The ease of formation and the stability of these polysilicate anions are what most likely enhance absorption and accumulation in tissue.[134]

Dietary Sources

In the FDA's Total Diet Study, the major contributors of Si to the U.S. diet were beverages (55%), primarily beer, coffee, and water, followed by grain and grain products (14%) and vegetables (8%).[135] Carrots, beetroot, and radish contain high concentrations of Si, but their bioavailability is lower than that of other vegetables.[136]

Recommended Dietary Allowances

The Institute of Medicine has not established a DRI for Si. Mean intakes of Si in adult men and women from the Total Diet Study were 40 and 19 mg/day, respectively.[135] The magnitude of difference between intakes on low-fiber and high-fiber diets is approximately 21 mg/day vs 46 mg/day.[137] No evidence indicates that dietary Si has any adverse effect. The use of antacids containing magnesium trisilicate has been associated with the development of urolithiasis resulting from the formation of Si-containing stones.[138]

Sites of Intestinal Absorption, Blood Transport, and Intracellular Forms

The mean uptake of Si in adult men and women from the Total Diet Study was 40 and 19 mg/day, respectively.[135] Si in food is broken down into the monomeric form in the gastrointestinal tract and is then absorbed. Serum Si concentration reaches a maximum 100 to 120 minutes following ingestion.[136] The mean plasma silicon in a Polish population was 152 ug/L.[137] Si is freely transported in the blood, possibly in a polymeric form.[139] It is moved into tissue from plasma and is readily excreted. The excretion of Si in humans increases as dietary Si intake increases.[140]

Urinary Si represents 41 ± 36% of total intake, a percentage suggesting that, for many foods, Si content can be used as an approximate indicator of food absorption.[136] This method cannot be used for the Si from root vegetables and bananas, however, because they are poorly absorbed. Individuals given an oral dose of orthosilicic acid ($Si(OH)_4$) excreted 96.3 ± 5.8% of the ingested Si in 24 h proportioned between urine 57 ± 9.5% and feces 39 ± 9.4%.[141]

Functions in Metabolism and Biology

Si has been reported to increase bone mineralization, but the mechanism of action remains unknown. Carlisle[128] suggested that Si is involved with phosphorus in events leading to calcification and that its primary effect is on connective tissue components, but she was never able to elucidate the biologic process. When she used an electron microprobe to determine the site of Si, however, Si was localized to the active growth areas of bone in growing mice and rats.[142,143] The localized concentration reached a maximum during the final stages of calcification and then dissipated. Eckhert[144] suggested from this, and from what is known from diatoms, that the mechanism of the effect of Si on bone could lie in proton buffering.

Assessment of Nutrient Status

Silicon balance has been studied in a small number of healthy adult men and women.[145] Serum Si values of 1,325 healthy adults 18 to 91 years old were measured using atomic absorption spectrometry.[146] In men 18 to 59 years of age, the median was 9.5 μmol/L, and it decreased to 8.5 μmol/L at 60 to 74 years of age. The median concentration in women increased from 10.0 μmol/L at 18 to 29 years of age to 11.1 μmol/L from 30 to 44 years and decreased to 9.2 μmol/L between 45 and 59 years. In adults 74 years old and older, the median was 7.7 μmol/L for men and 8.0 μmol/L for women.

Specific Causes and Manifestations of Deficiency and Excess

Human deficiencies of Si have not been reported. Studies in chickens, rats, and mice have suggested that Si may be important for the growth of bone.[136] Intramuscular injections of monomethyltrisilanol and in osteoporotic patients were shown to improve trabecular bone volume and femoral bone mass density.[131] Excessive intakes of Si were reported to cause urolithiasis, a deposition of calculi or uroliths containing monosilicic acid in the kidney, bladder, and urethra of grazing animals.[147] In humans, adverse effects are primarily limited to silicosis, a lung disease resulting from the inhalation of silica particles. No dose-response data are available to establish a UL for Si.[5]

Vanadium

Historical Overview

In the 1970s, V was reported by several laboratories to improve the growth of chickens, rats, and goats, but the results were inconsistent. In 1986 and 1989, Anke[148] used better-formulated diets and compared supplemented goats (2 μg/g diet) with V-deprived goats (10 ng/g diet). V-deficient goats exhibited higher rates of spontaneous abortion, and of offspring that developed convulsions and succumbed to death, at a rate of 41% during the first 3 months of life. Serum creatinine and β-lipoprotein concentrations were elevated, and serum glucose depressed in V-deprived goats. In rat adipocytes, pervanadate peroxides mimic insulin action by activation of the insulin receptor.[149] V is a possible carcinogen, and no evidence indicates that it is essential for humans, although pharmacologic concentrations promote the action of insulin and reduce blood glucose concentration.[150]

Terminology, Chemistry, Metabolic Roles, Interactions with Other Compounds, and Basic Importance in Normal Functions

V is a transition metal with six oxidations, of which three are biologically relevant: +III, +IV, and +V.[122] Foods contain the tetravalent vanadyl [VO_2^+] and pentavalent [VO_3^-] forms. Common compounds include V pentoxide (V_5O_5), sodium metavanadate ($NaVO_3$), sodium orthovanadate (Na_3VO_4), vanadyl sulfate ($VOSO_4$), and ammonium vanadate (NH_4VO_3).[151] The V ion is an enzyme cofactor and has been found to occur in certain tunicates.[152] Vanadate(V) ions compete with phosphate ions and inhibit Na^+ adenosine triphosphatase (ATPase). The *ortho*-Vanadate ion, VO_4^{3-}, resembles the *orthophosphate* ion, PO_4^{3-}. Unlike the phosphate ion, V(V) is easily reduced to IV and III with biologic reductants such as glutathione.[151]

Dietary Sources

Grains and grain products contribute 13% to 30% of dietary V. Marine organisms, such as tunicates, brown algae, lichen and mushrooms are rich sources of V. Eighty-eight percent of the foods evaluated in the FDA's Total Diet Study contained less than 2 μg/100 g of V.[102]

Recommended Dietary Allowances

Dietary intake of V ranges from 6 to 18 μg/day.[102] The Institute of Medicine has not assigned a DRI to V or determined a UL.

Sites of Intestinal Absorption, Blood Transport, and Intracellular Forms

Less than 5% of ingested V is absorbed.[152,153] Vanadyl sulfate and sodium metavanadate have been used as supplements. V pentoxide (V_2O_3) from occupational exposures is reduced from V(V) to (IV) in humans and other animals. Most V is in the form of the Vanadyl ion VO^{2+}(IV) in the stomach. Absorption occurs in the duodenum and upper gastrointestinal tract.[154] The vanadate anion(V) is absorbed three to five times more readily than the vanadyl ion(IV). The vanadate anion(V) enters cells through nonspecific anionic channels and is reduced by glutathione.[154,155] V is rapidly cleared from plasma and accumulates in kidney, liver, testes, bone, and spleen. The V anion(V) binds to the iron binding proteins, lactoferrin, transferrin, and ferritin, and tissue levels increase in rats fed V-supplemented diets. V can cross the placenta and is considered a reproductive toxin. V is excreted primarily through the kidney, with a small amount through bile. Tissues with the highest concentrations include lung, teeth, thyroid, and bone.

Functions in Metabolism and Biology

A functional role for V has not been identified in humans or other vertebrates. Vanadyl(IV) complexes potentiate the effect of insulin, but the underlying mechanism remains unknown.[151] Oral doses of vanadyl sulfate (100 mg/day) improved both hepatic and skeletal muscle insulin sensitivity in subjects with non–insulin-dependent diabetes mellitus (NIDDM), in part by enhancing insulin's inhibitory effect on lipolysis, but they did not alter insulin sensitivity in nondiabetic subjects.[156] Crystallographic analysis showed that V as VO_4^{3-} can be incorporated into bone matrix, thus replacing PO_4^{3-}.[157] Vanadate(V) anions inhibit phosphate-dependent enzymes and Na^+/potassium (K^+)-ATPase hydrolysis. However, rats fed low dietary V pentoxide, 1.8 to 18 mg/kg body weight, had higher alkaline phosphatase activity and DNA content in the femoral diaphysis, a finding indicating that it was beneficial, but diets containing 27 mg/kg body weight were inhibitory.[158] Thus, the range between beneficial and toxic appears to be exceedingly small.

In the in vitro cell culture gilthead sea bream (*Sparus aurata* L.) model of vertebrate skeletal development, 7.5 μM vanadate increased cell proliferation, but it decreased extracellular matrix mineralization in 20% of controls.[159] V is particularly toxic to macrophages.[160] Vanadium causes oxidation of thiols, including glutathione and cysteine, and introduces thiyl radicals; V toxicity is mediated, in part, by oxygen-derived free radicals. V amplifies the initial generation of singlet oxygen generated by the reduced form of NAD phosphate (NADPH) oxidase. The genotoxicity of V pentoxide results from oxidative damage to DNA that causes DNA strand breakage.

Assessment of Nutrient Status

No methods are available to evaluate V nutritional status.

Specific Causes and Manifestations of Deficiency and Excess

No human cases of V deficiency have been reported. V deficiency in goats was reported to increase rates of abortion, convulsions, bone malformations, and early death. V as V pentoxide is an industrial pollutant that is toxic,[161,162] but it is not present in foods.[162] How interconvertible various forms of V are in the stomach and when acted on by the intestinal microflora is not clear. V toxicity symptoms include abdominal cramps, diarrhea, hemolysis, increased blood pressure, and fatigue. V is a reproductive toxin affecting males more than females. It crosses the blood–placenta barrier, is teratogenic in rodents, and affects prepubertal animals. V in breast milk caused behavioral changes and deficits in myelin in neonatal rats.[163] The International Agency for Research on Cancer (IRAC) lists V as a possible carcinogen based on inhalation studies of V pentoxide in animals.[161] The US Environmental Protection Agency oral reference dose = NOAEL/UF = 0.009 mg/kg-day.[164] New methods for quantitation of V in serum and plasma should enable further studies of V biology in the future.[165]

References

1. Kaim W, Schwederski B, Klein A. In: Bioinorganic chemistry: Inorganic elements in the chemistry of life. 2nd edition, John Wiley; 1994. ISSN: 1939–5175.
2. Copi CJ, Schramm DN, Turner MS. Science 1995;267:192–9.
3. Jones DE, Ledingham K. Nature 1982;299:626–7.
4. Nriagu J. In: Environmental chemistry of arsenic. vol 1. Frankenberger WT, Jr., ed. CRC Press. 2002. https://doi.org/10.1201/9781482271102
5. Institute of Medicine. *Dietary Reference Intakes for Vitamin A, Vitamin K, Arsenic, Boron, Chromium, Copper, Iodine, Iron, Manganese, Molybdenum, Nickel, Silicon, Vanadium, and Zinc*. The National Academies Press; Washington, 2002.
6. Gunderson EL. J AOAC Int 1995;78:1353–63.
7. National Research Council. *Arsenic in Drinking Water: 2001 Update*. The National Academies Press; Washington, 2001.
8. Horner NS, Beauchemin D. Anal Chim Acta 2013;758:28–35.
9. Wester RC, Maibach HI, Sedik L, et al. Fund Appl Toxicol 1993;20:336–40.
10. Lu M, Wang H, Li X-F, et al. Chem Res Toxicol 2004;17:1733–42.
11. U.S. Department of Health and Human Services, Public Health Service. Toxicological Profile for Arsenic (2007). Accessed March 2023. www.atsdr.cdc.gov/toxprofiles/tp2.pdf
12. Abernathy CO, Liu YP, Longfellow D, et al. Environ Health Perspect 1999;107:593–7.
13. Bagla P, Kaiser J. Science 1996;274:174–5. Erratum in: Science 1996;274:1451.
14. Gamble MV, Liu X, Slavkovich V, et al. Am J Clin Nutr 2007;86:1202–9.
15. Rossman TG, Uddin AN, Burns FJ, et al. Environ Health Perspect 2002;110:749–52.
16. Costa M. Am J Clin Nutr 1995;61(suppl):666S–95.
17. Correia N, Carvalho C, Friões F, et al. Cases Journal 2009;2:7768.
18. Environmental Protection Agency, National Drinking Water Standards, 2006. Accessed March 2023. www.epa.gov/ground-water-and-drinking-water/national-primary-drinking-water-regulations
19. Ahmad SA, Kahn MH, Haque M. Risk Manag Healthc Policy 2018;11:251–61.
20. Stokstad E, Agricultural Pumping Linked to Arsenic. Science 2002;298:1535–7.
21. Michael HA. Science 2013;341:852–853.
22. Wittstein M, Apoger M. Cited in US Department of Agriculture Bulletin No. 797, 1857. Accessed March 2023. https://naldc.nal.usda.gov/download/CAT86200792/PDF
23. Bertrand G Cited in US Department of Agriculture Bulletin No. 797, 1912. Accessed March 2023. https://naldc.nal.usda.gov/download/CAT86200792/PDF
24. Wiley HW. Public Health Pap Rep 1906;32:186–93.
25. Beckley L. In: Indiana Historical Bureau Blog of Indiana History, pg 25, 2016. Accessed March 2023. https://blog.history.in.gov/dr-harvey-wiley-and-his-poison-squad/
26. Teresi JD, Hove E, Elvehjem CA, et al. Am J Physiol. 1944;140:513–8.
27. Downing RG, Strong PL. Exp Med 1999;12:205–12.
28. Follis RH. Am J Physiol 1947;150:520–2.
29. Hunt CD. Environ Health Perspect 1994;102:S35–43.

30. Nielsen FH, Hunt CD, Mullen LH, et al. FASEB J 1987;1:394–7.
31. Penland JG. Biol Trace Elem Res 1998;66:299–317.
32. Abdelnour SA, Abd El-Hack ME, Swelum AA, et al. J Trace Elem Med Biol 2018;50:296–304.
33. Nielsen FH, Meacham SL. Complemen Alt Med 2011. Accessed March 2023. https://doi.org/10.1177/2156587211407
34. Eckhert CD. J Nutr 1998;128:2488–93.
35. Rowe RI, Eckert CD. J Exp Biol 1999;202:1649–54.
36. Fort DJ, Propst TL, Stover EL, et al. Biol Trace Elem Res 1998;66:237–59.
37. Lanoue L, Taubeneck MW, Muniz J, et al. Biol Trace Elem Res 1998;66:271–98.
38. Eckhert CD, Rowe RI. J Trace Elements Exp Med 1999;12:213–9.
39. Ricardo A, Carrigan MA, Olcott AN, et al. Science 2004:303:196.
40. Woods WG. Environ Health Perspect 1994;102(Suppl 7):S5–11.
41. Loomis WD, Durst RW. Biofactors 1992;4:229–39.
42. Cui Y, Winton MI, Zhang Z-F, et al. Oncol Rep 2004;4:887–92.
43. Barranco W, Hudak P, Eckhert CD. Cancer Causes & Control, 2007;18, 71–7. Erratum: Cancer Causes & Control 2007;18:583–4.
44. Mahabir S, Spitz MR, Barrera SL, et al. Am J Epidemiol 2008;167:1070–80.
45. de Dardel F. Accessed March 2023. http://dardel.info/IX/other_info/history.html
46. Barranco WT, Eckert C. Br J Cancer 2006;94:884–90.
47. Smith KW, Johnson SL. Biochemistry 1976;15:560–5.
48. Kim DH, Marbois BN, Faull KF, et al. J Mass Spectrom 2003;38:632–40.
49. Kim DH, Faull KF, Norris AJ, et al. J Mass Spectrom 2004;39:743–51.
50. Kim DH, Hee SQ, Norris AJ, et al. J Chromatogr A 2006;1115:246–52.
51. Lee HC. Mol Med 2006;12:317–3.
52. Henderson K, Stella SL, Kobylewski SE, et al. PLOS One. 2009;4:e6009.
53. Yamada KE, Eckert CD. Biol Trace Elem Res 2019;188:2–10.
54. Yang CS, Brady JF, Hong J-K. FASEB J 1992;6:737–44.
55. Peng Q-Y, Wang Y-M, Chen C-X, et al. Brain Res 2018;1678:56–63.
56. Ince S, Kucukkurt I, Acaroz U, et al. J Biochem Mol Toxicol 2019;33:e22252.
57. Acaroz U, Ince S, Arslan-Acaroz D, et al. Toxicol Res (Camb) 2019;8:262–9.
58. National Center for Biotechnology Information. Accessed March 2023. www.ncbi.nlm.nih.gov/guide/genes-expression/
59. Onuh AF, Miwa K. Plant Cell Physiol 2021;62:590–9.
60. Naghii MR, Wall PM, Samman S. J Am Coll Nutr 1996;6:614–9.
61. Kim M-H, Bae Y-J, Lee YS, et al. Biol Trace Elem Res 2008;125:213–222.
62. Müezzinoğlu T, Korkmaz M, Nese M, et al. Biol Trace Elem Res 2011;144:49–57.
63. Pahl MV, Culver BD, Strong PL, et al. Toxicol Sci 2001;60:252–6.
64. Ku WW, Chapin RE, Moseman RF, et al. Toxicol Appl Pharmacol 1991;111:145–51.
65. Robbins WA, Wei F, Elashoff DA, et al. J Androl 2008;29:115–21.
66. Vendeville A, Winzer K, Heurlier KR, et al. Nat Rev Microbiol 2005;3:383–96.
67. Barranco WT, Eckert CD. Cancer Lett 2004;216:21–9.
68. Gallardo-Williams MT, Chapin RE, King PE, et al. Toxicol Pathol 2004;32:73–8.
69. Gonzalez A, Peters U, Lampe JW, et al. Cancer Causes Control 2007;18:1131–40.
70. Korkmaz M, Uzgören E, Bakirdere S, et al. Environ Toxicol 2007;22:17–25.
71. Scorei R, Cuibar R, Ciofrangeanu CM, et al. Biol Trace Elem Res 2008;122:197–205.
72. Acerbo AS, Miller LM. Analyst 2009;134:1669–74.
73. Hakki SS, Dundar N, Kayis SA, et al. J Trace Elem Med Biol 2012;27:148–153.
74. Nielsen FH, Stoecker BJ. J Trace Elem Med Biol 2009;23:195–203.
75. Armstrong TA, Spears JW, Crenshaw TD, et al. J Nutr 2000;130:2575–81.
76. Chapin RE, Ku WW, Kenney MA, et al. Fundam Appl Toxicol 1997;35:205–15.
77. U.S. Department of Health and Human Services, Public Health Service. Toxicological profile for Boron, 2010. Accessed March 2023. www.atsdr.cdc.gov/toxprofiles/tp26.pdf
78. National Library of Medicine. PubChem. Accessed March 2023. https://pubchem.ncbi.nlm.nih.gov/compound/Boric-acid.
79. European Food Safety Authority. EFSA J 2013;11:3407. Accessed March 2023. www.efsa.europa.eu/en/efsajournal/pub/3407
80. Bendel P, Wittig A, Basilico F, et al. J Pharm Biomed Anal 2010;51:284–7.
81. Schwarz K, Mertz W. Arch Biochem Biophys 1959;85:292–5.
82. Mertz W. J Am Coll Nutr 1998;17:544–7.
83. Vincent JB. Acc Chem Res 2000;33:503–10.
84. Wang H, Kruszewski A, Brautigan DL. Biochemistry 2005;44:8167–75.
85. Vincent JB. Curr Opin Clin Nutr Metab Care 2019;22:483–9.
86. Parr PD, Taylor FG Jr. Environ Exp Botany 1980;20:157–60.
87. Welch RM, Cary EE. J Agric Food Chem 1975;23:479–482.
88. Anderson RA, Bryden NA, Polansky MM. Biol Trace Elem Res 1992;32:117–21.
89. European Food Safety Authority. EFSA Journal 2014;12:3845. Accessed March 2023. https://efsa.onlinelibrary.wiley.com/doi/epdf/10.2903/j.efsa.2014.3845
90. Harris, DC. Biochemistry 1977;16:560–4.
91. Vincent JB. Curr Opin Clin Nutr Metab Care. 2019;22:443–89.
92. Polak L. In: Chromium, 1st ed. Burrows DE, ed. Chapman and Hall/CRC Press; New York, 1983.
93. Kirpnick-Sobol Z, Reliene R, Schiestl RH. Cancer Res 2006;66:3480–4.
94. Richert DA, Westerfield WW. J Biol Chem 1953;203:915–23.
95. Cohen HJ, Fridovich I, Rajagopalan KV. J Biol Chem 1971;246:374–82.
96. Misko A, Mahani K, Abbott J, et al. Gene Reviews 2023 (Internet). Accessed March 2023. www.ncbi.nlm.nih.gov/books/NBK575630/
97. National Research Council. *Geochemistry and the Environment: Volume I: The Relation of Selected Trace Elements to Health and Disease*. National Academies Press; Washington, 1974.
98. Novotny JA, Peterson CA. Adv Nutr 2018;9:272–3.
99. Schroeder HA, Balassa JJ, Tipton IH. J Chronic Dis 1970;23:481–99.
100. Mendel RR. Biofactors 2009;35:429–434.
101. Stone LR, Erdman JA, Feder GL, et al. J Range Management 1983;36:280–5.
102. Pennington JA, Jones JW. J Am Diet Assoc 1987;87:1644–50.
103. Tsongas TA, Meglen RR, Walravens PA, et al. Am J Clin Nutr 1980;33:1103–7.
104. Sievers E. J Nutr 2003;133:236–7.
105. Tejada-Jiménez M, Galván A, Fernández E. Proc Natl Acad Sci U S A. 2011;108:6420–5.
106. Thompson KH, Turnland JH. J Nutr 1996;126:963–72.
107. Turnlund JR, Keyes WR, Peiffer GL, et al. Am J Clin Nutr 1995;61:1102–9.
108. Cardin CJ, Mason J. Biochim Biophys Acta 1976;455:937–46.

109. Mills C, Bremner I. In: *Molybdenum and molybdenum-containing enzymes*. Couglan MP, ed. Elsevier; 1980. ISBN 978-0-08-024398-6
110. Ladefoged O, Stürup S. Vet Human Toxicol 1995;37:63–5.
111. Turnlund JR, Keyes WR, Peiffer GL. Am J Clin Nutr 1995;62: 790–6.
112. Allaway WH, Kubota J, Losee F, et al. Arch Environ Health 1968;16:342–8.
113. Turnlund JR, Keyes WR. J Nutr Biochem 2004;15:90–5.
114. Spiegel R, Schwahn BC, Squires L, et al. J Inherit Metab Dis 2022;45:456–69.
115. Abumrad NN, Schneider AJ, Steel D, et al. Am J Clin Nutr 1981;34:2552–9.
116. Kiersztan A, Winiarska K, Drozak J, et al. Mol Cell Biochem 2004;261:9–21.
117. Bartha R, Ordal EJ. J Bacteriol 1965;89:1015–9.
118. Schnegg A KM. [The essentiality of nickel for animal growth]. *Zur Essentialität von Nickel für das tierische Wachstum* 1975;36(2):63–74. Accessed March 2023. https://doi.org/10.1111/j.1439-0396.1975.tb01118.x
119. Nielsen FH, Myron DR, Givand SH, et al. J Nutr 1975;105: 1620–30.
120. Anke M, Grun M, Ditrich G, et al. *In: Trace Element Metabolism in Animals-2*, Hoekstra, WC, et al., Eds., University Park Press, Baltimore, 715–7, 1974.
121. Spears JW, Hatfield EE, Forbes RM, et al. J Nutr 1978;108: 313–20.
122. Kaim W, Schwederski B. In: Inorganic Elements in the Chemistry of Life. West Sussex, UK: John Wiley, 1994:1–38.
123. Nielsen F. Adv Nutr 2021;12:281–2.
124. Stojanović D, Nikić D, Lazarević K. Cent Eur J Public Health 2004;12:187–9.
125. Solomons NW, Viteri F, Shuler TR, et al. J Nutr 1982;112:39–50.
126. Sunderman FW Jr, Hopfer SM, Sweeney KR, et al. Proc Soc Exp Biol Med 1989;191:5–11.
127. Patriarca M, Lyon TD, Fell GS. Am J Clin Nutr 1997;66:616–21.
128. Carlisle EM. Science 1972;178:619–21.
129. Schwarz K, Milne D. Nature 1972;239:333–4.
130. Sripanyakorn S, Jugdaosingh R, Thompson RPH, et al. Nutrition Bulletin 2005;30:222–30.
131. Jugdaohsingh R, Tucker KL, Qiao N, et al. J Bone Miner Res 2004;19:297–307.
132. Hildebrand M, Volcani BE, Gassmann W, et al. Nature 1997;385:688–9.
133. Coradin T, Lopez PJ. ChemBioChem 2003;4:251–9.
134. Kinrade SD, Del Nin J, Schach AS, et al. Science 1999;285: 1542–5.
135. Pennington JA. Food Addit Contam 1991;8:97–118.
136. Jugdaohsingh R, Anderson SHC, Tucker KL, et al. Am J Clin Nutr 2002;75:887–93.
137. Kelsay JL, Behal KM, Prather ES. Am J Clin Nutr 1979;32: 1876–80.
138. Haddad FS, Kouyoumdjian A. Urol Int 1986;41:70–76.
139. Prescha A, Zablocka-Slowinska K, Grajeta H. Nutrients 2019;11:980.
140. Policard A, Collet H, Daniel Moussard H, et al. J Biophysical Biochem Cytol 1961;9:236–8.
141. Holt PF, Yates DM. Biochem J 1953;54:300–5.
142. Carlisle EM. Nutr Rev 1982;40:193–8.
143. Carlisle EM. *Silicon Biochemistry: Silicon as an Essential Trace Element in Animal Nutrition*. vol:. Ciba Foundation Symposium 121. Novartis Foundation Symposia; 2007.
144. Eckhert CD. *Other Trace Elements In: Modern nutrition in health and disease*, 10 ed. Lippincott Williams & Wilkins; 2006.
145. Pruksa S, Siripinyanond A, Powell JJ, et al. Nutr Metab (Lond) 2014;11:4.
146. Milligan AJ, Morel FM. Science 2002;297:1848.
147. Farrer JH, Rajfer J. J Urol 1984;132:739–740.
148. Anke M, Illing-Günther H, Gürtler H, et al. In: Trace Elements in Man and Animals 10, Springer; 2002. ISBN: 978-0-306-47466-8
149. Fantus I, Kadota S, Deragon G, et al. Biochemistry 1989;31: 8864–71.
150. Sekar N, Li J, Shechter Y. Crit Rev Biochem Mol Biol 1996;31:339–59.
151. Liochev SI, Fridovich I. Arch Biochem Biophys 1990;279:1–7.
152. Byrne AR, Kosta L. Sci Total Environ 1978;10:17–30.
153. Myron DR, Zimmerman TJ, Shuler TR, et al. Am J Clin Nutr 1978;31:527–31.
154. Nechay BR. Annu Rev Pharmacol Toxicol 1984;24:501–24.
155. Boyd DW, Kustin K. Adv Inorg Biochem 1984;6:311–5.
156. Crans DC, Smee JJ, Gaidamauskas E, et al. Chem Rev 2004;104:849–902.
157. Halberstam M, Cohen N, Shlimovich P, et al. Diabetes 1996;45:659–66.
158. Etcheverry SB, Apelia M, Baran EJ. J Inorg Biochem 1984;20: 269–74.
159. Yamaguchi M, Oishi H, Suketa Y. Res Exp Med (Berl) 1989;189:47–53.
160. Laizé V, Tiago DM, Aureliano M, et al. Cell Mol Life Sci 2009;66:3831–6.
161. International Agency for Research on Cancer. IRAC Monographs on the Evaluation of Carcinogenic Risks to Humans Volume 86, 2006.
162. Byczkowski JZ, Kulkarni AP. US Airforce Armstrong Laboratory report. AL/OE-TR-1996-0126, 1996. Accessed March 2023. https://apps.dtic.mil/sti/pdfs/ADA360746.pdf
163. Soazo M, Garcia GB. Neurotoxicol Teratol 2007;29:503–10.
164. Environmental Protection Agency. Integrated Risk Information System (IRIS). Accessed March 2023. https://iris.epa.gov/static/pdfs/0125_summary.pdf
165. Harrington JM, Haines LG, Essader AS, et al. Anal Lett 2021;54:2777–88.

C. VITAMINS

CHAPTER 17

Vitamin A

A. Catharine Ross

CHAPTER OUTLINE

Abbreviations

CRABP, cellular retinoic acid-binding protein; **CRALBP**, cellular retinal-binding protein; **CRBP**, cellular retinol-binding protein; **DR**, direct repeat; **EAR**, estimated average requirement; **HSC**, hepatic stellate cell; **IOM**, Institute of Medicine; **IRBP**, interstitial retinoid-binding protein; **LRAT**, lecithin: retinol acyltransferase; **NHANES**, National Health and Nutrition Examination Survey; **RA**, retinoic acid; **RAE**, retinol activity equivalent; **RALDH**, retinal dehydrogenase; **RAR**, retinoic acid receptor; **RARE**, retinoic acid response element; **RBP**, retinol-binding protein; **RDA**, recommended dietary allowance; **RDR**, relative dose response; **RE**, retinyl ester; **RPE**, retinal pigment epithelium; **RXR**, retinoid X receptor; **Stra6**, stimulated by retinoic acid gene 6; **TTR**, transthyretin; **UL**, tolerable upper intake level; **WHO**, World Health Organization

Historical Overview

The properties of vitamin A may have been recognized 2 millennia ago when Egyptian and Greek physicians proposed that liver, now known as a concentrated source of vitamin A, could be a cure for the condition now known as night blindness.[1] The modern era of vitamin A research began in 1913 with the independent discoveries by Osborne and Mendel and McCollum and Davis of "fat-soluble A," later named vitamin A. These investigators observed that young rats fed diets containing lard or olive oil as the only fat grew for only a short time, then lost weight and died unless they were fed small amounts of certain "lipins," which these investigators had extracted from foods such as butter, eggs, and cod liver oil. Discoveries in the next decades linked a dietary deficiency of vitamin A to abnormalities of the eye (xerophthalmia), tissue differentiation, reproduction, and immune function. In the 1930s, vitamin A (retinol) was first synthesized de novo by Karrer in the 1930s. In the 1940s, Arens and van Dorp synthesized the carboxylic acid form of vitamin A, retinoic acid (RA), and showed that it supported all of the known functions of vitamin A, except in vision. In the 1950s and 1960s, Wald, Hubbard, and others made key discoveries on the role of vitamin A in vision, for which "retinene" (retinal) is required as the essential light-absorbing component of rhodopsin.[2] These basic discoveries paved the way for understanding that vitamin A is a precursor for the synthesis of two main active metabolites: 11-*cis*-retinal and all-*trans*-RA, now recognized as the main bioactive forms of vitamin A for vision and cell differentiation, respectively, and still today is under intensive research.

In the 1980s, several research groups elucidated a mechanism of action of RA, showing that RA binds to nuclear receptors (RARs), which combine with retinoid X receptors (RXRs)

to regulate the expression of target genes.[3,4] Numerous synthetic analogs have been developed as therapeutic agents, mainly for the prevention or treatment of skin disorders and treatment of certain leukemias. The term retinoid is now used for both the naturally occurring forms of vitamin A and its synthetic analogues. In the area of public health, vitamin A supplementation has been effective in reducing young-child mortality.[5,6] Vitamin A is essential for maintaining almost all of the body's functions—reproduction, the immune system, the eyes, skin, intestine, and other organ systems—from early development and continuing throughout the life cycle.

Terminology, Chemistry, and Analysis

The term vitamin A is used specifically for retinol and related compounds present in foods and supplements. Dietary vitamin A is subdivided into two classes: preformed vitamin A (retinol) and provitamin A, namely, β-carotene and certain other carotenoids, which serve as precursors (see chapter 31). The fundamental components of the vitamin A molecule are 1) β-ionone ring, 2) conjugated polyene side chain, and 3) terminal functional group. When the functional group is a hydroxyl group, as in all-*trans*-retinol (**Figure 17.1A**), the hydroxyl group may be unesterified or esterified with a fatty acid to form retinyl esters (RE). Long-chain fatty acid esters of retinol are the most abundant form of vitamin A within tissues and, due to stability, a common form in supplements.

A few variants of retinol are known, which have "partial" vitamin A activity, for example, α-retinol in which the double bond is between carbons 4 and 5 instead of carbons 5 and 6, is present in tropical oils and carrots and possesses about half the bioactivity of all-*trans*-β-retinol. The variant known as vitamin A_2 (chemically 3,4-didehydroretinol) is present in fresh-water fish, and has about 40% the bioactivity of all-*trans*-β-retinol. It is also produced in skin as a metabolite of vitamin A.

Retinoids undergo three major kinds of transformation: 1) esterification and hydrolysis; 2) oxidation and reduction; and 3) isomerization of the carbon-carbon double bonds (*trans* to *cis* and reverse). As a result, retinoids exist in numerous forms. In most tissues, REs are the predominant storage form, whereas retinol is the predominant transport form in plasma; 11-*cis*-retinal is the predominant form in the retina, and all-*trans*-RA is the principal metabolite, present in low concentration in numerous tissues and capable of regulating gene expression. All these forms have very low solubility in aqueous solution and, thus, are nearly always bound to proteins. Both retinol and RA can be converted by conjugation with water-soluble molecules into water-miscible molecules, such as retinoyl-β-glucuronic acid.

A few of the many synthetic retinoid analogs of vitamin A with pharmacologic activity are illustrated in **Figure 17.1B**, including Am80/580, hydroxyphenylretinamide (fenretinide), and acitretin.

Most methods of retinoid analysis require a solvent extraction to release the retinoids from proteins and other lipids in the sample, followed by chromatographic separation and a method for the detection of individual retinoid species.[7] Prior to and during processing, samples and their extracts must be protected from light to prevent isomerization and oxygen to prevent spontaneous oxidation. For quantification of total retinol in tissues where most of the retinol is esterified, saponification is commonly used to convert the RE to retinol prior to chromatography and detection. Common methods of retinoid analysis employ high-performance liquid chromatography (HPLC) with detection by ultraviolet absorption at a wavelength optimized for each compound. Increasingly, single, or tandem mass spectrometry (LC-MS-MS) is being used, especially for low-abundance acidic retinoids.[7]

Dietary Sources and Units

All vertebrates require vitamin A, but none can synthesize it de novo and; therefore, a dietary source is essential. Most vertebrates can use either preformed vitamin A (i.e., retinol and RE present

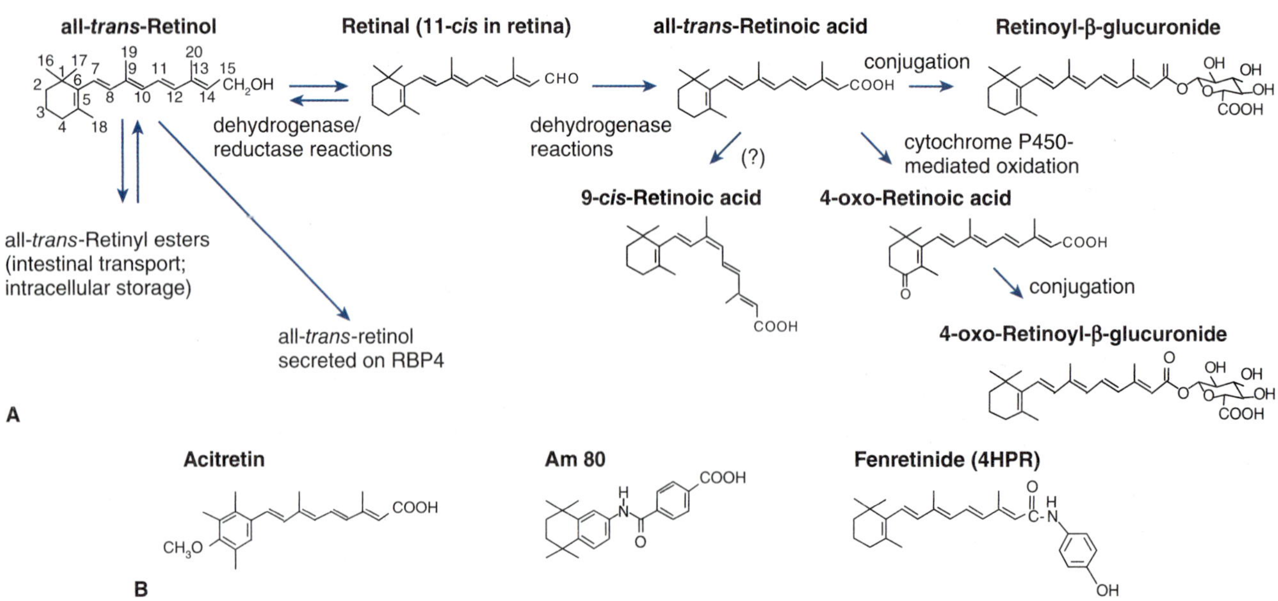

Figure 17.1 (**A**) Structures and major metabolic transformations of naturally occurring retinoids. (**B**) Examples of pharmacologic retinoids structurally related to all-*trans*-retinoic acid. *RBP*, retinol-binding protein.

in the tissues of animals that have metabolized carotenoid precursors) or provitamin A carotenoids themselves, which are synthesized exclusively by plants, fungi, and certain bacteria. Some exceptions to this generalization have been reported, for example, cats are able to absorb β-carotene but cannot convert it to retinol[8] so need preformed dietary vitamin A.

Foods with the highest concentration of preformed vitamin A are liver, fish liver oils, and organ meats, while the concentrations in milk and eggs are lower but contribute significantly to vitamin A intake. Fortified milk, margarine, and breakfast cereals are also significant sources.[9] In the United States and other higher-income countries where foods of animal origin are affordable, approximately two-thirds of the dietary vitamin A is consumed preformed in foods and nutritional supplements. In contrast, in the developing world, or for anyone who chooses a plant-based diet, most vitamin A is consumed as provitamin A carotenoids (see chapter 31).

Nutritional Units

Because vitamin A in foods present in different forms, and because the bioavailability of carotenoids and their bioconversion to vitamin A differs substantially due to food source and precise chemical form, it has been necessary for nutritionists to express dietary vitamin A in terms of equivalents. Over time, the international unit, IU, where 1 IU = 0.30 μg all-*trans*-retinol or 0.6 μg all-*trans*-β-carotene) has been largely replaced by the retinol activity equivalent, RAE,[10] which was established by the Institute of Medicine in 2001 to recognize that not all forms of provitamin A are equally bioavailable and bioactive. For RAE units, 1 μg RAE equals 1 μg of pure all-*trans*-retinol (in alcohol or ester form), 2 μg of pure all-*trans*-β-carotene in oil (a highly absorbable form), 12 μg of food-based all-*trans*-β-carotene in foods (from which absorption is less), and 24 μg of other, food-based, all-*trans*-provitamin A carotenoids. In practice, a person can obtain sufficient vitamin A from any mixture of preformed vitamin A and provitamin A carotenoids that provides an adequate dietary intake.[10]

Dietary Reference Intakes

DRI values for vitamin A are expressed in micrograms RAE/day (**Table 17.1**). The Estimated Average Requirement (EAR) was set for adults using a factorial method that takes into account the proportion of vitamin A stores that are lost per day (0.5%), an estimation of minimum acceptable liver vitamin A reserves of 20 μg/g liver,[11] and the efficiency of storage of vitamin A after ingestion. The EAR for pregnant women includes an additional amount to cover the transfer of vitamin A to the growing fetus, whereas the EAR for lactation includes an additional amount for the secretion of vitamin A in breast milk.[10] The EAR for children and adolescents was scaled down from adult values, based on children's body weight.[12] The RDAs for vitamin A were calculated from EARs by

Table 17.1 Dietary Reference Intake Values for Vitamin A by Life Stage Group

Life Stage Group	RDA (μg RAE/d)[a]		AI[b]	
	MALE	FEMALE	(μG/D)	UL[c] (μg preformed retinol/D)
0–6 mo			400	600
7–12 mo			500	600
1–3 y	300	300		600
4–8 y	400	400		900
9–13 y	600	600		1,700
14–18 y	900	700		2,800
>18 y	900	700		2,800/3,000[d]
Pregnancy				
14–18 y	750			2,800
19–50 y	770			3,000
Lactation				
14–18 y	1,200			2,800
19–50 y	1,300			3,000

[a]RDA, recommended dietary allowance, is the intake that meets the nutrient need of almost all (97% to 98%) individuals in a group. RAE, retinol activity equivalent.

[b]AI, adequate intake, is the observed average or experimentally determined intake by a defined population or subgroup that appears to sustain a defined nutritional status, such as growth rate, normal circulating nutrient values, or other functional indicators of health. The AI is used if sufficient scientific evidence is not available to derive an estimated average requirement (EAR). For healthy infants receiving human milk, the AI is the mean intake. The AI is not equivalent to an RDA.

[c]UL, tolerable upper intake limit, is the highest level of daily nutrient intake that is likely to pose no risk of adverse health effects for almost all individuals in the specified life stage group.

[d]UL equals 2,800 μg/day for women and 3,000 μg/day for men more than 18 years old.

Data from Food and Nutrition Board, Institute of Medicine. Dietary Reference Intakes for Vitamin A, Vitamin K, Arsenic, Boron, Chromium, Copper, Iodine, Iron, Manganese, Molybdenum, Nickel, Silicon, Vanadium, and Zinc. Washington, DC: National Academy Press, 2001.

adding a coefficient of variation of 20%. The DRI values for infants 0 to 6 and 7 to 12 months of age are expressed as an Adequate Intake (AI), which is calculated by multiplying the average concentration of vitamin A in breast milk by an average value for milk volume for each age group. Upper Tolerable Intake Levels (UL) apply only to preformed vitamin A, as only this form is associated with risk of vitamin A toxicity.[9,10] The critical indicators used by the IOM to specify the UL were birth defects in women of reproductive age and liver abnormalities for all other age-sex groups. It is worth noting that the difference between meeting the RDA and reaching the UL is quite narrow for vitamin A, especially for the younger age groups, e.g., an RDA of 300 μg RAE (from any source) and a UL of 600 μg (preformed retinol only) in children 1–3 y of age.

Reference intakes for vitamin A have also been established by other nations and by the Food and Agriculture Organization/World Health Organization (FAO/WHO); these values tend to be lower than the RDAs.[13] For adult men and women, respectively, the Scientific Advisory Committee on Nutrition of the United Kingdom[14] recommends 700 and 600 μg retinol/d and the EFSA Panel on Dietetic Products, Nutrition, and Allergies[15] recommends 570 and 490 μg/d.

Biochemistry and Metabolism

Vitamin A metabolism is shown biochemically in **Figure 17.2**. However, it should be appreciated that not all of the biochemical processes shown will take place in every organ. To a significant extent, the tissue-specific metabolism and cellular content of retinoids is specified by the specific retinoid-binding proteins present within various tissues, as well as the types of enzymes each cell type expresses.

Three categories of chaperone proteins facilitate vitamin A metabolism: 1) proteins that escort retinoids during circulation in plasma; 2) protein chaperones within cells; and 3) nuclear proteins that act as ligand-activated receptors and signaling proteins. The retinoid-binding proteins as well as nuclear retinoid receptors are critical for the normal metabolism of vitamin A, because they not only confer aqueous solubility to lipophilic molecules but also serve as chaperones during retinoid transport, metabolism, and signaling.[16]

The major plasma transport proteins are Retinol-binding Protein, RBP (which is sometimes referred to by its gene designation *RBP4*), and transthyretin, TTR. Each molecule of RBP binds one molecule of retinol in its hydrophobic binding pocket. The terms holo-RBP and apo-RBP apply to the RBP-retinol complex and to the empty protein, respectively.[17] Because RBP has a relatively low molecular weight (21 kDa) for a plasma protein, it can be easily lost in urine. This is largely prevented by the association of apoRBP with another plasma protein, TTR (previously known as prealbumin), forming a larger complex of about 75 kDa. That is not only less susceptible to oxidation but also reduces the rate of urinary loss. RBP can bind certain retinol analogs such as 4-HPR, but the complex interacts weakly with TTR compared with the complex with retinol bound, and, as a result, more of the circulating RBP is lost in urine.[18] α-Retinol, apparently due to its more planar overall structure compared with retinol with a β-ionone ring, does not bind effectively to RBP.[19] RBP is present in plasma at relatively low concentration of approximately 1 to 3 μM. It has a relatively short half-life, approximately 0.5 days for holo-RBP and 4 hours for apo-RBP.[20]

The *RBP4* gene is most highly expressed in hepatocytes, but also in adipose tissue and kidney, at approximately 3% to 10% of the level in the liver.[17] The expression of RBP4 in extrahepatic tissues suggests that these organs may also synthesize RBP protein and, in fact, adipose-derived RBP has been proposed to function as an adipokine in glucose homeostasis.[21] The liver is also the main, but not only, site of TTR synthesis.[22] As the molar concentration of TTR in plasma is higher than that of RBP, most TTR circulates as the tetramer without RBP. The TTR gene is highly polymorphic and some of the variants affect either the binding of thyroxine to TTR or the association of TTR with RBP; however, most variants are associated with familial amyloidotic polyneuropathies.[22,23]

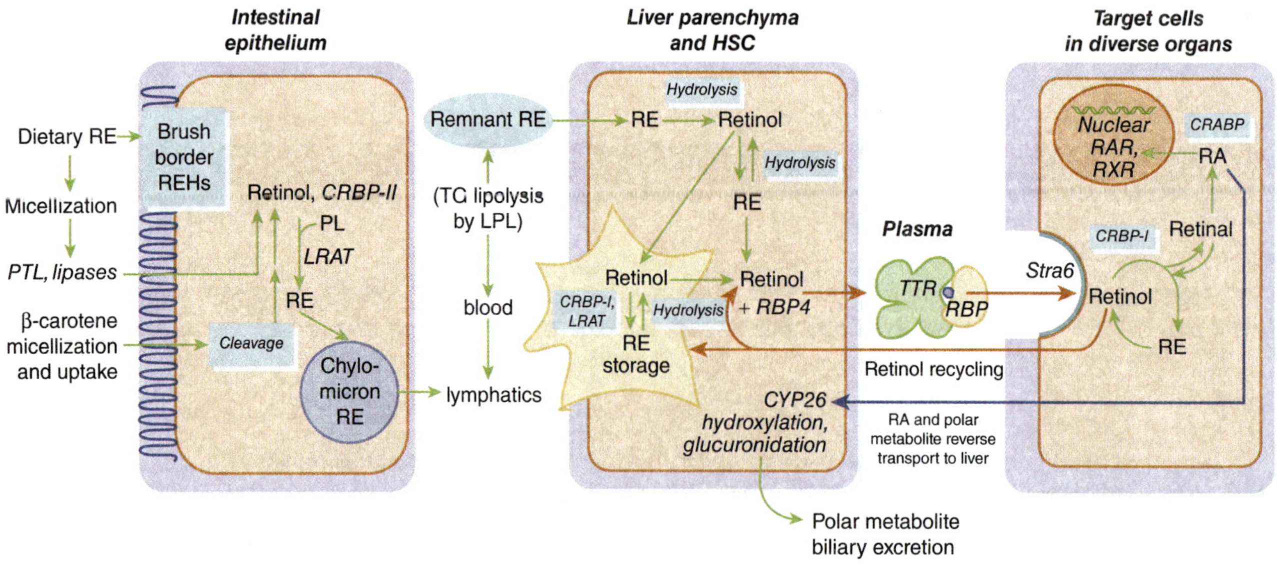

Figure 17.2 Major reactions in retinoid metabolism in the intestine, liver, and diverse target organs where RA is produced and utilized. *CRABP*, cellular retinoic acid–binding protein; *CRBP*, cellular retinol-binding protein; *HSC*, hepatic stellate cell; *LRAT*, lecithin: retinol acyltransferase; *PL*, phospholipid; *RA*, retinoic acid; *RAR*, retinoic acid receptor; *RBP*, retinol-binding protein; *RE*, retinyl ester; *RXR*, retinoid X receptor; *Stra6*, stimulated by RA gene 6; *TTR*, transthyretin.

Cellular retinoid-binding proteins are the major carriers of retinoids within cells; they are present in several forms in a variety of tissues. These intracellular proteins function as chaperones, mainly for retinol, retinal, or RA.[16,24] The cellular RBP family (CRBPs) -I through -IV belong to the larger gene family of fatty acid-binding protein (FABP)/CRBP, all of which have similar molecular weights of approximately 14.6 kDa and an overall structure described as a β-clam, containing a hydrophobic pocket that binds one molecule of the protein's retinoid cargo. Structural modeling of CRBP-I showed that retinol is oriented with its hydroxyl group inward. CRBP-I prefers to bind all-*trans*-retinol and is the most abundant form of CRBP in liver, kidney, testes, and other tissues, whereas CRBP-II, which binds either all-*trans*-retinol or retinal, is abundant in enterocytes. Recently, CRBP-II has been shown to also be a binding protein for 2-monoacylglycerols, an interesting result linking vitamin A and fat absorption.[25] CRBP-III and CRBP-IV are present in heart, skeletal muscle, kidney, and some other tissues.[26]

The subgroup of the FABP/CRBP family that binds all-*trans*-RA is known as CRABP-I and CRABP-II, each of which is expressed in its own tissue-specific patterns and generally at lower levels than the CRBPs. The two CRABPs also have different patterns of tissue expression, suggesting they may perform similar but yet distinct functions.[27]

Two other cellular retinoid-binding proteins are expressed almost exclusively in the eye (see the later section on ocular retinoid metabolism). They are cellular retinal-binding protein (CRALBP) and interstitial retinoid-binding protein (IRBP).

The nuclear retinoid receptors can also be considered a specialized class of retinoid-binding proteins with roles in the nucleus. The nuclear retinoid receptors are divided into two subfamilies, the RAR and RXR, each of which comprises three genes, α, β, and γ. There is strong structural similarity among them, especially within the ligand-binding domain and DNA-binding domain of each subgroup.[4] Although they are ligand-activated receptors, sometimes they function in a ligand-independent manner. Whereas the RARs tightly bind all-*trans*-RA (or its analogs) nearly exclusively, the RXRs binds all-*trans* and, at least in vitro, 9-*cis* retinoids as well as synthetic retinoids known as rexinoids.[28] For their function, the RXR and ligand-activated RAR proteins bind to each other as heterodimeric pairs, which then bind to specific DNA sequences known as retinoid-response elements (RARE) in retinoid-responsive genes, as described later. The binding of ligand to the ligand-binding domain of the receptor induces a conformational change in the receptor complex that influences its interaction with other nuclear proteins, including coactivator or corepressor proteins that, in turn, determine whether the target gene is induced or repressed, and with enzymes that modify histones, basal transcription factors, and the RNA polymerases responsible for gene transcription.[3,4,29] Nuclear receptor signaling is further regulated in a very complex manner by the rate of gene transcription, posttranscriptional protein modification, protein trafficking, and proteolysis.[30,31] The RXRs not only heterodimerize with RARs but also with the vitamin D receptor, peroxisome proliferator-activated receptor (PPAR), farnesoid X receptor (FXR), liver X receptor (LXR), and receptors for certain drugs and xenobiotics. Thus, they participate in a variety of regulatory networks.[31]

The canonical DNA sequence that makes an RARE consists of a direct repeat (DR) of the hexanucleotide sequence (A/G)/(G/T) GTCA with an intervening sequence of either five or two nucleotides, referred to as DR-5 or DR-2 elements. Most RAREs are in the 5′-regulatory region upstream of retinoid-responsive genes, although some lie in introns or outside of genes. RAREs are found in genes involved in many biological processes,[32] including the metabolism of vitamin A itself, for example, CRABP-II, RAR-β, and CYP26A1 each possess one or more RAREs (discussed in the section on metabolism), and are, therefore, able to sense and respond to RA in the tissues that express them.[33,34] Although some 500 genes have been reported to be physiologic targets of regulation by RA,[35] many of them appear to be regulated indirectly as no classical RARE has yet been identified within their regulatory sequence.

Characteristic Metabolic Processes

The processes that characterize vitamin A metabolism at the whole-body level include extensive interorgan recycling of retinol and extensive esterification of retinol and hydrolysis of RE, before the retinoid molecule is irreversibly oxidized and lost from the body. Kinetic modeling studies have shown that the average retinol molecule undergoes numerous cycles of uptake by various tissues, intracellular esterification of retinol and RE storage and then, as needed, hydrolysis of RE to replenish retinol, which then returns to the plasma pool from which it is taken up again; this recycling occurs from 2 to 3 times in humans to >100 times in animal models, before retinol is irreversibly degraded.[36,37] Another characteristic feature of retinol metabolism is its well-regulated oxidation to form retinal and RA. Retinoid metabolism and homeostasis are controlled to a significant extent by RA itself, through the autoregulation by RA of a number of genes that regulate and carry out the metabolism of vitamin A.[34]

Retinol Absorption and Processing in the Intestine

Efficient Isotope kinetic studies have shown that approximately 70% to 90% of a physiologic dose of vitamin A is absorbed.[38] The process of retinol absorption is relatively unregulated, and fractional absorption is high, even when the dose is very large,[10] a situation that may contribute to the development of hypervitaminosis A and toxicity (see later section). Despite wide variations in usual dietary intake, the concentration of retinol in plasma is well regulated to be nearly constant. Plasma retinol falls in states of vitamin A deficiency and RE concentrations rise when intake is significantly excessive over long periods. The steady level of holo-RBP in plasma has been referred to as "homeostatic regulation" (see later) which must ultimately be a function of the rates of release of holo-RBP from liver, retinol uptake by tissues, retinol recovery from the kidneys after filtration, and the irreversible disposal of retinol from the body.[11]

The vitamin A absorptive process closely resembles that of triacylglycerols. Like fats, the REs in foods must be released from the food matrix by acids and digestive enzymes and then emulsified with fatty acids and bile salts for incorporation into mixed lipid micelles. Pancreatic Triglyceride Lipase (PTL), a colipase-dependent hydrolase, hydrolyzes RE[39] while, additionally, microvillus-associated RE hydrolases have been described. It can be expected that conditions that interfere with digestion and lipid emulsification, including low dietary fat (less than approximately 5%), can reduce the efficiency of vitamin A absorption.[10]

After the uptake of unesterified retinol by duodenal and jejunal enterocytes, retinol is bound to CRBP-II and then esterified by the membrane-bound enzyme lecithin: retinol acyltransferase

(LRAT), which transfers the sn-1 fatty acid of membrane-associated phosphatidyl choline (lecithin) to retinol, forming RE. Therefore, the composition of sn-1 fatty acids in lecithin dictates that LRAT determines the mixture of RE species, with retinyl palmitate usually the predominant species.[40] Approximately 90% to 95% of retinol undergoes esterification. That LRAT is an obligatory enzyme that was shown by studies in mice that lack LRAT and, therefore, were unable to accumulate RE in their tissues.[41] The newly formed REs, along with newly formed triglycerides and cholesteryl esters, are packaged into the lipid core of the nascent chylomicron. Because amounts of dietary vitamin A differ substantially with dietary patterns and even with specific meals, the quantity of RE carried per chylomicron particle also varies widely and in direct proportion to the amount of vitamin A being absorbed at the time, from none after a meal free of vitamin A up to several milligrams RE per gram of lipid after ingestion of a very high–vitamin A meal or vitamin A supplement.[42]

The trafficking of chylomicron bound RE is mainly determined by the metabolism of the chylomicron itself. Chylomicrons enter the lymphatic system and then the venous circulation, reaching a peak plasma concentration approximately 2 to 6 hours after a meal. The triglyceride component of the chylomicrons is rapidly hydrolyzed by lipoprotein lipase, while the RE component mostly stays with the chylomicron remnant particle. These remnants are rapidly taken up by tissues, mostly by hepatocytes. Due to the rapid clearance of chylomicrons, the half-life of dietary REs in plasma is short, less than 20 minutes in normal individuals,[43] and limited to the absorptive phase. In contrast, fasting plasma contains very little RE, <5% of the total plasma retinol. If, however, chylomicron clearance is impaired, or the intake of vitamin A and rate of absorption of RE is extremely high, REs can be found in plasma even in the fasting state. Whereas the liver clears some 60% to 80% of chylomicron RE, smaller fractions are taken up into other tissues through lipoprotein receptors, during lipolysis by lipoprotein lipase, or are exchanged with other plasma lipoproteins.[26] Additionally, a small fraction of newly absorbed vitamin A is oxidized to polar retinoids within the intestine during the absorptive process, such that some RA is produced locally. RA is absorbed bound to albumin via the portal circulation. The concentration of portal blood RA was shown to increase after ingestion of β-carotene[44] and this most likely also occurs after ingestion of preformed vitamin A.

Polar retinoids such as hydroxylated retinoids and their glucuronide adducts are formed in the liver and possibly in other tissues, and present at low concentrations in plasma, from which they may be taken up into tissues. Little is known of their bioactivities. They are secreted from the liver into bile and thus enter the intestine, either for reabsorption or excretion in feces.

Retinoid Processing and Metabolism in the Liver

The liver plays a central role in whole-body retinoid homeostasis. As noted above, chylomicron remnant REs are taken up mostly by hepatocytes. They are rapidly hydrolyzed to retinol. This process is not sensitive to vitamin A status; however, the next steps in hepatic metabolism do depend on the vitamin A status of the individual.[33] In a study that traced the metabolism of chylomicron ^{3}H-RE in vitamin A–adequate rats, most of the ^{3}H-RE that was initially taken up by hepatocytes was transferred within 2 hours into hepatic stellate cells (HSCs),[45] where it reappeared as ^{3}H-RE. HSCs are known to contain CRBP-I and LRAT and, thus, are able to synthesize RE. These newly synthesized REs are stored within large cytoplasmic lipid droplets that characterize the HSC.[46] In the well-nourished state, REs in HSCs constitute approximately 50% to 85% of total body vitamin A.[47] In contrast, tracing the metabolism of chylomicron ^{3}H-RE in vitamin A–deficient rats showed that very little newly absorbed ^{3}H-RE was transferred to HSCs.[45] Instead, ^{3}H-retinol rapidly appeared in the plasma compartment, suggesting the retinol produced by the rapid hydrolysis of RE was prioritized for forming holo-RBP for release into plasma. Other studies showed that, in the situation of vitamin A deficiency, LRAT expression and activity decline dramatically,[48] which is likely part of a sparing mechanism that directs what little retinol remains into other uses, such as for secretion of holo-RBP or the conversion of retinol to RA to maintain signaling. Also, as vitamin A deficiency develops, the proportion of CRBP that is present as apo-CRBP increases, and apo-CRBP has been shown to stimulate the hydrolysis of RE by retinol ester hydrolases.[49] The result of the inhibition of retinol esterification and the stimulation of RE hydrolysis is that essentially all the vitamin A in the liver can be mobilized. Conversely, when a bolus dose of vitamin A was administered to vitamin A-deficient rats, holo-RBP secretion increased very rapidly,[50] and hepatic LRAT expression was induced[48], thus restoring the normal balance of plasma retinol secretion and RE storage within a few hours. Although most studies have been conducted in mice and rats, human retinol metabolism is likely to be similar, based on the presence of homologous genes, a similar range of vitamin A concentrations in human liver specimens,[51] and a similar rapid rise in plasma retinol soon after humans with vitamin A deficiency are given a supplement of vitamin A.[33]

In hepatocytes, RBP is synthesized as a 24-kDa preprotein that is cleaved during translation to form the mature 21-kDa protein,[17] which then moves through the secretory pathway, where it combines with retinol, resulting in secretion of holo-RBP.[50] However, in vitamin A deficiency, RBP mRNA remains relatively constant and is, apparently, transcribed because RBP protein accumulates within the hepatocytes as apo-RBP. Once retinol is available, holo-RBP is formed and rapidly released. In a study in vitamin A-deficient rats, plasma retinol and RBP rose from nearly zero to higher than normal within approximately 5 hours after the animals were supplemented with vitamin A, then stabilized at a normal concentration.[45,51] This finding of a rapid release of holo-RBP after vitamin A treatment if the individuals are deficient has provided the rationale for a useful clinical test, the relative dose response (RDR) test,[52,53] described later, for detecting whether hepatic vitamin A stores are sufficient (homeostatic regulation and no response) or inadequate (a rapid increase in holo-RBP).

Retinoids in Plasma

In the fasting state, plasma vitamin A is mainly (>95%) in the form of unesterified retinol. Plasma REs will have been elevated transiently after vitamin A–containing meals as noted earlier, but rapidly cleared. If RE constitutes 5% to 10% of the total retinol in fasting plasma, this suggests an abnormally slow clearance of chylomicrons, or it may suggest an excessive intake of dietary vitamin A (hypervitaminosis A; see later). In mammals other than humans, there is considerable variation in plasma vitamin A transport, with most rodents (those used in laboratory research) resembling humans, but in the great apes and several species of carnivores, including domestic dogs, vitamin A is mostly in the form of lipoprotein-bound RE.[54,55]

Human plasma retinol concentration normally ranges from approximately 1 to 3 μmol/L (equivalent to 28 to 86 μg/dL) in adults, with low day-to-day variation. The molar ratio of plasma retinol to RBP is approximately 0.82, indicating that a small fraction of RBP is normally present as apo-RBP. In the National Health and Nutrition Examination Survey (NHANES), plasma retinol has been shown to be age related, being lower in young children than in adolescents, higher in adult men than in premenopausal women,[56] and 15% to 35% higher in women <50 years using oral contraceptives.[57] Values in men and women over 50 years were similar. In studies of newborns, premature infants had lower concentrations than full-term infants.[58]

For interpretation of plasma retinol adequacy, the cut-off values commonly used are: <0.35 μmol/L for severe deficiency, 0.35–0.70 μmol/L for marginal deficiency, 0.71–1.05 μmol/L for low but subclinical vitamin A status, and >1.05 μmol/L for adequacy.[51] In NHANES from 1988 to 1994, serum retinol lower than 0.70 μmol/L was infrequent in all strata of the U.S. population.[57] However, the prevalence of serum retinol lower than 1.05 μmol/L was higher in non-Hispanic Black and Mexican American children than in non-Hispanic White children, even after controlling for covariates. The WHO and other organizations use plasma retinol as one criterion for deciding whether vitamin A is a public health problem in the region or country of interest.[59]

The relatively short half-life of plasma RBP and TTR, approximately 0.5 day and 2 to 3 days, respectively, necessitates a high rate of protein synthesis for replacement and, therefore, they are sensitive to deficiencies in protein and energy.[17] RBP and TTR are also reduced in states of infection and inflammation,[60] and trauma.[59] As the synthesis of RBP and the concentration of plasma holo-RBP generally respond rapidly during the recovery phase and with nutritional support, the monitoring of RBP is a useful indicator of visceral protein status and the response to nutritional support.[61,62]

Low plasma retinol and RBP are found in diseases of the liver, kidney, and thyroid gland.[63,64] Several aspects of retinol storage and transport are perturbed by inflammation, including a loss of RE from HSCs, which develop a myofibroblastic phenotype,[65] and reduced synthesis of both RBP and TTR in the liver.[66] Patients with inflammation had low plasma RBP, retinol, all-*trans*- and 13-*cis*-RA, all negatively correlated with plasma C-reactive protein, CRP.[67]

Genetic deficiencies of RBP are rare. Mutations affecting plasma RBP concentration were first identified in a detailed study, published in 1999, of two teenage sisters in Germany who presented at clinic with impaired vision (night blindness).[68] Two mutated alleles of the RBP gene were found to be responsible for two different, single amino-acid substitutions in the RBP protein. Despite the siblings having plasma retinol and RBP concentrations of 0.19 μmol/L and <0.60 μmol/L, respectively, their growth and development appeared normal. Their plasma RA was also normal, which may not be surprising given that only a small proportion of plasma retinol is actually used for RA production. It appeared that their dietary intake of vitamin A (e.g., chylomicron RE) was adequate to maintain the health of most of their tissues, excepting the eyes. A similar situation was observed in *RBP4*-null mice, in which low plasma retinol and abnormal electroretinograms were found, and which improved after several months on a vitamin A–adequate diet.[69] These findings suggest that the eyes are relatively more sensitive to a deficiency of circulating holo-RBP than are most other tissues. Mutations in the TTR gene have been identified that affect TTR binding to holo-RBP and that are also associated with low plasma RBP and loss of RBP in urine.[70]

The relationship of plasma vitamin A concentration to liver vitamin A stores is an important issue that has been addressed in many laboratories over many decades. In the mid-1980s, Olson[11] compiled data from which he proposed that plasma retinol is held at a nearly constant level across a wide range of liver vitamin A concentrations. Plasma retinol began to fall only when the liver was nearly depleted of vitamin A, around a concentration of 20 μg/g liver. This value is often cited as a cutoff point for inadequate liver vitamin A reserves. When liver vitamin A was elevated above approximately 300 to 500 μg retinol/g, plasma vitamin A began to rise, nearly entirely due to the presence of REs in lipoproteins. His analysis showed that, overall, plasma retinol is remarkably steady and said to be "homeostatically regulated" across a wide range of normal liver vitamin A concentrations.

The uptake of vitamin A from plasma into cells is now understood to be mediated by a plasma membrane channel protein called Stra6 (Stimulated by RA gene 6), which encodes a receptor for RBP.[71,72] Stra6 is abundant on the surface of retinal pigment epithelium (RPE) cells (see section on uptake of vitamin A from plasma into cells), providing a means for uptake of plasma retinol into the retina, where is it metabolized for production of visual pigments. The uptake of retinol through the Stra6 channel is aided by the presence of CRBP-I, which can immediately take up the retinol and which, in turn, couples to the esterification of retinol by LRAT, forming a RE storage pool in the RPE.[71,73] Other tissues express Stra6 at lower concentration. In Matthew-Wood syndrome, a genetic condition due to mutations in Stra6, vitamin A homeostasis is significantly disrupted.[73]

Many cell types can store RE, usually at concentrations well below those in the liver and RPE. Extrahepatic stellate cells analogous to the HSCs are present in lung, intestine, kidney, pancreas, and probably many other tissues. The local stores of RE are believed to be drawn upon to produce RA locally, or to release retinol into plasma for recycling and uptake by other tissues.[74]

Retinol Metabolism in the Kidney

Apo-RBP, as well as is any holo-RBP not bound to TTR, is readily filtered in the glomerulus. Megalin (gp330), a multiligand receptor on the apical surface of proximal tubule cells, can bind both TTR and RBP, as well as several other nutrient-related transport proteins. Megalin aids in the endocytic reuptake of RBP and retinol.[75] Some portion of the RBP that is taken up is degraded intracellularly, but another fraction undergoes transcytosis across the renal tubule cells, thereby recycling retinol back to plasma. In experiments conducted in mice null for the megalin gene, RBP uptake into the renal proximal tubules was defective, and retinol and RBP were lost in urine.[76] In humans, a rare genetic condition causing megalin deficiency affects multiorgan systems, including the eye.[77]

The clearance of RBP in the kidneys was estimated to be equivalent to approximately 7/8 L of plasma per day in a 70-kg human.[78] Kidney diseases that affect glomerular filtration rate are typically associated with higher plasma concentrations of retinol and RBP,[63] as well as TTR.[79] It is thought that apo-RBP, which normally is rapidly filtered from plasma in the glomerulus because of its poor binding to TTR, may provide a signal to the liver to stimulate the output of holo-RBP and, thus, to maintain plasma retinol homeostasis.[80] Retinol is lost from the kidneys in conditions of diarrhea and severe infections with proteinuria[81-83] in the proximal tubules.[75,84]

Another mechanism for the uptake of vitamin A into cells is receptor-mediated uptake of lipoproteins which, as noted earlier, can acquire REs from chylomicrons through exchange. Uptake of RE via receptors for low-density lipoproteins (LDL or scavenger receptors) has been demonstrated in cultured cells.[85]

The recycling of retinol is believed to help regulate plasma retinol concentration at a relatively steady state while, at the same time, maintaining adequate tissue concentrations through the redistribution of vitamin A. In studies with rats, using computer-based compartmental modeling to trace the retinol molecule, each molecule of retinol recycled an average of 9 to 11 times between plasma and the liver, kidneys, and other tissues before it was irreversibly degraded.[38] A similar compartmental analysis of data from a healthy young man who had consumed 105 μmol of ^{13}C-labeled retinyl palmitate showed that 50 μmol of retinol passed through plasma each day, whereas only 4 μmol/day was degraded,[37] implying that multiple cycles of each retinol molecule took place before it met with degradation. In contrast to retinol, RBP does not appear to be recycled, which suggests that new RBP must be synthesized in various tissues for retinol recycling. As noted above, RBP mRNA is expressed in several extrahepatic tissues. Conditions that alter the rate of retinol recycling include retinoid treatment, inflammation, and exposure to hepatotoxic agents.[86] Age may also be a factor as, in a study of normally nourished neonatal rats, retinol recycled more than 100 times between liver, kidney, lungs, and intestine before undergoing irreversible degradation.[87]

Besides retinol itself, numerous other retinoids are present in plasma. Typically, their concentrations are much lower than for retinol. RA and related metabolites are typically in the low nanomolar range. In a study where low concentrations of 13-*cis*-RA and 13-*cis*-4-oxo-RA were measured in plasma of healthy human adults, these concentrations increased two- to fourfold after ingestion of high intakes of retinyl palmitate,[88] or of liver high in vitamin A.[89] Plasma and bile both contain small amounts of more polar oxidation products and conjugated retinoids, which are partially or even fully aqueous soluble.

The concentration of RA in tissues is usually higher than its concentration in plasma, perhaps because cytoplasmic CRABPs bind and sequester RA intracellularly. Plasma uptake vs local production was shown to differ among tissues in a kinetic study in rats, where approximately 80% of RA in the liver was derived by uptake from plasma, whereas most of the RA present in the testis was produced locally.[90]

As RA is the main functional form of vitamin A in most cells, the processes that maintain the concentration and distribution of RA are crucial to the proper functioning of vitamin A. Intracellular RA concentration is regulated by both biosynthesis and oxidation. The first step in RA production, the oxidation of retinol to retinal, is reversible. Several enzymes may participate, including members of the families of alcohol dehydrogenase, ADH, retinol dehydrogenases named RDHs, and members of the short-chain dehydrogenases/reductase genes named SDRs or DHRSs.[91-96] Some of these enzymes have relatively broad substrate specificities, including for certain steroids as well as retinoids. Some RDHs preferentially oxidize all-*trans*-retinol and others the *cis* isomers of retinol. Additionally, some members of the alcohol dehydrogenase family also oxidize retinol to retinal. The oxidation of retinol to retinal is generally considered rate limiting for RA production, and the concentration of retinal within most tissues is low, for example, <2% as compared with unesterified retinol in a study of mouse liver.[94]

The second step in RA formation, the oxidation of retinal to RA, is irreversible. Again, multiple enzymes have been implicated, including members of the retinal dehydrogenase (RALDH) family and the cytochrome P-450 gene family. The gene most firmly established as being critical for RA production is *RALDH2*, because deletion of *RALDH2* was lethal to developing mouse embryos while administering RA to the dam rescued the normal phenotype.[97]

Several *cis* isomers of RA, such as 9-cis and 13-cis-RA, have been studied extensively, and used clinically, but their actions are still not well understood.[98] 13-*cis*-RA is a natural metabolite of vitamin A in human plasma. In some assays, it possesses bioactivity similar to that of all-*trans*-RA, but it is not a ligand for the RARs; it may function as a "prodrug" that undergoes slow isomerization to all-*trans*-RA or be formed enzymatically from retinol.[99] The significance of 9-*cis*-RA has been debated because, although its functioning in vitro as ligand for the RXRs has been documented, its physiologic presence in vivo is uncertain.[98] Because CRABP-I and CRABP-II preferentially bind the all-*trans*-RA isomer of RA, these proteins may stabilize this isomer within cells. CRABP-I was shown to facilitate, likely as an escort protein, the oxidation of all-*trans*-RA to more polar products,[16,91] and CRABP-II to shuttle all-*trans*-RA to the nucleus, where it interacted with RAR-α.[100]

The oxidation processes that metabolize RA mainly use enzymes of the microsomal cytochrome P-450 family of which the CYP26 family is best characterized. CYP26A1, which catalyzes the oxidative hydroxylation of RA, is of particular interest because of its strong inducibility by RA.[101] The CYP26 family may contribute importantly to sensing the build-up of RA and metabolizing it within tissues, thus serving a function that is detoxifying and regulatory. CYP26A1 is the form most abundant in liver, intestine, and reproductive organs.[101] Its promoter region contains multiple RAREs, which cooperate to induce a rapid and strong response.[102,103] CYP26B1 is present in several tissues, is also readily induced by RA, and may be important in patterning the developing brain.[104] CYP26C1 is also expressed in embryonic brain; it seems to be the only CYP26 isoform that can oxidize both *cis* and *trans* RA.[105] Products of RA oxidation may undergo conjugation such as glucuronidation (see Figures 17.1 and 17.2), resulting in water-soluble metabolites, which are readily excreted.

Functions

Retinoids in the Eye

Vitamin A plays two distinct roles in vision: in the retina, 11-*cis*-retinal is the visual pigment that absorbs light,[106,107] while in the cornea, similar to most other tissues, RA is required for the normal differentiation and integrity of cells, such as the corneal conjunctival membranes.

The retina's rod photoreceptor cells, which are specialized for vision in dim light and sensing motion, contain abundant membrane disks packed with the protein rhodopsin. Each rhodopsin molecule binds one molecule of 11-*cis*-retinal through a specific Schiff base. As **Figure 17.3** depicts, the absorption of a photon of light by rhodopsin-bound 11-*cis*-retinal triggers its photoisomerization as well as conformational changes in the rhodopsin protein that initiate the process of signal transduction. Signals that are generated simultaneously by multiple rod cells are integrated by nearby ganglion cells and then transmitted via the optic nerve to the brain's visual cortex.

The photoisomerization of 11-*cis*-retinal (also known as photobleaching) occurs in fractions of a second and results in intermediates that then form all-*trans*-retinal, which is released from the opsin molecule.

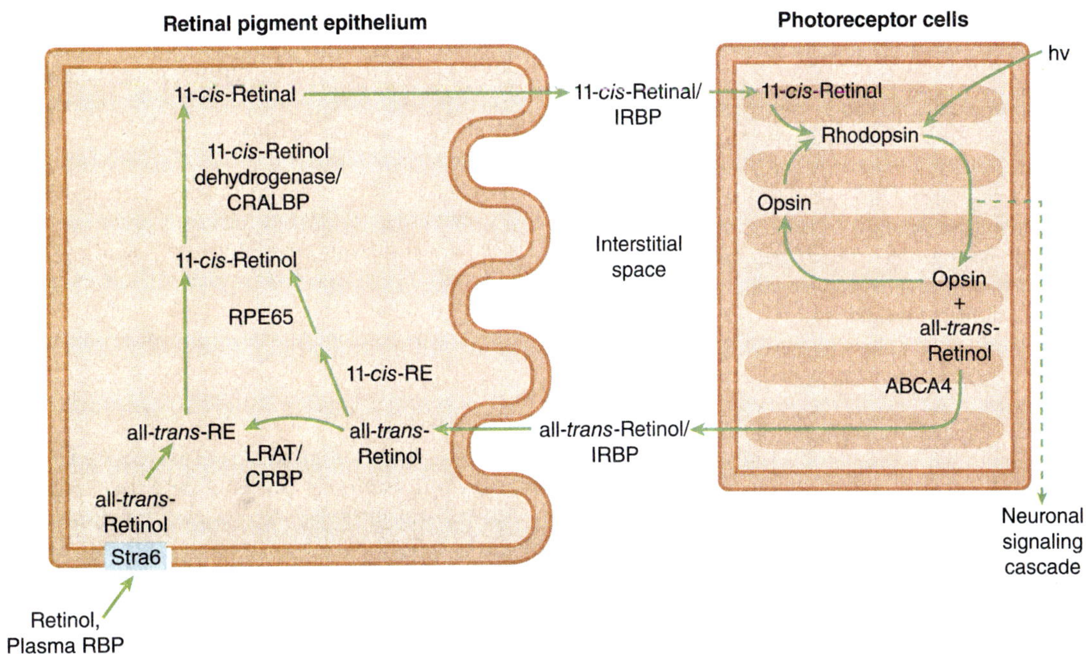

Figure 17.3 Retinoid metabolism in vision. *CRALBP*, cellular retinal-binding protein; *CRBP*, cellular retinol-binding protein; *IRBP*, interstitial retinoid-binding protein; *LRAT*, lecithin: retinol acyltransferase; *RBP*, retinol-binding protein; *RE*, retinyl ester; *RPE*, retinal pigment epithelium; *Stra6*, stimulated by retinoic acid.

For vision to continue without loss of acuity after photobleaching, 11-*cis*-retinal must be regenerated. This takes place in a series of reactions known as the visual photocycle.[106,108] Most of the enzymatic reactions that form 11-*cis*-retinal take place in the RPE, separated from the photoreceptor cells by the interphotoreceptor space. In brief, all-*trans*-retinal is first reduced enzymatically to all-*trans*-retinol in the photoreceptor cell outer segments. The all-trans-retinol is then moved out of the disk membrane by a transporter of the ABCA family, ABCA4, and subsequently shuttled across the interphotoreceptor space by the protein IRBP, returning retinol to the RPE.[109]

In the RPE, most of the retinol is esterified by LRAT and a palmitoylated, membrane-bound form of the protein RPE65 that form a local pool of RE.[107] Plasma retinol also enters this pool after uptake by Stra6. The cycle continues when RE is hydrolyzed and isomerized by the RPE-specific protein RPE65 to regenerate 11-*cis*-retinol, followed by oxidation resulting in the regeneration of 11-*cis*-retinal.[108] The latter reaction is facilitated by CRALBP in the RPE cells.[110] After 11-*cis*-retinal is transported back to the photoreceptor cells, it can react again with opsin to regenerate rhodopsin.[108,111] When the contents of the RE pool in the RPE becomes depleted, as occurs in vitamin A deficiency, the regeneration of 11-*cis*-retinal and the reformation of rhodopsin is slowed, resulting in a loss of vision, especially in dim light, known as night blindness (poor dark adaptation after exposure to bright light), which is further discussed later.

A similar cycle occurs in the color-sensing cone cells, which are concentrated in the fovea. Each cone cell expresses either a red-, green-, or blue-specific opsin that responds to a corresponding segment of the visible light spectrum.[112] The cone visual cycle rapidly regenerates 11-*cis*-retinal for binding to cone opsins. More of the cone cycle reactions take place within the cone cells, compared with the classical rod cycle, whereas the Müller cells adjacent to cones function more like the RPE. IRBP is likely functioning to shuttle retinoids across the space between the photoreceptor and accessory cells.

Xerophthalmia refers to the spectrum of ocular disease caused by severe vitamin A deficiency.[113] Most night blindness is the result of a lack of dietary vitamin A. The condition is reversed after adequate vitamin A is provided and more rhodopsin is synthesized. At least one other form of night blindness, Sorsby fundus dystrophy, is an autosomal dominant genetic disorder causing retinal degeneration. The condition was reported to improve in patients with early-stage disease within a week after receiving 50,000 IU (~15,000 μg) of vitamin A.[114,115]

Mutations in genes for the vitamin A–metabolizing proteins including LRAT, RDHs, and RPE65 are associated with diseases of the eye including retinal dystrophy, fundus albipunctatus, and cone degeneration.[116,117] Stargardt disease is an inherited degenerative condition of the retina in which excessive byproducts of retinal metabolism buildup.[118,119] SGDT-1 is due to an autosomal recessive mutation of the ATP-binding cassette, sub-family A, member 4, ABCA4, which flips and transports the 11-*cis* and all-*trans* isomers of retinal bound to phospholipid across the membranes of the phospholipid-rich photoreceptor discs during the visual cycle, allowing for movement of retinoid molecules through the photocycle and preventing the accumulation of potentially toxic retinoid adducts in photoreceptor and RPE cells. When ABCA4 is mutated as in SGDT-1, excess retinal bound to phospholipids, like phosphatidylethanolamine and other retinoid adducts, accumulates in and damages the retina.[119]

Vitamin A plays an essential role in the cornea and conjunctival membranes by promoting normal cell differentiation and structural integrity. Although the cornea is avascular, it obtains vitamin A in tear fluid. RBP is produced by the lacrimal gland.[120] As vitamin A deficiency develops, mucus-producing goblet cells of the conjunctival membranes are reduced, leading to dryness (xerosis). In a rabbit model of xerosis, treatment with eye drops containing retinyl palmitate increased the number of mucus-secreting conjunctival Goblet cells, demonstrating a direct link between vitamin A and Goblet cell integrity.[121]

Pre- and Postnatal Development

The role of vitamin A compounds in embryonic development has been studied since the 1930s. Early work in animal models showed that both vitamin A deficiency and vitamin A excess, at critical periods of development, result in severe malformations, especially of craniofacial structures, limbs, and visceral organs of the fetus.[122] It is now understood that retinoid signaling begins soon after gastrulation, when many of the regulatory factors, such as nuclear receptors and critical enzymes, are already expressed.[4] An appropriate concentration of RA is critical for the normal formation of structures derived from neural crest cells and somites and, later, during the period of organogenesis for the correct development of the heart, lungs, eyes, gonads, urogenital tract, and other organs.[122,123] When high doses of RA are administered by oral gavage to pregnant mice at embryonic day 9.5, limb development is abnormal, with shortening and an abnormal number of digits. Interestingly, vitamin A deficiency and retinoid excess cause a similar spectrum of defects. Recently, insufficient vitamin A has been linked to the development of the diaphragm and the condition of congenital diaphragmatic hernia.[124]

The embryo is thought to be able to produce its own RA from maternally derived retinol based on the finding that several genes involved in RA production, signaling, and catabolism are expressed early in development, when the body pattern is forming. These include RALDH2 (referred to as Aldh1a2 in mice), RARβ, and CYP26A1, which are expressed in temporally and spatially regulated patterns, often in adjacent cells or layers, but seldom within the same cells. Based on such observations, it is proposed that the local production of RA results in the diffusion of RA, forming a concentration gradient, to which nearby cells are exposed in a concentration-dependent manner. The expression of genes within these cells is sensitive to the concentration of RA as well as other regulatory signals.[122] Based on regional patterns of gene expression, it appears that RA production is turned on, evidenced by expression of RALDH2, and then is extinguished, again in specific patterns in different regions of the embryo, as evidenced by expression of CYP26 genes. Investigators have debated extensively whether RA itself is an endogenous morphogen that controls vertebrate development, or whether it acts as an inducer of other primary morphogenetic signals. Either way, retinoid signaling is part of a complex network that involves other proteins of the *Hox*, *Hedgehog*, fibroblast growth factor (*Fgf*), and *Wnt* signaling pathways, which also regulate the timing of development of the body's structures.[125]

Early in the postnatal period, the lungs are of special interest because they undergo extensive postnatal alveolarization, and vitamin A, in the form of RA, has been shown to accelerate this maturation process, even in the presence of inhibitory factors like hyperoxia or dexamethasone treatment.[126] Vitamin A, either given alone or in combination with RA, significantly increased the content of lung RE in the postnatal rat lung while, at the same time, regulating the expression of retinoid homeostatic genes, including *LRAT, CYP26B1*, and *Stra6*.[127] A Cochrane Database Review of clinical studies conducted in very-low-birth-weight infants concluded that supplementation of infants who weighed less than 1,500 g with vitamin A (5,000 IU, or 2.5 mg three times weekly) is associated with reductions in mortality and in the requirement for oxygen at 1 month.[128]

Tissue Repair

RA is one of the principal regulators of cell differentiation and tissue repair. In 1925, Wolbach and Howe[129] reported that vitamin A deficiency results in squamous metaplasia in which the epithelial lining of many tissues become flattened (squamous), dry, and keratinized. These seminal findings stimulated the investigation of retinoids as therapeutic agents in numerous conditions that involve cell differentiation. Vitamin A and cancer prevention is discussed below. As two other examples, out of many, RA has been shown to improve lung alveolar repair[130] and to improve the intestinal adaptive response, following tissue resection.[131]

Immunity

Poor resistance to infection was one of the earliest recognized pathologic features of vitamin A deficiency, and an original reason that vitamin A was referred to as "the anti-infective agent".[132] As elaborated in the chapter on Nutrition and the Immune System, it is now well established that vitamin A-deficient animals and humans are more susceptible to natural infections, or that they respond poorly to immunologic challenges.[133-135] Both vitamin A and its metabolite, RA, are important regulators of innate, adaptive, and mucosal immunity. RA is required for T-cell differentiation, influences the balance of Th1 to Th2 T-helper cells,[135-137] and the balance between FoxP3-positive T-regulatory cells and Th17 (generally inflammatory) cells.[138] RA also regulates intestinal homeostasis and prevention of autoimmune disorders.[139] Dendritic cells in intestinal mucosal lymph nodes produce or respond to RA in a way that favors the induction of T-regulatory over Th17 cells.[140-142] In addition, RA regulates B-cell maturation,[143] the magnitude of the antibody response,[135] and various aspects of innate and mucosal immunity.[135]

Manifestations of Vitamin A Deficiency and Excess

Deficiency

The symptoms and signs of vitamin A deficiency have been studied in greater detail than have those of any other nutritional deficiency disorder.[144] The eye is primarily involved, and xerophthalmia predominantly affects young children. Impaired dark adaptation or night blindness is an early symptom and can be elicited by a careful history and some simple tests in a poorly illuminated room.[145] Photopic and color vision, mediated by the retinal cones, is usually unaffected. Although a dietary deficiency is the primary cause, vitamin A deficiency may also result secondarily from disorders causing significant malabsorption. Extraocular manifestations of vitamin A deficiency include perifollicular hyperkeratosis, which may also be the result of other micronutrient deficiencies. Other changes associated with vitamin A deficiency may include impaired taste, anorexia, vestibular disturbance, bone changes with pressure on cranial nerves, increased intracranial pressure, infertility, and congenital malformations.

A nutritional deficiency of vitamin A is still a problem of public health proportion in many low- and middle-income countries. While deaths attributed to vitamin A deficiency had significantly declined in east and southeast Asia and Oceania, Latin America, and the Caribbean, in 2013, the prevalence of deficiency remained high in sub-Saharan Africa and south Asia.[146] Vitamin A supplementation is promoted by WHO/UNICEF for reducing the risk of mortality in children 6 months to 5 years of age in regions where vitamin A deficiency is still a public health problem. Typical doses are provided in the form of capsules containing 15 to 60 mg (50,000 to 200,000 IU) of retinol, depending on age, administered orally at 3- to 6-month intervals.[147] Pregnant and postpartum women are another target group for vitamin A supplementation. A Cochrane review of antenatal vitamin A supplementation, based on three large studies in low-income countries, concluded that there is good evidence that antenatal vitamin A supplementation reduces maternal night blindness and maternal anemia for women in areas where vitamin A deficiency is common or who are HIV-positive.[148]

Hypervitaminosis A and Adverse Effects

Hypervitaminosis A can be induced acutely by high intake of vitamin A,[144] but it is most often the result of long-term intake of lower, but still excessive, amounts of vitamin A, particularly the excessive use of vitamin A–containing supplements, food faddism such as an excessive consumption of liver, or self-medication with vitamin A preparations.[9] The severity of adverse effects depends on the dosage and includes severe headache, nausea, skin irritation, pain in bones and joints, coma, and death. Both prescription retinoids and excessive intakes of over-the-counter vitamin A supplements can pose similar risks.[149]

Most of the features of vitamin A toxicity are related to a rise in intracranial pressure: nausea, vomiting, headache, vertigo, irritability, stupor, fontanel bulging (in infants), papilledema, and pseudotumor cerebri (mimicking brain tumor).[150] Pyrexia and peeling of the skin also occur.

Chronic poisoning produces a bizarre clinical picture that is often misdiagnosed because of failure to consider excessive vitamin A intake.[150] It is characterized by anorexia, weight loss, headache, blurred vision, diplopia, dry and scaling pruritic skin, alopecia, coarsening of the hair, hepatomegaly, splenomegaly, anemia, subperiosteal new bone growth, cortical thickening (hands and feet and long bones of the legs), and gingival discoloration. Cranial sutures are widened in the young child.

Teratogenesis

Vitamin A and other retinoids are powerful teratogens in both pregnant experimental animals and women.[150,151] These agents cause fetal malformations (exencephaly, craniofacial malformations, eye defects, and cardiac abnormalities) that are quite similar across animal species and in humans.[152] The Teratology Society reviewed the animal and human literature in 1987 and concluded that at least seven case reports had been published of adverse pregnancy outcome associated with a daily intake of vitamin A of 25,000 IU or more.[153] In a study of the relationship between dietary vitamin A intake and birth defects in more than 22,000 pregnant U.S. women,[154] the investigators concluded that the risk of birth defects was significantly higher in women who consumed more than 10,000 IU/day (3,000 μg/day) of retinol during the periconceptual period. Concerns also have been raised about the frequent consumption of foods that are very high in vitamin A such as liver (>100,000 IU [~33,000 μg retinol] per 100 g) by women who either are or could become pregnant.[155] The WHO Micronutrient Initiative has published on the safe use of vitamin A in pregnant and lactating women.[156]

For determination of the UL, the IOM committee[10] reviewed case reports of liver abnormalities associated with elevated long-term intakes of vitamin A, including evidence of excessive vitamin A in perisinusoidal stellate cells, perisinusoidal fibrosis, hyperplasia, and hypertrophy of HSCs, for intakes of 1,500 to >14,000 μg/day over periods from 1 to 30 years. Accordingly, based on these data and the data above on birth defects, the UL for vitamin A was set at 3,000 μg retinol (preformed)/day.[10] Additionally, high intake of vitamin A over a long time may result in loss of bone mineral density. Clinical data from studies in older Swedish men and women, as well as in women in the United States, indicate that long-term intake of doses of vitamin A supplements >300 μg retinol/day is associated with an increased risk of bone fractures.[157]

Assessment of Vitamin A Deficiency and Excess

Vitamin A status exists as a continuum of states for which deficient, marginal, adequate, excessive, and toxic represent convenient descriptors.[51] Some of the plasma and liver changes indicative of these conditions are listed in **Table 17.2**. Investigators have been especially interested in ways to assess vitamin A deficiency and marginal vitamin A status in noninvasive or minimally invasive ways. Biomarkers of vitamin A have been reviewed, along with discussion of the impact of inflammation on the assessment of vitamin A status.[51]

Indicators and tests of vitamin A status comprise a number of biochemical, physical, and population measurements, discussed in detail in Rosales et al.[158] A low concentration of vitamin A or RBP in plasma, breast milk, or tear fluid may indicate vitamin A deficiency. Despite the limitations of plasma retinol concentration for assessing the vitamin A status of individuals, it is still useful for screening the status of large populations.[113] Plasma RBP is usually well correlated with serum retinol and is, therefore, a good surrogate, but values of both are likely to be low when either protein or energy are insufficient for the maintenance of RBP synthesis, as well as during inflammation,[158] when RBP transcription and translation are impaired.[66] Based on NHANES data, low plasma retinol is inversely related to CRP, a positive acute phase protein that is used as an indicator of inflammation. This suggests that persons with adequate vitamin A status and mild inflammation, as may exist in the generally healthy US population sampled by NHANES, could be misclassified as having low (inadequate) vitamin A based on their lower serum retinol values, when the true cause of the reduction is inflammation.[159]

Although liver total retinol is considered the "gold standard," it is rarely obtainable in human studies and, thus, indirect assays of liver vitamin A reserves have been sought.[51] The RDR test is based on the rapid release of stored apo-RBP from liver if the liver's vitamin A is not adequate. An increase in plasma retinol concentration of >20%, compared with baseline, 5 hours after administration of a small oral dose of vitamin A, is interpreted as a positive response indicating that hepatic vitamin A reserves are not adequate. A modified RDR (MRDR) test is similar in principle but uses an oral dose of vitamin A_2 (3,4-didehydroretinol). These dose-response tests are useful in clinical research settings and have been applied in some population-based studies.[51]

Bitôt spots are a classic sign of severe vitamin A deficiency.[160-163] Bitôt spots are comprised of desquamated cellular

Table 17.2 Typical Findings Associated with Categories of Vitamin A Status

Range	Plasma Retinol	Clinical Signs	Vulnerable Groups/Most Common Situations
Deficient	<0.35 μmol/L[a]	Night blindness; other ocular manifestations common	Preschool-age children and pregnant or lactating women with low vitamin A intakes; inflammation and poor nutritional status
Marginal	0.35–0.70[b] μmol/L	None or minimal (positive plasma response to vitamin A)[c]	Children, pregnant women in vulnerable populations, often with high rates of infection[d]
Adequate	>1.05–3.00 μmol/L	None	Typical of a well-nourished general population
Excessive	Upper normal to >3 μmol/L[e]	Not apparent or mild; may have elevated liver enzymes in plasma indicative of liver damage	Long-term supplement use; frequent intake of foods (e.g., liver) high in preformed vitamin A
Toxic	Like above, with circulating retinyl esters in fasting plasma	Headache; bone or joint pain; elevated liver enzymes and clinical signs of liver disease; very high vitamin A in liver and increased levels in extrahepatic tissues	Food faddists and users of high-dose vitamin A supplements; patients treated with retinoids[f]

[a]Rarely, plasma retinol may be low because of hereditary familial low retinol-binding protein (see text); 0.35 μmol/L = 10 μg retinol/dL.

[b]The range 1.05 to 0.70 μmol/L is sometimes used to indicate marginal status, and <0.70 μmol/L to indicate vitamin A deficiency. These ranges may be more appropriate for adults, in whom median plasma retinol levels are higher than those in children (see text).

[c]Positive relative dose response (RDR) or modified RDR test.

[d]Low plasma retinol in inflammation may indicate an acute phase response associated with reduced RBP production, rather than deficient vitamin A storage (see text).

[e]Retinol and RBP are normal, whereas total retinol is elevated because of retinyl esters, and liver total retinol exceeds approximately 300 μg/g.

[f]Retinoid administration typically reduces, not increases, plasma retinol.

debris and bacteria, often having a foamy appearance and are most commonly seen in the interpalpebral fissure on the temporal aspect of the conjunctiva.[161] Changes in the eye to this point can be reversed if vitamin A is administered.[163] A prevalence >0.5% in WHO classification XIB in young children is used to identify significant prevalence of vitamin A deficiency.[164,165] Blindness results when corneal involvement proceeds to xerosis and varying degrees of liquefaction (keratomalacia). If, however, vitamin A deficiency continues and the cornea becomes soft (keratomalacia) and ulcerated, the damage is irreversible. The WHO estimates that 250,000–500,000 children who are vitamin A-deficient become blind every year, and half of them die within 12 months of losing their sight.[113] Tests of night blindness or histologic evaluation of conjunctival cells have been applied as screening tools in regions where vitamin A intake is known to be low.

Dietary data are of greatest value in understanding the food habits of populations. Because vitamin A is concentrated in relatively few foods, which, in addition, may be consumed infrequently, it is necessary to either collect several days of dietary data or to use a relevant Food Frequency Questionnaire for screening.[166] Especially in higher income settings, it is important to assess the use of vitamin A–containing supplements.

Uses of Retinoids Other Than As Vitamin A

Aside from vitamin A as an essential nutrient, the vitamin A family includes compounds that have either known or potential therapeutic value as drugs. In dermatology, all-*trans*-RA, 13-*cis*-RA (isotretinoin), and the analog etretinate, are used as drugs, delivered either systemically or topically, used to treat skin conditions including severe cystic acne due to their comedolytic and anticomedogenic activity. Topical retinoids are a first line of treatment,[167,168] and systemic exposure is greatly reduced.[169] Topical retinoids are also used for reversal of photodamaged skin.[170,171] Caution has been expressed regarding nonprescription preparations sold over the counter or online as acne supplements, some of which contain vitamin A at levels higher than the UL.[172] In the United States, the prescription of acidic retinoids approved as systemic drugs by the Food and Drug Administration is under strict regulation, due to the risk of birth defects.[173] Even after discontinuation of the drug, the risk of teratogenesis may persist for months.[174] Some patients have reported that adverse events include depression and other psychiatric conditions.[175]

Due to the inherent role of vitamin A in maintaining cell differentiation, as mentioned above, vitamin A and related retinoids have long been of interest with respect to cancer prevention and treatment. Low vitamin A is linked to an increased risk of certain cancers, especially those of epithelial origin. Both natural and synthetic retinoids have been studied for several decades for possible roles in cancer chemoprevention [reviewed in].[32,176,177] Evidence does not support any benefit for cancer prevention of consuming more dietary vitamin A than the usual RDA.[177] In contrast, for therapeutic use, high-dose all-*trans*-RA is a proven front-line treatment for acute promyelocytic leukemia,[178,179] which has remarkably resulted in a high rate of cure. Synthetic analogs of RA are still being developed and tested clinically for a range of therapeutic benefits.[180]

References

1. Sommer A. J Nutr 1989;119:96–100.
2. Wald G. Science 1968;162:230–9.
3. Chambon P. FASEB J 1996;10:940–54.
4. Ghyselinck NB, Duester G. Development 2019;146:167502.
5. West KP, Jr. J Nutr 2002;132:2857S–66S.
6. Beaton GH, Martorell R, Aronson KA, et al. Food Nutr Bull 1994;15:282–9.
7. Kane MA, Napoli JL. Methods Mol Biol 2010;652:1–54.
8. Schweigert FJ, Raila J, Wichert B, et al. J Nutr 2002;132:1610S–2S.
9. Office of Dietary Supplements NIH. Dietary Supplement Fact Sheet: Vitamin A NIH, Washington, 2023. Accessed March 2023. https://ods.od.nih.gov/factsheets/VitaminA-HealthProfessional/
10. Institute of Medicine. *Dietary Reference Intakes for Vitamin A, Vitamin K, Arsenic, Boron, Chromium, Copper, Iodine, Iron, Manganese, Molybdenum, Nickel, Silicon, Vanadium, and Zinc*, National Academy Press, Washington, 2001.
11. Olson JA. J Natl Cancer Inst 1984;73:1439–44.
12. Ross AC, Moran NE. Curr Dev Nutr 2020;4:nzaa096.
13. FAO/WHO. Human Vitamin and Mineral Requirements. Report of a Joint FAO/WHO Expert Consultation. Vitamin A (Chapter 7). Bangkok, 2001. Accessed March 2023. www.fao.org/DOCREP/004/Y2809E/y2809e0d.htm
14. Gov.UK. SACN Review of Dietary Advice on Vitamin A. 2005. Accessed March 2023. www.gov.uk/government/publications/sacn-review-of-dietary-advice-on-vitamin-a
15. EFSA Panel on Dietetic Products, Nutrition, and Allergies [NDA] EFSA J 2015;13:4028. Accessed March 2023. www.efsa.europa.eu/en/efsajournal/pub/4028
16. Noy N. Biochem J 2000;348:481–95.
17. Soprano DR, Blaner WS. Plasma retinol-binding protein. In *The Retinoids: Biology, Chemistry and Medicine* (Sporn MB, Roberts AB, Goodman DS, eds) pp. 257–81, Raven Press, New York, 1994.
18. Berni R, Clerici M, Malpeli G, et al. FASEB J 1993;7:1179–84.
19. Dever JT, Surles RL, Davis CR, et al. J Nutr 2011;141:42–7.
20. Goodman DS. Plasma retinol-binding protein. In *The Retinoids, vol. 2* (Sporn MB, Roberts AB, and Goodman DS, eds) pp. 42–88, Academic Press, Orlando, FL, 1984.
21. Yang Q, Graham TE, Mody N, et al. Nature 2005;436:356–62.
22. Robbins J. Clin Chem Lab Med 2002;40:1183–90.
23. Plante-Bordeneuve V. J Neurol 2018;265:976–83.
24. Ong DE, Newcomer ME, Chytil F. Cellular retinoid-binding proteins. In *The Retinoids: Biology, Chemistry and Medicine* (Sporn MB, Roberts AB, Goodman DS, eds) pp. 283–317, Raven Press, New York, 1994.
25. Plau J, Golczak M, Paik J, et al. Biochim Biophys Acta Mol Cell Biol Lipids 2022;1867:159179.
26. O'Bryne SM, Blaner WS. Introduction to retinoids. In *Carotenoids and Retinoids: Molecular Aspects and Health Issues* (Packer L, Kraemer K, Obermüller-Jevic U, and Sies H, eds) pp. 1–22, AOCS Press, Champaign, IL, 2005.
27. Dong D, Ruuska SE, Levinthal DJ, et al. J Biol Chem 1999;274:23695–8.
28. Dawson MI, Zhang XK. Curr Med Chem 2002;9:623–37.
29. Idres N, Marill J, Flexor MA, et al. J Biol Chem 2002;277:31491–8.
30. Wei LN. Annu Rev Pharm Toxicol 2003;43:47–72.
31. Bastien J, Rochette-Egly C. Gene 2004;328:1–16.
32. di Masi A, Leboffe L, De Marinis E, et al. Mol Aspects Med 2015;41:1–115.
33. Ross AC. J Nutr 2003;133:291S–6S.
34. O'Connor C, Varshosaz P, Moise AR. Nutrients 2022;14:10.3390/nu14061312.
35. Balmer JE, Blomhoff R. J Lipid Res 2002;43:1773–808.
36. Green MH, Green JB. Adv Food Nutr Res 1996;40:3–23.
37. Von Reinersdorff D, Green MH, Green JB. Adv Exp Med Biol 1998;445:207–23.
38. Cifelli CJ, Green JB, Green MH. J Nutr 2005;135:746–52.
39. Harrison EH. Biochim Biophys Acta 2012;1821:70–7.
40. MacDonald PN, Ong DE. J Biol Chem 1988;263:12478–82.
41. O'Byrne S M, Wongsiriroj N, Libien JM, et al. J Biol Chem 2005;280:35647–57.
42. Ross AC, Zolfaghari R. J Nutr 2004;134:269S–75S.
43. Berr F. J Lipid Res 1992;33:915–30.
44. Wang XD, Krinsky NI. Subcell Biochem 1998;30:159–80.
45. Blomhoff R, Helgerud P, Rasmussen M, et al. Proc Natl Acad Sci U S A 1982;79:7326–30.
46. Matsuura T, Gad MZ, Harrison ER, et al. J Nutr 1997;127:218–24.
47. Sato M, Suzuki S, Senoo H. Cell Struct Funct 2003;28:105–12.
48. Randolph RK, Ross AC. J Biol Chem 1991;266:16453–7.
49. Napoli JL. Biochimica et Biophysica Acta 1999;1440:139–62.
50. Muto Y, Smith JE, Milch PO, et al. J Biol Chem 1972;247:2542–50.
51. Tanumihardjo SA, Russell RM, Stephensen CB, et al. J Nutr 2016;146:1816S–48S.
52. Makdani D, Sowell AL, Nelson JD, et al. J Am Coll Nutr 1996;15:439–49.
53. Tanumihardjo SA, Cheng JC, Permaesih D, et al. Am J Clin Nutr 1996;64:966–71.
54. Wilson DE, Hejazi J, Elstad NL, et al. Biochim Biophys Acta 1987;922:247–58.
55. Raila J, Buchholz I, Aupperle H, et al. Vet Res 2000;31:541–51.
56. Sowell A, Briefel R, Huff D, et al. FASEB J 1996;10:A813.
57. Ballew C, Bowman BA, Sowell AL, et al. Am J Clin Nutr 2001;73:586–93.
58. Shenai JP, Rush MG, Stahlman MT, et al. J Pediatr 1990;116:607–14.
59. Felding P, Fex G. Acta Physiol Scand 1985;123:477–83.
60. Zhao T, Liu S, Zhang R, et al. Nutrients 2022;14:950.
61. Raguso CA, Dupertuis YM, Pichard C. Curr Opin Clin Nutr Metab Care 2003;6:211–6.
62. Bernstein LH, Ingenbleek Y. Clin Chem Lab Med 2002;40:1344–8.
63. Smith FR, Goodman DS. J Clin Invest 1971;50:2426–36.
64. Floreani A, Baragiotta A, Martines D, et al. Aliment Pharmacol Ther 2000;14:353–8.
65. Safadi R, Friedman SL. Med Gen. Med 2002;4:27.
66. Rosales FJ, Ritter SJ, Zolfaghari R, et al. J Lipid Res 1996;37:962–71.
67. Fex GA, Larsson K, Nilsson-Ehle I. J Nutr Biochem 1996;7:162–5.
68. Biesalski HK, Frank J, Beck SC, et al. Am J Clin Nutr 1999;69:931–6.
69. Vogel S, Piantedosi R, O'Byrne SM, et al. Biochemistry 2002;41:15360–8.
70. Waits RP, Yamada T, Uemichi T, et al. Clin Chem 1995;41:1288–91.
71. Kawaguchi R, Yu J, Honda J, et al. Science 2007;315:820–5.
72. Zhong M, Kawaguchi R, Costabile B, et al. Proc Natl Acad Sci U S A 2020;117:9857–64.
73. Isken A, Golczak M, Oberhauser V, et al. Cell Metab 2008;7:258–68.
74. Senoo H, Yoshikawa K, Morii M, et al. Cell Biol Int 2010;34:1247–72.
75. Christensen EI, Moskaug JO, Vorum H, et al. J Am Soc Nephrol 1999;10:685–95.
76. Leheste JR, Rolinski B, Vorum H, et al. Am J Pathol 1999;155:1361–70.
77. Storm T, Heegaard S, Christensen EI, et al. Cell Tissue Res 2014;358:99–107.
78. Goodman DS. Retinol-binding Protein. In *The Retinoids, vol. 2* (Sporn MB, Roberts AB, and Goodman DS, eds) pp. 177–208, Academic Press, Orlando, FL, 1984.
79. Cano NJ. Clin Chem Lab Med 2002;40:1313–9.
80. Gerlach TH, Zile MH. FASEB J 1991;5:86–92.
81. Alvarez JO, Salazar-Lindo E, Kohatsu J, et al. Am J Clin Nutr 1995;61:1273–6.

82. Stephensen CB, Alvarez JO, Kohatsu J, et al. Am J Clin Nutr 1994;60:388–92.
83. Gerlach TH, Zile MH. J Lipid Res 1991;32:515–20.
84. Storm T, Tranebjaerg L, Frykholm C, et al. Nephrol Dial Transplant 2013;28:585–91.
85. Blomhoff R, Skrede B, Norum KR. J Intern Med 1990;228: 207–10.
86. Kelley SK, Nilsson CB, Green MH, et al. Toxicol Sci 2000;55: 478–84.
87. Tan L, Wray AE, Green MH, et al. J Lipid Res 2014;55:1077–86.
88. Chen C, Mistry G, Jensen B, et al. J Clin Pharmacol 1996;36: 799–808.
89. Arnhold T, Tzimas G, Wittfoht W, et al. Life Sci 1996;59: PL169–PL77.
90. Kurlandsky SB, Gamble MV, Ramakrishnan R, et al. J Biol Chem 1995;270:17850–7.
91. Napoli JL. Biochim Biophys Acta 2012;1821:152–67.
92. Pares X, Farres J, Kedishvili N, et al. Cell Mol Life Sci 2008;65:3936–49.
93. Belyaeva OV, Wu L, Shmarakov I, et al. J Biol Chem 2018;293:6996–7007.
94. Kane MA, Folias AE, Wang C, et al. Anal Chem 2008;80:1702–8.
95. Belyaeva OV, Chang C, Berlett MC, et al. Chem Biol Interact 2015;234:135–43.
96. Billings SE, Pierzchalski K, Butler Tjaden NE, et al. FASEB J 2013;27:4877–89.
97. Niederreither K, Vermot J, Le Roux I, et al. Development 2003;130:2525–34.
98. Blomhoff R, Blomhoff HK. J Neurobiol 2006;66:606–30.
99. Takahashi Y, Moiseyev G, Chen Y, et al. FEBS J 2011;278:973–87.
100. Budhu AS, Noy N. Mol Cell Biol 2002;22:2632–41.
101. Ross AC, Zolfaghari R. Annu Rev Nutr 2011;31:65–87.
102. Loudig O, Maclean GA, Dore NL, et al. Biochem J 2005;392:241–8.
103. Zhang Y, Zolfaghari R, Ross AC. Gene 2010;464:32–43.
104. Stoney PN, Fragoso YD, Saeed RB, et al. Brain Struct Funct 2016;221:3315–26.
105. Taimi M, Helvig C, Wisniewski J, et al. J Biol Chem 2004;279: 77–85.
106. Saari JC. Subcell Biochem 2016;81:231–59.
107. Saari JC. Invest Ophthalmol Vis Sci 2000;41:337–48.
108. Wright CB, Redmond TM, Nickerson JM. Prog Mol Biol Transl Sci 2015;134:433–48.
109. Gonzalez-Fernandez F. Vision Res 2003;43:3021–36.
110. Saari JC, Nawrot M, Kennedy BN, et al. Neuron 2001;29: 739–48.
111. Redmond TM. Proc Natl Acad Sci U S A 2017;114:10818–20.
112. Wang JS, Kefalov VJ. Prog Retin Eye Res 2011;30:115–28.
113. World Health Organization. Vitamin A deficiency. In *Nutrition Landscape Information System (NLiS)*. 2023. Accessed March 2023. www.who.int/data/nutrition/nlis/info/vitamin-a-deficiency
114. Cideciyan AV, Haeseleer F, Fariss RN, et al. Vis Neurosci 2000;17:667–78.
115. Jacobson SG, Cideciyan AV, Regunath G, et al. Nature Genet 1995;11:27–32.
116. Thompson DA, Gal A. Prog Retin Eye Res 2003;22:683–703.
117. Redmond TM. Exp Eye Res 2009;88:846–7.
118. Molday RS, Garces FA, Scortecci JF, et al. Prog Retin Eye Res 2022;89:101036.
119. Thirunavukarasu AJ, Ross AC, Gilbert RM. Front Nutr 2022;9:914457.
120. Ubels JL, Harkema JR. Invest Ophthalmol Vis Sci 1994;35: 1249–53.
121. Odaka A, Toshida H, Ohta T, et al. Clin Ophthalmol 2012;6:1585–93.
122. Clagett-Dame M, DeLuca HF. Annu Rev Nutr 2002;22:347–81.
123. McCaffery PJ, Adams J, Maden M, et al. Eur J Neurosci 2003;18:457–72.
124. Gilbert RM, Gleghorn JP. Am J Physiol Lung Cell Mol Physiol 2023;324:L456–67.
125. Zhao X, Duester G. Gene Expr Patterns 2009;9:430–5.
126. Massaro D, Massaro GD. Am J Physiol Lung Cell Mol Physiol 2002;282:L345–L58.
127. Wu L, Ross AC. J Lipid Res 2010;51:378–87.
128. Darlow BA, Graham PJ. Cochrane Database Syst Rev 2002;CD000501.
129. Wolbach SB, Howe PR. J Exp. Med 1925;42:753–77.
130. McGowan SE. Chest 2002;121:206S–8S.
131. Swartz-Basile DA, Wang LH, Tang YZ, et al. Am J Physiol Gastro Liv Physiol 2003;285:G424–G32.
132. Green HN, Mellanby E. Br Med J 1928;2:691–6.
133. Wiseman EM, Bar-El Dadon S, Reifen R. Crit Rev Food Sci Nutr 57:3703–14.
134. Stephensen CB, Lietz G. Br J Nutr 2021;126:1663–72.
135. Ross AC, Chen Q, Ma Y. Vitam Horm 2011;86:103–26.
136. Brown CC, Esterhazy D, Sarde A, et al. Immunity 2015;42: 499–511.
137. Amimo JO, Michael H, Chepngeno J, et al. Nutrients 2022; 14:5038.
138. Ziegler SF, Buckner JH. Microbes Infect 2009;11:594–8.
139. Mucida D, Cheroutre H. Cell Adh Migr 2007;1:142–4.
140. Yokota-Nakatsuma A, Takeuchi H, Ohoka Y, et al. Mucosal Immunol 2014;7:786–801.
141. Iwata M, Yokota A. Vitam Horm 2011;86:127–52.
142. Iwata M, Hirakiyama A, Eshima Y, et al. Immunity 2004;21: 527–38.
143. Ertesvag A, Naderi S, Blomhoff HK. Semin Immunol 2009;21:36–41.
144. Russell RM. Am J Clin Nutr 2000;71:878–84.
145. Duncan B, Canfield L, Barber B, et al. J Trop Pediatr 2000;46: 30–5.
146. Stevens GA, Bennett JE, Hennocq Q, et al. Lancet Glob Health 2015;3:e528–36.
147. World Health Organization. Vitamin A supplementation. 2023. Accessed March 2023. www.who.int/teams/immunization-vaccines-and-biologicals/essential-programme-on-immunization/integration/linking-with-other-health-interventions/vitamin-a
148. McCauley ME, van den Broek N, Dou L, et al. Cochrane Database Syst Rev 2015;CD008666.
149. Zamil DH, Burns EK, Perez-Sanchez A, et al. AMA J Ethics 2022;24:E376–81.
150. Hathcock JN, Hattan DG, Jenkins MY, et al. Am J Clin Nutr 1990;52:183–202.
151. Soprano DR, Soprano KJ. Annu Rev Nutr 1995;15:111–32.
152. Collins MD, Mao GE. Annu Rev Pharmacol Toxicol 1999;39: 399–430.
153. Teratology Society. 1987. Accessed March 2023. www.teratology.org/pubs/vitamina.htm
154. Rothman KJ, Moore LL, Singer MR, et al. N Engl J Med 1995;333:1369–73.
155. Blomhoff R, Alexander J. Tidsskr Nor Laegeforen 1993;113: 3037–9.
156. World Health Organization. (1998). Accessed March 2023. https://apps.who.int/iris/bitstream/handle/10665/63838/WHO_NUT_98.4_eng.pdf;jsessionid=8039457F2A622D3F16BBBD15EBD17774?sequence=1
157. Ribaya-Mercado JD, Solon FS, Fermin LS, et al. Am J Clin Nutr 2004;79:633–41.
158. Rosales FJ, Topping JD, Smith JE, et al. Am J Clin Nutr 2000;71:1582–8.
159. Stephensen CB, Gildengorin G. Am J Clin Nutr 2000;72: 1170–8.
160. Mishra K, Jandial A, Sandal R, et al. QJM 2019;112:225.
161. Madan S, Beri S. CMAJ 2017;189:E1264.
162. Kapil U, Bhadoria AS, Sareen N. J Trop Pediatr 2015;61:131–4.

163. McLaughlin S, Welch J, MacDonald E, et al. Eye (Lond) 2014;28:621–3.
164. Sommer A, West KP, Jr. *Vitamin A Deficiency: Health, Survival, and Vision*, Oxford University Press, Inc., New York, 1996.
165. McLaren DS, ed. *The Control of Xerophthalmia. A century of contributions and lessons*, Task Force Sight and Life, Basel, Switzerland, 2004.
166. Ambrosini GL, de Klerk NH, Musk AW, et al. Public Health Nutr 2001;4:255–64.
167. Dreno B, Kang S, Leyden J, et al. J Drugs Dermatol 2022;21: 734–40.
168. Baldwin HE, Nighland M, Kendall C, et al. J Drugs Dermatol 2013;12:638–42.
169. Clewell HJ, 3rd, Andersen ME, Wills RJ, et al. J Am Acad Dermatol 1997;36:S77–85.
170. Yaar M, Gilchrest BA. Br J Dermatol 2007;157:874–87.
171. Riahi RR, Bush AE, Cohen PR. Am J Clin Dermatol 2016;17: 265–76.
172. Zamil DH, Burns EK, Perez-Sanchez A, et al. Dermatol Pract Concept 2021;11:e2021075.
173. Food and Drug Administration. iPLEDGE Risk Evaluation and Mitigation Strategy (REMS). 2022. Vol. 2023, updated Jan 14, 2022. Accessed March 2023. www.fda.gov/bbs/topics/NEWS/2005/NEW01218.html
174. Nau H. J Am Acad Dermatol 2001;45:S183–7.
175. O'Donnell J. Am J Ther 2003;10:148–59.
176. Tang XH, Gudas LJ. Annu Rev Pathol 2011;6:345–64.
177. Ross AC. Vitamin A. In *Bioactive Compounds and Cancer* (Milner J, and Romagnolo D, eds), Springer, New York, 2010.
178. Pitha-Rowe I, Petty WJ, Kitareewan S, et al. Leukemia 2003;17:1723–30.
179. Meani N, Minardi S, Licciulli S, et al. Oncogene 2005;24:3358–68.
180. Uray IP, Dmitrovsky E, Brown PH. Semin Oncol 2016;43:49–64.

C. VITAMINS

CHAPTER 18

Vitamin D

Glenville Jones

CHAPTER OUTLINE

Abbreviations

1α,25-(OH)2D, 1α,25-dihydroxy vitamin D; **1α,25-(OH)$_2$D$_3$**, 1α,25-dihydroxy vitamin D_3; **25-OH-D**, 25-hydroxyvitamin D; **BMI**, body mass index; **Ca^{2+}**, calcium ion; **CYP**, cytochrome P-450; **DBP**, vitamin d-binding protein; **DRI**, dietary reference intake; **FGF-23**, fibroblast growth factor 23; **IOM**, Institute of Medicine; **NHANES**, National Health and Nutrition Examination Survey; **PCR**, polymerase chain reaction; **PTH**, parathyroid hormone; **RDA**, recommended dietary allowance; **UV**, ultraviolet; **VDR**, vitamin d receptor; **VDRE**, vitamin d-responsive element

History of Rickets and Vitamin D as an Antirachitic Factor

Although vitamin D was discovered less than 100 years ago, its deficiency diseases (rickets, and its adult-onset counterpart, osteomalacia) were clearly recognized by physicians, such as Daniel Whistler (1645) in the Netherlands and Francis Glisson (1650) in England, much earlier.[1,2] Rickets is characterized by a "bone matrix, which is insufficiently mineralized or calcified," usually as a result of vitamin D deficiency. Rickets results in a deformed and misshaped skeleton, particularly bending or bowing of the long bones and enlargement of the epiphyses of the joints of the rib cage, arms, legs, and neck. These defects result in painful breathing and difficulties holding up the head, and in women, they even give rise to problems with childbirth in later life, if rachitic deformities of the pelvis persist. By the turn of the twentieth century, rickets had reached almost epidemic proportions, particularly in the industrialized cities of Northern Europe, where the excessive degree of air pollution that blocked ultraviolet (UV) exposure and the sweatshop labor practices that kept workers indoors during the daylight hours combined to prevent skin synthesis of vitamin D.

During the nineteenth and early twentieth centuries, the following physicians and scientists investigated the cause of rickets and made key observations that led to the discovery of vitamin D: Sniadecki (1822) noted that rickets was common in city-dwelling children but not rural dwellers[3]; Palm (1890) noted the importance of latitude in the incidence of rickets in China and concluded that sunshine is the main etiologic factor in rickets[4]; Percival (1789) discussed the medicinal uses of cod liver oil in treating rickets[5]; Raczynski (1913), Huldschinsky (1919), and Hess and Unger (1922) showed that laboratory animals and children with rickets were cured when exposed to sunlight or mercury (UV) lamps[6-8]; Mellanby (1919) fed a semisynthetic oatmeal diet to dogs to induce rickets, which was reversible by administering cod liver oil[9]; McCollum et al., (1922) demonstrated that the antirachitic factor in cod liver oil was distinct from vitamin A and named it vitamin D[10]; Hess and Weinstock (1924) and Steenbock and Black (1924) were able to reconcile the apparently disparate concepts of sunlight and dietary factors by showing that irradiation of certain foods (e.g., plant oils or yeast) increased vitamin D activity[11,12] (this plant-derived antirachitic factor is now known as vitamin D_2); and finally, Windaus et al., (1936) received the Nobel Prize in 1928, principally for elucidating the structure of sterols, including vitamin D.[13]

Terminology and Glossary of Terms

The terminology used in the vitamin D field is confusing to many experts and nonexperts alike. A glossary of relevant terms is provided here to help the reader:

Vitamin D: Nutritional term coined in the 1920s to mean a substance that possesses the full antirachitic activity of the parent molecule vitamin D_3. The term is now often employed as a short form for the class of compounds with the biologic activity of 1α,25-dihydroxyvitamin D, abbreviated as 1α,25-$(OH)_2D$ and also known as calcitriol. It is also occasionally used to mean a summation of both vitamin D_2 and vitamin D_3,[14] especially in the reporting of clinical assay results where serum 25-hydroxyvitamin D (25-OH-D) is used to mean the sum of 25-hydroxyvitamin D_2 (25-OH-D_2) and 25-hydroxyvitamin D_3 (25-OH-D_3).

Vitamin D_3: This natural derivative of 7-dehydrocholesterol is made in the skin (see **Figure 18.1** for structure).

Vitamin D_2: Artificial or plant-derived form of vitamin D often used in food fortification, in dietary supplements, and in high-potency pharmaceutical preparations. It is metabolized to an active form, 1α,25-$(OH)_2D_2$, in a manner similar to the natural vitamin D_3. Vitamins D_2 and D_3 are considered biologically equivalent in terms of their ability to cure rickets.

Renal 1α-hydroxylase: Proximal tubular form of the enzyme responsible for the final 1α-hydroxylation step of vitamin D activation, comprising three proteins, the key one being the cytochrome P-450 (CYP) protein CYP27B1. Renal 1α-hydroxylase is regulated by calcium (Ca^{2+}) and phosphate (PO_4^{3-}) ions through the hormones parathyroid hormone (PTH) and fibroblast growth factor 23 (FGF-23).

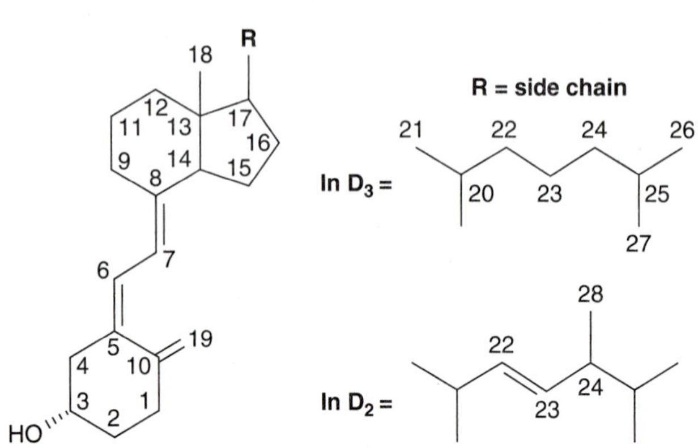

Figure 18.1 Structure of vitamins D_2 and D_3.

Reproduced from Makin HLJ, Jones G, Kaufmann M et al. Analysis of vitamins D, their metabolites, and analogues. In: Makin HLJ, Gower DB, eds. Steroid Analysis. New York: Springer, 2010:967–1096.https://doi.org/10.1023/b135931_11

Extrarenal 1α-hydroxylase: 1α-Hydroxylase, as evidenced by CYP27B1 protein, expressed in nonrenal tissues, especially where it locally produces 1α,25-$(OH)_2D_3$, which is believed to act in an autocrine or paracrine manner. It is regulated by cytokines and not by PTH and FGF-23.

Vitamin D receptor (VDR): Target cell protein that binds 1α,25-$(OH)_2D_3$, as well as specific sequences (vitamin D–responsive elements [VDREs]) in the genome to regulate gene expression of vitamin D–dependent genes at the transcriptional level. VDR does not bind any other metabolite with strong affinity, thus making it unlikely that vitamin D and 25-OH-D have any direct effect on gene expression under normal circumstances.

Dietary Sources of Vitamin D

As already pointed out, vitamin D can be derived from either endogenous synthesis in the skin or from the diet. In the strictest use of the term, vitamin D is, therefore, not a vitamin, at least during the summer months. Consequently, it should be referred to as a prohormone. Dietary sources become critical during the winter months (above 43 degrees north between October and April) when the zenith angle of the sun is such that UVB light does not penetrate the atmosphere and synthesis of vitamin D_3 in the skin is insignificant. Unfortunately, dietary sources of vitamin D are few. The only significant sources of vitamin D (D_2 or D_3) are animal liver, fatty fish (e.g., salmon, halibut, cod), egg yolk, and fish oil. Unfortified cow's milk is not a rich source. Because human milk is a poor source of vitamin D, breastfed infants require a vitamin D supplement. Most grains, lean meat, vegetables, and fruits are virtually devoid of measurable amounts of vitamin D. Although vitamin D_2 can be derived from the plant sterol ergosterol, the likelihood that this provitamin D_2 will be UV irradiated naturally seems low, although some cultures sun-dry vegetables. The artificial irradiation of shitake mushrooms, a rich source of ergosterol, has increased the possibility of finding vitamin D_2 in the diet.

Food fortification with either synthetic vitamin D_2 or, later, vitamin D_3 was pioneered and patented in the United States in the 1930s by Steenbock. His concept was to fortify staples such as breakfast cereals, milk, and margarine with vitamin D, initially in the form of irradiated ergosterol, to ensure delivery of this scarce nutrient, and to overcome the seasonal variability in potency found in natural sources (e.g., fish oils). This public health initiative to fortify certain foods with vitamin D eradicated rickets in the United States and virtually everywhere else in the world where it was introduced. In Quebec, Canada, which was the last province or state in North America to introduce vitamin D fortification, statistics showed a dramatic decline in the annual incidence of rickets at a downtown Montreal hospital (Ste Justine pour les Enfants) from 130 per 1,000 to virtually 0 per 1,000 over an 8-year span between 1968 and 1976; this period coincided with provincial legislation, making it mandatory for dairies to fortify milk.[15] A few countries resisted vitamin D fortification programs, not because they believed that they would be unsuccessful in the eradication of rickets, but because of financial questions about whether government or industry would pay for the programs. or out of concern about vitamin D intoxication in the neonatal period—fears that were proved largely unfounded.

Recommended Dietary Allowances of Vitamin D

The dietary reference intakes (DRIs) were reviewed in a report published by a panel of the Institute of Medicine (IOM, now the National Academy of Medicine) at the National Academy of Sciences in Washington, DC.[16] The chosen parameters for reporting optimal vitamin D intake are currently adequate intake (AI) for the newborn to 1-year life stage and recommended dietary allowance (RDA) for all other life stages. Excessive intake is defined as the tolerable upper intake level (UL) for all age groups. The DRI values for the various life stage groups are provided in **Table 18.1**.

Table 18.1 Vitamin D Dietary Reference Intakes by Life Stage (Amount/Day)

Life Stage Group	AI	EAR	RDA	UL
Infants				
0–6 mo	400 IU (10 μg)	—	—	1,000 (25 μg)
7–12 mo	400 IU (10 μg)	—	—	1,500 (38 μg)
Children				
1–3 y	—	400 IU (10 μg)	600 IU (15 μg)	2,500 IU (63 μg)
4–8 y	—	400 IU (10 μg)	600 IU (15 μg)	3,000 IU (75 μg)
Males				
9–13 y	—	400 IU (10 μg)	600 IU (15 μg)	4,000 IU (100 μg)
14–18 y	—	400 IU (10 μg)	600 IU (15 μg)	4,000 IU (100 μg)
19–30 y	—	400 IU (10 μg)	600 IU (15 μg)	4,000 IU (100 μg)
31–50 y	—	400 IU (10 μg)	600 IU (15 μg)	4,000 IU (100 μg)
51–70 y	—	400 IU (10 μg)	600 IU (15 μg)	4,000 IU (100 μg)
>71 y	—	400 IU (10 μg)	800 IU (20 μg)	4,000 IU (100 μg)
Females				
9–13 y	—	400 IU (10 μg)	600 IU (15 μg)	4,000 IU (100 μg)
14–18 y	—	400 IU (10 μg)	600 IU (15 μg)	4,000 IU (100 μg)
19–30 y	—	400 IU (10 μg)	600 IU (15 μg)	4,000 IU (100 μg)
31–50 y	—	400 IU (10 μg)	600 IU (15 μg)	4,000 IU (100 μg)
51–70 y	—	400 IU (10 μg)	600 IU (15 μg)	4,000 IU (100 μg)
>71 y	—	400 IU (10 μg)	800 IU (20 μg)	4,000 IU (100 μg)
Pregnancy				
14–18 y	—	400 IU (10 μg)	600 IU (15 μg)	4,000 IU (100 μg)
19–31 y	—	400 IU (10 μg)	600 IU (15 μg)	4,000 IU (100 μg)
31–50 y	—	400 IU (10 μg)	600 IU (15 μg)	4,000 IU (100 μg)
Lactation				
14–18 y	—	400 IU (10 μg)	600 IU (15 μg)	4,000 IU (100 μg)
19–31 y	—	400 IU (10 μg)	600 IU (15 μg)	4,000 IU (100 μg)
31–50 y	—	400 IU (10 μg)	600 IU (15 μg)	4,000 IU (100 μg)

AI, adequate intake; EAR, estimated average requirement; IU, international unit; RDA, recommended dietary allowance; UL, tolerable upper intake level.

Modified from Food and Nutrition Board, Institute of Medicine. Dietary Reference Intakes for Calcium and Vitamin D. Washington, DC: National Academy Press, 2011.

Current Understanding About Vitamin D Activation and Inactivation

Vitamin D_3 is synthesized from the sterol 7-dehydrocholesterol by a process involving UVB light in the wavelengths 290 to 315 nm[17] (**Figure 18.2**). UV irradiation opens the 9,10-bond of the provitamin to give a previtamin D_3 intermediate in the upper layers of the skin before it is isomerized nonenzymatically by heat to give vitamin D_3 in the lower layers. Transport of vitamin D_3 is by a specific plasma protein, vitamin D–binding protein (DBP), from skin to storage tissues, or to the liver, for the first step of activation. From the diet, both vitamin D_3 and vitamin D_2 are transported to storage depots or the liver on chylomicrons, although some evidence indicates that transfer from chylomicrons to DBP may also occur during transit. Vitamin D from skin or dietary sources does not circulate for long in the bloodstream, as it is taken up within hours by adipose tissue or liver for storage or activation.[18]

Ultimately, vitamin D_3 undergoes its first step of activation (**Figure 18.3**), namely, 25-hydroxylation in the liver. Some controversy has existed over whether 25-hydroxylation of vitamin D_3 is carried out by one or more enzymes, and whether this cytochrome P-450–based enzyme is found in mitochondrial or microsomal fractions of liver.[18] Biochemical research has established that one human mitochondrial enzyme (CYP27A1) and several microsomal cytochrome P-450 enzymes (including CYP2R1, CYP3A4, and CYP2J3) are able to carry out the 25-hydroxylation of vitamin D_2 or vitamin D_3 (see Figure 18.3). The physiologic relevance of one of these enzymes, CYP2R1, is particularly pertinent because there are several reports of mutations within the CYP2R1 gene in individuals with rickets[19-21]; the enzyme is also a 1α-OH-D_2-25-hydroxylase with high affinity for its vitamin D_2 substrate.[22] Investigations have provided a crystal structure of CYP2R1 with several of the known vitamin D substrates bound in the active site.[23] Furthermore,

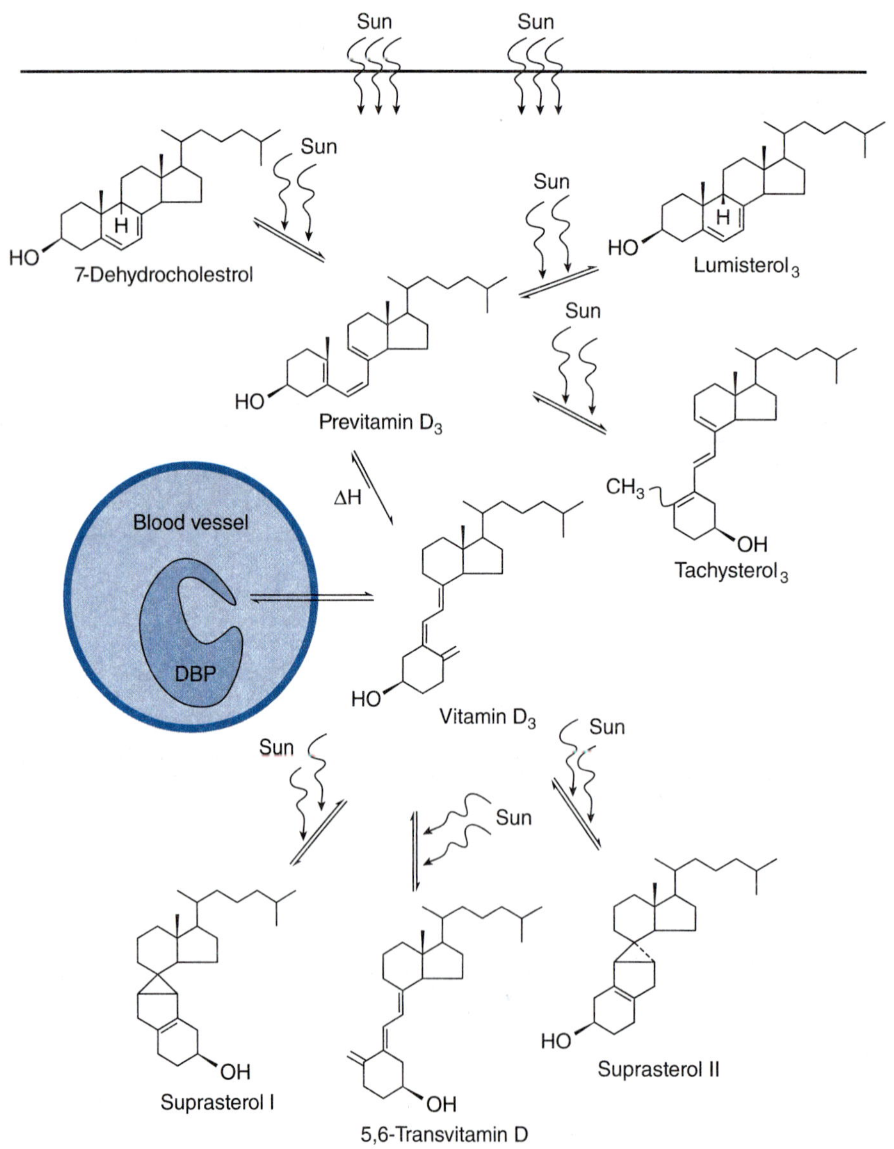

Figure 18.2 Photochemical events that lead to the production and regulation of vitamin D_3 in the skin. *DBP*, vitamin D–binding protein; *Sun*, ultraviolet B rays as a component of sunshine.

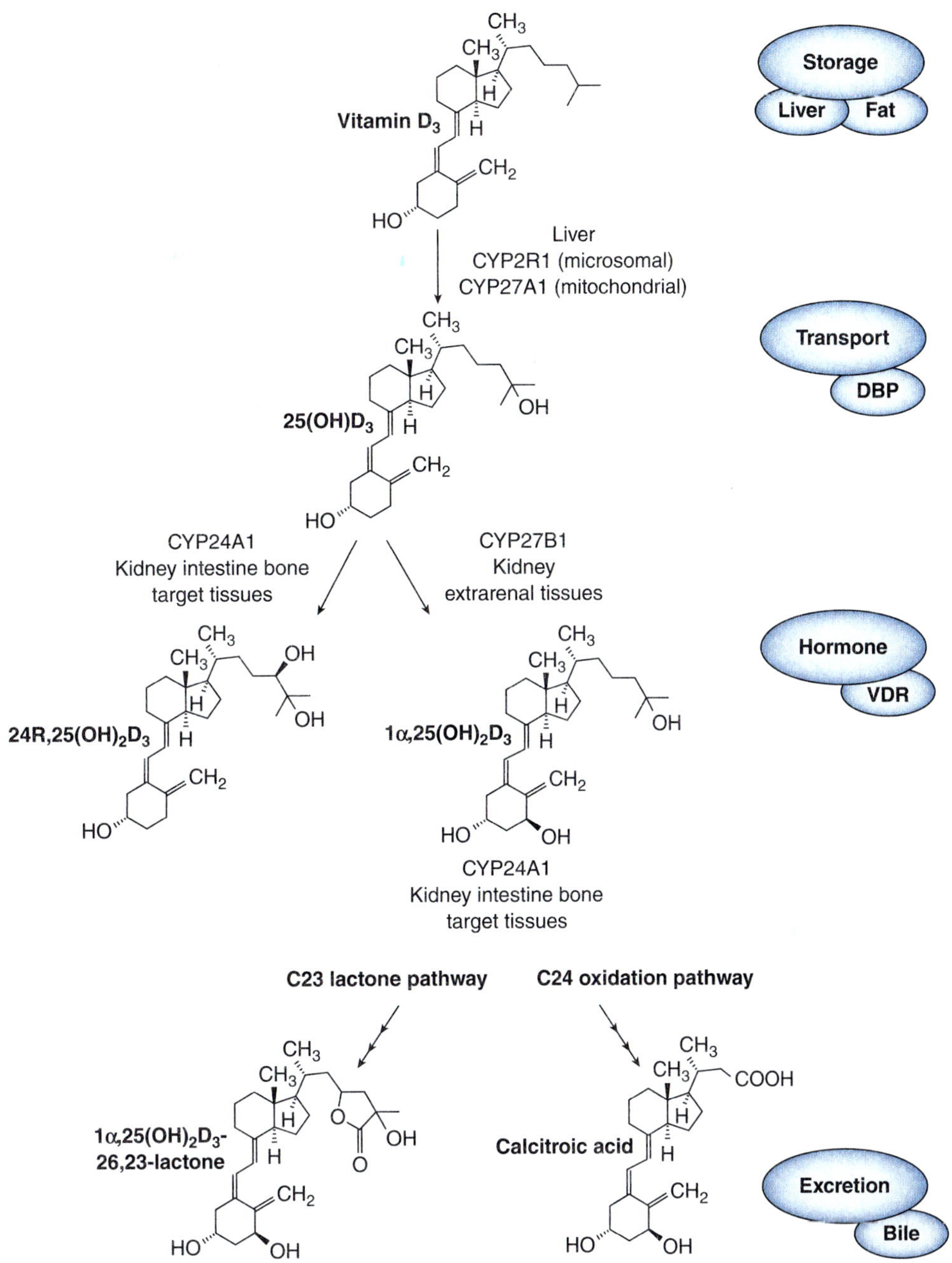

Figure 18.3 Metabolism of vitamin D_3. Current knowledge of the key metabolites in vitamin D metabolism in conjunction with the enzymes (all cytochrome P-450s) involved in their production. The hormonal form, 1α,25-$(OH)_2D_3$, is the sole "active" form through a transcriptional mechanism involving target cell machinery, in particular, the vitamin D receptor (VDR). The regulation of the enzyme system involves induction of CYP27B1 by parathyroid hormone and fibroblast growth factor 23 during a shortage of 1α,25-$(OH)_2D_3$ and induction of the CYP24A1 by the hormone itself during excess. Calcitroic acid is the biliary excretory product of 1α,25-$(OH)_2D_3$ generated by CYP24A1. A similar pathway exists for vitamin D_2. *DBP*, vitamin D–binding protein.

a genome-wide association study of the genetic determinants of serum 25-hydroxyvitamin D concentration[24] concluded that the chromosomal locus for CYP2R1 (11p15) showed the second strongest association of only a handful of sites in the whole genome; the others were DBP (or Gc), CYP24A1, and 7-dehydrocholesterol reductase (DHR7). Notably, variants of the other 25-hydroxylases, such as CYP27A1 or CYP3A4, were not associated with serum 25-OH-D concentration. This finding suggests that CYP2R1 is the most physiologically relevant 25-hydroxylase.

The product of the 25-hydroxylation step, 25-OH-D_3, is the major circulating form of vitamin D_3 and, in humans, is present in plasma at concentrations ranging from 10 to 80 ng/mL (25 to 200 nmol/L).[25] The main reason for the extended plasma half-life of 25-OH-D_3 is its strong affinity for DBP; the DBP-XO mouse shows accelerated rates of clearance and low 25-OH-D_3.[26] Serum 25-OH-D_3, therefore, represents a measure of the vitamin D status in vivo.

The circulating metabolite, 25-OH-D_3, is converted to the active form of vitamin D, known as calcitriol or 1α,25-$(OH)_2D_3$. The second step of activation, 1α-hydroxylation, occurs primarily in the kidney.[18] The synthesis of *circulating* 1α,25-$(OH)_2D_3$ in the normal, nonpregnant mammal appears to be the exclusive domain of the kidney. A specific mechanism appears to involve the cell surface receptors megalin and cubilin to provide uptake of the substrate, in the form of the 25-OH-D/DBP complex, by renal proximal cells (see Figure 18.3).[27] Megalin knockout mice show reduced vitamin D metabolite concentration and vitamin D deficiency. Additional evidence for renal synthesis of circulating calcitriol stems from clinical medicine. Patients with chronic kidney disease exhibit reduced 1α,25-$(OH)_2D_3$ and frank rickets or osteomalacia, resulting from deficiency of 1α,25-$(OH)_2D_3$ caused by lack of renal-1α-hydroxylase, a situation that can be reversed by 1α,25-$(OH)_2D_3$ hormone replacement therapy.[28] The cytochrome P-450 enzyme, CYP27B1, representing the 1α-hydroxylase enzyme, was cloned from several species, including rat, human, and mouse.[29-32] Investigators had known for some time that the kidney mitochondrial 1α-hydroxylase enzyme comprises three proteins—a cytochrome P-450, a ferredoxin, and a ferredoxin reductase for activity—and that it is strongly downregulated by 1α,25-$(OH)_2D_3$ and upregulated by PTH, as part of the calcium homeostatic loop.[33] Investigators showed that a similar phosphate homeostatic loop also exists involving FGF-23, the long-postulated phosphate-regulating hormone that downregulates CYP27B1 enzyme activity at the transcriptional level.[34] The promoter for the CYP27B1 gene is now known to contain the necessary regulatory elements (cyclic AMP response elements [CREs] and negative VDREs) necessary to explain the observed physiologic regulations of PTH, FGF23 and 1α,25-$(OH)_2D_3$, respectively, at the transcriptional level.[35] Human CYP27B1 gene mutations result in vitamin D dependency rickets type 1 (VDDR-I),[36] a disease state first proposed, in 1973, to result from a genetic defect of the 1α-hydroxylase enzyme.[37] Two independent groups generated the analogous mouse CYP27B1 knockout model by deletion of the cyp27B1 gene, and the 1α,25-$(OH)_2D_3$–deficient model revealed further insights into the regulation of the gene by different stimuli and subtleties in the roles of 1α,25-$(OH)_2D_3$.[38,39]

The final piece of the vitamin D metabolic machinery has been elucidation of the catabolism of 25-OH-D_3 and 1,25-$(OH)_2D_3$ in the body. This involves another cytochrome P-450 enzyme named CYP24A1, originally known as the 25-OH-D-24-hydroxylase. CYP24A1 inactivates both 25-OH-D_3 and 1α,25-$(OH)_2D_3$ by subjecting these metabolites to 24-hydroxylation, giving rise initially to 24R,25-$(OH)_2D_3$ and then to 1α,24,25-$(OH)_3D_3$, respectively[40,41] (see Figure 18.3).

Currently, evidence suggests that 24-hydroxylated metabolites may play a unique biologic role in bone fracture repair,[42] but most of the evidence favors the concept that 24-hydroxylation is primarily an inactivating step. The enzyme 24-hydroxylates both 25-OH-D_3 and 1α,25-$(OH)_2D_3$, the latter with a 10-fold higher efficiency.[43,44] Because circulating 25-OH-D_3 is approximately 1,000 times higher than that of 1α,25-$(OH)_2D_3$, the clearance of products of 25-OH-D_3 by CYP24A1 (e.g., 24R,25-$(OH)_2D_3$) is readily evident in the bloodstream. The enzyme, particularly the renal form, appears to be expressed at high amounts in the normal animal, and is probably involved in the inactivation and clearance of excess 25-OH-D_3 from the circulation.

Conversely, the extrarenal 24-hydroxylase appears to be primarily involved in target cell destruction of 1α,25-$(OH)_2D_3$.[45] Indeed, CYP24A1 has now been shown to be expressed almost ubiquitously wherever the VDR is expressed. CYP24A1 enzyme activity has been demonstrated in various cell lines representing specific vitamin D target organs (intestine, CaCo2 cells; osteosarcoma, UMR-106 cells; kidney, LLC-PK1 cells; keratinocyte, HPK1A and HPK1A-ras). Researchers have shown that 24-hydroxylation is the first step in the C-24 oxidation pathway, a five-step, vitamin D–inducible, ketoconazole-sensitive pathway that changes the hydroxylated vitamins D molecules to water-soluble truncated products, such as the biliary form, calcitroic acid.[46,47]

Although with human, rat, and mouse CYP24A1, the formation of calcitroic acid is the main pathway, other CYP24A1 isoforms, in particular guinea pig and opossum analogs, predominantly carry out a 23-hydroxylation pathway to a 26,23-lactone product.[48] The biologic value for the synthesis of the 26,23-lactone remains unknown, but the switch to the 23-hydroxylation pathway can be brought about by a single Ala326Gly mutation in human CYP24A1.[48] In most biologic assays, the intermediates and truncated products of these 23- and 24-hydroxylation pathways possess lower or negligible biologic activity. Furthermore, many of these compounds have little or no affinity for DBP, making their survival in plasma tenuous. Polymerase chain reaction (PCR) studies of CYP24A1 led to the detection of CYP24A1 mRNA in a wide range of tissues, corroborating earlier studies reporting widespread 24-hydroxylase enzyme activity in most, if not all, calcitriol target cells.

Additional studies showed that mRNA transcripts for CYP24A1 are virtually undetectable in naive target cells not exposed to 1α,25-$(OH)_2D_3$, but are dramatically induced by a VDR-mediated mechanism within hours of exposure to 1α,25-$(OH)_2D_3$.[49] In fact, the promoters of both human and rat CYP24A1 genes possess a double VDRE, which has been shown to mediate the calcitriol-dependent induction of CYP24 enzyme in both species. It is, therefore, attractive to propose that not only is 24-hydroxylation an important step in the inactivation of excess 25-OH-D_3 in the circulation but it is also involved in the inactivation of 1α,25-$(OH)_2D_3$ inside target cells. As such, one can hypothesize that C-24 oxidation is a target cell attenuation or desensitization process that constitutes a molecular switch to turn off calcitriol responses inside target cells.[50]

Loss-of-function mutations of human CYP24A1 result in the condition known as idiopathic infantile hypercalcemia, now renamed infantile hypercalcemia type 1, (IH)[51] characterized by hypercalcemia, hypercalciuria, nephrolithiasis and nephrocalcinosis, which supports a counter-regulatory role for CYP24A1 in vitamin D metabolism. Affected female patients with IH can exhibit episodes of hypercalcemia during pregnancy because of increased placental synthesis of 1,25-$(OH)_2D_3$.[52] The CYP24A1-knockout mouse exhibits a similar hypercalcemic phenotype and, in 50% of animals, is severe enough to cause death at around the time of weaning. Conversely, surviving animals have unexplained changes in bone morphology involving excess unmineralized osteoid that could suggest an alternative role for 24-hydroxylase in bone mineralization, although double knockouts lacking both CYP24A1 and VDR do not exhibit this bone phenotype.[53] Surviving CYP24A1 null animals have been shown to possess reduced ability to clear a bolus dose of [1β-^{3}H]1α,25-$(OH)_2D_3$ from their circulation, compared with normal, wild-type littermates.[54] Thus, not much evidence exists for efficient non–CYP24A1-mediated backup excretory systems for calcitriol catabolism at physiologic vitamin D concentrations.

Calcitroic acid, the final water-soluble biliary product of 1α,25-$(OH)_2D_3$ catabolism, is probably not synthesized in liver, because C-24 oxidation does not occur in hepatoma cells and presumably must, therefore, be transferred from target cells to the liver by some plasma carrier. Although calcitroic acid has been found in various tissues in vivo,[55] details of its transfer to bile have not been elucidated. Some emerging evidence indicates that high concentrations of vitamin D compounds can be metabolized by the general-purpose liver cytochrome P-450 enzyme, CYP3A4, which is induced in intestine by calcitriol.[56-58] As CYP3A4 is induced by several common drugs, such as anticonvulsants and rifampin, co-administration of drugs can cause accelerated destruction of vitamin D compounds by 4-hydroxylation to inactive forms and result in a type of rickets, termed vitamin D-dependent rickets type 3.[59] However, this deleterious phenomenon can be used to advantage when hypercalcemia is present, as in infantile hypercalcemia type 1, to lower 25-OH-D and 1,25-$(OH)_2D$ concentrations by administration of a CYP3A4-metabolized drug like rifampin.[60]

Role of Calcitriol in the Regulation of Vitamin D–Dependent Gene Expression

Current dogma holds that vitamin D, through its hormonal form 1α,25-$(OH)_2D_3$ (calcitriol), is a central regulator of calcium and phosphate homeostasis and that it promotes cellular differentiation and limits cell division of selected cell types.[18] Calcitriol achieves these functions through a VDR-mediated mechanism in which the hormone directly regulates gene expression at the transcriptional level of many different vitamin D–dependent genes in vitamin D target cells,[61] coding for proteins that, in turn, regulate cellular events such as intestinal calcium transport and cell division[18] (**Figure 18.4**). As in the classical steroid hormone model, 1α,25-$(OH)_2D_3$ enters the cell by traversing the plasma membrane in a free form and binds strongly to the VDR inside the nucleus (dissociation constant $[K_d] = 2 \times 10^{-10}$ M). The liganded or occupied VDR specifically targets only vitamin D–dependent genes by interacting with a specific sequence found upstream of the vitamin D–dependent gene. The sequence, known as a VDRE, is a tandem repeating oligonucleotide of six base pairs containing a 3-nucleotide spacer that is situated normally upstream of the 5′ end of the vitamin D–responsive gene. A consensus VDRE (AGGTCAnnnAGGTCA) is found in rat and human osteocalcin genes, the rat calbindin-9K gene, and the mouse osteopontin gene, whereas more complex elements are found in the collagen type I gene and the pre-pro-PTH gene, where these elements play a negative or suppressive role. Further research showed that VDR requires a heterodimeric partner, called the *retinoid X receptor* (RXR), and a plethora of other transactivators, termed a *DRIP complex*, to transactivate genes.[62]

The current model suggests that occupation of the ligand-binding domain of the VDR component triggers a protein conformational change in the AF-2 domain of the C-terminus of the VDR,[62] allowing recruitment of positive transcription factors and shedding of transcriptional inhibitory factors that lead to increased formation of a transcription initiation complex and an

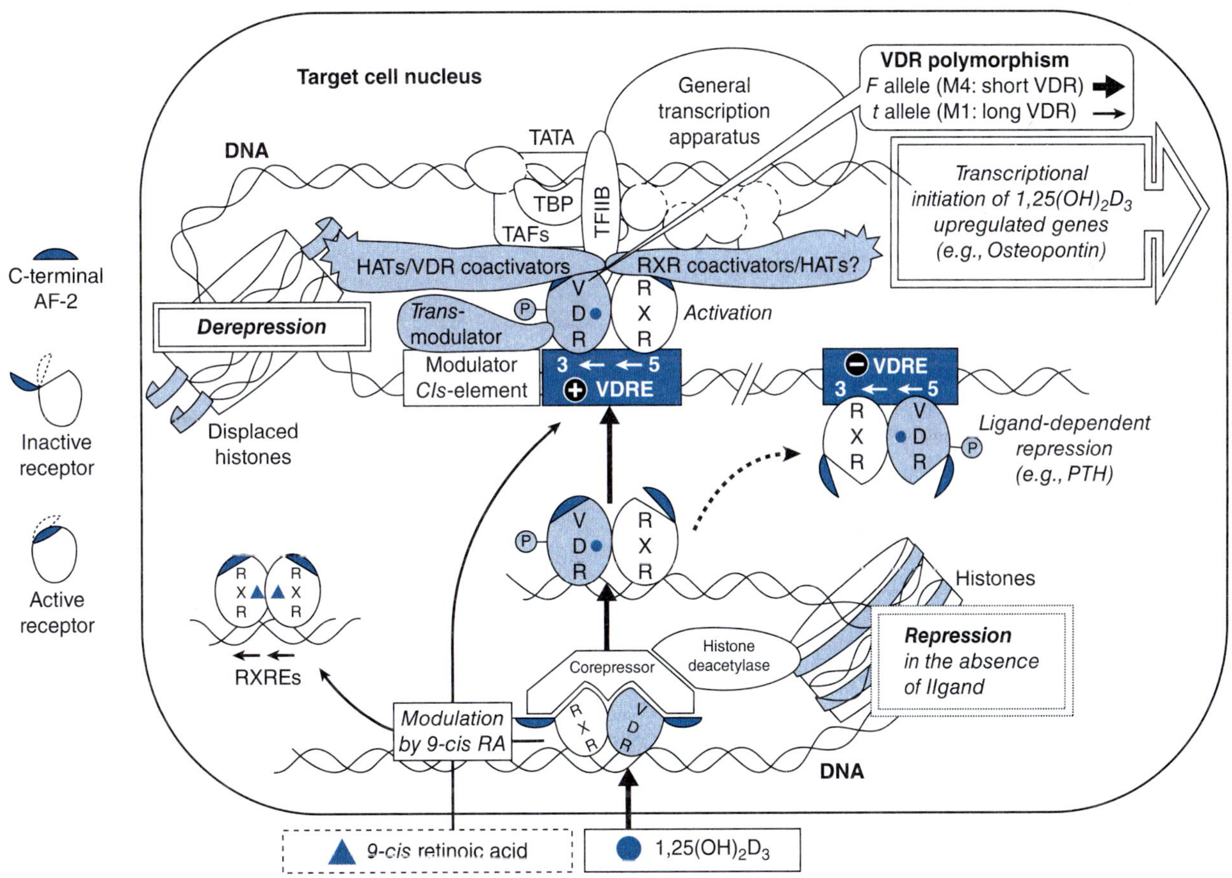

Figure 18.4 Mechanism of action of 1α,25-$(OH)_2D_3$ in gene expression at the transcriptional level. The simple target cell takes up 1α,25-$(OH)_2D_3$ as the free ligand originally ferried to the target cell bound to vitamin D-binding protein. The illustration shows the key elements of the transcriptional machinery including the vitamin D receptor (VDR), retinoid X receptor (RXR), and various coactivators in regulation of this machinery. *HAT*, histone acetyl transferase; *PTH*, parathyroid hormone; *RA*, retinoic acid; *RXRE*, retinoid X receptor response element; *TAF*, *TBP*, and *TFIIB*, transcription factors; *VDRE*, vitamin D–responsive element.

increased rate of gene transcription. The complexity of this mechanism and the number of specific and general transcription factors involved are impressive.

Calcemic and Noncalcemic Functions for Vitamin D

From autoradiographic studies using radiolabeled calcitriol,[63] VDR protein distribution studies,[64] gene chip arrays,[65] and data from various gene knockout mice,[38,39,54,66] investigators now recognize that vitamin D, in the form of calcitriol, acts on many different cells around the body. The spectrum of vitamin D–dependent genes is not limited to a handful of specific calcium and phosphate transport–related genes but is broad and probably in the hundreds.[67] The vitamin D–dependent genes point to several physiologic functions of 1α,25-$(OH)_2D_3$,[18] which are usually divided into calcemic and noncalcemic roles. *Calcemic roles* include the regulation of blood calcium and phosphate concentrations by actions at intestine, bone, parathyroid, and kidney. *Noncalcemic roles* include cell differentiation and antiproliferative actions in various cell types, such as bone marrow (osteoclast precursors and lymphocytes), immune system, skin, breast and prostate epithelial cells, muscle, and intestine.

Although most of the calcemic actions of calcitriol have been known since the early twentieth century, when dietary vitamin D deficiency was first demonstrated, the noncalcemic roles have emerged only from more subtle studies involving experiments probing the mechanism of action of calcitriol at the molecular level, and from studies of the VDR knockout mouse.[68]

The calcemic roles of calcitriol are mediated through regulation of a series of calcium-related genes, such as the calcium channel proteins (TRPV5 and 6), the cellular calcium-binding proteins (calbindins 9K and 28K), and the calcium pumps (e.g., calcium-dependent adenosine triphosphatases: $PMCA_{1b}$ and the sodium/Ca^{2+} exchanger, NCX1) which, together, explain the movement of Ca^{2+} across membranes of intestinal and kidney cells.[68] Although various VDR knockout mouse models showed a 60% decrease in intestinal calcium absorption and reduced expression of all of these proteins,[68] in the process demonstrating their vitamin D dependence, somewhat paradoxically, TRPV6 knockout mice[69,70] and calbindin-9K knockout mice[71,72] provided mainly negative or inconclusive data that these proteins are involved in calcium absorption. Explanations for these observations are that TRPV6 and calbindin-9k knockout mice have unknown compensatory vitamin D–dependent calcium transport mechanisms, and that the calcium pumps are the key vitamin D–dependent components.[68] Given the known effects of calcitriol on phosphate transport in intestine, a similar set of genes, including the type 2b sodium–phosphate cotransporter, must also exist for phosphate.[34,73]

The noncalcemic actions of calcitriol, especially on cell division and differentiation, are the result of effects on the cell cycle at the G_0 to G_1 transition. 1α,25-$(OH)_2D_3$ has been shown to have the following actions: it upregulates p21 and p27 cyclin-dependent kinase inhibitors, a process that results in the dephosphorylation of the retinoblastoma protein, which inhibits the transcription of the mammalian transcription factors E2F required for cell cycle progression[74,75]; it upregulates the transcription factor homeobox gene HOXA10[76]; and it regulates the dephosphorylation of p70S6 kinase, implicated in the cell cycle arrest in the G_1 to S phase transition in breast and colon cancer cells, respectively.[77] Alteration to growth factor signaling is another route by which 1α,25-$(OH)_2D_3$ and analogs have been observed to inhibit cell growth.[78] Several mechanisms whereby 1α,25-$(OH)_2D_3$ can induce apoptosis in cancer cells have been identified and include the following: (a) modulation of the relative amounts of antiapoptotic Bcl-2 and proapoptotic Bax[79,80]; (b) activation of proapoptotic μ-calpain by increasing cellular calcium concentration[81]; and (c) interaction with other signaling pathways that may lead to apoptosis, including through insulinlike growth factor and tumor necrosis factor.[82,83] Results obtained from the VDR knockout mouse suggest that 1α,25-$(OH)_2D_3$ could play a role in the differentiation of specialized cell types. For example, keratinocytes from VDR knockout animals exhibited uncontrolled hyperplasia as well as a reduction in the expression of the keratinocyte differentiation markers, including involucrin and loricrin, and these animals were more prone to carcinogen-induced tumorigenesis, compared with the wild-type animals.[84-86] Markers of terminal keratinocyte differentiation were also reduced in the CYP27B1 knockout animals, implying a role for 1α,25-$(OH)_2D_3$ in keratinocyte differentiation.

These discoveries in the basic science of vitamin D formed a basis to search for new functions or applications for vitamin D and, in some cases, are further supported by convincing epidemiologic data.[81] The actions of calcitriol in the immune system, skin, muscle, pancreas, kidney, and brain led to claims that vitamin D deficiency is associated with the pathogenesis of psoriasis, certain types of cancer, and autoimmune diseases such as multiple sclerosis and diabetes. These actions also point to roles for calcitriol in the synthesis of antimicrobial peptides, blood pressure regulation, and muscle cell differentiation. Consequently, these noncalcemic applications of vitamin D can now be potentially added to the calcemic roles in calcium- and phosphate-related diseases, such as rickets and osteomalacia, hyperparathyroidism, and osteoporosis[18,81] (**Figure 18.5**).

Emergence of Extrarenal CYP27B1 and Importance of Serum 25-Hydroxyvitamin D

Epidemiologic data placed increased emphasis on monitoring serum concentration of 25-OH-D, the circulating precursor to 1α,25-$(OH)_2D$, because serum 25-OH-D values correlate well, and better than serum 1α,25-$(OH)_2D$ values, with certain clinical parameters (e.g., bone mineral density).[87-89] Why may serum 25-OH-D be a better health indicator than serum 1α,25-$(OH)_2D$? This is best addressed in a newly emerging concept of an extrarenal 1α-hydroxylase that is expressed in many sites outside the kidney, and that augments the kidney-produced serum "endocrine" 1α,25-$(OH)_2D$ with locally produced "autocrine" or "paracrine" 1α,25-$(OH)_2D$ to promote additional roles for vitamin D outside the classical functions in calcium and phosphate homeostasis. This hypothesis had its roots in ideas developed in prostate disease,[90] but it has been broadened to explain the wide distribution of the 1α-hydroxylase and VDR in skin, immune system, and intestine.[91-93] In fact, the original concept of the extrarenal 1α-hydroxylase enzyme activity is already 2 decades old. Investigators suggested that an *extrarenal* 1α-hydroxylase may manifest in several physiologic or pharmacologic situations.

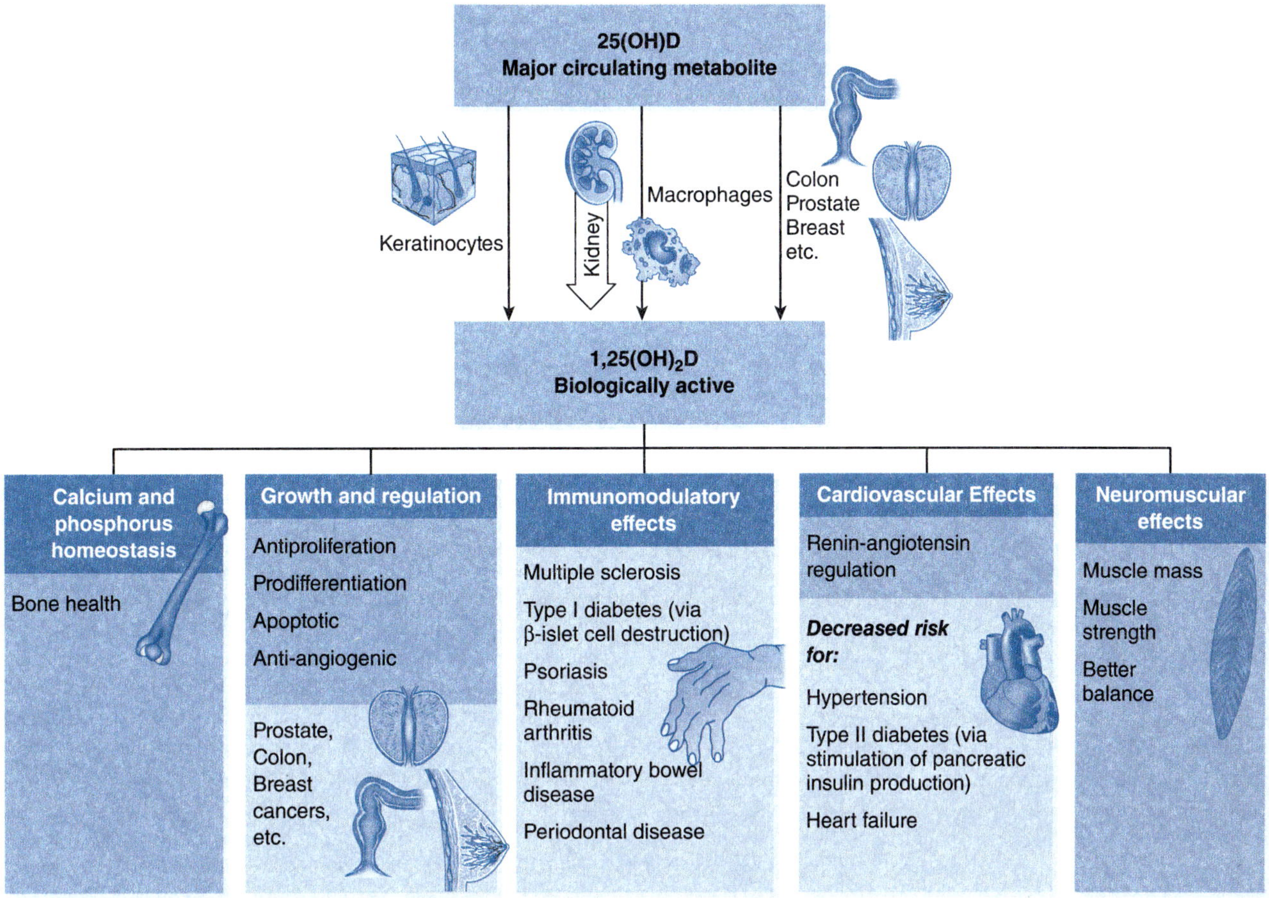

Figure 18.5 Calcemic and noncalcemic actions of calcitriol. The figure provides the range of biologic effects of calcitriol around the body through renally produced hormone by the local synthesis by the extrarenal 1α-hydroxylase (CYP27B1). Disease states associated with vitamin D deficiency are also indicated.

A placental 1α-hydroxylase was reported in the late 1970s and proved to be difficult to purify. Since cloning of the renal enzyme, however, its presence in placental tissue has been confirmed by real-time PCR of CYP27B1 mRNA and by specific antibody studies of the immunodetectable protein of CYP27B1.[94,95]

Before the use of calcitriol and its analogs in clinical medicine, occasional reports surfaced of anephric individuals who had been given large doses of vitamin D or 25-OH-D_3 and who had measurable blood concentration of a specific metabolite. This metabolite displaced 1α,25-$(OH)_2D_3$ in receptor binding assays,[96,97] a finding suggesting the existence of a significant extrarenal source of the 1α-hydroxylase. The concept was given additional momentum with work demonstrating the existence of loosely regulated 25-OH-D_3–1α-hydroxylase activity in sarcoid tissue that can cause elevated plasma 1α,25-$(OH)_2D_3$ and, in turn, hypercalciuria and hypercalcemia in patients with sarcoidosis.[98] The induction of extrarenal 1α-hydroxylase in macrophages by cytokines (e.g., interferon-γ) and growth factors as part of the inflammatory response was confirmed using molecular probes[99,100] and, even its role in antimicrobial peptide cathelicidin synthesis has been elucidated.

CYP27B1 mRNA and protein have been detected in many other extrarenal locations.[101] This knowledge has given rise to the concept that extrarenal CYP27B1 augments circulating 1α,25-$(OH)_2D_3$ with local production of 1α,25-$(OH)_2D_3$.[102,103] The locally high concentration of 1α,25-$(OH)_2D_3$ in certain sites, such as skin, prostate, and breast, are believed to give tissue-specific patterns of gene expression that, in turn, limit cell growth and lead to tissue-specific differentiation of specific cell types. Occasionally, as in sarcoidosis, locally high extrarenally synthesized calcitriol leaves the tissue and leaks into the main circulation.[98]

An important physiologic role for extrarenal 1α-hydroxylase is also associated with the important role of circulating 25-OH-D, which provides the substrate for this extrarenal enzyme, as well as for the renal enzyme. Serum 25-OH-D concentration is now considered an excellent predictor or biomarker of the noncalcemic actions of the D vitamins in the health of the immune system, skin, bone, certain epithelial cells, and muscle,[87] and is superior even to the serum 1α,25-$(OH)_2$D concentration.

Assessment of Nutrient Status: Serum 25-Hydroxyvitamin D as a Potential Biomarker for Health

Because serum 25-OH-D serves as a substrate not only for the kidney, but also wherever 1α-hydroxylase is found, the new theory has suggested wider use of the serum 25-OH-D parameter for monitoring vitamin D nutritional status. This theory has also led to a re-evaluation of the optimal circulating 25-OH-D concentration. The IOM report defined the different nutritional categories[16] as follows: (a) vitamin D deficiency remains defined as a 25-OH-D lower than 20 ng/mL; (b) vitamin D sufficiency as 25-OH-D between 20 and 50 ng/mL; and (c) vitamin D toxicity as 25-OH-D higher than 50 ng/mL.

Other investigators defined another category between deficiency and sufficiency that is known as *insufficiency*.[25] These investigators used a higher threshold of more than 30 ng/mL between inadequacy and adequacy. Some clinical chemists also use a much higher toxicity threshold of more than 80 ng/mL. Observed serum 25-OH-D values in U.S. residents from 2003 to 2006 in National Health and Nutrition Examination Survey (NHANES) data ranged from 5 to 50 ng/mL.

Central to controversies over the optimal ranges for serum 25-OH-D is that different health agencies[104] embrace different values for the important threshold between deficiency and sufficiency. Some vitamin D experts, like the IOM, set this value lower, at approximately 20 ng/mL, based on the blood concentration that results in normalization of bone health in 97.5% of the population (RDA).[105] Other experts endorsed the concept of an insufficiency range and suggested that the normal range starts somewhere between 20 and 60 ng/mL. Within this group, some experts emphasized the 25-OH-D threshold as the inflection point (at ~32 ng/mL) in a serum PTH/25-OH-D plot above which PTH values are normalized. Yet another group of bone experts used data emerging from NHANES studies of bone density or muscle strength parameters in broad populations of physiologically normal US women,[106,107] or in large epidemiologic correlations of serum 25-OH-D with breast, colon, and prostate cancer incidence, to suggest that the sufficiency threshold may be set at 40 ng/mL or even higher.

The IOM report[16] evaluated two Agency for Healthcare Research and Quality (AHRQ) reports and more than 1,000 studies in reaching the conclusion that vitamin D status could be based only on bone health, because data from nonskeletal health outcomes were inconclusive, contradictory, or negative. Time will tell whether investigators will need to reevaluate this perspective and set a higher normal range and, thus, new thresholds for vitamin D supplementation. Thus far, the IOM recommendations seem to be withstanding the test of time, as several recent high profile randomized controlled clinical trials with supplemental vitamin D have failed to show any additional benefit from augmenting vitamin D intake.[108,109]

In fact, a recent high dose Calgary-based trial featuring 10,000 IU/day vitamin D_3 found that the highest dose resulted in a deleterious effect, involving loss of bone density.[109] Accordingly, the 2011 IOM recommendations should be used as an appropriate guide for all North Americans because these guidelines are based on a thorough analysis of the best data currently available.

Serum 25-OH-D methodology uses both antibody-based and liquid chromatography tandem mass spectrometry–based approaches and provides reasonably reliable values for the physiologically relevant parameter, which is total 25-OH-D (sum of 25-OH-D_2 and 25-OH-D_3). Current controversies surrounding vitamin D methodology were reviewed,[110] and most vitamin D analysts voluntarily subscribed to a global vitamin D external quality assessment scheme (DEQAS) that oversees assay and analyst performance four times each year. If serum 25-OH-D is found to be lower than 20 ng/mL, clear justification exists for supplementation with appropriate doses of vitamin D for periods of approximately 6 weeks. Because of the time delay between vitamin D ingestion and a rise in the serum 25-OH-D concentration, and because 25-OH-D has a half-life of 15 to 20 days in the blood, no reason exists to measure serum 25-OH-D until 4 months after treatment begins. Most clinicians recommend monitoring serum 25-OH-D annually.

In evaluation of serum 25-OH-D, many otherwise normal individuals living at northern latitudes (>42 degrees), where skin synthesis is compromised for 6 months of the year, show 25-OH-D in the range of 10 to 40 ng/mL and are, thus, classified by some experts as insufficient, as least for part of the year.[25]

Bioequivalence of Vitamin D_2 vs Vitamin D_3

Qualitatively, vitamins D_2 and D_3 exhibit identical sets of biologic responses around the body, primarily through the VDR-mediated regulation of gene expression described earlier for calcitriol.[18,62] Physiologic responses to vitamin D_2 and D_3 hormones include regulation of calcium and phosphate homeostasis and regulation of cell proliferation and cell differentiation of specific cell types.[18] Quantitatively, considerable biochemical evidence indicates that most of the individual steps involved in the metabolism and actions of vitamin D_2 and D_3 are identical.[18] With the discovery of the critical importance of metabolism in vitamin D action, a series of vitamin D_3 metabolites were isolated and identified in the late 1960s and early 1970s; these included 25-OH-D_3, 1α,25-$(OH)_2D_3$, and 24R,25-$(OH)_2D_3$.[18] This discovery was followed by the identification of their vitamin D_2 counterparts: 25-OH-D_2, 1α,25-$(OH)_2D_2$, and 24R,25-$(OH)_2D_2$.[18] Noteworthy here was that the structural features unique to the vitamin D_2 side chain did not preclude either the 25-hydroxylation or 1α-hydroxylation steps in activation of the molecule or the first step of inactivation, namely, 24-hydroxylation.

Studies relating to the biochemistry of vitamin D since the late 1970s also documented that none of the steps in the specific vitamin D signal transduction cascade appeared to discriminate discernibly between the two vitamin D homologs at the molecular level in humans. These steps include the following: transport of vitamin D by DBP; 25-hydroxylation by CYP2R1; 1α-hydroxylation by CYP27B1; binding of 25-OH-D to the transport protein DBP; binding of 1,25-$(OH)_2D$ to the VDR; 24-hydroxylation of 25-OH-D or 1,25-$(OH)_2D$ by CYP24A1; and metabolic clearance of 1,25-$(OH)_2D_3$, all of which are similar for both vitamins D_2 and D_3.[18] The implication is that specific signal transduction systems designed to respond to vitamin D_3 also respond to physiologic doses of vitamin D_2 equally well.

Although reports suggested that certain species (e.g., avian species and new world monkeys)[111,112] discriminate against vitamin D_2, in humans, investigators have long assumed that the two vitamins are essentially equipotent. In 1940, Park[113] reviewed biopotency comparisons of viosterol (vitamin D_2) and cod liver oil (vitamin D_3) from more than 40 rickets treatment studies, and, although he suggested that many of the studies were of "poor quality," he concluded, "For practical purposes, the vitamin D in viosterol may be regarded as being equal to the vitamin D of cod liver oil. If viosterol is inferior to cod liver oil, rat unit for rat unit, the differences cannot be great."[113] Thus, the historical viewpoint in the medical literature that vitamin D_2 and vitamin D_3 are equipotent in treating rickets seems to have been reinforced by subsequent comparisons in rodents,[114,115] and by frequent reports that the two vitamins have similar effects in humans.[116]

Since 2000, however, the dogma that vitamins D_2 and D_3 are bioequivalent has been challenged,[117] leading to numerous efforts to reexamine the biopotency of the two forms in humans. Most, if not all, of these studies have been based on a surrogate, non-functional marker for biologic activity, namely, comparison of the plasma 25-OH-D concentration achieved after equivalent dosing of vitamin D_2 and vitamin D_3.

Multiple studies have provided evidence to suggest that vitamin D_2 is several-fold less effective in raising or maintaining 25-OH-D, compared with vitamin D_3,[118-121] whereas other studies have not been able to demonstrate a discrimination against vitamin D_2 and have argued that similar rises in 25-OH-D can be achieved with both forms.[122-124] Part of the apparent conflict between these different studies is the result of the significant variation in study design, with differences in the size, frequency, and duration of dosing (ranging from daily 1,000 IU/day doses for several months or years to 50,000 IU in a single dose), formulation, mode of administration, and baseline serum 25-OH-D (i.e., degree of deficiency). Whereas Armas et al.[119] used single doses of 50,000 IU and found markedly accelerated clearance of 25-OH-D_2, compared with 25-OH-D_3, Holick et al.,[123] used daily doses of 1,000 IU for 11 weeks and found the same rises in 25-OH-D and similar absolute 25-OH-D concentration at the end of the dosing period. Nevertheless, the complexity of the factors affecting the rise in serum vitamin D_3 and serum 25-OH-D_3 following various orally administered doses of vitamin D_3 cannot be overstated and were highlighted in the pharmacokinetic work of Heaney et al.[125]

This continuing debate about the relative biopotency of vitamin D_2 and vitamin D_3 can also be put into the context of frequent claims that vitamin D_2 compounds are less toxic than their vitamin D_3 counterparts in numerous mammalian species, from rodents to primates.[114,115,121] The implication of these diverse studies is that vitamin D_2 compounds may show differences in pharmacokinetics, particularly with pharmacologic doses (e.g., 50,000 IU/dose),[119] which are reflected in terms of the lower toxicity of vitamin D_2 but not necessarily a lower potency to cure rickets using physiologic doses (e.g., 1,000 IU/day).[123]

The challenge of the observed findings in dietary studies is to explain the in vivo discrimination against vitamin D_2 at higher doses,[118-121] discrimination consistent with the reports of lower toxicity,[114,115,126] whereas at lower doses, relative equipotency exists between the two vitamin forms. Among the possible explanations for this concentration-dependent phenomenon are the reports that vitamin D_2 and vitamin D_3 could be differentially susceptible to nonspecific inactivating modifications, such as those occurring to various drugs in the liver. These enzymes could include any of the liver CYPs that are known to metabolize vitamin D compounds differently, such as CYP27A1, which 25-hydroxylates vitamin D_3 and 24-hydroxylates vitamin D_2[18]; and CYP3A4, which 24- and 25-hydroxylates vitamin D_2 substrates more efficiently than vitamin D_3 substrates[127,128] and 23R- and 24S-hydroxylates 1α,25-$(OH)_2D_3$.[56] CYP3A4 has been shown to be selectively induced by 1α,25-$(OH)_2$D in the intestine[56,57] and by other drugs, e.g., rifampin, in the liver.[59,60]

Both CYP27A1 and CYP3A4 are known to have significantly lower Michaelis-Menten constant (K_m) values for vitamin D compounds[18] in the micromolar range, a property that questions their physiologic but not their pharmacologic relevance. Investigators[129] have shown that both human intestinal microsomes and recombinant CYP3A4 break down 1α,25-$(OH)_2D_2$ at a significantly faster rate than 1α,25-$(OH)_2D_3$. This finding suggests that this nonspecific cytochrome P-450 enzyme may preferentially limit vitamin D_2 action in target cells where it is expressed,[57,129] especially in the pharmacologic dose range. Thus, one explanation for the discrimination against vitamin D_2 could be the selective catabolism of vitamin D_2 by nonspecific cytochrome P-450 enzymes (e.g., CYP3A4) in liver and intestine. The same type of mechanism involving differential induction of nonspecific CYPs may also underlie the long-standing reports that coadministered drug classes, such as anticonvulsants,[130,131] cause accelerated degradation of vitamin D_2.

Populations at Risk for Vitamin D Deficiency

Because human breast milk is a poor source of vitamin D, breastfed infants constitute an important group to target for vitamin D supplementation. At particular risk are breastfed, dark-skinned children of African or Asian descent, especially those living in northern climates, such as Canada or Scandinavia. Over the past few decades, sporadic reports have noted persistent vitamin D deficiency rickets, almost always in breastfed, dark-skinned Canadians of African or Asian descent, but the total number of cases, even in a major metropolitan center such as Toronto, is small (17 over a 5-year period from 1988 to 1993).[132] Nevertheless, both Canadian and US pediatric societies have emphasized vitamin D supplementation programs and have recommended that daily intakes of vitamin D be doubled to 400 IU or 10 μg in this life stage.

Other life stages or groups at risk of vitamin D deficiency include individuals living at high latitude (Inuit), especially those who have forsaken their traditional high vitamin D–rich seafood diets, and individuals with dark skin, including those of African, Indian, Pakistani, or Sri Lankan descent, who are living at northerly latitudes.[133] These individuals have limited exposure to UV light or require longer exposure to UV light to make adequate amounts of vitamin D. Reports suggest that these groups have lower serum 25-OH-D than their lighter-skinned neighbors, a characteristic that may make them more susceptible to vitamin D deficiency during at least part of the year.

Another group at risk of vitamin D deficiency consists of individuals with high body mass index (BMI). With the increase in obesity to epidemic proportions in the United States and Western countries, much interest has been shown in the association of increasing BMI and vitamin D deficiency.[134,135] The widely held belief is that the vitamin D deficiency is the result of the sequestration of dietary vitamin D by adipose tissue. Because vitamin D is lipophilic, when it is absorbed from the gut, it enters the circulation, first on chylomicrons, and then is only partially transferred to DBP, and at a very slow rate.[5] Vitamin D has a relatively low affinity for DBP; reviews estimate this at between 10−5M and 10−7M.[136] Transport of dietary vitamin D contrasts significantly with that of vitamin D_3 made during skin synthesis, which is mainly DBP bound.[1] The consequences of chylomicron transport of dietary vitamin D include the possibility of uptake by peripheral tissues, such as adipose tissue and muscle, because of the action of lipoprotein lipase[137] and a short plasma half-life of 4 to 6 hours.[137] In contrast, the half-life of vitamin D stores is 2 months. The nutritional outcome of the sequestration of dietary vitamin D by adipose tissues is increased variability in the rise of serum 25-OH-D in response to vitamin D supplementation with increasing BMI. In contrast, weight loss studies show that even modest weight losses result in rises in serum 25-OH-D, presumably because of mobilization of adipose tissue stores that parallels adipose tissue depletion.

One group at particular risk of vitamin D deficiency is that of patients with chronic kidney disease, 80% to 100% of whom have extremely low 25-OH-D concentration.[138] This situation is further complicated by high concentration of FGF-23 secondary to phosphate retention, which results in CYP27B1 downregulation as well as increased 25-OH-D catabolism caused by CYP24A1 upregulation.[139] Correction of serum 25-OH-D into the sufficient range results in some improvement in most vitamin D–related outcomes in patients with chronic kidney disease and, in patients undergoing dialysis, vitamin D supplementation improves survival.[139,140]

Acute Toxicity Caused by Vitamin D

Numerous animal studies involving systematic vitamin D intoxication have been conducted since the late 1970s in various species, including rats, cows, pigs, rabbits, dogs, and horses.[136] Vitamin D_3 intoxication results in elevation of the blood concentrations of several metabolites, including vitamin D_3, $25(OH)D_3$, $24,25(OH)_2D_3$, $25,26(OH)_2D_3$, and $25(OH)D_3$-26,23-lactone, although it rarely raises plasma $1\alpha,25(OH)_2D_3$. Results of studies by Horst et al., (reviewed in reference 136) in pigs and dairy cows were definitive in suggesting that renal CYP27B1 is effectively turned off.

Focus, therefore, shifted to other vitamin D metabolites correlated with toxicity, especially the plasma threshold of 25-OH-D that must be exceeded to cause hypercalcemia. Shephard and DeLuca[141] acutely intoxicated rats with graded oral doses of vitamin D_3 (0.65 to 6500 ng/day for 14 days) or 25-OH-D_3 (0.46 to 4600 ng/day for 14 days) and used high-performance liquid chromatography and competitive binding assays to measure vitamin D metabolites postmortem. Findings showed that serum vitamin D_3 and $25(OH)D_3$ rose to micromolar concentrations in the plasma of rats given the highest intakes of vitamin D_3 and resulted in marked hypercalcemia. Dihydroxylated metabolites, including 24,25-$(OH)_2D_3$, 25,26-$(OH)_2D_3$, and $25(OH)D_3$-26,23-lactone, also rose to higher than 40 ng/mL, but plasma 1α,25-$(OH)_2D_3$ remained within the normal range. A dose of 460 ng/day of 25-OH-D_3 resulted in serum $25(OH)D_3$ concentration of 436 ± 53 ng/mL with normocalcemia, but such animals had no vitamin D burden. Based on this finding and on various studies in several animal species, it appears that serum 25-OH-D associated with acute toxicity is more than 200 ng/mL.

For ethical reasons, no systematic studies of vitamin intoxication have been conducted in humans. Numerous anecdotal reports collected over the years have described accidental vitamin D intoxication with either vitamin D_3 or vitamin D_2, however.[136,142] Because many of these studies involved measuring 25-OH-D and sometimes 1α,25-$(OH)_2$D, it is worth reviewing the vitamin D metabolite values that correlate with overt toxicity symptoms. Although an occasional report did find evidence of modest elevation of 1α,25-$(OH)_2$D,[136] all reported that 25-OH-D was well above the normal range, between 284 and 635 ng/mL. In a study of 35 hypervitaminotic patients with hypercalcemia resulting from chronic ingestion of overfortified milk,[143] the average 25-OH-D concentration was 224 ng/mL (range, 56 to 596 ng/mL). In an extended family group accidentally intoxicated with a veterinary vitamin D concentrate (peanut oil solution containing 2×10^6 IU cholecalciferol), Pettifor et al.,[144] showed that plasma 25-OH-D ranged from 339 to 661 ng/mL in intoxicated family members, whereas plasma 1α,25-$(OH)_2$D values were within the normal range in 8 of 11 patients.

A perusal of these data led reviewers in this field to the same conclusion,[136,142] namely, that hypercalcemia results only when 25-OH-D_3 concentration has been consistently higher than 200 ng/mL. The mechanism of vitamin D toxicity is unknown but is thought to involve vitamin D metabolites exceeding the vitamin D–carrying capacity of the DBP of the plasma. Displacement of one of the elevated metabolites, 1α,25$(OH)_2$D or 25-OH-D, into the target cell results in increases in unregulated biologic effects. Recent studies with the CYP27B1-knockout mouse, which is unable to make 1α,25-$(OH)_2D_3$ from 25-OH-D_3, suggests that 25-OH-D_3 and not 1α,25-$(OH)_2D_3$ is the toxic form,[145] as both knockout animals and their normal wild-type littermates were intoxicated at the same dietary vitamin D intakes.

With the increasingly popular call in the scientific literature[102,105] and even in the lay press for higher vitamin D supplementation, the question of vitamin D toxicity has switched to this: What are the long-term risks of moderately high doses of vitamin D? An honest answer to this important question is that we do not know whether vitamin D supplementation that produces serum 25-OH-D concentration of 50 to 100 ng/mL is safe or whether it causes any long-term side effects. Individuals with serum 25-OH-D in this range (e.g., lifeguards, outdoor workers) do not have acute toxicity (e.g., hypercalcemia), but whether they suffer more subtle side effects (e.g., more renal stones, higher cancer rates, higher mortality rates) has yet to be determined. Cancer epidemiologic data suggest that certain cancers show a U-shaped curve in response to vitamin D supplementation: low serum 25-OH-D is strongly associated with increased cancer risk, and moderately high serum 25-OH-D, greater than 50 ng/mL, also shows a risk, albeit at a lower level than in deficiency.[146] Concern has been expressed that long-term ingestion of moderately high vitamin D intake may cause changes in vascular calcification.

Experimentation using the hypercholesterolemic, low-density lipoprotein receptor knockout mouse model, rendered uremic by partial nephrectomy with low endogenous 1,25-$(OH)_2$D production, exhibited accelerated vascular calcification, but this was reversed by graded doses of various vitamin D analogs.[147] This model suggests that, contrary to popular belief, vitamin D compounds protect the vasculature from calcification, and only at very high doses do they trigger adverse effects through increasing serum $Ca^{2}+$ and PO_4^{3-}. Results of epidemiologic studies are consistent with these animal data.[148] Consequently, vitamin D supplementation that does not result in changes in calcium and phosphate homeostasis may be safe, and the serum 25-OH-D threshold for acute vitamin D toxicity (>200 ng/mL) may also be predictive for long-term risk. Currently, the normal range for serum 25-OH-D used by clinical chemistry laboratories is between 20 and 80 ng/mL, or between 20 and 100 ng/mL,[25] although NHANES data suggest that most US residents have values between 5 and 50 ng/mL.[16] This range may change significantly because the IOM Committee recommendations suggest a narrower normal range of 20 to 50 ng/mL, in view of concerns about long-term effects of chronic vitamin D dosing.[16,109] However, North Americans should be reassured that the ULs selected for different age groups (4,000 IU for 9 to more than 71 years) still include a large safety margin, at least based on acute toxicity symptoms (hypercalcemia).

References

1. Whistler D. De morbo puerili Anglorum, quem patrio idiomate indigenae vocant. The rickets MD thesis, University of Leiden, Leiden, Netherlands, 1645.
2. Glisson F. De Rachitide sive morbo puerili qui vulgo. The rickets dicitur. London, 1650.
3. Sniadecki J. Cited by Mozolowski W. Nature 1840;1939;143:121.
4. Palm TA. Practitioner 1890;45:270–9.
5. Percival T. Essays Medical, Philosophical and Experimental on the Medical Use of Cod-Liver Oil, vol 2. London, 1789.
6. Raczynski J. C R Assoc Int Pediatr 1913;308.
7. Huldschinsky K. Dtsch Med Wochenschr 1919;45:712–3.
8. Hess AF, Unger JF, Pappenheimer AM. J Exp Med 1922;36:427–46.
9. Mellanby E. Lancet 1919;1:407–12.
10. McCollum EV, Simmonds N, Becker JE, et al. J Biol Chem 1922;53:293–312.
11. Hess AF, Weinstock M. J Biol Chem 1924;62:301–13.
12. Steenbock H, Black A. J Biol Chem 1924;61:408–22.
13. Windaus A, Schenck F, van Werder F. Hoppe Seylers Z Physiol Chem 1936;241:100–3.
14. Makin HLJ, Jones G, Kaufmann M, et al. Analysis of vitamins D, their metabolites and analogues. In: Makin HLJ, Gower DB, eds. Steroid Analysis. New York: Springer, 2010:967–1096.
15. Delvin EE, Glorieux FH, Dussault M, et al. Med Biol 1978;57: 165–70.
16. Food and Nutrition Board, Institute of Medicine. Dietary Reference Intakes for Calcium and Vitamin D. Washington, DC: National Academy Press, 2011.
17. Holick MF. Photobiology of vitamin D. In: Feldman D, Pike JW, Glorieux FH, eds. Vitamin D. 2nd ed. New York: Elsevier, 2005:37–46.
18. Jones G, Strugnell SA, DeLuca HF. Physiol Rev 1998;78:1193–231.
19. Prosser DE, Jones G. Trends Biochem Sci 2004;29:664–73.
20. Cheng JB, Levine MA, Bell NH, et al. Proc Natl Acad Sci U S A 2004;101:7711–5.
21. Molin A, Wiedemann A, Demers N, et al J Bone Mineral Res. 2017;32:1589–96.
22. Jones G, Byford V, West S, et al. Anticancer Res 2006;26:2589–96.
23. Strushkevich N, Usanov SA, Plotnikov AN, et al. J Mol Biol 2008;380:95–106.
24. Wang TJ, Zhang F, Richards JB, et al. Lancet 2010;376:180–8.
25. Hollis BW: Detection of vitamin D and its major metabolites. In: Feldman D, Pike JW, Glorieux FH, eds. Vitamin D. 2nd ed. New York: Elsevier, 2005:931–50.
26. Safadi FF, Thornton P, Magiera H, et al. J Clin Invest 1999;103: 239–51.
27. Willnow TE, Nykjaer A. Endocytic pathways for 25-hydroxyvitamin D_3. In: Feldman D, Pike JW, Glorieux FH, eds. Vitamin D. 2nd ed. New York: Elsevier, 2005:153–63.
28. Martinez I, Saracho R, Montenegro J, et al. Nephrol Dial Transplant 1996;11:22–8.
29. St-Arnaud RH, Messerlian SH, Moir JM, et al. J Bone Miner Res 1997;12:1552–9.
30. Takeyama K, Kitanaka S, Sato T, et al. Science 1997;277:1827–30.
31. Monkawa T, Yoshida T, Wakino S, et al. Biochem Biophys Res Commun 1997;239:527–33.
32. Fu GK, Lin D, Zhang MYH, et al. Mol Endocrinol 1997;11:1961–70.
33. Armbrecht HJ, Hodam TL, Boltz MA. Arch Biochem Biophys 2003;409:298–304.
34. Quarles LD. Am J Physiol 2003;285:E1–9.
35. Meyer MB, Benkusky NA, Kaufmann M, et al. J Biol Chem 2019;294:9518–35.
36. Yamamoto K, Uchida E, Urushino N, et al. J Biol Chem 2005;280:30511–6.
37. Fraser D, Kooh SW, Kind P, et al. N Engl J Med 1973;289:817–22.
38. Dardenne O, Prud'homme J, Arabian A, et al. Endocrinology 2001;142:3135–41.
39. Panda DK, Miao D, Bolivar I, et al. J Biol Chem 2004;279:16754–66.
40. Holick MF, Schnoes HK, DeLuca HF, et al. Biochemistry 1972;11:4251–5.
41. Holick MF, Kleiner-Bossaller A, Schnoes HK, et al. J Biol Chem 1973;248:6691–6.
42. Martineau C, Naja RP, Husseini A, et al. J Clin Invest 2018;128:3546–57.
43. Ohyama Y, Okuda K. J Biol Chem 1991;266:8690–5.
44. Tenenhouse HS, Yip A, Jones G. J Clin Invest 1988;81:461–5.
45. Jones G, Vriezen D, Lohnes D, et al. Steroids 1987;49:29–53.
46. Makin G, Lohnes D, Byford V, et al. Biochem J 1989;262:173–80.
47. Reddy GS, Tserng KY. Biochemistry 1989;28:1763–9.
48. Prosser D, Kaufmann M, O'Leary B, et al. Proc Natl Acad Sci U S A 2007;104:12673–8.
49. Shinki T, Jin CH, Nishimura A, et al. J Biol Chem 1992;267: 13757–62.
50. Lohnes D, Jones, G. J Nutr Sci Vitaminol (Tokyo) 1992;Spec No:75–8.
51. Schlingmann KP, Kaufmann M, Weber S, et al. New Engl J Med 2011;365:410–21.
52. Shah AD, Hsiao EC, O'Donnell B, et al. J Clin Endocrinol Metab 2015;100:2832–6.
53. St-Arnaud R, Arabian A, Travers R, et al. Endocrinology 2000;141:2658–66.
54. Masuda S, Byford V, Arabian A, et al. Endocrinology 2005;146:825–34.
55. Esvelt RP, Schnoes HK, DeLuca HF. Biochemistry 1979;18: 3977–83.
56. Xue Y, Hashizume T, Shuhart MC, et al. Mol Pharmacol 2006;69:56–65.
57. Thummel KE, Brimer C, Yasuda K, et al. Mol Pharmacol 2001;60:1399–406.
58. Thompson PD, Jurutka PW, Whitfield GK, et al. Biochem Biophys Res Commun 2002;299:730–8.
59. Hawkes CP, Li D, Hakonarson H, et al. J Clin Endocrinol Metab 2017;102:1440–1446.
60. Roizen JD, Li D, O'Lear L, et al. J Clin Invest 2018;128:1913–18.
61. Haussler MR, Whitfield GK, Haussler CA, et al. J Bone Miner Res 1998;13:325–49.
62. Whitfield GK, Jurutka PW, Haussler C, et al. Nuclear receptor: structure-function, molecular control of gene transcription and novel bioactions. In: Feldman D, Pike JW, Glorieux FH, eds. Vitamin D. 2nd ed. New York: Elsevier, 2005:219–62.
63. Stumpf WE. Histochem Cell Biol 1995;104:417–27.
64. Pike JW. Annu Rev Nutr 1991;11:189–216.
65. White JH. J Steroid Biochem Mol Biol 2004;89–90:239–44.
66. Yoshizawa T, Handa Y, Uematsu Y, et al. Nat Genet 1997;16: 391–6.
67. Pike JW, Zella LA, Meyer MB, et al. J Bone Miner Res 2007;22: V16–9.
68. Bouillon R, Carmeliet G, Verlinden L, et al. Endocr Rev 2008;29:726–76.
69. Benn BS, Ajibade D, Porta A, et al. Endocrinology 2008;149: 3196–205.
70. Bianco SD, Peng JB, Takanaga H, et al. J Bone Miner Res 2007;22:274–85.
71. Kutuzova GD, Akhter S, Christakos S, et al. Proc Natl Acad Sci U S A 2006;103:12377–81.
72. Lee GS, Lee KY, Choi KC, et al. J Bone Miner Res 2007;22: 1968–78.
73. Xu H, Bai L, Collins JF, et al. Am J Physiol 2002;282:C487–93.
74. Liu M, Lee MH, Cohen M, et al. Genes Dev 1996;10:142–53.

75. Wang Q, Jones JB, Studzinski GP. Cancer Res 1996;56:264–7.
76. Rots NY, Liu M, Anderson EC, et al. Mol Cell Biol 1998;18:1918.
77. Bettoun DJ, Buck DW, Lu JF, et al. J Biol Chem 2002;277: 24847–50.
78. Masuda S, Jones G. Mol Cancer Ther 2006;5:797–808.
79. Wagner N, Wagner KD, Schley G, et al. Exp Eye Res 2003;77: 1–9.
80. James SY, Mackay AG, Colston KW, et al. J Steroid Biochem Mol Biol 1996;58:395–401.
81. Mathiasen IS, Sergev IN, Bastholm L, et al. J Biol Chem 2002;277:30738–45.
82. Xie SP, James SY, Colston KW. J Endocrinol 1997;154:495–504.
83. McGuire TF, Trump DL, Johnson CS. J Biol Chem 2001;276: 26365–73.
84. Xie ZJ, Komuves L, Yu QC, et al. J Invest Dermatol 2001;276: 11–6.
85. Sakai Y, Demay MB. Endocrinology 2000;141:2043–9.
86. Zinser GM, Sundberg JP, Welsh J. Carcinogenesis 2002;23: 2103–9.
87. Holick MF. Am J Clin Nutr 2004;80(Suppl):1678S–88S.
88. Grant WB, Garland CF. Nutr Cancer 2004;48:115–23.
89. Grant WB, Garland CF. J Intern Med 2002;252:178–9.
90. Schwartz GG, Hulka BS. Anticancer Res 1990;10:1307–11.
91. Bikle DD, Chang S, Crumrine D, et al. J Invest Dermatol 2004;122:984–92.
92. Bises G, Kallay E, Weiland T, et al. J Histochem Cytochem 2004;52:985–9.
93. Jones G, Ramshaw H, Zhang A, et al. Endocrinology 1999;140:3303–10.
94. Zeohnder D, Bland R, Williams MC, et al. J Clin Endocrinol Metab 2001;86:888–94.
95. Somjen D, Katzburg S, Stern N, et al. J Steroid Biochem Mol Biol 2007;107:238–44.
96. Barbour GL, Coburn JW, Slatopolsky E, et al. N Engl J Med 1981;305:440–3.
97. Dusso AS, Finch J, Brown A, et al. J Clin Endocrinol Metab 1991;72:157–64.
98. Adams JS, Gacad MA, Singer FR, et al. Ann N Y Acad Sci 1986,465:587–94.
99. Dusso AS, Kamimura S, Gallieni M, et al. J Clin Endocrinol Metab 1997;82:2222–32.
100. Stoffels K, Overbergh L, Giulietti A, et al. J Bone Miner Res 2006;21:37–47.
101. Hewison M, Adams JS. Extra-renal 1α-hydroxylase activity and human disease. In: Feldman D, Pike JW, Glorieux FH, eds. Vitamin D. 2nd ed. New York: Elsevier, 2005:1379–402.
102. Holick MF. N Engl J Med 2007;357:266–81.
103. Jones G. Semin Dial 2007;20:316–24.
104. Kidney Disease Outcomes Quality Initiative of the National Kidney Foundation. Am J Kidney Dis 2003;42(Suppl 3):S1–202.
105. Hollis BW. J Nutr 2005;135:317–22.
106. Bischoff-Ferrari HA, Dietrich T, Orav EJ, et al. Am J Med 2004;116:634–9.
107. Bischoff-Ferrari HA, Dietrich T, Orav EJ, et al. Am J Clin Nutr 2004;80:752–8.
108. Manson JE, Cook NR, Lee IM, et al VITAL Research Group. N Engl J Med. 2019;380:33–44.
109. Burt LA, Billington EO, Rose MS, et al. JAMA 2019;322:736–45.
110. Jones G, Horst RL, Carter G, et al. J Bone Miner Res 2007;22: V11–5.
111. Chen PS, Bosmann HB. J Nutr 1964;83:133–9.
112. Marx SJ, Jones G, Weinstein RS, et al. J Clin Endocrinol Metab 1989;69:1282–90.
113. Park EA. JAMA 1940;115:370–9.
114. Roborgh JR, de Man T. Biochem Pharmacol 1960;2:1–6.
115. Roborgh JR, de Man T. Biochem Pharmacol 1960;3:277–82.
116. Whyte MP, Haddad JG, Walters DD, et al. J Clin Endocrinol Metab 1979;48:906–11.
117. Houghton LA, Vieth R. Am J Clin Nutr 2006;84:694–7.
118. Trang HM, Cole DE, Rubin LA, et al. Am J Clin Nutr 1998;68:854–8.
119. Armas LA, Hollis BW, Heaney RP. J Clin Endocrinol Metab 2004;89:5387–91.
120. Romagnoli E, Mascia ML, Cipriani C, et al. J Clin Endocrinol Metab 93:3015–20.
121. Leventis P, Kiely PD. Scand J Rheumatol 2009;38:149–53.
122. Rapuri PB, Gallagher JC, Haynatzki G. Calcif Tissue Int 2004;74:150–6.
123. Holick MF, Biancuzzo RM, Chen TC, et al. J Clin Endocrinol Metab 2008;93:677–81.
124. Thacher TD, Obadofin MO, O'Brien KO, et al. J Clin Endocrinol Metab 2009;94:3314–21.
125. Heaney RP, Armas LA, Shary JR, et al. Am J Clin Nutr 2008;87:1738–42.
126. Hunt RD, Garcia FG, Walsh RJ. J Nutr 1972;102:975–86.
127. Gupta RP, Hollis BW, Patel SB, et al. J Bone Miner Res 2004;19:680–8.
128. Gupta RP, He YA, Patrick KS, et al. J Clin Endocrinol Metab 2005;90:1210–9.
129. Jones G, Byford V, Helvig C, et al. Abstract presented at the 14th International Vitamin D Workshop, Brugge, Belgium, October 4–8, 2009.
130. Tjellesen L, Gotfredsen A, Christiansen C. Calcif Tissue Int 1985;37:218–22.
131. Hosseinpour F, Ellfolk M, Norlin M, et al. Biochem Biophys Res Commun 2007;357:603–7.
132. Binet A, Kooh SW. Can J Public Health 1996;87:227–30.
133. Wu H, Gozdzik A, Barta JL, et al. Nutr Res 2009;29:255–61.
134. Wortsman J, Matsuoka LY, Chen TC, et al. Am J Clin Nutr 2000;72:690–3.
135. Reinehr T, de Sousa G, Alexy U, et al. Eur J Endocrinol 2007;157:225–32.
136. Jones G. Am J Clin Nutr 2008;88(Suppl):582S–6S.
137. Haddad JG, Matsuoka LY, Hollis BW, et al. J Clin Invest 1993;91:2552–5.
138. Gonzalez EA, Sachdeva A, Oliver DA, et al. Am J Nephrol 2004;24:503–10.
139. Judd SE, Tangpricha V. Am J Med Sci 2009;338:40–4.
140. Drechsler C, Pilz S, Obermayer-Pietsch B, et al. Eur Heart J 2010;31:2253–61.
141. Shephard RM, DeLuca HF. Arch Biochem Biophys 1980;202:43–53.
142. Vieth R. Bone Miner 1990;11:267–72.
143. Blank S, Scanlon KS, Sinks T, et al. Am J Public Health 1995;85:656–9.
144. Pettifor JM, Bikle DD, Cavaleros M, et al. Ann Intern Med 1995;122:511–3.
145. DeLuca HF, Prahl JM, Plum LA. Arch Biochem Biophys 2011;505:226–30.
146. Grant WG. Dermatoendocrinology 2009;1:289–93.
147. Mathew S, Lund RJ, Chaudhary LR, et al. J Am Soc Nephrol 2008;19:1509–19.
148. Giovannucci E. Curr Atheroscler Rep 2009;11:456–61.

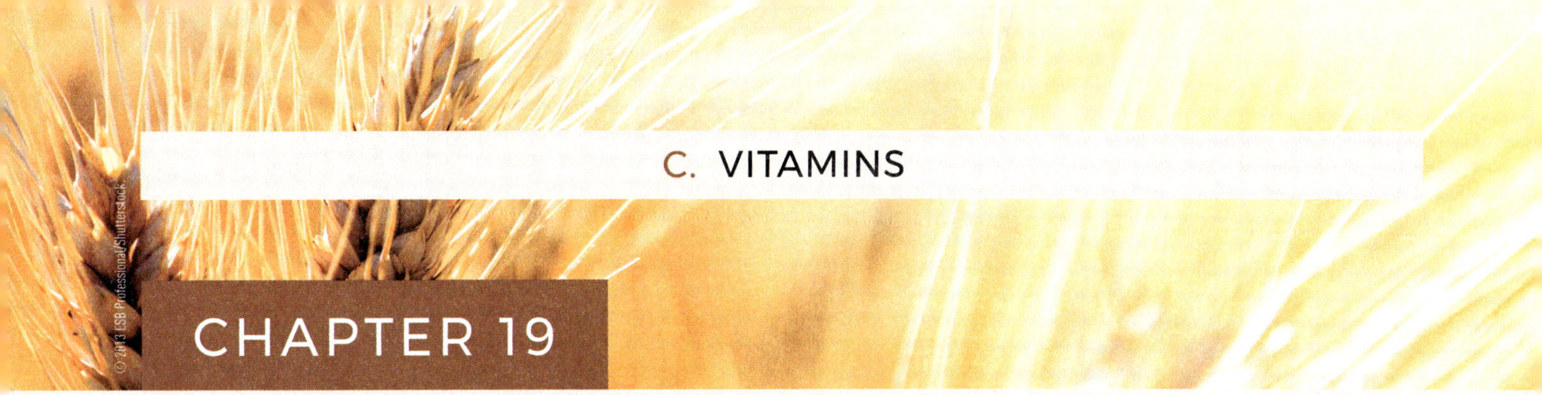

C. VITAMINS

CHAPTER 19

Vitamin E

Maret G. Traber

CHAPTER OUTLINE

Abbreviations

ABC, ATP-binding cassette; **α-CEHC**, 2,5,7,8-tetramethyl-2-(2′-carboxyethyl)-6-hydroxychroman; **apo**, apolipoprotein; **α-TTP**, α-tocopherol transfer protein; **AVED**, ataxia and vitamin E deficiency; **CHD**, coronary heart disease; **DRI**, dietary reference intakes; **EAR**, estimated average requirement; **γ-CEHC**, 2,7,8-trimethyl-2-(2′carboxyethyl)-6-hydroxychroman; **IOM FNB**, Institute of Medicine Food and Nutrition Board; **HDL**, high-density lipoprotein; **IU**, international unit; **LDL**, low-density lipoprotein; **LPL**, lipoprotein lipase; **MDR**, multi-drug resistance gene or p-glycoprotein; **NGT**, 2,7,8-trimethyl-2-(4,8,12-trimethyldecyl)-5-nitro-6-chromanol; **PLTP**, phospholipid transfer protein; **PUFA**, polyunsaturated fatty acid; **ROO**, peroxyl radical; **TPN**, total parenteral nutrition

Historical Overview

In 1922, Evans and Bishop,[1] during their investigations of infertility, first described fetal resorption as a symptom of vitamin E deficiency in rats fed "rancid lard." In 1936, Evans et al.,[2] isolated a factor from wheat germ and named it "α-tocopherol," a name derived from the Greek "tokos" (offspring) and "pherein" (to bear) with an "ol" to indicate that it was an alcohol. Two other tocopherols, β- and γ-, with lower biologic activities, were isolated from vegetable oils.[3] These early observations formed the foundation for defining the "biologic activity" of vitamin E, which is based on its ability to prevent or reverse specific vitamin E deficiency symptoms.[4] Necrotizing myopathy, fetal death and resorption, anemia, and accumulation of lipofuscin (a fluorescent pigment of "aging") in tissues are symptoms in animals fed experimentally vitamin E-deficient diets. A progressive, peripheral, sensory neuropathy is the first sign in humans of vitamin E deficiency, when caused by a genetic defect in the α-tocopherol regulatory protein.[5]

It was not until the mid-1960s that vitamin E deficiency was described in children with fat malabsorption syndromes.[6] Subsequently, vitamin E-deficient patients with peripheral neuropathies, but without fat malabsorption, were described.[6,7] Studies in such patients opened new avenues in vitamin E investigations because these latter patients were found to have a genetic defect in the hepatic α-tocopherol transfer protein (α-TTP).[8]

Terminology

The Institute of Medicine (IOM) Food and Nutrition Board,[9] defined that only α-tocopherol meets human vitamin E requirements. Molecules with α-tocopherol antioxidant activity include: four tocopherols and four tocotrienols (**Figure 19.1**). These molecules have similar chromanol structures: trimethyl (α-), dimethyl (β- or γ-), and monomethyl (δ-); tocopherols have a phytyl side chain, while tocotrienols have an unsaturated side chain. These various forms are not interconvertible by animals.

There are eight naturally occurring forms of vitamin E. Shown are *RRR*-α-tocopherol with naturally occurring stereochemistry in the three chiral centers at positions 2, 4′ and 8′ and α-tocotrienol with an unsaturated, isoprenoid tail. As indicated, both tocopherols and tocotrienols have similar chromanol head structures: trimethyl (α-), dimethyl (β- or γ-), and monomethyl (δ-).

α-Tocopherol, synthesized chemically by condensation of trimethyl hydroquinone with racemic isophytol,[10] contains eight stereoisomers (arising from the three chiral centers: 2, 4′, and 8′, specifically: *RRR*, *RSR*, *RSS*, *RRS*, *SRR*, *SSR*, *SRS*, *SSS*), and is designated *all-rac*-α-tocopherol (incorrectly called *d,l*-α-tocopherol). The naturally occurring *RRR*-α-tocopherol (formerly called *d*-α-tocopherol) is only one of the eight stereoisomers present in *all rac*-α-tocopherol. The IOM[9] determined that human vitamin E requirements are met only by 2*R*-α-tocopherols, which are preferentially retained by the human body (**Figure 19.2**). These 2*R*-α-tocopherols make up half of the stereoisomers in *all-rac*-α-tocopherol. Previously, α-tocopherol and other vitamin E forms were included as sources of vitamin E; these latter forms no longer are included because of a lack of evidence in humans showing that they have health benefits, or that they reverse vitamin E deficiency.[9]

The three chiral centers in natural α-tocopherol give rise to eight different stereoisomers in synthetic α-tocopherol (*all rac*-α-tocopherol). These are: *RRR*-, *RRS*-, *RSR*-, *RSS*-, *SRR*-, *SSR*-, *SRS*-, *SSS*-α-tocopherols. Shown are *SRR*-α- and *RRR*-α-tocopherols. *SRR*- is an example of 2*S*-α-tocopherols and its dramatic structural difference is readily apparent. The mittens indicate that the α-tocopherol transfer protein (α-TTP) is "right" handed and 2*R*-α-tocopherols only "fit" the α-TTP. This illustration explains why only 2*R*-α-tocopherols meet the human vitamin E requirement.

The IOM vitamin E definition has led to confusion about vitamin E units. α-Tocopherol supplements often contain esters, such as α-tocopheryl acetate, succinate, or nicotinate. Previously, the vitamin E International unit (IU) was defined as 1 IU = 1 mg

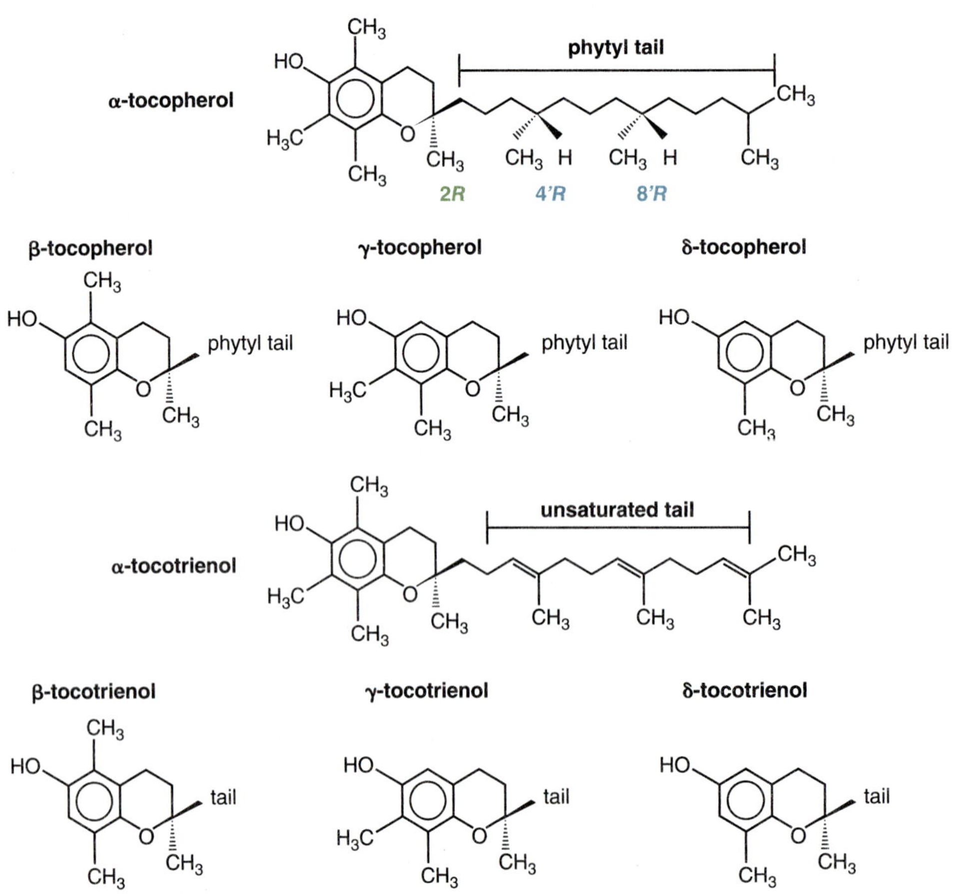

Figure 19.1 Tocopherols and Tocotrienol Structures.

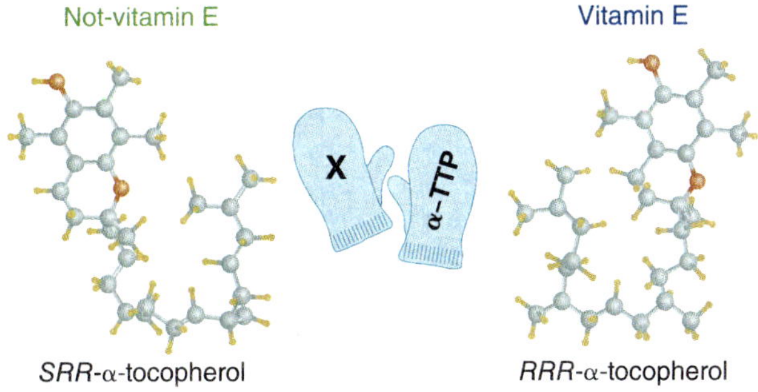

Figure 19.2 *RRR* and *SRR*-α-Tocopherol Structures.

all rac α-tocopheryl acetate or 0.67 mg *RRR*-α-tocopherol, or 0.74 mg *RRR*-α-tocopheryl acetate. However, the IOM (Table 6.1 in reference)[9] redefined the vitamin E requirement in milligrams of 2*R*-α-tocopherol and provided conversion factors, such that 1 mg *all rac*-α-tocopherol is equal to 0.5 mg *RRR*-α-tocopherol. Vitamin E milligrams are equivalent to 2*R*-α-tocopherol milligrams to meet the rules for labeling on Nutrition and Supplements Facts labels set by the US Food and Drug Administration (FDA).[11]

Vitamin E as defined[9,11]:

IU *all-rac*-α-tocopherol or its esters
= 0.45 mg 2*R*-α-tocopherol

IU *RRR*-α-tocopherol or its esters
= 0.67 mg 2*R*-α-tocopherol

Chemistry

Antioxidant Activity

Vitamin E functions in vivo as a lipid soluble, chain-breaking antioxidant.[12] Any functions as a signaling molecule are likely dependent on lipid oxidation products that arise when insufficient α-tocopherol is available for their protection.[13] For example, various polyunsaturated fatty acids can give rise to oxylipins, which are signaling molecules.[14]

Vitamin E is a potent peroxyl radical scavenger and especially protects polyunsaturated fatty acids (PUFAs). When peroxyl radicals (ROO·) are formed, these react 1,000-times faster with vitamin E (Vit E–OH) than with PUFA (RH), and form the tocopheroxyl radical (Vit E–O·):

In the presence of vitamin E: ROO· + Vit E–OH →
ROOH + Vit E–O·

In the absence of vitamin E: ROO· + RH →
ROOH + R·

$$R\cdot + O_2 \rightarrow ROO\cdot$$

In this way, vitamin E prevents further lipid autoxidation.

The two-electron oxidation product of α-tocopherol is α-tocopheryl quinone, but its relevance in vivo is difficult to document[15] because the one electron oxidation product can be reduced by other antioxidants (see next section). The regeneration of tocopherol from the tocopheroxyl radical in the erythrocyte (RBC) membrane is especially relevant to the discussion of oxidative stress causing RBC destruction and ultimately anemia, a vitamin E deficiency symptom. A number of antioxidant enzymes, including superoxide dismutase, glutathione peroxidase, and catalase, are also involved in the protection of RBCs against oxidative stress and the sparing of vitamin E, especially in the case of Wilson disease (e.g., copper overload).[16]

Vitamin E Antioxidant Network

The tocopheroxyl radical (Vit E–O·) formed in membranes emerges from the lipid bilayer into the aqueous domain.[17] It is here that the tocopheroxyl radical reacts with vitamin C (or other reductants serving as hydrogen donors, AH), thereby oxidizes the latter and returns vitamin E to its reduced state.

Vit E–O· + AH → Vit E–OH + A·

Biologically important hydrogen donors include ascorbate (vitamin C) and thiols, especially glutathione. This phenomenon has led to the idea of vitamin E recycling in which the antioxidant function of the vitamin E radical is continuously restored by other antioxidants, as well as the metabolic activity of cells.[18] Regeneration of tocopherol from its radical by vitamin C appears to be a physiologically relevant mechanism, based on studies in humans (see next section), guinea pigs[19] and rodents.[20] Studies in vitamin E-deficient zebrafish embryos suggest that dysregulation of choline and 1-carbon metabolism may also be caused by the metabolic consequences of increased lipid peroxidation and dysregulation and depletion of key metabolites in the interwoven metabolic cycles that are dependent on adequate protection by vitamin E (**Figure 19.3**).[21]

Increased Utilization of Vitamin E by Humans Under Oxidative Stress

Oxidative stress caused by ultramarathon running increased rates of plasma α-tocopherol disappearance in humans.[22] Moreover, prior vitamin E and C supplementation decreased markers of lipid peroxidation in runners.[23] Chronic oxidative stress and inflammation caused by cigarette smoking also increased plasma α-tocopherol disappearance rates in cigarette smokers compared with nonsmokers.[24] Moreover, the smokers with the lowest plasma ascorbic acid concentrations had the fastest α-tocopherol disappearance rates.[18] In a subsequent study, Bruno et al.[25] showed that these rates could be normalized by prior vitamin C supplementation. Importantly, both α- and γ-tocopherols were similarly affected by vitamin C status, a finding suggesting that oxidation of the tocopherols is the mechanism for the faster vitamin E disappearance in the presence of low vitamin C status.[25] Lower α-tocopherol bioavailability in persons with the Metabolic Syndrome was proposed to derive, in part, from their low vitamin C status and high levels of endotoxin, which caused increased oxidative distress.[26]

Structure—Function Relationships of non-α-Tocopherols Forms

Numerous reports have noted health benefits of non-α-tocopherols as anti-inflammatory agents, antioxidants, and antiangiogenic compounds in various chronic diseases.[27] One mechanism in

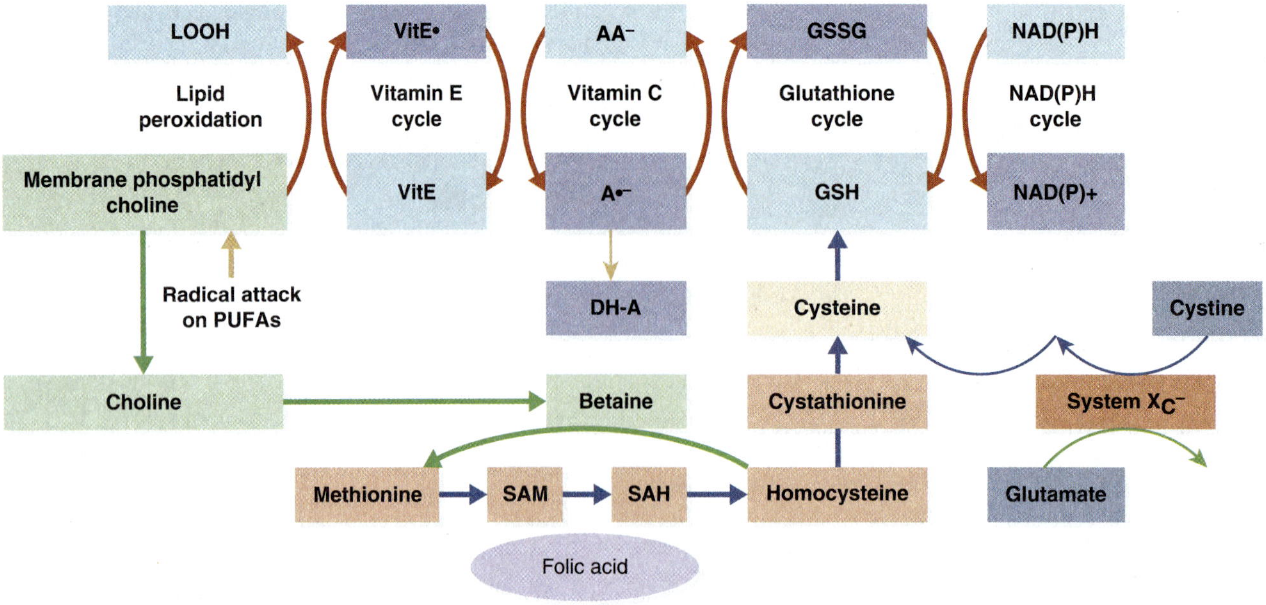

Figure 19.3 Lipid Peroxidation and the Antioxidant Cycles Impact Cellular Metabolism. Radical attack on polyunsaturated fatty acids in membrane phospholipids in the presence of oxygen results in the formation of the lipid peroxyl radical (LOO·), which can be reduced to the lipid hydroperoxide (LOOH) by the action of vitamin E. The vitamin E radical (VitE·) generated in this reaction can be reversibly reduced to vitamin E by the action of vitamin C (AA-). During this reaction, AA is oxidized to the ascorbyl radical, then to dehydroascorbic acid (DH-A). The ascorbyl radical can be regenerated to the reduced form by glutathione (GSH), leading to GSH oxidation (GSSG). GSSG then can be enzymatically reduced to GSH by the action of glutathione reductase employing NADPH as a cofactor. GSH can be regenerated either by the reduction of the oxidized glutathione (GSSG) through NADPH-mediated enzymatic process (glutathione cycle), or by GSH de novo synthesis from cysteine. Cysteine is obtained either through the function of the system Xc^- or from metabolism of cystathionine. Notably, phosphatidyl choline is a major source of choline, which is oxidized to betaine, and is a major source of one-carbon unit. By recruiting intermediate metabolites from other intersecting pathways, protection from lipid peroxidation becomes dependent on sulfur-containing amino acid metabolism and on one-carbon metabolism. Thus, α-tocopherol deficiency decreases thiol status and betaine, the oxidation product of choline has a critical role in maintaining both 1-carbon metabolism and generating thiols. By recruiting intermediate metabolites from adjacent pathways to sustain effective lipid hydroperoxide elimination, vitamin E action is linked to NADPH metabolism and formation, sulfur-containing amino acid metabolism, and to single-carbon metabolism.

Adapted from Head B, Traber MG. Free Radic Biol Med 2021;176:80-91. doi: 10.1016/j.freeradbiomed.2021.09.016

which α-tocopherol cannot participate is scavenging reactive nitrogen species. In vitro, β-, γ- or δ-tocopherols can be nitrated.[28,29] Hoglen et al.,[30] demonstrated that 5-nitro-γ-tocopherol (2,7,8-trimethyl-2-(4,8,12-trimethyldecyl)-5-nitro-6-chromanol; NGT) is the major reactive product between peroxynitrite and γ-tocopherol. NGT has been quantitated in human plasma,[15] in the plasma of patients with coronary artery disease,[31] from cigarette smokers,[32,33] and in brains collected post-mortem from patients with Alzheimer disease.[34] Indeed, lower tocopherols (both α- and γ-tocopherols) and higher oxidation products are associated with Alzheimer disease.[35]

Dietary Sources

The richest food sources of vitamin E are nuts, seeds, and edible vegetable oils,[36] which contain varying proportions of the eight homologs: α-, β-, γ- and δ-tocopherols or tocotrienols. *RRR*-α-tocopherol is especially high in wheat germ oil, safflower oil, and sunflower oil, whereas soybean and corn oils contain predominantly γ-tocopherol, as well as some tocotrienols. Foods that have been fortified with *all rac*-α-tocopheryl acetate include some breakfast cereals, tomato juice, orange juice, milk, etc.

Dietary Reference Intakes

The dietary reference intakes (DRIs) for vitamin C, vitamin E, selenium, and carotenoids were published in 2000.[9] The DRIs distinguish between natural and synthetic α-tocopherols (*RRR*- and *all rac*-α-tocopherols) because the structures are physically different and they have different fates with respect to transport and catabolism.[9]

The estimated average requirement (EAR) was based on the amount of 2*R*-α-tocopherol intake that reversed in vitro peroxide-induced erythrocyte hemolysis in men who were vitamin E deficient as a result of consuming a vitamin E deficient diet for 5 years.[9] The RDA values for α-tocopherol by life stage are given in **Table 19.1**. The EAR of 12 mg 2*R*-α-tocopherol was chosen because intakes at this level and above resulted in plasma α-tocopherol concentrations that prevented in vitro hydrogen peroxide-induced erythrocyte hemolysis. The assumption was made that men and women would have similar requirements because women, despite their lower body weight, have a larger percentage of body fat needing antioxidant protection. The 2000 RDA for adults (both men and women ≥19 years old), defined as 2*R*-α-tocopherol, is 15 mg/day.

Table 19.1 Dietary Reference Intake Values for Vitamin E by Life Stage Group[a]

Life Stage Group	RDA[b] (mg/d)	AI[c] (mg/d)
0–6 mo		4
7–12 mo		5
1–3 y	6	
4–8 y	7	
9–13 y	11	
14–18 y	15	
>18 y	15	
Pregnancy		
≤18 y	15	
19–50 y	15	
Lactation		
≤18 y	19	
19–50 y	19	

[a]In units of mg 2*R*-α-tocopherol.

[b]RDA, recommended dietary allowance. The intake that meets the nutrient needs of almost all (97%–98%) of individuals in a group.

[c]AI, adequate intake. The observed average or experimentally determined intake by a defined population or subgroup that appears to sustain a defined nutritional status, such as growth rate, normal circulating nutrient values, or other functional indicators of health. The AI is used if sufficient scientific evidence is not available to derive an estimated average requirement. For healthy infants receiving human milk, the AI is the mean intake. The AI is not equivalent to an RDA.

Data from Food and Nutrition Board, Institute of Medicine. Dietary Reference Intakes for Vitamin C, Vitamin E, Selenium, and Carotenoids. Washington, DC: National Academy Press, 2000.

The tolerable upper intake level (UL) was set at 1,000 mg/day for any form of supplemental α-tocopherol because all forms are absorbed.[9] This was one of the few UL that was set using data from rats because sufficient and appropriate quantitative data assessing long-term adverse effects of vitamin E supplements in humans were not available.

The α-tocopherol amounts consumed by most U.S. adults are sufficient to prevent overt symptoms of deficiency,[37] but the actual quantities consumed by U.S. adults are closer to 8 mg, as assessed by various surveys. Thus, 92% of men and 97% of women in the United States do not consume 12 mg vitamin E daily.[38] The report from the 2010 Dietary Guidelines Committee acknowledged this discrepancy between intake and recommendations, the 2015 Dietary Guidelines also recognized vitamin E as an under-consumed nutrient of concern,[39] and it remained a concern in the 2020 Guidelines.

Most people do not consume 15 mg α-tocopherol daily and; therefore, they may be at increased risk for various chronic diseases. A follow-up report from the Alpha-Tocopherol Beta-Carotene cancer prevention trial assessed baseline vitamin E status and dietary intakes of the 29,092 men who had been followed for 19 years since the study's initiation, during which time 13,380 deaths ensued. The men at baseline in the highest, compared with the lowest, serum α-tocopherol quintiles had significantly lower risks of total and cause-specific mortality, including cardiovascular disease and cancer.[40] The optimum relative reductions in mortality were associated with an estimated dietary vitamin E intake of 12 mg α-tocopherol,[40] a dietary value not different from the EAR (12 mg).[9] This finding suggests a health benefit of obtaining the RDA amount of vitamin E from the diet.

Health effects of high dietary vitamin E intakes also include beneficial effects on the microbiome. In a cross-sectional study of 1,709 members of a multiethnic cohort, higher serum α-tocopherol concentration was associated with higher bacterial diversity,[41] which is generally believed to be beneficial to the host. As these are associations, it is not clear whether the α-tocopherol concentrations are reflective of certain dietary patterns, socioeconomic status or other factors.

Physiologic Factors Influencing Utilization

Digestion and Intestinal Absorption

Vitamin E absorption measured in healthy women using multiple, different techniques was estimated to be 55%,[42] while the remaining ingested vitamin E was excreted in the feces.[42] Dietary fat is not necessary for α-tocopherol absorption,[42,43] but fat is required for absorption of vitamin E from supplements or fortificants containing vitamin E-esters.[44,45] Thus, hydrolysis of the vitamin E-ester by pancreatic enzymes is needed for uptake into enterocytes.[46] Bile acids are also required for vitamin E absorption, as children with cholestatic liver disease who have defects in secretion of bile become vitamin E deficient.[47] The key steps in vitamin E absorption are micellization, entry into the enterocyte, followed by packaging into chylomicrons (**Figure 19.4**).[37] Chylomicrons contain triglycerides, free and esterified cholesterol, phospholipids, and apolipoproteins (especially apolipoprotein (apo) B48). In addition, fat-soluble vitamins, carotenoids, and other fat-soluble dietary components are incorporated into chylomicrons.[48] Discrimination among forms of vitamin E does not occur during their intestinal absorption and secretion in chylomicrons.

The mechanisms for trafficking of vitamin E into and through the intestinal absorptive cells remains under investigation. Vitamin E uptake into enterocytes is facilitated by membrane transporters, such as Niemann-Pick C1-like 1 (NPC1-L1)[49] and by the scavenger receptor B1 (SR B-1).[50,51] Vitamin E is retained in the enterocytes until sufficient fat is available to promote chylomicron secretion.[42,52] Lipid droplets accumulate in enterocytes,[52,53] suggesting that these droplets allow metering of fat secretion into the circulation.[54,55] Similarly, various dietary vitamin E forms can accumulate in enterocyte lipid droplets, these droplets then coalesce with nascent chylomicrons to enrich their lipid-contents with vitamin E and, in the process, vitamin E is incorporated into chylomicrons.[42] A similar process has been shown for cholesterol[56] and for carotenoids.[51]

In summary, vitamin E absorption is a multistep process requiring micelle formation in the intestinal lumen, presentation to enterocytes, uptake into the cells and subsequent secretion in association with chylomicrons. Vitamin E is not appreciably secreted into plasma in persons lacking the ability to secrete chylomicrons (e.g., apoB48).

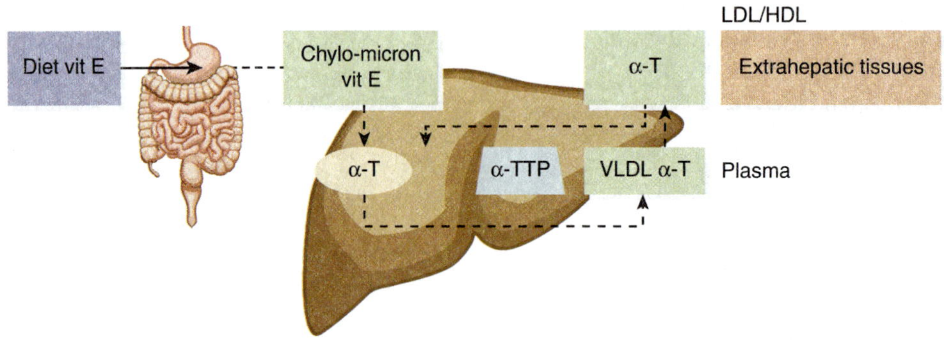

Figure 19.4 Pathways for Absorption of Vitamin E and Its Delivery to Tissues. Vitamin E absorption requires bile acids (secreted from the liver) and fatty acids and monoglycerides (released from dietary fat by pancreatic enzymes) for micelle formation. Following uptake into enterocytes of the intestine, all forms of dietary vitamin E are incorporated into chylomicrons. These triglyceride-rich lipoproteins are secreted into the circulation where lipolysis takes place and the resultant chylomicron remnants are taken up by the liver. In the liver, the α-tocopherol transfer protein (α-TTP) transfers α-tocopherol to the plasma membrane, where α-tocopherol is exchanged for phosphatidyl inositol phosphates (PIPs). The α-tocopherol in the plasma membrane is then exported by ABCA1 to an acceptor, such as nascent, very low-density lipoproteins (VLDL). Following VLDL secretion into plasma, lipolysis of VLDL by lipases results in the preferential enrichment of circulating lipoproteins with *RRR*-α-tocopherol. The uptake of these lipoproteins results in the delivery of *RRR*-α-tocopherol to extrahepatic tissues.

Data from Traber MG, Mah E, Leonard SW, Bobe G, Bruno RS. Am J Clin Nutr 2017;105(3):571-9. doi: 10.3945/ajcn.116.138495

Plasma Transport

Unlike other fat-soluble vitamins, which have their own specific plasma transport proteins, vitamin E is transported nonspecifically in plasma lipoproteins. During chylomicron catabolism in the circulation and its delipidation by lipoprotein lipase, some of the newly absorbed vitamin E is nonspecifically transferred to circulating lipoproteins.[37] During, this process, vitamin E is also transferred to HDLs,[42] which readily transfer vitamin E to other circulating lipoproteins.

The liver is the site of preferential selection of α-tocopherol from the various dietary vitamin E forms (Figure 19.4). Following partial delipidation of chylomicrons, the remnants are taken up by the liver, and fats are repackaged in very low density lipoproteins (VLDLs).[37] It is at this stage that 2*R*-α-tocopherol is selectively secreted into plasma (see next section). During the VLDL delipidation cascade, low-density lipoproteins (LDL) are formed. Some α-tocopherol remains with LDL, and some is transferred to HDL. Thus, the major plasma lipoproteins, LDLs and HDLs, become enriched in the circulation with α-tocopherol.

Hepatic α-Tocopherol Transfer Protein

α-TTP (30–35 kDa) has been isolated from humans and a variety of animals. α-TTP is expressed largely by hepatocytes,[57] but α-TTP mRNA has also been detected at low levels in rat brain, spleen, lung, and kidney.[58] Various mutations in the α-TTP gene (*TTPA*) result in a dysfunctional or absent α-TTP, thereby causing human vitamin E deficiency.[59]

TTPA belongs to a family of hydrophobic ligand-binding proteins that have a *cis*-retinal binding motif sequence (CRAL_TRIO). Panagablko et al.[60] showed that, of the CRAL_TRIO members, only α-TTP appears to have sufficient affinity to serve as a physiologic mediator of α-tocopherol transfer. α-TTP regulation is not modified by vitamin E deficiency,[61] or excess; however, the protein itself appears to be protected when bound to α-tocopherol.[62]

The hepatic mechanism for α-tocopherol selection and transfer by α-TTP is a multi-step process.[63] In the hepatocyte, the uptake of lipoprotein remnants and their hydrolysis in recycling endosomes leads to endosomal α-tocopherol enrichment. α-TTP selects the α-tocopherol from the endosomal membrane and, in the process, α-TTP undergoes a conformational change.[64] α-TTP then transfers α-tocopherol to the inner plasma membrane,[64] where the complex binds,[65] and α-tocopherol is exchanged for phospholipid inositol phosphates (PIPs).[63] The α-tocopherol is subsequently transferred from the membrane by ATP-binding cassette A1 (ABCA1) to an external acceptor, such as nascent VLDL or HDL, outside of the hepatocyte.[66] This process is well described in the hepatocyte, but α-TTP appears also to be important for transferring α-tocopherol within other tissues. α-TTP, in Müller cells of the retina, traffics vitamin E in the eyes.[67,68] Studies describing the mechanism of α-tocopherol delivery from specific astrocytes (glial fibrillary acidic protein-positive) to neighboring neurons show that this process is also α-TTP-facilitated.[69]

α-TTP is also critical during pregnancy; it is increased at the site of implantation,[70,71] expressed in the syncytiotrophoblast and trophoblast cells of the human placenta[71-73] and is additionally expressed by the human yolk sac.[74] Thus, it is likely that α-tocopherol is needed both by the fetus and by the mother to protect her from the oxidative stress of the rapidly growing fetus. Indeed, low plasma α-tocopherol concentration has been associated with doubled risk of miscarriage in Bangladeshi women.[75] Early failure of pregnancy is associated with lipid peroxidation with resultant damage to the syncytiotrophoblast.[76] Studies in zebrafish embryos have shown that α-TTP is needed for formation the developing brain and that its absence is lethal.[21] Downstream effects of inadequate vitamin E in the embryo include malformation of the nervous system, and dysregulation of energy and one-carbon metabolism. All these effects appear to be dependent on increased lipid peroxidation.[21]

Distribution to Tissues

Vitamin E is transported in plasma lipoproteins nonspecifically and mechanisms of lipoprotein metabolism facilitate the delivery of vitamin E to tissues.[37] Tissues likely acquire vitamin E by at least four major routes: 1) Vitamin E is delivered to tissues during triglyceride-rich lipoprotein catabolism mediated by lipoprotein lipase. 2) Vitamin E is delivered via lipoprotein uptake by the LDL receptor as well as by other lipoprotein receptors. 3) Vitamin E is delivered by HDL via the SR-B1 or CD36. 4) Vitamin E rapidly exchanges between lipoproteins, as well as between lipoproteins and membranes and; therefore, exchange mechanisms may enrich membranes with vitamin E.

Remarkably, the mechanisms of vitamin E efflux from tissues are not well characterized. More than 90% of the human body pool of α-tocopherol is located in the adipose tissue, with more than 90% of adipose tissue α-tocopherol in fat droplets.[77] Thus, the analysis of adipose tissue α-tocopherol content provides a useful estimate of long-term vitamin E intake.[78] In α-tocopherol-deficient humans, adipose tissue α-tocopherol content is lower than in healthy individuals, although it is not clear if this is due to decreased delivery, increased export, or increased utilization.[79] Studies in pediatric patients after burn injury suggest that adipose tissue does release α-tocopherol during oxidative distress.[80] Further studies are needed to establish mechanisms of intracellular α-tocopherol trafficking and tissue efflux.

Vitamin E Pharmacokinetics

Deuterated α-tocopherol has been used to assess the kinetics and distribution of α-tocopherol into various tissues in rats,[81] guinea pigs,[82] and humans[83]. From these studies, it is apparent that a group of tissues, such as erythrocytes, liver, and spleen, are in rapid equilibrium with the plasma α-tocopherol pool and readily replace "old" with "new" α-tocopherol.[84] Other tissues, such as heart, muscle, and spinal cord, have slower α-tocopherol turnover rates. Brain shows the slowest α-tocopherol turnover rate, perhaps because it expresses α-TTP.[69] By contrast, the peripheral nerves[5] are the most susceptible tissues in humans to α-tocopherol deficiency.[85]

A pharmacokinetic model of vitamin E transport in plasma has been developed using data from studies with deuteriumlabeled stereoisomers of α-tocopherol (*RRR*- and *SRR*-).[77] In control individuals, the fractional disappearance rates of deuterium labeled *RRR*-α-tocopherol (0.4 ± 0.1 pools per day) were significantly ($P < 0.01$) slower than for *SRR*-α-tocopherol (1.2 ± 0.6) The apparent half-life of *RRR*-α-tocopherol in normal individuals was approximately 48 hours, consistent with the "slow" disappearance of plasma *RRR*-α-tocopherol.[77] Because *RRR*-α-tocopherol is returned to the plasma, its apparent turnover is slow. This hepatic recirculation of *RRR*-α-tocopherol results in the daily replacement of nearly all of the circulating *RRR*-α-tocopherol.

Novotny et al.[86] used an extremely small dose of radioactive *RRR*-α-tocopherol (1.81 nmol [5-$^{14}CH_3$]-*RRR*-α-tocopherol; each dose had 3.70 kBq of ^{14}C) in healthy human adults to study its absorption, storage, and elimination. Complete urine and fecal collections were obtained over 21 days, and blood samples for 70 days, from dosing. A comprehensive, multicompartment, kinetic model of *RRR*-α-tocopherol metabolism and disposition was based on analyses of plasma, urine, and feces. According to the model, α-tocopherol absorption was estimated as 81%, based on recovery of the administered dose. Because the radioactive α-tocopherol could be followed for a long time, it was possible to show that the slowest turning over tissue was adipose tissue, with a half-life of 184 ± 48 days. Based on the results of this model, the authors suggested that the dietary requirements of vitamin E, under normal conditions, may be less than the current recommended dietary allowance (RDA) of 15 mg/day, and possibly as low as 4 mg.[86]

α-Tocopherol absorption was also quantified in healthy women using an intravenous (IV) bolus containing deuterium-labeled α-tocopherol (d_6-α-T, 30 mg), given as an oil: water emulsion to mimic chylomicrons, while simultaneously given an oral dose of a differently labeled α-tocopherol (d_3-α-T, 30 mg).[42] The IV d_6-α-T was rapidly cleared during the first 20 minutes following injection, then reappeared in the plasma; by contrast, the d_3-α-T took several hours to appear in the plasma. Absorption was quantified by estimating the ratio of the oral to the IV dose, by quantifying the amount of the oral dose excreted in feces, and separately modeling the two doses and estimating absorption.[42] These studies all found similar estimates of absorption, approximately 55%.

Catabolism and Excretion

Vitamin E is not accumulated in the liver to toxic levels, suggesting that the xenobiotic mechanisms of catabolism and excretion are important in preventing adverse vitamin E effects. Cytochrome P450 4F2 (CYP 4F2) initiates vitamin E catabolism.[87,88] During catabolism, all vitamin E forms undergo ϖ-oxidation,[89] followed by β-oxidation, conjugation and excretion in urine[48] or bile.[90] CYP 4F2 ω-hydroxylates the sidechain of both tocopherols and tocotrienols to form the long chain, 13′-hydroxy (13′-OH) catabolites.[87,89] The catabolite undergoes ω-oxidation to form the 13′-carboxy (13′-COOH) catabolite, which then undergoes several rounds of β-oxidation to form CEHC (**Figure 19.5**).[89]

Vitamin E catabolites, for example α-CEHC (2,5,7,8-tetramethyl-2-(2′-carboxyethyl)-6-hydroxychroman, Figure 19.5) and γ-CEHC (2,7,8-trimethyl-2-(2′carboxyethyl)-6-hydroxychroman)) are derived from α-tocopherols and α-tocotrienols and from γ-tocopherols and γ-tocotrienols, respectively.[91] High α-tocopherol intakes, e.g., most vitamin E supplements, lead to increases of plasma α-tocopherol, decreases of γ-tocopherol[92] and increases both α-CEHC and γ-CEHC excretion.[93,94]

The oxidized form of α-CEHC is α-tocopheronolactonyl quinone (α-TLQ, i.e., Simon's metabolite) and was first described in the 1950s.[95] α-TLQ can be generated both by α-CEHC oxidation in vitro and also appears to be generated in vivo from α-tocopheryl-quinone through the same catabolic process that generates α-CEHC from α-tocopherol (Figure 19.5).[96] Measurable plasma α-tocopheryl-quinone in humans also indicates that some α-tocopherol oxidation to the two-electron oxidation product occurs in vivo.[97] Importantly, α-TLQ is elevated tenfold in urine from children with diabetes compared with controls.[98] α-TLQ has also been reported in the urine of overweight people.[99] These findings suggest that α-TLQ may be a useful marker of in vivo oxidative stress and increased lipid peroxidation.

Although the tissues involved in vitamin E catabolism are not well identified, CYP 4F2 is expressed in liver and kidney,[100] as well as intestine.[101] The potential functions of vitamin E catabolites are under active investigation because they retain their antioxidant activity and are more water soluble than the parent compounds.[102] Health benefits[103] and vitamin E's anti-inflammatory activities have been attributed to the activity of its catabolites,[104] especially the 13′-COOH catabolite[105] because it inhibits 5-lipoxygenase—a key enzyme in the biosynthesis of leukotrienes from arachidonic acid.[106]

CYP 4F2 is not only involved in vitamin E catabolism but is also involved in 20-hydroxyeicosatetraenoic acid (20-HETE) production,[107,108] as well as catabolism of leukotrienes[109] and

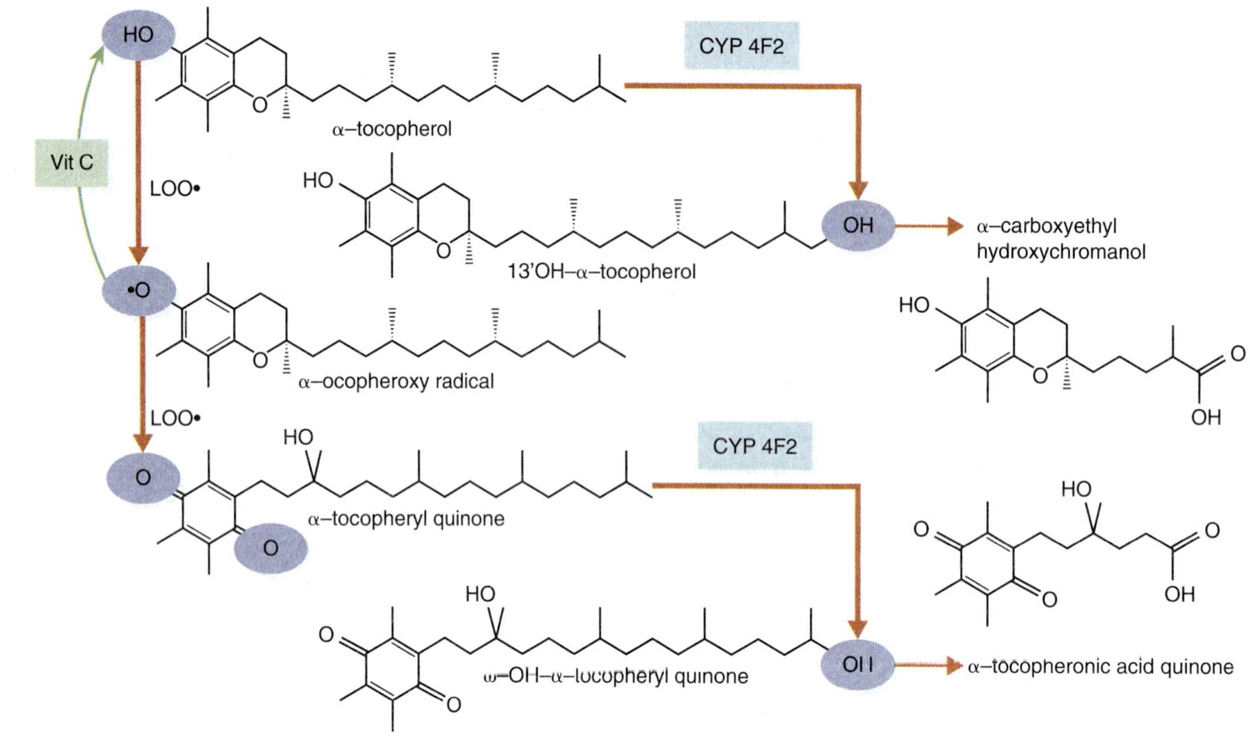

Figure 19.5 Pathways for Formation of Vitamin E Catabolites. "Extra" α-tocopherol is catabolized to α-carboxyethyl hydroxychromanol (α-CEHC). The first catabolic step depends on cytochrome P450 4F2 (CYP 4F2), which ω-hydroxylates the phytyl tail to permit β-oxidation to form α-CEHC. During lipid peroxidation, α-tocopherol quenches the peroxyl radical (LOO·) and becomes an α-tocopheroxyl-radical. Vitamin C (ascorbic acid, VitC) is necessary to "recycle" the α-tocopheroxyl-radical to its nonoxidized α-tocopherol form. If inadequate vitamin C is available, LOO· can further oxidize the α-tocopheroxyl-radical to α-tocopheryl-quinone. CYP 4F2 also initiates catabolism of the VitE-quinone, which, following β-oxidation, forms α-tocopheronic acid quinone, α-TLQ. Potentially, α-CEHC is a marker of vitamin E adequacy and α-TLQ is a marker of oxidative damage (α-tocopherol oxidized to the quinone).

long chain (16-26 carbon) saturated, unsaturated, and branched chain fatty acids.[110] CYP 4F2 is also a vitamin K oxidase that initiates vitamin K catabolism.[111,112] Vitamin E supplements (1,000 IU) in humans increased a biomarker of inadequate vitamin K status, under-γ-carboxylated prothrombin.[113] In contrast, 600 IU of natural source vitamin E on alternate days decreased thromboembolism in women participating in The Women's Health Study (39,876 women, 45+ years of age),[114] suggesting that inadequate vitamin K supported clotting mechanisms (see chapter 20).

The production of vitamin E catabolites is a xenobiotic process. Like other xenobiotics, CEHCs are sulfated or glucuronidated to increase their water solubility and excretion.[115,116] Xenobiotic transporters are likely candidates for mediating hepatic CEHC excretion because CEHCs are found in plasma, urine, and bile.[117] High liver α-tocopherol concentrations upregulated the rat multidrug resistance gene product, MDR1 (p-glycoprotein, or ABCB1)[117] and the mouse hepatic mdr1 gene.[118] These conjugates and transporters are potential candidates for promoting vitamin E catabolite excretion, but others are also possible.[119,120]

Assessment of α-Tocopherol Status

The IOM determined that a plasma concentration of lower than 12 μmol α-tocopherol/L was associated with evidence of α-tocopherol inadequacy; plasma concentrations in physiologically healthy individuals are approximately 20 μmol/L[9] or up to three times higher.[121] Several parameters can be measured in patients who may be α-tocopherol-deficient (**Table 19.2**). Although low serum or plasma α-tocopherol concentrations are indicative of vitamin E deficiency, measurement of plasma concentration is insufficient for patients with various forms of lipid malabsorption. Patients with elevated cholesterol or triglyceride concentrations may have α-tocopherol in the "normal" range, but this may not be sufficient to protect tissues.[122] Calculation of effective plasma α-tocopherol concentration needs to take into account plasma lipids by dividing the plasma α-tocopherol by the sum of plasma total cholesterol and triglycerides.[123]

Patients with peripheral neuropathies or retinitis pigmentosa of unknown causes should be evaluated to assess whether they are α-tocopherol-deficient. Plasma α-tocopherol concentration should be measured in all patients with ataxia to rule out α-tocopherol deficiency, because Friedreich ataxia is remarkably similar to that seen in patients with ataxia with vitamin E deficiency (AVED) (for Friedreich ataxia, see chapter 10).

During hyperlipidemia, elevated blood lipids "trap" α-tocopherol in the circulation, yielding "normal" plasma α-tocopherol concentration, while tissue α-tocopherol concentrations are inadequate and may permit oxidative injury.[122] An alternative marker for assessment of α-tocopherol status is urinary α-CEHC.[124] When α-tocopherol intakes are adequate, then urinary α-CEHC is excreted at low concentration. However, α-CEHC excretion is increased when dietary α-tocopherol intakes are increased,[125] and further increased in those persons taking

Table 19.2 Techniques for Assessment of α-Tocopherol Status

Measurements of plasma vitamin E concentration

- Normal values are ≥ 2 μmol α-tocopherol/L (μM) or 5 μg/mL
- Normal values are ≥ 8 mg α-tocopherol/g total lipid (cholesterol plus triglycerides) or 2.8 mg/g cholesterol

Measurements of adipose tissue vitamin E concentration

- ≥ 100 μg α-tocopherol/mg triglyceride

Functional signs of vitamin E deficiency

- Increased red cell hemolysis
- Increased lipid peroxidation (e.g., F_2-isoprostanes)
- Increased expired ethane or pentane

Clinical signs of vitamin E deficiency

- Neurologic testing: abnormal sensory nerve function
- Histopathology of peripheral nerves
- Electrophysiologic measurements

Gene testing

- Polymorphisms in α-tocopherol transfer protein (TTP)

Table 19.3 Disorders Requiring Administration of Supplemental α-Tocopherol to Prevent Vitamin E Deficiency

Genetic abnormalities

- α-Tocopherol transfer protein (AVED, ataxia with vitamin E deficiency)
- Apolipoprotein B (homozygous hypobetalipoproteinemia)
- Microsomal triglyceride transfer protein (abetalipoproteinemia)

Fat malabsorption syndromes

- Chronic cholestasis in children and adults
  - Idiopathic neonatal hepatitis
  - Familial cholestatic syndrome
  - Alagille syndrome
  - Paucity of interlobular bile ducts
  - Extrahepatic biliary atresia
  - Primary biliary cirrhosis
- Cystic fibrosis and pancreatic insufficiency
- Short bowel syndromes
- Crohn's disease
  - Mesenteric vascular thrombosis
  - Intestinal pseudo obstruction
- Chronic steatorrhea
  - Blind loop syndrome
  - Intestinal lymphangiectasia
  - Celiac disease
  - Chronic pancreatitis

Total parenteral nutrition

α-tocopherol supplements.[124] Thus, increases in urinary α-CEHC excretion reflect increases in α-tocopherol intake. By contrast, urinary α-CEHC excretion was lower in persons with lower α-tocopherol status.[43,126]

The problem with using α-CEHC as a biomarker of α-tocopherol status is that the regulation of α-tocopherol catabolism to α-CEHC is not well understood. This problem was studied in healthy women using IV deuterium-labeled α-tocopherol (d_6-α-T, 30 mg) and an oral dose of a differently labeled α-tocopherol (d_3-α-T, 30 mg), described above.[42,127] The concentrations of plasma and urinary labeled-α-CEHC suggest that α-tocopherol catabolism occurs during the first 24 h following dosing with the apparent peak in α-CEHC from the labeled doses peaking at approximately 12–18 h post-dosing.[127] More than 90% of the α-tocopherol catabolism to urinary α-CEHC took place in the liver. Labeled α-CEHC appears to leave the plasma and is excreted in urine in a linear manner, suggesting constant kidney clearance. The fecal long chain (11′-COOH- and 13′-COOH-α-T) catabolites derived from both the oral and the IV doses suggest that both the liver and the intestine are sources of these catabolites.[127]

Causes of Deficiency

Overt α-tocopherol deficiency occurs only rarely in humans.[121] Vitamin E deficiency symptoms include anemia[128]; an inverse correlation between vitamin E deficiency and hemolysis has been reported.[128,129] The earliest sign in humans is a sensory neuropathy,[6,130] which is usually seen in children, but adult onset has been observed.[131,132]

Vitamin E deficiency is usually caused by genetic defects in α-TTP[8,133,134] or genetic defects in lipoprotein metabolism, especially defects in synthesis and secretion lipoproteins containing apoB, including defects in the microsomal triglyceride transfer protein.[135] Vitamin E deficiency may be also caused by fat malabsorption syndromes (such as biliary obstruction, cholestatic liver disease, pancreatitis, or cystic fibrosis),[37,47] a result of severe malnutrition,[136] increased use of peroxidized oils,[137] and potentially increased consumption of ultra-processed foods with low micronutrient content.[138] Vitamin E absorption may be impaired by pharmaceutical agents designed to impair fat absorption,[139,140] or bariatric surgery to treat obesity.[141] (**Table 19.3**).

Genetic Defects in the α-Tocopherol Transfer Protein

The disorder "Ataxia with Vitamin E Deficiency" (AVED, MIM #277460), is characterized by extraordinarily low plasma α-tocopherol concentration[142] and progressive peripheral neuropathy and spinocerebellar ataxia[143] caused by a defect(s) in TTPA.[59] Both retinitis pigmentosa[144,145] and macular degeneration[146] have been reported in AVED. Vitamin E supplementation can prevent or halt progression of the disorder.[133,142,147] A dose of 800 to 1,200 mg/day is usually sufficient to prevent further deterioration of neurologic function, and in some cases, improvements have been noted.

Genetic Defects in Lipoprotein Synthesis

Vitamin E deficiency can be caused by defects in lipoprotein synthesis and secretion, including genetic defects in: 1) the microsomal triglyceride transfer protein (MTTP, abetalipoproteinemia,

OMIM 200100),[148] 2) in apoB (hypobetalipoproteinemia, OMIM 615558).[149,150] Long-term, daily α-tocopherol supplementation (>100 mg/kg body weight) can prevent symptom progression,[151] oxidative damage[152] and allows normal pregnancy.[153] Although plasma α-tocopherol concentrations never reach normal levels, adipose tissue α-tocopherol concentrations do reach normal levels in some affected people.[154,155]

Studies of persons with hypobetalipoproteinemia or abetalipoproteinemia (low to nondetectable circulating chylomicrons, VLDL, or LDL) have demonstrated that lipoproteins containing apoB are necessary for effective absorption and plasma transport of vitamin E.[37] Clinical features include retarded growth, acanthocytosis, retinitis pigmentosa, and a chronic progressive neurologic disorder with ataxia. These symptoms can be reversed with vitamin E supplementation if begun at a very young age.[156]

Fat Malabsorption Syndromes

Failure of intestinal micellar solubilization and malabsorption of dietary lipids leads to vitamin E deficiency. Indeed, any disorder that causes fat malabsorption can lead to vitamin E deficiency. The list of disorders associated with acquired vitamin E deficiency, assembled by Sokol,[5] includes cystic fibrosis, chronic dysfunction or resection of the small bowel, Crohn's disease, mesenteric vascular thrombosis or intestinal pseudo obstruction, blind loop syndrome, intestinal lymphangiectasia, celiac disease, and chronic pancreatitis. Serum vitamin E concentration may fall within 1 to 2 years of acquired lipid malabsorption in adolescents and adults; however, a 10 to 20 year interval between the identification of biochemical vitamin E deficiency and the onset of neurologic symptoms is generally observed in adults (5).[5] The prolonged time for onset of symptoms results from the prior accumulation of vitamin E in most tissues and its relatively slow depletion from nervous tissues (see Pharmacokinetics section).

Vitamin E supplementation in patients with fat malabsorption syndromes is difficult to achieve because they also malabsorb vitamin E. Treatment with tocopherol polyethylene glycol succinate (TPGS)[157,158] has improved serum α-tocopherol concentration[159] and neurologic function.[160]

Total Parenteral Nutrition

Patients receiving total parenteral nutrition (TPN) are provided, ideally, with their required nutrients—vitamin E (9–10 mg/day) is the level currently recommended by the European Society for Parenteral and Enteral Nutrition (ESPEN).[161] However, evaluation of the vitamin E status of TPN patients receiving lipid emulsions made with soybean oil have elevated markers of lipid peroxidation in vivo,[162] and adipose tissue α-tocopherol concentrations which suggest tissue stores are being depleted.[79] Furthermore, soybean oil-lipid emulsions are rich in omega-6 fatty acids and low in α-tocopherol. Some newly developed lipid emulsions contain fish oil and are rich in both omega-3 fatty acids and α-tocopherol.[163] These fish-oil emulsions have decreased fatty liver in mouse models.[164,165] A systematic review suggested that fish oil lipid emulsions may have benefit—and that they improved liver function in adults on long-term TPN.[166] Evaluation of safety and efficacy in a 1-year, open label study in children found no adverse effects from fish-oil emulsions.[167] In 2018, the US FDA approved a fish oil lipid emulsion containing vitamin E for use in pediatric patients with nutrition-associated cholestasis.[168] The mechanism of action of these novel lipid solutions remains under investigation.

Pathology of Human α-Tocopherol Deficiency

The primary manifestations of human α-tocopherol deficiency include spinocerebellar ataxia, skeletal myopathy, and pigmented retinopathy.[37,133] Additionally, hemolytic anemia is observed in premature infants fed deficient formula,[169] and is suggested to occur due to fast growth and inadequate α-tocopherol intake.[170,171]

The progression of neurologic symptoms resulting from α-tocopherol deficiency in humans follows a distinct pattern. Hypo- or areflexia is the earliest symptom observed. By the end of the first decade of life, untreated patients with chronic cholestatic hepatobiliary disease have a combination of spinocerebellar ataxia, neuropathy, and ophthalmoplegia. The progression of neurologic symptoms appears to depend on the level of oxidative stress accompanying the α-tocopherol deficiency.

Deficiency in children and adults results in a progressive peripheral neuropathy with a dying back of the large caliber axons in the sensory neurons.[142] Axonal dystrophy has been observed in the posterior columns of the spinal cord and the dorsal and ventral spinocerebellar tracts.[142] The large-caliber, myelinated axons of peripheral sensory nerves are the predominant targets in human α-tocopherol deficiency. Thus, axonal degeneration, rather than demyelination, is the primary sensory nerve abnormality.

Vitamin E Supplements

Epidemiologic studies and some intervention trials have suggested a beneficial role of supplemental α-tocopherol in decreasing chronic disease risk. Studies of the effects of vitamin E on heart attack risk have resulted in conflicting outcomes; however, neither benefit nor harm are supported by the clinical intervention studies.[172-174]

α-Tocopherol supplements may contribute to the amelioration of some chronic diseases, primarily in tissue requiring protection against lipid peroxidation. α-Tocopherol supplements have been shown as a promising therapy in reducing symptoms of fatty liver disease and are increasingly associated with improved brain and eye health in aging. Notably, these tissues (liver and brain) also express α-TTP,[175] again supporting the idea that these tissues have a higher vitamin E requirement.

Vitamin E Intervention Studies in Fatty Liver Disease

α-Tocopherol evidently plays a crucial role in liver health and adequate intakes may prevent progression of nonalcoholic fatty liver disease (NAFLD) to nonalcoholic steatohepatitis (NASH). A meta-analysis of five randomized trials concluded that α-tocopherol treatment consistently reduced biochemical measures of aspartate transaminase (AST), alanine aminotransferase (ALT), alkaline phosphatase (ALP), and steatosis, relative to control groups, with further improvement of fibrosis in NASH patients.[176] Vitamin E supplements may be a useful treatment in fatty liver disease.

The Role of Vitamin E in Alzheimer's Disease and Other Age-Related Dementias

Inadequate α-tocopherol intake in humans is linked to Alzheimer's disease (AD) incidence and progression,[177,178] and a recent meta-analysis showed that low serum α-tocopherol was positively associated with AD prevalence.[179] Farina et al., reported in a Cochrane Systematic Review that α-tocopherol supplements did not prevent dementia, although there was moderate quality evidence from a single study that vitamin E supplements may slow functional decline.[180]

The Role of Vitamin E in Age-Related Macular Degeneration

Age-related macular degeneration (AMD) appears to be responsive to vitamin E supplements. The Age-Related Eye Disease Study (AREDS), a large randomized trial, reported in 2001 that participants at high risk of developing advanced stage AMD reduced their risk by 25% by taking a daily supplement containing α-tocopherol (400 IU *dl*-α-tocopheryl-acetate), beta-carotene, vitamin C, zinc, and copper.[181] Mechanistically, α-tocopherol likely protects the eye in disease states by preventing lipid peroxidation in the sensitive phospholipids and carotenoids[182] of the macula, thus delaying AMD[183] and reducing the oxidative stress burden in the eye.

Vitamin E Safety Concerns

Meta-analysis of antioxidant intervention trials in humans suggests that the doses of vitamin E supplements (400 or 800 IU) given in many clinical trials are not associated with adverse effects.[184,185] However, a few trials have shown increased risk of death with use of supplements >400 IU/day.[186,187] To-date, no studies have documented the mechanism for adverse effects of vitamin E, other than its tendency to increase bleeding, which may be related to decreases in vitamin K status during vitamin E supplementation. It may be that the benefit arising from vitamin E supplementation in heart disease is not its antioxidant function, but a result of the ability of vitamin E to interfere with clot formation by interfering with vitamin K status.[114] Certainly, the upper level set by the IOM of 1000 mg α-tocopherol (1100 IU *dl*- or 1500 IU *d*-α-tocopherol) should not be exceeded by supplement users.[9]

Acknowledgments

The author was supported in part by the Linus Pauling Institute, Oregon State University.

The author has no conflicts of interest to report.

References

1. Evans HM, Bishop KS. Science 1922;56:650–1.
2. Evans HM, Emerson OH, Emerson GA. J Biol Chem 1936;113:319–32.
3. Emerson OH, Emerson GA, Mohammad A, et al. J Biol Chem 1937;122:99–107.
4. Machlin LJ. Vitamin E. In *Handbook of Vitamins* (Machlin LJ, ed) Vol. Second Edition pp. 99–144, Marcel Dekker, New York, 1991.
5. Sokol RJ. Vitamin E deficiency and neurological disorders. In *Vitamin E in Health and Disease* (Packer L, and Fuchs J, eds) pp. 815–49, Marcel Dekker, Inc., New York, 1993.
6. Sokol RJ, Kayden HJ, Bettis DB, et al. J Lab Clin Med 1988;111:548–59.
7. Traber MG, Sokol RJ, Kohlschutter A, et al. J Lipid Res 1993;34:201–10.
8. Ouahchi K, Arita M, Kayden H, et al. Nat Genet 1995;9:141–5.
9. Food and Nutrition Board, and Institute of Medicine. *Dietary Reference Intakes for Vitamin C, Vitamin E, Selenium, and Carotenoids*, National Academy Press, Washington, D.C., 2000.
10. Kasparek S. Chemistry of tocopherols and tocotrienols. In *Vitamin E: A Comprehensive Treatise* (Machlin LJ, ed) pp. 7–65, Marcel Dekker, New York, 1980.
11. Food and Drug Administration. Food labeling: revision of the nutrition and supplement facts labels. Final rule. (Department of Health and Human Services, ed) Vol. 81 pp. 33742–999, Federal Register, Washington, 2016.
12. Traber MG, Atkinson J. Free Radic Biol Med 2007;43:4–15.
13. Blaner WS, Shmarakov IO, Traber MG. Annu Rev Nutr 2021;41: 105–31.
14. Dyall SC, Balas L, Bazan NG, et al. Prog Lipid Res 2022;86: 101165.
15. Niki E, Noguchi N. Free Radic Res 2021;55:352–63.
16. Attri S, Sharma N, Jahagirdar S, et al. Pediatr Res 2006;59: 593–7.
17. Atkinson J, Marquardt D, DiPasquale M, et al. Free Radic Biol Med 2021;176:73–9.
18. Buettner GR. Arch Biochem Biophys 1993;300:535–43.
19. Burk RF, Christensen JM, Maguire MJ, et al. J Nutr 2006;136: 1576–81.
20. Babaev VR, Li L, Shah S, et al. Arterioscler Thromb Vasc Biol 2010;30:1751–7.
21. Head B, Traber MG. Free Radic Biol Med 2021;176:80–91.
22. Mastaloudis A, Leonard SW, Traber MG. Free Radic Biol Med 2001;31:911–22.
23. Mastaloudis A, Morrow JD, Hopkins DW, et al. Free Radic Biol Med 2004;36:1329–41.
24. Bruno RS, Ramakrishnan R, Montine TJ, et al. Am J Clin Nutr 2005;81:95–103.
25. Bruno RS, Leonard SW, Atkinson J, et al. Free Radic Biol Med 2006;40:689–97.
26. Traber MG, Buettner GR, Bruno RS. Redox Biol 2019;21:101091.
27. Jiang Q. Free Radic Biol Med 2022;179:375–87.
28. Cooney RV, Franke AA, Harwood PJ, et al. Proc Natl Acad Sci U S A 1993;90:1771–5.
29. Christen S, Woodall AA, Shigenaga MK, et al. Proc Natl Acad Sci USA 1997;94:3217–22.
30. Hoglen NC, Waller SC, Sipes IG, et al. Chem Res Toxicol 1997; 10:401–7.

31. Morton LW, Ward NC, Croft KD, et al. Biochem J 2002;364: 625–8.
32. Leonard SW, Bruno RS, Paterson E, et al. Free Radic Biol Med 2003;38:813–9.
33. Pei R, Mah E, Leonard SW, et al. Free Radic Res 2015;49: 1114–21.
34. Williamson KS, Gabbita SP, Mou S, et al. Nitric Oxide 2002;6: 221–7.
35. Casati M, Boccardi V, Ferri E, et al. Aging Clin Exp Res 2020;32: 459–64.
36. Sheppard AJ, Pennington JAT, Weihrauch JL. Analysis and distribution of vitamin E in vegetable oils and foods. In *Vitamin E in Health and Disease* (Packer L, and Fuchs J, eds) pp. 9–31, Marcel Dekker, Inc., New York, 1993.
37. Traber MG. Vitamin E. In *Modern Nutrition in Health and Disease* (Shils ME, Olson JA, Shike M, and Ross AC, eds) Vol. 9th ed. pp. 347–62, Williams & Wilkins, Baltimore, 1999.
38. Maras JE, Bermudez OI, Qiao N, et al. J Am Diet Assoc 2004;104:567–75.
39. US Department of Health and Human Services, and US Department of Agriculture. 2015–2020 Dietary Guidelines for Americans. Last accessed: May 2022. http://health.gov/dietaryguidelines/2015/guidelines/
40. Wright ME, Lawson KA, Weinstein SJ, et al. Am J Clin Nutr 2006;84:1200–7.
41. Frankenfeld CL, Hullar MAJ, Maskarinec G, et al. (2022) J Acad Nutr Diet 122, 78–98.
42. Traber MG, Leonard SW, Ebenuwa I, et al. Am J Clin Nutr 2019;110:1148–67.
43. Mah E, Sapper TN, Chitchumroonchokchai C, et al. Am J Clin Nutr 2015;102:1070–80.
44. Bruno RS, Leonard SW, Park SI, et al. Am J Clin Nutr 2006;83: 299–304.
45. Leonard SW, Good CK, Gugger ET, et al. Am J Clin Nutr 2004;79:86–92.
46. Cuerq C, Bordat C, Halimi C, et al. Nutrients 2020;13:129.
47. Sokol RJ, Heubi JE, Iannaccone S, et al. Gastroenterology 1983;85:1172–82.
48. Brigelius-Flohe R, Traber MG. FASEB J 1999;13:1145–55.
49. Narushima K, Takada T, Yamanashi Y, et al. Mol Pharmacol 2008;74:42–9.
50. Reboul E, Klein A, Bietrix F, et al. J Biol Chem 2005;281: 4739–45.
51. Reboul E. Antioxidants (Basel) 2017;6:95.
52. Young SG, Cham CM, Pitas RE, et al. J Clin Invest 1995;96: 2932–46.
53. Di Filippo M, Varret M, Boehm V, et al. J Clin Lipid 2019;13: 201–12.
54. Soayfane Z, Terce F, Cantiello M, et al. Nutr Metab 2016;13:48.
55. Khalifeh-Soltani A, Gupta D, Ha A, et al. JCI Insight 2016;1: e87418.
56. Beaumier-Gallon G, Dubois C, Senft M, et al. Am J Clin Nutr 2001;73:870–7.
57. Yoshida H, Yusin M, Ren I, et al. J Lipid Res 1992;33:343–50.
58. Hosomi A, Goto K, Kondo H, et al. Neurosci Lett 1998;256: 159–62.
59. Schuelke M. Ataxia with vitamin E deficiency. In *GeneReviews® [Internet]* (Adam MP, Ardinger HH, Pagon RA, Bird TC, Dolan CR, and Stephens K, eds) Vol. Available from www.ncbi.nlm.nih.gov/books/NBK1241/, University of Washington, Seattle, Seattle, WA, 2016.
60. Panagabko C, Morley S, Hernandez M, et al. Biochemistry 2003;42:6467–74.
61. Watt AT, Head B, Leonard SW, et al. J Nutr Biochem 2021;97:108801.
62. Thakur V, Morley S, Manor D. Biochemistry 2010;49:9339–44.
63. Arai H, Kono N. Free Radic Biol Med 2021;176:162–75.
64. Chung S, Ghelfi M, Atkinson J, et al. J Biol Chem 2016; 291:17028–39.
65. Lamprakis C, Stocker A, Cascella M. Front Mol Biosci 2015;2:36.
66. Shichiri M, Takanezawa Y, Rotzoll DE, et al. J Nutr Biochem 2010;21:451–6.
67. Tanito M, Yoshida Y, Kaidzu S, et al. Invest Ophthalmol Vis Sci 2007;48:396–404.
68. Shichiri M, Kono N, Shimanaka Y, et al. J Biol Chem 2012;287:2926–34.
69. Ulatowski L, Ghelfi M, West R, et al. J Biol Chem 2022;298:101712.
70. Jishage K, Arita M, Igarashi K, et al. J Biol Chem 2001;276: 1669–72.
71. Kaempf-Rotzoll DE, Igarashi K, Aoki J, et al. Biol Reprod 2002;67:599–604.
72. Kaempf-Rotzoll DE, Horiguchi M, Hashiguchi K, et al. Placenta 2003;24:439–44.
73. Rotzoll DE, Scherling R, Etzl R, et al. Eur J Obstet Gynecol Reprod Biol 2008;140:183–91.
74. Jauniaux E, Cindrova-Davies T, Johns J, et al. J Clin Endocrinol Metab 2004;89:1452–8.
75. Shamim AA, Schulze K, Merrill RD, et al. Am J Clin Nutr 2015;101:294–301.
76. Hempstock J, Jauniaux E, Greenwold N, et al. Hum Pathol 2003;34:1265–75.
77. Traber MG, Ramakrishnan R, Kayden HJ. Proc Natl Acad Sci U S A 1994;91:10005–8.
78. El-Sohemy A, Baylin A, Ascherio A, et al. Am J Clin Nutr 2001;74:356–63.
79. Steephen AC, Traber MG, Ito Y, et al. J Parenter Enteral Nutr 1991;15:642–52.
80. Traber MG, Leonard SW, Traber DL, et al. Am J Clin Nutr 2010;92:1378–84.
81. Ingold KU, Burton GW, Foster DO, et al. Lipids 1987;22: 163–72.
82. Burton GW, Wronska U, Stone L, et al. Lipids 1990;25:199–210.
83. Burton GW, Traber MG, Acuff RV, et al. Am J Clin Nutr 1998;67:669–84.
84. Burton GW, Traber MG. Annu Rev Nutr 1990;10:357–82.
85. Traber MG, Sokol RJ, Ringel SP, et al. N Engl J Med 1987;317: 262–5.
86. Novotny JA, Fadel JG, Holstege DM, et al. J Nutr 2012;142: 2105–11.
87. Sontag TJ, Parker RS. J Biol Chem 2002;277:25290–6.
88. Bardowell SA, Stec DE, Parker RS. J Nutr 2010;140:1901–6.
89. Sontag TJ, Parker RS. J Lipid Res 2007;48:1090–8.
90. Kiyose C, Saito H, Kaneko K, et al. Lipids 2001;36:467–72.
91. Birringer M, Pfluger P, Kluth D, et al. J Nutr 2002;132:3113–8.
92. Handelman GJ, Machlin LJ, Fitch K, et al. J Nutr 1985;115: 807–13.
93. Schultz M, Leist M, Elsner A, et al. Methods Enzymol 1997; 282:297–310.
94. Smith KS, Lee CL, Ridlington JW, et al. Lipids 2003;38:813–9.
95. Simon EJ, Eisengart A, Sundheim L, et al. J Biol Chem 1956;221:807–17.
96. Taylor L, Krueger N, Malysheva O, et al. Bioorg Med Chem 2018;26:5555–65.
97. Torquato P, Bartolini D, Giusepponi D, et al. Free Radic Biol Med 2019;131:115–25.
98. Sharma G, Muller DP, O'Riordan SM, et al. Free Radic Biol Med 2013;55:54–62.
99. Meulmeester FL, Luo J, Martens LG, et al. Nutr Metab Cardiovasc Dis 2021;31:2407–15.
100. Hirani V, Yarovoy A, Kozeska A, et al. Arch Biochem Biophys 2008;478:59–68.

101. Uehara S, Murayama N, Nakanishi Y, et al. Xenobiotica 2015;45:124–30.
102. Yoshida Y, Niki E. Biofactors 2002;16:93–103.
103. Galli F, Azzi A, Birringer M, et al. Free Radic Biol Med 2017;102:16–36.
104. Grammas P, Hamdheydari L, Benaksas EJ, et al. Biochem Biophys Res Commun 2004;319:1047–52.
105. Jiang Q. Free Radic Biol Med 2014;72:76–90.
106. Pein H, Ville A, Pace S, et al. Nat Commun 2018;9:3834.
107. Lasker JM, Chen WB, Wolf I, et al. J Biol Chem 2000;275:4118–26.
108. Stec DE, Roman RJ, Flasch A, et al. Physiol Genomics 2007;30:74–81.
109. Kikuta Y, Kusunose E, Kusunose M. J Biochem 2000;127:1047–52.
110. Hardwick JP. Biochem Pharmacol 2008;75:2263–75.
111. McDonald MG, Rieder MJ, Nakano M, et al. Mol Pharmacol 2009;75:1337–46.
112. Edson KZ, Prasad B, Unadkat JD, et al. Biochemistry 2013;52:8276–85.
113. Booth SL, Golly I, Sacheck JM, et al. Am J Clin Nutr 2004;80:143–8.
114. Glynn RJ, Ridker PM, Goldhaber SZ, et al. Circulation 2007;116:1497–503.
115. Stahl W, Graf P, Brigelius-Flohe R, et al. Anal Biochem 1999;275:254–9.
116. Pope SA, Burtin GE, Clayton PT, et al. Free Radic Biol Med 2002;33:807–17.
117. Mustacich DJ, Leonard SW, Devereaux MW, et al. Free Radic Biol Med 2006;41:1069–78.
118. Mustacich DJ, Gohil K, Bruno RS, et al. J Nutr Biochem 2009;20:469–76.
119. Cho JY, Kang DW, Ma X, et al. J Lipid Res 2009;50:924–37.
120. Bartolini D, De Franco F, Torquato P, et al. J Med Chem 2020;63:3701–12.
121. Peter S, Friedel A, Roos FF, et al. Int J Vitam Nutr Res 2015;85:261–81.
122. Sokol RJ, Heubi JE, Iannaccone ST, et al. N Engl J Med 1984;310:1209–12.
123. Traber MG, Jialal I. Lancet 2000;355:2013–4.
124. Lebold KM, Ang A, Traber MG, et al. Am J Clin Nutr 2012;96:801–9.
125. Michels AJ, Leonard SW, Uesugi SL, et al. J Nutr 2018;148:1924–30.
126. Traber MG, Mah E, Leonard SW, et al. Am J Clin Nutr 2017;105:571–9.
127. Traber MG, Leonard SW, Ebenuwa I, et al. Am J Clin Nutr 2021;113:92–103.
128. Farrell PM, Bieri JG, Fratantoni JF, et al. J Clin Invest 1977;60:233–41.
129. Cornblath M, Gordon HH, Nitowsky HM. AMA J Dis Child 1956;92:164–74.
130. El Euch-Fayache G, Bouhlal Y, Amouri R, et al. Brain 2014;137:402–10.
131. Gotoda T, Arita M, Arai H, et al. N Engl J Med 1995;333:1313–8.
132. Chan KH, O'Sullivan M, Farouji I, et al. Cureus 2021;13:e13389.
133. Di Donato I, Bianchi S, Federico A. Neurol Sci 2010;31:511–5.
134. Cavalier L, Ouahchi K, Kayden HJ, et al. Am J Hum Genet 1998;62:301–10.
135. Burnett JR, Hooper AJ, Hegele RA. Abetalipoproteinemia. In *GeneReviews(R)* (Adam MP, Everman DB, Mirzaa GM, Pagon RA, Wallace SE, Bean LJH, et al., eds), Seattle, WA, 1993.
136. Ghone RA, Suryakar AN, Kulhalli PM, et al. J Clin Diagn Res 2013;7:2146–8.
137. Pignitter M, Stolze K, Gartner S, et al. J Agric Food Chem 2014;62:2297–305.
138. Falcao R, Lyra CO, Morais CMM, et al. PLoS One 2019;14:e0224984.
139. Melia AT, Koss-Twardy SG, Zhi J. J Clin Pharmacol 1996;36:647–53.
140. McDuffie JR, Calis KA, Booth SL, et al. Pharmacotherapy 2002;22:814–22.
141. Sherf-Dagan S, Buch A, Ben-Porat T, et al. Surg Obes Relat Dis 2021;17:816–30.
142. Sokol RJ. Annu Rev Nutr 1988;8:351–73.
143. Ben Hamida C, Doerflinger N, Belal S, et al. Nat Genet 1993;5:195–200.
144. Yokota T, Shiojiri T, Gotoda T, et al. Ann Neurol 1997;41:826–32.
145. Yokota T, Uchihara T, Kumagai J, et al. J Neurol Neurosurg Psychiatry. 2000;68:521–5.
146. Iwasa K, Shima K, Komai K, et al. J Neurol Sci 2014;345:228–30.
147. Kohlschutter A, Finckh B, Nickel M, et al. Neurodegener Dis 2020;20:35–8.
148. Wetterau JR, Aggerbeck LP, Bouma ME, et al. Science 1992;258:999–1001.
149. Ross RS, Gregg RE, Law SW, et al. J Clin Invest 1988;81:590–5.
150. Malloy MJ, Kane JP, Hardman DA, et al. J Clin Invest 1981;67:1441–50.
151. Zamel R, Khan R, Pollex RL, et al. Orphanet J Rare Dis 2008;3:19.
152. Granot E, Kohen R. Am J Clin Nutr 2004;79:226–30.
153. Ferreira F, Patel V, Matts S. BMJ Case Rep 2014;2014.
154. Traber MG, Rader D, Acuff RV, et al. Atherosclerosis 1994;108:27–37.
155. Cuerq C, Restier L, Drai J, et al. Orphanet J Rare Dis 2016;11:114.
156. Buonuomo PS, Ruggiero A, Valeriani M, et al. Curr Opin Pediatr 2009;21:824–7.
157. Traber MG, Kayden HJ, Green JB, et al. Am J Clin Nutr 1986;44:914–23.
158. Traber MG, Thellman CA, Rindler MJ, et al. Am J Clin Nutr 1988;48:605–11.
159. Sokol RJ, Heubi JE, Butler-Simon N, et al. Gastroenterology 1987;93:975–85.
160. Sokol RJ, Butler-Simon N, Conner C, et al. Gastroenterology 1993;104:1727–35.
161. Berger MM, Shenkin A, Schweinlin A, et al. Clin Nutr 2022;41:1357–424.
162. Lemoyne M, Van Gossum A, Kurian R, et al. Am J Clin Nutr 1987;46:267–72.
163. Carpentier YA, Dupont IE. World J Surg 2000;24:1493–7.
164. Fell GL, Anez-Bustillos L, Dao DT, et al. PLoS One 2019;14:e0217155.
165. Baker MA, Cho BS, Anez-Bustillos L, et al. Am J Clin Nutr 2019;109:1038–50.
166. Kirk C, Haigh L, Thompson NP, et al. Clin Nutr 2022;41:80–90.
167. Raphael BP, Mitchell PD, Carey A, et al. J Pediatr Gastroenterol Nutr 2021;72:451–5.
168. U.S. Food and Drug Administration. Drug Approval Package: Omegaven (fish oil triglycerides). Accessed June 2022. www.accessdata.fda.gov/drugsatfda_docs/nda/2018/210589Orig1s000TOC.cfm
169. Ritchie JH, Fish MB, McMasters V, et al. N Engl J Med 1968;279:1185–90.
170. Kumar RK, Edwards KN, Bury G. Indian J Pediatr 2000;67:537–8.
171. Carey AN, Duggan C. J Pediatr 2017;181:162.
172. Fortmann SP, Burda BU, Senger CA, et al. Ann Intern Med 2013;159:824–34.

173. Bjelakovic G, Nikolova D, Gluud LL, et al. Cochrane Database Syst Rev 2012, CD007176.
174. Bjelakovic G, Nikolova D, Gluud C. PLoS One 2013;8: e74558.
175. Kono N, Arai H. Traffic 2015;16:19–34.
176. Sato K, Gosho M, Yamamoto T, et al. Nutrition 2015;31: 923–30.
177. Zandi PP, Anthony JC, Khachaturian AS, et al. Arch Neurol 2004;61:82–8.
178. Boothby LA, Doering PL. Ann Pharmacother 2005;39:2073–80.
179. Lloret A, Esteve D, Monllor P, et al. Int J Mol Sci 2019;20:879.
180. Farina N, Llewellyn D, Isaac MG, et al. Cochrane Database Syst Rev 2017;1:CD002854.
181. Age-Related Eye Disease Study Research G. Arch Ophthalmol 2001;119:1417–36.
182. Johra FT, Bepari AK, Bristy AT, et al. Antioxidants (Basel) 2020;9:1046.
183. Evans JR, Lawrenson JG. Cochrane Database Syst Rev 2017;7: CD000254.
184. Vivekananthan DP, Penn MS, Sapp SK, et al. Lancet 2003;361: 2017–23.
185. Berry D, Wathen JK, Newell M. Clin Trials 2009;6:28–41.
186. Miller ER, 3rd, Pastor-Barriuso R, Dalal D, et al. Ann Intern Med 2005;142:37–46.
187. Bjelakovic G, Nikolova D, Gluud LL, et al. JAMA 2007;297: 842–57.

C. VITAMINS

CHAPTER 20

Vitamin K

Sarah L. Booth, Russell Debose-Boyd, and Martin J Shearer

CHAPTER OUTLINE

Abbreviations

AI, adequate intake; **ApoE**, apolipoprotein E; **DRI**, dietary reference intake; **CVD**, cardiovascular disease; **EAR**, estimated average requirement; **ERAD**, ER-associated degradation; **FPP**, farnesyl pyrophosphate; **GGPP**, geranylgeranyl pyrophosphate; **Gla**, γ-carboxyglutamic acid; **GRP**, Gla-rich protein; **HCO_3^-**, biocarbonate; **HMGCR**, HMG CoA reductase; **INR**, international normalized ratio; **Km**, Michaelis-Menton constant; **MGP**, matrix Gla protein; **MK**, menaquinone; **NHANES**, National Health and Nutrition Examination Survey; **OC**, osteocalcin; **PT**, prothrombin time; **PXR**, pregnane X receptor; **RDA**, recommended dietary allowance; **SXR**, steroid and xenobiotic receptor; **UBIAD1**, UbiA Prenyltransferase Domain Containing Protein-1; **ucOC**, undercarboxylated osteocalcin; **Vitamin KH_2**, hydronaphthoquinone; **VKDB**, vitamin K deficiency bleeding; **VKORC1**, vitamin K epoxide reductase complex subunit 1

Introduction

Vitamin K was discovered in 1929 by Henrik Dam[1] when he noted that chicks ingesting diets that had been extracted with nonpolar solvents to remove cholesterol developed subdural or muscular hemorrhages, and that blood taken from these animals clotted slowly. By 1935, Dam had obtained sufficient evidence to propose the existence of a new fat-soluble antihemorrhagic factor, which he called vitamin K.[2] The chemical isolation and identification of vitamin K from natural sources was achieved by several groups, notably by Dam and Karrer's group at Zurich University and by Doisy's group at St Louis University. By 1939, both groups had isolated the vitamin from alfalfa as a yellow oil, while Doisy's group went on to characterize the active antihemorrhagic agent as 2-methyl-3-phytyl-1,4-naphthoquinone.[3] The Doisy team also isolated a structurally different vitamer from putrefied fish meal that contained an unsaturated polyprenyl side chain at the 3-position of the naphthoquinone ring. To distinguish between these vitamers, Doisy's group called the form with the phytyl side chain vitamin K_1, and the form with the polyprenyl side chain vitamin K_2. Early investigators recognized that the activity of some sources of the vitamin, such as putrefied fish meal, was the result of bacterial synthesis, and that several vitamers existed that differed in the number of prenyl groups at the 3-position. Doisy's group originally reported that the vitamin K_2 isolated from putrefied fish meal was 2-methyl-3-farnesylfarnesyl-1,4 naphthoquinone. However, in 1958, the organic chemist Otto Isler showed, by mixed melting point experiments, that this compound was actually 2-methyl-3-farnesylgeranylgeranyl-1,4 naphthoquinone with a sidechain of seven, not six prenyl units.[3]

At the time that vitamin K was isolated and characterized, the only plasma proteins known to be involved in blood coagulation were prothrombin and fibrinogen. Dam et al., isolated a crude prothrombin fraction from chick plasma and demonstrated that the activity of this fraction was decreased when it was obtained from a vitamin K–deficient chick.[4] The hemorrhagic condition resulting from biliary disease-related obstructive jaundice was also shown to be caused by poor utilization of vitamin K, and these bleeding episodes were initially specifically attributed to a lack of prothrombin. A real understanding of thrombus formation and of the various soluble and cellular factors involved in regulating the generation of thrombin from prothrombin did not begin until the mid-1950s. As factors VII, IX, and X were discovered through the study of patients with clotting disorders, these factors were shown to depend on vitamin K for synthesis. For a considerable time, these three factors and prothrombin were the only proteins known to require vitamin K for their synthesis.

Chemical Structure and Nomenclature

The term *vitamin K* is used as a generic descriptor of 2-methyl-1,4-naphthoquinone (menadione, historically called vitamin K_3), and all derivatives of this compound that exhibit an antihemorrhagic activity in animals fed a vitamin K–deficient diet (**Figure 20.1**). The plant-based source of vitamin K is preferably called phylloquinone (USP phytonadione), although historically, it was referred to as vitamin K_1. The major vitamin K_2 compound in putrefied fish meal, 2-methyl-3-farnesylgeranylgeranyl-1,4-naphthoquinone (originally abbreviated vitamin $K_{2(35)}$), is one of a series of vitamin K compounds with unsaturated side chains called menaquinones (MK), the majority of which are produced by a limited number of anaerobic bacteria and are present in large quantities in the lower bowel. Under modern nomenclature, MK are named according to the number of prenyl units in the side chain rather than the number of carbon atoms. Thus, the compound 2-methyl-farnesylgeranylgeranyl-1,4-naphthoquinone (vitamin $K_{2(35)}$) becomes menaquinone-7 (MK-7). The vitamins of the MK series with up to 13 prenyl groups have been identified. MK-4 (2-methyl-3-geranylgeranyl-1,4-naphthoquinone) can be formed in animal tissues by alkylation of menadione.[5]

Sources and Utilization of Vitamin K

Analysis, Food Content, and Bioaccessibility

Standardized high-performance liquid chromatography and tandem mass spectrometry procedures are now available for the assay of the phylloquinone and menaquinone content of foods. The United States Department of Agriculture's FoodData Central is a database that contains >40,000 values for phylloquinone and >1,250 values for MK-4[6] and is continually updated with new values for phylloquinone and, to a lesser extent, MK-4. Green leafy vegetables and certain plant oils, such as olive and soybean, are the primary dietary sources of phylloquinone, although in the United States, convenience foods also contribute significant phylloquinone in the diet given the addition of plant oils during food preparation.[7] MK-4 is found in meat and meat products obtained from animals fed menadione, which is the most common form of vitamin K in commercial animal feed. There is also some MK-4 present in dairy products, albeit in small amounts.

In contrast to phylloquinone and MK-4, the menaquinone content of foods is poorly characterized. The limited available data are from single foods obtained from a few geographic regions, with no systematic food sampling plan consistent with best practices, hence not comprehensive for use in dietary assessment. There has been considerable focus on a fermented soybean product, natto that is consumed mainly in the Japanese market and contains MK-7. However, more recent studies indicate that multiple menaquinones are abundant in the food supply.[8] Certain bacteria have unique menaquinone signatures, so MK-5 to -13 concentrations in any given food can vary widely based on the microbial species used in the fermentation process, the microbial content of the highly specialized ruminant digestive system, and/or the microbes present in the environment. This is best illustrated in a comparison of menaquinones in various cheeses produced in France and the United States, which reveal marked differences in the type and content of individual menaquinones for the same type of cheeses.[9]

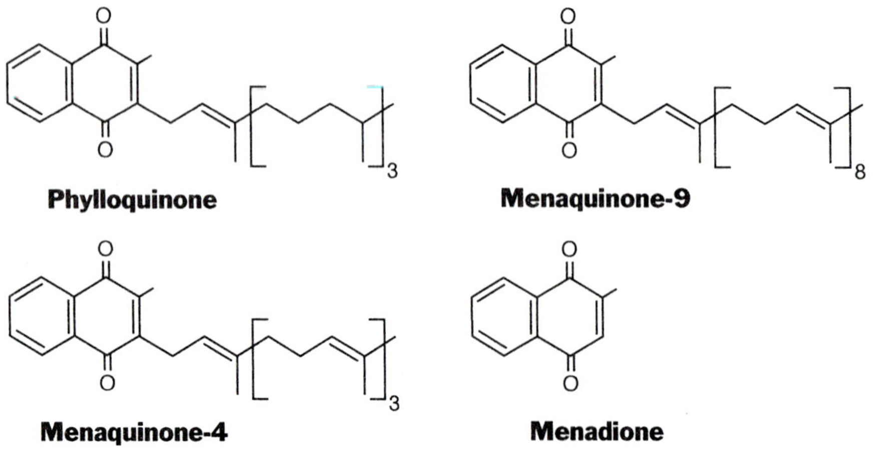

Figure 20.1 Structures of vitamin K active compounds. Phylloquinone (vitamin K_1) is the plant-based form of vitamin K. Menaquinone-9 is one of a series of menaquinones (vitamin K_2) produced by intestinal bacteria, and menadione (vitamin K_3) is a synthetic compound that can be converted to menaquinone-4 by animal tissues.

The relative bioavailability of phylloquinone from various foods has been difficult to assess. Initial studies in humans compared the increase in plasma phylloquinone from the consumption of green vegetables with that of pure phylloquinone. These limited studies suggested that the phylloquinone from various vegetable sources should be considered 15% to 20% bioavailable, compared with phylloquinone consumed as a supplement. The bioavailability of phylloquinone obtained from vegetable oil was found to be about twice that obtained from broccoli. There are no known bioavailability data comparing the individual menaquinones with phylloquinone obtained from various food sources.

Absorption and Transport of Vitamin K

Phylloquinone, the predominant dietary form of the vitamin, is absorbed from the proximal intestine through the lymphatic system,[10] and absorption is decreased in patients with various lipid malabsorption syndromes, particularly biliary disorders. Intestinal uptake of phylloquinone occurs via a saturable, energy-requiring process in which the cholesterol transporter, Niemann-Pick C1-like 1, also functions as a physiologic phylloquinone importer.[11] Phylloquinone in plasma is predominantly carried by the triglyceride-rich lipoprotein fraction containing very-low-density lipoproteins and chylomicron remnants, although some is located in the low-density lipoprotein and high-density lipoprotein fractions.[12] Plasma phylloquinone concentration in a healthy U.S. population has been shown to range widely from 0.3 to 2.6 nmol/L, with a mean of ~1.0 nmol/L (0.45 ng/mL).[13] As expected from this route of transport, plasma phylloquinone concentration is strongly correlated with plasma lipid concentrations.[14]

The uptake of lipoprotein-associated phylloquinone by tissues such as liver and bone is a complex process, involving different apoproteins on the surface of lipoproteins (e.g., ApoE), cell surface low affinity binding sites of heparan sulfate proteoglycans, and high-affinity lipoprotein receptors of the LDLR family that mediate internalization of the lipoprotein particles.[15] Studies designed to determine how common polymorphisms of ApoE (E2, E3, and E4) might differentially affect the uptake of phylloquinone by hepatocytes and osteoblasts, hence affecting the efficiency of γ-carboxylation of γ-carboxyglutamic acid (Gla) proteins, have yielded contradictory results.[15]

The total human body pool of phylloquinone is estimated to be ~100 μg, and turnover is rapid, as mirrored by the plasma clearance kinetics of oral or intravenous phylloquinone, which resolves into two exponential components with half-lives of ~15 minutes and ~2.5 hours, respectively.[16] Although the total vitamin K content of human liver is ~0.06 μg of phylloquinone equivalent per gram of liver, long-chain MKs, rather than phylloquinone, account for ~90% of stores.[3] Data based on liver biopsies of patients fed diets very low in vitamin K before surgery indicate that approximately two-thirds of hepatic phylloquinone is lost in 3 days, but nevertheless concentrations of MKs remain stable.[17] These findings are consistent with a small pool size of phylloquinone that turns over rapidly, and a large pool of MKs that turns over slowly.

The major route of ingested phylloquinone excretion is through the feces, in the form of biliary metabolites; very little unmetabolized vitamin is excreted. Many details of the catabolic pathways of vitamin K forms are currently lacking, but it is known that the side chains of phylloquinone and MK-4 are shortened to seven or five carbon atoms yielding a carboxylic acid group at the end.[10,18] These ^{5}C and ^{7}C-aglycones, which are the major metabolites of phylloquinone, are excreted in the urine at concentrations related to the intake of the vitamin.[19] Studies have also shown that glucuronides of menadione are excreted in urine at an amount that is positively related to phylloquinone.[20] The mechanism by which menadione is cleaved from various sources of vitamin K or its metabolites is not known. There are numerous other unidentified metabolites. For example, anticoagulant treatment with warfarin results in a substantial conversion of the body pool of phylloquinone to phylloquinone-2,3-epoxide, leading to the generation of new metabolites.[21]

Production and Use of Menaquinones

Substantial amounts of vitamin K in the form of long-chain MKs are produced by gut bacteria, which serve as electron carriers in microbial respiration.[22] In addition, MKs secreted by certain human gut bacterial species act as a growth factor for neighboring bacteria with incomplete MK synthesis pathways. This suggests that MKs are an important commensal factor in the human gut microbiota, and gut-derived menaquinones may influence human health indirectly through modulation of the gut microbial composition.[22] The relevance of gut-derived menaquinones to the vitamin K requirements of the host is still speculative.

In a mouse model, dietary vitamin K deficiency altered the intestinal menaquinone profile, but dietary supplementation with various vitamin K forms did not.[22] These findings suggest that the amount, rather than the form, of vitamin K in the diet influences intestinal microbial composition. In humans, there appear to be two dominant MK profiles in the adult gut microbiome, namely those with a greater abundance of either (1) Prevotella spp. and predominance of MK5 and MK11-13; or (2) Bacteroides spp. and predominance of MK9-10[23] These MK profiles do not appear to change in response to dietary interventions, which were not designed to alter vitamin K intake, and the implications of any MK grouping with respect to overall vitamin K status is not known. MK fecal concentration in infants measured at 6 weeks postpartum varies by the mode of delivery and feeding practice.[24] Specifically, fecal MK6, MK7, MK12, and MK13 were higher in formula-fed than breastfed infants, whereas MK8 predominated in breastfed and was 5-fold higher than in formula-fed infants. Total MK concentration was greater in vaginally delivered than in cesarean section infants. Interestingly, there was no association between infant and mother fecal MK concentrations. It is still unclear if the fecal menaquinone status of newborns has important implications for infant vitamin K status.

Vitamin K–Dependent Proteins

Hemostasis-Related Plasma Proteins

Prothrombin, the circulating zymogen of the procoagulant thrombin, was the first protein shown to depend on vitamin K for its synthesis. Prothrombin was also the first protein demonstrated to contain Gla residues. Plasma clotting factors VII, IX, and X were all initially identified because their activity was decreased in the plasma of a patient with a hereditary bleeding disorder,[25] and these factors were subsequently shown to depend on vitamin K_1 for their synthesis. Until the mid-1970s, these four vitamin K–dependent clotting factors were the only proteins known to require this vitamin for their synthesis.

A complex series of events (**Figure 20.2**), which lead to the generation of thrombin by proteolytic activation of protease zymogens,[26,27] is essential for hemostasis. The vitamin K–dependent clotting factors are involved in these activation and propagation events through formation of membrane-associated complexes with each other and with accessory proteins. The amino terminal Gla domains of these proteins are homologous: the region contains multiple Gla residues (10–13 each) that are positioned similarly to those in prothrombin.

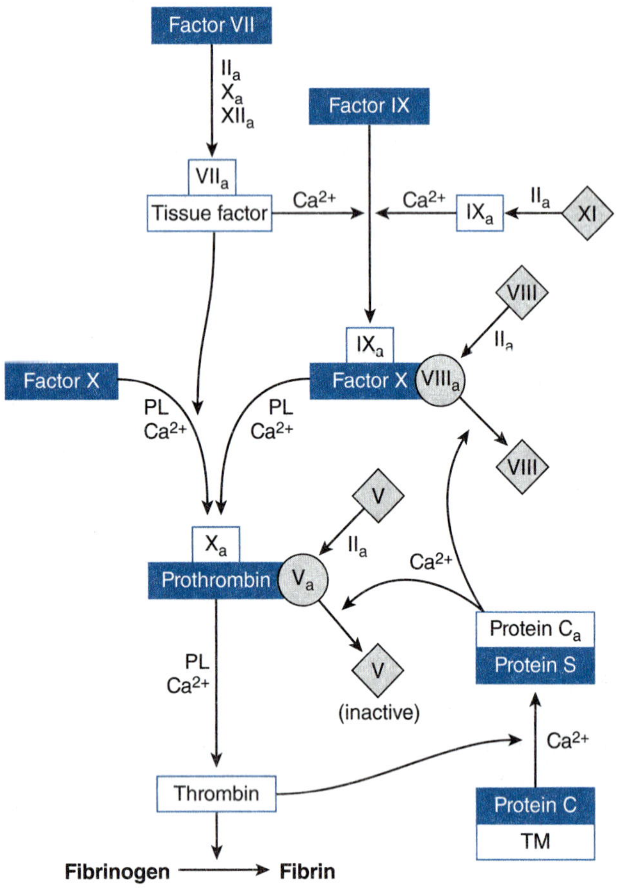

Figure 20.2 Vitamin K–dependent clotting factors involved in blood coagulation. The vitamin K–dependent procoagulants (prothrombin and factors VII, IX, and X) circulate as zymogens of serine proteases until converted to their active (subscript a) forms. Initiation of this process occurs when vascular injury exposes tissue factor to blood (extrinsic pathway). The product of the activation of one factor can activate a second zymogen, and this cascade can affect results in the rapid activation of prothrombin to thrombin and the subsequent conversion of soluble fibrinogen to the insoluble fibrin clot. Some of the steps in this activation involve an active protease, a second vitamin K–dependent protein substrate, and an additional plasma protein cofactor *(circles)* to form a calcium (Ca^{2+})-mediated association with a phospholipid (PL) surface. The formation of activated factor X can also occur through thrombin activation of factor XI and subsequently factor IX (intrinsic pathway). The other two vitamin K–dependent proteins participate in hemostatic control as anticoagulants, not procoagulants. Protein C is activated by thrombin (factor II_a) in the presence of an endothelial cell protein called thrombomodulin (TM). Activated protein C functions in a complex with protein S to inactivate factors V_a and VIII_a and to limit clot formation.

In addition to the classic vitamin K–dependent proteins, three more Gla-containing plasma proteins with similar homology have been discovered. Protein C and protein S participate in thrombin-initiated inactivation of factor V and, therefore, play an anticoagulant rather than a procoagulant role in normal hemostasis.[28] In addition to the approximately 40-amino acid Gla domain, the vitamin K–dependent proteins have other common features. The Gla domain of prothrombin is followed by two kringle domains, which are also found in plasminogen, and a serine protease domain. Factors VII, IX, and X and protein C contain two epidermal growth factor domains and a serine protease domain, whereas protein S contains four epidermal growth factor domains but is not a serineprotease. The function of the seventh Gla-containing plasma protein (protein Z), which is not a protease zymogen, has been shown to have an anticoagulant function under some conditions.[29] Because these proteins play a critical role in hemostasis, they have been extensively studied. They are primarily produced in the liver, the cDNA and genomic organization of each of them is well-documented, and many genetic variants of these proteins have been identified as risk factors in coagulation disorders.[30]

Extrahepatic Vitamin K–Dependent Proteins

A limited number of other mammalian proteins produced by tissues other than the liver have been found to contain Gla residues and, therefore, depend on vitamin K for their synthesis. The first extra-hepatic vitamin K–dependent Gla-containing protein discovered was isolated from bone.[31,32] This 49-residue protein containing 3 Gla residues was called osteocalcin (OC), and it had limited structural homology with the vitamin K–dependent plasma proteins. Although OC is the second most abundant protein in bone, its function is still not yet clearly defined. Rats maintained on a protocol of anticoagulant treatment and vitamin K administration to prevent bleeding problems developed fusion of the proximal tibia growth plate.[33] This finding suggests that OC is involved in some manner in the control of tissue mineralization or skeletal turnover. The original OC gene knockout mouse was reported to produce more dense bone rather than a defect in bone formation.[34] However, through use of CRISPR technology, mice that do not express OC have alterations in the collagen maturity and carbonate to phosphate ratio in the cortical bone, but no alterations in bone strength or mass.[35]

OC produced in bone appears in plasma at concentrations that are high in young children and approach adult levels at puberty, and its concentration is increased in Paget disease and other conditions of rapid bone turnover. Bone also contains, albeit in much smaller quantities, a second low-molecular-weight (79 residue) protein with 5 Gla residues, called matrix Gla protein (MGP). This protein has a structural relationship with OC but is synthesized in cartilage and many other soft tissues.[36] The protein has been difficult to study because of its hydrophobic nature, relative insolubility, and tendency to aggregate. As with OC, details of its physiologic role are unclear, but in studies with MGP knockout mice, death ensued from spontaneous calcification of arteries and cartilage.[37] Arterial calcification was also demonstrated in a warfarin-treated rat model.[38]

Gla-rich protein (GRP), the most extensively γ-carboxylated protein identified to date,[39] was initially found in sturgeon cartilage and was subsequently demonstrated to be expressed and accumulated in the soft tissue of rats and humans. Although the metabolic role of this protein has not been established, it appears to play a role in connective tissue calcification.[40]

The protein encoded by Growth arrest specific-gene 6 (Gas6) is a ligand for the tyrosine kinase Axl[41], which appears to be a growth factor for mesangial and epithelial cells. Two proline-rich Gla proteins (PRGP-1 and PRGP-2) were discovered as integral membrane proteins with an extracellular amino terminal domain that is rich in Gla residues.[42] Subsequently, two other members of this transmembrane Gla protein family (TMG-3 and TMG-4) were cloned.[43] The specifics of the role of these cell surface receptors are not yet known. Vitamin K–dependent proteins are not confined to vertebrates, and many toxic venom peptides secreted by marine *Conus* snails are rich in Gla residues.[44] Vitamin K–dependent proteins have also been found in snake venom,[45] and the carboxylase has been cloned from certain vertebrates, the *Conus* snail, a tunicate, and drosophila.[46] The strong sequence homology of the carboxylase enzyme from these phylogenetic systems suggests that this posttranslational modification of glutamic acid is of ancient evolutionary origin, and that numerous vitamin K–dependent proteins are yet to be discovered.

Biochemical Role of Vitamin K

A period of approximately 40 years elapsed between the discovery of vitamin K and the determination of its metabolic role. Early theories that vitamin K controlled the production of specific proteins at a transcriptional level could not be proven. Abnormalities of clotting times in anticoagulated patients suggested that a circulating inactive, but immunochemically similar, form of prothrombin, referred to as "abnormal prothrombin," was present in increased concentration in the plasma of these patients.

Characterization of the abnormal prothrombin isolated from the plasma of cows fed the anticoagulant dicumarol led directly to an understanding of the metabolic role of vitamin K. This protein lacked the specific calcium-binding sites present in normal prothrombin and did not demonstrate the calcium-dependent association with negatively charged phospholipid surfaces that was known to be essential for prothrombin activation. Acidic peptides were obtained by proteolytic enzyme digestion of prothrombin and were shown to contain Gla, a previously unrecognized acidic amino acid.[47,48] Gla residues could not be obtained by proteolysis of the abnormal prothrombin. All 10 of the glutamic acid residues in the first 42 residues of bovine prothrombin were subsequently shown to be post-translationally γ-carboxylated to form these effective calcium-binding groups.

Vitamin K–Dependent Carboxylase

The discovery of Gla residues in prothrombin led to the demonstration that crude rat liver microsomal preparations contained enzymatic activity (the vitamin K–dependent carboxylase) that promoted a vitamin K–dependent incorporation of ^{14}C bicarbonate ($H^{14}CO_3^-$) into endogenous precursors of vitamin K–dependent proteins present in these preparations.[49] Small peptides containing adjacent Glu-Glu sequences, such as Phe-Leu-Glu-Glu-Val, were substrates for the enzyme. The rough microsomal fraction of liver was highly enriched in carboxylase activity, and the carboxylation event was located to the luminal side of the rough endoplasmic reticulum. The vitamin K–dependent carboxylation reaction does not require adenosine triphosphate, and the energy to drive this reaction is derived from the oxidation of the reduced, hydronaphthoquinone, form of vitamin K (vitamin KH_2) by O_2 to form vitamin K-2,3-epoxide (**Figure 20.3**). Although the general scheme, shown in Figure 20.3, is consistent with the available data, the details of this mechanism are not yet clear.

Studies of substrate specificity at the vitamin K–binding site of the enzyme have shown that, although some differences in biologic activity can be measured, phylloquinone, MK-4, and multiple menaquinones are all effective substrates.

Very few of the proteins secreted by the liver to the plasma are vitamin K dependent, so an efficient mechanism for recognizing the precursors of the vitamin K–dependent proteins is an essential prerequisite for efficient carboxylation[50]. The primary gene products of these proteins contain a homologous domain between the amino terminus of the mature protein and the signal sequence that targets the polypeptide for the secretory pathway. This pro-peptide region appears to be both a docking or recognition site for the enzyme[51] and a modulator of the activity of the enzyme, by decreasing the apparent K_m of the Glu site substrate.[52] All vitamin K–dependent proteins contain this approximately 18-residue sequence, which is cleaved before secretion of the protein.

Vitamin K Epoxide Reductase

As vitamin K–dependent proteins are degraded, the Gla residues released are not used to form new proteins or metabolized, but are excreted in the urine.[53] The amount of Gla excreted by an adult human is in the range of 50 μmol/day, so a similar amount must be formed each day. The dietary requirement of vitamin K is only approximately 0.2 μmol/day, and tissue stores are low. A mole of vitamin is oxidized for each mole of Gla formed, and the vitamin K-2,3-epoxide generated by the carboxylase is actively recycled by an enzyme called vitamin K epoxide reductase.

The hepatic ratio of the epoxide relative to that of the vitamin was increased in animals administered the 4-hydroxycoumarin anticoagulant warfarin.[54] This finding led to the theory that warfarin inhibition of vitamin K action is indirect, through inhibition of the 2,3-epoxide reductase. Blocking of this enzyme prevents the reduction of the epoxide to both the quinone form of the vitamin and the carboxylase substrate, vitamin KH_2. The study of the activity of the epoxide reductase in livers of warfarin-resistant rats[55] was key to understanding the details of the vitamin K cycle[56,57] (**Figure 20.4**).

Three forms of vitamin K—the quinone (K), the hydronaphthoquinone (KH_2), and the 2,3-epoxide (KO)—can feed into this liver vitamin K cycle. In normal liver, the ratio of vitamin K-2, 3-epoxide to the less oxidized forms of the vitamin is approximately 1:10, but it can increase to a majority of epoxide in an anticoagulated animal. The epoxide reductase is now known to be a small, 163-amino acid, single-chain integral membrane protein-vitamin K epoxide reductase complex subunit 1 (VKORC1).[58,59] Although other cellular reductases can reduce the quinone form of vitamin K[60], investigators have shown that VKORC1 knockout mice die soon after birth of a blood clotting factor deficiency.[61] The enzyme contains a cysteine-based redox center, but regeneration of VKORC1 following vitamin K epoxide reduction requires a redox protein that has not yet been identified.

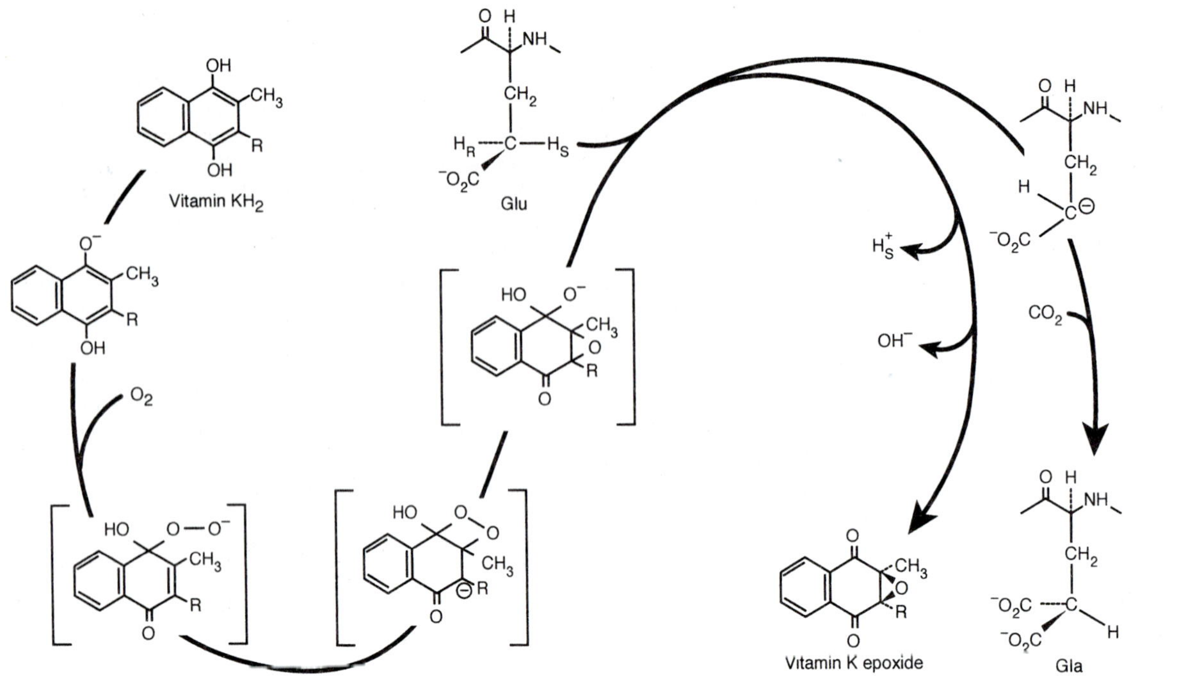

Figure 20.3 The vitamin K–dependent γ-glutamyl carboxylase. The available data support an interaction of oxygen (O_2) with vitamin KH_2, the reduced (hydronaphthoquinone) form of vitamin K, to form an oxygenated intermediate that is sufficiently basic to abstract the γ-hydrogen of the glutamyl residue. The products of this reaction are vitamin K-2,3-epoxide and a glutamyl carbanion. Attack of CO_2 on the carbanion leads to the formation of a γ-carboxyglutamyl residue (Gla). The bracketed peroxy, dioxetane, and alkoxide intermediates have not been identified in the enzyme catalyzed reaction but are postulated based on model organic reactions, and the available data are consistent with their presence.

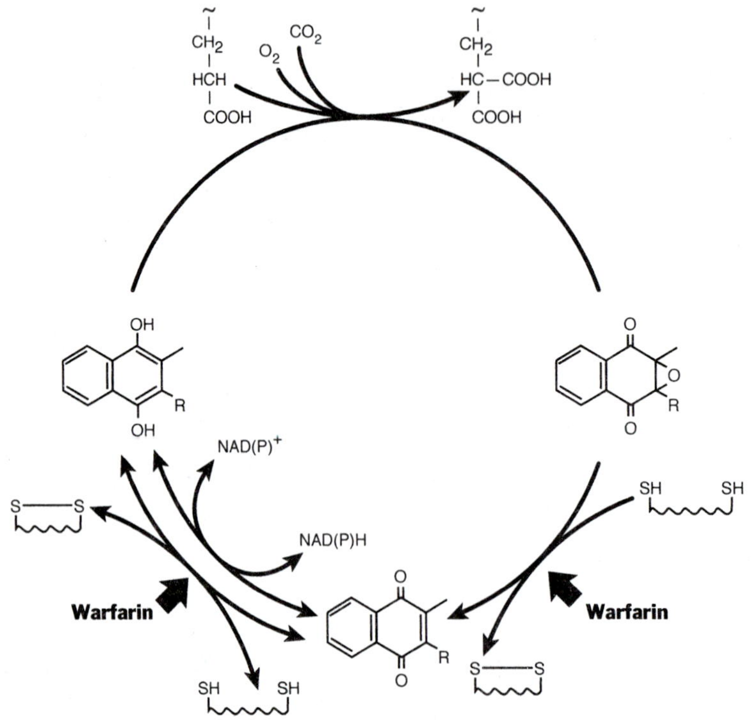

Figure 20.4 Tissue metabolism of vitamin K. Vitamin K epoxide formed in the carboxylation reaction is reduced to the quinone form of the vitamin by a warfarin-sensitive pathway, vitamin K epoxide reductase (VKORC1), which is driven by a reduced dithiol. In the absence of warfarin, the naphthoquinone form of the vitamin can be reduced to the hydronaphthoquinone form (KH_2) by VKORC1, or by one or more of the hepatic reduced nicotinamide adenine dinucleotide (NADH) or NADH phosphate (NADPH)–linked quinone reductases. A new discovery is that a NADP(H)-dependent reductase called ferroptosis suppressor protein 1 (FSP1) acts as the rescue enzyme to overcome warfarin inhibition to supply KH_2 to the carboxylase (Mishima E, Ito J, Wu Z, et al. Nature. 2022;608:778–83).

Synthesis and Function of Menaquinone-4 (MK-4)

MK-4 exhibits tissue-specific accumulation patterns, with relative concentration of the vitamer particularly high in the brain, kidneys, and pancreas, and comparatively low in heart and liver tissues. MK-4 is unusual among the menaquinones in that it is not a major product of bacterial synthesis. Instead, it is endogenously formed from conversion of other dietary forms of vitamin K.[62] Much of our understanding of this conversion is based on phylloquinone as the substrate. The phytyl sidechain of phylloquinone is removed by an unknown cleavage enzyme, generating menadione as an intermediate. Menadione is then prenylated with a geranylgeranyl sidechain by the enzyme UbiA prenyltransferase domain containing protein-1 (UBIAD1).[63] MK-4 accumulates in tissues when given orally or enterally, but not when administered parenterally,[64] and menadione is released in urine after oral, but not parenteral, administration.[65] These observations implicate the intestine as the location of the cleavage reaction, although this has yet to be clearly demonstrated.

It seems unlikely that a metabolic pathway leading to MK-4 evolved unless this vitamer has some specific role. This role is unlikely to involve the vitamin K–dependent carboxylase, because phylloquinone and MK-4 have similar activity as a substrate for this enzymatic activity. These findings suggest that MK-4 may be a control element for certain cellular functions[66,67], and they also suggest the possibility of a role for MK-4 that is completely different from the essentiality of vitamin K for Gla protein synthesis. MK-4 has been demonstrated to act as an endogenous ligand for the steroid and xenobiotic receptor (SXR) and its rodent ortholog, pregnane X receptor (PXR).[66] SXR/PXR is a nuclear receptor that dimerizes with the retinoid X receptor to regulate gene expression. While MK-4 is a potent ligand for SXR/PXR,[67] phylloquinone appears to be less active.[68,69] UBIAD1 also plays a role in regulation of cholesterol metabolism. Endoplasmic reticulum (ER)-localized 3-hydroxy-3-methylglutaryl coenzyme A HMG CoA reductase (HMGCR) is the rate-controlling enzyme in the mevalonate pathway, which produces cholesterol and the nonsterol isoprenoids farnesyl pyrophosphate (FPP) and geranylgeranyl pyrophosphate (GGPP). FPP and GGPP are used to prenylate several cellular proteins and to synthesize other nonsterol isoprenoids, including ubiquinone-10, heme, and dolichol, that play key roles in a variety of cellular functions. GGPP is the sidechain substrate used by UBIAD1 to synthesize MK4.[63,65] Several feedback control mechanisms converge on HMGCR to ensure that cells maintain a constant supply of nonsterol isoprenoids but avoid the overaccumulation of cholesterol and its sterol precursors.[70] One mechanism for feedback control of HMGCR involves its sterol-induced ubiquitination, which marks it for ER-associated degradation (ERAD) by 26S proteasomes. Sterols also cause a subset of HMGCR molecule to bind to UBIAD1. This binding inhibits the ERAD of HMGCR, which allows continued synthesis of GGPP and other nonsterol isoprenoids, even though sterols are abundant. Once GGPP accumulates in ER membranes, the isoprene binds to UBIAD1, triggering its dissociation from HMGCR. As a result, HMGCR becomes completely degraded and UBIAD1 translocates from the ER to the medial/trans Golgi. The physiologic significance of UBIAD1-mediated inhibition of HMGCR ERAD is highlighted by the finding that the embryonic lethality associated with *Ubiad1* deficiency in mice is rescued by ERAD-resistant HMGCR.[71] Despite their viability, these *Ubiad1*-deficient mice fail to thrive, which may be related to the complete absence and inability to synthesize MK-4 for carboxylation independent activity.[71]

Consequences of Vitamin K Deficiency

Anticoagulant Therapy

The most common deficiency of the vitamin K–dependent clotting factors is that acquired by treatment with oral anticoagulants. A naturally occurring antagonist of vitamin K was responsible for the hemorrhagic disease of cattle consuming improperly cured sweet clover hay that was prevalent in the upper midwestern United States and in western Canada in the 1920s. The cause of the prolonged clotting times, there was a decrease in the prothrombin activity of blood, and investigators attempted to isolate the compound from spoiled sweet clover. It was first isolated, characterized as 3-3′-methylbis-4-hydroxycoumarin by Link's group at the University of Wisconsin,[72] and called dicumarol (**Figure 20.5**). Analogs of dicumarol were synthesized, and the compound first used as both a rodenticide and a therapy for thrombotic disease was warfarin.

Although warfarin has a favorable pharmacologic profile and is essentially the only coumarin derivative prescribed in North America, structurally related compounds such as acenocoumarol, phenprocoumon, and ticlomarol are widely used in Europe. The pharmacology and clinical uses of these drugs are similar to those of warfarin.

Warfarin acts as an inhibitor of the vitamin K epoxide reductase. The results of this action are an acquired deficiency of vitamin K at the tissue level, and the secretion to the plasma of vitamin K–dependent proteins lacking all or a portion of the normal number of Gla residues. Although the activities of all of the vitamin K–dependent clotting factors are altered by warfarin treatment, the available evidence suggests that efficacy of treatment is best correlated with changes in prothrombin activity.

The magnitude of the anticoagulant effect produced by a given dose of warfarin varies substantially among and within individual patients over time. Some drugs have been shown to displace warfarin from its plasma albumin carrier, induce the hepatic cytochrome P-450 isoenzyme CYP2C9 that metabolizes warfarin, interfere with warfarin clearance, or bind warfarin in the gut. Alterations of vitamin K intake or absorption can also

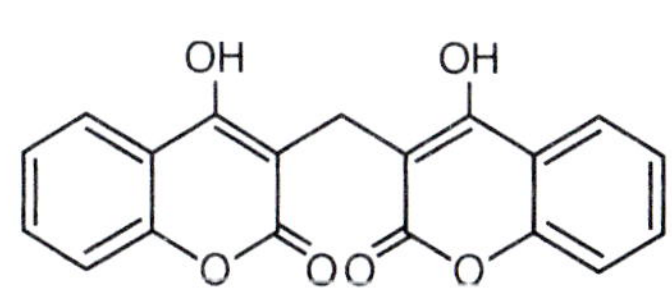

Dicumarol

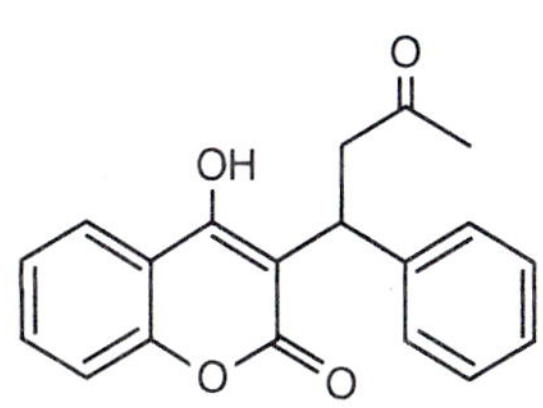

Warfarin

Figure 20.5 Structure of dicumarol and warfarin. Dicumarol was the compound isolated from sweet clover as a toxic hemorrhagic factor, and warfarin is the most commonly used of several 4-hydroxycoumarin anticoagulants.

alter warfarin efficacy, and genetic variability is also important. The amount of warfarin needed to stabilize anticoagulation is substantially affected by polymorphisms within the VKOR1 gene, the CYP2C9 gene, and the CYP4F2 gene, a vitamin K oxidase.[73,74] Pharmacogenetic testing of these alleles is being used to assist in determining the appropriate amount of warfarin needed for patients.[75]

The anticoagulant effect of warfarin therapy is monitored by measuring the PT, a measure of combined procoagulant status rather than a true measure of prothrombin activity. The international normalized ratio (INR) is used as a standardized method for reporting PT results. The INR allows interconversion of PT ratios (patient PT/mean normal PT) by use of an international sensitivity index that corrects for differences in thromboplastin sensitivities. The goal of anticoagulant therapy is the achievement of steady-state concentrations of vitamin K–dependent procoagulants in the range of 10% to 30% of normal (INR of 2 to 3). The most common complication of anticoagulant therapy, bleeding, is directly related to the INR; few episodes of bleeding occur at a stable INR of less than 4.0, and a relatively high incidence occurs with INRs greater than 7.0. Excessive anticoagulation can be brought back to the desired level by lowering the warfarin dose or, if levels are severely out of range, by oral, subcutaneous, or even slow intravenous infusion of phylloquinone.

Vitamin K-Deficiency Bleeding (VKDB) in Early Infancy

Vitamin K deficiency bleeding (VKDB) in early infancy (previously called hemorrhagic disease of the newborn) is the classic example of a human nutritional vitamin K deficiency.[76] The most common, "late onset" form of VKDB peaks between 3 and 8 weeks of age with an incidence, without prophylaxis, of about five per 100,000 births in Western European countries and up to 15 times higher in some Asian countries, such as Thailand, Japan, and China.[77] Worryingly, global surveys of late-onset VKDB show that 40% to 80% of cases first present as intracranial bleeding that may be lethal or neurodevelopmentally devastating.[77] Therefore, there is near-global consensus that all newborn infants should receive vitamin K supplementation to prevent VKDB.[77]

The main epidemiologic risk factor is exclusive feeding with breastmilk, which is attributed to its low content of vitamin K,[77] mainly comprising phylloquinone with some MK-4.[78] In some infants, underlying cholestatic pathologies that cause malabsorption of vitamin K may be identified, but a substantial proportion have no predisposing risk factor. Other known contributory risk factors are chronic diarrhea and persistent emesis. A vitamin K-specific risk factor that could explain the vulnerability to deficiency during early infancy is that vitamin K reserves in neonates lack the large pool of long-chain, bacterially-derived MKs that comprise the bulk of liver reserves in adults.[79] This means that neonates rely almost entirely on dietary phylloquinone, as their median hepatic concentrations are about one-fifth of the value in adults.[77]

The current recommendations of the American Academy of Pediatrics are that vitamin K (phylloquinone) should be given as single, intramuscular doses of 1 mg to term infants and 0.3 to 0.5 mg to preterm infants.[80] In some European countries, oral regimens are more popular whereas in the United Kingdom, the choice and dose of the prophylactic regimen is determined at the local level.[15]

Adult Deficiencies

Reports of uncomplicated adult deficiencies of vitamin K are rare, and most diets contain an adequate amount of vitamin K. The historical indication of a vitamin K deficiency, hypoprothrombinemia that responded to vitamin K administration, depended on the relatively insensitive PTs to assess adequacy of the vitamin K–dependent clotting factors.

Vitamin K deficiency has been reported in patients subjected to long-term total parenteral nutrition, and vitamin K supplementation is advised under these circumstances. Low lipid intake, or impaired lipid absorption resulting from the lack of bile salts, adversely affects vitamin K absorption, as will malabsorption syndromes and other gastrointestinal disorders (e.g., cystic fibrosis, sprue, celiac disease, ulcerative colitis, regional ileitis, *Ascaris* infection, and short bowel syndrome).

These reports and numerous cases of vitamin K–responsive hemorrhagic events in patients receiving antibiotics have been reviewed.[81] Most of these cases are assumed to result from decreased gut menaquinone utilization by these patients, but many may represent low dietary intake alone. The second- and third-generation cephalosporins have been implicated in many hypoprothrombinemic episodes,[82] and it is likely that these drugs exert weak carboxylase inhibition or a coumarin-like response that may be more important than an influence on the gut bacterial population.

Experimentally-induced vitamin K deficiencies that are sufficiently severe to reduce PT measurements are rare. An often-cited study[83] investigated the vitamin K requirement of starved, intravenously fed debilitated patients who were given antibiotics to decrease intestinal vitamin K synthesis. A significant degree of vitamin K–responsive hypoprothrombinemia was clearly established in these patients. More recently, controlled studies using diets containing approximately 10 μg/day or less of phylloquinone demonstrated alterations using more sensitive markers of vitamin K status,[84,85] but a clinically significant decrease in PT was not seen.

Role in Skeletal Health

Although MGPs, protein S, and Gla rich proteins are also present in bone or cartilage, it is the large amount of OC in bone has attracted attention as a possible factor in bone health. Small amounts of this protein enter circulation following their synthesis in osteoblasts and are used as a clinical measure of bone formation. The synthesis of OC is not regulated by vitamin K, which has a post-translational role in γ-carboxylation of the protein, hence measurement of the total amount of the protein in circulation provides little insight into vitamin K's role in bone. Instead, assessments of vitamin K status and bone in individuals within the physiologically normal population should be made by separate measurements of under γ-carboxylated species of OC (ucOC), expressing this fraction as a proportion (usually a percentage) of total circulating OC88.[86] The original assays defined ucOC as the fraction that does not adsorb to hydroxyapatite under standard conditions. Depending on assay conditions and specific epitopes detected by the assay kits used, the fraction of ucOC reported in normal populations has ranged from 60% to less than 10%. More recent studies using mass spectrometry have suggested that the fraction of ucOC is greater than 50% in normal populations.[87]

Habitual dietary intakes of vitamin K are not sufficient to maximally γ-carboxylate OC. Instead, supplementation with 1 mg phylloquinone per day (~10 × the current dietary reference intake (DRI)) has been proposed to achieve maximal γ-carboxylation, although this varies with the assay used.[86] The question that remains to be answered is what are, if any, the health consequences of not achieving maximal carboxylation? Reports linking this apparent marker of vitamin K insufficiency with bone health include epidemiologic observations that low vitamin K intake, primarily in the form of phylloquinone, is associated with increased hip fracture risk[88] and that higher values of either ucOC[89] or %ucOC[88] correlated with low bone mass.[89] These associations do not necessarily imply causation, and they may be surrogate markers of general nutrient deficiencies. Further, the proportion of circulating OC that is uncarboxylated reflects status of all forms of vitamin K, and not exclusively phylloquinone or individual forms of menaquinones. Unfortunately, the erroneous assumption that %ucOC can be used as a biomarker of bone health for individual molecular forms of vitamin K is still in the literature.

Multiple placebo-controlled randomized clinical trials have focused on the impact of phylloquinone in daily doses of 200 μg to 5 mg on age-related bone loss.[90] Many small trials of daily doses up to 45 mg MK4 supplementation assessing bone mineral density or fracture rates of postmenopausal osteoporotic women have also been conducted.[90] These studies used primary outcome measures of change in bone mineral density at various anatomical sites, or of bone turnover markers. A recent meta-analysis of these studies concluded that, for post-menopausal or osteoporotic patients, there is no evidence that vitamin K increases bone mineral density or reduces vertebral fractures.[90] One study that used a dose of 5 mg/d of phylloquinone for up to 4 years, did report a significant decrease in overall fracture among postmenopausal women. However, the study was not designed to test this question, nor was it statistically powered, so it remains unclear if this was a spurious or a true finding. Other caveats to the interpretation of existing studies are that none screened for individuals with low vitamin K status, the group that would most benefit from vitamin K, and most studies provided vitamin K in the context of adequate calcium and vitamin D status, thereby only testing the public health question of whether vitamin K has a protective role in bone health above that achieved by calcium and vitamin D.

Matrix Gla Protein and Vascular Calcification

Early studies of MGP knockout mice found that these animals died of massive calcification of large arteries within 8 weeks of birth,[37,38] and that other strategies to block MGP carboxylation in rats led to rapid calcification of the elastic lamellae of arteries and heart valves. These findings led to human studies relating low MGP activity to various aspects of soft tissue and blood vessel calcification.

Multiple forms of MGP are detectable in circulation that differ with respect to the carboxylation and phosphorylation status of the protein.[91] Carboxylation of MGP is required to bind to calcium, whereas the role of phosphorylation is less clear. Unlike osteocalcin, there are no assays available that determine the proportion of MGP that is carboxylated. Available assays either measure total MGP, which does not respond to vitamin K supplementation, or absolute amounts of uncarboxylated MGP (ucMGP), which does not account for the total amount of MGP available to be carboxylated. Total MGP varies with many factors related to vascular calcification, including age, all of which are independent of vitamin K.[92] Therefore, it is not clear if ucMGP is measuring the vitamin K-independent factors influencing total MGP or vitamin K status. As such, the utility of this measure as a biomarker of vitamin K status is uncertain.

Regardless of the uncertainty of the biomarker ucMGP as a measure of vitamin K status, there are many studies using ucMGP as an exclusive biomarker of vitamin K status and cardiovascular disease.[91] However, the overall evidence of the relative importance of phylloquinone and menaquinone to CVD in humans is equivocal. In some observational studies, higher dietary menaquinone intake, but not phylloquinone intake, was associated with less coronary artery calcification (a subclinical manifestation of CVD) and lower risk for clinical CVD events. These findings have led to claims that menaquinones have unique cardiovascular health benefits, compared with phylloquinone. However, this claim is not supported by the results of the limited number of intervention trials conducted to date. Another problem with the hypothesis of vitamer "uniqueness" is that it fails to take into account evidence that a fraction of multiple dietary forms of vitamin K can be converted during intestinal absorption into MK-4 via the UBIAD1 pathway.[62]

If vitamin K status does have an influence on vascular calcification, patients who have been treated with warfarin for years should be particularly susceptible to coronary calcification. Studies have had mixed results,[93,94] and more data will be needed to define more clearly the risk of vascular calcification that may be related to warfarin therapy. A genetic disorder, pseudoxanthoma elasticum, is also linked to the control of ectopic calcification by MGP[95,96], given that undercarboxylated MGP is increased in patients with this disease. Although the biological role of MGP in regulating vascular calcification appears to be established, the clinical value of supplemental vitamin K has not yet been determined.[91]

Dietary Requirement

There is a 2-fold range in current recommended vitamin K intakes when compared across countries, which reflects the critical gap in knowledge about optimal requirements for vitamin K. Ample data have established that essentially all individuals do not consume sufficient vitamin K to γ-carboxylate their circulating OC or MGP maximally. Supplementation with phylloquinone increases the proportion of the carboxylated forms of these circulating vitamin K-dependent proteins but, as described above, there is currently no established clinical significance of these apparent deficiencies. In this context, these biomarkers of vitamin K status cannot be used to set a reference value for vitamin K dietary requirements. The only clinically significant indicator of vitamin K status is still PT, and alterations in the PT by changes in dietary intake alone are uncommon. Because circulating phylloquinone concentration depends highly on the previous day's intake, it is also not a robust biomarker of an adequate intake at the individual level. Dietary menaquinones are not usually detected in circulation unless provided in supplemental form, hence cannot be used as a biomarker of intake.

In lieu of sufficient data in which to establish an Estimated Average Requirement for vitamin K, in North America, the reference values for vitamin K intake are referred to as Adequate

Intakes (AIs). These are based on the highest median intake for each age group reported by the third National Health and Nutrition Examination Survey (NHANES III)[7] (**Table 20.1**).

The AIs for infants are based on the phylloquinone content of human milk and assume that infants also receive prophylactic vitamin K at birth. Based on the NHANES III data, the intakes of pregnant or lactating women do not differ from those of the general population.

Because indications of toxicity following the ingestion of large amounts of vitamin K are not available, the DRI process was unable to define a tolerable upper intake limit (UL) for vitamin K.

Table 20.1 Adequate Intakes of Vitamin K[a]

Population	μg/d of Vitamin K
0–6-mo-old infants	2.0
7–12-mo-old infants	2.5
1–3-y-old children	30.0
4–8-y-old children	55.0
9–13-y-old boys and girls	60.0
14–18-y-old boys and girls[b]	75.0
19–>70-y-old men	120.0
19–>70-y-old women[b]	90.0

[a]Dietary reference intakes.
[b]No alteration of intake for pregnancy or lactation.

Data from Food and Nutrition Board, Institute of Medicine. Dietary Reference Intakes for Vitamin A, Vitamin K, Arsenic, Boron, Chromium, Copper, Iodine, Iron, Manganese, Molybdenum, Nickel, Silicon, Vanadium, and Zinc. Washington DC: National Academy Press, 2001.

Acknowledgments

The authors would like to acknowledge John W Suttie for contributions to the previous version of this chapter.

References

1. Dam H. Biochem Z 1929;215:475–92.
2. Dam H. Nature 1935;135:652–53.
3. Suttie JW. Vitamin K in Health and Disease. Boca Raton FL: CRC Press, Taylor and Francis Group; 2009.
4. Dam H, Schonheyder F. Biochem J 1936;30:897–901.
5. Shearer MJ. J Nutr 2022;152:917–19.
6. United States Department of Agriculture. FoodData Central 2022 (fdc.nal.usda.gov).
7. Harshman SG, Finnan EG, Barger KJ, et al. J Nutr 2017;147:1308–13.
8. Fu X, Shen X, Finnan EG, et al. J Agric Food Chem 2016;64:4531–5.
9. Fu X, Harshman SG, Shen X, et al. C Develop Nutr 2017;1.
10. Shearer MJ, Newman P. Thromb Haemost 2008;100:530 47.
11. Shearer MJ, Okano T. Annu Rev Nutr 2018;38:127–51.
12. Lamon-Fava S, Sadowski JA, Davidson KW, et al. Am J Clin Nutr 1998;67:1226–31.
13. Sadowski JA, Hood SJ, Dallal GE, et al. Am J Clin Nutr 1989;50:100–8.
14. Saupe J, Shearer MJ, Kohlmeier M. Am J Clin Nutr 1993;58:204–8.
15. Shearer MJ, Fu X, Booth SL. Adv Nutr 2012;3:182–95.
16. Jones KS, Bluck LJ, Wang LY, et al. Br J Nutr 2009;102:1195–202.
17. Usui Y, Tanimura H, Nishimura N, et al. Am J Clin Nutr 1990;51:846–52.
18. Wiss O, Gloor H. Vitam Horm 1966;24:575–86.
19. Harrington DJ, Booth SL, Card DJ, et al. J Nutr 2007;137:1763–8.
20. Thijssen HH, Vervoort LM, Schurgers LJ, et al. Br J Nutr 2006;95:260–6.
21. Shearer MJ, McBurney A, Breckenridge AM, et al. Clin Sci Mol Med 1977;52:621–30.
22. Ellis JL, Karl JP, Oliverio AM, et al. Gut Microbes 2021;13:1–16.
23. Karl JP, Fu X, Wang X, et al. Am J Clin Nutr 2015;102:84–93.
24. Ellis JL, Wang M, Fu X, et al. Curr Dev Nutr 2022;6:nzac019.
25. Giangrande PL. Br J Haematol 2003;121:703–12.
26. Dahlback B. Lancet 2000;355:1627–32.
27. Mann KG. Chest 2003;124:4S–10S.
28. Esmon CT. Chest 2003;124:26S–32S.
29. Yin ZF, Huang ZF, Cui J et al. Proc Natl Acad Sci U S A 2000;97:6734–8.
30. Endler G, Mannhalter C. Clin Chim Acta 2003;330:31–55.
31. Hauschka PV, Lian JB, Gallop PM. Proc Natl Acad Sci U S A 1975;72:3925–9.
32. Price PA, Otsuka AA, Poser JW, et al. Proc Natl Acad Sci U S A 1976;73:1447 51.
33. Price PA, Williamson MK, Haba T, et al. Proc Natl Acad Sci U S A 1982;79:7734–8.
34. Ducy P, Desbois C, Boyce B, et al. Nature 1996;382:448–52.
35. Diegel CR, Hann S, Ayturk UM, et al. PLoS Genet 2020;16:e1008361.
36. Fraser JD, Price PA. J Biol Chem 1988;263:11033–6.
37. Luo G, Ducy P, McKee MD, et al. Nature 1997;386:78–81.
38. Price PA, Faus SA, Williamson MK. Arterioscler Thromb Vasc Biol 1998;18:1400–7.
39. Viegas CS, Simes DC, Laize V, et al. J Biol Chem 2008;283:36655–64.
40. Viegas CS, Cavaco S, Neves PL, et al. Am J Pathol 2009;175:2288–98.
41. Manfioletti G, Brancolini C, Avanzi G, et al. Mol Cell Biol 1993;13:4976–85.
42. Kulman JD, Harris JE, Haldeman BA, et al. Proc Natl Acad Sci U S A 1997;94:9058–62.

43. Kulman JD, Harris JE, Xie L, et al. Proc Natl Acad Sci U S A 2001;98:1370–5.
44. McIntosh JM, Olivera BM, Cruz LJ, et al. J Biol Chem 1984;259:14343–6.
45. Brown MA, Hambe B, Furie B, et al. Toxicon 2002;40:447–53.
46. Bandyopadhyay PK. Vitam Horm 2008;78:157–84.
47. Nelsestuen GL, Zytkovicz TH, Howard JB. J Biol Chem 1974;249:6347–50.
48. Stenflo J, Fernlund P, Egan W, et al. Proc Natl Acad Sci U S A 1974;71:2730–3.
49. Esmon CT, Sadowski JA, Suttie JW. J Biol Chem 1975;250: 4744–8.
50. Rishavy MA, Hallgren KW, Wilson LA, et al. Blood 2022;140: 1710–22.
51. Furie B, Furie BC. N Engl J Med 1992;326:800–6.
52. Knobloch JE, Suttie JW. J Biol Chem 1987;262:15334–7.
53. Shah DV, Tews JK, Harper AE, et al. Biochim Biophys Acta 1978;539:209–17.
54. Bell RG, Matschiner JT. Nature 1972;237:32–3.
55. Lund M. Nature 1964;203:778.
56. Hildebrandt EF, Suttie JW. Biochemistry 1982;21:2406–11.
57. Zimmermann A, Matschiner JT. Biochem Pharmacol 1974; 23:1033–40.
58. Li T, Chang CY, Jin DY, et al. Nature 2004;427:541–4.
59. Rost S, Fregin A, Ivaskevicius V, et al. Nature 2004;427:537–41.
60. Tie JK, Stafford DW. J Thromb Haemost 2016;14:236–47.
61. Spohn G, Kleinridders A, Wunderlich FT, et al. Thromb Haemost 2009;101:1044–50.
62. Ellis JL, Fu X, Karl JP, et al. J Nutr 2022;152:981–93.
63. Nakagawa K, Hirota Y, Sawada N, et al. Nature 2010;468:117–21.
64. Okano T, Shimomura Y, Yamane M, et al. J Biol Chem 2008; 283:11270–9.
65. Hirota Y, Tsugawa N, Nakagawa K, et al. J Biol Chem 2013; 288:33071–80.
66. Tabb MM, Sun A, Zhou C, et al. J Biol Chem 2003;278:43919–27.
67. Yoshida T, Miyazawa K, Kasuga I, et al. Int J Oncol 2003:23: 627–32.
68. Landes N, Birringer M, Brigelius-Flohe R. Mol Aspects Med 2003;24:337–44.
69. Suhara Y, Hanada N, Okitsu T, et al. J Med Chem 2012;55: 1553–8.
70. Schumacher MM, Jun DJ, Johnson BM, et al. J Biol Chem 2018;293:312–23.
71. Jo Y, Kim SS, Garland K, et al. Elife 2020;9.
72. Link KP. Circulation 1959;19:97–107.
73. Flockhart DA, O'Kane D, Williams MS, et al. Genet Med 2008; 10:139–50.
74. McDonald MG, Rieder MJ, Nakano M, et al. Mol Pharmacol 2009;75:1337–46.
75. Gage BF, Eby C, Johnson JA, et al. Clin Pharmacol Ther 2008;84:326–31.
76. Lane PA, Hathaway WE. J Pediatr 1985;106:351–9.
77. Shearer MJ. Blood Rev 2009;23:49–59.
78. Indyk HE, Woollard DC. Analyst 1997;122:465–9.
79. Shearer MJ. Blood Rev 1992;6:92–104.
80. Hand I, Noble L, Abrams SA. Pediatrics 2022;149.
81. Savage D, Lindenbaum J. Clinical and experimental human vitamin K deficiency. In: Nutrition in Hematology. edn. Edited by Lindenbaum J. New York: Churchill Livingstone. 1983;271–320.
82. Weitekamp MR, Aber RC. JAMA 1983;249:69–71.
83. Frick PG, Riedler G, Brogli H. J Appl Physiol 1967;23:387–9.
84. Allison PM, Mummah-Schendel LL, Kindberg CG, et al. J Lab Clin Med 1987;110:180–8.
85. Booth SL, O'Brien-Morse ME, Dallal GE, et al. Am J Clin Nutr 1999;70:368–77.
86. Gundberg CM, Lian JB, Booth SL. Adv Nutr 2012;3:149–57.
87. Rehder DS, Gundberg CM, Booth SL, et al. Mol Cell Proteomics 2015;14:1546–55.
88. Booth SL, Tucker KL, Chen H, et al. Am J Clin Nutr 2000;71: 1201–8.
89. Vergnaud P, Garnero P, Meunier PJ, et al. J Clin Endocrinol Metab 1997;82:719–24.
90. Mott A, Bradley T, Wright K, et al. Osteoporos Int 2019;30: 1543–59.
91. Shea MK, Berkner KL, Ferland G, et al. Adv Nutr 2021;12: 632–646.
92. O'Donnell CJ, Shea MK, Price PA, et al. Arterioscler Thromb Vasc Biol 2006;26:2769–74.
93. Rennenberg RJ, van Varik BJ, Schurgers LJ, et al. Blood 2010;115:5121–3.
94. Villines TC, O'Malley PG, Feuerstein IM et al. Calcif Tissue Int 2009;85:494–500.
95. Hendig D, Zarbock R, Szliska C, et al. Clin Biochem 2008;41: 407–12.
96. Gheduzzi D, Boraldi F, Annovi G, et al. Lab Invest 2007;87: 998–1008.

C. VITAMINS

CHAPTER 21

Thiamine

Megan W. Bourassa, Chantal Bémeur, and Roger F. Butterworth

CHAPTER OUTLINE

Abbreviations

α-KTG, α-ketoglutarate dehydrogenase; **BBB**, blood-brain barrier; **eNOS**, endothelial nitric oxide synthase; **ETKAC**, erythrocyte transketolase activation coefficient; **HPLC**, high-performance liquid chromatography; **IgG**, immunoglobulin G, **KP**, Korsakoff psychosis; **NMDA**, N-methyl-D-aspartate; **ThDP**, thiamine diphosphate; **ThMP**, thiamine monophosphate; **ThTP**, thiamine triphosphate; **WE**, Wernicke encephalopathy

Historical Overview

Thiamine deficiency has been known for thousands of years, but only in the last two centuries has the cause been identified. Chinese medical texts referred to the condition known as beriberi as early as 2700 BC, but it was not until 1884 AD that Kanehiro Takaki, a surgeon general in the Japanese navy, showed that the disease was the consequence of a dietary inadequacy. In 1910, Umetaro Suzuki isolated vitamin B1 from rice bran and named it aberic acid.[1] Unfortunately, when his publication was translated from Japanese into German, it neglected to mention that he had isolated the vitamin. As a result, much of the credit goes to Christian Eijkman, a military doctor in the Dutch East Indies. He noticed that fowl fed a diet of cooked, polished rice developed paralysis that he attributed to a nerve poison in the endosperm of the grain. Eijkman would go on to win the Nobel Prize in Physiology or Medicine in 1929 for the discovery of the antineuritic vitamin (thiamine). However, it was Eijkman's colleague, Gerrit Grijns, who later correctly interpreted the connection between excessive consumption of polished rice and beriberi. He concluded that rice contained an essential nutrient in the outer layers of the grain that was removed in polishing.[2] In 1911, Cashmir Funk isolated the "antineuritic substance" from rice bran that he called a "vitamine" because of its vital role and amine functional group.[3] The "e" was later dropped from vitamin when it was realized that not all the subsequently isolated vitamins would contain an amine. Dutch chemists went on to isolate and crystallize the active agent whose structure (**Figure 21.1**) was determined in 1934 by Robert Williams, a U.S. chemist. Thiamine was synthesized in 1936.

Biochemistry

Thiamine, or vitamin B1, is a water-soluble vitamin. Aqueous solutions of thiamine are stable at acid pH but are unstable in alkaline solutions or when exposed to ultraviolet light. Both the pyrimidine and thiazole moieties (see Figure 21.1) are

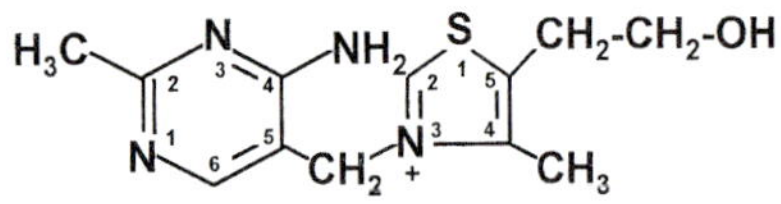

Thiamin

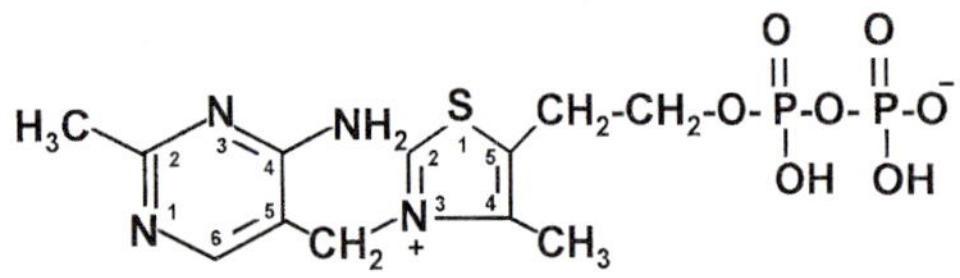

Thiamin pyrophosphate

Figure 21.1 The thiamine molecule consists of a pyrimidine ring and a thiazole moiety, which are linked by a methylene (CH_2) bridge. Thiamine is a water-soluble white crystalline solid.

required for biologic activity.[3] Thiamine is readily cleaved at the methylene bridge by sulfite treatment at pH 6.0. Thiamine is absorbed in its unphosphorylated form and, then, successive phosphorylations result in thiamine monophosphate (ThMP), thiamine diphosphate (ThDP) (or pyrophosphate), and thiamine triphosphate (ThTP). Approximately 80% of total body thiamine is in the metabolically active form of ThDP.[4] Prokaryotes, plants, and yeast cells are able to synthesize thiamine for their own use, but animals are dependent on dietary intake of the vitamin.[5]

Dietary Sources and Recommended Dietary Allowances

Thiamine concentrations are highest in yeast and in the pericarp and germ of cereals.[6,7] **Table 21.1** summarizes major dietary sources of thiamine.[8] Today, many cereals and breads are fortified with thiamine. Conversely, milk and dairy products, seafood, and most fruits are poor sources. Thiamine is sensitive to high temperatures, and prolonged cooking of foods may result in a loss of thiamine content. Baking bread, for example, leads to a 20% to 30% reduction in thiamine, and pasteurization of milk may also result in thiamine losses of up to 20%. By contrast, freezing foods does not result in significant reductions of thiamine. Because thiamine is a water-soluble vitamin, significant amounts are also lost in discarded cooking water. Thiamine is also destroyed by ultraviolet irradiation of food stuffs.[6,7]

Dietary reference intake values for thiamine by life stage group[7] are shown in **Table 21.2**. There are several lifestyle factors that may increase thiamine requirement, including diets high in carbohydrates, those who live in warm climates, those who weigh more, or those with high levels of physical activity.[9,10] It has also been proposed that these requirements should be increased based on modern lifestyles and diets.[11] Indeed, the recommended dietary allowance (RDA) for thiamine set by the Indian Food Safety and Standards Authority are based on activity levels and range from 1.4–2.3 mg/d for an adult male, higher than those set by the U.S. National Academies of Medicine (formerly Institute of Medicine).[12] There is no established UL for thiamine.

Table 21.1 Thiamine Content by Food Group

Food group	Food item	Thiamine content (mg) per 100g
Cereal	Wheat flour (unenriched)	
	■ White	0.08
	■ Whole grain*	0.50
	■ Wheat germ*	1.90
	Rice (unenriched)	
	■ White, raw	0.07
	■ White parboiled, raw	0.22
	■ Brown, raw*	0.54
	Oatmeal, unenriched*	0.73
	Sorghum, unenriched	0.33
Meat and fish	Meats	
	■ Pork, loin, lean only, roasted*	1.02
	■ Pork, cured (ham), lean only, canned*	0.89
	■ Chicken breast, meat only, roasted	0.07
	■ Beef, tenderloin, lean only, cooked	0.06
	Fish	
	■ Salmon, cooked	0.09
	■ Sardines, canned, in oil, drained	0.08
Pulses, nuts and seeds	Peas, boiled and drained*	0.26
	Lentils, boiled and drained*	0.17
	Soybeans (green), boiled and drained*	0.26 0.64
	Peanuts, fresh*	0.15
	Peanuts, dry roasted	1.48
	Sunflower seeds, dried*	
	Sunflower seeds, dry roasted	0.11
Dairy products	Full fat milk, pasteurized, unenriched	0.05
	Cheese, cow's milk	0.03
Vegetables	Broccoli, raw	0.07
	Cabbage	0.06
	Carrots	0.07
	Tomatoes	0.04
Fruits	Banana	0.03
	Papaya	0.02
	Pineapple	0.08
	Tamarinds, raw*	0.43
Tubers/ starchy roots	Cassava, fresh	0.09
	Potato, boiled	0.11
Other	Yeast	11

*Indicates good dietary sources of thiamine.

Modified from The New York Academy of Sciences (2019). National control and prevention programs for thiamine deficiency disorders: Technical Reference Materials. New York.

Table 21.2 Criteria and Dietary Reference Intake Values for Thiamine by Life Stage Group

		EAR[a] (mg/d)		RDA[b] (mg/d)		
Life Stage Group	**Criterion**	**Male**	**Female**	**Male**	**Female**	**AI[c] (Mg/D)**
0–6 mo	Average thiamine intake from human milk					0.2
7–12 mo	Extrapolation from adult requirements					0.3
1–3 y	Extrapolation from adult EAR	0.4	0.4	0.5	0.5	
4–8 y	Extrapolation from adult EAR	0.5	0.5	0.6	0.6	
9–13 y	Extrapolation from adult EAR	0.7	0.7	0.9	0.9	
14–18 y	Extrapolation from adult EAR	1.0	0.9	1.2	1.0	
18→70 y	Depletion/repletion studies; erythrocyte transketolase activity	1.0	0.8	1.2	1.1	
Pregnancy						
14–50 y	Adult female EAR plus estimated daily thiamine accumulation by fetus		1.2		1.4	
Lactation						
14–50 y	Adolescent female EAR plus average amount of thiamine secreted in human milk		1.2		1.4	

[a]EAR, estimated average requirement, the intake that meets the estimated nutrient needs of half of the individuals in a group.

[b]RDA, recommended dietary allowance, the intake that meets the estimated nutrient needs of almost all (97% to 98%) individuals in a group.

[c]AI, adequate intake, the observed average of experimentally determined intake by a defined population or subgroup that appears to sustain a defined nutritional status, such as growth rate, normal circulating nutrient values, or other functional indicators of health. The AI is used if sufficient scientific evidence is not available to derive an EAR. For healthy infants receiving human milk, the AI is the mean intake. The AI is not equivalent to an RDA.

Data from Food and Nutrition Board, Institute of Medicine. Dietary Reference Intakes for Thiamine, Riboflavin, Niacin, Vitamin B6, Folate, Vitamin B_{12}, Biotin, and Choline. Washington, DC: National Academy Press, 1998:58–86.

Thiaminases and Antithiamine Compounds in Foods

Certain foods contain thiaminases—thermolabile enzymes that rapidly degrade thiamine.[6] Thiaminase I is encountered in some raw fish, shellfish, and ferns, as well as in microorganisms such as *Clostridium thiaminolyticus*. Thiaminase II, which has an action distinct from that of thiaminase I, is found in other organisms such as *Candida aneurinolytica*. Thiaminases act during food storage or during passage through the gastrointestinal tract. Consequently, regular consumption of raw fish (with or without fermentation), raw shellfish, and ferns is a risk factor for the development of thiamine deficiency. Antithiamine compounds are thermostable and have been identified in some ferns, teas, and betel nut, in which the toxic agents were found to be analogs of polyphenolic compounds such as tannic acid (tannin). As a result, consuming raw or fermented fish or chewing betel nut have been known to contribute to thiamine deficiency.[13,14]

Absorption, Transport, and Excretion

Thiamine is absorbed by the small intestine by two distinct mechanisms, namely, active transport (at concentrations <2 μmol/L) and passive diffusion (at higher concentrations).[6] Active thiamine transport is greatest in the jejunum and ileum. Intestinal transport of thiamine is rate limiting in humans. Following uptake from the gastrointestinal tract, thiamine is transported by the portal blood to the liver.

In the normal adult human body, total thiamine concentrations are approximately 25 to 30 mg. Skeletal muscle, heart, liver, kidney, and brain contain relatively high thiamine concentrations and are where turnover rates are highest. Given these relatively fast turnover rates, and because thiamine is not stored to any large extent in tissue, a continuous dietary supply is necessary. Thiamine and its acid metabolites (2-methyl-4-amino-5-pyrimidine carboxylic acid, 4-methylthiazole-5-acetic acid, and thiamine acetic acid) are excreted principally in the urine.[6]

Assessment of Thiamine Status

Thiamine Diphosphate

There are two biochemical indicators of thiamine status: a direct measurement of ThDP in whole blood or red blood cells[14,15] and an indirect measurement that assays the activation of the thiamine diphosphate (ThDP)-dependent enzyme, transketolase in red blood cells.[16,17]

Erythrocyte Transketolase Activation Assay

Today, the erythrocyte transketolase (ETK) activation assay is less commonly used than the measurement of ThDP. It is an indirect measurement based on the transketolase activity in hemolysates of red blood cells (erythrocytes), using a UV spectrophotometer to detect changes in activity.[4] ETK activity is usually expressed as the ETK activity coefficient (ETKAC), which is a ratio of the basal transketolase activity and the activated levels of transketolase activity from exogenously added ThDP.[16] ETKAC values greater than 1.25 are generally considered an indicator of risk of thiamine deficiency, and values below 1.15 are sufficient.[7,18] Clinical symptoms of beriberi are associated with values around 1.4.[16]

Functions of Thiamine in Metabolism

After uptake into the cell, thiamine is rapidly phosphorylated to its diphosphate ester (ThDP), previously referred to as thiamine pyrophosphate. ThDP is an essential cofactor for enzymes involved in glucose and amino acid metabolism.[19-21] Such enzymes include the following: transketolase, a key component of the pentose shunt pathway; pyruvate dehydrogenase complex, an enzyme complex situated at the point of entry of pyruvate carbon into the tricarboxylic acid cycle; α-ketoglutarate dehydrogenase (α-KGDH), a rate-limiting enzyme and constituent of the tricarboxylic acid cycle; and branched-chain keto acid dehydrogenases. The first three of these ThDP-dependent enzymes are implicated in glucose and energy metabolism by the cell, as shown in the simplified schematic form in **Figure 21.2**.

Not surprisingly, given the mitochondrial localization of the dehydrogenases and the importance of the pentose phosphate pathway in cellular glucose metabolism, thiamine deficiency has a plethora of metabolic consequences, including a reduction of tricarboxylic acid intermediates, reduced synthesis of high-energy phosphates,[22] and the accumulation of alanine[21] and lactate=.[23] Alterations of brain pH resulting from focal lactic acidosis could contribute to the pathogenesis of thalamic neuronal damage and consequent cerebral dysfunction in thiamine deficiency.[24] In the brain, where the tricarboxylic acid cycle is essential for the synthesis of neurotransmitters such as acetylcholine and γ-aminobutyric acid, thiamine deficiency also results in a decrease in their synthesis[21,25] (see Figure 21.2). The addition of thiamine to thiamine-deprived cellular preparations *in vitro*[26] or to thiamine-deficient animals[21] results in rapid normalization in activities of ThDP-dependent enzymes and their associated metabolites and neurotransmitters. This reversible metabolic phenomenon has been referred to as *the biochemical lesion* in thiamine deficiency.

Causes of Thiamine Deficiency

Thiamine deficiency may result from inadequate dietary intake of the vitamin as well as from decreased absorption, defective transport, increased requirement, or enhanced losses.[7,27] Eating a staple diet of polished rice or ingesting foods containing thiaminases or antithiamine compounds with limited dietary diversity continue to be the major causes of thiamine deficiency in many parts of the world.[4,28] As described above, white rice often lacks nutrient enrichment, including thiamine.[29]

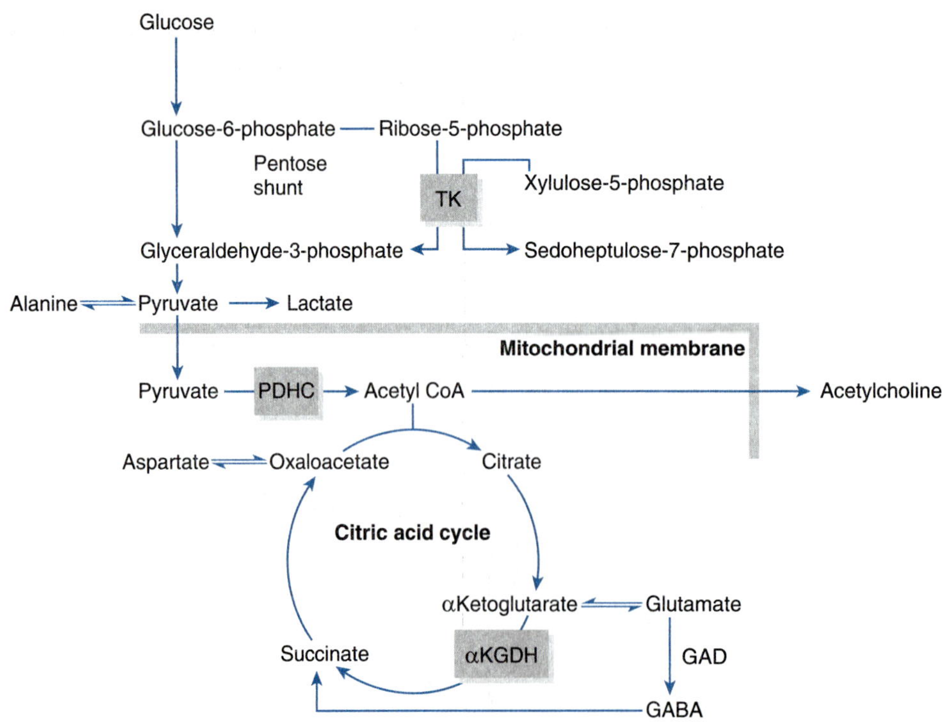

Figure 21.2 Thiamine diphosphate–dependent enzymes. *CoA*, coenzyme A; *GABA*, γ-aminobutyric acid; *GAD*, glutamic acid decarboxylase; *αKGDH*, α-ketoglutarate dehydrogenase; *PDHC*, pyruvate dehydrogenase complex; *TK*, transketolase.

Thiamine deficiency is increasing recognized in higher income settings in adults and children, often because of an underlying health condition or medical procedure.[27,30] Populations at particularly high risk of thiamine deficiency include those with alcoholism, eating disorders, human immunodeficiency virus and acquired immunodeficiency syndrome (HIV/AIDS),[31] those who have undergone gastric bypass surgery, as well as those receiving parenteral nutrition[25] when thiamine is omitted from the formula in error or when the thiamine is destroyed by prolonged contact with the amino acid solution. Certain drugs such as the antihyperglycemic agent tolazamide may also cause thiamine deficiency.[32]

Human Thiamine Deficiency Disorders

Due to thiamine's role in energy metabolism, clinical thiamine deficiency can present in several different forms, including metabolic, neurologic, cardiovascular, gastrointestinal, respiratory, and musculoskeletal symptoms.[33] The diversity of symptoms can make a diagnosis resulting from thiamine deficiency challenging.[34-37] Furthermore, abnormalities of ThDP-dependent enzymes have been reported in many different neurodegenerative and inherited metabolic disorders.[38,39] **Table 21.3** describes the major organ systems involved, the symptoms, reported age groups, and predisposing factors for thiamine deficiency.[33]

Beriberi

Historically, beriberi has been referred to as either "wet" or "dry" and is typically associated with infants. *Wet beriberi* is associated with edema, tachycardia, cardiomegaly, and congestive heart failure in addition to peripheral neuropathy. Hemodynamic changes in wet beriberi include high cardiac output and low peripheral resistance. *Dry beriberi* is characterized principally by peripheral neuropathy consisting of symmetric impairment of sensory, motor, and reflex functions that affects distal more than proximal limb segments and causes calf muscle tenderness. Rarely, patients have fulminant or "shoshin" beriberi, the major features of which are tachycardia and circulatory collapse.

The developing brain is more sensitive than the adult brain to the deleterious effects of thiamine deficiency.[40,41] For example, it is well established that infantile beriberi occurs in infants breastfed by mothers who themselves may be asymptomatic. Reports from many world populations continue to describe a high prevalence of thiamine deficiency and its complications in pregnant and lactating women, a population known to have an increased thiamine requirement.[4] Increased sequestration of thiamine by the fetus and placenta during the third semester of pregnancy is thought to result in increased thiamine requirement.[42] Thiamine concentrations are higher in umbilical cord blood compared with maternal blood; a finding consistent with preferential delivery of thiamine to the developing infant.

Other causes of maternal thiamine deficiency include alcohol abuse, gastrointestinal disease, hyperemesis gravidarum, and HIV/AIDS. Investigators have shown that maternal thiamine deficiency contributes to intrauterine growth retardation,[40] a condition that results in delayed myelination of the brain that may be related to reduced activity of ThDP-dependent enzymes. Maternal thiamine deficiency may also contribute to the pathogenesis of fetal alcohol syndrome.

While beriberi is most associated with infants, it is important to recognize that these symptoms can occur in older children and adults. To move away from the association between infants and beriberi and to recognize the array of symptoms, the term thiamine-deficiency disorder has been coined.[4,33]

Table 21.3 Thiamine Deficiency Symptoms

Organ system	Symptoms	Reported age groups	Possible predisposing factors
Metabolic	Acidosis—high lactate	Infants, adults	Genetic causes, alcoholism
Cardiorespiratory	Heart Failure	Infants, adults	Breastfed infants of women with a thiamine-insufficient diet. Consumption of anti-thiamine factors
	Dysphonia	Infants, adults	
	Tachypnea	Infants, adults	Breastfed infants of women with a thiamine-insufficient diet
Neurologic	Encephalopathy	Infants, children, adults	Alcoholism
	Developmental delay	Infants	Infants fed thiamine deficient formula
	Peripheral neuropathy	Children, adults	Bariatric surgery, obesity, pregnancy
	Ataxia	Children, Adults	Consumption of antithiamine factors
	Nystagmus or ophthalmoplegia	Infants, adults	Breastfed infants of women with a thiamine-insufficient diet. Bariatric surgery
Musculoskeletal	Weakness and atrophy	Children, adults	Pregnancy
Gastrointestinal	Vomiting	Infants, adults	Hyperemesis gravidarum, drug-induced hyperlactatemia, and disease-related malnutrition

Note: This table provides representative examples of key information. It is neither restrictive nor exhaustive.

Data from Smith TJ, Johnson CR, Koshy R, et al. Ann NY Acad Sci 2021;1498(1):9–28.

Wernicke Encephalopathy (Wernicke-Korsakoff Syndrome)

Wernicke encephalopathy (WE) is a common neuropsychiatric complication of chronic alcoholism.[19] It is also encountered in patients with severe gastrointestinal disease, those with HIV/AIDS, and patients who receive injudicious administration of parenteral glucose or hyperalimentation without adequate B-vitamin supplementation.[43] In brain tissue obtained at autopsy from patients with WE, activities of all three ThDP-dependent enzymes are reduced.[19]

In patients with alcoholism, thiamine deficiency results from inadequate dietary intake of the vitamin, reduced absorption resulting from gastrointestinal disease, and reduced liver thiamine stores, caused by hepatic steatosis or fibrosis.[19] In addition, ethanol itself inhibits thiamine transport in the gastrointestinal system and blocks brain phosphorylation of thiamine to its active form (ThDP).[44]

The diagnosis of WE is based generally on the acute appearance of ocular palsies, nystagmus, and gait ataxia as well as disorders of mentation.[2] In addition, more than 80% of patients with WE show signs of peripheral neuropathy. However, these diagnostic criteria are nonspecific, and the diagnosis of WE is missed in many patients with alcoholism[45] as well as in those with HIV/AIDS.[43] The reason for the high degree of underdiagnosis rests with the overzealous use of the classic triad of symptoms (ophthalmoplegia, ataxia; confusion) espoused by many textbooks. In practice, many cases of WE confirmed at autopsy do not manifest this triad of symptoms, and patients may show only psychomotor slowing or apathy.[43] In the meantime, thiamine deficiency should be suspected in all patients with grossly impaired nutritional status associated with chronic diseases, with particular attention paid to patients with chronic alcoholism, gastrointestinal diseases, HIV/AIDS, and persistent vomiting. Thiamine should be administered parenterally in a timely manner. It is essential to administer thiamine to all patients before infusions of glucose or parenteral nutrition are given.

Korsakoff psychosis (KP) is generally considered to occur with deterioration of brain function in patients initially diagnosed with WE.[2] However, KP may be present at the time of diagnosis of WE or even, in a few cases, present without WE symptoms. KP is an amnestic-confabulatory syndrome characterized by retrograde and anterograde amnesia, impairment of conceptual functions, and decreased spontaneity and initiative.

Thiamine Diphosphate–Dependent Enzymes in Brain in Neurodegenerative Diseases

Reduced activities of ThDP-dependent enzymes are found in brain tissue obtained at autopsy from patients with Alzheimer's disease.[46,47] In animal models of Alzheimer's disease, thiamine deficiency increases the formation of plaques and phosphorylation of the tau protein.[48-50] Treatment with benfotiamine, a thiamine derivative, has been shown to reverse these effects in mouse models of Alzheimer's disease.[51] As a result, benfotiamine has been tested in humans with mild cognitive impairment or Alzheimer's disease, with some promise in improving cognitive outcomes.[52,53] Activities of α-KGDH are significantly reduced in both genetic and sporadic forms of the disease. Decreased activities of α-KGDH have also been described in the brains of patients with other neurodegenerative diseases, including Parkinson's disease and progressive subnuclear palsy.[20] The most plausible explanation for the selective loss of α-KGDH activity in these diseases may relate to the deleterious effects of oxidative stress resulting from the cell death cascade mechanisms in these disorders.[20]

Other Thiamine Deficiency–Related Disorders

Other disorders in which a putative role for thiamine deficiency has been implicated include subacute necrotizing encephalomyelopathy, opsoclonic cerebellopathy (a paraneoplastic syndrome), and Nigerian seasonal ataxia.[54] In addition, several inherited disorders of ThDP-dependent enzymes have been reported.[55] Some of these inherited disorders may respond to thiamine treatment. End-stage chronic liver failure results in thiamine deficiency caused principally by depletion of liver thiamine stores. Consequently, brain dysfunction secondary to thiamine deficiency in chronic liver failure may occur.[56]

Clinical Response to Thiamine Administration

Parenteral thiamine should be administered promptly to patients suspected of having beriberi or WE. Doses in the 50- to 100-mg range are initially administered intravenously or intramuscularly to replenish cellular thiamine stores (particularly liver).[33] Parenteral rather than oral thiamine administration is particularly important in patients with gastrointestinal disease or alcoholism, in whom thiamine absorption is likely to be impaired.

In wet beriberi, rapid improvement consisting of reduction in heart rate, respiratory rate, and clearing of pulmonary congestion occurs generally within 24 hours.[6] Rapid improvements are also seen in infants with dry beriberi. Recovery of impaired sensation and motor weakness, in contrast, may take several weeks or months. There is also growing evidence that cognitive damage from thiamine deficiency in infants can be permanent. The most well-documented example of this comes from a group of infants in Israel who were fed infant formula which, inadvertently, was thiamine deficient.[57] Many of these otherwise well-nourished children went on to have severe cognitive and motor difficulties.[58] Even children who consumed the thiamine-deficient formula with subclinical thiamine deficiency had prolonged motor and learning disabilities well into childhood.[59,60]

The response to thiamine administration in patients with WE is variable, depending on the symptoms and the degree of neuronal cell loss. Ophthalmoplegia (nystagmus, ptosis) generally improves rapidly (within 24 hours), a finding suggesting that these symptoms are the result of biochemical (metabolic) lesions in oculomotor and vestibular nuclei. Gait ataxia, conversely, responds more sluggishly to thiamine administration because, in most cases, loss of cerebellar neurons is significant.[61] Similarly, the amnestic deficit that is generally thought to result from lesions in the medial-dorsal nucleus of the thalamus shows variable response to thiamine administration, and most patients have some residual memory deficit. Improvements of peripheral neuropathy in both beriberi and WE syndrome may require several months of thiamine treatment.[61] In addition, complete recovery from WE has been demonstrated following aggressive thiamine treatment.[62]

Selective Neuronal Cell Death Resulting from Thiamine Deficiency

Investigators have proposed that thiamine deficiency leads to two distinct types of neuropathologic lesions. The first type, characterized by neuronal disintegration, mild endothelial swelling, and sparing of the neuropil, is generally confined to the thalamus and inferior olives. Conversely, destruction of the neuropil, endothelial swelling, and neuronal sparing occur in mammillary bodies and periventricular brainstem nuclei.[63] Several mechanisms have been proposed to explain the phenomenon of selective neuronal cell damage and death resulting from thiamine deficiency. These mechanisms include cellular energy failure, oxidative and nitrosative stress, *N*-methyl-D-aspartate (NMDA) receptor–mediated excitotoxicity, and blood–brain barrier (BBB) breakdown.

Cellular Energy Failure

As discussed earlier, thiamine deficiency is characterized by decreased brain concentration of ThDP and reduced activity of ThDP-dependent enzymes.[21] Attention has been focused particularly on the role of decreased α-KGDH in the pathogenesis of neuronal cell death resulting from thiamine deficiency, because it is well established that α-KGDH is a rate-limiting enzyme in the tricarboxylic acid cycle responsible for cellular energy production.

Prolonged reductions in activity of α-KGDH resulting from thiamine deficiency leads to decreased glucose (pyruvate) oxidation and increased brain concentrations of alanine and lactate. Studies of oxidative metabolism in mitochondria isolated from the brains of thiamine-deficient animals revealed decreased respiration using α-ketoglutarate as substrate, but no such changes in respiration using succinate,[64] a finding consistent with decreased activities of α-KGDH (see Figure 21.2). High-energy phosphates in thiamine deficiency are decreased in the brainstem.[22] Decreased activity of α-KGDH resulting from thiamine deficiency also reduces synthesis of amino acid neurotransmitters, such as glutamate and γ-aminobutyric acid.[21] Focal accumulation of lactate leading to reduced pH has been described,[23] and disintegration of mitochondria has also been reported in degenerating diencephalic neurons of thiamine-deficient animals.[20] Furthermore, investigators have suggested that impaired branched-chain ketoacid metabolism could contribute to the neuronal dysfunction and ultimate thalamic neuronal cell death observed in thiamine deficiency.[24]

Oxidative and Nitrosative Stress

Accumulation of reactive oxygen species has been reported in the brain in thiamine deficiency.[65] Other indicators consistent with oxidative and nitrosative stress in the brain as a result of thiamine deficiency include reports of early activation of microglia[20,66] and increased expression of inducible nitric oxide synthase, leading to increased nitrotyrosine immunoreactivity in vulnerable brain regions,[67] as well as reports of increased expression of heme oxygenase-1, intercellular adhesion molecule-1, *S*-nitrosocysteine and cyclooxygenase-2.[20,68,69] Evidence suggests that vascular factors also contribute to thiamine deficiency–related brain damage. Such factors include increases of endothelial nitric oxide synthase (eNOS).[70] Moreover, targeted disruption (knockdown) of the eNOS gene attenuates the neuronal cell death in thiamine-deficient mice.[20,68] eNOS knockdown, but not knockdown of inducible NOS or neuronal NOS, leads to a reduction in protein tyrosine nitration. This finding suggests a major role of eNOS as the source of nitric oxide–related nitrosative stress in thiamine deficiency. Furthermore, treatment of thiamine-deficient rats with the antioxidant *N*-acetylcysteine prevents the downregulation of astrocytic glutamate transporter or excitatory amino acid transporter (EAAT-2), and increases neuronal survival.[71] Reactive oxygen species production results in decreased expression of astrocytic glutamate transporters and reduced activities of α-KGDH, with the potential to result in an amplification of cell death mechanisms in thiamine deficiency[72] (see also chapter 45).

NMDA Receptor–Mediated Excitotoxicity

The nature of the neuropathologic damage resulting from thiamine deficiency is similar, to some degree, to that encountered in excitotoxic brain injury (i.e., brain injury resulting from excessive stimulation of NMDA receptors by glutamate, a process known as *excitotoxicity* and shown to result in excessive accumulation of intracellular calcium, leading to the activation of cell death mechanisms). Evidence consistent with a role of excitotoxicity in the pathogenesis of thiamine deficiency–related brain damage includes the finding of increased extracellular glutamate in brain regions that are particularly vulnerable to thiamine deficiency,[73] and the report that pretreatment with the NMDA receptor antagonist MK801 leads to significant neuroprotection.[74] One possible explanation for the increased extracellular brain concentration of glutamate in thiamine deficiency is the reported loss in expression of high-affinity glutamate transporters in vulnerable brain regions.[71,75]

Blood–Brain Barrier Breakdown

The hemorrhagic lesions characteristic of experimental thiamine deficiency and of WE in humans are indicative of a breakdown of the BBB. Studies using immunoglobulin G (IgG) as an indicator of BBB integrity in thiamine-deficient animals revealed increased IgG immunoreactivity in inferior colliculus and inferior olive before the onset of cell death in these regions as well as in the medial thalamus.[76-78] Microglial activation leading to the release of reactive oxygen species and cytokines is an early cellular event responsible for BBB breakdown resulting from thiamine deficiency.[66] Investigators have also demonstrated that thiamine deficiency is associated with alterations in tight junction proteins (occludin, zona occludens-1, and zona occludens-2) and matrix metalloproteinase-9.[78] These alterations could be responsible for the initiation of changes to BBB integrity in thiamine deficiency.[78]

References

1. Umetaro Suzuki [Internet]. [cited 2022 Dec 23]. Available from: https://en.wikipedia.org/wiki/Umetaro_Suzuki
2. McCollum E v. A History of Nutrition. Cambridge, MA: Riverside Press, Houghton Mifflin; 1957.
3. Carpenter KJ. [Internet]. 2004 [cited 2022 Aug 5]. Available from: www.nobelprize.org/prizes/themes/the-nobel-prize-and-the-discovery-of-vitamins/
4. Whitfield KC, Bourassa MW, Adamolekun B, et al. Ann N Y Acad Sci 2018;1430:3.
5. Bettendorff L, Wins P. FEBS J 2009;276:2917–25.
6. Tanphaichitr V. Thiamin. In: Shils ME, Olsen JA, Shike M, et al., editors. Modern Nutrition in Health and Disease. 9th ed. Burlington: Jones & Bartlett Learning, LLS; 1999.
7. Food and Nutrition Board I of M. Washington, DC: Washington, DC: National Academies Press; 1999.
8. Gomes F, Bourassa MW, Bergeron G. National control and prevention programs for thiamine deficiency disorders: Technical Reference Materials. New York; 2019.
9. Whitfield KC, Smith TJ, Rohner F, et al. Ann N Y Acad Sci 2021;1498:29–45.
10. World Health Organization. Thiamine deficiency and its prevention and control in major emergencies 1999.
11. Marrs C, Lonsdale D. MDPI 2021;10.
12. ICMR-NIN Expert Group of Nutrient Requirement for Indians. Recommended Dietary Allowances (RDA) and Estimated Average Requirements (EAR). Hyderbad; 2020.
13. Vimokesant SL, Hilker DM, Nakornchai S, et al. Am J Clin Nutr 1975;28:1458–63.
14. Gallant J, Chan K, Green TJ, et al. Am J Clin Nutr 2021;114: 90–100.
15. Lu J, Frank EL. Clin Chem 2008;54:901–6.
16. Jones KS, Parkington DA, Cox LJ, et al. Ann N Y Acad Sci 2021;1498:77–84.
17. Dreyfus PM. N Engl J Med 1962;267:596–8.
18. Turck D, Bresson JL, Burlingame B, et al. Dietary Reference Values for thiamin EFSA Panel on Dietetic Products, Nutrition and Allergies (NDA) [Internet]. 2016. Available from: www.efsa.europa.eu/efsajournal
19. Butterworth RF, Kril JJ, Harper CG. Alcohol Clin Exp Res 1993;17:1084–8.
20. Gibson GE, Zhang H. Neurochem Int 2002;40(6):493–504.
21. Butterworth RF, Héroux M. J Neurochem 1989;52:1079–84.
22. Aikawa H, Watanabe IS, Furuse T, et al. J Neuropathol Exp Neurol 1984;43:276–87.
23. Hakim AM. Ann Neurol 1984;16:673–9.
24. Navarro D, Zwingmann C, Butterworth RF. Metab Brain Dis 2008;23:445–55.
25. Harper CG. Aust N Z J Med 1980;10:230–5.
26. Pannunzio P, Hazell AS, Pannunzio M, et al. J Neurosci Res 2000;62: 286–92.
27. Gomes F, Bergeron G, Bourassa MW, et al. Ann N Y Acad Sci 2021;1498:46–56.
28. Johnson CR, Fischer PR, Thacher TD, et al. Nutr Health 2019;25: 127–51.
29. Guerrero RT, Gebhardt SE, Holden J, et al. J Am Diet Assoc. 2009;109:1738–43.
30. Rakotoambinina B, Hiffler L, Gomes F. Ann N Y Acad Sci 2021;1498:57–76.
31. Mouly S. AIDS. 1996;10:931.
32. Kwee IL, Nakada T. N Engl J Med. 1983;309:599–600.
33. Smith TJ, Johnson CR, Koshy R, et al. Ann N Y Acad Sci 2021; 1498:9–28.
34. Koshy RM, Thankaraj S, Ismavel VA, et al. Ann N Y Acad Sci 2021; 1498:96–107.
35. Nazir M, Lone R, Charoo BA. Indian Pediatr 2019;56(8):673–81.
36. Bhat JI, Rather HA, Ahangar AA, et al. Indian Heart J 2017;69:24–7.
37. Adamolekun B, Hiffler L. Ann N Y Acad Sci 2017;1408:15–9.
38. Brown G. J Inherit Metab Dis 2014;37:577–85.
39. Ortigoza-Escobar JD, Alfadhel M, Molero-Luis M, et al. Ann Neurol 2017;82:317–30.
40. Butterworth RF. Am J Clin Nutr 2001;74:712–3.
41. Fournier H, Butterworth RF. Metab Brain Dis 1990;5:77–84.
42. Hiffler L, Rakotoambinina B, Lafferty N, et al. Front Nutr 2016;3:16.
43. Butterworth RF, Gaudreau C, Vincelette J, et al. Metab Brain Dis 1991;6:207–12.
44. Rindi G, Imarisio L, Patrini C. Biochem Pharmacol. 1986;35:3903–8.
45. Harper C. J Neurol Neurosurg Psychiatry 1979;42:226–31.
46. Gibson GE, Sheu KFR, Blass JP, et al. Arch Neurol 1988;45: 836–40.
47. Butterworth RF, Besnard AM. Metab Brain Dis 1990;5:179–84.
48. Zhao J, Sun X, Yu Z, et al. Neurotox Res. 2011;19:575–83.
49. Zhang Q, Yang G, Li W, et al. Neurobiol Aging 2011;32:42–53.
50. Karuppagounder SS, Xu H, Shi Q, et al. Neurobiol Aging 2009;30:1587–600.
51. Pan X, Gong N, Zhao J, et al. Brain 2010;133:1342–51.
52. Gibson GE, Luchsinger JA, Cirio R, et al. J of Alz Dis. 2020;78: 989–1010.
53. Pan X, Chen Z, Fei G, et al. Neurosci Bull. 2016;32:591–6.
54. Adamolekun B, Hiffler L. Ann N Y Acad Sci 2017;1408:15–9.
55. Blass JP. Inborn errors of pyruvate metabolism. Metabolic basis of inherited disease/[edited by] John B Stanbury. [et al]. 1983.
56. Butterworth RF. Metab Brain Dis 2009;24:189–96.
57. Fattal-Valevski A, Kesler A, Sela BA, et al. Pediatrics 2005;115: e233–8.
58. Mimouni-Bloch A, Goldberg-Stern H, Strausberg R, et al. Pediatr Neurol 2014;51:311–6.
59. Harel Y, Zuk L, Guindy M, et al. Matern Child Nutr 2017;13:e12397.
60. Fattal I, Friedmann N, Fattal-Valevski A. Brain 2011;134:1720–39.
61. Victor M. Contemp Neurol Series 1989;30.
62. Paparrigopoulos T, Tzavellas E, Karaiskos D, et al. In Vivo (Brooklyn) 2010;24:231–3.
63. Torvik A. Neuropathol Appl Neurobiol 1985;11:179–90.
64. Parker WD, Haas R, Stumpf DA, et al. Neurology 1984;34:1477.
65. Langlais PJ, Anderson G, Guo SX, et al. Metab Brain Dis 1997;12:137–43.
66. Todd KG, Butterworth RF. Glia 1999;25:190–8.
67. Calingasan NY, Park LCH, Calo LL, et al. Am J Pathol 1998;153:599–610.
68. Beauchesne É, Desjardins P, Hazell AS, et al. J Neurochem 2009;111:452–9.
69. Gu B, Desjardins P, Butterworth RF. Metab Brain Dis 2008;23:175–87.
70. Kruse M, Navarro D, Desjardins P, et al. Neurochem Int 2004;45:49–56.
71. Hazell AS, Sheedy D, Oanea R, et al. Glia 2010;58:148–56.
72. Desjardins P, Butterworth RF. Mol Neurobiol 2005;31:17–25.
73. Hazell AS, Butterworth RF, Hakim AM. J Neurochem 1993;61:1155–8.
74. Langlais PJ, Mair RG. J Neurosci. 1990;10:1664–74.
75. Hazell AS, Rao KVR, Danbolt NC, et al. J Neurochem 2001;78:560–8.
76. Calingasan NY, Baker H, Sheu KFR, et al. Exp Neurol 1995;134:64–72.
77. Harata N, Iwasaki Y. Metab Brain Dis 1995;10:159–74.
78. Beauchesne É, Desjardins P, Hazell AS, et al. Neurochem Int 2009;55:275–81.

CHAPTER 22

Riboflavin

A. Catharine Ross

CHAPTER OUTLINE

Abbreviations

EGRAC, erythrocyte glutathione reductase activity coefficient; **FAD**, flavin adenine dinucleotide; **FMN**, flavin mononucleotide; **G6PD**, glucose-6-phosphate dehydrogenase; **Na^+**, sodium; **RDA**, recommended dietary allowance; **RFT**, riboflavin transporter

Introduction

Riboflavin (vitamin B_2) was initially isolated from milk whey in a nonpure form in 1879,[1] followed by purification and determination of its structure and identification of its major coenzyme forms, flavin mononucleotide (FMN), and flavin adenine dinucleotide (FAD).[2-5] These coenzymes are both donors and acceptors of electrons and hydrogen that play essential roles in normal cellular functions, growth, and development. Specifically, FMN and FAD act as donors and acceptors of hydrogen and thereby function as cofactors for certain enzymes (flavoproteins) involved in electron transfer reactions, which take part in a wide range of biological processes, including: energy production reactions; metabolic conversion of essential micronutrients, such as folate, vitamin B_6, niacin; the metabolism of drugs and detoxification of toxins; and electron scavenging pathways. A nutritional deficiency of riboflavin, called ariboflavinosis, leads to degenerative changes in the nervous system, endocrine dysfunction, anemia, and skin disorders. Other symptoms, commonly used as clinical indicators of riboflavin deficiency, include inflammation (redness and swelling) of the lining of the mouth, tongue, and throat; cracks at the corners of the mouth (angular cheilitis); and red and itchy eyes (resulting from corneal vascularization). Symptomatic riboflavin deficiency, which is rare in the U.S.,[6] and suboptimal levels most commonly occur in persons with alcoholism; inflammatory bowel syndrome; and diabetes, as well as in elderly or other persons in whom diet or nutrient absorption may be insufficient. Four groups considered at risk of inadequate riboflavin intake include: vegetarian athletes, in which exercise increases metabolic pathways that use riboflavin; pregnant and lactating women and their infants, especially those women who rarely consume meats or dairy products; people who are vegan and/or consume little milk; and people with riboflavin transporter deficiency (see further below).[6]

Chemistry, Biochemistry, and Function of Riboflavin and Derivatives

Riboflavin is composed of two main components: a planar isoalloxazine ring system composed of carbon, nitrogen, and hydrogen, to which the sugar alcohol ribitol is conjugated as a side chain (**Figure 22.1**). The chemical name is 7,8-dimethyl-10-(1′-D-ribityl)-isoalloxazine, with a molecular weight of 376.4. Free riboflavin lacks biologic activity, it serves as a precursor for two coenzymes forms, often called "nucleotides;" flavin mononucleotide (FMN), which is generated in cells by the addition of a phosphate group derived from ATP; and flavin adenine dinucleotide,

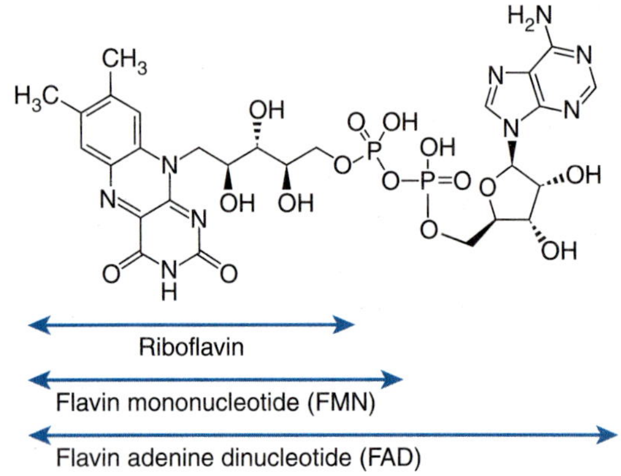

Figure 22.1 Structure of riboflavin and its coenzymes flavin mononucleotide (FMN) and flavin adenine dinucleotide (FAD).

in which an additional ATP donates adenosine and one phosphate group, forming a pyrophosphate link to riboflavin. Riboflavin itself is synthesized only by plants and microbes, while mammalian tissues have the enzymes required to produce the two coenzyme forms. The free vitamin has modest water solubility, which limits its use in parenteral and oral aqueous preparations. In aqueous solution, riboflavin is brightly fluorescent (yellow) and, as the riboflavin molecule readily absorbs light, it is photosensitive and needs to be protected from light. In alkaline solutions it degrades to lumiflavin (7,8,10-trimethyl-isoalloxazine) and in neutral-acidic solutions to lumichrome (7,8-dimethyl-alloxazine), both of which are biologically inactive compounds, yet capable of competing with riboflavin for uptake by different cells. For this reason, prolonged phototherapy in neonates with jaundice and in patients with certain skin disorders may negatively affect the normal whole-body and cellular homeostasis of riboflavin.

In the diet, riboflavin exists mostly in the form of FAD and FMN, with little in the free form. FMN is produced enzymatically by phosphorylation of the 5′-hydroxymethyl terminus of the ribityl side chain of the riboflavin molecule. This reaction is catalyzed by the enzyme flavokinase (**Figure 22.2**). When FMN is further modified by the addition of a pyrophosphate-bridged adenyl moiety, a reaction that is catalyzed by FAD synthetase (also known as FAD pyrophosphorylase), the more abundant FAD form of the vitamin is produced (see Figure 22.2).

Conversion of free riboflavin to its coenzyme forms within cells takes places mainly in the cytoplasm, although some conversion has also been reported to take place in the mitochondria.[7-10] The conversion of FMN to FAD appears to be affected by thyroid hormones, an effect that appears to be mediated by flavokinase activation.[11-13] Riboflavin can be regenerated from FMN and FAD in reactions that involve several phosphatases (see Figure 22.2).

More enzymes and cellular flavoproteins use FAD than FMN as a coenzyme. The cofactor forms are, in addition to being more biologically active forms of the vitamin, also trapped by interaction

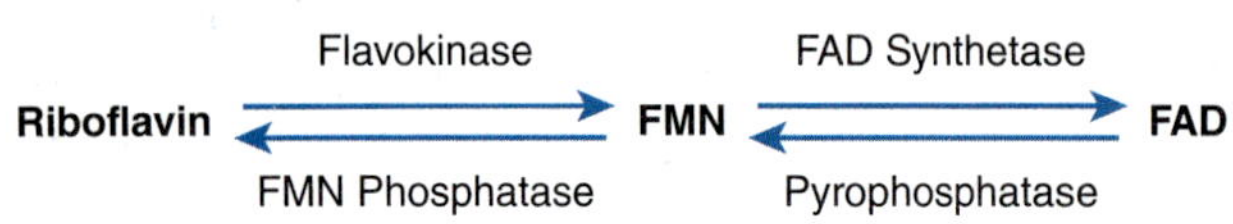

Figure 22.2 Interconversion of riboflavin to its coenzymes flavin mononucleotide (FMN) and flavin adenine dinucleotide (FAD).

with the proteins to which they bind, which serves as a mechanism for retaining this essential micronutrient inside cells. Most of the intracellular flavoproteins are located in the mitochondria; FAD (or riboflavin) is imported into this organelle by a mechanism that is different from that involved in the mitochondrial uptake of flavoproteins.[10,13-16] Transport of FAD and riboflavin into the mitochondria appears to occur by a specific carrier-mediated system in the mitochondrial membrane.[15] Little is known about how this system is regulated at the cellular and molecular levels, or about the factors that affect its function, however, see more below under Physiology of Riboflavin.

Assessment of Riboflavin Status

Most flavins in the plasma exist in the form of free riboflavin, although some FMN and FAD are also present. All plasma flavins are associated with plasma proteins, and mostly with albumin. Two main methods are commonly used to assess riboflavin nutritional status. The first method is based on the determination of activity known as the erythrocyte glutathione reductase activity coefficient, abbreviated as EGRAC,[17,18] and the second is based on fluorometric measurement of urinary excretion of riboflavin over a period of 24 hours (expressed as the total amount of riboflavin excreted in relation to creatinine excretion). A newer method involving estimation of the activity of erythrocyte pyridoxine phosphate oxidase has been described.[19] This method appears to be especially suited for use in populations with a high prevalence of glucose-6-phosphate dehydrogenase (G6PD) deficiency,[19] which is present in about 10% of African Americans, whereas estimation of riboflavin status in patients with G6PD deficiency by means of the EGRAC method may mask a riboflavin deficiency because G6PD deficiency is known to be associated with the enhanced binding of FAD to erythrocyte glutathione reductase.[20] Other analytical methods available to measure the levels of riboflavin and its derivatives in biologic samples include high-performance liquid chromatography and binding of riboflavin and its derivative to specific flavoproteins.[21]

Physiology of Riboflavin

Intestinal Absorption and Riboflavin Transporters

Mammals, including humans, cannot synthesize riboflavin and, thus, they must obtain the vitamin from exogenous sources through intestinal absorption. The intestine, therefore, plays a central role in regulating and maintaining normal riboflavin body homeostasis. The intestine is exposed to two sources of riboflavin: a dietary source, which is processed and absorbed in the small intestine; and a bacterial source, in which riboflavin is generated by the normal microflora in the large intestine and absorbed in that region of the gut.[22] Riboflavin in the diet exists mainly in the form of FMN and FAD, which are noncovalently bound to proteins; free riboflavin exists only in small amounts. The first step in the processing of dietary FMN and FAD is their release from proteins by the combined action of gastric acid and intestine-associated hydrolases. The released FMN and FAD molecules are then hydrolyzed to free riboflavin in the intestinal lumen and surface through the action of alkaline phosphatases before absorption.[23]

The mechanism of intestinal absorption of free riboflavin in the small intestine has been the subject of extensive investigations using various human and animal intestinal preparations. These preparations have ranged from intact intestinal tissue to purified vesicles isolated from the individual membrane domains of the polarized intestinal absorptive cells (i.e., from the apical brush-border membrane and from the basolateral membrane).[23-31] Collectively, these investigations have shown that absorption of free riboflavin occurs mainly in the proximal part of the small intestine and involves a specific sodium (Na^+)-independent, carrier-mediated system, referred to as riboflavin transporter (RFT), and subsequently further identified by gene names (see below). This RFT system is inhibited in a competitive manner by riboflavin structural analogs, such as lumiflavin and lumichrome, and by amiloride, an inhibitor of the Na^+/hydrogen [H^+] exchange process.[25] The intestinal uptake of riboflavin is also inhibited by chlorpromazine, a tricyclic phenothiazine drug that shares structural similarities with riboflavin.[26] Although some of the internalized riboflavin is phosphorylated inside the intestinal absorptive cells,[32] only free riboflavin exits across the basolateral membrane to the circulation. The latter process again involves a specific electroneutral, carrier-mediated mechanism.[26] Newer insights have been gained into the molecular identity of the systems involved in the intestinal riboflavin uptake process.[33,34] Two potential transport systems have been identified as riboflavin transporter-1 (RFT-1) and riboflavin transporter-2 (RFT-2).[33,34] RFT-2 appears to be a more promising candidate than RFT-1, because it transports riboflavin with much greater efficiency. Both systems, however, are expressed in the intestine.[33,34] Further studies are needed to determine which of these uptake system contributes mostly to riboflavin absorption in the native intestine in vivo.

A deeper understanding of intestinal riboflavin transport has been gained through the molecular cloning and characterization of specific riboflavin transporters, reviewed in[35] and summarized in **Table 22.1**. The riboflavin transporter known as RFVT-1, encoded by the human *SLC52A1* gene, is a vectorial plasma membrane protein that transports riboflavin into cells and, thus, plays a role in controlling cellular homeostasis of riboflavin in those tissues that express it (Table 22.1).[36] However, in the small intestine and colon, the transporter known as RFVT-3 (gene *SLC52A3*) is the primary means of riboflavin uptake across the plasma membrane. Studies by Subramanian et al.[37] showed that knocking out RFVT-3 in the mouse intestine resulted in riboflavin deficiency in mice fed an adequate diet, premature death of offspring, or growth retardation of surviving neonates and increased oxidative stress. As supplementation with pharmacologic doses of riboflavin could overcome these defects, the study showed not only that RFVT-3 is a normally required saturable transporter required for riboflavin uptake but also the importance of adequate riboflavin in overall health, growth, and development. Other studies in which RFVT-2 was deleted demonstrated embryonic lethality in the absence of this protein.[38] Another *SLC* gene family member, *SLC22A14*, has been identified as a riboflavin transporter located in the mitochondrial inner membrane.[39] Certain genetic defects and heritable diseases associated with mutations in riboflavin transporters are now being identified and studied further.[40,41]

The amount of riboflavin generated by the normal microflora of the large intestine depends on the type of the ingested diet. Higher amounts of riboflavin are produced following ingestion of a vegetable-based diet, compared with a meat-based diet.[42] In addition, considerable amounts of bacterially produced riboflavin exist in the large intestinal lumen, in the form of free riboflavin,[42,43] and are available for absorption, now understood to be by the RFVT-3 transporter mechanism above. Indeed, studies have shown that the large intestine is capable of absorbing luminally introduced free riboflavin.[44,45] The mechanism involved in riboflavin uptake by colonocytes has been characterized and involves an efficient and specific carrier-mediated mechanism like the one described in the small intestine.[46,47] Considering the time that the luminal content resides in the large intestine, it is reasonable to assume that this source of riboflavin contributes to the host's overall riboflavin nutrition, especially to the cellular nutrition of the localized colonocytes. These colonocytes are known to depend on luminal content for other nutrients (e.g., short-chain fatty acids produced by bacteria are used by the colonocytes for energy production). Further studies are needed to determine the exact level of contribution of this source of riboflavin to whole body riboflavin nutrition, and the way in which environmental factors affect such contribution. Recent work has identified specific bacteria including *Lactobaccilus plantarum* strains, which are capable of riboflavin production.[48] The topic of colon-targeted vitamins, including riboflavin, for disease amelioration is an emerging area,[49,50] with data suggesting that riboflavin supplementation can alter the metabolic output of gut microbiota.[51] Riboflavin has been suggested as a candidate prebiotic for functional foods[52]; further research is needed to understand the potential of riboflavin for improving gut health.

Fluctuation in the dietary intake of riboflavin plays a role in regulating the intestinal riboflavin uptake process. Investigators showed that riboflavin deficiency was associated with significant and specific upregulation in intestinal riboflavin uptake, whereas oversupplementation with pharmacologic

Table 22.1 **Characteristics of Riboflavin Transporters Present in Plasma and Mitochondrial Membranes**

			Ion Dependence			
Protein	**Gene**	**Synonyms**	**Na^+**	**Cl^-**	**H^+**	**Tissues where expressed**
RFVT1	*SLC52A1*	RFT1, GPR172B, PAR2	–	–	–	placenta, small intestine
RFVT2	*SLC52A2*	RFT3, GPR172A, PAR1	–	–	–	brain, salivary grand, liver, pancreas, *etc.*
RFVT3	*SLC52A3*	RFT2, C20orf54	–	–	+	testis, small intestine, kidney, placenta
	SLC22A14	MT RFT				spermatozoa

+: involved; –: not involved; MT, mitochondrial; RFT, riboflavin transporter

Data from Jin C, Yonezawa A. Pharmacol & Ther 2022;233:108023; Jin C, Matsui Y, Yonezawa A, et al. Biol Pharmacol Bull 2021;44(2):283–286.

concentrations of riboflavin resulted in significant and specific downregulation in riboflavin uptake.[25,46,53] These adaptive changes in the intestinal riboflavin uptake process appeared to be mediated by transcriptional mechanisms.[25,41,48] The intestinal riboflavin uptake process was also found to be under the regulation of pathways mediated by intracellular protein kinase A and calcium/calmodulin.[29,46] Furthermore, the intestinal riboflavin absorption process has been shown to undergo ontogenic regulation, with higher uptake during the suckling period compared with adulthood.[54]

Excretion and Reabsorption in the Kidney

The kidney also plays an important role in regulating and maintaining normal body homeostasis of riboflavin by controlling the amount of the vitamin that is lost in the urine. Under the normal condition of sufficient riboflavin intake, the amount of riboflavin that appears in the urine per day is approximately 120 μg, with free riboflavin representing between 60% and 70% of the total urinary flavin. Other flavin metabolites identified in the urine include 7- and 8-hydroxymethylflavins, lumichrome, 10 formylmethylflavin, 10-(2′-hydroxyethyl)flavin, 8α-flavin peptides, and 5′ riboflavinyl peptide ester.[18,54-60]

In the kidney, physiologic levels of riboflavin are filtered through the glomeruli and are then reabsorbed in the proximal tubule through an efficient and specific carrier-mediated process.[61-66] This process is, again, adaptively upregulated in riboflavin deficiency and is under the regulation of specific intracellular protein kinase–mediated pathways.[61-66] When plasma riboflavin is high (following ingestion of high doses of riboflavin), tubular secretion of riboflavin also occurs to enhance excretion of the vitamin from the body.[61-66]

Transport in Other Epithelia

In the placenta, riboflavin uptake has been studied for several decades, but molecular details are still vague. The RFVT-1 form of the riboflavin transporter is expressed,[36] but detailed uptake studies are needed. Studies of cell culture models and membrane vesicles isolated from the apical (maternal-facing) and basal (fetal-facing) membranes of the syncytiotrophoblast from full-term human placentas have shown the involvement of a specific and regulated carrier-mediated process.[8,62-64]

Studies have also characterized riboflavin transport in the liver,[67,68] which plays an important role in normal riboflavin metabolism and is the site of maximal utilization of the vitamin. Results have shown the involvement of a carrier-mediated process that is regulated by extracellular riboflavin concentrations and by a specific intracellular regulatory pathway.[69-72]

Uptake of riboflavin by human retinal pigment epithelium, which provides riboflavin to the metabolically active retina, has also been examined. Results have shown the involvement of a specific carrier-mediated mechanism that is upregulated in riboflavin deficiency and is under the control of specific intracellular regulatory pathways.[73]

A rare genetic and neurologic condition, referred to as riboflavin transporter deficiency (RTD),[74] is caused by pathogenic variants in the genes SLC52A2 and SLC52A3. This condition usually appears in infancy or childhood and is characterized by progressive peripheral and cranial changes that result in muscle weakness, including of respiratory muscles, and may be fatal.[75] The relationship of neuromuscular diseases to aberrations in riboflavin transport and cofactor metabolism is an emerging area,[76] which is likely to be fruitful given the known relationships of riboflavin compounds to cellular energetics.

Secretion in Milk

Both riboflavin and FAD are found in human and cow's milk; concentrations in human milk are higher than in cow's milk.[77,78] The concentration of flavin in human milk depends on maternal dietary intake of the vitamin.[79,80] Secretion of free riboflavin and FAD into milk appears to involve two separate mechanisms, with transport of riboflavin in the mammary gland mediated by the multidrug transport breast cancer resistance protein (BCRP/ABCG2), an ABC transporter.[81] Other flavin metabolites found in the milk include 10-(2′-hydroxyethyl) flavin, 7- and 8-hydroxymethylriboflavin, 10-formylmethylflavin, and lumichrome.[72,77,78]

Riboflavin Deficiency

Deficiency of riboflavin, ariboflavinosis, is often accompanied by other nutrient deficits. Clinical signs and symptoms include lesions on the outside of lips (cheilosis) and corners of the mouth (angular stomatitis), inflammation of the tongue (glossitis), redness or bloody (hyperemia) and swollen (edema) mouth or oral cavity, the inflammatory skin condition seborrheic dermatitis, anemia, and peripheral nerve dysfunction (neuropathy), among other signs (see chapter 57). People at risk for deficiency include those with congenital heart disease, some cancers, and excessive alcohol intake. The conversion of riboflavin into FAD and FMN is impaired in hypothyroidism and adrenal insufficiency.[12,56,82] Excretion of riboflavin is enhanced by diabetes, trauma, stress, and oral contraceptive use. When riboflavin intake is high, excess riboflavin is excreted in urine.

Because of the participation of FAD in intermediary metabolism, fatty acid oxidation is impaired in riboflavin deficiency. Riboflavin acts in concert with thiamin, niacin, and pantothenic acid, in reactions that include pyruvate dehydrogenase and α-ketoglutarate dehydrogenase, and in the metabolism of vitamin B_6 (conversion of pyridoxine or pyridoxamine phosphates to pyridoxal phosphate is catalyzed by a flavoprotein). Marginal riboflavin status may be associated with increased plasma homocysteine concentration, due to the requirement for riboflavin for 5,10-methylene tetrahydrofolate reductase, a key enzyme of folate metabolism. A hierarchy of cellular metabolic changes during the onset of riboflavin deficiency has been proposed: the core electron transfer chain required for adenosine triphosphate synthesis is preserved, whereas the enzymes required for the first step of fatty acid β-oxidation are diminished.[83]

Riboflavin deficiency ranges from mild to severe. Mild deficiency is detectable only by biochemical assays (elevated EGRAC or reduced erythrocyte glutathione reductase activity). Riboflavin supplementation results in changes that are detectable biochemically. In a systematic review of numerous riboflavin supplementation studies that used EGRAC or basal glutathione reductase activity to determine riboflavin status, EGRAC (14 studies) and glutathione reductase activity (five studies) showed highly significant association with riboflavin intake, although substantial study-to-study heterogeneity was noted.[84] Changes in EGRAC or glutathione reductase appear to be suitable biomarkers for changes in riboflavin intake in populations with severe to normal baseline intakes.[84]

Investigators have known, or assumed for some time, that riboflavin deficiency is prevalent in parts of the developing world with limited intakes of foods of animal origin, principally milk, eggs, and meats, which contain higher concentrations of the vitamin. Children and pregnant women are most likely to be affected. A 24-hour dietary recall study conducted in urban women of reproductive age living in Mali identified riboflavin as one of the four micronutrients for which the probability of adequacy was lowest.[85] Other studies evaluated multiple micronutrient supplementation as a means to improve diet quality and various health-related biomarkers, but riboflavin itself was not specifically tested. A study conducted in Poland that used EGRAC to assess riboflavin status in 20- to 25-year-old men and women found biochemical evidence of riboflavin deficiency in 33.7% of women and 25% of men.[86] The investigators associated these findings with lower intakes of riboflavin during a 7-day period in which dietary intake was recorded.[86]

Investigators have also shown interest in whether mild-to-moderate riboflavin deficiency is prevalent in affluent countries, and whether it may be part of a syndrome of suboptimal health, or affect the use of other micronutrients. Powers et al.[87] reported results of a randomized clinical trial of riboflavin supplementation in women in the United Kingdom who were 19 to 25 years old and were low milk consumers; improvements in hematologic status served as the primary endpoint. The study group had elevated EGRAC values of more than 1.4 at baseline and was randomized to receive 2 or 4 mg of riboflavin or a placebo for 8 weeks. Riboflavin status, assessed by reduction in EGRAC values, improved dose dependently. Hemoglobin status improved significantly and was greatest for the riboflavin-supplemented women with baseline EGRAC values of greater than 1.65. The investigators suggested that consideration should be given to raising the currently accepted EGRAC threshold for deficiency.

Sources and Recommended Dietary Allowances

Riboflavin, which occurs largely as a component of digestible coenzymes, is present in most plant and animal tissues. Especially good sources are eggs, organ meats (liver and kidney), lean meats, and milk. The US Department of Agriculture Continuing Survey of Food Intakes by Individuals found that food groups that supply more than 5% of total riboflavin intake are milk and milk drinks, bread and bread products, mixed foods (including sandwiches with meat, poultry, or fish as main ingredients), ready-to-eat cereals, and mixed foods with grain as the main ingredient.[88] Among vegetables, Brussels sprouts and broccoli contain more riboflavin per weight and energy unit than most other vegetables or fruits. Whole grains contain more riboflavin than milled, refined grains. Vitamin B–enriched breads and cereals are significant sources of riboflavin in the United States and other countries with similar nutrient fortification policies. Losses occur during cooking as a result of leaching of the heat-stable, light-sensitive flavins into water. Bioavailability is estimated at approximately 95% of food flavin, up to approximately 27 mg per single meal or dose.[88]

Table 22.2 Dietary Reference Intake Values for Riboflavin by Life-Stage Group

Life-Stage Group	RDA (mg/d)[a]		AI[b] (mg/d)
	Male	Female	
0–6 mo			0.3
7–12 mo			0.4
1–3 y	0.5	0.5	
4–8 y	0.6	0.6	
9–13 y	0.9	0.9	
14–18 y	1.3	1.0	
19→70 y	1.3	1.1	
Pregnancy	1.4		
Lactation	1.6		

[a]RDA, recommended dietary allowance, is the intake that meets the nutrient need of almost all (97%–98%) individuals in a group.

[b]AI is adequate intake. For healthy infants receiving human milk, the AI is the mean intake.

Data from Food and Nutrition Board, Institute of Medicine. Riboflavin. In: Dietary Reference Intakes: Thiamin, Riboflavin, Niacin, Vitamin B6, Vitamin B12, Pantothenic Acid, Biotin, and Choline. Washington, DC: National Academy Press, 1998:87–122.

For riboflavin, in contrast to that for thiamin, reference values have been set on the basis that the requirement does not increase when energy use is increased.[88] [See reference,[6] suggesting that high levels of physical activity may increase use such that marginal intakes pose a risk of insufficiency.] Recommended dietary allowances (RDAs) (**Table 22.2**) for riboflavin are expressed in milligrams/day. For conversion, 1 μmol riboflavin equals 0.376 mg, or inversely, 1 mg riboflavin equals 2.66 μmol. Clinical signs of deficiency in adults can be prevented with intakes of riboflavin greater than 0.4 mg/1,000 kcal, but more than 0.5 mg/1,000 kcal may be required to maintain tissue reserves in adults and children, as is reflected in urinary excretion, red cell riboflavin, and erythrocyte glutathione reductase.

For infants, an adequate intake, based on the content of riboflavin in human milk (0.35 mg/L) and the volume that is consumed, is now suggested as 0.3 mg/day for those 0 to 6 months of age and 0.4 mg/day for those 7 to 12 months of age (see Table 22.2). The RDA values for riboflavin[88] for children progressively increase on the basis of body weight, from 0.5 to 1.3 mg/day for the age ranges of 1 to 3 years, up to 18 years, with slightly greater amounts recommended for boys than for girls. Pregnancy and lactation impose extra demands, and an additional 0.3 mg/day and 0.5 mg/day, respectively, are deemed appropriate, added to the RDA for adult women.

The current DRI values were established in 1998. Recently, riboflavin has been the subject of an evidence scan by the National Academies of Sciences, Medicine and Engineering[89] to determine "whether there is new and relevant knowledge available that may merit a formal reexamination of DRIs for riboflavin." The report concluded, "The committee was not charged to determine whether the evidence scan results are sufficient to support a formal DRI review. However, this report will inform the federal agencies about the type and quality of new evidence and assist them in assigning a priority to a comprehensive systematic review for riboflavin."

In the European Community, Dietary Reference Values for Riboflavin are determined by EFSA.[90] For adults, the average requirement, AR, and population reference intake, PRI, are set at 1.3 and 1.6 mg/day, respectively.

Riboflavin is a common component of dietary supplements.[91] Given the close relationship of B-vitamins in one-carbon metabolism, riboflavin may be of benefit when other B-vitamins or their metabolism are suboptimal. An example is the emerging evidence that riboflavin, as the cofactor for methyl tetrahydrofolate reductase, MTHFR, may have a genotype-specific effect on blood pressure in persons with the MTHFR 677C→T polymorphism, which has been shown to increase the risk of hypertension and cardiovascular disease.[92]

Safety and Adverse Effects

When supplementation or therapy with riboflavin is warranted, oral administration of 5 to 10 times the RDA is usually satisfactory.

No toxic or adverse effects of high riboflavin intake in humans are known. A research study that administered 60 mg of supplemental riboflavin and 11.6 mg of riboflavin as a bolus intravenous dose reported no adverse effects.[93] The Food and Nutrition Board of the Institute of Medicine did not establish a tolerable upper intake level (UL) when the RDAs were revised in 1998.[88]

Acknowledgment

The author acknowledges the contributions of Hamid M. Said to the previous version of this chapter.

References

1. McCollum EV, Kennedy C. J Biol Chem 1916;24:491.
2. Stern KG, Holiday ER. Ber Dtsch Chem Ges 1934;67:1104.
3. Theorell H. Biochem Z 1934;272:155.
4. Theorell H. Biochem Z 1937;290:293.
5. Warburg O, Christian V. Biochem Z 1938;295:261.
6. Office of Dietary Supplements, NIH. Accessed November 18, 2022. Available at https://ods.od.nih.gov/factsheets/Riboflavin-HealthProfessional/#h7
7. Merrill AH Jr, Lambeth JD, Edmondson DE, et al. Annu Rev Nutr 1981;1:281–317.
8. Huang SN, Swaan PW. J Pharmacol Exp Ther 2001;298:264–71.
9. McCormick DB. Riboflavin. In: Brown ML, ed. Present Knowledge in Nutrition. 6th ed. Washington, DC: International Life Sciences Institute Press, 1990:146–54.
10. Barile M, Brizio C, Valenti D, et al. Eur J Biochem 2000;267:4888–4900.
11. Rivlin RS. Riboflavin. In: Bowman BA, Russel RN, eds. Present Knowledge in Nutrition. 8th ed. Washington, DC: International Life Sciences Institute Press, 2001;191–8.
12. Lees SS, McCormick DB. Arch Biochem Biophys 1985;237:197–201.
13. Cimino JA, Jhangiani S, Schwartz E, et al. Proc Soc Exp Biol Med 1987;184:151–3.
14. Spaan AN, Ijlst L, van Roermund CWT, et al. Mol Genet Metab 2005;86:441–7.
15. Nagao M, Tanaka K. J Biol Chem 1992;267:17925–32.
16. Barile M, Passarella S, Bertoldi A, et al. Arch Biochem Biophys 1993;305:442–7.
17. Sauberlich HE, Judd WH. Am J Clin Nutr 1972;25:756–62.
18. Briggs M, ed. Vitamins in Human Biology and Medicine. Boca Raton, FL: CRC Press, 1981.
19. Mushtaq S, Su H, Hill MH, et al. Am J Clin Nutr 2009;90:1151–9.
20. Flatz G. Nature 1970;226:755.
21. Kodentsova VM, Vrzhesinskaya OA, Spirichev VB. Ann Nutr Metab 1995;39:455–60.
22. Wrong OM, Edmonds CJ, Chadwich WS. Vitamins. In: The Large Intestine: Its Role in Mammalian Nutrition and Homeostasis. New York: Wiley, 1981:157–66.
23. Daniel H, Binninger E, Rehner G. Int J Vitam Nutr Res 1983;53:109–14.
24. Daniel H, Wille U, Rehner G. J Nutr 1983;113:636–43.
25. Said HM, Ma TY. Am J Physiol 1994;266:G15–21.
26. Said HM, Hollander D, Khani R. Biochim Biophys Acta 1993;1148:263–8.
27. Tomei S, Yuasa H, Inoue K. Drug Deliv 2001;8:119–24.
28. Said HM, Khani R, McCloud E. Proc Soc Exp Biol Med 1993;202:428–34.
29. Said HM, Ma TY, Grant K. Am J Physiol 1994;267:G955–9.
30. Hegazy E, Schwnk M. J Nutr 1983;113:1702–7.
31. Middleton HM. J Nutr 1985;120:588–93.
32. Gastaldi G, Ferrari G, Verri A, et al. J Nutr 2000;130:2556–61.
33. Yonezawa A, Masuda S, Katsura T, et al. Am J Physiol 2008;295:C632–41.
34. Yamamoto S, Inoue K, Ohta KY, et al. J Biochem 2009;145:437–43.
35. Jin C, Yonezawa A. Pharmacol Ther 2022;233;108023.
36. Sabui S, Ghosal A, Said HM. Gene 2014;553:49–56.
37. Subramanian VS, Lambrecht N, Lytle C, et al. Am J Physiol Gastrointest Liver Physiol 2016;310:G285–93.
38. Jin C, Matsui Y, Yonezawa A, et al. Biol Pharm Bull 2021;44:283–6.
39. Kuang W, Zhang J, Lan Z, et al. Cell Rep 2021;35:109025.
40. Yee SW, Giacomini KM. Drug Metab Dispos 2021;50:1193–1210.
41. Mosegaard S, Dipace G, Bross P, et al. Int J Mol Sci 2020;21:3847
42. Iinuma S. J Vitam 1955;2:6–13.
43. Ocese O, Pearson PB, Schwiegert BS. J Nutr 1948;35:577–90.
44. Sorrell MP, Frank O, Thomson AD, et al. Am J Clin Nutr 1971;24:924–9.
45. Kasper H. Am J Protocol 1970;21:341–5.
46. Said HM, Ortiz A, Moyer MP, et al. Am J Physiol 2000;278:C270–6.
47. Yuasa H, Hirobe M, Tomei SA, et al. Biopharm Drug Dispos 2000;21:77–82.
48. Mohedano ML, Hernandez-Recio S, Yepez A, et al. Front Microbiol 2019;10:1748.
49. Pham VT, Fehlbaum S, Seifert N, et al. Gut Microbes 2021;13:1–20.
50. LeBlanc JG, Levit R, Savoy de Giori G, et al. Appl Microbiol Biotechnol 2020;104:3331–7.
51. Liu L, Sadaghian Sadabad M, Gabarrini G, et al. Antioxid Redox Signal 2022.__(Online ahead of print, doi: 10.1089/ars.2022.0033)

52. Steinert RE, Sadaghian Sadabad M, Harmsen HJ, et al. Eur J Clin Nutr 2016;70:1348–53.
53. Said HM, Khani R. Gastroenterology 1993;105:1294–8.
54. Said HM, Ghishan FK, Greene HL, et al. Pediatr Res 1985;19: 1175–8.
55. Ohkawa H, Ohishi N, Yagi K. J Biol Chem 1983;258:5623–8.
56. McCormick DB. Riboflavin. In: Shils ME, Shike M, Olson JA et al, eds. Modern Nutrition in Health and Disease. 9th ed. Lippincott Williams and Wilkins, Baltimore, 1999:391–9.
57. Chastain JL, McCormick DB. J Nutr 1987;117:468–75.
58. Chia CP, Addison R, McCormick DB. J Nutr 1978;108:373–81.
59. Chastain JL, McCormick DB. Biochim Biophys Acta 1988;967: 131–4.
60. Chastain JL, McCormick DB. Am J Clin Nutr 1987;46:830–4.
61. Yanagawa N, Shih RN, Jo OD, et al. Am J Physiol 2000;279: C1782–6.
62. Yanagawa N, Jo OD, Said HM. Biochim Biophys Acta 1998;1415:56–62.
63. Kumar CK, Yanagawa N, Ortiz A, et al. Am J Physiol 1998;274:F104–10.
64. Yanagawa N, Jo OD, Said HM. Biochim Biophys Acta 1997;1330:172–8.
65. Lowy RJ, Spring KR. J Membr Biol 1990;117:91–9.
66. Spector R. J Pharmacol Exp Ther 1982;221:394–8.
67. Yao Y, Yonezawa A, Yoshimatsu H, et al. Eur J Pharmacol 2013;714:281–7.
68. Zempleni J, Link G, Kubler W. Int J Vitam Nutr Res 1992;62: 165–72.
69. Dancis J, Lehanka J, Levitz M. Pediatr Res 1985;19:1143–6.
70. Said HM, McCloud E, Yanagawa N. Biochim Biophys Acta 1995;1236:244–8.
71. Said HM, Ortiz A, Ma TY, et al. J Cell Physiol 1998;176: 588–94.
72. Aw YT, Jones DP, McCormick DB. J Nutr 1983;113:1249–54.
73. Said HM, Wang S, Ma TY. J Physiol 2005;566:369–77.
74. Balasubramaniam S, Christodoulou J, Rahman S. J Inherit Metab Dis 2019;42:608–19.
75. Cali E, Dominik N, Manole A, et al. (1993) In: *GeneReviews ((R))* (Adam MP, Everman DB, Mirzaa GM, et al., eds), Seattle (WA). In: GeneReviews® [Internet]. Seattle (WA): University of Washington, Seattle; 1993. 2015 Jun 11 [updated 2021 Apr 8].
76. Tolomeo M, Nisco A, Barile M. Methods Mol Biol 2021;2280: 275–95.
77. Roughead ZK, McCormick DB. J Nutr 1990;120:382–8.
78. Roughead ZK, McCormick DB. Am J Clin Nutr 1990;52:854–7.
79. Ortega RM, Quintas ME, Martinez RM, et al. J Am Coll Nutr 1999;18:324–9.
80. Allen LH. J Nutr 2003;133:3000S–7S.
81. van Herwaarden AE, Wagenaar E, Merino G, et al. Mol Cell Biol 2007;27:1247–53.
82. Powers HJ. Proc Nutr Soc 1999;58:435–40.
83. Ross NS, Hansen TP. Biofactors 1992;3:185–90.
84. Hoey L, McNulty H, Strain JJ. Am J Clin Nutr 2009;89:1960S–80S.
85. Kennedy G, Fanou-Fogny N, Seghieri C, et al. J Nutr 2010;140:2070S–8S.
86. Szczuko M, Seidler T, Mierzwa M, et al. Int J Food Sci Nutr 2011;62:431–8.
87. Powers HJ, Hill MH, Mushtag S, et al. Am J Clin Nutr 2011;93:1274–84.
88. Food and Nutrition Board, Institute of Medicine. Riboflavin. In: Dietary Reference Intakes: Thiamin, Riboflavin, Niacin, Vitamin B_6, Vitamin B_{12}, Pantothenic Acid, Biotin, and Choline. Washington, DC: National Academy Press, 1998:87–122.
89. Scanning for New Evidence on Riboflavin to Support a Dietary Reference Intake Review. In: Vorosmarti A, Yaktine AL, Schneeman BO, eds. Washington, DC, 2021.
90. Efsa Panel on Dietetic Products, Nutrition and Allergies, Turck D, Bresson JL, Burlingame B, et al. EFSA J 2017;15:e04919.
91. Jun S, Gahche JJ, Potischman N, et al. Obstet Gynecol 2020;135:623–33.
92. McNulty H, Strain JJ, Hughes CF, et al. Mol Aspects Med 2017;53:2–9.
93. Zempleni J, Galloway JR, McCormick DB. Am J Clin Nutr 1996;63:54–66.

C. VITAMINS

CHAPTER 23

Niacin

Mirella L. Meyer-Ficca, Ralph G. Meyer, and James B. Kirkland

CHAPTER OUTLINE

Abbreviations

ACMS, 2-amino-3-carboxymuconic-6-semialdehyde; **ACMSD**, 2-amino-3-carboxymuconic-6-semialdehyde decarboxylase; **ADP**, adenosine diphosphate; **ART**, mono-ADP-ribosyltransferase; **ATP**, adenosine triphosphate; **BER**, base excision repair; **CICR**, calcium-induced calcium release; **DRI**, dietary reference intake; **GI**, gastrointestinal; **GRP**, glucose-regulated-protein; **IP3**, inositol triphosphate; **NAADP**, nicotinic acid adenine dinucleotide phosphate; **NAD**, nicotinamide adenine dinucleotide; **NADH**, reduced nicotinamide adenine dinucleotide; **NADP**, nicotinamide adenine dinucleotide phosphate; **NADPH**, reduced nicotinamide adenine dinucleotide phosphate; **NER**, nucleotide excision repair; **NFκB**, nuclear factor κB; **PARP**, poly(ADP-ribose) polymerase; **RDA**, recommended dietary allowance; **TCA**, tricarboxylic acid; **UL**, tolerable upper intake level

Historical Overview

Pellagra is the clinical disease of niacin deficiency in humans. It is mainly caused by dependence on corn as a staple food. Although pellagra likely occurred at low incidence throughout history, it reached epidemic proportions in the southern United States and Europe as corn-based agriculture spread.[1] Corn does contain niacin, but in tightly bound structures; this binding is heat-stable, but is sensitive to alkaline treatment.[2] Native Americans developed various alkaline processing techniques that release the existing niacin, and the importance of this processing was not recognized when Columbus brought corn back to Europe.[1] Currently, in developed countries, pellagra mainly occurs in patients belonging to specific risk groups, while it remains a serious public health challenge in African and East Asian countries and India, with pellagra affecting up to one-third of the female population in some countries.[3-5]

The disease pellagra was first reported in eighteenth century Spain, and soon after, the term *pellagra* was derived from the Italian name for the condition, meaning "rough skin." Clinically, pellagra is characterized by the three "Ds" of dermatitis (sun sensitive), dementia, and diarrhea. Diarrhea is the least unique of these, but it leads to a vicious cycle of worsening status of niacin and other nutrients. Anorexia also sets in as the deficiency progresses. If left untreated, pellagra usually leads to the death of the patient, which is considered a fourth "D" of pellagra by some. The development of the more unique signs of dermatitis and dementia can be unpredictable from patient to patient, thus making pellagra difficult to diagnose in many cases. The historic epidemic in the southern United States occurred largely in people working outdoors, and sun-induced skin lesions were a clinical focus.[6] Similar outbreaks occurred in Spain, Italy, and Egypt during the 1700s and 1800s. The reported incidence in northern Europe was much lower,[6] but cooler weather and indoor work environments may have caused pellagra to manifest as poorly diagnosed dementia, for which unfortunate patients were often confined to an asylum and fed a disease-perpetuating corn-based diet. Even in the southern United States in the 1900s, outbreaks of pellagra were described in asylum populations.[6] Women were much more likely to develop pellagra than men, possibly because of unequal access to food resources.[1,7]

Remarkably, it took several hundred years for the dietary reliance on corn to be accepted as the cause of pellagra. Corn consumption was originally proposed to carry a disease or toxin.[8] Starting in 1915, Dr. Joseph Goldberger conducted clinical trials in which pellagra was induced in prison populations and was cured or prevented by balanced diets or yeast supplements.[9,10] Although nicotinic acid was first isolated in 1867, its role as the active vitamin was not identified until 1937, when Dr. Conrad Elvehjem cured black tongue, the manifestation of pellagra in dogs, with nicotinic acid.[11,12] A host of publications in 1937–1939 demonstrated that nicotinic acid cured pellagra in humans,[6,13-16] and *Time* magazine named Drs. D. Tom Spies, Marion Blankenhorn, and Clark Cooper "1938 Men of Comprehensive Science" for their contributions.

Nicotinamide adenine dinucleotide (NAD) was first identified in yeast extracts in 1906,[17] but its redox capabilities were not described until 1936,[18] followed by the connection of reduced NAD (NADH) formation with adenosine triphosphate (ATP) production in 1949.[19] For several decades, research focused on the extensive redox roles of NAD and NAD phosphate (NADP) in animal, plant, and microbial metabolism. An important advance was made in 1966, with the first publication on adenosine diphosphate (ADP)-ribose formation.[20] This advance expanded the view beyond NAD's role in redox reactions and led up to the discovery of NAD-dependent poly- and mono-ADP-ribosylation of proteins,[21-23] formation of cyclic ADP-ribose[24-26] and sirtuin activity,[27,28] which allowed a much better understanding of the unique metabolic origin of pellagra.

Terminology and Chemistry

The term *niacin* can have broad or narrow meaning. In the broader sense, as in the "niacin content of a diet," it refers to the combination of various dietary precursors of nicotinamide adenine dinucleotide (NAD). These molecules comprise nicotinic acid, free nicotinamide and nucleotide-bound nicotinamide, nicotinamide riboside (and its reduced form dihydronicotinamide riboside), all of which would directly contribute to NAD content or niacin status, with tryptophan as an additional non-niacin NAD precursor. Except for tryptophan, these molecules are also sometimes collectively designated as vitamin B_3. In its narrow meaning, niacin refers only to nicotinic acid, from which the name originally was derived (nicotinic acid vitamin); the term niacin is used in this narrow meaning in the literature on the pharmacologic use of nicotinic acid in the treatment of dyslipidemias and other conditions.

From an ecological perspective, niacin is introduced into the food chain, predominantly by plants, as nicotinic acid, nicotinamide, and the amino acid tryptophan (**Figure 23.1**). Plants often synthesize provitamin metabolites for purposes quite distinct from those of human cells. Plants use nicotinic acid to form pyridine nucleotides, but they also use nicotinic acid for nicotinic acid N- and O-glycosylation reactions, and for the synthesis of trigonelline (N'-methylnicotinate) and large amounts of alkaloids, such as nicotine,[29-32] for purposes such as pest resistance, growth regulation, and nicotinate detoxification. Some nicotinamide is formed in plants from nicotinic acid during pyridine nucleotide synthesis and may be released during plant cellular metabolism or during the digestion of plant matter in the human gastrointestinal (GI) tract.

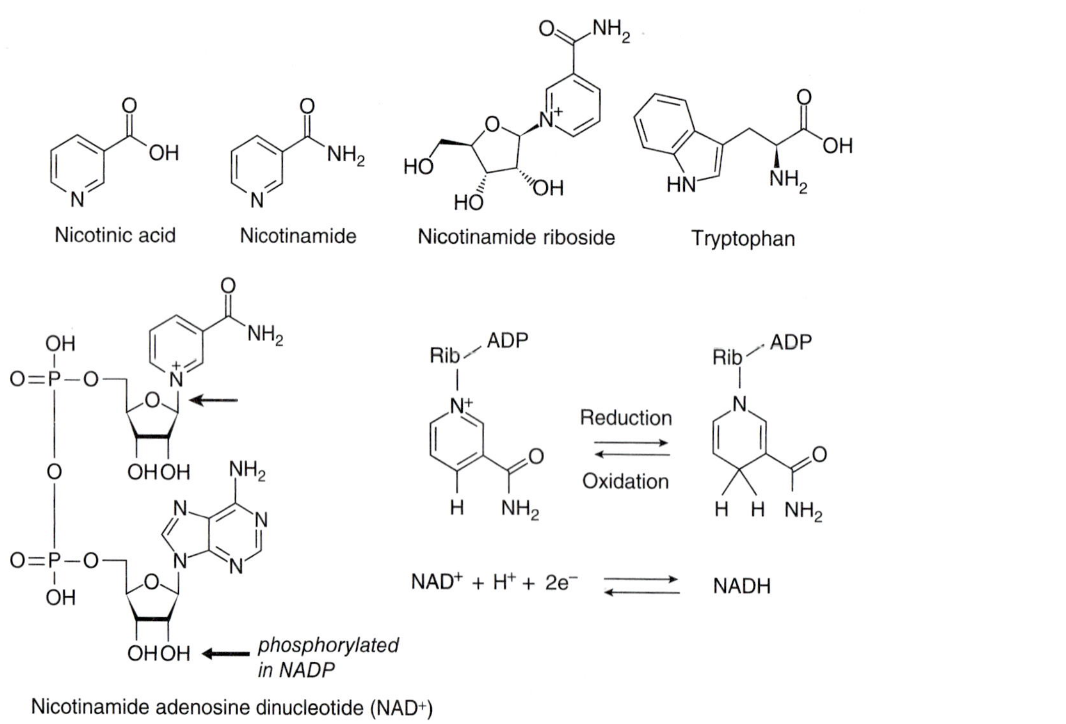

Figure 23.1 Dietary precursors for nicotinamide adenine dinucleotide (NAD) synthesis are shown in the *top row*. The *lower diagrams* show the structures of NAD on the left and the change in ring structure during reduction on the right. The site of phosphorylation to form nicotinamide adenine dinucleotide phosphate (NADP) is indicated by a block arrow, an arrow with open arrowhead marks the high-energy glycosidic bond.

Nicotinic acid and nicotinamide (niacinamide) are position 3 derivatives of the pyridine ring structure (carboxylic acid in the former, carboxamide in the latter structure), and nicotinamide riboside is a nicotinamide nucleoside (see Figure 23.1). Nicotinamide riboside and its reduced form, dihydronicotinamide riboside, are also considered NAD precursors; note that dihydronicotinamide riboside is a precursor for the direct synthesis of NADH that can subsequently be oxidized to NAD^+.[33,34] Tryptophan is an essential amino acid in animals, synthesized in plants as a derivative of an indole structure. Despite the differences in ring structure, tryptophan is used to form niacin in many plants,[35] and it is used to form NAD in the liver of animals, with variable efficiency with respect to niacin status.[36,37]

The biologically active forms of niacin compounds are the NAD and NADP coenzymes (see bottom right in Figure 23.1). The C-4 position on the pyridine ring of the nicotinamide moiety participates in oxidation and reduction reactions. Because of the electronegativity of the amide group and the nitrogen at position 1 on this ring, hydride ions can readily reduce the oxidized C-4 position. This reversible hydrogen transfer is the basis for the enzymatic redox reactions that are ubiquitous among organisms. With respect to the non-redox functions of NAD, the glycosidic linkage between nicotinamide and ribose is a high-energy bond (open arrow in Figure 23.1), and cleavage of this bond drives all types of ADP-ribose transfer reactions in the forward direction.

The oxidized and reduced forms of the coenzymes are designated NAD^+ or $NADP^+$ and NADH or NADPH, respectively. The designations NAD and NADP are used to describe the total pools. This nomenclature is often necessary if the method of quantification does not distinguish between oxidized and reduced forms or if a general statement about the nucleotide pool is made. The total pool of all four forms may be referred to as NAD(P). NAD and NADP have strong ultraviolet absorption at 340 nm in their reduced forms, which is often used to monitor the oxidation or reduction of these cofactors in enzyme assays.

Dietary Sources

Several categories of foods are good sources of niacin. Starting with plant-based foods, nuts, legumes, and grains have approximately 2 to 5 mg per 100 g serving, and they are important sources, given the consumption level of these staple foods. Remarkably, peanut butter and yeast extract spreads contain exceptionally high amounts of niacin, with 128 and 12–15 mg, respectively. Niacin in these foods is largely present in the form of nicotinic acid, in some cases bound in poorly available structures as seen in corn. Muscle-based foods, such as poultry, beef, and fish, provide approximately 6 to 19 mg per 100 g serving, mainly in the form of preformed nucleotides, which release nicotinamide during digestion. A third category of niacin-rich foods is created through fortification, usually of flour, cornmeal, and cereal products. In Canada and the United States, these products are fortified with approximately 5 to 6 mg/100 g flour. The eventual niacin content of ready-to-eat breakfast cereals can range up to 90 mg/100 g of dry cereal, according to US Department of Agriculture Food Data Central.[38]

The NAD precursor nicotinamide riboside has been approved as a supplement in the United States and Europe, and is present as a trace nutrient in milk, albeit at lower levels than nicotinamide.[39-41] The last category of niacin-rich foods consists of high-protein foods that provide tryptophan, converted at low efficiency in the liver to NAD. The contribution of tryptophan is not generally included in the niacin content of a food, but it is included in a calculation of niacin equivalents (1 NE = 1 mg niacin = 60 mg tryptophan, or milligrams niacin + milligrams tryptophan/60). The efficiency of tryptophan conversion is not easily predictable because it will be less efficient with low tryptophan intakes.[36,37]

Niacin in plant products is present mainly in the form of nicotinic acid, although much of it exists in poorly understood bound forms. These bound forms have been studied in wheat bran, corn, and other grains, and they are heterogeneous mixtures of polysaccharides and glycopeptides to which nicotinic acid is esterified.[2] In corn, most nicotinic acid is bound, and the tryptophan content is low, which makes pellagra a likely outcome when corn is consumed as a staple grain without alkaline processing. These conditions still occur in developing countries, and cause periodic outbreaks of pellagra.[42] Conversely, in the United States, the average daily intake of niacin climbed from approximately 16 mg in the 1930s to approximately 32 mg in 2004 as a result of fortification and increased intake of cereal products.[43] Thus, the incidence of clinically obvious pellagra in developed countries is extremely low. However, some evidence still points to subclinical deficiencies in developed countries, based on low blood NAD/NADP ratios.[44] Niacin deficiency and clinical signs of pellagra may appear in combination with other health conditions like anorexia nervosa,[45-48] alcoholism,[49-51] acquired immunodeficiency syndrome,[52-54] Crohn's disease,[55] cancer,[56,57] and chemotherapy.[57,58]

Recommended Dietary Allowances and Tolerable Upper Intake Limits

The Dietary Reference Intake (DRI) values adopted by the United States and Canada include recommended dietary allowances (RDAs), ranging from 2 to 6 mg/day in infants and toddlers, 8 mg in children (4 to 8 years), 12 to 16 mg in adolescents (9–18 years of age), 14 mg/day in women and 16 mg/day in men (**Table 23.1**), with tolerable upper intake levels (ULs) from 10 to 20 mg/day in children up to 35 mg/day in adults. The UL values only apply to niacin supplements plus niacin fortification and are based on the nicotinic acid–induced skin flush response. The skin flush is uncomfortable, but it is not directly related to any lasting health problems. Very few people have persistent skin flush responses to this level of niacin, and most niacin supplements and all B-50, B-75, and B-100 complex formulations greatly exceed the stated UL. Niacin supplements up to 500 mg are freely available for purchase.

Physicians prescribe nicotinic acid up to 3,000 mg/day to treat dyslipidemias. Niacin supplementation effectively decreases low-density lipoprotein cholesterol and lipoprotein a and is intended to reduce cardiovascular risk.[60-62] Although initial clinical trials have not yet been conclusive regarding cardiovascular benefits.[63-66] The strong skin flush responses decrease over time and can be modulated with cyclo-oxygenase inhibitors, but patients do struggle with compliance. Besides effects like skin flush and nausea, intake of pharmacologic niacin doses (at or above 500 mg daily) can temporarily affect the liver with transient elevation of liver enzymes and, in rare cases, lead to serious hepatotoxicity.[67-70]

Table 23.1 Recommended Dietary Allowances for Niacin[a]

Age (y)	0–0.5	0.5–1	1–3	4–8	9–13	≥14	Pregnancy	Lactation
RDA (mg)	2	4	6	8	12	Female: 14	18	17
						Male: 16		
UL (mg)	ND	ND	10	15	20	30–35	30–35	30–35

RDA, recommended dietary allowance.
[a]Values are niacin equivalents (NEs), except for values in infants younger than 6 months, which are expressed as adequate intake based on observations.

Data from Linus Pauling Institute's Micronutrient Information Center, https://lpi.oregonstate.edu/mic/vitamins/niacin#RDA, accessed July 10, 2022; and from Otten JJ, Hellwig JP, Meyers LD. Nat Acad Press 2006; Washington, DC[59].

The doubling in niacin intake between 1930 and 2005 in the United States preceded the increase in obesity and diabetes in children, and intervention trials showed that large doses of nicotinamide can impair glucose tolerance and might contribute to the ongoing increase in obesity.[43,71] The relevance of these results to normal variation in dietary niacin status is still uncertain.

The pharmacologic effects of high doses of nicotinic acid and nicotinamide occur by some common and some distinct molecular mechanisms,[72] and their deleterious effects need to be researched and assessed separately. Even single effects, such as inhibition of poly(ADP-ribose) polymerases, may have both beneficial and harmful effects on health.[73] Very high niacin intake can stress methyl donor status[74] and increase blood homocysteine concentration.[75-77] Current UL values are not being enforced, and the potential exists for toxicity of higher doses of nicotinic acid or nicotinamide supplements. Further work should be done to define more valid upper limits, which can then be applied in the marketplace.

Sites of Intestinal Absorption, Blood Transport, and Intracellular Forms

Preformed nicotinamide and nicotinic acid can be absorbed slowly through the stomach lining, but absorption in the small intestine is more rapid. Intact nucleotides are degraded in the upper small intestine to form free nicotinamide. The same might be true for nicotinamide riboside.[78,79] While exact mechanisms of intestinal niacin absorption are not yet fully clear, several mechanisms may contribute. Low concentrations of nicotinic acid and nicotinamide may be transported by sodium-dependent and -independent monocarboxylate transporters SMCT1/SMCT2 and MCT1, respectively, or by proton cotransporters or anion antiporters,[79-81] and several transporters involved in these processes are present in the small intestine.[82-84] Higher concentrations of both forms appear to be absorbed by passive diffusion, which will come into play with supplement use.

Once absorbed from the lumen into the intestinal mucosa, nicotinamide may be converted to NAD (**Figure 23.2**, reactions 4 and 5) or released into the portal circulation. Conversely, physiologic levels of nicotinic acid are largely converted through the Preiss-Handler pathway to NAD[7,80] (see Figure 23.2, reactions 1, 2, and 3). NAD glycohydrolases release nicotinamide into the portal circulation (see Figure 23.2, reaction 7). The liver then takes up and converts most of the remaining nicotinic acid in the portal blood to NAD, which it cleaves to release nicotinamide, as needed, into the systemic circulation. Red blood cells also take up nicotinic acid and nicotinamide, thus forming a circulating reserve pool of pyridine nucleotides.[85,86]

The liver is a central player in niacin metabolism. The liver receives nicotinamide and some nicotinic acid through the portal circulation as well as nicotinamide released from other extrahepatic tissues. In the liver, nicotinic acid and nicotinamide are metabolized to NAD or to yield compounds for urinary excretion, depending on niacin status. The liver is also the site of tryptophan conversion to NAD. The liver has very high basal levels of NAD, which are further increased by extra dietary niacin and create a medium-term storage pool that can be used to maintain blood nicotinamide concentration.[80] The liver also produces various methylated and hydroxylated products of both nicotinic acid and nicotinamide for urinary excretion. In humans, nicotinamide is primarily methylated to produce N^1-methylnicotinamide, whereas nicotinic acid is conjugated with glycine to form nicotinuric acid. High dietary intake can lead to unmodified nicotinic acid and nicotinamide in the urine[80] because methyl donor capacity may become limiting.[74]

Plants and microorganisms form nicotinic acid, nicotinamide, and tryptophan, which act as the dietary sources for the pyridine ring structure in mammals. Preiss and Handler[87-89] initially described the pathway for conversion of nicotinic acid to NAD in animal cells (Figure 23.2, reactions 1-3). Dietrich et al[90] showed that nicotinamide is salvaged by combining with phosphoribosyl pyrophosphate and then ATP to produce NAD directly (Figure 23.2, reactions 4 and 5). Nicotinamide is not demethylated to form nicotinic acid in humans, except by bacteria in the GI tract, which can occur at high levels of nicotinamide intake.[72]

A small proportion of tryptophan catabolized in the liver results in NAD formation, thus supporting niacin status. Most tryptophan is fully catabolized through 2-amino-3-carboxymuconic-6-semialdehyde (ACMS), (Figure 23.2, reaction 6) and finally to acetyl coenzyme when ACMS is catabolized by ACMS decarboxylase (ACMSD) (reaction 9). If ACMS accumulates, some of it rearranges spontaneously to quinolinic acid (see Figure 23.2, reaction 7), to allow formation of NAD. Thus, the production of NAD from tryptophan is favored by high activity of tryptophan or indoleamine 2,3-dioxygenase, low activity of ACMSD, and high activity of quinolinate phosphoribosyltransferase (reaction 8), all of which lead to a wide range in the efficiency of tryptophan to niacin conversion among species and individuals.[39-41] Increased ACMSD activity decreases tryptophan-dependent NAD synthesis and increases dependence on dietary nicotinic acid or nicotinamide.[91] This pathway appears to be regulated, in part, to

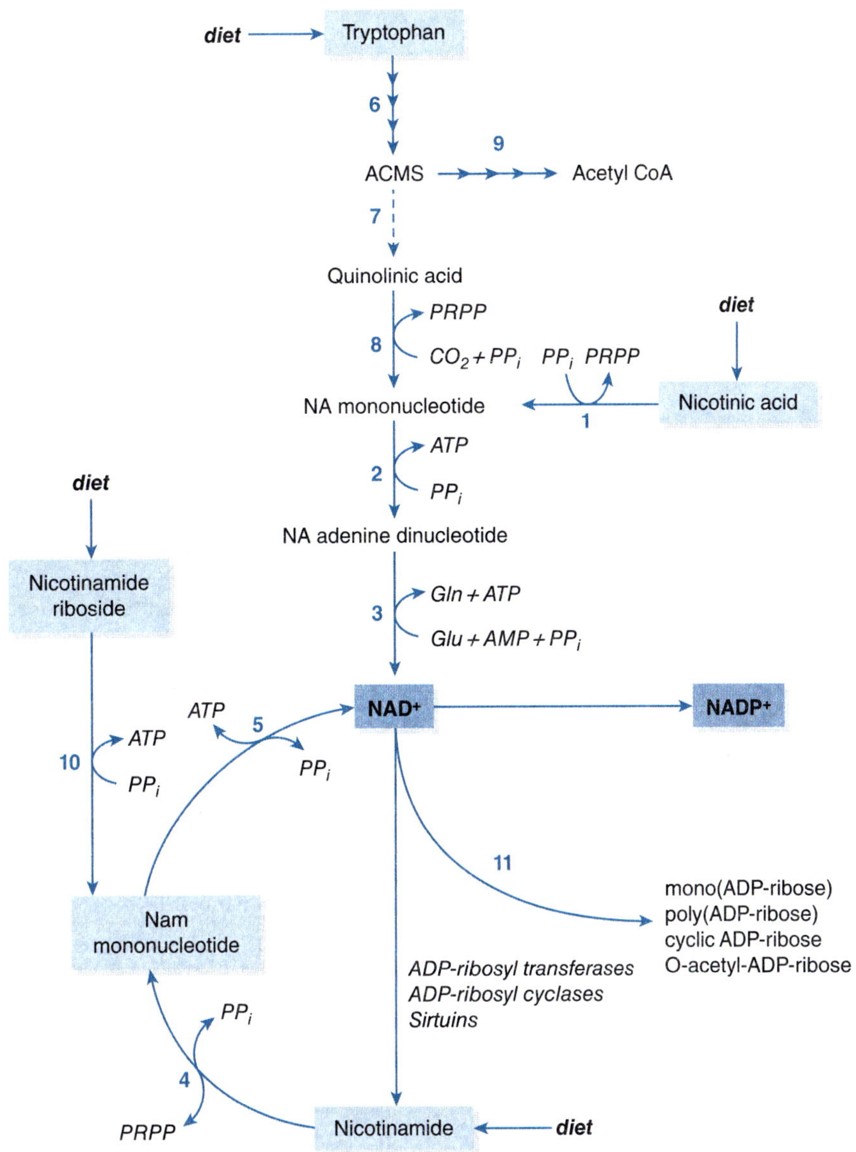

Figure 23.2 The synthesis and nonredox reactions of pyridine nucleotides. Reactions 1 to 3 constitute the Preiss-Handler pathway for de novo synthesis of nicotinamide adenine dinucleotide (NAD^+). Reactions 4 and 5 are used to convert dietary or endogenous nicotinamide into NAD^+. Reactions 6 to 8 describe steps required for the formation of NAD^+ from tryptophan. Reaction 9 represents the metabolism of ACMS to Acetyl CoA by ACMS decarboxylase (ACMSD), which is a prevalent pathway in humans. Reaction 10 illustrates the position where nicotinamide riboside enters the pathway to NAD. At position 11 are a large family of varied NAD^+ consuming ADP-(ribosyl)ation and NAD glycohydrolase reactions. Open boxes mark NAD precursors that can be taken up in food. *ACMS*, 2-amino-3-carboxymuconic-6-semialdehyde; *AMP*, adenosine monophosphate; *ATP*, adenosine triphosphate; *CoA*, coenzyme; *Gln*, glutamine; *Glu*, glutamate; *NA*, nicotinic acid; *Nam*, nicotinamide; PP_i, pyrophosphate; *PRPP*, phosphoribosyl pyrophosphate.

minimize quinolinate neurotoxicity during high protein intake, starvation, and ketosis.[92,93]

The traditional estimate of efficiency for tryptophan to NAD conversion is 1/60. This leads to the concept of niacin equivalents (1 NE = 1 mg niacin = 60 mg tryptophan). Significant individual variation exists, however.[94] Additionally, low levels of tryptophan intake cause a lack of tryptophan to NAD conversion.[37] In studies, young men consuming a diet containing 6 NE/day (RDA = 16 NE) for 5 weeks started receiving an additional 240 mg/day of tryptophan. This addition of 4 NE/day had no effect on blood NAD, and it appears that protein turnover takes precedence over niacin synthesis when tryptophan levels are low. This finding was also reported in animal models.[36] At the same time, high-protein diets and tryptophan supplements cure pellagra, while a genetic defect in tryptophan absorption, known as Hartnup disease, can cause pellagra in persons with marginal diets.

Assessment of Nutrient Status

At normal-to-low niacin intakes, most urinary excretion is of nicotinamide metabolites because nicotinic acid is efficiently converted into metabolically active nucleotides. As an individual becomes niacin-deficient, the urinary excretion of

N-methyl-2-pyridone-5-carboxamide decreases to a greater extent than that of *N*-methyl-nicotinamide does, with a ratio of less than 1, suggesting niacin deficiency.[95] Human studies by Fu et. al.,[37] subsequently showed that red blood cell NAD decreases during niacin deficiency, although the NADPH pool is very stable. These findings led to the use of (NAD/[NAD + NADP]) × 100, referred to as the niacin number, as an easily obtained index of human niacin deficiency.[44,57,96] This result was reproduced in other cell types in culture[97] and animal models[91,98] and showed a specific loss of the NAD^+ pool during niacin deficiency. This topic is considered later when the mechanisms of cellular disruption during niacin deficiency are discussed.

Functions in Metabolism

NAD and NADP are both indispensable cofactors for electron transfer in numerous redox reactions that are central to cellular energy metabolism and that do not affect total NAD concentration. Beyond that, they are substrates for numerous NAD^+ consuming nonredox enzymes with regulatory and signaling functions that involve permanent NAD degradation.[99-101]

Redox Reactions

The most critical function of the pyridine nucleotides is support of the oxidation and reduction reactions that are found throughout metabolism in all organisms. The oxidized nicotinamide ring in NAD^+ or $NADP^+$ can accept one electron at the positively charged nitrogen and a second electron (with associated proton) at the C-4 carbon (see Figure 23.1). The formation of a separate NADP pool is critical to the support of both oxidative and reductive processes, and the NAD kinase responsible for this is highly conserved among all levels of organisms.[102-106] The phosphorylation itself does not affect the redox properties of the cofactor, but it allows for enzyme specificity between the NAD and NADP pools. Consequently, the NAD pool is maintained in a largely oxidized state (as NAD^+), primarily by components of the electron transport chain.

Conversely, the NADP pool is maintained largely in a reduced state (as NADPH), primarily by the hexose monophosphate shunt (pentose phosphate pathway). The oxidizing NAD^+/NADH redox couple can then be linked to enzymes that oxidize substrates (e.g., glycolytic reactions, oxidative decarboxylation of pyruvate, oxidation of acetate in the tricarboxylic acid [TCA] cycle, oxidation of alcohol, β-oxidation of fatty acids, i.e., mostly catabolic reactions) and drive them forward by mass action. The reducing NADPH/$NADP^+$ redox couple can be linked to enzymes that reduce substrates (e.g., fatty acid and cholesterol synthesis, manufacture of deoxyribonucleotides, detoxification of hydrogen peroxide, i.e., mostly anabolic reactions). The critical nature of these reactions can be seen from the central pathways that depend completely on their function, including glycolysis, the TCA cycle, the electron transport chain, fatty acid synthesis, and β-oxidation. Changes in NAD and NADP pools and their redox states have significant consequences for human health, and how these separate NAD and NADP pools with their specific redox states are established and maintained is, therefore, subject to intense research.[100,107]

Poly(ADP-Ribose) Formation

Poly(ADP-ribose) (PAR) was discovered in 1966 by Paul Mandel's group.[20] Poly(ADP-ribose) polymerase-1 (PARP1) was the first PAR-synthesizing enzyme identified, because it is an abundant protein that affords the majority of cellular PARP activity. Eventually, PARP1 knockout mice were found to still synthesize 5% to 10% of control PAR levels, and a search for related enzymes led to the discovery of a large family of structurally related enzymes with 17 members in humans, of which only PARP1, PARP2, and tankyrases can produce PAR.[21,22,108,109]

PARP1 contains two zinc fingers that allow the enzyme to bind specifically to strand breaks in DNA and signal the catalytic portion of the protein to initiate poly(ADP-ribose) synthesis.[110,111] More than 30 nuclear proteins may act as acceptors, but most of the poly(ADP-ribose) is synthesized on PARP1 itself. This "automodification" of PARP1 is critical to its function in DNA repair. Automodification of PARP1 occurs by homodimerization of PARP1 or heterodimerization of PARP1 and PARP2.[112,113] As PARP1 becomes more poly(ADP-ribosyl)ated, it takes on a negative charge, which eventually causes it to be repelled from DNA and lose catalytic activity.[114,115]

PARP2 is like PARP1, with an abbreviated DNA-binding domain. PARP1 and PARP2 both interact with XRCC1 in the regulation of base excision repair (BER).[115-118] Both PARP1 and PARP2 null mice survive and reproduce, although they display genomic instability; double knockouts die *in utero*, thus demonstrating a redundancy of activity between these two enzymes.[118,119]

PARP3 lacks nick-sensing ability, but it can heterodimerize with PARP1 and tends to localize to the centrosome.[21,115] Tankyrase and tankyrase 2 are found around telomeres, the repetitive sequences at the end of mammalian chromosomes.[120] Their PARP activity relaxes the folded telomere tips and allows telomerase to access and elongate the DNA terminus. Telomerase is required by dividing cells to prevent erosion and instability at the ends of the chromosome, making tankyrase inhibitors prospective cancer therapeutics.[121]

It is beyond the scope of this chapter to describe the function of all members of the PARP enzyme superfamily in detail. Mechanistically, poly(ADP-ribose) has a strong negative charge similar to DNA. Proteins that are covalently modified with polymer are then repelled from DNA or other negatively charged binding partners. Additionally, the covalent modification can directly alter the activity of a protein.[122] Clouds of negatively charged polymer around sites of DNA damage can repel other strands of DNA and can help prevent deleterious translocation events. Finally, many proteins have specific, noncovalent, high-affinity binding sites for poly(ADP-ribose) and are thereby drawn to sites of polymer formation. These varied mechanisms have been most thoroughly studied in relation to the functions of PARP1 and PARP2 (PARP1/2) in DNA repair. Briefly, activation of PARP1/2 in response to DNA damage leads to a cascade of events. Activated PARP1/2, bound to a DNA strand break, will covalently modify nearby histones, thus causing them to be repelled from DNA and leading to local chromatin relaxation.

Additionally, histones have high-affinity polymer-binding sites[123] and are drawn out of nearby chromatin to bind to the cloud of poly(ADP-ribose) attached to PARP1/2. After DNA damage, this chromatin relaxation allows assembly of a complex of repair enzymes that is further aided by attraction of specific proteins with high-affinity, polymer-binding motifs such as XRCC1, p53, XPA, ATM, DEC, topoisomerase, DNA ligase, and DNA polymerase.[124] After the repair complex has been fully assembled, the automodified PARP1/2 can be repelled from the DNA to allow the repair process to be completed. Inhibition of PARP1 activity, or very low cellular NAD, can cause PARP1 to remain stuck to the site of DNA damage and can prevent repair from proceeding.[72] This "PARP trapping" inhibits DNA damage repair and is the underlying strategy of PARP inhibitors as cancer therapeutics.[125,126] Overall, the view is emerging that NAD is required as the biochemical fuel that allows PARP enzymes to preserve DNA integrity.[127,128] The importance of PAR and niacin for DNA damage repair is further seen in patients with xeroderma pigmentosum, a condition that is caused by mutation in DNA repair pathways. Those patients benefit clinically from niacin supplementation when they experience pellagra-like symptoms.[129]

Another area of interest is the role of PARP enzymes in controlling gene expression, likely using physical properties like those described earlier. PARP1 also binds to DNA in the absence of strand breaks and can be catalytically activated by certain secondary structures in DNA.[130] Thus, PARP1 can regulate chromatin structure and cause assembly of transcription factors in the absence of DNA damage. PARP1 activity can change gene expression indirectly by controlling epigenetic marks indirectly controlling gene expression.[131]

As a result, PARP1 is required for neural plasticity and learning,[132] processes thought to be driven more by regulation of gene expression than by DNA strand breaks. Similarly, interest has grown in the role of PARP1 in the regulation of nuclear factor κB (NFκB) signaling and inflammation.[133] Although this pathway likely has positive effects in responding to infection, these signals worsen tissue injury in many acute models, such as heart attack and stroke, organ transplantation, and septic shock, as well as in chronic models such as diabetes and cardiovascular disease.[73] Many investigators have demonstrated that PARP1 inhibition can greatly diminish the severity of these disease processes. This finding has caused quite a controversy in the literature on PARP1 and health. Many publications show the need for PARP activity to maintain genomic stability and long-term health, whereas many others demonstrate the negative impact of PARP1 activity on the progression of numerous important human health issues.[73] To make the most effective clinical use of PARP modifiers, including catalytic inhibitors and niacin supplementation, the positive and negative roles of PARP enzymes in each stage of each disease process must be defined.

Mono ADP-Ribosylation Reactions

In this class of reactions, a single ADP-ribose unit is moved from NAD^+ to an amino acid residue on an acceptor protein.[134-136] The bacterial toxins of the cholera, pertussis, and diphtheria pathogens and *Pseudomonas* ADP-ribosylate host G-proteins that disrupt host cellular function. Mammalian cells are now known to contain numerous endogenous mono-ADP-ribosyltransferases (ARTs), which act on various amino acid side chains as ADP-ribose acceptors and are classified either as choleratoxin-like ART (ARTC) or diphtheria-toxin like ART (ARTD).[22,137] Ecto-ARTs are secreted or expressed membrane-bound on the outside of cells, whereas endo-ARTs work inside cells. Ecto-ARTs include ARTC1 (ADP-ribosylates integrins and controls myogenesis) and ARTC2 (induces apoptosis through ADP-ribosylation of an ATP-gated ion channel).[138] Normally, extracellular NAD^+ concentration is very low, and the source of substrate for ecto-ARTs may involve NAD channels in the plasma membrane or NAD released from injured cells. These ideas suggest that NAD itself may be used as a signal of the metabolic state of a cell or as a signal for the death of nearby cells, thus leading to signaling events that may be paracrine or autocrine in nature.

Endo-ARTs act inside the cell. G-proteins are important components of cell signaling and have been shown to be substrates for arginine-specific ADP-ribosylation.[134,137] This process could control various pathways, depending on what the G-protein is controlling. Substrates of mono(ADP-ribosyl)ated G-proteins include elongation factor 2[139] as well as the 78-kDa glucose-regulated protein (GRP78), a molecular chaperone that aids in the correct folding of secreted proteins in the lumen of the endoplasmic reticulum. During metabolic or environmental stress, GRP78 is mono(ADP-ribosyl)ated, which results in decreased protein translation while at the same time preventing a total shutdown that would eventually kill the cell.[134,137]

Many of the enzymes that were originally classified as PARPs turned out to only possess mono(ADP-ribose) transferase activities.[22] These enzymes are involved in a range of physiologic processes, and the specific functions often are not yet well understood. PARP-4, or VPARP, associates with vault particles, which are massive ribonucleoprotein particles present in the cytosol of mammalian cells.[21,109] PARP-7 and PARP10 are likely involved in controlling gene expression control in response to environmental factors, even though specific mechanisms are not yet clear.[109,140] PARP11 enzyme is expressed in a tissue-specific manner, and is important for the rigidity of cellular membranes, immune response to cancer, as well as for male fertility.[141,142] Both PARP11 and PARP12 contribute to antiviral functions of the immune system.[143,144]

Cyclic ADP-ribose, Linear ADP-ribose, O-Acetyl-ADP-ribose, Nicotinic Acid Adenine Dinucleotide Phosphate, and Calcium Signaling

In 1993, cyclic ADP-ribose (cADPr) was identified as a metabolite of NAD^+ with potent intracellular calcium-mobilization properties.[148] Calcium concentrations are approximately 10,000-fold higher outside cells relative to the cytosol. Transient increases in intracellular calcium, arriving through the plasma membrane or released from intracellular stores (e.g., endoplasmic reticulum, mitochondria, lysosomes), regulate processes from neurotransmission to insulin release by β cells, muscle cell contraction, and T-lymphocyte activation, among others.[149] Cyclic ADP-ribose participates in the process of calcium-induced calcium release (CICR) and can be considered a second messenger molecule.[26] For example, an impulse traveling along a nerve axon arrives at a synapse in which voltage-gated channels allow a certain amount of calcium to cross the plasma membrane. This calcium causes the formation of inositol triphosphate (IP3) and cyclic ADP-ribose, which bind to ryanodine and IP3 receptors, respectively, thus causing more calcium release from intracellular stores. If intracellular calcium reaches a certain threshold, signals for the release of neurotransmitters will be sufficient to cause the impulse to be propagated across the synapse. Similar calcium release events occur in both the presynaptic and postsynaptic boutons, which enhance or dampen the strength of synapses and are involved in essentially all aspects of nervous system function.

More recently, a contaminant of commercial $NADP^+$ was found to mobilize calcium and was identified as nicotinic acid adenine dinucleotide phosphate ($NAADP^+$). $NAADP^+$ has been found in cultured cells and whole tissues and has been shown to respond to physiologic events.[150] Both cADPr and $NAADP^+$ are made by the same enzyme, termed CD38.[151] $NAADP^+$ causes calcium release through two-pore channels, which may initiate or amplify CICR.[152] Finally, this same class of enzymes forms linear ADP-ribose, either directly from NAD^+ or by hydrolyzing cyclic ADP-ribose.[153] Linear ADP-ribose is also formed through the formation and turnover of poly ADP-ribose. Linear ADP-ribose also causes calcium release, through TRPM2 channels.[154] TRPM2 channels can also release calcium in response to another NAD^+ metabolite, *O*-acetyl-ADP-ribose, which results from sirtuin activity. The emerging picture is that control of intracellular calcium release is a result of overlapping signals from IP3, cyclic ADP-ribose, linear ADP-ribose, *O*-acetyl-ADP-ribose, and $NAADP^+$. Clearly, calcium signaling is heavily integrated with pyridine nucleotide metabolism, cellular energy status, and neuronal and hormonal regulation. There even appears to be a link between ADP ribosyl cyclases and social skills.[155] Deregulation of the cADPr synthetase CD38 appears to be a major NAD sink that contributes to many pathophysiologies.[24,25]

Sirtuin Function

Another role for NAD^+ is as a substrate for sirtuin enzymes, a family of NAD-dependent protein deacetylases. These enzymes transfer an acetyl group from various proteins onto ADP-ribose, with the release of nicotinamide. The mammalian sirtuin family has seven members, the most thoroughly researched of which is SIRT1 (mammals)/Sir2 (yeast, worms, flies).[27,156]

SIRT1 functions as a protein deacetylase, whereas SIRT2 through SIRT7 have a mixture of deacetylase and ADP-ribosyl transferase activities.[27] Early work established that SIRT1 acted to deacetylate histones and p53. Deacetylation of histones leads to a more compact chromatin structure and gene silencing. Niacin deficiency could then, in theory, lead to a more open DNA structure, with more active gene expression and greater DNA sensitivity to damage and translocation events. SIRT1 further links control of chromatin structure to cellular energy status.

A relatively direct connection between SIRT1 and control of cellular energy status exists by transcription control,[156] where substrates that are deacetylated by SIRT1, besides histones and p53, include FOXO1, Ku70, p300, Rb, NFκB, and PGC-1α. Transcription factors like FOXO1 and PGC-1α directly control metabolic processes; p53 protein controls cell cycle checkpoints, DNA repair, and apoptosis, and acetylation appears to enhance p53 stability and accumulation. Thus, SIRT1 action appears to hinder p53 function and act as a tumor promoter.[157] Many other substrates will have to be studied to determine a healthy balance of SIRT1 actions with respect to apoptosis, genomic stability, and cancer.

SIRT1 may control the chromatin microenvironment around sites of DNA damage. Activation of PARP1/2 at strand breaks creates a localized depression in NAD^+ and an increase in nicotinamide. This inhibits SIRT1 activity, thus allowing acetylation of histones and leading to chromatin relaxation,[71] which aids access of repair enzymes (**Figure 23.3**).

Sir2 was found to mediate the life extension effects of caloric restriction in yeast, worm, fly and mouse models, which has

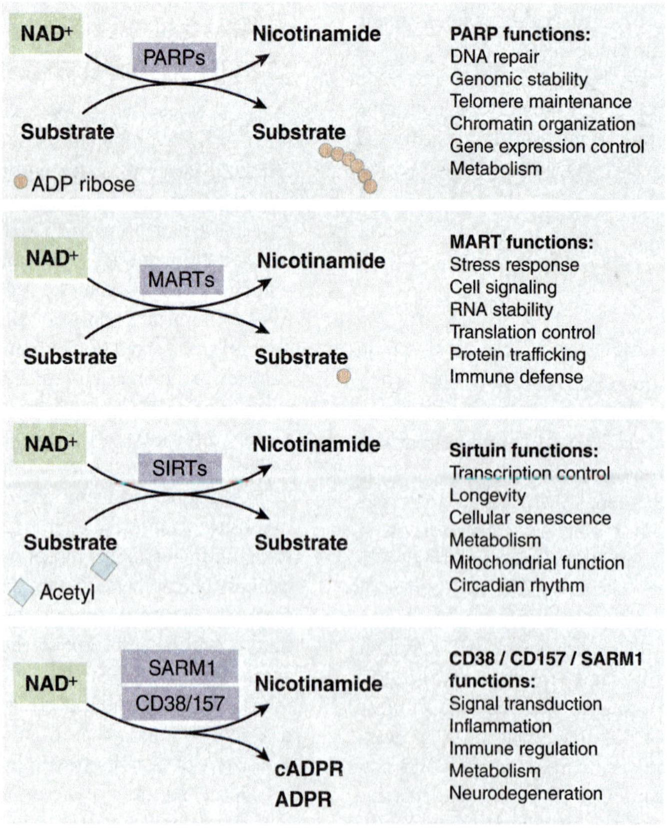

Figure 23.3 Overview of NAD^+ consuming enzyme groups, their reactions, and general functions.[21,145-147] *PARP*, Poly(ADP-ribose) polymerase; *MART*, Mono(ADP-ribose)transferase; *SIRT*, sirtuin; *cADPr*, cyclic ADP-ribose; *ADPr*, ADP-ribose; *SARM1*, Sterile alpha; *TIR*, motif containing 1.

Data from Gasparrini M, Sorci L, Raffaelli N. Cell Mo Life Sci 2021;78:3317–3331; Hogan KA, Chini CCS, Chinni EN. Cancer Aging Metab Dis. Front Immunol 2019;10:1187; Icso JD, Thompson PR. Curr Opin Chem Biol 2022;69:102176; Meyer-Ficca ML, Meyer RG, Jacobson EL, et al. Int J Biochem Cell Biol 2005;37:920–926.

brought sirtuin modulators to the forefront of aging and longevity research[158-160] and has been fueling an intense search for compounds that raise NAD$^+$ or activate sirtuins.[161,162] Resveratrol, a polyphenol found in grape products, was found to activate Sir2 and extend lifespan in the absence of caloric restriction, thereby boosting interest in red wine and health.

In summary, SIRT1 potentially extends longevity through chromatin structure and genomic stability, but it is also extensively involved in the regulation of gene expression related to control of energy metabolism in critical tissues such as liver, skeletal muscle, adipose tissue, and pancreas. SIRT1 activation, on balance, appears to maintain insulin sensitivity and decrease the risk of type 2 diabetes.[158,163] The exact mechanism of life span extension and whether sirtuin activation in humans functions in the same ways in animal models is still controversial and much remains to learn about sirtuins and human health and life span extension.[164]

Specific Causes and Manifestations of Deficiency and Excess States

Early researchers probably attributed the pathologic features associated with niacin deficiency to disruptions in redox cycling because this was the only metabolic role known for niacin at that time. However, the distinctive clinical signs of pellagra (dementia, sun-sensitive dermatitis) are better explained in relation to the ADP-ribosylation functions of NAD. The redox functions of pyridine nucleotides are too critical to lose and may be preserved as long as possible during niacin deficiency through high enzyme affinities and subcellular compartmentalization.[1,7] During moderate niacin deficiency, it is mainly the NAD$^+$ pool that declines, whereas NADH, NADP$^+$, and NADPH are maintained, and the GSH/GSSG (reduced and oxidized glutathione, respectively) couple is not as severely impaired.[98] A severe decrease in NAD/NADH is associated with decreased glutathione (GSH).[165] NAD is concentrated in the mitochondria, where it can serve multiple redox functions while being protected from use by most ADP-ribosyltransferases in the cell.[7,100] Recent metabolomics and molecule tracking methods have given us a better idea of the flux of NAD and of the responsiveness of redox reactions to niacin status in different tissues in whole animal models.[166]

The sun sensitivity of pellagra is dramatic and is not found with deficiencies of related redox nutrients such as riboflavin or iron. From experience with familial sun sensitivity disorders, such as xeroderma pigmentosum, sun sensitivity is generally recognized to reflect problems in DNA repair pathways such as nucleotide excision repair (NER). No major genetic deficiencies naturally occur in base excision repair (BER) genes, likely because they are critical for survival. Low niacin status and decreased PARP activity could impair NER, BER, and other pathways for DNA repair, given the broad participation of PARP enzymes in these processes.[127] Niacin deficiency increases the incidence of ultraviolet-induced skin cancer in animal models, and pharmacologic doses of niacin decrease skin cancer risk to less than that seen with adequate intake in both animal models and humans.[167,168] Cell culture models showed that poly(ADP-ribose) formation is very sensitive to niacin status, and inadequate NAD content decreases the ability to form polymers, which correlates with increased sensitivity to DNA damage.[169-171] Similar results were found in rat bone marrow cells in which niacin deficiency impaired PARP1- catalyzed poly(ADP-ribose) formation,[172] blocked nucleotide excision repair, and dramatically increased genomic instability,[173] thus increasing the development of nitrosourea-induced leukemias in rats.[174] In addition to decreased DNA repair, increased proinflammatory processes were also found in animal models of pellagra.[175]

These findings raise the question of niacin status and cancer risk in humans. Data do not appear to be available on long-term skin cancer risk in niacin-deficient human populations. The native population of the Transkei region in South Africa has a high risk of esophageal cancer.[176] A maize-based, low-protein diet is the staple for these people, and pellagra is common. Esophageal ulcerations and esophagitis, frequent in patients with pellagra, have been associated with development of carcinoma of the esophagus.[56] Studies of other populations have linked corn consumption with esophageal cancer risk.[177-179] In summary, the sun sensitivity of pellagra and other forms of genomic instability are probably related to low poly(ADP-ribose) synthesis, definitely by PARP1, and possibly by other members of the PARP family.

The other unique response to niacin deficiency is pellagrous dementia, which can progress from general depression to profound disruption of neural function, similar to schizophrenia. Patients may have auditory and visual hallucinations and display paranoid, suicidal, and aggressive behaviors.[180] Throughout the history of pellagrous epidemics, many patients in asylums for the insane were probably merely niacin deficient, even when skin lesions were not present. This situation is illustrated by a report of 11 cases admitted to the Georgia State Sanitarium in the early 1900s. Most of these patients with dementia showed no skin lesions initially and were not immediately diagnosed with pellagra. Dr. Little successfully treated 10 of the 11 patients with a nutrient-rich diet, starting with a combination of raw eggs in sweet milk.

Starting in 1937, numerous clinical reports of the response of patients with pellagra to nicotinic acid therapy were published. Clinicians reported that the most rapid and dramatic improvements were in neural function. The signs of pellagrous dementia often disappeared virtually overnight. Patients were able to recall their mental disturbances and marvel at their disappearance. This shows that niacin deficiency is disrupting a short-term process, such as cell signaling and neural transmission, as opposed to causing a degeneration of brain structure. Like sun sensitivity, dementia is not observed with deficiencies of other redox active nutrients such as riboflavin and iron. Pellagra generally involves deficiencies of protein, energy, and micronutrients in addition to niacin, such as riboflavin and thiamin. Following niacin therapy, riboflavin supplementation may be needed to resolve the oral lesions, and thiamin may be needed to treat peripheral nerve problems. Some researchers have focused on decreased tryptophan metabolites, including quinolinic acid and serotonin,[181] to explain pellagrous dementia, but the rapid improvement in central nervous system function following niacin supplementation is more supportive of the role of NAD metabolites.

It is now apparent that neural function can be altered by changes in various ADP-ribosylation reactions. PARP1 activity is required for long-term potentiation of synapses in several models of learning and memory.[132] Mono(ADP-ribosyl)ation of G-proteins could regulate synaptic sensitivity. The most likely link between niacin status and neural function involves calcium signaling, however. As mentioned earlier, cyclic ADP-ribose, linear ADP-ribose, NAADP, and *O*-acetyl-ADP-ribose can increase cytosolic calcium, thus leading to initiation of calcium spikes or enhancement through CICR processes. These processes are acting in presynaptic and postsynaptic boutons of all types of neurons with all types of neurotransmitters, so the disruption of neural function can be widespread and complex. Although niacin supplementation does not appear to be widely effective in the treatment of schizophrenia,[182] the similarity between this poorly understood condition and pellagrous dementia may lead to improved functional knowledge of both conditions in the future.

Increasing evidence identifies NAD deficiency and (epi) genetic and metabolic changes affecting NAD-generating pathways, as risk factors for psychoses, mood disorders like depression, and even neurodegenerative disorders like Alzheimer's disease or Parkinson's disease.[183,184]

Our understanding of the direct impact of niacin status on neural function is still incomplete. Levels of NAD^+ and cyclic ADP-ribose were altered in the brains of niacin-deficient, control, and pharmacologically supplemented rats, and these changes were associated with differences in learning and behavior.[185] CD38-null mice had lower levels of brain cyclic ADP-ribose and also displayed altered behavior,[185] but much work remains to be done to define the roles of cyclic and linear ADP-ribose, NAADP, and *O*-acetyl-ADP-ribose in the origin of pellagrous dementia. It appears likely that the profound neural changes and their rapid resolution will be related to changes in these signaling molecules.

Recent evidence further indicates that increased activity of the pathway converting tryptophan to NAD (kynurenine pathway), which is likely more active when dietary niacin is low, contributes to neuroinflammation, to neurodegenerative processes like Alzheimer's disease and Parkinson's disease and plays a role in mood disorders, depression, and suicidal behavior.[184,186,187] Molecular changes that occur in such a situation include decreased serotonin and increased quinolinic acid production. Quinolinic acid is a potent agonist for certain neuroreceptor types and can lead to overexcitation. In addition, it is a potent pro-oxidative molecule. Both mechanisms cause neurodegeneration and neuroinflammation, and have been observed in Alzheimer's and Parkinson's diseases.[184]

Diarrhea, the third D of pellagra, is less unique and occurs in other micronutrient and macronutrient deficiencies. The GI tract requires much energy to maintain, and this tissue atrophies during various forms of malnutrition. As the small intestinal villi shorten, surface area is lost, and undigested nutrients move to the colon, where they ferment and lead to diarrhea. The specific metabolic function of niacin that fails in the GI tract as diarrhea sets in is not clear but could be caused by reduced intestinal cell proliferation and increased mucosal inflammation. This is in line with the frequent occurrence of glossitis and esophageal lesions in pellagra, and it is the start of a vicious cycle of nutrient loss and worsening malnutrition that accelerates the occurrence of the final D of pellagra (death).

Pellagra is a fascinating condition that reflects the complex roles of niacin in whole body metabolism, from essential redox reactions to genomic stability, cell communication, and control of gene expression. Further work will be required to determine which niacin-dependent processes progressively fail as the signs and symptoms of pellagra develop. The focus of recent research has shifted from the extreme niacin deficiency causing pellagra to a more subtle and sometimes subclinical niacin deficiency seen in metabolic disorders and in aging.[100] The link between declining NAD and aging has caused a surge in pharmacologic and therapeutic approaches to supplement NAD precursors. While all supplements ultimately aim at increasing NAD, it appears as if different precursors are surprisingly not (entirely) redundant and different precursors might work better for certain cell or tissue types.[188] More research will be needed to further our understanding of NAD metabolism and to guide supplementation strategies.

References

1. Kirkland J. Niacin. In *Handbook of Vitamins, Fifth Edition* (Zempleni J, Suttie J, Iii J, Stover P, eds) pp. 149–190, CRC Press, 2013.
2. Mason JB., Gibson N, Kodicek E. Br J Nutr 1973;30:297–311.
3. Matapandeu G, Dunn SH, Pagels P. Am J Trop Med Hyg 2017;96:1244–7.
4. Seal AJ, Creeke PI, Dibari F, et al. Am J Clin Nutr 2007;85: 218–24.
5. Hołubiec P, Leończyk M, Staszewski F, et al. Folia Med Cracov 2021;61:125–37.
6. Carpenter KJ. Fed Proc 1981;40:1531–5.
7. Kirkland JB, Meyer-Ficca ML. Adv Food Nutr Res 2018;83: 83–149.
8. Johnson RB. J Okla State Med Assoc 1981;74:396–8.
9. Klevay LM. J Am Coll Nutr 1997;16:290–2.
10. Morabia A. JRSM 2008;101:566–8.
11. Elvehjem CA, Madden RJ, Strong FM, et al. J Am Chem Soc 1937;59:1767–8.
12. Elvehjem C.A, Madden RJ, Strong FM, et al. J Biol Chem 2002; 277:e22.
13. Spies TD. Exp Biol Med 1931;29:83–6.
14. Spies TD, Walker AA, Woods AW. JAMA 1939;113:1481.
15. Spies TD. JAMA 1938;110:622.
16. Bean WB. JAMA 1942;118:1176.
17. Harden A, Young WJ. Proc R Soc Lond B 1906;78:369–75.
18. Warburg O, Christian JW Biochemische Zeitschrift 1936;287:1.
19. Friedkin M, Lehninger AL. J Biol Chem 1949;178:611–44.
20. Chambon P, Weill JD, Doly J, et al. Biochem Biophys Res Comm 1966;25:638–43.
21. Meyer-Ficca ML, Meyer RG, Jacobson EL, et al. Int J Biochem Cell Biol 2005;37:920–6.
22. Lüscher B, Ahel I, Altmeyer M, et al. FEBS J 2021;289: 7399–410.
23. Kraus WL. Mol Cell 2015;58:902–10.
24. Takasawa S. Int J Mol Sci 2022;23:4306.
25. Zeidler JD, Hogan KA, Agorrody G, et al. Am J Physiol-Cell Physiol 2022;00451.
26. Ferrero E, Lo Buono N, Horenstein AL, et al. Front Biosci (Landmark Ed) 2014;19:986–1002.
27. Imai S, Guarente L. Trends Pharmacol Sci 2010;31:212–20.
28. Fiorentino F, Mautone N, Menna M, D'Acunzo F, et al. Future Med Chem 2022;14:915–39.
29. Kajikawa M, Hirai N, Hashimoto T. Plant Mol Biol 2009;69: 287–98.
30. Leete E. The Biosynthesis of Nicotine and Related Alkaloids in Intact Plants, Isolated Plant Parts, Tissue Cultures, and Cell-Free Systems. In *Secondary-Metabolite Biosynthesis and Metabolism* (Petroski, R. J. and McCormick, S. P., eds) pp. 121–139, Springer US, Boston, MA, 1992.
31. Alkaloids Derived from Nicotinic Acid. In *Alkaloids* pp. 181–92, Elsevier, 2015.
32. Li W, Zhang F, Wu R, et al. Plant Physiol 2017;174:1492–504.
33. Ziegler M, Nikiforov AA. Nat Metab 2020;2:291–2.

34. Yang Y, Zhang N, Zhang G, et al. Nat Metab 2020;2:364–79.
35. Tarr JB, Arditti J. Plant Physiol 1982;69:553–6.
36. Shibata K, Shimada H, Kondo T. Biosci Biotechnol Biochem 1996;60:1660–6.
37. Fu CS, Swendseid ME, Jacob RA, et al. J Nutr 1989;119:1949–55.
38. U.S. Department of Agriculture, Agricultural Research Service. FoodData Central, 2019. fdc.nal.usda.gov.
39. Ummarino S, Mozzon M, Zamporlini F, et al. Food Chem 2017;221:161–8.
40. EFSA Panel on Nutrition, Novel foods and Food allergens (NDA), Turck D, Castenmiller J, et al. 2002;17:EFS2.
41. Trammell SA, Yu L, Redpath P, et al. *J Nutr* 2016;146:957–63.
42. Mousa TY, Mousa OY. In *StatPearls* Publishing, Treasure Island, FL.
43. Li D, Sun W-P, Zhou Y-M, et al. World J. Gastroenterol 2010;16:2378–87.
44. Jacobson EL. J Am Coll Nutr 1993;12:412–16.
45. Portale S, Sculati M, Stanford FC, et al. Eat Weight Disord 2020;25:1493–6.
46. Jagielska G, Tomaszewicz-Libudzic EC, Brzozowska A. Eur Child Adolesc Psychiatry 2007;16:417–20.
47. MacDonald A, Forsyth A. Clin Exp Dermatol 2005;30:388–90.
48. Strumia R. Am J Clin Dermatol 2005;6:165–73.
49. Badawy AA-B. Alcohol and Alcoholism 2014;49:238–50.
50. Stevens HP, Ostlere LS, Begent RH, et al. Br J Dermatol 1993; 128:578–80.
51. Boileau M, Azib S, Staumont-Sallé D, et al. Annales de Dermatologie et de Vénéréologie 2022;S0151963822000370.
52. Bishnoi A, Ashraf R, Banerjee N, et al. Indian J Dermatol Venereol Leprol 2019;85:499–502.
53. Monteiro JP, da Cunha DF, Filho DC, et al. Nutrition 2004;20: 778–82.
54. Kabengele C, M'hango H, Mweemba D, et al. Pan Afr Med J 2021;39:73.
55. Hui S, Heng L, Shaodong W, et al. An Bras Dermatol 2017;92: 879–81.
56. Nogueira Dermatol Online J 2009;15:8.
57. Shah GM, Shah RG, Veillette H, et al. Am J Gastroenterol 2005; 100:2307–14.
58. Zhao C, Liu T, Ma C, et al. Mult Scler Relat Disord 2018;25:104–7.
59. Otten JJ, Hellwig JP, Meyers LD. DRI, dietary reference intakes: the essential guide to nutrient requirements. National Academies Press, Washington, D.C., 2006;102.
60. Carlson LA. J Intern Med 2005;258:94–14.
61. Goldie C, Taylor AJ, Nguyen P, et al. Heart 102:198–203.
62. Grundy SM, Feingold KR. In: Feingold KR, Anawalt B, Boyce A, eds. Guidelines for the Management of High Blood Cholesterol. South Dartmouth: MDText.com, Inc., 2000.
63. Schwartz GG, Ballantyne CM. Atherosclerosis 2022;349: 110–22.
64. The AIM-HIGH Investigators. N Engl J Med 2011;365:2255–67.
65. The HPS2-THRIVE Collaborative Group. N Engl J Med 2014;371:203–12.
66. Abdellatif M, Sedej S, Kroemer G. Circulation 2021;144: 1795–817.
67. Habibe MN, Kellar JZ. Niacin Toxicity. In *StatPearls* Publishing, Treasure Island, FL, 2022.
68. Leung K, Quezada M, Chen Z, et al. Hepatology Comm 2018;2: 1293–8.
69. Ellsworth MA, Anderson KR, Hall DJ, et al. Case Rep Pediatr 2014;2014:1–3.
70. Patterson DJ, Dew EW, Gyorkey F, et al. South Med J 1983;76: 239–40.
71. Zhou S-S, Li D, Zhou Y-M, et al. BMC Public Health 2010;10:746.
72. Kirkland JB. Curr Pharm Des 2009;15:3–11.
73. Kirkland JB. Exp Biol Med (Maywood) 2010;235:561–8.
74. ApSimon MM, Rawling JM, Kirkland JB. J Nutr 1995;125: 1826–32.
75. Stead LM, Jacobs RL, Brosnan ME, et al. Adv Enzyme Regul 2004;44:321–33.
76. Stead LM, Brosnan JT, Brosnan ME, et al. Am J Clin Nutr 2006;83:5–10.
77. Adiels M, Chapman MJ, Robillard P, et al. J Clin Lipidol 2018;12:810–821.e1.
78. Mehmel M, Jovanović N, Spitz U. Nutrients 2020;12:E1616.
79. Nabokina SM, Kashyap ML, Said HM. Am J Physiol Cell Physiol 2005;289:C97–103.
80. Henderson LM. Annu Rev Nutr 1983;3:289–307.
81. Takanaga H, Maeda H, Yabuuchi H, et al. J Pharm Pharmacol 1996;48:1073–7.
82. Gopal E, Miyauchi S, Martin PM, et al. Pharm Res 2007;24: 575–84.
83. Bongarzone S, Barbon E, Ferocino A, et al. Nucl Med Biol 2020;88–89:24–33.
84. Kumar JS, Subramanian VS, Kapadia R, et al. Am J Physiol Gastrointest Liver Physiol 2013;305:G207–13.
85. Lan SJ, Henderson LM. J Biol Chem 1968;243:3388–94.
86. Micheli V, Simmonds HA, Sestini S, et al. Arch Biochem Biophys 1990;283:40–45.
87. Preiss J, Handler P. J Biol Chem 1957;225:759–70.
88. Preiss J, Handler P. J Biol Chem 1958;233:488–92.
89. Preiss J, Handler P. J Biol Chem 1958;233:493–500.
90. Dietrich LS, Fuller L, Yero IL, et al. J Biol Chem 1966;241:188–91.
91. Palzer L, Bader JJ, Angel F, et al. Cell Rep 2018;25:1359–70.e4.
92. Kimura N, Fukuwatari T, Sasaki R, et al. Biosci Biotechnol Biochem 2005;69:273–9.
93. Fukuoka S-I, Ishiguro K, Yanagihara K, et al. J Biol Chem 2002;277:35162–7.
94. Horwitt MK, Harper AE, Henderson LM. Am J Clin Nutr 1981;34:423–7.
95. Dolphin D, Avramović O, Poulson R, eds. Pyridine nucleotide coenzymes: chemical, biochemical, and medical aspects. Wiley, New York, 1987.
96. Jacobson EL, Jacobson MK. Methods Enzymol 1997;280: 221–30.
97. Benavente CA, Jacobson EL. Free Radic Biol Med 2008;44: 527–37.
98. Tang K, Sham H, Hui E, et al. J Nutr Biochem 2008;19:746–53.
99. Xie N, Zhang L, Gao W, et al. Sig Transduct Target Ther 2020;5:227.
100. Strømland Ø, Niere M, Nikiforov AA, et al. Biochem Soc Trans 2019;47:119–30.
101. Xiao W, Wang R-S, Handy DE, et al. Antioxidants & Redox Signaling 2018;28:251–72.
102. Agledal L, Niere M, Ziegler M. Redox Rep 2010;15:2–10.
103. Pollak N, Niere M, Ziegler M. J Biol Chem 2007;282: 33562–71.
104. Love NR, Pollak N, Dölle C, et al. Proc Natl Acad Sci U S A 2015;112:1386–91.
105. Gossmann TI, Ziegler M, Puntervoll P, et al. FEBS J 2012;279: 3355–63.
106. Vickman O, Erives A. PLoS One 2019;14:e0220447.
107. Bockwoldt M, Houry D, Niere M, et al. Proc Natl Acad Sci U S A 2019;116:15957–66.
108. Meyer RG, Meyer-Ficca ML, Jacobson EL, et al. In: Burkle A, ed. Enzymes in poly(ADP-ribose) metabolism. Poly(ADP-ribosyl) ation, Landes Bioscience. 2004.
109. Burkle A. FEBS J 2005;272:4576–89.
110. Langelier M-F, Eisemann T, Riccio AA, et al. *Curr Opin Struct Biol* 2018;53:187–98.
111. Ikejima M, Noguchi S, Yamashita R, et al. J Biol Chem 1990;265:21907–13.

112. Mendoza-Alvarez H, Alvarez-Gonzalez R. J Biol Chem 1993;268:22575–80.
113. Langelier MF, Planck JL, Roy S, et al. Science 2012;336:728–32.
114. Althaus FR. J Cell Sci 1992;102:663–70.
115. van Beek L, McClay É, Patel S, et al. IJMS 2021;22:5112.
116. Ame JC, Rolli V, Schreiber V, et al. J Biol Chem 1999;274: 17860–8.
117. Schreiber, V, Amé J-C, Dollé P, et al. J Biol Chem 2002;277: 23028–36.
118. Menissier de Murcia J, Ricoul M, Tartier L, et al. EMBO J 2003;22:2255–63.
119. Huber A, Bai P, de Murcia JM, et al. DNA Repair (Amst) 2004; 3:1103–8.
120. Hsiao SJ, Smith S. Biochimie 2008;90:83–92.
121. Yu M, Yang Y, Sykes M, et al. J Med Chem 2022;65:5244–73.
122. Creissen D, Shall S. Nature 1982;296:271–2.
123. Karch KR, Langelier M-F, Pascal JM, et al. Mol Biosyst 2017;13: 2660–71.
124. Fahrer J, Popp O, Malanga M, et al. Biochemistry 2010;49: 7119–30.
125. Padella A, Ghelli Luserna Di Rorà A, Marconi G, et al. J Hematol Oncol 2022;15:10.
126. Poltronieri P, Miwa M, Masutani M. IJMS 2021;22:10829.
127. Ruszkiewicz JA, Bürkle A, Mangerich A. J Biol Chem 2022;298:102037.
128. Hanzlikova H, Caldecott KW. Trends Genet 2019;35:412–22.
129. Hijazi H, Salih MA, Hamad MH, et al. 2015; Clin Genet 2015;87:56–61.
130. Lonskaya I, Potaman VN, Shlyakhtenko LS, et al. J Biol Chem 2005;280:17076–83.
131. Sanchez A, Buck-Koehntop BA, Miller KM. Bioessays 2022;44:e2200015.
132. Goldberg S, Visochek L, Giladi E, et al. J Neurochem 2009;111: 72–9.
133. Boehi F, Manetsch P, Hottiger, MO. Cell Discov 2021;7:104.
134. Di Girolamo M, Dani N, Stilla A, et al. FEBS J 2005;272: 4565–75.
135. Yates SP, Jørgensen R, Andersen GR, et al. Trends Biochem Sci 2006;31:123–33.
136. Gan Y, Sha H, Zou R, et al. Front Cell Dev Biol 2022;10:864101.
137. Di Girolamo M, Fabrizio G. Biochem Pharmacol 2019;167: 86–96.
138. Ohlrogge W, Haag F, Löhler J, et al. Mol Cell Biol 2002;22: 7535–42.
139. Ergen K, Bektaş M, Gökçe S, et al. Biocell 2007;31:61–6.
140. Feijs KL, Žaja R. Biosci Rep 2022;42:BSR20212489.
141. Meyer-Ficca ML, Ihara M, Bader JJ, et al. Biol Reprod 2015;92:80.
142. Zhang H, Yu P, Tomar VS, et al. Nat Cancer 2022;3:808–20.
143. Guo T, Liu J, Chen X, et al. Nat Microbiol 2019;4:1771–3.
144. Li L, Shi Y, Li S, et al. *Cell Biosci* 2021;11:116.
145. Hogan KA, Chini CC, Chini EN. Front Immunol 2019;10:1187.
146. Gasparrini M, Sorci L, Raffaelli N. Cell Mol Life Sci 2021;78: 3317–31.
147. Icso JD, Thompson PR. Curr Opin Chem Biol 2022;69:102176.
148. Kim H, Jacobson EL, Jacobson MK. Biochem Biophys Res Commun 1993;194:1143–7.
149. Uhlén P, Fritz N. Biochem Biophys Res Commun 2010;396: 28–32.
150. Mándi M, Bak J. J Recept Signal Transduct Res 2008;28: 163–84.
151. Chini EN, Chini CC, Kato I, et al. Biochem J 2002;362:125–130.
152. Patel S, Marchant JS, Brailoiu E. Cell Calcium 2010;47: 480–90.
153. Malavasi F, Deaglio S, Funaro A, Physiol Rev 2008;88: 841–86.
154. Olah ME, Jackson MF., Li H, J Physiol 2009;587:965–79.
155. Chong A, Malavasi F, Israel S, et al. Psychoneuroendocrinology 2017;78:185–92.
156. Shahgaldi S, Kahmini FR. Life Sci 2021;282:119803.
157. Yi J Luo J. Biochim Biophys Acta 2010;1804:1684–9.
158. Imai S, Guarente L. Aging Mech Dis 2016;2.
159. Satoh A, Imai S, Guarente L. Nat Rev Neurosci 2017;18: 362–74.
160. Vidović T, Ewald CY. Front Aging 2022;3:935220.
161. Bonkowski MS, Sinclair DA. Nat Rev Mol Cell Biol 2016;17: 679–90.
162. Dai H, Sinclair DA, Ellis JL, et al. Pharmacol Ther 2018;188: 140–154.
163. Camins A, Sureda FX, Junyent F, et al. Biochim Biophys Acta 2010;1799:740–749.
164. Stacchiotti A, Favero G, Rezzani R. In: Badria AF, ed. Resveratrol and SIRT1 Activators for the Treatment of Aging and Age-Related Diseases. *Resveratrol - Adding Life to Years, Not Adding Years to Life* IntechOpen, 2019.
165. Yoon, J-Y, Daneshgar N, Chu Y, et al. Clin Transl Med 2022;12:e954.
166. Liu L, Su X, Quinn WJ, et al. Cell Metabolism 2018;27: 1067–80.e5.
167. Kirkland JB. Mutat 2012,733:14–20.
168. Park SM, Li T, Wu S, et al. Int J Cancer 2017;140:2023–31.
169. Jacobson EL, Nunbhakdi-Craig V, Smith DG, et al. In: Poirier GG, Moreau P, eds. ADP-ribose Polymer Metabolism: Implications for Human Nutrition. ADP-Ribosylation Reactions, New York, NY: Springer New York. 1992: 153–62.
170. Whitacre CM, Hashimoto H, Tsai ML, et al. Cancer Res 1995; 55:3697–701.
171. Benavente CA, Schnell SA, Jacobson EL. PLoS One 2012;7:e42276.
172. Boyonoski AC, Spronck JC, Gallacher LM, et al. J Nutr 2002;132:108–14.
173. Kostecki LM, Thomas M, Linford G, et al. Mutat Res 2007;625:50–61.
174. Bartleman AP, Jacobs R, Kirkland JB. Nutr Cancer 2008;60: 251–8.
175. Sugita K, Ikenouchi-Sugita A, Nakayama Y, et al. Sci Rep 2013;3:2973.
176. Warwick GP, Harington JS. In *Advances in Cancer Research* 17, pp. 81–229, Elsevier, 1973.
177. Wahrendorf J, Chang-Claude J, Liang QS, et al. Lancet 1989; 2:1239–41.
178. Franceschi S, Bidoli E, Barón AE, et al. J Natl Cancer Inst 1990; 82:1407–11.
179. Marshall JR, Graham S, Haughey BP, et al. Eur J Cancer B Oral Oncol 1992;28B:9–15.
180. Roberts SR. Pellagra: History, Distribution, Diagnosis, Prognosis, Treamtent. Etiology. C.V. Mosby Company, St. Louis, MO, 1912.
181. Dickerson JW, Wiryanti J. Proc Nutr Soc 1978;37:167–71.
182. Ramsay RA, Ban TA, Lehmann HE, et al. Can Med Assoc J 1970; 102:939–42.
183. Miller CL. Biomedicines 2022;10:561.
184. Hestad K, Alexander J, Rootwelt H, et al. Biomolecules 2022;12:998.
185. Young GS, Kirkland JB. Nutr Res Rev 2008;21:42–55.
186. Brundin L, Sellgren CM, Lim CK, et al. Transl Psychiatry 2016;6:e865.
187. Sha Q, Madaj Z, Keaton S, et al. Transl Psychiatry 2022;12:35.
188. Canto C. Metabolites 2022;12:630.

C. VITAMINS

CHAPTER 24

Vitamin B_6

Jesse F. Gregory III

CHAPTER OUTLINE

Abbreviations

ALAS, δ-aminolevulinate synthase; **CRP**, C-reactive protein; **FAD**, flavin adenine dinucleotide; **FMN**, flavin mononucleotide; **HPLC**, high-performance liquid chromatography; **NAD**, nicotinamide adenine dinucleotide; **4-PA**, 4-pyridoxic acid; **PL**, pyridoxal; **PLP**, pyridoxal 5′-phosphate; **PM**, pyridoxamine; **PMP**, pyridoxamine 5′-phosphate; **PN**, pyridoxine; **PNG**, pyridoxine 5′-β-D-glucoside; **PNP**, pyridoxine 5′-phosphate; **RDA**, recommended dietary allowance; **SHMT**, serine hydroxymethyltransferase; **TNAP**, tissue-nonspecific phosphatase; **VD**, vascular disease

Introduction

Since the 1930s, when evidence of vitamin B_6 was first reported, our understanding of its properties, metabolic function, and role in maintaining health has expanded enormously. Despite these advances, areas of uncertainty remain, including the optimal intake of the vitamin, the consequences of inadequacy, the best way to assess nutritional status, and the effects of supplementation on health. Since the previous edition of this book,[1] important advances have occurred regarding the assessment of vitamin B_6 nutritional status, metabolic effects of vitamin B_6 insufficiency, and the influence of vitamin B_6 status on the incidence of chronic disease. These topics are given special consideration here.

History

Evidence of a water-soluble nutritional factor later identified as vitamin B_6 was first reported in 1934.[2] Five laboratories reported the independent isolation and crystallization of pyridoxine (PN) in 1938.[3-7] The proposed structure was confirmed through successful synthesis the following year. Studies of the nutritional requirements of lactic acid bacteria led to the recognition of pyridoxal (PL) and pyridoxamine (PM).[8] The coenzyme form of vitamin B_6 was demonstrated to be a phosphorylated derivative,[9] and eventually it was identified as the 5′-phosphate.

Chemistry and Nomenclature

The term *vitamin B_6* is the preferred generic descriptor[10] for the family of 2-methyl,3-hydroxy,5-hydroxymethylpyridine derivatives that exhibit the nutritional activity of PN. "Pyridoxine" has been used as a generic term, especially in clinical contexts; however, it is strongly recommended that the consistent use of the generic "vitamin B_6" rather than "pyridoxine" be adopted to reduce confusion in vitamin B_6 nomenclature.

Vitamin B_6 exists as three main derivatives of a 2-methyl, 3-hydroxy,5-hydroxymethylpyridine nucleus that differ with respect to their substituent at the 4 position of the pyridine

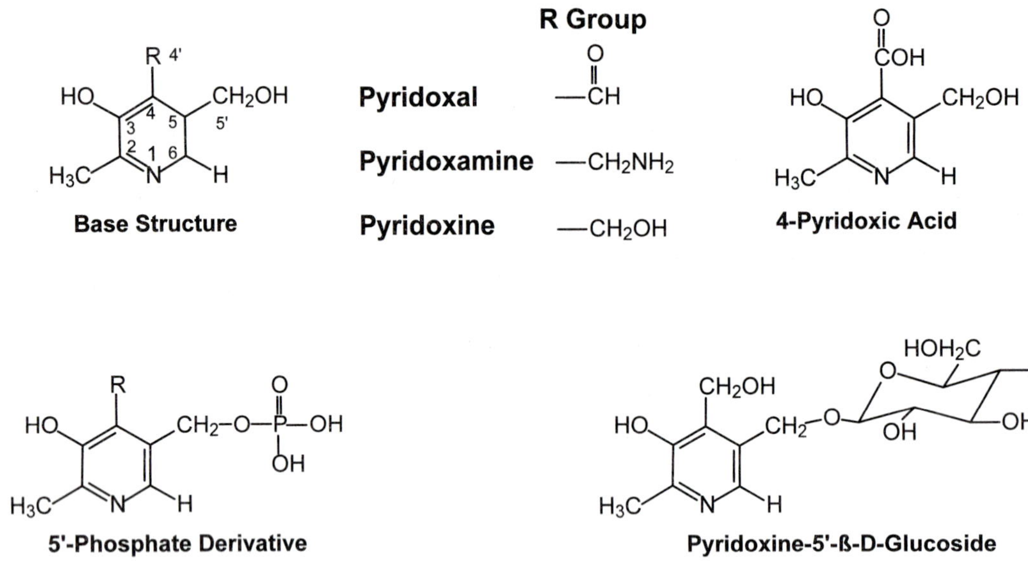

Figure 24.1 Chemical structures of vitamin B$_6$ forms.

ring (**Figure 24.1**). For the C-4 substituent, PN has a hydroxymethyl group, PL is an aldehyde, whereas PM has an aminomethyl group. Because PN is an alcohol, PN has been termed "pyridoxol" intermittently; this designator is obsolete and should be discontinued. PN, PL, and PM all can exist with a phosphate group esterified at the C-5′ position (i.e., PNP, PLP, and PMP). Pyridoxal 5′-phosphate (PLP) and pyridoxamine 5′-phosphate (PMP) are the coenzyme forms of vitamin B$_6$ and are interconverted as they function in the actions of the aminotransferase family of enzymes. Although pyridoxine 5′-phosphate (PNP) is not a coenzyme, it is an important intermediate in the metabolic pathway by which PLP is formed from dietary PN (**Figure 24.2**). 4-Pyridoxic acid (4-PA), which is the major metabolically inactive catabolic product of vitamin B$_6$ metabolism, has a carboxyl group at C-4 and is nutritionally and metabolically inactive. Also shown in Figure 24.1 is pyridoxine 5′-β-D-glucoside (PNG), a glycosylated form of vitamin B$_6$ commonly found in foods of plant origin.

Absorption and Bioavailability

Intestinal absorption of vitamin B$_6$ has been thought to take place in the jejunum by nonsaturable, passive diffusion of the nonphosphorylated forms.[11] However, evidence from in vitro studies using eukaryotic cells suggests that vitamin B$_6$ absorption is pH dependent and exhibits both saturable and nonsaturable components that are concentration dependent.[12] In this model, absorption appears to occur through a carrier-mediated pathway that involves proton exchange.[12] SLC19A2 and SLC19A3, also known as thiamin transporters 1 and 2, were shown recently to transport PN in several mammalian cell models. The acidic pH optimum of PN transport observed was consistent with transport in the acidic microclimate of the small intestinal mucosa.[13] PL and PM inhibited the uptake of radiolabeled PN, which suggests that they may also undergo uptake by these thiamin transporters.

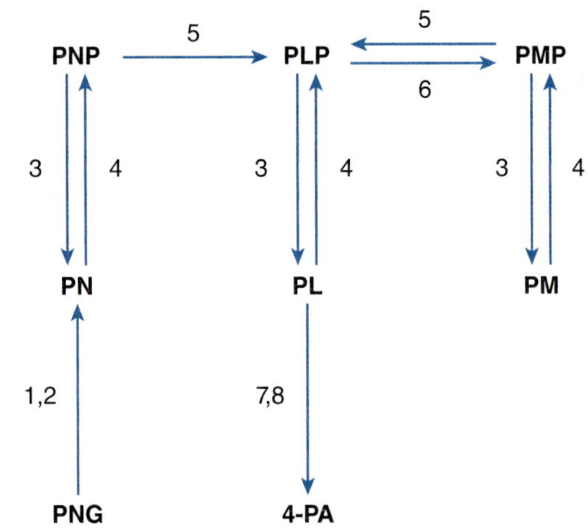

1: Lactase-phlorizin hydrolase
2: Pyridoxine-5′-β-D-glucoside hydrolase
3: Phosphatases
4: Pyridoxal kinase
5: Pyridoxine (pyridoxamine) phosphate oxidase
6: Aminotransferase
7: Aldehyde oxidase
8: Aldehyde dehydrogenase

Figure 24.2 Overview of vitamin B$_6$ metabolism. 4-PA, 4-pyridoxic acid; PL, pyridoxal; PLP, pyridoxal-5′-phosphate; PM, pyridoxamine; PMP, pyridoxamine-5′-phosphate; PN, pyridoxine; PNG, pyridoxine-5′-β-D-glucoside; PNP, pyridoxine-5′-phosphate.

Dietary PLP, PMP, and PNP are enzymatically dephosphorylated at the brush-border membrane by alkaline phosphatase before absorption.[1] Once absorbed, PN, PL, and PM are phosphorylated by pyridoxal kinase for the purpose of metabolic trapping. The addition of the phosphate group at the 5′ position of the pyridine ring creates a negative charge on the molecule that prevents the vitamers from diffusing through the cellular membranes in intestinal mucosal cells and in other tissues. To cross the basolateral membrane and enter into portal circulation, PN, PL, and PM are converted back to a nonphosphorylated form.

The bioavailability of nutrients in foods and supplements is an important issue in evaluating the adequacy of diets and the efficacy of supplements in meeting nutritional requirements and resolving inadequate status. The bioavailability of vitamin B_6 in humans consuming a mixed diet is approximately 75%,[14] and data from pigs indicate that the digestibility of vitamin B_6 from animal sources is approximately 10% greater than from plant sources.[15] As has been reviewed,[16] vitamin B_6 bioavailability is likely to be a function of the degree of entrapment in the food matrix (i.e., nondigestible residue) and the extent of utilization of glycosylated forms of vitamin B_6. PNG, the major glycosylated form of vitamin B_6 in the human diet, provides an average of approximately 15% of total daily vitamin B_6 intake,[17] although this percentage varies widely, depending on food selection. The bioavailability of purified PNG was approximately 30% in rats[18,19] and approximately 50% in humans, compared with free PN.[20,21] Rat[18] and human[20,21] studies using isotopically labeled PNG found that PNG was effectively absorbed, but it was not completely hydrolyzed in the small intestine to glucose and PN. Intestinal hydrolysis of PNG is catalyzed by two β-glucosidases: a cytoplasmic enzyme designated PNG hydrolase[22]; and the brush-border membrane enzyme lactase-phlorizin hydrolase.[23] PNG also can be absorbed intact, potentially hydrolyzed by glucosidase activity in the kidney, or excreted unchanged in the urine.[18,20,21]

Transport and Metabolism

Vitamin B_6, mostly as PL, enters the portal circulation and is bound to albumin in the plasma for transport.[24] Albumin binds both PL and PLP, and most of the plasma PLP in the general circulation is secreted from the liver presumably bound to albumin. PL and PLP constitute 75% to 80% of the circulating forms of vitamin B_6.[25] Erythrocytes can take up both PN and PL[26]; however, erythrocyte vitamin B_6 is probably not available for direct uptake from the circulation into tissues.

The liver is the primary site of vitamin B_6 metabolism and where PLP is generated for hepatic use and for export to extrahepatic tissues. The phosphorylation of PN, PL, and PM to yield PNP, PLP, and PMP, respectively, is catalyzed by pyridoxal kinase.[27] Conversion of PNP and PMP to PLP is catalyzed by the flavin mononucleotide (FMN)–dependent pyridoxamine-(pyridoxine-)5′-phosphate oxidase in the liver.[27] This reaction is critical to the metabolism of dietary vitamin B_6 because most extrahepatic tissues have comparatively little oxidase activity. Pyridoxamine-(pyridoxine-)5′-phosphate oxidase is subject to strong product inhibition that serves to minimize the production of an excessive amounts of PLP.[28] Dephosphorylation of PLP and PMP in liver and other tissues is catalyzed by tissue-nonspecific phosphatase (TNAP)[29] activity, as well as by a vitamin B_6–specific form of erythrocyte alkaline phosphatase.[30] Two hepatic enzymes, riboflavin (flavin adenine dinucleotide [FAD])–dependent aldehyde oxidase and nicotinamide adenine dinucleotide (NAD)–dependent aldehyde dehydrogenase, oxidize excess PL in tissues to 4-PA, the major catabolic product of vitamin B_6.[28]

As stated earlier, PLP and PL are the predominant circulating forms of vitamin B_6. The turnover of vitamin B_6, as PLP, has been described using a compartmental model that includes five body pools: muscle, liver, plasma, erythrocytes, and one to combine all other pools.[31] Total body concentrations of vitamin B_6 are estimated to be 15 nmol/g, which corresponds to approximately 1,000 μmol in an adult human.[32] PLP in the muscle pool represents 75% to 80% of total body vitamin B_6. Muscle PLP is largely bound in coenzyme form bound to glycogen phosphorylase.[31]

Tissue uptake of vitamin B_6 from circulation requires dephosphorylation. After enzymatic removal of the 5′-phosphate group by plasma membrane TNAP, vitamin B_6 can cross cellular membranes by a carrier-mediated transport system.[12,13,33] Vitamin B_6 in tissues is retained by phosphorylation and concentrated to the mitochondria and cytoplasm.

Functions

Vitamin B_6 functions as a coenzyme in various enzymatic reactions in the metabolism of amino acids, one-carbon units, lipids, and the pathways of gluconeogenesis, heme, and neurotransmitter biosynthesis. PLP is the most common vitamin B_6 coenzyme. The structures of PLP and PMP, and their respective nonphosphorylated forms, accommodate the formation of Schiff base linkages with other amines (PL/PLP) and aldehydes (PM/PMP). The structures of these vitamers make them well suited to serve as coenzymes for more than 160 different enzymes.

Amino Acids

Nearly all amino acids require at least one PLP-dependent enzyme in their metabolism. PLP is a coenzyme for aminotransaminases that catalyze reversible conversions of amino acids to their corresponding α-keto acids with simultaneous transfer of the amino group to yield PMP. Amino acids can also be modified by PLP-dependent decarboxylation and desulfuration reactions. Metabolism of several amino acids involved in one-carbon metabolism are catalyzed by PLP-dependent reactions, as described later. PLP-dependent decarboxylation reactions are important in the biosynthesis of neurotransmitters (γ-aminobutyric acid, dopamine, and norepinephrine), including the conversion of L-aromatic amino acids to active neurotransmitters (e.g., tryptophan conversion to serotonin).[34]

One-Carbon Units

PLP is a coenzyme for four enzymes in one-carbon metabolism and transsulfuration. Serine hydroxymethyltransferase isoforms (SHMT1 and SHMT2) reversibly transfer a one-carbon unit from serine to tetrahydrofolate to form 5,10-methylenetetrahydrofolate and glycine, and glycine decarboxylase transfers one-carbon units to tetrahydrofolate from glycine (**Figure 24.3**, reactions 1 and 2). These enzymatic reactions provide most of the one-carbon groups used for purine and thymidine synthesis, as well as the methyl group for remethylation of homocysteine to methionine.[35,36] The methyl group coupled with S-adenosylhomocysteine to form S-adenosylmethionine (SAM) may be used for SAM-dependent transmethylation reactions involved in the metabolism of creatine, DNA, RNA, lipids, proteins, and other molecules. The transsulfuration pathway is comprised of the PLP-dependent enzymes cystathionine β-synthase and cystathionine γ-lyase (see Figure 24.3, reactions 3 and 4). In this pathway, homocysteine is condensed with serine to produce

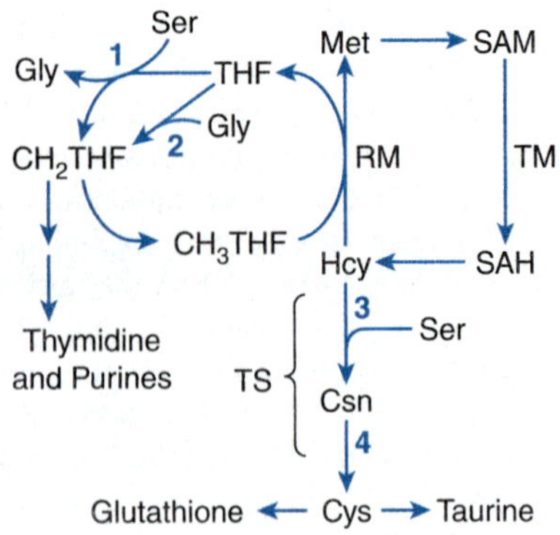

Figure 24.3 Pyridoxal-5′-phosphate (PLP) dependence of homocysteine (Hcy) and other one-carbon cycle reactions: *(1)* serine (Ser) hydroxymethyltransferase; *(2)* glycine (Gly) decarboxylase of the glycine cleavage system; *(3)* cystathionine (Csn) β-synthase; *(4)* cystathione γ-lyase. CH_2THF, 5,10-methylenetetrahydrofolate; CH_3THF, 5-methyltetrahydrofolate; *Cys*, cysteine; Met, methionine; RM, remethylation; SAH, *S*-adenosylhomocysteine; SAM, *S*-adenosylmethionine; TM, transmethylation; TS, transsulfuration.

cystathionine, which is subsequently cleaved to produce cysteine and α-aminobutyrate. Transsulfuration and generation of one-carbon units are impaired in severely vitamin B_6-deficient rats.[37,38] Elevated concentrations of glycine and cystathionine in plasma from young adult humans with low vitamin B_6 status suggest similar effects in humans during moderate vitamin B_6 restriction. Cysteine concentration in plasma and red blood cells and the rate of stable, isotopically determined generation of one-carbon units were not significantly affected by marginal vitamin B_6 status (20 to 30 nmol PLP/L) in healthy young adults, however.[39-42] The level of vitamin B_6 deficiency that is required to affect the function of these pathways in humans remains unclear.

Lipids

The role of vitamin B_6 in lipid metabolism is not clearly defined. Vitamin B_6–deficient rats had altered fatty acid profiles of tissue lipids, with reduced concentrations of arachidonic acid and increased linoleic acid,[43] whereas plasma cholesterol and triglyceride concentrations were increased.[43] Similarly, men and women fed a short-term vitamin B_6-restricted diet exhibited a reduction in plasma n-3 and n-6 fatty acids and a small increase in their n-6/n-3 ratio.[44] The biochemical mechanisms involved are uncertain; however, these observations may be explained by aberrations in the PLP-dependent enzymatic pathways involved in the transfer of methyl groups, which could yield lower concentrations of methylated phospholipids. The biosynthetic pathway of carnitine, which is essential for intramitochondrial transport of long-chain fatty acids, requires the activity of PLP-dependent 3-hydroxytrimethyllysine aldolase. Vitamin B_6 deficiency was shown to reduce plasma carnitine concentrations in rats,[45] but not in humans.[46]

Glycogenolysis and Gluconeogenesis

Vitamin B_6, as PLP, plays a dual role in the synthesis of glucose. Glycogen phosphorylase relies on PLP as a coenzyme in the enzymatic cleavage of glycogen that sequentially releases glucose-1-phosphate units. The 5′-phosphate group of PLP, rather than the 4′-formyl group (as in the aminotransferase reactions), is required for general acid catalysis in glycogenolysis. Both glycogen phosphorylase activity and muscle concentrations of vitamin B_6 are resilient to effects of dietary B_6 restriction.[47,48] PLP-dependent transaminases convert gluconeogenic amino acids to α-keto acids to create substrates to produce glucose. The effect of vitamin B_6 status on gluconeogenesis remains unclear.

Heme Biosynthesis

Heme biosynthesis relies on the activity of PLP-dependent δ-aminolevulinate synthase (ALAS). This enzyme catalyzes the condensation of succinyl coenzyme A and glycine to form δ-aminolevulinate, which is the precursor for the porphyrin ring. Chronic vitamin B_6 deficiency can precipitate microcytic hypochromic anemia in which the hemoglobin concentration of erythrocytes is reduced. Sideroblastic anemia is an inherited form of ALAS deficiency. This type of anemia often can be successfully treated with PN supplementation; however, some mutations alter the PLP binding sites on the ALAS enzyme[49] and thus diminish the efficacy of PN supplementation.

Interactions with Other Nutrients

The interconversion and metabolism of vitamin B_6 depend on riboflavin, niacin, and zinc (see Figure 24.2). Both pyridoxine (pyridoxamine) phosphate oxidase and aldehyde oxidase require riboflavin in the forms of FMN and FAD, respectively. Niacin, as NAD, serves as coenzyme for aldehyde dehydrogenase. The phosphorylation of vitamin B_6 is catalyzed by pyridoxal kinase, which requires zinc as cofactor. Insufficient dietary intake of these nutrients may adversely affect the metabolic utilization of vitamin B_6.

Niacin, folate, and carnitine require vitamin B_6 for their biosynthesis and metabolism. Biosynthesis of niacin from tryptophan is catalyzed by PLP-dependent kynureninase via the kynurenine pathway. As discussed earlier, PLP-dependent SHMT and glycine decarboxylase are essential for normal folate metabolism (see Figure 24.3, reactions 1 and 2). Carnitine is synthesized from lysine and methionine in a multistep process that requires PLP.

Vitamin B_6 in Foods and Supplements

Vitamin B_6 traditionally has been measured in foods and biologic materials by microbiologic assays using the yeast *Saccharomyces uvarum*.[50] A variety of high-performance liquid chromatographic (HPLC) and liquid chromatography-tandem mass spectrometry (LC-MS/MS) approaches now have largely replaced microbiologic assays for the determination of the various forms of vitamin B_6, including the glycosylated form of PN.[50-52]

Vitamin B_6 is widely distributed throughout the food supply. Foods of animal origin such as meat, fish, eggs, and, to a lesser extent, dairy products, are rich in vitamin B_6, mostly as PL and PM and their phosphorylated forms. Many vegetables and whole grain cereal products are good sources of the vitamin. The predominant forms of vitamin B_6 in foods of plant origin are PN and glycosylated PN.[11] Plant tissues may contain up to 75% of their vitamin B_6 as PNG,[52-54] which is presumed to be a storage form of the vitamin. According to the USDA Continuing Survey of Food Intakes by Individuals,[55] the adult U.S. population obtains most of its dietary vitamin B_6 from ready-to-eat fortified cereals, meat, fish, poultry, starchy vegetables, and noncitrus fruits. There is no added vitamin B_6 in enriched grain products. PN is currently used in the fortification of many breakfast cereals, meal replacement bars and many energy drinks. Analyses of U.S.D.A food composition data and food intake patterns from NHANES 2003–2006 have indicated the important role of such fortified food sources and use of dietary supplements in the total vitamin B_6 intake in the United States.[56]

Relatively little loss of vitamin B_6 occurs during food storage and handling, aside from losses occurring during the milling of grain, but losses of PL, PLP, PM, and PMP during cooking and commercial thermal processing of foods can be significant.[57] The nutritional impact of vitamin B_6 loss was acutely observed in the 1950s when infants were fed formula for which a change in thermal processing conditions led to excessive destruction of vitamin B_6 in the unfortified formula.[58] Certain infants consuming this formula developed convulsive seizures that were alleviated by PN supplementation. These incidents prompted the routine fortification of infant formulas with PN. Because of its greater stability than the other B_6 vitamers, PN hydrochloride is used in all types of food fortification and in most nutritional supplements. PN α-ketoglutarate also has been used as a vitamin B_6 supplement to enhance exercise performance; however, the evidence supporting this benefit is equivocal.[59] Several studies have suggested that high-dose supplemental pyridoxamine (PM) may reduce oxidative damage in conditions such as obesity, heart and other vascular disease, and renal failure, although confirmation is needed.

Assessment of Nutritional Status

Vitamin B_6 nutritional status can be evaluated using direct measurement of vitamin B_6 forms in blood or urine or by indicators based on the biochemical function of vitamin B_6.[60-61] The various indicators of vitamin B_6 status and their generally accepted cutoff values are summarized in **Table 24.1**. The reader is referred to a review of direct and functional methods for the assessment of vitamin B_6 nutritional status.[61]

The most used direct measure of vitamin B_6 status is the concentration of PLP in plasma, which can be readily measured by HPLC or LC-MS/MS methods. Plasma PLP concentration has been shown to correlate with tissue PLP concentration in rats[62] and with vitamin B_6 intake in controlled dietary studies with humans.[11,54] A plasma PLP concentration of more than 30 nmol/L traditionally has been considered to indicate adequate status in adult humans,[11,54] whereas a concentration higher than 20 nmol/L has been used as a more conservative cutoff value.[53] If the 20-nmol/L cutoff is used, we recommend that values in the 20- to 30-nmol/L range be interpreted as indicative of marginal status.

Certain genetic and physiologic conditions influence plasma PLP values[61]; thus, conclusions regarding vitamin B_6 status based on plasma PLP should be considered presumptive until confirmed by an alternate method of status assessment. Inflammatory conditions are associated with a reduction of plasma PLP that is independent of vitamin B_6 nutritional status.[63] This phenomenon has been hypothesized to be due to mobilization of PLP to sites

Table 24.1 Indices for Assessment of Vitamin B_6 Status and Suggested Minimal Values for Adequate Status

Index	Adequate Status
Direct	
Plasma pyridoxal phosphate	>30 nmol/L[a]
Plasma total vitamin B_6	>40 nmol/L
Urinary 4-pyridoxic acid	>3 μmol/d
Urinary total vitamin B_6	>0.5 μmol/d
Indirect	
Erythrocyte alanine aminotransferase coenzyme stimulation index[b]	<1.25
Erythrocyte aspartate aminotransferase coenzyme stimulation index[b]	<1.80
2-g L-Tryptophan load, urinary xanthurenic acid	<65 μmol/d
3-g L-Methionine load, urinary cystathionine	<350 μmol/d
Dietary Intake	
Vitamin B_6 intake, weekly average	>1.25–1.5 mg/d
Vitamin B_6/protein ratio	≥0.016 mg/g

[a]A plasma pyridoxal phosphate value lower than 20 nmol/L is considered indicative of deficiency.

[b]The coenzyme stimulation index is the ratio of enzyme activity values measured with and without preincubation of the erythrocyte hemolysate with added pyridoxal phosphate. This index is proportional to the fraction of enzyme in apoenzyme form.

Reproduced from Leklem JE. J Nutr 1990;120(Suppl):S1503–7.

of inflammation[64] and/or greater catabolism of PLP to 4-PA.[65] Erythrocyte PLP is a potential indicator of vitamin B_6 status, but consensus is lacking regarding its diagnostic utility, due largely to unresolved methodologic issues.

Urinary 4-PA excretion (>3 μmol/day) is indicative of vitamin B_6 adequacy.[60] Whereas urinary 4-PA is easily determined by HPLC, it must be considered a secondary status indicator because a complete 24-hour urine collection is usually required, and 4-PA excretion is strongly affected by recent vitamin B_6 intake. The urinary 4-PA/creatinine ratio of spot urine[66] samples may be an alternative to a 24h collection, but this indicator has not been fully validated.

The measurement of 4-PA in plasma relative to plasma PLP and PL (i.e., 4-PA/(PL+PLP)), designated the PA ratio (PAr), constitutes a biomarker of systemic inflammation based on the acceleration of vitamin B_6 catabolism during inflammation.[67] An example of the utility of the PAr was the fact that it provided greater efficacy than several other risk factors and inflammatory markers in predicting long-term mortality risk in a large-scale study of cardiovascular disease patients.[68]

Functional indicators of vitamin B_6 status are based on measures of PLP-dependent processes either in vivo or in blood cells. The urinary excretion of xanthurenic acid, either basal or following a tryptophan load, was the first functional indicator. The tryptophan load test has not been widely used since an incident of toxicity occurred in the 1980s resulting from the presence of a toxic impurity in a batch of tryptophan. The methionine load test involves the measurement of the rise in plasma homocysteine following an oral methionine load.[69] Vitamin B_6 deficiency has little effect on fasting plasma homocysteine concentration (unlike folate deficiency), but it yields a higher postmethionine load homocysteine concentration because of impairment of the transsulfuration pathway. The many variations in the protocol of methionine load tests (i.e., doses and blood sampling times) complicate interpretation and comparison of published findings. As mentioned earlier, plasma cystathionine concentration is a sensitive and very responsive biomarker of insufficient vitamin B_6 status[70-72]; however, reference ranges have not yet been established. If using elevated cystathionine as an indicator of vitamin B_6 deficiency, it is important to rule out deficiency of folate and B_{12}, which also cause elevated cystathionine. Finally, the in vitro measurement of erythrocyte aspartate aminotransferase or alanine aminotransferase in the presence and absence of added PLP allows calculation of an activation coefficient, which is an indirect measure of the degree of deficiency through assessment of the proportion of the enzyme in apoenzyme form.

Tryptophan metabolism through the kynurenine pathway has two PLP-dependent enzymes, each of which catalyze two reactions. Kynureninase can catalyze the conversion of kynurenine to anthranilic acid (AA) and 3-hydroxykynurenine (HK) to 3-hydroxyanthranilic acid (HAA). In addition, kynurenine aminotransferase catalyzes the conversion of kynurenine to kynurenic acid (KA) and HK to xanthurenic acid (XA). The development of sensitive and specific LC-MS/MS methods for the measurement of multiple constituents of the kynurenine pathway has extended our view of these constituents and their substrate/product ratios as potential biomarkers. Because kynureninase is more sensitive to cellular PLP insufficiency than kynurenine aminotransferase, evaluation of individual kynurenines and their ratios provides diagnostic evidence of vitamin B_6 insufficiency. HK increases nonlinearly with decreasing plasma PLP, which provides a functional biomarker of vitamin B_6 deficiency that does not require a tryptophan load.[73] The substrate/product ratios HK/XA and HK/HAA showed better ability to discriminate vitamin B_6 insufficiency indicated by low plasma PLP and less effect of inflammation and other confounders.[74] This approach has been extended by the evaluation of a substrate/product ratio HK/(KA+XA+HAA+AA), designated as the HK ratio (HKr) that provides an inverse marker of cellular vitamin B_6 status.[75]

Table 24.2 Recommended Vitamin B_6 Intakes by Age and Gender (mg/d)[a]

	Age (y)	Male	Female	Tolerable Upper Intake Level[b]
Infants	0–0.5	0.1[c]	0.1[c]	ND
	0.5–1	0.3[c]	0.3[c]	ND
Children	1–3	0.5	0.5	30
	4–8	0.6	0.6	40
Adolescents	9–13	1.0	1.0	60
	14–18	1.3	1.2	80
Adults	19–30	1.3	1.3	100
	31–50	1.3	1.3	100
	51–70	1.7	1.5	100
	>70	1.7	1.5	100
Pregnancy			1.9	100
Lactation			2.0	100

[a]1 mg of pyridoxine = 5.92 μmol.
[b]The tolerable upper intake level (UL) is the highest level of daily nutrient intake that is likely to pose no risk of adverse health effects to almost all individuals in the general population. The ULs for vitamin B_6 are the same for males and females.
[c]Adequate intakes; all other values are presented as recommended dietary allowances.
ND, Not determinable because of a lack of data on adverse effects in this age group and concern with regard to a lack of ability to handle excess amounts. The source of intake should be from food only, to prevent high levels of intake.

Data from Food and Nutrition Board, Institute of Medicine. Dietary Reference Intakes for Thiamin, Riboflavin, Niacin, Vitamin B_6, Folate, Vitamin B_{12}, Pantothenic Acid, Biotin, and Choline. Washington, DC: National Academy Press, 1998.

Requirements

The Food and Nutrition Board[55] revised the dietary reference intakes for vitamin B_6 in 1998 (**Table 24.2**). An estimated average requirement (EAR) of 1.1 mg/day and a recommended dietary allowance (RDA) of 1.3 mg/day for men and women (19 to 50 years old) were established as a result of this reassessment. These RDAs were reduced from the 1989 recommendations. To determine the Estimated Average Requirement and RDA values, plasma PLP concentration (<20 nmol/L) was used as the major vitamin B_6 status indicator because it best represents tissue stores.[55] As reviewed in the report,[55] vitamin B_6 requirements were based on data from controlled investigations examining dietary vitamin B_6 intake in combination with synthetic PN. Average vitamin B_6 intake, as estimated by nationally representative nutrient intake surveys in the United States, is approximately 1.5 mg/day for women and 2 mg/day for men. Although several studies showed that indicators of vitamin B_6 status declined with increasing protein consumption, this effect has not been consistently observed.[55] Thus, the current RDA is not expressed as a function of protein intake.[55] Some controversy exists regarding the RDA for vitamin B_6, especially considering findings suggesting that an optimal intake is greater than the current RDA of 1.3 mg/day[76,77] which, in some populations, is not enough to sustain circulating PLP levels in the adequate range.[78] Further research is required to answer this question. It is noteworthy that, in 2016, the European Food Safety Authority (EFSA) reported Population Reference Values for vitamin B_6 of 1.7 mg and 1.6 mg for adult men and women, respectively.[79]

Limited data exist regarding the requirements of infants and children for vitamin B_6. The Adequate Intake (AI) established for infants up to 11 months of age was primarily derived from the vitamin B_6 content of human milk (0.13 to 0.24 mg/L) and milk intake from healthy, exclusively breastfed infants.[55] Although insufficient dietary intake and consequent inadequate nutritional status of vitamin B_6 are not prevalent in the general population, significant portions of the population may be at risk for suboptimal vitamin B_6 intake and status. These groups include pregnant and lactating women,[17,80] smokers,[81,82] and elderly persons.[78] Women taking oral contraceptives (OC) have been reported to be at risk of vitamin B_6 insufficiency,[78,82] but any such effect on the vitamin B_6 requirement has been suggested to be modest.[83] Furthermore, studies of OC users and nonusers employing targeted metabolite profiling and functional biomarkers of vitamin B_6 status including plasma cystathionine, plasma HK, and the ratios HK/XA and HK/HAA, suggested that OC use had little aggrevating effect on vitamin B_6 nutritional status.[84] In addition, OC use did not interfere with the metabolic effects of PN supplementation as assessed by targeted metabolomics, functional biomarkers, and in vivo kinetic analysis.[85] Plasma PLP was modestly lower in OC users in a large observational study of Irish young adults, but PLP concentrations were in the adequate range, and kynurenine metabolite patterns were not indicative of vitamin B_6 deficiency.[86]

Vitamin B_6 in Health and Disease

The relationship between vitamin B_6 nutrition and vascular health and disease remains an active area of research. An increased risk of cardiovascular disease was first reported in animal studies, when vitamin B_6 deficiency led to the development of vascular lesions.[87,88] Elevated plasma homocysteine is associated with increased risk of stroke, coronary heart disease, and venous thrombosis.[89] Plasma homocysteine primarily is elevated in fasting conditions when vitamin B_6 intake and plasma PLP concentrations are very low.[90]

Several epidemiologic studies have reported an association between low plasma PLP and cardiovascular disease[91-93] that was independent of plasma homocysteine[94-96] and of plasma C-reactive protein (CRP).[63,91] This finding is not unexpected because the association of fasting plasma total homocysteine concentration with vitamin B_6 status is inconsistent and often weak in free-living subjects and in studies of vitamin B_6 depletion and supplementation.[96]

The association of low vitamin B_6 status with vascular disease (VD) tends to be strongest in studies involving populations already diagnosed with VD. This finding suggests an effect of VD progression on vitamin B_6 status, rather than an effect of vitamin B_6 status on VD risk. Intervention studies have largely focused on the plasma homocysteine-lowering action of B vitamins and the prevention of secondary incidence of vascular events. The results of these studies do not support the use of B vitamins to prevent recurrence of cardiovascular disease.[97-101] Secondary analysis of two trials, however, showed that B-vitamin supplementation lowered the incidence of stroke.[102,103] Concurrent folate and vitamin B_{12} supplementation complicated assessment of the effects of vitamin B_6 alone in these studies. Nevertheless, large-scale randomized trials that attempted to examine the effect of vitamin B_6 supplementation on secondary prevention have yielded largely negative results.[104]

Several hypotheses may explain how vitamin B_6 deficiency affects VD. In addition to effects on multiple pathways of homocysteine metabolism,[96] vitamin B_6 status may affect VD through effects on lipid metabolism,[44,105] endothelial function,[106] thrombogenesis,[105] and inflammation.[63] Low plasma PLP or low vitamin B_6 intake has been associated with elevated plasma CRP, an indicator of inflammation.[63,107,108] In addition, controlled dietary vitamin B_6 restriction did not affect plasma CRP in healthy young subjects.[70] Investigators have suggested that the requirement for vitamin B_6 increases in inflammation.[107] As noted previously, inflammation leads to increased vitamin B_6 catabolism. Plasma cystathionine was associated with increased risk of acute myocardial infarction[109] and mortality,[110] which suggests potential roles of altered transsulfuration associated with PLP availability associated with inadequate vitamin B_6 intake, catabolism due to inflammation, and oxidative stress.

The importance of adequate vitamin B_6 status for proper immune function, particularly cell-mediated and, to a lesser degree, humoral immunity, has been known for many years. Lymphoid tissue atrophy, reduced lymphocyte content of lymph tissues, extended allograft survival, and depressed antibody production, as well as reductions in lymphocyte proliferation, phagocytic activity of macrophages, and T-cell–mediated cytotoxicity in vitro, are all characteristic of vitamin B_6-deficient animals.[111] In human studies, lymphocytes isolated from vitamin B_6–deficient subjects display reduced proliferation and reduced interleukin-2 production in response to mitogens,[77,112] in addition to reduced antibody production in response to immunization.[111] Supplementation of vitamin B_6 has been reported to ameliorate immune responses in inflammatory conditions.[113] Critically ill patients supplemented with high doses of vitamin B_6 showed improved cell-mediated immune responses after 14 days.[114] Depressed lymphocyte proliferation in vitamin B_6 deficiency may stem from impaired DNA synthesis resulting from reduced activity of SHMT, a key enzyme in de novo synthesis of purines and thymidine (see Figure 24.3, reaction 1). The immune systems of elderly persons appear particularly sensitive to inadequate vitamin B_6 status,[112] and this population may benefit from vitamin B_6 supplementation.[115] Vitamin B_6 depletion-repletion studies in young and old subjects suggest that intake of vitamin B_6 equal to the current RDA may be insufficient to maximize immunocompetence.[77,112]

A role for vitamin B_6 in cancer was suggested by perturbations of vitamin B_6 metabolism observed in patients with breast cancer and within tumors.[116] The level of exposure to vitamin B_6 is inversely proportional to cell proliferation in experimental models of cancer, as reviewed in Komatsu et al.[116] Several epidemiologic studies reported an inverse relationship between vitamin B_6 intake or plasma PLP and colorectal cancer risk.[117] This protective effect was suggested to be independent of effects of vitamin B_6 status on one-carbon metabolites and inflammatory biomarkers.[118] Higher levels of plasma PLP have also been associated with lower risk of lung cancer.[119,120] This association has been suggested in other cancers as well, but the results are inconsistent.[121-123] Few clinical trial data are available to assess causality for low vitamin B_6 status in cancer. Potential protective effects of vitamin B_6 include modulation of steroid hormone action,[124] maintenance of one-carbon metabolism, and maintenance of immune function.[125] The PAr index is associated with risk of lung cancer, which suggests that greater vitamin B_6 catabolism during inflammation and immune activation may be at least partly responsible for the association of plasma PLP and lung cancer risk and potentially other types of cancer risk.[126]

The nervous system relies on certain PLP-dependent enzymes for neurotransmitter synthesis, as explained earlier. Serotonin production and γ-aminobutyric acid production are particularly sensitive to vitamin B_6 status in rats, and this sensitivity may account for changes in thyroid hormone levels and seizure activity observed in vitamin B_6–deficient animals.[34] In addition, these studies suggest a strong role for vitamin B_6 in cognitive development. PN-dependent seizure is a rare inherited condition in humans that begins before or within days of birth, ceases immediately on intravenous PN administration, and is controllable through daily PN supplementation (~0.2 to 3 mg/kg body weight).[127] The disease is caused by mutations in the gene encoding for α-aminoadipic semialdehyde dehydrogenase.[128] Loss of activity of this dehydrogenase results in accumulation of piperideine-6-carboxylate, which condenses with and inactivates PLP.[128] The resulting reduction in PLP-dependent glutamate decarboxylase activity is presumed to cause the seizures.[129] As discussed earlier, seizures also occurred in infants consuming infant formula deficient in vitamin B_6.[58] Abnormal brain wave activity was observed in these infants, as well as in some adults examined in vitamin B_6 deficiency studies.[130] Attempts to correct suspected neurotransmitter abnormalities with PN treatment in conditions such as headache, chronic pain, behavioral disorders, depression, autism, Down syndrome, schizophrenia, and various neuropathies have met with limited success.[131] Although the positive association between plasma total homocysteine and cognitive decline, dementia, and Alzheimer's disease in elderly persons implies a potential link between vitamin B_6 status and these conditions, few studies have found any association between vitamin B_6 status and cognitive function.[132-134] Both low plasma PLP and low dietary vitamin B_6 intake have been associated with higher likelihood of depressive symptoms.[135,136]

Multiple links exist between vitamin B_6 status and diabetes.[137] The roles of vitamin B_6 in gluconeogenesis and glycogenolysis are described in the earlier section on vitamin B_6 function. Low concentrations of plasma B_6 vitamers observed in type 1 or type 2 diabetes may be related to increased plasma alkaline phosphatase activity or the acute suppressive effect of an oral glucose load on plasma PLP concentration. Supraphysiologic concentrations of PLP inhibit advanced glycation reactions in vitro,[138] but the practical relevance of this observation is unclear. Similarly, high-dose vitamin B_6 administration markedly inhibited accumulation of advanced glycation end products in streptozotocin-induced diabetic rats[139] and improved diabetic neuropathy.[140] Claims regarding PM treatment have not been suitably evaluated clinically. Biochemical and functional indices indicate that declining vitamin B_6 status occurs with aging in both animals and humans.[141,142] Although the cause of these observations is uncertain, reduced dietary intake, impaired renal function, and effects of inflammation and the acute phase response on vitamin B_6 metabolism are possible contributors.[142] These indices, as well as immune function, are improved by PN supplementation.[115,143] In controlled nutritional studies, vitamin B_6 intakes needed to restore biochemical, functional, and immunologic indices of B_6 status in elderly persons to levels considered normal for younger populations were greater than the current RDA for this age group.[122,144]

Vitamin B_6 status is also perturbed in individuals with impaired renal function.[145,146] Low vitamin B_6 status is associated with elevated post–methionine load homocysteine concentration in renal transplant recipients, which can be ameliorated with supplemental PN.[147] Hemodialysis-induced peripheral neuropathy and sensory abnormalities also respond to supplemental PN.[148] Vitamin B_6 deficiency and hyperhomocysteinemia, which are independent risk factors for VD,[94] are prevalent in patients with renal disease and are coincident with elevated VD risk in this population.[146] Patients with renal disease who were supplemented with B vitamins had lower plasma homocysteine concentration than patients treated with a placebo, but the treatment did not improve survival or reduce the incidence of cardiovascular events.[149]

Vitamin B_6 metabolism is perturbed in patients with rheumatoid arthritis.[150] Patients have significantly lower plasma PLP concentration,[151] even at similar vitamin B_6 intake.[152] Depletion of vitamin B_6 was shown to be tissue specific in a rodent model of arthritis, a finding that implies redistribution of PLP to tissues that may have a higher demand for the coenzyme.[153] Supplementation with 100 mg PN/day, but not at 50 mg PN/day, suppressed production of proinflammatory cytokines in patients with rheumatoid arthritis.[113,154]

Pharmacologic Pyridoxine Therapy and Pyridoxine Toxicity

In addition to the uses of supplemental PN mentioned earlier, congenital homocystinuria has been successfully treated with 250 to 500 mg PN/day.[155] PN also improves hematopoiesis in patients with specific forms of sideroblastic anemia.[156] Pharmacologic doses of vitamin B_6 have also been used, with little proof of efficacy, to alleviate the symptoms of dysmenorrhea, morning sickness, asthma, carpal tunnel syndrome, and hyperoxaluria, among other conditions.[157] Those studies that showed benefit of supplemental PN often were small and poorly controlled. Trials of PN therapy for premenstrual syndrome, the treatment that may represent the most frequent use of large PN doses, yielded equivocal results.[158]

Although efficacy is questionable in many cases, PN treatment at pharmacologic doses continues to be used, either prescribed or self-administered, as stand-alone or adjunct therapy for many of the conditions mentioned earlier. Persistent PN use is based, in part, on the low perceived toxicity of PN supplementation compared with traditional medical approaches. However, long-term intake of pharmacologic doses of PN (>500 mg/day) is associated with a risk of sensory neuropathies, which are reversed on withdrawal of the PN supplements.[55] The Food and Nutrition Board of the Institute of Medicine established a tolerable upper intake level of vitamin B_6 at 100 mg/day, to prevent neuropathy.[55] This quantity of vitamin B_6 intake is not approachable by dietary means other than supplementation.

Vitamin B_6–Drug Interactions

Certain drugs, including cycloserine, hydralazine, phenelzine, gentamycin, penicillamine, isoniazid, and L-dopa, antagonize vitamin B_6 status by covalently binding to the carbonyl group of PLP or PL; this process reduces the availability of the PLP coenzyme.[159] The asthma drug theophylline interferes with PLP production by inhibition of PL kinase.[160] Vitamin B_6 status is usually recovered through PN supplementation without reducing drug efficacy.[159] PN supplementation was previously contraindicated in patients treated with L-dopa for Parkinson's disease because PN enhances peripheral metabolism of the drug.[157] Concurrent treatment with a peripheral decarboxylase inhibitor can be used to preserve the effectiveness of L-dopa during PN supplementation, however. Alcohol intake antagonizes vitamin B_6 status through production of acetaldehyde, which competes with PLP for binding sites of PLP-dependent enzymes.[161] Because chronic alcoholism likely increases vitamin B_6 catabolism through this mechanism, PN supplementation may be advisable for patients affected by this disease. Alcoholic liver disease is associated with reduced plasma PLP and elevated cystathionine.[162]

1-Amino-D-proline is a naturally occurring constituent of flax seed that exerts antivitamin B_6 activity by readily complexing with the carbonyl of PL and PLP,[163] which leads to the inhibition of several PLP-dependent enzymes. Incorporation of either purified 1-amino-D-proline or flax extract into the diets of moderately vitamin B_6-deficient rats led to reduced growth, depressed plasma levels of B_6 vitamers, and pronounced elevation of cystathionine.[164] Whether this phenomenon is relevant to humans consuming flax seed is presently unclear.

Acknowledgments

The author would like to acknowledge Vanessa R. Da Silva, Amy D. Mackey, and Steven R. Davis for contributions to the previous version of this chapter.

References

1. Da Silva V, Mackey AD, Davis SR, Gregory JF. Vitamin B_6. In: Ross AC, Caballero B, Cousins RJ et al., eds. Modern Nutrition in Health and Disease, 10th ed. () Baltimore: Lippincott Williams & Wilkins, 2014:341–350.
2. Gyorgy P. Nature 1934;133:448–9.
3. Lepkovsky S. Science 1938;87:169–70.
4. Keresztesy JC, Stevens JR. Proc Soc Exp Biol Med 1938;38:64–5.
5. Gyorgy P. J Am Chem Soc 1938;60:983–4.
6. Kuhn R, Wendt G. Ber Dtsch Chem Ges 1938;71B:780–2.
7. Ichiba A, Michi K. Sci Papers Inst Phys Chem Res 1938;34: 623–6.
8. Snell EE. Annu Rev Nutr 1989;9:1–19.
9. Gunsalus IC, Bellamy WD, Umbreit WW. J Biol Chem 1944;155:85–6.
10. American Institute of Nutrition. J Nutr 1990;120:12–9.
11. Leklem JE, Machlin LJ. *Vitamin B-6*. p 341, New York: Marcel Dekker, 1991.
12. Said HM. Annu Rev Physiol 2004;66:419–46.
13. Yamashiro T, Yasujima TY, Said HM, et al. J Biol Chem 2020;295:16998–17008.
14. Tarr JB, Tamura T, Stokstad ELR. Am J Clin Nutr 1981;34: 1328–37.
15. Roth-Maier DA, Kettler SI, Kirchgessner M. Int J Food Sci Nutr 2002;53:171–9.
16. Gregory JF III. Eur J Clin Nutr 1997;51:S43–8.
17. Andon MB, Reynolds RD, Moser-Veillon PB, et al. Am J Clin Nutr 1989;50:1050–8.
18. Ink SL, Gregory JF III, Sartain DB. J Agric Food Chem 1986;34:857–62.
19. Trumbo PR, Gregory JF III, Sartain DB. J Nutr 1988;118:170–5.
20. Nakano H, McMahon LG, Gregory JF III. J Nutr 1997;127: 1508–13.
21. Gregory JF III, Trumbo PR, Bailey LB, et al. J Nutr 1991;121: 177–86.
22. McMahon LG, Nakano H, Levy MD, et al. J Biol Chem 1997;272:320–25.
23. Mackey AD, Henderson GN, Gregory JF III. J Biol Chem 2002;277:26858–64.
24. Ink SL, Mehansho H, Henderson LM. J Biol Chem 1982;257: 4753–7.
25. Coburn SP, Mahuren JD, Kennedy MS, et al. Biofactors 1988;1:307–12.
26. Mehansho H, Henderson LM. J Biol Chem 1980;255:11901–7.
27. McCormick DB, Chen H. J Nutr 1999;129:325–7.
28. Merrill AH, Henderson JM. Ann N Y Acad Sci 1990;585: 110–117.
29. Van Hoof VO, De Broe ME. Crit Rev Clin Lab Sci 1994;31: 197–293.
30. Fonda ML. J Biol Chem 1992;267:159–78.
31. Coburn SP. Ann N Y Acad Sci 1990;585:76–85.
32. Coburn SP, Lewis DL, Fink WJ, et al. Am J Clin Nutr 1988;48: 291–4.
33. Zhang Z, Gregory JF III, McCormick DB. J Nutr 1993;123:85–9.
34. Dakshinamurti K, Paulose CS, Viswanathan M, et al. Ann N Y Acad Sci 1990;585:128–44.
35. Davis SR, Stacpoole PW, Williamson J, et al. Am J Physiol 2004;286:E272–9.
36. Lamers Y, Williamson J, Gilbert LR, et al. J Nutr 2007;137: 2647–52.
37. Lima CP, Davis SR, Mackey AD, et al. J Nutr 2006;136: 2141–7.
38. Scheer JB, Mackey AD, Gregory JF. J Nutr 2005;135:233–8.
39. Davis SR, Quinlivan EP, Shelnutt KP, et al. J Nutr 2005;135: 1045–50.
40. Davis SR, Scheer JB, Quinlivan EP, et al. Am J Clin Nutr 2005;81:648–55.
41. Lamers Y, O'Rourke B, Gilbert LR, et al. Am J Clin Nutr 2009;90:336–43.
42. Lamers Y, Williamson J, Theriaque DW, et al. J Nutr 2009;139: 666–71.
43. Bergami R, Maranesi M, Marchetti M, et al. Int J Vitam Nutr Res 1999;69:315–21.
44. Zhao M, Lamers Y, Ralat MA, et al. J Nutr 2012;142:1791–7.
45. Cho YO, Leklem JE. J Nutr 1990;120:258–65.
46. Gregory JF, Park Y, Lamers Y, et al. PLOS One 2013;8:e63544.
47. Black AL, Guirard BM, Snell EE. J Nutr 1978;108:670–7.
48. Coburn SP, Ziegler PJ, Costill DL, et al. Am J Clin Nutr 1991;53:1436–42.
49. Hurford MT, Marshall-Taylor C, Vicki SL, et al. Clin Chim Acta 2002;321:49–53.
50. Gregory JF. J Food Comp Anal 1988;1:105–23.
51. Midttun O, Hustad S, Ueland PM. Rapid Commun Mass Spectrom 2009;23:1371–9.
52. Bachman T, Maurer A, Rychlik M. Anal Bioanal Chem 2020;412:7237–52.
53. Gregory JF, Sartain DB. J Agric Food Chem 1991;39:899–905.
54. Kabir H, Leklem JE, Miller LT. J Food Sci 1983;48:422–5.
55. Food and Nutrition Board, Institute of Medicine. Dietary Reference Intakes for Thiamin, Riboflavin, Niacin, Vitamin B_6, Folate, Vitamin B12, Pantothenic Acid, Biotin, and Choline. Washington, DC: National Academy Press, 1998:592.
56. Fulgoni VL III, Keast DR, Bailey RL, et al. J Nutr 2011;141: 1847–54.
57. Gregory JF. Vitamins. In: Fennema's Food Chemistry, 5th ed. Damodaran S, Parkin KL, eds. Boca Raton, FL: Taylor and Francis/ CRC Press, 2017:439–522.

58. Coursin DB. JAMA 1954;154:406–8.
59. Gregory JF. Vitamin B_6 Deficiency. In: Carmel R, Jacobsen DW, eds. Homocysteine in Health and Disease, () Cambridge, United Kingdom: Cambridge University Press, 2001:307–320.
60. Leklem JE. J Nutr 1990;120:S1503–7.
61. Ueland PM, Ulvik A, Rios-Avila L, et al. Annu Rev Nutr 2015;35:33–70.
62. Lumeng L, Ryan MP, Li TK. J Nutr 1978;108:545–53.
63. Friso S, Jacques PF, Wilson PW, et al. Circulation 2001;103: 2788–91.
64. Paul L, Ueland PM, Selhub J. Nutr Rev 2013;71:239–41.
65. Ueland PM, McCann A, Midttun O, et al. Molec Aspects Med 2017;53:10–27.
66. Schuster K, Bailey LB, Cerda JJ, et al. Am J Clin Nut 1984;39: 466–470.
67. Ulvik A, Midttun O, Pedersen ER, et al. Am J Clin Nutr 2014;100:250–5.
68. Ulvik A, Pedersen ER, Svingen GFT, et al. Am J Clin Nutr 2016;103:1417–25.
69. Ubbink JB, van der Merwe A, Delport R, et al. J Clin Invest 1996;98:177–84.
70. Davis SR, Quinlivan EP, Stacpoole PW, et al. J Nutr 2006;136: 373–8.
71. Lamers Y, Williamson J, Ralat M, et al. J Nutr 2009;139:452–60.
72. Stabler SP, Sampson DA, Wang L-P, et al. J Nutr Biochem 1997;8:279–89.
73. Middtun O, Ulvik A, Pedersen ER, et al. J Nutr 2011;141:611–7.
74. Ulvik A, Theofylaktopoulou D, Midttun O, et al. Am J Clin Nutr 2013;98:934–40.
75. Ulvik A, Midttun O, McCann A, et al. Am J Clin Nutr 2020;111:178–186.
76. Hansen CM, Shultz TD, Kwak HK, et al. J Nutr 2001;131: 1777–86.
77. Kwak HK, Hansen CM, Leklem JE, et al. J Nutr 2002;132: 330–8.
78. Morris MS, Picciano MF, Jacques PF, et al. Am J Clin Nutr 2008;87:1446–54.
79. EFSA NDA Panel (EFSA Panel on Dietetic Products, Nutrition and Allergies), Scientific opinion on Dietary Reference Values for vitamin B6. EFSA J 2016;14:4485.
80. Reynolds RD, Polansky M, Moser PB. J Am Diet Assoc 1984;84:1339–44.
81. Ulvik A, Ebbing M, Hustad S, et al. Clin Chem 2010;56: 755–63.
82. Lussana F, Zighetti ML, Bucciarelli P, et al. Thromb Res 2003;112:37–41.
83. Leklem JE, Brown RR, Rose DP, et al. Am J Clin Nutr 1975;28: 535–41.
84. Rios-Avila L, Coats B, Chi Y-Y, et al. J Nutr 2015,145:87–95.
85. Rios-Avila L, Coats B, Ralat M, et al. J Nutr 2015;102:616–25.
86. Deac OM, Mills JL, Shane B, et al. J Nutr 2015;145:701–7.
87. Rhinehart JF, Greenberg LD. Am J Pathol 1949;25:481–91.
88. Smolin LA, Crenshaw TD, Kurtycz D, et al. J Nutr 1983;113: 2122–33.
89. Clarke R, Smulders Y, Fowler B, et al. Semin Vasc Med 2005;5: 75–6.
90. Selhub J, Jacques PF, Wilson PWF, et al. JAMA 1993;270: 2693–8.
91. Cheng CH, Lin PT, Liaw YP, et al. Nutrition 2008;2:239–44.
92. Hron G, Lombardi R, Eichinger S, et al. Haematologica 2007;92:1250–3.
93. Page JH, Ma J, Chiuve SE, et al. Circulation 2009;120:649–55.
94. Kelly PJ, Shih VE, Kistler JP, et al. Stroke 2003;34:e51–4.
95. Vanuzzo D, Pilotto L, Lornbardi R, et al. Eur Heart J 2007;28: 484–91.
96. McKinley MC. Proc Nutr Soc 2000;59:221–37.
97. Albert CM, Cook NR, Gaziano JM, et al. JAMA 2008;299: 2027–36.
98. Galan P, Kesse-Guyot E, Czernichow S, et al. BMJ 2010;341:c6273.
99. Hankey GJ, Eikelboom JW, Baker RI, et al. Lancet Neurol 2010;9:855–65.
100. Lonn E, Yusuf S, Arnold MJ, et al. N Engl J Med 2006;354: 1567–77.
101. Toole JF, Malinow MR, Chambless LE, et al. JAMA 2004;291: 565–75.
102. Saposnik G, Ray JG, Sheridan P, et al. Stroke 2009;40:1365–72.
103. Spence JD, Bang H, Chambless LE, et al. Stroke 2005;36: 2404–9.
104. Ebbing M, Bonaa KH, Arnesen E, et al. J Intern Med 2010;268:367–82.
105. Brattstrom L, Stavenow L, Galvard H. Scand J Clin Lab Invest 1990;50:873–7.
106. Miner SE, Cole DE, Evrovski J, et al. J Heart Lung Transplant 2001;20:964–9.
107. Morris MS, Sakakeeny L, Jacques PF, et al. J Nutr 2010;140: 103–10.
108. Shen J, Lai CQ, Mattei J, et al. Am J Clin Nutr 2010;91: 337–42.
109. Dhar I, Svingen GFT, Pedersen ER, et al. Int J Cardiol 2018;266:24–30.
110. Dhar I, Lysne V, Svingen GFT, et al. Am J Clin Nutr 2019;109:1546–1554.
111. Chandra RK, Sudhakaran L. Ann N Y Acad Sci 1990;585: 404–23.
112. Meydani SN, Ribaya-Mercado JD, Russel RM, et al. Am J Clin Nutr 1991;53:1275–80.
113. Huang SC, Wei JC, Wu DJ, et al. Eur J Clin Nutr 2010;64: 1007–13.
114. Cheng CH, Chang SJ, Lee BJ, et al. Eur J Clin Nutr 2006;60: 1207–13.
115. Talbott MC, Miller LT, Kerkvliet NI. Am J Clin Nutr 1987;46: 659–64.
116. Komatsu S, Yanaka N, Matsubara K, et al. Biochim Biophys Acta 2003;1647:127–30.
117. Larsson SC, Orsini N, Wolk A. JAMA 2010;303:1077–83.
118. Lee JE, Li HJ, Giovannucci E, et al. Cancer Epidemiol Biomarkers Prev 2009;18:1197–1202.
119. Johansson M, Relton C, Ueland PM, et al. JAMA 2010;303: 2377–85.
120. Theofylaktopoulou D, Middtun O, Ueland PM, et al. Int J Cancer 2018;142:2425–2434.
121. Ames BN, Wakimoto P. Nat Rev Cancer 2002;2:694–704.
122. Gibson TM, Weinstein SJ, Mayne ST, et al. Cancer Cause Control 2010;21:1061–9.
123. Stevens VL, McCullough ML, Sun J, et al. Am J Clin Nutr 2010;91:1708–15.
124. Allgood AE, Cidlowski JA. J Biol Chem 1992;267:3819–24.
125. Brown RR, Leklem JE, Reynolds RD. In: Leklem J, Reynolds, RD. Possible Role for Vitamin B-6 in Cancer Prevention and Treatment. Clinical and Physiological Applications of Vitamin B-6, New York: Alan R. Liss, 1988:279–301.
126. Zuo H, Ueland PM, Midttun O, et al. Cancer Res 2017;78: 302–8.
127. Gupta VK, Mishra D, Mathur I, et al. J Paediatr Child Health 2001;37:592–6.
128. Mills PB, Struys E, Jakobs C, et al. Nature Med 2006;12:307–9.
129. Baxter P. Biochim Biophys Acta 2003;1647:36–41.
130. Kretsch MJ, Sauberlich HE, Newburn E. Am J Clin Nutr 1991;53:1266–74.
131. Pfeiffer SI, Norton J, Nelson L, et al. J Autism Dev Disord 1995;25:481–93.

132. Balk EM, Raman G, Tatsioni A, et al. Arch Intern Med 2007;167:21–30.
133. Brady CB, Gaziano JM, Cxypoliski RA, et al. Am J Kidney Dis 2009;54:440–9.
134. Ford AH, Flicker L, Alfonso H, et al. Neurology 2010;75: 1540–7.
135. Merete C, Falcon LM, Tucker KL. J Am Coll Nutr 2008;27: 421–7.
136. Skarupski KA, Tangney C, Li H, et al. Am J Clin Nutr 2010;92:330–5.
137. Leklem JE. Ann N Y Acad Sci 1992;669:34–43.
138. Bender DA. Br J Nutr 1999;81:7–20.
139. Nakamura S, Li H, Adijiang A, et al. Nephrol Dial Transplant 2007;22:2165–74.
140. Jolivalt CG, Mizisin LM, Nelson A, et al. Eur J Pharmacol 2009;612:41–7.
141. van den Berg H, Bode W, Mocking JA, et al. Ann N Y Acad Sci 1990;585:96–104.
142. Bates CJ, Pentieva KD, Prentice A, et al. Br J Nutr 1999;81: 191–201.
143. Schrijver J, Westermarck T, Tolonen M, et al. In: Leklem J, Reynolds, RD, eds. Vitamin B-6 status and the effect of supplementation in Finnish and Dutch elderly. Clinical and Physiological Applications of Vitamin B-6. New York: Alan R. Liss, 1988:127–135.
144. Ribaya-Mercado JD, Russel RM, Sahyoun N, et al. J Nutr 1991;121:1062–71.
145. Lindner A, Bankson DD, Stehman-Breen C, et al. Am J Kidney Dis 2002;39:134–45.
146. Robinson K, Gupta A, Dennis V, et al. Circulation 1996;94: 2743–8.
147. Bostom AG, Gohh RY, Beaulieu AJ, et al. Ann Intern Med 1997;127:1089–92.
148. Okada H, Moriwaki K, Kanno Y, et al. Nephrol Dial Transplant 2000;15:1410–3.
149. Jamison RL, Hartigan P, Kaufman JS, et al. JAMA 2007;298: 1163–70.
150. Roubenoff R, Roubenoff RA, Selhub J, et al. Arthritis Rheum 1995;38:105–9.
151. Chiang EP, Bagley PJ, Roubenoff R, et al. J Nutr 2003;133: 1056–9.
152. Woolf K, Manore MM. J Am Diet Assoc 2008;108:443–53.
153. Chiang EP, Smith DE, Selhub J, et al. Arthritis Res Ther 2005;7:R1254–62.
154. Chiang EPI, Selhub J, Bagley PJ, et al. Arthritis Res Ther 2005;7:R1404–R11.
155. Barber GW, Spaeth GL. J Pediatr 1969;75:463–78.
156. Mason DY, Emerson PM. Br Med J 1973;1:389–90.
157. Bernstein AL. Ann N Y Acad Sci 1990;585:250–60.
158. Wyatt KM, Dimmock PW, Jones PW, et al. BMJ 1999;318: 1375–81.
159. Bhagavan BM. Curr Concepts Nutr 1983;12:1–12.
160. Ubbink JB, Delport R, Becker PJ, et al. J Lab Clin Med 1989;113:15–22.
161. Lumeng L. J Clin Invest 1978;62:286–93.
162. Medici V, Peerson JM, Stabler SP, et al. J Hepatol 2010;53: 551–7.
163. Klosterman HJ. J Agric Food Chem 1974;22:13–16.
164. Mayengbam S, Raposo S, Aliani M, et al. J Nutr 2016;146: 14–20.

CHAPTER 25

Pantothenic Acid

Paula R. Trumbo

CHAPTER OUTLINE

Abbreviations

AI, adequate intake; **ATP**, adenosinetriphosphate; **CoA**, coenzyme A; **DRIs**, dietary reference intakes; **EAR**, estimated average requirement; **ESADDI**, estimated safe and adequate daily intake; **GRAS**, Generally Recognized as Safe; **NBIA**, neurodegeneration with brain ironaccumulation; **PKAN**, pantothenate kinase–associated neurodegeneration; **RDA**, recommended dietary allowance; **TCA**, tricarboxylic acid

Historical Overview

Pantothenic acid belongs to the group of B vitamins. The name is a Greek derivation meaning "from everywhere." Earlier names used for pantothenic acid include vitamin B_5, chick antidermatitis factor, antidermatosis vitamin, and chick antipellagra factor. Pantothenic acid was isolated by R. J. Williams et al., in 1931,[1] and this isolate was shown to be a single acidic substance essential for the growth of yeast in 1933.[2] The structure of pantothenic acid was later determined in 1939.[3] In 1940, Williams et al., successfully synthesized pantothenic acid,[4] thus showing its relationship with inositol, thiamin, biotin, and vitamin B_6 as to the growth of yeast.[5] These researchers also developed assays for the isolation and measurement of pantothenic acid.[6] In 1947, Lipmann et al., identified pantothenic acid as one of the components of coenzyme A (CoA). The accepted biochemical structure of CoA was published in 1953.[7] It was not until 1954 that Bean and Hodges[8] reported that pantothenic acid is essential in human nutrition. Pantothenic acid–containing CoA has since been shown to be essential to the respiratory tricarboxylic acid (TCA) cycle, fatty acid synthesis and degradation, and many other metabolic and regulatory processes.

Terminology, Chemistry, and Biochemistry

Pantothenic acid is water soluble, exists as a yellow viscous oil, and is unstable to acids, bases, and heat. Pantothenic acid, d(+)-α-(-dihydroxy-β,β-dimethylbutyryl-β-alanine), is synthesized by microorganisms through an amide linkage of β-alanine and pantoic acid (**Figure 25.1**). Pantetheine consists of a β-mercaptoethylamine group added to pantothenate in humans. CoA is composed of 4′-phosphopantetheine linked by an anhydride bond to adenosine 5′-monophosphate, modified by a 3′-hydroxyl phosphate. In addition to serving as a component of CoA, 4′-phosphopantetheine is linked to certain proteins. 4′-Phosphopantetheine has been shown to be an essential cofactor in the biosynthesis of fatty acids (e.g., fatty acid synthetase), peptides (e.g., antibiotics), and polyketides.[9] Pantethine is a dimeric form of pantetheine, linked by a disulfide bridge, which is produced from pantothenic acid by the addition of cysteamine. Pantetheine is an intermediate in the production of coenzyme A by the body.

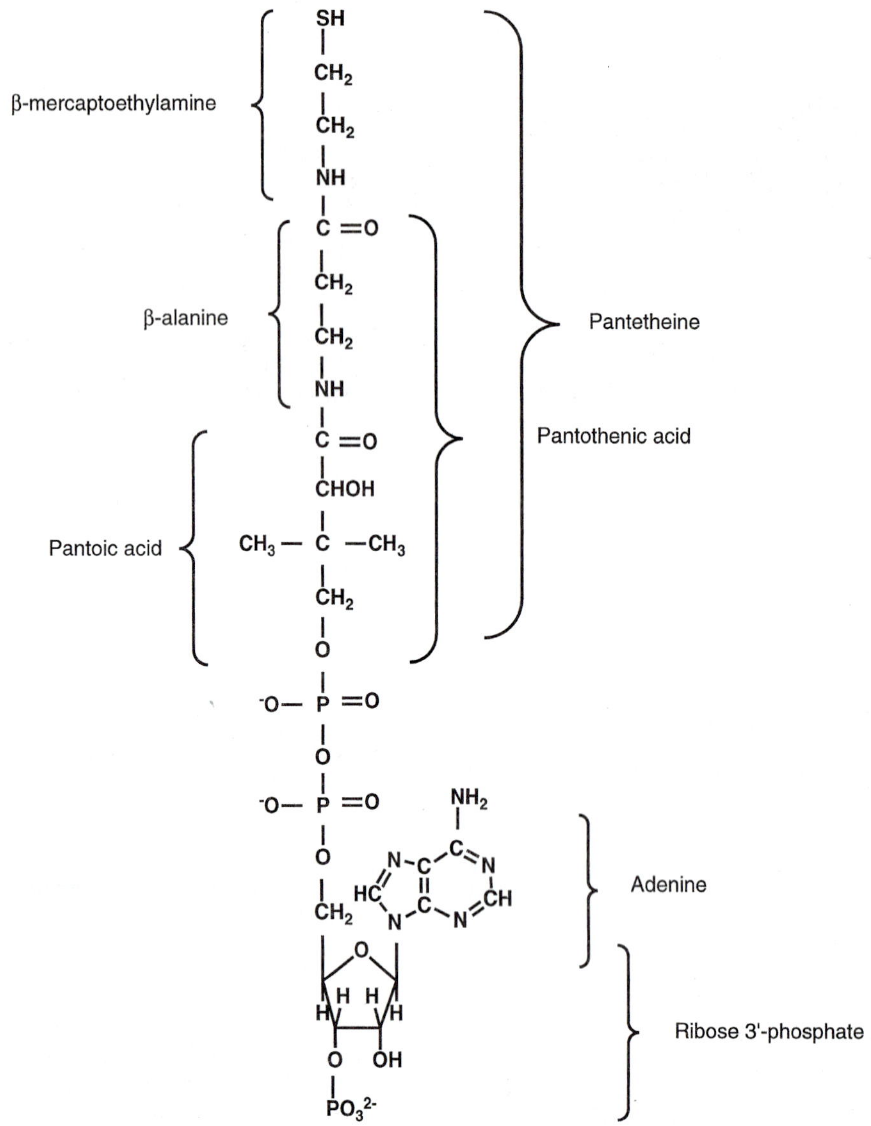

Figure 25.1 Coenzyme A and intermediaries.

Pantothenate, in the form of CoA, performs multiple roles in cellular metabolism. CoA facilitates the transfer of acetyl or acyl groups. β-Oxidation of fatty acids and the oxidative degradation of amino acids depend on CoA and thereby make the catabolic products available to the TCA cycle. Furthermore, acetyl CoA provides acetyl groups to oxaloacetic acid for the formation of citrate in the TCA cycle. The condensation of three acetyl CoA molecules yields 3-hydroxy-3-methylglutaryl-CoA (HMG CoA), an intermediate in cholesterol synthesis.

Dietary Sources

Free and conjugated pantothenic acid is found in various plant and animal foods. Approximately 85% of dietary pantothenic acid exists as CoA or phosphopantetheine.[10] Major sources of pantothenic acid include beef, chicken, liver, eggs, tomato products, broccoli, potatoes, and whole grains (**Table 25.1**).[11,12] Pantothenic acid, in the form of calcium pentothenate or sodium pantothenate, is considered to be Generally Recognized as Safe (GRAS) for the addition to various foods, such as breakfast cereals, beverages, and baby foods.[13,14] Fruit products and corn-based or presweetened cereals are among the poorest sources of pantothenic acid.

A review of studies in North America and the United Kingdom reported that the average pantothenic acid concentration in mature breast milk ranged from 2.2 to 2.5 mg/L.[15] The pantothenic acid concentration in human milk has been reported to be 6.7 mg/L, with no change occurring from 1 to 6 months postpartum.[16] The pantothenic acid content in human milk correlates with maternal intake of the vitamin.[17] The average pantothenate concentrations in fore and hind samples of preterm milk (3.31 and 3.72 mg/mL, respectively), were significantly higher than in term milk (2.64 and 2.48 mg/L, respectively).[17] In one report, the concentration of pantothenic acid in breast milk increased from 0.48 to 2.45 mg/L within 4 days after parturition.[18] Findings of another study reported that in women with relatively higher early postpartum milk, pantothenic acid concentrations exhibited decreases in their concentrations throughout lactation.[19]

U.S. regulations are provided for the nutrient content of infant formulas.[20] An infant formula should contain 300 μg pantothenic acid for each 100 kilocalories of the infant formula

Table 25.1 Pantothenic Acid Content of Foods

Food	Pantothenic Acid (Mg/100g Edible Portion)
Cashew nuts	8.86
Liver, chicken, cooked	6.67
Mushrooms, portabella, raw	1.14
Chicken, roasted	1.13
Egg, hard-boiled	1.40
Beef, ground, 93% lean, cooked	0.65
Broccoli, raw	0.57
Rice, brown	0.39
Milk, 2%	0.38
Potato, baked	0.38
Tomato, red, raw	0.09

Table 25.2 Adequate Intake for Pantothenic Acid

Age (Male and Female)	Adequate Intake (mg/d)[a]
0–6 mo[b]	1.7
7–12 mo	1.8
1–3 y	2
4–8 y	3
9–13 y	4
14–18 y	5
≥19 y	5
Pregnant women	6
Lactating women	7

[a]Based on usual intakes for groups 1 year old and older (see text).
[b]Based on the average contents of pantothenic acid in human milk consumed by infants (see text).

in the form prepared for consumption and as directed on the container. Calcium pantothenate and D-pantothenamide are sources of pantothenic acid that can be added to foods for special dietary uses, such as infant formula.[21]

Besides pantothenic acid, dietary supplements containing pantethine are commercially available. A study conducted in rats showed that the biologic efficiency in recovering from pantothenic acid deficiency was equivalent between the consumption of pantothenic acid and pantethine.[22]

Dietary Reference Intakes and Daily Values

In 1989, an estimated safe and adequate daily intake (ESADDI) of 4 to 7 mg/day of pantothenic acid was set because subjects were shown to excrete this level through the urine and feces.[23] Based on a scientific review of the data for setting dietary reference intakes (DRIs), it was determined that the information for setting an estimated average requirement (EAR) was insufficient, and, therefore, a recommended dietary allowance (RDA) could not be set. Therefore, an adequate intake (AI) value was set for pantothenic acid for all life stage and sex groups (**Table 25.2**).[24] The AI for infants 0 to 6 months old represents the average daily intake of pantothenic acid for infants exclusively fed human milk. The AIs for men and for nonpregnant and pregnant women are based on usual pantothenic acid intakes by adults in the United States. The AI for children 1 to 18 years old were set by extrapolating from the adult AI, based on body weight and growth factors. The AI during lactation is 7 mg/day because approximately 2 mg of pantothenic acid is secreted in human milk daily.[24]

The US Food and Drug Administration (FDA) amended the regulations, including the nutrient Daily Values (DVs), for the labeling of conventional foods and dietary supplements.[25] For each vitamin and mineral, one DV was set for males and females, 4 years of age and older, excluding pregnant and lactating women. The DV was set using the highest RDA or AI for each nutrient and, therefore, a DV of 5 mg was used for pantothenic acid. Using the AI for infants (up to 12 months of age), toddlers (1–3 years of age), and pregnant/lactating women, the FDA also set DVs for conventional foods and dietary supplements purported to be consumed specifically for these groups. The DVs for infants, toddlers, pregnant, and lactating women are 1.8, 2, 6, and 7 mg, respectively. Pantothenic acid is not required to be listed in the food label, unless added as an ingredient to a food or dietary supplement.

Dietary Intake

United States nutrition surveys do not estimate intakes of pantothenic acid. The mean pantothenic acid intake of men and women in Quebec in 1990 fell from approximately 6 to 3 mg/day with advancing age.[24] The average intake of pantothenic acid was 5.5 and 4.0 g/day for the male and female population, respectively, of New Brunswick, Canada.[26] The average intake of adults living in the Duluth and Minneapolis metropolitan area was 5.7 mg/day for nonsmokers and was significantly lower, at 4.6 mg/day, for smokers.[27] Usual intakes of approximately 4 to 7 mg/day have been reported for adolescents and adults.[24] The average dietary intake estimated for pregnant and lactating women has been reported at 2.8 mg/1,000 kcal[28] or 7.6 mg/day for lactating women[16]; for pregnant women living in Boston, the estimated intake was 6.6 mg/day[29]; and for elderly persons, it was 5.9 mg/day.[30] The average daily intake of pantothenic acid from multivitamin/mineral supplements, among supplement users, has been estimated at 10 mg/day.[31]

Metabolic Aspects

Digestion, Absorption, and Excretion

Dietary CoA is hydrolyzed in the intestinal lumen to dephospho-CoA, pantetheine, and phosphopantetheine. Pantetheine is further hydrolyzed by pantetheinase to pantothenic acid. Although pantothenic acid may be absorbed by passive diffusion, it is absorbed into the bloodstream of animals by a saturable, sodium-dependent, active transport mechanism.[32] Studies in mice indicated that the kinetics for this active transport system is not affected by variation in dietary intake levels of the vitamin.[33] Although animal studies demonstrated that intestinal microflora synthesize pantothenic acid,[33] the contribution to absorbed pantothenic acid in humans is unknown.

Absorbed pantothenic acid is transported by the red blood cells throughout the body.[34] The vitamin is also transported in the free acid form in the plasma at a concentration of approximately 1 μg/mL.[35] (The molecular weight of pantothenic acid is 219.24 g/mol; 1 mg equals 4.56 μmol; 1 μg/mL equals 4.56 μmolar [μM]). Concentrations in red blood cells are higher than in plasma. Maximum pantothenate concentrations occurred 3 minutes following intravenous injection and subsequently decreased, suggesting that the vitamin is rapidly taken up by red blood cells and other tissues.[36] A large increase in red blood cell pantothenate concentration was observed in men following the injection of a multivitamin mixture that included 45 mg D-panthenol.[37]

Animal studies showed that, following the intraluminal administration of radiolabeled pantothenate or CoA, approximately 40% was located in muscle, 10% in the liver, and 10% in the intestine.[38] Because pantothenic acid is an essential component for the biosynthesis of CoA, most tissues transport the vitamin through an active sodium cotransport mechanism.[39,40] Most of the pantothenic acid in tissues is present as CoA, with lesser amounts present as acyl carrier protein or free pantothenic acid.

Pantothenic acid is not synthesized in the brain and, therefore, must be supplied from blood across the blood-brain barrier to maintain brain homeostasis.[41] Brain concentrations of pantothenic acid are about 50-fold greater than in the plasma,[41] suggesting that an active transport mechanism exists at the blood-brain barrier. SLC5A6 is an Na^+-dependent multivitamin transporter that mediates cellular uptake of biotin, pantothenic acid, and lipoic acid.[42,43] It was shown that SLC5A6 is a major contributor to luminal uptake of pantothenic acid at the human blood-brain barrier.[44]

Before urinary excretion, CoA is hydrolyzed to pantothenate in a multistep reaction. Pantothenic acid is excreted in urine and is typically measured by microbiologic assay. When subjects were given 100 mg/day of pantothenic acid, urinary excretion was only 60 mg/day. This finding suggests that the vitamin can be stored when intake levels are high or the fractional bioavailability is relatively low.

Bioavailability

Data on the bioavailability of dietary pantothenic acid are limited. One study reported that dietary pantothenic acid was, on average, 50% (40% to 61%) bioavailable compared with the pure form of the vitamin (calcium pantothenate) given in a formula diet.[45] Approximately 60% of ingested pantothenic acid was excreted in urine when subjects were fed three different experimental diets.[46]

Functions in Metabolism

Synthesis of Coenzyme A

The first step in the synthesis of CoA is phosphorylation of pantothenic acid, which is catalyzed by pantothenate kinase. Following this, an adenosine triphosphate (ATP)–dependent condensation of 4′-phosphopantothenic acid with cysteine yields 4′-phosphopantothenoylcysteine, which is decarboxylated to 4′-phosphopantetheine.[47] CoA is formed by a series of transfers of adenosine monophosphate and phosphate from ATP to 4′-phosphopantetheine. Approximately 95% of CoA is located in the mitochondria. Because CoA does not cross the mitochondrial membrane, the final site of CoA synthesis is thought to be within the mitochondria.[48]

Cellular Metabolism and Physiology

Pantothenate, usually in the form of CoA, performs multiple roles in cellular metabolism.[49] Acetyl-CoA is central to the energy-yielding oxidation of glycolytic products and other metabolites through the mitochondrial TCA cycle. The first step of the TCA cycle involves the condensation of acetyl-CoA with oxaloacetate to yield citrate and subsequently succinyl-CoA, which provides energy for guanosine diphosphate phosphorylation. The β-oxidation of fatty acids and the oxidative degradation of amino acids are also CoA-dependent processes; the catabolic products of these processes become available to the respiratory TCA cycle for further degradation and energy production.

Pantothenic acid is required for the synthesis, by biosynthetically competent species, of several essential molecules including sphingolipids, leucine, arginine, and methionine. CoA is also required for the synthesis of isoprenoid derivatives, such as farnesol, cholesterol, steroid hormones, vitamin A, vitamin D, and heme A. Some of the isoprenoids are further bound to certain proteins, such as viral Ras proteins. Succinyl-CoA is required for the synthesis of δ-amino-levulinic acid, which is the precursor of the porphyrin rings in hemoglobin and the cytochromes and the corrin ring of vitamin B_{12}. CoA provides the essential acetyl group to the neurotransmitter acetylcholine, to serotonin in its conversion to melatonin, and to the acetylated sugars present in glycoproteins and glycolipids (*N*-acetylglucosamine, *N*-acetylgalactosamine, and *N*-acetylneuramic acid). CoA also plays a regulatory role in brain health through cerebral myelin biosynthesis.[50] Because of acetyl-CoA's role in energy metabolism and biosynthesis, the European Food and Safety Authority permits claims on food products sold in the European Union to say, "Pantothenic acid contributes to normal energy-yielding metabolism" and "Pantothenic acid contributes to normal synthesis and metabolism of steroid hormones, vitamin D and some neurotransmitters."[51]

While the mechanism is unclear, pantethine intake has been shown to reduce cholesterol concentrations. In a double-blind study, supplementation with 1,200 mg pantethine per day to individuals with type IIB hyperlipoproteinemia resulted in significant lowering of plasma total and low-density lipoprotein cholesterol (−13.5% for both parameters).[52] Switching from pantethine to placebo resulted in a rapid return to the baseline cholesterol concentrations. Compared with the placebo, subjects at low to moderate risk of cardiovascular disease, who consumed 600 to 900 mg/day pantethine, showed significant decreases in total and LDL cholesterol.[53] A number of pantethine dietary supplements are commercially available that bear structure/function claims related to maintaining healthy cholesterol levels.

Protein Acetylation

Most soluble proteins are *N*-terminally acetylated by CoA. *N*-Acetylation appears to alter the structure of certain proteins, thereby altering function or metabolism. Peptide hormones are acetylated, a process that alters their hormone activity. For example, acetylation results in the activation of α-melanocyte–stimulating hormone and the inactivation of β-endorphin. Acetylation of histones alters the conformation of chromatin and changes its sensitivity to nucleases. Two classes of proteins are internally acetylated: histones and α-tubulin. Acetylation of histones neutralizes the charge of acetylated lysine residues and thereby weakens interactions between the nucleosomes that depend on histone *N*-terminal tails. Histones that are highly acetylated tend to be associated with newly synthesized DNA or DNA that is transcriptionally active. Acetylated chromatin has a more unfolded conformation, as indicated by its increased sensitivity to nucleases.[54] Hyperacetylation of histones H3 and H4 has been shown to decrease the supercoiling within the nucleosome.[55]

Acetylation and deacetylation regulate the assembly and disassembly of microtubules. Microtubules, which are essential components of the cellular cytoskeleton, are assembled from α- and β-tubulin dimers that polymerize and depolymerize continuously. Acetylation of α-tubulin occurs in the assembled microtubule and appears to stabilize microtubules.[56] Deacetylation appears to be associated with depolymerization of microtubules.[57]

Protein Acylation

Many different cellular proteins are covalently modified with long-chain fatty acids donated by fatty acyl CoA. The fatty acids, myristic acid and palmitic acid, are commonly added to proteins by acylation. This modification affects the location and activity of many proteins, including those involved with signal transduction. Myristolated proteins are located in the cytoplasm, plasma membrane, endoplasmic reticulum, and nuclear membrane. Myristolation of proteins is irreversible and often combines with other protein modifications for regulation of the protein's activity or localization. Because of the reversibility of palmitoylation, this process has a regulatory function. Guanosine triphosphate–binding proteins, Src and Src-related tyrosine kinases, and most viral and cellular Ras proteins comprise an extensive group of proteins that become modified by addition of myristate or palmitate. Palmitoylation is required for oncogenic viral Ras proteins to bind to the plasma membrane and transform cells and for vesicular transport through the Golgi stacks.[58] Transmembrane receptors that are palmitoylated include the insulin receptor and the iron-transferrin receptor. Palmitate is acylated to certain membrane proteins involved with the cytoskeleton, including fibronectin and gap junction proteins.[59] Several neuronal proteins are modified with palmitate. Acetylcholinesterase, which degrades the neurotransmitter acetylcholine, is bound to the cell membrane by palmitoylation.[60] The reversible palmitoylation of proteins is involved with neural development by influencing neural motility and outgrowth in the developing brain.[61]

Neurodegenerative Diseases

Pantothenate kinase-associated neurodegeneration (PKAN) is a rare autosomal recessive disorder that largely affects children and young adults.[62] PKAN is a major cause of neurodegeneration with brain iron accumulation (NBIA). NBIA is a rare, inherited neurologic movement disorder in which a mutation of the *PANK2* gene results in a deficiency of pantothenate kinase and, therefore, inadequate synthesis of CoA. NBIA is characterized by dystonia, parkinsonism, and iron accumulation in the brain.[63] While iron accumulation is a prominent symptom of PKAN, evidence suggests that the pathogenesis of PKAN occurs because of a metabolic imbalance rather than because of the iron accumulation.[64] Approximately 25% of affected individuals have atypical presentations (e.g., prominent speech defects and psychiatric disturbances) with late onset (age >10 years).[65] *PANK2* is one of four known pantothenate kinases and is the only known gene to be associated with PKAN.[65] Acanthocytosis and a defect in plasma lipoproteins have also been noted in association with the *PANK2* mutations.[66] Approaches to treat PKAN have included supplementation with pantethine and 4′-phosphopatetheine to compensate for the reduced concentrations of the metabolic intermediates of CoA caused by altered Pank2 activity. In a pilot study on 15 children with PKAN, 24 weeks of pantethine treatment did not improve motor function.[67] Feeding mice 4-phosphopantetheine normalized CoA-, iron-, and dopamine-related biomarkers, as well as activities of mitochondrial enzymes. Human cell changes were also recovered with 4′-phosphopantetheine.[68] Several other neurodegenerative diseases are associated with relatively low pantothenic acid in the brain, affecting CoA synthesis. Huntington's disease is a slowly progressive, autosomal dominant neurodegenerative disorder with broad but distinct phenotypes characterized by cognitive decline, personality changes, psychiatric manifestations, incoordination, and involuntary movements.[69,70] Pantothenic acid concentration has been shown to be the most significantly reduced nutrient in various regions of Huntington's disease human brains, including the cerebellum, compared with controls.[71] Cerebral deficiency of pantothenic acid also occurs in Alzheimer's disease.[72] This deficiency has been shown to be worse in those regions of the brain known to undergo severe damage with Alzheimer's diseases, including the cerebellum, entorhinal cortex, middle temporal gyrus, and hippocampus. In brains of individuals with Parkinson's disease, pantothenic acid has been shown to be significantly lower in the cerebellum, substantia nigra, and medulla.[73] These findings mirror the significant decreases in similar regions of the brains in Alzheimer's disease or Huntington's disease cases. The most notable commonality among Huntington's disease, Alzheimer's disease, and Parkinson's disease was as an approximately 40% to 50% lower concentration of pantothenic acid in the cerebellum, compared with healthy controls.

Assessment of Status

Analytical Methods

Blood, urine, and tissue pantothenic acid concentrations are often measured by microbiologic growth assays, animal bioassay, or radioimmunoassay. Microbiologic assays for measuring pantothenic acid are highly sensitive and specific, but they are slow and tedious to perform. Results of radioimmunoassay compare favorably with those obtained by microbiologic methods. Samples containing bound vitamin (biologic sources excluding urine) must be hydrolyzed with enzymes or chemicals to release the pantothenate component of CoA before being assayed. Enzymes used to release pantothenic acid include papain, mylase-P, diastase, and clarase. High-performance liquid chromatography methods, coupled with mass spectrometry or fluorometric detection, have also been shown to be sufficiently sensitive for measuring pantothenic acid in urine.[74,75] Quantification of pantothenic acid in biologic samples has been developed using stable isotope dilution assay[76] and enzyme-linked immunosorbent assay.[77]

The conventional method for measuring pantothenic acid has been a microbiologic assay using *Lactobacillus plantarum* and this is an official AOAC method (AOAC 992.07) for measuring pantothenic acid in infant formula.[78] An ultra-performance liquid chromatography tandem mass spectrometry method has more recently been validated for measuring pantothenic acid in infant formulas and adult nutritional formulas (AOAC 2012.16).[79] An indirect competitive enzyme-linked immunosorbent assay and a gold nanoparticle-based lateral flow immunoassay have been established for measuring pantothenic acid in food products, based on an anti-pantothenic acid monoclonal antibody.[80]

Blood Concentrations

Blood concentrations of pantothenic acid that are lower than 100 μg/dL have been suggested to indicate inadequate intake.[81] Whole blood pantothenic acid concentrations have been reported to be significantly correlated with intake.[82] Whole blood concentrations fell from 8.9 to 6.4 μmol/L when men were fed a diet devoid of pantothenic acid for up to 9 weeks.[83] When the men were supplemented with 10 mg/day of pantothenic acid for 9 weeks, no difference in whole blood concentration was noted, compared with baseline. Dietary pantothenic acid intake did not correlate with blood concentration in a group of elderly individuals; however, when a supplement containing pantothenic acid was taken, blood concentration increased markedly.[84] Pantothenic acid intake and serum concentration were reported to correlate in adolescents, whereas no correlation was observed in adults.[85] Plasma concentrations did not reflect changes in intake or status of pantothenic acid.[37] The correlation between dietary and erythrocyte pantothenic acid in a group of well-nourished adolescents was 0.4, and average erythrocyte concentration was 1.5 μmol/L (334 ng/L).[82] At parturition, the pantothenic acid concentration was five times lower in the mother's serum compared with the infant's.[18]

Urinary Excretion

A dose-response relationship between dietary intake and urinary excretion of pantothenic acid was shown in women.[86,87] This correlation was also confirmed in adolescents.[85] Urinary pantothenic acid concentration in women fed a diet low in pantothenic acid (2.8 mg/day) exceeded the intake; suggesting that body stores were being lost or that synthesis was occurring.[86] The body has been shown to conserve pantothenic acid.[88]

Causes and Manifestations of Deficiency and Excess

Information on the causes of deficiency or overexposure of pantothenic acid in humans is limited. Because pantothenic acid is present, to some extent, in all foods, a deficiency in humans is rare except in severe malnutrition, and it most likely occurs in conjunction with other nutrient deficiencies. During World War II, prisoners of war in Japan, the Philippines, and Burma experienced numbness in their toes and painful burning sensations in their feet. These symptoms were relieved with pantothenic acid supplementation, but not when other B-complex vitamins were given.[89] When given a diet completely devoid of pantothenic acid[85] or when given an antagonist of pantothenic acid metabolism,[90,91] subjects experienced irritability, restlessness, sleep disturbances, numbness, and gastrointestinal disturbances. The European Food Safety Authority permits food products sold in the European Union to make the statements "Pantothenic acid contributes to normal mental performance" and "Pantothenic acid contributes to the reduction of tiredness and fatigue.[51]"

Pantothenic acid has been suggested to aid in wound healing[92,93] by increasing the number of migrating dermal fibroblasts[94] through modulation of gene expression.[95] Culturing human keratinocytes in a pantothenic acid-deficient medium inhibits proliferation and expression of keratogenic growth factor mRNA, suggesting that pantothenic acid promotes keratogenic growth factor synthesis and keratinocyte proliferation.[96] Low blood concentrations of pantothenic acid have been reported in patients with rheumatoid arthritis.[97] One randomized, double-blind study showed that daily supplementation with 1 g calcium pantothenate resulted in a significant reduction in pain in patients with rheumatoid arthritis[98] (see also the chapter on rheumatic and arthritis diseases). Evidence demonstrated that a class of pantothenic acid analogs represses the proliferation of the human malaria parasite, *Plasmodium falciparum*.[99] Prospective observational studies have reported a significant positive association between pantothenic acid intake and birth weight[100] and birth length.[29]

Administration of pantothenate analogs to humans has occurred with unintentionally deleterious side effects. Hopantenate is an analog of pantothenic acid, in which γ-amino butyric acid (GABA) replaces β-alanine. Hopantenate was used in Japan in individuals with emotional disturbance secondary to cerebrovascular disease and to alleviate tardive dyskinesia symptoms induced by tranquilizers. As a result, the patients exhibited severe side effects, including lactic acidosis, hypoglycemia, hyperammonemia, and eventually acute encephalopathy.[101] These effects were also demonstrated in dogs and were shown to be caused by pantothenic acid deficiency. Dogs given an equivalent amount of pantothenic acid and calcium hopantenate did not develop the disorder.[102]

No adverse effects of overconsumption of pantothenic acid in humans have been reported. For this reason, the Institute of Medicine did not set a tolerable upper intake level (UL).[21] When patients were treated with up to 15 g/day of pantothenic acid; however, symptoms of lupus erythematosus, nausea, and gastrointestinal distress were reported.[103,104] The toxic oral dose (LD_{50}) for mice was determined to be 10 g/kg, which led to death by respiratory failure.[105] A study in which rats were provided with up to 3% of their diet as calcium pantothenate for 29 days demonstrated that adverse effects (enlargement of testis, diarrhea, and hair damage) were observed at 3% but not at 1%.[106] The investigators suggested that the lowest-observed-adverse-effect level (LOAEL) and no-observed-adverse-effect level (NOAEL) for pantothenic acid should be 3% and 1%, respectively. This information, along with an uncertainty factor that takes into consideration the use of animal data, can be used for setting a UL.

References

1. Williams RJ, Bradway EM. J Am Chem Soc 1931;53:783.
2. Williams RJ, Lyman CM, Goodyear GH, et al. J Am Chem Soc 1933;55:2912–27.
3. Williams RJ, Weinstock HH, Rohrmann E, et al. J Am Chem Soc 1939:89:199–206.
4. William RJ, Mitchell HK, Weinstock HH, et al. J Am Chem Soc 1940;62:1784–5.
5. Williams RJ, Eakin RE, Snell EE. J Am Chem Soc 1940;62: 1204–7.
6. Pennington D, Snell EE, William RJ. J Biol Chem 1940;135: 213–22.
7. Baddiley J, Thain EM, Novelli GD, et al. Nature 1953;171:76–9.
8. Bean WB, Hodges RE. Proc Soc Exp Biol Med 1954;86:693–9.
9. Plesofsky-Vig N, Brambl R. Annu Rev Nutr 1988;8:461–82.
10. Bender DA, Bender AE, eds. Nutrition: A Reference Handbook. Oxford: Oxford University Press, 1997.
11. Walsh JH, Wyse BW, Hansen RG. J Am Diet Assoc 1981;78: 140–4.
12. US Department of Agriculture, Agricultural Research Service. FoodData Central. 2019. Available at: https://fdc.nal.usda.gov.
13. Code of Federal Regulations. Substances Generally Recognized as Safe. 21 CFR 182.5772 21 CFR 184.1212. 2022, Washington, DC: Office of the Federal Register, 2022.
14. Code of Federal Regulations. Direct Food Substances Affirmed as Generally Recognized as Safe. 21 CFR 184.1212. Calcium Pantothenate. Washington, DC: Office of the Federal Register, 2022.
15. Picciano MF. Water-soluble vitamins in milk. In: Jensen RG, ed. Handbook of Milk Composition. San Diego: Academic Press, 1995.
16. Johnston L, Vaughan L, Fox HM. Am J Clin Nutr 1981;34: 2205–9.
17. Song WO, Chan GM, Wyse BW, et al. Am J Clin Nutr 1984;40:317–24.
18. Robinson AF, Folkers K, eds. Vitamins and Coenzymes. New York: John Wiley, 1964.
19. Batalha MA, Ferreira ALL, Freitas-Costa NC, et al. Am J Clin Nutr 2021:114;1560–1573.
20. Code of Federal Regulation. Infant Formula. Subpart D–Nutrient Requirements. Nutrient Specifications. 21 CFR 107.100. Washington, DC: Office of the Federal Register, 2022.
21. Code of Federal Regulations. Food Additives Permitted for Direct Addition to Food for Human Consumption. 21 CFR 172.335 (D-Pantothenamide) and 172.330 (Calcium Pantothenate). Washington, DC: Office of the Federal Register, 2022.
22. Shibata K, Kaneko M, Fukuwatari T. J Nutr Sci Vitaminol 2013;59:93–99.
23. National Research Council. Recommended Dietary Allowances. Washington, DC: National Academy Press, 1989:169–73.
24. Food and Nutrition Board, Institute of Medicine. Dietary Reference Intakes for Thiamin, Riboflavin, Niacin, Vitamin B6, Folate, Pantothenic Acid, Biotin, and Choline. Washington, DC: National Academy Press, 1998;357–73.
25. Code of Federal Regulations. Nutrition Labeling of Food. Food Labeling and Nutrition Labeling of Dietary Supplements. 21 CFR 101.9 and 21 CFR 101.36.
26. New Brunswick Department of Health and Wellness. Appendix E. In: New Brunswick Nutrition Survey. Fredericton, New Brunswick, Canada: New Brunswick Department of Health and Wellness, 1997.
27. Raatz SK, Jahns L, Johnson LK, et al. Nutr Res 2017;45:30–37.
28. Song WO, Wyse BW, Hansen RG. J Am Diet Assoc 1985;85: 192–8.
29. Lagiou P, Mucci L, Tamimi R, et al. Eur J Nutr 2005;44:52–9.
30. Srinivasan V, Christensen N, Wyse BW, et al. Am J Clin Nutr 1981;34:1736–42.
31. Park SY, Murphy SP, Martin CL, et al. J Am Diet Assoc 2008;108:529–33.
32. Fenstermacher DK, Rose DC. Am J Physiol 1986;250:G155–60.
33. Stein ED, Diamond JM. J Nutr 1989;119:1973–83.
34. Eissenstat BR, Wyse BW, Hansen RG. Am J Clin Nutr 1986;44: 931–37.
35. Fox HM. Pantothenic acid. In: Machlin LJ, ed. Handbook of Vitamins. New York: Marcel Dekker, 1984:437.
36. Tahiliani AB, Beinlich CH. Vitam Horm 1991;46:165–28.
37. Baker H, Frank O, Thomson AD, et al. Am J Clin Nutr 1969;22:1469–75.
38. Shibata K, Gross CJ, Henderson LM. J Nutr 1983;113:2107–15.
39. Barbarat B, Podevin RA. J Biol Chem 1986;261:14455–60.
40. Prasad PD, Ramamoorthy S, Leibach FH, et al. Placenta 1997;18:527–33.
41. Spector R, Johanson CE. J Neurochem 2007;103:425–438.
42. de Carvalho FD, Quick M. J Biol Chem 2011;286:131–37.
43. Prasad PD, Wang H, Kekuda R. J Biol Chem 1998;273:7501–06.
44. Uchida Y, Ito K, Ohtsuki S, et al. J Neurochem 2105;134:97–112.
45. Tarr JB, Tamura T, Stocksatd EL. Am J Clin Nutr 1981;34: 1328–37.
46. Yu BH, Kies C. Plant Foods Hum Nutr 1993;43:87–95.
47. Brown G. J Biol Chem 1959;234:370–8.
48. Robishaw JD, Berkick D, Neely JR. J Biol Chem 1982;257: 10967–72.
49. Combs GF. The Vitamins: Fundamental Aspects in Nutrition and Health. New York: Academic Press, 1992:352.
50. Ismail N, Kureishy SJ, Church M, et al. Biochem Biophys Res Commun 202;220e225.
51. European Food Safety Authority Panel on Dietetic Products, Nutrition, and Allergies. EFSA Journal 2009;7:1218.
52. Gaddi A, Descovich GC, Noseda G, et al. Atherosclerosis 1984;50:73–83.
53. Evans M, Rumberger JA, Azumano I, et al. Vascular Health and Risk Management 2014:10;89–100.
54. Ridsdale JA, Hendzel MJ, Delcuve GP, et al. J Biol Chem 1990;265:5150–6.
55. Norton VG, Marvin KW, Yau P et al. J Biol Chem 1990;265: 19848–52.
56. Plesofsky-Vig N, Brambl R. Annu Rev Nutr 1988;8:461–82.
57. Lim SS, Sammak PJ, Borisy GG. J Cell Biol 1989;109:253–63.
58. Pfanner N, Orci L, Glick BS, et al. Cell 1989;59:95–102.
59. Maneti S, Dunia I, Benedetti EL. FEBS Lett 1990;262:356–8.
60. Randall WR. J Biol Chem 1994;269:12367–74.
61. Hess DT, Patterson SI, Smith DS, et al. Nature 1993;366: 562–5.
62. Gregory A, Hayflick SJ. Folia Neuropathol 2005;43:286–296.
63. Hayflick SJ, Westaway SK, Levinson B et al. N Engl J Med 2003;348:33–40.
64. Munshi MI, Yao SJ, Mamoun CB. J Biol Chem 2022;298:101577.
65. Gregory A, Hayflick SJ. Pantothenate kinase-associated neurodegeneration. In: Pagon RA, Bird TD, Dolan CR, Stephens K, eds. GeneReviews. Seattle: University of Washington, 2008.
66. Ching KH, Westaway SK, Levinson B, et al. Neurology 2002;58:1673–4.
67. Chang X, Zhang J, Jiang Y, et al. Orphanet J Rare Dis 202;15:248.
68. Jeong SY, Hogarth P, Placzek A, et al. EMBO Mol Med 2019;1:e10489.
69. Walker FO. Lancet 2007;369:218–28.

70. Tabrizi SJ. Langbehn DR, Leavitt BR, et al. Lancet Neurol 2009;8:791–801.
71. Xu J, Patassini S, Begley P, et al. Biochem Biophys Res Commun 2020;527:676–681.
72. Patassini S, Begley P, Xu J, et al. Metabolites 2019;9:113.
73. Scholefield M, Church SJ, Xu J, et al. Metabolites 2021;11:569.
74. Heudi O, Fontannaz P. J Sep Sci 2005;28:669–72.
75. Takahashi K, Fukuwatari T, Shibata K. J Chrom B 2009;877: 2168–72.
76. Rychlik M. J Agr Food Chem 2000;48:1175–81.
77. Gonthier A, Boullanger P, Fayol V, et al. J Immunoassay 1998;19:167–94.
78. Official Methods of Analysis, 21st ed., 992.07, Pantothenic Acid in Milk-based Infant Formula – 1995. Rockville, MD, AOAC International, 2019.
79. Official Methods of Analysis, 21st ed, 2012.16, Pantothenic acid (Vitamin B5) in Infant Formula and Adult/Pediatric Nutritional Formula, AOAC International, Rockville, MD, 2019.
80. Zeng L, Guo L Wang Z, et al. Food Chem 2021:348;129114.
81. Sauberlich HE, Skala JH. Laboratory Tests for the Assessment of Nutritional Status. Cleveland: CRC Press, 1974:88.
82. Eissenstat BR, Wyse BW, Hansen RG. Am J Clin Nutr 1986;44: 931–7.
83. Fry PC, Fox HM, Tao HG. J Nutr Sci Vitaminol (Tokyo) 1976;22:399–46.
84. Wyse BW, Hansen RG. Fed Proc 1977;36:1169.
85. Kathman JV, Kies C. Nutr Res 1984;4:245–50.
86. Fox HM, Linkswiler H. J Nutr 1961;75:451–4.
87. Tsuiji T, Fukuwatari T, Sasaki S, et al. Nutr Res 2010;30:171–8.
88. Karnitz LM, Gross CJ, Henderson LM. Biochim Biophys Acta 1984;769:486–92.
89. Glusman M. Am J Med 1947;3:211–23.
90. Hodges RE, Ohlson MA, Bean WB. J Clin Invest 1958;37: 1642–57.
91. Hodges RE, Bean WB, Ohlson MA, et al. J Clin Invest 1959;38:1421–25.
92. Aprahamian M, Dentiger A, Stock-Damage C, et al. Am J Clin Nutr 1985;41:578–89.
93. Lacroix B, Didier E, Grenier JF. Int J Vit Nutr Res 1988;58: 407–13.
94. Weimann BI, Hermann D. Int J Vit Nutr Res 1999;69:113–9.
95. Widerholt T, Heise R, Skazik C, et al. Exp Dermatol 2009;18: 969–78.
96. Kobayashi D, Kusama M, Onda M, et al. J Pharmacol Sci 2011;115:230–4.
97. Barton-Wright EC, Elliot WA. Lancet 1963;26:862–3.
98. Einstein P, Scheiner SA, eds. Overcoming the Pain of Inflammatory Arthritis: The Pain-Free Promise of Pantothenic Acid. Garden City, NY: Avery, 1999.
99. Lehane SM, Marchetti RV, Spry C, et al. J Biol Chem 2007;282:25395–405.
100. Haggarty P, Campbell DM, Duthie S, et al. Br J Nutr 2009;102:1487–97.
101. Otsuka M, Akiba T, Okita Y, et al. Jpn J Med 1990;29:324–8.
102. Noda S, Haratake J, Sasaki A, et al. Liver 1991,11:134–42.
103. Welsh AL. Arch Dermatol 1952;65:137–48.
104. Welsh AL. Arch Dermatol 1954;70:181–98.
105. Unna K, Greslin JG. J Pharmacol Exp Ther 1941;73:85–90.
106. Shibata K, Takahashi C, Fukuwatari T, et al. J Nutr Sci Vitaminol (Tokyo) 2005;51:385–91.

C. VITAMINS

CHAPTER 26

Folic Acid

Patrick J. Stover

CHAPTER OUTLINE

Abbreviations

AdoHcy, *S*-adenosylhomocysteine; **AdoMet**, *S*-adenosylmethionine; **AICARFT**, phosphoribosylaminoimidazolecarboxamide formyltransferase; **DFE**, dietary folate equivalent; **DHF**, dihydrofolate; **DHFR**, dihydrofolate reductase; **DNMT**, DNA methyltransferase; **GARFT**, phosphoribosylglycinamide formyltransferase; **GNMT**, glycine *N*-methyltransferase; **LINE-1**, long interspersed nuclear element-1; **MTHFD**, methylenetetrahydrofolate dehydrogenase; **MTHFR**, methylenetetrahydrofolate reductase; **MTR**, methionine synthase; **NADPH**, reduced nicotinamide adenine dinucleotide phosphate; **NTD**, neural tube defect; **PCFT**, proton-coupled folate transporter; **RDA**, recommended dietary allowance; **SHMT**, serine hydroxymethyltransferase; **THF**, tetrahydrofolate; **TYMS**, thymidylate synthase

Historical Background

Also known as vitamin B_9, vitamin B_c, vitamin M, *Lactobacillus casei* factor, folacin, and pteroyl-L-glutamic acid, folic acid was first discovered by Wills and Mehta in 1931 as a factor in yeast ("Wills' factor") that corrected the macrocytic anemia of pregnant Hindu women in India.[1] The factor was later isolated from spinach leaves and was given the name folic acid (Latin *folium*, "leaf") by Mitchell et al., in 1941, who demonstrated that it was required for growth of *Streptococcus lactis* R (*Streptococcus faecalis*).[2] In 1945, the chemical synthesis of pure crystalline folic acid was reported in the journal *Science*.[3] Synthetic folic acid was effective in reversing megaloblastic anemia that was refractory to treatment with liver extracts, but this agent was not able to prevent or improve neurologic damage that progressed from anemia that is now known to be caused by vitamin B_{12} deficiency.

Shortly after the discovery of folic acid as a growth-promoting factor, the development of folate antagonists as chemotherapeutics was undertaken by the Nobel Laureates Hitchings and Elion. In 1948, the folate antagonists aminopterin and, shortly thereafter, methotrexate were developed and administered to patients with childhood acute lymphoblastic leukemia and found to be effective treatments.[4] This success led to the development of numerous anticancer and antimicrobial agents over the following 50 years that targeted folate-requiring enzymes. Beginning in the 1950s to the present time, folate-dependent enzymes were purified to homogeneity, biochemical pathways were elucidated and, later, their genes were cloned and structures determined. Beginning in the 1980s, the importance of folic acid in prevention of chronic diseases, certain cancers, and birth defects gained appreciation. This knowledge led to fortification of the food supply with folic acid in the United States, Canada, and other countries to prevent a class of common birth defects known as neural tube defects (NTDs).[5] An excellent review of the history of folic acid was published in 2001 by Hoffbrand and Weir.[1]

Overview of Folate and Folic Acid

Folate is a generic term that refers to a family of water-soluble B vitamins that are found in natural food and in biologic organisms (**Figure 26.1**). Folates function as a family of enzyme cofactors that carry and chemically activate single carbons (referred to as one-carbons) for biosynthetic reactions. Folate is required for the biosynthesis of ribonucleotides and deoxyribonucleotide precursors for DNA synthesis. It is also required for amino acid metabolism, including the remethylation of homocysteine to methionine and, therefore, functions in the regulation of gene expression by methylation. Hence, folate cofactors are found in virtually all forms of life. Tetrahydrofolate (THF), which is the fully reduced form of the vitamin, carries one-carbons at one of three different oxidation levels, ranging from methanol to formate.[6-8] The one-carbons are covalently bound to the N5 or N10 position of THF. In the cell, five different one-carbon substituted forms of THF are present: 10-formyl-THF; 5-formyl-THF; 5,10-methenyl-THF; 5,10-methylene-THF; and 5-methyl-THF, and each of these forms is interconverted in the cell through enzyme-mediated catalysis. Folates are also modified through the addition of a glutamate polypeptide that is polymerized through unusual γ-linked peptide bonds.[9] The polyglutamate polypeptide increases the affinity of folate cofactors for folate-dependent enzymes and is required to retain folates within the cell and subcellular organelles. Folic acid (see Figure 26.1) is not a biologically active form of folate, but it can serve as a provitamin because it is converted to the reduced, natural form of folate once transported into cells. It is an oxidized form of folate generated during the oxidative degradation of folate, and normally does not accumulate in cells, although most degradation of THF is irreversible with degradation products that include oxidized pterin and *para*-aminobenzoylglutamate.[10] Folic acid is also a synthetic form of folate present in fortified foods and in dietary supplements.

Dietary Sources

Folate is a vitamin and, therefore, must be acquired from the diet. Folate nutritional status is supported by the intake of the vitamin found in natural foods, as well as dietary supplements and fortified foods.[11] The best dietary sources of natural folate include fresh fruits, leafy green vegetables, yeast, liver, and legumes.[12] Natural folates found in food are chemically labile and readily undergo irreversible oxidative degradation during food preparation and cooking. 5-Methyl-THF and formyl-substituted THF are the primary forms of folate present in natural foods, and they are also among the more stable forms of the vitamin. Folic acid, the synthetic, fully oxidized, and stable provitamin, is present in dietary supplements and fortified food (see Figure 26.1). Folic acid has greater bioavailability than natural food folate because of its chemical stability and lack of a polyglutamate moiety, which impairs absorption across the intestinal epithelium.[13] Once transported into the cell, folic acid is reduced to dihydrofolate (DHF) and subsequently THF by the enzyme DHF reductase (DHFR). Once fully reduced, folic acid is indistinguishable from natural

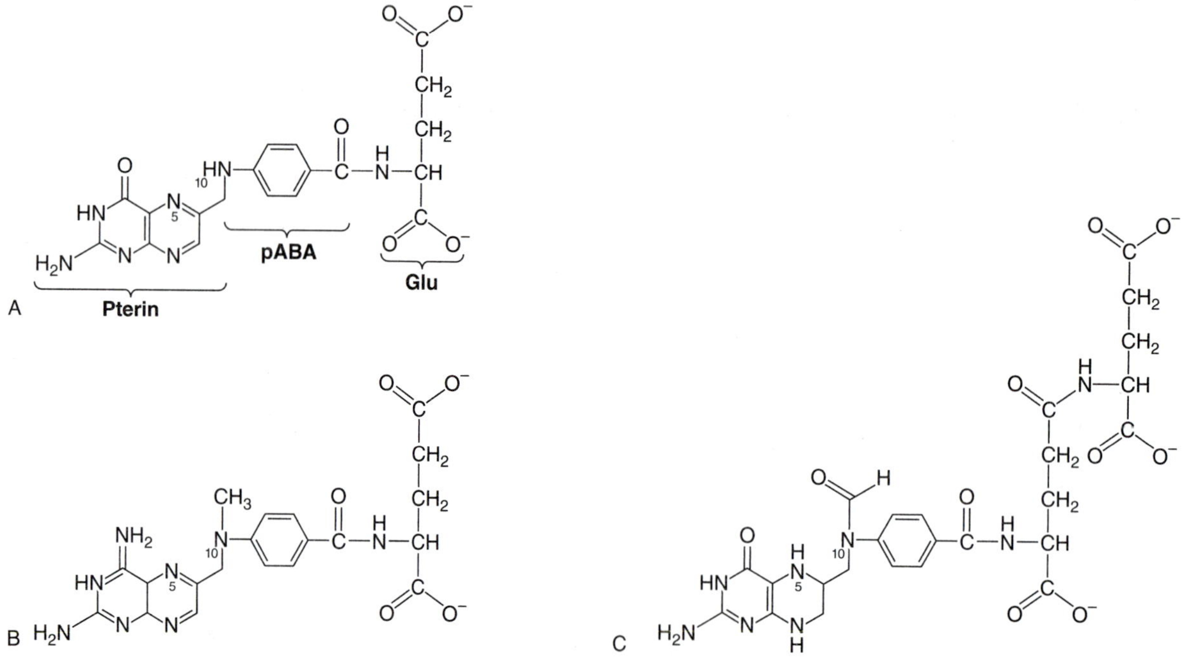

Figure 26.1 The chemical structure of folic acid (**A**), methotrexate (**B**), and 10-formyl-tetrahydrofolate diglutamate (**C**). Folic acid contains a pterin ring that is bridged to *para*-aminobenzoic acid (PABA) through a methylene group to form pteroic acid. The addition of the glutamate residue (Glu) through a peptide linkage results in the formation of folic acid. Methotrexate (4-amino-10-methylpteroylglutamic acid) (**B**) is a folate analog, antagonist, and pharmaceutical agent that inhibits the activity of DHFR. Once transported into the cell, folic acid is reduced to tetrahydrofolate and is modified by the addition of a glutamate polypeptide containing up to nine glutamate residues linked by unusual γ-peptide linkages. THF is also modified by the addition of single carbons at the N5 or N10 position or that bridge the N5 and N10 positions. The carbon moieties are carried at the oxidation states of formate, formaldehyde, or methanol. The structure of 10-formyl-tetrahydrofolate diglutamate is shown in **C**.

food folate. Low DHFR expression may result in the appearance of folic acid in the serum of individuals with high folic acid intake.[14] Evidence indicates that total DHFR activity is highly variable among individuals—which suggests a variable capacity to metabolize folic acid.[15]

Recommended Dietary Allowances and Folic Acid Fortification

The recommended intakes of folate established by the Food and Nutrition Board of the Institute of Medicine are shown in **Table 26.1**.[16] Dietary folate requirements are expressed as dietary folate equivalents (DFEs) because of the need to adjust for the increased bioavailability of folic acid, compared with natural food folate.[12] Folic acid is estimated to be 1.7 times more bioavailable than natural food folate. The recommended dietary allowance (RDA) for both men and women is 400 μg/day DFEs. The requirement for women of childbearing age is 400 μg of folic acid from fortified foods and supplements, in addition to food folate consumption from a varied diet.[16] The tolerable upper intake level for adults was set at 1,000 μg/day of folic acid exclusive of food folate, based on concerns that elevated intake of folic acid would exacerbate neurologic consequences of vitamin B_{12} deficiency.

The United States and Canada mandated the addition of folic acid at a level of 140 μg/100 g of product to enriched flour in 1998 to achieve a predicted intake of 100 μg/day folic acid to reduce the incidence of NTDs.[17] Before folic acid fortification, the median folate intake from food was estimated to be 250 μg/day. Population intake levels increased by 529 μg DFE/day between the interval 1993 to 1994 (before fortification) and 1999 to 2000 (after fortification), and then decreased by 135 mg in 2003 to 2004.[17] Folic acid fortification increased serum and red blood cell folate concentrations in the United States and decreased total plasma homocysteine by 6% to 13%.[18,19]

Table 26.1 Dietary Reference Intakes for Folate[a]

Group	Intake (μg of DFE/D)
Infants (mo)	*Adequate Intake*
0–5	65
6–11	80
Children and Adolescents (y)	*Recommended Dietary Allowance*
1–3	150
4–8	200
9–13	300
14–18	400
Adults (y)	*Recommended Dietary Allowance*
≥19	400
Pregnant Women	600
Lactating Women	500

DFE, dietary folate equivalent.

[a]Requirements are expressed as dietary folate equivalents, to account for the increased bioavailability of folic acid compared with natural food folate.

Data from Food and Nutrition Board, Institute of Medicine. Folate. In: Dietary Reference Intakes for Thiamin, Riboflavin, Niacin, Vitamin B_6, Folate, Vitamin B_{12}, Pantothenic Acid, Biotin, and Choline. Washington, DC: National Academy Press, 1998:196–305; and Bailey LB, Gregory JF III. J Nutr 1999;129:779–82.

Sites of Intestinal Absorption

Folate absorption across the intestinal epithelium occurs in the acidic environment of the upper small intestine through the proton-coupled folate transporter (PCFT),[20] which was originally discovered as a heme transporter. Loss of PCFT function is associated with severe folate malabsorption, indicating that it functions as the primary folate transporter in the gut. Only folate monoglutamates are bioavailable and absorbed. During digestion, the γ-glutamyl polypeptide of natural food folate is hydrolyzed to generate folate monoglutamate forms through a reaction catalyzed by the enzyme γ-glutamyl hydrolase. Folates circulate in serum as monoglutamate derivatives, primarily in the form of 5-methyl-THF. In circulating red blood cells, 5-methyl-THF polyglutamates are the primary form of folates, although individuals with polymorphisms in the methylenetetrahydrofolate reductase (MTHFR) gene accumulate 10-formyl-THF in red blood cells.[21] Transport into cells occurs primarily through the reduced folate carrier. Once transported into cells, folate monoglutamate derivatives are either converted to their polyglutamate forms by the addition of a γ-glutamyl polypeptide, usually consisting of five to nine glutamate residues in the cytoplasm, or transported into mitochondria as monoglutamate derivatives and converted to polyglutamate forms in that compartment. The glutamate polypeptide serves to retain the vitamin within mitochondria and within the cell.

Biologic Roles of Folate

THF polyglutamates function as coenzymes that donate or accept one-carbons in an integrated network of biosynthetic and catabolic reactions involved in nucleotide and amino acid metabolism. Collectively, the network is referred to as folate-mediated one-carbon metabolism. Folate metabolism is compartmentalized in the cytoplasm and nucleus (**Figure 26.2A**) and mitochondria (**Figure 26.2B**).[22] Each of these intracellular compartments is associated with specific metabolic pathways, and the compartments are interdependent through the exchange of common intermediates, including formate, serine, and glycine (see Figure 26.2B).[22] Folate-mediated one-carbon metabolism also requires the water-soluble vitamins riboflavin (vitamin B_2), niacin (vitamin B_3), choline, pantothenic acid (vitamin B_5), pyridoxal phosphate (vitamin B_6), and cobalamin (vitamin B_{12}) for its function[6] (see Figure 26.2).

Cytoplasm

The de novo biosynthesis of purine and thymidylate nucleotides and the remethylation of homocysteine to methionine occur in the cytoplasm. The cytoplasm is the only compartment in the network that involves all the one-carbon substituted forms of

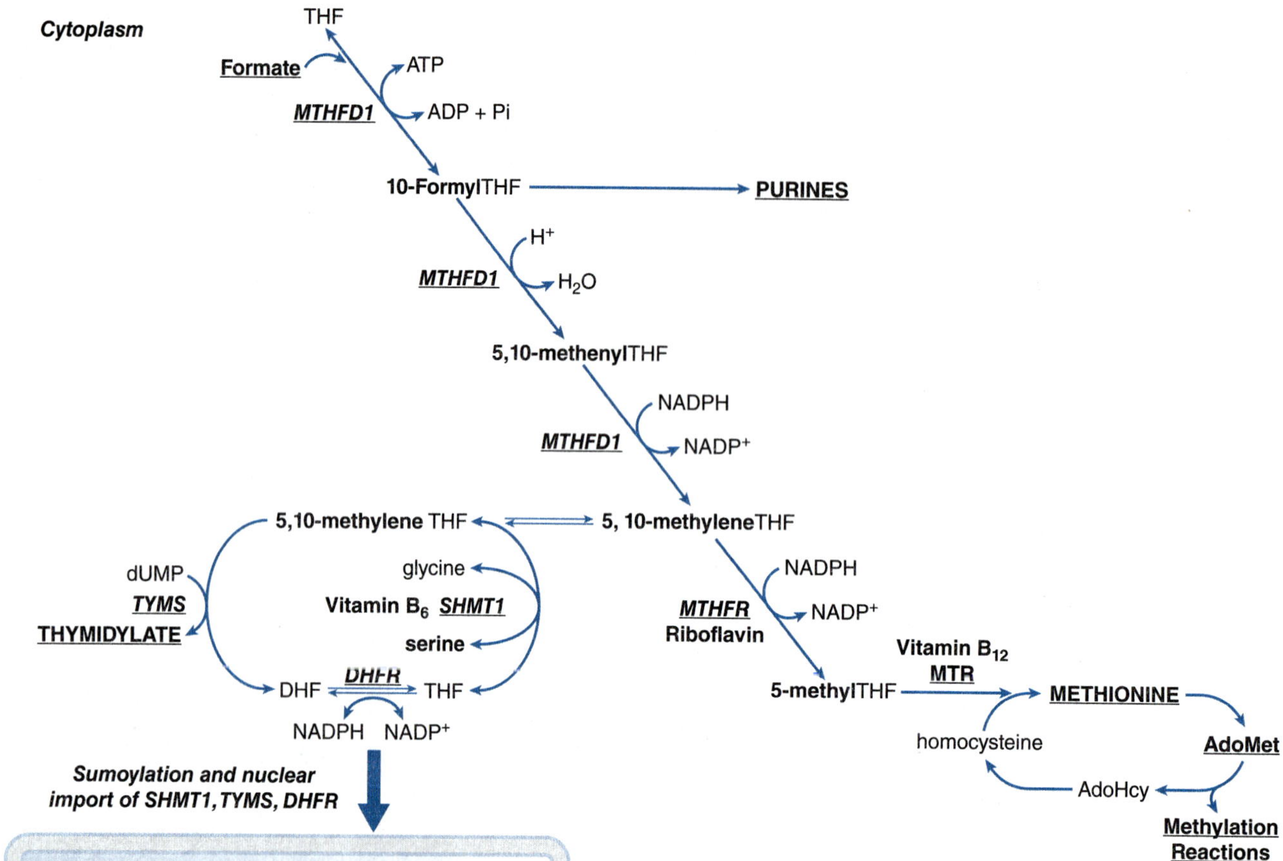

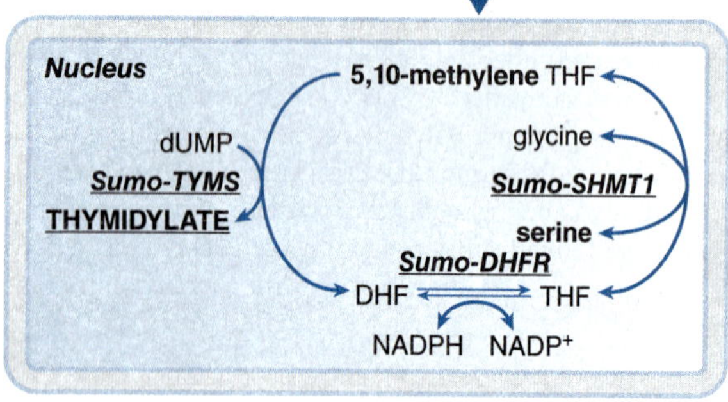

A

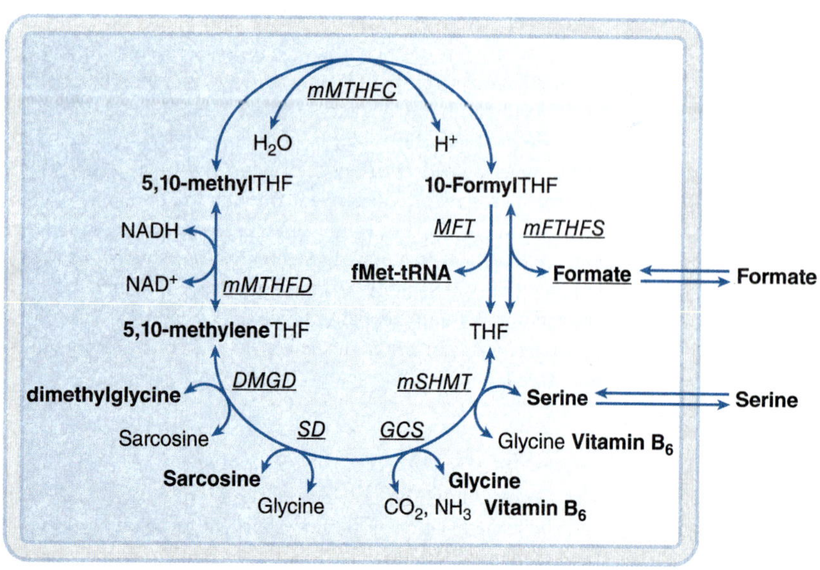

B

Figure 26.2 Compartmentation of folate-mediated one-carbon metabolism in the cytoplasm, mitochondria, and nucleus. (**A**) One-carbon metabolism in the cytoplasm is required for the de novo synthesis of purines and thymidylate and for the remethylation of homocysteine to methionine. One-carbon metabolism in the nucleus synthesizes thymidylate from uridylate and serine, and it occurs during the S phase of the cell cycle. (**B**) One-carbon metabolism in the mitochondria is required to generate formate for one-carbon metabolism in the cytoplasm. The folate and amino acid carriers of the one-carbon unit are indicated by **bold type**. *AdoHcy*, *S*-adenosylhomocysteine; *AdoMet*, *S*-adenosylmethionine; *ADP*, adenosine diphosphate; *ATP*, adenosine triphosphate; *DHF*, dihydrofolate; *DHFR*, dihydrofolate reductase, *DMGD*, dimethylglycine dehydrogenase; *dUMP*, deoxyuridine monophosphate; *GCS*, glycine cleavage system; *mFTHFS*, mitochondrial formyltetrahydrofolate synthetase; *mMTHFC*, mitochondrial methenyltetrahydrofolate cyclohydrolase; *mMTHFD*, mitochondrial methylenetetrahydrofolate dehydrogenase; *MTHFR*, methylenetetrahydrofolate reductase; *MTR*, methionine synthase; *M-tRNA-FT*, methionyl-tRNA formyltransferase; *NAD*, nicotinamide adenine dinucleotide; *NADP*, nicotinamide adenine dinucleotide phosphate; *NADPH*, reduced nicotinamide adenine dinucleotide phosphate; *Pi*, inorganic phosphate; *SD*, sarcosine dehydrogenase; *SHMT1*, cytoplasmic serine hydroxymethyltransferase; *THF*, tetrahydrofolate; *TYMS*, thymidylate synthase.

THF.[22] Formate serves as the primary source of one-carbon units for cytoplasmic one-carbon transfer reactions and is derived from amino acid catabolism in mitochondria.[23] In an adenosine triphosphate-dependent reaction, formate condenses with THF to form 10-formyl-THF, catalyzed by the 10-formyl-THF synthetase activity of the multifunctional enzyme methylenetetrahydrofolate dehydrogenase 1 (MTHFD1). Therefore, MTHFD1 is the primary entry point of one-carbons into the one-carbon metabolic network in the cytoplasm. The folate-dependent de novo synthesis of purine nucleotides involves 10 reactions and occurs through the formation of a multiple-enzyme complex termed a *purineosome*, which assembles when exogenous sources of purines are not available.[24,25] The activated formyl moiety of 10-formyl-THF is incorporated into the number 2 and number 8 positions of the purine ring. In the third reaction of de novo purine biosynthesis, phosphoribosylglycinamide formyltransferase (GARFT) catalyzes the 10-formyl-THF–dependent conversion of glycinamide ribotide (GAR) to form formylglycinamide ribonucleotide (FGAR) and THF (see Figure 26.2). In the ninth reaction, phosphoribosylaminoimidazolecarboxamide formyltransferase (AICARFT) catalyzes the 10-formyl-THF–dependent conversion of aminoimidazolecarboxamide ribotide (AICAR) to formylaminoimidazolecarboxamide ribonucleotide (FAICAR) and THF. Transformed cells are dependent on de novo purine biosynthesis, which accounts for the effectiveness of chemotherapeutic antifolates that target GARFT or AICARFT, including 6-*R*-dideazatetrahydrofolate (DDATHF; lometrexol that specifically targets GARFT).[26,27] Methotrexate (4-amino-10-methylpteroylglutamic acid) inhibits several folate-dependent enzymes, including both GARFT and AICARFT, by depleting 10-formyl-THF.

Alternatively, the one-carbon of 10-formyl-THF can be enzymatically reduced to 5,10-methylene-THF through the cyclohydrolase and reduced nicotinamide adenine dinucleotide phosphate (NADPH)–dependent dehydrogenase activities of MTHFD1. The de novo synthesis of thymidylate requires 5,10-methylene-THF as the one-carbon donating cofactor. 5,10-methylene-THF and uridylate are converted to thymidylate and DHF in a reaction catalyzed by the enzyme thymidylate synthase (TYMS). For this reaction, 5, 10-methylene-THF serves both as a one-carbon donor and as a source of two electrons through the oxidation of THF to DHF. THF is regenerated from DHF in a reaction catalyzed by the NADPH-dependent enzyme DHFR. To complete the de novo thymidylate synthesis cycle, THF is converted to 5,10-methylene-THF by the three catalytic activities of MTHFD1, as described earlier, or, alternatively, by the vitamin B_6–dependent enzyme serine hydroxymethyltransferase (SHMT1 and SHMT2α). The SHMT isozymes catalyze the conversion of serine to glycine to generate 5,10-methylene-THF from THF (see Figure 26.2).[22] Several chemotherapeutic agents that target TYMS have been developed, including the fluoropyrimidines 5-fluorouracil (5-FU) and 5-fluoro-2-deoxyuridine (FdUrd), and the antifolates raltitrexed, pemetrexed, and methotrexate. These agents have been proven effective in the treatment of head, neck, breast, stomach, and colon cancers.[28-30] They decrease TYMS catalytic function, while also increasing cellular TYMS concentration by preventing TYMS from binding to its mRNA, or by decreasing the rate of ubiquitin-independent enzyme degradation.[31-33]

The remethylation of homocysteine to methionine occurs through folate-dependent and folate-independent pathways. For the folate-dependent pathway, 5,10-methylene-THF is reduced to 5-methyl-THF in a reaction catalyzed by the NADPH- and flavin adenine dinucleotide (FAD)–dependent enzyme MTHFR. 5-Methyl-THF is a cofactor for homocysteine remethylation to methionine, which is catalyzed by methionine synthase (MTR) in a vitamin B_{12}–dependent reaction that converts 5-methyl-THF and homocysteine to methionine and THF. Homocysteine can be converted to methionine in a folate-independent reaction catalyzed by the enzyme betaine homocysteine methyltransferase, a reaction in which betaine serves as the one-carbon donor. Once formed, methionine can be adenosylated to form *S*-adenosylmethionine (AdoMet), which is a cofactor and one-carbon donor for numerous other methylation reactions.[34] *S*-Adenosylhomocysteine (AdoHcy) is a product of AdoMet-dependent transmethylation reactions and is cleaved to form adenosine and homocysteine, which completes the homocysteine remethylation pathway.

These three metabolic pathways in the cytoplasm are highly interconnected and interdependent. Folate-dependent enzymes bind folate polyglutamate cofactors tightly with binding constants in the low micromolar or nanomolar range. The cellular concentration of folate-binding proteins exceeds that of folate derivatives (which are present at 25 to 35 μM) and, therefore, the concentration of free folate in the cell is negligible.[10,35-37] Consequently, independent of their origin, metabolic impairments of one-carbon metabolism rarely affect a single pathway but, rather, influence the entire network. This primarily occurs because the folate-dependent pathways compete for a limiting pool of folate cofactors in the cytoplasm.[10,38]

Mitochondria

Mitochondria contain as much as 40% of total cellular folate, and folate polyglutamates in mitochondria are a distinct pool that does not exchange with folate polyglutamates in the cytoplasm.[39] In this compartment, one-carbon metabolism is required to formylate Met-tRNA to form fMet-tRNA, which is used for the initiation of mitochondrial protein synthesis. A primary role of one-carbon metabolism in mitochondria is to generate formate for one-carbon metabolism in the cytoplasm. Both pathways require 10-formyl-THF (see Figure 26.2B). Formate can be generated from the THF-dependent catabolism of the amino acids glycine, serine, dimethylglycine, and sarcosine, although serine and glycine are the primary sources of formate.[23,40] These amino acids donate a one-carbon to THF, thus generating 5,10-methylene-THF, which is subsequently oxidized to form 10-formyl-THF through the activities of MTHFD2 and MTHFD1L.[41] Formate derived in mitochondria traverses to the cytoplasm for use in cytoplasmic one-carbon metabolism.[23,42]

Nucleus

Approximately 10% of total liver folates reside in the nuclear compartment.[43] Nuclear one-carbon metabolism functions to generate thymidylate from uridylate during the S phase of the cell cycle and during ultraviolet radiation-induced DNA damage.[7,44] The enzymes that constitute the entire thymidylate synthesis pathway, including SHMT1, MTHFD1, SHMT2α, TYMS, and DHFR, are modified by the small ubiquitin-like modifier (SUMO), which facilitates the nuclear translocation of the entire pathway.[22] MTHFD1 and SHMT are the sources of one-carbons for nuclear thymidylate synthesis; mice lacking SHMT1 exhibit impaired thymidylate synthesis, and elevated uracil in nuclear DNA.[45,46] The necessity for redundancy in the compartmentation of de novo thymidylate synthesis in the cytoplasm and nucleus is not known.

Regulation of the Cellular Methylation Potential and Impact of Vitamin B_{12} Deficiency

The cellular methylation potential is highly regulated, primarily through the control of AdoMet synthesis and use. AdoMet synthesis occurs only when cellular AdoMet concentration is depleted, and excess AdoMet is consumed when its concentration is elevated. Both regulatory processes occur through a single mechanism that involves interactions among the metabolites AdoMet, 5-methyl-THF, MTHFR, and glycine *N*-methyltransferase (GNMT), an enzyme that catalyzes the AdoMet-dependent methylation of glycine to sarcosine for the purpose of disposing excess cellular AdoMet.[47] Because AdoMet is an allosteric inhibitor of MTHFR, 5-methyl-THF, which is the cofactor for the folate-dependent remethylation of homocysteine to methionine, it is synthesized only when AdoMet concentration is depleted. This feedback inhibition of MTHFR by AdoMet ensures that folate cofactors are available for nucleotide synthesis when the methylation potential is adequate to support cellular methylation reactions.

AdoMet concentration is also maintained through the allosteric inhibition of GNMT by 5-methyl-THF.[48] The depletion of 5-methyl-THF that results from the AdoMet inhibition of MTHFR activates GNMT which, in turn, lowers AdoMet through the conversion of glycine to sarcosine. GNMT prevents the cellular accumulation of AdoMet. Mice lacking GNMT exhibit a 36-fold elevation in AdoMet concentration and a 100-fold increase in the AdoMet/AdoHcy ratio. They also exhibit fatty liver.[49] Humans with mutations in GNMT have been identified with similar metabolic disruptions as seen in the *Gnmt* null mice.[49] GNMT expression is regulated by vitamin A[50] and glucocorticoids,[51] thus providing a mechanism whereby signals and nutrients unrelated to one-carbon metabolism can influence cellular methylation reactions and potentially epigenetic processes.[48]

Vitamin B_{12} deficiency has a major impact on the folate-mediated, one-carbon network and disrupts both homocysteine remethylation and nucleotide biosynthesis. This disruption is the metabolic origin of megaloblastic anemia. Vitamin B_{12} deficiency, whether from nutritional deficiency or excessive exposure to nitrous oxide, impairs MTR activity. Lack of MTR activity disrupts the homocysteine remethylation pathway and depletes AdoMet concentration, thereby activating MTHFR. MTHFR activation causes cellular folate to accumulate as 5-methyl-THF, a condition often referred to as the folate "methyl trap." Because the MTHFR-catalyzed generation of 5-methyl-THF is essentially irreversible in vivo, accumulation of 5-methyl-THF can impair purine and thymidylate de novo biosynthesis[52], as occurs in severe vitamin B_{12} deficiency.[53]

Analytic Methods and Biomarkers of Impaired One-Carbon Metabolism

Certain clinically useful biomarkers are responsive to dietary folate intake and can be used to assess folate status and disruptions in folate-mediated one-carbon metabolism. Most of the functional metabolic biomarkers that report on folate status lack specificity, however, because they are also responsive to nutritional status for other B vitamins as well as for common variants in genes that encode folate-dependent enzymes.[13] Impairments in the folate metabolic network can result from primary dietary folate deficiency and from deficiencies of other nutrients that function in one-carbon metabolism, including vitamin B_6, vitamin B_{12}, and riboflavin, from excessive alcohol intake, or from genetic variation that influences the activity or expression of folate-dependent enzymes (see Figure 26.2).[13]

Folate concentration is measured in serum or red blood cells to assess folate status in individuals as well as in epidemiologic and population studies.[54] Red blood cell folate concentration is the preferred indicator of long-term folate status.[55] The reason is that folate enters the red blood cells only during their development in the bone marrow and, therefore, these values reflect the average folate status over the life span of the adult red blood cell, which is 120 days. A value of 140 ng/mL of red blood cell folate is considered the lower limit of folate sufficiency. Serum folate represents both long-term status and recent folate intake and, thus, should be measured repeatedly when folate status is assessed. Plasma concentration lower than 7 nM indicates negative folate balance. The persistence of this condition usually progresses to megaloblastic anemia.

Measuring folate is particularly challenging because of the multiple chemical forms of the vitamin and its chemical liability.[10] Microbiologic assay has been the method of choice historically, in part because it measures growth of *Lactobacillus casei* when exposed to all forms of folate monoglutamates and can tolerate the high quantities of ascorbate that are required to prevent folate oxidation during sample preparation and assay. Radioisotope and chemiluminescent, and high-performance chromatography/mass spectrometry methods have also been widely used.[56,57] The high-performance chromatography methods have the advantage of distinguishing among folate one-carbon and polyglutamate forms.[55] Similarly, mass spectrometry methods are capable of resolving and quantifying the one-carbon forms of folate.[58] The distribution of the one-carbon forms of folate in serum or erythrocytes may serve as a more robust biomarker of folate status or functional indicator of impaired one-carbon metabolism. All of the existing methods are prone to errors of accuracy and precision, as discussed elsewhere.[59]

Two sensitive biomarkers of impaired folate metabolism are uracil content in nuclear DNA[60,61] and elevations in homocysteine in plasma and tissue.[55] Decreased rates of deoxythymidine triphosphate (dTTP) synthesis result in the incorporation of deoxyuridine triphosphate (dUTP) into DNA, because DNA polymerases do not discriminate between dUTP and dTTP.[62] Uracil content in white blood cell DNA may not be a robust surrogate for uracil content in tissue; however, so its utility for clinical and population assessment of folate status remains uncertain.[63] As mentioned earlier, total plasma homocysteine is responsive to folate status and has been used to assess folate status in the clinic and in population studies. This is because decreased rates of folate-dependent homocysteine remethylation result in elevations in cellular and plasma homocysteine.[64] Total plasma homocysteine is also responsive to vitamin B_6 and vitamin B_{12} status as well as genetic variation in the *MTHFR* gene.[65] Elevated homocysteine also results in elevated AdoHcy concentration because the equilibrium for the hydrolysis AdoHcy to adenosine and homocysteine favors AdoHcy synthesis.[34] AdoHcy is a potent inhibitor of AdoMet-dependent methylases, including DNA and protein methyltransferases,[66] and AdoHcy accumulation causes hypomethylated DNA.[67] AdoMet, AdoHcy, and the AdoMet/AdoHcy ratio have also been explored as functional indicators of folate status.[68] Other biomarkers of whole-body folate deficiency include DNA hypomethylation, elevated formiminoglutamate in urine (an intermediate in folate-dependent histidine catabolism), and neutrophil hypersegmentation.[13]

The biomarkers that were used to determine the RDA for dietary folate intake by the Institute of Medicine include erythrocyte folate concentration and plasma homocysteine and folate concentrations.[55] Each of these is responsive to inadequate dietary intake of folate. As stated earlier, however, secondary nutrient deficiencies including vitamin B_6 and B_{12}, genetics, and sex can influence metabolic biomarkers such as homocysteine that are used to assess whole body folate status.[69] Men tend to exhibit higher homocysteine than women.[55] The reference interval for total plasma homocysteine varies among research laboratories. A value of 10 μM total plasma homocysteine has been suggested to be a cutoff to assess folate status in populations. Erythrocyte folate concentration <140 ng/mL indicates folate deficiency, but this concentration is also influenced by genetic variation, including the common 677 C → T polymorphism in the *MTHFR* gene. Increasing evidence indicates that dietary folate requirement may differ by *MTHFR* genotype.[70]

Folate and Epigenetics

Epigenetics commonly refers to the inheritance of traits independent of DNA primary sequence, and this term is often used to describe the transmission of DNA methylation patterns and potentially other covalent chromatin modifications that may be heritable.[71] The term is often associated with various biologic phenomena including X chromosome inactivation and metabolic and nutritional imprinting. The influence of maternal nutrition during gestation and the suckling period on fetal and neonatal traits is referred to as *metabolic imprinting*.[72] Experimental animal models support the concept that maternal nutrition can influence the developing fetus and neonate by altering DNA methylation and gene expression patterns. These genomic changes persist throughout the lifetime of the animal and result in risk phenotypes, such as obesity and metabolic syndrome that increase the risk of adult-onset disease, including cardiovascular disease and certain cancers.[72] Dietary exposures associated with metabolic programming phenomena include energy undernutrition and overnutrition, as well as intakes of specific dietary components, including protein, fatty acids, folate, choline, methionine, and combinations of these B vitamins and methyl donors.[73,74]

Dietary folate and intake of other B vitamins and metabolites of one-carbon metabolism can induce alterations in DNA methylation patterns[75] that are potentially heritable.[76] The only established mechanism whereby folate influences epigenetic processes is through the homocysteine remethylation pathway. Cellular AdoMet and AdoHcy concentrations influence the activity of AdoMet-dependent methyltransferases, albeit by distinct mechanisms. AdoMet is the substrate for transmethylation reactions catalyzed by methyltransferases, including DNA and histone methyltransferases. AdoHcy, the product of AdoMet-dependent methyltransferase reactions, binds tightly and inhibits many AdoMet-dependent enzymes through product inhibition and, therefore, is a physiologically relevant inhibitor of chromatin methylation. The ratio of cellular AdoMet to AdoHcy concentration has often been referred to as the *cellular methylation potential*.[34] AdoHcy, which accumulates when homocysteine accumulates (see Figure 26.1), is the more important determinant of cellular methylation capacity and global DNA methylation, compared with AdoMet concentration, in human lymphocytes. In cystathionine β-synthase–deficient mice, which exhibit elevated homocysteine and AdoHcy, AdoHcy, but not AdoMet, predicts global DNA hypomethylation.[77] Similarly, vitamin B_{12} deficiency causes elevated homocysteine and DNA hypomethylation in rodents.[78]

Many examples of the impact of folate and other B-vitamin dietary deficiencies on global DNA CpG methylation have been reported. Mice fed a folate-deficient diet for 32 weeks exhibited 60% elevations in serum homocysteine and global DNA hypomethylation in splenocytes (reduced 9.1%) and colonic epithelial cells (reduced 7.2%), without changes in allelic-specific methylation at the mouse B1 element or the genetically imprinted genes *H19* or *Oct4*.[61] These effects may be mediated through elevations in homocysteine. In an endothelial cell culture model, exposure to homocysteine (to 50 μm) was sufficient to reduce DNA methyltransferase 1 (DNMT1) activity by 30%, without affecting DNMT1 protein concentration. Homocysteine exposure diminished DNA CpG methylation within the repressive cyclin-dependent element of the cyclin A promoter, thus leading to depressed *cyclin A* transcription. These results are consistent with a role for homocysteine-induced elevation in AdoHcy ratio in the regulation of DNMT1 activity.[79] In human studies, genetic impairments that elevate homocysteine are also associated with global DNA hypomethylation. A common polymorphism in the *MTHFR* gene, C677T, is associated with reduced MTHFR enzyme activity and elevated homocysteine[80] as well as reduced methylation potential and DNA hypomethylation in lymphocytes.[21] Folate nutritional status and global DNA CpG methylation do not exhibit a linear dose-response relationship, however. Folic acid supplementation in humans at 1 mg/day was shown not to alter long interspersed nuclear element-1 (LINE-1) methylation density, a proxy for DNA global methylation, in normal colonic mucosa cells.[81]

More severe elevations in plasma homocysteine, at concentrations shown in patients with homocystinuria (plasma homocysteine >50 μM), exhibited DNA CpG hypomethylation and biallelic expression of both sex-linked and imprinted genes. The magnitude of the shift from monoallelic expression to biallelic expression depended on homocysteine concentration. Supplemental folic acid corrected the DNA hypomethylation and restored imprinted patterns of gene expression.[82] However, in the absence of severe genetic disruptions, as seen in patients with homocystinuria, or severe dietary deficiencies, no evidence indicates that one-carbon metabolism influences gene silencing associated with classic parent-of-origin-specific genetic imprinting.

In transformed tissue, folate deficiency or genetic impairment in one-carbon metabolism affects both global and allelic-specific CpG methylation differently from that observed in nontransformed cells. Folate nutritional status and LINE-1 methylation do not correlate in normal colonic mucosa, but some evidence indicates that folate status does affect LINE-1 methylation once neoplasia develops.[83] Similarly, the common C677T polymorphism in *MTHFR* is associated with increased promoter methylation in colon cancer.[84] A study by de Vogel et al.[85] showed that the variants of MTR (A2756) and MTR reductase (A66G) genes may reduce mutL homolog 1 (*MLH1*) promoter hypermethylation in colorectal cancer. In non–small cell lung cancers, folate concentration in tumors correlated with global methylation, using LINE-1 methylation as a surrogate and allelic-specific methylation at the promoters of CDH13, RUNX3, but not MYOD1, RASSF1P16, APC, RARB. This study supports the concept that folate influences both global and some allelic-specific methylation in transformed cells.[86]

Although alterations in one-carbon metabolism and the cellular methylation potential can alter DNA methylation patterns and affect gene expression, the ability of maternal dietary folate and other metabolites of one-carbon metabolism to establish and then memorize specific changes in chromatin methylation appears to be possible only within specific developmental windows.[87] Harnessing the diet-epigenetic relationship will require further advances in our understanding of methylation targeting and limits of epigenetic plasticity in embryonic and adult stem cells as well as more differentiated cell lineages.

Folate and Neural Tube Defects

NTDs are neurodevelopmental anomalies that result from a failure of the neural tube to close during early embryonic development.[88] They are among the most common congenital birth defects in humans, with a worldwide prevalence ranging from 0.5 to 60/10,000 births.[5,89] The most common and severe NTDs include spina bifida, which results from a failure of posterior neural tube to close and results in exposure of the spinal cord and life-long paralysis, and anencephaly, which is lethal and defined by absence of the cranial vault and brain secondary to failure of the anterior neural tube to close. Maternal folic acid supplementation is the most effective intervention known to prevent NTDs, and it may prevent up to 70% of them. In the United States and Canada, folic acid fortification of enriched flour was initiated in 1998 to lower the incidence of NTDs, and this public health intervention has been successful.[90] Human genetic variation that contributes to a woman's risk of having an NTD-affected pregnancy includes genes that encode the folate-dependent enzymes *MTHFR, SHMT1*[91] and *MTHFD1*.[92] Both maternal and fetal *MTHFR* variants contribute to risk, whereas *MTHFD1* risk is exclusively maternal.

The impaired metabolic pathway responsible for NTDs has not been established. Homocysteine is cytotoxic at high concentration and induces oxidative stress, but mouse models of inborn errors of metabolism that exhibit severe hyperhomocysteinemia, including MTHFR deletion, do not develop NTDs. Similarly, elevated homocysteine in fetal culture medium does not induce NTDs in developing embryos.[93] Impairments in AdoMet-dependent methylation reactions, including genomic methylation, have also been proposed to underlie the origin of NTDs. Decreased chromatin methylation may affect neural tube closure by affecting cellular differentiation[94] or cellular migration processes,[95,96] both of which are critical for neural tube formation. In support of this notion, mice with targeted deletion of the de novo DNMT enzyme Dnmt3b exhibit altered differentiation capacity in ES cells,[97] and embryos have NTDs. These findings confirm the essentiality of de novo methylation and cell differentiation in neural tube closure. Targeted deletion of genes that mediate DNA methylation-mediated suppression of gene expression also result in NTDs.[98] The relevance of these mouse models to human NTDs, if any, is not known, however, nor is it known whether these NTDs can be prevented with folic acid.

Human embryos with NTDs have been shown to exhibit impaired de novo thymidylate synthesis,[99] a finding indicating a potential causal correlation between impaired thymidylate biosynthesis and NTDs. The rapid proliferation of the neuroepithelium during neural tube formation requires robust de novo nucleotide biosynthesis to maintain rates of cell division and limit uracil accumulation in DNA. Impairments in thymidylate biosynthesis during DNA replication and repair decrease rates of cell division during the critical period of neural tube closure.[100] Mouse models of genomic instability also exhibit NTDs, and genomic instability resulting from increased uracil accumulation in DNA has been demonstrated.[45,101] Disruption of the murine *Pax3* gene, which encodes a homeobox transcription factor, causes 100% penetrant spina bifida and impaired de novo thymidylate biosynthesis.[102,103] Maternal in utero folic acid supplementation or supplementation of culture media with either thymidine or folic acid prevented NTDs in homozygous *Pax3* null embryos, whereas methionine supplementation exacerbated the NTD phenotype. Collectively, the literature indicates that thymidylate biosynthesis is a strong candidate for the causal biosynthetic pathway involved in folate responsive NTD pathogenesis. However, human epidemiologic and murine fetal culture models have also identified choline as a modifier of NTD risk.[104]

Choline interacts with folate-mediated, one-carbon metabolism by two distinct mechanisms. Choline degradation is a source of one-carbon units for one-carbon metabolism in the cytoplasm. Choline biosynthesis from glycine is folate dependent, requiring three equivalents of AdoMet. The developing embryo may also be at risk for developmental anomalies resulting from vitamin B_{12} deficiency. Increasing evidence from cross-sectional studies indicates that NTD-affected pregnancies that are not prevented by maternal folic acid supplementation or wheat flour folic acid fortification may result from vitamin B_{12} deficiency, although no randomized controlled trials have been conducted.[5,105] Additional research is required to demonstrate conclusively which disruptions in one-carbon metabolism are causal and which are bystanders in the etiology of folate-responsive developmental anomalies. The potential role of other nutrients in the origin of NTD should be considered.

Folate in Cancer and Chronic Disease

Impairments in one-carbon metabolism, as indicated primarily by elevated plasma homocysteine or low circulating folate concentrations, have been shown to be associated with cardiovascular disease, cancers, and cognitive decline.[106-111] Gene–diet interactions are thought to be fundamental to the origins of virtually all folate-associated chronic disease. Genetic variation in the one-carbon metabolic network has been shown to be associated with cancer risk; the *MTHFR* 677C → T polymorphism is associated with increased risk of NTDs, but decreased risk of colon cancer.[112] Although mechanisms have yet to be established, proposed mechanisms underlying these disorders include the modification of cellular proteins by homocysteine that leads to loss of function,[113,114] alterations in genome methylation and gene expression profiles, uracil accumulation in DNA, and subsequent genome instability.[115] Proposed roles of low folate status in carcinogenesis have been the subject of several excellent reviews.[115,116] Low folate status increases uracil content in DNA,[22] which can lead to double strand breaks and altered DNA methylation patterns, all of which contribute to carcinogenesis. Folate has also been proposed to be a double-edged sword in relation to cancer risk. Whereas folate deficiency may increase the risk of cancer initiation, it has also been proposed to accelerate the growth of established cancers.

Randomized placebo-controlled clinical trials have not conclusively validated observational studies, indicating a preventive role for folate in cardiovascular disease and cancer. Secondary prevention trials have failed to demonstrate an effect of homocysteine lowering on cardiovascular disease outcomes.[117] Similarly, randomized clinical trials do not support a role for folic acid supplementation in the prevention of colon cancer.[118] This last finding indicates that cancer risk may be associated only with overt folate deficiency.[119] Given the role of folate in nucleotide biosynthesis, investigators have suggested that elevated folate status may accelerate cellular transformation or tumor growth in colon cancer, but definitive evidence has not fully supported this hypothesis to date.[119,120]

Folate-mediated one-carbon metabolism remains an attractive target for nutritional intervention to prevent or manage chronic disease, but a more comprehensive understanding of the causal pathways, their regulation, and mechanism of pathogenesis is required. Genome wide association studies are indicating a role for one-carbon metabolism in mitochondria in vascular disease,[121] but virtually nothing is known about the regulation of one-carbon metabolism in this compartment, including whether formate production is limiting in the one-carbon network. A more comprehensive understanding of folate metabolism and its regulation should provide a better mechanistic appreciation of the role of folate in human disease and should lead to the design of more effective therapies and preventive strategies.

References

1. Hoffbrand AV, Weir DG. Br J Haematol 2001;113:579–89.
2. Mitchell HK, Snell EE, Williams RJ. J Am Chem Soc 1941;63.
3. Angier RB, Boothe JH, Hutchings BL, et al. Science 1945;102: 227–8.
4. Farber S, Diamond LK. N Engl J Med 1948;238:787–93.
5. Crider KS, Qi YP, Yeung LF, et al. Annu Rev Nutr 2022;42: 423–52.
6. Fox JT, Stover PJ. Vitam Horm 2008;79:1–44.
7. Lan X, Field MS, Stover PJ. Wiley Interdiscip Rev Syst Biol Med 2018;10:e1426.
8. Mazat JP. Biochem Soc Trans 2021;49:1–15.
9. Moran RG. Semin Oncol 1999;26:24–32.
10. Suh JR, Herbig AK, Stover PJ. Annu Rev Nutr 2001;21:255–82.
11. Yang Q, Cogswell ME, Hamner HC, et al. Am J Clin Nutr 2010;91:64–72.
12. Bailey LB. Nutr Rev 1998;56:294–9.
13. Bailey LB, Stover PJ, McNulty H, et al. J Nutr 2015;145:1636S–80S.
14. Pfeiffer CM, Sternberg MR, Fazili Z, et al. J Nutr 2015;145: 520–31.
15. Bailey SW, Ayling JE. Proc Natl Acad Sci U S A 2009;106: 15424–9.
16. Dietary Reference Intakes for Thiamin, Riboflavin, Niacin, Vitamin B6, Folate, Vitamin B12, Pantothenic Acid, Biotin, and Choline, Washington, DC, 1998.
17. Quinlivan EP, Gregory JF. Am J Clin Nutr 2007;86:1773–9.
18. Ganji V, Kafai MR. J Nutr 2009;139:345–52.
19. Pfeiffer CM, Osterloh JD, Kennedy-Stephenson J, et al. Clin Chem 2008;54:801–13.
20. Hou Z, Gangjee A, Matherly LH. Faseb J 2022;36:e22164.
21. Friso S, Choi S-W, Girelli D, et al. Proc Nat Acad Sci 2002;99: 5606–11.
22. Field MS, Kamynina E, Chon J, et al. Annu Rev Nutr 2018;38: 219–43.
23. Tibbetts AS, Appling DR. Annu Rev Nutr 2010;30:57–81.
24. An S, Kumar R, Sheets ED, et al. Science 2008;320:103–6.
25. Pedley AM, Benkovic SJ. Trends Biochem Sci 2017;42:141–54.
26. Kovalev IS, Zyryanov GV, Santra S, et al. Molecules 2022;27.
27. Beardsley GP, Moroson BA, Taylor EC, et al. J Biol Chem 1989;264:328–33.
28. Hernando-Cubero J, Matos-García I, Alonso-Orduña V, et al. J Gastrointest Cancer 2017;48:135–47.
29. Takemura Y, Jackman AL. Anti-Cancer Drugs 1997;8.
30. Ferrari S, Severi L, Pozzi C, et al. Vitam Horm 2018;107: 473–513.
31. Kitchens ME, Forsthoefel AM, Rafique Z, et al. J Biol Chem 1999; 274:12544–7.
32. Forsthoefel AM, Pena MM, Xing YY, et al. Biochemistry 2004;43:1972–9.
33. Stover PJ. Nutr Rev 2003;61:427–31.
34. Finkelstein JD. Clin Chem Lab Med 2007;45:1694–9.
35. Misselbeck K, Marchetti L, Field MS, et al. Sci Rep 2017;7:797.
36. Schirch V, Strong WB. Arch Biochem Biophys 1989;269:371–80.
37. Stover PJ, Field MS. Adv Nutr 2011;2:325–31.
38. Scott JM, Dinn JJ, Wilson P, et al. Lancet 1981;2:334–7.
39. Lin BF, Huang RF, Shane B. J Biol Chem 1993;268:21674–9.
40. Lamers Y, Williamson J, Theriaque DW, et al. J Nutr 2009; 139:666–71.
41. Christensen KE, MacKenzie RE. Bioessays 2006;28:595–605.
42. Brosnan ME, Brosnan JT. Annu Rev Nutr 2016;36:369–88.
43. Shin YS, Chan C, Vidal AJ, et al. Biochim Biophys Acta 1976;444:794–801.
44. Fox JT, Shin WK, Caudill MA, et al. J Biol Chem 2009;284: 31097–108.
45. Field MS, Kamynina E, Stover PJ. Biochimie 2016;126:27–30.
46. MacFarlane AJ, Anderson DD, Flodby P, et al. J Biol Chem 2011;286:44015–22.
47. Lu SC, Mato JM. Physiol Rev 2012;92:1515–42.
48. Luka Z, Mudd SH, Wagner C. J Biol Chem 2009;284:22507–11.
49. Luka Z, Capdevila A, Mato JM, et al. Transgenic Res 2006;15: 393–7.
50. Nieman KM, Rowling MJ, Garrow TA, et al. J Biol Chem 2004;279:45708–12.
51. Rowling MJ, Schalinske KL. J Nutr 2003;133:3392–8.
52. Palmer AM, Kamynina E, Field MS, et al. Proc Natl Acad Sci U S A 2017;114:E4095–E102.
53. Scott JM. Proc Nutr Soc 1999;58:441–8.
54. Pfeiffer CM, Sternberg MR, Fazili Z, et al. Br J Nutr 2015;113: 1965–77.
55. Selhub J, Jacques PF, Dallal G, et al. Food Nutr Bull 2008;29: S67–73.
56. Verstraete J, Kiekens F, Strobbe S, et al. Anal Bioanal Chem 2019;411:4383–99.
57. Pfeiffer CM, Zhang M, Jabbar S. Ann N Y Acad Sci 2018;1414: 96–108.
58. Hannisdal R, Gislefoss RE, Grimsrud TK, et al. J Nutr 2010;140:522–6.
59. Green R. Food Nutr Bull 2008;29:S52–63; discussion S4–6.
60. Blount BC, Mack MM, Wehr CM, et al. Proc Natl Acad Sci U S A 1997;94:3290–5.
61. Linhart HG, Troen A, Bell GW, et al. Gastroenterology 2009;136: 227–35.e3
62. Ames BN. Ann N Y Acad Sci 1999;889:87–106.

63. Hazra A, Selhub J, Chao WH, et al. Am J Clin Nutr 2010;91: 160–5.
64. Finkelstein JD. Semin Thromb Hemost 2000;26:219–25.
65. Barbosa PR, Stabler SP, Machado AL, et al. Eur J Clin Nutr 2008;62:1010–21.
66. Clarke S. Homocysteine Health Dis 2001;63–78.
67. Jaenisch R, Bird A. Nat Genet 2003;33:245–54.
68. Obeid R, Kostopoulos P, Knapp JP, et al. Clin Chem 2007;53: 326–33.
69. Bailey LB, Gregory JF, 3rd. J Nutr 1999;129:779–82.
70. Solis C, Veenema K, Ivanov AA, et al. J Nutr 2008;138:67–72.
71. Ptashne M. Curr Biol 2007;17:R233–6.
72. Waterland RA, Garza C. Am J Clin Nutr 1999;69:179–97.
73. Lorenzo PM, Izquierdo AG, Rodriguez-Carnero G, et al. Adv Nutr 2022;13:1725–47.
74. Carlberg C, Velleuer E. Curr Opin Clin Nutr Metab Care 2023;26: 259–265.
75. Bokor S, Vass RA, Funke S, et al. Life (Basel) 2022;12.
76. Waterland RA, Jirtle RL. Mol Cell Biol 2003;23:5293–300.
77. Caudill MA, Wang JC, Melnyk S, et al. J Nutr 2001;131:2811–8.
78. Brunaud L, Alberto JM, Ayav A, et al. Digestion 2003;68:133–40.
79. Jamaluddin MD, Chen I, Yang F, et al. Blood 2007;110:3648–55.
80. Frosst P, Blom HJ, Milos R, et al. Nat Genet 1995;10:111–3.
81. Figueiredo JC, Grau MV, Wallace K, et al. Cancer Epidemiol Biomarkers Prev 1995;18:1041–9.
82. Ingrosso D, Cimmino A, Perna AF, et al. Lancet 2003;361: 1693–9.
83. Schernhammer ES, Giovannucci E, Kawasaki T, et al. 2003 Gut 2003;59:794–9.
84. Oyama K, Kawakami K, Maeda K, et al. Anticancer Res 2004;24:649–54.
85. de Vogel S, Wouters KA, Gottschalk RW, et al. Cancer Epidemiol Biomarkers Prev 2009;18:3086–96.
86. Jin M, Kawakami K, Fukui Y, et al. Cancer Sci 2009;100: 2325–30.
87. Waterland RA. Epigenetics 2009;4:523–5.
88. Beaudin AE, Stover PJ. Birth Defects Res A Clin Mol Teratol 2009;85:274–84.
89. Crider KS, Devine O, Hao L, et al. BMJ 2014;349:g4554.
90. Cordero AM, Crider KS, Rogers LM, et al. MMWR Morb Mortal Wkly Rep 2015;64:421–3.
91. Almekkawi AK, AlJardali MW, Daadaa HM, et al. J Pers Med 2022;12.
92. Brody LC, Conley M, Cox C, et al. Am J Hum Genet 2002;71: 1207–15.
93. Watanabe M, Osada J, Aratani Y, et al. Proc Natl Acad Sci U S A 1995;92:1585–9.
94. Endesfelder D, Burrell RA, Kanu N, et al. Cancer Res 2014;74:4853–63.
95. Issaeva I, Zonis Y, Rozovskaia T, et al. Mol Cell Biol 2007;27: 1889–903.
96. Rahnama F, Shafiei F, Gluckman PD, et al. Endocrinology 2006;147:5275–83.
97. Jackson M, Krassowska A, Gilbert N, et al. Mol Cell Biol 2004;24:8862–71.
98. Kim JK, Huh SO, Choi H, et al. Mol Cell Biol 2001;21:7787–95.
99. Dunlevy LP, Chitty LS, Burren KA, et al. Brain 2007;130: 1043–9.
100. Keller-Peck CR, Mullen RJ. Brain Res Dev Brain Res 1997;102: 177–88.
101. Herrera E, Samper E, Blasco MA. EMBO J 1999;18:1172–81.
102. Fleming A, Copp AJ. Science 1998;280:2107–9.
103. Wlodarczyk BJ, Tang LS, Triplett A, et al. Toxicol Appl Pharmacol 2006;213:55–63.
104. Zeisel SH. Annu Rev Nutr 2006;26:229–50.
105. Thompson MD, Cole DE, Ray JG. Am J Clin Nutr 2009;89:697S–701S.
106. Colapinto CK, O'Connor DL, Sampson M, et al. J Public Health (Oxf) 2016;38:e84–97.
107. Yuan S, Chen J, Dan L, et al. Am J Clin Nutr 2022;116: 1595–609.
108. Malouf R, Grimley Evans J. Cochrane Database Syst Rev 2008;Cd004514.
109. Wang F, Wu K, Li Y, et al. Am J Clin Nutr 2021;114:49–58.
110. Dai WM, Yang B, Chu XY, et al. Chin Med J (Engl) 2013;126:1957–64.
111. Katsa ME, Gil AP. Endocr Metab Immune Disord Drug Targets 2022;22:979–84.
112. Ma J, Stampfer MJ, Giovannucci E, et al. Cancer Res 1997; 57:1098–102.
113. Hubmacher D, Cirulis JT, Miao M, et al. J Biol Chem 2010;285:1188–98.
114. Perla-Kajan J, Jakubowski H. FASEB J 2010;24:931–6.
115. Mason JB. Nutr Rev 2009;67:206–12.
116. Hubner RA, Houlston RS. Br J Cancer 2009;100:233–9.
117. Smulders YM, Blom HJ. J Inherit Metab Dis 2011;34:93–9.
118. Cole BF, Baron JA, Sandler RS, et al. JAMA 2007;297:2351–9.
119. Maruvada P, Stover PJ, Mason JB, et al. Am J Clin Nutr 2020;112:1390–403.
120. Field MS, Stover PJ. Ann N Y Acad Sci 2018;1414:59–71.
121. Samani NJ, Erdmann J, Hall AS, et al. N Engl J Med 2007;357:443–53.

C. VITAMINS

CHAPTER 27

Cobalamin (Vitamin B_{12})*

Joshua W. Miller and Ralph Green

CHAPTER OUTLINE

Abbreviations

cbl, cobalamin; **CoA**, coenzyme A; **holoHC**, holohaptocorrin; **holoTC**, holotranscobalamin; **IF**, intrinsic factor; **MCV**, mean corpuscular volume; **PA**, pernicious anemia; **SCCD**, subclinical cobalamin deficiency

Historical Background

The history of cobalamin is inextricably bound to the disease, pernicious anemia (PA), which provides the most common setting for its clinical deficiency, even though cobalamin deficiency can arise from many other causes. The reader is referred to excellent reviews of the dramatic scientific and clinical story.[1-4] In 1849, Addison reported several patients with a "remarkable form of anemia" that was accompanied by languor and restlessness, among other signs and symptoms. Although Addison mistakenly attributed the anemia to adrenal disease, his report is considered to be the first of the disease, whose often fatal course later led Biermer to name it "pernicious anemia." That name is less apt now because the easily treated disease is no longer pernicious and because the disease is defined by its underlying gastric defect and not its anemic manifestation, which is sometimes minimal or even absent. Indeed, the strikingly megaloblastic anemia, although characteristic, is not specific to PA, or even to cobalamin deficiency.

*Système Internationale units: 1,000 pg cobalamin = 0.738 pmol.

The classical experiments by Minot and Murphy[5] and Whipple[6] transformed the lethal course of PA by feeding affected patients large amounts of liver and documenting their hematologic improvement. For this work, they shared a Nobel Prize. The second important contribution was Castle's discovery that patients with PA responded effectively to an "extrinsic factor" in the ingested liver or meat when it was combined with an "intrinsic factor" (IF) in gastric juice.[7] This demonstration sealed the long-suspected connection of PA with atrophic gastritis. The third critical achievement was the identification of cobalamin as the extrinsic factor. The synthesis of cobalamin was achieved,[8,9] as was the elucidation of its structure, the latter by Hodgkin,[10] who was also awarded a Nobel Prize for her crystallographic work.

As biosynthetic fermentation made cyanocobalamin readily available, PA became easy to treat. The vitamin also became one of the most frequently given injections in the United States and acquired the dubious status of a frequently misused placebo and "energizer." Having lost its grim implications, cobalamin deficiency began to be viewed with complacency by some health professionals, at times to their patients' disadvantage.

The past several decades have extended methodologic advances in accurate and sensitive metabolic assays that allow identification of cobalamin deficiency at early stages of depletion. As a result, asymptomatic subclinical cobalamin deficiency (SCCD)[11,12] is now understood to be far more prevalent than the relatively rare state of clinical deficiency.[13,14] This subclinical expansion has had major epidemiologic ramifications. Major advances have also been made in the molecular understanding of cobalamin transport and metabolism and the varied conditions and disorders that affect these processes, including inborn errors of metabolism, genetic polymorphisms, acquired disorders, and deficiencies or excesses of other nutrients such as folate and iron.

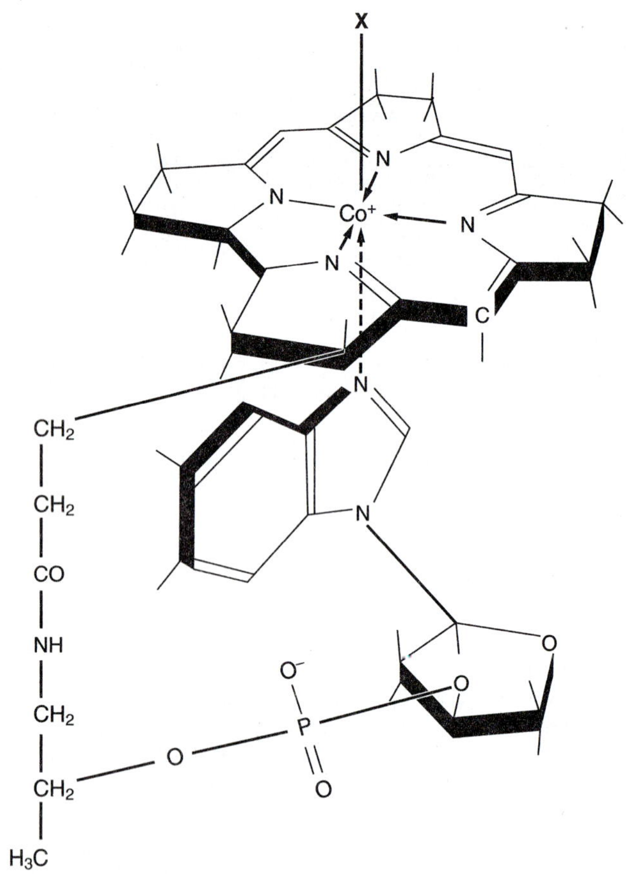

Figure 27.1 The structure of cobalamin. Attached to the central cobalt atom of the corrin tetrapyrrole and to one of the pyrrole rings is the α-ligand, the 5,6-dimethylbenzimidazole nucleotide, extending below the corrin plane. The β-ligand (marked as X in the figure) above the plane can be any of several moieties such as methyl, 5′-deoxyadenosyl, hydroxyl, or cyanide.

Chemistry

Cobalamin contains a planar tetrapyrrole (corrin), at whose center sits a cobalt atom, and has critically attached moieties (**Figure 27.1**). The cobalt fluctuates among monovalent, divalent, and trivalent states, with the reduced, monovalent cob(I)alamin being the active form. Linked to cobalt in the α position below the corrin plane is the 5,6-dimethylbenzimidazole nucleotide. Also linked to the cobalt atom but extending above the plane (β position) is any one of several interchangeable prosthetic moieties that lend their names to the cobalamin. The most important cobalamins are methylcobalamin and 5′-deoxyadenosylcobalamin, in which methyl and 5′-deoxyadenosine are the prosthetic moieties, respectively. Methylcobalamin predominates in cytoplasm and serves as a cofactor with 5-methyltetrahydrofolate in the methylation of homocysteine to methionine catalyzed by the enzyme, methionine synthase (**Figure 27.2**). 5′-deoxyadenosylcobalamin predominates in mitochondria, where it serves as a cofactor in the intramolecular rearrangement of L-methylmalonyl-coenzyme A (CoA) to succinyl-CoA by the enzyme methylmalonyl CoA mutase in propionate metabolism, odd-chain fatty acid metabolism, and catabolism of ketogenic amino acids (**Figure 27.3**). These two are the only known roles for cobalamin in humans.

Other cobalamins include hydroxocobalamin, which is stable and occurs widely; aquocobalamin; and sulfito cobalamin. Cyanocobalamin is a stable biosynthetic pharmaceutical that requires conversion to other cobalamins to become metabolically active; the term *vitamin B_{12}* refers specifically to cyanocobalamin,[8] but it often serves as a catch-all name for cobalamins as a whole. Altered corrinoids with structural deletions, referred to as cobalamin analogs, are nonfunctional in humans, but can find their way into tissues.[3,15-17] There are two major classes of analogs, the cobamides, in which ribose has been replaced by other moieties such as adenine, and the cobinamides, in which the nucleotide has been removed.[18] These analogs differ from the cobalamins in that they do not support the activities of the two cobalamin-dependent enzymes, methionine synthase and methylmalonyl CoA mutase, and do not support the growth of certain cobalamin-dependent bacteria such as *Euglena gracilis* and *Lactobacillus leichmannii*. The origin of analogs in humans is likely gut microbial metabolism of dietary cobalamin, but it is unclear how the analogs get into blood and tissues, as IF-mediated absorption of cobalamin [described in the "Absorption" section later in this chapter] likely excludes facilitated absorption of the analogs via this physiologic process. It is also unclear to what extent cobalamin analogs interfere with cobalamin-dependent reactions.

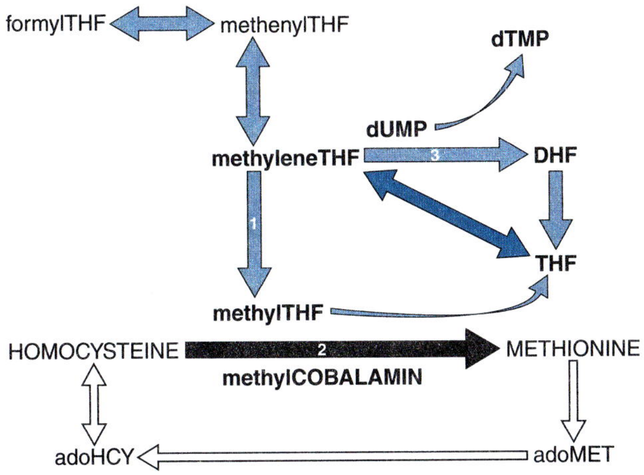

Figure 27.2 Schematic diagram of the intersection of cobalamin (*black arrow*) with folate metabolism (*blue arrows*) and the methionine cycle (*white arrows*). The direct role of folate in thymidylate synthesis (*reaction 3*) is also shown. *Reaction 1*: Reduction of 5,10-methylene tetrahydrofolate (THF) to 5-methylTHF by methyleneTHF reductase, which requires riboflavin. *Reaction 2*: Remethylation of homocysteine to methionine by methionine synthase, with methylTHF and methylcobalamin as substrate and cofactor, respectively; the THF that is produced is reused in the folate metabolic cycle. *Reaction 3*: Conversion of deoxyuridylate (dUMP) to deoxythymidylate (dTMP) by thymidylate synthase, in which 5,10-methyleneTHF is converted to dihydrofolate (DHF). (See the chapter on folate for fuller details of folate metabolism.) adoHCY, *S*-adenosylhomocysteine; adoMET, *S*-adenosylmethionine.

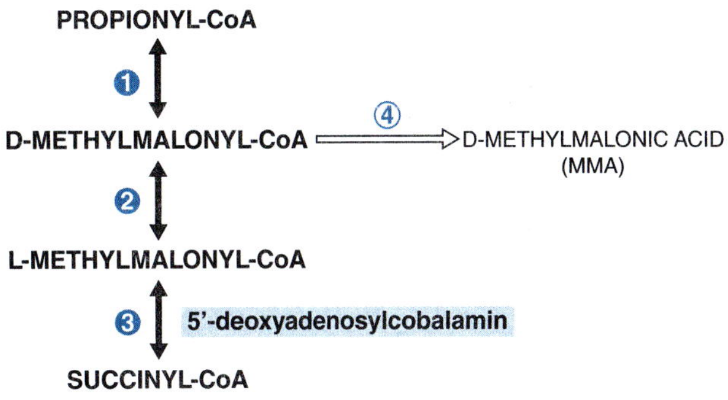

Figure 27.3 The sole function of 5′-deoxyadenosylcobalamin in humans. The mitochondrial conversion of propionyl-coenzyme A (CoA), which derives from diverse sources produces succinyl-CoA, which enters the tricarboxylic acid cycle and passes through three reversible reactions. Reaction 1: Carboxylation of propionyl-CoA by propionyl-CoA carboxylase, requiring adenosine triphosphate (ATP), biotin, and magnesium. Reaction 2: Racemization of d-methylmalonyl-CoA by methylmalonyl-CoA racemase. Reaction 3: Intramolecular rearrangement of l-methylmalonyl-CoA to succinyl-CoA by l-methylmalonyl-CoA mutase, which requires 5′-deoxyadenosylcobalamin. In addition, an irreversible side reaction converting d-methylmalonyl-CoA to methylmalonic acid, mediated by d-methylmalonyl-CoA hydrolase, produces methylmalonic acid (reaction 4). The metabolic fate of methylmalonic acid is unknown, but is largely excreted by the kidneys.

Dietary Sources and Nutritional Requirements

Certain bacteria and archaea synthesize cobalamin.[19] Animals acquire cobalamin by ingesting these microorganisms, by ingesting animal source foods containing cobalamin or, in the case of ruminants, by assimilating cobalamin produced by intestinal bacteria.[16,20,21] Animals and their products contain diverse amounts of cobalamin: from 139 μg/100 g in dark muscle of skipjack to 83 μg in cooked beef liver, 10 μg in shellfish, 3 to 8.9 μg in salmon and other fish, 0.9 to 1.4 μg in eggs, and 0.3 μg in milk.[21] Plants are negligible sources of cobalamin, although dried green and purple algae contain cobalamin that may be bioavailable.[21] Recently, a cultivated aquatic plant in the duckweed family, *Wolffia globosa* (Mankai), was found to contain substantial quantities of bioactive cobalamin.[22] As important as content in foods is bioavailability, which can vary 10-fold among different foods. Milk and cobalamin-fortified cereals are particularly efficient sources in the US diet, as is fish in the Norwegian diet. Bioavailability from meats is somewhat less than from milk and fortified cereals. Bioavailability from eggs is less than from meats, but eggs are still a significant dietary source of cobalamin.[21,23-26]

Nutritional requirements are influenced by two overarching principles. One is that the daily cobalamin turnover rate, relative to total body stores, renders most short-term changes in intake or assimilation irrelevant. The other principle is that in the long term, absorptive capacity affects cobalamin status more decisively than the amount of dietary intake.

Cobalamin malabsorption especially affects elderly persons (although not exclusively), who are most susceptible to cobalamin deficiency.[27-29] Their increased risk for deficiency largely arises from their greater susceptibility to atrophic gastritis, which affects absorption of cobalamin from foods, and pernicious anemia, which affects absorption of cobalamin from both dietary and supplement sources.[16,27,30-32] Dietary cobalamin intake does not appear unduly restricted in most older adults.[30,33-35]

Table 27.1 shows the adequate intakes and RDAs for adults, infants, children, and pregnant and lactating women, as well as intake estimates. For adults, the estimated daily loss of 1 μg cobalamin is minute compared with the body stores of approximately 2,500 μg.[16,20] This large disparity explains why depletion of body stores takes years after dietary intake is diminished or eliminated (e.g., upon initiating a vegetarian or vegan diet), and why clinically apparent cobalamin deficiency only infrequently arises because of poor dietary intake. Bioavailability approximates 50% at usual intake levels, and this explains the recommended dietary allowance (RDA) of 2.4 μg. The mean daily dietary intake, exclusive of supplements, was 5.3 μg in adults with normal cobalamin status in the 1999 to 2004 National Health and Nutrition Examination Survey data and 4.2 to 4.9 μg in persons with abnormal status.[36] Age and sex-stratified dietary intakes in the 1999 to 2000 National Health and Nutrition Examination Survey showed consistently lower intakes in women than in men[35] (Table 27.1). A Norwegian food frequency questionnaire survey showed higher intakes (including supplements), which differed more between sexes than between elderly and nonelderly persons.[24]

Table 27.1 Comparison of Recommended Dietary Allowances and Estimated Dietary Intakes of Cobalamin in Normal Population Subgroups[a]

Recommended Dietary Allowance			Estimated Dietary Intake		
				Mean/Median Intakes (μg/D)[b]	
Group	*Age (Y)*	*RDA (μg/d)*	*Age (Y)*	*Males*	*Females*
Infants	<1	0.4–0.5[c]	—	—	—
Children	1–3	0.9	<6	3.3/2.9	2.8/2.4
	4–8	1.2	6–11	3.9/3.1	3.7/3.0
	9–13	1.8	—	—	—
	14–18	2.4	12–19	5.4/4.3	3.4/2.6
Adults	19–70	2.4	20–59	6.1/4.4	4.0–4.1/2.7–3.1
Elderly adults	>70	2.4	>60	5.3/3.9	3.9/2.7
Pregnant women	—	2.6	—	—	—
Lactating women	—	2.8	—	—	—

RDA, recommended dietary allowance.

[a]Upper levels have not been determined for cobalamin intake and are, therefore, not shown in the table.

[b]The intake estimates are based on 24-hour dietary recall and do not include supplements. Mean and median values are shown separated by a forward slash.

[c]In place of RDA estimates, the adequate intake for infants was estimated from the cobalamin content of milk consumed.

Data from Food and Nutrition Board, Institute of Medicine. Dietary Reference Intakes: Thiamin, Riboflavin, Niacin, Vitamin B_6, Folate, Vitamin B_{12}, Pantothenic Acid, Biotin, and Choline. Washington, DC: National Academy Press, 1998:306–56. Daily intakes data are vitamin intake analyses from the 1999 to 2000 National Health and Nutrition Estimate Survey (From Ervin RB, Wright JD, Wang CY et al. Advance Data from Vital and Health Statistics. No. 339. Hyattsville MD: National Center for Health Statistics, 2004).

RDAs have traditionally been derived from amounts of parenteral cobalamin needed to prevent relapse in patients with pernicious anemia. Although such data have value, they bear a limited relationship to normal requirements. Judging intake adequacy by maintenance of normal serum cobalamin concentration and avoidance of hematologic manifestations is not ideal and has increasingly given way to metabolic end points. Surveys of various healthy subpopulations that matched intake to concentrations of metabolic biomarkers, such as serum methylmalonic acid and homocysteine [see 'Cellular Metabolism' and 'Analytic Methods' sections later in this chapter], suggest that blood biomarker concentrations may not plateau until daily intakes exceed 4.2 μg or even 6 to 7 μg, depending on the population studied.[21,37,30]

Cobalamin supplement intake, usually as part of a multivitamin preparation, has increased greatly, especially among elderly persons, and women, and strikingly often among those who need it least.[20,39] However, small supplement doses (e.g., 5 to 6 μg) are increasingly recognized to have suboptimal metabolic effect, not only in persons with severe malabsorption, such as PA, which is well known, but also, surprisingly, in the more frequent and milder malabsorption states with normal IF status. Many elderly persons with low cobalamin status but undetermined absorption status may not respond well until oral cobalamin doses exceed 50 μg, and sometimes not until they exceed 500 μg.[40-43]

In pregnancy, maternal cobalamin transfer favors the fetus. The transfer seems unaffected by the progressive decline in maternal serum cobalamin concentration throughout pregnancy, which usually reaches subnormal levels only in the last trimester.[16] Adequate intake of cobalamin in children has been estimated from comparisons based on the effectiveness of human milk of known cobalamin content,[20] but much depends on the criteria used to define deficiency in infants. Intakes of 0.4 to 0.5 μg are considered adequate in the first year of life, although transiently elevated serum methylmalonic acid levels of unknown mechanism or clinical relevance raise questions about cobalamin adequacy.

Absorption, Transport, and Metabolism

Absorption

IF-mediated absorption of cobalamin predominates in the ileum, where IF receptors are most abundant.[16,44] This efficient process effectively secures and concentrates true cobalamin, while simultaneously preventing excess absorption, perhaps to exclude nonfunctional or even harmful corrinoid analogs. The absorptive process is illustrated in **Figure 27.4**.

Food-bound cobalamin is released by gastric pepsin, whose activity becomes optimal at the acid pH of the normal stomach and degrades the food proteins that bind cobalamin[45] (see Figure 27.4, **panel 1**). The parietal cells that provide the acid also synthesize and secrete IF, a 48-kDa glycoprotein that binds cobalamin specifically. However, the released cobalamin is preferentially bound in the stomach at the low gastric pH (~1–2) by haptocorrin, a ~150-kDa glycoprotein (historically also known as R binder or transcobalamins I or III) secreted by the salivary gland epithelial cells.

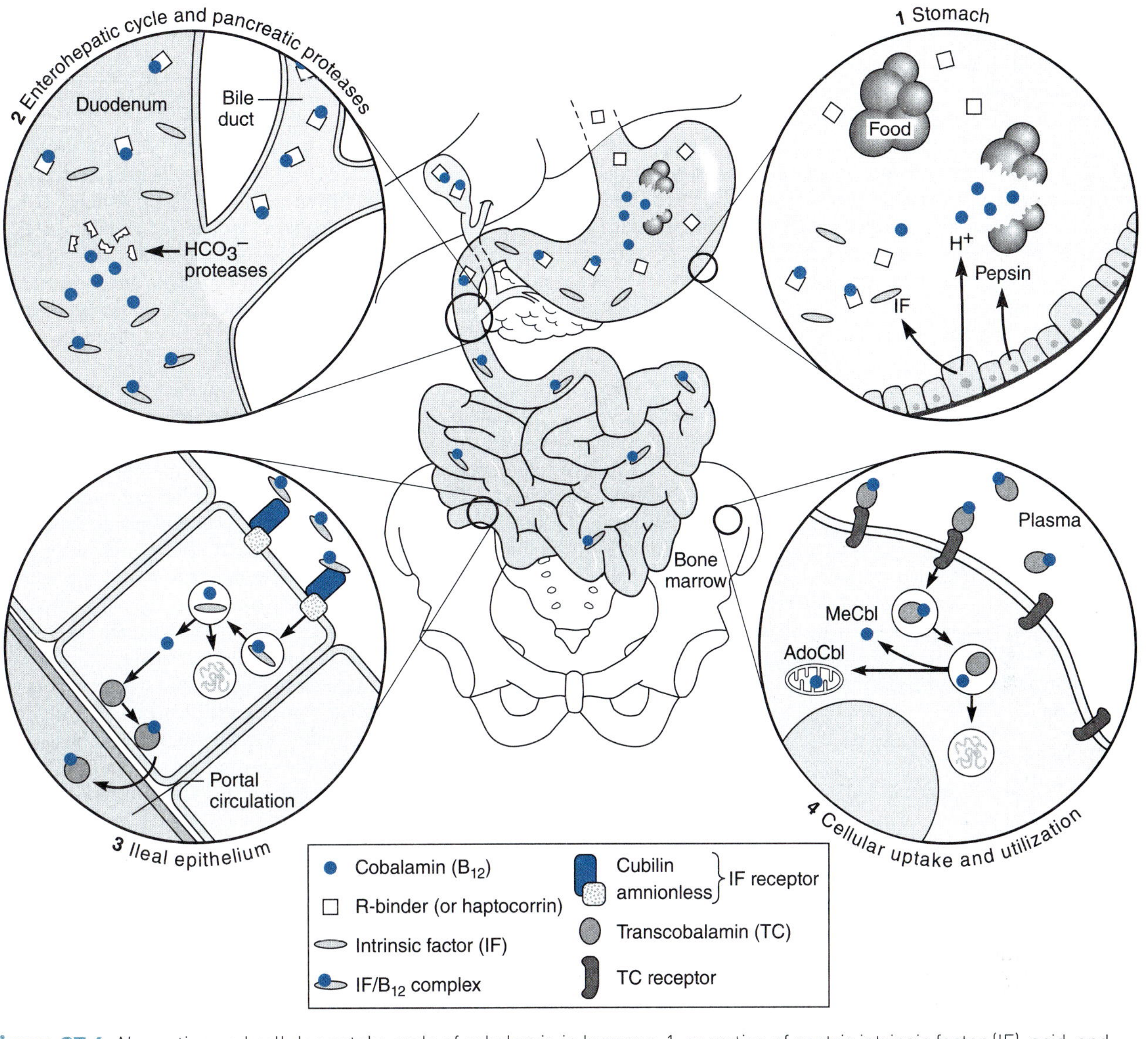

Figure 27.4 Absorption and cellular uptake cycle of cobalamin in humans. 1, secretion of gastric intrinsic factor (IF), acid, and pepsin and release of cobalamin from food and binding to R binder (haptocorrin); 2, biliary and pancreatic secretion and degradation of haptocorrin by pancreatic enzymes and binding of cobalamin to IF; 3, ileal cell uptake of IF-cobalamin by cubam (the cubilin-amnionless receptor), lysosomal processing, and transfer of transcobalamin (TC)–cobalamin (holoTC) to the portal circulation; 4, cellular uptake (e.g., in bone marrow) of plasma holoTC, lysosomal processing, and release of cobalamin for mitochondrial or cytoplasmic attachment to enzymes. AdoCbl, adenosylcobalamin; MeCbl, methylcobalamin.

Reproduced from Carmel R, Rosenblatt, DS. Disorders of cobalamin and folate metabolism. In: Handin RI, Lux SE, Stossel TP, eds. Blood: Principles and Practice of Hematology. 2nd ed. Philadelphia: Lippincott Williams & Wilkins, 2003. Originally adapted with permission from Carmel R. Cobalamin deficiency. In: Carmel R, Jacobsen DW, eds. Homocysteine in Health and Disease. Cambridge: Cambridge University Press, 2001.

Upon leaving the stomach, haptocorrin-bound cobalamin is exposed to increased pH (~6.0), due to alkalinization of the duodenal lumen by sodium bicarbonate released by the pancreas. The increased pH potentiates trypsin activity, which degrades the haptocorrin and releases the cobalamin.[46] The cobalamin then binds to the still-intact IF produced by the gastric parietal cells. In addition to dietary cobalamin, the IF also binds cobalamin released in the bile, thus promoting reabsorption and recycling of cobalamin via enterohepatic circulation (see Figure 27.4, **panel 2**). Hepatic clearance of cobalamin through bile approximates 1.4 μg daily; approximately 70% is normally reabsorbed via the physiologic process described below, whereas the remainder is lost in feces.[47] Inactive cobalamin analogs that may be produced by gut microbial metabolism of dietary cobalamin are excluded from IF-mediated absorption because IF, unlike haptocorrin, binds these analogs poorly.[48]

The IF-cobalamin complex travels to the ileum (see Figure 27.4, **panel 3**), where it is taken up by the IF receptor cubilin, a nontransmembrane, multiligand receptor on gut epithelial cells.[49] The membrane localization and function of the 460-kDa cubilin is supported by amnionless, a 45-kDa protein that provides transmembrane and cell signaling function in a complex called cubam.[50,51] Via endocytosis, the cubilin-IF-cobalamin complex is split in the ileal cell endosomes. The cobalamin eventually reaches the abluminal surface of the ileal cell, where it exits into the bloodstream ~3 to 4 hours after oral ingestion.[16,25,26,52] The cobalamin then travels in the portal circulation to the liver and beyond, bound to the ~45-kDa transport protein, transcobalamin (historically known as transcobalamin II).[53] There is some disagreement as to whether cobalamin is bound to transcobalamin within the enterocyte before entering the portal circulation, or if the cobalamin is transported unbound via multidrug resistance protein 1 into the portal circulation, where it encounters and binds to unsaturated transcobalamin.[54]

IF-mediated absorption of ingested cobalamin and reabsorption of much of biliary cobalamin is efficient but saturable. The only alternative pathway is nonspecific, passive diffusion. This inefficient process assumes importance only when the IF mechanism is damaged, as in PA, or overwhelmed by cobalamin doses larger than a few micrograms. Diffusion is not limited to the ileum and occurs at nongastrointestinal mucosal surfaces, such as the sublingual or nasal epithelium. The quantitative efficiencies of active and passive intestinal absorption are compared in **Table 27.2**.

Great disparities in efficiency exist between active and passive absorption, both of which begin with cobalamin release from food. More than 50% of the cobalamin in a typical meal will be absorbed actively if the IF-mediated absorptive system is intact. However, IF cannot accommodate much more than 2 μg cobalamin at a time (Table 27.2). Larger doses, such as those found in many supplements, exceed the capacity of the IF-cubam system. The excess cobalamin then becomes dependent on passive, nonspecific absorption, which is much less efficient (1% to 2% of the dose is absorbed) even though it is nonsaturable and linearly related to the amount of cobalamin presented. The absorption characteristics of free cobalamin in supplements are often assumed to remain unchanged when taken together with food, but the assumption appears unjustified in persons with PA[55] or food-bound cobalamin malabsorption.[40]

Transport

Cobalamin crosses membranes poorly and depends on several binding proteins for its transport throughout the body. The IF-mediated absorption process is limited to the gastrointestinal tract, although cubilin is abundant in the brush borders of renal tubular cells,[49] and IF fragments have been found in urine.[56] Once absorbed, cobalamin's transport and uptake depend on transcobalamin. Haptocorrin in the blood also binds cobalamin as well as corrinoid analogs, but its role appears to involve withholding cobalamin from microorganisms and, possibly more importantly, nonfunctional corrinoid analogs from cells.

Transcobalamin is encoded by the *TCN2* gene, which shares considerable homology with the *GIF* gene that encodes IF, although it is located on a different chromosome.[57] A common genetic variant substitutes proline for arginine in *TCN2* codon 259; its effects are uncertain, but it may impair transcobalamin function slightly.[58-60] Transcobalamin has a critical role in the blood, but small amounts also exist in milk, cerebrospinal fluid, semen, and elsewhere.[61] Transcobalamin-bound cobalamin (holotranscobalamin or holoTC) rapidly delivers cobalamin to all tissues (see Figure 27.4, **panel 4**), with a plasma half-life of only 90 minutes. A specific, calcium-dependent, 282-amino acid cell membrane receptor (CD320), whose gene has been identified,[62] is ubiquitous and seems to be regulated in synchrony with cell cycles.[63] The holoTC-CD320 complex is internalized by endocytosis. However, megalin, a separate calcium-dependent, 600-kDa, multiligand receptor that binds holoTC also exists in enterocytes, yolk sac, and other tissues.[49] Megalin has been studied best in the renal tubule,[64] where it may help conserve cobalamin by reabsorbing large amounts of filtered holoTC. The need for two receptor systems for holoTC awaits explanation.

Haptocorrin in serum originates in the specific granules of granulocyte precursors.[65] Its primary structure and immunologic behavior are identical to haptocorrin secreted by exocrine gland epithelial cells (e.g., the salivary glands), but glycosylation of the protein varies considerably.[61] Haptocorrin appears to have no specific cellular receptors, thus leaving the serum haptocorrin cobalamin complex (holoHC) to circulate with a half-life of ~9 to 10 days. Because of its longer half-life, at any given time, serum holoHC carries ~70% to 80% of the total serum

Table 27.2 Cobalamin Bioavailability from a Single Oral Dose Given to Persons with (A) Normal Absorption and (B) Abnormal Absorption without Intrinsic Factor (Pernicious Anemia)[a]

Oral Dose	(A) Amount (And %) Absorbed by Both Intrinsic Factor–Mediated and Passive Processes[b]		(B) Amount (And %) Absorbed by Passive Process Only[c]	
0.25 μg	0.19 μg	(75%)	—	—
1 μg	0.56 μg	(56%)	≈0.02 μg	(≈2%)
2 μg	0.92 μg	(46%)	—	—
3 μg	—	—	≈0.08 μg	(≈3%)
5 μg	1.4 μg	(28%)	—	—
10 μg	1.6 μg	(16%)	≈0.2 μg	(≈2%)
50 μg	1.5 μg	(3%)	≈0.5 μg	(≈1%)
100 μg	—	—	≈1.8 μg	(≈1.8%)
500 μg	—	—	≈6 μg	(≈1.2%)

[a]Data are not available for the partially compromised absorption in patients with food-bound cobalamin malabsorption, whose absorption efficiency is unknown but is presumed to be intermediate between (A) normal and (B) severely impaired absorption.

[b]In normal individuals, as is evident from comparing columns A and B, intrinsic factor–mediated absorption in normal individuals exceeds passive diffusion more than 10-fold until intake exceeds 10 μg.

[c]Intrinsic factor activity is lost in patients with pernicious anemia, so absorption is by passive diffusion only. The numbers in this column were obtained in a different study from those of normal individuals in the lefthand column. Comparability may, therefore, be limited, but the striking patterns illustrate the essential points.

Data from Chanarin I. The Megaloblastic Anaemias. 2nd ed. Oxford: Blackwell Scientific, 1979:94; Berlin H, Berlin R, Brante G. Acta Med Scand 1968;184:247–58.

cobalamin, compared with the shorter half-life holoTC, which carries ~20% to 30%.[66] Cross-species animal experiments suggest that plasma holoHC eventually becomes desialylated and is cleared by the nonspecific asialoglycoprotein receptors in the liver.[67] This process may initiate much of the enterohepatic recycling of cobalamin. Plasma haptocorrin also carries 100 to 380 pmol/L of corrinoid analogs,[68] and may divert unusable or even potentially harmful corrinoids from extrahepatic cells into the liver for ultimate release into the duodenum via bile excretions. Because IF specifically binds cobalamin and has little or no affinity for cobalamin analogs, this promotes the fecal excretion of such analogs. Haptocorrin exists in saliva, breast milk, tears, and other secretions, often in very high concentration. The haptocorrin presence in both secretions and granulocytes, and its ability to withhold cobalamin and analogs from both tissues and microorganisms, suggest that it may also serve an antibacterial role.[65]

Cellular Metabolism

After CD320-mediated uptake, cobalamin is released from holoTC within lysosomes and then enters the cytoplasm (see Figure 27.4, **panel 4**). At this point, the cobalt of cobalamin is in the +3-oxidation state and must be enzymatically reduced to the +1-oxidation state before it can be further metabolized into its active cofactor forms.

In the cytoplasm, some of the cobalamin acquires a methyl group from S-adenosylmethionine and then serves as a coenzyme for the enzyme, methionine synthase (see Figure 27.2). This methylation of the cobalamin essentially serves to prime the methionine synthase-methylcobalamin enzyme complex to catalyze the transfer of methyl groups to homocysteine to form methionine. The cobalamin then acquires subsequent methyl groups from methyltetrahydrofolate to continue the synthesis of methionine from homocysteine. In the process, methyltetrahydrofolate is converted to tetrahydrofolate, becoming available to acquire a new one-carbon unit from serine to form methylenetetrahydrofolate, which can then be reduced to methyltetrahydrofolate by the enzyme methylenetetrahydrofolate reductase. The methyltetrahydrofolate then donates the methyl group to homocysteine in another round of cobalamin-dependent methionine synthesis [for further details, see the chapter on folate in this volume].

Alternatively, some of the cytoplasmic cobalamin is transported into mitochondria where it is converted to 5′-deoxyadenosylcobalamin, which serves as a coenzyme for the conversion of methylmalonyl-CoA to succinyl-CoA catalyzed by the enzyme, methylmalonyl-CoA mutase (see Figure 27.3). This is an intermediate step in the metabolism of a variety of precursors, including propionate (as propionyl-CoA), odd-chain fatty acids, and ketogenic amino acids. It is important to note that, in addition to being converted to succinyl-CoA, methylmalonyl-CoA can be converted to methylmalonic acid. This is normally a minor side-reaction under conditions of adequate cobalamin status.

Causes of Deficiency

Identifying what caused a cobalamin deficiency is essential for patient care and also has important public health and research implications. It determines how the deficiency is best managed, what its course and prognosis are likely to be, and what associations and complications are probable. The causes are grouped in **Table 27.3** by categories of mechanisms, arranged in the sequence of events from cobalamin intake through cellular use, shown in Figure 27.4.

Table 27.3 Causes of Cobalamin Deficiency[a]

Inadequate dietary intake
Veganism or longstanding vegetarianism
Infants of vegan mothers (especially if also breast-fed by them)
Long-term, highly restricted diets (e.g., phenylketonuria diet)
Gastrointestinal malabsorption
Malabsorption of all cobalamin (free and food-bound)
Pernicious anemia: acquired or hereditary
Total gastrectomy
Partial gastrectomy (~30% of cases)
Ileal disease or damage (e.g., tropical sprue, ileal surgery)
Hereditary cobalamin malabsorption (Imerslund-Gräsbeck syndrome)
Bacterial overgrowth of the small bowel
Parasitic infestation (e.g., *Diphyllobothrium latum*)
Malabsorption limited to food-bound cobalamin
Atrophic gastritis
Partial gastrectomy (affects >50% of cases)
Other gastric surgery (e.g., gastric stapling, vagotomy)
Inhibitors of gastric acid secretion (H-blockers and proton-pump inhibitors)
Metabolic disorders
Acquired
Nitrous oxide toxicity
Hereditary
Transcobalamin deficiency
cbl mutations

[a]Claims for some unproven causes of cobalamin deficiency dot the literature. Examples are entities associated with cobalamin malabsorption but not yet convincingly linked to deficiency (e.g., pancreatic insufficiency, gastric acid-suppressing drugs), drugs associated with occasionally low cobalamin levels without proof of malabsorption or deficiency (e.g., metformin), and automatic equation of *Helicobacter pylori* infection with cobalamin deficiency.

Dietary Causes

Adult vegetarians, and even vegans, take many years to develop cobalamin deficiency because enterohepatic reabsorption of cobalamin released in the bile and retention of cobalamin via megalin-mediated reuptake of holoTC in the renal tubules are very efficient cobalamin recycling mechanisms.[47,64,69,70] As stated earlier (see "Dietary Sources and Nutritional Requirements" section), only about 1 μg of cobalamin is excreted per day and total body stores (highest amount found in liver) average around 2,500 μg, accounting for the extended time to develop deficiency. The consequences of dietary deficiency tend to be mild (e.g., borderline macrocytosis without anemia) or, most often, purely biochemical.[71,72] Notable clinical deficiency is uncommon. Still, the prevalence of low and marginal cobalamin status (based on serum cobalamin concentration) is common around the world in both adults and children.[14] This is most likely due to generally low availability and intake of cobalamin-containing animal source foods, although contributing gastrointestinal factors have not always been excluded in geographic studies.

Consequences may be greater when restricted intake begins in childhood, perhaps because of smaller body stores, the requirements of growth, and the greater vulnerability of the developing brain. Inadequacy in childhood may originate in utero if the mother has inadequate intake and status, leading to low initial tissue stores in the newborn and subsequent exposure to low dietary intake during breastfeeding and upon instituting solid foods that may be low in cobalamin.[73] An often-catastrophic syndrome affects infants born to vegan mothers or mothers with mild, undiagnosed PA with subnormal breast milk cobalamin content who breastfeed their babies.[74-77] These children develop severe neurologic complications, including seizures and developmental problems, whereas the mothers have only asymptomatic cobalamin deficiency. The frequency is unknown, but it may be the most common cause of clinical cobalamin deficiency in babies. Children may display growth retardation, impaired psychomotor development, cognitive problems, and poor school performance.[78]

Pernicious Anemia and Other Causes of Cobalamin Malabsorption

Severe malabsorption of cobalamin, which was diagnosable with the now-unavailable Schilling test[16,44] (see "Diagnostic Tests for Causes of Cobalamin Malabsorption" section below), was shown to cause 94% of clinically expressed cases of deficiency.[79] PA, in which IF secretion is lost irretrievably due to an autoimmune disorder, accounted for 76% of the cases. Its frequency varies; a survey of elderly community dwellers in Los Angeles discovered that 1.9% had PA in a mild, early, and often subclinical state of deficiency.[80]

In the slow evolution of classic acquired PA, atrophic gastritis, typically autoimmune and usually sparing the antrum, begins in late middle age. PA supervenes when parietal cell damage progresses to cause loss of IF.[31] However, chronic gastritis with achlorhydria may cause food-bound cobalamin malabsorption without progressing to PA. Once IF secretion fails, cobalamin depletion begins (or accelerates), leading to clinical deficiency several years later, typically in old age. PA sometimes affects young adults and even children,[16] especially among Black women and, to a lesser degree, Latin Americans.[81] The immune features of PA include two autoantibodies. The more prevalent of these is directed against the hydrogen, potassium-adenosine triphosphatase pump of parietal cells, but is not specific for PA. The less frequent but diagnostically more specific antibody is directed against IF.[44] Various autoimmune disorders coexist with PA, the most common being thyroid disorders.[82] Other immune disorders include vitiligo, myasthenia gravis, immune cytopenias, and agammaglobulinemia.[16,44] Iron deficiency, often (but not always) attributable to the iron malabsorption of achlorhydric gastritis, coexists in half of cases.[83] A worrisome complication in PA is an increased risk of gastric cancer and carcinoid tumors.[31,44,84,85]

A rarer form of PA results from the isolated loss of gastric IF secretion caused by mutations in the *GIF* gene.[86,87] In this situation, clinical cobalamin deficiency usually appears in the first few years of life.[88]

Partial gastric resection sometimes causes malabsorption of free, as well as food-bound, cobalamin. Loss of IF or significant cobalamin parasitization by bacteria colonizing blind loops in the upper small bowel resulting from the surgical procedure can be responsible, and clinical deficiency occurs in 15% to 30% of patients.[44,89] Most often, however, postgastrectomy malabsorption is limited to food-bound cobalamin, which usually causes SCCD only.[90] Bacterial overgrowth in the small bowel, resulting from blind loops, impaired gut motility, or giant diverticula, can parasitize ingested cobalamin and produce a malabsorptive picture and clinical deficiency. The tapeworm, *Diphyllobothrium latum*, resulting from raw fish ingestion, can do the same, but is infrequently seen today.

Intestinal causes of severe, IF-related malabsorption are disorders of the ileum, the major site for IF-mediated absorption. Acquired causes include tropical sprue; damage to the ileum by surgical bypass, resection, or radiation; and ileal bladder reserve pouches.[16,44,79] Other nutrients are often absorbed poorly as well. Hereditary malabsorption of cobalamin (Imerslund-Gräsbeck syndrome) causes isolated cobalamin malabsorption early in life.[88] It arises from mutations of the gene for cubilin[91,92] or of the gene for amnionless.[50] These children usually also have minor proteinuria, which reflects defective cubilin in the renal tubule.

Mild malabsorption limited to inadequate release of cobalamin from food and, thus, its decreased transfer to IF, was discovered in 1973 in cobalamin-deficient patients with normal Schilling test results.[90] This food-bound cobalamin malabsorption was tied to gastric surgery or to chronic gastritis, usually with intact IF secretion but diminished acid and pepsin secretion.[32] Other causes include bariatric gastric procedures[93,94] and manipulations to suppress acid secretion.[95] Food-bound cobalamin malabsorption affects 30% to 50% of persons with SCCD, but it also occurs in 10% to 15% of persons without deficiency, and in rare patients with severe deficiency.[31]

Helicobacter pylori infection was seen in 78% of patients with food-bound cobalamin malabsorption and in 44% of those without.[96] Histologic and functional gastric studies suggested that infected cobalamin-deficient patients with food-bound cobalamin malabsorption had mild gastritis and hypochlorhydria, whereas uninfected patients had severe atrophic gastritis and achlorhydria.[97] A study in a small subset showed that antibiotics reversed the food-bound cobalamin malabsorption in *H. pylori*–infected patients, but not in uninfected ones. The role of *H. pylori* remains unsettled,[40,98] despite provocative reports of improved cobalamin status after antibiotic therapy.

The food-bound cobalamin malabsorption data from its active period of investigation has been reviewed in detail,[32] but many studies after the late 1990s became unreliable, as absorption testing disappeared.[99] As discussed elsewhere,[99] some investigators substituted unproven diagnostic criteria whose misdiagnosis of food-bound cobalamin malabsorption cast doubt on data and conclusions, such as claims that persons with presumed food-bound cobalamin malabsorption respond to small oral cobalamin doses. Other assumptions and lack of absorption data cloud intriguing reports, suggesting that clinical improvement of cobalamin deficiency in *H. pylori*–infected individuals can be achieved with antibiotics alone.[40]

Metabolic Disorders

Clinical cobalamin deficiency develops much more quickly in disorders that disrupt cellular uptake and use of cobalamin than in malabsorptive conditions. Serum cobalamin concentration often remains normal, or even rises, in cellular disorders. The most common acquired metabolic disorder is long-term, recurrent exposure to nitrous oxide, which oxidatively destroys cobalamin and the methionine synthase to which it is attached.[100] Inhalant abuse of nitrous oxide is particularly widespread among young people,[101,102] and it can produce severe neurologic and mental changes. Although surgery provides too transient an exposure to produce clinical consequences, severe postoperative neurologic dysfunction can occur if a patient with unrecognized or incipient clinical cobalamin deficiency, such as PA, receives the nitrous oxide.[103,104]

Hereditary metabolic disorders are rare. These disorders include transcobalamin deficiency in which failure of cellular uptake of cobalamin causes megaloblastic anemia, and occasionally neurologic complications and immune dysfunction. Various cbl mutations affect methionine synthase or methionine synthase reductase activity, whose manifestations range from very mild and delayed to severe developmental, neurologic, and hematologic consequences that can be lethal in infancy, as often happens with cblC, the most common mutation. Other sources can be consulted for details of this complex area.[88,105,106]

Drugs

Unlike drugs that affect cobalamin metabolism directly (e.g., nitrous oxide), drugs with other actions, such as inhibitors of absorption, are likely to produce cobalamin deficiency only when they are taken uninterruptedly for many years. Thus, many drugs such as colchicine (interferes with IF-cobalamin receptor binding in the ileum),[107,108] proton-pump inhibitors and H_2-receptor antagonists (inhibit acid excretion in the stomach and thus reduce capacity to extract cobalamin from foods),[95] and alcohol (decreases secretion of pancreatic proteases that release cobalamin from haptocorrin in the duodenum)[109] can induce cobalamin malabsorption but rarely lead to cobalamin deficiency. Metformin, used to treat type 2 diabetes and polycystic ovary syndrome, also causes malabsorption, possibly by inhibiting Ca^{2+}-mediated IF-cobalamin receptor binding.[95] Alternatively, as was recently reported, metformin may reduce cobalamin bioavailability by inducing gut microbes to take up the vitamin by increasing their expression of cobalamin transporter genes.[110]

Cobalamin-Related Disorders that Do Not Cause Cobalamin Deficiency

Chronic pancreatic insufficiency can produce abnormal Schilling test results because pancreatic secretion is inadequate to degrade haptocorrin and release its cobalamin to IF in the gut (see Figure 27.4). This disorder allegedly sometimes causes cobalamin deficiency despite the absence of clinically convincing reports. Patients with hyperthyroidism and some malignant diseases may have increased cobalamin requirements, yet clinical consequences appear negligible. Patients with plasma cell neoplasms, like multiple myeloma, may have some degree of cobalamin deficiency, but treatment with cobalamin should be undertaken with caution because of the risk of accelerating the expansion of the tumor cell population.[111]

Hereditary deficiency of haptocorrin produces spuriously low serum cobalamin concentration. Because cellular cobalamin metabolism is unaffected, the low cobalamin concentration is typically discovered incidentally in adults.[112,113] The role of haptocorrin, which limits access of corrinoid analogs to cells and may withhold cobalamin from bacteria, is uncertain. Homozygous or compound heterozygous TCN1 mutations cause virtually undetectable haptocorrin concentration in serum, with cobalamin lower than 100 pg/mL, whereas simple heterozygosity causes modest serum haptocorrin and cobalamin reductions.[113,114] Originally thought rare, mild haptocorrin deficiency may explain 15% of cases of low cobalamin concentration,[112] and may often be mistaken for and treated as cobalamin deficiency.

Manifestations of Deficiency

Cobalamin deficiency is unique in several respects. With maximal disruption, namely the complete and persistent loss of IF absorption, including enterohepatic reabsorption, it still takes 2 to 5 years to deplete cobalamin stores sufficiently to induce clinical neurologic or hematologic manifestations.[16] Even the early biochemical changes that precede clinical manifestations may not appear until many months or several years after the process causing the deficiency has begun. The process is even more drawn out when the cause is dietary restriction because an intact absorption permits normal reabsorption of the sizeable daily output of biliary cobalamin. The same may be true of the development of deficiency in patients with food-bound cobalamin malabsorption, in which IF appears adequate for reabsorption of biliary cobalamin. In both of those conditions, depletion most often does not progress beyond the stage of SCCD.

Hematologic Features

The most frequent clinical expression of cobalamin deficiency is megaloblastic anemia, which may affect all blood cells, not just red blood cells.[14,16,44] Indeed, megaloblastic changes affect all dividing cells, such as epithelial cells in the gut, although the clinical consequences predominate in the blood system. (Curiously, megaloblastic anemia rarely accompanies cobalamin deficiency in most of the animal kingdom.) The cellular hallmarks are macrocytosis (large red blood cells), which is highly sensitive but not specific for megaloblastic anemia, and an abnormal nuclear maturation that is more specific than macrocytosis.

Macrocytosis is easily detected and quantified by the erythrocyte mean corpuscular volume (MCV), and cell sizing is routinely included in blood counts. In adults, the MCV is normally between 80 and 100 fl. As macrocytes replace normocytes that complete their 120-day life span, the MCV increases long before anemia becomes apparent.[115-117] This results in a gradual increase in the red cell distribution width (RDW), which is also now routinely included in a blood count and often precedes an increase in the MCV above the normal range.[118] As the anemia develops and worsens, the MCV continues to rise, sometimes to levels exceeding 120 fl. For that reason, normocytic anemia (i.e., normal MCV) should never be attributed to cobalamin deficiency except in persons with coexisting microcytosis, either from iron deficiency anemia or thalassemia, which can blunt or eliminate the expected macrocytosis in 7% of patients with PA.[44,119] However, macrocytosis can have many unrelated causes. For example, in many populations, alcohol abuse is a far more common cause of macrocytosis than cobalamin deficiency. Other causes include folate deficiency, liver disease, drugs (e.g., chemotherapeutic, antiviral, or immunosuppressive agents), hypothyroidism, copper deficiency (whose occasional myelopathy can also mimic cobalamin deficiency), and maturational hematopoietic disorders, such as the myelodysplastic syndrome.[44,120,121]

Megaloblastic changes consist of large nuclei with dispersed chromatin that impart an immature appearance to hematopoietic precursors in the bone marrow (**Figure 27.5**), but such changes may be difficult to recognize when mild. Mature granulocytes,

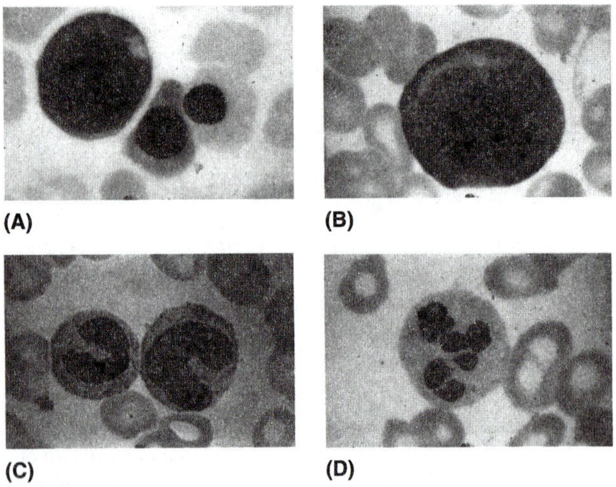

Figure 27.5 Megaloblastic changes in hematopoietic precursors in the bone marrow and their progeny in the peripheral blood, compared with normal cells. A. Normal erythroid precursor cells in bone marrow. Three nucleated precursors of increasing maturation are shown from left to right, with progressively smaller cell size and greater nuclear compactness; the cell on the right is ready to extrude its nucleus, become a mature erythrocyte, and exit into the bloodstream. B. Megaloblastic erythroid precursor cell at a maturation stage comparable to that of the normal large precursor cell on the left in A. Note the abnormal lacy appearance of the nuclear chromatin in the megaloblastic cell, compared with its normal equivalent. C. Peripheral blood containing a megaloblastic "giant" band cell (a nearly mature granulocyte) that can be contrasted with the normal band cell to its left; note the larger size of both the megaloblastic cell and its nucleus and the slightly more dispersed nuclear chromatin. D. Megaloblastic peripheral blood smear showing a hypersegmented neutrophilic granulocyte; the nucleus has more than six lobes, whereas normal neutrophils typically have four or fewer. Many mature red blood cells (no nuclei) tend to be unusually large when megaloblastic, some approaching the neutrophil in size, and some are oval rather than round; macro-ovalocytosis is typical of but not exclusive to megaloblastic anemia.

Reproduced from Carmel R. Folate deficiency. In: Carmel R, Jacobsen DW, eds. Homocysteine in Health and Disease. Cambridge: Cambridge University Press, 2001.

however, exhibit distinctive nuclear hypersegmentation in the peripheral blood. Hypersegmentation, diagnosed when 5% or more of neutrophils have five-lobed nuclei or when any neutrophil has six or more lobes (see Figure 27.5), is the earliest recognizable expression of megaloblastic anemia.[115] It also has greater specificity than macrocytosis for cobalamin or folate deficiency, but a few other conditions may display it, including iron deficiency and infection.[44,122-124]

Megaloblastosis produces ineffective hematopoiesis in which precursor cells are generated in abundance in the bone marrow, but do not survive to enter the bloodstream.[44] Biochemical markers of the massive, premature cell death, notably elevation in serum bilirubin and lactate dehydrogenase, become prominent as the anemia advances. The anemia and its mechanisms are identical in cobalamin and folate deficiencies, but the onset occurs later in the more slowly evolving cobalamin deficiency.[16,44,115]

In the past, hematologic manifestations were detectable in 73% to 87% of patients with PA, but nonanemic macrocytosis could persist for months before anemia supervenes.[44,116,117,125] More recently, hematologic manifestations have become less frequent in PA with an increase in neurologic manifestations.[14] The hematologic spectrum reflects a long progression: clinical signs of deficiency begin with asymptomatic macrocytosis without anemia; they pass through minimally symptomatic, mild macrocytic anemia; and they end with severe pancytopenia, as granulocyte and platelet counts fall along with the worsened anemia. The later stages feature symptoms resulting from impaired oxygen delivery, such as fatigue or shortness of breath,[44] but they can also be mitigated by the frequently more efficient oxygen delivery by the red blood cells in chronic anemia.[126] Clinical expression can be worsened by coexisting complications.

At the biochemical level, megaloblastic anemia of cobalamin deficiency arises from cobalamin's intersection with folate metabolism and is identical to folate deficiency anemia. The "methylfolate trap" hypothesis[127,128] provides a compelling focus on the methylation of homocysteine by methionine synthase, which requires both methyltetrahydrofolate and methylcobalamin (see Figure 27.2). This irreversible reaction is impaired in cobalamin deficiency, and methyltetrahydrofolate, though the most abundant folate, is unable to flow through the folate cycle via any other reaction and accumulates as other critical folates diminish. The reduced production of methionine and, therefore, also of *S*-adenosylmethionine, which fuels many critical methylation reactions (including methylations of DNA, RNA, histones, membrane phospholipids, neurotransmitters, and proteins), stimulates conversion of methylenetetrahydrofolate to methyltetrahydrofolate to generate more *S*-adenosylmethionine, but that only augments the trapping of folates as methyltetrahydrofolate. The trap also reduces methylenetetrahydrofolate availability for deoxyuridylate conversion to thymidylate. Excess uracil replaces thymidine incorporation into DNA, and active but futile excision repair cycles lead to strand breaks and ultimately to interphase arrest. This process appears to be a major contributor to megaloblastic conversion, but it may not be the entire explanation.[44,122]

Neurologic Features

Its potentially severe neurologic manifestations distinguish cobalamin deficiency from folate deficiency clinically, but the neurologic nuances and distinctions are debated.[129] The frequency of neurologic dysfunction has varied in part because it is less quantifiable than hematologic changes and its diagnosis, especially when changes are subtle, depends on the observers' expertise. Estimates suggest that more than half of patients with PA display neurologic findings.[129,130] Neurologic deficits can be the earliest and, in up to 27% of cases, even the sole clinical expression of cobalamin deficiency.[125,130,131] The severities of hematologic and neurologic manifestations tend to be inversely related in patients,[130,132] and the same expression tends to recur at relapse.[117,130,133] Just as many patients with PA present with only anemia as present with only neurologic changes. The pathophysiologic mechanisms underlying the differential hematologic and neurologic manifestations among patients are not understood. One hypothesis focused on common polymorphisms (C677T, A1298C) in methylenetetrahydrofolate reductase, which reduce this enzyme's activity, limit the supply of methyltetrahydrofolate for cobalamin-dependent methionine synthase, and increase availability of methylenetetrahydrofolate for thymidylate synthesis. In this way, it was proposed that the polymorphisms led to protection against the hematologic manifestations of cobalamin deficiency but promoted the neurologic manifestations. However, one study did not find empirical evidence for this hypothesis.[134]

The distribution and characteristics of the neuropathy and myelopathy tend to be stereotypical, but not specific to cobalamin deficiency. Histologically, myelin loss is followed by axonal degeneration and gliosis, and the larger, more myelinated fibers are preferentially affected.[133] The myelopathy affects the posterior and lateral columns, thus giving rise to "subacute combined degeneration" of the spinal cord. Symptoms tend to be symmetric and begin in the feet and later ascend to affect the legs, hands, and trunk.[16,130] The early clinical manifestations are diminished vibratory and position sense and paresthesias, but ataxia often follows, as can spasticity, incontinence, and other disabling manifestations. Motor function is largely spared, although gait disturbances and spasticity may become incapacitating. Cerebral manifestations can range from memory, mood, or personality changes to psychosis and occasionally delirium.[129-131,135] Autonomic dysfunction, optic neuritis, and visual changes can occur.

Magnetic resonance imaging may demonstrate surprisingly large patches of demyelination throughout the brain, in addition to classical upper spinal cord involvement. Electroencephalographic and other electrophysiologic abnormalities are common,[136-138] and they may occur even in asymptomatic patients.[12,139,140]

Neurologic abnormalities usually respond within a few weeks to months to cobalamin therapy, with complete response in 47% of cases and partial response in most of the rest.[141] Complete irreversibility occurs in only 6%, unlike the universal correction of the anemia (unless complicated by coexisting other anemias). The irreversibility is unpredictable but appears tied to the initial extent of the neurologic involvement and often to undue therapeutic delay in patients with a relentless cause of deficiency such as PA.[129,130,141,142] High folate intake or therapy has also been a suspected contributor; neurologically affected patients tend to have higher serum folate than unaffected patients,[132,143] but it is unclear if cobalamin-related metabolism or high folate intake explains the higher folate concentration. The partial response of cobalamin-deficiency anemia to folate sometimes delays recognition of the cobalamin deficiency.[16,129] Whether folate therapy simply delays recognition of cobalamin deficiency or can sometimes directly accelerate neurologic worsening remains unsettled. The amelioration of neurologic dysfunction resulting from cobalamin administration can be retarded in the presence of elevated serum folate.[144] [For more on this issue, see "Interactions–Folate" section later in this chapter.] Our continued ignorance about the basic mechanisms whereby cobalamin deficiency produces neurologic changes has stymied progress to date.

The pathogenic mechanism for the neurologic manifestations of cobalamin deficiency is unclear. Hypotheses have included abnormal myelinization resulting from impaired propionic acid metabolism, methylmalonic acid toxicity to neural cells, the accumulation of nonfunctional cobalamin analogs, disruption in the methylation cycle, and possible effects of cytokines. Many observations, including the limitation of classic neurologic deficits to genetic cobalamin disorders involving hyperhomocysteinemia alone and not those involving methylmalonic aciduria alone, favor the methionine synthase block as a linchpin, but the details are elusive. Cerebrospinal fluid studies in indirect models suggest that *S*-adenosylmethionine depletion may cause the neurologic dysfunction in cobalamin deficiency.[145] However, this does not explain the neurologic differences between cobalamin and folate deficiencies. Decreased cerebrospinal fluid *S*-adenosylmethionine concentration would be expected to result from both cobalamin and folate deficiencies. One possibility for this difference is that the brain may be protected from folate deficiency at the expense of peripheral tissues due to an abundance of folate-receptors in the choroid plexus.[146] Moreover, low plasma *S*-adenosylmethionine was reported to be a better predictor of anemia than of neurologic manifestations in patients with PA.[132] Plasma cysteine and folate concentrations were the most significant biochemical predictors of neurologic dysfunction in cobalamin deficiency.

Other Clinical Manifestations

Nonhematologic and non-neurologic abnormalities also occur in clinical deficiency and reverse with cobalamin therapy.[16,44] These include the following: occasional glossitis, sometimes severe enough to be the dominant symptom; unexplained weight loss; transient intestinal malabsorption; skin darkening, reddish hair, nail pigment changes, especially in darker skinned patients; and biochemical evidence of impaired bone formation.

Subclinical Deficiency and Public Health

SCCD was first described in 1985,[11,12,139] and its defining characteristics are shown in **Table 27.4**. Low cobalamin by itself is insufficient evidence of SCCD and requires metabolic support. The early, preclinical phases of PA satisfy the definition of SCCD[80,147] before they progress to the clinical stage, but most cases of SCCD are unrelated to PA or IF-related malabsorption.[12,13,27] Most causes of SCCD are unknown, and only 30% to 50% of cases are associated with food-bound cobalamin malabsorption.[13,27,32] SCCD progression to clinical deficiency is

Table 27.4 Criteria for Subclinical Cobalamin Deficiency

The diagnosis requires all four of the following criteria to be met:

1. Any two or more of the following test abnormalities must be present[a]:
   Elevated methylmalonic acid[b]
   Elevated plasma homocysteine[b]
   Low serum cobalamin[c]
   Low serum holotranscobalamin II
   Abnormal deoxyuridine suppression test
2. The abnormal methylmalonic acid and homocysteine results must respond to cobalamin therapy.[d,e]
3. Clinical signs of cobalamin deficiency must be absent.[f]
4. Pernicious anemia must be ruled out.[g]

[a]A lone abnormality is not reliable because each biomarker has limited sensitivity and specificity and because clinical confirmation, by definition, is unavailable in subclinical deficiency.

[b]The serum creatinine level must also be normal. Elevated creatinine suggests the likelihood of spuriously high metabolite values.

[c]The individual must not be pregnant or have haptocorrin deficiency, conditions that feature spuriously low cobalamin levels. These conditions nevertheless coexist with subclinical deficiency on occasion.

[d]The therapy must provide sufficient cobalamin: either a single injection of 100 to 1,000 μg or oral 1,000 μg doses daily for a week.

[e]Any rise in cobalamin and holotranscobalamin levels after cobalamin therapy simply reflects the influx of vitamin and has no diagnostic or metabolic implications or specificity.

[f]Subclinical deficiency is often encountered in persons undergoing evaluation for unrelated clinical findings that initially mimic cobalamin deficiency, such as anemia. The temptation to link subclinical deficiency to coexisting but nonspecific clinical findings such as nonmacrocytic anemia must be avoided unless there is a coexistent cause of microcytosis as described earlier.

[g]Anti-intrinsic factor antibody (which occurs in ~60% of PA patients) should be measured in every cobalamin-deficient patient to rule out pernicious anemia, which is the most common cause of clinical deficiency but is occasionally encountered at a preclinical stage before it inevitably and irreversibly progresses to a clinical stage.

not guaranteed, unlike the inevitable progression of PA. SCCD and its causes may be static, progress glacially, fluctuate, remit spontaneously, or occasionally accelerate, such as when chronic gastritis transforms into PA and cobalamin absorption plummets (**Figure 27.6**). As a result, prognosis in SCCD and PA differ, and so must management.

SCCD accounts for more than 80% of the cobalamin deficiency detected in population surveys.[148,149] The often-static or even transient course of SCCD[27,150,151] indicates that epidemiologic and clinical data are not interchangeable.[40,150,152] The malabsorption resulting from failure of the IF-mediated pathway that results in clinical deficiency usually predicts inevitable progression, if left untreated.

Like clinical deficiency, SCCD is more common in the aged population.[27,28] Low cobalamin concentration is seen in 5% to 15% of elderly persons, almost all of whom absorb free cobalamin adequately as shown with the Schilling test, and appear to have adequate cobalamin intake.[13,27,34] Cobalamin-related metabolic abnormalities accompany only 60% to 70% of those with low cobalamin concentration,[149,153] which suggests that one-third of such persons may not have SCCD or be cobalamin deficient at all. Moreover, isolated metabolic abnormalities often appear to be spurious. A longitudinal study (1 to 4 years) reported that 84% of isolated methylmalonic acid elevations reversed spontaneously or remained stationary,[151] a finding that supports older, limited observations that many persons with SCCD remain asymptomatic for many years.[154]

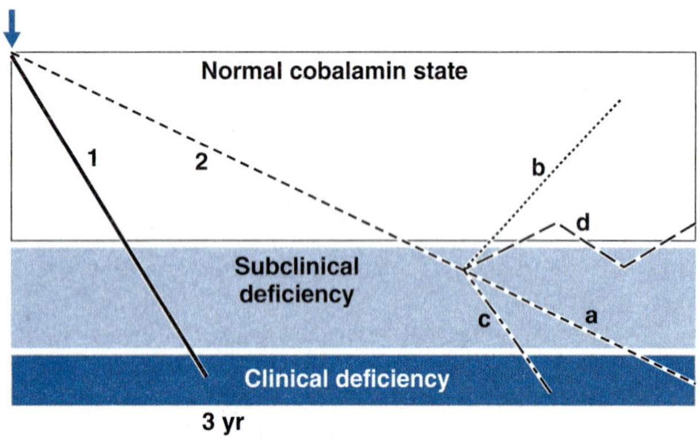

Figure 27.6 Schematic illustration of the diverse courses that cobalamin deficiency states may follow, depending on their underlying causes. The fields represent, from top to bottom, the normal cobalamin state, subclinical deficiency (mild metabolic abnormalities without clinical signs or symptoms), and clinical deficiency (mild and then progressively more severe hematologic or neurologic (or both) manifestations). The arrow (upper left) marks the onset of gradual cobalamin depletion whose progressions are arbitrarily represented as linear: line 1, the depletion produced by severe, permanent malabsorption typified by pernicious anemia; line 2, a less complete, less inexorable disruption of cobalamin balance (e.g., dietary insufficiency or a malabsorption limited to food-bound cobalamin). Based on published direct and indirect (but nonsystematic) observations, the diagram posits a slower course of unknown duration that increases the time spent transiting through subclinical deficiency, which may explain why subclinical deficiency is seen more often than clinical deficiency. At some point, this course may (line a) eventually progress sufficiently to produce clinical, symptomatic deficiency; (line b) remit completely for reasons that may or may not be known; (line c) accelerate and reach clinical deficiency more quickly (e.g., chronic gastritis transforms into pernicious anemia as intrinsic factor secretion disappears); or (line d) fluctuate indefinitely between normal and mildly subclinical deficiency states.

Modified from Carmel R. Am J Clin Nutr 2011;94(Suppl1):348S–58S.

The known neurologic consequences of clinical cobalamin deficiency explain the persisting disquiet about similar risks in SCCD. When tested, some patients with SCCD have demonstrated subtle cobalamin-responsive neurologic and electrophysiologic changes without apparent impact on health.[139,140] Nevertheless, the extensive literature on cognitive risks in SCCD is controversial. Observational studies cannot prove causation and disentangling the influence of cobalamin status from influences of folate, other B vitamins, and homocysteine has been difficult.[155-157] Initial clinical trials were equally inconclusive and often negative.[158-162]

One randomized clinical trial provided intriguing evidence that high daily doses of a combination of B vitamins (including 500 μg cobalamin, 800 μg folic acid, and 25 mg vitamin B_6) can reduce brain atrophy progression and delay cognitive decline in older adults with mild cognitive impairment.[163-165] Importantly, the effect of the B vitamins was strongest in those with elevated homocysteine at baseline, suggesting that the benefit of the intervention is confined to those patients with evidence of metabolic deficiency. The use of three B vitamins limits the ability to determine if the effects are specifically ascribable to one of the vitamins or some combination of two or all three, but it is noted that an *a posteriori* causal Bayesian network analysis indicated that cobalamin was primarily responsible for the decrease in homocysteine observed in the study.[164] Nonetheless, key issues remain to be addressed, including reproducing these findings in other cohorts, and whether older adults who are not experiencing mild cognitive impairment or cognitive decline, or those with advanced impairment (i.e., dementia, Alzheimer's disease), will benefit from cobalamin supplements with or without folic acid and vitamin B_6.

Other unresolved statistical associations with low or low-normal cobalamin concentration include: insulin resistance in children of mothers with marginal cobalamin status, depression, osteopenia, infertility, tinnitus, and some cancers. To lend perspective to this partial list, the data are often statistical exercises in which, for example, comparisons of the highest quartile of serum cobalamin to the lowest lend undue influence to outliers, which are rarely investigated in detail. Indeed, higher cobalamin is sometimes similarly associated with adverse outcomes,[166] but this may reflect haptocorrin changes more than cobalamin status.[61,166] Randomized clinical trials alone can determine whether cobalamin therapy modifies any of the many proposed risk associations with SCCD, which affects several million people in the United States. Studies must also confirm and clarify the effect of combined B vitamin therapy on cognitive decline.

Treatment of Deficiency

Even a 1 μg injection of cobalamin suffices to reverse the megaloblastic anemia of cobalamin deficiency temporarily, but the goals of treatment include repletion of stores and preventing relapse, not just reversal of manifestations. The use of a minimal effective dose is simply a demonstration of specific efficacy as used previously in therapeutic trials to distinguish cobalamin from folate deficiency.[16] Management requires ascertainment of the cause of the deficiency. Only knowing the cause can allow informed decisions about duration (ranging from a brief course to lifelong treatment), dose, and route of cobalamin therapy.[167]

Vegans, Vegetarians, and Other Patients with Intact Physiologic Cobalamin Absorption

The intact physiologic process of IF-mediated cobalamin absorption in vegans and vegetarians allows use of small oral supplements (e.g., 5 μg). Larger doses exceed the capacity of the IF-cubam system, and only 1% to 2% of the excess can be assimilated (see Table 27.2). It is prudent for vegans and vegetarians to take preventive cobalamin supplements, especially during pregnancy and while breast-feeding, and to avoid nitrous oxide anesthesia.

Patients Who Cannot Absorb Any Cobalamin

This category includes patients with PA, which is irreversible, and intestinal diseases with impaired IF-cobalamin absorption. Together, these disorders comprise more than 90% of clinically apparent cases of cobalamin deficiency. The optimal goal is to replete cobalamin stores. Larger parenteral doses produce greater absolute retention despite greater excretory losses.[16,44,55] After a short course of daily to weekly injections, monthly 100- or 1,000-μg injections of cyanocobalamin provide average retentions of 55 or 100 μg, respectively (**Table 27.5**). Patients can be taught to self-inject. For unknown reasons, perhaps related to clearance variation, occasional patients may require more frequent injections.[55,167]

Patients with PA also respond to oral cobalamin, provided that the dose is taken daily and is large enough (e.g., 1,000 μg) so that the ~1% to 2% bioavailability via diffusion provides enough absorbed vitamin.[55,168] Oral therapy prevents the discomfort, inconvenience, and cost of monthly injections, but it is not always effective in PA.[167] Clinical responses are sometimes suboptimal,[169] equivalent efficacy for severe neurologic symptoms is not fully proven, compliance with the usually lifelong requirement may wane, and relapse occurs sooner after discontinuance of oral therapy than discontinuance of parenteral therapy.[170] Noncompliance and relapse complicate all modes of cobalamin treatment[117] and they can be traced to poor understanding by patients and, sometimes, to complacency or misunderstanding among physicians about cobalamin therapy.

Patients with Malabsorption Limited to Food-Bound Cobalamin

Patients with food-bound cobalamin malabsorption absorb free, unbound cobalamin and theoretically should absorb cobalamin from supplements normally. However, that assumption may be premature.[40] Patients often respond incompletely to oral doses as high as 50 μg after gastric surgery, a known cause of food-bound cobalamin malabsorption,[93] and some elderly persons with SCCD (but unknown absorption status) had incomplete metabolic responses until doses reached 500 μg.[41-43] The optimal oral dose and the effect of meals on its bioavailability, which can be significant in patients with PA,[55] await formal study in elderly patients with and without documented food-bound cobalamin malabsorption.

Patients with Subclinical Cobalamin Deficiency

SCCD is several-times more common than clinical deficiency in the population, but the need to treat it is not yet clear. The cobalamin dose needed to improve metabolic status in SCCD itself can be surprisingly and unpredictably high.[41-43,93,94] Many surveys of supplementation have suggested that small oral doses are effective in the general study population, often without identifying the small SCCD subset at risk of non-responsiveness hidden by the normal, responsive majority. If supplementation is used for SCCD, the metabolic response must be monitored, and doses adjusted accordingly.[167] The duration of intervention, whether brief or lifelong, is also unclear in most persons with SCCD, because the causes of deficiency are often unknown.

Patients with Metabolic Disorders

Treatment of nitrous oxide toxicity must be started early and should be parenteral and extended because reversal can be incomplete. The optimal cobalamin form is uncertain. Prevention is

Table 27.5 Retention of Intramuscular Injection Doses of Cobalamin, with Comparison of Cyanocobalamin and Hydroxocobalamin

Intramuscular Injection Dose	Cyanocobalamin		Hydroxocobalamin	
	Amount Retained[a]	% Retained	Amount Retained	% Retained
10 μg	9.7 μg	97%	—	—
100 μg	55 μg	55%	92 μg	92%
500 μg	150 μg	30%	375 μg	75%
1,000 μg	150 μg	15%	710 μg	71%

[a]The amounts retained may be overestimated because they are based on losses by urinary excretion only. For example, losses by other routes (e.g., biliary) may conceivably differ between the two forms of cobalamin.

Data from Chanarin I. The Megaloblastic Anaemias. 2nd ed. Oxford: Blackwell Scientific, 1979:311.

important whenever possible, and cobalamin concentration and blood counts should be tested preoperatively if nitrous oxide use is anticipated. Most patients with hereditary disorders of metabolism require parenteral therapy and sometimes auxiliary measures as well. Other sources can be consulted for therapeutic details and rationales.[88,105,106]

Monitoring Response to Cobalamin Therapy

Monitoring response provides ultimate confirmation that the diagnosis was correct and allows early identification of nonresponsiveness or complications.[167] In clinically overt deficiency, reticulocyte counts show rises within 2 to 3 days and peak at 7 to 10 days after initiation of cobalamin supplementation.[16,44,167,171] Failure of full hematologic normalization by 8 weeks suggests that the diagnosis was incorrect or that another form of anemia, usually iron deficiency, coexists. Neurologic response, both clinical and electrophysiologic, also begins within the first few weeks, but its course and rate vary, and response can evolve over many months[130]; approximately 6% of patients have irreversible damage with greater risk occurring the longer diagnosis of deficiency and supplementation are delayed.[130]

Biochemical improvement, which is the only measurable response in asymptomatic SCCD, begins within a week. Methylmalonic acid and homocysteine concentrations reach normality after 1 to 2 weeks,[172,173] but they do not respond to folic acid given instead. Monitoring these metabolites is preferable to vitamin concentrations (cobalamin or holoTC), which rise whether therapy is effective or not. Maintenance therapy must continue for as long as the causative disorder persists.

Food Fortification

The conceptual and practical requirements for cobalamin fortification have been reviewed.[174,175] The case for fortification arises from a confluence of considerations: the potential to augment prevention of neural tube and perhaps other birth defects[166]; the high frequency of SCCD in elderly persons; and the possibility of mitigating neurologic risks that high folic acid intake may pose in persons with unsuspected clinical cobalamin deficiency [see "Interactions–Folate" section later in this chapter].

Those are important goals, but the decision to institute cobalamin fortification of foods must overcome the many information gaps.[174] Although folic acid fortification has been successful, it does not predict success for cobalamin fortification.[175-177] Two important differences exist. One is that bioavailability in the chief targeted subpopulation was never at issue with folic acid but is problematic with cobalamin. Although cobalamin added to flour and baked in bread has high bioavailability in older adults with intact cobalamin absorptive capacity,[25] its bioavailability is surprisingly poor, not just in PA, but in many elderly persons with SCCD or food-bound cobalamin malabsorption,[40-43,93,94] a shortcoming suggesting that small fortification doses may not suffice in an important target subpopulation. Moreover, bioavailability may be compromised when cobalamin is taken with meals,[55] which is the setting for fortification. Population studies of metabolic response to cobalamin supplements have generally provided limited information on the response of subsets at special risk. The first controlled study of fortified food (bread that provided a high daily amount of 9.6 μg) showed overall improvement in a small cohort.[178] However, elevated methylmalonic acid became normal in only seven of the 15 participants, and details about the nonresponders or about absorption status were unavailable.

Other issues that need resolution are whether SCCD indeed causes the neurocognitive abnormalities with which it is sometimes associated, and whether giving cobalamin in smaller doses can reverse or prevent them. Potential adverse effects of fortification also need consideration, guided by the experience from combined folic acid and high-dose cobalamin supplementation in patients with ischemic heart disease that detected increased risks of cancer[179] and reduction of renal function in diabetes patients.[180] Should adverse effects emerge, cobalamin accumulation may not be quickly dissipated, because cobalamin turnover is very slow.[174] Prospective clinical trials must address these important questions, including the crucial question of dose.

Characteristics of Cobalamin Supplements

Hydroxocobalamin is a suitable alternative to cyanocobalamin; its superior retention allows less frequent maintenance injections. Documentation of advantages for methylcobalamin is limited. Nasal and sublingual preparations have not been studied systematically.

Cobalamin has little toxicity even at high doses. However, the nonphysiologic injected form, cyanocobalamin, accumulates in red blood cells when doses reach 1,000 μg or more.[181] Allergic reactions can occur during routine therapy and can be severe.[182] Autoantibody to transcobalamin sometimes appears following injection of high-retention cobalamin preparations and produces very high serum cobalamin,[183] but harmful effects have not been noted. There is some evidence that cyanocobalamin supplementation in individuals with renal disease may lead to the accumulation of cyanide, as the supplement is converted into its active forms (methyl- and 5′-deoxyadenosylcobalamin) and the capacity to detoxify the freed cyanide is diminished.[184] Accordingly, it is suggested that hydroxocobalamin or methylcobalamin should be used as supplements in patients with glomerular filtration rates <50 mL/min/1.73 sq m.[185]

Analytic Methods

Analytic methods to diagnose cobalamin deficiency fall into two categories: assays of cobalamin amount, such as plasma or serum total cobalamin and the portion of total cobalamin bound to the transport protein transcobalamin or holotranscobalamin (holoTC); and measures of functional metabolic status, such as the plasma or serum metabolite biomarkers, methylmalonic acid and homocysteine. In addition, the deoxyuridine suppression test, which measures intracellular cobalamin-dependent metabolic pathways, can be used to assess cobalamin status; however, due to the complexity of the test, it is not currently being used in clinical settings. When clinical signs of deficiency are evident, one test usually suffices as confirmation,[167] but in research and epidemiologic surveys, the frequent absence of clinical identifiers usually requires the application of more than one diagnostic biomarker, i.e., multiple analyte testing.[150,186] At this time, no diagnostic gold standard exists and confirmation of cobalamin deficiency is often achieved empirically after treatment with cobalamin supplements (oral or intramuscular injections). The diagnostic approach in children must consider genetic disorders, as well as the entities that affect adults. Methylmalonic acid and homocysteine testing should always be included in children because cobalamin concentration may be normal in genetic disorders of cellular uptake and metabolism and because the two tests help narrow down the genetic possibilities and diagnostic focus. This complex approach is discussed elsewhere.[88] Low cobalamin in very early infancy suggests maternal cobalamin deficiency as their cause; the mother should also be tested.

Serum Total Cobalamin

Cobalamin exists in serum primarily as methylcobalamin (60%–70%) and adenosylcobalamin (20%–25%).[16] Supplement forms of cobalamin (cyanocobalamin and hydroxycobalamin) may constitute a small, transient amount of the total after supplementation. Serum total cobalamin is stable on long-term storage (although specific forms may be converted upon exposure to light) and can be assayed by various techniques. The earliest methods employed microorganisms, such as *Euglena gracilis* and *Lactobacillus leichmannii*, whose growth is proportional to the cobalamin content of the sample.[16] These assays, now automated, are still considered the standard by some reference laboratories. Radioisotopic methods rely on competitive binding of the sample's cobalamin by purified IF as the added cobalamin-binding protein; IF must not be contaminated with haptocorrin, which also binds nonfunctional corrinoids and causes falsely high cobalamin results in samples with such corrinoids. Immunoenzymatic-chemiluminescence techniques using anti-IF antibody to capture IF-complexed cobalamin now predominate in clinical laboratories for diagnostic use. These highly proprietary, automated assays have been less well defined and monitored than previous methods. They also appear susceptible to falsely normal results in some cobalamin-deficient sera,[187,188] probably by failing to inactivate endogenous anti-IF antibody in samples from patients with pernicious anemia.[188] This selective error does not seem to affect normal sera, which complicates detection of the error; it may explain puzzling reports of normal cobalamin concentration despite severe cobalamin deficiency.[150,189]

The cutoff point between deficient and adequate serum cobalamin concentration varies from method to method and laboratory to laboratory.[150] Most laboratories use traditional decision limits (cut-points) of 200 to 250 pg/mL (148 to 185 pmol/L) to define deficiency. The sensitivity of low serum cobalamin concentration for cobalamin deficiency has been questioned, but much depends on what kind of deficiency is considered, i.e., clinical or SCCD.[150,167] Sensitivity of serum total cobalamin exceeds 95% in patients who have obvious clinical manifestations of deficiency, such as megaloblastic anemia or neurologic abnormalities.[150,172,173,190] The lower the serum cobalamin concentration, the greater the likelihood of clinically severe deficiency,[16,191,192] but exceptions exist.[112,125] As with all biomarkers, diagnostic sensitivity decreases in subclinical conditions, and it is 38% to 39% in SCCD.[150] Low-normal cobalamin concentration with associated metabolic abnormalities is not unusual in population surveys,[30,148,149] but metabolic abnormalities can be spurious sometimes.[150]

The insensitivity of serum cobalamin in such surveys led some investigators to raise the cut-point for deficiency from the traditional 200 or 250 pg/mL to 350 pg/mL (258 pmol/L) or higher to ensure that no case of deficiency was missed.[148] This change, adopted by many laboratories, instantly increased the frequency of diagnosed deficiency. The overdiagnosis it creates is substantial. For example, the 5.3% to 24.8% frequency of "abnormal" cobalamin values in four surveys became 40.5% to 71.7% in study samples of older adults.[167] More relevantly, reanalysis showed that only one-third of the persons thus recategorized had elevated plasma or serum methylmalonic acid or homocysteine, thus making two-thirds of the new diagnoses metabolically suspect. Moreover, few of the one-third with elevated methylmalonic acid or homocysteine had clinical deficiency.[13,150] Furthermore, the advantages of overdiagnosis remain unrealized because the health risks of SCCD and the benefits of intervention in such cases remain unproven.

As mentioned, the specificity of low cobalamin concentration may be more limited than its sensitivity. **Table 27.6** lists the conditions associated with low serum cobalamin. The most notable causes of falsely low concentration include pregnancy and folate deficiency,[16,44] although true deficiency may accompany occasional cases. Substantive genetic influences on cobalamin concentration are also becoming evident. Mutations in the gene for the cobalamin binding protein, haptocorrin, underlie haptocorrin deficiency,[114,193] which may explain 15% of low serum cobalamin[112] and, thus, may make it a more frequent cause of low serum cobalamin than pernicious anemia.

Holotranscobalamin

Lindemans et al.[194] first suggested that measuring only the serum cobalamin attached to transcobalamin (i.e., holoTC) may enhance diagnostic specificity and sensitivity of cobalamin status testing. HoloTC originates either within the ileal cell or by rapid binding to unsaturated transcobalamin after entering the portal circulation from the ileal cell. Some holoTC may also have renal origins. Because holoTC is taken up by cells rapidly

Table 27.6 Conditions Frequently Associated with Lowered Serum Cobalamin Levels

Cobalamin deficiency
Clinically expressed deficiency[a]
Subclinical deficiency[b]
Normal cobalamin status
Pregnancy
Haptocorrin deficiency (severe or mild)[c]
Human immunodeficiency virus infection; acquired-immunodeficiency syndrome[d]
Folate deficiency
Multiple myeloma[d]
Aplastic anemia; myelodysplastic syndrome
Multiple sclerosis
Drugs (e.g., metformin, omeprazole)[e]

[a]Deficiency can be severe or very mild. The disorders usually arise from severe intrinsic factor-related malabsorption, such as pernicious anemia.
[b]Deficiency is expressed only biochemically. The causes of the deficiency are usually unknown; some arise from cobalamin malabsorption limited to food-bound cobalamin.
[c]Severe haptocorrin deficiency is rare but mild deficiency may be common.
[d]Most patients have no evidence of cobalamin deficiency or malabsorption, but some may have coexisting deficiency or malabsorption or coexisting pernicious anemia (described in some patients with multiple myeloma).
[e]Drugs taken briefly or irregularly rarely affect cobalamin status. Metformin decreases cobalamin levels by an unknown mechanism, and evidence of deficiency is weak. Omeprazole reduces cobalamin absorption, but low cobalamin levels have been few (it may require many years of constant use to induce deficiency).

(serum half-life ~90 minutes), only 20% to 30% of serum cobalamin is in holoTC at any given time. The remainder is carried by haptocorrin, which does not promote specific cellular uptake.[61]

HoloTC assays are now accurate, and a fully automated method is commercially available.[195] As with other cobalamin-related biomarkers, holoTC cut-points for deficiency/adequacy vary widely between 19 and 50 pmol/L.[150] Direct comparisons show a minor advantage for holoTC over total cobalamin; areas under the curve by receiver operator characteristic (ROC) curve analyses have varied between 0.75 and 0.90 for holoTC vs between 0.72 and 0.85 for total cobalamin.[150,196] However, most holoTC studies have involved large amorphous groups defined almost solely by their serum methylmalonic acid and/or their plasma total homocysteine concentrations, whose specificity for cobalamin status is not absolute [see below in this section]. Clinical or absorptive status is rarely assessed in these studies, and disagreements between holoTC and total cobalamin results are rarely explored.[150] One of the few clinical studies of patients with mild, clinically expressed cobalamin deficiency did not find holoTC to be superior to cobalamin in predicting response to therapy.[197] The claim that holoTC does not decline spuriously like total cobalamin in pregnancy and, therefore, permits better characterization of cobalamin status in pregnant women[198] may be premature; a previously unappreciated rise in maternal holoTC was observed following delivery.[150] Although prenatal concentration was not determined in this study, the finding suggests that holoTC probably does decline during pregnancy.

Controversy persists about holoTC because little is known about what other influences affect its serum concentration. Renal failure can produce striking holoTC elevation,[199] but preliminary reports of many cobalamin-unrelated influences on holoTC such as alcohol abuse, folate deficiency, myelodysplasia, and Gaucher disease await clarification.[150,200] Moreover, holoTC concentration lower than control values in cobalamin-repleted patients with PA[201] suggests that cobalamin absorption affects holoTC independently of metabolic status. Such dual influences could introduce diagnostic nonspecificity. For example, it seems possible that even transient dietary change or malabsorption (e.g., drug induced), lasting only several days or weeks and unaccompanied by and unlikely to evolve into cobalamin deficiency, might cause low holoTC concentration. These and other influences may explain isolated holoTC elevations that have heretofore been attributed to an unusual sensitivity of holoTC for SCCD. Exposure to excess intake of folic acid has been hypothesized to reduce serum holoTC, particularly in individuals with low or deficient cobalamin status, by diverting holoTC to support folic acid-induced reticulocyte synthesis, or by impairing reuptake of holoTC in the renal tubules.[202] This hypothesis remains to be tested.

Methylmalonic Acid

Methylmalonic acid accumulates in serum, and some methylmalonic acid is excreted in the urine, when 5′-deoxyadenosylcobalamin-dependent methylmalonyl-CoA mutase activity is reduced (see Figure 27.3). Methylmalonic acid can be measured reliably by liquid chromatography-mass spectrometry methods.[203] Defined cut points for methylmalonic acid vary among different research and clinical laboratories, with published values between 210 and 480 nmol/L.[150,190,204,205] Methylmalonic acid is elevated in 98% of patients with PA who have clinically expressed cobalamin deficiency, often to levels exceeding 1,000 nmol/L.[79,172,173,190] Methylmalonic acid elevation reverses soon after cobalamin therapy.[172,173]

Methylmalonic acid elevation is usually milder in SCCD because cobalamin stores are not severely depleted, and the sensitivity of high methylmalonic acid is probably lower than in clinical deficiency.[30,149,150,206] Specificity is clearly superior to that of homocysteine because folate status does not affect methylmalonic acid.[79,172,173] However, major known influences on methylmalonic acid include, in order of frequency, glomerular filtration (even a minimal decrease raises serum methylmalonic acid), cobalamin status, age, and perhaps, sex,[205,207] and they explain only 16% of the methylmalonic acid variation,[205] Another 12% of the variance is explained by common polymorphisms in the genes for 3-hydroxyisobutyryl-CoA hydrolase (*HIBCH*) and acyl-CoA synthetase family member 3 (*ACSF3*).[208] Asymptomatic babies can have moderate methylmalonic acid elevation that remits spontaneously after the first year of life.[209] The cause is unknown, but the association with mild homocysteine and cobalamin changes (neither of which usually reach abnormality) and their improvement after cobalamin treatment raise the possibility of relative cobalamin deficiency.[210]

Many experts consider methylmalonic acid the best metabolic test available to confirm cobalamin deficiency, and normal methylmalonic acid values provide strong evidence against deficiency. However, methylmalonic acid cannot be the diagnostic gold standard because its specificity is undefined.[150] It is particularly unclear what mild methylmalonic acid elevation without any other abnormalities mean.[36,150,211] The improvement of methylmalonic acid elevation after cobalamin is given suggests that these elevations represent mild SCCD.[149,173,211] However, normal methylmalonic acid concentration often declines after cobalamin therapy, a finding suggesting regression to the mean or supersaturation of methylmalonyl-CoA mutase by cobalamin as alternative explanations. A noteworthy longitudinal study of 432 patients left untreated over 1 to 3.9 years observed that 44% of isolated mild elevations of methylmalonic acid improved spontaneously, and only 16% progressed.[151] Antibiotics sometimes reduce cobalamin-unresponsive methylmalonic acid elevation,[173,212] suggesting that increased propionate metabolism by some intestinal bacteria may directly elevate methylmalonic acid without cobalamin deficiency. Also of note is the finding of markedly high serum methylmalonic acid concentration in school children living in a peri-urban area in Guatemala; mean serum methylmalonic acid concentration was 350 nmol/L in those with adequate plasma cobalamin concentration (>221 pmol/L), with many exceeding 500 nmol/L. Children with marginal (162–220 pmol/L) or low (<162 pmol/L) cobalamin status in this study had mean methylmalonic acid of 534 and 555 nmol/L, respectively.[213] These findings are contrasted with a study establishing reference intervals for methylmalonic acid in Danish children in which the mean methylmalonic acid was 170 nmol/L.[214] The difference in methylmalonic acid concentration between the Guatemalan and Danish cohorts is unexplained, but one possibility is that differences in gut microbiomes, particularly in propionate-producing intestinal bacteria, could be a contributing factor.

Homocysteine

Elevation of homocysteine because of impaired methionine synthase activity is nearly as sensitive as methylmalonic acid elevation for cobalamin deficiency. A variety of reliable assays for plasma total homocysteine (i.e., the concentration of protein-bound, free, and reduced homocysteine combined) are available, including high-pressure liquid chromatography, liquid chromatography-mass spectrometry, and enzymatic-based methods.[215] Sensitivity is 95% when cobalamin deficiency is clinically overt, and the homocysteine elevation is often striking.[76,173] However, hyperhomocysteinemia has many causes,[148,216,217] including preanalytic influences such as delayed blood sample processing or using serum instead of plasma, both of which promote artifactual release of blood cell homocysteine. Renal status and folate status affect homocysteine more than cobalamin status,[140,203,204] as observed best in countries that have not fortified the diet with folic acid.[207] The relative impact of cobalamin status on homocysteine rises in the aged, who have high rates of cobalamin deficiency. Other important influences on homocysteine include sex, menopausal status, hypothyroidism, genetic polymorphisms (especially in the enzyme, methylenetetrahydrofolate reductase, which produces methyltetrahydrofolate, the substrate for cobalamin-dependent conversion of homocysteine to methionine), drugs that interfere with one-carbon metabolism (e.g., methotrexate) or induce increased rates of S-adenosylmethionine-dependent methylation and increased homocysteine synthesis (e.g., L-dopa), alcohol abuse, lifestyle factors, and disorders of homocysteine transsulfuration.[216-219] Cut points for plasma homocysteine have varied widely, and this affects case definition. Plasma concentration lower than 10 μmol/L is considered optimal, but many laboratories use cut points of 12 to 14 μmol/L in adult men and 10 to 12 μmol/L in premenopausal women. Homocysteine is more reliable than cobalamin and is as reliable as methylmalonic acid in monitoring therapeutic response in cobalamin deficiency; both homocysteine and methylmalonic acid elevation respond to cobalamin, but only homocysteine responds to folic acid.[147,163,164]

Multiple Analyte Testing

In recent years, assessment of cobalamin status has shifted toward multiple analyte testing. This includes algorithmic sequential testing[147] or the use of two of the blood biomarkers described above sequentially or in parallel.[14,196,220,221] In clinical laboratories, this often consists of measuring serum total cobalamin and then subsequently measuring methylmalonic acid or homocysteine if the cobalamin value is low. Of note is that a new index of cobalamin status, termed "combined indicator of cobalamin status" or "c-B_{12}," has been introduced that uses a mathematical formula that combines two, three, or four of the cobalamin status blood biomarkers.[218] c-B_{12} has been shown to be more strongly associated with certain cobalamin-dependent outcomes than any of the single blood biomarkers alone, including hemoglobin in blood, cognitive function, and peripheral nerve conductivity.[144] The clinical utility of c B_{12} remains to be determined.

Diagnostic Tests for Causes of Cobalamin Malabsorption

Serum cobalamin and metabolite assays identify cobalamin deficiency, but not its cause. Identifying the cause has clinical and scientific importance, not just for diagnostic accuracy but also for its prognostic implications and management guidance on duration of therapy and disease complications.[99,167] Tests of cobalamin absorption have long been mainstays[16,44] because clinical deficiency is malabsorptive in 94% of cases.[79] The classical test for cobalamin malabsorption is the Schilling test.[222] Briefly, the test involves oral consumption of a cobalt-57-labeled cobalamin (57Co-B_{12}) followed about 2 hours later by a large, intramuscular dose of unlabeled cobalamin (1 mg). The unlabeled cobalamin saturates the serum B_{12} transport proteins (transcobalamin and haptocorrin), leading to the flushing of unbound 57Co-B_{12} absorbed from the gastrointestinal tract into the urine; low urinary excretion of the 57Co-B_{12} is diagnostic of cobalamin malabsorption. Subsequent repeating of the test with the addition of oral IF leads to a diagnosis of PA if the addition of the IF leads to increased urinary excretion of the 57Co-B_{12}. A modified version of the test using 57Co-B_{12} bound to proteins in egg yolk is used to diagnose food-bound cobalamin malabsorption such as that which occurs in atrophic gastritis patients.[12]

Notably, the Schilling test is now unavailable as a clinical diagnostic test due to a variety of issues, including exposure of patients to gamma radiation, reduced availability and high cost of the radioactive substrate and its disposal, risks associated with the use of animal source IF, requirement of normal renal function and complete 24-hour urine collection, and the fact that treatment of individuals with cobalamin supplementation may be unacceptably delayed because the supplements can confound the test results. This has created a major diagnostic vacuum.[99] Newer absorption tests, based on holoTC response to unlabeled oral cobalamin (the Cobasorb test)[223] or direct measure of carbon-14 or carbon-13 labeled cobalamin,[52,224,225] have not been extensively validated in clinical trials.

Surrogate tests for cobalamin malabsorption have limited value. Measuring serum antibody to IF is helpful because it is highly specific for diagnosing PA, unlike the parietal cell antibody.[226,227] However, only 50% to 70% of patients with PA have the anti-IF antibody, and the test provides no diagnostic information about other disorders. Serum gastrin is elevated and pepsinogen I is decreased in 80% to 90% of patients with PA, but both tests lack specificity.[228] Combining either test with IF antibody, along with concomitant measurement of cobalamin biomarkers discussed above (i.e., total cobalamin, holoTC, methylmalonic acid, and/or homocysteine) seems the best approach today in diagnosing PA. No reliable surrogate markers exist for the no longer available absorption tests for food-bound cobalamin malabsorption. Unproven indirect diagnostic criteria[229] predispose to misdiagnoses.[99]

Interactions

Folate

Cobalamin and folate are linked by close metabolic, clinical, and therapeutic connections [see also the chapter on folate]. Cobalamin's narrow dietary sources, slow rate of depletion, and highly specific, yet occasionally vulnerable, IF-dependent absorption processes explain why cobalamin deficiency, unlike folate deficiency, tends to be a drawn out malabsorptive deficiency state. As cobalamin deficiency progresses, it usually raises methyltetrahydrofolate concentration (and thus serum folate) as predicted by the methyltetrahydrofolate trap hypothesis [see "Manifestations of Deficiency—Hematologic Features" section above], whereas the poor cellular retention of methyltetrahydrofolate lowers red blood cell folate concentration. Folate deficiency lowers plasma cobalamin by unknown mechanisms, although intestinal malabsorption seems a likely cause, as concentration rebounds after folate therapy. Both vitamin deficiencies induce homocysteine elevation.

The anemia of cobalamin deficiency often responds to folate therapy (i.e., high-dose folic acid supplementation),[16] although the response can be partial and transient. Neurologic manifestations of cobalamin deficiency typically do not respond to folic acid[16,141,142] and may be exacerbated.[202] Data are insufficient to determine whether the sharply increased folate intake in the United States since 1997, due to folic acid fortification of cereals and grains accompanied by wider supplement use, has compromised the early hematologic diagnosis of cobalamin deficiency or worsened its neurologic complications. Cognitive findings related to the combination of high folate and low cobalamin status have conflicted in epidemiologic studies,[230-232] perhaps because the subsets at risk were all small and neurocognitive testing was too limited.[152] An unexpected, incidental finding was that the combination of low cobalamin and high folate was associated with greater methylmalonic acid abnormalities than when folate status was normal.[232,233] The nature of those metabolic associations is unclear. Indirect evidence[234] suggests that individuals with that metabolic combination had severe cobalamin deficiency (such as PA), which elevates serum folate, and did not have SCCD with unusually high folate intake.[152] However, as described above (see "Analytic Methods—Holotranscobalamin" section), the combination of low B_{12} and high folate was also associated with lower holoTC concentration in a cohort of older adults,[232] and it has recently been hypothesized that excess folic acid exposure diverts serum holoTC toward reticulocyte synthesis or excretion of holoTC in the urine.[202] The diversion of serum holoTC may further limit availability of cobalamin for other tissues which, in turn, may exacerbate the metabolic (i.e., elevations in serum methylmalonic acid and plasma homocysteine) and neurologic manifestations of cobalamin deficiency. This hypothesis remains to be tested.

Folate and cobalamin status interactions raise other issues in specialized medical settings. For example, patients with sickle cell diseases are routinely given folic acid because their chronic hemolytic anemia raises their folate requirements. However, reports now suggest that patients with sickle cell disease can develop PA despite their young age.[113,235] It seems prudent to screen such patients for clinical cobalamin deficiency periodically if they continue to take folic acid supplements.

Iron

More than half of patients with PA develop iron deficiency,[83] often because the atrophic gastritis underlying PA also compromises iron absorption. However, the increased risk of gastric cancer in patients with PA mandates a search for blood loss. When iron and cobalamin deficiency anemias coexist, the expected hematologic characteristics of either clinical entity may be blurred by the other[44,236]; MCV may be high, normal, or low, and markers of iron status sometimes can be masked by severe, untreated cobalamin deficiency.[16] The mixed anemia may fail to respond if only one of the two required hematinics is given.

Acknowledgments

The authors acknowledge Ralph Carmel, the previous author of this chapter, for his major contributions to the current version.

References

1. Kass L. Pernicious Anemia. Philadelphia: WB Saunders, 1976.
2. Castle WB. The conquest of pernicious anemia. In: Wintrobe MM, ed. Blood, Pure and Eloquent: A Story of Discovery, of People, and of Ideas. New York: McGraw Hill. 1980:284–317.
3. Smith EL. Vitamin B_{12}. New York: John Wiley, 1965.
4. Chanarin I. Br J Haematol 2000;111:407–15.
5. Minot GR, Murphy WP. JAMA 1926;87:470–6.
6. Robscheit-Robbins FS, Whipple GH. J Exp Med 1929;49: 215–27.
7. Castle WB. Am J Med Sci 1929;178:748–64.
8. Rickes EL, Brink NG, Koniuszy FR, et al. Science 1948;107: 396–7.
9. Smith EL, Parker LFJ. Biochem J 1948;43:viii–ix.
10. Hodgkin DC, Kamper J, Mackay M, et al. Nature 1956;178:64–6.
11. Carmel R, Karnaze DS. JAMA 1985;253:1284–7.
12. Carmel R, Sinow RM, Karnaze DS. J Lab Clin Med 1987;109: 454–63.
13. Carmel R. Annu Rev Med 2000;51:357–75.
14. Green R, Allen LH, Bjørke Monsen A-L, et al. Nature Rev Dis Primers 2017;3:17040.
15. Kondo H, Kolhouse JF, Allen RH. Proc Natl Acad Sci U S A 1980;77:817–21.
16. Chanarin I. The Megaloblastic Anemias. Oxford: Blackwell Scientific Publications, 1969.
17. Muir M, Landon M. Br J Haematol 1985;61:303–6.
18. Allen RH, Stabler SP. Am J Clin Nutr 2008;87:1324–35.
19. Martens AF, Barg H, Warren MJ. Appl Microbiol Biotechnol 2002;58:275–85.
20. Food and Nutrition Board, Institute of Medicine. Dietary Reference Intakes: Thiamin, Riboflavin, Niacin, Vitamin B_6, Folate, Vitamin B_{12}, Pantothenic Acid, Biotin, and Choline. Washington, DC: National Academy Press, 1998:306–56.
21. Watanabe F. Exp Biol Med 2007;232:1266–74.
22. Sela I, Yaskolka Meir A, Brandis A, et al. Nutrients 2020;12:3067.
23. Tucker KL, Rich S, Rosenberg I, et al. Am J Clin Nutr 2000;71: 514–22.
24. Vogiatzoglou A, Smith AD, Nurk E, et al. Am J Clin Nutr 2009;89:1078–87.
25. Garrod MG, Buchholz BA, Miller JW, et al. Nucl Instrum Methods Phys Res B 2019;438:136–40.
26. Garrod MG, Rossow HA, Calvert CC, et al. Nutrients 2019;11: 2148.
27. Carmel R. Am J Clin Nutr 1997;66:750–9.
28. Nilsson-Ehle H. Drugs Aging 1998;12:277–92.
29. Baik HW, Russell RM. Annu Rev Nutr 1999;19:357–77.
30. van Asselt DZ, de Groot LC, van Staveren WA, et al. Am J Clin Nutr 1998;68:328–34.
31. Carmel R. Pernicious Anemia: Definitions, Expressions, and the Long-Term Consequences of Atrophic Gastritis. In: Holt PR, Russell RM, eds. Chronic Gastritis and Hypochlorhydria in the Elderly. Boca Raton, FL: CRC Press, 1993:99–114.
32. Carmel R. Baillieres Clin Haematol 1995;8:639–55.
33. Selhub J, Jacques PF, Wilson PWF, et al. JAMA 1993;270:2693–8.
34. Howard JM, Azen C, Jacobsen DW, et al. Eur J Clin Nutr 1998;52:582–7.
35. Ervin RB, Wright JD, Wang CY, et al. Advance Data from Vital and Health Statistics. No. 339. Hyattsville, MD: National Center for Health Statistics, 2004.
36. Bailey R, Carmel R, Green R, et al. Am J Clin Nutr 2011;94: 552–61.
37. Bor MV, Lydeking-Olsen E, Moller J, et al. Am J Clin Nutr 2006;83:52–8.
38. Bor MV, von Castel-Roberts KM, Kauwell GPA, et al. Am J Clin Nutr 2010;91:571–7.

39. Kaufman DW, Kelly JP, Rosenberg L, et al. JAMA 2002;287:337–44.
40. Carmel R, Sarrai M. Curr Hematol Rep 2006;5:23–33.
41. Seal EC, Metz J, Flicker L, et al. J Am Geriatr Soc 2002;50:146–51.
42. Garcia A, Paris-Pombo A, Evans L, et al. J Am Geriatr Soc 2002;50:1401–4.
43. Rajan S, Wallace JI, Brodkin KI, et al. J Am Geriatr Soc 2002;50:1789–95.
44. Carmel R. Megaloblastic anemias: Disorders of Impaired DNA Synthesis. In: Greer JP, Foerster J, Rodgers GM et al, eds. Wintrobe's Clinical Hematology. 12th ed. Philadelphia: Lippincott Williams & Wilkins, 2009:1143–72.
45. del Corral A, Carmel R. Gastroenterology 1990;98:1460–6.
46. Allen RH, Seetharam B, Podell E, et al. J Clin Invest 1978;61:47–54.
47. el Kholty S, Guéant JL, Bressler L, et al. Gastroenterology 1991;101:1399–408.
48. Kolhouse JF, Kondo H, Allen NC, et al. N Engl J Med 1978;299:785–92.
49. Moestrup SK, Verroust PJ. Annu Rev Nutr 2001;21:407–28.
50. Fyfe JC, Madsen M, Hojrup P, et al. Blood 2004;103:1573–9.
51. Pedersen GA, Chakraborty S, Steinhauser AL, et al. Traffic 2010;11:706–20.
52. Carkeet C, Dueker SR, Lango J, et al. Proc Natl Acad Sci USA 2006;103:5694–9.
53. Quadros EV, Regec AL, Khan KM, et al. Am J Physiol 1999;277:G161–6.
54. Beedholm-Ebsen R, van de Wetering K, Hardlei T, et al. Blood 2010;115:1632–9.
55. Berlin H, Berlin R, Brante G. Acta Med Scand 1968;184:247–58.
56. Waters HM, Dawson DW. Clin Lab Haematol 1999;21:169–72.
57. Kalra S, Li N, Yammani RR, et al. Arch Biochem Biophys 2004;431:189–96.
58. Namour F, Olivier JL, Abdelmoutalleb I, et al. Blood 2001;97:1092–8.
59. McCaddon A, Blennow K, Hudson P, et al. Blood 2001;98:3497–9.
60. Miller JW, Ramos MI, Garrod MG, et al. Blood 2002;100:718–20.
61. Carmel R. Cobalamin-binding Proteins in Man. In: Silber R, Gordon AS, LoBue J et al, eds. Contemporary Hematology-Oncology, vol 2. New York: Plenum, 1981:79–129.
62. Quadros EV, Nakayama Y, Sequeira JM. Blood 2009;113:186–92.
63. Jiang W, Sequeira JM, Nakayama Y, et al. Gene 2010;466:49–55.
64. Birn H. Am J Physiol Renal Physiol 2006;291:F22–36.
65. Cowland JB, Borregaard N. J Leukoc Biol 1999;66:989–95.
66. Hall CA. Clin Sci Mol Med 1977;53:453–7.
67. Burger RL, Schneider RJ, Mehlman CS, et al. J Biol Chem 1975;250:7707–13.
68. Hardlei TF, Nexo E. Clin Chem 2009;55:1002–10.
69. Green R, Jacobsen DW, van Tonder SV, et al. Gastroenterology 1981;81:773–6.
70. Green R, Jacobsen DW, van Tonder SV, et al. J Lab Clin Med 1982;100:771–7.
71. Refsum H, Yajnik CS, Gadkari M, et al. Am J Clin Nutr 2001;74:233–41.
72. Carmel R, Mallidi PV, Vinarskiy S, et al. Am J Hematol 2002;70:107–14.
73. Jadhav M, Webb JK, Vaishnava S, et al. Lancet 1962;2(7262):903–7.
74. Higginbottom MC, Sweetman L, Nyhan WL. N Engl J Med 1978;299:317–23.
75. Graham SM, Arvela OM, Wise GA. J Pediatr 1992;121:710–4.
76. Grattan-Smith PJ, Wilcken B, Procopis PG, et al. Mov Disord 1997;12:39–46.
77. Centers for Disease Control and Prevention. MMWR Morb Mortal Wkly Rep 2003;52:61–4.
78. Dror DK, Allen LH. Nutr Rev 2008;66:250–5.
79. Savage DG, Lindenbaum J, Stabler SP, et al. Am J Med 1994;96:239–46.
80. Carmel R. Arch Intern Med 1996;156:1097–100.
81. Carmel R, Johnson CS, Weiner JM. Arch Intern Med 1987;147:1995–6.
82. Carmel R, Spencer CA. Arch Intern Med 1982;142:1465–9.
83. Carmel R, Weiner JM, Johnson CS. JAMA 1987;257:1081–3.
84. Borch K. Scand J Gastroenterol 1986;21:21–30.
85. Sjoblom SM, Sipponen P, Miettinen M, et al. Endoscopy 1988;20:52–6.
86. Yassin F, Rothenberg SP, Rao S, et al. Blood 2004;103:1515–7.
87. Tanner SM, Li Z, Perko JD, et al. Proc Natl Acad Sci U S A 2005;102:4130–3.
88. Watkins D, Rosenblatt DS. Inherited disorders of folate and cobalamin transport and metabolism. In: Valle D, Beaudet AL, Vogelstein B et al, eds. The Online Metabolic and Molecular Bases of Inherited Disease. New York: McGraw-Hill, 2011: part 17, chap 155.
89. Sumner AE, Chin MM, Abrahm JL, et al. Ann Intern Med 1996;124:469–76.
90. Doscherholmen A, Swaim WR. Gastroenterology 1973;64:913–9.
91. Aminoff M, Carter JE, Chadwick RB, et al. Nature Genet 1999;21:309–13.
92. Kristiansen M, Aminoff M, Jacobsen C, et al. Blood 2000;96:405–9.
93. Provenzale D, Reinhold RB, Golner B, et al. J Am Coll Nutr 1992;11:29–35.
94. Rhode BM, Arseneau P, Cooper BA, et al. Am J Clin Nutr 1996;63:103–9.
95. Miller JW. Adv Nutr 2018;9:511S–18S.
96. Carmel R, Aurangzeb I, Qian D. Am J Gastroenterol 2001;96:63–70.
97. Cohen H, Weinstein WM, Carmel R. Gut 2000;47:638–45.
98. Dierkes J, Ebert M, Malfertheiner P, et al. Dig Dis 2003;21:237–44.
99. Carmel R. J Nutr 2007;137:2481–4.
100. Guttormsen AB, Refsum H, Ueland PM. Acta Anaesthesiol Scand 1994;38:753–6.
101. Oussalah A, Julien M, Levy J, et al. J Clin Med 2019;8:551.
102. Ng J, Frith R. Lancet 2002;360(9330):384.
103. Schilling RF. JAMA 1986;255:1605–6.
104. Kinsella LJ, Green R. Neurology 1995;45:1608–10.
105. Quadros EV. Br J Haematol 2009;148:195–204.
106. Green R, Miller JW. Vitamin B_{12}. In: Zempleni J, Suttie JW, Gregory JF et al, eds. Handbook of Vitamins, 5th Edition. Boca Raton, FL: CRC Press, 2014:447–89.
107. Webb DI, Chodos RB, Mahar CQ, et al. N Engl J Med 1968;279:845–50.
108. Stopa EG, O'Brien R, Katz M. Gastroenterology 1979;76:309–14.
109. LeBauer E, Smith K, Greenberger NJ. Arch Intern Med 1968;122:423–5.
110. Yao L, Wang Y, Qin S, et al. Commun Biol 2023;6:96.
111. Baz R, Alemany C, Green R, et al. Cancer 2004;101:790–5.
112. Carmel R. Clin Chem 2003;49:1367–74.
113. Carmel R, Bellevue R, Kelman Z. Am J Hematol 2010;85:436–9.
114. Carmel R, Parker J, Kelman Z. Br J Haematol 2009;147:386–91.
115. Herbert V. Trans Assoc Am Physicians 1962;75:307–20.
116. Carmel R. Arch Intern Med 1979;139:47–50.
117. Savage D, Lindenbaum J. Am J Med 1983;74:765–72.
118. Green R, Dwyre DM. Semin Hematol 2015;52:279–86.

119. Green R, Kuhl W, Jacobson R, et al. N Engl J Med 1982;307:1322–5.
120. Carmel R. Sem Hematol 2008;45:224–34.
121. Green R, Datta Mitra A. Med Clin North Am 2017;101:297–317.
122. Green R, Datta Mitra A. Folate, Cobalamin, and Megaloblastic Anemias. In: Kaushansky K, Lichtman MA, Prchal JT, et al. Williams Hematology, 10th edition. New York: McGraw Hill, 2021.
123. Beard ME, Weintraub LR. Br J Haematol 1969;16:161–3.
124. Westerman DA, Evans D, Metz J. Br J Haematol 1999;107(3):512–5.
125. Carmel R. Arch Intern Med 1988;148:1712–4.
126. Fernandes-Costa FJ, Green R, Torrance JD. S Afr Med J 1978;53:709–12.
127. Noronha JM, Silverman M. On Folic Acid, Vitamin B_{12}, Methionine and Formiminoglutamic Acid Metabolism. In: Heinrich HC, ed. Vitamin B_{12} und Intrinsic Factor 2. Europäisches Symposion. Stuttgart: Enke, 1962:728–36.
128. Herbert V, Zalusky R. J Clin Invest 1962;41:1263–76.
129. Reynolds EH. Lancet Neurol 2006;5:949–60.
130. Healton EB, Savage DG, Brust JC, et al. Medicine (Baltimore) 1991;70:229–45.
131. Lindenbaum J, Healton EB, Savage DG, et al. N Engl J Med 1988;318:1720–8.
132. Carmel R, Melnyk S, James SJ. Blood 2003;101:3302–8.
133. Pant SS, Asbury AK, Richardson EP Jr. Acta Neurol Scand Suppl 1968;44:1–36.
134. Carmel R, Pullarkat V. Br J Haematol 2003;120:907–9.
135. Green R, Kinsella LJ. Neurology 1995;45:1435–40.
136. Fine EJ, Soria E, Paroski MW, et al. Muscle Nerve 1990;13:158–64.
137. Hemmer B, Glocker FX, Schumacher M, et al. J Neurol Neurosurg Psychiatry 1998;65:822–7.
138. Scherer K. N Engl J Med 2003;348:2208.
139. Karnaze DS, Carmel R. Arch Neurol 1990;47:1008–12.
140. Carmel R, Gott PS, Waters CH, et al. Eur J Haematol 1995;54:245–53.
141. Rundles RW. Blood 1946;1:209–19.
142. Ungley CC. Brain 1949;72:382–427.
143. Magnus EM. Eur J Haematol 1987;39:39–43.
144. Brito A, Verdugo R, Hertrampf E, et al. Am J Clin Nutr 2016;103:250–7.
145. Bottiglieri T, Hyland K, Reynolds EH. Drugs 1994;48:137–52.
146. Grapp M, Wrede A, Schweizer M, et al. Nat Commun 2013;4:2123.
147. Green R. Baillieres Clin Haematol 1995;8:533–66.
148. Lindenbaum J, Rosenberg IH, Wilson PWF, et al. Am J Clin Nutr 1994;60:2–11.
149. Carmel R, Green R, Jacobsen DW, et al. Am J Clin Nutr 1999;70:904–10.
150. Carmel R. Am J Clin Nutr 2011;94:348S–58S.
151. Hvas AM, Ellegaard J, Nexo E. Arch Intern Med 2001;161:1534–41.
152. Carmel R. Am J Clin Nutr 2009;90:1449–50.
153. Metz J, Bell AH, Flicker L, et al. J Am Geriatr Soc 1996;44:1355–61.
154. Waters WE, Withey JL, Kilpatrick GS, et al. Br J Haematol 1971;20:521–6.
155. Smith AD. Food Nutr Bull 2008;29:S143–72.
156. Ho RCM, Cheung MWL, Fu E, et al. Am J Geriatr Psychiatry 2011;19:607–17.
157. Carmel R. Curr Opin Gastroenterol 2012;28:151–8.
158. Malouf R, Areosa Sastre A. Cochrane Database Syst Rev 2003: CD004394.
159. McMahon JA, Green TJ, Skeaff CM, et al. N Engl J Med 2006;354:2564–72.
160. Balk EM, Raman G, Tatsioni A, et al. Arch Intern Med 2007;167:21–30.
161. Aisen PS, Schneider LS, Sano M, et al. JAMA 2008;300:1774–83.
162. Wald DS, Kasturiratne A, Simmonds M. Am J Med 2010;123:522–7.
163. Smith AD, Smith SM, de Jager CA, et al. PLoS One 2010;5:e12244.
164. Douaud G, Refsum H, de Jager CA, et al. Proc Natl Acad Sci U S A. 2013;110:9523–8.
165. de Jager CA, Oulhaj A, Jacoby R, et al. Int J Geriatr Psychiatry 2012;27:592–600.
166. Collin SM, Metcalfe C, Refsum H, et al. Cancer Epidemiol Biomarkers Prev 2010;19:1632–42.
167. Carmel R. Blood 2008;112:2214–21.
168. Kuzminski AM, Del Giacco EJ, Allen RH, et al. Blood 1998;92:1191–8.
169. Kondo H. Acta Haematol 1998;99:200–5.
170. Magnus EM. Scand J Haematol 1986;36:457–65.
171. Green R. Blood 2017;129:2603–11.
172. Allen RH, Stabler SP, Savage DG, et al. Am J Hematol 1990;34:90–8.
173. Lindenbaum J, Savage DG, Stabler SP, et al. Am J Hematol 1990;34:99–107.
174. Carmel R. Food Nutr Bull 2008;29:S177–87.
175. Green R. Am J Clin Nutr 2009;89:712S–6S.
176. Li F, Watkins D, Rosenblatt DS. Mol Genet Metab 2009;98:166–72.
177. Carmel R. J Inher Metab Dis 2011;34:67–73.
178. Winkels RM, Brouwer IA, Clarke R, et al. Am J Clin Nutr 2008;88:348–55.
179. Ebbing M, Bonaa KH, Nygard O, et al. JAMA 2009;302:2119–26.
180. House AA, Eliasziw M, Cattran D, et al. JAMA 2010;303:1603–9.
181. Gimsing P, Hippe E, Helleberg-Rasmussen I, et al. Scand J Haematol 1982;29:311–8.
182. Tordjman R, Genereau T, Guinnepain T, et al. Eur J Haematol 1998;60:269–70.
183. Skouby AP, Hippe E, Olesen H. Blood 1971;38:769–74.
184. Koyama K, Yoshida A, Takeda A, et al. Nephrol Dial Transplant 1997;12:1622–8.
185. Spence JD. Clin Chem Lab Med 2013;51:633–7.
186. Fedosov SN, Brito A, Miller JW, et al. Clin Chem Lab Med 2015;53:1215–25.
187. Carmel R, Brar S, Agrawal A, et al. Clin Chem 2000;46:2017–8.
188. Vlasveld LT, van't Wout JKW, Meuwissen P, et al. Clin Chem 2006;52:157–8.
189. Carmel R. Blood 2005;106:1136–7.
190. Stabler SP, Marcell PD, Podell ER, et al. J Clin Invest 1986;77:1606–12.
191. Adams J, Boddy K, Douglas A. Br J Haematol 1972;23:297–305.
192. Mollin DL, Anderson BB, Burman JF. Clin Haematol 1976;5:521–46.
193. Carmel R, Parker J, Kelman Z. Blood 2009;114:abstract 1989.
194. Lindemans J, Schoester M, van Kapel J. Clin Chim Acta 1983;132:53–61.
195. Brady J, McGregor L, Valente E, et al. Clin Chem 2008;54:567–73.
196. Miller JW, Garrod MG, Rockwood AL, et al. Clin Chem 2006;52:278–85.
197. Goringe A, Ellis R, McDowell I, et al. Haematologica 2006;91:231–4.
198. Morkbak AL, Hvas AM, Milman N, et al. Haematologica 2007;92:1711–2.

199. Carmel R, Vasireddy H, Aurangzeb I, et al. Clin Lab Haematol 2001;23:365–71.
200. Carmel R. Clin Chem 2002;48:407–9.
201. Chen X, Remacha AF, Sardà MP, et al. Am J Clin Nutr 2005;81:110–4.
202. Selhub J, Miller JW, Troen AM, et al. Adv Nutr 2022;13:16–33.
203. Kushnir MM, Komaromy-Hiller G, Shushan B, et al. J Chromatogr B Biomed Sci Appl 2000;741:231–41.
204. Rasmussen K, Moller J, Lyngbak M, et al. Clin Chem 1996;42:630–6.
205. Vogiatzoglou A, Oulhaj A, Smith AD, et al. Clin Chem 2009;55:2198–206.
206. Carmel R, Rasmussen K, Jacobsen DW, et al. Br J Haematol 1996;93:311–8.
207. Lewerin C, Ljungman S, Nilsson-Ehle H. J Intern Med 2007;261:65–73.
208. Molloy AM, Pangilinan F, Mills JL, et al. Am J Hum Genet. 2016;98:869–882.
209. Bjorke-Monsen AL, Refsum H, Markestad T, et al. Clin Chem 2003;49:2067–75.
210. Bjorke-Monsen AL, Torvik I, Saetran H, et al. Pediatrics 2008;122:83–91.
211. Hvas AM, Ellegaard J, Nexo E. Clin Chem 2001;47:1396–404.
212. Sentongo TA, Azzam R, Charrao J. J Pediatr Gastroenterol Nutr 2009;48:495–7.
213. Rogers LM, Boy E, Miller JW, et al. Am J Clin Nutr 2003;77:433–40.
214. Abildgard A, Knudson CS, Hoejskov CS, et al. Clin Chim Acta 2022;525:62–68.
215. Ueland PM, Refsum H, Stabler SP, et al. Clin Chem 1993;39:1764–79.
216. Carmel R, Jacobsen DW. Homocysteine in Health and Disease. Cambridge: Cambridge University Press, 2001.
217. Refsum H, Smith AD, Ueland PM, et al. Clin Chem 2004;50:3–32.
218. Hussein WI, Green R, Jacobsen DW, et al. Ann Intern Med 1999;131:348–51.
219. Miller JW, Selhub J, Nadeau MR, et al. Neurology 2003;60:1125–9.
220. Clarke R, Refsum H, Birks J, et al. Am J Clin Nutr 2003;77:1241–7.
221. Hvas AM, Nexo E. Haematologica 2006;91:1506–12.
222. Schilling R. J Lab Clin Med 1953;42:860–6.
223. Hardlei TF, Morkbak AL, Bor MV, et al. Clin Chem 2010;56:432–6.
224. Devi S, Pasanna RM, Shamshuddin Z, et al. Am J Clin Nutr 2020;112:1504–15.
225. Miller JW, Green R. Am J Clin Nutr 2020;112:1420–1.
226. Carmel R. Clin Exp Immunol 1992;89:74–7.
227. Waters HM, Dawson DW, Howarth JE, et al. J Clin Pathol 1993;46:45–7.
228. Carmel R. Am J Clin Pathol 1988;90:442–6.
229. Andres E, Kurtz JE, Perrin AE, et al. Am J Med 2001;111:126–9.
230. Morris MS, Jacques PF, Rosenberg IH, et al. Am J Clin Nutr 2007;85:193–200.
231. Clarke R, Sherliker P, Hin H, et al. Br J Nutr 2008;100:1054–9.
232. Miller JW, Garrod MG, Allen LH, et al. Am J Clin Nutr 2009;90:1586–92.
233. Selhub J, Morris MS, Jacques PF. Proc Natl Acad Sci U S A 2007;104:19995–20000.
234. Mills JL, Carter TC, Scott JM, et al. Am J Clin Nutr 2011;94:495–500.
235. Dhar M, Bellevue R, Carmel R. N Engl J Med 2003;348:2204–7.
236. Chan CW, Liu SY, Kho CS, et al. Int J Lab Haematol 2007;29:163–71.

C. VITAMINS

CHAPTER 28

Biotin

Toshinobu Kuroishi and Janos Zempleni

CHAPTER OUTLINE

Abbreviations

ACC, acetyl-CoA carboxylase; **AMP**, adenosine monophosphate; **CoA**, coenzyme A; **HLCS**, holocarboxylase synthetase; **MCC**, methylcrotonyl-CoA carboxylase; **Na^+**, sodium; **PC**, pyruvate carboxylase; **PCC**, propionyl-CoA carboxylase; **SMVT**, sodium-dependent multivitamin transporter

History of Discovery

Although a growth requirement for the "bios" fraction had been demonstrated in yeast, Boas first demonstrated the mammalian requirement for a factor—biotin—in rats fed egg-white protein. The severe dermatitis, hair loss, and neuromuscular dysfunction were termed "egg-white injury" and were cured by a factor present in liver. The critical event in this egg-white injury of both humans and rats is the highly specific and very tight binding (dissociation constant = 10^{-15} M) of biotin by avidin, a glycoprotein found in egg white. From an evolutionary standpoint, avidin probably serves as a bacteriostat in egg white; consistent with this hypothesis is the observation that avidin is resistant to a broad range of bacterial proteases. Because avidin is also resistant to pancreatic proteases, dietary avidin binds dietary biotin and biotin synthesized by intestinal microbes and, thus prevents absorption. Cooking denatures avidin and renders it susceptible to digestion and unable to interfere with absorption of biotin.

Structure, Chemistry, and Biochemistry

Structure

Biotin is a bicyclic compound (**Figure 28.1**). One ring contains a ureido group; the other contains sulfur and has a valeric acid side chain. The structure of biotin was elucidated independently by Kogl and by du Vigneaud in the early 1940s.[1] Eight stereoisomers exist, but only one (designated *d*-[+]-biotin or, simply, biotin) is found in nature and is enzymatically active.

Regulation

In mammals, biotin serves as an essential cofactor for five carboxylases, each of which catalyzes a critical step in intermediary metabolism. Biotin exists in free and bound pools within the cell

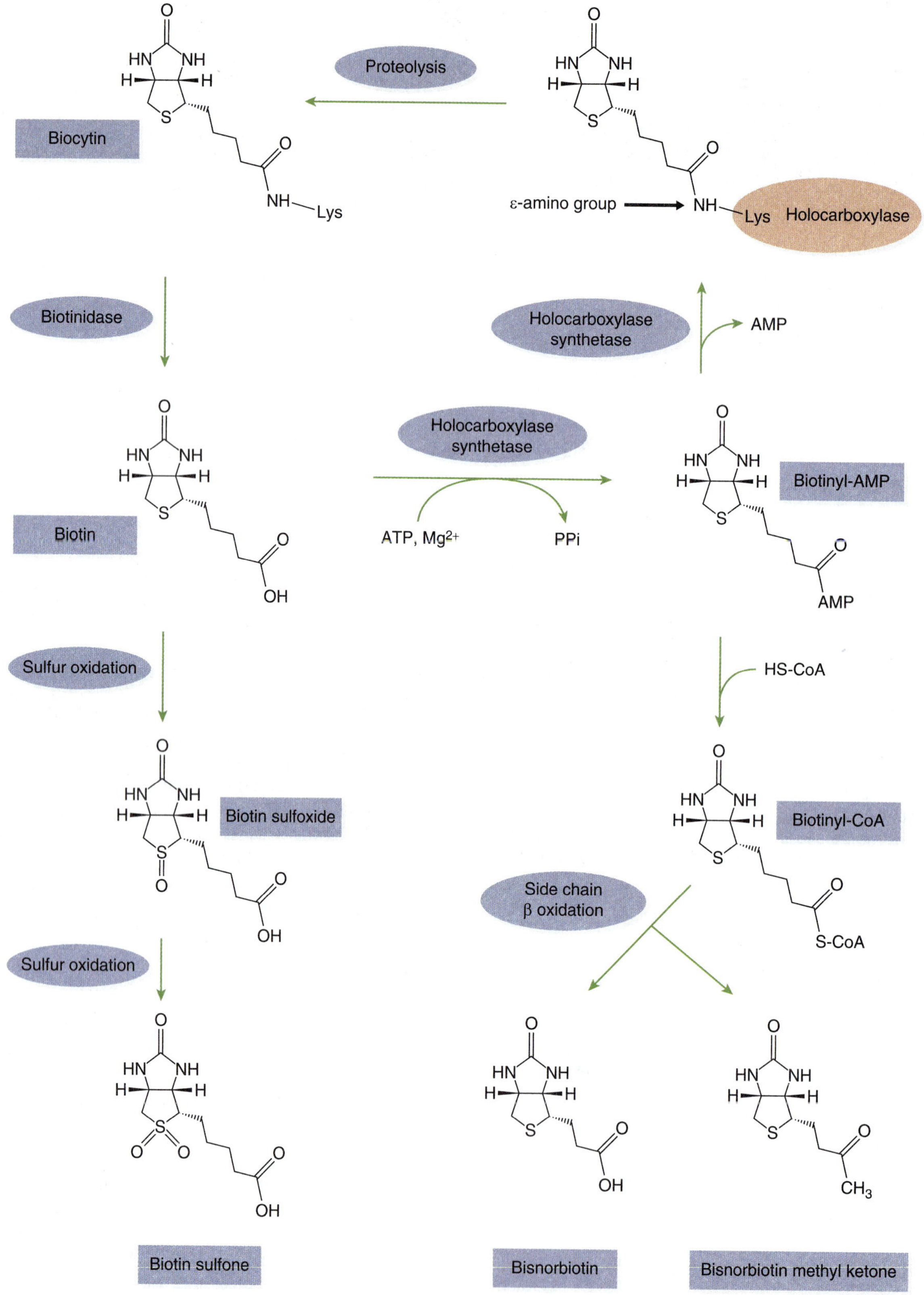

Figure 28.1 Biotin metabolism and degradation. *Ovals* denote enzymes or enzyme systems; *rectangles* denote biotin, intermediates, and metabolites. *AMP*, adenosine monophosphate; *ATP*, adenosine triphosphate; *CoA*, coenzyme A; *Pp$_i$*, pyrophosphate.

that are responsive to changes in biotin status.[2] The pool size is likely determined by a balance among cellular uptake, cellular release, incorporation into apocarboxylases and histones, release from these biotinylated proteins during turnover, and catabolism to inactive metabolites.

Attachment of biotin to the apocarboxylase (see Figure 28.1) is a condensation reaction catalyzed by holocarboxylase synthetase (HLCS). An amide bond is formed between the carboxyl group of the valeric acid side chain of biotin and the ε-amino group of a specific lysyl residue in the apocarboxylase; these apocarboxylase regions contain sequences of amino acids that tend to be highly conserved within and among species for the individual carboxylases.

Regulation of intracellular mammalian carboxylase activity by biotin remains to be elucidated. However, the interaction of biotin synthesis and production of holoacetyl–coenzyme A (CoA) carboxylase in *Escherichia coli* has been extensively studied. In the bacterial system, the availability of the apocarboxylase protein and of biotin (as the intermediate biotinyl–adenosine monophosphate [AMP]) act together to control the rate of biotin synthesis by direct interaction with promoter regions of the biotin operon which, in turn, control a cluster of genes that encode enzymes that catalyze the synthesis of biotin.

In the normal turnover of cellular proteins, holocarboxylases are degraded to biocytin (ε-*N*-biotinyl-L-lysine), or biotin linked to an oligopeptide containing, at most, a few amino acid residues (see Figure 28.1). Biotinidase (biotin amide hydrolase, EC 3.5.1.12) releases biotin for recycling. The clinical manifestations of biotinidase deficiency appear to result largely from a secondary biotin deficiency. The genes for HLCS and biotinidase have been cloned, sequenced, and characterized.[3] HLCS knockout was embryonic lethal in mice,[4] whereas mice with profound biotinidase deficiency developed normally and were asymptomatic if fed a biotin-supplemented diet.[5] When fed a biotin-deficient diet, these mice developed neurologic and cutaneous symptoms.

Chemistry

All five of the mammalian carboxylases catalyze the incorporation of bicarbonate as a carboxyl group into a substrate and employ a similar catalytic mechanism. In the carboxylase reaction, the carboxyl moiety is first attached to biotin at the ureido nitrogen opposite the side chain; then the carboxyl group is transferred to the substrate. The reaction is driven by the hydrolysis of adenosine triphosphate. Subsequent reactions in the pathways release carbon dioxide. Thus, these reaction sequences rearrange the substrates into more useful intermediates, but they do not violate the classic observation that mammalian metabolism does not result in the *net* fixation of carbon dioxide.[6]

Biotin-Dependent Carboxylases

The five biotin-dependent mammalian carboxylases are acetyl-CoA carboxylase (ACC; EC 6.4.1.2) isoforms I and II, pyruvate carboxylase (PC: EC 6.4.1.1), methylcrotonyl-CoA carboxylase (MCC; EC 6.4.1.4), and propionyl-CoA carboxylase (PCC; EC 6.4.1.3).

Both acetyl-CoA carboxylases catalyze the incorporation of bicarbonate into acetyl-CoA to form malonyl-CoA (**Figure 28.2**). Isoform I of ACC (ACC-1) is encoded by the *ACACA* gene and is located in the cytosol. The malonyl-CoA produced by ACC-1 is rate limiting in fatty acid synthesis (elongation). Isoform II of ACC (ACC-2) is encoded by the *ACACB* gene and is located on the outer mitochondrial membrane. ACC-2 controls fatty acid oxidation in mitochondria through the inhibition of carnitine palmitoyltransferase I by its product malonyl-CoA. Carnitine palmitoyltransferase I catalyzes the rate limiting step in fatty acid uptake into mitochondria and thus regulates the availability of fatty acids for oxidation. Thus, ACC-1 and ACC-2 are thought to have two very different roles in cellular metabolism; one controls fatty acid synthesis, and the other controls fatty acid oxidation. An inactive mitochondrial form of ACC may also serve as storage for biotin.[7]

The three remaining carboxylases are mitochondrial. PC catalyzes the incorporation of bicarbonate into pyruvate to form oxaloacetate, an intermediate in the Krebs tricarboxylic acid cycle (see Figure 28.2). In gluconeogenic tissues (i.e., liver and kidney), oxaloacetate can be converted to glucose. Deficiency of PC is probably the cause of lactic acidemia, central nervous system lactic acidosis, and abnormalities in glucose regulation observed in biotin deficiency and biotinidase deficiency.

MCC catalyzes an essential step in the degradation of the branched-chain amino acid leucine (see Figure 28.2). Deficient activity of MCC leads to metabolism of 3-methylcrotonyl-CoA to 3-hydroxyisovaleryl-CoA, 3-hydroxyisovaleryl-carnitine, and 3-hydroxyisovaleric acid by an alternate pathway.[1]

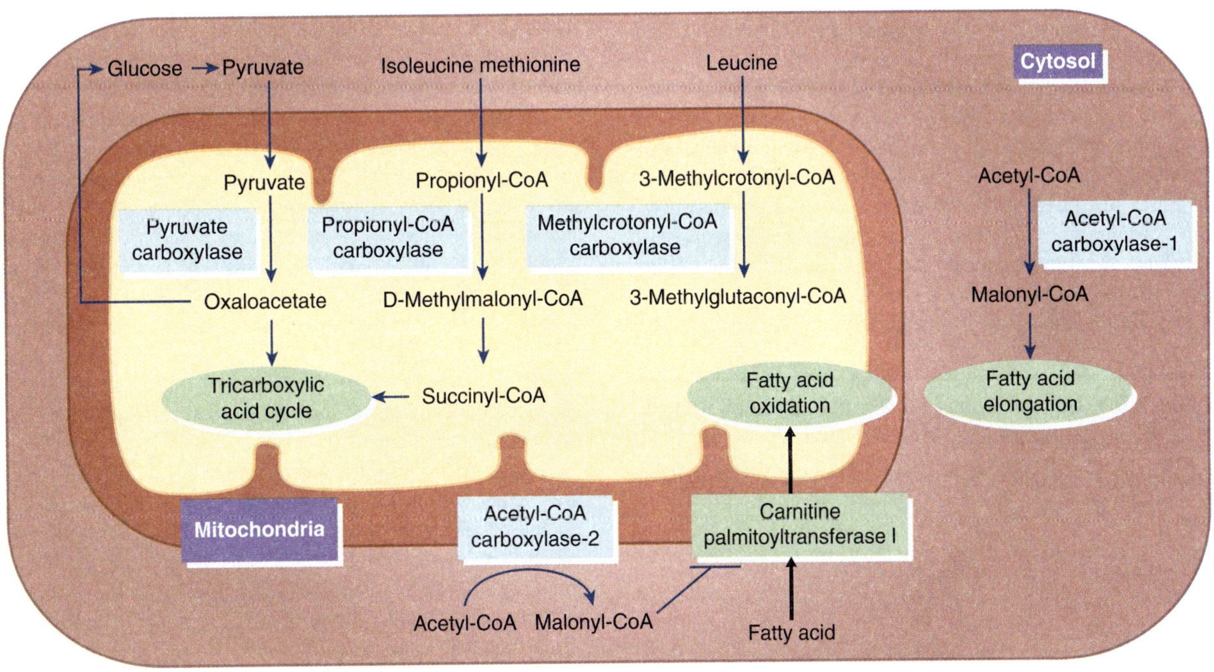

Figure 28.2 Interrelationship of pathways catalyzed by biotin-dependent enzymes *(gray boxes)*. *CoA*, coenzyme A.

Increased urinary excretions of 3-hydroxyisovaleryl-carnitine and 3-hydroxyisovaleric acid reflect deficient activity of MCC and are biomarkers of biotin deficiency.[1,8]

PCC catalyzes the incorporation of bicarbonate into propionyl-CoA to form methylmalonyl-CoA; methylmalonyl-CoA undergoes isomerization to succinyl-CoA and enters the tricarboxylic acid cycle (see Figure 28.2). In a fashion analogous to MCC deficiency, deficiency of PCC leads to increased urinary excretion of 3-hydroxypropionic acid and 3-methylcitric acid.

Metabolism

In humans, about half of biotin is catabolized to inactive metabolites before excretion in urine.[6] The two principal metabolites are bisnorbiotin and biotin sulfoxide. Bisnorbiotin is produced by β-oxidation of the valeric acid side chain. Biotin sulfoxide is produced from oxidation of the sulfur in the thiophane ring. Other minor metabolites result from continued β-oxidation of the side chain, further oxidation of the sulfur, or a combination.

On a molar basis, biotin accounts for approximately half of the total avidin-binding substances in human serum and urine (**Table 28.1**). During pregnancy and long-term anticonvulsant treatment, accelerated biotin catabolism is thought to contribute to biotin deficiency.

Measurement of Biotin and Metabolites

Various assays have been proposed to measure biotin at physiologic concentrations (i.e., 100 pmol/L to 100 nmol/L) and a few have been used to study biotin nutriture. For a more detailed review, please refer to the study by Mock.[9] Most published studies of biotin nutriture have used one of two basic types of biotin assays: bioassays or avidin-binding assays. Bioassays generally have adequate sensitivity to measure biotin in blood and urine, especially with modifications using injected agar plates or metabolic radiometry. However, the bacterial bioassays (and perhaps the eukaryotic bioassays as well) suffer interference from unrelated substances and variable growth response to biotin analogs. Bioassays give conflicting results if biotin is bound to protein.[9]

Avidin-binding assays generally measure the ability of biotin to compete with radiolabeled biotin for binding to avidin (isotope dilution assays), to bind to avidin coupled to a reporter and, thus, prevent the avidin from binding to a biotin linked to solid phase, or to prevent inhibition of a biotinylated enzyme by avidin. Various novel reporter systems have been described.[1,10] Avidin-binding assays generally detect all avidin-binding substances, although the relative detectabilities of biotin and analogs vary among analogs and among assays.[10] High-performance liquid chromatography separation of biotin analogs with subsequent avidin-binding assay of the chromatographic fractions appears to be both sensitive and chemically specific. Using avidin-binding assays, the fasting concentration of biotin in human plasma is approximately 250 pmol/L.

Recently, mass spectrometry coupled with ultra-high-performance liquid chromatography has been used to measure biotin in plasma,[11] milk,[12] and vegetables.[13]

Table 28.1 Normal Range for Biotin and Metabolites in Human Serum and Urine

Compound	Serum (pmol/L)	Urine (nmol/24 h)
Biotin	133–329	18–127
Bisnorbiotin	21–563	6–39
Biotin sulfoxide	0–120	5–19

Absorption

Digestion of Protein-Bound Biotin

The content of free biotin and protein-bound biotin in foods is variable, but most biotin in meats and cereals appears to be protein-bound by an amide bond between biotin and lysine.

Intestinal Absorption, Renal Reabsorption, and Uptake by Somatic Cells

At physiologic pH, biotin is at least modestly water soluble and requires a transporter to cross membranes of such cells as enterocytes, somatic cells, and renal tubule cells.

Intestinal Uptake

An excellent in-depth review of intestinal uptake of biotin has been published.[14] Human intestinal epithelial cells are highly specialized. Biotin transport must occur across two structurally and functionally different membrane domains: the brush-border membrane that faces the intestinal lumen and the basolateral membrane that faces the interstitium, which is in contact with the blood that perfuses the intestine.[14]

A biotin transporter is present in each of the membrane domains. In the brush-border membrane, transport occurs through a sodium (Na^+)-dependent, electroneutral, carrier-mediated mechanism that saturates at the micromolar range, and accounts for the overall limitation in nondiffusion transport.[14] In the presence of a Na^+ gradient, biotin transport occurs against a concentration gradient. This biotin transporter can also transport pantothenic acid and lipoic acid and hence has been named sodium-dependent multivitamin transporter (SMVT). Human SMVT is the product of the *SLC5A6* gene, which is located on chromosome 2p23. SMVT is exclusively targeted to the apical (brush-border) membrane. Intestinal knockout of the *SLC5A6* gene showed that SMVT is the sole intestinal biotin transporter in mice.[15]

Biotin transport across the basolateral membrane is also a carrier-mediated mechanism. However, this carrier is Na^+ independent and electrogenic, and it cannot accumulate biotin against a concentration gradient.[14]

Intestinal biotin transport upregulates in response to biotin deficiency in both humans and animal models. The mechanism likely primarily involves induction of SMVT mRNA synthesis and increased number of SMVT transporters per cell. The increase in SMVT is probably mediated by an induction in the activity of P1, which is one of two promoter regions upstream from the SMVT gene.[14]

In rats, biotin transport is upregulated with maturation and by biotin deficiency.[16] Although carrier-mediated transport of biotin is most active in the proximal small bowel of the rat, the absorption of biotin from the proximal colon is still significant—a finding supporting the potential nutritional significance of biotin synthesized and released by enteric flora. However, the quantitative importance of the contribution of enteric biotin synthesis to absorbed biotin remains to be determined.

Based on a study in which biotin was administered orally in pharmacologic amounts, bioavailability of biotin is approximately 100%. Thus, the pharmacologic doses of biotin given to treat biotin-dependent inborn errors of metabolism are likely to be well absorbed. Moreover, the finding of high bioavailability of biotin at pharmacologic doses provides at least some basis for predicting that bioavailability will also be high at the physiologic doses at which the biotin transporter mediates uptake.

Transport from the Intestine to Peripheral Tissues

Biotin is transported in blood from the site of absorption in the intestine to the peripheral tissues and the liver.[1] Biotin concentration in plasma is low, relative to other water-soluble vitamins. Most biotin in plasma is free and dissolved in the aqueous phase of plasma. However, approximately 7% is reversibly bound to plasma protein, and approximately 12% is covalently bound to plasma protein. Binding to human serum albumin likely accounts for the reversible binding. Biotinidase has been proposed as a biotin-binding protein or biotin-carrier protein for the transport into cells. Biotin transport of into erythrocytes is slow, consistent with passive diffusion.[17]

Uptake by the Liver

SMVT is widely expressed in human tissues. Studies by Subramanya and Said et al., using RNAi specific for SMVT, provided strong evidence that biotin uptake by liver (and likely many other somatic tissues) occurs through SMVT.[18] Metabolic trapping, (e.g., biotin bound covalently to intracellular proteins) is also important. After entering the hepatocyte, biotin diffuses into the mitochondria by a pH-dependent process, suggesting that biotin enters the mitochondria in the neutral, protonated form and dissociates into the anionic form in the alkaline mitochondrial environment, thus becoming trapped by the charge. Based on a study in rats, biliary excretion of biotin is quantitatively negligible.

Renal Handling

Because biotin is a small molecule (244 Da) and mainly not bound to plasma proteins, most plasma biotin appears in the glomerular filtrate. Thus, as with many other water-soluble vitamins, a specific system for the biotin reabsorption from the glomerular filtrate is needed to avoid substantial losses in urine. Said et al., identified the principal renal biotin transport system in brush-border membrane vesicles from human kidney cortex and in human-derived proximal tubular epithelial HK-2 cells.[19] These elegant studies, including gene silencing by siRNA specific to SMVT, provide conclusive proof that SMVT is the renal biotin transporter. Biotin uptake by SMVT is adaptively regulated by biotin deficiency—a finding consistent with previous studies demonstrating reduced biotin excretion early in experimentally induced biotin deficiency in human subjects. Subsequent egress of biotin from the tubular cells occurs through a basolateral membrane transport system that is not dependent on Na^+.

Transport into the Central Nervous System

Various animal and human studies suggest that biotin is transported across the blood–brain barrier.[1,20,21] The transporter is saturable and structurally specific for the carboxylate group on the valeric acid side chain. Transport into the neuron also appears to involve a specific transport system, as well as subsequent trapping of biotin by covalent binding to brain proteins, presumably apocarboxylases or histones.

Ozand et al., described several patients in Saudi Arabia with biotin-responsive basal ganglia disease.[22] Symptoms included confusion, lethargy, vomiting, seizures, dystonia, dysarthria, dysphagia, seventh-nerve paralysis, quadriparesis, ataxia, hypertension, chorea, and coma. The signs and symptoms recurred if biotin was discontinued. A defect in the biotin transporter system across the blood–brain barrier was postulated. Additional research identified a genetic defect in *SLC19A3*,[23] but it was shown conclusively that *SLC19A3* codes for THTR2, the thiamine transporter located in the apical membrane of intestinal, renal tubule, and hepatic cells.[24] THTR2 does not transport biotin; therefore, the biotin responsiveness of these patients is unexplained.

Placental Transport

Biotin concentrations are 3- to 17-fold greater in plasma from human fetuses compared with their mothers in the second trimester—a finding consistent with active placental transport.[25] SMVT is expressed in normal human placenta and, in fact, was originally discovered in human chorionic carcinoma cells. However, in the isolated, perfused, single cotyledon from placenta, transport of biotin across the placenta is relatively weak, thus potentially allowing greater fetal deficiency than maternal deficiency, as reported in mice.[26]

Transport into Human Milk

Greater than 95% of the biotin is free in the skim fraction of human milk.[27] The concentration of biotin in human milk varies substantially,[28] and exceeds the concentration in serum by one to two orders of magnitude—a finding suggesting a transport system into milk. Bisnorbiotin accounts for approximately 50% and biotin sulfoxide for approximately 10% of the total biotin, in addition to metabolites, in early and transitional human milk.[29] With postpartum maturation, the biotin concentration increases, but bisnorbiotin and biotin sulfoxide concentrations still account for 25% and 8% at 5 weeks postpartum. Current studies provide no evidence for a predominant trapping mechanism or for a soluble biotin-binding protein.

Biotin Transporter Deficiency

Mutations in the *SMVT* gene elicited a variety of phenotypes.[30] At age 15 months, an afflicted child failed to thrive, and presented with microcephaly and brain changes on MRI, cerebral palsy and developmental delay, variable immunodeficiency, and severe gastro-esophageal reflux requiring a gastrostomy tube/fundoplication, osteoporosis, and pathologic bone fractures. Supplemental administration of excess biotin, pantothenic acid, and lipoate improved clinical findings. Cells expressing the mutants showed impaired biotin uptake.

A child with biotin dependence resulting from defective biotin transport expressed in lymphocytes has been reported.[17]

This 18-month-old boy presented with sudden onset of coma; both the neurologic problems and a pattern of organic aciduria consistent with multiple carboxylase deficiency were responsive to biotin. Continued high-dose biotin supplementation was needed to prevent symptomatic relapse. The gene sequence of SMVT was normal. These investigators speculated that the defect in lymphocyte biotin transport also occurs in other tissues and mediates some critical aspect of biotin homeostasis.

An additional biotin transporter has been proposed; Daberkow and Zempleni et al., provided evidence that monocarboxylate transporter 1 (MCT1) is a biotin transporter in human lymphocytes.[31] MCT1 may also be responsible for biotin transport in keratinocytes.[32]

Biotin Deficiency

Circumstances Leading to Deficiency

The finding that normal humans have a requirement for biotin has been clearly documented in three situations: prolonged consumption of raw egg white, parenteral nutrition without biotin supplementation in patients with short-gut syndrome and other causes of malabsorption,[1] and infant feeding with an elemental formula devoid of biotin. Because biotin could not legally be added as a supplement to infant formulas in Japan until 2014, all reports related to infant formula have come from Japan.[33,34] These infants often required an elemental formula to treat intractable chronic diarrhea.

The clinical findings and biochemical abnormalities caused by biotinidase deficiency are quite similar to those of biotin deficiency. Common findings include periorificial dermatitis, conjunctivitis, alopecia, ataxia, and developmental delay. These clinical similarities suggest that the pathogenesis of biotinidase deficiency involves secondary biotin deficiency. However, the reported signs and symptoms of biotin deficiency and biotinidase deficiency are not identical. Seizures, irreversible neurosensory hearing loss, and optic atrophy have been observed in biotinidase deficiency, but these findings have not been reported in human biotin deficiency.

Velazquez et al., reported that biotin deficiency occurs in children with severe protein-energy malnutrition.[35] They speculated that the effects of biotin deficiency may be responsible for part of the syndrome of protein-energy malnutrition.

Long-term antiepileptic therapy can lead to biotin depletion in adults and children.[36-38] The mechanism may involve both accelerated biotin breakdown,[39,40] as well as impairment of biotin absorption caused by the anticonvulsants.[41]

Biotin deficiency has also been reported or inferred in several other circumstances, including:

1. Chronic alcoholism[1] and gastrointestinal diseases, perhaps through an effect on intestinal biotin uptake[18,42]
2. Leiner disease (a severe form of seborrheic dermatitis that occurs in infancy)[43-45]
3. Renal dialysis[46-49]

Clinical Findings of Frank Deficiency

Whether caused by egg-white feeding or omission of biotin from total parenteral nutrition, the clinical findings of frank biotin deficiency in adults, older children, and infants are similar. Typically, the findings appear gradually after weeks to several years of egg-white feeding or parenteral nutrition and include the following: thinning of hair and progression to loss of all hair including eyebrows and lashes; a scaly (seborrheic), red (eczematous) skin rash distributed around the eyes, nose, mouth, and perineal orifices; and neurologic findings such as depression, lethargy, hallucinations, and paresthesias of the extremities in adults; and hypotonia, lethargy, and developmental delay in infants. The cutaneous findings in conjunction with an unusual distribution of facial fat have been dubbed biotin deficiency facies.

Laboratory Findings

Indices of biotin status in humans were established primarily by experimentally inducing progressive biotin deficiency with feeding egg white.[6,50-56] In these studies, urinary excretion of biotin and lymphocyte PCC activity declined dramatically with time on the egg-white diet, and plasma concentration of 3-hydroxyisovaleryl-carnitine and urinary excretion of 3-hydroxyisovaleryl-carnitine and 3-hydroxyisovaleric acid increased steadily—findings providing evidence of decreased activity of MCC. Based on two studies totaling 12 subjects, urinary excretions of 3 hydroxyisovaleryl-carnitine and 3-hydroxyisovaleric acid, in response to a leucine challenge, were also sensitive indicators of marginal biotin deficiency. In contrast, plasma concentration of free biotin decreased to abnormal values in only half of the subjects.

Odd-chain fatty acid accumulation is also a marker of biotin deficiency. The accumulation of odd-chain fatty acid is thought to result from PCC deficiency (see Figure 28.2). Presumably, the accumulation of propionyl-CoA leads to the substitution of a propionyl-CoA moiety for acetyl-CoA in the ACC reaction and to the incorporation of a three (rather than two) carbon moiety during fatty acid elongation.

Biotin deficiency can be diagnosed using these indicators combined with resolution of the clinical and laboratory abnormalities in response to biotin supplementation. The clinical response to biotin has included resolution of hypotonia, lethargy, depression, and healing of the rash within a few weeks. Growth of hair was followed by accelerated mental and motor development in infants within months. Pharmacologic doses of biotin (e.g., 1 to 10 mg) have been used to treat most patients.

Biochemical Pathogenesis

The mechanisms by which biotin deficiency produces specific signs and symptoms remain to be completely delineated. The tacit assumption, as with most vitamins that act as required cofactors for essential enzymes, is that the clinical findings of biotin deficiency result directly or indirectly from deficient activities of the biotin-dependent carboxylases. Based on human and animal studies, the central nervous system effects of biotin deficiency (hypotonia, seizures, ataxia, and delayed development) are likely mediated through deficiency of brain PC and the attendant central nervous system lactic acidosis, rather than by disturbances in brain fatty acid composition.[57-59] Abnormalities in metabolism of fatty acids are likely important in the pathogenesis of the skin rash and hair loss.[60]

Several lines of research provided evidence for a potential role for biotin in gene expression: these findings likely will provide insights into the pathogenesis of biotin deficiency. Hymes and Wolf discovered that biotinidase can act as a biotinyl-transferase; biocytin serves as the source of biotin, and histones are specifically biotinylated.[3] Stanley et al., demonstrated that the abundance of biotinylated histones varies with the cell cycle, and that biotinylated histones are increased approximately twofold in activated, dividing lymphocytes, compared with quiescent lymphocytes.[61] These early observations provided initial evidence that biotinylation of histones may play a role in regulating DNA transcription and regulation as an additional element in the histone code. Initially, investigators believed that biotinylation of histones was catalyzed by biotinidase.[62] Indeed, approximately 25% of total cellular biotinidase activity localizes to the nucleus. Nuclear HLCS is associated with chromatin and contributes to histone biotinylation.[63-65] Histone biotinylation remains controversial and it has been proposed that biotinylation of histone H2A is an artifact.[66] Kuroishi et al., demonstrated that <0.001% of human histones H3 and H4 are biotinylated, raising questions as to how such a rare event can regulate gene expression.[67] It has been proposed that HLCS is a member of a multi-protein gene repression complex, and physically interacts with nuclear receptor co-repressor and histone deacetylase 1.[68] Studies conducted in rats and human patients with diabetes support an effect of biotin status on genes that affect carbohydrate metabolism, as well as nitric oxide, in the cyclic guanosine monophosphate signaling pathway.[69-71]

Table 28.2 Adequate Intake of Biotin[a]

Life-Stage Group	Age	Adequate Intake (mg/d)
Infants	0–6 mo	5
	7–12 mo	6
Children	1–3 y	8
	4–8 y	12
Males and females	9–13 y	20
	14–18 y	25
	≥19 y	30
Pregnancy	—	30
Lactation	—	35

[a]1 μg biotin = 4 nmol.

Requirements and Allowances

Data providing an accurate estimate of the dietary and parenteral biotin requirements for infants, children, and adults are lacking. However, recommendations for biotin supplementation have been formulated for oral intake in infants through adults, for oral and parenteral intake of biotin for preterm infants, and for parenteral intake in infants through adults[72] (**Table 28.2**). The Food and Drug Administration estimated that the daily intake of biotin is 30 μg in adults in the United States.[73] In Austria, Germany, and Switzerland, the estimated adequate intake for adults is 40 μg/day in adults.[74]

Dietary Sources

No published evidence indicates that mammals can synthesize biotin; thus, higher animals must derive biotin from other sources. The ultimate source of biotin appears to be de novo synthesis by bacteria; by primitive eukaryotic organisms such as yeast, molds, and algae; and by some plant species.

Most measurements of biotin content of foods have used bioassays.[75,76] These values likely contain substantial error.[77] However, some worthwhile generalizations can be made. Biotin is widely distributed in natural foodstuffs, but the absolute content of even the richest sources is low compared with the content of most other water-soluble vitamins. Foods relatively rich in biotin include egg yolk, liver, and some vegetables. Based on the data of Hardinge and Crooks,[75] the average dietary biotin intake in adults appears to meet or exceed intake recommendations in Switzerland (70 μg/day), Canada (60 μg/day),[78] the UK (35 μg/day),[79,80] and Japan (51 μg/day).[81]

Toxicity

Daily doses up to 200 mg orally and up to 20 mg intravenously have been given to treat biotin-responsive inborn errors of metabolism and acquired biotin deficiency. Toxicity has not been reported.

Acknowledgments

The authors acknowledge the author of the previous version of this chapter, Donald M. Mock for his contributions.

References

1. Mock DM. In: Ziegler EE, Filer LJ Jr, eds. Present Knowledge in Nutrition. 7th ed. Washington, DC: International Life Sciences Institutes, Nutrition Foundation, 1996:220–35.
2. Lewis B, Rathman S, McMahon R. J Nutr 2001;131:2310–5.
3. Wolf B. In: Scriver CR, Beaudet AL, Sly WS et al., eds. The Metabolic and Molecular Basis of Inherited Disease. 8th ed. New York: McGraw-Hill, 2001:3151–77.
4. Sadri M, Wang H, Kuroishi T, et al. Plos One 2022;17:e0265539.
5. Pindolia K, Jordan M, Guo C, et al. Mol Genet Metab. 2011;102:161–9.
6. Mock DM. In: Shils ME, Olson JA, Shike M, et al., eds. Modern Nutrition in Health and Disease. 9th ed., 1999:459–66.
7. Shriver BJ, Roman-Shriver C, Allred JB. J Nutr 1993;123:1140–9.
8. Stratton SL, Horvath TD, Bogusiewicz A, et al. Am J Clin Nutr 2010;92:1399–405.
9. Mock DM. In: Brown M, ed. Present Knowledge in Nutrition. 6th ed. Blacksburg, VA: International Life Sciences Institute, Nutrition Foundation, 1990:189–207.
10. Mock DM, Nyalala JO, Raguseo RM. J Nutr 2001;131:2208–14.

11. Yagi S, Nishizawa M, Ando I, et al. Biomed Chrom 2016;30:1285.
12. Shetty SA, Young MF, Taneja S, et al. J Chrom A 2020;1609:460452.
13. Zhao L, Zhao X, Xu Y, et al. Food Chem 2021;344:128712.
14. Said HM. J Nutr 2009;139:158–62.
15. Ghosal A, Lambrecht N, Subranmanya SB, et al. Am J Physiol Gastrointest Liver Physiol 2013;304:G64–71.
16. Balamurugan K, Ortiz A, Said HM. Am J Physiol Gastrointest Liver Physiol 2003;285:G73–7.
17. Mardach R, Zempleni J, Wolf B, et al. J Clin Invest 2002;109:1617–23.
18. Subramanya SB, Subramanian VS, Kumar JS, et al. Am J Physiol Gastrointest Liver Physiol 2011;300:G494–501.
19. Nabokina SM, Subramanian VS, Said HM. Am J Physiol Renal Physiol 2003;284:F737–42.
20. Spector R, Mock DM. J Neurochem 1987;48:400–4.
21. Spector R, Mock DM. Neurochem Res 1988;13:213–9.
22. Ozand PT, Gascon GG, Al Essa M, et al. Brain 1998;121:1267–79.
23. Zeng WQ, Al-Yamani E, Acierno JS Jr, et al. Am J Hum Genet 2005;77:16–26.
24. Subramanian VS, Marchant JS, Said HM. Am J Physiol Cell Physiol 2006;291:C851–9.
25. Mantagos S, Malamitsi-Puchner A, Antsaklis A, et al. Biol Neonate 1998;74:72–4.
26. Mock DM. J Nutr 2009;139:154–7.
27. Mock DM, Mock NI, Langbehn SE. J Nutr 1992;122:535–45.
28. Mock DM, Mock NI, Dankle JA. J Nutr 1992;122:546–52.
29. Mock DM, Stratton SL, Mock NI. J Pediatr 1997;131:456–8.
30. Subramanian VS, Constantinescu AR, Benke PJ, et al. Hum. Genet. 2017;136:253–261.
31. Daberkow RL, White BR, Cederberg RA, et al. J Nutr 2003;133:2703–6.
32. Grafe F, Wohlrab W, Neubert RH, et al. J Invest Dermatol 2003;120:428–33.
33. Fujimoto W, Inaoki M, Fukui T, et al. J Dermatol 2005;32:256–61.
34. Sato Y, Wakabayashi K, Ogawa E, et al. Pediatr Int 2016;58:867–871.
35. Velazquez A, Martin-del-Campo C, Baez A, et al. Eur J Clin Nutr 1988;43:169–73.
36. Krause KH, Berlit P, Bonjour JP. Ann Neurol 1982;12:485–6.
37. Krause KH, Berlit P, Bonjour JP. Int J Vitam Nutr Res 1982;52:375–85.
38. Mock DM, Dyken ME. Neurology 1997;49:1444–7.
39. Wang KS, Mock NI, Mock DM. J Nutr 1997;127:2212–6.
40. Mock DM, Mock NI, Lombard KA, et al. J Pediatr Gastroenterol Nutr 1998;26:245–50.
41. Said HM, Redha R, Nylander W. Am J Clin Nutr 1989;49:127–31.
42. Said HM, Sharifian A, Bagherzadeh A, et al. Am J Clin Nutr 1990;52:1083–6.
43. Nisenson A. J Pediatr 1957;51:537–48.
44. Nisenson A. Pediatrics 1969;44:1014–5.
45. Erlichman M, Goldstein R, Levi E, et al. Arch Dis Child 1981;567:560–2.
46. Livaniou E, Evangelatos GP, Ithakissios DS, et al. Nephron 1987;46:331–2.
47. Yatzidis H, Koutisicos D, Agroyannis B, et al. Nephron 1984;36:183–6.
48. Koutsikos D, Fourtounas C, Kapetanaki A, et al. Ren Fail 1996;18:131–7.
49. Descombes E, Hanck AB, Fellay G. Kidney Int 1993;43:1319–28.
50. Mock NI, Malik MI, Stumbo PJ, et al. Am J Clin Nutr 1997;65:951–8.
51. Mock DM, Henrich-Shell CL, Carnell N, et al. J Nutr 2004;134:317–20.
52. Mock DM, Henrich CL, Carnell N, et al. Am J Clin Nutr 2002;76:1061–8.
53. Mock DM, Henrich CL, Carnell N, et al. J Nutr Biochem 2002;13:462–70.
54. Horvath TD, Stratton SL, Bogusiewicz A, et al. Anal Chem 2010;82:4140–4.
55. Horvath TD, Stratton SL, Bogusiewicz A, et al. Anal Chem 2010;82:9543–8.
56. Stratton SL, Horvath TD, Bogusiewicz A, et al. J Nutr 2011;141:353–8.
57. Sander JE, Packman S, Townsend JJ. Neurology 1982;32:878–80.
58. Suchy SF, Rizzo WB, Wolf B. Am J Clin Nutr 1986;44:475–80.
59. Suchy SF, Wolf B. Am J Clin Nutr 1986;43:831–8.
60. Mock DM. J Pediatr Gastroenterol Nutr 1990;10:222–9.
61. Stanley JS, Mock DM, Griffin JB, et al. J Nutr 2002;132:1854–9.
62. Hymes J, Fleischhauer K, Wolf B. Clin Chim Acta 1995;233:39–45.
63. Gralla M, Camporeale G, Zempleni J. J Nutr Biochem 2008;19:400–8.
64. Chew YC, West JT, Kratzer SJ, et al. J Nutr 2008;138:2316–22.
65. Kobza K, Sarath G, Zempleni J. BMB Rep 2008;41:310–564.
66. Healy S, Perez-Cadahia B, Jia D, et al. Biochem Biophys Acta 2009;1789:719–33.
67. Kuroishi T, Rios-Avila L, Pestinger V, et al. Mol Genet Metab 2011;104:537–545.
68. Liu D, Zemplini J. Biochem J 2014;461:477–86.
69. Chauhan J, Dakshinamurti K. J Biol Chem 1991;266:10035–8.
70. Dakshinamurti K, Desjardins PR. Can J Biochem 1968;46:1261–7.
71. Collins JC, Paietta E, Green R, et al. J Biol Chem 1988;263:11280–3.
72. Greene HL, Hambridge KM, Schanler R, et al. Am J Clin Nutr 1988;48:1324–42.
73. McBurney MI, Hartunian-Sowa S, Matusheski NV. J Nutr 2017;147:1025–1030.
74. Jungert A, Elinger S, Watzl B, et al. Eur J Nutr 2022;61:1779–1787.
75. Hardinge MG, Crooks H. J Am Diet Assoc 1961;38:240–5.
76. Guilarte TR. Nutr Rep Int 1985;32:837–45.
77. Staggs CG, Sealey WM, McCabe BJ, et al. J Food Compost Anal 2004;17:767–76.
78. Hoppner K, Lampi B, Smith DC. Can Inst Food Sci Technol J 1978;11:71–4.
79. Bull NL, Buss DH. Hum Nutr Appl Nutr 1982;36A:125–9.
80. Lewis J, Buss DH. Br J Nutr 1988;60:413–24.
81. Watanabe T, Suemura K, Taniguchi A, et al. J Nut Sci Vitaminol 2010;56:279–86.

CHAPTER 29

Vitamin C

Mark Levine, Pierre-Christian Violet, and Ifechukwude Ebenuwa

CHAPTER OUTLINE

Abbreviations

DRI, dietary reference intake; **EAR**, estimated average requirement; **G6PD**, Glucose-6-phosphate dehydrogenase; **HPLC**, high-performance liquid chromatography; $\mathbf{K_m}$, Michaelis-Menten constant; **NHANES**, National Health and Nutrition Examination Survey; **NIH**, National Institutes of Health; **RDA**, recommended dietary allowance; **UL**, tolerable upper intake level; $\mathbf{V_{max}}$, maximal velocity

History

Scurvy, which we now know is caused by vitamin C deficiency, was described by Egyptians circa 3000 BC and by Hippocrates circa 500 BC.[1] Although sixteenth- and seventeenth-century sailor-explorers knew of scurvy, its fatal outcome, and its cure with fruits, lime, or plant products, the disease persisted among sailors and in northern latitudes when fruits and vegetables were scarce.

In 1753, James Lind published *A Treatise of the Scurvy*, a landmark controlled study showing that scurvy was easily treated.[2] In clinical experiments at sea, Lind divided 12 patients with severe scurvy into six groups. Each group received a different treatment, including cider, vinegar, seawater, or citrus fruit. The results unequivocally demonstrated that citrus fruit cured scurvy. Unfortunately, Lind included cold climate, dampness, lack of fresh air, and foggy weather as causative agents, thereby obscuring the clear outcome of his clinical trial. Only in 1795 did the British Royal Navy make it mandatory to issue an ounce of citrus juice (lemon, and later lime) daily to every sailor after 2 weeks at sea, but this rule was not enforced until 1804. Sailors in merchant navies continued to develop scurvy until the citrus fruit provision became mandatory following the Merchant Shipping Act in 1854.

Scurvy remained widespread during the American Civil War and World War I. After World War I, research intensified to identify the antiscorbutic principle. Using ox adrenal glands, oranges, and cabbages, Albert Szent-Gyorgyi, in 1928, isolated a six-carbon reducing substance. In 1932, the laboratories of Szent-Gyorgyi and C.G. King independently confirmed that this substance was the antiscorbutic principle.[3,4] Szent-Gyorgyi named it ascorbic acid, and he was awarded the Nobel Prize for this research in 1937.

Terminology, Chemistry, Metabolic Roles, Interactions with Other Compounds, and Basic Importance in Normal Functions

Terminology and Chemical Properties: Formation, Oxidation-Reduction, and Degradation

Vitamin C (L-ascorbic acid, ascorbate), a water-soluble micronutrient essential for humans, is a six-carbon α-ketolactone weak acid with a pH of 4.2 and a molecular weight of 176 (**Figure 29.1**). Plants use glucose and fructose to synthesize vitamin C. Vitamin C is abundant in plant leaves and in chloroplasts, and it may have a role in photosynthesis, stress resistance, plant growth, and development. Most mammals synthesize vitamin C from glucose in the liver, whereas some birds, reptiles, and amphibians synthesize the vitamin in the kidney.[5] Vitamin C is not synthesized by humans and nonhuman primates because of their lack of gulonolactone oxidase, the terminal enzyme in the biosynthetic pathway of vitamin C from glucose. The gulonolactone oxidase gene became nonfunctional in a common primate ancestor.[5] Guinea pigs, capybaras, bats, and some fish also do not synthesize ascorbate.[6,7] For all species unable to synthesize ascorbate, it is a vitamin by definition and must be obtained exogenously. Animals unable to synthesize vitamin C usually obtain sufficient amounts from plant diets, but they develop scurvy in captivity without adequate dietary supplementation.[6]

Vitamin C is an electron donor, or reducing agent (see Figure 29.1), and all of its known functions are attributable to this property. Vitamin C sequentially donates two electrons from the double bond between carbons two and three. When these electrons are lost, vitamin C is oxidized, and another compound is reduced, thereby forestalling oxidation of the reduced compound. Vitamin C is, therefore, commonly known as an antioxidant.

With loss of the first electron, vitamin C oxidizes to the ascorbate-free radical (semidehydroascorbic acid). In comparison with other free radicals, the ascorbate radical is relatively stable

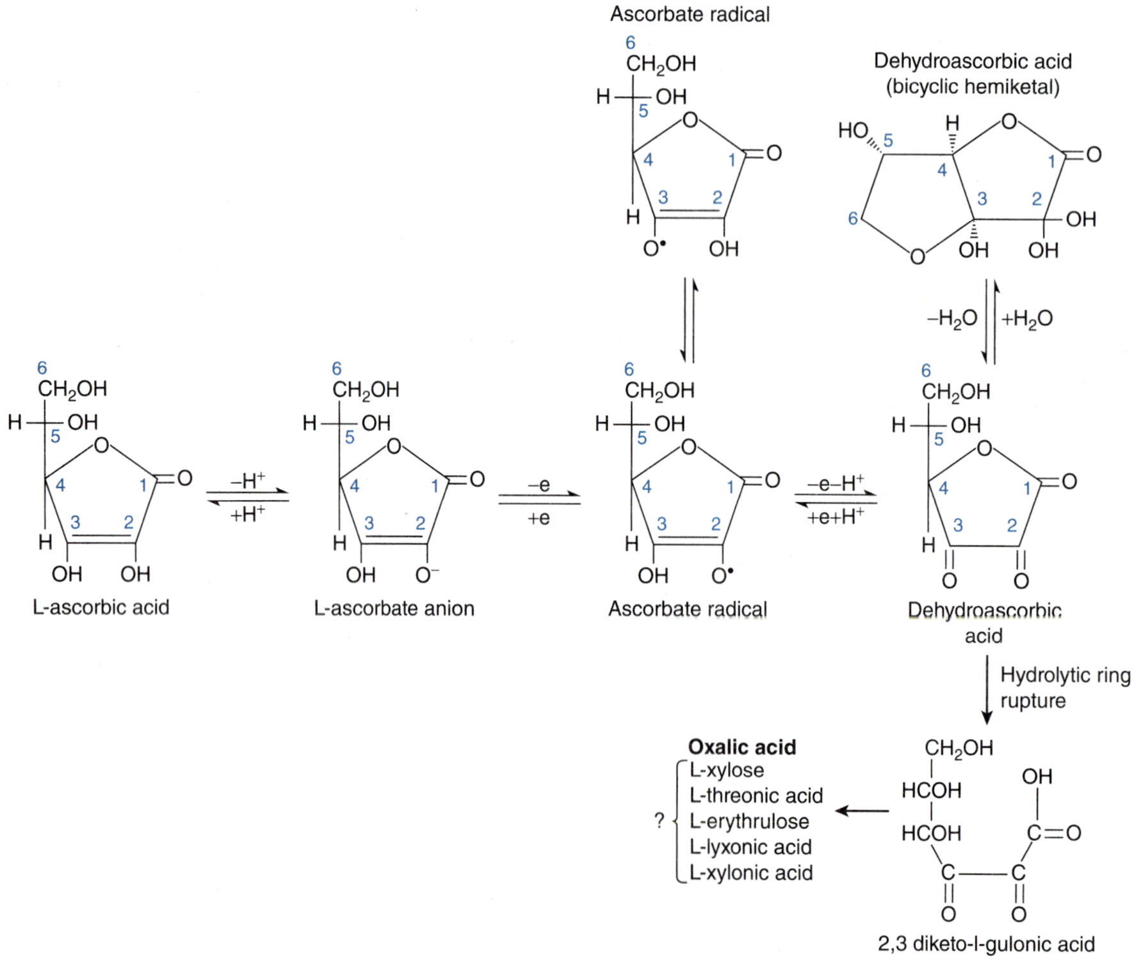

Figure 29.1 Ascorbic acid metabolism. Ascorbic acid and many of its metabolites exist in several resonant forms. These are not shown for simplicity, but two resonant forms of ascorbate radical are shown. Dehydroascorbic acid may exist in many structural forms. The nondehydrated form of dehydroascorbic acid and its hydrated bicyclic hemiketal forms are shown.[2,3] Diketo-l-gulonic acid undergoes further metabolism, resulting in several metabolites including the clinically significant product oxalate.

Modified from Washko PW, Welch RW, Dhariwal KR et al. Anal Biochem 1992;204:1-14.[8]

and unreactive. Reactive free radicals are reduced by vitamin C, and the less reactive ascorbate radical is formed in their place. This is the basis for characterizing vitamin C as a good free radical scavenger, or antioxidant.[9] Because of the short half-lives (i.e., $<10^{-3}$ seconds) of most free radicals, they cannot be measured directly and, instead, are measured indirectly by using other agents that form radical species with longer half-lives. The half-life of the ascorbate radical is long enough to be measured directly by electron paramagnetic resonance, however. Ascorbate radical half-life depends on concentration, the presence of trace metals, and oxygen, and it can vary from 10^{-3} seconds to minutes.

After formation, the ascorbate radical is either reversibly reduced to vitamin C or loses a second electron and is thereby oxidized to dehydroascorbic acid.[9] Although this substance is more stable than the ascorbate radical, the stability of dehydroascorbic acid depends on its concentration, temperature, and pH, and it is often stable for only minutes.[10] Dehydroascorbic acid may exist in one of several different structural forms (see Figure 29.1). Its dominant form in vivo is uncertain, but a good candidate is the hydrated hemiketal.[10,11] Because dehydroascorbic acid is probably not an acid in vivo, the designation "dehydroascorbate" is incorrect. Formation of both the ascorbate radical and dehydroascorbic acid from vitamin C in biologic systems is mediated by oxidants, such as molecular oxygen, with or without trace metals (iron and copper), superoxide, hydroxyl radical, hypochlorous acid, and reactive nitrogen species.

In biologic systems, dehydroascorbic acid has two fates. One is hydrolysis, with irreversible rupture of the ring to yield 2,3-diketogulonic acid. Although 2,3-diketogulonic acid metabolism is not well characterized, its metabolic products probably include oxalate, threonate, xylose, xylonic acid, and lynxonic acid.[10] Carbons from vitamin C were reported to be expired as carbon dioxide in animals, but this probably does not occur in humans.[12] Of vitamin C metabolites formed by dehydroascorbic acid hydrolysis, oxalate is an end-product with clinical significance (see the section "Manifestations of Vitamin C Deficiency and Excess").

The second fate of dehydroascorbic acid is reduction, either to the ascorbate radical by addition of one electron, or to vitamin C by addition of two electrons. Dehydroascorbic acid reduction in biologic tissues occurs chemically or is protein dependent.[5] Chemical reduction of dehydroascorbic acid is mediated in vivo by glutathione, with formation of glutathione disulfide. Enzymatic reduction of dehydroascorbic acid in vivo, with an electron donor, is often faster than by chemical reduction alone. Reduced nicotinamide adenine dinucleotide phosphate-dependent regenerating enzymes include 3-α-hydroxysteroid dehydrogenase and thioredoxin reductase. Glutathione-dependent regenerating enzymes include glutaredoxin (thioltransferase), protein disulfide isomerase, and dehydroascorbate (*sic*) reductase, with Michaelis-Menten constants (K_ms) for dehydroascorbic acid of 250 μM to several millimolars. Protein (enzyme)-mediated reduction results in ascorbate formation without ascorbate radical as an intermediate, as described for glutaredoxin.

Ascorbate radical can also be reduced to vitamin C. Although the reducing activities responsible have not been purified, several reducing activities have been reported in membranes of mitochondria, microsomes, and erythrocytes. The cytosolic enzyme thioredoxin reductase also reduces the ascorbate radical.[5]

In humans, ascorbate radical and dehydroascorbic acid reduction efficiency is incomplete. When vitamin C is removed from diets of healthy humans, deficiency occurs by 30 days, even if the individuals were initially saturated with vitamin C[13,14] (see the discussion about pharmacokinetics under the section "Physiology"). These data are a summed measure of both oxidation and reduction rates. The overall direction is vitamin C utilization, in which ascorbic acid is oxidized to dehydroascorbic acid, and dehydroascorbic acid undergoes irreversible hydrolysis.

Metabolic Roles, Biochemistry, and Importance in Normal Functions

General Principles of Vitamin C as an Electron Donor

Vitamin C is considered an outstanding antioxidant because of its reduction (redox) potential as an electron donor, taking into account anticipated concentrations in vivo. Under standard chemical conditions, the reduction potential of the couple dehydroascorbic acid/vitamin C is approximately +0.06 volts.[9] Reduction potentials are based on the Nernst equation:

$$E = E^o + \frac{RT}{nF} \ln \frac{[\text{electron acceptor}]}{[\text{electron donor}]}$$

Because vitamin C loses electrons sequentially, with ascorbate radical as an intermediate, the reduction potential for the dehydroascorbic acid/ascorbic acid couple is the sum of the dehydroascorbic acid/ascorbate radical and ascorbate radical/ascorbic acid couples. The redox potential of the couple ascorbate radical/ascorbic acid is approximately +0.3 volts under standard conditions.[8,9] Based on only this redox potential, ascorbic acid would not appear to be a good reducing agent. Standard reduction potentials assume that each member of the redox pair is at 1 M concentration, pH 7, at 25°C. Varying concentrations of each species are taken into account by the Nernst equation for calculating reduction potentials, and these can change when concentrations of electron donor and acceptor are different. Under physiologic conditions, predicted concentrations are ascorbic acid >> dehydroascorbic acid >> ascorbate radical, so that the summed redox potentials become favorable for reduction of many oxidizers.[9,10]

In addition to its redox potential, other properties of ascorbic acid make it an excellent biochemical electron donor. After one electron loss, the product ascorbate radical under physiologic conditions is relatively harmless and nonreactive, and it produces little superoxide because of poor reactivity with oxygen.[9] As noted earlier, some dehydroascorbic acid is reduced by cells to ascorbic acid, for reuse.[15,16]

Reductive Functions

Enzymatic Functions. Vitamin C is an electron donor for 21 enzymes, three of which are in fungi and are involved in reutilization pathways for pyrimidines or deoxynucleosides.[7] In mammals, vitamin C is a cofactor for 18 different enzymes that are either mono-oxygenases or dioxygenases (**Table 29.1**). The mono-oxygenases dopamine β-mono-oxygenase and peptidyl glycine α-mono-oxygenase incorporate a single oxygen molecule into a substrate, either dopamine for norepinephrine synthesis or a peptide with a terminal glycine for peptide amidation.[17,18] The remaining 16 mammalian enzymes are dioxygenases, which incorporate molecular oxygen (O_2) but with each oxygen atom incorporated in a different way. Nine dioxygenases add hydroxyl groups to proline

Table 29.1 Putative Enzymatic and Nonenzymatic Effects of Vitamin C in Humans

Cofactor for Enzymes		
Enzyme	*Function of Enzyme*	*Reference*
Dopamine β-monooxygenase	Norepinephrine biosynthesis	{Dhariwal 1989, Ref 17}
Peptidylglycine α-amidating monooxygenase	Amidation of peptide hormones	{Prigge 2000, Ref 18}
Prolyl 3-hydroxylase Lysyl hydroxylase Prolyl 4-hydroxylase	Collagen hydroxylation	{Prockop 1995, Ref 19}
Hypoxia-inducible factor 1, 2 and 3 (HIF) isoenzymes HIF-Prolyl 4-hydroxylase Asparaginyl hydroxylase or FIH-1 (factor inhibiting HIF)	Regulation of HIF	{Myllyharju 2008, Ref 20}
Trimethyllysine hydroxylase γ-Butyrobetaine hydroxylase	Carnitine biosynthesis	{Rebouche 1991, Ref 21}
4-Hydroxyphenylpyruvate dioxygenase	Tyrosine metabolism	{Lindblad 1970, Ref 22}
Flavin adenine dinucleotide-dependent amine oxidase (lysine-specific demethylase 1)	Histone demethylation	{Tsukada 2006, Ref 23}
Ten-eleven Translocation (Tet1-3)	DNA demethylation	{Tahiliani 2009, Ref 24; Minor 2013, Ref 25; Blaschke 2013, Ref 26; Hore, 2016, Ref 27}
Reducing Agent		
Site	*Action*	*Reference*
Small Intestine	Promote iron absorption	{Hallberg 1987, Ref 28; Cook 2001, Ref 29; Li 2020, Ref 30}
Antioxidant		
Site	*Action*	*Reference*
Cells	Regulate gene expression and mRNA translation, prevent oxidant damage to DNA and intracellular	{Toth 1995, Ref 31; Hitomi 1996, Ref 32; Padayatty 2003, Ref 33}
Plasma	Increase endothelium-dependent vasodilatation, reduce extracellular oxidants from neutrophils, reduce low-density lipoprotein oxidation, quench aqueous peroxyl radicals and lipid peroxidation products	{Polodori 2004, Ref 34; Steinberg 2009, Ref 35; Frei 2012, Ref 36}
Stomach	Prevent formation of N-nitroso compounds	{Helser 1992, Ref 37}
Prooxidant		
Target	*Effect*	*Reference*
DNA	DNA damage	{Podmore 1998, Ref 38}
Lipid hydroperoxidase	Decomposition of lipid peroxidase leading to DNA damage	{Lee 2001, Ref 39}
Ascorbate radical targets	Damage to some cancer cells	{Chen 2005, Ref 55; Chen 2008, Ref 41}

Data from Levine M, Rumsey SC, Daruwala R et al. Criteria and recommendations for vitamin C intake. JAMA 1999;281:1415–23.

or lysine. Of these, three prolyl 4-hydroxylase isoenzymes add hydroxyl groups to the amino acid proline in the collagen molecule to stabilize its triple helix structure.[19] Four prolyl 4-hydroxylases add hydroxyl groups to proline in hypoxia-inducible factor (HIF).[20] Two additional dioxygenases, prolyl-3-hydroxylase and lysyl hydroxylase, also modify collagen.[19] Beyond collagen synthesis, flavin adenine dinucleotide-dependent amine oxidase catalyzes histone demethylation coupled to oxidation of 2-oxoglutarate.[23] Three dioxygenases (TET1-3) convert cytosine to hydroxymethylcytosine, as part of a key pathway of DNA demethylation.[24-27] Of the remaining three mammalian dioxygenase enzymes, two participate in different steps of biosynthesis of carnitine, necessary for fatty acid transport into mitochondria for adenosine triphosphate synthesis,[21] and the remaining dioxygenase participates in tyrosine metabolism.[22] For some enzymes (i.e., dopamine beta-monooxygenase), ascorbate is a cosubstrate, whereas for others (i.e., TET1-3) ascorbate action as a cofactor may be conditionally dependent on concomitant Fe^{2+} availability and its oxidation. Scurvy is possibly the result of impaired function of ascorbate-dependent enzymes.

Nonenzymatic Reductive Functions: Vitamin C as an Antioxidant In Vitro. Vitamin C may have nonenzymatic functions resulting from its redox potential or free radical intermediate. In vitro evidence suggests that vitamin C has a role as a chemical reducing agent, both intracellularly and extracellularly. Intracellular vitamin C may prevent intracellular protein oxidation in tissues with millimolar ascorbate concentrations and high oxidant production or oxygen concentration, such as neutrophils, monocytes, macrophages, lung, and tissues of the eye that are exposed to light.[7,33]

In vitro, extracellular vitamin C may protect against oxidants and oxidant-mediated damage. Aqueous peroxyl radicals and lipid peroxidation products in isolated plasma are quenched by vitamin C,[34,42] which is preferentially oxidized before the plasma antioxidants uric acid, tocopherol, and bilirubin. In vitro, extracellular vitamin C affects several pathways involved in atherogenesis, including protection of low-density lipoprotein (LDL) from metal-catalyzed oxidation and regeneration of oxidized α-tocopherol (vitamin E) as a lipid-soluble antioxidant[34,42,43] (see chapter 19). Because α-tocopherol also prevents oxidation of LDL in vitro,[43] recycling of oxidized α-tocopherol by vitamin C was hypothesized to decrease atherosclerosis, as part of the oxidation modification hypothesis.[35] Unfortunately, vitamin C has minimal effects on markers of oxidation and endothelial activation in humans,[44] the oxidative modification hypothesis has not been supported by most clinical trials,[35,45,46] and the evidence is limited that α-tocopherol recycling occurs in vivo.[47,48]

Caution is necessary in extrapolating conclusions from in vitro experiments with vitamin C to in vivo conditions.[33,43] Reactions in vitro may not have a specific requirement for vitamin C as an antioxidant in vivo, and the type or concentration of the oxidant used in vitro may not be relevant in vivo. Oxidation in vitro is often induced by copper or iron, either added exogenously or as unintended trace contaminants in culture media. In vitro, metal-catalyzed LDL oxidation requires free copper or iron and relatively long lag periods for induction of oxidation. In vivo, both metals are tightly bound to proteins and may be unavailable to oxidize vitamin C at physiologic concentrations.

Extracellular vitamin C could have other effects as an antioxidant. For example, it may reduce oxidants from activated neutrophils[15] or macrophages that otherwise could damage collagen or fibroblasts.[49] Extracellular vitamin C in the intestinal lumen may keep iron reduced, facilitate iron absorption, and quench reactive oxidants in the stomach and duodenum (see section "Functional Consequences in Humans").

Other Cell Functions. In vitro, vitamin C may have other nonenzymatic, intracellular functions. Vitamin C has been reported to regulate gene transcription, mRNA stabilization, and signal transduction for certain genes. Examples include the following genes: collagen types I and III, elastin, acetylcholine receptor, fos-related antigen-1 (fra-1), activator protein-1 (AP-1), nuclear factor-κB (NF-κB), some forms of cytochrome P-450, tyrosine hydroxylase, collagen integrins, some ubiquitins, some osteoblastic marker proteins, and phosphatidylinositol transfer protein.[5,50,51] It is unknown whether and how vitamin C enters the cellular nucleus, or whether nuclear pores regulate entry of vitamin C as a charged molecule at physiologic pH. Vitamin C may stabilize intracellular tetrahydrobiopterin, perhaps enhancing endothelial nitric oxide synthesis/availability,[52,53] and also may regulate mRNA translation.[31] In red blood cells, vitamin C maintains the cytoskeletal protein beta-spectrin.[16]

Effects of vitamin C should be interpreted cautiously for many of these pathways when there is extrapolation to in vivo conditions based on in vitro data.[33] Often, control cells have no vitamin C. No corresponding in vivo condition exists, other than death. Sometimes added ascorbate concentrations are high enough to generate oxidants inadvertently or purposefully, and these oxidants are responsible for observed effects[41,50,54,55] (see pharmacokinetics, page 433).

Prooxidant Functions

Acting as an electron donor, it has been proposed that vitamin C could initiate pro-oxidant reactions, such as increasing 8-oxo-adenine in DNA or decomposition of lipid hydroperoxides.[34,35] The physiologic relevance of these systems is unclear, whether because vitamin C concentrations were not truly physiologic, in vitro conditions were not representative of in vivo physiology, or experimental artifacts may have complicated interpretation of the measurements. In vivo data do not support a pro-oxidant effect of vitamin C at physiologic concentrations.[14] Potential functions as a pro-oxidant at pharmacologic concentrations[41,55] are discussed in the sections "Physiology" (discussion on pharmacokinetics) and "Functional Consequences in Humans."

Dietary Sources and Intake

Food Sources of Vitamin C

The fruits and seeds of plants act as sink organs for synthesized ascorbate (**Table 29.2**). Because vitamin C is labile, its content in plant foods may vary depending on season, transportation, shelf time, storage, and cooking practices. Generally, 200 to 300 mg daily of vitamin C can be obtained from five servings of fruits and vegetables if a wide variety is consumed, whereas fruit and vegetable consumption restricted to a narrow selection may provide less.[56] Vitamin C is also available as a supplement in tablet and powder form, alone or in combination with other vitamins.[33]

Table 29.2 Food Sources of Vitamin C

Source (Portion Size)	Vit C (Mg)
Fruit	
Cantaloupe (¼ medium)	60
Fresh grapefruit (½ fruit)	40
Honeydew melon (⅛ medium)	40
Kiwi (1 medium)	75
Mango (1 cup, sliced)	45
Orange (1 medium)	70
Papaya (1 cup, cubes)	85
Strawberries (1 cup, sliced)	95
Tangerines or tangelos (1 medium)	25
Watermelon (1 cup)	15
Juice	
Grapefruit (½ cup)	35
Orange (½ cup)	50
Fortified Juice	
Apple (½ cup)	50
Cranberry juice cocktail (½ cup)	45
Grape (½ cup)	120
Vegetables	
Asparagus, cooked (½ cup)	10
Broccoli, cooked (½ cup)	60
Brussels sprouts, cooked (½ cup)	50
Cabbage	
Red raw, chopped (½ Cup)	20
Red, cooked (½ cup)	25
Raw, chopped (½ cup)	10
Cooked (½ cup)	15
Cauliflower, raw or cooked (½ cup)	25
Kale, cooked (½ c)	55
Mustard greens, cooked (1 cup)	35
Pepper, red or green	
Raw (½ cup)	65
Cooked (½ cup)	50
Plantain, sliced, cooked (½ cup)	15
Potato, baked (1 medium)	25
Snow peas	
Fresh, cooked (½ cup)	40
Frozen, cooked (½ cup)	20
Sweet potato	
Baked (1 medium)	30
Vacuum can (1 cup)	50
Canned, syrup-pack (1 cup)	20
Tomato	
Raw (½ cup)	15
Canned (½ cup)	35
Juice (6 fluid oz)	35

From Levine M, Rumsey SC, Daruwala R et al. JAMA 1999;281:1415-23, with permission of the American Medical Association.

Intake in the United State

Continuous National Health and Nutrition Examination Survey (NHANES) data are obtained from a representative sample of the U.S. population. From 1999–2018 NHANES data, mean vitamin C intake declined by nearly 25% for adults, from nearly 100 mg to less than 80 mg daily. Similar intake declines were noted for children.[57] The proportion of the population with intake below the Estimated Average Requirement (EAR) for vitamin C increased from 38% to 47% (pharmacokinetics data discussed later in the section "Physiology" can be used to convert plasma values to estimated intake). NHANES data limitations include reliance on dietary surveys, and exclusion of supplement intake. Mean serum vitamin C concentration decreased by approximately 5% when comparing data from 2003–2006 to 2017–2018, but the change was not significant. For 2017–2018 data, mean serum vitamin C concentrations were approximately 60 μM for children and adolescents, and 47 μM for adults.

More detailed analyses of vitamin C plasma data are available from NHANES 2003–2006.[58] Vitamin C deficiency, defined as plasma vitamin C concentration lower than 11.4 μM, occurred in 8.6% of males and 4.9% of females (see section "Assessment of Vitamin C Status"). Vitamin C deficiency was more common in some population subgroups, including low-income individuals and smokers. Vitamin C deficiency occurred in nearly 15% of smokers, in contrast to 3.3% of nonsmokers. To date, analyses of deficiency data from NHANES 2017–2018 are limited, but deficiency prevalence was unchanged.[59]

Dietary Reference Intakes

General Strategies for Deriving Recommendations

Ideally, recommendations for optimum intake of vitamin C should be based on intakes that produce good clinical outcomes. Without such data, other measures in combination can be used: dietary availability, steady-state concentrations in plasma or tissues in relation to dose, bioavailability, urinary excretion, adverse effects, biochemical and molecular functions in relation to concentration, beneficial effects in relation to dose, and prevention of deficiency.[60,61] Although data are available for some parameters,[13,14,62] clinical outcome data that describe the optimal intake of vitamin C in health and to prevent disease are lacking[63] (see section "Functional Consequences in Humans"). As described below, a successful approach for vitamin C has been the detailed characterization of normal vitamin physiology in healthy humans using pharmacokinetic principles.[13,14,62,64] With clarity of normal physiology, pathophysiology follows[62] (See renal reabsorption and excretion, page 435).

Dietary Reference Intake Values

DRI values for vitamin C were set by the Institute of Medicine.[61] Calculations of the EAR were based on neutrophil vitamin C concentrations in men, extrapolated vitamin C antioxidant action in neutrophils, and urinary vitamin C excretion in men, an approach reviewed elsewhere.[63] A limitation is use of in vitro data for neutrophil antioxidant function at arbitrary time points. The EAR for men 19 years and older was determined as 75 mg/day. Based on body weight differences by sex, requirements for women were extrapolated, and the EAR for women 19 years and older was set at 60 mg/day. EAR values were then used to calculate recommended dietary allowances (RDAs) for vitamin C in the United States and Canada; thus, RDAs were set at 90 mg/day for men and 75 mg/day for women (**Table 29.3**). Actual, rather than extrapolated, data for women became available only after the release of the DRI values,[14] and they have not yet been incorporated into these guidelines. Based on these newer pharmacokinetic data, other countries have set vitamin C intake recommendations at 100 to 110 mg daily.[14] Not included in the guidelines are pharmacokinetics data that describe the threshold for vitamin C renal excretion as a function of dose.[62]

Pregnancy

Plasma vitamin C concentration decreases during pregnancy, secondary to hemodilution, active transfer to the fetus, or increased renal loss. Vitamin C deficiency during pregnancy is associated with increased risk of infection, premature rupture of membranes, premature delivery, and eclampsia. It is unknown whether vitamin C deficiency contributes to these conditions or simply indicates poor general nutrition. An increase in intake from 75 mg daily in nonpregnant women to 85 mg/day during pregnancy was recommended, based on data that 7 mg/day of vitamin C prevents scurvy in infants.[61]

Use in Disease

Data were insufficient to recommend additional vitamin C other than in pregnant women, lactating women, and smokers.[61]

Upper Level

The tolerable upper intake level (UL) for vitamin C was set at 2 g/day, based on adverse gastrointestinal effects at higher doses.[61]

Table 29.3 Dietary Reference Intake Values for Vitamin C[a]

Life Stage	Sex	Age	Ear	RDA	AI	UL
Infants (mo)		0–6			40	[b]
		7–12			50	
Children (y)	Boys and girls	1–3	13	15		400
	Boys	4–8	22	25		650
		9–13	39	45		1,200
	Girls	14–18	63	75		1,800
		9–13	39	45		1,200
		14–18	56	65		1,800
Adults (y)	Men	19–30	75	90		2,000
		31–50	75	90		
		51–70	75	90		
		>70	75	90		
	Women	19–30	60	75		
		31–50	60	75		
		51–70	60	75		
		>70	60	75		
Pregnancy (y)		14–18	66	80		1,800
		19–30	70	85		2,000
		31–50	70	85		
Lactation (y)		14–18	96	115		1,800
		19–30	100	120		2,000
		31–50	100	120		
Smokers (y)	Men	>19	110	130[c]		
	Women	>19	95	115[c]		

AI, adequate intake; EAR, estimated average requirement; RDA, recommended dietary allowance; UL, tolerable upper intake level.

[a]Dietary reference intake values for vitamin C in milligrams, by life stage and gender.

[b]It is not possible to establish UL values for infants and children, for whom the source of vitamin C intake should be infant formula and food only.

[c]Whereas EARs were stated for smokers, RDAs for smokers were not explicitly documented. RDAs for smokers are calculated based on stated EAR × 1.2.

Data from Food and Nutrition Board, Institute of Medicine. Dietary Reference Intakes for Vitamin C, Vitamin E, Selenium, and Carotenoids. Washington, DC: National Academy Press, 2000. From Levine M, Padayatty SJ, Katz A et al. Dietary allowances for vitamin C: recommended dietary allowances and optimal nutrient ingestion. In: Asard H, May JM, Smirnoff N, eds. Vitamin C Function and Biochemistry in Animals and Plants. London: BIOS Scientific Publishers, 2004:291–316, with permission of BIOS Scientific Publishers, London.

Physiology

General Physiology and Tissue Distribution

Absorbed vitamin C reaches the liver through the hepatic portal venous system. Beyond the hepatic vein, vitamin C appears in the general circulation and is not protein bound. In blood, ascorbic acid is either the dominant chemical species or the only one.[65] In the kidney, vitamin C is freely filtered through glomeruli and is reabsorbed in proximal collecting tubules. When reabsorption becomes saturated, the remaining vitamin C is excreted unchanged in urine.[14,62]

Vitamin C distributes freely in the extracellular space as a water-soluble micronutrient,[40] and it is accumulated in almost all human tissues (**Table 29.4**). As an approximate conversion, 1 g of tissue equals 1 mL of internal volume. Concentration gradients range from a minimum of approximately 2- to 5-fold, to a maximum of approximately 100-fold for the pituitary and adrenal glands. Red blood cells are the only cells in which internal concentrations of vitamin C are similar to or less than plasma concentrations.[16] Because many measurements were performed on postmortem specimens before the advent of accurate high-performance liquid chromatography (HPLC) assays, it is likely that some literature values are underestimates.

It is uncertain why vitamin C is accumulated in millimolar concentration in many cells. For some types of cells, ascorbate may function as an enzyme cofactor. In the adrenal medulla, vitamin C is a cofactor for norepinephrine biosynthesis from dopamine. In the pituitary and perhaps the pancreas, vitamin C may be a cofactor for amidation of peptide hormones. In fibroblasts, osteoblasts, and chondrocytes, vitamin C is a cofactor for proline and lysine hydroxylation, and it perhaps regulates collagen or elastin gene transcription. In many cells, vitamin C may regulate DNA demethylation. Vitamin C has several postulated roles that are incompletely characterized, often involving antioxidant action, in neutrophils, monocytes, lens, retina, cornea, peripheral and central neurons, liver, pancreas, skeletal muscle, and endothelial cells. The purpose of accumulated vitamin C is uncertain in lymphocytes, platelets, adrenal cortex, testis, and ovary.

Transport and Accumulation Principles

Ascorbate is accumulated intracellularly by two distinct pathways: active transport as ascorbate and facilitated transport as dehydroascorbic acid through ascorbate recycling. In the former pathway, ascorbic acid itself is transported by one of two known sodium-dependent transporters, SLC23A1 and SLC23A2, which are also termed SVCT1 and SVCT2 (sodium-dependent vitamin C transporter).[68,69] The SVCTs are members of the nucleobase transporter superfamily and are dissimilar to other sodium-dependent transporters.[70] SVCT1 (SLC23A1) is localized to intestine, liver, and kidney and is a polarized epithelial cell transporter. SVCT1 has a Michaelis-Menten constant (K_m) of approximately 100 μM to 200 μM and a maximal velocity (V_{max}) of approximately 1 mM. These values are consistent with predicted vitamin C concentration in the intestinal lumen after oral ingestion, in the portal venous system and in the proximal renal tubule. SVCT2 (SLC23A2) is widely distributed among tissues; it has a K_m of approximately 5 μM to 10 μM and V_{max} of approximately 60 μM to 100 μM. These values are within the range of vitamin C concentrations found in human tissues, as described later. Both these transporters are sodium and energy dependent and do not transport dehydroascorbic acid.[11]

The second mechanism for ascorbate accumulation in cells is ascorbate recycling. In this pathway, external ascorbic acid is oxidized to dehydroascorbic acid, which is then transported by facilitated glucose transporters 1 to 4 and is immediately reduced to ascorbic acid intracellularly.[11,15,71] The affinity of dehydroascorbic acid for at least some facilitated glucose transporters (i.e., GLUT1), up to 1,000-fold, is greater than that of glucose.[72] Intracellular dehydroascorbic acid reduction is mediated by glutathione or reducing proteins. Dehydroascorbic acid is transported by activated neutrophils,[15,71] and is the only substrate transported by red blood cells to obtain their ascorbic acid.[16,73]

Vitamin C accumulation in vivo is driven primarily by sodium-dependent vitamin C transport. Specific cell types such as red blood cells and activated neutrophils use the dehydroascorbic acid–ascorbate recycling pathway. Although dehydroascorbic acid has been suggested to be the primary means for ascorbate

Table 29.4 Vitamin C Content of Human Tissues[a]

Organ/Tissue	Vitamin C Concentration	Organ/Tissue	Vitamin C Concentration
Pituitary gland	40–50	Lungs	7
Adrenal gland	30–40	Skeletal muscle	3–4
Eye lens	25–31	Testes	3
Liver	10–16	Thyroid	2
Brain	13–15	Cerebrospinal fluid	3.8
Pancreas	10–15	Plasma	0.3–1.0
Spleen	10–15	Saliva	0.07–0.09
Kidneys	5–15	Red Blood Cells	0.2–0.9

[a]Ascorbic acid content of human tissues (mg/100 g of wet tissue, mg/100 mL for fluids).[66,67] Values given are approximate and may vary with ascorbic acid intake, age, and possibly disease states.

Data from Hornig D. Ann N Y Acad Sci 1975;258:103–18, with permission of the New York Academy of Sciences, New York; Padayatty, S. J., and Levine, M. (2016) Oral Dis 22, 463-493; Tu, H., Li, H., Wang, Y. et al. (2015) EBioMedicine 2, 1735-1750.[7,16]

accumulation,[74,75] this is inconsistent with available data. Mouse experiments with genomic knockout of SVCT2 or chemical knockout of dehydroascorbic acid transport provide firm evidence that sodium-dependent vitamin C transport is the dominant mechanism in vivo.[73,76] Dehydroascorbic acid–ascorbate recycling depends on dehydroascorbic acid formation and availability. Using HPLC assays, only trace amounts of dehydroascorbic acid are found in blood and plasma,[65] but these measurements have limitations. For other vitamins with precursor and active forms (i.e., 25-OH vs 1,25-OH vitamin D), assays capture both forms distinctly, but this possibility is unavailable for the ascorbic acid dehydroascorbic acid couple in biologic samples. Assay development, specifically with separate measurement of ascorbic acid and dehydroascorbic acid, has great potential to advance vitamin C transport knowledge and biology in health and disease.

Other vitamin C transporters are predicted to exist. SVCT1 knockout mice absorb vitamin C, a finding suggesting that another absorptive transporter exists or is induced.[77] Because vitamin C is a charged molecule at physiologic pH and is not diffusible, transporters should mediate efflux as well as influx. Once transported into the intestinal epithelial cells, vitamin C must exit to reach the mesenteric vein. Similarly, absorbed vitamin C in the mesenteric vein must enter and exit the liver to reach the general circulation, and must enter and exit renal tubule cells, for reabsorption of vitamin C back into the circulation. In animals that synthesize ascorbate, it must exit hepatocytes. In addition, vitamin C is released by the adrenal, ovary, testes, stomach, and brain.[7,78-82] Although a candidate transporter is SLC23a3,[70] identification of vitamin C efflux transporters awaits future work.

Pharmacokinetics

Background

A key means to determine recommendations for any vitamin are concentration–function relationships. Graphically, these are *x*-axis/*y*-axis relationships, with the *x*-axis representing vitamin intake (or concentration) and the *y*-axis representing function. Before the establishment of DRIs, *y*-axis measures for vitamin C and other vitamins were based on prevention of deficiency, with added safety margins.[83] For DRIs, vitamin concentration–function relationships are based not only on prevention of deficiency but also on prevention of chronic disease.[61,83] Although such data are difficult to obtain, they are essential for ideal nutrient intake recommendations.[6,83,84]

Concentration (*x*-axis) data are provided by detailed physiology and pharmacokinetics experiments in humans, which describe vitamin physiology and how vitamin doses affect vitamin concentration. For vitamin C, depletion–repletion studies in humans are the pharmacokinetics study design of choice. Plasma vitamin C is measured because samples are readily available, measurements reflect extracellular concentration,[40] vitamin C is not protein bound, and dehydroascorbic acid is either not present or is present in amounts too small to detect using current assays.[65]

Depletion–Repletion Studies: Vitamin C Concentrations Are Tightly Controlled as a Function of Dose

Depletion–repletion studies for vitamin C have been conducted with both outpatients and inpatients. Most outpatient depletion–repletion data are limited by uncertainty about true vitamin C consumption, a problem solved by studying inpatients. In early inpatient depletion–repletion studies that were conducted in prisoners, body stores of vitamin C prevented the development of moderate to advanced scurvy for less than 6 weeks, and physical signs of scurvy were prevented by 10 mg/day of vitamin C.[85-87] Before the establishment of DRIs, these data were the basis of vitamin C RDAs. However, they are limited by an imprecise vitamin C assay, a diet that probably was deficient in other nutrients, a small number of individuals, and a narrow dose range.

Inpatient depletion–repletion studies that addressed these concerns were conducted at the National Institutes of Health (NIH) with healthy men and women.[13,14,62,64] These studies provided comprehensive dose concentration data using HPLC assays for analyses, and some results were used to calculate vitamin C DRIs. In these studies, seven healthy men and 15 women, each an inpatient for 5 to 7 months, received diets with a dietary intake of less than 5 mg/day of vitamin C and supplements to prevent other nutrient deficiencies. Vitamin C depletion occurred by the fourth week in all individuals, with plasma vitamin C concentrations falling to less than 7 μM to 8 μM. Repletion followed, in which participants were given a fixed daily dose of vitamin C, at escalating doses, until a fasting steady-state plasma vitamin C concentration was achieved. The fasting steady-state concentration was defined as five or more consecutive measurements, obtained over at least 7 days, for which the mean plasma vitamin C values had a mean (SD) of less than 10%. At steady state for each dose, bioavailability studies were performed (see absorption below); circulating blood cells were isolated for vitamin C measurements; timed urine samples matched to plasma values were collected to determine renal threshold; and 24-hour urine samples were collected for vitamin C and metabolites. When sample collections were completed, individuals were advanced to the next higher dose of vitamin C, a new steady state was reached, and the sampling sequence was repeated. Vitamin C doses were 30, 60, 100, 200, 400, 1,000, and 2,500 mg, administered in water as divided doses twice daily in the fasting state.

From these studies, extensive pharmacokinetics data for vitamin C were obtained. All plasma values from 15 women, and the depletion–repletion design, are shown (**Figure 29.2**). Every steady-state calculation for every individual is displayed as a function of dose for men and women (**Figure 29.3**). At plasma concentrations of up to 100 mg/day, a steep sigmoidal relationship was seen between vitamin C intake dose and vitamin C plasma concentration, and small changes in dose produced large changes in plasma. At doses of up to 100 mg/day, women achieved higher fasting steady-state plasma vitamin C concentration than men. At doses of 400 mg/day or higher, fasting steady-state plasma concentrations equaled 70 μM to 80 μM and increased little with higher doses. These data show that vitamin C plasma concentration is tightly controlled as a function of oral dose in both sexes. Underlying mechanisms include intake and intestinal absorption, tissue accumulation, utilization, and renal reabsorption and excretion.

Mechanisms of Tight Control

Absorption. Efficiency of intestinal absorption of vitamin C is assessed by bioavailability, with absolute bioavailability being the most accurate measurement. Data describing absolute bioavailability of vitamin C from foods are not available. Absolute bioavailability of pure vitamin C, however, as determined in the NIH study of healthy men at steady-state, was calculated using standard area-under-curve pharmacokinetics and a multicompartment model.[13,88] Bioavailability is expressed as a percentage, with 100% indicating complete absorption. Vitamin C bioavailability

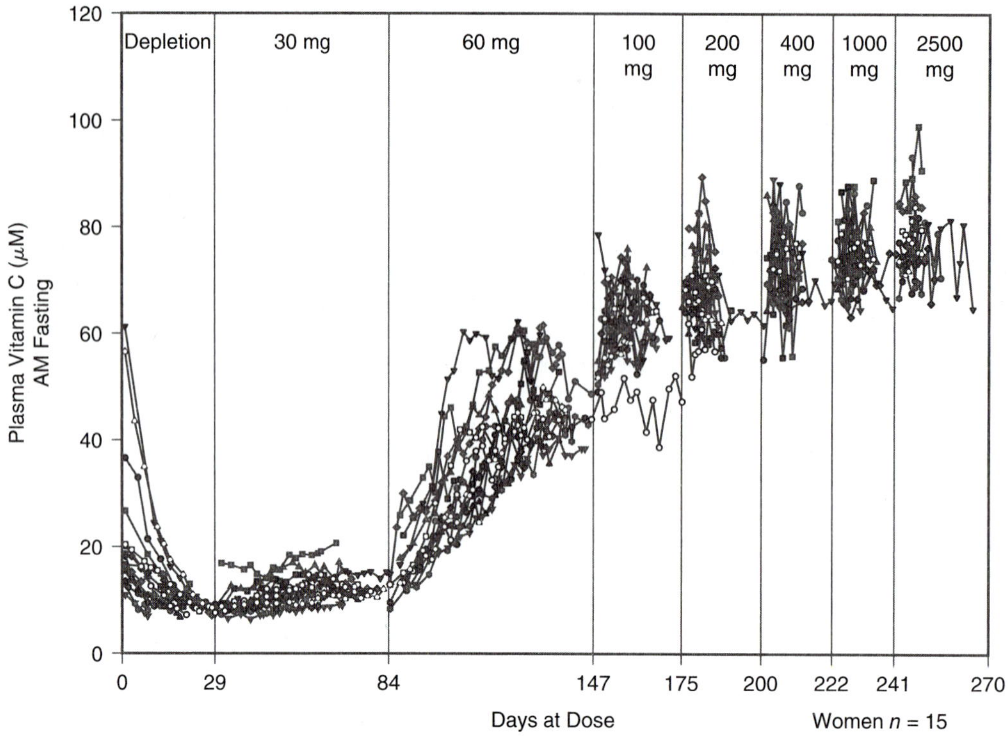

Figure 29.2 Fasting plasma vitamin C concentration as a function of dose for 15 healthy women. Individuals consumed a vitamin C–deficient diet, resulting in plasma and tissue vitamin C depletion. Vitamin C in solution was then administered by mouth at the doses shown until steady-state was reached for each dose.

Reproduced from Levine M, Wang Y, Padayatty SJ et al. Proc Natl Acad Sci U S A 2001;98:9842-6, with permission of the National Academy of Sciences, Washington, DC.

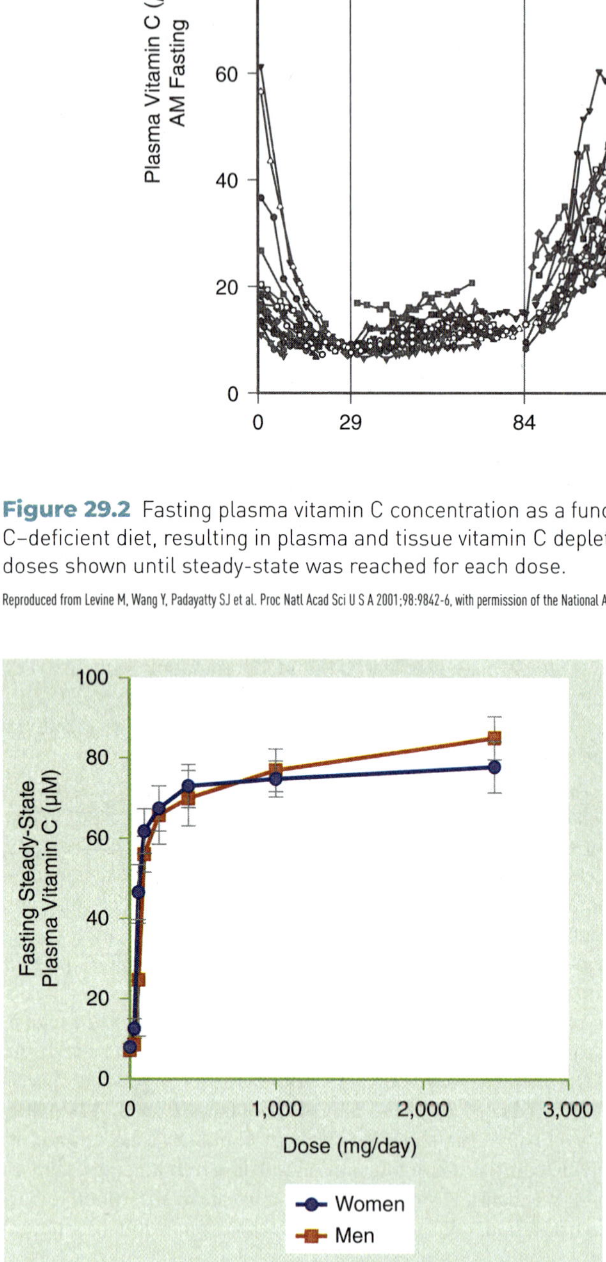

Figure 29.3 The relationship between oral doses of vitamin C and the mean fasting steady-state plasma ascorbic acid concentration in seven healthy men[13] and 15 healthy women.[14] The daily doses of vitamin C were 30, 60, 100, 200, 400, 1,000, and 2,500 mg. The dose-concentration curve is sigmoidal, with its steep portion between 30 and 100 mg of vitamin C daily. The figure shown is a composite of previously published dose concentration curves for men and women.[13,14]

Data from Levine M, Conry-Cantilena C, Wang Y et al. Proc Natl Acad Sci U S A 1996;93:3704-9; and Levine M, Wang Y, Padayatty SJ et al. Proc Natl Acad Sci U S A 2001;98:9842-6.

was 80% or greater for doses of 15 to 100 mg/day and decreased to less than 50% for 1,250 mg/day (**Table 29.5**), indicating that intestinal absorption contributes to tight control of vitamin C concentration.

Ascorbate is absorbed in the small intestine.[15] It is unclear whether ascorbate, dehydroascorbic acid, or both, is or are the species transported across the brush-border epithelium. Although SVCT1 is localized to the small intestine, SVCT1 knockout mice absorb an ascorbate analog that, when oxidized, is not transported by glucose transporters.[11,77] These data suggest that another sodium-dependent ascorbate transporter exits in the small intestine. That there is decreased bioavailability at higher vitamin C doses is best explained by saturable transport mechanisms. When single doses of ascorbic acid or dehydroascorbic acid are administered to humans, ascorbic acid concentrations increase in plasma similarly, but it is unknown whether dehydroascorbic acid is reduced to ascorbate in the stomach and/or intestinal lumen.[89]

Accumulation. Tight control of plasma vitamin C concentration results, in part, from concentration-dependent cell and tissue accumulation. In healthy humans, only limited tissue samples can be obtained, including the following: neutrophils, monocytes, lymphocytes, and platelets, all blood components; semen and seminal fluid from men; and urine. Because of their ready availability, circulating cells were used as a proxy for other tissues to determine vitamin C concentration in relation to dose over an 83-fold range (**Figure 29.4**). For the selected cell types, intracellular vitamin C concentration was always higher than plasma concentration and increased as the dose of vitamin C increased from 30 to 100 mg/day. As doses increased further, intracellular concentration remained constant. Cells achieved plateau concentration before plasma (Figures 29.3 and 29.4), a finding consistent

Table 29.5 Bioavailability of Vitamin C[a]

	Bioavailability of Ascorbic Acid	
	Method Using Area Under Curve Analyses	Method Using Multicompartment Mathematical Model
Dose (mg)	Mean (%) (SD)	Median (%)
15	—	89
30	—	87.3
50	—	85
100	—	80
200	112 (25)	72
500	73 (27)	63
1,250	49 (25)	46

[a]Bioavailability of vitamin C in healthy men at steady-state for each dose. Vitamin C bioavailability for three doses was calculated using an area under curve analyses.[13] This method could not be used for doses lower than 200 mg, when vitamin C did not have a constant volume of distribution or a constant rate of clearance.[88]

Data from Levine M, Conry-Cantilena C, Wang Y et al. Proc Natl Acad Sci U S A 1996;93: 3704–9; and Graumlich JF, Ludden TM, Conry-Cantilena C, et al. Pharmaceut Res 1997;14: 1133–1139.[82]

with SVCT2 transporter kinetics and the contribution of tissue accumulation to tight control. Ascorbic acid concentration in red blood cells was not determined because techniques were only developed more recently.[90]

Utilization. Vitamin C concentration may be affected by utilization rates, which can change steady-state values. Utilization rates may be affected by differences in transporter activity, recycling, enzyme efficiency, or conditions that accelerate utilization, such as oxidative stress. Although alternative explanations exist, increased utilization may account for low vitamin C concentrations in smokers and in disease, including critical illness, acute myocardial infarction, diabetes, and pancreatitis.[60,62,91-94] Utilization rates also differ even among healthy humans (see Figure 29.2).[6,14,86] Measurement of utilization rates will be possible when there are suitable assays for ascorbic acid and dehydroascorbic acid measurement, particularly those using liquid chromatography/mass spectrometry with administration of stable isotopes.

Renal Reabsorption and Excretion. With normal renal function, small molecules (i.e., glucose, amino acids) are filtered through glomeruli and are reabsorbed in the renal tubules. Based on tubular reabsorptive transport, individual nephrons have a maximal capacity to absorb specific substances, termed the *tubular maximum*. When the tubular maximum is within the range of plasma concentrations, the kidney has a central role in homeostasis. Plasma concentration corresponding to the tubular maximum is termed renal threshold. When plasma concentration exceeds renal threshold, reabsorption is saturated and urine excretion ensues.

Data from the NIH depletion repletion studies were used to determine a sex-specific renal threshold for vitamin C: 48±μM for healthy men and 58 μM± for healthy women[55] (see **Figure 29.5**). These values are consistent with the differences between men and women of steady-state plasma values at doses up to and including 100 mg daily (Figure 29.3). Plasma values at the 95% confidence limits of the renal threshold for each sex define a minimal elimination threshold, a concept that introduces new understanding of vitamin C in diseases that affect the kidneys. As expected, approximately 5% to 8% of healthy individuals have vitamin C in urine when the plasma concentration is below the minimal elimination threshold, while nearly ⅓ of individuals with diabetes have abnormal vitamin C urine loss.[62]

Renal reabsorption and excretion are an essential component of tight control of vitamin C concentration (Figure 29.3 and Figure 29.5). In the NIH depletion–repletion studies, no vitamin C

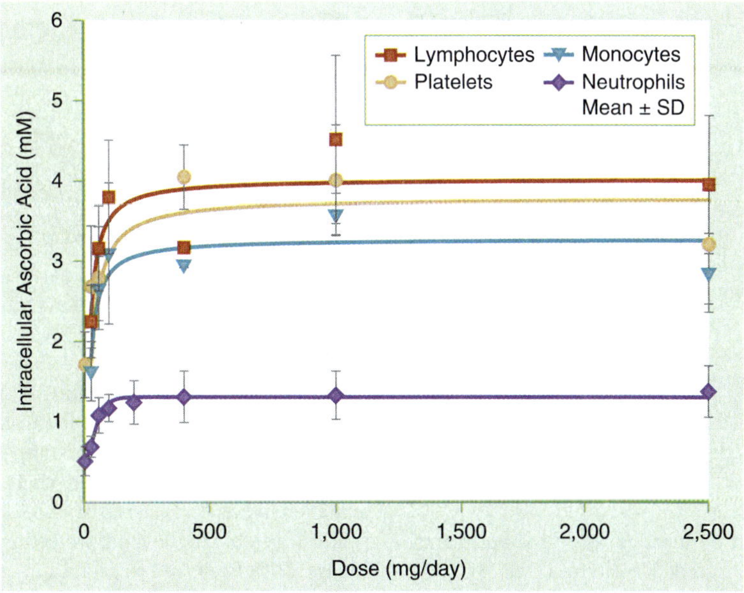

Figure 29.4 Intracellular vitamin C concentration in circulating cells as a function of dose in healthy women. Cells were isolated when steady-state was achieved for each dose.

Reproduced from Levine M, Wang Y, Padayatty SJ et al. Proc Natl Acad Sci U S A 2001;98:9842-6, with permission of the National Academy of Sciences, Washington, DC.

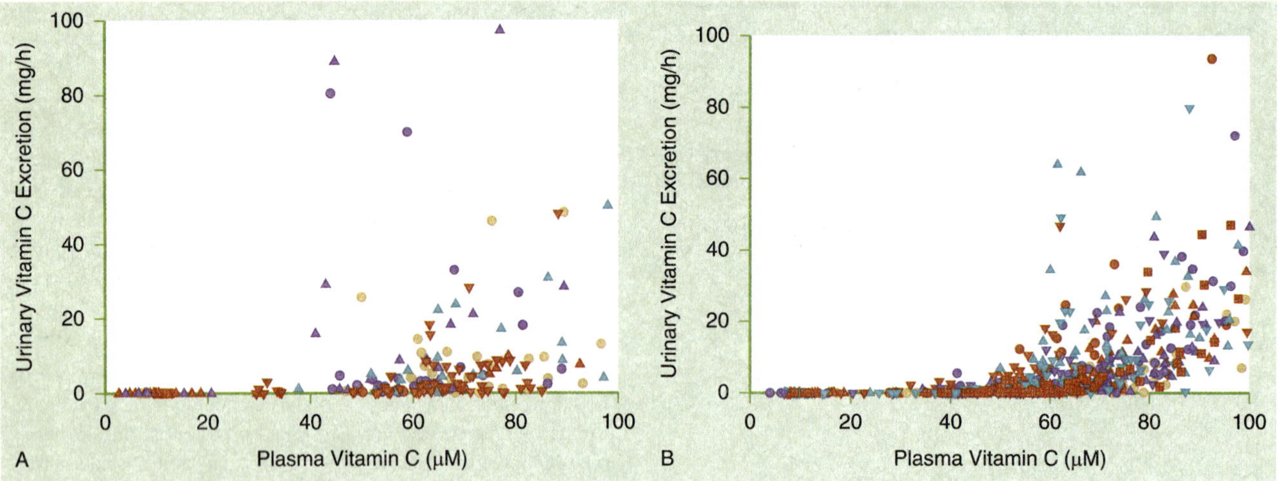

Figure 29.5 Urinary vitamin C excretion as a function of matched plasma vitamin C concentration in healthy men (A) and women (B).[62] Samples were collected as part of bioavailability studies (Figure 29.3).[13,14,88] Oral and intravenous doses (15–1,250 mg) were administered on successive days at steady-state for the dose. Blood samples were obtained at fixed intervals. All urine voided during the test period was collected, and each urine void formed a timed collection. The total amount of vitamin C in each urine sample is indicated on the *y*-axis in mg, which corrects for differing urine volumes. Each symbol shape represents a different individual. Data are shown from 10 to 14 pharmacokinetics studies (5–7 oral doses; 5–7 intravenous doses) for each participant. Participants were seven males and 10 females. See Reference #55 for additional information.[62]

was excreted at low doses, while at higher doses, all vitamin C administered intravenously or absorbed orally was excreted.[13,14] As an example, when 1,250 mg of vitamin C was given orally, approximately 600 mg was absorbed and was subsequently excreted in the urine. With administration of 1,250 mg intravenously, thereby removing an effect of intestinal absorption, virtually the entire administered dose of vitamin C was excreted.

In patients with end-stage renal disease, vitamin C is not excreted because no appreciable glomerular filtration occurs. Doses higher than 200 mg/day may accumulate and produce hyperoxalemia. Conversely, vitamin C is freely dialyzable and is lost during dialysis. Due to concern about hyperoxalemia from overreplacement of vitamin C, patients with end-stage renal disease who are undergoing dialysis often have chronically low plasma vitamin C concentration.[95]

Genetics. Because ascorbate transporters play a key role in tight control, genetic variation in transporter expression or activity could modify tight control of vitamin C concentration in healthy humans. For both known vitamin C transporters, single nucleotide polymorphisms occur in the respective genes, including those shown to decrease SVCT1 (SLC23A1) transport activity.[77] Decreased SVCT1 transport would decrease vitamin C renal reabsorption, thus leading to lower plasma vitamin C concentration.[77] Supportive data from population studies[96,97] suggest that genetic variations in vitamin C transporters may affect tight control.

Bypassing Tight Control: Pharmacology

Gram doses of ascorbate can be ingested as oral supplements, although doses higher than 3 g produce diarrhea (see "Manifestations of Vitamin C Deficiency and Excess"). With maximal oral dosing every few hours or with liposomal ascorbate as a supplement, plasma concentrations remained lower than 300 μM.[64,98] Plasma concentration is tightly controlled because of limited intestinal absorption coupled to renal excretion, due to saturation of both processes.[13,62,88] When ascorbate is administered intravenously (parenterally), limited intestinal absorption is bypassed, and millimolar concentrations of vitamin C occur within minutes in plasma. Over several hours, homeostasis is restored by glomerular filtration, saturation of tubular reabsorption, and renal excretion.[62,64,99] In humans, intravenous ascorbate produces peak plasma concentrations of 25 to 30 mM, which are several hundredfold greater than those from food ingestion.[41,99] Intravenous ascorbate use, as a drug, appears to be surprisingly safe,[99-101] with potential therapeutic application.

Functional Consequences in Humans

Potential Benefits of Vitamin C Consumption from Fruits and Vegetables

Vitamin C is found primarily in fruits and vegetables (Table 29.2). Recommendations for fruit and vegetable servings daily range from 4.5 (U.S.A.) to 8.5 (Australia). Higher intakes are associated with lower mortality, with effects achieving plateau at about five daily servings.[102] Although fruit and vegetable intake was previously believed to be inversely associated with occurrence of many cancers, these effects seem to be limited to certain specific cancer types, such as some kidney and breast malignancies.[102-105] Positive associations exist between fruit and vegetable intake and protection against cardiovascular disease.[103,106] Fruit and vegetable intake under controlled conditions was associated with decreases in blood pressure,[107] a risk factor for cardiovascular disease. However, nutrition prevention or clinical intervention trials confirming that fruits and vegetables themselves are protective are lacking.[108]

For prevention of cardiovascular disease, cancer, and other diseases, it is unknown whether associated benefits of fruit and vegetable intake result from vitamin C itself, from vitamin C in addition to other components of fruits and vegetables, or from fruit and vegetable components independent of vitamin C.[60,103] Vitamin C may simply be a surrogate marker for fruit and vegetable consumption or other healthy lifestyle practices. Fruit and vegetable consumption provides micronutrients, bulk, roughage, and satiety.

Outcome Studies

Vitamin C from foods and supplements has been investigated for prevention of cancer, cardiovascular disease, stroke, pre-eclampsia, and age-related eye diseases, often with conflicting results.[7,33,36] In one large-scale interventional study where vitamin C was partially obtained from foods, but was also consumed in combination with other antioxidant vitamins as a supplement, no benefit in preventing cancer or reducing vascular disease was observed.[109] In another, vitamins C and E slowed the progression of carotid atherosclerosis,[110] but a protective effect for atherosclerosis was not observed by other investigators.[33,111] Consistent with mouse studies,[77] vitamins C and E decreased pre-eclampsia or hypertension in pregnant women with low vitamin concentrations at study entry.[112] These findings were not confirmed in healthier populations, probably because individuals were near saturation with ascorbate.[113,114] For eye diseases, vitamin C supplements alone or in combination did not reduce cataract incidence.[115,116] In large placebo-controlled studies, combined supplements of vitamin C, vitamin E, β-carotene, and zinc reduced the odds of developing advanced age-related macular degeneration once the disease was present,[116] but evidence is lacking that antioxidant vitamins, including vitamin C, prevent macular degeneration.

Vitamin C supplementation was also tested for potential effects on outcome for hypertension, endothelial dysfunction, and respiratory diseases. Investigators noted a modest effect on lowering blood pressure in some but not all individuals, and no large-scale studies are available.[36,117,118] Oral vitamin C does not appear to have a sustained benefit on improving endothelial function in smokers.[119] There is evidence in humans that vitamin C diminishes endothelial vasomotor dysfunction and induces vasodilatation when administered arterially. However, intra-arterial (parenteral) administration bypasses tight control, and resulting plasma ascorbate concentration becomes higher than achievable orally. Vitamin C orally for 3 days potentiated nitroglycerin-induced vasodilatation in two small studies, but reproducibility has been inconsistent.[120-122] Vitamin C supplements probably do not prevent acute respiratory infections in healthy populations,[123] nor do they provide benefit in asthma treatment.[124]

Although vitamin C supplements were effective in treating pressure ulcers in a small study, other findings have been variable.[125,126] Original findings indicated that the control individuals were deficient in vitamin C in a population at risk, and supplementation corrected the deficiency. Despite scant data, vitamin C supplements are sometimes used in those at risk for pressure ulcers because of the low risk of treatment, the possibility that the treatment population has an ascorbate deficiency, and the difficulty in treating the condition.

No large-scale interventional studies of disease prevention have been reported with vitamin C as the sole supplement. Unfortunately, it remains a recurrent problem in many supplementation trials that there have not been inclusion criteria for low measured vitamin concentration at study entry. If individuals are near saturation, supplementation would be predicted to have minimal benefit.[7,33,36,127]

Effects of Vitamin C in the Gastrointestinal Tract

Vitamin C is secreted in gastric juice, and it achieves concentrations severalfold higher than plasma values.[81] Vitamin C content is low in the gastric juice of patients with atrophic gastritis and *Helicobacter pylori* infection, and eradication of the bacterium increases gastric vitamin C secretion.[128] Vitamin C can potentially quench reactive oxygen metabolites in the stomach and duodenum and can prevent formation of mutagenic N-nitroso compounds. Whether clinical benefit ensues is controversial.[129] Intake of foods high in vitamin C usually correlates with reduced risk of gastric cancer, but whether vitamin C itself or other substances in vitamin C–rich, plant-derived foods are responsible is unknown.[130] In a population at high risk for gastric cancer, vitamin C supplementation, with and without anti-*H. pylori* treatment, was associated with regression of precancerous lesions.[131]

In the small intestine, vitamin C reduces iron and is predicted to promote iron absorption (see chapter 10). A dose of 20 to 60 mg/day of vitamin C, found in vitamin C–rich foods, enhanced small intestinal iron absorption from test meals 1.5- to 10-fold, depending on iron status, vitamin C dose, and test meal type,[28] but absorption studies from single meals may overestimate effects.[29] Surprisingly, clinical findings indicate that effects of vitamin C with iron, compared with iron alone, are modest or absent when the outcome measure is raising hemoglobin concentration.[30] In practice, vitamin C is commonly administered with iron to increase its absorption, especially in pregnancy.

Effects of Pharmacologic Ascorbate

Pharmacologic ascorbate concentration, achieved only by parenteral administration, can produce hydrogen peroxide in the extracellular fluid, but not in blood, through reduction of molecular oxygen to form superoxide.[40,41,55] Hydrogen peroxide and pharmacologic ascorbate and trace metals produce reactive oxygen species that are selectively toxic to cancer cells in vitro; in animal models of cancer; and in small clinical trials.[41,55,101,132,133] Larger randomized clinical trials are necessary to learn whether pharmacologic ascorbate has efficacy in treating specific cancers in humans in combination with chemotherapy and/or immunotherapy.[101,134] Pharmacologic ascorbate has been tested as treatment in sepsis. In contrast to use in cancer, mechanism bases for pharmacologic ascorbate are imprecise, tested doses have been arbitrary, and clinical studies to date have shown minimal efficacy and inconsistent effects.[135]

Functions in Relation to Concentration In Vivo: Summary and Potential

An overarching function of vitamin C in humans, except to prevent scurvy, remains mysterious. Alternatively, requirement for vitamin C may be based on its multifunctionality. Nearly all tissues concentrate vitamin C. Perhaps this is due to the importance of regulating DNA demethylation, a process that can involve enzymes that are vitamin C dependent. However, epigenetic modification occurs in cultured cells, which have no vitamin C unless it is intentionally added. That vitamin C is concentrated in many tissues is perhaps a clue that this vitamin has other unrecognized functions. Current in vivo understanding of the 18 vitamin C–dependent enzymes and other nonenzymatic functions is incomplete. It is attractive to think of vitamin C as a critical antioxidant or electron donor in vivo, but here as well, conclusive evidence is absent.[33]

Currently, we lack definitive evidence that a particular vitamin C concentration or intake amount produces a clinically beneficial outcome, other than preventing deficiency.[7,33] Consuming five to nine servings of fruits and vegetables per day provides 200 to 400 mg of vitamin C and results in steady-state fasting plasma concentrations of 70 μM to 80 μM. We do not know whether such vitamin C concentrations optimize biochemical functions in vivo or improve clinical outcomes.

While vitamin C pharmacokinetics provide fundamental knowledge about concentrations (the *x*-axis), a paucity of data exists on the effect of these concentrations on function (the *y*-axis). Nevertheless, even without definitive *y*-axis data, *x*-axis data provide key insight to human outcome studies, which ideally should recognize tight control and the steep relationship between oral vitamin C dose and concentration. To determine whether vitamin C affects a given outcome, individuals with different vitamin C concentrations must be compared.[33,36,60,114] Unfortunately, many outcome studies have compared individuals with different estimated vitamin C intakes rather than different measured vitamin C concentrations at study entry. This common design flaw remains a limitation of vitamin C outcome studies, as well as those for other vitamins.[7,33,36,136] If individuals with the lowest intake are already beyond the steep portion of the pharmacokinetics curve (see Figure 29.3), increasing intake will not increase concentration, and outcomes should not differ. Future outcome studies should compare individuals with a range of vitamin C concentrations at baseline. The same *x*-*y* axis approach for physiology can reveal the possible benefit of pharmacologic ascorbate, when tight control is transiently bypassed with parenteral ascorbate administration.

Physiology and pharmacokinetics studies of vitamin C in healthy humans provide key tools to characterize aberrant physiology of vitamin C in disease: pathophysiology. An example of this promising path is that detailed understanding of normal vitamin C physiology in humans has unmasked new vitamin C pathophysiology in diabetes.[13,14,62,64]

Assessment of Vitamin C Status

In the absence of clinical scurvy, vitamin C status is based on white cell (leukocyte) or plasma ascorbate measurements; plasma ascorbate is used more often because of technical ease. Vitamin C deficiency is considered present when plasma concentration is less than 11.4 μM (0.2 mg/dL).[137,138] Marginal vitamin C status, with risk of developing deficiency, is indicated by plasma concentration between 11.4 μM and 28.4 μM (0.2 to 0.5 mg/dL).[137,138] Saturation occurs at approximately 70 μM and higher.[13,14]

Because no functional measure of vitamin C status exists other than clinical scurvy, values for deficiency and marginal status are arbitrary. For deficiency, values were obtained with an assay that overestimated vitamin C concentrations at low values.[85,86,139] In scurvy, the clinical findings of hemorrhage and hyperkeratosis do not occur until plasma concentration is less than 5 μM.[7,13,14] The first symptom of scurvy is fatigue which, unfortunately, is perhaps the most common general symptom in medicine. Fatigue occurs under controlled conditions when plasma vitamin C concentration is less than approximately 20 μM.[7,13,33] If vitamin C ingestion were to cease, a plasma vitamin C concentration of 28 μM represents an approximate 2- to 3-week reserve to prevent clinical scurvy. With intake at the RDA level for vitamin C for both men and women, plasma values are approximately 45 μM.[13,14]

Manifestations of Vitamin C Deficiency and Excess

Deficiency

Scurvy

Frank scurvy is now rare in industrialized countries. It occurs mainly in the following groups: malnourished populations; patients with cancer cachexia and malabsorption; persons with alcoholism, drug addicts, and poor or elderly persons with inadequate diets; institutionalized individuals; and those consuming idiosyncratic diets.[140] Scurvy occurs in war-torn areas and refugee camps. Subclinical vitamin C deficiency may be more common, but symptoms are nonspecific and not easily attributed to vitamin C deficiency.[141] Historically, Lind noted that the earliest symptoms of scurvy are weakness and lassitude.[2] Signs and symptoms of scurvy are further described in chapter 57. Diagnosis of scurvy is based on clinical findings and is confirmed by low plasma vitamin C. If untreated, scurvy is fatal, and treatment should not be delayed for laboratory confirmation.

Treatment and Prevention. Treatment should be initiated with vitamin C 100 mg three times daily. An initial intravenous dose of 60 to 100 mg of vitamin C may be given. Children may be given 100 to 200 mg/day either orally or parenterally. With prompt diagnosis and treatment, permanent damage can be prevented. Steady-state plasma concentration from vitamin C doses of 100 mg/day will forestall deficiency for approximately 1 month.[13,14,60]

Adverse Effects of Vitamin C Excess

Gastrointestinal Tract

Vitamin C is generally safe and well tolerated, with few dose-related side effects.[7,60,61] Ingestion of 3 g or more at once induces osmotic diarrhea and bloating and, accordingly, the UL has been set at 2 g daily. Because vitamin C might enhance iron absorption from the small intestine (discussed above), long-term vitamin C use could increase the risk of iron overload in susceptible people, including those with hemochromatosis, thalassemia major, sickle cell disease, sideroblastic anemia, and those who need multiple, frequent red blood cell transfusions.[142] Such people should avoid large doses of vitamin C but not fruits and vegetables.[143] In healthy individuals, vitamin C in doses as high as 2 g daily over 18 months did not induce iron overabsorption.[144]

Blood

Glucose-6-phosphate dehydrogenase (G6PD) deficiency is an X-linked inherited disease that may cause hemolytic crises, most often on exposure to oxidant stress. In people with G6PD deficiency, hemolysis was precipitated by vitamin C given intravenously and by single oral doses of at least 6 g.[100] Hemolysis was also precipitated by ascorbic acid, at estimated doses of 500 mg or more, in patients with the rare blood disease paroxysmal nocturnal hemoglobinuria.[7]

Kidney

Vitamin C doses of 3 g may cause transient hyperuricosuria, but this does not occur at doses lower than 1 g/day. Doses higher than 1 g/day may increase oxalate excretion in persons with known or occult hyperoxaluria and may precipitate oxalate kidney stone formation.[60,61] Incident kidney stones of unverified type were increased in men, but not women, from supplemental vitamin C more than 500 mg daily, but not from foods alone.[145] In patients with renal failure who were undergoing long-term hemodialysis, hyperoxalemia was induced by repeated intravenous vitamin C doses higher than 500 mg.[146] To prevent oxalosis, vitamin C intake in these patients should probably not exceed 200 mg/day.[60]

Miscellaneous

Vitamin C, at doses of 250 mg and higher, may cause false-negative results for stool occult blood with guaiac-based tests. Vitamin C intake should be reduced to less than 250 mg for several days before such testing. Several harmful effects have been erroneously attributed to vitamin C, including hypoglycemia, rebound scurvy, infertility, mutagenesis, and destruction of vitamin B_{12}.[60]

Acknowledgments

The authors acknowledge Sebastian J Padayatty for contributions to the previous version of this chapter.

References

1. Clemetson CA. Vitamin C, Volume I (Clemetson, CA, ed) Vol. I pp. 1–10, CRC Press, Boca Raton, FL, 1989.
2. Lind J. Treatise on Scurvy: A Bicentenary Volume Containing a Reprint of the First Edition of A Treatise of the Scurvy, Edinburgh University Press, Edinburgh, 1953.
3. Svirbely JL, Szent-Gyorgyi A. Biochem J 1932;26:865–870.
4. King CC, Waugh WA. Science 1932;75:357–358.
5. Linster CL, Van Schaftingen E. FEBS J 274:1–22.
6. Levine M. N Engl J Med 1986;314:892–902.
7. Padayatty SJ, Levine M. Oral Dis 2016;22:463–493.
8. Washko PW, Welch RW, Dhariwal KR, et al. Anal Biochem 1992;204:1–14.
9. Buettner GR. Arch Biochem Biophys 1993;300:535–543.
10. Lewin S. Vitamin C: Its Molecular Biology and Medical Potential, Academic Press, London, 1976.
11. Corpe CP, Lee JH, Kwon O, et al. J Biol Chem 2005;280: 5211–5220.
12. Baker EM, Halver JE, Johnsen DO, et al. Ann NY Acad Sci 1975;258:72–80.
13. Levine M, Conry-Cantilena C, Wang Y, et al. Proc Natl Acad Sci U S A 1996;93:3704–3709.
14. Levine M, Wang Y, Padayatty SJ, et al. Proc Natl Acad Sci U S A 2001;98:9842–9846.
15. Rumsey SC, Levine M. J Nutr Biochem 1998;9:116–130.
16. Tu H, Li H, Wang Y, et al. EBioMedicine 2015;2:1735–1750.
17. Dhariwal KR, Washko P, Hartzell WO, et al. J Biol Chem 1989;264:15404–15409.
18. Prigge ST, Mains RE, Eipper BA, et al. Cell Mol Life Sci 2000;57:1236–1259.
19. Prockop DJ, Kivirikko KI. Annu Rev Biochem 1995;64:403–434.
20. Myllyharju J. Ann Med 2008;40:402–417.
21. Rebouche CJ. Am J Clin Nutr 991;54:1147S–1152S.
22. Lindblad B, Lindstedt G, Lindstedt S. J Am Chem Soc 1970;92:7446–7449.
23. Tsukada Y, Fang J, Erdjument-Bromage H, et al. Nature 2006;439:811–816.
24. Tahiliani M, Koh KP, Shen Y, et al. Science 2009;324:930–935.
25. Minor EA, Court BL, Young JI, et al. J Biol Chem 2013;288: 13669–13674.
26. Blaschke K, Ebata KT, Karimi MM, et al. Nature 2013;500: 222–226.
27. Hore TA, von Meyenn F, Ravichandran M, et al. Proc Natl Acad Sci U S A 2016;113:12202–12207.
28. Hallberg L, Brune M, and Rossander-Hulthen L. Ann N Acad Sci 1987;498:324–32.
29. Cook JD, Reddy MB. Am J Clin Nutr 2001;73:93–98.
30. Li N, Zhao G, Wu W, et al. JAMA Network Open 2020;3:e2023644.
31. Toth I, Bridges KR. J Biol Chem 1995;270:19540–19544.
32. Hitomi K, Tsukagoshi N. Role of ascorbic acid in modulation of gene expression. Subcell Biochem. 1996;25:41–56.
33. Padayatty SJ, Katz A, Wang Y, et al. J Am Coll Nutr 2003;22: 18–35.
34. Polidori MC, Mecocci P, Levine M, et al. Arch Biochem Biophys 2004;423:109–115.
35. Steinberg D. J Lipid Res 2009;50:S376–381.
36. Frei B, Birlouez-Aragon I, Lykkesfeldt J. Crit Rev Food Sci Nutr 2012;52:815–829.
37. Helser MA, Hotchkiss JH, Roe DA. Influence of fruit and vegetable juices on the endogenous formation of N-nitrosoproline and N-nitrosothiazolidine-4-carboxylic acid in humans on controlled diets. Carcinogenesis. 1992;13:2277–80.
38. Podmore ID, Griffiths HR, Herbert KE, et al. Nature 1998;392: 559.
39. Lee SH, Oe T, Blair IA. Science 2001;292:2083–2086.
40. Chen Q, Espey MG, Sun AY, et al. Proc Natl Acad Sci U S A 2007;104:8749–8754.
41. Chen Q, Espey MG, Sun AY, et al. Proc Natl Acad Sci U S A 2008;105:11105–11109.
42. Carr A, Frei B. FASEB J 1999;13:1007–1024.
43. Jialal I, Fuller CJ. Can J Cardiol 1995;11:97G–103G.
44. Van Hoydonck PG, Schouten EG, Manuel YK, et al. Eur J Clin Nutr 2004;58:1587–1593.
45. Bjelakovic G, Nikolova D, Gluud LL, et al. Cochrane Database Syst Rev 2012;CD007176.
46. Fortmann SP, Burda BU, Senger CA, et al. Ann Intern Med 2013;159:824–834.
47. Jacob RA, Kutnink MA, Csallany AS, et al. J Nutr 1996;126: 2268–2277.

48. Bruno RS, Leonard SW, Atkinson J, et al. Free Radic Biol Med 2006;40:689–697.
49. Nualart FJ, Rivas CI, Montecinos VP, et al. J Biol Chem 2003;278:10128–10133.
50. Duarte TL, Lunec J. Free Radic Res 2005;39:671–686.
51. Griffiths HR, Willetts RS, Grant MM, et al. Br J Nutr 2009;101:1432–1439.
52. Heller R, Unbehaun A, Schellenberg B, et al. J Biol Chem 2001;276:40–47.
53. Mortensen A, Lykkesfeldt J. Nit Ox Biol Chem 2014;36:51–57.
54. Houglum KP, Brenner DA, Chojkier M. Am J Clin Nutr 1991;54:1141S–1143S.
55. Chen Q, Espey MG, Krishna MC, et al. Proc. Natl Acad Sci U S A 2005;102:13604–136041.
56. Johnston CS. JAMA 1999;282:2118.
57. Brauchla M, Dekker MJ, Rehm CD. Nutrients 2021;13:420.
58. Crook J, Horgas A, Yoon SJ, et al. Nutrients 2021;13:3910.
59. Narayanan S, Kumar SS, Manguvo A, et al. Curr Devel Nutr 2021;5:1067.
60. Levine M, Rumsey SC, Daruwala R, et al. JAMA 1999;281: 1415–1423.
61. Food and Nutrition Board. Vitamin C. In Dietary Reference Intakes for Vitamin C, Vitamin E, Selenium, and Carotenoids pp. 95–185; 181–120; 434–435; 442–443; National Academy Press, Washington DC, 2000.
62. Ebenuwa I, Violet PC, Padayatty S, et al. Am J Clin Nutr 2022;116:274–284.
63. Levine M, Padayatty SJ, Katz A, et al. In Vitamin C: Functions and biochemistry in animals and plants (Asard, H., May, J. M., and Smirnoff, N., eds) pp. 291–318, BIOS Scientific Publishers, London, 2004.
64. Padayatty SJ, Sun H, Wang Y, et al. Ann Intern Med 2004;140: 533–537.
65. Dhariwal KR, Hartzell WO, Levine M. Am J Clin Nutr 1991;54:712–716.
66. Hornig D. Ann NY Acad Sci 1975;258:103–118.
67. Voigt K, Kontush A, Stuerenburg HJ, et al. Free Radic Res 2002;36:735–739.
68. Tsukaguchi H, Tokui T, Mackenzie B, et al. Nature 1999;399: 70–75.
69. Daruwala R, Song J, Koh WS, et al. FEBS Lett 1999;460: 480–484.
70. Burzle M, Suzuki Y, Ackermann D, et al. Mol Aspects Med 2013;34:436–454.
71. Washko PW, Wang Y, Levine M. J Biol Chem 1993;268:15531–15535.
72. Rumsey SC, Kwon O, Xu GW, et al. J Biol Chem 1997;272: 18982–18989.
73. Tu H, Wang Y, Li H, Brinster LR, et al. EBioMedicine 2017;23: 125–135.
74. Kc S, Cárcamo JM, Golde DW. Faseb J 2005;19:1657–1667.
75. Heaney ML, Gardner JR, Karasavvas N, et al. Cancer Res 2008;68:8031–8038.
76. Sotiriou S, Gispert S, Cheng J, et al. Nat Med 2002;8: 514–517.
77. Corpe CP, Tu H, Eck P, et al. J Clin Invest 2010;120:1069–1083.
78. Koba H, Kawao K, Yamashita K. Tohoku J Exp Med 1971;104: 65–71.
79. Musicki B, Kodaman PH, Aten RF, et al. Biol Reprod 1996;54: 399–406.
80. Rebec GV, Wang Z. J Neurosci 2001;21:668–675.
81. Schorah CJ, Sobala GM, Sanderson M, et al. Am J Clin Nutr 1991;53:287S–293S.
82. Padayatty SJ, Doppman JL, Chang R, et al. Am J Clin Nutr 2007;86:145–149.
83. Levine MV, PC, Ebenuwa I, et al. In Vitamin C: New Biochemical and Functional Insights (Chen, Q. V., M.C.M., ed) Vol. 1 pp. 3–21, CRC Press, Taylor and Francis Group, Boca Raton, 2020.
84. Levine M, Dhariwal KR, Washko PW, et al. Am J Clin Nutr 1991;54:1157S–1162S.
85. Baker EM, Hodges RE, Hood J, et al. Am J Clin Nutr 1969;22: 549–558.
86. Baker EM, Hodges RE, Hood J, et al. Am J Clin Nutr 1971;24: 444–454.
87. Hodges RE, Hood J, Canham JE, et al. Am J Clin Nutr 1971;24:432–443.
88. Graumlich JF, Ludden TM, Conry-Cantilena C, et al. Pharmaceut Res 1997;14:1133–1139.
89. Tsujimura M, Higasa S, Nakayama K, et al. J Nutr Sci Vitaminol (Tokyo) 2008;54:315–320.
90. Li H, Tu H, Wang Y, et al. Anal Biochem 2012;426:109–117.
91. Bonham MJ, Abu-Zidan FM, Simovic MO, et al. Br J Surg 1999;86:1296–1301.
92. Padayatty SJ, Levine M. Am J Clin Nutr 2000;71:1027–1028.
93. Alberg A. Toxicology 2002;180:121–137.
94. Hager DN, Hooper MH, Bernard GR, et al. Trials 2019;20:197.
95. Sirover WD, Liu Y, Logan A, et al. J Renal Nutr 2015;25: 292–300.
96. Timpson NJ, Forouhi NG, Brion MJ, et al. Am J Clin Nutr 2010;92:375–382.
97. Duell EJ, Lujan-Barroso L, Llivina C, et al. Genes Nutr 2013;8:549–560.
98. Gopi S, Balakrishnan P. J Liposome Res 2021;31:356–364.
99. Hoffer LJ, Levine M, Assouline S, et al. Ann Oncol 2008;19: 1969–1974.
100. Padayatty SJ, Sun AY, Chen Q, et al. PLoS One 2010;5:e11414.
101. Nauman G, Gray JC, Parkinson R, et al. Antioxidants (Basel) 2018;7:89.
102. Wang DD, Li Y, Bhupathiraju SN, et al. Circulation 2021;143:1642–1654.
103. Willett WC, Stampfer MJ. Ann Rev Pub Health 2013;34: 77–95.
104. Willett WC. J Natl Cancer Inst 2010;102:510–511.
105. Boffetta P, Couto E, Wichmann J, et al. J Natl Cancer Inst 2010;102:529–537.
106. Dauchet L, Montaye M, Ruidavets JB, et al. Eur J Clin Nutr 2010;64:578–586.
107. Sacks FM Svetkey LP Vollmer WM, et al. N Engl J Med 2001;344:3–10.
108. Dauchet L, Amouyel P, Dallongeville Nat Rev Cardiol 2009;6: 599–608.
109. Heart Protection Study Collaborative Group. Lancet 2002; 360:23–33.
110. Salonen RM, Nyyssönen K, Kaikkonen J, et al. Circulation 2003;107:947–953.
111. Sesso HD, Buring JE, Christen WG, et al. JAMA 2008;300: 2123–2133.
112. Chappell LC, Seed PT, Briley AL, et al. Lancet 1999;354: 810–816.
113. Roberts JM, Myatt L, Spong CY, et al. N Engl J Med 2010;362:1282–1291.
114. Padayatty SJ, Levine M. N Engl J Med 2006;355:1065.
115. Lim JC, Caballero Arredondo M, Braakhuis AJ, et al. Nutrients 2020;12, 3142.
116. Fleckenstein, M., Keenan, T. D. L., Guymer, R. H. et al. Nat Rev Dis Primers 2021;7:31.
117. Harrison DG, Gongora M C. Med Clin North Am 2009;93: 621–635.
118. Guan Y, Dai P, Wang H. Medicine (Baltimore) 2020;99:e19274.
119. Raitakari OT, Adams MR, McCredie RJ, et al. J Am Coll Cardiol 2000;35:1616–1621.

120. Bassenge E, Fink N, Skatchkov M, et al. J Clin Invest 1998;102:67–71.
121. Watanabe H, Kakihana M, Ohtsuka S, et al. J Am Coll Cardiol 1998;31:1323–1329.
122. Münzel T, Daiber A, Gori T. Circulation 2011;123:2132–2144.
123. Hemilä H, Chalker E. Cochrane Database Syst Rev 2013 Jan 31;2013(1):CD000980.
124. Han YY, Forno E, Holguin F, et al. Curr Opin Allergy Clin Immunol 2015;15:369–374.
125. ter Riet G, Kessels AG, Knipschild PG. J Clin Epidemiol 1995;48:1453–1460.
126. Bechara N, Flood VM, Gunton JE. Antioxidants (Basel) 2022;11.
127. Padayatty SJ, Levine M. N Engl J Med 2006;355:1065.
128. Sobala GM, Schorah CJ, Shires S, et al. Gut 1993;34:1038–1041.
129. Vahid F, Davoodi SH. Nutr Cancer 2021;73:376–390.
130. Jenab M, Riboli E, Ferrari P, et al. Carcinogenesis 2006;27: 2250–2257.
131. Correa P, Fontham ET, Bravo JC, et al. J Natl Cancer Inst. 2000;92:1881–1888.
132. Schoenfeld JD, Sibenaller ZA, Mapuskar KA, et al. Cancer Cell 2017;32:268.
133. Shenoy N, Creagan E, Witzig T, et al. Cancer Cell 2018;34: 700–706.
134. Magri A, Germano G, Lorenzato A, et al. Sci Transl Med 2020;12.
135. Agarwal A, Oczkowski S, Sevransky J. Crit Care Med 2022;50:897–899.
136. Padayatty SJ, Levine M. JAMA 2013;309:545–546.
137. Schleicher RL, Carroll MD, Ford ES, et al. Am J Clin Nutr 2009;90:1252–1263.
138. Jacob RA, Skala JH, Omaye ST. Am J Clin Nutr 1987;46: 818–826.
139. Hodges RE. Am J Clin Nutr 1971;24:383–384.
140. Pichan C, Dhaliwal G, Cusick A, et al. N Engl J Med 2021;385:938–944.
141. Brown TM. J Acad Consult-Liaison Psychiatry 2021;62: 159–160.
142. Nienhuis AW. N Engl J Med 1981;304:170–171.
143. Barton JC, McDonnell SM, Adams PC, et al. Ann Intern Med 1998;129:932–939.
144. Cook JD, Watson SS, Simpson KM, et al. Blood 1984;64: 721–726.
145. Ferraro PM, Curhan GC, Gambaro G, et al. Am J Kidney Dis 2016;67:400–407.
146. Liu Y, Weisberg LS, Langman CB, et al. Clin Biochem 2016;49:1133–1139.

C. VITAMINS

CHAPTER 30

Choline

Isis Trujillo-Gonzalez and Steven H. Zeisel

CHAPTER OUTLINE

Abbreviations

Aβ, amyloid plaque; **Acho**, acetylcholine; **AD**, Alzheimer disease; **AdoHcy**, S-adenosyl-L-homocysteine; **AdoMet**, S-adenoslmethionine; **AI**, adequate intake; **ALT**, aminotransferase; ***ATP***, adenosine triphosphate; ***BHMT***, betaine-homocysteine methyl transferase; **CDP-*choline***, cytidine diphosphocholine; ***CoA***, coenzyme A; ***CoASH***, coenzyme A acetylcholine synthesis; **CHDH**, choline dehydrogenase; ***CTP***, cytidine triphosphate; **CVD**, cardiovascular disease; **DNMT**, DNA methyl transferase; **EGFR**, epidermal growth factor receptor; **ER**, estrogen recepton; **ERE**, estrogen-response element; **FMO**, flavin-dependent monooxygenase; **GPCho**, glycerophosphocholine; **Hcy**, homocysteine; **LTP**, long-term potentiation; ***MTHFD1***, methylenetetrahydrofolate dehydrogenase; ***MTHFR***, methylenetetrahydrofolate reductase; **NAFLD**, nonalcoholic fatty liver disease; **NHANES**, National Health and Nutrition Examination Survey; **PC**, phosphatidylcholine; **PEMT**, phosphatidylethanolamine N-methyl transferase; **SM**, Sphingomyelin; **SNP**, Single nucleotide polymorphism; **TMA**, trimethylamine; **TMAO**, trimethylamine N-oxide; ***THF***, tetrahydrofolate; **UL**, upper level; **VLDL**, very low-density lipoprotein

Introduction

Choline is an essential nutrient that is crucial for cellular function. Choline and its metabolites are vital for cell membrane formation and cholinergic transmission, and it is a methyl donor *via* the universal donor S-adenosylmethionine (AdoMet), among other functions. Because choline can be synthesized *de novo* through phosphatidylethanolamine N-methyl transferase (PEMT), forming phosphatidylcholine (PC), it was not considered an essential nutrient for humans. However, this nutrient needs to be consumed exogenously from the diet to meet the body's requirements and prevent organ dysfunction. In 1998, the recommended intake of choline was established by the Food and Nutrition Board of the United States.[1] Our knowledge about choline has increased significantly since then; now, we know that choline intake is essential for regulating our genes and hormonal status. In this chapter, we will discuss the multiple functions of choline, its metabolism, factors influencing personalized choline requirements, and the role of choline in disease.

Dietary Choline Intake and Sources

In 1998, the U.S. Institute of Medicine's Food and Nutrition Board established a recommended adequate intake (AI) for choline (**Table 30.1**).[1] The AI for choline was assessed by depletion-repletion studies in adult individuals and based on the choline intake level that prevents elevated plasma aminotransferase (ALT).[2,3] The AI for choline is approximately 550 mg/day for men and 425 mg/day for women. However, the requirement for women increases to 450 mg/day during pregnancy and to 550 mg/day during lactation.[1] There is no experimental data to calculate AI for infants; however, it is estimated from the calculated intake of human breast milk.[1]

The National Health and Nutrition Examination Survey (NHANES) 2013–2016 cohort, *What We Eat in America*, showed that the average dietary intake for choline is suboptimal for men and women, including women who are pregnant and lactating.[4] More recent studies have interrogated the notion that the current recommended choline intake may be too low. One such study demonstrated that the consumption of double the recommended intake for pregnant women, 930 mg/day, positively impacted the development of the fetus.[5]

The Institute of Medicine also established the tolerable upper level (UL) for choline intake in 1998. For adults, consumption of 3.5 g/day of choline was selected, based on several observed side effects including sweating, vomiting, and fishy body odor.[1] Other concerns about an excess of choline intake are related to the production of trimethylamine N-oxide (TMAO).[6] We will discuss later in this chapter the association between choline, TMAO, and cardiovascular disease.[6-8]

Choline has many functions: it is a methyl donor to make AdoMet through betaine, which is relevant for gene regulation. Betaine is also a critical osmolyte in the glomerulus of the kidneys that facilitates the reabsorption of water from the kidney tubule.[9] Choline contributes to cell membrane formation via PC production: PC is the major phospholipid of which cellular membranes are comprised.[10] Sphingomyelin (SM) resides mainly in the peripheral nervous system and promotes signal transduction and protein trafficking to the myelin.[11] Choline is part of cholinergic neurotransmission through acetylcholine (ACho), impacting synaptic transmission and behavior.[12] Although they represent a smaller proportion of the total choline pool, other important choline metabolites include lysophosphatidylcholine, phosphocholine, and glycerophosphocholine (GPCho).[13] In total, choline and its metabolites are important for homeostasis and cellular function (**Figure 30.1**).

Dietary Sources

Many foods eaten by humans contain significant amounts of choline and esters of choline.[14,15] Eggs and liver are excellent sources; one egg contains approximately 33% of the daily requirement (see the U.S. Department of Agriculture's website for a list of dietary sources of choline and betaine: www.nal.usda.gov/fnic/foodcomp/Data/Choline/Choline.html). In the 2013–16 cohort of the NHANES, only a few participants in all age groups in the United States ate diets achieving the recommended intake for choline (~550 mg/day/70 kg body weight).[4,16,17] Foods also contain the choline metabolite betaine,[10] which cannot be converted to choline but can be used as a methyl donor, thereby sparing some choline requirement. Plant-derived food sources can be rich sources of betaine, but only membrane-rich plant components (e.g., wheat germ) contain significant amounts of choline.

Human breast milk is rich in choline compounds.[14] The bioavailability of choline differs between human breast milk and infant formulas because they contain different amounts of the various choline compounds.[18,19] In 2007, most commercial infant formulas were modified to "humanize" their choline content to approximate the amount in human breast milk. Despite these

Table 30.1 Dietary Reference Intake Values for Choline

Population	Age	AI	UL
Infants	0–6 mo	125 mg/d, 18 mg/kg	Not possible to establish[a]
	6–12 mo	150 mg/d	
Children	1–3 y	200 mg/d	1,000 mg/d
	4–8 y	250 mg/d	1,000 mg/d
	9–13 y	375 mg/d	2,000 mg/d
Males	14–18 y	550 mg/d	3,000 mg/d
	≥19 y	550 mg/d	3,500 mg/d
Females	14–18 y	400 mg/d	3,000 mg/d
	≥19 y	425 mg/d	3,500 mg/d
Pregnancy	All ages	450 mg/d	Age-appropriate UL
Lactation	All ages	550 mg/d	Age-appropriate UL

[a]Source of intake should be food and formula only.

Data from Food and Nutrition Board, Institute of Medicine. Dietary Reference Intakes for Folate, Thiamin, Riboflavin, Niacin, Vitamin B_{12}, Pantothenic Acid, Biotin, and Choline. Washington DC: National Academy Press, 1998:390–422.

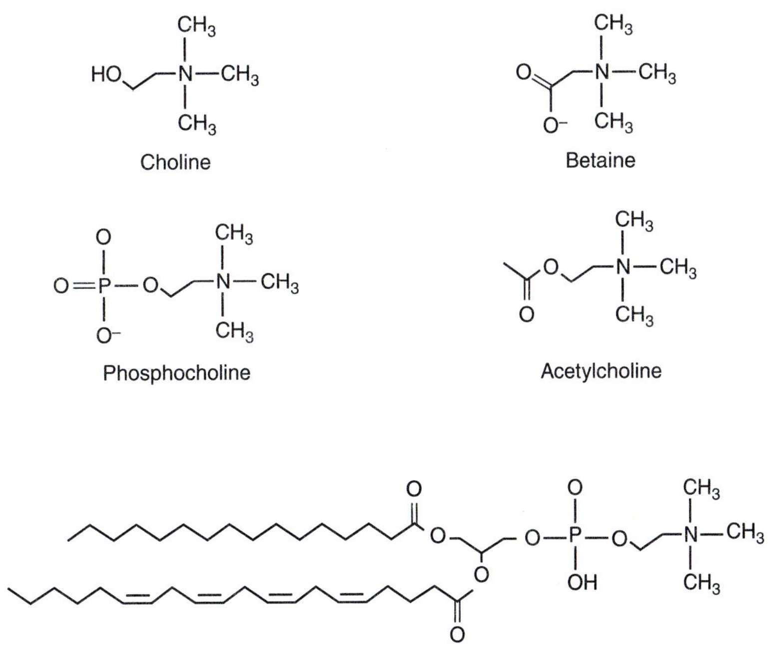

Figure 30.1 Choline and its metabolites. The structures of choline and some choline compounds are represented in this figure.

efforts, when comparing choline concentrations in infant formula with human breast milk fortifiers, a significant discrepancy still existed in choline concentrations. Specifically, both choline and PC are found to be less bioavailable in infant formula when compared with human breast milk.[20]

Mammary epithelial cells are capable of concentrative uptake of choline from maternal blood[21] and *de novo* biosynthesis of choline through PEMT.[22] The free choline content of human milk is high at the start of lactation and diminishes by 30 days postpartum.[23] Breast milk PC and choline concentrations are influenced by dietary choline intake. Dietary supplementation of PC can further increase breast milk choline, betaine, and PC concentrations.[24-26]

Digestion, Absorption, and Transport

As mentioned previously, choline can be obtained from *de novo* synthesis via PEMT and food sources. The extent to which dietary choline is bioavailable depends on the efficiency of its absorption in the intestine. Some variations in human choline requirements can be driven by differences in the gut microbiome.[27] The gut microbiome metabolizes choline to make methylamines, such as trimethylamine (TMA).[28] Following absorption, choline, betaine, PC, and GPCho enter the hepatic portal circulation and concentrate in the liver.[29] Once in the liver, choline is primarily exported as very low-density lipoprotein (VLDL) in the form of PC, VLDL being essential for packing and exporting triglycerides systemically.[30] Other choline metabolites, such as SM, circulate systemically and are taken up by several different tissues, eliciting a number of biologic effects.[31]

Choline uptake requires several different transporters to cross the plasma membrane. The choline transporter SCL44A1 is expressed in the brain, primarily in the cerebral cortex, hippocampus, spinal cord, and other tissues like colon and keratinocytes.[32-35] Cholinergic neurons depend on the high-affinity choline transporter, SLC5A7, to maintain cholinergic function, which is part of the rate-limiting step of ACho synthesis.[36]

During pregnancy, the developing fetus depends on the maternal blood supply to provide choline. Hence, the placenta maintains constitutive expression of the choline transporter SCL22A throughout pregnancy.[37-39] Choline is important for fetal brain development and normal neural tube closure.[40] In rodents, maternal choline influences the development of the hippocampus and the cerebral cortex.[41,42] This chapter will further discuss the relevance of choline and brain development.

Metabolism

A minor fraction of choline is acetylated by choline acetyltransferases to synthesize ACho.[43] As mentioned above, choline uptake in cholinergic neurons is mediated by the high-affinity choline transporter, SLC57A[36] (**Figure 30.2**). Cholinergic neurons are central to learning, mood, and memory; consequently, impaired cholinergic function is associated with neurologic diseases.[12,44,45]

In addition to ACho synthesis, choline is primarily used as a methyl donor for methylation reactions and for PC synthesis. Choline, methionine, and folate metabolism are highly interdependent (see Figure 30.2). Choline is important for the remethylation of homocysteine (Hcy) to methionine and the formation of AdoMet, a major substrate for methylation reactions in the cell. Initially, choline forms betaine-aldehyde, a major osmolyte of the liver and kidneys, via choline dehydrogenase (CHDH) in the inner mitochondrial membrane of the cell. Betaine-aldehyde undergoes an aldehyde dehydrogenation reaction forming betaine. Betaine, through betaine-homocysteine methyl transferase (BHMT), then donates its methyl group to

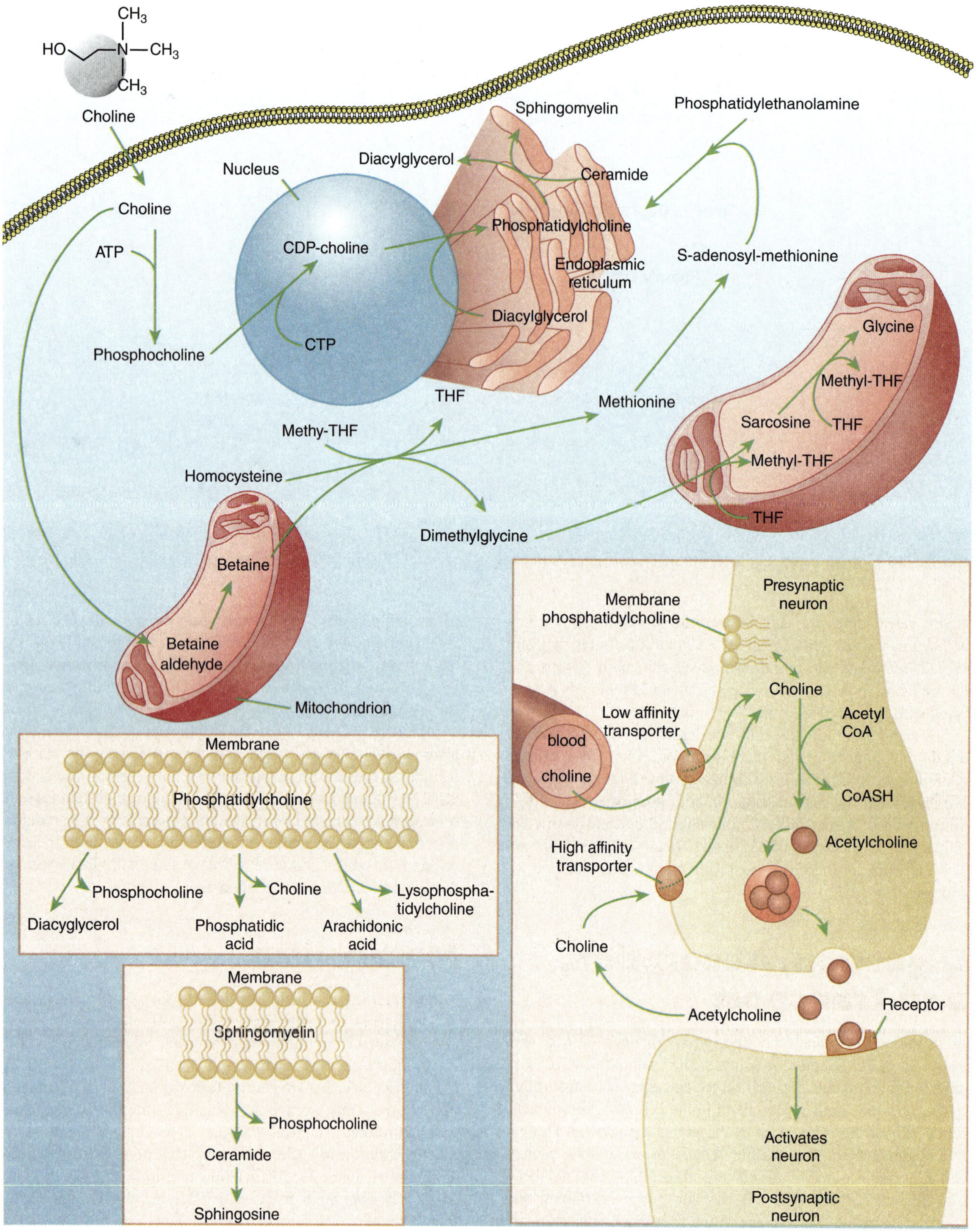

Figure 30.2 Metabolism of choline. The three primary metabolic uses for choline are as a precursor for PC biosynthesis, as a methyl donor, and as a precursor for acetylcholine biosynthesis. ATP, adenosine triphosphate; CDP-choline, cytidine diphosphocholine; CoA, coenzyme A; CoASH, coenzyme A acetylcholine synthesis; CTP, cytidine triphosphate; THF, tetrahydrofolate.

Hcy, forming methionine.[46] Folate also donates methyl groups for the remethylation of methionine via methionine synthase, which uses 5-methyltetrahydrofolate as its substrate. The remethylation of methionine is essential, as it forms AdoMet, the primary substrate for methylation reactions in the cell. Following a methylation reaction using AdoMet, S-adenosyl-L-homocysteine (AdoHcy) is formed.[47] The balance of Adomet and AdoHcy is essential for cellular homeostasis. This is due to AdoHcy being a competitive inhibitor with AdoMet, thereby inhibiting methylation enzymes. Thus, the AdoMet:AdoHcy ratio determines the activity of methyltransferases and is often referred to as the methylation potential.[48] Disruption of one of the methyl donor pathways forces compensatory mechanisms due to merging of these metabolic pathways.[46] Rodents, which lack the gene methylenetetrahydrofolate reductase (*MTHFR*), are sensitive to changes in choline/betaine intake and have low choline status.[49] These findings are important, as humans can have genetic variants in the *MTHFR* gene that increase their requirement for choline. Genetic variants of choline will be further discussed below.

It is important to mention that pregnant women with folate deficiency are more likely to have babies with neural tube defects.[50] Women consuming a low choline diet increased their risk of having a baby with cleft palate or a neural tube defect.[40,51]

PC biosynthesis occurs by two pathways: i) the Kennedy pathway, where choline is phosphorylated and converted to cytidine phosphocholine (CDP-choline); and ii) AdoMet is used via PEMT to sequentially methylate phosphatidylethanolamine. This pathway occurs primarily in the liver but has been identified in other tissues, including the brain.[52] PEMT activity in the brain is higher during the perinatal period and is needed for transporting secondary messengers such as docosahexaenoic acid (DHA).[53] Several single nucleotide polymorphisms (SNPs) have been identified in the *PEMT* gene and are associated with increased requirement for choline.[22] As mentioned before, genetic variants and hormonal status are major determinants of individual choline requirement.

Estrogen and the Requirement for Choline

De novo synthesis of choline is, in part, regulated via estrogen through expression of the *PEMT* gene. Estrogen binds to the estrogen receptor (ER) (ERα or ERβ), which then undergoes a conformational change and binds to DNA at estrogen-response elements (EREs); and can then drive gene expression, if located near the promoter of genes.[54] Multiple EREs are present in the promoter region of the *PEMT* gene[55] and estrogen causes marked upregulation in *PEMT* mRNA expression and enzyme activity in human hepatocytes.[55]

Thus, premenopausal women have enhanced capacity for *de novo* biosynthesis of the choline moiety, compared with postmenopausal women.[24] *De novo* biosynthesis is further upregulated during pregnancy as estradiol rises from <20–443 pg/mL to 28,036 pg/mL in the third trimester[56]; thereby activating the estrogen receptor and increasing the endogenous synthesis of choline via *PEMT*. This increase in endogenous production of choline during pregnancy is in response to the need to support fetal development and, later, lactation.

However, genetic variants in the PEMT promoter where the EREs reside are associated with a higher risk of developing organ dysfunction. Relevant genetic variants that individually modify the requirement of choline are discussed below.

Genetic Variants and Their Effect on Dietary Choline Requirement

Choline requirements are highly individualized by race, sex, hormonal status, and genetic variants.[3,24,57,58] Single nucleotide polymorphisms (SNPs) in the *PEMT* gene alter the endogenous synthesis of choline, while other choline metabolism SNPs influence dietary requirement by changing the efficiency by which the choline moiety is used. As discussed above, the endogenous formation of choline depends on the *PEMT* gene. One SNP in the promoter region of *PEMT* −744 G → C, rs12325817, increases the dietary choline requirement. Of 23 carriers of the C allele, 18 (78%) developed organ dysfunction when they were fed a low-choline diet.[59] This SNP is also in an ERE, underscoring the importance of estrogen in the regulation of *de novo* choline deficiency. This and other polymorphisms in the *PEMT* gene, including rs7946, are associated with greater risk of developing fatty liver.[60]

Conversely, a SNP in the coding region of the *CHDH* gene, an A → C substitution; rs9001, had a protective effect on susceptibility to low choline status, whereas a second *CHDH* variant (rs12676) was associated with increased susceptibility to low choline status.[59]

More research is needed to identify new genetic variants within different populations, to identify who may be at risk of developing organ dysfunction due to low choline intake. Genes associated with folate metabolism also contribute to an individual's choline requirement, due to the relative contributions of both folate and choline metabolism to the remethylation of Hcy.[46] Specifically, premenopausal women carrying a genetic variant in the methylenetetrahydrofolate dehydrogenase 1 (*MTHFD1*) gene, 1958G → A (rs2236225), are more than 15 times as likely to develop clinical signs of organ dysfunction when fed a low choline diet, compared with noncarriers. Genetic variants in the methylenetetrahydrofolate reductase (*MTHFR*) gene did not significantly affect people on a low choline diet.[57,59] However, consuming a diet high in choline benefits a subgroup of men with a SNP in the *MTHFR* gene.[61]

The *MTHFD1* 1958G → A SNP alters the delicately balanced flux between 5,10-methylene tetrahydrofolate and 10-formyltetrahydrofolate, thereby influencing the availability of 5-methyltetrahydrofolate for homocysteine remethylation.[62] This process increases the demand for choline as a methyl group donor. The risk of having a child with a neural tube defect increases in mothers with the 1958G → A SNP in *MTHFD1*.[63-65] These findings illustrate the complexity of the tight interrelation between folate and choline metabolism.

Gene Regulation and Epigenetics

As we previously discussed, choline is a major source of methyl groups through betaine. DNA and histone methylation are the most studied epigenetic modifications that alter chromatin and influence transcriptional regulation, thus influencing gene expression.[66] The addition of methyl groups to DNA occurs through an ATP-dependent reaction where methionine is converted to AdoMet, and DNA methyl transferases (DNMTs) attach methyl groups in a cytosine nucleotide, followed by guanine in the 5' → 3' direction (CpG).[67] Our knowledge about DNA methylation has evolved; we now know that DNA methylation regulates tissue-specific gene expression.[68] Histone methylation exists in either arginine (R) or lysine (K) residues in histone tails.[69] In the case of brain development in rodents, low choline status alters DNA methylation and modifies the expression of a histone methyltransferase (G9a) associated with repressive histone marks.[70] In pregnant women, choline supplementation with 930 mg/day led to increased methylation in the corticotropin promoter releasing hormone and glucocorticoid receptor, and increased histone methylation on the histone 3 K9me2 (H3K9me2).[71] In other organs, such as in mouse liver, deletion of *Bhmt* led to changes in CpG methylation across the genome, especially on chromosome 13.[72] Overall, choline has a clear role in epigenetic modifications that result in changes in gene expression.

Choline and Liver Health

As mentioned before, the majority of choline metabolism occurs in the liver. In humans and rodents, inadequate choline intake causes liver dysfunction. Choline, through its metabolite PC, is key to exporting lipids from the liver, as it is required for VLDL.[73,74] Lack of PC causes lipid accumulation in the liver, resulting in nonalcoholic fatty liver disease (NAFLD).[75] PEMT produces around 30% of PC in the liver. Mouse models lacking the *Pemt* gene develop fatty liver and liver failure when fed a low choline diet.[76,77]

In people, inadequate choline intake is associated with liver inflammation; however, individuals are affected differently. For instance, genetic mutations (such as insertions resulting in premature stop codons) disrupting the Kennedy pathway lead to severe fatty liver, low HDL cholesterol, and insulin resistance.[78] This indicates that other factors, like sex and genetics must be considered to assess the risk of developing NAFLD.[58,59,79] About 25% of adults in the United States have NAFLD, making it one of the most prevalent liver diseases.[80] Healthy humans with normal folate and vitamin B_{12} who consumed a low choline diet developed fatty liver (measured via elevated plasma alanine) and/or muscle damage (measured via elevated creatine phosphokinase), which was resolved when choline was restored to the diet.[58,81] Thus, identifying SNPs in the choline and folate metabolism can be helpful in predicting NAFLD development in humans.

Choline and the Developing Brain

The requirement for choline is increased during pregnancy. Adequate choline intake is crucial because it influences fetal development, particularly the brain. As discussed earlier, in mammals, the placenta regulates transport of choline to the fetus. In a mouse model, maternal choline supplementation modulates placental markers of inflammation and apoptosis.[82] Moreover, choline supplementation increases global DNA methylation in the placenta and improves fetal development.[83,84]

During the neonatal period, the capacity of the brain to extract choline from blood increases. This is coupled with increased enzymatic activity of PEMT in neonatal rat brains, thereby producing increased amounts of PC.[85] Furthermore, in the brains of newborn rats, AdoMet concentrations are elevated to 40 to 50 nmol/g[86]; a large pool of AdoMet would support a more enzymatically active form of PEMT. Postnatally, the choline is delivered in large quantities to the neonate via milk.

Low choline status during late pregnancy was associated with significant and irreversible changes in hippocampal function in the adult animal, including altered long-term potentiation (LTP).[87-89] Increased choline intake (approximately four times dietary levels) during neurogenesis (gestational days 11.5 to 17.5 in the rodent) increased hippocampal progenitor cell proliferation and decreased apoptosis in these cells.[41] On the other hand, low choline during gestational days 11.5 to 17.5 reduced global histone methylation in the developing brain.[70] This underscores that an adequate supply of choline is required for the developing hippocampus. The demand for choline in the developing brain also includes the cerebral cortex, a complex laminated structure important for perception, memory, and language.[42,90] Maternal choline intake regulates cerebral cortex development, at least in part, by governing cell proliferation, differentiation, and migration in neural progenitor cells.[42,91] The mechanism by which choline regulates cerebral cortex development is via epidermal growth factor receptor (EGFR), a pivotal receptor involved in cortical development,[92] through a specific microRNA, miR-129-5p, that is upregulated in low choline. MiR-129-5p inhibits the translation of *Egfr* mRNA, resulting in altered neuronal migration and cortical structure.[93] Hence, choline regulates cerebral cortex development partially by dictating microRNA expression.

Notably, the availability of choline during brain development has long-lasting effects. Rodents in utero exposed to high choline exhibited enhanced LTP when they were adults,[87,89] and enhanced visuospatial and auditory memory, by as much as 30%, in the adult animals throughout their lifespan.[86,88,89,91] Additionally, rodents with low choline displayed insensitivity to LTP when they were adult animals and decremented visuospatial and auditory memory.[91] Thus, low maternal choline during a critical period of pregnancy causes lifelong memory deficits in the offspring.

Significantly, in humans, higher intake of choline during the third trimester of pregnancy is associated with augmented processing speeds in children at age 4, 7, 10, and 13 months of age,[94] and children performed better in memory tests at 7 years old.[95]

Because brain and eye development share neural progenitor cells as precursors, it is not surprising to observe choline effects in the developing retina. The retinal structure comprises cell layers, including neurons, that take in light and send signals to the brain.[96] Low choline during retinal development led to breaks and disruptions in the retinal structure in mice. Moreover, vision changes observed in these mice persisted through adulthood.[97] In humans, the role of choline in eye health is a developing field. However, there is emerging evidence of favorable results of choline supplementation in dry eye syndrome.[98,99] The mechanism by which choline exerts its beneficial effects remains unclear; however, the cholinergic function seems critical, at least for amacrine cells.[100] Overall, evidence from rodent and human studies exemplifies the relevance of adequate choline intake during pregnancy for brain and eye function.

Choline and Neurological Disorders

The role of choline for brain function is not limited to development. Some neurologic disorders may be modulated by choline nutritional status.[53] Alzheimer disease (AD) is a challenging neuropathology where abnormal phospholipid metabolism is observed in postmortem brains.[101] In a mouse model of AD, persistent choline supplementation throughout life showed that choline protects the brain by blocking the production of amyloid plaque (Aβ) and attenuating microglial activation.[102] Moreover, choline supplementation through pregnancy and lactation reduces the amount of Aβ and improves hippocampal function.[103] Emerging evidence suggests that maternal choline supplementation in mice ameliorates AD trans-generationally by lowering Hcy concentration.[104] It is proposed that choline may be a possible adjuvant treatment for AD and Down Syndrome.[105-108] In pregnant women, impaired choline metabolism increased the risk of having a baby with Down Syndrome, and specifically, polymorphisms in the *BHMT* gene exacerbated the risk up to fivefold.[109]

Autism spectrum is a unique neurologic disorder characterized by language impairment, anxiety, and motor anomalies.[110] In two different mouse models, choline supplementation helped to improve social interaction and anxiety,[111,112] and in children, a combination of donepezil and choline improved receptive language skills.[113] Similar positive effects of choline supplementation were observed on learning and memory in rodent models of prenatal alcohol exposure.[114-116] In a double-blind, placebo-controlled clinical trial, maternal choline supplementation alleviated adverse effects of prenatal alcohol exposure.[117] How choline results in positive outcomes in neurologic disorders is current ongoing research. Some evidence indicates that it is through a combination of epigenetics and methylation potential,[46,70,93,118] phospholipid function,[119-121] and cholinergic function.[122] Everything considered, adequate choline intake during pregnancy, lactation, and throughout adult life is essential for brain development and cognition and improves neurodegenerative disease outcomes.

Choline and Cancer in Rodent Models and Humans

Mice fed a methionine and low choline diet lead to nonalcoholic steatohepatitis and formation of hepatic tumors.[123,124] Low choline status results in altered foci of hepatocytes. These hepatocytes subsequently express γ-glutamyltranspeptidase[125] and the placental form of glutathione s-transferase,[126] preceding the formation of adenomas and hepatocellular carcinomas.[127] Rats consuming a low choline diet are more prone to developing cancer when exposed to mutagens and carcinogens.[128] Interestingly, mice with a deletion of the *Bhmt* gene also develop hepatic cancer at 1 year old.[72,129] This is due, at least in part, to altered methylation.[72]

Recently, a meta-analysis of 11 epidemiologica studies showed that increased dietary intake of choline and betaine helps to reduce the risk of developing cancer.[130] In 2009, a population-based study showed that high intakes of choline and betaine reduce breast cancer mortality.[131] Additionally, in women with the *PEMT* rs12325817 SNP (impaired binding of the ER receptor, see above), who consumed more choline, reduced their risk of developing breast cancer by 24%.[132]

Although choline seems protective for rodents and humans for liver and breast cancer, higher choline intake has been associated with higher lethality in prostate cancer.[133]

Choline, Gut Microbiome, and Cardiovascular Disease

In humans, the intestinal microbiome metabolizes choline obtained from the diet to TMA, or choline is recycled by the gut.[134] TMA is then oxidized in the liver by flavin-dependent monooxygenases (FMOs) 1 and 3 to produce TMAO.[135] Some human diseases are associated with TMA production—for instance, the metabolic disorder trimethylaminuria is caused by mutations in the *FMO3* gene that perturb the oxidation of TMA to TMAO.[136] In the gut, choline cleavage is achieved by the *cut* gene cluster, which is widely distributed in human gut bacteria.[137] Relevant genes in the cluster are *cutC*, which encodes for the choline trimethylamine-lyase, and *cutD*, its activating enzyme.[138] Both *cutC* and *cutD* are needed for the anaerobic TMA production from choline and carnitine (**Figure 30.3**).[139]

In recent years, TMAO concentration has been associated with atherosclerosis and increased risk of stroke and myocardial infarction.[140,141] In 2011, an untargeted metabolomic analysis revealed that choline, betaine, and TMAO concentrations correlated with cardiovascular disease (CVD) risk.[142] Another precursor of TMAO formation is PC, which was found to be associated with increased mortality risk in women from the Nurses' Health Study (1980–2012) and men from the Health Professionals Follow-Up Study (1986–2012).[143] Other clinical studies have reported an association between TMAO and CVD risk and mortality in heart failure patients.[141,144,145]

On the other hand, the association between TMAO and CVD was not found in the Coronary Artery Risk Development In Young Adults Study (CARDIA), where one of the outcomes is coronary artery atherosclerosis.[146] Also, in postmenopausal Dutch women in the PROSPECT-EPIC cohort, betaine and choline were not associated with CVD.[147] In addition, the Jackson Heart Study of African Americans revealed that higher dietary choline was associated with reduced likelihood of ischemic stroke. In the same study, betaine intake was associated with a nonlinear higher risk of heart disease.[148]

People with weakened kidney function exhibit elevated plasma TMAO concentration.[149] Atherosclerosis is a complex disease and affects heart and kidneys; when glomerular filtration is impaired, TMAO increases.[150] This may explain why the association of CVD and TMAO is mostly in people with more severe CVD and in those who are older. The association between TMAO and CVD was not observed in less-severe atherosclerosis and younger people.

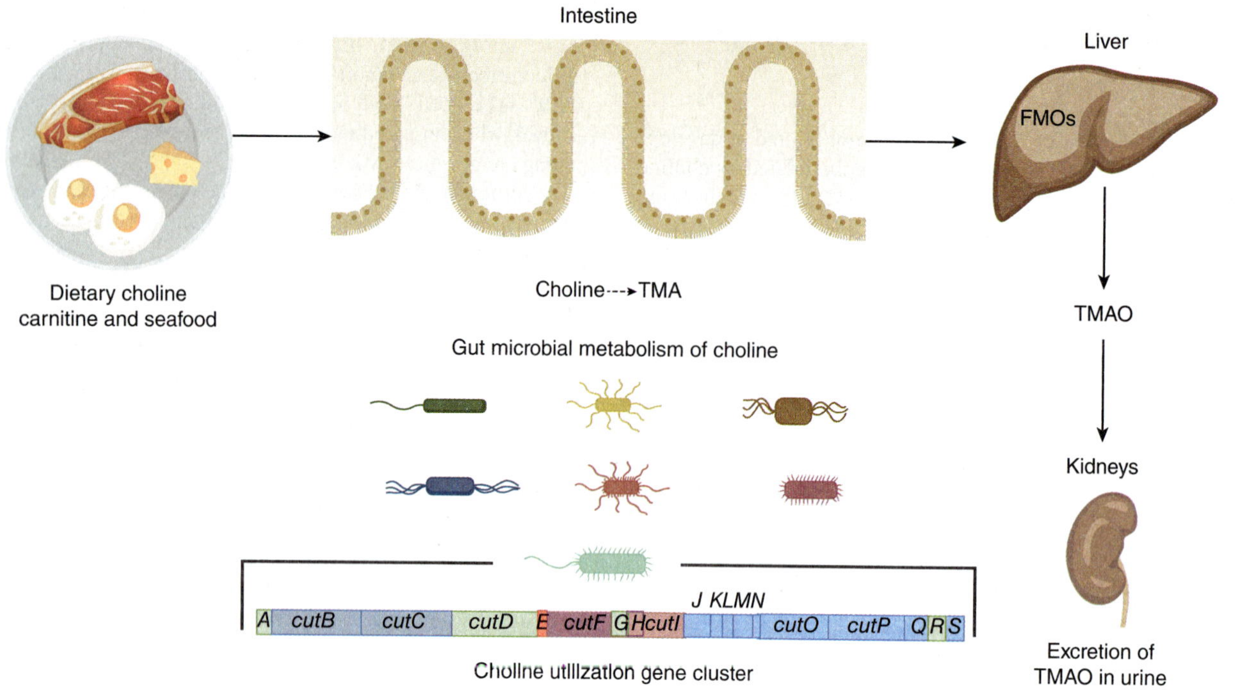

Figure 30.3 Gut microbial choline metabolism and trimethylamine N-oxide (TMAO) formation and excretion. Choline, carnitine, and choline-containing compounds are metabolized by bacteria expressing the choline utilization cluster[137] in the intestinal lumen. TMA is absorbed by the intestine and oxidized in the liver to form TMAO by flavin-containing mono-oxygenase enzymes (FMOs).[135] The kidney excretes TMAO into urine.[151]

Created with BioRender.com.

An Emerging Need for Accurate Biomarkers of Choline Status

In the last 30 years, a robust body of evidence has demonstrated the vital role of choline in many biologic functions. Yet, no precise biomarkers in humans adequately reflect choline status. One approach to calculating choline intake in a clinical setting is the use of food frequency questionnaires; however, there is no proven way to assess their accuracy, and sometimes, there is an underreporting of the actual choline intake.[152,153] Plasma free choline concentration is inaccurate due to the rapid degradation of choline. Most choline metabolism takes place in the liver; thus, choline depletion leads to fatty liver.[154] In a clinical setting, a liver biopsy would enable accurate choline concentration; however, this method is invasive and costly. A metabolomic approach showed that methionine and carnitine metabolism were promising predictors of low choline status[154]; these measurements are valuable but remain insufficient. Ongoing research is attempting to expand on these findings to accurately assess choline status in humans.

Summary

Choline is an essential nutrient that is important for many biological functions. It modulates cell signaling processes, neurotransmission, and is a methyl donor. Human requirements for choline are extremely individualized and depend on genetics, sex, and hormonal status. Choline is critical for brain development and has long-lasting effects in adults. Low choline status leads to disease states and its supplementation may be beneficial for neurodegenerative disorders.

Acknowledgments

Some of the work described in this review was supported by the National Institutes of Health (R01 DK115380). The authors thank David Horita, PhD, and Evan M. Paules, PhD, for the critical reading of this book chapter.

References

1. Institute of Medicine Standing Committee on the Scientific Evaluation of Dietary Reference I, its Panel on Folate OBV, and Choline. The National Academies Collection: Reports funded by National Institutes of Health. In Dietary Reference Intakes for Thiamin, Riboflavin, Niacin, Vitamin B(6), Folate, Vitamin B(12), Pantothenic Acid, Biotin, and Choline, National Academies Press (US). Copyright © 1998, National Academy of Sciences., Washington (DC), 1998.

2. Wallace TC, Blusztajn JK, Caudill MA, et al. Nutr Today 2018;53:240–53.
3. Zeisel SH, Da Costa KA, Franklin PD, et al. Faseb J 1991;5: 2093–8.
4. Service UAR. Usual Nutrient Intake from Food and Beverages, among Individuals Under 131% of Poverty Level, What We Eat in America, NHANES 2013–2016. 2019;52.
5. Bahnfleth CL, Strupp BJ, Caudill MA, et al. Faseb J 2022;36: e22054.
6. Zeisel SH, Warrier M. Annu Rev Nutr 2017;37:157–81.
7. Zeisel SH, Wishnok JS, Blusztajn JK. J Pharmacol Exp Ther 1983;225:320–4.
8. Kim RB, Morse BL, Djurdjev O, et al. Kidney Int 2016;89: 1144–52.
9. Craig SA. Am J Clin Nutr 2004;80:539–49.
10. van der Veen JN, Kennelly JP, Wan S, et al. Biochimica et Biophysica Acta (BBA) - Biomembranes 2017;1859:1558–72.
11. Poitelon Y, Kopec AM, Belin S. Cells 2020;9:812.
12. Picciotto MR, Higley MJ, Mineur YS. Neuron 2012;76:116–29.
13. Zeisel SH. Ann Rev Nutr 2006;26:229.
14. Howe JC, Williams JR, Holden JM, et al. Release One 2004.
15. Zeisel SH, Mar M-H, Howe JC, et al. J Nutr 2003;133:1302–7.
16. Wallace TC, Fulgoni VL. Nutrients 2017;9.
17. Wallace TC, Fulgoni VL, 3rd. J Am Coll Nutr 2016;35:108–12.
18. Holmes-McNary MQ, Cheng W-L, Mar M-H, et al. Am J Clin Nutr 1996;64:572–6.
19. Cheng W-L, Holmes-McNary MQ, Mar M-H, et al. J NutrBio 1996;7,457–64.
20. Shunova A, Böckmann KA, Minarski M, et al. Nutrients 2020;12:3815.
21. Chao C-K, Pomfret EA, Zeisel SH. Biochem J 1988;254:33–8.
22. Resseguie M, Song J, Niculescu MD, et al. Faseb J 2007;21: 2622–32.
23. Zeisel SH, Char D, Sheard NF. J Nutr 1986;116:50–8.
24. Fischer LM, da Costa KA, Galanko J, et al. Am J Clin Nutr 2010;92:336–46.
25. Davenport C, Yan J, Taesuwan S, et al. J Nutr Biochem 2015;26:903–11.
26. Lewis ED, Richard C, Goruk S, et al. Nutrients 2017;9:713.
27. Goh YQ, Cheam G, Wang Y. J Agricul Food Chem 2021;69: 10774–89.
28. Bodea S, Balskus EP. Purification and characterization of the choline trimethylamine-lyase (CutC)-activating protein CutD. In Methods in Enzymology Elsevier, 2018;606:73–94.
29. Zeisel SH, Story DL, Wurtman RJ, et al. Proc Natl Acad Sci 1980;77:4417–9.
30. Corbin KD, Zeisel SH. Curr Opin Gastro 2018;28:159.
31. Garner SC, Mar M-H, Zeisel SH. J Nutr 1995;125:2851–8.
32. Traiffort E, O'Regan S, Ruat M. Mol Aspects Med 2013;34: 646–54.
33. Brandon EP, Mellott T, Pizzo DP, et al. J Neuro 2004;24:5459–66.
34. Traiffort E, Ruat M, O'Regan S, et al. J Neurochem 2005;92: 1116–25.
35. Uchida Y, Inazu M, Takeda H, et al. J Pharmacol Sci 2009;109: 102–9.
36. Inazu M. Nutrients 2019;11:2265.
37. Sweiry J, Yudilevich D. J Physiol 1985;366:251–66.
38. Zeisel SH, Mar M-H, Zhou Z, et al. J Nut 1995;125:3049–54.
39. Baumgartner HK, Trinder KM, Galimanis CE, et al. Placenta 2015;36:1362–9.
40. Shaw GM, Carmichael SL, Laurent C, et al. Epidemiology 2006;285–91.
41. Albright CD, Tsai AY, Friedrich CB, et al. Dev Brain Res 1999;113:13–20.
42. Wang Y, Surzenko N, Friday WB, et al. FASEB J 2016;30: 1566–78.
43. Blusztajn JK, Wurtman RJ. Science 1983;221:614–20.
44. Bierer LM, Haroutunian V, Gabriel S, et al. J Neurochemistry 1995;64,749–60.
45. Dumas JA, Newhouse PA. PharmacolBiochem Behav 2011;99: 254–61.
46. Niculescu MD, Zeisel SH. J Nutr 2002;132:2333S–5S.
47. Froese DS, Fowler B, Baumgartner MR. J Inherited Metab Dis 2019:42:673–85.
48. Simile MM, Latte G, Feo CF, et al. Ann Gastro 2018;31:552.
49. Schwahn BC, Chen Z, Laryea MD, et al. FASEB J 2003;17,125.
50. Czeizel AE, Dudás I, Vereczkey A, et al. Nutrients 2013;5: 4760–75.
51. Shaw GM, Finnell RH, Blom HJ, et al. Epidemiology 2009; 714–9.
52. Blusztajn JK, Zeisel SH, Wurtman RJ. Brain Res 1979;179: 319–27.
53. Blusztajn JK, Slack BE, Mellott TJ. Nutrients 2017;9:815.
54. Klinge CM. Nucleic Acids Res 2001;29:2905–19.
55. Resseguie M, Song J, Niculescu MD, et al. FASEB J 2007;21: 2622–32.
56. Abbassi-Ghanavati M, Greer LG, Cunningham FG. Obstet Gynecol 2009;114:1326–31.
57. Kohlmeier M, da Costa KA, Fischer LM, et al. Proc Natl Acad Sci U S A 2005;102:16025–30.
58. Fischer LM, daCosta KA, Kwock L, et al. Am J Clin Nutr 2007;85:1275–85.
59. da Costa K-A, Kozyreva OG, Song J, et al. FASEB J 2006;20: 1336–44.
60. Zeisel SH. FASEB J 2006;20:2181–2.
61. Shin W, Yan J, Abratte CM, et al. J Nutr 2010;140:975–80.
62. Brody LC, Conley M, Cox C, et al. Am J Human Genet 2002;71:1207–15.
63. De Marco P, Merello E, Calevo MG, et al. J Human Genet 2006;51:98–103.
64. Meng J, Han L, Zhuang B. J Neur Sci 2015;348:188–94.
65. Carroll N, Pangilinan F, Molloy AM, et al. Human Genet 2009;125:247–56.
66. Portela A, Esteller M. Nat Biotech 2010;28:1057–68.
67. Anderson OS, Sant KE, Dolinoy DC. J Nutr Biochem 2012;23: 853–9.
68. Jones PA, Takai D. Science 2001;293:1068–70.
69. Tiffon C. Int J Molec Sci 2018;19:3425.
70. Mehedint MG, Niculescu MD, Craciunescu CN, et al. FASEB J 2010;24:184–95.
71. Jiang X, Yan J, West AA, et al. FASEB J 2012;26:3563–74.
72. Lupu DS, Orozco LD, Wang Y, et al. FASEB J 2017;31:2090–103.
73. Noga AA, Zhao Y, Vance DE. J Biol Chem 2002; 277, 42358–65.
74. Hörl G, Wagner A, Cole LK, et al. J Biol Chem 2011;286: 17338–50.
75. Arendt BM, Ma DW, Simons B, et al. Appl Physiol Nutr Metab 2013;38:334–40.
76. Li Z, Agellon LB, Vance DE. J Biol Chem 2005;280:37798–802.
77. Vance DE. Biochimica et Biophysica Acta (BBA)-Molecular and Cell Biology of Lipids 2013;1831:626–32.
78. Payne F, Lim K, Girousse A, et al. Proc Nat Acad Sci 2014;111:8901–6.
79. Song J, Da Costa KA, Fischer LM, et al. FASEB J 2005;19: 1266–71.
80. Cotter TG, Rinella M. Gastroenterol 2020;158:1851–64.
81. da Costa K-A, Badea M, Fischer LM, et al. Am J Clin Nutr 2004; 80:163–70.
82. King JH, Kwan ST, Yan J, et al. Nutrients 2019;11:374.
83. Kwan ST, King JH, Grenier JK, et al. Nutrients 2018;10:417.
84. King JH, Kwan ST, Bae S, et al. J Nutr Biochem 2019;72:108210.
85. Blusztajn JK, Zeisel SH, Wurtman RJ. Biochem J 1985;232: 505–11.

86. Pyapali GK, Turner DA, Williams CL, et al. J Neurophys 1998;79:1790–6.
87. DA M. Brain Res Dev Brain Res 2000;123:25–32.
88. Jones III J, Meck WH, Williams CL, et al. Dev Brain Res 1999;118:159–67.
89. Meck WH, Williams CL. Neuroreport 1997;8:3045–51.
90. Bystron I, Blakemore C, Rakic P. Nature Rev Neuro 2008;9: 110–22.
91. Albright CD, Friedrich CB, Brown EC, et al. Brain Res Dev Brain Res 1999;115:123–9.
92. Burrows RC, Wancio D, Levitt P, et al. Neuron 1997;19:251–67.
93. Trujillo-Gonzalez I, Wang Y, Friday WB, et al. FASEB J 2019;33:3601–12.
94. Caudill MA, Strupp BJ, Muscalu L, et al. *FASEB J* 2018;32: 2172–80.
95. Boeke CE, Gillman MW, Hughes MD, et al. Am J Epidem 2013;177:1338–47.
96. Hoon M, Okawa H, Della Santina L, et al. Prog Ret Eye Res 2014;42:44–84.
97. Trujillo-Gonzalez I, Friday WB, Munson CA, et al. FASEB J 2019;33:9194–209.
98. Choi JJ, Hwang JS, Shin YJ. Nutrients 2020;12:1526.
99. Hwang J-S, Shin Y-J. Int J Molec Sci 2021;22:4733.
100. Zhang C, Yu WQ, Hoshino A, et al. J Comp Neurol 2019;527:174–86.
101. Bartus RT, Dean RL, Goas JA, et al. Science 1980;209:301–3.
102. Velazquez R, Ferreira E, Knowles S, et al. Aging Cell 2019;18:e13037.
103. Mellott TJ, Huleatt OM, Shade BN, et al. PLoS One 2017;12:e0170450.
104. Velazquez R, Ferreira E, Winslow W, et al. Mol Psych 2020;25:2620–9.
105. J Strupp B, E Powers B, Velazquez R, et al. Cur Alz Res 2016;13:97–106.
106. Moon J, Chen M, Gandhy SU, et al. Behav Neuro 2010;124:346.
107. Kelley C, Ash J, Powers B, et al. Curr Alz Res 2016;13:84–96.
108. Alldred MJ, Chao HM, Lee SH, et al. Hippocampus 2018;28: 251–68.
109. Jaiswal S, Sukla K, Chauhan A, et al. Eur J Clin Nutr 2017;71: 45–50.
110. Sztainberg Y, Zoghbi HY. Nature Neuro 2016;19:1408–17.
111. Langley EA, Krykbaeva M, Blusztajn JK, et al. Behav Brain Res 2015;278:210–20.
112. Agam G, Taylor Z, Vainer E, et al. Trans Psych 2020;10:316.
113. Gabis LV, Ben-Hur R, Shefer S, et al. J Mol Neuro 2019;69: 224–34.
114. Idrus NM, Breit KR, Thomas JD. Neurotox Teratol 2017;59: 43–52.
115. Baker JA, Breit KR, Bodnar TS, et al. Nutrients 2022;14:2868
116. Thomas JD. Alcohol Clin Exp Res 2021;45:2465–7.
117. Jacobson SW, Carter RC, Molteno CD, et al. Alcohol Clin Exp Res 2018;42:1327–41.
118. Davison JM, Mellott TJ, Kovacheva VP, et al. J Biol Chem 2009;284:1982–9.
119. Kanfer J, Pettegrew J, Moossy J, et al. Neurochem Res 1993;18:331–4.
120. Ellison DW, Beal MF, Martin JB. Brain Res 1987;417:389–92.
121. Blusztajn JK, Gonzalez-Coviella IL, Logue M, et al. Brain Res 1990;536:240–4.
122. Mufson EJ, Counts SE, Perez SE, et al. Expert Rev Neurother 2008;8:1703–18.
123. Caballero F, Fernández A, Matías N, et al. J Biol Chem 2010;285:18528–36.
124. Yao Z, Vance DE. J Biol Chem 1988;263:2998–3004.
125. Yokota K, Singh U, Shinozuka H. Japanese J Cancer Res 1990; 81, 129–34.
126. Denda A, Kitayama W, Kishida H, et al. (2002) Japanese J Cancer Res 1990;93:125–32.
127. Ghoshal AK, Farber E. Carcinogenesis 1984;5:1367–70.
128. Zeisel SH, Costa K-Ad, Albright CD, et al. Diet Cancer 1995: 65–74.
129. Teng Y-W, Mehedint MG, Garrow TA, et al. J Biol Chem 2011; 286, 36258–67.
130. Sun S, Li X, Ren A, et al. Sci Rep 2016;6:1–11.
131. Xu X, Gammon MD, Zeisel SH, et al. FASEB J 2009;23:4022–8.
132. Xu X, Gammon MD, Zeisel SH, et al. FASEB J 2008;22:2045–52.
133. Richman EL, Kenfield SA, Stampfer MJ, et al. Am J Clin Nutr 2012;96:855–63.
134. Romano KA, Vivas EI, Amador-Noguez D, et al. MBio 2015;6:e02481-14
135. Bennett BJ, de Aguiar Vallim TQ, Wang Z, et al. Cell Metab 2013;17:49–60.
136. Christodoulou J. J Paed Child Health 2012;48:E153–E5.
137. Martínez-del Campo A, Bodea S, Hamer HA, et al. MBio 2015;6:e00042-15.
138. Craciun S, Balskus EP. Proc Natl Acad Sci 2012;109:21307–12.
139. Tang WW, Hazen SL. J Clin Invest 2014;124:4204–11.
140. Wang Z, Klipfell E, Bennett BJ, et al. Nature 2011;472:57–63.
141. Senthong V, Li XS, Hudec T, et al. J Am Coll Cardiol 2016;67:2620–8.
142. Wang Z, Klipfell E, Bennett BJ, et al. Nature 2011;472:57–63.
143. Zheng Y, Li Y, Rimm EB, et al. Am J Clin Nutr 2016;104:173–80.
144. Wang Z, Tang WW, Buffa JA, et al. Eur Heart J 2014;35: 904–10.
145. Tang WW, Wang Z, Shrestha K, et al. J Cardiac Fail 2015;21: 91–6.
146. Meyer KA, Benton TZ, Bennett BJ, et al. J Am Heart Assoc 2016;5:e003970.
147. Dalmeijer G, Olthof M, Verhoef P, et al. Eur J Clin Nutr 2008;62:386–94.
148. Millard HR, Musani SK, Dibaba DT, et al. Eur J Nutr 2018;57: 51–60.
149. Posada-Ayala M, Zubiri I, Martin-Lorenzo M, et al, Kidney Int 2014;85:103–11.
150. Missailidis C, Hällqvist J, Qureshi AR, et al. PloS One 2016;11:e0141738.
151. Ufnal M, Zadlo A, Ostaszewski R. Nutrition 2015;31:1317–23.
152. Cade J, Thompson R, Burley V, et al. Public Health Nutr 2002;5:567–87.
153. Fischer LM, Scearce JA, Mar MH, et al. J Nutr 2005;135:826–9.
154. Sha W, Costa KAD, Fischer LM, et al. FASEB J 2010;24: 2962–75.

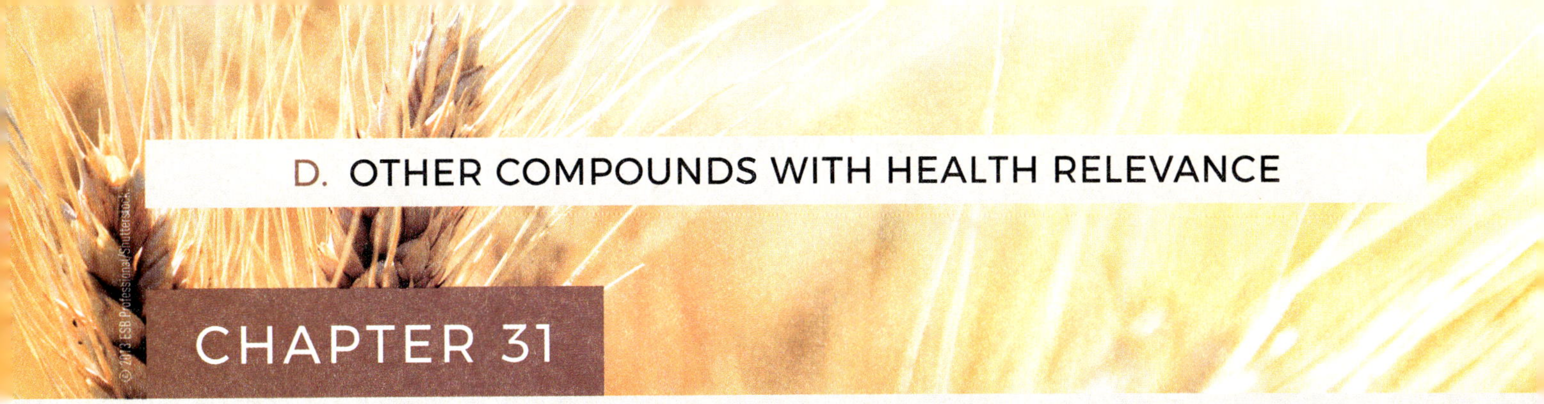

D. OTHER COMPOUNDS WITH HEALTH RELEVANCE

CHAPTER 31

Carotenoids

Xiang-Dong Wang

CHAPTER OUTLINE

Abbreviations

ARE, antioxidant-response element; **BCO1**, β-carotene-15,15′-oxygenase; **BCO2**, β-carotene-9′,10′-oxygenase; **Cx43**, connexin 43; **HDL**, high-density lipoprotein; **HPLC**, high-performance liquid chromatography; **IGF**, insulin-like growth factor; **IGFBP**, insulin-like growth factor binding protein; **ISX**, intestine-specific homeobox; **LDL**, low-density lipoprotein; **NHANES**, National Health and Nutrition Examination Survey; **NMR**, nuclear magnetic resonance; **Nrf2**, nuclear factor E2–related factor 2; **PPAR**, peroxisome proliferator-activated receptor; **PPRE**, peroxisome proliferator-response element; **RAR**, retinoic acid receptor; **RXR**, retinoid X receptor; **SNP**, single nucleotide polymorphism; **SR-B1**, scavenger receptor class B type 1 protein transporter; **CD36**, cluster determinant 36; **UV**, ultraviolet; **VLDL**, low-density lipoprotein

Historical Overview

Carotenoids are yellow, orange, and red lipophilic pigments that occur widely in plants, insects, fish, birds, algae, yeasts, and bacteria, and they carry out various important biologic functions. Humans do not synthesize carotenoids but obtain them primarily through dietary intake. The discovery of carotenoids had a fascinating history.[1] Studies on carotenoids began in 1831, when Wackenroder first isolated the crystalline yellow pigment *carotene* from carrots (*Daucus carota*), and studies continued in 1837, when Berzelius named the yellow pigments of autumn leaves *xanthophylls*.[2] In the early twentieth century, the technique of chromatography was discovered, and carotenoid analysis underwent great development. Researchers discovered a large family of carotenoids and found that these substances were isoprenoid derivatives. In 1913, very close to the discovery of vitamin A by McCollum and Davis (see chapter 17), Osborne and Mendel observed that the green parts of plants contained a relatively high amount of "fat-soluble A" activity. During the subsequent decade, Steenbock, Moore, and other investigators provided further information on the relationship between the fat-soluble A and yellow pigments from comparative studies of their growth-promoting activities in vivo.

In 1930 and 1931, Karrer et al., defined the chemical structures of both β-carotene and retinol purified from shark liver oil. These researchers determined that half of the chemical structure of β-carotene resembled the retinol structure. This finding led Karrer to suggest that the simple addition of two molecules of water at the central double bond of the carotene molecule should yield two molecules of retinol. However, it was not until 1965 that Goodman and Huang[3] and Olson and Hayaishi[4] independently demonstrated the enzymatic synthesis

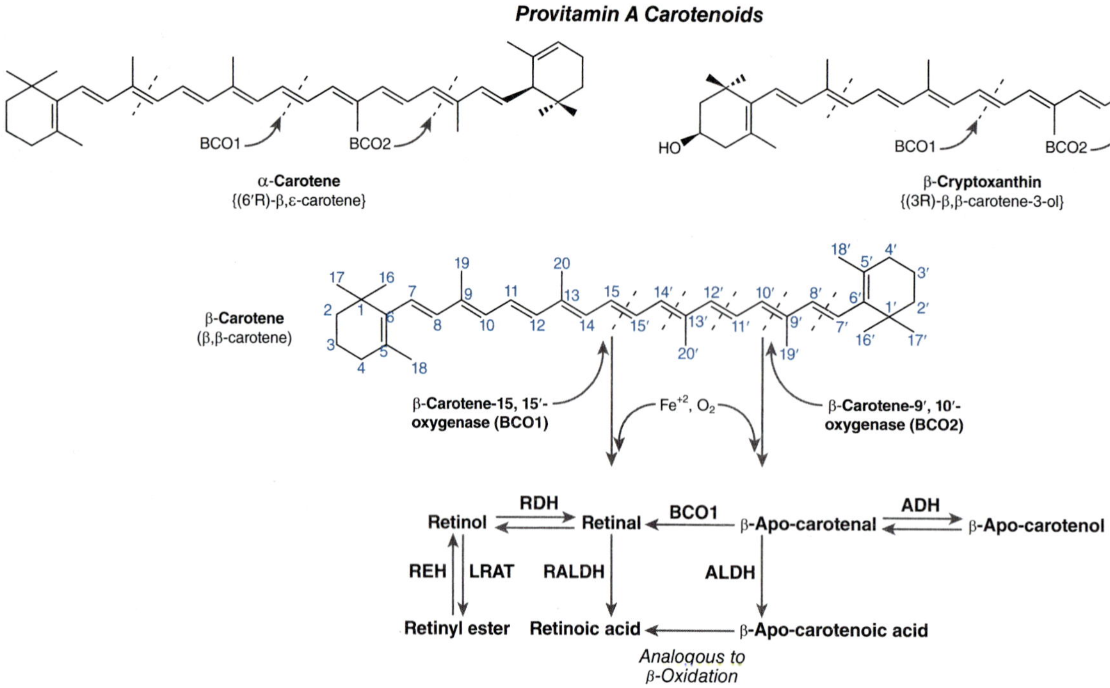

Figure 31.1 Metabolic pathway and chemical structures of the major provitamin A carotenoids (β-carotene, α-carotene, and β-cryptoxanthin) found in human plasma and tissues. Provitamin A carotenoids are symmetrically cleaved at the 15,15′ central double bond by β-carotene-15,15′-oxygenase (BCO1), which produces retinal (or β-apo-15,15′-carotenal) that can be oxidized to retinoic acid or reduced to retinol. Retinol can be converted to retinyl esters for storage. Provitamin A carotenoids can also be cleaved by β-carotene-9′,10′-oxygenase (BCO2) at either the 9,10 or 9′,10′ double bond, thus giving rise to β-apo-10′-carotenal and β-ionone (or 3-hydroxy β-apo-10′-carotenal and 3-hydroxy β-ionone). β-Apo-carotenals either can be precursors of vitamin A and retinoic acid by further BCO1 cleavage or can be reduced/oxidized to their corresponding β-apo-carotenol or β-apo-carotenoic acids. *ADH*, alcohol dehydrogenase; *ALDH*, aldehyde dehydrogenase; *LRAT*, lecithin:retinol acyltransferase; *RALDH*, retinal dehydrogenase; *RDH*, retinol dehydrogenase; *REH*, retinyl ester hydrolase.

of retinal (or apo-15,15′carotenal) from β-carotene in cell-free extracts of rat liver and intestine. These investigators did not detect any other cleavage products and, therefore, suggested that a process took place by means of cleavage symmetrically at the 15,15′ (C=C) central double bond of β-carotene that required molecular oxygen. These investigators named the enzyme β-carotene-15,15′-oxygenase (BCO1).

It was another 35 years until Wyss et al.,[5] and von Lintig and Vogt,[6] in different species, were able to achieve the molecular cloning of the *BCO1* gene; these investigators subsequently provided a biochemical and structural characterization.[7,8] These studies firmly established that the central cleavage of provitamin A carotenoids is the main pathway leading to the formation of vitamin A (**Figure 31.1**).

In 1954, Glover[9] proposed that β-carotene can undergo both central and eccentric cleavage. Eccentric cleavage is the asymmetric cleavage of carotenoids at positions outside of the central double bond (Figure 31.1). The existence of the eccentric cleavage pathway for β-carotene had been controversial among scientists since the 1970s,[10,11] however, and it was not confirmed until the molecular identification of β-carotene-9′,10′-oxygenase (BCO2) as a mitochondrial protein in humans and mice by Kiefer et al.[12] Investigators further showed that BCO2 cleaves provitamin A carotenoids and preferentially cleaves nonprovitamin A carotenoids (*cis*-lycopene, lutein, and zeaxanthin) (**Figure 31.2**).[12-15] Mice deficient for BCO2 accumulated carotenoid in many extrahepatic tissues[16] and apocarotenoids in their livers, indicating that the enzyme is a key component of apocarotenoid homeostasis in mice.[17]

Data indicate that carotenoid metabolites, called apocarotenoids in general, in addition to vitamin A, may participate in specific biologic activities in several important cellular signaling pathways and molecular targets.[18,19] These findings imply that carotenoid metabolites may have different biologic roles due to their different chemical structures, compared with their parent compounds in human health and disease. With the mapping of the human genome and the development of multi-omics analyses such as metabolomics, transcriptomics, and proteomics, greater understanding of carotenoid metabolism and activity is expected to provide new insight into the biologic functions of carotenoids.

Chemical Properties

More than 750 carotenoids with characterized chemical structures and key analytic data are listed in the *Carotenoids Handbook*[20] and new carotenoids continue to be identified. Among the carotenoids found in nature, approximately 40 to 50 occur in the human food chain, and 24 have been detected in human plasma and tissues.[21] The most abundant carotenoids in human plasma include β-carotene, α-carotene, β-cryptoxanthin, lutein, zeaxanthin, and lycopene. Carotenoids are divided into two major groups: xanthophylls, which are oxygenated carotenoids that include lutein, zeaxanthin, and β-cryptoxanthin and carotenes, which

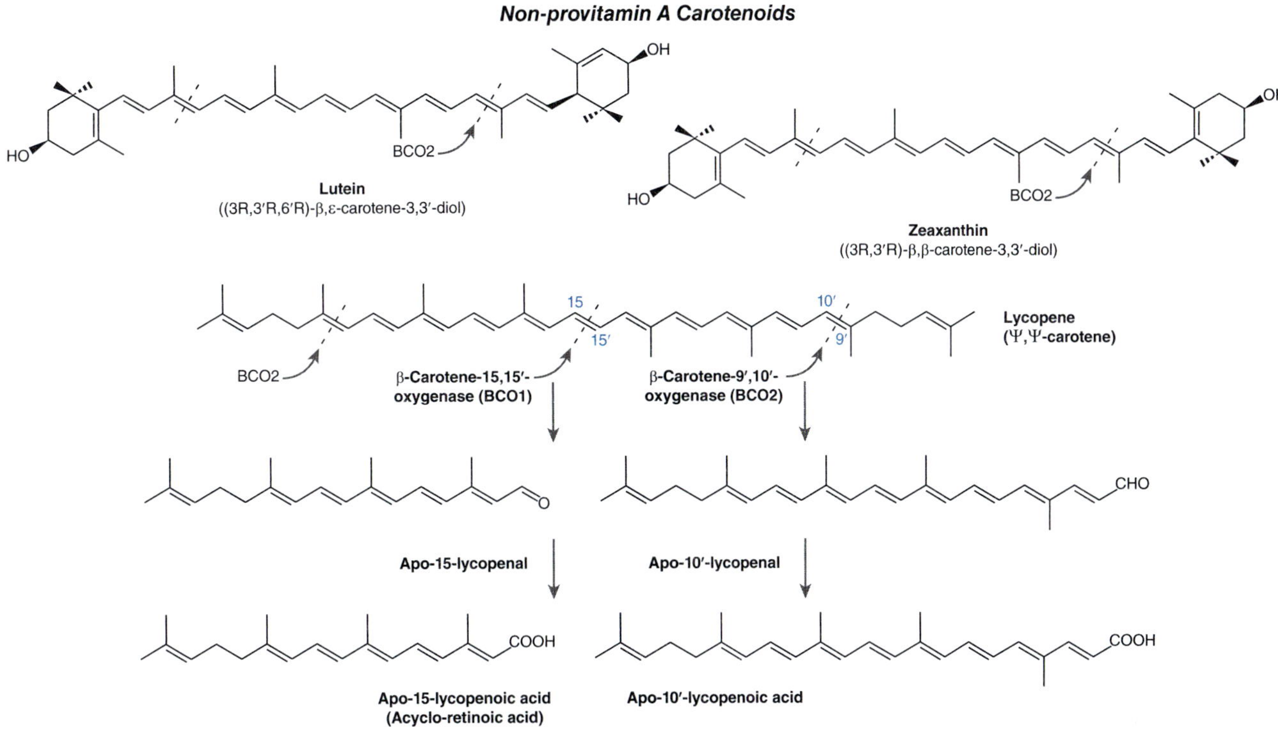

Figure 31.2 Proposed metabolic pathway and chemical structures of the major nonprovitamin A carotenoids (lutein, zeaxanthin, and lycopene) found in human plasma and tissues. Cleavage of xanthophylls (lutein and zeaxanthin) by β-carotene-9′,10′-oxygenase (BCO2) may occur at either the 9,10 or 9′,10′ double bond to produce 3-OH-β-apo-10′-carotenal and β-ionone, or 3-OH-β-ionone and β-apo-10′-carotenal, respectively. Cleavage of *cis*-lycopene by BCO2 may occur at either the 9,10 or 9′,10′ double bond to produce apo-10′-lycopenal, which can be oxidized to apo-10′-lycopenoic acid or reduced to apo-10′-lycopenol. *ADH*, alcohol dehydrogenase; *ALDH*, aldehyde dehydrogenase.

are hydrocarbon carotenoids that are either cyclized, such as α-carotene and β-carotene, or linear, such as lycopene.

The basic structure of a carotenoid is a 40-carbon atom conjugated polyene chain, sometimes terminated by ring structures (see Figure 31.1 and Figure 31.2).[20] The polyene chain, which typically contains a series of conjugated double bonds (e.g., –C=C-C=C–) in the central chain of the molecule, represents a chromophore that is responsible for the characteristic colors associated with carotenoids. It also causes instability in the carotenoids by making them susceptible to cleavage by oxidation, heat, light, acid, and isomerization from *trans* to *cis* forms. Conjugated systems with alternating single and double bonds result in a general delocalization of electrons across adjacent atoms that allows for resonance-stabilized structures. These structures, which impart both a capacity to act as biologic antioxidants and the ability to absorb and give off certain wavelengths of light, cause a compound to appear colored.

β-Carotene, α-carotene, and β-cryptoxanthin are important sources of vitamin A (see chapter 17). All provitamin A carotenoids have one or two β-ionone rings (Figure 31.1). β-Carotene has a symmetric chemical structure characterized by a long carbon chain with alternating double and single bonds, terminated at each end by a ring structure (Figure 31.1). As non-provitamin A carotenoids, lutein and zeaxanthin both have oxygen added to their ionone ring, but each differs from the other in the position of the double bond in one of the ionone rings (Figure 31.2). Lycopene has a symmetric open chain structure without a ring (Figure 31.2).

Major chemical data for these carotenoids are found in the *Carotenoids Handbook*.[20] Carotenoids usually occur in nature as all-*trans*-isomers, but the known exceptions include 9-*cis*-β-carotene in the alga *Dunaliella* and the 15-*cis*-phytoene, present in tomatoes and other organisms. The *cis-trans* isomerism of carbon-carbon double bonds is an important feature of the stereochemistry of carotenoids because these geometric isomers may have different biologic properties. Approximately 370 of the naturally occurring carotenoids are chiral, bearing 1 to 5 asymmetric carbon atoms; most individual carotenoids occur in nature in only a single configuration.

The term *apocarotenoids* or *apolycopenoids* refers to carotenoid derivatives in which the carbon skeleton is shortened by the removal of fragments from one or both ends of the carotenoid with the position of the point of cleavage indicated (e.g., β-apo-10′-carotenal from β-carotene [see Figure 31.1] or apo-10′-lycopenal from lycopene [see Figure 31.2]). The same applies to their further metabolites, such as alcohol and acid forms. Apocarotenoids are generated upon enzymatic cleavage of carotenoids both in plants and in mammals, and nonenzymatic cleavage in foods during food processing. Apocarotenoids are bioactive mediators in plants, in which they act as visual or volatile signals to attract pollinating and seed-dispersing agents. These substances are key players in allelopathic interactions, plant defense, and plant architecture.[22] Abscisic acid, formed by specific oxidative cleavage of the 11,12 carbon-carbon double bond of 9′-(Z)-neoxanthin, acts as a hormone in plants. In humans, the most important apocarotenoids are colorless retinoids (vitamin A and its derivatives (see chapter 17). Emerging evidence has shown that apocarotenoids other than retinoids also have a broad spectrum of biologic functions and play important roles for mammalian health.[19]

Dietary Sources

The major sources of carotenoids in the human diet are deeply pigmented yellow, orange to red fruits and vegetables (**Table 31.1**). β-Carotene is responsible for the orange color of carrots, lycopene for the red color of tomatoes and watermelon, and zeaxanthin for the pink color of salmon. In the green leaves of vegetables, the colors of the carotenoids are often masked by chlorophyll. A more extensive list of dietary sources of carotenoids is found in Table 31.1. Since the 1990s, great effort has been exerted on metabolic engineering of carotenoids into agricultural crops, such as golden rice, high–β-carotene maize, cassava roots, and potatoes. Scientists are evaluating these food products for their bioavailability to provide vitamin A and are assessing the potential of these products to combat vitamin A deficiency in certain developing countries.

Dietary Intake and Serum Concentrations

Information on the dietary intake and serum concentrations of carotenoids (Table 31.1) was obtained from the Third National Health and Nutrition Examination Survey (NHANES III)[23,24]. Carotenoids are not classified as essential nutrients, so values for their recommended daily intake have not been established. Although β-carotene and other provitamin A carotenoids are confirmed important sources of vitamin A, no specific recommendation exists on the percentage of total vitamin A that should be obtained from retinyl esters or carotenoids. There are no dietary reference intake recommendations (DRIs) for carotenoids due to insufficient evidence to make recommendations.[25] However, lutein has been proposed for DRI review, as it accumulates in the macula, and may exert beneficial effects against the risk of age-related macular degeneration.[25]

The reported values for the conversion of β-carotene to vitamin A (i.e., the number of β-carotene molecules that are nutritionally equivalent to one molecule of vitamin A) exhibit a wide range, from 2:1 for synthetic pure β-carotene in oil to 27:1 for β-carotene from vegetables.[26,27] Dietary β-carotene intake varies widely and is not normally distributed in the population. The majority of people consume 1–2 mg/day in the USA and the UK.[26]

Analysis

High-performance liquid chromatography (HPLC) is a powerful means of analyzing carotenoid composition and determining concentrations after the extraction of carotenoids from plasma, tissue, and food samples. The intrinsic ultraviolet (UV)-visible absorption spectrum provides the first criterion for identifying a carotenoid, and is the basis of quantitative analysis. Carotenoids have a characteristic absorption spectrum, and their concentrations can be calculated from specific extinction coefficients. The chemical data on major carotenoids, including UV-visible spectra and extinction coefficients, have been described by Britton et al.[20] The photodiode array detector allows simultaneous monitoring at a range of selected wavelengths and provides UV/V is spectra online for each component of a chromatogram as an aid to identification. Linked HPLC-mass spectrometry and HPLC-nuclear magnetic resonance (NMR) techniques are becoming more readily available.

The identification of carotenoids as known compounds should, at a minimum, be based on the following: (a) the UV-visible absorption spectrum (λmax) must be identical to that of an authentic sample; (b) the chromatographic properties must be identical to those of an authentic sample in HPLC, and cochromatography with an authentic sample should be demonstrated; and (c) if possible, a mass spectrum should be obtained that allows

Table 31.1 Dietary Intake, Serum Concentrations, and Natural Sources of Six Major Carotenoids

Carotenoids	Formula	Median Dietary Intake (μg/D) Reference Range (10th–90th Percentile)	Median Serum Concentration (μg/DL) Reference Range (10th–90th Percentile)	Major Natural Sources
β-Carotene	$C_{40}H_{56}$	1,665 (774–3,580)	14.7 (6.4–35.1)	Carrots, sweet potato, spinach, broccoli, kale, pumpkin, mango, apricot, and many yellow-orange fruits and green leafy vegetables
α-Carotene	$C_{40}H_{56}$	36 (2–1.184)	3.4 (1.3–9.2)	Carrots, pumpkin, some yellow-orange fruits and green vegetables
β-Cryptoxanthin	$C_{40}H_{56}O$	88 (24–319)	8.0 (4.0–16.4)	Tangerine, orange, sweet red pepper, peach, papaya; in small amounts, in yellow-orange fruits and in maize
Lutein + zeaxanthin	$C_{40}H_{56}O_2$	1,466 (714–3,021)	18.9 (11.1–33.0)	Egg yolk, broccoli, spinach, yellow pepper, kale, sweet corn, pumpkin, and all green fruits and vegetables
Lycopene	$C_{40}H_{56}$	8,031 (3,580–16,833)	22.4 (11.9–36.1)	Tomato and tomato products (e.g., sauce, pasta, soup, and juice) and watermelon

Data from the Third National Health and Nutrition Examination Survey (NHANES III), from 1988 to 1994, as discussed in Trumbo P, Yates AA, Schlicker S et al. J Am Diet Assoc 2001;101:294–301; Monsen ER. J Am Diet Assoc 2000;100:637–40, with permission.

for at least one confirmation of the molecular mass. For quantitative analysis by HPLC, an internal control, such as echinenone, is needed to assess the efficiency of extraction procedures. Full elucidation of the structure requires a fully assigned NMR spectrum and, for chiral compounds, comparison of a circular dichroism spectrum with that of an authentic reference sample.

Carotenoids are unstable and are vulnerable when exposed to oxygen, heat, light, and acid. Precautions and special procedures must be used to minimize the risk of degradation and the formation of artifacts. All analytic procedures should be carried out in an inert atmosphere (nitrogen or argon), at room temperature (~20°C), in the dark or in diffused light, under acid-free conditions, and with freshly purified, peroxide-free solvents. Geometric (*cis-trans*) isomerization occurs readily when carotenoids are exposed to factors such as light or heat and slowly occurs even in isolated or purified samples. Plasma or tissue samples should be stored at −80°C to minimize degradative reactions and isomerization.

Noninvasive resonance Raman spectroscopic and diffuse reflectance spectroscopic techniques show promise for measuring carotenoids in situ in the skin[28-30] and the retina.[31] Carotenoids in the skin of the palm of the hand, which correlate highly with serum concentration of carotenoids, can be measured using a portable Raman device. For example, using reflection spectroscopy, the lower skin levels of carotenoids were significantly associated with metabolic syndrome.[32] The lack of power to separate individual carotenoids is a limit to research, but great potential exists for monitoring carotenoid status from fruit and vegetable intake or the effect of carotenoid supplementation in human trials. Lutein, zeaxanthin, and meso-zeaxanthin, major macular pigments of the human retina with ocular health and functional benefits,[33] can also be measured using resonance Raman spectroscopy as a sensitive and specific objective method in the living human eye.[34] This noninvasive detection technique can be a screening method for macular pigment concentrations in the general population, and in patients with chronic disease, such as age-related visual loss or diabetes, and for potentially monitoring the effect of dietary carotenoids and supplement interventions in clinical trials.

Stable isotope techniques that use intrinsically labeled carotenoids have proven useful in determining carotenoid bioavailability, bioconversion, and bioefficacy from different human food sources.[35] Although these methods can distinguish between dosed and endogenous carotenoids and can determine the vitamin A equivalency of provitamin A carotenoids in healthy individuals, whether these methods are suitable for use in settings where there is a high prevalence of infection, iron deficiency, and liver and kidney disease needs further investigation.[36]

Absorption, Bioavailability, and Transport

Humans (along with monkeys, ferrets, and gerbils, but excluding rats, mice, and rabbits unless they are given very high doses) absorb a significant portion of intact carotenoids directly and circulate or accumulate them in their plasma, liver, and peripheral tissues.[27] A number of factors associated with carotenoid absorption and bioavailability, including age, body composition, hormonal fluctuations, and variation in genes have been well reviewed.[37] Carotenoids embedded in their food matrix cannot be absorbed efficiently. Food processing and cooking that cause the mechanical breakdown of the food matrix and the release of carotenoids can improve intestinal absorption. After release from the food matrix, ingested carotenoids must be emulsified and solubilized into micelles before they are absorbed into the intestinal mucosa (**Figure 31.3**).

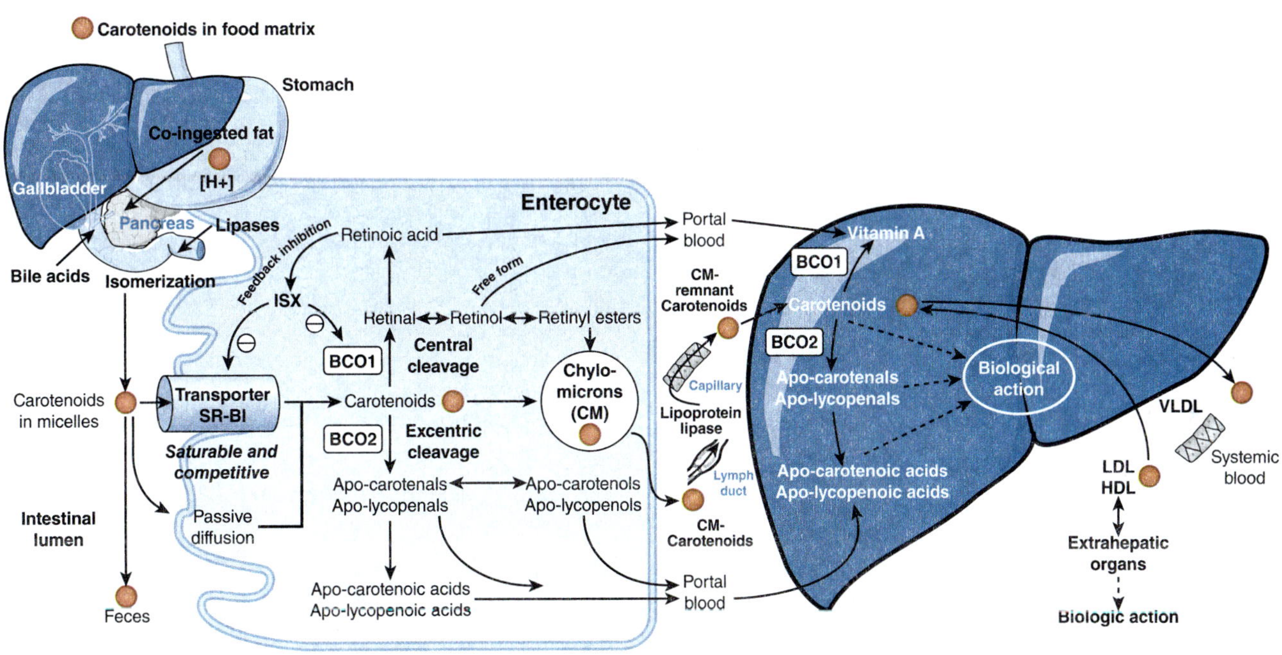

Figure 31.3 Simplified schematic illustration of absorption, metabolism, and transport of carotenoids. *BCO1*, β-carotene-15,15′-oxygenase; *BCO2*, β-carotene-9′,10′-oxygenase; *SR-B1*, scavenger receptor class B type 1 protein transporter; *CD36*, cluster determinant 36; *HDL*, high-density lipoprotein; *LDL*, low-density lipoprotein; *VLDL*, very-low-density lipoprotein. (See text for detailed information).

The efficiency of absorption of a moderate dose of β-carotene in oil is approximately 9% to 22%.[38,39] The bioavailability of β-carotene from vegetables is generally low (14% from mixed vegetables as compared with that of lutein (67% from mixed vegetables).[40] Several factors affect β-carotene bioavailability and conversion to vitamin A, such as the food matrix, host nutritional and health status (e.g., vitamin A status, malnutrition, parasitic infection, smoking, and alcohol intake) and genetic variations.[41] The mnemonic acronym SLAMENGHI lists the major contributors that affect carotenoid bioavailability.[42] SLAMENGHI stands for species of carotenoids, linkages at the molecular level, amount of carotenoid, matrix effectors, nutrient status, genetics, host-related factors, and interactions among these variables.[33,42] These factors are discussed in detail in the literature.[37,40]

The process of carotenoid absorption can occur via a passive diffusion through the brush border of the enterocytes, depending on the carotenoid concentration in the lumen.[38] Further studies indicated the involvement of an active process for the uptake of carotenoids through several apical membrane transporters, including scavenger receptor class B type 1 protein transporter (SR-B1)[7,38,43] and cluster determinant 36 (CD36), a surface membrane glycoprotein in the small intestine.[44] SR-B1 is found in the human small intestine, as well as the liver, adrenals, ovaries, placenta, kidneys, prostate, and brain. Therefore, SR-B1, which is also involved in the uptake of other fat-soluble vitamins (e.g., vitamins E and D), may be partially responsible for the transport of carotenoids from lipoprotein to tissues and from tissues to lipoproteins. A diet-responsive regulatory network involving the intestine-specific homeobox (ISX) transcription factor was shown to regulate intestinal β-carotene uptake and vitamin A production by a negative feedback regulatory mechanism.[7] ISX repressed both expression of intestinal BCO1 and SR-B1, which facilitates the absorption of dietary lipids and carotenoids. Because ISX is under the control of retinoic acid and retinoic acid receptor (RAR)–dependent mechanisms, during vitamin A insufficiency, both BCO1 expression and SR-B1 expression are induced to increase absorption and conversion of β-carotene to vitamin A (see Figure 31.3). Cleavage of β-carotene by BCO1 produces retinal, which can be oxidized to retinoic acid. Retinoic acid induces the expression of the ISX transcription factor and then represses expression of both BCO1 and SR-B1, to complete the dietary feedback mechanism (Figure 31.3). CD36, highly expressed at the brush border level of the duodenum and the jejunum, is not regulated by ISX; however, genetic variation in the CD36 gene modulated plasma concentrations of provitamin A carotenoids at a population level.[44] Although the same factors that influence the absorption of provitamin A carotenoids may affect other carotenoids similarly, more investigation of the absorption of individual carotenoids and their *cis* isomers is needed. For example, the Nieman Pick C1-like transporter 1, which is a major sterol transporter in the intestine, is involved with provitamin A carotenoids and lutein intestinal uptake, but not in that of lycopene, phytoene, or phytofluene.[38]

After β-carotene is taken up by the mucosa of the small intestine, it is either cleaved by BCO1 or BCO2 into vitamin A and other metabolites, or packaged into chylomicrons and secreted into the lymphatic system for transport to the liver and other peripheral tissues (see Figure 31.3). The differential absorption of carotenoids and their metabolites into lymph or portal blood seems to depend on the polarity of the metabolites involved. β-Carotene, retinyl esters, retinol, and the less polar metabolites are absorbed into the lymph.[45] Some polar metabolites in the enterocytes like, β-apocarotenals, can be directly transported into the liver through the portal blood system.[45] The polar metabolites of carotenoids could be further rapidly metabolized in the liver, which may explain the limited detection of apocarotenoids in systemic blood circulation. Chylomicrons in the bloodstream are partially degraded by lipoprotein lipase, a process that leaves chylomicron remnants that are quickly taken up by the liver (Figure 31.3, and see also chapter 4). Some carotenoids may be released from these lipoproteins and taken up directly by extrahepatic tissues. In the fed state, the liver stores or secretes the carotenoids in very-low-density lipoproteins (VLDLs) and (LDLs). In the fasting state, plasma carotenes are found mainly in LDL. Xanthophylls (lutein, zeaxanthin, and β-cryptoxanthin) are located mainly in both LDL and high-density lipoproteins (HDLs), and small proportions are located in VLDLs. LDL transport accounts for approximately 55%, HDL transport for 31%, and VLDL transport for 14% of total blood carotenoids. Specific factors that regulate tissue uptake, recycling of carotenoids back to the liver, and excretion are not yet understood.[37] The median concentrations of carotenoids were reported in NHANES III (Table 31.1). The half-life of plasma carotenoids ranges from up to 12 days for β-carotene, α-carotene, and cryptoxanthin; to 12 to 33 days for lycopene, and 33 to 61 days for zeaxanthin and lutein.[46]

Metabolism

Central Cleavage Pathway

For provitamin A carotenoids, central cleavage is the main pathway leading to the formation of vitamin A. β-Carotene, α-carotene, and β-cryptoxanthin are cleaved symmetrically at their central 15, 15′ double bond by BCO1, a cytoplasmic protein,[6,7] which is present in mammalian tissues (e.g., liver, lung, kidney, intestinal tract, and testis). Four conserved histidines and one conserved glutamate residue are essential for the catalytic mechanism of BCO1, presumably for the coordination of the iron cofactor required for catalytic activity. Purified recombinant human BCO1 enzyme cleaves β-carotene in vitro with a Michaelis-Menten constant (K_m) of 7 μM and maximal velocity (V_{max}) of 10 nmol retinal/mg/minute. Retinal formed from β-carotene can be subsequently reduced to retinol or oxidized further to form retinoic acid (Figure 31.1; see also the chapter 17). The open chain lycopene (Figure 31.2) can be converted by BCO1 into apo-15,15′-lycopenoid (acyclic retinoids) in vivo.[47,48] This pathway has been proposed to contribute retinoid-like activities under condition of severe vitamin A deficiency.[49] Other nonprovitamin A carotenoids, such as lutein and zeaxanthin, were not cleaved by purified recombinant mammalian BCO1.

Eccentric Cleavage Pathway

The eccentric cleavage of β-carotene leads to a series of homologous carbonyl cleavage products (see Figure 31.1).[10,11] The main enzyme of this pathway is β-carotene-9′,10′-oxygenase (BCO2), which is located in the inner mitochondrial membrane in humans, mice, and zebra fish.[12] BCO2 shares overall sequence homology with cytosolic BCO1, as well as the same conserved pattern of histidine residues and glutamate residues presumably involved in binding the iron cofactor in both proteins.[12] The activity of BCO2 is higher toward nonprovitamin A carotenoids, such as *cis*-lycopene isomers, lutein, and zeaxanthin, than toward provitamin A carotenoids as substrates.[13,15] BCO2 is highly expressed in the liver and testis and at lower levels in the intestine, kidney, lung, spleen, prostate, colon, brain, and adipose tissue.[12,15] The absence of BCO2 favors carotenoid accumulation in many extrahepatic tissues.[16] The almost ubiquitous expression of carotenoid oxygenases indicates that many tissues may contribute to their own metabolic homeostasis.[50] A recent study provided novel structural and functional insights into BCO2 catalysis and establishes the enzyme as a key component of apocarotenoid homeostasis in mice.[17] Although the exact contribution of BCO2 to vitamin A biosynthesis remains unknown, β-apocarotenals can be cleaved further by BCO1 to produce retinol and retinoic acid.[51,52] It has previously been shown that retinoic acid can be produced from eccentric cleavage pathway of β-carotene without retinal as an intermediate.[53,54] β-Apocarotenals can also rescue normal embryonic development in mice maintained on vitamin A free diet.[55] β-Apocarotenals can be oxidized to their corresponding apo-β-carotenoic acids, which may then undergo a process similar to β-oxidation of fatty acids to generate retinoic acid, until the further oxidation is blocked by the methyl group at the C13 position.[56] Whether the formation of other β-apocarotenals is the result of further metabolism of the apocarotenoid cleavage product or these substances are the primary cleavage products of additional carotene oxygenases remains to be discovered. Apocarotenoids do exist in plant foods due to a number of plant carotenoid cleavage oxygenases, and apocarotenoid formation can occur via nonspecific chemical oxidation of carotenoids (e.g., during storage, cooking, heat, acid, etc) and via site-specific cleavage of carotenoids by oxygenases in biologic systems.

Genetic Factors

Variability in carotenoid absorption and metabolism due to genetic variation has been documented in recent reviews.[37,57] Genetic variants of the BCO1/BCO2 genes have been associated with alterations in the status of human and animal carotenoid concentrations, and function of provitamin A carotenoids.[58] Single nucleotide polymorphisms (SNPs) in the human BCO1 gene are common and associated with reduced catalytic activity in the conversion of β-carotene to vitamin A.[59,60] Two SNPs (rs6564851 and rs12934922) in BCO1 were found to explain 5.6% of the variation in plasma lycopene concentration.[61] A SNP in BCO2 was strongly connected to age-related eye disease,[62] and to alterations in pro-inflammatory cytokine IL-18 expression, and fasting high-density lipoprotein cholesterol concentration.[63-65] Beyond carotenoid cleavage activities of BCO1/BCO2, several studies have found differential lipid, cholesterol, and oxidative stress-related physiologic abnormalities in individual BCO1 and BCO2 knock out (KO) mouse models.[66] Studies using systemic BCO1 KO mice indicate that BCO1 has a role in the development of hepatic steatosis and obesity,[67] lipid metabolism,[67] serum insulin, leptin, and cholesterol metabolism,[68] steroid metabolism,[69] and testosterone synthesis and prostatic cellular proliferation.[70] Similarly, systemic BCO2 ablation has been shown to result in mitochondrial dysfunction, oxidative stress,[6,66] anemia,[71] metabolic alteration, liver steatosis,[14] and prostate cancer progression.[72] Interestingly, when BCO1 and BCO2 are both systemically knocked out, mice developed hepatic steatosis and had significantly higher levels of hepatic and plasma triglycerides and total cholesterol, compared to wild type mice.[73] However, the precise roles of BCO1/BCO2 in the pathophysiology of most of these diseases are not presently clear.

Regulation

Knowledge of the molecular and regulatory framework of carotenoid metabolism is far from complete.[43] The difference in cellular compartmentalization may reflect the different substrate specificity and physiologic functions with respect to BCO1 and BCO2.[74] Molecular studies of the mouse and human BCO1 promoter demonstrate the presence of a peroxisome proliferator-response element (PPRE).[75,76] Peroxisome proliferator-activated receptor γ (PPARγ) and retinoid X receptor α (RXRα) agonists can transactivate the BCO1 promoter-reporter when they are cotransfected with the corresponding nuclear receptor.[76] Fatty acid binding protein-3, a PPARγ target gene, was downregulated in parallel with BCO1. Chronic alcohol consumption increased both expressions of BCO1 and PPARγ.[77] Dietary supplementation with various carotenoids may up- or downregulate tissue expressions of BCO1 and BCO2 in animal studies.[15,78,79] However, a molecular analysis failed to identify a PPRE within the mouse BCO2 promoter.[78] Some insight into BCO2 regulation is gained from studying BCO1-knockout mice which, compared with wild-type mice, have significantly elevated hepatic BCO2 expression.[69,80] This finding suggests the presence of concerted mechanisms governing the expression of BCO1 and BCO2 and needs further investigations.

Biologic Functions of Carotenoids and Their Metabolites

Early studies focused on the provitamin A carotenoids, in particular β-carotene, but research since 1980 has provided a framework for understanding other carotenoids and how their functions may benefit human health (See recent review).[37,81] β-Carotene, β-cryptoxanthin, lutein, zeaxanthin, and lycopene may have unique biologic roles in protecting against the development of several chronic and degenerative diseases, including vitamin A deficiency (see the chapter 17), age-related macular degeneration,[82,83] cognitive function and neurodegenerative disorders,[84,85] bone health,[86] cardiovascular disease,[87,88] diabetes,[89] obesity,[90,91] metabolic syndrome, fatty liver disease,[92] skin lesions,[93,94] and certain cancers.[94,95]

Although both cell culture and animal model studies provide strong evidence that carotenoids and their metabolites are active in several biologic activities[81] (see **Figure 31.4**), demonstrating these molecular effects in human systems, a process that involves multiple genetic and epigenetic events, is challenging. Specifically, the plasma values of carotenoids are biomarkers for consumption of diets rich in fruits and vegetables that contain other potentially bioactive nutrients, so association does not necessarily prove that carotenoids are the active compounds. As our understanding of carotenoid metabolism, molecular biologic properties, and their interactions with genetic and epigenetic factors improves, greater insight will be achieved into the role and application of carotenoids and their metabolites in human health and disease.

Retinoid Dependent- and Independent-Activities

The most clearly defined human biologic function of carotenoids is their vitamin A activity. Provitamin A carotenoids, through central cleavage, serve as precursors to vitamin A and

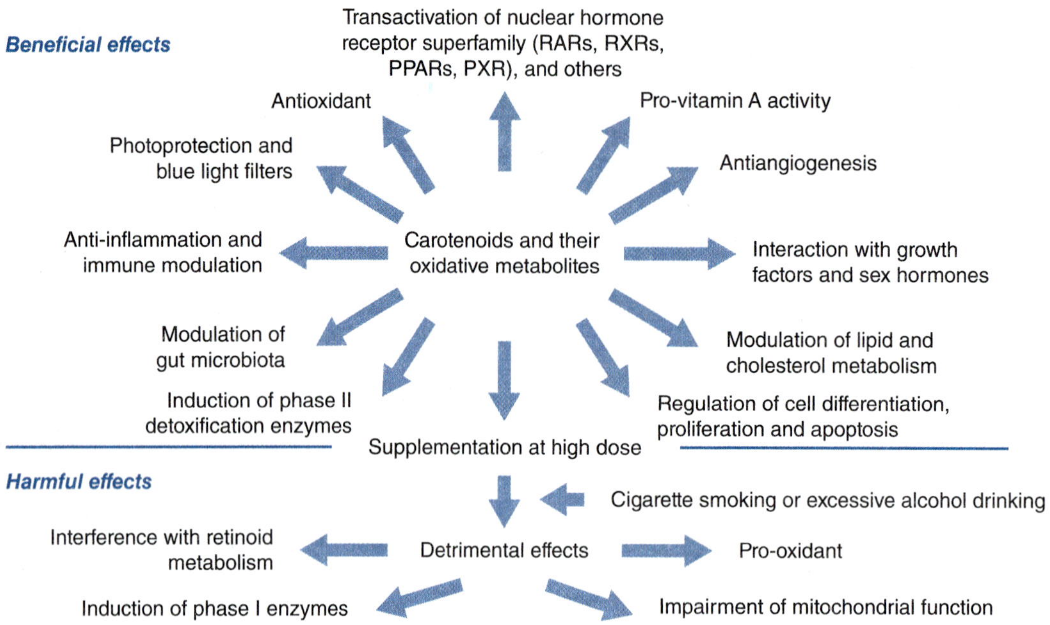

Figure 31.4 Schematic illustration of potential biologic effects, both beneficial and harmful, attributed to carotenoids and their metabolites to human health. Dietary carotenoids and their biologic active metabolites may offer protection against several chronic diseases through multiple molecular actions. Although small amounts of supplementation with specific carotenoids may be beneficial for certain disease, such as for lutein- and zeaxanthin-associated eye diseases, larger amounts through dietary supplementation may be harmful, especially when coupled with a highly oxidative environment (e.g., the lungs of cigarette smokers or liver of excessive alcohol drinkers). *PPAR*, peroxisome proliferator-activated receptor; *PXR*, pregnane X receptor; *RAR*, retinoic acid receptor; *RXR*, retinoid X receptor.

represent the major dietary sources of vitamin A for much of the world's population. Provitamin A carotenoids play important roles in protecting against vitamin A deficiency and its related health problems (see chapter 17). Through vitamin A action, provitamin A carotenoids exert effects on several critical life processes, including vision, reproduction, metabolism, differentiation, hematopoiesis, bone development, pattern formation during embryogenesis, and tumorigenesis. Provitamin A carotenoids can serve as direct precursors for retinoic acid,[53,54] which are ligands for retinoic acid receptor (RAR) and retinoid X receptor (RXR). Therefore, the molecular mode of the action of provitamin A carotenoids is likely to be mediated by retinoic acid through transcriptional activation of a series of genes.[81] Beyond participating in known retinoid signaling pathways, carotenoids may possibly be able to interact directly with transcription factors without their conversion into retinoids. β-Cryptoxanthin itself acts as a ligand (not through generating vitamin A or retinoic acid) for RAR transcription activation[96,97] and induces the CYP27A1 signaling pathway via RAR activation.[98,99] The discovery of the eccentric cleavage of carotenoids, and that its chemical structure assembles retinoids, heightened interest in carotenoid cleavage products and their possible biologic role in humans. Without being converted into retinoids, the nonvolatile apocarotenoids and apolycopenoids can inhibit cell growth, stimulate differentiation, transactivate nuclear receptors, or antagonize nuclear receptor activation. The volatile apocarotenoid β-ionone has also been shown to inhibit cell proliferation and induce apoptosis both in vitro and in vivo. Apocarotenoids and apolycopenoids[100,101] can activate other nuclear hormone receptor pathways, such as PPARs (see recent review).[81] However, whether these carotenoid metabolites function as nuclear hormone receptor ligands in vivo is still unclear, in part, due to the lack of their detection in biological systems.

Antioxidant Activities

Free radicals can cause cellular damage by reacting with proteins, lipids, carbohydrates, and DNA, and they are involved in the etiology of human diseases such as cancer, cardiovascular disease, and age-related diseases. Much of the biologic activity ascribed to carotenoids is attributed to their antioxidant capabilities, due to their polyene backbone consisting of a series of conjugated C=C bonds to interact with singlet oxygen and free radicals.[102] Carotenoids quench singlet oxygen ($^1O_2$) mostly through electron transfer to free radicals and formation of carotenoid radical cations. It is not fully understood if carotenoids scavenge other reactive oxidative species (ROS) in lipophilic environments, such as membranes and lipoproteins.[102] Another mechanism of carotenoids, including lutein and zeaxanthin, is as blue light filters that possibly prevent photo damage of the retina.[103] Indeed, the antioxidant properties of many carotenoids are well documented using in vitro and ex vivo systems where they are believed to play critical roles in protecting against chronic disease.[103] Accurate data regarding the antioxidant effects of carotenoids alone in in vivo biologic systems are limited, however.

Because most in vivo studies use fruit and vegetable products that contain various micronutrients and phytochemicals, including other carotenoids, polyphenols, vitamin C, and vitamin E, caution must be used when attributing the beneficial effects of vegetables and fruits to carotenoids or their antioxidant activity. Thus, although carotenoids demonstrate antioxidant activity in certain cell culture models and animal models, no definitive evidence indicates that carotenoids from diet and food sources act as in vivo antioxidants in human studies. In addition, a synergistic interaction between carotenoids and other antioxidants, such as vitamins E and C and other phytonutrients in fruits and vegetables, may play more important roles in the human antioxidant defense system.

Anti-Inflammatory Activities

Accumulating evidence shows that the beneficial effects of carotenoids may result from their action in modulating inflammatory signaling pathways.[81,104] NF-κB plays a prominent role to regulating a wide variety of genes, including inflammatory cytokines (e.g., IL-1β, IL-6, and TNF-α), chemokines (e.g., MCP1, IL-8), and inducible effector enzymes (e.g., iNOS, COX-2). Carotenoids can suppress the activation of the NF-κB and phosphatidylinositol 3-kinase/PTEN/Akt and mitogen-activated protein kinase pathways.[81,105] Carotenoids can also trigger the transcription factor nuclear factor E2–related factor 2 (Nrf2), which is a primary agent in the induction of antioxidant and detoxifying phase II enzymes,[81,105] such as heme oxygenase-1, glutathione S-transferases, and reduced nicotinamide adenine dinucleotide phosphate:quinone oxidoreductase. Under normal conditions, most of the Nrf2 is sequestered in the cytoplasm by Kelch-like erythroid Cap'n'Collar homolog-associated protein 1′ (Keap 1). Exposure to certain carotenoids leads to the dissociation of the Nrf2-Keap1 complex in the cytoplasm and the translocation of Nrf2 into the nucleus. The nuclear accumulation of Nrf2 subsequently activates target genes of phase II antioxidant enzymes. Not only β-carotene but also some nonprovitamin A carotenoids including lycopene, apo-lycopenoic acid, lutein, canthaxanthin, and astaxanthin, can induce several phase II enzymes both in vivo and in vitro.

Dietary insults, such as excess consumption of a high-fat diet, have been associated with microbial dysbiosis, an imbalance in both composition and function in the gut microbiome. Dysbiosis further influences metabolic pathways and inflammatory processes through the increased abundance of lipopolysaccharide (LPS) containing gram-negative bacteria, changes in both total bile acid pool and the generation of secondary bile acids, alterations in short chain fatty acid production, and disruption of gut permeability.[106,107] As the majority of ingested carotenoids transit to the small and large intestine, there is some recent evidence that carotenoids, including β-carotene,[108,109] lycopene,[110-112] astaxanthin,[113] and fucoxanthin,[114] can also contribute to health via interacting with the gut microbiome and gut immune responses, for instance acting as prebiotics. Interestingly, in moderately obese individuals accompanied by lycopene dose-dependent changes in the blood, liver metabolism, skeletal muscle, and skin parameters, there were also dose-dependent changes in the gut microbiota profile across lycopene groups (7 to 30 mg/daily) with an increase of relative abundance of, e.g., *Bifidobacterium adolescentis* and *Bifidobacterium longum*.[112] The colonic mucosal bacterial community was associated with serum carotenoid concentration at baseline, although it was not appreciably changed by dietary intervention.[115] These findings are noteworthy because they indicate that carotenoids may possess prebiotic effects, which have not been thoroughly explored. Additionally, the interactions among carotenoids, gut microbiota, and host genetics (e.g., BCO1 and BCO2) need further study. A study found that the deletion of BCO2 was associated with a significantly increased abundance of *Mucispirillum schaedleri* in mice, regardless of sex, and that supplemental astaxanthin, a non-provitamin A carotenoid, resulted in altered gut microbial abundances through increasing microbial diversity, reduced hallmarks of microbial dysbiosis, such as *Proteobacteria enrichments* (LPS-producing bacteria), and selectively increased the abundance of *Akkermansia muciniphila*, a bacterium associated with reduced serum LPS and increased gut barrier functionality.[113]

Modulation of Hormones and Growth Factors

Steroid hormones (e.g., androgens and estrogen) and insulin-like growth factor (IGF) signaling systems may play a role in the biologic action of carotenoids, in particular lycopene.[116,117] Lycopene reduced local prostatic androgen signaling, IGF-I expression, and basal inflammatory signals in normal prostate tissue.[116,118] Lycopene, phytoene, and phytofluene inhibited the estrogen-induced transactivation of the estrogen-response element bound by the nuclear estrogen receptors ERα and ERβ.[119] A higher dietary intake of lycopene may be associated with lower circulating IGF-I.[120,121] Although IGF-I concentration did not change, lycopene supplementation increased plasma insulin-like growth factor-binding protein 3 (IGFBP-3) concentration and was associated with inhibition of cigarette smoke–induced lung lesions in a ferret model.[122] However, a systematic review of the effects of lycopene on circulating IGF-1 and IGFBPs in humans showed no consistent modifying effect of lycopene supplementation on IGF-1 and IGFBPs.[123] Clearly, additional well-designed studies should be performed to elucidate whether lycopene may influence the IGF/IGFBPs system.

Gap Junction Communication

Several lines of in vitro evidence indicate that carotenoids (lycopene, β-carotene, and canthaxanthin) and their oxidative products and metabolites, can elevate gap junctional communication, as do the retinoids.[124] Gap junctions are cell-to-cell channels that enable connected cells to exchange nutrients, ions, waste products, and both electrical and biochemical signals between cells. Dysfunctional gap junctional communication has been linked to human disease such as cancer, skin disease, neuropathies, and cardiovascular dysfunctions. Each gap junction is derived from six connexin proteins, from each adjacent cell, for a total of 12 connexin proteins. The connexin 43 (Cx43) is the most widely expressed and is the form most often induced by retinoids and carotenoids.[124] Both provitamin A and nonprovitamin A carotenoids can inhibit carcinogen-induced neoplastic transformation and up-regulate Cx43 mRNA expression.[125,126] The lycopene and lycopene-5,6-epoxide, found in tomatoes, increased Cx43 expression in human keratinocytes, although the contribution of acycloretinoic acid to the activity of lycopene on gap junctional communication may be minimal.[126] However, RAR antagonists also inhibited upregulation of gap junctional communication by retinoids, but not carotenoids.[125] This finding, which warrants further study, suggested the possibility of two separate pathways of increasing gap junctional communication.

Effects Related to High Dosage

In the early 1980s, two key publications,[127,128] revealed that β-carotene could be an antioxidant and anticancer agent. This finding greatly stimulated the field of carotenoid research, and several intervention trials using pharmacologic doses of β-carotene (some 10 to 15 times more than the average dietary consumption) as a chemopreventive agent were performed. Unfortunately, the human trials concluded without evidence of a beneficial effect and actually showed an increased risk of lung cancer in heavy smokers and asbestos workers. These unexpected findings brought carotenoid researchers back to experimental research in animal and cell culture models in an attempt to find insight into this contradiction. The effects of dose responses, the antioxidant and pro-oxidant effect, the interaction with other micronutrients, and the coexistence of the central and eccentric cleavage pathways reveal the complexity of carotenoid metabolism in organisms and raise questions regarding the potential effects of interactions between exogenous factors (e.g., tobacco smoking and chronic alcohol consumption) and carotenoids and their metabolites.

Dosage and Undesirable Metabolites

The dosages of β-carotene used in two human intervention studies (the Alpha-Tocopherol, Beta-Carotene Cancer Prevention [ATBC] trial and the Beta-Carotene and Retinol Efficacy Trial [CARET]) were 20 to 30 mg/day for 2 to 8 years, more than tenfold higher than the intake of β-carotene in the typical US diet (~2 mg/day). This pharmacologic dose of β-carotene in humans is believed to have resulted in the accumulation of relatively high levels of β-carotene and its oxidative eccentric cleavage metabolites in lung tissue, especially after long periods of supplementation.

Research in animal and cell culture models suggested that β-carotene is unstable in the free radical–rich environment of lungs exposed to cigarette smoke, and that such an environment alters β-carotene metabolism and produces undesirable eccentric cleavage metabolites (see Figure 31.4). These metabolites have been shown to facilitate changes associated with the carcinogenic process, including induction of carcinogen-activating enzymes, binding of carcinogen metabolites to DNA, interference with vitamin A metabolism, downregulation of tumor suppressor genes, upregulation of oncogenes, induction of oxidative stress, and enhanced induction of cell transformation by carcinogens. Therefore, smokers and consumers of alcohol are warned to avoid high-dose carotenoid supplements. Given that β-carotene in the diet is less bioavailable than supplemental β-carotene, no current evidence indicates any harmful effect associated with high levels of dietary β-carotene from natural food sources, aside from the occasional appearance of carotenodermia. However, the Age-Related Eye Disease Study 2 (AREDS2) trial reported that nutritional supplementation with lutein (10 mg) and zeaxanthin (2 mg) can reduce risk of progression to advanced AMD.[129] Therefore, supplementation with individual carotenoids may be beneficial for specific purposes, such as for lutein- and zeaxanthin-associated eye diseases.[03]

Pro-Oxidant Effect

Some evidence indicates that carotenoids may behave as pro-oxidants under certain circumstances.[130] At higher oxygen concentrations, a carotenoid peroxy radical could be formed (e.g., Car-OO• or ROO-Car-OO•) that could then act as a pro-oxidant, promoting hydrogen abstraction and oxidation of unsaturated lipids and, hence, exacerbating membrane damage.[103] Based on the evidence presented from the large clinical trials of β-carotene supplementation in lung cancer, it appears that β-carotene may act as a protective antioxidant against cancer at physiologic concentrations, but it may lose its effectiveness or may even exert pro-oxidant effects at pharmacologic concentrations, especially in highly oxidative compartments of the body (see Figure 31.4).

Strong interactions among β-carotene, α-tocopherol, and ascorbic acid in vitro and the possible capability of these compounds to recycle each other have led researchers to speculate about the potential utility of combined antioxidant supplementation to eliminate potential pro-oxidant effects by a single agent. In animal studies, α-tocopherol and ascorbic acid decreased the production of undesirable oxidative metabolites and increased the formation of retinoids from β-carotene in lung tissue of smoke-exposed ferrets.[131]

Induction of Phase I Enzymes

In laboratory studies, tobacco smoking and chronic excessive alcohol consumption, especially when coupled with a high dose of carotenoids, induced the expression of cytochrome P-450 phase I enzymes (see Fig. 31.4).[132,133] These enzymes may activate procarcinogens present in alcoholic beverages, tobacco smoke, and diet and may lead to increased formation of carcinogen-DNA adducts. If not repaired or repaired incorrectly, these adducts may eventually lead to mutations and ultimately cancer, especially if the adducts are located in tumor suppressor genes. In addition, these same cytochrome P-450 enzymes can break down retinoic acid and lead to significantly decreased tissue retinoic acid concentration. These studies provide possible mechanistic explanations for the discordance between the results of observational epidemiologic studies and intervention trials using carotenoids as a potential beneficial agent.

Summary

Many epidemiologic studies have shown the benefit of carotenoid-rich fruits and vegetables on the risk of chronic diseases; however, clinical supplementation trials returned null findings or even evidence of harm in certain populations. Carotenoid supplementation is not recommended for the general population, and smokers and consumers of alcohol are warned to avoid high-dose carotenoid supplements. Assessment for individual variability of carotenoid bioavailability associated with genetic variation may provide useful information for personalized recommendations to prevent chronic diseases.[37] New research on carotenoid-vision and cognitive function and carotenoid–gut microbiota interactions and well-controlled human intervention studies are needed to establish future dietary recommendations.[134]

Despite numerous in vitro and animal studies that demonstrate the mechanisms of action of carotenoids, the metabolism and molecular biologic properties of many carotenoids remain to be determined.[81] The new activities and functions of carotenoid cleavage enzymes and carotenoid metabolites are expected to be discovered. Apo-carotenoids may be active in several important cellular signaling pathways and molecular targets, and they could have greater biologic roles than their parent compounds in human health and disease. As our understanding of carotenoid metabolism, molecular biologic properties, and their interactions with genetic and epigenetic factors improves, greater insight will be achieved into the role and application of carotenoids in human health and disease. Nevertheless, a prudent strategy to reduce the risk of chronic disease incidence and mortality is to increase consumption of carotenoid-rich vegetables and fruits as a part of a balanced diet.

Acknowledgments and Disclosures

This work is supported by the National Institutes of Health grant R01CA104932 and the U.S. Department of Agriculture, under agreement no. 58-1950-64S. Any opinions, findings, conclusions, or recommendations expressed in this publication are mine and do not necessarily reflect the views of the U.S. Department of Agriculture. Because of space limitations, not all excellent publications could be cited. I have no conflict of interest.

I thank Dr. Michael S. Daniels for his assistance in the preparation of this manuscript.

References

1. Britton G. Biochim Biophys Acta 2020;1865:158699.
2. Goodwin TW. The Biochemistry of the Carotenoids. 2nd edn. New york, NY, Methuen Inc.; 1980.
3. Goodman DS, Huang HS. Science 1965;149:879–80.
4. Olson JA, Hayaishi O. Proc Natl Acad Sci U S A 1965;54: 1364–70.
5. Wyss A, Wirtz G, Woggon W, et al. Biochem Biophys Res Commun 2000;271:334–6.
6. von Lintig J, Vogt K. J Biol Chem 2000;275:11915–20.
7. von Lintig J. Annu Rev Nutr 2010;30:35–56.
8. Harrison EH, Kopec RE. Biochim Biophys Acta 2020;1865: 158653.
9. Glover J. Vitam Horm 1960;18:371–86.
10. Ganguly J, Sastry PS. World Rev Nutr Diet 1985;45:199–220.
11. Wang XD, Tang GW, Fox JG, et al. Arch Biochem Biophys 1991;285:8–16.
12. Kiefer C, Hessel S, Lampert JM, et al. J Biol Chem 2001;276: 14110–6.
13. Mein JR, Dolnikowski GG, Ernst H, et al. Arch Biochem Biophys 2011;506:109–21.
14. Amengual J, Lobo GP, Golczak M, et al. FASEB J 2011;25: 948–59.
15. Hu KQ, Liu C, Ernst H, et al. J Biol Chem 2006;281:19327–38.
16. Arballo J, Amengual J, Black M, et al. Curr Dev Nutr 2021;5:69.
17. Bandara S, Thomas LD, Ramkumar S, et al. ACS Chem Biol 2021;16:480–90.
18. Wang XD. Biological activities of carotenoid metabolites. In: Britton G, Liaaen-Jensen, S. Pfander, H., ed. Carotenoids, Basel: Birkhauser, 2009;383–408.
19. Harrison EH, Quadro L. Annu Rev Nutr 2018;38:153–72.
20. Britton G, Liaaen-Jensen S, Pfander H. Carotenoids Handbook, Birkahäuser Verlag, Basel, Switerland, 2004.
21. Britton G, Khachik F. Carotenoids in Food. In: Britton G, Liaaen-Jensen, S. Pfander, H., ed. Carotenoids, Basel: Birkhauser, 2009; 45–66.
22. Moreno JC, Mi J, Alagoz Y, et al. Plant J 2021;105:351–75.
23. Trumbo P, Yates AA, Schlicker S, et al. J Am Diet Assoc 2001;101:294–301.
24. Monsen ER. J Am Diet Assoc 2000;100:637–40.
25. Erdman JW. Mol Aspects Med 2022;101116.
26. Grune T, Lietz G, Palou A, et al. J Nutr 2010;140:2268S–85S.
27. Böhm V, Lietz G, Olmedilla-Alonso B, et al. Nutr Rev 2020;79: 544–73.
28. Darvin ME, Lademann J, von Hagen J, et al. Antioxidants 2022;11:1451.
29. Mayne ST, Cartmel B, Scarmo S, et al. Arch Biochem Biophys 2013;539:163–70.
30. Ermakov IV, Ermakova M, Sharifzadeh M, et al. Arch Biochem Biophys 2018;646:46–54.
31. Hammond BR, Wooten BR. J Biomed Opt 2005;10:054002.
32. Takayanagi Y, Obana A, Muto S, et al. Antioxidants 2022;11:14.
33. Bernstein PS, Li B, Vachali PP, et al. Prog Retin Eye Res 2016;50:34–66.
34. Bernstein PS, Zhao D-Y, Sharifzadeh M, et al. Arch Biochem Biophys 2004;430:163–9.
35. Sheftel J, Loechl C, Mokhtar N, et al. Adv Nutr 2018;9:625–36.
36. Lietz G, Furr HC, Gannon BM, et al. Food Nutr Bull 2016;37: S87–S103.
37. Moran NE, Mohn ES, Hason N, et al. Adv Nutr 2018;9:465–92.
38. Reboul E. Nutrients 2019;11:838.
39. Bohn T, Desmarchelier C, El SN, et al. Proc Nutr Soc 2019;78: 68–87.
40. van het Hof KH, West CE, Weststrate JA, et al. J Nutr 2000;130:503–6.
41. Desmarchelier C, Borel P. Trends in Food Science & Technology 2017;69:270–80.
42. Castenmiller JJ, West CE. Annu Rev Nutr 1998;18:19–38.
43. von Lintig J, Moon J, Lee J, et al. Biochim Biophys Acta 2020;1865:158580.
44. Borel P, Lietz G, Goncalves A, et al. J Nutr 2013;143:448–56.
45. Wang XD, Krinsky NI, Marini RP, et al. Am J Physiol 1992;263:G480–6.
46. Rock CL, Swendseid ME, Jacob RA, et al. J Nutr 1992;122: 96–100.
47. Redmond TM, Gentleman S, Duncan T, et al. J Biol Chem 2001;276:6560–5.
48. dela Seña C, Narayanasamy S, Riedl KM, et al. J Biol Chem 2013;288:37094–103.
49. Aydemir G, Kasiri Y, Bartók E-M, et al. Mol Nutr Food Res 2016;60:2413–20.
50. Lobo GP, Isken A, Hoff S, et al. Development 2012;139: 2966–77.
51. Kelly ME, Ramkumar S, Sun W, et al. ACS Chem Biol 2018;13:2121–9.
52. Liu C, Wang XD, Russell RM. J Nutr Biochem 1997;8:652–7.
53. Napoli JL, Race KR. J Biol Chem 1988;263:17372–7.
54. Wang XD, Krinsky NI, Tang GW, et al. Arch Biochem Biophys 1992;293:298–304.
55. Spiegler E, Kim Y-K, Hoyos B, et al. Sci Rep 2018;8:8834.
56. Wang XD, Russell RM, Liu C, et al. J Biol Chem 1996;271: 26490–8.
57. Arballo J, Amengual J, Erdman JW. Antioxidants 2021; 10:342.
58. Lietz G, Oxley A, Boesch-Saadatmandi C, et al. Mol Nutr Food Res 2012;56:241–50.
59. Lietz G, Oxley A, Leung W, et al. J Nutr 2012;142:161S–5S.
60. Ferrucci L, Perry JR, Matteini A, et al. Am J Hum Genet 2009;84:123–33.
61. Moran NE, Thomas-Ahner JM, Fleming JL, et al. J Nutr 2019;149:381–97.
62. Meyers KJ, Mares JA, Igo RP, Jr., et al. Invest Ophthalmol Vis Sci 2014;55:587–99.
63. Borel P. Mol Nutr Food Res 2012;56:228–40.
64. Ford NA, Clinton SK, von Lintig J, et al. J Nutr 2010;140: 2134–8.
65. Tian R, Pitchford WS, Morris CA, et al. Anim Genet 2010;41: 253–9.
66. Wu L, Guo X, Lyu Y, et al. Sci Rep 2017;7:14624.
67. Hessel S, Eichinger A, Isken A, et al. J Biol Chem 2007;282: 33553–61.

68. Ford NA, Elsen AC, Erdman JW, Jr. Nutr Res 2013;33:733–42.
69. Amengual J, Gouranton E, van Helden YG, et al. PLoS One 2011;6:e20644.
70. Smith JW, Ford NA, Thomas-Ahner JM, et al. Am J Phys 2016;311:R1135–R48.
71. Lobo GP, Isken A, Hoff S, et al. Development 2012;139: 2966–77.
72. Gong X, Marisiddaiah R, Zaripheh S, et al. Mol Cancer Res 2016;14:966–75.
73. Lim JY, Liu C, Hu KQ, et al. Arch Biochem Biophys 2018;654: 1–9.
74. Wu L, Guo X, Wang W, et al. Exp Biol Med (Maywood) 2016;241:1879–87.
75. Boulanger A, McLemore P, Copeland NG, et al. FASEB J 2003;17:1304–6.
76. Gong X, Tsai SW, Yan B, et al. BMC Mol Biol 2006;7:7.
77. Luvizotto RA, Nascimento AF, Veeramachaneni S, et al. J Nutr 2010;140:1808–14.
78. Zaripheh S, Nara TY, Nakamura MT, et al. J Nutr 2006;136: 932–8.
79. Mustra Rakic J, Liu C, Veeramachaneni S, et al. J Nutr Biochem 2021;91:108596.
80. Praud C, Al Ahmadieh S, Voldoire E, et al. Exp Cell Res 2017;358:140–6.
81. Bohn T, Bonet ML, Borel P, et al. Nutr Res Rev 2021;1–27.
82. Arunkumar R, Gorusupudi A, Bernstein PS. Biochim Biophys Acta 2020;1865:158617.
83. Mrowicka M, Mrowicki J, Kucharska E, et al. Nutrients 2022;14:827.
84. Lakey-Beitia J, Kumar D. J, Hegde ML, et al. Int J Mol Sci 2019;20:5553.
85. Kabir MT, Rahman MH, Shah M, et al. Biomed Pharmacother 2022;146:112610.
86. Zia-Ul-Haq M, Riaz M, Modhi AO. Carotenoids and Bone Health. In: Zia-Ul-Haq M, Dewanjee S, and Riaz M, eds. Carotenoids: Structure and Function in the Human Body. Springer International Publishing, Cham, 2021;697–713.
87. Javed S, Mahmood S, Arshad M et al. Carotenoids and Cardiovascular Diseases. In: Zia-Ul-Haq M, Dewanjee S, and Riaz M, eds, Carotenoids: Structure and Function in the Human Body. Springer International Publishing, Cham, 2021;649–96.
88. Yao Y, Goh HM, Kim JE. Antioxidants 2021;10:1978.
89. Jiang y-W, Sun Z-H, Tong W-W, et al. Adv Nutr 2021;12: 1723–33.
90. Mounien L, Tourniaire F, Landrier J-F. Nutrients 2019;11:1562.
91. Bonet ML, Ribot J, Galmés S et al. Biochim Biophys Acta 2020;1865:158676.
92. Clugston RD. Biochim Biophys Acta 2020;1865:158597.
93. Baswan SM, Klosner AE, Weir C, et al. Photodermatol Photoimmunol Photomed 2021;37:490–504.
94. Darawsha A, Sharoni Y. Curr Dev Nutr 2022;6:8.
95. Rowles JL, Erdman JW. Biochim Biophys Acta 2020;1865:158613.
96. Lian F, Hu KQ, Russell RM, et al. Int J Cancer 2006;119:2084–9.
97. Matsumoto A, Mizukami H, Mizuno S, et al. Biochem Pharmacol 2007;74:256–64.
98. Fu H, Wu C, Riaz H, et al. Mol Nutr Food Res 2014;58:425–36.
99. Shirakura Y, Takayanagi K, Mukai K, et al. J Nutr Sci Vitaminol 2011;57:426–31.
100. Aydemir G, Kasiri Y, Birta E, et al. Mol Nutr Food Res 2013;57:739–47.
101. Cheng J, Miao B, Hu K-Q, et al. J Nutr Biochem 2018;56:26–34.
102. Edge R, Truscott TG. Antioxidants 2018;7:5.
103. Black HS, Boehm F, Edge R, et al. Antioxidants 2020;9:264.
104. Hajizadeh-Sharafabad F, Zahabi ES, Malekahmadi M, et al. Crit Rev Food Sci Nutr 2021;1–17.
105. Ben-Dor A, Steiner M, Gheber L, et al. Mol Cancer Ther 2005;4:177–86.
106. Tripathi A, Debelius J, Brenner DA, et al. Nat Rev 2018;15: 397–411.
107. Rinninella E, Raoul P, Cintoni M, et al. Microorganisms 2019;7:14.
108. Lyu Y, Wu L, Wang F, et al. Exp Biol Med (Maywood) 2018;243:613–20.
109. Honarbakhsh M, Malta K, Ericsson A, et al. Biochim Biophys Acta 2022;1867:159122.
110. Li CC, Liu C, Fu M, et al. Mol Nutr Food Res 2018;62:e1700738.
111. Xia H, Liu C, Li CC, et al. Cancer Prev Res 2018;11:797–810.
112. Wiese M, Bashmakov Y, Chalyk N, et al. Biomed Res Int 2019;2019:4625279.
113. Wu L, Lyu Y, Srinivasagan R, et al. J Nutr 2020;150:2687–98.
114. Terasaki M, Uehara O, Ogasa S, et al. Carcinogenesis 2020;42:210–9.
115. Djuric Z, Bassis CM, Plegue MA, et al. J Acad Nutr Diet 2018;118:606–16 e3.
116. Herzog A, Siler U, Spitzer V, et al. FASEB J 2005;19:272–4.
117. Wan L, Tan H-L, Thomas-Ahner JM, et al. Cancer Prev Res 2014;7:1228–39.
118. Siler U, Barella L, Spitzer V, et al. FASEB J 2004;18:1019–21.
119. Hirsch K, Atzmon A, Danilenko M, et al. Breast Cancer Res Treat 2007;104:221–30.
120. Kucuk O, Sarkar FH, Sakr W, et al. Cancer Epidemiol Biomarkers Prev 2001;10:861–8.
121. Mucci LA, Tamimi R, Lagiou P, et al. BJU Int 2001;87:814–20.
122. Liu C, Lian F, Smith DE, et al. Cancer Res 2003;63:3138–44.
123. Meshkini F, Ramezani-Jolfaie N, Sargazi S, et al. Phytother Res 2022;36:1633–43.
124. Bertram JS. Cancer chemoprevention by retinoids and carotenoids: proposed role of gap junctional communication. In: Jacobs MM ed. Vitamins and Minerals in the Prevention and Treatment of Cancer, CRC Press, 2018;31–50.
125. Hix L, Vine A, Lockwood S et al. Retinoids and carotenoids as cancer chemopreventive agents: role of upregulated gap junctional communication. In: Packer L, Obermuller-Jevic U, Kraemer K et al., eds. Carotenoids and Retinoids: Molecular Aspect and Health Issues. Champaign, IL: AOCS Press, 2005:182–203.
126. Stahl W, von Laar J, Martin H-D, et al. Arch Biochem Biophys 2000;373:271–4.
127. Peto R, Doll R, Buckley JD, et al. Nature 1981;290:201–8.
128. Burton GW, Ingold KU. Science 1984;224:569–73.
129. Age-Related Eye Disease Study 2 Research Group. JAMA 2013;309:2005–15.
130. Barros MP, Rodrigo MJ, Zacarias L. J Agric Food Chem 2018;66:5733–40.
131. Kim Y, Lian F, Yeum KJ, et al. Int J Cancer 2007;120:1847–54.
132. Wang XD. Carotenoid oxidative/degradative products and their biological activities. In: Krinsky NI, Mayne ST, and Sies H, eds. Carotenoids in Health and Disease. New York, Marcel Dekker, 2004;313–35.
133. Stice CP, Xia H, Wang XD. Chronic Dis Transl Med 2018;4: 211–24.
134. Martini D, Negrini L, Marino M, et al. Nutrients 2022;14:1191.

CHAPTER 32

Carnitine

Lu Qi

CHAPTER OUTLINE

Abbreviations

CAT, carnitine acetyltransferase; **CoA**, coenzyme A; **CPT**, carnitine palmitoyltransferase; **CVD**, cardiovascular disease; **ε**, epsilon; **NAD**, nicotinamide adenine dinucleotide; **TMAO**, trimethylamine-N-oxide

History of Carnitine

Carnitine (β-hydroxy-γ-*N*-trimethylaminobutyric acid), derived from an amino acid, is a conditionally essential nutrient playing a critical role in energy production and fatty acid metabolism. Carnitine exists as dextro (D-) and levo (L-) stereoisomers. The D-isomer is metabolically inert and of little interest.[1] Carnitine was first discovered in muscle extracts, independently by Gulewitsch and Krimberg and by Kutscher, in 1905, and the correct structure was assigned in 1927 by Tomita and Sendju.[2] Between 1948 and 1952, Fraenkel et al., demonstrated the essential nature of this compound for the mealworm, *Tenebrio molitor*, and assigned the term "vitamin B_T" to carnitine.[2] The role of carnitine in fatty acid oxidation was discovered independently by Bremer, and by Fritz and Yue, between 1962 and 1963.[3] The origin of the methyl groups of carnitine was identified by Wolf and Berger and by Bremer in 1961,[3] and the origin of the carbon chain of carnitine from the essential amino acid lysine was first reported by Tanphaichitr et al., in 1971.[4] Clinical syndromes associated with carnitine deficiency were first reported by Engel et al., in 1973[5] and 1975,[6] and primary systemic carnitine deficiency associated specifically with a defect in carnitine transport was identified by Treem et al., in 1988.[7] In 2013, Koeth et al., linked L-carnitine with gut microbiota metabolism and atherosclerosis.[8]

Chemistry and Nomenclature

L-Carnitine is a zwitterionic quaternary amino acid (**Figure 32.1**) with a molecular weight of 161.2 g/mol (inner salt). Only the L isomer is biologically active. L-Carnitine is present in biologic systems in both nonesterified and esterified forms. Short-chain (C_2–C_5) organic acids or medium (C_6–C_{12})- or long (C_{14}–C_{24})-chain fatty acids are transferred to and from coenzyme A (CoA) and the hydroxyl group of carnitine (Figure 32.1). These reversible reactions are catalyzed by a group of enzymes appropriately named carnitine acyltransferases.

Dietary Sources

Carnitine is widely distributed in food from animal sources, but there is limited availability in plants.[9] Among all dietary sources, red meats contain the highest concentrations of carnitine. Fruits,

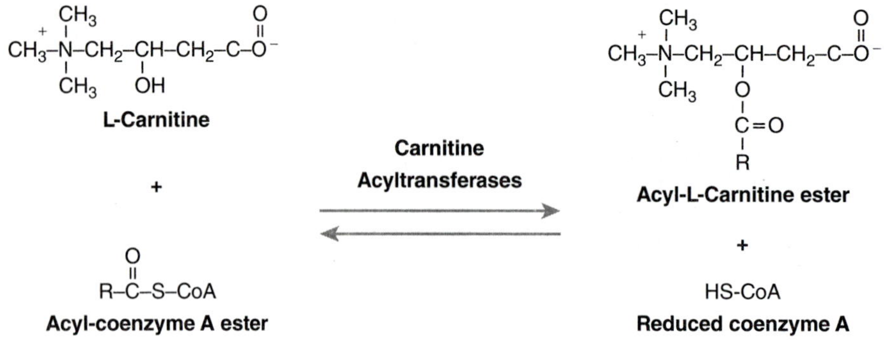

Figure 32.1 Carnitine structure and metabolic interconversions. *CoA*, coenzyme A; *HS-CoA*, coenzyme A.

vegetables, grains, and other plant-derived foods contain relatively little carnitine.[10] In humans, 75% of carnitine is obtained from the diet. A normal omnivorous diet provides approximately 2 to 12 μmol/kg body weight/day of carnitine, whereas a strict vegetarian diet contains approximately 0.1 μmol/kg body weight/day.[10]

Requirements and Recommended Intakes

Carnitine is not a required nutrient for children and adults. No recommended dietary allowance has been established. Population groups that may be most vulnerable to nutritional carnitine deficiency are strict vegetarians and newborn infants. Strict vegetarians (both adults and children) acquire very little carnitine in their diets, and most of the carnitine (>90%) is produced endogenously.[11] They have lower plasma carnitine concentration, compared with omnivores, but no indication of clinically relevant carnitine deficiency has been reported.[12]

Infants, particularly premature infants, are born with relatively low stores of carnitine, and rapid growth places a demand for carnitine accretion. In the past, but not currently, commercial soy protein–based infant formulas did not contain carnitine. Infants consuming these carnitine-free formulas grew at a normal rate and displayed no evidence of clinical carnitine deficiency, although some biochemical parameters relating to lipid metabolism (e.g., plasma free fatty acid concentration, rate of excretion of medium-chain dicarboxylic acids) were different, compared with infants consuming the same formulas but supplemented with carnitine.[13] An expert panel commissioned by the US Food and Drug Administration's Center for Food Safety and Applied Nutrition recommended a minimum carnitine content in infant formulas of 7.5 μmol/100 kcal, and a maximum level of 12.4 μmol/100 kcal, a value similar to the upper limit reported for human milk.[14] These recommendations were made based on reported biochemical differences when infants were fed carnitine-free diets, compared with similar diets with carnitine, and despite the lack of evidence that carnitine is essential for the full-term infant.

Homeostatic Mechanisms

Carnitine homeostasis in humans is maintained by the dynamic interactions of endogenous synthesis, acquisition from dietary sources, maintenance of concentration gradients across cell membranes, and regulation of renal reabsorption and excretion of carnitine.

Absorption and Bioavailability

Absorption of carnitine likely involves a combination of active transport and passive diffusion across the intestinal mucosal barrier. Evidence has accumulated from in vivo and in vitro studies, using several experimental preparations, to demonstrate that active transport of carnitine occurs across the apical brush-border membrane of enterocytes, but not across the basal membrane.[12] A passive component, at least at the serosal surface, is strongly suggested by experimental studies in rats[15,16] and Caco–2 cells[17] that demonstrated relatively rapid entry of carnitine into enterocytes from the luminal medium, but slow appearance in the serosal perfusate or medium.

Approximately 63% to 75% of carnitine is absorbed from the normal omnivorous diet.[18] The remainder is almost entirely degraded by bacteria in the large intestine. Primary organic degradation products of carnitine are trimethylamine (excreted in urine as trimethylamine oxide) and γ-butyrobetaine (excreted primarily in feces). Carnitine is not degraded by enzymes of animal origin.[19]

Biosynthesis and Metabolism

Carnitine is synthesized from the essential amino acids lysine and methionine (**Figure 32.2**).[20] Lysine provides the carbon chain and nitrogen atom, and three methionine molecules (as *S*-adenosyl-L-methionine) provide the methyl groups for one molecule of carnitine.[21] Methylation of the epsilon (ε) amino group of lysine is catalyzed by one or more protein:lysine methyltransferases. Lysine residues destined for carnitine synthesis must be peptide linked; no evidence indicates that free lysine is methylated enzymatically in mammals. ε-N-Trimethyllysine is released for carnitine synthesis through normal mechanisms of protein hydrolysis.

ε-N-trimethyllysine undergoes four sequential enzymatic reactions[21]: hydroxylation at position two of the carbon chain, catalyzed by ε-N-trimethyllysine hydroxylase (EC 1.14.11.8); aldol cleavage between carbons two and three of the carbon chain, catalyzed by serine hydroxymethyltransferase (EC 2.1.2.1); oxidation of the resulting aldehyde by any of several oxidized nicotinamide adenine dinucleotide (NAD1)–requiring aldehyde dehydrogenases (including one with high specificity for γ-N-trimethylaminobutyraldehyde); and a second hydroxylation, catalyzed by γ-butyrobetaine hydroxylase

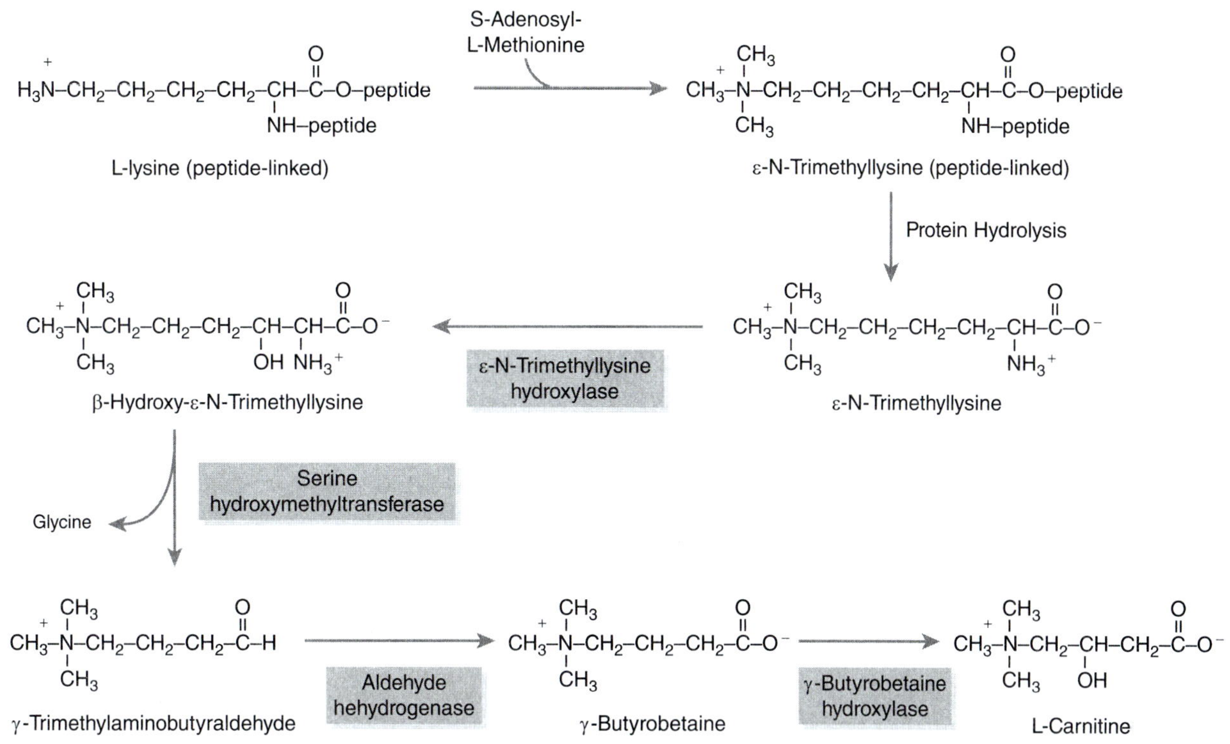

Figure 32.2 Pathway of carnitine biosynthesis in mammals.

Reproduced from Rebouche CJ. Am J Clin Nutr 1991;54(Suppl):1147S–52S.

(EC 1.14.11.1). cDNAs coding for each of these four enzymes have been cloned and sequenced.[22] All enzymes in the pathway except γ-butyrobetaine hydroxylase are ubiquitous in mammalian tissues. The last enzyme in the pathway is not found in cardiac or skeletal muscle.[21] γ-Butyrobetaine hydroxylase activity is highest in liver and testes. In some species, including humans, it is abundant in kidney.

The normal rate of carnitine synthesis in humans is approximately 1.2 μmol/kg body weight/day. This estimate was obtained from normal rates of urinary carnitine excretion by strict vegetarians, who acquire very little carnitine (~0.1 μmol/kg body weight/day) from dietary sources. Direct measurement of the rate of carnitine synthesis technically is not feasible.[10] The problems of isotope dilution (only a small percentage of the body lysine pool is used for carnitine synthesis) and uniform mixing within the body pool are overwhelming and preclude direct measurement of rates of carnitine synthesis from lysine. Direct determination of rates of carnitine synthesis from ε-N-trimethyllysine by isotopic conversion is also impractical because this precursor does not readily cross cell membranes and, thus, pools of intracellular free ε-N-trimethyllysine cannot be labeled uniformly.

The rate of carnitine synthesis in mammals is regulated by the availability of ε-N-trimethyllysine which, in turn, is determined by the extent of peptide-linked lysine methylation and by the rate of protein turnover.[10] ε-N-Trimethyllysine destined for carnitine synthesis is probably derived from the general protein pool, and not from any single or small group of proteins. Provision of excess lysine in the diet may increase carnitine synthesis modestly,[23] but the evidence is indirect, and the mechanism (e.g., increased flux through protein synthesis, methylation and turnover, or stimulation of a putative vestigial capability to methylate free lysine) has not been identified. The rate of carnitine biosynthesis does not appear to be affected by dietary carnitine intake or by changes in renal handling of carnitine.

Transport and Excretion

Carnitine is concentrated in most tissues of the body. In humans, the intracellular concentrations of carnitine in skeletal muscle and liver are approximately 76 and 50 times higher, respectively, than that in extracellular fluid (~50 μmol/L). Approximately 97% of all carnitine in the body is in skeletal muscle. Six carnitine transporters have been identified: three organic cation transporters OCTN1, OCTN2, and OCTN3; a carnitine transporter, CT2; an organic anion transporter, Oat9S; and an amino acid transporter, $ATB^{0,+}$.

OCTN1 is expressed in many tissues (although not in human adult liver),[24] but it has low affinity (translocation rate constant, K_t = 412 μM) and specificity for carnitine.[25] This pH-dependent, 63–kDa transporter may be responsible for secretion of carnitine and its short-chain esters across the renal epithelial brush-border membrane.[26,27] Carnitine is transported into most tissues by a high-affinity (K_t = 3 to 5 μM), sodium (Na^+) gradient–dependent organic cation transporter, OCTN2.[28, 29] This 63–kDa transporter is highly expressed in the heart, placenta, skeletal muscle, kidney, pancreas, testis, and epididymis[24,28] and weakly expressed in brain, lung, and liver.[30] OCTN2 binds carnitine, acetylcarnitine, and propionylcarnitine with comparable affinity.[31] It is likely to be quantitatively the most important carnitine transporter in all tissues except testis.

OCTN3 is expressed primarily and in the testes, and it has higher specificity for carnitine than OCTN1 or OCTN2.[24] Unlike with OCTN2, carnitine transport by OCTN3 is not driven by an inwardly directed Na^+ gradient. A murine Octn3–green fluorescent protein construct was expressed in HepG2 cells, where it localized in peroxisomes.[32] CT2 was found only in the luminal membrane of human epididymis and has high specificity for L-carnitine.[29] This protein may serve to secrete L-carnitine from epididymal epithelium into the lumen and may be relevant in the maturation of human spermatozoa. At the level of amino acid sequence, CT2 is distinct from OCT, OCTN, and OAT family members.[33]

Oat9S is expressed in the kidney and liver of mice.[34] This transporter is located at the apical side of the late portion of proximal tubules and at the sinusoidal side of hepatocytes. Oat9S has a high affinity for L-carnitine (K_t = 2.9 μM) when expressed in *Xenopus* oocytes. Its role in carnitine transport in humans has not been determined. The $ATB^{0,+}$ amino acid transporter cloned from mouse colon also transports carnitine.[35] This transporter is expressed primarily in the lung, mammary gland, and intestine. Carnitine binds with low affinity (K_t = 1 to 2 M) and low specificity, but the capacity of carnitine transport is high because of energization by transmembrane gradients of Na^+ and chloride ion (Cl^-) and by membrane potential.[35] This transporter may have a role in absorption of carnitine. Its tissue distribution and function as a carnitine transporter in humans have not been examined.

Carnitine and acylcarnitine esters are excreted by the kidney. At normal plasma carnitine concentration, the rate of carnitine excretion is very low, but it increases rapidly as the plasma carnitine concentration is elevated by increased oral consumption or by intravenous infusion of carnitine.[12] A threshold effect is observed at nearly normal plasma carnitine concentration, and above which the rate of carnitine excretion soon parallels the increase of filtered load (carnitine appearing in the glomerular filtrate).

Efficient reabsorption has a major role in maintenance of carnitine homeostasis. Approximately 95% of filtered carnitine is reabsorbed in healthy humans. The efficiency of carnitine reabsorption decreases as dietary intake of carnitine increases, independent of glomerular filtration rate and filtered load.[36] This adaptive response serves to maintain the circulating carnitine concentration in the presence of decreased input from dietary intake.

Carnitine transport across the renal brush-border membrane is mediated by OCTN2. γ-Butyrobetaine and short-chain acylcarnitine esters also are efficiently reabsorbed, probably by the same transporter. The mechanism by which intracellular carnitine is transferred across the serosal membrane of the renal epithelial cell is not completely understood. Mouse renal basolateral membranes transport carnitine by a high-affinity, Na^+-dependent process similar to renal brush-border membrane vesicles, but whereas Western blot analysis revealed OCTN2 in the brush-border membrane vesicles, it was not observed in basolateral membrane vesicles.[37] Carnitine, short-chain acylcarnitine esters, and γ-butyrobetaine are secreted from renal epithelial cells into the tubular lumen.[12] The brush-border membrane carrier responsible for this process has not been identified.

Functions in Metabolism

Mitochondrial Long-Chain Fatty Acid Oxidation

Long-chain fatty acids enter mitochondria only as acylcarnitine esters (**Figure 32.3**). Carnitine palmitoyltransferase 1 (EC 2.3.1.21),[38] located in the outer mitochondrial membrane,

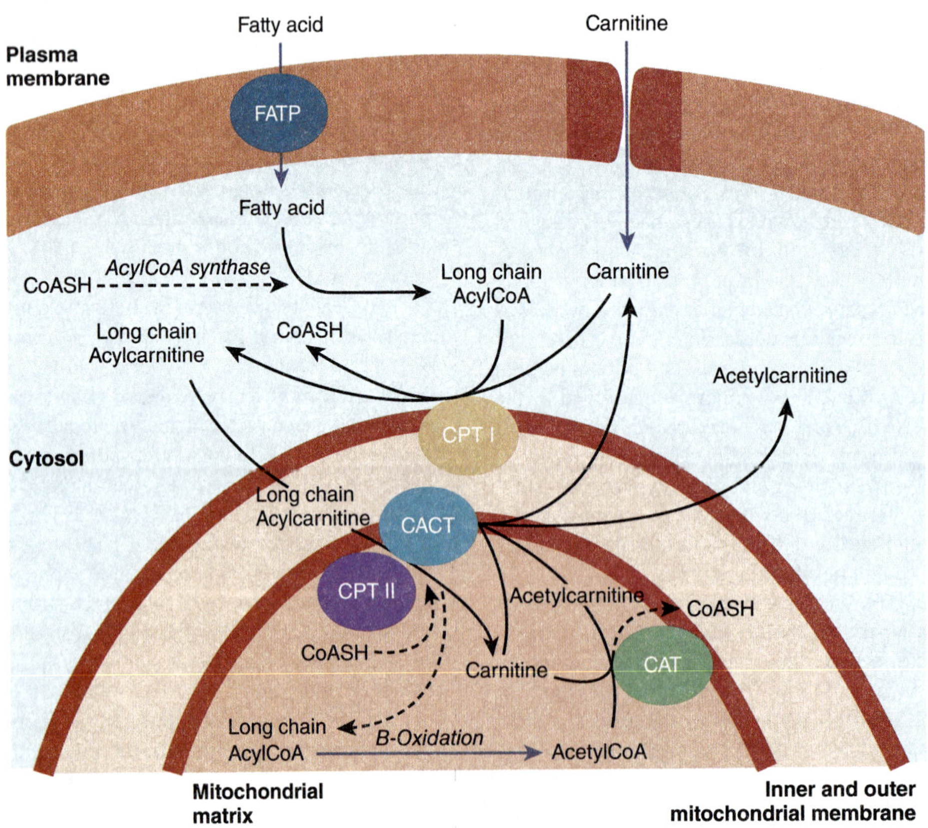

Figure 32.3 Carnitine function: facilitation of mitochondrial long-chain fatty acid oxidation and modulation of intramitochondrial acyl-coenzyme A (CoA)/CoA. *CAT*, carnitine acetyltransferase; *CPT I*, carnitine palmitoyltransferase I; *CPT II*, carnitine palmitoyltransferase II; *HS-CoA*, coenzyme A.

Reproduced from Flanagan JL, Simmons PA, Vehige J, Willcox MD, & Garrett Q. (2010). Role of carnitine in disease. Nutr Metab (Lond), 7, 1-14. https://doi.org/10.1186/1743-7075-7-30.

catalyzes transesterification of cytosolic long-chain fatty acids from CoA to carnitine. The acylcarnitine esters traverse the inner mitochondrial membrane by a carnitine-acylcarnitine translocase,[39,40] and the acyl moieties are transesterified to intramitochondrial CoA by the action of carnitine palmitoyltransferase 2, located on the matrix surface of the inner mitochondrial membrane.[29,41] Thus, carnitine is essential for mitochondrial use of long-chain fatty acids for energy production.

Modulation of Acyl-Coenzyme A/Coenzyme A Ratio

CoA is a required cofactor in many cellular reactions. If nonesterified CoA is not available in a cellular compartment (e.g., cytosol, mitochondria, peroxisomes) because it is completely esterified, the flux through pathways that require this cofactor will diminish. Carnitine is a reservoir for excess acyl residues, generated, for example, by high rates of β-oxidation in mitochondria, in which the acyl residue is transesterified from CoA to carnitine, thus freeing CoA to participate in other cellular reactions (see Figure 32.3). The acylcarnitine ester formed in this process may remain in the organelle or cell of origin, for use when needed, or it may be exported out of the cell for use by other cells or tissues, or for excretion. Unlike CoA and its esters, carnitine and its esters readily exchange across most membranes, facilitated by specific carriers and transporters.

This function has important implications for cellular energy metabolism. For example, carnitine facilitates oxidation of glucose in working hearts by relieving inhibition of pyruvate dehydrogenase by fatty acids.[42] The mechanism involves removal of acetyl groups generated from fatty acid β-oxidation by transesterification from acetyl-CoA to carnitine (catalyzed by carnitine acetyltransferase; EC 2.3.1.7), thus freeing CoA to participate in the pyruvate dehydrogenase reaction sequence. Carnitine may also increase the rate of glucose production and oxidation secondarily by facilitating the use of long-chain acyl-CoA in the hypothalamus. Inhibition of carnitine palmitoyltransferase I and the consequent increase in long-chain acyl-CoA concentration in the hypothalamus have been shown to promote anorexia and decreased hepatic glucose production.[43]

Other Functions in Cellular Metabolism

Carnitine and extra-mitochondrial carnitine palmitoyltransferase are important in long-chain fatty acid use for phospholipid remodeling and biosynthesis. Carnitine acts as a reservoir for long-chain fatty acids destined for incorporation into phospholipids as, for example, in erythrocytes during repair after oxidative insult,[44] as well as in lung alveolar cells in synthesis of dipalmitoylphosphatidylcholine, the major component of surfactant.[45]

Carnitine facilitates removal of chain-shortened fatty acid oxidation products from peroxisomes.[46] Peroxisomes oxidize very-long-chain fatty acids not metabolized in mitochondria. Chain-shortened products, primarily long-chain and medium-chain fatty acids esterified to CoA, are transesterified to carnitine (catalyzed by carnitine octanoyltransferase; EC 2.3.1.137) and are subsequently oxidized in mitochondria.

Antioxidant and radical scavenging activity of carnitine may facilitate maintenance of mitochondrial integrity and function by blocking formation or mitigating toxic effects of intracellular reactive oxygen species. Direct antioxidant effects of carnitine have been demonstrated in vitro.[47,48] Model reactions revealed concentration-dependent reducing power and hydrogen peroxide and radical scavenging activities of carnitine similar in magnitude to α-tocopherol and Trolox (a water-soluble analog of α-tocopherol). Carnitine protected erythrocytes from lysis by hypochlorite and 2,2′-azobis (2-amidinopooprane) dihydrochloride. Ferrous ion chelating ability of carnitine was also demonstrated. The chemical mechanisms for these activities are not known.

Assessment of Status

Carnitine status is most often reported as a function of circulating carnitine concentration and as the ratio of esterified to nonesterified carnitine. In general, a plasma-free carnitine concentration of 20 μmol/L or less, or a total carnitine concentration of 30 μmol/L or less, is considered abnormally low. However, these values demarcate only the lower range of normal plasma carnitine concentration. They do not reflect points at which functional carnitine deficiency is observed. A ratio of esterified to free carnitine of 0.4 or greater in plasma or serum (but not urine) is considered indicative of abnormal carnitine metabolism. This ratio is elevated primarily when mitochondrial energy metabolism is impaired, resulting in increased load of short-chain organic acids esterified to CoA, which are transesterified to carnitine for export from tissues into the circulation. Thus, in genetic disorders of fatty acid and organic acid oxidation (see the next section), the ratio of esterified to free carnitine in the circulation is often elevated. This elevated ratio is associated with carnitine depletion, resulting from either hyperexcretion of acylcarnitine esters or a decreased ability of the kidneys to reabsorb carnitine and its esters. Rates of carnitine excretion in urine do not provide a useful measure of carnitine status, because these rates vary considerably with dietary carnitine intake and other physiologic parameters. No validated tests or measures of functional carnitine deficiency are available to assess carnitine status in humans.

Causes and Effects of Deficiency and Depletion

Genetic and Acquired Diseases

Primary systemic carnitine deficiency of genetic origin results from mutations in the carnitine transporter OCTN2.[49] This autosomal recessive disorder is characterized by progressive cardiomyopathy, skeletal myopathy, hypoglycemia, and hyperammonemia.[50] The disease usually manifests within the first 5 years of life and is fatal if untreated. No primary carnitine deficiency resulting from a defect in carnitine biosynthesis has been identified.

Carnitine deficiency or depletion occurs secondary to many genetic and acquired disorders and conditions.[51] At least two basic mechanisms are responsible for these effects on carnitine status. In some disorders, the efficiency of carnitine reabsorption is impaired (e.g., medium-chain acyl-CoA dehydrogenase deficiency). In others, abnormal amounts of short-chain organic acids are produced and are removed from the body by urinary excretion as acylcarnitine esters (e.g., propionylcarnitine in propionyl-CoA carboxylase deficiency). In these disorders, the rate of excretion of carnitine as short-chain acylcarnitine esters exceeds the combined rates of endogenous synthesis and dietary carnitine intake, thus leading to a state of carnitine depletion.

Newborn screening for disorders of mitochondrial fatty acid oxidation disorders and other inborn errors of metabolism associated with primary and secondary carnitine deficiency is facilitated by tandem mass spectral analysis of specific acylcarnitine ester species in dried blood spots.[52] Quantitative analysis of plasma and urinary acylcarnitine esters, using tandem mass spectrometry, has also proven useful clinically and experimentally in the study of secondary carnitine deficiency diseases and conditions.

Nutrient and Drug Interactions

Valproic acid and pivalic acid–containing prodrugs negatively affect carnitine status in humans.[53,54] Valproic acid and its metabolites adversely affect mitochondrial β-oxidation and thus, contribute to hepatotoxicity and hyperammonemic encephalopathy. Valproic acid administration lowers circulating carnitine concentration in some patients. Although several possible mechanisms for valproic acid–induced carnitine depletion have been identified, the primary causes remain unclear. Supplementation with carnitine is recommended for young children treated with valproic acid.[53] Pivalic acid is conjugated to some antibiotics and antiretroviral drugs to improve their rates of absorption. In the intestinal mucosa, pivalic acid is cleaved by nonspecific esterases. Pivalic acid (as pivaloyl-CoA) is conjugated to carnitine, and is quantitatively excreted in urine as pivaloylcarnitine.[54] Prolonged treatment with these agents leads to depletion of the circulating carnitine pool and, presumably, to depletion of tissue carnitine pools as well.

Short-term treatment with the chemotherapeutic agents cisplatin[55] and ifosfamide[56] increased the rate of total carnitine excretion by approximately 10–fold and 30–fold, respectively, during the period of treatment. That plasma carnitine concentration that was modestly elevated during cisplatin treatment suggests a loss of carnitine from tissues. Elevated plasma carnitine concentration, as well as a diminished capacity to reabsorb carnitine, contributed to the increased rate of carnitine excretion. Conversely, metabolism of ifosfamide produces chloroacetaldehyde which, after oxidation, is esterified to carnitine and excreted in urine. Excretion of both nonesterified carnitine and acylcarnitine esters was elevated with ifosfamide treatment, a finding suggesting two mechanisms for increased urinary loss of carnitine: excretion or secretion of the abnormal acylcarnitine ester chloroacetylcarnitine (or both) and a diminished capacity to reabsorb carnitine.

Therapeutic Use and Supplements

L-Carnitine is used therapeutically (as replacement therapy) in patients with primary systemic carnitine deficiency, which results from mutations in the gene coding for the carnitine transporter OCTN2, and in patients with secondary carnitine deficiency, which is caused by genetic defects in organic acid metabolism.[51] L-Carnitine is also used in patients with end-stage renal disease who are undergoing long-term hemodialysis. These patients often exhibit an abnormally high ratio of esterified carnitine to free carnitine that is corrected by administration of L-carnitine. Studies have shown that carnitine administration improves lipid metabolism, antioxidant status, and anemia requiring erythropoietin, and it may reduce the incidence of intradialytic muscle cramps, hypotension, and cardiomyopathy.[57] The use of carnitine in these patients remains controversial, however.

L-Carnitine is included in parenteral and enteral nutrition solutions for hospitalized premature infants. L-Carnitine has been described as a "conditionally essential nutrient" for this population.[55] Studies show that it improves lipid metabolism and weight gain,[56] although available evidence does not provide conclusive proof of a requirement for carnitine.

L-Carnitine and acetyl-L-carnitine have been investigated as dietary supplements and marketed for various potential health benefits. These benefits include, but are not limited to, weight maintenance or reduction, performance enhancement and recovery from physical exertion, enhancement of male fertility and improvement in reproductive health, and maintenance of mental and physical function and reversal of decline with aging. These topics are reviewed elsewhere.[58] L-Carnitine and its esters, acetyl-L-carnitine and propionyl-L-carnitine, may be beneficial as dietary supplements in other medical conditions. These conditions include hyperthyroidism, human immunodeficiency virus infection and antiretroviral therapy, cancer chemotherapy, chronic fatigue syndrome, type 2 diabetes, chronic diabetic neuropathy, peripheral vascular disease, angina pectoris, and congestive heart failure.[58,59] Typical recommendations for the use of L-carnitine as a supplement are 0.5 to 4.0 g/day (oral administration) for children and adults. Toxicity has not been reported with use of these amounts of supplement, although occurrences of body odor ("fish odor syndrome") or diarrhea have been reported.

Relations with Chronic Diseases

Early studies, mostly from experimental settings, have shown beneficial effects of carnitine supplementation on chronic diseases, especially cardiometabolic disorders.[60] For instance, data have shown that carnitine supplementation may improve glucose tolerance and increase total energy expenditure in obese rodents.[61] In humans, treatment by acetyl-L-carnitine was found to improve glucose disposal in patients with type 2 diabetes,[62] and insulin resistance, polycystic ovaries, menstrual irregularities, and hypoadiponectinemia in women with polycystic ovary syndrome.[63] In 2013, a study reported that metabolism by intestinal microbiota of dietary L-carnitine, abundant in red meat, accelerated atherosclerosis in mice. Plasma L-carnitine concentration in adults undergoing cardiac evaluation predicted increased risks for both prevalent cardiovascular disease (CVD) and incident major adverse cardiac events (myocardial infarction, stroke or death). The associations between carnitine and cardiovascular events were dependent on trimethylamine-N-oxide (TMAO) concentration.[8] Further meta-analysis confirmed the adverse association between circulating L-carnitine and major adverse cardiovascular events/death.[64] However, circulating carnitine was not associated with type 2 diabetes[65] or colorectal cancer.[66] Although further studies, especially large-scale, prospective observational studies and clinical trials are needed to validate the findings and confirm the causality of L-carnitine in cardiovascular and metabolic disorders,[67] growing data have raised concerns for its use as a potential therapy and its widespread availability as a nutritional supplement.

Acknowledgments

The author acknowledges Charles J. Rebouche, author of the previous version of this chapter, for his contributions.

References

1. Liedtke AJ, Nellis SH, Whitesell LF, et al. Am J Physiol 1982;243:H691–7.
2. Fraenkel G, Friedman S. Vitam Horm 1957;15:73–118.
3. Bremer J. Physiol Rev 1983;63:1420–80.
4. Tanphaichitr V, Horne DW, Broquist HP. J Biol Chem 1971;246:6364–6.
5. Engel AG, Angelini C. Science 1973;179:899–902.
6. Karpati G, Carpenter S, Engel AG, et al. Neurology 1975;25:16–24.
7. Treem WR, Stanley CA, Finegold DN, et al. N Engl J Med 1988;319:1331–6.
8. Koeth RA, Wang Z, Levison BS, et al. Nat Med 2013;19:576–85.
9. Kendler BS. Prev Med 1986;15:373–90.
10. Rebouche CJ. FASEB J 1992;6:3379–86.
11. Dahash BA, Sankararaman S. Carnitine Deficiency. In StatPearls, Treasure Island, FL, 2022.
12. Rebouche CJ, Seim H. Annu Rev Nutr 1998;18:39–61.
13. Novak M. Biol Neonate 1990;58:89–92.
14. Klein CJ. J Nutr 2002;132:1395S–577S.
15. Gross CJ, Henderson LM. Biochim Biophys Acta 1984;772:209–19.
16. Gudjonsson H, Li BU, Shug AL, et al. Gastroenterology 1985;88:1880–7.
17. Rebouche CJ. J Nutr. Biochem 1998;9:228–35.
18. Rebouche CJ, Chenard CA. J Nutr 1991;121:539–46.
19. Rebouche CJ, Mack DL, Edmonson PF. Biochemistry 1984;23:6422–6.
20. Flanagan JL, Simmons PA, Vehige J, et al. Nutr Metab (Lond) 2010;7:30.
21. Rebouche CJ. Am J Clin Nutr 1991;54:1147S–52S.
22. Vaz FM, and Wanders RJ. (2002) Biochem J 361, 417–29.
23. Rebouche CJ, Bosch EP, Chenard CA, et al. J Nutr 1989;119:1907–13.
24. Tamai I, Yabuuchi H, Nezu J, et al. FEBS Lett 1997;419:107–11.
25. Xuan W, Lamhonwah AM, Librach C, et al. Biochem Biophys Res Commun 2003;306:121–8.
26. Yabuuchi H, Tamai I, Nezu J, et al. J Pharmacol Exp Ther 1999;289:768–73.
27. Tein I. J Inherit Metab Dis 2003;26:147–69.
28. Tamai I, Ohashi R, Nezu J, et al. J Biol Chem 1998;273:20378–82.
29. Ramsay RR, Gandour RD, van der Leij FR. Biochim Biophys Acta 2001;1546:21–43.
30. Wu X, Prasad PD, Leibach FH, et al. Biochem Biophys Res Commun 1998;246:589–95.
31. Wu X, Huang W, Prasad PD, et al. J Pharmacol Exp Ther 1999;290:1482–92.
32. Lamhonwah AM, Skaug J, Scherer SW, et al. Biochem Biophys Res Commun 2003;301:98–101.
33. Enomoto A, Wempe MF, Tsuchida H, et al. J Biol Chem 2002;277:36262–71.
34. Tsuchida H, Anzai N, Shin HJ, et al. Cell Physiol Biochem 2010;25:511–22.
35. Nakanishi T, Hatanaka T, Huang W, et al. J Physiol 2001;532:297–304.
36. Rebouche CJ, Lombard KA, Chenard CA. Am J Clin Nutr 1993;58:660–5.
37. Lahjouji K, Malo C, Mitchell GA, et al. Biochim Biophys Acta 2002;1558:82–93.
38. Zammit VA. IUBMB Life 2008;60:347–54.
39. Iacobazzi V, Naglieri MA, Stanley CA, et al. Biochem Biophys Res Commun 1998;252:770–4.
40. Pande SV, Murthy MS. Biochim Biophys Acta 1994;1226:269–76.
41. van der Leij FR, Huijkman NC, Boomsma C, et al. Mol Genet Metab 2000;71:139–53.
42. Broderick TL, Quinney HA, Lopaschuk GD. J Biol Chem 1992;267:3758–63.
43. Obici S, Feng Z, Arduini A, et al. Nat Med 2003;9:756–61.
44. Ravikumar S, Prabhu S, Vani R. Transfus Med 2020;30:215–25.
45. Arduini A, Zibellini G, Ferrari L, et al. Mol Cell Biochem 2001;218:81–6.
46. Ramsay RR. Am J Med Sci 1999;318:28–35.
47. Gulcin I. Life Sci 2006;78:803–11.
48. Solarska K, Lewinska A, Karowicz-Bilinska A, et al. Cell Mol Biol Lett 2010;15:90–7.
49. Li FY, El-Hattab AW, Bawle EV, et al. Hum Mutat 2010;31:E1632–51.
50. Nezu J, Tamai I, Oku A, et al. Nat Genet 1999;21:91–4.
51. Di Donato S. Disorders of Lipid Metabolism, McGraw-Hill, New York, 2004.
52. Chace DH, Kalas TA, Naylor EW. Annu Rev Genomics Hum Genet 2002;17–45.
53. Lheureux PE, Penaloza A, Zahir S, et al. Crit Care 2005;9:431–40.
54. Brass EP. Pharmacol Rev 2002;54:589–98.
55. Heuberger W, Berardi S, Jacky E, et al. Eur J Clin Pharmacol 1998;54:503–8.
56. Marthaler NP, Visarius T, Kupfer A, et al. Cancer Chemother Pharmacol 1999;44:170–2.
57. Stanley WC, Lopaschuk GD, Hall JL, et al. Cardiovasc Res 1997;33:243–57.
58. Rebouche C. L-Carnitine, Acetyl-L-Carnitine, and Propionyl-L-Carnitine, New York: Informa Healthcare, 2010.
59. Rebouche C. Carnitine, Washington, DC: International Life Sciences Institute, 2006.
60. Arsenian MA. Prog Cardiovasc Dis 1997;40:265–86.
61. Cave MC, Hurt RT, Frazier TH, et al. Nutr Clin Pract 2008;23:16–34.
62. Giancaterini A, De Gaetano A, Mingrone G, et al. Metabolism 2000;49:704–8.
63. Tauqir S, Israr M, Rauf B, et al. Adv Ther 2021;38:3842–56.
64. Heianza Y, Ma W, Manson JE, et al. J Am Heart Assoc 2017;6.
65. Lemaitre RN, Jensen PN, Wang Z, et al. JAMA Netw Open 2021;4:e2122844.
66. Guertin KA, Li XS, Graubard BI, et al. Cancer Epidemiol Biomarkers Prev 2017;26:945–52.
67. Johri AM, Hetu MF, Heyland DK, et al. Nutr Metab (Lond) 2022;19:26.

CHAPTER 33

Cysteine, Taurine, and Homocysteine

Sindhu Kashyap and Anura V Kurpad

CHAPTER OUTLINE

List of Compounds

Cysteine, 121.2 g/mol; **glutathione** (reduced form), 307.3 g/mol; **homocysteine**, 135.2 g/mol; **methionine**, 149.2 g/mol; **taurine**, 125.1 g/mol.

Abbreviations

Cys, cysteine (any form), with thiol and disulfide forms indicated as CySH, CySSCy, and CySSR; **cyst(e)ine**, Cys or cystine; **EAR**, estimated average requirement; **Glu**, glutamate; **Gly**, glycine; **GSH**, glutathione; **Hcy**, homocysteine (any form), with thiol and disulfide forms indicated as HcySH, HcySSHcy, and HcySSR; **homocyst(e)ine**, Hcy or homocystine; H_2S, hydrogen sulfide; K_m, Michaelis-Menten constant; **Met**, methionine; **NHANES III**, Third National Health and Nutrition Examination Survey; **RDA**, recommended dietary allowance; **SAA**, sulfur amino acid; **SAH**, *S*-adenosylhomocysteine; **SAM**, *S*-adenosylmethionine; **TauT**, taurine transporter; **tCys**, sum of all forms of plasma Cys including those present as thiol, half-disulfide, mixed disulfide, and protein-bound disulfide; **tHcy**, sum of all forms of Hcy; **THF**, tetrahydrofolate; **TPN**, total parenteral nutrition; **UL**, tolerable upper intake level

Historic Introduction

Cysteine (Cys), homocysteine (Hcy), and taurine are sulfur-containing amino acids, which are interrelated in their origin but have different biologic roles. Cystine, the oxidation product of two molecules of cysteine, was first isolated from urinary calculi by Wollaston in 1809.[1] The requirement of Cys for protein synthesis and growth was demonstrated by rapid growth restoration in rats fed a low casein diet (having low Cys) when supplemented with Cys.[2] It was later shown that Cys could be replaced by methionine (Met) to supplement a low Cys diet.[3] However, while Met could completely replace Cys and induce growth in rats, Cys alone without Met retarded growth.[4] Cys was, therefore, shown to be nonessential when adequate dietary Met was available.

Taurine, a product of cysteine catabolism, was first isolated from the bile (taurocholate) of the ox, *Box taurus*, in 1827.[5] Considered nonessential, it gained attention since 1975 when kittens fed a casein diet (without taurine) developed retinal degeneration.[6] This was followed by observation that preterm infants fed purified formulas had lower plasma and urine taurine concentrations than infants fed pooled human milk,[7] and the abnormal electroretinograms of children on parenteral nutrition with low plasma taurine became normal when they were supplemented with taurine.[8] These studies led to the addition of taurine to human infant formulas from the mid 1980s.

Hcy, a precursor of Cys, was first synthesized by du Vigneaud in 1932, during his research on intermediate sulfur metabolism.[9] Several subsequent studies showed the transfer of sulfur from Met to Cys through Hcy, by what is now known as the trans-sulfuration pathway.[9] It was later shown that choline or betaine acted as methyl donors to Hcy to form Met, by transmethylation.[9] Homocystinuria, an inborn error of metabolism, was identified in children presenting with mental retardation, vision impairment, and thrombovascular events. Several other conditions, such as cardiovascular disease and neural tube defects have also been associated with elevated Hcy concentration.[10-13]

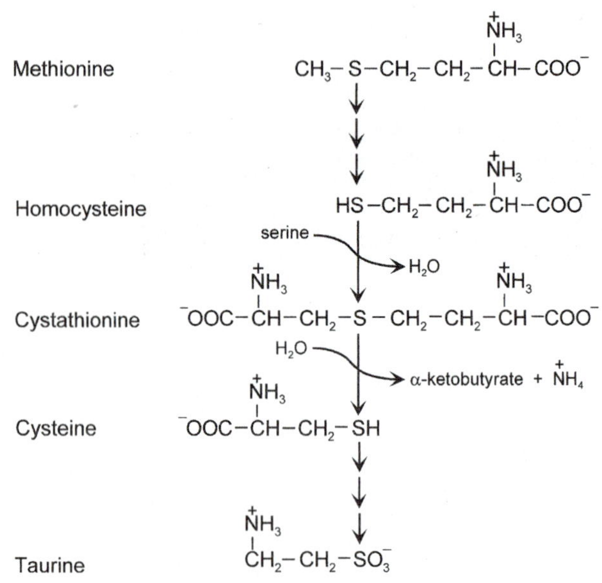

Figure 33.1 Structures and metabolic relations of sulfur amino acids.

Chemistry and Nomenclature

Cys and Hcy are α-amino acids with one asymmetric carbon atom and a free sulfhydryl group. Cys has the same backbone structure as serine (Ser) except for the hydroxyl group, which is replaced by sulfhydryl group. Hcy has one carbon less than Met. The L-isomers of Cys and Hcy are the biologically active forms. Taurine (2-aminoethane sulfonate) is a β-amino acid derived from Cys after removal of the carboxyl group and oxidation of the sulfhydryl group to sulfonic acid group. The carboxyl (pK_a ~1.7), sulfonic (pK_a ~1.5), sulfhydryl (pK_a ~8.3), and amino (pK_a ~9 to 11) groups all undergo ionization; the zwitterionic forms shown in **Figure 33.1** are the dominant species at physiologic pH. The structures of Cys, Hcy, and taurine and their relations with precursor amino acids (Met and serine (Ser)) are also shown in Figure 33.1.

Dietary Considerations, Typical Intakes, and Recommended Intakes

Cysteine

Cys can be synthesized in the body from Met. Since the intakes and requirements of both these amino acids (AA) are related, they are expressed in terms of total sulfur amino acids (SAA). Both Cys and Met are natural constituents of dietary protein, but the ratio of Cys to Met is higher in animal protein (**Table 33.1**), and; therefore, the intake of Cys and Met varies in different populations based on their habitual diet. In a Western diet, typical SAA intakes range from 2 to 4 g/day.[14] Based on the U.S. Third National Health and Nutrition Examination Survey (NHANES III; 1988 to 1994), the Cys intakes of men and women from 31 to 50 years old were 1.3 ± 0.02 and 0.89 ± 0.01 g/day (mean ± standard error (SE)), or 10.7 and 7.4 mmol/day. Met intakes for the same age group were 2.3 ± 0.04 and 1.6 ± 0.2 g/day (mean ± SE) or 15.4 and 10.7 mmol/day. The mean total SAA intake was 3.6 and 2.5 g/day or 26.1 and 18.1 mmol/day for men and women, respectively.

The current Estimated Average Requirement (EAR) and Recommended Dietary Allowance (RDA) of SAA for healthy adults are 15 mg/kg/day and 19 mg/kg/day.[14] The EAR matches with the requirements proposed by the 2007 FAO/WHO/UNU expert consultation.[15] The Cys and Met requirements are 4.1 mg/kg/day and 10.4 mg/kg/day, respectively,[15] and the normal Western diet has a similar Cys-to-Met ratio. In absence of Cys, the Met requirement has been found to be 15 mg/kg/day.[16] The EAR and RDA of SAA for different age groups are provided in **Table 33.2**.

The current EAR for protein intake in adults is 0.66 g/kg/day, and the RDA is 0.8/kg/day. Thus, a desirable amino acid pattern has at least 24 mg SAA/g of protein (16 mg/0.66 g = 24 mg/g). Mixtures of proteins consumed in the United States contain a higher proportion of SAAs, approximately 35 mg SAA/g of protein. Therefore, the RDA for SAAs (1.3 g/day for a 70-kg adult) is usually met in diets commonly consumed in the United States. Even the lowest SAA intakes reported in NHANES III (1st percentile; 1.87 g for men and 1.4 g for women in the 31- to 50-year age bracket) exceed the current RDA.[14] Although no tolerable upper limit has been set for Cys and Met, a no-observed-adverse-effect-level (NOAEL) and lowest-observed-adverse-effect-level (LOAEL) of supplemented Met, based on plasma Hcy concentration in healthy older adults, was shown to be 46.3 mg/kg/day and 91 mg/kg/day, respectively.[17]

An important consideration when calculating the sufficiency of SAA intake is their bioavailability or digestibility (which includes digestion and absorption) from dietary protein. The bioavailability/digestibility of SAA from animal protein (86%–99%) is greater than that from plant protein (52%–82%).[18] In addition, processing of dietary protein (heat or pressure) can lead to SAA oxidation, rendering them nutritionally unavailable.[19]

Despite high SAA in diets, individuals with inadequate protein intake or consuming SAA-deficient protein would not meet

Table 33.1 Methionine and Cysteine Content of Selected Foods

Food	Amount		Pattern	
	Methionine	Cysteine	Methionine	Cysteine
	(mg/100 g Edible Portion)		(mg/g Protein)	
Animal Foods				
Cheese, cheddar	652	125	26	5
Milk, whole	83	30	25	9
Egg, whole, chicken	392	289	32	24
Chicken, flesh only, cooked roasted	800	370	28	13
Beef, round, separable lean only	557	224	26	11
Plant Foods				
Wheat flour, whole meal	186	278	14	21
Corn grits, regular, dry	196	237	22	22
Oats, regular, dry	266	398	17	25
Peanut butter	292	365	10	13
Soybeans, green, cooked	150	113	12	9
Brown rice, dry	142	152	19	21
Lentil, raw	210	322	9	13

their requirement. An analysis of diets of long-term vegans living in California indicated an average protein intake of 64 g/day and an SAA intake of 1.04 g (7.6 mmol)/day[20]; this is equivalent to an intake of approximately 15 mg/kg/day of SAAs and an amino acid pattern of 16 mg SAA per gram of protein. This level of intake would meet the EAR, but not the RDA, for SAA. When corrected for bioavailability or digestibility, this would reduce to 10.5 mg/kg/day. Therefore, adults on a vegan diet, or those with higher-than-average requirements, would be at risk of inadequate SAA intake.

Table 33.2 Estimated Average Requirement and Recommended Dietary Allowance of Sulfur Amino Acids[14]

Age Group	EAR	RDA
Years	mg/kg/day	
0.7–1*	30	43
1–3*	22	28
4–8*	18	22
9–13 y, Boys	18	22
9–13 y, Girls	17	21
14–18 y, Boys	17	21
14–18 y, Girls	16	19
>19*	15	19
Pregnancy	20	25
Lactation	21	26

*Values are for both males and females.

Data from Food and Nutrition Board, Institute of Medicine. Dietary Reference Intakes for Energy, Carbohydrate, Fiber, Fat, Fatty Acids, Cholesterol, Protein, and Amino Acids. Washington, DC: National Academy Press, 2005.

Homocysteine

Very small quantities of Hcy are found in food. Analysis of Hcy in different foods with high or low levels of Met has shown that concentrations present are 150 to 10,000 times lower than the respective Met concentrations.[21,22] Hcy intake from foods, considering normal daily consumption, would be a maximum of 3 μmol, which will not significantly increase plasma Hcy concentration.

Taurine

Taurine is present in most animal foods, including meat and dairy products and is particularly high in seafood and shellfish. It is absent in most plant species but present in macroalgae such as seaweed. Food taurine content has not been widely determined, but data from several reports[23-26] are summarized in **Table 33.3**.

Like SAA, the intake of taurine varies with the type of diet being consumed. The analyzed taurine intake from a typical American omnivorous diet varied from 1,000 to 1,400 μmol/day (125.1 to 175.2 mg/day) whereas that of a lactovegetarian diet was 136 μmol/day; in the vegan diet, taurine was not detectable.[23] Taurine has been supplemented in protein-based beverages since the 1990s with 1,000 mg (8,000 μmol) per 240- or 250-mL can. These beverages, when consumed, can increase taurine intake by eight times or more. Adult vegetarians have lower

Table 33.3 Taurine Content of Selected Foods[23-26]

Food	Taurine Content
Animal Foods	
Poultry	89–2,245 μmol/100 g wet weight
Beef and Pork	307–489 μmol/100 g wet weight
Processed meats	251–981 μmol/100 g wet weight
Seafood	84–6,614 μmol/100 g wet weight
Cow's milk	18–20 μmol/100 mL
Yogurt, ice cream	15–62 μmol/100 mL
Cheese	Not detected
Plant Foods	
Most fruits, vegetables, seeds, cereals, grains, beans, peanuts	Not detected
Soybeans, chickpeas, black beans, pumpkin seeds, some nuts[a]	≤1–4 μmol/100 g wet weight
Seaweeds (marine algae)	1.5–100 μmol/100 g wet weight

[a]Low reported values should be regarded as upper limits because contamination of food or methodologic interference by compounds that co-elute with taurine could account for the observed low concentration of taurine.

Data from Laidlaw SA, Grosvenor M, Kopple JD. J Parenter Enteral Nutr 1990;14:183–8; Pasantes-Morales H, Quesada O, Alcocer L et al. Nutr Rep Int 1989;40:793–801; Roe DA, Weston MO. Nature 1965;203:287–8.

plasma taurine concentration and lower urinary taurine excretion compared with omnivores.

Infants normally receive taurine from their mother's milk. The taurine content of milk from lactating women was reported to be 52 ± 9 (SE) mg/day (413 ± 71 μmol/L) for early milk (1 to 7 days) and 42 ± 3.5 mg/day (337 ± 28 μmol/L) for later milk (>7 days).[27,28] The mean taurine content of milk of lactating vegan women is lower than that of lactating omnivores, but values between the two groups overlap considerably.[28] Children born to and nursed by vegan mothers appear to have normal growth and development.[28]

Absorption, Transport, Metabolism, and Excretion

Intestinal Absorption

Dietary Cys is absorbed as CySH, CySSCy, and Cys-containing peptides by a variety of L-amino acid and peptide transport systems. In intestinal mucosal cells, transport of Cys (CySH) is accomplished by the sodium (Na^+)–dependent neutral amino acid transporter system B (SLC6A19) in the apical membrane and by the Na^+-independent system ASC (SLC7A10+SLC3A2) in the basolateral membrane. Cystine (CySSCy) is transported by system $B^{0,+}$ (SLC7A9+SLC3A1) and x_c^- (SLC7A11+SLC3A2), both Na^+-independent systems that are present in the apical membranes of the intestinal mucosa.[29,30] System x_c^- is upregulated in response to oxidative stress or amino acid deficiency.[31]

Taurine absorption is facilitated by two apical membrane transporters: the β-amino acid or taurine transporter (TauT; SLC6A6), which is a Na^+ and chloride (Cl^-)–dependent carrier, with high affinity and low capacity that serves taurine, β-alanine, and γ-aminobutyric acid; and the hydrogen ion (H^+)–coupled transporter PAT1 (SLC36A1), which has high capacity but low affinity.[32] In the basolateral membrane, taurine absorption may also be mediated by TauT.[33] The expression of the TauT in the intestine does not respond to dietary SAA or taurine.[34]

Blood Transport and Intracellular and Extracellular Forms

After absorption, enterocytes use dietary Cys for protein and glutathione (GSH) synthesis.[35] The liver removes a substantial proportion of Cys from the portal circulation and uses it for synthesis of protein and GSH or for catabolism to taurine and sulfate. Cys then enters the circulating plasma to be removed by tissues. Plasma membranes of tissue cells have various transporters that transport Cys from the plasma. The disulfide forms of Cys (protein bound Cys, PSSCy and cystine, CySSCy) dominate in the more oxidized extracellular environment. Most of the intracellular Cys exists in peptide (GSH) or polypeptide/protein forms. For free Cys, the thiol form (CySH) dominates intracellularly.

Plasma Hcy concentration is normally low because small amounts are normally released from tissues into the plasma. Hcy may share the same transporters as Cys and has been found to be transported through the Cys transport system in vascular endothelium.[36] Intracellularly, low Hcy concentrations are present in free (HcySH) and protein-bound (PSSHcy) forms. Extracellularly, Hcy is present predominantly as mixed disulfides of Hcy with protein (PSSHcy) or Cys (HcySSCy).

Taurine is transported to different tissues and intracellularly through either TauT or PAT1, which are present in plasma membrane, nuclear membrane, and mitochondrial membrane. It is transported into endosome and lysosome through PAT1.[37]

Disorders of Cyst(e)ine Transport

Cystinuria is an inherited disorder of cystine and dibasic amino acid transport by the system $b^{0,+}$ transporter, which is expressed by the kidney and small intestine.[38] However, this results in elevated urinary concentrations of other amino acids like lysine, ornithine, and arginine, in addition to cystine, because of no renal tubular reabsorption.[39] Cystinuria results in the formation of kidney cystine stones, as cystine has low aqueous solubility (250 mg/L). The intestine is still able to absorb enough amino acids as other transport systems are not affected.

Cystinosis resulting from a loss-of-function mutation in cystinosin, a lysosomal cystine transporter[40] leads to accumulation of cystine in lysosomes, which causes tissue damage. Malfunctioning kidneys and corneal crystals are the main initial features of the disorder. Thiol cysteamine is used as a treatment. Cysteamine reacts with the cystine in the lysosome to form Cys and Cys- cysteamine disulfide, which can exit the lysosome through other transport systems.

Metabolism

Synthesis of Homocysteine and Cysteine and Taurine

Hcy exists at a critical intersection of either being remethylated to Met in the Met-Cycle or exiting the Met-Cycle to form Cys through trans-sulfuration. Due to the essentiality of Met, synthesis of Hcy and Cys depends on the intake of Met and regulation of Met transmethylation, Hcy remethylation, and Hcy trans-sulfuration (**Figure 33.2**).

Met-Cycle (Met Transmethylation and Remethylation Pathway). Dietary Met is either used for protein synthesis or converted to S-adenosyl Met (SAM) by Met adenosyltransferase. Met adenosyltransferase has three isoforms, two of which (MAT I and MAT III) are present in the liver encoded by *MAT1A* gene. The third isoform MAT II encoded by *MAT2* is expressed in both extrahepatic and hepatic tissues. While MAT I and MAT II have low Michaelis-Menten constant (K_m) and are inhibited by SAM, MAT III has high K_m and is activated by SAM. Hence, it responds to high plasma Met concentration and helps in its clearance. Therefore, under normal conditions, there is a constant concentration of SAM in extrahepatic tissues, while hepatic concentration can vary, depending on Met availability.[41]

SAM, a key methyl donor, donates its methyl group to acceptor molecules such as DNA, RNA, proteins, or other metabolites using different methyl transferases to form S-adenosyl Hcy (SAH). SAH is a feedback competitive inhibitor of methyl transferases; therefore, clearance of SAH is important. It is hydrolyzed to Hcy and adenosine by adenosyl homocystinase, which exists in all cells.[42] This is a reversible reaction with its equilibrium favoring the formation of SAH; however, the rapid clearance of Hcy and adenosine drives the reaction forward.[43] At higher concentrations, SAH exits the cells and is cleared by the kidneys.

Hcy is remethylated to Met by methionine synthase (methylfolate methyltransferase). The methyl group is donated by N^5-methyltetrahydrofolate (N^5-methyl-THF). This step requires methylcobalamin (vitamin B_{12}) as the cofactor.[41] Another enzyme, which links the folate cycle to the Met-cycle, and is necessary for synthesis of N^5-methyl-THF from $N^{5,10}$-methyl-THF, is methylene tetrahydrofolate reductase. These two enzymes are distributed in all cells.[44] The second pathway through which the Hcy is remethylated is by betaine Hcy methyltransferase, which is expressed in the liver and the kidney.[45] Betaine, a catabolic product of choline, is the methyl donor in this reaction. High dietary intake of Met and betaine/choline increases cellular concentration of SAM, which inhibits methylene tetrahydrofolate reductase and betaine Hcy methyltransferase, thereby reducing Hcy remethylation. Decreased Hcy remethylation increases Hcy concentration which, in turn, increases Hcy catabolism through trans-sulfuration.[46] Hcy can also be transferred from the site of origin to liver or other cells for remethylation or catabolism.

Vitamin B_9 and B_{12} status also influences the flux of Hcy to Met or Cys. In adult humans with folate deficiency, the concentration of Hcy was significantly higher than those with normal folate concentration, and the rate of Hcy remethylation was significantly lower, emphasizing the role of folate as a methyl donor

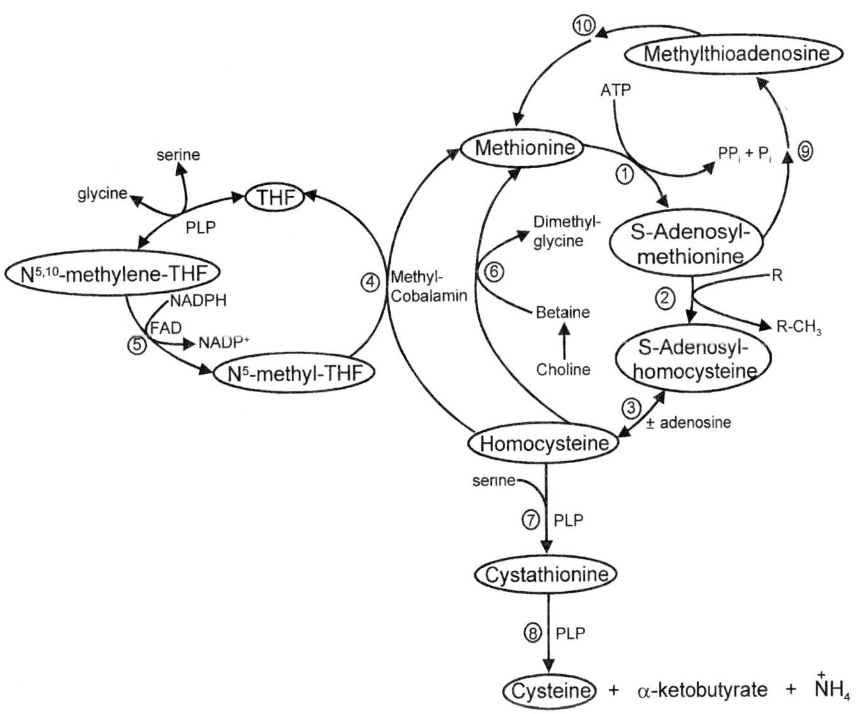

Figure 33.2 Methionine metabolism. Numbered reactions are catalyzed by the following enzymes: (1) Met adenosyltransferase (2) various methyltransferases (3) adenosylhomocysteinase (4) N^5-methyl-THF-homocysteine methyltransferase (5) $N^{5,10}$-methylene-THF reductase (6) betaine-homocysteine methyltransferase (7) cystathionine β-synthase (8) cystathionine γ-lyase (9) enzymes involved in polyamine synthesis and (10) enzymes involved in methylthioadenosine salvage pathway. FAD, flavin adenine dinucleotide; Pi, phosphate; PPi, free pyrophosphate; NADP+, nicotinamide adenine dinucleotide phosphate; NADPH, reduced nicotinamide adenine dinucleotide phosphate; NADPH, reduced nicotinamide adenine dinucleotide phosphate; PLP, pyridoxal 5′-phosphate; THF, tetrahydrofolate.

in this reaction.[47] In vitamin B_{12}-deficient adolescent females supplemented with vitamin B_{12}, those with high plasma B_{12} had a greater rate of Hcy remethylation and lower trans-sulfuration than those with low plasma B_{12} following a methionine loading test (MLT).[48] This observation is similar to the increased concentration of plasma Met in adults with hyperhomocysteinemia (high plasma concentration Hcy) when compared prior to, and following, vitamin therapy after a Hcy loading test.[49]

Trans-sulfuration Pathway. Met sulfur is transferred to Ser via Hcy to form Cys in a two pyridoxal 5′-phosphate-(PLP)-dependent enzyme catalyzed reactions. The first step is catalyzed by cystathionine β-synthase (CBS) and involves the condensation of Hcy and Ser to form cystathionine. This is an irreversible reaction and commits Hcy to its catabolism. Cystathionine γ-lyase (CHL) then hydrolyzes cystathionine to Cys, α-ketobutyrate, and ammonium.

SAM and SAH are allosteric activators of CBS. Both enzymes of Hcy catabolism have higher Km than the Met-Cycle enzymes, for instance the Km of CBS for Hcy is ~100 to 1,000 times higher than the remethylation enzymes.[41,50] Therefore, their activities are higher with greater availability of Met and Hcy. Rats fed diets containing high Met (3%) had relatively higher concentrations of SAM and SAH with a subsequent increase in Hcy, higher concentrations of cystathionine β-synthases, and lower serine concentration.[51]

CBS and CHL have limited expression in liver, kidney, small intestine, and pancreas, but only CBS is expressed in brain and adipose tissues in rats.[52] In humans, CBS is highly expressed in liver, pancreas, and brain, whereas CHL is expressed most in the liver, followed by the pancreas and kidney.[53,54] Tissues that do not express trans-sulfuration enzymes require Cys, which is exogenously supplied.

Trans-sulfuration enzymes are sensitive to vitamin B_6 status. Rats fed vitamin B_6-deficient diets had reduced trans-sulfuration enzyme activity, with CHL being more sensitive to variations in vitamin B_6 status.[55] Plasma cystathionine and Hcy concentrations were higher in vitamin B_6-deficient human adults with normal B_{12} and folate during MLT, which reduced significantly after B_6 supplementation.[56] Moderate vitamin B_6 restriction does not affect postprandial remethylation, transmethylation, or trans-sulfuration rates in the healthy population but does increase the rate of cystathionine synthesis.[57]

CBS enzyme is a heme protein and has higher activity as a metabolic response to oxidative stress, to facilitate glutathione synthesis (GSH), for which the availability of cysteine is rate limiting. In addition, hepatic CBS gene expression is increased by glucagon and glucocorticoids and decreased by insulin.[58,59] This regulation conserves Met for protein synthesis in the fed state and promotes its catabolism to produce α-ketobutyrate, a gluconeogenic substrate, in the fasted state.

At normal SAA intake, Hcy synthesis in men is ~19 mmol/day, and Cys synthesis from this Hcy is ~12 mmol/day. In men fed an SAA-free diet, Hcy synthesis was reduced to 6 mmol/day and Cys synthesis to 2 mmol/day.[60,61] The remaining Hcy is remethylated to Met. Cyst(e)ine is said to have a Met-sparing effect by reducing Met catabolism through the trans-sulfuration pathway, and this occurs with typical intakes in which the Met-to-Cys ratio ranges from ~1:1 to 2:1,[62] although other studies have suggested that the sparing is hard to quantify.[63] Maximal sparing of Met is about 64%, when excess cyst(e)ine and minimal Met diet is administered.[64] The action of supplemental cyst(e)ine when it is added to an SAA-free diet or to a low-Met diet may be explained at least partially by higher incorporation of Met into protein vs catabolism.[65]

Genetic Factors Influencing the Met-Cyle, Hcy Synthesis and Transsulfuration Pathway and Their Disorders.

Hyperhomocysteinemia and Homocystinuria. Elevated plasma tHCy can result from increased synthesis by transmethylation; decreased removal by remethylation or transmethylation; and decreased uptake, metabolism, or excretion from the kidneys. Several inborn errors of metabolism result in severe increases of tissue and plasma Hcy, leading to excretion of Hcy, homocystine, and mixed disulfides of Hcy, referred to as *homocystinuria*.

Lack of CBS activity causes homocystinuria and is associated with plasma tHcy concentration higher than 200 μmol/L and urinary Hcy >10 μmol/24 hours in untreated patients. It has an incidence of 1 in 335,000 births.[66] Another form of homocystinuria, resulting from lack of $N^{5,10}$-methylene-THF reductase activity, reduces the N^5-THF available for remethylation and affects folate metabolism. Disorders such as ocular abnormalities, cardiovascular thromboembolic episodes, and mental retardation have been attributed to homocystinuria, if untreated at birth. Standard treatment includes supplementation of vitamin B_6 or folic acid; vitamin B_{12} and betaine; or a Met-restricted, Cys-supplemented diet.[67]

Milder forms of homocysteinemia (e.g., >14 to 16 μmol/L) are common.[68] Mutations resulting from impaired activity of the functional enzymes account for mild to moderate cases of hyperhomocysteinemia, particularly in individuals with low folate or vitamin B_{12} intake. One common cause of hyperhomocysteinemia is the C677T (Ala222Val) variant of $N^{5,10}$-methylene-THF reductase, which results in an enzyme with reduced activity.[69] The occurrence of this homozygous polymorphism ranges from 1.5% in African Americans, 12% to 15% in North Americans and Asians, to 25% in Hispanics. Furthermore, ~30% to 40% of North Americans, Asians, and Hispanics are heterozygous.[70] This mutation is associated with lower plasma folate and higher plasma tHcy compared with controls.[71] Folate supplementation in these populations decreases the prevalence of hyperhomocysteinemia.[72] Mild-to-moderate hyperhomocysteinemia (i.e., 15 to 100 μmol/L) has also been observed in patients with renal disease, while severe hyperhomocysteinemia occurs in patients with renal failure.[73]

Other causes of hyperhomocysteinemia include the use of drugs that cause secondary vitamin deficiency. For example, theophylline is a vitamin B_6 antagonist, and valproate and carbamazepine have antifolate activity. Treatment of patients with metabolic syndrome or diabetes with fibrates resulted in abnormal renal excretion of betaine and an increase in plasma tHcy.[74] L-Dopa treatment in Parkinsonism also increased plasma tHcy.[75] These drugs can cause elevated plasma tHcy concentration because of decreased remethylation of Hcy.

Trisomy 21 is associated with overexpression of CBS, also present on chromosome 21. Children with this condition have significantly reduced plasma concentrations of Hcy, Met, SAH, and SAM and increased plasma cystathionine and Cys.[76]

Cysteine Metabolism to Taurine and Inorganic Sulfur. Taurine is an oxidation product of Cys through its major degradative pathway. Cys is dioxygenated by incorporating molecular oxygen to cysteine sulfinic acid (3-sulfinoalanine) in a reaction catalyzed by cysteine dioxygenase 1 (CDO1). The activity of CDO1 corresponds to its concentration and is rate limiting in this pathway. Concentration of the enzyme can change up to 10- or 45-fold in response to increased intake of SAA or protein.[77] CDO1 has high Km (~0.5 mmol/L), which is above the normal tissue concentration of Cys of ~0.2–0.3 mmol/kg in human liver.[78] It is expressed in liver, where much of the taurine synthesis occurs and in kidney, pancreas, adipose, and lungs in rodents. In humans, it is found in the liver, small intestine, and lungs.[79]

Cys sulfinic acid can either be decarboxylated to hypotaurine in a reaction catalyzed by cysteine sulfinic acid decarboxylase (CSD) or undergo transamination to form sulfinylpyruvate in a reaction catalyzed by glutamate oxaloacetic transaminase (GOT). Hypotaurine is then oxidized to taurine in an unknown reaction, which may, or may not, be enzymatic. Sulfinylpyruvate undergoes spontaneous decomposition to pyruvate and sulfite, which is then oxidized to sulfate. GOT has higher K_m (3–25 mmol/L) compared with CSD (0.04–0.17 mmol/L) and; therefore, the production of hypotaurine is preferred over transamination.[80,81] Activity of CSD, like the CDO1, depends on its concentration and is high in tissues, which have high CDO1 activity. In rats, the flux of partitioning between taurine and sulfate is 66% and 34%.[82] Taurine synthesis in adult humans, based on incorporation of ^{18}O (from inhaled $^{18}O_2$) into taurine, was estimated to be in the range of 200 to 400 μmol/day.[83] This is equivalent to 1% to 3% of the total SAA intake and compares well with the taurine excretion of approximately 250 μmol/day observed in strict vegans consuming a taurine-free diet.[20,28]

Cys is also catabolized by desulfhydration reactions, and catalyzed by CBS and CHL, to produce hydrogen sulfide (H_2S), pyruvate, and ammonia. The reaction is favored by high Cys, low Ser, and allosteric activation by SAM.[84] The K_m of CBS and CHL for Cys in high (~3–7 mmol/L) and; therefore, the synthesis of taurine is preferred over H_2S. CHL can also catalyze the desulfhydration reaction from Hcy to produce H_2S, 2-ketobutyrate, and ammonia. This reaction also occurs at higher concentration of Hcy, as the K_m of CHL for HCy is ~5 mmol/L.[85] H_2S is oxidized in mitochondria to sulfane sulfur and transferred to an acceptor molecule such as GSH or sulfite by sulfide quinone oxidoreductase (SQOR) in the presence of coenzyme Q to form glutathione persulfide (GSSH) or thiosulfate (SSO_3^{2-}).[86] Thiosulfate can also be formed by transfer of sulfane sulfur from GSSH to sulfite catalyzed by thiosulfate sulfur transferase. Ethylmalonic encephalopathy 1 (ETHE1) persulfide dioxygenase then catalyzes the cleavage of the S-S bond in GSSH to form sulfite (SO_3^{2-}).[87] Sulfite is further oxidized to sulfate (SO_4^{2-}) by sulfite oxidase[88] and excreted in the urine. These reactions may be critical for provision of reduced sulfur because mammals do not have the ability to reduce sulfate or sulfite to thiosulfate or sulfide. Metabolic pathways for Cys are shown in **Figure 33.3**.

Cys metabolism is influenced by its requirement for protein synthesis and GSH and by the demand for producing coenzyme A, taurine, and inorganic forms of sulfur. Tissue Cys concentration is maintained at levels adequate for protein and GSH synthesis. The regulation of tissue Cys concentration is by catabolism to taurine and sulfate through CDO1, the activity of which is sensitive to changes in Cys availability. Clearance of excess tissue Cys concentration is important, as it is both cytotoxic and neurotoxic. When SAA intake is marginal, preservation of Cys concentration occurs at the expense of taurine and sulfate production. GSH acts as a reservoir of Cys: with insufficient SAA intake, GSH hydrolysis preserves plasma Cys concentration and protects protein synthesis.

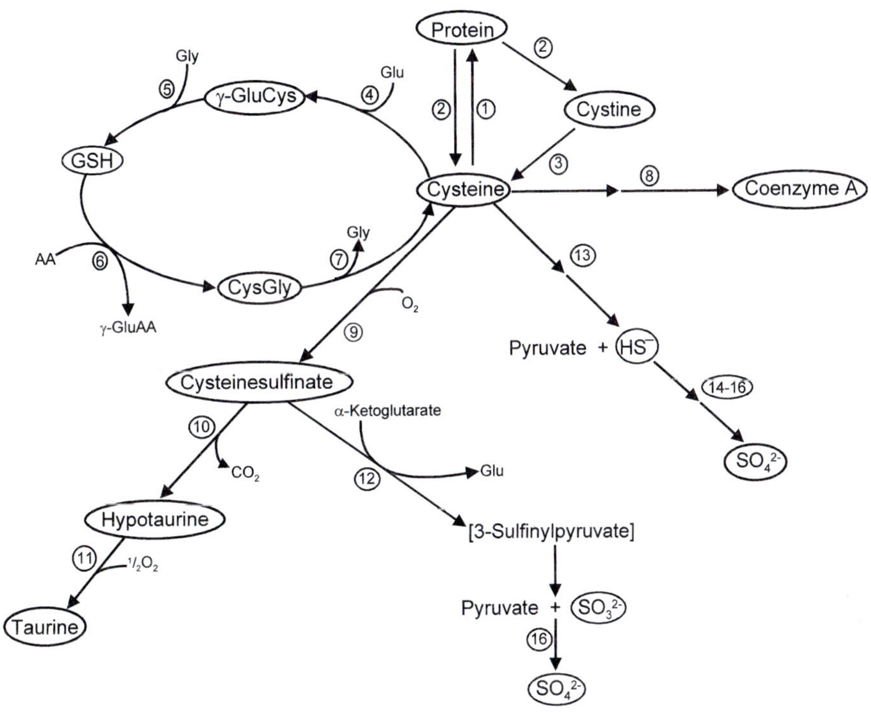

Figure 33.3 Pathways of cysteine metabolism. Numbered reactions are catalyzed by the following enzymes or pathways: (1) protein synthesis (2) protein degradation (3) GSH-thioltransferase or nonenzymatic thiol-disulfide exchange of cystine with GSH (4) γ-glutamylcysteine synthetase (5) GSH synthetase (6) GSH transpeptidase (7) dipeptidase (8) pathway of coenzyme A synthesis (9) cysteine dioxygenase (10) cysteinesulfinate decarboxylase (11) enzymatic or nonenzymatic oxidation of hypotaurine (12) aspartate (cysteinesulfinate) aminotransferase (13) cysteinesulfinate-independent or desulfhydration pathways of cyst(e)ine catabolism (14) sulfide oxidation by sulfide quinone oxidoreductase, sulfur dioxygenase, and thiosulfate sulfurtransferase (rhodanese) (15) GSH-dependent thiosulfate reductase and (16) sulfite oxidase. *GSH*, glutathione; *GLU*, glutamate; *HS*, hydrogen sulfide; *CYS*, cysteine.

Genetic Factors Affecting Cysteine Metabolism and Their Diseases. Neurologic conditions such as Parkinson's and Alzheimer's diseases are characterized by high plasma Cys,[89] indicative of low CDO1 activity. Clinical and biochemical features indicative of low CDO1 activity have also been reported with autoimmune, liver, and inflammatory diseases.[90,91] Polymorphisms in the CDO1 gene have been observed in patients with rheumatoid arthritis, suggesting a link between the low CDO1 and high cysteine:sulfate ratios observed in these patients compared with the control population.[91,92] Hallervorden-Spatz disease, a neurodegenerative disorder, is characterized by lower activity of CDO1 in the globus pallidus, leading to Cys accumulation, which chelates iron, causing the observed iron deposition there.[93]

Inborn errors in mitochondrial H_2S catabolism have been reported, but their prevalence remains unknown. Ethylmalonic encephalopathy, caused by a loss-of-function mutation of ETHE1, leads to high thiosulfate accumulation and excretion.[87] Mutations in sulphate oxidase leading to its deficiency result in elevated S-sulfocysteine, with low plasma Cys and high sulfite and thiosulfate concentrations in urine.[94] Treatment with a low Met and Cys diet has been shown to improve clinical outcomes in one patient with ETHE1 mutation[95] as well as the milder phenotype of sulphate oxidase mutation.[96]

Excretion

The renal proximal tubular epithelium has transport systems like those of the absorptive epithelium of the intestine. Renal reabsorption of Cys is normally high (>94%). Urinary cyst(e)ine excretion by adults has been reported as 63 to 285 μmol/day.[28,97] Typical urinary excretion of Cys accounts for 0.2% to 2% of average daily intake.

Urinary excretion of extracellular Hcy is limited, as Hcy exists bound to protein in plasma, which limits filtration, and free Hcy is readily reabsorbed. Only ~1% to 2% of Hcy filtered by the kidney is excreted and normal urinary Hcy excretion ranges from 3.5 to 9.8 μmol/day.[98] Hcy excreted in inborn errors of metabolism can range from 15 to 667 μmol/day.[99]

Unlike Cys and Hcy, taurine is not well reabsorbed, and excretion can vary over a wide range, depending on intake. Urinary taurine concentration of 250 μmol/day has been reported for adult vegans, whereas excretion of taurine by adult omnivores is usually greater than 600 μmol/day, and values greater than 1,000 μmol/day are common.[20,28,98] Low dietary taurine intake causes higher reabsorption of taurine from the filtrate due to enhanced TauT activity. Renal taurine concentration seems to be the signal for changes in TauT expression, activity, and subcellular localization.[100]

Functions of Cysteine and Taurine

Cys is essential for protein synthesis and other essential molecules, such as GSH, coenzyme A, taurine, H_2S, and inorganic sulfur (Figure 33.3). The functions of these molecules are discussed briefly here. The functions of coenzyme A are discussed in the chapter on pantothenic acid.

Glutathione

GSH is the major intracellular thiol owing to its cysteine residue. GSH is found in high concentration in most cells (1–10 mmol/L), whereas the extracellular concentration is typically low (2–20 μmol/L in plasma).[101,102] The intracellular ratio of reduced (GSH) to oxidized (GSSH) form is >100 in normal conditions, decreasing to <10 in oxidative stress.[103] One of the most important roles of GSH is to scavenge free radicals or reactive oxygen species (ROS) directly by donation of reducing equivalents or through enzymatic reactions, such as GSH peroxidases. GSH is converted to GSSH in such reactions and regenerated through the action of NADPH-dependent GSH reductase. GSH is necessary for the conversion of cystine to CySH, either enzymatically by thioltransferase or through thiol-disulfide exchange.

GSH is involved in excretion of xenobiotics and some endogenous compounds, such as estrogens, and prostaglandins by the formation of GSH conjugates by a group of enzymes called GSH S-transferases. These are degraded to form mercapturates, which are excreted in urine. GSH is also required for synthesis of prostaglandins, leukotrienes, and melanin polymer.[104] Endogenously produced formaldehyde is removed by GSH-dependent formaldehyde dehydrogenase. Protein glutathionylation, a disulfide linkage between Cys of protein and GSH, which may be a consequence of oxidative stress, has been shown to be regulatory and thought to be protective.[105]

Taurine

Taurine is present in high concentration in many human tissues and exerts may physiologic effects; however, its mechanism of action is not well understood.[106] Taurine is a substrate for formation of bile acids and taurocholate is an efficient bile acid.[106] The low pKa of taurocholate facilitates its ionization, detergent action, solubility, slower reabsorption, and higher intraluminal concentration. The ratio of glycocholate to taurocholate is 3:1, but varies substantially between individuals. This ratio changes with the body taurine pool.[107] Taurine supplementation has been shown to reduce cholestasis in rats[108] and to improve hepatic function parameters in cholestatic patients.[109] Taurine conjugates with xenobiotics, secondary bile acids, and all-*trans*-retinoic acid improves their solubility and excretion.

Taurine is an important osmolyte and is involved in cell volume regulation following osmotic stress.[110] Taurine efflux or influx pathways are modulated by cytoskeletal changes, protein phosphorylation, expression of TauT, reactive oxygen species (ROS) and calcium (Ca^{2+}) and calmodulin status.[110] In type 2 diabetes patients, several oxidative stress indices, such as increased activities of antioxidant enzyme and decreased lipid peroxidation products such as malonaldehyde, were reduced after taurine supplementation.[111]

Two modifications of mitochondrial (mt) tRNAs involving taurine conjugation (5-taurinomethyl-2-thiouridine in mt tRNAs for lysine, glutamine and Glu, and 5-taurinomethyl-uridine in mt tRNAs for tryptophan and leucine) at the wobble position in the t-RNAs are found to be essential.[112] Mutations affecting the conjugation have been implicated in mitochondrial myopathy, encephalopathy, lactic acidosis, and stroke (MELAS) and myoclonic epilepsy with ragged red fibers (MERRF).[113] Taurine supplementation has been shown to reduce stroke-like incidence in MELAS patients.[114]

Taurine has a role in the development of central nervous and visual systems in pre- and postnatal development. Human infants on taurine-free parenteral nutrition or formulas exhibited retinal abnormalities and immature brainstem auditory evoked responses.[106] Although unclear, taurine may affect neuronal progenitor cells via glycine and γ-aminobutyric acid–ergic (GABAergic) receptors.[115] It has been shown to influence neural stem cell proliferation, synapse development in vitro, and neuronal migration in vivo in mouse models.[116,117]

H_2S and Inorganic Sulfate

H_2S acts as a signaling molecule in vasodilation and smooth muscle relaxation, which may act by opening the potassium adenosine triphosphate (K_{ATP}) channels.[77] It has been shown to modify activity of proteins by sulfhydration with Cys residue. For example, the activity of glyceraldehyde-3-phosphate dehydrogenase increased because of sulfhydration.[118] Several roles have been attributed to H_2S, including angiogenic and anti-oxidative properties, and regulatory roles in the nervous system, and in glucose and lipid metabolism.[119,120]

Sulfur from Cys is required for synthesis of iron-sulfur clusters for iron-sulfur proteins, modification of uridine residues in tRNAs (thiouridine), and molybdopterin coenzyme biosynthesis.[121,122] Intracellular 3′-phospho-5′-phosphosulfate (PAPS) is a substrate for sulfotransferase reactions. Structural compounds such as proteoglycans and several integral membrane proteins are sulfated. Sulfoesters facilitate the excretion of many compounds, such as steroid hormones and the drug acetaminophen. Although inorganic sulfate is not considered essential, dietary inorganic sulfate may improve growth, feed efficiency, and sulfation of cartilage proteoglycans in animals.[123]

Assessment of Sulfur Amino Acid Status

SAA adequacy has generally been assessed by measures of nitrogen balance or growth. However, these are not good indicators of SAA sufficiency for optimal production of GSH, inorganic sulfur, or taurine. Adult humans remain in sulfur balance with approximately 14 to 28 mmol of sulfur equivalents excreted per day, primarily as inorganic sulfate. Free sulfate accounts for ∼77% to 92%, ester sulfate for ∼7% to 9%, taurine for ∼2% to 6%, and cyst(e)ine for ∼0.6% to 0.7% of the urinary sulfur excretion. Met, Hcy, cystathionine, *N*-acetylcysteine, mercaptolactate, mercaptoacetate, thiosulfate, and thiocyanate account for <0.2% of total sulfur excretion.[97] In young Japanese women, free sulfate correlated significantly with urea excretion, indicating that free sulfate excretion might be a good index of SAA intake.

Normal Plasma Concentrations of Cysteine, Homocysteine, and Taurine

Normal values for plasma total Cys (tCys), GSH, and related aminothiols are shown in **Table 33.4**. Cys is the major plasma thiol; tHcy and GSh are present at ≥10% and >1% of the tCys concentration. Both Cys and Hcy are predominantly present as protein-bound disulfides, with intermediate concentrations of disulfides (predominantly CySSCy and HcySSCy) and with very low concentrations of free thiols. The Cys-containing peptides derived from GSH turnover, cysteinylglycine (CysGly) and γ-glutamylcysteine (γGluCys), are also present in plasma and tissues.

Mean plasma tCys concentration in healthy adults ranges from 220 to 320 μmol/L.[124,125] Mean plasma tHcy in older Framingham Heart Study cohort survivors before folic acid fortification in the United States was 11.9 μmol/L, with a range of 3.5 to 66.8 μmol/L.[127] Mean plasma tHcy is lower for younger than for older adults and for women than for men.[68,128-130] For folate-supplemented populations, upper reference limits of 12 μmol/L for adults (15 to 65 years of age), 8 μmol/L for children and 16 μmol/L for elderly persons has been suggested. Hyperhomocysteinemia has been classified according to plasma tHcy concentration as moderate (15–30 μmol/L), intermediate (>30–100 μmol/L), or severe (>100 μmol/L).[130]

Table 33.4 Normal Values for Total Cysteine, Homocysteine, Glutathione, and Few Amino Thiols[124-126]

	Total	Reduced[1]		Oxidized[1]				Reduced/Oxidized Ratio	Reduced/Total Ratio
	μmol/L								
		Free				Protein Bound			
Cysteine	250	CySH	14	RSSCy	88	PSSCy	196	0.049	0.056
Cysteinylglycine	29	HSCyGly	4	RSSCyGly	5	PSSCyGly	18	0.174	0.138
Homocysteine	11	HcySH	0.05	RSSHcy	1	PSSHcy	10	0.005	0.005
Glutathione	6	GSH	4	RSSG	1.5	PSSG	1.6	1.290	0.667
γ-glutamylcysteine	3	γ-GluCySH	0.6	RSSCyGlu	2	PSSCyGlu	1	0.200	0.200

[1]RSH represents reduced the thiol form; RSSR or RSSR represents oxidized the disulfide of the thiol with itself or another thiol; PSSR represents protein-bound disulfides.
PS, sulfhydryl group of cysteinyl residue in protein; *RS*, unspecified thiol, usually CySH in plasma.

Data from Mansoor MA, Bergmark C, Svardal AM et al. Arterioscler Thromb Vasc Biol 1995;15:232–40; Andersson A, Isaksson A Brattstrom L et al. Clin Chem 1993;39:1590–7; Mansoor MA, Ueland PM, Svardal AM. Am J Clin Nutr 1994;59:631–5.

A wide range of plasma taurine concentration has been reported for human adults, depending on intake. Mean plasma and whole blood concentration of taurine in humans range from 39 to 116 μmol/L and 60 to 320 μmol/L.[131] Plasma taurine concentration is sensitive to taurine intake and differs from whole blood concentration. Plasma taurine concentration is lower in girls and women than in boys and men.[20,28]

Stable Isotopic Techniques for Sulfur Amino Acid Metabolism

Stable isotopes of Met, Hcy, Cystathionine, and Cys can be used in quantifying rates of appearance or disappearance of intermediates in the Met-Cycle, trans-sulfuration pathway, and Cys catabolism. This is a direct method of quantitatively assessing Met and Cys metabolism. The stable isotopic tracers labeled at the carbon, hydrogen, or sulfur moiety can be used for these studies; however, careful consideration of labeling positions of tracers is essential. Whole blood, plasma, expired breath, and urine samples can be used to measure the concentration and flux of these molecules. The classical study is to measure kinetics of Met-Cycle in fasted and fed conditions and employs [1-^{13}C] and methyl-[$^{2}H_3$] in a steady-state infusion. Here, the use of [1-^{13}C] label and methyl-[$^{2}H_3$] allows the measurement of rate of loss of Met via trans-sulfuration (oxidation), conservation of [1-^{13}C] Hcy via remethylation and the loss of [$^{2}H_3$] methyl group allows the measurement of transmethylation. $^{13}CO_2$, a product of methionine oxidation, is measured in expired breath and [1-^{13}C]; methyl-[$^{2}H_3$] Met and [1-^{13}C] Met enrichments are measured in plasma.[60] In later studies, the enrichment of plasma [1-^{13}C] Hcy is considered instead of [1-^{13}C] Met, as this more accurately represents intracellular [1-^{13}C] Met enrichment. These methods are used to understand SAA metabolism with differing intakes, or in vitamin B_9, B_{12}, B_6 deficiency, during pregnancy, in elderly adults, and in malnourished populations.[46-49,60,61]

Causes and Manifestations of Deficiency or Excess

Possible Causes of Deficiency of Cysteine or Taurine

Immaturity

Immaturity may be associated with a conditional requirement for Cys and taurine. Preterm infants (<32 weeks' gestation) have a low capacity for transsulfuration (low cystathionine γ-lyase activity), high plasma cystathionine and low Cys with low GSH synthesis rate from Met in erythrocytes.[132,133] Full-term, formula-fed infants also have high urinary cystathionine and low taurine concentrations, suggesting a limited capacity for trans-sulfuration.[134]

Cys or taurine is conditionally required for preterm infants, as they may have higher requirement for Cys due to more rapid growth and of taurine for development of the nervous and visual systems.[135,136] In addition, premature infants are born with lower stores of taurine; their fractional excretion of taurine ranged from 38% to 60% compared with 10% in term infants.[135]

Hepatic Dysfunction

The liver is the major site for transsulfuration and taurine synthesis: therefore, hepatic dysfunction can have adverse effects on SAA status. Patients with advanced liver dysfunction had low plasma taurine, Cys, and GSH concentrations; elevated plasma cystathionine concentration; decreased urinary taurine excretion; and increased urinary excretion of Cys and cystathionine.[137,138] They also had a decreased ability to metabolize Met to Cys with cystathionine accumulation; and decreased ability to metabolize Cys to taurine and inorganic sulfate with thiosulfate, Cys, and *N*-acetylcysteine accumulation.

Total Parenteral or Enteral Nutrition

Long-term total parenteral nutrition (TPN) is associated with adverse effects on SAA status, due to route of its administration and composition. TPN solutions contain little if any Cys, as it is rapidly converted to cystine, which has low aqueous solubility; taurine is not routinely added to adult TPN solutions. Hence, patients receiving TPN must synthesize both Cys and taurine from the Met in TPN. During parenteral administration, Cys and taurine synthesis from Met is limited, as first-pass metabolism by the liver is bypassed. This can be countered by oral administration.[139] In long-term, enterally tube fed patients, inadequate taurine supplementation can result in taurine deficiency.[140]

Drug Metabolism

Various drugs and toxins are excreted as conjugates of sulfate, GSH (mercapturate synthesis), or taurine. Acetaminophen is excreted mainly as glucuronide and sulfate conjugates; a smaller amount is excreted as mercapturate. In healthy adults administered different doses of acetaminophen (600 and 1,200 mg), the rate of turnover of the Cys pool available for GSH synthesis increased.[141] Therefore, consumption of high doses of drugs, which are metabolized by sulfate or GSH conjugation, could affect SAA status. N-acetyl cysteine, administered in acetaminophen toxicity, serves to replenish the lost Cys and GSH during acetaminophen excretion.

Possible Toxicity of Cysteine or Taurine

Large doses of Cys or cystine have been shown to be neuroexcitotoxic in several species. Cys injections (0.6 to 1.5 g/kg body weight (BW)) to 4-day-old rat pups led to permanent retinal dystrophy, brain atrophy, cortical neuron damage, and hyperactivity.[142,143] In 4 to 5-day old mice, Cys injections of >1.2 g/kg BW resulted in hypoglycemia and neurotoxicity.[144,145] Hippocampal brain damage and changes in brain morphology were also observed in survivors.[145] Dietary SAA administration has also been shown to influence lipid metabolism in rodents. Administration of 2% to 5% Cys by weight elevated hepatic cholesterol biosynthesis, plasma cholesterol concentration, and depressed plasma ceruloplasmin activity.[146] Daily Cys intakes in the United States are 33 to 83 times lower than intakes that produced adverse effects in animal studies. Evidence of adverse events is not associated with high taurine diets; however, possible adverse effects of energy drinks, of which taurine is a major constituent, in younger adults has been reported.[147]

Possible Adverse Effects of Hyperhomocysteinemia

Mild to moderate hyperhomocysteinemia has been associated with atherosclerotic and ischemic cardiovascular diseases, myocardial infarction, venous thromboses, and stroke. As a result, hyperhomocysteinemia is considered a risk factor for coronary, cerebral, and peripheral artery vascular disease.[10,11,128] Although most studies are observational, the effect of Hcy on endothelial dysfunction, potentially through inhibition of endothelial nitric oxide synthase, supports this observation.[148] Several other mechanisms of Hcy have been studied in this context, including increased production of ROS through decreased activity of antioxidant enzymes, such as glutathione peroxidase 1 (GPx-1) or superoxide dismutase (SOD). Homocysteinylation of proteins or incorporation of homocysteine thiolactone in place of methionine and its condensation with ε-NH_2 of lysine can impair protein function. In addition, high Hcy concentration could lead to elevation of the SAH:SAM ratio, and this change could cause altered methylation of DNA, which would result in endothelial dysfunction.[149] Although these mechanisms have been elucidated in vitro or in knockout mice models, the question of whether these mechanisms work synergistically needs to be assessed.

In severe hyperhomocysteinemia, caused by deficiency of cystathionine β-synthase activity, Met restriction, B-vitamin and betaine supplementation over a mean treatment time of 17.9 years lowered plasma tHcy from >150 μmol/L to an intermediate range of 30 to 100 μmol/L. This was associated with ~90% reduction of number of vascular events compared with that predicted for patients without treatment.[150] Improvement of cardiovascular risk after treatment of hyperhomocysteinemia with B vitamins remains uncertain. Long-term B-vitamin trials to lower plasma tHcy, for periods ranging from 2 to 7.3 years, did not show any benefit in cardiovascular endpoints but reduced tHcy in individuals with mild hyperhomocysteinemia.[151-154] These, and other studies, have shown that lowering of tHcy by B-vitamin supplementation had no effect on inflammatory markers, endothelial dysfunction, or hypercoagulability,[155,156] raising the question about whether increased Hcy concentration, itself, impacted the risk of cardiovascular diseases.

In addition to cardiovascular diseases, mild to moderate elevation in plasma tHcy is associated with neuropsychiatric disorders such as Alzheimer's disease and Parkinsonism, and pregnancy complications, such as pre-eclampsia, placental abruption or infarction and unexplained pregnancy loss.[157,158] The effect of tHcy on these pathologies remains unclear.

Acknowledgment

We gratefully acknowledge the substantial contribution by Martha H. Stipanuk, author of this chapter in the previous edition.

References

1. Vickery HB, Schmidt CL. Chem Rev 1931;9:169–318.
2. Osborne TB, Mendel LR. J Biol Chem 1915;20:351–78.
3. Jackson RW, Block RJ. J Biol Chem 1932;98:465–77.
4. Womack M, Kemmerer KS, Rose WC. J Biol Chem 1937;121:403–10.
5. Tiedemann F, Gmelin L. Ann Physik Chem 1827;9:326–37.
6. Hayes KC, Carey RE, Schmidt SY. Science 1975;188:949–51.
7. Sturman JA, Rassin DK, Gaull GE. Life Sci 1977;21:1–22.
8. Geggel HS, Ament ME, Heckenlively JR, et al. N Engl J Med 1985;312:142–6.
9. du Vigneaud VE. Trail of Research in Sulfur Chemistry and Metabolism and Related Fields. Ithaca, NY: Cornell University Press, 1952.
10. Clarke R, Daly L, Robinson K, et al. N Engl J Med 1991;324:1149–55.
11. Robinson K, Mayer E, Jacobsen DW. Cleve Clin J Med 1994;61:438–50.
12. Steegers-Theunissen RPM, Boers GHJ, Trijbels FJM, et al. Metabolism 1994;43:1475–80.
13. Wouters MCAJ, Boers GHJ, Blom HJ, et al. Fertil Steril 1993;60:820–5.
14. Food and Nutrition Board, Institute of Medicine. Dietary Reference Intakes for Energy, Carbohydrate, Fiber, Fat, Fatty Acids, Cholesterol, Protein, and Amino Acids. Washington, DC: National Academy Press, 2005.
15. Joint FAO, World Health Organization. Protein and amino acid requirements in human nutrition: report of a joint FAO/WHO/UNU expert consultation. World Health Organization; 2007.
16. Kurpad AV, Regan MM, Varalakshmi S, et al. Am J Clin Nutr 2003;77:1198–205.
17. Deutz NE, Simbo SY, Ligthart-Melis GC, et al. Am J Clin Nutr 2017;106:675–83.
18. Bandyopadhyay S, Kashyap S, Calvez J, et al. Adv Nutr 2022;13:1131–1143.
19. Rutherfurd SM, Moughan PJ. J AOAC Int 2008;91:907–13.
20. Laidlaw SA, Shultz TD, Cecchino JT, et al. Am J Clin Nutr 1988;47:660–3.
21. Pexa A, Fischer K, Deussen A, et al. Eur Food Res Technol 2008;226:933–5.
22. Sakamoto A, Nishimura Y, Ono H, et al. Pediatr Int 2002;44:409–13.
23. Laidlaw SA, Grosvenor M, Kopple JD. J Parenter Enteral Nutr 1990;14:183–8.
24. Pasantes-Morales H, Quesada O, Alcocer L, et al. Nutr Rep Int 1989;40:793–801.
25. Roe DA, Weston MO. Nature 1965;203:287–8.
26. Kataoka H, Ohnishi N. Agric Biol Chem 1986;50:1887–8.
27. Kim ES, Cho KH, Park MA, et al. Adv Exp Med Biol 1996;403:571–7.
28. Rana SK, Sanders TAB. Br J Nutr 1986;56:17–27.
29. Burdo J, Dargusch R, Schubert D. J Histochem Cytochem 2006;54:549–57.
30. Dave MH, Schulz N, Zecevic M, et al. J Physiol 2004;558:597–610.
31. Sato H, Nomura S, Maebara K, et al. Biochem Biophys Res Commun 2004;325:109–16.
32. Anderson CM, Howard A, Walters JR, et al. J Physiol 2009;587:731–44.
33. Roig-Pérez S, Ferrer C, Rafecas M, et al. J Membr Biol 2009;228:141–50.
34. Satsu H, Kobayashi Y, Yokoyama T, et al. Amino Acids 2002;23:447–52.
35. Bauchart-Thevret C, Stoll B, Chacko S, et al. Am J Physiol 2009;296:E1239–50.

36. Bŭdy B, O'Neill R, DiBello PM, et al. Arch Biochem Biophys 2006;446:119–30.
37. Baliou S, Kyriakopoulos AM, Goulielmaki M, et al. Mol Med Rep 2020;22:2163–73.
38. Mora C, Chillaron J, Calonge MJ, et al. J Biol Chem 1996;271:10569–76.
39. Sakhaee K. Miner Electrolyte Metab 1994;20:414–23.
40. Kalatzis V, Antignac C. Pediatr Nephrol 2003:18:207–15.
41. Finkelstein JD. J Nutr Biochem 1990;1:228–37.
42. de la Haba G, Cantoni GL. J Biol Chem 1959;234:603–608.
43. Duerre JA, Miller CH, Reams GG. J Biol Chem 1969;244: 107–111.
44. Banerjee RV, Matthews RG. FASEB J 1990;4:1450–9.
45. McKeever MP, Weir DG, Molloy A, et al. Clin Sci 1991;81: 551–6.
46. Hiramatsu T, Fukagawa NK, Marchini JS, et al. Am J Clin Nutr 1994;60:525–33.
47. Cuskelly GJ, Stacpoole PW, Williamson J, et al. Am J Physiol Endocrinol Metab 2001;281:E1182–90.
48. Bhat DS, Gruca LL, Bennett CD, et al. PLoS One 2018;13:e0196970.
49. Guttormsen AB, Schneede J, Ueland PM, et al. Am J Clin Nutr 1996;63:194–202.
50. Singh S, Padovani D, Leslie RA, et al. J Biol Chem 2009;284: 22457–66.
51. Finkelstein JD, Martin JJ. J Biol Chem 1986;261:1502–7.
52. Finkelstein JD. Am J Clin Nutr 2003;77:1094–5.
53. Bao L, Vlček Č, Pačes V, et al. Arch Biochem Biophys 1998;350: 95–103.
54. Ishii I, Akahoshi N, Yu XN, et al. Biochem J 2004;381:113–23.
55. Finkelstein JD, Chalmers FT. J Nutr 1970;100:467–9.
56. Ubbink JB, van der Merwe A, Delport R, et al. J Clin Invest 1996;98:177–-84.
57. Lamers Y, Coats B, Ralat M, et al. J Nutr 2011;141:835–42.
58. Jacobs RL, Stead LM, Brosnan ME, et al. J Biol Chem 2001;276:43740–47.
59. Ratnam S, Maclean KN, Jacobs RL, et al. J Biol Chem 2002;277:42912–8.
60. Storch KJ, Wagner DA, Burke JF, et al. Am J Physiol 1988;255: E322–31.
61. Storch KJ, Wagner DA, Burke JF, et al. Am J Physiol 1990;258:E790–8.
62. Di Buono M, Wykes LJ, Ball RO, et al. Am J Clin Nutr 2001;74:761–6.
63. Kurpad AV, Regan MM, Varalakshmi S, et al. Am J Clin Nutr 2004;80:1526–35.
64. Di Buono M, Wykes LJ, Cole DEC, et al. J Nutr 2003;133:733–9.
65. Stipanuk MH. Annu Rev Nutr 2004;24:539–77.
66. Yap S. J Inherit Metab Dis 2003;26:259–65.
67. Mudd SH, Skovby F, Levy HL, et al. Am J Hum Genet 1985;37: 1–31.
68. Dalery K, Lussier-Cacan S, Selhub J, et al. Am J Cardiol 1995;75:1107–11.
69. Jacques PF, Bostom AG, Williams RR, et al. Circulation 1996;93: 7–9.
70. Bailey LB, Gregory JF 3rd. J Nutr 1999;129:919–22.
71. Tsang BL, Devine OJ, Cordero AM, et al. Am J Clin N 2015;101:1286–94.
72. Tsai MY, Loria CM, Cao J, et al. Mol Genet Metab 2009;98: 181–6.
73. Arnadotti M, Hultberg B, Nilsson-Ehle P, et al. Scand J Clin Invest 1996;56:41–6.
74. Davidson MH. Cardiovasc Drugs Ther 2009;23:341–2.
75. Paul R, Borah A. Biochim Biophys Acta Gen Subj 2016;1860: 1989–97.
76. Finkelstein JD. Am J Clin Nutr 1998;68:224–5.
77. Stipanuk MH, Dominy Jr JE, Lee J-I, et al. J Nutr 2006;136: 1652S–9S.
78. Stipanuk MH, Ueki I. J Inherit Metab Dis 2011;34:17–32.
79. Uhlén M, Fagerberg L, Hallström BM, et al. Science 2015;347(6220):1260419. Accessed July 10, 2022. Available at www.proteinatlas.org/ENSG000001295CDO1/tissue.
80. Winge I, Teigen K, Fossbakk A, et al. Neurochem Int 2015;90:173–84.
81. Recasens M, Benezra R, Basset P, et al. Biochemistry 1980;19:4583–9.
82. Stipanuk MH, Rotter MA. J Nutr 1984;114:1426–37.
83. Irving CS, Marks L, Klein PD, et al. Life Sci 1986;38:491–5.
84. Majtan T, Krijt J, Sokolová J, et al. Antioxid Redox Signa 2018;28:311–23.
85. Chiku T, Padovani D, Zhu W, et al. J Biol Chem 2009;284: 11601–12.
86. Landry AP, Ballou DP, Banerjee R. Chembiochem 2021;22: 949–60.
87. Tiranti V, Viscomi C, Hildebrandt T, et al. Nat Med 2009;15: 200–5.
88. Goudarzi S, Babicz Jr JT, Kabil O, et al. J Am Chem Soc 2018;140:14887–902.
89. Heafield MT, Fearn S, Steventon GB, et al. Neurosci Lett 1990;110:216–20.
90. Davies MH, Ngong JM, Pean A, et al. J Hepatol 1995;22:551–60.
91. Bradley H, Gough A, Sokhi RS, et al. J Rheumatol 1994;21: 1192–6.
92. Waring RH, Ramsden DB, Wilkinson LJ, et al. World Intellectual Property Organisation; 2007;WO2007010258A2.
93. Perry TL, Norman MG, Yong VW, et al. Ann Neurol 1985;18: 482–9.
94. Tan WH, Eichler FS, Hoda S, et al. Pediatrics 2005;116:757–66.
95. Boyer M, Sowa M, Di Meo I, et al. Mol Genet Metab 2018;124: 57–63.
96. Del Rizzo M, Burlina AP, Sass JO, et al. Mol Genet Metab 2013;108:263–6.
97. Martensson J, Hermansson G. Metabolism 1984;33:425–8.
98. Refsum H, Helland S, Ueland PM. Clin Chem 1985;31:624–8.
99. Erbe RW. Inborn errors of folate metabolism. In: Blakley RL Whitehead VM, eds. Folates and Pterins, vol 3. New York: Wiley, 1986:413–65.
100. Sturman JA, Chesney RW. Pediatr Nutr 1995;42:879–97.
101. Griffith OW. Free Radic Biol Med 1999;27:922–35.
102. Lu SC. Mol Aspects Med 2009;30:42–59.
103. Jones DP. Meth Enzymol 2002;348:93–112.
104. Townsend DM, Tew KD, Tapiero H. Biomed Pharmacother 2003;57:145–55.
105. Ghezzi P. Biochim Biophys Acta Gen Subj 2013;1830:3165–72.
106. Huxtable RJ. Physiol Rev 1992;72:101–63.
107. Hardison WG, Grundy SM. Gastroenterology 1983;84:617–20.
108. van der Meer RO, Vonk RJ, Kuipers FO. Am J Physiol—Gastrointest Liver Physiol 1988;254:G644–9.
109. González-Contreras J, Gámez JV, Gómez-Sánchez AI, et al. Nutr Hosp 2012;27:1900–7.
110. Lambert IH. Neurochem Res 2004;29:27–63.
111. Maleki V, Mahdavi R, Hajizadeh-Sharafabad, et al. Diabetol Metab Syndr 2020;12:1–9.
112. Suzuki T, Suzuki T, Wada T, et al. Nucleic Acids Res Suppl 2001;1:257–8.
113. Yasukawa T, Kirino Y, Ishii N, et al. FEBS Lett 2005;579: 2948–52.
114. Ohsawa Y, Hagiwara H, Nishimatsu SI, et al. J Neurol Neurosurg Psychiatry 2019;90:529–36.
115. Albrecht J, Schousboe A. Neurochem Res 2005;30:1615–21.
116. Shivaraj MC, Marcy G, Low G, et al. PLoS One 2012;7(8):e42935.

117. Kilb W, Fukuda A. Front Cell Neurosci 2017;11:328.
118. Mustafa AK, Gadalla MM, Sen N, et al. Sci Signal 2009;2:ra72.
119. Zhou Y, Li XH, Zhang CC, et al. Am J Physiol Cell Physiol 2016;310:C305–17.
120. Carter RN, Morton NM. J Pathol 2016;238:321–32.
121. Shi R, Proteau A, Villarroya M, et al. PLoS Biol 2010;8:e1000354.
122. Noma A, Sakaguchi Y, Suzuki T. Nucleic Acids Res 2009;37:1335–52.
123. Anderson JO, Warnick RE, Dalai RK. Poult Sci 1975;54:1122–8.
124. Mansoor MA, Bergmark C, Svardal AM, et al. Arterioscler Thromb Vasc Biol 1995;15:232–40.
125. Andersson A, Isaksson A Brattstrom L, et al. Clin Chem 1993;39:1590–7.
126. Mansoor MA, Ueland PM, Svardal AM. Am J Clin Nutr 1994;59:631–5.
127. Selhub J, Jacques PF, Wilson PWF, et al. JAMA 1993;270: 2693–8.
128. Nygård O, Vollset SE, Refsum H, et al. JAMA 1995;274: 1526–33.
129. Rasmussen K, Moller J, Lyngbak M, et al. Clin Chem 1996;42:630–6.
130. Refsum H, Smith AD, Ueland PM, et al. Clin Chem 2004;50: 3–32.
131. Trautwein EA, Hayes KC. Am J Clin Nutr 1990;52:758–64.
132. Miller RG, Jahoor F, Jaksic T. J Pediatr Surg 1995;30:953–8.
133. Vina J, Vento M, Garcia-Sala F, et al. Am J Clin Nutr 1995;61:1067–9.
134. Martensson J, Finnstrom O. Early Hum Dev 1985;11:333–9.
135. Sturman JA. Physiol Rev 1993;73:119–47.
136. Helms RA, Christensen ML, Storm MC, et al. J Nutr Biochem 1995;6:462–6.
137. Chawla RK, Berry CJ, Kutner MH, et al. Am J Clin Nutr 1985;42:577–84.
138. Martensson J, Foberg U, Fryden A, et al. Scand J Gastroenterol 1992;27:405–11.
139. Stegink LD, den Besten L. Science 1972;178:514–6.
140. Cho KH, Kim ES, Chen JD. Adv Exp Med Biol 2000;483: 605–12.
141. Lauterburg BH, Mitchell JR. J Hepatol 1987;4:206–11.
142. Sandberg M, Orwar O, Hehmann A. J Neurochem 1991;57:S152.
143. Pedersen OO, Lund-Karlsen R. Invest Ophthalmol Vis Sci 1980;19:886–92.
144. Gazit V, Ben-Abraham R, Pick CG, et al. Pharmacol Biochem Behav 2003;75:795–9.
145. Gazit V, Ben-Abraham R, Coleman R, et al. Amino Acids 2004;26:163–8.
146. Yang B-S, Wan Q, Kato N. Biosci Biotechnol Biochem 1994;58:1177–8.
147. García A, Romero C, Arroyave C, et al. Eur J Nutr 2017;56(6):2081–91.
148. Lentz SR. J. Thromb. Haemost 2005;3:1646–54.
149. Xiao Y, Huang W, Zhang J et al. Arterioscler Thromb Vasc 2015;35:60–70.
150. Yap S, Boers GH, Wilcken B, et al. Arterioscler Thromb Vasc Biol 2001;21:2080–5.
151. Joseph J, Handy DE, Loscalzo J. Cardiovasc Toxicol 2009;9: 53–63.
152. Heinz J, Kropf S, Domröse U, et al. Circulation 2010;121: 1432–8.
153. Saposnik G, Ray JG, Sheridan P, et al. Stroke 2009;40:1365–72.
154. Løland KH, Bleie O, Blix AJ, et al. Am J Cardiol 2010;105: 1577–84.
155. Green TJ, Skeaff CM, McMahon JA, et al. Br J Nutr 2010;103:1629–34.
156. Khandanpour N, Armon MP, Jennings B, et al. Br J Surg 2009;96:990–8.
157. Gaiday A, Tussupkaliyev A, Bermagambetova S, et al. Hypertens Pregnancy 2021;40:118–23.
158. Kim J, Kim H, Roh H, et al. Arch Pharm Res 2018;41:372–383.

D. OTHER COMPOUNDS WITH HEALTH RELEVANCE

CHAPTER 34

Arginine, Citrulline, Nitric Oxide, and Glutamine

Guoyao Wu and J. Marc Rhoads

CHAPTER OUTLINE

Abbreviations

AA, amino acid; **ADMA**, asymmetric dimethylarginine; **ASPEN**, American Society for Parenteral and Enteral Nutrition; **BCAA**, branched-chain amino acid; **BH_4**, (6R)-5,6,7,8-tetrahydrobiopterin; **BW**, body weight; **CAT**, cationic amino acid transporter; **ESPEN**, European Society for Clinical Nutrition and Metabolism; **HIV**, human immunodeficiency virus; **ICU**, intensive care unit; **NMMA**, N^G-monomethyl arginine; **NEAA**, nutritionally nonessential amino acid; **NO**, nitric oxide; **NOS**, nitric oxide synthase; **PepT1**, peptide transporter 1; **SLC**, solute carrier

Introduction

Arginine (2-amino-5-guanidinovaleric acid), citrulline (α-amino-δ-carbamidovaleric acid), and glutamine (α-aminoglutaramic acid) were first isolated from etiolated lupine seedlings in 1886, from watermelon (the Latin word for *citrullus*) juice in 1930, and from sugar beet juice in 1883, respectively.[1] The side chains of arginine, citrulline, and glutamine possess guanidino, ureido, and ε-amide groups, respectively (**Figure 34.1**), and these three amino acids (AAs) are readily soluble in water (186, 152, and 48.1 g/L, respectively, at 25°C). At neutral pH, arginine has a positive charge but both citrulline and glutamine have no net charge. In physiologic fluid, arginine and citrulline are stable but glutamine spontaneously and slowly forms a brown cyclic product [pyroglutamate (a potential neurotoxin)]. In all members of the animal kingdom (including humans) and in plants, arginine, citrulline, and glutamine occur naturally as L-AAs.

Research over the past 50 years has revealed that arginine, citrulline, and glutamine are interconverted in most mammals (including humans) to meet physiologic needs and adapt to pathologic alterations.[2] Renewed interest in the roles of these AAs in the nutrition of humans arose from discoveries in the 1980s that (a) arginine is the physiologic nitrogenous precursor of nitric oxide [NO (a major vasodilator, a neurotransmitter, and a potent toxin for tumors)] in animals[3-5]; and (b) glutamine protects against intestinal mucosal injury,[5] while exerting an anabolic effect on skeletal muscle through stimulating protein synthesis and inhibiting proteolysis.[6] This chapter highlights the nutrition, metabolism, and clinical applications of the physiologic L-isomers of arginine, citrulline, and glutamine in humans.

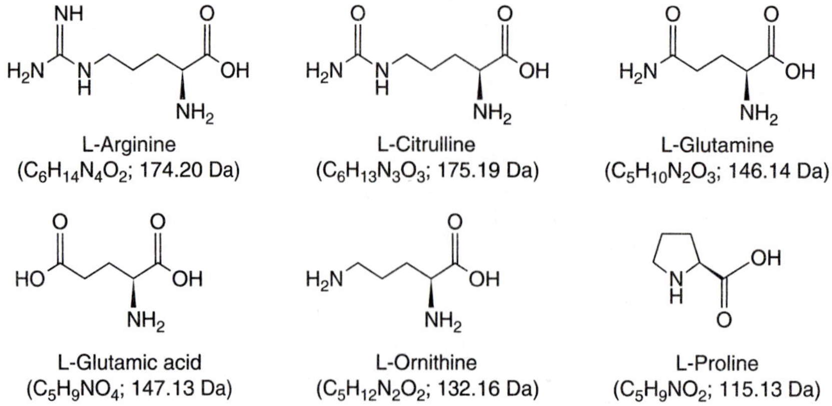

Figure 34.1 Chemical structures of L-arginine, L-citrulline, and L-glutamine, as well as related amino acids in unionized forms. Their composition and molecular weights are provided.

Composition in the Body and Physiologic Fluid

In healthy humans, glutamine is the most abundant free AA in the plasma (0.5–0.6 mM), the most abundant free α-AA in skeletal muscle (~20 mM), and accounts for 5.5% (g/g) of total AAs in tissue proteins.[7] Glutamine is an abundant free AA in human milk (0.6 mM) and accounts for 8.2% (g/g) of total AAs in its proteins.[1,8] By contrast, human milk contains only 0.02 mM free arginine, and arginine accounts for 3.9% (g/g) of milk proteins. Free arginine is present at much lower concentrations in the plasma (0.08–0.2 mM) and skeletal muscle (0.8–1.3 mM) of healthy humans than free glutamine.[1,8] However, arginine represents 7.1% of total AAs in whole-body proteins and carries the most nitrogen (14% of total AA nitrogen) among all AAs in the body.

Dietary Sources and Commercial Products

Dairy products are relatively low in total arginine but high in total glutamine and glutamate.[1] Watermelon juice is an excellent source of citrulline (i.e., 2.65 g/L) and arginine (0.56 g/L).[9] Meat (beef, pork, and poultry), eggs, fish, and shrimp not only provide large amounts of protein but also arginine, glutamine, and glutamate.[10] The mean intake of arginine by the U.S. adult population (consuming 80 g crude protein/day) was 4.4 g/day, with 25%, 20%, and 10% of people consuming daily <2.6 (suboptimal), 5 to 7.5, and >7.5 g/person, respectively.[11] According to data on the consumption of staple foods (consisting of 164 g wheat flour and 27.4 g white rice per day) by U.S. adults[12] and their AA content,[10] the mean intakes of arginine, glutamine, and glutamate by the U.S. adult population (consuming 87 g crude protein/person/day) were 5.7, 11.8, and 6.9 g/person/day, respectively. Because the content of glutamine in wheat flour is 472% greater than that in white rice,[13] the daily intake of glutamine by humans likely differs substantially among different regions of the world with diverse food patterns.

Commercially available arginine products for enteral and parenteral use in humans are arginine-HCl and arginine-α-ketoglutarate. Either citrulline or citrulline malate is marketed at U.S. health food stores as a dietary supplement, or can be used in parenteral nutrition. Free glutamine is commercially available only for oral administration. In hospitals, free glutamine can be freshly added to parenteral nutrition solution after filtration through 0.22-μm filters. By contrast, expensive but highly stable glutamine dipeptides (e.g., alanyl-glutamine [with a solubility of 568 g/L at 20°C] and glycyl-glutamine [with a solubility of 154 g/L at 20°C]) can be sterilized at high temperatures and high pressure before storage at 25°C for enteral and parenteral use.[14]

Digestion, Absorption, and Utilization by the Gastrointestinal Tract

Digestion

In the stomach (pH 2–3), dietary proteins undergo initial hydrolysis by active pepsins to large- and oligo-peptides.[14] Active rennin (chymosin) removes a glycopeptide segment from the *k*-casein of milk so that the remaining protein forms a clot or curd with calcium to slow the passage of milk in the intestinal tract. Undigested dietary proteins and their gastric digestion products enter the more alkaline duodenum (pH ~6.0; due to bile salts, pancreatic juice, and duodenal secretions), the jejunum (pH 6.5–7.0), and ileum (pH 7.0–7.5). In the lumen of the small intestine, active pancreatic proteases (including trypsin, chymotrypsin, elastase, carboxypeptidases A and B, and collagenase), along with enterocyte-derived aminopeptidase, carboxypeptidase O, and peptidases, hydrolyze dietary proteins as well as large- and oligopeptides to dipeptides, tripeptides, and free AAs. The ratio of di- and tripeptides to free AAs is about 80:20. Dietary citrulline is directly available for absorption by the small intestine. Arginine, glutamine, and citrulline are chemically stable in the lumen of the stomach, duodenum, jejunum, and ileum.[1]

Absorption by the Small Intestine

Dipeptides and tripeptides in the lumen of the small intestine are directly transported into enterocytes by H^+ gradient-driven peptide transporter 1 (PepT1; SLC15A1).[15] Free AAs or peptides containing ≥4 AA residues are not substrates for PepT1. Within the enterocytes, dipeptides and tripeptides are hydrolyzed rapidly by intracellular peptidases to form free AAs, including arginine and glutamine.[16] A small proportion of the small peptides may exit the enterocytes into the blood[1] or be taken up by intestinal microbes for utilization.[17]

Free AAs in the lumen of the small intestine are rapidly absorbed by enterocytes into the lamina propria (and then the blood) primarily via transmembrane transporters (**Figure 34.2**). Specifically, arginine is taken up by the enterocytes across their apical membrane through Na^+-independent transporters for basic AAs: y^+ [CAT-1 (SLC7A1) and CAT3 (SLC7A3)] and $b^{0,+}$ [$b^{0,+}$AT/rBAT (SLC6A14)] and, to a lesser extent, two Na^+-dependent transporters, with one being identified as ATB$^{0,+}$ (SLC6A14).[18] Glutamine and citrulline enter the enterocytes across their apical membranes via Na^+-dependent transporters, such as systems A [SNAT1 (SLC38A1), SNAT2 (SLC38A2), and SNAT4 (SLC38A4)]; N [SNAT5 (also known as SN2 (SLC38A5)]; NBB [B^{0}AT1 (SLC6A19)]; and ASC [ASCT2 (also known as ATB^0 (SLC1A5)], and $B^{0,+}$ [ATB$^{0,+}$ (SLC6A14)]. Arginine exits the enterocytes through their basolateral membranes into the lamina propria via Na^+-independent AA transport systems: y (CAT-1) and y^+L [y^+LAT1/4F2hc (SLC7A7/SLC3A2) and y^+LAT2/4F2hc (SLC7A6/SLC3A2)]. Because basic AAs share the same transporters,[18] improper ratios of arginine to lysine in diets (>3 for infants, g/g) may cause antagonism in protein nutrition.[19] By contrast, the export of glutamine and citrulline out of the enterocytes across their basolateral membrane into the lamina propria is mediated primarily by some Na^+-dependent AA transporters.[16] The latter include systems A (SNAT1, SNAT2, SNAT4) and N (SNAT5) and, to a lesser extent, Na^+-independent AA transports such as systems L1 [LAT1/4F2hc (SLC7A5/SLC3A2)], L2 [LAT2/4F2hc (SLC7A8/SLC3A2), y^+L [y^+LAT1/4F2hc (SLC7A7/SLC3A2), and y^+LAT2/4F2hc (SLC7A6/SLC3A2)].[16] y^+LAT1 and y^+LAT2 act only in conjunction with the membrane glycoprotein anchor 4F2hc and can also mediate the export of arginine out of the enterocytes across their basolateral membrane in exchange for equimolar glutamine from the lamina propria. The jejunum is the major site for the absorption of dipeptides, tripeptides, and free AAs, followed by the ileum and duodenum.[1] Due to their polarization, enterocytes take up glutamine, but no detectable amounts of other AAs (including arginine, citrulline, glutamate, ornithine, and proline) from the arterial blood.

In patients with lysinuric protein intolerance, mutations of the SLC7A7 gene result in the impaired export of arginine and other basic AAs from the enterocytes into the blood as well as the impaired reabsorption of these AAs from the lumen of renal tubules into the blood.[20] Arginine deficiency occurs in affected individuals, and this problem can be overcome by dietary supplementation with citrulline. To date, no inherited defects in PepT1 or major glutamine and citrulline transporters have been identified in humans.

Utilization of Citrulline and Arginine by the Small Intestine

Within enterocytes, dietary citrulline is converted into arginine in the presence of aspartate and ATP via argininosuccinate synthase and argininosuccinate lyase.[21] These cells utilize arginine via protein synthesis, NO production by NO synthase (NOS), and hydrolysis to ornithine plus urea by arginases I and II.[22] Protein synthesis accounts for about 10% of dietary AAs extracted by the small intestine.[1] Degradation of arginine by the NOS pathway is quantitatively limited in enterocytes of both infants and adults.[22] Likewise, arginase activity is negligible in enterocytes of preweaning infants but is induced by cortisol during weaning to degrade dietary arginine.[23] Thus, healthy adult humans extract a large amount (38%) of dietary arginine[24] via the arginase pathway to form ornithine and proline, with little oxidation of arginine carbons to CO_2.[22] In addition, small amounts of arginine and lysine are used for the synthesis of homoarginine by the small

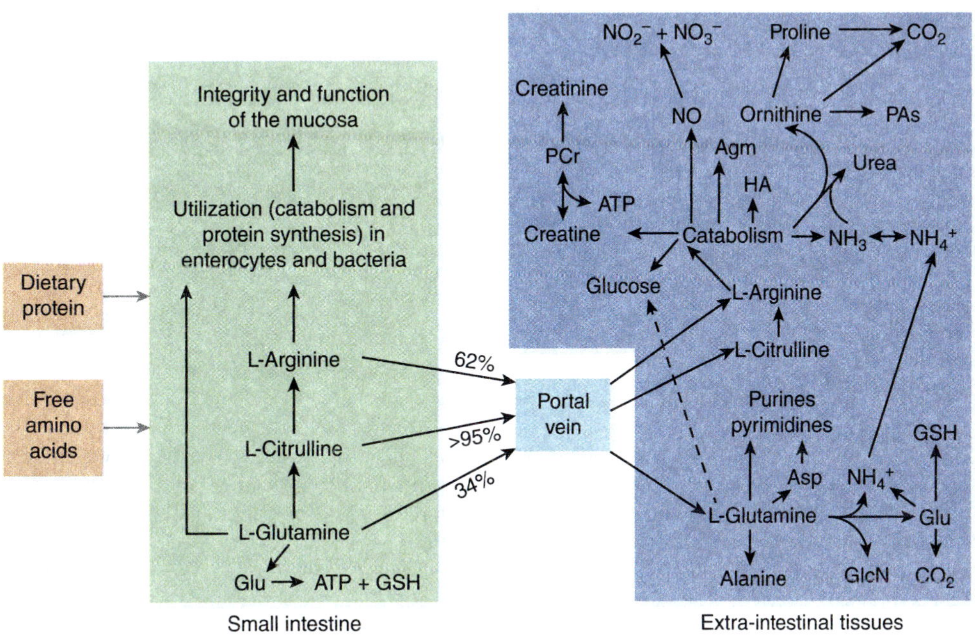

Figure 34.2 An overview of protein digestion and the use of its products in humans. Dietary protein is hydrolyzed by proteases and peptidases in the lumen of the stomach and the small intestine. *Agm*, agmatine; *Asp*, L-aspartate; *Cr*, creatine; *GlcN*, glucosamine-6-phosphate; *Glu*, L-glutamate; *GSH*, glutathione; *HA*, homoarginine; *NO*, nitric oxide; *PAs*, polyamines (putrescine, spermidine, and spermine); *PCr*, phosphocreatine.

intestine, possibly via arginine:glycine amidinotransferase.[25] In infants, almost all dietary arginine enters the portal vein, but in adults, 62% of dietary arginine is available for utilization by extraintestinal tissues.[24]

There is complex compartmentalization of citrulline synthesis in enterocytes. In these cells, arginase I (cytosol)- and II (mitochondria)-derived ornithine is preferentially channeled to the formation of proline and is not converted into citrulline.[22] Dietary or blood ornithine is a poor substrate for citrulline synthesis by enterocytes.[2] Rather, extracellular ornithine is preferentially metabolized to proline via ornithine aminotransferase and pyrroline-5-carboxylate reductase.[22] This explains why orally, or intravenously administered ornithine or intravenously administered proline cannot be used for citrulline production in humans or other mammals (e.g., swine and rats). By contrast, arginine is readily formed from citrulline in the cytoplasm of all cell types in mammals (including young and adult humans). Thus, citrulline, but not ornithine, must be supplemented to individuals with a deficiency of argininosuccinate synthase and/or argininosuccinate lyase to prevent arginine deficiency.

Citrulline is a useful biomarker for enterocyte mass and intestinal failure in infants and adults.[26-29] Its concentration in the plasma is reduced in: (a) short bowel syndrome, (b) active celiac disease, (c) a deficiency of intestinal ornithine carbamoyl transferase, (d) sepsis, (e) human immunodeficiency virus infection (HIV), and (e) following abdominal radiation. The plasma concentration of citrulline is usually proportional to the severity of disease. For example, in HIV enteropathy, a level of 22 μM predicts the need for parenteral nutrition.[27] Citrulline is currently the best biomarker for enteral autonomy (without parenteral support) in short gut syndrome in adults and children.[28] In preterm infants with a short gut, plasma citrulline concentration increased in those with bowel adaptation but did not increase in those remaining on parenteral nutrition, and was superior to bowel length in predicting which infants/children would remain on parenteral nutrition.[29] In addition, citrulline was useful to estimate bowel adaptation and mass in individuals with intestinal transplant,[30] and in the injured gut following chemotherapy.[31] Following intestinal transplantation, citrulline was a biomarker to estimate the time for recovery after acute cellular rejection.[32] Before the onset of severe graft-versus host disease, plasma citrulline concentration before stem cell transplant predicted the outcome.[33] In a group of adults with intestinal transplantation, plasma citrulline concentration was low shortly after the transplantation procedure; however, 31 to 45 days later, citrulline was similar to that in healthy controls; and it was 83% specific in detecting acute cellular rejection (but only 30% sensitive; it was 69% sensitive and 77% specific when adjusted for chronic renal failure).[34]

Utilization of Glutamine by the Small Intestine

Glutamine is extensively catabolized via the phosphate-activated glutaminase pathway in enterocytes of infants and adults as a major metabolic fuel, with ammonia, alanine, aspartate, ornithine, citrulline, arginine, and proline as nitrogenous products.[35,36] After weaning, some ammonia is converted into urea by enterocytes via a series of enzyme-catalyzed reactions analogous to the hepatic urea cycle,[37] and this intestinal pathway serves as the first line of defense against the toxicity of ammonia derived from the diet, intestinal lumen, and enterocyte metabolism. The sequential conversion of glutamine into ornithine and citrulline occurs only in the mitochondria of enterocytes (**Figure 34.3**). Most of the citrulline synthesized from glutamine, glutamate, and proline in enterocytes is converted locally into arginine in infants but is released into the blood for extraintestinal arginine synthesis in adults.[2] Besides citrulline and arginine synthesis, enterocytes also use glutamine to produce glucosamine-6-phosphate (the precursor for the synthesis of all glycoproteins) and glutamate by glutamine:fructose-6-phosphate transaminase, as well as purines and pyrimidines via a series of enzyme-catalyzed reactions.[1]

Most of dietary glutamine and glutamate are extensively extracted by the small intestine and other tissues of the portal-drained viscera in first-pass metabolism by infants and adults. For example, 65% of the enteral glutamine underwent first-pass splanchnic extraction in preterm infants with very low birth weight (0.73–1.34 kg), with the oxidation of glutamine to CO_2 in the splanchnic bed accounting for 57% of whole-body CO_2 production from glutamine.[38] Dietary glutamate was almost entirely utilized via first-pass metabolism in the splanchnic bed of premature infants,[39] with oxidation to CO_2 accounting for 86% of its metabolic fate.[40] Similarly, in healthy adult humans, 64% and 96% of the enterally delivered glutamine[41] and glutamate[42] were removed, respectively, in the first pass by the splanchnic bed, with 13% of the extracted glutamine being used for citrulline production.[43] Thus, most of the glutamine and glutamate extracted by the portal-drained viscera are oxidized to CO_2, and only ~10% of each AA is utilized for protein synthesis. Based on the finding that the fractions of plasma glutamate derived from plasma glutamine in adults were 34% and 76% during the intravenous and intragastric infusion of glutamine, respectively,[44] the conversion of the glutamine to glutamate occurred primarily in the splanchnic bed [most likely in enterocytes and in cells (e.g., immune cells and fibroblasts) of the lamina propria]. By contrast, glutamine synthesis in the intestinal mucosa is quantitatively insignificant due to low glutamine synthetase activity.[1] Furthermore, glutamine is essential for the synthesis of nucleotides, ATP, and amino sugars in the gut, while providing glutamate for glutathione production.[45]

Metabolism of Arginine and Citrulline by Extraintestinal Tissues

In the liver, periportal and perivenous hepatocytes express low levels of the y^+ system [CAT1, CAT2A and CAT2B (SLC7A2), and CAT3] for the uptake of arginine.[19] Thus, in mammals, only about 10% of arginine (e.g., 8.6% in humans[46] and 8.5% in pigs[1]) in the portal venous and hepatic arterial blood is extracted by the liver. These mammals do not take up citrulline from the blood due to a lack of transmembrane transporters, and there is no net flux of citrulline across the liver.[47] This is physiologically and nutritionally important to maximize the availability of enterally and intravenously delivered, as well as enterocyte-derived citrulline and arginine to extrahepatic tissues. Mammalian hepatocytes have exceedingly high arginase activity to degrade arginine. The intracellular concentration of arginine in hepatocytes is very low, ranging from 0.03 mM (e.g., in rats) to 0.1 mM (e.g., in pigs), depending on species, compared with 0.5 to 10 mM for other AAs.[19] Arginine is both an intermediate and an allosteric activator of N-acetylglutamate synthase of the urea cycle in periportal hepatocytes.[1] Thus, arginine is essential for the hepatic

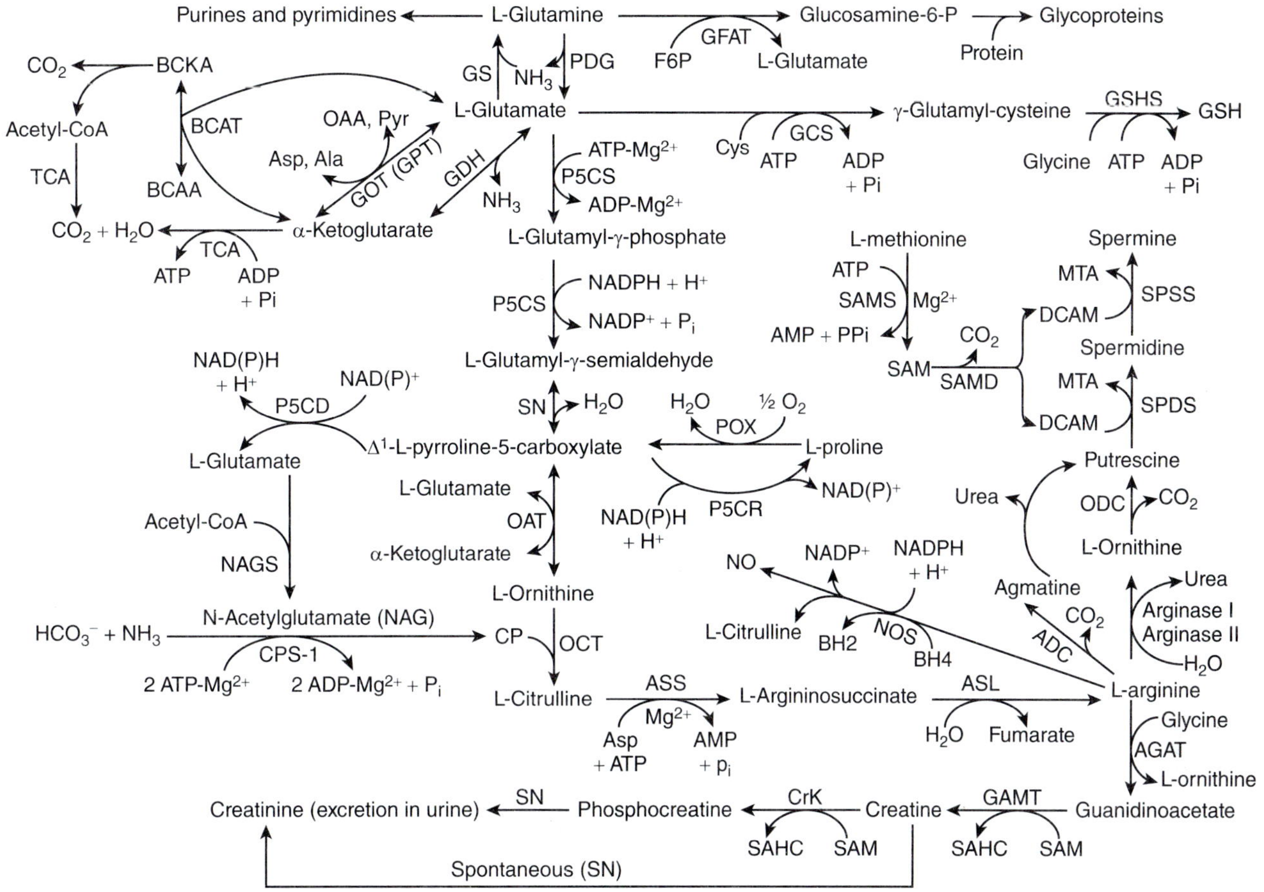

Figure 34.3 Metabolism of arginine, citrulline, and glutamine in humans. These pathways are cell-, tissue-, and age-specific, and are compartmentalized within given cells. *ADC*, arginine decarboxylase; *AGA*, agmatinase; *AGAT*, arginine:glycine amidinotransferase; *ASL*, argininosuccinate lyase; *ASS*, argininosuccinate synthase; *AS*, argininosuccinate; *Asp*, aspartate; *BCAA*, branched-chain amino acid; *BCAT*, branched-chain amino acid transaminase; *BCKA*, branched-chain α-ketoacid; *BH4*, (6*R*)-5,6,7,8-tetrahydro-L-biopterin; *BH2*, dihydro-L-biopterin; *CrK*, creatine kinase; *CP*, carbamoylphosphate; *CPS-I*, carbamoylphosphate synthetase-I (ammonia); *DCAM*, decarboxylated *S*-adenosylmethionine; *GDH*, glutamate dehydrogenase; *GA*, guanidinoacetate; *Glu*, glutamate; *Gln*, glutamine; *GMAT*, guanidinoacetate N-methyltransferase; *GCS*, γ-glutamyl-cysteine synthetase; *GOT*, glutamate-oxaloacetate transaminase; *GPT*, glutamate-pyruvate transaminase; *GSH*, glutathione; *GSHS*, glutathione synthetase; α-KG, α-ketoglutarate; *MTA*, methylthioadenosine; *NAG*, N-acetylglutamate; *NAGS*, N-acetylglutamate synthase; *NO*, nitric oxide; *NOS*, nitric oxide synthase; *OAA*, oxaloacetate; *OAT*, ornithine aminotransferase; *OCT*, ornithine carbamoyltransferase; *ODC*, ornithine decarboxylase; *POX*, proline oxidase; *P5CD*, pyrroline-5-carboxylate dehydrogenase; *P5CR*, pyrroline-5-carboxylate reductase; *P5CS*, pyrroline 5-carboxylate synthase; *SAM*, *S*-adenosylmethionine; *SAMD*, *S*-adenosylmethionine decarboxylase; *SAHC*, *S*-adenosylhomocysteine; *SPDS*, spermidine synthase; *TCA*, tricarboxylic acid cycle.

detoxification of ammonia as urea in mammals. Under physiologic conditions, there is no net synthesis of arginine by the liver.[2] Infants and adults with a plasma concentration of arginine at ≤32 μM are at high risk for hyperammonemia.[48]

Arginine and citrulline exit the liver via the hepatic vein into the systemic blood circulation.[1] All extra-hepatic and extra-intestinal cell types actively take up citrulline and arginine primarily via neutral (e.g., systems A and N) and basic (e.g., CAT1, CAT2A, CAT2B, y^+L, $b^{0,+}$, and $B^{0,+}$) AA transporters, respectively.[19] The kidneys are quantitatively the major organ for converting citrulline into arginine, and this metabolic pathway occurs in other cell types such as macrophages and endothelial cells.[49] The rate of de novo synthesis of arginine from citrulline represents only a small fraction (~10%) of whole-body arginine flux (plasma flux), which includes exogenous supply (e.g., diet; ~30%), and endogenous flux (mainly protein degradation; ~70%) in infants and adults.[2] Because most of the protein-derived arginine is recycled into the protein pool via protein synthesis and the remainder is catabolized via multiple pathways, intracellular protein turnover does not directly contribute net arginine to tissues in the state of nitrogen balance.[1] However, other AAs (e.g., glutamine and branched-chain AAs [BCAAs]) released from proteins play a role in the endogenous synthesis of arginine via the intestinal-renal axis.[49]

In humans, arginine degradation generates NO, ornithine, urea, polyamines, creatine, agmatine, and homoarginine, with ornithine being further converted into proline, glutamate, and glutamine (Figure 34.3). In a 70-kg adult person consuming 5.0 g arginine in protein (or 4.6 g digestible arginine) per day, 2.85 g arginine enters the portal circulation, and 2.60 g arginine is available to extrahepatic tissues. In the fed state, endogenous synthesis provides 1.52 g arginine/day in a 70-kg adult,[50] which is much less than the amount of arginine (2.3 g/day) required for creatine production.[2] Because most of dietary arginine enters the portal circulation, a single oral administration of arginine (e.g., 1.8 or 4.9 g) to adult humans increases the plasma concentration of arginine and whole-body NO synthesis in a dose-dependent manner.[51] Quantitatively, a negligible amount of arginine is catabolized to agmatine and homoarginine,[52] and ~1% of intravenously administered arginine is metabolized to NO.[50] The arginine:glycine amidinotransferase and other (primarily arginase) pathways contribute to the utilization of 56% and 44% of arginine in the extraintestinal tissues of adult humans, respectively.

Arginase I is expressed abundantly in hepatocytes and, to a limited extent, in extrahepatic cells, including the enterocytes of postweaning mammals (as noted previously), endothelial cells, mammary epithelial cells, macrophages, and red blood cells (only in primates).[53] By contrast, arginase II is widely expressed at relatively low concentration in virtually all mitochondria-containing extrahepatic cells (including neuronal, renal, vascular, and muscle cells) and plays an important role in regulating the synthesis of NO, proline, and polyamines.[19] Arginase released from cells and tissues is present in extracellular fluid (e.g., plasma and wounds) and hydrolyzes arginine to ornithine plus urea. Under conditions of inflammation and injury (particularly liver disease), a high activity of arginase in plasma results in a severe deficiency of arginine.[53] Likewise, arginase released from tumors can lead to arginine depletion in the surrounding microenvironment. Of note, extrahepatic cells do not convert arginase II-derived ornithine into citrulline, due to either the lack of carbamoylphosphate synthase-I and ornithine carbamoyltransferase or metabolic compartmentalization for proline formation, but can synthesize arginine from diet-, blood-, and NOS-derived citrulline.[2]

There are three isoforms of NOS: NOS1 (neuronal), NOS2 (inducible), and NOS3 (endothelial) in humans.[54] The NOS1 isoform was first discovered in neuronal tissues, the NOS2 isoform was originally found to be inducible under certain conditions in macrophages, and the NOS3 isoform was first identified in endothelial cells. Both NOS1 and NOS3 are Ca^{2+}-dependent and constitutively expressed, whereas NOS2 is Ca^{2+}-independent and expressed abundantly in response to immunologic challenges. The NOS isoforms can be present in the plasma membrane caveolae, cytoplasm, nucleus, rough endoplasmic reticulum, and mitochondria, depending on isoform and cell type. The NOS isoforms are encoded by three different genes, have 51% to 57% homology in nucleotide sequences, and require arginine, O_2, (6R)-5,6,7,8-tetrahydrobiopterin (BH4), NADPH, calmodulin (bound tightly to NOS2), FMN, and FAD for NO synthesis.[2] Arginine, BH4, and heme promote and stabilize the active dimeric form of all isoforms of the NOS. All isoforms of NOS are inhibited by asymmetric dimethylarginine (ADMA) and N^G-monomethyl arginine (NMMA), and symmetric dimethylarginine.[2] For example, the addition of 2 μM ADMA or NMMA to culture medium containing 0.2 mM arginine and physiologic concentrations of other AAs reduced NO synthesis in endothelial cells by 7%, compared with 1 μM ADMA or NMMA.[55]

The K_m values of NOS for arginine are 3–20 μM (≤10% of arginine concentration in extrahepatic cells), depending on isoform.[2] However, increasing extracellular concentration of arginine from 0.05 to 10 mM dose-dependently increases NO production by endothelial cells and activated macrophages.[56] For example, dietary supplementation with 10 g arginine/day (in three equally divided doses) for 1 week to healthy humans reduced systolic blood pressure and diastolic blood pressure by 6.2 and 6.8 mmHg, respectively, compared with control individuals consuming 4.1 g arginine/day.[57] Arginine promotes the translation of NOS1 mRNA in cytokine-stimulated astrocytes[58] and BH4 synthesis in endothelial cells,[56] therefore, enhancing NO production by NOS2 and NOS3, respectively. Under both physiologic and pathological conditions, arginase and NOS compete for arginine. Therefore, relative changes in their enzymatic activities serve as major determinants of NO and polyamine production in many cell types.[53] Both arginase-II and NOS2 are induced by proinflammatory cytokines (e.g., interleukin 1β and tumor necrosis factor-α) and lipopolysaccharide to promote arginine degradation.[53] Thus, a marked elevation in arginase activity reduces the availability of arginine for NOS2 and provides a mechanism responsible for the survival of parasites, bacteria, and tumors.[53]

In mammals, citrulline is effectively converted into arginine.[2] Thus, increasing the oral administration of citrulline to adult humans from 1 to 15 g/day enhanced plasma arginine concentration in a dose-dependent manner, with <5% of the administered citrulline being excreted in the urine.[59] In addition, oral administration of 6 g citrulline/day for 4 weeks to healthy young men (receiving the cold pressor test) decreased the brachial systolic blood pressure, aortic systolic blood pressure, and aortic pulse pressure by 6, 4, and 3 mmHg, respectively.[60] Because the half-life of citrulline in the blood is ~95% greater than that of arginine, enterally or intravenously delivered citrulline can increase plasma arginine concentration for a longer period of time than the same dose of arginine in mammals, including humans.[61]

Metabolism of Glutamine by Extraintestinal Tissues

Transmembrane transporters of glutamine in the liver and extrahepatic tissues include those expressed in the small intestine, as noted previously. The periportal hepatocytes take up glutamine, but not glutamate and aspartate, from the portal venous blood and the hepatic artery, whereas both glutamate and aspartate can enter the perivenous hepatocytes.[62] In the postabsorptive state, there is no net flux of glutamine across the liver in humans,[47] because periportal hepatocytes extract glutamine from the blood, but perivenous hepatocytes synthesize glutamine from glutamate and ammonia.[62] At elevated concentration (e.g., ≥1 mM) in the plasma, the liver has a net uptake of glutamine for oxidation to CO_2 and urea.[1] By contrast, in humans, skeletal muscle releases a large amount of glutamine in both postabsorptive and fed states.[63,64]

Because most dietary glutamine and glutamate are catabolized by the mucosa of the small intestine, these two AAs must be synthesized de novo by extraintestinal tissues. Their major precursors are BCAAs, which transaminate with α-ketoglutarate (formed mainly from glucose) to produce glutamate.[65] BCAAs are derived from diets and, under catabolic conditions, whole-body protein degradation. The degradation of many AAs also generates glutamate in a tissue-specific manner. Most cell types, mainly skeletal muscle but also heart, adipocytes, brain, placenta, and mammary glands, express glutamine synthetase to convert glutamate and ammonia into glutamine.[62] In the postabsorptive state (i.e., overnight fast), glutamine represents 25% (mol/mol) of total AAs released by human skeletal muscle.[63,64] In the fed state with a normal intake of protein, the skeletal muscle of humans continues to synthesize and export glutamine, which accounts for 99%, 83%, and 71% (mol/mol) of all released AAs at 1, 2, and 1 to 4 hours after meal ingestion, respectively.[64] This indicates the dominance of glutamine in the interorgan metabolism of AAs involving primarily skeletal muscle, kidneys, liver, small intestine, and white adipose tissue. The rate of release of glutamine by skeletal muscle, which is determined by its synthesis and catabolism,[1] is relatively constant after an overnight fast and during the postprandial period.[64] At the whole-body level, glutamine synthesis is very active, accounting for 82% and 70% of endogenous glutamine flux in preterm infants[66] and adult humans,[67] respectively. The physiologic significance of glutamine synthesis is epitomized by an autosomal recessive disorder resulting from mutations of the glutamine synthetase gene in humans.[68] Despite enteral or parenteral nutrition, affected newborn infants had a near absence of free glutamine in their serum (only 2 μM), exhibited hyperammonemia and severe neurologic problems, and died from multiple organ failure in the first month after birth.[68]

Renal ammoniagenesis from glutamine is vital to regulate the acid-base balance in humans in response to metabolic acidosis.[62] Under this condition, the liver releases glutamine, the use of glutamine by both the small intestine and immunocytes is substantially reduced, and the uptake of glutamine by the kidneys is greatly increased. During fasting, glutamine is a major precursor for hepatic and renal gluconeogenesis so that the liver and kidneys provide approximately equal amounts of glucose for use by the brain, red blood cells, retinal cells, renal medulla, and many other cell types including lymphocytes and macrophages.[69,70] In immunologically activated macrophages and lymphocytes, glutamine is a major metabolic fuel, in addition to its role in the phagocytosis, as well as the production of anti-inflammatory cytokines and antibodies.[14] Conversely, glucosamine, a product of glutamine metabolism, inhibits the expression of NOS2,[71] thereby alleviating autoimmune inflammatory injury in joints, such as arthritis in humans.

Glutathione, which is synthesized from glutamate, glycine, and cysteine, is abundant in red blood cells (RBCs) as a potent antioxidant. In humans, these cells contain glutamate in high concentration, but do not take up glutamate due to the absence of transmembrane transporters.[72] An exceedingly high enzymatic activity of glutamine: fructose-6-phosphate transaminase occurs in mammalian RBCs for the conversion of glutamine into glutamate.[2] This reaction may be crucial for the integrity and function of these cells.

In the postabsorptive state, uptake by the small intestine and other organs (primarily kidneys) accounts for about one-third and two-thirds of the plasma glutamine flux in healthy humans, respectively.[49] At the whole-body level, oxidation to CO_2 accounted for 87% of glutamine utilized by preterm infants, as determined from intravenously delivered [^{13}C]glutamine.[38] In adult humans, 73% and 58% of the enterally and intravenously delivered [^{13}C] glutamine was oxidized to CO_2, respectively, in adults, whereas 78% and 79% of the enterally and intravenously delivered [^{13}C] glutamate was oxidized to CO_2, respectively.[42,73] Glutamine in the blood readily enters the brain, but glutamate (neurotoxic at elevated concentrations) cannot cross the blood-brain barrier.[14] At elevated concentration (e.g., ≥2 mM as occurring in persons with inborn errors of urea-cycle defects), glutamine reduces blood flow to the brain through inhibiting the generation of NADPH (an essential cofactor of NOS) via the pentose cycle and, therefore, NO synthesis by endothelial cells.[1]

Use of Arginine, Citrulline, and Glutamine in Clinical Nutrition and Therapies

Arginine as a Nutritionally Essential AA in Humans

Neonates, particularly preterm infants, require dietary arginine for optimum growth and gut health.[48] Based on short-term nitrogen balance studies, arginine has long been considered a nutritionally nonessential (dispensable) AA (NEAA) for adult humans. However, as noted previously, endogenous synthesis alone is inadequate to meet the metabolic needs (e.g., creatine synthesis) of infants or adults for arginine. This notion is further supported by studies on fertility in males and females. For example, feeding an arginine-deficient diet to adult men for 9 days decreased both the number and motility of sperm cells by 90%.[74] Similarly, a deficiency of dietary arginine in young adult male rats over a period of 2 months resulted in progressive damage to testes, the absence of sperm production, as well as the filling of the lumina of the tubules with cellular debris, leukocytes, and macrophages.[74] Clearly, men cannot synthesize sufficient arginine to maintain their reproductive function such as spermatogenesis. In addition to males, females also require dietary arginine for the survival, development, and growth of embryos.[61] Furthermore, inadequate provision of arginine in diets impairs NO synthesis by endothelial cells, resulting in cardiac dysfunction in men and women.[56] Likewise, in humans with hyperammonemia resulting from arginine deficiency due to an insufficient synthesis of citrulline by the small intestine, administration of arginine is essential to prevent mortality,[48,52] further indicating the dietary essentiality of arginine. Thus, physiologic needs beyond nitrogen balance are important criteria for the classification of arginine as a nutritionally essential AA in humans.

Glutamine as a Conditionally Essential AA in Humans

Historically, little was known about the content of glutamine in food or animal-tissue proteins due to technical difficulties in its analysis. Thus, there were no nitrogen balance studies involving infants or adult humans fed a glutamine-free purified diet. Disappointingly, despite the lack of experimental absence, glutamine has long been classified as an NEAA for humans. Hou et al.[75] reported that although young rats could synthesize glutamine, its omission from a purified diet containing adequate amounts of all other AAs impaired their growth. Similar results have been reported for weanling pigs.[76] Clearly, glutamine cannot be synthesized sufficiently in young mammals and, therefore, must be present adequately in diets to sustain their growth. In addition, both infants and adults may need exogenous glutamine for improved health in catabolic states associated with its severe deficiency.[14]

Arginine and Glutamine as Functional AAs in Human Nutrition

Besides serving as building blocks of proteins, arginine and glutamine play important roles in regulating: the expression of genes (e.g., ornithine decarboxylase and heat shock proteins), cell signaling pathways (e.g., the mechanistic target of rapamycin [mTOR]), the digestion and absorption of dietary nutrients, DNA and protein synthesis, proteolysis, the metabolism of glucose and lipids, endocrine status (e.g., the secretion of insulin and glucocorticoids), male and female fertility, the acid–base balance, antioxidative responses, the detoxification of xenobiotics and endogenous metabolites, neurotransmission, inflammation, and immunity in mammals.[45,53,61,77] For example, glutamine is a substrate for pyrimidine synthesis needed for villus restitution, while also activating AMP-dependent protein kinase, mitogen-activated protein kinase, and c-Jun nuclear kinase signaling.[78] In addition, glutamine has powerful impacts on gut integrity, intestinal crypt cell proliferation, and anti-apoptosis; it upregulates heat shock protein expression as a response to injury and reduces circulating proinflammatory cytokines.[14] Furthermore, glutamine increases intestinal epithelial electroneutral NaCl absorption in viral or cryptosporidial enteritis—counterbalancing excessive fluid losses in the injured gut.[79] To summarize many studies, glutamine is physiologically and nutritionally essential for the integrity (e.g., the expression and distribution of tight-junction proteins), function (e.g., the absorption of water, glucose, and AAs), and immunity (e.g., the production of IgA) of the mucosa of the small intestine. These lines of evidence led to the new nutritional concept of functional AAs, which are defined as those AA that participate in and regulate key metabolic pathways to improve the health, survival, growth, development, lactation, and reproduction of humans and other animals.[1] Such biochemical processes include: (a) intracellular protein turnover (synthesis and degradation); (b) cell- and tissue-specific synthesis and catabolism of AAs; (c) generation of small peptides, nitrogenous metabolites, and sulfur-containing substances (e.g., H_2S); (d) urea cycle and uric acid synthesis; (e) lipid and glucose metabolism; (f) one-carbon unit metabolism and DNA synthesis; and (g) cellular redox signaling. Dietary requirements for arginine and glutamine are affected by a plethora of nutritional, physiologic, pathologic, and environmental factors.[1]

Reduced Availability of Arginine, Citrulline, and Glutamine in Stressed Patients

Concentrations of arginine, citrulline, and glutamine are commonly reduced in the plasma of patients during severe stress, such as starvation, sepsis, multiorgan system injury, or burns; and in these conditions, glutamine balance is substantially negative.[28,70] As a specific example, after abdominal surgery, intramuscular muscle glutamine concentration decreased by 25% to 60% and diminished to even lower concentration in sepsis.[7] Much of this decrease is mediated by glucocorticoid action on skeletal muscle.[14] Glutamine loss from the muscle is a major factor in the rapid wasting that can develop following surgical procedures. There is evidence that plasma glutamine concentration is the best index of glutamine depletion in humans.[80]

Use of Arginine, Citrulline, and Glutamine in Human Nutrition

Dietary supplementation with arginine, citrulline, or glutamine alone or in combination may be a necessary strategy to improve human health and well-being under many physiologic and pathologic conditions, such as exercise, diabetes, obesity, and vascular disorders. Patients with arginase deficiency or hepatic dysfunction have generally been excluded from clinical trials involving arginine and glutamine supplementation, respectively. Much work with animal models has shown the efficacy of enteral (e.g., 0.5%–1% in diet) or intravenous (e.g., 0.087–0.174 g/kg/day) administration of arginine on improving reproductive, cardiovascular, pulmonary, renal, gastrointestinal, liver, immune, and skin integrity and functions; ameliorating obesity, diabetes, and metabolic syndrome; and facilitating wound healing, enhancing insulin sensitivity, and maintaining tissue structure.[61] Likewise, extensive research involving animal models has identified beneficial effects of glutamine supplementation of enteral (e.g., 0.5%–2% in diet) or parenteral (0.3–0.5 g/kg/day) nutrition, alleviating or preventing intestinal dysfunction, shock, infection, inflammation, burns, trauma, infection, cancer with chemotherapy or irradiation, and other catabolic stresses.[14] As summarized in **Tables 34.1** and **34.2**, translational investigations have confirmed similar findings in humans under selected conditions for arginine[81-115] and glutamine[116-147] supplementation. Most of the clinical trials have involved adults, but some studies included critically ill neonates, as well as low birth weight and preterm infants. As of 2023, there have been hundreds of clinical trials involving arginine and glutamine, but only recently have we been able to identify whether and among whom supplementation can be advantageous.

Although there is no direct proof that normalization of plasma glutamine concentration is associated with improved clinical outcomes, it has been shown that individuals with adequate circulating glutamine have improvements in some health indices (Table 34.2). Of note, in a prospective study of premature infants in the United States, plasma glutamine and arginine concentrations were found to decrease substantially 1 week before the onset of necrotizing enterocolitis, whereas values gradually increased in healthy premature infants that did not develop the condition.[148] Interestingly, a Cochrane review of 12 randomized controlled trials involving a total of 2,877 preterm infants did not find evidence for improvements in mortality, necrotizing enterocolitis, or sepsis in response to glutamine supplementation.[149] Low plasma glutamine concentration at ICU admission and unfavorable outcome are also linked,[14] and this group of critically ill patients may be the most suitable candidates for glutamine supplementation. For example, meta-analysis of 4,671 critically ill adults indicated reductions in infections and the duration of mechanical ventilation with glutamine supplementation, but there was risk of bias and study heterogeneity.[150] Another meta-analysis, of 30 trials with 3,696 patients, found no significant difference in ICU or hospital mortality or the number of infectious complications with glutamine treatment, although authors recommended further randomized controlled trials to explore glutamine in more specific cohorts.[151] Finally, a large multicenter trial showed no benefit of glutamine and a trend toward increased mortality,[147] possibly related to inclusion of patients with liver or kidney disease, leading to American Society for Parenteral and Enteral Nutrition (ASPEN) guidelines

Table 34.1 Reports of Positive Effects of Arginine (Arg) or Citrulline (Cit) in Humans*

Condition	Amount of Arg or Cit	n	Effects	Ref.
Oral Administration				
Obesity	3 × 3 g Arg/day, 12 wk	20	White WAT ↓; Metabolic profiles ↑; LMB ↑	[81]
Type 2 diabetes	8.3 g Arg/day, 21 days	33	Blood glucose ↓; Metabolic profiles ↑	[82]
Endothelial dysfunction	7 g Arg × 3/day, 21 days	27	Endothelium-dependent relaxation in HCh ↑	[83]
	7 g Arg/day, 3 days	10	Endothelium-dependent relaxation in smokers ↑	[84]
	3 × 3 g Arg/day; 6 mo	26	Coronary endothelial dysfunction ↓	[85]
	2 × 3 g Arg/day; 10 days	7	Oral high fat-induced endothelial dysfunction ↓	[86]
Infertility in men	0.413 g Arg/day for 2–24 wk	6	Sperm counts ↑; Successful pregnancies	[87]
Erectile dysfunction	6 g Arg/day, 3 mo	51	Penile blood flow ↑; Erectile dysfunction ↓	[88]
Pre-eclampsia in women	3 g Arg/day, 4 wk	50	Maternal blood pressure ↓; Fetal well-being ↑; Birth weight ↑	[89]
	2 × 3.3 g Arg/day, 8–26 wk	228	Incidence of pre-eclampsia ↓	[90]
Preterm labor	3 g Arg/day, 7 days	37	Vascular resistance ↓; Fetal-placental blood flow ↑	[91]
SCD vasculopathy	2 × 0.05 g Cit/kg BW/day, 4 wk	5	Normalizing blood leukocyte counts; Well-being ↑	[92]
SCD (children)	3 × 0.1 g Arg/kg BW/day, 5 days or until discharged	28	Parenteral opioid use ↓; Pain crises ↓; Well-being ↑	[93]
Hypertension	2 × 3 g Arg/day, 6 wk	20	Systolic blood pressure ↓; Diastolic blood pressure ↓	[94]
NEC in preterm infants	0.261 g Arg/day, 28 days	75	Risk of NEC in infants with very low birth weight↓	[95]
	0.261 g Arg/day, 28–29 days	40	Risk of stage III NEC ↓; Mortality ↓	[96]
Intestinal injury[1]	10 g Cit (once before exercise)	10	Splanchnic perfusion ↑; Intestinal injury ↓	[97]
Liver injury	1 g Arg/day, 10 days	20	Liver injury ↓	[98]
Cystic fibrosis	1.3 g Arg (inhaled)	13	Pulmonary function ↑; Oxygen saturation ↑	[99]
Wound healing	4.5 or 9 g Arg/day, 3 wk	23	Severity of pressure ulcer ↓; Wound healing ↑	[100]
	20 g Arg/day, 3–6 wk[2]	36	Fistula formation ↓; Length of postoperative stay ↓	[101]
	10 mM Arg/day (SC)[3]	11	Healing of diabetic foot ulcers ↑	[102]
Immune response	8-20 g Arg/day, various wk	321	Ab titers ↑; Infectious risk ↓; ICU and hospital stay ↓	[103]
Immunity (COVID-19)	2 × 1.66 g Arg/day, 60 days	48	Immune responses ↑; Recovery from infections ↑	[104]
Blood flow to muscle	2 × 3 g Arg/day, 10 days	7	Blood flow to forearm muscle ↑	[86]
	6.6 g Arg once	15	Blood flow to forearm muscle ↑	[105]
Exercise performance	14.2 g Arg/day, 6 mo	11	Exercise endurance ↑; Muscle force generation ↑	[106]
	6 g Arg, followed by exercise	9	High-intensity exercise tolerance ↑	[107]
	3 g Arg, 15 days	12	Knee extension movement ↑; Work fatigue index ↑	[108]
Muscle Mit dysfunction	3 × 2.5 g Arg/day, 16 wk	5	Muscle mitochondrial function ↑	[109]
Hormone secretion	0.5 g Arg/kg BW over 30 min	42	Plasma insulin and growth hormone in children ↑	[110]
	30 g Arg over 30 min	7	Plasma insulin and growth hormone in adults ↑	[111]

(continues)

Table 34.1 Reports of Positive Effects of Arginine (Arg) or Citrulline (Cit) in Humans* *(continued)*

Condition	Amount of Arg or Cit	n	Effects	Ref.
Oral Administration				
Fetal IUGR	20 g Arg/day, 7 days	30	Fetal growth in women with IUGR ↑	[112]
End-stage renal failure	1.05 g Arg/h	12	NO-dependent venodilation in kidneys↑	[114]

*ASPEN guidelines: Immune-modulating formulas with arginine and essential fatty acids should be considered in severe trauma but not in sepsis.[115] The number (n) of study participants in the treatment group of each study is indicated.

[1]Exercise-induced intestinal injury.

[2]Patients with oral or laryngeal cancer underwent the surgical removal of tumors.

[3]The period of treatment ranged from 9 to 284 days.

↓, Decrease or alleviation; ↑, Increase or improvement.

Ab, antibody; *BW*, body weight; *HCh*, hypercholesterolemic patients; *ICU*, intensive care unit; *IUGR*, intrauterine growth restriction; *LMB*, lean body mass; *Mit*, mitochondria; *Mo*, months; *NEC*, necrotizing enterocolitis; *SC*, subcutaneously; *SCD*, sickle cell disease; *WAT*, white adipose tissue; *Wk*, weeks.

Table 34.2 Reports of Positive Effects of Glutamine (Gln) or Alanyl-glutamine (AG) in Humans*

Condition	Amount of Gln or AG	n	Effects	Ref.
Oral Administration				
Burn injury (adults)	26 g Gln/day, tube feeding	41	Blood bacteria counts ↓; Mortality ↓	[116]
	0.5 g Gln/kg BW/day, 14 days	47	Blood bacteria counts ↓; Mortality ↓	[117]
	AG (0.35 g Gln/kg BW/day)	40	Intestinal permeability ↓; Would healing ↑; Hospital stay ↓	[118]
Trauma	30.5 g Gln/100 g protein	72	Incidence of pneumonia ↓; Blood bacteria counts ↓; Clinical sepsis ↓	[119]
	0.5 g Gln/kg BW/day	10	Tolerance to enteral feeding ↑	[120]
ICU patients (adults)	26 g Gln/day	50	Hospital patient costs ↓	[121]
Critically ill infants	≤ 0.31 g Gln/kg BW/day	68	Incidence of hospital-acquired sepsis ↓; Morbidity ↓	[122]
VLBW infants	0.3 g Gln/kg BW/day, 27 days	54	Infectious morbidity ↓	[123]
NEC (preterm infants)	0.156–0.312 g Gln/kg BW/day; 28–29 days	25	Prevention of NEC (0% in Gln vs 9.2% in the control group); Mortality ↓	[124]
Diarrhea (children)	0.3 g Gln/kg BW/day, 7 days	63	Duration of diarrhea ↓	[125]
	16.2 g Gln/day, 10 days	53	Intestinal mucosal barrier function	[126]
SBS (adults)	30 g Gln/day, 28 days	47	Intestinal absorption of nutrients ↑; Fecal weight ↓	[127]
	30 g Gln/day, 4 wk	41	Tolerance to enteral feeding ↑	[128]
SCD (aged 5–58 yr)	2 × 0.3 g Gln/kg BW/day[1], 48 wk	152	Pain crises ↓; Hospitalizations ↓	[129]
COVID-19 infection	3 × 10 g Gln/day, 5 days	222	Serum IL-1β, TNF-α, hs-CRP, and MDA levels ↓; TAC ↑; Length of hospital stay ↓; Mortality ↓	[130]
HIV-infections	10 g Gln/day, 10 days	13	Nelfinavir-associated diarrhea ↓; Quality of life ↑	[131]
CRI oral mucositis	0.4 g Gln/kg BW/day, ≥14 days	24	Severity of mucositis ↓; Pain score ↑; Hospital stay ↓	[132]
CRI esophagitis	10 g Gln/8 h (children)	30	Severity of esophagitis ↓; Incidence of weight loss ↓	[133]
Exercise performance (adult)	2 × 5 g Gln/day, 30 days	23	Plasma GSH ↑; Strength and power of knee muscles ↑	[134]
	0.35 g Gln/kg BW/day, 8 wk	15	Upper and lower body strength ↑; Muscular power ↑	[135]
	0.3 g Gln/kg BW/day, 3 days	16	Muscle force recovery after eccentric exercise ↑; Muscle soreness after exercise ↓	[136]

Condition	Amount of Gln or AG	n	Effects	Ref.
Intravenous Administration				
Hormone secretion	30 g Gln over 2 min (adult)	8	Plasma GLP-1, GIP, glucagon, GH, and insulin ↑	[137]
Critically ill adults	0.57 g Gln/kg BW/day, 29–36 days	24	Nitrogen balance ↑; Hospital infections and stay ↓	[138]
ICU patients (adults)	25 g Gln/day, 6 mo	42	6-month survival ↑; Hospital infections ↓	[139]
	0.3 g AG/kg BW/day, ≥5 days	95	6-month survival ↑	[140]
	0.5 g AG/kg BW/day, ≤21 days	30	Hospital-acquired infections ↓	[141]
	AG (0.283 g Gln/kg BW/day), ≥3 days	413	Mortality ↓	[142]
	0.5 g AG/kg BW/day, ≥5 days	53	Incidence of hospital-acquired sepsis ↓	[143]
Patients requiring PN	20 g Gln/day, 6 mo	168	Length of hospital stay for surgical patients ↓	[144]
Burn patients (adults)	0.57 g Gln/kg BW/day, >7 days	12	Blood bacteria counts ↓	[145]
Acute pancreatitis	0.3 to 0.5 g AG/kg BW/day, 1–3 wk	505	Risk of mortality ↓; Infectious complications ↓	[146]

*ASPEN guidelines: Enteral glutamine should not be used routinely in critically ill patients.[115,147] The number (n) of study participants in the treatment group of each study is indicated.

[1] In 2017, the U.S. Food and Drug Administration approved the oral use of Endari (glutamine powder) for patients aged 5 years and older with sickle cell disease to reduce sickle cell-related acute pain events and hospitalizations.

↓, decrease or alleviation; ↑, increase or improvement.

CRI, chemotherapy and radiation-induced; *GH*, growth hormone; *GLP-1*, glucagon-like peptide 1; *GIP*, glucose-dependent insulinotropic polypeptide; *HIV*, human immunodeficiency virus; *hs-CRP*, high-sensitivity C-reactive protein; *ICU*, intensive care unit; IL-1β, interleukin-1β; *MDA*, malondialdehyde; *Mo*, months; *ORS, oral rehydration solution; PN*, parenteral nutrition; *SBS*, short-bowel syndrome; *SCD*, sickle cell disease; *TAC*, total antioxidant capacity; *TNF-α*, tumor necrosis factor-α; *VLBW*, very low birth weight; *Wk*, weeks; *Yr*, years.

against giving enteral glutamine.[115] In the multicenter pediatric CRISIS trial, 293 children received an immunonutrient solution containing enteral glutamine (0.3 g/kg/day), zinc, selenium, and metoclopramide, compared with a whey placebo (an unusual control), and no advantage was reported in the initial publication.[152] However, 9% of the study participants were immunocompromised and a subsequent 2×2 factor analysis of the immune compromised group showed protection from infection and sepsis in the treated group.[153]

In sickle cell disease, evidence has emerged from several trials that glutamine supplementation reduces the number of pain crises (individually or combined with hydroxyurea). Many studies have shown a 45% reduction in pain episodes and the duration of hospitalization.[129,154-156] However, barriers to treatment include denial of insurance coverage of glutamine (in up to 38% cases), high deductible charges (21%), and poor adherence (36%).[154]

In patients with extensive burn injury, recent European Society for Clinical Nutrition and Metabolism (ESPEN) guidelines recommend that glutamine should be given if burn surface area is >20%, or during the first 5 days after trauma (longer if there is delayed wound healing), but that no glutamine should be provided to ICU patients with liver or renal failure.[157] However, the U.S. RE-ENERGIZE trial of 0.5 g enteral glutamine/kg/day, vs placebo, showed no significant advantage in severe burn patients and was associated with increased adverse events.[158] In cancer patients exposed to chemotherapy and radiation therapy, mounting evidence indicates potential benefits (e.g., reduced oral ulcers and pain) of enteral glutamine supplementation.[159,160] Glutamine or alanyl-glutamine supplementation also may provide benefits in malnourished infants and children with enteropathy.[161,162]

Arginine can also be supplemented together with glutamine, other functional AAs, and their bioactive metabolites to humans for complementary or additive effects. For example, oral administration of a mixture of 0.5 g arginine, 1 g glutamine, 1 g monosodium glutamate, 1 g monosodium aspartate, 1 g glycine, and 0.5 g histidine, usually twice daily for 1 year, was reported to slow the progression of renal failure in uremic patients.[163] In addition, supplementation with a combination of arginine, glutamine, and β-hydroxy-β-methylbutyrate at various ratios (e.g., 14, 14, and 3 g daily, respectively) alleviated cachexia and morbid muscle loss in cancer- and acquired immunodeficiency syndrome patients,[164-166] promoted wound healing in people with pressure ulcers[167] or diabetic foot ulcers,[168] and enhanced blood flow to skeletal muscle[169] as well as muscle strength and health.[170]

As illustrated in the foregoing sections, the metabolism of arginine, citrulline, and glutamine in humans at the cell, tissue, and whole-body levels take place in the setting of complex compartmentalization. This means that the route of administration of these AAs can impact their availability and efficacy in the body. Not all clinical studies have shown consistent benefits in humans with arginine or glutamine administration, due to a plethora of factors, including doses and frequencies, nutritional intakes, endocrine status, pathologic states and, most importantly, the stage of disease. Inappropriate isonitrogenous controls involving the use of certain AAs (e.g., glycine and aspartate) may also explain why treatments may not confer clinical benefits. For example, glycine exerts multiple physiologic roles (including anti-inflammatory and antioxidative reactions, the stimulation of protein synthesis, and the enhancement of immunity), whereas aspartate is a major metabolic fuel for mammalian enterocytes and is required for arginine synthesis.[1]

In some clinical trials, incorrect calculations of nitrogen content in glutamine were made (e.g., references 126, 162), as glutamine contains 2, rather than 1, nitrogen atoms. Finally, in clinical, as in preclinical studies, negative results always raise the question of whether *absence of evidence indicates evidence of absence*.

Safety of Use of Arginine, Citrulline, and Glutamine in Clinical Nutrition and Therapies

Preterm infants and healthy adults can generally tolerate oral administration of 0.261 and 0.286 g supplemental arginine/day (in equally divided doses), respectively, beyond arginine intake from regular diets (Table 34.1). Healthy adults with normal gastrointestinal function can even tolerate oral administration of 30 g supplemental Arg (provided as arginine-HCl) per day (in equally divided doses) for at least 90 days.[171] Based on our knowledge of AA absorption by the small intestine, as well as the interorgan transport and metabolism of AAs,[1] the observed safe levels of citrulline for enteral and parenteral nutrition is expected to be like those of arginine. In support of this view, Miura et al.,[172] recently reported that a 70-kg adult tolerated oral administration of 24 g citrulline for at least 4 weeks. According to published work, infants and healthy adults can tolerate oral administration of 0.30 and 0.43 g supplemental glutamine/kg/day, respectively, beyond glutamine intake from regular diets. There is even evidence that pediatric oncology patients can tolerate oral administration of 0.65 g glutamine/kg/day.[173] However, there are concerns about hyperglutaminemia-induced increases in the risk of mortality in critically ill patients, often in the setting of liver disease.[174] Both infants and adults can tolerate intravenous administration of 0.5 g arginine (as its HCl salt) and 0.57 g glutamine or alanyl-glutamine/kg/day (Tables 34.1 and 34.2).[175,176]

In summary, humans can synthesize, degrade, and interconvert arginine and glutamine. Because of insufficient endogenous synthesis relative to utilization, young and adult humans have dietary requirements for arginine and, in catabolic states or infections, possibly also glutamine, to meet metabolic needs and improve health. When deficient in the blood and other tissues, supplementation with arginine, glutamine, or both may be beneficial for patients, such as in the settings of sickle cell disease, extensive burns, mucositis, muscle wasting, and male infertility; however, further well-designed trials are needed.

Acknowledgments

Work in our laboratories is supported by funds from Texas A&M AgriLife Research (H-8200) as well as grants R01 HD095158, P30 DK056338, and R01 5R01HD095158.

References

1. Wu G. Amino Acids: Biochemistry and Nutrition. CRC Press, Boca Raton, 2022.
2. Wu G, Morris SM Jr. Biochem J 1998;336:1–17.
3. Palmer RM, Ashton DS, Moncada S. Nature 1988;333:664–6.
4. Hibbs JB Jr, Taintor RR, Vavrin Z. Science 1987;235:473–6.
5. Fox AD, Kripke SA, De Paula J, et al. J Parenter Enteral Nutr 1988;12:325–31.
6. MacLennan PA, Smith K, Weryk B, et al. FEBS Lett 1988;237:133–6.
7. Furst P, Albers S, Stehle P. J Parenter Enteral Nutr 1990;14:118S–24S.
8. Baldeón ME, Mennella JA, Flores N, et al. SpringerPlus 2014;3:104.
9. Shanely RA, Zwetsloot JJ, Jurrissen TJ, et al. Nutr Res 202;76:9–19.
10. Li P, He WL, Wu G. Adv Exp Med Biol 2021;1332:189–210.
11. King DE, Mainous AG, Geesey ME. Nutr Res 2008;28:21–4.
12. U.S. Department of Agriculture. Food availability and consumption in the United States. Washington, DC. 2018.
13. Hou YQ, He WL, Hu SD, et al. Amino Acids 2019;51:1153–65.
14. Ziegler TR, Evans ME, Fernandez-Estivariz C, et al. Annu Rev Nutr 2003;23:229–61.
15. Killer M, Wald J, Pieprzyk J, et al. Sci Adv 2021;7:eabk3259.
16. Bröer S, Fairweather SJ. Compr Physiol 2019;9:343–73.
17. Garai P, Chandra K, Chakravortty D. Virulence 2017;8:297–309.
18. Closs EI, Boissel J-P, Habermeier A, et al. J Membr Biol 2006;213:67–77.
19. Wu G, Bazer FW, Davis TA, et al. Amino Acids 2009;37:153–68.
20. Sebastio G, Sperandeo MP, Andria G. Am J Med Genet 2011;157C:54–62.
21. Wu G, Knabe DA, Flynn NE. Biochem J 1994;299:115–21.
22. Wu G, Knabe DA, Flynn NE, et al. Am J Physiol 1996;271:G913–9.
23. Dillon EL, Wu G. Amino Acids 2021;53:1957–66.
24. Castillo L, Beaumier L, Ajami AM, et al. Proc Natl Acad Sci USA 1993;90:193–7.
25. Hou YQ, Hu SD, Jia SC, et al. Amino Acids 2016;48:993–1001.
26. Rhoads JM, Plunkett E, Galanko J, et al. J Pediatr 2005;146:542–7.
27. Papadia C, Kelly P, Caini S. Clin Nutr 2010;29:795–800.
28. Crenn P, Coudray-Lucas C, Thuillier F, et al. Gastroenterol 2000;119:1496–505.
29. Proli F, Faragalli A, Talbotec C, et al. Clin Nutr 2021;40:4941–7.
30. David AI, Selvaggi G, Ruiz P, et al. Transplantation 2007;84:1077–81.
31. Lutgens LC, Deutz N, Granzier-Peeters M, et al. Int J Radiat Oncol Biol Phys 2004;60:275–85.
32. Pappas PA, Tzakis AG, Saudubray J-M, et al. Transplant Proc 2004;36:345–7.
33. Hueso T, Gauthier J, Chevalier-Curt MJ, et al. Clin Gastroenterol Hepatol 2018;16:908–17.
34. Pironi L, Guidetti M, Lauro A, et al. Clin Transplant 2015;29:1039–46.
35. Windmueller HG. Adv Enzymol 1982;53:201–37.
36. Souba WW, Smith RJ, Wilmore DW. J Parenter Enteral Nutr 1985;9:608–17.
37. Wu G. Biochem J 1995;312:717–23.
38. van der Schoor S, Schierbeek H, Bet P, et al. Pediatr Res 2010;67:194–9.

39. Hays SP, Ordonez JM, Burrin DG, et al. Pediatr Res 2007;62: 353–6.
40. Riedijk MA, de Gast-Bakker DA, Wattimena JL, et al. Pediatr Res 2007;62:468–73.
41. Haisch M, Fukagawa NK, Matthews DE. Am J Physiol 2000;278:E593–602.
42. Battezzati A, Brillon DJ, Matthews DE. Am J Physiol 1995;269:E269–76.
43. van de Poll MC, Ligthart-Melis GC, Boelens PG, et al. J Physiol 2007;581:819–27.
44. Matthews DE. Clin Nutr 1990;9:50–1.
45. Rhoads JM, Wu G. Amino Acids 2009;37:111–22.
46. Siroen MP, van der Sijp JR, Teerlink T, et al. Hepatology 2005;41:559–65.
47. van de Poll MCG, Ligthart-Melis GC, Boelens PG. J Physiol 2007;581:819–27.
48. Wu G, Jaeger LA, Bazer FW, et al. J Nutr Biochem 2004;15: 442–51.
49. Luiking YC, Ten Have GAM, Wolfe RR, et al. Am J Physiol 2012;303:E1177–89.
50. Castillo L, Beaumier L, Ajami AM, et al. Proc Natl Acad Sci USA 1996;93:11460–5.
51. Mariotti F, Petzke KJ, Bonnet D, et al. Am J Clin Nutr 2013;97: 972–9.
52. Wu ZL, Hou YQ, Hu SD, et al. Amino Acids 2016;48:1541–52.
53. Morris SM Jr. Curr Opin Clin Nutr Metab Care 2012;15:64–70.
54. Soundararajan L, Dharmarajan A, Samji P. Cell Signal 2023;101: 110496.
55. Kohli R, Meininger CJ, Haynes TE, et al. J Nutr 2004;134:600–8.
56. Wu G, Meininger CJ. Annu Rev Nutr 2002;22:61–86.
57. Siani A, Pagano E, Iacone R, et al. Am J Hypertens 2000;13: 547–51.
58. Lee J, Ryu H, Ferrante RJ, et al. Proc Natl Acad Sci USA 2003;100:4843–8.
59. Moinard C, Nicolis I, Neveux N, et al. Br J Nutr 2008;99:855–62.
60. Figueroa A, Trivino JA, Sanchez-Gonzalez MA, et al. Am J Hypertens 2010;23:12–6.
61. Wu G, Meininger CJ, McNeal CJ, et al. Adv Exp Med Biol 2021;1332:167–87.
62. Curthoys NP, Watford M. Annu Rev Nutr 1995;15:133–59.
63. Marliss EB, Aoki TT, Pozefsky T, et al. J Clin Invest 1971;50:814–7.
64. Elia M, Folmer P, Schlatmann A, et al. Am J Clin Nutr 1989;49:1203–10.
65. Goldberg AL, Chang TW. Fed Proc 1978;37:2301–7.
66. Darmaun D, Roig JC, Auestad N, et al. Pediatr Res 1997;41: 391–6.
67. Darmaun D, Matthews DE, Bier DM. Am J Physiol 1986;251: E117–26.
68. Häberle J, Gorg B, Toutain A, et al. J Inherit Metab Dis 2006;29:352–8.
69. Kelly B, Pearce EL. Cell Metab 2020;32:154–75.
70. Deutz NE. Clin Nutr 2008;27:321–7.
71. Meininger CJ, Kelly KA, Li H, et al. Biochem Biophys Res Commun 2000;279:234–9.
72. Young JD, Wolowyk MW, Jones SM, et al. Biochem J 1983;216: 349–57.
73. Matthews DE, Marano MA, Campbell RG. Am J Physiol 1993;264:E848–54.
74. Holt LE Jr, Albanese AA. Trans Assoc Am Physicians 1944;58: 143–56.
75. Hou YQ, Yin YL, Wu G. Exp Biol Med 2015;240:997–1007.
76. Wu G, Bazer FW, Johnson GA, et al. J Anim Sci 2011;89:2017–30.
77. Efion DT, Barbul A. Curr Opin Clin Nutr Metab Care 1998;1: 531–8.
78. Rhoads JM, Argenzio RA, Chen W, et al. Gastroenterol 2000;118:90–100.
79. Rhoads JM, Keku EO, Quinn J, et al. Gastroenterol 1991;100:683–91.
80. Wernerman J. Ann Intensive Care 2011;1:25.
81. Hurt RT, Ebbert JO, Schroeder DR, et al. J Diet Suppl 2014;11: 40–52.
82. Lucotti P, Setola E, Monti LD, et al. Am J Physiol 2006;291: E906–12.
83. Clarkson P, Adams MR, Powe AJ, et al. J Clin Invest 1996;97: 1989–94.
84. Siasos G, Tousoulis D, Vlachopoulos C, et al. Int J Cardiol 2008;126:394–9.
85. Lerman A, Burnett JC, Higano T, et al. Circulation 1998;97:2123–8.
86. Marchesi S, Lupattelli G, Siepi D, et al. J Clin Pharma Ther 2001;26:343–9.
87. Tanimura J. Bull Osaka Med School 1967;13:84–9.
88. Menafra D, de Angelis C, Garifalos F, et al. J Endocrinol Invest 2022;45:941–61.
89. Pulido EEC, Benavides LG, Barón JGP, et al. Hypertens Pregnancy 2016;35:217–25.
90. Vadillo-Ortega F, Perichart-Perera O, Espino S, et al. BMJ 2011;342:d2901.
91. Rytlewski K, Olszanecki R, Lauterbach R, et al. Eur J Obstet Gynecol Reprod Biol 2008;138:23–8.
92. Waugh WH, Daeschner CW, Files BA, et al. J Natl Med Assoc 2001;93:363–71.
93. Morris CR, Kuypers FA, Lavrisha L, et al. Haematologica 2013;98:1375–82.
94. Pezza V, Bernardini F, Pezza E, et al. Am J Hypertens 1998;11: 1267–70.
95. Amin HJ, Zamora SA, McMillan DD, et al. J Pediatr 2002;140: 425–31.
96. Polycarpou E, Zachaki S, Tsolia M, et al. J Parenter Enteral Nutr 2013;37:617–22.
97. van Wijck K, Wijnands KA, Meesters DM, et al. Med Sci Sports Exerc 2014;46:2039–46.
98. Al-Khaqani MF. J Convent Knowl Holist Health 2019;3:198.
99. Grasemann H, Kurtz F, Ratjen F. Am J Respir Crit Care Med 2006;174:208–12.
100. Leigh B, Desneves K, Rafferty J, et al. J Wound Care 2012; 21:150.
101. De Luis DA, Izaola O, Cuellar L, et al. Eur Rev Med Pharmacol Sci 2009;13:279.
102. Arana V, Paz Y, Gonzalez A, et al. Biomed Pharmacother 2004;58:588–97.
103. Kang K, Shu X-L, Zhong J-X, et al. Asia Pac J Clin Nutr 2014;23:351–9.
104. Fiorentino G. EClinicalMedicine 2021;40:101125.
105. Monti LD, Casiraghi MC, Setola E, et al. Metabolism 2013;62:255–64.
106. Fricke O, Baecker N, Heer M, et al. Clin Physiol Funct Imaging 2008;28:307–11.
107. Bailey SJ, Winyard PG, Vanhatalo A, et al. J Appl Physiol 2010;109:1394–403.
108. Santos RS, Pacheco MTT, Martins RABL, et al. Isokinet Exerc Sci 2002;10:153–8.
109. Hafner P, Bonati U, Erne B, et al. PLoS ONE 2016;11:e0147634.
110. Root AW, Saenz-Rodriguez C, Bongiovanni AM, et al. J Pediatr 1969;74:187–97.
111. Pontiroli AE, Perfetti MG, Andreotti AC, et al. Metabolism 1993;42:1242–8.
112. Xiao XM, Li LP. Int J Gynecol Obstet 2005;88:15–8.
113. Heird WC, Nicholson JF, Driscoll JM, et al. J Pediatr 1972;81:162–5.
114. Hand MF, Haynes WG, Webb DJ. Kidney Int 1998;53:1068–77.
115. McClave SA, Taylor BE, Martindale RG, et al. J Parenter Enteral Nutr 2016;40:159–211.

116. Garrel D, Patenaude J, Nedelec B, et al. Crit Care Med 2003;31:2444–9.
117. Peng X, Yan H, You Z, et al. Burns 2004;30:135–9.
118. Zhou YP, Jiang ZM, Sun YH, et al. J Parenter Enteral Nutr. 2003;27:241–5.
119. Houdijk AP, Rijnsburger ER, Jansen J, et al. Lancet 1998;352: 772–6.
120. McQuiggan M, Kozar R, Sailors RM, et al. J Parenter Enteral Nutr 2008;32:28–35.
121. Jones C, Palmer TE, Griffiths RD. Nutrition 1999;15:108–15.
122. Neu J, Roig JC, Meetze WH, et al. J Pediatr 1997;131:691–9.
123. van den Berg A, van Elburg RM, Westerbeek EA. Am J Clin Nutr 2005;81:1397–404.
124. El-Shimi MS, Awad HA, Abdelwahed MA, et al. Int J Pediatr 2015;2015:856091.
125. Yalçin SS, Yurdakök K, Tezcan I, et al. J Pediatr Gastroenterol Nutr 2004;38:494–501.
126. Lima AA, Brito LF, Ribeiro HB, et al. J Pediatr Gastroenterol Nutr 2005;40:28–35.
127. Byrne TA, Persinger RL, Young LS, et al. Ann Surg 1995;222: 243–54.
128. Byrne TA, Wilmore DW, Iyer K, et al. Ann Surg 2005;242: 655–61.
129. Niihara Y, Miller ST, Kanter J, et al. N Engl J Med 2018;379: 226–35.
130. Mohajeri M, Horriatkhah E, Mohajery R. Inflammopharmacol 2021;29:1769–76.
131. Huffman FG, Walgren ME. HIV Clinical Trials 2003;4:324–9.
132. Widjaja NA, Pratama A, Prihaningtyas R, et al. Asian Pac J Cancer Prev 2020;21:2117–21.
133. Chang S-C, Lai Y-C, Hung J-C, et al. Medicine (Baltimore) 2019;98:e14463.
134. Amirato GR, Borges JO, Marques DL, et al. Nutrients 2021;13:1025.
135. Hakimi M, Mohamadi MA, Ghaderi Z. J Hum Sport Exerc 2012;7:770–82.
136. Legault Z, Bagnall N, Kimmerly DS. Int J Sport Nutr Exerc Metab 2015;25:417–26.
137. Greenfield JR, Farooqi IS, Keogh JM, et al. Am J Clin Nutr 2009;89:106–13.
138. Ziegler TR, Young LS, Benfell K, et al. Ann Intern Med 1992;116:821–8.
139. Griffiths RD, Allen KD, Andrews FJ, et al. Nutrition 2002;18: 546–52.
140. Goeters C, Wenn A, Mertes N, et al. Crit Care Med 2002;30: 2032–37.
141. Estivariz CF, Griffith DP, Luo M, et al. J Parenter Enteral Nutr 2008;32:389–402.
142. Wernerman J, Kirketeig T, Andersson, et al. Acta Anaesthesiol Scand 2011;55:812–8.
143. Grau T, Bonet A, Miñambres E, et al. Crit Care Med 2011;39:1263–8.
144. Powell-Tuck J, Jamieson CP, Bettany GE, et al. Gut 1999;45: 82–8.
145. Wischmeyer PE, Lynch J, Liedel J, et al. Crit Care Med 2001;29:2075–80.
146. Asrani V, Chang WK, Dong Z, et al. Pancreatology 2013;13: 468–74.
147. Heyland D, Muscedere J, Wischmeyer PE, et al. N Engl J Med 2013;368:1489–97.
148. Becker RM, Wu G, Galanko JA, et al. J Pediatr 2000;137: 785–93.
149. Moe-Byrne T, Brown JVE, McGuire W. Cochrane Database Syst Rev 2016;4:CD001457.
150. Tao K-M, Li X-Q, Yang L-Q, et al. Cochrane Database Syst Rev 2014;2014:CD010050.
151. Oldani M, Sandini M, Nespoli L, et al. Medicine (Baltimore) 2015;94:e1319.
152. Carcillo JA, Dean JM, Holubkov R, et al. Pediatr Crit Care Med 2012;13:165–73.
153. Carcillo JA, Dean JM, Holubkov R, et al. J Parenter Enteral Nutr 2017;41:1325–35.
154. Zaidi AU, Estepp J, Shah N, et al. Contemp Clin Trials 2021;110:106546.
155. Cieri-Hutcherson NE, Hutcherson TC, Conway-Habesh EE, et al. Pharmacotherapy 2019;39:1095–104.
156. Zhou Q, Verne ML, Fields JZ, et al. Gut 2019;68:996–1002.
157. Singer P, Blaser AR, Berger MM, et al. Clin Nutr 2019;38:48–79.
158. Rousseau A-F, Pantet O, Heyland DK. Curr Opin Clin Nutr Metab Care 2023;26:99–104.
159. Widjaja NA, Pratama A, Prihaningtyas R, et al. Asian Pac J Cancer Prev 2020;21:2117–21.
160. Tsujimoto T, Yamamoto Y, Wasa M, et al. Oncol Rep 2015;33: 33–9.
161. Moore SR, Quinn LA, Maier EA, et al. J Pediatr Gastroenterol Nutr 2020;71:393–400.
162. Lima AAM, Anstead GM, Zhang Q, et al. Clinics (Sao Paulo) 2014;69:225–33.
163. Yatzidis H. Int Urol Nephrol 2004;36:591–8.
164. Ellis AC, Hunter GR, Goss AM, et al. J Diet Suppl 2019;16: 281–93.
165. Pascoe J, Jackson A, Gaskell C, et al. BMC Cancer 2021;21:800.
166. Rathmacher JA, Nissen S, Panton L, et al. J Parenter Enter Nutr 2004;28:65–75.
167. Miu KYD, Lo KM, Lam KYE, et al. Aging Med Healthcare 2021;12:82–9.
168. Tattia P, Barber AE. J Endocrinol Metab 2012;2:26–31.
169. Ellis AC, Patterson M, Dudenbostel T, et al. Eur J Clin Nutr 2016;70:269–73.
170. Saka B, Şahin S, Kılıç D, et al. Clin Sci Nutr 2019;1:67–74.
171. McNeal CJ, Meininger CJ, Wilborn CD, et al. Amino Acids 2018;50:1215–29.
172. Miura N, Morishita K, Yasuda T, et al. Amino Acids 2023;55: 299–311.
173. Ward E, Picton S, Reid U, et al. Eur J Clin Nutr 2003;57:31–6.
174. Smedberg M, Helleberg J, Norberg Å, et al. Crit Care 2021;25:240.
175. Mohamad Ikram I, Quah B S, Noraida R, et al. Singapore Med J 2011;52:356–60.
176. Ziegler TR, May AK, Hebbar G, et al. Ann Surg 2016;263: 646–55.

D. OTHER COMPOUNDS WITH HEALTH RELEVANCE

CHAPTER 35

Functional Foods and Nutraceuticals in Health Promotion

Steven K. Clinton, Abigail A. Sommer, Maxine Z. Webb, Yael Vodovotz, and Elizabeth M. Grainger

CHAPTER OUTLINE

Abbreviations

DSHEA, Dietary Supplement Health and Education Act; **FDA**, U.S. Food and Drug Administration; **FDAMA**, Food and Drug Administration Modernization Act; **FD&C**, Federal Food, Drug, and Cosmetic Act; **FFSDU**, Foods for Special Dietary Uses; **FOSHU**, Foods for Specified Health Use; **FTC**, Federal Trade Commission; **GRAS**, generally recognized as safe; **NCD**, noncommunicable disease; **NDIN**, New Dietary Ingredient Notification; **NLEA**, Nutrition Labeling and Education Act; **SNP**, single nucleotide polymorphism

Introduction

Throughout thousands of years of human social and cultural evolution, societies have examined and used natural substances, including foods, and their components for medicinal purposes to promote health and treat disease. Written records dating from ancient China and South Asia, along with early Greek and Roman texts describe medicinal applications of plant and animal products.[1] The accelerated developments in science over recent centuries, particularly in chemical analysis and synthesis, led to the modern fields of pharmacognosy and pharmacology. These disciplines focus on the purification or synthesis of defined chemicals for medicinal purposes that are then evaluated in a stepwise process from preclinical models to human studies. In brief, in vitro and in vivo studies in experimental systems define putative mechanisms of action, toxicology, and perhaps utility in models of human disease. Promising compounds are then evaluated in a series of clinical trials defining safety and efficacy providing the foundation for large randomized, controlled clinical trials evaluating a new agent against the best standard-of-care treatment. Favorable results may then lead to a rigorous and objective evaluation by government agencies within nations, such as the U.S.

Food and Drug Administration (FDA) in the United States leading to approval for widespread use. The utilization and marketing of approved pharmaceuticals is then monitored to ensure quality and public safety. This paradigm, although at times contentious, has provided populations in nations around the globe with thousands of effective drugs with well-defined risks.

The homology between foods and medicine continues in the field of functional foods. To be clear, we are now focusing on chemically complex foods, not the isolation of pure bioactive compounds from a food. A key principle is that some foods may have health benefits above and beyond their supply of essential nutrients, such as vitamins and minerals that reverse specific deficiency syndromes. Foods may contain multiple bioactive compounds, often dozens with different mechanisms of action, with each bioactive provided at a modest dose. Thus, the advantages of a food are the modest risks of toxicity coupled with the possibility of additive or synergistic health benefits. In the area of functional foods, much of the emphasis is on phytochemicals that are not nutrients but exhibit bioactivities hypothesized to impact health. Such reasoning is supported by a vast preponderance of evidence showing that a phytochemical-rich, plant-based diet is beneficial for health and disease prevention.[2-4] Thus, encouraged by epidemiologic findings and a large literature repository regarding individual plant bioactives in preclinical studies, many investigators are pursuing translational research efforts to develop functional foods targeting specific disease processes.

Scientific progress in crop science and food technology is rapidly providing new opportunities to develop novel functional foods warranting testing for health promotion in a wide array of target populations. For example, plant genetics is providing new breeding and genetic tools to enhance patterns of putative health-promoting bioactives (and nutrients) in fruits and vegetables. Similarly, knowledge in horticulture and crop science can be applied to optimize bioactive patterns in crops to enhance potential benefits of functional foods. Food scientists have expanded capabilities to develop, test, and provide high quality foods that promote consumption and achieve desirable and consistent patterns of bioactives with favorable absorption and distribution characteristics. In parallel, modern biomedical science has rapidly expanded knowledge in genetics and biomarkers of future disease risk, providing increasingly more homogeneous cohorts that can be targeted for disease prevention by diet, pharmaceuticals, functional foods, or their combinations in phase I, II, or III clinical trials. The National Institutes of Health increasingly recognizes the concept of precision nutrition[5] and define optimal interventions to prevent or treat disease based on knowledge of genetics, race, gender, age, and other variables.

Thus, an opportunity exists for functional foods to serve in a research space between interventions focusing on whole dietary patterns and pure pharmaceuticals. Challenges in the field are several. Grant funding is limited, as many reviewers who have trained and practice in the modern era of reductionist science struggle to appreciate the potential value of a chemically complex food item in clinical research. Unlike the pharmaceutical industry that reaps great profits from new drugs, food companies do not see the potential for a similar return on investment, thus are reluctant to invest in costly clinical research to support health claims. We also have variable definitions for the diverse range of food products, which is further challenged by a complex regulatory process.

Definitions

Standardization of Terms

It is desirable to improve standardization of definitions and strategies for regulation and monitoring in the emerging field that is now global. Functional foods were first regulated in Japan in the 1980s, when a government approval process was established for Foods for Specified Health Use (FOSHU).[6] A Japanese manufacturer wishing to apply for governmental approval under FOSHU was required to tabulate and summarize all available preclinical and clinical research as well as internal reports dealing with the efficacy of the product and its ingredients.[7] This concept was subsequently embraced by many in the scientific community, and by multiple government agencies around the globe, to promote safety and quality in the marketplace. Ultimately, the Food and Agricultural Organization (FAO) of the United Nations provided The Codex Alimentarius, or "Food Code," as a collection of international standards and guidelines to protect consumer health and ensure fair practices in food trade.[8] Such standards enhance the ability to harmonize food safety regulations among nations. The Codex defines terms such as functional foods, nutraceuticals, and medical foods that are often used interchangeably, yet each term has come to represent a specific grouping of products that are often regulated differently.[8] Although the codex is a step forward, there remains great heterogeneity among nations in definitions and regulation of products within these categories of foods.

Functional Foods

Functional foods resemble conventional foods with either constitutive, enriched, or fortified bioactive compounds. Examples include a beverage enhanced with green tea extract rich in catechins or a bread with added wheat germ. In the United States, functional foods are not regulated as a separate product category, but rather fall under the regulation of foods, food additives, or dietary supplements. As such, many functional foods and their ingredients are regulated as foods since they resemble conventional food products. Many functional foods include components that are monitored by Federal Food, Drug, and Cosmetic (FD&C) Act. Any substance added to food is a "food additive" and must be approved by the FDA.[9] The key exceptions are substances that are generally recognized as safe (GRAS), a list of components that were used in the food industry, considered safe by experts and, thus, defined as exempt, a process initiated in 1958 with the Food Additives Amendment.[10] A listing of some GRAS substances may be found in the Code of Federal Regulations (21 USC § 182) and includes many spices, herbs, essential oils, and their extracts. To utilize the GRAS exception for a food ingredient not listed, the FDA has a voluntary notification procedure. The documentation includes ingredient identity, method of manufacture or production, ingredient specifications, previous dietary exposure, expected use levels, and evidence from preclinical and clinical safety studies.[11] The FDA reviews the documentation and decides whether the food ingredient meets the GRAS criteria.

Food Additive

If a food ingredient does not meet the requirements for an exemption, it is defined as a food additive and requires review by the FDA. Such substances may include color additives, pesticide residues in foods, new drugs found in animal products as a residue, or bioactive compounds added for the purpose of a health benefit to a functional food or nutraceutical. The FDA evaluates the safety of proposed food additives through consideration of the chemical composition and properties, amount typically consumed, health effects evaluated through multiple experimental models, human studies, and other factors.[12,13]

Nutraceuticals

Much like functional foods, there remains widespread inconsistencies and contradictions in the many definitions of nutraceuticals, including the terminology used by scientists, regulatory agencies, and those developing and marketing products.[14,15] In general, the term nutraceutical, derived from "nutrition" and "pharmaceutical," is applied to products that are often isolated from herbal products, fruits and vegetables, or dietary supplements, resulting in highly processed agents that are marketed as medicine.[6] While nutraceuticals are defined quite similarly to functional foods, they often resemble pharmaceuticals or supplements, frequently in the form of pills, powders, or liquid extracts. As these do not take the form of a conventional food product, nutraceuticals are generally regulated as dietary supplements.

Dietary Supplements

In the United States, the FDA is not obligated to evaluate safety or labeling of dietary supplements before marketing and sale, which is the result of the controversial Dietary Supplement Health and Education Act of 1994, known as DSHEA.[16] This is widely considered to be a disservice to the public by exempting the dietary and herbal supplement industry from stringent FDA regulations and monitoring, permitting them to be marketed without scientific evidence for safety or health and medical claims.[17,18] Thankfully, the FDA can act against postmarket adulterated or misbranded dietary supplements, but the FDA is simply not empowered with sufficient funds to effectively monitor this expansive industry.[15,19] Dietary supplements are not subject to a premarket approval system unless one or more ingredients are not considered GRAS or have been previously approved by the FDA. In this case, the manufacturer must submit a New Dietary Ingredient Notification (NDIN) to the FDA for review prior to product marketing. Regulation of functional foods as foods or dietary supplements requires that the product is not intended to prevent, treat, or cure disease, which would necessitate drug approval (**Figure 35.1**), leading to highly creative marketing strategies by the industry.

Foods for Special Dietary Use

While the broad term of functional foods is not specifically regulated in many countries, "Foods for Special Dietary Uses" (FFSDU) are more uniform. According to the Codex General Standard for the Labelling of and Claims for Prepackaged Foods for Special Dietary Uses, this term applies to foods that are processed or formulated to meet specific intake goals spawning from a physical or physiologic condition, disease, or disorder.[20] Similar definitions exist in the regulatory documents of The European Union, Korea, Japan, and the United States. In the United States, FFSDU can supply nutrients for conditions including pregnancy, lactation, allergies, food hypersensitivity, and being underweight or overweight.[20] Additionally, these foods can be applied to specific age groups, including infancy and childhood.[20] FFSDU also included foods fortified or enriched with vitamins, minerals, or other dietary components.[20] FFSDU do have the caveat that the product must be specifically processed or formulated and cannot include unprocessed foods with constitutive health-promoting components.[20] Functional foods may be either processed or unprocessed and are consequently broader in definition than FFSDU, although many functional foods many be considered as FFSDU as well.

Medical Foods

Under the category of FFSDU reside products termed medical foods. According to the Codex Alimentarius, medical foods are a category of foods for special dietary uses, "which are specifically processed or formulated and presented for the dietary management of patients and may be used only under medical

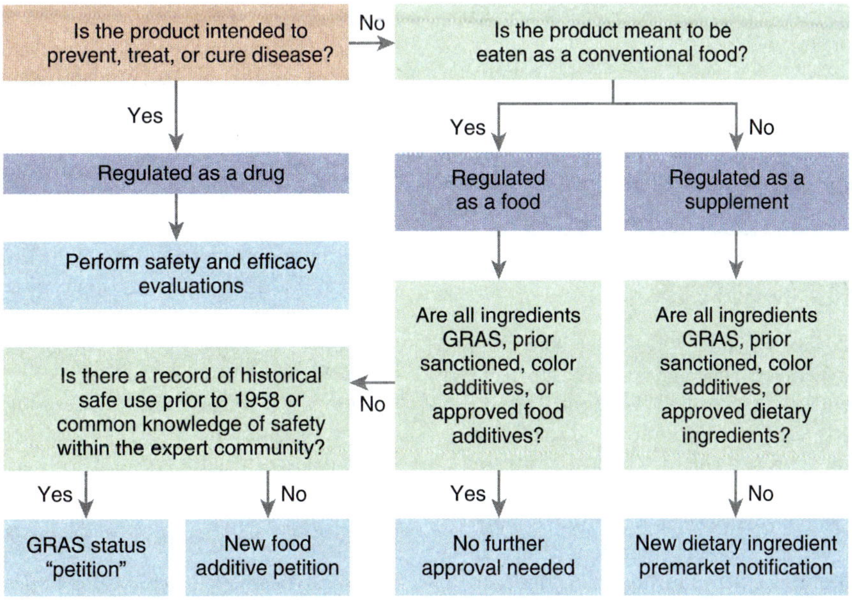

Figure 35.1 A summary of the regulation of functional foods and nutraceuticals.

supervision."[20] These foods are not recommended for health promotion in otherwise healthy individuals but rather, to help meet some dietary requirement or need due to an underlying disease or condition. The necessity of medical supervision is a unique aspect of medical foods. For example, products with low phenylalanine are those with phenylketonuria (PKU), a rare but severe genetic disease that results from an enzyme deficiency in phenylalanine degradation.

In the United States, medical foods have specific regulations under the Orphan Drug Act. Medical foods are administered under the supervision of a physician to manage a disease or condition (21 USC § 360ee). One aspect specific to medical foods is that they are exempt from the 21 CFR § 101.9 nutrition labeling requirements if a group of criteria are met. These criteria include that the product is formulated and processed for the partial or exclusive feeding of a patient who has limited or impaired capacity to ingest, absorb or metabolize foods or nutrients due to a therapeutic or chronic medical need, and that the product is used under recurring medical supervision.[21] Specific regulations for medical and similar food categories exist in Japan and in the European Union.

Fortified and Enriched Foods

While not "functional" foods *per se*, due to the focus on established nutrients, fortified and enriched foods warrant our review to clearly highlight the distinction. According to the General Principles of the Codex for the Addition of Essential Nutrients of Foods, fortification or enrichment adds one or more essential nutrients to a food in order to prevent or correct a deficiency.[20] Fortified and enriched foods are used around the world with great benefit to mediate nutrient deficiencies in general or specific populations. These include, for example, iodized salt products, B vitamins in foods made from grains, and many others.[20] Regulations of this category vary by country, with nutrient addition being either required or voluntary. A key part of the definition of fortification and enrichment is the term "nutrient." Nutrients refer only to essential components such as vitamins and minerals, where deficiency syndromes have been established and are reversed by administration of the pure nutrient. Other nonessential components that may also promote health can also be added to food products, although these would be considered functional foods and not fortified or enriched foods.

Regulation of Health Claims for Functional Foods, Nutraceuticals, and Supplements

If one considers that definitions and regulations in this discipline are complex and sometimes vague, the area of health claims will also leave most scientists unsatisfied or perplexed. The FAO defines nutrition claims as "any representation which states, suggests or implies that a food has particular nutritional properties, including but not limited to the energy value and to the content of protein, fat and carbohydrates, as well as vitamins and minerals".[8] Nutrition claims can include nutrient content, nutrient comparison, nonaddition, health, nutrient function, other function, and reduction of disease risk. Several types of compositional or health-related claims can be applied to functional food products. Many of these claims are regulated by governmental agencies and are expected to be accurate and not misleading. Additionally, government-endorsed claims should be evaluated periodically to account for new evidence.

Nutrient Content Claims

Nutrient content claims characterize the amount of a nutrient in the product. The claim can include an actual value (i.e., 100 mg sodium) or, in the case of comparative claims, a term such as "free," "high," "low," "reduced," or "lite." A non-addition claim states that an ingredient has not been added to a food. In the United States, several terms are allowed only in reference to specific amounts of nutrients. For example, "healthy" is a nutrient content claim that can be used only for products that contain a limited amount of total fat, saturated fat, cholesterol, and sodium. Conversely, products can be labeled as a "good source" of certain nutrients only if the nutrients are present at a level greater than 10% of the daily value per reference amount customarily consumed.

Health Claims

Health claims describe the relationship between the food product or dietary supplement and the risk of disease or health-related condition.[22] There are three types of health claims recognized in the United States: Nutrition Labeling and Education Act (NLEA) Authorized Health Claims, Health Claims Based on Authoritative Statements, and Qualified Health Claims.

NLEA authorized health claims characterize a relationship between a food or ingredient and the risk of a disease.[22] Upon submission of a health claim petition, the FDA evaluates if the claim is met by significant scientific agreement.[22] All proposed claims must be reviewed by the FDA. Approved claims are then published in the code of federal regulations 21 CFR Part 101. An example claim is "plant sterol/stanol esters may reduce the risk of heart disease."

While NLEA authorized health claims require FDA authorization, claims based on authoritative statements from certain scientific bodies only require a notification of the claim 120 days before introduction into interstate commerce. Health claims based on authoritative statements must still meet the "significant scientific agreement" standard. Currently, the regulation is only to be used for conventional foods and not dietary supplements, based on the wording on the FDA Modernization Act.[23]

The third type of health claim is used if the available evidence is not sufficiently robust to meet the standard of significant scientific agreement. The 2003 FDA Consumer Health Information for Better Nutrition Initiative allowed for use of a qualified health claim. The Initiative was intended to provide information to consumers during the development of the science if the strength and direction of the research were appropriately explained. Qualified health claims must include a disclaimer or other qualification to not mislead consumers. An example qualified health claim given by the FDA guidance is, "Scientific evidence suggests, but does not prove, that whole grains (three servings or 48 grams per day), as part of a low saturated fat, low cholesterol diet, may reduce the risk of diabetes mellitus type 2."[24]

Structure and Function Claims

The Dietary Supplement Health and Education Act introduced the term "structure and function claim" to describe the effect of supplements on the structure or function of the human body.[15] It may refer to how the component maintains or affects structure or function, contributes to general well-being, or helps with avoidance of a nutrient-deficiency disease. Although these originally applied only to dietary supplements, these claims may also be used for food products. Examples include: "fiber maintains bowel regularity" or "calcium builds strong bones." But a product may not be allowed to state, "prevents symptoms of Alzheimer's" but could perhaps have justification to state "helps maintain focus." For both foods and supplements, the claim must be accompanied by a phrase stating that the "statement has not been evaluated by the Food and Drug Administration. This product is not intended to diagnose, treat, cure, or prevent any disease."[9] For dietary supplements only, a notification must be sent to the FDA regarding the structure/function claim within 30 days of first marketing.[15] Food products with structure/function claims do not require FDA notification. However, under FDA regulations, misbranding is illegal, and all claims must be substantiated.

Claims in Marketing

While the FDA regulates claims on product packaging, the Federal Trade Commission (FTC) is primarily responsible for claims made in advertising including print, television, and Internet materials. If the advertising material includes a nutrient content claim, health claim, or structure and function claim, it is still required to meet FDA guidance. Additionally, companies are responsible for both express claims and reasonable implied claims.[9]

Public Perceptions and Desires

The above descriptions and regulatory policies are largely unknown and indecipherable to the public and by many in science, medicine, or the food industry. There is a critical need for improved clarity and enforcement of regulations regarding claims in this discipline. The consumer is inundated with communication by individuals, such as self-proclaimed experts (essentially unregulated), in various types of media regarding nutrients, non-nutrient supplements, and various foods to support health and treat disease, and these are increasingly prevalent in the modern world dominated by social media. The recent COVID crisis and profound impact of misinformation regarding the pandemic and vaccines is an example. Thus, it is a challenge to provide quality science-based guidance to the public, including those at risk of specific illness, in the areas of diet, nutrition, supplements, and the broad landscape of functional foods. Purveyors of products are more inclined to support media exposure than to support research. Companies touting the health benefits of products extensively push the legal limits on food labeling, in advertising, on websites, in social media networks, and at some scientific and health professional conferences, as well as traditional media outlets.

Consumers are concerned with health, physical performance, and chronic disease prevention.[25-27] Major drivers of functional food purchasing include actual or alleged health benefits along with food quality and acceptability, convenience, cost, and even the social reward of others presuming a healthy lifestyle.[28] However, the importance of each of these factors is limited by the others; for example, consumers are often not willing to compromise on taste in order to receive the purported health benefit.[29] This knowledge emphasizes that, for a food product that has proven evidence for an outcome, actual consumption and compliance will require a quality food product with taste and cost a priority.

The Role of Functional Foods for Public Health, Personalized Nutrition, or Precision Medicine

Quality basic and clinical science in functional foods for health will require research funding to achieve its potential to have impactful health benefits. Unlike the pharmaceutical industry, where massive profits foster vigorous reinvestment in research to develop new and more efficacious drugs; the food industry is very limited in its ability to support the types of preclinical and clinical research necessary to establish a health effect. Thus, research funds from government organizations are critical. To motivate such investment, scientists must consider the likely scenarios where functional foods may have their greatest impact on health.

For decades, the United States, many other nations, and various global organizations have provided guidelines regarding diet, nutrition, and fitness for health promotion and disease prevention in populations.[2-4] These guidelines are scientifically mature and based on extensive data with an increasingly objective and transparent analysis due to improved statistical tools of systematic reviews and meta-analysis.[2-4] Such guidelines are designed for health maintenance and prevention of disease across the population and provide a foundation for wise public policy. In contrast, the impact of functional foods is not likely to target the entire population but rather, more targeted subgroups at risk of disease or undergoing therapy.

One of the principles common to public health guidelines is that a plant-based diet rich in fruits, vegetables, whole grains, legumes, and fiber is recommended. Such knowledge, combined with decades of preclinical research evaluating plant bioactives to prevent disease, provides a conceptual foundation for the development and testing of novel functional foods. Designing foods with components showing bioactivities targeting pathways relevant to a specific disease process, either in a prevention or therapeutic paradigm, is an attractive approach. Such a strategy is supported by the concepts of personalized or precision nutrition, which emphasizes interventions for the individual, rather than the whole population. If we consider the issue of cardiovascular disease as an example, the major cause of mortality in the United States, it is appropriate for heart healthy dietary guidelines to target the entire population, given the prevalence of the disease, and, in parallel, to employ the principle of precision prevention by providing pharmaceuticals such as statins to those at greater risk due to dyslipidemia.

The strategy to target individuals raises the questions of what underlies person-to-person heterogeneity in disease risks, and whether we can improve our ability to identify high-risk subpopulations or individuals for effective interventions with functional foods. The importance of inherited germ-line DNA as key to disease risk is well established and continues to improve.[30] Increasingly, we appreciate the complex and dynamic impact of epigenetic modifications to modify the impact of genes. Environmental factors of many types, beginning with in utero exposure, and continuing over a lifetime, may impact epigenetic modifications that modulate the phenotype predicted from DNA alone. Epigenetic mechanisms include DNA methylation, covalent modifications of histones, and noncoding RNAs. Such modifications of the DNA have potential adaptive value for the host and thus, these processes are conserved over evolution. Epigenetic modifications may confer a Darwinian fitness advantage because they either adjust the phenotype to current circumstances or attempt to match an individual's responses to anticipated future exposures or experiences. When the phenotype is mismatched to the later environment (e.g., disease-promoting exposures, such as a poor diet), the risk of noncommunicable illness, such as cardiovascular disease, would be expected to increase. These mechanisms are also thought to have a role in processes such as aging and early onset of puberty, thus reinforcing a life-course perspective on such adaptive responses, especially the potential detrimental later effects of these responses.[31-35] Knowledge of an individual's genotype and epigenetic signatures may define those at risk of disease who can be targeted by functional foods to improve outcomes. For example, those with inherited dyslipidemias and inherited cancer syndromes may benefit from novel targeted functional foods.[36-38]

In parallel to defining target populations at disease risk, such individual genomic analysis will enhance our knowledge regarding why individuals vary in response to dietary patterns and bioactives in foods (as they do for pharmaceuticals) so as to improve outcomes.[39-44] Much genetic variation in response to drugs and bioactives is due to single nucleotide polymorphisms (SNPs). Such genetic variation is often linked to health or disease, such as the blood pressure sensitivity to salt, vitamin D receptor and bone development response to calcium, glutathione S-transferase detoxification and the anticancer response to cruciferous vegetables, and the estradiol receptor and sensitivity to soy, just to mention a few.[45-48] These efforts underly the emerging field of gene-nutrient interactions which, of course, goes beyond nutrients and includes dietary patterns, non-nutrient bioactives, and functional foods. The identification of genetic variants holds promise for identifying markers that may provide clues to individuals who will obtain benefits from consuming personalized dietary patterns or novel functional foods.[49,50]

Development and Evaluation of Functional Foods for Health

When developing a functional food to target a high-risk group, there are a multitude of considerations that will enhance the quality of science, promote development of food product with high desirability and compliance, and an impactful outcome.

Bioactive(s) Identification and Selection

A bioactive or multiple bioactives that may impact a disease process can be identified based on epidemiologic evidence, in vitro investigations, studies in rodent models, and clinical studies. For example, our development of a carotenoid and isoflavone-rich, tomato-soy juice product for clinical trials targeting prostate cancer includes all these strategies.[51-56] Foods often contain a portfolio of bioactives, with multiple mechanisms of action that may favorably target specific disease processes. For example, black raspberries (*Rubus occidentalis*) contain various phytochemicals including anthocyanins, ellagitannins, ferulic acid, flavonoids, fiber, vitamins, and minerals that, together, may provide a greater benefit than any pure compound. Black raspberry extracts and concentrates have been integrated into various food products that are undergoing clinical testing.[57-59]

Bioactive(s) Sourcing and Chemical Analysis

Foods such as tomatoes, berries, fruits, and vegetables show heterogeneity in bioactive content due to genetics, growing conditions (location, water, season, sunlight, etc.), harvesting and storage procedures. Collaboration with agricultural scientists with knowledge regarding how these processes impact the source of material you choose for a food product is valuable. Furthermore, such collaborations may allow investigators to purposely manipulate these processes, such as selective breeding, to enhance the phytochemical pattern desired. It is critical that the investigator establish analytic methodology, typically HPLC-MS or similar, to quantify the chemical profile, perhaps even to the level of specific isomers, in the components and the food product at each step during development.[60,61] Food products for human clinical research must have standardization of bioactive content over time and multiple trials to generate quality data.

Food Vehicle Selection and Testing

A food vehicle for bioactives of interest are diverse, such as confections/lozenges, bread products, juices, nectars, soups, or other products and may be chosen based on ease of integration into the daily diet, stability of the bioactives in the chosen matrix, or to maximize delivery of bioactives to a specific site. For example, choosing solid confections or nectar for delivery of black raspberry phytochemicals to the oral mucosa for cancer prevention.[58] The vehicle must be able to contain enough of the bioactive to have the biologic impact targeted. The food vehicle must be designed to protect the bioactives from degradation. For example, polyunsaturated lipids, such as eicosapentaenoic acid and docosahexaenoic acid in fish oil, are easily oxidized. The food matrix can be formulated to prevent this oxidation through addition of antioxidants or encapsulation. Additionally, the food matrix may influence the release characteristics of the bioactive compound(s) and can be formulated to control release throughout digestion and absorption. The compounds can be modified chemically using added enzymes or through processing such as fermentation or heat treatment. The food matrix can be utilized to perform this modification. For example, by adding almonds rich in beta glucosidase to soy bread to convert isoflavone glycosides to their aglycone equivalents.[62]

Food scientists employ several tools to help them design the optimal product for compliance and biologic impact. Methods, including microscopy, thermal analysis, and crystallography, give insight into the physical state of the bioactive component within the food product and how addition of that bioactive influences food physical properties that may impact taste and desirability. Through efforts to enhance the dose or profile of bioactives in an extract or concentrate and subsequently integrate into the food product, the physical properties of the food product may be impacted. This may include changes in texture, viscosity, color, aroma, and flavor. Instrumental testing of several types can be used to ensure that the food product maintains its identity and quality upon addition of the bioactive ingredient. Instrumental testing may also be used to optimize bioactive stability over time or evaluate release characteristics in vitro. Such efforts were key to developing a tomato-soy juice that was rich in bioactives but retained the desired taste qualities of tomato juice, while avoiding the undesirable taste and texture characteristics resulting from the addition of a soy powder.[52,54]

Sensory Evaluation

Sensory evaluation is imperative when developing a new product to ensure acceptability of the functional food and support daily integration into the diet. Sensory techniques are used including affective, difference, and threshold testing and standardized procedures have been developed.[63] Affective testing measures consumer acceptability of the product and its taste, aroma, texture, and appearance. Difference testing can be used to determine if the addition of a bioactive or various treatments change the product so that a difference can be detected from a control. Threshold testing determines the concentration at which a bioactive component is detected or rejected. Together, these efforts allow optimization of the product to ensure high compliance in clinical studies.

Scaling, Storage, Distribution, and Shelf Life

Small batches prepared using laboratory scale equipment can differ greatly from larger batches prepared using pilot or industrial equipment that will be necessary to provide sufficient product for clinical trials. The raw ingredients required are also much greater and potential variation between different batches must be anticipated, measured, and controlled to the degree possible. During scale up, the equipment changes may impact the food quality and stability of bioactives. Typically, additional instrumental and sensory testing is necessary after scaling up to ensure that the product was of similar quality to the product prepared in the laboratory.

Given that large batches of a food product may be necessary for a sufficiently powered clinical trial of many months or longer duration; and that cost concerns regarding production are an issue, it is desirable to prepare a small number of batches. Therefore, it is important to understand how storage and distribution of the product changes quality over the time course of a study. Shelf-life testing over time is used to monitor changes in product quality and bioactive concentration, often without refrigeration, based on the type of product.

Clinical Trials

The testing of novel functional foods can be considered in the same strategy that is applied to pharmaceuticals. In general, a phase I trial may evaluate various doses of a new product with an assessment of safety based on established laboratory tests and symptoms. A phase I trial of a functional food may involve normal healthy individuals or perhaps the target population for which the food was developed. Such trials will provide key data on compliance and characterization of putative biomarkers that can be employed in future studies. Biomarkers of internal exposure to bioactives may be quantified through analytical tools employing blood and urine and key dose-response relationships described that are helpful in designing future studies.[52,58] Biomarkers of impact related to a proposed health outcome can also be characterized through evaluation of tissue samples or blood fractions for hormonal and growth-factor alterations, gene expression, immunologic functions, and a multitude of proteomic or metabolomic signatures.

Phase II trials may include randomization to a control group, thus, at times, a placebo control food product may need to be designed. Such comparisons begin to focus on key biomarkers that are predictive of efficacy for a relevant health outcome under study. A phase II trial will concentrate on the target cohort of individuals. These could include, for example, patients with dyslipidemia, elevated hemoglobin A1C, a premalignant condition such as carcinoma-in-situ, those with a history of colon polyps, or a cohort at risk of a cancer due to genetic predisposition such as Lynch Syndrome.

Given that large batches of a food product may be necessary for a sufficiently powered clinical trial of many months or longer duration; and that cost concerns regarding production are an issue, it is desirable to prepare a small number of batches. Therefore, it is important to understand how storage and distribution of the product changes quality. Shelf-life testing over time is used to monitor changes in product quality and bioactive concentration, often with and without refrigeration based on the type of product (**Figure 35.2**).

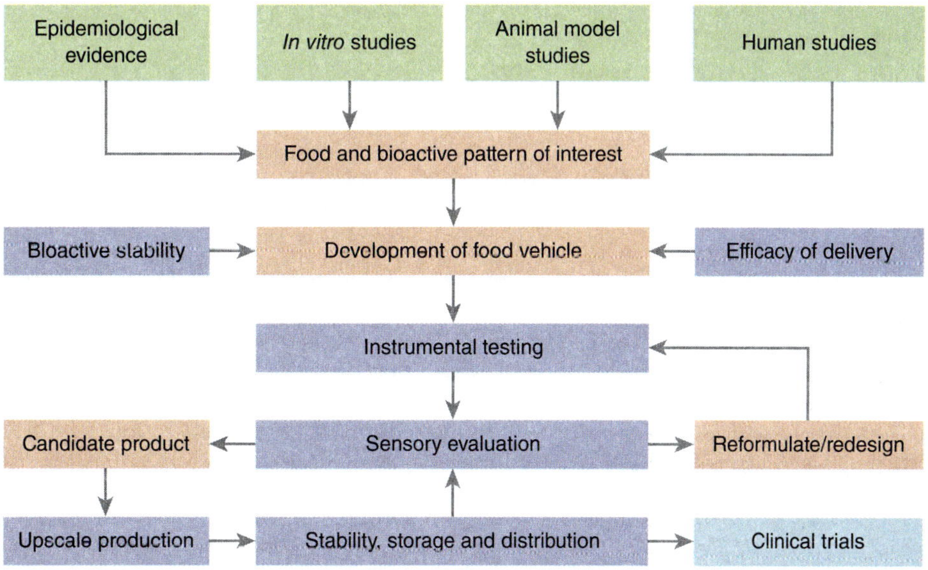

Figure 35.2 Summary of steps involved in developing a functional food for health and disease prevention. Evidence to support development of a bioactive-rich food product may be derived from multiple sources, including epidemiologic studies, in vitro cell biology studies, preclinical animal models, and prior clinical investigations. Once the food vehicle is chosen, analytic studies focusing on stability of bioactives and dissolution and delivery are needed. To optimize a food product for quality characteristics necessary to have high compliance in trials, the product is optimized for sensory characteristics and reformulated as needed. Finally, upscaling of production and defining stability, storage needs, and patient distribution issues are necessary prior to initiation of a clinical trial.

Conclusions and the Role of Research

Although the concept of developing specialized bioactive rich food products targeting specific individuals to promote health, prevent illness, enhance treatments of disease, and promote survivorship is appealing, there are significant challenges. The first issue is that current regulation and monitoring by government agencies is complex, and a multitude of enterprises can develop, market, and sell various supplements and functional foods for health outcomes with minimal scientific evidence and modest oversight.

Development and testing of functional foods, much like pharmaceuticals, is an expensive effort and research dollars directed to this topic have been modest by industry and funding agencies. Government and corporate funding is needed to see the potential fulfilled. This multidisciplinary research ranges from product development, analytical chemistry of bioactives and metabolism, characterization of biomarkers quantifying individual exposure and the understanding of person-to-person variability. In parallel, it is desired to characterize mechanisms of action for bioactives through biomarkers of physiologic and molecular impact on disease processes. Ultimately, large phase III trials to show disease impact are desired. Finally, integration of functional foods and supplements into public health strategies or into the clinics for individuals to enhance disease prevention, treatment, and survivorship will require careful messaging and oversight to insure impact and safety.

Acknowledgments

The authors thank John A. Milner, Cheryl Toner, and Cindy D. Davis, who authored the previous version of this chapter, which we have utilized extensively for the current update.

References

1. Hou Y, Jiang JG. Food Funct 2013;4:1727–41.
2. Diet, Nutrition, Physical Activity and Cancer: a Global Perspective. In Continuous Update Project Report 2018, World Cancer Research Fund/American Institute for Cancer Research.
3. Dietary Guidelines for Americans, 2020-2025. U.S. Department of Agriculture and U.S. Department of Health and Human Services.
4. Lichtenstein AH, Appel LJ, Vadiveloo M, et al. Circulation 2021;144:e472–e87.
5. US Department of Health and Human Services, National Institutes of Health. 2020–2030 Strategic Plan for NIH Nutrition Research: a report of the NIH Nutrition Research Task Force. 2020.
6. Food for Specified Health Uses (FOSHU). Accessed November 15, 2022. Available at www.mhlw.go.jp/english/topics/foodsafety/fhc/02.html.
7. Yamada K, Sato-Mito N, Nagata J, et al. J Nutr 2008;138: 1192S–8S.
8. Codex alimentarius. Food labelling, complete texts, Food and Agriculture Organization of the United Nations: World Health Organization.
9. Shield J, Kuler J, Gurnani A. In Bagchi D, and Nair S, eds. Developing New Functional Food and Nutraceutical Products San Diego: Academic Press. 2017;123–48.
10. Gaynor P. In Cianci S, ed. Food Safety Magazine The Target Group. 2005.
11. Mendes O, Chen J, Gowda R. In Bagchi D, ed. Nutraceutical and Functional Food Regulations in the United States and around the World, 3rd ed. Academic Press. 2019;107–20.
12. Safe Use of Chemical Additives in Foods. In: Press TNA, ed. Washington, DC: The National Academies Press, 1952: 28.
13. Overview of Food Ingredients, Additives & Colors. International Food Information Council (IFIC) and U.S. Food and Drug Administration (FDA), 2010.
14. Aronson JK. Br J Clin Pharmacol 2017;83:8–19.
15. Wong AY-T, Lai JMC, Chan AW-K. J Funct Foods 2015;17: 540–51.
16. Food and Drug Administration; Office of Dietary Supplement Programs. Dietary Supplements. 2022.
17. Hiltzik M. Column: Orrin Hatch is leaving the Senate, but his deadliest law will live on. In *Los Angeles Times*. 2018.
18. Novella S. Herbs Are Drugs. In Skeptical Inquirer 2013;37.
19. Guns ES, Goldenberg SL, Brown PN. Can J Urol 2002;9:1684-8; discussion, 9.
20. Kwak N-S, Jukes DJ. Food Control 2001;12:109–17.
21. Center for Food Safety and Applied Nutrition, Food and Drug Administration. Frequently Asked Questions About Medical Foods; 3rd ed. Guidance for Industry, 2023.
22. U.S. Food and Drug Administration. Label Claims for Conventional Foods and Dietary Supplements. 2022.
23. Guidance for Industry: Notification of a Health Claim or Nutrient Content Claim Based on an Authoritative Statement of a Scientific Body. Center for Food Safety and Applied Nutrition, 1998.
24. Questions and Answers on Health Claims in Food Labeling. Vol. 2022, U.S. Food and Drug Administration, 2017.
25. Bonnett M. US What/How America Eats Market Report. 2022. Mintel.
26. Formanski K. Supplements – US – October 2019. (Mintel, ed).
27. World Health Organization. Noncommunicable Diseases Progress Monitor 2022.
28. Topolska K, Florkiewicz A, Filipiak-Florkiewicz A. Int J Envir Res Pub Health 2021;18:5327.
29. Verbeke W. Food Qual Pref. 2006;17:126–31.
30. Antonarakis SE, Cooper DN. In: Rimoin D, Pyeritz R, and Korf B, eds. Emery and Rimoin's Principles and Practice of Medical Genetics, 6th ed. Oxford: Academic Press. 2013:1–48.
31. Braunschweig M, Jagannathan V, Gutzwiller A, et al. PloS One 2012;7:e30583.
32. Frantz EDC, Peixoto-Silva N, Pinheiro-Mulder A. Pancreas 2012;41:1–9.
33. McKay JA, Mathers JC. Acta Physiol 2011;202:103–18.
34. Ponzio BF, Carvalho MHC, Fortes ZB, et al. Life Sci 2012;90: 571–7.
35. vel Szic KS, Ndlovu MN, Haegeman G, et al. Biochem Pharmacol 2010;80:1816–32.
36. Bohn T, Blackwood M, Francis D, et al. Nutr Cancer 2013;65: 919–29.
37. Pigsborg K, Gurdeniz G, Rangel-Huerta OD, et al. Eur J Nutr 2022;61:2079–89.
38. Weghuber D, Widhalm K. Br J Nutr 2008;99:281–6.
39. Davis CD, Milner JA. J Nutrigenet Nutrigenomics 2011;4:1–11.
40. Dobbernack G, Meinl W, Schade N, et al. Carcinogenesis 2011;32:1734–40.

41. Glaser C, Lattka E, Rzehak P, et al. Matern Child Nutr 2011;7:27–40.
42. Mocchegiani E, Costarelli L, Giacconi R, et al. Age Res Rev 2012;11:297–319.
43. Perry GH, Dominy NJ, Claw KG, et al. Nature Genet 2007;39:1256–60.
44. Zeisel SH. Semin Cell Dev Biol 2011;22:624–8.
45. Dai Q, Xu W-H, Long J-R, et al. Pharmacogenet Genom 2007;17:161.
46. Rudnicki M, Mayer G. Pharmacogenomics 2009;10:463–76.
47. Seow A, Vainio H, Yu MC. Mutat Res 2005;592:58–67.
48. Zhang C, Wang C, Liang J, et al. Clinica Chimica Acta 2008;395:111–4.
49. Davis CD, Milner JA. Curr Cancer Drug Tar 2007;7:410–5.
50. Milner JA. Nutr Cancer 2006;56:216–24.
51. Giovannucci E. J Natl Cancer Inst 2007;99:1060–2.
52. Grainger EM, Moran NE, Francis DM, et al. J Nutr 2019;149:26–35.
53. Grainger EM, Schwartz SJ, Wang S, et al. Nutr Cancer 2008;60:145–54.
54. Moran NE, Thomas-Ahner JM, Fleming JL, et al. J Nutr 2019;149:381–97.
55. Moran NE, Thomas-Ahner JM, Wan L, et al. J Nutr 2022;152:1381–403.
56. Wang S, DeGroff VL, Clinton SK. J Nutr 2003;133:2367–76.
57. Gu J, Ahn-Jarvis JH, Riedl KM, et al. J Agric Food Chem 2014;62:3997–4006.
58. Roberts KM, Grainger EM, Thomas-Ahner JM, et al. Mol Nutr Food Res 2020;64:e1900800.
59. Knobloch TJ, Uhrig LK, Pearl DK, et al. Cancer Prev Res (Phila) 2016;9:159–71.
60. Cooperstone JL, Ralston RA, Riedl KM, et al. Mol Nutr Food Res 2015;59:658–69.
61. Unlu NZ, Bohn T, Francis DM, et al. Br J Nutr 2007;98:140–6.
62. Ahn-Jarvis JH, Riedl KM, Schwartz SJ, et al. J Agric Food Chem 2013;61:3111–20.
63. Lawless HT, Heymann H. Sensory Evaluation of Food: Principles and Practices. New York: Springer, 2010.

D. OTHER COMPOUNDS WITH HEALTH RELEVANCE

CHAPTER 36

Polyphenols: Focus on Flavonoids

Barbara Shukitt-Hale and Britt Burton-Freeman

CHAPTER OUTLINE

Abbreviations

C3G, cyanidin-3-glucoside; **DP**, degree of polymerization; **FMD**, flow mediated dilation; **MAPK**, mitogen-activated protein kinase; **PAC**, proanthocyanidin

Introduction

Flavonoids are a subclass of polyphenols, which are a subclass of phytochemicals (a.k.a, plant chemicals). Notably, flavonoids constitute a large class of bioactive components of food, aiding plant survival in many ways and, likewise, imparting beneficial effects on biologic processes when consumed by humans. Early research in the field of flavonoids was stimulated by the discovery of the French Paradox, that is, the low cardiovascular mortality rate observed in Mediterranean populations in association with red wine consumption, despite high saturated fat intake.[1] Further impetus for flavonoid research resulted from epidemiologic studies demonstrating relationships between dietary flavonoid intake and protection against various cancers[2,3] and cardiovascular diseases.[4,5]

Since the last update of this chapter, numerous books on phytochemical compounds have been published, including *Polyphenols: Prevention and Treatment of Human Disease*,[6] *Dietary Polyphenols: Their Metabolism and Health Effects*,[7] and others.[8-12] There have also been several review papers on the topics of polyphenols/flavonoids and health[13-16], including possible molecular mechanisms and signaling pathways behind their beneficial effects,[17] and more continue to be published.

Major Classes of Flavonoids Found in Foods

Flavonoids are perhaps the most important single group of phenolic compounds in plants, constituting about ⅔ of the polyphenols consumed in the diet. They can be divided by their basic structure into several different classes. More than 5,000, and up to as many as 15,000,[18] compounds have been described that have the same basic core structure, mainly in the form of various glycosides (i.e., glucoside, di-glucoside, rutinoside, rhamnosides, and more). Flavonoids are widely distributed in nature, albeit not uniformly. As a result, specific groups of foods are rich sources of one or more subclasses of these polyphenols. Flavonoids are often used by botanists for taxonomic classification. These substances may regulate plant growth, inhibit, or kill many bacterial strains, inhibit important viral enzymes, or destroy some pathogenic protozoans. However, their toxicity to animal cells is low.[19] The content of flavonoids in foods is strongly influenced by variations in plant type, breeding, growing conditions/farming practices, seasonal variation, storage/freshness, degree of ripeness, food preparation, and processing.[20] Plant stressors can also have major effects on the levels of flavonoids in the plant.

Flavonoid structure is based on a 2-phenyl-benzo[α]pyrane, or flavane, core (**Figure 36.1**). This core structure is defined by having a 15-carbon skeleton of two benzene rings (A and B), which are connected by an oxygen-containing pyrane ring (C). The flavonoids are further divided into subclasses (flavones, flavonols, flavanones, flavanols (flavan-3-ols), anthocyanins, and isoflavones) based on the connection of the aromatic B ring to the heterocyclic ring, as well as the oxidation state and functional groups of the heterocyclic ring. Within each subclass, individual compounds are characterized by specific hydroxylation and conjugation patterns and, in the case of flavan-3-ols, mono-, di-, tri- polymer structures. Because of limitations in space, the isoflavonoids are not considered in this chapter.

Flavones

The basic flavan nucleus is the structural feature of the flavone (see Figure 36.1). Among the flavones, apigenin and luteolin are present chiefly in grains, leafy vegetables, and herbs. Flavonols, which are discussed later, are 3-hydroxyflavones. Fewer research data are available on flavones, compared with the flavonols.

Apigenin, diosmetin, luteolin, acacetin, and chrysoeriol are the most common flavones, and they have been found in citrus fruits and juices, celery, parsley and other herbs, broccoli and more.[21,22] Apigenin has been shown to have strong cytostatic and antiangiogenic effects in vitro.[23] 2′,3′-dihydroxyflavone, fisetin, apigenin, and luteolin inhibit the proliferation of normal and tumor cells, as well as in vitro angiogenesis, at half-maximal

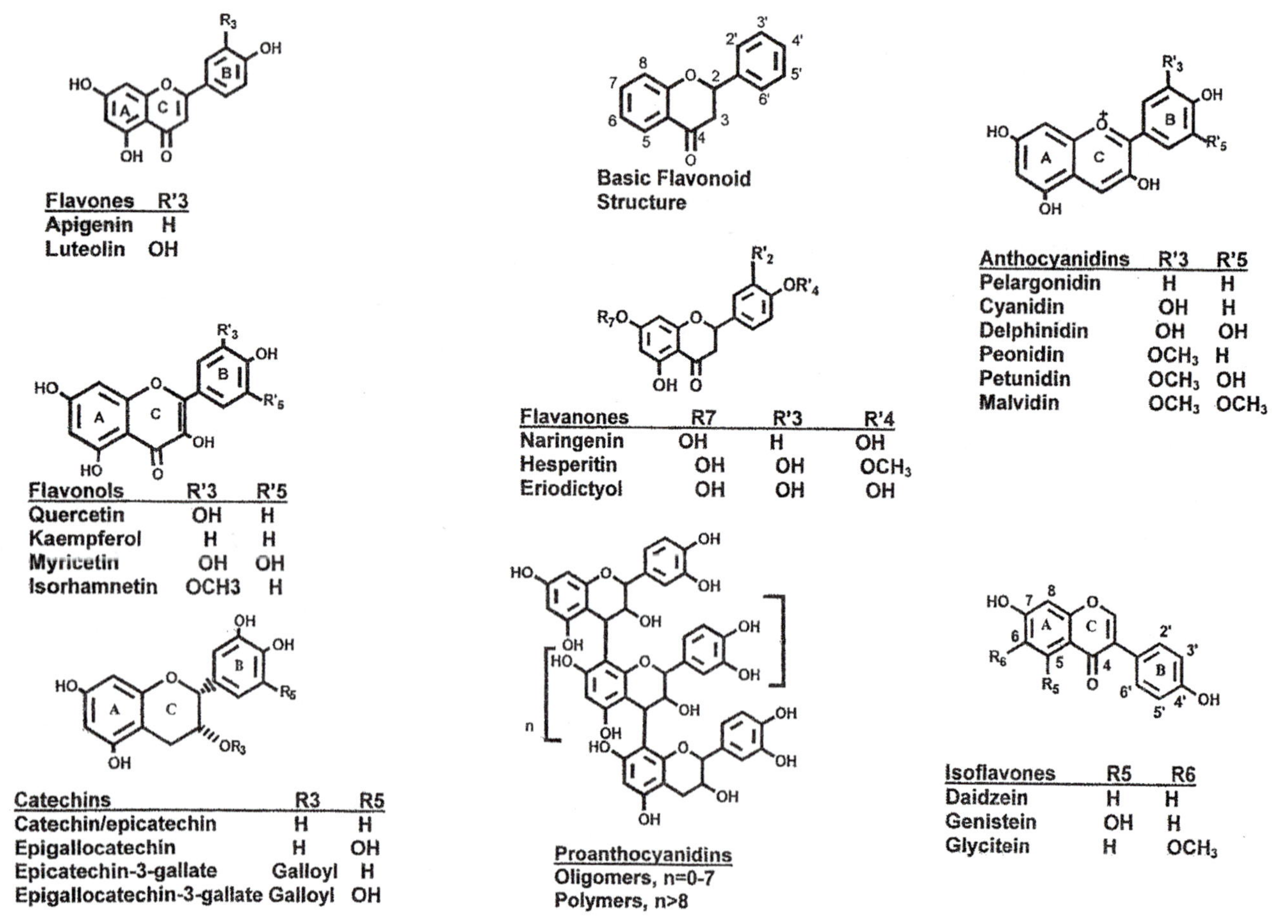

Figure 36.1 The basic flavonoid structure and structures of flavones, flavonols, flavanones, catechins and proanthocyanidins (flavan-3-ols), anthocyanidins, and isoflavones. Most naturally occurring flavonoids are glycosides (with sugars), but aglycones are shown in this figure. The flavanols or flavan-3-ols include catechins (monomer) and proanthocyanidins (polymer).

concentrations in the lower micromolar range.[24] Additionally, select flavones, such as luteolin, exert anticancer effects, both in cancer cell cultures and in xenografted mice.[25,26] Although flavonoids may show great potential as cancer chemopreventive agents in cell culture systems, this often does not translate well into in vivo activity.[27] When polyphenols are administered orally, only small quantities of these compounds appear in systemic circulation because of very high concentrations of uridine diphospho (UDP)-glucuronosyltransferases and sulfotransferases in the small intestine and liver, thus resulting in low oral bioavailability of intact flavonoids.[27] However, the case for superior anticancer activity for methoxylated flavones has been reviewed by Walle.[27]

A more recent review on flavones discussed the most common flavones in the diet, as well as their biologic roles in plants, bioavailability and metabolism, and biologic activity in animal models and human dietary intervention studies.[22] Like other flavonoids, flavones have low bioavailability; however, nuances of their chemistry may make them less well absorbed compared with other flavonoids.[22,28] Nonetheless, flavones have demonstrated many potentially beneficial activities such as lowered cholesterol and improved wound healing.[22] The beneficial physiologic and pharmacologic effects of citrus flavones was demonstrated by their antioxidant, anti-inflammatory, antimicrobial, and anticancer properties.[21] Consumption of Italian herbs containing varied flavonoids, including flavones, in a mixed macronutrient meal reduced postprandial plasma insulin and increased vessel flexibility, compared with the same meal without herbs.[29]

Flavonols

Chemistry

The flavonols (3-hydroxyflavones) are a subclass of flavonoids, which are diverse in their chemical structure and characteristics (see Figure 36.1). Quercetin, kaempferol, and myricetin are the most common flavonols. Quercetin (3,3′,4′,5,7-pentahydroxyflavone) is one of the most abundant and widely studied flavonoids. The availability of the pure compound made much of this research possible. The antioxidant property of quercetin is thought to be an important feature of its effect on human health, spurring research further. Fruits, vegetables, and beverages such as tea and red wine are major sources of flavonols in the human diet. Dietary quercetin is primarily present in plants in its glycoside form. Onions, elderberries, cranberries, herbs such as cilantro and dill weed, asparagus, kale, and apples with skin are among the most common dietary sources of quercetin.[30]

Sites of Absorption, Blood Transport, and Intracellular Forms

Early work suggested that myricetin and related compounds, such as quercetin, were absorbed from the small intestine following cleavage of the β-glucoside linkage.[31] Hollman et al.[32] found that humans absorb quercetin as the aglyconebut concluded that absorption was enhanced by conjugation with glucose. Although quercetin glycosides are subject to deglycosylation by enterobacteria for absorption in the large intestine, the small intestine can act as an effective absorption site for glucose-bound quercetin glucosides. Data suggest that β-glucosidases and also, membrane-bound, lactase-phlorizin hydrolase in the small intestine are capable of hydrolyzing quercetin glucosides, and that these compounds are thus taken up in the aglycone form preferentially.[33] Crespy et al.[34] found that quercetin, but not its glycosides, were absorbed from the rat stomach, an even earlier site of absorption.

Despite some discrepancies in the literature, it seems that the absorption of quercetin glycosides from the diet depends largely on the type of sugar groups attached to their phenolic group. Glucose-bound glycosides are likely to be much more absorbable than others that are sugar bound.[35] Hydrolysis to its aglycone form by enterocytes or enterobacteria seems to be important for the effective absorption of quercetin glycosides in the intestinal tract. Using a cell culture model for intestinal absorption, hydrolysis of quercetin glucosides accelerated quercetin absorption.[35]

Based on in vitro studies of Caco-2 cells, Walgren et al. demonstrated a complete lack of absorption of the glucosides of quercetin, mainly as a result of effective efflux by the multidrug resistance protein 2 transporter (MRP2),[36,37] but quercetin was readily absorbed. In subsequent human studies, Walle et al.[38] found that quercetin glucosides were hydrolyzed in the small intestine by bacterial enzymes.

The small intestine is also recognized as the site for metabolic conversion of quercetin and other flavonoids because it possesses enzymatic activity for glucuronidation and sulfation. The major circulating metabolites are glucuronide- and sulfate conjugates of isorhamnetin (3′-*O*-methyl quercetin) and of quercetin. Glucuronides, sulfates, and their *O*-methylated derivatives accumulate as quercetin metabolites in the circulation after intake of a quercetin-rich diet.[35,39] Modulation of the intestinal absorption and metabolism may be beneficial for regulating the biologic effects of dietary quercetin.[35]

Hollman et al.[40] found that absorption of quercetin in individuals with an ileostomy was 52%±5% for quercetin glucosides from onions, 17%±15% for quercetin rutinoside, and 24%±9% for quercetin aglycone. Excretion of quercetin or its conjugates in urine in four separate studies ranged from 0.07% to 17.4% of intake.[41] Gut microbiota can rapidly deconjugate the flavonols rutin (quercetin-3-rutinoside), isoquercetin, and a mixture of quercetin glucuronides. The main metabolite, 3,4-dihydroxyphenylacetic acid, appeared rapidly (within 2 hours) and was dehydroxylated to 3-hydroxyphenylacetic acid within 8 hours. Hydroxyphenylacetic acids were not methylated by colon microbes in vitro.[42]

Biologic Effects

Biologic effects of flavonols vary, depending on the flavonol metabolites produced. Following intragastric administration of quercetin (10 or 50 mg/kg), quercetin glucuronide or sulfate conjugates, or both, were present in the plasma, and the plasma was more resistant to copper sulfate–induced lipid peroxidation than the control plasma on the basis of the accumulation of cholesterol ester hydroperoxides and the consumption of α-tocopherol.[43] These results suggest that some conjugated metabolites of quercetin can act as effective antioxidants. Antioxidant activity of conjugated metabolites has also been observed in vitro.[39,44]

Numerous other in vitro effects have been reported.[45-49] However, one study suggested that quercetin supplementation in humans in doses of 500 to 1,000 mg/day for 12 weeks increased plasma quercetin concentration but did not alter several measures of oxidative stress or antioxidant capacity.[50] Studies have indicated that dietary quercetin may inhibit bone loss in ovariectomized mice,[51] and that it may increase mitochondrial biogenesis, with an associated increase in maximal endurance capacity and voluntary wheel-running activity in mice.[52] Although data from early clinical trials in humans were equivocal as to whether quercetin affects exercise outcomes,[53-55] a recent systematic review showed that polyphenols, and of note, quercetin, are viable supplements to improve performance in healthy individuals.[56,57]

A review of cohort and case-control epidemiologic studies focused on intake of flavonols, as well as flavones and flavanones; the studies revealed an inverse association with risk of some forms of cancer, cardiovascular disease, and other chronic conditions, suggesting that these flavonoids may beneficially affect disease etiology and pathophysiology.[58] Recent systematic reviews have shown that flavonols such as quercetin, kaempferol, myricetin and fisetin have positive effects on bone metabolism,[59] cancers (e.g., prostate and bladder cancers),[60] cardiovascular disease,[61] viral-induced lower respiratory tract infections,[62] blood pressure,[63] cholesterol,[64] and diabetes[65] in humans, as well as Alzheimer disease in animal models.[66]

Flavanones

Chemistry

Flavanones constitute most of all of flavonoids in citrus fruits, including oranges, tangerines, mandarins, grapefruit, lemons, and limes (see Figure 36.1). Overall, citrus flavonoids encompass a diverse set of structures, including numerous flavanone and flavone *O*- and *C*-glycosides and methoxylated flavones. Primary flavanones in citrus include hesperidin, naringin, narirutin, eriocitrin, neohesperidin, didymin, neoeriocitrin, and poncirin. After blood orange consumption, plasma naringenin and hesperetin increased, reaching peak concentrations at ~ 5 h, decreasing substantially by 8 h and were gone by 24 h.[67] Liquid chromatography–mass spectometry analyses of 12 dietary flavonoids in human urine showed that no flavonoid glycosides were excreted, and that citrus flavanones were excreted in higher amounts than flavonols.[68]

Biologic Effects

Citrus flavonoids exhibit several in vitro and in vivo anti-inflammatory and anticancer actions.[69] These biologic properties are consistent with their effects on the microvascular endothelial tissue. Evidence suggests that the biologic actions of the citrus flavonoids are possibly linked to their interactions with key regulatory enzymes involved in cell activation and receptor binding. The citrus flavonoids show little effect on normal, healthy cells and thus, typically exhibit remarkably low toxicity in animals. The citrus flavonoids extend their influence in vivo through their induction of hepatic phase I and II enzymes and through the biologic actions of their metabolites.[33,69]

Phytochemicals in grapefruit juice have some unique properties that can change the bioavailability of certain drugs and alter drug action. Naringin is likely one of the grapefruit components influencing drug metabolism. Grapefruit juice acts by inhibiting presystemic drug metabolism mediated by the cytochrome P-450 isoenzyme CYP3A4 in the small intestine. This interaction appears to be more relevant if the CYP3A4 content is high and the drug has a high degradation rate on first pass through the liver.[70,71]

Citrus flavanones have been shown to have anti-diabetes properties,[72] having several molecular targets in adipose, skeletal muscle, liver, and kidney and heart tissue aiding in the prevention and management of diabetes.[73] However, another review showed that hesperidin supplementation did not improve blood glucose control in humans.[74] Yet another review, focusing on the flavanones in citrus fruits, showed that they have a wide range of physiologic and pharmacologic effects, as shown in both in in vitro studies and in experimental animal investigations.[75] Although the authors point to the need for more human studies, they conclude that citrus flavanone compounds possess anticholesterolemic, antioxidant, anti-inflammatory and antimicrobial properties, as well as show potential anticancer as well as neuroprotective and cardioprotective activities.[75]

A recent systematic review of the flavanone farrerol discussed its anti-inflammatory, antioxidant, vasoactive, antitumor, and antimicrobial activities, as well as its underlying molecular mechanisms, as shown by in vitro and in vivo studies.[76] Naringin, a flavone glycoside, alone, or combined with other drugs, may be useful for treating cancers.[77]

Flavan-3-ols (Catechins)

Chemistry

Monomeric flavanols (aka flavan-3-ols) or catechins, which are precursors of proanthocyanidins (PACs,) are characterized by having a C6-C3-C6 skeleton with a hydroxyl group in the 3-postion of the C-ring (see Figure 36.1). Catechin and epicatechin are the most common of the flavan-3-ols. The (-)-epicatechin type is particularly high in cocoa. Galloyl-substituted derivatives, such as gallocatechin (GC), epigallocatechin (EGC), epicatechin gallate (ECG), and epigallocatechingallate (EGCG) are the forms usually found in select fruits, nuts, and, particularly, in teas. They are not usually found in their glycosylated from, unlike some other flavonoids such as anthocyanins.

Sites of Absorption, Blood Transport, and Intracellular Forms

Catechins are bioavailable. Plasma from humans was shown to contain up to 18 metabolites of (epi) catechin and (epi) gallocatechin, principally methylated, sulfated, and glucuronidated conjugates with peak plasma concentrations of 100 to 400 nM, which occurred within 0.8 to 2.3 hours of consumption, depending on the source and dose.[78-80] Blood and urine collected before and after 13 individuals consumed cocoa in skim milk containing 19.80 mg or 68.25 mg of flavanols showed increases in phase II derivatives of epicatechin peaking within 1.5 h of ingestion, consistent with small intestine absorption, but the most abundant metabolites were phase II derivatives peaking after 5 h, formed at the colonic level. 5-(4'-hydroxyphenyl)-γ-valeroalctone-sulfate may be a sensitive marker of cocoa intake.[81] (epi) Catechins are highly bioavailable, being absorbed and excreted to a much greater extent than most other flavonoids. Moreover, flavan-3-ol metabolites are rapidly turned over in the circulatory system, and, as a consequence, maximal concentration values are not an accurate quantitative indicator of the extent to which absorption occurs.[80]

Human intervention studies with green and black tea demonstrate a significant increase in plasma antioxidant capacity approximately 1 hour after consumption of moderate amounts of tea (1 to 6 cups/day).[82] The initial indications are that the enhanced blood antioxidant potential leads to reduced oxidative damage to macromolecules, such as DNA and lipids. The apparent bioavailability of the gallated catechins is lower than the non-gallated forms; however, absorbed gallated catechins are rapidly eliminated by excretion in the bile.[83] Catechins are excreted as free and conjugated in both intact and methylated forms. In animal studies, excretion of conjugated (epi) catechins was as high as 60% of the total consumed in some cases.[84] Most catechins and epicatechins excreted in the urine were in a methylated form at the 3′ or 4′ positions.[78,84,85] Excretion of epicatechins, including their methylated forms, ranged from 30% to 47% of the ingested amount, whereas that of catechins, including their methylated forms, ranged from 9% to 31%.[84] Urinary excretion of (epi) catechins was dose dependent and increased with the amount of (epi) catechins present in the diet. Based on excretory pattern of (epi) catechins in the urine, the bioavailability of epicatechins may be higher than catechins in rats.[84]

Biologic Effects

Tea remains the most widely consumed drink in the world after water. An accumulated number of population studies suggest that consumption of green and black tea beverages may bring positive health effects. One hypothesis explaining such effects is that the high levels of flavonoids in tea can protect cells and tissues from oxidative damage by scavenging oxygen free radicals. The effects of tea and green tea catechins on biomarkers of oxidative stress, especially oxidative DNA damage, appear very promising in animal models.[86,87] Human studies also show beneficial effects of epicatechin, based on biomarkers in pathologies involving inflammation, endothelial dysfunction/hypertension, and insulin resistance/diabetes.[87] However, the evidence relating oxidative stress biomarkers and flavonoid consumption to disease in humans is too limited to allow conclusive associations.[87]

Catechins may also function indirectly as antioxidants through the following: (a) inhibition of the redox-sensitive transcription factors, nuclear factor-κB and activator protein-1; (b) inhibition of "pro-oxidant" enzymes, such as inducible nitric oxide synthase, lipoxygenases, cyclo-oxygenases, and xanthine oxidase; and (c) induction of phase II and antioxidant enzymes, such as glutathione S-transferases and superoxide dismutases.[86,88] Chemically, the flavonoids found in green and black tea are effective radical scavengers. The tea flavonoids may, therefore, be active as antioxidants in the digestive tract or in other tissues after uptake.

Population studies of the association of high intakes of catechins with cancer incidence indicated that catechins derived primarily from fruits, (+)-catechin and (−)-epicatechin, tended to be inversely associated with upper digestive tract cancer, whereas catechins derived from tea were inversely associated with rectal cancer in postmenopausal women.[89] Many epidemiologic, case-control, and cohort studies have investigated the effects of tea consumption on human cancer incidence.[90-96] A brief synopsis[97] of 30 studies aimed at examining tea consumption as a factor in the incidence of colon and rectal cancers from 12 countries, and with data on consumption of both black and green tea, did not provide consistent evidence to support the theory, based on animal studies and basic research, that tea is a potent chemopreventive agent in humans. The assessment of tea consumption in most of these studies was based on a single question and, therefore, may be subject to significant measurement error, compared with more recent studies specifically aimed at assessing tea consumption.[97] Recent systematic reviews have shown that green tea catechins may reduce risk of prostate,[98] breast,[99] ovarian,[100] and lung[101] cancer risk, but there is less evidence for its treatment effects. Another review that assessed the association between green tea consumption and risk of developing cancer showed that the evidence was inconsistent, with some studies showing a beneficial effect, others indicating no effect, and some with even a slightly increased risk of cancer with green tea supplementation.[102] Although high intake of green tea seemed to slightly improve quality of life, compared with placebo, it was associated with some adverse side effects.[102] Overall, this review concluded that well conducted and adequately powered, randomized control trials would be needed to draw conclusions on the possible beneficial effects of green tea consumption on cancer risk.[102]

In animal models of atherosclerosis, green and black tea administration resulted in modest improvement in resistance of lipoproteins to ex vivo oxidation, and limited data suggest that green tea or green tea catechins inhibit atherogenesis.[86] One epidemiologic study indicated that high intake of tea and flavonoids may contribute to the primary prevention of ischemic heart disease.[86] Cardiovascular effects of monomeric catechins have been reviewed.[78] A systematic review of green tea catechin supplementation and biochemical measures associated with cardiovascular disease (CVD) risk showed significant reductions in body mass index (BMI), blood pressure, and plasma lipids, indicating that catechins from green tea possesses some potentially cardio- and vascular-protective properties, useful for reducing CVD.[103]

Cocoa flavan-3-ols are best known for their antioxidant, anti-inflammatory and vascular health promoting properties. Reviews of the science, including meta-analyses, relevant to cocoa flavanols and exercise, cardiometabolic and vascular health[104] largely support beneficial effects when consumed acutely, over the short-term, and longer term. Acute and short-term intake of cocoa flavanol reduced exercise-induced oxidative stress and altered fat and carbohydrate metabolism during exercise. Chronic intake improved mitochondrial efficiency in untrained individuals.[104] A meta-analysis including 19 randomized controlled trials, with 1,131 participants ingesting cocoa flavanols ranging from 166 to 2,110 mg/d over 2 to 52 weeks, showed modest but significant effects in lipid metabolism, insulin resistance, and systemic inflammation.[105] Vascular effects are indicated through a number of clinical trials assessing flow mediated dilation (FMD).[106] In a meta-analysis of 15 clinical trials with 730 participants, using interventions of at least 7 days with cocoa drink, chocolate, or both, FMD improved by 1.08%.[107] Others have shown similar results with an overall effect of 1.34% FMD (CI: 1.00%, 1.68%) for studies longer than 2 weeks (nine clinical trials, n = 545 participants).[108] FMD has been inversely associated with future cardiovascular events, with data supporting a 13% risk reduction of cardiovascular events with a 1% increase of FMD.[109,110] In a landmark clinical trial with more than 21,000 participants supplemented with a multivitamin and or cocoa extract (500 mg/d flavanols, 80 mg (-)-epicatechin), compared with placebo control, cardiovascular events were not significantly reduced among older adults, but cocoa extract supplementation reduced cardiovascular death by 27%.[111] In February of 2023, the United States Food and Drug administration (US FDA) granted a qualified health claim for high cocoa flavanol cocoa powder and risk reduction of CVD (www.fda.gov/media/165090/download, letter of enforcement discretion US FDA).

Proanthocyanidins

Chemistry

Proanthocyanidins (PACs), better known as condensed tannins, are oligomeric and polymeric flavan-3-ols (see Figure 36.1). They are ubiquitous and present as the second most abundant natural phenolic, after lignin in plants. The flavan-3-ol units are linked mainly through the C4→C8 bond, but the C4→C6 bond also exists (both called B-type) (see Figure 36.1). The flavan-3-ol units can also be doubly linked by an additional ether bond between C2→O7 (A-type). The size of the PAC molecules can be described by the degree of polymerization (DP).[112]

Sites of Absorption, Blood Transport, and Intracellular Forms

Unlike the smaller size oligomers, or monomeric catechins, PACs with a DP greater than 3 appear not to be absorbed directly in the gastrointestinal lumen.[113,114] They have been suggested to depolymerize into mixtures of epicatechin monomers and dimers in the acidic environment of the stomach, but this is highly dependent on the pH value of gastric juice, which is also affected by food intake. The resultant monomers and dimers are absorbed in the small intestine.[115] An observation in weaning pigs indicated that 8% to 15% of ingested polymers (DP >10) were depolymerized after 4 hours in the stomach, with a concurrent increase of monomers, dimers, and trimers. The depolymerization process is slow and, as such, most PACs may transit the small intestine intact.[116]

Major degradation of PACs takes place in the cecum and large intestine,[117] where the colonic microbiota play an important role in catabolizing PACs into a varied low molecular weight phenolic acids.[118] Déprez et al.[119] reported that incubation of polymeric procyanidins with human colonic microbes in vitro in anoxic conditions led to a complete degradation of procyanidins after 48 hours. The degradation products included phenylacetic, phenylpropionic, and phenylvaleric acids, with the monohydroxyl group mainly in the *meta*- or *para*- position. It has been suggested that these phenolic acids are the major metabolites of oligomeric and polymeric PACs in healthy humans.[116] Dimeric procyanidins can be detected in human plasma as early as 30 minutes after the consumption of a flavanol-rich food such as chocolate.[120,121] However, oligomers larger than the trimer have not been detected in plasma.

Free forms of dimers and trimers were detected in rat plasma, and they reached a maximum concentration 1 hour after oral intake of a grape seed extract.[122] Procyanidins A1 and A2 dimers were absorbed from the small intestine of rats, and they were absorbed more effectively than B2 dimers. Absorption of the A-type dimers was only 5% to 10% of monomeric epicatechin. Dimers were not conjugated or methylated, in contrast to epicatechin.[123] After oral administration of [^{14}C] procyanidin B2, 63% was excreted through the urine within 4 days. These data suggest that much of the procyanidin B2 was degraded by the gut microbiota before absorption.[124] Some of the human fecal microbial metabolites of procyanidins B2 have been characterized in an in vitro static model.[125]

A recent paper systematically reviewed the nutrikinetics (i.e., absorption, metabolism, distribution, and excretion) of up to 180 flavan-3-ol derived compounds following their intake by healthy subjects.[126] While the mean bioavailability of flavan-3-ols was 31%, the flavan-3-ol monomer, (−)-epicatechin, was more bioavailable, compared with ingested sources of PACs. In general, bioavailability is dependent on structure and inversely associated with molecular weight, influencing the metabolic efficiency in their biotransformation.[126]

Biologic Effects

PACs are of great interest in nutrition and medicine because they antagonize oxidative damage, enhance antioxidant capacity, and possibly have protective effects in reducing the risk of chronic diseases such as cancer and cardiovascular diseases.[127,128] In vitro studies have shown that procyanidins in chocolate inhibit human 5-lipoxygenase, decrease low-density lipoprotein oxidative susceptibility,[129,130] inhibit platelet function,[129] and promote transforming growth factor-β_1 homeostasis in peripheral blood mononuclear cells.[131] Procyanidins in grape seeds induce apoptotic death of human prostate carcinoma[132] and exhibit cytotoxicity toward MCF-7 breast cancer, A-427 lung cancer, and gastric adenocarcinoma cells, and they enhance the growth and viability of normal cells.[133]

PACs of different molecular size may differ in their physiologic effects. Mao et al.[134] studied the ability of the procyanidins to modulate interleukin-2 in vitro and found that higher oligomers inhibited interleukin-2 expression in stimulated peripheral blood mononuclear cells, whereas the monomer had no effect. The physiologic relevance of these findings is questioned because large PACs are not found in plasma. Tebib et al.[135] suggested that polymeric PACs were more effective than monomers in lowering blood cholesterol in rats fed hypercholesterolemic diets. The proposed mechanism is that PACs bind cholesterol in the intestine through hydrophobic association, which is more potent for PACs of higher DP. Different constituent flavan-3-ols and interflavan linkages in PACs may also influence their physiologic effects. In vitro studies have shown that procyanidins with the A-type interflavan linkage, isolated from cranberries, inhibit the adherence of uropathogenic *Escherichia coli* to the uroepithelial cell surfaces, whereas B-type procyanidins show no effects.[136] In addition, the A-type polymers from cinnamon have insulin-like biologic activity,[137] and cinnamon consumption (1 to 6 g) for 40 days in patients with type 2 diabetes lowered serum glucose, triglycerides, and total cholesterol.[138]

Oligomeric procyanidins from grapeseed extract have been shown to interact and induce the autophosphorylation of the insulin receptor, thus stimulating uptake of glucose.[139] However, the mechanism of activation differs from insulin activation and results in differences in downstream signaling. Grape seed procyanidins phosphorylated protein kinase B at Thr308 to a lesser degree than insulin; thus, insulin receptor activation was lower with procyanidins. However, grape seed procyanidins phosphorylated Akt at Ser473 to the same extent as insulin, and they phosphorylated p44/p42 and p38 mitogen-activated protein kinases (MAPKs) much more than insulin. These results point to Akt and MAPK proteins as key points for grape seed procyanidin-activated signaling pathways.[140] Treatment of hyperinsulinemic rats with grape seed procyanidins at a dose of 25 mg/kg body weight per day had a positive long-term effect on glucose homeostasis,[140] reflected by an improved Homeostasis Model Assessment—insulin resistance index, which was accompanied by downregulation of peroxisome proliferator-activated receptor γ2 (PPARγ2), glucose transporter type 4 (Glut4), and insulin receptor substrate 1 (IRS1) in mesenteric white adipose tissue.[140]

Thus, procyanidins seem to possess insulinomimetic properties by decreasing hyperglycemia in streptozotocin-diabetic rats and stimulating glucose uptake in insulin-sensitive cell lines. However, additional research is needed relative to procyanidin structure and concentrations that are effective in modulating insulin action.

Blood pressure regulation has been another clinical endpoint potentially influenced by consumption of PAC-rich foods/supplements. Grape seed extract, rich in monomeric and polymeric flavan-3-ols, have shown favorable outcomes for blood pressure lowering,[141-144] although not in all studies,[145] which may be a reflection of study population (healthy vs metabolic syndrome). One study design reporting decreased systolic blood pressure after 6 weeks of grape seed extract in a drink, also reported blood pressure increases back to baseline after discontinuation of grape seed extract, which coincided with a fall in circulating grape seed metabolites.[143] A more recent meta-analysis found that proanthocyanidin supplementation had significant positive effects on systolic and diastolic blood pressure, as well as mean arterial pressure, but no significant effect on pulse pressure.[146] Cranberry proanthocyanidins have been shown to be effective in preventing urinary tract infections (UTI) recurrence (by 26%) in generally healthy women.[147] A review paper found that proanthocyanidins can confer beneficial health effects in cardiovascular and metabolic disorders, and in some cancers.[148]

Anthocyanins

Chemistry

Anthocyanins are water-soluble plant secondary metabolites responsible for the blue, purple, and red color of many plant tissues (see Figure 36.1). They occur primarily as glycosides of their respective anthocyanidin-chromophores, with the sugar moiety attached at the 3-position on the C-ring or the 5-position on the A-ring. The common anthocyanidins (aglycones) are cyanidin, delphinidin, petunidin, peonidin, pelargonidin, and malvidin. The differences in chemical structure of these six common anthocyanidins occur at the 3′ and 5′ positions (see Figure 36.1). Aglycones are rarely found in fresh plant material. Several hundred anthocyanins are known, and they vary in the following ways: (a) the number and position of hydroxyl and methoxyl groups on the basic anthocyanidin skeleton; (b) the identity, number, and positions at which sugars are attached; and, (c) the extent of sugar acylation and identity of the acylating agent.[149] Common acylating agents are the cinnamic acids (caffeic, ρ-coumaric, ferulic, and sinapic). Acylated anthocyanins occur in some less common foodstuffs, such as red cabbage, red lettuce, garlic, red-skinned potato, and purple sweet potato.[150]

The chemistry and distribution of anthocyanins have been reviewed.[151] Like other flavonoids, anthocyanins and anthocyanidins (the aglycone form) have antioxidant properties.[152] The phenolic structure of anthocyanins (see Figure 36.1) conveys marked antioxidant activity in model systems by donation of electrons or transfer of hydrogen atoms from hydroxyl moieties to free radicals. The cyanidin glycosides tend to have higher antioxidant capacity than peonidin or malvidin glycosides,[152] likely because of the free hydroxyl groups on the 3′ and 4′ positions of cyanidin.

Cyanidin, the most common anthocyanidin, is present in 90% of fruits.[149,153] Anthocyanin concentrations (mg/100 g fresh weight) range from 0.25 mg/100 g in the pear to 500 mg/100 g in the blueberry,[153] and the fruits that are richest in anthocyanins (>20 mg/100 g fresh weight) are very strongly colored (deep purple or black) berries.

Sites of Absorption, Blood Transport, and Intracellular Forms

Anthocyanins are unique, compared with other flavonoids, in that they can be absorbed intact as glycosides. The mechanisms of absorption have recently been reviewed,[154,155] and they are much more bioavailable than previously thought, based on tracer studies illuminating their metabolic fate.[156] The intake of anthocyanins from the gastrointestinal lumen into the blood is likely to occur through the epithelium of the stomach, intestine and colon, although the mechanisms of absorption differ from site to site, and depend on the structure of the molecules that are absorbed.[154,155,157] Passamonti et al.[158] found that anthocyanins can serve as ligands for bilitranslocase, an organic anion membrane carrier found in the epithelial cells of the gastric mucosa—a finding suggesting that bilitranslocase could play a role in the bioavailability of anthocyanins from the stomach. The addition of sucrose to elderberry juice led to reduced and delayed excretion of anthocyanins[159]; this finding suggested that sugars may interfere with the anthocyanin transport mechanism.

At least 13 different anthocyanins from seven different food sources have been observed to be absorbed intact and to be present in plasma or urine.[149] In contrast to other flavonoids, the proportion of anthocyanins absorbed and excreted in the urine as a percentage of the intake seems to be small,[160] perhaps less than 0.1%. Maximum plasma concentrations of total anthocyanins are reported in the range of 1 to 120 nmol/L with doses of 0.7 to 10.9 mg/kg in human studies.[41,161-163] A summary of several studies indicated that, for a total anthocyanin intake of 0.05 to 1.9 g, the corresponding maximal concentration of 1 to 200 nM was reached (time of maximal concentration) within 0.5 to 4 hours after ingestion of the respective dose, and urinary excretion in no case exceeded 5% of the ingested dose.[164] The clearance of anthocyanins from the circulation is sufficiently rapid that, by 6 hours, very little is generally detected in the plasma.[161,165] Rats seem to differ from humans in that, in the rats, the aglycone (cyanidin) from cyanidin-3-glucoside (C3G) was detected in the jejunum,[166] and protocatechuic acid, which may be produced by degradation of cyanidin, was present in the plasma at concentrations eightfold higher than that of C3G.

In studies by Cao et al.[161] the two major anthocyanins in elderberry (C3G and cyanidin-3-sambubioside) were detected as glycosides in both plasma and urine of humans. Mulleder et al.[159] observed a greater urinary excretion of cyanidin-3-sambubioside than C3G (0.014% vs 0.004% of dose). The lower excretion of C3G may be the result of increased degradation, relative to cyanidin-3-sambubioside in the gastrointestinal tract.[167] A systematic review of human clinical trials of berry consumption identified 203 phenolic compounds in human samples, with C3G being the anthocyanin found most frequently in urine (58.1%) and plasma (69.5%), thereby leading the authors to suggest it as a potential biomarker of anthocyanin-rich berry intake in plasma and urine of healthy humans.[168]

The complexity of the glycosidic pattern does not seem to affect absorption noticeably. Food matrix or coexisting food components/ingredients will affect absorption.[169,170] As demonstrated with strawberries, peak plasma concentrations (C_{max}) of pelargonidin-3-O-glucoside (P3G) and pelargonidin-glucuronide (PG) were reduced after consuming a strawberry fruit beverage prepared with milk, compared with water.[170] Likewise, consuming a strawberry drink with a meal or alone 2 hours before a meal influenced absorption and pharmacokinetic patterns of anthocyanins.[171] Mazza et al.[172] suggested that acylated anthocyanins could be absorbed intact from blueberries, but they have not been detected in plasma or urine in other reports. Most likely, this is because these substances are present in low concentration, and methods were not sensitive enough to detect them. Most anthocyanins were excreted in urine during the first 4 hours after consumption.[160,161] However, total elderberry anthocyanin excretion in the first 4 hours accounted for only 0.077% of the dose.

The metabolism of anthocyanins in the gut is an area that has not been well understood, but has received increasing attention with interest in the reciprocal interactions with gut microbiota. Felgines et al.[173] were among the first to report the presence of anthocyanins in gut contents of rats after adaptation to a diet containing blackberry anthocyanins. Recovery of cyanidin plus C3G in the total cecal contents was approximately 0.25%, whereas larger amounts of malvidin-3-glucoside were recovered in the cecum (~1.3%). Later studies have demonstrated that anthocyanins can be metabolized by intestinal microbiota[78,174] or simply degraded chemically.[175,176] A review paper summarized the recent data on the absorption of anthocyanins in the upper gastrointestinal tract and the metabolism of anthocyanins by gut microbiota, concluding that consumption of a diet rich in anthocyanins may be beneficial to approaching a balanced gut microbial community.[177] Beneficial shifts in microbial communities were evident in people with prediabetes and insulin resistance after consuming red raspberries, a dietary source of anthocyanins and other polyphenols, for 4 weeks.[178] Correspondingly, altered plasma anthocyanin and metabolite profiles were evident after 4 weeks of red raspberry intake, suggesting that daily intake of anthocyanin- rich foods may influence host and microbial metabolic capacity for anthocyanins and other polyphenols potentially relevant in their biologic effects over time.[179] Similar findings were evident in older people consuming strawberries, a pelargonidin-rich source of anthocyanins, for 90 days, whereby marked increases in basal anthocyanin metabolites were evident along with enhanced postprandial anthocyanin metabolites.[180]

Tissue Anthocyanin Metabolism

Wu et al.[160] identified four anthocyanin metabolites from elderberry in the urine: peonidin-3-glucoside, peonidin 3-sambubioside, peonidin monoglucuronide, and C3G monoglucuronide. However, Miyazawa et al.[163] were not able to detect conjugated or methylated anthocyanins in human plasma, but did observe the presence of peonidin-3-glucoside in the liver of rats, following the consumption of red fruit anthocyanins (C3G).The formation of the peonidin metabolites likely takes place in the liver, through the catechol-O-methyl transferase reaction. Peonidin-3-glucoside was present in urine in rats fed a diet enriched with a lyophilized blackberry powder, and this likely resulted from hepatic methylation at the 3′-hydroxyl moiety of C3G.[173]

Delphinidin would be the only other anthocyanidin that could undergo this methylation reaction because malvidin and petunidin are already methylated in the 3′ position (see Figure 36.1). Urinary recovery of C3G in either intact or methylated forms was approximately 0.26% of the ingested amount, whereas that of malvidin 3-glucoside was 0.67%. This result suggests that the structure of the aglycone moiety of anthocyanins could play an important role in their metabolism.[173]

Biologic Effects

Vascular and Metabolic Effects

Diabetic retinopathy can lead to blindness, resulting from abnormally high synthesis of connective tissue to repair leaking capillaries and to form new capillaries. Adults with diabetes who were treated with 600 mg of anthocyanins per day for 2 months significantly decreased the biosynthesis of connective tissue, especially polymeric collagen and structural glycoproteins in gingival tissue.[181] Anthocyanins have been shown to have effects on cardiovascular health, including endothelial function and myocardium protection, based primarily on in vitro cell studies[78,182] and ex vivo rabbit aorta endothelial function tests.[183] Epidemiologic data examining dietary anthocyanin intake suggest a role in blood pressure regulation and reduced risk of myocardial infarction.[184,185] Data from systematic reviews showed correlations between anthocyanin consumption and improvements in key risk factors related to metabolic syndrome,[186] and that regular consumption of either purified anthocyanins or anthocyanin-rich berries could prevent cardiovascular diseases possibly via their lipid-lowering and anti-inflammatory properties.[187]

Diabetes is a growing problem and is a major risk factor for cardiovascular diseases. The association between anthocyanin intake and diabetes has gained increasing attention. In a meta-analysis investigating dietary anthocyanin and berry consumption, 15% and 18% reductions in type 2 diabetes (T2D) risk were associated with dietary anthocyanins and berry intake, respectively.[188] Data showing improvements in insulin sensitivity and inflammation with berry or anthocyanin intake are possible reasons for the observed risk reductions in T2D.[189] These results are supported by randomized controlled trial data in varied metabolically impaired populations showing improved insulin and inflammatory status responses with anthocyanin-rich berry intake, as well as improved insulin sensitivity using the gold-standard hyperinsulinemic-euglycemic clamp methodology.[190-192] Mechanisms of action for anthocyanins or their metabolites are likely due to both insulin dependent and independent pathways to improve glycemic control.[193,194]

Vision

Early reports and some anecdotal information exist on the association between anthocyanins and improved night vision. Zadok et al.[195] assessed the effect of anthocyanins on three night vision tests in a double-blind, placebo-controlled, crossover study in which normal volunteers were given either 12 or 24 mg anthocyanins, or a placebo, twice daily for 4 days. No significant effect was found on any of the night vision tests.

In a double-blind, placebo-controlled, crossover study with healthy humans, Nakaishi et al.[196] studied the effects of oral intake of a black currant anthocyanin concentrate on dark adaptation, video display terminal work-induced transient refractive alteration, and visual fatigue. Intake at three dose levels (12.5, 20, and 50 mg/person; $n = 12$) appeared to bring about a dose-dependent lowering of the dark adaptation threshold, with a significant difference at the 50-mg dose. In the assessment of subjective visual fatigue symptoms by questionnaire, significant improvement was recognized, based on statements regarding the eye and lower back, after black currant anthocyanin intake.

In a randomized, double-blind, placebo-controlled study, bilberry fruit extract (160 mg twice daily for 1 month) resulted in improvements in confirmed retinal abnormalities in 79% of the patients with either diabetic or hypertensive vascular retinopathy.[197] Patients with diabetic retinopathy who were given 480 mg of bilberry anthocyanins daily for 6 months showed improvement by the end of the trial period, as indicated by reduction of hemorrhage and alleviation of weeping exudates from the retina.[198] However, Muth et al.[199] failed to find an effect of bilberry anthocyanins on night visual acuity or night contrast sensitivity in individuals given 120 mg of anthocyanins daily for 21 days.

Overall, the response to anthocyanin exposure on vision is not consistent, based on the studies presented,[200,201] although anthocyanins did show some promise as an intervention for improving the outcomes of ocular disorders.[202] Dose and length of feeding are clearly factors affecting outcomes. Positive effects have been observed at intakes in the range of 300 to 600 mg/day over a period of several months. However, consumption of these levels of anthocyanins from foods is difficult unless one consistently consumes high anthocyanin-containing foods.

Other Effects

Fruits and berries may be protective through antioxidant mechanisms in preventing DNA damage, and they may also affect cell division, apoptosis, and angiogenesis.[203] Hou[204] summarized some of the molecular mechanisms whereby anthocyanins may have cancer chemoprevention properties, including those related to antioxidants and induction of apoptosis in tumor cells. In a preliminary human trial, participants with early memory changes were given blueberry juice. After 12 weeks, improved, paired associate learning and word list recall were observed.[205] Anthocyanins improved learning and memory in rats with estrogen deficit caused by ovariectomy.[206] Subsequent to these preliminary studies, studies have shown that anthocyanins may protect against age-related cognitive decline,[207] with the most evidence for the beneficial effects of blueberries.[208]

Purified anthocyanins from varying sources have been shown to decrease lipid deposition in rodent models of obesity.[209] In earlier studies in which anthocyanins were consumed as a part of the whole food or berry, antiobesity effects were not generally observed.[210,211] However, effects were observed with whole blueberries that protected against some of the disorders associated with obesity, including inflammation.[212] Anthocyanins have been shown to regulate adipocytokine gene expression and to ameliorate adipocyte dysfunction related to obesity and diabetes. Alterations in lipogenesis and lipolysis in adipose tissue, as well as in numerous adipokine and cytokine signaling pathways, have been suggested to explain the effects of anthocyanins on the development of obesity.[213] Anthocyanins can also improve the metabolic control involved in obesity by reducing lipogenesis, oxidative stress, and inflammation, thus boosting the speed of lipolysis and thermogenesis, regulating satiety, and reducing body fat accumulation.[214]

Dietary Sources and Intakes of Flavonoids

Data on the flavonoid content of selected foods can be found in the U.S. Department of Agriculture flavonoid database (https://data.nal.usda.gov/dataset/usda-database-flavonoid-content-selected-foods-release-33-march-2018), which may be the most widely used resource. However, additional database resources are also available, such as phenol explorer (http://phenol-explorer.eu], phytohub [https://phytohub.eu], and FooDB from Canada [https://foodb.ca]. According to data from the National Health and Nutrition Examination Survey (NHANES), the average intake in U.S. adults (aged ≥19 years) is between 200 and 250 mg/day, with 80% being flavan-3-ols, 8% for flavonols, 6% for flavanones, 5% for anthocyanidins, and ≤1% for isoflavones and flavones.[215,216] Other studies have also estimated daily consumption of flavonoids in various populations.[217-222]

Acknowledgments

The authors acknowledge Ronald L.Prior for contributions to the previous version of this chapter.

References

1. Nijveldt RJ, van Nood E, van Hoorn DE, et al. Am J Clin Nutr 2001;74:418–25.
2. Hertog MG, Hollman PC, Katan MB, et al. Nutr Cancer 1993;20:21–9.
3. Hollman PC, Hertog MG, Katan MB. Biochem Soc Trans 1996;24:785–9.
4. Hertog MG, Feskens EJ, Hollman PC, et al. Lancet 1993;342:1007–11.
5. Keli SO, Hertog MG, Feskens EJ, et al. Arch Intern Med 1996;156:637–42.
6. Watson RR. Polyphenols : Prevention and Treatment of Human Disease. San Diego, CA: Elsevier, 2018.
7. Tomas-Barberan FA, González-Sarrías A, García-Villalba Ro. Dietary Polyphenols : Their Metabolism and Health Effects. Hoboken, NJ: Wiley-Blackwell, 2021.
8. Rani V, Yadav UCS. Functional Food and Human Health. pp. 1 online resource (XXII, 694 pages 132 illustrations, 57 illustrations in color, Springer Singapore : Imprint: Singapore: Springer, 2018.
9. Saura-Calixto F, Pérez-Jiménez J, Royal Society of Chemistry (Great Britain). Non-Extractable Polyphenols and Carotenoids: Importance in Human Nutrition and Health, London: Royal Society of Chemistry, 2018.
10. Ullah MF, Ahmad A. Nutraceuticals and Natural Product Derivatives: Disease Prevention & Drug Discovery. Hoboken, NJ: John Wiley & Sons, Inc, 2018.
11. Upaganlawar AB. Health Benefits of Phenolic Antioxidants. New York: Nova Science Publishers, 2022.
12. Watson RR. Polyphenols : Mechanisms of Action in Human Health and Disease. San Diego, CA : Elsevier, , 2018.

13. Joseph SV, Edirisinghe I, Burton-Freeman BM. Crit Rev Food Sci Nutr 2016;56:419–44.
14. Zhang PY. Cell Biochem Biophys 2015;73:649–64.
15. Rodriguez-Garcia C, Sanchez-Quesada C, Gaforio JJ. Antioxidants (Basel) 2019;8:137.
16. Zhang Z, Li X, Sang S, et al. Foods 2022;11:2189.
17. Hamsalakshmi, Alex AM, Arehally Marappa M, et al. Inflammopharmacol 2022;30:111–36.
18. Xiao J. Crit Rev Food Sci Nutr 2017;57:1874–905.
19. Havsteen BH. Pharmacol Ther 2002;96:67–202.
20. Aherne SA, O'Brien NM. Nutrition 2002;18:75–81.
21. Barreca D, Mandalari G, Calderaro A, et al. Plants (Basel) 2020;9.
22. Hostetler GL, Ralston RA, Schwartz SJ. Adv Nutr 2017;8: 423–35.
23. Engelmann C, Blot E, Panis Y, et al. Phytomedicine 2002;9: 489–95.
24. Fotsis T, Pepper MS, Montesano R, et al. Baillieres Clin Endocrinol Metab 1998;12:649–66.
25. McCarty MF, Assanga SI, and Lujan LL. Med Hypotheses 141:109723.
26. Lin Y, Shi R, Wang X, et al. Curr Cancer Drug Targets 2008;8: 634–46.
27. Walle T. Semin Cancer Biol 2007;17:354–62.
28. Breiter T, Laue C, Kressel G, et al. Food Chem 2007;128; 338–47.
29. Huang Y, Tsai MF, Thorat RS, et al. Front Nutr 2022;9:811433.
30. Ay M, Charli A, Jin H, et al. Quercetin. In: Gupta RC, ed. Nutraceuticals: Efficacy, Safety and Toxicity. London: Academic Press, 2016;447–52.
31. Griffiths LA, Smith GE. Biochem J 1972;130:141–51.
32. Hollman PC, de Vries JH, van Leeuwen SD, et al. Am J Clin Nutr 1995;62:1276–82.
33. Rasmussen SE, Breinholt VM. Int J Vitam Nutr Res 2003;73: 101–11.
34. Crespy V, Morand C, Besson C, et al. J Agric Food Chem 2002;50:618–21.
35. Murota K, Terao J. Arch Biochem Biophys 2003;417:12–7.
36. Walgren RA, Karnaky KJ, Jr., Lindenmayer GE, et al. J Pharmacol Exp Ther 2000;294:830–6.
37. Walgren RA, Lin JT, Kinne RK, et al. J Pharmacol Exp Ther 2000;294:837–43.
38. Walle T, Otake Y, Walle UK, et al. J Nutr 2000;130:2658–61.
39. Morand C, Crespy V, Manach C, et al. Am J Physiol 1998;275:R212–9.
40. Hollman PC, van Trijp JM, Mengelers MJ, et al. Cancer Lett 1997;114:139–40.
41. Prior RL. Am J Clin Nutr 2003;78:570S–8S.
42. Aura AM, O'Leary KA, Williamson G, et al. J Agric Food Chem 2002;50:1725–30.
43. da Silva EL, Piskula MK, Yamamoto N, et al. FEBS Lett 1998;430:405–8.
44. Shirai M, Moon JH, Tsushida T, et al. J Agric Food Chem 2001;49:5602–8.
45. Graefe EU, Derendorf H, Veit M. Int J Clin Pharmacol Ther 1999;37:219–33.
46. Harasstani OA, Moin S, Tham CL, et al. Inflamm Res 2010;59: 711–21.
47. Liu S, Hou W, Yao P, et al. Toxicol In Vitro 2010;24:516–22.
48. Middleton E, Jr., Kandaswami C, Theoharides TC. Pharmacol Rev 2000;52:673–751.
49. Suri S, Liu XH, Rayment S, et al. Br J Pharmacol 2010;159: 566–75.
50. Shanely RA, Knab AM, Nieman DC, et al. Free Radic Res 2010;44:224–31.
51. Tsuji M, Yamamoto H, Sato T, et al. J Bone Miner Metab 2009;27:673–81.
52. Davis JM, Murphy EA, Carmichael MD, et al. Am J Physiol Regul Integr Comp Physiol 2009;296:R1071–7.
53. Dumke CL, Nieman DC, Utter AC, et al. Appl Physiol Nutr Metab 2009;34:993–1000.
54. Quindry JC, McAnulty SR, Hudson MB, et al. Int J Sport Nutr Exerc Metab 2008;18:601–16.
55. Utter AC, Nieman DC, Kang J, et al. Res Sports Med 2009;17: 71–83.
56. Somerville V, Bringans C, Braakhuis A. Sports Med 2017;47:1601.
57. Somerville V, Bringans C, and Braakhuis A. Sports Med 2017; 47:1589–99.
58. Graf BA, Milbury PE, Blumberg JB. J Med Food 2005;8:281–90.
59. Inchingolo AD, Inchingolo AM, Malcangi G, et al. Nutrients 2022;14.
60. Crocetto F, di Zazzo E, Buonerba C, et al. Nutrients 2021;13.
61. Prem PN, Sivakumar B, Boovarahan SR, et al. Phytomedicine 2022;101:154123.
62. Brito JCM, Lima WG, Cordeiro LPB, et al. Phytother Res 2021;35:4930–42.
63. Tamtaji OR, Milajerdi A, Dadgostar E, et al. Curr Pharm Des 2019;25:1372–84.
64. Tabrizi R, Tamtaji OR, Mirhosseini N, et al. Crit Rev Food Sci Nutr 2020;60:1855–68.
65. Bule M, Abdurahman A, Nikfar S, et al. Food Chem Toxicol 2019;125:494–502.
66. Zhang XW, Chen JY, Ouyang D, et al. Int J Mol Sci 2020;21.
67. Gardana C, Guarnieri S, Riso P, et al. Br J Nutr 2007;98:165–72.
68. Nielsen IL, Dragsted LO, Ravn-Haren G, et al. J Agric Food Chem 2003;51:2813–20.
69. Manthey JA, Grohmann K, Guthrie N. Curr Med Chem 2001;8: 135–53.
70. Ameer B, Weintraub RA. Clin Pharmacokinet 1997;33:103–21.
71. Lohezic-Le Devehat F, Marigny K, Doucet M, et al. Therapie 2002;57:432–45.
72. Azhar S, Sabaha R, Sajjad F, et al. Curr Diabetes Rev 2022;19:e070722206679.
73. Gandhi GR, Vasconcelos ABS, Wu D-T, et al. Nutrients 2020;12:2907.
74. Shams-Rad S, Mohammadi M, Ramezani-Jolfaie N, et al. Br J Clin Pharmacol 2020;86:13–22.
75. Barreca D, Gattuso G, Bellocco E, et al. Biofactors 2017;43: 495–506.
76. Qin X, Xu X, Hou X, et al. Pharm Biol 2022;60:9–16.
77. Ghanbari-Movahed M, Jackson G, Farzaei MH, et al. Front Pharmacol 2021;12:639840.
78. de Pascual-Teresa S, Moreno DA, Garcia-Viguera C. Int J Mol Sci 2010;11:1679–703.
79. Stalmach A, Mullen W, Steiling H, et al. Mol Nutr Food Res 2010;54:323–34.
80. Stalmach A, Troufflard S, Serafini M, et al. Mol Nutr Food Res 2009;53:S44–53.
81. Gomez-Juaristi M, Sarria B, Martinez-Lopez S, et al. Nutrients 2019;11.
82. Rietveld A, Wiseman S. J Nutr 2003;133:3285S–92S.
83. Manach C, Williamson G, Morand C, et al. Am J Clin Nutr 2005;81:230S–42S.
84. Khanal RC, Howard LR, Wilkes SE, et al. J Agric Food Chem 2010;58:11257–64.
85. Natsume M, Osakabe N, Oyama M, et al. Free Radic Biol Med 2003;34:840–9.
86. Frei B, Higdon JV. J Nutr 2003;133:3275S–84S.
87. Oteiza PI, Fraga CG, Galleano M. Redox Biol 2021;42:101914.
88. Higdon JV, Frei B. Crit Rev Food Sci Nutr 2003;43:89–143.
89. Arts IC, Jacobs DR, Jr., Gross M, et al. Cancer Causes Control 2002;13:373–82.
90. Katiyar SK, Mukhtar H. J Cell Biochem Suppl 1997;27:59–67.

91. Yang CS, Chung JY, Yang G, et al. J Nutr 2000;130:472S–8S.
92. Chung FL, Schwartz J, Herzog CR, et al. J Nutr 2003;133:3268S–74S.
93. Chen D, Milacic V, Chen MS, et al. Histol Histopathol 2008;23:487–96.
94. Inoue M, Robien K, Wang R, et al. Carcinogenesis 2008;29:1967–72.
95. Khan N, and Mukhtar H. Cancer Lett 2008;269, 269–80.
96. Kuo YC, Yu CL, Liu CY, et al. Cancer Causes Control 2009;20:57–65.
97. Arab L, Il'yasova D. J Nutr 2003;133:3310S–8S.
98. Guo Y, Zhi F, Chen P, et al. Medicine (Baltimore) 2017;96:e6426.
99. Yiannakopoulou E. Eur J Cancer Prev 2014;23:84–9.
100. Trudel D, Labbe DP, Bairati I, et al. Gynecol Oncol 2012;126:491–8.
101. Fritz H, Seely D, Kennedy DA, et al. Integr Cancer Ther 2013;12:7–24.
102. Filippini T, Malavolti M, Borrelli F, et al. Cochrane Database Syst Rev 2020;3:CD005004.
103. Lau SO, Georgousopoulou EN, Kellett J, et al. Beverages 2016;2:6.
104. Decroix L, Soares DD, Meeusen R, et al. Sports Med 2018;48:867–92.
105. Lin X, Zhang I, Li A, et al. J Nutr 2016;146:2325–33.
106. Vlachojannis J, Erne P, Zimmermann B, et al. Phytother Res 2016;30:1641–57.
107. Sun Y, Zimmermann D, De Castro CA, et al. Food Funct 2019;10:6322–30.
108. Hooper L, Kay C, Abdelhamid A, et al. Am J Clin Nutr 2012;95:740–51.
109. Inaba Y, Chen JA, Bergmann SR. Int J Cardiovasc Imaging 2010;26:631–40.
110. Ras RT, Streppel MT, Draijer R, et al. Int J Cardiol 2013;168:344–51.
111. Sesso HD, Manson JE, Aragaki AK, et al. Am J Clin Nutr 2022;115:1490–500.
112. Gu L, Kelm MA, Hammerstone JF, et al. J Agric Food Chem 2003;51:7513–21.
113. Deprez S, Mila I, Huneau JF, et al. Antioxid Redox Signal 2001;3:957–67.
114. Donovan JL, Manach C, Rios L, et al. Br J Nutr 2002;87:299–306.
115. Spencer JP, Chaudry F, Pannala AS, et al. Biochem Biophys Res Commun 2000;272:236–41.
116. Rios LY, Gonthier MP, Remesy C, et al. Am J Clin Nutr 2003;77:912–8.
117. Gu L, House SE, Rooney L, et al. J Agric Food Chem 2007;55:5326–34.
118. Zhang L, Wang Y, Li D, et al. Food Funct 2016;7:1273–81.
119. Deprez S, Brezillon C, Rabot S, et al. J Nutr 2000;130:2733–8.
120. Holt RR, Lazarus SA, Sullards MC, et al. Am J Clin Nutr 2002;76:798–804.
121. Urpi-Sarda M, Monagas M, Khan N, et al. Anal Bioanal Chem 2009;394:1545–56.
122. Serra A, Macia A, Romero MP, et al. Br J Nutr 2010;103:944–52.
123. Appeldoorn MM, Vincken JP, Gruppen H, et al. J Nutr 2009;139:1469–73.
124. Stoupi S, Williamson G, Viton F, et al. Drug Metab Dispos 2010;38:287–91.
125. Stoupi S, Williamson G, Drynan JW, et al. Arch Biochem Biophys 2010;501:73–8.
126. Di Pede G, Mena P, Bresciani L, et al. Mol Aspects Med 2022:101146.
127. Schewe T, Kuhn H, Sies H. J Nutr 2002;132:1825–9.
128. Li S, Xu M, Niu Q, et al. PLoS One 2015;10:e0139455.
129. Murphy KJ, Chronopoulos AK, Singh I, et al. Am J Clin Nutr 2003;77:1466–73.
130. Wan Y, Vinson JA, Etherton TD, et al. Am J Clin Nutr 2001;74:596–602.
131. Mao TK, Van De Water J, Keen CL, et al. Exp Biol Med (Maywood) 2003;228:93–9.
132. Tyagi A, Agarwal R, Agarwal C. Oncogene 2003;22:1302–16.
133. Ye X, Krohn RL, Liu W, et al. Mol Cell Biochem 1999;196:99–108.
134. Mao TK, Powell JJ, Van de Water J, et al. Int J Immunother 1999;15:23—9.
135. Tebib K, Besancon P, Rouanet JM. J Nutr 1994;124:2451–7.
136. Foo LY, Lu Y, Howell AB, et al. J Nat Prod 2000;63:1225–8.
137. Anderson RA, Broadhurst CL, Polansky MM, et al. J Agric Food Chem 2004;52:65–70.
138. Khan A, Safdar M, Ali Khan MM, et al. Diabetes Care 2003;26:3215–8.
139. Montagut G, Onnockx S, Vaque M, et al. J Nutr Biochem 2010;21:476–81.
140. Montagut G, Blade C, Blay M, et al. J Nutr Biochem 2010;21:961–7.
141. Feringa HH, Laskey DA, Dickson JE, et al. J Am Diet Assoc 2011;111:1173–81.
142. Kim JK, Kim KA, Choi HM, et al. J Med Food 2018;21:445–53.
143. Park E, Edirisinghe I, Choy YY, et al. Br J Nutr 2016;115:226–38.
144. Sivaprakasapillai B, Edirisinghe I, Randolph J, et al. Metab 2009;58:1743–6.
145. Ras RT, Zock PL, Zebregs YE, et al. Br J Nutr 2013;110:2234–41.
146. Ren J, An J, Chen M, et al. Pharmacol Res 2021;165:105329.
147. Fu Z, Liska D, Talan D, et al. J Nutr 2017;147:2282–8.
148. Blade C, Aragones G, Arola-Arnal A, et al. Biofactors 2016;42:5–12.
149. Prior RL. In: Meskin MS, Bidlack WR, Davies AJ, et al., eds. Absorption and Metabolism of Anthocyanins: Potential Health Effects. Phytochemicals: Mechanisms of Action , Boca Raton, FLCRC Press, 2004;1–19.
150. Clifford MN. J Sci Food Agric 2000;80:1063–72.
151. Strack D, Wray V. In: Harborne JB, ed. The Anthocyanins. The Flavonoids: Advances in Research Since 1986 London: Routledge, 1994;1–22.
152. Wang H, Cao GH, Prior RL. J Agr Food Chem 1997;45:304–9.
153. Macheix JJ, Fleuriet, A, Billot, J. In: Macheix JJ, ed. The Main Phenolics of Fruits. Fruit Phenolics Boca Raton, FL: CRC Press, 1990.
154. Lila MA, Burton-Freeman B, Grace M, et al. Annu Rev Food Sci Technol 2016;7:375–93.
155. Kay CD, Pereira-Caro G, Ludwig IA, et al. Annu Rev Food Sci Technol 2017;8:155–80.
156. Czank C, Cassidy A, Zhang Q, et al. Am J Clin Nutr 2013;97:995–1003.
157. Hribar U, Ulrih NP. Curr Drug Metab 2014;15:3–13.
158. Passamonti S, Vrhovsek U, Mattivi F. Biochem Biophys Res Commun 2002;296:631–6.
159. Mulleder U, Murkovic M, Pfannhauser W. J Biochem Biophys Methods 2002;53:61–6.
160. Wu X, Cao G, Prior RL. J Nutr 2002;132:1865–71.
161. Cao G, Muccitelli HU, Sanchez-Moreno C, et al. Am J Clin Nutr 2001;73:920–6.
162. Matsumoto H, Inaba H, Kishi M, et al. J Agric Food Chem 2001;49:1546–51.
163. Miyazawa T, Nakagawa K, Kudo M, et al. J Agric Food Chem 1999;47:1083–91.
164. McGhie TK, Walton MC. Mol Nutr Food Res 2007;51:702–13.

165. Bub A, Watzl B, Heeb D, et al. Eur J Nutr 2001;40:113–20.
166. Tsuda T, Horio F, Osawa T. FEBS Lett 1999;449:179–82.
167. Wu X, Pittman HE, 3rd, Prior RL. J Agric Food Chem 2006;54:583–9.
168. Sandoval-Ramirez BA, Catalan U, Fernandez-Castillejo S, et al. Nutr Rev 2020;78:597–610.
169. Yang M, Koo SI, Song WO, et al. Curr Med Chem 2011;18: 291–300.
170. Xiao D, Sandhu A, Huang Y, et al. Food Funct 2017;8: 3970–9.
171. Sandhu AK, Huang Y, Xiao D, et al. J Agric Food Chem 2016;64:4891–9.
172. Mazza G, Kay CD, Cottrell T, et al. J Agric Food Chem 2002;50:7731–7.
173. Felgines C, Texier O, Besson C, et al. J Nutr 2002;132: 1249–53.
174. Selma MV, Espin JC, Tomas-Barberan FA. J Agric Food Chem 2009;57:6485–501.
175. Kay CD, Kroon PA, Cassidy A. Mol Nutr Food Res 2009;53: S92–101.
176. Woodward G, Kroon P, Cassidy A, et al. J Agric Food Chem 2009;57:5271–8.
177. Tian L, Tan Y, Chen G, et al. Crit Rev Food Sci Nutr 2019;59: 982–91.
178. Zhang X, Zhao A, Sandhu AK, et al. J Nutr 2022;152:1438–49.
179. Zhang X, Sandhu A, Edirisinghe I, et al. Molecules 2020;25.
180. Sandhu AK, Miller MG, Thangthaeng N, et al. Food Funct 2018;9:96–106.
181. Boniface R, Robert AM. Klin Monbl Augenheilkd 1996;209: 368–72.
182. Xia M, Ling W, Zhu H, et al. Atherosclerosis 2009;202:41–7.
183. Edirisinghe I, Burton-Freeman B, Varelis P, et al. J Agric Food Chem 2008;56:9383–90.
184. Jennings A, Welch AA, Fairweather-Tait SJ, et al. Am J Clin Nutr 2012;96:781–8.
185. Cassidy A, Mukamal KJ, Liu L, et al. Circulation 2013;127: 188–96.
186. Wilken MR, Lambert MNT, Christensen CB, et al. Rev Diabet Stud 2022;18:42–57.
187. Xu L, Tian Z, Chen H, et al. Front Nutr 2021;8:747884.
188. Guo X, Yang B, Tan J, et al. Eur J Clin Nutr 2016;70:1360–7.
189. Jennings A, Welch AA, Spector T, et al. J Nutr 2014;144:202–8.
190. Basu A, Izuora K, Betts NM, et al. Nutrients 2021;13.
191. Edirisinghe I, Banaszewski K, Cappozzo J, et al. Br J Nutr 2011;106:913–22.
192. Stull AJ, Cash KC, Johnson WD, et al. J Nutr 2010;140:1764–8.
193. Fallah AA, Sarmast E, Jafari T. Food Res Int 2020;137:109379.
194. Burton-Freeman B, Brzezinski M, Park E, et al. Nutrients 2019;11.
195. Zadok D, Levy Y, Glovinsky Y. Eye (Lond) 1999;13:734–6.
196. Nakaishi H, Matsumoto H, Tominaga S, et al. Altern Med Rev 2000;5:553–62.
197. Perossini MC, Guidi, G, Siravo D. Ann Ottal Clini Ocul 1987;113:1173–90.
198. Orsucci PN, Sabbatini G, Menci S, et al. Clin Ocul 1983;5: 377–81.
199. Muth ER, Laurent JM, Jasper P. Altern Med Rev 2000;5: 164–73.
200. Kalt W, Hanneken A, Milbury P, et al. J Agric Food Chem 2010;58:4001–7.
201. Upton R. Bilberry Fruit, Vaccinium myrtillus L., Standards of Analysis, Quality Control, and Therapeutics. Santa Cruz, CA: American Herbal Pharmacopoeia, 2001.
202. Davinelli S, Ali S, Scapagnini G, et al. Front Nutr 2001;8:651441.
203. Baby B, Antony P, Vijayan R. Crit Rev Food Sci Nutr 2018;58:2491–507.
204. Hou DX. Curr Mol Med 2003;3:149–59.
205. Krikorian R, Shidler MD, Nash TA, et al. J Agric Food Chem 2010;58:3996–4000.
206. Varadinova MG, Docheva-Drenska DI, Boyadjieva NI. Menopause 2009;16:345–9.
207. Feng RC, Dong YH, Hong XL, et al. Nutr Rev. 2023;81: 287–303.
208. Travica N, D'Cunha NM, Naumovski N, et al. Brain Behav Immun 2020;85:96–105.
209. Prior RL. CAB Rev: Perspect Agric Vet Sci, Nutr Nat Resources 2010;5:1–9.
210. Prior RL, Wilkes SE, Rogers TR, et al. J Agric Food Chem 2010;58:3970–6.
211. Prior RL, Wu X, Gu L, et al. J Agric Food Chem 2008;56: 647–53.
212. DeFuria J, Bennett G, Strissel KJ, et al. J Nutr 2009;139: 1510–6.
213. Tsuda T, Ueno Y, Yoshikawa T, et al. Biochem Pharmacol 2006;71:1184–97.
214. Gomes JVP, Rigolon TCB, Souza M, et al. Nutr 2019;66: 192–202.
215. Kim K, Vance TM, Chun OK. Eur J Nutr 2016;55:833–43.
216. Sebastian RS, Wilkinson Enns C, Goldman JD, et al. J Nutr 2015;145:1239–48.
217. Zamora-Ros R, Andres-Lacueva C, Lamuela-Raventos RM, et al. J Am Diet Assoc 2010;110:390–8.
218. Murphy KJ, Walker KM, Dyer KA, et al. Nutr Res 2019;61: 64–81.
219. Chun OK, Chung SJ, Song WO. J Nutr 2007;137:1244–52.
220. Vogiatzoglou A, Mulligan AA, Lentjes MA, et al. PLoS One 2015;10:e0128132.
221. Lanuza F, Bondonno NP, Zamora-Ros R, et al. Front Nutr 2022;9:873774.
222. Jun S, Shin S, Joung H. Br J Nutr 2016;115:565.

D. OTHER COMPOUNDS WITH HEALTH RELEVANCE

CHAPTER 37

Biotics and Fermented Foods as Modulators of the Gut Microbiome

Hannah D. Holscher and Sharon M. Donovan

CHAPTER OUTLINE

Abbreviations

CRP, C-reactive protein; **CFU**, colony-forming units; **DP**, degrees of polymerization; **EFSA**, European Food Safety Authority; **FAO**, Food and Agriculture Organization of the United Nations; **FOS**, fructo-oligosaccharides; **FDA**, Food and Drug Administration; **GOS**, galacto-oligosaccharides; **HDL**, high-density lipoprotein; **HMOS**, human milk oligosaccharides; **ISAPP**, International Scientific Association for Probiotics and Prebiotics; **LDL**, low-density lipoprotein; **LTA**, lipoteichoic acids; **MAMPs**, microbial-associated molecular patterns; **NHANES**, National Health and Nutrition Examination Surveys; **PRRs**, pattern-recognition receptors; **RCT**, randomized controlled trials; **SCFA**, short-chain fatty acids; **TLR**, Toll-like receptors; **WHO**, World Health Organization

Introduction

Interest in dietary modulation of the gastrointestinal microbiome to support human health is an area of growing interest and research. There is no doubt that nutrition is critical to health. Indeed, specific nutrients, foods, and dietary patterns have been investigated for their roles in supporting health across the lifespan. With the advances in molecular and sequencing technologies, there is now a growing body of evidence indicating that diet affects the microbiome, and that changes in the gut microbiome and metabolome may mediate or moderate some diet-health relationships.[1,2]

The gut microbiome is the collection of trillions of microorganisms, including bacteria, archaea, fungi, and viruses, and their genes within the gastrointestinal tract. These microorganisms interact with nondigested food components (e.g., complex carbohydrates) to generate metabolites as well as modify host secretions, such as bile acids, in ways that affect health.[3] Considering the collective functional potential of the gut microbiome, the number of microbial genes outnumber that of the human genome by at least 100-fold.[4] This is particularly noteworthy when considering nondigestible carbohydrate—the carbohydrate lyases and hydrolyses contained within the microbiome is in the order of 16,000 genes,[5] while our human genome contains less than 20 of these types of genes, mostly targeting the glycosidic bonds in amylose and amylopectin, as well as the disaccharidases, sucrase, maltase, lactase, and isomaltase. In addition, the gut microbiome contributes to the biotransformation of phytonutrients, helping to facilitate their activation and bioaccessibility and, thus, ability to positively affect health status.[6]

The gut microbiome provides metabolic, immunologic, and protective functions for the host.[7] From the metabolic perspective, certain micro-organisms within the gut microbiota can metabolize nondigested dietary components, including dietary fiber and prebiotics, to generate short-chain fatty acids (SCFA).[8] Immunologic contributions include direct and indirect stimulation and education of the immune system.[9] Protective functions include the production of metabolites, like SCFA, that lower the pH of the gastrointestinal tract to limit the survival of pathogens and enhance intestinal barrier function.[10]

Consumption of probiotics, prebiotics, synbiotics, and postbiotics (collectively "biotics"), as well as fermented foods, can affect health (**Figure 37.1**). However, there are varying levels of evidence for health benefits related to the many different probiotic, prebiotic, synbiotic, and fermented food products and formulations. Each term has recognized scientific definitions, which include caveats related to issues such as the characterizations of the microorganisms present, the effects on the gut microbiota, and the need for a demonstrated health benefit in the host (**Table 37.1**). Therefore, it is important to be familiar with this field and to critically evaluate the evidence as it relates to different formulations and human health outcomes.

Probiotics

Definition and Criteria for Qualification as a Probiotic

In 2001, probiotics were defined by the Food and Agriculture Organization of the United Nations (FAO) and the World Health Organization (WHO) as "live microorganisms which, when administered in adequate amounts, confer a health benefit on the host".[11] Then, in 2002, the FAO & WHO published guidelines to assist regulators, scientists, and industry in interpreting the original document for applications to foods.[12] In 2014, a consensus panel of scientists organized by the International Scientific Association for Probiotics and Prebiotics (ISAPP) reaffirmed the major principles of the 2001 definition and published guidance on the criteria required for an organism to be deemed a probiotic, based on advances in identifying microbial strains, assessing impacts on the host and applications of probiotics beyond food sources (Table 37.1).[13]

Based on the criteria set out by the ISAPP consensus panel, evidence of a health benefit is required for a probiotic, at either a strain-specific or group level, depending on the nature of the benefit. Probiotics can have different means of administration, target host species (humans and animals), target populations, target sites (gut and beyond), efficacy endpoints, and regulatory categories. All probiotics must be safe for their intended use. Dead microbes, microbial products, and microbial components do not qualify as probiotics.[13] If a health benefit is observed, there is no requirement for proof that the probiotic colonizes the recipient, survive intestinal transit, demonstrates adherence or anti-pathogenic properties, or has the ability to balance the host microbiota.[13] Additionally, the source of the probiotic strain–human or otherwise–is not stipulated in the definition.[13] Fermented foods are not considered probiotics, as it is not always possible to distinguish the contribution of the live microbes from that of the food matrix, and most micro-organisms within fermented foods have not been characterized to the strain level. Additionally, although the commensal microbiota have been a source of probiotics, the panel recommended that only preparations of well-defined mixes of commensal microbes that provide a demonstrated health benefit would meet the criteria of a probiotic. An undefined fecal microbial transplant, however, would not meet the criteria, even if a benefit was shown.

Importance of Strain and Dose

Probiotics comprise both bacterial and fungal species. A probiotic must be described by their genus, species, and strain designations.[13] For example, for *Bifidobacterium lactis* BB-12, *Bifidobacterium* is the genus, *lactis* is the species, and BB-12 is the strain designation. All three components are necessary, because the genetics and functional capabilities of different strains of the same species may differ. For example, *Bifidobacterium longum longum*, which is common in formula-fed infants, does not metabolize human milk oligosaccharides (HMOs) to a similar degree as *Bifidobacterium longum infantis* ATCC15697, which is more common in breastfed infants.[14] This difference is attributed to a 43 kbp cluster encoding catabolic genes, extracellular solute binding proteins, and permeases predicted to be active on HMO in *B. infantis*.[15]

Sources

Probiotics are found as foods, dietary supplements, and pharmaceuticals and are marketed for humans, production animal species, and companion animals. Traditionally, lactobacilli, bifidobacteria, and other lactic acid-producing bacteria (LAB) have been used as probiotics, primarily isolated from fermented dairy products and the human fecal microbiome.[16] For example, *Roseburia intestinalis*, *Faecalibacterium prausnitzii*, *Eubacterium* spp., *Bacteroides* spp. and *Akkermansia muciniphila* have been isolated from the human gut and are under investigation for their probiotic potential, particularly as they produce butyrate, propionate, and other bioactives.[16]

Mechanisms of Action

Mechanisms of action of probiotics are complex, diverse, heterogeneous, and often strain-specific.[16] Probiotics interact with both the host and the commensal microbiome via molecular effectors present on the cell structure or secreted as metabolic products. Probiotic metabolites also affect the microbiota through cross-feeding interactions, changes in the gastrointestinal microenvironment (e.g., pH lowering), competition for nutrients and binding sites, and inhibition of growth via the production of strain-specific antibacterial compounds, including bacteriocins.[17,18]

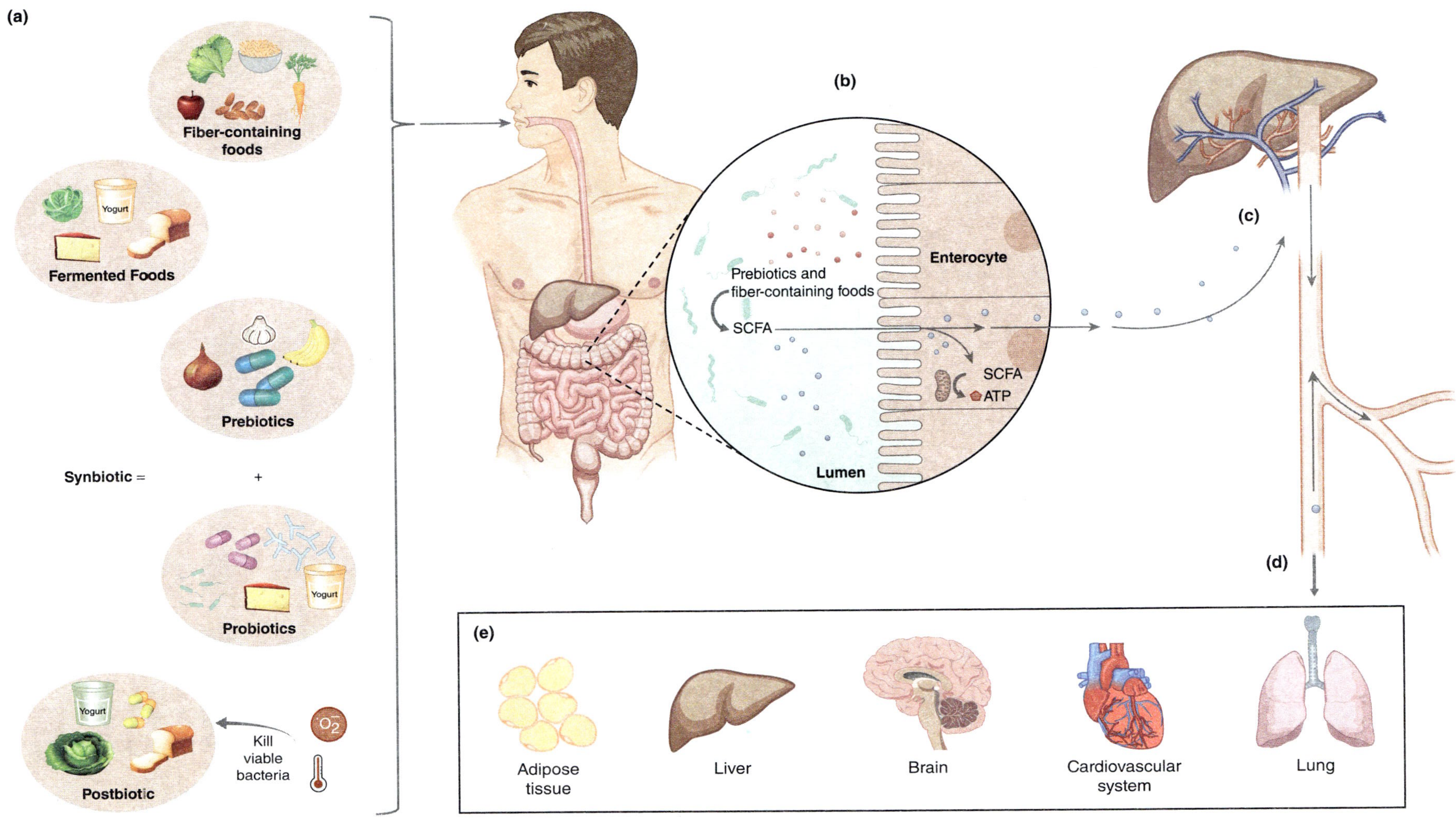

Figure 37.1 Sources and biologic activities of biotics and fermented foods. (**a**) Food and supplement sources of fermentable substrates include fiber-containing foods and prebiotics and complementary synbiotics (prebiotics+probiotics). Sources of live microbes include some fermented foods and probiotics. Nonviable bacteria and bacterial metabolites are contained in postbiotics. (**b**) Upon ingestion, resident microbiota or probiotic bacteria metabolize fermentable substrates to generate microbial metabolites, including SCFA and others. Metabolites generated in the lumen of the gut and those provided in fermented foods and postbiotics can be metabolized to generate ATP or interact with host epithelial, immune, and neural cells to signal to the host. MAMPS contained on cell walls of live microbes and microbial cells/fragments also bind to PRRs on host cells to signal to the innate immune system. (**c**) Some metabolites are absorbed into portal vein and (**d**) enter the circulation, where they (**e**) exert systemic effects on various organ systems throughout the body. **Abbreviations:** *ATP*, adenosine triphosphate; *MAMPS*, microbial-associated molecular patterns; *PRR*, pattern recognition receptors; *SCFA*, short chain fatty acids.

Created with BioRender.com.

Table 37.1 Definitions and Concepts Underlying Probiotics, Prebiotics, Synbiotics, Postbiotics, and Fermented Foods

	Probiotic	Prebiotic	Synbiotic	Postbiotic	Fermented Foods
Definition	Live microorganisms that, when administered in adequate amounts, confer a health benefit on the host (Hill et al., 2014)	A substrate that is selectively used by host microorganisms conferring a health benefit on the host (Gibson et al., 2017)	A mixture comprising live microorganisms and substrate(s) selectively used by host microorganisms that confers a health benefit on the host (Swanson et al., 2020) *Complementary synbiotic* is a mixture of probiotic + prebiotic *Synergistic synbiotic* contains a live microbe and a substrate that it can use for growth	Preparation of inanimate microorganisms and/or their components that confer a health benefit on the host (Salminen et al., 2021)	Foods made through desired microbial growth and enzymatic conversions of food components (Marco et al., 2020)
Live micro-organisms	Yes (strain designated, genome sequence available)	No	Yes (strain designated, genome sequence available)	No (Nonviable microbes and/or cell components with or without metabolites)	Sometimes; not required in final product (taxonomical characterization not required)
Examples	Bifidobacterium animalis subsp. lactis Bb12	Fructo-oligosaccharides (FOS), galacto-oligosaccharides (GOS)	Complementary synbiotic: inulin + Bifidobacterium animalis subsp. lactis Bb12	Some infant formulas	Yogurt, kefir, fermented vegetables (e.g., kimchee, sauerkraut), sourdough bread
Required demonstrated health benefit?	Yes	Yes	Yes	Yes	No
Health benefit linked to microbiota	No	Yes	Yes	Potentially, but not required	No

Data from Hill C, Guarner F, Reid G, et al. Nat Rev Gastroenterol Hepatol 2014;11:506-14.; Gibson GR, Hutkins R, Sanders ME, et al. Nat Rev Gastroenterol Hepatol 2017;14:491-502. ; Swanson KS, Gibson GR, Hutkins R, et al. Nat Rev Gastroenterol Hepatol 2020;17:687-701.; Salminen S, Collado MC, Endo A, et al. Nat Rev Gastroenterol Hepatol 2021;18:649-67. Marco ML, Hill C, Hutkins R, et al. J Nutr 2020;150:3061-7.

Probiotic effector molecules can interact directly with receptors in host intestinal epithelial, enteroendocrine, and immune cells, as well as vagal afferent fibers. These interactions influence intestinal barrier integrity and inflammation (e.g., via Toll-like receptors [TLR]), as well as systemic effects via host immune, endocrine, and nervous system mediators[17] (Figure 37.1). Probiotics can also perform enzymatic metabolism of host compounds, including dietary fiber, phytochemicals, bile salts, and xenobiotics, including drugs.[19] Surface-associated effector molecules or microbial-associated molecular patterns (MAMPs) can interact with host pattern-recognition receptors (PRRs), resulting in various immunomodulatory effects. These include exopolysaccharides, lipoteichoic acids (LTA), peptidoglycan, pili, and other surface-layer proteins, many of which are strain-specific.[19,20] TLRs are the best documented PRRs, being able to detect various probiotic MAMPs. As most probiotic taxa are Gram-positive bacteria, TLR2 is a key PRR, as it binds peptidoglycan and LTA.[20]

Although strain specificity of probiotic effects has been a core principle of probiotic science for decades, recent evidence suggests that some common mechanisms of action exist among probiotic microbes.[21] Hill and colleagues categorized the mechanisms of action of probiotics into three categories: those *widespread* among the studied probiotics, including colonization resistance, acid and SCFA production, regulation of intestinal transit, normalization of dysbiosis, increased epithelial cell turnover, and competitive exclusion of pathogens. Mechanisms that have been *frequently observed* at the species level include vitamin synthesis and enzymatic activity, direct antagonism, gut barrier reinforcement, bile salt metabolism, neutralization of carcinogens; and rare, species-specific effects, including neurologic, immunologic, and endocrine effects, as well as production of specific bioactives.[13]

Safety/Regulatory Considerations

There is no global harmonization of regulatory frameworks dictating manufacturing and claim substantiation requirements for probiotics. Thus, the quality of probiotic products, including reliability and accuracy of product labeling, can vary considerably across product categories and geographic regions. In the United States, the regulation of probiotics by the Food and Drug Administration (FDA) depends on the claim. Probiotics can be regulated as foods, dietary supplements, cosmetics, or drugs/biologics. For foods, the probiotic must be approved as a food additive or be Generally Regarded as Safe (GRAS). Many commonly exploited and currently available probiotic strains have GRAS status in the United States or belong to species with qualified presumption of safety (QPS) status with the European Food Safety Authority (EFSA).[16]

There are no probiotic health claims approved in the United States. In Europe, the only "probiotic" mechanism of action approved by EFSA is improved lactose digestion by live yogurt cultures containing at least 10^8 CFU/g *Lactobacillus delbrueckii* subsp. *bulgaricus* and *Streptococcus thermophilus,* because these bacteria are known to express the necessary enzymes (lactase or beta-galactosidase) to degrade lactose.[22]

Current Level of Evidence of Health Benefits

To determine whether causal relationships exist between an exposure and an outcome, appropriately powered, blinded, randomized controlled trials (RCT) are necessary. Optimally, the RCTs are further evaluated using a systematic review and meta-analysis process with evidence grading and risk of bias assessment. Blinded RCTs are often not possible for dietary interventions. However, the health effects of probiotics as supplements are supported by many well-conducted RCTs and high-quality systematic reviews and meta-analyses of infectious and chronic diseases across the lifespan. Although probiotic interventions often include different microbes and are administered at different doses for different durations, there are some common mechanisms of action among probiotic microbes[21] that allow data to be pooled for systematic reviews and meta-analyses. In early life, probiotic consumption has been associated with reduced necrotizing enterocolitis in preterm infants,[23] and antibiotic associated diarrhea.[24] A 2017 review reported that probiotics were associated with a reduction in *Clostridium difficile* infection in children and adults[25]; however, a 2020 Cochrane review of 82 studies involving 12,127 children and adults reported that "probiotics probably make little or no difference to the number of people who have diarrhea lasting 48 hours or longer, and we are uncertain whether probiotics reduce the duration of diarrhea".[26] The authors reported that risk of bias was high or unclear in many studies, and there was statistical heterogeneity when the primary outcomes were pooled in the meta-analysis. Effect size was similar in the sensitivity analysis and marked heterogeneity persisted.[26] It is likely that the protective effects of probiotics are both general and strain-specific, contributing to this heterogeneity in pooled analyses. This does not rule out that specific probiotics, such as *Lactobacillus rhamnoses* GG (LGG)[27] or *Saccharomyces boulardii*[28] may exert protection against diarrheal diseases or acute gastroenteritis. In addition, probiotic administration has shown promise for reducing the risk of atopic diseases. For atopic dermatitis, the use of LGG with or without other probiotics reduced the incidence of atopic eczema in pediatric patients at least to 7 years of age.[29] Furthermore, probiotics reduced allergic rhinitis severity scores, serum inflammatory cytokines, and the duration of cetirizine use in children.[30]

In adults, probiotics reduce the risk of infectious diarrhea,[31] antibiotic-associated diarrheal disease,[32] and *C. difficile* infection.[25] Probiotic administration improved systems in patients with ulcerative colitis but did not affect clinical outcomes in individuals with Crohn's disease, compared with the standard of care.[33,34] Cochrane reviews support the benefit of probiotics for reducing infections beyond the gastrointestinal tract, including respiratory tract infections[35] and bacterial vaginosis.[36] In the latter study, probiotic treatment for 1 to 3 months yielded superior beneficial outcomes and efficacy in the treatment of bacterial vaginosis than short-term probiotic treatment (<1 month).[36] Two systematic reviews revealed no benefit of probiotics for reducing urinary tract infections in children[37] or premenopausal adults.[38]

As metabolic diseases have become more common globally, investigators have focused on the potential for probiotics to aid in managing obesity and metabolic diseases. A systematic review of RCTs demonstrated that both probiotics and synbiotics, specifically certain strains of *Lactobacillus gasseri*; *L. rhamnosus*; L. plantarum; and *L. curvatus,* associated with other Lactobacillus species and/or with species from the *Bifidobacterium* genus, have the potential to aid in weight and fat mass loss in those with overweight and obesity.[39] However, given the heterogeneity in overweight and obesity, there is a need for more, longer-term clinical trials to inform accurate recommendations regarding strains, doses, and intervention durations. Probiotics have shown promise in reducing the symptoms of chronic metabolic diseases, including prediabetes and type 2 diabetes,[40] and inflammation and oxidative stress in diabetic kidney disease.[41] A recent umbrella meta-analysis reported that probiotics reduce fasting plasma glucose, HbA1c, HOMA-IR, and insulin concentrations. Probiotic supplementation for less than 8 weeks at moderate dosages (10^8 or 10^9 CFU) was an efficacious approach in improving glycemic parameters.[42]

These systematic reviews and meta-analyses support widespread beneficial effects of probiotics in the prevention and/or management of both acute infectious and long-term inflammatory and metabolic diseases. In addition, some RCTs show benefit of synbiotic administration with prebiotics,[39] which requires further investigation. While promising, it is important to note that most RCTs report concerns of risk of bias and heterogeneity. Thus, additional, well-controlled RCTs are warranted to develop broad-based recommendations for healthy populations interested in disease-risk reduction, which could fall under an FDA-approved health claim. Potential clinical interventions with probiotics would require additional regulatory approval as they imply claims related to diagnosis, cure, mitigation, or treatment of disease, which are beyond the current "authorized" and "qualified" health claims.[43]

Prebiotics

Definition and Criteria for Qualification as a Prebiotic

Prebiotics confer health benefits by acting as a nutrient source for host micro-organisms, thereby affecting either the composition or function of the microbiota. Prebiotics are defined by ISAPP as *"substrates that are selectively utilized by host microorganisms conferring a health benefit"* (Table 37.1).[44] At present, accepted prebiotics include readily fermented dietary fibers; galacto-oligosaccharides (GOS); and the inulin-type fructans: fructo-oligosaccharides (FOS) and inulin.[44] There is also growing interest in understanding how the health benefits of candidate prebiotics, including human milk oligosaccharides (HMOs)[44] and polyphenols,[45] may be imparting their benefits partially through the modulation of host microorganisms.

Fibers and prebiotics are similar but distinct concepts. Dietary fiber is defined as nondigestible carbohydrates and lignin that are intrinsic and intact in plants; isolated or synthetic, nondigestible carbohydrates determined by the U.S. FDA to benefit health.[46] Currently, the accepted prebiotics, inulin, FOS and GOS, are dietary fibers, but it's important to note that not all dietary fibers are prebiotics.

To distinguish what is considered a prebiotic, the substrate must be selectively utilized by host micro-organisms, as opposed to broadly affecting the microbiota.[44] The selective utilization is not limited to the classically studied micro-organisms (i.e., bifidobacteria and lactobaccili); it includes other taxa that have been shown to support health benefits, such as *Roseburia*, *Eubacterium*, and *Faecalibacterium*. The selective utilization implies that the micro-organism enriched uses the provided substrate as a nutrient source. Microbial fermentation of prebiotics can result in the production of beneficial metabolites, including the SCFA (acetate, probiotic, and butyrate).[8] SCFA have been shown to affect inflammation, neurotransmitters, and hormone section, thereby contributing to metabolic, gastrointestinal, and cognitive health benefits.[3,47]

Sources of Prebiotics

Inulin-type fructans and GOS can be found naturally intact within plants (i.e., fruits, vegetables, grains, legumes, pulses) or added into food or beverages. Plant sources that provide inulin-type fructans include chicory root, salsify, artichokes, garlic, leeks, onions, wheat, and bananas.[48,49] Inulin and oligofructose are also extracted from plants, mainly from chicory root, while FOS are chemically synthesized. GOS is naturally found within plant sources, including fenugreek, fava beans, lentils, and chickpeas, but can also be chemically synthesized.[50,51] Inulin-type fructans are made up of fructose polymers linked by β-2,1 linkages.[8,52] These fibers have varying degrees of polymerization (DP); Inulin has a DP of 3 to 60, oligofructose <10, and short-chain FOS 2–4. GOS are β-(1,4) inked galactose oligomers, attached to glucose by β-(1,4) bonds. GOS is typically between 2 and 8 units long.

Inulin-type fructans and GOS can be added to a range of foods and beverages, such as bakery products, bars, cereals, desserts, infant formula, sauces, and yogurt. To determine if a food/beverage contains added inulin-type fibers, check the ingredients list for the following terms: inulin, chicory, oligofructose, or FOS. To provide health benefits, at least 3 to 5 grams per day of prebiotics are needed,[44] regardless of whether it is found intrinsic and intact within plants or added to foods/beverages.

Current Level of Evidence of Health Benefits

Inulin-type fructans have been extensively studied, and the results are summarized within systematic reviews. A 2022 systematic review on prebiotic potential reported that consumption of inulin-type fructans by healthy adults enriches *Bifidobacterium*, lactobacilli, and *Faecalibacterium*.[52] The demonstrated health benefits in healthy adults include increasing intestinal absorption of calcium and magnesium, enhanced satiety, improved intestinal barrier function, laxation, lipid profiles and triglycerides, and insulin sensitivity.[52] Among adults and children with overweight and obesity, consumption of high amounts of inulin-type fructans has been shown to result in improvement in adiposity measures. For example, consumption of 21 g/d of oligofructose for 12 weeks reduced body weight, fat mass and trunk fat in adults with overweight and obesity,[53] while 8 g/d oligofructose enriched inulin for 16 weeks decreased body weight z-score, percent body fat, and percent truck fat in 7 to 12 year olds with overweight and obesity.[54] Meta-analyses have been conducted indicating the robustness of certain health outcomes related to the consumption of inulin-type fructans, including greater satiety,[55] glycemic control,[55] and lower low-density lipoprotein (LDL) concentration[56] in healthy adults; improved glycemic control,[56] increased high-density lipoprotein (HDL)-cholesterol concentration, reduced triglyceride and LDL concentrations in adults with type 2 diabetes[56,57]; and greater stool frequency, consistency, and reduced hardness of stool in adults with chronic constipation.[58]

The health effects of consumption of GOS are less studied but have shown benefits for psychological and metabolic health. For example, 5.5 g/d of GOS for 3 weeks reduced waking salivary cortisol and increased attentional vigilance in the processing of positive vs negative stimuli,[59] and 5.5 g/d for 12 weeks reduced total cholesterol concentrations, triglycerides, and insulin in adults with overweight,[60] and increased calcium absorption in adolescents and postmenopausal women.[61]

Meta-analyses of prebiotic health effects have reported that consumption of inulin-type fructans and GOS result in higher abundances of *Bifidobacterium* and Lactobacilli,[62] fewer acute infectious diseases in infants and children,[63] reduced total and LDL cholesterol in adults with overweight and obesity,[57] and reduced C-reactive protein (CRP) concentration.[64] Health benefits of certain prebiotics have also been reviewed and accepted by regulatory authorities. The EFSA supports the claim that "chicory inulin contributes to maintenance of normal defecation by increasing stool frequency."[65] Within the United States, the FDA has determined that evidence supports beneficial physiologic effects on bone mineral density and calcium absorption with the consumption of inulin-type fructans and GOS.[61]

Overall, there are many benefits of consuming prebiotics, including to intestinal, bone, and cardiometabolic health, as well as immunity. It is important to check food labels to look for relevant terms (inulin, chicory, oligofructose, FOS, and GOS) in adequate doses (>3g/d) to ensure that adequate amounts of the prebiotic are consumed to impart a health benefit.

Synbiotics

Definition and Criteria for Qualification as a Synbiotic

A synbiotic is defined by ISAPP as "*a mixture comprising live microorganisms and substrate(s) selectively utilized by host microorganisms that confers a health benefit on the host*" (Table 37.1).[66] Synbiotics may be formulated using two approaches: complementary or synergistic.[66]

A complimentary synbiotic is a mixture of probiotic(s) and prebiotic(s).[66] Each component works independently to achieve one or more health benefits in the host. For example, the probiotic may enhance vaccine-specific antibody levels and the prebiotic may be utilized by beneficial members of the host microbiota to reduce systemic inflammation. The combination thus provides a health benefit. The probiotic and prebiotic components of a complementary synbiotic must meet the criteria of their definitions and be provided at the same dosages with a demonstrated health benefit as an independent component. Consumption of the synbiotic mixture must result in a health benefit, but it is not necessary for the study to again demonstrate the selective utilization of the substrate in the synbiotic mixture.

A synergistic synbiotic is a mixture of a selectively utilized substrate and a live microbe chosen for its ability to deliver a health effect.[66] Components comprising synergistic synbiotics work together to confer health benefits in the target host. For a synergistic synbiotic, the health benefit and select utilization of the substrate by the live microbe contained in the synbiotic must be demonstrated in the same study. The substrate is utilized by the live microbe in the synbiotic mixture; the substrate may also be utilized by the host microbiota; both the live microbes present in the mixture and the enrichment of microorganisms in the host microbiota contribute to the health benefit in the host.

Current Level of Evidence of Health Benefits

The most studied live microorganisms within purported synbiotics include lactobacilli, *Bifidobacterium*, and *Streptococcus*.[66] Regarding the substrates, the accepted prebiotics, GOS, FOS, and inulin, at doses ranging from 100 mg to 15 g/d, have been studied. Although there have been numerous randomized controlled trials and even systematic reviews and meta-analyses of the health effects of putative synbiotics on disease, most do not meet the synbiotic definition as it stands because of the lack of concomitant evidence of the selective utilization by either the microbiota or the co-administer live microorganism.[66]

Synbiotics are generally formulated to be supplements in the form of capsules or sachets. However, synbiotics can also be provided within foods, such as yogurts.

Postbiotics

Definition and Criteria for Qualification as a Postbiotic

Postbiotics attempt to capture the health benefits beyond the inherent viability of probiotics.[67] Postbiotics were defined by ISAPP as a "*preparation of inanimate microorganisms and/or their components that confers a health benefit on the host*" (Table 37.1).[68] The term postbiotic was selected to refer to preparations resulting from living organisms (biotic) but after life or nonliving (post).[68] ISAPP further stipulated that inanimate strains typically belong to established probiotic taxa (*Lactobacilli, Bifidobacterium*); however, a postbiotic can also be derived from nonprobiotic species and *Akkermansia*, *Streptococcus*, and *Faecalibacterium*, as well as yeasts, have been reported as potential sources of postbiotics.[69] Viruses, bacteriophages, purified microbial components (e.g., exopolysaccharides) or purified microbial metabolites (e.g., SCFA) were considered not to be postbiotics by ISAPP.[68] However, this definition has been questioned by others,[70] who reference prior definitions of postbiotics as "any factor resulting from the metabolic activity of a probiotic or any released molecule capable of conferring beneficial effects to the host in a direct or indirect way." This definition focused on probiotic-produced soluble factors, not the complex matrices derived from microbial cells.[71,72] Under this definition, postbiotics include enzymes, secreted proteins, SCFA, vitamins, amino acids, peptides, organic acids, and others.[73] Inclusion of nonviable bacteria and their metabolic components along with the secreted components has been defined by others as "parabiotics," rather than postbiotics.[73] Nonviable bacterial cell walls provide peptidoglycans, teichoic acids, and surface proteins, which are ligands for host recognition receptors,[73] making it difficult to delineate the mechanism of action of postbiotics that include both nonviable bacteria and secreted factors.[69]

Sources

Sources of postbiotics include fermented foods that are thermally processed (typically pasteurization or autoclaving) to kill viable bacteria. Other nonthermal processes that reduce bacterial viability could also be applied, such as electric field, ultrasonication, high pressure, X-rays, ionizing radiation, spray-drying, freeze-drying, and exposure to oxygen.[74,75] However, various approaches could affect the level of microbial inactivation achieved and the functionality of a postbiotic.[76,77] Products include sauerkraut, kimchi, yogurt, kefir, kombucha, tempeh, and miso.[75]

Safety/Regulatory Considerations

Postbiotics must be safe for the intended use. The lack of viable bacteria in postbiotics has been stated as a safety advantage over probiotics.[78] To date, no regulatory framework specific to postbiotic-containing foods or food supplements has been established by the FDA. Thus, for a postbiotic to be considered for use as a food ingredient, it must either be determined to meet the criteria as generally regarded as safe (GRAS) by an expert panel or receive premarket approval as a food additive.[78]

Current Level of Evidence of Health Benefits

The site of action for postbiotics is not limited to the gut; they can be administered at a host surface, such as the mouth, nose, gut, skin, or urogenital tract; however, they cannot be administered parenterally.[68] There is growing evidence for the clinical benefits of postbiotics in managing highly prevalent conditions, including gastrointestinal, dermatologic, and neurological disorders, as well as respiratory infections and metabolic syndrome.[20,78] Some of these health benefits overlap with that of live probiotic strains; however, safety concerns about the use of live biotics in vulnerable populations (e.g., pediatric populations, the elderly, those with gastrointestinal or immune compromise), as well as challenges in maintaining probiotic viability in food products, has led to increased interest in postbiotics, as inactivated microbial cells or cell components are more stable.[78]

Mechanisms of Action

Because postbiotics are a heterogenous mixture of soluble components, intact nonviable bacteria, and bacterial cell wall fragments, they can influence health outcomes through several mechanisms.[79,80] For products of fermentation, such as SCFA, the mechanisms of action include increasing endogenous production and portal vein secretion of gut peptides (e.g., GLP-1, GLP2-, PYY), acting on specific G-protein-coupled receptors, controlling the anaerobic condition in the colon, and contributing to tight junction protein expression and assembly.[3,10] Similarly, the presence of nonviable bacteria or bacterial cell wall components will still allow for receptor-ligand interactions between MAMPs and PRRs that are associated with the host innate immune response,[79,80] similar to live probiotics.[20] The PPRs include TLRs, nucleotide-binding oligomerization domain (NOD) receptors, and C-type lectins, resulting in the expression of various cytokines and immune modulators.[79,80] Mosca and colleagues outlined five potential mechanisms of action for postbiotics, including (1) modulation of the resident microbiota; (2) enhancement of epithelial barrier functions; (3) modulation of local and systemic immune responses; (4) modulation of systemic metabolic responses; and (5) systemic signaling via the nervous system.[78]

In summary, postbiotics administered through foods or supplements show promise as a microbially directed means to support human health. They provide the benefits of fermented foods and probiotics but have greater stability and potentially fewer safety concerns. Clinical trial evidence is promising, but regulatory approval must be obtained before routine administration as a food ingredient or supplement.

Fermented Foods

Definition and Criteria for Qualification as a Fermented Food

Fermentation is a metabolic process in which micro-organisms convert food components into other substances. Fermented foods are foods made through designed microbial growth and enzymatic conversions of food components,[81] Table 37.1. For example, microbial fermentation of grapes results in the production of wine; milk, which can be utilized as a microbial substrate to make yogurt, kefir, and cheeses; and apples, which can be fermented to make apple cider vinegar. Most common foods and beverages require lactic acid bacteria, acetic acid bacteria, bacilli or other bacteria, yeasts, or filamentous fungi.[82] The most common micro-organisms involved in the creation of fermented foods are *Streptococcus*, *Lactobacillacea*, *Lactococcus*, and *Saccharomyces*.[83]

Sources of Fermented Foods

Fermented foods may or may not contain live microbes, as the live micro-organisms can be killed or removed during the production process through the actions of heating (e.g., sourdough bread), filtering (e.g., wine, beer), or roasting (e.g., chocolate, coffee) (**Table 37.2**).[81] Alternatively, some foods may resemble fermented food but are chemically produced (e.g., soy sauce), brined (e.g., pickles), or cured (e.g., sausage) rather than undergoing fermentation. Unfortunately, it can be difficult to discern if a food is fermented or not. So, reading the Nutrition Facts label and ingredients list and potentially contacting the manufacturer may be necessary to determine if the food is fermented.

Current Level of Evidence of Health Benefits

Several reviews have been published on the benefits related to consuming fermented foods,[82,84,85] which discuss putative mechanisms, including live microbes educating the immune system and providing enzymatic activity to metabolize lactose, enhanced microbially derived concentrations of vitamins, bioactive peptides, biogenic amines, conversion and activation of phytonutrients, and reducing anti-nutrients. Some fermented foods are very well researched, like fermented dairy. Others, like kefir, sauerkraut, natto, and sourdough bread, have been investigated in at least one RCT. Other fermented foods primarily rely on evidence from longitudinal trials or in vivo and in vitro research. Only recently have RCTs been conducted to investigate both the health benefits of consumption of fermented foods and the effects on the gut microbiota, including work on fermented foods in general[86,87] and fermented dairy.[88]

Table 37.2 Fermented Foods vs Nonfermented Foods

Fermented		Nonfermented
Live Micro-Organisms Present	***Live Micro-Organisms Absent***	
■ Yogurt ■ Sour cream ■ Kefir ■ Most cheeses ■ Miso ■ Natto ■ Tempeh ■ Fermented vegetables ■ Salami, pepperoni, fermented sausages ■ Fermented cereals, boza, bushera ■ Most kombuchas ■ Some beer	*Heated (baked, pasteurized)* ■ Bread ■ Shelf-stable vegetables ■ Sausage ■ Soy sauce ■ Vinegar ■ Some kombuchas *Filtered* ■ Wine ■ Most beers ■ Distilled spirits *Roasted* ■ Coffee ■ Chocolate beans	■ Chemically leavened bread ■ Fresh sausage ■ Vegetables pickled in brine or vinegar ■ Chemically produced soy sauce ■ Salted or cured meats and fish

Modified from Marco ML, Sanders ME, Gänzle M, et al. Nat Rev Gastroenterol Hepatol. 2021;18(3):196-208. doi:10.1038/s41575-020-00390-5.

Fermented milks, including yogurt and kefir, are the most studied fermented food group. Indeed, there are several systemic reviews and meta analyses on the topic. For example, a systematic review published in 2021 that summarized the health benefits of fermented milk in humans identified 108 studies, including 31 RCTs, on a range of health categories, including cardiometabolic health, bone health, obesity, and cancer.[89] Of the summarized research, 76 studies reported favorable outcomes related to the consumption of fermented milk products. Importantly, many of the studies used dairy products as controls, which is important to distinguish the effects of fermentation vs the effect of the nutrient content. The mechanisms underlying the health benefits include the live, active cultures improving lactose digestion and reducing lactose intolerance symptoms. Specific to cardiovascular health, bioactives within yogurts can provide anti-hypertensive properties that help to reduce blood pressure. However, some trials have demonstrated that blood pressure-lowering effects can be seen in both intervention and control conditions, suggesting that the milk itself, as opposed to fermented milk, may contribute the most robust benefit. Fermented milk consumption can also improve body weight, adiposity, and weight gain over time. There are also strong associations with reducing the risk for type 2 diabetes and metabolic syndrome.

Fermentation of grains to create sourdough bread results in the metabolism of proteins and certain carbohydrates, including gluten, amylase-trypsin inhibitors, fructans, and phytates.[90,91] Thus, the fermentation process has been shown to decrease the glycemic index of bread, which may be related to fermentation increasing the resistant starch and soluble fiber content within the bread and lowering alpha-amylase activity within the digestive tract. Sourdough bread provides increased antioxidant activity by releasing bound phenolic components. It decreases phytates, which allow for improved mineral absorption, especially iron.

Fermenting cabbage contributes to the biotransformation of phytochemicals into active forms, increases antioxidant activity, and inhibits nitric oxide production. Although the evidence remains limited, kimchi consumption has improved fasting blood glucose concentration, insulin sensitivity, blood lipids and blood pressure, and weight, BMI, and fat distribution.[92,93]

Miso and tempeh are foods created through soybean fermentation. Compared with nonfermented soybeans, fermented soybeans have increased digestibility, higher protein content availability, reduced antinutrients, and increased isoflavone content. Interestingly, despite the high sodium content of miso, it does not increase blood pressure compared with the equivalent salt intake—the purported mechanism relates to the microbial production of an angiotensin-converting enzyme inhibitor during the fermentation process.[94] Tempeh likely also provides health benefits beyond its nutrient content.[95] There are a moderate number of in vitro, ex vivo, in vivo, clinical, and population studies on the health benefits of tempeh. The suggested health benefits include enhancing IgA concentrations in the gastrointestinal tract to contribute to gut health and transforming isoflavone glycosides into more bioavailable forms to help reduce cancer risk.

Although there is growing research on the health benefits of consuming fermented foods, there are some potential risks. Making fermented foods at home creates the potential for improper production or storage that can lead to negative health effects. For example, consuming kombucha that is improperly brewed can cause lead poisoning because the low pH leaches heavy metals from containers. For this reason, it is imperative to make and store kombucha in glass containers. Fermented foods can also have negative effects on those that have histamine intolerances. The provision of live microbes may be contraindicated in immunocompromised individuals, such as those undergoing cancer treatments or with acquired immune deficiency syndrome. The high salt content and the presence of alcohol should be noted, as some individuals also need to limit or avoid the consumption of these components for health and/or personal reasons.

Conclusions

Fermented foods containing live or killed microbes (postbiotics) have been a part of the diet for thousands of years to extend food shelf-life and to enhance its organoleptic properties.[83,96] However, it is only recently that we have begun to appreciate the potential health benefits to the consumer. The United States lags behind other countries in the consumption of fermented foods and live microbes.[97] In 2022, the intake of live microbes in U.S. children/adolescents (aged 2–18 y) and adults ($\geq$19 y) [n = 74,466; 51.2% female] was assessed using cross-sectional data from nine cycles of National Health and Nutrition Examination Surveys (NHANES) (2001–2018).[98] The estimated level of live microbes (colony-forming units; CFU) per gram was set as: Low $<10^4$ CFU/g; Medium 10^4–10^7 CFU/g; or High $>10^7$ CFU/g. The High category was composed primarily of fermented dairy, whereas the Medium category included fermented fruits and vegetables containing live microbes. On average, 59% of children and 67% of adults consumed live microbes at the Medium/High level. Per capita intake of Medium/High foods was 85 g/d for children/adolescents, and 127 g/d for adults. Thus, microbe-containing foods are becoming a more widely accepted component of the U.S. diet. The evidence of health benefits of probiotics and, even postbiotics, has led to a call for a recommended intake of microbes in the United States,[99] which has already been incorporated in other countries.[97]

Our ever-expanding understanding of the human microbiome and host-diet-microbe interactions has enabled the development of new approaches to capitalize on probiotic, prebiotic, synbiotic, and postbiotics.[16] Technological advances in data collection and analysis will provide avenues for the development and optimization of biotics. For example, next-generation probiotics could be isolated from human commensal microbial niches (e.g., gut, nose, mouth), which could be optimized for site-specific administrations. These next-generation probiotics could be administered live in foods or supplements, in combination with prebiotics (synbiotics), which will be selected based on clinical evidence and predictive modeling based on artificial intelligence and machine learning to link prebiotics, probiotics/commensal bacteria, microbial metabolites, and host responses.[100] Next-generation prebiotics will be identified from new sources and optimized for prebiotic, synbiotic, or postbiotic administration. In addition, the potential to use synthetic biology to construct gut-engineered probiotics as live therapeutics is a promising avenue in the treatment of specific diseases, such as phenylketonuria and inflammatory bowel disease.[101] These opportunities will be enabled only with robust evaluation of safety and efficacy through clinical trial and population-based studies, regulatory oversight, and evidence-based translation into healthcare practice and policy.[16]

Acknowledgments

The authors acknowledge Sandra Tejero, Ian R. Rowland, Robert Rastall, and Glenn R Gibson for contributions to the previous version of this chapter.

References

1. Berry SE, Valdes AM, Drew DA, et al. Nat Med 2020;26:964–73.
2. Hjorth MF, Christensen L, Larsen TM, et al. Am J Clin Nutr 2020;111:1079–86.
3. de Vos WM, Tilg H, Van Hul M, et al. Gut 2020;71:1020–32.
4. Qin J, Li R, Raes J, et al. Nature 2010;464:59–65.
5. Wardman JF, Bains RK, Rahfeld P, et al. Nat Rev Microbiol 2022;20;542–56.
6. Cardona F, Andrés-Lacueva C, Tulipani S, et al. J Nutr Biochem 2013;24:1415–22.
7. Belkaid Y, Hand TW. Cell 2014;157:121–41.
8. Holscher HD. Gut Microbes 2017;8:172–84.
9. Goldszmid RS, Trinchieri G. Nat Immunol 2012;13:932–8.
10. Alexander C, Swanson KS, Fahey GC, et al. Adv Nutr 2019;10: 576–89.
11. Food and Agriculture Organization of the United Nations, Health and nutritional properties of probiotics in food including powder milk with live lactic acid bacteria, 2001. Accessed April 19, 2023. Available at www.fao.org/3/a0512e/a0512e.pdf.
12. Food and Agriculture Organization of the United Nations. Joint FAO/WHO working group report on drafting guidelines for the evaluation of probiotics in food, 2002. Accessed April 19, 2023. Available at www.fao.org/3/a0512e/a0512e.pdf.
13. Hill C, Guarner F, Reid G, et al. Nat Rev Gastroenterol Hepatol 2014;11:506–14.
14. Ward RE, Ninonuevo M, Mills DA, et al. Mol Nutr Food Res 2007;51:1398–1405.
15. Sela DA, Chapman J, Adeuya A, et al. Proc Natl Acad Sci USA 2008;105:18964–9.
16. Cunningham M, Azcarate-Peril MA, Barnard A, et al. Trends Microbiol 2021;29:667–85.
17. Plaza-Diaz J, Ruiz-Ojeda FJ, Gil-Campos M, et al. Adv Nutr 2019;10:S49–66.
18. Lebeer S, Bron PA, Marco ML, et al. Curr Opin Biotechnol 2018;49:217–23.
19. Monteagudo-Mera A, Rastall RA, Gibson GR. et al. Appl Microbiol Biotechnol 2019;103:6463–72.
20. Teame T, Wang A, Xie M, et al. Front Nutr 2020;7:570344. doi:10.3389/fnut.2020.570344
21. Sanders ME, Benson A, Lebeer S, et al. Curr Opin Biotechnol 2018;49:207–16.
22. Panel E, Nda A, Golly I. EFSA J 2010;8:1–18.
23. Alfaleh K, Anabrees J. Evidence-Based Child Heal 2014,9. 584–671.
24. Guo Q, Goldenberg JZ, Humphrey C, et al. Cochrane Database Syst Rev 2019;4(4):CD004827. doi:10.1002/14651858.CD004827.pub5
25. Goldenberg J, Yap C, Lytvyn L, et al. Cochrane Database Syst Rev 2017;12:1–154.
26. Collinson S, Deans A, Padua-Zamora A, et al. Cochrane Database Syst. Rev. 2020;12(12):CD003048. doi: 10.1002/14651858.CD003048.pub4
27. Sanklecha M, Verma L, Pai U, et al. Cureus 2022;14:e24594. doi:10.7759/cureus.24594
28. Fu H, Li J, Xu X, Comput Math Methods Med 2022:6234858. doi:10.1155/2022/6234858
29. Voigt J, Lele M. Am J Clin Dermatol 2022;23:801–11.
30. Wang X, Tan X, Zhou J. Int J Pediatr Otorhinolaryngol 2022;162:111300. doi:10.1016/j.ijporl.2022.111300
31. Wang F, Zhao T, Wang W, et al. Medicine (Baltimore) 2022;101:e30880. doi:10.1097/MD.0000000000030880
32. Zhang L, Zeng X, Guo D, et al. BMC Geriatr. 2022;22:1–9. doi:10.1186/s12877-022-03257-3
33. Müller L, Lorentz A. J Gastrointest Liver Dis 2022;31:74–84.
34. Chen M, Feng Y, Liu W. Ann Palliat Med 2021;10:11821–9.
35. Zhao Y, Dong BR, Hao Q. Cochrane Database Syst Rev 2022;8(8):CD006895. doi:10.1002/14651858.CD006895.pub4
36. Liu HF, Yi N. Eur Rev Med Pharmacol Sci 2022;26:90–8.
37. Meena J, Thomas CC, Kumar J, et al. Eur J Pediatr 2021;180: 3535–45.
38. Abdullatif VA, Sur RL, et al. Cureus 2021;13(10):e18843. doi:10.7759/cureus.18843
39. Álvarez-Arraño V, Martín-Peláez S. Nutrients 2021;13(10):3627. doi:10.3390/nu13103627
40. Naseri K, Saadati S, Ashtary-Larky D, et al. Pharmacol Res 2022;184:106399. doi:10.1016/j.phrs.2022.106399
41. Dai Y, Quan J, Xiong L, et al. Ren Fail 2022;44:862–80.
42. Zarezadeh M, Musazadeh V, Faghfouri AH, et al. Pharmacol Res 2022;183:106397. doi:10.1016/j.phrs.2022.106397
43. US Food and Drug Administration. Questions and Answers on Health Claims in Food Labeling, 2017. Accessed April 19, 2023. Available at www.fda.gov/food/food-labeling-nutrition/questions-and-answers-health-claims-food-labeling.
44. Gibson GR, Hutkins R, Sanders ME, et al. Nat Rev Gastroenterol Hepatol 2017;14:491–502.
45. Rodríguez-Daza MC, Pulido-Mateos EC, Lupien-Meilleur J, et al. Front Nutr 2021;8:689456. doi:10.3389/fnut.2021.689456
46. US Food and Drug Administration. Food labeling: Revision of the Nutrition and Supplements Facts Labels (21 CFR part 101). 2016. Accessed April 19, 2023. Available at www.fda.gov/media/134505/download.
47. Berding K, Vlckova K, Marx W, et al. Adv Nutr 2021;12: 1239–85.
48. Moshfegh AJ, Friday JE, Goldman JP, et al. J Nutr. 1999;129:1407S–11S.
49. Kalala G, Kambashi B, Everaert N, et al. Int J Food Sci Nutr 2018;69:682–9.
50. Torres DPM, Gonçalves MPF, Teixeira JA, et al. Compr Rev Food Sci Food Saf 2010;9:438–54.
51. Njoumi S, Josephe Amiot M, Rochette I, et al. Int J Food Sci Nutr 2019;70:551–61.
52. Hughes RL, Alvarado DA, Swanson KS, et al. Adv Nutr 2022;13:492–529.
53. Parnell JA, Reimer RA. Am J Clin Nutr 2009;89:1751–9.
54. Nicolucci AC, Hume MP, Martínez I, et al. Gastroenterol 2017;153:711–22.
55. Kellow NJ, Coughlan MT, Reid CM. Br J Nutr 2014;111: 1147–61.
56. Liu F, Prabhakar M, Ju J, et al. Eur J Clin Nutr. 2017;71:9–20.
57. Beserra BTS, Fernandes R, do Rosario VA, et al. Clin Nutr 2015;34:845–58.
58. Yurrita LC, Martín ISM, Ciudad-Cabañas MJ, et al. Nutr Hosp 2014;30:244–52.
59. Schmidt K, Cowen PJ, Harmer CJ, et al. Psychopharmacol (Berl) 2015;232:1793–1801.
60. Vulevic J, Juric A, Tzortzis G, et al. J Nutr 2013;143:324–31.

61. US Food and Drug Administration. Review of the Scientific Evidence on the Physiological Effects of Certain Non-Digestible Carbohydrates. 2018:1–52. Accessed April 19, 2023. Available at www.fda.gov/files/food/published/Review-of-the-Scientific-Evidence-on-the-Physiological-Effects-of-Certain-Non-Digestible-Carbohydrates-PDF.pdf.
62. So D, Whelan K, Rossi M, et al. Am J Clin Nutr 2018;107: 965–83.
63. Lohner S, Küllenberg D, Antes G, et al. Nutr Rev 2014;72: 523–31.
64. Mcloughlin RF, Berthon BS, Jensen ME, et al. Am J Clin Nutr 2017;106:930–45.
65. EFSA Panel on Dietetic Products, Nutrition and Allergies (NDA). EFSA J 2015;13(1):3951. doi:10.2903/j.efsa.2015.3951
66. Swanson KS, Gibson GR, Hutkins R, et al. Nat Rev Gastroenterol Hepatol 2020;17:687–701.
67. Cuevas-González PF, Liceaga AM, Aguilar-Toalá JE. Food Res Int 2020;136:109502. doi:10.1016/j.foodres.2020.109502
68. Salminen S, Collado MC, Endo A, et al. Nat Rev Gastroenterol Hepatol 2021;18:649–67.
69. Aguilar-Toalá JE, Hall FG, Urbizo-Reyes UC, et al. Proteins 2020;12:608–22.
70. Aguilar-Toalá JE, Arioli S, Behare P, et al. Nat Rev Gastroenterol Hepatol 2021;18:825–6.
71. Tsilingiri K, Rescigno M. Benef Microbes 2013;4:101–7.
72. Aguilar-Toalá JE, Garcia-Varela R, Garcia HS, et al. Trends Food Sci Technol 2018;75:105–14.
73. Nataraj BH, Ali SA, Behare PV, et al. Microb Cell Fact 2020;19(1):168. doi:10.1186/s12934-020-01426-w.–22
74. Moradi M, Molaei R, Guimarães JT. Enzyme Microb Technol 2021;143:109722. doi:10.1016/j.enzmictec.2020.109722
75. Thorakkattu P, Khanashyam AC, Shah K, et al. Foods 2022;11(19):3094. doi:10.3390/foods11193094
76. Tabanelli G, Burns P, Patrignani F, et al. Food Microbiol 2012;32:302–7.
77. Plovier H, Everard A, Druart C, et al. Nat Med 2017;23:107–13.
78. Mosca A, Abreu ATAY, Gwee KA, et al. Gut Microbes 2022;14(1):2117508. doi:10.1080/19490976.2022.2117508
79. Jastrząb R, Graczyk D, Siedlecki P. Int J Mol Sci 2021;22(24): 13475. doi:10.3390/ijms222413475
80. Peluzio MDCG, Martinez JA, Milagro FI. Trends Food Sci Technol 2021;108:11–26.
81. Marco ML, Sanders ME, Gänzle M, et al. Nat Rev Gastroenterol Hepatol 2021;18:196–208.
82. Marco ML, Heeney D, Binda S, et al. Curr Opin Biotechnol 2017;44:94–102.
83. Hutkins RW. Microbiology and Technology of Fermented Foods, 2nd edition. John Wiley & Sons, 2019.
84. Şanlier N, Gökcen BB, Sezgin AC. Crit Rev Food Sci Nutr 2019;59:506–27.
85. Dimidi E, Cox SR, Rossi M, et al. Nutrients 2019;11(8):1806. doi:10.3390/nu11081806
86. Wastyk H.C, Fragiadakis GK, Perelman D, et al. Cell 2021;184:4137–4153.e14
87. Berding K, Bastiaanssen TFS, Moloney GM, et al. Mol Psychiatry 2023;28:601–10.
88. Cannavale CN, Mysonhimer AR, Bailey MA, et al. Nutr Neurosci 2023;26:265–74.
89. Savaiano DA, Hutkins RW. Nutr Rev 2021;79:599–614.
90. Graça, C, Lima A, Raymundo A, et al. Fermentation 2021;7(4): 246. doi:10.3390/fermentation7040246
91. Arora K, Ameur H, Polo A, et al. Trends Food Sci Technol 2021;108:71–83.
92. Kim B, Mun EG, Kim D, et al. J Nutr Heal 2018;51(1):1–13. doi:10.4163/jnh.2018.51.1.1
93. Choi IH, Noh JS, Han JS, et al. J Med Food 2013;16:223–9.
94. Ito K. Environ Health Prev Med 2020;25(1):45. doi:10.1186/s12199-020-00883-4
95. Ahnan-Winarno AD, Cordeiro L, Winarno FG, et al. Compr Rev Food Sci Food Saf 2021;20:1717–67.
96. Verardo V, Gómez-Caravaca AM, Tabanelli G. Foods 2020;9(2):153. doi:10.3390/foods9020153
97. Chilton SN, Burton JP, Reid G. Nutrients 2015;7:390–404.
98. Marco ML, Hutkins R, Hill C, et al. J Nutr 2022;152: 1729–36.
99. Marco ML, Hill C, Hutkins R, et al. J Nutr 2020;150: 3061–7.
100. Oh KK, Gupta H, Min BH, et al. Cells 2022;11(18):2903. doi:10.3390/cells11182903
101. Huang Y, Lin X, Yu S, et al. ACS Synth Biol 2022;11:3134–53.

PART II

Nutritional Roles in Integrated Biologic Systems

A. NUTRIENT–GENE MECHANISMS

CHAPTER 38

Nutritional Genomics

Young Ah Seo and Edward A. Ruiz-Narváez

CHAPTER OUTLINE

Abbreviations

ATP, adenosine triphosphate; **ChIP**, chromatin immunoprecipitation; **CoA**, coenzyme A; **ELISA**, Enzyme-linked immunosorbent assay; **ES**, embryonic stem; **FAS**, fatty acid synthase; **GWAS**, genome-wide association studies; **HIF**, hypoxia-inducible factor; **miRNA**, microRNA; **PCR**, polymerase chain reaction; **siRNA**, small interfering RNA; **TF**, transcription factor; **USF1**, upstream transcription factor-1

Introduction

Gene expression is a multifaceted term that can have different interpretations, depending on the context in which it is used. In the context of health and disease, gene expression refers to the manifestation of specific genes that result in a particular phenotype. Conversely, gene expression can also refer to the mechanics and control factors of gene transcription and mRNA translation that influence which proteins are produced. The importance of nutritional influences on gene expression is increasingly recognized. Dietary conditions, including dietary patterns, can affect gene expression either directly, by interacting with transcription factors (TFs) or mRNA binding proteins, or indirectly, through exocrine means, thereby inducing changes that define the phenotypic expression. By contrast, variation in the consumption of nutrients in human populations can, in part, be explained by hereditary genetic variation. The technical methods outlined in this tutorial chapter are essential for all research in modern biology and are actively used in nutritional science.

Historical Perspective

In 1961, Nobel laureates François Jacob and Jacques Monod conducted classic experiments in bacteria, demonstrating that genes can influence the synthesis of enzymes involved in nutrient metabolism through an operon under nutrient control.[1] Following this study, researchers conducted experiments on mammalian systems using specific nutrients. Some notable experiments conducted with essential amino acids demonstrated that polyribosome formation depended on their presence in the diet. Additionally, metabolites of vitamin A and vitamin D were discovered to interact with nuclear receptors, resulting in physiologic effects.

Gene Regulation by Nutrients

Research on nutrient regulation of gene expression is a major focus in contemporary nutritional science. One key challenge is distinguishing between the direct effects of individual nutrients on gene expression, and the indirect effects mediated by physiologically controlled mediators and modulating molecules responsive to the diet. Therefore, cellular-level experiments are essential for identifying direct nutrient effects. However, to fully appreciate how dietary components and patterns influence gene expression in various tissues, cellular-level findings must be interpreted within an integrative context of the multicellular organ system. The interaction of the diet with hormones, cytokines, and growth factors to influence the differential expression of specific genes is now recognized as nutritional genomics, also known as nutrigenomics or nutrigenetics.[2] Nutritional genomics includes all genetic factors, including epigenetic events, as they modulate individual genes and gene networks. It is one of a growing number of terms in general use in the nutritional sciences literature (**Table 38.1**). This term has replaced the former term *nutrient–gene interactions*, which is a narrow term that implies a direct interaction of a nutrient with a gene. Examples of nutrient–gene interactions may include a nutrient binding to a

Table 38.1 Terms Frequently Used in Studies on Nutritional Regulation of Gene Expression

Term	Description	Term	Description
Acetylome	Reversible acetylation at the whole proteome level	Nuclear receptor	A transcription factor protein that requires a ligand (e.g., calcitriol for nuclear translocation and DNA binding)
Chromatin immunoprecipitation (ChIP)	Transcription factor antibodies precipitate DNA fragments to identify genes regulated by a specific transcription factor	Nutritional genomics	Genomic studies that relate nutritional factors in the regulation of genes that influence cellular processes genome wide
cis-Acting	DNA elements on the same strand as a structural gene, to which transcription factors bind and initiate transcription	Ortholog	A gene with a similar function to a gene in another species is evolutionarily related; ortholog comparisons help predict gene function
DNA array	Immobilized sequences of single-stranded DNA (probe) on a matrix that allows hybridization of mRNAs for quantitation of transcript abundance (also called gene chips or DNA chips)	PAGE	Polyacrylamide gel electrophoresis
Epigenetic	Nonmutational modification of a gene (e.g., by methylation and histone changes that influence the expression of a specific gene)	PCR	Polymerase chain reaction
Exon-intron junction	A junction between a block of coding sequence (exon) and an adjacent block of noncoding sequence (intron) present in DNA and precursor messenger RNA (pre-mRNA)	Polygenic	Disease or phenotypic characteristic caused by more than one gene
Functional genomics	Relationship of genes, proteins, and regulatory networks with the physiologic function	Protein array	Antibodies or other proteins immobilized on a matrix, allowing for the qualitative detection of specific proteins or the identification of interacting proteins
Genomics	Study of the singular or collective roles that genes play in cellular processes as influenced by external factors; prefixes, such as chemo-, epi-, pharmaco-, or toxico-, can define specialization in genomics	Proteomics	Proteome-wide analysis of protein structure, post-translational modification, interactions, and function
Haplotype	A set of DNA variations, or polymorphisms, usually inherited together; haplotype can refer to a combination of alleles or a set of single nucleotide polymorphisms found on the same chromosome	qPCR	Quantitative PCR in which the relative abundance of a sequence (mRNA-derived cDNA) is compared with a normalizing sequence
Homolog	A gene that has the same evolutionary origin and function in two or more species	Response element	The portion of a gene sequence that must be present for that gene to respond to a stimulus; response elements are binding sites for transcription factors
In silico	In or using computer simulation of complex biologic systems; term frequently used in microarray research in which extensive computational algorithms or comparisons are executed	RNA interference (RNAi or siRNA)	Use of short RNA molecules, frequently derived from double-stranded RNA, which, upon introduction into cells and complementary hybridization to specific mRNA, decrease gene expression

Term	Description	Term	Description
Metabolomics	Global analysis of all metabolites in a complex system	Single nucleotide polymorphism (SNP)	Single base substitution in the coding sequence of a gene; frequently determines phenotypic differences in a population (the human genome has ~10 million SNPs)
Micro RNA (miRNA)	Short regulatory form of RNA that binds to a target RNA molecule and generally suppresses its translation	Systems biology	Study of complex interactions of organ systems down to molecules
Monogenic	Disease or phenotypic characteristic produced by a single gene	*trans*-Acting factors	DNA-binding proteins (transcription factor) are *trans* because they are products of genes from other chromosomes that bind to regulatory elements; transcription factors that bind some nutrients are *trans*-acting factors
Noncoding RNAs	Segments of RNA that are not translated into amino acid sequences but may be involved in the regulation of gene expression	Transcription factor (TF)	Proteins that bind to regulatory regions of a gene and influence the transcription rate of the gene. Some bind to nutrients, vitamins, and minerals for activity
		Transcriptome	All transcribed mRNAs within a cell or tissue at a particular time

TF for subsequent interaction with a response element of a gene, the methylation of specific genes, nutrient-influenced acylation of TFs, or nutrient inhibition and activation of pathways that influence gene activation.

A generalized cell, showing different modes of gene regulation by nutrients, is illustrated in **Figure 38.1**. A "direct" effect of some nutrients, such as active metabolites of vitamins A and D, zinc, n-3 fatty acids, and sterols, on gene transcription involves a series of steps. First, the nutrient binds to a specific TF. This complex then undergoes translocation from the cytoplasm to the nucleus. Inside the nucleus, the TF interacts with a specific response element sequence located in the regulatory region of the gene. This interaction leads to a change in the transcription rate of the gene. Most situations involve a complex of multiple TFs and modifying proteins. Amino acid deprivation at the cellular level may activate transcription from specific defense genes through *cis*-acting nutrient-sensing response elements (see Chapter 46). The control of translation of specific mRNAs by iron is another example of a "direct" nutrient effect on gene expression—in this case, at the level of mRNA stability and translation efficiency—to increase protein abundance. Repression of mRNA by microRNA leading to mRNA degradation is shown in Figure 38.1.

Frequently, gene regulation by nutrients is complex. Multiple and interconnected factors, such as nutrient effects on signal transduction pathways, epigenetic effects on specific genes, gene polymorphisms, alternative mRNA splicing and translation, and post-translational modification, can influence the expression of a specific gene. These factors converge and define the indirect effects on gene expression. Studies have shown that nutritionally responsive TFs (e.g., sterol regulatory element–binding protein [SREBP]) can influence the activity of numerous promoters through TF isoforms and coregulatory nuclear proteins that regulate lipid metabolic genes.[3] Hypoxia-inducible factor (HIF), induced by iron deficiency, similarly regulates multiple genes of iron metabolism.[4] Activation of nuclear translocation and DNA binding of metal-response element binding transcription factor-1 (MTF1), by interaction with zinc, induces the expression of multiple zinc-regulatory and zinc-transporter proteins.[5] Moving further clockwise around the nuclear envelope, Figure 38.1 shows the influence of TF phosphorylation, which can be either activating or deactivating. Zinc-repressing phosphatase activity, with sustained TF activation, is given as an example.[6] Citrate, a product of intermediary metabolism, can diffuse into the nucleus, where adenosine triphosphate (ATP)–citrate lyase activity produces acetyl-coenzyme A (CoA). Nuclear acetyl-CoA functions in histone acetylation and activation of hexokinase 2 and other glucose-metabolizing enzymes and possibly leads to global changes in acetylation and gene expression.[7] An example of the complexity of nutrient gene regulation is the fatty acid synthase (FAS) gene.[8] During fasting, FAS is held in check by upstream transcription factors-1 and 2 (USF1 and USF2), but these TFs are deacetylated by histone deacetylases (HDACs), leading to FAS promoter inactivation. Upon feeding, USF1 is phosphorylated; this process generates interactions with numerous other interacting TFs (>7) to produce increased USF1 acetylation and FAS promoter activation.

Numerous dietary constituents, most notably folate and one-carbon donors, influence DNA methylation.[9] This process leads to the conversion of cytosine to thymidine through a methylation reaction. Methylation of the CpG sequences of gene promoters alters the affinity of the TF for the target gene and the transcription rate of the gene. These concepts and their effects on genetic variation are described in detail in Chapter 39 and Chapter 40.

Small RNA sequences are known to hybridize with mRNAs, resulting in translational repression or mRNA destabilization and degradation.[10] MicroRNAs (miRNAs), which are double-stranded RNAs of approximately 22 nucleotides, clearly regulate many physiologic responses in animals. These miRNAs can be transcribed by their promoters or from the intronic sequences of certain genes. The targets of miRNA are most commonly located in the 3′ untranslated region (UTR) of the target mRNA. An example of this process is the production of miRNA-33 from an intronic sequence within sterol-regulatory, element-binding factor-2, a TF controlling cholesterol synthesis. This miRNA inhibits the expression of the ATP-binding cassette

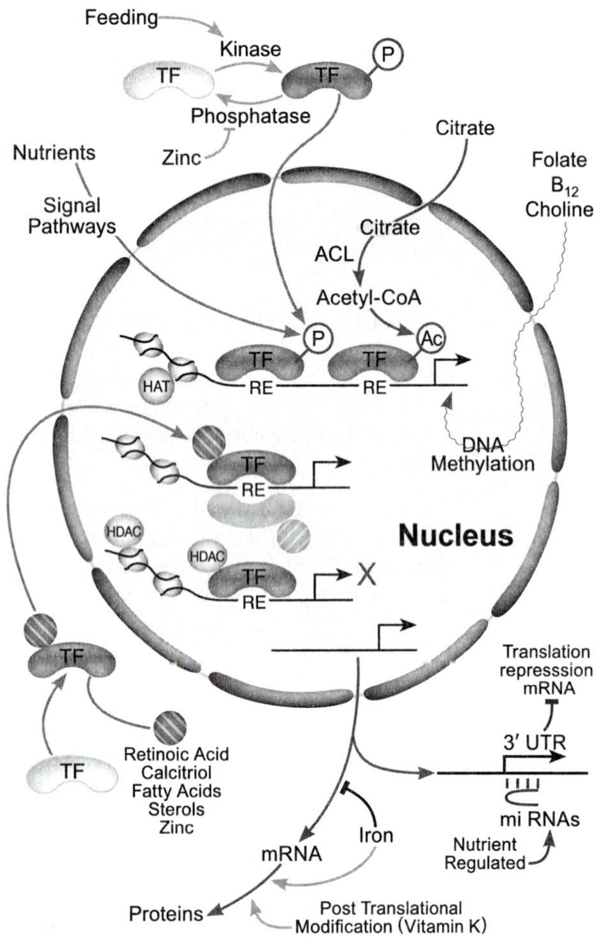

Figure 38.1 Generalized scheme illustrating the regulation of gene expression by nutrients. Metabolites of some fat-soluble vitamins (retinoic acid and calcitriol), fatty acids, sterols, and zinc bind to specific transcription factors (TF) and produce nuclear translocation and binding to specific DNA sequences (response elements) of target genes. TFs binding retinoic acid, calcitriol, and fatty acids are called nuclear receptors and bind DNA as heterodimers or homodimers. Histone deacetylases (HDAC) and histone acetyltransferases (HAT) regulate histone activity by acetylation and are components of larger DNA-binding complexes. Some nutrients activate transmembrane receptors, which activate intracellular signaling pathways to initiate or modify gene expression. Some nutrients influence TF phosphorylation and thus influence gene expression. Intranuclear ATP-citrate lyase can convert citrate to acetyl-coenzyme A (CoA), and acetylation influences gene expression through TF. Some nutrients, including folate and vitamin B_{12}, influence gene expression through DNA methylation. Some nutrients regulate genes that produce microRNAs (miRNAs) that repress gene expression primarily by repressing the translation of target mRNAs. Iron and other nutrients influence gene expression at the post-transcriptional level through control of degradation of some specific mRNAs and stabilization of other specific mRNAs. Specific nutrients (e.g., vitamin K) can produce post-translational modification of proteins. *Ac*, acetyl; *ACL*, acyl-CoAlyase; *P*, phosphoryl (group); *RE*, response element.

transporter G1 (ABCG1 transporter), thereby decreasing cholesterol efflux from cells.[11-13] Genome-wide miRNA-profiling methods reveal important roles for miRNAs in altering the effects of nutrients on specific genes.

Approaches to Study Genes Regulated by Individual Nutrients or Dietary Patterns

With the rapid advancement of technology in the field of biomedical sciences, specific methodologies may quickly become outdated, but the general process of evaluating dietary effects remains relevant. This discussion focuses on the process that investigators use to evaluate the effects of diet on the genome and proteome. A common approach adopted by researchers to analyze the effects of dietary components at the genome or protein level is illustrated in **Figure 38.2**. Vendors now offer products that facilitate sample acquisition and preservation. An example is the method introduced for obtaining blood cells that allows RNA stabilization, which is particularly important in clinical and field (intervention) studies where analysis may be delayed. Similarly, methods for protein analyses have also advanced, with mass spectral identification of proteins and metabolites now available. Commercial vendor websites can provide tutorials on applicable methods.

The approach outlined in Figure 38.2 shows how the abundance of a known transcript can be evaluated using the quantitative real-time polymerase chain reaction (qPCR). This technology has become the frontline method of choice for most research involving gene expression. Northern analysis has the advantage of providing an estimate of transcript size, but it has several limitations, including the requirement for a labeled DNA probe (usually with phosphorus-32 [^{32}P]), the ability to process only limited replicates, and lack of quantitation. In situ hybridization can identify a site of mRNA abundance within a cell and could establish a nutrient response within a given cell type or tissue, but this method is also not considered quantitative. Analysis at the genome-wide level most frequently uses DNA microarrays to obtain profiles of transcripts that increase or decrease in response to feeding a particular diet or specific nutrient. This approach detects potential associations and can identify previously unknown nutrient-responsive targets. PCR arrays are narrower in scope and are used to identify genes within a given process (e.g., oxidative stress, apoptosis, or a given cell signaling pathway). Chromatin immunoprecipitation (ChIP) with sequencing (ChIP-seq) is a technique that utilizes ChIP to immunoprecipitate a TF associated with DNA, followed by massively parallel DNA sequencing to identify the genes that the specific TF occupies. The specificity and selectivity of the identification rest with the selected antibody. This method enables researchers to identify novel loci for specific nutrition-related diseases and traits. An early example of this technology is the identification of vitamin D receptor–binding targets and the increase in DNA-binding targets produced by calcitriol stimulation.[14]

The flow chart in Figure 38.2 also provides an overview of techniques that answer questions related to nutritionally responsive processes at the proteome level. A blotting procedure typically estimates the abundance of a specific protein. The proteins are separated by size using PAGE, and then the protein of interest is detected using an immunologic method with a specific antibody. Sensitivity is increased using secondary antibodies that are linked to a reagent that produces luminescence before detection by X-ray film exposure. This process is frequently referred to as immunoblotting or western blotting. Immunoprecipitation is a less frequently used technique, but it can be a valuable aid for

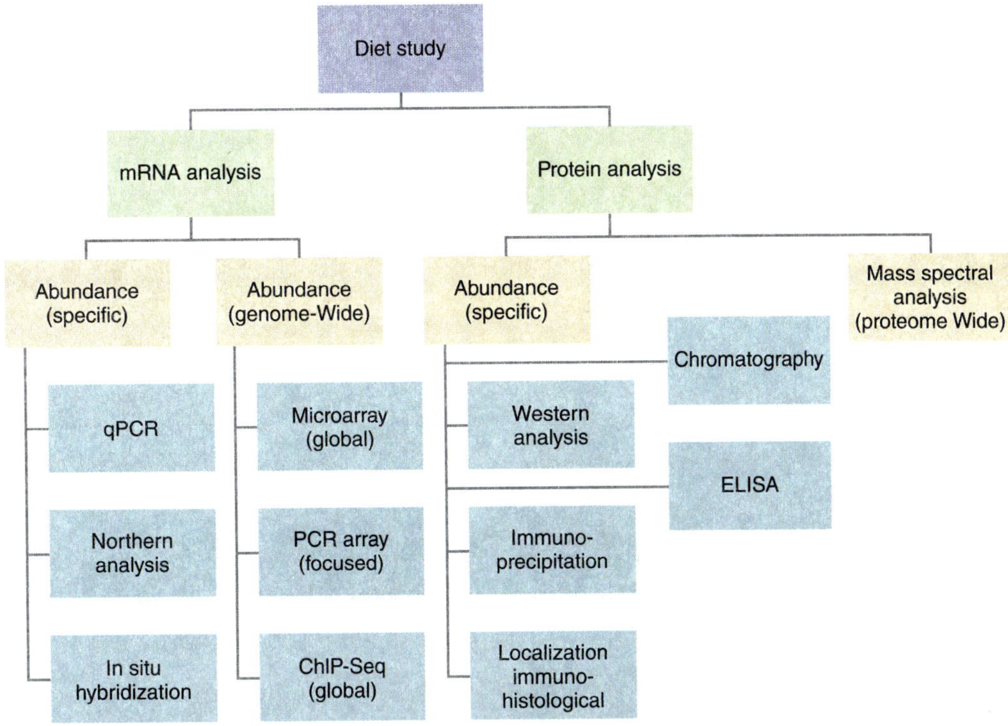

Figure 38.2 Flow diagram of some analytic approaches used to identify nutrient effects on gene expression. The methods are divided into those targeting the transcript and protein levels. *ChIP-seq*, chromatin immunoprecipitation with sequencing; *ELISA*, enzyme-linked immunosorbent assay; *PCR*, polymerase chain reaction; *qPCR*, quantitative real-time polymerase chain reaction.

sample enrichment before a further analytic procedure aimed at detecting a target protein. Antibodies are also used to detect specific proteins in histologic specimens. The method can accurately identify the location of a protein within a cell and determine whether the protein is subject to trafficking within the cell. For example, the endosomal recycling of nutrient transporters in response to nutrient availability can be visualized through these methods. Enzyme-linked immunosorbent assay (ELISA) methods are widely used in nutrition research and in clinically related studies to measure specific proteins of interest. Most frequently, ELISAs are used to detect small proteins and peptides, such as cytokines, found in serum samples. Chromatography is a less widely used method, but it leads to the identification of a specific protein with a nutritionally related process and can be an important first step in a protein purification assay (e.g., ion-exchange chromatography) before the detection of the abundance of the protein by a method with greater specificity.

The field of proteomics has shown exceptional promise as an analytic and research aid in nutrition-related studies.[15] The identification of specific proteins in an analytical sample typically uses matrix-assisted laser desorption/ionization mass spectrometry (MALDI-MS). This approach is useful because of the availability of extensive protein and peptide databases (see Figure 38.2). These methods are gaining popularity in identifying and measuring nutritionally responsive biomarkers.[16-18]

Figures 38.3 and **38.4** demonstrate the typical approach used in conducting studies related to nutritional processes. At the gene level, an initial aim is to establish whether the mode of nutritional regulation is transcriptional, post-transcriptional, or both. Subsequent analyses are directed at promoter activity, mRNA stability, and repression by miRNA. At the protein level, the studies are more analytically focused, based on abundance and cellular localization. Nevertheless, mechanistic studies aimed at nutritionally responsive acetylation and phosphorylation processes are important in evaluating post-translational modifications.

Approaches to Identify and Manipulate Genes Regulated by Individual Nutrients or Dietary Patterns

Transgenic Animals

The term *transgenic* refers to both overexpression and deletion of expression of a specific gene. However, the technique most associated with transgenesis involves the overexpression of a structural gene. Transgenic overexpression is achieved by producing a construct that includes a structural gene and a promoter, either the gene's normal promoter (homologous) or a different promoter (heterologous). The construct is then injected into fertilized eggs, typically from mice or pigs. If the constructed DNA integrates properly into the genome, transgenic animals can be produced after the eggs are returned to foster mothers for the full gestation period. Selective breeding can then produce homozygous lines of animals carrying transgenes.

Transgenic animals have been used to study various nutritional questions. A notable example is the overexpression of the glucose transporter (GLUT4) in mice using the aP2 fatty

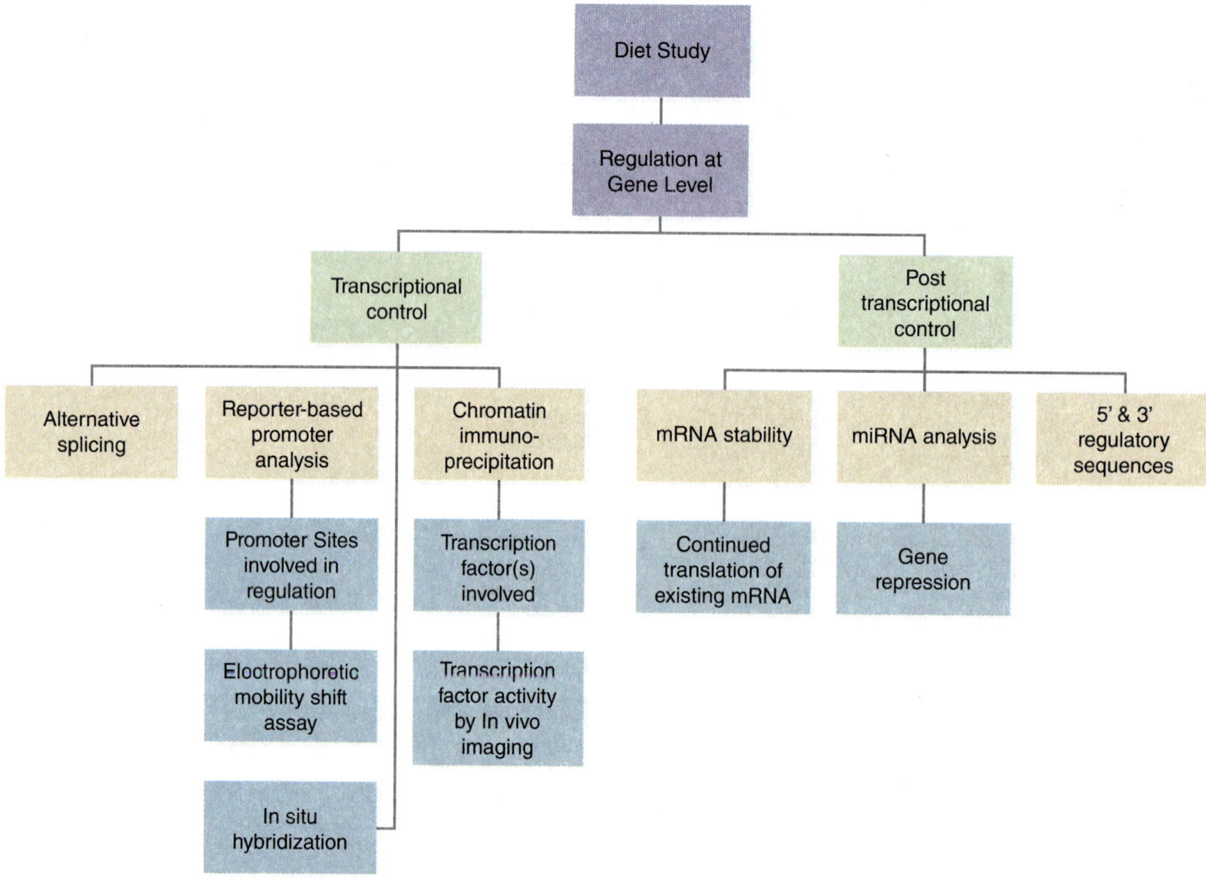

Figure 38.3 Flow diagram of research approaches to define nutrient effects on gene regulation. *miRNA*, microRNA.

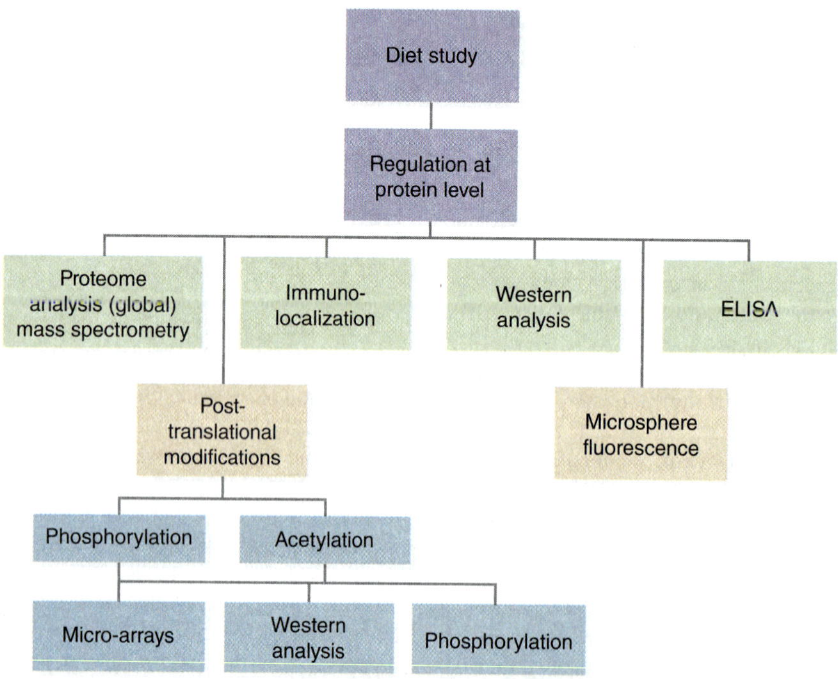

Figure 38.4 Flow diagram of research approaches used to define nutrient effects on regulation at the protein level. *ELISA*, enzyme-linked immunosorbent assay.

acid–binding protein promoter and a genomic DNA fragment containing the entire human *GLUT4* gene.[19] This overexpression resulted in higher glucose transport rates, lower glucose tolerance curves, and greater body fat. However, most transgenic mice strains do not exhibit such significant changes in phenotype, and the overexpression may even produce unanticipated outcomes. Nonetheless, the number of nutritionally relevant genes overexpressed in transgenic mice is now extensive. Moreover, many strains of transgenic mice are now commercially available through distributors of laboratory animals and mutant mouse research centers supported by the National Institutes of Health.

Gene Knockout (Null Mutation) Animals

Gene-knockout technology provides the opportunity to eliminate the expression of a specific gene (null mutation), thereby preventing the production of the normal gene product. The effects of null mutations can range from no apparent effect to embryonic lethality. Consequently, the technology is distinct from the spontaneous mutations that occur in laboratory animals and are propagated through selective breeding. These mutations usually result in the altered function of the gene product. An example of a spontaneous mutation is the *ob* gene mutation in mice.[20]

The technique of creating a knockout animal model or null mutation is more correctly called gene targeting by homologous recombination. This involves disrupting the targeted gene in one allele to produce heterozygotes with the null mutation. Two approaches are used to develop knockout mice.[21] The first approach is to isolate the murine gene under investigation, identify the exons by mapping, delete part or all of an exon, and replace it with the gene that produces a marker for selection. This construct then serves as the gene-targeting vector. The targeting vector is linearized and transfected into embryonic stem (ES) cells by microinjection or electroporation. The transfected cells are then injected into blastocysts that have been removed from pregnant mice, and the transfected cells are introduced into pseudopregnant mice. A second and more recent approach can provide cell type–specific targeting of gene deletion. The gene is engineered to have *loxP* sites on either side, and ES cell technology is then used to create a transgenic line that carries the target gene.

These methods have been extended to create conditional-knockout mutations.[22] In this approach, Cre recombinase is expressed with a tissue-specific promoter. This allows the production of mice whose genes are inactivated in a tissue-specific mode or during a specific period of development. Alternatively, a library of mutant ES cells covering much of the mouse genome is available through the International Gene Trap Consortium (www.genetrap.org). Gene trapping is a high-throughput method that uses vectors that produce *lacZ* fusion sequences with native endogenous gene transcripts that disrupt the normal transcription of that gene.[23,24] The flanked *lox* site insertion (Floxin) is a method for post-translational modification of specific gene trap vectors and can generate conditional loss-of-function modifications in gene trap ES cells.[25] Other methods, such as the use of proprietary zinc-finger nuclease technology, can be applied to produce targeted gene ablation for mutant mouse development.

Numerous knockout models have been used to study nutrient metabolism and function. Examples are the intestinal targets of the hormonal form of vitamin D (calcitriol) in controlling calcium absorption.[26,27] Frequently, complete ablation of a gene can lead to embryonic lethality or have no major phenotypic effect. Conditional knockout mice have been developed to prevent these phenotypic outcomes. For example, heart-specific knockout of the Ctr1 copper transporter reveals a systemic signaling mechanism in copper metabolism.[28] Another extension of knockout technology is crossbreeding with transgenic mice and knockout mice, which can produce valuable insights into metabolic pathways and phenotypes. For example, crossbreeding transgenic mice overexpressing apolipoprotein A-I with apolipoprotein E null mice resulted in an increased high-density lipoprotein concentration and increased atherosclerotic lesions in the offspring.[29]

Inhibition of Gene Expression by RNA Interference

Antisense RNA technology has been used to study a limited number of genes related to nutrition. This method operates under the principle that a small RNA sequence complementary (antisense) to a target mRNA can inhibit the translation of the target or stimulate its degradation. Initially, antisense RNA sequences for gene silencing employed short synthetic oligonucleotides for transient inhibition of the translation of specific mRNAs by hybridization. These oligonucleotides appear to be taken up by some tissues. In these cases, antisense DNA sequences can be introduced into specific areas of the brain.

The use of siRNA has become a widely accepted approach for gene silencing at the cell level.[30] Animals have an innate defense system that degrades double-stranded RNA by the action of an RNase (i.e., Dicer) to produce RNA sequences of 21 to 23 base pairs (bp) in length, called siRNA. These RNA strands bind to the target mRNA and promote its degradation. A synthetic oligonucleotide can replace the siRNA. In practice, double-stranded RNAs (200 to 1,000 bp) or short RNAs (20 to 25 bp), produced with commercially available reagents, expression vectors, and enzymes, are available to accomplish gene silencing. However, the siRNA approach has some disadvantages. For example, it only partially inhibits gene expression, and the inhibition achieved is frequently transient rather than stable. To address these issues, researchers have developed small hairpin RNA (shRNA) vectors, which provide gene silencing with greater stability and longer duration. This technology has been used to silence genes of nutritional interest. An advantage of siRNA is that it can knock down genes that would otherwise be lethal if knocked out in the embryo.

Dietary Patterns and Gene Expression in Humans

We will review here the evidence for the potential effects of the most common dietary patterns on gene expression in human populations. The Mediterranean diet (MedDiet) is one of the most studied dietary patterns in relation to diverse health outcomes. Traditional foods from countries around the Mediterranean Sea are the basis of the MedDiet dietary pattern. It is characterized by high consumption of nutrient-dense, plant-based foods and monounsaturated fats derived from olive oil. The diet also includes moderate consumption of fish, seafood, and fermented dairy products, and it emphasizes limiting the intake of red and processed meats, as well as sweets.[31] In observational studies, adherence to the MedDiet is assessed using *a priori* dietary patterns. Two of the most used scores are the Mediterranean Diet Score (MDS)[32] and the alternate Mediterranean Diet score (aMED).[33] Associations have been found between regular consumption of the MedDiet and a variety of health benefits, including lower risk of cardiovascular outcomes, diabetes, cancer, and mortality.[34-36] The landmark Spanish PREDIMED ("Prevención con Dieta Mediterránea") study is the largest intervention trial (N = 7,447 participants) that has evaluated the impact of the MedDiet on the incidence of cardiovascular disease (CVD) in a population at high risk of CVD.[37] Compared with a low-fat diet as a control, the MedDiet, supplemented either with extra virgin olive oil (EVOO) or nuts, showed associations with lower risk of CVD,[37] diabetes,[38] and remission of metabolic syndrome.[39] The MedDiet seems to affect health outcomes through the modulation of inflammatory and atherogenic mediators. A recent umbrella review of meta-analyses of observational studies and randomized trials found suggestive evidence that the MedDiet may be effective in increasing HDL and in reducing body weight, BMI, waist circumference, total cholesterol C-reactive protein (CRP), and Interleukin-6 (IL-6).[40] Using a candidate gene approach, intervention trials, including PREDIMED, have found that the MedDiet, particularly when supplemented with extra-virgin olive oil, was associated with reduced expression of several proatherogenic and inflammatory genes, such as *INFG* (interferon gamma), *ARHGAP15* (Rho GTPase activating protein 15), *IL7R* (interleukin-7 receptor), *ADRB2* (adrenoceptor beta 2), *POLK* (DNA polymerase kappa), *COX-2* (cytochrome c oxidase subunit II), *LRP1* (LDL receptor related protein 1), *CCL2* (C-C motif chemokine ligand 2; also known as *MCP1*), *RELA* (RELA proto-oncogene, NF-kB subunit; also known as *p65* subunit), *MMP9* (matrix metallopeptidase 9), and *TNF* (tumor necrosis factor) in peripheral blood mononuclear cells.[41-43] Transcriptome-wide analysis of 18,400 RNA transcripts, including 14,500 well-characterized genes, has shown that CVD biologic pathways are modulated by the MedDiet.[44] Another mechanism by which diet may affect gene expression is by modulating the content of exosomes. Exosomes are extracellular vesicles that play critical roles in intercellular communication by transporting diverse molecules (including nucleic acids) from cell to cell.[45] Recent evidence shows that the MedDiet may also influence exosomal RNAs.[46] Supplementation of the MedDiet with either extra-virgin olive oil or nuts modulated exosome RNA content, particularly miRNA content, after a 1-year intervention in the PREDIMED trial.[46]

In addition to the MedDiet, other dietary patterns have been associated with gene expression. The Dietary Guidelines for Americans Adherence Index (DGAI) is a score that measures adherence to the Dietary Guidelines for Americans (DGA) recommendations. Higher values of the DGAI score represent a healthier diet.[47] A transcriptome-wide association analysis in the Framingham Heart Study (FHS) Offspring and Third Generation found associations between the DGAI score and overexpression of 16 genes, and underexpression of three genes in blood.[48] Network analyses showed that the differentially expressed genes are involved in diverse biological pathways, such as insulin sensitivity, glucose tolerance, and energy consumption and metabolism.[48] Cross-sectional transcriptome-wide analyses have also reported gene-expression associations with other dietary patterns, such as the Western (i.e., rich in intake of red and processed meats as well as refined foods) and Prudent (i.e., high consumption of vegetables, fruits, legumes, and whole grains) patterns.[49,50] The Western dietary pattern was inversely associated with the expression of genes involved in the regulation of metabolic and inflammatory processes, and the Prudent pattern was associated with the expression of genes regulating DNA transcription.[49] In addition, both dietary patterns were associated with diverse biologic pathways, such as those involved in immune and inflammatory responses, as well as related to cancer and cardiovascular disease.[50]

Genetic Basis of Macronutrient Preferences

Food preferences vary significantly among individuals, and a variety of factors, including cultural, socioeconomic, and psychological factors, can influence these preferences. Additionally, hereditary genetic variation may also play a role in determining dietary preferences. Estimates of family-based heritability (i.e., the proportion of the variation of a phenotypic trait that is explained by genetic variation) of macronutrient intake range from 11% up to 65%.[51] Prior to the advent of genome-wide association studies (GWAS), only a few genomic regions were found linked to variations in energy and macronutrient intake.[52-55] GWAS are agnostic scans of the genome that identify common genetic hereditary variations associated with phenotypic traits. Because GWAS have no a priori hypothesis about the genomic location of gene variants associated with the phenotype of interest, they offer an unbiased examination of the whole genome.

The first two GWAS of macronutrient intake, each including more than 70,000 participants, were conducted in the DietGen, CHARGE (Cohorts for Heart and Aging Research in Genomic Epidemiology), and GLACIER (Gene-Lifestyle interactions And Complex traits Involved in Elevated disease Risk) consortia 10 years ago.[56,57] These two early GWAS identified two single nucleotide polymorphisms (SNPs) (rs838133 and rs838145) in the *FGF21* (fibroblast growth factor 21) gene associated with higher intake of carbohydrates and lower intake of protein. The *FGF21* gene codes for an endocrine factor that is primarily produced and secreted by the liver and regulates metabolism and energy homeostasis.[58] One SNP (rs1421085) in the *FTO* (Fat Mass and Obesity Associated) gene was associated with higher protein intake. *FTO* was the first GWAS-identified gene that carried variants associated with obesity.[59,60] *FTO* codes for an RNA demethylase[61] and is expressed in a wide variety of tissues, with high levels in the brain and hypothalamus.[62,63] *FTO* expression in the hypothalamus is involved in the regulation of food intake and energy homeostasis.[64,65]

More recent GWAS have expanded to 26 distinct genomic loci involved in macronutrient intake preferences.[66-68] The largest and most recent GWAS, which included 282,271 participants of European ancestry from the UK Biobank and the CHARGE consortium, estimated a genome-wide SNP-based heritability of macronutrient intake of 4.1% for carbohydrates, 3.5% for total fat, and 4.4% for protein.[68] The updated list of genes offers insight into the biologic mechanisms regulating the consumption of macronutrients. Variants in the *FGF21* locus had the strongest association with elevated intake of fat and protein and reduced consumption of carbohydrates, pointing to the key role of this gene in food intake regulation. Examination of gene expression in 54 tissues from the Genotype-Tissue Expression (GTEx) project[69] revealed that GWAS-identified genes were highly expressed in brain tissue, including cerebellum, cortex, frontal cortex, anterior cingulate cortex, and nucleus accumbens.[68] Single-cell analyses identified expression enrichment in GABAergic, dopaminergic, and glutamatergic neurons.[68] These results point to a critical role of the central nervous system in the regulation of body weight and energy metabolism.

In summary, available evidence highlights the existence of a genetic basis for preferences of macronutrient intake. However, causality of individual gene variants has not yet been established and caution must be taken in the interpretation of results, as residual confounding due to cultural and familial preferences is possible. Finally, it is worth noting that most of the participants in the discussed studies are from European ancestry. There is an urgent need to increase diversity in GWAS and whole-genome sequencing studies to better understand the genetic basis of macronutrient preferences in diverse human populations.

Conclusions

The field of nutritional genomics has advanced significantly in recent years, becoming a recognized research discipline in the nutritional sciences. The completion of animal and human genome sequencing, coupled with the technologies and approaches used in this field, has had a profound impact on our understanding of how diet and genetics influence the expression of traits. New technologies and approaches are still emerging, suggesting that nutritional genomics will continue to be an important area of research for years to come. The potential for further progress in this field is significant, and nutritional genomics holds great promise for improving human health and well-being.

Acknowledgments

The authors acknowledge Robert J. Cousins and Louis A. Lichten for contributions to the previous version of this chapter.

References

1. Jacob F, Monod J. J Mol Biol 1961;3:318–56.
2. Muller M, Kersten S. Nat Rev Genet 2003;4:315–22.
3. Bennett MK, Seo YK, Datta S, et al. J Biol Chem 2008;283: 15628–37.
4. Shah YM, Matsubara T, Ito S, et al. Cell Metab 2009;9:152–64.
5. Wang Y, Lorenzi I, Georgiev O, et al. Biol Chem 385, 623–32.
6. Eide DJ. (2020) Crit Rev Biochem Mol Biol 55, 88–110.
7. Wellen KE, Hatzivassiliou G, Sachdeva UM, et al. Science 2009;324:1076–80.
8. Wong RH, Chang I, Hudak CS, et al. Cell 2009;136:1056–72.
9. Tibbetts AS, Appling DR. Annu Rev Nutr 2010;30:57–81.
10. Bartel DP. Cell 2009;136:215–33.
11. Rayner KJ, Suarez Y, Davalos A, et al. Science 2010;328:1570–3.
12. Najafi-Shoushtari SH, Kristo F, Li Y, et al. Science 2010;328: 1566–9.
13. Marquart TJ, Allen RM, Ory DS, et al. Proc Natl Acad Sci 2010;107:12228–32.
14. Ramagopalan SV, Heger A, Berlanga AJ, et al. Genome Res 2010;20:1352–60.
15. Bantscheff M, Schirle M, Sweetman G, et al. Anal Bioanal Chem 2007;389:1017–31.
16. Moresco JJ, Dong MQ, Yates JR, 3rd. Am J Clin Nutr 2008;88: 597–604.
17. Kussmann M, Panchaud A, Affolter M. J Proteome Res 2010;9: 4876–87.
18. Linke T, Ross AC, Harrison EH. J Chromatogr A 2004;1043: 65–71.
19. Shepherd PR, Gnudi L, Tozzo E, et al. J Biol Chem 1993;268: 22243–6.
20. Ingalls AM, Dickie MM, Snell GD. J Hered 1950;41:317–8.
21. Leiter EH. Diabetologia 2002;45:296–308.
22. Liu P, Jenkins NA, Copeland NG. Genome Res 2003;13:476–84.
23. Yamamura K, Araki K. Cancer Sci 2008;99:1–6.
24. Lee T, Shah C, Xu EY. Mol Hum Reprod 2007;13:771–9.
25. Singla V, Hunkapiller J, Santos N, et al. Nat Methods 2010;7: 50–2.
26. Kutuzova GD, Akhter S, Christakos S, et al. Proc Natl Acad Sci 2006;103:12377–81.
27. Benn BS, Ajibade D, Porta A, et al. Endocrinol 2008;149: 3196–205.
28. Kim BE, Turski ML, Nose Y, et al. Cell Metab 2010;11:353–63.
29. Plump AS, Scott CJ, Breslow JL. Proc Natl Acad Sci 1994;91: 9607–11.
30. Morris KV, Chan SW, Jacobsen SE, et al. Science 2004;305: 1289–92.
31. Misirli G, Benetou V, Lagiou P, et al. Am J Epidemiol 2012;176: 1185–92.
32. Trichopoulou A, Kouris-Blazos A, Wahlqvist ML, et al. BMJ 1995;311:1457–60.
33. Fung TT, Rexrode KM, Mantzoros CS, et al. Circ 2009;119: 1093–100.
34. Martinez-Gonzalez MA, Gea A, Ruiz-Canela M. Circ Res 2019;124:779–98.
35. Salas-Salvado J, Becerra-Tomas N, Garcia-Gavilan JF, et al. Prog Cardiovasc Dis 2018;61:62–7.
36. Guasch-Ferre M, Willett WC. J Intern Med 2021;290:549–66.
37. Estruch R, Ros E, Salas-Salvado J, et al. N Engl J Med 2018; 378:e34.
38. Salas-Salvado J, Bullo M, Estruch R, et al. Ann Intern Med 2014;160:1–10.
39. Babio N, Toledo E, Estruch R, et al. CMAJ 2014;186:E649–57.
40. Dinu M, Pagliai G, Casini A, et al. Eur J Clin Nutr 2018;72:30–43.
41. Konstantinidou V, Covas MI, Munoz-Aguayo D, et al. FASEB J 2010;24:2546–57.
42. Llorente-Cortes V, Estruch R, Mena MP, et al. Atherosclerosis 2010;208:442–50.
43. Camargo A, Delgado-Lista J, Garcia-Rios A, et al. Br J Nutr 2012;108:500–8.
44. Castaner O, Corella D, Covas MI, et al. Am J Clin Nutr 2013;98:845–53.
45. Edgar JR. BMC Biol 2016;14:46.
46. Mantilla-Escalante DC, Lopez de Las Hazas MC, Crespo MC, et al. Eur J Nutr 2021;60:4279–93.
47. Sauder KA, Proctor DN, Chow M, et al. Br J Nutr 2015;113: 1773–81.
48. Lin H, Rogers GT, Lunetta KL, et al. Am J Clin Nutr 2019;110: 742–9.
49. Christensen JJ, Ulven SM, Thoresen M, et al. Nutr Metab Cardiovasc Dis 2020;30:2111–22.
50. Bouchard-Mercier A, Paradis AM, Rudkowska I, et al. Nutr J 2013;12:24.
51. Rankinen T, Bouchard C. Annu Rev Nutr 2006;26:413–34.
52. Collaku A, Rankinen T, Rice T, et al. Am J Clin Nutr 2004;79: 881–6.
53. Choquette AC, Lemieux S, Tremblay A, et al. Am J Clin Nutr 2008;88:1142–8.

54. Cai G, Cole SA, Bastarrachea RA, et al. Am J Clin Nutr 2004;80:1410–4.
55. Cai G, Cole SA, Butte N, et al. Obesity 2006;14:1596–604.
56. Tanaka T, Ngwa JS, van Rooij FJ, et al. Am J Clin Nutr 2013;97: 1395–402.
57. Chu AY, Workalemahu T, Paynter NP, et al. Hum Mol Genet 2013;22:1895–902.
58. Flippo KH, Potthoff MJ. Nat Metab 2021;3:309–17.
59. Frayling TM, Timpson NJ, Weedon MN, et al. Science 2007;316: 889–94.
60. Scuteri A, Sanna S, Chen WM, et al. PLoS Genet 2007;3:e115.
61. Jia G, Fu Y, Zhao X, et al. Nat Chem Biol 2011;7:885–7.
62. Gerken T, Girard CA, Tung YC, et al. Science 2007;318:1469–72.
63. Lein ES, Hawrylycz MJ, Ao N, et al. Nature 2007;445:168–76.
64. Olszewski PK, Fredriksson R, Olszewska AM, et al. BMC Neurosci 2009;10:129.
65. Karra E, O'Daly OG, Choudhury AI, et al. J Clin Invest 2013;123:3539–51.
66. Merino J, Dashti HS, Li SX, et al. Mol Psychiatry 2019;24: 1920–32.
67. Meddens SFW, de Vlaming R, Bowers P, et al. Mol Psychiatry 2021;26:2056–69.
68. Merino J, Dashti HS, Sarnowski C, et al. Nat Hum Behav 2022;6: 155–63.
69. Consortium GT. Science 2015;348:648–60.

CHAPTER 39

Genetic Variation: Effect on Nutrient Metabolism

Patrick J. Stover

CHAPTER OUTLINE

Abbreviations

ADH, alcohol dehydrogenase; **ALDH**, aldehyde dehydrogenase; **AMY1**, salivary amylase gene; **ApoE**, apolipoprotein E; **CEU**, Utah residents with Northern and Western European ancestry from the CEPH (Centre d'Etude du Polymorphisme Humain) collection; **CNV**, copy number variation, **GWAS**, genome-wide association study; **HapMap**, Human Haplotype Map Project; $\mathbf{k_{cat}}$, maximal rate of product formation at infinite substrate concentration; $\mathbf{K_m}$, Michaelis-Menten constant; **LCT**, lactase gene; **LD**, linkage disequilibrium; **LDL**, low-density lipoprotein; **LP**, lactase persistence; **MAF**, minor allele frequency; **meC**, methylcytosine; **MTHFR**, methylenetetrahydrofolate reductase gene; **PCSK9**, proprotein convertase subtilisin-like kexin type 9; **SNP**, single nucleotide polymorphism; **YRI**, Yoruba in Ibadan, Nigeria (West Africa)

Human Genetic Variation

Genetic variation contributes to individual phenotypic differences within and among human populations, including metabolic traits, nutrient and food needs, and differential susceptibility to common chronic and metabolic diseases.[1] Metabolic impairments are integral components of chronic diseases, developmental anomalies, cancers, neurologic disorders, and most other pathologic processes. Often, they precede anatomic and other signs of disease. Clinical investigations of inborn errors of metabolism provided some of the earliest and conclusive evidence that (a) metabolic impairments are heritable, (b) genes can modify nutrient use and metabolism, (c) metabolic impairments cause disease, and (d) the functional consequences of genetic mutations can be attenuated significantly by targeted nutritional therapies that compensate for and, less often, avoid genetically induced metabolic impairments.

Phenylketonuria provides a classic paradigm that demonstrates the potential effectiveness of diet in modifying deleterious phenotypes that result from genetic mutations that alter metabolism. Phenylalanine-restricted diets lessen and may even prevent severe cognitive deficits in children who carry mutations in the phenylalanine hydroxylase gene.[2] Inborn errors of metabolism are generally recessive, are relatively rare in most populations, and the initiation or progression of the associated disorders can be managed by diet or nutrition in some, but not all, cases.

Inborn errors of metabolism are typically monogenic disorders that follow Mendelian modes of inheritance and, therefore, are well characterized with respect to their molecular and genetic bases. However, the most prevalent human metabolic disorders are complex polygenic diseases with contributions from multiple low-penetrant susceptibility alleles. The risks associated with these alleles are modifiable by both lifestyle and environmental factors, including one or more dietary components.

The genetic and biochemical causes of many cancers and chronic diseases, including cardiovascular disease and type 2 diabetes, remain elusive. These disorders do not conform to classic Mendelian inheritance patterns and; therefore, genetic approaches based on "simple" linkage analyses are usually not possible. Genomic approaches that are enabled by the availability of complete genome sequences from several mammalian species and the generation of a comprehensive catalog of human genetic variation have been successful in identifying susceptibility genotypes that modify metabolism, alter nutritional requirements, and contribute to metabolic disease. Furthermore, through evolutionary genomics, the origins and consequences of human genetic variation are decipherable, and allelic variants and interacting environmental risk factors that impair metabolic pathways or modify optimal dietary requirements can be inferred.[1]

Origin of Human Genetic Variation

The pattern of human genetic variation is determined by interactions among different evolutionary forces.[3] The generation of primary sequence differences in DNA is a function of the DNA mutation rate; the expansion of the mutation within a population is a function of recombination, demographic history (e.g., fluctuations in effective population size, substructure, and migration), selection (the effect of the mutation on an organism's fitness), and random process (genetic drift).[4,5] Not all sequence variation has phenotypic consequences.[6] DNA sequences that do not affect function can mutate freely without consequences; whereas changes in DNA sequences that encode information or function may alter physiologic processes and; therefore, the propagation and expansion of such sequences will be more constrained.

Most human genetic variations present in noncoding regions, including those found in intronic and intergenic regions, are assumed to be selectively neutral and, therefore, a function of the DNA mutation rate,[7] which is estimated to be approximately 2.5×10^{-8} mutations per nucleotide per generation. However, this rate is not uniformly distributed across the whole genome.[8,9] The highest mutation rate for a human gene is approximately 1×10^{-5} per generation.[10]

Many factors contribute to DNA mutation rates. DNA replication and recombination do not occur with complete fidelity and, thereby, account for a significant portion of observable mutation rates. Polymerase error rates and DNA mutations are affected by nutrients, including iron, B vitamins, and antioxidants. For example, inhibition of folate-dependent deoxythymidine monophosphate synthesis results in the misincorporation of deoxyuridine triphosphate into DNA.[11] Purine and pyrimidine bases within DNA also undergo spontaneous chemical mutation; cytosine spontaneously deaminates to uracil with a frequency of 100 mutations/genome/day, and purine nucleotides undergo depurination mutations at a rate of 3,000 mutations/genome/day. DNA repair systems are effective in detecting and correcting most of these mutations.[12]

Genotoxic xenobiotics, both natural products and synthetic chemicals, are present in the food supply and can modify DNA chemically and increase mutation rates. One class of natural compounds, aflatoxins, can dramatically increase DNA mutation rates, trigger cancers in somatic cells, and lead to localized cancer epidemics.[13] DNA mutation rates are affected by dietary antioxidants[14] as well as by excesses in pro-oxidant nutrients, including iron.[15] However, only mutations that occur in the germ line contribute to a species' heritable genetic variation.

DNA mutation rates and polymorphism frequency vary throughout the human genome. Such region-specific differences within the genome have been attributed to the frequency of DNA recombination and to the mutagenic potential of specific nucleotide sequences. The most common genetic mutation within the human genome is the C to T transition.[16] The sequence CpG is enriched in the promoter regions of mammalian genes and is recognized by DNA methylases, which convert the cytosine base (C) to methylcytosine (meC). meC density within the genome is modifiable by dietary folate and one-carbon donors, and fetal methylation patterns established in utero can be metastable and can influence gene expression into adulthood.[17]

Cytosine methylation influences the transcription rates of genes by altering the affinity of DNA binding transcription factors, by enabling the recruitment of meC binding proteins that serve to silence gene transcription, or both. DNA methylation is usually associated with gene silencing and is critical for the inactivation of imprinted genes and X chromosome inactivation. Mutations at CpG dinucleotides occur approximately 10 times more frequently than at other loci, presumably because meC deaminates spontaneously to thymidine (T), whereas C deaminates to uracil. Uracil is recognized as foreign to DNA and is excised by the DNA repair enzymes, whereas T is not recognized as foreign. The sequence CpG is under-represented in the human genome, and its frequency has decreased throughout evolution, consistent with that inherent instability.[16]

DNA recombination rates also vary throughout the human genome. Recombination creates genetic variation by reshuffling existing genetic variation. The recombination rate is estimated to be 1 cM/Mb to approximately 1.33 cM/Mb; however, it is also heterogeneous across the human genome: approximately 33,000 "recombination hotspots" account for approximately 50% to 60% of the crossover events, but they occupy only approximately 6% of the human genome sequence.[18-23] Investigators have observed that genes that interact with the environment (e.g., immunity, cell adhesion, signaling) tend to locate at genomic regions with high recombination rates, whereas those genes that do not experience low recombination.[20] Recombination is also correlated with levels of genetic variation, a finding indicating that recombination itself may be mutagenic.[24]

Mutations that expand within a population contribute to genetic variation as polymorphisms, and this process is the basis for the molecular evolution of genomes. The expansion of mutations within a population occurs through the process of genetic drift or natural selection. Drift is a stochastic process that results from a chance assortment of chromosomes at meiosis. Only a few of all possible zygotes are generated or survive to reproduce[25]; therefore, mutations can expand in the absence of selection through random fluctuations in the transmission of alleles from one generation to the next, resulting from the random sampling of gametes. Because drift generally has a greater impact on allele frequencies in smaller populations, human demographic history has been a major force in shaping human genetic variation. Severe reductions in population size (bottleneck) can lead to a reduction in genetic variability, whereas rapid expansions can increase genetic variation.[18,26]

Migration and population admixture also affect allele frequency. Modern humans originated in Africa, and small subpopulations migrated to the rest of the world within the past 100,000 years.[6] Consequently, African populations have more genetic variations than other populations.[27] Investigators have shown that significantly more deleterious variations exist in European populations than in African populations, a finding indicating that genetic variation caused by demographic history has significant health consequences.[27,28] Specific diseases, such as breast cancer, Tay-Sachs disease, Gaucher disease, Niemann-Pick disease, and familial hypercholesterolemia within the Old Order Amish and Hutterite populations may be accounted for by demographic history.[29]

Selection is another important evolutionary force that shapes human genetic variation. Most substitutions in the genome are functionally neutral and do not have fitness consequences on their carriers. However, more and more genetic loci have been found to deviate from the neutral null model under various statistical tests, and the results suggest adaptive evolution. When a new mutation arises that affects the fitness in specific environmental contexts, (i.e., the capability to reproduce and propagate the genotype of its carriers), it will be subject to natural selection, which is defined as the differential contribution of genetic variants to future generations. The three general types of selection are positive, purifying, and balancing selection.

When a new mutation increases the fitness of its carriers, positive selection (adaptive evolution) drives the allele to high frequency in a population. Lactase tolerance provides a good example of positive selection.[6,30] Purifying selection, also called negative selection, drives deleterious alleles to low frequency or extinction. Balancing selection occurs when an allele has heterozygote advantage, or it is selected only when it reaches a specific frequency (frequency-dependent selection).[31] One of the best examples of balancing selection is illustrated by the variation in the hemoglobin gene, in which heterozygosity of a variant gene confers resistance against malaria infection, whereas homozygosity results in sickle cell anemia.

Because selection changes rates of molecular evolution at defined loci within the genome, not all genes are expected to evolve at the same rate. Comparison of mammalian genome sequences has permitted the identification of genes that have undergone accelerated evolution.[32,33] These rapidly evolving genes are assumed to enable adaptation and, thus, to have been positively selected because adaptive mutations expand within populations at accelerated rates relative to neutral mutations. The proportion of amino acid substitutions that result from positive selection is estimated to be 35% to 45%.[34] Specific examples of adaptive evolution include glucose-6-phosphate dehydrogenase (G6PD) in malaria,[35] the lactase gene *(LCT)* in lactase persistence,[36] amylase in starch digestion,[37] and C-C chemokine receptor 5 (CCR5) in immune defense.[38]

Comparison of mammalian genome sequences provides evidence that environmental exposures, including pathogens and dietary components, have been selective forces throughout evolution. These selective forces have influenced the generation of polymorphic alleles that alter the use and metabolism of dietary components and may be responsible for the generation of metabolic disease alleles across ethnically diverse human populations.[1,35] Variations that result from positive selection are expected to arise from region-specific selective factors. Therefore, the prevalence of specific functional polymorphisms may be associated with specific geographic or ethnic human populations to the degree that different selective pressures are operative across populations.

Specific allelic variants may be adaptive only in certain environments and neutral or less favored in others. For example, the relatively high prevalence of the E6V polymorphism in the β-globin gene is likely the result of an adaptation to the region-specific environmental challenge of the malaria parasite in African populations. This disease allele has high frequency in the population because of its enhanced fitness toward the region-specific environmental challenge of malaria in heterozygotes. Identifying and understanding the mechanism for the adaptive evolution of gene variants facilitate the discovery of human disease alleles. For example, a "thrifty gene" hypothesis was proposed to explain the epidemic of obesity and type 2 diabetes.[10] The putatively advantageous mutations may have resulted in more efficient adaptations to fasting conditions (e.g., more rapid decreases in basal metabolism) or physiologic responses that facilitate excessive intakes in times of plenty. Adaptive alleles may be recessive disease alleles, or they may become disease alleles, even in heterozygous, when the environmental conditions change profoundly, such as those brought about by the advent of civilization and agriculture, including alterations in the nature and abundance of the food supply.[1]

Classifying Human Genetic Variation

The primary sequence of the human genome contains approximately 3.2 billion nucleotide base pairs that are organized into chromosomes that range in size from 50 million to 250 million base pairs. The first human genome sequence was obtained from 5 to 10 persons of diverse ethnic and geographic backgrounds or ancestry.[39] The human genome, including both nuclear and mitochondrial DNA, contains an estimated 20,000 to 23,000 genes that serve as templates for 35,000 transcripts that encode information required for the synthesis of all cellular proteins, although a biologic function has not yet been determined for all human genes.[40,41] Other genes encode functional RNA molecules, including tRNAs, small nuclear RNAs, ribosomal RNA, and microRNA, which serve various roles in protein synthesis, mRNA processing, or gene expression regulation.[42]

Genes account for approximately 2% of the total human DNA primary sequence; the remaining DNA is termed *noncoding* and serves structural and/or regulatory or no known roles. The number of genes encoded within the genome does not limit the biologic complexity of the mammalian cell. A single gene can encode more than one RNA or protein product through post-transcriptional and post-translational processing reactions, including RNA editing, alternative splicing, protein splicing, and other modifications (e.g., differential phosphorylation).[43] As a result of such RNA and protein processing, and modification reactions, more than 100,000 proteins with distinct primary sequences can be derived from the human genome.

Human genetic variation is a product of complex and reciprocal interactions among the genome and environmental exposures and is manifested through the formation and propagation of primary sequence alterations in DNA.[44] Primary sequence variation among humans is a polymorphism and constitutes one of the molecular bases for human phenotypic variation, including variations in human behavior, morphology, and susceptibility to disease.[45]

Polymorphisms arise in populations through the independent and sequential processes of genetic mutation, followed by expansion of the mutant allele within the population. Environment can modify both these processes. Human genetic variation is estimated to be a 1% to 3% difference between any two sets of human chromosomes.[46] Human genetic variations are usually categorized into common and rare, according to the minor allele frequency (MAF, the frequency of the less-common allele) in human populations. Common variations, also called polymorphisms, have a MAF of at least 1% in human populations.[44] Genetic variants meeting the MAF threshold include single nucleotide changes and structural alterations, and they can result from mutations ranging from a single nucleotide base change to alterations of several hundred bases through deletions, insertions, translocations, inversions, and duplications.[20]

Single Nucleotide Polymorphisms

Single nucleotide polymorphisms (SNPs) are the simplest and most common type of polymorphism and are estimated to represent approximately 90% of all human DNA polymorphisms. SNPs differ from somatic mutations in that they are present in the germ line and are, therefore, heritable. SNPs are defined as nucleotide base pair differences in the primary sequence of DNA and can be single base pair insertions, deletions, or substitutions of one base pair for another. Nucleotide substitutions are the most common polymorphism; insertion or deletion mutations occur at one-tenth the frequency.[8,9]

The density of SNPs in the human genome varies within and among human chromosomes and ranges from 1 in 1,000 bases to 1 in 100 to 300 bases. Investigators have estimated that approximately 85 million SNPs exist in human genomes.[40] Nucleotide substitutions within protein coding regions of a gene can be classified as either nonsynonymous substitutions, which result in an amino acid replacement substitution within a protein, or synonymous (silent) substitutions, which do not change amino acid sequence resulting from degeneracy in the genetic code. Nonsynonymous SNPs in coding regions are more functionally relevant because they change the amino acid sequence of the encoded proteins, and they subsequently have the potential to affect virtually every aspect of protein function, including protein folding and stability, enzymatic functions, allosteric regulation, and post-translational modification. However, synonymous substitution can also have important functional consequences by altering mRNA splicing and protein translation efficiency. SNPs in introns, promoters, and intergenic regions may also be involved in regulating gene expression.

SNPs contribute to susceptibility for common diseases and developmental anomalies, and polymorphic alleles have been identified that increase the risk of common disorders, including neural tube defects, cardiovascular disease, cancers, hypertension, and obesity.[45,47] SNPs also influence physiologic responses to environmental exposures including diet,[48] pharmaceuticals,[49] pathogens, and toxins[50] and; therefore, many SNPs have diagnostic value. High-density human SNP maps facilitate the identification of disease risk alleles through gene mapping studies of complex disease, including low-penetrant alleles that make relatively small contributions to the initiation and/or progression of the disorder.

Haplotypes

Genetic variants across the human genome are not always independent of each other. SNPs that are physically close with respect to primary DNA sequence usually do not segregate. As a result of meiotic recombination, DNA sequence and the variation within this sequence are inherited in "blocks." SNPs that are captured within these blocks are said to be in linkage disequilibrium (LD), which is defined as the nonrandom association of alleles at a nearby locus. Inherited blocks of genetic variation are referred to as haplotypes. The size of the haplotype block depends on the number of meiotic recombination events that have occurred historically within a population. Therefore, the average haplotype block length varies among populations because of human evolutionary history: approximately 22 kb for European and Asian populations and approximately 11 kb for African populations.[47,51] However, the pattern of LD is not evenly distributed across the genome. Because genetic variants in the same haplotype tend to be redundant in defining unique genetic variation, investigators have estimated that approximately 1 million SNPs can capture most human genetic variation.[47]

Structural Change and Copy Number Variation

Structural variations are broadly defined as all genomic alterations that are not single nucleotide substitutions, such as insertions, deletions, inversions, block substitutions, duplications, translocations, and copy number variations (CNVs).[52] Retrotransposons are the most abundant transposable elements. Retrotransposons are classified by their sizes into long interspersed nuclear elements (*LINE*), which encode all necessary genetic components for movement within the genome, and to integrate within DNA and short interspersed nuclear elements (*SINE*), which require other transposable elements for mobility. The most abundant *SINE* is 280 base pair *Alu* element. An estimated 1.4 million copies are present in the human genome, and they occupy approximately 10% of the human genomic sequence. More than 1,200 *Alu* elements in the human genome integrated after early human migrations; a new *Alu* insertion occurs with every 200 births.[53] Therefore, current human populations are polymorphic for the presence or absence of these insertions.

Insertion of transposable elements can have significant functional consequences by disrupting genes, altering gene regulation, and contributing to the coding region of nearby genes. New *Alu* insertions directly cause approximately 0.1% of human genetic disorders, including Apert syndrome, cholinesterase deficiency, and breast cancer. Approximately 0.3% of human genetic disease results from *Alu*-mediated unequal homologous recombination events, resulting in other inherited disorders, including type 2 insulin-resistant diabetes and familial hypercholesterolemia.[53] *Alu*-mediated unequal homologous recombination events are inhibited by CpG methylation of the element.

CNV represents a copy number change involving a DNA segment that is approximately 1 kb or larger,[4] excluding those arising from insertion or deletion of transposable elements[37] CNV represents another major source of genetic variation, and it affects more nucleotides per genome than SNPs, with different estimates ranging from 12% to 30% of the genome.[4,54] The estimated genome-wide mutation rate of CNV ranges from 1.7×10^{-6} to 1.0×10^{-4} per locus per generation, which is 100 to 10,000 times higher than nucleotide substitution rates.[54] CNVs may exert their functional impacts through several mechanisms, such as modifying gene dosage, disrupting coding regions, interfering with proper splicing, and altering regulation of a nearby gene. Therefore, CNVs are subject to selection.[4]

When different functional categories are observed, exonic CNVs are subject to the strongest purifying selection, followed by intronic CNVs, and finally, intergenic CNVs.[4] CNVs can be subject to positive selection by contributing to regional adaptation, and they are enriched in genes that function in the immune system and muscle development.[4] Association studies have identified hundreds of CNVs contributing to phenotypic diversity, disease, and drug sensitivity. CNVs are also involved in starch digestion,[55] steroid hormone and xenobiotic metabolism, prostate cancer,[56] nicotine metabolism, regulation of food intake and body weight, neurodevelopment and neurologic disorders, colonic Crohn disease, toxin resistance, coronary heart disease risk, Alzheimer's disease, human immunodeficiency virus infection, and progression of acquired immunodeficiency syndrome.[37,52,56,57]

Functional Consequences of Genetic Variation

The metabolism of individual dietary components is affected by the activity, expression, and stability of protein transporters and enzymes. Polymorphisms affect gene expression as well as the physical and kinetic properties of cellular proteins, and thereby influence the flux through metabolic pathways and the steady-state concentration of reaction intermediates.

Gene Expression

High-throughput gene expression profiling approaches have identified many eQTLs (expression quantitative trait loci) that affect gene expression.[58] eQTLs have been identified in distinct tissues and cell types and contribute to complex traits and diseases. Both *cis-* and *trans*-regulatory polymorphisms influence gene expression differences within or among human populations.[59,60] Both SNPs and CNVs have dramatic influences on gene expression. To compare their relative importance, a study investigated the association between expression of 14,925 genes and genome-wide SNPs and CNVs in HapMap individuals. The results indicate that SNPs and CNVs contributed 83.6% and 17.7% to gene expression variations among these individuals, respectively.[61]

Polymorphisms in the insulin 5' promoter decrease insulin expression and increase the risk of type 1 diabetes; the risk of type 2 diabetes is associated with polymorphisms in the calpain-10 gene promoter.[62] Polymorphisms have been identified that affect the transcription of many metabolic and nutrient transport proteins, including alcohol dehydrogenase (ADH), apolipoproteins, catalase, cytochrome P-450 family members, glucokinase, lipase, and the vitamin D receptor.[63] Retroviral polymorphisms have been identified that influence gene expression by altering promoter methylation status in mice, and the degree of silencing depends on dietary folate and other one-carbon donors.[17]

Global differential gene expression patterns among human populations have also been investigated.[64-67] Genes in the inflammatory pathway and antimicrobial hormonal response are more likely to change expression from population to population; this finding indicates that gene expression differences among human populations could have resulted from local adaptation during human evolution.[1] Another study identified 356 transcript clusters that have differential expression between the CEU and YRI samples.[68] Twenty-seven genes show signals for adaptive evolution in at least one population. Among these 27 genes, certain metabolism-related molecular functions (e.g., lipid binding, metal ion binding, and transcription factor activity) are enriched, thus supporting the idea that differential gene expression among populations could have played important roles in adaptation to regionally specific dietary intake.[46]

Protein Function

The rate of enzyme-catalyzed reactions is determined by the concentration of both enzyme (E) and substrate (S), and by the intrinsic Michaelis-Menten kinetic properties (Michaelis-Menten constant [K_m] and catalytic constant [k_{cat}]) of the enzyme (or transport protein).

$$E + S \rightarrow ES \rightarrow E + P$$

The Michaelis-Menten constant, K_m, refers to the affinity of E for S and is defined as the substrate concentration required for the enzyme to achieve half-maximal velocity. Formation of the ES complex requires productive collisions between the enzyme and substrate and is governed by the law of mass action. Therefore, the rate of an enzyme-catalyzed reaction is usually directly proportional to the molecular concentrations of the reacting substances (both E and S). Breakdown of the ES complex to product (P) is determined by k_{cat}, which refers to the maximal rate of product formation at infinite substrate concentration (all enzymes are present as an ES complex).

Genetic variation influences both the formation of the ES complex and the rate of P generation. Polymorphisms affect the formation of the ES complex by influencing the concentration of E or the affinity of E for S (K_m). SNPs influence the concentration of E by altering its rate of synthesis (gene expression or mRNA stability) or rate of degradation (protein stability and turnover). Nonsynonymous substitutions that affect the K_m alter the concentration of substrate required to drive the formation of ES.

Therefore, SNPs that increase K_m result in the accumulation of metabolic intermediates in cells. SNPs also can influence k_{cat}, which is the rate of formation of P, by affecting the maximum rate of catalysis (conversion of S to P, at infinite S concentration). Alterations in k_{cat} can influence rates of nutrient uptake or clearance of metabolic intermediates and overall net flux through a metabolic pathway in a substrate-independent manner.

Similarly, genetic variation can affect the expression level and function of nutrient transporters and receptors. Functional effects include alterations in the affinity of transporters and receptors for nutrients, which can influence intracellular and plasma nutrient levels. Changes in transporter or receptor activity or abundance of these proteins in the membrane can affect rates of nutrient uptake and clearance. Genetic variation can also influence nutrient uptake and use indirectly by altering the expression and function of signaling peptides and hormones that regulate metabolic pathways.

Identification of Genetic Variation that Affects Nutrient Metabolism and Use

Genomes that confer either nutrient requirements that cannot be met by the mother, or severe metabolic disruptions that impair basic physiologic processes, will be selected against in large part because of fetal loss or the failure to survive to reproduce. Some common SNPs in genes that encode metabolic enzymes are not in Hardy–Weinberg equilibrium (alleles are not inherited at the expected frequency) because the homozygous state reduces fetal viability.[69] Nearly 62% of all human conceptuses are not viable and do not survive to the 12th week of gestation.[70,71] Genomes that survive gestation but confer atypical nutrient requirements or inefficient metabolism may encode one or more disease-causing alleles, and the penetrance of the disease allele may be modifiable by diet. High-dose vitamin therapy can rescue impaired metabolic reactions that result from genetic mutations and polymorphisms that decrease the affinity of substrates and cofactors for the encoded enzyme (K_m). Polymorphic risk alleles that affect nutrient metabolism and use have been identified with candidate gene approaches, and more recently, they have been inferred from unbiased comparative genomic analyses.

Locating Genes of Interest

Linkage Analysis and Genome-Wide Association Study

Linkage analyses and association studies are two commonly used approaches to map causal alleles underlying human traits, nutrient metabolism, and diseases.[72] Linkage analysis investigates candidate regions underlying the trait of interest in normal and affected individuals from the same family and determines whether genetic markers across the genome are inherited together with the trait. The tool has its limited power for research in complex diseases because sample sizes are usually small. The resolution can be so low that it is difficult to narrow candidate regions.[73,74]

Association studies investigate the coinheritance of genetic markers and the trait of interest in large population studies.[44,73,75] The candidate gene approach, which is frequently used in epidemiologic studies, is a type of direct association study that tests correlations between each candidate causal variant with the trait of interest. This methodology has better resolution than linkage analysis, but it can be greatly limited by the knowledge on the trait of interest. Candidate genes are selected based on knowledge of metabolic pathways and predictions that their impairment results in metabolic phenotypes that either mirror a particular disease state or affect the concentration of a biomarker associated with chronic disease. The candidate gene approach has been successful in identifying many disease susceptibility alleles,[76] but it is limited by incomplete knowledge of transcriptional and metabolic networks and interacting environmental factors that lead to inconsistent findings among studies. Furthermore, because many SNPs are in LD, it is not always possible to determine with certainty whether an individual SNP or allele is functional and disease causing or is linked to a causal polymorphism through genetic hitchhiking.

The genome-wide association study (GWAS) is an indirect approach that does not require prior knowledge of candidate genes underlying the trait of interest. The method uses a set of genetic markers, currently more than 1 million SNPs across the whole human genome, to detect associations between a particular genomic region and the trait of interest by using thousands of or even tens of thousands of normal and affected individuals.[54] Facilitated by SNPs identified through the HapMap project and the development of large-scale genotyping platforms, GWAS has been widely used and has generated candidate loci that may be causal for various complex diseases.[77-79] The list of traits examined by GWAS provide novel hypotheses for disease initiation and progression.

Adaptive Evolution and Genome-Wide Selection Scans

Adaptive evolution may have played important roles in determining human-specific traits that are different from other closely related primate species and human population-specific traits, such as appearance, disease susceptibility, and dietary response. Detecting positively selected alleles provides another approach to facilitate identifications of genes that play important roles in determining these traits.[80] Genetic adaptation during evolution can lead to unique features in a genome that are different from neutral expectation.

Statistical methods have been developed to identify these adaptive signals. The methods to detect adaptive evolution can be grouped into species comparison, using divergence data, and population comparison, using polymorphism data.[80] Methods for interspecies comparison include the following: the dN/dS or Ka/Ks test,[18,81] which searches for a significantly elevated ratio of nonsynonymous to synonymous changes in gene coding regions; and the Hudson, Kreitman, and Agaude (HKA)[82] and McDonald–Kreitman (MK)[6] tests, which identify significantly different distributions of genetic polymorphism within species compared with divergence among species.

Population-based methods can also be grouped into two categories: "frequency-spectrum based" and "haplotype based." Both positive selection and negative selection reduce genetic variation in selected regions: positive selection increases the frequency of advantageous alleles, and negative selection removes deleterious mutations. Different selection tests, such as Tajima's D test,[83] Fu and Li's test,[84] and Fay and Wu's H test,[85] have been developed to detect such reduction in genetic variation that differs from neutral expectation. As human ancestors migrated out of Africa and colonized different locations, human populations evolved in isolation, and allele frequencies became uniquely distributed in different populations, either from random drift or from local adaptation. The F_{st} test, designed to detect such population differences, provides potential target loci that were subject to regional adaptation.[86] Furthermore, the previously mentioned MK test for species comparison can also be modified to compare polymorphism data among populations.

The haplotype-based approaches to detect positive selection were made possible by the success of the HapMap project and further development of large-scale genotyping capability. Adaptive evolution increases the frequency of alleles faster than the neutral expectation. Therefore, the genetic variants in the same haplotype with the selected alleles will also increase in frequency. During this process, recombination does not have sufficient time to break down the haplotype as efficiently as it does under neutral expectation. As a result, adaptive evolution can lead to significantly longer haplotypes than the neutral expectation in the genome with high frequency in a population. Different haplotype-based methods, such as extended haplotype homozygosity (EHH) and relative EHH (REHH),[19] integrated haplotype score (iHS),[31] Cross Population EHH (XP-EHH),[19] and LD decay (LDD) test[87] were developed. These methods successfully identified hundreds of genes that may have gone through adaptive evolution in different human populations,[5,24,32,87-93] with many involved in nutrient metabolism.

Genetic Variation and Nutrient Metabolism

One-Carbon Metabolism (Folate Metabolism)

Folate-mediated, one-carbon metabolism is essential for the de novo biosynthesis of purines and thymidylate and the remethylation of homocysteine to methionine. The pathway is important for DNA synthesis and genome methylation.[94] Genetic variants of enzymes in the folate metabolic pathway, including the methylenetetrahydrofolate reductase gene (*MTHFR*)[95] and the methylenetetrahydrofolate dehydrogenase gene (*MTHFD1*),[96] are associated with altered metabolism and increased risk of birth defects in folate-deficient individuals. These deleterious variants may also be beneficial under certain environments. For example, individuals with C677T in *MTHFR* show a decreased risk of developing colon cancer.[97] Both deleterious and beneficial effects of C677T in *MTHFR* are influenced by dietary intake of folate and alcohol, a finding indicating that interactions between genetics and environment are critical for defining disease status of the genetic variants. This example illustrates the role that dietary interventions can play in modifying risks associated with potentially deleterious gene variants. The evolutionary mechanisms leading to distribution of genetic variants in one-carbon metabolism in different human populations remain to be illustrated.

Starch Digestion

CNVs can modify gene dosage and change levels of gene expression. A study of the salivary amylase gene (*AMY1*) illustrated that CNV may have played an important role in dietary adaptation.[98] The gene has extensive variation in copy number, which was found to be positively correlated with AMY1 protein, both among individuals and among human populations. Populations that consume high-starch diets exhibited a higher copy number of *AMY1* genes than populations that consumed low levels of starch. Comparison with other closely related primate species indicates that the increase in the *AMY1* gene copy number occurred in the human lineage. Indeed, the low level of nucleotide divergence among different *AMY1* gene copies indicates a very recent origin of *AMY1* gene duplication (~200,000 years ago). Taken together, these results indicate that adaptation to a regional food system may have played an important role in modulating the human genome and causing genetic variation among human populations.

Alcohol Metabolism

The metabolism of ethanol varies widely among human ethnic populations. Ethanol is oxidized to acetaldehyde by the enzyme ADH; acetaldehyde is subsequently oxidized to acetic acid by aldehyde dehydrogenase (ALDH). Three genes encode the class I ADH (ADH1) isozymes. The active enzyme is a homodimer or heterodimer composed of subunits encoded by ADH1A, ADH1B, and ADH1C. ADH1B and ADH1C are highly polymorphic, and variations in ADH1B show the greatest functional effects with respect to catalytic activity, affinity of the protein for ethanol, and rate of alcohol clearance from tissues. The ADH1B*1 variant predominates in non-Hispanic White and African-American populations, whereas the ADH1B*2 variant predominates in Japanese and Chinese populations. The origin and spread of this protective allele in East Asia were shown to coincide with the emergence and expansion of rice domestication and alcohol production, a finding indicating the role of Neolithic dietary shift in shaping the human genome.[99] The ADH1B*3 variant is mostly restricted to persons of African descent.

The second enzyme in the pathway, ALDH, is also polymorphic. Populations of Asian descent carry a common dominant null allelic variant (E487K) and develop a "flush" reaction when they consume alcohol that results from accumulation of the metabolic intermediate acetaldehyde. Individuals with high-activity ADH or low-activity ALDH2 are at lower risk of alcoholism than other individuals.[89,100,101]

Lactose Tolerance

Lactase hydrolyzes lactose, the main carbohydrate in milk, into glucose and galactose. After weaning, most mammals, including humans, lose the ability to digest lactose because of reduced lactase expression. However, northern European and African pastoralist populations maintain the ability to digest milk lactose into adulthood (lactase persistence [LP]).[102] Two SNPs (C/T-13910 and G/A-22018) identified in the *cis*-regulatory elements of the lactase gene (*LCT*) were shown to be important for the LP phenotype in European populations.[90]

Three other SNPs (G/C-14010, T/G-13915, and C/G-13907) in the regulatory region of the *LCT* gene were identified to be significantly associated with LP phenotype in African populations, a finding indicating that the LP trait evolved independently during human evolution.[91] Haplotype-based natural selection tests show that the adaptive sweep of different lactase variants in European and African populations occurred approximately within the past 7,000 years, consistent with the time when humans established cattle domestication. This classic example of dietary adaptation indicates that human cultural elements, in this case cattle domestication and adult milk consumption, played an important role in shaping the modern human genome.

Iron Metabolism

Hereditary hemochromatosis is a recessive iron storage disease common in populations of European descent, with an incidence of 1 in 300 persons. A common polymorphism in the *HFE* gene (C282Y), which encodes a protein that regulates iron concentration, is associated with the disease phenotype in 60% to 100% of Europeans, although mutations in other genes are also associated with the phenotype. Iron storage diseases exist in Asia and Africa, where the *HFE* C282Y allele is essentially absent. The penetrance of the C282Y *HFE* allele for the iron overload phenotype varies widely among homozygotes, with some persons being asymptomatic. The *HFE* C282Y polymorphism expanded in human populations relatively recently and may have conferred unidentified selective advantages.[92]

Lipid Metabolism

Apolipoprotein E (Apo-E) functions in lipid metabolism and cholesterol transport. The frequencies of three major Apo-E isoforms (E2, E3, and E4) vary in different human populations. These protein isoforms differ in their affinity both for lipoprotein particles and for low-density lipoprotein (LDL) receptors. Investigators estimated that Apo-E allelic variation could account for approximately 7% of variation in cholesterol concentrations in human populations.[93] In controlled dietary studies of low-fat and high-cholesterol diets, serum cholesterol increased in E4/E4 individuals but not in those with E3/E2 and E2/E2 genotypes. In human population studies, carriers of the E2 allele tend to display lower levels of plasma cholesterol than doE4 carriers. Furthermore, the E4 allele is associated with hypercholesterolemia and increased risk of late-onset Alzheimer disease.

Proprotein convertase subtilisin-like kexin type 9 (PCSK9) is a serine protease that regulates the plasma concentration of LDL.[103] Loss of function mutations lower plasma LDL and are associated with reduced risk of cardiovascular diseases.[104] Loss of function mutations, including two nonsense (Y142X, C679X)[105] and two missense (L253F, A443T)[103] mutations, inactivate PCSK9 in some African American individuals and are associated with an approximately 35% reduction in plasma LDL concentration. Another missense mutation, R46L, which also inactivates PCSK9, is common in European Americans.

Conclusion

Individuals respond differently to dietary exposures in the diet-disease relationship, which has led to initiatives to bring more precision into dietary guidance based on individual biologic responses to food and nutrients, and to lower healthcare costs related to diet-related chronic diseases. Food and nutrition have been powerful selective pressures in human evolution, contributing to diet-related chronic disease risk.[1] Early human populations that thrived and expanded adapted to the local food supply, in part through genetic selection. The human migration and globalization of the nature and abundance of the food supply has resulted in individual differences in how individuals respond to foods and diets. Understanding the complex intersections among human genetics and food environments is essential to address the increasing prevalence of diet-related chronic disease.

Acknowledgments

The author acknowledges Zhenglong Gu for contributions to the previous version of this chapter.

References

1. James WPT, Johnson RJ, Speakman JR, et al. J Intern Med 2019;285:533–49.
2. van Spronsen FJ. Mol Genet Metab 2010;100:107–10.
3. Johri P, Pfeifer SP, Jensen JD. bioRxiv 023;04.11.536488.
4. Conrad DF, Pinto D, Redon R, et al. Nature 464:704–12.
5. Bustamante CD, Fledel-Alon A, Williamson S, et al. Nature 2005;437:1153–7.
6. Sabeti PC, Reich DE, Higgins JM, et al. Nature 2002;419:832–7.
7. Campbell JN, Macosko EZ, Fenselau H, et al. Nat Neurosci 2017;20:484–96.
8. Nachman MW. J Genet 2004;83:231–3.
9. Nachman MW, Crowell SL. Genetics 2000;156:297–304.
10. Diamond J. Nature 2003;423:599–602.
11. Blount BC, Mack MM, Wehr CM, et al. Proc Natl Acad Sci U S A 1997;94:3290–5.
12. Linhart HG, Troen A, Bell GW, et al. Gastroenterol 2009;136: 227–35. e3.
13. Chen T, Heflich RH, Moore MM, et al. Environ Mol Mutagen 2010;51:156–63.
14. Moore SR, Hill KA, Heinmoller PW, et al. Environ Mol Mutagen 1999;34:195–200.
15. Shigenaga MK, Ames BN. Antimutagen anticarcin mechan 1993;419–36.
16. Walser J-C, Furano AV. Genome Res 20, 875–82.
17. Waterland RA. (2009) Epigenetics 4, 523–5.
18. Nielsen R, Hubisz MJ, Hellmann I, et al. Genome Res 2009;19: 838–49.
19. Sabeti PC, Varilly P, Fry B, et al. Nature 2007;449:913–8.
20. Frazer KA, Ballinger DG, Cox DR, et al. Nature 2007;449: 851–61.
21. Coop G, Wen X, Ober C, et al. Science 2008;319:1395–8.
22. Myers S, Bottolo L, Freeman C, et al. Science 2005;310:321–4.
23. Logsdon GA, Rozanski AN, Ryabov F, et al. bioRxiv 2023:11.11.566714.
24. Lercher MJ, Hurst LD. Trends Genet 2002;18:337–40.
25. Tishkoff SA, Verrelli BC. Ann Rev Genom Human Gen 2003;4: 293–340.
26. Castro ESMA, Ferraz T, Couto-Silva CM, et al. Mol Biol Evol 2022;39.
27. Fan S, Spence JP, Feng Y, et al. Cell 2023;186:923–39.e14.
28. Lohmueller KE, Indap AR, Schmidt S, et al. Nature 2008;451: 994–7.
29. Scheinfeldt LB, Tishkoff SA. Nat Rev Genet 2013;14:692–702.
30. Černý V, Priehodová E, Fortes-Lima C. Genes (Basel) 2023;14.
31. Voight BF, Kudaravalli S, Wen X, et al. PLoS Biol 2006;4:e72.
32. Nielsen R, Williamson S, Kim Y, et al. Genome Res 2005;15: 1566–75.
33. Keough KC, Whalen S, Inoue F, et al. Science 2023;3801696.
34. Wolfe KH, and Li WH. (2003) Nat Genet 33 Suppl, 255–65.
35. Tishkoff SA, Varkonyi R, Cahinhinan N, et al. Science 2001;293: 455–62.
36. Ranciaro A, Campbell MC, Hirbo JB, et al. Am J Human Genet 2014;94:496–510.
37. Freeman JL, Perry GH, Feuk L, et al. Genome Res 2006;16: 949–61.
38. Murray CJL, Abbafati C, Abbas KM, et al. Lancet 2020;396: 1135–59.
39. Campbell MC, Tishkoff SA. Ann Rev Genom Human Genet 2008;9:403.
40. Auton A, Brooks LD, Durbin RM, et al. Nature 2015;526:68–74.
41. Willyard C. Nature 2018;558:354–5.
42. Alles J, Fehlmann T, Fischer U, et al. Nucleic Acids Res 2019;47:3353–64.
43. Keren H, Lev-Maor G, Ast G. Nat Rev Genet 11:345–55.
44. Feero WG, Guttmacher AE, Collins FS. N Engl J Med 2010;362:2001–11.
45. Taliun D, Harris DN, Kessler MD, et al. Nature 2021;590: 290–9.
46. Zhang W, Dolan ME. Evol Bioinform Online 2008;4:171–9.
47. Manolio TA, Brooks LD, Collins FS. J Clin Invest 2008;118: 1590–605.
48. Stover PJ, King JC. J Nutr 2020;150:3058–60.
49. McCarthy JJ, Hilfiker R. Nat Biotechnol 2000;18:505–8.
50. Clark MA, Springmann M, Hill J, et al. Proc Natl Acad Sci U S A 2019;116:23357–62.
51. Gabriel SB, Schaffner SF, Nguyen H, et al. Science 2002;296: 2225–9.
52. Tuzun E, Sharp AJ, Bailey JA, et al. Nat Genet 2005;37:727–32.

53. Batzer MA, Deininger PL. Nat Rev Genet 2002;3:370–9.
54. Altshuler DM, Gibbs RA, Peltonen L, et al. Nature 2010;467:52–8.
55. Perry GH, Dominy NJ, Claw KG, et al. Nature Genetics 39:1256–60.
56. Xue Y, Sun D, Daly A, et al. Am J Hum Genet 2008;83:337–46.
57. Sebat J, Lakshmi B, Troge J, et al. Science 2004;305:525–8.
58. Zhang J, Zhao H. J Genet Genomics 2023:doi.org/10.1016/j.jgg.2023.05.003
59. Pickrell JK, Marioni JC, Pai AA, et al. Nature 2010;464:768–72.
60. Wittkopp PJ, Haerum BK, Clark AG. Nature 2004;430:85–8.
61. Stranger BE, Forrest MS, Dunning M, et al. Science 2007;315:848–53.
62. Guttmacher AE, Collins FS. N Engl J Med 2002;347:1512–20.
63. Rockman MV, Wray GA. Mol Biol Evol 2002;19:1991–2004.
64. Storey JD, Madeoy J, Strout JL, et al. Am J Hum Genet 2007;80:502–9.
65. Spielman RS, Bastone LA, Burdick JT, et al. Nat Genet 2007;39:226–31.
66. Degtyareva AO, Antontseva EV, Merkulova TI. Int J Mol Sci 2021;22.
67. Chiliński M, Sengupta K, Plewczynski D. Semin Cell Dev Biol 2022;121:171–85.
68. Duan S, Huang RS, Zhang W, et al. Pharmacogenomics 2009;10:549–63.
69. Stover PJ. Food Nutr Bull 2007;28:S101–15.
70. Edmonds DK, Lindsay KS, Miller JF, et al. Fertil Steril 1982;38:447–53.
71. Edwards RG. Int J Dev Biol 1997;41:255–62.
72. Abouzid M, Główka F, Kagan L, et al. Curr Drug Metab 2022;23:630–51.
73. Witte JS. Annu Rev Public Health 2010;31:9–20:4.
74. Risch N, Merikangas K. Science 1996;273:1516–7.
75. Borecki IB, Province MA. Adv Genet 2008;60:51–74.
76. Brandes N, Weissbrod O, Linial M. Genome Biol 2022;23:131.
77. Zhang HF, Qiu LX, Chen Y, et al. Hum Genet 2009;125:627–31.
78. McPherson R, Pertsemlidis A, Kavaslar N, et al. Science 2007;316:1488–91.
79. Samani NJ, Erdmann J, Hall AS, et al. N Engl J Med 2007;357:443–53.
80. Kelley JL, Swanson WJ. Annu Rev Genomics Hum Genet 2008;9:143–60.
81. Yang Z, Nielsen R. J Mol Evol 1998;46:409–18.
82. Hudson RR, Kreitman M, Aguade M. Genetics 1987;116:153–9.
83. Tajima F. Genetics 1989;123:585–95.
84. Fu YX. Genetics 1997;147:915–25.
85. Fay JC, Wu CI. Genetics 2000;155:1405–13.
86. Weir BS, Hill WG. Annu Rev Genet 2002;36:721–50.
87. Wang ET, Kodama G, Baldi P, et al. Proc Natl Acad Sci U S A 2006;103:135–40.
88. Carlson CS, Thomas DJ, Eberle MA, et al. Genome Res 2005;15:1553–65.
89. Crabb DW, Matsumoto M, Chang D, et al. Proc Nutr Soc 2004;63:49–63.
90. Enattah NS, Sahi T, Savilahti E, et al. Nat Genet 2002;30:233–7.
91. Tishkoff SA, Reed FA, Ranciaro A, et al. Nature Genet 2007;39:31–40.
92. Toomajian C, Kreitman M. Genetics 2002;161:1609–23.
93. Inoue K, Lupski JR. Annu Rev Genomics Hum Genet 2002;3:199–242.
94. Chon J, Stover PJ, Field MS. Mol Aspects Med 2002;53:48–56.
95. Christensen KE, Rohlicek CV, Andelfinger GU, et al. Hum Mutat 2009;30:212–20.
96. Brody LC, Conley M, Cox C, et al. Am J Hum Genet 2002;71:1207–15.
97. Mills PB, Camuzeaux SSM, Footitt EJ, et al. Brain 2014;137:1350–60.
98. Perry GH, Dominy NJ, Claw KG, et al. Nat Genet 2007;39:1256–60.
99. Peng Y, Shi H, Qi XB, et al. BMC Evol Biol 2010;10:15.
100. Bosron WF, Li TK. Hepatol 1986;6:502–10.
101. Loew M, Boeing H, Sturmer T, et al. Alcohol 2003;29:131–5.
102. Itan Y, Jones BL, Ingram CJ, et al. BMC Evol Biol 2010;10:1–11.
103. Kotowski IK, Pertsemlidis A, Luke A, et al. Am J Hum Genet 2006;78:410–22.
104. Horton JD, Cohen JC, Hobbs HH. Trends Biochem Sci 2007;32:71–7.
105. Cohen JC, Boerwinkle E, Mosley TH, Jr., et al. N Engl J Med 2006;354:1264–72.

A. NUTRIENT–GENE MECHANISMS

CHAPTER 40

Epigenetics

Jaclyn M. Goodrich and Dana C. Dolinoy

CHAPTER OUTLINE

Abbreviations

IGF-2, insulin-like growth factor-II gene; **SINE**, short interspersed nuclear element; **LINE1**, long interspersed nuclear elements; **IAP**, the intracisternal A particle; ***BRCA1***, breast cancer 1 early onset gene; **CpG**, cytosine-phosphate-guanine; **TET**, ten-eleven translocation methylcytosine dioxygenase; **DNMT**, DNA methyltransferase; **ncRNA**, noncoding RNA

Introduction to the Epigenome

The human genome contains information that is not fully described by the DNA sequence alone. Epigenetics encompasses a collection of mechanisms that influence gene expression without affecting the underlying genetic sequence.[1] This epigenetic information (from the Greek ἐπί, meaning "on" or "over"), or epigenome, is encoded in molecular modifications to DNA and chromatin. It fundamentally affects the way in which the information encoded in the DNA sequence is used. The epigenome is essential to the identity and the healthy functioning of cells. Epigenetic processes have been implicated in a wide range of health outcomes including cancer, cognition, cardiovascular disease, diabetes, and reproductive function. Our understanding of the effect of environmental factors—such as diet, the social and built environments, and chemical exposures—on the epigenome is growing rapidly.

In molecular terms, the epigenome represents a range of mechanisms including DNA methylation, histone modification, remodeling of nucleosomes and higher order chromatin reorganization, and regulation by noncoding RNAs.[1,2] A key characteristic of the epigenetic signal is that it is heritable and can be passed from somatic cell to daughter cell during mitosis.[1-4] There is some evidence that certain epigenetic signatures escape reprogramming following meiosis and are passed on to subsequent generations, a phenomenon called transgenerational inheritance.[5] While understanding of the epigenetic regulation of individual genes has increased greatly, coordinated epigenetic control of the genome on a much larger scale may be even more important. The human genome is made up of accessible regions of euchromatin and poorly accessible regions of heterochromatin, and these regions determine the ability of the transcriptional machinery of the cell to access genetic information. These regions may span many genes, and epigenetic regulation is critical to the transition between these states.[4]

DNA methylation is the most widely studied epigenetic mechanism in relation to nutrition. Methylation in mammalian cells typically takes place at a cytosine located 5′ to a guanosine (CpG site). A significant component of the global methylation signature (average level of methylation across the entire genome) is accounted for by the transposable elements that make up approximately 45% of the entire genome and are usually heavily methylated (~90%). The transposons include the long, interspersed nuclear elements (*LINE1*), the intracisternal A particle (*IAP*), short interspersed nuclear elements (*SINE*), and the *Alu* family of human *SINE* elements characterized by the action of the Alu restriction endonuclease.[6] Some classes of transposon can move around the genome and can cause abnormal function and disease if they are inserted into an important conserved sequence.[4,6] Within the genes that code for proteins, the most striking epigenetic distinction is between the imprinted and the nonimprinted genes. Most autosomal genes are expressed equally from both parental alleles, but imprinted genes are an exception.

Genomic imprinting refers to the epigenetic marking of genes in a manner specific to the parent of origin within the germ cells such that the subsequent expression pattern depends on the parent from whom the allele was derived.[1,3,4] Imprinted genes are particularly important in prenatal growth, placental function, and brain function and behavior.[7,8] Imprinted genes are usually found downstream of regions of DNA that have a high density of CpG sites.[4] Approximately 80% of imprinted genes are found in clusters with other imprinted genes, and this arrangement reflects coordinated regulation of the genes within a chromosomal domain.[4]

Regions rich in CpG sites, known as CpG islands, are found in gene bodies, endogenous repeats, and transposable elements, and are thought to be important in transcriptional repression.[2,9] The process of demethylation is, in many ways, as important to epigenetic regulation as methylation. Demethylation can occur passively or actively through ten-eleven translocation (TET) enzymes. It is now known that hydroxymethylation, an intermediate in the demethylation pathway, is also a stable epigenetic modification.[10,11] While hydroxymethylation is less abundant than methylation, it is a stable DNA modification, with detectable levels in the mammalian brain, liver, kidney, testes, placenta, colon, blood, and embryonic stem cells.[12-16] The functional roles of hydroxymethylation and methylation in gene regulation seem to be distinct.[17]

Post-translational histone tail modifications, also referred to as the histone code, make up another key category of epigenetic marks that dictate chromatin packaging.[18,19] The histone code works in concert with DNA methylation to regulate gene expression by altering the accessibility of transcriptional machinery to a given gene. We are learning more about nucleosome dynamics each day, with dozens of different types of chemical modifications known so far.[20]

Noncoding RNAs (ncRNA) are also important mechanisms of epigenetic regulation. ncRNA are highly conserved, regulating gene expression at the levels of transcription, translation, and splicing and serving as regulators for transposable elements.[21-23] Dysregulation in ncRNA systems has been implicated in cardiovascular disease, neurologic disorders, and cancer.[24-26] Additionally, small ncRNA serves as biomarkers for a variety of environmental factors that relate to human health outcomes later in life.[27-29] RNA interference (RNAi) encompasses two major mechanisms that regulate gene expression at the epigenetic and messenger RNA level, respectively: RNA-induced transcriptional silencing (RITS) and the RNA-induced silencing complex (RISC). RITS mechanisms have been routinely associated with increased DNA methylation and repressive histone modifications.[30-32] While RITS often induces chromatin modifications to regulate transcription, the RISC complex targets mRNA transcripts for cleavage and degradation, preventing translation.[33,34] Both RITS and RISC mechanisms are carried out by Argonaute proteins, which have been shown to induce epigenetic modifications in heterochromatin, as well as deadenylate and cleave target mRNAs to induce degradation.[35]

Epigenetic status varies among individuals[36] and is influenced by factors such as age, biological sex, socioeconomic status, and more.[37-39] The epigenome even differs between genetically identical monozygotic twins,[40] and these differences become more pronounced over the lifespan as twins' exposures and circumstances diverge. Much research has been carried out to identify the proportion of epigenetic variation that is influenced by nutrition. The objective of this chapter is to introduce epigenetics and its role in health and disease. We then discuss the broad mechanisms whereby nutrition can impact the establishment and maintenance of the epigenome. We provide examples from experimental and epidemiologic studies showcasing nutrient or diet-induced alterations to the epigenome. We also introduce nutriepigenomics—how nutrients impact the epigenome—and its promise for advancing precision health in years to come.

The Role of Epigenetics in Health and Disease

Epigenetic change has been implicated in all the major chronic diseases affecting humans, and there is evidence that epigenetic perturbations as early in life as gestation contribute to disease.[41] Historically, cancer is the disease in which epigenetics has been studied most extensively. A common observation in human tumors is epigenetic change, including altered methylation of DNA[42,43] and the histones associated with DNA.[44] Broad hypomethylation in tumor cells is thought to be an early trigger that predisposes cells to genomic instability, along with hypermethylation of tumor suppressor gene promoters. There is an epigenetic contribution to nearly every hallmark of cancer; epigenetic modification is thought to be one of the "key characteristics" by which carcinogenic chemicals exert their effects.[45-47] Advances in technology have allowed for exquisite profiling of tumor epigenomes, which has revealed common features across cancers and also unique epigenetic features that distinguish between cancer types and subtypes.[48,49] The role of epigenetic perturbation in carcinogenesis has led to the design of therapeutic interventions that target epigenetic processes and specific diets that may be chemopreventive, due to their influence on the epigenome.[50,51]

Imprinted genes play a role in many diseases. Certain imprinted genes are known tumor suppressors involved in cell proliferation. Loss of imprinting (gain or loss of DNA methylation or loss of allele-specific gene expression) is a common characteristic of cancer types, including breast, lung, colon, liver, and ovary.[52] Imprinting syndromes, in which the imprint is disrupted or absent, are associated with transient neonatal diabetes,[53] and cancer risk,[54] in addition to impairment of normal function that leads to obesity and impaired cognitive development.[55] There are currently 129 known imprinted genes in humans with >100 more predicted to be imprinted; many of these genes play key roles in development and disease.[56]

Epigenetic mechanisms contribute to a variety of diseases, and developing pharmacologic interventions to target epigenetic changes is an active area of research for many disease types. Epigenetic mechanisms have been implicated in neurodegenerative disorders (i.e., Alzheimer's disease) and neuropsychiatric conditions.[57-59] Epigenetic alterations associated with cardiovascular and metabolic diseases are well-established.[60] In addition to associations with specific genes with disease-relevant functions, broad epigenetic aging is associated with many diseases. The epigenome changes with age as physiologic function changes and declines over the lifespan. The study of DNA methylation across the lifespan has led to the development of epigenetic age biomarkers, referred to as epigenetic clocks.[61] Epigenetic clocks can be used to infer whether epigenetic age appears accelerated compared with chronologic age. Importantly, recent evidence suggests that these clocks can accurately estimate an individual's biological age.[62] Epigenetic age acceleration is associated with cancers, mortality, cardiovascular disease, and additional health endpoints in multiple large cohort studies.[63-65] Identifying environmental drivers of accelerated aging and preventive measures—including dietary interventions—that protect against biologic aging are important areas of research.

Epigenetic Reprogramming and Multi- and Transgenerational Inheritance

Two distinct waves of epigenetic reprogramming occur during early development, making this period vulnerable to nutritional and environmental perturbations.[66,67] The first wave of reprogramming occurs postfertilization, in which parental epigenetic marks are erased and the somatic epigenetic marks of the developing embryo are established.[68,69] The second wave of reprogramming occurs in the fetal primordial germ cells, in which sex and parent of origin-specific epigenetic patterning are established.[70] Nutritional and environmental disruptions in the normal epigenetic patterning in germ cells that are not corrected during the development of the future generation may result in multi- and transgenerational effects.[71] In a pregnant (F0) individual, environmentally induced epigenetic changes may occur in their somatic and germ cells, as well as in the somatic and primordial germ cells of the developing fetus.[71] Exposure-induced effects occurring in the offspring (F1) and grand offspring (F2) generations are termed multigenerational effects, as F0 and F1 generations receive direct exposure during the pregnancy, and the primordial germ cells of the F2 generation are also directly exposed.[71] In contrast, transgenerational epigenetic inheritance refers to the transmission of nutritional and/or other environmental effects to future generations that did not receive the exposure directly through embryonic development or germ cells, i.e., F3 generations and beyond.[71] Transgenerational effects of environmental exposures have been reported in nematodes, plants, fruit flies, and mammals.[71-74] However, the prevalence of this phenomenon and its relevance to health, in particular human health, is the subject of significant debate and ongoing study.[75,76] Most transgenerational studies have focused on chemical exposures and not dietary components.

Nutritional Effects on the Epigenome

The epigenome can be influenced by various kinds of exposures. Overall nutrition and specific dietary components are part of the "exposures" that can influence the epigenome. There are various mechanisms by which dietary and nutritional factors can influence the epigenome. In general, there are several major mechanisms that nutrient availability and dietary composition can work through. Nutrition can influence the availability of substrate used to mark DNA and histones epigenetically (i.e., methyl donors). Dietary components can have direct effects on the cellular machinery (i.e., methylating enzymes) involved in setting and interpreting epigenetic marks. They can also have direct effects on the structure and function of the genome.

The ultimate methyl donor for epigenetic methylation reactions is the folate-methylation cycle, and specifically the metabolite *S*-adenosylmethionine (SAM). Nutritional and genetic factors that affect the activity of this cycle also influence epigenetic marking. Poor folate status and elevated homocysteine have been linked to human DNA lymphocyte hypomethylation.[77] Mutation in the methylenetetrahydrofolate gene, which is involved in the provision of methyl groups, interacts with folate status to influence DNA methylation.[78,79] Direct effects of folate on the structure and function of the genome related to epigenetics are also possible. The human genome has more than 20 folate-sensitive fragile sites, which are regions of chromatin that fail to compact normally during mitosis in the presence of folate and thymidine deficiency.[80]

Other B vitamins have been implicated in epigenetic regulation. Examples include niacin and chromatin structure and function,[81] as well as biotin binding to histones and its effect on retrotransposons.[82] Acetylation of histones is under the control of histone deacetylase, which is inhibited by sulforaphane, a compound found in cruciferous vegetables.[83] Polyphenols in green tea, coffee, and soybeans are thought to influence epigenetic status by a direct effect on the DNA methyltransferases that add the methyl group to DNA.[50]

Windows of Sensitivity to Epigenetic Perturbation by Nutrition

Much of what we know about nutrition and the epigenome comes from studies of life stages vulnerable to the impact of external factors on epigenetic programming and maintenance. Many epigenetic events are restricted to specific phases of development, cellular differentiation, and cell division. Epigenetic regulation is central to the coordinated development of human gametes, the early embryo, and the fetus. The entire period before birth is marked by intense epigenetic activity, including a global demethylation event followed by remethylation and epigenetic resetting in the first trimester.[1,84] Extensive research in the field of nutritional epigenetics has focused on the long-term health consequences of nutritional exposure before birth due to the long-term impact that early life exposures can have on widespread epigenetic programming of the offspring.

Numerous studies in pregnant rodents have demonstrated that epigenetic regulation of specific genes in the offspring was influenced by the maternal intake of methyl donors such as folic acid, choline, betaine,[82,83] folic acid and low protein,[85] and phytoestrogens[86] during pregnancy. In human pregnancy, higher levels of methylation of the insulin-like growth factor-II gene *(IGF-2),* an imprinted gene important for fetal growth, have been observed in the umbilical cord blood DNA in babies of mothers who took folic acid supplements during pregnancy.[51] One of the most striking examples of the potential lasting impact of nutrition on the epigenome comes from a study of individuals exposed to the Dutch Hunger Winter famine of 1944–1945. Compared with same sex-siblings born outside of the famine window, altered *IGF-2* methylation was observed in the blood of individuals 60 years after early gestation exposure to the famine, but individuals exposed during late pregnancy did not exhibit the same effect. Additional differentially methylated regions in metabolically relevant genes were also identified, and the famine was associated with increased risk for breast cancer, coronary heart disease, high cholesterol, and more, in adulthood.[87-89]

Windows of epigenetic sensitivity to nutrition are not restricted to the period before birth but may occur throughout the lifespan.[90,91] The nutritionally programmed epigenome could become fixed and propagated during mitosis or meiosis in several ways (**Figure 40.1**).

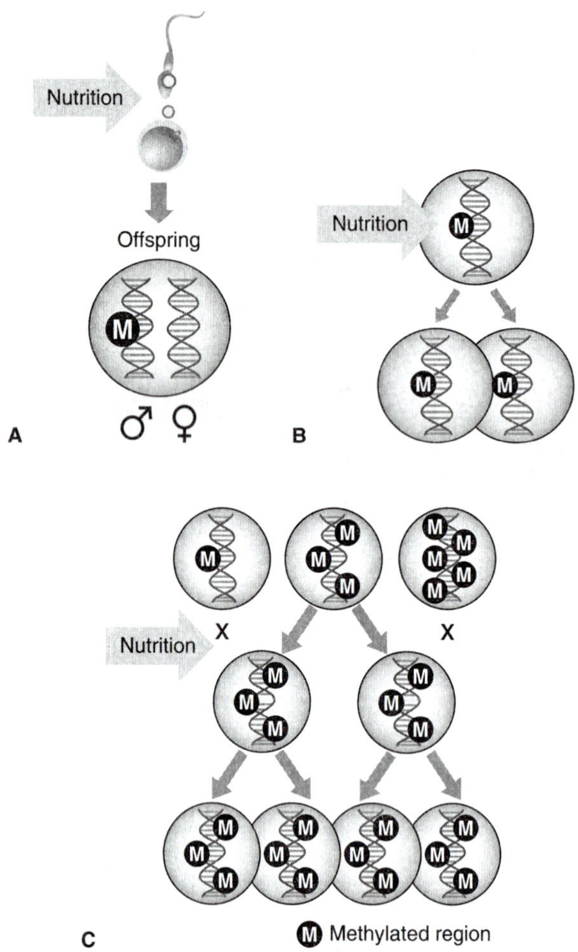

Figure 40.1 Potential mechanisms by which nutritionally programmed epigenetic status may be fixed and propagated. A. Exposure during irreversible marking events (e.g., imprinting). B. Exposure during transcriptional activity or key epigenetic processes (e.g., mitosis). C. Clonal selection following the generation of a range of cellular epigenotypes in response to nutritional exposure.[52]

Nutrition and Epigenetics: An Epidemiologic Perspective

It is important to understand whether epigenetic change is on the causal pathway to the development of disease and whether this process is influenced by nutrition in humans. Both harmful and beneficial diets or dietary components are necessary to elucidate as these can provide direction for encouraging avoidance of certain foods and promoting others. Causal relationships are relatively easy to establish in animal models, but species differences in epigenetic regulation and the origin of disease limit the usefulness of such models when investigating the determinants of human health. On the other hand, establishing causality in human nutritional studies presents its own challenges.

Humans, like all multicellular organisms, have multiple epigenomes. The epigenome varies based on the tissue type and stage of life. In most nutritional studies, investigators can sample only peripheral blood or buccal cell DNA. The critical tissues and organs that regulate and determine health and disease (liver, pancreas, heart, colon, brain, etc.) can be sampled only in the most invasive of protocols or in very specific study designs (e.g., detection of the epigenetic signature within minute traces of tumor tissue DNA released into the peripheral blood in cancer studies).[92] The rationale for blood and buccal cell sampling is that epigenetic status within these cells is a) indicative of key epigenetic events in the tissues and organs of interest, b) a biomarker of the impacts of exposure, and/or c) a biomarker predicting disease.

Epigenetic Biomarkers

Ideal epigenetic biomarkers in the field of nutrition will be responsive to dietary component(s) and predict future health and disease risk. It is preferable if the biologic basis for the biomarker response to nutrition and the mechanism linking this biomarker to health are known, but this knowledge is not essential. Epigenetic signals within blood and buccal cells may be useful biomarkers if they can be shown to predict disease or detect covert disease, irrespective of whether the mechanism has been established.

For some, but not all, genes the epigenetic signal in peripheral blood and buccal cells reflects epigenetic status in other tissues of interest. Researchers have observed associations between DNA methylation in blood with disease states, and some loci replicate in relevant target tissues, including adipose, skeletal muscle, and brain.[93-95] Loci that replicate across tissues hold the most promise for understanding mechanisms of disease. In genes that differ across tissues, associations between dietary components and blood DNA methylation can instead provide insight into whether an exposure impacts the epigenome. Epigenetic biomarkers in blood can also be predictive of disease. For example, DNA methylation of the breast cancer 1, early onset gene *(BRCA1)* is altered in tumor cells; changes have been detected in apparently normal epithelium adjacent to breast cancer as well as in peripheral blood and buccal cells of women with the disease or at increased risk.[96,97]

We expect some nutrients to have a broad impact on an epigenetic mark throughout the genome. For example, disruptions within the one-carbon metabolism cycle could lead to an inadequate amount of methyl donors for maintenance of DNA and histone methylation levels across tissues. Global DNA methylation has been demonstrated to be responsive to the intake of nutrients that influence the availability of methyl groups, and these global changes can be detected in human blood.[77,79] Global methylation may influence health by effects on repression of repetitive elements,[6] CpG islands,[9] and overall genome stability.[80] However, the high degree of covariance among folate status, homocysteine, B-vitamin–related genotype, and global methylation makes it difficult to identify causal mechanisms, linking nutrition to disease based on average DNA methylation levels. Thus, evaluating the association between nutrients and global methylation can be a first pass to assess the nutrients' broad impact, but more information is needed to link this finding to health and disease.

To identify specific loci/gene biomarkers associated with nutrition and/or disease, there are multiple gene-specific and genome-wide methods available to profile DNA methylation, noncoding RNAs, and histone modifications. We focus on DNA methylation here. Quantitative methods for DNA methylation at targeted genes include pyrosequencing[98] and the mass spectrometry based EpiTYPER assay,[99] which can be used to interrogate short regions of DNA. Recent advancements in the field include targeted bisulfite sequencing, which uses next-generation sequencing after amplification of genes of interest to provide coverage of longer segments of genes simultaneously.[100,101] Due to decreasing costs and rapid advancement of technologies for DNA methylome-wide screening, more epidemiologic studies are opting for a discovery-based approach. The most common tools used in such epigenome-wide association studies are probe-based arrays from the Illumina Infinium series. The most recent version, the EPIC array, quantifies DNA methylation at >850,000 CpG sites throughout the genome.[102] Typically, human studies run these arrays using DNA isolated from blood leukocytes or buccal cells.

Recent studies evaluating the impact of certain dietary components on adult DNA methylation using the Infinium arrays have provided novel insights. For example, in a study of overall diet quality, 24 loci were identified where the Alternative Healthy Eating Index was associated with DNA methylation levels. Importantly, these loci replicated across two cohorts, and the significant loci included genes important for adiposity and inflammation.[103] An analysis across three cohorts of middle-aged adults reported associations between multiple dietary components indicative of a Western diet (i.e., milk, cream, plant oils, butter, and alcohol) and DNA methylation at specific loci.[104] These and other studies point to the potential for a vast array of dietary components—not only nutrients that directly feed into the one-carbon cycle—to influence the epigenome. Links between diet and epigenetic aging have also been made. For example, adherence to the Dietary Approaches to Stop Hypertension (DASH) diet was associated with decelerated epigenetic aging, and this mediated the link between DASH and all-cause mortality among former and current smokers.[105]

Echoes of Early Exposure in Cohort Studies

Some epigenetic marks set very early in development are passed down through the somatic cell lineage such that, many divisions later, different cell types still carry the original signal. Thus, nutritional exposures during very early life have the unique potential to influence epigenetic programming across all developing tissues. The most striking example of this is found within imprinted genes and metastable epialleles. In both cases, epigenetic regulation is set at early stages of development and may be retained in multiple tissues throughout the entire life span. As described previously, famine conditions around the time of conception have been associated with imprinted gene DNA methylation measured decades later in blood; in a study conducted in the Gambia, methylation at metastable epialleles, measured in postnatal blood was also associated with famine conditions.[87,106]

Beyond imprinted genes and metastable epialleles, epigenome-wide association studies using Infinium arrays have identified many differentially methylated genes that are associated with gestational dietary exposures. In a study focused on allergy risk, perinatal probiotic and/or omega-3 fatty acid supplementation led to hypermethylation of multiple genes in neonatal T helper cells extracted from cord blood.[107] Maternal vitamin D supplementation throughout pregnancy and lactation was associated with altered methylation at hundreds of genes in infant blood leukocytes.[108] These and other epidemiologic studies showcase the importance of the gestational nutrition on establishment of the offspring epigenome.

Nutritional Epigenetics of the Future

Advances in epigenetics have given us an appreciation of the epigenome's role in development and disease, and the potential for many environmental factors—including specific nutrients and dietary patterns—to modify it. Many studies to date have been descriptive, characterizing the changes to the epigenome (and primarily to DNA methylation) that are associated with specific nutrients or dietary patterns. The future of the field, "nutriepigenomics," bridges nutritional sciences and epigenomics and a precision health paradigm to build upon and translate knowledge to improve health.[109] This field has enormous potential to lead to safe interventions that will target the epigenome and reverse progression of diseases like cancer.

Natural products—or therapeutics consisting primarily of compounds found in food and in nature—can modify epigenetic processes. Because of this, natural products are being developed to stop or prevent carcinogenic and other disease processes via epigenetic modification. For example, curcumin, resveratrol, genistein, and others are thought to regulate microRNAs that inhibit cancer cell growth. Therapies based on these compounds are in development.[110] In a cell model, docosahexaenoic acid repressed histone deacetylases and reduced expression of inflammatory genes in macrophages.[111] If replicated in vivo, docosahexaenoic acid supplementation could protect against inflammatory diseases. In addition to targeting specific diseases, "epigenetic diets" have been proposed that would serve as preventive diets—reducing cancer risk and promoting healthy aging by decreasing damage to the epigenome. Bioactive components believed to have a positive influence on the epigenome include epicatechins from green tea, genistein, selenium, folate, curcumin, and resveratrol. A diet naturally high in these components could reduce cancer risk and risks for other diseases.[50,112] While evidence is building for these protective diets, they are not yet recommended in standard clinical practice. Advances in nutriepigenomics should focus on solidifying the evidence base for health promotion through dietary epigenetic modifiers and translating these findings to patients and populations.

Acknowledgments

The authors acknowledge Paul Haggerty for his contributions to the previous version of this chapter.

References

1. Sasaki H, Matsui Y. Nat Rev Genet 2008;9:129–40.
2. Illingworth R, Bird A. Fed Eur Biochem Soc 2009;583:1713–20.
3. Jirtle RL, Skinner MK. Nat Rev Genet 2007;8:253–62.
4. Reik W, Walter J. Nature Reviews Genetics 2001;2:21–32.
5. Ben Maamar M, Nilsson EE, Skinner MK. Biol Reprod 2021;105: 570–92.
6. Waterland RA, Jirtle RL. Mol Cell Biol 2003;23:5293–300.
7. Tycko B, Morison I. J Cell Physiol 2002;192:245–58.
8. Wilkinson L, Davies W, Isles A. Nat Rev Neurosci 2007;8: 832–43.
9. Deaton AM, Bird A. Genes Dev 2011;25:1010–22.
10. Sadakierska-Chudy A, Kostrzewa RM, Filip M. Neurotox Res 2015;27:84–97.
11. Bachman M, Uribe-Lewis S, Yang X, et al. Nat Chem 2014;6: 1049–55.
12. Nestor CE, Ottaviano R, Reddington J, et al. Genome Res 2012; 22:467–77.
13. Globisch D, Munzel M, Muller M, et al. PLoS One 2010;5: e15367.
14. Szwagierczak A, Bultmann S, Schmidt CS, et al. Nucleic Acids Res 2010;38:e181.
15. Wu H, D'Alessio AC, Ito S, et al. Genes Dev 2011;25:679–84.
16. Stroud H, Feng S, Morey Kinney S, et al. Genome Biol 2011; 12:R54.
17. Lopez V, Fernandez AF, Fraga MF. Ageing Res Rev 2017;37: 28–38.
18. Strahl BD, Allis CD. Nature 2000;403:41–5.
19. Jayani RS, Ramanujam PL, Galande S. Methods Cell Biol 2010;98: 35–56.
20. Zentner GE, Henikoff S. Nat Struct Mol Biol 2013;20:259–66.
21. Chen L, Dahlstrom JE, Lee SH, et al. Epigenetics 2012;7:758–71.
22. Kalsotra A, Wang K, Li PF, et al. Genes Dev 2010;24:653–8.
23. Watanabe T, Totoki Y, Toyoda A, et al. Nature 2008;453:539–43.
24. Eulalio A, Mano M, Dal Ferro M, et al. Nature 2012;492:376–81.
25. Chang WS, Wang YH, Zhu XT, et al. Med Sci Monit 2017;23: 2721–31.
26. Fuziwara CS, Kimura ET. Front Med (Lausanne) 2015;2:64.
27. Martinez RM, Hauser R, Liang L, et al. Environ Int 2019;123: 20–8.
28. Marczylo EL, Amoako AA, Konje JC, et al. Epigenetics 2012;7: 432–9.
29. de Araujo ML, Gomes BC, Devoz PP, et al. Front Genet 2021;12:620744.
30. Buhler M, Verdel A, Moazed D. Cell 2006;125:873–86.
31. Iida T, Nakayama J, Moazed D. Mol Cell 2008;31:178–89.
32. Ishida M, Shimojo H, Hayashi A, et al. Mol Cell 2012;47:228–41.
33. Gregory RI, Chendrimada TP, Cooch N, et al. Cell 2005;123: 631–40.
34. Lessel D, Zeitler DM, Reijnders MRF, et al. Nat Commun 2020;11:5797.
35. Hutvagner G, Simard MJ. Nat Rev Mol Cell Biol 2008;9:22–32.
36. Bjornsson H, Sigurdsson M, Fallin M, et al. JAMA 2008;299: 2877–83.
37. Bell CG, Lowe R, Adams PD, et al. Genome Biol 2019;20:249.
38. Solomon O, Huen K, Yousefi P, et al. Mutat Res Rev Mutat Res 2022;789:108415.
39. Needham BL, Smith JA, Zhao W, et al. Epigenetics 2015;10: 958–69.
40. Fraga MF, Ballestar E, Paz MF, et al. Proc Natl Acad Sci U S A 2005;102:10604–9.
41. Bianco-Miotto T, Craig JM, Gasser YP, et al. J Dev Orig Health Dis 2017;8:513–9.
42. Vasilatos SN, Broadwater G, Barry WT, et al. Cancer Epidemiol Biomarkers Prev 2009;18:901–14.
43. Feinberg AP, Levchenko A. Science 2023;379:eaaw3835.
44. Fraga M, Ballestar E, Villar-Garea A, et al. Nature Genetics 2005;37:391–400.
45. Darwiche N. Am J Cancer Res 2020;10:1954–78.
46. Smith MT, Guyton KZ, Gibbons CF, et al. Environ Health Perspect 2016;124:713–21.
47. Smith MT, Guyton KZ, Kleinstreuer N, et al. Cancer Epidemiol Biomarkers Prev 2020;29:1887–903.
48. Cancer Genome Atlas Research N. Cell 2015;163:1011–25.
49. Figueroa ME, Lugthart S, Li Y, et al. Cancer Cell 2010;17:13–27.
50. Hardy TM, Tollefsbol TO. Epigenomics 2011;3:503–18.
51. Bates SE. N Engl J Med 2020;383:650–63.
52. Feinberg AP, Ohlsson R, Henikoff S. Nat Rev Genet 2006;7: 21–33.
53. Temple I, Shield J. Med Genet 2002;39:872–5.
54. Lim DH, Maher ER. Adv Genet 2010;70:145–75.
55. Ma VK, Mao R, Toth JN, et al. Appl Clin Genet 2023;16:41–52.
56. Jima DD, Skaar DA, Planchart A, et al. Epigenetics 2022;17: 1920–43.
57. Bakulski KM, Dolinoy DC, Sartor MA, et al. J Alzheimers Dis 2012;29:571–88.
58. Basavarajappa BS, Subbanna S. Int J Mol Sci 2021;22.
59. Rizzardi LF, Hickey PF, Rodriguez DiBlasi V, et al. Nat Neurosci 2019;22:307–16.
60. Wu YL, Lin ZJ, Li CC, et al. Signal Transduct Target Ther 2023; 8:98.
61. Topart C, Werner E, Arimondo PB. Clin Epigenetics 2020;12:97.
62. Fahy GM, Brooke RT, Watson JP, et al. Aging Cell 2019;18: e13028.
63. Wang C, Ni W, Yao Y, et al. EBioMedicine 2021;63:103151.
64. Chen BH, Marioni RE, Colicino E, et al. Aging (Albany NY) 2016;8:1844–65.
65. Lu AT, Quach A, Wilson JG, et al. Aging (Albany NY) 2019;11: 303–27.
66. Kafri T, Ariel M, Brandeis M, et al. Genes Dev 1992;6:705–14.
67. Monk M, Boubelik M, Lehnert S. Development 1987;99:371–82.
68. Rougier N, Bourc'his D, Gomes DM, et al. Genes Dev 1998;12: 2108–13.
69. Guo F, Li X, Liang D, et al. Cell Stem Cell 2014;15:447–59.
70. Seisenberger S, Andrews S, Krueger F, et al. Mol Cell 2012;48: 849–62.
71. Nilsson EE, Ben Maamar M, Skinner MK. Environ Epigenet 2022;8:dvac001.
72. Weigel D, Colot V. Genome Biol 2012;13:249.
73. Li Z, Yu Z, Yin D. Chemosphere 2021;280:130666.
74. Seong KH, Li D, Shimizu H, et al. Cell 2011;145:1049–61.
75. Horsthemke B. Nat Commun 2018;9:2973.
76. Martos SN, Tang WY, Wang Z. Prog Biophys Mol Biol 2015;118: 44–54.
77. Yi P, Melnyk S, Pogribna M, et al. J Biol Chem 2000;275: 29318–23.
78. Friso S, Choi SW, Girelli D, et al. Proc Natl Acad Sci U S A 2002;99:5606–11.
79. Bakulski KM, Dou JF, Feinberg JI, et al. Int J Environ Res Public Health 2020;17.
80. Robertson KD. Nat Rev Genet 2005;6:597–610.

81. Kirkland J. J Nutr 2009;139:2397–401.
82. Zempleni J, Chew YC, Bao B, et al. J Nutr 2009;139:2389–92.
83. Ho E, Clarke JD, Dashwood RH. J Nutr 2009;139:2393–6.
84. Guibert S, Forne T, Weber M. Genome Res 2012;22:633–41.
85. Lillycrop KA, Phillips ES, Jackson AA, et al. J Nutr 2005;135:1382–6.
86. Dolinoy DC, Weidman JR, Waterland RA, et al. Environ Health Perspect 2006;114:567–72.
87. Heijmans BT, Tobi EW, Stein AD, et al. Proc Natl Acad Sci U S A 2008;105:17046–9.
88. Bell CG, Lowe R, Adams PD, et al. Genome Biol 2019;20:249.
89. De Rooij SR, Bleker LS, Painter RC, et al. Int J Environ Health Res 2022;32:1432–46.
90. Haggarty P. Proc Nutr Soc 2007;66:539–47.
91. Lai CQ, Parnell LD, Smith CE, et al. Am J Clin Nutr 2020;112:1200–11.
92. Nassar FJ, Msheik ZS, Nasr RR, et al. Clin Epigenetics 2021;13:111.
93. Day SE, Coletta RL, Kim JY, et al. Epigenetics 2017;12:254–63.
94. Dick KJ, Nelson CP, Tsaprouni L, et al. Lancet 2014;383:1990–8.
95. Edgar RD, Jones MJ, Meaney MJ, et al. Transl Psychiatry 2017;7:e1187.
96. Snell C, Krypuy M, Wong E, et al. Breast Can Res 2008;10.
97. Widschwendter M, Apostolidou S, Raum E, et al. PLoS One 2008;3:e2656.
98. Busato F, Dejeux E, El Abdalaoui H, et al. Methods Mol Biol 2018;1708:427–45.
99. Suchiman HE, Slieker RC, Kremer D, et al. Front Genet 2015;6:287.
100. Moser DA, Muller S, Hummel EM, et al. Psychoneuroendocrinol 2020;120:104784.
101. King DE, Sparling AC, Joglekar R, et al. MicroPubl Biol 2021;2021.
102. Moran S, Arribas C, Esteller M. Epigenomics 2016;8:389–99.
103. Do WL, Whitsel EA, Costeira R, et al. Int J Epidemiol 2021;50:675–84.
104. Hellbach F, Sinke L, Costeira R, et al. Eur J Nutr 2023;62:1357–75.
105. Kim Y, Huan T, Joehanes R, et al. Am J Clin Nutr 2022;115:163–70.
106. Dominguez-Salas P, Moore SE, Baker MS, et al. Nat Commun 2014;5:3746.
107. Huoman J, Martinez-Enguita D, Olsson E, et al. Clin Epigenetics 2021;13:135.
108. Anderson C, Gillespie S, Thiele D, et al. Breastfeeding Machine 2018;13:371–80.
109. Bordoni L, Gabbianelli R. Biochimie 2019;160:156–71.
110. Zhang B, Tian L, Xie, J, et al. Biomed Pharmacother 2020;130.
111. Pham T, Bae M, Lee Y, et al. J Nutr Biochem 2018;57:162–9.
112. Lewis KA, Tollefsbol TO. Epigenomics 2017;9:1153–5.

B. DIGESTIVE, ENDOCRINE, IMMUNE, AND NEURAL MECHANISMS

CHAPTER 41

Nutritional Physiology of the Alimentary Tract

Ayman A Usmani, Sara O'Brien, and David L Burns

CHAPTER OUTLINE

Abbreviations

AgRP, Agouti-related peptide; **α-MSH**, α-melanocyte-stimulating hormone; **ATPase**, adenosine triphosphatase; **CART**, cocaine and amphetamine-regulated transcript; **CCqaK**, cholecystokinin; **Cl**, chloride; **CNS**, central nervous system; **EC**, enterochromaffin cells; **ECL**, enterochromaffinlike; **ENS**, enteric nervous system; **GALT**, gut-associated lymphoid tissue; **GI**, gastrointestinal; **GIP**, glucose-dependent insulinotropic peptide; **GLP**, glucagon-like peptide; **GLUT**, glucose transporter; **GRP**, gastrin-releasing peptide; **H^+**, hydrogen; **HCO_3^-**, bicarbon; **Ig**, immunoglobulin; **IGF-I**, insulinlike growth factor-I; **ILF**, isolated lymphoid follicle; **K**, potassium; **LCT**, long-chain triglyceride; **MC4R**, melanocortin-4 receptor; **MCT**, medium-chain triglyceride; **MMC**, migrating motor complex; **Na**, sodium; **NO**, nitric oxide; **NPY**, neuropeptide Y; **OXM**, oxyntomodulin; **POMC**, pro-opiomelanocortin; **PP**, pancreatic polypeptide; **PRR**, pattern recognition receptor; **PYY**, peptide YY; **SCFA**, short-chain fatty acid; **SGLT1**, sodium-glucose cotransporter-1; **VIP**, vasoactive intestinal polypeptide

Introduction

The alimentary tract is a tubular structure that extends from the posterior oropharynx to the anus. Its primary function is to digest and absorb ingested nutrients. The purpose of this chapter is to review the structural and functional components of the alimentary tract and describe the interactions of these components in response to a meal. The gastrointestinal (GI) tract flora and GI immune system are also reviewed briefly because of their importance in overall gut function.

Gastrointestinal Tract Structure

Substructures and Cells

The GI tract consists of four contiguous segments: the esophagus, stomach, small intestine, and colon (**Figure 41.1**). The wall of each segment contains four distinctive layers: the mucosa, submucosa, muscularis propria, and serosa or adventitia (**Figure 41.2**). The mucosa is composed of three distinct layers: the epithelium, lamina propria, and muscularis mucosae. The epithelial layer forms a barrier between the lumen and the underlying tissues. Region-specific secretory, absorptive, and barrier functions of the alimentary tract are accounted for by differences in the type of epithelial cell populations along the gut. The lamina propria is a connective tissue space between the epithelium and the thin layer of muscle fibers, the muscularis mucosae, which forms the lower boundary of the mucosa. The lamina propria contains many cells involved in immunologic functions, including immunoglobulin (Ig)-secreting plasma cells, macrophages, and lymphocytes. In addition, abundant lymphoid nodules are present, often extending through the muscularis mucosae into the underlying submucosa. Subepithelial fibroblasts produce collagen and many other extracellular matrix components that underlie the basal lamina of the epithelium. These fibroblasts and the extracellular matrix that they secrete have an important role in regulating cell proliferation and differentiation events within the overlying epithelium.

The mucosal epithelium contains numerous enteroendocrine cells, in addition to cells that serve secretory, absorptive, and barrier functions. The enteroendocrine cells found in gastric, intestinal, and colonic epithelium are characterized by their polygonal shape, broad base, and numerous basilar membrane–bound secretory granules. Enteroendocrine cells are joined to other adjacent cells in the epithelium via junctional complexes located near the apical pole. The regulatory peptides or bioamine products stored in the basally located secretory granules are secreted through the basolateral membrane and act through paracrine or endocrine mechanisms as mediators of GI secretion, absorptive function, and motility in response to luminally and/or basolaterally derived signals. Nutrients absorbed by the

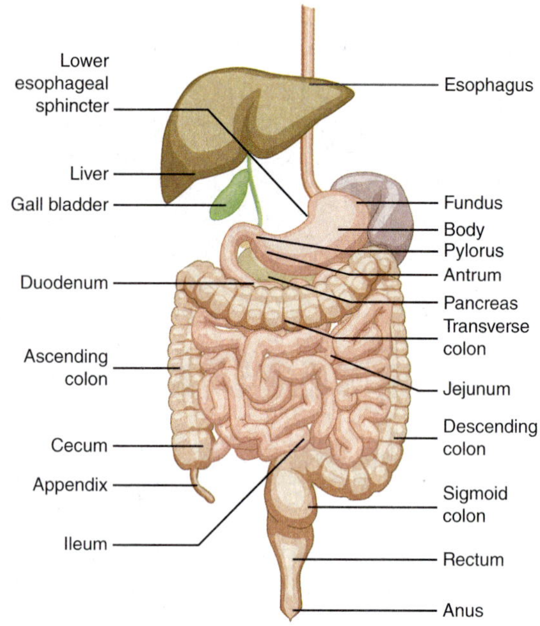

Figure 41.1 Anatomy of the stomach, small intestine, and large intestine. The duodenum is located in the retroperitoneal space and bends around the head of the pancreas. The jejunum lies within the peritoneal cavity and begins at the ligament of Trietz. Jejunal bowel loops are predominantly located in the left and middle upper abdomen. The proximal ileum lies in the middle abdominal region. The distal ileum lies in the right lower quadrant and joins the colon at the ileal cecal valve. The cut out reveals the duodenum and the ligament of Trietz that lie behind the transverse colon.

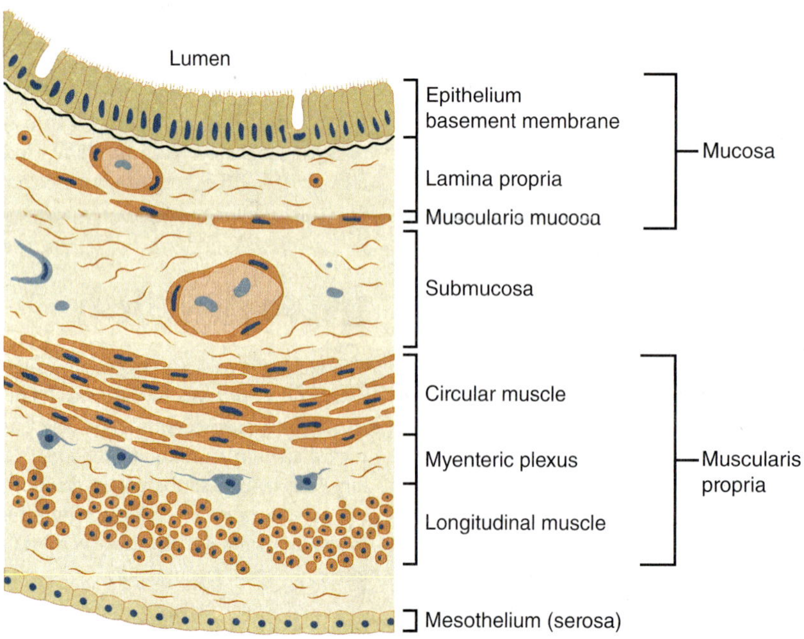

Figure 41.2 Schematic organization of the wall of the gastrointestinal tract.

alimentary tract are transported through blood and lymphatic vessels to other body tissues.

The submucosa extends from the mucosa to the muscularis externa and contains numerous small to moderate-sized veins, arteries, and lymphatic channels surrounded by connective tissue. Ganglion cells and autonomic nerve fibers of the Meissner plexus are also found in the submucosa. Fibers of this submucosal plexus, together with the myenteric plexus, form the enteric nervous system (ENS), which regulates and coordinates numerous intestinal functions, including motility. Additionally, scattered lymphoid aggregates or nodules may be found in this layer of the gut wall.

The muscularis propria is organized into two layers of muscle: an inner circular layer, in which muscle cells circle the intestine, and an outer longitudinal layer, in which muscle cells run parallel with the long axis of the intestine. In the upper esophagus, skeletal muscle fibers interdigitate with smooth muscle fibers, whereas the muscularis of the remaining alimentary tract is composed entirely of smooth muscle. Skeletal muscle in the proximal esophagus allows for volitional control of the initiation of swallowing.

Esophagus

The adult esophagus is approximately 25 cm long and extends from the posterior oropharynx at the level of the cricoid cartilage to just below the diaphragmatic hiatus, where it enters the stomach at the esophagogastric junction. The esophageal mucosa is lined with a thick, incompletely keratinized, stratified squamous epithelium that provides protection against abrasion during passage of a swallowed food bolus, and against refluxed stomach acid. The lamina propria contains occasional lymphoid aggregates and mucosal glands that secrete neutral mucus. Submucosal glands that secrete acidic mucus extend through the lamina propria and muscularis mucosae and are most abundant in the upper half of the esophagus.

In the upper esophagus, skeletal muscle fibers blend with the smooth muscle fibers found throughout the mid and distal esophagus. The upper esophageal sphincter consists of a thickened band of oblique muscle. A thickened band of circular smooth muscle adjacent to the esophagogastric junction forms the lower esophageal sphincter.

The arterial and venous vasculature of the esophagus is anatomically divided into thirds. The cervical esophagus receives its blood supply from the inferior thyroid artery and is drained by the inferior thyroid vein. The thoracic esophagus is supplied by bronchial branches of the aorta and drained by the azygos veins. The abdominal esophagus is supplied by branches of the left gastric and inferior phrenic arteries and drained by the left gastric vein, which is the only venous tributary of the esophagus stemming from the portal venous system.

The skeletal muscle fibers of the upper esophagus are under voluntary control and are involved in regulating the initial passage of a swallowed bolus into the upper esophagus. The remaining smooth muscle of the muscularis is innervated by parasympathetic fibers originating from the vagus nerve. Contraction of this specialized region of smooth muscle, coupled with the abrupt angulation of the esophagus as it passes through the diaphragmatic hiatus, where it joins the gastric cardia comprising the lower esophageal sphincter, provides a mechanism for preventing reflux of the acid contents of the stomach into the esophagus.

Stomach

The stomach is an asymmetric organ that extends from the gastroesophageal junction in the cardia to the duodenum (**Figure 41.3**). The portion of the stomach laying 1 to 2 cm below the gastroesophageal junction is referred to as the cardia. The upper portion of the stomach that lies under the left hemidiaphragm is called the fundus. The gastric body comprises the largest portion of the stomach and extends to the angularis, where the stomach abruptly bends. The gastric antrum lies between the angularis and the pylorus. The pyloric sphincter is a round band of muscle that forms the opening of the stomach into the duodenum. The flat glandular mucosa of the stomach changes to the villus epithelium seen in the duodenum at the pylorus.

The entire stomach is lined by a simple columnar epithelium. The mucosa contains numerous invaginating gastric pits, or foveolae, that form glands at their bases. Each glandular unit is composed of three regions: the upper pit region, lined by surface mucus-secreting cells; a narrow isthmus or neck, containing the proliferative zone; and many immature undifferentiated cells and mucous neck cells. The basilar gland contains parietal cells, chief cells, and enteroendocrine cells. Most of the gastric body and fundus is lined by oxyntic mucosa, consisting of fundic-type glands responsible for secretion of acid (H^+), pepsinogens, and intrinsic factor. These glands contain abundant parietal cells in the upper half of the gland. Chief cells predominate near the base of the glands in fundic-type mucosa. The cardiac glands, found in the first 3 to 4 cm adjacent to the esophagogastric junction, are primarily mucus-secreting glands with few parietal or chief cells. Pyloric glands in the prepyloric antrum are coiled and are remarkable for long foveolae and an increased population of enteroendocrine cells.

Surface mucous cells form a uniform population of columnar epithelial cells lining the surface mucosa and gastric pits. These cells secrete a glycoprotein-rich neutral mucous layer that protects the epithelium from acids, digestive enzymes, and pathologic micro-organisms.[1] Surface mucous cells are constantly shed into the gastric lumen and are replaced by replication of undifferentiated cells within the neck or isthmus region of each gastric gland that differentiate during migration up the foveola and onto the gastric mucosal surface. Mucous neck cells are also present in the neck of the gland. Mucous neck cells differ from the surface mucous cells in that their mucous granules are larger, they contain acidic glycoproteins, compared with the neutral glycoproteins of the surface mucous cells, and although they secrete mucous, they derive surface mucous, parietal, chief, and

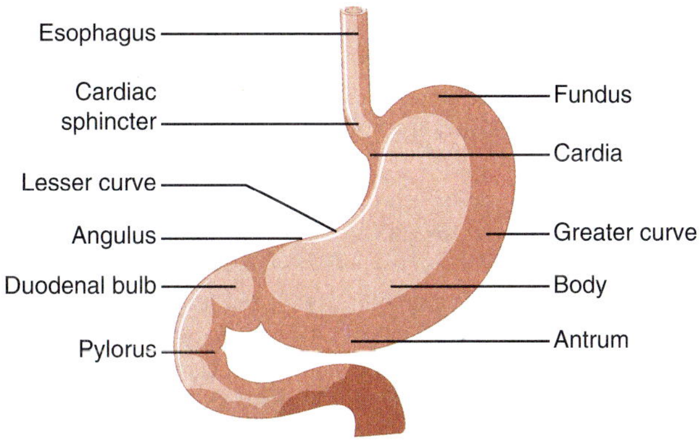

Figure 41.3 Regional organization of the stomach and proximal duodenum.

Reprinted with permission from Yamada T, Alpers DH, Owyang C et al, eds. Textbook of Gastroenterology. 2nd ed. Philadelphia: JB Lippincott, 1991:1304.

endocrine cells from the stem cell precursors[2] and likely respond to signals from the mesenchyme, possibly from myofibroblasts.[3]

The stomach has four tissue layers. The first is the mucosa, which is lined by the epithelial cells outlined above, and also contains lamina propria and a thin muscular layer called the muscularis mucosae. The layer beneath the mucosa is the submucosa. This is a connective tissue layer that contains blood vessels, lymphatics, and nerves. The next layer is the muscularis propria, which is comprised of three muscle layers: the oblique muscle, the circular muscle layer (which becomes the pyloric sphincter), and an outer longitudinal muscle. The final layer is the serosa.

The stomach's arterial supply arises mainly from the celiac trunk, with some contributing vessels of other origins. The left gastric artery arises directly from the celiac trunk. The short gastric, left gastro-omental and posterior gastric arteries arise from the splenic artery. The right gastric artery arises from the proper hepatic artery, while the right gastro-omental artery arises from the gastroduodenal artery. The innervation of the stomach is supplied via parasympathetics and sympathetic nerves. The parasympathetic nerve supply to the stomach is through the vagus nerve, through the left vagal and right posterior trunks. The sympathetic nerve supply arises from T6-T9 of the spinal cord.

Hormones and Secretory Products

Parietal cells secrete hydrochloric acid and are in the middle and basilar portions of the gastric glands. These cells are large, with clear or acidophilic cytoplasm and abundant mitochondria. They have well-developed intracellular canaliculi containing a microvillus border that greatly expands the apical surface available for acid secretion. Receptors for histamine, gastrin, and acetylcholine are located at the basolateral surface and regulate parietal cell secretory function. Hydrogen/potassium (H^+/K^+)-adenosine triphosphatase (ATPase), the enzyme that secretes H^+ into the lumen, is located on the canalicular membrane. Intrinsic factor, a binding protein for vitamin B_{12}, is secreted by parietal cells. In addition, parietal cells play a role in the regulation of gastric mucosa cell lineage differentiation. Transforming growth factor (TGF)-α, heparin-binding epidermal growth factor-like growth factor, amphiregulin and the morphogen sonic hedgehog (a peptide involved in gastric cellular growth and differentiation) are produced by parietal cells.[2]

Chief cells or zymogenic cells are found near the base of the gastric glands. These cells contain an extensive basilar rough endoplasmic reticulum and supranuclear zymogen granules, reflecting their role in the production of pepsinogens and other proteases. Pepsinogens are synthesized and secreted by these cells into the gastric lumen. Hydrochloric acid in the lumen catalyzes the conversion of the proenzyme pepsinogen to the active pepsins that begin breakdown of proteins to lower-molecular-weight polypeptides.

Enteroendocrine cells are most abundant in the prepyloric antrum and secrete many different neuropeptides and regulatory molecules. These cells are classified as open cells, which have apical membranes that are in contact with the lumen, or as closed cells, which are not in contact with the lumen. Gastrin-secreting G cells predominate in the antrum and are an example of an open endocrine cell. Enterochromaffin cells (ECs) are found throughout the gastric mucosa and produce serotonin, substance P, and motilin. Glucagon-secreting A cells are found in the proximal third of the stomach, and somatostatin-secreting D cells (an example of a closed endocrine cell) can be found in both the upper third of the stomach and the antrum. This complex web of enteroendocrine signals is important in integrating responses to both luminal conditions and basolateral signals.

GRP is released from the gastric mucosa in response to food intake, which stimulates the release of gastrin and somatostatin. These hormones further stimulate the secretion of histamine from oxyntic glands, which stimulates secretion of hydrochloric acid. Leptin is secreted from adipose tissue and chief cells. It acts to suppress hunger signals in the lateral hypothalamus.[4] Conversely, ghrelin is secreted from oxyntic glands in the fundus in the setting of fasting to stimulate hunger signals.[5]

Duodenum

The small intestine extends from the gastric pylorus to the ileocecal valve and is divided into three regions: the duodenum, the jejunum, and the ileum. The duodenum is approximately 30 cm in length and is fixed in place, moulded around the head of the pancreas. Histologically, the duodenum is characterized by the presence of abundant submucosal Brunner glands that secrete alkaline mucus. The first portion of the duodenum, known as the bulb, is attached to a mesentery that is folded onto the posterior wall of the peritoneal cavity. The second (descending), third (transverse), and fourth (ascending) portions of the duodenum are retroperitoneal in location. Bile and pancreatic secretions enter the second portion of the duodenum from the common bile duct at the ampulla (papilla) of Vater. The junction of the duodenum and the jejunum is defined by the position of the ligament of Treitz, where the small intestine re-enters the peritoneal cavity. No change occurs in the histologic appearance of the small intestine at this transition.

Jejunum and Ileum

The jejunum and ileum are mobile because of their attachment to an extensive mesentery. The proximal two fifths of the small intestine beyond the ligament of Treitz are defined as jejunum, whereas the distal three-fifths are ileum. The jejunum has a larger diameter, more prominent folds, and longer villi than the ileum. The ileum is characterized by the presence of abundant lymphoid follicles (Peyer patches) in the submucosa.

The length of the jejunum and ileum in adults is variable and ranges from 300 to 850 cm. Several structural features of the small intestine amplify the mucosal surface area available for nutrient absorption to more than 200 m^2, which is larger than a doubles tennis court (**Figure 41.4**). Magnification of the surface area is achieved by a series of folds and invaginations. First, the cylinder of the intestine is heaped up into circular folds (plicae circulares) involving both submucosa and mucosa. These folds are particularly prominent in the jejunum. Second, the mucosal surface is further expanded by the presence of numerous villi, long, finger-like projections of mucosa containing an arteriole, vein, and central draining lacteal. Third, the apical surface of each small intestinal epithelial cell along the villi is covered by microvilli, providing thousands of hills and valleys for surface expansion. The presence of folds, villi, and microvilli increase the surface area to 600 times that of the surface area present in a simple cylinder.

The simple columnar epithelium that lines the small intestine is composed of four principal differentiated cell types—absorptive enterocytes, goblet cells, Paneth cells, and enteroendocrine cells. Cells are joined to adjacent cells by junctional complexes that regulate the paracellular movement of fluid and macromolecules (see fluid and electrolytes section). Absorptive enterocytes are responsible for digestion of dipeptides, tripeptides, and disaccharides,

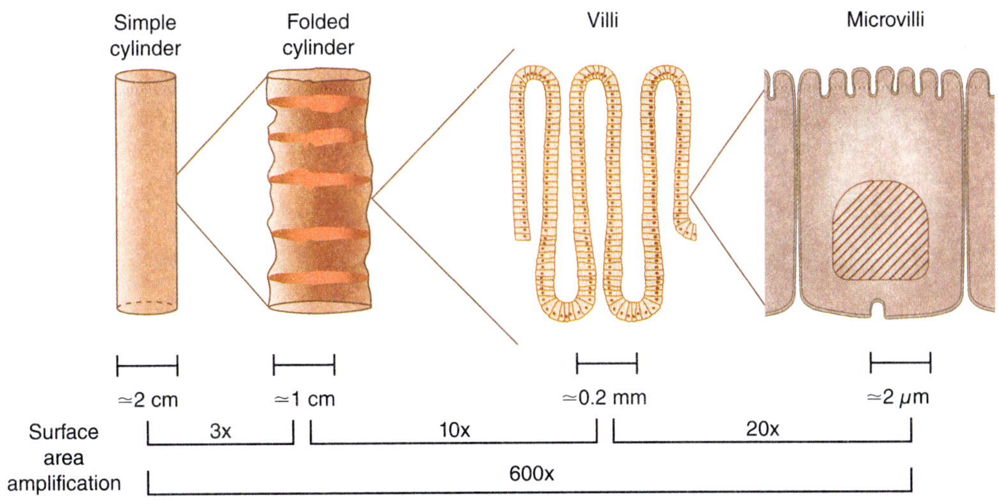

Figure 41.4 The intestinal surface area is expanded by the presence of intestinal folds aka plicae circulares and villi. Microvilli further expand the surface area of epithelial cells in contact with luminal contents. These structural features taken together expand the surface area of the small intestine by approximately 600-fold.

Reprinted with permission from Yamada T, Alpers DH, Owyang C et al, eds. Textbook of Gastroenterology. 2nd ed. Philadelphia: JB Lippincott, 1991:327.

and for nutrient absorption. The microvilli of absorptive enterocytes are supported by a central core of actin filaments that join with a dense terminal web of actin and myosin filaments oriented parallel to the apical surface of the enterocyte. The apical surface is covered by a glycoprotein-rich glycocalyx. Many enterocyte-encoded proteins that are important for digestive function are present at the apical surface, including dipeptidases, disaccharidases, enterokinase, and intestinal alkaline phosphatase. Goblet cells are flask-shaped cells with large apical vesicles that store and secrete mucus. Mucus secreted by the goblet cell forms a viscous gel that functions both as a lubricant and to protect the surface epithelium against adherence of invading pathogens. Goblet cells also secrete small cysteine-rich proteins (trefoil factors) that participate in host defense. Paneth cells reside at the base of the intestinal crypts and produce proteins involved in anti-bacterial actions, including lysozyme and a variety of defensins.

The vascular supply of the duodenum mainly arises from the superior and inferior pancreaticoduodenal arteries, with additional supply from the gastroduodenal, right gastric, common hepatic and right gastro-omental arteries. The vascular supply of the jejunum and ileum mainly sources from branches of the superior mesenteric artery. The small intestine receives parasympathetic innervation via branches of the vagus nerve, while the sympathetic nerve supply arises from branches of the superior mesenteric-celiac plexus.

Hormones and Secretory Products

Most enteroendocrine cells contain many neuroendocrine mediators (see the section on GI hormones) (**Table 41.1**). The distribution of individual enteroendocrine cell subpopulations within the epithelium differs along the length of the small intestine.

Table 41.1 Gastrointestinal Hormones

Peptide	Action	Site of Release	Stimulant
Endocrine			
Gastrin	Stimulates: Gastric acid secretion Histamine release Growth of gastric oxyntic gland mucosa Inhibits apoptosis Inhibits somatostatin	Antrum (duodenum)	Peptides Amino acids Distention Vagal stimulation GRP, PACAP, NPY
CCK	Stimulates: Gallbladder contraction Pancreatic enzyme secretion Somatostatin secretion Pancreatic bicarbonate secretion Growth of exocrine pancreas Inhibits gastric emptying Inhibits gastric acid production Appetite suppressant	Duodenum Jejunum	Peptides Amino acids Fatty acids >8C in length, monitor peptide, diazepam-binding inhibitor, CCK-releasing peptide

(continues)

Table 41.1 Gastrointestinal Hormones *(continued)*

Peptide	Action	Site of Release	Stimulant
Endocrine			
Secretin	Stimulates: Pancreatic bicarbonate secretion Biliary bicarbonate secretion Somatostatin secretion Growth of exocrine pancreas Pepsin secretion Inhibits: Gastric acid secretion Trophic effect of gastrin	Duodenum	Acid Pancreatic phospholipase A_2 Possibly bile and fatty acids
GIP	Stimulates insulin release Inhibits gastric acid secretion Inhibits gastric acid secretion	Duodenum Jejunum (K cells)	Glucose Amino acids Fatty acids
Peptide YY	Ileal brake Appetite suppressant May inhibit pancreatic secretion	Ileum Colon	Fatty acids Glucose
Motilin	Stimulates gastric and duodenal motility	Duodenum Jejunum	Ach, 5-HT_3
Oxyntomodulin	Inhibit gastric emptying Inhibit exocrine pancreatic secretions Appetite suppressant	Ileum Colon	Carbohydrate, protein, fat
Pancreatic polypeptide[a]	Inhibits: Pancreatic bicarbonate secretion Pancreatic enzyme secretion Gastric motility Appetite suppressant	Pancreas Colon	Protein Vagal stimulation
Enteroglucagon[a]	Inhibits gastric emptying	Ileum	Glucose Fat
Amylin (islet amyloid polypeptide)	Inhibit gastric emptying Inhibit glucagon secretion Appetite suppressant	Pancreas	Nutrient ingestion
Neurocrine			
VIP	Relaxes sphincters Relaxes gut circular muscle Stimulates intestinal secretion Stimulates pancreatic secretion Stimulate somatostatin release	Mucosa and smooth muscle of GI tract	Released from neurons and immune cells
GRP (bombesin)	Stimulates gastrin release Stimulates somatostatin release May stimulate exocrine pancreatic secretions	Gastric mucosa	Nutrient ingestion
Substance P	Mediates pain reflexes	Spinal afferent neurons	Afferent nerve input
Enkephalins, endomorphins, dynorphins	Stimulates smooth muscle contraction Inhibits intestinal secretion	Mucosa and smooth muscle of GI tract	Unknown,? Trpm5 cation channel
Paracrine			
Somatostatin	Inhibits: Gastrin release Other peptide hormone release Gastric acid secretion Exocrine pancreatic secretions	Gastric antrum and fundus Pancreatic islets	Acid Gastrin, GRP, VIP, PACAP, secretin, ANP, β_2/β_3-adrenergic agonists adrenomedullin, amylin, adenosine, CGRP Vagus, histamine, and interferon-γ inhibits release

Peptide	Action	Site of Release	Stimulant
Paracrine			
GLP-1, GLP-2	Stimulates insulin secretion, Increased cell proliferation, Inhibits apoptosis Inhibits motility (ileal brake) Appetite suppressant	Distal ileum through colon (L cells)	Nutrient ingestion
Insulin-like growth factor I	Increase cell proliferation	Mainly liver, less so gut mucosal cells	Nutrient ingestion
Histamine[b]	Stimulates gastric acid secretion	Oxyntic gland mucosa ECL cell	Gastrin
Epidermal growth factor	Stimulates cell proliferation Stimulates pepsinogen secretion Decrease gastric acid, increase gland cells	Salivary gland	Possible damage to the mucosa (ulceration or resection)
Leptin	Regulates food intake at hypothalamus, decreases NPY release	Adipose tissue, chief cells	CCK, gastric volume, glucose, cytokines
Ghrelin	Stimulates food intake, GH release	Gastric endocrine cells in the fundus	Fasting

Ach, acetylcholine; ANP, atrial natriuretic peptide; CCK, cholecystokinin; ECL, enterochromaffinlike; CGRP, calcitonin gene-related peptide; GH, growth hormone; GI, gastrointestinal; GIP; glucose-dependent insulinotropic peptide; GLP, glucagon-like peptide; GRP, gastrin-releasing peptide; 5-HT_3, 5-hydroxytryptamine; NPY, neuropeptide Y; PACAP, pituitary adenylate–cyclase activation peptide; VIP, vasoactive intestinal polypeptide.

[a]Unknown physiologic function.

[b]Histamine is an amine, not a peptide.

Data from Furness JB, Clerc N, Vogalis F et al. The enteric nervous system and its extrinsic connections. In: Yamada T, Alpers DH, Kaplowitz N et al, eds. Textbook of Gastroenterology. 5th ed. Philadelphia: Lippincott Williams & Wilkins, 2009:15–39; and from Hasler WL. Motility of the small intestine and colon. In: Yamada T, Alpers DH, Kaplowitz N et al, eds. Textbook of Gastroenterology. 5th ed. Philadelphia: Lippincott Williams & Wilkins, 2009:207–30.

Although enteroendocrine cells arise from the same stem cell as the other differentiated cell types found in the small intestine, they have a much longer half-life than enterocytes or goblet cells. Thus, their migration onto and along the intestinal villi is uncoupled from the migration of the other epithelial cell types in the intestine. Hormones produced in the small intestine include CCK, secretin, GIP, peptide YY, motilin, VIP, GLP-1, GLP-2, and enteroglucagon, all of which help to regulate response to food intake.

A small population of enteroendocrine cells differ from other enteroendocrine cells in that they do not contain secretory granules. An example is the brush cell in mice, which produces endogenous opioids as well as uroguanylin (a trypsin-resistant hormone that may act to increase bicarbonate [HCO_3^-] secretion) and express TRPM5, a Ca^{2+}-activated channel that mediates signaling in taste and may play a role in insulin regulation.[6]

Under normal physiologic circumstances, cells within the intestinal epithelium are continuously replaced by migration of cells onto the villus from several adjacent crypts of Lieberkühn or intestinal glands (**Figure 41.5**). The principal differentiated cell types of the small intestinal epithelium are all derived from multipotent stem cell(s) located near the base of each intestinal crypt.[7] These crypt stem cells divide rarely to produce a daughter stem cell (self-renewal) as well as a more rapidly replicating transit cell.[8] Transit cells, in turn, undergo four to six rapid cell divisions in the proliferative zone located in the lower half of each crypt, and their progeny subsequently differentiate during a bipolar migration away from this zone. Goblet cells and enterocytes undergo terminal differentiation as they are rapidly translocated upward from the zone of proliferation to the apical extrusion zone (a process lasting 48 to 72 hours) located adjacent to the villus tip, where they undergo apoptosis and are sloughed into the lumen. Paneth cells arise during downward migration to the crypt base, and enteroendocrine cells differentiate during migration from the zone of proliferation in either direction. Cell renewal, migration, and differentiation are interrelated processes that are regulated at multiple levels.

Colon

The colon is approximately 100 to 150 cm in length, extending from the ileocecal valve to the proximal rectum (see Figure 41.1). The colon consists of the cecum, ascending colon, hepatic flexure, transverse colon, splenic flexure, descending colon, and sigmoid colon. The terminal ileum enters the cecum on its posteromedial border at the ileocecal valve. The cecum is a large, blind pouch approximately 7.5 to 8.5 cm in diameter that projects from the antimesenteric side of the ascending colon. The appendix extends from a narrow opening in the base of the cecum. The diameter of the colon diminishes progressively; the sigmoid colon is approximately 2.5 cm in diameter and is the narrowest portion. The omentum is attached to the transverse colon on its anterior superior edge. The ascending colon, descending colon, rectum, and posterior surface of the hepatic and splenic flexures are fixed retroperitoneal structures and, therefore, lack a complete serosal layer. The cecum, transverse, and sigmoid colon are intraperitoneal and are covered by a complete serosal layer.

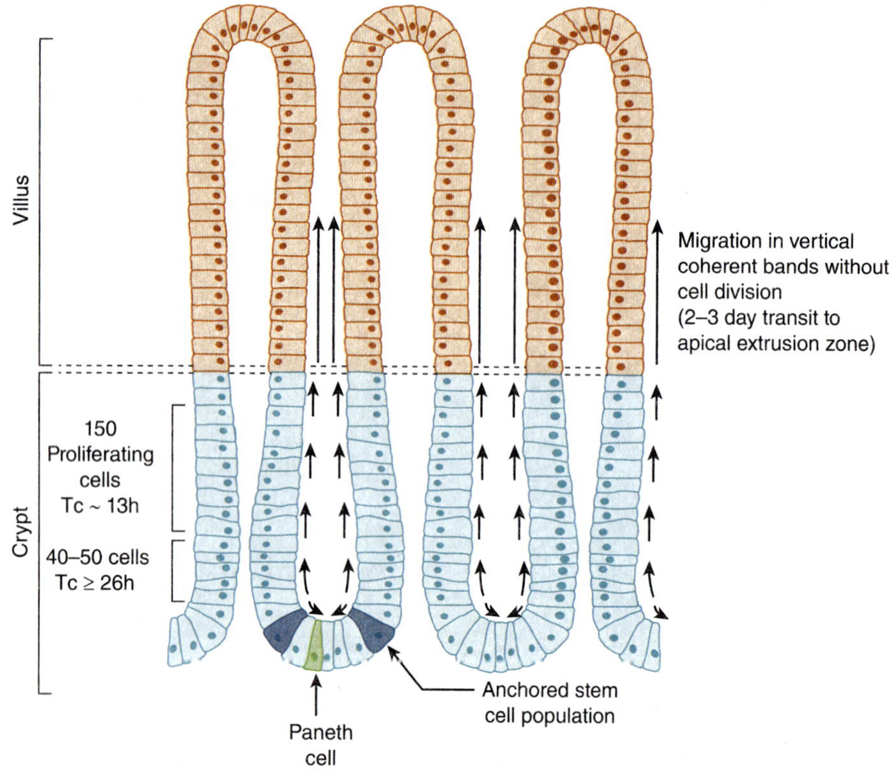

Figure 41.5 Schematic organization of the epithelium in the adult mouse small intestine. The small intestinal crypt contains approximately 250 cells. The lower five cell positions contain 40 to 50 cells that have an average cycle time (Tc) of 26 hours or longer. This region includes Paneth cells and is postulated to include undifferentiated, anchored stem cells at the fifth cell position above the base. The undifferentiated cells divide asymmetrically to give rise to proliferating transit cells (Tc ~13 hours) that migrate upward toward the villus and subsequently differentiate into enterocytes, goblet cells, and enteroendocrine cells. Paneth cells differentiate during downward translocation to the crypt base. Senescent cells are extruded near the villus tips.

Reprinted with permission from Yamada T, Alpers DH, Owyang C et al, eds. Textbook of Gastroenterology. 2nd ed. Philadelphia: JB Lippincott, 1991:1561.

Three principal differentiated epithelial cell types are present in the adult colonic epithelium: absorptive colonocytes, goblet cells, and enteroendocrine cells. As in the small intestine, all of these cell lineages appear to be derived from a common epithelial stem cell precursor. Undifferentiated cells, replicating cells, and enteroendocrine cells predominate near the base of each colonic gland (crypt). Cells belonging to each of the principal cell lineages differentiate as they migrate away from the zone of proliferation toward the surface epithelium. The average lifespan of goblet cells and absorptive cells, from their birth deep in the crypt to the point at which they are sloughed into the lumen, is approximately 6 days. As in the small intestine, some enteroendocrine cell subtypes appear to have a much longer life span than goblet cells or absorptive colonocytes.

As absorptive colonocytes differentiate during their migration up the crypt, they develop short microvilli and clear apically oriented vesicles containing a fibrillar glycoprotein-rich secretory product that may contribute to a glycocalyx. These apical vesicles are lost, and microvilli elongate and increase in number as the maturing absorptive cells emerge onto the surface epithelium. At this point, alkaline phosphatase activity appears on the brush border and the basolateral membranes have acquired a considerable amount of sodium (Na^+)/K^+-ATPase activity, reflecting their function in water and electrolyte transport.

The inner circular muscle fibers form a continuous layer around the colon. The outer longitudinal smooth muscle fibers are condensed into three bands (taeniae coli) equidistant around the circumference of the colon. Haustra are the bulging sacculations that form between adjacent taeniae coli. The serosa is a mesothelially derived cell layer that covers the peritoneal aspects of the colonic wall. Therefore, regions of the ascending colon, the descending colon, and the rectum that do not lie within the peritoneal cavity have no outer serosal layer.

The colon is supplied mainly by the superior mesenteric and inferior mesenteric arteries. The superior mesenteric artery branches into the ileocolic, right colic, and middle colic arteries, which supplies the proximal to midtransverse colon. The middle colic artery supplies from the mid-transverse colony to the distal descending colon. The inferior mesenteric artery branches into the left colic and sigmoid arteries, which provide the majority of blood supply to the left colon. The innervation of the colon is contingent on the embryologic origin of the structures. The midgut-derived ascending and proximal two thirds of the transverse colon receive their parasympathetic and sympathetic innervation from the superior mesenteric plexus, while the remainder of the hind-gut derived transverse, descending, and sigmoid colon receive their nerve supply from the inferior mesenteric plexus. See Table 41.1 for hormones secreted from the colon.

Rectum

The rectum is approximately 12 to 15 cm in length and extends from the sigmoid colon to the anal canal following the curve of the sacrum (see Figure 41.1). The rectal wall consists of mucosal, submucosal, inner circular, and outer longitudinal muscular layers. No serosal layer exists in the rectum. The anal canal is approximately 3 cm long. The anal verge is the junction between anal and perianal skin. Anal epithelium (anoderm) lacks hair follicles, sebaceous glands, and sweat glands. The dentate line is the true mucocutaneous junction located just above the anal verge. A 6- to 12-mm transitional zone exists above the dentate line, where the squamous epithelium of the anoderm becomes cuboidal and then columnar epithelium.

The rectum is supplied by three main arteries. The superior rectal artery arises from the inferior mesenteric artery, the middle rectal artery from the internal iliac artery, and the inferior rectal artery from the interior pudendal artery. The rectum receives a sympathetic nerve supply from the superior and inferior hypogastric plexuses along with lumbar splanchnic nerves, while parasympathetic supply arises from the pelvic splanchnic nerves and inferior hypogastric plexus.

Vascular Anatomy of the Alimentary Tract

Blood and lymphatic vessels provide the transportation system for delivering absorbed nutrients to other body tissues.[9] In addition, the arterial blood supply provides nutrients to the alimentary tract itself. In the small intestine, each villus contains a single arteriole that breaks into a capillary network at the villus tip before joining with a draining venule. Each villus contains a lymphatic vessel (lacteal) that drains into a submucosal plexus connected to larger lymphatics. In the colon, arterioles pass between crypts to the epithelial cell surface and form a network of capillaries around the crypts. Lymphatic vessels in the colon do not extend higher than the base of the crypts.

Blood from the small intestine and colon drains into the portal vein that delivers absorbed water-soluble nutrients directly to the liver, where these nutrients can be metabolized or released directly into the hepatic veins and ultimately, into the systemic circulation.[10] Enteric bile salts are actively absorbed in the terminal ileum and travel through the portal vein to the liver. Bile salts are then taken up by hepatocytes to produce bile and secreted back into the small intestine, providing an enterohepatic circulation for bile salts. This enterohepatic circulation of bile salts is very efficient with absorption of over 90% of the total body bile to maintain normal bile salt homeostasis and fat absorption. Intestinal lymphatic vessels, which are closely associated with arteries supplying the alimentary tract, carry absorbed fat-soluble nutrients to the thoracic duct, which drains into the left subclavian vein and the systemic circulation.

Adequate intestinal blood flow is critical because it provides the oxygen necessary for intestinal cell survival. Therefore, GI tract blood flow is carefully regulated by metabolic, vascular, and hormonal factors to ensure adequate tissue oxygenation.[11] Food ingestion increases intestinal blood flow and oxygen requirements.[12]

Enteric Nervous System and Motility

The Enteric Nervous System (ENS) is an extensive network of the peripheral nervous system within the GI tract that receives supply from both the parasympathetic and sympathetic nervous systems. The ENS can regulate complex and diffuse motility functions through its vast network throughout the GI tract. The ENS consists of approximately 100 million nerve cell bodies (neurons) and their processes that are embedded in the wall of the GI tract (**Figure 41.6**). These neurons lie in clusters (ganglia) and are segregated largely into two layers: (a) the myenteric ganglia, which form a continuous plexus between the circular and longitudinal muscle layers of the muscularis propria and extend from the upper esophagus to the internal anal sphincter; and (b) the submucous plexus, which is located in the submucosa and is especially prominent in the small and large intestines. The processes from these ganglia form dense networks

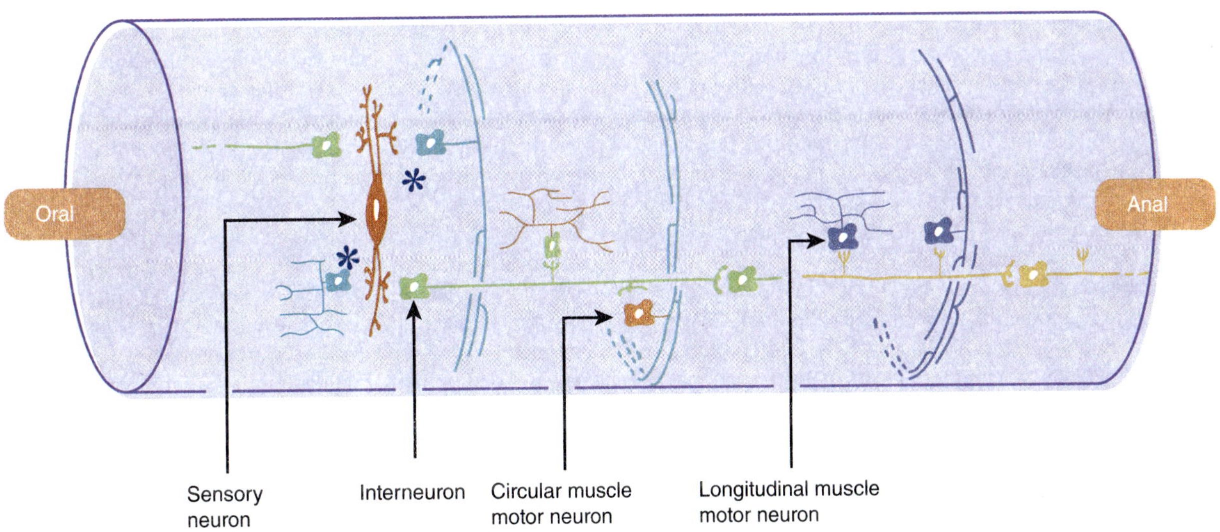

Figure 41.6 The pathways for propulsive reflexes in the intestine. A short segment of intestine is represented, on which the descending inhibitory reflex pathway and the first connections of the ascending pathway are depicted. They provide outputs to ascending and descending interneurons and monosynaptic connections to motor neurons *(asterisk)*. The interneurons form descending and ascending chains and provide outputs to motor neurons. In the descending pathway, some neurons excite the longitudinal muscle, and some neurons inhibit the circular muscle. Ascending reflex pathways supply inputs to excitatory longitudinal muscle motor neurons and excitatory circular muscle motor neurons.

Reprinted with permission from Yamada T, Alpers DH, Owyang C et al, eds. Textbook of Gastroenterology. 2nd ed. Philadelphia: JB Lippincott, 1991:15.

Table 41.2 Types of Neurons in the Enteric Nervous System

Location	Function	Chemical Transmitter
Circular muscle	Excitatory motor neurons Inhibitor motor neurons	Ach, tachykinins NO, ATP, VIP, PACAP
Longitudinal muscle	Excitatory motor neurons Inhibitory motor neurons	Ach, tachykinins GABA, VIP, ATP, PACAP
Muscle layers	Primary sensory neurons Interneurons (sensorimotor reflex, MMC) Secretomotor (gut endocrine cells, gastric glands)	Ach, CGRP, tachykinins ChAT, 5-HT, somatostatin Varies, GRP for nerves innervating G cells
Submucosa	Secretomotor/vasodilator Secretomotor (not vasodilator) Intrinsic primary afferent neurons	Ach, VIP/GAL (type 2) Ach Tachykinins (presumed)

Ach, acetylcholine; *ATP*, adenosine triphosphate; *CGRP*, calcitonin gene-related peptide; *GABA*, γ-aminobutyric acid; *GAL*, galanin; *GRP*, gastrin-releasing peptide; *5-HT*, 5-hydroxytryptamine; *MMC*, migrating motor complex; *NO*, nitric oxide; *PACAP*, pituitary adenylate cyclase–activating peptide; *VIP*, vasoactive intestinal polypeptide.

Data from Furness JB, Clerc N, Vogalis F et al. The enteric nervous system and its extrinsic connections. In: Yamada T, Alpers DH, Kaplowitz N et al, eds. Textbook of Gastroenterology. 5th ed. Philadelphia: Lippincott Williams & Wilkins, 2009:15–39.

and innervate the muscularis propria, muscularis mucosae, epithelium, and other structures. Also, nonganglionated plexuses supply all of the layers of the tubular GI tract accompanying the arteries that supply the gut wall.

The ENS has many different types of neurons (**Table 41.2**). Moreover, these neurons may differ in function in different regions of the digestive tract. Excitatory motor neurons innervate the longitudinal and circular smooth muscle and muscularis mucosae. In addition, they innervate enteric endocrine cells and Peyer patches. Secretomotor neurons in the small and large intestine and gallbladder regulate water and electrolyte secretion. In the stomach, they stimulate acid secretion. Interneurons are present in all regions of the gut, but their characteristics vary more than those of other neuron types. They are present as a chain within the myenteric plexus that runs from mouth to anus. Intrinsic reflex pathways that control gut movement, blood flow, and secretion are activated by sensory neurons that respond to mechanical and chemical stimuli and to distention. These neurons are now known as intrinsic primary afferent neurons. Intrinsic primary afferent neurons are multiaxonal and connect to other intrinsic primary afferent neurons, motor neurons, and interneurons. They differ from extrinsic sensory neurons because their responses can be modified by synapses at the cell body.

In recent years, substantial research has been conducted on communications between the ENS and CNS, a relationship colloquially referred to as the "brain-gut axis."[13] The ENS is connected to the central nervous system (CNS) by transmission along axons in both directions, from GI tract to brain and from brain to ENS. The connections are largely through the vagus nerve and pathways leaving the spinal cord. Most vagal fibers (75% to 90%) are afferent fibers that interact with neurons in the nucleus tractus solitarius in the midbrain. Because relatively few efferent vagal fibers exist, compared with the large number of ENS neurons, the vagus functions more to initiate activity of the integrated circuits in the ENS than to coordinate gut function by direct signaling. Efferent centers in the spinal cord can receive efferent signals from the CNS, which are relayed to the ENS. In addition, the spinal centers can process afferent signals from the gut.[14] Given the close relationship between the ENS and CNS, the GI tract is theorized to be an area of specific interest in the development of neurodegenerative pathologies such as Alzheimer's and Parkinson's, with future research focused on potential therapeutic targets focusing on this connection.[15]

The vagal and spinal components comprise the extrinsic branches of the autonomic nervous system, including the parasympathetic and sympathetic systems (**Figure 41.7**). The striated muscles in the upper esophagus and external anal sphincter are directly innervated by cholinergic fibers, whereas the remaining gut is innervated by a variety of neural mediators, including acetylcholine, gut peptides, and NO. These preganglionic fibers form synapses with the enteric plexuses, which, in turn, are connected with smooth muscle, secretory, and endocrine cells. The sympathetic nervous system contains preganglionic connections between prevertebral ganglia and the spinal cord, but the gut itself is innervated by postganglionic connections, mediated largely by epinephrine and norepinephrine. These postganglionic fibers innervate the plexuses of the ENS, as do the parasympathetic fibers, but the sympathetic fibers also directly innervate blood vessels, smooth muscle layers, and mucosal cells.

The sympathetic nervous system affects intestinal secretion, blood flow, and motility. The sensory fibers that accompany the sympathetic nerves (intestinofugal neurons) are primary sensory neurons that are not part of the autonomic nervous system and are not really "sympathetic" sensory nerves. Sympathetic efferent neurons inhibit motility by decreasing contractile activity and constricting sphincters. These various effects can be relayed along the gut to other regions before returning to the region of the initial stimulus by means of prevertebral ganglia connections. Examples of these inhibitory reflexes include slowing of gastric emptying by acidity or hypertonicity in the upper small intestine.

Intestinal smooth muscle is of the unitary type and is characterized by spontaneous activity, including active tension to stretching, and activity that is not initiated by nerves, but modulated by them. The circular muscle is innervated by both excitatory and inhibitory motor neurons and forms a thick syncytium surrounding the submucosa. Contraction shortens the radius but increases the length of each fiber and, in turn, of the syncytium.

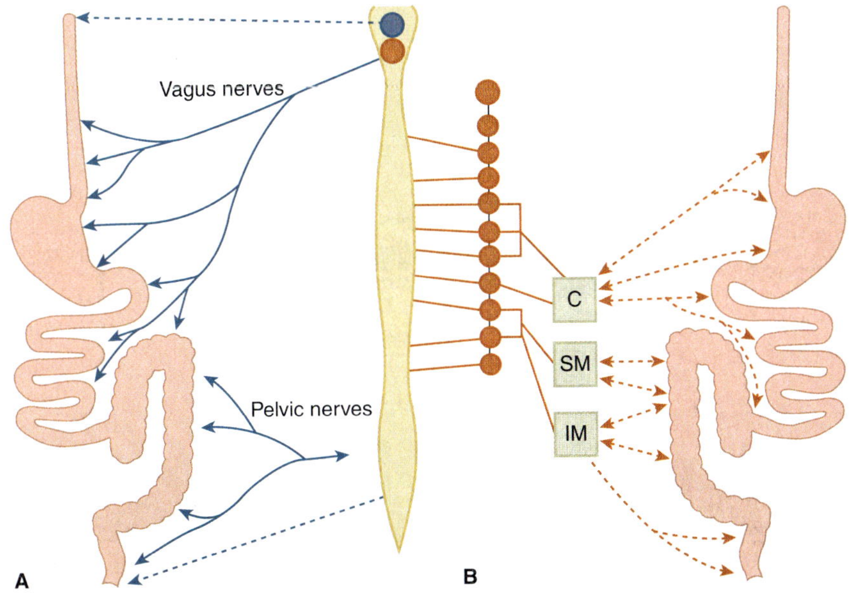

Figure 41.7 The extrinsic branches of the autonomic nervous system. (A) Parasympathetic. *Dashed lines* indicate cholinergic innervation of the striated muscle in the esophagus and external anal sphincter. *Solid lines* indicate afferent and preganglionic innervation of the remaining gastrointestinal tract. (B) Sympathetic. *Solid lines* denote the afferent and preganglionic efferent pathways between the spinal cord and the prevertebral ganglia. *Dotted lines* indicate the afferent and postganglionic efferent innervation. *C*, celiac; *IM*, inferior mesenteric; *SM*, superior mesenteric.

Reproduced from Johnson LR, Alpers DH, Jacobson ED et al, eds. Physiology of the Gastrointestinal Tract, vol 1. 3rd ed. New York: Raven Press, 1994:451.

In contrast, the longitudinal muscle layer surrounding the circular muscle is thin, is shortened by contraction (with enlarged radius), and is only innervated by excitatory neurons. Electrical slow waves are derived from the muscle cells and trigger action potentials, leading to contractile activity. Action potentials in intestinal smooth muscle are propagated through gap junctions from cell to cell, creating an electrical syncytium.

The regulation of peristalsis, the smallest unit of the propulsive reflex, is one of the simplest of the programmed motor activities of the ENS but is still quite complex (**Figure 41.8**). Two components of the reflex, orad contraction and caudad relaxation, combine to move intestinal contents in a caudad direction. The propulsive movement is the end result of contractions and relaxations of the longitudinal and circular external muscles and the muscularis mucosae. The circular muscle has the major role in mixing and propulsion by circumferential contractions that decrease the diameter of the intestine, whereas the longitudinal muscle creates shortening of the segment by sleeve contractions with little alteration in luminal diameter. Excitatory and inhibitory motor neurons supply the muscle, and inhibitory reflexes modulate these activities by monitoring luminal contents. Multiple chemical mediators are involved in this reflex (see Figure 41.8 and Table 41.1).

Many GI tract motor patterns have been described and involve complicated interactions between a series of stimulatory and inhibitory impulses from the ENS to GI smooth muscle. Intestinal smooth muscle consists of circular and longitudinal muscle layers, so the interaction of muscular contraction between layers determines the pattern of motility. The two most important motility patterns are the migrating motor complex (MMC) and peristalsis, which are programmed by the ENS.[16]

The MMC, also known as the "housekeeper of the gut," is the major complex fasting motility pattern in mammals. It is cyclic and passes from the stomach to the terminal ileum.[17,18] The migrating myoelectric complex consists of coordinated activity that empties the stomach and sweeps down the intestine, lasting 84 to 112 minutes, and is separated into three phases. In phase I, very little motor activity occurs with only a small amount of forward propulsion. In phase II, irregular contractions occur and duodenal cross-sectional diameter increases. This may be to accommodate biliary secretions, which are produced during this phase. Gut propulsion occurs during this phase, with rapid transition from phase II to phase III. Phase III lasts only for 5 to 10 minutes of the MMC cycle but propagates contractions over much longer distances than in phase II, with contractions starting in the gastric body. In addition, a few of the contractions in phase III are retrograde, which reflux duodenal contents and HCO_3^- into the antrum of the stomach. This increases the pH in this region of the stomach and may act to protect the mucosa in the fasting state. The maximal frequency of contractions is determined by the slow wave frequency (myocyte cell membrane potential fluctuations, which occur with a certain frequency and along the length of the bowel), which is 11 to 12 contractions per minute in the duodenum and seven to eight contractions per minute in the ileum. As "the housekeeper," the functional role of these interdigestive movements is to clear the gut for the next meal. During fasting, the fundus of the stomach is in a state of partial contraction. The pressure created by this partial contraction decreases with a food bolus in response to receptive relaxation (stimulated swallowing) and gastric accommodation (stimulated by gastric distention).[19,20]

In the fed pattern, the contractions in the stomach mimic those of phase II of the MMC. This lasts until the stomach is empty and starts 5 to 10 minutes after eating commences. These contractions propel the food distally and proximally in the stomach, ultimately mixing and grinding the food. The amount of time food is in the stomach depends on the amount of energy and fat consumed.[19] In the small intestine and colon, the MMC is replaced by a fed pattern of contraction. This is characterized by intermittent phasic contractions throughout the small intestine.

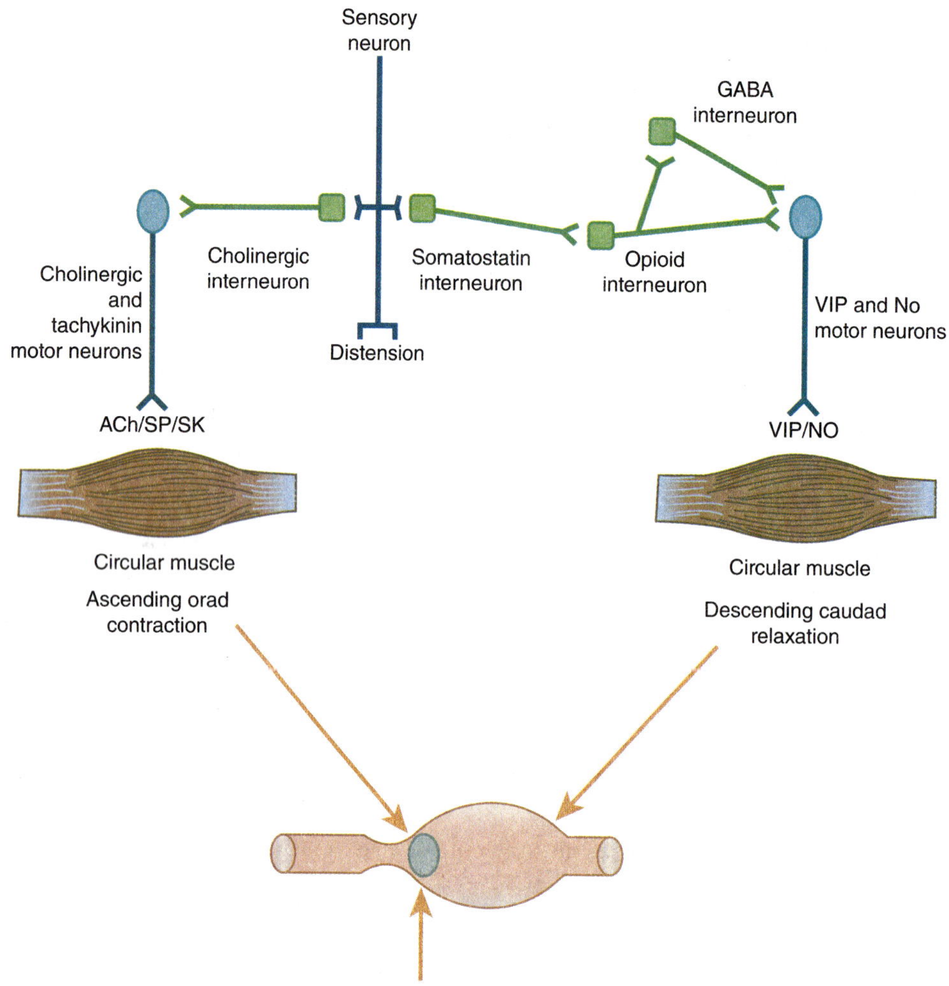

Figure 41.8 Regulation of peristaltic reflex by neurons of the myenteric plexus. The reflex has two components: ascending or orad contraction and descending or caudad relaxation. The stimulus (i.e., distention or mucosal stimulation) is relayed by sensory neurons to cholinergic interneurons coupled to vasoactive intestinal peptide (VIP) and nitric oxide synthase (NOS) neurons caudad and acetylcholine (Ach) and tachykinin (SP, SK) neurons orad. Somatostatin, opioid, and γ-aminobutyric acid (GABA) neurons exert a modulatory influence on VIP and NOS neurons.

Again, greater fat content increases the amount of time spent in this fed motor pattern. The return of the MMC after feeding is not well understood, but the first MMC contractions may start more distally in the small bowel. It is unclear what signals are involved.[18]

The presence of luminal nutrients can increase absorption by feedback regulation of intestinal motility, called the ileal brake. Fat or carbohydrate in the ileum stimulates the release of PYY, glucagon-like peptide-1 (GLP-1), and possibly oxyntomodulin (OXM) from ileal endocrine cells.[21] These hormones then enter the systemic circulation and inhibit gastric emptying, thereby slowing down small intestinal transit. Thus, this ileal brake mechanism enhances absorption by increasing the contact time between luminal nutrients and intestinal mucosa. Although luminal nutrients in the colon also have some effect on PYY and GLP-1 secretion, they have not been shown to affect small bowel transit times in humans at doses able to alter transit times when infused in the ileum.[18] However, small studies have shown that ileal infusions of high-dose lipids, carbohydrates and proteins may have the potential to induce satiety and reduce food intake.[22]

Endocrinology of the Gastrointestinal Tract

The largest endocrine organ in the human body is the GI tract, and the first hormone ever discovered was secretin, a peptide hormone produced in the GI tract. The mucosa of the GI tract differs from other endocrine organs in that the endocrine and peptidergic neurons are found diffusely throughout the GI tract.[23] These endocrine and peptidergic neurons produce regulatory substances that are critical for the precise coordination of activities necessary to assimilate a meal. These substances are mostly peptides that communicate by endocrine, neurocrine, and paracrine pathways (**Figure 41.9**; see Table 41.1), and some of the GI tract hormones can act via more than one pathway. Endocrine peptides are hormones released from sensory cells in the intestine in response to mechanical or chemical stimuli and enter the bloodstream to act on a distant target organ. Gut neurocrine peptides are produced within the ENS and are located in nerves within the gut itself. Many of these peptides (and their receptors) are also produced by the brain and represent a gut–brain axis.

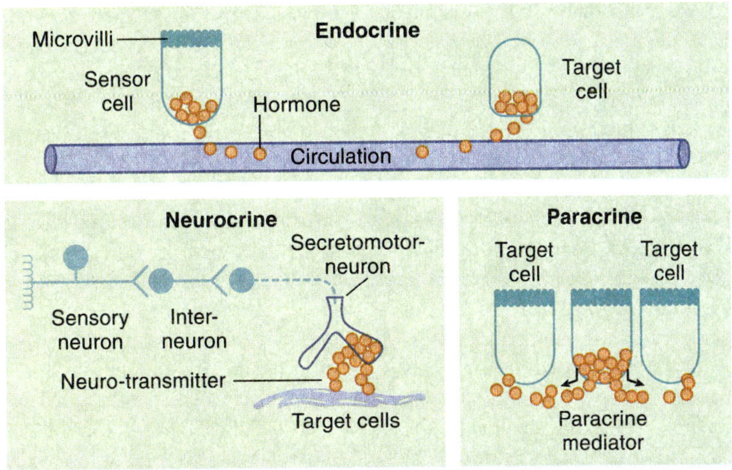

Figure 41.9 Three mechanisms of communication mediate responses in the gastrointestinal (GI) tract. The three mechanisms of communication that mediate responses are endocrine, neurocrine, and paracrine. For the endocrine mechanism, sensory cells respond to stimuli by releasing transmitters that travel by way of the blood to their target cells or tissues. Many examples of endocrine sensory cells through the GI tract respond to either mechanical or chemical stimuli to release their hormones. Some types of endocrine cells respond to changes in pH or osmolality, whereas others respond to changes in specific nutrients. For these neurocrine mechanisms, the sensing and transmissions to the target tissue are mediated by nerves and neurotransmitters. Nerves sense stimuli, such as nutrients, pH, and osmolality in the luminal contents, as well as movement of the contents and distention of the gut lumen.

Reproduced from Raybould H, Pandol SJ. Integrated Response to a Meal. Undergraduate Teaching Project, Unit 29. Bethesda, MD: American Gastroenterological Association, 1995.

Paracrine peptides (and histamine, an amine) are produced by intestinal cells and act on adjacent or nearby cells. This can occur either by direct cellular extension to other cells, or by release of the peptide (or histamine) into the mucosa (e.g., somatostatin, histamine) or into the intestinal lumen (e.g., monitor peptide, cholecystokinin [CCK]-releasing peptide, trefoil peptides).

The GI tract hormones have multiple effects, both short term in response to a meal, and more delayed on the growth and differentiation of enteric cells. Table 41.1 lists the better-characterized GI tract hormones. Many of the GI tract hormones that are important in response to a meal are produced by cells in the upper GI tract and act on the components of the upper GI tract (e.g., gastrin, CCK, secretin, motilin, glucose-dependent insulinotropic peptide [GIP], somatostatin, neuropeptide Y [NPY], leptin, ghrelin). All three macronutrients (protein, carbohydrate, fat) are responsible for the release of these substances. Coordination of function in the upper intestinal tract is complex, involving the stomach, duodenum, pancreas, and gallbladder. Many of these organized processes are mediated by local release of GI hormones.

The specificity and coordination of action of GI hormones depend on three major factors: the multiple functions of each hormone, the paracrine actions between neuroendocrine and mucosal cells, and the regulatory functions of the ENS. Most GI hormones have multiple actions and mediate both stimulatory and inhibitory functions (e.g., gastrin, CCK, secretin, GIP, vasoactive intestinal polypeptide [VIP], enkephalins) (see Table 41.1). Other GI hormones or amines are solely stimulatory (e.g., histamine, motilin, gastrin-releasing peptide [GRP], monitor peptide, CCK-releasing peptide), or inhibitory (e.g., somatostatin, PP). Thus, release of these hormones has the potential to create multiple effects on GI organs coordinated in time. The presence of multiple cells in the mucosa, each possessing receptors to many of the GI hormones, also helps to create specificity of response. For example, in isolated cell systems, CCK stimulates acid production. However, CCK injected into the intact animal does not stimulate acid production because of the greater effect of CCK on the D cell–producing somatostatin, an inhibitor of acid secretion, than on the parietal cell that produces acid. Conversely, gastrin has the reverse effects on those two mucosal cells, stimulating gastric acid secretion from the parietal cell. Thus, the variability of mucosal-specific cells adds a layer of complexity and feedback control to the presence of multiple hormones present in the mucosa. The ENS, with its many neuronal connections to mucosal cells, integrates the stimuli controlling GI hormone release. Both preganglionic parasympathetic cholinergic nerves and postganglionic fibers, mediated by neurocrine peptides, are important regulators of the GI response to feeding. Chemosensory neurons detect intraluminal events and regulate mucosal function by intrinsic mucosal reflexes.

Peptide hormones are also involved in appetite regulation (see Chapter 43). Hormones such as GLP-1, CCK, PYY, PP, OXM, amylin, insulin, glucagon, and ghrelin interact with the brain in both the hypothalamus and in the brainstem (area postrema) by passage through the blood–brain barrier, acting via vagal-brainstem-hypothalamic pathways, or both. All the hormones listed, with the exception of ghrelin, result in decreased energy intake and are considered appetite suppressing.

Some peptide hormones, such as GLP-1, GLP-2 and IGF-1 are important mitogens for the cells of the intestinal tract. Gastrin stimulates the growth of gastric oxyntic gland mucosa. GLP-1 and GLP-2 are produced in gut endocrine cells. These peptides are released by nutrient ingestion and regulate cell proliferation and differentiation in the intestine, in addition to their role in energy disposal. Insulin-like growth factor-I (IGF-I) is largely synthesized in the liver and gut mucosa. It serves as a potent trophic factor for intestinal mucosa, principally targeting epithelial cells, endothelial cells, and fibroblasts.[24] GLP-1 and GLP-2 also have antiapoptotic activity, thereby enhancing their effect on mucosal growth.[25]

Massive intestinal resection from ischemia, trauma, inflammatory bowel disease, malignancy, or congenital anomaly can result in patients with short bowel syndrome. This is characterized by weight loss, diarrhea and malabsorption with electrolyte disturbance, dehydration, and malnutrition. Specialized nutrition support, generally parenterally, is required for patients to maintain adequate nutrition and hydration. Short bowel syndrome occurs from loss of small bowel surface area in conjunction with impaired production of excitatory, inhibitory, and trophic gut hormones. This loss of hormone production affects acid production and neutralization, gastric emptying, production and secretion of bile, bicarbonate and pancreatic enzymes, gut motility and mucosal integrity. GLP-2 is one of several mucosal hormones that have trophic effects, stimulating intestinal mucosal repair and growth, with increased villus height and crypt depth. GLP-2 analogs are in use for therapy of short bowel syndrome, stimulating adaptation of the small intestine with bowel lengthening and dilation with increased absorptive surface area. This can result in decreased or eliminated need for parenteral supplementation.[26,27]

Integrated Response to a Meal

The integrated response of the GI tract to a meal represents a coordinated series of events that include the regulation of food intake, stimuli-evoked responses in anticipation of the meal, ingestion and transfer of the meal to the stomach, digestion and absorption of the meal, and elimination of waste products, bringing into play all of the individual regulatory controls reviewed earlier.

Regulation of Food Intake

The GI tract is involved in the earliest part of feeding, beginning with the control of nutrient ingestion. Peptide hormones and other neurotransmitters in the gut have been implicated in the short-term regulation of energy intake, although reasons for intake of food by humans is complex and includes both short-term "satiety signals" and long-term regulatory factors or "signals of adiposity."[28] Intake of food is also influenced by olfactory and visual signals, taste, mood, social situation, and degree of physical activity. The hypothalamic and brainstem centers are main sites where these signals converge and are integrated to control food intake. Leptin is the best-studied peripheral regulator of energy intake and is produced in adipose tissue; although adiponectin, resistin, and interleukin-6, also produced by adipose tissue, also likely play a role in modifying food intake. Neurotransmitters involved in the CNS include dopamine, serotonin, opiates (enkephalin, β-endorphin, and dynorphins), endocannabinoids and γ-aminobutyric acid[29] as well as the neuropeptides NPY, Agouti-related peptide (AgRP), pro-opiomelanocortin (POMC) and cocaine and amphetamine-regulated transcript (CART), α-melanocyte-stimulating hormone (α-MSH), and melanocortin-4 receptor (MC4R). Gut hormones that have been suggested as regulators of food intake include GLP-1, CCK, PYY, PP, OXM, amylin, insulin, glucagon, orexin, bestatin, and ghrelin. This section focuses on GLP-1, PYY, CCK, insulin, and ghrelin and their effects on neurotransmitters and neuropeptides.

In the CNS, both the hypothalamus and the brainstem integrate signals from the periphery to regulate food intake. Multiple nuclei in the hypothalamus are involved. The lateral hypothalamus, the ventromedial hypothalamic areas, the paraventricular nucleus, and the arcuate nucleus are likely the most important. Signals come from the periphery through neural pathways as well as endocrine mechanisms (possibly through a leaky blood–brain barrier near the median eminence of the hypothalamus or the area postrema of the brainstem).[30]

GLP-1, PYY, and CCK, act as satiety factors through neural pathways. Receptors for these hormones are found on vagal afferent nerve fibers in the splanchnic region.[31,32] Peripheral infusions of GLP-1 and PYY have led to decreases in food intake at meals, although oral administration has led to mixed results because of their short half-lives. CCK also reduces meal size; however, long-term administration leads to an increase in meal frequency, compensating for the decrease in meal size. Leptin and insulin are thought to act similarly through endocrine mechanisms. Both leptin and insulin stimulate neurons in the hypothalamus that produce POMC and CART. POMC/CART stimulate α-MSH production, which binds to MC4R in the paraventricular nucleus to suppress food intake. They also inhibit neurons that produce NPY and AgRP, which are potent stimulators of food intake.[28,30] The phenotype of leptin deficiency in *ob/ob* mice and in human congenital leptin deficiency is similar, including early-onset obesity, increased food intake, hypometabolism, hyperinsulinemia, and defective function of the hypothalamopituitary-thyroidal axis. Replacement of leptin in deficient humans has dramatic effects on intake of food, but none on the basal metabolic rate, even in the face of weight loss. Deletion of the NPY gene partially reverses the phenotype in *ob/ob* mice, confirming the balance between the actions of leptin and NPY. Although no monogenic disorders of NPY overexpression have been reported in humans, leptin receptor deficiency also presents with severe hyperphagia and weight gain. In addition, mild growth retardation and altered IGF-I secretion in these children suggest that the leptin receptor may interact with other hormonal systems.

Ghrelin, so named because of its action on the hypothalamus as a growth hormone–releasing peptide, is produced predominantly by parietal cells of the stomach, as well as in the intestines and pancreas in smaller amounts. Plasma ghrelin is increased by fasting and decreased by feeding, and levels are low in patients with obesity. Levels increase after energy restriction, but not as much after gastric bypass surgery, suggesting that ghrelin may be, in part, responsible for the inability to maintain energy restriction long term but may help to explain the reduced appetite seen in some patients after gastric bypass surgery. Ghrelin reaches the brain both by vagal innervations and through the blood–brain barrier. Ghrelin has the opposite effect of leptin in the CNS in that it stimulates food intake by activating neurons that produce NPY and AgRP and inhibits neurons that express POMC/CART.

Stimuli-Evoked Responses

The anticipatory responses to a meal are mediated by the CNS. Visual, olfactory, and auditory senses as well as the presence of food in the mouth can activate secretory responses from the salivary glands, stomach, and pancreas, and can initiate relaxation in the stomach, contraction of the gallbladder, and relaxation of the sphincter of Oddi. These actions prepare the GI tract to initiate digestion when the meal arrives. This preparation is important because digestive products of foodstuffs (e.g., amino acids, free fatty acids) are important stimuli in creating the maximum responses necessary to digest and absorb a meal. Thus, these nutrient products must be produced early in the meal. This cephalic phase of the meal is mediated through various brain centers, but the efferent signals all reach the gut through the vagus nerve. Once the meal enters the GI tract, the ENS becomes activated and works in concert with the CNS. For example, distention of the esophagus and/or stomach causes a contractile response mediated entirely by the ENS.

The most documented anticipatory CNS-mediated response is the cephalic phase of gastric secretion. Sensory input from the eyes, nose, ears, and mouth send afferent signals to the dorsal vagal complex in the midbrain, where they are integrated and transmitted to GI organs by vagal efferent nerves. In the stomach, the response is to produce acid and pepsin. Acetylcholine release from the vagus nerve stimulates pepsinogen release into the lumen of the stomach. In the distal stomach, the vagal efferent nerves activate the ENS to produce GRP and acetylcholine to release gastrin, stimulating acid and pepsinogen production. Thus, when food enters the stomach, some of the protein is rapidly converted to oligopeptides by the action of pepsin, produced from pepsinogen and activated in the presence of low pH. These oligopeptides stimulate the release of more gastrin to perpetuate the digestive process. In this process, as well as in other anticipatory responses, appetizing meals stimulate more response than bland or unappetizing meals. Thus, the higher centers of the CNS are important in regulating the initial response of the GI tract.

Although these anticipatory responses are clearly present at each meal, it is not certain to what degree they are essential for the assimilation of nutrients. For example, the stomach can be removed and digestion and absorption can proceed fairly completely. Anticipatory responses to a meal may be more important in determining the amount of food eaten at a meal than the absorption of nutrients. The loss of anticipatory relaxation of the proximal stomach allows only small volumes to be consumed at one time, and maintenance of sufficient food intake to maintain weight becomes difficult. Although this deficit can be overcome by cognitive training, the response to a meal is impaired. Impairment in the senses of sight, taste, and/or smell affects the cognitive drive that creates the desire to eat.

Mouth

Chewing and salivary secretion form the food into a round and smooth portion that can be swallowed. The mouth serves as the receptacle for these two functions: secretion and motility. Secretion into the oral cavity originates from the salivary glands and consists of fluid, electrolytes, and proteins. The structure and function of salivary glands, composed of acini that secrete their products through ducts, are analogous to the pancreas. Chloride (Cl) enters the lumen of the salivary gland through Cl channels, and Na enters paracellularly to maintain electroneutrality. In the ducts, the fluid is modified as Na and Cl leave the lumen; some Na is exchanged for K and some Cl is exchanged for HCO_3^-, thus producing a final salivary secretion rich in HCO_3^-. Stimulation of the parasympathetic nerves is the major factor in regulating salivary secretion by direct acinar and duct cell innervation and by altering the blood supply. However, vasoactive peptides are also released to regulate blood flow. Sympathetic nerve input also stimulates secretion, but to a much lesser extent.

Whole saliva is a complex solution of proteins, peptides, enzymes, hormones, sugars, lipids, and other compounds. It contains both salivary and nonsalivary components. The nonsalivary components include nasal and bronchial secretions, blood components, epithelial linings, food components, micro-organisms, and gingival crevicular fluid. Salivary glands (parotid, submandibular glands, sublingual glands, and minor salivary glands) produce salivary secretions from acinar cells. Proteins present in salivary secretions are important during the initial stages of nutrient assimilation. The influence of ptyalin, or salivary amylase, on starch digestion in the mouth and esophagus is small because of the short residence time of food in the mouth. However, in the stomach, the attachment of ptyalin to its substrate protects the enzyme from inactivation at the slightly acid environment (pH of 5 to 6) of the stomach when it is buffered by food. Thus, the enzyme achieves a significant initial hydrolysis of dietary starch while still in the stomach. In addition, a non–bile salt-dependent triglyceride lipase is produced by Ebner glands at the base of the tongue. The amount of digestion of triglycerides by this lipase is small, and the best dietary substrates for this enzyme are triglycerides that contain medium-chain fatty acids. The salivary glands also secrete haptocorrin (also known as R protein), a carrier protein that protects vitamin B_{12} from acid-peptic digestion in the stomach.[33] Many other proteins are found in whole saliva, with some estimates of at least 2,290 different proteins and significant overlap (27%) of proteins that are found in plasma. However, although the 22 most abundant proteins in plasma account for 99% of plasma's total protein content, only 40% of saliva's total protein content comes from the top 20 most abundant proteins.[34] These proteins serve a variety of functions in addition to digestion, including protecting against demineralization, aiding in remineralization, wound healing, immune defense, and protection from attack by oral micro-organisms. The most abundant classes of salivary proteins include mucins, amylase, basic proline-rich proteins, acidic proline-rich proteins, glycosylated proline-rich proteins, "S" cystatin, histatin, IgA, IgG, and statherin.[35] Saliva also may contain biomarkers of disease in the head and neck region, celiac disease, and possibly even Alzheimer's disease. However, the exogenous influence of food and oral hygiene products limit the ability to extract true biomarkers of disease processes.[36,37]

The taste of food is an important regulator of food intake and is governed both by input from smell (the olfactory bulb) and by taste receptors in the tongue. Seven transmembrane receptors (7TM) are now known to occur not only in the tongue but also in the GI tract, endocrine glands, and adipose tissue. These 7TMs are activated by amino acids, peptides, carbohydrates, or free fatty acids.[38] Taste tissue in the tongue contains receptors responding to amino acids and peptides (T1R1/T1R3, GPRC6A, CaR), monosaccharides (T1R2/T1R3), and FFAs (GPR120, FFA1). The FFA transporter CD36 also functions as a taste receptor and appetite regulator on the tongue.[39]

The motility functions of the oral cavity are coordinated with the upper esophageal sphincter to propel the food bolus into the esophagus. This action requires the coordination of extrinsic muscles to modify the shape of the pharyngeal cavity and close the airways and of intrinsic muscles to propel the bolus caudally. These two groups work in succession so that food does not reflux into the nose or larynx. These muscular units work in reverse order during the act of vomiting, again with the purpose of preventing the luminal contents from entering the airways.

Esophagus

The esophagus carries the food bolus from the mouth to the proximal stomach. Relaxation of the upper esophageal sphincter occurs immediately after swallowing, along with increased pharyngeal pressure. These pressure changes move the bolus into the esophagus. The esophagus is the first gut organ in which the phenomenon of peristalsis is encountered. Peristalsis along the length of the esophagus (primary peristalsis) is enhanced by distention in the esophagus produced by the food bolus (secondary peristalsis). The coordinated caudal movement of contraction and relaxation waves moves the food bolus along the length of the esophagus. The act of swallowing initiates both pharyngeal and esophageal peristalsis and relaxation of the lower esophageal sphincter, allowing the swallowed bolus to enter the proximal stomach. Immediately after a swallow, the lower esophageal sphincter pressure also decreases to that of the stomach and remains low until the swallow is completed. At the end of the swallow, the lower esophageal sphincter contracts, stripping the end of the esophagus of any remaining food contents. The most important neurotransmitters for the motility pattern in the esophagus are acetylcholine (contraction) and VIP/NO (relaxation). Although the esophagus is often depicted as an open tube, the walls of the esophagus are approximated to each other during fasting conditions and in areas not distended by a food bolus during feeding. Thus, the bolus cannot travel down the esophagus in the absence of peristalsis. Surprisingly, gravity is not a significant factor in the function of the esophagus.

Stomach

The food bolus enters the stomach as large particles after chewing action in the mouth. In the stomach, the food is mixed and ground with secreted fluid and enzymes and converted to a suspension of particles small enough to pass the pylorus into the duodenum. In addition, fats are converted into an emulsion by mixing action, and small amounts of fatty acids and monoglycerides are formed. Protein and starch digestion also proceeds to create monomeric and oligomeric nutrients that can act further in the duodenum to potentiate the intestinal response to a meal. The two major components responsible for these overall actions of the stomach are motility and acid or peptic secretion.

The anticipatory cephalic phase and distention of the stomach by a meal both lead to receptive relaxation of the proximal stomach, thus accommodating the meal without increasing gastric pressure. Vagal afferent fibers in the gastric wall respond to changes in tension in the muscular coat of the stomach. These responses are processed in the dorsal vagal nucleus in the medulla and create vagal efferent responses that not only relax the proximal stomach but also increase gastrin, acid, and pepsinogen secretion; initiate antral and gallbladder contraction; relax the sphincter of Oddi; and stimulate pancreatic secretion. These vagal reflexes are important in the coordinated function of the organs of the upper GI tract (stomach, duodenum, gallbladder, and pancreas), and are part of the reason that these organs are considered as a cluster unit. The likely neural mediators of these reflexes are VIP and NO. Although the functions of the four upper GI organs are being considered separately, it is important to realize that these functions do not proceed in isolation but are part of a carefully programmed response involving the entire cluster unit.

Antral (distal stomach) contractions are initiated by distention of the stomach. The propulsion, grinding, and retropulsion action in the distal stomach serves to grind the meal into small pieces and mix it with gastric secretions rich in acid and pepsin. The food bolus is ground until the particle size becomes less than 2 mm, so it can pass through the pylorus during the propulsive component. Peristalsis in the stomach is slow, at a frequency of approximately three cycles per minute, mediated in large part by vagal and intrinsic gastric wall cholinergic neurons.

Gastric emptying is a closely regulated phenomenon and is modulated by factors other than particle size. The fastest rate of gastric emptying occurs with isotonic solutions. Most solid meals produce hypertonic solutions, and most liquids are either hypotonic or hypertonic. Thus, most meals will not be emptied at the fastest possible rate. The rate of gastric emptying after a meal is normally approximately 2 mL/minute. At this rate, the digestive and absorptive functions of the upper small intestine are not overwhelmed. Other inhibitory mechanisms affecting the rate of gastric emptying involve H^+ ion concentration and energy load delivered to the duodenum.

Another major function of the stomach is to produce secretions rich in H^+ and pepsinogen. The parietal and chief cells are most responsible for the products entering the gastric lumen in response to a meal (**Table 41.3**), and occur in both the cephalic and gastric phases of gastric secretion, as mentioned. Postprandially (gastric phase), the volume of gastric secretion increases and the ion concentration changes, almost entirely the result of parietal cell secretion. Nonparietal cell secretion from mucous and chief cells contributes HCO_3^--rich fluid in the fasting state. After a meal, H^+ is exchanged for Na^+, and Cl^- replaces HCO_3^- secretion. Most of these secretory changes occur during the gastric phase of acid secretion, which occurs maximally about 60 to 90 minutes after ingestion of food. In healthy persons, the predominant proteins secreted into the stomach include pepsins A and C as well as gastric lipase and gastric intrinsic factor. However, in disease states, such as gastric cancer or chronic gastritis, the range of secreted proteins become more complex and includes not only the pepsins and gastric lipase but also albumin, transthyretin, IgG, IgA, calgranulin A, and α_1-antitrypsin.[40,41] Multiple studies in gastric cancer proteomics have identified proteins that are upregulated in this patient population, including heat-shock 27-kDa protein, nicotinamide N-methyltransferase, and annexin 2.[42]

Table 41.3 Gastric Cell Secretory Products and Function

Cell Type	Product	Function
Surface cells	Mucus	Lubrication
Neck cells	Bicarbonate Trefoil peptides	Protection
Parietal cells	Hydrogen ion Intrinsic factor	Protein digestion Binding of cobalamin (vitamin B_{12})
Chief cells	Pepsinogen Gastric lipase	Protein digestion when activated Triglyceride digestion, not requiring bile salt, MCT >LCT
Endocrine cells	Gastrin Histamine Somatostatin	Release of histamine Stimulation of acid secretion Inhibition of acid secretion

LCT, long-chain triglyceride; *MCT*, medium-chain triglyceride.

The control of parietal cell secretion in the gastric phase involves multiple cell types: parietal cells, enterochromaffin-like (ECL) cells, somatostatin (D) cells, ECs, and gastrin (G) cells. These cells are distributed in two different anatomic portions of the stomach; the parietal, ECL, and fundic D cells in the fundus, and the antral D cell and G cells in the antrum. Gastrin, acetylcholine, and histamine are the major stimulators of parietal cell acid secretion via receptors on the basolateral surface of the parietal cell. Although the histamine pathway is mediated by a different messenger than the acetylcholine and gastrin pathway, they all result in stimulation of the proton pump on the apical membrane. In addition, histamine is the main determinant of gastric acid secretion, possibly through a potentiating effect on acetylcholine and gastrin. Gastrin also influences the production of histamine from ECL cells. Gastrin is released from the G cells in response to a meal by multiple mechanisms; acetylcholine and GRP acting through vagal and intrinsic nerves release gastrin during the cephalic and gastric phases of secretion, and luminal amino acids released by pepsin also stimulate gastrin release during the gastric phase. Gastrin acts in an endocrine fashion by binding to CCK-B receptors on the ECL cell, causing histamine release by exocytosis. Somewhat later, synthesis of histidine decarboxylase is activated, causing additional histamine production. Finally, gastrin stimulates ECL cell growth. As a result of these three effects, ECL cell histamine production is increased and drives parietal cell activation and secretion. Gastrin accounts for approximately 70% of the stimulated histamine release, the remainder being driven by acetylcholine via muscarinic receptors, epinephrine via adrenergic receptors, and gastrin directly via CCK-B receptors. Acetylcholine stimulates acid secretion directly via M3 receptors on parietal cells and, indirectly, by binding to M2 and M4 receptors on D cells, which inhibit somatostatin secretion. Ghrelin also stimulates acid secretion and glutamate indirectly by inhibiting somatostatin secretion.[5]

Gastric acid secretion is further regulated by feedback inhibition, mediated largely by somatostatin released from specialized endocrine (D) cells in the antrum and fundus.[43] D cells typically act locally, either through cytoplasmic processes or the local circulation; therefore, fundic D cells are more important than antral D cells in regulating histamine secretion from ECL cells and directly inhibiting acid secretion in the parietal cells, whereas antral D cells exert their effects mainly on G cells and EC cells in the antrum. Different conditions may mediate the release of somatostatin from D cells in these two locations, and a host of factors are involved in the control of somatostatin secretion, including gastrin, GRP, VIP, pituitary adenylate–cyclase activation peptide (PACAP), β2/β3-adrenergic agonists, secretin, atrial natriuretic peptide (ANP), adrenomedullin, amylin, adenosine, and calcitonin gene–related peptide (CGRP), which stimulate secretion of acetylcholine, histamine, and interferon-γ inhibit somatostatin. Gastric pH is also involved in the control of acid secretion. When the pH in the antral lumen falls to less than 3.0, somatostatin is released from antral D cells, thus inhibiting gastrin release from G cells by paracrine mechanisms. Moreover, luminal acid directly decreases gastrin release from G cells. This example of the regulation of gastric acid secretion after a meal is provided as one of the best-known examples of the intricate and complex coordination of GI function, using elements of the CNS, ENS, and GI hormones.[43]

Gastric parietal cells also produce intrinsic factor, a carrier protein necessary for ileal absorption of vitamin B_{12}. Vitamin B_{12} fits into a hydrophobic pit of the intrinsic factor protein, forming the intrinsic factor–vitamin B_{12} complex that is the obligatory ligand for receptor-mediated absorption in the terminal ileum.

Duodenum

The duodenum is at the center of another elaborately coordinated regulatory process, integrating the functions of gastric emptying, bile formation, gallbladder and duodenal motility, and pancreatic and biliary secretion. For this reason, the concept of the duodenal cluster unit has been developed. This concept is also embryologically cogent. Each of the organs of the duodenal cluster unit (stomach, duodenum, liver, common bile duct, gallbladder, and pancreas) are derived from closely related structures at an early stage of fetal development. The liver, gallbladder, common bile duct, and ventral pancreas bud off together from the antimesenteric side of the duodenum, whereas the dorsal pancreatic bud develops from the mesenteric surface. The ventral pancreas then rotates to join the dorsal pancreas. It is not surprising, therefore, that sensors in the duodenum can regulate function in the other organs of the cluster unit.

The duodenum acts as both a simple mixing chamber and a regulatory center by containing cells and nerve endings that sense nutrient content, pH, osmolarity, and distention. The major hormones involved in regulating the duodenal cluster unit are CCK and secretin, although their effects are not exclusive. Moreover, the GI hormones that act in the duodenal cluster unit may act via an endocrine mechanism (through the bloodstream) or via paracrine mechanisms (locally within the intestinal mucosa). An acidic pH leads to release of secretin and activation of extrinsic and intrinsic nerves to increase pancreatic and biliary secretion of water and HCO_3^-.[44] The presence of digestive products of nutrients (amino acids, fatty acids, monosaccharides) leads to release of CCK and activation of extrinsic and intrinsic nerves, which inhibit gastric emptying and acid secretion, stimulate gallbladder contraction, stimulate pancreatic enzyme secretion, and initiate the small bowel motility pattern of the fed state. In addition, taste receptors have been found in the intestinal mucosa, where they act to trigger responses to nutrients.[45,46] G-protein–coupled receptors (GPCRs) for sweet, bitter, and umami (amino acid) taste have been found on enteroendocrine L cells and intestinal brush cells. The sweet sensors may stimulate secretion of GLP-1 and GIP. They may also enhance glucose absorption via upregulation of mRNA encoding sodium-glucose cotransporter-1 (SGLT1) and T1R2/T1R3 sweet receptor-stimulated translocation of glucose transporter 2 (GLUT2) to the apical membrane of enterocytes, although further research in intact animals is needed to fully understand these interactions. Bitter taste receptors, or taste 2 receptors (TAS2Rs) are primarily expressed on the tongue to protect against ingestion of hazardous compounds. However, TAS2Rs have been found to be expressed throughout the gut, and are hypothesized to trigger antibacterial mechanisms on stimulation by bitter compounds.[47] Much less is known about the intestinal receptors involved in lipid sensing, although research suggests that T2R receptors on enteroendocrine cells may sense lipids, either in addition to sensing bitter, or by using bitter as a surrogate marker for lipids, leading to release of CCK. Glutamate is involved in the sensing of amino acids, and one receptor for glutamate, mGluR1 has been found on the apical membranes of chief cells. In the duodenum, glutamate stimulates the appearance of PepT1 (an oligopeptide transporter), with resultant rapid internalization of T1R1, T1R3 and α-gustducin; however, the significance of these findings is unclear.[46]

Gastric acid secretion can be inhibited by neural or hormonal systems originating in the duodenum (**Table 41.4**). This process is distinguished from inhibition by antral somatostatin in that the regulation of gastric acid secretion by duodenal acid,

Table 41.4 Negative Regulation of Gastric Acid Secretion

Region	Stimulus	Mediation	Inhibits Gastrin	Directly Inhibits Acid
			Release	Secretion
Oxyntic gland area	Acid, CGRP, secretin, VIP	Somatostatin (endocrine)	Yes	Yes
Antrum	(pH <3.0)	Somatostatin (paracrine)	Yes	Yes
	Acid	Nervous reflex	No	Yes
		Decreases gastrin release from G cells	Yes	No
Duodenum	Hyperosmotic	Unidentified, might include CCK, peptide-YY, secretin, neurotensin, GLP-1 (peptides with enterogastrone activity)		
	Solutions, nutrients		No	Yes
Duodenum and jejunum	Fatty acids	GIP GLP-1, delayed gastric emptying	Yes	Yes

CCK, cholecystokinin; *CGRP*, calcitonin gene related peptide; *GIP*, glucose dependent insulinotropic peptide; *GLP*, glucagon like peptide; *VIP*, vasoactive intestinal polypeptide.

hyperosmolarity, and fatty acids also leads to inhibition of gastric emptying. In this way, the duodenal mucosa is doubly protected from an excessive influx of acid. GIP, formerly called gastric inhibitory polypeptide, released by the duodenum, inhibits gastric acid secretion, and stimulates insulin release from pancreatic β cells.

The release of CCK from duodenal endocrine I cells after a meal is of crucial importance for meal digestion. CCK acts as a hormone to stimulate pancreatic secretion and increase antral, pyloric, and duodenal contractions. Moreover, CCK, acting as a neurocrine peptide, stimulates vagal afferent fibers that form part of the vagal efferent outflow after a meal, with subsequent effects on proximal gastric relaxation, increased acid output, gut motility, and pancreatic secretion. In fact, most of the effects of CCK after a meal may be through its role as a neurocrine peptide. CCK is important in regulation of the biliary system and its components. This action is mediated by both the hormonal and the neurocrine functions of CCK. Food intake stimulates the release of CCK, which, in turn, acts humorally on CCK-A receptors in the gallbladder to stimulate gallbladder contraction. Moreover, in response to sensory afferent nerves activated by CCK, vagal efferent nerves mediated by acetylcholine contract the gallbladder, and vagal afferent nerves releasing VIP/NO relax the sphincter of Oddi. The result is that CCK supports concentrated bile entry into the duodenum.

A complex system regulates the release of CCK from endocrine I (CCK-secreting) cells in the duodenum. Luminal nutrients, especially protein, amino acids, and free fatty acids, initiate the signal. Protein is involved in stimulating the release of three peptides which, in turn, release CCK-monitor peptide produced in pancreatic acinar cells, diazepam-binding inhibitor from porcine intestine, and CCK-releasing peptide produced in rat duodenal mucosal cells. Pancreatic phospholipase A_2 from pancreatic juice may also act as a secretin-releasing peptide (**Table 41.5**). Release of both peptides is mediated by parasympathetic (vagal) efferent nerves. Between meals, these peptides are degraded by luminal trypsin that is highly concentrated. Thus, little CCK secretion occurs during fasting. However, as large quantities of protein enter the gut after a meal, the amount overwhelms luminal trypsin activity, and most of the putative regulatory peptides escape degradation. In this way, the ingestion of protein regulates the release of CCK which, in turn, stimulates the release of proteolytic enzymes from the pancreas in conjunction with vagal efferent stimulation.

Another important role of the duodenal cluster unit is to neutralize the gastric acid delivered to the proximal duodenum and maintain a constant intraluminal pH. Multiple organs are involved in this regulation, including the duodenal mucosa, biliary system, and pancreas. The meal itself provides buffers, mostly

Table 41.5 Phases of Pancreatic Secretion After a Meal

Phases	Pancreatic Response (%)	Stimulants	Vagal-Cholinergic Pathways
Cephalic	25	Sight, smell, taste, eating	Cholecystokinin, secretin
Gastric	10	Distention	Enteropancreatic reflexes
Intestinal	50–75	Amino acids	Other hormones (?)
		Fatty acids	
		Calcium and hydrogen ions	Other hormones (?)

in the form of peptides and fatty acids. Most of the neutralization comes from HCO_3^- secreted from the pancreas, biliary ducts, and duodenal mucosa. Secretin mediates the biliary and pancreatic response, whereas the ENS mediates the mucosal response. The major mucosal sensor is the endocrine secretin (S) cell, which is activated to release secretin when the luminal pH falls to less than 4.5. A low intraluminal pH stimulates duodenal HCO_3^- secretion via both central and enteric nerves, local prostaglandin production and hormones. Mechanisms invoked can be cyclic adenosine monophosphate mediated (dopamine agonists, enteropeptide receptor agonists, VIP), cyclic guanosine monophosphate mediated (guanylin, uroguanylin), calcium mediated (muscarinic M_3 agonists, CCK_A agonists), or inhibition by neurotransmitters (α_2-adrenoceptor agonists, NPY receptor agonists, NO).

A final important role for the duodenum is to produce and maintain isotonicity of luminal contents, thereby avoiding large shifts of fluid across the semipermeable membrane of the gut. This function is performed by the duodenal mucosa alone, without other organs in the cluster unit. Most meals are either hypertonic or hypotonic. Thus, the duodenum must either add or absorb fluid and electrolytes. Under normal circumstances, the maximum rate of gastric emptying is about 2 mL/minute, so the proximal duodenum is not presented with larger volumes than it can accommodate for isotonic adjustment.

Enterocytes also produce brush-border membrane hydrolases (**Table 41.6**). These hydrolases are glycoproteins and are secreted from the cell and inserted into the brush-border membrane; the hydrophobic end attaches to the membrane, whereas the oligosaccharidase component projects into the lumen. Brush-border hydrolases are only expressed in villous enterocytes, predominantly in the duodenum and jejunum, with lower expression distally. Enzyme expression and activity are regulated by transcriptional, translational, and post-translational processes, which are modified by dietary intake, pancreatic enzyme activity, trophic factors, and GI diseases.

Thus, passage through the duodenum changes the physical properties of the meal because of the contributions of the organs in the duodenal cluster unit. Large amounts of pancreatic hydrolases and bile salts are added, digesting nearly all ingested macromolecules (except dietary fiber) to oligomers or monomers, solubilized in a form compatible with absorption. Intestinal fluid leaving the duodenum is more isosmotic, and the pH is made more neutral.

Liver and Biliary System

Bile is composed of both bile salts and excretory endogenous and exogenous compounds. Bile salts are crucial for solubilization and absorption of lipid-soluble nutrients. They are synthesized and secreted by the liver, conjugated to either taurine or glycine to improve solubility, stored and concentrated in the gallbladder, and delivered to the duodenal lumen in response to a meal. Bile is an aqueous solution made up of about 95% water with bile salts, fatty acids, cholesterol, protein, vitamins, bilirubin, phospholipids and other compounds (e.g., drugs, environmental chemicals).[48] Bile salts are cofactors for pancreatic lipase and make lipids soluble by forming micelles. Between meals, the gallbladder stores and concentrates the bile salts that are extracted by the liver from the blood. Two factors regulate the supply of bile salts after a meal: CCK-stimulated contraction of the gallbladder and relaxation of the sphincter of Oddi releases the gallbladder contents into the upper duodenum. This provides the first and immediate load of bile salts to enhance digestion by pancreatic lipase and fatty acid/monoglyceride and cholesterol solubilization. Second, bile salts move down the small intestine to the ileum, where they are absorbed by a receptor-mediated mechanism and returned to the liver via the bloodstream. The enterohepatic circulation (reabsorption in the ileum, uptake by the liver, and secretion back into the intestine) preserves bile salts and diminishes the need for new synthesis in the 1 to 2 hours after a meal. The entire body pool of bile salts (~3 to 4 g) is recirculated two to four times after each meal, providing 6 to 16 g of bile salts to the upper duodenum during the first hours after a meal. With a total luminal volume from diet and secretions of 2 to 3 L after each meal, this provides a large margin of safety for maintaining a luminal concentration above the critical micellar concentration of 2 mM needed for lipid solubilization and activation of pancreatic lipase.[49]

Proteomic analysis of bile has improved throughout the years. There have been greater than 2,500 proteins identified in bile, including potential biomarkers for biliary and pancreatic diseases.[50]

Table 41.6 Intestinal Brush-Border Membrane Hydrolase Activity in Normal Human Biopsy Specimens

Hydrolase	Approximate Activity (Units/G Protein)
Glucoamylase	250
Sucrase	100
α-Dextrinase	100
Lactase	45

Pancreas

Three phases of pancreatic secretion occur after a meal: cephalic, gastric, and intestinal (see Table 41.5). These phases classify the multitude of events that occur postprandially. As seen in the other digestive organs, pancreatic secretion is mediated by neural (vagal) efferent responses and by gut hormones.[51] In humans, the cephalic phase of secretion is mediated by the vagus nerve. In this phase and in the gastric phase, the pancreas secretes mostly water and HCO_3^-. PP, located in specific cells in the pancreatic islets, acts as a negative feedback mechanism for the vagally stimulated portion of pancreatic secretion. It is released in response to vagal efferent stimulation, and inhibits the vagal effect on the pancreas.

In the intestinal phase, pancreatic enzymes are added to the large volume of fluid secreted. As noted, products of proteolysis and lipolysis stimulate the CCK (endocrine I) cells to release CCK, which likely acts neurally (through vagal stimulation of afferent pathways) and humorally on the pancreatic acinar cells to produce enzymes. Although the mechanism by which CCK senses these products remains unclear, it may involve a trypan-sensitive, CCK-releasing peptide, which leads to the release of 5-hydroxytryptamine from EC cell, which activates substance P neurons in the submucosa, turning the signal into a cholinergic signal. This signal may then be transmitted to CCK-, peptide-producing cells, thus leading to the release of CCK. In fact, both CCK and 5-HT stimulate pancreatic secretion through vagal pathways and are the primary stimulators of postprandial pancreatic enzyme secretion.[40] H^+ ions stimulate the S cell to release secretin, which acts on the pancreatic duct cells to secrete a HCO_3^--rich fluid to neutralize gastric acid and allow pancreatic enzymes to be effective. Enteropancreatic reflexes within the ENS, sensitive to distention, osmolarity, and various nutrients, stimulate pancreatic enzyme secretion mediated by gastrin, GRP, VIP, and NO.[51] Other neuropeptides that may increase pancreatic secretions include substance P, neurotensin, serotonin, and calcitonin gene–related peptide. In addition, insulin may modulate pancreatic secretion by potentiating the response of secretin and CCK.

Less is known about the inhibition of pancreatic secretion, compared with the stimulation of pancreatic secretion. Hyperglycemia and amino acid infusions can decrease pancreatic secretion, and it has been suggested that this effect is mediated by glucagon; however, this remains unclear. Somatostatin may also inhibit pancreatic secretions, possibly by blocking the effect of CCK at a central vagal site. CCK secretion and, therefore, pancreatic secretion can also be inhibited by nutrients in the colon. PYY (produced in the distal small bowel and colon) may also be involved in inhibiting pancreatic secretion via cholinergic and hormonal routes.[52,53] GLP-1 also inhibits pancreatic secretion after ileal perfusion, which appears to be mediated through a central vagal mechanism.[53] PP, localized to the islet of Langerhans, is also involved in inhibiting pancreatic secretions and likely modulates vagal efferent output to the pancreas via the CNS. Ghrelin and leptin (a hormone produced by adipose tissue) also inhibit pancreatic secretion by a neurohormonal mechanism.[52] A feedback mechanism exists, with pancreatic enzymes and CCK secretion, possibly through a CCK-releasing factor protein, but this feedback mechanism is not completely understood in humans.[52,53] Bile and bile salts have also been postulated to be involved in a pancreatic secretion negative feedback loop; however, this remains controversial.[52]

About 0.7% to 10% of the pancreatic juice is protein. Most proteins secreted are enzymes and proenzymes, inactive precursor forms of enzymes that are cleaved to their active forms in the duodenal lumen. Most of the enzymes secreted by the pancreas are proteases and are secreted in an inactive precursor form to prevent digestion within the pancreas (**Table 41.7**).

Table 41.7 Pancreatic Proteases

Protease	Function
Endopeptidases	
Trypsin	Cleaves internal bonds at lysine or arginine residues and cleaves other pancreatic proenzymes
Chymotrypsin	Cleaves bonds at aromatic side chains or neutral amino acid residues
Elastase	Cleaves bonds at aliphatic amino acid residues
Exopeptidases	
Carboxypeptidase A	Cleaves aromatic amino acids from carboxy terminal end of protein and peptides
Carboxypeptidase B	Cleaves arginine or lysine from carboxy terminal end of proteins and peptides

Trypsinogen accounts for 40% of the pancreatic protein secreted. In the intestinal lumen, trypsinogen is activated to trypsin by the enzyme enterokinase, produced by duodenal enterocytes. Trypsin, in turn, converts trypsinogen and all the other proenzymes to their active form, and initiates the intraluminal phase of intestinal digestion. Both CCK and insulin stimulate the production of pancreatic enzymes, which is thought to be mediated through an increase in translation, because mRNA concentrations in the acinar cells do not increase after stimulation with CCK and insulin.

Pancreatic insulin secretion in response to a meal is enhanced by the release of GIP and GLP-1 from enteroendocrine cells in the duodenum (K and L cells, respectively). Although GIP was first recognized for its ability to inhibit gastric acid secretion, it was later found that the major function of this peptide is to mediate meal-stimulated insulin release from the pancreas. This observation led to changing the name of GIP from gastric inhibitory polypeptide to glucose-dependent insulinotropic polypeptide. Intraluminal glucose stimulates GIP and GLP-1 release, which act humorally to augment the glucose-mediated release of insulin from β cells in pancreatic islets. This action of GIP helps to maintain blood glucose concentration within a reasonable range after a meal, and provides another example of the redundancy characteristic of the regulation of GI function after a meal. GLP-1 may also act on neurons in a similar manner as CCK. GLP-1 also has a second phase of secretion, when nutrients reach L cells in the distal small bowel, which is likely responsible for the other effects of GLP-1, including the ileal brake and appetite control.[54,55]

Nutrient Absorption

Fluid and Electrolytes

The GI tract absorbs large volumes of fluid each day. Approximately 9 L of water are delivered to the upper small intestine daily from dietary intake (2,000 mL), saliva (1,500 mL), gastric secretions (2,500 mL), bile (500 mL), pancreatic secretions (1,500 mL), and small intestinal secretions (1,000 mL). Ninety-eight percent of the daily fluid load is absorbed, whereas only 100 to 200 mL/day is excreted in stool; approximately 85% (7.5 L) of water is absorbed in the jejunum and ileum and 13% (1.4 L) in the colon.

Water is absorbed passively throughout the intestine and is regulated primarily by active electrolyte absorption.[56] Specific features of epithelial cells are important in regulating fluid and electrolyte absorption. First, the apical (luminal) membrane contains specific electrolyte transporters and channels.[57] Second, the basolateral (serosal) membrane contains a Na^+ pump that provides the drive for electrolyte absorption. Third, intestinal epithelial cells are linked to each other by tight junctions that are located close to the apical surface.[58] The permeability of the intestinal epithelium depends on the number of tight junctions. The permeability of these intercellular junctions to solute, ion, and water movement decreases distally through the intestine. Therefore, the jejunum is more permeable than the ileum, which is more permeable than the cecum, which is more permeable than the rest of the colon.

Fluid and electrolytes are absorbed from the intestinal lumen directly through (transcellular pathway) or between (paracellular pathway) epithelial cells. Passive transport does not require energy and can occur transcellularly or paracellularly.[59] The lipid content of the epithelial cell membrane prevents passive diffusion of charged electrolytes. Specialized proteins present in the apical membrane form channels or pores, which permit electrolyte transport. Passive transport through membrane channels is regulated by concentration and electrochemical gradients across the membrane. Ion channels are usually specific and can be opened or closed by cellular messages. In the open state, more than 1 million ions can pass through per second, but no ions pass when the channel is closed.[60] Passive transport can also occur via carriers, which are proteins located in the cell membrane. Carriers are specific for certain solutes or ions and facilitate their passive movement along a concentration or electrochemical gradient across the cell membrane. Carrier-mediated transport is much slower than movement through channels.

Active transport requires energy and permits the movement of a solute or ion against a concentration or electrochemical gradient. Active transport only occurs transcellularly and is mediated by a "pump" that moves ions in and out of the cell. The most important epithelial cell pump is the Na pump (also known as Na/K-ATPase) and moves three Na ions across the basolateral membrane in exchange for two K ions (**Figure 41.10**). Therefore, the Na pump makes the intracellular Na concentration low, and the intracellular potential difference negative, compared with the extracellular environment.

Secondary active transport represents transport that combines both passive and active processes. For example, the negative intracellular voltage of epithelial cells enhances cation entry and anion exit from the cell. Thus, ions may move passively against their concentration gradients because of the electrical potential difference across the cell generated by the active Na pump. The use of oral rehydration therapy in patients with severe diarrhea, such as those with cholera or short bowel syndrome, takes advantage of secondary active transport and the Na-glucose cotransporter, SGLT1 protein, in small intestinal epithelium (see Figure 41.10).[61] This transporter, present in the apical membrane, binds both Na and glucose. Glucose is transported across the cell membrane into the cell against its concentration gradient because of the low Na concentration and the negative potential difference present in the cell. As glucose accumulates in the cell, it moves along its concentration gradient across the basolateral membrane via a specific transport carrier. Similar Na cotransport mechanisms also facilitate absorption of amino acids, vitamins, and bile salts.[62] The Na pump also drives passive absorptive or secretory transport of H^+, Cl, K, and HCO_3^- (see Figure 41.10). Transport regulation can occur at the channel, carrier, or pump levels.

Water is absorbed passively throughout the GI tract and follows the absorption of electrolytes and other osmotically active nutrients. As noted, water moves both transcellularly and paracellularly in response to an increase in osmolarity of the intracellular and subepithelial spaces. Na absorption is the most important factor in regulating water absorption. The Na-nutrient cotransporter and electroneutral NaCl exchange transporter are responsible for most water absorption. Furthermore, water absorbed between epithelial cells can increase the absorption of solutes present in water, a process known as solvent drag (**Figure 41.11**). Movement of both Na and water in response to an osmotic gradient is much greater in the jejunum than in the ileum because of the increased permeability of the junctions between epithelial cells in the jejunum, compared with those in the ileum. In the jejunum, Na is primarily absorbed by uptake via the Na-nutrient cotransporter and solvent drag. Therefore, ingestion of fluids or a meal with low Na content

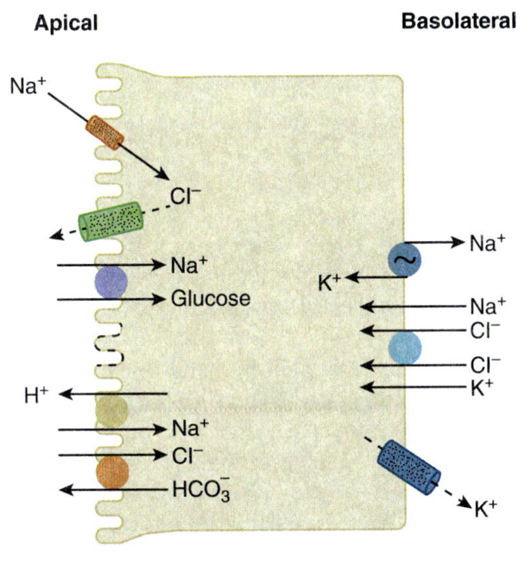

Figure 41.10 Electrolyte and solute absorption. Sodium can travel from the intestinal lumen into the epithelial cell by (a) an ion channel (apical side top), (b) the sodium (Na^+)-glucose cotransporter (apical side middle), or (c) an Na^+-hydrogen (H) exchanger (apical side bottom). The release of H^+ creates a favorable gradient for bicarbonate (HCO_3) exit, which facilitates chloride (Cl) entry through the Cl/HCO_3 exchanger. The Na/potassium (K)/Cl cotransporter in the basolateral membrane also increases Cl uptake. Electrogenic Cl secretion occurs via a Cl channel on the apical membrane. Intracellular glucose accumulation favors glucose transport across the basolateral membrane via a specific carrier protein. The Na pump (Na/K-adenosine triphosphatase [ATPase]) provides the energy for these processes by generating low intracellular Na concentrations and a transmembrane electrochemical gradient.

Reproduced from Sleisinger MH, Fordtran JS, Scharschmidt BF et al, eds. Gastrointestinal Disease. 5th ed. Philadelphia: WB Saunders, 1993:954–76.

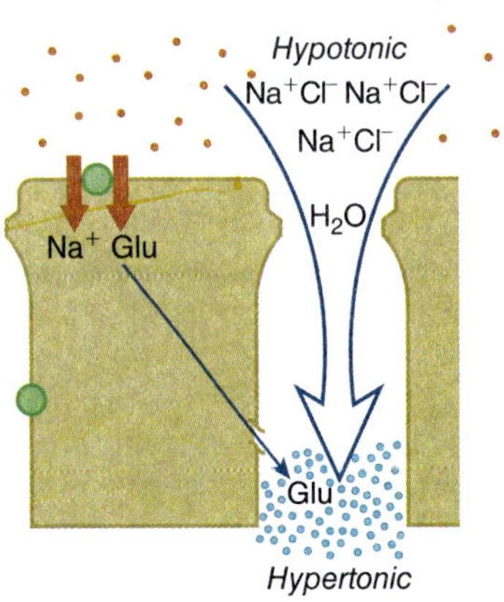

Figure 41.11 Electrolyte and water absorption in the jejunum. The sodium (Na)–glucose (Glu) cotransporter, SGLT1, present in the small intestine binds both sodium and glucose and transports them across the epithelial cell membrane. As glucose accumulates in the cell, it moves along its concentration gradient across the basolateral membrane via a specific transport carrier. Water is absorbed passively by both transcellular and paracellular routes in response to an increase in osmolarity of the intracellular and subepithelial spaces. The Na–nutrient cotransporter shown in this figure and the electroneutral Na chloride (NaCl) exchange transporter are responsible for most water absorption. Water absorbed between epithelial cells can increase the absorption of solutes present in water by "solvent drag."

decreases the osmolality in the upper small intestine and causes net secretion of water and Na into the lumen. Patients with a jejunostomy and less than 100 cm of jejunum have difficulty maintaining fluid and electrolyte balance; thus, longer lengths of small intestine are required for optimal fluid and electrolyte absorption. Balance studies performed after liquid ingestion in patients with a very short bowel ending in a jejunostomy show that drinking solutions with Na concentration less than 75mmol/L leads to net Na and water losses, whereas drinking a solution with at least 75 mmol/L causes net Na and fluid absorption.[61] Although most water is absorbed in the small intestine, approximately 1 to 1.5 L enters the colon each day. Ninety-five percent of the fluid that enters the colon is absorbed. The colon has the capacity to absorb up to approximately 5 L of fluid per day.[63]

Lipid

Approximately 100 g of fat, equivalent to approximately 40% of total energy intake, is consumed daily in an adult Western diet. The majority (92%–96%) of fat intake consists of long-chain triglycerides (LCTs) whereas the remainder includes cell membrane phospholipids, sphingolipids, galactolipids, cholesterol, other sterols, and fat-soluble vitamins. In addition, about 20 g of phospholipids and 1 to 2 g of cholesterol in bile enter the duodenum each day.[64] The upper limit of normal fecal fat output while consuming a 100-g fat diet is less than 7 g per day. Therefore, at least 95% of fat delivered to the intestine is usually absorbed. Most dietary fat is absorbed before the fat contained in a meal reaches the ileum. However, even when no dietary fat is ingested, a small amount of fat can still be detected in stool because of the contribution from endogenous sources.[64]

Assimilation of dietary fat provides a good general index of intestinal absorptive function because it is the most complicated macronutrient to digest and absorb. Triglycerides are particularly difficult to digest and absorb because they are insoluble in water. Therefore, absorption requires (a) the breakdown of ingested fat into an emulsion, which enhances contact between lipolytic enzymes and triglycerides; (b) enzymatic hydrolysis of triglycerides; (c) water-soluble micelle formation, which permits transport across the unstirred water layer to intestinal epithelial cells; (d) uptake of fatty acids by epithelial cells; (e) repackaging of fatty acids into water-soluble chylomicrons within the epithelial cell; and (f) secretion of chylomicrons into the systemic circulation by lymphatic vessels.

The stomach is important in initiating fat digestion. Approximately 20% of ingested triglycerides are hydrolyzed in the stomach by gastric lipase. Gastric lipase is produced by chief cells in an acidic environment and is resistant to denaturation by gastric pepsin. The stomach aids food digestion via gastric muscular contractions breaking down food into small particles. Then action by gastric acid, lipase, and pepsin facilitate release of dietary lipids. In the duodenum, the lipids are further stabilized by the addition of bile salts and phospholipids secreted by the gallbladder, creating an emulsion. The presence of gastric acid in the duodenum stimulates secretin release from duodenal mucosa. Secretin enters the portal circulation and stimulates the pancreas to secrete HCO_3^-, which raises the intraluminal pH to greater than 6. The presence of fatty acids and amino acids in the duodenum stimulates CCK release from duodenal mucosa, which then enters the portal circulation and stimulates the pancreas to secrete lipase, colipase, and other digestive enzymes, and stimulates gallbladder contraction and bile flow into the duodenum.[65] Lipase and colipase are secreted by the pancreas in a 1:1 molar ratio and act at the surface of the emulsion particles to hydrolyze triglycerides to monoglycerides and fatty acids.[64] The near-neutral pH of the duodenum maximizes lipase and colipase activity, as pancreatic lipase is not functional in an acidic environment. Colipase is a critical cofactor for lipolysis by acting as a link between pancreatic lipase and triglycerides. Pancreatic lipase is unable to gain access to triglycerides within the emulsion without colipase because of interference from bile salts and phospholipids coating the emulsion particles. Pancreatic lipase is responsible for the majority of intestinal triglyceride lipolysis. The pancreas also secretes bile salt–activated lipase, which hydrolyzes ester linkages in cholesterol, phospholipids, and fat-soluble vitamins. Fat digestion by gastric and pancreatic lipases is effective and most ingested triglycerides are hydrolyzed within the first 100 cm of jejunum.[66]

Bile salts contain both a hydrophilic and hydrophobic portion. Cholesterol, phospholipids, and bile salts form water-soluble mixed micelles. These mixed micelles have a hydrophobic interior and hydrophilic exterior. Dietary lipids enter the hydrophobic center of the micelle and the hydrophilic exterior allows the micelle to emulsify in an aqueous environment.[67] Thus, bile salts make fatty acids, monoglycerides, cholesterol, and other intraluminal lipids soluble in water by "hiding" these hydrophobic lipid products inside of mixed micelles (**Figure 41.12**). There is a critical micellar concentration of bile salts (2 mmol/L) to allow micelle formation.[67] Inadequate bile salts result in fat malabsorption and steatorrhea.

The products of triglyceride digestion by pancreatic lipase can also coalesce to form vesicles. Lipid within these vesicles is usually transferred to micelles but the vesicles can also transport lipid directly to the mucosa.[68,69] It is believed that the formation of vesicles permits absorption of more than half of ingested triglycerides when bile salts are absent, such as in patients with severe cholestasis. Dietary lipid soluble vitamins D, E, and K require micelle formation for adequate absorption.

Mixed micelles must pass through the unstirred water layer located at the surface of the intestinal epithelium to deliver their contents to the apical portion of the enterocytes. The quantitative diffusion of fatty acids through the unstirred water layer is enhanced more than 100-fold when fatty acids are carried within micelles rather than as monomeric fatty acids. Fatty acid and lipid uptake across the epithelial brush-border membrane occurs by passive diffusion, facilitated diffusion, and active transport. Members of the ATP-binding cassette transporter superfamily (ABCG) identified in the human small intestine may transport fatty acids, monoglycerides, and cholesterol across the enterocyte apical membrane.[70] Defects in the ABCG5 and ABCG8 genes are associated with the rare autosomal-recessive disorder sitosterolemia, which results in the inability to secrete plant sterols, with accumulation in the blood and arteries. CD36, a scavenger receptor is found throughout the small bowel, decreasing in concentration from the proximal to the distal end. CD36 is required for uptake of very-long-chain fatty acids.[71] Cholesterol may also be absorbed via the scavenger receptor-B1 (SR-B1) pathway or with CD36 at the enterocyte brush border.[72]

After fatty acids and lipolytic products enter the intestinal epithelial cell, they are bound to cytosolic fatty-acid binding proteins. These binding proteins are found predominantly in jejunal enterocytes and their expression declines progressively down the GI tract. Fatty acid–binding proteins are important for intracellular trafficking by directing fatty acids from the cell membrane to the smooth endoplasmic reticulum for triglyceride synthesis. The intracellular fatty acid transport system enhances

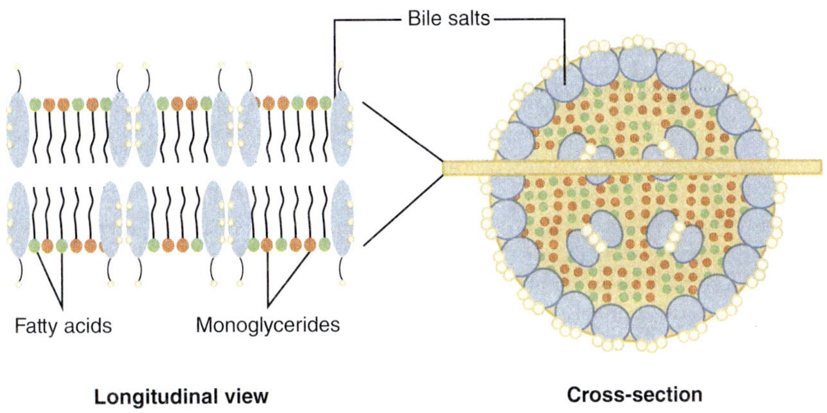

Figure 41.12 Structure of a mixed lipid–bile salt micelle. The products of lipolysis are solubilized in the interior of the particle. The bile salt molecules orient with their hydroxyl groups (black circles) facing the aqueous phase or when they are in the interior of the micelle, facing each other. Fatty acids and monoglycerides orient in the micelle with their polar head groups in contact with the aqueous phase and their hydrocarbon tails in the interior of the micelle.

Reprinted with permission from Chang EB, Sitrin MD, Black DD, eds. Gastrointestinal, Hepatobiliary, and Nutritional Physiology. Philadelphia: Lippincott-Raven, 1996:147.

fatty acid uptake by maintaining a fatty acid concentration gradient, preventing potentially toxic accumulation of fatty acids.

Fatty acids and monoglycerides present in smooth endoplasmic reticulum are used to produce triglycerides and phospholipids. Triglyceride, phospholipid, cholesterol, and fat-soluble vitamins are joined by apolipoproteins, made in the rough endoplasmic reticulum, to form chylomicrons. Chylomicrons consist of a core of triglyceride, cholesterol esters, fat-soluble vitamins, and other lipids and a surface of phospholipids, free cholesterol, and apolipoproteins (apolipoproteins B-48, A-IV, and A-I) (**Figure 41.13**). These nascent chylomicrons are transferred to the Golgi apparatus and are incorporated into secretory vesicles, which fuse with the basolateral membrane of the epithelial cells, and are released by exocytosis into the extracellular space. These chylomicrons move through the lamina propria into the villous core, which contains a network of capillaries and a single lymph lacteal. Chylomicrons cannot enter the bloodstream directly because they are too large to pass through the fenestrations between capillary endothelial cells. Fat absorption stimulates lacteal distention, which produces gaps between endothelial cells and facilitates chylomicron uptake by the lymphatic system and ultimate delivery into the systemic circulation. Newly formed circulating chylomicrons interact with other circulating lipoproteins and exchange components, thereby acquiring additional apolipoproteins, including apolipoproteins C-II and E, which serve important functions in chylomicron metabolism.[73]

Medium-chain triglycerides (MCTs) are fatty acids with a chain length of 6 to 12 carbon atoms. A typical American diet contains predominantly long chain triglyceride (LCT) and does not contain appreciable amounts of MCTs., Absorption of MCTs differs markedly from that of LCTs. MCTs are hydrolyzed more rapidly by lipases than LCTs, do not require bile salts for absorption because MCTs are water soluble, and can be absorbed as an intact triglyceride. Once inside the intestinal epithelial cell, MCTs and medium-chain monoglycerides are rapidly hydrolyzed to medium-chain fatty acids by specific cellular lipases. Medium-chain fatty acids do not bind to fatty-acid binding proteins, are not re-esterified to triglycerides, and are not packaged in chylomicrons. After leaving the enterocyte, medium-chain fatty acids enter the portal system, where they become bound to albumin and are transported to the liver. Given the complexity of LCT

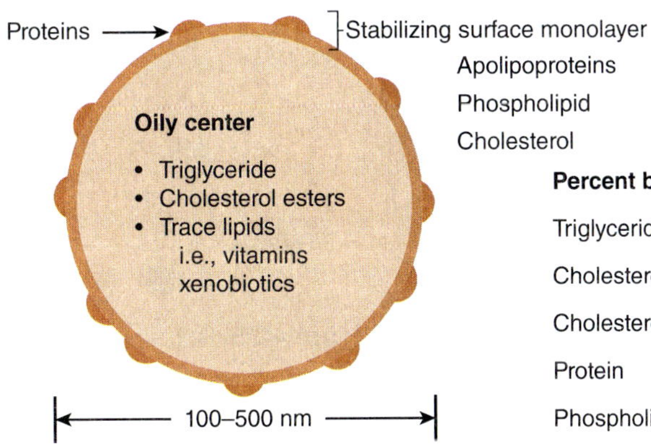

Percent by weight	
Triglyceride	90%
Cholesterol ester	1–3%
Cholesterol	1%
Protein	1.5%
Phospholipid	6.5%

Figure 41.13 Chylomicrons are fat droplets that are coated with a monolayer of phospholipid and cholesterol. Dispersed in the monolayer are apoproteins (Apo) A-1, apoA-IV, and Apo B and probably also some Apo C-11 and Apo C-111. These proteins help direct the tissue uptake and catabolism of the chylomicrons. In the circulation, chylomicrons also acquire additional apoproteins. Chylomicrons predominantly carry triglyceride and cholesterol, fat-soluble vitamins, and small amounts of many other trace lipophilic molecules.

Reprinted with permission from Patton JS, Hoffman AF. Lipid Digestion. Undergraduate Teaching Project, Unit 19. Bethesda, MD: American Gastroenterological Association, 1986.

assimilation and absorption vs the simpler process of MCT metabolism, specialized diets can be prescribed for patients who have fat malabsorption. These diets are generally low in fats (LCT) and include supplementation with MCT oil or MCT-enriched liquid formulas.

Carbohydrate

A typical Western diet contains 200 to 300 g/day of carbohydrate (45% of total energy intake), which includes starch derived from cereals and plants (amylose, amylopectin); sugars derived from fruits and vegetables (glucose, fructose, sucrose), milk (lactose), and refined processed foods (sucrose, fructose, oligosaccharides, polysaccharides); and fiber derived from plant wall polysaccharides and lignin. Starch consists of long chains of glucose molecules joined together by α-1,4 linear linkages (amylose), or by both α-1,4 linear and -1,6 branched linkages (amylopectin) (**Figure 41.14**). Ingested sugars consist of monosaccharides (glucose, fructose) and disaccharides (sucrose: glucose linked with fructose, lactose: glucose linked with galactose). Complex carbohydrates also include dietary fiber. Approximately 10 to 20 g of dietary fiber is ingested daily in a typical Western diet, consisting mostly of celluloses and hemicelluloses and also pectin, gums, and lignin. Cellulose is glucose molecules joined together by β-1,4 linear linkages. Hemicellulose consists of pentose and hexose monomers joined together by both β-1,4 linear and branched linkages. Most dietary carbohydrates are digested and absorbed in the jejunum. However, dietary fiber cannot be digested in the small intestine, in part because the β-1,4 bond is resistant to the action of amylase.[74]

Amylase, secreted by the salivary glands and pancreas, cleaves the α-1,4 bond but not the α-1,6 bonds of starch, generating linear oligosaccharides, branched α limit dextrins, maltotriose, and maltose (see Figure 41.14). Pancreatic amylase is responsible for most starch digestion. The contribution from salivary amylase depends on the duration and amount of contact between salivary amylase and ingested starch; slow and careful chewing would increase starch digestion by salivary amylase. Furthermore, the physical interaction between salivary amylase and its substrate provides some protection from acid denaturation after ingested carbohydrates and amylase enter the stomach.

Brush-border membrane hydrolases, found predominately in the jejunum, glucoamylase (maltase), sucrase-α dextrinase (sucrase-isomaltase), and lactose-phlorizin hydrolase (lactase) are required to hydrolyze dietary disaccharides and the products of amylase starch digestion before their total absorption can occur. Glucoamylase cleaves α-1,4 bonds releasing one glucose molecule at a time from oligosaccharides. Sucrase-α dextrinase represents two enzyme subunits with distinct properties. Sucrase hydrolyzes sucrose to glucose and fructose and short-chain α-1,4–linked oligosaccharides to glucose. α-Dextrinase also hydrolyzes short-chain α-1,4–linked oligosaccharides to glucose and can hydrolyze α-1,6–linked α limit dextrins. Lactase hydrolyzes lactose to glucose and galactose. Digestion of disaccharides, trisaccharides, and oligosaccharides at the surface brush-border membrane usually exceeds the capacity of monosaccharide enterocyte transport. However, hydrolysis of lactose is the rate-limiting step for absorption because lactase activity is lower than all other brush-border hydrolases, even in persons who have complete lactase activity (see Table 41.6). When there are undigested oligosaccharides, they are osmotically active and can lead to diarrhea, while undigested disaccharides lead to the formation of gas and resultant flatulence.

Transport proteins, known as glucose transporters, present in the apical and basolateral cell membranes, facilitate monosaccharide absorption (**Figure 41.15**).[75] These transporters are expressed on small-intestine enterocytes. Glucose and galactose absorption occur principally by a Na-monosaccharide cotransporter, SGLT1, which delivers two Na molecules for every monosaccharide across the cell membrane. GLUT5 facilitates Na-independent fructose absorption, but fructose is not as well absorbed as glucose.[76] Glucose and fructose exit the enterocyte through the basolateral membrane into the portal circulation via the Na-independent GLUT2 transporter.

Starch and dietary fiber not absorbed in the small intestine enter the colon, where colonic bacteria metabolize these

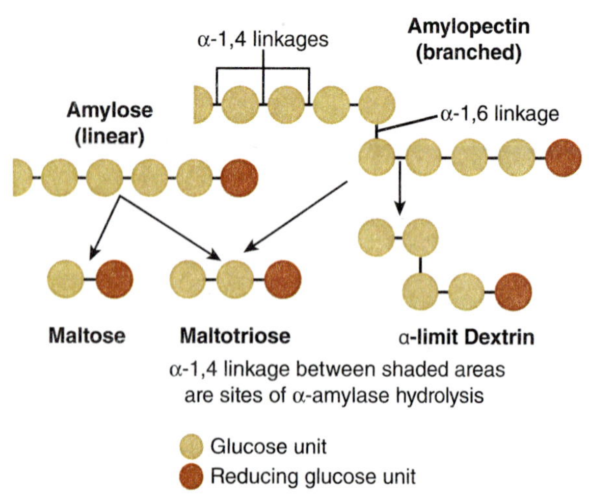

Figure 41.14 Starch (amylose and amylopectin) digestion by pancreatic amylase produces maltose, maltotriose, and α-limit dextrans.

Reprinted with permission from Chang EB, Sitrin MD, Black DD, eds. Gastrointestinal, Hepatobiliary, and Nutritional Physiology. Philadelphia: Lippincott-Raven, 1996:122.

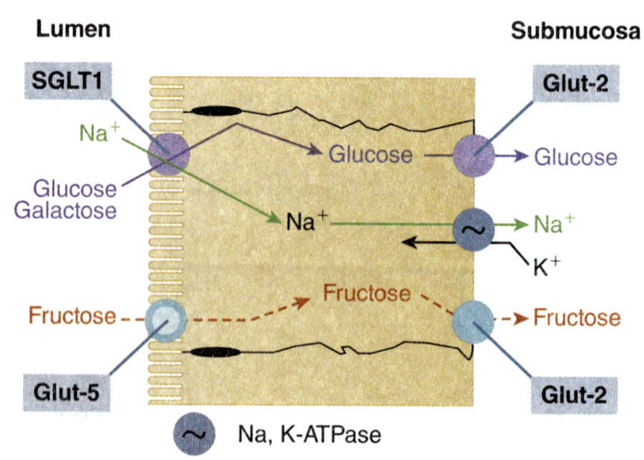

Figure 41.15 Monosaccharide absorption by the enterocyte occurs by active and passive processes. Glucose and galactose are absorbed with a sodium (Na)-dependent glucose/galactose transporter (SGLT1), driven by an Na^+ gradient generated by Na^+/potassium (K^+)/adenosine triphosphatase (ATPase) at the basolateral membrane of the enterocyte. Fructose is absorbed by facilitated diffusion using a glucose transporter called GLUT5. All monosaccharides exit the enterocyte by facilitated diffusion via a carrier protein called GLUT2.

Reprinted with permission from Chang EB, Sitrin MD, Black DD, eds. Gastrointestinal, Hepatobiliary, and Nutritional Physiology. Philadelphia: Lippincott-Raven, 1996:125.

carbohydrates to short-chain fatty acids (SCFAs) (acetate, propionate, and butyrate), carbon dioxide, and H^+. Absorption of SCFAs occurs via the monocarboxylate transporter, MCT1, allowing the colon to salvage a considerable amount of energy that would otherwise be lost in stool.[77] Butyrate is a preferred large intestine fuel and provides approximately 70% of daily colonic fuel requirements, propionate may have important effects on hepatic metabolism, and acetate provides an important systemic fuel. Further, SCFA absorption enhances colonic Na and water absorption.

Protein

A typical Western diet contains approximately 70 to 100 g of protein, representing approximately 15% of total daily energy intake. Additional proteins enter the GI tract from salivary, gastric, biliary, pancreatic, and intestinal secretions (~35 g/day), desquamated intestinal cells (~30 g/day), and plasma protein (~2 g/day). Normally, more than 95% of the total protein load delivered to the gut is absorbed.

Protein digestion begins in the stomach, where a family of proteolytic enzymes (pepsins) hydrolyzes peptide bonds.[78] Pepsins are generated from pepsinogens, which are inactive proenzymes produced mostly by chief cells. When exposed to the acidic environment of the stomach, pepsinogen undergoes a conformational change, with loss of a terminal peptide to its active pepsin form. Pepsin is active at low pH and is inactivated in an alkaline environment. The stomach is not essential for protein digestion, and patients with atrophic gastritis and even a total gastrectomy can absorb protein normally. However, the release of amino acids in the stomach triggers part of the initial GI response to a meal: gastric acid secretion, CCK secretion, gastrin secretion, and gastric emptying.

Most protein digestion occurs in the duodenum; 60% is digested by the time it reaches the proximal jejunum. Several proteases (see Table 41.7), in the form of inactive proenzymes, are secreted into the duodenal lumen by the pancreas. Enterokinase, a brush-border enzyme released into the lumen by bile acids, cleaves the N-terminal peptide from trypsinogen to form trypsin. Trypsin activates additional trypsinogen molecules as well as the other pancreatic proenzymes. Pancreatic proteases act as either endopeptidases (trypsin, chymotrypsin, and elastase) or exopeptidases (carboxypeptidase A and B). Endopeptidases and exopeptidases work efficiently in concert to degrade protein into smaller subunits. However, proline-containing peptides are resistant to cleavage by pancreatic proteases. After pancreatic hydrolysis of proteins is complete, approximately 70% of amino nitrogen is present as oligopeptides containing two to six amino acids and 30% as free amino acids.

The mucosal brush-border membrane contains approximately 20 peptidases that cleave specific amino acids present in dipeptides, tripeptides, and oligopeptides, thereby generating free amino acids, dipeptides, and tripeptides.[79,80] These peptidases are produced by enterocytes, released at the cell surface, and anchored to the cell membrane with the active site projecting into the lumen. Most brush-border peptidases are aminopeptidases, which sequentially cleave the N-terminal amino acid from oligopeptides. Several specific peptidases are able to hydrolyze proline-containing peptides, thus compensating for the inability of pancreatic proteases to cleave the proline-amino acid bond.

Amino acids, dipeptides, and tripeptides generated by intraluminal and brush-border protein hydrolysis, are transported across the enterocyte apical cell membrane by specific transport mechanisms.[81] Amino acid transport on the apical cell membrane is facilitated by several transport systems.[82] Some amino acids can use different carriers because of overlapping specificity between systems. Amino acid transport in most systems is coupled to Na uptake (Na dependent). However, amino acid uptake can also occur in Na-independent processes by facilitated or passive diffusion. Dipeptides and tripeptides are absorbed intact by intestinal epithelia with a Na-independent process that involves H^+-peptide cotransport along a H^+ gradient. Human dipeptide/tripeptide transporters belong to the proton-dependent oligopeptide transporter family and include hPepT1, expressed only in the intestine, and hPepT2, expressed in both kidney and intestine. Peptide transport represents an important mechanism for amino acid absorption; in the jejunum, most amino acids are absorbed faster as peptides than as free amino acids.

Peptides present in the enterocyte are hydrolyzed to individual amino acids by several cytosolic peptidases. Dipeptidases and tripeptidases are more abundant inside the enterocytes than the brush-border membrane. Intracellular amino acids are moved out of the enterocyte through the basolateral membrane by active transport, facilitated diffusion, and simple diffusion. During meals, most amino acid transport out of the cell occurs by facilitated or simple diffusion because of the large amino acid concentration gradient across the cell membrane. Several amino acid transport systems have been identified. Passive diffusion and the L, y^+L, A, GLY, and y^+ facilitated carrier systems are involved in amino acid exit from the enterocyte from the basolateral membrane, whereas the active Na^+-dependent A and ASC systems and the Na^+-independent ASC, b^o and y^+ systems are principally involved in amino acid uptake across the apical membrane.[82]

Absorbed amino acids can have several fates: Some provide fuel for the small intestine itself (particularly glutamate and glutamine) and some are used for protein synthesis. Most amino acids are transported into the portal circulation for metabolism in the liver or for subsequent delivery to peripheral tissues via the bloodstream. Despite the presence of intracellular peptidases, approximately 10% of portal blood amino nitrogen is in the form of peptides that have escaped intracellular hydrolysis. Villus cells receive their amino acid requirements topically from absorption of luminal proteins. In contrast, crypt cells receive most of their amino acids from the bloodstream.

Minerals

Dietary mineral absorption involves three steps: (a) intraluminal events that transform ingested minerals into absorbable forms, (b) mucosal events that govern mineral uptake by intestinal epithelium, and (c) postmucosal events that regulate the mineral transport into the mesenteric and portal circulation for subsequent delivery to the liver and peripheral tissues. Although some general comments regarding intestinal mineral absorption are made in this section, the specific absorptive processes for each mineral are reviewed elsewhere in this volume.

Minerals ingested in the diet are frequently bound to proteins within a matrix of organic molecules. Assimilation by mastication, dispersion, acidification, and digestion by pancreatic enzymes convert ingested minerals into forms necessary for effective absorption. Unlike other nutrients, intestinal absorption of some minerals is regulated by body stores to prevent excessive uptake and toxicity. Further, the absorption of one mineral can decrease the absorption of another.

For example, absorptive interactions occur between calcium and magnesium as well as iron, zinc, and copper. These interactions can be used therapeutically; oral zinc supplementation inhibits copper absorption in patients with Wilson's disease who have excessive tissue copper loads.

Mineral absorption can be complicated because some minerals are released into the lumen as charged ions, whereas others are a component of an organic complex. For example, iron is ingested as a component of heme (animal sources) and nonheme (animal and plant sources) compounds. Dietary nonheme iron is usually present in the ferric (Fe^{3+}) form, which is soluble in the acid pH of the stomach but insoluble at pH greater than 3. Other dietary compounds and intestinal secretions can either enhance iron absorption by making iron more soluble (by forming unstable chelates or reducing iron to the more soluble ferrous [Fe^{2+}] form) or decrease iron absorption by making iron less soluble (by precipitating iron or forming stable chelates). Heme iron is soluble at the alkaline pH of the small intestine and is more efficiently absorbed than nonheme iron. Most iron is absorbed in the duodenum, whereas other minerals are absorbed throughout the small intestine.

Vitamins

Water-soluble vitamins (thiamin, riboflavin, niacin, pyridoxine, biotin, pantothenate, folate, cobalamin, and ascorbic acid) are usually present in foods as part of a coenzyme system and are often associated with proteins. This complex arrangement must be digested to a simpler form before the vitamins can be transported across the apical epithelial cell membrane. Vitamins are usually present in the diet in low concentrations and require active carrier systems for adequate absorption. However, water-soluble vitamins are also absorbed by passive diffusion. Therefore, oral vitamin supplementation with large doses can often overcome defects in normal vitamin transport by achieving high intraluminal concentrations. All water-soluble vitamins are absorbed primarily in the upper small intestine with the exception of vitamin B_{12}, which is absorbed in the terminal ileum. The specific mechanisms involved in the absorption of each water-soluble vitamin are reviewed in chapters on those vitamins.

Absorption of fat-soluble vitamins (vitamins A, D, E, and K) requires bile salts for solubilization within micelles, which enhance their delivery through the unstirred water layer to the enterocyte apical membrane. Therefore, the absence of bile salts can seriously impair fat-soluble vitamin absorption, particularly the highly insoluble vitamins D and K. Vitamin K is unique in that the body stores reflect absorption of both vitamin K_1 (phylloquinone), ingested in the diet, and vitamin K_2 (menaquinone), produced by intestinal bacteria. Vitamin K of bacterial origin comes predominantly from vitamin K synthesized by small intestinal bacteria or colonic bacteria that refluxed into the small intestine because absorption by the colon is limited. Once inside the enterocyte, fat-soluble vitamins become incorporated within the core of chylomicrons for transport into intestinal lymphatics. Most ingested fat-soluble vitamins are absorbed in the proximal small intestine, although often less than 50% of total dietary intake is absorbed. Steatorrhea, potentially resultant from pancreatic insufficiency, surgical small bowel resection, or Roux-en-Y resection, will lead to malabsorption of vitamins D, E, A and K. The specific mechanisms involved in the absorption of each fat-soluble vitamin are reviewed in the chapters on those vitamins.

Gut Microbiota

The human GI tract contains approximately 100 trillion bacteria, and has been suggested to be a "microbial organ" within the host[83] or "virtual organ."[84] The gut microbiota not only contains bacteria but other classes of microbes as well, including fungi and viruses; however, this section will focus on the bacteria present in the gut microbiota.

The diversity of this virtual organ has only relatively recently been discovered with advances in detection methods that do not require culturing organisms, as many of the gut microbiota are difficult to culture. Furthermore, the status of a person's gut microbiome is dynamic, depending on changes in environment, diet and overall health. The GI tract is sterile prior to birth with microbial colonization occurring at the time of delivery with variations in the specific microbiomes of infants contingent on their mode of delivery.[85] **Figure 41.16** demonstrates both the microbial diversity in the human gut and the proportion of organisms that are able to be cultured, vs those that are detected only by other methods (in this case, small subunit rRNA-based testing).[86] The organisms in the gut ecosystem have the capacity for interacting with the host and communicating with one another. They are involved with energy metabolism of themselves and the host, and they mediate chemical reactions for the host. For example, anaerobic bacteria in the colon produce SCFAs, especially butyric acid, a preferred substrate for the colonocyte. Subsets of bacteria of a host's microbiota have a synergistic relationship with the overall health of the host. Studies have shown that higher concentrations of Bifidobacterium in infants at age 6 months are associated with lower adiposity at 18 months.[87] However, pathogens can adapt to the conditions provided by the host, as shown by the elaboration by the gut mucosa of a unique respiratory electron acceptor that Salmonella have then mutated to provide a growth advantage.[88] Thus, this relationship is not always mutually beneficial, because pathogens can cause infections and some organisms can produce substances that are carcinogenic.

Although the gut microbiota has one of the highest densities of organisms for any microbial ecosystem, it is not an extremely diverse population. For example, nine phyla are represented in the colon, compared with 20 phyla represented in a soil sample.[89] Significant interindividual variation exists in the microbiota, but intraindividual microbiota populations are consistent within the gut ecosystem compartments.[90] One mechanism whereby bacteria can be modified is the direct transfer of genetic material from bacteria contained in food to the gut microbiota. One study demonstrated this communication of marine bacteria from food to certain Japanese people, allowing them to subsequently metabolize carbohydrates in seaweed.[91] Another mechanism appears to involve the alteration of microbial diversity by carbohydrate-rich diets.[92] It appears that a core microbiota exists that is stable through life, although the bacteria that make up the core may vary from person to person.[93] However, the microbial genes contained within the microbiota (referred to as the microbiome) may be similar; therefore, the shared "core" may be at the gene (core microbiome) rather than organism (core microbiota) level.[94]

The gut ecosystem can be compartmentalized into sections: the oral cavity, esophagus, stomach, small intestine, and colon. The oral cavity shares a similar composition of microbiota as the colon. Per the expanded Human Oral Microbiome Database (eHOMD), 774 oral bacterial species have been identified, with 58% named and cultivated, 16% cultivated but not yet named,

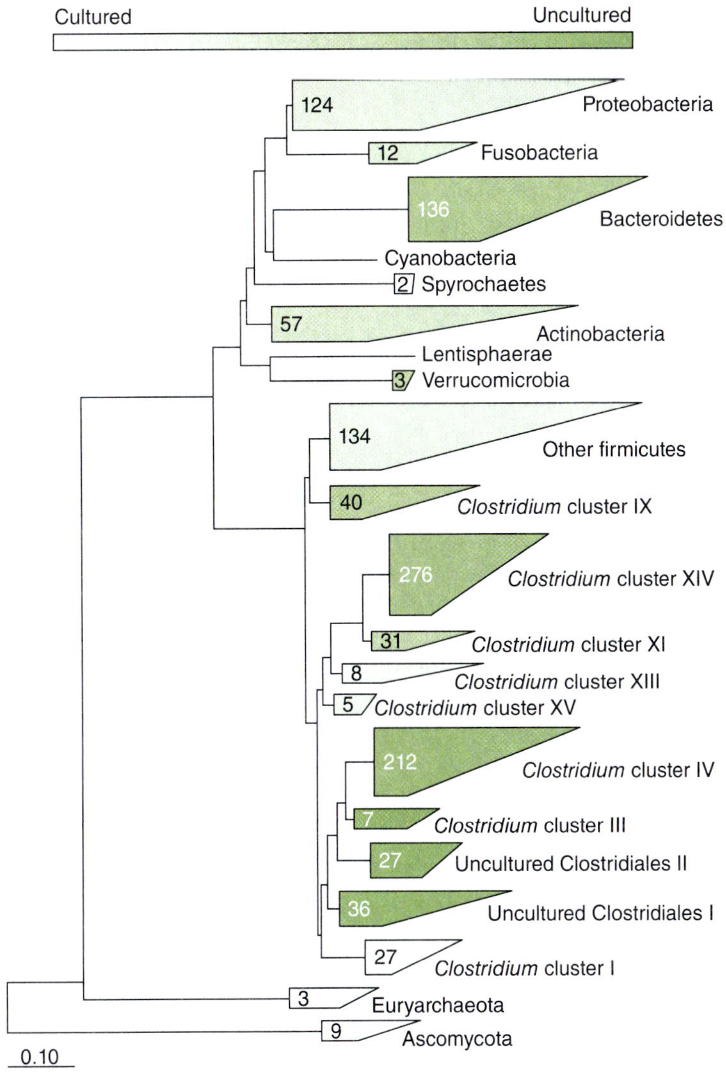

Figure 41.16 Small subunit (SSU) rRNA-based phylogenetic tree of the distinct phylotypes that have been found in the human gastrointestinal tract. The relative proportion of the phylotypes that correspond to cultured representatives is indicated by the darkness of the filling.

and 26% only known as uncultivated phylotypes.[95] However, the distribution of the oral flora is not uniform and bacterial composition and density vary with location. The most densely populated areas are the gingival crevices. Poor oral hygiene and immunologic variations permit overgrowth of subgingival organisms, leading to gingivitis. The esophagus has not been studied as extensively as the colon or oral cavity; however, one study found diversity of phyla similar to the colon and at least 166 species.[96] The data from this study have also raised the possibility of two different types of microbiomes, one associated with gastroesophageal reflux disease. A study of 26 patients compared 16 with either Barrett's Esophagus or distal esophageal/esophagogastric junction cancer to 10 control patients and found significant differences in the esophageal microbiota.[97] Previously, the stomach was thought to be inhospitable to most microbes because of the acidic environment, with survival of a few and colonization only with *Helicobacter pylori* (an important cause of gastritis and ulcer disease). However, more recent research has demonstrated the presence of at least 5 phyla and 102 phyllodes, which differ from those present in the oral cavity or colon, although when *H. pylori* were present, there was less diversity in the microbiota.[98] Using a different technique, 13 phyla and 262 phylotypes have been found in the stomach.[99] The microbiota of the small intestine have not been studied extensively because of the difficulty with obtaining samples, but research suggests that the microbiota diversity from the duodenum and jejunum vary from that found in the terminal ileum, with the most abundant genera coming from Streptococcus, Veillonella, and Clostridium in the proximal small bowel[93] and Bacteroides, Firmicutes, and Proteobacteria dominating the terminal ileum,[100] as well as density of microbes increasing from proximal to distal small bowel.[101] Furthermore, studies have shown that the microbiota of small bowel effluents from patients with ileostomies vary, contingent on carbohydrate intake of the host and what time of day the sample was collected.[101] In the colon, the number of micro-organisms increases 100,000-fold, consisting of nine different phyla, with Firmicutes and Bacteroides dominating.[89] The ileocecal valve represents a physical barrier between the small and large intestine. Resection of the ileocecal valve permits translocation of bacteria flora from the colon to the remaining ileum, where the bacterial population becomes similar to that of the colon.

The interaction between enteric microflora and the host is complex. The presence of enteric organisms enhances defense against pathogenic bacteria by stimulating antibody production, increasing cell-mediated immunity, and preventing the overgrowth of more pathogenic organisms. The mucosal barrier is a combined physicochemical complex function composed, in part, of mucus secretion, mucin glycoproteins, trefoil peptides, and surfactant phospholipids. It forms a separation of luminal contents from the mucosa and creates a framework for host–bacterial interactions. The innate immune system of the GI tract provides another set of defense mechanisms against pathogenic bacteria and parasites (see immune system section). Pattern recognition receptors (PRRs; also called Toll-like receptors) are present in some epithelial cells, and are more broadly expressed in macrophages and other immune cells. These receptors sense the presence of various bacterial macromolecules and initiate a nonspecific inflammatory response. Endogenous antimicrobial peptides, known as defensins, are produced by Paneth cells at the base of intestinal crypts and offer a broad spectrum of antimicrobial activity. Goblet cells also produce small cysteine-rich proteins, which have antihelminthic activity. Normal flora effectively compete for intraluminal fuels and adhere better to the intestinal wall, preventing pathogenic bacteria from establishing residence. The importance of this defense mechanism is illustrated by germ-free animals, who cannot survive exposure to hostile microbes.

The gut microbiota is influenced by diet and interacts with the host in nutrient metabolism. Research suggests that carnivores, omnivores, and herbivores have different microbiota, and that human microbiota resemble that of other omnivorous primates.[102] Data from murine models demonstrated that a switch to a high-fat diet was associated with a decrease in Bacteroides and an increase in both Firmicutes and Proteobacteria. Intestinal bacteria also have important metabolic and nutritional functions, including hydrolysis of cholesterol esters, androgens, estrogens, and bile salts; use of carbohydrate, lipid, and protein; and consumption (vitamin B_{12} and folate) and production (biotin, folate, and vitamin K) of vitamins. All compounds that enter the alimentary tract by ingestion or intestinal secretion are potential substrates for bacterial metabolism (**Table 41.8**). Furthermore, the microbiota may influence and be influenced by obesity. Obese individuals have less diversity in their microbiota and a higher ratio of Firmicutes to Bacteroides,[103] but also may have gut microbiota that are more efficient in nutrient harvest, degrading foodstuffs that would otherwise be indigestible to humans.[94]

Recent research has revealed that the gut microbiota also interacts with noninfectious host disease processes, with studies suggesting that certain bacterial species grow in higher concentrations in patients with colon cancer and inflammatory bowel disease than in control patients.[104,105] Other studies have shown that gut microbiota metabolites may play a role as etiologic factors in the development of irritable bowel syndrome.[106] Research on how this may dictate new modalities of therapeutic intervention are ongoing.

Given the multitude of interactions between the gut microbiota and the host, the gut microbiome has implications on human health that we can manipulate to improve the health of the host. Antibiotics have revolutionized our treatment of infectious disease and have been used to treat pathogenic infections in the GI tract. They have also been used to treat small intestine bacterial overgrowth. Historically, Crohn's disease has also been treated with antibiotics. However, the efficacy of antibiotics in inducing remission is inconsistent, and the relationship between the microbiome and Crohn's disease is not clearly defined.[107] Antibiotics have also been used to treat pouchitis in post colon resection with ileal pouch-anal anastomosis.[108] However, antibiotics can alter the gut microbiota in negative ways as well, leading to a state of dysbiosis (disruption of the gut microbiome).[109] The adult gut microbiota largely rebounds from the effects of antibiotics within 4 weeks, but antibiotic-associated diarrhea can occur and the most common causative agent is *Clostridium difficile*. However, subtle changes in the gut microbiota may persist after antibiotic administration, the effects of which are subjective to the individual and largely unknown.[109,110]

Table 41.8 Biochemical Reactions by Intestinal Bacteria

Reaction	Representative Substrate
Hydrolysis	
Glucuronides	Estradiol-3-glucoronide
Glycosides	Cycasin
Sulfamates	Cyclamate, amygdalin
Amides	Methotrexate
Esters	Acetyldigoxin
Nitrates	Pentaerythritol trinitrate
Dehydroxylation	
C-Hydroxy groups	Bile acids
N-Hydroxyl groups	N-Hydroxyfluorenylacetamide
Decarboxylation	Amino acids
D-Demethylation	Biochanin
Deamination	Amino acids
Dehydroxygenase	Cholesterol, bile acids
Dehalogenase	DDT
Reduction	
Nitro groups	P-Nitrobenzoid acid
Double bonds	Unsaturated fatty acids
Azo groups	Food dyes
Aldehydes	Benzaldehydes
Alcohols	Benzyl alcohols
N-Oxides	4-Nutroquinolone-1-oxide
Nitrosamine formation	Dimethylnitrosamine
Aromatization	Quinic acid
Acetylation	Histamine
Esterification	Gallic acid

Data from Kim YS, Erickson RH. Gastroenterology 1985;88:1071–3.

Other ways to influence the gut microbiota include probiotics and prebiotics (see the chapter 37). Probiotics are live micro-organisms that have beneficial effects on the GI tract, including downregulating inflammatory cytokines, increasing IgA production, improving mucosal barrier function, and inhibiting pathogenic mucosal adherence.[109] Probiotics have been successfully used in inflammatory bowel disease, irritable bowel syndrome, and antibiotic-associated diarrhea. Prebiotics are substances that are not digested by humans but are fermented by gut microbiota and stimulate the growth or activity of the gut microbiota. Prebiotics are also resistant to gastric acid, hydrolysis by host enzymes, or absorption from the host GI tract. It is thought that prebiotics promote the growth of beneficial gut microbes, and they have been shown in mouse models to decrease proinflammatory cytokines.[111] Newer studies have been conducted on the effectiveness of fecal microbiota transplantation for the restitution of microbiota diversity in those who have received antibiotics, with encouraging results.[112] Ongoing research is active on the effects of dysbiosis of the gut microbiome and its influence on multiple disease processes, such as Alzheimer's, Parkinson's, and breast cancer.[113,114]

Our knowledge of gut microbiota and host interactions as well as the effects on human health is still in its infancy. Further research into these interactions and the manipulation of the gut microbiome are challenging but will further our ability to use the gut microbiome to improve human health.

Immune System

The alimentary tract houses a major portion of the body's immune system directed toward defending the host against bacterial, viral, parasitic, and food antigens that are constantly present in the intestinal lumen. The intestinal immune system consists of two arms, the innate and the adaptive immune system. These two systems share some components, and they work in concert to protect the GI tract from pathogens. Components of the innate immune system include specialized epithelial cells, including Paneth cells, and goblet cells, that secrete defensive peptides, as well as antigen presenting cells. This arm of the immune system is more primitive and orchestrates a complex system of both defense and activation of the adaptive immune system as well as tolerance to protect but avoid unnecessary inflammation. This system also includes nonimmunologic components, including gastric acid, digestive enzymes, mucus, bile acids, and peristalsis; all of which make the GI tract less hospitable for many microbes. Moreover, mucosal cells are part of the innate system, including NK-T cells (shared with adaptive), phagocytes, mast cells, and myeloid cells. This system recognizes lipopolysaccharides, peptidoglycans, and lipoteichoic acids from pathogens. Receptors for the innate system are expressed on monocytes, macrophages, dendritic cells, and B cells, as well as epithelial cells. These PRRs include TLRs and macrophage mannose receptors. The number of receptors are limited, as these recognize patterns, unlike adaptive immune cells that produce specific responses. The innate system responds rapidly, within hours. The adaptive system takes days.

Paneth cells reside in the base of crypts near the multipotent stem cells. Paneth cells not only protect the crypt by producing both antimicrobial peptides and enzymes (α-defensins HD-5 and HD-6, group IIA phospholipase A_2, and lysozymes) but also produce inflammatory cytokines that attract antigen-presenting cells and lymphocytes, activating the adaptive immune response. Paneth cells are typically only found in the small intestine but can be seen in the colon when it is inflamed.[115] Current research has found a complex relationship between Paneth cells and intestinal stem cell differentiation, with Paneth cells regulating the secretion of multiple signaling pathways and growth factors.[116]

Nonspecialized epithelial cells can also produce antimicrobial peptides and enzymes, as well as inflammatory cytokines, for activating the adaptive immune response; however, the types of peptides and enzymes are different (β-defensins, athelicidin, bactericidal/permeability-increasing protein). In addition, epithelial cells provide a barrier to pathogens and their toxins through intercellular junctions (tight junctions, adherens junctions, and desmosomes).[117]

Goblet cells secrete mucus, which is thought to provide a barrier at the mucosal surface of the intestine and contains numerous antimicrobial compounds, as well as compounds that may play a protective role in colitis, such as trefoil factor-3.[118]

Antigen-presenting cells are also an integral part of both the innate and adaptive immune response, essentially bridging the two components of the immune system. Multiple cell types can present antigens, such as macrophages, B lymphocytes cells, basophils as well as epithelial cells. However, accumulated data place importance on a specialized macrophage called a dendritic cell, both with activation of an immune response as well as tolerance.[119] It is unclear how these cells acquire antigen. Investigators have hypothesized that these cells may extend protrusions into the lumen to acquire antigen, recruit lymphocytes, induce Ig class switch, and stimulate Tregs,[99] which play an important role in tolerance.[119] Newer investigations have revealed that goblet cells in the small intestine deliver antigen to CD103(+) dendritic cells in the lamina propria.[120]

The adaptive immune system regulates antigen-specific immune responses and consists of the following components: (a) T lymphocytes, (b) B lymphocytes, (c) natural killer cells, (d) myelomonocytic cells (monocytes, neutrophils, eosinophils, and basophils), (e) cytokines, (f) antibodies (IgG, IgM, and secretory IgA), and (g) gut-associated lymphoid tissue (GALT). Adaptive immunity also has two components, humoral and cellular. Antibodies from B lymphocytes defend against extracellular events. Cellular immunity from T cells protects against intracellular processes. Clonally distinct T cells maintain response to specific epitopes, determined by immunoglobulin receptors for B cells and T-cell receptors. Each cell can detect self from nonself. Adaptive immunity depends on tolerance, otherwise autoimmunity occurs. Autoimmunity in the intestine is not common, suggesting that GALT can prevent activation of pathogenic clones. On repeated exposure, the adaptive system responds more rapidly than on previous exposures. This system includes inductive sites where antigen is presented and effector sites in the mucosa where either stimulation or tolerance occurs, depending on the antigens presented to the inductive site.

The secretion of the dimeric Ig, IgA, is an important GI tract protective mechanism. Secretory IgA, the predominant intestinal Ig, is produced by B lymphocytes in the lamina propria. Secretory IgA binds dietary antigens, thereby preventing their absorption, and can bind to pathogenic micro-organisms, thereby preventing epithelial cell adherence and intestinal colonization.

GALT contains anatomically organized and nonorganized compartments within the submucosa, lamina propria, and epithelium to provide specialized host defense functions (**Figure 41.17**). The nonorganized components include lymphocytes, plasma cells, macrophages in the lamina propria and epithelium, and mucosal and submucosal mast cells. The more

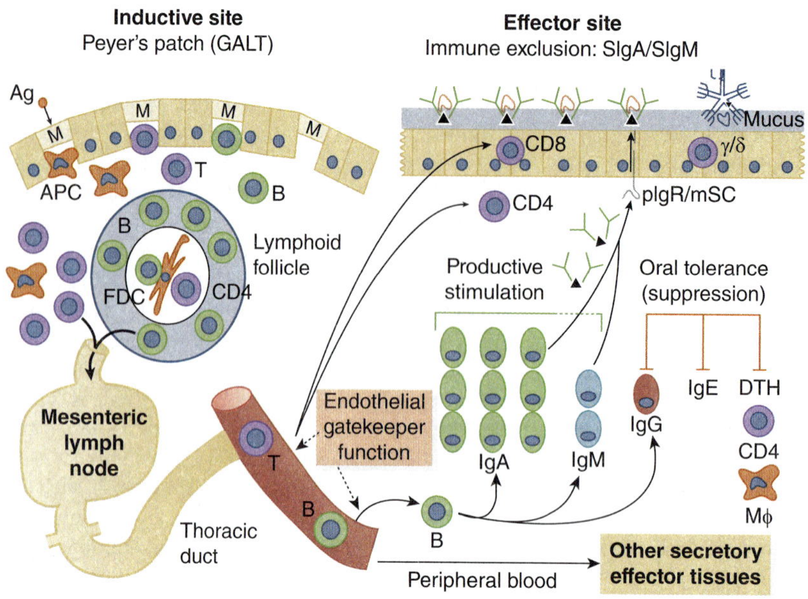

Figure 41.17 Schematic depiction of the intestinal immune system. Inductive sites for mucosal T and B cells are constituted by gut-associated lymphoid tissue (GALT) such as Peyer patches with B-cell follicles and M cell (M)–containing follicle associated epithelium through which exogenous antigens (Ag) are actively transported to reach professional antigen-presenting cells (APC), including dendritic cells, macrophages (MΠ), B cells, and follicular dendritic cells (FDCs). After being primed, naive T and B cells become memory/effector cells and migrate from GALT to mesenteric lymph nodes via efferent lymph and then via the thoracic duct to peripheral blood for subsequent extravasation at mucosal effector sites. This process is directed by the profile of adhesion molecules and chemokines expressed on the microvasculature—the endothelial cells thus exert a local "gatekeeper" function for mucosal immunity. The mucosal lamina propria (effector site) is illustrated with its various immune cells, including B cells (B), Ig-producing plasma cells, and CD4$^+$ T cells. The distribution of intraepithelial lymphocytes (mainly T-cell receptor α/β^+CD8$^+$ and some γ/δ^+ T cells) is also schematically depicted. Additional features are the generation of secretory IgA (SIgA) and secretory IgM (SIgM) via pIgR/membrane secretory component (mSC)–mediated epithelial export. The combined effect of oral tolerance mechanisms, mainly the action of regulatory T cells (not shown), provides a suppressive tone in the gut, normally keeping inflammation driven by IgG and IgE antibodies as well as cell-mediated (CD4$^+$ T cells and MΠ) delayed-type hypersensitivity (DTH) under control.

Reproduced from Brandtzaeg P. Scand J Immunol 2009;70:505–15.

organized structures include Peyer patches, isolated lymphoid follicles (ILFs), cryptopatches, and mesenteric lymph nodes.

Peyer patches are secondary lymphoid tissue in the GI tract. They develop during the prenatal period and are composed of three or more clusters of lymphoid aggregates that release lymphocytes after antigen processing.[121,122] Peyer patches do not contain afferent lymphatics; instead, they sample antigens with an overlying follicle-associated epithelium that contains M cells. M cells provide a selective site for sampling intraluminal antigens by permitting the transport of large molecules and microorganisms. These antigens come into contact with lymphocytes and macrophages located within an indented space below the M cell before entering Peyer patches. Activated lymphocytes from Peyer patches migrate to mesenteric lymph nodes, the systemic circulation, and back to specific mucosal sites, where they provide protective immunity from the offending antigen.

ILFs are non–Peyer patch B-lymphocyte containing aggregates. In their mature form, they look much like Peyer patches except that they lack a discrete T-cell zone. This is thought to be because of how they form; ILFs are thought to be formed from cryptopatches and, therefore, can develop or regress in response to changes in the intestinal flora.[122]

Cryptopatches are small clusters of cells that include dendritic cells, immature hematopoietic cells, very few T or B lymphocytes, and vascular cell adhesion molecule (VCAM) 1+ stromal cells. Some debate remains on the role of cryptopatches, but it is thought that cryptopatches are the precursor lymphoid aggregate that leads to the development of ILFs and may also generate lymphocytes.[123] Furthermore, cryptopatches have yet to be found in humans; however, this may be because of the high prevalence of ILFs in the colon, where most of the biopsies studied from humans have been taken. Like Peyer patches, cryptopatches are thought to be formed early in life. Although the number of cryptopatches and ILFs may change throughout life, the sum of both cryptopatches and ILFs remains constant throughout life.[122]

Mesenteric lymph nodes are found within the mesentery of the small bowel. They are activated within a few hours of oral antigen exposure. Although the mechanism by which they are activated is not completely understood, it likely involves activated T cells migrating to the mesenteric lymph node, dendritic cells migrating to the mesenteric lymph node–presenting antigen, and free antigen reaching the mesenteric lymph node.[121] More recently, the roles, and regulation of Toll-like receptors and NOD-like receptors, collectively referred to as PRRs, which interact with microbial components (e.g., flagellin and lipopolysaccharide) and other ligands, have become increasingly understood in the intestine.[124,125] Intestinal PRRs appear to play key roles in host immunity and inflammatory responses and also interact with the host microbiome to help maintain homeostasis between the intestine and its microbiota. Dysregulation of PRR signaling may be implicated in the development of inflammation and subsequent complications, such as tumorigenesis.[126]

Acknowledgments

The authors acknowledge Shelby Sullivan, David Alpers and Samuel Klein for their contributions to the previous version of this chapter.

References

1. Galura GM, Chavez LO, Robles A, et al. Curr Gastro Rep 2019;21:34.
2. Geibel J. In: Wang TC, Camilleri M, Lebwohl B, et al., eds. Yamada's Textbook of Gastroenterology, John Wiley & Sons, Ltd. 2022;313–33.
3. Dimaline R, Varro A. Exper Physiol 2007;29:591–601.
4. Baskin DG. Dagogo-Jack S, ed. Springer Cham. 2015;25–37.
5. Muller TD, Nogueiras R, Andermann ML, et al. Ghrelin 2015;4: 437–460.
6. Liman ER. TRPM5. In: Nilius, B, Flockerzi, V, eds. Mammalian Transient Receptor Potential (TRP) Cation Channels. 2014;222: 489–502.
7. Fink J, Koo BK. In: Jansen, M, Wright, N, eds. Stem Cells, Pre-neoplasia, and Early Cancer of the Upper Gastrointestinal Tract. 2016;11–25.
8. Van der Flier LG, Clevers H. Annu Rev Physiol 2009;71: 241–60.
9. Granger DN, Richardson PD, Kvietys PR, et al. Gastroenterol 1980;78:837–63.
10. Geboes K, Geboes KP, Maleux G. Best Pract Res Clin Gastro 2001;15:1–14.
11. Hodsfield BD, Markel TA. In Wang TC, Camilleri M, Lebwohl B, et al., eds. Yamada's Textbook of Gastroenterology 2022;168–90.
12. Chou CC. Fed Proc 1983;42:1658–61.
13. Chanpong A, Borrelli O, Thapar N. Facul Rev 2022;11. DOI: 10.12703/r/11-7
14. Farmer AD, Aziz Q In: Podolsky DK, Camilleri M, Fitz JG et al., eds. Yamada's Textbook of Gastroenterology. 2015;227–38.
15. Fried S, Wemelle E, Cani PD, et al. Neuropharmacol 2021;197: 108721.
16. Spencer NJ, Hu H. Nat Rev Gastroenterol Hepatol 2020;17: 338–51.
17. Deloose E, Janssen P, Depoortere I, et al. Nat Rev Gastroenterol Hepatol 2012;9:271–85.
18. Hasler W. In: Yamada T, Alpers DH, Kalloo AN et al., eds. Yamada's Textbook of Gastroenterology. Wiley-Blackwell. 2009;213–63.
19. Hasler W. In: Yamada T, Alpers DH, Kalloo AN et al., eds. Yamada's Textbook of Gastroenterology. Wiley-Blackwell. 2009;207–30.
20. Goyal RK, Guo Y, Mashimo H. Neurogastroenterol Motil 2019;31: 13546.
21. Maljaars PWJ, Peters HPF, Mela DJ, et al. Physiol Behav 2008;95: 271–81.
22. Van Avesaat M, Troost F, Ripken D, et al. Int J Obes 2015;39: 235–43.
23. Miller LJ. In: Yamada T, Alpers DH, Kalloo AN et al., eds. Yamada's Textbook of Gastroenterology. Wiley-Blackwell. 2009;207–30.
24. Kuemmerle JF. Endocrinol Metabol Clin 2012;41:409–23.
25. Baggio LL, Drucker DJ. Gastroenterol 2007;132:2131–57.
26. Jeppesen PB, Pertkiewicz M, Messing B, et al. Gastroenterol 2012;143:1473–81.
27. Kim ES, Keam SJ. Drugs 2017;77:345–52.
28. Boguszewski CL, Paz-Filho G, Velloso LA. Endokyronologia Polska 2010;61:194–206.
29. Fulton S. Front Neuroendocrinol 2010;31:85–103.
30. Suzuki K, Simpson KA, Minnion JS, et al. Endocrine J 2010;57: 359–72.
31. Zhang X, Shi T, Holmberg, et al. Proc Natl Acad Sci U. S. A. 1997;94:729–34.
32. Browning KN, Verheijden S, Boeckxstaens GE. Gastroenterol 2017;152:730–44.
33. Blanco A, Blanco G. In: Blanco A, Blanco G, eds. Medical Biochemistry. 2017;251–273.
34. Loo JA, Yan W, Ramachandran P, et al. J Dent Res 2010;89: 1016–23.
35. Messana I, Inzitari R, Fanali C, et al. J Separat Sci 2008;31: 1948–63.
36. Pawlik P, Błochowiak K. Diagnostics 2021;11:371.
37. Milanowski M, Pomastowski P, Ligor T, et al. 2017;47:251–66.
38. Wellendorph P, Johansen LD, Bräuner-Osborne H. Vit Horm 2010;84:151–84.
39. Niot I, Besnard P. Biochimie 2017;136:27–32.
40. Lee K, Kye M, Jang JS, et al. Proteomics 2004;4:3343–52.
41. Liang CR, Tan S, Tan HT, et al. Proteomics 2010;10:3928–31.
42. Leal MF, Wisnieski F, de Oliveira Gigek C, et al. Tumor Biol 2010;37:9991–10010.
43. Schubert ML, Peura DA. Gastroenterol 2008;134:1842–60.
44. Nayeb-Hashemi H, Kaunitz JD. Curr Opin Gastroenterol 2009; 25:537–43.
45. Hyde R, Taylor PM, Hundal HS. Biochem J 2003;373:1–18.
46. Alpers DH. Curr Opin Gastroenterol 2010;26:134–9.
47. Liszt KI, Wang Q, Farhadipour M, et al. J Clin Invest 2021;132, 1–16.
48. Boyer J. Comprehen Physiol 2013;3:1035–78.
49. Wang HH, Portincasa P, Wang DQ Front Biosci 2008;13: 401–23.
50. Barbhuiya M, Sahasrabuddhe N, Pinto SM, et al. Proteomics 2011;11:4443–53.
51. Williams JA. Curr Opin Gastroenterol 2010;26:478–83.
52. Morisset J. Pancreas 2008;37:1–12.
53. Chandra R, Liddle R. Curr Opin Gastroenterol 2009;25:441–6.
54. Parker HE, Reimann F, Gribble FM. Exp Rev Mol Med 2010; 12:236–42.
55. Zhang T, Perkins MH, Chang H, et al. Cell 2022;185:2478–94.
56. Kiela PR, Ghishan FK. Best Pract Res Clin Gastroenterol 2016;30:145–59.
57. Kiela PR, Xu H, Ghishan FK. J Physiol Pharmacol 2006;57: 51–79.
58. Barbara G, Barbaro MR, Fuschi D, et al. Front Nutr 2021,13. 718356.
59. Pappenheimer JR, Reiss KZ. J Membrane Biol 1987;100:123–36.
60. Alberts B, Johnson A, Lewis J, et al. In: Dries, DJ ed. Molecular Biology of the Cell, 4th ed. 2002;616–88.
61. Binder HJ. Digest Dis Sci 2019;65:349–54.
62. Alpers DH, Stenson WF, Taylor BE, et al. Manual of Nutritional Therapeutics, 6th ed. Lippincott Williams & Wilkins. 2015.
63. Debongnie JC, Phillips SF. Gastroenterol 1978;74:698–703.
64. Whitcomb DC, Lowe ME. Digest Dis Sci 2007;52:1–17.
65. Williams JA. Curr Opin Gastroenterol 2010;26:478–83.
66. Singh H, Ye A, Horne D. Prog Lipid Res 2009;48:92–100.
67. Wang HH, Portincasa P, Wang DQ. Frontiers Biosci 2008;13: 401–23.
68. Hernell O, Staggers JE, Carey MC. Biochem 1990;29:2041–56.
69. Rigotti A. Molec Aspects Med 2007;28:423–36.
70. Sun Y, Wang J, Long T, et al. Proc Natl Acad Sci 2021;118:1.
71. Su X, Abumrad NA. Trends Endocrin Metab 2009;20:72–7.
72. Nguyen DV, Drover VA, Knopfel M, et al. J Lipid Res 2009;50: 2235–44.
73. Dash S, Xiao C, Morgantin C, et al. Ann Rev Nutr 2015;35: 265–94.
74. Elia M, Cummings JH. Eur J Clin Nutr 2007;61:S40–74.

75. Elferink H, Bruekers JPJ, Veeneman GH, et al. Cell Molec Life Sci 2020;77:4799–826.
76. Douard V, Ferraris RP. Am J Physiol Endocrinol Metab 2008;295: 227–37.
77. Gomes SD, Oliveira CS, Azevedo-Silva J, et al. Curr Med Chem 2020;27:4087–108.
78. Bergen WG, Wu G. J Nutr 2009;139:821–5.
79. Kim YS, Erickson RH. Gastroenterol 1985;88:1071–3.
80. Tobey N, Heizer W, Yeh R, et al. Gastroenterol 1985;88:913–26.
81. Brandsch M, Knütter I, Bosse-Doenecke E. J Pharmacy Pharmacol 2008;60:543–85.
82. Bröer S, Fairweather SJ. Compr Physiol 2018;13:343–73.
83. Bäckhed F, Ley RE, Sonnenburg JL, et al. Science 2005;307: 1915–20.
84. O'Hara AM, Shanahan F. EMBO Rep 2006;7:688–93.
85. Ghaisas S, Maher J, Kanthasamy A. Pharmacol Thera 2016;158: 52–62.
86. Rajilić-Stojanović M, Smidt H, de Vos WM. Environ Microbiol 2007;9:2125–36.
87. Dogra S, Sakwinska O, Soh SE, et al. mBio 2015;6:321–5.
88. Winter SE, Thiennimitr P, Winter MG, et al. Nature 2010;467: 426–9.
89. Ley RE, Peterson DA, Gordon JI. Cell 2006;124:837–48.
90. Marchesi JR. (2010) Adv Appl Microbiol 2006;72:43–62.
91. Hehemann JH, Correc G, Barbeyron T, et al. Nature 2010;464: 908–12.
92. De Filippo C, Cavalieri D, DI Paola M, et al. Proc Natl Acad Sci U. S. A, 2010;107:14691–6.
93. Rajilić-Stojanović M, Heilig, HG, Molenaar D, et al. Environ Microbiol 2009;11:1736–51.
94. Turnbaugh PJ, Gordon JI. J Physiol 2009;587:4153–8.
95. Escapa IF, Chen T, Huang Y, et al. mSystems 2018;3:e00187–18.
96. Yang L, Lu X, Nossa CW, et al. Gastroenterol 2009;137: 588–597.
97. Lv J, Guo L, Liu JJ, et al. World J Gastroenterol 2019;25: 2149–61.
98. Engstrand L, Lindberg M. Best Pract Res Clin Gastroenterol 2013;27:39–45.
99. Andersson AF, Lindberg M, Jakobsson H, et al. PloS One 2008;3:e2836.
100. Wang X, Heazlewood SP, Krause DO, et al. J Appl Microbiol 2003;95:508–20.
101. Kastl Jr AJ, Terry NA, Wu GD, et al. Cell Molec Gastroenterol Hepatol 2020;9:33–45.
102. Ley RE, Hamady M, Lozupone C, et al. Science 2008;320: 1647–51.
103. Ley RE, Bäckhed F, Turnbaugh P, et al. Proc Natl Acad Sci 2005;102:11070–75.
104. He T, Cheng X, Xing C. Bioengineered 2021;12:7046–60.
105. Alam MT, Amos GCA, Murphy ARJ, et al. Gut Pathol 2020;12:1.
106. Xiao L, Liu Q, Luo M, et al. Front Cell Infect Microbiol 2021;11: 729346.
107. Su JW, Ma JJ, Zhang HJ. J Digest Dis 2015;16:58–66.
108. Navaneethan U, Shen B. Curr Gastroenterol Rep 2010;12: 485–94.
109. Fujimara KE, Slusher NA, Cabana MD, et al. Expert Rev Anti-Infect Ther 8, 435–54.
110. Valdes AM, Walter J, Segal E, et al. BMJ 2018;361:k2179.
111. Kanauchi O, Oshima T, Andoh A, et al. Scand J Gastroenterol 2008;43:1346–52.
112. Taur Y, Coyte K, Schluter J, et al. Sci Trans Med 2018;10: eaap9489.
113. McKee AM, Hall LJ, Robinson SD. Br Cancer Manag 8:BMT29.
114. Dutta SK, Verma S, Jain V, et al. J Neurogastroenterol Motil 2019;25:363–76.
115. Müller CA, Autenrieth IB, Peschel A. Cell Mol Life Sci 2005;62: 1296–1307.
116. Mei X, Gu M, Li M. Stem Cell Res Ther 2020;11:p 349.
117. Mahida YR. Best Pract Res Clin Gastroenterol 2004;18:241–53.
118. Artis D, Grencis RK. Muc Immunol 2008;1:252–64.
119. Rescigno M. Curr Opin Immunol 2010;22:131–6.
120. McDole JR, Wheeler LW, McDonald KG, et al. Nature 2012;483:345–349.
121. Newberry RD, Lorenz RG. Immunol Rev 2005;206:6–21.
122. Newberry RD. Curr Opin Gastroenterol 24, 121–8.
123. Eberl G, Sawa S. Trend Immunol 2010;31:50–5.
124. Marques R, Boneca IG. Cell Mol Life Sci 68, 3661–73.
125. Carvalho FA, Aitken JD, Vijay-Kumar M, et al. Ann Rev Physiol 2012;74:177–98.
126. Fukata M, Arditi M. Muc Immunol 2013;6:451–463.

B. DIGESTIVE, ENDOCRINE, IMMUNE, AND NEURAL MECHANISMS

CHAPTER 42

Sensory Nutrition

Valerie B. Duffy and Shristi Rawal

CHAPTER OUTLINE

Abbreviations

ACC, accumbens; **AM**, amygdala; **ATP**, adenosine triphosphate; **AVI**, anterior ventral insular cortex; **CCK**, cholecystokinin; **CN**, cranial nerve; **CNS**, central nervous system; **CTN**, chorda tympani nerve; **DI**, dorsal insular cortex; **ENaC**, epithelial sodium channel; **FGF**, fibroblast growth factor; **GI**, gastrointestinal; **GLP**, glucagon-like peptide; **GLUT**, glucose transporter; **GMP**, guanosine monophosphate; **GPCR**, G-protein coupled receptor; **IMP**, inosine monophosphate; **IP**, inositol trisphosphate; **LH**, lateral hypothalamus; **LOFC**, lateral orbitofrontal cortex; **MOFC**, medial orbitofrontal cortex; **MSG**, monosodium glutamate; **MUC**, transmembrane mucin; **NaCl**, sodium chloride; **NHANES**, National Health and Nutrition Examination Survey; **NST**, nucleus of the solitary tract; **OB**, olfactory bulb; **OC**, olfactory cortex; **OE**, olfactory epithelium; **OR**, olfactory receptor; **OSN**, olfactory sensory neuron; **PPC**, posterior parietal cortex; **PROP**, propylthiouracil; **PTC**, phenylthiocarbamide; **SGLT**, sodium-driven glucose transporter; SNP, single nucleotide polymorphism; **SOM**, somatosensory cortex; **TAS2R**, taste type 2 receptor; **TRP**, transient receptor potential; **VC**, visual cortex; **VPM**, ventral posteromedial thalamic nucleus; **VFD**, venus flytrap–like domain

Introduction

Foods and beverages provide sustenance and diverse pleasurable experiences. Specific chemosensory organ systems respond to chemicals in foods, eliciting neurologic, biobehavioral, and metabolic responses that stimulate complex emotional, pleasurable, and memory responses. Odors emanating from restaurants can stimulate our appetites and entice us to stop and eat. At the table, chewing releases and pumps food volatiles to olfactory receptors, which lie behind the bridge of the nose. No single word in the English language fully captures the perceptual experiences of food. Olfactory sensations blend with true taste (salty, sweet, sour, bitter, meaty/savory sensations) and oral somatosensory sensations (mouthfeel, texture, temperature, astringency, irritation) to form a unique and integrated flavor message.[1]

Food flavor, colloquially referred to as taste, is an important driver of food choice, yet individuals experience different flavor worlds because of physiologic variation in perceptual responses to food and beverage chemicals. Over more than a decade, most adults in nationally representative samples have reported that taste was the primary driver of food purchases, surpassing price, health, convenience, and environmental sustainability.[2]

Individual differences in sensory systems are present at birth, and sensory response changes throughout life with maturation and environmental interactions. Interindividual taste differences were reported as early as 1888.[3] In the 1960s, Fischer et al., connected variation in taste with dietary preferences, smoking, and body weight.[4] Since then, accumulating evidence has indicated that chemosensory factors have influences parallel to metabolic controls on food intake and body weight, acting through receptors, nervous signals, and gut-brain signaling. Today, the 2020–2030 Strategic Plan for the National Institutes of Health (NIH) Nutrition Research recognizes the need to "Improve Understanding of Sensory Nutrition and Ingestive Behaviors" as an important part of developing precision nutrition strategies for optimal health.[5] This chapter reviews the chemosensory systems that comprise sensory nutrition, their variations with genetics and disease, as well as how these variations explain differences in what we like and choose to consume, thus ultimately influencing risk for diet-related diseases. The chapter ends with approaches to translate sensory nutrition into tailored nutrition education interventions.

Background

Chemosensation includes detection and response to chemicals in the external world and signals from the gastrointestinal and respiratory systems. Chemosensory perception is the conscious experience that arises from these chemicals.

Sensation

To elicit a chemosensory response, most chemicals (odors; sweet, bitter, umami [i.e., (meaty/savory] stimuli; irritants) bind to specific receptors. Transduction of chemosensory events commonly involves G-protein–coupled receptors (GPCRs)—seven-transmembrane proteins that trigger G-protein–based signaling cascades when activated with chemical (ligand) binding. Linda Buck and Richard Axel received the 2004 Nobel Prize for their work on the genetic basis of olfactory receptors.[6] Receptors may be finely tuned, responding to a few specific chemicals, or broadly tuned, responding to a repertoire of compounds. True tastes may be exemplified by simple prototypical tastants (e.g., quinine is bitter), whereas most odors are complex mixtures of chemicals (e.g., coffee aroma requires upward of 50 different volatiles).[7] Receptor cells generally are bipolar neurons; chemical signals are transduced through an elaborate electrochemical cascade mechanism that eventually triggers the action potential of the surrounding nerve cells and signals to the central nervous system (CNS) that a tastant has been received. Receptor binding is not necessary to transduce salty or sour tastes, which permeate through ion channels to stimulate taste receptor cells. Various sensors respond to somatosensory sensations, including oral touch for mouthfeel, creaminess, and texture sensations.

There is neuroplasticity in the chemosensory system, with neurogenesis in response to extrinsic and intrinsic factors. The olfactory system can generate throughout life beyond critical developmental periods.[8] There has been great interest in neurogenesis of the olfactory sensory neurons and supportive cells and recovery of smell loss with COVID-19 infection. Taste receptors are continually replaced and functionally change in response to chemical environments. Continuous, constant exposure causes adaptation (e.g., inability to smell the perfume or cologne you wear) or desensitization (e.g., less burn from long-term chili ingestion).

How do we perceive sweet vs bitter, garlic vs basil, Riesling vs Pinot Grigio? How information is transmitted from periphery to the CNS (i.e., taste coding) for sweet or bitter quality, for example, is still debated with opposing beliefs of a labeled line (chemical-specific neurons and fibers) vs unit pattern theory (distribution of responsive cells).[9] Odors are coded as patterns of receptor stimulation that are reflected in olfactory spatiotemporal representations and sequentially processed through the olfactory pathways in the brain.[10] Although reaching the CNS through separate pathways, there are multiple brain stem and cortex areas of integration of chemosensory signals into unique flavor percepts.[1]

Perception

Psychophysicists study how perceptions vary with the physical world—how saltiness varies with salt concentration, smell with odor exposure, and creaminess with level of fat. The following is a brief overview of selected psychophysical techniques, with detailed information available for taste[11] and smell.[12]

Threshold is the lowest physical concentration required for detection or recognition of a taste, smell, or irritant. An elevated threshold (low sensitivity) means a higher concentration is required for detection and recognition. Threshold procedures require substantial control to discern functional ability from chance and biases (e.g., subtle differences in stimulus delivery, diluent, temperature). Threshold may fail to correspond to perception of concentrated stimuli. An individual with a low salt threshold (high sensitivity) may perceive concentrated salt as less intense than one with a higher threshold. Thus, characterizing individuals by threshold may not help explain differences in dietary intake patterns.[11]

Suprathreshold reflects the ability to perceive odors, taste, and irritants at levels common to foods, Identification tasks are the most common suprathreshold tests. For olfaction, participants are asked to smell the stimuli, one at a time, and identify the correct response from a listing of distractors in a forced-choice procedure.[12] The task must include familiar odors to minimize cognitive challenges. "Scratch-and-sniff" measures, such as The University of Pennsylvania Smell Identification Test have norms, based on age and sex, to identify normal, hyposmia, and anosmia. New technologies exist to imbed odorants into portable cards for improved odor delivery for rapid testing of olfaction.[13,14] Identification tasks for taste are less informative, with only five qualities and common confusion between sour and bitter.[11]

Identification tasks assess olfactory dysfunction, not acuity. Individuals can show a range of olfactory abilities and still identify an odor correctly. With good stimulus control, as with an olfactometer or embedded odors,[13,14] such as in the Lift'nSmell® technology, adding intensity judgments (discussed on the next page) to identification tasks provides a measure of dysfunction and acuity. Retronasal tests are important for olfaction and diet studies. As a screener (or demonstration), have participants pinch their nostrils closed, place a gourmet jellybean in the mouth, chew, and then open the nostrils. With the nostrils plugged, the sensation is minimally sweet and little else. Opening the nostrils allows retronasal olfaction and increases sweetness resulting from flavor integration (described later). Retronasal impairment is defined if participants fail to notice differences between fully plugged and unplugged. There are new measures of retronasal olfaction in clinical use.[15,16]

Another suprathreshold measure is direct scaling of perceived intensity or degree of liking/disliking, such as how perceived intensity grows with increasing concentration (slope). The liking function usually forms an inverted U: low and high concentrations are less pleasant than those in between. Direct scaling methods aim to make the intensity ratings objective and permit comparison of these ratings across individuals.

An understanding of absolute meaning is achieved by expressing the intensity of interest relative to another sensory modality (magnitude matching is the gold standard for perceived intensity measures).[17] Cross-modal references can be actual (e.g., 1,000-Hz white noise) or remembered (e.g., brightness of the sun) sensations, assuming the other sensory modality does not vary systematically with the intensity of interest.[18] Importantly, participants must use the same scale to make intensity judgments within the context of all sensations. For example, participants rate the intensity of sweetness on the same scale as intensity of tones or lights. One participant may rate sweetness of tea about one-half as intense as a 72-dB, 1,000-Hz tone whereas another may find the sweetness equally intense.

Scales with adjective/adverb labels (e.g., weak, strong) are commonly used to obtain intensity ratings. The context for the labels is necessary as illustrated by Stevens: "Mice may be called large or small, and so may elephants, and it is quite understandable when someone says it was a large mouse that ran up the trunk of the small elephant."[19] The size judgment was within the context of mice (large relative to mice) or elephants (small relative to elephants). A common pitfall is incorrectly assuming the same context for labels, falsely equating the top of the scale, and, according to the scale length, determining that the mouse is *larger* than the elephant (**Figure 42.1**). Correct comparisons of absolute size are possible if the scale is generalized to all sizes with the top understandable to all (e.g., city block).

Because of taste receptor gene polymorphisms and differences in receptor density, we know that individuals vary in the range of their orosensory scale, from nontasters to supertasters (see below). Incorrect ratings of oral perceived intensity result from failing to provide a context for the top of the scale or applying the top only to oral sensation. Falsely equated (e.g., equating mice and elephant scales) scales may blunt or reverse orosensory differences between nontasters and supertasters. Generalizing the scale top to strongest sensation of any kind[20] or something tangible like the brightness of the sun (assuming brightness does not vary systematically with orosensory variation), allows the correct differentiation of oral sensations between those who perceive less intense and more intense oral sensations. Similarly, generalizing hedonic scales outside food and beverage experiences can identify individuals for whom food is more pleasurable than any other pleasurable activity.[11]

In summary, intensity measures test chemosensory function across a range of concentrations, providing more information than thresholds about genotype-phenotype-diet-health connections.[21] In the presence of good stimulus control, adding intensity to odor identification tasks tests for dysfunction and acuity. Intensity scales should be generalized to all sensations (not just chemosensations) or, for hedonic assessment, all pleasurable and unpleasurable activities. Participants should rate intensity of chemosensations relative to sensory modalities of comparison in a practice session to determine if they are able to correctly order the comparison series (e.g., weakest to brightest light). Experimenters can use the comparison intensity ratings to normalize the chemosensory ratings[11,22] or to treat as covariates in statistical analyses.[23]

Taste

Any chemical that is soluble in the oral cavity watery medium (saliva, mucus) can stimulate taste—perceptual qualities of sweet, salty, sour, bitter, and umami (meaty/savory) via activating taste receptor cells. In general, sugars, alcohols, and some peptides are sweet; salts are salty; organic/inorganic acids are sour; many plant alkaloids, terpenoids, flavonoids, and some salts and peptides are bitter; and certain amino acids are meaty/savory. Some have characterized a response to macronutrients and gut sensing called alimentary taste[24] that would include umami, fatty acids, and carbohydrate sensing. Pleasure responses to sweetness and disliking for bitterness (and probably strong sour and umami) are present at birth and not learned, whereas response to saltiness develops during the first year of life.[25]

Taste chemicals dissolve in saliva and mucus layers to reach the taste receptors within taste buds in the oral cavity. Taste buds are found on the soft palate, pharynx, larynx, and epiglottis and within gustatory papillae on the tongue (**Figure 42.2**). Within taste buds, taste receptor cells are discrete, ovoid structures composed of 50 to 150 specialized cells arising from the epithelium.

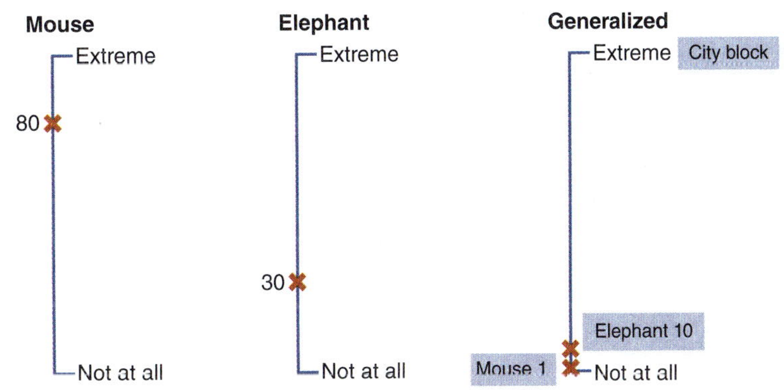

Figure 42.1 Perceived size ratings when the context of the label is defined only by what is being asked, from only mice (left) to only elephant (center); if comparing the absolute ratings (80 vs 30), the experimenter would incorrectly conclude that the mouse was more than twice as large as the elephant. The right scale allows correct judgments of the absolute size with telling the participant that the top of the scale was labeled to a size generally understood and broader than the size of animals.

Reproduced from Duffy VB, Rawal S, and Hayes JE. (2021). Measurement of Gustation: From Clinical to Population-Based Methods. In Sensory Science and Chronic Diseases: Clinical Implications and Disease Management (Joseph PV, and Duffy VB, eds) pp. 65-102, Springer International Publishing.

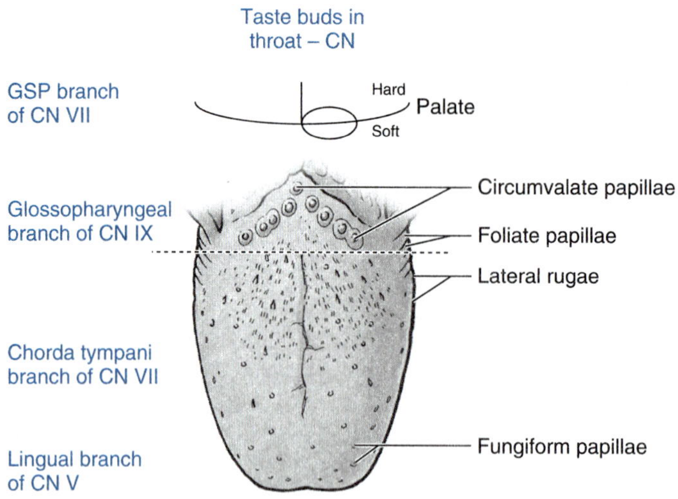

Figure 42.2 Drawing showing cranial taste and trigeminal innervation of the tongue and throat as well as the taste papillae on the tongue. *CN*, cranial nerve.

These cells are organized by function and structure, with three types of elongated cells (type I, type II, and type III) with microvilli that contain the taste receptors and extend through a pore into the oral cavity, as well as Type IV (basal cells precursor cells). Type I cells are most abundant and perform glia-like functions to maintain the taste bud integrity. Transduction functions occur in Type II (expressing GPRC for sweet, bitter, and umami) and Type III (sensory receptors for sour taste). Salt taste responses appear separate from these specialized cells.

After chemical activation and depolarization of receptor cells, afferent taste fibers within branches of three cranial nerves (CNs) transmit taste signals to the rostral nucleus of the solitary tract (gustatory NST), which is also involved in control of digestive, cardiovascular, and respiratory systems. The chorda tympani nerve (CTN), CN VII, innervates fungiform papillae on the tongue tip (Figure 42.2). Foliate papillae on the posterior lateral tongue are innervated by the CTN (anterior papillae) and lingual nerve (CN IX). The lingual nerve innervates circumvallate papillae (posterior tongue in a rearward facing V). The superficial petrosal nerve (CN VII) innervates soft palate taste buds, and the vagal nerve superior branch (CN X) innervates the epiglottis. All taste qualities are perceivable on all areas of CN innervation unless there is taste damage to a single CN, making the "taste map" concept incorrect. Afferent taste fibers ending in the nucleus of the solitary tract synapse into second order neurons to the ventrobasal thalamus, then to the gustatory cortex, orbitofrontal cortex, amygdala, and lateral hypothalamus. Glutamate, an excitatory neurotransmitter, modulates information traveling from peripheral taste receptors to the brain; others likely regulate information carried from the brain to the peripheral taste system. Mechanoreception is an important component of oral sensations for fat and textures in food as well as the overall flavor of foods and beverages. The mechanosensory innervation of the fungiform and filiform papillae as well as the hard palate has recently been described in healthy human tissue.[26]

Taste receptors are expressed beyond the oral cavity, where they do not mediate perception but are involved with the gut-brain access, immune function, and metabolism. GPCR are present in the enteroendocrine cells in the GI tract, the genitourinary system, cardiovascular system, central nervous system, as well as upper and lower airways.[27] Genetic variation in taste receptors could explain differential responses to nutrients, health, and disease risk, with or without influencing perceptual experiences.[28] As reviewed,[29] enteroendocrine cells in the gut sensory epithelium respond to nutrients, mechanical stretching, and bacteria via GPCRs, transporters, ion channels, and mechano-receptors. In response and for gut-brain signaling, some enteroendocrine cells produce hormones for slower, sustained signaling of satiety and some are synapsed with afferent nerves in the mucosa (including the vagal nerve) for rapid neurotransmission signaling of pleasure and reward.[29]

The density of fungiform papillae and their taste buds varies across individuals.[30] Some research finds associations with oral sensations[31-33] and others do not.[34] Densities can decouple from taste intensity related to environmental effects on taste and papilla morphology[35,36] such that fungiform papillae can provide a separate phenotype from taste intensity for explaining variation in diet.[37,38] It is challenging to assess density of fungiform papillae on the tongue tip, including blue food coloring staining and video microscopy[31] and digital photography[34,39] to discern pink fungiform papilla from blue-stained filiform papilla. Using digital pictures with augmentation and automated counting based on a deep learning approach called Convolution Neural Networks,[40] may be less biased than manual counting by accounting for fungiform papillae of different sizes and shapes.[30] Taste buds on fungiform papilla can be assessed longitudinally with confocal laser scanning microscopy.[41]

Sweet

Diverse chemicals in structure and weight stimulate a singular perceptual experience that is qualitatively similar—carbohydrate sweeteners, polyols, amino acids, peptides, inorganic salts, natural sweeteners[42] and synthetic noncaloric sweeteners stimulate the receptors for sweetness with some compounds as perceptual mixtures of sweetness and bitterness[43] such as saccharin.[44] Multiple lines of psychophysical evidence exist for multiple sweet transduction mechanisms,[45] including lack of cross-adaptation (sweeteners sharing similar binding would cross-adapt), inability to block sweetness of all sweeteners, and that sweetener combinations can produce greater-than-expected sweetness (i.e., synergy). In food, the level of sugars shows strong correlation with the sweetness.[46,47] The food supply is significant for added sugars, presenting a concern for diet quality and diet-related chronic diseases.[46] However, sweetness per se does not translate to poor diet quality. Diets higher in sweetness from fruits and healthier foods can be higher in diet quality.[48]

The primary sweet taste receptor, a heterodimer of two seven-transmembrane domain proteins, T1R2 and T1R3 (T1R2/T1R3),[49] has at least six chemical binding sites. Humans carry three *TAS1R* taste receptor genes in a single cluster on chromosome 1. These proteins are part of class C GPCRs, which have large N-terminal Venus flytrap–like domains (VFD). Most sweeteners bind to the VFD subunit T1R2, sucrose to T1R3, and others bind to the transmembrane domain.[50] Sweet taste receptors are in the oral cavity and in other organs, including gut, pancreas, bladder, bone, and adipose tissues.[51] After receptor binding, sweet and bitter tastes share similar transduction pathways in the taste bud in a canonical (i.e., established) pathway involving heterotrimeric G protein complex, activation of enzyme phospholipase Cβ_2 and generation of a second messenger Inositol trisphosphate (IP_3) and release of calcium ions, and opening of the cation channels TRPM4 and TRPM5 to allow sodium ions into the cell for depolarization and release of adenosine triphosphate (ATP) for activation of the afferent nerve fibers and signal to the CNS (as reviewed).[52] Transporters of glucose across the gastrointestinal tract also may be involved in transduction of sugar sweeteners, independent from receptors, including glucose transporters (GLUTs) and sodium-driven glucose symporters (SGLTs) that communicate to gustatory nerves through separate pathways, including glucagon-like peptide-1 (GLP-1), ultimately involved in cephalic phase insulin response.[52]

Differential binding to multiple taste receptors and different transduction responses partially explain flavor profile differences between sugar-based and artificial sweeteners.[53] Sweet receptors in the brain are involved in reward, but the effects could be different with responses to sugar sweeteners and low-calorie sweeteners having differential effects on appetite (as reviewed).[53] Artificial sweeteners also stimulate bitter receptors, making them less pleasant,[54] especially to those with greater propensity to experience bitterness.[44]

Bitter

Bitter can be a sign of poison but many nutritionally important compounds in foods are bitter in taste[55] with the level of bitterness suggestive of nutritional vs pharmaceutical application.[56] Multiple mechanisms and receptors respond to many chemicals with diverse structures that are bitter,[57] with more than 1,000 identified in the database of bitter stimuli (http://bitterdb.agri.huji.ac.il/dbbitter.php). The human bitter taste receptors, taste type 2 receptors (hTAS2Rs),[58,59] are 25 proteins within the GPCR superfamily, mostly on chromosomes 7 and 12. Bitter receptor genes are expressed in papillae in the oral cavity and extraorally, including the lung tissue, reproductive tract, skin, brain, heart, gastrointestinal tract, and pancreas (as reviewed)[60] where they respond to endogenously produced chemicals and those acquired exogenously.[61]

Most bitters stimulate multiple bitter receptors (i.e., broadly tuned) and most hTAS2Rs respond to multiple compounds. Some hTAS2Rs are tuned more narrowly to bitter chemicals (i.e., agonists), whereas some receptors bind a wide spectrum. The most studied, hTAS2R38, responds broadly to isothiocyanates, including those with thiourea groups, phenylthiocarbamide (PTC) and propylthiouracil (PROP).[21,62] hTAS2R16 binds to β-glycopyranosides (i.e., bitter sugars), plus has a dual binding region to respond to a wide number of ligands.[61] Broadly tuned receptors include hTAS2R10, hTAS2R14, and hTAS2R4, responding to many bitter chemicals,[57] with hTAS2R14 the most promiscuous. There are specific food bitters that bind to hTAS2R.[56] Bitter chemicals quinine and caffeine, for example, bind multiple bitter receptors. Bitter taste transduction follows the pathway described with sweet.

The hTAS2Rs likely evolved first as nutrient sensors and then to respond as taste receptors in response to ingestion of plants,[63] and they show high degrees of allelic variation across the world to influence bitter perception.[64] Evolutionary adaptation to local plant environments may account for which alleles are favored.[63] Variation in bitter taste perception and TAS2Rs has implications for acceptance and preference for important foods and for human health (see discussion below) and, thus, are of immense interest to nutritional and pharmaceutical interventions.

Salty

Primarily added to food as sodium chloride (NaCl), salt is important for saltiness, bitter blocking, flavor enhancement, and functional purposes (e.g., preservative). Salt evokes different qualities, depending on concentration—weak concentrations taste sweet, higher concentrations are salty, and the highest are irritating.[65]

Taste transduction for salty and sour occurs via ion channels. In rodents, low concentration of sodium stimulates sodium-specific and amiloride-sensitive epithelial sodium channel (ENaC) on the anterior tongue[66]; salt taste is blocked with the application of amiloride. The application to humans is unclear, in part because the ENaC responses are localized to the fungiform papillae in rodents, and taste is perceived with the whole mouth. Further, salt stimulates irritation and saltiness at higher NaCl concentrations, as reviewed.[67] Recent work in mice shows that sodium ions in the oral cavity stimulate taste receptor cells through ENaC and recruitment of CALHM1/3 channels, both part of amiloride-sensitive transduction of salty taste.[68] There also appears to be salt responses independent of ENaC but reliant on CALHM1/3 channels to drive attraction toward salt.[68] The application of this recent work to humans is still challenged,[69] especially in the understanding of our affinity toward saltiness. However, the transient receptor potential cation subfamily V member 1 (TRPV1) channel also appears to be involved in salt taste.[70] Variation in salt taste associates with single nucleotide polymorphisms in these channels[70,71] and risk of being hypertensive.[71]

Sour

Organic acids (lactic acid, citric acid, malic acid, and acetic acid) and inorganic acids (hydrochloric acid, nitric acid, and sulfuric acid) stimulate sour taste.[72] Acids elicit sour taste by stimulating acid-seeking cells within taste buds and, if strong enough, elicit rejection. Prototypical sour tastes are also perceived as irritating, bitter, and astringent. A putative sour proton-entry channel, OTOP1, has been identified recently.[73] Higher concentrations of acid would stimulate trigeminal fibers and aversive responses.[74] A twin study suggested that genetics may explain the 50% variation in citric-acid intensity.[75]

Umami

Many classify umami, described as savory or meaty, as the fifth basic taste. Monosodium glutamate (MSG), the sodium salt of free L-glutamate (glutamic acid) and first identified umami stimuli, is the most used prototypical umami stimuli in psychophysical testing.[76] In addition to glutamates, compounds eliciting or enhancing umami taste include certain amino acids, small organic acids such as succinic and gluconic acid, peptides, and nucleotides, particularly guanosine monophosphate (GMP), and inosine monophosphate (IMP), which strongly enhance umami taste in the presence of glutamate.[76] Umami-inducing compounds are found naturally in a wide range of foods including meat, seafood, cheese, and certain vegetables such as mushrooms and tomatoes.[76]

Like sweet taste, there are multiple umami receptors[76,77]—the GPCR heterodimer of two seven-transmembrane domain proteins, TAS1R1 and TAS1R3, is the predominant umami taste receptor. TAS1R1+TAS1R3 heterodimer, responds to glutamate, aspartate, and l-2-amino-4-phosphonobutyrate, with signal potentiation by purinic ribonucleotides IMP and GMP.[76,77] Four taste-specific metabotropic glutamate receptors, GRM1 to GRM4, (previously called mGLUR1 to 4) are activated by glutamate and analogs.[76]

Glutamate taste receptor heterodimers are present throughout the gastrointestinal (GI) tract, where stimulation of these receptors have been suggested to play a role in nutrient absorption through regulation of satiety hormones such as cholecystokinin (CCK).[24] Although findings in the literature are mixed, a few behavioral studies have also shown that glutamate consumption may increase satiety and decrease food intake in adults and infants.[24] Single nucleotide polymorphisms (SNPs) in TAS1R1 and TAS1R3 receptor genes are associated with variation in sensitivity to umami, with a subset of individuals considered to be MSG/umami hypotasters or nontasters.[77] In vitro studies also demonstrate functional variation in ability to bind MSG with amino acid substitutions in TAS1R1 and TAS1R3 receptor genes.[77]

Fat and Taste

Texture is the primary oral sensory cue for fat, especially for fungiform papillae, which act as mechanical sensors as the tongue moves during food manipulation and licking. Fungiform papilla number correlates with fat sensation.[78] Mounting evidence supports fat as a basic or alimentary taste. Receptors have been identified in human fungiform papillae for fat taste sensing, including GPCRs and ion channels, such as CD36, FFAR4, FFAR2, GPR84, and certain types of DRK channels like KCNA2,[79] with the main candidate genes CD36 and G-protein coupled receptor 120. The CD36 gene, coding for a receptor that binds cell adhesion molecules and functions in the transport of fatty acid into cells, associates with fat taste detection threshold[80] and perceived fattiness of potato chips.[81] Change in response to a free fatty acid receptor expressed in the fungiform papilla was examined in twins on a low-fat or high-fat diet. After 8 weeks, the low-fat diet was associated with increased expression of free fatty acid receptor 4, which also associated with change in fat-taste sensitivity.[82]

Smell

Odors must be volatile, hydrophobic, single, or complex chemicals with relatively low molecular weight. The sense of smell, a dual sensory process, comprises transport of odors orthonasally via the nostrils and retronasally via the nasopharynx to olfactory receptors in the olfactory epithelium (**Figure 42.3**). Brain activation differs with route of delivery.[83] Hedonic responses to odors are not innate, but rather are learned through positive (e.g., pairing odors with energy, repetitive exposure) and negative (e.g., flavor aversions) conditioning.

There are two routes to olfactory perception. Orthonasal olfaction occurs passively with breathing—we may not think of eating until breathing in food smells. Sniffing increases the quantity and quality of odors reaching the olfactory epithelium with large numbers of olfactory sensory neurons (OSNs). The olfactory epithelium sits behind the bridge of the nose, in the dorsal nasal cavity near the septum, and from superior to anterior middle turbinate to stimulate neural activity throughout the olfactory system. Retronasal olfaction is an active process in which the mouth, tongue, and swallowing movements work in synchrony to release and warm the volatiles and create a pressure differential that pumps them up through the oropharynx and nasopharynx to the olfactory epithelium.[84] Food volatiles are integrated with taste and somatosensory sensations into a unitary percept in the insular[85] and orbitofrontal[86] cortex (Figure 42.3).

OSNs contain olfactory receptor cells for transduction. Olfactory receptor cells are bipolar neurons (giving rise to a dendrite on one side, an axon on the other) associated in the olfactory epithelium with supporting (mucus producing) and basal (for generating new neurons) cells. By diffusion and active transport via binding proteins, odors cross the mucus layer before binding with transmembrane GPCRs on long cilia of olfactory receptor neurons, dendritic side.

The olfactory sensory neurons express GPCRs on long hair-like extensions (i.e., cilia), each neuron responding to a single odorant (i.e., agonist) and communicating a pattern of input (an odor image, similar to patterning in the visual system[87]) to the olfactory bulb and CNS for discrimination of an odor percept comprising of few to multiple odorants.[88] In humans, there are ~400 different olfactory receptor genes expressing intact open reading frame[89] and these form the largest gene family in mammals. Expression of specific OSNs in the olfactory epithelium are upregulated with odor exposure and downregulated with odor adaptation (i.e., explains inability to smell your own body odors).[88] Ectopic olfactory receptors are expressed in the gut, lungs, kidneys, eyes, muscle, and sperm, where they respond to bioactive food components,[90] endogenously produced odorants involved in cell division, growth, and death, potentially linked with carcinogenesis and cancer treatment,[91] including immunotherapy.[92]

Odor-receptor binding on the cilia initiates the transduction cascade—activation of Gaolf, then adenylyl cyclase activation, and nucleotide-gated channels.[93] A single CN (CN I) carries olfactory messages from the peripheral nervous system to the CNS. cAMP depolarizes the olfactory neuron; the action potential is carried by an unmyelinated axon through the cribriform plate of the ethmoid bone, where the axon synapses on second-order neurons located in olfactory bulb glomeruli.[87] Olfactory bulb microcircuits further enhance and sharpen odor images. The odor image is further modified with travel to the olfactory cortex, which refines, stores, and coordinates this image with complex behavioral responses. Through two more synapses, odor images are compared with templates of past experiences in the piriform cortex (e.g., this ice cream has coconut and almond flavors), integrated in the orbitofrontal cortex with taste, somatosensory, visual, and auditory sensations into a flavor percept (e.g., Almond Joy Ice Cream), and processed in the hippocampus and amygdala for odor memory (e.g., this ice cream reminds me of the beach) and hypothalamus as the feeding center (e.g., I want more). The odor image is

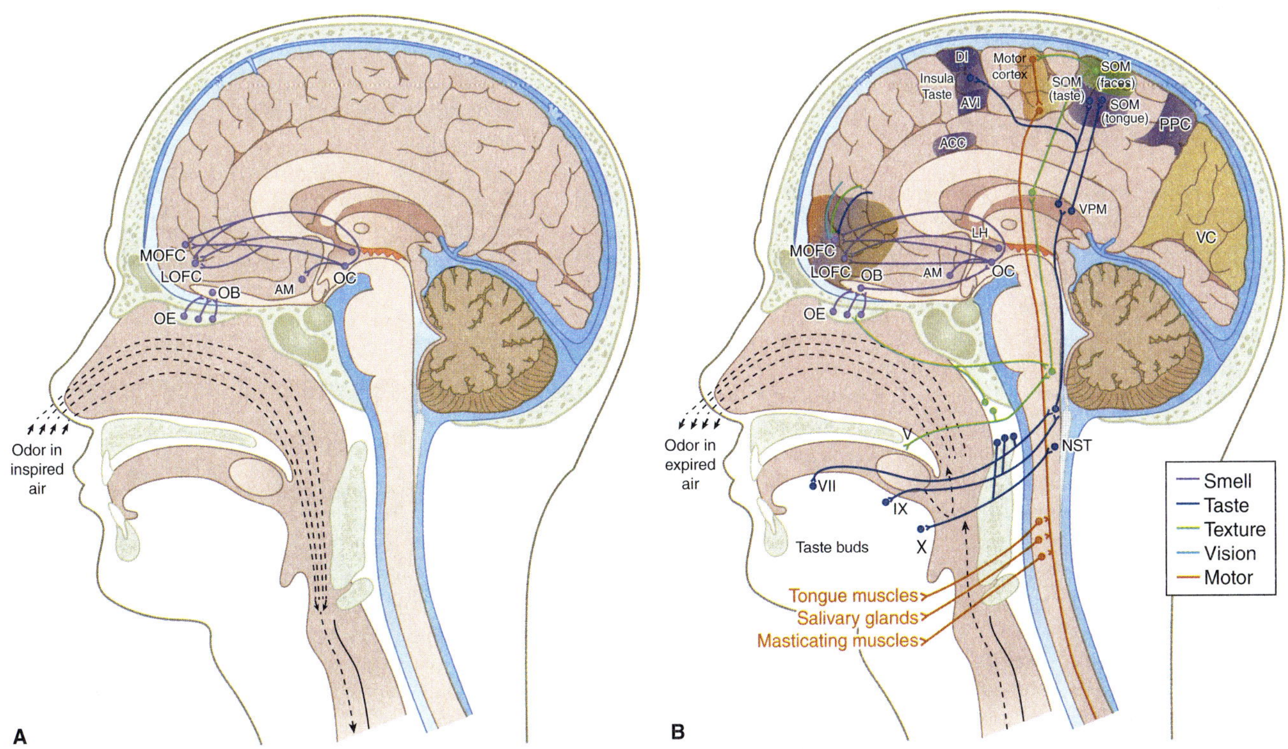

Figure 42.3 A. Brain systems involved in smell perception during orthonasal olfaction (sniffing in). B. Brain systems involved in smell perception during retronasal olfaction (breathing out), with food in the oral cavity. Air flows indicated by *dashed and dotted lines; dotted lines* indicate air carrying odor molecules. *ACC*, accumbens; *AM*, amygdala; *AVI*, anterior ventral insular cortex; *DI*, dorsal insular cortex; *LH*, lateral hypothalamus; *LOFC*, lateral orbitofrontal cortex; *MOFC*, medial orbitofrontal cortex; *NST*, nucleus of the solitary tract; *OB*, olfactory bulb; *OC*, olfactory cortex; *OE*, olfactory epithelium; *PPC*, posterior parietal cortex; *SOM*, somatosensory cortex; *V*, *VII*, *IX*, *X*, cranial nerves; *VC*, primary visual cortex; *VPM*, ventral posteromedial thalamic nucleus.

modulated from the receptor level upward, in which odor adaptation desensitizes olfactory receptors, or from the top downward, in which hunger state, for example, influences the awareness and hedonic response to odors.[87]

Olfactory dysfunction occurs with loss of functional olfactory receptors, damage to olfactory neuron axons as they pass through the cribriform plate, and olfactory bulb reduction.[94] Dysfunction is severe if the anatomic structures fail to regenerate. Odor quality is distorted if regeneration is incorrect. Olfactory ability also could vary with genetics. Understanding how polymorphisms in olfactory receptor genes influence olfactory perception is complicated by the existence of hundreds of OSNs involved in coding of the olfactory message. Parallel to phenotypic variation with taste receptor gene polymorphisms, olfactory acuity could differ with variation in olfactory receptor genes. The study of behavioral-receptor genotype relationships in olfaction is in its infancy and involves understanding functional genes from pseudogenes and copy-number variation with deletion alleles, particularly on chromosome 11.[95] An extensive genotyping and phenotyping of over 300 individuals showed associations between genetic variation in a single odor receptor gene and odor intensity,[95] suggesting that certain olfactory receptors are more important than others to olfactory coding.

Somatosensory Input

Touch, temperature, and chemesthesis are terms for somatosensation and are important constituents of food flavor. Mechanoreceptors mediate touch and texture (e.g., particle size, mouthfeel, creaminess) by stimulating touch afferents in the trigeminal nerve, including those within fungiform papillae.[96] Thermoreceptors respond to food temperature. On the posterior tongue, the glossopharyngeal nerve carries sensory fibers for pain and temperature. Astringency is a dry, rough feeling caused when acids and polyphenols hinder salivary protein lubrication in the oral cavity,[97] possibly through a transmembrane mucin (MUC1) mechanism, and/or by activating trigeminal chemoreceptors specifically tuned to a group of compounds.[98] Foods high in polyphenols, such as tea and wine, can have astringency, which influences their acceptance.[47]

Chemesthesis is the response to chemicals that elicit burning, stinging, tingling covers or anesthetic sensations in the oral cavity as well as across the body.[99] In the oral cavity and nose, these chemicals, concentrated odors, and tastants, as well as noxious temperatures, stimulate receptors of transient receptor potential (TRP) superfamily, chemonociceptor neurons, and thermosensitive neurons in the trigeminal nerve, glossopharyngeal nerve, and vagal nerve. Because chemesthesis is polymodal, activation of the TRPV1 (vanilloid–1) receptor on trigeminal neurons gives rise to hot painful burning sensations, regardless of whether the stimulus is the capsaicin in chili peppers or soup that is too hot.

Repeated application of irritants, such as capsaicin, results in sensitization (building burning or pain) whereas application of an irritant and removal for 2 to 3 minutes results in desensitization (reapplication of the irritant produces much less burning or pain).[100] Desensitization offers potential for analgesia of peripheral oral pain[101]; the challenge is to minimize the pain of the capsaicin application and test with rigorous trials.[102]

Integration of Smell, Taste, and Somatosensation

In 1825, Brillat-Savarin wrote, "smell and taste are in fact but a single composite sense, whose laboratory is the mouth and its chimney the nose." The CNS integrates sensory information into a flavor percept that elicits hedonic and dietary responses. Analysis of functional magnetic resonance imaging of the brain helps to understand the functional connectivity of brain regions in time and space for behavioral response to chemosensation.[103]

From an orthonasal smell perspective, it is difficult to imagine eating a cheese with a pungent odor. Oral sampling, however, balances strong smell, salt-bitter-sour taste, and creaminess into a pleasant flavor. Coupling sweet taste with sweet odor (e.g., strawberry) synergizes sweetness, whereas increasing creaminess perceptually suppresses sweetness. Somatosensory sensations have tight integration with smell and taste sensations. Odor compounds, if concentrated enough, cause intranasal stimulation of olfactory and trigeminal nerves. Concentrated or noxious odors irritate the eye and affect change in respiratory and circulatory systems to warn that harmful substances are present. Individuals with anosmia may distinguish some odors through intranasal stimulation of the trigeminal nerve. Similarly, some concentrated sour and salty tastes also produce irritation.

The balanced flavor percept and behavioral response are influenced by receptor density and functionality, as well as by variability and modulation of chemosensation throughout the perceptual pathway from receptor to CNS, including effects of the gut-brain axis.

Genetic Variation in Taste: Are There Supertasters?

Observations of individual variation in taste date to the nineteenth century.[3] In the 1930s, Fox reported individual variation in ability to taste PTC bitterness,[104] which was attributed to a genetic trait.[105] Variation in PTC/PROP detection threshold has been studied hundreds of times with thousands of participants. PROP is preferred as a safer and purer taste stimulus. In the 1960s, Fischer et al.,[4] connected quinine and PROP sensitivity to dietary preferences, smoking, and body weight, an observation suggesting that these bitter compounds captured broader differences in oral sensation. Genetic variation in the ability to taste PTC or PROP is most studied but has been extended to other phenotypes (taste qualities, taste papilla) and genetic markers.

Bartoshuk, who first used the supertaster label, noticed that nontasters had a homogeneous inability to taste PROP. Tasters were much more variable, including supertasters, who tasted PROP as intensely bitter. Suprathreshold identifies PROP supertasters,[31] not threshold, originally by the ratio of perceived intensity of PROP to NaCl. Subsequent work indicated that supertasters tasted both NaCl and PROP as more intense.[106] Additionally, PROP tasting phenotype reflected more than ability to sense the unique thiourea moiety of PROP/PTC and a heightened taste ability beyond that explained by TAS2R38 or other receptor proteins.[21] Supertasters, defined by PROP bitterness alone or PROP/NaCl ratio, report greater intensities from other basic tastes, retronasal odors, and somatosensory stimuli like irritation and the texture of fat.[78,96,107-109] An association between taste genetics and lifespan was reported in a genetically isolated geographic region. There was a greater proportion of centenarians who were homozygous tasters for the *TAS2R8* gene than homozygous nontasters,[108] which researchers proposed could be explained by healthier diet and nondiet factors, such as better immune functioning.

Bitterness of PROP does not need to define supertasting. Bitterness of quinine, for example, shows greater variability across adults than intensity of prototypical sour, salty, and sweet stimuli (**Figure 42.4**), yet PROP shows the most variance across all prototypical tastes.

Chemosensory Changes with Aging and Environmental Insults

Genetic chemosensory predispositions are modified across life with maturation and exposure to pathogens and the environment. Females generally outperform males in taste and smell testing.[110] These senses may be heightened in women during childbearing age to ensure healthy pregnancies. In a longitudinal study across pregnancy, bitter perception rises across pregnancy to its first trimester peak and falls to its third trimester nadir.[111] Aging is associated with chemosensory change with exposures to pathogens, chronic conditions, and environmental insults. Causes of chemosensory impairments have been reviewed.[112]

For the first time, smell and taste assessments were included in the 2013–2014 National Health and Nutrition Examination Survey (NHANES). NHANES is a continuous, cross-sectional, multifaceted, nationally representative assessment of the health and diet of the US population. The NHANES Taste and Smell protocol provides an opportunity to examine the associations of taste and smell with a broad array of diseases and conditions (e.g., cardiovascular, diabetes, kidney disease, obesity, oral health, respiratory), environmental exposures, and behaviors (e.g., dietary, physical activity) collected in a home visit and mobile examination center. The SARS-CoV-2 increased attention to chemosensation, with infection causing losses to smell, taste, and chemesthesis.[113]

Total taste loss (ageusia) is rare, and most "taste" disturbances are olfactory, including with COVID-19.[114] Taste loss was a true symptom of the SARS-CoV-2; a meta-analysis of nearly 139,000 patients who tested positive for the infection, 4 of 10 had some type of measured taste impairment, with greater prevalence among individuals of ages 18 to 65 years, as well as females versus males.[113] Importantly, the self-report of taste loss mirrored the findings from the measured taste function.[113] The cause of taste loss is not fully understood, potentially altering taste receptors, cells supporting the taste receptor cells, taste nerves, and central processing.[114,115] More research into COVID-related taste loss is needed.[113] Aging associates with elevations in taste threshold[116] that may not correspond to reductions in suprathreshold taste intensity as those experienced during eating.[117]

In the NHANES, ≈19% reported a taste problem (either taste problems in the past 12 months; worse ability to taste sweet, sour, salty, or bitter; or to perceive food flavors compared with when 25 years old, or dysgeusia—taste or other sensation in the mouth in the past 12 months that does not go away), with increased risk for individuals of lower education, xerostomia,

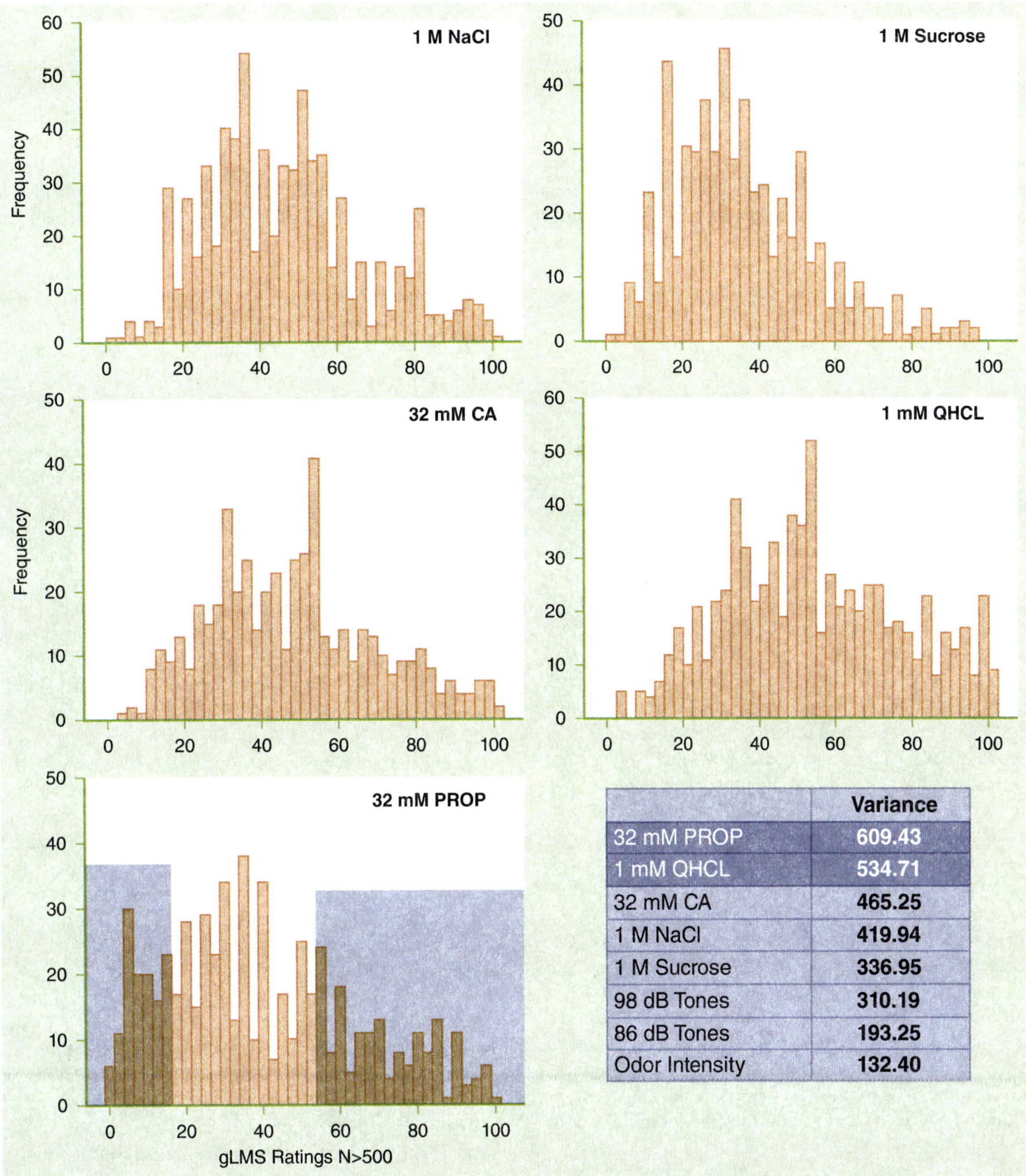

	Variance
32 mM PROP	609.43
1 mM QHCL	534.71
32 mM CA	465.25
1 M NaCl	419.94
1 M Sucrose	336.95
98 dB Tones	310.19
86 dB Tones	193.25
Odor Intensity	132.40

Figure 42.4 Perceived intensity of tones, citric acid, quinine, and propylthiouracil (PROP) showing greatest variance in perceived intensity of PROP.

self-reported fair/poor health, and nose/facial injury.[118] Further analysis from 2011–2014 NHANES data revealed a 5.2% prevalence of dysgeusia (described as a metallic taste 30.4%, bad/foul 24.2%, bitter 22.8%, sour 9.9%, burning/tingling 8.1%), with higher rates in women, those with xerostomia or poor oral health, and of lowest income.[119] Other large cohort studies provide support for taste changes with aging using taste testing.[120] In a longitudinal study of more than 1,000 older adults, the number of fungiform papilla decreased over an average of 4 years,[121] which may reflect deterioration of the form, size, and shape of the papilla.

Individuals can suffer from taste damage to a single CN without a change in ability to taste with the whole mouth, as normally occurs with eating. However, change in input from a single taste nerve may modify the integrated flavor percept associated with differences in food preference and dietary behaviors. Most studied is taste alteration from the CTN, which changes the balance of taste, oral somatosensory, and retronasal olfactory input. A nineteenth-century example comes from Brillat-Savarin. As punishment, a prisoner had his anterior tongue cut off; after recovering, however, the prisoner did not complain of taste loss, but of intense, painful sensations from sour or bitter. The punishment eliminated CTN input (CN VII) to the CNS and released inhibition of inputs from other taste (CN IX and X) and trigeminal (CN V) nerves to maintain whole mouth taste and heighten pain from concentrated tastes.[122]

Middle ear surgical procedures and infection as well as dental procedures can change orosensations via alterations of the CTN.[41,123,124] Patients of middle ear surgery can show improvements after 6 months.[125] Anesthesia for dental procedures[126] or experimental procedures[127,128] provides insight into clinical manifestations and impact on flavor perception. For example, anesthesia abolishes anterior tongue taste while intensifying taste (especially bitterness) on the posterior tongue, thus diminishing retronasal intensity and suggesting a mechanism for dysgeusia[127,128] and oral pain phantoms. Oral pain phantoms are associated with CTN taste damage.[129] CTN depression of taste can heighten somatosensory sensations such as creaminess because papilla density and their trigeminal innervation are retained.[18] Olfactory functioning can be depressed (hyposmia) or absent (anosmia) for some (specific anosmia) or all odors (total anosmia) and altered in quality (parosmia). Age-related olfactory dysfunction is more common than whole mouth taste dysfunction; olfactory information is carried by one CN vs three for taste. Oral sensory integration makes it difficult to distinguish taste from smell dysfunction. Careful questions and testing can characterize the type of chemosensory dysfunction.[118,130]

Olfactory dysfunction results from alterations in peripheral and/or central processes,[94] and the most studied is smell through the nostrils (orthonasal olfaction). In recent population-based studies, prevalence estimates for measured olfactory dysfunction based on odor identification tasks have ranged from 13.6% to 32.9% in adults 40 years and older, with rates increasing with age.[131-134] From self-report in the NHANES 2011, ≈23% of the population reported a smell problem (either smell problems in the past 12 months, worse smell ability compared with when age 25 years, or phantom odor perception—smelling an unpleasant/bad/burning odor when nothing is there), with increased risk among individuals who reported sinonasal symptoms, heavy drinking, xerostomia, history of loss of consciousness from head injury, or lower income.[118] Further analysis from the 2011–2014 NHANES data revealed a 6.5% prevalence of phantom odor perception with similar risk factors; those with smell dysfunction had an overall poor self-rating of health, and few sought medical care for this disorder.[135] From an 8-item identification task (four food, four household odors),[136] olfactory dysfunction prevalence was 12.4%, with age-related increases (4.2%, 40-49 years; 39.4% 80+ years). Between 20% and 31% of those ≥70 years misidentified smoke and natural gas, the two warning odors important from a public safety perspective.

The primary causes of olfactory dysfunction have been described in an evidence-based review[137]—neurodegenerative disorders (e.g., Alzheimer's disease, Parkinson's disease) and cognitive decline, postviral loss, sinonasal inflammatory disease, head trauma, or toxic exposure, including smoking. Individuals with severely elevated weight have a greater risk of olfactory dysfunction.[137] These diseases diminish olfaction by (a) reducing odor transport to receptors, (b) damaging receptors that receive and transduce olfactory messages, or (c) damaging peripheral or central neurophysiologic systems. With respect to COVID-induced olfactory loss, most evidence points to the SARS-CoV-2 virus targeting the supporting cells of the olfactory neuroepithelium,[115] with many experiencing parosmia with return of olfactory function.[138] Genetics may interact with the environment to influence risk of sensory alterations or dysfunction (e.g.,)[139] but further research is needed.[140] Impaired oral health can decrease retronasal perception even if the olfactory system is intact (e.g., poorly fitting dentures, Sjögren syndrome).

Systemic diseases, such as liver disease, kidney disease, diabetes, or diseases that influence mucus secretions (e.g., cystic fibrosis), can alter chemosensory function, with impairment related to the disease severity and complications, buildup of toxic metabolites, impaired nutritional status, and medication side effects (as reviewed).[112] Some medication effects are pronounced and have well-documented effects, particularly on taste (e.g., angiotensin-converting enzyme inhibitors),[141] with implications for drug toxicity.[142] Cancer therapies can directly impair chemosensory processes, thus hindering stimulus transport (e.g., diminished mucus secretions) and changing transduction mechanisms.[143] Except for head and neck cancer, cancer therapies (not cancer itself) impair chemosensation. Most frequently documented are mouth blindness, dysgeusia, and oral pain, presumably secondary to peripheral damage to taste nerves, and intensifying orosensation in the CNS. Clinicians can address nutritional gaps and quality of life by providing practical coping strategies and recommendations to patients with cancer who have chemosensory complaints.[144]

Treatments for chemosensory disorders are sparse with the call across an international panel of experts to support research that integrate basic science and clinical assessments, as well as to increase awareness in this area.[145] Taste alterations, including dysgeusia, have the most negative impact on those afflicted and require a thorough assessment and management plan.[129] Some types of olfactory dysfunction recover spontaneously (e.g., after viral loss or head trauma), depending on severity of insult, age, and the duration of dysfunction.[137] Nasal or sinus disease is one of the few treatable causes of olfactory dysfunction. Treatment includes controlling the cause (e.g., allergens, nasal infections) and level of inflammation (e.g., short cores of systemic or intranasal corticosteroids).[137] Olfactory training, purposely smelling odors daily to stimulate neurogenesis, has been tested for olfactory dysfunction related to viral loss with some promise including in parosmia,[146] yet it requires rigorous evidence-based testing.[147]

A special note on zinc is required. Zinc supplementation fails to correct smell and taste dysfunction in normally nourished adults[148] and likely will be ineffective for the older adult,[149] unless there is zinc deficiency.[150] Zinc supplementation may improve taste perception with zinc deficiency of chronic diseases.[151] The benefit of zinc supplementation to olfactory functioning is questionable and potentially dangerous. Specifically, intranasal zinc gluconate, an over-the-counter, alternative therapy for common cold prevention and treatment, causes hyposmia and anosmia, based on clinical, biologic, and experimental data.[152]

Chemosensory Variation, Nutrition, and Health

Sensory nutrition is the study of how chemosensation influences nutrition and health.[153] As taste drives liking and consumption which, in turn, influences our health, the question of interest becomes: can taste serve as a biomarker of susceptibility to chronic diseases?

The rest of this section follows the conceptual model shown in **Figure 42.5**, reviewing the associations between orosensory

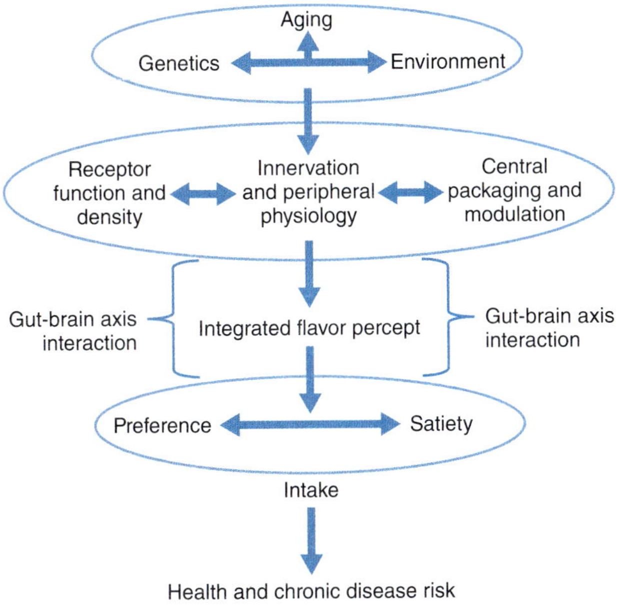

Figure 42.5 Hypothesized pathways of associations among factors that explain variation in chemosensation to influence flavor perception, dietary behaviors, intake, and health outcomes.

variation and food preference, dietary behaviors, and chronic disease risk with a focus on adults. Although early studies had inconsistent findings, evidence for these associations has grown through methodologic advances in characterizing chemosensory and hedonic experiences, as well as orosensory phenotypes and genotypes. Novel approaches show promise for capturing taste phenotype across taste qualities,[154,155] spatial patterns of taste function[156] as well as potential for polygenic taste scores[157,158] to advance understanding of sensory nutrition, diet, and health. Implications of genetic variation in olfactory receptors on olfactory acuity, diet, and health are largely unknown. A family-based genome-wide analysis found some increased odds of early onset, extreme obesity linked with chromosome 11q11 deletion, a region of three olfactory receptor genes (OR4P4, OR4S2, and OR4C11), potentially related to energy metabolism.[159] Nutritional implications of age-related changes in olfaction and orosensation are reviewed elsewhere.[117]

Food sensations could influence food consumption and metabolism through cephalic-phase responses and communication with the CNS through parasympathetic and/or sympathetic pathways, as well as through gut-brain connections.[153] The CNS also modulates orosensory sensations in fed and fasted states to encourage satiety and desire to eat, respectively, with research needed to identify the specific neural circuitry.[160]

Eating behaviors, including the rate of eating, bite size, and chewing, interact with the physical properties of the food (texture, textural complexity) to influence satiety.[161,162] Smelling external odors (orthonasal olfaction) can stimulate appetite, whereas taste interacts with texture[162] and retronasal olfaction[163] to promote satiation with slower eating and savoring the flavor of what is eaten. Interactions between food sensations and intake are operationalized in the theory of sensory-specific satiety (SSS)—how liking flavors of eaten food decreases relative to foods unconsumed.[164] There is desire to eat dessert even when full because of limited sweet satiation during the meal. Buffets promote overeating because of difficulty satiating on the diverse sensations, and monotonous diets limit flavor diversity to promote less intake.[165] Regarding gut-brain connections, genetic variation in taste and receptors have been shown to associate with variation in orosensations, gut receptor expression,[166] as well as to be modulated in metabolic syndrome.[167]

Dietary Bitters

Consumption of dietary bitters, such as vegetables, alcohol, and caffeine are of interest because of their important health implications. Many important phytonutrients are bitter that interact with other flavor components in nutritionally important foods and beverages.[55] Most evidence associates PROP phenotype (i.e., taste responsiveness) or genotype with vegetable consumption, initially focused on vegetables with the N-C=S moiety, particularly Brassicaceae such as broccoli and kale.[107] PROP tasters by phenotype or genotype report these vegetables as more bitter and consume them less frequently.[37,168] Those who taste PROP as more bitter report sampled vegetables as more bitter, less sweet, and having a stronger retronasal flavor, which decreases preference for sampled vegetables and generalizes to less consumption of all vegetables.[169] Individuals vary in multiple genes or genetic-environmental influences that can mask single SNPs, diet, and health associations. The TAS2R38 haplotype does not explain all the variability observed with the PROP phenotype. The influence of PROP phenotype or TAS2R38 genotype on bitter liking and intake may also depend on sex, age, fungiform papillae density,[37] or eating traits such as food neophobia[170] and dietary restraint.[171] Environments with less access to healthy foods may heighten the bitter effects on less-healthy diets and increased body weight in children.[172] Poverty, challenged living conditions, and access to health care can increase risk of otitis media, which can alter the bitter and sweetness from vegetables[173] to associate with lower preference for vegetables.[173,174] Modifying vegetable tastes and balancing bitterness and sweetness could hold promise in developing liking for sour- or bitter-tasting foods in children,[168,175,176] especially children with supertaster characteristics. Farm to table approaches can maximize sensory characteristics of bitter foods and beverages[177] within an approach that considers the sociocultural and economic influences on vegetable availability.[178]

Taste genetics likely interact with environmental and metabolic factors to influence behaviors toward alcoholic beverages and nicotine. PROP supertasters taste alcoholic beverages as most bitter and least sweet, and consume them least frequently.[179] Those who taste heightened bitterness may have sensory hindrance to starting smoking. In analysis of the 2013/2014 NHANES, adults with reduced bitterness of quinine on the tongue tip were more likely to be a chronic and dependent smoker.[156] The taste effects may show interactions with race/ethnicity[180,181] and use of menthol cigarettes,[156,182,183] which can attenuate the nicotine harshness with the minty/cooling sensations of menthol. TAS2R38 has been associated with alcohol intake and/or dependence,[184] as well as nicotine dependence.[180] TAS2R16 SNPs also have been associated with alcohol intake[185] and dependence.[186]

Causal relationships between bitter taste, diet, and health have been tested in Mendelian randomization studies. This design simulates randomization in controlled trials, based on the assumption that traits follow independent assortment (Mendel's law) and random segregation from other traits to form exposed and control groups with equal distribution of confounding variables (measured and unmeasured) between the two groups.[187] Multivariate models[187] assess if measured variables in the population-based studies follow direct or indirect pathways of causal association with outcomes of interest. Causal relationships between taste genetics have shown effects of PROP bitterness or TAS2R38 receptor gene in Mendelian randomization design on children's body weight,[188] consumption of bitter beverages (tea, coffee, alcohol),[189] and dietary risks of cancer.[190] Describing phenotype by multiple bitter markers can increase the ability to examine links between taste, diet, and health outcomes.[38,191] Taste or chemosensory scores composed of multiple taste qualities can further enhance the ability to explain health outcomes.[191,192]

Sweet Intensity and Preference

Humans vary in perceived intensity of sweetness from sucrose,[78] up to 33% attributable to heritability.[193] TAS1R2 (vs TAS1R1 and TAS1R3) is highly polymorphic,[194] showing associations with sucrose sensitivity and consumption of sugars,[195-197] as well as sweet receptor function in the gastrointestinal tract with impacts on glucose homeostasis.[198] Individuals who taste PROP as more bitter report greater sweetness from sucrose in water[199] as well as in milks that varied in fat.[78]

The liking for sweet is innate and pronounced in childhood, possibly cueing an energy source to sustain growth. A topic of debate has been whether exposure to highly sweetened foods early in life promotes greater desire for sweetness throughout life. Critical reviews of scientific evidence found insufficient evidence to support this supposition.[46,200] There also is not support for exposure to sweetness and body weight in adults from a scoping review.[201]

Rodent work shows that separate gut-brain signal pathways support greater preference for sugar over noncaloric sweetener.[202] Preferred level of sweetness also varies; 10% sucrose is optimal for some and 20% for others.[38] Liking for sweetness can be characterized by response to increasing concentration of sucrose, producing sweet dislikers (liking decreases), sweet likers (liking increases, peaks, and decreases), and extreme likers (liking increases).[203]

Twin studies show a significant heritable component to sweet preference.[193,204] Variation in sweet taste preference could reflect differences in receptor responses, CNS processing of this information to drive hedonic responses and reward, as well as gut-brain interactions. Polymorphisms of TAS1R2 associate with consumption of sugars in two populations of overweight and individuals with obesity[205] and, from the Amish Family Diabetes Study, polymorphisms of three TAS2R9 SNPs, not TASR1 or TASR3, associate with glucose homeostasis mechanisms.[206] Variation in brain reward responses to sweet stimuli have been examined in groups who vary in sweet preference and consumption.[207] Genetic effects on sweet taste and preference are beyond the peripheral receptor variation, including in the CNS; preference for and consumption of sugar intake was associated with regions of the FTO gene[208] and the fibroblast growth factor (FGF21) regulating glucose uptake in adipose tissue[208] and the liver.[209] The oral microbiome could interact with genetic variation in taste, sugar intake, and salivary flow to influence risk of dental caries.[210]

Salt and Sour Preference

Most sodium in the food supply is added during processing, and this may habituate us to higher levels. Some foods are consumed for saltiness (e.g., snacks). For others, salt is a flavor enhancer (e.g., cheese). Because sodium is vital for body fluid homeostasis, cell volume, and transmembrane potential, sodium appetite is controlled based on need, with excessive salt ingestion resulting from food processing, flavor enhancing, cultural food patterns, and preference for salty taste.[211] The ability to taste salt develops during the first year of life[212] and preference for salty taste may be based on early loss of sodium, dietary exposure, and growth.[211] Children of mothers who experienced dehydration during pregnancy report greater salt preference during infancy[213] and adulthood.[211] Salt appetite in adults is not responsive to sodium loss, diet, or deficiency.[211] Females and males differ in affinity for salt,[106] possibly because of sex hormones, as shown in pregnancy, when increasing salt preference is associated with the need to expand blood volume.[111]

Saltiness does not drive preference per se, except when it is missing and unpleasant sensations such as bitterness are heightened in food products. In soups, added sodium varies greatly, by more than twofold across regular sodium soups. Perceived saltiness and liking for high-sodium foods interacts with sex and markers of genetic variation in taste.[106] The NaCl added to soups could be reduced and still be within the concentration level that is most preferred by males and females and across PROP phenotype groups.[106] TRPV1-rs8065080 genotype failed to associate with variation in salt preference, intake, or blood pressure in older adults.[214] However, consuming high-sodium diets over a longer term can blunt the aversive responses to concentrated sodium by downregulating the bitter taste sensor, transient receptor potential channel M5 (TRPM5), as seen in mouse models and patients with hypertension.[215] There also could be interactions between salt-sensitive blood pressure, salt preference, and discretionary salt use.[216]

Consumption of sodium in US adults generally exceeds recommended levels.[217] Consumers appear to be using less discretionary salt,[218] while most sodium consumed is from processed foods.[217,219] Salt preference is malleable, but it can take months for adults to learn to like lower sodium foods.[220] Children will consume lower-sodium foods if that is what is available in the home, supporting the role families have in promoting healthier diets.[221] NaCl levels can be reduced modestly in some foods (potato products, bread) without negative impact on liking, yet voluntary industry reductions in sodium in the food supply has slowed since 2012.[222] Use of savory seasonings[223] and monosodium glutamate[224] can improve palatability of lower sodium products, with attention to nutrition education[225] and a total diet and behavioral approach to control blood pressure and cardiovascular risk. Because high-sodium diets are associated with hypertension and potassium may benefit blood pressure control, research is ongoing on salt substitutes, which can have unwanted, unpleasant tastes.[226]

A preference for sour taste has been maintained across evolution and may reflect the need for digestible energy and nutrients in fermented foods.[227] There is phenotypic variation in sour intensity and preference. A minority of adults and children[228] are sour likers in response to citric acid and berry products.[47,229] Sour likers could experience different qualities in the sour stimuli and have an adventuresome eating personality.[47,230] Sour taste intensity and liking associate with fruit consumption in infants[231] and children.[232] Those liking an intense sour taste had the highest salivary flow rates to buffer the sourness.[230]

Fat Preference

Preference for high-fat foods is variable, showing associations with a range of demographic, biologic, and psychological factors.[233] Some variability in liking and/or consumption for fat is associated with PROP tasting, fungiform papilla density taste phenotype[78,234] as well as polymorphisms in CD36, a fatty acid translocase.[235,236] Fat, by itself, is disliked, yet, in mixtures with sweet or salt, it can be pleasurable[78,237] and provide strong reward value of eating.[238] Change in level of fat closely relates with preference for high fat sweetened milks.[38] Reported liking for high-fat foods associates with consumption of high-fat foods[239,240] but can show a tighter relationship to adiposity than reported fat intake—individuals are more willing to tell you what they like (than what they eat) and reported preferences reflect habitual dietary behaviors tied into nutritional status endpoints such as adiposity.[241] Individuals who report greater fat preference have higher adiposity.[241-243]

Chemosensory Variation and Adiposity

Fischer was the first to associate variation in oral sensation with body type.[4] Sheldonian ectomorphs (thin) were the most sensitive to PROP and quinine taste, while endomorphs (heavier) were the least.[4] There is likely a bidirectional interaction between taste and obesity, with certain taste phenotypes/genotypes overconsuming energy dense, palatable foods and adipose tissue compounds influencing taste perception through hormones, signaling proteins, and taste receptors in the brain.[244]

Diminished taste intensity, either due to clinical or genetic variation, has been associated with elevated adiposity in some[171,188,192,245] but not all studies. Risk factors of CTN taste damage such as otitis media have also been associated with excessive adiposity in adults and children.[174,192] A longitudinal observational study conducted among 1900+ adults, however, found greater weight gain over 5 years among those with higher taste intensity at baseline.[154] In experimental studies, female PROP nontasters consumed more energy at a laboratory-based buffet meal than supertasters, following a high-fat soup preload.[246] In an observational study, which modeled direct and indirect associations using structural equation models, a preference-based diet quality index partially mediated the association of bitter taste phenotype on central adiposity, suggesting that lower central adiposity among those with greater bitter sensitivity is partly due to liking for foods with healthier diet quality.[191]

Greater BMI has been associated with lower olfactory acuity or olfactory dysfunction in smaller and population-based studies.[247,248] Data from NHANES and others suggest that the influence of olfactory dysfunction on BMI may vary by age, sex, as well as the type and severity of the impairment.[132,245,249] In NHANES, olfactory dysfunction was associated with lower BMI in older men (≥65 years), but with higher BMI and higher waist circumference in middle-age women (40–64 years), even after adjusting for potential confounders such as smoking and socioeconomic status.[132]

Sensory Nutrition—Considering Food Preferences

The sensory characteristics of food, but even more so, how much they are liked, drives what people consume. Advances in nutritional science hinge on the ability to understand what people usually eat, yet measures of dietary intake are challenged with bias, are tedious for individuals to complete, and current biomarkers may be invasive, impractical, and reflect only part of the diet.[250] Data support that asking people their level of liking/disliking of foods and beverages is a proxy of usual consumption which, in turn, shows links with chemosensation and health.[11] Combining survey-reported liking with consumption identifies individuals who are restraining their consumption (not eating what they like)[191] or eating less liked but healthier foods.[191,251] Survey reported liking correlates with biomarkers of dietary consumption and health in children[251] and adults.[23,252] Liking survey responses can form an index of diet quality (like the Healthy Eating Index[253]) to explain nutrition and health biomarkers in children[251] and adults.[23,191,254]

Interventions personalized or tailored to an individual's dietary behaviors can improve the level of response and outcome.[255] Survey liking, as a proxy of usual consumption, is simple for participants and researchers to complete and can be easily measured online via smartphones in children and adults for tailoring health information and messages to motivate healthier behaviors.[256,257] People chose low-fat or low-carbohydrate diets for weight loss that is consistent with their food preferences.[258] However, those allowed to choose their preferred macronutrient composition diet (i.e., low fat vs low carbohydrate) did not achieve greater weight loss from baseline to 24 weeks than those who were randomized into the diet.[259] Furthermore, those who learned to like the diet, especially the low carbohydrate diet, showed greater weight loss.[259] Personalizing the macronutrient composition of the diet to taste genetic phenotype also may explain differential weight loss in a lifestyle intervention.[260] Finally, improving preference for a higher-quality diet may explain a greater level of weight loss in a behavioral weight loss program[261] and across bariatric surgery.[262]

Summary

Sensory nutrition aims to understand how taste, smell, and oral somatosensation influence what we like and choose to eat, and ultimately, our health and risk of chronic diseases and conditions. Functional variation in taste, smell, and oral somatosensation, and the integrated flavor percept arises from genetics, development, environmental exposures, and aging. The variation may be narrowly tuned to a single receptor genotype or broadly tuned to generalized supertasting. Losses to these sensory inputs also may be quality specific or generalized, including loss of smell and taste with viruses such as the SARS-CoV-2. The importance of investigation on sensory nutrition was highlighted in the 2020–2030 Strategic Plan for NIH Nutrition Research. Future research will likely identify new biomarkers of chemosensory variation that best predict differences in specific diet-related health outcomes. Attention to chemosensation and preference could help tailor interventions to lower the dietary risk of chronic diseases and conditions.

References

1. Linne BM, Simons CT. Integration of Taste, Smell, and Chemesthesis: Clinical Implications. In: Joseph PV, and Duffy VB, *eds*. Sensory Science and Chronic Diseases: Clinical Implications and Disease Management. Springer International Publishing, Cham, 2021;105–24.
2. IFIC. 2022 Food and Health Survey. International Food Information Council Foundation, 2022.
3. Bailey EH, Nichols EL. Science 1888;11:145–6.
4. Fischer R, Griffin F, Rockey MA. Perspect Biol Med 1966;9: 549–77.
5. Research NOoN. 2020-2030 Strategic Plan for NIH Nutrition Research. 2020;2021.
6. Buck L, Axel R. Cell 1991;65:175–87.
7. Sunarharum WB, Williams DJ, Smyth HE. Food Res Intl 2014;62:315–25.
8. Coppola DM, White LE. J Bioenerg Biomembr 2019;51:53–63.
9. Lorenzo PMD. Nutrients 2021;13.
10. Kato M, Okumura T, Tsubo Y, et al. Proc Natl Acad Sci U S A 2022;119:e2114966119.
11. Duffy VB, Rawal S, Hayes JE. In: Joseph PV, and Duffy VB, eds. Measurement of Gustation: From Clinical to Population-Based Methods. Sensory Science and Chronic Diseases: Clinical Implications and Disease Management 2021;65–102, Springer International Publishing.
12. Parma V, Boesveldt S. In: Joseph PV, and Duffy VB, eds. Measurement of Olfaction: Screening and Assessment. Sensory Science and Chronic Diseases: Clinical Implications and Disease Management Springer International Publishing. 2021;45–63.
13. Parma V, Hannum ME, O'Leary M, et al. Chem Senses 2021;46:bjab012.
14. Weir EM, Hannum ME, Reed DR, et al. Chem Senses 2022;47: bjac036.
15. Pieniak M, Oleszkiewicz A, Klockow M, et al. Eur Arch Otorhinolaryngol 2022;279:779–84.
16. Yoshino A, Pellegrino R, Luckett CR, et al. Eur Arch Otorhinolaryngol 2021;278:3847–56.
17. Marks L, Stevens J, Bartoshuk L, et al. Chem Senses 1988;13: 63–87.
18. Bartoshuk LM, Duffy VB, Hayes JE, et al. Philos Trans R Soc Lond B Biol Sci 2006;361:1137–48.
19. Stevens, S. S. Am J Psychol. 1958;71:633–46.
20. Bartoshuk LM, Duffy VB, Green BG, et al. Physiol Behav 2004;82:109–14.
21. Hayes JE, Bartoshuk LM, Kidd JR, et al. Chem Senses 2008;33:255–65.
22. Duffy VB, Peterson J, Bartoshuk LM. Physiol Behav 2004;82: 435–45.
23. Xu R, Blanchard BE, McCaffrey JM, et al. Nutrients 2020;12.
24. Hartley IE, Liem DG, Keast R. Nutrients 2019;11:182.
25. Mennella JA. Am J Clin Nutr 2014;99:704s–11s.
26. Moayedi Y, Michlig S, Park M, et al. J Comp Neurol 2021;529: 3046–61.
27. Martens K, Steelant B, Bullens DMA. Cells 2021;10.
28. Turner A, Veysey M, Keely S, et al. Nutrients 2021;13.
29. Kaelberer MM, Rupprecht LE, Liu WW, et al. Annu Rev Neurosci 2020;43:337–53.
30. Miller I, Reedy F. Physiol Behav 1990;47:1213–9.
31. Bartoshuk LM, Duffy VB, Miller IJ, Jr. Physiol Behav 1994;56: 1165–71.
32. Bangcuyo RG, Simons CT. Exp Brain Res 2017;235,2679–88.
33. Sobek G, Jagielski P. Nutrients 2022;14.
34. Dinnella C, Monteleone E, Piochi M, et al. Chem Senses 2018;43:697–710.
35. Kaufman A, Kim J, Noel C, et al. Int J Obes (Lond) 2020;44: 739–43.
36. Lieder B, C˘onka J, Reiner AT, et al. Nutrients 2022;14.
37. Duffy VB, Hayes JE, Davidson AC, et al. Chemosens Percept 2010;3:137–48.
38. Hayes JE, Duffy VB. Physiol Behav 2008;95:77–87. Epub May 2, 2008. doi:10.1016/j.physbeh.2008.04.023.
39. Fischer ME, Cruickshanks KJ, Schubert CR, et al. Chem Senses 2013;38:669–77.
40. Cattaneo C, Liu J, Wang C, et al. Sci Rep 2020;10:18694.
41. Saito T, Ito T, Ito Y, et al. Eur J Oral Sci 2017;125:44–8.
42. çiçek SS. Molecules 2020;25.
43. Di Pizio A, Ben Shoshan-Galeczki Y, Hayes JE, et al. Neurosci Lett 2019;700:56–63.
44. Bartoshuk LM. Science 1979;205:934–5.
45. Hayes JE. Chemosens Percept 2008;1:48–57.
46. Trumbo PR, Appleton KM, de Graaf K, et al. Adv Nutr 2021;12:343–54.
47. Duffy VB, Rawal S, Park J, et al. Appetite 2016;107:116–25.
48. van Langeveld AWB, Teo PS, de Vries JHM, et al. Br J Nutr 2018;119:1195–206.
49. Nelson G, Hoon MA, Chandrashekar J, et al. Cell 2001;106: 381–90.
50. DuBois GE. Physiol Behav 2016;164:453–63.
51. Laffitte A, Neiers F, Briand L. Curr Opin Clin Nutr Metab Care 2014;17:379–85.
52. von Molitor E, Riedel K, Krohn M, et al. Front Hum Neurosci 2021;15:667709.
53. Hunter SR, Reister EJ, Cheon E, et al. Nutrients 2019;11.
54. Kamerud JK, Delwiche JF. Chem Senses 2007;32:803–10.
55. Xu L, Zang E, Sun S, et al. Crit Rev Food Sci Nutr 2022;1–21.
56. Bayer S, Mayer AI, Borgonovo G, et al. J Agric Food Chem 2021;69:13916–24.
57. Dagan-Wiener A, Di Pizio A, Nissim I, et al. Nucleic Acids Res 2019;47:D1179–D85.
58. Chandrashekar J, Mueller KL, Hoon MA, et al. Cell 2000;100: 703–11.
59. Matsunami H, Montmayeur JP, Buck LB. Nature 2000;404: 601–4.
60. Behrens M, Lang T. Front Nutr 2022;9:881177.
61. Fierro F, Giorgetti A, Carloni P, et al. Sci Rep 2019;9:8437.
62. Kim UK, Jorgenson E, Coon H, et al. Science 2003;299: 1221–5.
63. Wooding SP, Ramirez VA, Behrens M. Evol Med Public Health 2021;9:431–47.
64. Kim U, Wooding S, Ricci D, et al. Hum Mutat 2005;26: 199–204.
65. Green B, Gelhard B. Chem Senses 1989;14:259–71.
66. Heck GL, Mierson S, DeSimone JA. Science 1984;223:403–5
67. Bigiani A. Nutrients 2020;12.
68. Nomura K, Nakanishi M, Ishidate F, et al. Neuron 2020;106: 816–29.e6.
69. Liman ER. Neuron 2020;106:709–11.
70. Dias AG, Rousseau D, Duizer L, et al. Chem Senses 2013;38: 137–45.
71. Tapanee P, Tidwell DK, Schilling MW, et al. Int J Hypertens 2021;2021:5559831.
72. Ganzevles PGJ, Kroeze JHA. Chem Senses 1987;12:563–76.
73. Turner HN, Liman ER. Annu Rev Physiol 2022;84:41–58.
74. Zhang J, Jin H, Zhang W, et al. Cell 2019;179:392–402.e15.
75. Wise PM, Hansen JL, Reed DR, et al. Chem Senses 2007;32: 749–54.
76. Diepeveen J, Moerdijk-Poortvliet TCW, van der Leij FR. J Food Sci 2022;87:1449–65.
77. Wu B, Eldeghaidy S, Ayed C, et al. Crit Rev Food Sci Nutr 2022;62:7015–24.
78. Hayes JE, Duffy VB. Chem Senses 2007;32:225–36.

79. Liu D, Costanzo A, Evans MD, et al. Br J Nutr 2018;120: 64–73.
80. Pepino MY, Love-Gregory L, Klein S, et al. J Lipid Res 2012;53:561–6.
81. Lin C, Colquitt L, Wise P, et al. Chem Senses 2020;45:467–81.
82. Costanzo A, Liu D, Nowson C, et al. Br J Nutr 2019;122: 1212–20.
83. Small DM, Gerber JC, Mak YE, et al. Neuron 2005;47: 593–605.
84. Burdach KJ, Doty RL. Physiol Behav 1987;41:353–6.
85. Gogolla N. Curr Biol 2017;27:R580–R6.
86. Rolls ET. Prog Neurobiol 2015;127--128, 64–90.
87. Shepherd G. Nature 2006;444:316–21.
88. Horgue LF, Assens A, Fodoulian L, et al. Nat Commun 2022;13:2929.
89. Malnic B, Godfrey PA, Buck LB. Proc Natl Acad Sci U S A 2004; 101:2584–9.
90. Raka RN, Wu H, Xiao J, et al. Crit Rev Food Sci Nutr 2022;62:5424–43.
91. Maßberg D, Hatt H. Physiol Rev 2018;98:1739–63.
92. Martin AL, Anadon CM, Biswas S, et al. Mol Cancer Ther 2022;21:1184–94.
93. McClintock TS, Khan N, Xie C, et al. Chem Senses 2020;45: 805–22.
94. Boesveldt S, Postma EM, Boak D, et al. Chem Senses.2017.
95. Trimmer C, Keller A, Murphy NR, et al. Proc Natl Acad Sci U S A 2019;116:9475–80.
96. Prutkin JM, Duffy VB, Etter L, et al. Physiol Behav 2000;61: 161–73.
97. Huang R, Xu C. Compr Rev Food Sci Food Saf 2021;20: 1036–74.
98. Schobel N, Radtke D, Kyereme J, et al. Chem Senses 2014;39: 471–87.
99. Simons CT, Carstens E. In: Fritzsch B, ed. 3.20 - Oral Chemesthesis and Taste. The Senses: A Comprehensive Reference (Second Edition). Elsevier, Oxford. 2020;398–422.
100. Green B. Neurosci Lett 1989;107:173.
101. Berger A, Bartoshuk L, Duffy V, et al. In: DeVita V, Hellman S, and Rosenberg S, eds. Capsaicin for the treatment of oral mucositis pain. Principles and Practice of Oncology-Update. Philadelphia: J.B. Lipponcott Company, 1995; 243–8.
102. Mena M, Dalbah L, Levi L, et al. J Dent Anesth Pain Med 2020;20:337–56.
103. Farruggia MC, Pellegrino R, Scheinost D. Front Syst Neurosci 2022;16:865929.
104. Fox AL. Sci News Lett 1931;9:249.
105. Blakeslee AF. Proc Natl Acad Sci 1932;18:120–30.
106. Hayes JE, Sullivan BS, Duffy VB. Physiol Behav 2010;100: 369–80.
107. Tepper BJ. Annu Rev Nutr 2008;28:367–88.
108. Melis M, Errigo A, Crnjar R, et al. Sci Rep 2019;9:18047.
109. Nolden AA, McGeary JE, Hayes JE. Chem Senses 2020;45: 383–90.
110. Doty RL, Cameron EL. Physiol Behav 2009;97:213–28.
111. Duffy VB, Bartoshuk LM, Striegel-Moore R, et al. Ann N Y Acad Sci 1998;855:805–9.
112. Duffy VB. In: Meiselman HR, R. S., ed. Causes of Smell, Taste And Oral Somatosensory Disorders Affecting Eating And Drinking. Handbook of Eating and Drinking: Interdisciplinary Perspectives Springer Nature, https://link.springer.com/referencework/10.1007/978-3-319-75388-1, 2020.
113. Hannum ME, Koch RJ, Ramirez VA, et al. Chem Senses 2022;47.
114. Hintschich CA, Niv MY, Hummel T. Int Forum Allergy Rhinol 2022;12:210–6.
115. Boscolo-Rizzo P, Polesel J, Vaira LA. BMJ 2022;378:o1653.
116. Trius-Soler M, Santillán-Alarcón DA, Martínez-Huélamo M, et al. Compr Rev Food Sci Food Saf 2020;19:3755–73.
117. Duffy VB, Hayes JE. In: Chernoff R, ed. Smell, taste, and oral somatosensation: age-related changes and nutritional implications In Geriatric Nutrition pp. 115–64, Jones & Bartlett, 2014.
118. Rawal S, Hoffman HJ, Bainbridge KE, et al. Chem Senses 2016;41:69–76.
119. Hoffman H, Bartoshuk L, Losonczy K, et al. Chem Senses (in press).
120. Barragan R, Coltell O, Portoles O, et al. Nutrients 2018;10.
121. Karikkineth AC, Tang EY, Kuo PL, et al. Aging (Albany NY) 2021;13:24989–5003.
122. Bartoshuk LM, Snyder DJ, Grushka M, et al. Chem Senses 2005;30:i218–i9.
123. Bull TR. J Laryngol Otol 1965;79:479–93.
124. Kveton J, Bartoshuk L. Laryngoscope 1994;104:25–9.
125. Choi N, Ahn J, Cho YS. Otol Neurotol 2018;39:591–6.
126. Lecor PA, Touré B, Boucher Y. Clin Oral Investig 2018;22: 1093–102.
127. Lehman CD, Bartoshuk LM, Catalanotto FC, et al. Physiol Behav 1995;57:943–51.
128. Yanagisawa K, Bartoshuk LM, Catalanotto FA, et al. Physiol Behav 1997;63:329–35.
129. Grushka M, Su N. In: Joseph PV, and Duffy VB, eds. Oral Health and Chemosensory Problems: Clinical Implication and Disease Management. Sensory Science and Chronic Diseases: Clinical Implications and Disease Management. pp. Springer International Publishing, Cham, 2021; 203–24.
130. Rawal S, Duffy VB, Berube L, et al. Nutrients 2021;13.
131. Murphy C, Schubert CR, Cruickshanks KJ, et al. JAMA 2002;288:2307–12.
132. Gallo S, Byham-Gray L, Duffy V, et al. Physiol Behav 2020;215: 112702. doi: 10.1016/j.physbeh.2019. Epub 2019 Oct 17
133. Bramerson A, Johansson L, Ek L, et al. Laryngoscope 2004;114:733–7.
134. Schubert CR, Cruickshanks KJ, Fischer ME, et al. Chem Senses 2012;37:325–34.
135. Bainbridge KE, Byrd-Clark D, Leopold D. JAMA Otolaryngol Head Neck Surg 2018;144:807–14.
136. Hoffman HJ, Rawal S, Li CM, et al. Rev Endocr Metab Disord 2016;17:221–40.
137. Patel ZM, Holbrook EH, Turner JH, et al. Int Forum Allergy Rhinol 2022;12:327–680.
138. Lerner DK, Garvey KL, Arrighi-Allisan AE, et al. Laryngoscope 2022;132:633–9.
139. Dong J, Wyss A, Yang J, et al. Mol Neurobiol 2017;54: 8021–32.
140. Kousathanas A, Pairo-Castineira E, Rawlik K, et al. Nature 2022;607:97–103.
141. Schiffman SS. World J Otorhinolaryngol Head Neck Surg 2018;4:84–91.
142. Wang T, Glendinning J, Grushka M, et al. Toxicol Sci 2017;156:315–24.
143. Nolden AA. In: Loss of Taste and Smell Function in Cancer Patients. Sensory Science and Chronic Diseases Springer, 2021;181–201.
144. Kiss N, Symons K, Hewitt J, et al. J Acad Nutr Diet 2021;121: 278–304.
145. Mainland JD, Barlow LA, Munger SD, et al. Chem Senses 2020;45:493–502.
146. Altundag A, Yilmaz E, Kesimli MC. Laryngoscope 2022;132:1433–8.
147. Nag AK, Saltagi AK, Saltagi MZ, et al. Ann Otol Rhinol Laryngol 2022;34894221118186.

148. Henkin R, Schechter P, Friedwald W, et al. Am J Med Sci 1976;272:285–99.
149. Greger J, Geissler A. Am J Clin Nutr 1978;31:633–7.
150. Stewart-Knox BJ, Simpson EE, Parr H, et al. Br J Nutr 2008;99:129–36.
151. Hewlings S, Kalman D. Nutrients 2020;12.
152. Davidson TM, Smith WM. Arch Otolaryngol Head Neck Surg 2010;136:673–6.
153. Reed DR, Alhadeff AL, Beauchamp GK, et al. Am J Clin Nutr 2021;113(1):232–245.
154. Fischer ME, Cruickshanks KJ, Schubert CR, et al. J Acad Nutr Diet 2014;114:1195–202.
155. Gervis JE, Chui KKH, Ma J, et al. J Nutr 2021;151:2843–51.
156. Berube L, Duffy VB, Hayes JE, et al. Physiol Behav 2021;240:113554.
157. Gervis J, Ma J, Chui K, et al. Curr Dev Nutr 2022;6:1112.
158. Robino A, Concas MP, Catamo E, et al. Nutrients 2019;11.
159. Diels S, Huybreghts S, Van Hoorenbeeck K, et al. Mol Genet Metab Rep 2020;25:100656.
160. Fu O, Minokoshi Y, Nakajima KI. Front Neural Circuits 2021;15:609824.
161. Krop EM, Hetherington MM, Nekitsing C, et al. Appetite 2018;125:253–69.
162. Boesveldt S, de Graaf K. Perception 2017;46:307–19.
163. Yin W, Hewson L, Linforth R, et al. Appetite 2017;114: 265–74.
164. Rolls B. Nutr Rev 1986;44:93–101.
165. Brown CL, Halvorson EE, Cohen GM, et al. Pediatr Clin North Am 2015;62:1241–61.
166. Robino A, Rosso N, Guerra M, et al. Appetite 2021;166: 105595.
167. Bertran L, Portillo-Carrasquer M, Martínez S, et al. Nutrients 2021;13.
168. Fisher JO, Mennella JA, Hughes SO, et al. J Acad Nutr Diet 2012;112:235–45.
169. Duffy VB, Hayes JE, Sharafi M. J Agr Food Res 2, https://doi.org/10.1016/j.jafr.2020.100044
170. De Toffoli A, Spinelli S, Monteleone E, et al. Nutrients 11.
171. Tepper BJ, Koelliker Y, Zhao L, et al. Obesity (Silver Spring) 2008;16:2289–95.
172. Burd C, Senerat A, Chambers E, et al. Obesity (Silver Spring) 2013;21:786–94.
173. Dinehart ME, Hayes JE, Bartoshuk LM, et al. Physiol Behav 2006;87:304–13.
174. Peracchio HL, Henebery KE, Sharafi M, et al. Physiol Behav 2012;106:264–71.
175. Sharafi M, Hayes JE, Duffy VB. Chemosens Percept 2013; 6:8–19.
176. Gajari D, Rumbak I, Ranilovic´ J, et al. Appetite 2022;173: 105996.
177. Mennella JA, Colquhoun TA, Bobowski NK, et al. J Food Sci 2017;82:1713–9.
178. Chawner LR, Hetherington MM. Physiol Behav 2021;238:113493.
179. Lanier SA, Hayes JE, Duffy VB. Physiol Behav 2005;83: 821–31.
180. Mangold JE, Payne TJ, Ma JZ, et al. J Med Genet 2008;45: 578–82.
181. Risso DS, Kozlitina J, Sainz E, et al. PLoS One 2016;11: e0164157.
182. Oncken C, Feinn R, Covault J, et al. Nicotine Tob Res 2015;17:1416–20.
183. Duffy VB, Glennon SG, Larsen BA, et al. Physiol Behav 2019;15:111–22.
184. Wang JC, Hinrichs AL, Bertelsen S, et al. Alcohol Clin Exp Res 2007;31:209–15.
185. Hayes JE, Wallace MR, Knopik VS, et al. Chem Senses 2011;36:311–9.
186. Hinrichs AL, Wang JC, Bufe B, et al. Am J Hum Genet 2006;78:103–11.
187. Gupta V, Walia GK, Sachdeva MP. Pub Health 2017;145: 113–9.
188. Bouthoorn SH, van Lenthe FJ, Kiefte-de Jong JC, et al. Int J Obes (Lond) 2014;38:1005–10.
189. Ong JS, Hwang LD, Zhong VW, et al. Sci Rep 2018;8:16414.
190. Sacerdote C, Guarrera S, Smith GD, et al. Am J Epidemiol 2007;166:576–81.
191. Sharafi M, Rawal S, Fernandez ML, et al. Physiol Behav 2018;194:103–12.
192. Rawal S, Huedo-Medina TB, Hoffman HJ, et al. Obesity (Silver Spring) 2017;25:781–7.
193. Keskitalo K, Knaapila A, Kallela M, et al. Am J Clin Nutr 2007;86:55–63.
194. Valente C, Alvarez L, Marques PI, et al. Genome Biol Evol 2018;10:1139–52.
195. Dias AG, Eny KM, Cockburn M, et al. J Nutrigenet Nutrigenomics 2015;8:81–90.
196. Han P, Keast RSJ, Roura E. Br J Nutr 2017;118:763–70.
197. Chamoun E, Carroll NA, Duizer LM, et al. Nutrients 2018;10.
198. Serrano J, Seflova J, Park J, et al. Mol Metab 2021;54:101339.
199. Melis M, Tomassini Barbarossa I. Nutrients 2017;9.
200. Appleton KM, Tuorila H, Bertenshaw EJ, et al. Am J Clin Nutr 2018;107:405–19.
201. Higgins KA, Rawal R, Baer DJ, et al. Adv Nutr 2022;13: 2341–56.
202. Buchanan KL, Rupprecht LE, Kaelberer MM, et al. Nat Neurosci 2022;25:191–200.
203. Armitage RM, Iatridi V, Yeomans MR. Physiol Behav 2021;235:113398.
204. Bretz WA, Corby PM, Melo MR, et al. Arch Oral Biol 2006;51:1156–60.
205. Eny KM, Wolever TM, Corey PN, et al. Am J Clin Nutr 2010;92:1501–10.
206. Dotson CD, Zhang L, Xu H, et al. PLoS One 2008;3:e3974.
207. Szajer J, Jacobson A, Green E, et al. Brain Res 2017;1674: 101–10.
208. Hwang LD, Lin C, Gharahkhani P, et al. Am J Clin Nutr 2019;109:1724–37.
209. Søberg S, Sandholt CH, Jespersen NZ, et al. Cell Metab 2017;25:1045–53.e6.
210. Esberg A, Haworth S, Hasslöf P, et al. Nutrients 2020;12.
211. Leshem M. Br J Nutr 2020;123:1312–20.
212. Beauchamp GK, Cowart BJ, Mennella JA, et al. Dev Psychobiol 1994;27:353–65.
213. Crystal SR, Bernstein IL. Appetite 1998;30:297–307.
214. Ferraris C, Turner A, Kaur K, et al. Nutrients 2020;12.
215. Cui Y, Wu H, Li Q, et al. Hypertension 2019;74:1021–32.
216. Pilic L, Mavrommatis Y. Br J Nutr 2018;120:721–31.
217. U.S. Department of Agriculture and US Department of Health and Human Services. 2020. available at DietaryGuidelines.gov.
218. Quader ZS, Patel S, Gillespie C, et al. Public Health Nutr 2016;19:2195–203.
219. Bhat S, Marklund M, Henry ME, et al. Adv Nutr 2020;11: 677–86.
220. Blais CA, Pangborn RM, Borhani NO, et al. Am J Clin Nutr 1986;44:232–43.
221. Bobowski N, Mennella JA. J Nutr 2019;149:870–6.
222. Moran AJ, Wang J, Sharkey AL, et al. Am J Public Health 2022;112:325–33.
223. Dougkas A, Vannereux M, Giboreau A. Nutrients 2019;11.
224. Halim J, Bouzari A, Felder D, et al. J Food Sci 2020;85: 2902–14.

225. Silva-Santos T, Moreira P, Rodrigues M, et al. Nutrients 2021;14.
226. Xiao H, Yan Y, Gu Y, et al. Food Res Int 2022;156:110822.
227. Frank HER, Amato K, Trautwein M, et al. Proc Biol Sci 2022;289: 20211918.
228. Kildegaard H, Tønning E, Thybo AK. Food Qual Prefer 2011;22:620–7.
229. Laaksonen O, Ahola J, Sandell M. Appetite 2013;61:85–96.
230. Liem DG, Westerbeek A, Wolterink S, et al. Chem Senses 2004;29: 713–20.
231. Blossfeld I, Collins A, Boland S, et al. Br J Nutr 2007;98: 1084–91.
232. Liem DG, Bogers RP, Dagnelie PC, et al. Appetite 2006;46: 93–6.
233. Lampuré A, Deglaire A, Schlich P, et al. Br J Nutr 2014;112: 1353–63.
234. Tepper BJ, Nurse RJ. Ann N Y Acad Sci 1998;855:802–4.
235. Keller KL, Liang LC, Sakimura J, et al. Obesity (Silver Spring) 2012;20:1066–73.
236. Graham CA, Pilic L, McGrigor E, et al. Chem Senses 2021;46.
237. Drewnowski A, Brunzell JD, Sande K, et al. Physiol Behav 1985;35:617–22.
238. DiFeliceantonio AG, Coppin G, Rigoux L, et al. Cell Metab 2018;28:33–44.e3.
239. Lampure A, Castetbon K, Deglaire A, et al. Int J Behav Nutr Phys Act 2016;13:74.
240. Chmurzynska A, Mlodzik-Czyzewska MA, Malinowska AM, et al. Appetite 2021;159:105053.
241. Duffy VB, Lanier SA, Hutchins HL, et al. J Am Diet Assoc 2007;107:237–45.
242. Lampuré A, Adriouch S, Castetbon K, et al. Eur J Nutr 2020;59:249–61.
243. Park H, Shin Y, Kwon O, et al. Nutrients 2018;10, pii: E877.
244. Rohde K, Schamarek I, Blüher M. Diabetes Metab J 2020;44: 509–28.
245. Simchen U, Koebnick C, Hoyer S, et al. Eur J Clin Nutr 2006;60:698–705.
246. Tepper BJ, Melis M, Koelliker Y, et al. Nutrients 2017;9.
247. Pastor A, Fernández-Aranda F, Fitó M, et al. PLoS One 2016;11: e0148734.
248. Hwang SH, Kang JM, Seo JH, et al. PLoS One 2016;11: e0148813.
249. Mattes R, Cowart B, Schiavo M, et al. Am J Clin Nutr 1990;51:233–40.
250. Labonte ME, Kirkpatrick SI, Bell RC, et al. Appl Physiol Nutr Metab 2016;41:1096–9.
251. Sharafi M, Perrachio H, Scarmo S, et al. Child Obes 2015;11: 530–40.
252. Pallister T, Sharafi M, Lachance G, et al. Twin Res Hum Genet 2015;18:793–805.
253. Krebs-Smith SM, Pannucci TE, Subar AF, et al. J Acad Nutr Diet 2018;118, 1591–602.
254. Sharafi M, Duffy VB, Miller RJ, et al. Appetite 2016;99:157–67.
255. Celis-Morales C, Livingstone KM, Marsaux CF, et al. Int J Epidemiol 2017;46:578–88.
256. Hildrey R, Karner H, Serrao J, et al. Foods 2021;10.
257. Hubert PA, Fiorenti H, Duffy VB. Nutrients 2022;14.
258. McVay MA, Voils CI, Coffman CJ, et al. Appetite 2014;83: 117–24.
259. McVay MA, Voils CI, Geiselman PJ, et al. Appetite 2016;103: 336–43.
260. Burgess B, Raynor HA, Tepper BJ. Obesity (Silver Spring) 2017;25:1682–90.
261. Sharafi M, Faghri P, Huedo-Medina TB, et al. Nutrients 2021;13.
262. Hubert P, Papasavas P, Stone A, et al. Nutrients 2019;11:804.

CHAPTER 43

Regulation of Appetite and Food Intake

Paul A.M. Smeets and Monica Mars

CHAPTER OUTLINE

Abbreviations

AgRP, agouti-related peptide; **ARC**, arcuate nucleus of the hypothalamus; **CCK**, cholecystokinin; **CPR**, cephalic phase response; **DVC**, dorsal vagal complex; **GI**, glycemic index; **GLP-1**, glucagon-like peptide 1; **IQR**, interquartile range; **NAc**, nucleus accumbens; **NPY**, neuropeptide Y; **NTS**, nucleus of the solitary tract; **OFC**, orbitofrontal cortex; **PBN**, parabrachial nuclei; **POMC**, pro-opiomelanocortin; **PP**, pancreatic polypeptide; **PVN**, paraventricular nucleus; **PYY**, peptide YY; **RYGB**, Roux-en-Y gastric-bypass; **SCFA**, short-chain fatty acid; **SN**, substantia nigra; **SSS**, sensory-specific satiety; **VMH**, ventromedial hypothalamus; **vmPFC**, ventromedial prefrontal cortex; **VTA**, ventral tegmental area

Introduction

Food intake is a physiologic necessity and; therefore, eating is highly rewarding. Many animals, including humans, eat in brief periods, which allows them to engage in other activities. Food intake is the counterpart of energy expenditure. If energy intake and energy needs are in balance over longer periods (multiple days or longer), body weight remains relatively stable. Excess food intake (overconsumption) and low energy expenditure, promoted by an "obesogenic" environment, are widely viewed as the most important cause of the obesity epidemic.[1,2]

The regulation of food intake is complex as it entails concurrent short- and longer-term regulation of, among other things, blood glucose, glycogen stores, and fat stores, which must be managed together with physiologic processes critical for daily functioning, such as growth, physical activity, sleep, stress, and immune function. There has been much debate about what exactly is being regulated, e.g., body temperature (thermostat), blood glucose (glucostat), fat mass (lipostat), protein needs (protein leverage), and all of these putative drivers of food intake have received some support.[3,4] However, none of them provides a complete or satisfactory explanation of what is driving daily food intake, which contests the complexity of the myriad physiological processes involved. A very basic distinction is that between "homeostatic eating," driven by physiologic need and "hedonic eating" (also termed "nonhomeostatic eating" or "eating in the absence of hunger").[5] The modern viewpoint is that these processes are not so distinct as initially proposed; homeostatic and hedonic processes interact to produce eating behavior. The motivation to eat specific foods can vary, depending on different factors, like their sensory characteristics (sight, smell, taste), eating context (environmental, social), and personal characteristics (e.g., reward sensitivity, dietary restraint, health beliefs). A well-known model that tries to capture the different levels of signaling and their timing is the so-called "satiety cascade" put forward by Blundell et al.[6-8] Two main phases distinguished in this model are the phase of (mainly) oral and preabsorptive signals that lead to satiation (meal termination) and the subsequent phase of satiety, which is mainly characterized by oral postabsorptive signals. This chapter aims to outline the underlying mechanisms of daily appetite and food intake regulation.

The Eating Cycle

Unlike energy expenditure, eating is episodic; people consume meals or snacks followed by periods of fasting. The eating cycle (**Figure 43.1**) starts with the process of anticipation and is characterized by the start of consumption (meal initiation), its end (meal termination), and the time in between eating episodes. The process that leads to meal termination is called satiation and determines meal size. What follows is a state of satiety, which is characterized by a feeling of fullness and a lack of eating motivation, which persists until appetite rises again and a next eating episode is started.[8] The length of the intermeal interval determines meal frequency, i.e., the number of meals per day. What and how much is ingested is determined by conscious and subconscious decisions about what to eat, when to start, and when to stop.

Ingestive behavior is strongly linked to digestive processes. Four phases can be distinguished: the cephalic, oral, gastric, and intestinal phase (Figure 43.1). In all of these phases, the presence of food is detected by chemical or physical stimulation, which elicits physiologic anticipatory (feedforward) signals or feedback signals. These processes are collectively referred to as the "satiety cascade".[8]

During the cephalic phase, food intake is anticipated, and the first sensory signals are perceived (sight, orthonasal smell). These trigger multiple innate and learned physiologic responses that serve to prepare the body for the influx of nutrients and optimize digestion, so as to minimize the disturbance of homeostasis.[9] During the oral phase, i.e., when the food is chewed and swallowed, there is continued sensory stimulation, among which are taste, but also trigeminal sensations (texture, temperature) and retronasal smell. In this phase, sensory stimulation can decrease appetite, specifically for the type of food consumed.[10] The gastric phase involves neural and endocrine signaling, triggered by the stretching of the stomach and the presence of food, notably in the duodenum. As soon as macronutrients enter the duodenum the intestinal phase starts. This triggers the release of several intestinal hormones by enteroendocrine cells, such as cholecystokinin (CCK), glucagon-like peptide 1 (GLP-1), and peptide YY (PYY), which reflect macronutrient intake and contribute to satiety. In the coming paragraphs, the physiologic factors affecting appetite and food intake will be discussed in more detail.

Food Anticipation (Cephalic Phase)

Cephalic phase responses (CPRs) are mediated by the vagus nerve and are part of the complex regulation of appetite, digestion, and metabolism. They consist of anticipatory physiologic responses that prepare the digestive tract for biochemical and mechanical breakdown of food and optimal nutrient absorption. They also prepare other organs such as the liver and adipose tissue for the metabolization and efficient storage of the absorbed nutrients.[12-14] CPRs are elicited in response to the thought, sight, smell, and taste of food. Probably the most well-known CPR is the conditioned increase in saliva secretion ("mouth-watering") upon exposure to a food cue. This response can also be observed when merely thinking or talking about a food.[15] In addition, there are many other exocrine and endocrine secretions, such as gastric acid and digestive enzymes, secreted throughout the digestive tract, that show anticipatory responses. This includes pancreatic hormones, such as insulin and pancreatic polypeptide. Furthermore, there are other anticipatory physiologic responses, like changes in heart rate and gastric and intestinal motility.

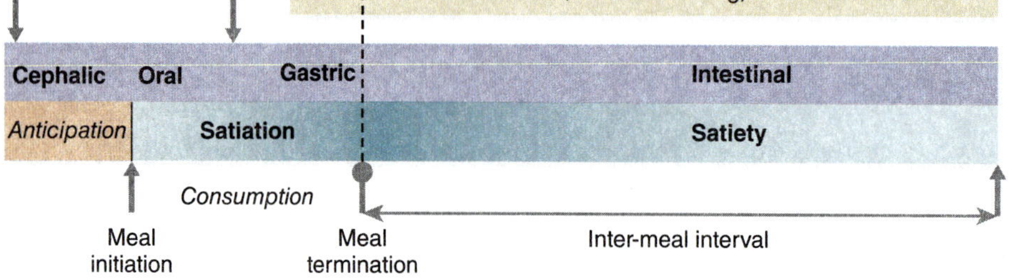

Figure 43.1 Overview of the eating cycle and the different phases of eating and digestion. A darker shade represents a greater degree of satiety.

Data from Blundell JE, Tremblay A. Nutr Res Rev 1995;8:225–42. doi:10.1079/NRR19950014 and Smeets PA, Erkner A, de Graaf C. Nutr Rev 2010;68:643–55. doi:10.1111/j.1753-4887.2010.00334.x[8,11]

Most research on endocrine CPRs has been done on insulin and pancreatic polypeptide (PP), which are secreted by the pancreas. Insulin is a key hormone involved in glucose homeostasis. PP has various known physiologic effects, such as inhibition of pancreatic exocrine secretion and bile duct motility after digestion; however, the specific role of cephalic phase PP release is unclear. Nonetheless, cephalic-phase PP release is viewed as a marker of vagal activation by food cues. While cephalic phase insulin and PP responses are consistently found in rodents, human studies are less conclusive. A meta-analysis of Lasschuijt, et al.[16] found a median (IQR, interquartile range) insulin increase of 2.5 (1.6–4.5) μIU/mL, 30% above baseline at 5±3 min after food cue onset (based on study treatments that induced ≥1 μIU/mL insulin increase). In only 22% of 119 treatments, this constituted a significant increase from baseline. In a less-detailed meta-analysis, overall support for the mere occurrence of cephalic-phase insulin release was found.[17] In humans, solid foods that stimulate multiple sensory modalities are the most potent triggers for cephalic-phase insulin release (as well as other CPRs). However, it should be noted that the magnitude of this response is small compared with spontaneous fluctuations.

For PP, Lasschuijt, et al.[16] report a median (IQR) increase of 99 (26–156) pg/mL, 68% above baseline at 9±4 min after food cue onset. In 21% of 42 treatments, the PP increase was significant from baseline. Both cephalic phase PP and insulin responses show substantial variation in magnitude and onset time within and between studies and individuals. There is little evidence for effects of these two cephalic responses on satiation and subsequent glucose homeostasis.[16] Thus, the physiologic relevance of these endocrine CPRs outside experimental settings is still not well established.

The cephalic phase is typically characterized by an increase in appetite, i.e., a desire to start eating. This has also been labeled the "appetizer effect."[18] However, there is no link between the magnitude of endocrine CPRs and subsequent meal size.[11]

Factors Affecting Satiation and Meal Termination

Oral Phase

During consumption, sensory signals provide crucial information about what is ingested and contribute to satiation. This mainly involves stimulation of receptors in the mouth by taste, trigeminal stimulation (texture, temperature) and orthonasal odor, which all contribute to oro-sensory stimulation. Taste has a clear role as a nutrient sensor. Together with texture, taste affects eating rate, and thereby determines the duration of orosensory exposure, which contributes to satiation.[16] More orosensory stimulation, either through longer or more intense stimulation, leads to faster satiation and, thus, a smaller meal.[16] This is called sensory satiation.[10,19] Eating rate is an important factor, which influences satiation (meal size) as do other aspects of food oral processing. Oral processing is determined by the interaction of personal physiologic characteristics (e.g., size of the oral cavity, chewing frequency) and mechanical and rheologic properties of the food, such as its texture.[20]

It is also well established that in the absence of overall satiety, sensory-specific satiety (SSS) can occur. This is defined as a reduction in liking for an eaten food, relative to an uneaten one.[21] A common example is that, after a savory meal, there may still be appetite for a sweet dessert. However, SSS has also been shown to occur for specific aromas, tastes, and even visual characteristics of food.[22,23] It is thought to promote variety in the diet.[24]

Apart from basic gustatory stimulation of taste buds on the tongue (sweet, salty, sour, bitter), there is mounting evidence for more direct sensing of nutrients throughout the alimentary tract, including in the oral cavity. There is ongoing work on the sensing of fat (fatty acids, for example),[25,26] nonsweet carbohydrates (oligosaccharides, for example[27,28]) and protein (umami taste; glutamate, kokumi taste; γ-glutamyl peptides).[29] More detail about the chemical senses and nutrient sensing can be found elsewhere in this book (*Chapter 42,* and *Chapter 46*).

Gastric Phase

Several signals from the stomach and the intestines are involved in the regulation of food intake and appetite. It is thought that the main function of these signals is to foster efficient, optimal and safe digestion. By regulating appetite and satiety, the inflow of nutrients into the digestive system is also optimized. We can distinguish between pre- and postabsorptive signals, i.e., signals that arise before and after nutrients (glucose, fatty acids amino acids) enter the bloodstream.

While ingesting food, the upper part of the stomach relaxes to accommodate the meal. Simultaneously, mechanoreceptors from the stomach wall sense the tension and stretch caused by the food's volume and weight, and this information is signaled to the brain via vagal and spinal visceral nerves.[30] For liquids, stomach emptying starts almost simultaneously with ingestion. More energy-dense foods, semisolids, and solids empty more slowly from the stomach.[31] The process of gastric emptying can take up to 4 to 5 hours, depending on the food's structure, nutrient content, and size of the meal.

Increasing the degree of stomach filling lowers food intake. For example, Lieber and colleagues have shown that an intragastric balloon filled with over 400 mL water lowered subsequent food intake.[32] However, gradual balloon distention is perceived as painful and unpleasant compared with intragastric infusion of a nutrient-containing drink with the same rate and net volume, indicating that nutrient sensing is important for gastric accommodation of food and associated feelings of satiation.[33]

Next to gastric volume, oral exposure to food can also decrease subsequent energy intake, although the effects are not additive.[34] While higher energy density lowers gastric emptying rate and increases experienced satiety, prolonging the oral exposure to food does not result in additional effects on gastric emptying or satiety.[35] This may be explained by a ceiling effect. Another study confirmed that higher energy density of meal shakes decreases gastric emptying, and that increased viscosity has a similar but smaller effect. In contrast, experienced fullness was lowered more by increased viscosity[36]. This underscores the lack of satiating efficiency of energy in quickly ingested drinks, such as sugar-sweetened beverages. Related to this, the degree of stomach filling and associated satiety feelings can be affected by what has been termed "gastric sieving"; low-calorie liquids can pass through the stomach quickly, while larger particles from the same meal are retained much longer. Blending a mixed liquid and solid meal into a homogeneous soup delays gastric emptying and increases hormonal responses, which may enhance postprandial satiety.[37]

Factors Affecting Satiety

Gastrointestinal Hormones

Several hormones are responsible for establishing a safe and efficient inflow of nutrients into the digestive system. These hormones are often referred to as "satiety hormones."[38] They are released from different parts of the digestive tract and show differential responses to certain nutrients.[39] These hormones and their agonists are extensively investigated as they are of potential interest in the development of medications to treat obesity. Many hormones are involved in the regulation of food intake; in the following paragraphs, we discuss those that have been investigated most frequently in relation to appetite and food intake.

Ghrelin

Ghrelin is the only known hunger or so-called orexigenic hormone; unlike the other hormones, its circulating level decreases after eating has started.[40] It is produced predominantly by closed-type endocrine cells in the oxyntic glands of the fundus of the stomach.[41,42] Infusion of ghrelin has been shown to enhance appetite and increase food intake in humans.[43]

The mechanisms regulating circulating ghrelin concentration are poorly understood. Ghrelin has a strong diurnal pattern.[40,44] Plasma concentration rises just before a meal and drops after eating, an opposite pattern compared with insulin.[40] Ghrelin secretion may be affected by nutrients in the stomach and by other hormones and metabolites from the circulation, as ghrelin-producing cells express nutrient-sensitive receptors on both their basal and lateral cell membrane.[39]

Macronutrients differentially affect circulating ghrelin concentration; fat suppresses ghrelin less effectively than carbohydrate or protein.[45] Moreover, after a glucose drink, ghrelin rebounded; the concentration started increasing again after 140 minutes, up to concentrations of 5% to 10% above baseline, while after the protein and lipid-rich drinks, ghrelin concentration returned to baseline after 6 hours. Of note, these findings were based on pure glucose and may not be applicable to other carbohydrates, especially those that are not fully broken down into glucose. Basal ghrelin concentration is associated with weight status, and drops after acute energy restriction.[46] Moreover, basal ghrelin and meal-induced ghrelin responses show substantial between and within individual variability.[47,48]

CCK

Cholecystokinin (CCK) was the first hormone shown to affect food intake. It has been studied since the 1970s in animals and the 1980s in humans.[49-51] Multiple studies have shown that intravenous infusion of physiologic doses of CCK lowers food intake in subsequent meals.[52] CCK is predominantly produced by I cells in the duodenum.[53] It is especially released when proteins and lipids are present[54] and primarily induces gallbladder contraction and slows gastric emptying, presumably mediated by CCK-1 receptors expressed by smooth muscle cells and enteric nerves. In addition, receptors on vagal afferent neurons mediate CCK stimulation of pancreatic enzyme secretion and inhibit food intake. There are also CCK1-1 receptors in parts of the midbrain (see the section on *Central regulation*).

GLP-1

Glucagon-like peptide-1 (GLP-1) is a so-called incretin hormone, which is secreted from the L cells in the small intestine after food intake. GLP-1 stimulates insulin secretion in a glucose-dependent manner, resulting in reduced glucose fluctuation after food intake. GLP-1 is also produced in parts of the nucleus of the solitary tract, a brainstem region that receives various inputs from the GI tract. In addition to its incretin function, GLP-1 also acts on receptors in the stomach to delay gastric emptying and inhibit gastric secretions. It also has a role in the so-called ileal brake, a feedback mechanism in the ileum that controls the transit of undigested nutrients in the GI tract.[55,56] Next to this, GLP-1 acts directly on receptors in the brain to reduce appetite for more on the many functions of GLP-1.[56,57]

Incretin hormones such as GLP-1 are interesting targets for managing obesity and metabolic disorders.[58] There is ongoing research on the use of long-acting GLP-1 receptor agonists, or so-called GLP-1 mimetics, to promote weight loss.[59,60] Unlike endogenously produced GLP-1, GLP-1 mimetics are stable against degradation by peptidases, giving them a half-life of up to 13 to 15 hours (liraglutide) or even a week (semaglutide). Liraglutide has been shown to reduce body weight in patients with prediabetes and with obesity,[61] and promoted weight loss in adolescents with obesity when added to lifestyle therapy.[62] It has been approved to support chronic weight management in combination with a reduced-energy diet and increased physical activity for adults with overweight or obesity and, in the United States, for children >12y with obesity.[63-65] Similarly, semaglutide, primarily developed for the management of type 2 diabetes,[66] can also reduce body weight in adults with obesity.[67] Semaglutide has been approved to support chronic weight management in adults in the United States and Europe.[68-70]

Peptide YY (PYY)

Peptide YY, or PYY in short, is a member of the pancreatic polypeptide (PP) family. It is produced by the L cells of the distal ileum and colon and acts on the brainstem and hypothalamus to reduce appetite. Fasting concentration of PYY is negatively associated with BMI.[71] Plasma PYY rises after eating[72,73] proportional to ingested energy.[74,75] Next to having a direct effect on the hypothalamus, PYY is also thought to be part of the ileal break mechanism that involves GLP-1.[55]

Exogenous infusion of PYY significantly reduces appetite and 24-h food intake in normal weight people, and in people with obesity.[71,76] However, such significant effects on satiety require supra-physiologic plasma PYY concentrations that are two to three times above those induced by a high energy meal, i.e., 1,500 kcal.[77] Likewise, two studies that compared PYY infusions and meals concluded that infusions are only effective in lowering appetite sensations or reducing energy intake when circulating blood concentration is much higher than those observed following the consumption of a large meal.[52,78]

Postabsorptive Factors

Postabsorptive factors, such as circulating nutrients and their metabolites, play an important role in satiety and the initiation of the next meal. Changes in plasma nutrient concentrations are sensed by hypothalamic neurons in the brain involved in appetite regulation. In addition, pancreatic hormones released postabsorption, such as insulin and PP, play a role in the regulation appetite and food intake. For example, peripheral PP infusion delays gastric emptying and reduces food intake in humans.[79,80] As mentioned previously, these pancreatic hormones are also involved in cephalic-phase responses to prepare the body for incoming nutrients.

A large body of literature is focused on the role of circulating glucose on appetite and food intake. Glucose dynamics are the result of carbohydrate ingestion. Carbohydrates, including polysaccharides such as starch, and disaccharides (e.g., sucrose, lactose, maltose), are broken down by enzymes into monosaccharides (glucose, fructose, and galactose) before they are absorbed. Subsequently, blood glucose rises and insulin is secreted by the pancreas so that glucose can enter the cells to be used or stored in the form of glycogen. In healthy individuals, glucose concentration stays within a relatively narrow physiologic range, due to the interplay between insulin and glucagon.

The dynamics of circulating glucose have a role in meal initiation. For example, Melanson and colleagues showed, in 10 healthy, time-blinded volunteers, that transient and dynamic declines in blood glucose were frequently followed by meal requests and accompanied by higher subjective hunger.[81] This finding is also consistent with the idea that low glycemic index foods enhance satiety. The glycemic index (GI) is a measure that ranks carbohydrate-containing foods according to the amount by which they raise blood glucose after consumption, compared with a reference food (75 g pure glucose or 75 g white bread).[82] Thus, low glycemic foods are hypothesized to prolong satiety. It has also been shown that glucose and sucrose reduce subsequent food intake in a dose-dependent manner.[83]

Although there are reports of associations between glucose homeostasis and appetite and food intake, the exact biologic mechanism is not clear.[83] It has been suggested that the absolute concentration of glucose in the blood is not the critical factor but, instead, reduced glucose utilization or intracellular glucose concentrations are thought to stimulate meal initiation.[83] Therefore, other factors like insulin and (incretin) gastrointestinal hormones may be involved. This is supported by the fact that experimental studies in healthy volunteers with insulin infusions showed both an increase and a decrease in food intake after hyperglycemia.[84,85] Also, when glucose is infused in the ileum, larger satiating effects are observed compared with infusion in the duodenum.[86] This suggests a role for peripheral glucose sensing in the intestines, presumably through the release of GLP-1 and PYY.[86]

Central Regulation and Integration

Eating Decisions

Eating can occur for various reasons, which are commonly dichotomized into eating out of physiologic need (homeostatic eating) and eating in the absence of this (hedonic eating). Obviously, eating out of need is rewarding, as food is a primary reward. However, there may be other physiologic and psychological reasons for eating (or not) that may pertain to specific types of foods. These include physiologic or psychological stress, illness, and considerations or beliefs about foods, for example, with regard to their healthiness or sustainability. Eating decisions are made in the brain by virtue of the integration of a multitude of neural and hormonal signals that reflect one's internal state and the environment and, thus, rely on the ability of the brain to weigh all of these different factors in order to make eating decisions.[87] These decisions determine the nutrient composition of the diet, eating frequency, and portion size. Understanding what drives eating decisions is crucial for understanding maladaptive patterns of eating behavior and fostering their prevention and remediation.

Eating decisions come about in the brain and are governed by the same neurocircuitry as other types of decisions. Our understanding of how the many factors that influence eating behavior act on this circuitry is far from complete but has advanced with the help of neuroimaging techniques and other increasingly sophisticated methods to study different populations of neurons and how they interact. There are three types of neural control systems that govern food choices, as well as other decisions: Pavlovian, habitual, and goal-directed.[88] The Pavlovian system is based on classical conditioning, where specific stimuli (food cues) are coupled to a response, such as salivation or approaching food. The habitual system allows for more flexible behavioral responses, based on previous (rewarding) experiences; it works well as long as the learned association between a certain action and its physiologic outcome does not change. Crucially, habitual actions may not be optimal when they have longer-term (adverse) effects or in a setting where reward contingencies change rapidly. The goal-directed system is most flexible and tries to select the best action (eating decision) by considering both basic and immediate outcomes (e.g., taste) as well as abstract and delayed ones (e.g., health effects).[88]

Research using comparable behavioral tasks in both humans and rats has implicated homologous cortical regions (medial prefrontal cortex/medial orbital cortex in humans and prelimbic cortex in rats) and the dorsal striatum (anterior caudate in humans and dorsomedial striatum in rats) in goal-directed action and in the control of habitual actions.[89] For goal-directed decisions, the human brain "weighs" all relevant aspects by assigning them values, which are integrated in a net value signal in the ventromedial prefrontal cortex (vmPFC).[90]

In humans, all aspects of feeding can be controlled by the goal-directed system, i.e., we can exert cognitive control over our eating behavior. However, there may be competition with the other two systems and the resulting eating behavior may not be optimal. Commonly, there is a conflict between immediate gratification (eat something tasty now) and longer-term health-related consequences.[91] Which decision is made depends on many factors, and some individuals may be better equipped to make healthier decisions than others, depending on factors like stress.[92]

Homeostasis: Hindbrain and Hypothalamus

A key brain area involved in integrating multiple neural and endocrine signals related to homeostatic processes is the hypothalamus, a cherry-sized area at the base of the brain. It is the most investigated brain area involved in the regulation of food intake and energy balance but intimately interacts with the hindbrain (caudal brainstem) and areas in the basal forebrain.[93] Most hormones involved in the short- and long-term regulation of food intake, as well as vagus nerve signals from the gut, act on the brain stem, which is a first site of integration of such signals. Notably, a study in decerebrate rats showed that the presence of the hindbrain is sufficient to produce meal termination for a known food.[94] The hypothalamic circuits that modulate food intake and energy balance target parts of the hindbrain's dorsal vagal complex (DVC), which integrates sensory information from the gastrointestinal tract with rhythmic signals that orchestrate food consumption; see Cheng et al.[95] for a detailed overview of the relevant hindbrain circuits. This includes vagal afferent and taste signaling to the nucleus of the solitary tract (NTS). Together with the DVC, the parabrachial nuclei (PBN) are involved in the integration of energy-related peripheral signals and processing of taste and gut signals. These brainstem areas receive satiation-related input at three levels[16]: First, neurons sense circulating metabolites and hormones such as glucose and CCK. Second, the brainstem receives vagal input from the sensory system and the gastrointestinal tract, such as taste (NTS, PBN) and gastric distention signals (PBN). Third, the brainstem receives input from the mid- and forebrain, which is integrated together with energy-related signals in the area postrema. The brainstem then sends signals to hepatic and gastric efferent nerves.

The hypothalamus is part of the limbic system, which is a set of interconnected deep brain structures involved in emotional responses, motivation, and various autonomic functions. It is composed of many histologically distinct nuclei that process and integrate myriad neural and hormonal inputs and outputs.[5,96] Neural hypothalamic inputs (afferents) include olfactory, visceral, limbic, and cortical signals. Visceral inputs come mainly from the vagus nerve, which receives information from mechano- and chemoreceptors in the gastrointestinal tract. The vagus nerve projects to the hypothalamus, the amygdala, and the thalamus via the brainstem (NTS, PBN).

Classic lesion experiments by Hetherington and Ranson[97,98] led to the dual center hypothesis, which posits that the lateral hypothalamus (LH) is responsible for hunger and subsequent feeding, while the ventromedial hypothalamus (VMH) is responsible for satiation and satiety[99,100]; stimulation of the LH increases food intake, while its destruction attenuates feeding and causes weight loss. Stimulation of the VMH inhibits food intake, while its destruction causes overeating and weight gain. However, apart from the VMH and the LH, several other hypothalamic regions, including the arcuate nucleus, paraventricular nucleus (PVN), and dorsomedial nucleus, have been implicated in food intake and body weight regulation. The current view is that several neuronal populations in the hypothalamus express specific neurotransmitters that mediate effects on food intake and/or energy expenditure, and that they are regulated by specific signals of nutritional state.[101] Specifically, there are two distinct neuronal populations in the arcuate nucleus of the hypothalamus (ARC): appetite-suppressing, pro-opiomelanocortin (POMC) neurons and appetite-increasing neuropeptide Y (NPY)/agouti-related peptide (AgRP) neurons. These neurons are well-positioned to integrate peripheral and central signals and are in turn modulated by multiple neurotransmitters and/or hormones, such as ghrelin. ARC neurons also project to other hypothalamic areas, including the PVN (catabolic) and the LH (anabolic), which enhance and suppress satiation signals, respectively. For a comprehensive overview of the central regulation of short- and longer-term fuel homeostasis, the reader is referred to Myers et al.[102]

Reward

The act of eating is rewarding. In the cephalic phase, when food intake is anticipated, appetite for a food is linked to its perceived reward value (**Figure 43.2**) Apart from a learning component, food reward has a motivational and a hedonic component, referred to as "wanting" and "liking," respectively.[103,104] Wanting is the motivation to eat a food (incentive salience), which is normally related to metabolic state (degree of nutrient depletion) and reflected in circulating hormone concentrations. "Liking" is the hedonic reaction to the pleasure of a reward, i.e., experienced pleasantness. However, it is hard to separate this from the associated food preference, which is a result of learning. Wanting is mediated by a distributed network of dopamine-related mesocorticolimbic brain areas, including brain stem areas (PBN), the lateral hypothalamus, amygdala, striatal areas (nucleus accumbens (NAc), ventral pallidum, dorsal striatum), parts of the insula, and the orbitofrontal cortex (OFC). Liking is mediated by a coordinated network of small brain areas referred to as "hedonic hotspots," which are in proximity of several of the wanting-related brain areas. When one is appropriately stimulated, e.g., with opioid- or endocannabinoid-stimulating drugs, they interact to enhance "liking" reactions.[104] These hedonic hotspots exist within the OFC, insula, NAc, ventral pallidum, and the PBN. Wanting and liking are mediated by interlinked brain mechanisms and normally arise together. However, in some cases, such as addiction, they can become segregated because their neural mechanisms are separable.[103,104] Not surprisingly, the brain areas involved in reward form a major part of what has been termed the "appetitive brain network.[87]

Effects of Bariatric Surgery

Roux-en-Y gastric-bypass (RYGB) and other bariatric surgeries, like sleeve gastrectomy and gastric banding, are increasingly used in the management of obesity and type 2 diabetes, as they have relatively high efficacy.[106] These procedures have substantial but incompletely understood effects on gastrointestinal hormone signaling and energy intake regulation, which may differ, depending on the type of surgery and the individual.[107] Although malabsorption, to some extent, may occur due to the surgery, the weight loss observed after bariatric surgery mainly results from a reduced intake of energy. The exact drivers of this are poorly understood as they relate to changes in multiple aspects relevant for eating behavior, including food preferences, food reward and appetite.[108] The evidence so far indicates that only a reduction in overall energy intake and an increase in postprandial satiety are maintained, whereas relative macronutrient intake and premeal hunger remain unchanged beyond 6-months postsurgery.[109] Research on this topic is complicated by the fact that dietary intake may be hard to measure accurately. For example, work on changes in food preferences has not yielded conclusive results.[110] Further work using objective measurement of changes in appetite and dietary intake and assessment of food perceptions and preferences, is needed to better determine the mechanisms contributing to weight regulation in the longer term.[109]

Role of the Microbiome

The effects of the intestinal microbiota on both gut and gut-brain signaling have received recent attention. This is putatively achieved through the production of short-chain fatty acids (SCFAs) and neurotransmitters, such as serotonin, dopamine, and GABA, by the microbiome.[111] These compounds are thought to signal to the brain via enteroendocrine cells, e.g., GLP-1 and PYY releasing cells, and vagal afferents.[112] The composition of the microbiome is influenced by diet, especially the amount and type of dietary fiber. These undigested nutrients act as a selective culture medium for the microbiota, hereby influencing the number

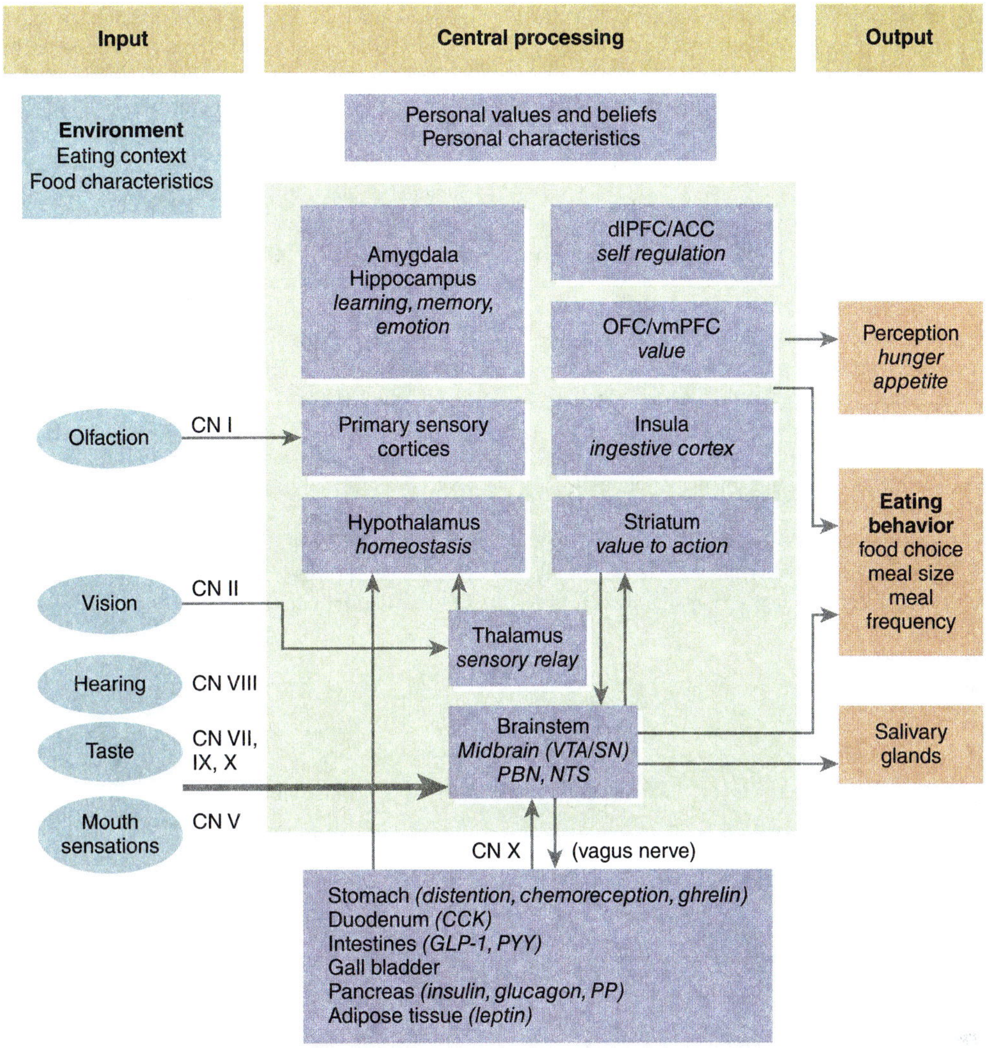

Figure 43.2 Overview of the central regulation of eating behavior. *NTS*, nucleus of the solitary tract; *OFC*, orbitofrontal cortex; *PBN*, parabrachial nuclei; *SN*, substantia nigra; *vmPFC*, ventromedial prefrontal cortex; *VTA*, ventral tegmental area.

Data from Neseliler S, Han JE, Dagher A. In: Harris RBS, ed. The Use of Functional Magnetic Resonance Imaging in the Study of Appetiteand Obesity. Appetite and Food Intake: Central Control, Boca Raton (FL). 2017;117–34. doi:10.1201/9781315120171-6[105]

and type of bacteria and indirectly the secretion of their bioactive fermentation products. Ample studies have investigated the effect of dietary fibers and prebiotics on satiety and food intake .[113-115] However, studies on the composition of the microbiome and its fermentation products on satiety and food intake regulation are still limited (see Bastings et al. for a recent review[111]).

Other Factors Affecting Appetite and Food Intake Regulation

In addition to the physiologic factors mentioned so far, there are several other factors that affect the regulation of appetite and food intake. For example, meal size and composition can be strongly influenced by state and trait physiologic factors related to age, sex,[116,117] adiposity, fasting duration, sleep deprivation, circadian rhythms, and stress (both acute and chronic). Food selection and consumption are also influenced by personal psychological factors, such as trait reward sensitivity, inhibitory control, dietary restraint, health beliefs, and religious beliefs, by cognitive effects related to previous experiences with the food (learning), and by contextual factors, such as the eating context (home vs restaurant setting, presence of other foods, serving size) and the social context (e.g., number and type of people present). Several of these factors are genetically determined, as they relate to personality traits like reward sensitivity and genetic predispositions that affect the propensity to obesity,[118] possibly through indirect effects on eating behavior.[119]

What makes the system even more complex and robust is that the short-term processes underlying daily food intake interact with overarching processes related to longer-term regulation of food intake and body composition, such as adipose tissue-related signals like leptin and insulin.[120]

Implications and Outlook

As we have seen, the regulation of food intake is immensely complex. Even on a basic physiologic level, there are numerous neural and hormonal signals that interact. Circulating physiologic concentrations of so-called satiety hormones have been shown to affect appetite and satiety but may be compensated or nullified by other signals. It is, therefore, a robust system, which is not easily perturbed. Targeting this system, for example to lower food intake to induce weight loss is, therefore, a challenge. In particular, the effects of dietary interventions may be too subtle to affect the system, or habituation may occur.[121]

Animal studies aimed at resolving the neurobiologic and physiologic mechanisms underlying ingestive behavior may give the false impression that manipulating a single factor can influence eating behavior. For example, the discovery of leptin in 1994 sparked hopes of an effective obesity treatment, but these proved to be idle; due to many feedback mechanisms and interactions, a "golden bullet" intervention applicable to humans is unlikely. Many single-molecule approaches have shown low efficacy with weight loss of <10%, which has led to ongoing work on more advanced compounds, so-called polyagonists, that target multiple systems at once.[122]

There is wide consensus that a multi-disciplinary approach is necessary to change daily eating behavior and, thereby, body adiposity and health. The most effective treatments to date are surgical in nature and target the stomach and intestines and, thus, many of the satiety factors discussed. These include the bariatric surgical treatments used in the management of severe obesity.[106] However, their exact mechanisms of action have not yet been resolved. Promising pharmacologic agents, such as GLP-1 mimetics, are emerging as supportive treatments. Pharmaceutical treatments and something as radical as bariatric surgery may change the balance between the multitude of signals involved, resulting in a shift in food intake. However, it is currently unclear how long such perturbations may persist and whether a new dynamic equilibrium is reached.

We do know that interventions accompanied by changes in lifestyle are an important prerequisite for longer-term success. Thus, there is still a need to identify both central and peripheral targets for pharmacologic interventions to either decrease or increase satiety and food intake. We anticipate that personalized multidisciplinary approaches aimed at influencing energy intake as part of larger lifestyle programs will ultimately prove most effective for prevention and treatment of overweight and obesity, given the large environmental and interindividual variability.

Acknowledgments

The authors acknowledge Syed Dufyan Hussain, Akila de Silva, and Stephen Robert Bloom for their contributions to the previous version of this chapter.

References

1. Swinburn B, Egger G. BMJ 2004;329:736–9.
2. Swinburn BA, Sacks G, Hall KD, et al. Lancet 2011;378:804–14.
3. Simpson SJ, Raubenheimer D. Obes Rev 2005;6:133–42.
4. Levitsky DA, Barre L, Michael JJ, et al. Front Nutr 2022;9:826334.
5. Saper CB, Chou TC, Elmquist JK. Neuron 2002;36:199–211.
6. Blundell J, Halford J. Proc Nutr Soc 1994;53:407–18.
7. Fetissov SO. Nat Rev Endocrinol 2017;13:11–25.
8. Blundell JE, Tremblay A. Nutr Res Rev 1995;8:225–42.
9. Woods SC. Psychol Rev 1991;98:488–505.
10. Graaf C. Physiol Behav 2012;107:496–501.
11. Smeets PA, Erkner A, de Graaf C. Nutr Rev 2010;68:643–55.
12. Zafra MA, Molina F, Puerto A. Neurosci Biobehav Rev 2006;30: 1032–44.
13. Power ML, Schulkin J. Appetite 2008;50:194–206.
14. Skvortsova A, Veldhuijzen DS, Kloosterman IE, et al. Neurosci Biobehav Rev 2021;126:447–64.
15. Feldman M, Richardson CT. Gastroenterol 1986;90:428–33.
16. Lasschuijt MP, Mars M, de Graaf C, et al. Adv Nutr 2020;11: 1364–83.
17. Wiedemann SJ, Rachid L, Illigens B, et al. Appetite 2020;155: 104792.
18. Yeomans MR. Appetite 1996;27:119–33.
19. Weijzen PL, Smeets PA, de Graaf C. Br J Nutr 2009;102:1091–7.
20. Chen J. Food Hydrocolloids 2009;23:1–25.
21. Rolls BJ, Rolls ET, Rowe EA, et al. Physiol Behav 1981;27: 137–42.
22. Rolls BJ, Rowe EA, Rolls ET. Physiol Behav 1982;29:409–17.
23. Rolls ET, Rolls BJ, Rowe EA. Physiol Behav 1983;30:185–92.
24. de Graaf C, Kok FJ. Nat Rev Endocrinol 2010;6:290–3.
25. Chale-Rush A, Burgess JR, Mattes RD. Chem Senses 2007;32: 423–31.
26. Running CA, Craig BA, Mattes RD. Chem Senses 2015;40: 507–16.
27. Lapis TJ, Penner MH, Lim J. Chem Senses 2016;41:755–62.
28. Pullicin AJ, Penner MH, Lim J. PloS One 2017;12:e0183008.
29. Keast R, Costanzo A, Hartley I. Nutrients 2021;13:667.
30. Cifuentes L, Camilleri M, Acosta A. Nutrients 2021;13:1158.
31. Kunz P, Feinle C, Schwizer W, et al. J Magn Reson Imaging 1999;9: 75–80.
32. Geliebter A, Westreich S, Gage D. Am J Clin Nutr 1988;48:592.
33. Ly HG, Dupont P, Van Laere K, et al. Neuroimage 2017;144: 101–12.
34. Wijlens AG, Erkner A, Alexander E, et al. Obesity 2012;20: 2226–32.
35. Wijlens AG, Erkner A, Mars M, et al. J Nutr 2015;145:365–71.
36. Camps G, Mars M, De Graaf C, et al. Am J Clin Nutr 2016; 104:73–80.
37. Marciani L, Hall N, Pritchard SE, et al. J Nutr 2012;142:1253–8.
38. Cummings DE, Overduin J. J Clin Invest 2007;117:13–23.
39. Steinert RE, Feinle-Bisset C, Asarian L, et al. Physiol Rev 2016;97:411.
40. Cummings DE, Purnell JQ, Frayo RS, et al. Diabetes 2001;50:1714.
41. Date Y, Kojima M, Hosoda H, et al. Endocrinol 2000;141: 4255–61.
42. Kojima M, Hosoda H, Date Y, et al. Nature 1999;402:656–660.
43. Wren AM, Seal LJ, Cohen MA, et al. J Clin Endo Metab 2001: 86:5992.
44. Templeman I, Smith HA, Walhin J-P, et al. J Appl Physiol 2021;130:1534–43.
45. Foster Schubert KE, Overduin J, Prudom CE, et al. J Clin Endo Metab 2008;93:1971–9.
46. Blom WA, Mars M, Hendriks HF, et al. Obesity 2006;14:838–46.
47. King JA, Deighton K, Broom DR, et al. Med Sci Sports Exerc 2017;49:1219–28.
48. Goltz FR, Thackray AE, Atkinson G, et al. J Nutr 2019;149: 1159–69.

49. Gibbs J, Young RC, Smith GP. J Comp Physiol Psychol 1973;84:488–95.
50. Kissileff HR, Pi-Sunyer FX, Thornton J, et al. Am J Clin Nutr 1981;34:154–60.
51. Pi-Sunyer X, Kissileff HR, Thornton J, et al. Physiol Behav 1982;29:627–30.
52. Mars M, Stafleu A, de Graaf C. Physiol Behav 2012;105: 483–8.
53. Liddle RA. J Gastroenterol 2000;35:181–7.
54. Pilichiewicz AN, Little TJ, Brennan IM, et al. Am J Physiol Regul Integr Comp Physiol 2006;290:R668–77.
55. Maljaars PWJ, Peters HPF, Mela DJ, et al. Physiol Behav 2008;95:271–81.
56. Holst JJ. Dig Dis Sci 2022;67:2716–20.
57. Muller TD, Finan B, Bloom SR, et al. Mol Metab 2019;30: 72–130.
58. Bays HE, Fitch A, Christensen S, et al. Obes Pillars 2022;2: 100018.
59. Del Prato S, Gallwitz B, Holst JJ, et al. Obes Rev 2022;23:e13372.
60. Mahapatra MK, Karuppasamy M, Sahoo BM. Pharm Res 2022;39:1233–48.
61. Le Roux CW, Astrup A, Fujioka K, et al. Lancet 2017;389: 1399–409.
62. Kelly AS, Auerbach P, Barrientos-Perez M, et al. N Engl J Med 2020;382:2117–28.
63. U.S. Food and Drug Administration. FDA approves weight management drug for patients aged 12 and older. Accessed February 28, 2023. Available at www.fda.gov/drugs/news-events-human-drugs/fda-approves-weight-management-drug-patients-aged-12-and-older.
64. European Medicines Agency. Saxenda recommended for approval in weight management in adults. Accessed February 28, 2023. Available at www.ema.europa.eu/en/news/saxenda-recommended-approval-weight-management-adults.
65. National Institute for Health and Care Excellence. Liraglutide for managing overweight and obesity. Accessed March 29, 2023. Available at www.nice.org.uk/guidance/ta664.
66. Mahapatra MK, Karuppasamy M, Sahoo BM. Rev Endocr Metab Disord 2022;23:521–39.
67. Wadden TA, Bailey TS, Billings LK, et al. JAMA 2021;325: 1403–13.
68. U.S. Food and Drug Administration. FDA Approves New Drug Treatment for Chronic Weight Management, First Since 2014. Accessed February 28, 2023. Available at www.fda.gov/news-events/press-announcements/fda-approves-new-drug-treatment-chronic-weight-management-first-2014.
69. European Medicines Agency. Wegovy. Accessed February 28, 2023. Available at www.ema.europa.eu/en/medicines/human/EPAR/wegovy.
70. National Institute for Health and Care Excellence. NICE recommended weight-loss drug to be made available in specialist NHS services. Accessed February 28, 2023. Available at www.nice.org.uk/news/article/nice-recommended-weight-loss-drug-to-be-made-available-in-specialist-nhs-services.
71. Batterham RL, Cohen MA, Ellis SM, et al. N Engl J Med 2003;349: 941–8.
72. Adrian T, Ferri G, Bacarese-Hamilton A, et al. Gastroenterol 1985;89:1070–7.
73. Chan J, Stoyneva V, Kelesidis T, et al. Diabetologia 2006;49: 169–73.
74. Degen L, Oesch S, Casanova M, et al. Gastroenterol 2005;129: 1430–6.
75. Le Roux C, Batterham R, Aylwin S, et al. Endocrinol 2006; 147:3–8.
76. Batterham RL, Cowley MA, Small CJ, et al. Nature 2002;418: 650–4.
77. Beglinger C, Degen L. Physiol Behav 2006;89:460–4.
78. Lim JJ, Poppitt SD. Nutrients 2019;11:1517.
79. Jesudason DR, Monteiro MP, McGowan BM, et al. Br J Nutr 2007;97:426–9.
80. Batterham R, Le Roux C, Cohen M, et al. J Clin Endo Metab 2003;88:3989–92.
81. Melanson KJ, Westerterp-Plantenga MS, Saris WH, et al. Am J Physiol 1999;277:R337–R45.
82. Wolever T. Mechanisms by which different carbohydrates elicit different glycaemic responses. In The Glycaemic Index: a physiological classification of dietary carbohydrate. CABI Wallingford UK: 2006; 64-82.
83. Anderson GH, Woodend D. Nutr Rev 2003;61:S17–S26.
84. Rodin J, Wack J, Ferrannini E, et al. Metabolism 1985;34: 826–31.
85. Gielkens HA, Verkijk M, Lam WF, et al. Metab 1998;47: 321–4.
86. Poppitt SD, Shin HS, McGill A-T, et al. Am J Clin Nutr 2017;106:725–35.
87. Dagher A, Neseliler S, Han J-E. Appetite as motivated choice: Hormonal and environmental influences. In Decision Neuroscience Elsevier, 2017; 397–409.
88. Rangel A. Nat Neurosci 2013;16:1717–24.
89. Balleine BW, O'doherty JP. Neuropsychopharmacol 2010;35: 48–69.
90. Levy DJ, Glimcher PW. J Neurosci 2011;31:14693–707.
91. Hare TA, Malmaud J, Rangel A. J Neurosci 2011;31:11077–87.
92. Maier SU, Makwana AB, Hare TA. Neuron 2015;87:621–31.
93. Grill HJ, Kaplan JM. Front Neuroendocrinol 2002;23:2–40.
94. Grill HJ, Norgren R. Science 1978;201:267–9.
95. Cheng W, Gordian D, Ludwig MQ, et al. Nat Metab 2022;4: 826–35.
96. Berthoud H-R. Neurosci & Biobehav Rev 2002;26:393–428.
97. Hetherington A, Ranson S. Anat Rec 1940;78:149–72.
98. Hetherington A, Ranson SW. Am J Physiol Leg Cont 1942;136: 609–17.
99. Anand BK, Brobeck JR. Yale J Biol Med 1951;24:123–40.
100. Stellar E. Psychol Rev 1954;61:5–22.
101. Williams G, Bing C, Cai XJ, et al. Physiol Behav 2001;74: 683–701.
102. Myers Jr MG, Affinati AH, Richardson N, et al. Nat Metab 2021; 3:737–50.
103. Berridge KC. Physiol Behav 2009;97:537–50.
104. Nguyen D, Naffziger EE, Berridge KC. Curr Opin Behav Sci 2021;39:72–8.
105. Neseliler S, Han JE, Dagher A. In: Harris RBS, ed. The Use of Functional Magnetic Resonance Imaging in the Study of Appetite and Obesity. Appetite and Food Intake: Central Control, Boca Raton (FL). 2017;117–34.
106. Arterburn DE, Telem DA, Kushner RF, et al. JAMA 2020;324: 879–87.
107. Steinert RE, Feinle-Bisset C, Asarian L, et al. Physiol Rev 2017; 97:411–63.
108. Al-Najim W, Docherty NG, le Roux CW. Physiol Rev 2018;98: 1113–41
109. Redpath TL, Livingstone MBE, Dunne AA, et al. Obes Rev 2021;22:e13202.
110. Gero D, Steinert RE, le Roux CW, et al. Curr Ather Rep 2017;19:1–8.
111. Bastings JJ, Venema K, Blaak EE, et al. Trends Endocrinol Metab 2023;34:243–255.
112. Dalile B, Van Oudenhove L, Vervliet B, et al. Nat Rev Gastroenterol Hepatol 2019;16:461–78.
113. Clark MJ, Slavin JL. J Am Coll Nutr 2013;32:200–11.
114. Wanders AJ, van den Borne JJ, de Graaf C, et al. Obes Rev 2011;12:724–39.

115. Poutanen KS, Dussort P, Erkner A, et al. Am J Clin Nutr 2017;106:747–54.
116. Hurtado MD, Cifuentes L, Campos A, et al. Intl J Obes 2022;46:2156–62.
117. Vien S, Fard S, El Khoury D, et al. J Nutr 2021;151:2161–74.
118. Loos RJ. Best Pract Res Clin Endocrinol Metabo 2012;26: 211–26.
119. Herle M, Smith AD, Kininmonth A, et al. Curr Obes Rep 2020;9:512–21.
120. Schwartz MW, Woods SC, Porte Jr D, et al. Nature 2000;404: 661–71.
121. Halford JCG, Masic U, Marsaux CFM, et al. Obes Rev 2018;19: 1329–39.
122. Müller T, Clemmensen C, Finan B, et al. Pharmacol Rev 2018; 70:712–46.

CHAPTER 44

Nutrition and the Immune System

Charles B. Stephensen and Ryan G. Snodgrass

CHAPTER OUTLINE

Abbreviations

AA, arachidonic acid; **APC**, antigen-presenting cell; **BCR**, B-cell receptor; **CLR**, C-type lectin receptor; **CD**, cluster of differentiation; **COX**, cyclo-oxygenase, **CRP**, C-reactive protein; **CTL**, cytotoxic T lymphocyte; **CYP**, cytochrome P450 oxidase; **DC**, dendritic cell; **DHA**, docosahexaenoic acid; **DTH**, delayed-type hypersensitivity; **EPA**, eicosapentaenoic acid; **HIV**, Human Immunodeficiency Virus; **ICAM-1**, intercellular adhesion molecule-1; **IFN**, interferon; **Ig**, immunoglobulin; **IL**, interleukin; **LOX**, Lipoxygenase; **LPS**, lipopolysaccharide; **LTB4**, leukotriene B4; **LTE4**, leukotriene E4; **MBL**, mannose-binding lectin; **MHC**, major histocompatibility complex; **NADPH**, nicotinamide dinucleotide phosphate; **NET**, neutrophil extracellular trap; **NF-kB**, nuclear factor-kB; **NK cell**, Natural Killer cell; **NKT-cell**, Natural Killer T-cell; **NO**, nitric oxide; **NLR**, nucleotide-binding and oligomerization domain (NOD)-like receptors; **PAMP**, pathogen-associated molecular pattern; **PGE2**, prostaglandin E2; **PUFA**, polyunsaturated fatty acid; **pIgR**, polymeric immunoglobulin receptor; **PRR**, pattern recognition receptor; **RA**, retinoic acid; **RBP**, retinol-binding protein; **RNA**, ribonucleic acid; **ROR**, retinoic acid orphan receptor; **SPM**, specialized pro-resolving mediator; **TCR**, T-cell receptor; **TGF**, transforming growth factor; **Tfh**, T-follicular-helper cell; **Th cell**, T-helper cell; **TLR**, Toll-like receptor; **TNF**, tumor necrosis factor; **Treg cell**, regulatory T-cell; **VDR**, vitamin D receptor

Introduction

The principal function of the immune system is to protect the host from death and disability caused by infections.[1] Most pathogenic micro-organisms (viruses, bacteria, protozoa, or helminthic parasites) have evolved methods of evading the innate immune response and must be cleared by adaptive immunity. Some evade adaptive immunity as well (e.g., malaria or Human Immunodeficiency Virus [HIV]). "Opportunistic" pathogens can cause disease when the immune system is compromised by malnutrition, other infections, or advancing age. Commensal organisms, primarily bacteria, colonize the skin, intestine, respiratory tract, and urogenital tracts, and are benign or beneficial to the host. However, these organisms might also be harmful under certain circumstances and thus, are also subject to control, but not elimination, by the immune system.[2]

In addition to infections, the immune system can be activated by sterile injury that causes tissue damage but does not involve micro-organisms.[3] Such "sterile inflammation" is an important factor in the development of many chronic inflammatory diseases (e.g., coronary artery disease),[4] discussed elsewhere in this book.

Innate and Adaptive Immunity

The immune system comprises two major subsystems referred to as "innate" and "adaptive,"[5] which work together as an integrated whole. The "innate" system is evolutionarily older, and it is fully functional at birth. Innate immune cells use a diverse group of receptors to recognize and respond to signature molecules from classes of micro-organisms (e.g., flagella from some bacteria, cell-wall carbohydrate from yeast, RNA from viral genomes). These responses are essentially the same for all individuals within a species. The "adaptive" system is different in that the host's response adapts to a specific pathogen (e.g., Measles virus specifically and not RNA viruses in general) to develop "immunologic memory" that will respond more quickly and more efficiently the next time the same pathogen is encountered. Thus, individuals have different levels of adaptive immunity, depending on their exposure history. The adaptive nature of this response explains why the first encounter with a childhood pathogen (e.g., Measles) can make a child quite ill, but subsequent infections may go unnoticed.

Passive Protection of Infants

Infants have a full complement of innate immune cells at birth, although these cells respond less vigorously to micro-organisms than the same cell types from adults.[6] In contrast, infants have not yet developed adaptive immunologic memory. However, infants transiently acquire some components of adaptive immunity from their mothers. For example, serum immunoglobulin G (IgG) antibody is transferred across the placenta, giving infants protection against infections such as Measles for up to 9 months.[7] In addition, breastfed infants receive secretory IgA antibody and many antimicrobial factors from colostrum and breast milk.[8] This maternally derived protection for infants is important because the infant's adaptive immune system responds less robustly to pathogens than the adult system.[9] This attenuated response may be beneficial because colonization of the gut and other epithelial surfaces with commensal microflora present a major challenge to the developing immune system, and over-responding could be detrimental. However, colonization of the gut of the breastfed infant is influenced toward beneficial flora, particularly *Bifidobacterium infantis*, by the human milk oligosaccharides that are not absorbed in the small intestine and are preferentially utilized by *B. infantis*, giving it an advantage in colonizing the infant gut.[10] One benefit of this organism appears to be supporting development of better adaptive immunity, as suggested by the greater vaccine responses seen in infants with higher levels of *B. infantis*.[11]

Organization of the Immune System

The immune system in humans and other mammals is made up of organs and tissues located strategically throughout the body to protect against invasion by microorganisms.[1,5] "Primary" organs, where immune cells develop, include the bone marrow and thymus. All white blood cells (i.e., leukocytes; see **Table 44.1**) originate in the bone marrow, but one subset of lymphocytes, T lymphocytes, (also known as "T-cells") needs an additional maturation step in the thymus. In mammals, B lymphocytes (B-cells) mature in the bone marrow. The lymph nodes, spleen, and mucosa-associated lymphoid tissue are "secondary" organs and tissues. These secondary sites are meeting places for immune cells that are connected by the blood and lymphatic systems to allow transmission of information from the innate to the adaptive immune system.

The lymph nodes are located regionally (e.g., along lymphatic vessels draining specific regions of the body) and this information transfer occurs when an antigen-presenting cell (APC), after an encounter with invading micro-organisms, travels via lymphatic vessels from peripheral tissues (e.g., skin, respiratory mucosa, gut) to enter the closest draining lymph node.[1,5] Lymphatic vessels drain all tissues of the body, thus, this APC-based surveillance system can deliver information from any site of infection to a regional lymph node. "APC" is a functional definition and antigen presentation can be made by several cell types, including dendritic cells, macrophages, and B-cells.

The spleen, like the lymph nodes, provides a site for APCs to transfer information to lymphocytes. The spleen also filters the blood. In the case of a breach of peripheral defenses, bloodborne micro-organisms, or infected erythrocytes (e.g., in the case of malaria) are removed from the blood by phagocytic cells in the spleen, particularly macrophages.

Intercellular Communication in the Immune System

Cells of the immune system aggregate in secondary lymphoid tissues and at sites of inflammation. These cells communicate with one another, via cell-cell contact and soluble mediators, to trigger changes in activity (e.g., chemotaxis) and gene expression. Cytokines, including interleukins, and chemokines are protein mediators produced by immune and other cells that trigger a variety of responses in cells bearing the appropriate receptors. One large family of chemokines has a standard Cysteine-Cysteine or C-C motif, while a second family has a C-X-C motif. These chemokines are known as "CC" and "CXC" chemokines, respectively. Bioactive "lipid mediators" also regulate immunity. In addition to providing structural integrity to cell membranes through incorporation into phospholipids, polyunsaturated fatty acids (PUFAs), including arachidonic acid (AA, the main n-6 precursor) and eicosapentaenoic acid (EPA, the main n-3 precursor), function as substrates for the generation of lipid signaling molecules. These lipid mediators regulate a wide array of biologic properties and are synthesized by three types of oxidation reactions involving cyclo-oxygenase (COX), lipoxygenase (LOX), and cytochrome P450 oxidase (CYP) or epoxygenase enzymes. The COX pathway generates prostanoids, including prostaglandins (PGs) and thromboxanes (Txs) while the LOX pathway produces leukotrienes (LTs) as well as a diverse group of specialized pro-resolving mediators (SPMs; e.g., resolvins, protectins, and maresins) that exhibit inflammation-resolving properties.[12]

Innate Immunity

Epithelial Surfaces and Barrier Defenses

The innate immune system protects "portal-of-entry" sites used by pathogens to cause infections, including the skin, conjunctiva, respiratory tract, gut, and urogenital tracts.[1] Tissues at these

Table 44.1 Cells of the Immune System

Myeloid Cells
Common myeloid progenitor—found in bone marrow; progenitor of all myeloid cells, including monocyte lineage and granulocyte lineage cells
Monocyte Lineage
Monocyte—Found in blood, can differentiate into macrophage upon entering tissues
Macrophage—Phagocytic cell found in tissues involved in defense against micro-organisms and in "sterile" inflammation initiated by tissue damage (e.g., wound or plaque in coronary artery)
Immature dendritic cell—Found in blood, differentiates to dendritic cell in tissues
Dendritic cell—Function as "antigen-presenting cells"; deliver antigen from periphery to lymphocytes in draining lymph nodes
Myeloid-derived suppressor cells (MDSCs) —Immature myeloid cells at varied stages of differentiation that have immune-suppressive activity
Granulocyte Lineage
Neutrophil—Principal phagocytic cell in blood; enters tissue in response to inflammation to kill invading bacteria by phagocytosis (ingestion), oxidative metabolism, and secretion of anti-bacterial peptides
Eosinophil—Found in blood; enters tissue to mediate inflammation in response to parasitic infections and allergies, including asthma
Basophil—Found in blood; enters tissue in response to parasitic infection
Mast cell—Found in tissues primarily at submucosal sites; responds to some antigens, including allergens, via IgE molecules on surface of masT-cell; this activation triggers release of mediators that triggers local and systemic inflammation, including anaphylaxis
Lymphoid Cells
T-cell—Normally found in blood and lymph nodes, as well as in tissue at sites of inflammation; cell-surface T-cell receptor (TCR) recognizes peptide antigens; $CD8^+$ "killer" T-cells recognize and kill virus-infected hosT-cells; $CD4^+$ helper T-cells produce cytokines that stimulate development of $CD8^+$ T cells, B cells and stimulate protective responses of some myeloid cells, including macrophages
B-cell—Normally found in blood and lymph nodes; cell-surface B-cell receptor (BCR) is a membrane-anchored immunoglobulin that recognizes foreign antigens: after antigenic stimulation, B cells develop into antibody-secreting plasma cells found in bone marrow and at submucosal surfaces
Innate lymphoid cells (ILCs) —Include cytotoxic natural killer (NK) cells and groups of cytokine-producing (or helper-like) ILCs; mainly reside at barrier surfaces and regulate tissue homeostasis and immunity; do not have antigen-specific cell-surface receptor
Natural killer (NK) cell—Found in blood and tissues; does not have antigen-specific cell-surface receptor; recognizes and kills virus-infected and other "stressed" or damaged cells via change in expression of cell surface receptors
NK T cell—Unconventional T-cell that responds to nonpeptide "antigens" (typically lipid; presented via CD1 rather than MHC) via TCR of limited diversity; can be cytotoxic or regulatory
Mucosal-associated invariant T (MAIT) cell—Unconventional T-cell whose TCR responds to nonpeptide, pyrimidine antigens, derived from intermediates in microbial riboflavin metabolism; abundant in mucosal tissues such as lung and gut lamina propria
Other Cells
Megakaryocyte—Found in bone marrow; precursor to small, non-nucleated platelets that are found in blood and mediate blood clotting
Erythroid progenitor cells—Found in bone marrow; progenitors for red blood cells

portals are designed to protect against infection using a variety of common mechanisms. These sites have a surface layer of epithelial cells interspersed with a few lymphoid or myeloid immune cells. The subepithelial tissue provides structure and contains blood vessels to provide entry into the epithelium for immune cells, when needed, and lymphatic drainage to allow egress of APCs to the draining lymph node. Two interesting examples to consider are the skin and the intestine.

Skin consists of two cellular layers, epidermis and dermis.[13] The epidermis consists of four layers of keratinocytes interspersed with melanocytes and Langerhans cells, a professional APC, and the principal immune cell of the uninfected epidermis. Some commensal micro-organisms adhere to the epithelial surface and are adapted to persist in this niche.[14] Pathogens, including strains of *Staphylococcus aureus*, may penetrate the skin using special virulence factors (e.g., enzymes to break down extracellular matrix) to cause deeper infections which, if the local immune response is not sufficient, may become systemic.[1,15] The dermis contains blood capillaries and lymphatic drainage as well as a variety of immune cells, the number and type varying depending on the immunologic challenge. All such challenges do not come from micro-organisms. Inflammation in the skin may be triggered by irritants (e.g., chemicals, ultraviolet [UV] light), to which a person may become sensitized (e.g., poison ivy, which elicits an adaptive immune response).

The mucosal epithelium of the intestine consists of a single layer of absorptive epithelial cells interspersed with a number of other cells, including: a) goblet cells that secrete a protective layer of mucus, b) M cells that collect particulate antigen from the lumen for delivery to mucosa-associated APC in underlying lymphoid aggregates, c) interdigitating dendritic cells (DC, a type of APC) that send cytoplasmic arms between epithelial cells to directly sample antigen from the gut lumen,[2] and d) Paneth cells in intestinal crypts that secrete anti-fungal α-defensins. The lamina propria underlying the gut epithelium contains abundant immune cells, particularly lymphocytes. Unlike the dermis, there are many lymph nodes in the lamina propria (termed Peyer's patches). A number of factors, including peristalsis, the mucus barrier, the relatively rapid turnover of epithelial cells, and secreted factors (e.g., IgA, antimicrobial peptides), help protect this epithelial barrier from micro-organisms.[16,17] IgA and IgM are transported across intestinal epithelial cells and into the gut lumen via the polymeric immunoglobulin receptor (pIgR). The extensive network of APCs, in concert with regulatory T (Treg) cells in the lamina propria, help the body differentiate commensal organisms from pathogens.[18]

Other mucosal sites include the mouth, nasopharynx, trachea, esophagus, stomach, and urogenital tracts. These sites have similar organizational features and functions.[1] The lungs present a unique challenge in that alveoli are gas exchange surfaces and due to limits of gas diffusion, cannot be organized into multicellular layers. The final line of defense in the lungs is formed by the alveolar macrophages, which engulf and clear tiny particles and micro-organisms (e.g., *Mycobacterium tuberculosis*).[19]

Recognition of Pathogens by Innate Immune Cells

Innate immune cells, including monocytes and macrophages as well as "immune-like" epithelial cells, recognize and respond to pathogens and, thus, are an integral part of the response to infection.[5] Recognition of micro-organisms occurs via pattern-recognition receptors (PRRs) that recognize "signature" pathogen-associated molecular patterns (PAMPs) found in macromolecules common to groups of micro-organisms, but not typically found in mammals. PRR families include Toll-like receptors (TLRs), C-type lectin receptors (CLRs), nucleotide oligomerization domain-like receptors (NLRs), and RIG-I-like receptors (RLRs).[20] The TLR family is the best studied, and distinguishes PAMPs from different classes of bacteria, yeast, and viruses.[20,21] For example, bacterial lipopolysaccharide (LPS) is recognized by TLR4, bacterial flagellin by TLR5, single-stranded RNA by TLR7, and repeated DNA sequences of the bases C and G (common in bacterial but not mammalian genomes) by TLR9.

Specific PRRs, termed "inflammasomes," function as signaling hubs that serve as activation platforms for inflammatory cytokine production.[22] In response to distinct stimuli, these receptors assemble into a multiprotein complex in the cytoplasm that results in cleavage of pro-interleukin (IL)-1β and pro-IL-18 to produce the active cytokines that are key to host defense against pathogens. Inflammasomes can also be activated by nonmicrobial tissue "irritants" such as uric acid crystals, which accumulate in tissues of patients with gout, and the adjuvant alum, which is used in many human vaccines.

Local Inflammation

Except for inflammasomes, binding of PAMPs to their cognate receptors activates cytoplasmic signal transduction pathways that initiate gene transcription in the nucleus. For example, the transcription of many proinflammatory cytokine and chemokines genes is regulated by the transcription factor nuclear factor-κB (NF-κB).[23] Genes induced by NF-κB include tumor-necrosis factor (TNF)-α, IL-6, cyclooxygenase-2, and 5-lipoxygenase. Keratinocytes in the skin express TLRs that are activated during infections causing production of chemokines that attract T-cells (e.g., CCL20 and CXCL9, 10, and 11), neutrophils (CXCL1 and 8) and other immene cells,[24] produce cationic antimicrobial peptides (AMP), such as cathelicidin and β-defensin,[25] that mediate killing of invading bacteria and, thus, protect epithelial surfaces from infection.

The innate immune response can also protect against viral infections.[21] Recognition of virus-derived nucleic acids by nucleic acid sensors induce transcription of interferon (IFN)-α and IFNβ. Pathogen-derived nucleic acid sensors include TLR3 and retinoic acid-inducible gene (RIG)-1, which sense double-stranded viral RNA, TLR7/8, which recognize single-stranded viral RNA and cyclic GMP–AMP synthase (cGAS), which recognizes double-stranded viral DNA.[21,26] Secreted interferons bind to cell-surface receptors on the same and adjacent cells and induce protective factors that degrade viral nucleic acids or otherwise interfere with viral replication.[27] IFNα and IFNβ also activate NK cells to kill target cells.

These initial responses to infection trigger a local inflammatory response involving cells already at the site and cells recruited to the site by soluble mediators.[1,5] Many tissues contain resident macrophages that also respond to infection by producing chemokines (including CXCL8), cytokines (including IL-12, IL-1β, TNF-α and IL-6), leukotrienes (including LTB4 and LTE4), prostaglandins (including PGE2), and platelet-activating factor, which mediate inflammation. The goal of this inflammation is to eliminate the pathogen or to minimize the spread of the pathogen until adaptive immunity can produce a pathogen-specific response. The key events in inflammation include a) release of preformed mediators, rapid enzymatic production of lipid mediators, followed by transcription and translation of chemokine and cytokine genes; b) induction of cell-adhesion molecules (e.g., ICAM-1) in the vascular endothelium in adjacent capillaries, which slows the progress of leukocytes; c) loosening of tight junctions between epithelial cells to allow egress of leukocytes along a chemokines gradient; d) stimulation of blood clotting by activation of platelets to minimize "escape" of pathogens; and e) killing of microorganisms or infected cells by the leukocytes attracted to the site. Later, f) a recovery phase stimulates repair of damage caused by pathogens or the responding leukocytes.

Killing of Bacteria by Macrophages and Neutrophils

Monocytes from the blood differentiate into macrophages following extravasation.[28] Macrophages ingest invading micro-organisms into phagocytic vesicles, the phagosome, using several cell-surface receptors. The phagosome fuses with lysosomes containing antibacterial peptides and enzymes (e.g., lysozyme). Following fusion, a respiratory burst involving NADPH oxidase acidifies the phagolysosome and injects reactive oxygen species, which kill ingested micro-organisms. Neutrophils are the most common white blood cell but are not found in healthy tissue.

Their numbers at sites of inflammation increase rapidly during bacterial infection. In addition to killing engulfed bacteria, neutrophils can expel neutrophil extracellular traps (NETs) during a process called "NETosis".[29] NETs are web like DNA structures decorated with granular proteins, which contribute to pathogen killing. The lifespan of the neutrophil is short, and these cells typically die after one round of phagocytosis and granule discharge. Macrophages live longer, have more cellular transcription machinery, and can regenerate phagosomes. Macrophages play a prominent role in responses to intracellular pathogens, such as viruses and *M. tuberculosis*.

Opsonization and complement-mediated killing. Some serum proteins (e.g,, mannose-binding lectin [MBL]) and C-reactive protein (CRP), and secretory proteins such as surfactant proteins A and D produced in the lungs, bind to PAMPs on the surface of bacteria and enhance their uptake and killing by phagocytic cells.[30] This activity is termed "opsonization." The complement system of serum proteins opsonizes bacteria by binding to the bacterial surface directly or to MBL or antibody bound to the bacteria. Complement proteins undergo conformational change and enzymatic activation upon binding, and a cascade of such events leads to formation of biologically active components, such as C3a and C5a, which are chemoattractants for phagocytes, and C3b, which is an opsonin. In addition, accumulation of several terminal complement components, known as a "membrane attack complex," on the surface of a bacterial cell, forms a pore that disrupts membrane integrity and kills the bacteria.[5]

Systemic Inflammation and the Acute-phase Response

When production of TNFα, IL-1β, and IL-6 at a site of inflammation is high, serum concentrations of these cytokines increase, and systemic effects are triggered, including activation of the acute phase response.[31] These include fever, malaise, muscle aches, and decreased appetite. Fever is induced by PGE2 acting on the hypothalamus. One early name for TNFα was "cachexin" because it decreases appetite, an effect also mediated through the central nervous system. These cytokines also act on hepatocytes to increase synthesis of positive acute-phase proteins, including ferritin, CRP, and MBL, and to decrease synthesis of negative acute-phase proteins, including albumin and retinol-binding protein, the serum transport protein for vitamin A. The positive acute-phase proteins typically have a protective role in the innate immune response. They increase within a few hours and reach a peak at 10- to >100-fold their initial concentration within a few days. CRP, for example, increases from about 1 milligram/L to >100 milligram/L during bacterial pneumonia and binds to cell wall polysaccharides, acting as an opsonin. The reason for decreased concentrations of negative acute phase proteins, which may drop by 25% to 50%, is not clear. Serum iron, which is bound to the transport protein transferrin, also decreases during the acute phase response because of increased hepcidin synthesis (also see chapter 10). Hepcidin blocks normal recycling of transferrin-bound iron through macrophages, resulting in increased intracellular and decreased serum iron concentrations.[32] Increased ferritin synthesis may facilitate intracellular iron storage. This sequestration of iron decreases its availability to opportunistic pathogens. Chronic inflammation may result in "anemia of chronic disease" by decreasing the availability of iron for erythropoiesis. Serum zinc also decreases during the acute phase response, perhaps to inhibit bacterial zinc acquisition. Macronutrient metabolism is also altered during the acute phase response with elevated serum triglycerides, decreased β-oxidation of fatty acids, and increased gluconeogenesis. Neutrophils are also mobilized from bone marrow to increase availability at sites of inflammation, and TNFα stimulates activation of APC and their migration to lymph nodes.

APC Functions: Linking Innate to Adaptive Immunity

The role of the APC is to stimulate an adaptive immune response by transferring information about a specific microorganism from the site of infection to the draining lymph node for presentation to T-cells.[5] At least three types of "information" are transferred: First, unique peptides from microbial proteins, or antigens, are displayed on the APC surface by major histocompatibility complex (MHC) molecules for "presentation" to naïve T-cells in the draining lymph node. This presentation leads to the formation of peptide-specific memory T-cells that will respond to a specific pathogen, as discussed below. Some antigens stimulate development of memory B-cells and an antibody response without T-cell help. Such "T-cell independent antigens" include bacterial polysaccharides with repeating epitopes that bind to multiple antigen-specific immunoglobulin molecules on the surface of naïve B-cells, thus stimulating the development of memory B-cells and antibody-producing plasma cells without help from T-cells. Memory B and T-cells leave the lymph node via an efferent lymphatic vessel and eventually reach the bloodstream via the thoracic duct. From the bloodstream these cells may recirculate to the initial site of infection, to specific tissues (e.g., mucosal sites or skin) or to lymph nodes.

The second type of information (or "signal") transferred from APC to T-cell is termed "co-stimulation" and involves activation of cell-surface receptors on the T-cell such as CD28 (using the cluster of differentiation [CD] naming convention)[28] by corresponding co-stimulatory molecules on the surface of the activated APC (e.g., CD80 and CD86). This co-stimulation enhances T-cell proliferation and survival and is provided only by an APC, thus preventing unintended development of adaptive immunity by chance encounters of T and B-cells with antigen in other contexts. The third type of information transmitted to the T-cell by the APC is the differentiation signal, which consists principally of soluble mediators, typically cytokines, that help direct T-cell differentiation into a particular helper phenotype, as discussed below.

Adaptive Immunity

T-cells and B-cells are the principal cellular components of the adaptive immune system. The adaptive immune response develops more slowly after initial infection than the innate response but eventually has a greater capacity to defend against the targeted pathogen. All mammals have similar adaptive immune systems and laboratory mice are the favored model for research because of the availability of antibody reagents to characterize murine cells and molecules and because genetic manipulation of mice allows mechanistic studies in which specific genes can be targeted for decreased or enhanced expression.

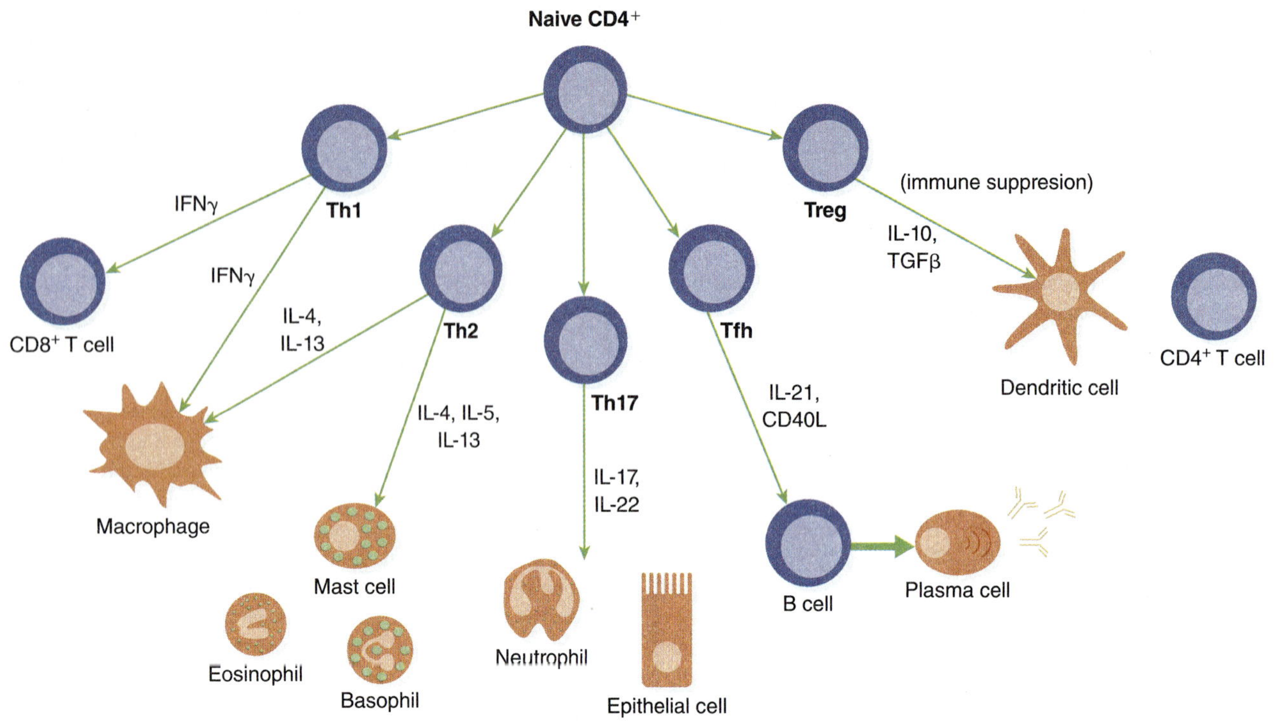

Figure 44.1 $CD4^+$ helper and regulatory T-cells. The figure shows the major subtypes described in the text, including major effector cytokines, cell types acted on by these cytokines, and the overall effect of the cell types, activation of pathogen clearance mechanisms or suppression of cell-mediated responses.

T-helper Lymphocytes

The T lymphocyte pool (identified by expression of CD3 on the cell surface) is composed of helper, cytotoxic, and regulatory cells (**Figure 44.1**). The "helper" designation derives from the ability of these cells to mediate development of T- and B-cell responses, but they also help promote and maintain "effector" functions of innate immune cells and thus, participate directly in the clearance of invading pathogens at sites of infection. Mature T helper (Th) cells express CD4 whereas cytotoxic T lymphocytes (CTL) express CD8. Naïve $CD4^+$ T-cells differentiate into Th1, Th2, Th17, T follicular-helper (Tfh) and regulatory (Treg) subtypes upon activation and these subtypes each have different effector functions (Figure 44.1). These subtypes form persistent lineages represented in the memory response to different pathogens, although there is some plasticity to this commitment.[18,33-36]

Th1 cells support development of type 1 immunity involving CTL and macrophages to protect against intracellular pathogens (e.g., viruses and some bacteria). Th2 cells support type 2, antiparasite immunity involving epithelial defenses (e.g., mucus production), mast cells, basophils, and eosinophils. Th1 and Th2 lymphocytes develop in response to specific, differentiation-inducing cytokines (IL-12 and IFNγ for Th1 and IL-4 for Th2) produced by APC or other cell types during initial exposure of a naïve T-cell to antigen. Mature Th1 and Th2 cells express signature transcription factors [T-box expressed in T-cells (Tbet) for Th1 and GATA-binding factor 3 (Gata3) for Th2], which help mediate and maintain their unique phenotypes, including distinct patterns of cytokine production. The signature effector cytokine of Th1 cells is IFNγ, which promotes CTL development and supports macrophage function in inflamed tissues. Th2 cells produce IL-4, -5, -13 to support type 2 immunity, including development of eosinophils in bone marrow and production of mucus by goblet cells at epithelial surfaces. Macrophages exposed to Th1 effector cytokines adopt a pro-inflammatory phenotype to facilitate clearance of intracellular pathogens, while macrophages exposed to Th2 effector cytokines contribute to anti-helminth immunity.[37] IFNγ produced by Th1 cells inhibits the generation of Th2 cells, whereas IL-4 produced by Th2 cells inhibits Th1 development. Th17 cells (so-called due to their production of IL-17) support type 3 immunity to protect against extracellular pathogens, which may involve epithelial defenses (e.g., antimicrobial peptide production) and promotion of neutrophilic inflammation. Th17 cell development often occurs in the intestinal lymphoid tissue in response to bacterial pathogens or to some commensals that adhere to intestinal epithelial surfaces and, thus, stimulate the intestinal immune system. Th17 development is stimulated by the cytokines IL-6, transforming growth factor (TGF)-β, and IL-23, and express the signature transcription factor retinoic acid orphan receptor (ROR)-γT (38). Th1 and Th17 cells bridge the innate and adaptive immune responses by secreting IFNγ and IL-17, respectively, which enhance innate effector mechanisms at sites of infection. IFNγ activates macrophage-mediated killing of intracellular pathogens, while IL-17 induces neutrophil recruitment and the production of proinflammatory cytokines, chemokines, and metalloproteinases by epithelial cells to enhance resistance to extracellular bacteria.[18,38]

Cytotoxic T-Lymphocytes

CTL are the principal type of adaptive immune cells that mediate the killing of cells infected with viruses (e.g., influenza, hepatitis B, and severe acute respiratory syndrome coronavirus 2) and intracellular bacteria (e.g., *M. tuberculosis* and *Salmonella typhimerium*)[5,39] although evidence suggests a cytotoxic role for some $CD4^+$ T-cells as well.[40] CTL also respond to intracellular parasitic infections caused by *Plasmodium* sp. (malaria),

Toxoplasma gondii (toxoplasmosis), and *Trypanosoma cruzi* (Chagas disease). CTL kill infected cells by the CD95 and perforin/granzyme-mediated lytic pathways.[41,42] Natural killer (NK) cells of the innate immune system also kill cells exhibiting specific markers of stress by these mechanisms.[43]

Regulatory T Lymphocytes

Treg cells may develop in the thymus or from naïve T-cells in peripheral lymphoid tissue in a manner like the development of Th1, Th2, and Th17 cells. Treg cells play a critical role in the induction of self-tolerance and, thereby, significantly contribute to preventing autoimmune reactions, including organ-specific autoimmunity, systemic autoimmunity, and colitis.[44,45] Treg cells that develop in the thymus are important in controlling development of autoimmunity in response to self-antigens. Treg cells also play a role in immune homeostasis by suppressing excessive immune responses that may develop in response to infection and be damaging to the host. Treg cells also develop in response to exposure to commensal bacteria, which may be "seen" by the immune system but are nonpathogenic. Treg cells can be $CD4^+$ or $CD8^+$, although $CD8^+$ Treg cells have not been as extensively characterized.[45,46] The major class of $CD4^+$ Treg cells are characterized and identified by the expression of the signature transcription factor forkhead box P3 (FoxP3) and the surface marker CD25, which is the α chain of the IL-2 receptor. Expression of this chain of the receptor confers high-affinity IL-2 receptor activity. The ability of Treg cells to constitutively use IL-2, an important growth cytokine for developing T-cells, via this receptor, can inhibit development of Th1, Th2, or Th17 cells via depletion of IL-2 in the lymph node micro-environment. Treg cells also typically produce the inhibitory cytokines IL-10 and TGFβ, which can dampen production of effector cytokines and decrease antigen presentation by APC. Treg cells may develop in intestinal and other peripheral immune tissues in response to antigens presented in a "non-inflammatory" manner (e.g., without IL-12, IL-4, or IL-6 present, which might drive Th1, Th2, or Th17 development). Treg development requires the presence of TGFβ and is promoted by the vitamin A-derived metabolite retinoic acid (RA), which is produced by dendritic cells in the intestinal lymphoid tissue. This is a principal pathway for development of Treg cells in intestinal immune tissue.[5,36] Thus, development of Treg cells at mucosal surfaces appears to be an important aspect of preventing an over-response of the adaptive immune system to commensal bacteria.

T Follicular Helper Cells

Tfh cells develop in peripheral lymph nodes after exposure to antigen by APC and the cytokine IL-6.[36] The developing Tfh cell then expresses its signature transcription factor Bcl-6 and produces the cytokine IL-21, which acts in an autocrine manner to reinforce Tfh development. However, development of Tfh cells also requires direct interaction of maturing Tfh cells with B-cells, necessitating migration from the T-cell zone of the lymph node (paracortex) to an adjacent germinal center containing B-cells via Tfh expression of the chemokine receptor CXCR5, which allows migration in response to the chemokine CXCL13 produced in the germinal centers. After arrival, the developing Tfh cell receives further costimulatory activity, IL-6 exposure, and continuing antigenic stimulation (to the same antigen expressed by the initiating APC) from a developing memory B-cell. Mature Tfh cells promote the survival and differentiation of these antigen-specific B-cells into memory B-cells and antibody-producing plasma cells by providing stimulation and survival signals via the stimulatory molecule CD40 ligand (CD40L) and the cytokines IL-4 and IL-21, produced by Tfh cells. Tfh cells develop during immune responses to antigen that produce Th1, Th2, and Th17 cells, as all such protective responses to pathogens typically require antibody production as well as effector T-cells.

Unconventional T-cells

In contrast to conventional $CD4^+$ and $CD8^+$ T-cells, unconventional T-cells, which comprise approximately 10% of circulating T-cells, typically express TCRs that are of limited diversity and recognize nonpeptide antigens, such as lipids, vitamin B metabolites, and phosphoantigens.[47] Cell subsets include natural killer T (NKT) cells, mucosal-associated invariant T (MAIT) cells, and γδ T-cells. Unconventional T-cells are thought to comprise most T-cells in peripheral tissues, including the liver and gut mucosa, and mount rapid effector responses after activation.

B Lymphocytes

The primary role of B-cells is to produce antibodies that are specific for a foreign antigen.[5] Naïve B-cells develop into two different cell types following antigen-specific stimulation: (1) memory B-cells that provide a more rapid response following second exposure to the same antigen and; (2) antibody-producing plasma cells. Antibodies are immunoglobulins (Ig) that have a basic unit shaped like a Y consisting of two identical heavy (H) and two identical light (L) polypeptide chains (**Figure 44.2**). The C-terminus consists of the two heavy chains and denotes the constant (C) region. The two variable (V) regions formed at the N-terminal fork of the Y structure are composed of one heavy and one light chain each. It is the variable region that is responsible for binding to foreign antigens. There are five classes of Igs: IgM, IgD, IgG, IgA, and IgE. IgM exists as a pentamer of Ig units held together by a J chain peptide. IgA can form either a monomer or dimer joined by the J chain. Antibodies bind directly to antigen via the two antigen-binding sites on each molecule. In solution, this binding has many protective effects. Antibody binds to soluble, foreign antigens (e.g., to "neutralize" bacterial toxins), to virus particles in blood, or at mucosal surfaces to neutralize infectivity, and to surfaces of micro-organisms to enhance uptake and killing by phagocytic cells of the innate immune system.

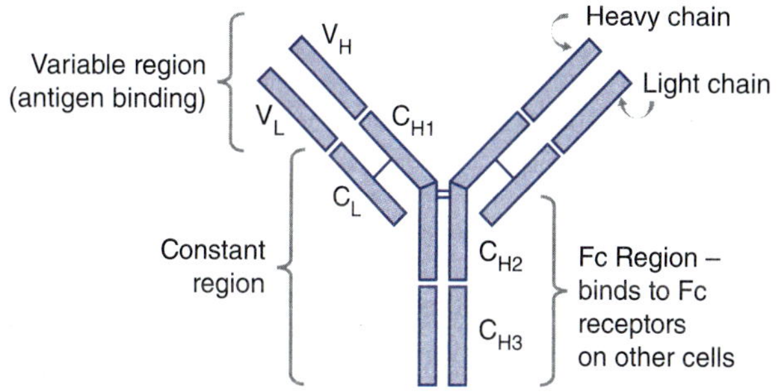

Figure 44.2 Immunoglobulin structure. Structure of immunoglobulin (Ig) unit. The variable region binds to antigen. The heavy chain determines the class of Ig—IgM, D, G, A, or E. The Fc region can bind to Fc receptors and activate other immune cells to kill pathogens or induce internalization of the Ig-pathogen immune complex by phagocytosis.

Impact of Nutrition on Immunity

Vitamin A

Vitamin A and its active metabolite, retinoic acid (RA), are involved in a wide range of immunologic processes through binding to nuclear retinoic acid receptors to regulate gene expression. Vitamin A plays a critical role in embryonic life during development and differentiation of intestinal epithelial cells, the thymus, and secondary lymphoid organs, including fetal lymph nodes and Peyer's patches.[48] Vitamin A deficiency causes squamous metaplasia postnatally on epithelial surfaces, impairing epithelial barrier function, particularly in the respiratory tract.[49] In animal models, the availability of RA *in utero* was shown to control the fitness of the immune system in the adult offspring.[50] In addition to its role in embryonic development and lymphopoiesis, vitamin A controls postnatal immune function, especially in sub-mucosal lymphoid tissues where cells of the innate immune system produce RA. In particular, vitamin A directs lymphocyte homing to mucosal immune tissues by enhancing expression of $\alpha 4\beta 7$ integrin and CCR9 on lymphocytes that undergo antigenic stimulation and develop at these sites.[51] These molecules allow trafficking back to the gut for mature effector T-cells, B-cells, and IgA-producing plasma cells.[52] Vascular endothelium in the gut expresses mucosal vascular addressing cell adhesion molecule 1 (MAdCAM-1), to which $\alpha 4\beta 7$ binds, thus allowing extravasation. Epithelial and other cells in the gut express CCL25, which attracts CCR9-expressing cells. RA produced by some cells of the immune system, including APCs, also act to promote development of Treg cells in the intestine, which play a significant role in maintaining "tolerant" rather than "inflammatory" responses toward gut flora.[53] Vitamin A deficient (VAD) animals show impaired generation of peripheral Treg cells in response to diet, resulting in impaired oral tolerance.[54] In contrast to promoting Treg development, RA suppresses differentiation of Th17 cells, which defend against extracellular bacteria and fungi and are implicated in the development of autoimmune diseases. RA also modulates B-cell development and function. VAD mice infected with influenza A virus exhibited lower influenza-specific salivary IgA response than control mice[55] while VAD rats had a marked decrease of lamina propria IgA+ plasma cells, coinciding with increased susceptibility to local infections.[56]

Vitamin A deficiency increases the risk of death for infants and young children living in areas where malnutrition is common and with a high burden of infectious diseases. Community-based trials have shown that prevention of vitamin A deficiency with high-dose vitamin A capsules reduces infant and child mortality from infectious diseases when administered from 6 m to 5 y of age.[57] In similar trials, neonatal vitamin A supplementation did not reduce mortality prior to 6 m of age.[58] However, high-dose vitamin A supplementation can sometimes have adverse effects, such as increasing the severity of pneumonia when such supplements are given to acutely ill children,[59] perhaps due to unexpected effects on the immune response to infection.

B Vitamins

As B vitamins affect many aspects of basic cellular metabolism, one would expect deficiencies in B vitamins to impair immune function. Nutrition literature from the mid-20th century does show that B vitamin deficiencies impair host responses to infections in animal models, suggesting deficits in immunity, according to a comprehensive review published in 1967.[60] More recent mechanistic studies are less abundant,[61] but some examples are worth noting. Vitamins B_6, B_{12}, and folate play critical roles in 1-carbon metabolism and are essential for the synthesis of nucleic acids and proteins, and deficiencies in these nutrients impair both T-cell and B-cell function.[62] Vitamin B_6 deficiency impairs antibody responses and cellular immunity in animal studies, with some similar findings in humans.[63] Vitamin B_{12} is required for erythrocyte development and is needed for nucleic acid synthesis. In women immunized with pandemic influenza A (H1N1) vaccine during gestation, vitamin B_{12} supplementation during pregnancy and lactation increased influenza vaccine-specific IgA responses in plasma and colostrum.[64] There is also a developing research focus on the role of B vitamins produced by commensal intestinal bacteria on intestinal immunity, as recently reviewed,[65] and one interesting example is the newly described role of folic acid in development and survival of intestinal Treg cells and control of intestinal inflammation in mice.[66] This and other similar studies suggest a role for folic acid, and other B vitamins, in intestinal health.

Vitamin C

Vitamin C deficiency affects both innate and adaptive immunity. Humans fed a vitamin C deficient diet had decreased delayed type hypersensitivity (DTH) skin responses, which are mediated by the Th1 cytokine IFNγ.[67] Supplementation of these individuals with vitamin C normalized the DTH response, indicating that vitamin C is involved in maintenance of Th1 function. In elderly adults, vitamin C supplementation for 1 month increased the ex vivo proliferative responses of T-cells to mitogen.[68] Studies in a mouse model for asthma showed that high dose supplementation with vitamin C increased the ratio of IFNγ to IL-5 in bronchoalveolar fluid, again indicating that vitamin C promotes Th1 function.[69] In type 2 diabetes patients with poor glycemic control, vitamin C supplementation for 6 weeks increased ex vivo polymorphonuclear cell phagocytosis and oxidative burst.[70] Neutrophils have a high cytoplasmic level of vitamin C and rapid regeneration of ascorbate,[71] presumably to protect the host cell against the oxidative stress associated with bacterial killing. It is possible that improved phagocyte function, as well as better protection against oxidative damage, reduces the symptoms of infections such as the common cold and, thus, accounts for the somewhat shorter duration of cold symptoms seen in regular users of vitamin C supplements.[72]

Vitamin D

Vitamin D signaling is active in both the innate and adaptive arms of the immune system. In the innate system, the active metabolite of vitamin D, calcitriol, can be produced by 1-α-hydroxylase-expressing innate immune cells of the monocytic lineage, including monocytes, macrophages, and dendritic cells upon pathogen activation via appropriate PRR3.[73] In macrophages, for example, synthesized calcitriol can then act in an autocrine or paracrine fashion to increase expression of the antimicrobial peptides cathelicidin and $\beta 2$ defensin via the nuclear vitamin D receptor, thus enhancing protection against bacteria and viruses. Vitamin D has been shown to affect development and function of various innate immune cells, including neutrophils, eosinophils, mast cells, as well as NK T-cells, which are a subset of T-cells that lack immunologic memory.[73] In contrast to the enhancement of protection implicit in increased production of anti-microbial peptides, vitamin D can dampen immune responses as well by decreasing the antigen-presenting ability of dendritic cells, for example.[74] As a demonstration of this, vitamin D decreased airway eosinophilia in a murine model of allergic airway diseases,[75] and reduced eosinophilic airway inflammation in patients with nonatopic asthma and severe eosinophilic airway inflammation.[76]

Calcitriol from innate immune cells can act directly on developing T-cells to influence effector phenotype development. In both human and mouse studies, calcitriol can inhibit Th1 cell differentiation.[75,77-79] Knockout of the VDR in mice enhances development of inflammatory bowel disease,[80] which is mediated by an inflammatory T-cell response. Vitamin D_3 treatment was also shown to inhibit the production of TNFα and IFNγ by Th1 cells, and enhance expression of IL-4 by Th2 cells in a mouse model for experimental colitis.[78] However, the enhancement of Th2 responses remains controversial, as others have shown that vitamin D_3 inhibits IL-4 synthesis in mice and inhibits proliferation and synthesis of immunoglobulins, while promoting apoptosis, in human B-cells.[81] Vitamin D_3 inhibits the generation of Th17 cells in vitro and impairs development of Th17 cells in vivo.[82] Vitamin D_3 increases the development of Treg cells, resulting in increased FoxP3 expression and IL-10 and TGFβ production.[78] Vitamin D deficiency also increases severity of experimental autoimmune encephalomyelitis,[83] a mouse model of multiple sclerosis. The increase in Treg cell activity, as well as number, suggests that vitamin D_3 may have an overall immunosuppressive activity toward the adaptive immune response and a causal link has been postulated between vitamin D deficiency and the risk of autoimmune inflammatory disease in humans, including multiple sclerosis.[73] However, intervention trials treating multiple sclerosis with vitamin D have not yet shown consistent benefits.[84]

In addition to treatment of autoimmune disease, vitamin D has been used for the treatment of respiratory infections and asthma. Tuberculosis was thought to be a particularly good target because vitamin D deficiency is common in tuberculosis patients and the causative agent, *M. tuberculosis*, infects macrophages, which are responsive to vitamin D. In a recent meta-analysis of such trials using high-dose vitamin D as an adjunct to standard-of-care antimicrobial therapy,[85] vitamin D did not hasten the recovery in patients overall, compared with placebo, but recovery was faster in patients who had contracted infection with a multidrug resistant strain of *M. tuberculosis*, suggesting that vitamin D may be beneficial in the absence of an effective antibiotic regimen. In addition, such trials have been carried out to prevent and treat acute upper respiratory infections, which are often caused by viruses. Vitamin D supplementation had a small but statistically significant benefit on reducing the risk of developing the common cold.[86] Asthma is not an infectious disease but exacerbations of asthma symptoms requiring therapy with systemic corticosteroids in patients with moderate to severe asthma can be triggered by viral respiratory infections. In a recent meta-analysis,[87] vitamin D reduced the rate of such exacerbations, relative to placebo, suggesting that vitamin D may be of modest therapeutic benefit in asthma, perhaps related to its anti-microbial properties.

Vitamin E

Vitamin E affects both innate and adaptive immunity. In the innate system, vitamin E has anti-inflammatory effects related to its antioxidant activity and has been shown to decrease COX2 activity, production of PGE2 and production of nitric oxide (NO) by innate immune cells.[88] In the adaptive system, vitamin E promotes Th1 development from naïve $CD4^+$ T-cells.[89] Increases in DTH skin responses have been shown with vitamin E supplementation. In purified $CD4^+$ T-cells from young and old mice, vitamin E enhanced the formation of immune synapses between the TCRs and APCs.[90] Many of the human studies with vitamin E have been performed on elderly adults and these data suggest that vitamin E supplementation may be important for improving the declining innate and adaptive immune responses in the aged and decreasing the risk of some infections.[91]

Zinc

Research on the effect of zinc on immunity has advanced in recent years through the examination of differential expression of zinc transporters in specific immune cell types including T-cells, B-cells and macrophages.[92] Studies with humans have shown that deficiency of dietary zinc resulted in thymic atrophy, decreased numbers of peripheral T-cells, and reduced IL-2 and IFNγ production by T-cells.[93,94] Human T-cells express the ZIP6 and ZIP8 transporters with ZIP6, regulating zinc flux during activation of T-cells. Zinc-deficient individuals have a decreased DTH response due to the reduction in IFNγ production, apparently due to specific effects on Th1 development and function. Zinc deficiency also affects B-cells, which express the ZIP7 transporter (in humans), resulting in reduced B-cell activation and downstream antibody production by plasma cells. Perhaps as a result of these specific mechanisms, infants born to mothers at risk of deficiency who received zinc supplements during pregnancy and postpartum had enhanced antibody responses to hepatitis B vaccination.[95] Zinc deficiency is common among young children in resource-poor areas around the world and has been associated with increased risk of morbidity and mortality from infectious diseases.[96] As a result, clinical and community trials have examined the effects of zinc supplementation on treating diarrhea and other infections, as well as on decreasing overall mortality. A recent meta-analysis showed a benefit of zinc supplementation given to treat diarrhea in reducing the duration of symptoms, particularly in children with signs of malnutrition.[97] Similarly, a meta-analysis of community trials in children under 5 y examining the effects of zinc supplementation on child survival showed a decrease in all-cause mortality, including deaths apparently due to pneumonia.[98] Older adults in the United States are also at risk of zinc deficiency and zinc supplementation in elderly residents of a residential care facility that increased stimulated T-cell proliferation and the number of peripheral T-cells.[99] Zinc supplementation is also of interest in decreasing morbidity in older adults but supplementation trials, including involving treatment of pneumonia,[100] have not shown benefit.

Iron

Iron is required for the growth of micro-organisms and pathogens are specifically adapted to acquire iron in the relatively iron-poor environment of the human host.[101] This need for iron by pathogens suggests that the decrease in serum iron seen during the acute phase is an attempt by the host to restrict iron availability to pathogens. This may explain the association of hemochromatosis (which results in increased tissue iron concentration) with increased severity of invasive bacterial infections[102] and the increased risk of infectious diarrhea with the use of iron supplements.[103] As iron is an essential element for both host and pathogen, iron concentration at the tissue level is exquisitely governed by the immune system, with macrophages playing a vital role in systemic iron homeostasis.[104] Macrophages mediate the recycling of iron through engulfment of senescent red blood cells and by sequestering iron during injury and infection to diminish iron-induced oxidative stress and mediate bacteriostasis. Iron deficiency can also impair adaptive immunity; a recent study in a birth cohort found that iron deficiency was associated with lower antibody responses to vaccination, while iron supplementation was associated with higher responses,[105] suggesting enhancement of adaptive immunity.

Omega-3 and Omega-6 Fatty Acids

Arachidonic acid (AA; C20:4, n-6) released by phospholipase A2 activity from the membrane of macrophages, granulocytes and, at times, lymphocytes, is used as a precursor for synthesis of lipid mediators, including LTs, PGs, and epoxides by LOX, COX, CYP pathways, respectively.[12] AA-derived eicosanoids including 2-series prostaglandins (e.g., PGE2) and 4-series leukotrienes (e.g., LTB4) are potent pro-inflammatory mediators, which regulate the cardinal signs of inflammation, including local capillary blood flow, heat, swelling, and pain.[106,107] PGE2, the most abundant prostaglandin, modulates many biologic processes due to its activation of numerous receptors widely expressed throughout the body. PGE2 mediates recruitment of neutrophils, macrophages, and mast cells,[108-110] enhances the secretion of proinflammatory cytokines in dendritic cells and macrophages,[111-113] but also exhibits numerous anti-inflammatory properties.

Omega-3 fatty acids, including EPA (C20:5, n-3) and docosahexaenoic acid (DHA; C22:6, n-3), also serve as substrates for lipid mediator production. In contrast to AA-derived eicosanoids, EPA-derived eicosanoids including 3-series prostaglandins and 5-series leukotrienes are generally less biologically active.[114] Furthermore, both EPA and DHA are substrates for biosynthesis of SPMs, which, over the last decade, have been increasingly recognized for their potent anti-inflammatory properties.[107,115] SPMs play fundamental roles in resolving acute inflammation by governing neutrophil trafficking, enhancing macrophage efferocytosis of apoptotic neutrophils, and downregulating pro-inflammatory mediators.[116]

Diets in the United States are generally low in omega-3 fatty acids, but the use of supplements or consumption of omega-3-rich foods increases concentrations of n-3 PUFAs in the membranes of monocytes and granulocytes, resulting in greater production of EPA- and DHA-derived lipid mediators and changes in immune function.[117] Many studies in healthy individuals or in patients with chronic inflammatory conditions have shown reduced production of AA-derived eicosanoids following supplementation with EPA and DHA.[114]

DHA has also been shown to block TLR-mediated signaling initiated by various PAMPs.[118,119] Other polyunsaturated fatty acids (PUFA) have similar but lesser effects. Interestingly, saturated fatty acids can stimulate TLR mediated signaling and mimic, to a degree, the effect of TLR ligands such as LPS.[120] The TLR-blocking effect of DHA and other PUFA is independent of eicosanoid and SPM production and may be mediated by affecting lipid raft formation, which would affect TLR dimerization (in the case of TLR4), which could then induce signal transduction. A similar mechanism has been proposed for DHA-mediated decreases in TCR activation and T-cell proliferation.[121]

Acknowledgments

The authors acknowledge Susana J Zunino for her contributions to the previous version of this chapter.

References

1. Mims CA, Nash A, Dalziel RG, et al. Mims' Pathogenesis of infectious disease. Elsevier, Amsterdam: Academic Press, 2015.
2. Kayama H, Okumura R, Takeda K. Annu Rev Immunol 2020;38: 23–48.
3. Zindel J, Kubes P. Annu Rev Pathol 2020;15:493–518.
4. Nahrendorf M, Pittet MJ, Swirski FK. Circulation 2010;121: 2437–45.
5. Murphy K, Weaver C. Janeway's Immunobiology. New York, NY: Garland Science/Taylor & Francis Group, LLC, 2016.
6. Maródi L. Clin Immunol 2006118:137–44.
7. Hasselquist D, Nilsson JA. Philos Trans R Soc Lond B Biol Sci 2009;364:51–60.
8. Lackey KA, Fehrenkamp BD, Pace RM, et al. Annu Rev Nutr 2021;41:283–308.
9. Wells JM, Gao Y, de Groot N, et al. Annu Rev Nutr 2022;42: 165–200.
10. Taft DH, Lewis ZT, Nguyen N, et al. Nutrients 2022;14:1423.
11. Huda MN, Ahmad SM, Alam MJ, et al. Pediatrics 2019;143:e20181489.
12. Christie WW, and Harwood JL. Essays Biochem 2020;64: 401–21.
13. Kashem SW, Haniffa M, Kaplan DH. Annu Rev Immunol 2017;35:469–99.
14. Harris-Tryon TA, Grice EA. Science (New York, N.Y.) 2022; 376:940–5.
15. Feng Y, Chen CJ, Su LH, et al. FEMS Microbiol Rev 2008;32: 23–37.
16. Chairatana P, Nolan EM. Crit Rev Biochem Mol Biol 2017;52: 45–56.
17. Mestecky J. Mucosal immunology. Elsevier, Amsterdam: Elsevier/AP, Academic Press, 2015.
18. Ivanov II, Tuganbaev T, Skelly AN, et al. Annu Rev Immunol 2022;40:559–87.
19. Evren E, Ringqvist E, Willinger T. Immunol 2020;160: 126–38.
20. Brubaker SW, Bonham KS, Zanoni I, et al. Annu Rev Immunol 2015;33:257–90.
21. Tan X, Sun L, Chen J, et al. Annu Rev Microbiol 2018;72: 447–78.
22. Ross C, Chan AH, von Pein JB, et al. Annu Rev Immunol 2022;40:249–69.
23. Taniguchi K, Karin M. Nat Rev Immunol 2018;18:309–24.
24. Nestle FO, Di Meglio P, Qin JZ, et al. Nat Rev Immunol 2009;9:679–91.
25. Yang D, Biragyn A, Hoover DM, et al. Ann Rev Immunol 2004;22:181–215.
26. Ma Z, Ni G, Damania B. Annu Rev Virol 2018;5:341–62.
27. Wang W, Xu L, Su J, et al. Trends Microbiol 2017;25:573–84.
28. Guilliams M, Mildner A, Yona S. Immunity 2018;49: 595–613.

29. Thiam HR, Wong SL, Wagner DD, et al. Annu Rev Cell Dev Biol 2020;36:191–218.
30. Bottazzi B, Doni A, Garlanda C, et al. Annu Rev Immunol 2010;28:157–83.
31. Gabay C, Kushner I. N Engl J Med 1999;340:448–54.
32. Golonka R, Yeoh BS, Vijay-Kumar M. J Innate Immun 2019;11:249–62.
33. Dong C. Annu Rev Immunol 2021;39:51–76.
34. Ruterbusch M, Pruner KB, Shehata L, et al. Annu Rev Immunol 2020;38:705–25.
35. Taniuchi I. Annu Rev Immunol 2018;36:579–601.
36. Yin X, Chen S, Eisenbarth SC. Annu Rev Immunol 2021;39: 759–90.
37. Murray PJ. Annu Rev Physiol 2017;79:541–66.
38. Burkett PR, Meyer zu Horste G, Kuchroo VK. J Clin Invest 2015;125:2211–9.
39. Teijaro JR, Farber DL. Nat Rev Immunol 2021;21:195–7.
40. Cheroutre H, Husain MM. Semin Immunol 2013;25:273–81.
41. Chowdhury D, Lieberman J. Annu Rev Immunol 2008;26: 389–420.
42. Golstein P, Griffiths GM. Nat Rev Immunol 2018;18:527–35.
43. Mujal AM, Delconte RB, Sun JC. Annu Rev Immunol 2021;39:417–47.
44. Muñoz-Rojas AR, Mathis D. Nat Rev Immunol 2021;21: 597–611.
45. Savage PA, Klawon DEJ, Miller CH. Annu Rev Immunol 2020;38:421–53.
46. Niederlova V, Tsyklauri O, Chadimova T, et al. Eur J Immunol 2021;51:512–30.
47. Pellicci DG, Koay H-F, Berzins SP. Nat Rev Immunol 2020;20: 756–70.
48. Larange A, Cheroutre H. Annu Rev Immunol 2016;34:369–94.
49. Smith AD, Panickar KS, Urban JF, et al. Annu Rev Anim Biosci 2018;6:227–54.
50. van de Pavert SA, Ferreira M, Domingues RG, et al. Nature 2014;508:123–7.
51. Iwata M, Hirakiyama A, Eshima Y, et al. Immunity 2004;21: 527–38.
52. Mora JR, von Andrian UH. Semin Immunol 2009;21:28–35.
53. Coombes JL, Siddiqui KR, Arancibia-Carcamo CV, et al. J Exp Med 2007;204:1757–64.
54. Hall JA, Cannons JL, Grainger JR, et al. Immunity 2011;34: 435–47.
55. Gangopadhyay NN, Moldoveanu Z, Stephensen CB. J Nutr 1996;126:2960–7.
56. Bjersing JL, Telemo E, Dahlgren U, et al. Clin Exp Immunol 2002;130:404–8.
57. Imdad A, Mayo-Wilson E, Haykal MR, et al. Cochrane Database Syst Rev 2022;3:CD008524.
58. Haider BA, Sharma R, Bhutta ZA. Cochrane Database Syst Rev 2017; 2:CD006980.
59. Stephensen CB, Franchi LM, Hernandez H, et al. Pediatrics 1998;101:E3.
60. Scrimshaw NS, Taylor CE, Gordon JE, et al. Inter Nutr Infect. World Health Organization, 1968.
61. Yoshii K, Hosomi K, Sawane K, et al. Front Nutr 2019;6:48.
62. Wintergerst ES, Maggini S, Hornig DH. Ann Nutr Metab 2007;51:301–23.
63. Meydani SN, Hayek M, Coleman L. Ann N Y Acad Sci 1992;669:125–39; discussion 39–40.
64. Siddiqua TJ, Ahmad SM, Ahsan KB, et al. Eur J Nutr 2016;55: 281–93.
65. Uebanso T, Shimohata T, Mawatari K, et al. Mol Nutr Food Res 2020;64:e2000426.
66. Kinoshita M, Kayama H, Kusu T, et al. J Immunol 2012;189: 2869–78.
67. Jacob RA, Kelley DS, Pianalto FS, et al. Am J Clin Nutr 1991;54:1302S–9S.
68. Kennes B, Dumont I, Brohee D, et al. Gerontol 1983;29: 305–10.
69. Chang H-H, Chen C-S, Lin J-Y. J Agric Food Chem 2009;57:10471–6.
70. Chuangchot N, Boonthongkaew C, Phoksawat W, et al. Nutr Res 2020;79:50–9.
71. Washko PW, Wang Y, Levine M. J Biol Chem 1993;268: 15531–5.
72. Hemilä H, Chalker E. Cochrane Database Syst Rev 2013;CD000980.
73. L Bishop E, Ismailova A, Dimeloe S, et al. JBMR 2021;5:e10405.
74. Griffin MD, Lutz W, Phan VA, et al. Proc Natl Acad Sci U S A 2001;98:6800–5
75. Matheu V, Bäck O, Mondoc E, et al. J Allergy Clin Immunol 2003;112:585–92.
76. de Groot JC, van Roon ENH, Storm H, et al. J Allergy Clin Immunol 2015;135:670–5.e3.
77. Boonstra A, Barrat FJ, Crain C, et al. J Immunol 2001;167: 4974–80.
78. Daniel C, Sartory NA, Zahn N, et al. J Pharmacol Exp Ther 2008;324:23–33.
79. Sloka S, Silva C, Wang J, et al. J Neuroinflamm 2011;8:56
80. Froicu M, Weaver V, Wynn TA, et al. Mol Endocrinol 2003;17:2386–92.
81. Chen S, Sims GP, Chen XX, et al. J Immunol 2007;179:1634–47.
82. Tang J, Zhou R, Luger D, et al. J Immunol 2009;182:4624–32.
83. Nashold FE, Spach KM, Spanier JA, et al. J Immunol 2009;183:3672–81.
84. Jagannath VA, Filippini G, Di Pietrantonj C, et al. Cochrane Database Syst Rev 2018;9:CD008422.
85. Jolliffe DA, Ganmaa D, Wejse C, et al. Eur Respir J 2019;53.
86. Jolliffe DA, Camargo CA, Sluyter JD, et al. Lancet Diabetes Endocrinol 2021;9:276–92.
87. Jolliffe DA, Greenberg L, Hooper RL, et al. Lancet Respir Med 2017;5:881–90.
88. Lee GY, Han SN. Nutrients 2018;10:E1614.
89. Meydani SN, Han SN, Wu D. Immunol Rev 2005;205: 269–84.
90. Marko MG, Ahmed T, Bunnell SC, et al. J Immunol 2007;178:1443–9.
91. Meydani SN, Leka LS, Fine BC, et al. JAMA 2004;292:828–36.
92. Wessels I, Fischer HJ, Rink L. Semin Cell Dev Biol 2021;115: 62–9.
93. Overbeck S, Rink L, Haase H. Arch Immunol Ther Exp (Warsz) 2008;56:15–30.
94. Prasad AS. Curr Opin Clin Nutr Metab Care 2009;12:646–52.
95. Ahmad SM, Hossain MB, Monirujjaman M, et al. *Eur J Nutr* 2016;55:1823–9.
96. Fischer Walker C, Black RE. Ann Rev Nutr 2004;24:255–75.
97. Lazzerini M, Wanzira H. Cochrane Database Syst Rev 2016;12:CD005436.
98. Rouhani P, Rezaei Kelishadi M, Saneei P. Eur J Nutr 2022;61: 37–54.
99. Barnett JB, Dao MC, Hamer DH, et al. Am J Clin Nutr 2016;103:942–51.
100. Sharafi S, Allami A. Int J Tuberc Lung Dis 2016;20:685–8.
101. Monteith AJ, Skaar EP. Trends Endocrinol Metab 2021;32: 916–28.
102. Khan FA, Fisher MA, Khakoo RA. Int J Infect Dis 2007;11: 482–7.
103. Gera T, Sachdev HPS. BMJ 2002;325:1142.
104. Winn NC, Volk KM, Hasty AH. JCI Insight 2020;5.
105. Stoffel NU, Uyoga MA, Mutuku FM, et al. Front Immunol 2020;11:1313.

106. Kanaoka Y, Boyce JA. Allergy, Asthma & Immunol Res 2014;6:288–95.
107. Serhan CN, Libreros S, Nshimiyimana R. Sem Immunol 2022;101597.
108. Treutlein E-M, Kern K, Weigert A, et al. J Biol Chem 2018;293:9685–95.
109. Weller CL, Collington SJ, Hartnell A, et al. Proc Natl Acad Sci U S A 2007;104:11712–7.
110. Yu Y, Chadee K. J Immunol 1998;161:3746–52.
111. Raucci F, Saviano A, Casillo GM, et al. Br J Pharmacol 2022;179:1857–73.
112. Schiffmann S, Weigert A, Männich J, et al. Biochem Pharmacol 2014;87:625–35.
113. Sheibanie AF, Yen J-H, Khayrullina T, et al. J Immunol 2007;178:8138–47.
114. Ruscica M, Sirtori CR, Carugo S, et al. Pharmacol Res 2022;182:106342.
115. Serhan CN, Chiang N, Van Dyke TE. Nat Rev Immunol 2008;8:349–61.
116. Buckley CD, Gilroy DW, Serhan CN. Immunity 2014;40: 315–27.
117. Adkins Y, Kelley DS. J Nutr Biochem 2010;21:781–92.
118. Lee JY, Zhao L, Hwang DH. Nutr Rev 2010;68:38–61.
119. Snodgrass RG, Huang S, Choi IW, et al. J Immunol 2013;191:4337–47.
120. Huang S, Rutkowsky JM, Snodgrass RG, et al. J Lipid Res 2012;53:2002–13.
121. Kim W, Khan NA, McMurray DN, et al. Prog Lipid Res 2010;49:250–61.

B. DIGESTIVE, ENDOCRINE, IMMUNE, AND NEURAL MECHANISMS

CHAPTER 45

Defenses Against Oxidative Stress

A. Catharine Ross

CHAPTER OUTLINE

Abbreviations

AMD, age-related macular degeneration; **ATP**, adenosine triphosphate; **Cd**, cadmium; **CYP**, cytochrome P-450; **Cys**, cysteine; **CySSG**, cysteine-glutathione disulfide; **DRI**, dietary reference intake; **ER**, endoplasmic reticulum; **Fe^{2+}**, ferrous iron; **Fe^{3+}**, ferric iron; **G6P**, glucose 6-phosphate; **GGT**, γ-glutamyltransferase; **GPX**, glutathione peroxidase; **GSH**, glutathione; **GSSG**, glutathione disulfide; **GST**, glutathione transfer; **H_2O_2**, hydrogen peroxide; **HNE**, 4-hydroxynonenal; **MDA**, malondialdehyde; **Met**, methio(nine); **NAD^+**, oxidized nicotinamide adenine dinucleotide; **NADH**, reduced nicotinamide adenine dinucleotide; **$NADP^+$**, oxidized nicotinamide adenine dinucleotide phosphate; **NADPH**, nicotinamide adenine dinucleotide phosphate; **NO**, nitric oxide; **NOS**, nitric oxide synthase; **Nox**, nitric oxide; **NOS**, nitric oxide synthase; **NRF-2**, nuclear factor-2; **OH**, hydroxyl radical; **PDI**, protein disulfide isomerase; **Prx**, peroxiredoxin; **PUFA**, polyunsaturated fatty acid; **RNS**, reactive nitrogen species; **ROS**, reactive oxygen species; **Sec**, selenocysteine; **SOD**, superoxide dismutase; **Trx**, thioredoxin; **UV**, ultraviolet

Overview

Oxidative metabolism is central to macronutrient utilization, as well as to a wide variety of transformations of bioactive molecules. Thus, oxidation, and redox biology, is central to all nutritional biochemistry. The formation of free radicals (molecules with an unpaired electron) is a natural consequence of oxidative metabolism. To maintain life and health, oxidative processes must be regulated and kept in proper balance. Numerous processes related to aging and metabolic derangements are linked to "oxidative stress," a term derived from life in an aerobic environment, where the carbon-, hydrogen-, nitrogen- and sulfur-containing chemicals are oxidized while O_2 serves as the electron acceptor, ultimately being reduced to water. However, reductive stress is also possible. It is generally agreed that natural processes, such as aging, occur side by side with, and perhaps are causally related to, the buildup of various oxidized substances, such as free radicals, so-called "age pigments" like malondialdehyde (MDA), lipofusceins, and macular pigments in the retina, to name a few. These are often the products of oxidative or condensation reactions and, as the products formed are not easily cleared, they accumulate in cells and tissues. Oxidized products of nucleic acid metabolism may underlie aspects of mutagenesis. Therefore, antioxidant defenses are essential to maintain appropriate balance for all of metabolism.

Nature has evolved a rich variety of antioxidant defense mechanism that, in various ways, helps to control and regulate the rates and extent of biologic oxidations. It is the concept that antioxidants, especially those related to diet, are central to the regulation of metabolism. That is the focus of this chapter. The idea that certain foods or specific nutrients possess and exert antioxidant properties has fascinated both scientists and the public for a very long time. One result of interest in antioxidants has been the conduct of controlled trials to test whether supplementation with specific forms, such as micronutrients or small diet-derived molecules, can have positive effects on processes such as aging, cancer, or other acute or chronic diseases. As has been noted,[1] antioxidant defenses against oxidative stress have a rich history of scientific investigation, influenced by medicine, public health, commerce, and politics. This chapter focuses on mechanisms by which antioxidants act to balance and regulate the body's oxidative metabolism. The chapter surveys the sources and types of oxidative stress and the multiple systems that interact to protect against excessive oxidative stress. Oxidative stress is highly complex, and not adequately described as a simple imbalance between pro-oxidants and antioxidants. Other chapters describe in detail the properties and redox characteristics of other specific nutrients, including metals (e.g., iron, zinc, copper, selenium), vitamins (β-carotene, vitamin E, vitamin C), amino acids (e.g., the sulfur amino acids methionine [Met] and cysteine [Cys], and glutamine), and bioactive compounds such as glutathione (GSH). Overall, diet and nutrition are central to the function of the body's redox signaling systems, both directly by maintaining the redox signaling components (e.g., enzymes, transporters, transcription factors) and indirectly by optimizing gene expression and epigenetic control of protective and repair systems. Increasingly, research on nutrition and defenses against oxidative stress is employing approaches that provide better spatial and temporal resolution of oxidative reactions within cells, including at the single cell level, and is integrating redox systems biology with omics studies of the genome, epigenome, proteome, and metabolome.

Definition of Oxidative Stress

Oxidative stress is defined as an imbalance in pro-oxidant and antioxidant reactions that causes macromolecular damage or disrupts biologic redox signaling and control.[2,3] **Figure 45.1** illustrates the spectrum of oxidative stress. A common feature of all these processes is electron transfer, or "redox," reactions, in which one or more electrons is lost (termed *oxidation*) from a donor chemical, and one or more electrons is gained (termed

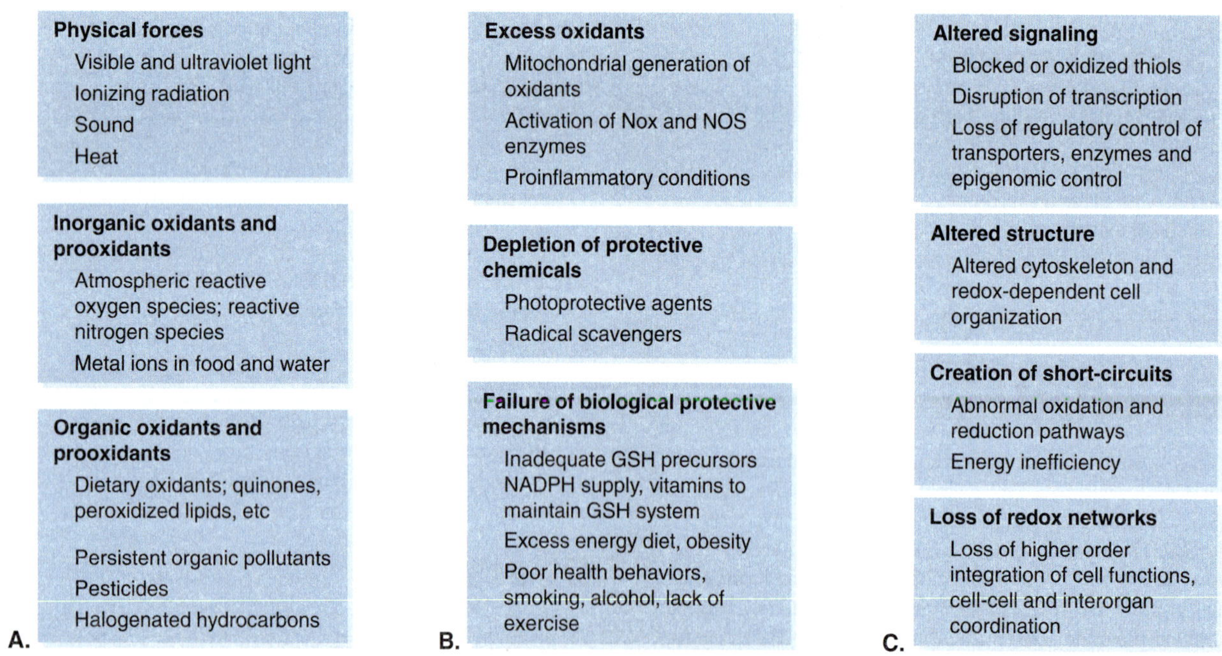

Figure 45.1 The spectrum of oxidative stress. A. Direct-acting environmental and dietary agents contribute to oxidative stress, including physical, inorganic, and organic agents. B. Endogenous oxidant generation by cellular metabolism and failure of the metabolic systems to protect against oxidants create imbalances in specific pro-oxidant and antioxidant reactions that result in oxidative stress. C. More subtle mechanisms of oxidative stress involve disruption of redox signaling and control mechanisms, such as occur by nonradical oxidants; these mechanisms may represent the most critical aspect of oxidative stress in chronic and age-related disease. *GSH*, glutathione; *NADPH*, nicotinamide adenine dinucleotide phosphate; *NOS*, nitric oxide synthase; *Nox*, nicotinamide adenine dinucleotide phosphate oxidase.

reduction) by an acceptor chemical. Conservation of matter requires that these processes be coupled (i.e., whenever something is oxidized, something else becomes reduced).

Pro-oxidants and Antioxidants

The reactions of pro-oxidants and antioxidants have been extensively studied and provide a foundation to understand the biochemistry and biology of oxidative stress. Pro-oxidants are agents that stimulate aberrant electron transfer in biologic systems and cause oxidative stress; these include both free radical oxidants and nonradical oxidant agents that initiate radical reactions, oxidize biologic components, or interfere with normal reductive and antioxidant functions. Free radicals (or more simply, radicals) are organic molecules or ions with an unpaired electron that are often reactive and function as pro-oxidants. Radicals support a unique chemistry in which chain reactions occur, such as those involved in polymerization of plastics, or in biologic systems the polymerization (adduct formation) of lipids such as polyunsaturated fatty acids. Nonradical oxidants are chemicals that participate in oxidation reactions not involving radical mechanisms. Biologic antioxidants (**Figure 45.2**) are defined as agents that function at low concentrations to stop oxidative stress. These include radical-scavenging chemicals that accept or donate electrons to terminate radical chain reactions (see later) and chemicals and enzyme systems that eliminate or protect against oxidants, some of which are listed in **Figure 45.3**. As stressed later, these enzyme systems are crucial to the formation as well as the disposition of oxidants in vivo.

The term *antioxidant* is loosely defined. It includes agents that inactivate radical initiation and catalysis, such as metal ion chelators, agents that block radiation-induced oxidation, and agents that counter oxidation of a substance with reduction. The term includes agents that protect against radical and nonradical oxidative mechanisms, and more specific terms such as "radical-scavenging antioxidant" or "thiol antioxidant" are used to discriminate these.

Antioxidants: agents that function at low concentrations to stop oxidative stress

1. Radical-scavenging chemicals that accept or donate electrons to terminate radical chain reactions (e.g., vitamin C)
2. Chemicals and enzyme systems that eliminate or protect against oxidants (e.g., GSH peroxidases)
3. Agents that prevent radical initiation (e.g., inhibitors of Cyp enzymes, radiation shields)
4. Agents that bind catalysts of oxidation (e.g., metal ion chelators)
5. Agents that counter oxidation of a substance with reduction (e.g., thiol antioxidants)
6. Agents that stabilize macromolecules against oxidation (e.g., Zn^{2+})

Figure 45.2 Antioxidants refer to agents that act at low concentrations to stop oxidative stress. The term is loosely defined and can refer to radical scavengers as well as agents, such as zinc (Zn^{2+}), which interacts with thiols to decrease tendency toward oxidation. *Cyp*, cytochrome P-450; *GSH*, glutathione.

Challenges to Define Antioxidant Requirements and Dietary Recommendations

Fundamental knowledge about oxidative chemistry is derived mostly from studies with purified chemical systems. It remains a great challenge to extrapolate this chemistry to biology and nutrition because the simple in vitro models do not effectively describe complex in vivo systems. For most vitamins and minerals, knowledge about their roles in supporting specific proteins and biologic functions is used to infer the amounts that are needed for those respective functions. For chemical reactions with high-rate constants, such as those involved in free radical reactions associated with oxidative stress, one must extrapolate data from homogeneous chemical systems to nonhomogeneous biologic systems. In all, estimates of oxidation and antioxidant defenses in vivo are based on extrapolation or surrogate indications.

Oxidation and reduction reactions in cells are typically highly compartmentalized, a primary example being the oxidative phosphorylation of ADP to form ATP in the electron transport system, located on the inner mitochondrial membrane. Within such a compartment, the rate of reaction at a specific subcellular site is critical. Antioxidants that block radical reactions are consumed

Pro-oxidant Radicals		Nonradical oxidants	
Superoxide	$O_2^{-}\cdot$	Hydrogen peroxide	H_2O_2
Hydroxyl radical	$\cdot OH$	Hypochlorous acid	HOCl
Alkoxyl radicals	$RO\cdot$	Singlet oxygen	$^1O_2$
Peroxyl radical	$ROO\cdot$	Ozone	O_3
Nitric oxide	$NO\cdot$	Hydroperoxide	ROOH
Carbon-centered radical	$\cdot CCl_3$	Peroxynitrite	$ONOO^-$
		Alkylperoxynitrite	ROONO
		Nitrous acid	HNO_2
Stable radicals		Dinitrogen tetraoxide	N_2O_4
Ubisemiquinone		Disulfide	RSSR′
Semidehydroascorbate		Sulfenate	RSO^-
Vitamin E radical		Sulfinate	RSO_2^-

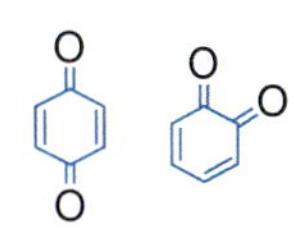
Quinones

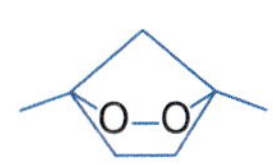
Endoperoxide

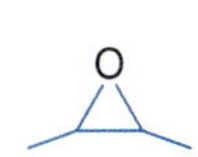
Epoxide

Figure 45.3 Pro-oxidant radicals and nonradical oxidants in biologic systems. (See text for descriptions.)

at such sites, with the result that the subcellular compartment becomes a "sink" for the antioxidant. Diffusion of an antioxidant to a sink varies as a function of the concentration difference (ΔC) between the source (e.g., blood) and the sink, according to the Fick equation (flux = diffusion coefficient $\times$ ΔC). In principle, there is no upper limit to the amount of antioxidant that could be beneficial. This is fundamentally different from the case for vitamins and minerals that saturate protein-dependent systems, and it represents an ongoing challenge to the ability to define antioxidant requirements. Furthermore, beyond the problem of defining requirements for delivery to specific subcellular sites, defining antioxidant requirements is limited by overlapping and redundant activities. Multiple antioxidants can and often do block the same oxidative reactions. Thus, even though an overall antioxidant capacity may be needed, assignment of that requirement to an individual chemical is not possible.

Vitamin C and E Are the Only Antioxidants with Dietary Reference Intakes

Many naturally occurring antioxidant chemicals block free radical reactions, but only two, vitamins C (ascorbate) and E (specifically α-tocopherol), have been found to have sufficient specificity for elimination of disease to warrant their inclusion as vitamins (see chapters 19 and 29). Other antioxidants support the same types of radical-scavenging activities as vitamin C and E but, because there is not the same biochemical evidence to define a requirement, there is no DRI for them. As indicated earlier, other chemicals have overlapping and redundant antioxidant activities so that requirements are difficult to establish and justify scientifically.

Contemporary Refinement in Definition

Around 2000, an advance in the definition of oxidative stress came about because of research findings[2,4-6] showing that the principal thiol-disulfide systems controlling the oxidation-reduction states of proteins are not in a near-equilibrium state but rather are kinetically limited.[7] This discovery, coupled with rapidly developing knowledge of redox-signaling nicotinamide adenine dinucleotide phosphate (NADPH) oxidases, provided a new framework to consider thiol redox circuits as a global system for the regulation of cell and organ system function.[8,9] Concomitantly, the publication of results from large-scale, double-blind interventional studies that tested supplementation with radical-scavenging antioxidants[10-17] showed little or no benefit on human health outcomes. In consequence of these new results, the original definition of oxidative stress has been qualified in the sense that an imbalance is now considered only in terms of specific reactions or pathways and not in terms of a "global" imbalance. In the current conceptualization, an imbalance in a specific reaction or pathway can cause damage even if there is not an overall imbalance in the system, and, conversely, a change in overall balance of pro-oxidants and antioxidants can occur without causing damage. This concept focuses attention on reactions, pathways, and specific end-effects. At the same time, advances in the development of "-omic" technologies and systems biology have facilitated a more integrated "network" approach. Older criteria of oxidative stress were limited to gross measures, such as cell death[18] or macromolecular damage, including lipid peroxidation,[19] global protein carbonylation,[20] or DNA damage.[21] Modern methods support measures of expression of specific genes,[22] modification of specific lipids,[23] oxidation of specific proteins,[24] and global effects on metabolic pathways.[25,26] These newer methods are enabling detailed understanding of the nutritional contributions to both radical and nonradical mechanisms of oxidative stress.[27]

One of the major significant advances has been the understanding of the role of the transcription factor Nuclear Factor E2-Related Factor 2 (Nrf2), which lies at the center of a complex regulatory network, regulating the transcription of components of the glutathione (GSH) and thioredoxin (Trx) antioxidant systems, as well as enzymes involved in phase I and phase II detoxification of exogenous and endogenous products, the regeneration of NADPH, and heme metabolism.[28-30] This regulatory system is further discussed later in this chapter.

Spectrum of Oxidative Stress

The spectrum of oxidative stress can be considered in terms of oxidants and pro-oxidants obtained directly from the diet and environment (see Figure 45.1A); oxidants generated endogenously with a failure of counterbalance by endogenous protective systems (see Figure 45.1B); and more subtle disruptions of redox signaling and control mechanisms (see Figure 45.1C), or in common words: too much, too little, or the wrong mix. For some of these situations, nutrition provides limited to no protection. For others, antioxidant defenses are clearly important to human health, are affected by nutrition, and will be considered further below. It should be noted that oxidative stresses also affect the food supply (e.g., exposures to light, temperature, and radiation), and may be factors in food safety and acceptability, which indirectly affect human health.

Environmental Causes of Oxidative Stress

Some examples of oxidative stress that are due to environment, and in which antioxidant defenses play an important role, are damages due to light. While light in the visible range (400nm–700nm) can cause oxidative damage, the more intense blue light causes photodamage that has been linked to disease. In the retina, blue light damage to the photoreceptor and retinal pigment epithelial cells in the sensory retina contributes to age-related macular degeneration (AMD).[31] Ultraviolet (UV) light (100nm–400nm) from the sun causes more extensive oxidative stress, such as to the skin, resulting in sunburn and skin cancers.[32,33] The evolution of humans as diurnal rather than nocturnal mammals, with relatively poor protection by hair, has resulted in nutritional requirements that are distinct from requirements for other species. Ionizing radiation causes oxidative DNA damage and contributes to leukemias and other cancers.[34,35] Sound waves cause oxidative damage in lithotripsy of kidney stones and damage cochlear cells in hearing loss.[36,37] Heat increases oxidative reactions with resulting oxidative damage in thermal injury. Thus, exposures in the physical world cause oxidative stress, some of which is avoidable by proper protection.

Although few nutritional means are available to protect against these physical forces, some exceptions exist. Some carotenoids, in particular lutein and zeaxanthin, accumulate regionally in the macula (fovea) of the retina and are believed to act as a filter to protect this central region of the retina from blue light-induced photo-oxidation. Light may also be a catalyst for the formation of "bis"-retinoids, adducts of retinaldehyde and phospholipids that accumulate in the retina with age.[38] Nutritional supplements that may provide protection in persons with AMD have been studied in clinical trials by the National Eye Institute.[39] In the first trial, termed the Age-Related Eye Disease Study (AREDS), β-carotene was included in the formulation as an antioxidant. In the AREDS-2 trial, β-carotene was replaced by lutein and zeathanthin. Both the AREDS and AREDS-2 clinical trials demonstrated some slowing of the progression of AMD in participants with advanced disease.[31] Based on their known chemistry, carotenoids in the AREDS-2 formula (which is now over the counter) function as intraocular filters to capture and dissipate light energy and scavenge singlet oxygen, a reactive form of molecular oxygen generated by photochemical activation.[40] In the skin, UV radiation is most effectively blocked by melanin, a natural polymer derived from tyrosine oxidation.[41] Nutrition is not very useful for enhancing protection from UV radiation, and the use of sunscreens (ointments or protective clothing) is by far more effective. Photodamaged skin, having experienced erythema from UV radiation, perhaps because of activation of endogenous antioxidant systems,[42] may receive some benefit in the repair processes from sulfur amino acids, lysine, zinc, vitamin C, vitamin A, and polyunsaturated fatty acids (PUFAs). However, avoidance of sun exposure and use of commercial sunscreens is still the recommended action.

Ionizing radiation from radioactive decay is like UV light in that nutrition is important in inflammatory signaling and repair, but not by providing an ability to filter the radiation. The chapter on iodine refers to supplemental iodine in displacing and facilitating excretion of radioactive iodine from accidental nuclear radiation exposure. Sound nutritional strategies are essential for recovery following radiation therapy, for the building blocks to generate tissue repair.

The Chemical Basis of Antioxidant Defenses

Antioxidant defenses can be classified in different ways. Here, we first classify them as inorganic chemical, and organic compounds, each with specific ways of defending against oxidant stress.

Direct-Acting Inorganic Chemicals and Xenobiotics

Direct-acting inorganic chemicals cause oxidative stress, especially affecting the respiratory and gastrointestinal systems. Atmospheric pollution contains certain reactive oxygen species (ROSs) and reactive nitrogen species (RNSs) that contribute to oxidative damage in the airways.[43,44] The roles of ROSs and RNSs in physiologic and pathologic conditions has been reviewed.[45] ROSs are oxygen-containing chemicals that are reactive with organic molecules and include a superoxide anion radical, hydrogen peroxide (H_2O_2), lipid hydroperoxides, ozone, and related oxygen-centered radicals (**Figure 45.4**). RNSs include nitric oxide (NO·), peroxynitrite, and other oxides of nitrogen. The collective terms *ROS* and *RNS* are commonly used because the chemistry is too complex, and the analytic methods insufficient to identify specific reactive species.

Many transition metals catalyze electron transfer and are important environmental sources of oxidative stress. Environmental dispersal of oxidative transition metals such as iron[46] and cadmium (Cd)[47-49] can cause acute toxicities. Cd is especially important as a contaminant in the diet because it is a common pollutant associated with industrialization and natural recycling of wastes as fertilizers. Pharmaceuticals,[50] drugs of abuse, and chemicals from commercial products and industrial wastes[51,52] are also exogenous sources of oxidative stress. Ames[53] emphasized that natural toxicants in food pose a much greater risk than manufactured sources. Although this may be generally true, the apparent increase in foodborne toxicants, which cannot be compensated for by better nutrition, indicates an important need for nutrition and food scientists to find ways to limit sources of such contaminants.

Oxidant Production in Energy Metabolism

The most actively studied areas of nutrition and oxidative stress have involved endogenous oxidant generation and the failure of endogenous antioxidant defenses (see Figure 45.1). This includes damaging processes that result from central oxidative metabolism processes in cells. The major macromolecules of cells (protein, nucleic acids, lipids, carbohydrates) are composed mostly of carbon, hydrogen, oxygen, nitrogen, phosphorus, and sulfur—elements that form stable chemical structures in which pairs of electrons are shared between elemental nuclei. The interconversion of these biochemicals mostly involves conservation of the pairs of electrons, with hydride ion transfer via reduced nicotinamide adenine dinucleotide/oxidized nicotinamide adenine dinucleotide ($NADH/NAD^+$) or NADH/oxidized nicotinamide adenine dinucleotide phosphate ($NADP^+$) (see Figure 45.4), providing a common mechanism for 2-electron (2-e^-) transfer. These electron transfers occur in highly specific reactions, many of which have relatively high rates to support critical needs for energy metabolism, anabolism and repair, detoxification, and elimination of wastes.

Two and One Electron Transfers

The 2-e^--transfer reactions linked to $NADH/NAD^+$ and $NADPH/NADP^+$ are connected to one-electron (1-e^-) transfer reactions in metabolic pathways through chemical structures that exist in interconvertible fully oxidized, 1-e^--reduced and 2-e^--reduced forms (see Figure 45.4). These include hydroquinone/semiquinone/quinone (e.g., coenzyme Q) and flavoprotein systems (e.g., the flavin mononucleotide (FMN) of mitochondrial NADH dehydrogenase). Living organisms use 1-electron chemistry in central energy capture and transfer reactions of photosynthesis and mitochondrial respiration. These 1-e^- transfers are physically contained (compartmentalized) within special membrane structures of chloroplasts and mitochondria. These features support the oxidation of macromolecular energy sources (fat, carbohydrate, and protein) through 2-e^- transfers with

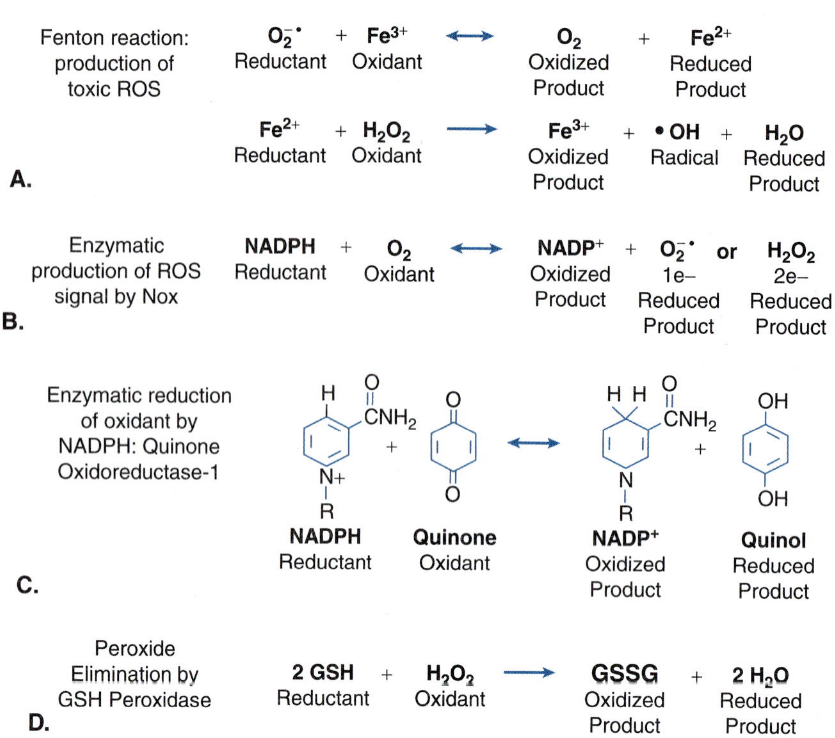

Figure 45.4 Different types of nonenzymatic and enzymatic oxidation-reduction reactions are relevant to oxidative stress. A. The Fenton reaction is an iron-catalyzed Haber-Weiss reaction that generates the highly reactive hydroxyl radical (•OH). B. Flavin-containing enzymes, such as redox signaling proteins, nicotinamide adenine dinucleotide phosphate (NADPH) oxidase (Nox), catalyze both 1-electron and 2-electron transfers to O_2, thereby forming both $O_2^{-\bullet}$ and hydrogen peroxide (H_2O_2). The most active NADPH oxidase, Nox-2, is present in phagocytic cells and activated to kill invading microorganisms. C. Two-electron transfer, shown for the detoxification enzyme NADPH:quinone reductase-1 (NQO1), is used to support chemical interconversions in intermediary metabolism. The niacinamide group shown for NADPH is identical to that in NADH and serves to transfer a pair of electrons as a hydride ion (H^-) without radical formation. D. Thiols are used as reductants to eliminate H_2O_2. In the reaction shown, GSH peroxidase catalyzes reduction of H_2O_2 to water without forming radicals as intermediates. The oxidized product formed from two GSH molecules is glutathione disulfide, GSSG. It is reduced back to GSH by an NADPH-dependent reductase (not shown), thus completing a detoxification cycle for peroxides. *Fe*[21], ferrous iron; *Fe*[31], ferric iron; *ROS*, reactive oxygen species.

coupling to 1-e⁻ transfer pathways for production of adenosine triphosphate (ATP). The 1-e⁻ systems of mitochondria and chloroplasts are important in oxidative stress because 1) these systems have very high electron transfer rates; 2) they are sensitive to visible and UV light, trace metals, oxidants, and reactive electrophiles; and 3) they cause extensive destruction to other biologic components.

The fact that photosynthetic plants are rich sources of radical scavenging antioxidants is likely because of their own need to control radical injury from sunlight while at the same time using the light-derived energy to drive photosynthesis and O_2 production in their chloroplasts. In contrast, mitochondria and the electron transport system therein have high rates of electron transfer and require O_2 for ATP production. In both chloroplasts and mitochondria, damaged or malfunctioning electron transfer pathways represent a major source of endogenous oxidative stress.

Disruption of mitochondrial electron transfer has been extensively studied as a contributing mechanism in aging and disease. Damaged mitochondria have increased rates of 1-e⁻ transfer to O_2 to produce the radical, superoxide anion ($O_2^{-\bullet}$). Hundreds of studies have shown that disruption of mitochondrial function results in injury and disease symptoms. Copper deficiency results in aberrant megamitochondria.[54] The anticancer drug doxorubicin and antiretroviral drugs used for human immunodeficiency virus (HIV) infection treatment specifically cause mitochondrial damage.[55,56] Such conditions often damage mitochondrial DNA, and increase ROS generation, so such conditions can be self-perpetuating and autocatalytic in destruction of mitochondria. However, evidence that one can protect against mitochondrial oxidative stress by nutritional or dietary antioxidant supplementation in humans is limited.

Protective Mechanisms in Mitochondria

Mitochondria have multiple defense systems to protect against excessive oxidant generation, and these may be more critically dependent on the adequacy of precursors for NADPH supply than on supplementation with radical scavengers. Endogenous defenses include the electron transfer intermediate, coenzyme Q,[57] which exists in the mitochondrial inner membrane in a stable, radical form (semiquinone) that protects against radical reactions. Mitochondria also contain specific GSH and Trx systems.[58,59] (See section below on *Thioredoxin-Dependent Systems.*)

Pharmacologic forms of coenzyme Q show some promise of benefit in protecting mitochondria, but nutritional forms of coenzyme Q do not appear to be effectively delivered to mitochondrial inner membranes. Both GSH and Trx systems depend on adequate dietary selenium and riboflavin, although excesses of these are considered likely to be harmful, rather than to enhance activity further.

In addition to the effective antioxidant systems within mitochondria, a turnover mechanism exists to eliminate damaged mitochondria by autophagy[60] and maintain mitochondrial number by biogenesis.[61] When these mechanisms are functioning properly, oxidative stress may be apparent only transiently under relatively severe conditions. Mitochondria produce $O_2^{-\bullet}$ under all aerobic conditions, and it is possible that physiologically produced $O_2^{-\bullet}$ functions as a biologic signal to maintain mitochondrial homeostasis or mitochondrial–cytoplasmic communication.[4] Most of the experimental evidence for mitochondrial injury resulting from oxidative stress can also be interpreted in terms of disruption of homeostatic mechanisms signaled by $O_2^{-\bullet}$ or H_2O_2. Because of this, it may be necessary to understand the signaling functions of these agents before developing nutritional approaches for protecting mitochondrial function. Ongoing research on mitochondrial dysfunction related to chronic excess energy intake, obesity, and metabolic syndrome[62] is expected to provide a basis for interventional studies in this important area.

Superoxide Anion Radical and Hydrogen Peroxide

Although considerable research has focused on $O_2^{-\bullet}$ as a highly deleterious compound, damage from $O_2^{-\bullet}$ is self-limiting because $O_2^{-\bullet}$ reacts with itself in a spontaneous dismutation reaction to form O_2 and H_2O_2. Additionally, superoxide dismutases (SODs) exist in the cellular cytoplasm, in mitochondria, and in plasma, to support efficient elimination of $O_2^{-\bullet}$. Thus, $O_2^{-\bullet}$ damage typically occurs in combination with another reactant, such as H_2O_2 or NO. In the Haber-Weiss reaction, $O_2^{-\bullet}$ reacts with H_2O_2 to form a highly destructive hydroxyl radical (•OH). In biologic systems, the reaction is catalyzed by transition metals and is termed the *Fenton reaction*. In this reaction, $O_2^{-\bullet}$ serves as a reductant to ferric iron (Fe^{3+}), producing ferrous iron (Fe^{2+}), which transfers an electron to H_2O_2 to form water and •OH (see Figure 45.4, and chapter 10). The OH reacts at diffusion-controlled rates with all macromolecules in living systems. OH is produced in biologic systems exposed to ionizing radiation, and this reaction is used therapeutically to kill cancer cells in radiation therapy.

Loss of SOD in knockout mice is lethal, but heterozygotic knockout mice with decreased SOD activity, as well as transgenic mice with increased activity, have a longer life span. At present, these observations have no simple interpretation, but several lines of evidence suggest that the H_2O_2 produced by the SOD reaction is central to oxidative stress, especially in the presence of iron. H_2O_2 is efficiently eliminated by GSH peroxidases, a family of Trx-dependent peroxidases (peroxiredoxins [Prxs]), and catalase. One can anticipate that conversion of $O_2^{-\bullet}$ to H_2O_2 under conditions of limited H_2O_2 elimination can create conditions of OH generation. However, H_2O_2 itself is toxic, and the failure of radical-scavenging antioxidants to reduce disease in human clinical trials suggests that excess H_2O_2 production may be more relevant to oxidative stress in humans. Superoxide is produced by a family of NADPH oxidases (Noxs) that function in phagocytic killing of bacteria and in cell signaling.[63] These enzymes support both 1^- and 2-e^- reduction of O_2, and both $O_2^{-\bullet}$ to H_2O_2 produced by these enzymes appears to contribute to toxic reactions (see Figure 45.4). $O_2^{-\bullet}$ reacts with the vasoreactive radical NO•, produced by NO· synthases (NOSs), to form a highly reactive and toxic product, peroxynitrite, and H_2O_2 serves as a substrate for myeloperoxidase in the production of reactive, bacteriocidal, hypochlorous acid.[64]

Other Endogenous Sources of Reactive Oxygen Species

In addition to the Nox family enzymes, at least 25 other enzymes produce H_2O_2 as a normal biologic product; and many flavoproteins, hemoproteins, and other iron-containing proteins can catalyze 1-e^- transfer to generate ROS. The NOSs have been extensively studied as a pathologic source of $O_2^{-\bullet}$ in a nonphysiologic uncoupled side reaction.[65] Other biologic sources of ROS include cytochrome P-450 (Cyp) enzymes that function in elimination of dietary and hormonal molecules, as well as environmental and therapeutic xenobiotics. These enzymes activate O_2 to rid the body of the foreign chemicals; they are abundant in the endoplasmic reticulum (ER) of liver, and in the process of normal function, $O_2^{-\bullet}$ or H_2O_2 can be released into the cytoplasm. Some chemicals facilitate generation of ROS by accepting an electron from reactive hemes or flavins and transfer the electron to O_2 in a catalytic process known as redox cycling.[66] These reactions can be a major source of toxic oxidant generation by initiating lipid peroxidation (see later). In addition, flavins and flavoproteins are photoreactive, absorbing visible light that activates electron transfer to other chemicals. This reaction results in destruction of the flavins, making it important that nutritional supplements and other preparations containing riboflavin be protected from light.

Defenses Against Peroxides

Peroxisomes and Catalase

Peroxisomes are specialized organelles containing several enzymes that produce H_2O_2. These are important for nutrition because they support oxidation and elimination of nonnutritive dietary components, such as D-amino acids and α-hydroxyl acid. With supply of such chemicals to isolated hepatocytes, up to 40% of cellular O_2 can be converted to H_2O_2 in these organelles. The organelles contain a sufficiently high concentration of catalase, which converts H_2O_2 to nontoxic O_2 and H_2O, to eliminate the H_2O_2 without causing acute cell death. Many peroxisomal enzymes are flavoproteins, and a decrease in hepatic peroxisome content occurs in riboflavin deficiency. Red blood cells and immune cells often have cytoplasmic catalase, serving as a low-affinity, high-capacity system to eliminate H_2O_2 when activities of GSH peroxidases and Prxs are insufficient. Some of these processes are regulated by Forkhead Box class O (FoxO) transcription facts, which stimulate the transcription of numerous genes coding for several antioxidant proteins, such as in mitochondria (i.e., superoxide dismutase-2, peroxiredoxins 3 and 5) and peroxisomes (catalase).[67]

Glutathione-Dependent Antioxidant Systems

GSH is the most extensively studied antioxidant system and, with an extensive literature, one can anticipate considerable complexity, as well as misinterpretations and contradictions.

Salient features are that GSH supports critical reactions in the elimination of peroxides, detoxification of reactive electrophiles produced by radical reactions, and regulation of proteins through maintenance of sulfhydryl groups and via covalent modification (*S*-glutathionylation).[68] GSH is a reductant for selenium-dependent GSH peroxidases (GPXs) and selenium-independent Prx-6, and some selenium-independent GSH transferases (GSTs) (see chapter 14). The peroxidases act on H_2O_2 and lipid hydroperoxides. The product of these oxidative reactions, glutathione disulfide (GSSG), the oxidized product of GSH donating its H to other substrates, is reduced back to GSH by GSSG reductase in most tissues.

Rate limitation in peroxide elimination has been studied most extensively in liver. Activities of the peroxidases greatly exceed the rate of NADPH supply for GSSG reduction.[69] Additionally, GSSG reductase has a high Michaelis-Menten constant (K_m) for GSSG relative to the steady-state concentration of GSSG. Therefore, stimulated peroxide metabolism results in increased GSSG concentration and stimulates its export, which takes place in an ATP-dependent manner. These rates exceed the rates of GSH synthesis and result in decreased cellular GSH concentrations. Although endogenous peroxide generation in vivo rarely approaches the maximal rates of metabolism, the effect of NADPH supply limitation is of considerable nutritional importance (see chapter 23). Cells have limited mechanisms for NADPH supply, and NADPH is required for detoxification reactions by Cyp enzymes in the endoplasm, and for anabolic processes, such as fatty acid biosynthesis. Much of the cytoplasmic NADPH is produced from glucose 6-phosphate (G6P) via the pentose phosphate pathway. G6P becomes limiting when glycolysis is stimulated, and this can limit NADPH supply. Conversely, in individuals with excess energy intake in the form of carbohydrate and protein, the demand for NADPH to convert these precursors to fatty acids approaches the maximal rates of NADPH generation. Thus, either energy insufficiency or energy excess can adversely affect the capacity to eliminate peroxides. These findings emphasize the importance of balance in biologic oxidation and antioxidant defenses.

GSH synthesis. GSH is synthesized intracellularly in a two-step pathway by glutamate Cys ligase, an enzyme composed of catalytic and regulatory subunits, and GSH synthetase.[70] The rates of synthesis are slow relative to the maximal rates of peroxide elimination and NADPH supply, as described earlier. Regulation of cellular concentrations has often been described simply in terms of feedback regulation, but this description cannot account for the variation in GSH concentration, which ranges from about 0.1 mM in small intestinal mucosa under fasting conditions to 10 mM in liver and kidney. The abundance of both enzymes is increased in response to activators of the transcription factor Nrf2 (see later), but additional regulation of cytoplasmic and mitochondrial concentrations occurs through transport, mediated by the mitochondrial dicarboxylate carrier and a monocarboxylate carrier.[71] GSH is transported out of cells by several multidrug resistance proteins (MRPs) and the chloride channel transmembrane conductance regulator CFTR, which is mutated in cystic fibrosis. GSH is transported into the cisternae of the ER,[72] but the molecular nature of the transporter is not known.

Additional control of GSH synthesis occurs through the supply of sulfur amino acid precursors. Excess Cys is oxidized by Cys dioxygenase, but a fraction of Cys is also oxidized by an unknown mechanism to the disulfide cystine. Cystine distributes into the total body water in a sufficient amount to provide a 24-hour reservoir of Cys. Cystine uptake by cells occurs by x_c^-, a transporter under transcriptional regulation by Nrf2, the same transcription factor that controls glutamate–Cys ligase expression. Nrf2 is activated by many phytochemicals, so diets rich in these activators can be important in maintaining GSH-dependent antioxidant systems.

Other aspects of GSH metabolism indirectly affect the maintenance of GSH-dependent antioxidant functions. The major route of turnover involves the export of GSH from cells. In the plasma, GSH reacts nonenzymatically with cystine to produce a Cys–GSH disulfide (CySSG).[70] Both GSH and CySSG are degraded in extracellular compartments by γ-glutamyltransferase (GGT). GGT is present in high abundance at the surface of the brush border of the kidney, small intestine, and several other tissues and is also present in the cisternae of the secretory pathway in some cells.[73] GSH has several functions in the extracellular space, including detoxification of dietary electrophiles catalyzed by GSH *S*-transferase associated with mucus in the small intestine.[74] GSH is present in many foods,[75] and it is also supplied to the intestinal lumen via bile.[76] Both dietary and biliary sources of GSH protect against intestinal epithelial oxidative stress by supporting elimination of dietary lipid peroxides.[77,78]

GSH is used by GSTs to detoxify reactive electrophiles, such as those generated by lipid peroxidation. These enzymes include microsomal and nonmicrosomal forms, some of which act on biosynthetic intermediates for prostaglandins and leukotrienes. A fraction of GSH is present as *S*-nitroso-GSH, a transnitrosylating agent generated from NO or its metabolites.[79] GSH functions in metabolism as a coenzyme for formaldehyde dehydrogenase, glyoxylase, and other metabolic reactions.[79,80] In these reactions, GSH is cyclically removed by one reaction and regenerated in a second reaction. Several thiol transferases, also known as glutaredoxins, catalyze the introduction and removal of GSH in reactions that function in cell signaling and control.[81,82] Several proteins are regulated by this glutathionylation mechanism, and many others undergo glutathionylation under oxidative stress conditions.[70,83,84]

Thioredoxin-Dependent Systems

Enzymes of the Trx family provide an important complement to the GSH system in peroxide elimination and protection against oxidative stress.[85] Trxs are small proteins that have two Cys residues at the active site. This thiol pair serves as a reductant in the synthesis of deoxyribose for DNA synthesis as well as for protection against oxidative stress by reducing peroxides, protein disulfides, protein sulfenic acids, and protein Met sulfoxides. These reactions convert the Trx active site to a disulfide, which is recycled back to the dithiol form by Trx reductases. Distinct Trx systems are present in the nuclear and cytoplasmic (Trx-1) and mitochondrial (Trx-2) compartments. Reduced Trx-1 or Trx-2 also protect against oxidant-induced apoptosis by binding to apoptosis signal-regulating kinase-1 (ASK1), thus inhibiting its function. The activity of Trx-1 is controlled in part by binding the vitamin D_3–binding protein, TXNIP.[86] In cell nuclei, Trx-1 also maintains the reduced DNA-binding forms of several important transcription factors, including nuclear factor (NF)-κB, Nrf-2, activator protein-1 (AP-1), P53, glucocorticoid receptor (GR), estrogen receptor, and hypoxia inducible factor-1α (HIF-1α).

Of perhaps greatest importance as an antioxidant, Trxs support six Prxs that have heterogeneous subcellular distributions, but a common activity to eliminate peroxides.[87] Prx-3 and Prx-5 are present in mitochondria, whereas Prx-1 and Prx-2 are widely distributed as cytoplasmic and nuclear forms. Of note, kinetic studies have shown that these systems can be quantitatively more important that the GSH systems in elimination of peroxides. Both Trx and GSH systems depend on NADPH; however, so NADPH supply can be an important target for nutritional support. NADPH is used as a cofactor by glutathione reductase to reduce oxidized glutathione (GSSG/2GSH) and, likewise, it is used by thioredoxin reductase to reduce oxidized thioredoxin, both of which contribute to defense against oxidative stress.[88]

Extracellular Defense Systems

Although most oxidative stress research has focused on cellular defenses, extracellular spaces represent about 30% of the body water, and all cells have extracellular surfaces with proteins that are subject to oxidative damage. Soluble proteins include enzymes, antibodies, clotting factors, and enzymes required for maintenance of basement membranes; and cell surfaces include transporters, receptors, adhesion sites, and structural elements. To provide protection against oxidative stress, extracellular fluids contain secreted forms of many antioxidant proteins, such as SOD, GPX, and Trx-1. In addition, the amino acid Cys and its disulfide cystine interact with the circulating pool of GSH and GSSG to maintain a dynamic redox communication between cellular and extracellular compartments and among organ systems.[69,85] Certain extracellular fluids, such as the lung's alveolar lining fluid and bile, have relatively high GSH concentrations. In most other fluids; however, Cys is the predominant low-molecular-weight thiol. In nonenzymatic reactions, Cys is about 10 times more reactive than GSH. Consequently, the continuous supply of Cys to extracellular spaces provides a chemical system to support many of the same functions as provided by GSH within cells.

In the small intestine, Cys is part of a redox control system in which transporters shuttle Cys out and cystine in to maintain luminal thiol-disulfide redox states.[89] The transport systems differ on the apical and basal surfaces of cells, thus allowing compartment-specific redox regulation. GSH transport occurs via different transporters, so the Cys–cystine and GSH–GSSG redox couples are not equilibrated. This allows different protein systems on cell surfaces and in the extracellular space to be independently controlled. Additionally, protein disulfide isomerase (PDI) is associated with cell surfaces and functions in regulation of receptors and structural proteins.

Summary of Oxidant Generation and Detoxification

The most coherent conclusions from the extensive scientific literature about endogenous generation of oxidants are as follows: (a) $O_2^{-\bullet}$ and H_2O_2 are ubiquitous products of aerobic metabolism; (b) high $O_2^{-\bullet}$ and H_2O_2 production, especially in the presence of Fe^{3+} or other redox-active transition metal, is damaging to biologic systems; (c) $O_2^{-\bullet}$ and H_2O_2 are naturally produced by enzymes with evolved catalytic activities; and (d) aberrant production of $O_2^{-\bullet}$ and H_2O_2 occurs by many mechanisms. The most salient features of the defense systems against oxidants are as follows: (a) multiple systems with overlapping activities are present; (b) the systems have relatively high capacities to eliminate $O_2^{-\bullet}$, H_2O_2, and other peroxides; (c) importantly, and not always well appreciated, the major peroxide-metabolizing systems are ultimately limited by NADPH supply, which is linked to cellular energy metabolism; and (d) compartmentation of antioxidant systems is of considerable importance in providing protection.

Disruption of Redox Signaling and Control

Redox signaling and control mechanisms are central to cellular communication and regulation. These mechanisms depend on highly reactive thiols, which sense oxidative signals and operate at relatively low rates of electron transfer. Hence, disruption of these redox signaling and control mechanisms represents a more subtle form of oxidative stress than that caused by exogenous oxidants or an imbalance of endogenous oxidant and antioxidant defenses (see Figure 45.1). The discovery of the Noxs, enzymes that generate $O_2^{-\bullet}$ and H_2O_2 as signaling molecules,[90,91] showed that it is not possible to simply equate ROS with harmful chemistry. The Nox enzymes are ubiquitously distributed and have signaling functions that are integrated with the extensive kinase-phosphatase pathways governing most aspects of cellular regulation. Noxs produce both $O_2^{-\bullet}$ and H_2O_2, and both of these could have signaling functions. However, the redox potentials of redox couples in biologic systems generally support reactions in which $O_2^{-\bullet}$ serves as a one-electron donor and H_2O_2 as a two-electron acceptor. Consequently, the redox characteristics raise the possibility that $O_2^{-\bullet}$ supports different signaling functions than H_2O_2. Available data, however, show that H_2O_2 provides a relevant redox signal in many systems.

In biochemical systems in which the rates of 1-e^--transfer and 2-e^--transfer to O_2 have been compared, the predominant product is H_2O_2, not superoxide anion.[69,92] Radical and nonradical oxidant production is, therefore, partitioned, as depicted in **Figure 45.5**. Redox proteins that generate $O_2^{-\bullet}$ often show rapid, sequential electron transfer to produce H_2O_2. Further, $O_2^{-\bullet}$ (two molecules) is rapidly dismutated to H_2O_2 and O_2. Consequently, the 2-e^-, nonradical oxidant, H_2O_2, is produced at a higher rate and appears to represent the major oxidant burden. In mammalian systems, this rate has been estimated to be 1% to 4% of the rate of O_2 consumption,[93] but precise rates are not known and probably vary considerably among cell types. Within cells, rates are higher in peroxisomes[94] and mitochondria[69] and lower in the cytoplasm or nuclei.[95]

Summary of the Spectrum of Oxidative Stress and Defense Systems

Humans experience oxidative stress from external and internal sources that are diverse and unavoidable. For many of these, prudent nutrition to maintain health appears to be the best strategy for protection. For others, such as light-induced damage to the retina and skin, phytochemicals that filter light or induce endogenous protective systems appear to be beneficial. No specific recommendations are available, but these issues would appear to be adequately covered by public health recommendations concerning consumption of diets rich in fruits and vegetables. DRIs exist for radical scavenging antioxidants vitamins C and E and for selenium to support antioxidant enzymes. Other nutritional factors affect oxidant defenses in different ways, especially related to sulfur amino acid metabolism and regulation of the GSH system. Contemporary research has led to the recognition that radical and nonradical mechanisms both contribute to oxidative stress in health and disease. The following sections address these distinct mechanisms in more detail, specifically with the goal to help direct nutrition scientists to the important needs to understand complex redox systems.

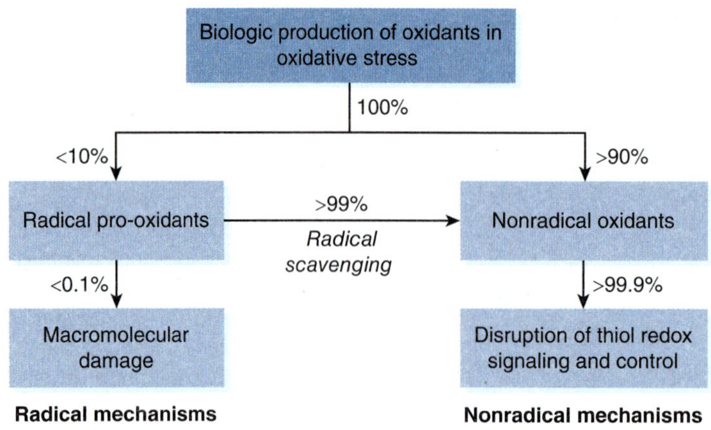

Figure 45.5 Oxidant production in biologic systems is partitioned between radical and nonradical pathways. In this schematic representation, two-electron transfers predominate at biologic active sites that reduce O_2 to superoxide and hydrogen peroxide (H_2O_2), so radicals are always produced at lower rates (e.g., <10% of total) than two-electron products, such as H_2O_2 (>90% of total). Radical scavenging mechanisms such as superoxide dismutase are highly efficient in converting radicals to nonradical species. Consequently, more than 99% of the oxidants generated are nonradical in nature. The remaining radicals cause a low level of radical-mediated macromolecular damage. Nonradical oxidants preferentially target reactive thiols of redox signaling and control pathways, resulting in a disruption of signaling and control.

Radical Mechanisms of Oxidative Stress

A Brief History of Discoveries

Research on radicals in biology was popularized by the discovery of lipid peroxidation (**Figure 45.6**), the antioxidant activity of vitamin E (see chapter 19), and the selenium-dependent activity of GSH peroxidase.[96,97] Study of a commonly used household spot remover, carbon tetrachloride (see Figure 45.6), showed that radical reactions of lipid peroxidation caused hepatotoxicity[98,99] and raised awareness of this toxic mechanism. Additionally, Denham Harman hypothesized that radicals contributed to aging,[100] thus bringing the concepts of oxidative stress to a very broad audience.

Although this constellation of findings and conceptual developments advanced the biochemical study of free radicals, the unanticipated finding of an enzyme that detoxifies $O_2^{-\bullet}$ attracted the attention of the broad spectrum of respiratory biologists and invigorated a new field of free radical biology and medicine. Study of the O_2-dependent reduction of cytochrome c by xanthine oxidase led to the discovery that the abundant blood protein, erythrocuprein, was an enzyme that converted the 1-e^- reduction product of O_2 (i.e., the superoxide anion radical, $O_2^{-\bullet}$), to O_2 plus H_2O_2.[101] The discovery of SOD[102] was seminal because it provided evidence that biologic systems had evolved an enzyme to eliminate radicals. This provided clear evidence that radicals must be commonly produced, and are a threat to biologic systems.

Radical Chemistry of Lipid Peroxidation

Chemical studies of PUFA showed that rancidity of oils and fats occurred by a radical chain reaction now known as lipid peroxidation.[103] In this reaction process, an initiation event forms a radical, and subsequent hydrogen abstraction propagates formation of additional radicals until finally terminated by reactions that eliminate the radicals. In the process, hundreds of molecules of PUFA can be oxidized because of a single initiating event. Quinones and other chemicals that interact with flavoproteins and redox cycle are a common source of initiation (see Figure 45.6, top left), as are halogenated hydrocarbons and other environmental chemicals activated to radicals by Cyp proteins. In biologic membranes and depot fat, radical propagation occurs with abstraction of a hydrogen atom from PUFA (see Figure 45.6, middle). This results in rearrangement of the newly formed fatty acid radical to a more stable form with conjugated double bonds. Molecular oxygen (O_2) reacts with this intermediate to form a peroxyl radical, which then propagates the radical chain reaction by hydrogen atom abstraction from another PUFA. In the presence of trace amounts of Fe^{2+}, the process is amplified by creation of a new radical from the lipid hydroperoxide (see Figure 45.6, lower left, "Propagation") in a Fenton-type reaction. Vitamins C and E terminate the process by reducing the reactive intermediates and breaking the chain reaction (see Figure 45.6, bottom).

Lipid Peroxidation in Biologic Systems

The lipid peroxidation mechanisms described earlier are more complex in living organisms because of the high concentrations of protein, the abundance of free radical scavenging antioxidants, and the processes of biologic turnover that eliminate and replace damaged macromolecules and cells. Because of the high concentration of protein in cells, the number of propagation reactions tends to be small, thus emphasizing the importance of initiation events in toxicity. In pure lipid systems, free radical chain reactions can occur in which an initiation event causes modification of 200 to 400 molecules of fatty acid before being terminated by radicals reacting with each other. However, in biologic systems, such a chain reaction does not occur, for several reasons. One is that initiation events are prevented by keeping free metal ion concentrations very low. Another is that the concentration of protein is so high that H is often abstracted from proteins rather than other polyunsaturated lipids, thereby blocking propagation. Biologic

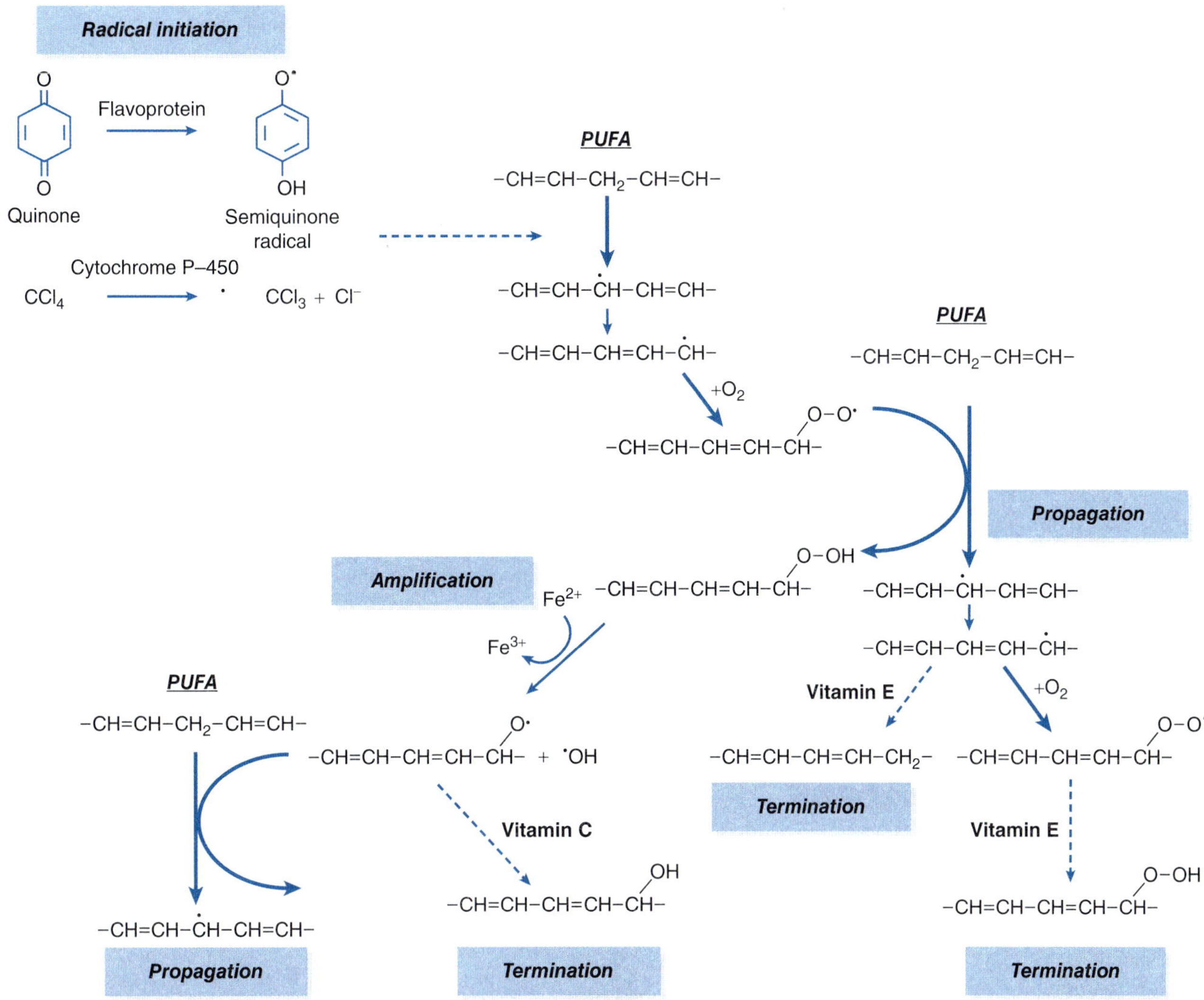

Figure 45.6 Lipid peroxidation is a radical chain reaction that occurs in biologic systems. Radical chain reactions occur through initiation, propagation, and termination steps. Initiation (top left) often occurs in biologic systems by redox cycling of quinone compounds and reductive activation of halogenated hydrocarbons (e.g., CCl_4). In the redox cycling reactions, quinones accept an electron from a reduced flavoprotein and are converted to a radical. The radical abstracts a hydrogen atom from a polyunsaturated fatty acid (PUFA) (top middle). This is most energetically favorable at a carbon adjacent to carbons with double bonds. The PUFA radical rearranges to form a conjugated double bond with an adjacent carbon-centered radical. O_2 rapidly adds to form a peroxyl radical (center). The peroxyl radical reacts with a second PUFA (right middle) to propagate the radical chain reaction. The peroxyl radical is reduced to a lipid hydroperoxide, which is normally reduced by glutathione (GSH) peroxidases or peroxiredoxin-6 (not shown). When activities of the latter systems are impaired or free iron is present, the process can amplify lipid peroxidation by ferrous iron (Fe^{2+})–dependent initiation of another radical chain reaction (bottom left), as occurs in the Fenton reaction. In the presence of radical scavengers such as vitamin C and vitamin E, the chain reactions are terminated. Vitamins C and E are most active with different types of radicals and complement each other.

systems also contain high concentrations of chain-terminating antioxidants, such as coenzyme Q, vitamin C and vitamin E. Amplification is also effectively prevented by the removal of lipid hydroperoxides by GSH peroxidases and Prx-6.

Reactive Aldehyde Products of Lipid Peroxidation

The propagation reactions described above can create multiple rearrangement products and eliminate other reactive species, especially conjugated aldehydes such as 4-hydroxynonenal (HNE) that react with proteins and DNA. The reactions generating HNE also generate carbon-centered radicals that propagate chain reactions. Other products are formed, including epoxides as shown earlier in Figure 45.3, which can be important as more stable oxidants with greater specificity in reaction with macromolecules; and isoprostanes,[104] which are rearrangement products that are useful as biomarkers of this chemical reaction sequence. Conjugated aldehydes react with thiols and the amine groups on proteins to create modified proteins and degradation products that are universally detectable in biologic systems. Thus, lipid peroxidation is an ongoing process in biologic systems that can be measured as protein carbonyls.[20] Malondialdehyde (MDA) is one such dicarbonyl, frequently used for assessment of lipid oxidation, and it can form adducts with proteins.[105]

Summary of Radical Reactions in Nutrition

Radical mechanisms are important in nutrition and food science because foods become rancid by these mechanisms. This limits food storage and preservation methods and can result in diminished nutritional value of the food. GSH-dependent systems are present in the gastrointestinal tract to detoxify reactive electrophiles and lipid hydroperoxides in food, thereby protecting against their absorption. Although radical reactions are important in vivo in some toxicologic processes, low concentrations of isoprostanes, protein carbonyls, and other well-documented oxidation products are found in normal cells, and healthy organisms are well protected against these reactions. Substantial evidence shows that acute physical and chemical exposures cause radical reactions in biologic systems, but when these occur, their intensity overwhelms defenses. Based on extensive double-blind interventional studies with radical scavengers in humans, supplementation at reasonable levels does not contribute significantly to long-term health in humans.[10-17] Thus, in everyday life, i.e., in the absence of high concentrations of acute environmental exposure, the importance of radical mechanisms as causative processes in human disease has probably been exaggerated since the mid-twentieth century.[69]

Nonradical Mechanisms of Oxidative Stress

Nonradical oxidants are quantitatively more important than radicals and may be more relevant to chronic toxicity through the disruption of redox signaling and control, whether macromolecular damage occurs or not. Important nonradical oxidants include H_2O_2, lipid hydroperoxides, quinones, disulfides, and peroxynitrite.[106,107] Although the free radical hypotheses of Harman, mentioned earlier, have been extensively studied, relatively less attention has been given to the nonradical mechanisms. The concepts have been formalized in the redox hypothesis of oxidative stress,[69] which provides four postulates to guide investigation into detailed mechanisms, ways to detect oxidative stress, and strategies to intervene to avoid or minimize oxidative stress. These postulates point to a need to incorporate systems biology, modern -omic technologies, and bioinformatics into nutrition research on oxidative stress. These four postulates are as follows:

1. All biologic systems contain redox elements (e.g., redox-sensitive viscinal Cys, Cys, residues), which function in cell signaling, macromolecular trafficking, and physiologic regulation.
2. Organization and coordination of the redox activity of these elements occur through redox circuits dependent on common control nodes (e.g., Trx, GSH).
3. The redox-sensitive elements are spatially and kinetically insulated so that "gated" redox circuits can be activated by translocation or aggregation and catalytic mechanisms.
4. Oxidative stress is a disruption of the function of these redox circuits caused by specific reaction with the redox-sensitive thiol elements, altered pathways of electron transfer, or interruption of the gating mechanisms controlling the flux through these pathways.

Targets of Nonradical Oxidants

Three functional groups of proteins undergo reversible oxidation: the thiol moiety in Cys, the thioether moiety in Met, and the selenol moiety in selenocysteine (Sec). The oxidation states of the sulfur in Cys include the thiol (-SH), disulfide (-SS-), sulfenate ($-SO^-$), sulfinate ($-SO_2^-$), and sulfonate ($-SO_3^-$) forms. Thiyl radicals (-RS·) generated in the presence of oxygen-centered radicals,[108] as well as other reactive sulfur species,[109] can also be considered as toxic species in oxidative stress, but these rapidly react to form disulfides.[110] Sulfenates are relatively unstable and are converted to disulfides in the presence of thiols; sulfenates are stabilized in some protein structures as sulfenamides.[111,112] Higher oxidation states, sulfonates, and sulfonates are typically not reversible in mammalian systems. Sulfiredoxin reduces sulfinate in Prx[113] and could be important in redox signaling.[114,115]

Met oxidation[116,117] and Sec[118,119] can also be important in toxicologic mechanisms. Met is oxidized to Met sulfoxide in oxidative stress and aging.[117,120] Two types of Met sulfoxide reductases are important in protecting against distinct *S*- and *R*-sulfoxides in Met.[116,121,122] These reductases depend on Trxs[123] and are also associated with longevity.[124-127] This less-frequently studied type of oxidation involves both Met and Sec and is of nutritional importance because of the variable availability of Met and Sec. The selenol moiety of Sec is critical for the catalytic functions of Trx reductases and selenium dependent GSH peroxidases,[118,119] enzymes present at key positions in both Trx and GSH pathways and circuits that protect against nonradical oxidative stress.

Subcellular Redox Compartmentalization

Methods have been developed to allow study of oxidative stress within specific subcellular compartments.[95,128] The genome of mammalian cells contains about 214,000 encoded Cys residues, and redox biochemical approaches are beginning to provide an understanding of the organization of the redox network structure.[7,129,130] This research has revealed that oxidative stress is not uniform within cells but, rather, selectivity affects specific pathways in compartments. The extracellular compartments of plasma and interstitial space are generally more oxidized than cellular compartments,[95] have fewer antioxidant systems, and are more vulnerable to oxidants. The lung alveoli, oral mucosa, and intestinal lumen are supplied with the thiol antioxidant, GSH, and have protective enzymes associated their respective mucous lining layers. Although some evidence for direct protection by dietary and orally supplied GSH and GSH precursors is available, no rigorously designed studies have established the effectiveness of nutritional means to support these tissue pools.

Within cells, different redox systems function in the organelles. The ER and secretory pathway use PDI and an oxidase system (EROS) to oxidize proteins during processing for secretion.[131] Disruption of this oxidative pathway activates cell death through an ER stress mechanism.[131,132] In comparison, nuclei and mitochondria are more reduced,[94] and each has specific redox-sensitive proteins. Mitochondria have a unique Trx2 and glutaredoxin-2, Grx2, whereas Trx-1 and glutaredoxin-1 are found in the cytoplasm. Trx-1 is translocated into nuclei during oxidative stress,[133] and nuclei also contain Grx-2, at least when it is overexpressed in cells.[113,129]

Integration of Antioxidant Defenses by the NRF2 Transcription Factor

It has already been emphasized that components of redox biology and the defenses against oxidative stress are highly integrated. A main "hub" of regulation is the NRF-2 transcription factor and the downstream genes it regulates. NRf-2 is evolutionarily conserved and present in many tissues. When NRF-2 resides in the cytoplasm, it is itself regulated in an inactive form by protein-protein association with the protein known as Kelch-like ECH-associated protein 1 (Keap1), which is itself an adapter substrate of the Cullin 3 (Cul3) ubiquitin E3 ligase complex. By sequestering Nrf2 in the cytoplasm under homeostatic conditions, NRF2 is repressed by KEAP1 through its role in protein ubiquitination and proteosomal degradation. Two mechanisms of Nrf2 activation, described as canonical and noncanonical, have been described.[28] In the first-described canonical mechanism, electrophilic compounds, and oxidative stress result in the oxidation of some of the cysteine residues in Keap1, resulting in a decrease in Nrf2 ubiquitination and thereby promoting its nuclear translocation and activation. Once inside the nucleus, Nrf2 induces a variety of genes that have an antioxidant response element (ARE) in their own regulatory regions. The more recently described noncanonical mechanism of Nrf2 activation involved the disruption of the direct interaction of Nrf2 and Keap1 by other proteins such as p62, p21, dipeptidyl peptidase III (DPP3), Wilms tumor gene on X chromosome (WTX), and others by direct interaction with Keap1, thereby decreasing Nrf2 ubiquitination and increasing its nuclear translocation and activation. As a result, Nrf2 protein is available for translocation and, subsequently, genes dependent on Nrf2 become activated.[29] A model of the regulation of Nrf2 and some of its target genes are shown in **Figure 45.7**.

The Nrf-2/Keap-1 system is expressed in various cell types and is regulated in a highly complex manner.[28-30] Several inflammatory, xenobiotic-regulated, hormonal, and proliferation-regulating pathways impinge on Nrf2 activation.[28] Depending on activating signals and on the cell's expression of downstream genes, networks of defense mechanisms can be activated to protect various types of cells and tissues. Based on studies in endothelial cells (ECs), agents with protective effects against oxidative stress mainly include phenylpropanoids, flavonoids, terpenoids, and alkaloids. Most of these agents alleviate cell apoptosis due to oxidative stress, and the mechanisms are related to Nrf2 signaling activation and heme oxygenase-1 (HO-1) expression.[134] Nrf2-induced expression of ARE-containing cytoprotective genes forms a network of cooperating enzymes involved in phase I, II, and III drug detoxification reaction and elimination of pro-oxidants to maintain cellular homeostasis.[135]

Disruption of Redox Circuits

Accumulating evidence indicates that most agents that cause oxidative stress do so by disrupting specific redox control pathways associated with specific organelles. Very little nutrition-related research has been focused on these compartment-specific functions, although some in vitro evidence indicates that the nuclear compartment is relatively resistant to oxidation compared with cytoplasmic and mitochondrial compartments.

Key points of oxidative stress, according to the redox hypothesis, include altered function of redox circuits caused by specific reaction with redox-sensitive thiol elements.[136] This differs from earlier concepts of macromolecular damage in oxidative stress by encompassing adaptability of an organism without overt toxicity. In systems controlled by redox circuits, antioxidants or other chemicals that create new pathways for electron transfer can result in pathway failure resulting from creation of short circuits. Additionally, low-flux regulatory circuits are likely to control high-flux systems such as energy pathways providing ATP.[4] Exposures that disrupt these low-flux pathways can, therefore, contribute indirectly to many disease processes. Mass spectrometry-based redox proteomics methods provide a systematic means to examine the redox dependence of protein redox circuits. With

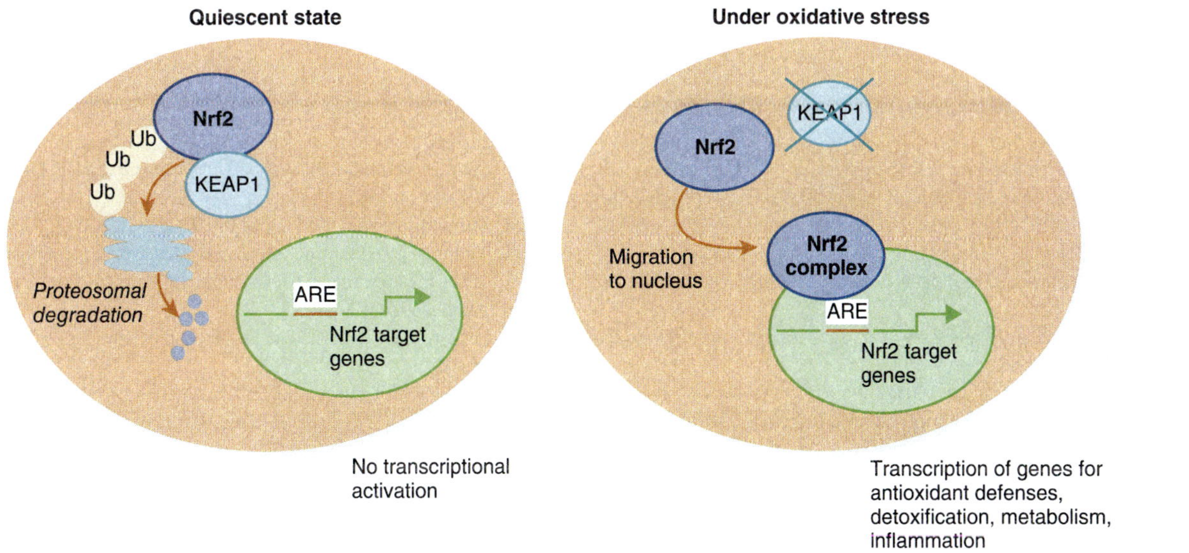

Figure 45.7 Schematic of Nrf2 regulation. Under normal homeostatic conditions, Nrf2 is sequestered in the cytoplasm by KEAP1 and, therefore, subject to ubiquitination and proteolysis, keeping its level low. Under conditions of oxidative stress, KEAP1 is modified and Nrf2 is free to enter the nucleus. Therein, it combines with other nuclear factors to bind to the ARE elements in the regulatory regions of a variety of genes (see text).

Data from He F, Ru X, Wen T. Int J Mol Sci 2020:21:4777. doi:10.3390/ijms21134777[30]

currently available methods, the percentage of oxidation of specific Cys residues can be measured in hundreds of proteins,[130] thereby allowing detailed studies of nutritional effects on these critical subcellular systems.[136]

Perspectives on Nutrition and Defenses Against Oxidative Stress

The concepts of oxidative stress have been and are continuing to undergo revision from the earlier definition as an imbalance of pro-oxidants and antioxidants that cause macromolecular damage, to one that also includes oxidative stress as a disruption of vital redox signaling and control processes within complex biologic systems.[1,2,6,88,136] In human health, accumulating data suggest that the latter is more relevant with respect to chronic disease, whereas the former is likely to be important under some acute conditions. Research is needed on the timely identification of radical processes contributing to disease under real-life conditions so that aggressive interventions to minimize injury and facilitate recovery can be instituted. A critical issue is that complex systems, such as humans consuming complex diets, respond to challenges with adaptation to maintain function. Because subclinical disease conditions can occur without overt signs, efforts are needed to study controlled variations in nutrition systematically to determine subclinical effects on redox circuits so that improved antioxidant strategies can be developed. Although such approaches may only confirm current nutritional principles for healthy individuals, they will provide knowledge that could lead to improved therapeutics to corrected aberrant redox control during disease.

Acknowledgments

The author acknowledges the author of the previous version of this chapter, Dean P. Jones, for his contributions.

References

1. Jones DP. In: Ross et al., eds. Modern Nutrition in Health and Disease, 11th ed. Walters Kluwer Lippincott Williams & Williams, 2014;46:611–625.
2. Sies H, Jones DP. In: Fink G, ed. Encyclopedia of Stress. 2nd ed. New York: Academic Press, 2007:45–8.
3. Murphy MP, Smith RA. Annu Rev Pharmacol Toxicol 2007;47:629–56.
4. Jones DP. Chem Biol Interact 2006;163:38–53.
5. Jones DP. Rejuv Res 2006;9:169–81.
6. Jones DP. Antioxid Redox Signal 2006;8:1865–79.
7. Kemp M, Go YM, Jones DP. Free Radic Biol Med 2008;44: 921–37.
8. Jones DP. J Intern Med 2010;268:432–48.
9. Jones DP, Go YM, Anderson CL, et al. FASEB J 2004;18:1246–8.
10. Anonymous. N Engl J Med 1994;330:1029–35.
11. Anonymous. Lancet 1999;354:447–55.
12. Age-Related Eye Disease Study Research Group. Arch Ophthalmol 2001;119:1417–36.
13. de Gaetano G. Lancet 2001;357:89–95.
14. Hennekens CH, Buring JE, Manson JE, et al. N Engl J Med 1996;334:1145–9.
15. Lonn E, Bosch J, Yusuf S, et al. JAMA 2005;293:1338–47.
16. Yusuf S, Dagenais G, Pogue J, et al. N Engl J Med 2000;342: 154–60.
17. Zureik M, Galan P, Bertrais S, et al. Arterioscler Thromb Vasc Biol 2004;24:1485–91.
18. Orrenius S. Toxicol Lett 2004;149:19–23.
19. Morrow JD, Roberts LJ 2nd. Methods Enzymol 1999;300:3–12.
20. Levine RL, Garland D, Oliver CN, et al. Methods Enzymol 1990;186:464–78.
21. Wallace DC. Science 1999;283:1482–8.
22. Bammler T, Beyer RP, Bhattacharya S, et al. Nat Methods 2005;2:351–6.
23. Kagan VE, Borisenko GG, Tyurina YY, et al. Free Radic Biol Med 2004;37:1963–85.
24. Go YM, Park H, Koval M, et al. Free Radic Biol Med 2010;48: 275–83.
25. Deo RC, Hunter L, Lewis GD, et al. PLoS Comput Biol 2010;6:e1000692.
26. Hiller K, Metallo CM, Kelleher JK, et al. Anal Chem 2010;82: 6621–8.
27. Dennis KK, Go Y-M, Jones DP. J Nutr 2019;149:553–65.
28. Tonelli C, Chio IIC, Tuveson DA. Antioxid Redox Signal 2018;29:1727–45.
29. Silva-Islas CA, Maldonado PD. Pharmacol Res 2018;134:92–9.
30. He F, Ru X, Wen T. Int J Mol Sci 2020:21:4777.
31. Rosenthal JM, Kim J, de Monasterio F, et al. Invest Ophthalmol Vis Sci 2006;47:5227–33.
32. Hideg E, Vass I. Plant Science 1996;115:251–60.
33. Jurkiewicz BA, Buettner GR. Photochem Photobiol 1996;64: 918–22.
34. Bernhard WA. In: Greenberg MM, ed. Radical and Radical Ion Reactivity in Nucleic Acid Chemistry. New York: Wiley, 2009: 41–68.
35. Jackson SP, Bartek J. Nature 2009;461:1071–8.
36. Kopke R, Bielefeld E, Liu J, et al. Acta Otolaryngol 2005;125: 235–43.
37. Suhr D, Brummer F, Hulser DF. Ultrasound Med Biol 1991;17: 761–8.
38. Kim, HJ, Sparrow JR. J Lipid Res 2021;62:100042.
39. National Eye Institute. Accessed May 2023. Available at www.nei.nih.gov/research/clinical-trials/age-related-eye-disease-studies-aredsareds2.
40. Li B, Ahmed F, Bernstein PS. Arch Biochem Biophys 2010;504: 56–60.
41. Solano F. Molecules 2020;25:1537.
42. Heinrich U, Neukam K, Tronnier H, et al. J Nutr 2006;136: 1565–9.
43. Ruidavets JB, Cournot M, Cassadou S, et al. Circulation 2005;111:563–9.
44. Seinfeld JH, Pandis SN. Atmospheric Chemistry and Physics: From Air Pollution to Climate Change. 2nd ed. New York: John Wiley & Sons, 2006.
45. Di Meo S, Reed TT, Venditti P, et al. Oxid Med Cell Longev 2016;1245049.
46. Becana M, Moran JF, Iturbe-Ormaetxe I. Plant Soil 1998;201: 137–47.
47. Das P, Samantaray S, Rout GR. Environ Pollut 1997;98:29–36.

48. Klaassen CD, Liu J, Choudhuri S. Annu Rev Pharmacol Toxicol 1999;39:267–94.
49. Ragunathan N, Dairou J, Sanfins E, et al. Environ Health Perspect 2010;118:1685–91.
50. Tafazoli S, Spehar DD, O'Brien PJ. Drug Metab Rev 2005;37: 311–25.
51. Araujo JA, Barajas B, Kleinman M, et al. Circ Res 2008;102: 589–96.
52. Siddique HR, Gupta SC, Mitra K, et al. J Appl Toxicol 2008;28:734–48.
53. Ames BN. Science 1983;221:1256–64.
54. Wakabayashi T, Asano M, Kurono C. Acta Pathol Jpn 1975;25: 15–37.
55. Lewis W, Gonzalez B, Chomyn A, et al. J Clin Invest 1992;89: 1354–60.
56. Wallace KB. Cardiovasc Toxicol 2007;7:101–7.
57. Shults CW, Oakes D, Kieburtz K, et al. Arch Neurol 2002;59: 1541–50.
58. Arner ES, Holmgren A. Eur J Biochem 2000;267:6102–9.
59. Holmgren A. J Biol Chem 1989;264:13963–6.
60. Las G, Shirihai OS. Diabetes Obes Metab 2010;12:15–9.
61. Chan DC. Cell 2006;125:1241–52.
62. Hernández-Aguilera A, Rull A, Rodríguez-Gallego E, et al. Mediators Inflam 2013;2013:135698.
63. Nauseef WM. Curr Opin Immunol 2019;60:130–40.
64. Weiss SJ, Klein R, Slivka A, et al. J Clin Invest 1982;70:598–607.
65. Xia Y, Roman LJ, Masters BS, et al. J Biol Chem 1998;273: 22635–9.
66. Kappus H, Sies H. Experientia 1981;37:1233–41.
67. Klotz L-O, Sánchez-Ramos C, Prieto-Arroyo I, et al. Redox Biol. 2015;6:51–72.
68. Hayes JD, McLellan LI. Free Radic Res 1999;31:273–300.
69. Lu SC. Biochim Biophys Acta 2013;1830:3143–53.
70. Jones DP. Am J Physiol Cell Physiol 2008;295:C849–68.
71. Lash LH. Chem Biol Interact 2006;163:54–67.
72. Banhegyi G, Lusini L, Puskas F, et al. J Biol Chem 1999;274: 12213–6.
73. Pompella A, Corti A, Paolicchi A, et al. Curr Opin Pharmacol 2007;7:360–6.
74. Samiec PS, Dahm LJ, Jones DP. Toxicol Sci 2000;54:52–9.
75. Jones DP, Coates RJ, Flagg EW, et al. Nutr Cancer 1992;17: 57–75.
76. Sies H, Graf P. Biochem J 1985;226:545–9.
77. Aw TY, Williams MW. Am J Physiol 1992;263:G665–72.
78. Aw TY, Williams MW, Gray L. Am J Physiol 1992;262:G99–106.
79. Staab CA, Alander J, Brandt M, et al. Biochem J 2008;413: 493–504.
80. Ahmed U, Dobler D, Larkin SJ, et al. Ann N Y Acad Sci 2008;1126:262–4.
81. Lillig CH, Holmgren A. Antioxid Redox Signal 2007;9:25–47.
82. Lofgren S, Fernando MR, Xing KY, et al. Invest Ophthalmol Vis Sci 2008;49:4497–505.
83. Fratelli M, Demol H, Puype M, et al. Proc Natl Acad Sci U S A 2002;99:3505–10.
84. Klatt P, Lamas S. Eur J Biochem 2000;267:4928–44.
85. Lu J, Holmgren A. Free Rad Biol Med. 2014; 66:75–87.
86. Nishiyama A, Matsui M, Iwata S, et al. J Biol Chem 1999;274: 21645–50.
87. Rhee SG, Yang KS, Kang SW, et al. Antioxid Redox Signal 2005;7:619–26.
88. Sies H, Berndt C, Jones DL. Annu Rev Biochem 2017;86; 715–748.
89. Circu ML, Aw TY. Free Radic Res. 2011;45:1245–66.
90. Griendling KK. Antioxid Redox Signal 2006;8:1443–5.
91. Lambeth JD. Nat Rev Immunol 2004;4:181–9.
92. Fridovich I. J Biol Chem 1970;245:4053–7.
93. Chance B, Sies H, Boveris A. Physiol Rev 1979;59:527–605.
94. Jones DP, Eklow L, Thor H, et al. Arch Biochem Biophys 1981; 210:505–16.
95. Go YM, Jones DP. Biochim Biophys Acta 2008;1780:1273–90.
96. Mills GC. J Biol Chem 1957;229:189–97.
97. Rotruck JT, Pope AL, Ganther HE, et al. Science 1973;179: 588–90.
98. Recknagel RO, Ghoshal AK. Nature 1966;210:1162–3.
99. Recknagel RO, Ghoshal AK. Lab Invest 1966;15:132–48.
100. Harman D. J Gerontol 1956;11:298–300.
101. Fridovich I. J Biol Chem 1997;272:18515–7.
102. McCord JM, Fridovich I. J Biol Chem 1969;244:6049–55.
103. Tribble DL, Aw TY, Jones DP. Hepatol 1987;7:377–86.
104. Morrow JD, Roberts LJ 2nd. Methods Mol Biol 2002;186:57–66.
105. Estévez M, Padilla P, Carvallo L, et al. Redox Biol 2019;26:101277.
106. Thomas DD, Liu X, Kantrow SP, et al. Proc Natl Acad Sci U S A 2001;98:355–60.
107. Thomas DD, Ridnour LA, Espey MG, et al. J Biol Chem 2006;281:25984–93.
108. Winterbourn CC, Metodiewa D. Arch Biochem Biophys 1994;314:284–90.
109. Giles GI, Tasker KM, Jacob C. Free Radic Biol Med 2001;31: 1279–83.
110. Starke DW, Chock PB, Mieyal JJ. J Biol Chem 2003;278: 14607–13.
111. Salmeen A, Andersen JN, Myers MP, et al. Nature 2003;423: 769–73.
112. van Montfort RL, Congreve M, Tisi D, et al. Nature 2003; 423:773–7.
113. Biteau B, Labarre J, Toledano MB. Nature 2003;425:980–4.
114. Woo HA, Chae HZ, Hwang SC, et al. Science 2003;300:653–6.
115. Woo HA, Kang SW, Kim HK, et al. J Biol Chem 2003;278: 47361–4.
116. Kim HY, Gladyshev VN. Mol Biol Cell 2004;15:1055–64.
117. Stadtman ER, Van Remmen H, Richardson A, et al. Biochim Biophys Acta 2005;1703:135–40.
118. Gladyshev VN, Hatfield DL. J Biomed Sci 1999;6:151–60.
119. Kryukov GV, Castellano S, Novoselov SV, et al. Science 2003;300:1439–43.
120. Stadtman ER, Levine RL. Amino Acids 2003;25:207–18.
121. Carp H, Miller F, Hoidal JR, et al. Proc Natl Acad Sci U S A 1982;79:2041–5.
122. Hansel A, Kuschel L, Hehl S, et al. FASEB J 2002;16:911–3.
123. Weissbach H, Etienne F, Hoshi T, et al. Arch Biochem Biophys 2002;397:172–8.
124. Miller RA, Buehner G, Chang Y, et al. Aging Cell 2005;4:119–25.
125. Moskovitz J, Bar-Noy S, Williams WM, et al. Proc Natl Acad Sci U S A 2001;98:12920–5.
126. Richie JP Jr, Leutzinger Y, Parthasarathy S, et al. FASEB J 1994;8:1302–7.
127. Sanz A, Caro P, Ayala V, et al. FASEB J 2006;20:1064–73.
128. Hansen JM, Go YM, Jones DP. Annu Rev Pharmacol Toxicol 2006;46:215–34.
129. Gallogly MM, Starke DW, Mieyal JJ. Antioxid Redox Signal 2009;11:1059–81.
130. Jones DP, Go YM. Curr Opin Chem Biol 2010;15:1–10.
131. Cenci S, Sitia R. FEBS Lett 2007;581:3652–7.
132. Todd DJ, Lee AH, Glimcher LH. Nat Rev Immunol 2008;8: 663–71.
133. Go YM, Jones DP. Antioxid Redox Signal 2010;13:489–509.
134. Zhang Q, Liu J, Duan H, et al. J Adv Res. 2021;34:43–63.
135. Hayes JD, Dinkova-Kostova AT. Trends Biochem Sci. 2014;39: 199–218.
136. Jones DP. In: Meyers RA, ed. Encyclopedia of Sustainability Science and Technology. New York: Springer Science+Business Media, 2012.

B. DIGESTIVE, ENDOCRINE, IMMUNE, AND NEURAL MECHANISMS

CHAPTER 46

Mechanisms of Nutrient Sensing

Douglas G. Burrin and Teresa A. Davis

CHAPTER OUTLINE

Abbreviations

AMP, adenosine monophosphate; **AMPK**, adenosine monophosphate–activated protein kinase; **ATF4**, activating transcription factor 4; **ATP**, adenosine triphosphate; **CASTOR1**, cytosolic arginine sensor for mTORC1 subunit 1; **CCK**, cholecystokinin; **ChREBP**, carbohydrate-response element-binding protein; **4EBP1**, eukaryotic initiation factor 4E–binding protein 1; **EE**, enteroendocrine; **eEF2**, eukaryotic elongation factor 2; **eIF**, eukaryotic initiation factor; **GATOR,** GAP activity toward Rags; **GβL**, G-protein β–like protein; **GCN2**, general control nonderepressible 2; **GI**, gastrointestinal; **GLP**, glucagon-like peptides; **GPCR**, G-protein–coupled receptor; **IGF**, insulin-like growth factors; **IRS-1/2**, insulin receptor substrate-1/2; **LAT1**, L-type amino acid transporter 1; **LC-PUFA**, long-chain polyunsaturated fatty acid; **LXR**, liver X receptor; **LRH-1**, lipid receptor homolog 1; **met-tRNA$_i$**, initiator methionyl-transfer RNA; **mTOR**, mammalian target of rapamycin; **mTORC**, mammalian target of rapamycin complex; **MuRF1**, muscle-specific ring finger protein 1; **PGC-1α**, peroxisome proliferator-activated receptor-γ coactivator-1α; **PI3K**, phosphoinositide 3-kinase; **PKB**, protein kinase B; **PPAR**, peroxisome proliferator-activated receptor; **PPRE**, peroxisome proliferator response element; **PepT1**, peptide transporter 1; **Rag**, Ras-related GTP-binding protein; **Rheb**, ras homolog enriched in brain; **S6K1**, 70-kDa ribosomal protein S6 kinase; **SGLT1**, sodium/glucose cotransporter 1; **SIRT1**, silent information regulator T1; **SLC38A9**, solute carrier family 28 member 9; **SNAT2**, system A sodium-dependent neutral amino acid transporter 2; **TSC1/2**, tuberous sclerosis complex 1/2; **X5P**, xylulose 5-phosphate

Introduction

The term *nutrient sensing* has emerged to describe the molecular mechanisms by which nutrients and their metabolites interact with various cell surface receptors, intracellular signaling proteins, and nuclear receptors, and modulate the activity of a complex network of signaling pathways that regulate the growth and function of cells. Nutrients also trigger the release of hormones and neurotransmitters that act on neighboring or distant cells through paracrine or endocrine mechanisms to regulate cellular growth and function. This chapter discusses some of the key nutrient-sensing mechanisms in different tissues within the body.

Nutrient Sensing by the Gut

Nutrient sensing in humans, as in most mammals, begins in the gastrointestinal (GI) tract, including the oral cavity. Nutrient sensing begins with taste buds that line the tongue and palate epithelium. This vital sensory process of tasting foods functions to recognize foods that are nutritious, giving a sweet taste, as well as ingested compounds that may be poisonous or harmful, giving a bitter taste (see Chapter 42). Specialized cells and receptors presented on taste buds mediate the five general taste sensations of sweet, sour, salty, bitter, and savory (also called umami).[1,2] These taste sensations are produced within the taste buds by specialized taste receptor cells that express numerous surface receptors that recognize different chemicals.

The taste sensations of sweet and umami are mediated by a family of three G-protein–coupled receptors (GPCRs) that function as taste receptors, namely, T1R1, T1R2, and T1R3, which form homodimeric or heterodimeric receptor complexes. Taste cells expressing the combination of T1R2 + T1R3 recognize sugars, artificial sweeteners, and some D-amino acids, and the result is a sweet taste (**Figure 46.1**). Taste cells expressing the heterodimeric T1R1+T1R3 receptor recognize the D-amino acids, glutamate, and aspartate, and the result is the savory or umami taste. Other metabotropic receptors are found in taste cells and may mediate some of the umami taste.[3-5] The bitter taste sensation is mediated by specific taste cells that express another family of GPCRs, namely T2R, which recognizes an array of bitter tastants, such as denatonium and quinine. Studies showed that the salty taste sensation is mediated by the epithelial sodium channel expressed on taste cells.[6] Finally, the sour taste sensation is mediated by a member of the transient receptor potential ion channel family, PKD2L1, that functions as an acid receptor in taste cells.

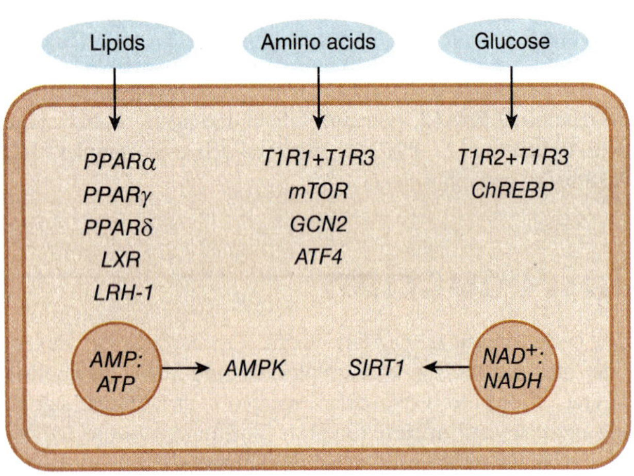

Figure 46.1 Lipid, amino acid, and glucose sensing. Cellular sensing of fatty acids involves the peroxisome proliferator-activated receptor (PPAR) family: PPARα, PPARγ, and PPARδ. Amino acids are sensed in the gut by T1R1+T1R3 receptors and in various tissues by yet unknown mechanisms to modulate intracellular signaling pathways involving mammalian target of rapamycin (mTOR), general control non-derepressible 2 (GCN2), and activating transcription factor 4 (ATF4). Glucose is sensed in the gut by T1R2+T1R3 receptors (described in text). Increased intracellular glucose concentrations increase carbohydrate-response element-binding protein (ChREBP). Changes in the nicotinamide adenine dinucleotide-to-reduced nicotinamide adenine dinucleotide (NAD^+/NADH) and adenosine monophosphate-to-adenosine triphosphate (AMP/ATP) ratios are detected by the energy-sensing molecules, silent information regulator T1 (SIRT1), and adenosine monophosphate–activated protein kinase (AMPK).

Nutrient sensing beyond the oral cavity of the GI tract is accomplished by specialized, differentiated epithelial cells that sense the presence of nutrients in the stomach and intestinal lumen as they are processed after meal ingestion. These cells, called enteroendocrine (EE) cells, are one of five different cell lineages derived from stem cells that reside in the inner layer of the mucosal lining referred to as the crypt.[7-9] The EE cells function as key nutrient sensors within the mucosal wall that recognize carbohydrates, triglycerides, and protein in the gut lumen. The EE cells function to coordinate the recognition of luminal nutrients with activation of gut physiologic functions, such as motility, fluid secretion, and blood flow, by secreting hormones and neurotransmitters. More than 20 different EE cell types are recognized, and they differ in their gut location and type of hormones secreted.[10] The EE cells differ from most other epithelial cells, such as parietal cell or enterocytes, which are programmed to produce acid and digestive enzymes to digest food components into their simplest, constituent units, namely, sugars, fatty acids, and amino acids. Tuft cells are another intestinal secretory cell involved in sensory function that contain the taste receptors mentioned above, as well as receptors for succinate and gamma-amino butyric acid.[11]

Some examples of specific nutrient sensing in EE cells include their responses to carbohydrate, lipids, and protein. Glucose absorption from the intestine occurs through the sodium/glucose cotransporter 1 (SGLT1) expressed mainly on absorptive enterocytes. In vivo animal studies showed that the SGLT1 transporter is upregulated by the presence of glucose, as well as by nonmetabolizable glucose analogs, in the intestinal lumen.[10,12] The mechanism by which glucose increases the expression of SGLT1 in enterocytes is thought to involve EE cells because they also express the same taste receptors (T1R2+T1R3) present in taste buds (see Figure 46.1). Glucose also activates the EE cell release of incretin hormones involved in insulin secretion and peripheral glucose uptake, namely, glucose-dependent insulinotropic peptide (GIP) and the glucagonlike peptides, GLP-1 and GLP-2. The current theory is that glucose-dependent activation of T1R in EE cells triggers the release of hormones that eventually lead to increased SGLT1 expression and glucose uptake. Luminal glucose can also activate EE cell release of 5-hydroxytryptamine (5-HT) or serotonin, which regulates gastric emptying and pancreatic exocrine and intestinal fluid secretion by interaction with vagal afferent neural circuits.[10,13]

The EE cells also respond to luminal lipids by secretion of cholecystokinin (CCK), which regulates several physiologic GI functions and appetite by activation of vagal nerves.[10] The mechanism for EE cell sensing has been ascribed to the presence of several seven-transmembrane GPCRs that recognize fatty acids, namely, GPR120, FFAR1, FFAR2, and FFAR3. These cell receptors are expressed on EE cells that colocalize with GLP-1 and peptide YY (PYY).[14]

Products of protein hydrolysis also activate release of CCK by EE cells, with physiologic effects like those of lipids.[13] Peptides are taken up by the intestinal enterocyte apical membrane by peptide transporter 1 (PepT1), a cotransporter specific for dipeptides and tripeptides. Studies suggested that peptidomimetic compounds specific for PepT1 induce CCK release and the expected inhibition of gastric motility. Emerging reports indicate that additional amino acid sensing receptors, including T1R3, metabotropic glutamate receptor 1 to 4, and calcium-sensing receptor (CaSr), are also expressed on EE cells and other epithelial cells in the gut.[15,16] The physiologic significance of these cell receptors in nutrient sensing awaits further study.

Intracellular Nutrient Sensors

Glucose, Amino Acid, and Lipid Sensing

Once nutrients are absorbed from the intestine into the bloodstream, they are sensed by somatic cells through various cellular mechanisms. Several intracellular or nuclear receptors are regulated by changes in the cellular availability of nutrients such as glucose, amino acids, and fatty acids.

Mechanisms Related to Carbohydrate-Responsive Element-Binding Protein and Liver X Receptor

An important cellular glucose sensing mechanism involves the transcription factor carbohydrate-response element-binding protein (ChREBP), which is activated in response to increased cellular glucose concentration (Figure 46.1).[17,18] ChREBP is mainly expressed not only in the liver but also in other glucose-responsive tissues such as adipose, brain, and pancreas. In fasting conditions of low glucose concentration, ChREBP is located in the cytosol in its phosphorylated form associated with the 14-3-3 protein. After a meal, however, the increased cellular influx of glucose results in increased production of xylulose 5-phosphate (X5P) through the pentose phosphate pathway. Increased cellular X5P concentration leads to activation of protein phosphatase 2A, which dephosphorylates ChREBP and enables its translocation in the nucleus. Once in the nucleus, ChREBP interacts with a binding partner, Max-like protein X, which then binds to the carbohydrate-response element of multiple target genes, and this increases their transcription. Many of the target genes activated by ChREBP are enzymes involved in lipogenesis and glucose metabolism. Another transcription factor reported to bind glucose is the nuclear receptor liver X receptor (LXR), whose primary ligands are oxysterols, such as cholesterol.[19] Ligand activation of LXR by oxysterols induces heterodimerization with retinoid X receptor and binding of target gene promoter sequences, processes that lead to activation of lipogenesis in liver and adipose tissue. The relative importance of ChREBP, compared with LXR, in glucose sensing has not been identified.[20] The liver receptor homolog-1 (LRH-1) is activated by phospholipids and plays a regulatory role in multiple metabolic pathways involved with bile acid and cholesterol homeostasis and glucose and lipid processing.[21]

Mechanisms Related to Activating Transcription Factor 4

A key cellular mechanism for sensing amino acids, especially in conditions of amino acid deficiency or imbalance, involves stimulation of activating transcription factor 4 (ATF4) (Figure 46.1).[22] In conditions of food deprivation or dietary protein restriction, the level of activated or charged transfer RNA (tRNA) bound to amino acids decreases. This results in an increase in uncharged tRNAs, which bind to the general control non-derepressible 2 (GCN2) kinase and, in turn, increase the phosphorylation of the eukaryotic initiation factor 2α (eIF2α) (**Figure 46.2**). The phosphorylated form of eIF2α suppresses general protein synthesis by inhibiting translation initiation factor, eIF2B, which is critical for ribosomal assembly. Paradoxically, the amino acid–limiting conditions alter ribosomal processing and result in increased translation of ATF4, which then binds to a conserved promoter region in many genes involved in amino acid transport (cationic amino acid transporter [CAT-1], and system A sodium-dependent neutral amino acid transporter [SNAT2]), metabolism (asparagine synthetase [ASNS]), and cell death (C/EBP homologous protein [CHOP], and tribbles homolog 3 [TRB3]). The ATF4 signaling pathway senses amino acid stress and functions as a counterpart to the mammalian target of the rapamycin (mTOR) pathway (discussed later) that responds to amino acid sufficiency to promote cell anabolism and growth.

Mechanisms Related to Peroxisome Proliferator-Activated Receptor

An important class of nuclear receptors that function in cellular sensing of fatty acids is the peroxisome proliferator-activated receptor (PPAR) family, including PPARα, PPARγ, and PPARδ (see Figure 46.1).[24-27] Lipid products and fatty acids play a critical role in metabolic control, and PPAR family members have emerged as central transcriptional regulators of lipid and carbohydrate metabolism. Saturated and unsaturated long-chain fatty acids and their eicosanoid derivatives are natural activators of this subclass of nuclear receptors. The relative impact of dietary sources compared with endogenous production of ligands on PPAR activation has not been well characterized, however. Investigators believe that many cellular lipid-modifying enzymes are involved whereby cyclo-oxygenases (COX), lipoxygenases (LO), epoxygenases/cytochrome P-450 enzymes, and lipase-use fatty acids, triglycerides, or phospholipids as substrates to generate endogenous PPAR ligands.

At the cellular level, the actions of these bioactive fatty acids are thought to be mediated by their cellular uptake by fatty acid–binding protein, which acts as a chaperone to facilitate molecular binding and activation of nuclear receptors. The PPAR family translates these lipid signals into responses, which control energy homeostasis and cell function. When activated by a ligand, such as long-chain polyunsaturated fatty acids (LC-PUFAs), the PPAR proteins heterodimerize with the retinoid X receptor and bind to a specific DNA sequence element called a peroxisome proliferator response element (PPRE) in target genes. Activation of PPRE increases the transcription of various target genes that control lipid and glucose metabolism, as well as inflammation. The PPARα gene is highly expressed in tissues with active fatty acid catabolism, such as liver, heart, kidney, brown adipose tissue, muscle, and small intestine. PPARα exerts a dominant role in fatty acid catabolism and ketone body synthesis in the liver by upregulation of both mitochondrial and peroxisomal oxidative enzymes.

One of the important mechanisms by which LC-PUFAs, especially the n-3 forms, docosahexaenoic acid and eicosapentaenoic acid, affect metabolic function is by modulating the expression of genes involved in fat and glucose metabolism. The beneficial health effects of fish oil are believed to be mediated by these bioactive n-3 LC-PUFAs. The n-3 LC-PUFAs act to suppress the transcription of genes encoding transcription factors (e.g., sterol regulatory element binding protein [SREBP-1] and ChREBP) and specific lipogenic enzymes, and induce the expression of gene encoding enzymes involved in peroxisomal and microsomal fatty acid oxidation. The transcriptional induction of PPARα-responsive genes promotes hepatic intracellular

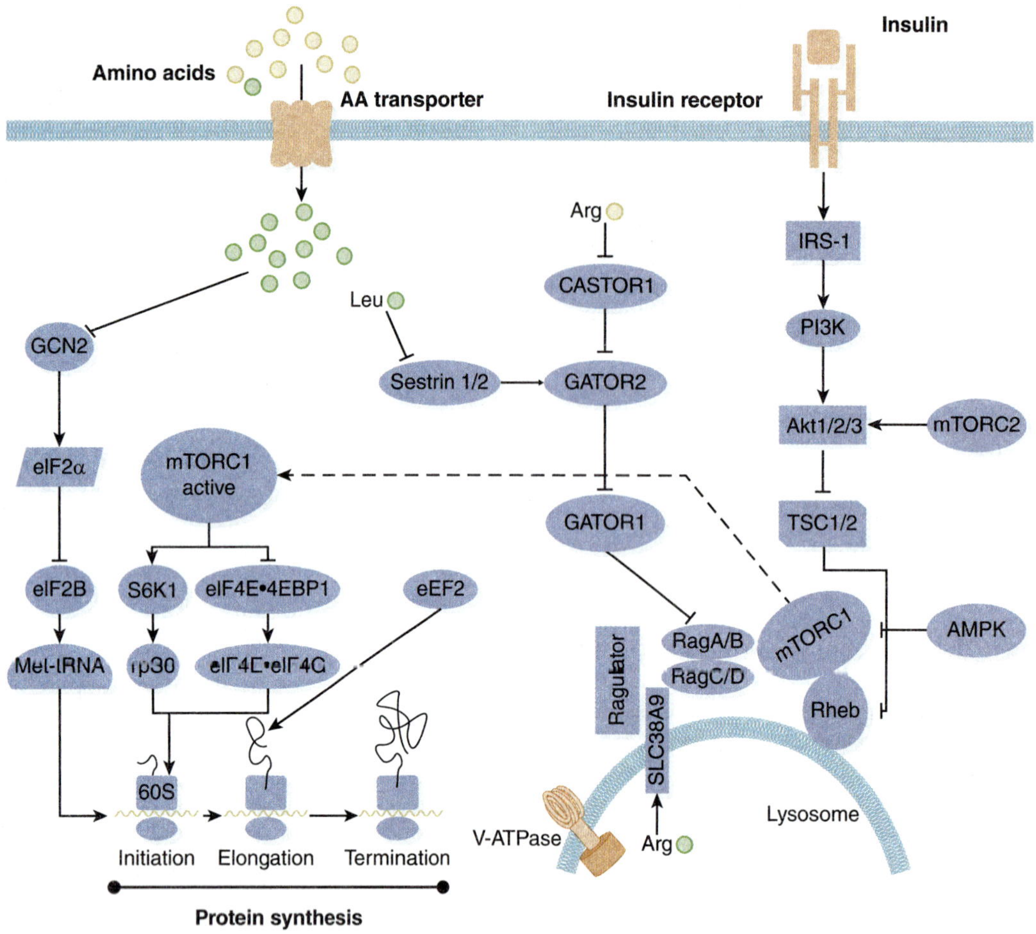

Figure 46.2 Nutrient sensing by mammalian target of rapamycin (mTOR). Insulin activates the insulin receptor and insulin receptor substrate-1 (IRS-1), followed by activation of phosphoinositide-3 kinase (PI3K), and Akt. Activation of Akt inactivates tuberous sclerosis complex 1 and 2 (TSC1/2), inducing the activation of ras homolog enriched in brain (Rheb) and mammalian target of rapamycin (mTOR). Adenosine monophosphate–activated protein kinase (AMPK) activation increases TSC1/2 activation and decreases mTOR activation. Both amino acids and insulin activate mTOR, which exists in a complex (mTORC1) with raptor and G-protein β-like protein (GβL). Activated mTOR phosphorylates ribosomal protein S6 kinase 1 (S6K1) and eukaryotic initiation factor 4E (eIF4E)–binding protein-1 (4EBP1). Phosphorylation of S6K1 activates ribosomal subunit S6 (rpS6). Phosphorylated 4EBP1 releases eIF4E from an inactive complex with 4EBP1 and thus enables the formation of the active eIF4E-eIF4G complex that mediates mRNA binding to the ribosome. The binding of initiator methionyl-tRNA (met-$tRNA_i$) to the 40S ribosomal subunit to form the 43S preinitiation complex is mediated by eIF2B, which can be inhibited by phosphorylation of the α subunit of eIF2 in response to amino acid deprivation sensing by general control non-derepressible 2 (GCN2). Eukaryotic elongation factor 2 (eEF2) is regulated by eEF2 kinase.

fatty acid uptake, the conversion of fatty acids to their acyl-coenzyme A derivatives, and channeling toward mitochondrial/peroxisomal oxidation.

In contrast to PPARα, a dominant effect of PPARγ is to control fatty acid storage and glucose metabolism (see Figure 46.1).[28,29] PPARγ is a pivotal transcription factor in adipose tissue differentiation and survival as well as in maintaining adipocyte specific functions, such as lipid storage in white adipose tissue. In addition, PPARγ is involved in glucose metabolism through an improvement of insulin sensitivity and thus, represents a molecular link between lipid and carbohydrate metabolism. The extent to which dietary and endogenous fatty acid ligands activate PPARγ is poorly understood. Much of what is known about this role of PPARγ is based on studies using synthetic, thiazolidinedione, and antidiabetic drugs, such as rosiglitazone, which are high-affinity agonist ligands for PPARγ.

A powerful coactivator of PPARγ is PPARγ coactivator-1α (PGC-1α), which is highly expressed in brown adipose tissue and serves to augment mitochondrial biogenesis expression, thermogenesis, and increased cellular respiration. PGC-1α is also regulated by protein kinase cascades, such as adenosine monophosphate–activated protein kinase (AMPK) and protein kinase B (PKB)/Akt either to increase oxidative metabolism in brown fat or to suppress hepatic glucose production. A more recently identified member of the PPAR family is PPARδ, which is expressed in various tissues, including skeletal muscle, fat, and the heart.[27] Activation of PPARδ stimulates fatty acid catabolism and peroxisomal metabolism and leads to decreased triglyceride stores, improved endurance performance, and enhanced cardiac function. PPARδ receptor activation in the liver also suppresses hepatic glucose output and contributes to improved glucose homeostasis.

Nutrient Signaling Pathways

Nutrient Sensing by Mammalian Target of Rapamycin

Food ingestion stimulates the synthesis of tissue proteins and, in skeletal muscle, this response is more profound in the young, rapidly growing individual.[30-33] The stimulation of protein synthesis after a meal is triggered by the postprandial rise in amino acids, particularly the branched-chain amino acid leucine, as well as the hormone insulin,[34-37] which is secreted by the pancreas largely in response to the rise in circulating glucose.[38] Although remarkable progress has been made in understanding the intracellular signaling pathway by which insulin regulates protein synthesis, the mechanism by which amino acids are sensed and modulate protein synthesis is less well understood. Investigators have widely accepted,[39] however, that amino acids and insulin induce their anabolic action to stimulate protein synthesis by activating independent signaling pathways that converge at the protein kinase mechanistic target of rapamycin complex 1 (mTORC1). Indeed, mTORC1 plays a central role in sensing the availability of nutrients, such as amino acids, and in integrating this information with other signals arising from hormones, exercise, and cell stress to regulate cellular processes including metabolism, gene expression, and protein synthesis.[40]

The nutritional status of an individual is monitored in a coordinated fashion at both the systemic and the cellular levels by mTORC1.[41-44] Systemic nutrient sensing involves the insulin signaling pathway. The binding of insulin to its cell surface receptors induces the autophosphorylation of the receptors on tyrosine residues, followed by the activation of the tyrosine kinase activity of the receptors (see Figure 46.2).[45] Insulin receptor binding leads to the recruitment and activation of insulin receptor substrate-1/2 (IRS-1/2) proteins. The resulting activation of phosphoinositide 3-kinase (PI3K) results in the phosphorylation and activation of Akt (also known as protein kinase B [PKB]), which phosphorylates and inactivates an inhibitor of cell growth, called tuberous sclerosis complex 1 and 2 (TSC1/2).[46,47] Inhibition of the function of the TSC1/2 complex results in the activation of ras homolog enriched in brain (Rheb), followed by the activation of mTORC1.

The protein kinase mTOR is composed of two independently regulated complexes: mTORC1 (Figure 46.2) and mTORC2.[48] mTORC1 consists of mTOR, raptor, and the G-protein β–like protein (GβL), whereas mTORC2 is composed of mTOR, rictor, and GβL. mTORC1 is regulated by insulin as well as by amino acids, whereas mTORC2 does not appear to be activated by amino acids.

The mechanisms by which amino acids activate mTORC1 have not been as well defined as that for insulin and is an area of intense research. Amino acid transport into cells is an essential step for mTORC1 activation. Amino acids such as leucine and arginine, through their transporters, L-type amino acid transporter 1 (LAT1) and sodium-coupled neutral amino acid transporter member 2 (SNAT2) and solute carrier family 28 member 9 (SLC38A9), activate mTORC1 via nutrient-sensing components in the cytosol or lysosomal membrane.[49-52] The most accepted model by which amino acids activate mTORC1 is that amino acids signal to the Ras-related GTP-binding protein A/B (RagA/B) and RagC/D complexes independently of Akt to induce the recruitment of mTORC1 to the lysosomal surface where it becomes activated. This activation involves the association of mTORC1 with the lysosomal Ragulator complex and GTP-loaded Rheb. Sensors for specific amino acids in the cytosol have been identified and the sensors for leucine (stress response proteins, Sestrins) and arginine (cytosolic arginine sensor for mTORC1 subunit 1, CASTOR1) bind to the amino acids, causing the release of these sensors from GAP activity toward Rags (GATOR)2, which leads to the inhibition of GATOR1 by GATOR2 and the subsequent activation of mTORC1. The signaling pathways activated by amino acids downstream of mTORC1 are like those of insulin.[53]

Nutrient Signaling to Translation

Activation of mTOR, through signaling from either insulin or amino acids, stimulates the phosphorylation of the 70-kDa ribosomal protein (rp) S6 kinase (S6K1) and eIF4E–binding protein 1 (4EBP1), which are important regulatory components of mRNA translation (Figure 46.2).[54-58] Translation of mRNA into protein consists of three distinct stages: (a) initiation, in which the initiator tRNA and mRNA bind to the ribosomal subunit; (b) elongation, whereby tRNA-bound amino acids are incorporated into the growing peptide chain; and (c) termination, which results in release of the completed protein from the ribosome.[40,56] mTOR controls several components involved in the initiation and elongation stages of translation.

Translation initiation involves two essential steps: first, the binding of initiator methionyl-tRNA (met-tRNA$_i$) to the 40S ribosomal subunit to form the 43S preinitiation complex; and second, the binding of mRNA to the 43S preinitiation complex.[40,54-58] The binding of initiator met-tRNA$_i$ to the 40S ribosomal subunit is mediated by eIF2 (Figure 46.2). Phosphorylation of the α subunit of eIF2 reduces the activity of eIF2B and thereby decreases initiator met-tRNA$_i$ binding to the ribosome. The activity of eIF2B may be diminished by starvation and amino-acid depletion.

The binding of mRNA to the 43S preinitiation complex is facilitated by a group of initiation factors, including eIF4E, the protein that binds to the mRNA, and eIF4G, a scaffolding protein that binds to eIF4E and the 43S preinitiation complex (Figure 46.2).[40,54-58] The eIF4E-eIF4G complex, which is highly regulated by 4EBP1, is crucial for the 43S preinitiation complex binding with mRNA. When unphosphorylated, 4EBP1 forms an inactive complex with eIF4E, with resulting inhibition of translation of mRNA. In the hyperphosphorylated state, 4EBP1 is no longer able to bind to eIF4E and thus allows eIF4E and eIF4G to form an active complex. Both insulin and amino acids, particularly leucine, increase the activity of mTORC1; the results are increased phosphorylation of 4EBP1 and subsequent eIF4E-eIF4G complex formation.

The selection of mRNAs for translation involves modulation of the activity of S6K1 (Figure 46.2).[40,54-58] Its target, rpS6, is associated with the regulation of translation of a subset of mRNAs that code for proteins involved in the protein synthetic machinery, such as ribosomal proteins. Amino acids, particularly leucine, and insulin induce phosphorylation and thus activation of S6K1 by augmenting mTORC1 activity.

During the elongation process, eukaryotic elongation factor 2 (eEF2) mediates the translocation of the ribosome relative to the mRNA after addition of each amino acid to the nascent chain (Figure 46.2).[40,54-58] The activity of eEF2 is regulated by eEF2 kinase. Although insulin and amino acids can regulate the activation of eEF2 in an mTORC1-dependent manner, the elongation process is not limiting for protein synthesis under normal physiologic conditions.

Nutrient Signaling to Protein Degradation

Nutrient depletion stimulates autophagy, a process whereby cytoplasmic constituents are engulfed in vesicles and are degraded in the lysosomes into amino acids and other molecules to provide nutrients for the cell.[41,59-61] mTOR acts as a master regulator of cell growth vs autophagy. Activation of mTORC1, in response to stimulation by nutrients, such as amino acids or insulin, inhibits autophagy. Stimulation of protein degradation in response to nutrient deprivation also involves the ubiquitin proteasome pathway. The most common measure of the activation of this pathway is the increased expression of atrogin-1 and muscle-specific ring finger protein 1 (MuRF1). Transcription of atrogin-1 and MuRF1 can be activated by the family of forkhead O–box (FOXO) transcription factors and is repressed by mTORC1 in response to alterations in insulin and/or amino acid signaling.

Energy Sensing by Adenosine Monophosphate–Activated Protein Kinase and Silent Information Regulator T1

AMPK serves as an energy sensor that monitors cellular energy status.[41,62-64] AMPK activity is increased under conditions of cellular stress, characterized by a rise in adenosine monophosphate (AMP) and a decline in adenosine triphosphate (ATP) and thus, an increase in the AMP-to-ATP ratio in the cell (see Figure 46.1). AMPK activates pathways that generate ATP and inhibits pathways that consume ATP. AMPK activation leads to an increase in TSC2 activity which, in turn, inhibits mTORC1 activity and leads to a reduction in protein synthesis and an increase in protein degradation (Figure 46.2). AMPK activation in skeletal muscle stimulates the translocation of the glucose transporter, GLUT4, to the plasma membrane and thereby increases glucose uptake. Increases in glycolytic flux and fatty acid oxidation and reductions in glucose and triglyceride biosynthesis in response to AMPK activation also enhance energy production and reduce energy utilization by the cell. Activation of AMPK in response to food deprivation activates a signaling process that involves PGC-1α to reduce energy usage. Although AMPK is activated during nutrient depletion and in response to muscle contraction, whether AMPK activity is altered during the normal fasting and feeding cycle is still debated.[65]

The histone/protein deacetylase silent information regulator T1 (Sirtuins) have also been recognized as a fuel-sensing molecule (Figure 46.1).[66-68] Although SIRT1 is best known for its role in mediating the increase in longevity with energy restriction, studies suggested that SIRT1 interacts with AMPK to regulate energy metabolism. The link between the two fuel sensors involves alterations in the ratio of nicotinamide adenine dinucleotide to reduced nicotinamide adenine dinucleotide (NAD^+/NADH ratio), as occurs in response to starvation and exercise. SIRT1, like AMPK, promotes the utilization of substrates for energy generation through PGC-1α in response to reduced fuel (glucose or fatty acid) availability.

Acknowledgments and Disclosure

Drs. Davis and Burrin are recipients of research support from the National Institutes of Health HD085573, HD099080, DK094616, GM108940, and U.S. Department of Agriculture/Agricultural Research Service (USDA/ARS) 3092-51000-060-01S, and they have no disclosures to report. This work is a publication of the USDA/ARS Children's Nutrition Research Center, Department of Pediatrics, Baylor College of Medicine, Houston. The contents of this publication do not necessarily reflect the views or policies of the U.S.D.A., nor does the mention of trade names, commercial products, or organizations imply endorsement by the U.S. Government.

References

1. Chandrashekar J, Hoon MA, Ryba NJ, et al. Nature 2006;444:288–94.
2. Yarmolinsky DA, Zuker CS, Ryba NJ. Cell 2009;139:234–44.
3. Yasumatsu K, Horio N, Murata Y, et al. Am J Clin Nutr 2009;90:747S–52S.
4. Nakamura E, Hasumura M, San Gabriel A, et al. J Pharmacol Sci 2010;112:13–8.
5. San Gabriel A, Maekawa T, Uneyama H, et al. Am J Clin Nutr 2009;90:743S–6S.
6. Chandrashekar J, Kuhn C, Oka Y, et al. Nature 2010;464:297–301.
7. May CL, Kaestner KH. Mol Cell Endocrinol 2010;323:70–5.
8. Burman A, Kaji I. Nutrients. 2021;22;13(11):3712.
9. Lu VB, Gribble FM, Reimann F. Nutrients. 2021;9;13(3):883.
10. Raybould HE. Auton Neurosci 2010;153:41–6.
11. Hendel SK, Kellermann L, Hausmann A, Bindslev N, Jensen KB, Nielsen OH. Front Immunol 2022;14;13:822867.
12. Dyer J, Daly K, Salmon KS, et al. Biochem Soc Trans 2007;35:1191–4.
13. Raybould HE, Glatzle J, Freeman SL, et al. Auton Neurosci 2006;125:28–33.
14. Miyauchi S, Hirasawa A, Ichimura A, et al. J Pharmacol Sci 2010;112:19–24.
15. Conigrave AD, Hampson DR. Trends Endocrinol Metab 2006;17:398–407.
16. Wellendorph P, Brauner-Osborne H. Br J Pharmacol 2009;156:869–84.
17. Postic C, Dentin R, Denechaud PD, et al. Annu Rev Nutr 2007;27:179–92.
18. Uyeda K, Repa JJ. Cell Metab 2006;4:107–10.
19. Mitro N, Mak PA, Vargas L. Nature 2007;445:219–23.
20. Denechaud PD, Bossard P, Lobaccaro JM, et al. J Clin Invest 2008;118:956–64.
21. Sun Y, Demagny H, Schoonjans K. Biochim Biophys Acta Mol Basis Dis 2021;1;1867(8):166145.
22. Kilberg MS, Shan J, Su N. Trends Endocrinol Metab 2009;20:436–43.
23. Suryawan A, El-Kadi SW, Nguyen HV, et al. J Nutr 2021;151:2636–45.
24. Pyper SR, Viswakarma N, Yu S, et al. Nucl Recept Signal 2010;8:e002.
25. Lefebvre P, Chinetti G, Fruchart JC, et al. J Clin Invest 2006;116:571–80.
26. Michalik L, Auwerx J, Berger JP, et al. Pharmacol Rev 2006;58:726–41.
27. Barish GD, Narkar VA, Evans RM. J Clin Invest 2006;116:590–7.

28. Lehrke M, Lazar MA. Cell 2005;123:993–9.
29. Tontonoz P, Spiegelman BM. Annu Rev Biochem 2008;77: 289–312.
30. Hoffer LJ, Yang RD, Matthews DE, et al. Br J Nutr 1985;53:31–8.
31. Davis TA, Burrin DG, Fiorotto ML, et al. Am J Physiol 1996;270:E802–9.
32. Denne SC, Rossi EM, Kalhan SC. Pediatr Res 1991;30:23–27.
33. Davis TA, Fiorotto ML. Curr Opin Clin Nutr Metab Care 2009;12:78–85.
34. O'Connor PM, Bush JA, Suryawan A, et al. Am J Physiol 2003;284:E110–9.
35. Suryawan A, O'Connor PMJ, Bush JA, et al. Amino Acids 2009;37:97–104.
36. Escobar J, Frank JW, Suryawan A, et al. Am J Physiol 2005;288:E914–21.
37. Columbus DA, Fiorotto ML, Davis TA. Amino Acids 2015;47: 259–70.
38. Davis TA, Fiorotto ML, Burrin DG, et al. Am J Physiol 2002;282:E880–90.
39. Laplante M, Sabatini DM. J Cell Sci 2013;126:1713–19.
40. Kimball SR. Am J Clin Nutr 2014;99:237S–42S.
41. Hietakangas V, Cohen SM. Annu Rev Genet 2009;43:389–410.
42. Ma XM, Blenis J. Nat Rev Mol Cell Biol 2009;10:307–18.
43. Liu GY, Sabatini DM. Nature Rev Mol Cell Biol 2020;21:183–203.
44. Saxton RA, Sabatini DM. Cell 2017;168:960–76.
45. Boura-Halfon S, Zick Y. AJP Endocrinol Metab 2009;296: E581–91.
46. Manning BD, Token A. Cell 2017;169:381–405.
47. Avruch J, Hara K, Lin Y. Oncogene 2006;25:6361–72.
48. Jhanwar-Uniyal M, Wainwright JV, Mohan AL, et al. Adv Biol Regul 2019;72:51–62.
49. Heng L, Zhang W, Zhou Y, et al. Int J Mol Sci 2016;17:1636–51.
50. Wolfson RL, Sabatini DM. Cell Metab 2017;26:301–9.
51. Hu X, Guo F. Endo Rev 2021;42:56–76.
52. Suryawan A, Davis TA. J Nutr 2018;825–33.
53. Wang X, Proud CG. Physiol 2006;21:362–9.
54. Leplante M, Sabatini DM. J Cell Sci 2013;126:1713–19
55. Saxton RA, Sabatini DM. Cell 2017;168:960–76
56. Proud CG. Cold Spring Harb Perspect Biol 2019;11:a033050.
57. Condon KJ, Sabatini DM. J Cell Sci 2019;132:jcs222570.
58. Jackson RJ, Hellen CU, Pestova TV. Nat Rev Mol Cell Biol 2010;11:113–27.
59. Rudar M, Fiorotto ML, Davis TA. Annu Rev Anim Biosci 2019;7:309–35.
60. Sandri M. Int J Biochem Cell Biol 2013;45:2121–29.
61. Neufeld TP. Curr Opin Cell Biol 2010;22:157–68.
62. Herzig S, Shaw RJ. Nature Rev Mol Cell Bio 2018;19:121–35.
63. Lin SC, Hardie DG. AMPK Cell Metab. 2018 Feb 6;27(2): 299–313
64. González A, Hall MN, Lin SC, Hardie DG. Cell Metab 2020;31:472–92
65. Suryawan A, Orellana RA, Nguyen H. Am J Physiol Endocrinol Metab 2007;293:E1597–605.
66. Chang HC, Guarente L. Trends Endocrinol Metab 2014 25: 138–145.
67. Zhou S, Tang X, Chen XC. Front Endocrinol 2018;9:748.
68. Solon-Biet SM, Mitchell SJ, de Cabo R. J Endocrinol. 2015;226:R17–28.

PART III

Nutritional Assessment and Needs During the Life Cycle

CHAPTER 47

Body Composition

Diana Thomas and Steven Heymsfield

CHAPTER OUTLINE

Abbreviations

3-MH, 3-methylhistidine; **ADP**, air displacement plethysmography; **aLST**, appendicular lean soft tissue; **AT**, adipose tissue; **BCM**, body cell mass; **BD**, body density; **BIA**, bioelectric impedance analysis; **BM**, bone mineral; **BMI**, body mass index; **BV**, body volume; **CT**, computed tomography; **DXA**, dual-energy x-ray absorptiometry; **ECW**, extracellular water; **FFM**, fat-free mass; **FM**, fat mass; **HT**, height; **IBW**, ideal body weight; **ICW**, intracellular water; **LST**, lean soft tissue; **MRI**, magnetic resonance imaging; **NCAA,** National Collegiate Athletic Association; **NHANES**, National Health and Nutrition Examination Survey; **PBF**, percent body fat; **PCA,** principal component analysis; **SAT**, subcutaneous adipose tissue; **SM**, skeletal muscle; **TBK**, total body potassium; **TBN**, total body nitrogen; **TBP**, total body protein; **TBW**, total body water; **UBW**, usual body weight; **VAT**, visceral adipose tissue

Introduction

A person's body composition reflects their net lifetime accumulation of nutrients and effects of physical activities. These components, ranging from elements to tissues and organs, are the building blocks that give mass, shape, and confer function to all living things. Body composition assessment methods allow scientists to measure and describe how these components adapt with age, growth, and metabolic state. Clinicians rely on body composition measurements for diagnosis, determining disease risk and optimizing efficacy of therapies to improve clinical outcomes. Body composition measuring techniques can be divided into simple anthropometric measurements and laboratory-based measures. Simple anthropometric measures height, weight, circumferences (waist, hip, arm), while laboratory-based measures include techniques such as dual-energy x-ray absorptiometry (DXA), bioelectric impedance (BIA), and densitometry. Simple anthropometric measurements are described in more detail in another chapter.

Assessment of body composition is important to understand normal changes in the human body during growth, maturation, and senescence, but also for understanding disease-associated changes. Body composition assessment is often challenging, as there is large variation within and among healthy individuals. In this chapter, a physiologic background of body composition measures will be provided, with in-depth description of methods used to assess body composition.

Body Composition

In the simplest of terms, body composition can be divided into two compartments, fat mass and fat-free mass (FFM). FFM can additionally be divided into three compartments: bone, muscle mass, and intra/extracellular fluid.

Body Fat and Adipose Tissue

FM is the most variable component of body composition, ranging from 6% to 60% of body WT. Baumgartner described age trends.[1] Infants, on average, have approximately 10% to 15% fat at birth, increasing to approximately 30% by 6 months of age, when a gradual decline begins. Between 5 and 8 years of age, a preadolescent adiposity rebound occurs, which continues during adolescence at a rate of approximately 1.4 kg/year in girls and approximately 0.6 kg/year in boys. Percent fat increases from an average of approximately 20% to approximately 26% in girls from age 9 to 20 years; conversely, in boys, it decreases from approximately 17% to approximately 13% after age 13 years, as FFM rapidly increases. Earlier body fat percentage rebound is generally associated with greater adolescent and adult disease risk.[2]

Total body fat increases slowly with age during young and middle adulthood. The rate of increase varies by sex and possibly, race and ethnicity, although comprehensive data are lacking. Estimates vary from approximately 0.37 kg/year to 0.52 kg/year in women and 0.37 kg/year to 0.57 kg/year in men. These estimates are, undoubtedly, affected by methodologic differences, and some data suggest a lower rate in middle-aged men (45 to 66 years) compared with younger men (18 to 45 years). Maximum body fat is reached between 50 and 65 years of age, and thereafter, body fat is maintained until losses occur in old age.[3] Healthy ranges have been proposed based on reviews of the literature, expert opinion,[4] and estimation of percent fat values corresponding to BMI-based definitions of overweight and obesity in adults, based on the BMI–% fat relationship (**Table 47.1**).

Laurson et al[5] developed percent fat standards for children and youth based on the association between body fat and chronic disease risk (**Table 47.2**). Similar analyses are needed in adults.

Fat Distribution

Fat distribution refers to the relative amounts of fat in primary compartments where AT and fat are stored. Main depots include the subcutaneous and intra-abdominal compartments, with small amounts in intermuscular and intramuscular stores.[6] Changes in subcutaneous AT (SAT) and fat are described best according to their accessibility, although the emergence of methods such as CT and MRI made it possible to measure internal fat compartments such as intra-abdominal AT and fat as well as fat in non-AT locations, such as liver and perivascular and intramuscular locations.

Assessment of fat distribution often entails measuring one site or variable in relation to another so that a dichotomous fat distribution "type" can be identified, as in gynoid or android distribution. The term *fat pattern* has been used to refer to SAT distribution to distinguish it from the accumulation of internal fat. Age, sex, and ethnic or racial differences in fat patterns have been studied frequently, using anthropometric measures and various indexes and ratios.[7] In general, SAT increases on the trunk in boys during adolescence, and gluteal fat increases in girls, thus leading to distinct male (android) and female (gynoid) fat patterns characteristic of adulthood. These changes are associated with sexual maturity, sex hormone levels, and changes in plasma lipid and lipoprotein cholesterol concentrations. In adults of either sex, an android fat pattern is associated with a spectrum of metabolic risk factors for chronic disease, including hypercortisolism, hypercholesterolemia, hypertension, and insulin resistance. The risk for metabolic disorders is more related to the

Table 47.1 Percent Body Fat Ranges for Adults

	Men			Women		
Age[a] (y)	**Minimal**	**Recommended**	**Obese**	**Minimal**	**Recommended**	**Obese**
<34	<5	8–22	>25	10–12	20–35	>35
35–55	<5	10–25	>28	10–12	23–38	>40
>56	<5	10–25	>28	10–12	25–38	>40
Age[b] (y)	**Men**			**Women**		
20–39	8	9–19	25	21	22–33	39
40–59	11	12–21	28	23	24–34	40
60–79	13	19–24	30	24	25–36	42

[a]Data from Lohman TG. Advances in Human Body Composition. Champaign, IL: Human Kinetics; 1992, based on literature review, expert opinion.
[b]Data from Gallagher D, Heymsfield SB, Heo M et al. Am J Clin Nutr 2000;72(3):694–701, derived from regression of percent body fat on BMI.

Table 47.2 Health-Related Percent Body Fat Standards for Children and Youth

	Girls			Boys		
Age (y)	HFZ	NI: Some Risk	NI: High Risk	HFZ	NI: Some Risk	NI: High Risk
5	9.8–20.8	20.9	≥28.4	8.9–18.8	18.9	≥27.0
6	9.9–20.8	20.9	≥28.4	8.5–18.8	18.9	≥27.0
7	10.1–20.8	20.9	≥28.4	8.3–18.8	18.9	≥27.0
8	10.5–20.8	20.9	≥28.4	8.4–18.8	18.9	≥27.0
9	11.0–22.6	22.7	≥30.8	8.7–20.6	20.7	≥30.1
10	11.6–24.3	24.4	≥33.0	8.9–22.4	22.5	≥33.2
11	12.2–25.7	25.8	≥34.5	8.8–23.6	23.7	≥35.4
12	12.7–26.7	27.8	≥35.5	8.4–23.6	23.7	≥35.9
13	13.4–27.7	26.8	≥36.3	7.8–22.8	22.9	≥35.0
14	14.0–28.5	29.2	≥36.8	7.1–21.3	21.4	≥33.2
15	14.6–29.1	29.8	≥37.1	6.6–20.1	20.2	≥31.5
16	15.3–29.7	30.5	≥37.4	6.5–20.1	20.2	≥31.6
17	15.9–30.4	30.5	≥37.9	6.7–20.9	21.0	≥33.0
>17	16.5–31.3	31.4	≥38.6	7.0–22.2	22.3	≥35.1

HFZ, Healthy Fitness Zone; NI: Some risk, needs improvement, some risk; NI: High risk, needs improvement, high risk.

Data from Laurson KR, Eisenmann JC, Welk GJ. Am JPrevent Med 2011;41: S93–9. doi:10.1016/j.amepre.2011.07.003, with permission; criterion-referenced (presence or absence of metabolic syndrome) body fat standards used in Fitnessgram, health-related fitness assessment (www.cooperinstitute.org/youthfitnessgram)[5]

intra-abdominal or VAT, compared with an SAT pattern. Many studies have used waist, waist-to-hip, or other circumference ratios as surrogates to describe VAT,[8] although longitudinal data on representative samples are lacking. Some data suggest that waist circumference increases more in women (0.28 cm/year) than in men (0.18 cm/year) over their lifespan, although men have greater waist-to-hip ratios relative to total body fat than women.

Waist circumference and waist-to-hip ratio are fairly insensitive measures of visceral adiposity, especially for detecting changes and in elderly populations. Nevertheless, the few studies that have used imaging techniques to describe age-related changes in VAT have generally supported patterns described using their anthropometric surrogates.[1] VAT increases with age, especially during middle age. Women have a marked increase during menopause. During old age, absolute amounts of VAT remain fairly stable, although VAT may increase relative to total body fat because the latter declines during senescence. Despite its limitations, waist circumference is widely used as a surrogate for VAT, and reference data have been developed. In combination with BMI, waist circumference predicts disease risk better than either measure alone (see Table 47.2).

As obesity develops, the increase in size and number of adipocytes leads to lipid deposits within and around organs and tissues involved in energy metabolism, as excess lipids are redirected to the liver, skeletal muscle (SM), heart, blood vessels, and the pancreatic β cells. This type of ectopic fatty infiltration contributes to increased risk of diabetes and cardiovascular disease through the secretion of relaxing factors, proatherogenic cytokines, and smooth muscle cell growth factors and by increasing vascular stiffness, thereby altering the flow of blood and lymph.[1]

Fat Free Mass

FFM is a heterogeneous compartment on the chemical level of analysis. Its primary constituents of intracellular and extracellular fluid, protein, and osseous and non-osseous minerals can be combined to form various models on which assessment methods are based (see Table 47.1). FFM can be projected from body density (BD) estimated by underwater weighing,[9] TBK estimated by whole body counting,[10] and total body water (TBW) estimated by hydrometry.[11] Each approach relies on a conversion factor based on an assumption of a constant relationship between the measured component and FFM. In healthy young adults, the assumption of chemical constancy introduces relatively little error. However, significant changes in FFM components with growth and maturation, aging, and illness are well described[1,12-14] and introduce significant error unless appropriate adjustments are made. Sex and race or ethnicity differences are known[15] as well as effects of physical training.[16] It is imperative that use of constants and equations be restricted to the groups for which they were developed, unless their validity in another group has been shown. Alternatively, and especially when the condition of steady state is not met, application of multiple component models improves accuracy,[17] although the requirement for more measures increases cost and patient burden, thus limiting their use.

Skeletal Muscle Mass

SM comprises approximately 30% to 40% of weight in a healthy woman, and 40% to 50% of weight in a healthy man.[18] In adults, most SM is found in the legs, with lesser amounts in the head, trunk, and arms.[18] Assessment methods historically have emphasized body fat, reflecting interest in a compositional variable related to chronic disease risk, especially heart disease and type 2 diabetes. Interest in SM has increased as greater appreciation for its importance for health and physical function emerged. Clinicians require estimates of SM to evaluate the progression of catabolic disease, prognosis, and therapy. Geriatricians require longitudinal assessments of SM to monitor muscle loss with aging, its functional effects, and the efficacy of programs aimed at maintaining SM, mobility, and quality of life for elders.

Imaging methods such as computed tomography (CT) and magnetic resonance imaging (MRI) are considered the most accurate means available for in vivo quantification at the tissue–organ level.[19] Protocols have been developed for estimating whole body or regional SM from multiple CT or MRI images. Because acquisition and analysis of contiguous images over a given region are expensive and time consuming, axial images are typically collected with interslice gaps (~20 to 40 mm). Volumes are then calculated using geometric models based on tissue areas and the distance between images. Because densities for tissues and organs are reasonably constant from person to person, CT and MRI volume measures can be converted to mass by multiplying the volume by the density for the tissue of interest.[19]

Some caution is needed when using CT to assess SM in patients. Normal muscle attenuation varies considerably, depending on which muscle is examined.[20,21] Muscle density is assumed constant at 1.04 g/cm^3.[22] Individual variation is small in healthy individuals. However, clinical conditions that change muscle density can confound conversion of muscle volume to mass. Appendicular LST (aLST), measured by DXA, is composed mainly of SM, skin, and soft connective tissue. Research has shown that aLST closely approximates the SM of arms and legs, and this method has been proposed for screening for sarcopenia in older individuals.[23] Longitudinal observations of change in aLST in healthy and clinical populations and its relationship with function are needed. DXA aLST has significant potential for following changes in appendicular SM leading to sarcopenia, impaired function, and mobility disability.

On the molecular level, endogenous components or metabolites of SM metabolism are used to estimate whole body SM. Two metabolites, creatinine and 3-methylhistidine (3-MH), have been used. Use of either marker assumes that it is found only in SM, the size of the marker pool is constant, turnover rate is relatively unchanged over long periods of time, and the compound is not further metabolized. These assumptions are not strictly valid. Data from different studies suggest that 1 g of creatinine during a 24-hour period is derived from approximately 18 to 20 kg of muscle.[24,25] This range undoubtedly reflects differences in muscle sampling and other methodologic variation between studies. There is large intraindividual variability in daily urinary creatinine (11% to 30%) for individuals consuming self-selected diets. Diet affects the creatinine pool, and urinary creatinine excretion may be somewhat independent of body composition. The requirements for a controlled, meat-free diet and repeated urine collections are other practical concerns. Similar concerns exist for estimation of SM from 3-MH excretion. Although SM has the greatest 3-MH concentration (3 to 4 μMol/g fat-free, dry WT), intermediate concentrations exist in cardiac and some smooth muscle tissues (1 to 2 μMol/g), and low concentrations (<1 μMol/g) are found in spleen, liver, and kidney.[26] The 3-MH concentration in muscle appears relatively constant (3 to 4 μMol/g) between ages 4 and 65 years.[27] Thereafter, it decreases with age, thus reflecting either a decreased 3-MH turnover with age or decreased SM.[28] Many studies have shown that urinary 3-MH reflects SM assessed directly with nuclear methods or indirectly with a surrogate indicator,[29] although concern has been raised over the potential influence of non-SM protein turnover (e.g., from skin and gastrointestinal tract) on excretion of 3-MH.[30]

Skeletal Muscle Tissue Quality

Changes in SM quality with aging, along with muscle atrophy, contribute to the functional consequences of sarcopenia and add to risk of chronic disease. Both CT and MRI can be used to measure SM tissue composition. X-ray attenuation depends on molecular composition. Because lipid has a lower density than water and protein, attenuation for AT is lower than for lean SM.[31,32] Consequently, marbled or intermuscular AT can be separated from lean SM, based on the differences in attenuation characteristics. Also, the average attenuation for AT-free SM voxels can be used as an index of SM lipid content. With the application of chemical shift methods, MRI can also be used to measure quality of muscle[33,34] although, as with CT, it is not possible to separate the lipid measurement into intramyocellular and extramyocellular compartments. Partitioning the lipid signal into separate compartments can be accomplished using proton magnetic resonance spectroscopy,[35,36] and this method has been used to describe SM lipid distribution (e.g., intramyocellular and extramyocellular lipid) and its relation to insulin resistance.[37-39] CT and MRI have also been used to show the replacement of muscle with adipose and connective tissue with aging. Investigators have recognized for some time that age-related changes in SM may interact with changes in other body composition elements, particularly body fat. Changes in fat and FFM are correlated and generally occur in a constant proportional relationship with weight change (~70% fat to ~30% FFM).[40] However, with aging, this relationship may become dysregulated, resulting in discordant changes in lean and other soft tissue components, leading to sarcopenic obesity[41,42] or low muscle mass in the presence of high body fat.

Bone Mineral Mass, Content, and Density

The skeleton provides structure, mobility, support, and organ protection, and acts as a reservoir for essential minerals. More than 80% of BM is found in the skeleton, and low bone mass, due to aging or other reasons, has significant functional consequences. Historically, osteoporosis, low bone mass leading to fragility fractures, has been a concern in middle-aged and elderly persons; and slowing bone loss with aging has been the main prevention strategy. The important role of optimal bone development on future fracture risk is now better appreciated. More than 90% of BM is accrued during childhood and adolescence.[43] Attainment of peak bone strength during adolescence is arguably the best prevention for osteoporosis and fracture later in life.

Bone, muscle, and fat are interdependent. Total body mass influences bone through independent effects of fat and LST mass. Although muscle's positive effect on bone is clear, the relationship between adiposity and bone remains uncertain. Mechanistically, fat and bone are linked by multiple pathways[44] and, conceivably, fat serves the function of modifying the skeleton to support increases in body mass. However, more recent studies have shown that VAT and intramuscular fat are unique pathogenic fat depots that are inversely associated with bone strength.[45,46] Patients with type 2 diabetes have more fragility fractures,[47] suggesting that impaired glucose tolerance and insulin resistance have negative consequences for the skeleton.

Historically, skeletal assessment was limited, given no verified in vivo method for quantifying bone tissues. Development of newer technologies, especially DXA, have facilitated description of bone as a component of body composition and made it possible to follow changes in skeletal mineral status.

Body Composition Measuring Methods

Anthropometric Measures

Body composition methods can traditionally be divided into anthropometric measurements and laboratory methods. Anthropometric measurements are noninvasive methods to assess body composition, while laboratory methods are typically more invasive. However, each category has their advantages and disadvantages. With anthropometric measures, these are often more feasible with portable and inexpensive equipment; however, laboratory methods provide higher validity and more accurate measurement. Anthropometric measures typically include height, weight, waist circumference, hip circumference, skinfold thickness, and indices across these variables, such as Body Mass Index, Waist-to-Hip Ratio, Waist-to-Height ratio, and more. In this chapter, anthropometric measures will be briefly reviewed. More in-depth use and interpretation of these measures can be found in Chapter 48. Laboratory methods include densitometry, hydrometry, dual-energy, MRI, CT, and BIA.

Height

Height measures an individual's linear growth and stature. Height attainment is influenced by genetic and environmental factors. Measuring height is important to ensure proper growth and development of children, detecting malnutrition, but also to help predict future disease risk in adults.[48,49] Height is often used in adults to monitor height loss, which can be predictive of osteoporotic fractures, with consistent height monitoring, diagnosis can occur earlier, allowing for better outcomes.

Height can be used to calculate ideal body weight (IBW), which was developed by Actuarial Society of America and the Association of Life Insurance Medical Directors of America based on mortality of insured persons from 1885 to 1908.[50] Since then, it has received much scrutiny in the healthcare setting for the methods used to develop the measure and population used. However, in the clinical setting, it may still be used to create feeding plans, avoiding malnutrition, identifying transplant eligibility, and is used in bariatric surgery settings. When using IBW, it is important its use is accompanied with clinical judgment and evaluation of a variety of factors.[50]

Body Weight

Body weight is the measure of the mass of an individual's body. Weight is impacted by a variety of factors, including body frame, sex, genetics, medical history, and lifestyle habits. Body weight typically varies less than $\pm$1–2 kg/day in weight-stable adults.[51] Weight fluctuations are due to short-term impacts like sleep, water retention, diet, exercise, and illness. IBW may be useful in certain settings; however, an individual's usual body weight (UBW) (rather than IBW) may provide additional information that is useful for evaluating an individual's nutritional status. The difference between current and UBW or IBW may be compared with clinical parameters to determine risk of morbidity and mortality. A clinically significant rate of weight loss as an indicator of moderate malnutrition is considered to be 1% to 2% over 1 week, 5% over 1 month, 7.5% over 3 months, or 10% or greater over 6 months.[52] Severity of weight loss may also be evaluated by the absolute weight reduction, which also has prognostic value. An absolute weight of 85% to 95% of UBW (or 80% to 90% of IBW) indicates mild malnutrition, 75% to 84% of UBW (or 70% to 79% of IBW) indicates moderate malnutrition, and 75% or less of UBW (or $\leq$69% of IBW) indicates severe malnutrition.[52]

Absolute weight reduction to less than 55% to 60% of IBW places an individual at the limits of starvation.[53] This is based on unintentional loss and can be assessed for malnutrition-related weight loss that occurs over a period longer than 6 months. In infirm individuals, a weight loss between 10% and 20% of preillness weight over 6 months has been associated with functional abnormalities,[54] whereas a loss of more than 20% of preillness weight suggests significant protein-energy malnutrition.[54] The minimum survivable body weight in humans is between 48% and 55% of IBW or a BMI of approximately 13 kg/m^2. Overconsumption of energy relative to expenditure results in a positive energy balance which, if sustained, leads to weight gain and excess adiposity. The maximum survivable body weight is approximately 500 kg (a BMI of ~150 kg/m^2).[55]

When using weight as an estimate of energy and protein needs, clinicians must consider factors that affect weight fluctuations or otherwise confound the assumption that weight is a surrogate of energy stores, such as rapid fluid shifts (intracellular to extracellular or intravascular to extravascular spaces) and accumulation of fluid secondary to inflammation. Edema and ascites and the medications used to treat them may cause fluid to shift to extracellular spaces, masking body composition changes and artificially increasing weight. Tumor growth or abnormal organ enlargement in disease states may cause an increase in weight and mask tissue loss (i.e., loss of fat or FFM). Individuals who are attempting weight loss must also be monitored, as rapid loss may increase risk of health complications due to severe protein-energy malnutrition and semistarvation. Finally, physical activity or diet-induced changes in energy intake and expenditure affect glycogen mass (which is bound to water), and body sodium, which is associated with fluid readjustment and weight fluctuations.

Body Mass Index

Ratios of weight to height have a long history in studies of body habitus. BMI (kg/m^2) is the most commonly used index because height squared minimizes the relationship between height and weight, at least in adults. Although not a direct measure of adiposity, BMI is a widely used surrogate for composition, based on the weak assumption that excess weight results from body fat. Although BMI and body fat are correlated, use of BMI as an "adiposity" index is confounded by differences in body proportions (e.g., trunk-to-leg length ratio), fat distribution, and composition, relative to height. Individuals with greater than average muscularity, for example, may be misclassified as overweight or obese. In addition, composition and location of excess weight vary with sex, race, and age, which is information not captured by BMI.[56] Despite these limitations, BMI predicts disease risk and standard definitions of overweight and obesity are in use (Table 47.2).

Laboratory Methods of Body Composition Assessment

Densitometry

Historically, hydrodensitometry (underwater weighing) was used to estimate body volume (BV) and BD, which was converted to estimates of PBF and FFM.[57] For young children, the elderly, those with certain disabilities, and other special populations, complete submersion in water is very difficult, if not impossible. An alternative approach, air displacement plethysmography (ADP), uses pressure–volume relationships to estimate BV and BD. The most recent form of ADP, the Bod Pod (COSMED USA, Inc. [formerly Life Measurements, Inc.], Concord, CA), provides a reliable means of determining BV[58,59] and eliminates the need for submersion in water. The procedure can be performed by children and adults, although it does require a breathing maneuver to measure thoracic gas volume that may be difficult for young children and patients with pulmonary disease.

A major source of error in densitometry is the model used to convert BD to composition. In the classic two-component model, the densities of fat and FFM are assumed to be 0.9 and 1.1 g/mL, respectively. Using these densities, it is possible to derive an equation for estimating percent fat from BD (see Table 47.1). The density of FFM is derived from its primary constituents, water, protein, and mineral, as well as their respective fractions and densities (**Table 47.3**). The more closely the FFM components and their densities fit the individual being measured, the more accurate the result will be.

Many studies have demonstrated considerable variation in FFM composition and thus density attributed to growth and maturation,[65] aging,[66] and specialized training.[67] Sex and racial differences also exist and, even within a population, considerable interindividual variation[68] invalidates the assumption of FFM chemical constancy. Consequently, three-component and four-component models (see Table 47.1), which require fewer assumptions because more components are measured, are more accurate than the two-component model. In children and in patients with edema, combining a measure of TBW along with BD significantly improves estimation of FFM; similarly, in elderly patients and in patients with significant bone loss, combining a measure of body mineral with BD gives a more accurate estimate of FFM. When a multicomponent model is not feasible, accuracy can be improved by using a population-specific equation that has been adjusted for the anticipated changes that occur with growth, maturation, and aging (**Table 47.4**).

Table 47.3 Fat Free Body Composition and Density in Children and Youth

	Male			Female		
Age (y)	**TBW/FFM (%)**	**M_o/FFM (%)**	**DFFM (g/cc)**	**TBW/FFM (%)**	**M_o/FFM (%)**	**DFFM (g/cc)**
1	79.0	3.7	1.068	78.8	3.7	1.069
1–2	78.6	4.0	1.071	78.5	3.9	1.071
2–4	77.8	4.3	1.075	78.3	4.2	1.073
5–6	77.0	4.8	1.079	78.0	4.6	1.075
7–8	76.8	5.1	1.081	77.6	4.9	1.079
9–10	76.2	5.4	1.084	77.0	5.2	1.082
11–12	75.4	5.7	1.087	76.6	5.5	1.086
13–14	74.7	6.2	1.094	75.5	5.9	1.092
15–16	74.2	6.5	1.096	75.0	6.1	1.094
17–20	74.0	—	1.098	74.8	—	1.095
20–25	73.8	—	1.100	74.5	—	1.096

DFFM, density of fat free mass; FFM, fat free mass; M_o, osseous mineral as a fraction of weight; TBW, total body water.

Data from Boileau RA, Lohman TG, Slaughter MH. Scand J Sports Sci 1987:17.[60]; Fomon SJ, Haschke F, Ziegler EE et al. Am J Clin Nutr 1982;35:1169–75.[52]; Haschke F, Fomon SJ, Ziegler EE. Pediatr Res 1981;15:847–9.; Haschke F. Acta Paediatr Scand 1983;307(Suppl):11.[61,62]; Lohman TG, Boileau RA, Slaughter MH. Body composition in children and youth. In: Boileau RA, ed. Advances in Pediatric Sport Sciences. Champaign, IL: Human Kinetics, 1984:29–57.; Lohman TG, Slaughter MH, Boileau RA et al. Hum Biol 1984;56:667–79. with permission, with some modifications of the estimates of Fomon SJ, Haschke F, Ziegler EE et al. Am J Clin Nutr 1982;35:1169–75. to provide for linear changes in body water and bone mineral with age.[63,64]

Table 47.4 Equations[a] for Estimating Percentage of Fat from Body Density in Children and Youth

	Males		Females	
Age (y)	C_1	C_2	C_1	C_2
1	5.72	5.36	5.69	5.33
1–2	5.64	5.26	5.65	5.26
3–4	5.53	5.14	5.58	5.20
5–6	5.43	5.03	5.53	5.14
7–8	5.38	4.97	5.43	5.03
9–10	5.30	4.89	5.35	4.95
11–12	5.23	4.81	5.25	4.84
13–14	5.07	4.64	5.12	4.69
15–16	5.03	4.59	5.07	4.64
18	4.95	4.50	5.05	4.62

[a]Adjusted for average changes in water, protein, and mineral fractions of fat-free mass with growth and maturation (see Table 47.3). The C_1 and C_2 terms are derived from the general equation for estimating the percentage of fat from body density:
$\%Fat = 1/BD\,[(d\downarrow1\ d\downarrow2)/(d\downarrow1 - d\downarrow2)]–[d\downarrow2/(d1 - d2)] \times 100$ where BD = body density, $d\downarrow1$ = density of fat-free mass, and $d\downarrow2$ = density of fat = $0.90\downarrow$(g/cc) for all age groups,
$\%Fat = [C\downarrow1/BD - C2] \times 100$

Whole Body Counting and Total Body Potassium

Potassium is mainly an intracellular ion that can be measured by whole body potassium counting.[10] Reproducibility of measurements is good, even in small children.[69] TBK is useful for estimating BCM, TBP,[70-72] and SM mass,[70,71] although it is more commonly used to estimate FFM using a two-component model,[56] in which TBK resides in FFM. This model assumes a stable TBK/FFM ratio; however, as the SM fraction of FFM increases during growth, the TBK/FFM ratio increases. These changes lead to complexities in developing appropriate model coefficients. In young, healthy adults, the TBK/FFM ratios in men (2.66 g, K/kg FFM) and women (2.55 g, K/kg FFM) are well established and reasonably stable. In children, use of the adult TBK/FFM ratio underestimates FFM.[72] In older adults, and the infirm with muscle loss (sarcopenia), a similar problem of underestimation occurs.

Hydrometry and Total Body Water (TBW)

On the molecular level, the water compartment consists of a singular molecular species, hydrogen oxide, which lends itself to the use of the isotope dilution principle for assessment of TBW. Although several tracers have been used, isotopes of water (radioactive tritium oxide, deuterium oxide, and oxygen-18 hydride) provide the most precise and accurate estimates of TBW.[14] TBW is used in models to estimate body composition on the molecular, cellular, and tissue levels (see Table 47.1), although estimation of FFM based on the two-component model that restricts all water to FFM is the most common component estimated from TBW. Its calculation assumes constant hydration of FFM. This assumption is incorrect in individuals who either are dehydrated or have abnormal water metabolism leading to edema. Among healthy adults, TBW is relatively constant and FFM is estimated assuming its water proportion is 73% (FFM = TBW/0.73). Infants and children have higher TBW/FFM ratios, and age-appropriate hydration constants must be used to estimate FFM (see Table 47.3). Malnourished patients with severe protein depletion may have hydration factors as high as 75%,[73] and disease states that alter water metabolism producing edema also result in higher hydration constants.[74] Some metabolically healthy groups also have higher TBW/FFM; for example, bodybuilders with expanded SM compartments have hydration constants elevated by 2% to 3%.[75] This occurs not because of greater hydration but because of a larger SM fraction of FFM. Pregnancy also results in an increase in TBW that depends on the trimester.[76]

Dual-Energy X-Ray Absorptiometry (DXA)

DXA, a way of measuring bone density using x-ray energy beams, is often available in clinical settings and easily performed on most persons, thus making it an attractive method for clinical studies. The scans take about 10 to 20 minutes and radiation exposure is low; thus it is able to be used in children, although very young children may need to be sedated. The main limitations are the weight limits of the scanners, errors related to larger patient size, and the cost of the machine to implement.[67] Also, hardware and software differences exist across scanners, even from the same manufacturer, and longitudinal studies must be done using the same scanner and software.[67]

The DXA method provides estimates of three main components: fat, lean soft tissue (LST), and BM. The LST includes two primary components at the cellular level, BCM, and extracellular fluid. FFM is estimated as the sum of LST and BM. Both LST and BCM increase with age, although at different rates, such that the BCM component increases relative to LST with greater age during development.[77,78] Thus, LST is not metabolically homogeneous with respect to age in children, and results must be interpreted accordingly.

Like all indirect methods, DXA relies on assumptions of tissue constancy that may not always be accurate. Estimation of soft tissue composition and BM depends on attenuation ratios (R values) that are assumed stable for specific components, such as fat and FFM. In vitro studies with homogeneous materials and in vivo comparisons against multicomponent models demonstrate that correlation values may systematically change, as bone thickness or depth varies.[79-81] Consequently, percent fat may be overestimated in patients with higher percent fat and underestimated in those with lower percent fat.[78] Variation in body water from the assumed TBW/FFM ratio may also confound DXA estimates of FFM and percent fat, although considerable deviation is necessary for influence because an increase or decrease in hydration of 5% biases DXA estimates of percent fat by only 1% to 2.5%.[76,82]

Bioelectric Impedance Analysis (BIA)

Impedance is a measure of the opposition of electrical flow. In the body, water is the conductor of electrical flow. As different compartments of the body have different amounts of water, BIA can be used to analyze the overall body composition based on its conductivity. The electric current flows differently through the extracellular water (ECW) and intracellular water (ICW) as a function of the current frequency. At low frequencies, the current flows through the ECW; and at high frequencies, the current penetrates all tissues. Thus, using different frequencies, it is possible to estimate different fluid spaces. Both single and multiple frequency analyzers have been developed.

Single-frequency analyzers use a weak, alternating current (e.g., 800 μA, 50 kHz) that penetrates the TBW space. Measurements of resistance and reactance can be done at bedside with the patient lying supine. Handheld devices and weighing scales that measure impedance are also available. The underlying theory is that resistance (R) is directly proportional to the length (L) of the conductor, and inversely proportional to its cross-sectional area (A). Application of BIA assumes that the conductor is a perfect cylinder and that volume resistivity is constant. Neither is true. Resistivity varies from tissue to tissue, and the arms and legs contribute most of the resistance of the body. Variation in specific resistivities among tissues and segments and among individuals, because of differences in tissue composition, account for some of the inter-individual differences and predictive errors. Other variables that affect the measurements include body position, hydration status, consumption of food and beverages, ambient air and skin temperatures, recent physical activity, and bladder activity. Standard protocols controlling for these factors must be followed.[83]

Bioelectric impedance measures do not directly measure any biologic quantity of interest. Instead, the resistance index, HT^2/R, is used as a predictor in regression equations. These equations describe statistical relationships found for a particular population, and each equation is useful only for individuals who are a close match to the reference population. Numerous prediction equations for estimating TBW, FFM, and PBF from BIA have been published[83,84] (**Table 47.5**).

Estimation of TBW using single-frequency BIA is reasonably accurate,[85] but its use to estimate FFM and percent fat depends on a constant TBW/FFM ratio (73%). In part to counter errors related to invalid assumptions, HT^2/R is used with anthropometric measures to predict body composition.[84] Overall, published equations provide reasonably accurate body composition estimates for groups, but their accuracy for individuals depends on numerous factors specific to the construction of the equations.[87,88]

The clinical uses of body composition, as assessed by BIA, frequently concerns conditions where water distribution is disturbed,[89] such as in cancer,[90] human immunodeficiency virus infection,[13] and dialysis.[88] Disturbances of ICW are characteristic of protein-energy malnutrition, and direct or indirect measures of TBW do not reliably reflect FFM in these conditions.[91,92] Single-frequency BIA is likely invalidated as an assessment of response to parenteral and enteral nutrition in these patients. Also, the ability to predict body fat percentage in severely obese persons remains a problem because a greater proportion of body mass and water in the trunk, greater FFM hydration, and an increased ECW/ICW ratio. Acute weight changes resulting from dieting or infusion, and acute loss attributable to protein-energy malnutrition are also not reliably detected by single-frequency BIA.

Table 47.5 Equations for Predicting Body Composition from Bioelectric Impedance and Anthropometry

Males 12–80 y
TBW = 1.203 + 0.176 (WT) + 0.449 (HT²/R)
FFM = −10.678 + 0.262 (WT) + 0.652 (HT²/R)
Females 12–80 y
TBW = 3.747 + 0.113 (WT) + 0.45 (HT²/R) + 0.015 (R)
FFM = −9.529 + 0.168 (WT) + 0.668 (HT²/R) + 0.016 (R)
Adults 18–86 y
SMM = 5.102 + 0.401 (HT²/R) + 3.825 (gender) − 0.07 (age)

FFM, fat free mass; HT, height (cm); R, resistance (ohms); SMM, total skeletal muscle mass; TBF, total body fat=weight−FFM; TBW, total body water; WT, weight (kg).

Data from Sun SS, Chumlea WC, Heymsfield SB, et al. Am J Clin Nutr 2003;77:331–40[85]; Janssen I, Heymsfield SB, Baumgartner RN, et al. J Appl Physiol 2000;89:465–71,[86] with permission.

Total Body Water (TBW)

Water is essential, serving as a solvent for biochemical reactions and as a transport media. A 15% decrease in body water attributable to dehydration is life threatening. Even a small change in TBW can produce a measurable change in body weight and, thus, determining TBW is central to measuring body composition.

Water is an important constituent of models at the molecular, cellular, and tissue levels of models describing body composition.[93] On the molecular level, water consists of a single molecular species, hydrogen oxide. On the cellular level, water is found in two compartments: the BCM, which is approximately 73% water and 27% solids[94]; and the extracellular fluid compartment, which is approximately 94% water and 6% solids.[93] On the tissue level, water resides in five compartments: ICW, in the cytoplasm and nucleus of every tissue; plasma; interstitial water, in the lymphatic system; connective tissue water, which includes the water found in bone, cartilage, and other dense connective tissues; and transcellular water, a diverse collection of largely excretory extracellular fluids such as bile, gastrointestinal secretions, mucus, cerebrospinal fluid, and other minor components.[95]

As noted earlier, multiple frequency impedance is capable of discriminating fluid compartments and has the potential clinically for providing estimates of TBW, ECW, and ICW.[88]

Advances in Body Composition Measurements Accessible to a Wide Community

Despite body composition measurements being a staple of nutrition and obesity studies, body composition is rarely measured in clinical settings. Even simple methods, such as measuring circumferences with a tape measure or skinfold assessments, are rarely applied in primary care settings even though the connection between body composition and cardio-metabolic health is well established.[96] In addition, funded research of large cohort studies like the Nutrition for Precision Health Program[97] benefit from a balance between accuracy and easily obtainable body composition measurements. A similar need arises from tracking percent body fat in elite athletes quickly and efficiently on site.[98,99] Finally, public consumption of personal health metrics to gauge health and fitness status[100,101] have generated comparable demands.

Unfortunately, research grade measurements using DXA, MRI, CT, Whole Body Counting and Total Body Potassium, and hydrometry require specialized expensive equipment and licensed technical operators, and add a labor-intensive burden to staff and participants or patients, which are unfeasible for measurement in large cohort studies or for public use. This gap calls for practical, cost-effective, and scalable body composition measurement technologies, which has led to new state-of-the-art methods for measuring body composition.

Digital Photography

The Pew Research Center estimates that 71% of the U.S. population owns a smartphone,[102] which translates to widespread readily available camera access. Both smartphone and standalone cameras have been used to estimate body composition with good, comparable accuracy to DXA.[103,104]

Recently,[103] a principal components analysis (PCA) was applied to 3D optical scans and the components were applied to predict DXA measured body composition. Participants were then photographed in the front using a 2D camera. The PCA results were matched to the 2D camera silhouettes, which then were used to transitively predict body composition. An independently validated 2D camera predicted PBF against DXA measurements and resulted in a coefficient of determination of $R^2 = 0.81$ for males ($n = 31$) and $R^2 = 0.74$ for females ($n = 39$).

Another recent study[104] photographed 97 children and 226 adults twice; one photograph from the front and one from the back. A 3D rendered body volume and body shape were developed from the dual 2D photographs and volume and body shape features were combined with age, race, sex, and BMI in a support vector machine model to predict DXA measured body composition. High correlations between 2D-predicted PBF and DXA measured body fat were observed in both adults and children.

These small-scale studies demonstrate that camera-based methods combined with machine learning predictions have good agreement with DXA measured body composition. More research is needed to test this emerging technology in larger, more heterogeneous populations; however, commercial products like the Amazon Halo™[105] are already disseminating camera-based body composition technology to a large consumer base.

3D Body Image Scanners

Three-dimensional body image scanners use different technologies to render 3D avatars within seconds. The avatars can then be distilled into anthropometric measurements, which are being used to predict several body shape measurements, including body composition. Advances in 3D body image scanners have resulted in commercially available devices that are affordable for individual use. These 3D rendering technologies can be classified into three groups: (1) optical methods using passive stereo, structured light, or time-of-flight technologies, (2) laser triangulation technology, as previously described, and (3) 2D photographs with reconstruction methods, including feature extraction, to create a 3D human avatar.

Several different 3D body shape scanner models have been used in academic research,[60-63,106-110] the clothing industry,[64,111,112] and the fitness industry.[113] While some models, like the Human Solutions scanner[112] are stationary, expensive, and require set-aside space, newer models, like the Fit3D,[61] have reasonable costs and are portable.

A comparison between manually measured, clinically based anthropometrics and 3D body shape scanners has demonstrated good agreement with 3D body scanning device measurements.[114-116] Several comparisons have been made between DXA measured body composition and 3D body shape scanning measures. The ShapeUp! study ($n = 188$) found a small mean difference between FFM (1.2±3.4 kg) and FM (1.3±3.4 kg). Another study[106] compared four different commercially available scanners to body composition measured using the four-compartment method.[117] All four scanners had low root mean square error coefficients of variation, ranging from 2.5–4.3% for PBF and 0.7–1.4% for FFM.

Newer 3D body scanning models leverage smartphone camera technology to render 3D avatars. One recent study[118] compared two smartphone 3D body scanning apps to four compartment measured body composition measurements. One of the apps, Amazon HALO, had good agreement of PBF with the four compartment measurements (mean differences <±1.0 kg for FM and FFM). The coefficient of determination for PBF between these apps and DXA was $R^2 = 0.96$.

Smartwatch BIA Technology

Smartwatch BIA technology collects BIA measurements when fingers touch electrodes on the side of the watch. With this technology, body composition measurements are estimated within 15 seconds. Like other new body composition technologies, smartwatch BIA technology is fairly new and not well studied. One study[119] compared smartwatch BIA measurements to standard BIA and DXA ($n = 75$). Compared with DXA measurements, smartwatch-measured FFM correlated well ($R^2 = 0.96$). For PBF, there were similar findings ($R^2 = 0.87$). There were significant differences in FFM and PBF measurements from the smartwatch, compared with DXA, and the study found that the agreement was not as precise as clinical research grade measurements.

A second study[120] evaluated smartwatch bioimpedance and standard bioimpedance against four compartment model measurements[117] in 186 participants. Total body water estimates from the smartwatch and standard BIA were compared with measurements from bioimpedance spectroscopy. While both the smartwatch and the standard BIA showed agreement with the four-compartment method, the smartwatch overestimated FM, and the standard BIA measurement and underestimated FM. The study authors cautioned clinicians to be aware "that there may be large errors relative to clinical measures".[120]

Portable Ultrasound

Portable ultrasound as a method of measuring body composition, primarily in athletes, was introduced over the past few decades.[121-123] These measurements were proposed not only to capture static body composition but also to measure changes in body composition, particularly for athletes performing in weight category sports.[122]

Portable ultrasound devices use a handheld transducer, which is placed on the skin.[122] The transducer sends out ultrasound waves and then receives the waves as echoes. A resulting

Table 47.6 Summary of Novel Feasible, Scalable, and Affordable Methods to Measure Body Composition

Measurement Type	Description	Strengths and Weaknesses
Digital Photography[124,125]	Body shape information is extracted from digital photographs and used to predict DXA-measured body composition.	Strengths: Good correlation between actual and predicted measurements and digital photography is low cost and widely accessible. Weaknesses: Existing studies have been small and regional.
3D Body Image Scanners	Rendering of 3D avatars within seconds using different technologies. Anthropometry from the rendered images are used to predict body composition.	Strengths: Good agreement between PBF and FFM measured with DXA or the four-compartment method. Weakness: Devices are affordable but still expensive for individual purchase. Not all are portable. More validation research is needed for smartphone applications.
Smartwatch BIA technology	BIA capturing technology relying on electrodes on the watch estimate body composition within seconds.	Strengths: The smartwatch is affordable and convenient for personal use and the agreement with DXA is adequate. Weaknesses: Body composition measurements are not as precise as DXA and there is systematic bias, which can be correctable.
Portable Ultrasound	Body composition is measured using depths estimated by the ultrasound, which can quantify the thickness of fat between the skin and the muscle. Measurements taken at specific body sites are then placed into a model that predicts PBF.	Strengths: The device is portable and convenient with high interrater reliability. Weaknesses: Body composition measurements, especially in females, are not as accurate as other methods.

image is produced from waves bouncing off of changes in body tissue at different depths. The depths, which are measured by the ultrasound, can quantify the thickness of fat. Measurements taken at specific body sites are then placed in a model that predicts PBF.

Studies on the accuracy and reliability of portable ultrasound systems found mixed results. In a study of 93 athletes,[122] PBF was well correlated to DXA measurements ($r = 0.97$ females, 0.98 males) and showed good agreement in a Bland Altman analysis. Another study[121] in a more heterogeneous population ($n = 144$), performed six consecutive measurements at eight body sites with two raters for error reduction. The study found that while there was good agreement with ultrasound-measured body composition and DXA, there was also a source of variability due to the rater applying the ultrasound. Moreover, the study found that the measurements were more accurate in males. In another recent study, body composition was measured in 45 National Collegiate Athletic Association (NCAA) athletes via air displacement plethysmography, portable ultrasound, and skinfolds.[123] Significant differences were observed in the measurements obtained from all three methods, with a larger mean difference observed in PBF measurements from females (5%) compared with males (1.5%). There was also higher interrater reliability for the portable ultrasound, compared with skinfolds. The authors concluded that the portable ultrasound is an adequate measure of body composition but should be used with caution, especially in females.

Summary

Emerging portable, reduced-cost body composition methods that do not require specialized licensed technologies are providing exciting new opportunities in the field of body composition measurement (Summary in **Table 47.6**). While the gold standards of body composition methods for research focus on accuracy, these new technologies allow for mass measurement of heterogeneous populations that can provide new insights into nutrition and obesity research. Most importantly; however, the new methods allow clinicians and healthcare professionals to use devices to effectively measure and track body composition to guide patient care.

Acknowledgments

The authors acknowledge the authors of the previous version of this chapter, Scott Going, Melanie Hingle, and Joshua Farr, for their contributions.

References

1. Baumgartner R. Age. In: Human Body Composition Champaign, IL: Human Kinetics, 2005; 259–69.
2. Roubenoff R. Eur J Clin Nutr 2000;54:S40–7.
3. Mott JW, Wang J, Thornton JC, et al. Am J Clin Nutr 1999;69: 1007–13.
4. Lohman TG. Advances in Body Composition Assessment. In: Current Issues in Exercise Science Series: Monograph No. 3. Champaign, IL: Human Kinetics, 1992.
5. Laurson KR, Eisenmann JC, Welk GJ. Am JPrevent Med 2011;41: S93–9.

CHAPTER 48

Use and Interpretation of Anthropometry

Karen E. Peterson, Aria Grabowski, and Suzanne M. Cole

CHAPTER OUTLINE

Abbreviations

AC, Arm circumference; **BIA**, bioelectrical impedance analysis; **BMI**, body mass index; **%BF**, Body fat percentage; **CDC**, Centers for Disease Control and Prevention; **CVD**, cardiovascular disease; **DXA**, dual x-ray absorptiometry; **FFM**, fat-free mass; **NHES**, National Health Examination Survey; **HC**, Head circumference; **IASO**, International Association for the Study of Obesity; **IDF**, International Diabetes Federation; **IOTF**, International Obesity Task Force; **IUGR**, intrauterine growth restriction; **INTERGROWTH-21**, International Fetal and Newborn Growth Consortium for the 21st Century; **LMS**, least means squared; **MGRS**, Multicenter Growth Reference Study; **MUAC**, mid-upper arm circumference; **NCSS**, Newborn Cross-Sectional Study; **NCHS**, National Center for Health Statistics; **NICE**, National Institute for Health and Care Excellence; **NIH**, National Institute of Health; **NHANES**, National Health and Nutrition Examination Survey; **PI**, Ponderal Index; **SD**, standard deviation; **SGA**, small for gestational age; **UNICEF**, United Nations Children's Fund; **RFM**, relative fat mass; **RFMp**, relative fat mass pediatric; **WC**, waist circumference; **WHO**, World Health Organization; **WHR**, waist-hip ratio

Introduction

Anthropometry is defined as the measurement of humans for the purposes of understanding physical variation, using noninvasive techniques.[1] Changes in lifestyle behaviors such as diet and physical activity, and in health status, can lead to changes in anthropometry. Anthropometric measures are widely used to assess growth and development in infants and children and health status and disease risk in adults. The World Health Organization (WHO) has developed guidelines and growth references to provide guidance on the use and interpretation of anthropometric measures.[1] In this chapter, the most used anthropometric measurements are presented, as well as the use and interpretations of these measures.

Anthropometric Measures and Indices

The most used anthropometric measures in adults and children include weight, stature (length or height), and waist and hip circumference. These measures or a combination, e.g., an index,

Table 48.1 Strengths and Limitations of Commonly Used Anthropometric Measures

	Definition	Strengths	Limitations
Weight	The mass of the entire human body, ideally without clothes or other items	■ Easy to use, inexpensive ■ Indicates short-term nutrition status; readily impacted by changes in lifestyle and health status	■ Impacted by numerous factors, including dehydration, edema, and other acute and chronic diseases
Stature: Length or Height	The distance from the base of the heels to the top of the head, without shoes	■ Easy to use, inexpensive ■ Indicates long-term nutrition status	■ Not suitable for individuals who are unable to stand. Alternatives include arm span, knee height, or tibia length ■ Infant length may be subject to measurement error due to moving infant
BMI (kg/m²)	Weight-for-height index, calculated as weight (kg)/height (m)^2	■ Easy to use, inexpensive ■ Commonly used, allowing for comparability across studies ■ High correlation with percent body fat in population studies ■ Cut points have been developed in adults and children ■ Associated with risk of morbidity and mortality	■ Assumes fat-to-lean ratio is the same for all ■ Does not provide information on distribution of body fat ■ Should not be used in pregnant individuals ■ May have different relationships with total body fat and fat distribution and health risk in different populations
Waist circumference	The distance around the waist, above the hips and below the rib cage Measured using a non stretch tape	■ Portable and inexpensive to measure ■ Correlated with abdominal adiposity ■ Can be a better predictor of obesity-related conditions than BMI	■ No consensus on measuring protocol regarding tape positioning, WHO and NIH approaches are recommended. Can be subject to measurement error by staff and/or type of measuring tape
Skinfold Thickness	A measure of thickness of subcutaneous fat at multiple sites	■ Portable and inexpensive to measure ■ Can be used to calculate body density and percent body fat	■ Population-specific predictive equations must be used to estimate body density and body fat ■ Can be subject to measurement error by staff or poorly calibrated calipers

such as body mass index (BMI), are often used to assess body size and composition. The ideal measure of body size and composition should meet several requirements. First and foremost, the measure must be useful and accessible in terms of simplicity, cost, and ease of use, in both clinical settings and epidemiologic studies, with small measurement error. Additionally, it should predict risks of health consequences and allow cut-points to classify individuals in terms of health risk.[2] An in-depth presentation of body composition methods is provided in Chapter 47. The strengths and limitations of the most used anthropometric measures and indices are summarized in **Table 48.1**.

Adults

Body Mass Index

BMI is a simple index calculated as weight (kg)/height (m^2). BMI is commonly used worldwide to classify underweight, overweight, and obesity. Although BMI has numerous limitations, it is readily used due to its relative ease, low cost, and a strong association with body fat and health risks in populations.[3-5] At the individual level, the relationship between BMI and body fat has been shown to vary due to a variety of factors, such as age, sex, body shape, race/ethnicity, and fitness level.[6-10]

BMI Cut-Points

An Expert Consultation on Obesity convened by the WHO in 1993 recognized obesity as a disease and determined cut-points to classify BMI for international use.[1] These cut-points for underweight, overweight, and obesity in adults (**Table 48.2**), were established based on data relating BMI to mortality. The BMI cut-points allow for comparisons within and between populations and can be used to classify groups and individuals at higher risk of morbidity and mortality, and to prioritize those who may benefit from intervention. It is important to note that BMI should be used as a screener rather than a diagnostic tool. Health history, other clinical and anthropometric measures and presence of other risk factors should be considered when using BMI in a clinical setting.[11]

Research has suggested racial or ethnic differences in associations among BMI, percent body fat (%BF), fat distribution, and health risks.[12,13] Since the late 1990s, there has been debate on whether population- or ethnicity-specific BMI cut-points should be used for the classification of obesity.[14]

A 2000 multi-agency report by the WHO, the International Obesity Task Force (IOTF), and the International Association for the Study of Obesity (IASO) presented Asian-specific BMI cut-point values for overweight (BMI >23.0 kg/m^2) and obese (BMI >25 kg/m^2) categories. Recommended cut-points were based

Table 48.2 World Health Organization Body Mass Index Cut-Points for Adults

Classification	BMI (kg/m²)
Underweight	<18.50
Severe thinness	*<16.00*
Moderate thinness	*16.00–16.99*
Mild thinness	*17.00–18.49*
Healthy Weight Range	18.50–24.99
Overweight	25.00–29.99
Obese	≥30.00
Obese Class I	*30.00–34.99*
Obese Class 2	*35.00–39.99*
Obese Class 3	*≥40.00*

Data from World Health Organization. Obesity: preventing and managing the global epidemic. Report of a WHO consultation. Geneva: World Health Organization, 2000:1–253. Technical Report Series No. 894. World Health Organization. Appropriate body-mass index for Asian populations and its implications for policy and intervention strategies. Lancet 2004;363:157–63.

on studies among Asian populations that found increased risk of conditions such as diabetes and cardiovascular disease (CVD) at lower BMI cut-points, compared with the WHO international BMI classifications.[15] Variability in the relationship between BMI, %BF, and disease risk across different Asian populations and rising rates of type 2 diabetes and CVD in Asia led to a 2002 WHO Expert Consultation to review the evidence regarding Asian-specific BMI thresholds.[14,16,17] The Expert Consultation group found that disease risk-associated BMI cut-points varied across Asian populations, with observed risk and high risk ranging from 22–25 kg/m² and 26–31 kg/m², respectively.[14] They suggested additional BMI cut-points including 23 (risk) and 27.5 kg/m² (high risk) for possible public health action in Asian populations.[14] Nevertheless, they emphasized establishing specific overweight and obese BMI cut-points for all Asian populations was not possible given the variability in the associations between BMI, adiposity, and disease risk, but suggested additional BMI cut-points that could be used by Asian countries for public health assessment of disease risk in their respective populations. Subsequent studies generally support lower disease-risk BMI cut-points for Asian populations, although the cutoff values may vary depending on the population and the disease.[18-21]

The relationship between BMI and obesity-related disease risk also differs in other racial or ethnic groups. A large population-based study in the United Kingdom found that BMI cut-points associated with diabetes risk were lower in Arab, Black, Chinese, and South Asian populations, compared with the BMI cutoff values used to assess diabetes risk in non-Hispanic-White populations.[22] Similarly, a U.S. study reported lower diabetes-risk BMI cut-points in African American, Chinese American, and Hispanic adult participants >45 years of age, compared with the BMI cut-point associated with diabetes risk in non-Hispanic White participants.[23] This evidence has led some to advocate for racial/ethnic-specific BMI cut-points, as the utility of population-specific BMI cut-points may be especially relevant for disease screening.[24,25]

Relative Fat Mass Index

Relative fat mass (RFM) was developed in 2018 as an alternative to BMI.[26] Using data from multiple NHANES cohorts, the new index equations were developed to estimate whole-body fat mass:

$$RFM = 64 - 20 \times (height/waist)) + (12 \times sex),$$

where *sex* = 0 for male and 1 for female.
Height and waist circumference are in the same units.
U.S. units or metric units may be used.

Compared with BMI, RFM had a stronger relationship with whole-body fat percentage measured by dual x-ray absorptiometry (DXA) among both men and women and improved accuracy and precision. RFM also was shown to be more accurate than BMI across racial or ethnic groups, age categories, and body fat quintiles.[26] The RFM has been externally validated in Mexico as a better predictor of fat mass percent than BMI across all body composition methods measured (DXA, BIA, and air-displacement).[27] Using NHANES data, the RFM index has also been shown to be better than BMI to classify risk of obesity-related conditions, e.g., diabetes, high blood pressure, hospitalization, and arthritis.[28] The RFM is still a new index and further evaluation of its usefulness in a variety of settings will need to be explored.

Waist Circumference and Ratios

Waist circumference (WC) has been recommended by the National Institute of Health (NIH), WHO, and the International Diabetes Federation (IDF) as a measure of abdominal adiposity.[29,30] WC is associated with mortality and chronic diseases such as cardiovascular disease and type 2 diabetes,[31-34] and has been shown to be a better predictor of obesity-related health risk compared with BMI alone.[35,36] In 1997, the WHO reported cut-points to indicate concern for WC risk; however, these cut-points were presented as examples only.[11,37,38] Example cut-points indicated that individuals with a WC >102 cm for men or >88 cm for women may be at increased risk of metabolic complications. In 1998, the NIH adopted these as recommendations.[37] In 2008, the WHO held an expert consultation on WC and Waist-Hip Ratio (WHR) to discuss recommendations and the use of these measures. A consensus was not determined but, rather, recommendations were made for future steps to be taken and the creation of a workgroup to address these steps. At the time of this publication, no update had been provided.[30] In 2006, the IDF released recommendations on categorizing metabolic syndrome risk by sex and ethnicity, including WC as the measure of obesity instead of BMI, shown in **Table 48.3**.

Subsequent reviews by expert panels have considered WC and WHR.[30,39] However, due to gaps in knowledge on diverse populations in terms of race/ethnicity and differences in body composition, no consensus recommendations have yet been developed.

Two ratios using WC have been proposed to assess the distribution of adipose tissue: the WHR and waist-to-height ratio. The WHR is waist circumference divided by hip circumference.[30] WHR had the highest precision in predicting type 2 diabetes in a multi-ethnic study, compared with BMI, WC, and %BF[40]; whereas a meta-analysis found that BMI, WC, and WHR showed similar relative risk of diabetes incidence.[41]

Table 48.3 Waist-Circumference Cut-Points Based on Country/Ethnic Group and Sex

Country/Ethnic Group	Sex	WC Cut-Points (cm)
Europids	Male	≥94 cm
	Female	≥80 cm
South Asian, Chinese, Japanese	Male	≥90 cm
	Female	≥80 cm
Ethnic South and Central Americans	Use South Asian recommendations until more data are available	
Sub-Saharan Africans and Eastern Mediterranean and Middle East (Arab) populations	Use European data until more specific data are available	

Reproduced from Alberti KG, Zimmet P, Shaw J. Diabet Med 2006;23:469–80. doi:10.1111/j.1464-5491.2006.01858.xv[29]

Waist-to-height ratio is calculated by dividing the waist circumference by height (using the same units of measurement). Waist-to-height ratio in a meta-analysis was shown to improve detection of cardiometabolic risk factors, compared with BMI and WC,[42] and to be a better predictor of mortality than BMI.[43] A waist-to-height ratio cut-off of 0.5 has been proposed as a potential indicator of risk,[44] creating a simple public health message "keep your waist circumference to less than half your height." The U.K. National Institute for Health and Care Excellence (NICE) has recommended cut-points of 0.4–0.49 to indicate no increased health risk, 0.5–0.59 to indicate increased health risk and >0.6 to indicate further increased health risks.[45] Currently, no consensus has been established for adult cut-points for these ratios; this and their added benefit in clinical and research settings remains a matter of debate. Further research is needed to establish guidelines for these ratios.

Skinfold Thickness

Skinfold thickness measurements are commonly used in the field, e.g., nonresearch settings, as an inexpensive and portable tool to estimate adiposity. Skinfold thickness measurements have been used since the 1900s to estimate the amount of subcutaneous adipose tissue in pinched skin, using calipers, and can be used to provide an estimate of total body fat. Measurements are typically taken for chest, axilla, triceps, biceps, subscapular, abdomen, suprailia, thigh, and calf, using standardized protocols described by Lohman et al.[46,47] Measurements can be used as a sum of all components, or they can be used to estimate body density and %BF.

Several assumptions underlie the use of skinfold thickness measurements to estimate body density and body fat. The first is that the subcutaneous fat in a fold of skin represents total body fat, while in reality, 50% to 70% of total body fat is found subcutaneously.[48] The validity of this assumption increases if multiple sites are measured. Additionally, it is assumed that the proportions of water, mineral, and protein in fat-free mass (FFM) are consistent across populations, assuming that FFM is made up of 73.8% water, 6.8% mineral, and 19.4% protein.[49,50]

To calculate body density using skinfold measures, numerous sex- and age-specific equations have been developed. Among these, the most used are from Durnin & Womersley or Jackson & Pollock (**Table 48.4**).[51-54] The Durnin & Womersley equations have also been revised for use in specific racial/ethnic populations.[55] Once body density is estimated, this value can be used to calculate total body fat, using additional equations, commonly those from Siri and Brozek (Table 48.4).[49,56] Due to limitations and assumptions of the equations, which lead to a wide range in %BF estimations, it has been proposed that using the sum of skinfold sites be used, instead of using the measurements to estimate body density and body fat.[57-59]

In addition to the limitations and assumptions of skinfold thickness measurements and predictive equations, measurement error is a concern. It has been reported that untrained individuals averaged 2 mm higher values on the same individuals, compared with individuals trained in skinfold thickness measurements.[60] In a study comparing students to a trained

Table 48.4 Equations to Estimate Body Density and Subsequent Body Fat from Skinfold Measures

Body Density (BD) Predictive Equations		
Durnin & Womersley, ages 16–68[A]	Male	$BD = 1.1765 - (0.0744 \times \log \Sigma 4ST)^{C}$
	Female	$BD = 1.1567 - (0.0717 \times \log \Sigma 4ST)^{C}$
Jackson & Pollock, 7 site[B]	Male	$BD = 1.112 - (0.00043499 \times \Sigma 7ST)^{D} + (0.00000055 \times \Sigma(7ST)^2 - (0.0002574 \times age^{E})$
	Female	$BD = 1.097 - (0.00046971 \times \Sigma 7ST)^{D} + (0.00000056 \times \Sigma(7ST)^2 - (0.00012828 \times age^{E})$
%BF Predictive Equations		
Siri		$\%BF = ((0.4950/BD) - 4.500) \times 100\%$
Brozek		$\%BF = ((0.4570/BD) - 4.142) \times 100\%$

Footnote:

A. The equations by Durnin and Womersley reflect summarized equations for ages 16–68 y; however they also generated age-specific equations for 16–19, 20–29, 30–39, 40–49, and 50+

B. The equations by Jackson & Pollock are for seven sites, predictive equations for three sites are available

C. Σ4ST = sum of triceps, subscapular, suprailiac, and biceps, in mm

D. Σ7ST = sum of chest, axilla, triceps, subscapular, abdominal, suprailiac, and thigh skinfolds, in mm

E. Age in years

professional, measurement error ranged from 0.5 to 4.9 mm, depending on the site and student's sex.[61] Individuals performing skinfold thickness measurements should be trained periodically and standardized to ensure accuracy and precision of measures for use in clinical and research settings.

Infants and Children

Anthropometry of infants and children poses unique challenges and limitations compared to anthropometry of adults. Infants and children grow and gain lean body mass and adipose tissue at different rates, and there are large differences in and between populations, as well as interindividual and intraindividual variations. Children's maturation status and growth patterns also affect their body composition.[62] Additionally, children experience secular changes in height growth and in body composition, which complicate the interpretation of anthropometric measurements.[63-65]

Growth References and Standards

Since the early part of the last century, healthcare providers serving infants and children have sought standards of normal growth against which to compare and assess the growth performance of individual cases and populations. Reference charts were created to address this need, but these charts relied primarily on data from small, nonrepresentative samples. Due to a need for growth charts that are more comprehensive, in the 1970s, the US National Center for Health Statistics (NCHS) compiled growth values based on a large sample of children from several U.S. cohorts and generated the 1977 NCHS charts.[66] In 1978, the WHO adopted and recommended the use of these charts for growth monitoring and health promotion in primary care.[67] Since then, the WHO updated growth standards for children (ages 0–5) in 2006 and subsequent growth references for school-aged children and adolescents (ages 5–19).[68,69] Pediatric growth charts have been widely used globally by researchers, pediatricians, nurses, and parents to assess children's growth and nutritional status. Of note, growth charts should not constitute the sole diagnostic instrument but contribute to forming an overall clinical impression for the child being measured.[70] There are several versions of growth references recommended for international use to assess children's physical growth, summarized below in **Table 48.5**.

1978 WHO/NCHS Growth References

The 1978 WHO/NCHS reference growth charts were used in the United States and worldwide for more than 2 decades in primary care and research. These reference curves were initially developed in 1977, based on serial, cross-sectional data from the National Health Examination Survey (NHES) and from the National Health and Nutrition Examination Survey (NHANES), representative of the U.S. population in the 1960s and 1970s.[66] Reference growth data for infants and children from 0–36 months were based on data from the Fels Longitudinal Study of human growth and body composition, initiated in 1929 to follow 818 infants.[71] In 1978, the WHO produced a normalized version with Z-scores based on the 1977 NCHS curves and advocated their use worldwide, despite some limitations.[67,72-75] The growth reference for children 0 to 36 months was developed based on retrospective data on infants who were primarily formula fed, and from one area in the Midwest region of the United States. This was of concern because the growth pattern of breastfed infants differs from that of formula-fed infants, and also would reflect the social and environmental conditions only in the midwestern United States. Furthermore, follow-up intervals provided insufficient data to describe the rapid and changing rate of growth in early infancy.[76] Nevertheless, the WHO recommended adoption of the 1977 NCHS reference curves, given the representative data on a healthy population and the large sample size, which permitted smoothing of the outer percentiles used to classify under/over nutrition.

Table 48.5 Anthropometric Measures and Indicators Provided in WHO and CDC Growth References and Standards

	2000 CDC Growth Reference		2006 WHO Growth Standards		2007 WHO Growth Reference
Applicable age range (y)	0–3[a]	2–20	0–2	2–5	5–19
Length/height/stature-for-age	x	x	x	x	x
Weight-for-age	x	x	x	—	x
Weight-for-length/height	x	x	x	x	—
BMI-for-age	—	x	x	x	x
Head circumference-for-age	x	—	x	—	—
Arm circumference-for-age	—	—	x[b]	x	—
Subscapular skinfold-for-age	—	—	x[b]	x	—
Triceps skinfold-for-age	—	—	x[b]	x	—

BMI, body mass index; *CDC*, Centers for Disease Control and Prevention; *WHO*, World Health Organization; *x*, available measures; —, not provided.

[a]Not recommended for children 0 to 2 years.

[b]From 3 months to 5 years.

Data fom World Health Organization. The WHO Child Growth Standards. 2006. Available at: www.who.int/childgrowth/en/. Accessed August 10, 2012.; Centers for Disease Control and Prevention. CDC GrowthCharts. 2000. Available at: www.cdc.gov/growthcharts/cdc_charts.htm. Accessed August 10, 2012.

In 1995, a WHO Expert Committee reviewed existing growth references and research findings and recommended the continued use of the 1978 WHO/NCHS growth charts on a provisional basis until better references became available.[67]

2000 CDC Growth Charts

In 2000, the U.S. Centers for Disease Control and Prevention (CDC) developed growth charts to replace the 1977 NCHS charts,[77] based on data from the NHES and NHANES and supplemental data from other sources. Other reference data included birth length from birth certificates in two states, and information on length from the CDC Pediatric Nutrition Surveillance System. The 2000 CDC Charts comprise a series of sex-specific percentile curves for weight-for-age, length-for-age, weight-for-length, and head circumference-for-age, from birth to 36 months, and from 2 to 20 years. The growth curves are available in two layouts: Individual Growth Charts and Clinical Growth Charts. The clinical charts modify the presentation of reference curves, allowing two individual charts to appear on a single page with grids scaled to metric and English units.[78]

In 2010, the CDC released a recommendation for health care providers to use the WHO growth charts for children 0 to 2 years of age.[79] This recommendation highlighted limitations of the NCHS reference infant growth data, which were not based on breastfed infants and used a small sample size in the first 6 months of life.

An extended BMI-for-age growth chart was released by the CDC in 2022, given the increases in prevalence of severe obesity in children in the United States. This extension allows for visual growth and tracking for children with severe obesity and is designed to help providers optimize care for children with a BMI above the 97th percentile.[80]

2006/2007 WHO Growth Charts

In 2006, the WHO released new growth standards for children 0 to 5 years, developed using data from the prospective, longitudinal Multicentre Growth Reference Study (MGRS).[81] MGRS participants were exclusively breastfed, healthy infants and children, whose mothers did not smoke during pregnancy or after delivery, and who lived in households with socioeconomic conditions favorable to growth. The MGRS was conducted in Brazil, Ghana, India, Norway, Oman, and the United States. The data showed clear similarities in growth across countries,[82] and demonstrated that preschool children worldwide have the same growth potential, if raised in an optimal environment. The 2006 WHO charts combined growth data from the six countries and include sex-specific reference standards for weight-for-age, length-for-age, weight-for-length, BMI-for-age, head circumference-for-age, arm circumference-for-age, triceps skinfold-for-age, and subscapular skinfold-for-age.[83]

In 2007, the WHO released a new, cross-sectional growth reference for children 5 to 19 years, based on data from the original 1977 NCHS growth reference population.[69] These data were supplemented with information from the WHO 2006 reference populations to allow for a smooth transition in reference curves from 0 to 5 years to beyond 5 years.[69] The 2007 charts for children and adolescents provide sex-specific reference curves for height-for-age, weight-for-age, and BMI-for-age.

Utilizing Growth Charts

Percentiles and Z-Scores

Healthy growth patterns in infants and children vary by age and sex. Thus, percentiles and/or Z-scores are required to assess physical growth performance, relative to the reference standard or population and to permit comparisons across age and sex. Percentiles and Z-scores shown on reference growth charts are interconvertible (**Table 48.6**). For example, Z-scores of +2 and –2 correspond to the 97.7th and 2.3rd percentiles of the reference population at a specific given age and sex. Z-scores are more useful in research, whereas percentiles are commonly used in clinical settings.

Percentiles. A percentile is the value of a variable below which a certain percentage of observations (or the population) falls. Therefore, the percentile refers to the position of an individual relative to the reference distribution. A potential limitation of using percentiles is that the same interval of percentile values corresponds to different ranges in absolute values for different measures across ages, and in males and females. Even within the distribution of one measure for a specific age and sex, the same increments at different percentiles could correspond to different changes in absolute measures. Quantifying the change in percentile value near the extremes of the reference distribution yields very small values, thus Z-scores are usually considered better for this purpose.

Z-Scores. The Z score is the number of standard deviations (SD) away from the mean. When the distribution is normal, the Z score indicates how far and in what direction (positive vs negative) a measured value deviates from the mean for a given age and sex in the reference population, expressed in units of the population SD. It is derived from dividing the difference between individual value (x) and the population mean (μ) by the population SD (σ).

$$Z = \frac{x - \mu}{\sigma}$$

The 2006/2007 WHO and 2000 CDC growth charts both used the least means squared (LMS) method to normalize percentile distributions.[84] Sex- and age-specific LMS parameters allow users to calculate the Z-score corresponding to each individual child's measured value, using the following formula, where y is

Table 48.6 The Corresponding Values Between Percentiles and Z-Scores

Percentiles	0.2nd	2.3rd	2.5th	5th	15th	16th	50th (median)	84th	85th	95th	97.5th	97.7th	99.8th
Z-scores	−3	−2	−1.96	−1.64	−1.04	−1	0	+1	+1.04	+1.64	+1.96	+2	+3

the individual observation, whereas the LMS parameters for the individual's age and sex need to be applied.

$$Z = \frac{\left(\frac{y}{M}\right)^{L} - 1}{SL}$$

The use of Z-scores to assess growth performance is recommended in research settings for several reasons. As standardized measures, Z-scores are comparable across ages, sexes, and anthropometric measures. A group of Z-scores can be subject to summary statistics, such as mean and SD and can be studied as a continuous variable. In addition, Z-score values provide a readily interpretable approach to quantify the growth performance of individuals and groups outside of the percentile ranges.[1]

Interpretation

Growth charts are used in primary care settings worldwide to track infant and children's growth over time. The first step to using growth charts for clinical use is choosing the right growth chart (e.g., WHO, CDC) best suited to one's purpose and context.[85] To monitor growth of individuals, select the growth chart appropriate for the child's sex and age to plot growth measurements. When plotting anthropometric measures from two or more visits on a graph, connecting adjacent points allows for better trend observation. In general, healthy growth patterns will be relatively consistent over time, in terms of a particular Z-score or percentile on reference growth charts. Growth that crosses major Z-score lines or percentiles may indicate risk and requires further assessment. Crossing upwards 2 or more weight-for-length percentiles in the first 24 months has been associated with increased risk of obesity at age 5,[86] whereas crossing percentiles downward may indicate acute wasting in response to undernutrition or infection.

The WHO has established cut-points for Z-scores indicating risk and severity of overweight/obesity, stunting, and wasting (**Table 48.7**). Cutoffs of ±2 SD on WHO charts correspond to the 2.3rd and 97.7th percentiles on the CDC reference growth distribution. To facilitate growth monitoring of children 0 to 24 months using WHO growth charts that have been modified by the CDC for use in the United States,[79] the 2nd and 98th percentile are shown as cutoffs to identify low and high growth performance.[87]

Weight-for-Age Charts

Weight-for-age charts can be used to classify the infant or child as being underweight or severely underweight. The WHO 2006 and 2007 growth charts use a cut-point of Z-score of −2 to indicate underweight and −3 for severely underweight, and the CDC provides the 2nd and 98th percentiles on WHO charts modified for use in young children. Weight-for-age cannot, however, be utilized to classify overweight and obesity; therefore, infants and children with weights at above the median should also be assessed using the weight-for-length or height or BMI-for-age growth charts. Similarly, weight-for-age will not distinguish wasting, which requires use of assessment of weight-for-length/height-for-age and additional clinical assessment.

Newborns. A newborn's growth measures can be plotted on WHO and CDC growth charts. The distribution of birth weight varies, however, by gestational age, and reference growth charts apply to term infants, born at or after 37 weeks of gestation. A general classification of low birth weight, independent of

Table 48.7 Summary of WHO Classification of Growth Problems Based on Z-Scores

	Growth Indicators			
Z-Score	**Length/Height-for-Age**	**Weight-for-Age**	**Weight-for-length/height**	**BMI-for-age**
Above 3	Note 1	Note 2	Obese	Obese
Above 2			Overweight	Overweight
Above 1			Possible risk of overweight (Note 3)	Possible risk of overweight (Note 3)
0 (median)				
Below −1				
Below −2	Stunted (Note 4)	Underweight	Wasted	Wasted
Below −3	Severely Stunted (Note 4)	Severely Underweight (Note 5)	Severely wasted	Severely wasted

Measurements in the shaded boxes are in the normal range.

Notes:

1. A child in this range is very tall. Tallness is rarely a problem unless it is so excessive that it may indicate an endocrine disorder. Refer a child in this range for assessment if parents of normal height have a child who is excessively tall for his or her age.
2. A child whose weight-for-age falls in this range may have a growth problem, but this would be better assessed using weight-for-length/height or BMI-for-age.
3. A plotted point above 1 shows possible risk. A trend toward the 2 z-score line shows definite risk.
4. It is possible for a stunted or severely stunted child to become overweight.
5. Referred to as very low weight in the Integrated Management of Childhood Illness (IMCI), In-service training modules. WHO, Geneva, 1997.

Reproduced from World Health Organization (2008) WHO Child Growth Standards: Training Course on Child Growth Assessment. Geneva: World Health Organization.[85]

gestational age, provided by UNICEF and WHO, classifies low birth weight (microsomia) as <2,500 grams or 5.5 pounds.[88] High birth weight, also known as macrosomia, does not have a standardized definition but is generally accepted as >4,000 g or >4,500 g.[89-91] Low birth weight can result from intrauterine growth restriction (IUGR), prematurity, or both.[92-94] Low birth weight due to IUGR has been related to maternal undernutrition and short stature, socioeconomic status, infections such as malaria, among other risk factors.[95-98] Maternal gestational diabetes, obesity, and weight gain during pregnancy have been associated with increased incidence of high birth weight.[89-91,99] Macrosomia has been associated with increased risk of neonate hypoglycemia, shoulder dystocia, and caesarean section delivery.[89-91,99]

Cut-points of <10th percentile and >90th percentile of fetal weight for gestational age have been widely used to indicate small-for-gestational age or large-for-gestational age, respectively.[100-102] These percentiles were originally generated using data from 5,635 newborns ranging from 24 to 42 weeks of gestational age. However, these cut-points were based on a predominantly non-Hispanic White reference population in 1960, living at an altitude of over 5,000 feet above sea level, and may not reflect risk in other populations. More recent reference data on U.S. live births have been used to classify the term, small-for-gestational age infants for international comparison.[93,103] Oken et al.,[104] argued that a cutoff at the 10th percentile to classify small-for-gestational age birth weights does not fully illuminate causes and consequences of fetal growth. These authors used U.S. natality datasets with nearly 7 million infants born from 1999 to 2000 to compute percentiles for birth weight for gestational age from 22 through 44 weeks on firstborn and non-firstborn infants, providing more continuous reference data for research purposes.

In 2009, the International Fetal and Newborn Growth Consortium for the 21st century (INTERGROWTH-21) launched a multi-country project to complement the MRGS initiative that led to the 2006/2007 WHO standards.[105] INTERGROWTH-21 conducted the Newborn Cross-Sectional Study (NCSS) to create growth standards for weight, length, and head circumference at birth, based on infants born to 20,486 women with adequate antenatal care, living at or below 1,600 meters, absent known environmental exposures (pollution, tobacco, wood smoke, etc.). Like the MGRS, the NCSS was considered a prescriptive population of women with pregnancies at low risk for fetal growth impairment. INTERGROWTH-21 standards present tables and charts for 3rd, 5th, 10th, 50th, 90th, and 97th weights for weeks' gestation based on fetal ultrasound.[105]

Length/Height-for-Age Charts

Linear growth is a key marker of healthy growth. As such, length- or height-for-age charts are used to identify children with short stature on WHO or CDC growth charts, e.g., classified as stunted, relative to children of the same age and sex in the reference population. Stunting is defined as impaired growth, reflected in short stature, that often results from chronic undernutrition and repeated infections, among other risk factors.[106] Stunting in early childhood has been associated with short stature in adulthood, delayed cognitive, educational performance, and low adult wages and productivity. The WHO 2006 and 2007 growth charts use a cut-point of Z-score of −2 to indicate the child being stunted and −3 for severely stunted (Table 48.7). For children under 24 months monitored on CDC modified WHO growth charts, the 2nd percentile is used to identify those with stunting (Figure 48.1).

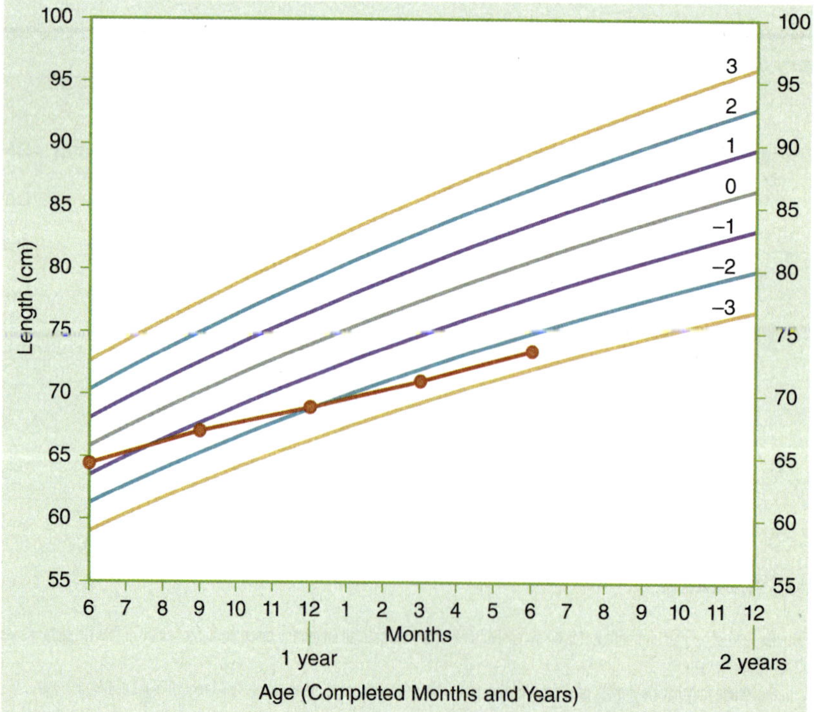

Figure 48.1 How to apply a growth chart to assess a child's growth trajectory and health status. This case shows a child whose linear growth has slowed over the past year, compared with expected (crossing two z-score lines); she is now considered stunted.

Reproduced from World Health Organization (2008) WHO Child Growth Standards: Training Course on Child Growth Assessment. WHO, Geneva.[15]

For older children, height-for-age less than the 5th percentile on CDC growth charts is used as an indicator of short stature.[107]

Weight-for-Length/Height Charts and BMI-for-Age

Both weight-for-length or height and BMI-for-age are used to indicate wasting and overweight and obesity. Weight-for-length or height can be useful in situations where children's ages are unknown.[85] Wasting usually indicates recent or severe weight loss due to inadequate dietary intake and/or frequent or prolonged illness.[108] Wasting is often considered a sign of acute malnutrition, while stunting largely reflects chronic exposure to undernutrition and disease.

Cut-points for weight-for-length or height and BMI-for-age on WHO growth standards have the same values and interpretation. *Z*-scores below −2 indicate wasting and below −3 indicate severe wasting. *Z*-scores between 1 and 2 indicate risk of being overweight, between 2 and 3 overweight, and >3 indicates obesity. Of note, the BMI-for-age *Z*-score equal to 1 at 19 years of age in the 2007 WHO reference population was 25.4 for boys and 25.0 for girls, which is close to the WHO BMI cut-point of 25 used to classify overweight in adults. Given this correspondence, a *Z*-score for BMI-for-age equal to 1 was recommended to classify overweight and *Z*-score greater than 2 to classify obesity on WHO charts. Among children 0 to 24 months, the 2nd percentile and 98th percentile are recommended to indicate low and high weight-for-length, respectively. BMI-for-age values plotted on CDC charts for children from 2 to 20 years <5th percentile and ≥95th percentile classify underweight and obesity, respectively. A BMI-for-age cut-point of >85th percentile and <95th percentile indicates risk of overweight.[109]

2000 International Obesity Task Force Cut-Points. The International Obesity Task Force (IOTF) endorsed a series of sex- and age-specific BMI cut-points to classify overweight and obesity among children aged 2 to 18 years in 2000 (**Table 48.8**), based on data from multinational (Brazil, United Kingdom, Hong Kong, Netherlands, Singapore, and United States) surveys.[110] These cut-points were developed from sex-specific, BMI-age curves that pass through a BMI of 25 for overweight and 30 for obesity at 18 years of age.[110] These classifications were considered more meaningful than percentiles or *Z*-scores based on extremes of the reference growth data distribution.

The IOTF reference has some unique strengths for international use, e.g., it is based on large data sets, covers different races/ethnicities, and BMI cut-points are linked to adult cut-points for overweight and obesity. However, there are also potential limitations.[111] For example, there is wide variation in the prevalence of obesity across the countries on which the IOTF reference population was based. The IOTF initially did not provide cut-points to classify underweight. The procedure used to generate high BMI cut-points was subsequently applied using the same data, generating the cut-points for "thinness" in children, using the WHO-recommended cut-point for adults of BMI <17 for grade 2 thinness.[112] Notably, use of the term "thinness" is more precise than "underweight," as it accounts for weight in relation to stature, e.g., BMI.

Table 48.8 International Obesity Task Force Body Mass Index Cut-Points for Overweight and Obesity by Sex between 2 and 18 Years

	Overweight		Obesity	
Age(y)	Males	Females	Males	Females
2	18.41	18.02	20.09	19.81
2.5	18.13	17.76	19.80	19.55
3	17.89	17.56	19.57	19.36
3.5	17.69	17.40	19.39	19.23
4	17.55	17.28	19.29	19.15
4.5	17.47	17.19	19.26	19.12
5	17.42	17.15	19.30	19.17
5.5	17.45	17.20	19.47	19.34
6	17.55	17.34	19.78	19.65
6.5	17.71	17.53	20.23	20.08
7	17.92	17.75	20.63	20.51
7.5	18.16	18.03	21.09	21.01
8	18.44	18.35	21.60	21.57
8.5	18.76	18.69	22.17	22.18
9	19.10	19.07	22.77	22.81
9.5	19.46	19.45	23.39	23.46
10	19.84	19.86	24.00	24.11
10.5	20.20	20.29	24.57	24.77
11	20.55	20.74	25.10	25.42
11.5	20.89	21.20	25.58	26.05
12	21.22	21.68	26.02	26.67
12.5	21.56	22.14	26.43	27.24
13	21.91	22.58	26.84	27.76
13.5	22.27	22.98	27.25	28.20
14	22.62	23.34	27.63	28.57
14.5	22.96	23.66	27.98	28.87
15	23.29	23.94	28.30	29.11
15.5	23.60	24.17	28.60	29.29
16	23.90	24.37	28.88	29.43
16.5	24.19	24.54	29.14	29.56
17	24.46	24.70	29.41	29.69
17.5	24.73	24.85	29.70	29.84
18	25	25	30	30

Reproduced from Cole TJ, Bellizzi MC, Flegal KM et al. BMJ 2000;320:1240–3.

Relative Fat Mass. The relative fat mass developed for adults may also be used for adolescents, ages 15–19. For children 8 to 14 years, sex-specific RFM pediatric (RFMp) equations have been developed and validated using NHANES data[113]:

$$RFMp = 74-(22 \times (height/waist))+(5 \times sex),$$
$$\text{where } sex = 0 \text{ for male and 1 for female.}$$

Height and waist circumference are in the same units and English or metric units may be used (e.g., inches, cm). The RFMp was more accurate in estimating whole-body fat percentage measured by DXA among males and females than BMI. It also was shown to be a better predictor of body fat than BMI across ethnic groups.[113]

Ponderal (Rohrer's) Index. The Ponderal or Rohrer's Index is like BMI, but mass is normalized with the third power of body height, rather than the second. For adults, the Ponderal index (PI) is calculated as weight (kg)/height (m^3); for infants and newborns, weight is recorded in grams and length in centimeters. The PI can be used to assess weight status relative to height for very short or very tall individuals and is primarily used in pediatrics. A normal range for newborns is 2.2 to 3.0, irrespective of gestational age and sex. A common use of the Ponderal Index is to classify small size for gestational age (SGA) at birth, considered an indicator of intrauterine growth retardation.[114] Asymmetric SGA, indicated by PI <2, can reflect fetal growth delay later in pregnancy, when weight velocity is highest, whereas symmetric SGA (PI >2) has been associated with more chronic delays throughout fetal life.[114]

Head-Circumference-for-Age

Head circumference (HC) in children has been related to brain size. HC-for-age is used in primary care to screen for potential developmental or neurologic disabilities, although evidence for this relationship is mixed.[115,116] HC can also be impacted by infections like the Zika Virus, reflected in microcephaly.[117] The WHO provides sex-specific charts for *Z* scores and percentiles for HC-for-age for children 0 to 13 weeks, 0 to 2 years, and 0 to 5 years, but no specific recommendation to indicate low and high values of concern. The CDC provides HC reference growth charts for children 0 to 36 months and cut-points of <5th percentile and >95th percentile to indicate potential developmental problems.[109] Microcephaly is generally defined by pediatric practitioners as HC >2 SD below the mean, and severe microcephaly is >3 SD below the mean of HC reference curves.[115]

Arm Circumference-for-Age

Arm circumference measurement is used for assessing nutritional status when stature and weight are difficult to collect, for example, during famines, conflict, or environmental crises.[118] The WHO recommends evaluation of children with a mid-upper arm circumference (MUAC) <115 mm, who may be at high risk of severe acute malnutrition and death.[119] Sex-specific WHO reference charts displaying *Z* scores for children 3 months to 5 years can be used to identify children with high or low values of concern.[118] Additionally, the WHO provides tables with *Z* score cutoffs (±1, 2, 3) by sex and month of age in several formats, including a simplified field table, as well as tables that can be incorporated into national health cards. Reference percentiles for MUAC-for-age have been computed for the sample of 32,952 children 1 to 20 years used to create the 2000 CDC growth references for BMI-for-age.[120] A range of sex-specific percentiles (3rd, 5th, 10th, 25th, 50th, 75th, 85th, 90th, 97th) are provided at 6-month intervals, as well as L, M, S coefficients to permit calculation of *Z* scores. These authors note that a constant cutoff of 115mm or 125 mm to classify undernutrition did not correspond to a consistent percentile in children under 5 years of age, highlighting the potential limitations of this indicator in healthy children.

Skinfold Thickness-for-Age

Skinfold thickness is a widely used method of estimating body fat by measuring subcutaneous adipose tissue in standardized sites.[121] Skinfold measures are a useful addition to the battery of growth standards for assessing fat mass and distribution. Equations can be used to estimate body composition from skinfolds. Because the choice of equation impacts the validity of body composition estimates, it has been recommended to use skinfold thickness measures as indices, rather than as a measure of total body adiposity.[59] Subscapular and triceps skinfolds fluctuate by age, declining after about 1 year, then rebounding. The most marked changes in subscapular and triceps skinfolds occur during adolescence, coinciding with puberty, and the timing and magnitude of these changes varies by sex, with girls showing increases and boys demonstrating decreases in these measures during puberty.[122]

The 2006 WHO provides sex-specific reference growth charts and tables of *Z*-scores for subscapular skinfold thickness-for-age and triceps skinfold thickness-for-age for children aged 3 months to 5 years. Reference curves for triceps and subscapular skinfolds for U.S. children 1.5 to 19.99 years of age were developed in 2010 based on the same national samples used to generate the CDC 2000 BMI reference growth charts.[123] Sex-specific tables provide a full range of percentiles, (3rd, 5th, 10th, 25th, 50th, 75th, 85th, 90th, 97th) every 6 months of age created using the LMS method and L, M, S values are provided to permit calculation of *Z*-scores.

Waist Circumference

As with adults, waist circumference and related ratios have been explored to identify children at risk of abdominal obesity. In children, WC has wide variation across ethnic groups and life stages. Therefore, waist-to-height ratio is used in children as a sex- and age-independent variable.[124] Using data from NHANES, researchers created smoothed charts for WC and waist-hip ratio, for children 5 to 19 years old. Using these charts, they concluded that WC and waist-hip ratio were more strongly associated with cardiometabolic outcomes than BMI- *Z* score.[125-128] In a systematic review and a meta-analysis, a cut-off of 0.5 for waist-hip (the same as recommended for adults) has been proposed to indicate risk of excess adiposity and cardiometabolic sequelae.[44,128] However, universal guidelines have yet to be established.

Conclusions

Anthropometry provides a group of useful, inexpensive, and noninvasive measures and indicators to assess the size, shape, and composition of the human body as well as risk of adverse health outcomes, such as malnutrition and obesity in adults and children. They are indirect measures of body composition, whereas direct body composition assessment methods such as DXA and air displacement plethysmography can provide accurate measures (e.g., total body fat, fat distribution) but are more costly and invasive. In general, the commonly used anthropometric measures, such as BMI and WC, have good validity in measuring body fat and predicting future health risk. Ongoing research is needed to guide the appropriate applications of anthropometric measures and the use of international growth references and standards in different populations. Efforts should be made to develop innovative new anthropometric measures and technologies to meet the research and clinical needs in the biomedical field and public health settings.

Acknowledgments

The authors acknowledge Youfa Wang, Hyunjung Lim, and Benjamin Caballero, authors of the previous version of this chapter, for their contributions.

References

1. World Health Organization. Physical Status: The Use and Interpretation of anthropometry. Report of a WHO Expert Committee. World Health Organ Tech Rep Ser 1995;854:1–452.
2. Power C, Lake JK, Cole TJ. Int J Obes Relat Metab Disord 1997;21:507–26.
3. Jeong SM, Lee DH, Rezende LFM, et al. Sci Rep 2023;13:3472.
4. Flegal KM, Shepherd JA, Looker AC, et al. Am J Clin Nutr 2009;89:500–8.
5. Ranasinghe C, Gamage P, Katulanda P, et al. BMC Public Health 2013;13:797.
6. Deurenberg-Yap M, Chew SK, Deurenberg P. Obes Rev 2002;3:209–15.
7. Heymsfield SB, Peterson CM, Thomas DM, et al. Obes Rev 2016;17:262–75.
8. Deurenberg P, Deurenberg Yap M, Wang J, et al. Int J Obes Relat Metab Disord 1999;23:537–42.
9. Werkman A, Deurenberg-Yap M, Schmidt G, et al. Ann Nutr Metab 2000;44:235–42.
10. Romero-Corral A, Somers VK, Sierra-Johnson J, et al. Int J Obes (Lond) 2008;32:959–66.
11. World Health Organization. World Health Organ Tech Rep Ser 2000;894:i–xii, 1–253.
12. Deurenberg P, Yap M, van Staveren WA. Int J Obes Relat Metab Disord 1998;22:1164–71.
13. Wang J, Thornton JC, Burastero S, et al. Obes Res 1996;4:377–84.
14. WHO Consultation. Lancet 2004;363:157–163.
15. World Health Organization. Regional Office for the Western Pacific, International Association for the Study of Obesity, and International Obesity Taskforce.
16. Deurenberg-Yap M, Schmidt G, van Staveren WA, et al. Int J Obes Relat Metab Disord 2000;24:1011–7.
17. Zhou BF, Cooperative Meta-Analysis Group of the Working Group on Obesity in China. Biomed Environ Sci 2002;15:83–96.
18. Cheong KC, Yusoff AF, Ghazali SM, et al. Public Health Nutr 2013;16:453–9.
19. Haldar S, Chia SC, Henry CJ. Adv Food Nutr Res 2015;75:97–154.
20. Ding J, Chen X, Bao K, et al. J Diabetes 2020;12:372–84.
21. Misra A, Khurana L. Int J Obes (Lond) 2011;35:167–87.
22. Caleyachetty R, Barber TM, Mohammed NI, et al. Lancet Diabetes Endocrinol 2021;9:419–26.
23. Rodriguez LA, Bradshaw PT, Shiboski SC, et al. Diabet Med 2021;38:e14377.
24. Aggarwal R, Bibbins-Domingo K, Yeh RW, et al. Ann Intern Med 2022;175:765–73.
25. Stanford FC, Lee M, Hur C. Mayo Clin Proc 2019;94:362–3.
26. Woolcott OO, Bergman RN. Sci Rep 2018;8:10980.
27. Guzman-Leon AE, Velarde AG, Vidal-Salas M, et al. PLoS One 2019;14:e0226767.
28. Cichosz SL, Rasmussen NH, Vestergaard P, et al. Diabetes Metab Syndr 2022;16:102590.
29. Alberti KG, Zimmet P, Shaw J. Diabet Med 2006;23:469–80.
30. World Health Organization. Waist circumference and waist-hip ratio: report of a WHO expert consultation, Geneva, 8-11 December 2008, World Health Organization, Geneva, 2011.
31. Song X, Jousilahti P, Stehouwer CD, et al. Eur J Clin Nutr 2013;67:1298–302.
32. Cerhan JR, Moore SC, Jacobs EJ, et al. Mayo Clin Proc 2014;89:335–45.
33. Pischon T, Boeing H, Hoffmann K, et al. N Engl J Med 2008;359:2105–20.
34. Zhang C, Rexrode KM, van Dam RM, et al. Circ 2008;117:1658–67.
35. Janssen I, Katzmarzyk PT, Ross R. Am J Clin Nutr 2004;79:379–84.
36. Lofgren I, Herron K, Zern T, et al. J Nutr 2004;134:1071–6.
37. Clinical Guidelines on the Identification, Evaluation, and Treatment of Overweight and Obesity in Adults—The Evidence Report. National Institutes of Health. Obes Res 1998;6:51S–209S.
38. Lean ME, Han TS, Morrison CE. BMJ 1995;311:158–61.
39. Ross R, Neeland IJ, Yamashita S, et al. Nat Rev Endocrinol 2020;16:177–89.
40. Zethof M, Mosterd CM, Collard D, et al. Diabet Care 2021;44:1692–8.
41. Vazquez G, Duval S, Jacobs DR, Jr., et al. Epidemiol Rev 2007;29:115–28.
42. Ashwell M, Gunn P, and Gibson S. Obes Rev 2012;13:275–86.
43. Ashwell M, Mayhew L, Richardson J, et al. PLoS One 2014;9:e103483.
44. Browning LM, Hsieh SD, Ashwell M. Nutr Res Rev 2010;23:247–69.
45. National Institute for Health and Care Excellence Guidelines. Obesity: identification, assessment and management. In: Obesity: identification, assessment and management, London, 2023.
46. Lohman TG, Lohman TG, Roche AF, et al. Anthropometric Standardization Reference Manual, Human Kinetics Books, Champaign, IL, 1988.
47. Anthropometry Procedures Manual, Centers for Disease Control and Prevention, National Center for Health Statistics, 2021.
48. Lohman TG. Hum Biol 1981;53:181–225.
49. Brozek J, Grande F, Anderson JT, et al. Ann N Y Acad Sci 1963;110:113–40.
50. Wagner DR, Heyward VH. Res Q Exerc Sport 1999;70:135–49.
51. Durnin JV, Womersley J. Br J Nutr 1974;32:77–97.
52. Jackson AS, Pollock ML. Br J Nutr 1978;40:497–504.
53. Jackson AS, Pollock ML, Ward A. Med Sci Sports Exerc 1980;12:175–81.
54. Jackson AS, Pollock ML. Phys Sportsmed 1985;13:76–90.

55. Davidson LE, Wang J, Thornton JC, et al. Med Sci Sports Exerc 2011;43:542–9.
56. Brodie DA. Sports Med 1988;5:11–40.
57. Johnston FE. Hum Biol 1982;54:221–45.
58. Doran DA, Mc Geever S, Collins KD, et al. Int J Sports Med 2014;35:95–100.
59. Reilly JJ, Wilson J, Durnin JV. Arch Dis Child 1995;73:305–10.
60. Burkinshaw L, Jones PR, Krupowicz DW. Hum Biol 1973;45: 273–9.
61. Barreira T. Int J Exerc Sci 2013;6:20–8.
62. Ong KK, Ahmed ML, Dunger DB. Mol Cell Endocrinol 2006; 254–255:8–12.
63. Freedman DS, Khan LK, Serdula MK, et al. Arch Pediatr Adolesc Med 2000;154:155–61.
64. Hauspie RC, Vercauteren M, Susanne C. Horm Res 1996;45 Suppl 2:8–17.
65. Moreno LA, Fleta J, Sarria A, et al. Int J Obes Relat Metab Disord 2001;25:1656–60.
66. Hamill PV, Drizd TA, Johnson CL, et al. Vital Health Stat 1977;11:i-iv, 1–74.
67. World Health Organization. A growth chart for international use in maternal and child health care: guidelines for primary health care personnel, WHO Publications Centre, Geneva, 1978.
68. World Health Organization. World Health Organization releases new Child Growth Standards. World Health Organization, Online, 2006.
69. de Onis M, Onyango AW, Borghi E, et al. Bull World Health Organ 2007;85:660–7.
70. World Health Organization. Paediatr Child Health 2010;15: 84–98.
71. Roche AF. Grrowth, Maturation and Body Composition: the Fels Longitudinal Study 1928-1991. Cambridge University Press, 1992.
72. Wang Y, Moreno LA, Caballero B, et al. Food Nutr Bull 2006;27: S175–88.
73. Victora CG, Morris SS, Barros FC, et al. J Nutr 1998;128: 1134–8.
74. de Onis M, Yip R. Bibl Nutr Dieta 1996;74–89.
75. Dibley MJ, Staehling N, Nieburg P, et al. Am J Clin Nutr 1987;46: 749–62.
76. Ogden CL, Kuczmarski RJ, Flegal KM, et al. Pediatrics 2002;109:45–60.
77. Kuczmarski RJ, Ogden CL, Guo SS, et al. Vital Health Stat 2002;11:1–190.
78. Centers for Disease Control and Prevention. Clin Grow Charts 2023, Available at CDC.gov.
79. Grummer-Strawn LM, Reinold C, Krebs NF, et al. MMWR Recomm Rep 2010;59:1–15.
80. Centers for Disease Control and Prevention. 2022;2023. Available at CDC, cdc.gov.
81. de Onis M, Garza C, Victora CG, et al. Food Nutr Bull 2004;25:S15–26.
82. WHOMGRS Group. Acta Paediatr Suppl 2006;450:7–15.
83. World Health Organization. WHO Child Growth Standards: Length/Height-for-Age, Weight-for-Age, Weight-for-Length, Weight-for-Height and Body Mass Index-for-Age : Methods and Development, World Health Organization, Geneva, 2006.
84. Martinez-Millana A, Hulst JM, Boon M, et al. PLoS One 2018;13: e0208362.
85. World Health Organization. WHO Child Growth Standards: Training Course on Child Growth Assessment. (Safety, N. a. F., ed), World Health Organization, 2008.
86. Taveras EM, Rifas-Shiman SL, Sherry B, et al. Arch Pediatr Adolesc Med 2011;165:993–8.
87. Centers for Disease Control and Prevention. Assessing Growth Using the WHO Growth Charts. In Growth Chart Training : Using the WHO Growth Charts 2022;2023.
88. Bramer GR. World Health Stat Q 1988;41:32–6.
89. Chatfield J. Am Fam Physician 2001.
90. Berard J, Dufour P, Vinatier D, et al. Eur J Obstet Gynecol Reprod Biol 1998;77:51–9.
91. Gregory KD, Henry OA, Ramicone E, et al. 1998 Obstet Gynecol 1998;92:507–13.
92. Black RE. Nestle Nutr Inst Workshop Ser 2015;81:1–7.
93. Lee AC, Katz J, Blencowe H, et al. Lancet Glob Health 2013;1: e26–36.
94. Zohdi V, Sutherland MR, Lim K, et al. Int J Nephrol 2012; 136942.
95. Kramer MS. Bull World Health Organ 1987;65:663–737.
96. Kramer MS. J Nutr 2003;133:1592S–6S.
97. Rasmussen KM, Yaktine AL. In: Rasmussen KM, and Yaktine AL, eds. Weight Gain During Pregnancy: Reexamining the Guidelines. Weight Gain During Pregnancy: Reexamining the Guidelines. Washington, DC: 2009.
98. WHO, UNICEF. Low Birthweight Estimates: Levels and Trends 2000–2015 New Global, Regional and National Estimates of Low Birthweight, UNICEF, 2019. Available at unicef.org.
99. Mohammadbeigi A, Farhadifar F, Soufi Zadeh N, et al. Ann Med Health Sci Res 2013;3:546–50.
100. Lubchenco LO, Hansman C, Dressler M, et al. Pediatrics 1963;32:793–800.
101. Battaglia FC, Lubchenco LO. J Pediatr 1967;71:159–63.
102. Lee PA, Chernausek SD, Hokken-Koelega AC, et al. Pediatrics 2003;111:1253–61.
103. Alexander GR, Himes JH, Kaufman RB, et al. Obstet Gynecol 1996;87:163–8.
104. Oken E, Kleinman KP, Rich-Edwards J, et al. BMC Pediatr 2003;3:6.
105. Villar J, Cheikh Ismail L, Victora CG, et al. Lancet 2014;384: 857–68.
106. World Health Organization. Stunting in a Nutshell. WHO Publications Centre, 2015.
107. National Center for Chronic Disease Prevention and Health Promotion (U.S.). Division of Nutrition PA, & Obesity. Use and Interpretation of the WHO and CDC Growth Charts for Children from Birth to 20 Years in the United States. Centers For Disease Control, 2013.
108. World Health Organization. Malnutrition. WHO Publications Centre, 2021.
109. National Center for Chronic Disease Prevention and Health Promotion (U.S.). Division of Nutrition PA, & Obesity. Growth Chart Training: Overview of the CDC Growth Charts for Use in the United States Among Children and Teens Aged 2 Years to 20 Years. Centers for Disease Control, 2022.
110. Cole TJ, Bellizzi MC, Flegal KM, et al. BMJ 2000;320:1240–3.
111. Wang Y. Int J Obes Relat Metab Disord 2004;28 Suppl 3:S21–8.
112. Cole TJ, Flegal KM, Nicholls D, et al. BMJ 2007;335:194.
113. Woolcott OO, Bergman RN. Sci Rep 2019;9:15279.
114. Osuchukwu OO, Reed DJ. Small for Gestational Age. In: StatPearls, Treasure Island, FL, 2023.
115. Harris SR. Can Fam Physician 2015;61:680–4.
116. Nicolaou L, Ahmed T, Bhutta ZA, et al. BMJ Glob Health 2020;5.
117. Centers for Disease Control and Prevention. Microcephaly & Other Birth Defects. Centers for Disease Control, 2019.
118. World Health Organization. WHO child growth standards : head circumference-for-age, arm circumference-for-age, triceps skinfold-for-age and subscapular skinfold-for-age : methods and development. WHO Publications Centre, 2007.

119. World Health Organization. WHO Child Growth Standards and the Identification of Severe Acute Malnutrition in Infants and Children: A Joint Statement by the World Health Organization and the United Nations Children's Fund. In: WHO Child Growth Standards and the Identification of Severe Acute Malnutrition in Infants and Children: A Joint Statement by the World Health Organization and the United Nations Children's Fund. Geneva: 2009.
120. Addo OY, Himes JH, Zemel BS. Am J Clin Nutr 2017;105: 111–20.
121. Duren DL, Sherwood RJ, Czerwinski SA, et al. J Diabetes Sci Technol 2008;2:1139–46.
122. Tanner JM. Foetus into Man: Physical Growth from Conception to Maturity. Harvard University Press. 1989.
123. Addo OY, Himes JH. Am J Clin Nutr 2010;91:635–42.
124. Fredriksen PM, Skar A, Mamen A. Scand J Public Health 2018;46:12–20.
125. Bacopoulou F, Efthymiou V, Landis G, et al. BMC Pediatr 2015;15:50.
126. Sharma AK, Metzger DL, Daymont C, et al. Pediatr Res 2015;78:723–9.
127. Savva SC, Tornaritis M, Savva ME, et al. Int J Obes Relat Metab Disord 2000;24:1453–8.
128. Munoz-Hernando J, Escribano J, Ferre N, et al. Clin Nutr 2022;41:508–16.

CHAPTER 49

Dietary Assessment Methods

Sharon I. Kirkpatrick and Katherine L. Tucker

CHAPTER OUTLINE

Abbreviations

24HDR, 24-hour dietary recall; AMPM, Automated Multiple-Pass Method; ASA24, Automated Self-Administered 24-hour Dietary Assessment Tool; **DHQ**, Diet History Questionnaire; **FFQ**, Food frequency questionnaire; **NCI**, National Cancer Institute; **NHANES**, National Health and Nutrition Examination Survey; **SFFQ**, Semi-Quantitative Food Frequency Questionnaire; **WWEIA**, What We Eat in America

Introduction

This chapter provides an overview of methods used to collect information on consumption of foods, beverages, and dietary supplements, with a focus on self-report dietary assessment methods, as they are most commonly used in various types of research. The chapter starts with an overview of the importance of and complexities in dietary assessment, followed by a discussion of short-term and long-term dietary assessment methods. Technological advances and biomarkers for assessing dietary intake are then briefly discussed. The focus is generally on assessing dietary intake among populations and population subgroups for research needs rather than among individuals for clinical purposes.

Importance of Dietary Assessment

Reasonably accurate and reliable information on what people eat and drink, including consumption of foods, beverages and, often, vitamin and mineral supplements, is fundamental to nutrition research and practice. Dietary intake data are used in surveillance and monitoring to assess how closely the intake of a population, population subgroup, or a given individual aligns with food group recommendations and nutrient requirement estimates; as well as in epidemiologic research to examine associations between dietary exposures and health and disease outcomes. In turn, surveillance and epidemiologic research relying on dietary intake data are critical to developing and updating dietary guidance intended to promote dietary patterns that are consistent with meeting nutrient needs and reducing chronic disease risk. Dietary intake data are also used to assess the influence of factors such as socio-economic status and characteristics of the food environment on dietary intake, and to inform and evaluate policies, programs, and other interventions aimed at supporting healthy eating. For example, estimates of the prevalence of the population with inadequate or excess intake are used to identify nutrients of concern,[1] with implications for dietary guidance and policies related to food assistance programs[2] and food labeling. Such information is also relevant to the food and supplement industry, for example, with respect to reformulation to reduce sodium in the food supply.[3] Given the implications of the food system and dietary intake for greenhouse gas emissions, land and water use, and other factors associated with planetary health,[4] dietary intake data are also increasingly used to evaluate the environmental sustainability of what we eat and drink,[5] as well as the interrelationship between healthy and climate-friendly patterns of dietary intake.[6] At the individual level, dietary assessment is important as a basis for clinical guidance.

Complexities in Dietary Assessment

Multidimensionality and Dynamism

To address these various needs for information on dietary intake, dietary assessment methods are typically designed to capture *what was consumed* (foods, drinks, and supplements), preferably including details such as methods of preparation, as well as the *quantities* consumed and/or the *frequency* with which a given food or beverage is consumed, among individuals and groups. Capturing these facets of dietary intake is complex. This is because humans consume a range of foods and nutrients and, potentially, supplements, in combinations that vary over time, making dietary intake both multidimensional and dynamic.[7] Relatedly, it is increasingly recognized that it is likely not consumption of a particular food or nutrient that influences health and disease outcomes but rather, the totality of the foods that we consume, with synergistic and antagonistic effects of different food combinations.[8] For this reason, the Dietary Guidelines for Americans emphasize overall dietary patterns that are associated with health,[9] and recommendations for sustainable eating also focus on overall dietary patterns.[10] Given the multidimensionality of dietary intake and of dietary guidance, dietary assessment methods often aim to capture the total diet, that is, all foods, beverages, and supplements consumed over a given period. This information is coded using food composition databases to estimate intake of nutrients, food groups, and potentially other components in foods, such as toxins (e.g., pesticide residues), as well as to assess overall diet quality and dietary patterns relative to dietary guidance for health[11] and the sustainability of intake.[12]

Dynamism is another important consideration when collecting and analyzing dietary intake data. Intake can vary greatly from day to day[13-15] with some foods, such as refined grain products, consumed most days by most members of the population, and others, such as fish and seafood, consumed infrequently, or episodically, by most members of the population. Dietary intake may fluctuate across seasons, although seasonality does not appear to be a major factor shaping dietary intake in the United States, given importation of foods grown and produced in other areas of the world throughout the year.[16] Dietary intake is also dynamic across the lifecycle, with sensitive periods, such as pregnancy, early life, adolescence, and later life.[7] For surveillance and monitoring, epidemiologic, intervention, and clinical research, interest is typical in capturing habitual, or usual, dietary intake over some period, not intake on a given day or a few days.[17] Because our bodies can store nutrients, food group recommendations and nutrient requirement estimates are typically intended to be met, on average, over time. Furthermore, it is habitual intake rather than intake on a given day that influences health and disease risk, as well as the overall environmental impact of our dietary intake. Accordingly, interventions ideally shift habitual intake to positively impact human and planetary health over the long term.

Assessing Usual Intake

Despite ongoing technological innovations and advances in dietary assessment, it is currently not possible to directly measure usual intake. For example, while observation can be used to assess intake with considerable accuracy, for example, in a school setting[18] or in the context of controlled feeding studies,[19] it is not typically feasible to conduct observation over long periods of time. Similarly, while a class of biomarkers called recovery biomarkers[20] can be used to estimate intake of energy and a small number of nutrients, including protein, potassium, and sodium,[21] with a relatively high degree of accuracy, such biomarkers are expensive and burdensome. Furthermore, it is typically feasible to use biomarkers only in small samples and with limited repeat administrations to assess long-term intake. Importantly, biomarkers neither provide insights into what people actually eat or drink nor contextual information about dietary intake,[22] whereas these details are often of interest.

Therefore, self-report dietary assessment methods remain the most commonly used methods across various types of research. Self-report methods generally comprise tools that capture intake over a short period of time, such as a day or a few days, and those that capture intake over a longer period, such as a month or year. Short-term methods include 24-hour dietary recalls (24HDR) and diet or food records, whereas long-term methods include food frequency questionnaires (FFQs) and screeners.

Considerations in Interpreting Information on the Validity of Dietary Assessment Methods

Like all self-report methods, the data collected are affected by measurement error, which may be random or systematic.[21] Random measurement error results in estimates that are not inaccurate but are imprecise, whereas with systematic error, measurements consistently deviate from truth in the same direction, leading to bias (e.g., underestimation of intake). Because short-term methods capture intake over a day or a few days, the data are affected to a greater extent by random error driven by within-person variation, whereas long-term methods reliant on preset food lists and on recalling and averaging intake over long periods of time are affected to a greater extent by systematic error or bias.[21] However, both short- and long-term self-report methods are affected by bias, which may come about due to forgetting some foods that were consumed (recall bias) or changing intake because it is being recorded (reactivity bias). Social desirability bias may lead to omission of some foods and beverages, with mixed evidence on whether foods and beverages that are considered less healthy are misreported to a greater extent than other foods.[23]

Random error, which mainly affects short-term dietary data, can be addressed through data collection and analytic strategies, as discussed subsequently. Unless a criterion marker of true intake, such as observational or recovery biomarker data, is available, it is not possible to analytically remove the systematic error or bias. Depending on the direction, bias may lead to underestimating or overestimating intake of a given dietary component, as well as misestimating overall diet quality based on intake of a number of food groups and/or nutrients. Because social desirability bias may lead to over-reporting of foods that are encouraged and under-reporting of foods to be consumed infrequently, the interpretation within surveillance research may be that intake of fruits and vegetables is even lower or intake of sugary beverages even higher than the data suggest, necessitating interventions to shift dietary patterns. However, misestimation of intake can make it challenging to identify targets for dietary guidance and intervention research. In intervention studies, exposure to the intervention itself may change reporting,[24,25] another source of bias. Further, bias may be differential among individuals with different characteristics, such as body weight status, complicating comparisons of intake among subgroups. Bias also has implications for epidemiologic research to examine diet-health relationships, typically resulting in attenuation of observed associations, as well as reduction of statistical power to detect such associations.[26]

Considering the types and extent of error that affect data collected using different dietary assessment methods is critical to determining the most appropriate approach to address a given research question, as well as in guiding data analysis and interpretation to mitigate the effects of error.[17,26-28] However, characterizing error in dietary assessment methods is challenging due to the paucity and burden of measures of true intake, as noted earlier. Studies using observation and recovery biomarkers are typically short term and/or involve small, select samples that may not be generalizable to the broader population. Consequently, for validation purposes, self-report methods are often compared with other self-report methods that are thought to be affected by less bias. For instance, FFQ data may be compared with data from a diet or food record. However, because the errors in different self-report methods may be correlated,[29,30] comparisons of methods to one another may lead to results that underestimate the bias and overestimate the validity of the method being evaluated.

Contextual Factors Related to Dietary Intake

Beyond characterizing the foods and beverages consumed in assessing dietary intake, we may seek to capture contextual information, such as the timing and patterning of eating occasions; with whom meals were eaten; the location of eating; the sources of the foods and beverages consumed (e.g., grocery store, restaurant, hunted or foraged, food pantry); and engagement in other activities, such as using a computer or other device, while eating. This contextual information is important for understanding how facets beyond what is consumed may influence health; for example, there is growing attention to how meal frequency and temporality may relate to diet quality and health.[31-33] Information on contextual factors can also inform interventions that address factors at various levels of the socioecological model[34,35] that shape dietary intake, including those that may contribute to dietary inequities. Self-report dietary assessment methods differ in their capacity to collect contextual details, such as where meals were eaten and with whom, and whether meals were consumed while engaging in activities such as the use of a computer or mobile device.

Methods

Self-Report Methods

A comparison of the key features of self-report methods is provided in **Table 49.1**.

Short-Term Self-Report Methods

Short-term methods of dietary assessment provide a snapshot in time, typically aiming to capture comprehensive information about the foods, beverages, and supplements consumed over a day or few days to allow quantification of intake with detail. Short-term methods often seek to capture not only the foods and beverages consumed but also the amounts, methods of preparation

Table 49.1 Key Features of Different Self-Report Dietary Assessment Methods

	Advantages	Disadvantages
24-Hour Dietary Recall	■ Aims to capture all foods and beverages, and potentially, supplements, consumed over the 24-hour period ■ Not limited to a preset food list and can be used with diverse populations ■ Mean usual intake for a population or population subgroup can be estimated using a single recall per person ■ Can collect contextual information, such as timing and location of eating occasions ■ Does not alter eating behaviors if unannounced ■ Estimated intakes are less affected by bias compared with data collected using long-term methods ■ Multiple-pass systems can enhance recall and accuracy ■ If interviewer administered, literacy requirements are minimal ■ Online self-administered systems eliminate the need for trained interviewers and coders	■ Paper-based 24HDRs require trained interviewers and coders ■ Affected by recall and social desirability bias ■ Requires at least two nonconsecutive recalls among a representative subsample to estimate distributions of usual intake ■ Depending on the dietary component of interest, a large number of nonconsecutive recalls is needed to estimate usual intake for a given individual ■ Some population subgroups, such as children, may not be able to independently complete, although proxy reporting can be used

(continues)

Table 49.1 Key Features of Different Self-Report Dietary Assessment Methods *(continued)*

	Advantages	Disadvantages
Diet or Food Record	■ Aims to capture all foods and beverages, and potentially, supplements, consumed during the days recorded ■ Not limited to a preset food list and can be used with diverse populations ■ Mean usual intake versus intakes for a population or population subgroup can be estimated using a single record per person ■ Can collect contextual information, such as timing and location of eating occasions ■ Amounts consumed can be weighed to improve accuracy ■ If completed in real time, not affected by recall error ■ Estimated intakes are less affected by bias compared with data collected using long-term methods ■ Mobile records leverage images to improve real-time recording ■ Online records offer automated coding and linkage to food composition databases	■ High participant burden ■ Paper-based records require a substantial coding effort to link with food composition databases ■ Affected by reactivity bias such that participants may change their intake in response to recording ■ Requires literacy and numeracy ■ Requires at least two nonconsecutive records among a representative subsample to estimate distributions of usual intake ■ Depending on the dietary component of interest, a large number of nonconsecutive records is needed to estimate usual intake for a given individual ■ Some population subgroups, such as children, may not be able to independently complete, although proxy reporting can be used
Food Frequency Questionnaire	■ Aims to capture usual intake over a period of time with a single administration ■ Does not alter eating behaviors ■ Can be self-administered in populations with literacy and numeracy skills ■ Online versions using automated skip patterns can reduce respondent burden	■ Food lists must be tailored to the populations of interest ■ May not capture portion size details ■ Affected by substantial bias compared with short-term methods ■ Does not provide contextual information, such as meal timing and location
Screener	■ Aims to capture usual intake over a period of time with a single administration ■ Does not alter eating behaviors ■ Can be self-administered in populations with literacy and numeracy skills ■ Online versions using automated skip patterns can reduce respondent burden	■ Does not capture all foods and beverages consumed ■ Affected by substantial bias compared with short-term methods ■ Does not provide portion size details or contextual information, such as meal timing and location

and, in some cases, contextual details, including the location of eating occasions, with whom meals or snacks were eaten, the sources of the foods consumed, and other activities engaged in while eating, such as using a mobile device. Intake over the short-term period may be recalled retrospectively, as in the case of 24HDRs, or recorded in real time, as in the case of diet or food records.

24-Hour Dietary Recall

The 24HDR uses a structured interview consisting of open-ended questions to query all foods and beverages, and usually supplements, consumed during the previous day. 24HDRs may be implemented using paper and pencil but are increasingly conducted using computerized systems, either by interviewer or by self-administration. Although in-person collection by a trained interviewer is ideal, telephone and online self-administered recalls have been shown to have reasonable accuracy.[36-38] Recalls are typically implemented using multiple passes to capture foods and beverages consumed at eating occasions during the previous day and details including preparation methods, the consumption of commercial vs homemade versions of foods, and amounts consumed. Recipes may be recorded and disaggregated by ingredient to improve the accuracy of estimates of intake of nutrients and other dietary components. Contextual information, such as the location of eating occasions and with whom meals were eaten, may also be collected. A recall may take 20 to 60 minutes to complete.[29]

Strengths of 24HDRs include the level of detail collected and the ability to report from among the great diversity of foods available, including foods specific to different cultures, as opposed to long-term methods that typically use a preset list of foods. However, because intake is recalled, some foods and beverages consumed may be forgotten. Alternatively, commonly consumed foods and beverages may be reported, even if they were not consumed on the recalled day. If the recall is not unannounced, there may be reactivity bias, such that individuals change their intake. A study that pooled data from several validation efforts using recovery biomarkers found that factors associated with misreporting of intake included age, educational status, and body weight status.[21] Additionally, the need to remember what was consumed and to estimate amounts consumed may limit the use of recalls in some groups, such as children who may not yet have fully formed cognitive skills. Proxy-reported or proxy-assisted recalls may be used for children and other individuals who are unable to complete the recall independently.

24HDR are often used in surveillance systems,[39,40] including in the United States. The What We Eat in America (WWEIA) dietary intake interview component of the National Health and Nutrition Examination Survey (NHANES)[41] provides data to monitor the dietary intake of the population and population subgroups.[42] To improve the accuracy of the data collected on foods and beverages consumed the prior day from midnight to midnight within WWEIA, the U.S. Department of Agriculture developed and evaluated the structured and computerized step-by-step Automated Multiple-Pass Method (AMPM) to conduct recalls in English and Spanish.[43] The steps include the quick list; the forgotten foods list; the time-and-occasion step; the detail cycle that prompts for detailed descriptions of foods, portion sizes, and additions (e.g., milk added to cereal or coffee); and a final probe that asks the participant to report anything else consumed during the 24-hour period that is the focus of the recall (**Table 49.2**). Portion size estimation is aided using the U.S. Department of Agriculture (USDA) Food Model Booklet,[44] as well as measuring cups and spoons. The method incorporates specialized training of interviewers, as well as coders who code the foods and beverages reported using USDA food composition databases. Proxy and proxy-assisted reporting are used for children under the age of 12 years and for other individuals who are unable to self-report their intake. Validation studies using recovery biomarkers and comparing 24HDR data collected using AMPM to data from food records and FFQs suggest reasonable validity, specifically indicating that the method can assess intake of energy and macronutrients, as well as sodium intake, with reasonable accuracy.[43,45-48]

Collection of 24HDRs has traditionally been costly and burdensome because of the need for trained interviewers and coders. To eliminate this need, automated self-administered recall interfaces have been developed.[49] In the United States, the Automated Self-Administered 24-hour Dietary Assessment Tool (ASA24) was developed in English and Spanish by the National Cancer Institute (NCI).[49] ASA24 has been adapted for use in Canada (in English and French) and Australia, with linkage to country-specific food composition databases. The ASA24 system includes a website that allows researchers and practitioners to identify the parameters for the recalls (e.g., number of recalls) and obtain nutrient and food group files; and a respondent website that prompts participants to enter the foods and beverages consumed. Recalls can be completed on computers or mobile devices. To facilitate self-administration, the recall is organized by eating occasion, using multiple passes adapted from the AMPM.[50] The passes include a meal-based quick list, meal gap review, detail pass, final review, forgotten-foods list, last chance to report foods and beverages, and finally, a usual intake question querying whether the amount consumed yesterday was more than usual, usual, or less than usual.[51] To report foods and beverages consumed, participants choose from food list terms or can use free-text entry if they cannot find the specific food or beverage. Consequently, similar to interviewer-administered recalls, it is possible to report a diversity of foods, although the codes assigned to free-text entries may require verification.[52] Portion size reporting is aided by digital images, informed by cognitive testing.[50,53] Foods and beverages reported are automatically coded using USDA food composition databases to yield estimates of energy, nutrient, and food group intakes. Researchers and practitioners can choose to have ASA24 provide participants with a Respondent Nutrition Report, comparing their intake to dietary guidance, after the completion of a recall.

Table 49.2 Steps in the Automated Multiple-Pass Method (AMPM)

Step	Approach
Quick List	Ask for a list of foods and beverages consumed on the previous day
Forgotten Foods	Ask about frequently forgotten foods (e.g., beverages, snacks)
Time & Occasion	Ask for the time of day and meal type for each eating occasion
Detail Cycle	For each food or beverage, ask for detailed description (recipe if needed), portion size, additions to the food (e.g., sugar in coffee?)
Final Probe	Ask about anything else that may have been consumed

Recalls can be completed using ASA24 for the prior day from midnight to midnight or for the 24-hours preceding the initiation of the recall (e.g., if the recall is initiated at noon on Monday, the beginning of the recall period queried is noon on Sunday). The NCI indicates that completing a recall takes 17 to 34 minutes, with an initial recall taking longer than subsequent recalls.[54] Researchers can opt to include modules capturing with whom meals were eaten, locations of eating occasions, sources of foods and beverages, and intake of supplements (participants can report supplements in any study using ASA24 but selecting the supplement module means that they will be explicitly prompted to report use of vitamin and mineral supplements). The U.S. version of ASA24 also includes a sleep module assessing sleep timing, quantity, and quality.[54]

Observational feeding studies have been conducted to compare ASA24 recall data to estimated true intake for select eating occasions, in addition to comparisons to 24HDR data collected using AMPM.[38] These studies suggest that ASA24 recall data have reasonable validity relative to true intake and are comparable to interviewer-administered AMPM 24HDRs for estimating intakes of nutrients and food groups among a diverse sample of adults, as well as women with low incomes.[37,38] ASA24 has also been evaluated relative to recovery biomarkers, indicating that similar to interviewer-administered recalls, estimates of absolute intake of energy, protein, potassium, and sodium from ASA24 are affected by more random error and less systematic error, compared with data from FFQs.[55] There is some indication that the use of digital images, which can be tailored to different foods and beverages, may improve portion size estimation compared with the U.S. Department of Agriculture Food Model Booklet.[56] However, accurate portion size estimation remains a major source of error in 24HDR and other self-report dietary intake data,[57] necessitating continued innovation.

Furthermore, self-administered 24HDRs may not be appropriate for all population groups. For example, ASA24 can be challenging for children under 12 years of age to complete independently[58] given incompletely developed cognitive skills. In some cases, a research assistant may aid participants in completing the recall or use the demonstration version to walk through the steps with the participant in advance of requesting that they complete their own recall. The NCI provides a variety of resources to aid researchers and practitioners, including quick start guides that can be shared with participants.[59] Piloting of the research protocol with the population of interest is encouraged in advance of launching a study using ASA24.[60]

Diet or Food Record

Diet or food records are used to prompt participants to record everything they eat and drink, and corresponding details such as brand names, preparation methods, and amounts consumed, as well as supplement use, in real time. Records may be kept by paper and pencil or using online interfaces. Similar to 24HDRs, food records are typically open ended, allowing individuals to report a diversity of foods and beverages that are not limited to predetermined food lists. As with 24HDRS, recipes may be recorded and disaggregated by ingredient to improve the accuracy of estimates of intake of nutrients and other dietary components. Contextual information related to eating occasions, such as location, may also be collected. Portion size may be estimated, potentially facilitated by food models or other aids, or participants may be asked to weigh the amounts of foods and beverages consumed. Because individuals record their intake in real time, and often use portable scales to weigh the exact amount, this method may reduce omissions and enhance complete recording and was once considered to be a gold standard for dietary assessment.[61]

Although records can enable collection of detailed intake data, they are burdensome for participants, who are often asked to keep records for consecutive days, such as 3, 4, 5, or 7 days, with recording generally requiring at least 15 minutes per day,[29,30] depending on the number and complexity of foods and meals consumed. Participants may receive training on the level of detail to record. Some participants may not record their intake consistently in real time, but rather, recall it at the end of the day, introducing recall bias due to imperfect memory. By contrast, real-time recording can introduce reactivity whereby participants change their intake, eating differently and less during the recording period. In some cases, observers may weigh and measure foods to ensure accuracy, but their presence may also change intake. However, having a trained research assistant review the record with the participant after completion may improve data quality.[62] Quality and completeness of records may decrease with increasing number of days recorded due to fatigue among participants, with the potential for selection bias in that those who complete the full record differ from those who do not. Some validation studies have shown that, among methods, total energy intake was lowest using food records.[63] More recent validation studies suggest that records, particularly when they are weighed, perform the best relative to 24HDRs and FFQs in comparison to recovery biomarkers based on attenuation factors and correlation coefficients.[55] Nonetheless, the literacy and numeracy demands of food records may preclude use with some subgroups. As with 24HDRs, food records may be completed by proxies, such as parents on behalf of children.

Diet or food records are used in some surveillance systems[39,40] and have been used in some large epidemiologic studies.[64] However, few large studies use records due to the cost and respondent burden, although this may change in the future, as similar to advances in 24HDR methodology, interfaces have been developed to enable online and mobile records. For example, the ASA24 system has been expanded to include the capacity to collect diet or food records.[60] While collecting paper-based records is not necessarily burdensome for researchers, depending on the level of support provided to participants, coding the completed records requires a substantial effort. ASA24 allows for automatic coding using USDA food composition databases, greatly reducing the effort needed by researchers. Similar to the use of ASA24 for 24HDRs, researchers and practitioners use the researcher website to set the parameters for data collection, including whether they wish participants to complete single-day or consecutive-day records. Participants are instructed to visit the respondent website to record the foods and beverages they consume along with their details, such as preparation methods and amounts consumed, in real time. After recording an eating occasion, participants are queried about whether it is their last meal for the day and if so, they are prompted to complete the final review, forgotten foods pass, and final question on whether their intake on the recorded day was usual. Participants can review and edit the information they have entered for up to 1 day after completing the record.[51]

ASA24 can be used to complete recalls or records on computers or mobile devices, such as tablets and smart phones. Other mobile food record applications may allow participants to take images of foods and beverages before and after eating occasions, as well as to enter foods and beverages consumed by typing in food descriptions, making voice recordings, capturing nutrition labels or barcodes, or selecting from lists of foods.[65] Such applications may reduce participant burden associated with recording foods and beverages as they are consumed. In systems using images, the images may be reviewed by trained humans, with verification by the participant, or image review may be automated using pattern recognition and volume estimation techniques.[65,66] Mobile records can also provide information on the timing and duration of eating.[66]

For instance, developed within the U.S. context, the technology-assisted dietary assessment system includes a mobile application that allows the user to enter information and capture images, and a cloud-like system that allows for food image analysis and integrates database servers.[65] Participants are instructed to include a fiducial marker (e.g., a checkerboard with known colors and dimensions) in their images to allow for color and spatial calibration and are alerted if the marker is not detected in the image. Pattern recognition techniques are used to classify foods by comparing the image with database images, and portion size is quantified based on the estimated volume.[65] As with other self-report methods, amounts of foods and beverages consumed are then converted to estimated intake of nutrients and food groups using food composition databases.

A recent review identified varied characteristics of smartphone-based dietary assessment methods, including food record applications, used in research,[66] identifying that available methods vary widely in their features and characteristics. The variation across methods may lead to differences in the information recorded, necessitating care in selecting a mobile application that matches the data needed to address a given research question. The authors also note that the validity of smartphone-based methods is rarely examined.[66]

Mitigating Measurement Error in Data from Short-term Methods

In some cases, it may be of interest to use data from short-term methods to characterize intake on a single day or over a few days; for example, with respect to binge drinking alcohol or exposure to pesticide residues or other toxins. However, as noted, interest in various types of research is typically in usual or habitual intake. Because they are focused on one or a few days, short-term dietary intake data are affected by considerable random measurement error, driven mainly by day-to-day, within-person variability,[13,67] and do not directly measure usual intake. Studies of day-to-day variability show large differences across nutrients and foods and estimates of the number of days needed for relatively accurate estimates can be quite high, particularly for nutrients like vitamin A that are found in a limited number of foods.[67] While many repeat administrations of short-term methods are consequently needed to make inferences about the usual intake of a given individual,[67] a single administration of a short-term method can be used to estimate mean usual intake of a nutrient or other dietary component among a population or population subgroup, particularly if the recalls are spread out across days of the week and seasons, if applicable, at the group level.[17]

In many cases, there may be interest in characteristics of the distribution of usual intake beyond the mean, for example, to enable estimation of the proportions of populations or population subgroups with usual intakes below or above cut points, such as food group recommendations and nutrient requirement estimates.[68] For these applications, it is not possible to use the observed distribution of intake based on data for one or a few days, because this distribution is too wide, relative to the distribution of usual intake. Because of the random error, the ends of the distribution of intake based on a single recall, or even the mean of two recalls, vs the distribution of usual intake, are extended and estimates of the proportion of the population with low or high intake are exaggerated. To estimate the distribution of usual intake, it is necessary to collect repeat administrations of the short-term method in at least a subsample, which should be representative of the full sample (**Figure 49.1**).[69] The repeat recalls can be used in usual intake modeling, which removes the within-person variation.[17,70] In other words, analytic methods are used to adjust for the random measurement error, resulting in pulling in the tails of the distribution of intake for a population or population subgroup. The repeat administrations should be nonconsecutive because intake on consecutive days may be correlated, potentially leading to underestimation of within-person variation.[13,71] This means that repeat administrations of multiple-consecutive-day diet or food records are recommended if there is interest in using record data to estimate distributions of usual intake to allow for estimates of the prevalence above or below a cut-point.

Several software implementations have been developed to enable usual intake modeling, mainly focused on 24HDRs given their use in surveillance, including in WWEIA. For example, the National Cancer Institute (NCI) method allows univariate modeling of a single dietary component, bivariate modeling for two correlated dietary components (e.g., saturated fat or added sugars intake as a proportion of energy intake), and multivariate modeling for multiple correlated dietary components, such as the several components included in diet quality indices.[72-74] In addition, the NCI method not only allows accounting for correlations among dietary components but also whether dietary components are consumed most days by most persons vs episodically consumed, along with other potential factors influencing intake, such as whether the recall was for a weekday vs weekend day and the sequence of the recalls. Covariates such as age and sex can be included to enable estimation of distributions of usual intake among subgroups used in dietary guidance. The NCI method has been used to estimate prevalence of inadequate or excessive intakes among population subgroups to inform the Dietary Guidelines for Americans,[75] as well as to estimate the distribution of Healthy Eating Index scores based on simulated usual intakes.[11,76] Ongoing method development is underway to improve the user friendliness of software implementations for usual intake modeling.[77]

Even with appropriate accounting for random measurement error in data from short-term methods, interpretation of results must consider that the data are also affected by bias.[21,27,55,78,79] The distribution of intake, including the mean, based on biased self-report data is shifted left (underestimation) or right (overestimation) compared with the distribution of true intake. Given availability of recovery biomarker data, it may be possible to

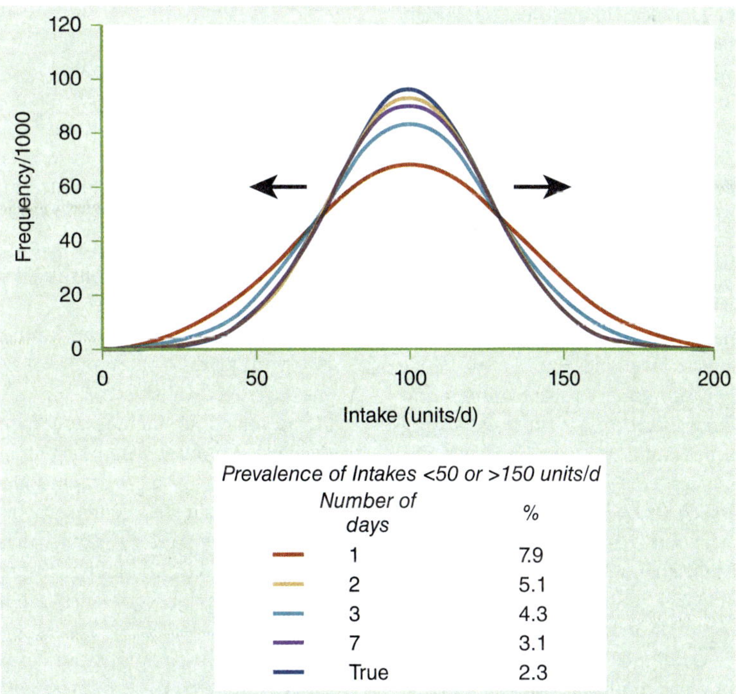

Figure 49.1 Simulation of impact of random error on observed intake distributions; also shown is the impact that days of data collected would have on the proportions of individuals with low (<50 units) or high (>150 units).

Reproduced from Beaton GH. Am J Clin Nutr. 1994;59(1 Suppl):253S-261S. doi:10.1093/ajcn/59.1.253S

calibrate short-term data to reference data to improve accuracy, including for epidemiologic purposes.[80] Although recovery biomarker data are typically not available, because short-term methods tend to collect intake data with less bias than long-term methods, expansion of their use, including to assess diet-health associations, may nonetheless improve nutrition research.[81] Expanding capacity to collect short-term data with low cost and burden has been the impetus for innovations such as ASA24 and mobile food record applications.

Long-Term Self-Report Methods

FFQs and screeners aim to capture usual intake, typically over a month or a year. They may capture the frequency with which particular foods and beverages are consumed, and possibly, typical portion sizes.

Food Frequency Questionnaires

FFQs query the foods, beverages, and often, supplements, consumed during the last month or year (or other specified time period), from a list of food items, and how frequently each was consumed. FFQs are typically self-administered using paper-based or online questionnaires. An FFQ may be interviewer-administered under some circumstances, such as when it is used with individuals with low literacy. To limit the length of the questionnaire and promote completion, a finite set of questions, or line items, usually queries groups of related foods, which limits the granularity of the possible data analysis and the ability to estimate intake of some nutrients. To capture the total diet, an FFQ will typically include line items specifying upward of 80 foods and beverages and require at least 30 minutes to complete.[29,30] For frequently consumed foods likely to be major contributors to nutrient intake, follow-up questions, such as type of milk used or type of breakfast cereal, may be included. To avoid lengthy FFQs, follow-up questions are limited, meaning that details on issues of current interest, such as level of processing and specific recipe used, may not be available. Furthermore, FFQ food lists become outdated as the food supply changes, which can be problematic when a questionnaire is used in multiple waves of data collection. Updates may be made to the questionnaire for use in subsequent waves of the same study, but the impact of the changes on comparability of estimates of nutrients and food groups over time should be assessed. Technological innovation in FFQs is limited compared with advances in 24HDRs and food records, though online versions enable automation of skip patterns, simplifying the completion of the questionnaire for participants.

A strength of FFQs is that they aim to collect information on usual intake with a single administration, therefore, collecting data affected by little random measurement error. Consequently, repeat administrations are not needed to enable estimation of distributions of usual intake, such that FFQs are cost-effective. Analysis is straightforward compared with short-term methods, as estimated usual intake of nutrients and food groups over a period of time based on an FFQ can be directly compared with food group recommendation and nutrient requirement estimates, although if portion size is not ascertained, it is only possible to estimate frequency of consumption of different foods. However, FFQ data are affected by considerable systematic bias, arising from recall bias, the cognitive challenges inherent in averaging intake of foods and beverages over a long period of time, omission of foods and beverages commonly consumed in the population(s) of interest within the food lists, and social desirability bias. Furthermore, FFQs do not collect information such as methods of food preparation, nor meal timing and other contextual factors that may be of interest.

Quantitative estimates from an FFQ must be interpreted in light of the substantial bias, which is why short-term methods are typically preferred for surveillance. However, given their feasibility, FFQs are widely used in large studies, including epidemiologic research. The bias in FFQ data is problematic in this context because it can attenuate observed relationships between diet and health.[26] Bias also causes a loss of statistical power, meaning that very large samples may be needed to study diet-health associations.[26] Although the administration of the FFQ itself may not be overly costly, recruiting a large sample and administering other study measures to that sample may be cost prohibitive.

Evaluated and commonly used FFQs in the U.S. context include the Harvard or Willett Semi-Quantitative Food Frequency Questionnaire (SFFQ), the Block FFQ, and the Diet History Questionnaire (DHQ). The Willett SFFQ FFQ was developed for use in the Nurse's Health Study[82] and originally had 61 items. It was expanded for use in the Male Health Professionals Study[83] with 131 items. Selection of the food list was based on the ability to rank individuals by nutrient intake, using regression methods, with input from nutrition experts. The SFFQ does not include portion size options, but rather, lists a standard portion size with the instructions to change the frequency if needed because a different portion size is used. Based on extensive testing in their nurse population, it was determined that the greatest variation was in frequency and that portion size was not needed. Rather, adjusting for total energy intake helps to restandardize intakes for good relative ranking.[84] In contrast, the Block FFQ was developed at the NCI using 24HR data from NHANES[85] and originally had 110 items. The food list was developed by ranking food sources of key nutrients to ensure that the most commonly consumed foods were included. Based on variation in portion sizes noted in the 24HR, the Block FFQ includes small, medium, and large portion sizes, which are then quantified and multiplied by the frequency to obtain estimates of usual intakes. Both FFQs have associated food composition databases that are used to convert frequencies to estimated intakes of nutrients and food groups.

Each of these FFQs have been used in numerous studies. However, they are not without limitations. The primary limitation is the food lists. In both cases, the list is based on the most commonly consumed foods in the United States. However, foods that are less frequently consumed in the population overall but that are frequently consumed by subsets of the population are not included. To address this limitation, tailored FFQs have been developed for different populations of interest. A multiethnic FFQ was developed for use in a large study in Hawaii and California and designed to include the foods commonly consumed by five major groups, non-Hispanic White, African American, Latino, Japanese American, and Native Hawaiian, based on diet records.[86,87] As it was recognized that not only foods but also portion sizes may vary by ethnic identity, photographs of portion sizes were used to aid in reporting accuracy.[88] Subsequently, many food frequency questionnaires tailored to particular populations have been developed.[89-92] The use of appropriately different FFQs with different populations can, however, complicate syntheses of evidence, for example, to inform dietary guidance.

Based on cognitive interviewing, the NCI created a new format for the FFQ, resulting in the DHQ I.[93] Individuals were asked to talk aloud as they completed different FFQ versions to identify potential areas of difficulty. Identified areas for improvement included changing from the grid format to sequential questions about frequency and portion size, clarifying when foods are eaten seasonally, and ordering foods to avoid overreporting or redundancy. The initial version of the DHQ included 124 food list items, along with portion size assessment, and dietary supplement questions. The current version, the DHQ III, is online only and is freely available and intended for use with adults 19 years of age and older.[94] It was developed using 24HR data from NHANES, in this case 2007-14. The DHQ III queries 135 food and 26 dietary supplement items, and is available in English and Spanish and with or without portion size prompts.

Efforts have been undertaken to assess the validity of the Harvard SFFQ, the Block FFQ, and the DHQ, suggesting that they perform relatively similarly and that, with energy adjustment, they are comparable for the assessment of diet-health associations.[95] As noted earlier, biomarker-based validation studies indicate that various versions of FFQs, including paper-based and online, are affected by considerable bias compared with 24HDRs and FRs.[21,55]

Brief Dietary Screeners

Brief dietary instruments, called screeners, are commonly used to rapidly capture usual intake over a period such as a month or a year. Screeners are often frequency-based and are typically self-administered, by paper and pencil or online. They often focus on a limited number of dietary components such as fat or fruits and vegetables, generally do not include portion size information, and may require less than 15 minutes to complete, thus lessening respondent burden compared with a full FFQ. Due to their brevity, screeners do not capture the total diet, necessitating care in the interpretation of the resulting data depending on the purpose. For instance, unless purposively designed to do so, screeners may not enable quantification of intake of a given dietary component, but rather, they can provide a crude indication of high or low intake of various foods, or the quality of the overall diet in the case of multidimensional screeners.

As with FFQs, the foods queried by a screener must be relevant to the population or population subgroup of interest, with the omission of major contributors of the dietary component(s) of interest lessening the screener's usefulness. Furthermore, to the extent that a screener requires participants to recall and average intake over a long period of time, such as a month or year, they can be cognitively challenging and may not be appropriate for use with all population subgroups.

Because screeners do not capture the total diet, they are not generally evaluated relative to recovery biomarkers but may be compared with other self-report methods to provide an indication of their comparative validity. As noted earlier, because of correlated errors between self-report methods, correlation coefficients describing associations between estimates from a screener and another method may be inflated and thus, overly optimistic.

Examples of screeners include the Center for Disease Control and Prevention and Surveillance's Behavioral Risk Factor Surveillance System Fruit and Vegetable Module[96] and the NCI's Dietary Screener Questionnaire (DSQ),[97] which has been used in NHANES.[98]

Mitigating Measurement Error in Data from Long-Term Methods

As noted previously, it is not possible to analytically remove the bias in self-report data unless a reference measure, such as a recovery biomarker, is available, which is rarely the case. Energy adjustment is recommended to mitigate error.[99] In biomarker-based validation research, FFQs perform better for nutrients expressed relative to energy intake compared with absolute nutrients.[55] This is important given interest in understanding the composition of the diet rather than only intake of isolated nutrients. In some cases, short- and long-term methods may be used in combination to leverage the strengths and mitigate the limitations of each method.[100,101] In particular, using a combination of short- and long-term methods has been recommended to improve our capacity to identify diet and health relationships in epidemiologic research.[100] This is because systematic error in FFQ data can distort the observed relationship between diet and health and disease outcomes, whereas the focus of short-term methods on a few days can overlook infrequently consumed foods that are important to diet-disease relationships. For example, within the context of a cohort study using FFQs as the main dietary assessment method, 24HDRs or records may be collected from a subsample in a calibration substudy and regression calibration[102] used to mitigate the impact of bias in the FFQ on observed diet and health associations.[26] Although the reference measure—24HDRs or records in this case—is not unbiased, its use to mitigate the impact of bias in the FFQ data may improve estimates of diet-health associations. Work is also underway to leverage concentration, predictive, and metabolomic-based markers to improve estimation of diet-health associations, as discussed briefly subsequently.

Screeners are not typically used in epidemiology, but they are used in surveillance, often in surveys that query an array of topics. When data are available from a brief screener along with a more comprehensive self-report method, it may be possible to develop algorithms to calibrate the screener data to the more accurate method. Thompson et al. used 24HDR data from WWEIA along with screener data from NHANES to develop scoring algorithms for the DSQ that generated estimates of mean intake and prevalences below thresholds that aligned closely with estimates from the 24HDRs.[98] A similar approach was used to estimate the prevalence of individuals meeting recommendations for fruit and vegetable intake based on data from the BRFSS Fruit and Vegetable Module,[103] expanding the utility of the screener. However, the analytic approach is intensive, making use of the NCI method for estimating distributions of usual intake using the 24HDR data, and the algorithms developed for the screener may not be applicable to other datasets or populations.[98]

Technological Innovations

In addition to technological advances to improve the feasibility and potentially the accuracy of traditional self-report methods, an array of other technology-based methods are under investigation. These include image-based methods such as wearable cameras, methods to measure the mass of foods using cutlery, wearable sensors that assess factors such as chewing frequency and bite size,[104] and motion sensors to detect eating activities.[105] These approaches differ in the aspects of dietary intake they capture, with implications for the research questions they are able to address. A range of novel methods for assessing different facets of intake will undoubtedly continue to emerge as researchers aim to address the limitations of self-report methods.

Biomarkers

The identification of recovery biomarkers that reflect usual intake of dietary components has been critical to the validation of self-report dietary assessment methods. However, few such biomarkers have been identified. Most available biomarkers are indirectly associated with intake, as they are absorbed, metabolized, and stored differentially in the body.

Recovery Biomarkers

Recovery biomarkers, as noted above, can be used as reference measures of true intake in validation studies. They are directly related to intake and are not affected by homeostasis or substantial differences in metabolism between persons.[20] Recovery biomarkers have been identified for energy, protein, potassium, and sodium. Doubly labeled water represents energy expenditure over a 10 to 14 day period, with the assumption that individuals are in energy balance.[106] Protein intake over 24 hours is based on 24-hour nitrogen level in urine,[107] with some systematic bias possible due to incomplete collection of 24-hour urines. Potassium and sodium are also based on 24-hour urine collection.[21] Similar to short-term dietary assessment methods, recovery biomarkers are affected by within-person variation.[106] Therefore, repeat administrations are needed to estimate distributions of usual intake among a group by characterizing and adjusting for within-person random error.

Although short-term studies to assess 24HDRs and food or diet records may suggest that they can estimate energy intake with reasonable validity, validation studies based on recovery biomarkers suggest that energy is poorly estimated by self-report methods.[21,78] This is not surprising because almost all foods and beverages contribute energy and small errors in the estimation of each food and beverage compound.[22] Consequently, it is recommended that self-report data, particularly from FFQs, not be used to estimate energy intake. Protein and potassium appear to be better estimated by self-report methods, whereas sodium is less well estimated,[21,78] potentially partially due to the variation in sodium content among foods that is not well captured by food composition databases.

The discovery of additional recovery biomarkers would make valuable contributions to our capacity to assess the error in self-report dietary assessment methods for estimating intake of a range of nutrients and for use to assess diet and health associations and evaluate interventions.

Concentration Biomarkers

Unlike recovery biomarkers, concentration biomarkers do not have a stable relationship with intake.[108] However, concentration biomarkers are correlated with intake.[109] For example, serum carotenoids may be correlated with fruit and vegetable intake. Concentration biomarkers have been used to provide insight into the validity of self-report dietary assessment methods[110,111] and work is ongoing to explore the potential to use concentration biomarkers to improve estimation in different types of research, including in epidemiologic and intervention studies.[112-114]

Predictive Biomarkers

Predictive biomarkers are similar to recovery biomarkers, with a stronger dose-response relationship to intake than concentration biomarkers,[108,115] but they do not provide unbiased estimates of true intake. Predictive biomarkers may be useful for identifying reporting errors and mitigating error in data from self-report methods. Examples include 24-hour urinary fructose and sucrose as markers of sugar intake.[116,117]

Metabolomics

Most recently, metabolomics has been identified as a potentially promising avenue for indicating intake of different foods and nutrients and for assessing the validity of self-report and other methods for estimating intake of some foods or nutrients.[118] Metabolomics techniques can identify hundreds of metabolic compounds in the blood with a single assessment, allowing for the identification of potential markers of food or nutrient intake, as well as clarification of metabolic pathways. Similar to efforts surrounding concentration biomarkers, research is underway to explore the use of metabolomic-based biomarkers in understanding diet and health associations.[119]

Summary

There has been substantial effort over recent decades to improve methods for collecting and analyzing dietary intake data. Ongoing innovations leveraging pattern recognition and other machine learning techniques, as well as capabilities of mobile devices and other technologies, are likely to shift the field of dietary assessment significantly in the coming years. However, given the complexity of dietary intake, including multidimensionality and dynamism, limitations remain in current methods, necessitating careful method selection and data analysis and interpretation, as well as continued improvements in dietary assessment. Given the critical need to pursue human health, health equity, and planetary health, as well as efforts to understand heterogeneity in dietary patterns and diet-health associations,[120] it is crucial to ensure that method development, evaluation, and application are responsive to the dietary intake, cognitive abilities, and other characteristics of diverse populations.

References

1. Bailey RL, Ard JD, Davis TA, et al. J Nutr 2021;151:1197–204.
2. NASEM. Review of WIC food packages: improving balance and choice: final report, The National Academies Press, Washington, D.C., 2017.
3. US Food & Drug Administration. Sodium Reduction. Accessed August 4, 2023. Available at www.fda.gov/food/food-additives-petitions/sodium-reduction.
4. Jennings R, Henderson AD, Phelps A, et al. Nutrients 2023;15.
5. Willits-Smith A, Lopez M, O'Malley K, et al. Curr Dev Nutr 2022;6:83.
6. Willits-Smith A, Tooze J, Heller M, et al. Curr Dev Nutr 2019;13.
7. Reedy J, Subar AF, George SM, et al. Nutrients 2018;10.
8. World Cancer Research Fund/American Institute for Cancer Research. Diet, Nutrition, Physical Activity and Cancer: a Global Perspective. Continuous Update Project Expert Report. Available at dietandcancerreport.org. 2018.
9. U.S. Department of Health and Human Services, and U.S. Department of Agriculture. Dietary Guidelines for Americans. Accessed August 4, 2023. Available at www.dietaryguidelines.gov/.
10. Willett W, Rockstrom J, Loken B, et al. Lancet 2019;393:447–92.
11. Reedy J, Lerman JL, Krebs-Smith SM, et al. J Acad Nutr Diet 2018;118:1622–33.

12. O'Malley K, Willits-Smith A, Rose D. Am J Clin Nutr 2023; 117:701–8.
13. Tarasuk V, and Beaton GH. Am J Clin Nutr 1991;54:464–70.
14. Beaton GH, Milner J, McGuire V, et al. (1983) Am J Clin Nutr 37, 986–95.
15. Beaton GH, Milner J, Corey P, et al. Am J Clin Nutr 1979;32: 2546–59.
16. Crane TE, Latif YA, Wertheim BC, et al. Am J Epidemiol 2019;188:1304–10.
17. Kirkpatrick SI, Guenther PM, Subar AF, et al. J Acad Nutr Diet 2022;122:1246–62.
18. Simons-Morton BG, Baranowski T. J Sch Health 1991;61:204–7.
19. Kirkpatrick SI, Baranowski T, Subar AF, et al. J Acad Nutr Diet 2019;119:1801–16.
20. Kaaks R, Ferrari P, Ciampi A, et al. Public Health Nutr 2002;5: 969–76.
21. Freedman LS, Commins JM, Moler JE, et al. Am J Epidemiol 2015;181:473–87.
22. Subar AF, Freedman LS, Tooze JA, et al. J Nutr 2015;145: 2639–45.
23. Hebert JR, Clemow L, Pbert L, et al. Int J Epidemiol 1995;24: 389–98.
24. Natarajan L, Pu M, Fan J, et al. Am J Epidemiol 2010;172: 819–27.
25. Neuhouser ML, Tinker L, Shaw PA, et al. Am J Epidemiol 2008;167:1247–59.
26. Freedman LS, Schatzkin A, Midthune D, et al. J Natl Cancer Inst 2011;103:1086–92.
27. Kipnis V, Subar AF, Midthune D, et al. Am J Epidemiol 2003;158:14–21; discussion 2–6.
28. Kirkpatrick SI, Collins CE, Keogh RH, et al. Nutrients 2018;10.
29. National Cancer Institute. Dietary Assessment Primer. Accessed August 4, 2023. Available at https://dietassessmentprimer.cancer.gov/.
30. Thompson FE, Kirkpatrick SI, Subar AF, et al. J Acad Nutr Diet 2015;115:1986–95.
31. Leech RM, Livingstone KM, Worsley A, et al. J Nutr 2016; 146:2027–34.
32. Leech RM, Worsley A, Timperio A, et al. Nutr Res Rev 2015;28: 1–21.
33. Aqeel MM, Guo J, Lin L, et al. J Nutr 2020;150:3259–68.
34. McLeroy KR, Bibeau D, Steckler A, et al. Health Educ Q 1988;15:351–77.
35. Centers for Disease Control and Prevention. The Social-Ecological Model: A Framework for Prevention. Accessed August 4, 2023. Available at www.cdc.gov/violenceprevention/about/social-ecologicalmodel.html.
36. Bogle M, Stuff J, Davis L, et al. J Am Diet Assoc 2001;101:216–22.
37. Kirkpatrick SI, Guenther PM, Douglass D, et al. J Nutr 2019;149:114–22.
38. Kirkpatrick SI, Subar AF, Douglass D, et al. Am J Clin Nutr 2014;100:233–40.
39. Huybrechts I, Aglago EK, Mullee A, et al. Proc Nutr Soc 2017;76:549–67.
40. De Keyzer W, Bracke T, McNaughton SA, et al. Nutrients 2015;7:3587–620.
41. Centers for Disease Control and Prevention. National Health and Nutrition Examination Survey. Accessed February 20, 2024. Available at www.cdc.gov/nchs/nhanes/index.htm.
42. Centers for Disease Control and Prevention. NHANES Dietary Analyses. Accessed August 5, 2023. Available at wwwn.cdc.gov/nchs/nhanes/tutorials/DietaryAnalyses.aspx.
43. Blanton CA, Moshfegh AJ, Baer DJ, et al. J Nutr 2006;136:2594–9.
44. Centers for Disease Control and Prevention. National Health and Nutrition Examination Survey Measuring Guides for the Dietary Recall Interview. Accessed August 5, 2023. Available at www.cdc.gov/nchs/nhanes/measuring_guides_dri/measuringguides.htm
45. Conway JM, Ingwersen LA, Vinyard BT, et al. Am J Clin Nutr 2003;77:1171–8.
46. Conway JM, Ingwersen LA, Moshfegh AJ. J Am Diet Assoc 2004; 104:595–603.
47. Moshfegh AJ, Rhodes DG, Baer DJ, et al. Am J Clin Nutr 2008; 88:324–32.
48. Rhodes DG, Murayi T, Clemens JC, et al. Am J Clin Nutr 2013; 97:958–64.
49. Subar AF, Kirkpatrick SI, Mittl B, et al. J Acad Nutr Diet 2012; 112:1134–7.
50. Zimmerman TP, Hull SG, McNutt S, et al. J Food Compost Anal 2009;22:S48–S51.
51. National Cancer Institute. ASA24® Respondent Website Methodology. Accessed August 5, 2023. Available at https://epi.grants.cancer.gov/asa24/respondent/methodology.html.
52. Zimmerman TP, Potischman N, Douglass D, et al. Procedia Food Science 2015;4:160–72.
53. Subar AF, Crafts J, Zimmerman TP, et al. J Am Diet Assoc 2010; 110:55–64.
54. National Cancer Institute. ASA24® Respondent Website Features. Accessed August 5, 2023. Available at https://epi.grants.cancer.gov/asa24/respondent/features.html.
55. Kirkpatrick SI, Troiano RP, Barrett B, et al. Am J Epidemiol 2022; 191:1125–39.
56. Kirkpatrick SI, Potischman N, Dodd KW, et al. J Nutr 2016; 146:2567–73.
57. Whitton C, Ramos-Garcia C, Kirkpatrick SI, et al. Adv Nutr 2022;13:2620–2665.
58. Raffoul A, Hobin EP, Sacco JE, et al. (2019) J Nutr 149, 1019–26.
59. National Cancer Institute. Downloadable ASA24® Resources. Accessed August 5, 2023. Available at https://epi.grants.cancer.gov/asa24/resources/.
60. National Cancer Institute. Automated Self-Administered 24-Hour (ASA24®) Dietary Assessment Tool. Accessed August 5, 2023. Available at https://epi.grants.cancer.gov/asa24/.
61. Carlsen MH, Lillegaard IT, Karlsen A, et al. Nutr J 2010;9:37.
62. Cantwell MM, Millen AE, Carroll R, et al. J Nutr 2006;136:440–5.
63. Sawaya AL, Tucker K, Tsay R, et al. Am J Clin Nutr 1996;63: 491–9.
64. Bingham SA, Welch AA, McTaggart A, et al. Public Health Nutr 2001;4:847–58.
65. Boushey CJ, Spoden M, Zhu FM, et al. Proc Nutr Soc 2017;76: 283–94.
66. Konig LM, Van Emmenis M, Nurmi J, et al. Health Psychol Rev 2022;16:526–50.
67. Basiotis PP, Welsh SO, Cronin FJ, et al. J Nutr 1987;117:1638–41.
68. Institute of Medicine (US) Standing Committee on the Scientific Evaluation of Dietary Reference Intakes. In: DRI Dietary Reference Intakes: Applications in Dietary Assessment, Washington, DC, 2000.
69. Beaton GH. Am J Clin Nutr 1994;59:253S–61S.
70. Dodd KW, Guenther PM, Freedman LS, et al. J Am Diet Assoc 2006;106:1640–50.
71. Larkin FA, Metzner HL, Guire KE. J Am Diet Assoc 1991;91: 1538–42.
72. Zhang S, Midthune D, Guenther PM, et al. Ann Appl Stat 2011;5:1456–87.
73. Zhang S, Krebs-Smith SM, Midthune D, et al. Int J Biostat 2011;7:1.
74. Tooze JA, Midthune D, Dodd KW, et al. J Am Diet Assoc 2006;106:1575–87.
75. Krebs-Smith SM, Guenther PM, Subar AF, et al. J Nutr 2010;140:1832–8.
76. Guenther PM, Kirkpatrick SI, Reedy J, et al. J Nutr 2014;144: 399–407.

77. Luo H, Dodd KW, Arnold CD, et al. J Nutr 2021;151:1329–40.
78. Freedman LS, Commins JM, Moler JE, et al. Am J Epidemiol 2014;180:172–88.
79. Subar AF, Kipnis V, Troiano RP, et al. Am J Epidemiol 2003;158: 1–13.
80. Prentice RL, Mossavar-Rahmani Y, Huang Y, et al. Am J Epidemiol 2011;174:591–603.
81. Schatzkin A, Subar AF, Moore S, et al. Cancer Epidemiol Biomarkers Prev 2009;18:1026–32.
82. Willett WC, Sampson L, Stampfer MJ, et al. Am J Epidemiol 1985;122:51–65.
83. Rimm EB, Giovannucci EL, Stampfer MJ, et al. Am J Epidemiol 1992;135:1114–26; discussion 27–36.
84. Willett W. Nutritional Epidemiology. Third Edition., Oxford University Press, Oxford; New York, 2013.
85. Block G, Hartman AM, Dresser CM, et al. Am J Epidemiol 1986;124:453–69.
86. Hankin JH, Wilkens LR. Am J Clin Nutr 1994;59:198S–200S.
87. Stram DO, Hankin JH, Wilkens LR, et al. Am J Epidemiol 2000;151:358–70.
88. Hankin JH, Wilkens LR, Kolonel LN, et al. Am J Epidemiol 1991;133:616–28.
89. Carithers TC, Talegawkar SA, Rowser ML, et al. J Am Diet Assoc 2009;109:1184–93.
90. Tucker KL, Bianchi LA, Maras J, et al. Am J Epidemiol 1998;148:507–18.
91. Mitchell DC, Tucker KL, Maras J, et al. J Nutr Health Aging 2012;16:667–72.
92. Denova-Gutierrez E, Tucker KL, Salmeron J, et al. Salud Publica Mex 2016;58:608–16.
93. Subar AF, Thompson FE, Smith AF, et al. J Am Diet Assoc 1995;95:781–8; quiz 9–90.
94. National Cancer Institute. Diet History Questionnaire III (DHQ III). Accessed September 27, 2022. Available at https://epi.grants.cancer.gov/dhq3/
95. Subar AF, Thompson FE, Kipnis V, et al. Am J Epidemiol 2001;154:1089–99.
96. Centers for Disease Control and Prevention. Using the New BRFSS Module. Accessed August 7, 2023. Available at www.cdc.gov/nutrition/data-statistics/using-the-new-BRFSS-modules.html.
97. National Cancer Institute. Short Dietary Assessment Instruments. Accessed August 8, 2023. Available at https://epi.grants.cancer.gov/diet/screeners/.
98. Thompson FE, Midthune D, Kahle L, et al. J Nutr 2017;147: 1226–33.
99. Rhee JJ, Cho E, Willett WC. Public Health Nutr 2014;17: 1054–60.
100. Carroll RJ, Midthune D, Subar AF, et al. Am J Epidemiol 2012;175:340–7.
101. Freedman LS, Midthune D, Arab L, et al. Am J Epidemiol 2018;187:2227–32.
102. Spiegelman D, McDermott A, Rosner B. Am J Clin Nutr 1997;65:1179S–86S.
103. Moore LV, Dodd KW, Thompson FE, et al. Am J Epidemiol 2015;181:979–88.
104. Doulah A, McCrory MA, Higgins JA, et al. IEEE Access 2019;7:49653–68.
105. Heydarian H, Adam M, Burrows T, et al. Nutrients 2019;11.
106. Schoeller DA, Hnilicka JM. J Nutr 1996;126:348S–54S.
107. Bingham SA. J Nutr 2003;133(Suppl 3):921S–4S.
108. Jenab M, Slimani N, Bictash M, et al. Hum Genet 2009;125: 507–25.
109. Lampe JW, Huang Y, Neuhouser ML, et al. Am J Clin Nutr 2017;105:466–75.
110. Bingham SA, Gill C, Welch A, et al. Int J Epidemiol 1997;26 (Suppl 1):S137–51.
111. Burrows TL, Warren JM, Colyvas K, et al. Obesity (Silver Spring) 2009;17:162–8.
112. Keogh RH, Carroll RJ, Tooze JA, et al. Stat Med 2016;35:4493–508.
113. Freedman LS, Midthune D, Carroll RJ, et al. Am J Epidemiol 2011;174:1238–45.
114. Freedman LS, Tasevska N, Kipnis V, et al. Am J Epidemiol 2010;172:836–42.
115. Tasevska N, Runswick SA, McTaggart A, et al. Cancer Epidemiol Biomarkers Prev 2005;14:1287–94.
116. Tasevska N. Nutrients 2015;7:5816–33.
117. Tasevska N, Midthune D, Potischman N, et al. Cancer Epidemiol Biomarkers Prev 2011;20:490–500.
118. Guasch-Ferre M, Bhupathiraju SN, Hu FB. Clin Chem 2018;64:82–98.
119. Zheng C, Pettinger M, Gowda GAN, et al. J Nutr 2022;152: 1711–20.
120. Livingstone KM, Ramos-Lopez O, Perusse L, et al. Trends Food SciTechnol 2022;128.

CHAPTER 50

Metabolic Consequences of Starvation

L. John Hoffer

CHAPTER OUTLINE

Abbreviations

BCM, body cell mass; **BMI**, body mass index; **CED**, chronic energy deficiency; **N**, nitrogen; **NDKA**, nondiabetic ketoacidosis; **PEM**, protein-energy malnutrition

Chapter Objectives

This chapter explains the metabolic and nutritional consequences of adult starvation, with a focus on its modification by metabolic adaptation and systemic inflammation, and its relationships to clinical topics such as obesity therapy, frailty, and physical activity. An underlying objective is to demonstrate the importance of physiologic reasoning in clinical nutrition.

Introduction

Starvation is the physiologic state that develops when food energy and/or protein consumption are less than energy expenditure and amino acid catabolism. Negative energy and protein balance trigger metabolic and behavioral adaptations that aim to minimize losses of body fat and protein. Brief periods of starvation are not uncommon and are without adverse consequence, but prolonged starvation leads to the disease called protein-energy malnutrition (PEM), which is the main topic of this chapter. Other names for PEM are starvation disease, hunger disease, and starvation-related malnutrition.

This chapter explains starvation both as a regulated physiologic phenomenon and as a disease. The important causes of starvation in the world are famine, war, social disruption, tyranny, and poverty. In peaceful societies that enact humane social and economic policies, starvation disease almost always develops as a complication or consequence of a medical, surgical, or neuropsychiatric disease. The main determinant of the clinical outcome of prolonged starvation is the starving person's ability to obtain, consume, and absorb enough food. Two other phenomena—metabolic adaptation and systemic inflammation—modify its characteristics and clinical trajectory. Metabolic adaptation enables survival during prolonged starvation, whereas sustained systemic inflammation—the constant companion of serious infection and tissue injury—prevents or reverses metabolic adaptation and worsens PEM.

The hallmark anatomic features of PEM are a generalized reduction of muscle mass and adipose tissue loss. The pathologic features that make PEM a disease emerge when the body's active protein mass has become depleted enough to impair specific physiologic functions. The loss of muscle mass and subcutaneous adipose tissue depletion that characterize PEM are easy to identify by physical examination, so one might assume that its diagnosis is straightforward. Too often, this is not the case. PEM frequently goes unnoticed in medical settings, partly because many health-care professionals are poorly trained in nutritional awareness, and partly because PEM merges with, and can be mistaken for, other common conditions associated with loss of muscle mass; these conditions are considered later in the chapter.

Basic Concepts

Principles of Energy and Protein Balance

Gain and loss of adipose tissue from the body may be determined by serial measurements of body weight combined with structured physical examinations, by body composition measurements, or by measuring energy balance—the difference between known energy intake and measured energy expenditure. Energy expenditure can be measured at a given moment in time by indirect calorimetry—a technique that uses the rate of oxygen uptake to infer the rate of carbon oxidation—or by means of sophisticated tracer-labeled water techniques. Energy balance roughly predicts the rate of adipose tissue gain or loss from the body, because the oxidation of intracellular triglycerides yields approximately 9.5 kcal of energy per/g (dietary fat, because it is slightly malabsorbed, provides 9.0 kcal/g), and adipose tissue is approximately 85% pure fat.[1]

For example, a person whose energy balance is −500 kcal/day for 1 week will catabolize 3,500 kcal of endogenous fat to make up the energy deficit and, therefore, will lose approximately 3,500/9.5=368 g fat and 368/0.85=433 g adipose tissue from their body.

Energy expenditure is conceptually divided into resting (basal) energy expenditure, energy expenditure above the basal rate due to formal physical exercise and external work, and non–exercise-associated thermogenesis.[2,3] The increased energy expenditure that occurs in hypermetabolic conditions, including fever, is a fourth category that comes into play in certain disease states.

Gain or loss of protein from the body may be assessed by repeated physical examinations, formal body composition measurements, or by measuring nitrogen (N) balance. N balance is the difference between dietary N consumption and loss from the body (almost all of it as urinary urea and ammonium). As N represents 16% of the mass of formed proteins, the amount of protein in a purified sample may be calculated by measuring its N content and multiplying the resulting value by 6.25.[4]

The total mass of metabolically active, protein-rich cells in the body is called the body cell mass (BCM). The BCM makes up 40% to 50% of the total body weight of a normal, idealized young adult. Skeletal muscle (the peripheral protein compartment) makes up approximately 80% of the BCM; the remaining 20% (the central protein compartment) consists of the fat-free cells of the liver, spleen, intestines, kidneys, heart, and other organs, and the bone marrow, blood cells, and immunocytes. The BCM is normally hydrated to the extent of approximately 4 g intracellular water to 1 g formed protein. This factor can be used to extrapolate a patient's rate of gain or loss of BCM from their N balance. For example, a person whose N balance is −5 g/day is losing approximately $5 \times 6.25 \times 5 = 156$ g BCM, and the same amount of body weight on that basis.[1,4]

Effects of Starvation on Energy and Protein Metabolism

Energy Metabolism

The body normally maintains a relatively constant adipose tissue mass through the activity of hormonal, neural, and behavioral regulatory systems that attempt to match energy expenditure to energy consumption. On the output side of this homeostatic equation, a sustained increase in daily energy expenditure (with its associated reduction in body fat) triggers an increase in food energy consumption. A sustained reduction in energy expenditure tends to reduce voluntary energy consumption—absent, that is, a metabolic predisposition to accumulate body fat and absent the modern obesogenic dietary and lifestyle environment.[5]

On the input side of the equation, a sustained reduction in food energy consumption triggers a reduction in energy expenditure. A sustained increase in food energy consumption triggers a compensatory increase in resting energy expenditure. The response to increased energy intake is more variable and less tightly regulated than the hypoenergetic response to dietary energy deficiency. The general consequence of these regulatory mechanisms is relative constancy of the body's major energy store—fat—with a bias more strongly oriented toward minimizing fat loss than preventing fat gain.

The negative energy balance created by starvation triggers three adaptive mechanisms which, when they are successful, re-establish zero energy balance, despite the persisting energy deficit. The first mechanism—governed by changes in thyroid hormone metabolism and sympathetic nervous system activity, detectable within a few days of energy deficiency—is reduced energy expenditure by the metabolically active tissues of the body. In prolonged or severe starvation, a second and more consequential mechanism develops: muscle loss. Skeletal muscle is responsible for a large proportion of total resting energy expenditure,[6] so a smaller muscle mass expends less energy. Finally, chronically starving people commonly reduce their total energy expenditure by doing less formal exercise and external work, and by limiting non–exercise-associated thermogenesis.[7] When adaptation is successful, the reduction in resting and total daily energy expenditure matches the reduced rate of energy intake, re-establishing zero energy balance and halting the loss of body fat.[1,8]

Protein Metabolism

As with energy homeostasis, protein consumption and amino acid catabolism are normally closely matched.[9] Very high tissue amino acid concentrations are toxic, so the body has the capacity to increase amino acid catabolism sufficiently to match high protein intake. A reduction in habitual protein consumption triggers a commensurate reduction in amino acid catabolism. The lowest rate to which the body can reduce amino acid catabolism (through a combination of reduced endogenous amino acid catabolism and increased conservation of the amino acids entering from the diet) determines a person's minimum protein requirement. The outcome of this homeostatic process is dynamic constancy of the BCM for a protein intake that meets or exceeds the minimum protein requirement. Anabolic states (as occur in growth, pregnancy, and recovery from starvation) increase hunger (and hence protein and energy consumption) and increase the efficiency of endogenous amino acid recycling and dietary amino acid retention. Catabolic states have the opposite effect.[1]

There are two reasons why protein is an essential nutrient. First, because the body is unable to reduce endogenous amino acid catabolism below a certain minimum rate—the so-called protein minimum or obligatory N excretion rate—dietary protein is continually required to replace this obligatory loss. Second, dietary protein digestion, amino acid absorption into the circulation, and their incorporation into newly synthesized endogenous proteins, are also inherently inefficient. The minimum dietary protein requirement is determined by a combination of endogenous obligatory N excretion and the maximum efficiency with which the body can process the protein entering from the diet.[4] When protein intake is below the minimum requirement level, amino acid catabolism exceeds protein intake and the BCM diminishes. Under these conditions, two mechanisms minimize the loss of muscle mass that would otherwise occur.

The first mechanism is muscle loss itself. The rate at which amino acids exit muscle is proportional to the existing amount of muscle protein, in a first-order kinetic process in which the rate of muscle loss decreases as total muscle mass diminishes. The second mechanism is a metabolic adaptation that increases the efficiency of endogenous protein turnover and the efficiency with which amino acids are captured from the diet and deposited into endogenous protein. As starvation progresses, the body's increasingly depleted muscle mass requires less dietary protein to maintain its existing protein mass, incoming dietary protein is captured more efficiently, and endogenous amino acids are recycled more efficiently. When adaptation is successful, progressive muscle loss has reduced the amount of dietary protein required to maintain the existing muscle mass, and endogenous and dietary protein retention increases to the point at which the reduced rate of protein loss equals the increased retention of dietary protein; consequently, zero N balance is re-established.[1]

Protein and Energy Metabolism Interact

At any given level of protein intake, N balance is worsened by a reduction in energy intake and improved by an increase in energy intake. Energy balance, rather than energy intake, is probably the key physiologic variable that, when negative, worsens N balance and, when positive, improves it. Deficient protein intake and negative energy balance independently induce negative N balance. As explained above, muscle loss is the body's backup strategy for reducing energy expenditure; energy deficiency, therefore, triggers negative N balance even when the level of protein intake is conventionally adequate. The muscle loss induced by energy deficiency can, however, be minimized or prevented, at least temporarily, by increasing protein provision to a level well above the normal minimum requirement.[4]

Metabolic and Clinical Cost of Adaptation

Adaptation is successful when zero energy and N balance have been re-established through a process of controlled muscle loss that stabilizes muscle mass before its adverse consequences become intolerable. The benefit of successful adaptation is survival, but its metabolic and functional costs are high—indeed, they account for most of the clinical features of PEM. The most apparent cost is muscle loss, with its associated loss of physical power. A hypometabolic state develops that shares features with, but is not the same as, hypothyroidism. Starving patients commonly feel chilly and are at risk of hypothermia, partly because their BCM is hypometabolic and partly because loss of subcutaneous fat reduces their thermal insulation. The loss of muscle mass reduces options for protein remodeling in response to changing metabolic needs. Starving people mount a blunted increase in protein turnover and a smaller catabolic response to metabolic stress.

In physiology, the term *adaptation* usually refers to homeostatic adjustment within the normal range. Because adaptation to prolonged starvation comes at a high cost, the terms *accommodation* and *pathological adaptation* are more precise descriptors.[1] In addition to suffering muscle loss and its consequences, people with PEM suffer from deficits in central protein mass and function that include anemia, reduced heart mass and function, decreased pulmonary mechanical function and a diminished response to stimuli to breathe and cough, altered gut anatomy (with possibly mildly impaired absorptive function), impaired wound healing, bone loss, altered metabolism of many drugs,[10] and impaired barrier immunity. Concurrent micronutrient deficiencies combine to impair immunity in other ways. In general, however, the starving body protects its central protein mass by providing it with amino acids that are made available by ongoing net muscle catabolism.[1]

The Minnesota Experiment

During the later years of the Second World War, 32 young American men who objected on moral grounds to engaging in combat volunteered to participate in a study of the physiology of prolonged adult starvation. The practical aim of the experiment was to determine the best composition of starvation rations to send famine victims in Europe. These healthy young men subsisted for 6 months on a diet that provided approximately 1,500 kcal/day (approximately 60% of their normal maintenance energy requirement) and 50 g protein/day, close to the current recommended safe intake of 0.80 g/kg/day. Carried out on the campus of the University of Minnesota, this landmark study is known as the Minnesota experiment.[11,12]

Table 50.1 indicates the average values and percentage losses of body weight of the Minnesota volunteers due to changes in BCM, body fat, and extracellular fluid volume after 6 months of starvation. These values illustrate important characteristics of prolonged starvation. First, large losses of fat (70%) and BCM (25%) reduced average body weight by 23% and reduced the body mass index (BMI; kg/m^2) from approximately 23 to 17.5. Note that BCM—a larger and denser body compartment than fat—made the most important contribution to total weight loss. Second, weight loss closely tracked that of BCM, illustrating the clinical rule that serial body weight measurements indicate the severity of BCM depletion in nonobese starving people. Third, extracellular fluid volume increased. Because of this, total weight loss was less than the sum of BCM and fat loss, and, therefore, underestimated it. Finally, the Minnesota experiment demonstrated the phenomenon of pathologic adaptation to starvation. After 6 months of continuous starvation, the Minnesota volunteers stopped losing weight. They had re-established zero energy and N balance, so their body composition remained stable despite continuing starvation.[1]

Table 50.1 The Minnesota Experiment. Effects of 24 Weeks of Starvation on Body Composition

Component	Zero Time (%)	24 wk (%)	% Change
Body weight	100	100	−23
Fat	14	5	−71
Body cell mass	57	56	−24
Extracellular fluid	24	34	+4

Obesity Therapy Involves Modified Starvation

Excess body fat is lost by creating and sustaining negative energy balance. Frustratingly for obese people, the adaptations that are lifesaving for underweight starving people confound efforts to eliminate excessive body fat. Weight-loss diets trigger metabolic adaptations and behavioral signals directed at minimizing fat loss, and they potentially include muscle loss. As shown earlier in this chapter, an energy balance of −3,500 kcal will diminish a person's adipose tissue mass by approximately 1 lb; this calculation is known as the "3,500 kcal per pound rule".[3] Although this rule is accurate as stated, it is frequently misinterpreted as indicating that someone who reduces their habitual energy consumption by 500 kcal/day will reliably lose 1 lb of adipose tissue every week. This will not happen, because adaptive mechanisms kick in to reduce energy expenditure and make the energy balance less negative.

Can this adaptive process be safely counteracted? To be effective, a weight-loss regimen must sustain negative energy balance; hence the importance of preserving muscle mass during dieting. Doing so keeps the resting energy expenditure rate high—helping sustain the negative energy balance—and it preserves physical well-being, strength, and the ability to exercise. The muscle loss induced by negative energy balance can be prevented or minimized by increasing protein intake above the usual recommendation of 0.8 g/kg normal body weight.[13] Muscle loss is further prevented by following a program of formal exercise and generally increased physical activity.[14]

Severe obesity imposes an exercise load on the body. The muscle mass of otherwise healthy, obese people is greater than that of otherwise comparable nonobese people and greater than the muscle mass they will have after they have eliminated their excess fat. Muscle loss of this kind is normal and inevitable after successful weight reduction.

Isolated Protein Deficiency is Starvation

Protein-deficiency diseases hardly ever occur without energy deficiency, because people eat food, not nutrients. Protein-deficient diets are rarely rich in energy, and energy-sufficient diets are rarely deficient in protein. It was long speculated that the pediatric disease, kwashiorkor, is caused by such a diet but, as will be explained later, isolated protein deficiency does not appear to be its sole or major cause. Pure protein deficiency disease—manifested by loss of BCM but not fat—has been observed in people with chronic renal disease who were prescribed a low-protein, adequate-energy diet to preserve their renal function.[1] The most common situation in which pure protein deficiency develops is when highly protein-catabolic, critically ill patients are prescribed energy-rich but amino acid- or protein-deficient nutritional support regimens.[15,16]

Outcome of PEM

Survival

The disease experienced by the Minnesota volunteers is the simplest form of PEM. Clinical settings in which this uncomplicated form of PEM occur include anorexia nervosa and certain other neuropsychiatric diseases, nonmalignant partial obstruction of the esophagus, gastric outlet or intestinal tract, and intestinal angina. In principle, uncomplicated PEM is easy to prevent and cure: simply provide food and allow the patient to eat it or eliminate any barriers preventing them from eating enough. Indeed, this is usually possible, but challenging situations frequently arise in hospital settings. In such cases, clinicians are required to determine what severity of PEM justifies the risks and complications involved in providing artificial nutrition support.

The risk of PEM-related complications, including death, increases as BCM depletion worsens. A 50% depletion of the BCM (or a BMI <13) is barely compatible with survival. (BMI as low as 10 has been reported in some cases of uncomplicated PEM in otherwise healthy adolescents or young adults.[17] When all other practical measures for improving the patient's nutritional status have failed, a reasonable tipping point for intervening with artificial nutrition (along with other pertinent factors and including, as a last resort, forced feeding of psychiatrically incompetent patients) is a 25% decrease in BCM or a BMI as low as 15 or 16 in the setting of uncomplicated PEM.[1]

The immediate causes of morbidity and death from PEM—so-called "malnutrition-related complications"—are infectious pneumonia (related to decreased ventilatory mechanical function and drive, lung stasis, and ineffective cough); delirium; injurious falls; skin thinning and breakdown due to physical immobility with consequent local and systemic infection; sepsis from intravenous infusion and other catheters; diarrhea with volume depletion and dehydration; and synergistic worsening of the patient's primary disease. Contributing to all these causes is starvation-induced immunodeficiency, which results from decreased active protein stores, hypothermia, and micronutrient deficiencies. For some patients, death results from a cardiac arrhythmia precipitated by the loss of cardiac muscle and electrolyte derangements.

Successful and Failed Adaptation

Adaptation (or better, pathologic adaptation) is successful when the starving person re-establishes zero energy and N balance, despite their continuing unfavorable nutritional environment. But adaptation can fail. The cardinal indicator of failed adaptation is the resumption of weight loss in a previously weight-stable starving patient or failure to gain weight despite the development of edema. Either situation suggests that muscle and fat loss have resumed. Factors that impair adaptation in a starvation-adapted person include a further diminution of food intake, worsening of the primary disease (or the development of one of its complications), treatments that affect protein or energy metabolism, certain micronutrient deficiencies, or the development of a new disease that imposes a new energetic or protein-catabolic stress, especially systemic inflammation.

Reasons for Failed Adaptation

Food Restriction. Food deprivation may be too severe for adaptation to be possible.

Micronutrient Deficiencies. Tissue deficiencies of potassium, phosphorus, zinc, and magnesium prevent effective adaptation and an anabolic response to refeeding. Thiamine deficiency causes muscle loss. It is important to bear in mind that starving people consume deficient amounts of these micronutrients.[1,18]

Metabolic Disease, or the Administration of Hormones or Antimetabolites. Certain endocrine diseases (such as uncontrolled diabetes or states of glucocorticoid excess) and high-dose glucocorticoid and cytotoxic chemotherapy pathologically reduce muscle mass.

Systemic Inflammation. The acute-phase response to severe systemic infection and tissue injury prevents or reverses adaptation. Chronic systemic inflammation of lesser intensity, as occurs in chronic heart, lung, renal, or gastrointestinal disease; inflammatory arthritis; vasculitis; and some forms of cancer, has a similar, but less dramatic, effect. The cytokines and stress hormones released as part of the acute-phase response stimulate net muscle protein breakdown and tend to increase energy expenditure, with resulting muscle and fat loss.

Prolonged Fasting

Fasting is the most extreme form of starvation. Here it is defined as the consumption of a diet devoid of protein, carbohydrate, and fat.[1,19] The defining metabolic characteristic of prolonged fasting is ketosis. It must be emphasized that ketosis does not necessarily indicate starvation, nor does its absence rule starvation out. Thin young adults in normal health may excrete ketone bodies in their urine (ketonuria) after an overnight fast, and mild ketosis is expected and normal during the consumption of a severely carbohydrate-restricted diet. As carbohydrate intakes as low as 100 g per day abolish ketosis, very few starving people demonstrate it.

The fed state ends (and fasting begins) after the last meal has been absorbed and the body has switched from exogenous to endogenous fuel consumption. The condition that exists after the overnight fast (before "breakfast") is termed the "basal" or "postabsorptive" state. The postabsorptive state is characterized by a physiologically low (but non-zero) circulating insulin concentration that creates an environment of fatty acid oxidation and net release of glucose from liver glycogen, and of amino acids from muscle. Under usual conditions, glucose is the brain's only fuel. Reduction of glucose delivery to the brain below a critical rate impairs consciousness and, if prolonged, will cause neuron death. Given the brain's fixed and high energy requirement—approximately half the body's total glucose production—the liver is obliged to deliver adequate amounts of glucose into the bloodstream at every moment. As fasting continues, insulin decreases even more, and the liver gradually releases its entire glycogen supply into the bloodstream. Under these conditions, the only source of new glucose molecules within the body is gluconeogenesis from amino acids (derived from muscle protein breakdown) and glycerol, released when triglycerides that have been hydrolyzed to free fatty acids. Some of the fatty acids reaching the liver are fully oxidized (they represent its main energy supply), but most are oxidized only as far as acetoacetic acid, which spontaneously interconverts with β-hydroxybutyric acid and, to a lesser extent, is decarboxylated to acetone. These three molecules are known as the *ketone bodies*.

In prolonged fasting, the liver becomes a factory that absorbs the fatty acids delivered to it from adipose tissue, converts most of them to ketone bodies, and exports them into the bloodstream. Glycogen depletion markedly reduces hepatic glucose release, but the serum glucose concentration decreases only modestly, because tissue glucose uptake and metabolism also decrease. Only part of this reduction in glucose metabolism is due to reduced terminal glucose oxidation in muscle and adipose tissue. The most important reason why glucose use decreases in early fasting, and the main reason why it decreases further in prolonged fasting, is that the brain progressively switches its source of energy from glucose to ketone bodies, which readily traverse the blood-brain barrier.[1,19,20]

Ketoacidosis and Ketoacidemia

Ketosis is an abnormally high blood ketone body concentration. By the second or third day of a total fast, circulating β-hydroxybutyrate and acetoacetate concentrations rise from their usual micromolar concentrations into the low millimolar range. The characteristic sweet odor of acetone is detectable in the breath by day 3 or 4 of fasting.

The terms *ketosis* and *ketoacidosis* (ketosis severe enough to lower the serum bicarbonate concentration but still within its normal buffering capacity) occur in cases of diabetes, when severe insulin deficiency mobilizes fatty acids and primes the liver for ketone body production and gluconeogenesis, just as in total fasting. In the complete absence of endogenous insulin, glucose derived from ingested carbohydrate is not transported into fat and muscle cells. The resulting hyperglycemia overloads the reabsorptive capacity of the renal tubules, resulting in glycosuria and an osmotic diuresis that depletes the body of extracellular fluid and intracellular free water. Ketone body concentration seldom rises above 6 to 8 mmol/L in prolonged total fasting but can rise much higher than this in diabetic ketoacidosis. The resulting acid load is too great for the body's buffering system to absorb, and a dangerous fall in pH occurs: this is *ketoacidemia*.

Why is the ketosis of simple fasting mild and clinically benign, while the ketoacidosis of diabetes commonly escalates into life-threatening ketoacidemia? In simple fasting, a very low but residual insulin concentration regulates and restrains ketogenesis, and the hypometabolic state induced by starvation places an upper limit on its maximum possible rate, as ketogenesis is energetically costly.[1,20] The combination of severe hypoinsulinemia and increased energy expenditure, as may be induced by glucagon, cortisol, and catecholamines, stimulates ketogenesis and increases its capacity, even in the absence of hyperglycemia. Clinical insight into this process is provided by the uncommon syndrome of nondiabetic ketoacidosis (NDKA). NDKA typically occurs after a period of protracted vomiting and volume depletion in persons embarked on a prolonged alcoholic binge during which no food is consumed. It also occurs rarely in pregnancy, and in people treated with sodium-glucose cotransporter inhibitors.[21] The ketoacidemia of NDKA can be as severe as diabetic ketoacidemia despite the normal glucose concentration. Alcoholic, gestational, and drug-induced NDKA occur in settings of severe hypoinsulinemia complicated by volume depletion and metabolic stress.

Survival in Prolonged Fasting

Brief, prudent, controlled fasts are usually innocuous, and may confer health benefits,[22-24] but prolonged fasting can be lethal. Survival in prolonged fasting is mostly determined by the body's initial fat store. The historic literature suggests that nonobese fasting adults who consume adequate amounts of water commonly die after approximately 60 days of continuous fasting. This period is consistent with complete exhaustion of body fat but only about one-third of the BCM. In contrast, some extremely obese adults have survived prolonged fasts of many months' duration, albeit with the unacceptable consequence of severe muscle loss. The body needs a certain minimum amount of body fat to survive prolonged fasting. The prolonged fasting brain is almost entirely dependent on ketone bodies for its energy, and the prolonged fasting liver needs fatty acids delivered to it to synthesize ketone bodies and to provide the energy required to drive gluconeogenesis.[1]

Substrate Metabolism in Inflammatory Disease States

Severe tissue injury (as occurs in trauma or sepsis) precipitates a systemic inflammatory response (also called the catabolic response) that increases energy expenditure to varying degrees. However, more importantly, it dramatically accelerates amino acid release from muscle protein into the general circulation. Upon entering the free amino acid pool, these amino acids provide the substrate for increased protein synthesis at sites of tissue injury and within the central protein compartment to optimize immune function and regulate the inflammatory response. Rapid body protein remodeling is an extremely inefficient process. Muscle protein loss vastly exceeds gains of protein elsewhere in the body, so body protein loss increases dramatically.[4]

For example, critically ill patients commonly excrete 15 g N/day or more in their urine in the absence of exogenous protein provision. This rate of body N loss implies a loss of $15 \times 6.25 = 94$ g of muscle protein (0.47 kg of muscle tissue) every day. Given our current understanding of the physiology of protein requirements and the fact that critically ill patients are unable to eat, there is an impelling physiologic argument for promptly providing such patients with sufficiently generous amounts of dietary protein to, if only partially, mitigate their massive loss of BCM.[16]

Energy expenditure and body protein remodeling increase in direct proportion to the severity of the injury and the body's catabolic response to it, although with wide and poorly predictable variability. Low-grade chronic inflammation also increases energy expenditure and muscle protein catabolism, although much less than occurs in response to major injury or sepsis.

Systemic inflammation and starvation interact. The purpose of adaptation—to minimize or halt fat and protein loss during starvation—is opposite to the rapid protein remodeling of systemic inflammation. Systemic inflammation, thus, interferes with successful adaptation and magnifies fat and protein loss in a starving person. Various interactions between inflammation and starvation create specific categories of starvation-related diseases, as explained in the next section.

Clinical Relevance

Categories of Malnutrition and Related Diseases

What is Malnutrition?

Malnutrition is a blanket term that is commonly used as a convenient shortcut-synonym for PEM but is often used in a variety of other ways, such as to describe any starvation-related disease or any diet that, if continued, could eventually result in such a disease. The problems created by this ambiguity are compounded by the publication of checklist tools that define malnutrition as a numeric score, or as an indicator of the risk of adverse clinical outcomes that may partly be due to PEM but have other contributing causes.[25,26]

Many clinical guidelines and clinical trial enrollment rules define malnutrition—intended here as a synonym for PEM—as BMI less than 16 or 17. There can be no doubt that a person with a BMI as low 16 has PEM—the only way such a BMI can come about is by denuding the body of much of its BCM and adipose tissue. But there is a problem with "diagnosis by BMI"—it fosters the incorrect assumption that a BMI greater than 16, 17, 20, or 25 rules PEM out. The increased body weight caused by extracellular fluid accumulation, and the obesity epidemic now engulfing the world, invalidate this assumption. Patients with normal or high BMI and important PEM occupy the waiting rooms of chronic disease clinics and the beds of modern acute care hospitals.[27-30]

It is a problem that many malnutrition scoring systems do not clearly distinguish between diagnosis and screening.[25,26,31] Diagnosis is the process of identifying a specific pathologic entity—PEM in this case—by reviewing the medical history, pertinent findings on physical examination, and laboratory and imaging reports.[32] Complete diagnosis also involves an estimation of the probability that the diagnosis is correct and a judgment about its severity. By contrast, screening is the application of a test that identifies people at sufficiently high risk of having a target disease to warrant carrying out definitive procedures to establish the diagnosis or rule it out, or identifies people at high enough risk of developing the target disease to warrant specific preventive interventions.[33] Screening tools and risk predictors are extremely useful, but they should not be confused with clinical diagnosis. PEM is not a number; it is a disease.

Recognition of the harmful confusion created by the widespread vague and ambiguous use of "malnutrition" in the scientific and clinical literature has spurred a concerted effort to develop clear, pathophysiologic-based diagnostic terminology.[32,34,35] Currently recommended terms are summarized below.

Starvation-Related Malnutrition

Starvation-related malnutrition refers to PEM solely due to prolonged starvation, as illustrated by the Minnesota experiment or uncomplicated anorexia nervosa. Starvation-related malnutrition is partially or successfully adapted PEM. It is characterized by generalized muscle loss, adipose tissue diminution, and the clinical picture of hypometabolism described earlier. The history, physical examination, and pathophysiologic features (including strength and physical function) will dictate the correct diagnosis and its severity, which is categorized as A (absent), B (possible or minor), or C (definite and consequential).[32,36]

Systemic inflammation is absent; consequently, the serum albumin concentration is normal or slightly increased above normal because of hemoconcentration.

Chronic Disease-Related Malnutrition and Cachexia Syndrome

Chronic disease-related malnutrition and cachexia syndrome refer to much the same entity: PEM complicated by systemic inflammation.[37,38] Chronic low-grade systemic inflammation, as occurs in chronic infection; inflammatory autoimmune diseases; chronic renal, heart, and often pulmonary failure; and some inflammatory cancers, creates a pernicious synergistic combination of mild protein catabolism that promotes muscle loss, fatigue, and anorexia. Chronic disease-related malnutrition can be difficult to reverse. Fortunately, the deficit on the input side of the balance equation (inadequate food consumption due to anorexia) is usually a stronger driver of the patient's PEM than deficits on the output side (increased protein catabolism and mildly increased resting energy expenditure), and this makes the disease amenable to nutritional intervention. Serum albumin concentration is typically in the low-normal or subnormal range.

Acute Disease-Related Malnutrition

Acute disease-related malnutrition refers not to a disease entity but rather to a metabolic-nutritional environment that creates an extremely high risk of PEM.[37] Severe systemic inflammation (as occurs in severe tissue injury or sepsis) precipitates rapid muscle loss and increases energy expenditure in patients who are simultaneously unable to eat (and would refuse food even if they were able to eat).[39] People with acute disease–related malnutrition may not have PEM at present but will rapidly develop it if their primary disease is not effectively treated and appropriate nutritional support not promptly provided. A patient with already-existing PEM (or important muscle loss for any reason) will rapidly experience even more serious, and potentially lethal, loss of muscle unless appropriate nutritional support is provided promptly.

Hypoalbuminemia

Hypoalbuminemia is frequently misinterpreted as equivalent to "malnutrition," but it is only a risk factor for it. Serum albumin is a negative acute-phase protein whose concentration decreases as part of the acute-phase response to systemic inflammation. A patient with hypoalbuminemia may or may not have PEM, and they may never develop it.

The acute-phase response increases capillary permeability, allowing large intravascular proteins (including albumin) and immunocytes to migrate into the interstitial fluid to combat infection and facilitate healing. Most hypoalbuminemia of systemic inflammation is due to redistribution of the body's albumin pool from the intravascular into the interstitial space, but it is also partly due to a combination of increased albumin catabolism and reduced hepatic synthesis, the relative importance of which depends on the specific disease process.[40]

Hypoalbuminemia is an important predictor of death and morbidity.[41] Although hypoalbuminemia is not "malnutrition," it is, like other acute-phase proteins, a red flag that should alert the clinician to the need for a diagnostic nutritional assessment and consideration of a preventive or therapeutic nutritional intervention. Dietary protein deficiency, starvation, and muscle loss worsen systemic inflammation-induced hypoalbuminemia by denying the liver the amino acids it needs to synthesize albumin. Hypoalbuminemia will not completely renormalize while systemic inflammation persists, no matter how much protein the patient eats. The body's circulating albumin pool has a turnover half-life of approximately 3 weeks. Consequently, approximately 2 or 3 months may be required for hypoalbuminemia to renormalize even under optimum conditions of resolved systemic inflammation and appropriate nutrition.

It is important to emphasize that the serum albumin concentration is normal in simple, starvation-related malnutrition. A normal serum albumin concentration in an adequately hydrated patient rules against important systemic inflammation and, in that respect, signifies a more hopeful prognosis, but it does not exclude the possibility of dangerous starvation. The most ominous combination is systemic inflammation-induced hypoalbuminemia in the setting of severe muscle loss and persisting starvation.

Marasmus and Kwashiorkor

Marasmus is adapted PEM in a young child. Kwashiorkor is a syndrome of uncertain etiology characterized by hypoalbuminemia, abnormal hair pluckability, skin or wound breakdown, irritability, and otherwise unexplained edema in a starving (and micronutrient-deficient) child, usually in the setting of infection or another cause of systemic inflammation. The widely repeated theory that kwashiorkor is caused by a protein-deficient diet is no longer tenable. Also erroneous is the notion that marasmus and kwashiorkor represent opposite ends of a continuous pathophysiologic spectrum, with marasmic kwashiorkor in the middle.[42,43]

Chronic Energy Deficiency

Starvation and PEM are easy to diagnose in wealthy societies but are so common in impoverished ones that there is a risk of perceiving them as normal.[44,45] Appreciating this situation, the World Health Organization recognizes a form of adapted starvation called adult chronic energy deficiency (CED): starvation on the cusp of PEM. CED is compatible with gainful employment, pregnancy, and other aspects of normal daily life.[8,46] CED is classified into three grades of severity: grade I, BMI 17.0 to 18.4; grade II, BMI 16.0 to 16.9; and grade III, BMI less than 16. In otherwise healthy adults, grade II and III CED are associated with an increasing probability of days of illness, reduced physical work capacity, poorer reproductive function, and poorer lactation performance.[47] Voluntary physical activity is decreased in grade III CED, which is equivalent to uncomplicated PEM (adapted starvation-related malnutrition).

Interactions Among PEM, Systemic Inflammation, Obesity, Old-Age Muscle Loss, Physical Inactivity, and Frailty

Why do healthcare professionals fail to diagnose and respond to PEM? Three factors are involved. The first is the lack of clinically relevant nutritional content in undergraduate and postgraduate medical education, which fosters a medical culture that subconsciously discounts malnutrition as irrelevant to the care they provide. Second, in-hospital starvation and PEM confront hospital administrations with time-consuming and financially demanding challenges they are disinclined to deal with. Finally, PEM in hospitalized and chronically ill patients often merges with, and is often confused with, other common conditions that cause muscle loss.

PEM so invariably accompanies serious advanced diseases that clinicians sometimes mistakenly assume that their patient's loss of muscle mass and fat depletion are usual, or even defining, characteristics of their primary disease, when the actual cause is starvation, a potentially remediable complication of the primary disease. The majority of patients with inexorably advancing untreatable diseases will have PEM and associated micronutrient deficiencies. Should the PEM be treated? Doing so will not change the trajectory of the primary disease, although correcting the micronutrient deficiencies may improve quality of life. As a rule, nutritional intervention is justified when the pace of debility and death from PEM is greater than the rate of inexorable progression of the primary disease.

Systemic Inflammation, Acute and Chronic Disease-Related Malnutrition, and Cachexia

Chronic systemic inflammation is easy to diagnose and often to treat effectively, but clinicians should also attend to its commonly associated loss of muscle mass, which is often largely driven by anorexia and inadequate food consumption.

Hormone Deficiency or Excess

High-dose glucocorticoid therapy and cortisol, growth hormone, and testosterone deficiency are important causes of muscle loss.

Frailty and Sarcopenia

Frailty is a condition of reduced vigor, diminished physiologic reserve, and vulnerability to stressors due to impairments in multiple integrated systems in the body. The frailty of advanced old age comes about as a combination of decreased sensory acuity; cognitive ability; pulmonary, cardiovascular, and renal function; physical coordination; speed and agility; bone mass; joint and tendon flexibility and range of movement; skin thickness and resiliency; immune system responsiveness and regulation; and decreased muscle mass.[48] Muscle loss that is solely due to advanced old age was traditionally named "sarcopenia," but many authors now give this term other meanings, sometimes using it as a precise synonym for "muscle loss" and sometimes to refer to the muscle loss associated with cancer or systemic inflammation.

There is a tendency to regard frailty as purely an age-related phenomenon, but many elderly people are robust and high functioning, and some young people—those suffering from a serious chronic disease—are frail. An old person with muscle loss will likely have some degree of old-age sarcopenia and may be frail, but the presence of old-age muscle loss does not rule out the possibility of concurrent PEM.

Disuse Muscle Loss

PEM and systemic inflammation combine to induce muscle loss and reduce voluntary physical activity, which then causes disuse muscle loss (disuse muscle atrophy). The frailty of old age causes, and is worsened by, reduced physical activity, with consequent deconditioning and disuse muscle loss.

Neuromuscular Diseases

Certain primary myopathic and neurologic diseases reduce muscle mass.

Central Importance of Muscle Loss

The severity of muscle loss is an important predictor of a patient's clinical outcome. Although adipose tissue loss is also important, except in extreme situations, it is much less consequential than muscle loss. In most cases, various combinations of six causes—PEM, systemic inflammation, hormone excess or deficiency, old-age sarcopenia, physical inactivity, and a neuronal or myopathic disease—converge to create and worsen the patient's muscle loss. The impact of each should be considered when evaluating the patient and formulating a treatment plan.[30,33,49-56] Old-age sarcopenia, by definition, cannot be reversed, but it can be minimized and delayed; PEM can be halted and reversed by an appropriate nutritional intervention; and disuse muscle atrophy is amenable to physical rehabilitation. The muscle loss caused by starvation will not respond to nutrition therapy alone, nor will muscular exercise without appropriate nutrition reverse PEM: nutritional and physical rehabilitation are both required

Summary

Starvation is the physiologic state that develops when food energy and protein consumption consistently fail to match energy expenditure and amino acid catabolism. Prolonged and severe starvation progresses to the disease called PEM. The hallmark anatomic features of PEM are generalized muscle loss and adipose tissue depletion. The pathologic features that make it a disease emerge when the body's active protein mass has become depleted enough to impair specific physiologic functions.

When it is solely caused by starvation, PEM may be compatible with prolonged survival. When PEM is complicated by other factors, especially chronic moderate systemic inflammation (chronic disease-related malnutrition or cachexia) or acute severe inflammation, rapid and dangerous muscle loss and often severe adipose tissue depletion may occur. PEM always requires treatment, but complicated PEM requires more urgent and intensive nutritional intervention while the patient's medical and surgical complications are being diagnosed and treated.

The cardinal feature of PEM that indicates its severity and prognosis is the extent of generalized muscle loss. Fat depletion is also important, but the modern obesity epidemic has created a subset of obese patients suffering from PEM whose muscle loss is concealed by ample residual adipose tissue. PEM interacts with and shares features with other common medical conditions—chronic systemic inflammation, old-age sarcopenia, disuse muscle atrophy, and frailty—and is frequently confused with them, an error that can delay its diagnosis and treatment.

References

1. Hoffer LJ. (2012) In: Ross AC, Caballero B, Cousins RJ, Tucker KL, Ziegler TR, eds. Metabolic Consequences of Starvation. Modern Nutrition in Health and Disease. Philadelphia, PA: Lippincott William & Wilkins. 2012;660–77.
2. Levine JA. Am J. Physiol Endocrinol Metab 2004;286:E675–E85.
3. Hall KD, Heymsfield SB, Kemnitz JW, et al. Am J Clin Nutr 2012;95:989–94.
4. Hoffer LJ. JPEN J Parenter Enteral Nutr 2016;40:460–74.
5. Muller MJ. Am J Clin Nutr 2019;110:542–3.
6. Zurlo F, Larson K, Bogardus C, et al. J Clin Invest 1990;86: 1423–7.
7. Martin CK, Das SK, Lindblad L, et al. J Appl Physiol 2011;110: 956–63.
8. Kurpad AV, Muthayya S, Vaz M. Public Health Nutr 2005;8: 1053–76.
9. Liu GY, Sabatini DM. Nat Rev Mol Cell Biol 2020;21:183–203.
10. Verrest L, Wilthagen EA, Beijnen JH, et al. Clin Pharmacokinet 2021;60:1149–69.
11. Keys A, Brozek J, Henschel A, et al. The Biology of Human Starvation. The University of Minnesota Press, Minneapolis, 1950.
12. Wischmeyer PE. Crit Care 2017;21:316.
13. Kim JE, O'Connor LE, Sands LP, et al. Nutr Rev 2016;74: 210–24.
14. Weinheimer EM, Sands LP, Campbell WW. Nutr Rev 2010;68: 375–88.
15. Hoffer LJ, Bistrian BR. F1000Res 2016;5:2531.
16. Hoffer LJ. Pract Gastroenterol 2021;45:10–23.
17. Rigaud D, Tallonneau I, Brindisi MC, et al. Clin Nutr 2012;31: 693–8.
18. Ververs M, Muriithi JW, Burton A, et al. MMWR Morb Mortal Wkly Rep 2019;68:72–5.
19. Cahill GF, Jr. Annu Rev Nutr 2006;26:1–22.
20. Halperin ML, Cheema-Dhadli S. Diabetes Metab Rev 1989;5: 321–36.
21. Qiu H, Novikov A, Vallon V. Diabet Metab Res Rev 2017;33:07.
22. Michalsen A. Current Pain Headache Rep 2010;14:80–7.
23. de Cabo R, Mattson MP. N Engl J Med 2019;381:2541–51.
24. Aoun A, Ghanem C, Hamod N, et al. Nutr Today 2020;55: 270–7.
25. Hoffer LJ, Bistrian BR. JPEN J Parenter Enteral Nutr 2013;37: 300–9.
26. Hoffer LJ. Nestle Nutr Workshop Ser Clin Perform Programme 2009;12:41–52.
27. Thibault R, Pichard C. Ann Nutr Metab 2012;60:6–16.
28. Gonzalez MC, Correia M, Heymsfield SB. Curr Opin Clin Nutr Metab Care 2017;20:314–21.
29. Ryan AM, Prado CM, Sullivan ES, et al. Nutrition 2019;67–68: 110539.
30. Prado CM, Purcell SA, Alish C, et al. Ann Med 2018;50: 675–93.
31. Gilbert R, Logan S, Moyer VA, et al. West J Med 2001;174: 405–9.
32. Cederholm T, Jensen GL, Correia M, et al. Clin Nutr 2019;38: 1–9.
33. Bellanti F, Lo Buglio A, Quiete S, et al. Nutrients 2022;14:21.
34. Jensen GL, Compher C, Sullivan DH, et al. JPEN J Parenter Enteral Nutr 2013;37:802–7.
35. Soeters P, Bozzetti F, Cynober L, et al. Clin Nutr 2017;36: 896–901.
36. Allard JP, Keller H, Gramlich L, et al. Clin Nutr 2020;39: 2771–7.
37. White JV, Guenter P, Jensen G, et al. JPEN J Parenter Enteral Nutr 2012;36:275–83.
38. Martin L. Curr Opin Clin Nutr Metab Care 2016;19:188–98.
39. Preiser JC, Ichai C, Orban JC, et al. Br J Anaesth 2014;113: 945–54.
40. Kaysen GA, Dubin JA, Muller HG, et al. Kidney Int 2002;61: 2240–9.
41. Tojek K, Kowalczyk G, Czerniak B, et al. Biomark Med 2019;13.
42. Seres DS, Resurreccion LB. Nutr Clin Pract 2003;18:297–301.
43. Bhutta ZA, Berkley JA, Bandsma RHJ, et al. Nat Rev Dis Prim 2017;3:17067.
44. James WP. Eur J Clin Nutr 1994;48 Suppl 3:S1–8.
45. Delisle H, Agueh V, Fayomi B. BMC Int Health Hum Rights 2011;11 Suppl 2:S10–S.
46. Subasinghe AK, Walker KZ, Evans RG, et al. PLoS one 2014;9:e87423.
47. James WPT, Ferro-Luzzi A, Waterlow JC. Eur J Clin Nutr 1988;42:969–81.
48. Nishikawa H, Fukunishi S, Asai A, et al. Int J Mol Med 2021;48: 1–8.
49. Jeejeebhoy KN. Curr Opin Clin Nutr Metab Care 2012;15: 213–9.
50. Tappenden KA, Quatrara B, Parkhurst ML, et al. JPEN J Parenter Enteral Nutr 2012;37:482–97.
51. Fischer M, JeVenn A, Hipskind P. Nutr Clin Pract 2015;30: 239–48.
52. Cohen S, Nathan JA, Goldberg AL. Nat Rev Drug Discov 2015;14: 58–74.
53. Landi F, Camprubi-Robles M, Bear DE, et al. Clin Nutr 2019;38:2113–20.
54. Deutz NEP, Ashurst I, Ballesteros MD, et al. J Am Med Dir Assoc 2019;20:22–7.
55. Schuetz P, Seres D, Lobo DN, et al. Lancet 2021;398:1927–38.
56. Reinders JJ, Hobbelen JSM, Tieland M, et al. J Multidiscip Healthc 2022;15:1247–60.

CHAPTER 51

Metabolic and Molecular Consequences of Dietary Restriction

Cara L. Green, Dudley W. Lamming, and Luigi Fontana

CHAPTER OUTLINE

Abbreviations

Aβ, amyloid-β; **AMPK**, AMP-activated protein kinase; **BCAA**, branched chain amino acid; **CALERIE**, comprehensive assessment of long-term effects of reducing intake of energy; **DASH**, dietary approaches to stop hypertension; **DR**, dietary restriction; **GCAT**, glycine c-acetyltransferase; **FGF21**, fibroblast growth factor 21; **FOXO**, class O of forkhead box transcription factors; **HDL**, high-density lipoprotein; **IF**, intermittent fasting; **IGF**, insulin-like growth factor; **IL**, interleukin; **LDL**, low density lipoprotein; **mTOR**, mammalian target of rapamycin; **NAD**, nicotinamide adenine dinucleotide; **NASH**, non-alcoholic steatohepatitis; **NIA**, National Institute on Aging; **PPF**, prolonged periodic fasting; **TGFβ**, transforming growth factor-β; **TNF-α**, tumor necrosis factor-α; **TRF**, time-restricted feeding; **UW**, University of Wisconsin

Introduction

Dietary restriction (DR) is defined as a chronic or intermittent reduction of food intake or of specific nutrients. DR without malnutrition is the most robust way to extend lifespan in model organisms, including rodents.[1] In rhesus monkeys, DR promotes longevity and protects against obesity, diabetes, cardiovascular disease, cancer, neurodegeneration, and frailty and, in humans, DR results in a wide range of metabolic, molecular, cellular, and immunological alterations that safeguard against multiple age-associated chronic conditions (**Figure 51.1**).[2-4] In animal studies, DR involves reducing *ad-libitum* food intake by 10% to 60% with adequate protein and vitamin and mineral intake. DR can be achieved by chronically reducing food intake or by changing meal frequency and timing. Selective forms of DR are isocaloric

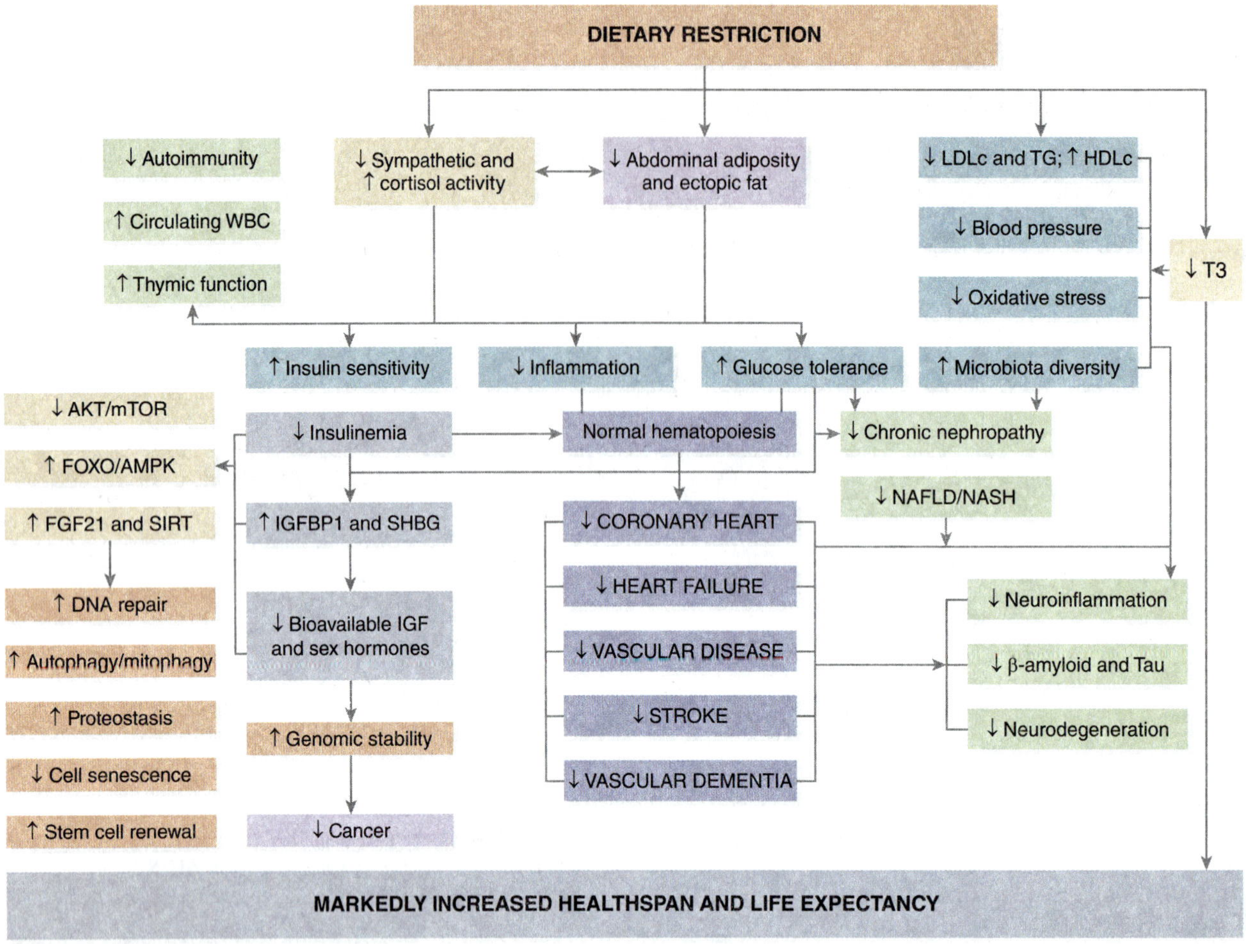

Figure 51.1 The Hallmarks of dietary restriction. Dietary restriction with adequate nutrition modulates important metabolic, hormonal, and molecular factors that play a crucial role in the prevention of multiple chronic diseases and in the promotion of healthy aging. *WBC*, white blood cells; *LDLc*, low-density lipoprotein cholesterol; *HDLc*, high-density lipoprotein cholesterol; *TG*, triglycerides; *T3*, tri-iodothyronine; *IGFBP-1*, insulin-like growth factor binding protein 1; *SHBG*, sex hormone-binding globulin; *NAFLD*, non-alcoholic fatty liver disease; *NASH*, non-alcoholic steatohepatitis; *AKT*, serine/threonine protein kinase; *mTOR*, mammalian target of rapamycin; *FOXO*, forkhead family of proteins; *AMPK*, AMP-activated protein kinase; *FGF21*, fibroblast growth factor 21; *SIRT*, sirtuin.

restriction of protein or specific amino acids. In humans, DR refers to a state in which energy or macronutrient intake is chronically or intermittently reduced below what is typical or habitual and, ideally, sufficiently low to achieve or maintain low-normal waist circumference without causing deprivation of essential nutrients. Combining regular endurance and resistance exercise training is essential to preserve adequate muscle and bone mass and physical function while practicing DR, especially in older adults.[5]

Dietary Restriction and Lifespan

In invertebrate animal models and rodents, lifespan rises as food intake is reduced, but then suddenly declines with excessive restriction (starvation). The degree of DR that maximizes longevity depends on genetic and epigenetic factors, as even in inbred strains of mice, the increase of lifespan in response to the same level of DR differs considerably. Many different gene mutations have been discovered in mice that extend longevity and have since been confirmed in knockout and transgenic mouse models.[2] Experiments in rhesus monkeys demonstrate that the longevity effects of DR are not limited to short-lived model organisms. In the University of Wisconsin (UW) primate study, monkeys randomized to a 30% semi-purified DR diet had an almost 50% decrease in mortality rate, compared with unrestricted controls,[4] and DR similarly reduced the methylation-based calculated biologic age by an average of 7 years.[6] Although the likely underpowered National Institute on Aging (NIA) primate study reported no difference in estimated survival, the average lifespan of DR animals consuming natural ingredient-based, minimally refined diets was extraordinarily high. While the median and 10% survival of rhesus monkeys in captivity is ~26 and ~35 years of age, respectively, about one-third of the NIA monkeys on late-onset DR have lived more than 40 years, with one male monkey dying at 44 years of age (the equivalent of 135 years for a human being).[4] These trials, together with accumulating data in model organisms and humans, underscore two key concepts: (i) the beneficial effects of DR on lifespan and, most importantly, on healthspan, appear conserved across species; and (ii) the interactive roles of energy restriction, meal timing, and diet composition in regulating life expectancy and disease development and progression.

While long-term randomized trials to assess the effects of DR on young, healthy individuals have not been performed, accumulating data show that in both young and older adults, moderate DR with adequate nutrition, especially when coupled with insulin-sensitizing endurance exercise training, exerts protective effects on multiple metabolic and molecular aging pathways, i.e., it protects against the accumulation of cellular and tissue damage, and it prevents or slows the development and clinical progression of multiple chronic conditions. As reviewed in depth elsewhere,[2] inhibition of AKT/mTOR and TGFβ signaling, coupled with activation of FOXO, AMPK, FGF21, and SIRT pathways, trigger cellular mechanisms that protect against aging, including increased protein homeostasis (proteostasis) and autophagy, augmented stem cell and mitochondrial function, and reduced accumulation of driver mutations and senescent cells.

Dietary Restriction, Obesity, and Diabetes

Excessive abdominal adiposity is a potent risk factor for common age-related illnesses, including type 2 diabetes, heart disease, stroke, kidney and liver disease, cancer, neurodegeneration, and cognitive decline. Chronic inflammation, insulin resistance, hyperinsulinemia, and other associated hormonal (e.g., high bioavailable IGF-1 and sex hormones), immunologic (e.g., clonal hematopoiesis), and associated molecular perturbations are of particular concern because they drive the accumulation of cellular damage, leading to accelerated aging of multiple organ systems.[7,8]

In rodents, DR has been shown to drastically reduce the occurrence of type 2 diabetes and non-alcoholic fatty liver disease.[9] Moreover, DR has been shown to improve glucose tolerance, pancreatic beta-cell function, and insulin sensitivity across many species, including lean and high-fat fed mice,[10,11] rhesus macaques[4] and even in non-obese healthy men and women.[12] Several compounds frequently used to control blood glucose in humans also extend lifespan in mice, including acarbose[13] and canagliflozin.[14] Low-dose metformin, one of the most widely used diabetes medications in humans, has also been investigated as a drug that targets biology to promote healthy aging (i.e., a geroprotector or geroprotective agent).[15,16] However, a nationwide prospective study in Denmark indicated that paternal metformin treatment was associated with birth defects,[17] emphasizing "caution" in the use CR-mimetics in healthy individuals.

Human and non-human primate DR trials have demonstrated that DR-induced change in weight is predominantly due to total body and abdominal fat loss.[18-20] In a U.K. weight loss trial, 86% of patients with type 2 diabetes who lost 15 kg or more achieved remission at 1 year.[21] In another trial, >80% of NASH patients who lost more than 10% of their baseline body weight achieved steatohepatitis resolution, and even fibrosis regression.[22] In the CRESO trial of diabetes patients with abdominal obesity, 6 months of moderate DR markedly reduced albuminuria and glomerular hyperfiltration, two strong predictors of end-stage renal disease.[23]

Dietary Restriction and Cancer

Cancer is the leading cause of death in rodents, and DR is the most potent and reproducible intervention for delaying and/or protecting against both spontaneous and experimentally induced (radiation or carcinogen) tumors.[24] However, the cancer preventive effects of DR are not universal, as cancers with constitutive activation of the PI3K pathway are resistant to DR.[25]

In both non-human primates and humans, DR exerts powerful cancer protective effects. In the NIH and UW monkey trials, 30% DR caused a 40% to 55% reduction in the incidence of adenocarcinomas.[4] In patients with morbid obesity, bariatric surgery-induced DR resulted in 29% reduction in the incidence of several obesity-associated cancers (breast, endometrial, and ovarian cancers), and in 23% lower cancer mortality.[26] Accordingly, overweight patients with type 2 diabetes who lost as little as 6.8 kg in the Look AHEAD trial experienced a 16% lower risk of obesity-related cancers.[27]

The precise biological mechanisms mediating the DR-induced protection against cancer initiation and progression are complex and may involve reduced stem cell proliferation, improved proteostatic and DNA repair functions, decreased oxidative stress and inflammation, and enhanced anti-cancer immune response.[2,28]

Dietary Restriction and Kidney Disease

Kidney disease is the second leading cause of death in rodents, and 40% DR consistently prevents many of the typical age-dependent glomerular, interstitial and tubular lesions leading to kidney failure.[29] In humans, type 2 diabetes and hypertension play crucial roles in the pathogenesis of end-stage kidney disease, leading to dialysis or transplantation. In healthy and unhealthy men and women, DR provides a strong protective effect against type 2 diabetes and hypertension.[30] In a recent trial of DR in people with type 2 diabetes,[23] individuals who lost 5 to 7 kg over 6 months showed marked improvement in glomerular filtration rate and albuminuria, with concomitant reductions in waist circumference, blood pressure, insulin resistance, glycated hemoglobin, LDL-to-HDL cholesterol ratio, C-reactive protein, and angiotensin II, a master regulator of blood pressure and aging. A study prior to kidney transplantation showed that DR improved kidney function for at least one month and reduced the incidence of rejection.[31]

Dietary Restriction and Cardiovascular Disease

DR protects against cardiomyopathic degenerative fibrotic disease in wild-type rodents[32] as well as protecting against coronary heart disease in genetic mouse models of atherosclerosis.[33] In UW rhesus monkeys, 30% DR reduced heart disease by ~50%.[4] Unlike in rodents, atherosclerotic cardiovascular disease is the main cause of disease and death in humans consuming Western diets, and DR with adequate nutrition causes super physiologic reductions of multiple cardiometabolic risk factors, including systolic and diastolic blood pressure, cholesterol-to-HDL ratio, triglycerides, high-sensitivity C-reactive protein, insulin resistance, and oxidative stress, as assessed by urinary F2-isoprostanes and DNA/RNA markers of oxidative damage.[12,34-36] As in DR rodents, people practicing DR have lower intimal media thickness of the common carotid arteries, improved left ventricular diastolic function, and improved heart-rate variability, strongly suggesting that DR retards human cardiovascular aging.[37]

Dietary Restriction and Neurodegeneration

Murine animal models show that DR has profound beneficial neuroprotective effects: it enhances balance, spatial and non-spatial reference, and working memory in most[38,39] but not all of the studies conducted so far.[40] DR has also reduced amyloid deposition,[41] and improved cognitive dysfunction[42] in genetic animal models of Alzheimer's disease, and enhanced motor function and neuronal survival in rodent models of Parkinson's and Huntington diseases.[43,44] Data from the UW and NIA monkey trials demonstrate that 30% DR lessened gray matter sclerosis.[18] In the squirrel monkey model of Alzheimer's disease, DR caused a significant reduction of Aβ1-40 and Aβ1-42 in the temporal cortex.[45]

Findings from the 2-yr CALERIE randomized clinical trial demonstrate that a 14% reduction in energy intake is sufficient to ameliorate working memory tests in non-obese young and middle-aged individuals.[46] In another trial of elderly men and women, DR increased memory performance, which correlated with improvements in insulin and inflammatory biomarkers.[47] Similarly, cognitive function was significantly improved in hypertensive patients randomized to 4 months of an energy-restricted DASH diet and exercise training.

Dietary Restriction and Autoimmune Diseases

In rodents, DR exerts a potent beneficial effect against allergic and autoimmune diseases. DR markedly increases survival and reduces glomerulonephritis and tubular lesions in autoimmune-prone mice.[48] Similar protective effects of DR were seen in animal models of experimental autoimmune colitis, Sjogren syndrome, multiple sclerosis, atopic dermatitis, systemic lupus erythematosus, uveoretinitis, and immune-mediated occlusive coronary vascular disease.[1,3] Moreover, DR strongly inhibits allergic and inflammatory responses to house dust mite antigen and ozone (O3) inhalation in rodents.[49,50]

The impact of DR on autoimmune diseases in humans is not fully understood. However, data from observational studies and trials demonstrate a powerful anti-inflammatory and immune modulating effect.[51,52] This occurs through several mechanisms, including reduced inflammatory cytokine production and changes in gut microbiota.[3] Some concerns about DR and its potential role in immunosuppression come from studies in monkeys, where there were detrimental effects on cell-mediated immunity,[53] and in rodents, where there was increased risk of deadly infections.[54] However, DR improved immunity in other mouse and monkey studies,[55,56] and a recent 2-year trial in non-obese volunteers showed that 14% DR lowered serum C-reactive protein, TNF-α and white blood cells by ~40%-50%, and increased thymic volume and function,[57] without impairing adaptive immunity.[36]

Impact of Age of Onset of Dietary Restriction on Longevity

In rodents, DR can be implemented for the entire lifespan, or in blocks of specified time. When DR starts shortly after weaning, life extension is maximized. However, it has been shown that DR can also significantly extend longevity when started in middle and old age, but its effects are weaker. When started very late in life, the effects of DR on lifespan may be minimal, reflective of nutritional memory, or may even possibly be detrimental.[58,59]

In the NIA primate study, DR was initiated at different ages to assess the impact of age of onset of food restriction on its biological effects. Both male and female monkeys who started DR at older ages weighed less than controls, and had lower plasma triglycerides, glucose, cholesterol and F2-isoprostanes (a marker of oxidative stress). In contrast, young-onset DR monkeys had fasting glucose concentration similar to controls until 23 years of age, after which values started to diverge, with higher concentration in control males.[4] Age-associated diseases occurred at half the rate in young-onset DR monkeys, compared with *ad libitum* fed animals, and cancer incidence was reduced.[4,19] Unfortunately, the first manifestation of age-associated disease could not be detected in the old-onset NIA monkeys because of their advanced age when the study began.

Data from randomized trials indicate that the extensive body composition, and beneficial metabolic and hormonal adaptations induced by moderate DR are similar in young (20–50 years) and older (50–60 years) non-obese adults.[60] Because most people consuming Western diets gain considerable weight (on average 0.5–1.0 kg per year, mostly fat) during early and middle adulthood, we recommend that health professionals prescribe DR (coupled with regular endurance and resistance exercise) to prevent the accumulation of excess abdominal and ectopic fat, well before people become obese. Indeed, epidemiologic studies suggest that small amounts of weight gain (≥2.5 kg to <10 kg), even in lean people, are associated with a significantly increased risk in major chronic diseases and unhealthy aging.[61] Excessive energy and protein intake during pubertal development and early adulthood, when many stem cells within the bone marrow and epithelial tissues undergo rapid cell division, overstimulates multiple nutrient sensing and mitogenic signaling pathways, and increases opportunities for mutations to occur and accumulate.[62] For example, in a retrospective cohort study, low energy intake prior to first birth, followed by a subsequent pregnancy in women with anorexia nervosa prior to age 40, was associated with a striking 76% reduction in breast cancer incidence.[59]

What To Eat: Protein and Specific Amino Acid Restriction

Protein Restriction

It is estimated that adults in many Western societies eat twice the amount of protein recommended by the U.S. Institute of Medicine, approximately 90 to 100 g, of which (~70%–85%) is animal protein.[63] High dietary intake of protein is associated with increased risk of obesity, type 2 diabetes, and cardiovascular disease mortality.[3] In humans between age 50 to 65 years, those in the lowest protein intake group (<10% energy from protein) had reduced incidence of cancer and lower mortality, while high protein intake was associated with diabetes in those aged 50 years or older.[64]

In rodents, lifelong experiments have shown that isocaloric protein restriction increases lifespan,[65] and that low protein diets protect against several cardiometabolic and neurodegenerative conditions.[66] In contrast, high-protein diets drive atherosclerosis by stimulating macrophage mTOR signaling and apoptosis in atherosclerotic plaques.[64] Mechanistic studies indicate that excessive protein intake causes insulin resistance and compensatory hyperinsulinemia[67] and induces other long-term pro-aging effects by overstimulating the AKT/mTOR pathway and inhibiting FGF21 signaling.[68]

Feeding high-protein diets for 6 to 18 weeks to overweight individuals with metabolic syndrome or type 2 diabetes reduced insulin-mediated glucose disposal.[69] Moreover, one carefully conducted randomized controlled trial found that protein supplementation (1.3 g/kg/day) in an energy-reduced weight loss diet did not affect muscle mass or strength[70] and, unlike with normal protein diets (0.8 g/kg/day), also prevented the improvement in insulin-mediated glucose disposal induced by 10% weight loss in women with obesity.[71] In addition, a small RCT found that short-term (4–6 weeks) isocaloric protein restriction significantly reduced fat mass and improved blood glucose, FGF21, and C-reactive protein concentrations in middle-aged, overweight and mildly obese males.[72]

The benefits of lower protein consumption may start in early life: newborns fed a lower protein-content formula (like that found in human milk) had lower risk of childhood obesity than infants consuming an isocaloric high protein-content formula.[73] Although the effect of protein restriction on human lifespan is unknown, the longest living population in the world, the Okinawans, have traditionally eaten a diet containing 9% energy from protein.

Methionine Restriction

In 1993, Orentreich and colleagues showed that lifelong ~80% methionine restriction resulted in a 30% increase in lifespan in male rats.[74] Many of the pro-longevity adaptations to methionine restriction have been attributed, at least in part, to lowering circulating IGF-1, insulin, and thyroid hormones,[75] decreased activity of the protein kinase mTORC1,[76] and to the pleiotropic actions of FGF21, which are upregulated by methionine restriction in both young and aged mice.[77,78] Some beneficial effects of methionine restriction may be due to the depletion of the antioxidant glutathione, which is generated from methionine via cysteine.[79] Methionine restriction results in lower concentration of the universal methyl donor SAM, and thus, has profound effects on histone methylation and gene expression.[80] Finally, both protein and methionine restriction promote the generation of hydrogen sulfide (H_2S), which has been shown to regulate longevity and health in multiple species.[81-84] Restricting methionine by 80% protects mice against diet-induced obesity, improving glucose tolerance and insulin sensitivity.[85]

Methionine restriction improves metabolic health not only in rodents, but in humans as well, by promoting reduced total and liver adiposity, improved glucose metabolism, and increased fat oxidation.[86] These data suggest the possibility that methionine restriction could be of use in both the prevention and treatment of obesity and diabetes. Due to the naturally lower levels of methionine in plant-based foods, people chronically consuming vegan diets have lower concentration of this amino acid in their blood, which might explain, in part, the reputed health benefits of veganism.[87]

Branched Chain Amino Acid Restriction

In mice, dietary branched chain amino acid (BCAA) restriction recapitulates some of the beneficial effects of protein restriction, including reduced adiposity, improved glucose tolerance, and elevated energy expenditure, but not increased FGF21.[72,88,89] Conversely, high BCAA diets are associated with hyperphagia, increased body weight, enlarged fat mass, and insulin resistance.[88,90] The extremely long lived Ames dwarf mouse has low circulating BCAAs; initiation of a 67% BCAA-restricted diet in young wild-type male mice reduced frailty and extended lifespan by more than 30%.[91] However, restriction of BCAAs by 50% or 80% at 12 weeks of age, or restriction of BCAAs by 67% starting at 16 months of age, improved body composition and glucose homeostasis but did not increase mouse lifespan.[90,91] Finally, and further supporting a negative effect of dietary BCAAs on mammalian healthspan and lifespan, dietary supplementation with additional BCAAs resulted in hyperphagia, obesity, and reduced lifespan of mice.[90]

Blood concentrations of branched-chain amino acids are correlated with obesity and diabetes in humans. The branched-chain amino acids leucine, isoleucine, and valine were first observed to be elevated in the blood of obese humans in the late 1960s[92] and, since then, several studies have observed increased serum concentrations of the three BCAAs in insulin-resistant humans and rodents.[65,93] The effects of one of the BCAAs, isoleucine, appears to be particularly potent. Restriction of dietary isoleucine is both necessary and sufficient for the metabolic benefits of a low-protein diet.[94] In humans, dietary isoleucine was positively associated with body mass index[94]; another recent study found that blood isoleucine was correlated with increased mortality in humans.[95]

Tryptophan Restriction

In studies in the 1970s and 1980s, tryptophan restriction in mice and rats increased average and maximum lifespan while delaying aging pathology, including cancers.[96] Despite this long history, it was only recently demonstrated that tryptophan restriction may be an evolutionarily conserved geroprotective intervention, with ibuprofen, an inhibitor of tryptophan uptake, extending the lifespan of yeast, worms, and flies.[97] Tryptophan is catabolized via the kynurenine pathway, and its metabolites include NAD^+, which is a key regulator of metabolism and an essential cofactor for the activity of the sirtuin family of enzymes.[98] One recent study found that serum tryptophan was associated with the onset of diabetes.[99] There was also a negative correlation between tryptophan and cognitive function in humans.[100] Finally, restriction of dietary tryptophan induces expression of FGF21 and recapitulates metabolic effects of a protein restriction diet in mice.[101] Much remains to be done to examine if tryptophan restriction can promote healthy aging in mice or in humans.

Threonine Restriction

A recent study surveyed the nine dietary essential amino acids to try and identify those responsible for the metabolic effects of a protein restriction diet, as modeled by a diet in which 5% of energy were derived from casein. Intriguingly, in this dietary setting, restriction of threonine mimicked the effects of a protein restriction diet, increasing energy expenditure, insulin sensitivity, and FGF21.[101] Concentrations of an enzyme in the threonine catabolic pathway, GCAT (glycine-C-acetyltransferase), decline with age in *C. elegans* and mice, and downregulation of this gene with RNAi extends the healthspan of *C. elegans*.[102] Overall, while threonine restriction may promote health and longevity, this remains to be formally tested in mice and humans.

When to Eat: Meal Frequency and Timing

Unlike most humans practicing DR, who typically reduce energy intake at every single meal, hungry DR rodents binge eat their once-a-day restricted food allotment in 1 to 4 hours (depending on the mouse strain) followed by a daily prolonged period of fasting. These observations have opened an exciting new field of translational research that is gaining momentum. Intermittent fasting (IF) in rodent models usually refers to an every other day 24-hr complete fast while, in humans, it refers to a variety of styles, such as complete fasting or severe DR (e.g., 500–600 kcal per day) on alternate days or 2 non-consecutive days per week (5:2 diet). Time-restricted feeding (TRF) involves consuming all daily food in a 4- to 12-h time window, whereas prolonged, periodic fasting (PPF), lasting for 2 to 7 consecutive days, and repeated cyclically, is an extreme form of DR that might have specific clinical indications.

However, one often overlooked caveat is that humans can fast for much longer than mice: a 24-hr fast-feed cycle in mice most likely equates to recurrent ~5 days fast-feed cycles in humans. Another problem is that laboratory rodents during the feast days consume a nutritionally balanced chow; in contrast, most people practicing different forms of IF, PPF, or TRF eat the unhealthy Western-like obesogenic food that has been shown to cause negative consequences on metabolic and gut microbiome health even during DR, and potentiate vitamin and mineral deficiencies.[103,104] Finally, unlike laboratory rodents, many obese adults take a range of medications (e.g., antidiabetes and antihypertensive agents) that could have serious negative, and potentially fatal consequences when coupled with fasting, including hypotension and severe hypoglycemia.

Intermittent Fasting

Fasting for 24-hours every other day or twice per week extends lifespan in mice and rats; IF started at 10 months of age caused a 15% change in mean and maximal lifespan of A/J inbred mice that are widely used to model cancer and for carcinogen testing, given their high susceptibility to carcinogen-induced tumors.[105] IF in rodents also reduces the incidence of a wide range of chronic diseases, including stroke, cardiomyopathy, hypertension, diabetes, and several neurodegenerative diseases, but the effects on cancer are controversial.[2] A 24-hour fast can also increase intestinal stem-cell function in mice through induction of fatty acid oxidation.[106]

Interestingly, as implemented in the laboratory, rodents on DR are typically fed once per day; as CR animals rapidly consume all their food within 1–4 h, CR-fed animals are subject to a prolonged collaterally imposed fast between meals.[107,108] We hypothesized that this imposed daily fast might contribute to the effects of DR and, using a series of dietary regimens, recently investigated the contribution of this imposed fast to DR. Surprisingly, we found that this prolonged fast between meals was necessary for the metabolic and geroprotective effects of a DR diet in C57BL/6J male mice, and that, in the absence of fasting, a reduction in energy intake alone did not extend lifespan or reduce frailty.[109] Furthermore, fasting alone—through imposition of a 21-hour per day interval between meals—was sufficient to recapitulate many of the metabolic and molecular effects of a CR diet.[109]

It is important to note that in adult C57BL/6J mice, plasma ketones begin to increase after just 4 to 7 hours of fasting, and peak at about 24 hours but, in humans, the production of ketone bodies after an overnight fast is negligible; plasma concentrations start to creep up only after 18 to 24 hours of fasting, and progressively increase (more rapidly in women than in men), peaking at 4 to 7 mM after 2 weeks fasting, which may be why IF is less effective in humans than mice.[2] IF in rodents exerts neuroprotective effects by increasing brain-derived neurotrophic factor (BDNF) concentration.[110] However, IF clinical trials reported significant reduction of circulating BDNF in humans.[111] Short-term randomized trials have shown potentially favorable effects of IF in humans. However, not all demonstrated beneficial effects and, unlike in rodents where some of the metabolic adaptations of IF are independent of food intake and weight loss, well-conducted human studies suggest that energy restriction is required to improve health.[112]

A 12-mo randomized trial comparing IF (25% of energy needs on fast days; 125% of energy needs on alternating "feast days") and chronic 25% DR, found no significant differences in body weight, blood pressure, triglycerides, HOMA-IR, or C-reactive protein at 6 or 12 months.[113] However, plasma LDL-cholesterol concentration was much higher (11.5 mg/dL) at the end of the study in the IF compared with the chronic DR group, further highlighting the importance of diet quality during feast days. Interestingly, 38% of the IF participants, vs 29% in the DR group, dropped out of the study, and many IF volunteers converted their diet to chronic DR as the trial progressed. Similar metabolic results have been reported for two other 1-yr trials comparing 5:2 fasting vs chronic DR in overweight adults.[111]

Prolonged, Periodic Fasting

In contrast to IF, which consists of frequent short periods without food, prolonged, periodic fasting lasts more than 24 hours and is repeated once or twice per month. Because water-only fasting for more than 48 hours is deadly in mice, an alternative is to feed them reduced portions of a diet low in protein and carbohydrate, and high in unsaturated fat, providing between 10% and 50% of normal AL intake. Feeding 16-mo old C57BL/6 mice such a fasting mimicking diet (FMD) for 4 days twice per month was enough to reduce weight and visceral fat accumulation, while preserving lean and bone mass. This FMD regimen reduced cancer burden, rejuvenated the immune system, improved motor and memory performance, and increased median, though not maximum, lifespan by 11%.[114] Weekly 3-day FMD in mouse models of multiple sclerosis reduced pro-inflammatory cytokines, suppressed autoimmunity, and promoted remyelination of axons.[115] Similarly, weekly 4-day FMD restored insulin sensitivity and glucose homeostasis in type 2 and type 1 diabetes mouse models.[116]

As reviewed elsewhere,[28] preclinical data have shown that both prolonged fasting and FMD can induce differential stress resistance responses in tumor-bearing mice, with the potential to maximize chemotherapy toxicity to cancer cells while protecting normal cells. Interestingly, in a recent trial of women with HER2-negative stage II/III breast cancer, 4 days of plant-based FMD each week during neoadjuvant chemotherapy indicated that the radiologic and pathologic tumor response was improved and chemotherapy-induced DNA damage to T-cells was reduced.[117] In this trial, insulin concentration was lowered in compliant patients and FMD significantly reduced circulating IGF-1 and increased ketone bodies.

Time-Restricted Feeding

Preclinical and epidemiologic data show that consuming food out of sync with the day/night cycle or over an extended time frame impairs metabolic health.[118] In mice, TRF (8-hour daily access to food) is thought to restore disrupted circadian clock rhythms and protect against obesity, fatty liver disease, insulin resistance, hyperinsulinemia, and inflammation independently of energy and fat intake.[119] These positive metabolic effects are attributed to fine tuning circadian clock genes in response to metabolic regulators, even when weekend *ad libitum* "cheat days" are permitted. Disrupting circadian rhythms by genetic manipulation of clock genes that encode circadian rhythmicity induces obesity, alters glucose metabolism, and reduces lifespan in mouse models.[120] A recent study showed that TRF is required for the metabolic, molecular, and geroprotective effects of DR to work in mice, and DR without a fasting period doesn't increase lifespan.[121]

Data on early TRF (feeding from 6am to 2pm) in non-human primates are limited to a study of geriatric vervet monkeys showing a significant increase in HDL-cholesterol and reverse cholesterol efflux, but no change in adiposity.[122] This is consistent with findings from a randomized trial of women affected by polycystic ovary syndrome showing that early meal timing (980 kcal breakfast, 640 kcal lunch, and 190 kcal dinner) was associated with more weight loss, higher insulin sensitivity, lower circulating testosterone, and increased ovulation rate, than controls eating isocaloric diets with a later meal pattern (190 kcal breakfast, 640 kcal lunch, and 980 kcal dinner).[123] A 5-week RCT in males with prediabetes found that, independent of weight loss, early time-restricted feeding (6-hr feeding period, with dinner before 3 pm) improved insulin sensitivity, β-cell responsiveness and blood pressure, but increased plasma triglycerides and total cholesterol.[124] A number of small short-term (2–4 month) pilot studies in overweight individuals suggest a beneficial effect of TRF (~8–12 hours per day) on body weight and composition, but the large TREAT RCT showed that overweight or obese men and women randomized to TRF (8 hours per day) did not experience weight loss or any cardiometabolic or glucose homeostatic improvements over 3 months.[125] Interestingly, unlike in rodents, the benefits of TRF do not appear to outweigh the benefits of energy restriction. A recent 12-month trial of obese men and women consuming a TRF (eating only between 8:00 a.m. and 4:00 p.m.) diet with energy restriction did not produce greater weight, visceral and liver fat loss, or improvements in blood pressure, glucose, insulin and lipid concentrations than the regimen of daily (time unrestricted) energy restriction alone.[126]

Summary and Conclusions

Although DR was discovered to extend lifespan and healthspan in rodents almost a century ago, the past few years have led to a surge of discoveries regarding the physiologic, dietary, temporal, and molecular mechanisms though which DR may function in rodents, as well as an increasing translation of DR-mimetic interventions in humans. It is important to note that rodents are not small humans—and caution must be taken as investigators seek to translate interventions and findings from mice to humans. Moreover, there is a growing awareness that humans are genetically heterogeneous, and multiple experiments by a range of laboratories have shown that both sex and genetic background play a critical role in the response to dietary interventions, including DR.[3,127-129] We still have much to learn about how diet interacts with other health interventions, including exercise training and medications. Due to all of these factors, it is clear that individuals considering adopting the types of dietary interventions discussed here should consult with their physician prior to implementing such regimens. As our understanding progresses, we anticipate that the clinical translation of these findings will use a personalized food-as-medicine approach, allowing each person to identify the dietary regimen and other factors that will maximize healthy aging.

Acknowledgments

The authors acknowledge Edward P. Weiss for his contributions to the previous version of this chapter.

References

1. Speakman JR, Mitchell SE. Mol Aspects Med 2011;32:159–221.
2. Green CL, Lamming DW, Fontana L. Nat Rev Mol Cell Biol 2022;23:56–73.
3. Green CL, Pak HH, Richardson NE, et al. Cell Metab 2022;34: 209–26 e5.
4. Mattison JA, Colman RJ, Beasley TM, et al. Nat Comm 2017;8:14063.
5. Villa E, Sahu U, O'Hara BP, et al. Mol Cell 2021;81:2076–93 e9.
6. Maegawa S, Lu Y, Tahara T, et al. Nat Comm 2017;8:539.
7. Longo VD, Fontana L. Trends Pharmacol Sci 2010;31:89–98.
8. Devesa A, Lobo-Gonzalez M, Martinez-Milla J, et al. Eur Heart J 2022;43:1809–28.
9. Kim KE, Jung Y, Min S, et al. Sci Rep 2016;6:30111.
10. Mitchell SE, Delville C, Konstantopedos P, et al. Oncotarget 2015;6:23213–37.
11. Yu D, Tomasiewicz JL, Yang SE, et al. Cell Rep 2019;29: 236–48 e3.
12. Fontana L, Meyer TE, Klein S, et al. Proc Natl Acad Sci U S A 2004;101:6659–63.
13. Strong R, Miller RA, Antebi A, et al. Aging Cell 2016;15: 872–84.
14. Miller RA, Harrison DE, Allison DB, et al. JCI Insight 2020;5.
15. Martin-Montalvo A, Mercken EM, Mitchell SJ, et al. Nat Comm 2013;4:2192.
16. Barzilai N, Appleby JC, Austad SN, et al. Aging Dis 2020;11: 725–9.
17. Wogatzky JB, Wirleitner A, Stecher P et al. Ann Internal Med 2022;175, 665–73.
18. Colman RJ, Anderson RM, Johnson SC, et al. Science 2009;325: 201–4.
19. Mattison JA, Roth GS, Beasley TM, et al. Nature 2012;489: 318–21.
20. Kraus WE, Bhapkar M, Huffman KM, et al. Lancet Diabetes Endocrinol.2019;7:673–83.
21. Lean MEJ, Leslie WS, Barnes AC, et al. Lancet Diabetes Endocrinol 2019;7:344–55.
22. Vilar-Gomez E, Martinez-Perez Y, Calzadilla-Bertot L, et al. Gastroenterology 2015;149:367–78.e5; quiz e14–5.

23. Ruggenenti P, Abbate M, Ruggiero B, et al. Diabetes 2017;66: 75–86.
24. Hursting SD, Smith SM, Lashinger LM, et al. Carcinog 2010;31: 83–9.
25. Kalaany NY, Sabatini DM. Nature 2009;458:725–31.
26. Carlsson LMS, Sjöholm K, Jacobson P, et al. N Engl J Med 2020;383:1535–43.
27. Look ARG, Yeh H-C, Bantle JP, et al. Obesity 2020;28:1678–86.
28. Clifton KK, Ma CX, Fontana L, et al. CA Cancer J Clin 2021;71:527–46.
29. Blackwell BN, Bucci TJ, Hart RW, et al. Tox Pathol 1995;23: 570–82.
30. Romagnani P, Remuzzi G, Glassock R, et al. Nat Rev Dis Prim 2017;3:17088.
31. Jongbloed F, de Bruin RWF, Steeg HV, et al. Aging (Albany NY) 2020;12:12441–67.
32. Ahmet I, Tae HJ, de Cabo R, et al. J Mol Cell Cardiol 2011;51: 263–71.
33. Guo Z, Mitchell-Raymundo F, Yang H, et al. Mech Ageing Dev 2002;123:1121–31.
34. Kraus WE, Bhapkar M, Huffman KM, et al. Lancet Diabetes Endocrinol 2019;7:673–83.
35. Il'yasova D, Fontana L, Bhapkar M, et al. Aging Cell 2018;17.
36. Meydani SN, Das SK, Pieper CF, et al. Aging (Albany NY) 2016;8:1416–31.
37. Riordan MM, Weiss EP, Meyer TE, et al. Am J Physiol-Heart Circul Physiol 2008;294:H1174–H82.
38. Ingram DK, Weindruch R, Spangler EL, et al. J Gerontol 1987;42:78–81.
39. Pitsikas N, Algeri S. Neurobiol Aging 1992;13:369–73.
40. Markowska AL. Neurobiol Aging 1999;20:177–89.
41. Patel NV, Gordon MN, Connor KE, et al. Neurobiol Aging 2005;26:995–1000.
42. Halagappa VK, Guo Z, Pearson M, et al. Neurobiol Dis 2007;26: 212–20.
43. Duan W, Mattson MP. J Neurosci Res 1999;57:195–206.
44. Bruce-Keller AJ, Umberger G, McFall R, et al. Annals of Neurol 1999;45:8–15.
45. Qin W, Chachich M, Lane M, et al. J Alzheimer's Dis 2006;10: 417–22.
46. Leclerc E, Trevizol AP, Grigolon RB, et al. CNS Spectrums 2020;25:2–8.
47. Witte AV, Fobker M, Gellner R, et al. Proc Natl Acad Sci 2009;106:1255.
48. Fernandes G, Good RA. Proc Natl Acad Sci U S A 1984;81: 6144–8.
49. Dong W, Kari FW, Selgrade MK, et al. Environ Health Perspect 2000;108:1125–31.
50. Kari F, Hatch G, Slade R, et al. Am J Respir Cell Mol Biol 1997;17:740–7.
51. Contreras NA, Fontana L, Tosti V, et al. GeroScience 2018;40: 279–91.
52. Meydani SN, Das SK, Pieper CF, et al. Aging 2016;8:1416–31.
53. Roecker EB, Kemnitz JW, Ershler WB, et al. J Gerontol Ser A Biol Sci Med Sci 1996;51:B276–9.
54. Ritz BW, Aktan I, Nogusa S, et al. J Nutr 2008;138:2269–75.
55. Yang H, Youm YH, Dixit VD. J Immunol 2009;183:3040–52.
56. Messaoudi I, Warner J, Fischer M, et al. Proc Natl Acad Sci U; S A 2006;103:19448–53.
57. Spadaro O, Youm Y, Shchukina I, et al. Science 2022;375: 671–7.
58. Cheney KE, Liu RK, Smith GS, et al. J Gerontol 1983;38: 420–30.
59. Hahn O, Drews LF, Nguyen A, et al. Nat Metab 2019;1: 1059–73.
60. Most J, Tosti V, Redman LM, et al. Ageing Res Rev 2017;39: 36–45.
61. Zheng Y, Manson JE, Yuan C, et al. JAMA 2017;318:255–69.
62. Mitchell E, Spencer Chapman M, Williams N, et al. Nature 2022; 606:343–50.
63. Gardner CD, Hartle JC, Garrett RD, et al. Nutr Rev 2019;77: 197–215.
64. Levine ME, Suarez JA, Brandhorst S, et al. Cell Metab 2014;19: 407–17.
65. Solon-Biet SM, McMahon AC, Ballard JW, et al. Cell Metab 2014; 19:418–30.
66. Wahl D, Solon-Biet SM, Wang QP, et al. Cell Rep 2018;25: 2234–43 e6.
67. Smith GI, Yoshino J, Stromsdorfer KL, et al. Diabetes 2015;64: 1555–63.
68. Mittendorfer B, Klein S, Fontana L. Nat Rev Endocrinol 2020;16: 59–66.
69. Sargrad KR, Homko C, Mozzoli M, et al. J Am Diet Assoc 2005;105:573–80.
70. Smith GI, Commean PK, Reeds DN, et al. Obesity (Silver Spring) 2018;26:854–61.
71. Smith GI, Yoshino J, Kelly SC, et al. Cell Rep 2016;17:849–61.
72. Fontana L, Cummings NE, Arriola Apelo SI, et al. Cell Rep 2016;16:520–30.
73. Weber M, Grote V, Closa-Monasterolo R, et al. Am J Clin Nutr 2014;99:1041–51.
74. Orentreich N, Matias JR, DeFelice A, et al. J Nutr 1993;123: 269–74.
75. Miller RA, Buehner G, Chang Y, et al. Aging Cell 2005;4: 119–25.
76. Simcox J, Lamming DW. Dev Cell 2022;57:691–706.
77. Lees EK, Krol E, Grant L, et al. Aging Cell 2014;13:817–27.
78. Yu D, Yang SE, Miller BR, et al. FASEB J 2018;32:3471–82.
79. Wanders D, Stone KP, Forney LA, et al. Diabetes 2016;65: 1499–510.
80. Haws SA, Yu D, Ye C, et al. Mol Cell 2020;78:210–23 e8.
81. Hine C, Harputlugil E, Zhang Y, et al. Cell 2015;160:132–44.
82. Longchamp A, Mirabella T, Arduini A, et al. Cell 2018;173: 117–29 e14.
83. Das A, Huang GX, Bonkowski MS, et al. Cell 2018;173: 74–89 e20.
84. Miller DL, Roth MB. Proc Natl Acad Sci U S A 2007;104: 20618–22.
85. Ables G, Perrone C, Orentreich D, et al. PloS One 2012;7: e51357.
86. Plaisance EP, Greenway FL, Boudreau A, et al. J Clin Endocrinol Metab 2011;96:E836–40.
87. Schmidt JA, Rinaldi S, Scalbert A, et al. Eur J Clin Nutr 2016;70:306–12.
88. Cummings NE, Williams EM, Kasza I, et al. J Physiol 2018;596:623–45.
89. Trautman ME, Richardson NE, Lamming DW. Aging Cell 2022;e13626.
90. Solon-Biet SM, Cogger VC, Pulpitel T, et al. Nat Metab 2019;1:532–45.
91. Richardson NE, Konon EN, Schuster HS, et al. Nat Aging 2021;1:73–86.
92. Felig P, Marliss E, Cahill GF, Jr. N Engl J Med 1969;281:811–6.
93. Ribeiro RV, Solon-Biet SM, Pulpitel T, et al. Nutrients 2019;11.
94. Yu D, Richardson NE, Green CL, et al. Cell Metab 2021;33: 905–22 e6.
95. Deelen J, Kettunen J, Fischer K, et al. Nat Comm 2019;10: 3346.
96. Ooka H, Segall PE, Timiras PS. Mech Ageing Dev 1988;43: 79–98.

97. He C, Tsuchiyama SK, Nguyen QT, et al. PLoS Genet 2014;10: e1004860.
98. Imai S, Armstrong CM, Kaeberlein M, et al. Nature 2000;403: 795–800.
99. Chen T, Zheng X, Ma X, et al. PLoS One 2016;11:e0162192.
100. Ramos-Chavez LA, Roldan-Roldan G, Garcia-Juarez B, et al. Oxid Med Cell Longev 2018;2018:8604718.
101. Yap YW, Rusu PM, Chan AY, et al. Nat Comm 2020;11:2894.
102. Ravichandran M, Priebe S, Grigolon G, et al. Cell Metab 2018;27:914–25 e5.
103. Griffin NW, Ahern PP, Cheng J, et al. Cell Host Microbe 2017;21:84–96.
104. Dey N, Wagner VE, Blanton LV, et al. Cell 2015;163:95–107.
105. Goodrick CL, Ingram DK, Reynolds MA, et al. Mech Ageing Dev 1990;55:69–87.
106. Mihaylova MM, Cheng CW, Cao AQ, et al. Cell Stem Cell 2018;22:769–78 e4.
107. Acosta-Rodríguez VA, Rijo-Ferreira F, Green CB, et al. Nat Comm 2021;12:2862.
108. Acosta-Rodriguez VA, de Groot MHM, Rijo-Ferreira F, et al. Cell Metab 2017;26:267–77 e2.
109. Pak HH, Haws SA, Green CL, et al. Nat Metab 2021;3:1327–41.
110. Mattson MP, Arumugam TV. Cell Metab 2018;27:1176–99.
111. Schübel R, Nattenmüller J, Sookthai D, et al. Am J Clin Nutr 2018;108:933–45.
112. Hutchison AT, Liu B, Wood RE, et al. Obesity (Silver Spring) 2019;27:50–8.
113. Trepanowski JF, Kroeger CM, Barnosky A, et al. JAMA Intern Med 2017;177:930–8.
114. Brandhorst S, Choi In Y, Wei M, et al. Cell Metab 2015;22: 86–99.
115. Choi IY, Piccio L, Childress P, et al. Cell Rep 2016;15: 2136–46.
116. Cheng C-W, Villani V, Buono R, et al. Cell 2017;168:775–88. e12.
117. de Groot S, Lugtenberg RT, Cohen D, et al. Nature Comm 2020;11:3083.
118. Mattson MP, Allison DB, Fontana L, et al. Proc Natl Acad Sci 2014;111:16647–53.
119. Hatori M, Vollmers C, Zarrinpar A, et al. Cell Metab 2012;15: 848–60.
120. Turek FW, Joshu C, Kohsaka A, et al. Science (New York, N.Y.) 2005;308:1043–5.
121. Yu D, Richardson NE, Green CL, et al. Cell Metab 2021;33: 905–22.e6.
122. Kavanagh K, Bashore AC, Davis M, et al. Innov Aging 2019;3:S104–S.
123. Jakubowicz D, Barnea M, Wainstein J, et al. Clin Sci (Lond) 2013;125:423–32.
124. Sutton EF, Beyl R, Early KS, et al. Cell Metab 2018;27: 1212–21.e3.
125. Lowe DA, Wu N, Rohdin-Bibby L, et al. JAMA Intern Med 2020;180:1491–9.
126. Liu D, Huang Y, Huang C, et al. N Engl J Med 2022;386: 1495–504.
127. Green CL, Lamming DW. Nat Metab 2021;3:1144–5.
128. Liao CY, Rikke BA, Johnson TE, et al. Aging Cell 2010;9: 92–5.
129. Mitchell SJ, Madrigal-Matute J, Scheibye-Knudsen M, et al. Cell Metab 2016;23:1093–112.

CHAPTER 52

Nutrition in Pregnancy

Christine D. Garner

CHAPTER OUTLINE

Abbreviations

ACOG, American College of Obstetricians and Gynecologists; **AI**, adequate intake; **BEE**, basal energy expenditure; **BMI**, body mass index; **DHA**, docosahexaenoic acid; **DRI**, dietary reference intake; **EAR**, estimated average requirement; **FAS**, fetal alcohol syndrome; **FGR**, fetal growth restriction; **GD**, gestational diabetes; **GWG**, gestational weight gain; **IOM**, Institute of Medicine; **LBW**, low birth weight; **LGA**, large for gestational age; **NTD**, neural tube defect; **PKU**, phenylketonuria; **RDA**, recommended dietary allowance; **SGA**, small for gestational age; **TEE**, total energy expenditure; **UL**, tolerable upper intake level; **USPSTF**, United States Preventive Services Taskforce

Introduction

Optimal nutrition is integral to a healthy pregnancy, which can be described as "without physical or psychological pathology in mother and results in delivery of a healthy baby".[1] Although the influence of poor nutritional status on adverse pregnancy outcome was documented early in the 20th century, retrospective studies considering the effects of food shortages in the Netherlands during World War II clearly identified the influence of diet on pregnancy outcome.[2-4] Nutrition can affect the mother's health and risk of pregnancy complications; it also affects the growth and development of the fetus, risk of birth defects, and health of the infant at delivery. Further studies have linked both undernutrition and overnutrition during pregnancy to increased risks of obesity, coronary heart disease, hypertension, diabetes, metabolic syndrome, and psychiatric disorders in the children; these findings suggest a persistent change in gene expression in response to the intrauterine environment.[5-9]

Box 52.1 Preconception Wellness Measures

Pregnancy intention
Access to care
Preconception folic acid use
Tobacco avoidance
Absence of uncontrolled depression
Healthy weight
Absence of sexually transmitted infection
Optimal glycemic control
Teratogen avoidance in chronic conditions

Reproduced from Frayne DJ, Verbiest S, Chelmow D, et al. Obstet Gynecol 2016;127:863–72.[14]

Current Public Health Objectives Related to Pregnancy and Neonatal Health

Maternal and infant health are important predictors of the future health of the world's citizens. The World Health Organization (WHO) states that women who remain healthy during pregnancy and after birth are more likely to stay healthy later in life and have better birth outcomes, influencing infancy, childhood, and adulthood.[10] In the United States, the overarching vision for the U.S. Department of Health and Human Services' *Healthy People 2030* is for "all people to achieve their full potential for health and well-being across the lifespan".[11] Public health issues related to maternal and child health include morbidity and mortality of pregnant and postpartum women; fetal, perinatal, and infant mortality; birth outcomes; prevention of birth defects; health behaviors before conception and during pregnancy; and access to preventive care. Progress has been made toward objectives related to fetal and infant deaths, prenatal care, preterm birth, and prevention of neural tube defects (NTDs); however, the occurrence of low birth weight (LBW), maternal deaths, and use of alcohol have increased.[12,13] Healthy People 2030 objectives continue to emphasize the importance of nutrition, prenatal care, and preconception health in improving the health of mothers and infants.

Preconception Health

Nutritional status before pregnancy is a key factor in overall maternal health and in the risk of birth defects. Women who are contemplating pregnancy can make dietary and lifestyle changes that will reduce the risk of poor pregnancy outcome. The Preconception Health and Health Care Initiative Clinical Workgroup, a public-private multiorganizational partnership in the United States, recommends that healthcare systems adopt and monitor nine preconception wellness measures (**Box 52.1**), many of which are directly or indirectly related to nutritional status.[14]

Folic acid supplementation is recommended from at least 1 month before pregnancy through the first 2 to 3 months of pregnancy to reduce the risk of NTDs and other birth defects. Ideally, all women of childbearing age should be consuming 400 μg/day of folic acid in addition to folate provided through foods because just under one-half of all pregnancies in the United States are unplanned.[15] Women following vegan or other strict vegetarian diets should also take supplemental vitamin B_{12} because the status of this vitamin is another risk factor for NTDs.[16]

Preconception iron status is important to reducing the risk during pregnancy of iron deficiency and anemia which, in turn, can lead to fetal growth restriction (FGR) and preterm birth.[17] Preconception care should include screening for iron deficiency anemia. Multivitamin and mineral supplementation may help to improve nutritional status in women who are following inappropriate diets, are avoiding numerous foods or groups of foods, are underweight, are trying to lose weight, or are abusing alcohol.

Achieving a healthy weight before pregnancy may improve chances of conception, pregnancy outcome, and lactation.[18] Women with obesity at the start of pregnancy are at greater risk of gestational diabetes (GD) and pre-eclampsia, and of experiencing induced labor and cesarean delivery. Women with obesity also tend to have lower intention to breastfeed and may have more difficulty initiating and sustaining breastfeeding.[19] Infants born to women who were obese before pregnancy are at increased risk of congenital abnormalities, NTDs, stillbirth, macrosomia, and obesity later in life.[18] Physical activity can help to improve weight and nutritional status; however, the amount of physical activity needed daily for weight management, chronic disease risk reduction, and enhanced physical fitness varies.[20]

Management of pre-existing chronic disease is another important element of preconception planning. Women with hypertension are at risk of maternal, fetal, and neonatal morbidity and mortality. Even slightly elevated blood pressure prior to conception and in early pregnancy are associated with development of hypertensive disorders of pregnancy.[21] The presence of pre-existing hypertension, severity of hypertension, and the presence of pre-eclampsia negatively affect pregnancy outcomes.[21-23]

Pregestational diabetes increases the risk of birth defects, especially defects of the heart and central nervous system, and it also increases the risk of spontaneous abortion.[24,25] Attaining good blood glucose control before conception and during organogenesis can substantially reduce risks.[26]

An estimated 3,000 to 6,000 women of childbearing age in the United States have phenylketonuria (PKU).[27] To prevent microcephaly, intellectual or developmental disabilities, facial dysmorphism and congenital heart disease in their infants, pregnant women with PKU must maintain a low-protein, amino acid–modified diet during pregnancy.[28] Strict dietary control and treatment of PKU is recommended before conception with continued tight control throughout pregnancy.

Maternal Physiologic Changes During Pregnancy

Numerous anatomic, biochemical, and physiologic changes occur during pregnancy to maintain a healthy environment for the growing fetus without compromising the mother's health.

Many of these changes begin in the early weeks of pregnancy, and together they regulate maternal metabolism, promote fetal growth, and prepare the mother for labor, birth, and lactation. A review of the physiologic changes during pregnancy sets the stage for understanding the changes in nutrient requirements that accompany pregnancy.

Maternal plasma volume begins to expand near the end of the first trimester, with a total increase in volume of 50% by 30 to 34 weeks of gestation. Red blood cell production is stimulated with a total increase in red blood cell mass of approximately 33%. As plasma volume expansion during pregnancy is greater than the red blood cell mass increase, hematocrit and hemoglobin decline until late pregnancy. Declining concentrations of plasma proteins and other nutrients are expected because of the expansion of blood volume. Poor plasma volume expansion predicts a poorly growing fetus and poor pregnancy outcome.[29]

Cardiac output increases approximately 30% to 50% during pregnancy. Elevated cardiac output occurs in response to increased tissue demands for oxygen and is accompanied by an increase in stroke volume. The size of the heart increases by approximately 12%, probably because of the increased blood volume and cardiac output. Systemic blood pressure declines slightly during pregnancy, with most of the change occurring in diastolic pressure (5 to 10 mm Hg). Diastolic pressure returns to prepregnancy levels near term.

Respiratory changes support increased maternal and fetal requirements for oxygen. As the uterus enlarges, the diaphragm is elevated, which reduces lung capacity by about 5%, and residual volume is reduced by approximately 20%.[30] Tidal volume increases as pregnancy progresses, resulting in increased alveolar ventilation and more efficient gas exchange, given that oxygen consumption increases by only 15% to 20%.[30] Respiration rate increases only slightly.

The kidneys increase slightly in both length and weight during pregnancy; and the ureters elongate, widen, and become more curved. Glomerular filtration rate increases by approximately 50%, and renal plasma flow rate increases by 25% to 50%. Renin increases early in the first trimester and continues to rise until term. Most pregnant women are resistant to the pressor effects of the resulting elevation of angiotensin II, but enhanced renin secretion may help to explain pre-eclampsia. A marked increase in excretion of glucose, amino acids, and water-soluble vitamins occurs, probably because the higher glomerular filtration rate presents higher concentrations of nutrients than the tubules can reabsorb.

Changes along the gastrointestinal tract support the increased demand for nutrients during pregnancy. Appetite increases, although initially this may be offset by nausea and vomiting. High progesterone decreases the production of motilin, a hormone that stimulates smooth muscle throughout the body, including in the gastrointestinal tract. The result is reduced muscle contraction throughout the gastrointestinal tract, including the lower esophageal sphincter, as well as elongation in gastrointestinal transit time, especially by the third trimester of pregnancy.[31] Gallbladder emptying time is also reduced and often incomplete. These changes can cause gastrointestinal symptoms, including heartburn and constipation.

The basal metabolic rate rises by the fourth month of gestation and has usually increased by 15% to 20% by term. An elevated basal metabolic rate reflects the increased demand for and consumption of oxygen. Most (50%–70%) of the energy needs of the fetus are provided by glucose, with approximately 20% coming from amino acids and the remainder derived from fat. Use of fatty acids for fuel is enhanced in the mother to conserve glucose for use by the fetus.

Prepregnancy Weight and Weight Gain

Gestational weight gain (GWG) and maternal body mass index (BMI) before pregnancy influence maternal and infant outcomes. In 2009, the U.S. Institute of Medicine (IOM, now known as the National Academy of Medicine) published recommendations for gestational weight gain that have been broadly used.[32] These recommendations reflect the GWG and rate of weight gain associated with the best pregnancy outcome for both mother and child based on prepregnancy BMI category (**Table 52.1**). Recommended ranges of GWG decrease as BMI increases. At the time when these recommendations were developed, insufficient data were available to provide GWG guidelines for each obesity class.

Since the 2009 IOM guidelines were released, further research has investigated optimal GWG, particularly for higher BMI groups. Results from The LifeCycle Project, which pooled data from 196,670 women in Europe and North America, suggest that the 2009 guidelines, which proposed a single range for all obesity subgroups because of insufficient data for heavier women, may be too high for optimal health for women with obesity grades 2 (BMI 35.0–39.9) and 3 (BMI >40.0).[33] Importantly, gestational weight gain had lower predictive value for pregnancy outcomes than prepregnancy BMI. Attaining a healthy weight prior to pregnancy, thus, appears to be key to minimizing risk for both mother and child.

Studies that examine GWG among obese individuals have produced conflicting results. Some research suggests that GWG

Table 52.1 Gestational Weight Gain Recommendations for Singleton Pregnancies

Weight Category	Prepregnancy Body Mass Index (Kg/M²)	Total Weight Gain (Kg)	Rate of Weight Gain[a] (mean; Kg/Wk)
Underweight	<18.5	12.5–18.0	0.51
Normal weight	18.5–24.9	11.5–16.0	0.42
Overweight	25.0–29.9	7–11.5	0.28
Obese	≥30.0	5–9	0.22

[a]Second and third trimesters.

Reproduced from Institute of Medicine, National Research Council. Weight Gain During Pregnancy: Reexamining the Guidelines. Washington, DC: National Academies Press, 2009.

below IOM guidelines may optimize outcomes among obese mothers.[33,34] Conversely, results from a study including more than 15.8 million mother-infant dyads that examined infant morbidity and mortality outcomes support the IOM guidelines and caution against weight gain below current recommendations even for women with obesity.[35] Notably, this study did not examine maternal outcomes, and research in this area continues.

Determinants of Gestational Weight Gain

A variety of factors influence the amount of weight gained during pregnancy, including environmental factors, maternal genetics and body size, medical and psychologic conditions, and behavioral factors.[32] Prepregnancy BMI appears to be the strongest independent predictor of GWG.[33,36] Evidence suggests that GWG among underweight (BMI <18.5 kg/m^2) and normal weight (BMI 18.5 to 24.9 kg/m^2) women is more often within the recommended range, whereas GWG among women with overweight (25.0 to 29.9 kg/m^2) and obesity (≥30 kg/m^2) tends to be higher than recommended.[32,37]

Additional predictors of GWG have been examined. Limited evidence from a systematic review suggests that dietary patterns higher in vegetables, fruits, nuts, legumes, and fish and lower in sugar and red and processed meats reduce the risk of excess GWG.[38] Additional modifiable predictors of excess GWG include planned weight gain above recommendations, frequent food cravings, fast food intake, eating in front of a screen, and >3 hours of screen time daily.[39,40] Nulliparity and depression also appear to be associated with excess GWG in some populations.

Effects on Fetal and Maternal Outcomes

Interactions between GWG and prepregnancy BMI on pregnancy outcomes are complex. Evidence suggests that maternal prepregnancy BMI is more strongly associated with adverse pregnancy outcomes than GWG.[33] Importantly, both appear to have significant impacts on many outcomes. Excessive and inadequate GWG are both associated with adverse outcomes. Across all prepregnancy BMI groups, GWG below recommendations increases the odds of small-for-gestational age (SGA) infant and preterm delivery.[41] Conversely, GWG above recommendations increases the odds of large-for-gestational age (LGA), macrosomia, and cesarean delivery.[42]

The relationship between GWG and birthweight appears to be strongest in the lower prepregnancy BMI groups.[32,42] Particularly among women beginning pregnancy with a low BMI, inadequate GWG increases the risk of SGA, while excess GWG increases the risk of LGA and macrosomia. Generally, poor weight gain is associated with poor fetal growth, LBW, SGA birth weight, and risk of preterm delivery. Carmichael and Abrams[43] found that a marked acceleration or deceleration of gain toward the end of pregnancy was associated with higher risk of spontaneous preterm delivery. Low GWG is also associated with failure to initiate breastfeeding.[32,44]

Excessive weight gain affects infant growth, increases the chance of LGA birth weight and cesarean delivery, and is associated with higher body fatness in childhood. Although nearly half (47%) of pregnant women in the United States gain above the guidelines, women who are overweight or obese are more likely than women of normal weight to gain excess weight,[37] and exceeding recommendations was found to be more likely among low-income women.[45] Excessive GWG is further associated with early breastfeeding cessation, postpartum weight retention, and future overweight or obesity.[32,44]

Weight gain recommendations should be individualized to promote best outcomes while reducing risk for excessive postpartum weight retention and reducing the risk of later chronic disease in the child. The most effective approach for reducing negative outcomes associated with GWG is for women to be within the normal range for BMI at the time of conception.[32,46] Even small weight gains between pregnancies increase the risk of complications in future pregnancies.[47] Women should receive counseling about appropriate GWG during preconception and prenatal care, with tracking of actual weight gain across pregnancy and referral to services for counseling on diet and physical activity as needed.[18,32]

Energy and Nutrient Needs

To support the growth of the fetus and the health of the mother, requirements for energy and for most vitamins and minerals are higher during pregnancy.

Energy

Energy is needed to support basal energy expenditure (BEE), physical activity, the thermic effect of food, and, in pregnant women, growth of the fetus, placenta, and deposition of maternal tissues (fat mass, breast tissue, and uterus). BEE increases because of the enhanced metabolism of the uterus and fetus and the increased work of the heart and lungs. Increased BEE represents the major component of increased energy requirements. The cumulative increase in BEE is estimated at 106 to 180 kcal/day, although variation among individuals is substantial.[48] Late in pregnancy, the fetus uses approximately 56 kcal/kg/day, which represents about one-half of the increment of BEE.

The theoretical energy cost of energy deposition can be estimated from the amount of protein and fat deposited.[48] Mean total energy deposition is 39,862 kcal (180 kcal/day).[32] Analysis of studies employing the doubly labeled water method shows a median change in total energy expenditure (TEE) of 8 kcal/gestational week. The estimated energy requirement for pregnancy is thus derived from the sum of TEE for a nonpregnant woman plus 8 kcal/gestational week plus 180 kcal/day for energy deposition. This increase in recommended energy intake is suggested only for the second and third trimesters because TEE changes little in the first trimester, as on average physical activity decreases slightly and weight gain is minimal.[49] Therefore, during the second trimester, an additional 340 kcal/day greater than nonpregnant energy requirements is recommended. This increase climbs to 452 kcal/day extra in the third trimester.

Data used in estimating energy requirements during pregnancy have left obese women and racial and ethnic minorities largely un- or under-represented. Recent models have been developed that take higher BMI into account and suggest that overweight women may require smaller increases in energy intake during pregnancy, estimating ranges of 260 to 330 and 270 to 360 kcal/day required in the second and third trimesters, respectively.[50] Energy needs for obese women during pregnancy remain controversial with indications that obese women may require even smaller increases in energy intake during pregnancy, or that mobilization of energy stores without increases in energy intake may support GWG consistent with recommendations.[49,50]

Ultimately, the best method for assessing the adequacy of energy intake is to monitor GWG. The recommended balance of energy sources during pregnancy is the same as for nonpregnant women: 10% to 35% of kcal as protein, 45% to 65% as carbohydrate, and 20% to 35% as fat.[51] The Dietary Guidelines for Americans recommendations for nutrient intake during pregnancy are summarized in **Table 52.2**.

Table 52.2 Daily Nutritional Goals and Upper Limits for Nutrients During Pregnancy

Nutrient	Source	Nutrient Goal	Upper Limit
Macronutrients			
Protein (g/d)	RDA	71	n/a
Protein (% kcal)	DGA	10–35	n/a
Carbohydrate (g/d)	RDA	175	n/a
Carbohydrate (% kcal)	DGA	45–65	n/a
Total fiber (g/d)	AI	28–36	n/a
Added sugars (% kcal)	DGA	<10	n/a
Total fat (g/d)	DGA	20–35	n/a
Saturated fatty acids (% kcal)	DGA	<10	n/a
18:2 Linoleic acid (g/d)	AI	13	n/a
18:3 Linolenic acid (g/d)	AI	1.4	n/a
Vitamins			
Vitamin A (μg/d)	RDA	770	3,000
Vitamin C (mg/d)	RDA	85	2,000
Vitamin D (IU)	RDA	600	4,000
Vitamin E (μg/d)	RDA	15	1,000
Vitamin K (μg/d)	AI	90	ND
Thiamin (mg/d)	RDA	1.4	ND
Riboflavin (mg/d)	RDA	1.4	ND
Niacin (mg/d)	RDA	18	35
Vitamin B_6 (mg/d)	RDA	1.9	100
Folate (μg/d)	RDA	600	1,000[a]
Vitamin B_{12} (μg/d)	RDA	2.6	ND
Pantothenic Acid (mg/d)	AI	6	ND
Biotin (μg/d)	AI	30	ND
Choline (mg/d)	AI	450	3,500
Minerals			
Calcium (mg/d)	RDA	1,000	2,500
Copper (μg/d)	RDA	1,000	10,000
Fluoride (mg/d)	AI	3	10
Iodine (μg/d)	RDA	220	1,100
Iron (mg/d)	RDA	27	45

(continues)

Table 52.2 Daily Nutritional Goals and Upper Limits for Nutrients During Pregnancy *(continued)*

Nutrient	Source	Nutrient Goal	Upper Limit
Magnesium (mg/d)	RDA	350	350[b]
Manganese (mg/d)	AI	2.0	11
Molybdenum (μg/d)	RDA	50	2,000
Phosphorus (mg/d)	RDA	700	3,500
Selenium (μg/d)	RDA	60	400
Zinc (mg/d)	RDA	11	40
Potassium (mg/d)	AI	2,900	ND
Sodium (mg/d)	CDRR	2,300	ND
Chloride (g/d)	AI	2.3	3.6
Macronutrients			
Protein (g/d)	RDA	71	n/a
Protein (% kcal)	DGA	10–35	n/a
Carbohydrate (g/d)	RDA	175	n/a
Carbohydrate (% kcal)	DGA	45–65	n/a
Total fiber (g/d)	AI	28–36	n/a
Added sugars (% kcal)	DGA	<10	n/a
Total fat (g/d)	DGA	20–35	n/a
Saturated fatty acids (% kcal)	DGA	<10	n/a
18:2 Linoleic acid (g/d)	AI	13	n/a
18:3 Linolenic acid (g/d)	AI	1.4	n/a

RDA = Recommended Dietary Allowance, AI = Adequate Intake, CDRR = Chronic Disease Risk Reduction Level, DGA = Dietary Guidelines for Americans, 2020–2025

ND = Not determinable due to lack of data.

[a]In the form of folic acid.

[b]As pharmacologic agents only.

Data from U.S. Department of Agriculture and U.S. Department of Health and Human Services. Dietary Guidelines for Americans, 2020-2025. 9th Edition. December 2020. Available at DietaryGuidelines .gov and Institute of Medicine 2006. Dietary Reference Intakes: The Essential Guide to Nutrient Requirements. Washington, DC: The National Academies Press. https://doi.org/10.17226/11537

Protein

During pregnancy, whole body protein turnover increases, and substantial amounts of protein are accumulated by the growth of the fetus, uterus, blood volume, placenta, amniotic fluid, and maternal skeletal muscle.[48] Considering protein deposition over the last two trimesters, the recommended dietary allowance (RDA) increases from 46 g/day to 71 g/day. For a reference woman weighing 57 kg, this is an additional 0.27 g/kg/day for a total of 1.1 g/kg/day during pregnancy. Protein intake should increase proportionally with energy intake.

Carbohydrate

The fetus uses glucose as its major energy source. The transfer of glucose from mother to fetus is estimated at 17 to 26 g/day. By the end of pregnancy, all of this glucose is thought to be used by the fetal brain.[48] The estimated average requirement (EAR) for carbohydrate increases from 100 g/day to 135 g/day, which translates to an RDA for carbohydrate for pregnant women of 175 g/day.

Fat

Fat is a major source of energy for the body and aids in the absorption of fat-soluble vitamins and carotenoids. The developing brain accumulates large amounts of docosahexaenoic acid (DHA) during prenatal and postnatal development. Maternal n-3 PUFA intake has been associated positively with cognitive and retina development. Interventions that increase maternal n-3 PUFA intake either through food or supplements may reduce preterm birth, likely due to anti-inflammatory effects.[52] The major dietary source of n-3 fatty acids is fish and seafood, of which most childbearing-age and pregnant women in the United States Consume inadequate amounts.[53] The adequate intake (AI) values for essential fatty acids during pregnancy are based on median intakes among pregnant women in the United States: 13 g/day for linoleic acid and 1.4 g/day for α-linolenic acid.

Fat-Soluble Vitamins

Vitamin A is important for regulation of gene expression and for cell differentiation and proliferation, particularly for the development of the vertebrae and spinal cord, the limbs, heart, eyes, and ears. The increase in maternal requirement of 70 μg/day as retinol activity equivalents supports growth, maternal metabolism, and accumulation for fetal liver reserves.[54] Maternal liver stores are considered sufficient to cover this requirement in non-deficient individuals.

Vitamin A deficiency affects 19 million pregnant women globally.[55] Maternal night blindness, a symptom of deficiency, is associated with infant mortality risk. Vitamin A supplementation in endemic regions has not decreased maternal or infant mortality but reduces maternal night blindness and maternal anemia.[56]

Excess retinol intake is a known human teratogen. The most critical period for damage appears to be in the first trimester, resulting in spontaneous abortions and birth defects affecting the cardiovascular system, central nervous system, craniofacial area, and thymus.[57] The threshold for risk remains controversial[57,58]; however, teratogenicity was used as the critical adverse effect for women of childbearing age in determining the tolerable upper intake level (UL) of 3,000 μg/day of preformed retinol. Supplements containing high doses of retinol should be avoided due to risk of toxicity; the beta-carotene form is preferred. Health authorities in some high-income countries recommend avoiding or limiting liver consumption during pregnancy due to its high vitamin A content.[59,60] Although limiting liver consumption may be prudent in the first trimester, avoiding it for the duration of pregnancy has not been supported by evidence. Use of the synthetic analog 13-*cis*-retinoic acid (isotretinoin or Accutane) is contraindicated during pregnancy and prior to conception.

Adequate vitamin D status is needed for fetal growth and development, skeletal development, and tooth enamel formation.[61] Extraskeletal effects of maternal vitamin D related to preeclampsia, gestational diabetes, and immune-related outcomes in the infant have also been suggested, but causal roles have not been confirmed.[61,62] Furthermore, consensus is lacking on optimal serum 25-hydroxyvitamin D status during pregnancy. Thus, the requirement for vitamin D in pregnancy is based on skeletal outcomes, and the RDA (600 IU/day) is not increased for pregnancy.[61] Although vitamin D deficiency is prevalent worldwide, screening of all pregnant women is not currently supported.[63] Risk factors of deficiency include pregnancy, dark skin, and low sun exposure.[64] Deficiency during pregnancy results in defects in skeletal and tooth formation that persist into childhood.[61] The UL for vitamin D is 4,000 IU/day for both pregnant and nonpregnant women. However, evidence has shown no adverse effects on a developing fetus from supplemental doses as high as 4,000 IU/day.[65]

Vitamin E is a chain-breaking antioxidant that prevents propagation of lipid peroxidation, especially for polyunsaturated fatty acids within membrane phospholipids and in plasma lipoproteins. Blood concentration of α-tocopherol increases during pregnancy, and the rate of placental transfer appears constant. Vitamin E supplementation in pregnancy has no effect on maternal or infant outcomes.[66] In the absence of reports of vitamin E deficiency during pregnancy, the RDA for pregnant women is the same as for nonpregnant women.[67]

Vitamin K is used as a coenzyme in the synthesis of certain proteins involved in blood coagulation and bone metabolism. Data on vitamin K status during pregnancy are limited, and no data exist on vitamin K content of fetal tissue.[54] Thus, the AI for vitamin K is based on median intake and is the same for pregnant and nonpregnant women.

Water-Soluble Vitamins and Choline

Thiamin participates as a coenzyme in the metabolism of carbohydrate and branched-chain amino acids. An increased requirement of approximately 30% during pregnancy is based on increased growth in maternal and fetal compartments, along with a small increase in energy utilization.[68]

Riboflavin acts as a coenzyme in numerous oxidation-reduction reactions. Riboflavin is also needed for biosynthesis of niacin-containing coenzymes, for formation of pyridoxal-5-phosphate, and for reduction of 5,10-methylenetetrahydrofolate. Additional requirements for riboflavin in pregnancy are based on increased growth and energy utilization, along with lower urinary excretion of riboflavin.[68]

Niacin is required for the formation of nicotinamide-adenine dinucleotide (NAD) and nicotinamide-adenine dinucleotide phosphate (NADP) and, as such, is involved with oxidation of energy sources and biosynthesis of fatty acids and steroids. A small increment in niacin intake is considered adequate to cover increased energy utilization and growth.[68]

The coenzyme forms of vitamin B_6 are involved in the metabolism of amino acids, glycogen, and sphingoid bases. Vitamin B_6 coenzymes catalyze the first step in heme synthesis and are involved in the *trans*-sulfuration pathway from homocysteine to cysteine. Indicators of vitamin B_6 status in plasma decrease throughout pregnancy; and during the second and third trimesters, fetal blood concentration of pyridoxal phosphate is higher than in the mother.[68] The fetus and placenta are estimated to accumulate approximately 25 mg of vitamin B_6, but this is only an increased need of 0.1 mg/day, averaged over the course of gestation. Vitamin B_6 supplementation is not associated with improvement in maternal or infant outcomes.[69] However, supplemental vitamin B_6 has been used for decades in the treatment of morning sickness during pregnancy with no evidence of side effects, even with high doses.[68]

Folate functions as a coenzyme for single-carbon transfers in the metabolism of nucleic and amino acids. DNA synthesis depends on a folate coenzyme (for pyrimidine nucleotide biosynthesis); thus, folate is required for normal cell division. Folate requirements increase substantially during pregnancy because of a higher rate of single-carbon transfer reactions, especially in nucleotide synthesis and cell division. Folate is also transferred to the fetus in substantial amounts. When intake is inadequate, maternal serum and erythrocyte folate concentrations decline, and megaloblastic changes may occur.[68] Folate deficiency during pregnancy increases occurrence or recurrence of NTDs in the fetus. The RDA for pregnancy was set at 600 μg/day of dietary folate equivalents, based on a controlled metabolic study, which indicated that this was adequate to maintain normal folate status.[70] Randomized trials of folic acid (synthetic folate) supplementation show decreases in neural tube defects by 70% to 93%.[71] Thus, 400 to 800 μg/day folic acid supplementation is recommended at least 1 month prior to conception and continuing through pregnancy.[72] Higher levels of supplementation are recommended for those with previous NTDs.

Vitamin B_{12} functions as a coenzyme in the metabolism of odd-chain-length fatty acids and in methyl transfer reactions, working closely with folate. Adequate vitamin B_{12} is essential for normal blood formation and neurologic function. During pregnancy, absorption may decrease, and serum total vitamin B_{12} concentration may decline in the first trimester more than would be expected solely from hemodilution.[68] The placenta appears to concentrate vitamin B_{12} and then transfer it to the fetus, so that serum vitamin B_{12} concentration in the newborn is about twice that found in the mother. Based on liver content, the fetus accumulates 0.1 to 0.2 μg/day, which necessitates a slight increase in the RDA.[68] Risk for vitamin B_{12} deficiency is increased among pregnant women who follow strict vegetarian or vegan diets.[73] Vitamin B_{12} deficiency during pregnancy increases risk for maternal and fetal megaloblastic anemia, fetal demyelination, and NTDs.[74] Observational data indicate associations with preterm birth and low birth weight; however, randomized trials are needed.[75]

Pantothenic acid is a component of coenzyme A and phosphopantetheine. Little information is available on pantothenic acid use and status during pregnancy. Usual intakes in the United States and Canada appear to support healthy outcomes and; therefore, the AI is set at 6 mg/day for pregnancy.[68]

Biotin coenzymes function in bicarbonate-dependent carboxylations. These reactions include the formation of malonyl-coenzyme A and carboxylation of pyruvate for the tricarboxylic acid cycle or glucose formation. Degradation of leucine and formation of D-methylmalonyl-coenzyme A also depend on biotin.[68] Insufficient evidence was available to set a biotin requirement for pregnancy, thus an AI level was set at 30 μg/day, consistent with the AI for all adults. Notably, animal studies suggest that biotin deficiency is teratogenic,[76] and a recent controlled feeding study in pregnant women suggests that the current AI is insufficient.[77] Further evidence is needed to determine optimal biotin intake during pregnancy.

Choline serves as a precursor to acetylcholine, phosphatidylcholine, and sphingomyelin, and thus, plays a critical role in brain development. Choline is an essential dietary component during pregnancy as choline is transferred in large amounts to the fetus. Research in animals and humans shows that choline supports brain development and mitigates negative effects of alcohol exposure on the fetal brain.[78,79] The AI for choline during pregnancy is 450 mg/day, although an estimated 90% of pregnant women consume less than this and choline is typically not present in prenatal supplements.[68,78] Pregnant women are encouraged to meet needs through food sources, including eggs, meats, and seafood.[51]

Vitamin C is a water-soluble antioxidant and a cofactor for enzymes in collagen, carnitine, and neurotransmitter synthesis. During pregnancy, additional maternal vitamin C intake is required for transfer to the fetus, thus, increasing the RDA to 85 mg/d.[67] Women who smoke cigarettes or use alcohol may have a higher requirement. Vitamin C supplementation in pregnancy has shown no beneficial or adverse effects.[80]

Water and Electrolytes

Water, the largest component of the human body, is essential for maintaining vascular volume, serves as the medium for transport of nutrients and waste, a solvent for biochemical reactions, and helps to control body temperature. The AI for nonpregnant, nonlactating women, 2.7 L/day, is based on a median intake of total water from fluids and foods.[81] Total water accumulation of 6 to 9 L occurs during pregnancy, with approximately 1.8 to 2.5 L as intracellular fluid.[81] Plasma osmolality decreases by 8 to 10 mOsm/kg during gestation and stays low until term. The AI of 3.0 L/day for total water intake during pregnancy is based on median intake during pregnancy, of which 2.3 L/day is assumed to be consumed from beverages, including drinking water.[81]

Sodium and chloride are required for maintenance of extracellular volume and serum osmolality. Sodium is the principal cation of extracellular fluid, and chloride is the principal anion in extracellular fluid. Chloride is also important for production of gastric acid. Although substantial changes in both intracellular and extracellular fluid volumes occur during pregnancy, high-quality evidence to support dietary recommendations specific to pregnancy are lacking.[82] The Chronic Disease Risk Reduction Intake (CDRR) for sodium was established as 2,300 mg/day for nonpregnant women and also applies during pregnancy.[82] The AI for chloride does not change with pregnancy.[81]

Potassium is the major extracellular cation in the body and has a great influence on neural transmission, muscle contraction, and vascular tone. Potassium accretion in the fetus is estimated to be 120 mg/day in the third trimester.[82] Hypokalemia in pregnancy is uncommon, as renal reabsorption of potassium increases during pregnancy, and progesterone may help to conserve potassium, contributing to homeostasis. In 2019, the AI for potassium in pregnancy was set at 2,900 mg/day based on highest median intake in a nationally representative sample of normotensive pregnant women.[82]

Macrominerals

Calcium is essential for bone mineralization, contributing to the strength of bones and teeth. It also mediates vascular contraction and vasodilation, muscle contraction, nerve transmission, and glandular secretion. Approximately 25 to 30 g of calcium is transferred to the fetus, with the majority of calcium accretion occurring in the third trimester.[61] Fetal demand for calcium is met by a twofold increase in intestinal absorption of calcium by the mother in response to increased maternal 1,25-dihydroxyvitamin D (1,25[OH]$_2$D).[61] If maternal intake is low, calcium is mobilized from maternal stores to meet this demand. Results of supplementation studies among mothers with low calcium intakes showed mixed results on maternal and offspring bone mineral density.[61,83] Conversely, calcium supplementation may reduce the risk of pre-eclampsia and hypertension, among women with low calcium diets.[84] In fact, to reduce the risk of pre-eclampsia, the World Health Organization (WHO) recommends 1,500 to 2,000 mg of daily calcium supplementation (provided in three daily doses) in populations with low dietary calcium intake or high pre-eclampsia risk.[85] Among women whose intake is sufficient for maximizing bone accretion when not pregnant, intake does not need to be increased in pregnancy due to the increased efficiency of calcium absorption during pregnancy. Notably, the RDA for calcium during adolescence is higher (1,300 mg/day) than in adulthood (1,000 mg/day), although neither increase during pregnancy.

Phosphorus is a major component of bones, which contain 85% of total body phosphorus, and is essential in all tissues for structural (phospholipids, nucleotides, nucleic acids) and regulatory functions. A term infant has approximately 17 g of phosphorus, the majority (88%) of which is in bone and water.[86] Maternal increases in 1,25(OH)$_2$D that enhance calcium absorption during pregnancy also increase phosphorus absorption and cover the increased requirement for phosphorus. Thus, the RDA for phosphorus during pregnancy is the same as for nonpregnant women. Excess phosphorus intake from inorganic phosphate additives has detrimental effects on bone.[87] However, evidence of effects of excess intake in pregnant individuals is lacking. The UL for nonpregnant adults is based on intakes associated with the normal range of serum phosphorus. Because efficiency of phosphorus absorption increases by 15% during pregnancy, the UL for pregnant women (3,500 mg/day) is less than for nonpregnant women (4,000 mg/day).[86]

Magnesium is a required cofactor for more than 300 different enzymes. A newborn contains approximately 750 mg of magnesium, 60% of which is in the skeleton.[86] Decreases in serum magnesium concentration appear to be due to hemodilution. Evidence does not support magnesium supplementation during pregnancy to improve outcomes.[88] Conversely, pharmacologic magnesium sulfate has been used for decades to prevent the risk of seizures in women with pre-eclampsia in the peripartum and postpartum periods.[89] The RDA for pregnancy rises by 40 mg/day to account for lean tissue accretion along with enhanced bioavailability.[86]

Trace Minerals

Chromium has been suggested to potentiate insulin action, thereby playing a role in macronutrient metabolism. In 2001, an AI was set based on available evidence. Subsequent research suggests that, while chromium may have benefits at pharmacologic doses, it is not an essential mineral, as deficiency of chromium does not produce abnormalities that can be reversed with chromium provision.[90] Thus, although an AI exists, a recommendation for chromium intake may be inappropriate.

Copper is a component of metalloenzymes that act as oxidases in the reduction of molecular oxygen, including ceruloplasmin, which plays a role in iron metabolism and superoxide dismutase. Serum copper and ceruloplasmin increase during pregnancy, which could mask copper deficiency. The EAR is based on estimates of the amount of copper that must be accumulated during pregnancy to account for copper in the fetus and products of pregnancy.[54] A full-term infant contains approximately 13.7 mg of copper, primarily in the liver. This, combined with 4.6 mg of copper accumulated in the placenta and maternal tissues, translates into an RDA of 1,000 μg/day (100 μg/day higher than in the nonpregnant state).[54]

Fluoride is mainly associated with calcified tissues. It also inhibits the initiation and progression of dental caries and stimulates new bone development. Fluoride crosses the placenta and is incorporated into primary teeth. Evidence does not support provision of fluoride supplements during pregnancy to prevent dental caries in offspring.[91] In fact, fluoride balance is maintained at intakes similar to in nonpregnant women, so no increase in AI is recommended.[86] Excess fluoride intake during pregnancy is not associated with increased susceptibility to fluorosis. Interestingly, emerging evidence from human and animal studies suggests a controversial possible negative impact on child neurodevelopment.[92,93]

Iodine is an essential component of thyroid hormones that regulate key biochemical reactions, including protein synthesis and enzymatic activity, thus greatly impacting regulation of growth and development. Thyroid hormone is important for myelination of the central nervous system and is most active in the perinatal periods. Lack of iodine is particularly damaging to the developing brain. Iodine deficiency disorders include intellectual and developmental disabilities, hypothyroidism, and goiter.[94] Cretinism is an extreme form of neurologic damage from fetal hypothyroidism that results in severely impaired neurologic function, and varying degrees of short stature, deaf/mutism, and spasticity.[54,95]

The requirement for iodine in pregnancy is based on the iodine content of the newborn thyroid gland (50 to 100 μg), an amount that almost completely turns over each day.[54] It is estimated that fetal uptake of iodine is 75 μg/day. Based on supplementation trials, it was determined that an average intake of 160 μg/day of iodine could prevent goiter in pregnancy, thus, the RDA for pregnancy was set at 220 μg/day to cover the needs of 97% of the population.[54] The WHO and the American Thyroid Association recommend higher daily iodine intake; 250 μg/day during pregnancy.[95,96] Women who do not regularly consume dairy, eggs, seafood, or iodized salt are less likely to meet increased needs during pregnancy.[51] The American Thyroid Association recommends supplementation with 150 μg/day 3 months prior to and during pregnancy, but states that universal salt iodization is the most cost-effective way to improve iodine delivery for maternal and infant health.[95] Supplementation trials in populations with mild to moderate iodine deficiency show a decrease in postpartum hyperthyroidism and an increase in digestive intolerance yet showed no clear benefit on other outcomes, in part due to poor-quality data.[97] Despite this, the importance of iodine during pregnancy is well established. Prenatal exposure to excess iodine more than double the UL, either from supplements or from foods such as seaweed or kombu (edible kelp), results in goiter and hypothyroidism in the newborn.[98,99] The UL for iodine during pregnancy is 1,100 μg/day.[54]

Iron is a component of proteins in four major classes: heme proteins, iron-sulfur proteins, proteins for iron storage and transport, and other iron-containing or iron-activated enzymes.[54] Nearly two-thirds of total body iron is found in hemoglobin and facilitates oxygen movement and delivery to tissues throughout the body. Pregnancy imposes a large demand for iron to support maternal erythrocyte expansion and fetal growth and brain development. Iron needs across pregnancy increase by an estimated 1,000 to 1,200 mg, 330-400 mg of which is for fetal growth.[100] Physiologic demand is lower in the first trimester and increases across gestation, reflecting fetal growth.[100] The RDA for iron (27 mg/day) assumes 25% dietary iron absorption and uses estimated needs in the third trimester to build iron stores during the first trimester of pregnancy.[54] Because bioavailability of nonheme iron from a vegetarian diet is substantially lower, requirements for vegetarians are estimated to be 1.8 times higher than for nonvegetarians. Notably, recommendations for iron requirement and supplementation in pregnancy vary worldwide and across health organizations.[100]

Like many nutrients, iron exhibits a U-shaped risk curve such that adverse outcomes are higher with both inadequate and excessive nutrient availability. Iron deficiency, which limits expansion of maternal red cell mass, is the most prevalent single nutrient deficiency worldwide, occurring in 18% to 35% of pregnant women in the United States and Europe and higher in developing countries.[101,102] In retrospective studies, iron deficiency during pregnancy has been associated with preterm birth, LBW, low iron stores in the neonate, and perinatal mortality.[100,103] Women with iron deficiency or living in regions where iron deficiency is prevalent are likely to benefit from iron supplementation, as recommended by the World Health Organization.[104,105] However, inadequate evidence exists to recommend universal iron supplementation during pregnancy. Trials show that iron supplementation reduces maternal anemia and iron deficiency in pregnancy, but benefits for other maternal or infant outcomes were less clear.[106] Further concern has emerged about use of iron supplements among iron-replete individuals, and comprehensive assessment of iron status may be required to assess appropriate iron supplementation at the individual level.[104,107] Additional considerations in iron supplementation include its gastrointestinal side effects and impairment of zinc and copper absorption. Intermittent oral iron supplementation has shown to be effective for improving iron status with fewer gastrointestinal side effects.[108]

Manganese is essential for the formation of bone and in the metabolism of amino acids, cholesterol, and carbohydrates. Data on manganese and pregnancy are limited. The manganese concentration in fetal tissues ranges from 0.35 to 9.27 μg/g dry weight.[54] Adverse effects associated with manganese deficiency during pregnancy in animals have not been observed in humans. Despite limited data, the AI for manganese during pregnancy is higher than for nonpregnant women to account for fetal bone and cartilage development.[109]

Molybdenum is a cofactor for the enzymes sulfite oxidase, xanthine oxidase, and aldehyde oxidase, which are involved in the catabolism of sulfur amino acids and heterocyclic compounds. No direct data are available on molybdenum requirements for pregnancy, so the RDA for pregnant women (50 μg/day) was extrapolated from values for nonpregnant women using a median weight gain of 16 kg.[54] The UL for adults, 2 mg/day, is based on adverse reproductive effects of excess molybdenum intake in animal studies.

Selenium-containing proteins defend against oxidative stress, regulate thyroid hormone action, and regulate the redox status of vitamin C and other molecules. During pregnancy, selenium is transferred to the infant to the detriment of maternal status, and research is exploring a relationship with outcomes.[110] Selenium intake during pregnancy should promote adequate accumulation by the fetus to saturate selenoproteins.[67] Using an estimated selenium content of 250 μg/kg body weight, a 4-kg fetus would contain 1,000 μg. This translates to an additional requirement of 4 μg/day.

Zinc has catalytic, structural, and regulatory functions. Nearly 100 enzymes depend on zinc, representing all six enzyme classes.[54] Maternal and fetal tissues are estimated to accumulate progressively more zinc over the course of pregnancy, with a value of 0.73 mg/day in the final quarter of pregnancy.[54] Based on the average fractional absorption of zinc of 27%, the EAR during pregnancy increases by 2.7 mg/day to 9.5 mg/d, resulting in the RDA of 11 mg/d.[54] The requirement for zinc may be up to 50% higher for vegetarians, especially when the major food staples have a high ratio of phytate to zinc (e.g., grains, legumes).[111] Insufficient data exist for determining a different UL for pregnant women.

Observational studies suggest a possible relationship between low maternal dietary zinc intake and infant LBW and pre-eclampsia.[112] An association with preterm birth and gestational diabetes have been explored, but evidence is lacking. Supplementation trials among women with low income have shown little evidence of improvement in maternal or infant outcomes.[113] A possible relationship exists between maternal zinc intake and infant birthweight and severe pre-eclampsia; however, further study is warranted.[112] Poor zinc status may instead reflect overall poor nutritional status, thus, strategies to improve overall diet quality among low-income women may result in greater improvements in maternal and infant outcomes than supplementation of individual nutrients.

Recommendations and Adequacy

Dietary Recommendations During Pregnancy

In the past, public policy guidelines did not include dietary recommendations for pregnant women. For the first time, the *2020–2025 Dietary Guidelines for Americans* expanded recommendations to include pregnant and lactating women, which serves to provide guidance to these important populations and their health professionals.[51] These guidelines support a healthy dietary pattern, which includes daily consumption of a variety of vegetables, fruits, whole grains, low-fat dairy or fortified dairy alternative, and healthy proteins (meats, poultry, eggs, seafood, nuts, seeds, and soy products).

Dietary choices of pregnant women need to support the increased nutritional needs for the developing fetus and maternal tissues, along with appropriate weight gain. Women who are pregnant are encouraged to follow a healthy dietary pattern and to slightly increase intake in each of the food groups as energy needs increase across pregnancy (**Table 52.3**).[51] Because the requirement of increased energy (~16% to 22% greater than nonpregnant needs) is less than the increased need for most nutrients, food choices must be nutrient dense. Nutrient needs increase the most for iron (50%), folate (50%), iodine (47%), and zinc (38%). Thus, to meet the higher nutrient requirements without exceeding energy needs, increased consumption of nutrient-dense foods is necessary. Nutritional needs should be met primarily through foods and beverages. However, certain nutrients of concern may require specific attention.

Adequacy of Maternal Diets

Limited data are available on dietary intakes of pregnant women. A systematic review, including 18 studies from 10 countries across the globe examined adherence to dietary guidelines during preconception and pregnancy.[114] Results indicated that most women during the preconception period and during pregnancy may not meet minimum recommendations for vegetable and grain intake. Vegetables are an important source of folate and grains are an important source of folate and iron—two nutrients of concern for pregnancy. In contrast, preconceptual and pregnant women may be likely to exceed recommended fat intake. Women with higher education level and nonsmoking status had greater adherence to dietary guidelines.

The most comprehensive data about usual dietary intake among pregnant women in the United States is provided by a single study that used seven survey cycles (2001–2014) of the National Health and Nutrition Examination Survey (NHANES), a nationally represented, cross-sectional survey, which included 1,003 pregnant women aged 20 to 40 years.[115] Some nutrients that are a concern in the general population were also consumed in inadequate amounts among most pregnant women, including vitamin D (92%), magnesium (53%), and calcium (21%). Nutrients that are of particular concern during pregnancy were also inadequately consumed. From foods alone, a substantial portion of pregnant women consumed inadequate amounts of iron (84%), folate (36%), and choline (93%). Among those who used dietary supplements, only 36% consumed inadequate iron, and 16% consumed inadequate folate, while choline intake was virtually unchanged, as it is typically absent or minimally present in prenatal supplements. Interestingly, among

Table 52.3 Food Groups and Subgroups for a Healthy U.S.-Style Dietary Pattern During Pregnancy and Lactation

Food Group or Subgroup[b]	Calorie Level[a]					
	1,800	2,000	2,200	2,400	2,600	2,800
Vegetables (cup eq/day)[c]	**2.5**	**2.5**	**3**	**3**	**3.5**	**3.5**
Vegetable subgroups in weekly amounts						
Dark green (cup eq/wk)	1.5	1.5	2	2	2.5	2.5
Red & orange (cup eq/wk)	5.5	5.5	6	6	7	7
Beans, peas, lentils (cup eq/wk)	1.5	1.5	2	2	2.5	2.5
Starchy (cup eq/wk)	5	5	6	6	7	7
Other vegetables (cup eq/wk)	4	4	5	5	5.5	5.5
Fruits (cup eq/day)[c]	**1.5**	**2**	**2**	**2**	**2**	**2.5**
Grains (ounce eq/day)[d]	**6**	**6**	**7**	**8**	**9**	**10**
Whole grains (ounce eq/day)	3	3	3.5	4	4.5	5
Refined grains (ounce eq/day)	3	3	3.5	4	4.5	5
Dairy (cup eq/day)[e]	**3**	**3**	**3**	**3**	**3**	**3**
Protein foods (ounce eq/day)[f]	**5**	**5.5**	**6**	**6.5**	**6.5**	**7**
Protein foods subgroups in weekly amounts						
Meats, poultry, eggs (ounce eq/wk)	23	26	28	31	31	33
Seafood (ounce eq/wk)[g]	8	8	9	10	10	10
Nuts, seeds, soy products (ounce eq/wk)	4	5	5	5	5	6
Oils (grams/day)	**24**	**27**	**29**	**31**	**34**	**36**
Limit on calories for other uses (kcal/day)[h]	**140**	**240**	**250**	**320**	**350**	**370**
Limit on calories for other uses (%/day)	8%	12%	11%	13%	13%	13%

Eq = equivalent; wk = week.

[a]Calorie level ranges include calorie estimates during the first trimester (when calorie needs generally do not increase compared with prepregnancy needs) as well as later trimesters when calorie needs increase. Calorie needs vary based on many factors, including age, sex, height, weight, activity level, and pregnancy trimester.

[b]All foods are assumed to be in nutrient-dense forms; lean or low-fat; and prepared with minimal added sugars, refined starches, saturated fat, or sodium.

[c]Vegetables, fruits (1 cup eq) = 1 cup raw or cooked vegetable or fruit; 1 cup vegetable or fruit juice; 2 cups leafy salad greens; ½ cup dried fruit or vegetable.

[d]Grains (1ounce eq) = ½ cup cooked rice, pasta, or cereal; 1ounce dry pasta or rice; 1 medium (1 ounce) slice bread, tortilla, or flatbread; 1 ounce of ready-to-eat cereal (about 1 cup of flaked cereal).

[e]Dairy (1 cup eq) = 1 cup milk, yogurt, or fortified soymilk; 1½ ounces natural cheese (e.g., cheddar) or 2 ounces processed cheese.

[f]Protein foods = 1 ounce lean meats, poultry, or seafood; 1 egg; ¼ cup cooked beans or tofu; 1 Tbsp nut or seed butter; ½ ounce nuts or seeds.

[g]Seafood choices should be options lowest in methylmercury. Refer to FDA and EPA websites for more information. Available at www.fda.gov/food/consumers/advice-about-eating-fish

[h]Calories up to the specified limit can be used for added sugars and/or saturated fat, or to eat more than the recommended amount of food in a food group.

Reproduced from U.S. Department of Agriculture and U.S. Department of Health and Human Services. Dietary Guidelines for Americans, 2020-2025. 9th Edition. December 2020. Available at DietaryGuidelines.gov

those who used dietary supplements, 33% of pregnant women consumed folate in excess of the UL, and 28% consumed iron in excess of the UL. Notably, 94% of pregnant women consumed sodium at a level greater than the UL. Although data were lacking to estimate iodine intakes, iodine intakes have been reported to be inadequate in 23% to 56% of pregnant women.[116,117]

Concern about inadequate consumption of certain nutrients that play particularly important roles during pregnancy lead many health professionals to recommend prenatal supplements. Among the 70% of pregnant women who report taking such supplements, supplements increased intake and adequacy of many nutrients (iron, folate, vitamin D, calcium, iron, zinc), except for choline.[115] Supplementation with folic acid and iron has been recommended universally for pregnant women, particularly in low- and middle-income countries.[105] Recent estimates of intake among pregnant women in the United States suggest that individualized guidance may be warranted to avoid excessive intake of these nutrients. Ideally, early prenatal visits should assess the adequacy of the maternal diet and make recommendations for supplementation as needed. Furthermore, if a woman begins pregnancy either underweight or obese, individualized strategies may be needed to ensure adequate nutrition and appropriate weight gain.

Dietary Patterns

Studies that investigate associations between single nutrients and maternal and infant outcomes are numerous. Although such research reveals important insights, it also fails to consider or incorporate the complexity of diets, interactions between foods and nutrients, and the cumulative impact on health and disease. Conversely, research on dietary patterns examines the totality of the diet and its relationship with outcomes.

Maternal outcomes have been examined based on maternal dietary pattern. A systematic review reported limited yet consistent evidence that certain healthy dietary patterns before pregnancy are associated with reduced risk of gestational diabetes.[118] Evidence also suggested that healthy dietary patterns before and during pregnancy are associated with reduced risk of hypertensive disorders of pregnancy.[119] Associations between maternal dietary pattern and birth outcomes have also been examined with limited, but consistent, evidence suggesting that healthy dietary patterns during pregnancy are associated with lower risk of preterm birth.[120] Protective healthy dietary patterns were higher in vegetables, fruits, whole grains, nuts, legumes, and fish. They were lower in red and processed meats (GD), meats and refined grains (hypertensive disorders of pregnancy), or red and processed meats and fried foods (preterm birth). Notably, these studies were primarily conducted among non-Hispanic White women with access to health care; substantial heterogeneity existed across studies, thus limiting conclusions. Although further research is needed to draw stronger conclusions, the protective dietary patterns described in these studies are consistent with recommended dietary patterns of the *2020–2025 Dietary Guidelines for Americans*.

Other Dietary and Lifestyle Factors

Obesity

The prevalence of obesity among U.S. women has grown to approximately 42%.[121] More women are beginning pregnancy with high BMI and are also gaining more weight than is recommended.[32,42] National surveys began collecting data about prepregnancy BMI in 2016. Subsequently, prepregnancy obesity prevalence has steadily increased across all maternal ages, race and ethnicity groups, and education levels.[122] Obesity is associated with low-grade inflammation, detrimental metabolic changes, and insulin resistance. Obesity has been linked to subfertility, infertility, and miscarriage.[123] Risk of GD, pre-eclampsia, and gestational hypertension increase with obesity, along with rates of preterm birth, cesarean delivery, and labor complications.[123] Children of mothers with obesity are more likely to have congenital defects, including NTDs, and are at higher risk of intrauterine death and LGA birth weight. Maternal overweight and obesity are associated with childhood overweight, and prepregnancy BMI is an important modifiable risk factor to reduce risk to both mother and child.[124] Additionally, children of obese mothers may be at higher risk of developing metabolic syndrome. Routine health care for women should focus on dietary and physical activity counseling to maximize the number of women who conceive at a normal BMI.

Exercise

Physical activity during pregnancy should be encouraged in the absence of contraindications.[125] Regular exercise during pregnancy helps improve physical fitness, prevent excessive weight gain, promotes faster delivery, and hastens recovery. Randomized trials reveal that exercise interventions reduce the risk of GD, pregnancy-induced hypertension, and pre-eclampsia.[126] Further evidence suggests that exercise reduces the severity of lower back pain and pelvic girdle pain, common discomforts of pregnancy, although it did not reduce the odds of developing such pain.[127] Evidence also suggests that physical activity has a small, yet significant, effect of reducing depressive symptoms.[128] Previous concerns arose from epidemiologic studies that suggested a link between strenuous physical activities and poor fetal growth and preterm delivery; however, these findings have not been substantiated in more recent cohort studies or randomized trials.[125]

Many different physical activities can be maintained during pregnancy. Those that have been studied the most extensively and found to be beneficial include walking, stationary cycling, dancing, resistance exercises, stretching exercises, and water aerobics.[129] Those to be avoided include sports with a high potential for contact, activities with a high risk of falling, vigorous racquet sports with a risk of abdominal trauma, scuba diving, and any exercise in the supine position after the first trimester.[129] It is recommended that pregnant women participate in at least 150 minutes (2 hours and 30 minutes) per week of moderate-intensity physical activity, spreading aerobic activity throughout the week.[20] Women who continue a regular exercise program must be sure to maintain adequate energy, nutrient, and fluid intake throughout pregnancy, along with adequate weight gain.

Food Safety

The risk of foodborne illness is elevated during pregnancy. Specific recommendations for reducing risk of foodborne illness include avoiding soft cheeses made with unpasteurized milk, reheating luncheon meats and frankfurters until steaming, and avoiding unpasteurized milk or juice, raw or partially cooked eggs, raw or undercooked meat and poultry, raw sprouts, and raw or undercooked fish and shellfish.[51] National and international guidelines recommend limiting or avoiding certain types of fish due to high content of methylmercury, which crosses the placenta and can cause significant neurodevelopmental abnormalities.[130] Specifically, pregnant women should avoid large predatory fish such as shark, swordfish, tilefish, and king mackerel.[131] Conversely, they should eat at least 8 and up to 12 ounces per week of a variety of other choices that are lower in methylmercury.[51] Pregnant women should check local advisories regarding locally caught seafood. Recent studies suggest positive neurodevelopmental outcomes among children whose mothers consumed moderate amounts of fish during pregnancy compared with those who did not consume fish, despite the potential risks posed by methylmercury intake.[132,133]

Vegetarian Diets

During pregnancy, nutrient needs for vegetarians are the same as for nonvegetarians, except for a higher recommendation for iron intake because of the lower bioavailability of nonheme iron from plant sources.[54] Studies suggest that pregnant vegetarians tend to consume lower levels of protein, vitamin B_{12}, choline, calcium, and zinc, iron, iodine, and long-chain essential fatty acids.[134] The risk of vitamin B_{12} and iron deficiencies are particularly high.[73] More severe restrictions of animal-source foods, as in a vegan diet, may increase risk of inadequacies of these nutrients. Studies examining the effects of vegetarian diets during pregnancy are limited in number and design; however, severe adverse outcomes or birth defects have not been observed, with the exception of an increase in hypospadias among children born to vegan mothers in one large study.[73] Vegetarian diets can be planned to meet requirements for most nutrients.[135] Particular attention should be paid to adequacy of vitamin B_{12}, choline, vitamin D, calcium, iron, zinc, and DHA/EPA, with supplementation as appropriate to meet an individual's needs.[51]

Caffeine

Caffeine is widely consumed. Eighty-five percent of U.S. adults consume at least one caffeinated beverage daily, primarily from coffee, tea, soft drinks, and energy drinks.[136] The need to restrict or eliminate caffeine intake during pregnancy remains controversial. Caffeine is metabolized more slowly in pregnant women and passes readily through the placenta to the fetus. Epidemiologic studies investigating a link between caffeine intake and risk of spontaneous abortion or LBW have been inconclusive. A systematic review of 381 studies concluded that up to 300 mg of caffeine per day was not associated with adverse pregnancy or fetal outcomes.[137] Conversely, a subsequent review of observational studies determined that caffeine was associated with negative outcomes in a dose-dependent manner and, thus, that even lower levels of caffeine consumption were unsafe.[138] Results are likely affected by challenges in estimating caffeine intake, an inverse relationship between caffeine consumption and pregnancy-related nausea, and confounders such as age, education, income, tobacco smoking, and alcohol consumption.[139] It is recommended by American College of Obstetricians and Gynecologists (ACOG) to limit caffeine consumption during pregnancy to less than 200 mg per day, this is also prudent, as most caffeine-containing foods are low in nutritional value.[140] Notably, single servings of some energy drinks may exceed this daily limit, and it is recommended to avoid energy drinks altogether during pregnancy.[1]

Herbal and Other Dietary Supplements

Although many pregnant women benefit from supplementation with vitamins and minerals to achieve the recommended nutrient intakes in pregnancy, less is known about the benefits or risks of herbal and other dietary supplements. Few studies have examined the efficacy and safety of alternative therapies during pregnancy[141]; therefore, it is most prudent to consider these remedies as suspect until proven safe. Remedies promoted to pregnant women are often for easing gastrointestinal distress. Substantial evidence shows ginger to be effective in reducing pregnancy-induced nausea with no adverse effects.[141,142] In two observational studies and one randomized trial, castor oil showed mixed results on induction of labor, and nausea was a reported side effect. Some botanicals, such as red raspberry, cranberry, peppermint, and echinacea have a few studies examining their efficacy, with mixed results and often with no obstetric or neonatal outcomes reported. Many botanical therapies have not been formally studied.

Many herbal products have been identified as potentially unsafe for use during pregnancy.[1] Safety issues range from potential embryotoxicity to the more likely hormonal effects and drug interactions.[142] Given the lack of premarket requirement for proof of safety and efficacy for dietary supplements, pregnant women should discuss such supplements with their healthcare providers before they begin or continue to use them. Unfortunately, the advice of the healthcare provider sometimes carries its own risks. Finkel and Zarlengo[143] reported a case of a woman advised by her obstetrician to drink a tea made from blue cohosh to induce labor. Two days after delivery, the infant suffered a stroke, and the cocaine metabolite benzoylecgonine was detected in the infant's urine and in the mother's bottle of blue cohosh.

Alcohol

In the United States, data from the Behavioral Risk Factor Surveillance System indicate that 13.5% of pregnant women report alcohol use and 5.2% reported binge drinking.[144] Data from the Pregnancy Risk Assessment Monitoring System (PRAMS) suggest similar prevalence of alcohol use during pregnancy. Among those who continued alcohol use after becoming pregnant, about half reduced their consumption while half did not.[145] Pregnant women who reported frequent mental distress and no usual healthcare provider had higher prevalence of alcohol use.[144]

The developing fetus lacks sufficient enzymes needed to convert ethanol to acetaldehyde and then to acetate. Cytotoxic ethanol remains in fetal circulation, crosses the blood-brain barrier, and causes impaired brain development.[146] Circulating acetaldehyde causes free radical damage to fetal cells, further affecting development.

Alcohol use during pregnancy can cause numerous adverse effects on the developing fetus, and a range of associated disabilities from mild to severe. Fetal alcohol spectrum disorders (FASD) comprise four categories: fetal alcohol syndrome (the most severe), partial fetal alcohol syndrome, alcohol-related neurodevelopmental disorders, and alcohol-related birth defects.[147] FASDs are the most common preventable disabilities in the world. Prevalence estimates in the United States range from 1.1% to 5% of children.[148] The highest prevalence has been documented at 11% in South Africa.[149] The specific amount of alcohol exposure required to cause FASD has not been determined. Dose, timing, duration of exposure, genetic factors, and protective factors are all contributors.

The diagnosis of fetal alcohol syndrome requires (a) confirmed maternal exposure, (b) the presence of a characteristic pattern of facial anomalies, (c) growth retardation (pre- and/or postnatal), and (d) central nervous system neurodevelopmental abnormalities.[147] The characteristic facial features include short palpebral fissures, epicanthal folds, midface hypoplasia, a flat nasal bridge, a long philtrum, and a thin vermilion border of the upper lip. Central nervous system abnormalities can include decreased head size, brain structure abnormalities, recurrent nonfebrile seizures, poor fine motor skills, hearing loss, poor tandem gait, and poor hand–eye coordination.[147] Further studies indicate long-term detrimental health effects, including insulin resistance, dyslipidemia, and increased adiposity, as well as child behavioral effects.[150,151] The only preventive measure for FASD is the complete abstinence from alcohol during pregnancy.

Pregnant women who drink alcohol may have poor diet quality and nutrient intake. Furthermore, alcohol can interfere with absorption of thiamine, riboflavin, pyridoxine, vitamin A, vitamin C, and folic acid, making these nutrients less available for the growing fetus.[152] Interestingly, certain nutrients moderate the effect of prenatal alcohol exposure on infant outcomes. In a prospective longitudinal birth cohort, mothers who consumed alcohol during pregnancy and had poorer nutrition, including poor weight gain; or poor energy, iron, or choline intake, had infants with more severe growth restriction compared with mothers who consumed alcohol but had better weight gain or nutritional intake.[153] Additional evidence indicates that iron deficiency contributes to the severity of FASD.[154] Animal studies suggest that supplementing a mother with antioxidants may help reduce the detrimental effects of alcohol on the fetus, but this has yet to be shown in humans.[146] Regardless, optimizing nutritional intake of a pregnant mother who consumes alcohol is prudent.

Smoking

Cigarette smoking during pregnancy is linked to LBW, preterm delivery, spontaneous abortion, and stillbirth. Infants exposed prenatally to both alcohol and cigarettes had a nearly 12-fold risk of SIDS.[155] Carbon monoxide and nicotine from cigarettes increase fetal carboxyhemoglobin and reduced placental blood flow, thus limiting oxygen delivery to the fetus. Metabolic changes like those observed in offspring of undernourished mothers have been observed in offspring of mothers who smoked; increased risk of overweight and obesity later in life may occur through a developmental origin of disease mechanism.[156] Fortunately, smoking prevalence during pregnancy is declining. In 2007, 16% of pregnant women reported smoking and in 2017, 6.9% reported smoking.[157] Smoking is highest (10.7%) among women in their early 20s. American Indian Alaska Native women (16.7%) have the highest prevalence of smoking during pregnancy, followed by White, non-Hispanic women (10.5%); those with a high school education or less also have higher smoking prevalence during pregnancy.[158]

Illicit Drugs and Cannabis

In addition to alcohol and tobacco, illicit drugs such as cannabis, cocaine, and heroin can have devastating effects on a developing fetus. Approximately 4.4% of pregnant women use illicit drugs.[159] Although it is often difficult to isolate the effects of an illicit drug from concurrent use of alcohol or tobacco and other confounding factors, cannabis and some illicit drugs have been linked to reduced fetal growth and increased risk of stillbirth.[160-162] Cannabis accounts for most drug use, and effects of cannabis use during pregnancy remains controversial.[163-165] However, evidence from both animal and human data indicate that cannabis use during pregnancy may cause harm, including in development of the placenta, heart, and pancreas, as well as fetal growth restriction, and may lead to future metabolic disease.[161,166] Cannabidiol (CBD), a non-psychoactive cannabinoid, has reportedly been used to address pain, anxiety, and depression, all of which are common during pregnancy.[167] Research is needed to understand effects of CBD on pregnancy and postnatal health.[162]

Exposure to opioids, including heroin, leads to neonatal abstinence syndrome that affects the central nervous, autonomic nervous, and gastrointestinal systems of the neonate.[168] Longer-term effects on behavior, language, and academic achievement have also been observed.[160] Use of drugs can interfere with both adequate intake of essential nutrients as well as absorption and utilization of certain nutrients.[152] Malnutrition can result, particularly of folic acid, zinc, choline, iron, copper, and selenium.[152,169] Vasoconstriction that occurs with heroin or cocaine use also reduces placental transport of nutrients to the fetus, further compromising fetal growth.

Nutrition-Related Complications and Problems

Gastrointestinal Problems

Among the most common problems experienced during pregnancy are nausea and vomiting (commonly called "morning sickness" although these symptoms can and do occur at any time of day). Nausea affects 50% to 80% of pregnant women and vomiting and retching affect approximately 50%.[170] Pregnant women primarily experience nausea and vomiting of pregnancy in the first trimester, but it can continue to 20 weeks or beyond in some women. The etiology of nausea and vomiting of pregnancy is largely unknown, although several theories exist. Early nausea is associated with gastric dysrhythmias and hormonal changes that slow gastrointestinal motility. Increases in human chorionic gonadotropin, produced by the placenta, and high estrogen concentrations are temporally associated with onset of nausea and vomiting of pregnancy and, thus, are purported causes.[171] The role of human chorionic gonadotropin in nausea and vomiting of pregnancy is supported by the known increased risk of hyperemesis with higher placental weight.

Treatment of symptoms is necessary for quality of life and mental health.[171] Beginning use of a multivitamin prior to or in early pregnancy is associated with less nausea and vomiting symptoms. Management of nausea and vomiting depends on the severity of symptoms; most women with mild episodes are helped by eating smaller, more frequent meals, avoiding offending odors, and drinking adequate fluids. Use of ginger, due to its aromatic characteristics and ability to increase gastrointestinal motility, has shown some benefit for reducing nausea symptoms and is considered safe.[171,172] Pharmacologic treatment with vitamin B_6 alone or in combination with another medication may be needed.[171] Early treatment of nausea and vomiting of pregnancy is recommended to prevent progression to hyperemesis gravidarum.

Hyperemesis gravidarum is a severe form of nausea and vomiting of pregnancy, occurring in 0.3% to 3% of pregnancies.[170] This clinical diagnosis includes persistent vomiting not related to other causes, a measure of acute starvation, and weight loss. Hyperemesis gravidarum is one of the most common reasons for hospitalization during pregnancy, second only to preterm labor.[171,173] Dehydration and nutrient deficiencies can occur, as well as severe morbidities, including Wernicke encephalopathy resulting from thiamine deficiency.[174] Rehydration and correction of nutrient deficiencies may be needed along with nutrition support.

Gastroesophageal reflux (commonly called "heartburn") is another common gastrointestinal complaint experienced by between 40% and 85% of pregnant women.[1,175] Increased progesterone and estrogen concentrations lead to relaxation of the lower esophageal sphincter. Heartburn is generally relieved by eating smaller, more frequent meals, avoiding lying down after eating, elevating the head while sleeping, and avoiding known irritants.

Constipation resulting from slowed gastrointestinal motility can be aggravated by high-dose iron supplements. Between 25% and 40% of pregnant women may experience constipation at some point during pregnancy. Including generous amounts of fiber, adequate fluids, and regular exercise can help to relieve constipation.[1]

Pica

Pica, the intentional consumption and craving of nonfood substances, affects pregnant women worldwide with prevalence varying between 4% and 77%.[176] Commonly craved or consumed nonfood items include ice, laundry starch, cornstarch, baking soda, chalk, clay, and dirt.[1] Pica is thought to occur as a result of both biologic and sociocultural circumstances. A meta-analysis of 70 studies worldwide identified the best predictors of pica to be anemia, low education, and geographic region; pica prevalence was highest in Africa and among African Americans.[176] Iron deficiency has been proposed as a cause of pica.[177] Conversely, nutrient deficiencies may result from pica, as substances such as clay may bind to and inhibit absorption of micronutrients.[178] Pica should be addressed during pregnancy by identifying foods or alternative behaviors to be substituted for consumption of such nonfood substances.

Gestational Diabetes

In 2020, GD affected 7.8% of all pregnant women, nearly double from that a decade prior.[179] The rate of GD per 100 births increases with maternal age and is lowest for non-Hispanic Black, and highest for Asian and American Indian or Alaska Native mothers.[179] Risk factors associated with increased incidence of GD include age, high prepregnancy BMI, family history of diabetes, and ethnicity. Maternal complications associated with GD include higher rates of hypertensive disorders, cesarean delivery, recurrent GD, and future development of type 2 diabetes. For the fetus, GD increases the risk of macrosomia, hyperbilirubinemia, hypoglycemia, and erythema. Macrosomia (typically defined as a birth weight >4,000 g) is the most common fetal complication and is associated with high prepregnancy BMI and previous GD.[180] Initial therapy includes nutrition and exercise counseling, along with intensive self-monitoring of blood glucose. If these fail to adequately control blood glucose, insulin therapy is the preferred pharmacologic method of managing GD, although oral medication is sometimes used.[181] Lifestyle modification studies among women with GD have demonstrated reduced LGA, macrosomia, and neonatal fat mass in infants born to mothers who received the interventions. More work is needed to identify the best specific approaches and optimal format for effective intervention.[182]

Hypertensive Disorders

Gestational hypertension is defined as hypertension (blood pressure ≥140 mm Hg systolic or 90 mm Hg diastolic) without proteinuria occurring after 20 weeks' gestation.[89] Approximately 25% of women with gestational hypertension will develop *pre-eclampsia,* which is defined as hypertension with proteinuria (>300 mg/24 hours) after 20 weeks' gestation.[89] The clinical definition has evolved to recognize the complex involvement of other organ systems. In the absence of proteinuria, alternative criteria may warrant a diagnosis of pre-eclampsia, including thrombocytopenia, renal insufficiency, impaired liver function, pulmonary edema, or headache not accounted for by alternative diagnoses.[89] Pre-eclampsia may progress to *eclampsia,* a condition marked by seizures that is life threatening to both mother and infant. Pre-eclampsia affects 3% to 5% of pregnancies in the United States and is associated with substantial risks including FGR, preterm delivery, and maternal renal failure, seizures, pulmonary edema, stroke, and death.[183] Currently, the cause of pre-eclampsia is unknown, and accurate screening tests are not available. Risk factors include history of pre-eclampsia, chronic hypertension, pregestational diabetes, obesity, multifetal pregnancy, primiparity, assisted reproductive technologies, kidney disease, and older age.

Pre-eclampsia is one of the largest contributors to maternal deaths. It is a disease state characterized by poor placental development and maternal vascular malperfusion to the placenta. As pregnancy progresses, the placenta secretes increasing amounts of factors that cause vascular inflammation, endothelial dysfunction, and maternal vascular injury.[183] In addition to these maternal features, reduced perfusion extends to virtually all organs and is caused by vasoconstriction, microthrombi formation, and reduced circulating plasma volume.

Because of an increased incidence of pre-eclampsia noted in women of low socioeconomic status, nutritional factors have long been suggested as contributing to the disorder. Multiple vitamins and minerals have been studied for causal or preventive roles; however, most of these have not provided conclusive links with pre-eclampsia, except for calcium. Calcium supplementation (1,000 mg split over 2 or 3 doses daily) may help prevent pre-eclampsia among those with chronically low dietary calcium intake and is recommended in regions where dietary calcium intake is low.[84,85] Pharmacologic magnesium sulfate is often used to reduce the risk of eclamptic seizure.[89] Women with a history of pre-eclampsia are at increased risk for developing future cardiovascular disease and should follow appropriate dietary recommendations to reduce these risks and related long-term complications.

Low Birth Weight and Fetal Growth Restriction

An estimated 20 million infants per year (15% to 20%) are born with low birthweight.[184] LBW (<2,500 g) can be divided into two categories: those born too early and those with FGR. In developed countries, a large portion of LBW infants are preterm, whereas in developing countries, most LBW infants are affected by FGR. In 2020, approximately 8% of live births in the United States were LBW, and approximately 10% were preterm, although the preterm rate for singletons was 8.4%.[185] LWB and preterm birth together are the second leading cause of infant mortality, after birth defects. Improving birth weight has a significant effect on infant mortality.

Poor nutrition is a known cause of LBW and FGR. Other contributing factors include smoking, infections, pregnancy-related hypertension (including pre-eclampsia), multiple gestation, periodontal disease, and environmental factors.[186] Low weight gain in either the second or third trimester increases the risk of FGR, as does low pre-pregnancy BMI and young age.[187] Longitudinal studies indicate that FGR may negatively affect neurodevelopment, including motor, cognitive, and language delay, and increase the future risk of chronic disease.[188,189] Consistent, early prenatal care can help improve maternal nutrition and identify patterns of weight gain that pose a risk of LBW.

Neural Tube Defects

NTDs are one of the most common major congenital malformations in the United States,[190] and result in improper closure of the neural tube. NTDs include varying degrees of disturbance in the neurulation process (anencephaly, spina bifida [meningomyelocele and meningocele], and craniorachischisis (the most severe form of NTD)) and can lead to a range of disabilities or death.

Neurulation is the first organogenetic process to be initiated and completed.[68] The process begins approximately 21 days after conception and is completed by day 28.

The etiology of NTDs includes heredity, probably related to multiple genes influenced by environmental factors. The relationship between folate and NTDs was first suggested by Hibbard in 1964.[68] Studies of folic acid supplementation have been shown to reduce NTDs by 70% to 93%.[71] Possible mechanisms of folate inadequacy include disruption of nucleotide synthesis and uracil misincorporation into DNA, and thus DNA instability, disruption of histone methylation altering protein synthesis, disruption of one-carbon metabolism in the cytosol, and disruption of homocysteine methylation, resulting in toxic concentrations in the developing embryonic neural cells.[191,192]

As a result of the accumulated evidence, in 1992, the U.S. Public Health Service first recommended universal supplementation of folic acid for all women planning or capable of pregnancy. In 1998, mandatory food fortification was implemented in the United States of 140 μg folic acid per 100 g of flour, rice, pasta, cornmeal, and other grain products with the label, "enriched."[193] Subsequent estimates of folic acid intake suggest that fortification provides 138 μg/day of folic acid in addition to folate intake from other sources.[194] It is estimated the NTDs decreased by 28% after food fortification was implemented.[195] Studies examining the effect of folic acid supplementation following food fortification suggested less benefit; however, they had substantial limitations, given their observational study designs.[196] Thus, in 2017, the United States Preventive Services Taskforce (USPSTF) reaffirmed its recommendation to supplement all women planning or capable of pregnancy with 400 to 800 μg/d of folic acid.[72] Notably, this recommendation, in conjunction with food fortification, may result in intake exceeding the upper limit, and in some instances, individual assessment may be necessary to assure that total folate intake is adequate but not excessive.[115,197]

Summary

Healthy outcomes for both mother and baby can result from preconception assessment, good nutrition, healthy lifestyle choices, adequate weight gain, and early prenatal care. Prenatal care is important for nutrition assessment, evaluation of risk factors, and follow-up to ensure optimal outcomes. Early screening can identify physiologic or psychologic problems and initiate appropriate therapy. As the world strives toward the WHO Sustainable Development Goals, improving maternal nutritional status remains a major priority for reducing maternal morbidity and mortality and improving child outcomes.[10] Researchers must continue to identify nutritional intervention strategies that are effective at improving pregnancy outcomes.

Acknowledgments

The author acknowledges R Elaine Turner, the author of the previous version of this chapter, for her contributions.

References

1. Kaiser LL, Campbell CG. J Acad Nutr Diet 2014;114:1447.
2. Stein Z, Susser M. Pediatr Res 1975;9:76–83.
3. Stein Z, Susser M. Pediatr Res 1975;9:70–6.
4. Stein AD, Ravelli AC, Lumey LH. Hum Biol 1995;67:135–50.
5. Painter RC, Roseboom TJ, Bleker OP. Reprod Toxicol 2005;20:345–52.
6. Barker DJ. J Intern Med 2007;261:412–7.
7. Lumey LH, Stein AD, Susser E. Annu Rev Public Health 2011;32:237–62.
8. Stephenson J, Heslehurst N, Hall J, et al. Lancet 2018;391:1830–41.
9. Fleming TP, Watkins AJ, Velazquez MA, et al. Lancet 2018; 391:1842–52.
10. World Health Organization. Maternal health fact sheet on sustainable development goals. World Health Organization Regional Office for Europe, Copenhagen, Denmark, 2017.
11. U.S. Department of Health and Human Services. Healthy People 2030 Framework, 2022. Accessed December 10, 2023. Available at https://health.gov/our-work/national-health-initiatives/healthy-people/healthy-people-2030.
12. Hoyert D. NCHS Health E-Stats, National Center for Health Statistics. Healthy People 2020 Final Review: Healthy People 2020 Progress Table, 2022. Accessed December 10, 2023. Available at www.cdc.gov/nchs/healthy_people/hp2020-final-review.htm.
13. National Center for Health Statistics. Healthy People 2020 Final Review: Healthy People 2020 Progress Table, 2022. Available at https://www.cdc.gov/nchs/healthy_people/hp2020-final-review.htm. Accessed December 10, 2023.
14. Frayne DJ, Verbiest S, Chelmow D, et al. Obstet Gynecol 2016; 127:863–72.
15. Finer LB, Zolna MR. N Engl J Med 2016;374:843–52.
16. Pepper MR, Black MM. Semin Cell Dev Biol 2011;22:619–23.
17. Georgieff MK. Am J Obstet Gynecol 2020;223:516–24.
18. Stang J, Huffman LG. J Acad Nutr Diet 2016;116:677–91.
19. Achike M, Akpinar-Elci M. Breastfeed Med 2021;16:678–86.
20. U.S. Department of Health and Human Services. Physical Activity Guidelines for Americans, 2nd edition. U.S. Department of Health and Human Services, Washington, DC, 2018.
21. Nobles CJ, Mendola P, Mumford SL, et al. Hypertens 2020; 76:922–9.
22. Lu Y, Chen R, Cai J, et al. Br Med Bull 2018;128:75–84.
23. Bramham K, Parnell B, Nelson-Piercy C, et al. Br Med J 2014; 348:g2301.
24. Malaza N, Masete M, Adam S, et al. Int J Environ Res Public Health 2022;19.
25. Papazoglou AS, Moysidis DV, Panagopoulos P, et al. J Matern Fetal Neonatal Med 2022;35:7685–94.
26. Alexopoulos AS, Blair R, Peters AL. JAMA 2019;321:1811–9.
27. Resta R. Am J Med Genet A 2012;158a:1118–23.
28. Prick BW, Hop WC, Duvekot JJ. Am J Clin Nutr 2012;95:374–82.
29. Sanghavi M, Rutherford JD. Circ 2014;130:1003–8.
30. De Swiet M. The respiratory system. In: Hytten F, and Chamberlain G, eds. Clinical Physiology in Obstetrics, 2nd ed. Oxford, Great Britain: Blackwell Scientific Publications, 1991.
31. Shin GH, Toto EL, Schey R. Am J Gastroenterol 2015;110:521–9.
32. Institute of Medicine and Nutrition Research Council. Weight Gain During Pregnancy: Reexamining the Guidelines. Washington, DC: National Academies Press, 2009.
33. LifeCycle Project-Maternal Obesity, Childhood Outcomes Study Group. JAMA 2019;321:1702–15.

34. Bodnar LM, Siega-Riz AM, Simhan HN, et al. Am J Clin Nutr 2010;91:1642–8.
35. Wang L, Zhang X, Chen T, et al. JAMA Netw Open 2021; 4:e2141498.
36. Santos S, Eekhout I, Voerman E, et al. BMC Medicine 2018;16:201.
37. Deputy NP, Sharma AJ, Kim SY. Morb Mortal Wkly Rep 2015; 64:1215–20.
38. Donovan S, Dewey K, Novotny R, et al. USDA Nutrition Evidence Systematic Reviews. Dietary Patterns during Pregnancy and Gestational Weight Gain: A Systematic Review. Alexandria,VA: USDA Nutrition Evidence Systematic Review, 2020.
39. Dolin CD, Gross RS, Deierlein AL, et al. Int J Environ Res Public Health 2020;17:352.
40. Yu ZM, Van Blyderveen S, Schmidt L, et al. J Womens Health 2021;30:1006–15.
41. Kim SY, Sharma AJ, Sappenfield W, et al. Obstet Gynecol 2014; 123:737–44.
42. Goldstein RF, Abell SK, Ranasinha S, et al. JAMA 2017;317: 2207–25.
43. Carmichael SL, Abrams B. Obstet Gynecol 1997;89:865–73.
44. Huang Y, Ouyang Y-Q, Redding SR. Breastfeeding Med 2019; 14:366–74.
45. Endres LK, Straub H, McKinney C, et al. Obstet Gynecol 2015; 125:144–52.
46. Marshall NE, Abrams B, Barbour LA, et al. Am J Obstet Gynecol 2022;226:607–32.
47. Bogaerts A, Van den Bergh BRH, Ameye L, et al. Obstet Gynecol 2013;122:999–1009.
48. Institute of Medicine. Dietary Reference Intakes for energy, carbohydrate, fiber, fat, fatty acids, cholesterol, protein, and amino acids. Washington, DC: The National Academies Press, 2005.
49. Most J, Dervis S, Haman F, et al. Nutrients 2019;11:1812.
50. Thomas DM, Navarro-Barrientos JE, Rivera DE, et al. Am J Clin Nutr 2012;95:115–22.
51. U.S. Department of Agriculture, and U.S. Department of Health and Human Services. Dietary Guidelines for Americans, 2020–2025.
52. Middleton P, Gomersall JC, Gould JF, et al. Cochrane Database Syst Rev 2018;11:Cd003402.
53. Zhang Z, Fulgoni VL, Kris-Etherton PM, et al. Nutrients 2018; 10:416.
54. Institute of Medicine. Dietary Reference Intakes for Vitamin A, Vitamin K, Arsenic, Boron, Chromium, Copper, Iodine, Iron, Manganese, Molybdenum, Nickel, Silicon, Vanadium, and Zinc. Washington, DC: National Academies Press, 2001.
55. World Health Organization. Global prevalence of vitamin A deficiency in populations at risk 1995–2005. WHO Global Database on Vitamin A Deficiency. Geneva, Switzerland: World Health Organization, 2009.
56. McCauley ME, van den Broek N, Dou L, et al. Vitamin A supplementation during pregnancy for maternal and newborn outcomes. Accessed March 9, 2020. Available at http://onlinelibrary.wiley.com/doi/10.1002/14651858.CD008666.pub3/abstract.
57. Simpson JL, Bailey LB, Pietrzik K, et al. J Matern Fetal Neonatal Med 2011;24:1–24.
58. Voyles LM, Turner RE, Lukowski MJ, et al. J Am Diet Assoc 2000;100:1068–70.
59. Finnish Food Authority. Consumption of liver and liver-based food during pregnancy. Accessed October 27, 2022. Available at www.ruokavirasto.fi/en/foodstuffs/instructions-for consumers/safe-use-of-foodstuffs/liver-and-liver-based-foods/consumption-of-liver-and-liver-based-food-during-pregnancy/.
60. National Health Service. Foods to Avoid in Pregnancy. Accessed: October, 27, 2022. Available at www.nhs.uk/pregnancy/keeping-well/foods-to-avoid/.
61. Institute of Medicine. Dietary Reference Intakes for Calcium and Vitamin D. Washington, DC: National Academy Press, 2011.
62. Curtis EM, Moon RJ, Harvey NC, et al. Br Med Bull 2018;126:57–77.
63. American College of Obstetricians and Gynecologists. Obstet Gynecol 2011;118:197–8.
64. van Schoor N, Lips P. Endocrinol Metab Clin North Am 2017;46:845–70.
65. Kiely ME, Wagner CL, Roth DE. J Steroid Biochem Mol Biol 2020;201:105669.
66. Rumbold A, Ota E, Hori H, et al. Vitamin E Supplementation in Pregnancy. Accessed April 13, 2022. Available at http://onlinelibrary.wiley.com/doi/10.1002/14651858.CD004069.pub3/abstract
67. Institute of Medicine. Dietary Reference Intakes for Citamin C, Vitamin E, Selenium, and Carotenoids. Washington, DC: National Academy Press, 2000.
68. Institute of Medicine. Dietary Reference Intakes for Thiamin, Riboflavin, Niacin, Vitamin B6, Folate, Vitamin B_{12}, Pantothenic Acid, Biotin, and Choline. Washington, DC: The National Academies Press, 1998.
69. Salam RA, Zuberi NF, Bhutta ZA. Cochrane Database Syst Rev 2015;Cd000179.
70. Caudill MA, Cruz AC, Gregory JF, 3rd, et al. J Nutr 1997;127: 2363–70.
71. De-Regil LM, Peña-Rosas JP, Fernández-Gaxiola AC, et al. Cochrane Database Syst Rev 2015;Cd007950.
72. US Preventive Services Task Force, Bibbins-Domingo K, Grossman DC, et al. JAMA 2017;317:183–9.
73. Piccoli GB, Clari R, Vigotti FN, et al. BJOG 2015;122:623–33.
74. Simpson JL, Bailey LB, Pietrzik K, et al. J Matern Fetal Neonatal Med 2010;23:1323–43.
75. Rogne T, Tielemans MJ, Chong MF, et al. Am J Epidemiol 2017; 185:212–23.
76. Mock DM. J Nutr 2017;147:1487–92.
77. Perry CA, West AA, Gayle A, et al. J Nutr 2014;144:1977–84.
78. Korsmo HW, Jiang X, Caudill MA. Nutrients 2019;11:1823.
79. Derbyshire E, Obeid R. Nutrients 2020;12:1731.
80. Rumbold A, Ota E, Nagata C, et al. Vitamin C Supplementation in Pregnancy. Accessed December 12, 2021. Available at http://onlinelibrary.wiley.com/doi/10.1002/14651858.CD004072.pub3/abstract.
81. Institute of Medicine. Dietary Reference Intakes: The Essential Guide To Nutrient Requirements. Washington, DC: The National Academies Press, 2006.
82. National Academies of Sciences Engineering and Medicine. Dietary Reference Intakes for Socium and Potassium. Washington, DC: The National Academies Press, 2019.
83. Tihtonen K, Korhonen P, Isojärvi J, et al. Ann N Y Acad Sci 2022;1509:23–36.
84. Hofmeyr GJ, Lawrie TA, Atallah Á, et al. Cochrane Database Syst Rev 2018;CD001059.
85. World Health Organization. WHO Recommendation: Calcium Supplementation During Pregnancy for The Prevention of Pre-Eclampsia and Its Complications. Geneva, Switzerland: World Health Organization, 2018.
86. Institute of Medicine. Dietary Reference Intakes for Calcium, Phosphorus, Magnesium, Vitamin D, and Fluoride. Washington, DC: The National Academies Press, 1997.
87. Vorland CJ, Stremke ER, Moorthi RN, et al. Curr Osteoporos Rep 2017;15:473–82.
88. Makrides M, Crosby DD, Bain E, et al. Cochrane Database Syst Rev 2014;Cd000937.
89. American College of Obstetricians and Gynecologists. Obstet Gynecol 2020;135:e237–e60.
90. EFSA Panel on Dietetic Products N and Allergies. EFSA J 2014; 12:3845.
91. Takahashi R, Ota E, Hoshi K, et al. Cochrane Database Syst Rev 2017;10:Cd011850.

92. Farmus L, Till C, Green R, et al. Environ Res 2021;200:111315.
93. Bashash M, Thomas D, Hu H, et al. Environ Health Perspect 2017;125:097017.
94. Pearce EN, Lazarus JH, Moreno-Reyes R, et al. Am J Clin Nutr 2016;104 Suppl 3:918s–23s.
95. Alexander EK, Pearce EN, Brent GA, et al. Thyroid 2017;27: 315–89.
96. World Health Organization and United Nations Childrens Fund. Joint Statement by the WHO and UNICEF Reaching optimal iodine nutrition in pregnant and lactating women and young children. Accessed April 30, 2023. Available at www.who.int/publications/m/item/WHO-statement-IDD-pregnantwomen-children
97. Harding KB, Peña-Rosas JP, Webster AC, et al. Cochrane Database Syst Rev 2017;3:Cd011761.
98. Connelly KJ, Boston BA, Pearce EN, et al. J Pediatr 2012;161: 760–2.
99. Nishiyama S, Mikeda T, Okada T, et al. Thyroid 2004;14:1077–83.
100. Brannon PM, Taylor CL. Nutrients 2017;9.
101. Mei Z, Cogswell ME, Looker AC, et al. Am J Clin Nutr 2011; 93:1312–20.
102. Milman N, Taylor CL, Merkel J, et al. Am J Clin Nutr 2017; 106:1655s–62s.
103. Allen LH. Am J Clin Nutr 2000;71:1280S–4S.
104. Georgieff MK, Krebs NF, Cusick SE. Annu Rev Nutr 2019;39: 121–46.
105. World Health Organization. Guideline: Daily iron and folic acid supplementation in pregnant women. Geneva, Switzerland: World Health Organization, 2012.
106. Peña-Rosas JP, De-Regil LM, Garcia-Casal MN, et al. Cochrane Database Syst Rev 2015;2015:CD004736.
107. Iqbal S, Ekmekcioglu C. J Matern Fetal Neonatal Med 2019; 32:1528–40.
108. Peña-Rosas JP, De-Regil LM, Gomez Malave H, et al. Cochrane Database Syst Rev 2015;CD009997.
109. Wood RJ. Nutr Rev 2009;67:416–20.
110. Schomburg L. Int J Mol Sci 2021;22:8532.
111. Hunt JR. Am J Clin Nutr 2003;78:633s–9s.
112. Wilson RL, Grieger JA, Bianco-Miotto T, et al. Nutrients 2016; 8:641.
113. Carducci B, Keats EC, Bhutta ZA. Cochrane Database Syst Rev 2021;3:Cd000230.
114. Caut C, Leach M, Steel A. Matern Child Nutr 2020;16:e12916.
115. Bailey RL, Pac SG, Fulgoni VL, 3rd, et al. JAMA Netw Open 2019;2:e195967.
116. Kerver JM, Pearce EN, Ma T, et al. Am J Obstet Gynecol 2021; 224:82.e1–e8.
117. Caldwell KL, Pan Y, Mortensen ME, et al. Thyroid 2013;23: 927–37.
118. Raghavan R, Dreibelbis C, Kingshipp BJ, et al. USDA Nutrition Evidence Systematic Reviews. Dietary Patterns before and during Pregnancy and Risk of Gestational Diabetes Mellitus: A Systematic Review. Alexandria, VA: USDA Nutrition Evidence Systematic Review, 2019.
119. Raghavan R, Dreibelbis C, Kingshipp BJ, et al. USDA Nutrition Evidence Systematic Reviews. Dietary Patterns before and during Pregnancy and Risk of Hypertensive Disorders of Pregnancy: A Systematic Review. Alexandria, VA: USDA Nutrition Evidence Systematic Review, 2019.
120. Raghavan R, Dreibelbis C, Kingshipp BJ, et al. USDA Nutrition Evidence Systematic Reviews. Dietary Patterns before and during Pregnancy and Gestational Age at Birth: A Systematic Review. Alexandria, VA: USDA Nutrition Evidence Systematic Review, 2019.
121. Stierman B, Afful J, Carroll MD, et al. In: National Center for Health S, ed. National Health and Nutrition Examination Survey 2017–March 2020 Prepandemic Data Files Development of Files and Prevalence Estimates for Selected Health Outcomes. National Health Statistics Reports. Accessed November 15, 2023. Available at http://dx.doi.org/10.15620/cdc:106273, Hyattsville, MD.
122. Driscoll A, Gregory E. Increases in prepregnancy obesity: United States, 2016–2019. NCHS Data Brief, no 392. Hyattsville, MD: National Center for Health Statistics, 2020.
123. Creanga AA, Catalano PM, Bateman BT. N Engl J Med 2022;387:248–59.
124. Voerman E, Santos S, Patro Golab B, et al. PLoS Med 2019;16:e1002744.
125. American College of Obstetricians and Gynecologists. Obstet Gynecol 2020;135:e178–e88.
126. Davenport MH, Ruchat SM, Poitras VJ, et al. Br J Sports Med 2018;52:1367–75.
127. Davenport MH, Marchand AA, Mottola MF, et al. Br J Sports Med 2019;53:90–8.
128. Marconcin P, Peralta M, Gouveia É R, et al. Biology (Basel) 2021;10:1331.
129. Berghella V, Saccone G. Am J Obstet Gynecol 2017;216:335–7.
130. Taylor CM, Emmett PM, Emond AM, et al. Public Health Nutr 2018;21:2149–59.
131. US Food and Drug Administration. Advice about eating fish: for those who might become or are pregnant or breastfeeding and children ages 1-11 years, 2022. Accessed December 10, 2023. Available at www.fda.gov/food/consumers/questions-answers-fdaepa-advice-about-eating-fish-those-who-might-become-or-are-pregnant-or
132. Snetselaar L, Bailey R, Sabaté J, et al. USDA Nutrition Evidence Systematic Reviews. Seafood Consumption during Pregnancy and Lactation and Neurocognitive Development in the Child: A Systematic Review. Alexandria, VA: USDA Nutrition Evidence Systematic Review, 2020.
133. Golding J, Taylor C, Iles-Caven Y, et al. Neurotoxicol 2022;91:22–30.
134. Sebastiani G, Herranz Barbero A, Borrás-Novell C, et al. Nutrients 2019;11:557.
135. Melina V, Craig W, Levin S. J Acad Nutr Diet 2016;116:1970–80.
136. Mitchell DC, Knight CA, Hockenberry J, et al. Food Chem Toxicol 2014;63:136–42.
137. Wikoff D, Welsh BT, Henderson R, et al. Food Chem Toxicol 2017;109:585–648.
138. James JE. BMJ Evid Based Med 2021;26:114–5.
139. Doepker C, Lieberman HR, Smith AP, et al. Annu Rev Food Sci Technol 2016;7:117–37.
140. American College of Obstetricians and Gynecologists. Obstet Gynecol 2010;116:467–8.
141. Dante G, Bellei G, Neri I, et al. Curr Opin Obstet Gynecol 2014;26:83–91.
142. Sarecka-Hujar B, Szulc-Musioł B. Pharmaceutics 2022;14:171.
143. Finkel RS, Zarlengo KM. N Engl J Med 2004;351:302–3.
144. Gosdin L, Deputy N, Kim S, et al. Morb Mortal Wkly Rep 2022;71:10–3.
145. Kitsantas P, Gaffney KF, Wu H, et al. Arch Gynecol Obstet 2014;289:771–9.
146. Ciafrè S, Ferraguti G, Greco A, et al. Neurosci Biobehav Rev 2020;118:654–68.
147. Hoyme HE, Kalberg WO, Elliott AJ, et al. Pediatr 2016;138:e20154256.
148. May PA, Chambers CD, Kalberg WO, et al. JAMA 2018;319: 474–82.
149. Lange S, Probst C, Gmel G, et al. JAMA Pediatr 2017;171:948–56.
150. Akison LK, Reid N, Wyllie M, et al. Alcohol Clin Exp Res 2019; 43:1324–43.
151. Flak AL, Su S, Bertrand J, et al. Alcohol Clin Exp Res 2014; 38:214–26.

152. Sebastiani G, Borrás-Novell C, Casanova MA, et al. Nutrients 2018;10:1008.
153. Carter RC, Senekal M, Duggan CP, et al. Am J Clin Nutr 2022; 116:460–9.
154. Bradley R, Lakpa KL, Burd M, et al. Nutrients 2022;14:4223.
155. Elliott AJ, Kinney HC, Haynes RL, et al. EClinicalMedicine 2020;19:100247.
156. Rogers JM. Birth Defects Res 2019;111:1259–69.
157. Azagba S, Manzione L, Shan L, et al. BMC Pregnancy Childbirth 2020;20:52.
158. Drake P, Driscoll A, Mathews T. Cigarette Smoking During Pregnancy: United States, 2016. NCHS Data Brief, no 305. Hyattsville, MD: National Center for Health Statistics, 2018.
159. Wendell AD. Clin Obstet Gynecol 2013;56:91–6.
160. Behnke M, Smith VC, Committee on Substance Abuse, et al. Pediatr 2013;131:e1009–e24.
161. Metz TD, Borgelt LM. Obstet Gynecol 2018;132:1198–210.
162. Nashed MG, Hardy DB, Laviolette SR. Front Psychiatry 2020;11:624275.
163. Torres CA, Hart CL. Am J Obstet Gynecol 2017;217:227.
164. Torres CA, Medina-Kirchner C, O'Malley KY, et al. Front Psychol 2020;11:816.
165. Chaput KH, Lebel C, McMorris CA. Front Psychol 2020;11:1891.
166. Lee K, Hardy DB. Int J Mol Sci 2021;22:9528.
167. Corroon J, Phillips JA. Cannabis Cannabinoid Res 2018;3:152–61.
168. Kocherlakota P. Pediatr 2014;134:e547–61
169. Hohman EE, Corr TE, Kawasaki S, et al. Nutrients 2023;15:1891.
170. Matthews A, Haas DM, O'Mathúna DP, et al. Cochrane Database Syst Rev 2015;2015:Cd007575.
171. American College of Obstetricians and Gynecologists. Obstet Gynecol 2018;131:e15–e30.
172. Vutyavanich T, Kraisarin T, Ruangsri R. Obstet Gynecol 2001;97:577–82.
173. Gazmararian JA, Petersen R, Jamieson DJ, et al. Obstet Gynecol 2002;100:94–100.
174. Giugale LE, Young OM, Streitman DC. Obstet Gynecol 2015;125:1150–2.
175. Body C, Christie JA. Gastroenterol Clin North Am 2016;45: 267–83.
176. Fawcett EJ, Fawcett JM, Mazmanian D. Int J Gynaecol Obstet 2016;133:277–83.
177. Borgna-Pignatti C, Zanella S. Expert Rev Hematol 2016;9: 1075–80.
178. Young SL. Annu Rev Nutr 2010;30:403–22.
179. Gregory E, Ely D. Trends and characteristics in gestational diabetes: United States, 2016-2020. National Vital Stattistics Reports; vol 71 no 3. Hyattsville, MD: National Center for Health Statistics, 2022.
180. Ye W, Luo C, Huang J, et al. BMJ 2022;377:e067946.
181. American College of Obstetricians and Gynecologists. Obstet Gynecol 2018;131:e49–e64.
182. Brown J, Alwan NA, West J, et al. Cochrane Database Syst Rev 2017;5:Cd011970.
183. Chappell LC, Cluver CA, Kingdom J, et al. Lancet 2021;398: 341–54.
184. World Health Organization. Global Nutrition Targets 2025: Low Birth Weight Policy Brief (WHO/NMH/NHD/14.5). Geneva, Switzerland: World Health Organization, 2014.
185. Osterman M, Hamilton B, Martin J, et al. Births: Final Data for 2020. National Vital Statistics Reports. Hyattsville, MD: National Center for Health Statistics, 2022.
186. American College of Obstetricians and Gynecologists. Obstet Gynecol 2019;133:e97–e109.
187. Strauss RS, Dietz WH. J Nutr 1999;129:988–93.
188. Levine TA, Grunau RE, McAuliffe FM, et al. Pediatr 2015;135:126–41.
189. Crispi F, Miranda J, Gratacós E. Am J Obstet Gynecol 2018; 218:S869–S79.
190. Centers for Disease Control and Prevention. Data and statistics on birth defects, 2022. Accessed December 10, 2023. Available at www.cdc.gov/ncbddd/birthdefects/data.html
191. Molloy AM, Pangilinan F, Brody LC. Annu Rev Nutr 2017; 37:269–91.
192. Field MS, Kamynina E, Chon J, et al. Annu Rev Nutr 2018; 38:219–43.
193. U.S. Food and Drug Administration, Department of Health and Human Services. Food Standards: Amendment to the Standards of Identity for Enriched Grain Products to Require Addition of Folic Acid. Federal Register Vol. 61, 1996.
194. Yang Q, Cogswell ME, Hamner HC, et al. Am J Clin Nutr 2010; 91:64–72.
195. Williams J, Mai CT, Mulinare J, et al. Morb Mortal Wkly Rep 2015;64:1–5.
196. Viswanathan M, Treiman KA, Kish-Doto J, et al. JAMA 2017; 317:190–203.
197. Field MS, Stover PJ. Ann N Y Acad Sci 2018;1414:59–71.

CHAPTER 53

Nutrition in Lactation

Deborah L. O'Connor and Michelle K. McGuire

CHAPTER OUTLINE

Abbreviations

AAP, American Academy of Pediatrics; **AHRQ**, Agency for Healthcare Research and Quality; **BMD**, bone mineral density; **DHA**, docosahexaenoic acid; **FSH**, follicle-stimulating hormone; **GnRH**, gonadotropin-releasing hormone; **HIV**, human immunodeficiency virus; **FSH**, follicle-stimulating hormone; **PROBIT**, Promotion of Breastfeeding Intervention Trial; **WHO**, World Health Organization; **WIC**, Special Supplemental Nutrition Program for Women, Infants, and Children

Introduction

Human milk is a unique food that provides the recipient infant with much more than nutrients. In addition to macronutrients and micronutrients, an impressive body of evidence indicates that human milk (now considered to be a biological system) contains a host of other components. These include anti-inflammatory substances, immunoglobulins, antimicrobial constituents, antioxidants, immune cells, microRNAs, oligosaccharides, microbes, cytokines, hormones, and growth factors—all of which likely have biological activities related to development, metabolic regulation, inflammation, immune maturation, and pathogenesis.[1-4] The combined effects of these and other biologically active components likely result in the protection that human milk provides breastfeeding infants against infectious diseases, allergic disorders, and chronic diseases with an immunologic and/or inflammatory basis.[5-7]

In recognition of the benefits of human milk and breastfeeding, the World Health Organization (WHO), the American Academy of Pediatrics (AAP), the Dietary Guidelines for Americans, and Health Canada all recommend exclusive breastfeeding for ~6 months after birth.[8-11] At ~6 months, it is advised that infants be introduced to nutrient-rich, age-appropriate foods and that breastfeeding be continued for the first 12 to 24 months, and beyond if desired. Exclusive breastfeeding is defined as not receiving any infant foods, solids, or liquids other than human milk.[8]

This chapter summarizes information on the prevalence of lactation, its physiologic aspects, and the variable composition of human milk. In addition, it highlights established and potential beneficial impacts of lactation on both the breastfed, full-term infant and the breastfeeding mother, briefly reviews dietary recommendations during lactation, and suggests directions for future research related to human milk composition, lactation, and health. We wish to acknowledge that we have used the terms women and breastfeeding throughout this chapter but recognize that not all persons who breastfeed identify as women, and some lactating individuals consider themselves chestfeeding rather than breastfeeding.

Prevalence of Breastfeeding

Throughout the World

In a cross-sectional pooled analysis of 487 nationally representative surveys from 113 countries, it was reported that 89% of infants received "any breastmilk" at 6 months in 2019; this did not differ from the 88% reported for 2000.[12] While the percentage of infants receiving any breastfeeding at 6 months (>90%) did not change appreciably in low- and lower-to-middle income countries between 2000 and 2019, there were significant improvements in both upper-middle (67%–78%) and high-income countries (31%–55%) over the same timeframe. Improvements were attributed to many unique country-specific initiatives to promote and support breastfeeding, such as allowing lactating persons to express milk during working hours (United States), peer support (United Kingdom), and extensive paid postpartum leave for parents (Sweden). In 2019, 43% of infants from high-income countries, 59% of infants from upper-middle income countries, 82% of infants from lower-middle countries, and 89% of infants from low-income countries were breastfed to at least 1 year of age.[12] Rates for any breastfeeding of infants at 6 months of age in 2019 were estimated to be highest in west and central Africa (97%) and in eastern and southern Africa (94%), and lowest in western Europe (45%) and North America (61%). A key metric of breastfeeding success at the population level is "exclusive" breastfeeding at 6 months. Because so few national data from high income countries were available, determination of rates of exclusive breastfeeding were restricted to low- and middle-income countries. In this analysis, exclusive breastfeeding at 6 months increased from 35% in 2000 to 49% in 2019. While the increases are encouraging, they remain far below the target of 70% set by the World Health Assembly for 2030.[13]

In North America

Breastfeeding rates in North America are currently reported though the Centers for Disease Control and Prevention Breastfeeding Score Card (every 2 years) and continuously through the National Immunization Survey in the United States and the Public Health Agency of Canada.[14-16] Early national data on breastfeeding were captured through the proprietary Ross Laboratories Mothers Survey, initiated in 1954.[17,18] Reported percentage of mothers who ever breastfed in the United States was at its lowest between the 1950s and 1990.[17,18] In 1990, for example, only 34% of participants of the Special Supplemental Nutrition Program for Women, Infants, and Children (WIC) program and 63% of non-WIC participants provided any breastmilk to their infant while still in the hospital. WIC is a U.S. federal program that provides supplemental foods, healthcare referrals, and nutrition education for low-income pregnant, breastfeeding, and nonbreastfeeding postpartum women, and to infants and children up to 5 years of age. At 6 months, only 8% and 24% of WIC and non-WIC participants, respectively, provided any human milk to their infants.

Breastfeeding rates have increased significantly in the United States since 1990. In 2019, 83% and 56% of U.S. infants received some breastmilk at birth and at 6 months, respectively[15]; although exclusive breastfeeding through 6 months remains low at only 25% (**Table 53.1**). Significant differences exist in breastfeeding rates by sociodemographics, where infants born to women who are less educated, or who are non-Hispanic Black or living in poverty, have lower likelihood of receiving any breastmilk or being breastfed exclusively. In Canada, 2017 to 2018 national data suggest that 91% of mothers initiated breastfeeding, and 62% continued to provide breastmilk for at least 6 months.[16] Fewer (35%) reported exclusively breastfeeding for at least 6 months. The U.S. Healthy People 2030 objectives (published in 2020) are to increase the proportion of U.S. mothers who are still exclusively breastfeeding at 6 months to 42% and to increase those who are still breastfeeding at 1 year to 54%.[19]

Multi-Level Equitable Approaches to Support Breastfeeding

As illustrated in **Figure 53.1**, protecting, promoting, and supporting breastfeeding involves strategies that address barriers and facilitators from a structural to an individual level.[20] Breastfeeding in a civil society is not the sole responsibility of the mother. Policies that protect breastfeeding include skilled prenatal and postpartum lactation counseling and support, universal paid maternity leave, the right of a woman to breastfeed in public, and universal workplace break time with appropriate facilities for expressing milk.[9,20]

Mammary Glands and Regulation of Milk Secretion

Milk synthesis and secretion are complex, involving specialized anatomic structures and a well-orchestrated series of physiologic functions. The mature mammary glands of a nonpregnant, nonlactating woman are made up of glandular tissue that can, when stimulated to do so, synthesize and store milk; ligaments that provide a framework and support; metabolically active adipose (fatty) tissue; blood vessels that deliver nutrients and other milk constituents to the gland; myoepithelial cells that contract to cause milk ejection; a treelike pattern of branching "secondary" milk ducts that transport milk from the glandular tissue to the main milk ducts; and 4 to 18 larger "main" milk ducts that deliver milk to the nipple.[21] Alveolar clusters (alveoli), made up of spheres of mammary epithelial (secretory) cells that synthesize milk constituents exist in a dynamic state, with growth and complexity increasing and decreasing in response to the hormonal changes of the menstrual cycle.[22,23] During pregnancy, lobular alveolar complexes expand dramatically in response to the hormones progesterone, prolactin, and placental lactogen. The complex shifts required to signal synthesis of milk constituents in the mammary gland occur about halfway through pregnancy, and are sometimes referred to as "stage I lactogenesis." Although small amounts of milk can sometimes be expressed during the very late stage of pregnancy, copious milk secretion

Table 53.1 Prevalence of any and Exclusive Breastfeeding by Sociodemographic Characteristics Among Children Born in The United States in 2019[15]

	Any Breastfeeding[a]			Exclusive Breastfeeding[a] Through 6 Months
	Ever Breastfed	Breastfed at 6 Months	Breastfed at 12 Months	
All Infants	83.2	55.8	35.9	24.9
Hispanic	83.0	51.4	33.2	23.5
Non-Hispanic White	85.3	59.9	39.4	26.9
Non-Hispanic Black	74.1	44.0	24.1	19.1
Non-Hispanic Asian	90.8	70.2	45.7	28.5
Less than high school	71.6	39.2	26.6	19.8
High school graduate	73.2	39.7	23.4	18.6
Some college or technical school	85.7	54.2	34.2	23.8
College graduate	91.5	71.9	47.7	31.2
Poverty Income Ratio <100[b]	74.0	41.1	24.9	19.3
100–199	80.1	49.9	31.5	22.3
200–399	85.3	59.0	39.1	28.5
400–599	90.7	68.4	46.4	28.3
600 or greater	91.6	70.4	44.3	29.0

[a]Data from 20,906 infants for any breastfeeding and 20,217 for exclusive breastfeeding.
[b]Poverty Income Ratio = Ratio of self-reported family income to the federal poverty threshold value, depending on the number of people in the household.

(sometimes referred to as "stage II lactogenesis" or "secretory activation") does not occur until the early postpartum period, as it is inhibited by elevated circulating progesterone levels from the placenta during pregnancy.[24,25]

Lactogenesis and lactation are regulated through complex endocrine system control mechanisms that coordinate the actions of various hormones, including the reproductive hormones prolactin, progesterone, placental lactogen, oxytocin, and estrogen.[25,26] Although it is clear that progesterone suppresses active milk secretion during stage I lactogenesis, hormonal regulation during this stage is not well understood.[26] After parturition, secretory activation is initiated through the rapid decline in circulating progesterone, combined with high levels of prolactin. This process results in secretion of antibody-rich colostrum ("early milk") and then milk. Initiation of secretory activation does not require infant suckling, but suckling must begin by 3 to 4 days postpartum to maintain milk secretion. The hormone prolactin, required to maintain milk production after lactation, is established (at least for a period of time) and released into the maternal circulation from the anterior pituitary gland in response to suckling or pumping. During lactation, release of prolactin is mediated by a transient decline in the secretion of dopamine, an inhibiting factor, from the hypothalamus. Because plasma prolactin concentration does not correlate with the rate of milk secretion, and can return to prepregnancy levels during late lactation, investigators have suggested that prolactin may be a permissive factor for milk secretion rather than a regulatory factor.[26]

A diagram of an alveolar complex, the milk-secreting unit of the mammary gland, is shown in **Figure 53.2**.[27] As described above, each alveolus consists of a spherical layer of secretory epithelial cells surrounded by various supporting structures including myoepithelial cells, blood vessels, and a supportive stroma that contains adipocytes, fibroblasts, and antibody producing immune cells. To produce milk, four integrated secretory processes take place in the alveolar complex, as follows.

- Exocytosis of milk protein, lactose, and other components of the aqueous phase via Golgi-derived secretory vesicles from inside the secretory epithelial cell into the lumen of the alveolus.
- Lipid synthesis and secretion in milk fat globules via a complex process whereby the milk fat globules are ultimately surrounded by a milk fat globule membrane comprised of a phospholipid monolayer from the endoplasmic reticulum and a phospholipid bilayer from the secretary epithelial cell membrane.
- Secretion of ions, water, and glucose via transport molecules.
- Transcytosis of antibodies and other substances from the interstitial spaces via paracellular pathways.

Milk is secreted into the inner lumen of the alveoli and stored there until it is forced into the secondary milk ducts by contraction of the myoepithelial cells.[24] Although milk production is a continuous process, the amount produced is regulated primarily by infant demand. It is noteworthy that, unlike the lactating cow, which can store large amounts of milk in structures called lactiferous sinuses within the mammary glands, very little milk can be stored in the ducts of a breastfeeding woman's mammary glands. As such, milk removal is critical for continued milk production.

Suckling or pumping causes neural impulses to be sent from the nipple to the hypothalamus. This triggers oxytocin release

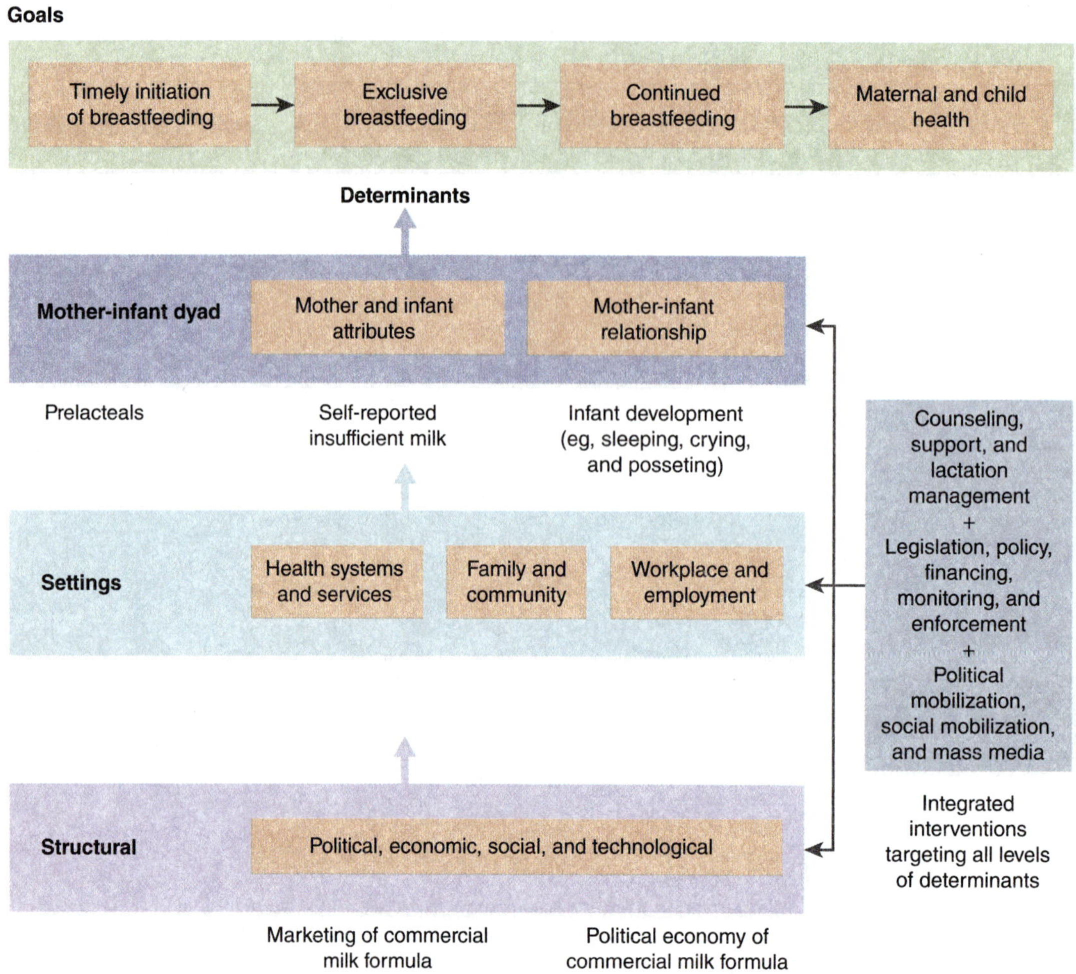

Figure 53.1 Conceptual framework illustrating the multiple levels interacting to facilitate or create barriers to breastfeeding.

Reproduced from Perez-Escamilla R, Tomori C, Hernandez-Cordero S, et al. Lancet 2023;401:472–485. doi:10.1016/S0140-6736(22)01932-8[28]

from the posterior pituitary gland. Oxytocin brings about contraction of the myoepithelial cells, which forces milk into the secondary milk ducts and eventually the main milk ducts that converge at the nipple. The coordinated actions of the infant's mouth and tongue (or pump) then remove the milk from the breast. It is noteworthy that this process (let-down) can also be triggered simply by seeing the infant or hearing the infant cry.

After initiation of breastfeeding (or pumping) after parturition, milk volume increases significantly within several days postpartum. During lactation, the typical daily volume of milk transferred to the infant increases from 0.50 mL on day 1, to 500, 650, and 750 mL by day 5, 1 month, and 3 months, respectively. Most women can produce considerably more milk than needed by a single infant as long as the milk is removed (as would occur with, for example, twins). When milk is not removed either by infant suckling or pumping, involution of the mammary epithelium occurs, and milk production stops within 1 to 2 days.

Composition of Human Milk

Human milk is a remarkably complex biologic fluid. It is composed of thousands of constituents that are dispersed throughout various phases, including an aqueous phase with true solutions (87%), colloidal dispersions of casein molecules (0.3%), emulsions of fat globules surrounded by fat globule membranes (4%), and living cells (both maternally derived and microbial). Milk composition changes substantially as early milk develops the characteristics of mature milk that are evident by day 10 of lactation. For example, lactose increases; sodium, potassium, and chloride decrease; total lipids increase; the immune factors lactoferrin and secretory immunoglobulin A decrease; and oligosaccharides decrease. New efforts are underway globally to more precisely define milk composition and the myriad maternal, environmental, infant, and behavioral factors that influence it.[29] Representative values for early and mature human milk constituents are listed in **Table 53.2**.[30,31] Both the composition and volume of human milk secreted are influenced to some degree by factors such as genetic individuality, maternal dietary or supplement intake (particularly fatty acids, vitamin B_{12}, thiamin, riboflavin, vitamin B_6, vitamin A, selenium, and iodine), adiposity (total fat) and time postpartum.[3,32-36]

The mammary gland can extract most nutrients actively from the circulation, independently of maternal regulatory systems, so milk may contain adequate nutrients even during inadequate maternal dietary intake. However, persistent maternal deficiencies may result in inadequate micronutrient concentrations in the milk. Although most constituents of human milk (including nutrients) are biologically active, nutritional components are

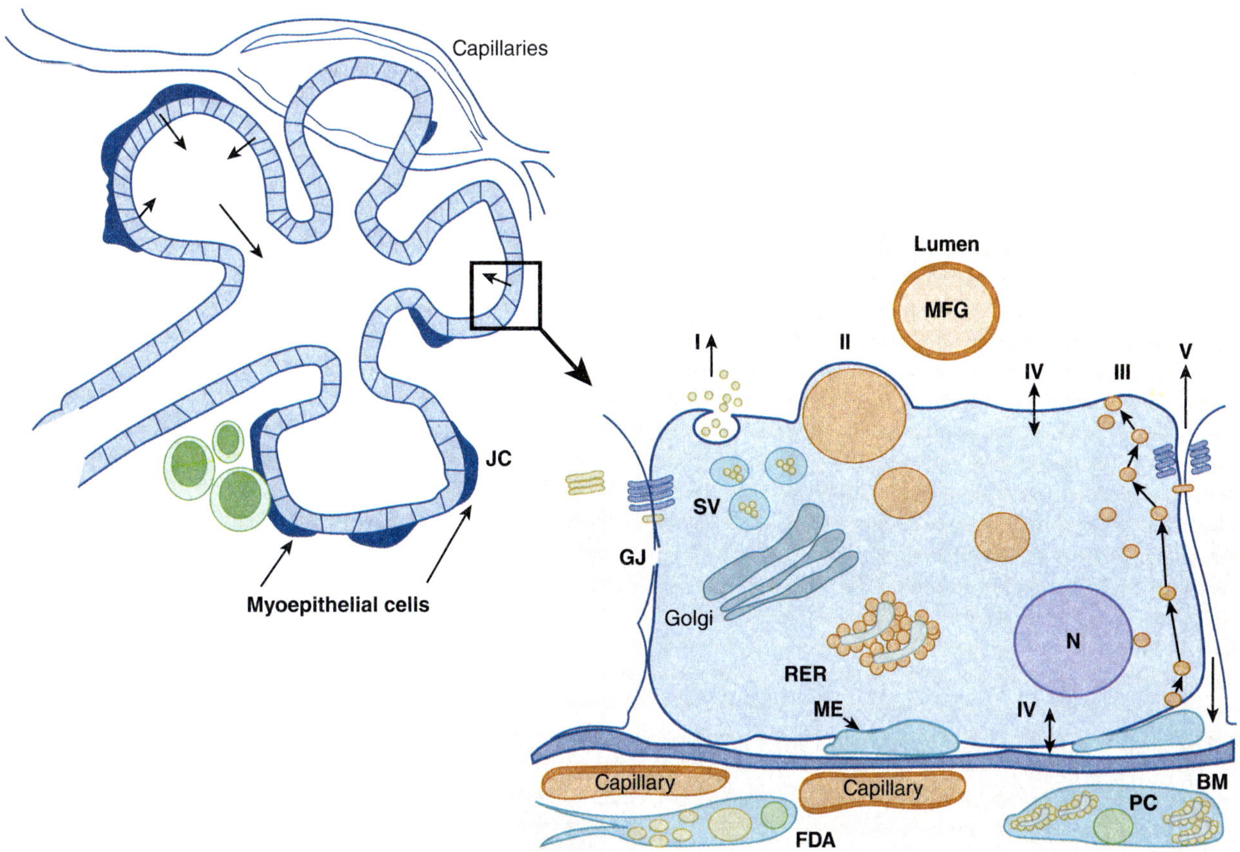

Figure 53.2 Diagram of the mammary alveolus and alveolar (secretory) epithelial cell showing pathways for milk secretion. Milk is secreted by alveolar epithelial cells into the alveolar lumen and is forced into the secondary milk ducts by contraction of myoepithelial cells (ME). The alveolus is surrounded by a well-developed vasculature and support framework (stroma) that includes extracellular matrix components, fibroblasts, and adipose tissue. The region inside the box is expanded to show key structural and transport properties of a single secretory epithelial (alveolar) cell. *I*, Exocytosis of milk proteins, lactose, calcium, and other aqueous-phase milk components. *II*, Formation of cytoplasmic lipid droplets that move to the apical membrane to be secreted as a membrane-bound milk fat globule (MFG). *III*, Vesicular transcytosis of proteins such as immunoglobulins (antibodies) from the interstitial space. *IV*, Transporters for the movement of monovalent ions, water, and glucose across the apical and basal membranes of the cell. *V*, Transport of plasma components and immune cells through the paracellular pathway (open only during pregnancy, involution, and in inflammatory states such as mastitis). *BM*, basement membrane; *FDA*, fat-depleted adipocyte; *GJ*, gap junction; *JC*, junctional complex; *N*, nucleus; *PC*, plasma B cell; *RER*, rough endoplasmic reticulum; *SV*, secretory vesicle.

Reproduced from McManaman JL, Neville MC. Adv Drug Deliv Rev 2003;55:629–41.[77]; Neville MC, Allen JC, Watters C. In: Neville MC, Neifert MR, eds. The Mechanisms of Milk Secretion. Lactation. Springer, 1983.[78]

grouped separately from "biologically active factors" in the following discussion.

Nutritional Components

Macronutrients

Protein constituents of human milk provide essential amino acids for growth, protective factors (e.g., immunoglobulins, lysozymes, lactoferrin), carriers for vitamins (e.g., folate-, vitamin D–, and vitamin B_{12}–binding proteins) and hormones (e.g., thyroxine- and corticosteroid-binding proteins), enzymatic activity (e.g., amylase, bile salt–stimulated lipase), and other biologic activities (e.g., insulin, epidermal growth factor).[37] Although the total protein content of human milk is low compared with that of most mammals, it is easily digestible and its incorporation into lean tissue in the infant is exceptionally high.[38] Human milk also contains a relatively high amount of nitrogen incorporated in nonprotein molecules (referred to as the nonprotein nitrogen fraction). This nonprotein nitrogen fraction comprises more than 200 compounds, including free amino acids, carnitine, taurine, amino sugars, nucleic acids, urea, nucleotides, and polyamines. There is substantial research ongoing to determine the purpose of these nonprotein, nitrogen-containing compounds for the infant. Maternal nutrition may alter both the total protein and nonprotein nitrogen constituents of human milk. However, healthy full-term exclusively breastfed infants do not, as a rule, show signs of protein deficiency, regardless of maternal protein intake.[32]

Human milk lipids, the major energy-yielding fraction (45%–55% of total kcal), are the most variable constituents in human milk. Circulating fatty acids, a reflection of maternal diet and adipose stores, are the main substrates for human milk fat—at least when the woman is

Table 53.2 Representative Values For Selected Constituents of Human Milk

Constituent (per liter)[a]	Early Milk	Mature Milk
Energy (kcal)		653–704
Carbohydrate		
Lactose (g)	20–30	67
Glucose (g)	0.2–1.0	0.2–0.3
Oligosaccharides (g)	22–24	12–14
Total nitrogen (g)	3.0	1.9
Nonprotein nitrogen (g)	0.5	0.45
Protein nitrogen (g)	2.5	1.45
Total protein (g)	16	9
Casein (g)	3.8	5.7
β-Casein (g)	2.6	4.4
κ-Casein (g)	1.2	1.3
α-Lactalbumin (g)	3.62	3.26
Lactoferrin (g)	3.53	1.94
Serum albumin (g)	0.39	0.41
Serum immunoglobulin A (g)	2.0	1.0
Immunoglobulin M (g)	0.12	0.2
Immunoglobulin G (g)	0.34	0.05
Total lipids (g)	20	35
Triglyceride (% total lipids)	97–98	97–98
Cholesterol[b] (% total lipids)	0.7–1.3	0.4–0.5
Phospholipids (% total lipids)	1.1	0.6–0.8
Fatty acids (weight %)	88	88
Total saturated	43–44	44–45
C12:0		5
C14:0		6
C16:0		20
C18:0		8
Monounsaturated		40
C18:1 ω-9	32	31
Polyunsaturated	13	14–15
Total ω-3	1.5	1.5
C18:3 ω-3	0.7	0.9
C22:5 ω-3	0.2	0.1
C22:6 ω-3	0.5	0.2
Total ω-6 (weight %)	11.6	13.06
C18:2 ω-6	8.9	11.3
C20:4 ω-6	0.7	0.5
C22:4 ω-6	0.2	0.1
Water-soluble vitamins		
Vitamin C (mg)		100
Thiamin (μg)	20	200
Riboflavin (μg)		400–600
Niacin (mg)	0.5	1.8–6.0
Vitamin B_6 (mg)		0.09–0.31
Folate (μg)		80–140
Vitamin B_{12} (μg)		0.5–1.0
Choline (total, μg)[c]	486–560	1378–1410
Pantothenic acid (mg)		2.0–2.5
Biotin (μg)		5–9
Fat-soluble vitamins		
Vitamin A (mg)	2	0.3–0.6
Carotenoids (mg)	2	0.2–0.6
Vitamin K (μg)	2–5	2–3
Vitamin D (μg)		0.33
Vitamin E (mg)	8–12	3–8
Minerals		
Major minerals		
Calcium (mg)	250	200–250
Magnesium (mg)	30–35	30–35
Phosphorus (mg)	120–160	120–140
Sodium (mg)	300–400	120–250
Potassium (mg)	600–700	400–550
Chloride (mg)	600–800	400–450
Trace minerals		
Iron (mg)	0.5–1.0	0.3–0.9
Zinc (mg)	8–12	1–3
Copper (mg)	0.5–0.8	0.2–0.4
Manganese (μg)	5–6	3
Selenium (μg)	40	7–33
Iodine (μg)		150
Fluoride (μg)		4–15

[a]All values are expressed as per liter of milk except for fatty acids, which are expressed as a percentage based on either milk volume or weight of total fatty acids identified.

[b]The cholesterol content of human milk ranges from 100 to 200 mg/L in most samples of human milk after day 21 of lactation.

[c]Chen MY, Northington R, Yan J. FASEB J 2018;31:lb392–lb392.

Data from Picciano MF. Ped Clin North Am 2001;48(1):263–4. doi:10.1016/s0031-3955(05)70299-6

well nourished and not fasting. Some shorter-chain fatty acids can also be made in the mammary gland, which can also elongate and desaturate some fatty acids. The complex characteristic features of human milk lipids are reviewed elsewhere.[37,39] Human milk is a rich source of linoleic acid and α-linolenic acid, both essential fatty acids, as well as their long-chain polyunsaturated fatty acid derivatives, arachidonic acid and docosahexaenoic acid (DHA). Because lipid digestion is not yet fully developed in the newborn, several enzymes join forces to aid in digestion of milk lipids. These include the following: lingual lipase, which initiates hydrolysis in the stomach; gastric lipase; pancreatic lipase; and bile salt–dependent lipase, which is a constituent of human milk.

Lactose, a disaccharide made from glucose and galactose, is a major human milk constituent and its primary digestible carbohydrate.[40] There is little variability in milk lactose concentration, and it is thought that most of the variability that exists results mainly from maternal individuality. Nonetheless, milk lactose concentration increases rapidly in early lactation. Glucose and fructose are also present in human milk, but in relatively small quantities. Human milk also contains amylase, an enzyme that can aid in carbohydrate digestion, nucleotide sugars, glycolipids, glycoproteins, and oligosaccharides, which inhibit the growth and function of certain pathogens and are likely important for immune development among other functions.[41]

Micronutrients

In general, the vitamin content of human milk is related to acute and chronic maternal dietary and supplement intake. If maternal vitamin status is low, milk vitamin concentrations also are low, but they increase if maternal intake increases. If maternal status is adequate, milk vitamin concentrations also are adequate, and are less affected by maternal intake.[3,40,42] Some vitamins (e.g., folate) are preferentially partitioned into milk and, thus, it is only in severe maternal folate deficiency that milk concentrations become inadequate for the infant. In contrast to vitamins, mineral concentrations in human milk generally do not correlate well with maternal intake, except for selenium and iodine.

Human milk contains the fat-soluble vitamins A, D, K, and E, as well as certain carotenoids (α-carotene, β-carotene, lutein, cryptoxanthin, lycopene) that have varying degrees of biologic activity. The vitamin A content of human milk is influenced more by maternal vitamin A intake than by overall maternal vitamin A status.[3,43] Retinyl esters in chylomicrons and plasma retinol-binding protein–retinol are the sources of vitamin A in human milk. The retinyl esters are directly related to maternal intake, whereas retinol-binding protein–retinol is relatively constant, regardless of maternal vitamin A liver stores.

Historically, a main source of vitamin D for humans has been cutaneous synthesis from cholesterol after exposure to ultraviolet B light. In the absence of adequate sun exposure (e.g., daily sunscreen use, clothing that covers most of the body) human milk alone will not provide sufficient vitamin D to prevent deficiency in infants, except when mothers consume high levels of supplemental vitamin D (4,000 to 6,400 IU/day) to increase milk concentration.[44-47] The tolerable upper limit (UL) for vitamin D for lactating women is 4,000 IU.[48] To avoid the risk of skin cancer and dehydration, health authorities recommend that infants not be exposed to direct sunlight.[49] Rickets attributable to inadequate vitamin D in infants continues to be a problem in North America and other Western countries, particularly among exclusively breastfed infants and those of dark pigmentation.[50] In North America, it is, therefore, recommended that breastfed infants be supplemented with 400 IU/day of vitamin D beginning the first few days of life.[8,51] The Canadian Pediatric Society further recommends providing 800 IU vitamin D, from all sources, for infants residing in communities above the fifty-fifth parallel (approximately Edmonton, Alberta).[52,53]

The vitamin K content of human milk does not correlate well with maternal dietary vitamin K intake. Some studies report, however, that maternal vitamin K supplementation with pharmacologic doses (5 or 20 mg/day) significantly increases vitamin K concentration in milk and improves vitamin K status of breastfed infants.[3,31] Infants are born with low tissue stores of vitamin K and, thus, normally receive a prophylactic dose at birth to reduce the risk of hemorrhagic disease.[54,55]

Most (83%) of the vitamin E in human milk is in the form of α-tocopherol, although small quantities of β-, γ-, and δ-tocopherols are present. Some data indicate that vitamin E concentration in human milk can be increased only through supplementation with large amounts of the vitamin.[3]

Water-soluble vitamins in human milk include vitamin C, thiamin (B_1), riboflavin (B_2), niacin, vitamin B_6 (pyridoxine and related compounds), vitamin B_{12} (cobalamin), folate, choline, and biotin.[3,30,34,35,42] Their concentrations in human milk all depend on maternal diet. Vitamin B_{12} deficiencies have been reported in infants nursed by mothers who follow a strict vegetarian diet without vitamin supplementation. Although vitamin B_{12} supplementation resolves deficiency-related abnormalities, some evidence suggests that lasting neurologic disability may result.[35,56] It appears that preferential use of folate by the mammary gland over the maternal hematopoietic system occurs, so the risk of suboptimal folate in breastmilk is minimal unless maternal insufficiency is severe.[35] While maternal supplementation with low doses of folic acid (400 μg) have little or no impact on milk folate concentration, higher doses of folic acid (e.g., 1,000 and 5,000 μg) are associated with an increase in total folate content, as well as shifting the forms of folate in milk (i.e., increased unmetabolized folic acid in milk).[57,58]

Human milk contains the major minerals calcium, phosphorus, magnesium, sodium, and potassium, as well as various trace minerals, including iron, copper, zinc, manganese, selenium, and iodine. The concentrations of major minerals in human milk do not generally correspond to their values in maternal serum. In fact, increased maternal bone resorption of calcium and phosphorus, accompanied by decreased urinary excretion, appear to provide the necessary amounts of these minerals for milk production, independent of maternal dietary intake.[48] A mother's bone mineral content can decline 5% to 10% during a 2- to 6-month time period of exclusive breastfeeding. However, it normally returns to baseline 6 to 12 months after weaning and, thus, does not appear to increase the risk of low bone mineral content or osteoporosis.

Although the amounts of iron, copper, and zinc in human milk are relatively low, regardless of maternal intake, high bioavailability of iron and zinc in human milk has been reported.[31] The iron provided by human milk, in addition to a full-term infant's stored iron, is adequate for breastfed infants for approximately the first 6 months after birth. Unlike most minerals, concentrations of selenium and iodine in human milk depend on maternal diet and can vary widely by geographic region; depending on soil concentration.[3,59] For example, the milk iodine content in iodine-sufficient areas is approximately 10 times that of milk in areas where iodine deficiency disorders, which include brain damage and intellectual disability, are prevalent. Maternal iodine supplementation can be effective in preventing iodine deficiency disorders.[60]

Biologically Active Components

Although most components of human milk are important for infant health (i.e., have biological activity), researchers tend to refer to those that are not essential nutrients as "biologically active (bioactive) components." Along with the nutrients, these biologically active components provide three overlapping functions for the infant.[61]

- Nourishment, supporting physical growth, development, and metabolism (e.g., digestive enzymes, microbes)
- Protection, shielding the infant from infection and supporting the infant during a time when the immune system is naïve (e.g., antibodies, oligosaccharides)
- Communication, relaying complex information as to how to best thrive in a particular environment and culture and with a specific genetic background (e.g., stem cells, miRNAs)

Some of the hundreds of known biologically active components of human milk and their possible functions are outlined in **Table 53.3**.[41,61-63] It is noteworthy that some components may impact infant health through more than one mechanism, and emerging evidence suggests that components might interact (both physiologically and physically) to exert their effects. For example, lactoferrin, a glycosylated protein present in greater amounts in early milk than in mature milk, has broad-spectrum and complex antimicrobial functions. It exhibits antiviral activity against herpes simplex virus, cytomegalovirus, and HIV. It also exhibits various immunomodulating activities, and possesses anti-inflammatory activity carried out through scavenging iron and other processes.[41] An example of possible complex interactions among milk components is

Table 53.3 Possible Functions of Selected Biologically Active (Non-Nutritive) Factors in Human Milk

Biologically Active Factors	Possible Function
Proteins and Peptides (Nonenzymatic)	
Secretory immunoglobulin A (sIgA)	Antimicrobial and antiviral activity by preventing bacteria and viruses from attaching to mucosal surfaces (immune exclusion), neutralizing microbial toxins, increasing viral excretion, shaping infant's GI microbiota
Lactoferrin	Antimicrobial activity by chelation of iron, immune exclusion, cytotoxicity of breakdown products (i.e., lactoferricin B)
Lysozyme	Antimicrobial activity by cell wall (β-1,4 linkages of *N*-acetylmuramic acid and *N*-acetylglucosamine) and cell membrane (proteoglycan matrix) destruction
κ-Casein	Antimicrobial activity
Cytokines (e.g., IL-10, TNF-α, interferon-γ)	Modulation of maturation and function of the immune system
sCD14	Antimicrobial activity by detection of bacterial lipopolysaccharide; regulation of T- and B-cell activation
Complement (C3, C4) and complement receptors	Antimicrobial activity by bacteriolysis, neutralization, immune exclusion, and enhanced phagocytosis
Human β-defensin-1	Antimicrobial activity
Toll-like receptors	Antimicrobial activity
Milk fat globule membrane proteins	Glycosylated proteins (e.g., MUC-1, butyrophilin, lactadherin) may enhance lipid digestion in infants; lactadherin, MUC-1, and xanthine oxidase may have antimicrobial properties
Carbohydrates	
Oligosaccharides	Antimicrobial activity by immune exclusion; shaping infant GI microbiota, modulation of immune system
Fructose	Possible modulator of infant growth and development
Glycoconjugates	Antimicrobial and antiviral activity
Free fatty acids (produced from triglycerides during digestion)	Antimicrobial, antiviral, and antiprotozoan activity by disruption of cell membranes and intracellular pH and immune exclusion; immune maturation and modulation of immune response (Th1, Th2)
Nonessential Nutrients and Other Dietary Components	
β-Carotene	Anti-inflammatory activity
Flavor or scent compounds	May prepare infant for flavors of cultural diet

Biologically Active Factors	Possible Function
Enzymes	
Bile salt–dependent lipase	Production of free fatty acids with antibacterial, antiviral, and antiprotozoan activity
Catalase	Anti-inflammatory activity
Glutathione peroxidase	Anti-inflammatory activity
PAF factor: acetylhydrolase	Protection against necrotizing enterocolitis
Antienzymes	
α_1-Antitrypsin	Inhibition of inflammatory proteases
α_1-Antichymotrypsin	Inhibition of inflammatory proteases
Prostaglandins	Possible cytoprotection in intestine of the newborn
Cells (e.g., macrophages, polymorphonuclear cells, lymphocytes)	Antimicrobial activity, production of lymphokines and cytokines, enhancement of other protective agents, and modulation of maturation and function of immune system
Microbes	Bacteria, viruses, and fungi might protect the infant and mammary gland against pathogens and train the infant's immune system to tolerate common environmental microbes; likely important colonizers of the infant gastrointestinal tract
Immune cells (e.g., macrophages, lymphocytes (e.g., T and B cells))	Antimicrobial activity by production of lymphokines and cytokines, and phagocytosis, and modulation of maturation and function of the immune system (via traversing the neonatal intestine); anti-inflammatory activity
Growth/Development-Related Substances	
Erythropoietin	Stimulation of erythropoiesis in newborns
Insulin	Neonatal glycemia effects
Prolactin	Help in regulation of development of neuroendocrine, reproductive, and immune system functioning
Adrenal steroids	Stimulation of organ maturation
Gonadotropin-releasing hormone (GnRH)	Increase in ovarian GnRH receptors in newborns
Growth hormone-releasing hormone (GHRH)	Possible regulation of GH secretion in newborns
Thyrotropin-releasing hormone (TRH)	Possible regulation of TSH secretion in newborns
Thyroid-stimulating hormone (TSH)	Possible regulation of T_3 (tri-iodothyronine) and T_4 (thyroxine) secretion in newborns
Epidermal growth factor (EGF)	Stimulation of gastrointestinal tract growth, acceleration of gastrointestinal tract closure
Transforming growth factor alpha (TGF-α)	Stimulation of gastrointestinal tract growth
Insulin-like growth factors (IGFs)	Stimulation of gastrointestinal tract growth, possible systemic growth effects
Nonprotein, Nitrogenous Compounds	
Urea	May contribute nitrogen to the infant via GI microbiota metabolism
Free amino acids	Protein synthesis, synthesis of nonessential amino acids, and energy (ATP) production
Ammonia	Nonessential amino acid synthesis
Nucleotides	Enhanced T-cell maturation, natural killer cell activity, and antibody response to certain vaccines
Polyamines	Cell division and proliferation, gastrointestinal development, immune development, shaping gastrointestinal microbiome
Micro-RNAs (miRNAs)	Communication to infant, immune modulation, regulation of energy metabolism

Data from McGuire MK, O'Connor DL, eds. Human Milk: Sampling and Measurement of Energy-Yielding Nutrients and Other Macromolecules. Academic Press, 2021[37]; Hamosh M. Pediatr Clin North Am 2001; 48(1):69–86[41] doi:10.1016/s0031-3955(05)70286-8; Smilowitz JT, Allen LH, Dallas DC, et al. Am J Clin Nutr 2023;117:S28–S42[61]; Grosvenor CE, Picciano MF, Baumrucker CR. Endocr Rev 1993;14:710–28[62] doi:10.1210/edrv-14-6-710; Hosea Blewett HJ, Cicalo MC, Holland CD, Field CJ. Adv Food Nutr Res 2008;54:45–80[63] doi: 10.1016/s1043-4526(07)00002-2. PMID: 18291304; Gregg B, Ellsworth L, Pavela G, et al. Pediatr Obes 2022;17(7):e12892[64] doi:10.1111/ijpo.12892

the fact that TGF-β (present in milk) remains inactive until it is exposed to low pH and protease activity (either gastric or milk proteases). Indeed, scientists now consider milk to be a biological system, characterized by the fact that its overall biological activity is greater than the sum of its parts.[2] Understanding how all of milk's constituents work in a coordinated fashion to influence health is a current focus of research in this area.

Many hormones and growth factors in human milk survive digestion in the gastrointestinal tract, are absorbed into the circulation of the infant, and likely carry out important functions. These biologically active substances include agents from numerous glands, organs, and organ systems, including the pituitary, pancreas, hypothalamus, thyroid, parathyroid, intestine, adrenal gland, and gonads.[62,63] Much remains to be learned about the functions of both known human milk constituents and constituents that likely are yet to be discovered, and this remains an exciting area of current investigation.

Special Considerations (Banked Donor Milk, and Informal Milk Sharing)

This chapter focuses on breastfeeding as it pertains to the healthy mother-infant dyad, but a few special considerations are worth mentioning here. First, there has been a resurgence over the past decade in donor human milk banking; the primary recipients in North America being preterm very low birth weight infants when parent milk is unavailable. Donor human milk, used as a supplement to parent milk for very low birth weight infants during in-hospitalization, has been shown to reduce risk of necrotizing enterocolitis, a serious bowel emergency.[65,66] By definition, a donor human milk bank is a service established to recruit and screen potential human milk donors, and to collect, screen, process, store, and distribute donated human milk; usually to a neonatal intensive care unit.[67] In most parts of the world, including in North America, donated human milk is heat-treated (i.e., pasteurized) to remove pathogens that could harm an infant. The majority of milk banks have some kind of external oversight—e.g., local public health authorities and milk banking associations that provide accreditation. There is significant interest in whether the use of donor milk should be extended to term-born infants when maternal milk is unavailable and a safe supply of donor human milk allows, for example on compassionate grounds (e.g., maternal death or illness). Due to the manner in which human milk is currently processed (e.g., freezer storage, freeze and thaw cycles, heat treatment) and the impact on its nutritional quality, it is likely that donated human milk alone may not meet the nutritional requirements of term-born infants.

Outside of the hospital setting, some families secure donated milk through informal milk sharing. This is the practice of donating and receiving expressed human milk without going through a donor human milk bank.[68,69] Often, informal milk sharing occurs via community peer-to-peer or over-the-Internet interaction. Informal milk sharing is not subject to the same regulation or safety controls as donor human milk banks and; therefore, donors may not be adequately screened for infectious disease (e.g., HIV) or chemical contaminants (illegal or a limited list of contraindicated prescription drugs). Furthermore, milk could be spoiled or contaminated with micro-organisms, if not handled properly. Families are advised to consult with their healthcare provider if they are considering informal milk sharing.

Impact of Lactation on the Infant

Nutritional Status

Human milk, which provides an appropriate balance of nutrients in forms that are readily digestible and bioavailable, provides optimal nutrition for nearly all healthy, full-term newborns. After the first 2 to 3 months of life, exclusively breastfed infants tend to have a slower rate of weight gain compared with formula-fed infants.[70] The pattern of growth of exclusively breastfed infants is thought to be the gold standard for all infants, regardless of feeding method and, as such, the WHO has developed prescriptive growth charts (standards) using an international sample of breastfed infants living in favorable economic and environmental conditions in countries around the world.[70] It is noteworthy that, regardless of international location and genetic background, healthy exclusively breastfed infants used to construct the WHO growth standards appear to have similar growth patterns during the first year of life.

When a mother is well nourished, exclusive breastfeeding can meet all of the infant's nutrient requirements for approximately 6 months, except for vitamin D and iron. As described previously, to prevent vitamin D deficiency, the AAP recommends that exclusively breastfed infants and those consuming less than 28 oz of infant formula each day receive vitamin D supplements (400 IU/d), beginning at hospital discharge and throughout breastfeeding.[71] This recommendation applies to both exclusively and partially breastfed infants. The Canadian Pediatric Society further recommends providing 800 IU vitamin D, from all sources, for infants residing in communities above the fifty-fifth parallel.[52,53] To prevent iron deficiency, the AAP recommends that exclusively breastfed term infants receive an iron supplementation of 1 mg/kg/d, starting at 4 months of age and continued until appropriate iron-containing complementary foods have been introduced. Beginning at 4 months of age, partially breastfed infants who receive more than half of their daily feedings as human milk and who are not receiving iron-containing complementary foods should also receive 1 mg/kg per day of supplemental iron.[72] The preterm infant ($<$37 weeks' gestation) who is fed human milk should receive a supplement of elemental iron at 2 mg/kg per day starting by 1 month of age and extending through 12 months of age. In Canada, routine iron supplementation is not recommended for healthy term, exclusively breastfed infants with no risk factors for iron deficiency.[73] Risk factors include preterm delivery or being born with $<$2,500 grams, low intake of iron-rich complementary foods such as meat, meat alternatives, and iron-fortified infant cereals after 6 months.

Because only animal-source foods contain vitamin B_{12}, special care must be taken to make sure that milk produced by vegan mothers has enough of this nutrient. When the mother is at risk of vitamin B_{12} deficiency, human milk may not provide sufficient vitamin B_{12}.[74,75] In these cases, the mother and/or infant fed human milk may require a vitamin B_{12} supplement. Parents, caregivers, and guardians should consult with a healthcare provider to determine whether supplementation is necessary.

Morbidity

In 2016, the Lancet published a series of two papers, which included a summary of the findings from 28 systematic reviews (22 specifically commissioned for this series) assessing the association between breastfeeding and health outcomes in both mother and child.[10] The authors reported that appropriate breastfeeding practices reduced child morbidity due to diarrhea, respiratory infections, and otitis media (ear infections). In low- and middle-income countries, where infectious diseases are common causes of death, the authors concluded that breastfeeding provides major protection, but even in high-income populations, breastfeeding was associated with lower mortality from causes such as necrotizing enterocolitis (preterm infants) and sudden infant death syndrome. The authors also concluded that breastfeeding was related to increased intelligence, and likely reduced the risk of becoming overweight or obese and developing diabetes in infants and young children. These outcomes are discussed in greater detail below.

Brain Development

One of the most consistently described benefits of breastfeeding in developed countries is the impact on cognitive development[10,76,77] Differences in cognitive development between human milk-fed and formula-fed infants are particularly pronounced for those born preterm with low birth weight.[78-80] However, because it is unethical to randomize infants to breastfeeding or formula feeding under normal circumstances, studies in this field are primarily observational and are frequently unable to control for important sociodemographic characteristics that could affect brain development. Sociodemographic differences (e.g., income, maternal age, maternal education, quality of the home environment) differ between families that provide breastmilk and those that provide formula. In a systematic review that excluded studies that did not adjust for infant stimulation or interaction in the home, Horta and colleagues demonstrated a small but statistically significant benefit of any breastfeeding, vs exclusive formula feeding, on infant intelligence (mean difference 2.62 points).[76]

The only randomized controlled trial included in the systematic review was the Promotion of Breastfeeding Intervention Trial (PROBIT).[77] In this trial, one set of maternity hospitals and their associated clinics in Belarus were randomized to a breastfeeding promotion program (experimental group), and a second set received the usual practices and policies with regard to breastfeeding (control). Belarus is an upper middle-income Eastern European country that had very low breastfeeding rates at the time PROBIT commenced. Approximately 17,000 term-born infants were followed. Maternity hospitals in Belarus randomized to receive the breastfeeding promotion intervention had a significant increase in exclusive breastfeeding to 3 months (43%), vs hospitals randomized not to receive the program (6%). A significant increase in the duration of breastfeeding was also observed. At 6.5 years of age, children at the intervention sites had higher scores on all the Wechsler Abbreviated Scales of Intelligence, with cluster-adjusted mean increases of 7.5, 2.9, and 5.9 points for verbal, performance, and full-scale IQ scores, respectively. Teachers' academic ratings for both reading and writing were also significantly higher in the intervention group.

Researchers in the field suggest that the positive link between breastfeeding and cognitive development may be partly the result of the long-chain polyunsaturated fatty acids DHA and arachidonic acid, which are present in high concentration in human milk but were not added to infant formulas in North America until approximately 25 years ago. However, results from clinical trials designed to evaluate whether preformed DHA and arachidonic acid need to be added to infant formula in addition to the precursor fatty acids (essential nutrients alpha-linolenic acid and linoleic acid, respectively) are mixed.[81] Some studies show a short-term benefit on either visual or cognitive development, whereas others show no benefit at all. DHA is especially important for development of the central nervous system, including the rapid growth of the brain and retina that takes place in the fetus during the last trimester and in the neonate during the first several postnatal months. Although humans can synthesize DHA from the precursor fatty acid alpha-linolenic acid, the capacity to do this appears to be low.

Because human milk fatty acid composition reflects maternal dietary fatty acid intake, considerable interest has been shown in whether supplementing lactating mothers with a source of DHA and other long-chain polyunsaturated fatty acids could improve the neurocognitive development of their children. A Cochrane Review concluded that available evidence did not appear to support this relationship.[82]

Overweight and Obesity

Investigators hypothesize that breastfeeding may indirectly reduce the risk of becoming overweight or obese by promoting development of satiety responsiveness and altering the gastrointestinal microbiome which, in turn, is associated with body composition.[83,84] Still, others suggest that it is not breastfeeding itself that is protective, but rather the more rapid postnatal growth induced by formula feeding that programs the infant for later developing obesity and metabolic syndrome. As education, socioeconomic status, and race are associated with both breastfeeding intensity and adiposity in high income countries, there is concern that reported associations between breastfeeding and the risk of overweight and obesity are inherently confounded by these known and unknown factors.[85-89]

In a recent update of their systematic review of the association between breastfeeding with overweight and obesity, Horta and colleagues reviewed 159 studies, most carried out with children 1–9 years of age.[84] These investigators reported that breastfeeding was associated with a 27% reduction in overweight or obesity. In the 19 studies assessed to have the lowest risk of publication bias, residual confounding, and misclassification, a 15% reduction in risk associated with breastfeeding was still observed. The only randomized trial that examined the impact of a breastfeeding promotion intervention on obesity rates was the PROBIT study, conducted by Kramer and colleagues and described earlier in this chapter.[90] Although the breastfeeding promotion intervention significantly increased the duration and exclusivity of breastfeeding, it did not reduce measures of adiposity in children at 6.5 years.

Transmission of Pathogens from Mother to Child via Breastfeeding

As described previously, human milk is not sterile. Although experts agree that most microbes in milk are likely commensal or protective in nature, sometimes they can be potentially pathogenic. "Vertical transmission" is a term used to describe mother-to-child transmission of a pathogen. Vertical transmission of a pathogen from mother to infant can occur during pregnancy, labor, delivery, and after birth (e.g., via breastfeeding). The AAP recognizes few contraindications to breastfeeding (including providing expressed milk) due to vertical transmission of disease. Mothers who are infected with untreated brucellosis are advised to temporarily not breastfeed and should not feed expressed milk until they are treated.[9] Infants born to women with hepatitis B should receive an initial dose of hepatitis B vaccine and hepatitis B immune globulin within 12 hours of birth, but there is no need to delay breastfeeding initiation until after the infant is immunized. Although hepatitis C virus can be detected in milk, transmission via breastfeeding has not been documented; as such neither the AAP nor the CDC considers maternal hepatitis infection a contraindication to breastfeeding. However, mothers with hepatitis C who have cracked or bleeding nipples are advised to refrain from breastfeeding from the affected breast or feeding expressed milk from that breast until the nipple has healed. It is noteworthy that avoidance of breastfeeding when the mother has HIV was previously advised by the AAP. However, this is no longer the case. In its most current guidelines on HIV and infant feeding released in 2016, the WHO recommends that all mothers living with HIV receive antiretroviral therapy (ART) to support their health and to ensure the well-being of their infants.[91] Furthermore, the WHO released guidelines in 2016 advises that, in countries that have opted to promote and support breastfeeding together with ART, mothers living with HIV who are on ART and adherent to therapy should breastfeed exclusively for the first 6 months and then add complementary feeding until 12 months of age. Breastfeeding with complementary feeding may continue until 24 months of age or beyond.

The recent COVID-19 pandemic quickly brought into question whether SARS-CoV-2 could be vertically transmitted from mother to infant via the act of breastfeeding and/or human milk, itself. Fortunately, researchers relatively quickly determined that breastfeeding was not only safe for infected women and their breastfeeding infants but also likely provided some protection to the latter.[92,93] Clearly, this is an area of great interest and one needing continued research—particularly when an outbreak, epidemic, or pandemic occurs.

Maternal Marijuana and Alcohol Use

The AAP has concluded that research is insufficient to assess the effects of exposure of infants to maternal marijuana use (including edible products) during breastfeeding. As a result, they advise women to abstain from all marijuana use during breastfeeding and to avoid secondhand marijuana smoke exposure.[94] As for alcohol, which rapidly becomes incorporated into milk when consumed by the breastfeeding woman, the AAP discourages consuming more than two standard alcoholic drinks daily and advises that the mother wait at least 2 hours after a single drink before nursing or expressing milk to be fed to the infant.[9] Health Canada advises avoiding alcohol if you are breastfeeding, but provides direction if a breastfeeding mother wishes to drink alcohol on a special occasion to limit intake to one or two drinks per occasion, and to drink after and not before breastfeeding.[8,95]

Impact of Lactation on the Mother

Just as for the infant, breastfeeding has several direct health benefits for the mother. Exclusive breastfeeding to 6 months, for example, is associated with a delay in postpartum resumption of menses and hence improved birth spacing.[96] As reviewed in greater detail below, there is evidence that breastfeeding reduces the risk of breast and ovarian cancer and has a beneficial effect on glucose metabolism and type 2 diabetes.[10,97] No relationship was found between a history of lactation and the risk of osteoporosis. Breastfeeding (any or exclusive) is associated with reduced risk of postpartum depression.[10,98,99] The direction of the association remains unclear. There is also some evidence that breastfeeding improves parental sleep duration and has a positive impact on mood and reaction to stress, again diminishing the chances of postpartum depression development.[100-102]

Fertility

Breastfeeding is accompanied by a period of amenorrhea (lack of menses) and infertility that results, at least in part, from suckling-induced suppression of ovarian activity. Because suckling interferes with the normal pulsatile secretion pattern of hypothalamic gonadotrophin-releasing hormone, breastfeeding also decreases secretion of the gonadotropins luteinizing hormone (LH) and follicle-stimulating hormone (FSH)—at least in the early postpartum period. Nonetheless, normal pulsatile secretion of FSH returns early in lactation, and ovarian follicles may develop; LH secretion, however, remains suppressed for a variable period of time. Because ovarian follicles do not produce normal levels of estradiol while LH remains suppressed, ovulation is also suppressed and lactational amenorrhea is the result.[103,104]

This estimate of contraceptive efficacy of breastfeeding was confirmed by prospective studies of lactating women in both developing and developed countries.[105,106] A study of more than 4,000 mother–infant pairs reported that, for full breastfeeding, cumulative pregnancy rates during lactational amenorrhea ranged from 0.9% to 1.2% in the first 6 months postpartum, but they increased to 6.6% to 7.4% at 12 months postpartum.[105] A decline in sucking stimulus appears to be a critical factor that determines when postpartum ovulation will resume.[103] Thus, supplementation with formula or solid foods during breastfeeding will likely hasten the return of fertility. Kramer and Kakuma, in their Cochrane systematic review, reported that data from two Honduran trials, and from observational studies from Bangladesh and Senegal, suggest that exclusive breastfeeding through 6 months is associated with delayed resumption of menses and greater maternal weight loss, at least in generally malnourished women with high breastfeeding frequency.[107] However, a U.S.-based study provided evidence that well nourished women with shorter durations of amenorrhea appear to lose substantially more body fat during the postpartum period than women with delayed return of ovulation. This relationship does not appear to be related to baseline body fat levels in the early postpartum period.[108] Substantially more research is needed in this area, particularly to understand potential differences amongst populations.

Weight Retention and Cardiometabolic Disease

In the United States, approximately one-third of women have obesity prior to becoming pregnant, and at least half will gain more weight during pregnancy than recommended.[109,110] Although many women express a desire to lose weight after childbirth and return to their prepregnancy weight, postpartum weight loss is highly variable. Weight gained during pregnancy and not lost postpartum contributes to overweight and obesity in women of childbearing age. A woman who is lactating has the same requirement for regulating body weight as one who is not lactating, except that she is producing a continuous supply of milk and, thus, creating higher energy output. The total energy cost to a woman who is exclusively breastfeeding an infant to 6 months postpartum is estimated to be 500 kcal/day. If energy intake and physical activity remain unchanged, theoretically, you could expect a 0.5 kg/week weight loss. The Agency for Healthcare Research and Quality (AHRQ) reviews concluded, however, that there was insufficient evidence as to whether breastfeeding is associated with postpartum weight change.[97,111] They noted that other factors likely have a greater effect on postpartum weight retention, including income, prepregnancy body mass index, ethnicity, gestational weight gain, and energy intake. In general, women included in the studies reviewed did not adhere to recommendations to breastfeed exclusively for 6 months, and it has been argued that breastfeeding of sufficient intensity and duration is necessary to promote accelerated postpartum weight loss.

One of the risk factors for unsuccessful initiation, intensity, and duration of breastfeeding is obesity, which complicates assessment of the possible impact of breastfeeding on postpartum weight retention.[112] Baker and colleagues[113] showed that breastfeeding had a significant independent contribution to postpartum weight loss among women from the Danish National Birth cohort (n = 36,030) whose pregnancy weight gains (~12 kg) were generally inline with recommendations and who breastfed as recommended (exclusive breastfeeding to 6 months and to any extent to 12 months). In fact, by 6 months, postpartum weight retention was eliminated.

Breastfeeding may also have a beneficial effect on maternal glucose and lipid metabolism.[97,111] In women with gestational diabetes, lactation has been associated with improved pancreatic β-cell function. AHRQ reviews concluded that both ever breastfeeding and longer duration of lifetime breastfeeding was associated with a reduced risk of developing type 2 diabetes. Similarly, longer duration of breastfeeding was associated with lower rates of maternal hypertension (low strength of evidence for both outcomes).

Breast and Ovarian Cancers

Systematic reviews of the literature demonstrate an inverse relationship between ever and duration of breastfeeding and risk of breast cancer.[10,111] In their recent review, the AHRQ report a 22% reduction in breast cancer risk for ever breastfeeding compared to never breastfeeding.[111] The principal mechanisms by which breastfeeding is thought to reduce the risk of breast cancer is the delay in re-establishing ovulation, which reduces long-term exposure to reproductive hormones. Additionally, breastfeeding supports differentiation of mammary cells following pregnancy, and differentiated cells are less likely to become cancerous. Finally, apoptosis at the end of lactation is also thought to decrease breast cancer risk through the elimination of cells with DNA damage and mutations.

In addition to a reduction in breast cancer, systematic reviews of the literature also suggest an association between ever breastfeeding and longer duration of breastfeeding and lower risk of ovarian cancer.[10,111] Epithelial ovarian cancers, which account for approximately 90% of all ovarian cancers, have significant histologic differences, suggesting that these cancers may have heterogeneous causes. While the biologic mechanisms of how breastfeeding reduces ovarian cancer risk are not well understood, the leading hypothesis is that ovulation suppression during breastfeeding inhibits epithelial cell division and proliferation, thereby reducing the opportunity to initiate or promote carcinogenesis. This may be especially important in the first few months postpartum when immune function and tumor surveillance mechanisms remain suppressed. It is hypothesized that breastfeeding may play a role in long-term modulation of inflammatory, immune, or metabolic pathways and, in doing so, reduces the risk of ovarian cancer. Some evidence indicates that reproduction-related risk factors are inversely related to risk for nonmucinous tumors (e.g., serous, endometrioid, clear cell) but not for mucinous tumors.[114-116]

Osteoporosis

Calcium resorbed from the bones of the lactating woman is the primary source of calcium in human milk[117,118] and evidence indicates that increasing calcium intake either through diet or supplementation does not prevent resorption.[48,119,120] However, studies consistently show that the skeletal calcium lost during lactation is rapidly regained after weaning.[117,118] The recovery of bone mineral density (BMD) after weaning appears to be influenced by the duration of lactation and postpartum amenorrhea, and recovery varies among skeletal sites.[117] For example, greater BMD increases have been found in the lumbar spine than in the femoral neck during the first 6 months after weaning.[117,121] Bone recovery is complete for most women and occurs even with multiple and closely spaced pregnancies and periods of lactation.[121,122] A study that measured BMDs of 30 Finnish-American women who had given birth to at least six children, and who breastfed each child for at least 6 months, found that repeated pregnancies and lactation, without a recovery interval, were not associated with either lowered BMD or osteoporosis or osteopenia in these women. Moreover, women with 10 or more children did not have lower BMD than women who had six to nine children.[122] In general, duration of lactation does not appear to be associated with increased fracture risk or with osteoporosis later in life.[117]

General Dietary Recommendations for Breastfeeding Women

In general, nutrient requirements increase during lactation to support the synthesis of milk. The amount of additional energy needed by the breastfeeding mother depends on whether she is exclusively breastfeeding or feeding a combination of human milk and infant formula. The National Academy of Medicine recommends an additional 330 kcal/day in the first 6 months postpartum and an additional 400 kcal/day in the second 6 months postpartum.[123]

Recommendations for micronutrients are generally higher than preconception, but similar to those during pregnancy, although some (such as vitamin A) are higher; and others (e.g., folate and iron) are lower. It is also important that lactating women consume enough fluids because a lack of fluid can decrease milk volume. The Adequate Intake (AI) for water is 3.8 liters (13 cups) per day during lactation, which includes fluids in both foods and beverages.

Future Research Directions

Evaluate Strategies to Increase Breastfeeding Initiation and Duration

Additional research effort should focus on evaluating promising strategies to increase breastfeeding rates generally and to address disparities in breastfeeding specifically.[124] While we have much more to learn, there has been more than ample evidence for a very long time to compel action in support of breastfeeding. Yet, less than half of infants globally are breastfed in accordance with WHO recommendations, and one in three infants in low- and middle-income countries receive prelacteal feeds (e.g., mostly water and animal milk)[20,125]; a strong predictor or early cessation of breastfeeding. Only half of infants born in the United States are exclusively breastfed at 3 months of age.[15] Research efforts will require investigation of ongoing and new promising novel policy and community level interventions delivered through healthcare facilities, public health, and community agencies.[126]

Study Human Milk as a Biological System

Breastfeeding clearly provides optimal nutrition and other health benefits for the newborn, as well as health benefits for the lactating mother. Human milk composition and its biologic functions are, however, more than the sum of their parts.[2] Nonetheless, much of the research to date has focused on the effect of a single nutrient or bioactive component in human milk one at a time. There is a need to understand the complexity of human milk composition as a biologic system. Beyond an understanding of the average concentration of key nutrients and biologically active components in human milk required by the infant, a more fulsome appreciation of the maternal and infant inputs that influence milk composition and biology is required.

Could, as posited by Christian et al.,[2] human milk be an exemplar of "precision nutrition" where the milk produced by a mother is specifically tailored for her growing infant in the context in which they both find themselves (e.g., geography, days postpartum, exposure to infectious disease, maternal nutritional status)? Alternatively, does milk composition merely reflect maternal genetics and environment, and the resultant milk variability has little impact on the nutritional status, growth, or development of the nursing baby? Understanding the entire ecosystem of the mother, environment, and infant and how it impacts milk composition is critical to understanding the most fundamental aspects of maternal and infant health, especially in compromised settings. Indeed, this path of investigation holds much promise to improve maternal and child health.

To this end, a variety of emerging multi-omics technologies, systems biology approaches, artificial intelligence methods, and statistical modeling techniques promise rich opportunities for multidisciplinary researchers to unravel the multifunctional components of human milk and their collective role in early child growth and development.

Acknowledgments

The authors acknowledge Mary Francis Picciano for her contributions to the previous version of this chapter.

References

1. Andreas NJ, Kampmann B, Mehring Le-Doare K. Early Hum Dev 2015;91(11):629–35.
2. Christian P, Smith ER, Lee SE, et al. Am J Clin Nutr 2021;113(5): 1063–1072.
3. Dror DK, Allen LH. Adv Nutr 2018;9(Suppl 1):278S–294S.
4. Hunt KM, Foster JA, Forney LJ, et al. PLoS One 2011;6(6): e21313.
5. Carr LE, Virmani MD, Rosa F, et al. Front Immunol 2021;12: 604080.
6. Poulsen KO, Sundekilde UK. Curr Nutr Rep 2021;10(1): 12–29.
7. Vizzari G, Morniroli D, Ceroni F, et al. Children (Basel) 2021; 8(10).
8. Health Canada. Breastfeeding Your Baby, 2022. Accessed March 17, 2023. Available at www.canada.ca/en/public-health/services/health-promotion/childhood-adolescence/stages-childhood/infancy-birth-two-years/breastfeeding-infant-nutrition.html
9. Meek JY, Noble L, Section on Breastfeeding. Pediatrics 2022;150(1).
10. Victora CG, Bahl R, Barros AJ, et al. Lancet 2016;387:475–90.
11. WHO Guideline for complementary feeding of infants and young children 6–23 months of age. Geneva: World Health Organization; 2023. Licence: CC BY-NC-SA 3.0 IGO.
12. Neves PAR, Vaz JS, Maia FS, et al. Lancet Child Adolesc Health 2021;5:619–630.
13. WHO/UNICEF Discussion Paper: The Extension of the 2025 Maternal, Infant and Young Child Nutrition Targets to 2030, 2019. Accessed March 16, 2023. Available at https://data.unicef.org/resources/who-unicef-discussion-paper-nutrition-targets/
14. Centers for Disease Control and Prevention. Breastfeeding Among U.S. Children Born 2012-2019, CDC National Immunization Survey, 2023. Accessed April 10, 2023. Available at www.cdc.gov/breastfeeding/data/nis_data/results.html
15. Centers for Disease Control and Prevention. Breastfeeding Report Card, 2022. Accessed March 17, 2023. Available at www.cdc.gov/breastfeeding/data/reportcard.htm/
16. Government of Canada. Canada's Breastfeeding Dashboard, 2022. Accessed March 17, 2023. Available at https://health-infobase.canada.ca/breastfeeding/
17. Ryan AS, Wenjun Z, Acosta A. Pediatrics 2002;110:1103–9.
18. Ryan AS, Zhou W. Pediatrics 2006;117:1136–46.
19. U.S. Department of Health and Human Services. Healthy People 2023, 2018. Accessed March 17, 2023. Available at https://health.gov/healthypeople
20. Perez-Escamilla R, Tomori C, Hernandez-Cordero S, et al. Lancet 2023;401:472–485.
21. Ramsay DT, Kent JC, Hartmann RA, Hartmann PE. J Anat 2005;206:525–34.
22. Alex A, Bhandary E, McGuire KP. Adv Exp Med Biol 2020; 1252:3–7.

23. Sriraman NK. Curr Probl Pediatr Adolesc Health Care 2017;47:305–310.
24. Neville MC. Pediatr Clin North Am 2001,48.13–34.
25. Neville MC, Morton J. J Nutr 2001;131:3005S–8S.
26. Neville MC, McFadden TB, Forsyth I. J Mammary Gland Biol Neoplasia 2002;7(1):49–66.
27. McManaman JL, Neville MC. Adv Drug Deliv Rev 2003;55:629–41.
28. Neville MC, Allen JC, Watters C. In: Neville MC, Neifert MR, eds. The Mechanisms of Milk Secretion. Lactation. Springer, 1983.
29. National Academies of Sciences Engineering and Medicine. Scanning for New Evidence on the Nutrient Content of Human Milk: A Model for the Derivation of Age-Specific Nutrient Requirements. The National Academies Press, 2020.
30. Chen MY, Northington R, Yan J. FASEB J 2018;31:lb392–lb392.
31. Picciano MF. Pediatr Clin North Am 2001;48:53–67.
32. Institute of Medicine. Dietary Reference Intakes for Energy, Carbohydrate, Fiber, Fat, Fatty Acids, Cholesterol, Protein, and Amino Acids. The National Academy Press, 2005.
33. Institute of Medicine. Dietary Reference Intakes for Vitmin A, Vitamin K, Arsenic, Boron, Chromium, Copper, Iodine, Iron, Manganese, Molybdenum, Nickel, Silicon, Vanadium, and Zinc. The National Academies Press, 2001.
34. Institute of Medicine. Dietary Reference Intakes for Vitamin C, Vitamin E, Selenium, and Carotenoids. The National Academies Press, 2000.
35. Institute of Medicine. Dietary Reference Intakes for Thiamin, Riboflavin, Niacin, Vitamin B6, Folate, Vitamin B12, Pantothenic Acid, Biotin, and Choline. The National Academies Press; 1998.
36. Daniel AI, Shama S, Ismail S, et al. Am J Clin Nutr 2021; 113:1009–1022.
37. McGuire MK, O'Connor DL, eds. Human Milk: Sampling and Measurement of Energy-Yielding Nutrients and Other Macromolecules. Academic Press, 2021.
38. Motil KJ, Sheng HP, Montandon CM, Wong WW. J Pediatr Gastroenterol Nutr 1997;24:10–7.
39. Jensen RG. Lipids 1999;34:1243–71.
40. Jensen RG. Handbook of Milk Composition. Academic Press, 1995.
41. Hamosh M. Pediatr Clin North Am 2001;48(1):69–86.
42. Kodentsova VM, Vrzhesinskaya OA. Bull Exp Biol Med 2006;141:323–7.
43. Haskell MJ, Brown KH. J Mammary Gland Biol Neoplasia 1999; 4:243–57.
44. Hollis BW, Wagner CL, Howard CR, et al. Pediatrics. 2015;136:625–34.
45. Kulie T, Groff A, Redmer J, et al. J Am Board Fam Med 2009;22(6):698–706.
46. Perumal N, Al Mahmud A, Baqui AH, et al. Public Health Nutr 2017;20:1865–1873.
47. Wagner CL, Hulsey TC, Fanning D, et al. Breastfeed Med 2006; 1:59–70.
48. Institute of Medicine. Dietary Reference Intakes: Calcium and Vitamin D. The National Academies Press, 2011.
49. Government of Canada. Sun safety tips for parents, 2022. Accessed Dec 10, 2023. Available at www.canada.ca/en/health-canada/services/sun-safety/sun-safety-tips-parents.html
50. Ward LM, Gaboury I, Ladhani M, Zlotkin S. CMAJ 2007; 177(2):161–6.
51. Golden NH, Abrams SA, Committee on Nutrition. Pediatrics 2014;134:e1229–43.
52. Canadian Paediatric Society. Preventing Symptomatic Vitamin D Deficiency and Rickets among Indigenous Infants and Children in Canada, 2022. Accessed March 16, 2023. Available at https://cps.ca/en/documents/position/vitamin-d-deficiency-and-rickets-among-indigenous-infants-and-children
53. Paediatr Child Health 2007;12:583–98.
54. Ng E, Loewy AD. Paediatr Child Health 2018;23:394–402.
55. Centers for Disease Control and Prevention. Vitamin K Deficiency Bleeding, 2021. Accessed April 17, 2023. Available at www.cdc.gov/ncbddd/vitamink/vitamin-k-fact-sheet-general.html
56. von Schenck U, Bender-Gotze C, Koletzko B. Arch Dis Child 1997; 77:137–9.
57. Houghton LA, Yang J, O'Connor DL. Am J Clin Nutr 2009; 89:216–20.
58. Page R, Robichaud A, Arbuckle TE, et al. Am J Clin Nutr 2017; 105:1101–1109.
59. Dror DK, Allen LH. Adv Nutr 2018;9:347S–357S.
60. Delange F. Proc Nutr Soc 2000;59(1):75–9.
61. Smilowitz JT, Allen LH, Dallas DC, et al. Am J Clin Nutr 2023; 117:S28–S42.
62. Grosvenor CE, Picciano MF, Baumrucker CR. Endocr Rev 1993;14:710–28.
63. Hosea Blewett HJ, Cicalo MC, Holland CD, et al. Adv Food Nutr Res 2008;54:45–80.
64. Gregg B, Ellsworth L, Pavela G, et al. Pediatr Obes 2022; 17(7):e12892.
65. O'Connor DL, Gibbins S, Kiss A, et al. JAMA 2016;316(18): 1897–1905.
66. Quigley M, Embleton ND, McGuire W. Cochrane Database Syst Rev 2018;6:CD002971.
67. Tyebally Fang M, Chatzixiros E, Grummer-Strawn L, et al. Bull World Health Organ. 2021;99:892–900.
68. Pound C, Unger S, Blair B. Paediatr Child Health 2020;25: 549–550.
69. FDA U.S. Food and Drug Administration. Use of Donor Human Milk, 2018. Accessed April 10, 2023. Available at www.fda.gov/science-research/pediatrics/use-donor-human-milk
70. WHO Child Growth Standards: Length/Height-for-Age, Weight-for-Age, Weight-for-Length, Weight-for-Height and Body Mass Index-for-Age Methods and Development (World Health Organization), 2006.
71. Wagner CL, Greer FR, American Academy of Pediatrics Section on B, American Academy of Pediatrics Committee on N. Pediatrics 2008;122:1142–52.
72. Baker RD, Greer FR, Committee on Nutrition American Academy of P. Pediatrics. 2010;126:1040–50.
73. Unger SL, Fenton TR, Jetty R, et al. Paediatr Child Health 2019;24:555–556.
74. Karcz K, Krolak-Olejnik B. Crit Rev Food Sci Nutr 2021;61: 1081–1098.
75. Pawlak R, Vos P, Shahab-Ferdows S, et al. Am J Clin Nutr 2018;108(3):525–531. doi:10.1093/ajcn/nqy104
76. Horta BL, Loret de Mola C, Victora CG. Acta Paediatr. 2015; 104:14–9.
77. Kramer MS, Aboud F, Mironova E, et al. Arch Gen Psychiatry 2008;65:578–84.
78. Anderson JW, Johnstone BM, Remley DT. Am J Clin Nutr 1999;70:525–35.
79. O'Connor DL, Jacobs J, Hall R, et al. J Pediatr Gastroenterol Nutr 2003;37:437–46.
80. Vohr BR, Poindexter BB, Dusick AM, et al. Pediatrics 2007;120:e953–9.
81. Jasani B, Simmer K, Patole SK, et al. Cochrane Database Syst Rev 2017;3:CD000376.
82. Delgado-Noguera MF, Calvache JA, Bonfill Cosp X, et al. Cochrane Database Syst Rev 2015;2015:CD007901.
83. Dewey KG, Gungor D, Donovan SM, et al. Am J Clin Nutr 2021;114:1774–1790.
84. Horta BL, Rollins N, Dias MS, et al. Acta Paediatr 2023;112: 34–41.
85. Beauregard JL, Hamner HC, Chen J, et al. MMWR Morb Mortal Wkly Rep 2019;68:745–748.

86. Gionet L, Statistics Canada. Breastfeeding trends in Canada, 2015. Health at a glance. Accessed March 17, 2023. Available at www.southeasthealthline.ca/healthlibrary_docs/Breastfeeding%20trends%20in%20Canada.pdf
87. Orr SK, Dachner N, Frank L, et al. CMAJ. 2018;190:E312–E319.
88. Satkunam M, Anderson LN, Carsley S, et al. Can J Public Health. 2018;109:489–497.
89. Woo Baidal JA, Locks LM, Cheng ER, et al. Am J Prev Med 2016;50:761–779.
90. Kramer MS, Matush L, Vanilovich I, et al. J Nutr 2009;139: 417S–21S.
91. World Health Organization. Guideline: Updates on HIV and Infant Feeding: The Duration of Breastfeeding, and Support from Health Services to Improve Feeding Practices Among Mothers Living with HIV. Geneva, Switzerland: World Health Organization, 2016. Accessed April 17, 2023. Available at www.ncbi.nlm.nih.gov/books/NBK379874/
92. Lackey KA, Pace RM, Williams JE, et al. Matern Child Nutr 2020;16(4):e13032.
93. Pace RM, Williams JE, Jarvinen KM, et al. mBio 2021;12
94. Ryan SA, Ammerman SD, O'Connor ME, Committee On Substance USE, Prevention, Section On B. Pediatrics 2018;142.
95. Health Nexus Sante. Mixing Alcohol and Breastfeeding. Resource for Mothers and Partners about Drinking Alcohol While Breastfeeding. Accessed April 10, 2023. Available at https://resources.beststart.org/wp-content/uploads/2018/11/A28-E.pdf
96. Calik-Ksepka A, Stradczuk M, Czarnecka K, et al. Int J Mol Sci. 2022;23.
97. Ip S, Chung M, Raman G, et al. Evid Rep Technol Assess (Full Rep) 2007;:1–186.
98. Alimi R, Azmoude E, Moradi M, et al. Breastfeed Med 2022;17: 290–296.
99. Dias CC, Figueiredo B. J Affect Disord 2015;171:142–54.
100. Gila-Diaz A, Herranz Carrillo G, Arribas SM, et al. Children (Basel) 2021;8.
101. Groer MW. Biol Res Nurs 2005;7:106–17.
102. Hamdan A, Tamim H. Int J Psychiatry Med 2012;43:243–59.
103. McNeilly AS. J Mammary Gland Biol Neoplasia 1997;2:291–8.
104. McNeilly AS. Adv Exp Med Biol 2002;503:199–205.
105. World Health Organization Task force on Methods for Natural Regulation of Fertility: Fertil Steril 1999;72:431–40.
106. Kennedy KI, Visness CM. Lancet 1992;339:227–30.
107. Kramer MS, Kakuma R. Cochrane Database Syst Rev 2012; 2012:CD003517.
108. Domer MC, Beerman KA, Ahmadzadeh A, et al. J Hum Lact. 2015;31:282–93.
109. Driscoll AK, Gregory ECW. NCHS Data Brief 2020;392:1–8.
110. Centers for Disease Control and Prevention. Weight gain during pregnancy, 2022. Accessed Dec 10, 2023. Available at www.cdc.gov/reproductivehealth/maternalinfanthealth/pregnancy-weight-gain.htm/
111. Feltner C, Weber RP, Stuebe A, Grodensky CA, Orr C, Viswanathan M. Breastfeeding Programs and Policies, Breastfeeding Uptake, and Maternal Health Outcomes in Developed Countries. AHRQ Comparative Effectiveness Reviews, 2018.
112. Achike M, Akpinar-Elci M. Breastfeed Med 2021;16:678–686.
113. Baker JL, Gamborg M, Heitmann BL, et al. Am J Clin Nutr 2008;88:1543–51.
114. Purdie DM, Siskind V, Bain CJ, et al. Am J Epidemiol. 2001;153:860–4.
115. Titus-Ernstoff L, Perez K, Cramer DW, et al. Br J Cancer. 2001;84:714–21.
116. Tung KH, Goodman MT, Wu AH, et al. Am J Epidemiol. 2003;158:629–38.
117. Kalkwarf HJ, Specker BL. Endocrine 2002;17:49–53.
118. Kovacs CS. J Clin Endocrinol Metab. 2001;86:2344–8.
119. Kalkwarf HJ, Specker BL, Ho M. J Clin Endocrinol Metab 1999;84:464–70.
120. Prentice A, Jarjou LM, Stirling DM, et al. J Clin Endocrinol Metab 1998;83:1059–66.
121. Karlsson C, Obrant KJ, Karlsson M. Osteoporos Int 2001;12: 828–34.
122. Henderson PH, 3rd, Sowers M, Kutzko KE, et al. Am J Obstet Gynecol 2000;182:1371–7.
123. Institute of Medicine. Dietary Reference Intakes. The Essential Guide to Nutrient Requirements. The National Academies Press, 2006.
124. O'Connor DL. Am J Clin Nutr 2021;114:1577–1578.
125. Rollins N, Piwoz E, Baker P, et al. Lancet 2023;401:486–502.
126. Segura-Perez S, Hromi-Fiedler A, Adnew M, et al. Int J Equity Health 2021;20:72

CHAPTER 54

Nutritional Requirements of Infants and Children

Olivia S. Anderson and Melissa J. Slotnick

CHAPTER OUTLINE

Abbreviations

AAP, American Association of Pediatrics; **AI**, adequate intake; **CACFP**, Child and Adult Care Food Program; **DGA**, Dietary Guidelines for Americans; **DRI**, dietary reference intake; **EAR**, estimated average requirement; **EER**, estimated energy requirement; **EPA**, Environmental Protection Agency; **FDA**, Food and Drug Administration; **FNB**, Food and Nutrition Board; **IOM**, Institute of Medicine; **LBW**, low birth weight; **LC-PUFA**, long-chain polyunsaturated fatty acid; **NASEM**, National Academies of Sciences, Engineering, and Medicine; **NSLP**, National School Lunch Program; **PUFA**, polyunsaturated fatty acid; **RDA**, recommended dietary allowance; **SBP**, School Breakfast Program; **SFSP**, Summer Food Service Program; **SNAP**, Supplemental Nutrition Assistance Program; **SSB**, Sugar-Sweetened Beverages; **UL**, tolerable upper intake level; **USDA**, United States Department of Agriculture; **WHO**, World Health Organization; **WIC**, Special Supplemental Nutrition Program for Women, Infants, and Children

Introduction

The nutritional requirements of infants and children reflect this population's unique nutrient needs for growth and development. Moreover, because the metabolic rate of infants and children is greater and the turnover of nutrients more rapid than adults, the nutrient needs for growth and development are superimposed on higher maintenance requirements. This chapter addresses such nutrient needs of infants and children and discusses how current dietary guidelines can help to shape healthy dietary patterns, beginning in infancy.

Dietary Reference Intakes for Infants and Children

Overview of Dietary Reference Intakes

The latest recommendations of the Food and Nutrition Board (FNB) of the National Academies of Sciences, Engineering and Medicine (NASEM),[1-8] are designated Dietary Reference Intakes (DRIs). The reference intakes include the Estimated Average Requirement (EAR) and the Recommended Dietary Allowance (RDA), as well as other "reference intakes" including Adequate Intake (AI) and Tolerable Upper Intake Level (UL).

Estimated Average Requirement (EAR) refers to the daily intake of a specific nutrient that results in some predetermined physiologic outcome. In infants, the outcomes include satisfactory growth rates and development, as well as prevention of specific nutrient deficiencies. The EAR meets the needs of roughly half of the population in which it was established, but not necessarily the needs of the other half. Of note, EARs are not set for infants 0 to 6 months of age.

The Recommended Dietary Allowance (RDA) is the daily intake of a specific nutrient deemed by a scientifically knowledgeable group to meet the "requirement" of most healthy members of a population. In general, if the EAR of a specific population is known and is normally distributed, the RDA is set at the EAR plus two standard deviations. RDAs are useful for assessing the nutrient intakes of individuals; that is, the usual intake of a nutrient at its RDA has a low probability of being inadequate. As with EARs, RDAs are not set for infants 0 to 6 months of age. An EAR for a few nutrients, including iron, zinc, and protein, has been established for the 7- to 12-month old infant, and for several nutrients for children, either directly or by extrapolation from EARs of adults or older children. Hence, for these, an RDA can be (and has been) established.

The Adequate Intake (AI) of a specific nutrient is the daily intake of that nutrient by a group of healthy individuals and is based on the group's observed or approximated intake of that nutrient when sufficient scientific evidence is not available to set an EAR. In general, the content of specific nutrients in the average volume of human breastmilk consumed by healthy, normally growing, exclusively breastfed infants is considered the AI for most nutrients for infants less than 6 months of age. This definition is consistent with national and international recommendations for exclusive breastfeeding for the first 6 months of life.[9,10] For the 7- to 12-month-old infant, the AI of many nutrients is set at the amount of the nutrients in the average volume of breastmilk plus the average amount of complementary foods consumed by healthy, normally growing 7- to 12-month-old infants. The AIs of other nutrients for the 7- to 12-month-old infant are extrapolated from that of the 0- to 6-month-old infant or that of the older child or adult.

The upper tolerable intake level (UL) is the highest daily intake of a specific nutrient that is likely to pose no risk. It is not a recommended level of intake but rather a guide for avoiding excessive intakes and adverse effects secondary to such intakes.

The most recent reference intakes of various nutrients proposed by the FNB for infants and children less than 8 years of age are summarized in **Table 54.1**. The DRIs of some nutrients for the 0- to 6-month old infant, the 7- to 12-month-old infant, the 1- to 3-year-old child, and the 4- to 8-year-old child are discussed briefly in the sections that follow.

Energy

Per unit of body weight, an infant or young child requires at least twice as much energy as the adult (i.e., 80 to 100 kcal/kg/day vs 30 to 40 kcal/kg/day). This greater need reflects the infants' and children's needs for tissue deposition and other special needs for growth and development.

The estimated energy requirement (EER) of the infant and young child proposed by the FNB,[7] that is, the energy intake predicted to maintain energy balance (which is not the same as EAR), is based on analysis of total energy expenditure data obtained by the doubly labeled water method (Total Energy Expenditure (TEE) = 89 × weight (kg) − 100) plus an allowance for energy deposition incident to growth as determined from measurements of weight gain and body composition of normally growing infants and young children.[11]

Equations for predicting the EER (kcal/day) of infants and children less than 3 years of age are as follows:

0 to 3 months (89 × weight of infant − 100) + 175 kcal
4 to 6 months (89 × weight of infant − 100) + 56 kcal
7 to 12 months (89 × weight of infant − 100) + 22 kcal
1 to 3 years (89 × weight of child − 100) + 20 kcal

The EER of the infant younger than 6 months of age determined in this way is close to the mean energy intake of exclusively breastfed infants.

The EER of the 3- to 8-year-old child is also based on total expenditure measured by the doubly labeled water method, plus an allowance for growth (20 kcal/day), and an adjustment for physical activity level. For this age group, the equation predicting TEE differed between boys and girls, and included age, height, and weight. This was adjusted for physical activity level (PA, 1.0 for sedentary to 1.42 [boys] or 1.56 [girls] if very active).

For 3- to 8-year-old boys, the equation for EER (kcal/day) is as follows:

$$\text{EER} = 88.5 - (61.9 \times \text{age [years]}) + \text{PA} \times (26.7 \times \text{weight [kg]} + 903 \times \text{height [m]}) + 20 \text{ kcal}$$

For girls, it is the following:

$$\text{EER} = 135.3 - (30.8 \times \text{age [years]}) + \text{PA} \times (10 \times \text{weight [kg]} + 934 \times \text{height [m]}) + 20 \text{ kcal}$$

Carbohydrates

The AI of carbohydrates proposed by the FNB for a 0- to 6-month-old infant (60 g/day [~10 g/kg/day]) is based on the carbohydrate content from the average intake of human milk.[7] The AI for the 7- to 12-month-old infant (95 g/day [~10.5 g/kg/day]) is based on the average consumption of carbohydrate from human milk plus the average intake of complementary foods. An EAR for carbohydrates for the older child was established by extrapolation from adult requirements. It is 100 g/day for both the 1- to 3-year-old child (8.3 g/kg/day) and the 4- to 8-year-old (5 g/kg/day) child. The RDA for carbohydrate is 130 g/day (10.8 and 6.5 g/kg/day, respectively, for the younger and older child).

Fats

The AI of fat proposed by the FNB for the 0- to 6-month-old (31 g/day [~5 g/kg/day]) is based on the fat content of average intake of human milk.[7] The AI for the 7- to 12-month-old infant (30 g/day [~3.3 g/kg/day],) is based on the average

Table 54.1 Dietary Reference Intakes (DRI) of Nutrients for Infants and Children[a]

	DRI by age group			
Nutrient	0–6 mo (6 kg)	7–12 mo (9 kg)	1–3 y (13 kg)	4–8 y (22 kg)
Energy (kcal [kJ]/24 h)	550 (2,310)	750 (3,013)	1,074 (4,494)	See text
Fat (g/24 h)	31	30	—	—
Linoleic acid (g/24 h)	4.4	4.6	7	10
α-Linolenic acid (g/24 h)	0.5	0.5	0.7	0.9
Carbohydrate (g/24 h)	60	95	130	130
Protein (g/24 h)	9.1	11[a]	13[a]	19[a]
Calcium (mg/24 h)	200	260	700[a]	1,000[a]
Phosphorus (mg/24 h)	100	275	460[a]	500[a]
Magnesium (mg/24 h)	30	75	80[a]	130[a]
Sodium (mg/24 h)	110	370	800	1,000
Chloride (g/24 h)	0.18	0.57	1.5	1.9
Potassium (mg/24 h)	400	860	2,000	2,300
Iron (mg/24 h)	0.27	11[a]	7[a]	10[a]
Zinc (mg/24 h)	2	3[a]	3[a]	5[a]
Copper (μg/24 h)	200	220	340[a]	440[a]
Iodine (μg/24 h)	110	130	90[a]	90[a]
Selenium (μg/24 h)	15	20	20[a]	30[a]
Manganese (mg/24 h)	0.003	0.6	1.2	1.5
Fluoride (mg/24 h)	0.01	0.5	0.7	1.0
Chromium (μg/24 h)	0.2	5.5	11	15
Molybdenum (μg/24 h)	2	3	17[a]	22[a]
Vitamin A (μg/24 h)	400	500	300[a]	400[a]
Vitamin D (μg/24 h)	10	10	15[a]	15[a]
Vitamin E (mg α-TE/24 h)	4	5	6[a]	7[a]
Vitamin K (μg/24 h)	2.0	2.5	30	55
Vitamin C (mg/24 h)	40	50	15[a]	25[a]
Thiamin (mg/24 h)	0.2	0.3	0.5[a]	0.6[a]
Riboflavin (mg/24 h)	0.3	0.4	0.5[a]	0.6[a]
Niacin (mg NE/24 h)	2	4	6[a]	8[a]
Vitamin B_6 (μg/24 h)	0.1	0.3	0.5[a]	0.6[a]
Folate (μg)	65	80	150[a]	200[a]
Vitamin B_{12} (μg/24 h)	0.4	0.5	0.9[a]	1.2[a]
Biotin (μg/24 h)	5	6	8	12
Pantothenic acid (mg/24 h)	1.7	1.8	2	3
Choline (mg/24 h)	125	150	200	250
Water (L/24 h)	0.7	0.8	1.3	1.7

α-*TE*, α-tocopherol equivalent; *NE*, niacin equivalent.
[a]Recommended dietary allowance (RDA); other intakes are adequate intake (AI).

consumption of fat from human milk plus average intake of complementary foods. AIs for fat beyond 1 year of age have not been determined.

The AIs of n-6 polyunsaturated fatty acids (PUFAs; primarily linoleic acid) and n-3 PUFAs (primarily α-linolenic acid) proposed for the 0- to 6-month-old, based on the average consumption of these fatty acids by exclusively breastfed infants, are 4.4 g/day (~0.73 g/kg/day) and 0.5 g/day (~83 mg/kg/day), respectively.[7] Those for the 7- to 12-month-old child, based on the average consumption of these fatty acids from human milk plus complementary foods, are 4.6 g/day (~0.5 g/kg/day) and 0.5 g/day (~56 mg/kg/day), respectively. AIs of these fatty acids for the 1- to 3-year-old child and the 4- to 8-year-old child are based on the median intakes of these fatty acids by children of these age groups reported by the U.S. Continuing Survey of Food Intake by Individuals (1994–96, 1998), and are 7 and 10 g/day (0.58 and 0.5 g/kg/day), respectively, for n-6 PUFAs and 0.7 and 0.9 g/day (58 mg/kg/day and 45 mg/kg/day), respectively, for n-3 PUFAs. On average, AIs of these two fatty acid groups account for 5% to 7% and 0.5% to 1.0% of the EER, respectively.

Concern exists that infants may also require a preformed intake of at least some of the longer-chain, more unsaturated derivatives of linoleic and α-linolenic acids (e.g., arachidonic and docosahexaenoic acids). These fatty acids are present in human milk but, until recently, were not present in formulas. Furthermore, the contents of these fatty acids in plasma and erythrocyte lipids are lower in infants fed unsupplemented formulas, vs breastfed infants or those fed formulas supplemented with these fatty acids.[12,13] However, the results of functional outcome studies of breastfed vs formula-fed infants and infants fed formulas with and without arachidonic and docosahexaenoic acid are inconclusive.[14] Overall, these studies provide little evidence that the absence of these fatty acids in infant formulas is problematic, provided intakes of both linoleic and α-linolenic acid are adequate.[14,15] There is also no convincing evidence that the amounts of long-chain PUFAs (LC-PUFAs) in available supplemented formulas pose safety concerns, and an argument can be made for the likelihood that some infants may benefit from the supplemented fatty acids.

In concert with the recommendation that the dietary fat intake of the general population be reduced to improve cardiovascular health, it has been suggested that this guideline be applied to children 2 years of age and older.[16] However, because fat is a major source of energy, as well as the only source of essential fatty acids, groups responsible for making recommendations for infants and young children have not endorsed this recommendation for those less than 2 years of age. The Acceptable Macronutrient Distribution Range of fat suggested for the 1- to 3-year-old child by the Panel on Macronutrients of the FNB is 30% to 40% of total energy per day.[7] The range suggested for the 4- to 8-year-old child is 25% to 35% of total energy per day (5% to 10% of n-6 and 0.6% to 1.2% as n-3 fatty acids).

Studies of low saturated fat and low cholesterol diets among children 7 months to 11 years of age in the United States indicate that long-term, lower-fat diets reap similar health benefits to that evidenced in adult studies.[17,18] Outcomes include lower LDL cholesterol and ability to make healthier food choices without compromising growth and development.

Protein

The protein needs of the infant and young child, per unit of body weight, are also greater than those of the adult, reflecting primarily the infant's and child's additional needs for growth. The AI of protein established by the FNB for the 0- to 6-month-old infant, 9.1 g/day or approximately 1.52 g/kg/day (assuming a mean weight of 6 kg), is based on the observed mean protein intake of infants fed principally with human breast milk.[7]

EARs for protein intake were established for the 7- to 12-month-old infant as well as the 1- to 3-year-old and 4- to 8-year-old children.[7] These values are based on maintenance protein needs plus the additional need for protein deposition as determined by measurements of body composition of infants and children for normal growth, assuming an efficiency of deposition of dietary protein intake of 56%. The EAR for the 7- to 12-month-old infant is 0.98 g/kg/day, for the 1- to 3-year-old child is 0.86 g/kg/day, and for the 4- to 8-year-old child is 0.76 g/kg/day. Because the calculated coefficient of variation is approximately 12%, RDAs are 1.24 × EAR: 1.2 g/kg/day for the 7- to 12-month-old infant (11 g/day), 1.05 g/kg/day for the 1- to 3-year-old child (13 g/day), and 0.95 g/kg/day for the 4- to 8-year-old child (19 g/day).

AIs of the essential amino acids for the 0- to 6-month-old infant are set at the amounts of each in the amount of human breast milk protein equal to the AI of protein. For the 7- to 12-month-old, 1- to 3-year-old child, and 4- to 8-year-old child, EARs of the essential amino acids are based on the pattern of these amino acids in body protein and the EAR of protein. The AIs of the essential amino acids for the 0- to 6-month-old infant and the EARs of the older infant and young child are shown in **Table 54.2**.

Minerals

Accretion of calcium during infancy and early childhood ranges from 60 to 100 mg/day in children 2 to 5 years of age to 100 to 160 mg/day for those 6 to 8 years of age. The AIs of calcium set by the FNB for the 0- to 6- and 7- to 12-month-old infants are based, respectively, on the amount of calcium in the average intake of principally breastfed infants and the average intakes of calcium from human milk plus complementary foods (200 and 260 mg/day, respectively).[4] Calcium absorption of formula-fed infants is less than that of those breastfed, but the calcium content of formulas is higher, thus, calcium retention of breastfed and formula-fed infants differs minimally. The EARs established

Table 54.2 Dietary Reference Intakes (mg/kg/d) of Essential Amino Acids for Infants and Children

Amino Acid	0–6 mo[a]	7–12 mo[b]	1–3 y[b]	4–8 y[b]
Aromatic amino acids	120	61	46	38
Isoleucine	78	36	28	25
Leucine	139	71	56	47
Lysine	95	66	51	43
Sulfur amino acids	52	32	25	21
Threonine	65	36	27	22
Tryptophan	25	10	7	6
Valine	77	42	32	27

[a]Adequate intake (AI).
[b]Recommended dietary allowance (RDA).

for young children are based on balance studies of calcium leading to sufficient accretion. The EAR of calcium for the 4- to 8-year-old child is 700 mg/day. The EAR for the 1- to 3-year-old child is 500 mg/day and is based on extrapolation from the EAR of the 4- to 8-year-old child. The respective RDAs for calcium are 1,000 and 700 mg/day.

The AI of phosphorus is 100 mg/day for the 0- to 6-month-old infant and 275 mg/day for the 7- to 12-month-old infant.[5] These values are based on the average intake of the 0- to 6-month-old breastfed infant and the combined intake of breast milk and complementary foods of the 7- to 12-month-old infant. EARs of phosphorus were established for the 1- to 3- and 4- to 8-year-old children, based on factorial estimates; these are 380 and 405 mg/day, respectively. RDAs (EAR × 1.20) for phosphorus for the 1 to 3 and 4 to 8-year-old children are 460 and 500 mg/day, respectively.

Trace Minerals and Vitamins

DRIs have been established for all trace minerals except arsenic, boron, nickel, silicon, and vanadium, as well as for all vitamins.[1,6] These recommendations are summarized in Table 54.1. DRIs of major importance are iron, zinc, and vitamin D.

An infant born to a person with adequate iron intake and body levels should have sufficient stores of iron at birth to meet requirements for up to 4 to 6 months. Alternatively, iron deficiency during infancy can occur due to iron deficits of the birthing parent during pregnancy, and, thus, inadequate transfer for fetal storage. Despite the low iron content of human milk, the FNB set the AI of iron for the 0- to 6-month-old infant at the intake of iron by the principally breastfed infant, at 0.27 mg/day.[6] Moreover, the iron content of human milk is more bioavailable than that of formulas. For this reason, only iron-fortified formulas are recommended. The EARs of iron for the 7- to 12-month-old infant, the 1- to 3-year-old child, and the 4- to 8-year-old child are based on a factorial approach, accounting for obligatory losses as well as increases in hemoglobin mass, tissue iron, and storage iron. Assuming 10% bioavailability of iron for infants and children, EARs were set at 6.9, 3, and 4.1 mg/day for the 7- to 12-month-old infant, the 1- to 3-year-old child, and the 4- to 8-year-old child, respectively. RDAs of iron are 11, 7, and 10 mg/day, respectively.

Like other nutrients, the AI of zinc for 0- to 6-month-old infants is based on the mean zinc intake of exclusively breastfed infants.[6] Because the zinc concentration of human milk falls from approximately 4.0 mg/L at 2 weeks postpartum to approximately 1.0 mg/L at 6 months postpartum, the AI of 2 mg/day reflects an average intake of human milk of 0.78 L and a zinc concentration of 2.5 mg/L. EARs of zinc for the 7- to 12-month-old infant, the 1- to 3-year-old child, and the 4- to 8-year-old child are based on factorial analysis or extrapolation from the adult EAR, both of which are similar (2.5 mg/day for the 7- to 12-month-old infant and the 1- to 3-year-old child; 4 mg/day for the 4- to 8-year-old child). RDAs for zinc reflect 1.2 × EAR.

The major function of vitamin D is to maintain serum calcium and phosphorus concentrations within the normal range by enhancing their absorption from the small intestine and to prevent rickets in this population. Vitamin D is present in very few foods naturally; the majority of the body's active vitamin D is synthesized from sterols in skin following exposure to UVB rays from sunlight. Due to public health precautions against sunlight exposure across all life-stages for the prevention of skin cancers, dietary vitamin D reference intakes have increased over recent decades. The most recent AIs established by the FNB for vitamin D throughout infancy is 400 IU/day. Breastfed infants 0 to 6 months of age are recommended to take an oral vitamin D supplement, due to low transfer of vitamin D to breast milk, starting within the first few days of life.[19] Infant formulas typically provide as much as 10 μg/day. The EAR is established at 12.5 μg/day for the 1- to 3 and 4- to 8-year-old children, and is based on the assumption that minimal vitamin D is being obtained by exposure to sunlight.[5] These intakes maintain normal serum 25-hydroxyvitamin D (active vitamin D) values and are not associated with evidence of vitamin D deficiency. Thus, the RDA for children 1 to 8 years of age is set at 15 μg/day, or 1.2 standard deviations of the EAR.

Water and Electrolytes

The AI of water for an infant is based on the average fluid intake of the predominantly breastfed 0- to-6-month-old child (~700 mL/day).[8] Breastfed and formula-fed infants 0- to 6-months old receive enough water through breast milk and formula, respectively, and do not require supplemental water. Exceptions would only occur if vomiting or diarrhea is occurring, and a provision of 150 mL/kg/day would be recommended. The AI of water for the 7- to 12-month-old infant is based on the average fluid intake of the breastfed infant plus complementary foods (~800 mL/day). Intakes of electrolytes by the breastfed and formula-fed infant, as well as children between 1 and 8 years of age fed conventional foods, appear to approximate the DRIs of each (see Table 54.1).

Breast Milk and Infant Formula

It is necessary to meet specific nutrient needs for the infant, who experiences considerable growth as well as developmental advances. In the first 6 months of life, human breast milk or formula should be the infant's sole source of nutrients. We provide this section to explore the unique components of breast milk and infant formula, which are recommended as the sole nutrient source for the first 6 months of life.

Breastfeeding and Breast Milk Composition

Human breast milk is uniquely adapted to the infant's needs. In addition to nutrients, breast milk contains bioactive compounds, such as antimicrobial factors, hormones, growth factors, anti-inflammatory factors, digestive enzymes, and transporters that are protective of the infant's health and support adequate nutrient digestion and absorption in the immature gastrointestinal tract of an infant.[10] Thus, the World Health Organization (WHO) has pointed to breastfeeding as the gold standard for infant feeding and as the "single most effective public health intervention for infant mortality."[10] It is estimated that if optimal breastfeeding was achieved, 820,000 lives of children would be saved per year globally.[20] The WHO and AAP both recommend that the infant breastfeeds exclusively for the first 6 months of life and, after 6 months, up to 2 years of age with complimentary foods.[9,10]

The bioactive factors in breastmilk are well evidenced in the protection against gastrointestinal and respiratory tract infections, as well as otitis media (ear infections) across the globe.[20] Newer evidence shows support for breastfeeding against chronic illness, such as obesity or Type 2 Diabetes later in the infant's life.[20] There is also some evidence that breastfeeding results in higher IQ levels in childhood, which translates to higher earnings over one's life.[20]

Although nutrients in breast milk are adequate for the child to develop, there are a few of concern to discuss. The maternal transfer of vitamin D to breastmilk is low, which evolved as such due to UVB exposure to the sun and biosynthesis of active vitamin D. In the present day, with public health messaging to avoid sun exposure to minimize risk of skin cancer, vitamin D deficiencies are increasing in prevalence among infants and children.[21] Thus, it is recommended to supplement the breastfed infant with 400 IU/day (10 ug/day) through liquid droplets.[19] Iron supplements are recommended only if the birthing parent was diagnosed with iron deficiency anemia, as the iron stores of the infant may be too low to support its growth and development before complementary foods are introduced.[6]

Maternal contraindications to breastfeeding are rare. According to the AAP, the main contraindications are when the breastfeeding parent is HIV positive with a detectable viral load or if the infant is diagnosed with galactosemia.[9] Although in high-income countries it is considered a contraindication for an infant to breastfeed from a parent who is HIV positive, due to increased risk of the infant acquiring the infection, in low-income countries, the benefits of breastfeeding from an HIV positive parent for the infant may outweigh this risk.[22] An infant diagnosed with galactosemia is unable to break down galactose found in breast milk, leading to a dangerous buildup in the infant's blood. Other contraindications to be aware of include the breastfeeding parent having active tuberculosis, substance abuse, taking specific medications, or severe neuroses or psychoses.

Formula Feeding

Numerous, safe formulas that support normal growth and development are available for feeding the infant. Most are available in ready-to-use, concentrated liquid or powdered forms. Powdered formulas are lower in cost and are used with increasing frequency. Term infants require a basic formula modeled after breast milk composition, which universally contains about 20 kcal per ounce. Basic formulas will contain lactose as their carbohydrate source and protein from cow's-milk. There is no evidence to recommend one brand over another; most formulas are nutritionally interchangeable.[23]

Lactose-free bovine milk formulas are available. The most used soy protein formulas contain either sucrose or a glucose polymer. Thus, these formulas or lactose-free bovine milk protein formulas are useful for the infant with either transient or congenital lactase deficiency.

The fat content of both bovine milk and soy protein formula usually comprises approximately 50% of the nonprotein energy. In general, intestinal absorption of the blend of vegetable oils present in current formulas is at least 90%. Formulas supplemented with the LC-PUFAs, docosahexaenoic acid (DHA), and arachidonic acid (AA), have become available, as these fatty acids are found in breast milk but are not in conventional formula. The intent of supplementing formulas with LC-PUFAs is to further support the development of membrane constituents in the central nervous system, but clinical trials of the effects of AA and DHA on neural development are inconsistent and most randomized trials show no benefit.[14] Thus, recent systematic reviews conclude that formula DHA and AA supplementation should not be recommended based on current evidence.[14] Additionally, these supplemented formulas cost more than nonsupplemented formulas.

The electrolyte, mineral, and vitamin contents of most formulas are similar and, when formulas are fed in adequate amounts, they provide the DRIs for minerals and vitamins. Both iron-supplemented (~12 mg/L) and nonsupplemented (~1 mg/L) formulas are available, but it is recommended that all infants receive iron-fortified formula to prevent iron deficiency anemia. Some parents choose these non–iron-supplemented formulas with the belief that iron causes stomach upset. Family physicians should strongly counsel parents to opt for iron-supplemented formulas.

The nutrient content of infant formulas marketed in the United States is regulated by the Infant Formula Act and enforced by the U.S. Food and Drug Administration (FDA). Most industrialized, and many developing, countries have similar regulations. All formulas must contain minimum amounts of all nutrients known or thought to be required by infants, and increasing emphasis is placed on avoiding specified maximum amounts of each. The most recent recommendations for the minimum and maximum nutrient contents of term infant formulas marketed in the United States are shown in **Table 54.3**.[24] The minimum recommended amount of each nutrient is greater than the amount of that nutrient in human milk and, hence, greater than the DRI of that nutrient for infants less than 1 year old (see Table 54.1).

In addition to assuring the FDA that each marketed formula contains the minimum recommended amount and no more than the maximum amount of each nutrient for the intended shelf-life of the formula, manufacturers of infant formulas must also assure that the formula was manufactured safely and hygienically. Thus, each batch of manufactured formula is assayed continually throughout its shelf-life. Manufacturers also are responsible for assuring the FDA that each marketed formula, as the sole source of nutrition, supports normal growth and development of infants for at least the first 4 months of life. This is usually done by conducting growth studies with a new formula during the first 4 months of life in a sufficient number of infants to detect a 3 g/day difference in rate of weight gain. The efficacy and safety of substituting alternative sources of various nutrients also must be demonstrated by appropriate studies.

Cultivating Healthy Eating in Infants and Children

Healthy eating patterns and habits begin to develop early in a child's life, and self-feeding and self-regulation can begin as early as in infancy. For example, infants should be permitted to participate in feeding themselves as soon as they seem physically able to do so, usually before 1 year of age. By the end of the second year of life, toddlers should be largely responsible for feeding themselves. Parents, caregivers, and the child's food environment play large roles in shaping eating habits and diet quality. For example, feeding styles in which parents encourage self-feeding and self-regulation, have been associated with higher diet quality than other feeding styles.[25]

Other social aspects of the home food environment can influence nutrient intake and diet quality in children. For example, meal times that are family-centric, screen-free, and regularly-scheduled have been associated with healthier eating patterns.[25] However, modern lifestyles can make it difficult for

Table 54.3 Nutrient Specifications Per 100 Kilocalories of Infant Formulas According to the FDA

Nutrients (Unit of Measure)	Minimum Level	Maximum Level
Protein (g)	1.8	4.5
Fat (g)	3.3	6
Percent calories of fat (%)	30	54
Linoleic acid (mg)	300	
Percent calories of linoleic acid (%)	2.7	
Vitamins		
Vitamin A (IU)	250	750
Vitamin D (IU)	40	100
Vitamin E (IU)	0.7	
Vitamin K (μg)	4	
Thiamine (Vitamin B_1) (μg)	40	
Riboflavin (Vitamin B_2) (μg)	60	
Vitamin B_6 (μg)	35	
Vitamin B_{12} (μg)	0.15	
Niacin[1] (μg)	250	
Folic acid (Folacin) (μg)	4	
Pantothenic acid (μg)	300	
Biotin[2] (μg)	1.5	
Vitamin C (mg)	8	
Choline[2] (mg)	7	
Inositol[2] (mg)	4	
Minerals		
Calcium (mg)	60	
Phosphorus (mg)	30	
Magnesium (mg)	6	
Iron (mg)	0.15	3
Zinc (mg)	0.5	
Manganese (μg)	5	
Copper (μg)	60	
Iodine (μg)	5	75
Selenium (μg)	2	7
Sodium (mg)	20	60
Potassium (mg)	80	200
Chloride (mg)	55	150

Modified from United States Electronic Code for Federal Regulations National Archives Code of Federal Regulations. (2023) Nutrient Requirements. Vol. eCFR: 21 CFR Part 107 Subpart D.[24]

busy families to prioritize a healthy home food environment. Furthermore, sociodemographic factors can affect food accessibility, cooking at home, food insecurity, and many other aspects of the home food environment. Thus, guidelines, programs, and policies, such as those discussed below, can nurture healthy diets by providing strategies for healthy eating while enriching the home food environment.

A Shift to Dietary Patterns and Guidelines

Dietary patterns, representing what people eat and drink in total, may be better predictors and modulators of disease risk than individual nutrients alone.[26,27] Dietary patterns associated with lower risk of disease include those with higher intake of fruits and vegetables, legumes, low-fat dairy, whole grains, lean meats and poultry, seafood, nuts, and unsaturated vegetable oils; and lower intake of red and processed meats, sugar-sweetened beverages (SSBs), and refined grains.[27] Dietary guidelines today are largely based on scientific understanding of these healthy dietary patterns. This section focuses on dietary guidelines set forth by agencies in the United States, and touches upon some commonalities with international dietary guidelines for young children. Lastly, U.S. federal programs with the potential to improve dietary patterns for children are highlighted.

An overarching principle of the U.S.D.A. Dietary Guidelines for Americans is that everyone can benefit from shifts to healthier dietary patterns, and dietary patterns with nutrient-dense, healthy foods should be consumed at every stage of the lifespan to maintain healthy body weight and reduce disease risk.[27] Among children and adolescents living in the U.S., an estimated 7% have high cholesterol, over 200,000 have diabetes, and about 40% have either overweight or obesity,[27] highlighting the need for a shift to overall healthier dietary patterns across the lifespan.

Dietary Guidelines for Infants and Toddlers

While the U.S.D.A. Dietary Guidelines for Americans are updated every 5 years based on the current scientific evidence, the 2020–2025 Dietary Guidelines for Americans are the first to include recommendations for infants and toddlers. The following guidelines are summarized from these 2020–2025 USDA Dietary Guidelines for Americans titled "Make Every Bite Count."[27] During the first 11 months, U.S.D.A. dietary guidelines are not provided per se but, rather, it is recommended that actions are taken to set infants on a path to a healthy dietary pattern. For the first 6 months of life, breastmilk is recommended exclusively, when possible. If breastmilk is not available, infants should be fed a commercial formula that is iron fortified and formulated specifically for infants. Vitamin D supplementation is not needed for infants receiving only infant formula, but is recommended for infants receiving both breastmilk and formula or for those who are consuming breastmilk exclusively.

Between 4 and 6 months, infants start to show developmental readiness for eating solid foods (e.g., sitting up, exhibiting head and neck control, grasping small objects and bringing them to their mouth), and healthy solid foods can be introduced at this time.[27] A variety of nutrient-dense complementary foods with different tastes and textures should be offered repeatedly. These foods should include iron and zinc-rich foods (e.g., meats, beans, seafood), fruits and vegetables, whole grains, and low-fat dairy. Potentially-allergenic foods (e.g., peanuts, tree nuts, cow's milk products, eggs, wheat, shellfish, fish, and soy) should be introduced during this time. In particular, for infants at high risk of peanut allergy, the U.S. DGA state that peanut-containing foods could be introduced as early as 4 to 6 months, if approved by a healthcare provider, to reduce risk of developing a peanut allergy. Cow's milk, however, should not be introduced as a beverage until at least 12 months.

Continuing past 12 months, infant formula is no longer recommended, less breastmilk should be consumed, and a healthy dietary pattern should supply most of the energy and nutrients needed. Following the U.S. DGA, a healthy dietary pattern should contain a variety of nutrient-dense fruits and vegetables, whole grains, lean meats, seafood, nuts and seeds, dairy, and oils with minimal added sugars and sodium. Recommendations to limit saturated fats do not apply to children under 2, and; therefore, higher-fat dairy is included in guidelines for toddlers 12 to 23 months. The recommended U.S.D.A. healthy dietary pattern for children ages 12 to 23 months is shown in the table below (**Table 54.4**).[27] A healthy vegetarian dietary pattern may also be possible for children ages 12 to 23 months and is included in the

Table 54.4 Healthy U.S.-Style Dietary Pattern for Toddlers Aged 12 through 23 Months Who Are No Longer Receiving Human Milk or Infant Formula, with Daily or Weekly Amounts from Food Groups, Subgroups, and Components

Calorie Level of Pattern[a]	700	800	900	1,000
Food Group or Subgroup[b,c]	**Daily Amount of Food from Each Group[d]** (Vegetable and protein foods subgroup amounts are per week)			
Vegetables (cup eq/day)	**⅔**	**¾**	**1**	**1**
	Vegetable Subgroups in Weekly Amounts			
Dark-Green Vegetables (cup eq/wk)	1	⅓	½	½
Red and Orange Vegetables (cup eq/wk)	1	1¾	2½	2½
Beans, Peas, Lentils (cup eq/wk)	¾	⅓	½	½
Starchy Vegetables (cup eq/wk)	1	1½	2	2
Other Vegetables (cup eq/wk)	¾	1¼	1½	1½
Fruits (cup eq/day)	½	¾	**1**	**1**
Grains (ounce eq/day)	**1¾**	**2¼**	**2½**	**3**
Whole Grains (ounce eq/day)	1½	2	2	2
Refined Grains (ounce eq/day)	¼	¼	½	1
Dairy (cup eq/day)	**1⅔**	**1¾**	**2**	**2**
Protein Foods (ounce eq/day)	**2**	**2**	**2**	**2**
	Protein Foods Subgroups in Weekly Amounts			
Meats, Poultry (ounce eq/wk)	8¾	7	7	7¾
Eggs (ounce eq/wk)	2	2¾	2½	2½
Seafood (ounce eq/wk)[e]	2–3	2–3	2–3	2–3
Nuts, Seeds, Soy Products (ounce eq/wk)	1	1	1¼	1¼
Oils (grams/day)	**9**	**9**	**8**	**13**

[a]Calorie level ranges: Energy levels are calculated based on median length and body weight reference individuals. Calorie needs vary based on many factors. The DRI Calculator for Healthcare Professionals, available at usda.gov/fnic/dri-calculator, can be used to estimate calorie needs based on age, sex, and weight.

[b]Definitions for each food group and subgroup and quantity (i.e., cup or ounce equivalents) are provided in Chapter 1 and are compiled in Appendix 3.

[c]All foods are assumed to be in nutrient-dense forms and are prepared with minimal added sugars, refined starches, or sodium. Foods are also lean or in low-fat forms with the exception of dairy, which includes whole-fat fluid milk, reduced-fat plain yogurts, and reduced-fat cheese. There are no calories available for additional added sugars, saturated fat, or to eat more than the recommended amount of food in a food group.

[d]In some cases, food subgroup amounts are greatest at the lower calorie levels to help achieve nutrient adequacy when a relatively small number of calories are required.

[e]If consuming up to 2 ounces of seafood per week, children should only be fed cooked varieties from the "Best Choices" list in the FDA/EPA joint "Advice About Eating Fish," available at FDA.gov/fishadvice and EPA.gov/fishadvice. If consuming up to 3 ounces of seafood per week, children should only be fed cooked varieties from the "Best Choices" list that contain even lower methylmercury: flatfish (e.g., flounder), salmon, tilapia, shrimp, catfish, crab, trout, haddock, oysters, sardines, squid, pollock, anchovies, crawfish, mullet, scallops, whiting, clams, shad, and Atlantic mackerel. If consuming up to 3 ounces of seafood per week, many commonly consumed varieties of seafood should be avoided because they cannot be consumed at 3 ounces per week by children without the potential of exceeding safe methylmercury limits; examples that should not be consumed include: canned light tuna or white (albacore) tuna, cod, perch, and black sea bass. For a complete list, please see: FDA.gov/fishadvice and EPA.gov/fishadvice.

Reproduced from USDA. (2020) Dietary Guidelines for Americans, 2020–2025.[27]

U.S. DGA. Such diets may need additional nutrient supplement such as iron and vitamin B_{12}.

Dietary guidelines put forth by the American Association of Pediatrics (AAP) echo those stated above.[28] Specifically, it is recommended that an infant consume human milk exclusively for the first 6 months of life, and that "breastfeeding should continue for 2 years or longer, as mutually desired by both mom and baby."[9] If an infant is bottle fed, the bottle should not be propped and only breastmilk or formula should be given to an infant through the bottle. During the first year, a child should only drink breastmilk or formula, and juice should be avoided until the child is a toddler. At the very least, juice should not be introduced until 6 to 9 months, and intake should be limited to 4 to 6 ounces. Sugar-sweetened juices and other beverages should be avoided. The introduction of solid foods can begin at 6 months of age, and a wide assortment of healthy foods with a variety of tastes and textures should be offered. After 9 months, 2 to 3 healthy snacks should be consumed each day. At 1 to 2 years, children should be consuming approximately three meals and two snacks a day, and fats should not be restricted. Balanced snacks should include "a protein food, a fruit and/or vegetable, and/or a grain or starchy vegetable." Toddlers should continue to be offered healthy, nutrient-dense foods including protein-containing foods (such as meat, seafood, eggs, and nuts), dairy products, fruits and vegetables, and whole grains.

There are additional dietary guidelines for infants and toddlers set forth by agencies around the world. Many consistencies are observed in these guidelines and agree with those stated above, as have been detailed by the NASEM in 2020,[29] yet a few differences remain. As stated in the NASEM report, it is consistently recommended that breast milk be provided exclusively to infants until 6 months of age, and that breastfeeding should continue until at least 12 months, but to what age breastfeeding should continue to is inconsistent. Supplemental formula feeding for breastfed infants should be avoided. If breast milk is unavailable, commercial cows-milk based formula is recommended through 12 months of age. Under special circumstances, soy-based formula may be recommended. In the NASEM report,[29] it was also determined that guidelines consistently recommend introduction of plain water after 6 months, no juice in the first 12 months, and no more than 4 ounces of juice per day beyond 12 months. SSBs, coffee, tea, and caffeinated beverages should be avoided for young children, and plain whole cow's milk is consistently recommended in limited amounts after 12 months (although cow's milk recommendations for ages 9 to 12 months are inconsistent). When foods are introduced, a variety of foods, food groups, flavors, and textures should be offered as part of a carefully-planned diet that is low in added sugars and salts. Including iron-rich foods is important, although the age at which such foods are recommended is inconsistent. Lastly, it is consistently recommended to not delay introduction to potentially allergenic foods, although there is some variation as to when and how such foods should be introduced.

Finally, the WHO puts forth recommendations for a healthy diet to guard against malnutrition and noncommunicable diseases.[30] Like those above, these guidelines recommend exclusive breastfeeding during the first 6 months for infants, and recommend breastfeeding "until 2 years of age and beyond." As with other guidelines, the introduction of a variety of nutrient-dense complementary foods low in sugar and salts is recommended starting at 6 months of age.

Dietary Guidelines for Older Children (ages 2–10 years)

The AAP offers specific dietary guidelines for preschoolers that emphasize offering a variety of healthy foods, including fresh fruits and vegetables, low-fat dairy or dairy substitutes, lean proteins, and whole-grain cereals and breads.[28] The AAP recommends that preschoolers should be offered the same healthy meals as the family, but children should choose their own serving size and decide how much and what to eat. Two healthy snacks per day should be offered at regular times. Milk or water should be given as a beverage, and sugary drinks should be limited. The AAP states that it is better to offer whole fruit rather than juice, but if juice is offered, it should be 100% fruit juice, limited to ≤4 ounces per day and served with a meal. These recommendations state that children's diets should include a range of foods from all major food groups (vegetables; fruit; bread, cereal or pasta; protein foods; and dairy products). Diets for children older than 2 years old should be lower in fats and saturated fats, with no more than ⅓ of fat kilocalories coming from saturated fat, and dietary sugar and salt should be limited.

Similarly, the U.S. DGA continue to focus on nutrient-dense foods with little added sugar, saturated fat, and sodium, throughout childhood.[27] There is an emphasis on limiting SSBs, especially through age 8, and encouraging whole fruit intake rather than 100% fruit juice. The guidelines state that although fruit juice can be a part of a healthy dietary pattern, children should consume 4 to 10 ounces maximum per day, depending on their current energy intake level. All children should continue to be offered an assortment of nutrient-dense foods, served multiple ways; snacks present an opportunity to continue to introduce such foods. However it is important to note that overall household dietary patterns are important for children and incorporating nutrient-dense foods and healthy eating behaviors into regular household routines will encourage healthy dietary patterns for children of all ages.

The U.S. DGA state that children ages 2 to 4 should consume approximately 1,000 to 1,400 kilocalories per day for females and 1,000 to 1,600 kilocalories per day for males.[27] All types of physical activity (light to moderate to vigorous) should be encouraged in this age group, and while preschoolers should engage in approximately 3 hours of active play per day, energy intake requirements vary. While a focus on nutrient-dense foods is recommended for all age groups, energy requirements guide dietary recommendations for each age group.

Older children, ages 5 to 8, require approximately 1,200 to 1,800 kilocalories for females and 1,200 to 2,000 kilocalories for males.[27] It is recommended in the U.S. DGA that school-age children participate in a minimum of 60 minutes of "moderate-to-vigorous" physical activity per day, incorporating both aerobic and strengthening activities. Nutrient-dense foods should be continued to be emphasized for older children, but amounts consumed should increase proportionally to increased energy requirement. Healthy dietary patterns recommended by the U.S. DGA for children ages 2 to 8 are presented in **Table 54.5**.

Federal Programs and Dietary Guidelines

Federal nutrition assistance programs in the United States aimed at supporting healthy diets of infants and children[31] are aligned with the above dietary guidelines and are outlined in the U.S. DGA. For example, the Special Supplemental

Table 54.5 Healthy U.S.-Style Dietary Pattern for Children Ages 2 Through 8, With Daily or Weekly Amounts from Food Groups, Subgroups, and Components

Energy (kcal) Level of Pattern[a]	1,000	1,200	1,400	1,600	1,800	2,000
Food Group or Subgroup[b]	**Daily Amount of Food from Each Group** (Vegetable and protein foods subgroup amounts are per week.)					
Vegetables (cup eq/day)	**1**	**1½**	**1½**	**2**	**2½**	**2½**
	Vegetable Subgroups in Weekly Amounts					
Dark-Green Vegetables (cup eq/wk)	½	1	1	1½	1½	1½
Red and Orange Vegetables (cup eq/wk)	2½	3	3	4	5½	5½
Beans, Peas, Lentils (cup eq/wk)	½	½	½	1	1½	1½
Starchy Vegetables (cup eq/wk)	2	3½	3½	4	5	5
Other Vegetables (cup eq/wk)	1½	2½	2½	3½	4	4
Fruits (cup eq/day)	**1**	**1**	**1½**	**1½**	**1½**	**2**
Grains (ounce eq/day)	**3**	**4**	**5**	**5**	**6**	**6**
Whole Grains (ounce eq/day)	1½	2	2½	3	3	3
Refined Grains (ounce eq/day)	1½	2	2½	2	3	3
Dairy (cup eq/day)	**2**	**2½**	**2½**	**2½**	**2½**	**2½**
Protein Foods (ounce eq/day)	**2**	**3**	**4**	**5**	**5**	**5½**
	Protein Foods Subgroups in Weekly Amounts					
Meats, Poultry, Eggs (ounce eq/wk)	10	14	19	23	23	26
Seafood (ounce eq/wk)[c]	2–3[d]	4	6	8	8	8
Nuts, Seeds, Soy Products (ounce eq/wk)	2	2	3	4	4	5
Oils (grams/day)	**15**	**17**	**17**	**22**	**22**	**24**
Limit on Calories for Other Uses (kcal/day)[e]	**130**	**80**	**90**	**150**	**190**	**280**
Limit on Calories for Other Uses (%/day)	13%	7%	6%	9%	10%	14%

[a]Calorie level ranges: Ages 2 through 4, Females: 1,000 to 1,400 calories; Males: 1,000 to 1,600 calories. Ages 5 through 8, Females: 1,200 to 1,800 calories; Males: 1,200 to 2,000 calories. Energy levels are calculated based on reference height (median) and reference weight (healthy) corresponding with a healthy body mass index (BMI). Calorie needs vary based on many factors. The DRI Calculator for Healthcare Professionals, available at nal.usda.gov/fnic/dri-calculator, can be used to estimate calorie needs based on age, sex, height, weight, and activity level.

[b]Definitions for each food group and subgroup and quantity (i.e., cup or ounce equivalents) are provided in Chapter 1 and are compiled in Appendix 3.

[c]The U.S. Food and Drug Administration (FDA) and the EPA provide joint advice regarding seafood consumption to limit methylmercury exposure for children. Depending on body weight, some children should choose seafood lowest in methylmercury or eat less seafood than the amounts in the Healthy US-Style Dietary Pattern. More information is available on the FDA and EPA websites at FDA.gov/fishadvice and EPA.gov/fishadvice.

[d]If consuming up to 2 ounces of seafood per week, children should only be fed cooked varieties from the "Best Choices" list in the FDA/ EPA joint "Advice About Eating Fish," available at FDA.gov/fishadvice and EPA.gov/fishadvice. If consuming up to 3 ounces of seafood per week, children should only be fed cooked varieties from the "Best Choices" list that contain even lower methylmercury: flatfish (e.g., flounder), salmon, tilapia, shrimp, catfish, crab, trout, haddock, oysters, sardines, squid, pollock, anchovies, crawfish, mullet, scallops, whiting, clams, shad, and Atlantic mackerel. If consuming up to 3 ounces of seafood per week, many commonly consumed varieties of seafood should be avoided because they cannot be consumed at 3 ounces per week by children without the potential of exceeding safe methylmercury limits; examples that should not be consumed include: canned light tuna or white (albacore) tuna, cod, perch, black sea bass. For a complete list, please see: FDA.gov/fishadvice and EPA.gov/fishadvice.

[e]Foods are assumed to be in nutrient-dense forms; lean or low-fat; and prepared with minimal added sugars, refined starches, saturated fat, or sodium. If all food choices to meet food group recommendations are in nutrient-dense forms, a small number of calories remain within the overall limit of the pattern (i.e., limit on calories for other uses). The number of calories depends on the total calorie level of the pattern and the amounts of food from each food group required to meet nutritional goals. Calories up to the specified limit can be used for added sugars and/or saturated fat, or to eat more than the recommended amount of food in a food group.

Reproduced from USDA. (2020) Dietary Guidelines for Americans, 2020–2025.[27]

Nutrition Program for Women, Infants, and Children (WIC) supports healthy dietary guidelines by providing iron-fortified formula, nutrient-dense foods for older infants and toddlers, and nutrition education and counseling services. The Child and Adult Care Food Program (CACFP) administers reimbursements for nutritious meals and snacks to qualifying infants, toddlers, and children in childcare centers, after school programs, and emergency shelters. The Supplemental Nutrition Assistance

Program (SNAP) provides temporary monthly benefits to eligible families that can be used to access foods supporting health dietary patterns. The National School Lunch Program (NSLP) and School Breakfast Program (SBP) can shape developing dietary patterns, as they may provide up to half of daily energy intake for millions of children in the United States.[32] Additionally, the Summer Food Service Program (SFSP) helps to provide nutritious meals and healthy snacks during the summer months for children residing in areas of low income.

Most recently, corresponding with the National Strategy on Hunger, Nutrition, and Health, the 2022 US White House Conference on Hunger, Nutrition, and Health outlined the administration's goal of "ending hunger and increasing healthy eating and physical activity by 2030."[33,34] Recognizing a gap between the recommended U.S. DGA and actual dietary patterns of Americans, the National Strategy on Hunger, Nutrition, and Health aims to equitably increase food security through access and affordability, prioritize nutrition in overall health, promote healthy choices and physical activity, and further research to support equitable policies.[33] Prioritizing these issues has the potential to influence dietary patterns for children through actionable goals such as providing free school meals for all students, increasing access to federal nutrition assistance programs, improving breastfeeding support, and expanding nutritional support for low-income families.[33]

Lastly, detectable levels of lead, arsenic, mercury, and cadmium in baby foods[35,36] has led to a call for federal action in the United States to monitor and reduce heavy metals in foods that young children eat.[37,38] Subsequently, the US FDA launched a "Closer to Zero" action plan, setting a goal to reduce arsenic, lead, cadmium, and mercury exposure in infants and young children through regulatory standards. In collaboration with other federal agencies, the program also aims to increase research and review evidence related to exposure and potential mitigation strategies.

Heavy Metals and Childhood Nutrition

Children are vulnerable to environmental exposures at different developmental stages due to both behavior and physiology[39]; furthermore, exposure to toxicants in childhood can disrupt critical development.[40] Specifically, exposure to heavy metals during childhood has been associated with various neurologic, behavioral, growth, and developmental effects.[41] Although heavy metal exposure may occur through different pathways, tap water and diet are two possible exposure routes for children.[41] While public drinking water supplies are regulated by the U.S. EPA and routinely tested for many metals,[42] lead-laden plumbing components can contribute to lead exposure at the tap, particularly in socioeconomically disadvantaged communities with an aging drinking water infrastructure.[43] In the United States, private wells are not regulated and, therefore, should be routinely tested for contaminants by homeowners, as advised by the U.S. EPA.[42] The AAP also provides guidance on water testing for families on private wells and recommends that families contact their local drinking water supplier to learn more about local drinking water quality.[28]

Dietary exposure to arsenic, lead, cadmium, and mercury are of concern for children due to measured levels of these metals in foods that children consume.[35,36,44,45] The AAP provides guidelines for reducing infant and toddler dietary exposure to heavy metals.[28] Serving a variety of foods (fruits, vegetables, grains, and proteins), reading labels of commercial baby food blends to ensure that these foods contain a variety of ingredients, and washing fresh fruits and vegetables in cool water before eating is recommended by the AAP to minimize exposure. Arsenic levels can be high in some rice products due to groundwater arsenic sources.[46] Therefore, serving a variety of whole grain cereals (including oats, barley, bulger, and quinoa) while minimizing rice milk and brown rice syrup is recommended by the AAP.[28] To limit exposure to other heavy metals, AAP recommends that babies should be breastfed whenever possible, fruit juices should be avoided, and certain types of fish should be limited. In general, to avoid methylmercury exposure, fish lower on the food chain such as light tuna, salmon, cod should be consumed while avoiding larger predatory fish such as shark, orange roughy, and swordfish.[28,44] Local fish consumption guidelines should also be referred to when consuming locally sourced fish. These fish recommendations are echoed by the U.S. FDA and U.S. EPA[47] for breastfeeding women and children ages 1 to 11, and support guidelines in the U.S. DGA.[27]

A healthy dietary pattern may also help to modulate the toxicity of some heavy metals and other toxicants through anti-inflammatory pathways as well as mechanistic interactions.[48] For example, folic acid and other nutrients may help to facilitate arsenic excretion,[49] and several nutrient components of healthy diets (e.g., calcium, iron, zinc, selenium, vitamin C) may reduce susceptibility to lead exposure.[50] While more research is needed on the role of nutrition in modulating heavy metal exposure,[51,52] the U.S. EPA currently provides guidelines for WIC participants on the importance of a healthy diet for mitigating lead exposure.[53] These recommendations include eating foods high in vitamin C (e.g., fruits and peppers), calcium (e.g, milk products, green leafy vegetables) and iron (e.g., legumes, iron-fortified breakfast cereals).[53] Additionally, consistent food intake is important for minimizing lead exposure,[53,54] highlighting the importance of programs to increase food security and provide healthy diets for all children.

Conclusions

It is critical to give special attention to nutrient needs and dietary patterns of infants and children to ensure proper development, health, and growth during childhood while instilling healthy eating behaviors for life. Breastfeeding, when possible, alongside a varied diet rich in nutrient-dense foods that are low in added sugars and salt, can help to provide a solid nutritional foundation while minimizing exposure to toxicants. Governmental and institutional policies that support breastfeeding as well as equitable access to healthy foods for infants and children have the potential to improve dietary patterns during these critical developmental stages.

Acknowledgments

The authors acknowledge William C Heird, the author of the previous version of this chapter, for his contributions.

References

1. National Academies of Science Engineering and Medicine. Dietary Reference Intakes for Vitamin C, Vitamin E, Selenium, and Carotenoids. National Academies Press, 2000.
2. National Academies of Sciences E and Medicine. In: Stallings VA, Harrison M, and Oria M, eds. Dietary Reference Intakes for Sodium and Potassium. Washington, DC: The National Academies Press, 2019;594.
3. National Academies of Science. Dietary Reference Intakes for Thiamin, Riboflavin, Niacin, Vitamin B6, Folate, Vitamin B_{12}, Pantothenic Acid, Biotin, and Choline. | National Academies Press, 1998.
4. National Academies of Science. Dietary Reference Intakes for Calcium and Vitamin D. | National Academies Press, 2011.
5. National Academies of Science. Dietary Reference Intakes for Calcium, Phosphorus, Magnesium, Vitamin D, and Fluoride. The National Academies Press. National Academies Press, 1997.
6. National Academies of Science. Dietary Reference Intakes for Vitamin A, Vitamin K, Arsenic, Boron, Chromium, Copper, Iodine, Iron, Manganese, Molybdenum, Nickel, Silicon, Vanadium, and Zinc. National Academies Press, 2001.
7. National Academies of Science. Dietary Reference Intakes for Energy, Carbohydrate, Fiber, Fat, Fatty Acids, Cholesterol, Protein, and Amino Acids. The National Academies Press, 2005.
8. National Academies of Science. Dietary Reference Intakes for Water, Potassium, Sodium, Chloride, and Sulfate. National Academies Press, 2005.
9. Meek JY, Noble L. Pediatrics 2022;150.
10. World Health Organization. Breastfeeding. 2023.
11. Butte NF, Hopkinson JM, Wong WW, et al. Pediatr Res 2000;47:578–85
12. Ponder DL, Innis SM, Benson JD, et al. Pediatr Res 1992;32: 683–7.
13. Agostoni C, Riva E, Bellù R, et al. J Am Coll Nutr 1994;13: 658–64.
14. Simmer K, Patole SK, Rao SC. Cochrane Data Syst Rev 2011.
15. Gibson RA, Chen W, Makrides M. Lipids 2001;36:873–83.
16. Gidding SS, Dennison BA, Birch LL, et al. Circ 2005;112:2061–75.
17. Dorgan JF, McMahon RP, Friedman LA, et al. J Clin Endocrinol Metab 2006;91:3992–6.
18. Talvia S, Lagström H, Räsänen M, et al. Arch Pediatr Adolescent Med 2004;158:41–47.
19. Wagner CL, Greer FR, Breastfeeding So, et al. Pediatr 2008; 122:1142–52.
20. Victora CG, Bahl R, Barros AJ, et al. Lancet 2016;387:475–90.
21. Balk SJ, Health CoE, and Dermatology So. Pediatr 2011;127: e791–e817.
22. Eidelman AI, Schanler RJ, Johnston M, et al. Pediatr 2012;129: e827–e41.
23. O'Connor NR. Am Fam Physician 2009;79:565–70.
24. National Archives Code of Federal Regulations. Nutrient Requirements. eCFR: 21 CFR Part 107 Subpart D, 2023. www.accessdata.fda.gov/scripts/cdrh/cfdocs/cfcfr/cfrsearch.cfm?fr=107.100
25. Haines J, Haycraft E, Lytle L, et al. Appetite 2019;137:124–33.
26. Neuhouser ML. Nutr Res 2019;70:3–6.
27. U.S. Department of Agriculture. Dietary Guidelines for Americans, 2020-2025, 2020.
28. American Academy of Pediatrics. Healthy Children, 2023. Accessed December 10, 2023. Available at www.healthychildren.org/
29. National Academies of Sciences. Feeding Infants and Children from Birth to 24 Months: Summarizing Existing Guidance. Washington DC: The National Academies Press, 2020.
30. World Health Organization. Healthy Diet, 2020. Accessed December 10, 2023. Available at www.who.int/news-room/fact-sheets/detail/healthy-diet
31. U.S. Department of Agriculture. Food and Nutrition Service Nutrition Programs, 2023. Accessed December 10, 2023. Available at www.fns.usda.gov/programs
32. Cullen KW, Chen T-A. Prev Med Rep 2017;5:82–5.
33. Psota T, Maniscalco S, Kijek M, et al. Am J Clin Nutr S0002-9165 (23) 00540.
34. The White House. The Biden-Harris Administration National Strategy on Hunger, Nutrition, and Health. 2022.
35. Gardener H, Bowen J, Callan SP. Sci Total Environ 2019;651: 822–7.
36. Healthy Babies Bright Futures. What's in My Baby's Food. 2019.
37. US House of Representatives. Baby foods are tainted with dangerous levels of arsenic, lead, cadmium. and mercury. Staff Report, Subcommittee on Economic and Consumer Policy. 2021.
38. American Academy of Pediatrics. AAP: Babies should eat a varied diet to protect against heavy metals in commercial, homemade food. 2022.
39. Moya J, Bearer CF, Etzel RA. Pediatrics 2004;113:996–1006.
40. Vrijheid M, Casas M, Gascon M, et al. Int J Hyg Environ Health 2016;219:331–42.
41. Al Osman M, Yang F, Massey IY. Biometals 2019;32:563–73.
42. US Environmental Protection Agency. Ground Water and Drinking Water. 2023.
43. Roy S, Edwards MA. Curr Opin Environ Sci Health 2019;7: 34–44.
44. Counter SA, Buchanan LH. Toxicol Applied Pharmacol 2004;198:209–30.
45. Ventre S, Desai G, Roberson R, et al. Curr Probl Pediatr Adolesc Health Care, 101276, 2022.
46. Zavala YJ, Duxbury JM. Environ Sci Technol 2008;42:3856–60.
47. US Food and Drug Administration. Advice about Eating Fish. 2022.
48. Hoffman JB, Hennig B. Ann New York Acad Sci 2017;1398: 99–107.
49. Abuawad A, Bozack AK, Saxena R, et al. Toxicol 2021;457: 152803.
50. Ahamed M, Siddiqui MKJ. Clin Nutr 2007;26:400–8.
51. Mayne ST. Am J Clin Nutr 2023;S0002-9165 (23) 04150
52. Desai G, Anzman-Frasca S, Vernarelli JA, et al. Acad Pediatr 2021;21:471–9.
53. U.S. Department of Agriculture. Protect against Lead Exposure with WIC Foods, 2023. Available at https://wicworks.fns.usda.gov/resources/protect-against-lead-exposure-wic-foods
54. Liu J, McCauley L, Compher C, et al. Environ Health 2011;10: 1–8.

CHAPTER 55

Nutrition in Adolescence

Kathleen L. Keller and Jennifer L. Temple

CHAPTER OUTLINE

Abbreviations

AI, adequate intake; **BMD**, bone mineral density; **BMI**, body mass index; **CVD**, cardiovascular disease; **DGA**, Dietary Guidelines for Americans; **DRI**, dietary reference intake; **EAR**, estimated average requirement; **FFM**, fat-free mass; **FM**, fat mass; **FSH**, follicle-stimulating hormone; **GABA**, Gamma aminobutyric acid; **GnRh**, gonadotropin-releasing hormone; **HDL-C**, high-density lipoprotein cholesterol; **IGF**, insulin-like growth factor; **IR**, insulin resistance; **LH**, luteinizing hormone; **NHANES**, National Health and Nutrition Examination Survey; **OSFED**, Other Specified Feeding and Eating Disorder; **SAT**, subcutaneous adipose tissue; **SSB**, sugar-sweetened beverage; **T2D**, type 2 diabetes

General Definitions and Overview

What Is Adolescence?

Adolescence is the transition period between childhood and adulthood, typically ranging from ages 10 to 21 years. This period can be further divided into early adolescence (10–14 years), middle adolescence (15–19 years), and the transition from late adolescence to young adulthood (20–24 years).[1] The growth in physical size that occurs during this period, which is most profound in early adolescence, is second only to the amount of growth experienced in infancy. Adolescents gain approximately 20% of their adult height, 50% of their adult weight, and experience increases in bone mass by as much as 40%.[2] These changes in physical growth and development contribute to increased demand for both macro- and micronutrients that ideally are met by increases in intake of nutrient-dense foods. However, challenges to optimal nutrition often arise during this time due, in part, to the dramatic changes that occur in psychosocial, neurocognitive, and behavioral domains as youth develop greater autonomy from their family units. Adolescents are known to engage in risk-taking behaviors that can contribute to suboptimal nutrition. Of public health concern, metabolic disorders such as type 2 diabetes (T2D), which previously were found primarily in adults, are increasingly diagnosed among youth.[3] Adolescents need appropriate levels of support to help them learn to make healthful lifestyle choices, so they can reach their growth potential and avoid future chronic diseases. This chapter will cover the following learning objectives:

- Understand the primary physical, physiologic, and psychosocial changes that occur during adolescence and review the implications for each of these on the ability to meet nutritional requirements.

- Discuss the Dietary Reference Intakes (DRIs) and Dietary Guidelines for Americans (DGAs) for adolescents, and relate nutritional recommendations to physiologic changes.
- Define the major trends in eating behavior among adolescents and discuss how these behaviors affect nutritional status and growth during this period.
- Review special considerations for adolescents whose circumstances require unique nutritional approaches, including dieting, eating disorders, pregnancy, obesity, the teenage athlete, and sleep.

Early adolescence also encompasses the majority of physical changes that occur with *puberty*. During *puberty*, secondary sexual characteristics develop, and an individual becomes capable of sexual reproduction. Females typically enter puberty approximately 2.5 years earlier than males[4] although, for both sexes, the age at which puberty begins has decreased worldwide over the past 100 years.[5] There are also racial disparities in pubertal timing, with non-Hispanic Black females experiencing earlier development of breasts and pubic hair, compared with non-Hispanic White females.[6] The appearance of secondary sexual characteristics at an earlier age than expected, i.e., precocious puberty, varies across populations. Secular changes in pubertal development are thought to be driven, in part, by increases in adiposity that have occurred among adolescents at the population level.[7] While the sequence of events that occur during puberty is consistent, the pattern and timing of physical changes varies widely for males and females, primarily because of the differential effects of the hormones, estrogen, and testosterone. The implications of these variations for nutrition are that physical growth and development are better indicators of nutrient needs than chronologic age.

Physiologic Changes that Occur with Puberty

The timing of puberty has important implications for nutrition, as precocious puberty has adverse effects on both physical growth[8] and psychosocial development.[9,10] Similarly, nutritional status is one of the most important determinants of pubertal timing.[11] The changes that signal the onset of puberty involve complex coordination of somatotrophic peptides (e.g., growth hormone and insulin-like growth factors [IGFs]), gonadotrophic hormones (e.g., estrogen and testosterone) and adipostatic hormones (e.g., leptin) in the hypothalamus (**Figure 55.1**).[12] During the majority of childhood, the neurotransmitter Gamma aminobutyric acid (GABA) in the central nervous system suppresses activity of the hypothalamic-pituitary-gonadal axis. At puberty, excitatory neurotransmitters signal the release of gonadotropin-releasing hormone (GnRh) in the hypothalamus, which subsequently stimulates secretion of the gonadotropins, luteinizing hormone (LH), and follicle-stimulating hormone (FSH). Increased concentrations of LH and FSH stimulate maturation of the gonads and production of the sex steroid hormones, testosterone (responsible for the development of secondary sex characteristics that occur in males) and estrogen (responsible for the development of secondary sex characteristics in females). Estrogen and progesterone (synthesized in the corpus luteum and released in response to LH) control the menstrual cycle and development of secondary sex characteristics in females.

Neurons from the hypothalamus release gonadotropin-releasing hormone (GnRH), which stimulates release of gonadotropin hormones, luteinizing hormone (LH), and follicle stimulating

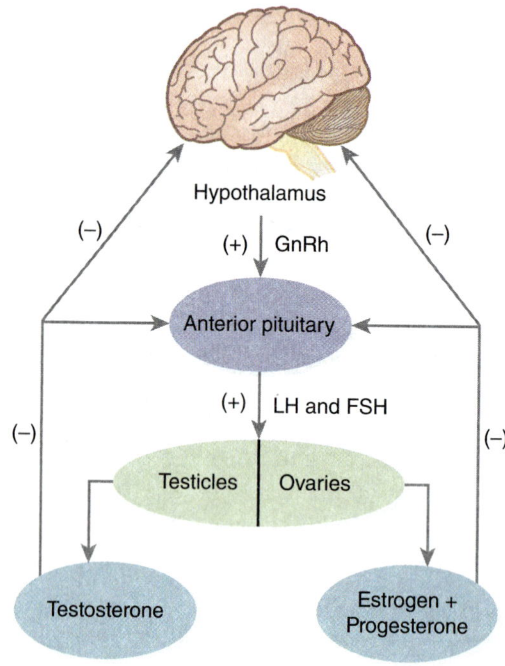

Figure 55.1 The mature hypothalamic-pituitary-gonadal axis.

hormone (FSH) from the anterior pituitary. FSH stimulates the ovaries to release estrogen and progesterone and the seminiferous tubules in the testicles to produce testosterone. Sex steroids (estrogen and testosterone) exert negative feedback signals on the anterior pituitary and hypothalamus. During puberty, linear growth of the skeleton occurs through the process of endochondral ossification. This process starts when insulin-like growth factor produced by both bone and liver stimulates cartilage-producing cells, called chondrocytes, to proliferate within the growth plate. These chondrocytes continue increases in height and promote invasion of blood vessels that produce an extracellular matrix rich in collagen and cytokines. In addition, growth factors within this matrix promote mineralization, as the cartilage is eventually replaced by new bone.[12] Because half of peak bone mass is acquired during adolescence, both calcium requirement and absorption increase during this time.[13]

Undernutrition has been recognized as a key regulator of sexual maturation in both males and females. The discovery of leptin, a peptide hormone secreted from adipocytes, has shed light on this relationship.[14] Animals and humans with a defect in the gene that secretes leptin experience both extreme obesity and infertility. This is, in part, because leptin is necessary for the function of the GnRh pulse generator responsible for pulsatile secretion of sex hormones.[15] Undernourished males and females experience delays in sexual maturation that can be partly explained by reduction in leptin as a response to low body fat. Similarly, overnutrition associated with increased adiposity gain in childhood, has been associated with early sexual maturation.[7] The extent to which puberty onset is the result of obesity itself, or to the effects of hormones released from adipocytes, is unclear. Multiple population-based studies have demonstrated a link between childhood adiposity gain and early sexual maturation in females.[16] Some studies have also demonstrated a similar inverse association between adiposity and pubertal timing in males, although the findings have not been as straightforward as they are in females.[17-19] Because the prevalence of obesity has increased among youth over the past four decades,[20] it has been hypothesized that increasing adiposity may influence the onset of puberty in both males and females by stimulating the hypothalamic-pituitary axis.

Sexual Maturity Ratings: Tanner Stages

Although the timing of major pubertal milestones in females and males varies substantially, the sequence of events that occurs during puberty is consistently observed. To illustrate the importance of this variation for determining nutrient needs, one can compare two females, both 12 years of age, where female A has completed her growth spurt, but female B is still in the prepubertal stage. Female A would likely require less energy for growth, but she may require additional micronutrients, such as iron, to account for blood losses during menstruation, compared with female B. As one quickly realizes, chronologic age is irrelevant because of the substantial variation in onset and timing of pubertal events. During puberty, sexual maturation is more important in assessing nutrient needs, growth, and development than chronologic age.

The gold standard method for assessing pubertal development by health professionals is by evaluating Tanner stages, named for James Tanner, the pediatrician who first described these stages.[21] Trained evaluators can use a series of photographs, called Tanner's Sexual Maturation Scale (SMS)[22,23] that correspond to the Tanner stages in **Table 55.1**. These scales rate pubertal development based on secondary sex characteristics: testicular and penile development and appearance of pubic hair in males, and breast development and appearance of pubic hair in females. Tanner stage 1 signifies the prepubertal stage, Tanner stages 2 to 5 depict various stages of puberty, with Tanner stage 5 signifying the completion of puberty (Table 55.1, Figure 55.1). Typically, Tanner staging is conducted by health-care professionals in a pediatrician's office, but self- and parental assessment tools have been validated for use in research when hiring medical personnel is less feasible.[24]

The onset and length of the Tanner stages vary by race and ethnicity, particularly in females. Non-Hispanic Black females enter puberty at an earlier age than non-Hispanic White females.[25] Nearly 50% of non-Hispanic Black females have been reported to be in Tanner stage 2 by 8 years of age. However, by menarche, the age of the first menstrual period, the differences between non-Hispanic Black and non-Hispanic White females are less pronounced.[25] Early puberty is a risk factor for future development of insulin resistance (IR), cardiovascular disease (CVD), and other chronic diseases.[26] Therefore, determining the reasons for early puberty has clinical relevance. Increasing prevalence of obesity can have implications for accuracy of determination pubertal onset. When obesity coincides with accumulation of fat in the breasts, self-assessment of the Tanner stage can be compromised.[27]

Table 55.1 Tanner Staging in Adolescent Females and Males

Females		
Stage	**Breast Development**	**Pubic Hair Growth**
1	Prepubertal; nipple elevation only	Prepubertal; no pubic hair
2	Small, raised breast bud	Sparse growth of hair along labia
3	General enlargement of/raising of breast and areola	Pigmentation, coarsening and curling, with an increase in amount
4	Further enlargement with projection of areola and nipple as secondary mound	Hair resembles adult type, but not spread to medial thighs
5	Mature, adult contour, with areola in same contour as breast and only nipple projecting	Adult type and quantity, spread to medial thighs
Males		
Stage	**Genital Development**	**Pubic Hair Growth**
1	Prepubertal; no change in size or proportion of testes, scrotum, and penis from early childhood	Prepubertal; no pubic hair
2	Enlargement of scrotum and testes; reddening and change in texture in skin of scrotum; little or no penis enlargement	Sparse growth of hair at base of penis
3	Increase first in length, then width of penis; growth of testes and scrotum	Darkening, coarsening and curling, increase in amount
4	Enlargement of penis with growth in breadth and development of glands; further growth of testes and scrotum, darkening of scrotal skin	Hair resembles adult type, but not spread to medial thighs
5	Adult size and shape genitalia	Adult type and quantity, spread to medial thighs

Reproduced from Tanner JM. Growth at Adolescence. 2nd ed. Oxford: Blackwell Scientific, 1962.

Body Composition Changes

Marked changes occur in both height and body composition during adolescence, both of which have major implications for determining energy and nutrient requirements and may subsequently affect body image and food choices. During this time, muscle mass in males increases and shoulders broaden, whereas females increase body fat and develop rounder hips and smaller waists. The pattern and rate of development in body composition differ in males and females. Females attain peak height growth velocity at 11.5 years of age, vs 13.5 years of age among males,[28] but males attain greater maximal height growth velocity and experience increases in linear growth for a longer duration, primarily due to a more prolonged prepubertal interval, of approximately 2 additional years.

Females gain fat mass (FM) steadily through age 16. Males have an initial increase in FM between ages 8 and 14 years, then a decline between ages 14 and 16 years, followed by a plateau.[29] The distribution of FM also changes; in males, increased deposition of subcutaneous adipose tissue (SAT) occurs in the trunk area, whereas in girls, SAT is deposited in the gluteal–femoral region. This results in differing body composition patterns, in which males have more upper body fat and females have wider hips and more lower body fat. FFM increases in females until age 15 years, and FFM increases in males through age 18 years, with the most rapid increases in males occurring between 12 and 15 years.[29] The composition of FFM also changes during this time, from 80% water in young childhood, to approximately 73% water by ages 10 to 15 years.[29] The rise in density of FFM is caused by accretion of protein and minerals in the FFM compartment during growth.

For females, a negative relationship has been observed among age at menarche, body mass index (BMI), and body fatness. Females who are more pubertally advanced tend to be taller and have more FM, bone mineral content, and FFM, compared with same-age females at a lower stage of pubertal development. Adolescent females with early growth spurts reach Tanner stage 2 and menarche earlier than adolescent females with average and late growth spurts.[30] Furthermore, females with early menarche obtain greater adiposity by the end of puberty, than females with late menarche. This is of concern because tracking of FM into adulthood is strong: females in the highest FM category have a 55% chance of staying within that category 10 years later, whereas females in the lowest FM category in adolescence have a 77% chance of staying within their category.[31]

For males, the relationship between body composition and puberty onset is equivocal. Some studies have reported that increased adiposity is associated with delayed puberty in males,[32-34] but others have found the opposite, with higher adiposity associated with earlier puberty onset.[35-38] One explanation for the potential sex difference is that, unlike menarche, which is easy to observe and can be ascribed to a narrow time window, the onset of puberty in males is harder to pinpoint and takes place over a longer period. Another challenge is the high degree of variability in assessment tools used for adiposity, with some studies relying on BMI and others using more precise measures, such as skin folds, waist circumference, and bioelectrical impedance. A recent study by O'Keefe and colleagues[7] addressed these prior limitations in a longitudinal cohort study of 4,167 adolescent males and females, aged 9 to 18 years, using multiple measures of adiposity, and observational measures of puberty and fat mass, collected over time. The authors found negative associations between adiposity and age of puberty onset in females, confirming prior research and, additionally, found similar associations between adiposity and puberty in males.

Another aspect of growth that is highly dynamic during puberty is bone remodeling. Bone's response to forces and its capacity for growth are greatest during adolescence. Endogenous estrogens and androgens independently exert effects on bone acquisition and maintenance. Estrogen lowers the bone remodeling threshold, and females experience greater gains in bone mass during puberty than males.[39] Bone acquisition and metabolism are influenced by hormonal, dietary, and lifestyle factors. In addition to the sex hormones, growth hormone, IGF-I, cortisol, thyroid hormones, parathyroid hormone, vitamin D, and leptin influence bone metabolism during puberty. Physical activity during puberty has positive effects on bone accrual and turnover. Increased lean mass improves bone mass strength, and bone metabolism is influenced by dietary intakes of high-quality protein, calcium, magnesium, phosphorus, and vitamins D, K, and C.[40]

Changes in body composition that occur during adolescence guide nutrient recommendations: growth increases energy demands and protein requirements, and bone accrual requires protein, minerals, and vitamins. The adolescent period, and the changes in body composition that occur, can pose emotional and psychological distress, which can lead to unhealthy eating patterns, affect subsequent health in adulthood, and set the stage for increased metabolic risk.

Daily Recommended Intakes for Adolescents

Dietary Guidelines for Americans

The Dietary Guidelines for Americans (DGA) are developed by the U.S. Department of Agriculture (USDA) in collaboration with the United States Department of Health and Human Services. By law, the Guidelines are updated every 5 years. The most recent version of the DGAs, 2020–2025, takes a lifespan approach to recommending healthy dietary patterns for all age groups, including adolescents.[41] This is the first version of the Guidelines to present healthy dietary recommendations broken down by stage of life, making it particularly relevant to adolescents. While this age group tends to meet needs for total grains, their diets lack sufficient fruits, vegetables, dairy, and high-quality protein. For children and adolescents aged 9 to 13 years, added sugars, saturated fat, and sodium are well above recommended levels. For children 14 to 18 years, the discrepancies between actual and recommended intake for food groups, such as fruits, vegetables, and dairy, are wider than for any other age group, placing them at relatively higher risk for dietary inadequacy. Adolescent males, on average, have inadequate intakes of phosphorus, magnesium, and choline. Overall, females in this age group have inadequate intakes of iron, folate, Vitamin B_6, and Vitamin B_{12}. Considering these dietary concerns with the increased physiologic demand for both macro- and micronutrients needed to support growth and maturation, there is a need to find effective methods of increasing consumption of nutrient-dense foods in this age group.

Dietary Reference Intakes for Adolescents

The Dietary Reference Intakes (DRIs)[42] are established and published by the National Academy of Medicine (NAM, formerly the Institute of Medicine (IOM). Committees consisting of U.S. and Canadian experts on specific nutrients review the scientific literature, consider the roles of nutrients in reducing disease risk, evaluate indicators of adequacy, and estimate average requirements for each nutrient. This information is interpreted considering current intakes by various North American population groups.

The DRIs consist of four types of reference values. The estimated average requirement (EAR) is the amount of a nutrient that would meet the requirements of 50% of healthy individuals of different sex and age groups. This value is used for micronutrient and macronutrient recommendations. The recommended dietary allowance (RDA) is calculated from the EAR to meet the needs of 97% to 98% of healthy individuals. The adequate intake (AI) is established when an EAR cannot be determined from available data. It is based on experimental data or determined from estimated intakes of a group of healthy individuals. The underlying assumption is that the amount of the nutrient consumed by these individuals is adequate to sustain health. The tolerable upper intake level (UL) is the highest amount of a nutrient that can be consumed without posing a risk of adverse side effects for almost all individuals.

For setting nutrient intake guidelines, the IOM, (now NAM), defines adolescence as ages 9 to 18 years. DRIs for adolescents account for variability in requirements related to growth rates. **Table 55.2** shows the nutritional goals set forth by the DGA based on DRI and dietary guidelines.

Table 55.2 Nutritional Goals for Age-Sex Groups Based on Select Dietary Reference Intakes and Dietary Guideline Recommendations

Nutrient (units)	Source of Goal	Female 9–13 y	Male 9–13 y	Female 14–18 y	Male 14–18 y
Protein (g)	RDA	34	34	46	52
Protein (% energy)	AMDR	10–30	10–30	10–30	10–30
Carbohydrate (g)	RDA	130	130	130	130
Carbohydrate (% energy)	AMDR	45–65	45–65	45–65	45–65
Fiber (g)	IOM	26	31	26	38
Fat (% energy)	AMDR	25–35	25–35	25–35	25–35
Saturated fat (% energy)	DG	<10	<10	<10	<10
Linoleic acid (g)	AI	10	12	11	16
Linoleic acid (% energy)	AMDR	5–10	5–10	5–10	5–10
α-Linolenic acid (g)	AI	1	1.2	1.1	1.6
α-Linolenic acid (% energy)	AMDR	0.6–1.2	0.6–1.2	0.6–1.2	0.6–1.2
Calcium (mg)	RDA	1,300	1,300	1,300	1,300
Iron (mg)	RDA	8	8	15	11
Magnesium (mg)	RDA	240	240	360	410
Phosphorus (mg)	RDA	1,250	1,250	1,250	1,250
Potassium (mg)	AI	2,300	2,500	2,300	3,000
Sodium (mg)	DG	<1,800	<1,800	<2,300	<2,300
Chloride (g)	RDA	2.3	2.3	2.3	2.3
Fluoride (mg)	RDA	2	2	3	3
Zinc (mg)	RDA	8	8	9	11
Copper (μg)	RDA	700	700	890	890
Selenium (μg)	RDA	40	40	55	55
Iodine (μg)	RDA	120	120	150	150

(continues)

Table 55.2 Nutritional Goals for Age-Sex Groups Based on Select Dietary Reference Intakes and Dietary Guideline Recommendations *(continued)*

Nutrient (units)	Source of Goal	Female 9–13 y	Male 9–13 y	Female 14–18 y	Male 14–18 y
Vitamin A (μg RAE)	RDA	600	600	700	900
Vitamin D (μg)	RDA	15	15	15	15
Vitamin E (mg AT)	RDA	11	11	15	15
Vitamin C (mg)	RDA	45	45	65	75
Thiamin (mg)	RDA	0.9	0.9	1	1.2
Riboflavin (mg)	RDA	0.9	0.9	1	1.3
Niacin (mg)	RDA	12	12	14	16
Folate (μg)	RDA	300	300	400	400
Vitamin B_6 (mg)	RDA	1	1	1.2	1.3
Vitamin B_{12} (μg)	RDA	1.8	1.8	2.4	2.4
Pantothenic Acid (mg)	RDA	4	4	5	5
Biotin (μg)	RDA	20	20	25	25
Choline (mg)	AI	375	375	400	550
Vitamin K (μg)	AI	60	60	75	75
Water (l)	AI	2.1	2.4	2.3	3.3

AI, adequate intake; AMDR, acceptable macronutrient distribution range; *AT*, α-tocopherol; *DG*, dietary guidelines; *IOM*, Institute of Medicine; *RAE*, retinoic acid equivalents; *RDA*, recommended dietary allowance.

Modified from US Departments of Agriculture and of Health and Human Services. Report of the Dietary Guidelines Advisory Committee on the Dietary Guidelines for Americans, 2010. Washington, DC: US Government Printing Office, 2010.

Dietary Behaviors of Adolescents

Dieting Behavior

Adolescence is the developmental period where dieting behaviors begin to emerge, with more than 37% of those aged 16 to 19 reporting weight loss attempts in the past year. Females are more likely than males to report weight loss attempts, with percentages equal to 45.3% and 30.1%, respectively.[43] Dieting behavior typically begins earlier in adolescence, often coinciding with the onset of puberty,[44] and earlier onset of dieting behavior is associated with greater weight gain over time. Dieting behavior emerges for a variety of reasons that can be largely categorized as family, peers, and the media.[45] Furthermore, social media is increasingly seen as a driver of dieting behavior, with positive associations between smartphone usage and body image distortion, weight loss attempts, and inappropriate weight loss strategies observed for both teen boys and girls.[46]

Weight loss attempts include a variety of dieting behaviors that can be broadly classified as healthy or unhealthy. Healthy strategies include increased intake of fruits and vegetables and engagement in physical activity. Unhealthy strategies include meal skipping and use of appetite suppressants or laxatives. Many of these behaviors are considered typical and subclinical, but clinical eating disorders also tend to develop in adolescence, which will be discussed below.

In addition to adoption of dieting behaviors, adolescents also shift dietary patterns and engage in meal skipping, especially breakfast. Consuming breakfast provides adolescents with an opportunity to obtain much needed nutrients that are lacking in their diets. Breakfast consumption is recommended by the American Academy of Pediatrics for both dietary and cognitive outcomes,[47] but it tends to decrease with age during adolescence.[48] According to data from the National Health and Nutrition Examination Survey (NHANES), between 2015 and 2018, breakfast was reportedly consumed on a given day by only 73% of male adolescents and 72.8% of female adolescents. Breakfast consumption was lowest among non-Hispanic Black adolescents, with only 67.1% reporting consumption on a given day. Among adolescents who consumed breakfast, only 31.5% reported consuming milk.[49] This has implications for the health of adolescents, because low meal frequency and breakfast skipping have been identified as factors associated with obesity.[50]

Several epidemiologic studies have reported that breakfast consumption is associated with protection against obesity.[48,51,52] In the School Nutrition Dietary Assessment Study, BMI decreased by 0.15 points for every additional breakfast meal consumed per week.[53] This effect was strongest among non-Hispanic White children and was not observed among Hispanic children. Furthermore, a systematic review in 16,130 children and adolescents found poorer cardiometabolic outcomes associated with more frequent breakfast skipping.[54] It is important to note, however, that most studies that demonstrate a protective role of breakfast intake on weight status have been observational. A systematic review of the effects of ready-to-eat-cereal on body weight in children and adolescents found that few randomized controlled trials had been done, and they were unable to find support for a direct role for breakfast cereal on body weight.[55] Independent of body weight; however, breakfast consumption has an important impact on nutrient intakes, particularly fiber and calcium[48]; and adolescents who skip breakfast have lower intakes of most vitamins and minerals, including B vitamins, folate, calcium, phosphorus, magnesium, iron, and zinc, than those who eat ready-to-eat cereal for breakfast.[51]

Beverage Consumption

Another dietary behavior that has important nutritional implications for adolescents is beverage consumption. Consumption of sugar-sweetened beverages (SSBs) is high among adolescents, but has declined in recent decades. According to data from the NHANES 2011–2016, two-thirds of youth reported consuming a sugar-sweetened beverage on a given day, with soda being the most popular beverage choice among adolescents.[56] In 2011–2014, sugar-sweetened beverages made up approximately 9.3% and 9.7% of total energy intake for 12 to 19 year-old males and females, respectively.[57] Patterns of beverage intake also change dramatically between childhood and adolescence. Intakes of both milk and fruit juice decline by approximately 30% over the 10-year period spanning childhood to middle adolescence.[58] Furthermore, the proportion of children consuming soda remains stable, but the amount of soda consumed increases,[58] whereas daily milk consumption continues to decline by approximately 0.5 servings per day in adolescents from ages 15 to 20 years.[59]

Trends in beverage consumption affect the nutritional adequacy of adolescent diets. Soda consumers at age 5 to 7 years have been shown to have lower percentage of total dietary energy intake from protein, lower fiber, calcium, magnesium, phosphorus, vitamin K, and vitamin D, and higher intake of added sugars, than non-soda consumers through mid-adolescence.[58] In Project EAT (Eating Among Teens), more than 72% of 15-year-old females and 55% of males had calcium intake below the AI for their age.[59] Over the 5-year follow-up, intakes decreased such that 68% of adult females and 53% of adult males consumed less than the AI, even though the AI is 300 mg lower for calcium at age 20 years than at age 15 years. Low intake of dietary calcium can result from low milk consumption, and this may increase risk for both low bone mineral content and density.[60] Furthermore, greater intake of SSBs is also associated with negative metabolic consequences, including increased body weight[61] and hypertension[62] among both children and adolescents.

Caffeinated beverage consumption also increases during adolescence. According to data from NHANES, 83% of teens reported consuming one or more servings of caffeine per week.[63] The primary source of caffeine intake in teens is soda, but energy drink use is increasing, with the highest rates of youth energy drink consumption reported in 13-19-year old teens.[64] Caffeine use among adolescents has been associated with poor sleep,[65] obesity,[66] and with high intake of snack foods and fast foods.[67] Because of the negative impacts on health, The American Academy of Pediatrics recommends limiting caffeine consumption among children of all ages.[68]

Diet Nutrient Density

The current food environment has been described as "obesogenic" because of the ready availability of large portions; the high quantity and variety of energy-dense snacks, sweets, and fast foods; and the low accessibility of whole grains and fresh fruits and vegetables. Adolescents are eating excess amounts of energy-dense, nutrient-deplete foods, and their diets are lacking in whole grains, fruits, and vegetables. However, data from NHANES comparing diets for youth in 1999–2000 and 2015–2016 show some improvements in overall dietary quality, based on the American Heart Association dietary score. Still, adolescents were more likely than other age groups to report poor diet quality, with 66.6% of 12-19-year olds demonstrating poor diet quality (compared with 52.5% and 39.8% in 6-11-year olds and 2-5-year olds, respectively.[69] This underscores other population-based data, which shows declines in diet quality as children transition into adolescents. Intervening during this time is important because the dietary patterns of adolescence often track into adulthood.[70]

The National Heart, Lung, and Blood Institute Growth and Health Study followed a cohort of 2,371 non-Hispanic Black and White 9-10-year-old girls for 10 years. Dietary patterns that were most common during this time showed high intakes of processed grains, lunch meats, pizza, fried potatoes, sweet breads, and fruit, and low intake of vegetables. Only 12% of girls met the requirements of a "healthy" diet: rich in fruits, green salads, and other vegetables; unprocessed whole grains; baked or steamed meat and poultry; and low-fat, unflavored dairy.[71] High intake of fast food is a common dietary pattern, particularly problematic in late adolescence, because this is when individuals have more autonomy and money to purchase food.[72] These dietary patterns are of concern because they contain excess fat, sodium, and added sugars, and lack dietary fiber, folic acid, calcium, and potassium.

Snacking has increased in adolescents since the 1970s. In 1977 to 1978, 61% of adolescents reported snacking on any given day whereas, from 2005 to 2006, 83% did. In addition, snacking frequency has increased; the percentage of adolescents who report consuming three or more snacks daily has more than doubled. Snack foods provide 23% of total daily energy requirement, approximately 526 kcal, for adolescents. In general, snack foods for this age group are high in sugars, solid fats, or both, and contain fewer vitamins and minerals than foods consumed at meal times.[73] Although snacking has been associated with increased total energy intake, the relationship with BMI has been inconsistent, with some studies finding a protective role for snacking on weight status, and other studies finding no association.[74,75] However, many prior studies did not account for potential biases in reporting of dietary data. A more recent analysis of NHANES data found that snacking was more frequent among 12-19-year olds with obesity, compared with those with overweight or normal weight status. Furthermore, while adolescents with overweight or obesity consumed more added sugars, saturated fat, and sodium from snacks, the energy density of snacks consumed by these groups was lower than for adolescents with normal weight.[76] Also, because fruits, such as apples, oranges, bananas, and orange juice, are common snacks, snacking fulfills more than 25% of vitamin E and C requirements and 23% of magnesium requirement. In addition, 20% of daily calcium requirement is reportedly achieved through snacking. Clearly, cutting out snacks entirely is not advisable, but improving the quality of these snacks by reducing "empty calorie" sources and increasing dairy, high-quality protein, fiber and whole grains, and nutrient-rich fruits and vegetables may be beneficial.[73]

Adolescents who follow healthier eating patterns share certain characteristics. First, eating dinners together has been associated with increased consumption of fruits and vegetables and reduced intakes of SSBs and fried foods.[77] Having healthy foods available in the home and modeling healthy eating behaviors have been associated with improvements in adolescent eating behavior.[78] Reducing the amount of unhealthy foods that are available and easily accessible at home can also improve adolescents' diets.[79] Even though adolescents may strive for independence at this time, they still need continued parental involvement and support to assist with the development of healthful, lifelong eating behaviors.

Special Considerations

Eating Disorders

Eating disorders are the third most common chronic disease in adolescence, with only obesity and asthma occurring at higher rates.[80] Eating disorders were once thought to affect primarily non-Hispanic White females of higher socioeconomic status, but more recent evidence suggests that the prevalence is increasing among males[81] minority populations,[82] and those who identify as sexual minorities.[83]

Eating disorders share two common features: disturbed eating (e.g., undereating, overeating) and distorted body image (e.g., feeling fat, extreme fear of weight gain). The best-known eating disorders are anorexia nervosa and bulimia nervosa. According to the *Diagnostic and Statistical Manual of Mental Disorders*, 5th edition,[84] *anorexia nervosa* is defined as a restricted energy intake leading to significantly low body weight, an intense fear of gaining weight, and an altered perception of one's weight or shape. Subtypes of anorexia nervosa include both adolescents who purely restrict (i.e., fast, diet, and exercise) or those who engage in periods of binge-eating and purging to control weight. In *bulimia nervosa,* eating patterns are disrupted such that individuals show repeated patterns of bingeing, followed by compensatory behaviors, such as excessive exercise or purging. *Binge-eating disorder*, which is thought to be the most common eating disorder, refers to recurrent episodes of binge eating that are not accompanied by attempts to compensate for the excess energy consumed, along with feelings of mental anguish associated with binge episodes. While prevalence of anorexia and bulimia among adolescents is relatively low, with lifetime prevalence of 0.3% and 0.9%, respectively,[85] symptoms related to these disorders are common, with up to 14% of adolescents engaging in symptoms such as fasting, bingeing, and purging, without meeting all of the diagnostic criteria, and some individuals present with symptoms of both anorexia and bulimia. Eating disorder symptomatology was reportedly worsened from the isolation and social distancing required because of the COVID-19 pandemic lockdown.[86]

The term "orthorexia" has been used to describe individuals who adopt extremely restrictive dietary behaviors intended to improve overall nutrition. While orthorexia is not classified on its own as an eating disorder (it is classified under Other Specified Feeding and Eating Disorders [OSFED]), it is associated with anorexia nervosa and obsessive-compulsive disorder.[87] Although some adolescents adopt restrictive diets due to legitimate cultural and/or health concerns (e.g., vegetarianism), overly rigid dietary changes are often a red flag that may be masking a more serious underlying eating disorder pathology.[88]

Adolescents are vulnerable to developing eating disorders for several reasons. First, the physical changes that occur during puberty may often be accompanied by feelings of dissatisfaction with one's body and, as a result, adolescents may attempt to diet. Dieting is an independent risk factor for developing eating disorders and, consequently, childhood overweight is a risk factor for developing eating disorders later in life.[89] Media use in adolescents may predispose adolescents to eating disorders. Adolescents spend more than 7 hours/day watching television, reading magazines, and surfing the Internet,[90] and those who read fashion magazines are more likely to develop distorted images about their own bodies.[91] In addition, the use of photo-based social media apps, like Instagram and TikTok, has been associated with body dissatisfaction and disordered eating, although prospective studies are needed to strengthen this evidence.[92,93] The number of pro-anorexia and pro-bulimia websites on the Internet that contain strategies for hiding eating disorders from parents, pictures of famous celebrities with exceedingly thin figures, and unhealthy tips for teenagers to manage body weight have also increased.[94] Exposure to these websites has been associated with low self-esteem, poor body image, and increased concern about body weight, compared with individuals who viewed control websites.[95] A third reason why eating disorders often appear during adolescence has been attributed to an epigenetic effect, or an interaction between genes and the environment that influences the onset of disordered eating traits. Twin studies have demonstrated moderate levels of heritability of eating disorders.[96] However, genetic effects on the development of eating disorders vary by age: eating disorders that develop before adolescence have low heritability, whereas those that develop in early adolescence through adulthood are more likely to be heritable.[97] The exact mechanisms by which the events of puberty trigger the onset of eating disorders have not been determined.

Although eating disorders are primarily considered psychological disorders, nutrition plays an integral role in determining their medical complications and treatment outcomes. The medical complications of eating disorders can be widespread and affect all body systems.[98] However, depending on the severity and timing of the energy restriction, adolescents suffer from some complications that are thought to be irreversible, such as loss of bone mineral density (BMD) and growth retardation.[99] In addition, puberty can often be delayed in individuals with eating disorders, particularly in those who lose significant amounts of body fat.[100]

Adolescent Athletes

Adolescent participation in organized sports in the United States has increased, with approximately 60% of youth ages 6 to 17 years reporting participating in sports in the previous 12 months.[101] Young athletes have added nutritional requirements based on the increased energy expenditure incurred in sports participation. In general, fluid intake should be 0.5 to 1 L/day greater than baseline requirements to compensate for added fluid lost in sweat,[102] and energy intakes should be increased above and beyond the needs for normal growth and basal metabolic requirements. If energy intake is adequate to support growth and development while compensating for the added expenditure from sports participation, adolescents should meet all of their nutrient needs. In other words, although absolute protein and carbohydrate needs are higher in teenage athletes than in their nonathletic counterparts, the recommendations are the same in terms of percentage of energy intake: 12% to 15% of energy for protein and at least 50% for carbohydrates. Because of their greater food intake, athletes generally achieve or approach daily recommended intakes for vitamins and minerals more readily than nonathletic teenagers.[103]

Nutritional issues can arise if the teenage athlete adopts a vegetarian diet or does not consume adequate energy or hydration. Voluntary dehydration is used in some sports to meet a weight category requirement and can cause an athlete to enter a competition in a dehydrated state. This can lead to hyponatremia and impaired performance. Vegetarian or plant-based diets offer many health benefits, but they can also pose nutritional issues if they are not well planned or are overly restrictive.[104] Iron absorption from plant foods is lower than that from animal foods and can lead to low iron status. Vitamin B_{12} is found only in animal products, and deficiency of this nutrient can cause macrocytic anemia. Finally, because of the high bulk, or low-energy density, of vegetarian diets, it may be important for some adolescent athletes to eat more frequent snacks and meals to ensure adequate energy intake.[104]

Extreme cases of undernutrition in athletes may lead to a syndrome known as relative energy deficiency in sport (RED-S), which is associated with both declines in health and athletic performance, as energy and hydration needs of athletes are chronically unmet.[105] Symptoms of RED-S include fatigue, rapid weight loss, missed periods and delayed puberty (females), low libido (males), frequent illness, hair loss, inability to focus, bone loss, irritability, and depression (**Figure 55.2**). It has been estimated that 22% to 58% of athletes have RED-S, but assessment tools are unreliable, so the problem may be even more widespread than it appears.[106] A subset of athletes with RED-S may also meet the criteria for the female athlete triad, described in an American College of Sports Medicine position paper as an interrelationship of low energy availability, amenorrhea, and osteoporosis (represented by the triangle in Figure 55.2). This may occur in the absence of disordered eating, but it provides inadequate available energy for use in cellular maintenance, thermoregulation, growth, and reproduction.[106] *Amenorrhea* is the absence of menstrual cycles for more than 3 months, and osteoporosis is reflected by a BMD Z-score lower than −2. In female athletes, chronic low energy intake results in low FM, which leads to low circulating leptin. Low leptin is proposed to reduce LH pulse frequency and increase pulse amplitude, thus resulting in disturbance or loss of menstruation. Low energy intake can also reduce bone formation.[107] Athletes at greatest risk are those who restrict their intake, exercise for long periods, or adopt restrictive vegetarian diets. Dieting, psychological predisposition, low self-esteem, family dysfunction, abuse, and biologic and genetic factors have been identified as contributors to the female athlete triad.[106]

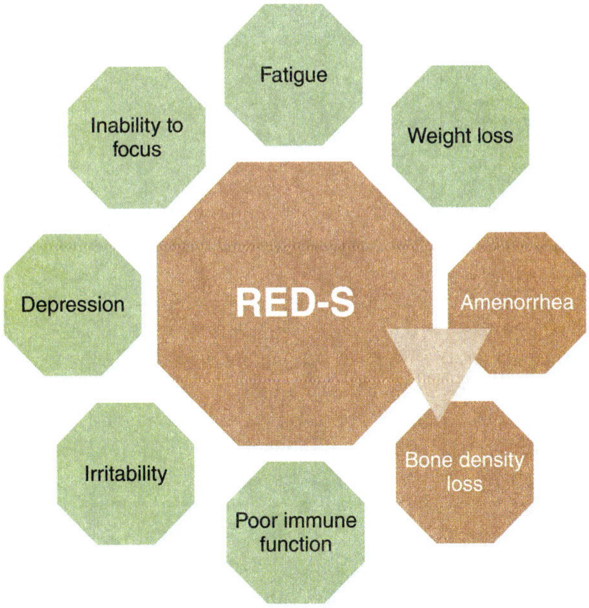

Figure 55.2 Relative energy deficiency in Sport. This figure shows the health consequences of Relative Energy Deficiency in Sport (RED-S), including fatigue, weight loss, amenorrhea, bone density loss, poor immune function, irritability, depression, and inability to focus. The symptoms outlined in red and denoted by connections with the red triangle are part of the Female Athlete Triad, which includes energy deficiency, amenorrhea, and bone density loss. Female Athlete Triad increases risk for fractures and osteoporosis.

Although exercise in adolescence is healthy and can lead to the establishment of lifelong physical activity habits, some associated behaviors and dietary habits must be monitored. Adequate energy intake for growth and the added demands of exercise is essential for both optimal performance and growth and maturation.

Pregnancy

Birth rates among adolescents in the United States have been declining since 1991, with estimates from 2019 reporting that only 16.7 out of 1,000 live births were attributed to teenagers. However, birth rates among Hispanic and non-Hispanic Black teenagers are more than two times higher than they are for non-Hispanic White teenagers.[108] These declines in adolescent pregnancy rates are beneficial for public health, because of the physical, psychological, economic, and social demands associated with raising children. Pregnancy places additional nutritional demands on adolescents, who already have increased energy requirements to meet the demands of their own rapid growth. Younger mothers are also at higher risk of pregnancy complications, such as pregnancy-induced hypertension, abnormally high weight gain, anemia, and renal disease.[109]

Reviews of dietary surveys suggest that pregnant adolescents consume inadequate amounts of a variety of nutrients, including total energy, iron, folate, calcium, vitamin E, and magnesium.[110] These nutrients all play critical roles in fetal growth. Pregnant non-Hispanic Black females from low-income households are at greater risk than other ethnic groups of consuming poor-quality diets.[111] In addition, dietary patterns that are common in adolescence, such as skipping meals and high SSB intake, are particularly detrimental during pregnancy. Support for pregnant adolescents must be multidisciplinary to address the social, behavioral, medical, and nutritional issues faced during this time. Nutrition education to help adolescents achieve appropriate weight gain for their BMI is crucial for ensuring the healthy birth weight of the infant, as well as the optimal growth and development of the adolescent mother.[112] Low-income mothers can gain nutritional support and receive supplemental sources of protein, vitamins, and minerals by joining the Special Supplemental Nutrition Program for Women, Infants, and Children (WIC).

Obesity and Metabolic Disorders

Obesity rates among US adolescents continue to rise. Data from the NHANES 2011 to 2020 for 14,967 children and adolescents found that the prevalence of obesity increased for adolescent males and females during this time. Among 12-19-year olds, prevalence of obesity, defined as a body mass index (BMI)-for-age percentile ≥95 increased from 20.1% in 2011 to 25.6% in 2020. Compared with non-Hispanic White adolescents, prevalence of obesity in 2020 was even higher among Mexican American (27.9) and non-Hispanic Black adolescents (29.8).[113] Of additional concern is the sharp rise in the prevalence of severe obesity, BMI at the 99th percentile or greater for sex and age, among children and adolescents. A 300% increase in the prevalence of severe obesity occurred in 2-19-year olds between NHANES 1976 to 1980 and 1999 to 2004,[114] and, in 2016, 7.7% of 12-19-year olds had severe obesity,[115] defined as a BMI at or above 120% of the sex-specific 95th percentile cut-off on the US Centers for Disease Control and Prevention growth charts. Recent increases in obesity among adolescents have been hastened due to the social isolation required during the COVID-19 pandemic.

The high rates of overweight and obesity among adolescents are of concern for several reasons. First, body composition in adolescence tracks into adulthood, thereby increasing the odds of remaining overweight or obese as adults. Second, adult obesity is associated with a range of metabolic disorders, including T2D, CVD, hypertension, cancers, sleep disorders, osteoarthritis, and breathing problems. Third, obesity in adolescence also increases the risk of developing metabolic syndrome and T2D. *Metabolic syndrome* is a constellation of risk factors for CVD and T2D that include large waist circumference or obesity, dyslipidemia (low high-density lipoprotein cholesterol [HDL-C] and/or high triglycerides), elevated fasting glucose or insulin resistance (IR), elevated blood pressure, and, in some cases, inflammation, microalbuminemia, and thrombosis.[116]

High body fatness in adolescence is a clear risk factor for T2D and CVD. In adolescents, progression of IR to T2D occurs more quickly than in adults, for whom the period of IR can persist for decades before T2D develops.[117] In addition, adverse CVD risk profiles are observed in adolescents. Data from the Bogalusa Heart Study showed that, among 5-17-year olds, the prevalence of having two metabolic syndrome risk factors, whether high sum of skinfolds (adiposity), triglycerides, low-density lipoprotein cholesterol, fasting insulin, blood pressure, or low HDL-C, was 59% in those with severe obesity, whereas it was 5% in those with BMI below the 25th percentile.[118] The prevalence of having three or more metabolic risk factors was present in 7% of obese and 33% of severely obese adolescents.[114]

Puberty itself has been shown to lead to a state of IR. Insulin-stimulated glucose uptake is reported to be approximately 30% lower in adolescents in Tanner stages 2 to 4, compared with those in Tanner stage 1, and insulin sensitivity is reduced by 25% to 30%.[117] The timing of lowest insulin sensitivity seems to be in Tanner stage 3, and it tends to recover in Tanner stage 5. The rise in fasting plasma glucose, fasting plasma insulin, and the acute insulin response to glucose during this period is similar across sex and ethnic groups, and the relative changes are equivalent between lean and obese adolescents.[117] However, controversy exists concerning the role of change in body fat on the transient IR observed during puberty. Some investigators reported that it was not associated with changes in body fat, visceral fat, sex steroid hormones, or IGF-I,[117] whereas others suggested that it may be caused by changes in total and distribution of body fat and hormone release.[77] Nonetheless, it is agreed that transient IR, during puberty, is a natural occurrence[117] that may help promote growth.[119]

Sleep

Reduced sleep duration and quality is another factor that may impact adolescent health. Sleep needs vary with age and, during adolescence, recommended sleep duration ranges between 8.5 and 9.25 hours/night. Trends in U.S. adolescents show that an increasing number of 12-17-year olds get insufficient sleep, with two separate cohorts finding that >40% of respondents reported <7 hours per night.[120] Factors associated with insufficient sleep include social technology,[121] screen time,[122] >10 hours per week of homework and working a paid job.[120] Adolescence is a developmental time period where natural shifts in circadian rhythms favor later sleep-wake times, which are habitually at odds with actual sleep duration, due to early school start times or other social obligations. This phenomenon, known as social jet lag, creates sleep debt that may not be "repaid" by longer sleep hours on the weekends or during holidays.

Social jet lag, sleep debt, and insufficient sleep may have serious health consequences in adolescence. A study of 16-19-year olds reported that sleep time <8 hours/night, was associated with high energy intake from snacks.[123] This may explain why there is an inverse association between sleep duration and prevalence of obesity in both children and adults,[124-126] and a positive relationship between social jet lag and excess weight gain in adolescents.[127] Social jet lag is also associated with poorer mental health[128,129] and lower academic performance.[130]

Summary

Adolescence is a critical period of development because of the rapid changes in physical, cognitive, and psychological growth that occur. During this time, eating behaviors change as children grow and begin taking more autonomy over their feeding. Offering adolescents appropriate levels of support can improve their chances of maintaining healthful dietary behaviors into adulthood. This is particularly critical for adolescents who have additional needs, including athletes, pregnant teens, and those who have disordered eating behaviors. Finally, the prevalence of obesity and metabolic diseases has been increasing throughout the population, including adolescents. Comprehensive nutrition support for this age group needs to include a strong focus on excess weight gain prevention.

Acknowledgments

The authors acknowledge Marie-Pierre St Onge for her contributions to the previous version of this chapter.

References

1. Black RE, Victora CG, Walker SP, et al. Lancet 2013;382:427–51.
2. Baxter-Jones AD, Faulkner RA, Forwood MR, et al. J Bone Miner Res 2011;26:1729–39.
3. Andes LJ, Cheng YJ, Rolka DB, et al. JAMA Pediatr 2020;174: e194498.
4. Barnes HV. Med Clin North Am 1975;59:1305–17.

5. InterLACE Study Team. Hum Reprod 2019;34:881–93.
6. Rosenfield RL, Lipton RB, Drum ML. Pediatrics 2009;123: 84–8.
7. O'Keeffe LM, Frysz M, et al. Human Reprod 2020;35:2784–92.
8. Carel JC, Lahlou N, et al. Human Reprod Update 2004;10: 135–47.
9. Rudolph KD, Troop-Gordon W, Lambert SF, et al. Dev Psychopathol 2014;26:1423–44.
10. Lynne-Landsman SD, Graber JA, Andrews JA. Dev Psychol 2010;46:853–68.
11. Soliman A, De Sanctis V, Elalaily R. Indian J Endocrinol Metab 2014;18:S39–47.
12. Wood CL, Lane LC, Cheetham T. Best Pract Res Clin Endocrinol Metab 2019;33:101265.
13. Weaver CM. J Bone Miner Res 2007;22:45–9.
14. Vermeulen A. Environ Health Persp Suppl 1993;101:91–100.
15. Apter D, Bützow TL, et al. J Clin Endocrinol Metab 1993;76: 940–9.
16. Huang A, Reinehr T, Roth CL. Curr Opin Endocr Metab Res 14:160–68.
17. Pereira A, Busch AS, Solares F et al. J Clin Endocrinol Metab 2021;106:1352–61.
18. Brix N, Ernst A, Lauridsen LLB, et al. Paediatr Perinat Epidemiol 2019;33:70–78.
19. Brix N, Ernst A, Lauridsen LLB, et al. Int J Epidemiol 2020;49: 834–44.
20. Tsoi MF, Li HL, Feng Q, et al. Obesity Facts 2022;15:560–69.
21. Tanner JM. Growth at Adolescence. 2nd ed. Oxford: Blackwell Scientific, 1962.
22. Marshall WA, Tanner JM. Arch Dis Child 1969;44:291–303.
23. Marshall WA, Tanner JM. Arch Dis Child 1970;45:13–23.
24. Koopman-Verhoeff ME, Gredvig-Ardito C, Barker DH, et al. J Adolesc Health 2020;66:597–602.
25. Herman-Giddens ME, Slora EJ, Wasserman RC, et al. Pediatrics 1997;99:505–12.
26. Solorzano CMB, McCartney CR. Reproduction 2010;140: 399–410.
27. Bonat S, Pathomvanich A, Keil MF, et al. Pediatrics 2002;110: 743–7.
28. Veldhuis JD, Roemmich JN, Richmond EJ, et al. Endocr Rev 2005;26:114–46.
29. Siervogel RM, Demerath EW, Schubert C, et al. Horm Res 2003;60:36–45.
30. Buyken AE, Bolzenius K, Karaolis-Danckert N, et al. Am J Hum Biol 2011;23:216–24.
31. Vink EE, van Coeverden SC, van Mil EG, et al. Obesity 2010;18:1247–51.
32. Crocker MK, Stern EA, Sedaka NM, et al. J Clin Endocrinol Metab 2014;99:E1519–29.
33. Lee JM, Wasserman R, Kaciroti N, et al. Pediatrics 2016;137: e20150164.
34. Zhai L, Zhao J, Bai Y, et al. J Pediatr Endocrinol Metab 2013;26:895–901.
35. Busch AS, Hollis B, Day FR, et al. Human Reprod 2019;34: 1514–22.
36. Busch AS, Højgaard B, Hagen CP, et al. J Clin Endocrinol Metab 2020;105:e1667–e1672.
37. Li W, Liu Q, Deng X, et al. Am J Hum Biol 2018;30:e23195.
38. Aghaee S, Deardorff J, Quesenberry CP, et al. Am J Epidemiol 2022;191:2026–36
39. Schiessl H, Frost HM, Jee WS. Bone 1998;22:1–6.
40. Pérez-López FR, Chedraui P, Cuadros-López JL. Curr Med Chem 2010;17:453–66.
41. U.S. Department of Agriculture and U.S. Department of Health and Human Services. Dietary Guidelines for Americans, 2020-2025. 9th ed, 2020.
42. Institute of Medicine of the National Academies. Dietary Reference Intakes: The Essential Guide to Nutrient Requirements. Washington, DC: The National Academies Press, 2006.
43. McDow KB, Nguyen DT, Herrick KA, et al. Attempts to lose weight among adolescents aged 16-19 in the United States, 2013-2016. In: Centers for Disease Control and Prevention NCHS, ed. Hyattsville, MD, 2019.
44. Harnois-Leblanc S, Van Hulst A, Sabiston CM, et al. Int J Eat Disord 2021;54:764–72.
45. Balantekin KN, Birch LL, Savage JS. Eat Weight Disord 2018;23:215–23.
46. Kwon S, Kim R, Lee JT, et al. JAMA Netw Open 2022;5: e2213237.
47. HealthyChildren.org Available at www.healthychildren.org /English/healthy-living/nutrition/Pages/Breakfast-for-Learning.aspx
48. Affenito SG, Thompson DR, Barton BA, et al. J Am Diet Assoc 2005;105:938–45.
49. Terry AL, Wambogo E, Ansai N, et al. Breakfast intake among children and adolescents: United States, 2015-2018. Hyattsville, MD: Centers for Disease Control, National Center for Health Statistics, 2020.
50. Moreno LA, Rodriguez G, Fleta J, et al. Crit Rev Food Sci Nutr 2010;50:106–12
51. Deshmukh-Taskar PR, Nicklas TA, O'Neil CE et al. J Am Diet Assoc 2010;110:869–78.
52. Merten MJ, Williams AL, Shriver LH. J Am Diet Assoc 2009;109:1384–91.
53. Gleason PM, Dodd AH. J Am Diet Assoc 2009;109:S118-28.
54. Monzani A, Ricotti R, Caputo M, et al. Nutrients 2019;11:387.
55. Sanders LM, Dicklin MR, Zhu Y, et al. Adv Nutr 2022.
56. Russo RG, Northridge ME, Wu B, et al. J Racial Ethn Health Disparities 2020;7:1100–16.
57. Rosinger A, Herrick KA, Gahche J, et al. Sugar-Sweetened Beverage Consumption Among U.S. Youth, 2011–2014. Hyattsville, MD: National Center for Health Statistics, 2017.
58. Fiorito LM, Marini M, Mitchell DC, et al. J Am Diet Assoc 2010;110:543–50,
59. Larson NI, Neumark-Sztainer D, Harnack L, et al. J Nutr Educ Behav 2009;41:254–60.
60. Black RE, Williams SM, Jones IE, et al. Am J Clin Nutr 2002;76:675–80.
61. Malik VS, Pan A, Willett WC, et al. Am J Clin Nutr 2013;98: 1084–102.
62. Farhangi MA, Nikniaz L, Khodarahmi M. J Translat Med 2020;18:344.
63. Benson SM, Unice KM, Glynn ME. Food Chem Toxicol 2019;125:271–78.
64. Ruiz LD, Scherr RE. Am J Lifestyle Med 2019;13:22–25.
65. Tomanic M, Paunovic K, Lackovic M, et al. Nutrients 2022;14: 3813.
66. Yu L, Mei H, Shi D, et al. Environ Sci Pollut Res 2022;29: 57618–28.
67. Nuss T, Morley B, Scully M, et al. Nutr J 2021;20:64.
68. Committee on Nutrition and the Council on Sports Medicine and Fitness Pediatrics 2011;127:1182–9.
69. Liu J, Rehm CD, Onopa J, et al. . JAMA 2020;323:1161–74.
70. Mikkilä V, Räsänen L, Raitakari OT, Br J Nutr 2005;93:923–31.
71. Ritchie LD, Spector P, Stevens MJ, et al. J Nutr 2007;137: 399–406.
72. Cutler GJ, Flood A, Hannan P, et al. J Nutr 2009;139:323–8.
73. Sebastian RS, Goldman JD, Wilkison Enns C. Snacking patterns of U.S. Adolescents: What we eat in America, NHANES 2005-2006. In: Group FSR, ed.,2010.
74. Keast DR, Nicklas TA, O'Neil CE. Am J Clin Nutr 2010;92: 428–35.
75. Larson N, Story M. Child Obes 2013;9:104–15.

76. Tripicchio GL, Kachurak A, Davey A, et al. Nutrients 2019; 1:1486.
77. Gillman MW, Rifas-Shiman SL, Frazier AL, et al. Arch Fam Med 2000;9:235–40.
78. Cutler GJ, Flood A, Hannan P, et al. J Am Diet Assoc 2011;111: 230–40.
79. Vereecken C, Haerens L, De Bourdeaudhuij I, et al. Pub Health Nutr 13, 1729–35.
80. Reijonen JH, Pratt HD, Patel DR, Greydanus DE. J Adolesc Res 2003;13:209–22.
81. Mitchison D, Mond J, Bussey K, et al. Psychol Med 2020;50: 981–90.
82. Austin SB, Spadano-Gasbarro J, Greaney ML, et al. J Adolesc Health 2011;48:109–12.
83. Cao Z, Cini E, Pellegrini D, Fragkos KC. Eur Eat Disord Rev 2023;31:46–64.
84. American Psychiatric Association. Diagnostic and Statistical Manual of Mental Disorders. 5th ed. Arlington, VA; 2013.
85. Swanson SA, Crow SJ, Le Grange D, et al. Arch Gen Psychiatry 2011;68:714–23.
86. D JD, Han A, Anderson A, et al. Int J Eat Disord 2023;56:5–25.
87. Costa CB, Hardan-Khalil K, Gibbs K. Issues in Mental Health Nursing 2017;38:980–88.
88. Robinson-O'Brien R, Perry CL, Wall MM, et al. J Am Diet Assoc 2009;109:648–55.
89. Kotler LA, Cohen P, Davies M, et al. J Am Acad Child Adolesc Psychiatry 2001;40:1434–40.
90. Strasburger VC, Jordan AB, Donnerstein E. Pediatrics 2010; 125:756–67.
91. Hogan MJ, Strasburger VC. Adolesc Med State Art Rev 2008;19: 521–46.
92. Holland G, Tiggemann M. Body Image 2016;17:100–10.
93. Borzekowski DL, Schenk S, Wilson JL, et al. Am J Public Health 2010;100:1526–34.
94. Bardone-Cone AM, Cass KM. Int J Eat Disord 2007;40: 537–48.
95. Bulik CM, Blake L, Austin J. Psychiatr Clin N Am 2019;42: 59–73.
96. Klump KL, Burt SA, Spanos A, et al. Int J Eat Disord 2010;43: 679–88.
97. Rosen DS. Pediatrics 2010;126:1240–53.
98. Katzman DK. Int J Eat Disord 2005;37:S52–9.
99. Misra M, Aggarwal A, Miller KK, et al. Pediatrics 2004;114: 1574–83.
100. Burns RD, Bai Y, Brusseau TA. J Phys Act Health 2020;17: 1197–204.
101. Bonci L. Pediatr Ann 2010;39:300–6.
102. Petrie HJ, Stover EA, Horswill CA. Nutrition 2004;20:620–31.
103. Barr SI, Rideout CA. Nutrition 2004;20:696–703.
104. Mountjoy M, Sundgot-Borgen J, Burke L, et al. Brit J Sports Med 2014;48:491–97.
105. Logue DM, Madigan SM, Melin A, et al. Nutrients 2020;12:835.
106. Nattiv A, Loucks AB, Manore MM, et al. Med Sci Sports Exerc 2007;39:1867–82.
107. Warren MP, Chua AT. Ann N Y Acad Sci 2008;113:244–52.
108. Martin JA, Hamilton BE, Osterman MJK, et al., Natl Vital Stat 2021;70:1–51.
109. Lenders CM, McElrath TF, Scholl TO. Curr Opin Pediatr 2000;12:291–6.
110. Moran VH. Br J Nutr 2007;97:411–25.
111. Grunbaum JA, Kann L, Kinchen SA, et al. MMWR Surveill Summ 2002;51:1–62.
112. Anderson AS, Wrieden WL. In: Ramsay MM, Symonds ME, eds. Teenage pregnancies. Maternal-Fetal Nutrition During Pregnancy and Lactation. Cambridge: Cambridge University Press, 2010.
113. Hu K, Staiano AE. JAMA Pediatr 2022;176:1037–39.
114. Skelton JA, Cook SR, Auinger P, et al. Acad Pediatr 2009;9: 322–9.
115. Hales CM, Fryar CD, Carroll MD, et al. JAMA 2018;319: 1723–25.
116. Levesque J, Lamarche B. J Nutrigenet Nutrigenomics 2008;1: 100–8.
117. Goran MI, Ball GDC, Cruz ML. J Clin Endocrinol Metab 2003;88:1417–27.
118. Freedman DS, Mei Z, Srinivasan SR, et al. J Pediatr 2007;150: 12–17.
119. Roemmich JN, Clark PA, Lusk M, et al. Int J Obes Relat Metab Disord 2002;26:701–9.
120. Twenge JM, Krizan Z, Hisler G. Sleep Med 2017;39:47–53.
121. Charmaraman L, Richer AM, Ben-Joseph EP, et al. J Adolesc Health 2021;69:162–65.
122. Hisler GC, Hasler BP, Franzen PL, et al. Sleep Health 2020;6: 731–42.
123. Weiss A, Xu F, Storfer-Isser A, et al. Sleep 2010;33:1201–9.
124. Patel S, Hu FB. Obesity 2008;16:643–53.
125. Chen X, Beydoun MA, Wang Y. Obesity 2008;16:265–74.
126. de Oliveira IDR, Maciel NMS, da Costa BT, et al. J Pediatr 2023;99:45–52.
127. Hayes JF, Schumacher LM, Lanoye A, et al. J Behav Med 2022;45:794–803.
128. Mathew GM, Li X, Hale L, Chang AM. Chronobiol Int 2019;36:461–69.
129. de Souza CM, Hidalgo MP. Chronobiol Int 2014;31:199–205.
130. Tamura N, Komada Y, Inoue Y, et al. Chronobiol Int 2022;39:311–22.

CHAPTER 56

Nutrition In Older Adults

Connie Watkins Bales and Mary Ann Johnson

CHAPTER OUTLINE

Abbreviations

AD, Alzheimer disease; **AI**, adequate intake; **CDDR**, Chronic Disease Risk Reduction Intake; **CHD**, coronary heart disease; **CKD**, chronic kidney disease; **CVD**, cardiovascular disease; **DRI**, dietary reference intake; **EAR**, estimated average requirement; **H2RA**, histamine H_2 receptor antagonist; **IDA**, iron deficiency anemia; **NHANES**, National Health and Nutrition Examination Survey; **RDA**, recommended dietary allowance; **USDA**, United States Department of Agriculture

Overview

Almost all countries are experiencing growth in their older adult populations.[1,2] The "graying" of the world's population brings considerable burden of chronic diseases and many, if not most, of these diseases have a strong nutritional component. Thus, this chapter reviews the impact of diet on chronic health conditions and selected nutrients of concern, as well as the infrastructure in the United States for meeting the food and nutrient needs of older adults in the continuum of care in community and long-term care settings. Although knowledge of how nutrition supports health throughout life is growing, much remains to be learned and applied in: (a) the behavioral sciences, concerning ways to improve eating habits and thus lessen the burden of chronic disease: (b) the policy arena, to ensure that all older people have access to safe and nutritious foods at all times: and (c) the basic and clinical sciences, to further delineate the role of specific foods and nutrients in maximizing health and minimizing the adverse consequences of sarcopenia, weight loss, nutritional frailty, and other age- and nutrition-related concerns.

Current and Future Demographics of Aging

In 2018, for the first time in history, persons aged 65 years or over worldwide outnumbered children under age 5. Projections indicate that by 2050, there will be more than twice as many persons above 65 as children under five.[2] In 2018, 16% of the U.S. population was at least 65 years of age.[1] On average, approximately 5% of older adults lived in institutional settings (long-term care facilities or community housing with services), but this number increased with age from 2% for those 65 to 74 years old, to 5% for those 75 to 84 years old, and to 20% for those 85 years old or older. About 1.1 million people aged 65 and older were residents of long-term care facilities (2016). Approximately 31% of older women and 19% of older men lived alone. Among those aged 65 to 74 and 75 and over, the prevalence of poverty in men was 7.7% and 8.8%, respectively; and in women, it was 9.6% and 13.2%[1], respectively.

The number of individuals in the United States who are 65 years old and older will nearly double from 52.4 million in 2018 to 994.7 million in 2060, whereas those aged 85 years old and older will increase by more than fourfold to 19.9 million and will become 5% of the U.S. population.[1] Ethnic and racial diversity are also increasing. Between 2018 and 2060, the number of older non-Hispanic White adults is expected to increase from 40.1 million to 52.2 million, non-Hispanic Blacks from 4.8 million to 12.1 million, Asians from 2.4 million to 7.9 million, and Hispanics or Latinos (any race) from 4.4 million to 19.9 million.[1] Under mortality conditions of 2018, people who survived to age 65 could expect to live an average of 19.5 additional years (18.1 years for men and 20.7 years for women), whereas the life expectancy of people who survive to age 85 was 7.0 years for women and 6.0 years for men.[1] Although life expectancy is increasing in most countries (prepandemic), the United States ranks only 46th in life expectancy worldwide.[3]

In summary, major demographic shifts are occurring that will be associated with particularly large increases in the oldest old (>85 years of age) and of racial and ethnic minorities. These trends bring new challenges in health care, especially for preventive and therapeutic nutritional care for older adults.

Physiologic and Other Changes Affecting Nutritional Risk

Certain physiologic and metabolic changes inherent in the aging process have the potential to increase nutritional risk. The requirements for some, but not all, nutrients can be altered by these changes. Some of these changes and their potential influences on nutrient needs and intakes are shown in **Table 56.1**. Additionally, medical comorbidities and other factors, including economic, geographic, and psychosocial concerns, can also affect diet behaviors and thus nutritional status.

Assessment of Nutritional Status

Inadequate nutritional status is an independent risk factor for poor health outcomes and loss of independence. Early identification of older adults with malnutrition (and risk of malnutrition) is critical for mounting a successful nutrition intervention.[4] Detection of malnutrition and risk of malnutrition involves a two-step process, screening followed by assessment. Screening tools are used to detect the likelihood that malnutrition exists. Assessment tools are subsequently used to determine the existence of malnutrition and to search for contributing elements, so that appropriate interventions can be implemented.

Table 56.1 Potential Physiologic and Metabolic Determinants of Nutrient Needs and Intakes in Older Adults

	Factor or Condition	Effect on Dietary Requirements
Physiologic changes	Decreased total energy expenditure and reduced physical activity	Decreased energy requirement: increased importance of nutrient-dense diet
	Decreased muscle mass and strength	Possible increased protein requirement: functional impairments could limit food access
	Decreased immune competence	Possible increased requirement for iron, zinc, other nutrients
	Detrimental oral changes	Decreased amount and/or quality of nutrient intake
	Gastrointestinal: atrophic gastritis	Increased requirements for folate, calcium, vitamin K, vitamin B_{12}, and iron
	Menopause	Decreased requirement for iron
Metabolic changes	Reduced skin synthesis of previtamin D_3: impaired renal activation of and reduced gut response to 1,25-dihydroxy vitamin D	Increased requirements for vitamin D and calcium
	Increased retention of vitamin A: altered hepatic metabolism	Decreased requirement for vitamin A
	Decreased ability to regulate fluid balance	Fluid needs possibly increased or decreased: fluid monitoring required

Nutrition Screening

Nutrition screening is most effective when it is conducted as the first step in a defined clinical protocol tied to a direct plan of action when the need for intervention is identified. Screening tools should be easy to use, quick, and economical. Screening tools should also be standardized and validated, as well as being both sensitive and specific.[4] Screening tools confirmed to have moderate validity for identifying malnutrition include the Nutrition Risk Screening-2002 (NRS-2002) and the Mini Nutritional Assessment–Short Form (MNA-SF). The MNA-SF has been especially well studied in older adults.[4,5]

Nutritional Assessment

While screening tools identify risk, assessments seek to determine the actual nutrition status. The goal is to determine if there is a true nutrition problem and to assess its severity, if so. The assessment parameters may be similar to those collected during screening, but they are explored in more depth. Broadly speaking, a nutrition assessment can include any of an array of subjective and objective measures—including diet and medical history, physical examination and anthropometric measurements, functional assessment, laboratory values, and quality of life assessment.[4] Not surprisingly, the approach(es) for nutritional assessment are varied and the identification of the best approach to use is more complex in older adults because their care settings cover such a wide continuum of need. For example, their care needs can range from those of a home-bound older adult to older patients experiencing critical-care hospitalization or end of life care in a skilled nursing home. A variety of age-related physiologic changes and chronic illnesses affect these assessments. For example, both critical and chronic illness cause inflammation, requiring a reprioritization of protein synthesis that leads to lower serum concentrations of albumin and prealbumin. Thus, changes in visceral proteins like serum albumin and prealbumin characterize inflammation rather than malnutrition.[6] Likewise, standardized semiquantitative methods of assessing food intake, such as 24-hour recalls and food frequency questionnaires can be problematic in older adults because sensory deficits or memory problems may diminish accuracy. At present, there is no one "gold standard" assessment tool for the evaluation of nutritional status in older adults.[7]

Over the past decade, assessment of nutritional status has evolved to depend less on anthropometric measures and laboratory (biochemical) values and more on physical (clinical) indicators. The American Academy of Nutrition and Dietetics (AND) and the American Society for Parenteral and Enteral Nutrition (ASPEN) collaborated to provide recommendations for evaluating six criteria to indicate the presence and severity of malnutrition. The criteria are (1) Inadequate energy intake, (2) Weight loss, (3) Loss of muscle mass, (4) Loss of subcutaneous fat stores, (5) Fluid retention that may mask weight loss, and (6) Diminished functional status, as measured by handgrip strength.[8] Reflecting these criteria, the nutrition-focused physical examination (NFPE) uses a full body exam to assess muscle mass, fat stores, fluid retention, micronutrient deficiencies, and functional capacity. While more evidence in various populations is needed to confirm the use of the NFPE to accurately diagnose malnutrition as defined by the AND/ASPEN criteria, the approach has gained wide acceptance and there are a variety of resources for acquiring training and expertise to conduct a standardized NFPE.[9]

Dietary Reference Intakes for Older Adults

Dietary recommendations for intakes of essential nutrients[10] by age, sex, ethnicity, and other demographic indices are set by the National Academies of Sciences, Engineering, and Medicine (NASEM). These recommendations, along with typical intakes of older adults[11] are shown in **Table 56.2**. Some dietary reference intake (DRI) recommendations are higher for men compared with women, including for

Table 56.2 Recommendations and Intakes of Selected Nutrients for Older Adults

	RDA (EAR) or AI or CDRR[a]				Intakes from Food from 2015–2018[b] (unless otherwise indicated)			
	Men 50–70 y	**Men >70 y**	**Women 50–70 y**	**Women >70 y**	**Men 50–70 y**	**Men >70 y**	**Women 50–70 y**	**Women >70 y**
Energy (kcal)					2,456	2,200	1,800	1,674
Protein (g)	56 (46)	56 (46)	46 (38)	46 (38)	95.7	82.0	69.2	62.9
Dietary fiber (g)	30	30	21	21	19.1	18.5	15.8	15.4
Sodium (mg)	2,300	2,300	2,300	2,300	4,060	3,561	2,916	2,702
Potassium (mg)	3,400	3,400	2,600	2,600	3,086	2,930	2,403	2,280
Calcium (mg)	1,000 (800)	1,200 (1,000)	1,200 (1,000)	1,200 (1,000)	1,046	969	838	779
Calcium (% taking supplement)					46%	47%	53%	61%
Vitamin D (μg)[c]	15 (10)	20 (10)	15 (10)	20 (10)	5.1	6.0	4.0	4.3
Vitamin D (% taking supplement)					45%	55%	53%	70%
Magnesium (mg)	420 (350)	420 (350)	320 (265)	320 (265)	353	318	277	250
Iron (mg)	8 (6)	8 (6)	8 (5)	8 (5)	16.2	16.1	12.2	11.8

(continues)

Table 56.2 Recommendations and Intakes of Selected Nutrients for Older Adults *(continued)*

	RDA (EAR) or AI or CDRR[a]				Intakes from Food from 2015–2018[b] (unless otherwise indicated)			
	Men 50–70 y	Men >70 y	Women 50–70 y	Women >70 y	Men 50–70 y	Men >70 y	Women 50–70 y	Women >70 y
Zinc (mg)	11 (9.4)	11 (9.4)	8 (6.8)	8 (6.8)	13.2	11.9	9.4	8.6
Folate (μg DFE)	400 (320)	400 (320)	400 (320)	400 (320)	572	554	426	418
Folic acid (% taking supplement)					38%	39%	35%	45%
Vitamin B_{12} (μg)	2.4 (2.0)	2.4 (2.0)	2.4 (2.0)	2.4 (2.0)	5.70	5.45	4.04	3.95
Vitamin B_{12} (% taking supplement)					41%	45%	41%	55%
Vitamin B_6 (mg)	1.7 (1.4)	1.7 (1.4)	1.5 (1.3)	1.5 (1.3)	2.47	2.25	1.70	1.64
Vitamin A (μg RAE)	900 (625)	900 (625)	700 (500)	700 (500)	710	805	598	703
Vitamin E (mg alpha-tocopherol)	15 (12)	15 (12)	15 (12)	15 (12)	10.5	8.7	8.4	8.0
Vitamin K (μg)	120	120	90	90	130	122	119	123

AI, adequate intake; *CDDR*, Chronic Disease Risk Reduction Intake; *DFE*, dietary folate equivalent; *EAR*, estimated average requirement; *NHANES*, National Health and Nutrition Examination Survey; *RAE*, retinol activity equivalents; *RDA*, recommended dietary allowance.[10]

[a]Recommendations for intake are from the Dietary Reference Intakes as cited by DGA 2020–2025.[10]

[b]Intakes are from Agricultural Research Service.[11]

[c]Multiply micrograms of vitamin D by 40 to obtain international units.

Data from Agricultural Research Service, US Department of Agriculture, National Health and Nutrition Examination Survey, 2007 to 2008. What we Eat in America. Nutrient Intakes from Food: Mean Amounts Consumed per Individuals, One Day, 2007–2008. Available at: www.ars.usda.gov/Services/docs.htm?docid=18349. Accessed April 16, 2011.; Data from Bailey RL, Dodd KW, Glodman JA, et al. J Nutr 2010;140:817–22; Data from Bailey RL, Dodd KW, Gahche JJ et al. Am J Clin Nutr 2010;91:231–7.

protein, fiber, magnesium, zinc, vitamin B_6, vitamin A, and vitamin K. DRI recommendations increase with age for vitamin D.[10]

As already indicated (Table 56.1), a host of physiologic and psychosocial factors can influence food intake and determine whether diets consumed by older adults meet nutritional needs. As illustrated in Table 56.2, results of National Health and Nutrition Examination Survey (NHANES) food intake assessments show that mean intakes from diet alone exceeded recommendations for protein, sodium, iron, zinc, folate, vitamin B_{12}, vitamin B_6, and vitamin K. Nutrients for which dietary intakes were generally lower than recommendations included fiber, potassium, magnesium, calcium, and vitamins A, D, and E. The intakes of most nutrients were consistently higher in persons 50 to 70 years old compared with those 70 years old or older, except for intakes of vitamin D and vitamin A.

Dietary Guidelines for Older Americans

Along with the DRIs for nutrients, the Dietary Guidelines for Americans uses a food-based based approach to assist with meal planning for congregate and home-delivered meals and in institutional settings, as well as for general dietary guidance.[10] Food-based recommendations at various energy intakes facilitate meal planning (e.g., the recommended servings of fruit, vegetables, whole grains, meat equivalents, and milk products). Specific recommendations relevant to older adults emphasize consuming vitamin B_{12} from fortified foods such as breakfast cereals. Across the life span, the nutrients of greatest concern were identified as vitamin D, calcium, potassium, and dietary fiber.

Nutrient-Specific Concerns in Older Adults

Energy, Protein, Fiber, and Fluid

Energy requirements as well as intakes decrease with advancing age, such that the reduction in energy requirements are 500 to 700 kcal/day between early adulthood and late life in healthy men and women.[12] These decreases in energy requirement and intake create challenges in meeting requirements for protein, as well as for some vitamins and minerals.[12]

Protein intakes decrease throughout adulthood, with the lowest intakes among those >70 years of age. The current recommended dietary allowance (RDA) for protein does not change for men (56 g) or women (46 g) from 19 to 51+ years of age[10] and 21% of women and 13% of men aged 70 and older consume less than the RDA for protein. Protein intake recommendations (RDA) are also expressed as 0.80 g/kg of body weight/day but emerging evidence suggests that protein intakes of 1.0 to 1.2 g/kg of body weight may be optimal for older adults to prevent muscle wasting through mechanisms involving improved update of dietary amino acids and reduced anabolic signaling of protein synthesis.[12] Higher protein intakes may also be involved in prevention and treatment of musculoskeletal disorders such as frailty, falls, and osteoporosis.[12] Most community-dwelling individuals are not at high risk for protein or protein-energy malnutrition but home-bound[13] and hospitalized older adults (see subsequent sections) as well as nursing home residents are all at risk of protein insufficiency. Reduced food intake resulting from the anorexia of aging may also jeopardize the adequacy of protein and other essential nutrients. Frailty secondary to poor nutritional intake is addressed in a subsequent section.

Fiber

The adequate intake for total fiber is based on energy intake and not on age.[10] Intake of fiber is less than recommended (Table 56.2). Dietary sources of fiber include whole grain foods, fruits, and vegetables and may be soluble and insoluble with different functional properties.[10,14] Meta-analyses of prospective cohort studies reported associations of whole-grain consumption with lower risk of heart disease, stroke, and cardiovascular disease, as well as decreased overall mortality and mortality due to CVD, cancer, diabetes, and other chronic conditions.[14] There is also interest in discovering the mechanisms and health impacts of the interrelationships of immune function, the microbiota and dietary factors, such as fiber.[15]

Dietary strategies to address chronic constipation include consuming the recommended intake of fiber, including coarse wheat bran fiber, as well as adequate intake of water and related fluids.[12] The prevalence of constipation increases with age.[16] Expert consensus on the assessment and treatment of constipation in older adults include dietary strategies, such as consuming 25 to 30 g dietary fiber daily,[14,16] while keeping in mind declines in masticatory function in advanced age that may require texture modified foods.[17] Prebiotic fibers such as inulin (e.g., from chicory) incorporated into foods may be appropriate for individuals with dysphagia and/or who require tube feeding.[17] Fiber-containing oral nutritional supplements or enteral nutrition will likely have metabolic and intestinal benefits.[16]

Fluid

Comprehensive assessment of nutritional status includes determination of fluid intake and hydration status.[16] Recommendations to identify, treat and prevent dehydration in older persons in general, and from a geriatric perspective, were developed by the European Society for Clinical Nutrition and Metabolism,[16] including recommendations related to low-intake dehydration and volume depletion (e.g., excess losses of fluids and electrolytes, especially sodium). Drinks or beverages account for about 70% to 80% of fluid intake, leading to recommendations to consume 1.6 L of drinks per day for women and 2.0 L for men. These recommendations apply to conditions of moderate environmental temperatures and moderate physical activity levels, and should be modified in higher summer temperatures and/or with greater physical activity. Excessive fluid losses due to fever, diarrhea, blood loss, and related problems are other indications for increasing fluid intake. Some conditions, such as heart failure and renal failure, may require restriction of fluid intake. It is a misconception that caffeinated drinks are "dehydrating" and it is important to offer a variety of drinks that are preferred by older adults so that hydrating will be more enjoyable compared with always drinking water.[16]

Alcohol

Alcohol use has increased in those 60 years of age and older over the past few decades, especially among women.[18] Approximately 20% of adults aged 60 to 64 years and 11% of those over age 65 years report current binge drinking.[18] Routine office visits and the Medicare Annual Wellness Visit may be opportunities to screen for problematic alcohol use with validated tools.[19]

Specific health and nutritional concerns include age-related declines in tolerance to alcohol with heavy drinkers being at risk for B-vitamin deficiencies such as thiamine,[20] folate, and vitamin B_{12}, increased antioxidant nutrient requirements (from increased oxidative stress with heavier alcohol consumption), and malnutrition.[21] Duration, amount, and source of alcohol consumption (red wine compared with other sources) also matter.[20] Chronic alcohol consumption in alcoholics is related to negative impacts on brain structures, including increases in volume of cerebral spinal fluid and decreases in volume of gray and white matter.[20] Heavy alcohol consumption is associated with alcohol-related dementia in the aging brain, while mild to moderate intake is associated with neuroprotective effects and prevention of cognitive decline and dementia.[20] Heavy drinking also increases risk of cardiovascular disease and damage to the cardiovascular system (that may be exacerbated by malnutrition and thiamine deficiencies), while mild alcohol consumption may help prevent damage to the cardiovascular system and decrease the risk of myocardial function.[20]

Numerous terms and definitions of unhealthy alcohol use from national and international organizations are used,[22] including DSM-5 criteria for alcohol use disorder and applications to older adults.[19] Alcohol intake guidelines relevant to adults of all ages includes, "Adults of legal drinking age can choose not to drink, or to drink in moderation by limiting intake to 2 drinks or less in a day for men and 1 drink or less in a day for women, when alcohol is consumed. Drinking less is better for health than drinking more.[10]" The National Institute on Alcohol Abuse and Alcoholism defines "risky use" for older adults as three drinks per day or seven drinks per week (42 g/d or 98 g/week of alcohol) for all adult women of any age and men 65 years or older.[22]

In contrast to population level recommendations,[10,22] geriatric recommendations and perspectives are often individualized. Examples include, "Alcohol use may be contraindicated in older adults due to pre-existing health conditions, heavier use of OTC and prescribed medications, greater risk of balance problems, frailty and falls, and/or cognitive dysfunction,[21]" "the interplay of ethanol with physiological changes of aging, comorbidity and polypharmacy increases risks for falls, traffic accidents, drug interactions and victimization,[19]" and "the risk in "at risk" drinking includes the drinker and a spouse [or other care-recipient] who is also at risk due to disability, frailty, and cognitive impairment.[19]"

Vitamin D and Calcium

Calcium and vitamin D are well known for their roles in bone health and have many other functions, such as differentiation, growth, immunity, insulin secretion, blood pressure, and neuromuscular function.[23] Intakes of vitamin D and calcium are below recommendations (Table 56.2). Adequate calcium and vitamin D intake throughout life protects skeletal health. Relationships of vitamin D, calcium and osteoporosis are discussed in an upcoming section. Menopause leads to bone loss through decreased estrogen that increases bone resorption and decreases calcium absorption.

Although there is no clinical biomarker for calcium status, vitamin D status can be assessed through serum/plasma 25-hydroxyvitamin D[24]. A 25-hydroxyvitamin D concentration of 50 or more nmol/L is generally considered adequate.[24] It is generally considered that 25-hydroxyvitamin D concentration less than 2,530 nmol/L (10–12 ng/mL) is indicative of vitamin D deficiency and should be treated.[24] Although guidelines and recommendations are numerous and inconsistent, it is reasonable for health practitioners to recommend a dietary supplement of 800 IU/day of vitamin D (preferably as the more bioavailable form, D_3) in those individuals with measured serum 25-hydroxyvitamin D <50 nmol/L and in individuals with osteoporosis.[24] Correction of vitamin D deficiency likely requires consideration of disease states and risk factors when recommending higher doses of vitamin D along with follow-up measurements of 25-hydroxyvitamin D.[24]

In NHANES 2001 to 2006, the prevalence of serum 25-hydroxyvitamin D ([OH]D)[25] concentration lower than 30 nmol/L (risk of deficiency) was 6%, 7%, 11%, and 11% in men 51 to 70 years, men more than 70 years, women 51 to 70 years, and women more than 70 years of age, respectively, whereas the prevalence of serum[25] (OH)D concentration of 30 to 49 nmol/L (risk of inadequacy) was 25%, 24%, 28%, and 27% in these groups, respectively.[25] After 80 years of age, risk factors such as advanced age, race (Black vs White), season, and nonuse of dietary supplements continued to be associated with poor vitamin D status.[26]

Vitamin B_{12}, Vitamin B_6, and Folate

Vitamin B_{12} is required for two enzymes and supports development, myelination, and function of the central nervous system, healthy red blood cell formation, and DNA synthesis.[27] Vitamin B_{12} intake in older adults is well above recommendations (Table 56.2) and the increased prevalence of deficiency with aging is attributed mainly to acquired malabsorption.[27,28] Given concerns about malabsorption of protein-bound vitamin B_{12}, it is recommended that older adults consume vitamin B_{12} from fortified foods, such as breakfast cereals, and dietary supplements.[10,28]

Pernicious anemia is an autoimmune disease with loss of intrinsic factor and is the cause of severe malabsorption of vitamin B_{12} in about 1% to 2% of older adults.[28] Other causes of malabsorption are chronic autoimmune atrophic gastritis that develops with antibodies to gastric parietal cells (and to intrinsic factor in 50% of affected individuals), bacterial overgrowth in the gastrointestinal tract, some infections (e.g., *Helicobacter pylori*, the most common cause of atrophic gastritis), bariatric surgery, total or partial gastric resection or bypass, ileal disease or resection, and medications including nitrous oxide, metformin, and stomach acid blockers (histamine H_2 receptor antagonists and proton pump inhibitors that lead to achlorhydria).[28,29] Lack of intake of foods of animal origin and malabsorption of protein-bound vitamin B_{12} (related to gastritis) are other causes of deficiency.[28] Even between the ages of 80 and 100+ years, risk factors such as advanced age, atrophic gastritis, nonuse of dietary supplements, and race (White race) were associated with poor vitamin B_{12} status.[30]

Serum vitamin B_{12} and holotranscobalamin are biomarkers of circulating vitamin B_{12} concentrations.[31] Serum MMA and homocysteine are functional biomarkers of vitamin B-12 status,[31] but may also be elevated by impaired renal function. While elevated MMA is specific for low vitamin-B_{12} status, elevated homocysteine also occurs in response to inadequate folate, riboflavin, or vitamin B_6 status. Commonly used cutoffs for serum vitamin B_{12} concentration to identify deficiency, low, moderate, and other categories vary.[31] Examples include: deficient is <100 pg/mL (74 pmol/L and requires intramuscular or high oral supplements): suboptimal/indeterminate/low/marginal is between 100 and 350 pg/mL (74–258 pmol/L): and adequate is >350 pg/mL (258 pmol/L).[27-29,31]

Using multiple measures of vitamin B_{12} status is costly and potentially difficult to interpret, so strategies could include ensuring adequate concentration of serum vitamin B_{12}[28] of at least 300 to 350 pg/mL or higher through fortified foods and/or dietary supplements. Also, high-dose dietary supplements of 1,000 to 2,000 μg/day to normalize serum vitamin B_{12} to concentrations similar to that from intramuscular vitamin B_{12}.[27] A strategy for prevention of vitamin B_{12} deficiency could involve food fortification[28] as has been implemented in the United States and numerous other countries to enhance folate status.

The recommended intakes for vitamin B_6 are higher for older, compared with younger, people. Average daily intake among older adults is higher than recommendations (Table 56.2). A large portion of vitamin B_6 intake is from fortified foods and supplements.[29] In the United States, insufficient intake, or low concentrations of vitamin B_6 biomarkers, generally occur with poverty, chronic alcoholism, malabsorption syndromes, and drug-nutrient interactions.[29] Inflammation is associated with lower vitamin B_6 biomarkers.[29] At-risk populations for deficiency are older adults, residents of nursing homes, and those with alcoholism independently from liver disease, rheumatoid arthritis, status epilepticus, chronic infections (including HIV), and taking anti-tuberculosis medications (isoniazid) or anti-Parkinson medications.[29] Vitamin B_6 can be toxic in intakes of 300 to 500 mg/day that are sometimes recommended as nonevidence-based therapies: symptoms are sensory ataxic neuropathy that are generally, but not always, reversible.[29]

The EAR and RDA for folate are similar for older and younger adults, except that no specific recommendation exists for older women to consume folic acid.[10] Dietary intake alone (including folic acid from fortified foods) is much higher than the recommendations (see Table 56.2). Adults aged 51 to 70 years are more likely than other age groups to take supplements containing folic acid.[32] There is ongoing research and interest in the health benefits and risks of higher intakes of folic acid from supplements, such as potentially decreasing the risk of stroke, decreased cognition and/or dementia, as well as potentially increasing the risk of some cancers.[32] As this research continues, it is prudent for older adults who choose to take a dietary supplement to select a supplement with up to 400 μg of folic acid/day and keep their overall intake of folic acid below the upper level of 1,000 μg folic acid/day.

Iron

Aging decreases iron requirements for older women (cessation of menstruation) such that iron recommendations are the same for older men and women.[33] Average intakes of iron among older adults in the United States generally exceed recommendations (Table 56.2).

Iron deficiency anemia occurs in about 10% of people aged 65 years and older, and about 20% in those aged 85 years and older.[34] Iron deficiency is only one of many contributors to anemia. In older adults, anemia is generally associated with impaired physical and cognitive function, decreased quality of life, increased hospital admissions, prolonged hospital stays, and increased mortality.[35]

The three iron compartments are 1) functional hemoglobin, myoglobin, and respiratory enzymes: 2) iron reserves in the liver as ferritin: and 3) transferrin.[34] The two types of iron deficiency are 1) absolute iron deficiency with low serum ferritin and low transferrin associated with the decrease in iron reserves caused by gastrointestinal tract bleeding, ulcers, malignancies, infections, some medications, inflammatory bowel disease, and other disorders and 2) functional iron deficiency in which iron reserves are high or normal due to inflammation associated with a number of chronic conditions prevalent in older adults.[34] Inflammation also leads to increases in hepcidin that decreases intestinal absorption of iron and promotes sequestration of iron (increasing ferritin and decreasing serum iron). With chronic kidney disease, endogenous erythropoietin is inadequate to support red blood cell production. Differentiating these types of iron deficiency alongside hormonal changes (hepcidin and erythropoietin) and multiple comorbidities is essential in identifying causes and treatments of iron deficiency.[34,35]

Strategies in older adults to replenish iron stores continue to evolve and include using multiple biomarkers of iron status and inflammation, addressing general malnutrition if it co-exists with iron deficiency, correcting erythropoietin deficiency, providing relatively high dose iron supplements, and/or parenteral iron, as appropriate to the cause and degree of iron deficiency and associated comorbidities.[34,35] Perspectives on oral and parenteral iron administration are of particular interest.[34,35]

Vitamins A, E, and K

Vitamin A intakes from food are generally higher than the EAR but lower than the RDA in older adults (see Table 56.2). Although vitamin A intake recommendations do not change with age, age-associated or alcohol-associated impairments in renal and/or liver function[36] may increase the risk of vitamin A toxicity, e.g., from high-dose dietary supplements. Although there are no specific guidelines for older adults, it seems reasonable to follow guidance for those who are or might be pregnant and those who are lactating to not take high doses (no more than 3,000 μg RAE, 10,000 IU daily) of vitamin A supplements.[37]

Mean intakes of vitamin E are less than the EAR (see Table 56.2). Although vitamin E is not a nutrient of concern according to the Dietary Guidelines for Americans,[10] others propose that recommendations should be increased for older adults, based on evidence suggesting benefits of higher intake of vitamin E on immune function, inflammatory processes, and resistance to infection.[38] For example, evidence suggests that 200 IU/day (or 200 mg/day in some studies) is effective in improving certain aspects of immune function in older adults.[38] There is also evidence from two intervention trials that long-term use of high intakes of vitamin E as alpha-tocopherol may increase risk of hemorrhagic stroke (50 mg/day in Finnish male smokers: 180 mg/day in male physicians in the US).[39] Identifying and balancing potential risks and benefits of supplemental vitamin E through clinical trials is needed.

The recommended adequate intake (AI) for vitamin K decreases with aging. While mean intakes appear adequate (see Table 56.2), more than 50% of those aged 71 years and older have intakes less than the AI.[40] Vitamin K intake is primarily from green leafy vegetables and vegetable oils, as well as mixed dishes.[40] Normal coagulation depends on several vitamin K-dependent proteins.[40,41] Vitamin K can interfere with the efficacy of coumarin medications (warfarin) to prevent clotting,[41] which may lead some individuals to avoid vitamin K-rich foods such as leafy greens. However, there is evidence that anticoagulation control is less variable and more stable among those with higher, compared with lower, intakes of vitamin K.[41] Communications with patients and routine monitoring of clotting time are useful in assessing the impact of any changes in vitamin K intake (from diet or supplements).

Other vitamin K–dependent proteins are found in cartilage, bone, and vascular tissue and vitamin K may be involved in cardiovascular disease, joint disease, inflammation, brain, and endocrine function.[40,41] Although research is ongoing, it is premature to determine whether higher intakes of vitamin K may be protective against age-associated disorders.[40,41]

Magnesium

Magnesium dietary recommendations remain the same for adults after the age of 30 years, and men have higher recommendations than women.[42] Magnesium intakes were lower than the EAR in older adults (see Table 56.2). Several medications can affect magnesium status by increasing or decreasing magnesium absorption, excretion, or translocation.[43] Although older adults take medications that may interfere with magnesium absorption, such as proton pump inhibitors, neither age nor sex affected the relationship between controlled intakes of magnesium and losses (in feces and urine).[43,44]

Results from cross-sectional and prospective studies are inconsistent regarding relationships of lower magnesium status (intake and/or blood concentration) with chronic conditions, including inflammation, cardiovascular disease, hypertension, stroke, metabolic syndrome, type 2 diabetes, bone health, osteoporosis, and osteoporosis.[43] Randomized controlled trials are needed to better understand the clinical significance of higher magnesium intakes and disease-related outcomes. As cited by others,[43] RCT for hypertension are promising and show reductions in blood pressure. Also, magnesium is one of the few minerals that has been shown in RCT to improve muscle function.[45,46] Among older women enrolled in a weekly exercise program, magnesium supplements significantly enhanced muscle strength and physical performance (e.g., chair stands, walking speed), especially among those with lower baseline physical performance or lower intakes of magnesium.[46]

Zinc

Zinc is required for the activity of hundreds of enzymes and has roles in immune function, protein, and DNA synthesis, cell signaling and division, and in bone structure.[47,48] Dysregulation of immune function in zinc deficiency includes increased inflammatory cytokines and decreased antioxidant status that may increase susceptibility to infection, which is important given the known decrease in immune function with aging.[48] However, zinc supplementation is not consistently associated with reductions in duration or severity of the common cold.[47] Supplements including zinc and certain antioxidants were associated with slowing the progress of age-related macular degeneration; however, the unique role of zinc is unclear.[47] Also, the role of zinc supplements to prevent or ameliorate other age-related disorders, such as type 2 diabetes and osteoporosis, remains unclear.[47]

The interaction of aging with zinc requirements is poorly understood. Moreover, dependable markers of true zinc status are lacking. Mean zinc intakes are higher than the RDA in community-dwelling older adults (see Table 56.2). A systematic review of zinc status provided evidence that potential zinc inadequacy was more prevalent in those dwelling in institutions, compared with those living in the community.[48]

Micronutrient Supplements, Cardiovascular Disease, and Cancer

Cancer and heart disease are the top two causes of death in the United States. Regarding use of vitamins and/or multivitamins for prevention of cardiovascular disease and cancer in community-dwelling, nonpregnant adults, the 2022 United States Preventive Services Task Force (USPSTF)[49]:

- Recommends against the use of beta carotene or vitamin E supplements, Grade D, meaning do not use.

- Concludes that the current evidence is insufficient to assess the balance of benefits of harms of the use of multivitamin supplements, Grade I, meaning insufficient evidence.
- Concludes that the current evidence is insufficient to assess the balance of benefits and harms of the use of single or paired nutrient supplements (other than beta carotene and vitamin E), with Grade I, meaning insufficient evidence.

The summary for clinicians notes that an important harm of increased lung cancer incidence was reported with the use of beta carotene by persons who smoke tobacco or have occupational exposure to asbestos; excessive doses of vitamin supplements can cause adverse effects; it is uncertain whether there might be heterogeneity across populations by baseline nutrient level, or by socioeconomic factors, such as food insecurity, in the effects of vitamin, mineral, and multivitamin supplementation on cardiovascular disease and cancer outcomes.[49] Moreover, the USPSTF states that "clinical decisions involve more considerations than evidence alone. Clinicians should understand the evidence but individualize decision making to the specific patient or situation".[49] USPSTF[49] also acknowledges other evidence-based recommendations from the American Heart Association and US DHHS[10] that the preference is that nutrient needs to be met primarily from nutrient-dense foods and beverages.

Nutrition-Related Health Concerns and Community-Based Services

Physical Activity and Obesity

Age-related declines in physical activity occur due to a combination of reduced opportunity, interest, and functional ability to participate in exercise and other active pursuits. A leading cause of this physical decline is loss of lean muscle mass over time, a reduction from about 50% of body weight during young adulthood to about 25% at 75 to 80 years.[50] This marked change in body composition leads to slower movement, declines in strength and power, and increased risk for falls. Fear of falling is another element of declining physical activity. A cycle of inactivity and further functional decline elevates the risk for chronic conditions, including cardiovascular disease, hypertension, certain cancers, and type 2 diabetes, as well as for premature mortality.

While reduced physical activity plays a role, age-associated decreases in energy requirement and a gradual accumulation of body fat mass are also important contributors to the increase in adiposity and body weight experienced by many older adults. More than 40% of adults 60 and older are obese,[51] having a body mass index (BMI) $\geq$30 kg/m^2. Obesity accelerates chronic diseases and conditions already prevalent in older adults, including heart disease, type 2 diabetes, and osteoarthritis. Of particular concern is a common type of geriatric obesity wherein excess fat mass is accompanied by sarcopenia (age-related loss of muscle mass and function), a condition known as sarcopenic obesity. Definitions and diagnostic criteria for sarcopenic obesity were recently published by the European Society for Clinical Nutrition and Metabolism (ESPEN) and the European Association for the Study of Obesity (EASO) and the cumulative risks of the combination of the two conditions were emphasized.[52] Additionally, with continued study, it appears that sarcopenic obesity may also affect bone health and risk of falling and fracture.[53]

Unfortunately, the treatment of obesity and sarcopenic obesity in older adults is not straightforward. Obesity can be especially difficult to treat in older adults, who have lower basal metabolic rate, energy requirement, and physical activity. Additionally, there are well-recognized concerns about weight loss in older adults, including concomitant losses of lean muscle and bone mass.[54] Moreover, there is evidence that an overweight BMI is preferable to either a normal BMI or an obese BMI in terms of survival for older adults and increases in BMI have been linked with lower mortality risks compared with decreases in BMI.[55] Nonetheless, obesity in older adults is clearly linked with detrimental effects on both health and function, and intervention trials have shown clinically significant benefits of weight reduction with regard to osteoarthritis, physical function, diabetes, and cardiovascular disease.[56] When exercise is combined with weight loss, it may promote conservation of lean body mass and bone mass, and improve cardiorespiratory function and balance. Clinical studies continue to explore the safest and most effective intervention strategies for obese older adults and to seek more individualized interventions, based on risk factors and quality-of-life priorities.

Osteoporosis

Osteoporosis is the most common bone disease affecting older adults. Combining the cases of osteoporosis and low bone mass (at the femoral neck or lumbar spine), an estimated 53.6 million U.S. adults, aged 50 years and older, were affected by age-related bone disease in 2010.[57] Although many dietary factors influence bone health throughout life, the most critical bone-supporting nutrients for older adults are calcium and vitamin D. In addition to being essential for healthy bones, vitamin D has also been shown to reduce the risk of falls.[58] While this should reduce the risk of fractures, adequate calcium status is likely also required to reduce this risk.[59] As noted earlier, intakes of both calcium and vitamin D tend to be lower in older adults. Thus, assessment of both calcium intake and vitamin D status (via the blood biomarker 25-hydroxyvitamin D) may be warranted in those known to have low bone mass or at high risk for it. Not only are these nutrients essential for bone mass maintenance, they are an important cornerstone of pharmacologic management of osteoporosis.[60]

Diabetes

Among adults 65 years old or older in the United States, approximately 25% have diabetes and 50% have prediabetes,[61] with these numbers expected to increase in the coming decades. Older adults with diabetes have higher rates of related conditions such as hypertension, coronary heart disease, and stroke, than those without diabetes. Along with higher rates of accelerated muscle loss and functional disability, they also have higher rates of common geriatric syndromes such as cognitive impairment, depression, visual impairments, urinary incontinence, injurious falls, and chronic pain and have higher likelihood of premature death.[61]

There are unique considerations and challenges in the treatment of older adults with diabetes that are not traditionally associated with diabetes management. Clinical presentation, psychosocial environment, and available resources are variable in the older population. Those with a greater number of existing comorbidities and/or those in a more disadvantaged living situation will have different treatment goals than those who are mostly healthy, free living and socially supported. Living situation (e.g., community, assisted living, nursing home) and available social support must be taken into account for both treatment goals and methods of management.

Hypoglycemia can be a serious concern in older adults. "Simplification" (making it easier for the patient to adhere to) and/or "deintensification" (reducing the potency of the treatment regimen) may be recommended to reduce this risk.[62] Detailed guidance on treatment goals for glucose and lipid control has been recently published and detailed elsewhere.[61,63] At more advanced ages treatment goals tend to shift away from tight glycemic control and focus more on control of other cardiovascular risk factors. Generally speaking, older adults who are mostly healthy, with few coexisting chronic illnesses and intact cognitive function and functional status, have more stringent glycemic goals (e.g., A1C <7.5%), while those with multiple coexisting chronic illnesses, cognitive impairment, or functional dependence have less-stringent glycemic goals (e.g., A1C 8.0%–8.5%). Special recommendations apply for those with advanced diabetes complications, life-limiting comorbid illnesses, and at the end of life.[61]

Cardiovascular Disease, Dyslipidemia, and Hypertension

Based on data collected in 2015 to 2017 in the Behavioral Risk Factor Surveillance System,[64] 21.7%% of community dwelling adults aged 65 years and older report having coronary heart disease (CHD) and/or a stroke. Dyslipidemia and hypertension are major risk factors for these conditions, yet older adults are often undertreated due to concerns about comorbidities, drug interactions, and other considerations. The relationship of total cholesterol and LDL-cholesterol (LDL-C) with mortality tends to be U-shaped after the 7th decade of life, complicating therapies to address dyslipidemia in older adults.[65] Yet, LDL-C remains associated with CHD events in later life and there is strong evidence supporting the benefits of lipid-lowering medications for secondary prevention of CHD through about age 85 years. While statins are the first-line therapy for atherosclerotic CHD, they are underprescribed in older adults and may have lower adherence rates in this population, thus emphasizing that increased physical activity and following heart healthy dietary guidelines (daily diet with <10% of energy from saturated fats, <10% from added sugars, and <2,300 mg sodium/day) are also fundamental components of therapy.[66]

Hypertension is the most common modifiable cardiovascular risk factor in older adults, with a prevalence of about 70% in adults aged 75 years and over. It is a major risk factor for CHD, peripheral artery disease, incident myocardial infarction, stroke, and heart failure.[66] Unfortunately, blood pressure control is often difficult to achieve in older adults, due to factors such as a widened pulse pressure, the need for ≥2 medications to achieve adequate control (and concomitant problems with side effects, polypharmacy), more frequent use of medications that increase blood pressure (e.g., nonsteroidal anti-inflammatory agents), and higher sodium intake due to increased consumption of prepared foods and/or changes in taste ability that increase the preference for salt.[66,67] Whenever possible, nonpharmacologic treatments are preferred and may be sufficient alone in milder cases of hypertension (e.g., reduction of excessive body weight, lowered sodium and alcohol intake, increased physical activity, and following a Dietary Approaches to Stop Hypertension [DASH]- type diet that is rich in fruits, vegetables, and low-fat dairy foods and low in saturated and total fat). Even if not sufficient to fully achieve normal blood pressure, these lifestyle measures can still reduce the number and dose of antihypertensive medications.[67]

Stroke

Cerebrovascular accident (CVA: stroke) is a major cause of debilitation and death in the United States, and the risk of stroke increases with age, such that the prevalence is 15% in those aged 80+ years.[66] The majority of strokes are ischemic in origin, and the leading risk factors for ischemic stroke are hypertension, dyslipidemia, diabetes, and smoking. There is strong evidence that control of hypertension and dyslipidemia can reduce new and recurrent strokes although, as already noted, both of these conditions can become increasingly difficult to manage with increasing age.[66]

For those experiencing a CVA, one of the most common and life-threatening symptoms of neurologic disorders is oral pharyngeal dysphagia (OD), occurring in at least 50% of patients with ischemic or hemorrhagic stroke.[68] OD greatly increases the risk of malnutrition and dehydration, as well as the danger of aspiration pneumonia. Based on instrumental swallow evaluation (fiberoptic endoscopic or videofluoroscopic), patients with swallowing disorders may be prescribed texture-modified diets (e.g., soft, chopped, pureed, minced) with or without thickened fluids, as well as receive counseling to consume higher amounts of energy and protein. Whether or not patients with stroke are malnourished at the time of hospital admission, they are at high risk of becoming malnourished during their recovery, not only because of OD but also because of other physical impairments, such as paralysis and neuropsychological disturbances. There is considerable evidence that malnutrition in patients with stroke is associated with longer length of hospital stay, greater hospitalization costs, and increased mortality. Available evidence also supports that all stroke patients should be screened for risk of malnutrition on admission to hospital (within 48 hours) to identify those most likely to benefit from medical nutrition therapy. Utilization of screening and assessment tools validated for the stroke population should be emphasized.[68] Severe swallowing disorders that are expected to preclude adequate oral intake for a week or more indicate the need for early enteral nutrition via feeding tube (at least within 72 hours). While it may not be feasible to start the feeding in uncertain situations (e.g., cerebral hemorrhage or need for ventilation), an early start of enteral nutrition in acute disease does have advantages, including keeping the barrier function of the gut mucosa intact and, thus, reducing bacterial translocation of gut-bacteria into the systemic blood flow.[69]

Renal Disease

Chronic kidney disease (CKD) is common in older adults and carries significant health and economic burdens, with major precipitating conditions including obesity, diabetes, and hypertension. Additionally, malnutrition may lead to increased risk of kidney function decline and CKD, emphasizing the need to address malnutrition risk, even in those older adults without advanced kidney disease.[70] There are six different stages of renal disease, Stage 1 mild CKD (eGFR >90mL/min), Stage 2 mild (eGFR 60 to 89 mL/min), Stage 3a moderate (eGFR 45 to 59 mL/min), Stage 3b moderate (eGFR 30 to 44 mL/min), Stage 4 severe (15 to 29 mL/min) and Stage 5 kidney failure (eGFR <15 mL/min: also called end-stage renal disease: ESRD), making diet therapy complex, because it varies by stage. Recommendations for intakes of protein, sodium, phosphorus, potassium, and fluids must be carefully individualized and also adapted over time if there is disease progression. In particular, a diet low in protein (0.6 to 0.8 g/kg/day) is recommended in all but ESRD to delay the progression of CKD and help mitigate metabolic derangements such as metabolic acidosis and hyperphosphatemia.[71] Only in the case of ESRD and ongoing dialysis treatment is the protein restriction lifted, because hypermetabolism increases the protein requirement and dialysis removes protein from the blood. Recommendations are for a much higher protein intake (1.2–1.4 g/kg of ideal body weight/day) in ESRD patients being treated with dialysis.[72]

An additional concern comes from the increasing recognition that older adults with CKD are predisposed to conditions where protein wasting plays an important role, including sarcopenia, protein-energy wasting, and frailty. This emphasizes the need to harmonize kidney-preserving dietary protein restriction with the need to maintain sufficient protein intake to assure adequate energy and nitrogen balance.[73] It has been suggested that a plant-dominant, low-protein diet (composed of >50% plant-based protein sources) may be helpful for relieving uremic burden and metabolic complications in CKD compared with animal-dominant protein consumption.[71] It also follows that CKD patients should be routinely monitored for signs of nutritional concerns, including protein-energy wasting.

Food Insecurity

In the United States, food insecurity is said to exist when a household (of one or more persons) answers affirmatively to these three questions, "We worried whether our food would run out before we got money to buy more." "The food we bought just didn't last, and we didn't have money to get more." and "We couldn't afford to eat balanced meals.[74]" For 2020, USDA reported the national prevalence of food insecurity as 6.9% among households with an older adult, and 8.3% among older adults living alone.[74]" Food insecurity in older adults is associated with numerous problems, including poor food and nutrient intakes, adverse physical and mental health outcomes, medication management problems, and weight-related disability.[75-77]

Food and Nutrition Assistance Programs and Home and Community-Based Services

The Academy of Nutrition and Dietetics and the Society for Nutrition Education and Behavior issued a position statement that stated, "older adults should have access to evidence-based food and nutrition programs that ensure the availability of safe and adequate food to promote optimal nutrition, health, functionality, and quality of life" and "The rapidly growing older population increases demand for integrated continuous support systems, and the rising cost of health care underscores the need for these programs.[78]" A network of food and nutrition assistance programs across the nation helps meet these needs[78] with funding from the U.S. Department of Health and Human Services (DHHS), USDA food and nutrition programs, and state, local, and charitable sources. USDA typically refers to their programs as assistance programs, e.g., Supplemental Nutrition Assistance Program (SNAP). DHHS coordinates with states and local providers to provide congregate meals, home delivered meals, and health promotion programs, among several other home and community-based services. DHHS is well positioned to integrate nutrition services with many other assistance program areas for older adults.[79]

Several of these programs and their target audience, eligibility, services, participation, and expected outcomes offer opportunities for involvement by registered dietitians, nutritionists, and nutrition and dietetic technicians.[78] It is recommended that program outcomes include: decrease risk of malnutrition: prevent or reverse unintended weight loss: improve dietary alignment with the Dietary Guidelines for Americans, as determined by validated screening and assessment tools: improve food security: decrease avoidable admissions to hospitals, nursing homes, and other care settings associated with poor nutrition: and reduce hospital readmissions through integrated services and recognition of malnutrition risk during transitions of care.[78] Achieving these outcomes requires food assistance and meal programs, nutritional screening and assessment, nutrition education, medical nutrition therapy, monitoring, evaluation, and documentation of these evidence-based outcomes. To enhance the relevance and increase the funding of these programs, it also is recommended to design and implement uniform outcome data analysis for continuous quality improvement: to better target the programs to those most in need: to communicate and advocate for improved collaboration with healthcare systems, the continuum of care, policymakers and payers: and to provide and engage in education to enhance understanding of geriatric nutrition and aging.

Hospitalized Older Adults

Hospitalization is associated with substantial nutritional risk in older adults. Many are malnourished at the time of admission (about one in three[80]) and many more become malnourished during their hospital stay for reasons including pain, sedation, medication side effects, and restrictions for medical testing. Other risk factors include increased energy requirements for healing, poor appetite, and reduced ability to chew and/or swallow.[81] Malnutrition, an ICD10-AM-coded condition, is strongly and independently predictive of adverse healthcare outcomes, increased mortality, and elevated treatment costs, especially in older adults with multiple comorbidities.[80] An obvious first step for addressing this is early identification of malnutrition and prompt initiation of appropriate interventions to prevent further nutritional decline during hospitalization. However, even when identified, nutrition support prescriptions are often delayed or neglected.[80] Impediments to improving nutritional status in hospitalized older adults include both system and process barriers, yet there is evidence that early intervention for malnutrition will shorten length of hospital stay and reduce complications and readmission rates, mortality, and medical care costs. An interdisciplinary intervention is the most likely formula for success, which should be applied both in the hospital and during the acute post-discharge period.[82]

Late Life Concerns

Nutrition Concerns During Transitions of Care

The ideal preference of many older adults is to "age in place," transitioning from one stage of care to another without needing to move from their home. However, for most individuals, several changes in dwelling place will occur during the latter part of life, including hospital admissions and relocations to assisted living, rehabilitation, and/or skilled nursing care. Whenever transitions occur, there is a risk that gaps in continuity of nutrition care/support will occur. In particular, as already noted, the worsening of malnutrition during and after hospitalization is common, due to pressures for rapid discharge and low rates of uptake/implementation of post discharge nutrition care.[83] Even for those who are able to age in place, home care needs will evolve over time, creating the need to assess and address different kinds of nutritional concerns. Keller et al.[84] created two nutrition care pathways that offer guidance to health care and service providers: one addresses malnourished older adults transitioning from hospital to home and the other provides guidance for reducing malnutrition in community-dwelling older adults. Nutritional concerns in long-term care settings are particularly challenging due to the unique needs of these patients, who have multiple comorbidities and may be facing terminal illness (see End of Life).

The Role of Nutrition in Frailty, Sarcopenia, and Functional Decline

An overarching definition of frailty characterizes it as "a clinically recognizable state of increased vulnerability resulting from aging-associated decline in reserve and function across multiple physiologic systems.[85]" The classic operational definition from Fried et al.[86] defines the frailty phenotype as one meeting at least three of the five following criteria: low grip strength, low energy, slowed walking speed, low physical activity, and unintentional weight loss. The classification of "pre-frail," in which one or two of criteria exist, indicates high likelihood of progressing to frailty. Physical frailty, which is characterized by diminished strength and endurance and is associated with functional decline, falls, and fractures and reduced quality of life, accounts for many of the attributes included in the frailty criteria. A major contributor to physical frailty is sarcopenia, an age-related decline in muscle mass and strength. While genetics and physiologic characteristics help determine sarcopenia risk, nutrition is one of the few potentially modifiable etiologic factors. This is based on evidence of a protective role for higher protein intake that helps preserve lean body mass and prevent frailty, which are "accepted intervention targets for reducing the risk of sarcopenia."[87] Higher protein intake promotes muscle protein synthesis by increasing availability of plasma amino acids and can lead to improved muscle mass, strength, and function over time.[88] Older adults need more protein to optimally stimulate muscle protein synthesis due to age-related blunting of the muscle protein synthesis response to a given dose of protein/amino acids, or to exercise stimuli (anabolic resistance).[89] Protein energy malnutrition resulting from inadequate food intake can also be a major contributor to frailty. Malnutrition has been independently associated with physical pre-frail and frail conditions in community-dwelling older adults.[90] Further study is needed to understand the combined effects of negative energy balance and protein intake on lean mass and skeletal muscle protein turnover, risk of sarcopenia, and other components of frailty.[91] Causes of poor food intake can include loss of appetite, alterations in taste and smell, poor oral health, gastrointestinal changes, and a reduced ability to regulate appetite. In these instances, the best practice is to recommend the intake of high nutrient density foods to avoid weight loss, to achieve adequate intake of protein, and micronutrients, and to promote regular physical activity. For obese and frail older adults who will clearly benefit from weight reduction, a modest reduction of 200 to 500 kcal per day with a target of 0.5 to 1 kg loss per week, protein intake of at least 1 g/kg body weight per day, and a program of physical activity/exercise may be recommended.[92]

Nutritional Issues at the End of Life

Difficult decisions about nutrition and hydration are required during the management of any terminal illness, regardless of patient age,[93] and the recommendations differ from management among older adults in general.[16] Guidelines published by the European Society for Clinical Nutrition (ESPEN) in 2016 provide valuable, detailed information on a variety of issues concerning artificial nutrition and hydration and include sections related to the older patient and age-related conditions such as dementia.[93] End-of-life care for geriatric patients (patients over age 65 years, but often over age 80 years), otherwise known as Geriatric Palliative Care (GPC), presents specific challenges and complicating complexities, including reduced decisional capacity on the part of the patient, problems with communication and behavior, complex social considerations, and an extended trajectory of disease.[94] The extended trajectory of disease occurs when an older patient experiences a slow deterioration due to debilitating disease conditions, such that the point at which they become a "dying patient" may not be easily distinguishable.[93]

For this situation, there is sparse clinical evidence available for guidance.

It is important that physicians, nutritionists, and other healthcare providers be well informed so that they can assist families and other caregivers with decisions regarding nutrition and hydration at the end of life. For example, artificial nutrition support is discouraged for patients with advanced dementia when they refuse to eat or have dysphagia because there is no evidence that it improves quality or duration of life,[95] and it may even increase mortality.[96]

COVID-19: A Current Threat to the Health and Nutritional Wellbeing of Older Adults

Unvaccinated older adults have shown extreme vulnerability to COVID-19, both in terms of susceptibility and severity of complications when they become infected.[97] Even with the availability of vaccines and boosters, older adults remain especially vulnerable to the effects of the disease. Their susceptibility to malnutrition may be increased while experiencing the typical symptoms of COVID-19, including fever, bodily aches, vomiting, diarrhea, and loss of taste.[81] Guidelines are available for the nutritional care of patients with COVID-19: this includes an emphasis on hydration, and protein and energy intake for those at home and early feeding via an enteral route for hospitalized patients. Older patients in particular need early screening for underlying malnutrition.[98]

In addition to those directly affected by COVID-19 infection, an even larger population of older adults is experiencing negative health effects resulting from COVID-19-related isolation. With repeated periods of long-term isolation, many older adults have lost consistent access to their support networks and healthcare providers, with the result being social isolation and delays in seeking essential medical treatments and follow-up.[99] Especially with regards to social engagement, interruptions in activities such as congregate meals at Senior Centers, religious services, exercise classes, fitness center activities, and volunteer commitments, have compounded the negative effects of isolation. In some instances, reduced access to nutritious foods due to loss of income and pandemic-related inflation of food and other basic costs has intensified the risk of malnutrition and unintentional weight loss, leading to marginal nutritional deficiencies, reduced immunity, and increased mortality. In other cases, promotion of geriatric obesity (another form of malnutrition) has been reported, as isolation leads to reduced physical activity, overeating, and weight gain due to increased stress and reduced access to healthy food options and opportunities to exercise safely. This increased obesity exacerbates chronic conditions like type 2 diabetes and hypertension and enhances the likelihood of complications and death if COVID-19 infection occurs.[81] The optimal goal for the older adult population is to achieve a balance between protection of the vulnerable and a disproportionate level of detrimental social isolation.

References

1. Federal Interagency Forum on Aging. U.S. Government Printing Office, 2020. Accessed June 13, 2022. Available at https://agingstats.gov/docs/LatestReport/OA20_508_10142020.pdf
2. United Nations, Department of Economic and Social Affairs. World Population Prospects 2019. Accessed June 13, 2022. Available at https://population.un.org/wpp/Publications/Files/WPP2019_Highlights.pdf
3. US Central Intelligence Agency. Accessed June 13, 2022. Available at www.cia.gov/the-world-factbook/countries/
4. Reber E, Gomes F, Vasiloglou MF, et al. J Clin Med 2019;8:1–19.
5. El Chaar D, Mattar L, Fakih El Khoury C. JPEN 2022;46: 1061–70.
6. Evans DC, Corkins MR, Malone A, et al. Nutr Clin Pract 2021;36:22–8.
7. Marshall S, Craven D, Kelly J, et al. Clin Nutr 2018;37:1902–12.
8. White JV, Guenter P, Jensen G, et al. JPEN 2012;36:275–83.
9. Hummel AC, Cummings M. Nutr Clin Pract 2021;37:41–9.
10. U.S. Department of Agriculture and U.S. Department of Health and Human Services. Dietary Guidelines for Americans. 9th Edition. Accessed June 20, 2022. Available at www.dietaryguidelines.gov/sites/default/files/2020-12/Dietary_Guidelines_for_Americans_2020-2025.pdf
11. Agricultural Research Service, U.S. Department of Agriculture. What we Eat in America, Table A. 2015-2018. Accessed June 20, 2022. Available at www.ars.usda.gov/ARSUserFiles/80400530/pdf/usual/Usual_Intake_gender_WWEIA_2015_2018.pdf
12. Roberts SB, Silver RE, Das SK, et al. Adv Nutr 2021;12:1438–48.
13. Ritchie CS, Burgio KL, Locher JL, et al. Am J Clin Nutr 1997; 66:815–8.
14. Dominguez LJ, Veronese N, Baiamonte E, et al. Nutrients 2022; 14:1–22.
15. Clements SJ, Carding RS. Crit Rev Food Sci Nut 2018;58:651–61.
16. Volkert D, Beck AM, Cederholm T, et al. Clin Nutr 2019;38:10–47.
17. Zheng S, Yao J. Aging Med (Milton). 2018;1:8–17.
18. National Institute on Alcohol Abuse and Alcoholism. Accessed June 28, 2022. Available at www.niaaa.nih.gov/alcohols-effects-health/special-populations-co-occurring-disorders/older-adults
19. Rodin MB. Clin Geriatr Med. 2022;38:145–58.
20. Mende MA. Front Neurosci 2017;3:713.
21. Heuberger RA. J Nutr. Elder 2009;28:203–235.
22. U.S. Preventive Services Task Force. JAMA 2018;320:1899–1909.
23. National Institutes of Health, Office of Dietary Supplements. Vitamin D. Accessed June 28, 2022. Available at https://ods.od.nih.gov/factsheets/VitaminD-HealthProfessional/
24. Pilz S, Zittermann A, Trummer C, et al. Endocr Connect 2019; 8:R27–R43.
25. National Center for Health Statistics. NCHS data brief, no 59. 2011. Accessed September 11, 2022. Available at www.cdc.gov/nchs/data/databriefs/db59.pdf
26. Johnson MA, Davey A, Park S, et al. J Nutr Health Aging 2008; 12:690–5.
27. National Institutes of Health, Office of Dietary Supplements. Vitamin B_{12}. Accessed June 28, 2022. Available at https://ods.od.nih.gov/factsheets/VitaminB12-HealthProfessional/
28. Stabler SP. Vitamin B_{12}. In: Marriott BP, Birt DF, Stallings VA, et al., eds. Present Knowledge in Nutrition, 11th ed. International Life Sciences Institute: Elsevier, Inc., 2020:257–71.
29. Suter PM. The B-Vitamins. In: Prasad AS, Brewer GJ, eds. Essential and Toxic Trace Elements and Vitamins in Human Health. Academic Press: Elsevier Inc., 2020;217.
30. Johnson MA, Hausman DB, Davey A, et al. J Nutr Health Aging 2010;14:339–45.
31. Pfeiffer CM, Lacher DA, Schleicher RL, et al. Adv Nutr 2017; 8:290–307.
32. National Institutes of Health, Office of Dietary Supplements, Folate. Accessed June 28, 2022. Available at https://ods.od.nih.gov/factsheets/Folate-HealthProfessional/
33. National Institutes of Health, Iron. Accessed June 28, 2022. Available at https://ods.od.nih.gov/factsheets/Iron-HealthProfessional/
34. Manckoundia P, Konaté A, Hacquin A, et al. Clin Interv Aging 2020;15:1927–1938.
35. Auerbach M, Spivak J. Clin Geriatr Med 2019;35:307–317.
36. Carazo A, Macáková K, Matoušová K, et al. Nutrients 2021; 13:1703.
37. National Institutes of Health, Office of Dietary Supplements, Vitamin A and Carotenoids. Accessed June 28, 2022. Available at https://ods.od.nih.gov/factsheets/VitaminA-HealthProfessional/#en1
38. Meydani SN, Lewis ED. Adv Nutr 2018;9:533–543.
39. National Institutes of Health, Office of Dietary Supplements, Vitamin E. Accessed June 28, 2022. Available at https://ods.od.nih.gov/factsheets/VitaminE-HealthProfessional/
40. Shea MK, Berkner KL, Ferland G, et al. Adv Nutr 2021;12: 632–646.
41. Ferland G. Vitamin K. In: Marriott BP, Birt DF, Stallings VA, et al., eds. Present Knowledge in Nutrition, 11th ed. International Life Sciences Institute: Elsevier, Inc., 2020;137–53.
42. National Institutes of Health, Office of Dietary Supplements, Magnesium. Accessed July 28, 2022. Available at https://ods.od.nih.gov/factsheets/Magnesium-HealthProfessional/
43. Costello RB, Rosanoff A. Magnesium. In: In: Marriott BP, Birt DF, Stallings VA, et al., eds. Present Knowledge in Nutrition, 11th ed. International Life Sciences Institute: Elsevier, Inc., 2020; 349–73.
44. Hunt CD, Johnson LK. Am J Clin Nutr 2006;84:843–52.
45. van Dronkelaar C, van Velzen A, Abdelrazek M, et al. J Am Med Dir Assoc 2018;19:6–11.
46. Veronese N, Berton L, Carraro S, et al. Am J Clin Nutr 2014;100:974–81.
47. National Institutes of Health, Office of Dietary Supplements, Zinc. Accessed September 11, 2022. Available at https://ods.od.nih.gov/factsheets/Zinc-HealthProfessional/
48. Vural Z, Avery A, Kalogiros DI, et al. Nutrients 2020;12:1072.
49. United States Preventive Services Task Force, Final Recommendation Statement, Vitamin, Mineral, and Multivitamin Supplementation. Accessed: July 26, 2022. Available at www.uspreventiveservicestaskforce.org/uspstf/recommendation/vitamin-supplementation-to-prevent-cvd-and-cancer-preventive-medication#bootstrap-panel--10

50. Kalyani RR, Corriere M, Ferrucci L. Lancet Diabetes Endocrinol 2014;2:819–829.
51. Hales CM, Carroll MD, Fryar CD, et al. NCHS Data Brief 2020;360:1–8.
52. Donini LM, Busetto L, Bischoff SC, et al. Clin Nutr 2022;41: 990–1000.
53. Scott D, Shore-Lorenti C, McMillan L, et al. Arch Gerontol Geriatrics 2018;75:125–131.
54. Villareal DT, Apovian CM, Kushner RF, et al. Am J Clin Nutr 2005;825:923–34.
55. Javed AA, Aljied R, Allison DJ, et al. Obes Rev 2020;218:1–12.
56. DiMilia PR, Mittman AC, Batsis JA. Current Diabetes Rep 2019;19:114.
57. Wright NC, Looker AC, Saag KG, et al. J Bone Min Res 2014;29:2520–26.
58. Cameron ID, Murray GR, Gillespie LD, et al. Coch Data Syst Rev 2010:CD005465.
59. Thanapluetiwong S, Chewcharat A, Takkavatakarn K, et al. Medicine (Baltimore) 2020;99:1–12.
60. Eastell R, Rosen CJ, Black DM, et al. J Clin Endocrinol Metab 2019;104:1595–1622.
61. American Diabetes Association. Diabet Care 2020;43(Suppl 1): S152–62.
62. Alexopoulos AS, Kahkoska AR, Pate V, et al. JAMA Netw Open 2021;4:1–13.
63. Leung E, Wongrakpanich S, Munshi MN. Diabetes Spectr 2018;31:245–253.
64. Centers for Disease Control and Prevention. Behavioral Risk Factor Surveillance System website. Accessed September 13, 2022. Available at www.cdc.gov/brfss
65. Liu Y, Liu F, Zhang L, et al. Sci Rep 2021;11:22111.
66. Fleg JL, Forman DE, Berra K, et al. Circ 2013;128:24222446.
67. Aronow WS, Fleg JL, Pepine CJ, et al. Circ 2011;123:2434–2506.
68. Burgos R, Breton I, Cereda E, et al. Clin Nutr 2018;37:354–96.
69. Wirth R, Smoliner C, Jager M, et al. Exp Transl Stroke Med 2013;5:14.
70. Lu Y, Nyunt MSZ, Gao Q, et al. J Ren Nutr 2022;32:560–68.
71. Ko GJ, Kalantar-Zadeh K. Korean J Intern Med 2021;36:795–806.
72. Ko GJ, Obi Y, Tortorici AR, et al. Curr Opin Clin Nutr Metab Care 2017;20:77–85.
73. Narasaki Y, Rhee CM, Kramer H, et al. Curr Opin Clin Nutr Metab Care 2021;24:10–17.
74. U.S. Department of Agriculture, Economic Research Service. Household Food Security in the United States in 2020, ERR 298, U.S. Accessed July 28, 2022. Available at www.ers.usda.gov/webdocs/publications/102076/err-298.pdf?v=1461.4
75. Report for Feeding America. An Annual Report, 2019. Accessed September 11, 2022. Available at www.feedingamerica.org/research/senior-hunger-research, www.feedingamerica.org/sites/default/files/202205/State%20of%20Senior%20Hunger%202019.pdf
76. Bengle R, Sinnett S, Johnson T, et al. J Nutr Elder 2010;29: 170–91.
77. Brewer DP, Catlett CS, Porter KN, et al. J Nutr Elder 2010;29: 150–69.
78. Saffel-Shrier S, Johnson MA, Francis SL. J Acad Nutr Diet 2019; 119:1188–1204.
79. Administration for Community Living, Nutrition Services. Accessed July 20, 2022. Available at https://acl.gov/programs/health-wellness/nutrition-services
80. Holst M, Geirsdóttir ÓG, Bell JJ. In: Geirsdóttir ÓG, Bell JJ, eds. Nutrition Support in Older Adults. Interdisciplinary Nutritional Management and Care for Older Adults: An Evidence-Based Practical Guide for Nurses. Cham: Springer International Publishing: 2021;65–77.
81. Porter Starr KN, Borack MS, Rincker JC, Bales CW. In: Wilson T, Temple NJ, Bray GA, eds. Nutrition and Aging: Meeting the Unique Needs of Older Adults. Nutrition Guide for Physicians and Related Healthcare Professions. Cham: Springer International Publishing: 2022;59–69.
82. Tappenden KA, Quatrara B, Parkhurst ML, et al. JPEN 2013; 37:482–497.
83. Laur C, Curtis L, Dubin J, et al. Healthcare (Basel) 2018;6.1–13.
84. Keller H, Donnelly R, Laur C, et al. JPEN 2022;46:141–52.
85. Xue QL. Clin Geriatr Med 2011;27:1–15.
86. Fried LP, Tangen CM, Walston J, et al. J Gerontol A Biol Sci Med Sci 2001;56:M146–156.
87. Beasley JM, Shikany JM, and Thomson CA. Nutr Clin Practice 2013:28:684-90
88. Wolfe, RR. Br J Nutr 2012;108:S88–93.
89. Aragon AA, Tipton KD, Schoenfeld BJ. Nutr Rev 2022; Aug:1–14.
90. Buhl SF, Beck AM, Olsen PO, et al. Clin Nutr ESPEN 2022;49: 449–458.
91. Lorenzo-Lopez L, Maseda A, de Labra C, et al. BMC Geriatrics 2017;17:108.
92. Goisser S, Guyonnet S, Volkert D. J Frailty Aging 2016;5:74–7.
93. Druml C, Ballmer PE, Druml Clin Nutr 2016;35:545–56.
94. Voumard R, Rubli Truchard E, Benaroyo L, et al. BMC Geriatrics 2018;18:220.
95. Mitchell SL, Mor V, Gozalo PL, et al. JAMA 2016;316:769–70.
96. Ticinesi A, Nouvenne A, Lauretani F, et al. Clin Nutr 2016; 35:1512–16.
97. Shahid Z, Kalayanamitra R, McClafferty B, et al. J Am Geriatr Soc 2020;68:926–29.
98. Detopoulou P, Tsouma C, Papamikos V. Top Clin Nutr 2022; 37:187–202.
99. Berg-Weger M, Morley JE. J Nutr Health Aging 2020;24:456–58.

CHAPTER 57

Clinical Manifestations of Nutrient Deficiencies and Toxicities

Sherry Tarleton, Michelle DiBaise

CHAPTER OUTLINE

Abbreviations

ATP, adenosine triphosphate; **DRI**, dietary reference intake; **HDN**, hemolytic disease of the newborn; **CPN**, central parenteral nutrition

Introduction

Nutritional disorders result from imbalances among the body's requirements for nutrients and energy sources and the supplies of these substrates of metabolism. These imbalances may take the form of either deficiency or excess and may be attributable to inappropriate intake, to defective utilization or, frequently, a combination of both.

Despite our extensive understanding of human nutritional requirements for maintenance of health, malnutrition continues to be one of the main causes of morbidity and mortality in low- and middle-income countries, especially in young children.[1] In resource-rich countries, undernutrition continues to occur in hospitalized patients and other at-risk groups, including individuals with chronic medical disorders, long-term alcohol or drug use, premature infants, long-term use of parenteral nutrition, and those who have restricted diets or decreased access to food.[2,3] The chronic medical conditions that lead to deficiency states include malabsorptive conditions (inflammatory bowel disease, short bowel syndrome, celiac disease, chronic diarrhea, chronic cholestatic liver disease), pancreatic insufficiency (including cystic fibrosis), and end-stage renal disease. Deficiency due to severely restricted diets can occur among vegans, persons with eating disorders such as anorexia nervosa (restricting and binge-purge type)[4,5] and Avoidant Restrictive Food Intake Disorder (ARFID), commonly seen in but not limited to Autism Spectrum Disorder,[6-9] and persons with food allergies and sensitivities, including lactose intolerance.[3,10] Individuals experiencing poverty, food insecurity, or homelessness also have restricted diets.[3,11]

Vigilance is needed to detect secondary undernutrition resulting from malabsorption; failure in transport, storage, or cellular utilization; excessive losses; or inactivation by genetic mutations of essential metabolic pathways that increase needs. The improper use of nutrient supplements, often because of ignorance about proper dosage or by failure of excretion through renal failure with continued nutrient intake, has been a major cause of toxicity.[12]

This chapter addresses clinical manifestations of nutritional disorders related to vitamins, minerals, and essential fatty acids. Descriptions of clinical symptoms of deficiency of each nutrient are followed by brief consideration of who is likely to be at risk of deficiency and, if relevant, who is likely to be at risk of toxic levels and the appropriate workup to confirm the diagnosis. The micronutrient reference chart (**Table 57.1**) provides examples of common food sources, lab tests, and supplementation regimens, as well as the dietary reference intake (DRI) and possible risk of toxicity with high-dose supplementation. The electrolytes have been strategically left off this chart due to the considerable clinical correlation and interpretation required in assessing abnormal concentrations. Iodine, fluoride, chromium, molybdenum, and manganese have also been purposefully left off, as their monitoring and supplementation are not routinely part of nutrition care. Supplementation recommendations used in the chart are found in recent data, however, may not denote a definite approach. A detailed review of medication use, past medical and surgical and social history combined with a nutrition-focused physical exam (NFPE), and the clinical interpretation of appropriate biochemical tests, evaluated with a multidisciplinary approach, offers the best approach to correcting abnormal micronutrient status.

Vitamins

Vitamins are required for normal growth and metabolism and, except for vitamins D and K, cannot be synthesized by the human body, necessitating dietary intake. Vitamin D can be acquired through exposure to sunlight and from dietary intake. Vitamin K can be produced by bacteria in the gut. Vitamins are either fat-soluble (A, D, E, and K) or water-soluble (B vitamins and C). Fat-soluble vitamins are stored in the liver and adipose tissue and, with excess intake, pose a risk for toxicity. The fat-soluble vitamins require adequate fat for digestion and absorption; dyscrasias in this process may lead to deficiencies. Excess water-soluble vitamins are uncommon, as they are minimally stored in tissues and are rapidly excreted in urine. Patients with chronic kidney disease are, however, at risk of toxicity of water-soluble vitamins due to excretion defects.[83]

Vitamin A (Retinol, Retinal, Retinyl Esters) and Provitamin A (Carotenoids)

Deficiency

Vitamin A is important for vision, particularly the development of the cornea and conjunctiva, and is a component of the protein rhodopsin in the retina.[2,3,13,84] Deficiency of vitamin A leads to xerophthalmia (severely dry cornea and conjunctiva) that can progress to squamous cell proliferation of the cornea (Bitot spots) (**Figure 57.1A**). Corneal involvement may start as a superficial punctate keratopathy and proceed to xerosis and varying degrees of "ulceration" and liquefaction (keratomalacia) (see **Figure 57.1B**), which frequently results in blindness.[85] Impaired dark adaptation or night blindness (i.e., decreased vision in dim light) is an early symptom and can be elicited by a careful history and some simple tests in a poorly illuminated room.[86]

Extraocular manifestations of deficiency include perifollicular hyperkeratosis (phrynoderma), an accumulation of keratinized skin around hair follicles most commonly seen on the lateral aspects of the upper arms and the thighs. Phrynoderma has also been linked to deficiencies of the B vitamins, vitamin E, and essential fatty acids.[87,88] Other changes may occur, including decreased immunity and bone growth abnormalities.[2,89]

Vitamin A is also important in wound healing and optimizing negative effects of corticosteroids, due to its vital role in epithelial growth. Oral vitamin A supplementation has been shown to improve the inflammatory response, with an increase in IL-10 and a decrease in TNF-alpha, leading to accelerated resolution of acute wounds.[14] Small studies have demonstrated a benefit of oral vitamin A on chronic wound healing as well.[14] The use of long-term corticosteroids will disrupt healing by interfering with fibroblast proliferation and collagen deposition. The use of vitamin A supplementation reverses these effects in patients on long-term corticosteroids.[14] There is risk of long-term supplementation toxicity, particularly for patients with renal or hepatic failure. Therefore, some clinics and hospitals have protocols to provide vitamin A supplementation (topical or oral) in those with chronic wounds or on corticosteroids, with close monitoring of plasma vitamin A concentration.[14]

Vitamin A is absorbed in the small intestine, requiring pancreatic enzymes and bile salts. Therefore, deficiency can occur with malabsorptive disorders, pancreatic insufficiency, severe cholestatic diseases, and other interference of fat absorption (i.e., medication, bariatric surgery).[2,10,90]

Serum concentration of vitamin A may be suppressed in the setting of zinc deficiency, malnutrition, and decreased liver function, as the synthesis and mobilization of retinol-binding protein (RBP) is decreased in these disorders.[91] They are best tested through the measurement of plasma retinol and carotenoid concentrations.[13]

Short-term vitamin A supplementation should be considered in persons who are known to be deficient in vitamin A, who are on glucocorticoids for more than 1 month, have diabetic ulcers, radiation-induced injuries, burns, immunosuppression, or who are on cyclophosphamide or vincristine. Depending on the degree of deficiency and clinical presentation, a common oral dose would range from 15,000 to 20,000 IU/day for 2 to 3 weeks.[14]

The treatment for xerophthalmia is immediate use of oil-based oral vitamin A of 200,000 IU, with a repeat dose on

Table 57.1 Micronutrient Replacement Considerations

Micronutrient	Food Sources	DRIs (Amount per day)	Lab Test(s)	Supplementation Recommendations for Deficiency (Combine with clinical interpretation, NFPE, and multidisciplinary approach)	Potential Risks with High Dose Supplementation
Vitamin A	Beef; poultry; vegetable, such as sweet potato, carrots, spinach, pumpkin, butternut squash	**RDA:** Men: 3,000 IU (900 μg retinol) Women: 2,300 IU (700 μg retinol) **UL:** 10,000 IU (3,000 μg)	Plasma retinol and carotenoid concentrations	■ A common short-term oral dose is 15,000–20,000 IU/day for 2–3 weeks for: Deficiency, on glucocorticoids (>1 month), diabetic wounds, radiation-induced injuries, burns, immunosuppression or on cyclophosphamide or vincristine ■ Xerophthalmia: • Immediate use of oil-based oral vitamin A of 200,000 IU, repeat dose on days 2 & 14 • Additional 200,000 IU every 4–6 months • Use water-miscible vitamin A for fat malabsorption	**Acute toxicity:** Nausea, vomiting, headache, vertigo, irritability, stupor, fontanel bulging (in infants), papilledema, and pseudotumor cerebri (mimicking brain tumor) **Chronic toxicity:** Ataxia, blurred vision, diplopia, dry and scaling pruritic skin, alopecia, hepatomegaly, splenomegaly, and bone or muscle pain ■ Potentially teratogenic 1st trimester of pregnancy at doses above the RDA ■ Bone fractures with long term high dose retinal in older population
Vitamin D	Fatty fish, fortified milk, mushrooms	**RDA:** Expressed in μg of cholecalciferol Men: 15 μg Women: 15 μg **UL:** 100 μg 1 μg cholecalciferol = 40 IU vitamin D	25(OH)D concentration	Supplement according to 25(OH)D concentration: ■ <12 ng/mL: • 50,000 IU 1 ×/week for 6–8 weeks • Followed by 800 IU daily long term ■ 12 to 20 ng/mL: • 800–1000 IU ■ 20 to 30 ng/mL: • 600–800 IU ■ Cholecalciferol (vitamin D3) is preferred ■ Repeat lab 12–16 weeks after supplementation with 25(OH)D concentration <20 ng/mL	**Acute toxicity:** Anorexia, nausea, vomiting, confusion, muscle weakness, hypotonia, stupor, and hypertension **Chronic toxicity:** Renal colic resulting from stone formation, bone demineralization, and pain
Vitamin E	Nuts, green leafy vegetables, fortified cereals, vegetable oils	**RDA:** Men and Women: 15 mg of dietary α-tocopherol **UL:** 1,000 mg	■ Serum α-tocopherol concentrations when serum lipid concentrations are normal ■ In hyperlipidemia, the α-tocopherol concentration must be divided by the total lipids (cholesterol + triglycerides)	■ 200–2000 mg/day, depending on deficiency ■ Water miscible vitamin E (tocopheryl polyethylene glycol succinate) for fat malabsorption	■ Increase in all-cause mortality ■ Mixed data re: >1,200 mg/day & competition with vitamin K (increased bleeding, decrease in platelet adhesion with warfarin)

(continues)

Table 57.1 Micronutrient Replacement Considerations *(continued)*

Micronutrient	Food Sources	DRIs (Amount per day)	Lab Test(s)	Supplementation Recommendations for Deficiency (Combine with clinical interpretation, NFPE, and multidisciplinary approach)	Potential Risks with High Dose Supplementation
Vitamin K	Liver, green leafy vegetables, broccoli, peas, and green beans	**AI:** Men: 120 μg Women: 90 μg	▪ Mild deficiency: PT/INR as will affect Factor VII first ▪ Severe deficiency: both PT and PTT can be affected ▪ Measurement of vitamin K-dependent proteins (Factor II [prothrombin], VII, IX, X, and/or protein C) will guide the diagnosis	▪ Supplementation regimens range from 2.5 to 10 mg, twice weekly to daily with oral or injected forms ▪ Extra care should be taken to monitor dosing with injectable forms due to inconsistency in formulations	Once the goal INR is met, daily vitamin K intake should remain consistent with warfarin due to risk of clotting
Thiamin	Whole grains, meat, fish	**RDA:** Men: 1.2 mg Women: 1.1 mg	▪ Serum thiamin preferred ▪ Erythrocyte thiamin transketolase activity (EKTA) test is also available	▪ Acute infantile beriberi: • 100–150 mg IV ▪ At risk for WE: • 100–250 mg/day for 3–5 days ▪ Suspected to have WE require immediate treatment: • 500 mg IV over 30 minutes 3 times/day for 2 days • 250 mg (IV or IM) once/day for 5 days • 100/day mg oral until no further risk of WE ▪ IV Thiamin preferred over oral in setting of possible WE to optimize absorption ▪ Consider 1 g magnesium sulfate with IV thiamin ▪ Give IV thiamin prior to dextrose to decrease risk of worsening WE	Rare reports of anaphylactic reaction
Riboflavin (B2)	Dairy, eggs, fish, meat, poultry, fortified cereals	**RDA:** Men: 1.3 mg Women: 1.1 mg	▪ Erythrocyte glutathione reductase activity coefficient (EGRAC) 24-hour urine for riboflavin for those with G6PD deficiency ▪ Serum riboflavin concentration (less sensitive than EGRAC)	▪ No consensus guidelines ▪ Injections available in the setting of oral intolerance	Negative effects on ocular lens proteins and retina with >100 mg/day long term

Micronutrient	Food Sources	DRIs (Amount per day)	Lab Test(s)	Supplementation Recommendations for Deficiency (Combine with clinical interpretation, NFPE, and multidisciplinary approach)	Potential Risks with High Dose Supplementation
Niacin (B3)	Animal-based food, grains, nuts, legumes Niacin is converted from Trp (nuts, seeds, legumes, soybeans, Beef, poultry)	**RDA:** Men: 16 mg Women: 14 mg **UL:** 35 mg	■ Urinary N1-methylnicotinamide ■ Erythrocyte NAD: NADP ratio	■ Oral nicotinamide 10–150 mg/day, depending on deficiency ■ Intake of 1 g of Trp (=100 g protein/day) ■ Treatment of pellagra: • 100–300 mg/day SQ niacinamide in 3 doses	1-3 g/day may cause: Flushing, pruritus, nausea, vomiting, constipation, headache, elevated LFTs, insulin resistance (in diabetes patients), increased uric acid concentration
Pyridoxine (B6)	Whole grains, vegetables, nuts, meats	**RDA:** Men: 1.3–1.7 mg Women: 1.3–1.5 mg **UL:** 100 mg	Plasma pyridoxal-5-phosphate (PLP)	■ 25–600 mg/day depending on symptom severity ■ IV dosing reserved for seizures caused by deficiency	Sensory neuropathies and movement disorders
Biotin	Organ meats, eggs, fish, meat, sunflower seeds, peanuts, sweet potatoes	**AI:** 30 μg	■ White blood cell biotinylated methylcrotonyl CoA carboxylase and propionyl-CoA carboxylase ■ Serum or urine biotin concentration	No consensus guidelines	■ Falsely elevated T3 and T4 concentrations (mimicking Grave's disease) ■ Falsely low troponin concentration
Vitamin B12 (Cobalamin)	Animal products, particularly meat, fish, dairy products, and eggs, fortified cereals	**RDA:** Men: 2.4 μg Women: 2.4 μg	■ Serum or plasma vitamin B12 concentration ■ Methylmalonic acid (MMA) ■ Homocysteine	■ A common regimen is: 1,000 μg cyanocobalamin IM once/week until deficiency is corrected ■ Pernicious anemia: • 100–1,000 μg cyanocobalamin IM for 5 days • 100–1,000 μg/month	No reports of adverse effects of vitamin B_{12} supplementation have been noted in the literature
Folate (Folic Acid)	Dark green vegetables, grains, cereals	**RDA:** Men: 400 μg Women: 400 μg **UL:** 1,000 μg	Serum folate	■ 1–5 mg of oral folic acid, depending on deficiency ■ IV dosing is reserved for rapid repletion or oral intolerance	≥1,000 μg/day may have risk of colon, prostate, and breast cancers

(continues)

Table 57.1 Micronutrient Replacement Considerations *(continued)*

Micronutrient	Food Sources	DRIs (Amount per day)	Lab Test(s)	Supplementation Recommendations for Deficiency (Combine with clinical interpretation, NFPE, and multidisciplinary approach)	Potential Risks with High Dose Supplementation
Vitamin C (Ascorbic Acid)	Citrus fruits, peppers, tomato sauce, fortified juices, broccoli, spinach, potatoes strawberries	**RDA:** Men: 90 mg Women: 75 mg **UL:** 200 mg	Serum or leukocyte vitamin C concentration	■ Up to 300 mg/day in pediatrics ■ Up to 1,000 mg/day in adults ■ 300 mg/day for 1 month will decrease symptoms ■ 300 mg/day for 3 months for full recovery from scurvy	**Test inference:** ■ Falsely elevate blood glucose concentration in point-of-care glucose monitors ■ False negative stool guaiac ■ Excess iron absorption ■ Oxalate nephropathy ■ Diarrhea
Choline	Eggs, organ meats, fish	**AI:** Men: 500 mg Women: 425 mg **UL:** 3.5 g	Plasma choline	No consensus guidelines	Vomiting, hypotension, increased risk of cardiovascular disease, and a "fishy" body odor
Essential Fatty Acids (Linoleic and α-Linolenic Acid)	Oils including corn, soy, and safflower	**% Energy in adults:** n-6 PUFAs (linoleic acid): 5–10% n-3 PUFAs (α-linolenic acid): 0.6–1.2%	Serum triene-tetraene ratio	■ Oils high in EFA can be taken orally–need to consider absorption capability (i.e.: fat malabsorption) • Ensure appropriate pancreatic enzyme replacement in patients with pancreatic insufficiency ■ Mixed data on topical oil (sunflower and safflower) application ■ Intravenous lipid emulsions in patients who are on CPN can be used as a source of EFA. Amounts will vary with different formulations.	N/A
Iron	Heme: Meat, poultry, and fish Non-heme: Vegetables, fruits, fortified products	**RDA:** Men: 8 mg Women: 18 mg (18–50 years old) 8 mg (51 & older) **UL:** 45 mg	Serum iron concentration, total iron binding capacity, transferrin, serum ferritin, and transferrin saturation	■ 40–200 mg orally in the AM on alternate days ■ Iron infusions are available for oral iron intolerance	**Acute toxicity:** Vomiting, upper abdominal pain, and diarrhea **Chronic toxicity:** Damage to pancreas, liver, and heart

Micronutrient	Food Sources	DRIs (Amount per day)	Lab Test(s)	Supplementation Recommendations for Deficiency (Combine with clinical interpretation, NFPE, and multidisciplinary approach)	Potential Risks with High Dose Supplementation
Copper	Beef, soy, nuts, seeds, beans, shellfish, mushrooms	**RDA:** Men: 900 μg Women: 900 μg **UL:** 10,000 μg	Serum copper and ceruloplasmin concentrations	Recommendations for copper supplementation in Roux-en-Y gastric bypass: ■ Mild to moderate deficiency • Oral copper 3–8 mg/day until levels are corrected ■ Severe symptomatic deficiency: • 2–4 mg/day of intravenous copper for 6 days • Switch to oral therapy	■ Vomiting, diarrhea, hepatic, cardiac and/or renal failure ■ Zinc concentration should be monitored during long-term copper repletion
Zinc	Beef, fish, oysters, liver, beans, and soybeans.	**RDA:** Men: 11 mg Women: 8 mg **UL:** 40 mg	Plasma or serum zinc concentration	■ Oral replacement is 2–3 mg/kg/day of elemental zinc, will alleviate all clinical manifestations within 1–2 weeks ■ Acrodermatitis enteropathica: • Oral supplementation of 1–2 mg/kg/day for life	■ Gastric irritation with nausea, vomiting, and gastric hemorrhage ■ Copper concentration should be monitored during long-term zinc repletion
Selenium	Poultry, nuts, seeds, wheat, barley, beef, and potatoes	**RDA:** Men: 55 μg Women: 55 μg **UL:** 400 μg	■ Serum or plasma selenium concentration ■ Selenium levels in hair and nails can provide a more accurate long-term assessment of intake	■ No consensus guidelines ■ A common dose of oral selenium for adults is 100 μg/day	Hair loss, nail changes, nausea, vomiting, diarrhea, vision changes, and polyneuritis

References [2,3,13-82]

AI, Adequate Intakes; *DRI*, Dietary Reference Intakes; *EFA*, essential fatty acids; *IM*, intramuscular; *IU*, International Unit; *IV*, intravenous; *LFTs*, liver function tests; *μg*, microgram; *mg*, milligram; *NFPE*, Nutrition Focused Physical Exam; *PT/INR*, prothrombin time/International normalized ratio; *PTT*, partial thromboplastin time; *PUFAs*, polyunsaturated fats; *RDA*, Recommended Dietary Allowance; *SQ*, subcutaneous; *Trp*, tryptophan; *UL*, Tolerable Upper Intake Levels; *WE*, Wernicke encephalopathy

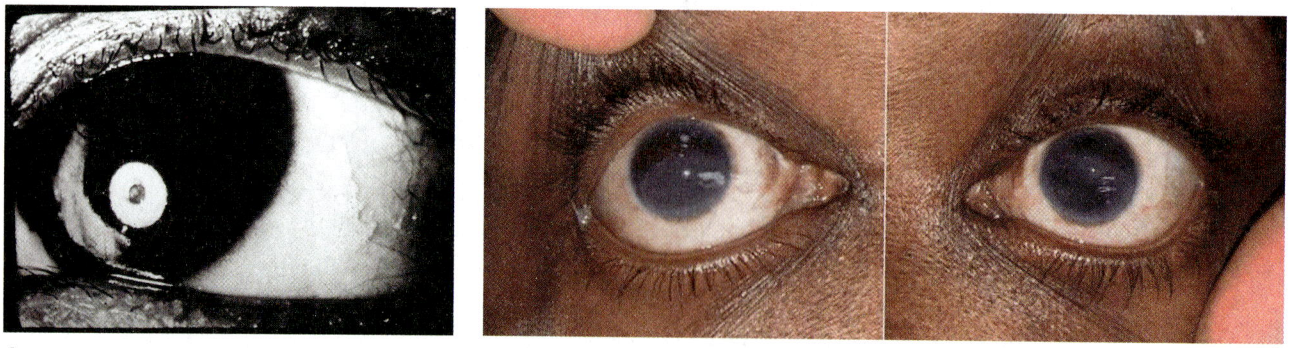

A B

Figure 57.1 A. Vitamin A deficiency. Bitot spot in the temporal interpalpebral fissure. B. Vitamin A deficiency. Keratomalacia.

A. Courtesy of Centers for Disease Control and Prevention; B. Reproduced from Bijan S, Filutowski O R, Safari S (August 01, 2022) Bilateral Keratomalacia From Vitamin A Deficiency in Pancreatic Insufficiency. Cureus 14(8): e27569. doi:10.7759/cureus.27569

days 2 and 14. Follow-up supplementation with 200,000 IU is recommended every 4 to 6 months.[92] For those with fat malabsorption, water-miscible vitamin A supplementation is available.[25]

Toxicity (Hypervitaminosis A)

Acute toxicity of vitamin A presents with nausea, vomiting, headache, vertigo, irritability, stupor, fontanel bulging (in infants), papilledema, and pseudotumor cerebri (mimicking brain tumor).[16,93] Chronic toxicity is characterized by ataxia, blurred vision, diplopia, dry and scaling pruritic skin, alopecia, hepatomegaly, splenomegaly, and bone or muscle pain.[16]

Vitamin A and other retinoids are teratogenic in pregnant women during the first trimester.[16,17] Spontaneous abortions, microcephaly, and cardiac anomalies have been reported with the use of isotretinoin (to treat acne) and with supplementation of more than 10,000 IU/day before the 7th week of gestation.[94] Despite the teratogenicity of vitamin A, it is still necessary for adequate fetal development. Vitamin A deficiency in pregnancy is seen more commonly in the latter part of pregnancy due to hypervolemia with increasing development of the fetus but may also occur in the setting of gestational diabetes, restriction of foods high in vitamin A and/or maternal infections.[95] Significant evidence indicates that long-term intake of high-dose supplements of retinol is associated with increased risk of bone fractures in older Swedish men[18] and women[19], as well as in women in the United States.[20]

Hypercarotenosis

Excessive intake of carotenoids can cause hypercarotenosis. Yellow or orange discoloration of the skin (xanthosis cutis, carotenoderma) that affects areas where sebum secretion is greatest; in the nasolabial folds, forehead, axillae, and groin and, keratinized surfaces such as the palms and soles (see **Figure 57.1C**). The sclerae and buccal membranes are not affected, which is a feature that distinguishes this condition from jaundice, in which these tissues are yellowed. No toxicity is apparent, and the discoloration gradually disappears with reduction of intake.

Vitamin D (Calciferol)

Deficiency

Vitamin D is important in bone metabolism and can be produced from ultraviolet activation within the skin or obtained through ingestion. Activation of vitamin D requires 25-hydroxylation in the liver (25-hydroxy-vitamin D [25OHD]) and subsequent 1 α-hydroxylation in the kidney (1,25-dihydroxy-vitamin D [$1,25(OH)_2D$]). The best measure of vitamin D sufficiency is a

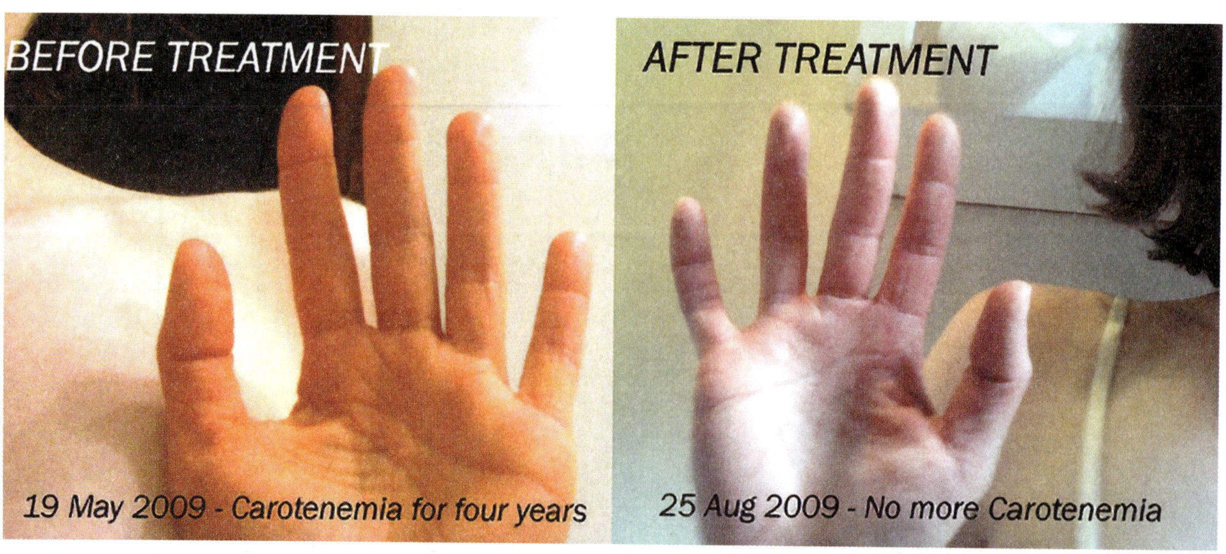

Figure 57.1 C. Palm of a woman with hypercarotenosis (left), compared with the palm of a person without hypercarotenosis (right).

25OHD concentration, as this is the major circulating variant; however, $1,25(OH)_2D$ is the active hormonal variant.[96] Vitamin D deficiency can occur in areas with limited exposure to or continuous protection from the sun, with an intake of foods low in vitamin D, in exclusively breastfed infants, in individuals with chronic renal disease, after gastrectomy or bariatric surgery, and in individuals with malabsorption such as in celiac disease.[97-99] Individuals on medications may be at risk for vitamin D deficiency, particularly antiseizure medications, isoniazid, and rifampin.[100] Vitamin D deficiency may be manifested as hypocalcemia with or without hypophosphatemia. The severity of disease is related to the duration of the deficiency. Prolonged deficiency with hypocalcemia leads to secondary hyperparathyroidism, which subsequently causes bone demineralization in an attempt to normalize blood calcium. Prolonged bone demineralization will progress to rickets in children and osteomalacia in adults, presenting with bone pain, muscle weakness, and risk of fracture.

An infant with rickets may be restless and sleep poorly due to pain. Craniotabes, softening of the bones of the skull, is often the earliest sign of rickets in addition to frontal bossing and delayed closure of the fontanels. Sitting, crawling, and walking are generally delayed. Weight-bearing results in bowing of the arms, knock-knees (genu valgum), or outward bowing of the lower legs (genu varum) (**Figure 57.1D**). Low calcium concentration may manifest, with hypoplastic dental enamel, decreased muscle tone, and seizures.[101]

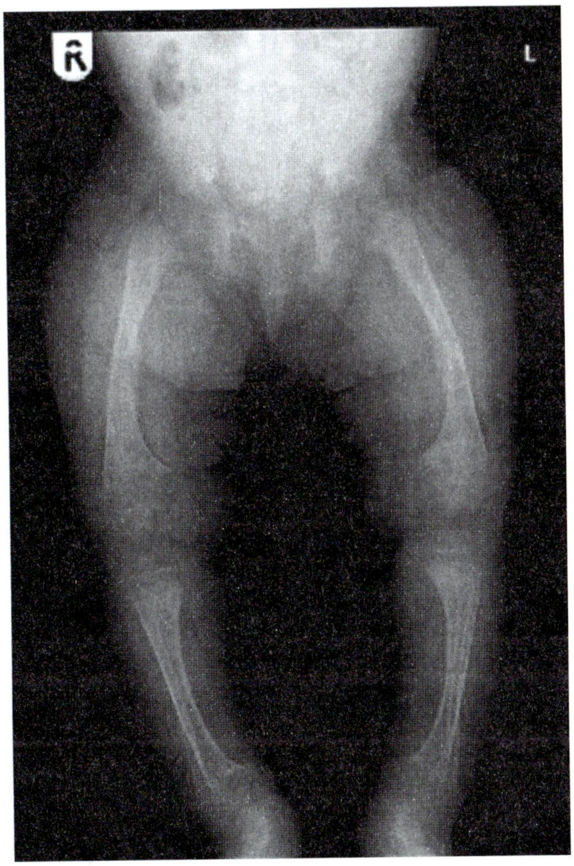

Figure 57.1 D. Rickets. Genu varum in a 30-month-old girl with nutritional rickets and progressive bowing of the legs since she began walking at the age of 11 months.

Michael L. Richardson, M.D.

Table 57.2 Dosing Regimens for Vitamin D

25(OH)D Concentration	Initial Supplement Dose
<12 ng/mL (30 nmol/L)	50,000 IU of vitamin D_2 or D_3 orally, once weekly for 6–8 weeks, then 800 IU of vitamin D2 or D3 per day long term
12 to 20 ng/mL (30–50 nmol/L)	800–1,000 IU of vitamin D_2 or D_3, per day long term
20 to 30 ng/mL (50–75 nmol/L)	600–800 IU vitamin D_2 or D_3 per day long term

Data from Dawson-Hughes, B., Vitamin D deficiency in adults: Definition, clinical manifestations, and treatment: Summary and recommendations. Vitamin D deficiency in adults: Definition, clinical manifestations, and treatment—UpToDate Last accessed: May 15, 2022.[21]

The main features of osteomalacia are bone pain and tenderness, muscle spasms, difficulty walking, fractures, and weakness of the proximal muscles.[102] Looser zones are radiotranslucent zones, sometimes termed *pseudofractures*, which usually occur in the long bones, pelvis, ribs, and around the scapulae in a bilaterally symmetric pattern. Osteomalacia may result in a waddling gait and in difficulty going up and down the stairs.

The evaluation of vitamin D deficiency includes 25(OH)D concentration, serum calcium, phosphorus, parathyroid hormone, alkaline phosphatase, BUN, creatinine, and antibodies for celiac disease (anti-tissue transglutaminase). Radiography of the affected area is required when individuals exhibit bone pain to rule out other causes of bone pathology.[103] Anyone with serum 25(OH)D <20 ng/mL (50 nmol/L) should have repeat lab testing done 12 to 16 weeks after starting supplementation to ensure improvement and assessment of dosing changes.

Vitamin D supplementation recommendations (**Table 57.2**) depend on the severity of deficiency and clinical picture, such as a history of malabsorption. Cholecalciferol (vitamin D_3) is preferred (if available), rather than ergocalciferol (vitamin D_2), and calcium intake should accompany vitamin D supplementation. Dosing regimens are based on baseline vitamin D concentration.[21]

To avoid deficiency, breastfed infants need 400 IU per day from the first days of life.[104]

Toxicity (Hypervitaminosis D)

The symptoms and signs in acute toxicity are related to hypercalcemia and include anorexia, nausea, vomiting, and confusion. Muscle weakness, hypotonia, stupor, and hypertension are less common. Polyuria and polydipsia are caused by hypercalciuria. Chronic toxicity can lead to renal colic resulting from stone formation, bone demineralization, and pain.[22] Serum concentration of 25(OH)D should be measured in the presence of hypercalcemia. In addition, malignancy, sarcoidosis and primary hyperparathyroidism need to be ruled out as the etiology of hypercalcemia.

Vitamin E (Tocopherol)

Deficiency

Vitamin E predominantly serves to protect cells from oxidative stress. Individuals at risk of vitamin E deficiency include those with malabsorption disorders, including pancreatic insufficiency and cystic fibrosis, as well as cholestatic disorders. Deficiency may also be seen in severe protein-energy malnutrition.[105] Genetic disorders can also lead to vitamin E deficiency manifested by ataxia from a mutation in the α-tocopherol transfer protein (α-TTP) gene or with abetalipoproteinemia, due to a mutation in the gene coding for microsomal triglyceride transfer protein.[106] In severe vitamin E deficiency, individuals have progressive ataxia of gait, dysarthria, areflexia, extensor plantar signs, and impaired vibratory and positional sense. In abetalipoproteinemia, patients have acanthocytes (erythrocytes with spiny projections of the membrane), retinitis pigmentosa-like changes in the retina, steatorrhea, ataxia, and failure to thrive.[106]

In addition to neuromuscular dysfunction, vitamin E deficiency can lead to hemolysis and has also been associated with the deposition of lipofuscin, a ceroid pigment, within the bowel, muscle tissue,[23] and the retina.[107] Testing for vitamin E deficiency is best measured with serum α-tocopherol concentration with normal cholesterol and adjusted by dividing the concentration by total cholesterol plus triglyceride values in the presence of hyperlipidemia.[108]

There are no consensus guidelines on vitamin E supplementation in the setting of deficiency. Depending on clinical presentation along with the degree of deficiency, common adult treatment with oral vitamin E may range from 200 to 2,000 mg/day.[109] Over-the-counter water-miscible vitamin E (tocopheryl polyethylene glycol succinate) supplements are available in the United States for those with fat malabsorption.[110]

Toxicity

Toxicity is seen primarily with over-supplementation. There are reports that low birth weight infants receiving pharmacologic doses of vitamin E have shown a high incidence of sepsis and necrotizing enterocolitis.[111,112] A meta-analysis reporting increased all-cause mortality in adults taking daily vitamin E supplements of 400 mg or more has prompted discussion of whether the tolerable upper limit should be reduced from its current level of 1,000 mg/day.[25,113,114] There are concerns that high doses of vitamin E (>1,200 mg/day) may increase the risk of competition with vitamin K, leading to increased bleeding for patients undergoing surgery and those taking warfarin.[24,115,116]

Vitamin K (Phylloquinone)

Deficiency

Vitamin K is a cofactor for several proteins required for coagulation. Deficiency is uncommon, as vitamin K is found in a variety of foods and is produced by bacteria in the gut.[26] The risk of vitamin K deficiency is most common in newborns, individuals with pancreatic insufficiency, cholestatic disorders, fat malabsorption disorders, and with the use of medications that either interfere with fat absorption or that destroy the gut bacteria.[117] Deficiency of vitamin K in the newborn is due to an immature liver and poor utilization of vitamin K in the first month of life.[27,118] Vitamin K deficiency bleeding (VKDB) of early onset (within 24 hours of birth) is most commonly seen in mothers who were on warfarin, cephalosporins, and other antibiotics or anticonvulsants that block vitamin K. The classic variant of VKDB of early onset occurs between days 2 and 7 after birth, whereas the late variant can present from 3 weeks up to 8 months after birth. Vitamin K deficiency bleeding may present with bruising of the skin and mucosa, GI tract, site of circumcision, or as intracranial hemorrhage. Vitamin K deficiency bleeding in the newborn led to the widespread intramuscular injection of phylloquinone at a recommended dose of 0.5 to 1.0 mg for all infants at birth.[119]

In adults with vitamin K deficiency, the bleeding may present as easy bruising of the skin and mucosal surfaces and blood in the stool and urine (melena and hematuria). Testing of vitamin K deficiency is first evaluated by a prothrombin time (PT) and international normalized ratio (INR) to assess for prolonged bleeding. A partial thromboplastin time (PTT) will be positive only after prolonged or severe deficiency. Confirmation of vitamin K deficiency requires measurement of the vitamin K-dependent clotting factors: Factors II, VII, IX, X, and protein C.[120]

There is no agreed-upon supplementation regimen in the literature for vitamin K deficiency. Vitamin K can be provided in oral and injectable forms with doses ranging from 2.5 to 10 mg to be given twice weekly to daily. Extra discretion should be used when using lipid injectable forms of vitamin K, as amounts may vary, depending on the manufacturing process.[117] Daily vitamin K intake consistency is imperative once the INR goal is met, as there could be a risk of clotting with further increase in vitamin K without adjusting warfarin dose.[121]

Toxicity

The DRI report on vitamin K states that a "search of the literature revealed no evidence of toxicity associated with the intake of either phylloquinone or menaquinone forms of vitamin K." A synthetic form of vitamin K, menadione, has been associated with liver damage and, therefore, is no longer used therapeutically.[122]

Thiamin (Vitamin B_1)

Deficiency

Thiamin is involved in a variety of metabolic activities, particularly propagation of nerve impulses. Thiamin deficiency can occur with intake of diets low in thiamin or foods that block or degrade thiamin (milled rice, raw shellfish, and foods high in sulfites), in starvation diets, chronic alcohol use disorder, in patients after gastric bypass surgery, and in parenteral nutrition therapy with inadequate thiamin supplementation.[123-126] Depleted magnesium concentration can also cause thiamin deficiency, as magnesium is needed to convert thiamin into thiamin pyrophosphate.[28,127,128]

Deficiency can present as either beriberi or Wernicke-Korsakoff syndrome. Beriberi has two presentations: infantile and adult. The infantile variant presents around 2 to 3 months of age, predominantly in breastfed infants whose mothers have a diet deficient in thiamin. The symptoms are cardiovascular (cardiomegaly, cyanosis, pulmonary hypertension), dyspnea, vomiting and neurologic (meningeal signs, nystagmus, seizure, altered consciousness). Treatment of acute infantile beriberi requires parenterally administered thiamin, using doses in infants of 100 to 150 mg, typically leading to rapid improvement of symptoms.[29]

The adult variants are subsequently subdivided into "wet" beriberi or predominantly cardiovascular issues, and "dry" beriberi or predominantly neurologic symptoms.

Cardiovascular (wet) beriberi usually manifests as chronic high-output right- and left-sided heart failure with tachycardia, rapid circulation time, elevated peripheral venous pressure, sodium retention, and edema.[129] Individuals may also present with stocking-glove cyanosis, extreme tachycardia, cardiomegaly, hepatomegaly, and neck-vein distention. Edema is usually absent.[130] Neurologic (dry) beriberi presents with a symmetrical peripheral neuropathy including symmetric footdrop, associated with marked tenderness of the calf muscles, and a mild disturbance of sensation over the outer aspects of the legs and thighs and in patches over the abdomen, chest, and forearms. Ataxia with loss of position and vibration sense, burning paresthesias in the feet, and amblyopia are less common.[130]

Wernicke–Korsakoff Syndrome includes a spectrum of disease manifestations. Wernicke's encephalopathy, commonly seen with low thiamin intake and chronic alcohol use disorder or after gastric bypass surgery, presents with nystagmus, ataxia, and confusion and is a medical emergency.[131,132] Korsakoff syndrome stems from Wernicke's encephalopathy and is manifested by anterograde amnesia (the inability to store new memories) and retrograde amnesia (the inability to recall past memories) with confabulation (filling gaps in memory with fabricated events).[133,134]

The preferred method of testing is a serum thiamin concentration, although a functional test, the erythrocyte thiamin transketolase activity (EKTA) test, is also available.[30] There are no consensus guidelines on dose and frequency of thiamin supplementation for persons at risk for the development of Wernicke's encephalopathy,[31,32,135-138] but 100–250 mg per day for 3 to 5 days are common doses.[31,32] It is also important to remember that thiamin supplementation may not be optimized if magnesium concentration is depleted. The clinician may want to consider providing an initial dose of 1,000 mg magnesium sulfate with the IV thiamin.[28]

For suspected Wernicke encephalopathy, treatment should occur immediately, due to high risk of death. A common regimen is 500 mg IV thiamin delivered over 30 minutes three time per day for 2 days, followed by 250 mg (IV or IM) once per day for 5 days.[32] Oral thiamin is discouraged due to risk of poor absorption with underlying malnutrition and/or alcohol use disorder. Ideally, IV thiamin should be provided prior to glucose administration, as glucose can exacerbate Wernicke encephalopathy. After the full IV thiamin regimen is completed, the patient should be placed on 100 mg thiamin per day orally until the risk of Wernicke encephalopathy is resolved.[33]

Possible Toxicity

Large doses of thiamin are routinely given to patients with alcoholism. Adverse effects, including sensitization of an anaphylactic nature, have been reported only rarely.[35,39] No anaphylactic reactions have been reported with multivitamins containing thiamin used in central parenteral nutrition (CPN) solutions.

Riboflavin (Vitamin B_2)

Deficiency

Riboflavin is readily available in dairy, eggs, and meat, in addition to being produced by bacteria in the large intestine.[36] A deficiency can occur in individuals who avoid these foods but also in alcohol use disorder, malabsorptive disorders and in states of starvation, including anorexia nervosa.[2] The most common manifestations of riboflavin deficiency are changes to the skin and oral mucosa: glossitis with a magenta tongue (see **Figure 57.1E**), angular stomatitis (fissures in the angle of the lips) and cheilitis (fissures on the vermillion surface) (see **Figure 57.1F**). These fissures may become infected with *Candida albicans*, thus giving rise to the appearance known as perlèche. These mucocutaneous changes may also be seen in other nutrient deficiencies or in persons with edentulism with angular stomatitis.

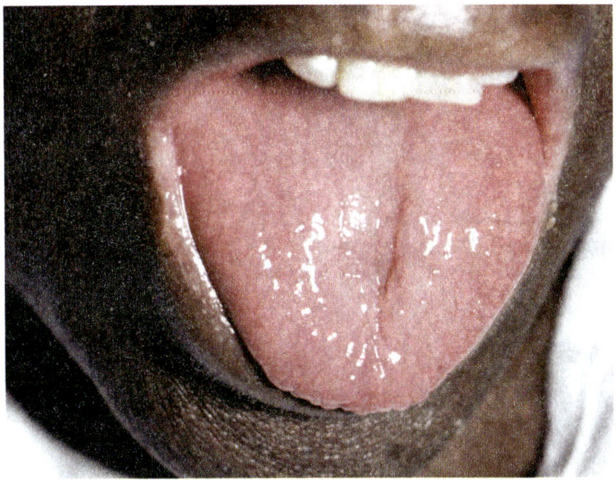

Figure 57.1 E. Riboflavin deficiency. Magenta tongue.

Courtesy of Visual Dx.

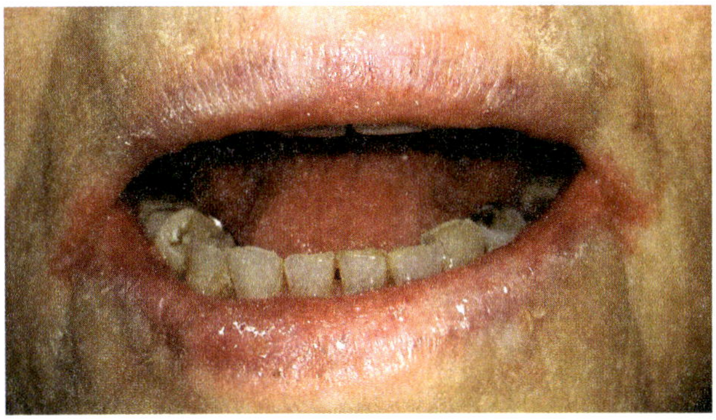

Figure 57.1 F. Riboflavin deficiency. Cheilosis and angular stomatitis.

Matthew Ferguson 57 (https://commons.wikimedia.org/wiki/File:Angular_cheilitis1.jpg), „Angular cheilitis1", https://creativecommons.org/licenses/by-sa/3.0/legalcode

Individuals can also have extensive seborrheic dermatitis that may be resistant to treatment. Areas of skin involvement mainly include the nasolabial folds, external ears, eyelids, scrotum in boys and men, and labia majora in girls and women (see **Figure 57.2A**). These areas become reddened, scaly, greasy, painful, and/or pruritic. Photophobia, lacrimation, and conjunctival infection may also be present. Because vitamin deficiencies are often multiple, it is rarely possible in clinical practice to demonstrate the precise cause. Normocytic normochromic anemia, leukopenia, thrombocytopenia from marrow hypoplasia, and peripheral neuropathies with hyperesthesia, altered temperature sensation, and pain have also been reported with riboflavin deficiency.[37]

Testing is best measured with the erythrocyte glutathione reductase activity coefficient (EGRAC). It cannot be used in individuals with glucose-6-phosphate dehydrogenase (G6PD) deficiency. The alternative test is a 24-hour urine for riboflavin.[139] There is no agreed-upon supplementation regimen for riboflavin deficiency. Injectable forms are available for oral intolerance.

Toxicity

Reports of administration of single oral doses up to 38 times the recommended dietary allowance demonstrated no adverse effect.[39] Despite this, it is advised to use caution with long-term high-dose supplementation (>100 mg/day), due to reports of light-induced damage on the retina and ocular lens proteins.[40,41]

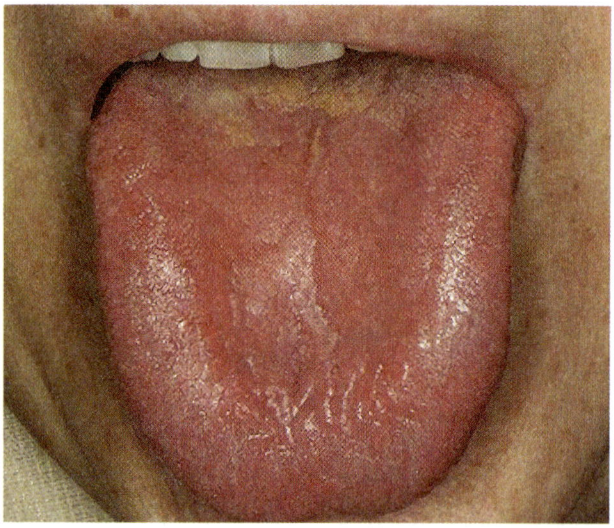

Figure 57.2 A. Bright red and atrophic tongue.

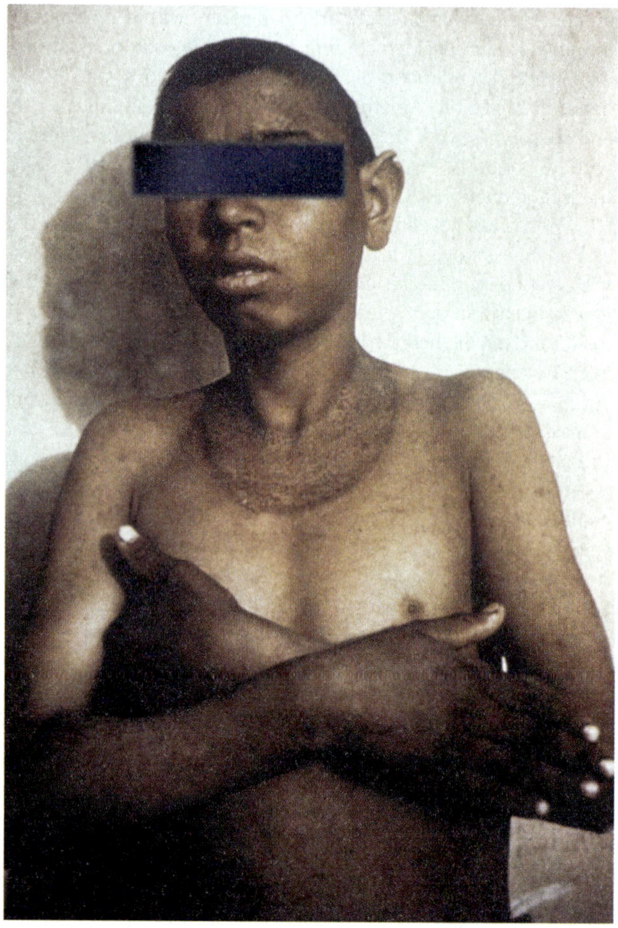

Figure 57.2 B. Pellagra. Casal necklace, a broad band or collar of dermatitis, induced by exposure to sunlight, is a classic sign of pellagra. This boy was suffering from a skin rash associated with a case of pellagra, brought on by a niacin, vitamin B_3, deficiency.

Courtesy of Centers for Disease Control and Prevention.

Niacin (Vitamin B_3)

Deficiency

Niacin is not only found in animal foods and grains but can also be converted from Tryptophan (Trp). A deficiency of niacin is seen after bariatric surgery, in chronic alcohol use disorder, and in malabsorptive disorders.[42] Tryptophan deficiency can occur in individuals with a vitamin B_6 deficiency or in the presence of carcinoid tumor.[43] The deficiency presents as pellagra and produces the "four Ds" of dermatitis, diarrhea, dementia, and ultimately death. Dermatitis is usually the earliest and most prominent manifestation. It is symmetric and appears on parts exposed to sunlight or trauma with erythema that progresses to keratosis and scaling with pigmentation. The back of the hands, wrists, forearms, face, and neck (Casal necklace) are typically affected[42] (see **Figure 57.2B**). The tongue often has a "raw beef" appearance and is bright red, swollen, and painful. Symptoms of gastritis, bouts of diarrhea, and signs of malabsorption suggest similar changes in the gastrointestinal tract. Nervous system involvement includes periods of depression with insomnia, headaches, and dizziness and, if undiagnosed, dementia and encephalopathy leading to death.[44] Testing for deficiency is best measured with urinary N1-methylnicotinamide or an erythrocyte NAD:NADP ratio.[39]

Niacin needs can be met by consuming 1 g of Trp, which equals 100 g of protein per day, regardless of niacin intake. Oral niacin, in the form of nicotinamide, can be used to treat deficiency, with dosing ranges of 10 to 150 mg/day.[45] Pellagra requires doses of 100 to 300 mg subcutaneous niacinamide daily for 3 days.[46]

Toxicity

The most notable side effects of megadoses of niacin (e.g., 1 to 3 g/day), include flushing, pruritus, nausea, vomiting, constipation, and headache.[47] Liver dysfunction manifesting as elevated serum liver enzymes is common, and liver failure can occur. Patients with diabetes should monitor their blood glucose, as supplemental niacin may worsen insulin resistance. Patients with gout should supplement with caution, as niacin can elevate uric acid concentration.

Pyridoxine (Vitamin B_6)

Deficiency

Pyridoxine deficiency may occur in individuals with malabsorptive diseases, alcohol use disorder, chronic renal insufficiency, autoimmune disorders, such as rheumatoid arthritis, and the use of isoniazid, levo/carbidopa, and anticonvulsants. The deficiency can also occur with genetic disorders such as homocystinuria.[48] Pyridoxine deficiency induced by poor intake in adults is rarely severe enough to produce signs or symptoms. The most common presentation of deficiency is similar to riboflavin, with seborrheic dermatitis along the nasolabial folds, cheeks, neck, and perineum, glossitis, angular stomatitis, blepharitis, and peripheral neuropathy. Pyridoxine deficiency can also manifest as microcytic anemia and seizures, particularly in infants.[2,87] Severe deficiency can present as pellagra with impairment of the conversion of tryptophan to niacin.[2] Testing for pyridoxine deficiency can be measured through plasma pyridoxal-5-phosphate (PLP).[39]

Dosing and route of supplementation will depend on the severity of symptoms and degree of deficiency, with oral doses ranging from 25 mg to 600 mg/day.[49] Intravenous dosing is indicated for individuals with seizures that are caused by a vitamin B_6 deficiency.

Toxicity

The highest amount of pyridoxine that could be consumed daily that is likely to avoid toxicity, known as the tolerable upper intake level (UL), is 100 mg/day. Megadoses of >250 mg/day can cause sensory neuropathy, including progressive sensory ataxia and profound lower limb impairment of position and vibration sense.[50] Touch, temperature, and pain perception are less affected.

Biotin

Deficiency

Biotin deficiency has occasionally been induced in patients who consumed large amounts of raw egg whites over a prolonged period. Egg white contains avidin, which antagonizes the action of biotin.[38] In the past, biotin deficiencies occurred in children and adults maintained on long-term CPN before biotin was included in commercial vitamin formulations and is uncommon today. A varied diet should provide adequate biotin for the average person.[38]

There are two variants of an inherited biotin metabolism defect entitled multiple carboxylase deficiency, tested in newborn screening panels in many states. One variant presents in the first week after birth and manifests with vomiting, lethargy and hypotonia. The other variant presents later in life and manifests with ketoacidosis, seizures, an erythematous rash, and neurologic findings such as nystagmus.[140] The classic presentation in patients with malabsorption disorders, alcohol use disorder, and excess egg white ingestion is alopecia, brittle nails, facial dermatitis that resembles seborrheic dermatitis, and in severe deficiency paresthesias, myalgias, lethargy, and mental status changes.[51] Testing for decreased biotin concentration in the urine can be an early sign of deficiency.[52] Serum concentration can be measured but is not reliable with mild deficiency. The best assessment is white blood cell biotinylated methylcrotonyl CoA carboxylase (MCC) and propionyl-CoA carboxylase.[53] There are currently no consensus guidelines for biotin supplementation in the setting of deficiency.[38]

Toxicity

No reports of adverse effects of biotin intake have been published; however, high supplementation doses may interfere with thyroid testing (T4 and TSH) with a false appearance of Grave's disease and can also interfere with troponin concentration.[54-56]

Vitamin B_{12} (Cobalamin)

Deficiency

Vitamin B_{12} is found in animal products and is added to cereals.[33] Absorption requires vitamin B_{12} to bind with intrinsic factor, a protein produced by the parietal cells of the stomach, which is then absorbed from the terminal ileum. A vitamin B_{12} deficiency can, therefore, occur with disorders of the stomach (achlorhydria, atrophic gastritis, after bariatric surgery) and the ileum (inflammatory bowel disease, resection of the distal ileum). Deficiency is also seen in individuals on metformin for longer than 4 months or proton pump inhibitors or H2 blockers for longer than 12 months. An autoimmune disease, pernicious anemia, occurs when antibodies are produced against intrinsic factor, thereby inhibiting B_{12} absorption. Other population groups at risk of vitamin B_{12} deficiency include vegans, strict vegetarians, and adults over 75 years of age.[57] Caution should be exercised in the use of nitrous oxide anesthesia, or when it is used recreationally, because of the possibility of precipitating abrupt symptoms of B_{12} deficiency.[141]

Pernicious anemia usually manifests after middle age, especially in persons with prematurely gray hair and blue eyes. A slight female preponderance exists. The most common presentation includes signs of anemia (fatigue, exercise intolerance, palpitations, skin pallor, and shortness of breath), glossitis, cheilitis, anorexia, weight loss, progressive neurologic findings (stocking-glove sensory loss, symmetric paresthesias, ataxia, weakness, impaired position and vibratory sense, impaired deep tendon reflexes, and paraplegia), and neurocognitive changes (depression, forgetfulness, insomnia, dementia, and psychosis).[58] When dietary lack or malabsorption is the cause of the deficiency, megaloblastic anemia is usually the most prominent feature, but glossitis and neurologic changes have also been described.

On testing with a complete blood count (CBC) with differential and peripheral smear, individuals may have megaloblastic anemia (low hemoglobin with elevated MCV and MCH, leukopenia, thrombocytopenia, nucleated RBCs). The two vitamin deficiencies to consider with megaloblastic anemia include B12 and folate deficiency. The workup should include tests for deficiency of both vitamins. Testing for B_{12} deficiency is best measured with serum cobalamin.[59] A serum folate test will rule out folate deficiency. Low serum cobalamin concentration requires confirmatory testing with methylmalonic acid (MMA) and homocysteine. Both tests will be elevated in vitamin B_{12} deficiency.[60]

Neurologic impairment in the setting of vitamin B_{12} deficiency can be permanent if not treated.[142,143] Data suggest that high-dose oral vitamin B_{12} supplementation (1,000–2,000 μg/day) can be comparable to injections.[61,142,143] However, there may not be an improvement in MMA concentration as there is with IM subcutaneous dosing.[57] A common dosing regimen is 1,000 μg cyanocobalamin IM injection once per week until the deficiency is corrected. The British Society for Hematology recommends using oral supplementation for those with mild deficiency, no absorption issues, and who are known to be compliant.[57] Pernicious anemia dosing ranges are 100 to 1,000 μg IM injections of cyanocobalamin for 5 days followed by 100 to 1,000 μg monthly.[144] Vegans and strict vegetarians should commit to consuming fortified cereals on a regular basis and/or take a vitamin B_{12} supplement to decrease the risk of deficiency.[57]

Toxicity

No reports of adverse effects of vitamin B_{12} supplementation have been noted in the literature.

Folate (Folic Acid)

Deficiency

Folate is found naturally in foods such as fruits and vegetables. Folic acid is the synthetic form used in supplements and fortified foods and is better absorbed than folate found in food.[145] Folic acid is readily available in fortified grains and cereals. Therefore, deficiency is seen in individuals who have low intake (alcohol use disorder, anorexia nervosa), increased requirement (pregnancy, hemolytic anemia, exfoliative dermatitis), decreased absorption (bariatric surgery, celiac disease), loss from hemodialysis without replacement of folic acid, and interference from medications (methotrexate, sulfasalazine). The megaloblastic anemia of folic acid deficiency has morphologic features indistinguishable from those of vitamin B_{12} deficiency, but it develops more rapidly. The tongue may be red and painful in the acute stage. In chronic deficiency, the tongue papillae atrophy, leaving a shiny, smooth surface. Hyperpigmentation of the skin is occasionally seen predominantly in the hands and feet.[145-147] Neurocognitive dysfunction can develop in addition to symptoms of depression.

Folic acid therapy before conception is now accepted as protective against neural tube defects.[148] Low plasma folate concentration is associated with increased risk of early spontaneous abortion, placental abruption, spina bifida, and anencephaly.[149] Emerging research suggests that maternal folic acid supplementation may also be protective in reducing the risk of autism spectrum disorder.[150-152]

Similar to a vitamin B_{12} workup, a complete blood count (CBC) will demonstrate macrocytic anemia (low hemoglobin with elevated MCV and MCH, leukopenia, thrombocytopenia). Testing for folate deficiency is best measured with serum folate concentration.[34] Low serum folate requires confirmatory testing with homocysteine, which will be elevated. A methylmalonic acid (MMA) test will be normal in the absence of vitamin B_{12} deficiency.[60]

Oral or IV folic acid is available for supplementation in the setting of deficiency, although IV dosing is reserved for those who need rapid repletion and/or cannot tolerate oral supplementation. A folate deficiency is treated with 1 to 5 mg of folic acid daily. This can be given for a period of four months to lifelong, depending on the cause of deficiency.[57]

Toxicity

The required addition of 400 μg per serving of ready-to-eat cereals has raised the question of excess. Using data from the U.S. Food and Drug Administration, the Food and Nutrition Board of the National Academy of Medicine stated, "It is unlikely that intake of folate added to foods or as supplements would regularly exceed 1,000 μg for any of the life stage or gender groups".[34] This is important as studies have shown an increase in colon, prostate, and breast cancer at doses of 1,000 μg and higher in some individuals.[62,153,154]

Administration of folate for megaloblastic anemia should be given only after ruling out cobalamin deficiency as the primary cause because folate administration may improve the hematologic manifestations of vitamin B_{12} deficiency without arresting its neurologic effects.

Vitamin C (Ascorbic Acid)

Deficiency

Vitamin C is predominantly found in fresh fruits and vegetables, particularly citrus fruits, broccoli, spinach, and potatoes.[36] Lack of access to these foods or poor intake can lead to a condition known as scurvy. Individuals most at risk include the elderly with restricted diets (tea and toast, fast-food only) and individuals with food insecurity who have limited access to fresh fruits and vegetables. Individuals at risk for poor intake include infants on exclusive cow's milk feedings, those with anorexia nervosa, anorexia due to treatment for cancer or other health disorders, alcohol use disorder, and children with food aversions, including those with ARFID.[7-9] Other causes of deficiency include increased requirements, as in pregnancy and lactation, cigarette smokers, type I diabetes, malabsorption, particularly of the small intestine or after bariatric surgery, and iron overload disorders.[2,155,156]

Vitamin C is essential for collagen and connective tissue formation and is important in wound healing. A deficiency will manifest with perifollicular hyperkeratosis, hemorrhage, and corkscrew hairs as early as 3 months without proper intake. Individuals may complain of fatigue and muscle and joint pain and exhibit spongy, friable gingiva that bleed with trauma, interdental inflammation, tooth loss, and poor wound healing. The classic oral mucosal changes may be exacerbated by poor dental hygiene. In addition to the previously noted symptoms, infants and children may demonstrate tenderness and swelling, most marked at the knees or ankles, with limping if ambulatory. These signs result from characteristic bone changes demonstrable by radiograph. The infant often adopts the "pithed frog" position of maximum comfort, with the legs flexed at the knees and the hips partially flexed and externally rotated.

Bleeding may not only occur anywhere in the skin (the orbit is a frequent site) or mucous membranes but also in the renal tract. In infancy, intracranial hemorrhages are rapidly progressive if treatment is delayed, and death may occur. Microcytic hypochromic anemia is common, whereas a normochromic normocytic picture is less so. Without treatment, individuals can develop shortness of breath, hypotension, jaundice, neuropathy, seizures, and death[157] (see **Figure 57.2C** and **Figure 57.2D**). The best measure of deficiency is with serum or leukocyte vitamin C concentration.[2,156]

Vitamin C repletion in the setting of deficiency can be given up to 300 mg per day in pediatrics, and up to 1,000 mg per day in adults.[63] Symptoms should be improved within 1 month with 300 mg per day, allowing recovery from scurvy within 3 months.[64,65] One published adult supplementation regimen is 2 g vitamin C daily for 3 days, followed by 500 mg daily for 1 week, followed by 100 mg daily for 3 months.[2]

Toxicity

Long-term supplementation of vitamin C in excess of the tolerable upper limit can cause diarrhea, kidney stones, and excess iron absorption.[63] A false-negative stool guaiac can be seen with large doses.

Choline

Choline is needed for phospholipid synthesis, and choline deficiency has been linked to intestinal failure-associated liver disease (IFALD). Individuals on long-term parenteral nutrition (PN) were noted to develop choline deficiency with liver enzyme abnormalities and evidence of liver steatosis.[158,159] Choline is not included in IV multi-vitamin injection (MVI) for PN but is incorporated into the lipid component in the form of phosphatidylcholine.[160] The American Society for Parenteral and Enteral Nutrition recommends that IV choline be available as a separate additive in PN.[161] A low-choline diet ingested for 3 weeks by healthy men resulted in decreased plasma choline concentration and some abnormal liver function, as indicated by a liver function test.[66] Studies have shown benefit of added choline to parenteral formulas in reversing liver changes associated with long-term PN and as prevention against IFALD.[158,159]

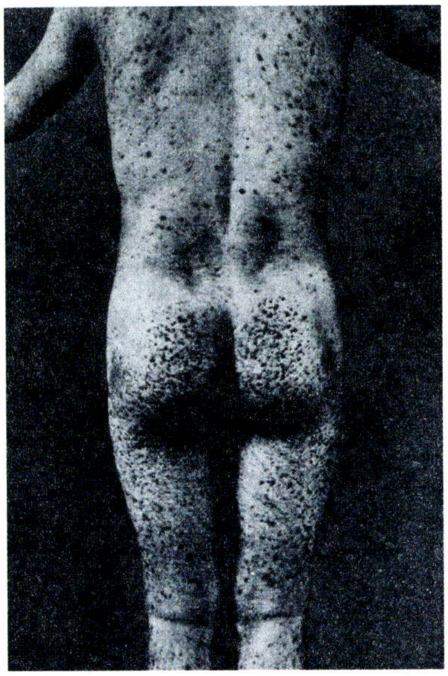

C

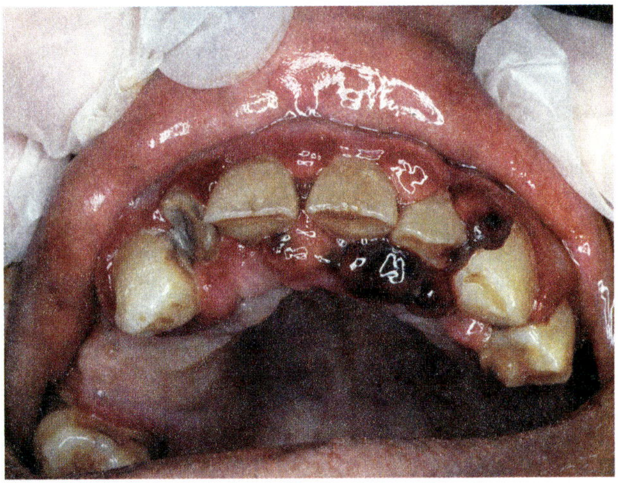

D

Figure 57.2 C. Scurvy. Back view of male scurvy victim, showing characteristic hemorrhages under the skin. D. Gingival bleeding and missing teeth due to scurvy.

C. National Library of Medicine; D. Courtesy of Visual Dx.

Toxicity

High doses of choline have been linked to vomiting, hypotension, an increased risk of cardiovascular disease, and a "fishy" body odor.[34]

Essential Fatty Acids (Linoleic Acid and α-Linolenic Acid)

Deficiency

Linoleic acid and α-linolenic acid are found in ingested fats and oils and cannot be synthesized by humans.[4,67] Deficiency can occur in individuals with malabsorptive disorders, after surgical procedures of the gut (bariatric surgery, resection of small bowel), in pancreatic insufficiency, cholestatic disorders, and in individuals on long-term PN without added fat emulsion.[4] Individuals with essential fatty acid (EFA) deficiency present with alopecia; dry, flaky skin of the trunk; poor wound healing; increased susceptibility to infection; and, in children, growth retardation.[4]

Individuals who are deficient in linoleic acid and α-linolenic acid show an increase in the synthesis of mead acid (a triene). Conversely, there is a decrease in the production of arachidonic acid (a tetraene).[68] Measurement requires a serum triene to tetraene ratio, which will be elevated >0.2 in the presence of EFA deficiency.[4]

Prevention of EFA deficiency in people on PN is accomplished through the provision of at least 10% of total energy from fat, of which 2% to 4% are from linoleic acid. Dosing of IV lipid emulsion (ILE) (100 g soybean oil-based ILE) per week should be adequate to prevent deficiency, but higher doses may be needed in patients with a preexisting deficiency.[10,162] A variety of other lipid emulsions are available containing different fatty acid compositions, such as fish oil and olive oil. These provide decreased risk of intestinal failure-associated liver disease but require higher dosing to meet essential fatty acid requirements.[163,164] Some individuals cannot receive ILE secondary to allergy or elevated triglycerides.

Oils that are a good source of EFA can be taken orally if GI absorption is adequate.[69] Pancreatic enzyme replacement is imperative in pancreatic insufficiency to avoid worsening fat malabsorption with EFA replacement.[3] There are limited and conflicting data regarding the use of topical sunflower or safflower oil application.[69] Caution is warranted when using topical therapy, and patients should be monitored closely.

Minerals

Minerals are inorganic elements that start off in water and soil, get absorbed by plants, and are, therefore, found in our food. They are necessary for many essential roles in our bodies. Minerals that have an electric charge and dissociate in solution are referred to as electrolytes and are responsible for maintaining homeostasis. Minerals needed in amounts less than 100 mg per day are referred to as trace elements. In this section of the chapter, the electrolytes: sodium, calcium, phosphorus, and magnesium, will be discussed as well as trace elements: iodine, iron, copper, zinc, fluoride, selenium, chromium, molybdenum, and manganese.

Sodium

Sodium is used to flavor and preserve food, is the major cation in the extracellular fluid in the body, and helps regulate extracellular fluid volume and water distribution. Normal sodium concentration is 135 to 145 mEq/L.[165]

Hyponatremia

Sodium is regulated through the kidneys predominantly under the control of the renin-angiotensin-aldosterone system. Total body water is controlled primarily through the excretion or suppression of antidiuretic hormone (ADH). Dysfunction in water control presents as hypo—or hypernatremia, whereas dysfunction in sodium control manifests as hypo- or hypervolemia.[166] Water normally diffuses freely across the cell membrane to maintain an equilibrium of osmolality (concentration of a solution) in the intracellular and extracellular fluid. The three major solutes in serum are sodium, glucose, and urea. When serum osmolality increases, water flows into the bloodstream and increases fluid volume, and conversely, when osmolality is low, fluid leaves the bloodstream, leading to decreased volume.

Hyponatremia has three categories, based on the osmolality of the blood: hypertonic (>295 mOsm/L), isotonic (275–295 mOsm/L), and hypotonic (<275 mOsm/L).[166,167] Hypertonic hyponatremia is caused by osmotically active substances in the serum, such as mannitol, or excessive glucose, such as in diabetes. Sodium concentration must be corrected with hyperglycemia: Corrected Na+ = Serum Na+ + [0.016 × (Serum Glucose) − 100].

Isotonic hyponatremia is seen in patients with hyperlipidemia, particularly hypertriglyceridemia or obstructive jaundice, which drives total serum cholesterol concentrations. Individuals with hyperproteinemia, as in multiple myeloma, can also present with isotonic hyponatremia. Hypotonic hyponatremia is the most common electrolyte imbalance, occurring in up to 25% of patients in the hospital[166,167] and is due to multiple etiologies including but not limited to primary polydipsia, beer potomania, tea-and-toast diet, diuretic use, volume depletion from any cause, congestive heart failure, cirrhosis, syndrome of inappropriate antidiuretic hormone (SIADH), Addison's disease, and exercise-associated hyponatremia.[166-169] Hypotonic hyponatremia can be further categorized based on the patient's fluid status: hypovolemic, hypervolemic, and euvolemic. Causes of hypovolemic hypotonic hyponatremia include copious fluid loss from diarrhea, fistulas, drains, burns, and excessive sweating. Euvolemic hypotonic hyponatremia can occur in the presence of SIADH and Addison's disease.[169]

Hypervolemic hypotonic hyponatremia is seen in fluid overload states such as heart failure, cirrhosis, and nephrotic syndrome. Exercise-associated hyponatremia (EAH) is a complex condition with multiple causes, including overhydration, sodium losses through sweat coupled with inappropriate ADH stimulation, and decreased renal diluting ability.[168] Volume status with EAH can range from hypovolemic to hypervolemic.

Signs and symptoms of hyponatremia are dependent on the severity of the deficiency and the rapidity of occurrence. The earliest findings are nausea, vomiting, and headache, progressing to cerebral edema with neurologic symptoms (altered mental status, seizures, obtundation, coma, and death). Diagnosis is made through a comprehensive history and physical exam and a serum metabolic panel to include electrolytes, glucose, and creatinine. Treatment options depend on the etiology and fluid volume status.

Hypernatremia

Hypernatremia is classified as serum sodium greater than 145 mEq/L.[167] Symptoms of hypernatremia depend on the speed with which the sodium increase occurs. In chronic hypernatremia, individuals experience dry mucous membranes, orthostatic blood pressure changes, and tachycardia, but can progress to cognitive and neurologic dysfunction as hypernatremia worsens. In the abrupt onset of hypernatremia, such as IV administration of hypertonic saline, patients can develop demyelinating brain lesions, hemorrhage, and neurologic changes, including seizures, coma, and death. Individuals at risk for hypernatremia include those over 65 years of age, impaired thirst or inability to access water, diarrhea, excessive sweating, diuretic use, diabetes insipidus, excess mineralocorticoid secretion, and excessive infusion of isotonic or hypertonic sodium solution.

Calcium

Hypocalcemia

Calcium homeostasis involves the interplay of multiple factors, including vitamin D, parathyroid hormone, phosphate concentration, magnesium concentration, and adequate renal function. When calcium begins to decrease, parathyroid hormone is secreted, which has three effects. Calcium is reabsorbed in the distal tubules of the kidney with a resultant decrease in urinary calcium excretion. Vitamin D production is increased in the kidneys which, in turn, increases calcium absorption from the intestines. Calcium is released from the bone (increased bone resorption).[170]

There are several causes of hypocalcemia; they can be divided based on whether the parathyroid hormone is low or high. Causes of hypocalcemia with low parathyroid hormone relate to the destruction of the parathyroid glands or genetic mutations that affect the development of the parathyroid gland or the production/structure of parathyroid hormone. Hypoparathyroidism is most commonly due to damage to the parathyroid glands after thyroid surgery but may occur after irradiation of the neck area or damage from some other cause.[171] Causes of hypocalcemia with high parathyroid hormone include vitamin D deficiency, end-stage renal disease, hyperphosphatemia (excess phosphate intake, rhabdomyolysis, acute renal failure, or tumor lysis syndrome), acute pancreatitis, or in critically ill or septic patients.[172] Hypocalcemia can also occur with low magnesium concentration, which is typically seen in individuals with chronic alcohol use disorder and malabsorption.

The most characteristic sign in hypocalcemia is tetany, consisting of the following: (a) paresthesias about the lips, tongue, fingers, and feet; (b) carpopedal spasm (see **Figure 57.2E**); (c) generalized muscle aches and (d) spasm of the facial muscles. Chvostek sign is a contraction of the facial muscles on light tapping of the facial nerve. Trousseau sign is a carpopedal spasm induced by restriction of the blood supply to a limb by a blood pressure cuff applied for 3 minutes or fewer. Hypocalcemia may also present with depression, irritability, anxiety, psychosis, generalized seizures, hypotension, and congestive heart failure. Cases of papilledema and optic neuritis have been reported in severe hypocalcemia.[173]

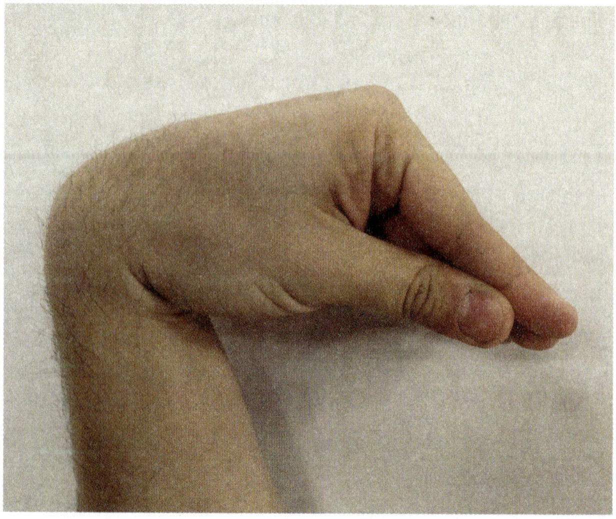

Figure 57.2 E. Hypocalcemia. The characteristic contraction of the hands (tetany) associated with the presence of marked hypocalcemia often secondary to magnesium depletion.

Courtesy of Tmdswan.

Calcium insufficiency, especially during growth when bone mass is developing, and in later life, is a risk factor for osteomalacia, rickets, and osteoporosis. Osteoporosis is common in postmenopausal women and men over the age of 70. Patients have bone deformity, localized pain, and fractures. Osteomalacia may coexist. The most common deformity is loss of height caused by vertebral collapse, which accounts for most of the pain. Fractures of the neck, femur, or distal radius fracture above the wrist (Colles fracture) are most commonly precipitated by trauma, which may be trivial, in elderly persons with osteoporosis. True rickets can be produced by dietary calcium deficiency in the presence of normal vitamin D status.[174] Such cases respond better to calcium therapy alone than to vitamin D supplementation alone.[175]

Testing is best assessed with serum calcium concentration. Nearly half of all calcium is transported bound to plasma proteins, predominantly albumin. In the presence of low albumin, serum calcium needs to be corrected using the formula:

$$\text{Calcium} = \text{serum calcium} + [0.8 \times (\text{Normal Albumin} - \text{Patient's albumin})].$$

Caution is needed when applying this formula in critically ill patients with acid-base disorders, as the calculation will be inaccurate.[176] Alternatively, ionized calcium can be ordered, which will provide a direct measurement of the metabolically active calcium.

Hypercalcemia

Hypercalcemia has various causes including hyperparathyroidism, malignant disease, thyrotoxicosis, excess calcium intake in the presence of chronic kidney disease, hypervitaminosis D, or prolonged immobilization. Individuals on lithium are also at risk of developing mild hypercalcemia. The most common signs and symptoms of hypercalcemia are covered in the mnemonic "stones, bones, abdominal groans, and psychic moans." Renal system involvement produces polyuria, nocturia, polydipsia, stone formation, and sometimes hypertension and signs and symptoms of uremia. Muscle weakness, myopathy, and bone pain can occur. Gastrointestinal symptoms include anorexia, nausea, vomiting, constipation, abdominal pain, and ileus. Severe hypercalcemia, which causes psychosis, delirium, stupor, and coma, may be fatal. Hypercalcemia caused by excessive calcium intake is uncommon, even in those ingesting large doses of calcium supplements. However, the combination of calcium supplements with sodium bicarbonate or foods containing oxalic acid (rhubarb, spinach, almonds) increases the risk of nephrolithiasis.

Phosphorus

Hypophosphatemia

Hypophosphatemia is seen primarily in hospitalized patients, particularly in those with sepsis or who have suffered trauma, in addition to individuals with alcohol use disorder. Acute hypophosphatemia, without total body phosphate depletion, occurs in any situation stimulating anaerobic glycolysis, such as infusion of hypertonic glucose (e.g., CPN), especially in cachectic patients without adequate phosphate replacement. This results in a rapid shift of phosphate into cells, thereby reducing serum phosphate and potentially depleting adenosine triphosphate (ATP) and impairing multiple phosphate-requiring metabolic processes, including glycolysis. Marked hypophosphatemia, for which patients with cachexia are particularly at risk, can cause the refeeding syndrome, which consists of dramatic hyperglycemia, weakness, muscle paralysis, cardiorespiratory failure and, unless treated promptly, death.[177]

Total body phosphate depletion occurs with total body nitrogen loss as the result of various diseases that lead to excessive loss in the stool (e.g., malabsorption, vitamin D deficiency) or in the urine (e.g., hyperparathyroidism, congenital or drug-induced renal tubular acidosis, severe potassium depletion). Poor intake of phosphate-rich foods in the presence of chronic diarrhea can lead to phosphate depletion. In the management of advanced renal disease, the administration of phosphate-binding gels intended to reduce phosphate absorption in association with restricted phosphate in the diet may lead to symptomatic phosphate deficiency.[178,179] Vitamin D deficiency will decrease GI absorption of phosphate and can lead to secondary hyperparathyroidism. Parathyroid hormone will increase urinary phosphate excretion; therefore, hypophosphatemia will occur in both primary and secondary hyperparathyroidism.[180] Chronic hypophosphatemia leads to increased bone resorption and hypercalciuria and places individuals at increased risk of stone formation, flank pain, and hematuria, in addition to osteomalacia and fractures.

When serum phosphorus concentration is low, unless the cause of the decreased value is readily evident, a 24-hour urine to measure phosphate excretion is necessary. When urinary phosphate excretion is low in the presence of hypophosphatemia, the cause is redistribution of phosphorus such as in refeeding syndrome or chronic diarrhea. When urinary phosphate excretion is elevated, the cause is most commonly hyperparathyroidism or vitamin D deficiency.

Treatment of hypophosphatemia is dependent on the severity of the deficit and GI tract function. Oral and IV formulations are available, with the latter used in severe deficits. Oral repletion may cause diarrhea in some individuals. Multiple repletion regimens exist and can be weight-based or based on the severity of the lab value.[181]

Hyperphosphatemia

Chronic hyperphosphatemia is common in acute renal failure and end-stage renal disease due to impaired urinary excretion. An increase in phosphate can also occur in tissue release, as seen in tumor lysis syndrome (hyperkalemia, hyperuricemia, and hyperphosphatemia), particularly with lymphoma; in rhabdomyolysis, and with high phosphate intake, such as with phosphate-containing laxatives. Hypoparathyroidism, vitamin D toxicity, and acromegaly have also been associated with hyperphosphatemia. These conditions are potentially serious because of the development of nephrocalcinosis and calcification of soft tissues.

Potassium

Hypokalemia

Potassium is the major cation in the intracellular fluid.[165] Hypokalemia can occur with low intake of potassium, increased losses from urine (diuretic use, hyperaldosteronism, renal tubular acidosis, diabetes insipidus), sweat, or the gut (vomiting, diarrhea, nasogastric secretions), and with renal dialysis.[182] It can also be the result of redistribution of potassium into cells due to increased insulin, increased catecholamine release (stress, use of epinephrine, alcohol withdrawal, use of beta-agonists and sympathomimetic agents, acute myocardial infarct), respiratory or metabolic alkalosis, hypokalemic periodic paralysis, hypothermia, and overdose (barium, cesium, and chloroquine).[182]

Like hypocalcemia, low magnesium concentration can lead to hypokalemia, which is typically seen in individuals with chronic alcohol use disorder and malabsorption.

Symptoms of hypokalemia generally present when serum potassium falls below 30 mEq/L and include muscle weakness, cramps, paralysis, rhabdomyolysis, respiratory failure, paralytic ileus, and tetany. Cardiac effects include hypotension and arrhythmias. The electrocardiogram is characteristic of ST-segment depression, increased U-wave amplitude, and T-wave amplitude less than that of U-wave amplitude in the same lead. Premature ventricular and atrial contractions and ventricular and atrial tachyarrhythmias occur. When serum potassium is low, serum magnesium should also be assessed. With potassium concentration less than 2.5 mEq/L, continuous ECG monitoring should be considered while potassium is corrected and the cause is evaluated.[183]

Hypokalemia can be treated orally (via tablet, liquid, or powder mixed with water), depending on the severity and adequacy of GI function. The formulations for IV potassium supplementation include potassium chloride, potassium acetate, and potassium phosphate, and are measured in milliequivalents (mEq). The form used depends on the clinical profile of the patient and other laboratory values. Oral potassium supplements are commonly ordered for patients on loop or thiazide diuretics.[184] Individuals with chronic kidney disease should limit high potassium foods. The patient will need thorough diet education, as potassium is found in an abundance of food items.

Hyperkalemia

Hyperkalemia can occur in the presence of respiratory and metabolic acidosis, in uncontrolled diabetes, rhabdomyolysis, tumor lysis syndrome, excessive ingestion or infusion, or digitalis overdose. Hyperkalemia can also occur with a decrease in urinary potassium excretion due to decreased aldosterone secretion, acute and chronic kidney disease, or the use of potassium-sparing diuretics.[182] Cardiac toxicity, of serious import, starts with shortening of the QT interval of the electrocardiogram and tall, peaked T waves. Progressive toxicity with serum potassium concentration >7.0 mEq/L causes nodal and ventricular arrhythmias, widening of the QRS complex, PR interval prolongation, and disappearance of the P wave and finally, degeneration of the QRS complex with ventricular asystole or fibrillation and death.

Magnesium

Hypomagnesemia

Hypomagnesemia can occur due to losses from the gut (diarrhea, malabsorption, acute pancreatitis, use of proton pump inhibitors) or from the kidney (diuretic use, primary hyperaldosteronism, chronic alcohol use). Low magnesium has also been seen in individuals on ketogenic diets and with refeeding syndrome.

The early symptoms and signs of deficiency are primarily neuromuscular: Trousseau and Chvostek signs, muscle fasciculations, tremor, muscle spasm, seizures, with later personality changes, anorexia, nausea, and vomiting. Despite the hypocalcemia present in severe magnesium deficiency, deep tendon reflexes are normal or depressed.[185] Low dietary intake of magnesium has been associated with impaired lung function and wheezing.[186] Electrocardiogram changes occur with widened QRS and peaked T waves, prolonged PR interval, and atrial and ventricular premature beats or development of atrial fibrillation.[187] Because hypomagnesemia is often accompanied by hypocalcemia and hypokalemia, symptoms associated with low calcium and potassium are likely to develop.

When serum magnesium concentration is low, unless the cause of the decreased value is readily evident, a 24-hour urine to measure magnesium excretion is necessary. Urinary magnesium excretion will be higher with renal losses and lower when the cause is due to GI losses.

Both oral (pill and liquid forms) and IV formulations are available for magnesium supplementation. Magnesium is measured in milligrams, grams, and milliequivalents. One gram of magnesium equals 1,000 milligrams or 8 mEq. In acute hypomagnesemia, the IV route is preferred due to more rapid onset of action and a decrease in GI intolerance.[176,188-190]

Hypermagnesemia

Hypermagnesemia occurs in individuals with end-stage kidney disease and in women provided magnesium infusion for eclampsia or pre-eclampsia. Nausea, vomiting, headache, and lethargy may appear in persons with serum magnesium concentration between 4 to 6 mEq/L. At concentrations higher than 6 mEq/L, deep tendon reflexes disappear, hypotension, and electrocardio graphic abnormalities (prolonged PR interval, widening of QRS complex, and heart block) occur. Respiratory depression, muscle paralysis, and cardiac arrest may occur with concentrations greater than 10 mEq/L.

Iodine

Deficiency

People who do not consume iodized salt, dairy, seafood, and/or eggs may be at risk for iodine deficiency.[191] Iodine is incorporated into the thyroid hormones tri-iodothyronine (T3) and thyroxine (T4). Iodine function is diminished in the setting of selenium deficiency, as selenoprotein deiodinases are needed in the conversion of thyroid hormone.[192] Enlargement of the thyroid gland (goiter) is the most common clinical sign of iodine deficiency. With the advent of iodized salt, goiter from iodine deficiency is rarely seen in the United States and other developed countries.[144] The consumption of cassava root, common in parts of Africa and South America, produces thiocyanate, a metabolite that has antithyroidal properties.[193] Early in iodine deficiency, the enlargement of the thyroid is soft, symmetric, and smooth; later, multiple nodules and cysts may appear. Low thyroid hormone concentration leads to hypothyroidism. Iodine deficiency in pregnancy can lead to stillbirth, spontaneous abortion, congenital malformations, and intellectual disability in the neonate.[194] Severe endemic goiter is often accompanied by cretinism. Endemic cretinism occurs in two distinct forms, the myxedematous and the neurologic, which may coexist.[195] In most areas of the world, the neurologic form is most common. Other than goiter, individuals who develop hypothyroidism will present with symptoms associated with a slowing of all metabolic processes, including but not limited to: complaints of feeling cold when others are feeling hot, dry, coarse skin and hair, hair loss, enlarged tongue, edema particularly around the eyes and the pretibial region, constipation, shortness of breath particularly with exertion, sleep apnea, irregular menses, infertility issues, and erectile dysfunction.

Iodine deficiency can be assessed through a 24-hour urine specimen for iodine excretion. Alternatively, a urine iodine to creatinine ratio can be obtained. Thyroid studies may also reflect iodine deficiency, including serum thyroglobulin concentration.

Toxicity

Excess iodine intake can occur from over-the-counter supplements and medications, prescription medications, and with the use of radiocontrast media. Individuals with dysfunctional thyroid glands may be unable to handle the excess iodine load, which can lead to hypothyroidism. This is especially true in patients with pre-existing Hashimoto's thyroiditis. In these cases, individuals will present with signs of hypothyroidism.

Iron

Deficiency

Iron is incorporated into the hemoglobin molecule and is essential for tissue oxygenation. Iron deficiency leads to a decrease in hemoglobin production and a distinct microcytic, hypochromic anemia. Iron deficiency anemia can occur with blood loss in menstruating women; acute blood loss from trauma; in frequent blood donors; in individuals with malignancies, particularly colon cancer; or after GI surgeries. It can also occur due to poor intake, especially in preterm or low-birth-weight infants and individuals following vegan and vegetarian diets. Relative iron deficiency anemia can occur with increased fluid states, such as in pregnant women and individuals with chronic heart failure.[195]

The clinical presentation of iron deficiency anemia depends on the rapidity of development of the anemia and on its severity. An insidious onset generally manifests as increasing fatigue, dyspnea, particularly with exertion, tachycardia, palpitations, and slight pallor (best seen in the mucous membranes), while more acute onset or severe anemia can present with cardiorespiratory signs and symptoms including angina, claudication, night cramps, cardiac bruits, reversible cardiac enlargement and, if cardiac failure occurs, basal crepitations, peripheral edema, and ascites.[70] Children may present with developmental delays. Plummer-Vinson syndrome is the association of iron deficiency anemia, koilonychia (spoon-shaped nails), dysphagia, and post-cricoid esophageal webs usually seen in middle-aged women.[196] Neuromuscular involvement is evidenced by headache, tinnitus, vertigo, faintness, increased cold sensitivity, and retinal hemorrhage. Gastrointestinal symptoms include anorexia, nausea, constipation, and diarrhea. Iron deficiency itself has certain characteristics not usually associated with other forms of anemia. Nonspecific glossitis with almost complete loss of filiform papillae is common. Angular stomatitis is less frequent. Pica (geophagia) is an occasional feature.[197]

When iron deficiency is suspected, the first test to obtain is a complete blood count, which will demonstrate microcytic hypochromic anemia.[70] This finding is nonspecific, therefore, follow-up studies are required, including: serum iron, total iron binding capacity (TIBC), serum ferritin, and transferrin saturation.[70] In iron deficiency anemia, serum iron, serum ferritin, and transferrin saturation will be low, and TIBC will be increased.

Oral iron supplementation is the first-line treatment for iron deficiency; however, efficacy may be decreased due to GI side effects, which can lead to decreased compliance. Iron supplementation causes a 24-hour increase in plasma hepcidin concentration, which acts as a regulator of iron balance. Iron supplementation taken during hepcidin elevation will inhibit iron absorption. Multiple recent studies have demonstrated that single-dose supplements on alternate days may increase iron absorption better than daily dosing.[71,197-200]

Iron sulfate is most frequently used; however, iron gluconate and iron fumarate are good alternatives. Recent recommended dosing for adults favors 40 to 200 mg in the morning on alternate days,[71,198] preferably with vitamin C, which is known to optimize iron absorption. If oral iron supplementation cannot be tolerated, IM or IV iron formulations should be initiated.[72] There are six IV formulations available in the United States: INFeD (iron dextran), Venofer (iron sucrose), Injectafer (ferric carboxymaltose), Ferrlecit (sodium ferric gluconate), Feraheme (ferumoxytol), and Monoferric (ferric derisomaltose).[201-206] Hypersensitivity reactions, including anaphylaxis, are possible with each of these preparations. It is, therefore, recommended that patients are closely monitored with appropriate resuscitation equipment and personnel available.[201-206] There are currently no forms of iron suitable for addition to PN in the United States.

Toxicity

Excess iron in the body is stored in the liver, spleen, and bone marrow as either ferritin or hemosiderin and is stored as myoglobin in muscle tissue.[207] Excess can occur from intake, repeated blood transfusions (secondary hemochromatosis), or hereditary hemochromatosis, an autosomal recessive disorder that leads to iron overload. Acute iron poisoning is most commonly seen from unintentional ingestion among children under the age of 6.[208] Acute iron poisoning causes vomiting, upper abdominal pain, and diarrhea. Shock can develop with pallor, cyanosis, a high-anion gap metabolic acidosis, and hypotension. Death occurs from shock or liver failure, as iron overload can lead to hepatic necrosis.[208]

Chronic toxicity (hemochromatosis) affects many tissues. "Bronze diabetes" occurs in about half of all patients with hereditary hemochromatosis; so named from iron deposition in the pancreas damaging beta cells, and deposition in the skin leading to discoloration.[209] In addition, individuals present with hepatomegaly, which can lead to cirrhosis and hepatocellular cancer. Individuals are at risk for the development of cardiomyopathy and heart failure, joint pain, lethargy, and ECG abnormalities. Pituitary failure may cause testicular atrophy and loss of libido. Focal hemosiderosis can also damage the lungs and kidneys.[209]

Copper

Deficiency

Copper is incorporated into ceruloplasmin, a transport protein that moves copper from the liver to tissues and is also important in iron metabolism and in a number of enzymes that aid in skin pigmentation, clotting, and the development of neurotransmitters.[73] Copper absorption can be inhibited by increased intake of zinc, iron, and vitamin C. Zinc is found in over-the-counter cold lozenges and many denture adhesives.[210] Because zinc and copper compete for absorption, overuse of these products can lead to copper deficiency. Copper deficiency can be acquired or hereditary. Individuals at risk for acquired deficiency include gastric bypass or gastrectomy, malabsorptive disorders and chronic diarrhea, premature infants provided long-term parenteral nutrition or milk formulas without sufficient copper added, chronic excess zinc ingestion, or overtreatment of Wilson's disease.[73] A hereditary form of copper deficiency is Menkes disease, an x-linked disorder that results in a defect of intestinal absorption of copper. The resultant copper deficiency leads to "kinky hair syndrome," loss of skin pigmentation, growth failure, and progressive neurologic changes (ataxia, intellectual developmental delay, seizures).[211]

Testing for deficiency is assessed by measurement of serum copper and ceruloplasmin concentrations. Menkes disease diagnosis requires genetic testing for the ATP7A gene mutation.

No consensus guidelines exist for copper supplementation in the setting of deficiency; however, the American Society for Metabolic and Bariatric Surgery does recommend routine copper supplementation (1–2 mg/day) for post-Roux-en-Y gastric bypass and after biliopancreatic diversion (2 mg/day), regardless of copper status.[74] Both IV and oral formulations are available.[212,213] With mild-to-moderate deficiency, the dosing recommendation is 3 to 8 mg/day of oral copper until concentrations are corrected, and with severe deficiency, 2 to −4 mg/day of IV copper for 6 days, then continued oral supplementation until concentrations are corrected. Hematologic abnormalities should reverse within 4 to 12 weeks of therapy.[74,212,213] Treatment of Menkes disease consists of parenteral administration of a copper histidine complex.[75]

Toxicity

Acute poisoning has resulted from ingestion of solutions of copper salts, contaminated water, food supplies, or dialysis fluid.[214,215] Individuals present with vomiting and diarrhea and, in severe cases, evidence of hepatic, cardiac, and/or renal failure.

Wilson's disease is an autosomal recessive condition that leads to a deficiency of ceruloplasmin, causing copper deposition and damage to various organs, including the liver (cirrhosis), pancreas (diabetes), heart, and neurologic system. Copper deposition also leads to blue nailbeds (azure lunula) and deposition in the eye (Kayser-Fleischer rings).[215]

Testing for toxicity requires ceruloplasmin and copper concentrations, which will be increased. The diagnosis of Wilson's disease requires an ophthalmologic exam looking for Kayser-Fleischer rings and a 24-hour urinary copper excretion test. If the diagnosis remains uncertain, a liver biopsy or genetic testing for the ATP7B gene can be completed.[76] Patients with Wilsons's disease require copper chelation and should avoid foods high in copper.[214] As zinc competes with copper for absorption, zinc intake in the early stages of Wilson's disease may help delay onset of symptoms.[212,213]

Zinc

Deficiency

Zinc deficiency can be acquired or hereditary. Individuals at risk for zinc deficiency include those with malabsorptive disorders, short bowel syndrome, alcohol use disorder, long-term PN without added zinc, chronic liver disease, chronic renal disease, chronic diarrhea, sickle cell disease, diabetes, GI surgery, pregnancy and lactation, or in exclusively breastfed infants after 6 months of age.[2] Inherited zinc deficiency is due to an autosomal recessive disorder that interferes with zinc absorption. The presenting condition is termed acrodermatitis enteropathica and presents in infancy.[2] Regardless of whether the zinc deficiency is acquired or congenital, individuals present with alopecia, dysgeusia, anorexia, a vesicular or pustular perioral, perineal and acral dermatitis, diarrhea, impaired immune function, infertility issues, and growth retardation[2] (see **Figure 57.2F**), (see **Figure 57.2G**). Deficiency can be measured with plasma or serum zinc concentration; but these may not always be accurate; therefore, correlation with clinical presentation and risk factors is suggested.[2]

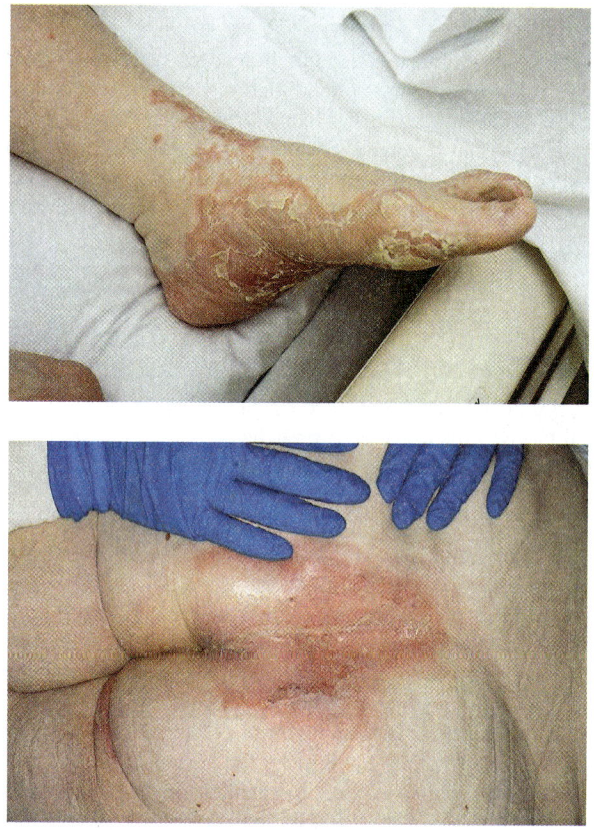

F

G

Figure 57.2 F and G. Dermatitis in a patient with malabsorption and acrodermatitis enteropathica on the feet and gluteal cleft. The patient responded to zinc administration.

Courtesy of Department of Dermatology, Mayo Clinic.

Oral supplementation replacement of 2 to 3 mg/kg of elemental zinc per day should eradicate clinical symptoms within 1 to 2 weeks. Individuals with acrodermatitis enteropathica require lifelong oral supplementation of 1 to 2 mg/kg/day. High doses of zinc have been shown to cause gastric irritation with nausea, vomiting, and gastric hemorrhage. Copper concentration should be monitored in the setting of long-term zinc supplementation due to zinc competition with copper absorption.[63]

Toxicity

Ingestion of large amounts of zinc, usually from an acid food or drink from a galvanized container, or long-term consumption of high doses of zinc supplements, has caused vomiting and diarrhea.[77,78] High doses of zinc will competitively inhibit copper leading to copper deficiency.

Fluoride

Deficiency

Fluoride is found in ground water, but concentrations vary by region. Fluoride is important in bone mineralization and the hardening of tooth enamel. Areas with low fluoride content in the water supply have high rates of dental caries.[216] Fluoridation of the water or use of supplemented toothpaste is associated with significantly lower incidence of dental caries. Testing for fluoride concentration is not routinely done.

Toxicity (Fluorosis)

Fluoride toxicity can occur from ingestion of water with high concentration (>10 ppm), fluoride supplemented products, or from skin contact or inhalation of chemicals containing fluorine. The most common cause of fluoride toxicity is the use of fluoride-containing toothpaste in children younger than 6 years of age.[217] Chronic exposure to fluoride leads to fluorosis, associated with discoloration of the permanent teeth. The earliest changes, chalky white, irregularly distributed patches on the surface of the enamel, become infiltrated by yellow or brown staining, giving rise to the characteristic "mottled" appearance (see **Figure 57.2H**). More severe fluorosis also causes pitting of the enamel. Long-term ingestion of large amounts of fluoride (>5 mg/day) for years may lead to crippling skeletal fluorosis, progressing from occasional stiffness or joint pain to chronic pain and osteoporosis of long bones. This rare condition is associated with drinking high-fluoride well water.[218]

Selenium

Deficiency

Selenium is found in soil and in foods grown in selenium-rich soil. Selenium deficiency is more common in areas with selenium-deficient soil, particularly among individuals on vegetarian or vegan diets. Selenium deficiency is also associated with long-term hemodialysis; long-term PN; chylous loss; phenylketonuria, malabsorption disorders; and HIV disease.[79,80,219,220] In areas of China where the soil is deficient in selenium, Keshan disease, consisting of highly fatal cardiomyopathy that mainly affects young children and women of childbearing age, was identified. A good response to selenium supplementation has been reported.[221] Selenium deficiency can present with whitened nailbeds, impaired immune function, skeletal myopathy, and cardiomyopathy.[80,220] Testing is best assessed with serum or plasma selenium concentration, but selenium levels in hair and nails can provide a more accurate long-term assessment of intake.[79]

A routine dose for oral adult selenium supplementation is 100 μg/day.[80] The amount of selenium in trace element injection formulations for PN varies by age group: neonate and pediatric formulations are weight-based, whereas adult recommendations are a standard 60 μg/day.[203]

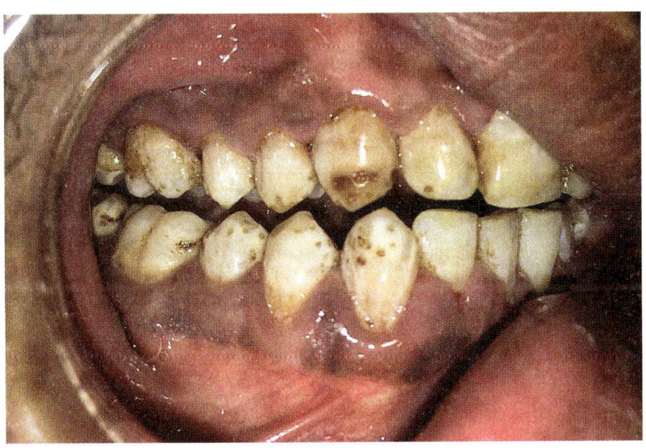

Figure 57.2 H. Fluorosis. Early stage with brown mottling that is most marked on upper central incisors.

Courtesy of D.C. Heimburger.

Toxicity

Selenium toxicity occurs with excess ingestion from foods high in selenium or through supplementation. The most frequently observed signs are loss of hair and nail changes, but nausea, vomiting, diarrhea, vision changes, and polyneuritis have been reported.[81]

Chromium

Deficiency

Chromium deficiency is associated with malnutrition in individuals with increased nutritional demands, such as those with burns or trauma, and in those with short bowel syndrome.[222,223] The most common presentation of deficiency includes weight loss, peripheral neuropathy, and glucose intolerance. Chromium has recently been removed from the standard trace element preparation used in CPN.[81] It had been discovered that patients were receiving 10% to 100% more chromium in the PN solution via contamination.[224] The amount provided by contaminants should be adequate for most patients. Although rare, chromium deficiency can still be possible. There have been anecdotal cases of chromium deficiency, with patients developing glucose intolerance that responded favorably when they received chromium. This is difficult to quantify as not all cases are published and testing of chromium concentration is not routinely done, as serum and plasma concentrations may be inaccurate due to lack of equilibrium between tissue and plasma concentrations. Plasma concentration may also be falsely decreased in the setting of illness.[224]

Toxicity

Toxicity has not been reported with dietary chromium. There have been reports of toxicity with pentavalent and hexavalent chromium, which are chemicals used in various industries, such as in steelworkers. Exposure to these variants of chromium can cause skin ulcers on the hands and perforation of the nasal septum and have been linked to the development of lung cancer.[225]

Molybdenum

Deficiency

The autosomal recessive disorder, molybdenum cofactor deficiency, is one of several inborn errors of metabolism. This genetic mutation results in increased concentrations of sulfites in blood and urine within the first few days of life.[226] Neonates present with increased startle, seizures, and severe brain damage. About half of those diagnosed fail to survive beyond early infancy. A urine dipstick for increased sulfite concentration and genetic testing are used for diagnostic purposes.

Only one clear case of deficiency related to prolonged CPN has been reported, involving tachycardia, tachypnea, headache, night blindness, central scotomas, nausea, vomiting, lethargy, disorientation, and coma.[227] These manifestations were reversed by administration of molybdenum, and the urinary excretion of abnormal amounts of methionine was dramatically decreased.

Toxicity

Excess molybdenum can occur from ingestion of contaminated water. Elevated urine molybdenum concentration was associated with hyperuricemia and a gout-like syndrome with joint pain from contaminated food and water.[227] Another study demonstrated pneumoconiosis that developed in workers exposed to molybdenite dust.[227]

Manganese

Deficiency

Manganese deficiency is rare but has been described in very restricted diets.[221,228] Clinical signs of deficiency include transient dermatitis, impaired growth, bony defects, infertility, and impaired glucose tolerance.[228]

Toxicity

Manganese toxicity from food intake has not been reported. Toxicity generally occurs from exposure through contaminated water supplies, soy-based infant formulas, CPN, exposure to airborne manganese in exhaust fumes, or among welders and workers in factories that use manganese (i.e., battery manufacturers).[221,228-230] Patients on CPN are at the highest risk for manganese toxicity as 100% of IV manganese in the PN solution is absorbed, compared with 5% when taken orally.[231] Manganese in PN solutions occurs from both trace element injection and additional manganese present from contamination.[222,231] Discussions between ASPEN and parenteral formula companies are ongoing to develop recommendations for decreasing manganese with long-term CPN.[161] Manganese is eliminated via the hepatobiliary system, and with cholestatic liver disease, individuals can rapidly develop toxicity. Manganese and copper toxicity should be monitored and treated in patients with cholestasis, biliary dysfunction, or cirrhosis.[203]

Excess manganese can accumulate in any organ but has an affinity for brain tissue. Initial signs include insomnia, depression, and delusions, followed by anorexia, arthralgias, dystonia, and Parkinson-like symptoms.[232] Manganese may also accumulate in the liver, causing liver cirrhosis, or in the heart leading to abnormal ECG findings.[232] Manganese toxicity has been linked to an increased incidence of attention deficit disorder and infant mortality.[229]

Acknowledgments

The author(s) acknowledge Douglas C Heimburger, the author of the previous version of this chapter, for his contributions.

References

1. Black RE, Allen LH, Bhutta ZA, et al. Lancet 2008;371:243–60.
2. Galimberti F, Mesinkovska NA. Cleve Clin J Med 2016;83:731–9.
3. Shashidhar H. Accessed May 3, 2022. Available at https://emedicine.medscape.com/article/985140-overview
4. Hanachi M, Dicembre M, Rives-Lange C, et al. Nutrients 2019;11.
5. Achamrah N, Coeffier M, Rimbert A, et al. Nutrients 2017;9.
6. Chiu M, Watson S. BMJ Case Rep 2015.
7. Yule S, Wanik J, Holm EM, et al. J Acad Nutr Diet 2021;121;467–92.
8. Schmidt R, Hiemisch A, Kiess W, et al. Nutrients 2021;13:400.
9. Liuzzo Scorpo M, Corsello G, Maggio MC. Am J Case Rep 2021;22:e930583.
10. Mogensen KM. Pract Gastroenterol 2017;41:37–44.
11. Shifler Bowers K, Francis E, Kraschnewski JL. Prev Med Rep 2018;12:294–7.
12. Hathcock JN. Am J Clin Nutr 1997;66:427–37.
13. Blaner WS. Vitamin A and provitamin A carotenoids. In Present Knowledge in Nutrition pp. 73–91, Elsevier, 2020.
14. Zinder R, Cooley R, Vlad LG, et al. Nutr Clin Pract 2019;34:839–49.
15. Maqbool A, Stallings VA. Curr Opin Pulm Med 2008;14:574–81.
16. Hathcock JN, Hattan DG, Jenkins MY, et al. Am J Clin Nutr 1990;52:183–202.
17. Geubel AP, De Galocsy C, Alves N, et al. Gastroenterol 1991;100:1701–9.
18. Michaëlsson K, Lithell H, Vessby B, et al. N Engl J Med 2003;348:287–94.
19. Melhus H, Michaëlsson K, Kindmark A, et al. Ann Intern Med 1998;129:770–8.
20. Feskanich D, Singh V, Willett WC, et al. JAMA 2002;287:47–54.
21. Dawson-Hughes B. Vitamin D Deficiency in Adults: Definition, Clinical Manifestations, and Treatment: Summary and Recommendations. Vitamin D Deficiency in Adults: Definition, Clinical Manifestations, and Treatment—UpToDate Accessed May 15, 2022.
22. Kimball S, Vieth R. Ann Clin Biochem 2008;45:106–10.
23. Alkurdi A, Rubin D, Seelhoff A, et al. Case Rep Gastroenterol 2021;15:960–5.
24. Mueller CM, Lord L, Marian M, et al, Eds. Vitamins and Trace Elements, in The ASPEN Adult Nutrition Support Core Curriculum, American Society for Parenteral and Enteral Nutrition, 2017;139–82.
25. Miller ER, 3rd, Pastor-Barriuso R, Dalal D, et al. Ann Intern Med 2005;142:37–46.
26. Conly JM, Stein K, Worobetz L, et al. Am J Gastroenterol 1994;89:915–23.
27. Shearer MJ. Blood Rev 2009;23:49–59.
28. Giacalone M, Martinelli R, Abramo A, et al. Nutr Clin Pract 2015;30:104–10.
29. Rao SN, Chandak GR. J Trop Pediatr 2010;56:284–5.
30. Talwar D, Davidson H, Cooney J, et al. Clin Chem 2000;46:704–10.
31. Agabio R. Alcohol Alcohol 2005;40:155–6.
32. Cook CC, Hallwood PM, Thomson AD. Alcohol Alcohol 1998;33:317–36.
33. Martel JL, Kerndt CC, Doshi H, et al. Vitamin B1 (Thiamine). In: StatPearls, StatPearls Publishing. Copyright © 2023, StatPearls Publishing LLC., Treasure Island, FL: 2023.
34. Institute of Medicine (US) Standing Committee on the Scientific Evaluation of Dietary Reference Intakes and its Panel on Folate OBV and Choline. Dietary reference intakes for thiamin, riboflavin, niacin, vitamin B6, folate, vitamin B12, pantothenic acid, biotin, and choline. Washington (DC): National Academies Press; 1998.

35. Juel J, Pareek M, Langfrits CS, et al. BMJ Case Rep, 2013.
36. U.S. Department of Agriculture ARS, Nutrient Data Laboratory. USDA Branded Food Products Database; FoodData Central. Accessed May 29, 2022. Available at https://fdc.nal.usda.gov/fdc-app.html#/?component=0
37. Lopez F, Cole HS, Montoya MF, et al. J Pediatr 1975;87:420–2.
38. Said HM. Am J Clin Nutr 2002;75:179–80.
39. Institute of Medicine Standing Committee on the Scientific Evaluation of Dietary Reference I; Its Panel on Folate OBV, and Choline. The National Academies Collection: Reports funded by National Institutes of Health. In: Dietary Reference Intakes for Thiamin, Riboflavin, Niacin, Vitamin B(6), Folate, Vitamin B(12), Pantothenic Acid, Biotin, and Choline. National Academies Press, United States. Copyright © 1998, Washington (DC): National Academy of Sciences, 1998.
40. Institute of Medicine Committee to Review Dietary Reference Intakes for Vitamin D, and Calcium. In: Ross AC, Taylor CL, Yaktine AL, and Del Valle HB, eds. The National Academies Collection: Reports Funded by National Institutes of Health. Dietary Reference Intakes for Calcium and Vitamin D. National Academies Press, United States. Copyright © 2011, National Academy of Sciences., Washington, D.C., 2011.
41. Pinto JT, Zempleni J. Adv Nutr 2016;7:973–5.
42. Savvidou S. Clin Pract 2014;4:637.
43. Li R, Yu K, Wang Q, et al. Nutr Clin Pract 2016;31:785–9.
44. Redzic S, Hashmi MF, Gupta V. Niacin Deficiency. In: StatPearls, StatPearls Publishing. Copyright © 2023, StatPearls Publishing LLC., Treasure Island, FL, 2023.
45. Badawy AA. Alcohol Alcohol 2014;49:238–50.
46. Papadakis MA, Rabow MW. Medical Diagnosis & Treatment. San Francisco, CA: McGraw Hill, 2019.
47. Bonjour J M. In: Machlin LJ, ed. Nutritional, Biochemical, and Clinical Aspects, 1984.
48. McCormick D. Present Knowledge in Nutrition 1, 269–77, 2006.
49. Brown MJ, Ameer MA, Beier K. Vitamin B6 Deficiency. In *StatPearls*, StatPearls Publishing. Copyright © 2023, StatPearls Publishing LLC., Treasure Island, FL, 2023.
50. Schaumburg H, Kaplan J, Windebank A, et al. N Engl J Med 1983;309:445–8.
51. Ross C, Vitamin A, Coates P. London and New York: Informa Healthcare 2010;778–91.
52. Zempleni J, Wijeratne SS, Hassan YI. Biofactors 2009;35:36–46.
53. Eng WK, Giraud D, Schlegel VL, et al. Br J Nutr 2013;110:321–9.
54. Elston MS, Sehgal S, Du Toit S, et al. J Clin Endocrinol Metab 2016;101:3251–5.
55. Kummer S, Hermsen D, Distelmaier F. N Engl J Med 2016;375:704–6.
56. U.S. Food and Drug Administration. The FDA Warns that Biotin May Interfere with Lab Tests: FDA Safety Communication, 2019. Accessed May 3, 2022. Available at https://public4.pagefreezer.com/content/FDA/16-06-2022T13:39/www.fda.gov/medical-devices/safety-communications/update-fda-warns-biotin-may-interfere-lab-tests-fda-safety-communication
57. Devalia V, Hamilton MS, Molloy AM. Br J Haematol 2014;166: 496–513.
58. Rojas Hernandez CM, Oo TH. Discov Med 2015;19:159–68.
59. Carmel R. Vitamin B12. In: Ross AC CB, Cousins RJ, Tucker KL, Ziegler TR, eds. Modern Nutrition in Health and Disease. Baltimore, MD: Lippincott Williams & Wilkins, 369–89.
60. Allen LH, Miller JW, de Groot L, et al. J Nutr 2018;148:1995s–2027s.
61. Abboud Leon C, Georges P, Zaeeter W. N Engl J Med 2013;368:2041.
62. Figueiredo JC, Grau MV, Haile RW, et al. J Natl Cancer Inst 2009; 101:432–5.
63. Abdullah M, Jamil RT, Attia FN. Vitamin C (ascorbic acid). In StatPearls, StatPearls Publishing. Copyright © 2023, StatPearls Publishing LLC., Treasure Island, FL, 2023.
64. Hodges RE. Monogr Ser World Health Organ 1976;120–5.
65. Baradhi KM, Vallabhaneni S, Koya S. Proc (Bayl Univ Med Cent) 2018;31:227–8.
66. Zeisel SH, Da Costa KA, Franklin PD, et al. FAESB J 1991;5: 2093–8.
67. Ernst KD. J Perinatol 2017;37:695–7.
68. Gramlich L, Meddings L, Alberda C, et al. JPEN J Parenter Enteral Nutr 2015;39:61s–6s.
69. Gramlich L, Ireton-Jones C, Miles JM, et al. JPEN J Parenter Enteral Nutr 2019;43:697–707.
70. Clark SF. Nutr Clin Pract 2008;23:128–41.
71. Stoffel NU, Cercamondi CI, Brittenham G, et al. Lancet Haematol 2017;4:e524–e33.
72. Camaschella C. N Engl J Med 2015;372:1832–43.
73. Danks DM. Annu Rev Nutr 1988;8:235–57.
74. Parrott J, Frank L, Rabena R, et al. Surg Obes Relat Dis 2017; 13:727–41.
75. Nastoulis E, Karakasi MV, Couvaris CM, et al. Forensic Sci Rev 2017;29:77–91.
76. Steindl P, Ferenci P, Dienes HP, et al. Gastroenterol 1997; 113:212–8.
77. Turnlund JR, Durkin N, Costa F, et al. J Nutr 1986;116:1239–47.
78. Prasad AS, Brewer GJ, Schoomaker EB, et al. JAMA 1978;240:2166–8.
79. Freitas RG, Nogueira RJ, Antonio MA, et al. Rev Paul Pediatr 2014;32:126–35.
80. Hasunuma N, Umebayashi Y, Manabe M. JAMA Dermatol 2014;150:779–80.
81. Centers for Disease Control (CDC). MMWR Morb Mortal Wkly Rep 1984;33157-8.
82. US Department of Agriculture, Agricultural Research Service, Food and Nutrition Information Center. DRI Nutrient Reports. Accessed May 3, 2022. Available at www.nal.usda.gov/legacy/fnic/dri-nutrient-reports
83. Jankowska M, Rutkowski B, Dębska-Ślizień A. Nutrients 2017;9.
84. Saari JC. Annu Rev Nutr 2012;32:125–45.
85. Sommer A, Emran N, Tamba T. Am J Ophthalmol 1979;87:330–3.
86. Sommer A, Hussaini G, Muhilal, et al. Am J Clin Nutr 1980; 33:887–91.
87. Bessey OA, Adam DJ, Hansen AE. Pediatrics 1957;20:33–44.
88. Weintraub LR, Conrad ME, Crosby WH. N Engl J Med 1966;275:169–76.
89. Sommer A, and Davidson FR. J Nutr 2002;132:2845s–50s90.
90. Slater GH, Ren CJ, Siegel N, et al. J Gastrointest Surg 2004;8:48–55; discussion 4–5.
91. McClave SA, Taylor BE, Martindale RG, et al. JPEN J Parenter Enteral Nutr 2016;40:159–211.
92. World Health Organization. Vitamin A supplements: a guide to their use in the treatment and prevention of vitamin A deficiency and xerophthalmia, 2nd Ed. 1997; Geneva.
93. Olson JA. J Nutr 1989;119:1820–4.
94. Rothman KJ, Moore LL, Singer MR, et al. N Engl J Med 1995;333: 1369–73.
95. Bastos Maia S, Rolland Souza AS, Costa Caminha MF, et al. Nutrients 2019;11.
96. Bikle DD. In: Feingold KR, Anawalt B, Boyce A, et al, eds. Vitamin D: Production, Metabolism and Mechanisms of Action. Endotext. MDText.com, Inc. Copyright © 2000–2022, MDText.com, Inc., South Dartmouth, MA, 2000.
97. LaClair RE, Hellman RN, Karp SL, et al. Am J Kidney Dis 2005; 45:1026–33.
98. Shaker JL, Brickner RC, Findling JW, et al. Arch Intern Med 1997;157:1013–6.
99. Peterson LA, Zeng X, Caufield-Noll CP, et al. Surg Obes Relat Dis 2016;12:693–702.
100. Meier C, Kraenzlin ME. Ther Adv Musculoskelet Dis 2011; 3:235–43.

101. Misra M, Pacaud D, Petryk A, et al. Pediatrics 2008;122:398–417.
102. Basha B, Rao DS, Han ZH, et al. Am J Med 2000;108:296–300.
103. Bingham CT, Fitzpatrick LA. Am J Med 1993;95:519–23.
104. Chang SW, Lee HC. Pediatr Neonatol 2019;60:237–44.
105. Kalra V, Grover J, Ahuja GK, et al. J Trop Pediatr 1998;44:291–5.
106. Rosenberg RN. N Engl J Med 1995;333:1351–3.
107. Robison WG, Kuwabara T, Bieri JG. Invest Ophthalmol Vis Sci 1979;18:683–90.
108. Albahrani AA, Greaves RF. Clin Biochem Rev 2016;37:27–47.
109. McKeever L. In: Mueller CM LL, Marian M, et al, Eds. Vitamins and Trace Elements. The ASPEN Adult Nutrition Support Core Curriculum, American Society for Parenteral and Enteral Nutrition. American Society for Parenteral and Enteral Nutrition, 2017;139–82.
110. Thébaut A, Nemeth A, Le Mouhaër J, et al. J Pediatr Gastroenterol Nutr 2016;63:610–5.
111. Johnson L, Bowen FW, Jr., Abbasi S, et al. Pediatrics 1985;75:619–38.
112. Finer NN, Peters KL, Hayek Z, et al. Pediatrics 1984;73: 387–93.
113. Kitagawa M, Mino M. J Nutr Sci Vitaminol (Tokyo) 35, 133–42.
114. Dereska NH, McLemore EC, Tessier DJ, et al. J Surg Res 2006;132:121–9.
115. Biesalski HK. Gastroenterol 2009;137:S92–104.
116. Bjelakovic G, Nikolova D, Gluud LL, et al. Cochrane Database Syst Rev 2012;2012:Cd007176.
117. Shearer MJ. Gastroenterol 2009;137:S105–18.
118. Motohara K, Endo F, Matsuda I. Lancet 1985;2:242–4.
119. Lippi G, Franchini M. Blood Transfus 2011;9:4–9.
120. Vermeer C, De Boer-Van den Berg MA. Haematologia (Budap) 1985;18:71–97.
121. American Academy of Pediatrics Committee on Fetus and Newborn. Pediatrics 2003;112:191–2.
122. Institute of Medicine Panel on M. Dietary Reference Intakes for Vitamin A, Vitamin K, Arsenic, Boron, Chromium, Copper, Iodine, Iron, Manganese, Molybdenum, Nickel, Silicon, Vanadium, and Zinc. National Academies Press, United States, Copyright 2001 by the National Academy of Sciences. All rights reserved., Washington, DC:, 2001.
123. Masumoto K, Esumi G, Teshiba R, et al. JPEN J Parenter Enteral Nutr 2009;33:417–22.
124. Matrana MR, Davis WE. South Med J 2009;102:1025–31.
125. Ahmed A, Daida Y, Novotny R. Clin Med Res 2011;9:165.
126. Francini-Pesenti F, Brocadello F, Manara R, et al. Nutrition 2009;25:142–6.
127. Dyckner T, Ek B, Nyhlin H, et al. Acta Med Scand 1985;218: 129–31.
128. McLean J, Manchip S. Lancet 1999;353:1768.
129. Campbell CH. Lancet 1984;2:446–9.
130. Dabar G, Harmouche C, Habr B, et al. Nutr J 2015;14:51.
131. Foster J. J Neurol Neurosurg Psychiatr 1989;52:1217.
132. Seehra H, MacDermott N, Lascelles RG, et al. BMJ 1996;312:434.
133. Singh S, Kumar A. Neurol 2007;68:807–11.
134. Haas RH. Annu Rev Nutr 1988;8:483–515.
135. Ambrose ML, Bowden SC, Whelan G. Alcohol Clin Exp Res 2001;25:112–6.
136. Day E, Bentham P, Callaghan R, et al. Cochrane Database Syst Rev 2004;Cd004033.
137. Kumar N. Neurol Clin 2010;28:107–70.
138. Mirtallo J, Canada T, Johnson D, et al. JPEN J Parenter Enteral Nutr 2004;28:S39–70.
139. Said AR. Riboflavin. In: Ross AC, Caballero B, Cousins RJ, et al, Eds. Modern Nutrition in Health and Disease, 2014. Baltimore, MD: Lippincott Williams & Wilkins, 325–30.
140. Baumgartner ER, Suormala T. Int J Vitam Nutr Res 1997;67: 377–84.
141. Garakani A, Jaffe RJ, Savla D, et al. Am J Addict 2016;25: 358–69.
142. Graells J, Ojeda RM, Muniesa C, et al. J Am Acad Dermatol 2009;60:498–500.
143. Shipton MJ, Thachil J. Clin Med (Lond) 2015;15:145–50.
144. Butler CC, Vidal-Alaball J, Cannings-John R, et al. Fam Pract 2006;23:279–85.
145. Khan KM, Jialal I. Folic Acid Deficiency. StatPearls Publishing LLC, 2023, Treasure Island, Florida.
146. Stabler SP. N Engl J Med 2013;368:149–60.
147. Green R, Datta Mitra A. Med Clin North Am 2017;101:297–317.
148. Centers for Disease Control and Prevention. Folic Acid. Accessed May 3, 2022. Available at www.cdc.gov/ncbddd/folicacid/about .html
149. Wolff T, Witkop CT, Miller T, et al. Ann Intern Med 2009;150: 632–9.
150. Hoxha B, Hoxha M, Domi E, et al. Cells 2021;10:1976.
151. Sampaio AC, Matos FFN, Lopes LL, et al. Rev Bras Ginecol Obstet 2021;43:775–81.
152. Liu X, Zou M, Sun C, et al. J Autism Dev Disord 2022;52:522–39.
153. Lubecka-Pietruszewska K, Kaufman-Szymczyk A, Stefanska B, et al. Biochem Biophys Res Commun 2013;430:623–8.
154. Hirsch S, Sanchez H, Albala C, et al. Eur J Gastroenterol Hepatol 2009;21:436–9.
155. Jacob RA, Sotoudeh G. Nutr Clin Care 5, 66–74.
156. Weinstein M, Babyn P, Zlotkin S. Pediatrics 2001;108:E55.
157. Mutgi KA, Ghahramani G, Wanat K, et al. J Fam Pract 2016;65:927–30.
158. Protara Therapeutics. Protara Therapeutics Receives Fast Track Designation from U.S. FDA for Intravenous Choline Chloride for the Treatment of Intestinal Failure Associated Liver Disease. Accessed May 3, 2022. Available at www.globenewswire.com /news-release/2020/05/26/2038632/0/en/Protara-Therapeutics -Receives-Fast-Track-Designation-from-U-S-FDA-for-Intravenous -Choline-Chloride-for-the-Treatment-of-Intestinal-Failure -Associated-Liver-Disease.html
159. Kaufman SS, Pehlivanova M, Fennelly EM, et al. J Pediatr 2010;156:580–5.e1.
160. Nilsson AK, Pedersen A, Malmodin D, et al. Eur J Nutr 2021; 60:1081–9.
161. Vanek VW, Borum P, Buchman A, et al. Nutr Clin Pract 2015; 30:559–69.
162. McCray S, Parrish CR. Pract Gastroenterol 2011;94:12–32.
163. Memon N, Hussein K, Hegyi T, et al. JPEN J Parenter Enteral Nutr 2019;43:438–41.
164. Jones CJ, Calder PC. Clin Nutr 2018;37:285–91.
165. Terry J. J Intraven Nurs 1994;17:240–7.
166. Buffington MA, Abreo K. J Intensive Care Med 2016;31:223–36.
167. Hamilton C. In: Mueller CM, McClave SA, and Kuhn JM, eds. Fluids, Electrolytes, and Acid-Base Disorders. ASPEN adult nutrition support core curriculum, American Society for Parenteral and Enteral Nutrition, 3rd ed. American Society for Parenteral and Enteral Nutrition, 2017;113–37.
168. Hew-Butler T, Loi V, Pani A, et al. Front Med (Lausanne) 2017;4:21.
169. Yasir M, Mechanic OJ. Syndrome of Inappropriate Antidiuretic Hormone Secretion. In *StatPearls*, StatPearls Publishing. Copyright © 2023. Treasure Island, FL: StatPearls Publishing LLC., 2023.
170. Bradley EW, Westendorf JJ, Van Wijnen AJ, et al. Primer on the Metabolic Bone Diseases and Disorders of Mineral Metabolism 2018;31–7.
171. Favus MJ, Bushinsky DA, Lemann Jr J. Primer on the Metabolic Bone Diseases and Disorders of Mineral Metabolism 2006;6: 76–83.
172. Hannan FM, Thakker RV. BMJ 2013;346:f2213.

173. Arshad MF, Karim R, Paling D, et al. BMJ Case Rep 2018;11.
174. Bishop N. N Engl J Med 1999;341:602–4.
175. Thacher TD, Fischer PR, Pettifor JM, et al. N Engl J Med 1999; 341:563–8.
176. Dickerson RN, Alexander KH, Minard G, et al. JPEN J Parenter Enteral Nutr 2004;28:133–41.
177. Weinsier RL, Krumdieck CL. Am J Clin Nutr 1981;34:393–9.
178. Knochel JP. N Engl J Med 1985;313:447–9.
179. Berner YN, Shike M. Annu Rev Nutr 1988;8:121–48.
180. Murer H, Lötscher M, Kaissling B, et al. Kidney Int 1996;49: 1769–73.
181. Gaasbeek A, Meinders AE. Am J Med 2005;118:1094–101.
182. Gennari FJ. Crit Care Clin 2002;18:273–88, vi.
183. Drew BJ, Califf RM, Funk M, et al. Circ 2004;110:2721–46.
184. Cohn JN, Kowey PR, Whelton PK, et al. Arch Intern Med 2000;160:2429–36.
185. Cohen L, Kitzes R. Isr J Med Sci 1987;23:1238–41.
186. Britton J, Pavord I, Richards K, et al. Lancet 1994;344:357–62.
187. Agus MS, Agus ZS. Crit Care Clin 2001;17:175–86.
188. Oster JR, Epstein M. Am J Nephrol 1988;8:349–54.
189. Sacks GS, Brown RO, Dickerson RN, et al. Nutrition 1997; 13:303–8.
190. Hébert P, Mehta N, Wang J, et al. Crit Care Med 1997;25: 749–55.
191. Pearce EN. AACE Clinical Case Reports 2015;1:e81–e2.
192. Veisa V, Kalere I, Zake T, et al. Medicina (Kaunas) 2021;57.
193. Bourdoux P, Delange F, Gerard M, et al. J Clin Endocrinol Metab 1978;46:613–21.
194. Hetzel BS, Dunn JT. Annu Rev Nutr 1989;9:21–38.
195. Hetzel BS, Hay ID. Clin Endocrinol (Oxf) 1979;11:445–60.
196. Novacek G. Orphanet J Rare Dis 2006;1:36.
197. Andersen HT, Barkve H. Scand J Clin Lab Invest Suppl 1970;114:1–62.
198. Moretti D, Goede JS, Zeder C, et al. Blood 2015;126:1981–9.
199. Stoffel NU, Zeder C, Brittenham GM, et al. Haematologica 2020;105:1232–9.
200. Uyoga MA, Mikulic N, Paganini D, et al. Am J Clin Nutr 2020;112:1132–41.
201. Pharmacosmos Therapeutics, Inc. Monferric Package Insert. Accessed May 3, 2022. Available at file:///C:/Users/M065711/Downloads/20220208_55859d2d-0456-4fa9-b41f-f535accc97db%20(1).pdf
202. Allergan. INFeD Package Insert. Accessed: May 3, 2022. Available at www.allergan.com/assets/pdf/infed_pi
203. American Regent, Inc. Venofer Package Insert. Accessed: May 3, 2022. Available at www.venofer.com/Indications_Dosage
204. Daiichi Sankyo. Injectafer Package Insert. Accessed May 3, 2022. Available at https://daiichisankyo.us/prescribing-information-portlet/getDocument?product=IF&inline=true
205. Sanofi-Aventis U.S. LLC. Ferrlecit Package Insert. Accessed: May 3, 2022. Available at www.accessdata.fda.gov/drugsatfda_docs/label/2011/020955s013s015lbl.pdf
206. AMAG Pharmaceuticals, Inc. Feraheme Package Insert. Accessed: May 3, 2022. Available at https://injectafer.com/prescribing-informationportlet/getDocument?product=IF&inline=true
207. Bacon BR, Adams PC, Kowdley KV, et al. Hepatol 2011;54: 328–43.
208. Morris CC. South Med J 2000;93:352–8.
209. Niederau C, Strohmeyer G, Stremmel W. Adv Exp Med Biol 1994;356:293–302.
210. Nations SP, Boyer PJ, Love LA, et al. Neurol 2008;71:639–43.
211. Verrotti A, Carelli A, Coppola G. J Child Neurol 2014;29: 1757–64.
212. Chhetri SK, Mills RJ, Shaunak S, et al. BMJ 2014;348:g3691.
213. Mechanick JI, Youdim A, Jones DB, et al. Surg Obes Relat Dis 2013;9:159–91.
214. Viveiros B, Caron G, Barkley J, et al. MMWR Morb Mortal Wkly Rep 2021;70:1501–4.
215. Rodriguez-Castro KI, Hevia-Urrutia FJ, Sturniolo GC. World J Hepatol 2015;7:2859–70.
216. Council NR. Fluoride in drinking water: a scientific review of EPA's standards. 2007.
217. Martínez-Mier EA. J Evidence-Based Compl AltMed 2012;17: 28–32.
218. Meena L, Gupta R. N Engl J Med 2021;385:1510.
219. Etani Y, Nishimoto Y, Kawamoto K, et al. J Trace Elem Med Biol 2014;28:409–13.
220. de Berranger E, Colinet S, Michaud L, et al. JPEN J Parenter Enteral Nutr 2006;30:173–4.
221. Chen X, Yang G, Chen J, et al. Biol Trace Elem Res 1980;2: 91–107.
222. Jeejeebhoy KN, Chu RC, Marliss EB, et al. Am J Clin Nutr 1977;30:531–8.
223. Verhage AH, Cheong WK, Jeejeebhoy KN. JPEN J Parenter Enteral Nutr 1996;20:123–7.
224. Moukarzel A. Gastroenterol 2009;137:S18–28.
225. Stearns DM, Wetterhahn KE. Chem Res Toxicol 1994;7:219–30.
226. Schwahn BC, Van Spronsen FJ, Belaidi AA, et al. Lancet 2015;386:1955–63.
227. Barceloux DG. J Toxicol Clin Toxicol 1999;37:231–7.
228. Aschner M, Erikson K. Adv Nutr 2017;8:520–1.
229. O'Neal SL, Zheng W. Curr Environ Health Rep 2015;2:315–28.
230. Evans GR, Masullo LN. In: © SPC, ed. Manganese Toxicity. Treasure Island, FL: StatPearls Publishing LLC, 2022.
231. Dobson AW, Erikson KM, Aschner M. Ann N Y Acad Sci 2004;1012:115–28.
232. Chen P, Bornhorst J, Aschner M. Front Biosci (Landmark Ed) 2018;23:1655–79.

CHAPTER 58

Nutrition Assessment

Laura Matarese, Rebecca Brody, and Riva Touger-Decker

CHAPTER OUTLINE

Abbreviations

ACE, elderly in acute care; **ADL**, activities of daily living; **AND/ASPEN**, Academy of Nutrition and Dietetics/American Society for Parenteral and Enteral Nutrition; **CN**, cranial nerves; **GLIM**, Global Leadership Initiative on Malnutrition; **LTCE**, elderly in long-term care; **MNA**, mini-nutritional assessment; **MNA-SF**, mini-nutritional assessment–short form; **MST**, Malnutrition Screening Tool; **MUST**, malnutrition universal screening tool; **NA**, not available; **NCPM**, Nutrition Care Process and Model; **NFPE**, Nutrition Focused Physical Exam; **NRS**, nutritional risk screening; **NST/BAPEN4**, nutrition screening tool/British Association of Parenteral and Enteral Nutrition; **PEG**, Percutaneous Endoscopic Gastrostomy; **PG-SGA**, patient-generated subjective global assessment; **SCREEN-II AB**, seniors in the community: Risk Evaluation for Eating and Nutrition, Version II, abbreviated version; **SGA**, subjective global assessment; **SLP**, speech language pathologist, **SNAQ**, short nutritional assessment questionnaire

Introduction

"The Skeleton in the Hospital Closet" brought attention to the need to address the nutritional status of hospitalized patients and the prevalence of hospital malnutrition or what Butterworth called "physician-induced malnutrition".[1] This landmark paper led to a shift in nutrition care practices and the advent of nutrition screening and assessment tools in hospitals, other healthcare settings, and the community. This chapter addresses nutrition screening and assessment, nutrition-focused physical examination, and dysphagia screening as part of the assessment. Other chapters in this text provide greater depth in approaches to dietary assessment.

Nutrition Screening

Malnutrition has been linked to poor outcomes. Most would agree that no disease state benefits from malnutrition. Thus, evaluating nutritional status is essential to determining who may benefit from nutrition intervention. There are various methods used to assess nutritional status. The components of these tools and the degree to which they determine the nutritional state of an individual vary. There is a distinct difference between nutrition screening and assessment. The goals of each and the tools used are different. There are various definitions of "screening".[2] Generally, a screening tool is a simple tool used to identify individuals who may have nutrition problems and, therefore, require a more comprehensive nutrition assessment and possibly nutrition intervention.

The individuals performing the nutrition screen will vary from institution to institution. It may be a dietetic technician, nurse, clerk, or another healthcare provider. The foremost issue is using a *validated* and *reliable* screening tool.[2] A *valid* nutrition screening tool is one that accurately identifies the nutrition problem. Tool *reliability* refers to the ability of a tool to produce the same results each time it is used. The screening tool should have a high level of *sensitivity*; that is, it is more likely to correctly

Table 58.1 Validated Screening Tools (adapted from Skipper et al.)[3]

Screening Tool	Population	Reference Standard	Sensitivity %	Specificity %	Positive Predictive Value	Negative Predictive Value
NRS-2002	Hospitalized medical/surgical	SGA	62	93	85	79
NRS-2002	Acute hospitalized	MNA	39	83	85	37
NRS-2002	Acute hospitalized	SGA	70	85	79	78
MNA-SF	Ambulatory	MNA	100	69.5	92.8	78.5
MNA-SF	Subacute and ambulatory	MNA	97.9	100	98.7	NA
MST	Acute hospitalized	SGA	74	76	27.9	96
MST	Acute hospitalized	SGA	93	93	98	73
MST	Oncology outpatient	PG-SGA	100	92	80	100
MST	Oncology outpatient	SGA	100	81	40	100
MUST	Medical	SGA	NA	NA	NA	NA
MUST	Medical/surgical hospitalized	SGA	61	76	65	76
NST/BAPEN	Acute hospitalized	Dietitian assessment	86	95	80	96
Simple Two-part Tool	Acute hospitalized	SGA	63	97	70.2	95.5
NRS	Acute care, medical/surgical	Dietitian assessment	NA	NA	NA	NA
SCREEN II-AB	Community seniors and Geriatric clinic	Dietitian assessment	84	58	NA	59
Rapid Screen	Subacute care (rehabilitation center)	Standardized nutrition assessment	78.6	97.3	95.7	85.7
Tool #1	ACE; LTCE	Dietitian assessment	ACE: 75; LTCE:63; total sample: 70	ACE: 75; LTCE:63; total sample: 70	NA	NA
SNAQ	Hospitalized medical/surgical	Cross validation	76–86	83–91	62–79	89-96

ACE, elderly in acute care; *LTCE*, elderly in long-term care; *MNA*, mini-nutritional assessment; *MNA-SF*, mini-nutritional assessment–short form; *MST*, malnutrition screening tool; *MUST*, malnutrition universal screening tool; *NA*, not available; *NRS*, nutritional risk screening; *NST/BAPEN4*, nutrition screening tool/British Association of Parenteral and Enteral Nutrition; *PG-SGA*, patient-generated subjective global assessment; *SCREEN-II AB*, seniors in the community: Risk Evaluation for Eating and Nutrition, Version II, abbreviated version; *SGA*, subjective global assessment; *SNAQ*, short nutritional assessment questionnaire.

Data from Skipper A, Ferguson M, Thompson K, Castellanos VH, Porcari J. JPEN J Parenter Enteral Nutr. 2012;36(3):292-8.

identify patients with nutrition problems, resulting in a low percentage of false-negative results. Equally important, the tool should have high *specificity*, resulting in a greater level of identification of those patients who do not have a nutrition problem and, thus, a lower level of false-positive tests. Although significant, sensitivity and specificity are not enough to determine a reliable screening tool. The *predictive value* of the screening test, that is, the proportion of people with a given test result (positive or negative) who do or do not have a nutrition problem, is also required. A screening tool with low positive predictive value would correctly identify malnutrition in nearly all individuals with the condition but would also result in a high percentage of false-positive results. Multiple validated nutrition screening tools are available (**Table 58.1**).[3] Although all of these tools evaluate unintentional weight loss, the other components of these tools vary (**Table 58.2**).[3,4] Based on a systematic review of the literature,[4] the Academy of Nutrition and Dietetics' position is that the Malnutrition Screening Tool (MST) should be used

Table 58.2 Components of Currently Validated Screening Tools[3,4]

Tool	Recent Unintentional Weight Loss	Appetite	BMI	Disease Severity
Malnutrition Screening Tool (MST)	X	X		
Malnutrition Universal Screening Tool (MUST)	X		X	X
Mini Nutrition Assessment Short Form (MNA-SF)	X	X		X
Mini Nutrition Assessment Short Form BMI	X	X	X	X
Nutrition Risk Screening 2002 (NRS—2002)	X		X	X
Short Nutrition Assessment Questionnaire (SNAQ)	X	X		

to screen adults for malnutrition (undernutrition) regardless of their age, medical history, or setting.[5] Their position is that clinicians should abandon all nonvalidated malnutrition screening tools, including tools that were validated, then modified without rigorous revalidation against a standard definition of malnutrition. Adding items, modifying questions, or interpreting scores differently than intended by the tool's authors should be avoided, as these changes invalidate the tool.

Nutrition evaluation is not intuitive and requires training. It is possible that the experienced clinician would be able to identify individuals at nutritional risk simply based on clinical judgment. In one study, clinicians agreed on the nutritional status of patients electively admitted to a general surgical ward in 81% of the instances.[6] But these were experienced physicians, and the interobserver reproducibility, although high, potentially missed a determination of malnutrition in almost 20% of the cases. For most clinicians without this level of experience, a validated screening tool that is easy to use may help to identify those individuals who may need a more in-depth nutrition assessment.

There may be instances where the nutrition screen is skipped, and a complete nutrition assessment is conducted. For example, patients in an intensive care unit can be classified as "at risk" simply because they are critically ill. The assessment may evaluate the degree of weight loss, lean muscle mass, nutrient requirements, and mode of nutrition intervention. Patients in hospice or palliative care most likely do not need a nutrition screen. They will need an assessment to determine palatable foods and beverages that can be easily consumed. The goal in this instance is not to correct malnutrition but to provide comfort and improve the remaining quality of life.

Nutrition Assessment

Nutrition assessment differs from nutrition screening primarily in the level and intensity of the investigation.[2] Nutrition assessment continues the data gathering process initiated during the screen. It is the next step.[7] The data collected in the nutrition assessment are often like data collected in the screening process, but are more in-depth. The purpose is to detect the degree and type of malnutrition, nutrient deficiencies or toxicities, and other diet and nutrition problems such as diet quality and quantity of food intake, food access, factors affecting intake, digestion, absorption, mechanical and functional issues, and any sensory deficits. The nutrition assessment is also performed to determine if and what interventions are required. This may include diet, enteral or parenteral nutrition, counseling, education, food insecurity initiatives, referral for bariatric surgery or to other health professionals such as physical therapy, occupational therapy, speech-language pathology, dentistry, behavioral health, physician, or dietitian specialists.

The use of a validated nutrition assessment tool is essential. There are two components to validation. First, how well does it identify patients who will respond to nutrition intervention, and second, how well does the tool predict patient outcomes? As with screening tools, the importance of validation cannot be overstated, and there are risks to using a nonvalidated assessment tool.[8]

Without solid validation studies, there is no indication that an assessment tool will accurately identify malnutrition. This means that malnutrition could be missed in those who require nutrition intervention, or falsely labeling someone as malnourished, leading to unnecessary nutrition interventions that increase risk. There are also potential financial consequences. Labeling someone as having malnutrition when they do not and coding for malnutrition can lead to audits and substantial fines to the institution.[9] Additionally, any research study that uses a nonvalidated assessment tool as a component is at risk for false, erroneous, or inaccurate conclusions. Several assessment tools have been validated and recommended because of their high sensitivity and specificity in predicting nutritional status (**Table 58.3**). There are similarities in the approach taken to this complex diagnosis.

The Subjective Global Assessment (SGA) is the most highly validated tool to identify malnutrition in many patient populations. To date, it remains the gold standard by which other assessment tools are developed and validated. The SGA is based on the clinical history and physical exam. It includes five nutritionally relevant features of the clinical history, including decreased nutrient intake, unintentional weight loss, symptoms affecting oral intake, functional capacity, and metabolic demand. The physical examination focuses on subcutaneous fat loss, muscle wasting, and fluid accumulation.[10] Since the SGA evaluates intake, gastrointestinal symptoms, and the physical exam, it is poised to capture potential micronutrient deficiencies. There is good intrarater and interrater reliability.[6] It has been validated by predicting patient outcomes[5] and controlled clinical studies demonstrating that the SGA identifies patients who respond to nutrition intervention.[13,14] The SGA has been validated in numerous settings, including acute care,[10,15,16] rehabilitation and community settings,[17,18] as well as long-term care.[19] It has also been validated among numerous patient populations, including surgical,[10] geriatrics,[17-20] oncology,[16] and renal.[15] "Subjectivity" is sometimes cited as a weakness of the SGA. However, "subjectivity" can be considered a strength

Table 58.3 Validated Nutrition Assessment Tools[10-12]

Assessment Tool	Population	Setting	Parameters
Subjective Global Assessment (SGA)	Surgery, Geriatrics, Oncology, Renal	Acute Care, Rehabilitation, Community, Long-term Care	Medical History (weight loss, intake, GI symptoms, functional capacity) and physical exam to evaluate adipose stores, lean muscle mass, fluid retention, ascites
Patient Generated Subjective Global Assessment (PG-SGA)	Oncology, Renal, Stroke	Acute Care	Medical History (weight, intake, symptoms, functional capacity, metabolic demand) and physical exam
Mini-Nutritional Assessment (MNA)	Geriatrics	Acute Care, Community, Rehabilitation, Long-term Care	Diet history, anthropometry (weight history, height, mid-arm circumference, calf circumference), medical and functional status

of the SGA, requiring thought, reasoning, and interpretation.[8] Rarely are health conditions diagnosed using only "objective" data. Any objective data must be interpreted based on the patient's unique clinical presentation.

The Patient-Generated SGA (PG-SGA) includes all of the components of the SGA in addition to metabolic demand.[11] It has been validated in the acute care setting[21,22] and for patients in oncology settings,[21] with kidney disease[22] and stroke.[23]

The Mini Nutritional Assessment (MNA) includes both screening and assessment components.[12] This tool includes diet history, anthropometry (weight history, height, mid-upper arm circumference, and calf circumference), and medical and functional status. It has been validated in acute care, community, and rehabilitation settings, as well as long-term care for use in the geriatric population aged 65 and older.[12]

Two other nutrition assessment tools have been proposed, the Academy of Nutrition and Dietetics/American Society for Parenteral and Enteral Nutrition (AND/ASPEN) Characteristics Recommended for the Identification and Documentation of Adult Malnutrition (Undernutrition)[24] and the Global Leadership Initiative on Malnutrition (GLIM).[25] Each of these is in various stages of validation testing. Both are based on the SGA. The AND/ASPEN consensus statement uses inflammation as a surrogate marker of malnutrition and is etiology based. The tool includes six clinical characteristics of malnutrition: 1: insufficient food and nutrition intake compared with nutrition requirements, 2: weight loss over time, 3: loss of muscle mass, 4: loss of fat mass, 5: fluid accumulation, and 6: measurably diminished grip strength, with specific criteria defining severe and nonsevere malnutrition.[24] Two components are required to identify malnutrition. However, it may be that only one of the components is necessary to identify malnutrition; or it may require three components. It is also possible that a certain combination of factors (i.e., involuntary weight loss, inadequate diet intake) has greater sensitivity and specificity than others, especially in specific clinical situations. The ability of the tool to capture micronutrient deficiencies is unknown.

The GLIM consensus tool introduced a diagnostic framework based on phenotypic and etiologic criteria. Obesity or overweight status is not a GLIM criterion although these individuals often are malnourished. Of note, a validated screening tool is required as an initial step to identify patients at risk of malnutrition.[25] The presence of one phenotypic (weight loss, low BMI, reduced muscle mass) and one etiologic criterion (reduced food intake or assimilation, inflammation) is proposed to be sufficient to diagnose a patient with malnutrition. As with the AND/ASPEN criteria, the tool's ability to capture micronutrient deficiencies is unknown. Although the GLIM criteria have been found to have a high degree of *criterion* validity and reliability, in one study, different combinations of diagnostic criteria had different levels of validity.[26] Thus, as with the AND/ASPEN criteria, various combinations of diagnostic criteria and cutoff points are most likely to be required. The AND/ASPEN and GLIM criteria are works in progress and likely to change as more evidence-based information from well-designed, prospective trials becomes available.[27-30]

More in-depth body composition evaluations, such as bioelectrical impedance, are generally not used routinely in a hospital setting. In some instances, they may be used in an outpatient clinic setting where the patients may be more stable.

Once malnutrition has been identified and an intervention has been used, it is crucial to monitor the patient and reassess their status to determine that the nutrition intervention was effective. This may help to decide if weaning or other changes to the intervention are indicated or, in some instances, treatment should be escalated.

Patients at High Risk of Malnutrition: History and Previous Diagnoses

The importance of knowing the history of the patient cannot be overstated. It is the vital first step.[8] Certain patients will require further investigation by virtue of their current and previous diagnoses and history. For example, someone with a known diagnosis of short bowel syndrome is, by nature of the diagnosis, at risk of having or developing malnutrition. The patient history combined with diagnosis is often the initial step in the assessment process. The physical exam confirms or refutes the clinician's suspicions of the presence and degree of malnutrition.

Nutrition assessment incorporates a review of the patient's health record, starting with their history and current diagnoses and symptoms. With that information, the clinician is prepared to continue the nutrition assessment with the patient interview and physical exam. The combination of history, clinical data, patient presentation, and exam provides the clinician with the necessary information to assess the patient's nutritional status.[31] This includes identifying if other diagnostic tests are needed to evaluate micronutrient deficiencies or toxicities, determining the presence or absence of malnutrition, the route of feeding, necessary referrals, and other special needs the patient may have regarding diet management.

Nutrition-focused Physical Exam (NFPE)

The physical exam has been used as a component of nutritional status assessment since the 1980s[6,10,32] when the Subjective Global Assessment (SGA) was introduced. The term nutrition-focused physical examination (NFPE) evolved as part of nutrition care as dietitians sought to distinguish their approach to physical examination from that of physicians, nurses, or other health professionals, and to integrate NFPE into the nutrition assessment process.[33] With the advent of the Academy of Nutrition and Dietetics' (hence termed Academy) Nutrition Care Process and Model (NCPM) in 2003, examination and assessment of "nutrition-focused physical findings" became a standard part of nutrition assessment for dietitians.[7,34] Subsequently, competency in the assessment of these signs and symptoms was integrated into the Standards of Practice for general and specialist practice dietitians in 2012.[35]

The goal of NFPE is to identify clinical factors that may impact nutritional status and the ability to consume foods and fluids, and to reflect on patient nutrition problems, including signs and symptoms of malnutrition and micronutrient deficiencies. NFPE should take a head-to-toe approach to thoroughly evaluate the patient and provide the clinician with physical findings that reflect patients' mechanical and sensory abilities to eat and drink, activities of daily living (ADL), signs and symptoms of micronutrient deficiencies, and fat and muscle stores. Although problem-focused NFPE may be used in monitoring patients' response to treatment, the initial encounter should include an NFPE that considers the whole patient and not focus solely on examination for conditions such as malnutrition. NFPE alone is not diagnostic; however, the findings from the exam can confirm or contribute to nutrition and medical diagnoses.

A comprehensive NFPE includes measurement of body composition (height and weight, at a minimum but may include skinfold measures and bioelectrical impedance analysis), and clinical examination through a nutrition lens of the skin, head, neck, face (facial symmetry, cranial nerves, muscles of mastication, temporomandibular joints), mouth (extra- and intra-oral exam including dentition, saliva, and soft tissues) and some or all of the cranial nerves[33,36-42] along with the arms, hands, fingers, abdomen, lower limbs, and feet. In addition to providing diagnostic clues, the NFPE can provide insights into diet modifications that may be needed due to physical exam findings. **Table 58.4** provides an approach to NFPE that includes most of the components of a comprehensive physical examination from a nutrition lens. Clinical implications of the exam findings can be found in related chapters of this text. In combination with the patient's history, and subjective and objective data, the NFPE provides a wealth of information to inform the nutrition assessment and subsequent care plan.

NFPE starts with "observation" and "inspection" during the initial patient encounter.[33,43] These first impressions when the clinician or the patient enters the room provide insight into the patient's demeanor, degree of alertness and orientation, posture, gait, hygiene, and overall appearance. The power of observation afforded by the NFPE cannot be understated.[44] The inspection also includes observation for clinical signs of micro- and macronutrient deficiencies or excesses[45] and subcutaneous fat and muscle depletion, hydration status, functional status, disease-specific clinical signs, and any other physical symptoms that may impact functional or sensory ability to consume foods

Table 58.4 Components of the Physical Examination with a Nutrition Lens[10,36,39,40,42]

Anatomic Area	Anatomic Structure	Exam Components	Purpose
Head and Neck			
Face	Eyes	Inspect eyeballs, eyelids, & peripheral vision; observe for symmetry	Observation of micronutrient deficiencies
	Muscles of Mastication	Palpate for tenderness; ask about pain with biting & chewing	Ability to bite and chew
	Temporalis muscle (also a muscle of mastication)	Palpate for tenderness; inspect for signs of muscle wasting; ask about pain with biting & chewing	Ability to bite and chew Observation of muscle wasting
	Muscles of facial expression	Cranial Nerve (CN) VII (Facial)	Dysphagia risk
	Temporomandibular joint	Ask about joint sounds & pain with opening, biting & chewing; palpate for tenderness; inspect mandibular opening jaw strength and movement; CN V (Trigeminal)	Mouth opening, ability to bite and chew
	Nasolabial folds	Observe for symmetry CN VII (Facial)	Dysphagia risk
	Cranial Nerves	I (Olfactory), III (Oculomotor), V (Trigeminal), VII (Facial)	Ability to smell and see, dysphagia risk
	Skin	Inspect color; inspect texture & temperature	Hydration, observation of macro- and micronutrient deficiencies, skin breakdown
	Lips	Inspect for moisture, texture, cracks (especially at the corners), fissures, lesions & ulcers	Hydration, observation of micronutrient deficiencies
	Cheeks	Observe for parotid gland swelling	Salivary gland disorder(s)

(continues)

Table 58.4 Components of the Physical Examination with a Nutrition Lens[10,36,39,40,42] *(continued)*

Anatomic Area	Anatomic Structure	Exam Components	Purpose
Head and Neck			
Mouth	Overall Wetness	Inspect salivary flow and appearance	Salivary gland disorder(s), ability to chew and swallow
	Labial Mucosa	Evert lips, inspect color, moisture and texture	Hydration, observation of micronutrient deficiencies
	Tongue	Inspect dorsal, ventral, lateral borders, range of motion, strength, pigmentation, erythema, growths, lesions and any fissures; ask about pain, burning, taste changes; CN XII (Hypoglossal)	Observation of micronutrient deficiencies, dysphagia risk
	Gingiva	Inspect color and texture, lesions, bleeding, swelling, & ulcerations; ask about pain interfering with eating	Hydration, ability to eat
	Buccal Mucosa	Inspect color, wetness, and texture, Stenson's duct	Salivary gland disorders
	Floor of the mouth	Inspect color and texture, saliva pooling; Wharton's duct	Salivary gland disorders
	Hard Palate	Inspect color, texture, lesions, and ulcerations	Observation of micronutrient deficiencies and infection
	Soft Palate and Oropharnyx	Observe uvula and tonsils; inspect color, lesions, ulcerations, and texture, swallow, and elevation of the soft palate; CN IX (Glossopharyngeal), X (Vagus); ask about difficulty swallowing	Dysphagia risk, observation of infection
	Dentition	Observe extent and location of edentulism, occlusion, full or partial dentures, broken teeth; if tooth loss, ask about any difficulty eating; if dentures ask about use for eating, use of adhesives	Ability to bite and chew
Neck	Cranial nerves (CN)	Posture, CN XI (Accessory)	Ability to hold the head upright for eating, dysphagia risk
Skin and Nails			
Skin	Along all sections of the body	Inspect color, turgor, temperature, lesions, rashes, ulcerations, wounds, edema, and dryness	Hydration, observation of macro- and micronutrient deficiencies, skin breakdown
Nails		Inspect color of nail beds	Observation of micronutrient deficiencies
Trunk			
Arms	Triceps, Biceps	Inspect for fat wasting	Sarcopenia, malnutrition
Hands	Fingers	Observe range of motion, strength of handshake; inspect color, turgor, temperature; ask about numbness or pain with movement	Muscle wasting, ability to hold eating utensils
Upper Extremity	Clavicles Chest Deltoids	Inspect for fat and muscle wasting	Sarcopenia, malnutrition
Abdomen	Abdomen	Inspect for Ascites; distension, soft, non-tender, ostomy or PEG tube, prior surgical scars	Fluid status and GI function
Lower Extremity	Quadriceps Gastrocnemius	Inspect for muscle wasting (both);	Sarcopenia, malnutrition
Ankles		Palpate for pedal edema	Hydration
Feet	Toes	Inspect for ulcerations and color; ask about paresthesias	Observe for micronutrient deficiencies and diabetes sequela

Data from Detsky AS, McLaughlin JR, and Baker JP. JPEN J Parenter Enteral Nutr 1987;11:8–13; Brody RA, Touger-Decker R, Radler DR, et al. Topics in Clinical Nutrition 2014;29:57–68; Zechariah S, Lehman L, Waller JL, et al. Healthcare 2021;9:17; Zechariah S, Waller JL, De Leo G, et al. Healthcare 2021;9:20; Zechariah S, Waller JL, Stallings J, et al. Healthcare 2022;10:21.

and fluids. In addition to inspection, the NFPE may include palpation, auscultation, and percussion.[46,47] A thorough physical examination is noninvasive, simple to perform, and provides diagnostic clues that may not be obvious from laboratory and radiologic examinations.[48]

Definitions of NFPE and its components included in nutrition assessment tools vary.[10,12,24,49] The SGA, PG-SGA, and Academy/ASPEN tools, which focus on the diagnosis of malnutrition, are covered in detail in the nutrition assessment section of this chapter. They include several physical exam components, such as weight and weight change, muscle (in the quadriceps and deltoids), and fat loss (triceps and mid-axillary region), hydration status (edema in the ankles and sacrum and ascites) and functional capacity.[10,11,24] The AND/ASPEN tool includes adipose tissue, muscle, handgrip strength, and fluid status.[24]

Clinicians working in specialty areas may incorporate specific components of the physical exam based on the population served. The examination may vary based on clinical practice guidelines, standards of practice of various professions and disciplines, and research identifying abnormal physical findings in different disease states or conditions. Those working with patients with lipid disorders may focus on xanthomas; clinicians practicing bariatric surgery may focus more on clinical signs and symptoms of micronutrient deficiencies. Clinicians working with patients with head and neck cancers may include a more in-depth orofacial examination to note signs and symptoms of side effects of radiation and chemotherapy and the impact of head and neck surgeries on sensory and functional changes in the mouth. Similarly, those working in dental settings will include a more in-depth orofacial examination, including signs of micronutrient deficiencies. The chapter on oral health in this text contains tables on orofacial nutrition risk factors to consider in physical examination. Providers working in geriatrics, neurology, or neurosurgery settings may do a more comprehensive cranial nerve exam and include cognitive evaluation and dysphagia screening (addressed elsewhere in this chapter). Clinicians practicing in gastroenterology settings may perform a more in-depth abdominal examination.

The Physical Exam with a Nutrition Lens

Despite the advent of artificial intelligence, electronic health records, and computer technology in health care, the value of a physical examination with a nutrition lens to clinicians continues to be immeasurable.[8,43-46] A thorough physical examination, in combination with diagnostic tests and patient history, can lead to earlier diagnosis, intervention, and referrals to other disciplines and professions, as needed.[31] Touch and palpation provide a physical connection between the provider and the patient and another source of clinical information. A benefit of the inspection component of the physical exam has been described by Gupta et al.[43] as "inspection allows us to humanize our patients, to see them as people who have lives outside of their illnesses." They say, "remembering to actually look at our patients, we see what is hiding in plain sight." The benefits of seeing the patient's dentition, posture, arm and hand movements, and ability to hold up their head provide insights into their eating ability. The patient's physical appearance and dress provide insight into self-care and hygiene practices.[44] Physical exam findings may help explain patient symptoms (Table 58.4). Muscle wasting, which may be apparent on observation or evaluated using palpation, may reflect malnutrition but could also reflect non-nutrition diagnoses like sarcopenia, cancer cachexia, or muscle atrophy from nonuse. However, when the finding is combined with the rest of the NFPE, history, and diagnostic tests, an accurate diagnosis can be determined.

Dysphagia Screening as a Component of Nutrition Assessment

The risk of dysphagia, a swallowing disorder characterized by difficulty moving food, liquids, secretions, or medications from the mouth to the stomach,[50] can be identified through dysphagia screening as part of a comprehensive nutrition assessment and NFPE.[37] The goal of dysphagia screening is to identify those at risk, make referrals to the Speech-Language Pathologist (SLP) and other appropriate healthcare team members for treatment or additional instrumental testing, and start interventions such as diet modification and swallowing therapy early.[37,51,52] While dysphagia screening is recognized as necessary for populations at risk for dysphagia, bedside screening tools with high sensitivity and specificity in predicting aspiration risk are lacking.[53] Dysphagia screening tools vary, and the literature describes many methods, including patient interviews, self-administered questionnaires, medical record reviews, eating observations, oral-motor examination, or a brief swallow test.[37,51,52,54] Screening can be administered by nursing professionals, SLPs, dietitians, or other healthcare professionals.[51] Clinical findings obtained from dysphagia screening inform the nutrition assessment and interventions, such as diet or liquid modifications or enteral nutrition initiation. Physical examination includes sensory and motor evaluation of the five cranial nerves involved in swallowing (CN V trigeminal, CN VII facial, CN IX glossopharyngeal, CN X vagus, and CN X11 hypoglossal nerves).[52] The accessory nerve (CN XI) is also examined for its innervation with muscles required for proper head and neck position and posture, which is needed for a safe swallow (**Table 58.5**). Listening for voice changes and hoarseness after eating or drinking may provide evidence of aspiration. Cognitive assessment is incorporated into dysphagia screening because if patients cannot follow commands and stay awake during a meal, they are at higher risk of dysphagia. These individuals may forget they are eating, pocket food, or poorly clear it from the oral cavity, resulting in aspiration. Having a patient demonstrate their ability to cough on demand may indicate the individual's capacity to clear food from their airway in the event of aspiration. A weak cough is a risk factor for dysphagia.[55]

There is a bidirectional association between malnutrition and dysphagia. Dysphagia can lead to malnutrition because of an inability to safely and adequately consume macro- and micronutrients.[56,57] The resulting muscle wasting associated with malnutrition is theorized to contribute to developing or worsening dysphagia as muscles critical to protecting the airway are compromised.[58] Individuals have less ability to clear their airway of aspirated food or fluids and to maintain good oral hygiene, due to their decompensated condition.[59] As muscle function is reduced, functional status, including the ability to perform ADLs or even sit up straight to maintain a safe posture for swallowing, can be impaired. Whole-body skeletal muscle depletion, including the muscles involved in swallowing, was first described in 2012 as sarcopenic dysphagia.[60] This interplay between malnutrition and dysphagia warrants immediate screening and nutrition assessment to identify dysphagia risk and determine appropriate interventions.[37]

Table 58.5 Cranial Nerve Assessment Related to Swallowing[52]

Cranial Nerve	Function	Examination	Presentation of Cranial Nerve Dysfunction
V—Trigeminal	(S) Bilateral sensations to face, mouth, teeth to sharp and soft touch along three nerve branches (ophthalmic, maxillary, and mandibular). Sensory perception of mucocutaneous structures of the face, mouth, nose, and teeth. (M) Masticatory muscles: jaw movement and strength.	(S) Test face along both sides of the face along the mandible, maxilla, and forehead with sharp (pin prick) and light touch (cotton swab). (M) Have individual open and close jaw, move side to side, and clench teeth. Place hand under chin of individual and have them resist attempts of the examiner to close their jaw.	(S) Impaired/altered oral sensation. - Drooling - Pocketing of food - Food residue in oral cavity (M) Weak jaw; asymmetry on jaw opening/closing; impaired mastication. - Easily fatigues with eating - Food leakage
VII—Facial	(S) Taste anterior two-thirds of tongue. (M) Facial expressions related to muscles of lips and cheeks. (A) Secretion of lacrimal, sublingual, and submandibular glands.	(S) Test taste sensation of sweet, salt, and sour. (M) Observe bilateral facial expressions and examine symmetry. Have individual smile to show teeth, frown, blow out cheeks, purse lips, and raise eyebrows. Observe if nasolabial folds are symmetrical.	(S) Altered/impaired or absent taste (M) Facial asymmetry, unequal nasolabial folds, poor lip seal. - Drooling - Pocketing of food (A) Decreased saliva - Increased oral preparatory stage
IX—Glossopharyngeal	(S) Taste posterior one-third of tongue; sensation to pharynx, tonsils, soft palate. (M) Elevation of soft palate; swallowing. (A) Secretion of parotid gland.	(S) Test taste sensation of bitter. (M) Have individual say "AH." Using a penlight, look for bilateral soft palate elevation and midline uvula.	(S) Altered/impaired or absent taste posterior one-third of tongue. No pharyngeal elevation. Altered or impaired swallowing. - increased risk of aspiration. (M) Limited palate elevation. Altered or impaired swallowing. - pocketing - prolonged eating time - Regurgitation of food into nasopharynx
X—Vagus	(M) Movement of soft palate, pharynx, and larynx; swallowing.	Test IX and X together (see above)	(See IX above)
XII-Hypoglossal	(M) Tongue movement and range of motion.	(M) Have individual protrude tongue (should be midline), move it side to side, and press against cheek to demonstrate tongue strength.	(M) Limited tongue movement or deviation; oral phase dysphagia.
XI—Accessory	(M) Strength of the sternocleidomastoid (SCM) and trapezius muscles.	(M) Have individual shrug shoulders against examiner's pressure on shoulders. Observe posture and head positioning.	(M) Limited strength of the SCM and trapezius muscles may limit ability to maintain proper posture and head positioning while eating. - Increased risk of aspiration

(S): Sensory Response; (M): Motor Response; (A): Autonomic Response.

Data from Speyer R, Cordier R, Farneti D, et al. Dysphagia 2022;37:333–49.

Dysphagia Etiologies and Presentation

Up to 80% of individuals who have a stroke develop poststroke dysphagia, although the prevalence ranges widely, based on study design, population, and method of diagnosing dysphagia.[61] Those with poststroke dysphagia are four times more likely to be at risk of pneumonia than those with stroke but no dysphagia.[62] Oropharyngeal dysphagia and aspiration pneumonia are positively correlated in older adults with stroke.[63] Neurogenic-related diagnoses, including traumatic brain and cervical spinal cord injuries, multiple sclerosis, and amyotrophic lateral sclerosis as examples, place individuals at high risk for dysphagia.[50] Similarly, myogenic disorders that affect the muscles associated with swallowing increase dysphagia risk. These may include muscular dystrophy, myasthenia gravis, and aging due to muscle atrophy.[50] Obstructive disorders block or disrupt the passage of food or liquids during the swallowing process and may include candidiasis, fistulas, and head and neck cancers.[50] Iatrogenic causes of dysphagia involve side effects from chemo- or radiotherapy and postsurgical complications where anatomical changes impact swallowing.[50]

Specific populations are at greater risk of dysphagia than others; early identification of risk is critical to symptom management and prevention of nutritional decline. Physical changes associated with dysphagia in older adults include edentulism and ill-fitting dentures, which impact the ability to masticate and move the food bolus in the oral cavity.[64] Xerostomia and hyposalivation are of concern, as many medications can cause the sensation of dry mouth or reduced saliva production. The residue left in the oral cavity due to xerostomia or hyposalivation can increase bacterial growth aspiration of food particles.[64] Additionally, if the muscles required for swallowing are atrophied, a disordered or slow swallow can occur. The strength of muscle contractions that induce peristalsis is also reduced.[64]

Older adults with dysphagia are prone to psychosocial consequences. These individuals may be socially isolated as they do not want to eat with others due to their signs and symptoms of dysphagia, such as drooling, regurgitation of food, and choking.[51] They also may have a reduced quality of life, with evidence pointing toward adverse effects on self-esteem, enjoyment of life, socialization, and decreased ability to perform activities of daily living.[65] Many patients with dysphagia suffer from anxiety due to fear of food sticking in their throat or feeling like they are choking. In long-term care settings, up to 60% of residents have dysphagia.[66,67]

Of all of the cancer diagnoses, those with head and neck cancer are most at risk for dysphagia due to the location and extent of cancer and related treatment in the anatomic regions associated with swallowing.[68] Radiation therapy to the head and neck region, often given in tandem with chemotherapy, can cause significant treatment effects that can worsen pre-existing dysphagia or may cause dysphagia when it does not exist pre-treatment.[68] Oral sequela include mucositis, xerostomia, hyposalivation, thickened saliva, dysgeusia, odynophagia, or pain with swallowing and pharyngeal fibrosis.[68] Individually, these problems can negatively affect nutritional well-being; patients with head and neck cancer often experience more than one of these symptoms, thus increasing their risk. Similarly, chemotherapy treatment effects can exacerbate dysphagia due to mucositis, stomatitis, or ulcerations in the structures involved in swallowing.[68] Surgical resections due to head and neck cancer also increase dysphagia risk. Dysphagia can impede adequate intake and contribute to malnutrition during head and neck cancer treatment, resulting in delays or interruptions in treatment.[68]

The primary negative outcome of dysphagia is aspiration pneumonia, an infection caused by the inhalation of oral secretions or food particles colonized by pathogenic bacteria.[51,69] Other notable health consequences related to dysphagia are dehydration due to suboptimal intake of fluids for fear of choking or aspiration or prescribed liquid restrictions due to aspiration risk.[51]

Dysphagia presents in the oral, pharyngeal, and esophageal phases of the swallow (**Table 58.6**). Oral-phase dysphagia is associated with neurogenic and cognitive disorders, hyposalivation, and oropharyngeal lesions. It can present with weakness of the lips, causing food or liquids to leak from the oral cavity, and speech disorders. Decreased oral sensitivity from cranial nerve dysfunction reduces the feeling that food is left in the oral cavity, thus increasing the likelihood of aspiration. This can also occur when there is poor tone in the cheeks or lip region, causing food residue to build up. Tongue weakness limits bolus formation, as the tongue is necessary for shaping and moving the bolus.[50,70]

Pharyngeal dysphagia is related to a disorganization of the sequencing of events that protect the airway and move the bolus through the pharynx. Food particles can remain in the pharynx, making aspiration most likely to occur during the pharyngeal phase of swallowing if the airway is not protected adequately. It is most often associated with neuromuscular disorders.

Table 58.6 Etiologies and Signs and Symptoms by Phase of Dysphagia[52,70]

Type of Dysphagia	Etiologies	Signs and Symptoms
Oral phase dysphagia: *Difficulty moving food or liquids in the oral cavity to prepare the bolus for transport from the mouth to the pharynx.*	▪ Neurogenic disorders ▪ Decreased salivation ▪ Oropharyngeal lesions ▪ Cognitive disorders	▪ Weakness of lips ▪ Decreased oral sensitivity ▪ Poor bolus formation due to tongue weakness or range of motion ▪ Food falls into the anterior or lateral sulcus due to poor labial or buccal tone
Pharyngeal Phase Dysphagia: *Difficulty moving food or liquids through the pharynx into the esophagus.*	▪ Neuromuscular disorders ▪ Pharyngeal muscle function is deranged due to nerve or muscle disease ▪ Must confirm by instrument testing	▪ Difficult passage of bolus above the suprasternal notch ▪ Choking, coughing ▪ Regurgitation of food or liquid into the nasopharynx
Esophageal Phase Dysphagia: *Difficulty transporting food or liquids through the esophagus into the stomach.*	▪ Obstructive disorders ▪ Motility disorders ▪ Motor dysfunction ▪ Must confirm with instrument testing	▪ Food "sticking" in throat or chest ▪ Neck or chest pain, heartburn ▪ Regurgitation of food or pills ▪ Coughing when lying down after meals ▪ Dysphagia to solids, can progress to liquids

Data from Speyer R, Cordier R, Farneti D, et al. Dysphagia 2022;37:333–49; Siwiec RM, and Babaei A. (2020). Oropharyngeal Dysphagia. In Evaluation and Management of Dysphagia : An Evidence-Based Approach (Patel DA, Kavitt RT, and Vaezi MF, eds) pp. 43-62, Springer International Publishing, Cham, Switzerland.

Individuals with pharyngeal dysphagia often complain of the difficult or delayed passage of the bolus above the suprasternal notch, the indentation in the throat region between the clavicles. Choking, coughing, and regurgitation of food or liquids into the nasal cavity occur because of improper elevation of the soft palate to protect food or liquids from entering the nasopharynx. Pharyngeal dysphagia must be confirmed by instrument testing.[50,70]

Esophageal phase dysphagia results from impaired esophageal function with retention of liquids and food in the esophagus after swallowing. Obstructive, motor, or motility disorders impede the passage of food through the esophagus. Individuals typically first complain of difficulty swallowing solid foods, which progresses to soft consistencies and eventually liquids due to a progressive narrowing or blockage of the esophagus. As with pharyngeal dysphagia, esophageal dysphagia must be confirmed with instrument testing.[50]

Early identification of dysphagia and an interprofessional approach to its management positively impact patient care. Dysphagia identification can significantly reduce the risk of aspiration pneumonia, length of hospital stay, dependency, and mortality in hospitalized patients.[71,72]

Special Considerations in Nutrition Assessment

Nutrition assessment practices will vary based on many factors, including the lifespan stage, care setting, and disease state. There are tools and techniques specific to age groups, such as those designed for pediatrics or geriatrics.[12] Likewise, there are disease-specific considerations for patients with chronic diseases such as chronic kidney disease[73] or cirrhosis.[74] Patients with cognitive impairment may be unable to give accurate histories or fully cooperate during an NFPE; involving a caregiver or family member during the assessment process can provide insight and result in greater patient cooperation. While some assessment tools, like the SGA, have been validated in various settings and populations, others, like the MNA, are specific to the setting and patient population.

Standards used for nutrition assessment, including those used for body weight status and composition and determination of energy and nutrient needs, can vary. Reference standards for determining body fatness, energy, some nutrient needs, and selecting laboratory data, are based on biological sex. Linsenmeyer et al., suggest that in the absence of evidence-based guidelines or standards for nutrition assessment of individuals who are transgender, the decision of which sex to use as the reference standard should be based on whether the individual is undergoing or has undergone hormonal therapy to complete a medical transition to a different sex.[75] There are also racial and ethnic differences in body adiposity[76] that merit consideration when determining body composition. Chapters in this text on body composition and anthropometry provide greater detail on these topics.

Cultural competency is essential in all aspects of healthcare. Culture, race, ethnicity, and religion-based guidance systems must be considered when completing a nutrition assessment.[77] It is essential for the providers performing all facets of the nutrition assessment to consider the patient's race, ethnicity, culture, and religion and their impact on the patient's presentation, food patterns, and diet history, and practice sensitivity in the conduct of the physical examination.

Summary

Nutrition screening and assessment are essential in all care settings to determine nutritional status and possible interventions. Validated tools should be used to determine nutritional status accurately. A comprehensive nutrition assessment includes a physical examination, dysphagia screening, evaluation of oral intake, unintentional weight loss, functional capacity, and metabolic demand to determine clinical signs and symptoms that can impact a patient's nutritional status and response to treatment.

References

1. Butterworth CE. Nutrition Today 1974;9:4–8.
2. Charney P. Nutr Clin Pract 2008;23:366–72.
3. Skipper A, Ferguson M, Thompson K, et al. JPEN J Parenter Enteral Nutr 2012;36:292–8.
4. Skipper A, Coltman A, Tomesko J, et al. J Acad Nutr Diet 2020;120:669–708.
5. Academy of Nutrition and Dietetics. J Acad Nutr Diet 2020;120:709–13.
6. Baker JP. N Engl J Med 1982;306:969–72.
7. Lacey K, Pritchett E. J Am Diet Assoc 2003;103:1061–72.
8. Matarese LE, Charney P. Nutr Clin Pract 2017;32:11–4.
9. Department of Health and Human Services. Hospitals Overbilled Medicare $1 Billion By Incorrectly Assigning Severe Malnutrition Codes To Inpatient Hospital Claims. Office of Inspector General, 2020.
10. Detsky AS, McLaughlin JR, Baker JP. JPEN J Parenter Enteral Nutr 1987;11:8–13.
11. Ottery F. In: McCallum P, and Polisena C, eds. Patient Generated Subjective Global Assessment. The Clinical Guide to Oncology Nutrition. Chicago, IL: American Dietetic Association, 2005; 11–23.
12. Guigoz Y, Vellas B, Garry PJ. Facts Res in Gerontol 1994; Suppl 2:15–59.
13. Norman K, Kirchner H, Freudenreich M, et al. Clin Nutr (Edinburgh, Scotland) 2007;27:48–56.
14. Norman K, Pirlich M, Smoliner C, et al. Eur J Clin Nutr 2011;65:735–42.
15. Cooper BA, Bartlett LH, Aslani A, et al. Amer J Kid Dis 2001;40:126–32.
16. Thoresen L, Fjeldstad I, Krogstad K, et al. Palliat Med 2002;16: 33–42.
17. Christensson L, Unosson M, Ek AC. Eur J Clin Nutr 2002;56: 810–8.
18. Duerksen DR, Yeo TA, Siemens JL, et al. Nutr 2000;16: 740–4.
19. Sacks GA, Dearman K, Replogle WH, et al. J Amer Coll Nutr 2000;19:570–7.
20. Persson MD, Brismar KE, Katzarski KS, et al. J Amer Geri Soc 2002;50:1996–2002.
21. Bauer J, Capra S, Ferguson M. Eur J Clin Nutr 2002;56:779–85.
22. Desbrow B, Bauer J, Blum C, et al. J Renal Nutr 2005;24: 1073–7.

23. Martineau J, Bauer JD, Isenring E, et al. Clin Nutr 2005;24, 1073–7.
24. White JV, Guenter P, Jensen G, et al. J Acad Nutr Diet 2012;112:730–8.
25. Jensen GL, Cerderholm T, Correia MITD, et al. JPEN J Parenter Enteral Nutr 2019;43:32–40.
26. El Chaar D, El Khoury C, Mattar L. JPEN J Parenter Enteral Nutr 2022;6:563.
27. Allard JP, Keller H, Gramlich L, et al. Clin Nutr (Edinburgh, Scotland) 2020;39:2771–7.
28. Balci C, Bolayir B, Eşme M, et al. JPEN J Parenter Enteral Nutr 2021;45:1172–80.
29. Brito JE, Burgel CF, Lima J, et al. Clin Nutr 2021;40:4366–72.
30. Hand RK, Murphy WJ, Field LB, et al. J Acad Nutr Diet 2016;116:856–64.
31. Abou Ziki MD, Podell DN, Schiliro DM. JAMA Intern Med 2015;175:1901–2.
32. Duerksen DR, Laporte M, Jeejeebhoy K. Nutr Clin Pract 2021;36:942–56.
33. Touger-Decker R. Topics Clin Nutr 2006;21:190–8.
34. Swan WI, Pertel DG, Hotson B, et al. J Acad Nutr Diet 2019;119:840–55.
35. Academy Quality Management Committee. J Acad Nutr Diet 2013;113:S56–S71.
36. Brody RA, Touger-Decker R, Radler DR, et al. Topics Clin Nutr 2014;29:57–68.
37. Brody RA, Touger-Decker R, VonHagen S, et al. J Am Diet Assoc 2000;100:1029–37.
38. Touger-Decker R, Mobley C, Epstein JB. Nutr Oral Med, Springer, 2014.
39. Zechariah S, Lehman L, Waller JL, et al. Healthcare 2021;9:17.
40. Zechariah S, Waller JL, De Leo G, et al. Healthcare 2021;9:20.
41. Zechariah S, Waller JL, Stallings J, et al. Healthcare 2022;10:13.
42. Zechariah S, Waller JL, Stallings J, et al. Healthcare 2022;10:21.
43. Gupta S, Saint S, Detsky AS.) JAMA Intern Med 2017;177: 757–8.
44. Fitzgerald FT, Tierney Jr LM. West J Med 1982;137:169.
45. Esper DH. Nutr Clin Pract 2015;30:194–202.
46. Hummell AC, Cummings M. Nutr Clin Pract 2022;37:41–9.
47. Hirschtick RE. JAMA 2016;316:1363–4.
48. Bickley L, Szilagyi PG. Bates' Guide to Physical Examination and History-Taking. Philadelphia, PA: Lippincott Williams & Wilkins, 2012.
49. Tomesko J, Brody R, Chen ST, et al. Topics Clin Nutr 2022;37: 2–14.
50. Lynch KL, Katzka DA. In: Patel DA, Kavitt RT, and Vaezi MF, eds. Dysphagia: How to Recognize and Narrow the Differential. Evaluation and Management of Dysphagia: An Evidence-Based Approach Cham, Switzerland: Springer International Publishing 2020;1–12.
51. McGinnis CM, Homan K, Solomon M, et al. Nutr Clin Pract 2019;34:80–95.
52. Speyer R, Cordier R, Farneti D, et al. Dysphagia 2022;37: 333–49.
53. Virvidaki IE, Nasios G, Kosmidou M, et al. J Clin Neurol 2018;14:265–74.
54. Belafsky PC, Mouadeb DA, Rees CJ, et al. Ann Otol Rhinol Laryngol 2008;117:919–24.
55. Field M, Wenke R, Sabet A, et al. Dysphagia 2018;33:827–39.
56. Brody R. Nutr Clin Pract 1999;14:S47–S51.
57. Blanař V, Hödl M, Lohrmann C, et al. J Adv Nurs 2019;75: 3566–76.
58. Veldee MS, Peth LD. Dysphagia 1992;7:86–101.
59. Ortega O, Martín A, Clavé P. J Am Med Dir Assoc 2017;18: 576–82.
60. Kuroda Y, Kuroda R. J Am Ger Soc 2012;60:1785–6.
61. Takizawa C, Gemmell E, Kenworthy J, et al. Dysphagia 2016;31:434–41.
62. Banda KJ, Chu H, Kang XL, et al. BMC Geriatrics 2022;22:420.
63. van der Maarel-Wierink CD, Vanobbergen JN, Bronkhorst EM, et al. J Dent Res 2011;90:1398–404.
64. Ney DM, Weiss JM, Kind AJ, et al. Nutr Clin Pract 2009;24: 395–413.
65. Azzolino D, Passarelli PC, De Angelis P, et al. Nutrients 2019;11: 2898.
66. Namasivayam AM, Steele CM. J Nutr Gerontol Ger 2015; 34:1–21.
67. Shanley C, O'Loughlin G. J Gerontol Nurs 26, 35–48.
68. Baijens LW, Walshe M, Aaltonen L-M, et al. Euro Arch Oto-Rhino-Laryngol 2021;278:577–616.
69. Teramoto S. Respiratory Investigation 2022;60:45-55.
70. Siwiec RM, Babaei A. Oropharyngeal Dysphagia. In: Patel DA, Kavitt RT, and Vaezi MF, eds. Evaluation and Management of Dysphagia: An Evidence-Based Approach. Cham, Switzerland: Springer International Publishing, 2020; 43–62..
71. Sherman V, Greco E, Martino R. J Am Heart Assoc 2021;10: e018753.
72. Eltringham SA, Kilner K, Gee M, et al. Cerebro Dis 2018;46: 97–105.
73. Ikizler TA, Burrowes JD, Byham-Gray LD, et al. Am J Kidney Dis 2020;76:S1–S107.
74. Johnson TM, Overgard EB, Cohen AE, et al. Nutr Clin Pract 2013;28:15–29.
75. Linsenmeyer W, Garwood S, Waters J. J Acad Nutr Diet 2022;122: 1081–6.
76. Heymsfield SB, Peterson CM, Thomas DM, et al. Obes Rev 2016;17:262–75.
77. Gbareen M, Barnoy S, Theilla M. BMC Nurs 2021;20:1–10.

CHAPTER 59

Interprofessional Teams and Nutrition

Rebecca Kuang, David G. Binion, and Ruy Cruz

CHAPTER OUTLINE

Abbreviations

ASPEN, American Society for Parenteral and Enteral Nutrition; **EN**, enteral nutrition; **IPE**, interprofessional education; **IPEC**, Interprofessional Education Collaborative; **LCME**, Liaison Committee on Medical Education; **NP**, nurse practitioner; **NST**, nutrition support team; **PA**, physician's assistant; **PN**, parenteral nutrition; **RD**, registered dietitian; **RN**, registered nurse, **TBC**, team-based care; **UPMC**, University of Pittsburgh Medical Center

Introduction

The field of nutrition has dramatically evolved over the past 5 decades. This explosion of research interest and resultant knowledge is evidenced by the 25-fold increase in nutrition publications spanning the most recent 5 decades. In 1971, there were 2,408 nutrition publications in PUBMED, which rose to 67,113 nutrition-indexed publications in 2021. Societal and scientific factors have triggered this rise in nutrition research and knowledge. These include the growing appreciation of diet as a key environmental factor playing a critical role in human health and disease, the emergence of the obesity epidemic and diseases linked to metabolic syndrome in the United States And throughout the world, and an explosion of research into the microbiome and how diet plays a critical "prebiotic" role in shaping the architecture of the gut microflora.

This rise in nutrition science has paradoxically witnessed a reduction in the hours of nutrition instruction included in the average medical student curriculum. The Liaison Committee on Medical Education (LCME), the accrediting organization for U.S. medical schools offering the M.D. degree, recommends a minimum of 25 hours of nutrition education across the 4-year medical school curriculum, which is consistent with the landmark 1985 report on nutrition education in U.S. medical schools from the National Academy of Sciences.[1] However, nutrition education remains extremely limited in medical schools across the country and many do not meet these recommendations.[2] A Lancet systematic review (2019) found that "medical students are not supported to provide high-quality, effective nutrition care."[3] This mismatch in nutrition expertise and a working knowledge of nutrition in caring for patients with acute and chronic illnesses, combined with the shift across the professions to interprofessional education to create patient-centered interprofessional team care, has

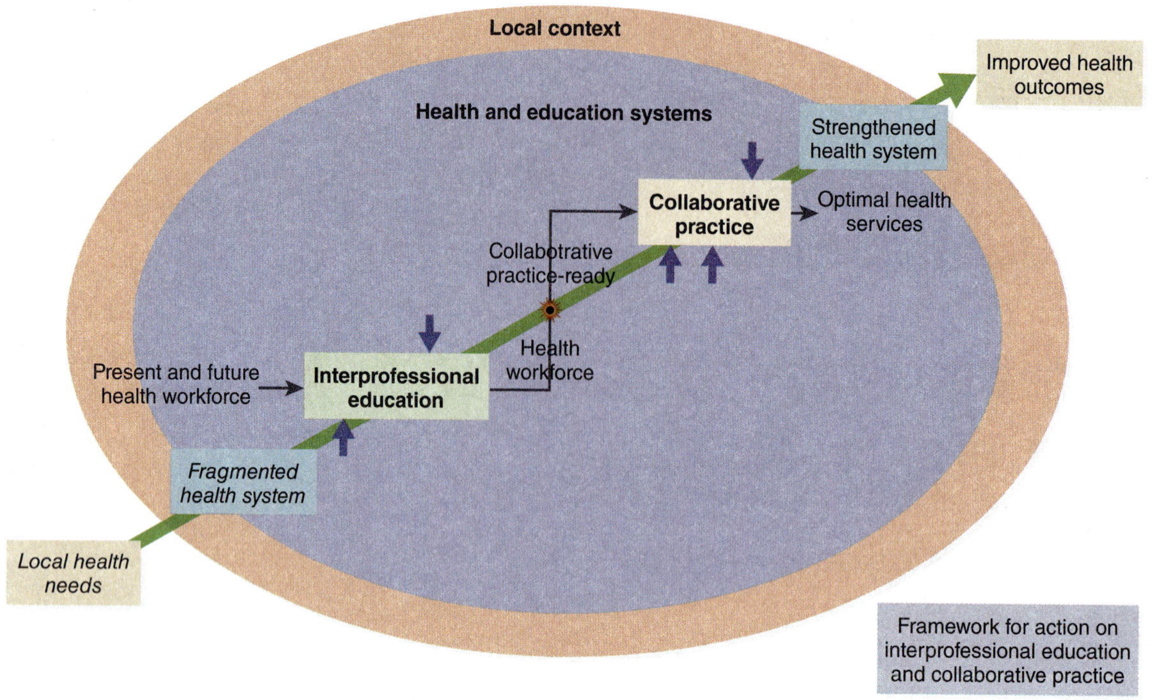

Figure 59.1 Health and education systems.

Reproduced from World Health Organization. Framework for Action on Interprofessional Education and Collaborative Practice (No. WHO/HRH/HPN/10.3). World Health Organization, Geneva, 2010.

created an impetus for greater collaboration among all healthcare providers, including dietitians. Formalization of this collaboration as interprofessional teams has been pioneered in several areas, including nutrition support with parenteral nutrition (PN), transplant medicine, intestinal rehabilitation, diabetes management, inflammatory bowel disease, and celiac disease.

This chapter reviews the history behind the growth in interprofessional education and teams. The goals, scope, and team member roles of a nutrition support team (NST) for administering parenteral (PN) and enteral nutrition (EN) as an example of an interprofessional team are described. The role of interprofessional nutrition teams in diabetes management and education, intestinal rehabilitation, and small intestinal and multivisceral transplant are discussed. New models of care where telemedicine has facilitated interprofessional expert nutrition team management and the integration of nutrition in patient management are addressed. Finally, examples of interprofessional nutrition management at the University of Pittsburgh Medical Center are provided along with a discussion on the roles of different healthcare team members and the synergistic relationships between the nutrition service, small bowel transplant team, and intestinal rehabilitation team.

The Paradigm of Interprofessional, Collaborative Practice

Interprofessional or collaborative practice occurs when "multiple health workers from different professional backgrounds work together with patients, families, and communities to deliver the highest quality of care" and allow "health workers to engage any individual whose skills can help achieve local health goals."[4] Over the past half-century, there has been significant global interest in interprofessional collaboration and team-based healthcare delivery to improve the quality of care, population health, cost of care, patient safety, and patient/provider satisfaction.[5] There have been several influential reports on interprofessional education since the beginning of the 21st century from international health organizations.

In 2010, the World Health Organization (WHO) published The Framework for Action on Interprofessional Education and Collaborative Practice (**Figure 59.1**). The Framework's authors state that "collaborative practice strengthens health systems and improves health outcomes." Health leaders "who champion interprofessional education (IPE) and collaborative practice position their health system to facilitate achievement of the health-related Millennium Development Goals."[4] The report argues that greater emphasis on and investment in IPE is crucial to realizing the benefits of interprofessional collaboration.

The 2010 Lancet Commission report on health professionals for a new century, which examined the 100-year history of health education since the publication of the influential 1910 Flexner report on the education of health professionals, promoted the importance of interprofessional education that "breaks down professional silos while enhancing collaborative and non-hierarchical relationships in effective teams" as one of 10 proposed reforms for health education in the new century.[6] The U.S. Interprofessional Education Collaborative (IPEC) 2011 core competencies for effective interprofessional collaborative practice were updated in 2016 to reflect two major changes that occurred in American health care during the interim: the implementation of the Patient Protection and Affordable Care Act and the increased focus on the Triple Aim goals developed by the Institute for Healthcare Improvement.[7] The four core competencies outlined by IPEC and supported by many health professions focus on values and ethnics for interprofessional practice, roles, and responsibilities of health professionals, interprofessional communication and teams, and teamwork.[7] These are outlined in **Table 59.1**.

Table 59.1 2016 Update to the Core Competencies for Effective Interprofessional Collaborative Practice

1. Work with individuals of other professions to maintain a climate of mutual respect and shared values. (Values/Ethics for Interprofessional Practice)
2. Use the knowledge of one's own role and those of other professions to appropriately assess and address the healthcare needs of patients and to promote and advance the health of populations. (Roles/Responsibilities)
3. Communicate with patients, families, communities, and professionals in health and other fields in a responsive and responsible manner that supports a team approach to the promotion and maintenance of health and the prevention and treatment of disease. (Interprofessional Communication)
4. Apply relationship-building values and the principles of team dynamics to perform effectively in different team roles to plan, deliver, and evaluate patient/population centered care and population health programs and policies that are safe, timely, efficient, effective, and equitable. (Teams and Teamwork)

Reproduced from Interprofessional Education Collaborative. Core Competencies for Interprofessional collaborative practice: 2016 update. Washington, DC: Interprofessional Education Collaborative, 2016.

A landmark 2015 report published by The National Academies of Sciences, Engineering and Medicine, titled "Measuring the Impact of Interprofessional Education on Collaborative Practice and Patient Outcomes" concluded that despite the consensus that the delivery of health care can be improved by integrating IPE into health professions education programs, "definitive evidence linking IPE to desirable immediate and final outcomes does not exist."[5] This is mainly due to a lack of engagement between education and healthcare delivery systems, a conceptual model of IPE intervention and impact, and purposeful and thoughtfully designed studies.[5] Despite this, the report made three major conclusions and two recommendations to "shed additional light on the value of collaboration between educators and practitioners and patients, families, and communities...to improve lives through treatment and palliation, disease prevention, and wellness interventions."[5]

There has been rapid growth in IPE in health professions education curricula. According to a 2021 report from the *American Journal of Pharmaceutical Education*, the percentage of institutions in medicine requiring IPE activities has risen to 92%, and the accrediting bodies for the majority of the health professions require IPE in their curricula.[8] In the healthcare workforce, the importance of interprofessional teamwork is reflected by programs such as TeamSTEPPS. TeamSTEPPS was developed by the *Agency for Healthcare Research and Quality* and the Department of Defense as an evidence-based interprofessional team approach "to improve communication and teamwork skills among healthcare professionals."[9]

The continued emphasis on and growth of IPE is critical to the success of future healthcare professionals because of the myriad of interprofessional teams in practice in the American healthcare system today. These teams specialize in different areas and provide care to patients with complex needs, including those requiring nutrition support. The American Society for Parenteral and Enteral Nutrition (ASPEN) defines nutrition support therapy as the "provision of enteral or parenteral nutrients to treat or prevent malnutrition"[10] to patients who have or are at risk for malnutrition and cannot meet their energy and nutrient needs by diet alone.[10] This includes patients with compromised gastrointestinal tract function due to various causes, including diseases of the GI mucosa (e.g., celiac, inflammatory bowel disease) that impair absorption, inadequate motility of the intestines and surgical alterations of GI anatomy. The prevalence of malnutrition among hospitalized patients depends on many factors and has been reported to vary from 8.9% in the 2018 Healthcare Cost and Utilization Project (HCUP)[11] to as high as 30% to 60%.[12-14]

Traditional Nutrition Support Teams (NST)

History

In the early to mid-1970s, the concept of a specialized, multidisciplinary team for the management of nutrition support gained popularity in North America.[15] The impetus for developing hospital-based nutrition support teams (NSTs) stemmed from the widespread use of PN in the care of patients with complex medical and surgical conditions, where compromised GI function prevented the use of routine enteral nutrition (EN). Given the complexity of PN administration, there was a concern that PN would be associated with higher rates of infectious, metabolic, and mechanical complications that would increase morbidity, mortality, and costs. The purpose of a specialized NST was to train a team of physicians, dietitians, pharmacists, and nurses with knowledge of EN and PN and to harness the multidisciplinary knowledge of its members to standardize treatment protocols for procedures. Such procedures included those needed for safe and effective PN delivery, including central catheter placement and dressing changes, assessment of energy and nutrient needs, ordering proper amounts and components of formulations, and using appropriate laboratory tests with testing frequency and acceptable result ranges. This interprofessional team with expertise in PN understood common complications and how to avoid and manage them.

The development of NSTs was highly successful in the late 1970s and early 1980s and led to improved outcomes, including reduction of morbidity and mortality and cost savings in hospital care.[16,17] However, a shift in emphasis toward cost-cutting in the late 1980s and early 1990s resulted in the reduction and ultimate demise of NSTs in several large academic and community medical centers. According to John Wesley, former president of ASPEN, "the rationale was that nurses, dietitians, and pharmacists working in various parts of the hospital where patients receive nutrition support were already conversant with all of the issues associated with the delivery of sophisticated and safe nutrition support and that a specialized team was no longer needed."[18] Recent studies have demonstrated the clear benefits of NSTs in patient care. Reber et al., studied the efficacy and efficiency of NSTs. They found that NSTs are associated with decreased complications, reduced mortality, improved monitoring and dietary intake, and the use of PN for the correct indications.[19] Stidham et al., conducted a systematic review examining PN usage in hospitalized adult patients from 2004 to 2020; their findings indicated that NSTs were associated with decreased incidence of inappropriate PN use.[20]

While designated NSTs are less prevalent now, there has been an increase in other types of interprofessional teams involved in the nutrition management of patients. With the rapid advancements in modern medicine, especially in trauma, surgery, and oncology, there are increasing numbers of patients with medically complex conditions who are living longer, due to new drug therapies and surgical technology advancements that were not seen 50 years ago when NSTs first gained prominence. These patients often require significant interprofessional support, including nutrition therapy. While many different types of interprofessional teams involved in patient nutrition care exist in the healthcare system today, the goals, scope, and operation of the traditional NST provide a historical and foundational model for thinking about a team-based approach to nutrition care.

Team Goals

Historically, the primary goals of an NST were to 1) identify patients in need of nutrition support who could benefit from such services, 2) assess the nutrition status of such patients, and 3) provide safe and effective nutrition support.[21]

Over the last 3 decades, the goals of an NST have expanded to include "minimizing, auditing, and managing complications of EN and PN, minimizing resource costs of nutrition-related complications, and providing psychological support for patients on PN and their caregivers."[22] This is particularly important for patients on long-term home PN, as it may affect their quality of life and impact their ability to attend work or school. Teaching is an important aspect of a well-functioning NST. This includes training other providers on screening and assessing patients for risk and presence of malnutrition and treating malnutrition. Similarly, given the constantly evolving nature of medicine, it is advisable for the NST to periodically review methods and search systematically and periodically for improvements in PN safety and efficacy. To accomplish this, additional goals of an NST are to collect data on the effectiveness and side effects of EN and PN, conduct research, and develop clinical guidelines.

Finally, one of the goals of the NST must be to consider the institutional and broader healthcare environment in which it exists and to demonstrate its value within that system. NSTs have been challenged to justify their existence since their inception. In an expert review, Barrocas stated that the "rise or demise" of NSTs will depend on many factors, but it is vitally important to understand what those who "control the purse strings are seeking in terms of salutary cost and quality outcomes in the current value-based system."[23]

Scope and Operation of the NST

The types of patients seen by an NST can vary among institutions but generally includes patients considered for PN and EN support.[24] The same team will manage both PN and EN. Generally, there are separate teams for adult and pediatric populations. The team is responsible for the daily monitoring of patients and team rounds. In addition, a successful NST will provide training and educational programs, such as lectures, for other medical, nursing, diet and nutrition, and pharmacy staff, and community education programs, and develop educational materials for professional and patient use.[23] The team must also aid in developing and monitoring strict adherence to hospital guidelines, policies, and procedures for PN and EN therapy to maintain the standards of care. The NST participates in discharge planning for patients on the service, including patient education and training, referrals for posthospital care, such as extended care facilities, and ambulatory care follow-up.[23]

In terms of operation, the NST may have complete autonomy over a patient's nutrition care, or function as a consultation service in which it advises the primary medical team on nutrition management. There are pros and cons to both approaches. Functioning as a primary team, the NST has total control to assess and prescribe the nutrition and fluid requirements of the patient, but this may not be practical, especially with many patients. As a consult service, the NST may see more patients. However, effective care delivery requires good communication with the primary team, who will ultimately need to act on the recommendations.

Team Roles

The members of an NST can vary between institutions and consist of professionals from different specialties and professions. At its core, the NST should have a physician, a registered dietitian (RD), a pharmacist, and a nurse.[15,25] While no specific certification is required to practice as part of an NST, the Certified Nutrition Support Clinician (CNSC®) certificate is a widely recognized and accepted nutrition support credential that all members of the NST can obtain. As an interprofessional team, it is important to note that while each of the professionals has a role to play, there is significant overlap in knowledge, skills, and practice between the providers from these different professions. The transition from silos of roles to a knowledge-sharing, interprofessional, and collaborative approach improves patient care.

The physician normally has overall responsibility and authority for all patient care rendered by the NST. To do so effectively, they must be knowledgeable in the pathophysiology and clinical manifestations of malnutrition, the different options for nutrition interventions and their potential complications, and nutrient metabolism in health and disease. Gastroenterologists with expertise in intestinal failure and nutrition are the prototypical physician leader of the NST, but experts in inflammatory bowel disease have also taken on leadership roles in this context. Surgical leadership of the NST is also well represented, as surgical nutrition expertise and expertise with intestinal transplantation will lend itself naturally to this role.

With the increasing number of nurse practitioners (NPs) and physician assistants (PAs) entering the healthcare workforce, teams may also include these healthcare professionals in a physician extender role. The complexity of the patient such as those with intestinal failure, and the administration of PN, lends itself to developing NP/PA providers who are experts in this role. On-the-job experience and training with the daily supervision of PN administration typically prepares NP and PA providers to function at extremely high levels.

The multifaceted role of the RD is a central component of the NST. It is multifaceted and has advanced considerably since the inception of NSTs.[26] Dietitians possess a deeper understanding of food and nutrition science and how this contributes to health, disease, and recovery. The RD is typically the expert on the NST in core concepts of nutrition, including energy and nutrient intake and needs, and the ultimate transition to EN and oral feeding. Thorough nutrition assessment of the patient and documentation of nutrition diagnoses, the RD makes recommendations and, in some institutions, orders for the route of feeding, energy, and macro and micronutrient needs. They will also carefully monitor and evaluate the patient's response to nutrition therapy using laboratory results, fluid status, and nutrition-focused physical exam, and adjust the nutrition prescription accordingly. In many hospital settings, RDs with expertise in parenteral support function with pharmacists to provide daily nutrition support, particularly in institutions that lack designated physician supervisors.

The role of the pharmacist involves working with the primary physician and the RD to optimize the composition of enteral and parenteral formulations. Because PN bypasses the gut and its defense system against food's microbial and toxic elements, PN formulations should be prepared with a similar level of care required in drug preparations. The pharmacist will work together with the nurse and RD to ensure the safe delivery of PN. In addition, the pharmacist has the greatest amount of training in pharmacokinetics, drug metabolism, drug-drug, and drug-nutrient interactions. Pharmacists often advise on the compatibility and stability of drug/nutrient interactions after carefully reviewing patient medications with PN formulations.

The NST nurse specialist works closely with all other members of the NST. Their roles include teaching and training the patient or caregiver in the care of central venous catheters (CVCs), tubes, stomas, home PN administration, and recognizing and managing complications of CVCs. The nurse specialist will have the most contact with the patient and can relay the daily realities of the patient from first-hand observations of patient reactions to disease and treatment to the rest of the team. In some instances, they may also serve as a case manager and assist the patient in all aspects of their nutrition care, including communication with the insurance company, coordinating transport, etc.

The roles and responsibilities of the NST members can be formalized by creating collaborative practice agreements among the supervising physician and the various team members. This process will typically depend on the local, governing State Medical Board. NPs and PAs will typically function under the license of a supervising physician. Collaborative Practice Agreements can also be created for pharmacists and other providers.

Other Interprofessional Teams in the Management of Patient Nutrition

While the NST is an example of an interprofessional team approach to providing PN and EN, many other potential interprofessional teams are involved in managing patient nutrition in health care today. These include teams focused on diabetes management, transplant, intestinal rehabilitation, and others. Similar to NSTs, these interprofessional teams are composed of healthcare professionals from different backgrounds and areas of expertise across multiple specialties to provide the highest level of nutrition care to patients.

Diabetes Management Team

Diabetes is a chronic metabolic disease affecting the body's ability to produce and respond to insulin. The Centers for Disease Control and Prevention (CDC) estimates the current prevalence of diabetes in the United States to be 37.3 million people, or 11.3% of the population,[27] with projections for disease prevalence to increase.[28] As with other chronic diseases, the use of team-based care (TBC) is common in many provider networks to pursue the "Triple Aim" of decreasing costs and improving health outcomes and the patient experience.[29] This is consistent with the shift in health professions education and training from silos of roles to interprofessional practice where overlap in knowledge, skills, and practices among team members allow for collaborative practice. Ideally, the team-based approach will make broad interprofessional expertise available to the patient, resulting in better patient outcomes and experiences and lower costs.

According to the American Diabetes Association (ADA), diabetes management teams may consist of members, including a primary care provider (i.e., physician, NP, etc.), an endocrinologist, an RD, and a registered nurse (RN) or NP with the Certified Diabetes Care and Education Specialist (CDCES) credential, along with an ophthalmologist, a podiatrist, a pharmacist, a dentist, and a mental health professional.[30] Team composition depends on the unique purpose, setting, and resource availability of the institution that it functions within. The five broad principles of effective TBC "include shared goals between the team and the patient, clear roles and expectations for all team members, mutual trust among team members, effective communication, and measurable processes and outcomes."[31]

The goals of diabetes-specific TBC are to facilitate patients' transitions toward self-management and treatment adherence through education; timely and appropriate testing and examinations; promote the adoption of healthy lifestyle choices through diet, physical activity, and smoking cessation; prevent diabetes-related complications; and improve quality of life.[32] The meta-analysis of Levengood et al. revealed that TBC significantly improved patient blood glucose, blood pressure, and lipid concentrations, compared with controls.[32] Miller-Rosales et al., found that patients with access to interprofessional care teams with expertise in different self-management skills reported better experiences with care.[33]

Good diabetes care necessitates adequate blood sugar management, which is inextricably linked to diet and nutrition. A comprehensive team effort is needed to ensure that the patient has an adequate intake of nutrients while controlling overall and especially postprandial glucose concentration. The ADA recommends that all patients with diabetes be assessed and referred for medical nutrition therapy, education in diabetes self-management, and emotional health at the time of diagnosis and then annually or when new complicating factors arise or transitions in care occur.[34] The diabetes-certified RD provides medical nutrition therapy (MNT), which includes nutrition assessment and diagnosis, intervention/treatment, and treatment monitoring with as-needed adjustments to the treatment plan relative to diet and medication management of blood glucose.[35] The RD may also provide instructions for self-monitoring of blood glucose, insulin administration, adjusting diabetes medication, and education/training on continuous glucose monitoring devices and continuous subcutaneous insulin infusions.[35]

The team pharmacist provides important expertise on diabetes medication selection in patients who are both noninsulin and insulin dependent. They will consider the patient's current medications, medication adherence, intolerances and side effects, dietary habits, and nutrition status.[30] Either the RN or RD diabetes care and education specialist can guide patients on the use of continuous glucose monitors, how to give insulin injections, and signs and symptoms of common complications to watch out for. The mental health professional will assist in assessing for and treating mental health conditions such as depression, substance use, suicidal ideation, etc. In addition, they may work with other team members to help patients achieve lifestyle changes such as smoking cessation, weight loss, and increased physical activity. Other specialists, such as dentists, ophthalmologists, and podiatrists, will be focused on screening for and caring for related comorbid conditions and complications of diabetes.[30] A primary care provider, either a physician or an NP/PA provider, oversees this entire effort through a comprehensive diabetes medical evaluation at initial, follow-up, and annual visits and will be the point person for the patient.[30]

Celiac Disease: Nutrition Management Team

Celiac disease, or gluten-sensitive enteropathy, is a chronic, autoimmune, enteropathic, inflammatory disease with a global incidence of about 1%.[36] Individuals with celiac disease develop an inflammatory mucosal injury in response to gluten, the protein in wheat, rye, and barley. First described by the Dutch pediatrician Willem Dicke following World War II, a gluten-free diet is essential to treat celiac disease. The RD will assess the patient's nutrition status, determine their nutrition diagnoses, energy, macro, and micronutrient needs and provide MNT on a gluten-free diet. This includes a diet prescription and diet counseling on gluten avoidance and prescribing any dietary supplements needed to meet the patient's micronutrient needs. In addition, the RD will monitor the patient's response to treatment and adjust the diet accordingly.

Solid Organ Transplant Team

The origins of modern transplantation started in 1869; the Swiss surgeon Jacques-Louis Reverdin[37] performed the first skin transplant. In 1959, the first successful kidney transplant was performed in the United States in Louisiana between individuals who were not genetically identical.[37] By the late 1960s, organ transplantation had become a viable treatment option.[38] Today, the field of transplant medicine is separated mainly into solid organ transplants (i.e., kidney, liver, etc.) and tissue transplants (i.e., bones, skin, tendons, etc.). This section will focus on solid organ transplants and the interprofessional teams that care for these patients.

According to the United Network for Organ Sharing (UNOS), there were 41,354 total solid organ transplants in 2021, which is a 5.9% increase from the year prior, and sets the record for an all-time high.[39] The most commonly transplanted organ is the kidney, followed by the liver and the heart.[39] Interprofessional transplant teams exist in all major transplant centers across the country and typically consist of a transplant physician, surgeon, pharmacist, social worker, dietitian, psychologist, financial counselor, and nursing coordinator.[40,41]

While the team's overarching goal is successful organ transplantation, nutrition therapy is integral to transplant care. More than 20% to 50% of patients with cirrhosis planning to undergo liver transplantation have malnutrition.[42] This can be due to many causes, such as anorexia and cachexia from the catabolic effects of chronic illness, frequent fasting periods for medical exams, diet counseling to avoid high protein sources in patients at high risk for encephalopathy, autonomic neuropathy that alters gustatory sensation, gastroparesis and slowed transit time, etc.[43] Patients needing kidney transplants are similarly affected with malnutrition and muscle mass loss, with estimated prevalence ranging from 11.1% to 49.6% of recipients of kidney transplants.[44] Because malnutrition, as evidenced by sarcopenia, has been linked to increased morbidity and mortality in the transplant population,[45,46] optimizing nutrition status is a key part of the transplant team's work.

The transplant physician and surgeon generally supervise the surgical and medical care before, during, and after the transplant. This includes monitoring nutrition status. Working with the physician and other team members, the RD assesses the patient's nutrition status and determines risk for and presence of malnutrition, determines nutrition diagnosis(es), and provides a feeding prescription for macro- and micronutrient needs and route. The pharmacist works with the RD and physician in "optimizing pharmacotherapy from the intensive care unit to the ambulatory care clinic"[47] and TPN formulations if indicated for the patient. The social worker will provide psychosocial support during the transplantation process, including issues like access to food and affording medications. The psychologist will help the patient with stress and mental health issues, while the nursing coordinator will oversee and assist the patient with all aspects of the transplantation process. It is important to note that each health professional is not exclusively limited to the general role descriptions stated above. Most professions involved in interprofessional teams have overlapping knowledge, skills, and practices and may be involved in all aspects of a patient's care to differing degrees.

Small Bowel Transplant and Intestinal Rehabilitation Team

Intestinal failure is defined as "the inability of the gastrointestinal tract to sustain life autonomously"[48] due to a failure of the small intestine to digest and absorb adequate fluids and nutrients. The many etiologies of small intestine failure include short bowel syndrome (SBS), mucosal abnormalities, and motility disorders that lead to malabsorption. SBS describes an anatomical state where the small intestine is <30% normal or a small bowel length of <75cm in children or <180cm in adults.[49] This can occur after iatrogenic resection or congenitally. Mucosal abnormalities can be due to disorders like inflammatory bowel disease (IBD), infection, celiac disease, etc. Almost all patients with intestinal failure require PN support, which is costly, comes with a host of potential complications, and can lead to a reduced quality of life.[50]

Fortunately, with advances in transplant medicine, small bowel transplants are an additional option for these patients. Like other solid organ transplants, successful small bowel transplants and intestinal rehabilitation require a concerted interprofessional team effort. Team members typically include the transplant surgeon, gastroenterologist(s), RDs, NPs, and pharmacists.[48] The team's goal is to "provide all modalities of intestinal rehabilitation" and, if this fails, offer transplantation to those who develop severe complications of PN.[51] Given that individuals who are candidates for small intestinal transplantation have intestinal failure by definition, they will need PN support in essentially all cases. The ability of surgical small bowel transplant teams to intervene with either bowel lengthening procedures or transplantation to achieve autonomy from PN support typically takes the form of a multimonth transition. The complexity and heterogeneity of this patient population will necessitate an interprofessional team approach, where RDs and pharmacists experienced in PN support work closely with surgeons, gastroenterologists, and associated NP/PA providers to provide routine care and adjustments of PN support during interventions.

The UPMC Experience with Interprofessional Teams for Nutrition

The composition and specific roles of the NST are most commonly defined by the institution it serves. The University of Pittsburgh Medical Center (UPMC) is an integrated healthcare delivery network that includes over 40 hospitals and over 600 outpatient clinics centered in Pittsburgh, with coverage across western Pennsylvania and neighboring states. This healthcare delivery network includes multiple level 1 trauma centers and solid organ transplant programs (including small intestine and multivisceral transplant), a National Cancer Institute-designated cancer center, and a tertiary referral inflammatory bowel disease center, among the clinical service lines that have used nutrition support most heavily. In response to this clinical need, UPMC has developed multiple NSTs to serve a growing population of patients with malnutrition and intestinal failure. This institutional experience provides a perspective on the growth and expansion of this interprofessional team approach to expert nutrition care to address the changing landscape of complex medical and surgical issues that require specialized nutrition care.

The initial NST developed at UPMC was guided by senior surgeons and hospital leadership, who saw a need for standardization of PN use among surgical teams and inpatients. This initial formation of the NST included a senior gastroenterologist, a pharmacist, critical care RD, and nursing support, who were tasked with overseeing the inpatient use of PN and following these patients in the outpatient setting. Specialty pharmacy support was provided through a new entity, which was a joint effort with UPMC. This early effort grew from the early 2000s to the present time, quadrupling the average number of home PN support patients it manages (**Figure 59.2**). The UPMC hub academic hospital has restricted the use of PN to the dedicated NST, which functions initially as a consultative service.

During the COVID-19 pandemic, the widespread implementation and use of telemedicine allowed the NST at the hub hospital to expand to function as a service line for patients requiring nutrition support throughout the UPMC System, which substantially increased the utilization of this service. Small intestinal and multivisceral transplantation has maintained a smaller independent NST throughout this process, due to the complexity of their patient population and the need for more direct control of nutrition support decision making.

The function of the main UPMC hub NST is to stabilize patients presenting with acute intestinal failure and those for whom EN is contraindicated, with the goal of transitioning these patients to an outpatient setting for management. Acute care and trauma surgery are the primary services requiring the NST, followed by pancreaticobiliary surgery, thoracic surgery, GI oncology surgery, colorectal surgery, transplant surgery, and gastroenterology. The independent small intestinal/multivisceral transplant NST is dedicated to serving a selective patient population eligible for surgical management of intestinal rehabilitation when surgical options become available, including those who are candidates for both small intestinal and multivisceral transplantation and who require nutrition support during their pre- and post-transplant care.

There is constant crosstalk between the two NSTs. The UPMC hub NST receives referrals for all patients requiring nutrition support throughout the UPMC system and will medically manage issues requiring both acute and long-term intestinal rehabilitation. Patients who are eligible for complex surgical management will subsequently transition to the small intestinal/multivisceral transplant team. A core group of 10 RDs rotate on the inpatient NSTs and provide outpatient management. This core group is supported by six NP/PAs, two inpatient pharmacists, two outpatient specialty pharmacy pharmacists, and four full-time nurse clinicians. Three gastroenterologists and three transplant surgeons supervise care. The role of each professional on the team is as described in the traditional NST section. Each professional's overlapping layers of expertise allows for comprehensive patient care. Other medical and surgical services welcome this type of dedicated nutrition support service. This is partly due to the complexity of the patient population requiring nutrition support, who often require multiple admissions.

Normally, patients are admitted to the primary service from which they receive care, pancreaticobiliary, thoracic, GI oncology, colorectal or transplant surgery, or general GI. With a dedicated NST, however, nutrition support patients are no longer shouldered

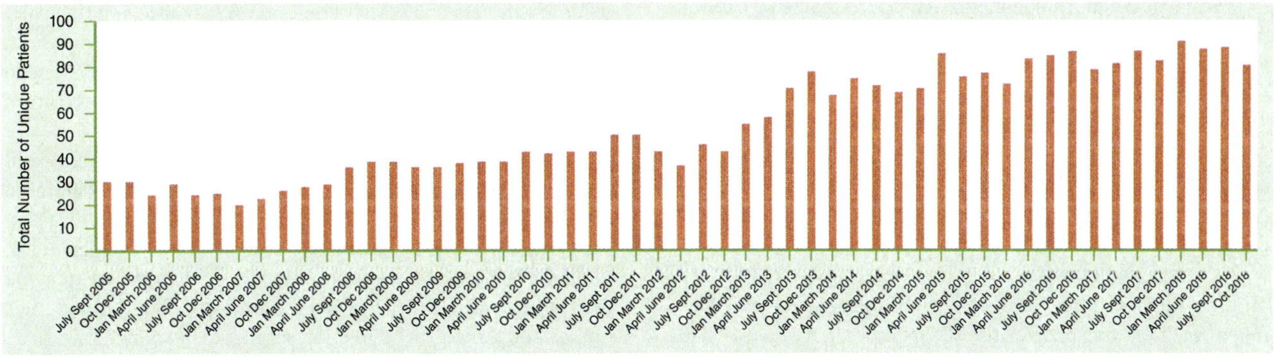

Figure 59.2 Utilization of home parenteral nutrition support in the UPMC Presbyterian-Montefiore hospital during a 14-year time period following the inception of a nutrition support service.

Kuang R., Binion D. G., Cruz R. (2020) Unpublished data.

by individual services. In addition to relieving pressure on other services, the availability of a dedicated NST allows UPMC to more effectively standardize and coordinate nutrition support care, leading to improved patient outcomes. The homegrown expertise at UPMC, and the workflow of the main NST in conjunction with the transplant/rehabilitation NST, allows for the highest levels of nutrition support in patient management.

COVID-19 and Telehealth: Expanding Access for the NST

Telehealth has been broadly defined as "the use of and access to multiple technologies that have the capacity to share information and provide clinical care, education, public health, and administrative services at a distance."[52] This may include videoconferencing, image exchange, electronic monitoring, and wireless communication. Since the onset of the COVID-19 pandemic, telehealth has become a central component of patient healthcare delivery, as many of the barriers to the routine use of telehealth were temporarily lifted to allow for remote care and decentralization of patients congregating in clinics, which might pose a risk of infection transmission.[53] According to the U.S. Department of Health and Human Services (HHS), the use of telehealth by Medicare beneficiaries increased significantly during the pandemic.[54] Previously, telehealth was primarily used for primary care needs and made up only 1% of all visits for Medicare patients.[54] It is now increasingly used for specialized and urgent care delivery,[53] with the greatest growth in behavioral health. This is partly due to the myriad of benefits telehealth brings, including patient convenience, greater scheduling flexibility, remote triaging, electronic intensive care unit monitoring,[55] improved access to information, and reduced healthcare costs.[56]

The efficacy of telehealth-based interventions for chronic illnesses is supported by research. In a single-arm, prospective, longitudinal pilot study, McKenzie et al., demonstrated the utility of using a continuous remote care model for diabetes prevention and normalization of glycemia by delivering nutrition therapy through telemedicine.[57] Kelly et al., found that telehealth interventions can improve patient diet quality through increased fruit/vegetable intake and decreased sodium intake.[58] In a systematic review, Huang et al., found that telehealth interventions of 6 months or longer were effective in lowering the body mass index (BMI) of people with overweight and obesity.[59]

Social distancing guidelines largely spurred the move toward telehealth in nutrition support to prevent the spread of infectious agents during the COVID-19 pandemic. Visits to healthcare providers were delayed or canceled. This included appointments for many patients who received outpatient care for EN or PN, and patients receiving medical nutrition therapy for diagnoses such as obesity, cancer, or malnutrition. For these patients, telehealth can keep patients and providers safe while allowing for continuity of care.

Current uses of telehealth in nutrition support include one-on-one outpatient appointments with interprofessional team members, including physicians, RDs, and other providers.[60] This type of visit provides greater convenience for patients and providers. Patients save time traveling to and from appointments while providers are free from waiting for in-person registration and room turnover processes. It also allows geographically isolated patients who lack provider resources to access specialty expertise, especially for services usually performed in large academic centers.

In a systematic review and meta-analysis, Marx et al. found that telehealth was effective in managing malnutrition-related interventions with older adults living at home.[61] Providers can also gain valuable information from visualizing the patient's home environment. This is especially relevant for patients receiving PN, as it allows healthcare professionals to see how supplies and equipment are set up and stored at home and, if needed, assess potential improvements to in-home care to prevent central line-associated bloodstream infections and other risks for readmission. In addition, telehealth allows for easier coordination of group visits,[60] such as bowel transplant multidisciplinary clinics where interprofessional team members from different specialties may come together to discuss care, including nutrition support. As interprofessional clinics have been shown to improve treatment outcomes[62,63], telehealth is a convenient option for patients and providers in these settings.

Potential limitations of telehealth include patient access to technology, specifically high-speed broadband service and good connection to a wireless network, and organizational and bureaucratic difficulties.[56] Although most U.S. adults have broadband Internet service at home, it is unevenly distributed. Racial minorities, individuals in rural areas, and those with lower incomes are less likely to have broadband services,[64] contributing to greater healthcare inequity. In addition, pediatric and geriatric populations and individuals who are hearing or sight impaired may require greater instruction and adaptive services to use telehealth services. There may be regulatory barriers and insurance reimbursement challenges with telehealth.[52] Although the broad adoption of telehealth services during COVID-19 has provided some relief from these specific barriers, they merit consideration.

The COVID-19 pandemic continues to progress and change, with new variants discovered periodically. While social distancing restrictions have relaxed, COVID-19 is just one of many new viruses within a larger pattern of rising infectious outbreaks over the past few decades.[65] It is unknown what the future holds, but telehealth is here to stay.

Conclusions

The increasing complexity of medical and surgical challenges facing patients and their healthcare providers at present has created a growing need for nutrition experts to be integrated into patient care. The NST structure, effectively combining the expertise of RDs, pharmacists, and medical and surgical providers, is a well-established and successful interprofessional team paradigm for achieving state-of-the-art nutrition guidance in the care of patients requiring specialized EN and PN.

There are multiple pressures facing medicine that benefit from nutrition expertise. The growing geriatric population and the parallel growth in the demand for successful medical and surgical approaches in caring for patients with complex conditions have provided a strong rationale for expanding nutrition in routine care. The NST is an efficient, cost-effective, and, most importantly, highly effective interprofessional expert care structure to meet this increasing demand. With nutrition education still lagging in most medical education curricula, there is a need to reaffirm the importance and need for these expert teams now more than ever.

References

1. National Research Council Committee on Nutrition in Medical Education. Nutrition Education in U.S. Medical Schools. Washington, DC: National Academies Press, 1985.
2. Van Horn L, Lenders CM, Pratt CA, et al. Adv Nutr 2019;10: 1181–1200.
3. Crowley J, Ball L, Hiddink GJ. Lancet Planet Health 2019;3: e379–e389.
4. World Health Organization Framework for Action on Interprofessional Education and Collaborative Practice (No. WHO/HRH/HPN/10.3). Geneva, Switzerland: World Health Organization, 2010.
5. Institute of Medicine; Board on Global Health; Committee on Measuring the Impact of Interprofessional Education on Collaborative Practice and Patient Outcomes. The National Academies Press, 2015.
6. Frenk J, Chen L, Bhutta ZA, et al. Lancet 2010;376:1923–58.
7. Interprofessional Education Collaborative. Core Competencies for Interprofessional Collaborative Practice: 2016 update. Washington, DC: Interprofessional Education Collaborative, 2016.
8. Brennan LF, McBride A, Akinola M, et al. Am J Pharm Educ 2021;85:848116.
9. Agency for Healthcare Research and Quality, Rockville, MD. About TeamSTEPPS. Accessed November 19, 2022. Available at www.ahrq.gov/teamstepps/about-teamstepps/index.html
10. ASPEN. What Is Nutrition Support Therapy. Accessed October 20, 2022. Available at www.nutritioncare.org/About_Clinical_Nutrition/What_Is_Nutrition_Support_Therapy_/
11. Guenter P, Abdelhadi R, Anthony P, et al. Nutr Clin Pract 2021;36:957–969.
12. Bedock D, Bel Lassen P, Mathian A, et al. Clin Nutr ESPEN2020;40:214–219.
13. van Vliet IMY, Gomes-Neto AW, de Jong MFC, et al. Nutr 2020; 77:110814.
14. Correia MITD, Perman MI, Waitzberg DL. Clin Nutr 2017;36: 958–967.
15. Wesley JR. Nutr Clin Pract 1995;10:219–28.
16. Faubion WC, Wesley JR, Khalidi N, et al. JPEN J Parenter Enteral Nutr 1986;10:642–5.
17. Trujillo EB, Young LS, Chertow GM, et al. JPEN J Parenter Enteral Nutr 1999;23:109–13.
18. Wesley, John R. Introduction. Nutrition Support Teams: An Anthology. Silver Spring, MD: ASPEN 2004;VII–VIII.
19. Reber E, Strahm R, Bally L, et al. J Clin Med Aug 22;8:1281.
20. Stidham MA, Douglas JW. JPEN J Parenter Enteral Nutr 2020;44:1447–1460.
21. Hamaoui E. JPEN J Parenter Enteral Nutr 1987;11:412–21.
22. Staun M, Pironi L, Bozzetti F, et al. Clin Nutr 2009;28:467–79.
23. Barrocas A. Nutr Clin Pract 2019;34:806–821.
24. Nightingale J. Frontline Gastroenterol 2010;1:171–177.
25. McShane CM, Fox HM. JPEN J Parenter Enteral Nutr 1985;9: 263–8.
26. Corrigan ML, Bobo E, Rollins C, et al. Nutr Clin Pract 2021;36(6):1126–1143.
27. Centers for Disease Control and Prevention. National Diabetes Statistics Report. Accessed October 20, 2022. Available at www.cdc.gov/diabetes/data/statistics-report/index.html
28. Deshpande AD, Harris-Hayes M, Schootman M. Phys Ther 2008;88:1254–64.
29. Berwick DM, Nolan TW, Whittington J. Health Aff (Millwood) 2008;27:759–69.
30. Draznin B, Aroda VR, Bakris G, et al. Diabet Care 2022;45: S244–53.
31. Wynia MK, Von Kohorn I, Mitchell PH. JAMA 2012;308:1327–8.
32. Levengood TW, Peng Y, Xiong KZ, et al. Am J Prev Med 2019;57:e17–e26.
33. Miller-Rosales C, Rodriguez HP. J Am Board Fam Med 2021;34:151–161.
34. Powers MA, Bardsley J, Cypress M, et al. Clin Diabetes 2016;34:70–80.
35. Academy Quality Management Committee. J Acad Nutr Diet 2018;118:132–140.e15.
36. See J, Murray JA. Nutr Clin Pract 2006;21:1–15.
37. Nordham KD, Ninokawa S. Baylor University Medical Center Proceedings 2021;35:1, 124–128.
38. Schlich T. Lancet 2011;378:1372–3.
39. UNOS. All-Time Records Again Set in 2021 for Organ Transplants, Organ Donation from Deceased Donors. Accessed October 20, 2022. Available at https://unos.org/news/2021-all-time-records-organ-transplants-deceased-donor-donation
40. National Kidney Foundation. Your Transplant Care Team. Accessed October 20, 2022. Available at www.kidney.org/atoz/content/txcareteam
41. Klesney-Tait J, Eberlein M, Geist L, et al. Chest 2015;147: 1435–1443.
42. Haj Ali S, Abu Sneineh A, Hasweh R. World J Hepatol 2022;14:1694–1703.
43. Anastácio LR, Davisson Correia MI. World J Gastroenterol 2016;22(4):1513–22.
44. Kosoku A, Uchida J, Nishide S, et al. Sci Rep 2020;10:266.
45. Nolte Fong JV, Moore LW. Front Med (Lausanne) 2018;5:302.
46. Mazurak VC, Tandon P, Montano-Loza AJ. Liver Transpl 2017;23: 1451–1464.
47. Alloway RR, Dupuis R, Gabardi S, et al. Am J Transplant 2011;11:1576–83.
48. Kocoshis SA, Beath SV, Booth IW, et al. J Pediatr Gastroenterol Nutr 2004;39:S655–61.
49. Weale AR, Edwards AG, Bailey M, et al. Postgrad Med J2005;81: 178–84.
50. Sudan D. Gastroenterol 2006;130:S158–62.
51. Weseman RA, Gilroy R. Nutr Clin Pract 2005;20:509–16.
52. Knotowicz H, Haas A, Coe S, et al. Gastroenterol 2019;157: 594–597.
53. Colbert GB, Venegas-Vera AV, Lerma EV. Rev Cardiovasc Med 2020;21:583–587.
54. ASPE. Medicare Beneficiaries' Use of Telehealth in 2020: Trends by Beneficiary Characteristics and Location. Accessed October 20, 2022. Available at https://aspe.hhs.gov
55. Hollander JE, Carr BG. N Engl J Med 2020;382:1679–1681.
56. Hjelm NM. J Telemed Telecare 2005;11:60–70.
57. McKenzie AL, Athinarayanan SJ, McCue JJ, et al. Nutr 2021; 13:749.
58. Kelly JT, Reidlinger DP, Hoffmann TC, et al. Am J Clin Nutr 2016;104:1693–702.
59. Huang JW, Lin YY, Wu NY. J Telemed Telecare 2019;25:389–401.
60. Mehta P, Stahl MG, Germone MM, et al. J Acad Nutr Diet 2020;120:1953–7.
61. Marx W, Kelly JT, Crichton M, et al. Maturitas 2018;111: 31–46.
62. Kozak VN, Khorana AA, Amarnath S, et al. Clin Colorectal Cancer 2017;16:366–371.
63. Erskine KE, Griffith E, Degroat N, et al. Per Med 2013;10: 73–80.
64. Pew Research Center. Internet, Science & Technology. Internet/Broadband Fact Sheet. Accessed October 7, 2022. Available at www.pewresearch.org/internet/fact-sheet/internet-broadband
65. Smith KF, Goldberg M, Rosenthal S, et al. J R Soc Interface. 2014;11:20140950.

PART IV

Prevention and Management of Disease

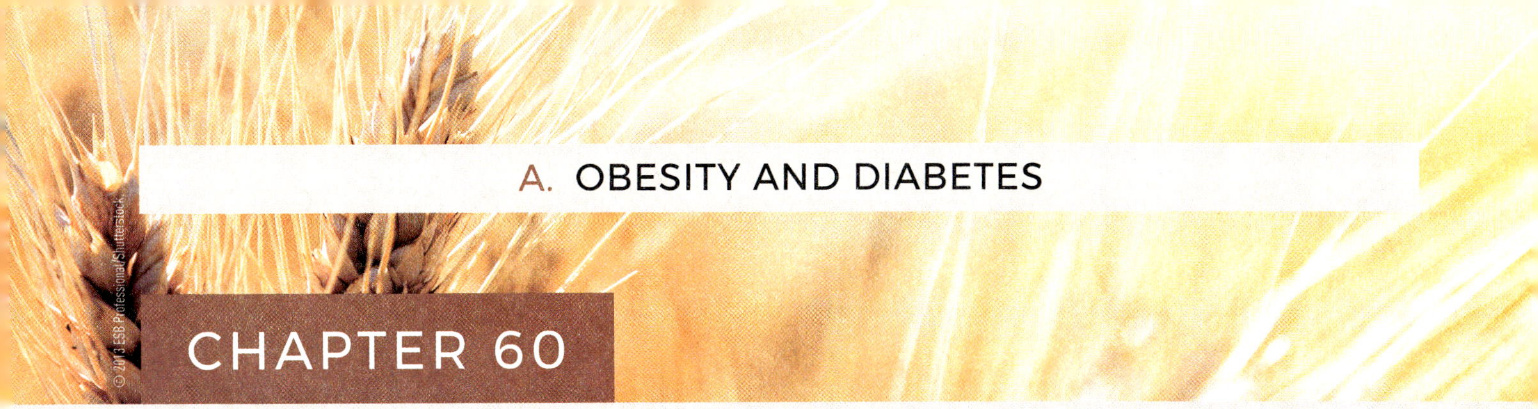

A. OBESITY AND DIABETES

CHAPTER 60

Obesity: Epidemiology, Etiology, and Prevention

Augusto César Ferreira de Moraes, Marcus V. Nascimento-Ferreira, Francisco Leonardo Torres-Leal, Alexandre Chiavegatto Filho, and Deanna M. Hoelscher

CHAPTER OUTLINE

Abbreviations

AD-36, adenovirus-36; **BMI**, body mass index; **BMR**, basal metabolic rate; **HFCS**, high-fructose corn syrup; **MCR4**, melanocortin-4 receptor; **NHANES**, National Health and Nutrition Examination Survey; **PAEE**, physical activity–related energy expenditure; **PWS**, Prader-Willi syndrome; **REE**, resting energy expenditure; **TEE**, total energy expenditure; **TEF**, thermic effect of food; **WHO**, World Health Organization

Introduction

The development of obesity depends on complex interactions among genetic, environmental, and behavioral factors that act on energy balance (i.e., energy intake, energy expenditure, energy storage). Obesity has existed in the population throughout recorded history, but only in more recent generations has prevalence increased to the extent that United States[1] and global[2] public health experts describe trends as an epidemic.[3] The cumulative and challenging increase in the prevalence of obesity over the last 50 years has not only captured the interest of healthcare professionals but also of the media, employers, schools, private industry, and policy makers. Obesity is a major contributor to preventable death in the United States and has been recognized as a chronic disease by the American Medical Association.[4,5] Because of the association between obesity and many other chronic diseases, obesity has become one of the most important public health issues of our time. This chapter provides an overview of obesity development within individuals and societies.

Epidemiology of Obesity

Consequences of an Ancient Metabolism in a Modern World

Genes involved in the regulation of body weight are estimated to have evolved 200,000 to 1 million years ago, at a time when environmental factors controlling habitual physical activity and food acquisition were dramatically different.[2] The consistent increase in the global prevalence of obesity occurred over a period too short for a change in the gene pool, suggesting that factors in the environment are promoting weight gain. Humans evolved in an environment of paucity in which high levels of physical activity were necessary for survival and food acquisition and starvation was a serious threat. Thus, physiologic mechanisms were useful to prevent weight loss but not weight gain. The capacity to conserve and store energy were critical for survival; inefficient energy storage and excessive nonessential physical activity were maladaptive.

Over time, the modern environment changed to one with a range of inexpensive, plentiful, palatable, energy-dense foods and technologic advances designed to decrease physical activity. As a result, our current environment has a strong, sustained propensity to promote positive energy balance and obesity. Our physiology has not developed processes or abilities to oppose these environmental pressures, so the relatively recent increase in the prevalence of obesity in the population can be attributed to a mismatch between our physiology and our environment.

Worldwide Rates of Obesity

The World Health Organization (WHO) estimated that numbers of adults with obesity worldwide have more than doubled from 1980 to 2018, to approximately 650 million.[6] An estimated 38.2 million children younger than 5 years old were classified as overweight worldwide in 2019.[6,7] The relationship between economic development and obesity is apparent. Less-developed countries show increases in the prevalence of obesity as they become more affluent.[8] The WHO categorizes countries according to economic development, and the prevalence of obesity in the population is related to the degree of economic development. As economies grow from "least developed" to "developing" to "economy in transition" to "developed market economy," the obesity prevalence increases along a continuum.[9] Previously viewed as an issue primarily affecting affluent nations, excessive weight gain and obesity are now becoming increasingly prevalent in lower- and middle-income countries, especially within urban environments. In Africa, for example, there has been a 24% increase in the number of overweight children under the age of 5 since the year 2000. In 2019, close to 50% of all under-5 children identified as overweight or obese resided in Asia.[6,7]

Changes in United States: Obesity Rates Over Time

Since the mid-1980s, the prevalence of obesity has increased steadily and markedly in both Westernized and non-Westernized countries. Recent U.S. data suggest that obesity prevalence may be leveling off but remains high. Over two-thirds of U.S. adults have high BMI; as such, they have substantially increased morbidity and mortality from hypertension, stroke, coronary artery disease, dyslipidemia, type 2 diabetes (T2D), sleep apnea, and numerous other conditions.[10-12]

Representative data about changes over time in overweight and obesity in the United States come from the National Health and Nutrition Examination Survey (NHANES). The NHANES program of the National Center for Health Statistics, Centers for Disease Control and Prevention, includes a series of nationally represented, cross-sectional surveys beginning in 1960, enabling an examination of trends over time. Previous national surveys include the National Health Examination Survey (NHES I, 1960 to 1962) and the first, second, and third NHANES surveys (NHANES I, 1971 to 1974; NHANES II, 1976 to 1980; and NHANES III, 1988 to 1994). Beginning in 1999, NHANES became a continuous survey with data released in 2-year cycles. Data for the first 12 years of the continuous NHANES (1999 to 2010) were published in 2012.[10]

NHANES data reveal that the prevalence of obesity was relatively constant from 1960 to 1980 and then increased in the 1990s. The 2000s saw significant increases in obesity.[12] Up to 2010, a significant increase in obesity was noted mainly among men, African American women, and Mexican American women.[12] Approximately 68% of US adults had overweight (BMI 25 to 29.9) or obesity (BMI ≥30) in the 2007 to 2008 period, 12% higher than the overweight and obesity estimates obtained from the 1990s and 21% higher than in the 1970s.[13] Among adults aged 20 to 74 years, the estimated prevalence of obesity (BMI ≥30) doubled between NHANES II (1976–1980) to NHANES 1999 to 2000, from approximately 15% to 31%, with further, though less dramatic, increases since then.[14]

Self-reported data from the Behavioral Risk Factor Surveillance System showed that the prevalence of clinically severe obesity in the United States increased more than the prevalence of mild obesity between 1986 to 2000.[13] During this period, the prevalence of a BMI of 40 or greater quadrupled, from 1 in 200 adults in the United States, to 1 in 50; the prevalence of a BMI of 50 or greater increased by a factor of 5, from approximately 1 in 2000 to 1 in 400. In contrast, obesity doubled during the same period, from approximately 1 in 10 to 1 in 5. Thus, increases in prevalence of overweight and obesity may underestimate the consequences for the healthcare system because obesity-related comorbidities are much higher among persons with severe obesity.

Sociodemographic Characteristics and Obesity Rates

Sex and Race/Ethnicity

Overweight and obesity prevalence vary greatly across racial and ethnic groups. In data from NHANES 2017,[11] the prevalence of obesity for adults aged ≥20 years, adjusted for age, was 41.9%, while severe obesity stood at 9.2%. Obesity prevalence was highest in adults identifying as non-Hispanic Black (49.9%), followed by adults identifying as Hispanic (45.6%), non-Hispanic White (41.4%), and non-Hispanic Asian (16.1%).

Obesity affects both males and females, although with some notable differences. Body fat distribution differs between the sexes, with males more predisposed to visceral (abdominal) obesity.[16] In NHANES data from 2017 to March 2020, more males than females were overweight, but more females than males had obesity.[11] The sex differences vary with race and ethnicity. While obesity rates were similar in males (43.1%) and females (39.6%) identifying as non-Hispanic White, and in males (45.3%) and females (45.7%) identifying as Hispanic, obesity prevalence was markedly higher in females identifying as African-American women (57.9%) than in males identifying as African-American men (40.4%).

Among U.S. males, obesity prevalence from 2017 to 2020 was lowest among those identifying as non-Hispanic Asian (17.6%) and highest among those identifying as Hispanic (45.2%). (**Figure 60.1**)[11] Among females, those identifying as non-Hispanic Black had the highest prevalence of obesity

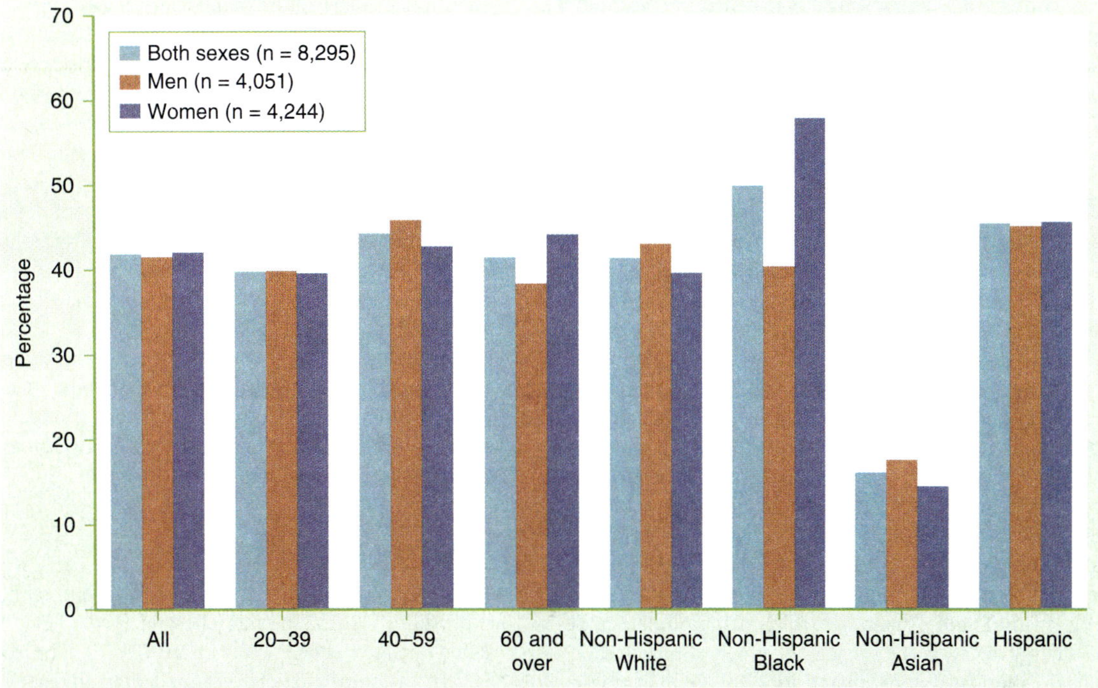

Figure 60.1 Prevalence of overweight and obesity for adults aged 20 years or older by age, sex, and racial or ethnic group: United States 2017–2020 (pre-pandemic).

Data from Stierman B, Afful J, Carroll MD, et al. (2021). National Health Statistics Reports No.158; 1-21. http://dx.doi.org/10.15620/cdc:106273

(57.9%), while those identifying as non-Hispanic Asian (14.5%) had the lowest prevalence. Among younger males (2 to 19 years), obesity prevalence was higher in those identifying as Hispanic compared with non-Hispanic White or non-Hispanic Black.[9] Younger females (aged 2 to 19 years) identifying as non-Hispanic Black had the highest prevalence of obesity compared with those in other NHANES race and ethnic groups.[11]

Socioeconomic and Environmental Factors

In the United States, an inconsistent relationship exists between social determinants of health (SDOH) and obesity.[17] The SDOH framework groups individual SDOH into distinct domains: healthcare access and quality, education access and quality, social and community context, economic stability, food insecurity, and neighborhood and built environment.[17] Among U.S. adults participating in the National Health Interview Survey (NHIS) between 2013 and 2017, the prevalence of obesity was higher for each incremental quartile of SDOH burden. After adjusting for relevant factors, individuals in the highest quartile of SDOH burden had 15% higher relative prevalence of overweight, 50% higher prevalence of obesity class 1 and 2, and 70% higher prevalence of obesity class 3, compared with those in the lowest quartile. These findings highlight the impact of SDOH on weight-related health outcomes and emphasize the need to address social factors in efforts to lower the prevalence of obesity.

Childhood Obesity

Obesity among schoolage children and adolescents, defined as BMI for age ≥95th percentile, tripled beginning in 1980, but plateaued from 1999–2000 to 2007–2008.[18] Recent data from NHANES 2017–2020 show increasing prevalence of obesity with age; 12.7% of children <2 to 5 years, 20.7% of children 6 to 11 years, and 22.2% of children 6 to 19 years, were living with obesity.[11] Researchers have estimated that 30% of adults became obese during childhood, and that approximately 80% of obese adolescents become obese adults.[19] Obesity-related disorders, including T2D, hypertension, gallbladder disease, hyperlipidemia, orthopedic complications, sleep apnea, and nonalcoholic steatohepatitis, are now seen with increasing frequency in the pediatric population.[20]

Obesity as a Disorder of Energy Balance

Obesity can develop only because of an imbalance between energy intake and energy expenditure (i.e., positive energy balance). Understanding the etiology of obesity requires understanding the complex ways in which positive energy balance can occur. Both energy intake and energy expenditure are influenced by genetics and many factors within the environment in which we live. Furthermore, changes in energy expenditure can influence energy intake and vice versa. Because of this complexity, the development of obesity cannot be attributed simply to excessive energy intake or low energy expenditure.

Energy balance can be illustrated in the following equation, often called the energy balance equation. The equation states that any change in body weight (ΔBody weight) must be caused by a difference between energy intake (E_{in}) and energy expenditure (E_{out}):

$$E_{in} - E_{out} = \Delta\text{Body weight}$$

The first law of thermodynamics states that energy can be neither created nor destroyed. Energy balance in the body occurs when the energy content of food consumed is matched by the amount of energy expended. When intake is lower than expenditure, negative energy balance occurs, and body energy stores are reduced. When energy intake exceeds expenditure, positive energy balance occurs, and body energy stores are increased.

Hypothalamus and Energy Balance

The hypothalamus plays a critical role in controlling appetite and energy balance through complex signaling pathways that involve various hormones and neural circuits.[21] One of the key hormones involved in appetite and energy balance regulation is leptin, which is produced by fat cells and acts on the hypothalamus to suppress appetite and increase energy expenditure. When leptin concentration is high, the hypothalamus signals a decrease in hunger and an increase in metabolism.[22-25]

Conversely, when leptin is low, the hypothalamus signals an increase in hunger and a decrease in metabolism to conserve energy. This feedback loop helps maintain a balance between energy intake and expenditure. The hypothalamus also responds to other hormones and neurotransmitters, such as ghrelin, insulin, and dopamine, which can modulate appetite and energy balance. For example, ghrelin, produced by the stomach, stimulates appetite and food intake by acting on receptors in the hypothalamus.[26]

In addition to hormonal signals, the hypothalamus receives input from neural circuits that respond to external cues, such as the sight and smell of food. These cues can trigger the release of neurotransmitters, such as neuropeptide Y and agouti-related protein, which can stimulate appetite and food intake.[26] Overall, the complex interplay between hormonal and neural signals in the hypothalamus helps regulate appetite and energy balance, and disruptions in these signaling pathways can lead to conditions such as obesity and metabolic disorders.[27]

In individuals with obesity, the hypothalamus may become less sensitive to signals that indicate fullness, such as the hormone leptin, leading to overeating and weight gain.[23] This can create a vicious cycle, as increased body fat can further decrease sensitivity to leptin.[24] Additionally, obesity can cause inflammation in the hypothalamus, which can lead to damage to the neurons that regulate appetite and metabolism. This damage can further disrupt the normal signaling processes that control eating behavior and energy expenditure.[28] In some cases, obesity-related damage to the hypothalamus can lead to conditions such as hypothalamic obesity, characterized by severe, treatment-resistant weight gain, and disrupted metabolic function.[29] This condition can be challenging to manage and may require specialized medical care. Overall, the effects of obesity on the hypothalamus underscore the importance of maintaining a healthy weight and lifestyle to support optimal brain function and overall health.[30]

Components of Energy Balance

Energy Intake

The macronutrient sources of dietary energy are fat, carbohydrate, protein, and alcohol. Humans regulate their food intake in a complex way that is not fully understood. After consumption of food, satiety signals are generated in the periphery from the mouth, from the rest of the gastrointestinal system, and as a consequence of peripheral metabolic processes involved in digestion and absorption of nutrients. Signals from the periphery are monitored by a sophisticated neural system. Many hormones and peptides, deriving both from the periphery and from the central nervous system, appear to be involved in the food intake regulatory system. Neural peptides such as α-melanocyte–stimulating hormone, Agouti-related peptide, neuropeptide Y, and melanocyte-concentrating hormone are known to influence eating behavior.[31] In addition, gut peptides such as cholecystokinin, gastrin-releasing peptide, glucagon-like peptide-1, bombesin, insulin, peptide YY, and ghrelin can also modify feeding behaviors.[31,32] Longer range signals of satiety may be related to body energy stores (e.g., body fat, glycogen). Leptin is a hormone secreted from adipocytes as fat cell size increases and may serve as a signal within the brain to decrease food intake and increase energy expenditure.[33] Circulating nutrients, such as free fatty acids, glucose, and triglycerides, affect food intake as well.[34-36]

Energy Expenditure

Total energy expenditure (TEE) is the sum of resting energy expenditure (REE), thermic effect of food (TEF), and physical activity–related energy expenditure (PAEE). The components of TEE and a comparison of TEE in sedentary and active persons are shown in **Figure 60.2**.

Resting Energy Expenditure. The majority of human energy expenditure is through the body's metabolism at rest (REE). REE comprises 60% to 80% of TEE in most people. REE is the energy required by the body to maintain basic physiologic functions, such as pumping blood, making hormones, and maintaining body temperature. Basal metabolic rate (BMR) is the theoretical minimum level of energy expended by the body to sustain life. REE is the body's energy expenditure measured during rest in the fasted state. REE is slightly (~3%) higher than BMR because of the energy required for arousal. REE is, in general, related to lean

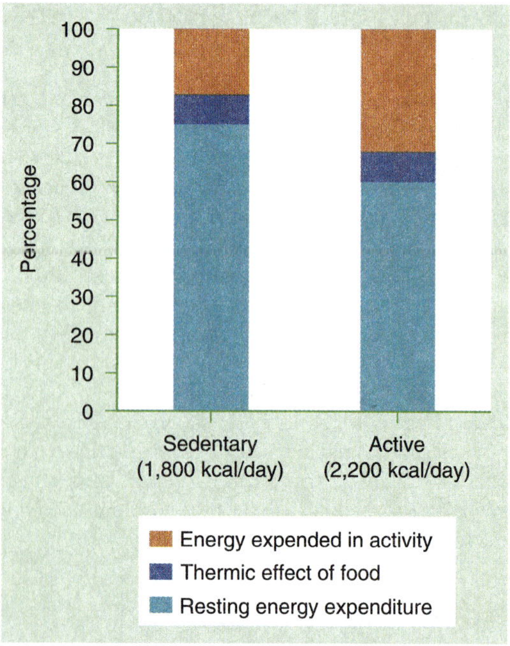

Figure 60.2 Components of total energy expenditure (TEE) in sedentary and active persons.

Table 60.1 Contribution of Different Organs and Tissues to Energy Expenditure

	Weight		Metabolic Rate	
Organ or Tissue	**kg**	**(% of Total)**	**kcal/kg/d**	**(% of Total)**
Kidneys	0.3	(0.5)	440	(8)
Brain	1.4	(2.0)	240	(20)
Liver	1.8	(2.6)	200	(21)
Heart	0.3	(0.5)	440	(9)
Muscle	28.0	(40.0)	13	(22)
Adipose tissue	15.0	(21.4)	4	(4)
Other (e.g., skin, gut, bone)	23.2	(33.0)	12	(16)
Total	70.0	(100)	—	(100)

Data from Matthews DE. Proteins and amino acids. In: Shils ME, Olson JA, Shike M et al, eds. Modern Nutrition in Health and Disease. 9th ed. Burlington: Jones & Bartlett Learning, LLC, 1999:11–48.

body mass, primarily organ and muscle mass. The resting energy requirements of different organs and tissues differ dramatically (**Table 60.1**). In adults, lean organs account for almost 75% of REE, although they constitute only 10% of body weight. Skeletal muscle consumes approximately 20% of REE and comprises 40% of body weight. Adipose tissue normally comprises 20% of body weight but consumes only 5% of REE. REE is generally higher in obese persons than in lean persons because of an increase in lean body mass (organ mass and muscle mass), in addition to the increase in adipose tissue.[37]

Physical Activity–Related Energy Expenditure. PAEE is the component of energy expenditure that is most under voluntary control because it is influenced most strongly by physical activity. PAEE is the most variable component of energy expenditure and can range from 10% of TEE in sedentary persons to 40% of TEE in highly active persons. PAEE includes volitional activities, such as daily activities and exercise, and nonvolitional behaviors, such as spontaneous muscle contractions, maintenance of posture, and fidgeting. Although obese and lean persons expend the same amount of energy during activities in which the body weight is supported, persons with obesity expend more energy than lean persons during weight-bearing activities because of the increased work involved in carrying more weight. Physical activity provides the greatest source of flexibility in the energy expenditure system, and it is the component through which large changes in energy expenditure can be achieved.

Thermic Effect of Food. TEF is the increase in energy expenditure associated with digestion, absorption, and storage of ingested macronutrients, usually 7% to 10% of the total energy content of the meal. The energy cost of a meal is associated with the macronutrient composition of food consumed. The TEF is higher for carbohydrate and protein than for fat. This is because the process for energy storage of ingested fat is very efficient, whereas for carbohydrate and protein, additional energy is needed for conversion to an appropriate storage form (e.g., glucose to glycogen, amino acids to protein). Whether persons with obesity have a systematically lower TEF than lean persons is a matter of controversy. Data suggest that this is the case, but supposed differences are very small and of questionable importance in body weight regulation. If such differences are present, it is not clear whether they existed before the development of obesity, and thus may have contributed to weight gain, or whether they arise as a consequence of the obese state.[38] A reduction in TEF in the person with obesity could be related to the increased insulin resistance and blunted sympathetic nervous system activity that is often associated with obesity.[39]

Energy Stored in the Body. The body stores energy as protein, carbohydrate, and fat. The body has limited storage capacity for both protein (in muscle and organs) and carbohydrate (as glucose and glycogen). The capacity of the body to store fat in adipose tissue depots is virtually unlimited. Because of their high energy density and hydrophobic nature, triglycerides are a better fuel per unit mass than glycogen. A lean adult has approximately 35 billion adipocytes, each containing 0.4μ to 0.6 μg of triglyceride. Triglycerides liberate 9.3 kcal/g when oxidized; by comparison, glycogen stored in the liver and muscle produces 4.1 kcal/g when oxidized. Triglycerides are stored compactly inside the fat cell, thereby accounting for 85% of its weight. Thus, total storage capacity of adipose tissue in lean persons is approximately 80,000 to 130,000 kcal. In persons with obesity, triglyceride stores can increase tremendously because of both increased adipocyte size and increased adipocyte cell number. The total body stores of glycogen and protein (as muscle) in an average 70-kg man are approximately 1,800 and 110,000 kcal, respectively, in a person of normal weight. However, the body can mobilize only approximately half of its protein stores for fuel before life-threatening loss of lean tissue occurs. Thus, adipose tissue represents an effective mechanism for storing fuel and permits survival during periods of food deprivation. The duration of survival during starvation depends on the amount of body fat mass; in lean men, death occurs after approximately 60 to 90 days. In contrast, obese persons have undergone prolonged therapeutic fasts by ingesting only noncaloric fluids, vitamins, and minerals for more than a year without consequences.[40]

Consequences of Energy Imbalance

The average person consumes close to 1 million kcal per year, yet many persons are able to maintain a remarkable state of equilibrium. Even a slight perturbation in energy balance could lead to dramatic weight gain or loss. An error in matching intake to expenditure of only 5% would result in a change of 15 kg over the course of a year. Short-term overfeeding and underfeeding studies suggest that energy expenditure is affected when energy intake is altered. During food restriction, energy expenditure declines, attenuating the loss of body weight that results from the negative energy balance.[41] During overfeeding, some increase in energy expenditure occurs to mitigate the increase in body weight that would occur because of positive energy balance.[42] The changes in energy expenditure are much greater with underfeeding than with overfeeding, suggesting that the body has a strong ability to defend against loss of body weight and a weaker ability to defend against body weight gain.

Negative Energy Balance Results in Weight Loss

An energy deficit of approximately 3,500 kcal is needed to lose 1 lb of body weight. Approximately 75% to 85% of weight lost by dieting is composed of fat, and 15% to 25% is fat-free mass. There is regional heterogeneity in the distribution of fat loss, with loss of subcutaneous adipose tissue generally preceding the loss of muscle mass and visceral fat.[43] Most of the loss in fat is caused by a decrease in lipid content of existing adipocytes; however, long-term fat loss may decrease the number of fat cells as well.

Positive Energy Balance Results in Weight Gain

When energy intake exceeds energy expenditure, a state of positive energy balance is reached, and excess energy is stored in the body. The composition of weight gain during positive energy balance is predominantly fat (~70%–80%), with some gain in lean body mass (20%–30%). Not all excess energy is stored in the body during overfeeding, as efficiency of storage is not 100%. It is generally accepted that efficiency of storage of excess nutrients is somewhere between 60% and 90%.[44,45] Efficiency of storage of excess energy seems to be influenced by characteristics of the individual (i.e., genetics) and the composition of the diet over-consumed. Bouchard et al.[46] found that twins responded with similar amounts of body weight gain when they were overfed, a finding suggesting that genes influence the efficiency of storage during overfeeding. Horton et al.[47] observed that storage efficiency of fat is better than carbohydrates, with storage of excess energy at 90% to 95% for fat, compared with 75% to 85% for carbohydrates.

Etiology of Obesity

Body fat mass is determined by the equilibrium established between energy intake and energy expenditure. Therefore, obesity within an individual is a result of a long-term imbalance between energy intake and expenditure, such that energy intake is either too high for that person's energy expenditure, or energy expenditure is too low for that person's energy intake.[48] These determinants of energy balance (energy intake and expenditure) are themselves influenced by numerous biologic and environmental factors, which makes the etiology of obesity more complex than it may first appear. For example, inter-individual variations in body weight and body composition can be explained, in part, by genetic factors that affect energy balance,[49] but also by environmental influences.

Biological Influences on Obesity

Evidence from family studies shows strong genetic influences on body weight. Rare single gene defects contribute to the development of obesity in some individuals; however, common obesity is likely a highly polygenic and complex disorder. Congenital or acquired medical disorders can also contribute to the development of obesity.

Genetic Heritability of Body Weight

In family studies, BMI is correlated between first-degree family members,[50] and having an overweight parent increases the risk that the children will become overweight.[51,52] In adoption studies, genetic factors account for 20% to 60% of the variation in BMI.[33] The BMI of the biologic, not adoptive, parent is more strongly correlated with the adult weight of the adoptive child.[53] In data from numerous twin studies (>25,000 pairs), genetic factors explain 50% to 90% of the variance in BMI, especially in identical twins,[54] regardless of whether twins were raised separately or apart.[55]

Genetic Disorders Causing Obesity

Susceptibility to obesity appears to be a polygenic trait (related to more than one gene) in most people[56,57]; however, several rare monogenic (single gene) disorders have been identified where severe, early-onset obesity is often the predominant factor. Mutations or deficiency in genes that regulate body fat, body weight, or satiety have been associated with severe, early-onset obesity and include leptin,[58] the leptin receptor,[59] the melanocortin-4 receptor (MCR4),[60] pro-opiomelanocortin (POMC),[60,61] and prohormone convertase 1 (PC1).[62] MCR4 mutations are the most common, accounting for up to 6% of cases.[63]

Pleiotropic syndromes have multiple, widespread effects resulting from a single gene defect. The most common pleiotropic obesity syndrome is Prader-Willi syndrome (PWS), occurring in 1 in 25,000 births.[64] PWS results from an abnormality in chromosome 15q11.2 and produces infantile myotonia, cognitive delay, hypogonadism, overeating, and early-onset obesity.[65] Other obesity syndromes include Bardet-Biedl syndrome, Albright hereditary osteodystrophy, fragile X syndrome, Börjeson-Forssman-Lehmann syndrome, Cohen syndrome, and Alström syndrome.[65]

Genetic Influences on Common Susceptibility to Obesity

Numerous genes have been identified that appear to be associated, either directly or indirectly, with the regulation of body weight.[66] These susceptibility genes likely code for metabolic and hormonal factors that regulate aspects of energy intake, energy use, and energy expenditure. Common variants (or polymorphisms) of these genes could affect individual susceptibility to obesity. Through natural selection, these genes may have become more common because of the evolutionary advantage that they offer by promoting energy storage to survive periods of food deprivation. Within our current environment, these genes are now associated with increased risk of obesity and associated metabolic diseases, such as T2D.[67] Identification of new genes, their role in weight gain, and their interactions with the environment is a rapidly progressing area of research. Polymorphisms within the fat mass and obesity-associated gene (*FTO*) on chromosome 16 became the first to be associated reproducibly with the risk of being overweight or obese in multiple populations.[68] Studies have yet to identify the exact mechanism of action, but they suggest that the FTO gene product may be involved in controlling food intake.

The field of epigenetics, described in Chapter 40 of this textbook, sheds light on some of the impact of genes on the development of obesity. Epigenetics is the study of heritable alterations in gene function, induced without DNA sequence modifications.[69] Epigenetic marks can affect different biologic processes, such as imprinting. Imprinted disorders, such as PWS (discussed above) often include obesity among their clinical characteristics.[69]

Medical Conditions That Can Contribute to Development of Obesity

Numerous medical conditions are associated with obesity. Endocrine disorders associated with weight gain include Cushing's syndrome, hypothyroidism, adult growth hormone deficiency, and polycystic ovarian syndrome.[70] Psychiatric conditions include binge eating disorder,[71] night eating syndrome, and depression.[72] Injury to the ventromedial or paraventricular regions of the hypothalamus or the amygdala leads to hyperphagia and obesity. Iatrogenic weight gain is often caused by certain medications, such as steroid hormones, antidepressants, and antidiabetic agents (**Table 60.2**). Weight gain with these drugs is generally modest, except with high-dose corticosteroids, which can produce true obesity.[70] Recently, semaglutide, a GLP-I (glucagon-like peptide receptor-I) agonist shown to reduce hunger and improve satiety, has been FDA approved for chronic weight management in patients with both high BMI (≥ 30) and weight-related comorbidities.[73] Current research suggests that semaglutide is among the most effective and safe pharmacotherapeutic options for achieving weight loss among currently overweight and obese adults.

Table 60.2 Drugs That May Promote Weight Gain and Therapeutic Alternatives

Drugs That May Promote Weight Gain	Drugs That Are Weight Neutral or Promote Weight Loss
Psychiatric/neurologic medications	Alternative psychiatric/neurologic medications
Antipsychotic drugs	Antipsychotic drugs
Olanzapine, clozapine	Ziprasidone, risperidone, quetiapine
Antidepressants	Antidepressants
SSRIs, TCAs, MAOIs	Bupropion, nefazodone
Antiepileptic drugs	Antiepileptic drugs
Gabapentin, valproate, carbamazepine	Topiramate, lamotrigine
Lithium	Alternatives to steroid hormones
Steroid hormones	Barrier contraceptive methods
Hormonal contraceptives	NSAIDs
Corticosteroids	Weight loss
Progestational steroids	Alternative antidiabetic agents
Antidiabetic agents	Metformin
Insulin	Acarbose, miglitol
Sulfonylureas	Orlistat, sibutramine
Thiazolidinediones	Decongestants, inhalers
Antihistamines	Alternative antihypertensive agents
Antihypertensive agents	ACE inhibitors, calcium channel blockers
α- and β-Adrenergic blockers	
Protease inhibitors[a]	

ACE, angiotensin-converting enzyme; *MAOI*, monoamine oxidase inhibitor; *NSAID*, nonsteroidal anti-inflammatory drug; *SSRI*, selective serotonin reuptake inhibitor; *TCA*, tricyclic antidepressant

[a]May cause weight gain, but less than the drugs they replace.

Data from Aronne LJ, Segal KR. J Clin Psychiatry 2003;64:22–9.

Environmental Influences on Energy Intake

The environment can influence health behaviors that, in turn, affect the amount and composition of the food we eat.[74] Factors related to diet composition, portion size, dietary variety, and cost and convenience of food may all affect energy intake and, thus, have the potential to contribute to positive energy balance and the development of obesity.

Dietary Fat

Diets high in fat have been suggested to increase the risk of obesity. Sedentary animals given unrestricted access to high-fat diets gain weight and became obese, compared with those given unrestricted access to a low-fat diet.[75] Because humans also tend to eat a constant weight of food on both high- and low-fat diets, high-fat diets seem to increase the risk for excessive energy intake.[75] A comprehensive review of the epidemiologic literature on the relationship between dietary fat intake and body weight[76] concluded that, although the data were not entirely consistent, the more dietary fat humans consumed, the greater their body weight. However, more recent analyses unveil only weak evidence associating low-fat diets with weight loss,[77] showing that different types of dietary fats differ in directionality of association with total mortality and weight gain.[78,79] Specifically, polyunsaturated fats and monosaturated fats have been associated with lower risk of total mortality, while saturated fats and trans fats have been associated with higher mortality risk[78] and weight gain.[79] These findings are reflected in the 2020–2025 U.S. Dietary Guidelines (DGA), which recommend that Americans consume healthy oils low in saturated fats (such as olive oil, canola oil, and corn oil) and limit saturated fats (such as those present in full-fat meat and dairy products).[80]

Energy Density

Energy density is the "amount of energy (kilocalories or joules) in a given weight (grams) of food (kcal/g or kJ/g)."[81] Because of the high energy content of fat (9 kcal/g), foods with high fat content have relatively high energy density. Water content reduces energy density because it contributes to the weight of a food, but not its energy. Thus, foods higher in water content (e.g., fruits, vegetables, whole grains) typically have lower energy density. Research suggests that because people tend to eat a consistent weight of food, when the energy density of a food is decreased, total energy intake is reduced.[82] Population studies also suggest that dietary energy density could influence body weight. For example, adults of normal weight report consuming diets with lower energy density than obese individuals,[83] and dietary energy density has been associated with weight gained over time.[84] Clinical trials also suggest that encouraging intake of food lower in energy density improves weight loss.[82]

Sugar-Sweetened Beverages

The rise in obesity over the past several decades has occurred at the same time as an increase in the consumption of sugar-sweetened beverages. Investigators have suggested that sugar-sweetened beverages may promote weight gain because the energy consumed in these beverages is not compensated for in subsequent meals.[85] In a large prospective study, with >50,000 women followed over an 8-year period, higher consumption of sugar-sweetened beverages was associated with greater weight gain over time.[86] The current literature supports that sugar-sweetened beverage intake is a significant contributor to weight gain and obesity.[87-90]

High Fructose Corn Syrup

High fructose corn syrup (HFCS) was introduced into the food supply just before 1970. By 2000, HFCS represented greater than 40% of caloric sweeteners added to foods and beverages and was the sole caloric sweetener in soft drinks in the United States.[91] The marked rise in obesity prevalence from the 1970s to 1990s roughly corresponded to the period in which food processors increased use of HFCS, raising concerns about a link between the two.[91] Fructose is digested, absorbed, and metabolized differently from glucose,[91] and once inside the cell, fructose can enter the pathways that provide the glycerol backbone for triglyceride synthesis more efficiently than glucose.[92] Evidence from epidemiologic studies and longer term randomized controlled trials are inconclusive, and no large, long-term human studies to date have shown a unique link between HFCS and obesity.[93,94]

Portion Size

The 2020 DGA[80] urge people to pay careful attention to portion sizes while making nutrient-dense choices from a variety of foods. Portion sizes in the United States have increased significantly since 1990.[95] Americans associate quantity with quality, so restaurants and food manufacturers are providing larger and larger quantities as evidence of value. As a result, Americans are surrounded by larger portions sizes at relatively low prices. Young and Nestle[96] determined marketplace portion sizes, identified changes in these sizes with time, and compared these marketplace portions with federal standards as defined by the U.S. Department of Agriculture (USDA) and the U.S. Food and Drug Administration (FDA). These investigators found that most marketplace portions exceed standard serving sizes by at least a factor of two and sometimes eightfold. Research suggests that people eat more when they are presented with larger portion sizes.[97,98]

Dietary Variety

Dietary variety has been defined as occurring when a meal or diet is composed of foods that are different in at least one sensory characteristic (e.g., flavor, color, shape).[99] Both animal and human studies show that food intake is greater when there is more variety in a meal or diet, and that greater dietary variety is associated with higher body weight.[99,100] Investigators have suggested that the mechanism behind these findings is related to a process called *sensory specific satiety,* in which the hedonic (pleasurable) ratings of foods that are eaten decrease more than those not eaten, thus satiation (feeling of fullness) occurs only for the foods that have been eaten.[99] Thus, when a greater variety of foods is presented at a meal, it takes longer to become full, and overall, food intake increases. The increased variety of foods available in the current food supply (including numerous highly palatable snacks and desserts) may contribute to the development and maintenance of obesity.

Economic Factors Influencing Diet

The relationships between fat and sugar consumption, energy density of food, and cost of food have several implications for the obesity epidemic in the United States.[101] First, the highest rates of obesity are found among those with the lowest incomes and levels of education.[102,103] Second, an inverse relationship existed between energy density of foods and energy cost (the cost in U.S. dollars per kcal or MJ) such that energy-dense foods with refined grains, fats, and sugars represent some of the lowest cost options available to consumers. In contrast, more nutritious foods, such as lean meats, fish, fresh vegetables, and fruit generally cost more.[101] Third, studies suggest that foods higher in energy density are associated with increases in energy intake. Fourth, lower incomes are associated with spending less money on food, food insecurity, and diets lower in quality.[104,105] These factors suggest that the observed relationship between obesity and socioeconomic status may be related to dietary energy density and energy cost,[106] and that the higher cost of healthier diets may help explain why the highest rates of obesity are found among lower income groups.[106]

Environmental Influences on Energy Expenditure

Trends in Physical Activity Over Time

A review of trends in physical activity since the mid-twentieth century in the United States[107] found: relatively stable or slightly increasing levels of leisure-time physical activity, declining work-related activity, declining transportation activity, and declining activity in the home. Since the mid-twentieth century, the percentage of the work force in high-activity occupations declined, whereas those in low-activity occupations increased.[108] This shift in job categories, combined with increased automation and computer use, has led to a substantial decline in work-related physical activity. In contrast, sedentary activity (i.e., television and computer use) has increased dramatically.[107] The net result is an overall trend of declining total physical activity, putting the majority of the population at high risk of inactivity.[107]

Relationship Between Level of Activity and Body Weight

A low level of physical activity decreases TEE and, unless matched by a decrease in energy intake, causes weight gain. Restriction of activity in rodents produces weight gain, and animals in zoos tend to be heavier than animals in the wild. However, results comparing physical activity levels between obese and lean humans, have been conflicting. These data are confounded by declines in activity as individuals gain weight. However, a decrease in physical activity preceded the increase in the prevalence of obesity in the UK population.[109,110] Cross-sectional analysis of both baseline and follow-up data from NHANES I revealed that recreational physical activity was inversely related to body weight. In NHANES III, men and women in the lowest tertile of activity were almost four times as likely to gain weight.[111] In the Coronary Artery Disease Risk Development in Young Adults (CARDIA) study, maintaining a high level of activity was associated with less gain in BMI over 20 years of follow-up, compared with low activity, after adjustment for race, baseline BMI, age, education, smoking, alcohol use, and energy intake.[111] Although population-based studies of physical activity and obesity have limitations from restricted measurement tools for physical activity, reverse causation, and lifestyle confounding, the data implicate physical activity as a major player in body weight regulation.

Early-Life Environmental Factors

Prenatal Influences

Environmental factors very early in life may affect adult body weight. Exposure to an unfavorable environment either in utero or in the early postnatal period has been hypothesized to cause changes that increase risk of developing disease as an adult, including obesity.[112,113] Epigenetic variation from these environmental influences in utero may be contributing permanent metabolic changes in the children.[69] Several studies have found that adults born small for gestational age were more likely to have higher BMI, higher waist-to-hip circumference ratio, and higher likelihood of metabolic syndrome and coronary artery disease, than those of normal size at birth.[114-117] A meta-analysis of 14 studies estimated that maternal smoking during pregnancy increased the odds that children would be obese later in life (3 to 33 years old) by 50%.[118] Higher maternal BMI at conception[119] and the amount of weight gain during pregnancy both affect the body weight of the children.[120] Children born to women with diabetes also appear to be at increased risk of being overweight as children and adults.[121]

Breastfeeding

Evidence suggests that breastfeeding could protect against obesity.[122-129] A U.S. survey of more than 15,000 adolescents and their mothers found that children who had been only or mostly fed breast milk, compared with formula, were less likely to become overweight, after adjusting for confounders (age, sex, energy intake, time watching television, physical activity, maternal BMI, and other variables reflecting social, economic, and lifestyle factors).[129] Meta-analyses point to a protective effect of breastfeeding on the risk of overweight and obesity in childhood,[124,130] although other cohort studies[131,132] found no significant effect. The reasons for these contradictory findings are unclear but may relate to differences in methods used for determining exposure to breast milk or to assess and adjust for confounders, the selection of different endpoints for measurement of obesity, and the statistical power of the studies.[133] A recent review concluded that breastfeeding may be associated with lower risk of obesity after ages 2 and older, but further well-designed studies that minimize confounders are necessary, to understand the underlying mechanisms and determine a causal relationship between breastfeeding and obesity.[134] Despite uncertainty in the relationship between breastfeeding and obesity, breastfeeding is associated with many other benefits for both the mother and baby[134] and current guidelines recommend breastfeeding infants for at least 6 months.[80,135]

Other Environmental Influences on Obesity

Viruses

Infecto-obesity, obesity induced by an infectious agent, is an emerging field of study. The human adenovirus-36 (AD-36) may be contributing to the obesity pandemic.[136,137] AD-36 may increase adiposity by increasing glucose uptake and decreasing leptin secretion.[138] Human studies, including twin studies, have found positive correlations between AD-36 antibody positivity and body weight.[139,140] However, more studies are needed to establish a causal link.[130]

Toxins

Numerous environmental chemicals with hormone-like activity can induce negative health effects.[141-145] Exposure to these endocrine-disrupting chemicals, such as bisphenol A, organotins, and phytoestrogens, during development, may induce obesity.[146-152]

Smoking Cessation

Weight gain is common when people discontinue smoking. Flegal et al.[153] analyzed data from a national sample of more than 5,000 adults and found that weight gain over a 10-year period associated with the cessation of smoking was 4.4 kg for men and 5.0 kg for women. Smoking cessation increased the odds of obesity by at least twofold.

Sleep

Increasing evidence suggests that short sleep duration and poor sleep health play a role in the development of obesity. The reasons are unclear but may be related to decreased physical activity from a decrease in core body temperature and an increase in fatigue, and increased food intake from changes in hormones mediating hunger and satiety and increased time to eat.[154]

The biologic plausibility of the sleep time effect indicates that the sleep-wake cycle regulates the production and activation of cells responsible for regulating immunity, and that immune system cell blood concentrations are highest in the early evening and lowest in the morning.[155,156] Although the physiologic mechanism of these diurnal variations is unknown, studies have demonstrated[157] that disruption of the normal sleep-wake cycle via sleep deprivation can significantly affect immune system function. These mechanisms suggest that sleep plays an important role in energy homeostasis, and that sleep disturbances increase risk for obesity.[158] Epidemiologic and animal studies supporting these observations have shown an interplay between sleep and body weight regulation.[159] The association between body weight and sleep pattern or architecture[160] appears to be modulated by clock genes (e.g., CLOCK, BMAL1, RORa, REV-ERBa) related to circadian rhythm.[161,162] Humans evolved to predict light and dark cycles over a 24-hour period by developing an endogenous circadian (*circa*: about and *dies*: day) clock, which is synchronized to external time cues.[163]

A review of 71 original studies and seven reviews concluded that (a) short sleep duration is consistently associated with development of obesity in children and young adults, but less consistently in older adults; and (b) significant methodologic issues in the current literature make it difficult to interpret cause and effect.[164] From a public health perspective, the increasing prevalence of sleep deprivation in the general population and its association with the obesity trends merit further research.

Strategies to Reduce Obesity Rates

Behavior, Environment, and Culture

Both increases in energy intake and decreases in physical activity have influenced the high rates of obesity seen in the United States and many other countries. However, these behaviors are influenced by the physical and social environments in which we live.

Prevention vs Treatment

Obesity can be reduced by prevention and/or management. Both strategies can and should be pursued simultaneously but, from a public health point of view, it may be worth considering where to prioritize resources. It will require less behavior change to prevent excessive weight gain than to reverse obesity in those who are already obese.

Resting metabolic rate, TEF, and the energy cost of physical activity all decline with weight loss.[165,166] Increasing amounts of physical activity could counter this drop in energy requirement. Thus, after weight loss, people must either consume less energy or increase physical activity, compared with before weight loss. Hill[167] attempted to quantify the degree of reduction in energy intake and/or increase in energy expenditure required to maintain weight loss. Energy requirements of people formerly with obesity are likely lower than those of similar-sized people who have never been obese. This is because a period of maintaining an obese state may elicit permanent metabolic changes that serve to defend the obese state.[168,169] Work from the National Weight Control Registry showed that successfully reduced obese individuals show greater physical activity than individuals of similar weight who were never obese.[170]

In contrast, prevention of weight gain should require less effort than reversal of obesity. Hill et al.,[74] suggested that most weight gain in adults could be prevented with energy reduction equaling 100 kcal/d or fewer. Other investigators have provided similar estimates from non-U.S. populations.[171,172] In children, excessive weight gain appears to be preventable with energy reduction of approximately 150 kcal/day.[173] A small-changes approach[74] could be a feasible strategy to prevent excessive weight gain in populations at risk for weight gain.[174,175]

Where Should Obesity Prevention Occur?

Obesity prevention efforts will likely be required in many different settings, including home, school, work, and community.[176-179] Most prevention interventions have had limited success in preventing obesity through schools. The difficulty of addressing obesity in schools is illustrated by the Healthy Schools Study.[180] In this multicenter study, researchers addressed both behavior and environment in a comprehensive fashion in 42 schools in sixth through eighth graders. Although the combined prevalence of overweight and obesity declined overall, no difference was noted between control and experimental schools.[180]

Although interest in addressing obesity in the workplace and the home is growing, it is likely that multi-component strategies will be required. The Shape Up Somerville study[181] intervention addressed behaviors and environment in schools, at home, and in the community. Researchers demonstrated a smaller increase in measures of overweight and obesity in the intervention community, compared with a control community.

When Should Prevention Occur?

The optimum time to initiate obesity prevention efforts is debated, but successful efforts may start with young children. Obesity prevention effects are now being evaluated in children 5 years of age or younger. In a randomized controlled trial among infants born to mothers intending to breastfeed, Paul et al.,[182] demonstrated that the combination of helping parents discriminate between hunger-related fussiness and helping overcome rejection of healthy foods through repeated presentation resulted in lower weight-for-length percentiles at 1 year of age.

Addressing the Physical and Social Environments

Even small behavioral changes are difficult to sustain over the long term in a physical environment in which good tasting, energy-dense, inexpensive food is readily available and there is little need or opportunity for physical activity. Many researchers have pointed out the importance of addressing our physical environment to make healthy behavior more sustainable.[74] Although there is ongoing research to identify how to modify the physical environment[183-184] and interest in creating "healthier places," no clear strategy for environmental change currently exists. Christakis and Fowler[185] demonstrated that social networks can influence weight and obesity. Despite the current social pressures that promote weight gain, it may be possible to use social forces and social networks to help in prevention of weight gain and obesity. Given that U.S. society has successfully addressed other social issues such as smoking, recycling, and seat belts, it may be possible to learn from those experiences.[186]

Prediction Models to Prevent Obesity

Predictive models in medicine have the potential to improve patient outcomes by allowing physicians to identify high-risk patients, develop personalized treatment plans, and monitor patients for disease progression.[187] Predictive statistical models that analyze large sets of individual patient data to make predictions about the future development of obesity can also be used to identify patterns and relationships to predict the likelihood of a specific health disease after the onset of obesity.

Early identification of cardiovascular risk factors, including obesity, has important implications for long term health.[188] Nevertheless, complex extrinsic and intrinsic factors, such as sociodemographic, maternal, environmental, behavioral, and anthropometric variables known to be associated with obesity must also be considered.[189] Wang et al.,[190] proposed a nomogram for predicting metabolic syndrome in adults that incorporates reliable noninvasive anthropometric features to facilitate clinical practice or self-risk assessment. An individualized prognostic tool for metabolic syndrome based on statistical models is currently unavailable for children, but researchers have identified the predictive capacity of BMI for general and abdominal obesity.[191-193] The SAYCARE Risk Score, a diagnostic tool to assess risk of developing abdominal obesity in a multicenter study in South American adolescents, incorporates maternal and child anthropometry and key diet and physical activity behaviors (**Figure 60.3**). These authors call for prospective validation studies of such nomograms in other study populations to confirm their predictive ability to diagnose abdominal obesity. Going forward, the development of models to predict population health conditions and the outcome of treatment interventions has an important place in both clinical practice and epidemiologic research.[194]

Machine learning (ML) can be used to predict obesity and related health outcomes by analyzing data, identifying patterns, and making recommendations to improve clinical decisions.[195] In general, machine learning is a set of data-driven tools used to support decision making and has had recent applications in many scientific areas, especially in health care. Supervised learning algorithms, among the most frequent ML applications, learn to predict a specific outcome, such as the incidence of diseases or future patient prognosis.[196] Healthcare providers and researchers are using ML to predict and prevent obesity, providing sophisticated and elegant tools to describe, classify, and predict obesity-related risks and outcomes.[197,198]

Popular ML techniques for structured data have been described in a recent systematic review.[199] Briefly, to develop a ML prediction model, the sample is frequently divided into training and testing sets with a ratio of 70/30. In this process, the model is built from the available data and the hyperparameters are frequently selected using 10-fold cross validation with the GridSearch technique, which allows selection of the best combination of all possible hyperparameters.[198] The predictive performance of the algorithm is frequently assessed using Area Under the Receiver Operating Characteristics (AUROC) on the test set. Feature selection methods capable of handling high-dimensional, heterogeneous, censored data are also used.

A recent systematic review of computerized decision support and ML applications for prevention and treatment of childhood obesity found that ML algorithms, such as decision trees and artificial neural networks, can be used to derive knowledge from large amounts of demographic, clinical, and behavioral data to predict and diagnose childhood obesity.[200] Overall, ML has the potential to revolutionize the prediction, prevention, and management of obesity by providing personalized recommendations and interventions based on individual data and risk factors. However, it is important to note that ML algorithm models are only as good as the training data, and further research is needed to validate the accuracy and effectiveness of ML-based approaches to obesity prediction and prevention.

Summary

The current obesity epidemic can be attributed, in large part, to a "modern" environment that discourages physical activity and encourages the consumption of supersized portions of energy-dense foods. Traditional approaches that have emphasized education and individual responsibility alone are unlikely to be effective. Thus, individuals may know what they must do to prevent weight gain, or to lose weight and maintain weight loss, but may be unable to implement the needed lifestyle changes. Machine learning has the potential to improve the prevention, diagnosis, and management of obesity by providing personalized recommendations, identifying risk factors, and informing public health interventions. Thus, efforts toward the prevention of obesity must both aid individuals in managing their weight and target the environment for change. Successful prevention will require public policy initiatives to enhance safe and easy access for physical activity and facilitate low-energy food choices. Strategies for the prevention of obesity should involve public and private sector partnerships among community leaders, school administrators, employers, healthcare providers, and governmental agencies.

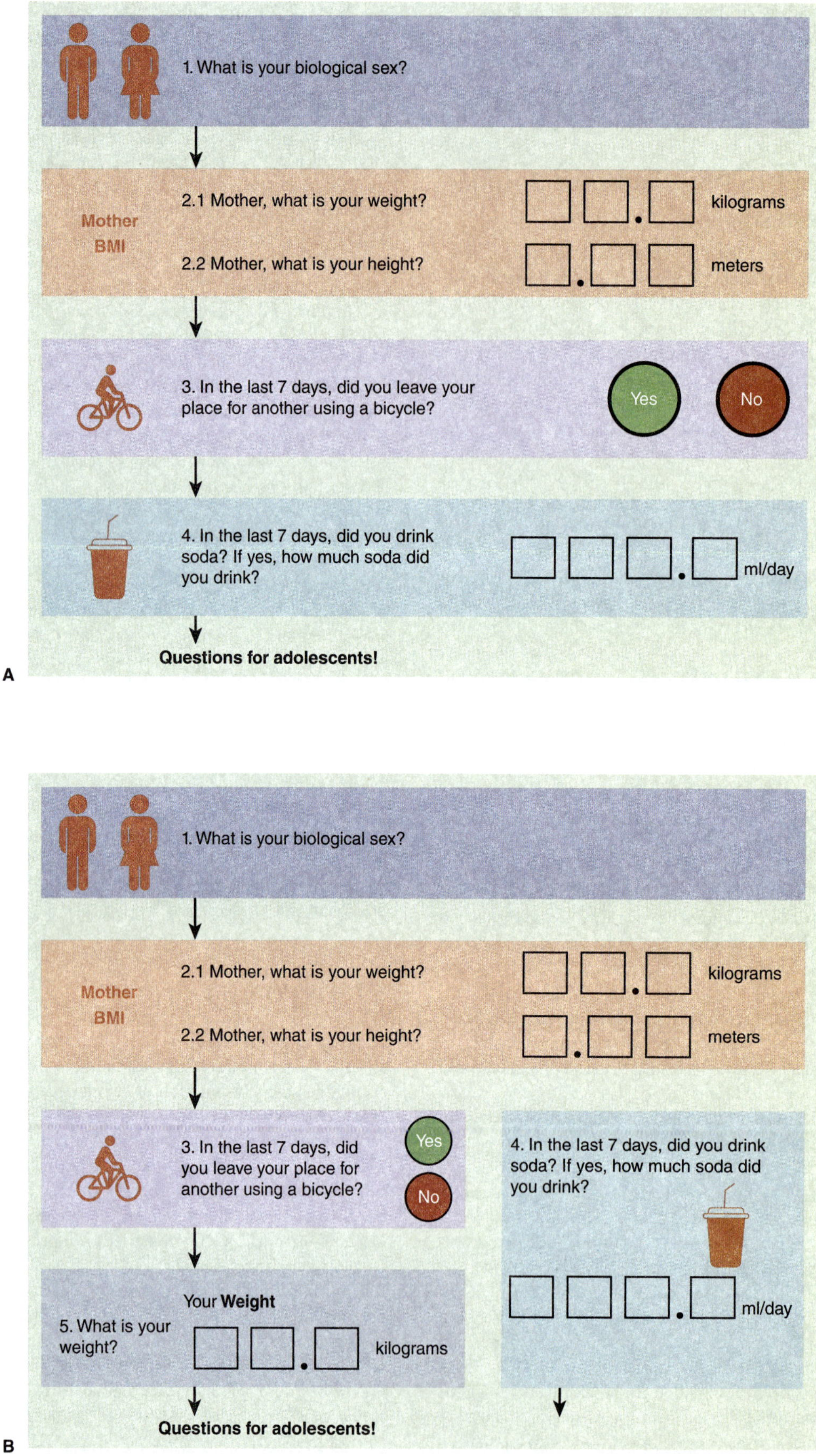

Figure 60.3 South American Youth Cardiovascular and Environmental study (SAYCARE) abdominal obesity score. Clinical risk score level I (A) and level II (B).

Acknowledgments

This work was supported in part by the Start Fund from The University of Texas Health Science Center at Houston (De Moraes); Brazilian Government from National Counsel of Technological and Scientific Development-CNPq; proc. 402391/2021-7 and by the Federal University of Tocantins-PROPESQ Universal No. 088/2022 (Nascimento-Ferreira). We have no disclosures to report.

The authors acknowledge Sarit Polsky, Victoria A Catenacci, Holly R Wyatt, and James O Hill, the authors for their contributions to the previous version of this chapter.

References

1. Hruby A, Hu FB. Pharmacoeconomics 2015;33:673–689.
2. Inoue Y, Qin B, Poti J, et al. Curr Obes Rep 2018;7:276–288.
3. Kyle TK, Dhurandhar EJ, Allison DB. Endocrinol Metab Clin North Am 2016;45:511–520q.
4. American Medical Association. JAMA 2013;420:A–13.
5. Sheikh AB, Nasrullah A, Haq S, et al. Cureus 2017;9:e1435.
6. World Health Organization. Obesity and overweight. Global prevalences. Available at www.who.int/news-room/fact-sheets/detail/obesity-and-overweight
7. Di Cesare M, Sorić M, Bovet P, et al. BMC Med 2019;17:212.
8. Calle EE, Thun MJ, Petrelli JM, et al. N Engl J Med 1999;341:1097–105.
9. Zimmerman R. Clin Fam Pract 2002;4:229–47.
10. Dai H, Alsalhe TA, Chalghaf N, et al. PLoS Med 2020;17:e1003198.
11. Stierman B, Afful J, Carroll MD, et al. National Health Statistics Reports No.158;1–21. Available at http://dx.doi.org/10.15620/cdc:106273
12. Flegal KM, Carroll MD, Kit BK, et al. JAMA 2012;307:491–7.
13. Flegal KM, Carroll MD, Ogden CL, et al. JAMA 2010;303:235–41.
14. Ogden CL, Yanovski SZ, Carroll MD, et al. Gastroenterol 2007;132:2087–102.
15. Sturm R. Arch Intern Med 2003;163:2146–8.
16. Streicher SA, Lim U, Park SL, et al. PLoS One 2023;18:e0279932.
17. Javed Z, Valero-Elizondo J, Maqsood MH, et al. Obesity 2022;30:491–502
18. Ogden CL, Carroll MD, Curtin LR, et al. JAMA 2010;303:242–9.
19. Pi Sunyer F. Obesity. In: Shils M, Olson J, Shike M et al, eds. Modern Nutrition in Health and Disease. 9th ed. Burlington, MA: Jones & Bartlett Learning, LLC, 1999:1395–418.
20. Barlow SE, Dietz WH. Pediatrics 1998;102:E29.
21. Clarke IJ. Compr Physiol 2015;5:217–253.
22. Leinninger GM, Jo YH, Leshan RI, et al. Cell Metab 2009;10:89–98.
23. Crujeiras AB, Carreira MC, Cabia B, et al. 2015 Life Sci 2009;140:57–63.
24. Liu H, Du T, Li C, Yang G. Nutr Metab 2021;18:39.
25. de Vrind VAJ, Rozeboom A, Wolterink-Donselaar IG, et al. Obesity 2019;27:1123–1132.
26. Cui H, López M, Rahmouni K. Nat Rev Endocrinol 2017;13:338–351.
27. Sonnefeld L, Rohmann N, Geisler C, et al. Eur J Endocrinol 2023;188:R37–R45.
28. Jais A, Brüning JC. J Clin Invest 2017;127:24–32.
29. Abuzzahab MJ, Roth CL, Shoemaker AH. Horm Res Paediatr 2019;91:128–136.
30. Della Guardia L, Codella R. Antioxidants (Basel) 2023;12:297.
31. Tritos N, Maratos-Flier E. In: Eckel R, ed. Neuroendocrine control of energy balance. Obesity: Mechanisms and Clinical Management. Baltimore, MD: Lippincott Williams & Wilkins, 2003:128–46.
32. Neary M, Batterham R. Pharmacol Ther 2009;124:44–56.
33. Eckel R. In: Eckel R, ed. Obesity: a disease or a physiologic adaptation for survival? Obesity: Mechanisms and Clinical Management. Baltimore, MD: Lippincott Williams & Wilkins, 2003:3–30.
34. Song Z, Levin BE, McArdle JJ, et al. Diabetes 2001;50:2673–81.
35. Urayama A, Banks WA. Endocrinol 2008;149:3592–7.
36. Wang H, Astarita G, Taussig MD, et al. Cell Metab 2011;13:105–13.
37. Ravussin E, Burnand B, Schutz Y, et al. Am J Clin Nutr 1982;35:566–73.
38. Segal KR, Lacayanga I, Dunaif A, et al. Am J Physiol 1989;256:E573–9.
39. de Jonge L, Bray GA. Obes Res 1997;5:622–31.
40. Stewart WK, Fleming LW. Postgrad Med J 1973;49:203–9.
41. Hill J, Donahoo W. In: Eckel RH, ed. Environmental influences on obesity. Obesity: Mechanisms and Clinical Management. Baltimore, MD: Lippincott Williams & Wilkins, 2003:75–90.
42. Levine JA, Eberhardt NL, Jensen MD. Science 1999;283:212–4.
43. Keys A, Brozek J, Henschel A, et al. Body fat. The Biology of Human Starvation, vol 1. Minneapolis: University of Minnesota Press, 1950:161–83.
44. Diaz EO, Prentice AM, Goldberg GR, et al. Am J Clin Nutr 1992;56:641–55.
45. Jebb SA, Prentice AM, Goldberg GR, et al. Am J Clin Nutr 1996;64:259–66.
46. Bouchard C, Tremblay A, Despres JP, et al. N Engl J Med 1990;322:1477–82.
47. Horton TJ, Drougas H, Brachey A, et al. Am J Clin Nutr 1995;62:19–29.
48. Hofbauer KG. J Int Assoc Stud Obes 2002;26:S18–27.
49. Bray GA, Bouchard C. In: Bouchard C PL, Rice T, Rao DC, eds. Handbook of obesity. Genetics of Human Obesity. New York: Marcel Dekker, 2004:157–200.
50. Bouchard C, Perusse L, Leblanc C, et al. Int J Obes 1988;12:205–15.
51. Agras WS, Hammer LD, McNicholas F, et al. J Pediatr 2004;145:20–5.
52. Agras WS, Mascola AJ. Curr Opin Pediatr 2005;17:648–52.
53. Price RA, Cadoret RJ, Stunkard AJ, et al. Am J Psychiatry 1987;144:1003–8.
54. Maes HH, Neale MC, Eaves LJ. Behav Genet 1997;27:325–51.
55. Stunkard AJ, Harris JR, Pedersen NL, et al. N Engl J Med 1990;322:1483–7.
56. Barsh GS, Farooqi IS, O'Rahilly S. Nature 2000;404:644–51.
57. Boutin P, and Froguel P. (2001) Best Pract Res Clin Endocrinol Metab 15, 391–404.
58. Montague CT, Farooqi IS, Whitehead JP, et al. Nature 1997;387:903–8.
59. Clement K, Vaisse C, Lahlou N, et al. Nature 1998;392:398–401.
60. Farooqi IS. Front Horm Res 2008;36:1–11.
61. Krude H, Biebermann H, Luck W, et al. Nat Genet 1998;19:155–7.
62. Jackson RS, Creemers JW, Ohagi S, et al. Nat Genet 1997;16:303–6.
63. Farooqi IS, Keogh JM, Yeo GS, et al. N Engl J Med 2003;348:1085–95.
64. Goldstone AP. Trends Endocrinol Metab 2004;15:12–20.
65. Farooqi IS. Best Pract Res Clin Endocrinol Metab 2005;19:359–74.

66. Korner A, Kiess W, Stumvoll M, et al. Front Horm Res 2008;36: 12–36.
67. Sabin MA, Werther GA, Kiess W. Best Pract Res Clin Endocrinol Metab 2011;25:207–20.
68. Fawcett KA, Barroso I. Trends Genet 2010;26:266–74.
69. Herrera BM, Keildson S, Lindgren CM. Maturitas 2011;69:41–9.
70. Bray GA. J Fam Pract 2002;4:249–75.
71. McGuire M, Jeffery R, French S. J Fam Pract 2002;4:319–31.
72. Luppino FS, de Wit LM, Bouvy PF, et al. Arch Gen Psychiatry 2010;67:220–9.
73. Singh DG, Krauthamer M, Bjalme-Evans M. J Investigative Med 2022;70:5-13.
74. Hill JO, Wyatt HR, Reed GW, et al. Science 2003;299:853–5.
75. Hill JO, Wyatt HR, Melanson EL. Med Clin North Am 2000;84:333–46.
76. Lissner L, Heitmann BL. Eur J Clin Nutr 1995;49:79–90.
77. Tobias DK, Chen M, Nanson JE, et al. Lancet Diabetes Endocrinol 2015;3:968–70.
78. Wang DD, Li Y, Chiuve SE, et al. JAMA Intern Med 2017;176:1134–45.
79. Field AE, Willett WC, Lissner L, et al. Obesity 2007;15:967–76.
80. US Department of Agriculture and U.S. Department of Health and Human Services. Dietary Guidelines for Americans, 2020-2025. 9th ed. December 2020.
81. Ello-Martin JA, Ledikwe JH, Rolls BJ. Am J Clin Nutr 2005;82: 236S–41S.
82. Rolls BJ. Physiol Behav 2009;97:609–15.
83. Ledikwe JH, Blanck HM, Kettel Khan L, et al. Am J Clin Nutr 2006;83:1362–8.
84. Bes-Rastrollo M, van Dam RM, Martinez-Gonzalez MA, et al. Am J Clin Nutr 2008;88:769–77.
85. Malik VS, Popkin BM, Bray GA, et al. Circ 2010;121:1356–64.
86. Schulze MB, Manson JE, Ludwig DS, et al. JAMA 2004;292: 927–34.
87. Malik VS, Hu FB. Nat Rev Endocrinol 2022;18:205–18.
88. Malik V, Pan A, Willett WC, et al. Am J Clin Nutr 2013;98: 1084–102.
89. Ebbeling CB, Feldman HA, Steltz SK, et al. J Am Heart Assoc 2020;9:e015668.
90. Hu FB. Obes Rev 2013;14:606–19.
91. Bray GA, Nielsen SJ, Popkin BM. Am J Clin Nutr 2004;79: 537–43.
92. Bray GA. Clin Dermatol 2004;22:281–8.
93. Forshee RA, Storey ML, Allison DB, et al. Crit Rev Food Sci Nutr 2007;47:561–82.
94. White J, Foreyt J, Melanson K, et al. Am J Life Med 2010;4: 515–20.
95. Nielsen SJ, Popkin BM. JAMA 2003;289:450–3.
96. Young LR, Nestle M. J Am Diet Assoc 2003,103.231–4.
97. Rolls BJ, Morris EL, Roe LS. Am J Clin Nutr 2002;76: 1207–13.
98. Diliberti N, Bordi PL, Conklin MT, et al. Obes Res 2004;12: 562–8.
99. Raynor HA, Epstein LH. Psychol Bull 2001;127:325–41.
100. McCrory MA, Fuss PJ, McCallum JE, et al. Am J Clin Nutr 1999;69:440–7.
101. Drewnowski A, Specter SE. Am J Clin Nutr 2004;79:6–16.
102. Goto R, Nianogo R, Okubo Y, et al. JAMA Pediatr 2022;176: 937–40.
103. He J, Zhu Z, Bundy DB, et al. JAMA 2021;326:1286–98.
104. Liu J, Rehm CD, Onapa J, et al. JAMA 2020;323:1161–74.
105. Leung CW, Epel ES, Ritchie LD, et al. J Acad Nutr Diet 2014;114:1943–53.
106. Drewnowski A, Darmon N. J Nutr 2005;135:900–4.
107. Brownson RC, Boehmer TK, Luke DA. Annu Rev Public Health 2005;26:421–43.
108. King GA, Fitzhugh EC, Bassett DR, Jr., et al. Int J Obes Relat Metab Disord 2001;25:606–12.
109. Prentice AM, Jebb SA. BMJ 1995;311:437–9.
110. Physical Activity and Health: A Report of the Surgeon General. Atlanta: US Department of Health and Human Services, 1996.
111. Hankinson AL, Daviglus ML, Bouchard C, et al. JAMA 2010;304:2603–10.
112. Gluckman PD, Hanson MA. Pediatr Res 2004;56:311–7.
113. Hanson M, Gluckman P, Bier D, et al. Pediatr Res 2004;55: 894–7.
114. Barker DJ, Winter PD, Osmond C, et al. Lancet 1989;2: 577–80.
115. Phillips DI, Barker DJ, Hales CN, et al. Diabetologia 1994;37: 150–4.
116. Valdez R, Athens MA, Thompson GH, et al. Diabetologia 1994;37:624–31.
117. Barker DJ, Hales CN, Fall CH, et al. Diabetologia 1993;36:62–7.
118. Oken E, Levitan EB, Gillman MW. Int J Obes (Lond) 2008;32: 201–10.
119. Ruager-Martin R, Hyde MJ, Modi N. Early Hum Dev 2010;86: 715–22.
120. Oken E, Taveras EM, Kleinman KP, et al. Am J Obstet Gynecol 2007;196:322 e1–8.
121. Dabelea D, Hanson RL, Lindsay RS, et al. Diabetes 2000;49: 2208–11.
122. von Kries R, Koletzko B, Sauerwald T, et al. Adv Exp Med Biol 2000;478:29–39.
123. Bergmann KE, Bergmann RL, Von Kries R, et al. Int J Obes Relat Metab Disord 2003;27:162–72.
124. Harder T, Bergmann R, Kallischnigg G, et al. Am J Epidemiol 2005;162:397–403.
125. Toschke AM, Vignerova J, Lhotska L, et al. J Pediatr 2002;141: 764–9.
126. Taveras EM, Rifas-Shiman SL, Scanlon KS, et al. Pediatrics 2006;118:2341–8.
127. Arenz S, von Kries R. Adv Exp Med Biol 2005;569:40–8.
128. Owen CG, Martin RM, Whincup PH, et al. Pediatrics 2005;115:1367–77.
129. Gillman MW, Rifas-Shiman SL, Camargo CA, Jr., et al. JAMA 2001;285:2461–7.
130. Horta BL, Rollins N, Dias MS, et al. Acta Paediatr 2023; 112:34–41.
131. Li R, Jewell S, Grummer-Strawn L. Am J Clin Nutr 2003;77: 931–6.
132. Hediger ML, Overpeck MD, Kuczmarski RJ, et al. JAMA 2001; 285:2453–60.
133. Clifford TJ. BMJ 2003;327:879–80.
134. Dewey KG, Gungor D, Donovan SM, et al. Am J Clin Nutr 2021;114:1774–90.
135. Meek YJ, Lawrence N, Section on Breastfeeding. Pediatr 2022; 150:e2022057988.
136. van Ginneken V, Sitnyakowsky L, Jeffery JE. Med Hypotheses 2009;72:383–8.
137. Atkinson RL. Mayo Clin Proc 2007;82:1192–8.
138. Mitra AK, Clarke K. Obes Rev 2010;11:289–96.
139. Atkinson RL, Dhurandhar NV, Allison DB, et al. Int J Obes (Lond) 2005;29:281–6.
140. Gabbert C, Donohue M, Arnold J, et al. Pediatr 2010;126: 721–6.
141. McLachlan J. Estrogens in the Environment II , New York, NY: Elsevier/North-Holland, 1995.
142. Colborn T, vom Saal FS, Soto AM. Environ Health Perspect 1993;101:378–84.
143. Colborn T, Clement C. Chemically Induced Alterations in Sexual and Functional Development: The Wildlife/Human Connection. Princeton, NJ: Princeton Scientific, 1992.

144. Colborn T, Dumanski D, Myers J. Our Stolen Future. New York, NY: Penguin, 1996.
145. Newbold RR, Padilla Banks E, Jefferson WN, et al. Int J Androl 2008;31:201–8.
146. Baillie-Hamilton PF. J Altern Complement Med 2002;8: 185–92.
147. Heindel JJ. Toxicol Sci 2003;76:247–9.
148. Heindel JJ, Levin E. Birth Defects Res A Clin Mol Teratol 2005;73:469.
149. Newbold RR, Padilla-Banks E, Snyder RJ, et al. Reprod Toxicol 2007;23:290–6.
150. Newbold RR, Padilla-Banks E, Jefferson WN. Endocrinol 2006; 147:S11–7.
151. Newbold RR, Padilla-Banks E, Snyder RJ, et al. Mol Nutr Food Res 2007;51:912–7.
152. Liu Y, Peterson KE. Curr Env Health Rep 2015;2:339–47.
153. Flegal KM, Troiano RP, Pamuk ER, et al. N Engl J Med 1995;333: 1165–70.
154. Patel SR, Hu FB. Jul 2008;16:643–53.
155. Besedovsky L, Lange T, Born J. Pflugers Arch 2012;463: 121–137.
156. Rico-Rosillo MG, Vega-Robledo GB. Ginecol Obstet Mex 2012; 80:332–340.
157. Mortensen RF. Immunol Res 2001;24:163–176.
158. Flegal KM, Carroll MD, Ogden CL, Curtin LR. JAMA 2010;303: 235–241.
159. Kalra M, Mannaa M, Fitz K, et al. Obes Surg 2008;18:675–679.
160. Mavanji V, Billington CJ, Kotz CM, Teske JA. Neurosci Biobehav Rev 2012;36:1015–1029.
161. Sookoian S, Gemma C, Gianotti TF, et al. Am J Clin Nutr 2008;87:1606–1615.
162. Nascimento Ferreira MV, Goumidi L, Carvalho HB, et al. Sleep Med 2018;46:56–60.
163. Froy O. ISRN Obes 2012;437198.
164. Nielsen LS, Danielsen KV, Sorensen TI. Obes Rev 2011;12: 78–92.
165. Hill J, Levine J, Saris W. Bourchard B, ed. Energy expenditure and physical activity. Handbook of Obesity New York, NY: Marcel Dekker, 2003; 631–54.
166. Hill JO. Endocr Rev 2006;27:750–61.
167. Hill JO. Am J Clin Nutr 2009;89:477–84.
168. MacLean PS, Higgins JA, Johnson GC, et al. Am J Physiol Regul Integr Comp Physiol 2004;287:R288–97.
169. Goldsmith R, Joanisse DR, Gallagher D, et al. Am J Physiol Regul Integr Comp Physiol 2010;298:R79–88.
170. Catenacci VA, Grunwald GK, Ingebrigtsen JP, et al. Obesity (Silver Spring) 2011;19:1163–70.
171. Brown WJ, Williams L, Ford JH, et al. Obes Res 2005;13: 1431–41.
172. Zhai FY, Wang HJ, Wang ZH, et al. Wei Sheng Yan Jiu 2006;35:72–6.
173. Wang YC, Gortmaker SL, Sobol AM, et al. Pediatr 2006;118: e1721–33.
174. Rodearmel SJ, Wyatt HR, Barry MJ, et al. Obesity (Silver Spring) 2006;14:1392–401.
175. Rodearmel SJ, Wyatt HR, Stroebele N, et al. Pediatr 2007;120: e869–79.
176. Kamath CC, Vickers KS, Ehrlich A, et al. J Clin Endocrinol Metab 2008;93:4606–15.
177. Harris KC, Kuramoto LK, Schulzer M, et al. CMAJ 2009;180: 719–26.
178. Doak CM, Visscher TL, Renders CM, et al. Obes Rev 2006;7: 111–36.
179. Lytle L, Hearst M. Curr Nutr Food Sci 2009;5:134–48.
180. Healthy Schools Group, Foster GD, Linder B, et al. N Engl J Med 2010;363.
181. Economos CD, Hyatt RR, Goldberg JP, et al. Obesity (Silver Spring) 2007;15:1325–36.
182. Hesketh KD, Campbell KJ. Obesity (Silver Spring) 2010;18: S27–35.
183. Paul IM, Savage JS, Anzman SL, et al. Obesity (Silver Spring) 2011;19:353–61.
184. Feng J, Glass TA, Curriero FC, et al. Health Place 2010;16: 175–90.
185. Christakis NA, Fowler JH. N Engl J Med 2007;357:370–9.
186. Economos CD, Brownson RC, DeAngelis MA, et al. Nutr Rev 2001;59:S40–56; discussion S7–65.
187. Takura T, Hirano Goto K, Honda A. BMC Med 2021;19:15.
188. Titmuss AT, Srinivasan S. J Paediatr Child Health 2016;52: 928–934.
189. Corvalán C, Garmendia ML, Jones-Smith J, et al. Obes Rev 2017;18:7–18.
190. Wang S, Jiang S, Ye Q. Eur J Cardiovasc Nurs 2020;19: 223–229.
191. Cole TJ, Lobstein T. Pediatr Obes 2012;7:284–294.
192. Chen G, Yan H, Hao Y, et al. BMC Pediatr 2019;19:127.
193. Nascimento-Ferreira MV, Berg G, González Zapata LI, De Moraes AC, Carvalho HB. Clin Nutr ESPEN 2021;45:333–340.
194. Szklo M, Javier Nieto F. Epidemiology: Beyond the Basics. Burlington, MA: Jones & Barlett Learning, 2018.
195. Bi Q, Goodman KE, Kaminsky J, Lessler J. Am J Epidemiol 2019;188;2222–2239.
196. Delpino FM, Costa Â, Farias SR, et al. Pub Health 2022;205: 14–25.
197. DeGregory KW, Kuiper P, DeSilvio T, et al. Obes Rev 2018;19: 668–685.
198. Fernandes, FT, de Oliveira TA, Teixeira et al. Sci Rep 2021;11: 3343.
199. Silva, GF, Fagundes TP, Teixeira BC, et al. Curr Hypertens Rep 2022;24:523–533.
200. Triantafyllidis A, Polychronidou E, Alexiadis A, et al. Artif Intell Med 2020;104:101844.

A. OBESITY AND DIABETES

CHAPTER 61

Obesity Management

Kavita H. Poddar, Lilian De Jonge, and Lawrence J. Cheskin

CHAPTER OUTLINE

Abbreviations

ACSM, American College of Sports Medicine; **BMI**, body mass index; **BMR**, basal metabolic rate; **FDA**, U.S. Food and Drug Administration; **FFQ**, food frequency questionnaire; **LCD**, low-calorie diet; **METs**, metabolic equivalents; **NHLBI**, National Heart, Lung, and Blood Institute; **VLCD**, very low-calorie diet

Introduction

The prevalence of obesity has nearly doubled in recent decades, with approximately 40% of U.S. adults now having obesity. If current trends continue, almost the entire U.S. population is projected to be overweight by 2030.[1] The increase in prevalence is even more dramatic among children and those with extreme obesity.[2] Obesity has truly reached epidemic proportions in the United States, coincident with unfavorable shifts in diet and exercise habits.[3] Largely due to the adoption of Western diet and exercise patterns, even developing nations are seeing an increase in obesity, with the projected number of those with obesity estimated to be in the billions by 2030.[4,5]

In the United States, although the percentage of energy intake derived from fat is decreasing (from a high of 40% to ~32% today), total daily energy and refined carbohydrate intakes have increased.[6] Furthermore, increasing numbers of adults and children engage in minimal physical activity.[7] The environment is a large contributor to this inactivity and, thus, to the epidemic of obesity in the United States.[8]

Obesity is a risk factor for diseases of virtually every organ system, including many cancers and is the most important risk factor in the development of diabetes and other health complications.[9] The risk of complicating medical conditions increases with the degree of obesity,[9] although for some complications—notably dyslipidemias, which are associated with cardiovascular risk factors, type 2 diabetes, and hypertension—the risk correlates best with the regional distribution of fat.[10] Central (visceral) deposition of fat (the "apple-shape" pattern), seen more commonly in men, increases risk, whereas excess fat in the lower body (thighs, hips, and buttocks), seen more commonly in women (the "pear-shape" pattern), is associated with a lesser risk of comorbidities.

Overweight and obesity increases overall mortality risk and shortens life expectancy in individuals with body mass index (BMI) greater than 25 kg/m^2; this life expectancy reduction can be dramatic for those who have a BMI greater than 35 kg/m^2.[11] In addition to the medical risks—and often more motivating for many people seeking to lose weight—are the untoward psychosocial consequences of obesity.[12-14] Prejudice against individuals with obesity is widespread. The resulting social effects and job discrimination contribute to low self-esteem and depression among people with obesity who seek treatment. Notable also is the greater social stigma borne by women with obesity, compared with men with obesity, in the United States and the higher prevalence of obesity among those of low socioeconomic status, Blacks, Latinos, and Native Americans. For example, more than 80% of middle-aged Black women in the United States are overweight or obese. There is compelling evidence of health benefits associated with even modest weight loss, especially for those with existing comorbidities.[15-18]

This chapter aims to provide an overview of the assessment and treatment of obesity, with the primary focus on dietary and exercise interventions. Pharmacologic and surgical therapies are a second-line option in the treatment of obesity and are reviewed in less depth.

Evaluating Obesity

Physical Assessment

Classification of obesity was proposed by an expert panel convened by the National Institutes of Health (NIH), National Heart, Lung, and Blood Institute (NHLBI) in 1998 after extensive review of health complications associated with this condition.[19] The panel classified obesity and overweight status based on BMI and waist circumference (see **Table 61.1**). Specifics on the panel's classifications are covered in Chapter 48 of this book. Broadly, a BMI of 25 to 30 kg/m^2 is defined as overweight, 30 to 35 kg/m^2 as obesity class I, 35 to 40 kg/m^2 obesity class II, and 40+kg/m^2 obesity class III.[20] For children, obesity is defined as BMI for age and sex ≥95th percentile, based on the 2000 Centers for Disease Control and Prevention (CDC) growth chart, & overweight as BMI ≥85th percentile to <95th percentile.

Waist circumference is the second step in the assessment of obesity. A waist circumference >88 cm (35 in) in women or >102 cm (40 in) in men constitutes abdominal or visceral obesity and is associated with increased risk of health complications.[20] In the case of abdominal fat deposition, even mild excess adiposity may pose medical complications, such as increased risk of hypertension, dyslipidemias, and type 2 diabetes.[10,20,21] Visceral obesity can exist even in the absence of overall obesity (i.e., at BMIs below the cutoff points for obesity or overweight).[10,20,21] With visceral obesity, even with normal BMI, weight loss can benefit health, especially if the patient already suffers from complicating medical conditions or has a strong family history of type 2 diabetes, cardiovascular, or cerebrovascular disease. For patients with cosmetic or trivial obesity, the benefits (and motivators) for successful weight loss are more psychosocial than medical. These patients should be encouraged to focus on a healthier diet (with low refined carbohydrate, low saturated fat, and high fiber) and increased physical fitness, rather than just the number on the scale.

Psychosocial and Behavioral Assessment

A psychosocial and behavioral assessment can provide considerable information about the patient's readiness to lose weight, as well as identify disordered eating behaviors. Individuals with obesity commonly exhibit depression, with the degree of severity often greater with severe obesity.[22-24] A behavioral psychologist or other skilled professional can inquire into depressive symptoms by asking about the patient's mood and related symptoms and signs or assessing depression with formal tests.[25] Individuals suffering from obesity with significant depression should be provided appropriate treatment (cognitive behavior therapy and/or pharmacotherapy) before or during weight reduction efforts.

Approximately 30% of obese individuals who seek weight reduction suffer from *binge eating disorder*.[26] Binge eating is characterized by consumption of large amounts of food until one is uncomfortably full and eating alone when not hungry. When purging (vomiting or using laxatives or diuretics) or compulsive exercise is used following binging to control weight, *bulimia nervosa,* rather than binge eating disorder, is the likely diagnosis. Although anorexia nervosa and bulimia nervosa are commonly

Table 61.1 Classification of Overweight and Obesity by Body Mass Index, Waist Circumference, and Associated Disease Risk

		Disease Risk Relative to Normal Weight and Waist Circumference		
Weight Category	**BMI**	**Obesity Class**	**Men ≤102 cm (40 in) Women ≤88 cm (35 in)**	**Men >102 cm (40 in) Women >88 cm (35 in)**
Underweight	<18			
Normal	18.5–24.9			
Overweight	25.0–29.9		Increased	High
Obesity	30.0–34.9	I	High	Very high
	35.0–39.9	II	Very high	Very high
	>40.0	III	Extremely high	Extremely high

BMI, body mass index.

Data from National Institutes of Health, National Heart, Lung, and Blood Institute. Clinical Guidelines on the Identification, Evaluation, and Treatment of Overweight and Obesity in Adults—The Evidence Report. National Institutes of Health.

recognized as serious eating disorders,[27] binge eating disorder is more common and often occurs in those with overweight or obesity, rather than those with normal weight or underweight. Eating disorders and their treatment are covered in greater detail in Chapter 98 of this book.

Evaluating Dietary and Activity Habits

A comprehensive assessment of dietary intake and activity is helpful in evaluating the contribution of these factors to weight gain.[28] These assessments can identify problem areas related to diet and exercise that may need modification to achieve weight loss. Taking current medications, including over-the-counter or prescription drugs, herbs, and vitamin or mineral supplements, into account is important, as they can significantly influence weight loss efforts. This information can help identify potential food–drug interactions, assess daily intake in relation to nutrient requirements, and assist in the evaluation of methods used to address weight and nutritional issues by the patient. Often, a person with obesity reveals more information to a dietitian when specifically probed to answer these questions, than when meeting with a primary healthcare provider. Any food allergies or intolerances (e.g., gluten, lactose) should also be covered.

The data obtained from any dietary assessment should be interpreted with caution, however, because under-reporting and over-reporting are common. FFQs are self-administered instruments with multiple questions about the usual frequency of consumption and portion size of foods over the prior 1, 3, or 12 months. Questions in FFQs also may include information on food purchasing and preparation methods. FFQs can help identify inadequacy or overconsumption of specific food groups, or patterns of specific foods or preparation methods. Like the food diary, the FFQ can offer a more real-life assessment of the patient's typical food choices because it can be completed outside of the provider's office. Dietary assessments and their use are covered more in depth in Chapter 49 of this book.

Reduced levels of physical activity can be a major factor in the etiology of obesity and may be a direct result of acute or chronic illness, job change or retirement, or sedentary lifestyle in general, such as more television viewing or screen time.[29-32] Depending on the level of obesity, joint strain from increased weight can become an underlying cause of low levels of physical activity. Food record forms can include a place to inventory physical activity, which can be helpful in discussing exercise habits and goals. However, the provider must recognize and communicate that exercise alone is not an effective method for losing weight. However, regular aerobic exercise and strength training will also improve cardiovascular fitness and promote growth of metabolically more active muscle tissue.

The exercise assessment, preferably done by an exercise physiologist or certified trainer, should record the current fitness level, limiting factors such as joint disease or previous injuries and types of activity the patient finds enjoyable. Physical activity level can be assessed broadly by inquiring about the amount of walking done in a day, flights of stairs climbed, and hours of television watched.[33] More formal assessment can be conducted using a pedometer to determine the number of steps walked daily, or an accelerometer, which also can assess the intensity of activity. Exercise interventions for weight loss are considered later in this chapter.

A rule of thumb in prescribing an exercise regimen is to use a phased-in approach. Most patients with obesity start out with limited capacity to exercise. Rather than suggesting a type or level of activity that is unlikely to maximize adherence, make sure that the plan fits the patient's current abilities, schedule, and lifestyle. The first phase often consists of increasing the amount of everyday physical activity, so-called lifestyle activity, without introducing a formal exercise regimen. Lifestyle activities include taking the stairs in gradually increasing fashion, parking the car farther away from the destination, walking to the mailbox, and the like. This phase alone may double the level of physical activity in a very sedentary person.

The next phase is a walking plan. People are most likely to comply with such a plan if the walk is scheduled during typically available times, such as a break or lunchtime during work hours. Scheduling exercise when the individual's daily energy level is the highest (e.g., early morning for many people) is often more effective than at the end of a long day. Having a companion to walk with and a place to walk indoors also are helpful in increasing adherence.

Thirty minutes is a good minimum time to recommend that a patient should make available for each session of exercise. An hour or longer is best for weight control. The distance walked is more important than the intensity of the exercise: walking at a leisurely pace for 1 hour is roughly equal to walking the same distance at a briskly pace for half an hour. Allow the patient to set the pace. Initially, it may be quite slow, but in the absence of severe pulmonary, cardiovascular, or joint disease, most patients soon find the going easier and faster. Goal setting can strengthen this reinforcement. Having the patient keep a log of the time spent walking and the distance covered after each session is useful. The patient can then see the progress being made and can set the goal a bit higher, as warranted.

In the next phase of a progressive exercise plan, the types of activities performed should be broadened. Walking or jogging can and should remain a feature at this stage, but with the addition of other forms of aerobic exercise. Other suggestions for exercise include aerobics classes, stationary or outdoor bicycling, swimming, a cross-country skiing machine, or just about anything else that will burn energy and be enjoyable. Team or racquet sports and golf can be suggested to provide social interaction, as well as to increase energy expenditure. Again, the most important criterion for a good exercise plan is one that the patient is likely to follow and be comfortable with as a lifelong habit.

Weight Loss Readiness

Successful weight loss can be achieved and maintained when the individual is determined and motivated. For this, the provider must assess and evaluate the state of readiness. It is essential for the individual to be motivated to make lasting lifestyle changes; however, internal self-motivation is more sustainable than external motivators, such as a significant other's demands or the anticipation of a special event. The stages of change model for behavior change intervention may be useful in assessing where an individual is regarding making behavior changes, and in helping the individual move along the continuum from precontemplation to action stages.[34] Although internal motivation is the key to successful weight loss and maintenance, external stimuli, such as support from friends and family, and environmental factors, such as easy access to healthful foods and safe places to walk and run, also play important roles. Garnering these support systems can assist the individual to be mentally ready to move along the path of the action stage of weight loss and then weight maintenance efforts.

The provider should assist the individual to set "smart" goals (i.e., that are specific, measurable, achievable, realistic, and timely), which are more likely to be achieved. Social support from family and friends also can improve the likelihood of success in making behavior changes. Online or local weight loss support groups also can provide opportunities for discussing challenges and giving or receiving support. These can improve a patient's outcome both during and after initial treatment

Selecting the Right Treatment

Initial weight loss should be achieved with a comprehensive program that includes dietary changes, increased physical activity, and behavior modification. The recommended rate of weight loss depends on the degree of obesity, the presence or absence of comorbid conditions, the results of behavioral assessment, and patient preference. According to NHLBI guidelines,[20] overweight individuals (BMI = 25 to 29.9 kg/m^2) without any associated risk factors should be encouraged to prevent further weight gain or lose weight with simple dietary modifications. However, if this condition is accompanied by one or more cardiovascular risk factors, lifestyle modification is recommended in which diet, exercise, and behavior therapy may be helpful. Similarly, for individuals with moderate obesity (BMI ≥30 kg/m^2) and no additional risk factors, lifestyle modifications are beneficial, aiming for a rate of weight loss of 1 to 1.5 lbs/wk (an energy deficit of 500 to 750 kcal/d), a rate that is generally safe and rapid enough to sustain motivation. However, for individuals with BMI ≥30 kg/m^2, or >27 with medically important comorbid conditions, if lifestyle modifications alone are not successful, pharmacologic interventions may be an option. For individuals with more severe classes of obesity (BMI ≥35 kg/m^2 with comorbidities, or ≥40 kg/m^2 without), more aggressive energy restriction under medical supervision may be preferable. In the presence of significant comorbid conditions for individuals suffering from severe obesity, bariatric surgery may be considered.

Dietary Intervention

Diet therapy remains the cornerstone of weight loss,[35] but consistent adherence to diet therapy is often difficult.[36,37] Standard heart-healthy diets, which contain 50% to 55% of energy from carbohydrates, ~30% from fat, and ~15% protein often result in low satiety, have palatability issues, and lack variety, resulting in low adherence.[36] Moreover, internal and social factors, including level of motivation, confidence in one's ability to make the suggested dietary changes (self-efficacy), type of diet prescribed, social support, accountability, and access to healthful foods all play an important role in the level of success with weight loss a person can achieve.[37,38]

Achieving a negative energy balance is the most important factor associated with the amount and rate of weight loss over time, according to the 2013 American Heart Association (AHA)/ American College of Cardiology (ACC)/The Obesity Society (TOS) guidelines for the management of overweight and obesity in adults.[35] To lose weight at the rate of 1 to 2 lbs/wk, which is considered a healthy rate of weight loss, per NHLBI guidelines,[39] an energy deficit of 500 to 1,000 kcal/d (3,500 to 7,000 kcal/wk) is required. Both dietary intake restriction and increased exercise can contribute to this energy deficit (3,500 kcal = 1 lb of fat).[39]

To achieve successful weight loss, an energy intake target of ~1,200 to 1,500 kcal/d in women or 1,500 to 1,800 kcal/d in men is often recommended. Dietary energy intake reduction is achieved either via prescribed energy intake or by using *ad libitum* approaches without formal energy-deficit targets. *Ad libitum* diets achieve lower energy intake by restricting or eliminating specific foods or food groups.[35] More detailed discussion of various approaches follows.

Diets That Create Energy Deficit

Moderate Energy-Deficit Diets

Moderate energy-deficit diets (1,500 to 1,800 kcal/d) follow the guidelines laid out by the regulatory agencies such as the AHA/ The Obesity Society[35] and the U.S. Department of Agriculture *Dietary Guidelines for Americans*.[40] A low-calorie, food-based diet can be designed that has either a *balanced deficit* (reducing total energy while keeping proportions from carbohydrate, fat, and protein roughly the same) or with most of the energy reduction resulting from restriction of fat or carbohydrate intake.[41-45] The fat-deficit approach may be preferable for individuals consuming too much fat, especially saturated fat. In addition, a greater volume of food can be eaten on a diet that emphasizes complex and vegetable-source carbohydrates and reduces fat to less than 30% of energy.

According to the most recent Dietary Guidelines for Americans (2020–2025), more than two-thirds of American adults exceed recommendations for added sugars, saturated fats, and sodium intake.[40] More than 85% of Americans consume diets that do not meet recommended intakes of fruits, vegetables, and whole grains, which contributes to low intake of dietary fiber as well.[40] Restricting dietary fat is an approach that can serve to minimize hunger while maximizing satiety by replacing dietary fat with complex carbohydrates such as whole grains and vegetables, which are less energy-dense, higher in fiber and water content,[46,47] and more filling than foods with higher energy density. High energy-density foods tend also to be higher in fat, sugar, and sodium content; are often ultra-processed; and are associated with weight gain.[48,49] For this reason, emphasis on consuming a combination of complex carbohydrates, particularly those with higher fiber content, and lean proteins, provides a larger volume of food and sensation of fullness for any given energy intake level. In addition, lowering fat intake can help boost total metabolism by increasing the thermic effect of food percentage, as less energy is used to convert dietary fat into energy or body fat, compared with carbohydrates and proteins. A relatively low-fat diet, especially lower in saturated fat, can also improve blood lipoprotein profiles and reduce risk for many chronic diseases, such as cardiovascular disease. Finally, these diets may be easier to adhere to because relatively smaller changes are required in eating habits without compromising the volume of food eaten.

Low-Calorie Diets

A low-calorie diet (LCD) is arbitrarily defined as one that provides 800 to 1,500 kcal/d.[50] Various diet strategies can be used to achieve an LCD[51] and these vary in protein, carbohydrate, and fat content.[51-54] For individuals with BMI between 25 and 35 kg/m^2, with comorbidities such as type 2 diabetes or hypertension, an LCD is an appropriate first-line approach. Research shows that for those with a starting BMI ≥30, these diets are helpful in producing a loss of more than 5% initial body weight (an amount of weight loss associated with significant improvements in health outcomes) over 3 to 6 months of treatment.[51-54] Studies also show improvement for body fat distribution, bone, cardiovascular, and renal health, pain, and overall quality of life.[52-54]

Very Low-Calorie Diets

A very low-calorie diet (VLCD) is arbitrarily defined as one containing <800 kcal/d[50] and is a strategy for achieving quick weight loss. A VCLD is most suitable for individuals with class 3, severe obesity (BMI ≥40), who will benefit from rapid weight loss (sometimes prescribed prior to major surgery/bariatric surgery) with relative preservation of lean body mass.[35,55,56] VLCDs often provide high protein (up to 1.5 grams of protein/kg of ideal body weight per day) to preserve lean body mass. VLCDs may be food based or may use meal replacements, containing vitamins and minerals to prevent micronutrient deficiencies (e.g., soy-, egg-, or milk-based high protein, low carbohydrate, low-fat shakes or bars). In addition to high protein content, meal replacement supplements often include low intakes of carbohydrates (up to 80 g/d) and fats (up to 15 g/d) with 100% of the recommended daily allowance for essential vitamins and minerals. They are generally viewed as safe to use under medical supervision.[57-61] This approach can be useful if it is monitored and accompanied by a comprehensive program of behavior modification and physical activity.[62-64] Initial, but usually temporary, side effects may include hunger, fatigue, and lightheadedness; later, constipation and intolerance to cold can occur.[60] Rapid weight loss (>1.5 kg/wk) is the hallmark of VLCDs[57-64] and is associated with risk of gallstone development; however, this may be transient and can be prevented.[65-67] The rate of weight loss reduces to approximately 0.8 kg/wk after the first 6 weeks, resulting in an average weight loss of 10% to 20% in VLCDs of up to 6 months.[57-63] Of note, the extent of weight regain after the cessation of VLCDs is high.[57-60] Thus, use of VLCDs is most effective in the context of a multidisciplinary approach to weight loss and long-term weight control, with careful attention to a weight maintenance program following the VLCD phase. VLCDs are most appropriate for individuals who have medical need for rapid weight loss and have failed weight loss attempts with LCDs and behavior changes.

Diets That Emphasize Different Macronutrient Compositions

Low-Carbohydrate and Very Low-Carbohydrate Ketogenic, High Protein, High Fat Diets

Interest in low carbohydrate (CHO) diets has grown in recent years for individuals with diabetes and prediabetes, to improve glycemic control[68,69] and other cardiometabolic risk factors.[70-73] There is mounting evidence that low CHO diets may have a therapeutic role in disease conditions ranging from polycystic ovary syndrome, to acne, neurologic diseases, and cancer.[73] A low CHO diet is defined as CHO intake of less than the recommended lower level of the acceptable macronutrient distribution range for healthy adults, which is 45% to 65% of total daily energy (TDE) intake.[40,74] There are numerous versions of low CHO diets. For example, a moderate CHO diet provides 26% to 44% TDE from CHO, such as the South Beach[75], the Zone,[76] and the paleolithic diets.[77] Then, there are low CHO diets that provide 10% to 25% TDE from CHO (130 g/d of CHO on a 2,000 kcal diet),[78] and very low CHO diets that provide <10% TDE from CHO (<50 g CHO/d), such as the Atkins diet. Atkins-style diets include moderate amounts of protein and fats along with very low CHO intake.[79] Contemporary very-low CHO diets, also known as ketogenic diets (KDs or "keto diets") also limit protein intake to moderate levels without restricting fat or total energy intake.[80] Low and moderate CHO diets can be moderate or high in fat and moderate or high in protein.[75-79] While there are conflicting views on the advantages of low CHO diets, there is some evidence that low CHO diets provide increased satiety, thereby reducing overall energy intake and contributing to weight loss.[81-83] Others promote the theory that low CHO diets moderate to high in protein intake have more energy utilization by the body, as use of energy from proteins is a metabolically "expensive" process when compared with other diets.[81-83] On the other hand, very low CHO diets, of 20 to 50 grams of CHO/d, produce a state of nutritional ketosis and provide high fat (70%-80% TDE).[73,84,85] Low carb and very low CHO ketogenic diet approaches hypothesize that decreased insulin secretion with low CHO intake produces a catabolic state in which a fat-burning mode (lipolysis) ensues.[86]. This increase in lipolysis causes production of ketone bodies in the blood, typically with concentration ≥0.3 mmol/L, but no more than 7 to 8 mmol/L.[73,85] A steady state of ketosis can be achieved within 2 to 3 weeks if dietary circumstances are sustained that promote ketogenesis (<50 g CHO/d). Ketosis may be important for the rapid initial weight loss seen in individuals who follow these diets and is accompanied by bodily fluid loss and diuresis.

Very low CHO diets typically recommend up to 20 g of CHO at the onset of the diet, with high intakes of protein and fat, and involve three or four phases. The first phase is called the induction phase, which is the weight loss initiation phase and includes no more than 10% of energy from CHO, 60% to 70% energy from fat, and the remaining from protein.[79,86] The second phase, the ongoing phase, is a continuation of weight loss, with the proportions of CHO (9%), protein (33%), and fat (58%) slightly liberalized. The third and fourth phases are pre-maintenance and maintenance phases, in which CHO intake increases to no more than 20% of total energy, with 25% to 27% protein, and approximately 52% fat.[79,87]

Any restrictive diet can cause diuresis in the initial 1 to 2 wks, resulting in 2% to 4% weight loss, largely because of water loss.[88,89] Low CHO and very low CHO diets have favorable metabolic effects on energy expenditure and food intake, along with positive effects on satiety and appetite control.[72,73,80-83,85,86] Despite beneficial effects on weight loss, studies show that weight loss may not be superior with low and very low CHO ketogenic diets, compared with conventional strategies.[43,90-94] Evidence from several studies indicates that low and very low CHO diets are initially (3–6 months) more effective than diets low in fat and energy for weight loss. However, no significant differences in weight loss were present between diet types after 6 months.[43,90-94] Diuresis may account for some of the more rapid weight loss seen with low CHO diets in the early phase. Although benefits of low CHO diets have been reported on markers of cardiovascular risk factors, concerns exist regarding high intake of fat in the diet.[43,91-94] High fat intake, particularly of saturated fatty acids (SFA), is associated with health problems such as cardiovascular (CV) disease and certain types of cancers.[95,96] Controlling SFA intake is critical to preventing significant increases in LDL-C and for achieving improved CV health. Other adverse health effects, including cognitive impairment, constipation, diarrhea, dizziness, halitosis, headaches, insomnia, kidney stones, and nausea, have also been reported in individuals on low CHO diets.[97]

Low-Fat, Low-Energy Density Diets with Moderate to High Carbohydrate Intake

The typical U.S. diet is moderately high in fat intake and in energy density.[40] It is well known that energy-dense food intake is associated with higher BMI and body weight.[49-51,98-100] Weight gain from high intake of energy-dense foods such as fats has been attributed to passive overconsumption, likely because of the relatively low satiety value and high palatability of such foods.[99] Moreover, U.S. adults are not meeting dietary recommendations for fruits, vegetables, beans, and whole grain intakes, each of which is associated with lower energy density and higher micronutrient content.[40,101-103] Inadequate intake of nutrient-dense foods across several food groups has resulted in underconsumption of calcium, potassium, dietary fiber, and vitamin D, all dietary components of public health concern.[40,101-103]

Reducing the energy density of the diet by replacing fatty foods with more fruit, vegetables, beans, and whole grains can be an effective strategy for weight control as fat, at 9 kcal/g, provides more than double the energy of carbohydrates (4 kcal/g) or proteins (4 kcal/g).[42] Moreover, low energy-dense foods, such as fruits, vegetables, and whole grains lower energy intake because of increased satiety of the diet,[104,105] which translates to lower BMI.[48,98-100,106]

Low fat, low-energy diets concentrate on reducing overall energy intake by reducing total fat intake. Studies using this approach have reported significant weight loss in obese individuals, compared with low-fat, high-CHO diets, which are not energy controlled.[107,108] Examples of low fat diets include the Dean Ornish diet and the Pritikin diet, which were originally developed to reverse heart disease in the 1970s.[109-111] Adhering to low-fat intake while increasing vegetables, fruits, legumes, and whole grains inevitably leads to overall energy reduction and weight loss.[109-111] Another example of a strict low-fat diet originally developed for reversal of heart disease includes the whole food plant-based diet plan developed by Dr. Caldwell B. Esselstyn at the Cleveland Clinic, which recommends avoidance of all foods that have fat[112] and has shown beneficial effects on weight loss.[113] Longer duration studies are needed to assess adherence to these diets and to understand if individuals compensate for lower energy intake later and regain weight.

Types of Controlled Carbohydrate Diets

Low Glycemic Index, Balanced Nutrient Diets

The *glycemic index (GI)* of foods refers to the effects produced by carbohydrate-rich foods on blood glucose and insulin concentrations.[114] GI is defined as the ratio of the area under the curve (AUC) of glucose 2 hours after consuming 50 g of carbohydrate in test food, vs that for a standard food.[114] Glucose or white bread serves as the standard food, defined as a GI of 100. Foods are divided into three categories based on GI values: high GI foods (>70), medium GI foods (55–70), and low GI foods (<55).[114] Ingestion of high GI foods such as potatoes and white bread produce rapid increases in blood glucose and insulin concentrations.[115-117] These changes result in short sustenance of satiety, low-fat oxidation, and subsequent weight gain, secondary to poor appetite regulation caused by the rapid rise and fall in blood glucose and insulin.[115-117] Diets that are constructed to be low GI contain more fruits and vegetables, which are high in fiber, and contain lower intakes of foods high in simple carbohydrates, such as sugars, highly refined foods, and starchy vegetables.[115-117]

A low-GI, energy-restricted diet may yield greater benefits than a reduced fat, energy-restricted diet for weight loss.[35,72,75,76,116-122] One study found that lowering the glycemic load of an energy-restricted diet significantly reduced body weight, increased fat oxidation, and decreased fat regain after weight loss.[116] Another study found that lowering the glycemic load of an energy-restricted diet significantly reduced body weight but had no effects on hunger, satiety, lipid profiles, or other inflammatory and metabolic risk markers.[117] A recent meta-analysis concluded that low-GI diets were more effective at controlling body weight than high-GI diets.[118]

Diet Pattern Based Approaches

Balanced Deficit/Portion-Controlled Diets

A balanced-deficit diet involves balanced restrictions in the major food groups from the USDA's MyPlate food guidelines.[123] These diet plans focus more on food patterns and less on nutrients, as promoted by the USDA 2020–2025 Dietary Guidelines for Americans.[40] Here, the focus is to highlight foods to eat more of (vegetables, fruits, whole grains, fat-free or low-fat dairy, legumes, nuts, seeds, lean meats, and oils) within a diet plan. It also emphasizes restricting fat content of the diet to 20% to 25% of total energy consumed by lowering saturated fat.[124] Overall, these diets do not deviate significantly from recommended macronutrient and micronutrient content,[125] unlike other popular diets.[75,76,110,126] One study evaluated the nutrient adequacy of the OmniHeart diet, a modified DASH diet, against five popular diets for weight loss (Atkins, Ornish, South Beach, Mediterranean, and Zone diets).[125] This report concluded that (on a 2,000 kcal/d diet), the Ornish diet included 75% of energy from CHO, 67 g fiber, had the lowest percent energy from fat (7%) and saturated fat (1%), and had only 6 mg/d cholesterol.[125] The Atkins diet included 9% energy from CHO, 20 g of fiber, 62% of energy from fat, 23% from saturated fat, and 731 mg/d total cholesterol. None of these diets met national guidelines for macronutrient intake.[75,76,110] The Zone diet had the highest protein content, at 35% total energy, and met national guidelines for fat and protein intake, but not for CHO intake. The other diets analyzed had moderate mixes of protein (16%–26%), CHO (33%–54%), and fat (28%–46%).[125] Another study showed that diet quality (as measured by the alternate healthy eating index score (AHEI)) of different popular weight loss diet plans varied, with the Ornish diet scoring 64.6, which was the highest score and Atkins diet (45 g/d of CHO) scoring 42.3, the lowest, with a maximum possible score of 70.[126]

Balanced, portion-controlled diet approaches generally allow less restriction in the volume of food consumed but reduce fatty foods that are more compact and energy dense while meeting guidelines for fat intake. For example, one can swap a small piece of cheese, which may contain 250 kcal, with a large plate of salad providing only 125 kcal. Even reducing serving sizes modestly from each food group can provide energy restriction of 300 to 500 kcal/d, resulting in weight loss of up to 1 lb/wk. However, weight loss may not be consistent and may frustrate individuals trying to lose weight quickly. In balanced-deficit diets, the goal is to achieve an energy deficit while leaving initial dietary choices unchanged. Adherence to such diets may also be better because it is based on personal and cultural preferences and habits rather than avoidance of specific foods or food groups that may be favorites. Foods rich in fats and sugar need not be replaced in a balanced-deficit diet but incorporated in a portion-controlled manner. The Healthy US-style,[40] the Healthy Mediterranean-Style,[126,127] Healthy Vegetarian diet pattern,[128,129] the Dietary Approaches to Stop Hypertension (DASH) diet,[125,130] and Vegan diet patterns are all associated with weight loss and reduced chronic disease risk.[131-133] These diets can be modified to incorporate personal and cultural preferences while taking an individual's medical history into account.[35,40]

U.S. diets today often include huge serving sizes, which contribute to high energy intake.[134] The size of meals in the United States increased by almost 75% between the 1970s and the 2000s, whereas bread slices grew by 24% likely reflecting popular perceptions of the size of meals at different stages in history.[135] Several studies showed that portion-controlled diets lead to mean weight loss of up to 10% of initial body weight.[136-140] Several studies have shown that patients on portion controlled diets lost more weight and fat mass during short-term interventions, compared with controls.[136-139] One study included a comprehensive lifestyle modification program for weight loss, with portion-controlled diets, and reported a dramatic 10% weight loss in 16 to 26 wks of treatment.[140]

Using a variety of tools to control portions (plates, cooking utensils, educational aids, bowls, forks, glasses) is better than using smaller plates alone.[141-143] While it may be possible for individuals to adhere better to portion-controlled diets, nutrition education efforts are also needed to help individuals understand portion sizes.[144] Individuals should be encouraged to use measuring cups and spoons, food scales, and/or familiar objects (e.g., a deck of cards or the palm of the hand), both to estimate their typical portions consumed and to serve themselves. For example, 3 ounces constitutes a serving of protein (meat, chicken, or fish), one half-cup a serving of cooked starches (rice, pasta, or potatoes), and one teaspoon (or thumb tip) is a serving of fat (oil).[145] Checking labels for serving size and number of servings per package is imperative to understanding exactly how much of a food is being consumed. One package does not necessarily equal one serving.[146]

Meal Replacement Diets

Meal replacement (MR) diets may be a good choice when conventional LCDs have not worked for those needing rapid initial weight loss. Good initial results can enhance motivation to continue losing weight, and MRs are a useful means of enforcing energy restraint. MR diets include one or more meals, which are replaced by nutritionally balanced, and macronutrient and energy controlled (800 to 1500 kcal/d) commercial dietary formulations.[147-152] MRs are portion-controlled products that are commonly packaged as powders that can be reconstituted as shakes or soups or are available as liquid shakes that can be used as is. Others are in the form of high protein bars, shelf-stable or frozen entrees, or snacks. The combination of reduced energy intake, smaller portions, and macro- and micronutrient balance contributes to the weight loss observed in patients on MR diets.[148-150] However, there is no conclusive evidence that these diets improve cardiometabolic risk markers (e.g., blood glucose).[149]

Several studies of MRs have reported greater initial weight loss compared with standard LCDs,[148-151] and that MRs have greater nutrient adequacy and superior diet quality.[148-152] While these diets are highly successful for short-term weight loss, one study found that those on the MR diet regained more during the 24-wk maintenance phase compared with controls.[150] A study evaluating the efficacy of the U.S. Army's education-based, weight-management program, "Weigh to Stay," with and without MRs among soldiers exceeding body fat standards concluded that those in the MR group lost more weight, more percent body fat, and more fat mass than those in the Weigh to Stay program alone.[152] Weight maintenance on MRs on a long-term basis, as with most dietary approaches to weight loss, has not been demonstrated[153] and sustaining compliance may be difficult.

Meal Timing and Frequency Patterns Influencing Weight

Eating behaviors have evolved in the last 4 decades in the United States. Total energy consumed in meals as opposed to snacks decreased from >80% in the 1970s to 77% in 2010; however, total energy consumed from snacks has risen from 18% to 23%.[154,155] Mealtimes among free-living individuals have also shifted; overall, Americans are eating the lunch meal approximately 30 minutes earlier and more frequently skipping breakfast.[156] While the number of people eating three standard meals has declined for both men and women, these variations reflect changes in eating pattern rather than eating frequency, which has increased.[154-156] Shift workers tend to have erratic eating patterns associated with higher energy and macronutrient intakes.[157-159] Eating patterns in healthy nonshift workers have also become more erratic. One study reported that >35% energy was being consumed after 6:00 pm in healthy nonshift working individuals, and <25% energy consumed before noon.[160] Clearly, adults in the United States are eating around the clock, which may lead to increased energy intake and elevated risk of obesity. The circadian clock of the body plays a critical role in many aspects of physiologic functions, including the sleep-wake cycle, blood pressure, body temperature, etc.[161-164] Disruption in synchronization can result in circadian misalignment and the development of metabolic disorders. Studies show that meal timing and frequency can, indeed, disrupt peripheral clock synchrony with the master clock and can increase the risk of developing cardiometabolic diseases.[161-164]

Higher Meal Frequency and Weight

Higher meal frequency, e.g., three meals and two to three snacks, is often encouraged for weight loss, compared with eating three large meals.[165-167] Moreover, higher meal frequency has been associated with better diet quality in both adults and children[168-170] and improved lipid and glycemic profiles.[171-173] Nevertheless, several studies report contradictory findings. Notably, A systematic review and meta-analysis of 22 studies showed that increased meal frequency (i.e., four to nine events daily compared with three) was associated with 200 kcal/d higher energy consumption among adults in the United States.[174] Similarly, the prevalence of daily snacking among children increased from >70% in 1977–78 to >90% in 2003–2006, and was associated with >160 kcal/d consumption increase.[175] In adolescents, the percentage of daily snackers increased from 61% in 1977–1978 to 83% in 2003–2006 and the daily energy provided by snacks increased from 300 to 526 kcal.[176] A positive association between higher meal frequency and body weight has been reported in children and adults.[177,178] For example, one prospective analysis showed that men consuming >3 meals/d had a higher risk of >5 kg weight gain over 10 years.[177] Another study, however, reported no significant difference in weight loss with higher meal frequency under isocaloric conditions.[179] Possible explanations for inconsistent results on the impact of meal frequency on weight include planned vs unplanned snacking, type of foods eaten (e.g., high energy vs lower energy food/beverage intake), which determine energy intake per ingestive event, and portion size of meals/snacks eaten. Larger randomized clinical trials of longer duration are needed to clarify the effect of meal frequency on body weight in humans.

Intermittent Fasting and Weight

Intermittent fasting as a weight-loss strategy has increased in popularity over the past decade and has been purported to have many health benefits.[180] Fasting is an age-old practice rooted in many cultural practices.[181] Intermittent fasting (IF) includes different approaches such as: 1) time-restricted eating (TRE) typically with an 8 to 10 hour window of eating; 2) alternate day fasting (ADF), which involves decreased energy intake of ~25% of total energy needs on the day of "fasting" and *ad libitum* food intake during other day; 3) periodic fasting, also known as 5:2 fasting, which involves decreased energy intake of ~25% of total energy needs at lunch for 2 consecutive days and ad libitum energy intake during the other 5 days; and 4) the *fasting mimicking diet,* which involves consuming a low energy (~30% of energy needs), high unsaturated fat diet for 5 days, once per month, or once every 3 or 4 months.[182] The two most common forms of IF are TRE and ADF. Several studies found that these approaches are promising strategies for weight loss.[183-189] Studies using the TRE method show that people with excess adiposity can lose up to 4% of their weight and, although not clinically meaningful (defined as 5% or more for cardiometabolic benefits), this is a promising method, as individuals lost weight even with *ad libitum* energy intake.[183-186] Similarly, the ADF approach has shown modest weight loss in individuals with no energy restriction during the "feast" days and restricted or no energy intake during the fasting days.[187-189] Studies also show that these eating approaches may be useful in improving risk factors besides weight, such as fasting insulin, insulin resistance, hemoglobin A1c, triglycerides, and blood pressure.[183-192] While studies show that TRE and ADF are effective strategies to lose weight and improve cardiometabolic health, it should be noted that these results are not superior to those provided by diets that offer balanced daily energy restriction.[188-191] No long-term, randomized, controlled trials beyond 52 wks have been conducted to assess the benefits and feasibility of these approaches, and if such benefits can be sustained.

Summary: Dietary Intervention

Body weight is determined by a complex interplay among behavior, environment, and biological factors. The trajectory of weight loss is nonlinear and does not reflect the targeted energy deficit consistently as metabolic adaptation and suboptimal dietary adherence. Maximum average weight loss can be achieved by 6 months. After this, smaller losses can be maintained for up to 2 years, as treatment and follow-up taper off. Of note, energy requirements decrease following successful weight loss in proportion to the amount of weight lost. Consequently, if targeted energy intake is not reduced, maintaining the weight loss is unlikely.

The evidence suggests that no single dietary approach or eating pattern offers clearly superior short- or long-term success. Any approach that reduces energy intake will result in weight loss. There are wide ranges of options available with established efficacy, from which healthcare practitioners can recommend evidence-based strategies to their patients who would benefit from weight loss. Most approaches used a multidisciplinary team approach in which highly trained medical, nutrition, and behavioral experts work with individuals on weight loss management. Other settings may not yield comparable success.

While several diets and eating approaches have been found to have advantages for achieving weight loss, the choice of a specific dietary approach needs to be based on an individual's goals, medical needs, and preferences. Many people are motivated to lose weight faster during the initial phases of the weight loss journey, and this can favor an initially more aggressive level of energy restriction, if medically appropriate. For others, long-term weight control is the more motivating factor, and then standard dietary approaches at a more modest level of energy restriction may be preferable. The general health of the individual with obesity, including the presence or absence of specific comorbid conditions, also plays an important role in selecting a dietary approach and/or eating pattern. The provider and the client must have an ongoing relationship in which a long-term, multidisciplinary approach can be used. Interventions that include increased levels of social support and self-regulatory methods, such as goal setting and self-monitoring, along with increased contact with providers, have been shown to be more effective, irrespective of the mode of intervention.[193]

Physical Activity for Weight Loss

As previously discussed, a sustained, daily 500-kcal deficit is required to lose one pound of fat per wk. Adhering to this level of energy deficit by diet alone can be challenging. Adding physical activity to a LCD can facilitate achieving this energy deficit and sustaining it in the long term. According to a joint panel of experts from the American College of Sports Medicine (ACSM) and the CDC,[194] healthy adults should engage in at least 30 minutes of moderate intensity (defined as three to six metabolic equivalents [METs]) physical activity, preferably every day. These recommendations were updated by a joint panel of experts from the ACSM and AHA by detailing the types of activity and amounts needed to maintain health and prevent weight gain.[195] Adherence to these recommendations should be phased in with introductions and increases in lifestyle activities and walking before moving onto higher intensity cardio and resistance training that patients enjoy.

These specifications are for a minimum of 30 minutes of moderate-intensity physical activity (e.g., brisk walking) for 5 d/wk or vigorous-intensity (>6 METs; e.g., jogging) exercise for 3 d/wk. However, for individuals with obesity, a review of evidence from research studies by the ACSM indicates that more than 60 minutes of moderate-intensity physical activity for at least 4 d/wk can help in achieving clinically significant weight loss without dietary energy restriction.[196] However, when moderate-intensity physical activity (between 30 and 60 min/d for 5 d/wk) is coupled with energy restriction from the diet, weight loss is easier to achieve.[196] For weight maintenance or prevention of weight regain, moderate-intensity physical activity of more than 60 min/d for at least 5 days is needed.[196]

Effect of Exercise on Weight Loss

The level of physical activity to lose weight recommended by the ACSM may seem like a daunting task for many individuals with overweight and obesity. When starting an exercise program, it is worthwhile to start with small goals and increase the level and intensity of exercise gradually to reach ACSM recommendations. One study reported that, at the end of 1 year, energy restriction plus more than 200 minutes of physical activity per week resulted in significantly greater weight loss among sedentary obese participants, along with improved cardiovascular fitness, compared with those who were physically active for less than 150 min/wk. This study reported no significant effect of the exercise intensity level on change in body weight.[197]

Although adding physical activity to a weight loss regimen can make achieving a desired level of energy restriction easier, another study reported no added effect of physical activity plus energy restriction on body weight loss, fat loss, and visceral fat loss after 6 months of intervention.[198] In this report, the control group (dietary energy restriction of 25%) had lost more than 18 lb, approximately 6 kg of fat mass, and approximately 1 kg of visceral fat, whereas the intervention group (12.5% reduced energy intake and 12.5% increased energy expenditure through exercise) weight loss was approximately 18 lb, approximately 6.5 kg, and approximately 1 kg, respectively. However, those who exercised along with dietary energy restriction showed improved cardiovascular fitness.[198] Like these results, Nicklas et al.,[199] reported no significant differences in body weight or visceral fat loss among those who used a weight loss regimen that included energy restriction with or without moderate- and vigorous-intensity exercise. However, those who lost weight and visceral fat on the energy restriction along with vigorous activity preserved lean body mass better.[199]

Thus, several studies showed beneficial effects of energy restriction plus exercise on body composition changes, but without additional weight loss benefits of added exercise.[197-200] It is important that exercise along with energy restriction appears to enhance cardiovascular health benefits and preserve lean body mass[196-200] and this approach is, therefore, advisable to include in a weight-loss regimen unless physical conditions contraindicate exercise.

Exercise and Weight Maintenance

Although losing weight is difficult, maintaining weight is even more challenging and weight regain is common and often rapid once a strict weight loss regimen is ended. This regain can be attributed to low adherence to maintenance levels of exercise and dietary compliance.[201] Many factors contribute to this relapse, including low motivation and diminished or terminated support from the provider. One study showed that after the initial weight loss phase (2 months; mean weight loss achieved ~14 kg), participants regained, on average, more than 60% of the weight lost by the end of an 8- to 31-month unsupervised weight maintenance phase. This weight regain was accompanied by significant gains in fat mass and waist circumference, along with decreased levels of physical activity.[201]

Although weight regain may be difficult to prevent entirely, continuation of some form of exercise may slow and diminish its degree.[201,202] It was shown that after a 12-wk weight loss regimen, those who continued walking at moderate intensity regained weight and fat at a slower pace than those who were sedentary during a 40-wk maintenance phase.[202] Those who exercised regained approximately 7 lb less in body weight and had a 3.8-cm smaller waist circumference, compared with those who did not exercise.[203] A supervised weight maintenance regimen can help in slowing weight and fat mass regain significantly. Borg et al.[202] reported the benefits of supervised walking or resistance training for 6 months after weight loss. Participants incorporating walking or resistance training regained less (~4 lb) or no weight and maintained fat mass.[202] However, other studies show that energy restriction along with high-intensity physical activity may be the most effective strategy to maintain weight loss on a long-term basis.[204,205]

Resistance Training for Weight Loss and Maintenance

While aerobic exercise helps in increasing energy expenditure, resistance (strength) training can be added to help preserve muscle mass and lower the decrease in basal metabolic rate (BMR).[194,195,201] Studies have shown that resistance training, along with aerobic exercise, in addition to an energy-restricted diet, results in weight loss along with muscle strength preservation and improvements in physical functioning.[205] One study reported that those who incorporated resistance training during the weight maintenance phase were likely to regain fat mass at a slower pace, or were less likely to regain fat mass, compared with those who did not, in spite of overall body weight regain.[201] However, it is important to note that resistance training may not correlate with body weight changes during the weight loss phase and may even result in weight gain due to muscle mass gain, which is denser than fat.[194,206]

Although aerobic training has been emphasized for losing body weight and fat, combining resistance training with an energy-restricted regimen for weight loss can indirectly aid in weight and visceral fat loss by improving muscle mass, strength, and endurance.[206,207] Moreover, these changes are associated with health benefits, such as increased insulin sensitivity in type 2 diabetes.[207,208] One study reported that resistance training improved insulin sensitivity and weight loss more than aerobic training in an African American population.[208] In addition, resistance training appears to decrease bone loss, which can be seen in those who lose weight by energy restriction alone[209] and improves BMR,[210] which is associated with more fat-free mass in individuals with obesity.[211]

Summary: Physical Activity for Weight Loss

Physical activity along with energy restriction has a more positive influence on weight loss and maintenance of weight loss than energy restriction alone. Additionally, energy restriction alone may lower BMR and fat-free mass and adversely affect bone parameters. Adding endurance and resistance training to energy restriction may mitigate these weight loss-associated issues. Exercise, paired with energy restriction, should be performed at least five times per week for more than 240 minutes to lose weight and, thereafter, at least five times per week for 150 to 240 minutes to help maintain weight loss. Although cutting back on energy and simultaneously increasing energy expenditure may seem intimidating, planning an exercise schedule that is based on enjoyable activities may help ease this set of behavior changes. Encouraging individuals to add more movement into their daily lives (e.g., taking the stairs rather than the elevator), brainstorming how they can exercise with family and friends, and helping them add variety to their routines to avoid boredom are all useful tools. Exercise should start slowly and be increased progressively, based on individuals' tolerance level to enhance adherence.

Metabolic Changes and Sleep as Related to Weight and Dieting

Along with the increased prevalence of obesity,[212] an increase in self-reported sleep deprivation has been noted. Average sleep time has fallen in the general population by an average of 2 hours per night.[213,214] It has been suggested that chronic sleep deprivation and obesity could be linked. Both conditions are most prevalent in lower socio-economic classes, women, and racial-ethnic minorities.[215] Chronic sleep deprivation has been associated with chronic conditions such as increased body weight, glucose intolerance, and cardiovascular disease.[216-219] On the other hand, sleep duration >8 hours has been associated with increased mortality, suggesting a U-shaped curve between sleep duration and morbidity.[220]

It is a widespread notion that there may be adaptations to chronic sleep deprivation that prevent its negative effects, but growing clinical evidence suggests otherwise. Mice with faulty body clocks became morbidly obese after only a few weeks.[221] In humans, a short course of sleep restriction-induced insulin resistance, hypercortisolism, leptin deficiency, and increases in interleukin-6 (IL-6) and other proinflammatory cytokines, which all have the potential for long-term weight gain.[222] In one study, 12 healthy young men were randomly assigned to 2 nights of sleep restriction or sleep extension in a controlled laboratory setting for 6 weeks. When volunteers remained in bed for no more than 4 hours, decreased leptin, increased ghrelin, and increased cortisol concentrations were observed. Additionally, hunger rating scores were higher after sleep restriction than sleep extension, proportional to the changes in the ghrelin-to-leptin ratio.[222]

Studies in young, healthy volunteers suggest that sleep deprivation is both a biologic and a physiologic stressor.[223] Chronic periods of short sleep were shown to increase appetite, blood pressure, and evening cortisol, insulin, and blood glucose. In addition, increased interleukin-1, tumor necrosis factor-α, and IL-6, have been shown to contribute to a diabetogenic environment.[224] Since the 2004 report by Hasler et al.,[225] increasing epidemiologic evidence points to the effects of sleep deprivation on body weight and related comorbidities. A meta-analysis showed that insufficient sleep is associated with increased risk of obesity in children (odds ratio (OR) 1.89; 1.46–2.43; $P < 0.0001$) and in adults (OR 1.55; 1.43–1.68; $P < 0.0001$).[226] However, self-reported sleep and body weight data were used for this analysis and it is not clear how much, and for how long, sleep deprivation must be present before an effect on weight is evident.[227] The association between short sleep and obesity has been primarily related to increased appetite and food intake,[227-230] whereas the repercussions of chronic sleep deprivation on resting energy expenditure (REE) and BMR have not been explored.

Sleep deprivation causes interconnected systemic endocrine and immune system dysregulation.[231-233] Sleep curtailment also affects the endocrine system in its entirety, alters sympathovagal balance, and leads to activation of both efferent limbs of the stress system, the hypothalamic-pituitary-adrenal (HPA) axis, and the sympathetic nervous system.[234,235] Concomitantly, the GH, gonadal, and thyroid axes are inhibited.[236] Sleep deprivation is associated with alterations in body temperature regulation,[237] heart rate,[215] and metabolic rate.[238] Given the known stimulatory effects of cortisol and catecholamines on energy expenditure, sleep deprivation may, in theory, lead to increased energy expenditure.[239] With the evidence of sleep on obesity risk, healthy sleep patterns should be a part of lifestyle behaviors recommended for treatment of overweight and obesity.

Behavior Therapy

Behavior therapy to treat obesity aims to identify events and stimuli that trigger inappropriate behaviors associated with excess energy intake, and to develop approaches to control these behaviors. Behavioral interventions focus on modifying cognitive, emotional, and social events that influence the individual's choices that result in weight gain. Studies show that behavior therapy as an adjunct to exercise and diet therapy to lose weight produces significantly better results than exercise and diet alone.[240,241]

Several components of behavior therapy can help the individual develop skills to achieve weight loss and successfully maintain it. These include setting goals, identifying the process to achieve these goals based on individual needs, monitoring progress, removing barriers, building support from friends and family, and controlling stimuli to inappropriate behaviors.[242] Weight loss and maintenance for obese individuals requires long-term lifestyle changes; thus, behavior modification should aim at setting realistic and measurable goals. Interventions that include group sessions rather than individualized sessions can be more effective treatment modalities for obesity.[243-245] Studies show that group treatment can produce significantly greater weight loss (~13%) than individual sessions (>11%)[244] and is more likely to help maintain weight loss.[245] The known benefits of social support may explain the superiority of group sessions over individual sessions.

Behavior modification can help address weight regain by focusing on strategies that promote higher adherence to lifestyle changes. One study showed that teaching individuals to take responsibility for weight maintenance by developing problem-solving skills to overcome setbacks improves weight loss maintenance.[246] These techniques and skills include providing incentives for motivation, learning how to preplan low-energy meals and control portions, providing additional support from peers, increasing provider and patient contact, encouraging self-monitoring, and others. Social support, along with these techniques, improves weight maintenance after initial weight loss.[246] Thus, several studies suggested that behavior therapy beyond the initial weight loss phase is important in maintaining weight loss.[245-248]

Because the long-term results of attempts at weight loss are often poor, it is important to introduce the individual at the beginning of treatment to attitudes and behaviors that are likely to foster long-term maintenance of weight loss including assessing readiness, allowing flexibility, and recording and rewarding successes.

Pharmacologic Treatment for Weight Loss

Lifestyle modifications discussed earlier form the cornerstone and first stage of any weight loss plan. Adjunctive anorectic medications may be useful if this approach (lifestyle modification including diet, exercise, and behavior therapy) alone does not result in weight loss. Pharmacotherapy may be particularly helpful when compliance to lifestyle modification begins to waver, or physical hunger becomes an issue during dieting. There is little doubt that such medications significantly increase weight loss during the period in which they are used, and they may help with maintaining weight loss (although weight tends to be regained even with continued drug use).[246] However, pharmacologic treatments are best suited for those who are severely obese (BMI >40 kg/m^2) or who have two or more significant medical comorbidities.[20]

The most used weight loss drug approved by the FDA is phentermine.[249] In addition to phentermine, several newer pharmacologic treatments have recently been approved to target the neurochemical pathways associated with energy homeostasis. The discovery of leptin unveiled several neurohormonal targets for pharmacologic treatments that include inhibition of neuropeptide Y, which stimulates food intake, and stimulation of the melanocortin-4 receptor, which inhibits food intake.[249]

Phentermine

Phentermine is a noradrenergic compound approved by the FDA in 1959 for "short-term" use, generally defined as a period of <12 wks.[249] One long-term study examined data from 300 patients treated with phentermine, 15 to 75 mg/d and found significant weight loss, and maintenance of more than 10% of initial weight loss for up to 8 years.[250] In general, participants did not report feeling hungry and reported better control over food intake and cravings; this control tended to diminish over time but could often be regained if doses were increased progressively. Common adverse effects reported included dry mouth and insomnia.[250]

One randomized controlled trial of a 24-wk intervention with a combination of pramlintide and phentermine (37.5 mg) reported weight loss of 11.3 %.[251] Increases in heart rates (4.5 beats/minute) and diastolic blood pressure (3.5 mm Hg) were also observed.[251] In combination with fenfluramine (removed from the market in 1997), phentermine use was shown to cause valvular heart disease.[252] However, phentermine alone has not been implicated. Phentermine use is common in the United States because of its generic availability at low cost. However, administration of phentermine for more than 12 wks is still considered "off-label" use in the United States; in Europe, the drug is not licensed for use at all.[249] It has been more recently available in combination with topiramate, an antimigraine/antiseizure medication such as Qysmia®. This combination shows greater efficacy than either agent alone.

Most recently, semaglutide (Ozempic®, and Wegovy®), a glucagon-like peptide receptor-I agonist (GLP-I RA), and Mounjaro®, a dual GLP-I and gastric inhibitory peptide (GIP) agonist were approved by the FDA for obesity, and diabetes, respectively. These agents are remarkably effective, with mean weight losses in the 14% to 22% range. Long-term experience and effectiveness are awaited.

In summary, pharmacologic therapy for the treatment of obesity holds promise and can be a useful adjunct to diet and lifestyle change. Moreover, when lifestyle modification has been ineffective or stalls, pharmacologic treatment as an adjunct therapy may be very helpful. However, side effects are associated with pharmacologic therapies and these drugs should be prescribed and monitored regularly by medical practitioners. Also, indiscriminate use by individuals who have no strong medical indication for weight loss leads to limited access to the short supply of the newest medications.

Surgical Treatment

Although lifestyle modifications that include combinations of diet, exercise, behavior therapy, and pharmacologic agents for severely obese people can produce changes in body weight, these changes may prove to be insignificant, with insufficient health benefits. In these patients, weight-loss surgery is a second-stage approach. The NIH consensus conference on gastrointestinal surgery for severe obesity concluded that weight-loss surgery may be an appropriate option for those who are severely obese (BMI >40 kg/m^2) or those who have a BMI >35 kg/m^2 with two or more obesity-related comorbidities.[20,253] Moreover, weight-loss surgery in such individuals has the potential to resolve associated medical comorbidities.[253] Bariatric surgery and its use in obesity treatment is covered in Chapter 62 of this book.

Conclusions

Obesity management is perhaps the greatest challenge faced by healthcare professionals today. With the rise in obesity rates comes a plethora of medical complications. Although increasing research in obesity has provided some clarity on the causes of the obesity epidemic, and has helped to develop increasingly effective treatment modalities. Further research is needed to understand the basis of behaviors adopted by affected individuals and ways to modify these behaviors effectively to achieve long-term weight management. Almost all adults who become obese do so not because of a specific medical or metabolic condition but because of lifestyle behaviors leading to increased food consumption and/or decreased energy expenditure.

Treatment of overweight and obesity involves a complex combination of dietary, physical activity, sleep, and potentially pharmaceutical interventions. There is no universal diet or treatment for overweight and obesity; thus patient-centered care and environmental changes are critical to reducing prevalence. Improving school meals and snacks, educating individuals on the importance of healthy foods, providing resources to improve physical activity, and educating parents on ways to enhance healthy eating and physical activity behaviors in their children are examples of measures that can reduce obesity rates. Although the obesity treatment modalities discussed in this chapter can be successfully used to assist individuals suffering from obesity, improving our environment may be the most important way to prevent and reverse this public health epidemic.

References

1. Wang Y, Beydoun MA, Liang L, et al. Obesity (Silver Spring) 2008;16:2323–30.
2. Ogden CL, Carroll MD, Curtin LR, et al. JAMA 2010;303: 242–9.
3. Hill JO. Endocr Rev 2006;27:750–61.
4. Kelly T, Yang W, Chen CS, et al. Int J Obes (Lond) 2008;32: 1431–7.
5. Popkin BM. Nutr Rev 2004;62:S140–3.
6. Du H, Feskens E. Acta Cardiol 2010;65:377–86.
7. Giskes K, van Lenthe F, Avendano-Pabon M, et al. Obes Rev 2011;12:e95–e106.
8. Feng J, Glass TA, Curriero FC, et al. Health Place 2010;16: 175–90.
9. Field AE, Coakley EH, Must A, et al. Arch Intern Med 2001;161: 1581–6.
10. Janssen I, Katzmarzyk PT, Ross R. Arch Intern Med 2002;162: 2074–9.
11. Finkelstein EA, Brown DS, Wrage LA, et al. Obesity (Silver Spring) 2010;18:333–9.
12. Puhl RM, Andreyeva T, Brownell KD. Int J Obes (Lond) 2008;32:992–1000.
13. Carr D, Jaffe KJ, Friedman MA. Obesity (Silver Spring) 2008;16: S60–8.
14. Latner JD, Stunkard AJ, Wilson GT. Obes Res 2005;13:1226–31.
15. Look ARG, Wing RR. Arch Intern Med 2010;170:1566–75.
16. Williamson DF, Thompson TJ, Thun M, et al. Diabet Care 2000;23:1499–504.
17. Gregg E, Williamson DF. The Relationship of Intentional Weight Loss to Disease Incidence and Mortality. New York, NY: Guilford Press, 2002.
18. Knowler WC, Barrett-Connor E, Fowler SE, et al. N Engl J Med 2002;346:393–403.
19. Rush EC, Goedecke JH, Jennings C, et al. Int J Obes (Lond) 2007;31:1232–9.
20. National Institutes of Health, National Heart, Lung, and Blood Institute. Obes Res 1998;6:51S–209S.
21. National Institutes of Health, National Heart, Lung, and Blood Institute. Obes Res 1998;6:51S–209S.
22. Ma J, Xiao L. Obesity 2010;18:347–53.
23. Onyike CU, Crum RM, Lee HB, et al. Am J Epidemiol 2003;158:1139–47.
24. Zhao G, Ford ES, Dhingra S, et al. Int J Obes (Lond) 2009;33: 257–66.
25. Steer RA, Brown GK, Beck AT, et al. Psychol Rep 2001;88: 1075–6.
26. de Zwaan M. Int J Obes Relat Metab Disord 2001;25:S51–5.
27. Sim LA, McAlpine DE, Grothe KB, et al. Mayo Clin Proc 2010;85:746–51.
28. Gomez-Rubalcava S, Stabbart K, Phelan S. In: Wadden TA, Bray GA eds. Behavioral Treatment of Obesity. Handbook of Obesity Treatment. New York, NY: Guilford Press, 2016; 336–49.
29. Abbott RA, Davies PS. Eur J Clin Nutr 2004;58:285–91.
30. Jackson DM, Djafarian K, Stewart J, et al. Am J Clin Nutr 2009;89:1031–6.
31. Centers for Disease Control and Prevention. Morb Mortal Wkly Rep 2011;60:614–8.
32. Slingerland AS, van Lenthe FJ, Jukema JW, et al. Am J Epidemiol 2007;165:1356–63.
33. Andersen RE, Wadden TA, Bartlett SJ, et al. JAMA 1999;281: 335–40.
34. Di Noia J, Prochaska JO. Am J Health Behav 2010;34:618–32.
35. Jensen MD, Ryan DH, Apovian CM, et al. Circ 2014;129: S102–38.
36. Abete I, Astrup A, Martinez JA, et al. Nutr Rev 2010;68:214–31.
37. Gibson AA, Sainsbury A. Behav Sci 2017;7:44.
38. Tay A, Hoeksema H, Murphy R. Nutrients 2023;15:1297.
39. Borecki IB, Higgins M, Schreiner PJ, et al. Obes Res 1998;6: 107–14.
40. U.S. Department of Agriculture Center for Nutrition Policy and Promotion. Dietary Guidelines for American 2020–2025. Washington, DC: US Government Printing Office, 2020.
41. Sacks FM, Bray GA, Carey VJ, et al. N Engl J Med 2009;360: 859–73.
42. Feidantsis K, Methenitis S, Ketselidi K, et al. Nutr 2021;91-92: 111365.
43. Ge L, Sadeghirad B, Ball GDC, et al. BMJ 2020;369:m696.
44. Razavi Zade M, Telkabadi MH, Bahmani F, et al. Liver Int 2016;36:563–71.
45. Nordmann AJ, Nordmann A, Briel M, et al. Arch Intern Med 2006;166:285–93.
46. Offringa LC, Hartle JC, Rigdon J, et al. Nutrients 2021;13:3625.
47. Iversen KN, Carlsson F, Andersson A, et al. Clin Nutr ESPEN 2021;45:155–69.
48. Cordova R, Kliemann N, Huybrechts I, et al. Clin Nutr 2021;40: 5079–88.
49. Harb AA, Shechter A, Koch PA, et al. Eur J Clin Nutr 2023;77, 619–627.
50. Clinical Guidelines on the Identification, Evaluation, and Treatment of Overweight and Obesity in Adults-The Evidence Report. National Institutes of Health [published correction appears in Obes Res 1998;6:51S–209S.
51. Karam G, Agarwal A, Sadeghirad B, et al. BMJ 2023;380: e072003.
52. Bischoff SC, Damms-Machado A, Betz C, et al. Int J Obes (Lond) 2012;36:614–24.
53. Lopez-Gomez JJ, Izaola-Jauregui O, Torres-Torres B, et al. Nutr Hosp 2018;35:71–7.
54. Noakes M, Keogh JB, Foster PR, et al. Am J Clin Nutr 2005;81:1298–306.
55. Holderbaum M, Casagrande DS, Sussenbach S, et al. Surg Obes Relat Dis 2018;14:237–44.
56. Gils Contreras A, Bonada Sanjaume A, Montero Jaime M, et al. Obes Surg 2018;28:3756–68.
57. Basciani S, Camajani E, Contini S, et al. J Clin Endocrinol Metab 2020;105:dgaa336.
58. Moreno B, Bellido D, Sajoux I, et al. Endocr 2014;47:793–805.
59. Gomez-Arbelaez D, Bellido D, Castro AI, et al. J Clin Endocrinol Metab 2017;102:488–98.
60. Goday A, Bellido D, Sajoux I, et al. Nutr Diabetes 2016;6:e230.
61. Sellahewa L, Khan C, Lakkunarajah S, et al. Curr Diabetes Rev 2017;13:35–46.
62. Ardavani A, Aziz H, Smith K, et al. Adv Ther 2021;38:149–63.
63. Saris WH. Obes Res 2001;9:295S–301S.
64. Atkinson RL, Fuchs A, Pastors JG, et al. Am J Clin Nutr 1992;56:199S–202S.
65. Stokes CS, Lammert F. Visc Med 2021;37:254–60.
66. Broomfield PH, Chopra R, Sheinbaum RC, et al. N Engl J Med 1988;319:1567–72.
67. Son SY, Song JH, Shin HJ, et al. J Metab Bariatr Surg 2022;11: 30–8.
68. Tay J, Luscombe-Marsh ND, Thompson CH, et al. Am J Clin Nutr 2015;102:780–90.
69. Tay J, Thompson CH, Luscombe-Marsh ND, et al. Diabet Obes Metab 2018;20:858–71.
70. Hyde PN, Sapper TN, Crabtree CD, et al. JCI Insight 2019;4:e128308.
71. Harvey C, Schofield GM, Zinn C, et al. Peer J 2019;7:e6273.
72. O'Neill B, Raggi P. Atherosclerosis 2020;292:119–26.

73. Paoli A, Rubini A, Volek JS, et al. Eur J Clin Nutr 2013;67: 789–96.
74. Trumbo P, Schlicker S, Yates AA, et al. J Am Diet Assoc 2002;102:1621–30.
75. Baron M, Health Care Food Nutr Focus 2004;21:10–11.
76. Baron M, Health Care Food Nutr Focus 2004;21:8–11.
77. Singh A, Singh D. Cureus 2023;15:e34214.
78. Oh R, Gilani B, Uppaluri KR. Low Carbohydrate Diet. Treasure Island, FL: *StatPearls*, 2023.
79. Atkins R. Dr Atkins' New Diet Revolution. New York, NY: Harper Collins, 2002.
80. Veldhorst M, Smeets A, Soenen S, et al. Physiol Behav 2008;94: 300–7.
81. Owen OE, Morgan AP, Kemp HG, et al. J Clin Invest 1967;46: 1589–95.
82. Westerterp-Plantenga MS, Nieuwenhuizen A, Tome D, et al. Annu Rev Nutr 2009;29:21–41.
83. Halton TL, Hu FB. J Am Coll Nutr 2004;23:373–85.
84. Feinman RD, Pogozelski WK, Astrup A, et al. Nutr 2015;31: 1–13.
85. Veech RL. Prostaglandins Leukot Essent Fatty Acids 2004;70:309–19.
86. Adam-Perrot A, Clifton P, Brouns F. Obes Rev 2006;7:49–58.
87. Tahreem A, Rakha A, Rabail R, et al. Front Nutr 2022;9:960922.
88. Shils M, Olson J, Shike M, et al. eds. Modern Nutrition in Health and Disease. Burlington, MA: Jones & Bartlett Learning, LLC, 1999.
89. Frigolet ME, Ramos Barragan VE, Tamez Gonzalez M. Ann Nutr Metab 2011;58:320–34.
90. Naude CE, Brand A, Schoonees A, et al. Cochrane Database Syst Rev 2022;1:CD013334.
91. Naude CE, Schoonees A, Senekal M, et al. PLoS One 2014;9:e100652.
92. Sainsbury E, Kizirian NV, Partridge SR, et al. Diabet Res Clin Pract 2018;139:239–52.
93. van Zuuren EJ, Fedorowicz Z, Kuijpers T, et al. Am J Clin Nutr 2018;108:300–31.
94. Schwingshackl L, Hoffmann G. J Acad Nutr Diet 2013;113: 1640–61.
95. Sacks FM, Lichtenstein AH, Wu JHY, et al. Circ 2017;136: e1–e23.
96. Wang DD, Hu FB. Annu Rev Nutr 2017;37:423–46.
97. Johnston CS, Tjonn SL, Swan PD, et al. Am J Clin Nutr 2006;83:1055–61.
98. Vernarelli JA, Mitchell DC, Rolls BJ, et al. Eur J Nutr 2018;57:351–61.
99. Westerterp KR. Physiol Behav 2006;89:62–5.
100. Bowman SA. Impact of Fruit and Vegetable Intakes on Food and Nutrient Intakes of Adults: What We Eat in America, NHANES 2017-2018. FSRG Dietary Data Briefs. Beltsville (MD): United States Department of Agriculture (USDA), 2022.
101. Du M, Mozaffarian D, Wong JB, et al. Am J Clin Nutr 2022;116:1704–14.
102. Hoy MK, Goldman JD, Moshfegh A. Potassium Intake of the U.S. Population: What We Eat in America, NHANES 2017-2018. FSRG Dietary Data Briefs. Beltsville, MD: United States Department of Agriculture (USDA), 2022.
103. Hoy MK, Goldman JD. Calcium Intake of the U.S. population: What We Eat in America, NHANES 2009-2010. FSRG Dietary Data Briefs. Beltsville (MD): United States Department of Agriculture (USDA), 2014.
104. Ledikwe JH, Blanck HM, Khan LK, et al. J Am Diet Assoc 2006;106:1172–80.
105. Rolls BJ, Roe LS, Meengs JS. Am J Clin Nutr 2010;91:913–22.
106. Hall KD, Ayuketah A, Brychta R, et al. Cell Metab 2019;30: 67–77.
107. Hays NP, Starling RD, Liu X, et al. Arch Intern Med 2004;164: 210–7.
108. Mueller-Cunningham WM, Quintana R, Kasim-Karakas SE. J Am Diet Assoc 2003;103:1600–6.
109. Gardner CD, Kiazand A, Alhassan S, et al. JAMA 2007;297: 969–77.
110. Li Z, Heber D. JAMA 2020;323:1104.
111. Lei L, Huang J, Zhang L, et al. Front Nutr 2022;9:935234.
112. Esselstyn CB, Jr., Gendy G, Doyle J, et al. J Fam Pract 2014;63:356–64b.
113. Friedman SM, Barnett CH, Franki R, et al. Am J Lifestyle Med 2022;16:374–81.
114. Jenkins DJ, Wolever TM, Taylor RH, et al. Am J Clin Nutr 1981;34:362–6.
115. De Rougemont A, Normand S, Nazare JA, et al. Br J Nutr 2007;98:1288–98.
116. Abete I, Parra D, Martinez JA. Clin Nutr 2008;27:545–51.
117. Juanola-Falgarona M, Salas-Salvado J, Ibarrola-Jurado N, et al. Am J Clin Nutr 2014;100:27–35.
118. Ni C, Jia Q, Ding G, et al. Nutrients 2022;14.
119. Pereira MA, Swain J, Goldfine AB, et al. JAMA 2004;292: 2482–90.
120. Ma Y, Pagoto SL, Griffith JA, et al. J Am Diet Assoc 2007;107: 1786–91.
121. Chiavaroli L, Viguiliouk E, Nishi SK, et al. Nutrients 2019; 11:338.
122. Van Horn L, Carson JA, Appel LJ, et al. Circ 2016;134: e505–e29.
123. Chrisman M, Diaz Rios LK. J Nutr Educ Behav 2019;51: 899–903.
124. Swain JF, McCarron PB, Hamilton EF, et al. J Am Diet Assoc 2008;108:257–65.
125. de Souza RJ, Swain JF, Appel LJ, et al. Am J Clin Nutr 2008;88: 1–11.
126. Esposito K, Maiorino MI, Ciotola M, et al. Ann Intern Med 2009;151:306–14.
127. Dellis D, Tsilingiris D, Eleftheriadou I, et al. Nutr 2020;71: 110578.
128. Sofi F, Dinu M, Pagliai G, et al. Circ 2018;137:1103–13.
129. Huang RY, Huang CC, Hu FB, et al. J Gen Intern Med 2016;31:109–16.
130. Sacks FM, Svetkey LP, Vollmer WM, et al. N Engl J Med 2001;344:3–10.
131. Crosby L, Rembert E, Levin S, et al. J Acad Nutr Diet 2022;122:1922–39 e0.
132. Ivanova S, Delattre C, Karcheva-Bahchevanska D, et al. Foods 2021;10:3052.
133. Kahleova H, Dort S, Holubkov R, et al. Nutrients 2018;10:1302.
134. Ello-Martin JA, Ledikwe JH, Rolls BJ. 2005;82:236S–41S.
135. Wansink B, Wansink CS. Int J Obes (Lond) 2010;34:943–4.
136. Gupta AK, Smith SR, Greenway FL, et al. Diabet Obes Metab 2009;11:330–7.
137. Faucher MA, Mobley J. J Midwifery Womens Health 2010;55: 60–4.
138. Pedersen SD, Kang J, Kline GA. Arch Intern Med 2007;167: 1277–83.
139. Hannum SM, Carson L, Evans EM, et al. Obes Res 2004;12: 538–46.
140. Wadden TA, Butryn ML, Wilson C. Gastroenterol 2007;132: 2226–38.
141. Vargas-Alvarez MA, Navas-Carretero S, Palla L, et al. Nutrients 2021;13:1978.
142. Jia SS, Liu Q, Allman-Farinelli M, et al. Nutrients 2022;14:892.
143. Almiron-Roig E, Majumdar A, Vaughan D, et al. Nutrients 2019;11:1095.
144. Christoph MJ, Ellison B. J Acad Nutr Diet 2017;117:1528–37.

145. U.S. Department of Agriculture. What is MyPlate? www.myplate.gov/eat-healthy/what-is-myplate#:
146. Kliemann N, Kraemer MVS, Scapin T, et al. Nutrients 2018;10:891.
147. Ditschuneit HH. Nestle Nutr Workshop Ser Clin Perform Programme 2006;11:171–81.
148. Ashley JM, Herzog H, Clodfelter S, et al. Nutr J 2007;6:12.
149. Konig D, Deibert P, Frey I, et al. Ann Nutr Metab 2008;52:74–8.
150. Davis LM, Coleman C, Kiel J, et al. Nutr J 2010;9:11.
151. Raynor HA, Anderson AM, Miller GD, et al. J Acad Nutr Diet 2015;115:731–42.
152. Smith TJ, Sigrist LD, Bathalon GP, et al. J Am Diet Assoc 2010;110:268–73.
153. Wadden TA, Butryn ML, Bryne KJ. Obes Res 2012;12:151S–62S.
154. Kant AK, Graubard BI. J Acad Nutr Diet 2015;115:50–63.
155. Kant AK, Graubard BI. Am J Clin Nutr 2019;109:1145–53.
156. Kant AK, Graubard BI. Am J Clin Nutr 2015;102:661–70.
157. Pereira Marot LP, Tibiletti Balieiro LC, do Vale Cardoso Lopes T, et al. Eur J Nutr 2023;62:1707–18.
158. Silva CM, Teixeira BS, Wright KP, Jr., et al. Nutrients 2022;14:2202.
159. Marot LP, Rosa DE, Lopes T, et al. J Am Coll Nutr 2021;40:624–31.
160. Gill S, Panda S. Cell Metab 2015;22:789–98.
161. Kovac J, Husse J, Oster H. Mol Cells 2009;28:75–80.
162. Haupt S, Eckstein ML, Wolf A, et al. Biomolecules 2021;11:516.
163. Cermakian N, Boivin DB. Obes Rev 2009;10:25–36.
164. Gnocchi D, Bruscalupi G. Biology (Basel) 2017;6:10.
165. Speechly DP, Rogers GG, Buffenstein R. Int J Obes Relat Metab Disord 1999;23:1151–9.
166. Speechly DP, Buffenstein R. Appetite 1999;33:285–97.
167. Ma Y, Bertone ER, Stanek EJ, 3rd, et al. Am J Epidemiol 2003;158:85–92.
168. Murakami K, Livingstone MB. J Acad Nutr Diet 2016;116:1101–13.
169. Murakami K, Livingstone MB. Public Health Nutr 2016;19:1635–44.
170. Bailey RL, Leidy HJ, Mattes RD, et al. Curr Dev Nutr 2022;6:nzac132.
171. Farshchi HR, Taylor MA, Macdonald IA. Eur J Clin Nutr 2004;58:1071–7.
172. Titan SM, Bingham S, Welch A, et al. BMJ 2001;323:1286–8.
173. Mekary RA, Giovannucci E, Willett WC, et al. Am J Clin Nutr 2012;95:1182–9.
174. Higgins KA, Hudson JL, Hayes AMR, et al. Adv Nutr 2022;13:248–68.
175. Piernas C, Popkin BM. J Nutr 2011;141:1159–64.
176. Sebastian RS, Goldman JD, Enns CW Snacking Patterns of U.S. Adolescents: What We Eat In America, NHANES 2005-2006. FSRG Dietary Data Briefs. Beltsville (MD): United States Department of Agriculture (USDA). 2010.
177. van der Heijden AA, Hu FB, Rimm EB, et al. Obesity 2007;15:2463–9.
178. Toschke AM, Kuchenhoff H, Koletzko B, et al. Obes Res 2005;13:1932–8.
179. Cameron JD, Cyr MJ, Doucet E. Br J Nutr 2010;103:1098–101.
180. Varady KA, Cienfuegos S, Ezpeleta M, et al. Annu Rev Nutr 2021;41:333–61.
181. Kerndt PR, Naughton JL, Driscoll CE, et al. West J Med 1982;137:379–99.
182. Di Francesco A, Di Germanio C, Bernier M, et al. Science 2018;362:770–5.
183. Chow LS, Manoogian ENC, Alvear A, et al. Obesity 2020;28:860–9.
184. Domaszewski P, Konieczny M, Dybek T, et al. Exp Gerontol 2023;174:112116.
185. Cienfuegos S, Gabel K, Kalam F, et al. Cell Metab 2020;32:366–78 e3.
186. Gabel K, Hoddy KK, Haggerty N, et al. Nutr Healthy Aging 2018;4:345–53.
187. Stekovic S, Hofer SJ, Tripolt N, et al. Cell Metab 2020;31:878–81.
188. Gabel K, Kroeger CM, Trepanowski JF, et al. Obesity (Silver Spring) 2019;27:1443–50.
189. Trepanowski JF, Kroeger CM, Barnosky A, et al. JAMA Intern Med 2017;177:930–8.
190. Lowe DA, Wu N, Rohdin-Bibby L, et al. JAMA Intern Med 2020;180:1491–9.
191. Carter S, Clifton PM, Keogh JB. JAMA Netw Open 2018;1:e180756.
192. Zhu S, Surampudi P, Rosharavan B, et al. Curr Opin Clin Nutr Metab Care 2020;23:387–94.
193. Greaves CJ, Sheppard KE, Abraham C, et al. BMC Public Health 2011;11:119.
194. Pate RR, Pratt M, Blair SN, et al. JAMA 1995;273:402–7.
195. Haskell WL, Lee IM, Pate RR, et al. Med Sci Sports Exerc 2007;39:1423–34.
196. Donnelly JE, Blair SN, Jakicic JM, et al. Med Sci Sports Exerc 2009;41:459–71.
197. Chambliss HO. Clin J Sport Med 2005;15:113–5.
198. Redman LM, Heilbronn LK, Martin CK, et al. J Clin Endocrinol Metab 2007;92:865–72.
199. Nicklas BJ, Wang X, You T, et al. Am J Clin Nutr 2009;89:1043–52.
200. Cox KL, Burke V, Morton AR, et al. Metab 2003;52:107–15.
201. Borg P, Kukkonen-Harjula K, Fogelholm M, et al. Int J Obes Relat Metab Disord 2002;26:676–83.
201. Fogelholm M, Kukkonen-Harjula K, Nenonen A, et al. Arch Intern Med 2000;160:2177–84.
203. Vogels N, Westerterp-Plantenga MS. Int J Obes (Lond) 2005;29:849–57.
204. Leser MS, Yanovski SZ, Yanovski JA. J Am Diet Assoc 2002;102:1252–6.
205. Anton SD, Manini TM, Milsom VA, et al. Clin Interv Aging 2011;6:141–9.
206. Hills AP, Shultz SP, Soares MJ, et al. Obes Rev 2010;11:740–9.
207. Walberg JL. Sports Med 1989;7:343–56.
208. Winnick JJ, Gaillard T, Schuster DP. Ethn Dis 2008;18:152–6.
209. Daly RM, Dunstan DW, Owen N, et al. Osteoporos Int 2005;16:1703–12.
210. Dolezal BA, Potteiger JA. J Appl Physiol (1985) 1998;85:695–700.
211. Lazzer S, Bedogni G, Lafortuna CL, et al. Obesity (Silver Spring) 2010;18:71–8.
212. Mokdad AH, Serdula MK, Dietz WH, et al. JAMA 1999;282:1519–22.
213. Webb W. Bull Psychon Soc 1975;6:47–8.
214. Patel SR, Hu FB. Obesity (Silver Spring) 2008;16:643–53.
215. Van Cauter E, Spiegel K. Ann N Y Acad Sci 1999;896:254–61.
216. Spiegel K, Leproult R, Van Cauter E. Lancet 1999;354:1435–39.
217. Irwin M. Brain Behav Immun 2002;16:503–12.
218. Arnedt JT, Owens J, Crouch M, et al. JAMA 2005;294:1025–33.
219. Taheri S, Lin L, Austin D, et al. PLoS Med 2004;1:e62.
220. Bliwise DL, Young TB. Sleep 2007;30:1614–5.
221. Turek FW, Joshu C, Kohsaka A, et al. Science 2005;308:1043–5.
222. Spiegel K, Tasali E, Penev P, et al. Ann Intern Med 2004;141:846–50.
223. McEwen BS. Metab 2006;55:S20–3.
224. Hotamisligil GS. Nature 2006;444:860–7.

225. Hasler G, Buysse DJ, Klaghofer R, et al. Sleep 2004;27:661–6.
226. Cappuccio FP, Taggart FM, Kandala NB, et al. Sleep 2008;31: 619–26.
227. Horne J. Sleep 2008;31:595–6.
228. Penev PD. Curr Opin Endocrinol Diabet Obes 2007;14: 374–381.
229. Spiegel K, Tasali E, Penev P, et al. Ann Intern Med 141:846–850.
230. St-Onge MP, Roberts AL, Chen J, et al. Am J Clin Nutr 2011;94:410–416.
231. Stamatakis KA, Punjabi NM. Chest 2010;137:95–101.
232. Schmid SM, Hallschmid M, Jauch-Chara K, et al. Sleep 2011;34:371–377.
233. Spiegel K, Leproult R, Van Cauter E. Lancet 1999;354: 1435–1439.
234. McEwen BS, Seeman T. Ann NY Acad Sci 1999;896:30–47.
235. Vgontzas AN, Mastorakos G, Bixler EO, et al. Clin Endocrinol 1999;51:205–215.
236. Chrousos GP, Gold PW. JAMA 267:1244–1252.
237. Lack LC, Gradisar M, Van Someren EJ, et al. Sleep Med Rev 2008;12:307–317.
238. Jung CM, Melanson EL, Frydendall EJ, et al. J Physiol 589: 235–244.
239. Tataranni PA, Larson DE, Snitker S, et al. Am J Physiol 1996;271:E317–E325.
240. Shaw K, O'Rourke P, Del Mar C, et al. Cochrane Database Syst Rev 2005;CD003818.
241. Avenell A, Broom J, Brown TJ, et al. Health Technol Assess 2004;8:iii-iv, 1–182.
242. Lang A, Froelicher ES. Eur J Cardiovasc Nurs 2006;5:102–14.
243. Renjilian DA, Perri MG, Nezu AM, et al. J Consult Clin Psychol 2001;69:717–21.
244. Miller WM, Franklin BA, Nori Janosz KE, et al. Metab Syndr Relat Disord 2009;7:441–6.
245. Cresci B, Tesi F, La Ferlita T, et al. Eat Weight Disord 2007;12:147–53.
246. Perri M, Corsica J. In: Wadden TA, ed. Improving the Maintenance of Weight Lost in Behavioral Treatment of Obesity. Handbook of Obesity. New York, NY: Guilford Press, 2002; 357–94.
247. Leermakers EA, Perri MG, Shigaki CL, et al. Addict Behav 1999;24:219–27.
248. Wadden TA, Vogt RA, Foster GD, et al. J Consult Clin Psychol 1998;66:429–33.
249. Vetter ML, Faulconbridge LF, Webb VL, et al. Nat Rev Endocrinol 2010;6:578–88.
250. Hendricks EJ, Greenway FL, Westman EC, et al. Obesity 2011;19:2351–60.
251. Aronne LJ, Halseth AE, Burns CM, et al. Obesity 2010;18: 739–46.
252. Connolly HM, Crary JL, McGoon MD, et al. N Engl J Med 1997;337:581–8.
253. Brolin RE. Nutr 1996;12:403–4.

A. OBESITY AND DIABETES

CHAPTER 62

Bariatric Surgery

Charis Ripley-Hager, Thomas Magnuson, and Michael Schweitzer

CHAPTER OUTLINE

Abbreviations

BMI, body mass index; **D-RYGB**, distal Roux-en-Y gastric bypass; **DS-BPD**, duodenal switch with biliopancreatic diversion; **IF**, intrinsic factor; **JIB**, jejunoileal bypass; **LAGB**, laparoscopic adjustable gastric band; **LVSG**, laparoscopic vertical sleeve gastrectomy; **RYGB**, Roux-en-Y gastric bypass; **TPN**, total parenteral nutrition

Introduction

Obesity, a disease that is created by a multitude of genetic and environmental factors, is a significant health concern in the United States. The consequences of obesity are as equally complex as its etiology, affecting every organ system in the human body, as well as imposing serious psychological stress—often associated with social isolation, depression, and numerous other psychologic comorbidities. Medical management has made significant advances in recent years with the introduction of GLP1 agonists for weight loss. These medications can achieve up to 15% total body weight loss (TBWL), the effect of which can be sustained over 2 years when combined with diet and exercise.[1] Despite these advances, bariatric surgical procedures remain the most effective means to achieve sustained weight loss, and provide durable treatment of obesity-associated morbidities.

Weight-loss surgery is not a simple "cure" for this very complex and debilitating disease. It does, however, provide a powerful tool for patients to achieve success. A successful long-term outcome depends on the patient's commitment to a lifetime of dietary and lifestyle changes. For this reason, there must be a multidisciplinary approach that includes surgeons, primary care physicians, psychologists, nurses, and dietitians, to provide critical instructions to help patients adhere to the dietary and lifestyle changes consistent with the surgery.

Types of bariatric surgery differ in expected outcomes, in terms of weight loss and the likelihood of predisposing patients to nutritional deficiencies postoperatively. To understand these deficiencies and their appropriate management fully, it is imperative to understand the origin of the deficit. The purpose of this chapter is to review the various surgical procedures currently offered and the potential nutritional deficiencies that may ensue. Healthcare professionals must be aware of these deficiencies and the practice guidelines that must be followed to prevent these deficiencies because some of them may have serious consequences.

Overview

The prevalence of obesity continues to increase at an alarming rate throughout the world. Based on the most recent CDC data (2020), obesity affects 42% of adults 20 years old and older in the

United States, amounting to more than 137 million people. The occurrence of obesity is similar between men and women, with the highest prevalence among adults between the ages of 40 and 59 years. More than 9% of adults are classified as morbidly obese, with a body mass index (BMI) greater than 40; this prevalence is higher in women than in men.[2]

Obesity is increasing across the globe in both adults and children and is a leading preventable cause of death worldwide. It is one of the most serious public health problems of the 21st century. Obesity is stigmatized in much of the modern world (particularly in the Western world), although it was widely perceived as a symbol of wealth and fertility at other times in history, and still is in some parts of the world.

Healthcare professionals need to be concerned about the prevalence of obesity because of the well-established relationships between excess body weight and serious medical conditions such as type 2 diabetes, hypertension, and heart disease, to name a few. These relationships have long been established in the adult obese population and more recently also have been observed at an increasing rate in the adolescent population.

Unfortunately, no single solution to prevent or treat obesity is beneficial for everyone. Treatment of obesity may include a combination of diet, exercise, behavior modification, and medications. For most patients, although these methods may provide a moderate amount of weight loss, the benefits are usually short lived. Hence, bariatric surgery has evolved over the past couple of decades and has been shown to be effective in reducing obesity-related comorbidities, improving the quality of life, and decreasing the number of sick days, monthly medication costs, and overall mortality. With the increasing rates of weight loss procedures, the quality, efficacy, and surgical outcomes have improved, particularly with the creation of Metabolic and Bariatric Surgery Accreditation and Quality Improvement Program (MBSAQIP) accredited centers through the American Society of Metabolic and Bariatric Surgery, and the American College of Surgeons. The benefits of bariatric procedures in morbidly obese patients outweigh the risks. With the advent of minimally invasive surgical procedures, bariatric surgery is a reasonable treatment option in those who strongly desire substantial weight loss and have life-threatening comorbid conditions.

Definition of Morbid Obesity

The definition and classification of obesity are based on a calculation of the BMI—weight in kilograms divided by height in meters squared. For most of the population, BMI provides a reliable indicator of the body fat composition. It is used to stratify patients into categories that may lead to health problems. Patients with a BMI of 30 to 35 kg/m^2 are considered to have class I obesity, 35 to 40 kg/m^2 is class II, and higher than 40 kg/m^2 is class III. Morbid obesity is defined as a BMI of 40 kg/m^2 or higher, or a BMI of 35 kg/m^2 or higher in patients with comorbidities. Patients are defined as suffering from super-obesity or mega-obesity if their BMI is higher than 50 or 70 kg/m^2, respectively.

Indications

The European Association for Endoscopic Surgery released clinical practice guidelines in 2020. These were endorsed by the International Federation for the Surgery of Obesity (IFSO) and highly correlated with the most recent ASMBS guidelines, which were supported by multiple medical and governmental groups, including the International Diabetes Federation[3] (**Table 62.1**). Patients are considered candidates for bariatric surgery if they have a BMI of 40 kg/m^2 or greater, a BMI between 35 and 40 kg/m^2 if a comorbidity exists that would be expected to improve with weight loss, or a BMI between 30–35 kg/m^2 with poorly controlled diabetes or hypertension despite maximum medical therapy. Relative contraindications include an inability to comply with postoperative requirements and follow-up, active alcohol or substance abuse, and uncontrolled psychiatric disease.

Table 62.1 Indications for Bariatric Surgery for Morbid Obesity

1. BMI ≥40 kg/m^2
2. BMI 35–40 kg/m^2 with associated comorbidities that are expected to improve with weight loss (hypertension, diabetes, sleep apnea)
3. BMI 30–35 kg/m^2 with poorly controlled diabetes or hypertension despite optimal medical therapy
4. BMI thresholds should be reduced by 2.5 kg/m^2 for Asian patients[4]
5. No medical contraindication to surgery

BMI, body mass index.

Preoperative Assessment of the Obese Patient

The evaluation of potential patients for bariatric surgery should involve a multidisciplinary team approach. This team should include a dietitian and a mental health professional familiar with bariatric surgery. They must obtain a complete past dietary and behavioral eating history, educate the patient on postoperative dietary expectations, examine the social support structure, and ensure that any psychiatric or behavioral disorders are optimally controlled. At the Johns Hopkins Center for Bariatric Surgery in Baltimore, all patients are required to attend a multidisciplinary preoperative education seminar. Postoperative participation in support groups is also encouraged.

Nutritional Deficits in the Obese Patient

Nutritional assessment must be an essential part of the preoperative evaluation of the obese patient. Despite the increased energy intake of the obese population, many suffer from various nutritional deficiencies, particularly the morbidly obese. Consumption of excess energy does not usually correlate with overconsumption of fresh fruits and vegetables or high-quality, nutrient-dense foods. Instead, this more likely correlates with the consumption of higher energy-dense processed foods of low nutritional quality, which are common in developed countries such as the United States. In fact, it is estimated that 27% to 30% of the daily energy intake of the average U.S. adult or child consists of these low-nutrient-dense food sources, with sweeteners and desserts contributing an estimated 18% to 24% of the total.[5,6] However, there have been positive trends in overall intake of higher quality nutritional foods among U.S. adults over the

21st century, with a recent decrease in proportion of energy consumption from low-quality carbohydrates.[7]

As the obesity epidemic continues to flourish, obesity should be recognized as a risk factor for many nutrient deficiencies. For example, obese individuals tend to have lower mean concentration of vitamin D, compared with lean subjects.[8] There are many theories behind this observation, including decreased consumption of vitamin D–fortified milk, sedentary lifestyle, reduced exposure to sunlight, and sequestration of the lipid-soluble vitamin in the excess adipose tissue, which can be verified by studies that show serum 25-hydroxyvitamin D (25[OH]D) concentration to be inversely proportional to increasing fat mass.[9,10] Low vitamin D can have deleterious effects on the immune system and may play a role in increasing the risk of cancers, diabetes, autoimmune diseases, and cardiovascular disease.[11] It is estimated that 25% to 80% of adult patients before bariatric surgery may have baseline vitamin D deficiency.[12,13] Studies reporting baseline nutritional deficits of adults presenting for bariatric surgery also showed low concentrations of the other fat-soluble vitamins, A, K, and E.[14,15]

Low vitamin B_{12} has been reported in up to 18% of severely obese adults,[16] and vitamin B_1 (thiamin) deficiency has been noted in up to 29% of patients undergoing bariatric surgery.[12] The prevalence of deficiencies of other B vitamins are not currently known because they are not frequently included in screening. Unfortunately, depending on the type of surgical procedure and compliance with postoperative supplementation, these deficiencies can become highly exacerbated.

Types of Procedures

The dramatic growth in bariatric surgical procedures performed over the past few decades can be attributed to many factors. Increased patient acceptance is one major factor, attributed in large part to the introduction of laparoscopic and minimally invasive surgical techniques. Significant advantages are offered with laparoscopic and minimally invasive surgery such as less pain, fewer wound complications, and early recovery with relatively low complication rates. Advances in anesthesia, critical care, and nutrition support are other milestones in the success of bariatric procedures.

The bariatric surgical options have traditionally been classified into the following three categories: restrictive procedures, malabsorptive procedures, and combined restrictive and malabsorptive procedures. These definitions are oversimplified, however, based on the complex interplay between hormonal and mechanical changes that are known to occur following bariatric surgical procedures. Restrictive procedures depend on restriction of the amount of food that enters the foregut and typically involve changes in gut hormones controlling hunger and satiety. In contrast, malabsorptive procedures depend on the malabsorption of nutrients by bypassing various segments of the small intestine. Combined restrictive and malabsorptive procedures are a combination of the two.

Restrictive Procedures

The laparoscopic adjustable gastric band (LAGB) received Food and Drug Administration approval in 2001 and has been in clinical use in the United States since that time, although its use has been decreasing since 2011, owing to weight regain and complications such as band slippage and erosion. It now accounts for less than 10% of primary bariatric operations performed in the United States. Overall rates of complications vary in the literature but have been reported up to 33% in some series, often requiring reoperation and contributing to a high (40%) failure rate over 5 years.[17] The LAGB is the only device that is adjustable after surgery, thus allowing for tightening or loosening of the band through a subcutaneous port placed for fluid injection. Other advantages of the band include relative ease of placement, lack of operative staple lines or need for bowel transection, and reversibility. The band does require, however, an average of five to six adjustments in the first year after surgery; and its success depends on patient compliance and close follow-up.

Dissection for the LAGB (**Figure 62.1**) is first performed bluntly at the angle of His, thus freeing up attachments for later insertion of the band. The gastrohepatic ligament is then opened, and the plane posterior to the gastroesophageal junction is bluntly dissected. The adjustable band is placed in the abdomen through a trocar in the left upper quadrant and is secured around the gastroesophageal junction, with a slightly diagonal orientation up toward the angle of His. One to four sutures are then placed from the fundus to the proximal stomach around the band, to secure the band in place and minimize the possibility of band migration or herniation. The band tubing is brought out through the left upper quadrant trocar site, where it is secured to the subcutaneous injection port. Fascia is cleared in this area, and the port is secured to the fascia, with care taken not to entrap or kink the band tubing. The band is left empty until 6 weeks postoperatively, when patients receive their first fill. Frequent office visits are usually necessary, especially over the first year, for fluid fills or removal to obtain an appropriate restriction with food ingestion to sustain an appropriate weight loss.

The laparoscopic vertical sleeve gastrectomy is currently the most performed operation in the United States, accounting for more than 70% of primary operations performed. Since its adoption, the LVSG has gained significant popularity, partially due to low complication rates and excellent weight loss and metabolic outcomes. While the LVSG is generally considered a

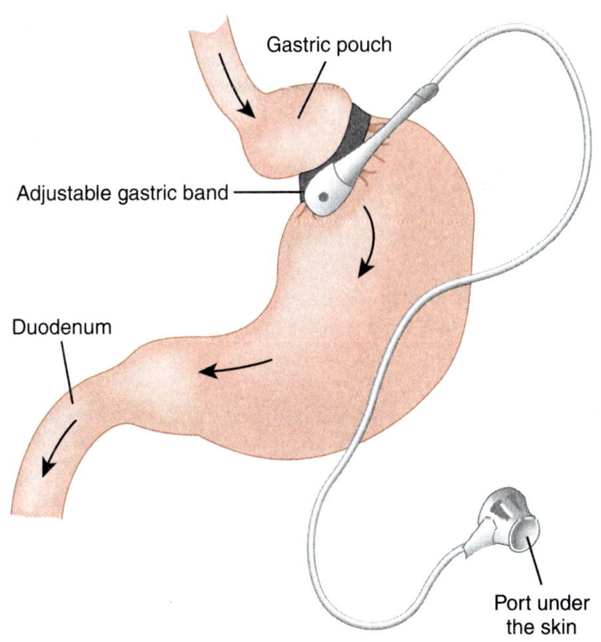

Figure 62.1 Laparoscopic adjustable gastric band.

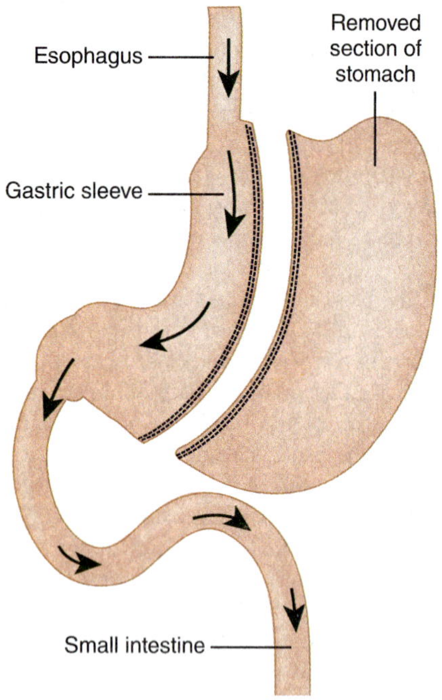

Figure 62.2 Laparoscopic vertical sleeve gastrectomy.

restrictive operation, it may also achieve weight loss by affecting satiety. Serum ghrelin, a proappetite hormone produced in the fundus, is reduced after the LVSG because that area of the stomach has been resected. The sleeve procedure is not reversible because a partial gastrectomy is performed, but it can be converted to a gastric bypass or duodenal switch later, if greater weight loss is desired.

The LVSG (**Figure 62.2**) is performed by first dividing the short gastric vessels along the greater curve of the stomach, beginning near the antrum and extending to the angle of His. A 40F bougie is placed in the stomach and is directed along the lesser curve. The stomach is then divided with the laparoscopic stapler by using the bougie as a guide, beginning 6 cm from the pylorus on the greater curve side and continuing up to the angle of His. The lateral stomach specimen is then removed from one of the trocar sites.

Malabsorptive Procedures

Jejunoileal bypass (JIB) is a purely malabsorptive procedure that was common in the 1960s and 1970s, despite lack of scientific study of its mechanism of action. This procedure bypasses approximately 90% of the small intestine. The proposed mechanism of action was induced weight loss by a surgically induced short gut syndrome. This procedure is based on canine studies performed in the mid-1950s that showed that 50% of the small intestine in dogs can be removed without apparent effects, with profound interference with fat absorption associated with weight loss.[18] However, many complications were associated with bypassing a large portion of the small intestine. Patients suffered frequent flatulence and diarrhea secondary to bypassing the site of bile acid reabsorption. Electrolyte deficiencies were common secondary to loss of potassium, calcium, and magnesium. Various vitamin deficiencies frequently led to neuropathies, bone demineralization, and protein malnutrition. Exposure of colonic mucosa to excessive bile salts and fatty acids created calcium oxalate renal stones. In addition, bacterial overgrowth in the bypassed small intestine led to hepatic decomposition and arthritis. It was later determined that the actual mechanism of action of weight loss from this procedure was learned behavior. The rectal complications and intense anal irritation from the diarrhea led to patients' changes in eating habits.[19] Patients learned very quickly that, to function in society, they had to consume minimal fat and nutrients before venturing away from their homes. For these reasons, the JIB procedure has long been abandoned, but it did pave the way for the more recent techniques of bariatric surgery.

The laparoscopic duodenal switch with biliopancreatic diversion (DS-BPD) is predominantly a malabsorptive operation that involves preservation of the gastric pylorus and creation of a short, 150-cm ileal "common channel," where food and biliopancreatic enzymes are allowed to mix. Because of the potential for malabsorption-related nutritional deficiencies and the complexity of the operation, DS-BPD is less commonly performed than either the LVSG or RYGB (approximately 10% of primary operations).

The first step in the DS-BPD (**Figure 62.3**) involves dividing the small bowel 300 cm from the ileocecal valve. The proximal end of the bowel is then anastomosed to the distal ileum 150 cm from its juncture with the cecum. A vertical sleeve gastrectomy is then performed over a 48F bougie to reduce the size of the stomach, thus providing some restriction as well. The duodenum is divided approximately 3 to 4 cm distal to the pylorus with a laparoscopic stapler. The Roux limb is then brought antecolic up to the end of the proximal duodenum, and a side-to-side anastomosis is performed. A simplified version of the DS-BPD termed the single-anastomosis duodenoiliostomy (SADI-S) involves the creation of a sleeve gastrectomy, followed by a single looped duodenoileostomy anastomosis 250–300 cm from the ileocecal valve.

Combined Restrictive and Malabsorptive Procedures

Roux-en-Y gastric bypass (RYGB) is the second most common bariatric procedure performed in the United States (just over 20%). It has been demonstrated in numerous reports to achieve durable long-term weight loss and remission of metabolic disease with a reasonably low (~3.6%) complication rate.[20] The procedure-related rate of remission of type 2 diabetes is among the highest of the bariatric procedures, second only to DS-BPD, but with fewer nutritional side effects: ~57% to 60%, depending on the preoperative severity and duration of the diabetes.[21,22] Normoglycemia often occurs within days after the operation, well before significant weight loss has occurred.[23] This finding suggests that the resolution of type 2 diabetes is related not only to restriction of energy intake but also to changes in gut peptide secretion secondary to bypassing a portion of the foregut. The exact mechanism remains to be elucidated, but it is an area of ongoing research.

For the RYGB (**Figure 62.4**), the jejunum is initially divided approximately 60 to 100 cm distal to the ligament of Treitz with a laparoscopic stapler. The proximal biliopancreatic limb of the jejunum is then anastomosed to the distal segment of jejunum 75 to 150 cm distal to the point of division. This anastomosis is performed in a side-to-side fashion. The mesenteric defect is closed with a running suture to help minimize the risk of internal hernia.

Next, dissection is performed at the angle of His, to expose the left crus, and at the gastrohepatic ligament, to gain access

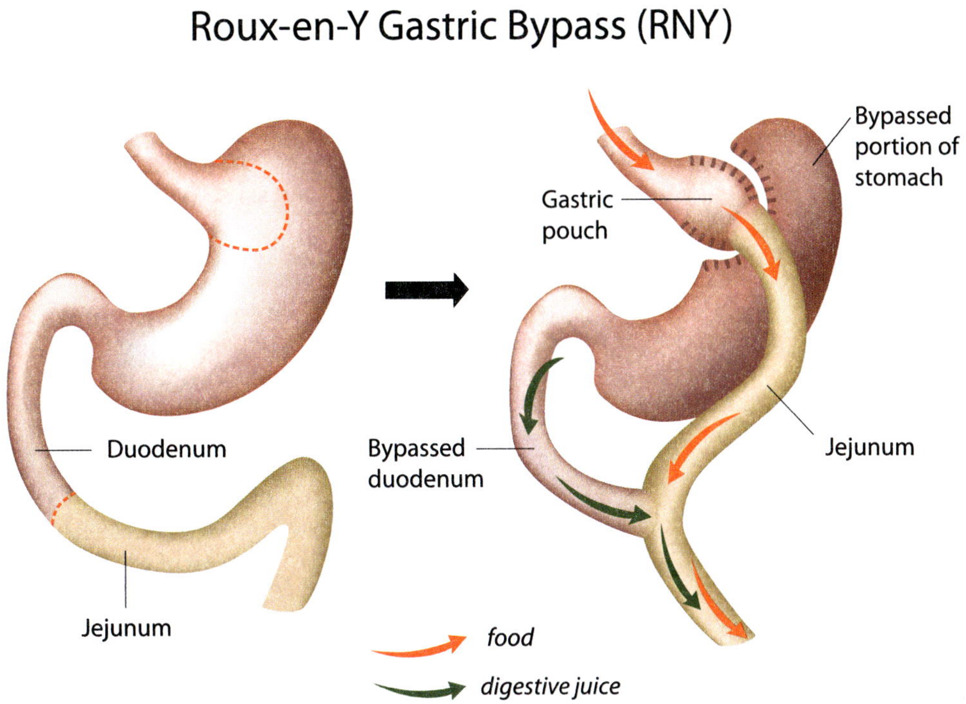

Figure 62.3 Duodenal switch with biliopancreatic diversion.

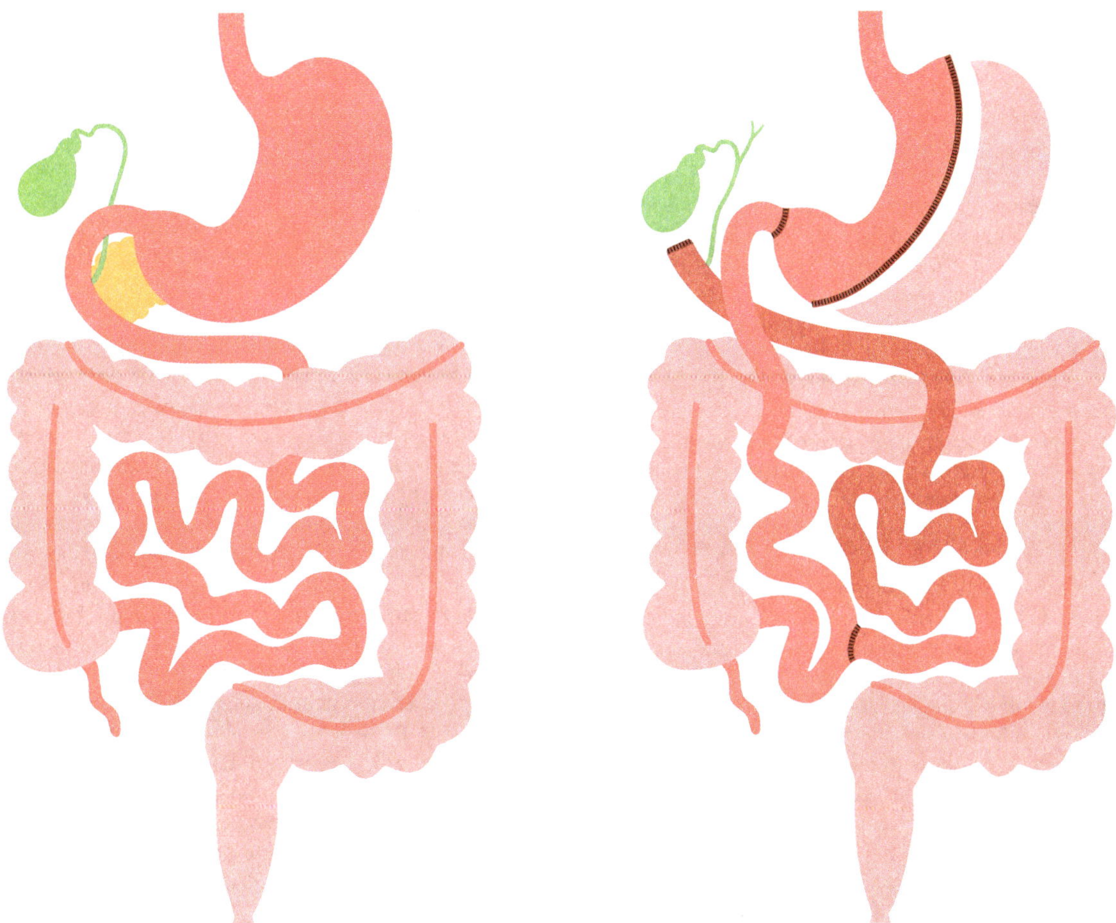

Figure 62.4 Roux-en-Y gastric bypass.

to the lesser sac. Multiple staple cartridges are then used to transect the stomach up to the angle of His, thus creating a vertically oriented, 20-mL proximal gastric pouch.

The Roux limb of the jejunum is routinely brought up to the gastric pouch in an antecolic-antegastric orientation. This seems to reduce the incidence of internal hernia and is simpler to perform than the retrocolic-retrogastric approach. The gastrojejunostomy is performed using a standard side-to-side technique.

Outcomes following the standard RYGB continue to provide good weight loss and resolution of comorbid conditions, and the procedure is considered by most surgeons to be the gold standard for weight-loss surgery. However, weight loss failure (BMI >35 or <50% EWL) can and still does occur, reportedly in 15% to 35% of cases.[24,25] This outcome is more commonly seen in superobese patients. It is common for these patients to seek additional surgery to attain their goals. One way to do this is to convert to a distal Roux-en-Y gastric bypass (D-RYGB). With this procedure, conversion is made to a 400 to 450 cm total alimentary limb length, therefore, promoting further malabsorption, which can result in 65.7% excess weight loss at 3 years.[26]

Potential Nutrient Deficits After Bariatric Surgery

As previously mentioned, early attempts at bariatric surgery, such as the JIB, shed light on the potential nutritional consequences of bypassing a significant portion of the small intestine. Even as the surgical techniques have evolved and improvements have been made, very specific guidelines still must be followed to prevent serious complications.

As the number of people who undergo bariatric surgery increases, more patients are being followed by general practitioners who need to be aware of these preventable complications. The type and frequency of the nutritional deficiencies are related to the type of operation performed. Restrictive procedures, such as the LAGB and LVSG, have the least impact on vitamin and mineral absorption because none of the small intestine is bypassed. DS-BPD is the procedure performed today with the greatest impact on nutrients because a large portion of the small bowel is bypassed, with only a short common channel for absorption. Regardless of the procedure, patients must constantly be monitored for the development of nutritional deficiencies and receive the appropriate supplementation.

Protein-Energy Malnutrition

Protein malnutrition, characterized by hypoalbuminemia, anemia, edema, and alopecia, can represent a serious complication of bariatric surgery, usually late in the postoperative period, with a peak incidence of 1 to 2 years after surgery.[27] The cause may be related to either excessive malabsorption from bypassing segments of small intestine in which protein is resorbed (more common in the malabsorptive procedures such as DS-BPD or distal-RYGB) or food limitation because of poor compliance, twist or stenosis of a sleeve gastrectomy, or stenosis of the gastric outlet in bypass procedures.[28]

Several studies have assessed protein deficiency after gastric bypass, showing that the length of the Roux limb is highly correlated with the chances of developing protein malnutrition. Brolin et al.,[29] performed a prospective randomized study of superobese patients (BMI >50); 13% of patients who underwent D-RYGB developed protein malnutrition over a 2-year follow-up. A similar study showed a 5.9% protein deficiency 20 months after D-RYGB[29]. Protein deficiency after standard RYGB is relatively uncommon, with a rate of up to 1.4% at 1 year.[27]

Protein malnutrition after DS-BPD was documented to occur in as many as 18% of patients in a retrospective study of 134 patients at 28 months postoperatively,[30] although the common channel that was created was only 75 cm long. Due to complications from intestinal malabsorption, most DS-BPD are currently formed with longer common channel lengths of at least 150 cm.

Regardless of the procedure, a few weeks of total parenteral nutrition (TPN) will likely correct the acute problem. Dietetic counseling and increased protein intake can help prevent recurrences in most patients.[27,28] If protein malnutrition persists despite counseling, reoperation with lengthening of the common channel may be required.

Patients should consume at least 60 g of protein daily after weight-loss surgery. This means selecting foods rich in protein and use of appropriate supplementation, such as protein shakes daily to meet this goal. This recommendation is important, to maintain lean body mass and prevent visceral protein wasting, especially in the initial 3 to 6 months after gastric bypass.[31]

Vitamin B_{12} and Folate Deficiency

Vitamin B_{12} deficiency, one of the most common deficiencies following gastric bypass surgery, occurs in up to one-third of patients 1 year after surgery.[32-34] Several mechanisms contribute to deficiency in vitamin B_{12} following RYGB. First, the amount of acid and pepsin produced in the new gastric pouch is inadequate to allow for the release of bound vitamin B_{12} from food.[35,36] Second, investigators have demonstrated that the gastric juice of the bypassed segment of stomach does not contain enough intrinsic factor (IF).[37] Therefore, there is a lack of vitamin B_{12}–IF complex available for uptake by the ileal enterocytes. Finally, the consumption of high B_{12} containing foods such as milk and meat is decreased following gastric bypass.[38-40]

Because of the alterations in the normal physiologic mechanisms by which vitamin B_{12} is absorbed after gastric bypass, most patients are not able to maintain normal vitamin B_{12} concentration with diet alone and require supplementation. Investigators have shown that normal serum vitamin B_{12} may be maintained by providing at least 350 to 500 μg of vitamin B_{11} daily, which is supported by the most recent ASMBS guidelines.[34,41] Intramuscular injection of 1 mg every month may also be given.[41] Other routes of vitamin B_{12} administration include sublingual (500 μg/day) and nasal spray (500 μg/week). Standard multivitamin supplements contain low amounts of vitamin B_{12} (6 to 25 μg/tablet) and are usually not sufficient to prevent the development of deficiency.

Vitamin B_{12} deficiency is usually subclinical in presentation and detected only by low serum concentration. However, occasionally, patients may present with megaloblastic anemia, thrombocytopenia, leukopenia, glossitis, and peripheral neuropathy. All are reversible with replacement therapy in the short run, but there is risk of irreversible neuropathy if deficiency is maintained over a long period of time.

Despite its absorption throughout the small intestine, folate deficiency also occurs after weight-loss surgery. Decreased folate concentration has been reported after restrictive procedures because nutritional intake is reduced despite the type of surgery.[42] Deficiency of folate may also be secondary to deficient vitamin B_{12} because the vitamin is needed to convert folate to its active form. Folate deficiency is associated with macrocytic anemia, leukopenia, thrombocytopenia, glossitis, megaloblastic marrow, and elevated homocysteine.[43] Folate supplementation of 400 μg/day is recommended because it is a water-soluble vitamin and is not stored in the body, adequate supplementation is present is most bariatric multivitamins.

Thiamin (Vitamin B_1) Deficiency

Thiamin (B_1) is another water-soluble vitamin that is not stored in the body. Investigators have shown that stores can be depleted within 10 to 20 days without adequate intake.[44] Asymptomatic thiamin deficiency is common and can occur in up to 18% of patients after an RYGB.[11] The development of peripheral neuropathy and Wernicke encephalopathy secondary to thiamin deficiency has been reported after all types of bariatric procedures and appears to be more common in patients who experience persistent vomiting, poor oral intake, or rapid weight loss after surgery.[45-47] Other predisposing factors include TPN deficient in thiamine, steroid ingestion, and malabsorption after bariatric surgery.[48] Because thiamin plays a role in carbohydrate metabolism, administration of intravenous dextrose fluid rehydration without concomitant administration of thiamin can further exacerbate brain cytotoxicity in bariatric patients.[49] Therefore, it should be common practice when postoperative patients present to the emergency department with significant dehydration associated with protracted vomiting or poor oral intake, that they be placed on an electrolyte and vitamin replacement protocol similar to patients who have alcoholism or eating disorders.

The classic triad of Wernicke encephalopathy including confusion, ataxia, and nystagmus may occur in patients deficient in thiamin. Other atypical neurologic features include third and sixth cranial nerve palsies, sensory and motor polyneuropathy, dysmetria, myoclonus, seizures, hearing loss, papilledema, paresis, and psychosis.[50] Symptomatic patients with signs and symptoms of Wernicke encephalopathy should have aggressive thiamine supplementation with up to 500 mg IV thiamine three times daily for the first 2 days of therapy, followed by 250 mg daily for an additional 5 days. Oral doses of 100 mg/day should then be continued until symptoms resolve.[51] For the majority of patients, this regimen will prevent the development of long-term, irreversible cognitive impairment. Asymptomatic patients may receive lower doses of thiamin at 100 mg IV daily until oral supplementation can be initiated. For patients who do not develop protracted vomiting and/or rapid weight loss, the administration of a daily multivitamin is generally sufficient to prevent deficiency.

Calcium and Vitamin D Deficiency

Calcium and vitamin D deficiencies are well-known risk factors in bariatric surgery. Vitamin D is absorbed mainly in the jejunum and ileum, whereas calcium is absorbed more proximally in the duodenum and proximal jejunum.[49] Malabsorption of these vitamins occurs secondary to bypass of these segments of the small intestine. In addition, the malabsorption of vitamin D contributes further to calcium malabsorption. With a decrease in serum calcium, the production of parathyroid hormone increases, resulting in bone resorption that ultimately leads to osteoporosis.

As mentioned earlier, many patients who present for bariatric surgery are already deficient in vitamin D. These patients must receive supplementation preoperatively, and all patients require supplementation postoperatively. Vitamin D and calcium deficiencies are more prevalent in patients who undergo DS-BPD and D-RYGB, which correlate with the shorter length of the common channel.[52] Despite supplementation, inadequate vitamin D and calcium can still occur after these procedures.

Iron Deficiency

Iron deficiency is one of the most common deficits after bariatric surgery, estimated to be between 30% and 50%.[50] This deficit is particularly prevalent in patients after RYGB secondary to bypassing the primary sites of absorption in the duodenum and jejunum. Inadequate intake of foods rich in iron, such as meats, is also a probable cause of deficiency. In addition, iron that is ingested is subjected to decreased acid from the small gastric pouch, which limits the ability to convert dietary ferric iron into the more absorbable ferrous form.[53]

Unfortunately, iron supplementation may not be well tolerated, especially early in the postoperative period secondary to the development of constipation. However, if patients have low iron saturation postoperatively, it is recommended that additional doses of oral ferrous sulfate (325 mg one to three times daily) be added.[41] Occasionally, patients may also require intravenous infusion of iron to replete stores. It is important to note that concomitant enteral administration of calcium with iron can inhibit iron absorption, so these supplements should be staggered in their administration.[54] Vitamin C supplementation may increase nonheme iron absorption but studies in human subjects have shown mixed results.[55,56] In patients with refractory iron deficiency anemia, copper deficiency should be considered, as copper is essential for iron absorption into cells.[57]

Other Nutritional Deficiencies

As mentioned previously, deficiencies in the fat-soluble vitamins (D, E, K, A) are common in the obese patient. These deficiencies can be exacerbated following surgery, especially after malabsorptive procedures, particularly DS-BPD. Symptomatic deficiencies, such as night blindness from vitamin A deficiency are uncommon, but have been described.[28,58]

Numerous reports in the bariatric literature have documented other micronutrient deficiencies such as zinc, magnesium, and selenium. Zinc deficiency may cause hair loss[49] and selenium deficiency is associated with cardiomyopathy,[30] but the overall significance of these deficiencies remains unclear because the bariatric surgery literature is limited to case reports at this time. Deficiencies are most often seen after malabsorptive procedures and usually in patients who do not follow the recommended postoperative multivitamin regimen.

Conclusions

Bariatric surgery is currently the most effective method of sustainable weight loss among morbidly obese patients. The types of surgery currently offered can be divided into three categories: restrictive, malabsorptive, and combined restrictive and malabsorptive. In general, patients who undergo restrictive procedures experience less weight loss and resolution of comorbidities, but they have the least risk for long-term nutritional complications. Malabsorptive procedures are associated with the greatest amount of weight loss and resolution of comorbidities, but they are associated with the greatest risk of nutritional complications. Most of these complications can be avoided by adhering to strict nutritional guidelines outlined in this chapter. In addition, the registered dietitian is a vital member of the bariatric team and provides critical instructions to help patients adhere to dietary changes before and after surgery. Given the increasing prevalence of obesity and bariatric surgery procedures, caring for patients who are contemplating or have undergone surgery will be an expanding role for the nutritionist and primary care physician alike.

Acknowledgments

The authors acknowledge Kevin Tymitz for his contributions to the previous version of this chapter.

References

1. Garvey WT, Batterham RL, Bhatta M, et al. Nat Med 2022;28:2083–91.
2. Adult obesity facts. Accessed: July 8, 2022. www.cdc.gov/obesity/data/adult.html
3. Di Lorenzo N, Antoniou SA, Batterham RL, et al. Surg Endosc 2020;34:2332–58.
4. American Diabetes Association. Diabet Care 2021;44:S100–S10.
5. Kant AK. Arch Pediatr Adolesc Med 2003;157:789–96.
6. Kant AK. Am J Clin Nutr 2000;72:929–36.
7. Shan Z, Rehm CD, Rogers G, et al. JAMA 2019;322:1178–87.
8. Yetley EA. Am J Clin Nutr 2008;88:558S–64S.
9. Buffington C, Walker B, Cowan GS, Jr., et al. Obes Surg 1993;3:421–4.
10. Vilarrasa N, Maravall J, Estepa A, et al. J Endocrinol Invest 2007;30:653–8.
11. Holick MF. N Engl J Med 2007;357:266–81.
12. Flancbaum L, Belsley S, Drake V, et al. J Gastrointest Surg 2006;10:1033–7.
13. Ernst B, Thurnheer M, Schmid SM, et al. Obes Surg 2009;19:66–73.
14. Pereira S, Saboya C, Chaves G, et al. Obes Surg 2009;19:738–44.
15. Clements RH, Katasani VG, Palepu R, et al. Am Surg 2006;72:1196–202; discussion 203–4.
16. Gemmel K, Santry HP, Prachand VN, et al. Surg Obes Relat Dis 2009;5:54–9.
17. Suter M, Calmes JM, Paroz A, et al. Obes Surg 2006;16:829–35.
18. Kremen AJ, Linner JH, Nelson CH. Ann Surg 1954;140:439–48.
19. Condon SC, Janes NJ, Wise L, et al. Gastroenterol 1978;74:34–7.
20. Chen SY, Stem M, Schweitzer MA, et al. Surgery 2015;158:777–86.
21. McTigue KM, Wellman R, Nauman E, et al. JAMA Surg 2020;155:e200087.
22. Vetter ML, Cardillo S, Rickels MR, et al. Ann Intern Med 2009;150:94–103.
23. Pories WJ, Albrecht RJ. World J Surg 2001;25:527–31.
24. Buchwald H, Avidor Y, Braunwald E, et al. JAMA 2004;292:1724–37.
25. Parikh M, Pomp A, Gagner M. Surg Obes Relat Dis 2007;3:611–8.
26. van der Burgh Y, Boerboom A, de Boer H, et al. Surg Obes Relat Dis 2020;16:381–8.
27. Skroubis G, Sakellaropoulos G, Pouggouras K, et al. Obes Surg 2002;12:551–8.
28. Scopinaro N, Adami GF, Marinari GM, et al. World J Surg 1998;22:936–46.
29. Brolin RE, LaMarca LB, Kenler HA, et al. J Gastrointest Surg 2002;6:195–203; discussion 4–5.
30. Dolan K, Hatzifotis M, Newbury L, et al. Ann Surg 2004;240:51–6.
31. Raymond JL, Schipke CA, Becker JM, et al. Surgery 1986;99:15–9.
32. Rhode BM, Arseneau P, Cooper BA, et al. Am J Clin Nutr 1996;63:103–9.
33. Rhode BM, Tamin H, Gilfix BM, et al. Obes Surg 1995;5:154–8.
34. Juhasz-Pocsine K, Rudnicki SA, Archer RL, et al. Neurol 2007;68:1843–50.
35. Smith CD, Herkes SB, Behrns KE, et al. Ann Surg 1993;218:91–6.
36. Behrns KE, Smith CD, Sarr MG. Dig Dis Sci 1994;39:315–20.
37. Marcuard SP, Sinar DR, Swanson MS, et al. Dig Dis Sci 1989;34:1238–42.
38. Moize V, Geliebter A, Gluck ME, et al. Obes Surg 2003;13:23–8.
39. Yue TP, Mohd Yusof BN, Nor Hanipah ZB, et al. Clin Nutr ESPEN 2022;48:321–8.
40. Avinoah E, Ovnat A, Charuzi I. Surgery 1992;111:137–42.
41. Parrott J, Frank L, Rabena R, et al. Surg Obes Relat Dis 2017;13:727–41.
42. Gasteyger C, Suter M, Calmes JM, et al. Obes Surg 2006;16:243–50.
43. Malone M. Ann Pharmacother 2008;42:1851–8.
44. Sechi G, Serra A. Lancet Neurol 2007;6:442–55.
45. Bozbora A, Coskun H, Ozarmagan S, et al. Obes Surg 2000;10:274–5.
46. Escalona A, Perez G, Leon F, et al. Obes Surg 2004;14:1135–7.
47. Makarewicz W, Kaska L, Kobiela J, et al. Obes Surg 2007;17:704–6.
48. Singh S, Kumar A. Neurol 2007;68:807–11.
49. Bloomberg RD, Fleishman A, Nalle JE. Obes Surg 2005;15:145–54.
50. Shuster MH, Vazquez JA. Crit Care Nurs Q 2005;28:227–60; quiz 61–2.
51. So Y. Wernicke encephalopathy. Accessed: July 29, 2022. www.uptodate.com/contents/wernicke-encephalopathy
52. Hamoui N, Kim K, Anthone G, et al. Arch Surg 2003;138:891–7.
53. Rhode BM, Shustik C, Christou NV, et al. Obes Surg 1999;9:17–21.
54. Cook JD, Dassenko SA, Lynch SR. Am J Clin Nutr 1991;54:717–22.
55. Li N, Zhao G, Wu W, et al. JAMA Netw Open 2020;3:e2023644.
56. Cook JD, Reddy MB. Am J Clin Nutr 2001;73:93–8.
57. Collins JF, Prohaska JR, Knutson MD. Nutr Rev 2010;68:133–47.
58. Meyer JH. Gastroenterol Clin North Am 1994;23:227–60.

A. OBESITY AND DIABETES

CHAPTER 63

Nutrition Management of Diabetes

Eugene Arnold, Shabina Roohi Ahmed, and Susan Oh

CHAPTER OUTLINE

Abbreviations

ADA, American Diabetes Association; **CDCES**, certified diabetes care and education specialist; **CKD**, chronic kidney disease; **CVD**, cardiovascular disease; **DCCT**, Diabetes Control and Complications Trial; **DKA**, diabetic ketoacidosis; **DPP-4**, dipeptidyl peptidase-4; **DSMES**, Diabetes Self-Management Education and Support; **FDA**, Food and Drug Administration; **FPG** fasting plasma glucose; **GAD**, glutamic acid decarboxylase; **GD**, gestational diabetes; **GI**, glycemic index; **GLP-1**, glucagon-like peptide-1; **GLUT4**, glucose transporter-4; **GRAS**, generally regarded as safe; **HNF**, hepatocyte nuclear factor; **HHS**, hyperglycemic hyperosmolar state; **IDF**, International Diabetes Federation; **IF**, Intermittent fasting; **IFG**, impaired fasting glucose; **IGT**, impaired glucose tolerance **IPF-1**, insulin promoter factor-1; **LC**, low carbohydrate; **LDL-C**, low-density lipoprotein cholesterol; **MNT**, medical nutrition therapy; **MODY**, maturity-onset diabetes of the young; **NeuroD1**, neurogenic differentiation factor-1; **OD**, other types of diabetes; **OGTT**, oral glucose tolerance test; **PUFA**, polyunsaturated fat; **RDA**, recommended dietary allowance; **RDN**, registered dietitian nutritionist; **RYGB**, Roux-en-Y gastric bypass; **SFA**, saturated fat; **T1D**, type 1 diabetes; **T2D**, type 2 diabetes; **TRE**, time-restricted eating; **UKPDS**, UK Prospective Diabetes Study; **VLC**, very low carbohydrate; **VLCD**, Very low-energy (calorie) diet

Introduction

Diabetes mellitus (diabetes) is a metabolic disorder primarily characterized by abnormally high concentrations of glucose (sugar) in the blood, with other multiple metabolic abnormalities often present. It is one of the most prominent epidemics worldwide. Diabetes afflicts 37.3 million people of all ages in the United States and 463 million worldwide, with the rate of diabetes-related deaths predicted to double between 2005 and 2030.[1-3] In 2014, diabetes was the direct cause of 1.5 million deaths, and 48% of all deaths due to diabetes occurred before the age of 70 years.[3] Diabetes and its complications have serious health and economic impacts. It is the seventh leading cause of death in the United States,[4] and a major cause of blindness, nontraumatic amputations, end-stage renal disease and dialysis, and peripheral neuropathy.[5]

Diabetes significantly increases the risk of heart disease and stroke.[5] Bloom et al., estimate that losses in GDP worldwide from 2011 to 2030, including both the direct and indirect costs of diabetes, will total U.S. 1.7 trillion dollars, comprising U.S. $900 billion for high-income countries and U.S. $800 billion for low- and middle-income countries.[5] With decreased physical activity and increased availability of energy-dense foods, the prevalence of diabetes, like obesity, is increasing globally, especially in developing countries and among adolescents. It has more than doubled in the last 20 years.

As demonstrated by the Diabetes Control and Complications Trial and the Diabetes Prevention Program,[6,7] lifestyle modification is the most successful strategy in treating and preventing diabetes. Because serum glucose depends on dietary intake, medical nutrition therapy (MNT) remains a cornerstone of treatment, and it can be effective in managing the disease. Physical activity is also essential to achieving lifestyle changes. Self-monitoring of blood glucose concentration and appropriate food intake, exercise, and medication adjustments can facilitate optimal glycemic control. In all cases, an integrated, interprofessional healthcare team is essential for persons with diabetes to maintain quality of life and longevity.

Historical Overview

Classic symptoms of diabetes, such as polyuria and polydipsia, have been observed and treated with dietary interventions for thousands of years in Egyptian, Indian, and Greek civilizations. One of the earliest reports, the Ebers Papyrus, written in 1500 BC (discovered in a tomb in the Thebes region in southern Egypt in 1862 and named after the Egyptologist Geary Ebers), describes a common symptom—polyuria of the "sugar disease."[8] Egyptians suggested various dietary remedies for this syndrome, including a diet of beer, fruits, grains, and honey.[8] Early Indians described diabetes-like symptoms and recommended freshly harvested cereals and bituminous preparations containing benzoates and silica as a remedy.[9] Aretaeus of Cappadocia (81 to 138 AD) first used the Greek word "diabetes," literally meaning "to run through" or siphon. He described the disease as the "melting down of flesh and limbs into the urine." He concluded that diabetes was a disease of the stomach and should be treated with milk, gruel, cereals, fruits, and sweet wines. Milk, water, wine, and beer were also used as the main fluids to relieve excessive thirst until the second century AD, when diabetes was thought to be a disease of the kidneys.[10] A London physician, Thomas Willis, added "mellitus," meaning honeylike, after noticing the sweet taste of urine.

In the late 1700s, a French physician began prescribing undernutrition or semistarvation diets interspersed with frequent fasting periods for patients with diabetes.[10] During the early 1900s, Dr. Frederick M. Allen developed his starvation diet in the United States. He was one of the first to customize or "individualize" diets to patients' preferences while providing only 1,000 kcal per day. Although many of his patients had malnutrition, Allen is credited with helping many survive before the introduction of insulin therapy in 1921.[10] The discovery of insulin in the 1920s dramatically increased survival for those afflicted with diabetes.

However, diet recommendations remained controversial. In the 1940s, it became evident that there was a need for consistent and standardized food values to develop a simple method for planning the diabetes diet. The Academy of Nutrition and Dietetics (formally called the American Dietetic Association), the American Diabetes Association, and the U.S. Public Health Service created a method known as the exchange system or "exchange scheme."[10] Six "convenient" sample meal plans were developed from these exchange lists that detailed the grams of carbohydrate, protein, and fat for various energy concentrations. In the early 1970s, these sample meal plans were extensively used without nutrition education or follow-up with a registered dietitian. Patients found it difficult to adhere to these rigid diets and quickly lost motivation. In the 1970s, official dietary recommendations emphasized a diet higher in complex carbohydrates and lower in fat and protein than most earlier diets. During this time, the American Diabetes Association began emphasizing the need for nutrition education about the exchange system. The exchange lists for meal planning emphasized carbohydrates. They reflected a move from the main focus of the lethal effects of short-term complications (ketoacidosis and hypoglycemia) to a new focus, concern about long-term complications.[10] The exchange system continued to be taught and used and was perceived as the diabetes diet until the 1980s and 1990s.

In 1994, shortly after clinical findings from the Diabetes Control and Complications Trial (DCCT) were published, the American Diabetes Association (ADA) published a revised set of nutrition guidelines to refocus on an "individualized approach to nutrition self-management that is appropriate for the personal lifestyle and diabetes management goals of the individual with diabetes."[11] This approach allows more flexibility by taking into account personal preferences and lifestyle. Although diet recommendations have evolved from ancient times to the present day, diet remains a vital cornerstone of diabetes self-management.

Classification

Diabetes is a clinically heterogeneous, complex metabolic group of disorders that share hyperglycemia in common, and that result from insufficient insulin secretion, resistance to insulin action, or both.[12] An appropriate system of classification is important for clinical management. The ADA and the WHO classify diabetes into four main clinical classifications: type 1 (T1D), type 2 (T2D), other specific types (OD), and gestational diabetes (GD).[2,13]

T1D is characterized by absolute insulin deficiency caused by autoimmune-mediated destruction of the pancreatic β cells, accounting for 5% to 10% of all cases.[13] T2D, which accounts for about 90% to 95% of cases, is characterized by two primary defects: insulin resistance (diminished peripheral tissue sensitivity to insulin) and relatively impaired β-cell function (delayed or inadequate insulin release).[4] GD is defined as abnormal glucose status first recognized during pregnancy. The other types (i.e., unusual genetic forms or secondary causes of diabetes) account for the remaining cases in the United States. The origin and classification can be found in **Table 63.1**.

The WHO states there have been recent calls to review and update the classification system for diabetes. This is because many people with diabetes do not fit into any single category, and because there have been recent advances in knowledge of pathophysiologic pathways and emerging technologies, thus a trend toward individualized treatment.[2]

Table 63.1 Etiologic Classification of Diabetes Mellitus

I. T1D
   A. Immune mediated: Complete or almost complete insulin deficiency, usually caused by autoimmunity (destruction of pancreatic β cells)
   B. Idiopathic: Unusual form of phenotypic t1D with a strong hereditary component and no evidence of autoimmunity; reported mainly in Africa and Asia
   C. Latent autoimmune diabetes of adulthood: Adult onset, slowly progressive, eventual insulin requirement but may respond to oral agents initially; positive GAD, IA2, and/or islet cell antibodies

II. T2D: Results from a progressive insulin secretory defect in the background of peripheral insulin resistance and may range from predominantly insulin resistance with a relative insulin deficiency to a predominantly secretory defect with insulin resistance

III. Other specific types, including the following:
   A. Genetic defects of β-cell function: Chromosome 12, HNF-1α (MODY3); chromosome 7, glucokinase (MODY2); chromosome 20, HNF-4α (MODY1); chromosome 13, IPF-1 (MODY4); chromosome 17, HNF-1α (MODY5); chromosome 2, NeuroD1 (MODY6); Chromosome 9, carboxyl ester lipase (MODY7); transient neonatal diabetes (Chromosome 6p22 or 6p24, ZAC encoding zinc finger protein); permanent neonatal diabetes (Chromosome 11p15, usually KCNJ11 encoding for KIR6.2 subunit of the beta-cell K_{ATP} channel; mitochondrial DNA, others
   B. Genetic defects in insulin action: Type A insulin resistance, leprechaunism, Rabson-Mendenhall syndrome, lipoatrophic diabetes, others
   C. Diseases of the exocrine pancreas: Pancreatitis, trauma or pancreatectomy, neoplasia, cystic fibrosis, hemochromatosis, fibrocalculous pancreatopathy, others
   D. Endocrinopathies: Acromegaly, Cushing syndrome, glucagonoma, pheochromocytoma, hyperthyroidism, somatostatinoma, others
   E. Drug or chemical induced: Vacor (*N*-3 pyridylmethyl-*N'*-4 nitrophenyl urea), pentamidine, nicotinic acid, glucocorticoids, thyroid hormone, diazoxide, β-adrenergic agonists, thiazides, phenytoin, antipsychotics, γ-interferon, infections (Congenital rubella, cytomegalovirus, others), others
   F. Uncommon forms of immune-mediated diabetes: "Stiff-man" syndrome, anti-insulin receptor antibodies, others
   G. Other genetic syndromes sometimes associated with diabetes: Down syndrome, Klinefelter syndrome, Turner syndrome, Wolfram syndrome, Friedreich ataxia, Huntington chorea, Laurence-Moon-Biedl syndrome, myotonic dystrophy, porphyria, Prader-Willi syndrome, others

IV. Gestational diabetes: Diabetes first diagnosed during pregnancy that is not clearly from earlier overt diabetes

GAD, glutamic acid decarboxylase; *HNF*, hepatocyte nuclear factor; *IPF-1*, insulin promoter factor-1; *MODY*, maturity-onset diabetes of the young; *NeuroD1*, neurogenic differentiation factor-1

Data from American Diabetes Association. Diabetes Care 2022;45:S17-S38; and World Health Organization. Classification of Diabetes Mellitus 2019. Available at: www.who.int/publications/i/item/classification-of-diabetes-mellitus

Epidemiology

The prevalence and incidence of diabetes have risen dramatically over the last several decades. In 2019, the global prevalence was estimated to be 9.3% (463 million people), with a projected rise to 10.2% (578 million) by 2030 and 10.9% (700 million) by 2045.[3] The rising incidence has been linked to the changing lifestyles seen around the globe, including unhealthy, energy-dense diets, reduced physical activity, and increased BMI, a risk factor for developing T2D.[14] In the United States, approximately two-thirds of the population is overweight or obese; in 2020, approximately 37.3 million people (11.3% of the population) in the United States were living with T2D.[1,15] Projections suggest that up to one-third of Americans will be diagnosed with diabetes by 2050.[16]

The economic impact of diabetes and its complications is extensive. In the United States, the cost of diabetes in 2017 was estimated at $327 billion, including $237 billion in direct medical costs and $90 billion in indirect costs. Between 2015 and 2017, Americans spent approximately $56 billion per year on diabetes prescription medications alone[17]; this and other factors, including the rising cost of insulin and its accessibility, can directly impact daily care for patients. Indirect costs include absenteeism from work, reduced work productivity, unemployment from disease-related comorbidity, and lost productive capacity from early mortality.[18] Further, total estimated global healthcare expenditures on diabetes were $760 billion in 2019, and are expected to grow to $825 billion in 2030 and $845 billion by 2045.[19]

Considerable geographic and genetic variation is noted in the incidence of diabetes. The global prevalence of T1D in 2017 was estimated to be just over 9 million[20] and comprises 5% to 10% of diabetes diagnoses.[21] There is considerable variation in the prevalence of T1D geographically; for example, in China, T1D represents less than 1% of all prevalent cases of diabetes while, in contrast, in the United Kingdom and Finland, it accounts for 8.6% and 17% of all cases, respectively.[22] The incidence of T1D is increasing by almost 2% to 3% per year worldwide,[21,22] a finding suggesting an environmental contribution to the disease.[23] Further supporting this hypothesis is that migrants develop the same risk of the disease as the native population of their new area of residence.[24] Epidemiologic studies link environmental risk factors early in life as possible contributors to the increasing incidence of T1D, including enteroviral infections in pregnant women,[25-27] older maternal age,[28] cesarean section delivery,[29] increased birth weight, increased rate of postnatal growth,[30,31] and childhood obesity.[32] There is a strong association between the gut microbiome and the risk of developing T1D, and further studies examining this relationship are warranted.[33-35] More recently, a possible link between Covid-19 infection and the development of T1D has been speculated, and trials are being set up to study this association.[36,37] Viruses have the potential to initiate autoimmunity toward islet cells and may cause β cell damage.[38] Vitamin D supplementation in early infancy may be protective.[39,40]

T2D affects 90% to 95% of people with diabetes globally. Onset is usually in adults older than 35 years old.[41] Risk factors include physical inactivity (exercising less than three times per week),[42] high-risk ethnicity (e.g., non-Hispanic Black, non-Hispanic Asian, and Hispanic),[43,44] delivery of a baby weighing more than 9 lbs (4 kg) or a diagnosis of GD,[45] metabolic syndrome,[46] smoking,[47] polycystic ovarian syndrome,[48] previously identified impaired fasting glucose (IFG) or impaired glucose tolerance (IGT),[49] clinical conditions associated with insulin resistance, and history of cardiovascular disease (CVD).[50] Having a first-degree relative with T2D increases the risk by two- threefold.[51] Abdominal obesity confers higher risk, and waist circumference cutoffs may vary by ethnicity.[52]

Other risk factors for T2D include older age,[53] greater parity,[54] low *or* excessive (vs moderate) alcohol use,[55-57] smoking,[47] stress,[58] quality and quantity of sleep,[59] low socioeconomic status,[60] Western diet (i.e., high in red and processed meats, low fiber, high fat dairy, and desserts),[61] low magnesium intake,[62] and sugar-sweetened beverages.[63] Strong urbanization trends associated with T2D are depicted in the following two cases. On the Pacific island of Nauru, Pima Indians took on a more Western lifestyle, leading to more obesity, and their diabetes prevalence increased from nearly 0% to about 50%.[64] This trend has also been seen in other populations of Pima Indians, such as those in Arizona and Mexico.[65] In South Asia, urban communities have up to twice the prevalence of T2D, compared with rural communities.[66]

Like T2D, the prevalence of prediabetes is also rising. Persons with prediabetes have an increased risk of developing T2D, CVD, and microvascular diseases, such as peripheral neuropathy. In 2019, based on fasting glucose or hemoglobin A1c concentrations, 96 million (or more than one in three) adults in the United States had prediabetes.[67]

GD affects an estimated 17% of pregnancies globally,[68] and 7.8% of pregnancies in the United States in 2020.[69] Risk factors for GD include prepregnancy BMI ≥30 kg/m^2, prior history of GD, excessive weight gain during the first two trimesters of pregnancy, and older maternal age (≥35 years).[70,71] Of particular concern, up to 50% of women with GD will develop T2D in their lifetime.[72]

Diagnosis

In the past, a diagnosis of diabetes was based only on fasting plasma glucose (FPG) or the 2-hour 75-g oral glucose tolerance test (OGTT). In 1997, the diagnostic criteria were revised after observing associations between FPG concentration and the presence of retinopathy as the key factor on which to base the threshold glucose value.[13] These analyses helped define the diagnostic cut point of FPG at greater than 126 mg/dL (7.0 mmol/L) and confirmed the long-standing diagnostic, 2-hour plasma glucose threshold of greater than 200 mg/dL (11.1 mmol/L), still used today.[13] The diagnostic criteria, based on a hemoglobin A1c cut point of 6.5% or greater, are similarly associated with an inflection point for retinopathy prevalence.[13] Generally, FPG, 2-h PG during 75-g OGTT, and A1C are equally appropriate for diagnostic screening. To be classified as having diabetes, a patient needs to satisfy at least one diagnostic criterion. **Table 63.2** contains a summary of the criteria.

Table 63.2 Diagnostic Criteria for the Diagnosis of Diabetes

FPG ≥126 mg/dL (7.0 mmol/L). Fasting is defined as no energy intake for ≥8 h.
OR
2-h PG ≥200 mg/dL (11.1 mmol/L) during OGTT. The test should be performed as described by WHO using a glucose load containing the equivalent of 75 g anhydrous glucose dissolved in water
OR
Hemoglobin A1c ≥6.5%. The test should be performed in a laboratory using a method that is NGSP certified and standardized to the DCCT assay
OR
In a patient with classic symptoms of hyperglycemia or hyperglycemic crisis, a random plasma glucose ≥200 mg/dL (11.1 mmol/L) on at least two occasions

DCCT, Diabetes Control and Complications Trial; *FPG*, fasting plasma glucose; *NGSP*, National Glycohemoglobin Standardization Program; *OGTT*, oral glucose tolerance test.

Data from American Diabetes Association. Diabet Care 2022;45:S17-S38

Diagnostic criteria also exist for persons with increased risk for diabetes but whose serum glucose concentration does not yet meet the criteria. The terms impaired glucose tolerance (*IGT*) and impaired fasting glucose (*IFG*) refer to a stage intermediate between normal glucose homeostasis and diabetes.[13] IFG is defined as FPG concentration between 100 mg/dL (5.6 mmol/L) and 125 mg/dL (6.9 mmol/L) by the ADA, or between 110 mg/dL (6.1 mmol/L) and 125 mg/dL (6.9 mmol/L) by the International Diabetes Federation (IDF). IGT refers to a 2-hour glucose concentration of 140 mg/dL (7.8 mmol/L) to 199 mg/dL (11.0 mmol/L). The ADA refers to these conditions as "prediabetes," whereas the WHO prefers "intermediate hyperglycaemia." A criterion for a "category of increased risk of diabetes" based on a hemoglobin A1c concentration of 5.7% to 6.0% has also been introduced.[73]

The diagnosis of GD is based on an OGTT performed during pregnancy. In the revised ADA guidelines,[13] all women, regardless of their risk factors, are recommended to undergo a 75-g OGTT during weeks 24 to 28 of pregnancy. In women at high risk, OGTT should also be considered at the beginning of pregnancy to diagnose overt diabetes.

Body Fuel Regulation

Physiology of Normal Blood Glucose Regulation

Carbohydrate metabolism, or glucose homeostasis, depends on the interaction of several hormones. Insulin is central in maintaining this homeostasis, as the only hormone that lowers blood glucose. However, glucagon, glucocorticoids, catecholamines, and growth hormone also have significant glucose-raising effects that are interactive with insulin.[74,75] Following the ingestion of a meal, nutrients are digested; broken down into glucose, amino acids, and fatty acids; and rapidly absorbed by the small intestine. Glucose is first carried to the liver by the portal vein, where a substantial portion (30% to 70%) enters the liver by facilitated (carrier-mediated) diffusion, driven by the concentration gradient that exists in the fed state.[75,76] The majority of this glucose is transformed into glycogen and stored, although some is converted to lipid or consumed by energy-generating pathways. A small portion of glucose in the liver is metabolized through glycolytic pathways to produce adenosine triphosphate. The remainder enters the peripheral circulation, where regulated insulin secretion and target skeletal tissue responses to insulin contribute to insulin-mediated glucose clearance and control of blood glucose concentration. Skeletal muscle represents the primary peripheral tissue site for removal of this circulating blood glucose.[75-77]

Plasma glucose concentration is the main regulator of insulin secretion by the pancreatic β cells. Intestinal factors, called incretins (e.g., gastrointestinal-inhibitory peptide [GIP] and glucagon-like peptide-1 [GLP-1]), and neural (vagal) factors augment insulin secretion such that the insulin secretory response to oral glucose greatly exceeds the response to an equivalent intravenous glucose infusion.[78,79] High insulin concentration stimulates the transport of glucose and amino acids into muscle and adipose tissue. Insulin also facilitates the conversion of glucose products to fatty acids for storage as triglycerides in fat cells.

The overall effects of the increase in insulin concentration, in response to increased glucose entry into the circulation, are suppression of glucose production by the liver and stimulation of glucose transport into the muscle and adipose tissue, where it is consumed as a metabolic fuel or stored. Insulin also inhibits the catabolism of the alternative energy sources, fat, and protein. This is an appropriate response to the abundance of circulating nutrients after meals.[76]

During the fasting state, low serum insulin permits the mobilization of fuel and energy from storage forms. For example, under stressful circumstances, hypoglycemia, or trauma, glucagon and other counterregulatory hormones, including catecholamines, glucocorticoids, and growth hormone, act specifically to decrease peripheral glucose uptake, to promote hepatic glucose production, and to mobilize fatty acids.[80]

During periods of starvation, maintaining glucose homeostasis is critically important (see chapter 50 in this book). The brain cannot synthesize glucose or store fuel as glycogen for more than a few minutes. Thus, the brain depends on a continuous supply of glucose from plasma. Only the liver and kidneys contain glucose-6-phosphatase, the enzyme necessary to release glucose into circulation. Starvation is associated with a decline in insulin and a rise in glucagon concentrations that result in increased gluconeogenesis. The change from a glucose-based to a lipid-based (free fatty acids and ketone bodies) energy supply in prolonged starvation helps minimize skeletal muscle protein catabolism by reducing the need for amino acid–derived gluconeogenesis.[80]

Figure 63.1 provides a brief overview of the impact of insulin on multiple tissues, and its specific actions that result in lowering blood glucose and inhibiting the mobilization of alternative metabolic fuels such as fat and protein. The disruption in glucose homeostasis and other metabolic derangements observed with diabetes can be explained by the loss of these actions of insulin. As previously mentioned, diabetes can be caused by either a relative or absolute deficiency of insulin and/or diminished tissue responsiveness to insulin that ultimately results in hyperglycemia.[76]

Pathophysiology of Diabetes

In the fed state, insulin is secreted from the pancreatic β cells in response to the increased circulating glucose concentration, thus promoting glycogen synthesis in the liver and muscle, lipid formation in adipocytes, and amino acid uptake and protein synthesis in most cells. In the postabsorptive state, during starvation, and in response to stress, decreased insulin and increased glucagon contribute to glycogen breakdown, lipolysis, hepatic ketogenesis, decreased synthesis, and increased protein degradation. This decline in insulin also results in an increased hepatic release of glucose into the systemic circulation to maintain glucose concentration.[76,77]

In diabetes, decreased insulin action poses a range of metabolic abnormalities that span from the effects of mild insulin deficiency, as seen in hyperglycemia, to the effects of insulinopenia, as seen in diabetic ketoacidosis (DKA), associated with fluid and electrolyte depletion.[78] In postabsorptive or fasted states, hyperglycemia does not resolve, and often worsens. Low insulin activity results in exaggerated counter-regulatory responses that normally protect against the development of hypoglycemia.[79]

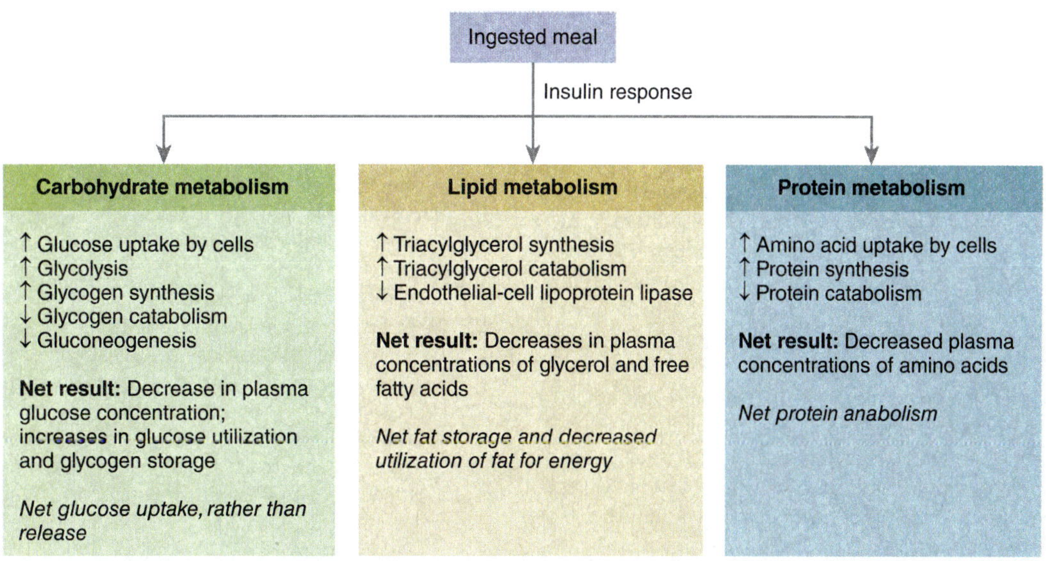

Figure 63.1 Metabolic effects of insulin on macronutrients.

The result of decreased insulin action and elevated counterregulatory hormones (glucagon, catecholamines and, to a lesser extent, growth hormone and cortisol) includes, initially, the conversion of stored glycogen to glucose. Glucagon is a potent activator of glycogenolysis and gluconeogenesis (in the liver) and can increase endogenous glucose production.

Relative or absolute insulin deficiency leads to a marked decrease in the glucose transporter-4 (GLUT4) activity, mainly due to decreased insulin-stimulated GLUT4 localization to the surface membranes of skeletal muscle. This results in a decrease in the normal postmeal flux of glucose into skeletal muscle and increased circulating plasma glucose.[75] Diabetes is also associated with increased activity of enzymes involved in gluconeogenesis, decreased activity of glycolytic and oxidative enzymes, and relative or absolute hyperglucagonemia, resulting from the loss of the suppressive effect of insulin on glucagon secretion by the pancreatic α cell.

Complications

Diabetes is a chronic disease that can cause complications leading to significant morbidity and premature death, particularly if not well treated. Diabetes is a major cause of heart disease and stroke and is the seventh leading cause of death in the United States.[1] In 2019, the WHO reported that diabetes and diabetes-related complications were the 9th leading cause of death globally.[3] Diabetes is the leading cause of end-stage renal disease, blindness, nontraumatic limb amputation, and cardiovascular morbidity and mortality. Medical costs, lost work, and wages for people with diagnosed diabetes total $327 billion annually.

Acute Complications

Symptoms of hyperglycemia include increased polyuria (frequent urination), polydipsia (excessive thirst), fatigue, irritability, blurred vision, and weight loss. Blurry vision results from osmolar shifts from hyperglycemia within the lens.[77] Polyuria and polydipsia occur when blood glucose increases above the urinary filtering threshold of 180 mg/dL. In normal glucose states, the tubules reabsorb all glucose filtered at the glomerulus. However, the elevated plasma glucose in diabetes leads to an increase in the filtered load of glucose that exceeds the maximum tubular reabsorptive capacity, and large amounts of glucose may be excreted[81].

For the same reason, ketones may appear in the urine. These urinary losses further deplete the body of nutrients and lead to weight loss.

DKA and the hyperglycemic hyperosmolar state (HHS) are acute complications of diabetes and contribute significantly to morbidity and mortality. DKA was formerly considered a hallmark of T1D, but it may also occur in T2D.[82] HHS is primarily seen in individuals with T2D. Both disorders are associated with absolute or relative insulin deficiency, volume depletion, and acid–base abnormalities. In DKA, the osmotic force exerted by unreabsorbed glucose and ketones leads to water retention in the tubule, thereby preventing its reabsorption and leading to fluid depletion. Sodium reabsorption is also retarded, and the net result is the marked excretion of sodium and water, which can lead, in worst cases, to hypotension, brain damage, and death if left untreated.[83] Hypoglycemia is the most frequent complication in individuals with diabetes who are treated with insulin. However, hypoglycemia can also occur in patients who are not taking insulin but are using hypoglycemic agents, such as insulin secretagogues.

Chronic Complications

Chronic complications occur over years or decades of hyperglycemia and are often difficult or impossible to reverse. Examples include microvascular complications (i.e., small vessel disease) such as retinopathy, neuropathy, and end-stage renal disease, or macrovascular complications (i.e., large vessel disease) such as coronary heart disease, peripheral vascular disease, or stroke. The pathophysiologic features of both microvascular and macrovascular complications are similar; both are secondary to oxidative damage from long-term, uncontrolled hyperglycemia, resulting in plaque formation and narrowing of small and large blood vessels , with ischemic damage to end organ tissues. Diabetes is the leading cause of end-stage renal disease. The risk of CVD is increased two to four times in patients with diabetes.[84]

Chronic microvascular complications of diabetes can be delayed or prevented by optimal glycemic control (i.e., hemoglobin A1c <7%), as demonstrated in several key studies, including the DCCT in T1D and the UK Prospective Diabetes Study (UKPDS) in T2D. These landmark trials concluded that achieving and maintaining serum glucose concentration in this range slowed the onset and progression of eye, kidney, and nerve diseases. Macrovascular complications, such as heart disease, can be reduced with optimal glycemic control, as demonstrated in long-term follow-up of studies such as the DCCT and UKPDS, and with cardiovascular risk factor modification of comorbidities, including hypertension and dyslipidemia.

Other common complications include poor wound healing, increased susceptibility to infections, periodontal disease,[85] erectile dysfunction, and gastroparesis. In addition, many comorbidities associated with diabetes may influence its management, including human immunodeficiency virus infection, cystic fibrosis, polycystic ovarian syndrome, postpancreatectomy diabetes, and Cushing syndrome. Sleep apnea and depression are also common (**Table 63.3**). The list in Table 63.3 provides the more frequent, but not all, complications.

Table 63.3 Chronic Complications of Diabetes Mellitus

- **I. Microvascular**
  - A. Retinopathy (nonproliferative or proliferative)
    - a. Macular edema
  - B. Neuropathy
    - a. Peripheral
    - b. Autonomic
  - C. Nephropathy
- **II. Macrovascular**
  - A. Cardiovascular disease
  - B. Peripheral vascular disease
  - C. Cerebrovascular disease
- **III. Other less-common complications**
  - A. Gastrointestinal diseases (e.g., gastroparesis, diarrhea)
  - B. Genitourinary diseases (e.g., uropathy, sexual dysfunction)
  - C. Dermatologic or musculoskeletal diseases (e.g., shin spots, osteoporosis)
  - D. Infectious diseases (e.g., osteomyelitis, zygomycoses)
  - E. Hematologic or malignant diseases (e.g., anemia, pancreatic cancer)
  - F. Neurologic or psychiatric diseases (e.g., dementia, depression)

Diabetes is manageable. Its complications are not inevitable with optimal glycemic control and cardiovascular risk factor management, and they can be treated if they occur.

Goals for Medical Nutrition Therapy

Medical Nutrition Therapy (MNT) is vital to all diabetes prevention and management plans. MNT varies with the type of diabetes and the age of the individual. It is provided by registered dietitian nutritionists (RDNs) and can be defined as the "treatment of a disease or condition through the modification of nutrient or whole-food intake".[86] MNT promotes healthy eating habits, helps control blood glucose and lipid concentrations and, as appropriate, weight management by facilitating lifestyle modifications. It is effective, with reported decreases in hemoglobin A1c of 1% to 2%, depending on the type and duration of diabetes.[87,88] MNT has the greatest impact at the initial diagnosis and continues to be an effective intervention at any time throughout the disease process.[88] Many RDNs have the certified diabetes care and education specialist (CDCES) credential, and they can provide comprehensive diabetes care and education beyond MNT.

Once a diagnosis of diabetes is established, treatment includes an MNT plan, pharmacologic therapy (oral agents, noninsulin injectables, insulin, or a combination of these), regular monitoring by the healthcare team (physicians, nurse practitioners, physician assistants, clinical pharmacists, registered nurses, RDNs, and licensed clinical social workers) and, most importantly, self-monitoring and continuing education of the patient or caregiver about managing their condition. MNT should be offered in several phases, based on the individual's comprehension and readiness to learn. During the initial treatment, basic principles, such as identifying sources of carbohydrates and prevention and treatment of hypoglycemia, may be introduced. During later sessions, in-depth self-management guidance, such as carbohydrate counting and adjusting insulin-to-carbohydrate ratios, is provided. Throughout the treatment process, individualization is key. **Table 63.4** depicts the ADA goals for MNT for various circumstances in which treatment should be tailored to achieve these goals.[89]

Table 63.4 American Diabetes Association Goals for Medical Nutrition Therapy

For those with diabetes or prediabetes:

- To promote and support healthful eating patterns by emphasizing a variety of nutrient-dense foods in appropriate portion sizes to improve overall health and more specifically:
  - Improve A1C, blood pressure, and cholesterol concentration (goals differ for individuals based on age, duration of diabetes, health history, and other present health conditions)
  - Achieve and maintain body weight goals
  - Delay or prevent complications of diabetes
- To address individual nutrition needs based on personal and cultural preferences, health literacy and numeracy, access to healthful food choices, willingness and ability to make behavioral changes, as well as barriers to change.
- To maintain the pleasure of eating by providing positive messages about food choices while only limiting foods when indicated by scientific evidence.
- To provide the individual with diabetes with practical tools for day-to-day meal planning.

Modified from Evert AB, Dennison M, Gardner CD, et al. Diabet Care 2019;42:731–754.[89]

Nutritional Plan

To achieve these goals, formal process steps include an initial nutrition assessment, which consists of a nutrition diagnosis and plans for intervention, monitoring, and evaluation (see chapter 58 in this book). Nutrition management includes monitoring diet, blood glucose, medications, physical activity, education, behavior modification, and evaluation of cardiovascular and renal status to ensure appropriate nutrition therapy. The following steps are modified from the Academy of Nutrition and Dietetics Nutrition Recommendations for Management of Diabetes[90] and the Nutrition Subcommittee of the Diabetes Care Advisory Committee of Diabetes UK.[91]

Nutrition Assessment

Nutrition assessment serves as the basis for the implementation of the nutrition prescription, goals, and intervention and should include the following:

1. Diet history: meal patterns, food choices, nutritional adequacy, beliefs or misconceptions, food intake with a focus on added sugar, sodium, fat, and refined carbohydrate commonly found in ultra-processed foods.
2. Clinical information: age, type of diabetes, and treatment, including but not limited to medications (insulin, oral hypoglycemic drugs, or diet alone), dietary supplements, metabolic control (hyperglycemia or hypoglycemia, lipids, and blood pressure), smoking and other CVD risk factors, nutrition-focused physical examination, anthropometric measurements, physical activity, relevant surgical history (particularly, the gastrointestinal tract), and other medical conditions (nephropathy).
3. Personal information: socioeconomic circumstances, ethnicity, literacy, ability, and willingness to change, numeracy, emotional state (if distressed by a new diagnosis of diabetes).

Once the preceding is established, the RDN can identify the nutrition-related problem(s) they are responsible for treating. This is known as the nutrition diagnosis.

Nutrition Diagnosis

The nutrition diagnosis should detail specific nutritional problems that may be ameliorated through nutrition intervention. The diagnosis contains three components: the nutrition problem, etiology, and signs/symptoms. Nutrition problems are distinct from medical diagnoses; for example, a nutrition diagnosis for a person with diabetes could be "excessive carbohydrate intake," whereas a medical diagnosis describing the same patient could be "uncontrolled T2D." For effective MNT, the RDN must identify the second piece of the diagnosis, the etiology of the problem, as this will guide the intervention. Identifying and documenting signs and symptoms of the nutritional problem will aid in measuring the effectiveness of the intervention.[92]

Nutrition Interventions

Nutrition interventions need to be individualized and jointly developed between the RDN and the patient to achieve nutrition therapy goals.[92] Components of the intervention include:

1. Encouragement to modify macronutrient intake based on the individual's health status. Factors to consider are metabolic goals (glycemic control, lipid profile, etc.), food access, and the ability to be physically active.[89] This will need to be customized to individuals based on their current health status. For example, patients with severe insulin resistance may benefit from consuming less carbohydrate than the RDA of 130 grams/day.
2. Implementation of nutrition education and counseling. Nutrition counseling must be sensitive to the individual's personal needs, cultural preferences, and willingness and ability to make changes. It is recommended that people with diabetes have at least three visits with an RDN within the first 6 months of beginning MNT. Additionally, at least one follow-up encounter is recommended annually.[89]
3. Weight management. Even modest amounts of weight loss (7%–10% of body weight) are highly effective in preventing and treating T2D in those with overweight or obesity (see also the chapter 61 in this book).[89] Clinicians should encourage clients to participate in at least 150 minutes of moderate intensity each week to help achieve and maintain weight loss.

Nutrition Monitoring and Evaluation

Components of nutrition monitoring and evaluation include:

1. Coordination of care with the interprofessional team.
2. Monitoring and evaluating food intake, medication, metabolic control (e.g., glycemia, lipids, and blood pressure), anthropometric measurements, and physical activity.
3. Use of blood glucose or continuous interstitial glucose monitoring results primarily in evaluating the achievement of goals and effectiveness of MNT. Glucose monitoring results help determine whether adjustments in foods or meals will be sufficient to achieve blood glucose goals or whether medication additions or adjustments need to be combined with MNT.

Over time, many attempts have been made to identify one approach to the nutritional treatment of diabetes. No one "diabetes diet" exists, just as no single medication or insulin regimen applies to all persons with diabetes. Instead, various interventions, such as reduced energy and fat intake, carbohydrate counting, simplified meal plans, healthy food choices, weight management, meal planning strategies, insulin-to-carbohydrate ratios, physical activity, and behavioral strategies,[90] are customized to each individual. Nutrition education and counseling must be sensitive to the individual's personal needs and cultural preferences and consider the person's willingness and ability to make changes.[88] Continual monitoring and follow-up encounters are also important to support these lifestyle changes, evaluate nutrition-related outcomes, and assess medication needs.[93]

Distribution of Energy Intake

Current recommendations for macronutrient proportions are to establish individualized goals for patients based on metabolic goals, personal preferences, cultural backgrounds, accessibility, and level of physical activity.[89]

Although numerous studies have attempted to identify the optimal mix of macronutrients for meal plans of people with diabetes, it is unlikely that one such combination of macronutrients exists. The best mix of carbohydrates, protein, and fat appears to vary, depending on individual circumstances. Balanced macronutrient (fat, carbohydrate, protein) intake is recommended, although some people do well with more extreme restrictions of fat or carbohydrate. Individualization of the macronutrient composition depends on the metabolic status of the patient (e.g., lipid profile, renal function) and food preferences.[93]

In 2019, an expert panel appointed by the American Diabetes Association issued a consensus report based on a review of updated evidence of nutrition therapy interventions.[93] The following reflects some of their updated findings and recommendations for macronutrients.

Carbohydrates

Carbohydrates have the greatest postprandial effect on serum glucose and are the most important nutrient to consider in treating diabetes. The long-standing recommended dietary allowance (RDA) for digestible carbohydrates is 130 g/day for adults.[94] This is based on the average amount of glucose used by the brain when an individual is not in nutritional ketosis. Based on the Institute of Medicine's 2002 report, the recommendation advises the consumption of carbohydrates in the form of whole grains, vegetables, fruits, pulses, nuts, seeds, and milk products. While there has been a decrease in carbohydrate intake over the last few decades in the United States, carbohydrates still represent about 50% of total energy intake, which is about 230 g/day in a typical 2,000 kcal diet. About 21% of carbohydrate intake is in the form of low-quality carbohydrates, such as refined carbohydrates, fruit juice, and potatoes, while 14% is in the form of added sugars to beverages.[95]

Several studies have demonstrated greater efficacy of low carbohydrate, compared with high carbohydrate, diets in both glycemic control and weight reduction.[96,97] These studies align with the ADA position statement that low carbohydrate eating patterns may lead to improved glycemia and a decrease in antihyperglycemic medications.[93] A dose-dependent relationship between the level of carbohydrate intake and the amount of weight loss, glycemia, and other factors, such as improvements in cholesterol and blood pressure, is evident. However, sustaining such changes are often not seen long term as patients often, over time, revert to their previous lifestyle habits.[98]

The Nutrition Practice Guidelines released in 2017 by the Academy of Nutrition and Dietetics made several recommendations on carbohydrate management strategies to improve glycemic control and limit hypoglycemia. Patients who learn carbohydrate counting (see Carbohydrate Counting section) should dose insulin based on an insulin-to-carbohydrate ratio. For patients on fixed-dose insulin or insulin secretagogues (medications like glimepiride, glipizide, nateglinide, etc.), carbohydrate intake should be consistent in amount and timing.[99]

Glycemic Response to Carbohydrates

The rate at which glucose is absorbed from the intestinal tract is another component controlling glucose concentration. The glycemic index of foods was developed to provide a numeric value that represents the effect of the food on postprandial glycemia concentration, compared with other food, usually white bread, or sugar.[100] Foods with high glycemic index values are digested and absorbed more rapidly and cause greater fluctuations in blood glucose per unit of carbohydrate than foods with lower glycemic index values.[100]

The Glycemic index is an equation calculated by the increase in postprandial glycemia concentration over the baseline during a 2-hour period after consumption of a specified amount of carbohydrate (usually 100 g), compared with an equal amount of carbohydrate as a reference food (white bread or glucose). A low glycemic index is considered lower than 55, and a value greater than 70 is regarded as high.[101] Many factors affect glycemic responses to foods. These include the type of carbohydrate (e.g., glucose, fructose, sucrose, lactose, amylase, resistant starch), the cooking method (longer cooking times produce more breakdown of the starch), the type of food processing, and other meal components such as fat and proteins.

An additional reference guide is the glycemic load, which considers both the quantity and quality of the carbohydrate in a meal and is calculated by multiplying the glycemic index by the grams of carbohydrates in a serving of the food. Chiavaroli et al. reported that improved glycemic control, reduced risk for coronary heart disease, and a modest reduction in hemoglobin A1c were seen when the glycemic index of the diet was lowered in individuals with moderately controlled T1D or T2D, who were also treated with antihyperglycemic medications.[102] Although the glycemic index is sometimes helpful for the conscientious patient, it may be better for the general public to consider it a broad guide to better carbohydrate food choices, because most processed and simple sugars have a high glycemic index. In contrast, whole grain products have a lower glycemic index.

Protein

Evidence is insufficient to suggest that usual protein intake (15% to 20% of energy) should be modified for persons with diabetes and normal renal function.[103] Although about one-fifth of protein does convert to glucose, in well-controlled diabetes, glucose from ingested protein does not appear in the general circulation and does not raise blood glucose concentration.[104] In persons with T2D, however, ingested protein can increase insulin response without increasing plasma glucose. Therefore, protein should not be used to treat acute, or to prevent nighttime, hypoglycemia.[105] Individuals with T1D are more susceptible to glycemic excursions related to dietary protein consumption. Intake of high protein meals leads to a rise in plasma glucagon, cortisol, and IGF-1 concentrations and a subsequent increase in glucose concentration. Furthermore, of the 20 amino acids, 18 are glucogenic, meaning they can be converted to glucose via gluconeogenesis. The consumption of a high protein meal in the presence of insufficient circulating insulin leads to an increase in amino acid uptake by the liver, increased gluconeogenesis, and, consequently, an increase in glucose concentration.[106]

Reduction of protein intake to 0.8 to 1.0 g/kg/day in patients with early stages of chronic kidney disease (CKD) and 0.8 g/kg/day in the later stages of CKD may improve measures of renal function (urine albumin excretion rate, glomerular filtration rate).[90] For patients with diabetic nephropathy, improvements in albumin excretion rate, but not in glomerular filtration rate, are reported with protein intake lower than 1 g/kg/day.[107-110] For persons with late-stage diabetic nephropathy (CKD stages 3 through 5), further restriction of protein intake beyond 0.8 g/kg may be warranted for select populations. The National Kidney Foundation recommends 0.6–0.8 g/kg for those with an energy intake of 30 kcal/kg and CKD stages 3–5. Further, the PROT-AGE study group recommended that individuals with CKD stages 3–5 and sarcopenia consume at least 0.8 g/kg.[111] There may be benefits from certain animal or plant proteins, but this needs to be further explored. Several studies suggest that whey protein may help augment GLP-1 concentration, thereby increasing insulin secretion and reducing postprandial hyperglycemia. Jakubowicz et al., found that a premeal shake consisting of 50g whey protein concentrate reduced postprandial glycemia by 28%, compared with 18% for nateglinide.[112] Protein sources, including eggs, white meats, and fish, have not been linked to worsening insulin sensitivity, while red meats and highly processed meats (sausages, cured meats, salami) are considered risk factors for T2D. Evidence on plant proteins is mixed with some studies showing similar benefits (weight loss, improved A1C, reduced uric acid concentration, blood pressure reduction, and cholesterol reduction) for equicaloric diets high in animal or plant protein.[113,114]

Fat

Recommendations for fat depend on treatment goals. Over the last 10 years, literature suggests that total dietary fat consumption is not associated with T2D risk.[115] Because persons with diabetes have a higher risk for atherosclerosis and CVD, they should aim to follow the Dietary Guidelines for Americans (2020–2025) and limit saturated fat (SFA) to less than 10% of total energy.[116] Total acceptable fat intake is around 20%–35% of total energy, according to the National Academy of Medicine.[94] Given that saturated fat and *trans*-fatty acids have a higher potential to increase total serum cholesterol and low-density lipoprotein cholesterol (LDL-C) concentrations, it may be reasonable to apply recent evidence suggesting that replacing 5% of dietary SFA with polyunsaturated fats (PUFA) can equate to a ~10% reduction in CHD risk.[117-119] As many animal-based foods contain fat and cholesterol, it is worth noting that dietary cholesterol intake, while associated with increased cholesterol concentrations, is not associated with an increase in cardiovascular events.[89] Lastly, the acute effects of dietary fat must be considered. Moderate to high fat meals (>20g) delay gastric emptying and, thus, reduce the early postprandial glucose response while augmenting the late postprandial response. Individuals with diabetes on multiple daily injections may need to give extra insulin before the meal and/or administer additional insulin over a longer period to maintain glycemic control following a high fat meal.[120]

Fiber

People with diabetes are encouraged to choose a variety of fiber-containing foods because they provide vitamins, minerals, and other important components for overall good health (see also chapter 3). The ADA recommendations for dietary fiber intake for persons with diabetes are the same for the general population, with a DRI of 14 g/1,000 kcal. The DRI recommendations for dietary fiber are based on protection against CVD, given consistent and strong data for the inverse relationship between lipid concentrations and fiber intake.[121,122] A landmark metanalysis by Brown et al.,[123] showed that daily intake of 2 to 10 g soluble fiber significantly lowered serum total cholesterol and LDL-C concentrations. Studies in participants without diabetes showed that diets high in total and soluble fiber, as part of cardioprotective nutrition therapy, further reduced total cholesterol by 2% to 3% and LDL-C by up to 7%.[124]

Mixed evidence indicates that increasing dietary fiber will improve glycemic outcomes in people with diabetes. However, a higher fiber diet, specifically soluble fiber, may have glycemic benefits that could warrant a higher fiber recommendation than the DRI. Many physiologic mechanisms may be involved. Juvonen et al., proposed a mechanism involving viscous and gel-forming fibers and positive changes in glucagon-like peptide 1 (GLP-1), peptide YY, and ghrelin, which modulate satiety.[125] Additionally, Weikert et al., found an enhanced early insulin response and an early increase in postprandial glucose-dependent insulinotropic polypeptide in response to insoluble dietary fiber.[126] Interestingly, the effect of the insoluble fiber persisted beyond the day of consumption, with improved postprandial glucose response noted 24 hours following the fiber-rich intervention.[127] Fiber-rich meals are processed more slowly, and nutrient absorption occurs over a longer period.[128] This slowing of digestion, which flattens the glucose and insulin response in persons without diabetes, and factors that influence glycemic index, such as cooking time, may contribute to the effect.

Franz et al.,[90] looked at five studies comparing high-fiber (40 to 60 g) with low-fiber (10 to 20 g) diets, with similar macronutrient percentages of energy. Two of these studies showed no significant differences between diets on hemoglobin A1c, although overall serum glucose concentrations were lower in the higher fiber groups.[128,129] Giacco et al., showed a 2% reduction in hemoglobin A1c only in participants compliant with the 50-g fiber diet.[130] In other studies, diets that included 30 to 50 g of fiber per day from whole foods, with 7 to 13 g as soluble fiber, appeared to produce lower serum glucose concentration, compared with a low-fiber diet.[90,128] Soluble fiber appears to reduce serum cholesterol and LDL-C. It can delay gastric emptying and increase transit time and thus, may help lower serum glucose concentration. Additional benefits include prolonging satiety from the meal for better control of appetite. Recommended soluble fiber sources include oats, beans, fruits, and vegetables, whereas insoluble fiber comes from whole wheat bread, bran cereal, legumes, brown rice, vegetables, and many fruits.[131] Some disadvantages should be noted in consuming very-high-fiber diets (>75 g/day). These include flatulence and abdominal fullness.[132,133]

Sweeteners

Sweeteners are an important aspect of the quality of life for people with diabetes. It is important to distinguish between those sweeteners with (nutritive) and those without (non-nutritive) significant energy content. **Table 63.5** and **63.6** provide comparative summaries of many available nutritive and non-nutritive sweeteners.[134,135]

All nutritive sweeteners, such as sucrose, fructose, and sugar alcohols (polyols) cause some degree of hyperglycemia. Sucrose, also known as table sugar, is a "simple" carbohydrate that produces a glycemic response equivalent to those of other carbohydrates. Although sucrose can be substituted isocalorically in place of other carbohydrates without eliciting poor glycemic control, Coulston et al.,[137] concluded that additive sucrose, rather than substitution, to the total intake results in increased hyperglycemia and serum lipid concentrations. Fructose requires no insulin to metabolize and causes a small effect on blood glucose concentration. However, less than 20% of total energy should be consumed as fructose, to avoid hypertriacylglycerolemia.[138]

Sugar alcohols (polyols) such as sorbitol, mannitol, and xylitol, are classified as hydrogenated monosaccharides, hydrogenated disaccharides, and oligosaccharides, and are absorbed at a slower rate. They contain 2 kcal/g but, because they are only partially absorbed through passive diffusion in

Table 63.5 Nutritive Sweeteners

Type	kcal	Percentage As Sweet As Sucrose	Description
Sucrose	16 kcal/tsp	100%	Sugar (glucose+fructose)
Fructose	11 kcal/tsp	110%–200%	Sweetest sugar
Sugar alcohols:	(2 kcal/g average)		
D-Tagatose	1.5 kcal/g	75%–92%	Sweetness synergizer; texturizer, stabilizer, humectant
Erythritol	0.2 kcal/g	60%–80%	Flavor enhancer, formulation aid, humectant, stabilizer and thickener, sequestrant, and texturizer
HSH	3 kcal/g	25%–50%	Also called hydrogenated starch hydrolysates; maltitol syrup
Isomalt	2 kcal/g	45%–65%	Bulking agent
Lactitol	2 kcal/g	30%–40%	Bulking agent
Maltitol	2.1 kcal/g	90%	Bulking agent
Mannitol	1.6 kcal/g	50%–70%	Possible laxative effect from load ≥20 g
Sorbitol	2.6 kcal/g	50%–70%	Possible laxative effect from load ≥50 g
Trehalose	4 kcal/g	45%	Texturizer, stabilizer, and humectant
Xylitol	2.4 kcal/g	100%	New forms have better free flowing abilities

Data from American Dietetic Association. Position of the American Dietetic Association. J Am Diet Assoc 2004;104:255–75.

Table 63.6 Nonnutritive Sweeteners Approved by the Food and Drug Administration

Type	Kcal	Common Names	Glycemic Response	Sweetening Power Reduced During Heating	Number of Times Sweeter Than Sucrose	Acceptable Daily Intake
Acesulfame-K	0 kcal/g	Sunett, Sweet 'n Safe, Sweet One	None	No	200	15 mg/kg body weight/day
Advantame			None	No	20,000	32.8 mg/kg body weight/day
Aspartame	4 kcal/g	NutraSweet, Equal, Sugar Twin (Blue box)	Limited	Decomposes during excessive heating	160–220	50 mg/kg body weight/day
Luo han guo extract	0 kcal/g	Nectresse, Monk Fruit in the Raw, PureLo	None	No	150-300	No ADI determined
Neotame	0 kcal/g		None	No	7,000–13,000	18 mg/kg body weight/day
Saccharin	0 kcal/g	Sweet'NLow, Sweet Twin, Sweet'NLow Brown, Necta Sweet	None	No	300	12 mg/kg body weight/day
Stevia rebaudiana Bertoni (steviol glycosides)	0kcal/g	Truvia PureVia Enliten	Limited	No	200–300	4 mg/kg body weight/day
Sucralose	0 kcal/g	Splenda	None	No	600	5 mg/kg body weight/day

Data from Fitch C, Keim KS. J of the Acad Nutr and Diet 2012;112:739–758.[136]; Bloomgarden Z. Diabet Care 2011;34:e46–51[135]; and Food and Drug Administration website. Available at www.fda.gov/food/food-additives-petitions/additional-information-about-high-intensity-sweeteners-permitted-use-food-united-states

the small intestine,[135] they have a reduced energy value per gram. However, they can still elicit a small glycemic response. One must take caution when consuming large quantities (e.g., >20 g/day of mannitol or >50 g/day of sorbitol) because sugar alcohols have also been known to have laxative effects and can cause diarrhea and other gastrointestinal distresses. Nutritive sweeteners may be used as replacement carbohydrate sources to minimize energy and carbohydrate consumption. Specific sugar alcohols such as isomalt, maltitol, lactitol, and xylitol can impact the microbiota composition in humans by increasing bifidobacteria in healthy individuals.[139] Despite numerous studies demonstrating a link between sugar alcohol intake and microbiome alterations, clinically relevant outcomes data are still lacking surrounding sugar alcohol consumption and diabetes.[140]

Non-nutritive sweeteners can benefit persons with diabetes by adding flavor without adding energy or eliciting a glycemic response. Non-nutritive sweeteners are derived from substances of several different chemical classes that interact with taste receptors. They typically exceed the sweetness of sucrose by a factor of 30 to 13,000 times.[141] Currently, six non-nutritive sweeteners are approved by the U.S. Food and Drug Administration (FDA) and regulated as food additives for use in diabetes, that report no effect on changes in glycemic response. Two additional sweeteners in Table 63.6, *Stevia rebaudiana* Bertoni and Luo Han Guo, have received the "generally regarded as safe" (GRAS) designation.[142] In addition, some are heat resistant and can be used for cooking and baking. See Table 63.6 for a comparative summary of individual non-nutritive sweeteners.

Results from some studies suggest that users of non-nutritive sweeteners may tend to consume more energy and even gain weight because of specific effects of the sweetener on appetite. However, the data on this issue are controversial,[143-145] and a definitive conclusion awaits further research.

Carbohydrate Counting

Carbohydrate counting and carbohydrate awareness are essential because carbohydrate intake is the primary nutrient affecting postprandial blood glucose concentration.[146] Carbohydrate counting also allows flexibility in food choices and helps promote glycemic control.[147] The exchange system and experience-based estimations are other methods for estimating carbohydrate intake.[148] Carbohydrate awareness is helpful for all persons with diabetes, but it is essential in treating T1D, so the patient knows the effect of the meal on blood glucose and can better match food intake to insulin doses.

The model of carbohydrate counting is a nutrition strategy that requires the patient to know carbohydrate amounts in food. It is highly dependent on the patient's ability to monitor blood glucose and perform mathematical conversions to determine the amount of carbohydrate in meals. The three levels of carbohydrate counting (basic, intermediate, and advanced) can be evolved by a

Table 63.7 Levels of Carbohydrate Counting

Level	Diabetes Type	Description
I: Basic	Insulin requiring T2D or GD	Key concept of carbohydrate counting, awareness of which foods contain carbohydrates, fixed carbohydrate portion sizes, avoidance of sweets, and sweetened beverages
II: Advanced	T1D	Teaching how to calculate carbohydrate-to-insulin ratios when using multiple daily injections or insulin infusion pumps to match short-acting insulin to carbohydrate
Both basic and advanced carbohydrate counting emphasize portion control.		

Data from Souto DL, Zajdenverg L, Rodacki M, et al. Nutrition 2014;30:286–290.[152]

motivated individual, once carbohydrate amounts and equivalents in foods are maintained at consistent blood glucose concentration (**Table 63.7**).

Carbohydrate counting uses a grouping method to place foods in categories of similar carbohydrate equivalency and estimate intake by total grams of carbohydrate or servings (**Table 63.8**). For example, one carbohydrate serving is equal to a 15-g carbohydrate serving of starch, grain, fruit, or milk. A patient's ability to follow this method will allow for a greater variety of food choices.[149] Clinicians should use discretion when recommending carbohydrate counting to a patient, given the potentially high degree of complexity involved. Brazeau et al., found a mean difference of 20.9±9.7% in total estimated carbohydrate grams between adults with diabetes and an RDN using a computerized program to calculate the carbohydrate.[150] Larger meals were more likely to be underestimated. Smart et al., 2009 found an overestimation of 10g on a 60g carb meal, a 17% error, but it did not result in a statistically significant difference in 2.5-hour postprandial glucose concentration.[151]

Micronutrients

Continual interest exists in using various vitamin, mineral, and trace element supplements in the treatment of diabetes. Trace elements and minerals, such as chromium, potassium, magnesium, vanadium, and zinc may have effects on blood glucose control. The evidence, however, is slim and unconvincing that supplementation with any of these trace elements has a beneficial effect, except perhaps in true deficiency. When an individual is in poor glucose control or taking diuretic medications, serum magnesium concentration may be low and testing is recommended to determine whether a deficiency exists. Chromium may positively affect glucose metabolism; however, studies have yielded conflicting results, and routine supplementation is not currently recommended.

Table 63.8 General Food Groups and Foods in a VLCD

Food Group	Foods
Proteins	Eggs, meats, seafood, nuts, cheese
Dairy	Heavy cream, sour cream, hard cheeses, cream cheese, plain yogurt/cottage cheese
Fats	Olive oil, butter, avocado oil, coconut oil, nuts and seeds, avocados
Vegetables	Colorful leafy vegetables, nonstarchy vegetables
Beverages	Water, unsweetened tea/coffee, unsweetened almond/coconut milk

Routine supplementation of vitamins E and C and carotene is not advised because of the lack of evidence of efficacy and concern related to long-term safety. Vitamin E supplementation was found to have no beneficial effects on cardiovascular outcomes, microvascular complications, or glycemic control in persons with diabetes and CVD.[153] Vitamins B_1, B_6, and B_{12} are sometimes used to treat diabetic peripheral neuropathy, but without strong supporting evidence of benefit. Metformin therapy is associated with lower B_{12} concentration, and supplementation may benefit a subset of the population who have been on metformin therapy for extended periods.[154] Aroda et al., concluded that a longer duration of metformin therapy increased the risk of low or borderline B_{12} concentration and B_{12} deficiency. The authors recommended routine B_{12} screening for individuals taking metformin.

Supplements are indicated for vitamins, minerals, and trace elements when a deficiency is suspected or likely. High-risk populations include elderly persons, pregnant or lactating women, strict vegetarians, or persons on an energy-restricted diet with poor glycemic control, or taking medications that alter micronutrient metabolism. Folate supplementation is well documented to improve the outcome of pregnancy, with or without diabetes.[155]

In addition, deficiencies in vitamins in many populations have been reported due to decreased sun exposure, aging, and/or lactose intolerance (decreased intake of milk fortified with vitamin D). The literature suggests an association between vitamin D insufficiency and glycemic control,[156] although further studies are needed. Calcium supplementation is indicated, particularly in elderly persons, if daily intake is less than 1.0 to 1.5 g.

In summary, evidence is weak that vitamin, mineral, or trace element supplementation benefits patients with diabetes without a true deficiency, and none of these supplements have clear benefits on glucose control.[157] If the diet is adequate, and there are no nutrient deficiencies, supplementation has little or no role in controlling diabetes, and general nutritional guidelines for vitamins and trace elements should be followed. Food group sources of important shortfall nutrients are in **Table 63.9**.

Table 63.9 Food Groups Containing Essential Micronutrients

Nutrient	Foods to Focus on
Calcium	Cow's milk, yogurt, cheese, fortified plant "milks," fortified cereals, calcium-set tofu, fortified juices, broccoli, dark leafy greens, beans (soybeans, chickpeas, black beans), almonds, almond butter
Iron	Fortified cereals, soybeans, spinach, swiss chard, beans, eggs
Protein	Beans, peas, lentils, whole grains, soy products, nuts and nut butters, dairy products, eggs
Vitamin B_{12}	Eggs, dairy products, B_{12} fortified foods (nutritional yeast, select meat substitutes, fortified cereals)
Vitamin D	Eggs, cow's milk, fortified plant based "milks," fortified orange juice, fortified cereals

Data from Klemm, S. (2021, October 15). Food sources of 5 important nutrients for vegetarians. EatRight. Retrieved September 7, 2022, from www.eatright.org/food/nutrition/vegetarian-and-special-diets/food-sources-of-important-nutrients-for-vegetarians[191]

Dietary Eating Patterns

The optimal diet to achieve weight loss and to improve glycemic control remains to be seen, due to the broad range of diets studied, often with conflicting results. The ADA currently recommends individualized meal plans focusing on nutrient-dense foods consumed in appropriate portions low in refined carbohydrates and saturated fats.[93]

Several dietary eating patterns have demonstrated reductions in A1C and weight, including Mediterranean, vegetarian/vegan, low carbohydrate, and very low carbohydrate diets.[158-162] Low-fat diets have also been shown to reduce A1c and weight, but often to a lesser degree.[163] High carbohydrate diets demonstrated improvements in these parameters if implemented with low glycemic index (GI) foods and high fiber. Reduced energy interventions, such as intermittent fasting, and portion control tools, such as the "diabetes plate method," have also shown efficacy in A1c reduction and weight loss.[164] Eating patterns should be tailored to personal and cultural food preferences, with appropriate energy needs for weight control and diabetes prevention and management. These eating patterns are described in more detail in the following sections.

Mediterranean Diet

The Mediterranean diet has been adopted by populations living near the Mediterranean Sea and is thought to be partly responsible for the good health observed in this region. A precise definition of the diet is difficult, given that the population in this region is composed of various countries, ethnicities, cultures, religious backgrounds, and economic statuses. However, it can broadly be defined as a diet rich in fruits, vegetables, whole grains, olive oil as the major source of fat, low to moderate amounts of fish and poultry and alcohol, and minimal amounts of red meat.[165]

The Mediterranean diet has long been shown to have various health benefits. This eating pattern has improved cardiovascular risk factors, such as waist circumference, lipids, blood pressure, and triglycerides, in individuals with and without diabetes.[166-169] Those at high risk for CVD were found to have a reduction in cardiovascular events and stroke when the diet was supplemented with olive oil or mixed nuts (including tree nuts and peanuts).[170] Individuals following an energy-restricted Mediterranean diet also improve glycemic control and weight loss.[166,171] These favorable outcomes are thought to be due to the intake of whole quality foods rich in antioxidants that induce a low-inflammatory state which, in turn, promotes peripheral insulin sensitivity[172] and improvements in endothelial function.[168] The general components of the Mediterranean diet are outlined in **Figure 63.2**.

Low Carbohydrate and Very Low Carbohydrate Diets

Low (LC) and very low carbohydrate (VLC) diets have gained popularity in recent decades. There is a growing body of evidence that this eating pattern may improve glycemic control and weight loss. This eating pattern is also referred to as the ketogenic diet, due to its mechanism of action in inducing a state of nutritional ketosis. Although it was popularized at the end of the 20th century by the Atkins Diet Revolution,[174] ketogenic diets have been employed since the late 1800s and early 1900s to treat diabetes before the availability of insulin,[175,176] and for pediatric epilepsy.[177]

In low carbohydrate eating patterns, carbohydrates are restricted to <130 grams per day for LC diets or <20–50 grams per day in VLC (ketogenic) diets, with similar amounts of protein (80–120 grams per day) and the rest of the energy coming from fat. It's postulated that by restricting carbohydrates, the body shifts its source of gluconeogenesis to ketone bodies (byproducts of lipolysis), with decreased lipogenesis mediated by lowered insulin concentration.[178] Coupled with increased satiety induced by the diet,[179] weight loss and improved glycemic control are observed.

Prior studies of VLC diets report a significant reduction in A1c of >1% over a period of 12 weeks to 1 year, with reductions in glucose-lowering drugs. The greatest area of uncertainty is the impact of such a dietary pattern on cardiovascular risk, as long-term studies with CVD endpoints are lacking. Improvements in blood pressure, HDL cholesterol, and triglycerides have been observed across studies. In most studies, LDL cholesterol remains unchanged or lowered.[180-185] LDL particle size may affect overall cardiovascular risk, and there is an observed trend toward larger, less atherogenic LDL particles with VLC diets.[186,187] Although further studies are needed to evaluate LCD and VLC diets about long-term efficacy and cardiovascular benefits, there is growing acceptance of this eating pattern within the medical community. To summarize, the 2022 update to the ADA Standards of Medical Care in Diabetes states, "For individuals with T2D not meeting glycemic targets, or for whom reducing glucose-lowering drugs is a priority, reducing overall carbohydrate intake with a low- or very-low-carbohydrate eating pattern is a viable option."[188]

Details regarding the composition of the diet are included in **Figure 63.3**.

Vegetarian and Vegan Diets

Vegetarian and Vegan diets, typically defined by elimination of some or all animal products, have been shown to provide health benefits. Several meta-analyses show that vegetarian dietary patterns reduce body weight and blood pressure while simultaneously improving glycemic control and blood lipids. Recent literature suggests that vegetarian and vegan dietary patterns can result in a HbA1C reduction of between 0.29% and 0.40%. Several mechanisms elucidate the benefits of omitting or partially omitting

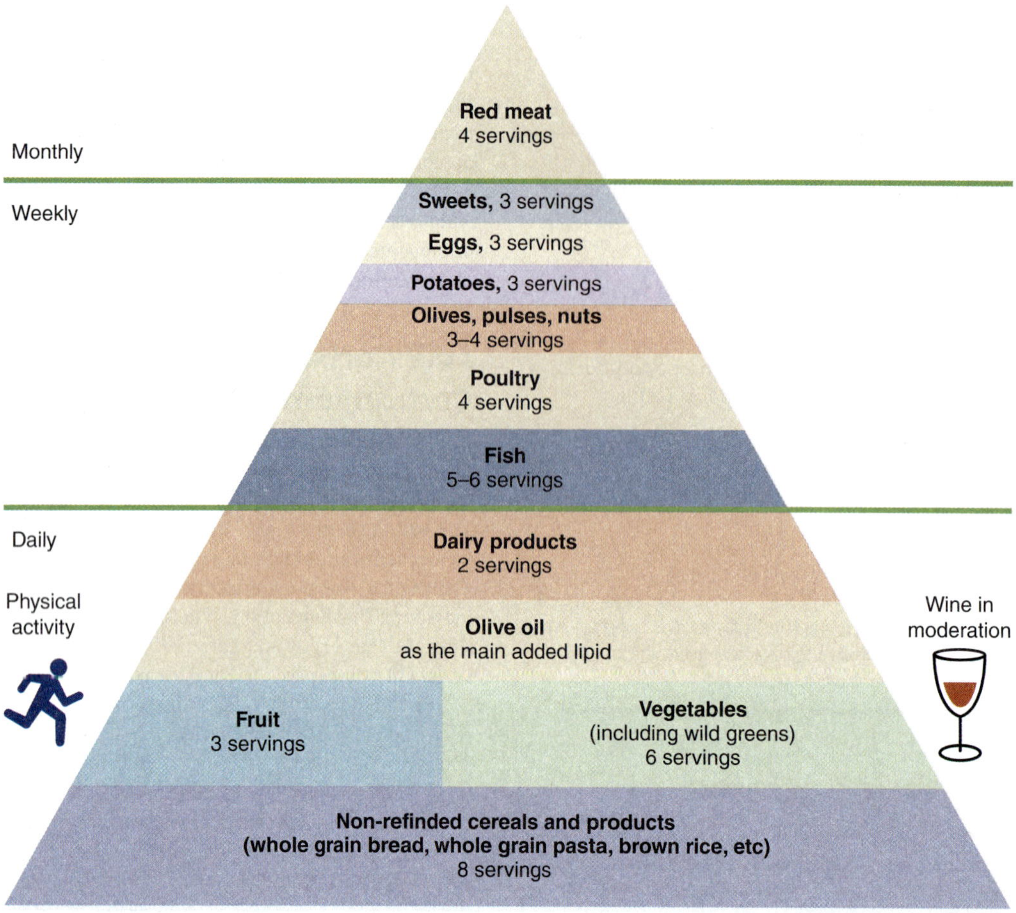

One serving equals approximately half of the portions as defined in Greek market regulations (portions served in restaurants)

Also remember to:
- Drink plenty of water
- Avoid salt and replace it by herbs (e.g. oregano, basil, thyme, etc)

Figure 63.2 Mediterranean diet food pyramid.

Reproduced from Ministry of Health and Welfare. Available at www.hhf-greece.gr/hydria-nhns.gr/adultdietary_eng.html[173]

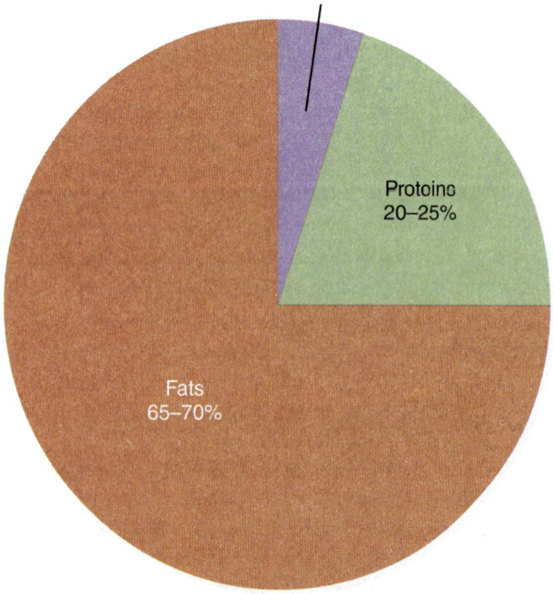

Figure 63.3 Proportion of macronutrients in a very low carbohydrate diet.

animal products. First, vegetarian and vegan diets typically lead to a reduction in energy intake, which can facilitate weight loss, resulting in improvement in glycemic control. Vegetarian and vegan diets provide approximately 10% less fat than standard eating patterns. Low-fat diets are associated with improving insulin sensitivity via a reduction in intramyocellular lipid concentrations. High intake of dietary fiber, a common characteristic of vegetarian and vegan eating patterns, is a well-known and accepted strategy to improve cardiometabolic health.[189,190] Animal foods are rich in iron, calcium, B_{12}, vitamin D, and protein; eliminating these foods can place people at risk for nutrient deficiencies. People following vegetarian and vegan diets should try to select fortified foods in alignment with their dietary preferences and consider supplementation, if necessary.

Energy and Fat Restricted Diets

Very low-energy diets (VLCD) can be defined as eating plans allowing <800 kcal per day. Such diets should be followed under medical supervision, as there is a relatively high risk of micronutrient deficiency and inadequate protein intake. There is considerable evidence to support VLCD in the management of diabetes, with several trials demonstrating significant weight loss, improved β-cell function, and insulin sensitivity.[192] A trial by Jackness et al., found that individuals following a VLCD of 500 kcal/d lost an equivalent amount of weight to those undergoing Roux-en-Y gastric bypass (RYGB) surgery during the 21-day trial period.[193] Additionally, insulin sensitivity improved similarly in both groups, as did insulin secretion, suggesting improved β-cell function. This trial suggests that VLCD are noninferior to RYBG in terms of improved glycemic control and weight loss in the short term. It is generally regarded that RYBG produces a stronger incretin effect and may be superior to VLCD in the long term.[193] Unfortunately, not all people with diabetes respond well to VLCD; Steven et al., found that longer duration of diabetes, being on more diabetes medications and having lower circulating insulin, were predictors of a poor response to VLDC, despite weight loss.[194] It is important to note that long-term adherence to VLCD is challenging and not sustainable for a large percentage of the population. Longer-term studies on VLCD have found similar weight loss effects compared with low-energy diets consisting of approximately 1,200 to 1,500 kcal/day.[192,195]

Intermittent Fasting

Intermittent fasting (IF) and time-restricted eating (TRE) are often used interchangeably to describe an eating pattern that focuses on the timing of energy intake rather than the types of foods or macronutrients consumed. There are several variations of IF. Two common approaches include restricting nutrient intake for 16 to 20 hours each day or alternate-day fasting.[196] IF, alternate-day fasting, and traditional energy restriction are all legitimate methods to improve insulin sensitivity and reduce body fat. A review by Barnosky et al., reported reductions in circulating insulin concentration between 20% and 31% for alternate day fasting or IF, and 11% to 41% for traditional energy restriction in individuals with prediabetes. The authors noted a relationship between the amount of weight lost and greater improvements in insulin.[197] IF is generally regarded as a safe and effective strategy to reduce body weight, diabetes medication burden, and leptin concentration in diabetes patients. A case series by Furmli et al., placed three patients on 24-hour fasts 3 days per week; each patient was able to discontinue insulin therapy within 1 month after beginning the fasting plan.[198] Beta cell loss is a component of diabetes; therefore, not all individuals who intermittently fast will be able to discontinue diabetes medications. This raises concerns for additional risk in terms of hypoglycemia frequency and severity for those who take insulin or secretagogues.[199] For any person with diabetes who is interested in starting a TRE or IF program, the importance of close contact with their clinical team cannot be understated. The risks of these eating patterns are not only limited to hypoglycemia but also include possible protein-energy malnutrition, micronutrient deficiencies, and dehydration, which can be mitigated by close follow-up with the medical team. Generally, most adults can participate in fasting safely, although certain populations should consider more moderate eating patterns, given the unique risks associated with fasting. Pregnant women, children, older adults, people with eating disorders, and individuals with immunosuppression should forgo IF.[200]

Plate Method

MyPlate replaced MyPyramid in 2011 as the official USDA nutrition guide and remains unchanged. There is literature dating as far back as 1988 describing a similar portioning methodology developed in Sweden for diabetes nutrition education known as the "plate method."[201] Each of the plate-based graphics are designed to simplify the meal-building process by encouraging portion control and relative portion sizes. Those following the plate method are instructed to assemble a meal using a 9-inch plate with approximately 1/2 consisting of non-starchy vegetables, ¼ lean protein, and ¼ starch or grain-based product. The plate method focuses on what one should eat and places less emphasis on foods to avoid without the requirement of tedious measuring or weighing of food.[202] A 2016 randomized controlled trial compared the A1C lowering effect of Diabetes Self-Management Education and Support (DSMES) using the plate method to carbohydrate counting with label reading and using suggested serving sizes. Both groups enjoyed similar benefits, regardless of the educational intervention. Notably, those in the plate method group reported higher treatment satisfaction. Also of note, participants of high numeracy benefited regardless of the intervention compared with those with lower numeracy, who only realized A1c improvement in the plate method. This suggests that the plate method is a viable alternative to carbohydrate counting and may be the preferred starting point for nutrition education and those with known low numeracy.[202]

Physical Activity

The American College of Sports Medicine (ACSM) defines physical activity as "bodily movement that is produced by the contraction of skeletal muscle and that substantially increases energy expenditure."[203] Regular physical activity is strongly encouraged in all persons with diabetes and should be incorporated into their daily lifestyle. The ADA and ACSM currently recommend 150 minutes/week of moderate to vigorous aerobic activity over at least 3 days/week and 2 to 3 days/week of moderate to vigorous resistance exercises, with no more than 2 consecutive days without exercise.[188] A 2022 position statement by the ADA specifically calls for a reduction in sedentary time, including activities such as desk work, television viewing, and video gaming. When possible, people with diabetes should aim to interrupt sitting or sedentary activities with light walking or other non-exercise activities at least every 30 minutes, as this improves glycemic control.[204] Exercise has beneficial effects on glycemic control, body composition, hypertension, hyperlipidemia, and obesity, as well as psychologic effects.[205-207] Snowling and Hopkins[206] reported an average hemoglobin A1c reduction of −0.8% after 130 to 270 minutes/week of exercise for 6 months, within the range to promote significant reductions in microvascular, macrovascular, and nonvascular complications in persons with diabetes. Physical activity also increases insulin sensitivity, so adjustments to prevent low blood glucose, such as exercise after meals, additional carbohydrate intake, or less insulin than usual (≤50%) are recommended, depending on the intensity and duration of exercise.

Special precautions must be considered in prescribing an exercise plan. Because patients with diabetes are also known to be at a higher risk of developing CVD, neuropathy, nephropathy, and retinopathy, further evaluation may be warranted to determine whether these complications exist and/or the degree of progression, to ascertain an appropriate physical activity program. For example, CVD is the major cause of mortality for people with diabetes; therefore, a careful cardiac assessment should be performed before any fitness program is initiated in high-risk individuals. This assessment may include a graded exercise stress test. Individuals with T1D are more prone to glycemic variability during and following exercise. Given the multitude of strategies one might employ to achieve stable glucose concentration,[204] it may be of great value for people with T1D to meet with a CDCES or endocrinologist to develop a personalized plan surrounding exercise.

Additional Considerations

Children

The primary goals of nutrition therapy for children with diabetes are to promote normal growth and development, achieve good glycemic control, prevent hypoglycemia, and decrease risk of complications. The nutritional management of T1D and T2D in children varies, because most children with t1D are thin at diagnosis, whereas most children with T2D are overweight.

T1D

Providing enough energy for growth has always been a concern when developing a nutrition plan for children with diabetes. Parents and children with T1D need to be guided in adjusting the insulin dose to the child's increasing energy needs, and be warned against withholding food to keep blood glucose concentration under control. Adequate growth should be monitored several times a year to ensure appropriate growth for the child's age and sex, especially in the first few years after diagnosis. As with adults, the nutrition plan should be customized to the child's needs and preferences, and carefully evaluated and readjusted as the child grows. The ADA recommends a personalized A1c goal for each patient, with <7% being appropriate for many children and even <6.5% during the early phase of T1D. Progressively less-stringent goals, such as <7.5% and up to <8%, may be appropriate for children with a history of severe hypoglycemia.[188] Care and education should be provided by a diabetes team that includes a pediatric endocrinologist, a nurse, an RDN, and a mental health counselor, who are all familiar with the normal stages of childhood and adolescent development and how they affect diabetes management.[208]

T2D

The increasing prevalence of T2D in youth is related to the increasing obesity epidemic in children. Current estimates suggest that there are approximately 5,000 new cases of T2D in youth each year, with an annual increase of 2.3%.[188] The increased incidence of T2D, especially in minority ethnic groups (e.g., African Americans), is related to the increased prevalence of obesity in children, which often occurs in the context of low socioeconomic status and challenging cultural environments.[188] Treatment involves medications to normalize glycemia, lifestyle modifications in food intake and physical activity to promote weight loss, and control of comorbidities.[209] Notably, T2D in children carries novel features, including a more rapid decline in β-cell function and earlier onset of diabetes complications. Currently, only insulin, metformin, and glucagon-like peptide 1 receptor agonists are approved for use in children by the FDA.[188]

Older Adults

T2D in persons more than 65 years old is a major public health concern. More than one-quarter of persons over 65 years old have diabetes, and almost one-half of older adults have prediabetes.[1] Normal aging is associated with impaired insulin sensitivity,[210-212] possibly because of the lower density of the glucose carrier GLUT4 in the muscle that could contribute to insulin resistance.[213] Biologic changes associated with aging may also contribute to impaired insulin sensitivity, including increased abdominal fat mass, decreased physical activity, mitochondrial dysfunction, hormonal changes, increased oxidative stress, and inflammation.[214] The presence of comorbidities, cognitive dysfunction, and functional disabilities affects the management of diabetes, especially in elderly persons. Older persons with diabetes are also at greater risk than other older adults for several geriatric syndromes, such as polypharmacy, cognitive impairment, depression, urinary incontinence, injuries, falls, chronic pain, sarcopenia, and frailty.[215] These conditions are associated with difficulties in self-management abilities and quality of life.[188] Older adults with diabetes are two to three times more likely to have functional disabilities, including difficulties walking a quarter mile, lifting heavy objects, doing housework, or participating in leisure activities, compared with their counterparts without diabetes.[216]

The population of older adults with diabetes is heterogenous in race/ethnicity, duration of diabetes, comorbidity, and functional status,[215] which should be taken into consideration when developing guidelines for treatment. Lifestyle interventions recommended for clinical treatment should be tailored to frailty status. Weight loss recommendations for older adults who are overweight and obese are appropriate. In contrast, nursing home residents with debilitation may not be good candidates for weight loss.[217] The ADA recommends a hemoglobin A1c goal of <7.0–7.5% (53–58 mmol/mol) for otherwise healthy older adults with T2D and a life expectancy of more than 5 years.[74] However, older adults with multiple comorbidities, functional disabilities, and/or limited expected life expectancy may benefit from less-stringent hemoglobin A1c goals (i.e., <8%), although more studies are needed.[218] Hypoglycemia is a serious complication of diabetes treatment in older patients. Factors associated with hypoglycemia include renal impairment, coadministration of insulin sensitizers or insulin, exercise, skipped meals, energy restriction, recent hospitalization, polypharmacy, and therapy with salicylates, sulfonamides, fibric acid derivatives, and warfarin.[219]

Pharmacology

Diet and exercise may be sufficient to control blood glucose for persons with T2D. However, for those unable to meet hemoglobin A1c goals, modern pharmacologic regimens lend themselves to various combinations tailored to an individual's needs. Currently, 11 classes of noninsulin medications are available for diabetes: sulfonylureas, biguanides, meglitinides, α-Glucosidase inhibitors, thiazolidinediones, dipeptidyl peptidase-4 (DPP-4) inhibitors, GLP-1 agonists, colesevelam, bromocriptine, and SGLT2 inhibitors (**Table 63.10**), as well as combination products. These classes are generally used in T2D, although biguanides and pramlintide may be used in select T1D patients. Insulin (**Table 63.11**) carries the greatest risk for hypoglycemia, followed by sulfonylureas. In contrast, the risk of experiencing the symptoms of hypoglycemia for a person not taking any diabetes medications is rare. Patients and health professionals must understand the effects of meals and medications on hypoglycemia and glycemic control.[188]

Table 63.10 Oral and Injectable Blood Glucose-Lowering Medications

Classification Medication	Route	Mechanism of Action	Time and Dose	Complications or Comments
Sulfonylureas ■ Glimepiride (Amaryl) ■ Glipizide (Glucotrol) ■ Glipizide ER (Glucotrol XL) ■ Glyburide (DiaBeta, Micronase)	Oral	Stimulate first-phase insulin secretion by binding to and causing closure of potassium-ATP channels on pancreatic β-cell membranes, with consequent membrane depolarization, calcium influx, and insulin exocytosis	One or two times a day	Contraindicated in DKA; not to be used for therapy in T1D; patients should avoid excessive alcohol use (increased risk for hypoglycemia); may cause weight gain, nausea, diarrhea, or heartburn; check liver function tests; should avoid use in the frail or elderly due to higher risk of hypoglycemia
Biguanides ■ Glucophage (Metformin) ■ Glucophage XR (Metformin XL)	Oral	Decreases hepatic glucose production, and decreases intestinal absorption of glucose	Two to three times a day; XR once a day	Considered first line for most patients with T2D. Well tolerated if taken with or after meals (reduces GI upset); used with caution in mild renal insufficiency with appropriate dose reductions, and should not be used in patients with stage 4 or 5 chronic kidney disease, excessive alcohol use (increases risk of lactic acidosis) should be avoided; may cause nausea, vomiting, diarrhea, flatulence, abdominal pain, cobalamin (vitamin B_{12}) deficiency, or asthenia (physical weakness or loss of strength)
α-Glucosidase inhibitors ■ Miglitol (Glyset) ■ Acarbose (Precose)	Oral	Competitively and reversibly inhibit enzymes (α-glucoside hydrolases) that break down complex sugars in the small intestinal brush border; cause delayed absorption of simple sugars from the gut, thus reducing postprandial hyperglycemia	Take before each meal; swallow with first bite of food	Contraindicated in GI conditions such as inflammatory bowel disease, intestinal obstruction or ileus, conditions potentially exacerbated by increased intestinal gas, conditions associated with digestion or absorption, or colonic ulcerations; GI disturbances (flatulence, diarrhea, bloating, abdominal pain) can occur (in ≤74%)
Thiazolidinediones ■ Pioglitazone (Actos)	Oral	Increase insulin-dependent glucose disposal primarily by decreasing insulin resistance in the periphery; also affect fatty acid metabolism	Once or twice daily with or without food	Liver damage possible; enzymes to be monitored carefully; can lead to weight gain or heart failure; not to be used in persons with congestive heart failure (NYHA class III or IV heart failure); rosiglitazone no longer available in the United States
Meglitinides ■ Repaglinide (Prandin) ■ Senaglinide (nateglinide; Starlix)	Oral	Increases insulin secretion by pancreatic β cells; short-acting	5–30 min before meals	Better control of postprandial hyperglycemia and associated with a lower risk of delayed hypoglycemic episodes; hypoglycemia, headache, nausea, vomiting, diarrhea, upset stomach, joint pain possible
DPP-IV inhibitors ■ Sitagliptin (Januvia) ■ Saxagliptin (Onglyza) ■ Linagliptin (Tradjenta)	Oral	Inhibit the degradation of incretins such as GLP-1 by inhibiting the enzyme IV (DPP-IV); prolonged incretin effect enhancing glycemic control through various mechanisms	Usually once a day	Possible nasopharyngitis or upper respiratory tract infections, headache, nausea, diarrhea, abdominal pain, urinary tract infections, peripheral edema; weight neutral

(continues)

Table 63.10 Oral and Injectable Blood Glucose-Lowering Medications *(continued)*

Classification Medication	Route	Mechanism of Action	Time and Dose	Complications or Comments
GLP-1 Agonists ■ Exenatide* (Byetta) ■ Liraglutide* (Victoza) ■ Lixisenatide* (Adlyxin) ■ Exenatide ER (Bydureon) ■ Dulaglutide (Trulicity) ■ Semaglutide (Ozempic, or oral form Rybelsus*) ■ Tirzepatide (Mounjaro) combo GIP agonist/GLP-1 ■ Agonist Short-acting, daily or twice daily dosing	Injectable	Stimulate glucose-dependent insulin secretion, slow gastric emptying; inhibit glucagon secretion; suppress appetite; tirzepatide has a GIP agonist, which works in an additive fashion to increase glucose-dependent insulin secretion and may enhance glucagon activity in hypoglycemic states	Once or twice a day, inject within an hour of meals; or once weekly	Nausea, usually improving over time; hypoglycemia, especially with sulfonylureas; CV benefits; associated with weight loss, seen most with long-acting forms, especially tirzepatide
Antihyperglycemic agent ■ Pramlintide (Symlin)	Injectable	Slows gastric emptying; suppresses an exaggerated rise in postprandial glucagon; induces satiety	Inject before major meals	Reduced insulin dosage needed with drug initiation to avoid severe hypoglycemia
■ Colesevelam (Welchol)	oral	Unclear mechanism of action for glucose lowering	Take before meals	Often causes constipation; not routinely used as a treatment for T2D
■ Bromocriptine (Cycloset, Parlodel)	oral	Uncertain mechanism, but may be related to dopaminergic action in the brain and inhibition of sympathetic tone	Take once daily	Nausea, diarrhea, orthostatic hypotension; not routinely used as a treatment for T2D
SGTL-2 Inhibitors ■ Canagliflozin (Invokana) ■ Dapagliflozin (Farxiga) ■ Empagliflozin (Jardiance) ■ Ertugliflozin (Steglatro)	oral	Promote renal excretion of glucose	Once daily	Several have shown to reduce CV-related hospitalization and mortality, prevent progression of CKD; may help with modest blood pressure and weight reduction; use with caution in CKD (each has specific GFR ranges); may lead to euglycemic DKA and should not be used in patients with prior DKA; may cause UTIs, mycotic genital infections

ATP, adenosine triphosphate; *DPP-IV*, dipeptidyl peptidase IV; *FDA*, Food and Drug Administration; *GI*, gastrointestinal; *GLP-1*, glucagon-like peptide-1; *NYHA*, New York Heart Association; *CV*, cardiovascular; *SGLT2*, sodium-glucose co-transporter 2; *CKD*, chronic kidney disease; *GFR*, glomerular filtration rate; *DKA*, diabetic ketoacidosis; *UTI*, urinary tract infection.

Insulin Therapy

Insulin therapy is indicated for everyone with diagnosed T1D, even in people whose glucose concentration is nearly normal (latent autoimmune diabetes of adulthood). If T1D is diagnosed, the current evidence recommends starting insulin immediately, both in anticipation of diminished β-cell function and to preserve some islet function. Given the lability of T1D, generally, both long-acting and short- or fast-acting insulin should be started. Frequent self-monitoring of blood glucose by the patient guides the therapy. Patients are recommended to test before meals (goal blood glucose, 70 to 120 mg/dL), occasionally also 2 hours after meals (goal blood glucose, 140 to 180 mg/dL), at bedtime (goal blood glucose, 100 to 140 mg/dL), with symptoms of hyperglycemia or hypoglycemia, and occasionally overnight. Goal glucose ranges may be more conservative for older adults and patients with frailty to prevent hypoglycemia. Most insulin dose adjustments should be in 10% to 20% increments, depending on the degree of glucose abnormality. Table 63.11 contains a summary of currently available insulins.[188]

Herbal Supplements

In the United States and globally, supplement use remains popular among consumers, with the hopes of improving or maintaining health, to correct a dietary deficiency or, therapeutically, for a specific health condition, despite the lack of evidence for most popular supplements on the market.[220] Herbal supplements should not be used in place of conventional medical therapy for diabetes.

There has been recurrent interest in whether the routine use of herbal and micronutrients, such as cinnamon[221] and curcumin,[221] improve glycemia in people with diabetes. However, there is insufficient evidence of efficacy and possibly concern over long-term safety. There are mixed results as to whether

Table 63.11 Types of Insulin

Type	Generic Name (Brand Name)	Onset	Peak	Duration	Comments
Rapid-acting insulin	■ Insulin aspart (NovoLog, Fiasp) ■ Insulin glulisine (Apidra) ■ Insulin lispro (Humalog, Ademalog, Lyumjev) ■ Inhaled insulin (Afrezza)	10–20 min	2 h	4 h	Fastest-acting insulins on the market; can be taken right before or during a meal; Fiasp is ultra-fast acting; inhaled insulin may cause bronchospasm, not widely prescribed
Short-acting insulin	Regular insulin (Humulin R, Novolin R)	30–60 min	2–4 h	6–8 h	Injected 30 min before meals to cover the sugars absorbed from food
Intermediate-acting insulin	NPH insulin (Humulin N, Novolin N), U-500 insulin	1–3 h	4–10 h	10–16 h	Often used in combination with rapid- or short-acting insulin; can also be given in combination with oral agents in T2D; usually given twice a day; U-500 insulin can be used in severely insulin resistant patients requiring >200 units/day, used infrequently
Long-acting insulin	■ Insulin glargine (Lantus, Basaglar) ■ Insulin detemir (Levemir)	1–3 h	Theoretically peak-less (glargine); 6–8 h (detemir)	20–24 h (glargine); 6–24 h (detemir)	Often used with a rapid- or short-acting insulin to cover the sugars absorbed from food at mealtimes; can also be given in combination with oral agents in T2D; given once or twice a day (detemir)
Ultra-long acting	■ Insulin glargine U-300 (Toujeo) ■ Insulin degludec (Tresiba, U-100 and U-200)	6 h (glargine); 1-3 hours (degludec)	No peak	>24 h	Often used with a rapid- or short-acting insulin to cover the sugars absorbed from food at mealtimes; can also be given in combination with oral agents in T2D; less hypoglycemia than long-acting agents; higher concentration allows for lower volume of injection; given once daily
Mixtures	Novolin 70/30, Humulin 70/30, NovoLog mix 70/30, Humalog mix 75/25, Humalog mix 50/50	Varies according to type	Varies according to type	Varies according to type	Convenient for people who use a mixture of short- or rapid-acting insulin and long-acting insulin by providing both in one syringe; helpful for those with poor dexterity or eyesight or for anyone who has problems drawing up insulin from two different bottles or reading the instructions and dosages on bottle labels

NPH, neutral protamine Hagedorn.

cinnamon (*Cinnamon cassia*) can enhance insulin signaling and increase glycogen synthase activity. Human trials investigated the use of 1 to 6 g/day. A modest effect on reducing fasting blood glucose (5%–24%) with short-term administration was reported; however, the results were mixed.[221] A meta-analysis conducted by Allen et al., concluded that, although there were statistically significant decreases in fasting plasma glucose, there was no significant effect on hemoglobin A1c.[221] Curcumin is a bioactive molecule present in the rhizome of the *Curcuma longa* plant, also known as turmeric. Curcumin has different pharmacologic and biologic effects that have been described by both in vitro and in vivo studies and include antioxidant, cardioprotective, anti-inflammatory, antimicrobial, nephron-protective, antineoplastic, hepatoprotective, immunomodulatory, hypoglycemic, and anti-rheumatic effects.[222]

Although the benefits of some of these compounds have been observed, there continues to be no clear evidence of benefit from herbal or nonherbal supplementation in the treatment of diabetes. In some cases, these remedies can have adverse effects. The 2022 ADA Lifestyle Management recommendations conclude that, after a review of the evidence, there is still not enough evidence to substantiate recommending the use of herbs or spices as treatment for T2D.[188] The ADA also states that the use of any herbal supplements, which are not regulated and vary in content, may provide more risk than benefit in that herbs may interact with other medications that are taken to control diabetes.[188]

Conclusion

Diabetes is a chronic disease with a global prevalence that continues to grow and is associated with significant individual and public health burden. Lifestyle modifications, including medical and nutritional therapies, remain the cornerstones of the successful management of diabetes, in conjunction with glucose-lowering medications, when indicated. In addition, education on the principles of physical activity, the need for self-monitoring of blood glucose, and the adjustment of appropriate medications during times of illness is vital to the person with diabetes. The interprofessional healthcare team should work together with the patient to achieve good glycemic control, attain optimal serum lipid concentrations and blood pressure, and maintain desirable body weight, as well as other risk factor modifications to prevent the development of long-term complications and reduce the morbidity and mortality associated with this chronic disease.

Acknowledgments

The authors acknowledge Rita Rastogi Kalyani and Adrian Dobs for their contributions to the previous version of this chapter.

References

1. Centers for Disease Control and Prevention. National Diabetes Statistics Report 2020. Reviewed June 29, 2022. Accessed September 5, 2022. Available at www.cdc.gov/diabetes/data/statistics-report/index.html
2. World Health Organization. Classification of Diabetes Mellitus 2019. Released April 21, 2019. Accessed on September 1, 2022. Available at www.who.int/publications/i/item/classification-of-diabetes-mellitus
3. Saeedi P, Petersohn I, Salpea P, et al. Diabet Res Clin Pract 2019;157:107843.
4. Centers for Disease Control and Prevention. Diabetes Fast Facts. Reviewed September 30, 2022. Accessed on October 1, 2022. Available at www.cdc.gov/diabetes/basics/quick-facts.html
5. World Health Organization. Global Report on Diabetes. WHO 2016,. Released April 21, 2016. Accessed on September 18, 2022. Available at www.who.int/publications/i/item/9789241565257
6. Knowler WC, Barrett-Connor E, Fowler SE et al. N Engl J Med 2002;346:393.
7. Diabetes Control and Complications Trial Research Group. N Engl J Med 1993;329:977–86.
8. Zajac J, Shrestha A, Patel P, et al. In: Poretsky L, ed. The main events in the history of diabetes mellitus. Principles of Diabetes Mellitus. 2nd ed. New York: Springer, 2010:3–16.
9. Ahmed A. Saudi Med J 2002;23:373–8.
10. Chalmers K. In: Kahn R, King G, Moses, A et al. Medical nutrition therapy. Joslin's Diabetes Mellitus. 14th ed. Baltimore, MD: Lippincott Williams & Wilkins, 2005:611–632.
11. American Diabetes Association. Diabetes Care 1994;17: 490–518.
12. Solis-Herrera C, Triplitt C, Reasner C, et al. In: Feingold KR, Anawalt B, Boyce A, eds.Classification of Diabetes Mellitus. South Dartmouth, MA: Endotext [Internet]. MDText.com, Inc.;2018. Available at www.ncbi.nlm.nih.gov/books/NBK279119/
13. American Diabetes Association. Professional Practice Committee. Diabet Care 2022;45: S17–S38.
14. Bays HE, Chapman RH, Grandy S; SHIELD Investigators' Group. Int J Clin Pract 2007;61:737–47. Erratum in: Int J Clin Pract 2007;61:1777–8.
15. Centers for Disease Control and Prevention. National Center for Chronic Disease Prevention and Health Promotion, Division of Nutrition, Physical Activity, and Obesity. Data, Trend and Maps. Reviewed October 4, 2022. Accessed on October 05, 2022. Available at www.cdc.gov/nccdphp/dnpao/data-trends-maps/index.html
16. Boyle JP, Thompson TJ, Gregg EW, et al. Popul Health Metr 2010;8:29.
17. Zhou X, Shrestha SS, Shao H, et al. Diabetes Care 2020;43:2396–2402.
18. American Diabetes Association. Diabetes Care 2018;41(5):917–928.
19. Williams R, Karuranga S, Malanda B, et al. Diabetes Res Clin Pract 2020;162:108072.
20. Green A, Hede SM, Patterson CC, et al. Diabetologia 2021;64, 2741–2750.
21. Mobasseri M, Shirmohammadi M, Amiri T, et al. Health Promot Perspect 2020;30;10:98–115.
22. Mayer-Davis EJ, Lawrence JM, Dabelea D, et al. N Engl J Med 2017;13;376:1419–1429.
23. Rewers M, Ludvigsson J. Lancet 2016;387:2340–2348.
24. Oilinki T, Otonkoski T, Ilonen J, et al. Pediatr Diabet 2012;13: 176–80.
25. Faulkner CL, Luo YX, Isaacs S, et al. Rev Med Virol 2020;e2209.
26. Vehik K, Lynch KF, Wong MC, et al. Nat Med 2019; 25:1865–72.
27. Hyoty H, Hiltunen M, Knip M, et al. Diabetes 1995;44:652–7.

28. Goldacre, R. Diabetologia 2018; 61:616–25.
29. Tanoey J, Gulati A, Patterson C, et al. Curr Diabetes Rep 2019; 19:124.
30. Harder T, Roepke K, Diller N, et al. Am J Epidemiol 2009;169: 1428–1436.
31. Yassouridis C, Leisch F, Winkler C, et al. Pediatr Diabetes 2017;18:103–10.
32. Verbeeten KC, Elks CE, Daneman D, et al. Diabetes Med 2011; 28:10–8.
33. Kostic AD, Gevers D, Siljander H, et al. Cell Host Microbe 2015; 17:260–73.
34. Zhao G, Vatanen T, Droit L, et al. Proc Natl Acad Sci USA 2017; 114:E6166–75.
35. Han H, Li Y, Fang J, et al. Int J Mol Sci 2018; 19:995.
36. Boddu SK, Aurangabadkar G, Kuchay MS. Diabetes Metab Syndr 2020;14:2211–7.
37. Unsworth R, Wallace S, Oliver NS, et al. Diabetes Care 2020;43: e170–e1.
38. Principi N, Berioli MG, Bianchini S, et al. J Clin Virol 2017;96:26–31.
39. Murri M, Leiva I, Gomez-Zumaquero JM, et al. BMC Med 2013;11:46.
40. Stene LC, Joner G. Am J Clin Nutr 2003;78:1128–34.
41. Centers for Disease Control and Prevention. Type 2 Diabetes. Reviewed December 16, 2021. Accessed September 9, 2022. Available at www.cdc.gov/diabetes/basics/type2.html
42. Lao XQ, Deng HB, Liu X, et al. Br J Sports Med 2019;53: 895–900.
43. Menke A, Casagrande S, Geiss L, et al. JAMA 2015;8;314:1021–9.
44. Shai I, Jiang R, Manson JE, et al. Diabetes Care 2006;29:1585–90.
45. Herath H, Herath R, Wickremasinghe R. PLoS One 2017;23;12:e0179647.
46. Sattar N, McConnachie A, Shaper AG, et al. Lancet 2008;7;371:1927–35.
47. Willi C, Bodenmann P, Ghali WA, et al. JAMA 2007;12;298: 2654–64.
48. Ehrmann DA, Barnes RB, Rosenfield RL, et al. Diabetes Care 1999;22:141–6.
49. Edelstein SL, Knowler WC, Bain RP, et al. Diabetes 1997;46: 701–10.
50. Tenenbaum A, Motro M, Fisman EZ, et al. Am J Med 2003; 114:271–5.
51. InterAct Consortium, Scott RA, Langenberg C, et al. Diabetologia 2013;56:60–9.
52. Jayedi A, Soltani S, Motlagh SZ, et al. BMJ 2022;18;376:e067516.
53. Bellary, S., Kyrou, I., Brown, J.E. et al. Nat Rev Endocrinol 2021;17:534–548.
54. Li P, Shan Z, Zhou L, et al. Eur J Endocrinol 2016;175:R231–45.
55. Kao W, Puddey I, Boland L, et al. Am J Epidemiol 2001; 154748–757.
56. Wei M, Gibbons L, Mitchell T, et al. Diabetes Care 2000;2318–22.
57. Wannamethee SG, Camargo CA, Manson JE, et al. Arch Intern Med 2003;163(11):1329–1336.
58. Harris ML, Oldmeadow C, Hure A, et al. PLoS One 2017;21;12(2):e0172126.
59. Cappuccio FP, D'Elia L, Strazzullo P, et al. Diabetes Care 2010; 33:414–20.
60. Suwannaphant K, Laohasiriwong W, Puttanapong N, et al. J Clin Diagn Res 2017;1:LC18–LC22.
61. van Dam RM, Rimm EB, Willett WC, et al. Ann Intern Med 2002;136:201–9.
62. Zhao B, Zeng L, Zhao J, et al. BMJ 2020;10:e032240.
63. Malik VS, Popkin BM, Bray GA, et al. Circ 2010;23;121:1356–64.
64. Zimmet P, Alberti KG, Shaw J. Nature 2001;13;414(6865): 782–7.
65. Schulz LO, Chaudhari LS. Curr Obes Rep 2015;4(1):92–8.
66. Cheema A, Adeloye D, Sidhu S, et al. J Glob Health 2014;4:010404.
67. Centers for Disease Control and Prevention. Diabetes Report Card: Prediabetes. Reviewed May 17, 2022. Accessed September 5, 2022. Available at www.cdc.gov/diabetes/library/reports/reportcard /prediabetes.html
68. Guariguata L, Linnenkamp U, Beagley J, et al. Diabet Res Clin Pract 2014;103(2):176–85.
69. Gregory EC, Ely D. Natl Vital Stat Rep 2022;71:1–15.
70. Solomon CG, Willett WC, Carey VJ, et al. JAMA 1997;278: 1078–83.
71. Hedderson MM, Gunderson EP, Ferrara A. Obstet Gynecol 2010; 115(3):597–604. Erratum in: Obstet Gynecol 2010;115:1092.
72. Vounzoulaki E, Khunti K, Abner SC. BMJ 2020;369:m1361.
73. World Health Organization & International Diabetes Federation. Definition and diagnosis of diabetes mellitus and intermediate hyperglycaemia: report of a WHO/IDF consultation. World Health Organization. Accessed on September 10, 2022. Available at https://apps.who.int/iris/handle/10665/43588
74. Hantzidiamantis PJ, Lappin SL. Physiology, Glucose. Updated 2021 Sep 20. In: StatPearls [Internet]. Treasure Island, FL: StatPearls Publishing; 2022. Accessed on September 5, 2022. Available at www.ncbi.nlm.nih.gov/books/NBK545201/
75. McCowen KC, Deaconess BI, Smith RJ. In: Caballero B, ed. Classification and chemical pathology: diabetes mellitus. Encyclopedia of Human Nutrition. 2nd ed. London:, U.K. Elsevier, 2005:543–51.
76. Shrayyef MZ, Gerich JE. In: Poretsky L, ed. Normal Glucose Homeostasis. Principles of diabetes mellitus. 2nd ed. New York: Springer, 2010:19–35.
77. Vander A, Sherman J, Luciano D. In: Nunes I, Schauck D, Bradley J, eds. Regulation of organic metabolism, growth, and energy balance. Human Physiology. 5th ed. New York, NY: McGraw-Hill, 1990:555–600.
78. Woerle HJ, Meyer C, Dostou JM, et al. Am J Physiol Endocr Metab 2003;284:E716–25.
79. Gosmanov NR, Szoke E, Israelian Z, et al. Diabetes Care 2005;28: 1124–31.
80. Ruder NB, Myers MG Jr, Chipkin SR, et al. In: Kahn C, King G, Moses A, et al, eds. Hormone-Fuel Interrelationships: Fed State, Starvation, and Diabetes Mellitus. Joslin's Diabetes Mellitus. 14th ed. Philadelphia: Lippincott Williams & Wilkins, 2005:127–44.
81. Fonseca V, Pendergrass M, McDuffie R. In: Fonseca V, Pendergrass M, McDuffie R, eds. Complications of Diabetes. Diabetes in Clinical Practice. London: Springer, 2010:41–57.
82. Umpierrez G, Casals M, Gebhart S, et al. Diabetes 1995;44: 790–5.
83. Feng Y, Fleckman A. In: Poretsky L, ed. Acute Hyperglycemia Syndromes: Diabetic Ketoacidosis and the Hyperosmolar State. Principles of Diabetes Mellitus. 2nd ed. New York, NY: Springer, 2010:281–95.
84. Taylor KS, Heneghan CJ, Farmer AJ, et al. Diabetes Care 2013;36(8):2366–71.
85. Preshaw PM, Alba AL, Herrera D, et al. Diabetologia. 2012; 55(1):21–31.
86. Institute of Medicine. The role of nutrition in maintaining health in the nation's elderly: evaluating coverage of nutrition services for the Medicare population [Internet], 1999. Accessed October 2, 2018. Available at www.nap.edu/catalog/9741/the-role-of-nutrition -in-maintaining-health-in-the-nations-elderly
87. Morris SF, Wylie-Rosett J. Clin Diabetes 2010;28:12–18.
88. Pastors J, Warshaw H, Daly A, et al. Diabetes Care 2002;25: 608–13.
89. Evert A.B, Dennison M, Gardner CD, et al. Diabetes Care 2019;42: 731–754.
90. Franz MJ, Powers MA, Leontos C, et al. J Am Diet Assoc 2010;110:1852–89.
91. Connor H, Annan F, Bunn E, et al. Diabetes Med 2003;20:786–807.

92. Swan WI, Vivanti A, Hakel-Smith NA, et al. J Acad Nutr and Diet 2017;117:2003–2014.
93. American Diabetes Association. Diabetes Care 2019;42:S46–S60.
94. Food and Nutrition Board, Institute of Medicine. Dietary Reference Intakes for Energy, Carbohydrate, Fiber, Fat, Fatty Acids, Cholesterol, Protein, and Amino Acids (Macronutrients). Washington, DC: National Academy Press, 2005.
95. Shan Z, Rehm CD, Rogers G, et al. *JAMA*. 2019;322:1178–1187.
96. Tay J, Luscombe-Marsh ND, Thompson CH, et al. Am J Clin Nutr. 2015;102:780–90.
97. Yamada Y, Uchida J, Izumi H, et al. Intern Med 2014;53: 13–9103.
98. Jayedi A, Zeraattalab-Motlagh S, Jabbarzadeh B, et al. Am J Clin Nutr 2022;116:40–56.
99. Evert A, Reppert A, Robinson M. J Acad Nut and Diet 2017; 117:1637–1658.
100. Brand-Miller JC, Stockmann K, Atkinson F, et al. Am J Clin Nutr 2009;87:97–105.
101. Foster-Powell K, Holt S, Brand-Miller J. Am J Clin Nutr 2002;76:5–56.
102. Chiavaroli L, Lee D, Ahmed A, et al. BMJ 2021;374:n1651.
103. American Diabetes Association. Diabetes Care 2011;34:S11–61.
104. Khan MA, Gannon MC, Nuttall FQ. J Am Coll Nut 1992;11: 701–706.
105. American Diabetes Association, Bantle J, Wylie-Rosett J, et al. Diabetes Care 2008;31:S61–78.
106. Paterson M, Bell KJ, O'Connell SM, et al. Curr Diabetes Rep 2015;15:61.
107. Hansen H, Christensen P, Tauer-Lassen E, et al. Kidney Int 1999;55:621–8.
108. Raal FJ, Kalk WJ, Lawson M, et al. Am J Clin Nutr 1994;60: 579–85.
109. Meloni C, Morosetti M, Suraci C, et al. J Renal Nutr 2002;12: 96–101.
110. Meloni C, Tatangelo P, Cipriani S. J Renal Nutr 2004;14:208–13.
111. Isaka, Y. Nutrients 2021;13:1205.
112. Jakubowicz D, Froy O, Ahrén B, et al. Diabetologia 2014;57:1807–1811.
113. Sucher S, Markova M, Hornemann S, et al. Diabetes Obes Metab 2017;19:944–952.
114. Moorthi RN, Vorland CJ, Gallant KM. Curr Diabetes Rep 2017; 17:1–8.
115. Rice Bradley, BH. Curr Nutr Rep 2018;7:214–226.
116. U.S. Department of Agriculture and U.S. Department of Health and Human Services. Dietary guidelines for Americans, 2020–2025. 9th ed.
117. Clifton PM, Keogh JB. Nutr Metab Cardiovasc Dis 2017;27: 1060–1080.
118. Micha R, Mozaffarian D. Lipids 2010;45:893–905.
119. Imamura F, Micha R, Wu J, et al. PLoS Med 2016;13:e1002087.
120. Bell K, Fio C, Twigg S, et al. Diabetes Care 2020;43:59–66.
121. Marlett JA. In: Cho SS, Dreher ML, eds. Dietary Fiber and Cardiovascular Disease. Handbook of Dietary Fiber. New York, NY: Marcel Dekker, 2001:17–30.
122. Jensen MK, Koh-Banerjee P, Ju FB, et al. Am J Clin Nutr 2004;80:1492–9.
123. Brown L, Rosner B, Willett W, et al. Am J Clin Nutr 1999;69: 30–42.
124. Van Horn L, McCoin M, Kris-Etherton P, et al. J Am Diet Assoc 2008;108:287–331.
125. Juvonen K, Purhonen A, Salmenkallio-Marttila M, et al. J of Nutr 2009;139:461–466.
126. Weickert M., Mohlig M., Koebnick C., et al. Diabetologia 2005;48:2343–2353.
127. Jenkins D, Kendal C, Augustin LS, et al. Am J Med 2002;113: 30S–7S.
128. Chandalia M, Garg A, Lutjohann, et al. N Engl J Med 2000;342: 1392–8.
129. Hagander B, Asp NG, Efendic S, et al. Am J Clin Nutr 1988;47: 852–8.
130. Giacco R, Parillo M, Rivellese A. Diabetes Care 2000;23:1461–6.
131. Slavin JL. J Am Diet Assoc 2008;109:1716–31.
132. Bolin TD, Stanton RA. Eur J Surg 1998;582:115–8.
133. Tomlin J, Lowis C, Read NW. Gut 1991;32:665–9.
134. American Dietetic Association. J Am Diet Assoc 2004;104: 255–75.
135. Bloomgarden ZT. Diabetes Care 2011;34:e46–51.
136. Fitch C, Keim KS. J Acad Nutr and Diet 2012;112:739–758.
137. Coulston A, Hollenbeck C, Donner C, et al. Metabolism 1985;34:962–6.
138. Chong MFF, Fielding BA, Frayn KN. Am J Clin Nutr 2007;85:1511–20.
139. Ruiz-Ojeda F, Plaza-Díaz J, Sáez-Lara M, et al. Adv Nutr, 2019;10:S31–S48.
140. Wiebe N, Padwal R, Field C, et al. BMC Med 2011;9:1–18.
141. Whitehouse C, Boullata J, McCauley, L. AAOHN J 2008;56: 251–9.
142. Office of the Commissioner. How Sweet It Is: All About Sugar Substitutes. U.S. Food and Drug Administration., Reviewed on November 30, 2021. Accessed on December 23, 2022. Available at www.fda.gov/consumers/consumer-updates/how-sweet-it-all-about-sugar-substitutes
143. Ludwig DS. JAMA 2009;302:2477–8.
144. Mattes, R, Popkin B. Am J Clin Nutr 2009;89:1–14.
145. Pepino M, Bourne C. Curr Opin Clin Nutr Metab Care 2011;14:391–5.
146. Sheard N, Clark N, Brand-Miller J, et al. Diabetes Care 2004;27:2266–71.
147. Chiesa G, Piscopo M, Rigamonti A, et al. Acta Biomed 2005;76:44–8.
148. Wheeler ML, Pi-Sunyer FX. J Am Diet Assoc 2008;108:S34–9.
149. Gillespie SJ, Kulkarni KD, Daly AE. J Am Diet Assoc 1998;98:897–905.
150. Brazeau A, Mircescu, H, Desjardins K, et al. Diabetes Res Clin Pract 2013;99:19–23.
151. Smart C, Ross K, Edge J, et al. Diabetes Med, 2009;26:279–285.
152. Souto D, Zajdenverg L, Rodacki M, et al. Nutr 2014;30: 286–290.
153. Lonn E, Dagenais G, Yusuf S, et al. Diabetes Care 2002;25: 1919–27.
154. Liu Q, Li S, Quan, H, Li J. PloS One 2014; 9:e100379.
155. Sebastiani G, Herranz Barbero A, Borrás-Novell C, et al. Nutrients 2019;11:557.
156. Pittas A, Lau N, Hu F, et al. J Clin Endocrinol Metab 2007;92:2017–29.
157. Valdés-Ramos R, Ana Laura GL, Beatriz Elina MC, et al. Endocr Metab Immune Dis Drug Targets (formerly Curr Drug Targets-Immune Endocr Metab Dis) 2015;15:54–63.
158. Shai I, Schwarzfuchs D, Henkin Y, et al. N Engl J Med 2008;359:229–241.
159. Barnard ND, Cohen J, Jenkins DJ, et al. Diabetes Care 2006;29(8):1777–83.
160. Yokoyama Y, Barnard ND, Levin SM, et al. Cardiovasc Diag Ther 2014;4(5):373–82.
161. Sainsbury E, Kizirian NV, Partridge SR, et al. Diabetes Res Clin Pract 2018;139:239–252.
162. Gardner C, Landry M, Perelman D, et al. Am J Clin Nutr 2022;116(3):640–652.
163. Schwingshackl L, Chaimani A, Hoffmann G, et al. Eur J Epidemiol. 2018;33:157–170.
164. Zhang Y, Han H, Chu L. Prim Care Diabetes 2022;16(3): 368–374.

165. Willett WC, Sacks F, Trichopoulou A, et al. Am J Clin Nutr 1995;61:1402S–1406S.
166. Elhayany A, Lustman A, Abel R, et al. Diabetes Obes Metab 2010;12:204–209.
167. Esposito K, Maiorino MI, Bellastella G, et al. Endocrine 2017;56:27–32.
168. Torres-Peña J, Garcia-Rios A, Delgado-Casado N, et al. Atherosclerosis2018;269:50–56.
169. Kastorini CM, Milionis H, Esposito K, et al. J Am Coll Cardiol 2011;57:1299–1313.
170. Estruch R, Ros E, Salas-Salvadó J, et al. N Engl J Med 2013;368:1279–1290.
171. Miller CK. Evid Based Nurs 2015;18:74.
172. Mirabelli M, Chiefari E, Arcidiacono B, et al. Nutrients 2020;12:1066.
173. Hellenic Ministry of Health. Mediterranean Diet Food Pyramid. Accessed September 29, 2022. Available at www.hhf-greece.gr/hydria-nhns.gr/adultdietary_eng.htlm
174. Atkins RC. Dr. Atkins' new diet revolution. New York, NY: M. Evans, 2002.
175. Osler, W. The Principles and Practice of Medicine. D. Appleton and Company, 1912.
176. Allen FM, Stillman E, Fitz, R. Total Dietary Regulation in the Treatment of Diabetes. New York, NY: Rockefeller Institute for Medical Research, 1919.
177. D'Andrea MI, Romão TT, Pires do Prado HJ, et al. Front Neurosci. 2019 Jan 29;13:5.
178. Noakes TD, Windt J. Brit J Sports Med 2017;51:133–139.
179. Nickols-Richardson SM, Coleman MD, Volpe JJ, et al. J Am Diet Assoc 2005;105:1433–7.
180. Hallberg SJ, McKenzie AL, Williams PT, et al. Diabetes Ther 2018;9:583–612.
181. Saslow LR, Daubenmier JJ, Moskowitz JT, et al. Nutr Diabetes 2017;7:304.
182. Krebs JD, Bell D, Hall R, et al. Diabetes Care 2009;32:1147–52.
183. Davis NJ, Tomuta N, Schechter C, et al. Diabetes Care 2009;32: 1147–52.
184. Tay J, Thompson CH, Luscombe-Marsh ND, et al. Diabetes Obes Metab 2018;20:858–71.
185. Westman EC, Yancy WS, Mavropoulos JC, et al. Nutr Metab 2008;5:36.
186. Guay V, Lamarche B, Charest A, et al. Metab 2012;61:76–83.
187. Volek JS, Fernandez ML, Feinman RD, et al. Prog Lipid Res 2008;47:307–18.
188. American Diabetes Association. Diabetes Care 2022;45.
189. Viguiliouk E, Kendall CW, Kahleová H, et al. Clin Nutr 2019;38:1133–1145.
190. Yokoyama Y, Barnard ND, Levin SM, et al. Cardiovasc Diag Ther 2014;4):373.
191. Klemm S, Academy of Nutrition and Dietetics. Food Sources of 5 Important Nutrients for Vegetarians. Released October 15, 2021. Accessed on September 7, 2022. Available at www.eatright.org/food/nutrition/vegetarian-and-special-diets/food-sources-of-important-nutrients-for-vegetarians
192. Tsai AG, Wadden TA. Obesity 2006;14:1283–1293.
193. Jackness C, Karmally W, Febres G, et al. Diabetes 2013;62: 3027–3032.
194. Steven S, Hollingsworth K, Al-Mrabeh A, et al. Diabetes Care 2016;39:808–815.
195. Blankenship JD, Wolfe BM. Dietary Management of Obesity. Surgical Management of Obesity. Elsevier Inc. 2007:67–72.
196. Mattson MP, Longo VD, Harvie M. Age Res Rev 2017;39,46–58.
197. Barnosky A, Hoddy K, Unterman T, et al. Trans Res 2014;164: 302–311.
198. Furmli S, Elmasry R, Ramos M, et al. BMJ Case Rep 2018;2018: bcr-2017-221854.
199. Corley B, Carroll R, Hall R, et al. Diabetes Med 2018;35(5): 588–594.
200. Grajower MM, Horne BD. Nutrients 2019;11(4):873.
201. Camelon K, Hadell K, Jamsen P, et al. J Am Diet Assoc, 1998;98:1155–1158.
202. Bowen M, Cavanaugh K, Wolff K, et al. Pat Edu Counsel 2016;99:1368–1376.
203. American College of Sports Medicine. Benefits and Risks Associated with Physical Activity. 10th ed. Accessed on December 28, 2022. Available at www.acsm.org/docs/default-source/publications-files/acsm-guidelines-download-10th-edabf32a97415a400e9b3be594a6cd7fbf.pdf?sfvrsn=aaa6d2b2_0
204. Colberg S, Sigal R, Yardley J. et al. Diabetes Care 2016;39: 2065–2079.
205. Marwick T, Hordern M, Miller T, et al. Circ 2009;119:3244–62.
206. Snowling NJ, Hopkins WG. Diabetes Care 2006;29:2518–62.
207. Williamson D, Rejeski J, Lang W, et al. Arch Intern Med 2009; 169:163–71.
208. Silverstein J, Klingensmith G, Copeland K, et al. Diabetes Care 2005;28:186–212.
209. Rosenbloom A, Silverstein J, Amemiya S, et al. Pediatr Diabetes 2009;10:17–32.
210. Røder M, Schwartz R, Prigeon R, et al. J Clin Endocrinol Metab 2000;85:2275–80.
211. DeFronzo RA. Diabetes Care 1981;4:493–501.
212. Elahi D, Muller D, Egan J, et al. Novartis Found Symp 2002;242:222–42.
213. Houmard J, Weidner M, Dolan P, et al. Diabetes 1995;44: 555–60.
214. Goulet E, Hassaine A, Dionne I, et al. Exp Gerontol 2009;44: 740–4.
215. Laiteerapong N, Huang ES. In: Cowi CC, Casagrande SS, Menke A, et al., eds. Diabetes in Older Adults. In Diabetes in America. 3rd ed. Bethesda, MD: National Institute of Diabetes and Digestive and Kidney Disease (US), 2018:33651542.
216. Kalyani R, Saudek C, Brancati F, et al. Diabetes Care 2010;33: 1055–60.
217. Wedick N, Barrett-Connor E, Knoke J, et al. J Am Geriatr Soc 2002;50:1810–5.
218. Brown A, Mangione C, Saliba, et al. J Am Geriatr Soc 2003;51:S265–80.
219. Neumiller J, Setter S. Am J Geriatr Pharmacother 2009;7: 324–42.
220. Hannon BA, Fairfield WD, Adams B, et al. Nutr Diabetes 2020;10.
221. Allen RW, Schwartzman E, Baker WL, et al. Ann Fam Med 2013;11:452–9.
222. Pivari F, Mingione A, Brasacchio C, et al. Nutrients 2019;11:1837.

A. OBESITY AND DIABETES

CHAPTER 64

Metabolic Syndrome: Definition, Pathophysiology, and Clinical Utility

Max C. Petersen and Dominic N. Reeds

CHAPTER OUTLINE

Abbreviations

AHA/NHLBI, American Heart Association/National Heart, Lung, and Blood Institute; **BMI**, body mass index; **BP**, blood pressure; **CHD**, coronary heart disease; **CETP**, cholesteryl ester transfer protein; **CVD**, cardiovascular disease; **DHA**, docosahexaenoic acid; **EPA**, eicosapentaenoic acid; **FA**, fatty acid; **FBG**, fasting blood glucose; **FFA**, free fatty acid; **HDL-C**, high-density lipoprotein cholesterol; **HIV**, human immunodeficiency virus; **HOMA-IR**, homeostasis assessment model assessment of insulin resistance; **HTN**, hypertension; **IAAT**, intra-abdominal adipose tissue; **IDF**, International Diabetes Federation; **IFG**, impaired fasting glucose; **IGT**, impaired glucose tolerance; **IHTG**, intrahepatic triglyceride; **IR**, insulin resistance; **LDL-C**, low-density lipoprotein cholesterol; **LPL**, lipoprotein lipase; **MS**, metabolic syndrome; **RAAS**, renin-angiotensin II-aldosterone system; **SNS**, sympathetic nervous system; **T2D**, type 2 diabetes; **TG**, triglyceride; **VLDL-TG**, very-low-density lipoprotein-triglyceride; **WC**, waist circumference; **WHO**, World Health Organization

Introduction

The term *metabolic syndrome* (MS) is used to describe a cluster of metabolic disorders: insulin resistance (IR) or hyperglycemia, abdominal obesity, dyslipidemia (high very-low-density lipoprotein-triglyceride [VLDL-TG] concentration and low plasma high-density lipoprotein cholesterol [HDL-C]), and essential hypertension (HTN). These factors are important because each component increases the risk of development of type 2 diabetes (T2D) and/or cardiovascular disease (CVD). That the components of MS are associated with both T2D and CVD has been appreciated since the 1930s. Progress in understanding the pathogenesis of the syndrome has been hindered by the challenge of understanding the complex relationships among features of MS, IR, or insulin sensitivity, pancreatic β-cell function, and host factors.

Historical Context and Definitions

Until the 1960s, the prevailing belief was that an absolute insulin deficiency was the primary metabolic defect in T2D. This belief persisted despite studies performed as early as the 1930s that indicated that resistance to insulin-mediated stimulation of glucose clearance was present in individuals with T2D.[1-5] The insulin radioimmunoassay, developed by Yalow and Berson in 1960, established that most patients with diabetes had greater plasma insulin concentration than healthy individuals.[6] The ability to measure both plasma glucose and insulin concentration facilitated the development of oral glucose tolerance testing[7] and the glucose clamp technique.[8] Various glucose clamp techniques exist; however, the most commonly performed is the hyperinsulinemic-euglycemic clamp. In this protocol, a constant infusion of insulin is administered intravenously to the individual to establish and maintain a stable level of hyperinsulinemia, in conjunction with a glucose infusion adjusted to maintain euglycemia. After a stable or "clamped" plasma glucose concentration is achieved, the rate of glucose infusion is equal to the rate of insulin stimulated glucose disposal (provided that the degree of hyperinsulinemia is sufficient to fully suppress hepatic glucose output). Concomitant infusion of stable isotope–labeled tracers of amino acids, glucose, and fatty acids (FAs) during clamps may be performed to allow for calculation of glucose production, amino acid deposition, VLDL-TG synthesis, and lipolysis, among other metabolic measures.[9-13] These methods have proven critical in dissecting the complex relationships between tissue-specific insulin-resistance (IR) and insulin secretion.

Subsequent studies showed that most individuals with T2D had resistance to insulin action in adipose tissue (impaired suppression of lipolysis), liver (impaired suppression of glucose production), and skeletal muscle (impaired stimulation of glucose disposal).[14-16] IR almost always precedes the development of T2D; this paradigm is supported by studies showing that IR is present at an early age in first-degree relatives of people with T2D,[17] and that IR increases the risk of developing T2D.[18-22]

The relationship between IR and insulin secretion is complex. In general, IR causes increased insulin secretion and reduced hepatic insulin clearance, resulting in systemic hyperinsulinemia. Although a focus is often placed on insulin as a regulator of blood glucose, insulin plays a key role in the regulation of lipid and protein metabolism, in addition to cellular growth and development. Seminal studies by Hollenbeck, Reaven, Yeni-Komshian, and others[23,24] systematically examined IR in individuals without underlying T2D, finding that insulin-stimulated glucose disposal surprisingly varied up to eightfold within a group of healthy individuals.

Further work revealed that insulin resistance was associated with higher plasma concentrations of insulin, VLDL-TG, and glucose during the oral glucose tolerance test.[25] In his 1988 Banting lecture, Reaven proposed that T2D was not the only adverse outcome associated with hyperinsulinemia. He hypothesized that activation of insulin-responsive pathways could also cause dyslipidemia and HTN.[26] He termed this collection of metabolic disturbances *syndrome X*. Subsequently, several studies described clinical associations among IR, dyslipidemia, HTN, elevated waist circumference (WC), and the risk of CVD and T2D.[27-31] The term *syndrome X* gave way in the 1990s to the term *metabolic syndrome* (MS).[32]

The first formal definition of MS was by the World Health Organization (WHO) in 1998 (**Table 64.1**).[32] This initial definition focused on IR as the main contributor to the syndrome and *required* the presence of IR in addition to two of the following: obesity, HTN, high plasma TG concentration, low plasma HDL-C concentration, or microalbuminuria. In 2001, the report of the Adult Treatment Panel III (ATP III) of the National Cholesterol Education Program also noted the relationship between IR and known CVD risk factors in proposing criteria for MS (**Table 64.2**).[33] In contrast to the 1998 WHO definition, the 2001 ATP III definition did not require the presence of insulin-resistant glucose metabolism but, rather, emphasized abdominal obesity as measured by waist circumference (WC), suggesting an additional risk posed by an accumulation of abdominal fat. The

Table 64.1 World Health Organization Metabolic Syndrome Criteria

Insulin resistance identified by one of the following:
- Type 2 diabetes
- FBG >110 mg/dL
- IGT >140 mg/dL at 2-hour time point of 75 g oral glucose tolerance test
- FBG <110 mg/dL but in lowest quartile of glucose disposal during hyperinsulinemic, euglycemic conditions

And any two of the following:
- TG ≥150 mg/dL
- HDL-C <35 mg/dL (men), <39 mg/dL (women)
- BMI >30 kg/m^2 and/or waist-to-hip ratio >0.9 (men), >0.85 (women)
- Urinary albumin excretion rate ≥20 μg/min or albumin-to-creatinine ratio ≥30 mg/g

BMI, body mass index; *FBG*, fasting blood glucose concentration; *HDL-C*, high-density lipoprotein cholesterol; *IGT*, impaired glucose tolerance; *TG*, triglycerides.

Data from Grundy SM. Circulation 2004;109:433–8.

Table 64.2 National Cholesterol Education Program-Adult Treatment Panel III (ATP-III) Definition of Metabolic Syndrome

	ATP III
Waist circumference (in)	>40 (men), >35 (women)
HDL-C (mg/dL)	<40 (men), <50 (women)
TG (mg/dL)	>150
BP (mm Hg)	Systolic ≥130 and/or diastolic ≥85
FBG (mg/dL)	≥110

Metabolic syndrome is defined as the presence of any three (or more) of the five criteria. *HDL-C*, high-density lipoprotein cholesterol; *TG*, triglyceride concentration; *BP*, blood pressure; *FBG*, fasting blood glucose

Data from Grundy SM. Circulation 2004;109:433–8.

Table 64.3 Harmonized Metabolic Syndrome Criteria

Measure	Category Cutoff Points
Elevated waist circumference	Population and country specific definitions
Plasma TG or drug treatment	≥150 mg/dL
Plasma HDL-C or drug treatment	<40 (men) mg/dL, <50 (women) mg/dL
BP (mm Hg) or drug treatment	Systolic ≥130 and/or diastolic ≥85
FBG or drug treatment	≥100 mg/dL

Metabolic syndrome is defined as the presence of any three (or more) of the five criteria. *BP*, blood pressure; *FBG*, fasting blood glucose; *HDL-C*, high-density lipoprotein cholesterol; *TG*, triglyceride.

Data from Grundy SM. Circulation 2009;120:1640–5.

International Diabetes Federation (IDF) and the American Heart Association/National Heart, Lung, and Blood Institute (AHA/NHLBI) each released updated MS criteria in 2005, which were harmonized in 2009 (**Table 64.3**). These harmonized criteria differ from the ATP III criteria only in the recommendation to use population- and country-specific WC cutoffs, and by a downward revision of the cutoff for elevated fasting glucose from 110 mg/dL to 100 mg/dL.

Since these initial definitions, MS has entered the clinical vernacular and is used to describe a clinical state associated with increased risk for the development of CVD and T2D. Reaven himself did not propose *syndrome X* for use as a diagnostic entity but, rather, to provide a conceptual framework for understanding the complex relationships among abdominal obesity, IR, and the adverse consequences of hyperinsulinemia. The rest of this chapter describes pathophysiologic and clinical aspects of each component of MS and discusses the utility of the diagnosis of MS in clinical practice.

Waist Circumference

Upper body obesity, particularly abdominal obesity, may confer a greater cardiometabolic risk than obesity alone.[34,35] Because precise measurement of abdominal fat requires expensive imaging techniques, WC is often used as a surrogate of both total and abdominal adiposity.[36-38] While there is no universal approach for measurement of WC, the reproducibility of WC measurement is high when performed by trained technicians.[39] The most commonly used sites for measurement of WC are the midpoint between the lowest rib and the iliac crest, the umbilicus, and the site of the narrowest measured WC.

WC cutoff values of 40 inches in men and 35 inches in women were used in the ATP III definition of MS because these values corresponded to a body mass index (BMI) of 30 kg/m^2 in a large Scottish study.[40] However, the use of population- and country-specific WC cutoffs was recommended in the 2005 IDF MS criteria and the 2009 IDF/AHA/NHLBI harmonized MS criteria.[41,42] This recommendation was based on evidence that the ATP III WC cutoffs have poor sensitivity for identifying individuals at risk for underlying HTN, dyslipidemia, and T2D in East Asian and South Asian populations.[43] As a result, population-specific WC cutoffs associated with increased cardiometabolic risk have now been developed for several Asian countries.[44] Whether different WC cutoff values should be used to define MS in other populations remains uncertain, however, because of challenges in specifically defining populations of interest and results that have largely been generated using cross-sectional rather than prospective longitudinal approaches.[44-47] The effect on WC of gender-affirming hormone therapy in transgender individuals is incompletely studied, but likely to be modest.[48]

The rationale for the use of WC in the criteria for MS, rather than BMI, is the superiority of WC over BMI in predicting CVD and T2D risk.[35,42,49-53] A consensus statement issued in 2007 by the Obesity Society, the American Society for Nutrition, and the American Diabetes Association stated that WC provides "incremental value in predicting diabetes mellitus, coronary heart disease, and mortality rate above and beyond that provided by BMI."[52] The regional distribution of adipose tissue affects CVD and T2D risk, such that greater amounts of lower-body (e.g., gluteofemoral subcutaneous) adipose tissue are associated with lesser, whereas greater amounts of intra-abdominal adipose tissue are associated with greater, CVD and T2D risk at any given BMI.[54] WC is a better correlate of intra-abdominal adipose tissue (IAAT), often referred to as visceral adipose tissue, than BMI, waist-to-hip ratio, or waist-to-height ratio.[55] IAAT should not be considered strictly synonymous with "visceral fat"; the latter term includes fat surrounding specific viscera (e.g., pericardiac and perinephric adipose depots), which may exert direct mechanical and/or paracrine effects on organ function, whereas IAAT refers specifically to adipose tissue located within the peritoneum (e.g., mesenteric, omental, and epiploic adipose depots).

The reasons for the close relationship between IAAT and cardiometabolic risk are not fully understood. One hypothesis posits that IAAT has a direct effect on insulin sensitivity, lipid metabolism, and blood pressure. The venous drainage of IAAT is to the hepatic portal vein, such that FA released by IAAT disproportionately contribute to hepatic FA delivery.[56] Hepatic FA delivery increases as IAAT mass increases.[56] FA delivered to the liver can be exported as VLDL-TG, oxidized, or stored. Failure to export or oxidize these FA thus promotes hepatic steatosis which, in turn, causes hepatic IR.[57] Despite the high relative rate of delivery of FA to the liver by IAAT, its absolute contribution to whole-body FA turnover is far smaller than that of subcutaneous adipose tissue, even in the setting of obesity (**Figure 64.1**).[56]

Because intrahepatic triglyceride (IHTG) and IAAT are strongly correlated, it is not clear whether increased WC (and by implication, increased IAAT) and increased IHTG are independent risk factors for dyslipidemia and IR. However, studies of adults with obesity stratified by IHTG and IAAT suggest that IHTG is a better predictor of IR (**Figure 64.2**) and VLDL-TG production than IAAT.[58] Insulin sensitivity in liver, muscle, and adipose tissue (measured by using the euglycemic-hyperinsulinemic clamp procedure) and VLDL-TG production (measured by using stable isotope labeled tracer methodology) was assessed in 31 volunteers with obesity who were then separated into groups, initially by matching on IAAT volume, and in a second analysis on IHTG content. In the groups matched for IAAT volume, VLDL-TG production rates were increased, and multiorgan insulin sensitivity was decreased, in individuals with high IHTG compared with low

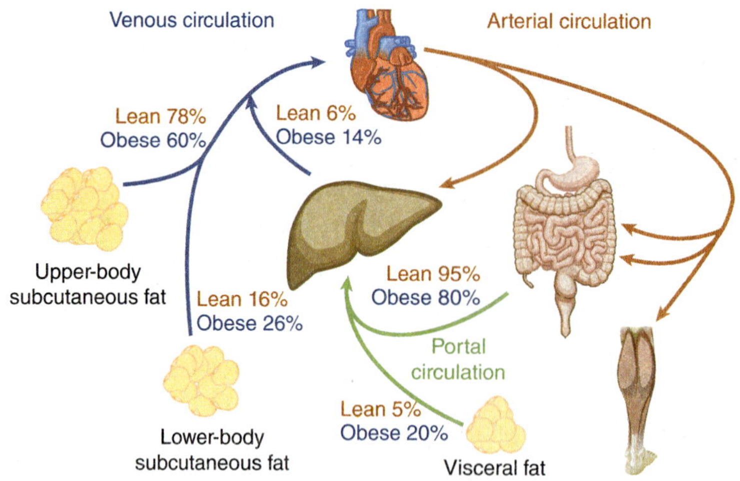

Figure 64.1 The relative contributions of different adipose depots to fatty acid (FA) delivery to the portal and systemic venous circulations. In lean people, approximately 95% of portal venous FA are derived from the systemic circulation and only about 5% come from intra-abdominal visceral adipose tissue (IAAT). The proportion from IAAT increases in obesity, but only to approximately 20% of hepatic FA delivery. Only about 6% to 14% of systemic FA are delivered from the hepatic veins; the remainder derive from subcutaneous adipose tissue depots.

Reproduced from Klein S. J Clin Invest 2004;113(11):1530-2. Data are from Nielsen S, Guo Z, Johnson CM et al. J Clin Invest 2004;113(11):1582-8.

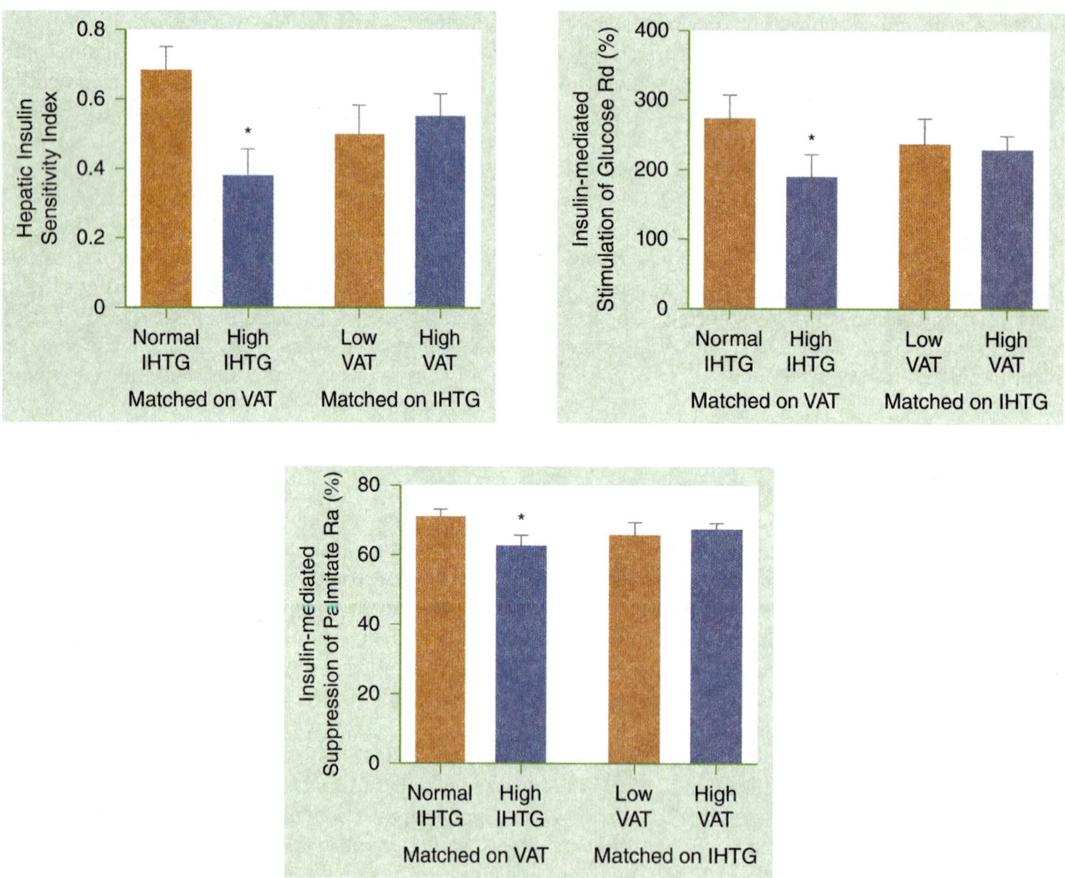

Figure 64.2 Hepatic (A), skeletal muscle (B), and adipose tissue (C) insulin sensitivity in individuals matched on intra-abdominal adipose tissue (IAAT) volume with either normal or high intrahepatic triglyceride (IHTG) content and individuals matched on IHTG content who had either low or high IAAT volume.

AU, arbitrary units.

Values are means ±SEM.,* value is significantly different from the corresponding value in the normal IHTG group, $p < 0.05$. Hepatic insulin sensitivity was determined by calculating the reciprocal of the product of basal endogenous glucose production rate, in micromoles per kilogram of fat-free mass (FFM) per minute, and fasting plasma insulin concentration in milliunits per liter). Ra, rate of appearance; Rd, rate of disappearance.

Reprinted with permission from Fabbrini E, Magkos F, Mohammed BS et al. Proc Natl Acad Sci U S A 2009;106:15430–5.

IHTG. In contrast, VLDL-TG production and insulin sensitivity were not different among groups when individuals were matched for IHTG and then divided into high and low IAAT volume. These findings strongly suggest that differences in hepatic FA handling (i.e., FFA oxidation or export vs storage) are critical in determining whether abdominal obesity causes metabolic abnormalities.

The precise mechanism(s) responsible for the relationship between increasing adiposity, greater IHTG content, and features of MS are unclear. Clearly, the ability to compensate for energetic oversupply and increased adiposity is necessary for the small proportion of people with obesity who do not have any other features of MS, often referred to as "metabolically healthy obesity."[59] It has been proposed that differences in the capacity of adipose tissue to expand by hypertrophy and hyperplasia, and/or differential development of fibrosis, hypoxia, and inflammation during adipose tissue expansion, determine whether overnutrition and/or obesity cause IR and hyperlipidemia. However, recent data suggest that adipose tissue macrophage infiltration can be dissociated from adipose tissue insulin resistance.[60,61] For example, during weight loss, insulin sensitivity improves before changes in adipose tissue inflammation are seen, suggesting that inflammation and impaired insulin sensitivity do not share a causal link.[61,62] Failure to adequately sequester fat in adipocytes and achieve healthy expansion of adipose tissue may promote increased plasma FFA concentrations and lipid deposition in ectopic sites such as liver and skeletal muscle (termed *lipotoxicity*).[63] Tissue FA uptake is generally proportionate to plasma FA concentration, and increases in FA utilization inhibit glucose uptake and oxidation, causing glucose intolerance.[64-66] Increased plasma FFA concentration and accumulation of bioactive lipids in liver and muscle are strongly associated with impaired insulin sensitivity in these organs.[57] This process may be further exacerbated by increased expression of CD36, a FA sensor/transporter, expressed broadly in target organ membranes and in Kupffer cells, macrophages, and immune cells.[58]

In conclusion, WC can be easily and reliably measured in a clinical setting. WC identifies people at higher risk of development of T2D, dyslipidemia, and CVD than BMI alone, although the magnitude of this difference is small. WC may be particularly helpful in studies conducted in differing racial and ethnic groups because of differences in normal BMI values and distributions among populations. Although WC largely reflects increased abdominal adiposity, IHTG content is a better predictor of IR than IAAT mass. The mechanism(s) affecting the impact of IAAT on IHTG and excess cardiometabolic risk are not fully understood.

Dyslipidemia

Dyslipidemia is common in patients with other features of MS and indicates IR in liver, adipose tissue, and skeletal muscle in many disease entities, from common obesity to lipodystrophy to HIV infection.[67] A schema for the relationships among visceral fat, skeletal muscle, and the liver with respect to lipid metabolism is shown in **Figure 64.3**. In individuals with IR, low plasma HDL-C concentration and hypertriglyceridemia are more common than high low-density lipoprotein cholesterol (LDL-C) plasma concentration. Although abundant data link low HDL-C and, to a lesser degree, hypertriglyceridemia, with subsequent development of CVD, pharmacologic therapies targeting these abnormalities have shown limited efficacy in reducing CVD risk. Instead, lowering of LDL-C is the focus of pharmacologic lipid modulation for CVD prevention. The dyslipidemia of IR is associated with elevated concentration of small, dense LDL particles, which are a better predictor of CVD risk than elevated LDL-C or TG alone.[68,69]

Hypertriglyceridemia

Plasma triglycerides circulate on TG-rich lipoproteins, mainly chylomicrons (carrying dietary TG) and VLDL particles exported from the liver. However, because the plasma residence time of chylomicrons is short, the fasting hypertriglyceridemia of MS primarily reflects high plasma VLDL-TG concentration. These elevations in VLDL-TG concentration result from increased VLDL-TG production, reduced clearance, or a combination of both factors. VLDL-TGs are synthesized from FFAs obtained either from hepatic de novo lipogenesis or circulating FFA released from subcutaneous adipose tissue, IAAT, or intrahepatic lipid stores. TG clearance occurs through the action of lipoprotein lipase (LPL) on the capillary endothelial cell surface, a process that is itself regulated by several circulating proteins, including apolipoprotein C-II, apolipoprotein C-III, apolipoprotein A5, angiopoietin-like 3, angiopoietin-like 4, glycosylphosphatidylinositol anchored high density lipoprotein binding protein 1 (GPIHBP1), and others. Some of these LPL regulators have themselves become therapeutic targets for pharmacologic lipid lowering.[70,71]

The hypertriglyceridemia of MS, like insulin resistance, is a consequence of chronic positive energy balance in a susceptible individual; however, as VLDL secretion and clearance are each controlled by multiple mechanisms, the extent of the elevation in fasting plasma triglycerides is correlated with, but not precisely predicted by, the degree of insulin resistance. Importantly, the average fasting triglyceride concentration at a given level of IR is lower in Americans of African ancestry than Americans of other ancestries, which has clinical implications for the screening of individuals at risk for T2D.[72,73] The mechanistic basis of this observation is incompletely understood but may relate to differences in TG content within secreted VLDL particles.[74,75]

Weight loss typically reduces VLDL-TG concentration, mainly because of a reduction in the contribution of "nonsystemic FA" (i.e., from IHTG stores, de novo lipogenesis, and IAAT) to VLDL-TG synthesis, rather than a reduction in whole-body fatty acid turnover.[76] Pharmacologic interventions that reduce lipolysis, such as acipimox, not only reduce VLDL-TG and raise HDL but also appear to improve insulin sensitivity in insulin-resistant individuals, emphasizing the close relationship between plasma FFA concentration and glucose and lipid metabolism.[77,78]

Data regarding the clinical utility of reducing TG to reduce CVD risk are not robust, although when plasma TG concentration is greater than 500 mg/dL despite correction of secondary causes of hypertriglyceridemia (especially uncontrolled diabetes and alcohol use), pharmacologic therapy is generally indicated to reduce the risk of pancreatitis. Commonly used agents for TG lowering include the peroxisome proliferator activator receptor-α (PPAR-α) modulators gemfibrozil and fenofibrate, as well as multiple omega-3 fatty acid preparations. The Helsinki Heart Study and the Veterans Affairs HDL Intervention Trial (VA-HIT) showed an approximate 34% relative risk reduction in CVD risk in patients treated with gemfibrozil; however, these benefits were most pronounced in those with low HDL-C rather than hypertriglyceridemia.[79,80] Those

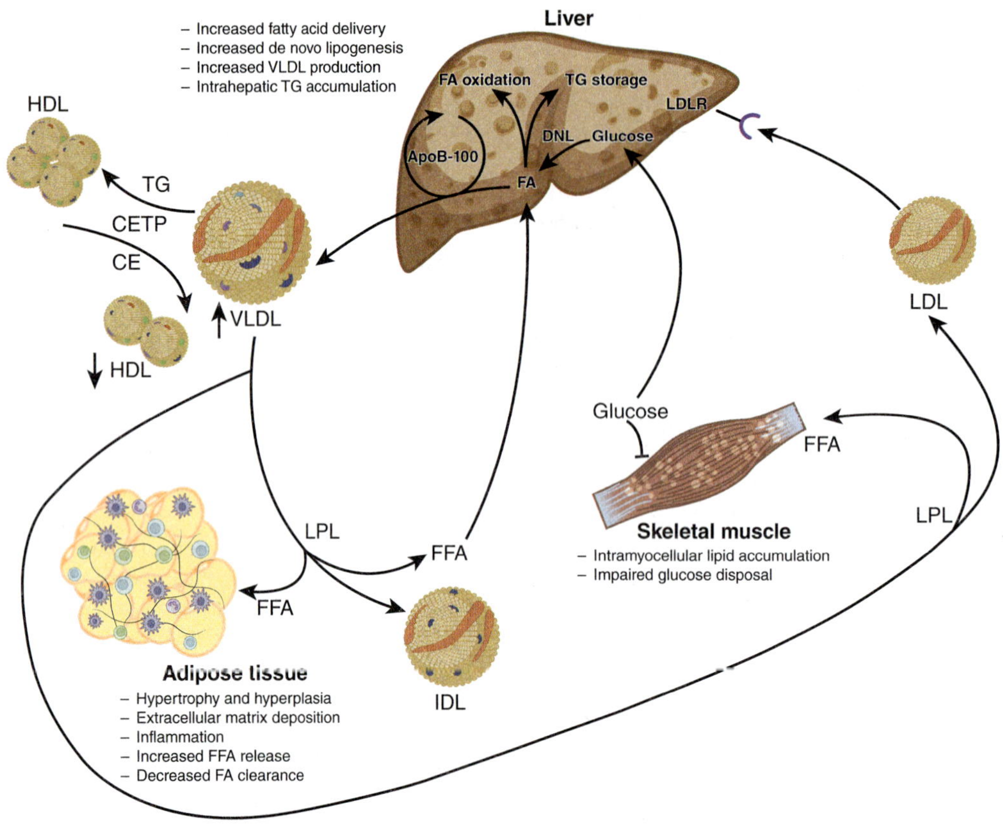

Figure 64.3 Interactions among adipose tissue, skeletal muscle, and liver in the development of metabolic syndrome. Chronic positive energy balance results in expansion of adipose tissue depots by hypertrophy and hyperplasia. In people with MS, this expansion is associated with extracellular matrix deposition, inflammation, increased FFA release, and decreased FA clearance. This results in increased FA delivery to liver and skeletal muscle. In skeletal muscle, increased FA delivery may contribute to intramyocellular lipid accumulation, which is associated with resistance to insulin-mediated glucose disposal. Impaired skeletal muscle glucose disposal supports hepatic de novo lipogenesis, which, together with increased hepatic FA delivery, results in intrahepatic triglyceride accumulation and increased very-low-density-lipoprotein-triglyceride (VLDL-TG) export. Increased VLDL-TG promotes transfer of TG to high-density-lipoprotein (HDL) particles via cholesteryl ester transfer protein (CETP), which accelerates HDL clearance and decreases HDL-C concentration. Increased tissue FA delivery from VLDL-TG by the action of lipoprotein lipase (LPL) perpetuates these processes.

Figure created using Biorender.com.

with the highest plasma insulin concentration at study entry had the greatest benefit. In contrast, the Fenofibrate Intervention and Event Lowering in Diabetes (FIELD) and the Action to Control Cardiovascular Risk in Diabetes (ACCORD) lipid study failed to show CVD risk reduction with the use of fenofibrate in patients with T2D,[81,82] although in extended follow-up of ACCORD study participants, the subgroup with hypertriglyceridemia and low HDL-C did display a 27% relative CVD risk reduction after fenofibrate therapy.[83] Of the omega-3 fatty acid preparations, only icosapent ethyl has been shown to reduce CVD risk, with a relative risk reduction of ~25% in the Reduction of Cardiovascular Events with Icosapent Ethyl-Intervention Trial (REDUCE-IT).[84] In this trial, icosapent ethyl was particularly effective in CVD risk reduction in those with TG ≥200 mg/dL and HDL-C ≤35 mg/dL. Among studied omega-3 fatty acid preparations, icosapent ethyl is atypical in that it contains only an eicosapentaenoic acid (EPA) derivative rather than the typical combination of EPA and docosahexaenoic acid (DHA) derivatives. The reason for the success of icosapent ethyl in CVD risk reduction, despite the failure of several other omega-3 fatty acid preparations that achieved a similar or greater degree of TG lowering,[85] is unclear.

High-Density Lipoprotein

HDL particles remove cholesterol from macrophages, inhibit lipid oxidation, promote endothelial nitric oxide synthesis, and are anti-inflammatory—all of which confer protection against atherosclerosis.[86] These functions of HDL particles, rather than the amount of cholesterol associated with HDL particles, are thought to mediate the link between HDL and CVD.[86] This is emphasized in that the cholesterol efflux capacity of HDL to macrophages has been found to be superior to HDL-C concentration as a predictor of CVD risk.[86] The anti-oxidant and anti-inflammatory functions of HDL are also impaired in MS,[87] as is the ability of HDL to promote nitric oxide synthesis.[88]

The pathophysiology of low HDL-C in MS is complex. One contributing factor is the transfer of TG from TG-rich lipoproteins to HDL particles via cholesteryl ester transfer protein (CETP), which accelerates HDL clearance.[89] MS is also associated with destabilizing modifications of apolipoprotein A-I and other key HDL-associated proteins which, in turn, accelerates HDL clearance.[86] The composition of HDL particles in MS is altered, with a decrease in large buoyant HDL particles and a shift toward small, TG-rich and cholesteryl ester-depleted HDL.[86]

Curiously, while low HDL-C is clearly associated with CVD risk in large epidemiologic data sets, CVD risk does not appear to be reduced by either genetic variants associated with higher HDL-C or by pharmacologic agents that raise HDL-C. Most notably, multiple CETP inhibitors failed to reduce CVD risk in large clinical trials despite large (40%–130%) increases in HDL-C.[90-92]

Glucose Metabolism and Insulin Resistance

Impaired fasting glucose (IFG) is common in patients with HTN, dyslipidemia, and abdominal obesity; and it increases the risk of developing T2D and CVD. The IFG threshold of 100 mg/dL used in the 2009 harmonized MS criteria and by the American Diabetes Association was chosen so that the prevalence of undiagnosed T2D would be similar in the setting of IFG and IGT.[93] The risk of conversion from IFG to T2D increases sharply as fasting glucose approaches the diabetic range and is also increased in people with MS, compared with those without MS (**Figure 64.4**).[94,95] Measurement of fasting glucose or oral glucose tolerance testing is more challenging to perform in clinical practice than measurement of random blood glucose. However, even modest elevations in random blood glucose are highly predictive of diabetes risk. Relative to individuals with two random blood glucose concentrations <110 mg/dL, those with random blood glucose values of 110 to 119 mg/dL or 120 to 129 mg/dL have odds ratios of 5.4 and 45.3, respectively, of receiving a diagnosis of T2D within the next year.[96] A similar pattern holds for hemoglobin A1c, the commonly used estimator of recent mean glycemia, which is more commonly used clinically than either IFG or IGT for the diagnosis of prediabetes.[97] Lifestyle intervention (i.e., weight loss and physical activity), biguanides, thiazolidinediones, and glucagon-like peptide-1 receptor agonists have all been shown to reduce the rate of progression to T2D in people with IGT or IFG.[98-101] Importantly, the Diabetes Prevention Program found that while metformin and lifestyle intervention each reduced progression to T2D, lifestyle intervention was nearly twice as effective as metformin.[98]

IFG partially reflects IR, but the relationship of IR to IFG and IGT is complex and IFG is an insensitive marker of IR.[102] IR alone is insufficient to cause T2D, nor is it sufficient to cause IFG/IGT. For example, insulin sensitivity was examined in a cohort of extremely obese women (mean BMI=49 kg/m^2) with evidence of IR (acanthosis nigricans).[103] Despite profound IR (fasting plasma insulin concentrations were sixfold higher and rates of glucose disposal during hyperinsulinemic-euglycemic clamp studies were 50% lower in individuals with acanthosis nigricans than in lean control individuals), the women with IR displayed normal FBG (87±5 mg/dL) and normal oral glucose tolerance. Furthermore, IR, to an extent comparable to that observed in T2D, is present in as many as 25% of otherwise healthy, nondiabetic people.[104] This finding reflects the wide variation in the capacity of β-cells and insulin target tissues to compensate for IR.[23]

Do these factors—the insensitivity of IFG for the presence of IR, the presence of IR in people with normal glycemia—indicate that clinical assessment of IR is not useful? On the contrary, IR is one of the earliest detectable risk factors for T2D.[105] Moreover, IR [as assessed by the homeostasis model assessment of IR, or HOMA-IR, which is calculated as (FBG in mmol/L x fasting plasma insulin in μU/mL/22.5)] is a better predictor of CVD risk than FBG alone.[106] However, HOMA-IR is limited in its clinical utility because it is strongly influenced by the fasting plasma insulin concentration, which is rarely measured clinically owing to persistent challenges with assay standardization[107] and its diminishing utility as the pathophysiology of T2D develops and progresses. In T2D, the capacity and durability of the β-cell response diminish and the inadequate β-cell response to hyperglycemia may not be captured by the fasting insulin concentration. Despite these limitations, IR clearly has prognostic implications for T2D and CVD risk.

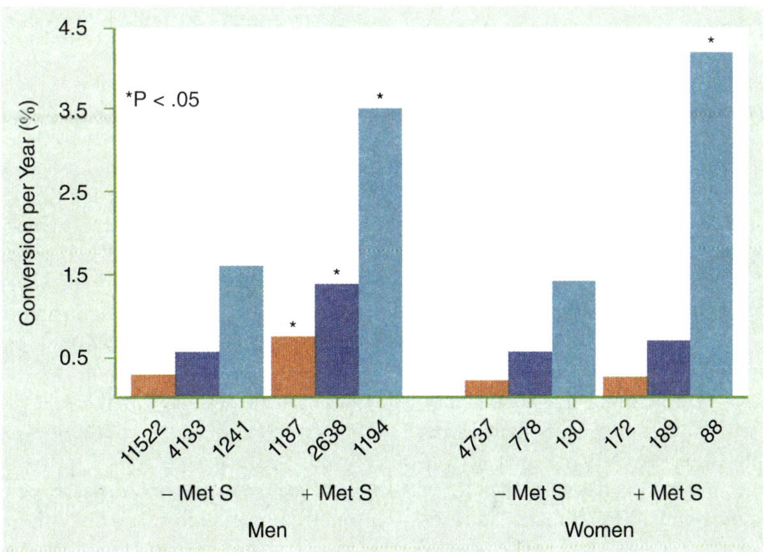

Figure 64.4 Interaction among fasting blood glucose (FBG), sex, and metabolic syndrome (MS) in determining type 2 diabetes (T2D) risk. Annualized conversion to T2D was calculated in a longitudinal study of 28,209 adults. Orange bars indicate FBG <101 mg/dL at baseline, dark blue bars indicate FBG 101 to 108 mg/dL at baseline, and light blue bars indicate FBG 108 to 126 mg/dL at baseline. Individuals were also stratified by sex and by presence of MS using the 2009 harmonized IDF/AHA/NHLBI criteria. P values denote comparison of +MetS and −MetS between people of the same sex and FBG category.

Reproduced from DeFina LF, Vega GL, Leonard D, Grundy SM. J Investig Med 2012;60(8):1164-8.

Hypertension

HTN is the most common component of the metabolic syndrome in men and is highly prevalent in the general population.[108] The presence of MS is associated with a near-doubling of CVD risk in adults with HTN.[109] HTN is probably most clinically relevant when it occurs in the setting of dyslipidemia. For example, in the Copenhagen Male Study of 3,000 men with HTN, CVD risk was not increased in participants with essential HTN and normal HDL-C and VLDL-TG concentration but was greatly increased in those with the lowest HDL-C and highest VLDL-TG concentrations.[110]

IR is clearly associated with HTN,[111] and several lines of evidence suggest that IR may contribute directly to the development of HTN. In the longitudinal Multi-Ethnic Study of Atherosclerosis (MESA), the presence of IFG or T2D at baseline increased the risk of developing HTN over the next 5 years.[112] Moreover, the risk of HTN was positively associated with both fasting glucose and fasting insulin concentrations across all ethnicities studied.[112] In a longitudinal analysis of participants in the Framingham Study, insulin sensitivity predicted HTN incidence and progression, particularly in younger and lower-weight individuals.[113]

However, there is also evidence that IR is not the major causal factor in the HTN of MS. In some studies, the association of IR with HTN disappears after adjustment for WC or BMI,[114,115] and varies by race/ethnicity.[116] Although insulin promotes renal sodium reabsorption and increases sympathetic nervous system (SNS) activity, it also has vasodilatory effects, and neither acute experimental hyperinsulinemia nor chronic hyperinsulinemia caused by insulinoma is sufficient to cause HTN in humans.[117] The HTN of MS is associated with chronic activation of the renin-angiotensin-aldosterone system (RAAS) and SNS, as well as increased perinephric adipose tissue, which has been proposed to contribute to HTN through mechanical compression of multiple components of the nephron.[118]

Lifestyle interventions such as weight loss and aerobic exercise improve blood pressure and all other features of MS, including IR.[119] Conversely, drugs that improve insulin action, such as thiazolidinediones or metformin, do not appear to improve blood pressure. Because of the other metabolic actions of these agents, it is difficult to determine whether potential blood pressure benefits from either agent are obscured by their side effects. Pharmacologic interventions to treat HTN clearly reduce CVD risk, although the specific blood pressure targets remain debated.

Clinical Utility of Metabolic Syndrome

People with MS are at increased risk for development of and death from CVD.[120] However, in examining the contribution of MS to CVD risk, people with T2D should be excluded from analysis. T2D so dramatically increases CVD risk that it overwhelms the value of any other indices for risk assessment. For example, data from the Second National Health and Nutrition Examination Survey (NHANES II) revealed that T2D conferred a hazard ratio of five for CHD (compared with a hazard ratio of 3.5 of MS).[121] In patients with known CVD, the addition of T2D increased CHD risk 11-fold.[121] However, when people with T2D are excluded from analysis, MS is still associated with greater CVD mortality compared with the general population.[122]

The increased CVD risk seen with MS is logical, given that the components of MS each individually increase the risk of CVD. However, whether these components exert synergistic effects, such that MS confers a greater risk for CVD than would be expected for any combination of its components, is doubtful. Several studies attempted to address whether the diagnosis of MS adds prognostic value beyond that conferred by individual MS components. Synergistic effects of multiple co-morbid MS features (specifically, hyperinsulinemia, HTN, and hypertriglyceridemia) on carotid intima-media thickness (a measure of atherosclerosis) were reported in a 2002 study of nearly 12,000 individuals without T2D, treated dyslipidemia, or CHD at baseline.[123] By contrast, the diagnosis of MS was found to not add prognostic value for CVD risk beyond the sum of its components in multiple prospective cohort studies: in women aged 60 to 79 years without baseline CHD,[124] adults aged 70 to 82 years at high CVD risk,[125] adults without diabetes aged 65 to 74 years,[126] and men aged 45 to 63.[127] Furthermore, not all permutations of three metabolic syndrome criteria confer a greater CVD risk than that conferred by certain combinations of two metabolic syndrome criteria.[128]

The utility of MS as a clinical diagnosis is further challenged by several factors, including: i) the use of dichotomous cutoff points to classify continuous variables; ii) incomplete understanding of the impact of ethnicity and genetic background on these cutoff points (as discussed above for WC and serum triglycerides); and iii) significant heterogeneity among individuals that could meet diagnostic criteria for MS (e.g., identical treatment of individuals with differing degrees of hypertension, IFG, or dyslipidemia).[129] One commonly proposed clinical use of MS is as a CVD risk prediction tool, but other algorithms, such as the Framingham Risk Score, are able to more accurately predict CVD risk because of the inclusion of other risk factors, such as serum cholesterol and tobacco use and the use of continuous rather than dichotomous variables.[130] Moreover, the treatment of MS remains rooted in the individual treatment of its component features. In light of these problems, MS is perhaps best thought of as a conceptual framework and/or epidemiologic construct that has limited direct clinical utility.

Summary

MS represents a cluster of strongly associated metabolic abnormalities that each increase the risk for development of CVD and T2D. IR is associated with all components of MS and is likely necessary, though not sufficient, for its pathogenesis. MS itself does not increase the risk of adverse outcomes beyond that predicted by each individual component of the syndrome. The main clinical utility of MS is as a pattern-recognition heuristic: the presence of one or more MS components should lead the clinician to evaluate for the other MS components, as well as other IR-associated comorbidities, such as nonalcoholic fatty liver disease and polycystic ovarian syndrome. With respect to treatment, a focus should clearly be placed on interventions that reduce progression of glucose intolerance to T2D, because T2D is a far greater risk factor for CVD than is MS alone. Even modest weight loss in combination with exercise dramatically reduces the risk of progression of IGT to T2D. Treatment of HTN also reduces CVD risk. Finally, treatment of the dyslipidemia of MS should usually focus on lowering LDL-C, not raising HDL-C or lowering TG.

References

1. Himsworth HP. Lancet 1939;2:118–22.
2. Himsworth HP. Lancet 1939;2:65–8.
3. Himsworth HP. Lancet 1939;2:1–6.
4. Himsworth HP. Lancet 1939;2:171–6.
5. Himsworth HP. Lancet 1949;256:464–73.
6. Yalow RS, Berson SA. J Clin Invest 1960;39:1157–75.
7. Shen SW, Reaven GM, Farquhar JW. J Clin Invest 1970;49: 2151–60.
8. DeFronzo RA, Tobin JD, Andres R. Am J Physiol 1979;237: E214–23.
9. Magkos F, Mittendorfer B. Clin Lipidol 2009;4:215–30.
10. Steele R. Ann N Y Acad Sci 1959;82:420–30.
11. Smith GI, Atherton P, Reeds DN, et al. J Appl Physiol 1985;107: 1308–15.
12. Sidossis LS, Magkos F, Mittendorfer B, et al. Clin Nutr 2004;23: 457–66.
13. DeFronzo RA, Ferrannini E. Diabetes 1982;31:683–8.
14. DeFronzo RA, Simonson D, Ferrannini E. Diabetologia 1982;23: 313–9.
15. Efendic S, Wajngot A, Vranic M. Proc Natl Acad Sci U S A 1985;82: 2965–9.
16. DeFronzo RA. Diabetes 1988;37:667–87.
17. Laws A, Stefanick ML, Reaven GM. J Clin Endocrinol Metab 1989; 69:343–7.
18. Ginsberg H, Kimmerling G, Olefsky JM, et al. J Clin Invest 1975; 55:454–61.
19. Liese AD, Mayer-Davis EJ, Haffner SM. Epidemiol Rev 1998; 20:157–72.
20. Warram JH, Martin BC, Krolewski AS, et al. Ann Intern Med 1990;113:909–15.
21. Meigs JB, D'Agostino RB, Sr., Wilson PW, et al. Diabetes 1997; 46:1594–600.
22. Ginsberg H, Olefsky JM, Reaven GM. Diabetes 1974;23:674–8.
23. Hollenbeck C, Reaven GM. J Clin Endocrinol Metab 1987; 64:1169–73.
24. Yeni-Komshian H, Carantoni M, Abbasi F, et al. Diabetes Care 2000;23:171–5.
25. Abbasi F, Brown BW, Jr., Lamendola C, et al. J Am Coll Cardiol 2002;40:937–43.
26. Reaven GM. Diabetes 1988;37:1595–607.
27. Zavaroni I, Bonora E, Pagliara M, et al. N Engl J Med 1989;320:702–6.
28. Mitchell BD, Haffner SM, Hazuda HP, et al. Am J Epidemiol 1992;136:12–22.
29. Haffner SM, Valdez RA, Hazuda HP, et al. Diabetes 1992;41: 715–22.
30. Schmidt MI, Duncan BB, Watson RL, et al. Diabetes Care 1996; 19:414–8.
31. Schmidt MI, Watson RL, Duncan BB, et al. Metabolism 1996; 45:699–706.
32. Alberti KG, Zimmet PZ. Diabetes Med 1998;15:539–53.
33. Expert Panel on Detection E, Treatment of High Blood Cholesterol in A. JAMA 2001;285:2486–97.
34. Hubert HB, Feinleib M, McNamara PM, et al. Circ 1983;67: 968–77.
35. Balkau B, Deanfield JE, Despres JP, et al. Circ 2007;116:1942–51.
36. Kissebah AH, Vydelingum N, Murray R, et al. J Clin Endocrinol Metab 1982;54:254–60.
37. Pouliot MC, Despres JP, Lemieux S, et al. Am J Cardiol 1994;73:460–8.
38. Pouliot MC, Despres JP, Nadeau A, et al. Diabetes 1992;41: 826–34.
39. Rimm EB, Stampfer MJ, Colditz GA, et al. Epidemiol 1990;1:466–73.
40. Lean ME, Han TS, Morrison CE. BMJ 1995;311:158–61.
41. Federation ID. The IDF consensus worldwide definition of the metabolic syndrome. Brussels, Belgium: 2006.
42. Alberti KG, Eckel RH, Grundy SM, et al. Circ 2009;120: 1640–5.
43. Lear SA, James PT, Ko GT, et al. Eur J Clin Nutr 2010;64: 42–61.
44. Ross R, Neeland IJ, Yamashita S, et al. Nat Rev Endocrinol 2020;16:177–89.
45. Sumner AE, Bagheri MH. Nat Rev Endocrinol 2020;16:1–3.
46. Katzmarzyk PT, Bray GA, Greenway FL, et al. Obesity (Silver Spring) 2011;19:1272–8.
47. Staiano AE, Bouchard C, Katzmarzyk PT. Obes Facts 2013;6: 317–24.
48. Klaver M, de Blok CJM, Wiepjes CM, et al. Eur J Endocrinol 2018;178:163–71.
49. Snijder MB, Zimmet PZ, Visser M, et al. Int J Obes Relat Metab Disord 2004;28:402–9.
50. Yusuf S, Hawken S, Ounpuu S, et al. Lancet 2005;366:1640–9.
51. Winter Y, Rohrmann S, Linseisen J, et al. Stroke 2008;39: 3145–51.
52. Klein S, Allison DB, Heymsfield SB, et al. Obesity (Silver Spring) 2007;15:1061–7.
53. Vega GL, Adams-Huet B, Peshock R, et al. J Clin Endocrinol Metab 2006;91:4459–66.
54. Karpe F, Pinnick KE. Nat Rev Endocrinol 2015;11:90–100.
55. Han TS, McNeill G, Seidell JC, et al. Int J Obes Relat Metab Disord 1997;21:587–93.
56. Nielsen S, Guo Z, Johnson CM, et al. J Clin Invest 2004;113: 1582–8.
57. Petersen MC, Shulman GI. Physiol Rev 2018;98:2133–223.
58. Fabbrini E, Magkos F, Mohammed BS, et al. Proc Natl Acad Sci U S A 2009;106:15430–5.
59. Smith GI, Mittendorfer B, Klein S. J Clin Invest 2019;129: 3978–89.
60. Espinosa De Ycaza AE, Sondergaard E, Morgan-Bathke M, et al. Diabetes 2022;71:381–93.
61. Jia Q, Morgan-Bathke ME, Jensen MD. Am J Physiol Endocrinol Metab 2020;319:E254–E64.
62. Magkos F, Fraterrigo G, Yoshino J, et al. Cell Metab 2016;23: 591–601.
63. McQuaid SE, Hodson L, Neville MJ, et al. Diabetes 2011;60: 47–55.
64. Randle PJ, Garland PB, Hales CN, et al. Lancet 1963;1:785–9.
65. Ferrannini E, Barrett EJ, Bevilacqua S, et al. J Clin Invest 1983; 72:1737–47.
66. Kelley DE, Mokan M, Simoneau JA, et al. J Clin Invest 1993;92: 91–8.
67. Reeds DN, Yarasheski KE, Fontana L, et al. Am J Physiol Endocrinol Metab 2006;290:E47–E53.
68. Ginsberg HN. Circ 2002;106:2137–42.
69. Ikezaki H, Lim E, Cupples LA, et al. J Am Heart Assoc 2021; 10:e019140.
70. Gaudet D, Alexander VJ, Baker BF, et al. N Engl J Med 2015;373:438–47.
71. Raal FJ, Rosenson RS, Reeskamp LF, et al. N Engl J Med 2020;383:711–20.
72. Sumner AE, Finley KB, Genovese DJ, et al. Arch Intern Med 165:1395–400.
73. Raygor V, Abbasi F, Lazzeroni LC, et al. Diab Vasc Dis Res 2019;16:153–9.
74. Miller BV, 3rd, Patterson BW, Okunade A, et al. J Lipid Res 2012;53:2767–72.

75. Chung ST, Cravalho CK, Meyers AG, et al. Circ Res 2020;126: 94–108.
76. Mittendorfer B, Patterson BW, Klein S. Am J Physiol Endocrinol Metab 2003;284:E549–56.
77. Vaag A, Skott P, Damsbo P, et al. J Clin Invest 1991;88:1282–90.
78. Cusi K, Kashyap S, Gastaldelli A, et al. Am J Physiol Endocrinol Metab 2007;292:E1775–81.
79. Manninen V, Tenkanen L, Koskinen P, et al. Circ 1992;85:37–45.
80. Rubins HB, Robins SJ, Collins D, et al. N Engl J Med 1999;341:410–8.
81. Keech A, Simes RJ, Barter P, et al. Lancet 2005;366:1849–61.
82. Group AS, Ginsberg HN, Elam MB, et al. N Engl J Med 2010;362:1563–74.
83. Elam MB, Ginsberg HN, Lovato LC, et al. JAMA Cardiol 2017;2:370–80.
84. Bhatt DL, Steg PG, Miller M, et al. N Engl J Med 2019;380: 11–22.
85. Aung T, Halsey J, Kromhout D, et al. JAMA Cardiol 2018;3: 225–34.
86. Rosenson RS, Brewer HB, Jr., Ansell BJ, et al. Nat Rev Cardiol 2016;13:48–60.
87. Hansel B, Giral P, Nobecourt E, et al. J Clin Endocrinol Metab 2004;89:4963–71.
88. Siervo M, Bluck LJ. Metab 2012;61:680–8.
89. Lamarche B, Uffelman KD, Carpentier A, et al. J Clin Invest 1999;103:1191–9.
90. Lincoff AM, Nicholls SJ, Riesmeyer JS, et al. N Engl J Med 2017;376:1933–42.
91. Group HTRC, Bowman L, Hopewell JC, et al. N Engl J Med 2017;377:1217–27.
92. Schwartz GG, Olsson AG, Abt M, et al. N Engl J Med 2012;367:2089–99.
93. Genuth S, Alberti KG, Bennett P, et al. Diabetes Care 2003;26: 3160–7.
94. Gabir MM, Hanson RL, Dabelea D, et al. Diabetes Care 2000;23:1108–12.
95. DeFina LF, Vega GL, Leonard D, et al. J Investig Med 2012;60:1164–8.
96. Rhee MK, Ho YL, Raghavan S, et al. PLoS One 2019;14:e0219964.
97. Zhang X, Gregg EW, Williamson DF, et al. Diabetes Care 2010;33:1665–73.
98. Knowler WC, Barrett-Connor E, Fowler SE, et al. N Engl J Med 2002;346:393–403.
99. DeFronzo RA, Tripathy D, Schwenke DC, et al. N Engl J Med 2011;364:1104–15.
100. le Roux CW, Astrup A, Fujioka K, et al. Lancet 2017;389: 1399–409.
101. American Diabetes Association Professional Practice C. Diabetes Care 2022;45:S39–S45.
102. Salazar MR, Carbajal HA, Espeche WG, et al. Diabetes Vasc Dis Res 2011;8:109–16.
103. Reeds DN, Stuart CA, Perez O, et al. Metab 2006;55: 1658–63.
104. Reaven GM, Hollenbeck CB, Chen YD. Diabetologia 1989;32: 52–5.
105. James DE, Stockli J, Birnbaum MJ. Nat Rev Mol Cell Biol 2021;22:751–71.
106. Gast KB, Tjeerdema N, Stijnen T, et al. PLoS One 2012;7:e52036.
107. Horber S, Achenbach P, Schleicher E, et al. Biotechnol Adv 2020;39:107359.
108. Ford ES, Giles WH, Dietz WH. JAMA 2002;287:356–9.
109. Schillaci G, Pirro M, Vaudo G, et al. J Am Coll Cardiol 2004;43:1817—22.
110. Jeppesen J, Hein HO, Suadicani P, et al. Arch Intern Med 2001;161:361–6
111. Ferrannini E, Buzzigoli G, Bonadonna R, et al. N Engl J Med 1987;317:350–7.
112. Levin G, Kestenbaum B, Ida Chen YD, et al. Am J Epidemiol 2010;172:1144–54.
113. Arnlov J, Pencina MJ, Nam BH, et al. Circ 2005;112:1719–27.
114. Arshi B, Tohidi M, Derakhshan A, et al. J Hum Hypertens 2015;29:260–7.
115. Poirier P, Lemieux I, Mauriege P, et al. Hypertens 2005;45:363–7.
116. Saad MF, Rewers M, Selby J, et al. Hypertens 2004;43: 1321–31.
117. da Silva AA, do Carmo JM, Li X, et al. Can J Cardiol 2020;36: 671–82.
118. Hall JE, do Carmo JM, da Silva AA, et al. Nat Rev Nephrol 2019;15:367–85.
119. Valenzuela PL, Carrera-Bastos P, Galvez BG, et al. Nat Rev Cardiol 2021;18:251–75.
120. Lakka HM, Laaksonen DE, Lakka TA, et al. JAMA 2002;288:2709–16.
121. Malik S, Wong ND, Franklin SS, et al. Circ 2004;110:1245–50.
122. Mottillo S, Filion KB, Genest J, et al. J Am Coll Cardiol 2010;56:1113–32.
123. Golden SH, Folsom AR, Coresh J, et al. Diabetes 2002;51: 3069–76.
124. Lawlor DA, Smith GD, Ebrahim S. Diabetologia 2006;49:41–8.
125. Sattar N, McConnachie A, Shaper AG, et al. Lancet 2008;371:1927–35.
126. Wang J, Ruotsalainen S, Moilanen L, et al. Eur Heart J 2007;28:857–64.
127. Yarnell JW, Patterson CC, Bainton D, et al. Heart 1998;79: 248–52.
128. Wilson PW, D'Agostino RB, Parise H, et al. Circ 2005;112: 3066–72.
129. Reaven GM. Am J Clin Nutr 2006;83:1237–47.
130. Wannamethee SG, Shaper AG, Lennon L, et al. Arch Intern Med 2005;165:2644–50.

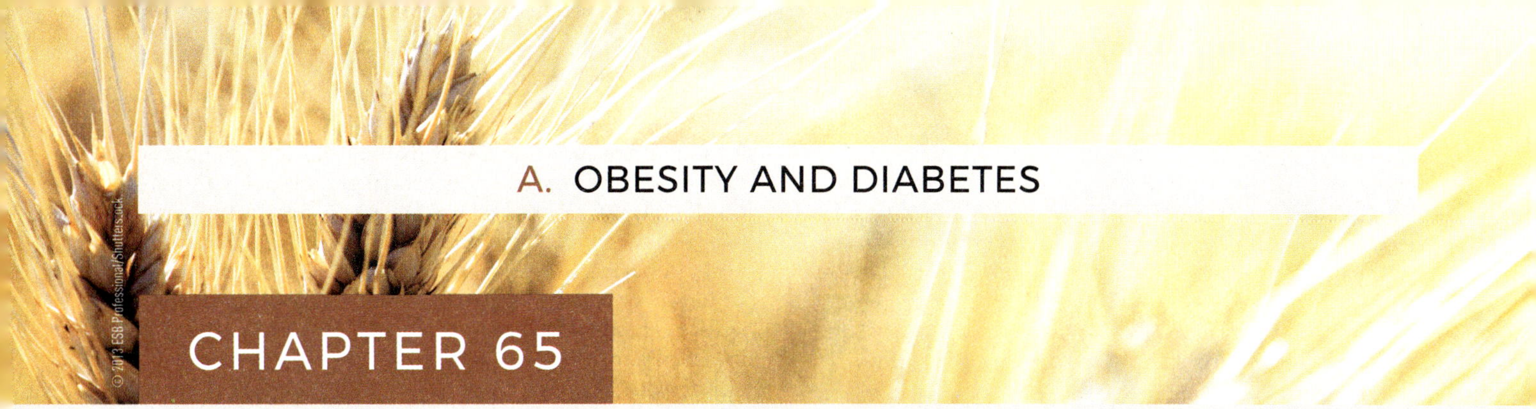

A. OBESITY AND DIABETES

CHAPTER 65

Nutrition and Inflammatory Processes

Philip C. Calder

CHAPTER OUTLINE

Abbreviations

AGE, advanced glycation end product; **AIDI**, anti-inflammatory dietary index; **AP-1**, activator protein 1; **CFU**, colony-forming units; **COX**, cyclo-oxygenase; **CRP**, C-reactive protein; **DHA**, docosahexaenoic acid; **DII**, dietary inflammation index; **DIS**, dietary inflammatory score; **EDIP**, empirical dietary inflammatory pattern score; **EPA**, eicosapentaenoic acid; **IBD**, inflammatory bowel disease; **IL-6**, interleukin 6; **iNOS**, inducible nitric oxide synthase; **LOX**, lipoxygenase; **MCP**, monocyte chemoattractant protein; **MMP**, matrix metalloproteinase; **NF-κB**, nuclear factor kappa-light-chain-enhancer of activated B cells; **PPAR**, peroxisome proliferator-activated receptor; **PRR**, pattern recognition receptor; **PUFA**, polyunsaturated fatty acid; **RAGE**, receptor for AGE; **SCFA**, short-chain fatty acid; **TLR**, toll-like receptor; **TNF-α**, tumor necrosis factor-α

Inflammation

Introductory Comments

Inflammation is a normal defense mechanism that protects the host from infection and other insults. It initiates pathogen killing, is involved in the removal of cellular debris as well as tissue repair processes, and helps to restore homeostasis at infected or damaged sites. Inflammation is typified by redness, swelling, heat, pain, and loss of function, although these may be transient. It involves interactions among many cell types, movement of cells between body compartments and the production of, and responses to, many chemical mediators.[1] Normally, the host is tolerant of microbes and other environmental components that do not pose a threat. This tolerance involves only a limited host response or an active response that is tightly controlled. When an inflammatory response does occur, it is normally well regulated and does not cause excessive damage to the host, is self-limiting, and resolves rapidly. This self-regulation involves the activation of negative feedback mechanisms such as the secretion of anti-inflammatory cytokines, inhibition of proinflammatory signaling cascades, shedding of receptors for inflammatory mediators, and activation of regulatory cells. As such, and controlled properly, regulated inflammatory responses are essential to remaining healthy and maintaining homeostasis.

Pathologic inflammation involves a loss of tolerance or a loss of regulatory processes. If this situation becomes excessive, irreparable damage to host tissues and disease can occur.[1] Typically, diseases or conditions with a well-recognized inflammatory component are treated with general or specific anti-inflammatory pharmaceuticals. However, because many dietary components may influence various aspects of inflammation, nutrition may play a role in predisposing to inflammatory conditions and altered nutrition may be useful in the prevention or therapy of such conditions. This chapter describes inflammation and the inflammatory response, considers the role of inflammation in various diseases and conditions, identifies common mechanisms and markers of inflammation, discusses evidence that diet and selected dietary components can influence inflammatory processes, and indicates the likely mechanisms of action of these dietary components.

General Aspects of the Inflammatory Process

Inflammation is largely orchestrated by leukocytes (neutrophils, monocytes/macrophages, eosinophils, and mast cells) designed to eliminate pathogens and to heal tissue injury using multiple chemical mediators (chemokines, cytokines, adhesion molecules, lipid mediators, growth factors, proteases). Inflammation may be classified as acute or chronic.[2] Acute inflammation is the initial response of the body to harmful stimuli and is achieved by the increased movement of plasma and leukocytes (especially granulocytes, which include neutrophils, eosinophils, and mast cells) from the blood into the injured tissues. A cascade of biochemical events propagates and matures the inflammatory response, involving the local vascular system, the immune system, and various cells within the injured tissue. Prolonged inflammation, known as chronic inflammation, leads to a progressive shift in the type of cells present at the site of inflammation (e.g., monocytes and macrophages become more involved) and is characterized by simultaneous destruction and healing of the tissue from the inflammatory process. Eventually, T and B lymphocytes can become involved; T cells shape and sustain the ongoing inflammatory response while B cells produce antibodies specific to the response. The features of acute and chronic inflammation are compared in **Table 65.1**.

Common to both forms of inflammation are an afferent phase, in which the presence of a "foreign material" is "sensed" by some cell types, and an efferent phase, in which an inflammatory response is generated to eliminate the perceived hostile intruder. The "sensing" of pathogens is carried out by pattern-recognition receptors (PRRs) on the leukocytes that recognize so-called microbe-associated molecular patterns; toll-like receptors (TLRs) are an example of PRRs. The purpose of the inflammatory response to harmful micro-organisms is obvious, and the response is beneficial and necessary to protect the integrity of the body, as long as it does not become unnecessarily destructive or long lasting. Inflammation caused by nonpathogenic agents can also be beneficial and can remove the foreign or damaged material, but it may also have negative health effects, especially if it has a long duration. Irrespective of the cause of the inflammation, the response involves four major events:

1. An increase in blood supply to the site of inflammation.
2. Increased capillary permeability caused by opening of junctions between endothelial cells. This permits plasma and larger molecules, not normally capable of crossing the endothelium, to do so and thus delivers soluble mediators to the site of inflammation.
3. Leukocyte migration from the capillaries into the surrounding tissue. This process is promoted by release of chemoattractants from the site of inflammation and by the upregulation of adhesion molecules on the endothelium. Once in the tissue, the leukocytes move to the site of inflammation.
4. Release of mediators from leukocytes at the site of inflammation. These may include lipid mediators (e.g., prostaglandins, leukotrienes), peptide mediators (e.g., cytokines, chemokines), reactive oxygen species (e.g., superoxide), amino acid derivatives (e.g., histamine), and enzymes (e.g., matrix proteases), depending on the cell type, the nature of the inflammatory stimulus, the anatomic site, and the stage during the inflammatory response. These mediators normally would play a role in host defense, but when they are produced inappropriately or in an unregulated fashion, they can damage host tissues and lead to disease. Several of these mediators may act to amplify the inflammatory process by acting, for example, as chemoattractants. Some of the inflammatory mediators may escape the inflammatory site into the circulation and, from there, exert systemic effects. For example, the cytokine interleukin (IL-6) induces hepatic synthesis of the acute phase protein C-reactive protein (CRP), whereas the cytokine tumor necrosis factor-α (TNF-α) elicits metabolic effects within skeletal muscle, adipose tissue, and bone.

Figure 65.1 provides an overview of the events that occur during the inflammatory response.

TABLE 65.1 Features of Acute and Chronic Inflammation

	Acute	Chronic
Causative agent	Pathogens, injured tissues	Persistent acute inflammation due to continued exposure to pathogens or loss of tolerance to harmless microbes, other environmental agents or self
Major cells involved	Neutrophils and other granulocytes, mononuclear cells (monocytes, macrophages)	Mononuclear cells (monocytes, macrophages, T lymphocytes, B lymphocytes)
Primary mediators	Vasoactive amines (e.g., histamine), bioactive lipids (e.g., prostaglandins, leukotrienes), chemokines	Cytokines, chemokines, bioactive lipids, growth factors, reactive oxygen species, hydrolytic enzymes
Onset	Immediate	Delayed
Duration	Hours to a few days	May be months or years; may be intermittent
Outcomes	Resolution, abscess formation, chronic inflammation	Tissue destruction, fibrosis, necrosis, disease

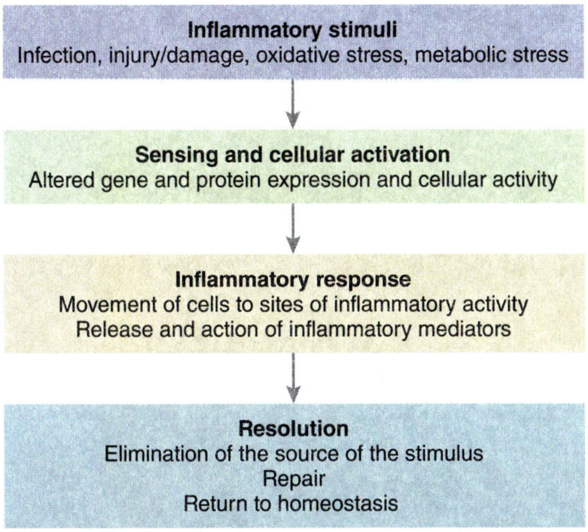

Figure 65.1 Overview of the inflammatory response. Inflammatory stimuli trigger an inflammatory response after they are "sensed" by receptors. This initiates cell signaling processes leading to cellular activation, some immediate cell responses and altered gene and protein expression. Cells move to sites of inflammatory action and there is release of, and responses to, a plethora of inflammatory mediators. These aim to eliminate the source of initial trigger and then, during resolution, to remove cell debris and to repair local tissue damage.

Characteristics of Inflammatory Conditions

Inflammation is a recognized contributor to the pathology of many conditions.[1,2] In some cases, such as rheumatoid arthritis, inflammatory bowel diseases (IBDs), asthma, and psoriasis, the central role of inflammation to the pathologic features is well recognized. Persons with these conditions have heavy infiltration of inflammatory cells at the site of disease activity (e.g., joints, intestinal mucosa, lungs, skin), and they have elevated concentrations of inflammatory mediators at those sites and in the systemic circulation. These conditions are treated with varying levels of success by anti-inflammatory drugs. In other cases, such as atherosclerosis and obesity, the role of inflammation has emerged more recently, and its contribution to the pathologic features alongside the many other relevant factors is less clear.[3,4] Persons with these conditions show infiltration of inflammatory cells at the site of disease activity (e.g., blood vessel wall, adipose tissue) and have moderately elevated concentrations of inflammatory mediators in the systemic circulation.[1,2,5]

Although inflammation-induced tissue damage occurs in an organ-specific manner (joints, gut, lungs, skin, blood vessel wall, adipose tissue) in different diseases or conditions, some commonality exists among the responses seen in the different organs.[1,2] In general, the inflammatory response observed is normal, but it occurs in the wrong context; this relates to loss of barrier function (epithelial or endothelial), inappropriate triggering (i.e., a response to a normally benign stimulus equivalent to a loss of tolerance), lack of downregulation to control the response, and tissue destruction with a loss of function. In some cases, the inflammation is the result of exogenous triggers such as allergens or microbes. In other cases, it is the result of loss of tolerance to self, as seen in autoimmune conditions, while in others it is secondary to tissue damage caused by endogenous molecules such as oxidized low-density lipoproteins.

Most, if not all, of the chronic inflammatory diseases are characterized by overproduction of cytokines (TNF-α, IL-1β, IL-6), chemokines (IL-8, monocyte chemoattractant protein (MCP)-1), eicosanoids (prostaglandin E_2, 4-series leukotrienes), and matrix metalloproteinases. Elevated concentrations of these mediators act to amplify the inflammatory process (e.g., by attracting further inflammatory cells to the site) and contribute to tissue destruction and to the clinical symptoms observed. Many of these mediators are positively regulated through the transcription factor Nuclear factor kappa-light-chain-enhancer of activated B cells (NF-κB)[6] or through cytoplasmic protein complexes called inflammasomes[7], and some are negatively regulated through peroxisome proliferator-activated receptors (PPARs).[8] **Figure 65.2** provides an overview of inflammatory signaling. Entry of inflammatory cells to sites of inflammatory activity is facilitated by upregulation of adhesion molecules on the endothelium, a process that is promoted by inflammatory cytokines and by a range of inflammatory triggers, frequently acting through NF-κB. The continuous process of tissue injury, healing, and repair, in response to the release of cytokines, chemokines, growth factors, and matrix metalloproteinases by infiltrating inflammatory cells, as well as resident tissue cells, results in tissue remodeling.

Why resolution of inflammation is absent or abnormal in so many pathophysiologic processes is largely unknown, although several mechanisms may be considered.[9] First, persistent insult (i.e., chronic infection, continued exposure to triggering stimuli) may provide a continued proinflammatory stimulus. Second, the inflammatory response results in tissue damage, and loss of barrier function may lead to exposure of antigens and loss of tolerance to autoantigens or components of the microbiota, which then provide a trigger to drive prolonged inflammation. Third, local overproduction of survival factors such as IL-5, granulocyte-macrophage colony-stimulating factor, and IL-1β may result in prolonged survival and activity of granulocytes. Finally, negative feedback (control) mechanisms that lead to loss of inflammatory control may be deficient. Although the relative importance may differ, these mechanisms seem to contribute to most inflammatory conditions.

Diet, Foods, and Inflammatory Processes

General Considerations

Irrespective of the nature of the inflammation, it is important to distinguish between dietary factors as direct causal factors (i.e., triggers or stimulators) of the inflammatory response and dietary factors as modifiers or regulators of the inflammatory response to another trigger or stimulant (**Figure 65.3**). Examples in which diet is a direct causal factor include allergy triggered by exposure to a food allergen (e.g., cow's milk protein, hen's egg protein, peanut protein)[10] or in celiac disease, an inappropriate adverse immunologic response to gluten and gluten-like proteins of some grains (wheat, rye, barley) that results in chronic inflammation of the small intestinal mucosa.[11] In these conditions, avoidance of foods that contain the trigger of the inflammation is the obvious key to treatment.

Other examples of dietary components that act as direct triggers of inflammation are emerging. In vitro experiments show that saturated fatty acids can activate inflammatory cells through TLR4 and the NF-κB pathway in the same way as bacterial lipopolysaccharide.[12] This finding raises the possibility that exposure of inflammatory cells to certain nonesterified saturated fatty acids

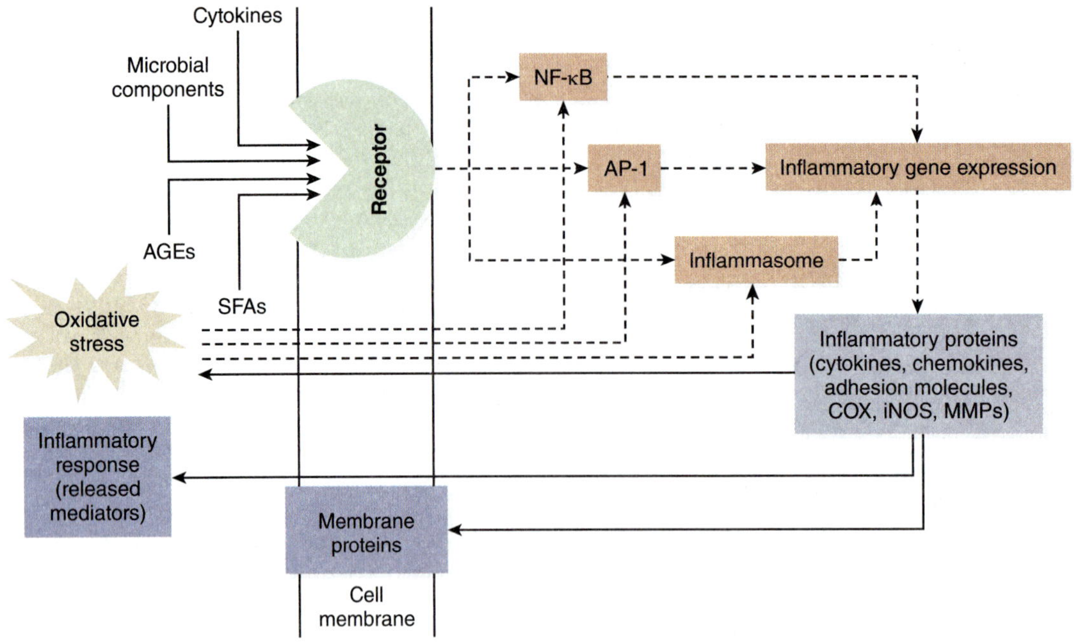

Figure 65.2 A generalized overview of inflammatory signaling. A range of inflammatory stimuli are sensed, often by receptors such as pattern recognition receptors, and initiate cell signaling pathways. These pathways lead to activation of various transcription factors, including nuclear factor kappa-light-chain-enhancer of activated B cells (NF-κB) and activator protein 1 (AP-1), and inflammasomes. Inflammatory gene expression is upregulated and inflammatory proteins (including cytokines, chemokines, adhesion molecules, cyclo-oxygenase (COX), inducible nitric oxide synthase (iNOS), and matrix metalloproteinases (MMPs)) are produced. Oxidative stress can activate these same inflammatory pathways while the inflammatory response induces production of oxidants, creating oxidative stress. AGEs, advanced glycation end products; SFAs, and saturated fatty acids.

could be an important factor in driving inflammation; this could be important in metabolic inflammation as seen in obesity, type 2 diabetes, nonalcoholic fatty liver disease, and atherosclerosis. Advanced glycation end products (AGEs) are formed by chemical reactions between glucose and amino acid residues during food processing and in cooking; they are also formed in vivo. Inflammatory cells express receptors for AGEs (RAGEs), and RAGEs induce inflammatory signaling.[13] Thus, AGEs from foods or formed endogenously may possibly trigger inflammation.

Breakdown of gut barrier function ("leaky gut") is a key factor in some inflammatory diseases such as IBDs and is also believed to contribute to the development of metabolic disease. An inadequate or inappropriate diet may contribute to a breakdown in the gastrointestinal barrier which, perhaps in the presence of other factors, could influence the initiation, progression, and severity of the disease. Furthermore, the composition of the gut microbiota is now recognized as a key "environmental" factor contributing to IBDs; diet clearly can influence the gut microbiota and, therefore may have an indirect influence on IBDs in this way. Thus, diet can be seen to be one step removed from direct causal factors (gut barrier breakdown; microbiota composition) of the initiation, progression, and severity of IBDs.

A strong interaction occurs between oxidative stress and inflammation. The generation of oxidants (e.g., superoxide radicals, hydrogen peroxide) is part of the host inflammatory response. Oxidants can damage components of host cells. In turn, oxidants, and oxidized cell components, acting through transcription factors such as NF-κB, induce production of inflammatory eicosanoids and cytokines, among other mediators. Thus, dietary components that contribute to oxidative stress (e.g., AGEs and oxidized lipids that result from heating cooking oils to high temperature) could promote inflammatory responses, whereas dietary components that inhibit or quench oxidative stress (various antioxidants) could decrease the strength of inflammatory responses.

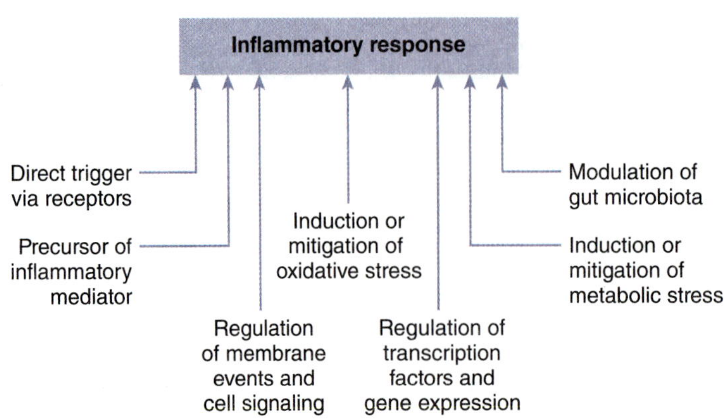

Figure 65.3 Overview of the ways in which dietary components can influence inflammation.

The nature of the diet can affect the concentrations of hormones (e.g., insulin, leptin, cortisol) that, in turn, affect inflammatory processes. Some dietary components act as substrates for biosynthesis of inflammatory mediators (e.g., arginine is the precursor of nitric oxide and the n-6 fatty acid arachidonic acid is the precursor of inflammatory prostaglandins). Thus, diet could play a key role in maintaining availability of such substrates to fuel the inflammatory response. Finally, some dietary components act as regulators of various aspects of inflammatory cell responses. For example, marine n-3 fatty acids influence signaling pathways in inflammatory cells through actions at the membrane, the cytosol, and the nucleus and therefore, they act to modify the responses initiated by various inflammatory triggers.

Dietary Approaches to Preventing or Ameliorating Inflammation

Energy Restriction and Weight Loss

Weight loss is accompanied by decreased concentrations of circulating markers of inflammation,[14] but whether this effect is secondary to the weight loss itself or to the nature of the diet used to induce weight loss is difficult to determine. However, it seems likely that reduced secretion of proinflammatory mediators from adipocytes or activated macrophages of adipose tissue contributes to the effect of weight loss. Conversely, energy restriction itself may play an anti-inflammatory role.[15] Key mediators of the effect are proteins of the sirtuin and FoxO families, which are induced or activated during states of limited energy supply. The sirtuins are oxidized nicotinamide adenine dinucleotide dependent deacetylases of substrates ranging from histones to transcriptional regulators. Therefore, metabolic efficiency is improved, cell defenses against stress are strengthened, and inflammatory activities are dampened, notably by decreasing the activation of NF-κB.[16,17] FoxO proteins are transcription factors that regulate the expression of genes involved in energy homeostasis, cell survival, and inflammatory responses including NF-κB.[18,19] Reducing energy intake appears to be more important than the nature of low-energy foods; decreased concentrations of inflammatory markers are observed with a low-energy, fat-rich diet as well as with a low-energy, carbohydrate-rich diet.[20] However, it is conceivable that some dietary components are better regulators of sirtuin or FoxO activity than others.

Dietary Patterns

Epidemiologic studies examined the associations between particular dietary patterns or intake of specific types of foods and measures of inflammation. These studies have usually focused on blood markers of inflammation (e.g., plasma concentrations of CRP or cytokines) and have been conducted largely in the context of the low-grade chronic inflammation associated with cardiovascular disease, insulin resistance, and overweight and obesity.

Greater adherence to the traditional Mediterranean diet (rich in fruits, vegetables, whole grains, legumes, nuts, fish, and low-fat dairy products, with moderate consumption of wine, and with olive oil as the principal source of fat) is associated with lower blood concentrations of inflammatory markers in healthy persons.[21,22] Intervention studies demonstrated that consuming a Mediterranean diet decreases inflammation in healthy persons, in obese persons, and in individuals with high cardiovascular risk.[23-26] Dietary patterns consistent with vegetarianism have been associated with lower concentrations of inflammatory markers in the bloodstream compared with nonvegetarian diets.[27,28] Healthy eating patterns, as captured by scoring systems such as the Healthy Eating Index, the Alternate Healthy Eating Index, the Diet Quality Score, or the Prudent Diet Score, have all been shown to be inversely associated with circulating markers of inflammation such as CRP and IL-6.[22,29,30] Using data from the Nurses' Health Study, a dietary pattern that was significantly associated with higher concentrations of several inflammatory markers was identified; this pattern was high in sugar-sweetened soft drinks, refined grains, diet soft drinks, and processed meat but was low in wine, coffee, cruciferous vegetables, and yellow vegetables.[31] In contrast to a prudent diet, which was associated with lower concentrations of several inflammatory markers, a Western diet pattern was associated with higher concentrations of these same markers.[32]

Specific Foods

Observational studies report inverse associations between intake of whole grains, nuts and seeds, fruits and vegetables, fish, and tea and certain blood biomarkers of inflammation, often including CRP and IL-6.[5,33] Regular consumption of small doses of dark chocolate decreased markers of inflammation in healthy adults.[34] Intervention studies with whole grain foods are inconsistent about the effect on inflammation.[5,33] Interventions with fruits and vegetables as a food group have been successful at reducing the blood concentrations of inflammatory markers.[5,33] However, studies focusing on a single variety of vegetable or fruit have been inconsistent. Soy protein appears not to affect circulating markers of inflammation. Intervention trials of drinking black tea, green tea, or coffee have not yielded a clear picture regarding a possible anti-inflammatory effect.[5]

Selected Dietary Components that May Prevent or Ameliorate Inflammation

Observations that healthy diets and their components (whole grains, nuts and seeds, fruits and vegetables, and fish) are associated with reduced inflammation have focused attention on the nutrients and other components provided by those diets and foods as having anti-inflammatory properties. Key among these nutrients are marine n-3 fatty acids, antioxidant micronutrients, and polyphenols.

Marine n-3 Polyunsaturated Fatty Acids

The key link between fatty acids and inflammation is that eicosanoids, which act as mediators and regulators of inflammation, are generated from 20-carbon polyunsaturated fatty acids (PUFAs) of the n-6 and n-3 families.[35,36] Because inflammatory cells typically contain a high proportion of the n-6 PUFA arachidonic acid and low proportions of other 20-carbon PUFAs, arachidonic acid is usually the major substrate for eicosanoid synthesis. Eicosanoids, which include prostaglandins, leukotrienes, and other oxidized PUFA derivatives, are generated from arachidonic acid by reactions catalyzed by cyclo-oxygenase (COX), lipoxygenase (LOX), and cytochrome P450 enzymes. At least two COX enzymes and several LOX and cytochrome P450 enzymes are expressed in different cell types, according to different conditions; between them, they produce a range of mediators involved in modulating the intensity and duration of inflammatory responses. These mediators have cell- and stimulus-specific sources and frequently have opposing effects. Thus, the overall physiologic (or pathophysiologic) outcome depends on the cells present, the nature of the stimulus, the timing of eicosanoid generation, the concentrations of different eicosanoids generated, and the sensitivity of target cells and tissues to the eicosanoids generated.

The amount of arachidonic acid in inflammatory cells can be decreased by increased consumption of marine n-3 PUFAs (eicosapentaenoic acid [EPA] and docosahexaenoic acid [DHA]) found in seafood, especially oily fish, and usually given as fish oil in experimental settings.[37] Thus, less substrate is available for synthesis of inflammatory eicosanoids from arachidonic acid. EPA is also able to act as a substrate for COX, LOX, and cytochrome P450 enzymes, and produces eicosanoids with a structure slightly different from that formed from arachidonic acid. The functional significance of this is that the mediators formed from EPA are typically less potent than those formed from arachidonic acid.[37] Furthermore, so-called specialized pro-resolving mediators (SPMs) are formed from both EPA and DHA.[30,39] SPMs include E- and D-series resolvins, protectins, and maresins. SPMs have been shown to exert potent anti-inflammatory and inflammation-resolving actions. Thus, one anti-inflammatory mechanism of action of marine n-3 PUFAs is antagonism of production of inflammatory eicosanoids from arachidonic acid coupled with the generation of less-potent EPA-derived eicosanoids and anti-inflammatory SPMs from EPA and DHA. Altered eicosanoid and SPM profiles have downstream effects, for example, altered production of inflammatory cytokines. However, lipid mediator-independent effects of marine n-3 PUFAs also seem likely.[37,40] These PUFAs have been shown to decrease activation of the proinflammatory transcription factor NF-κB and the inflammasome, to activate the anti-inflammatory transcription factor PPAR-γ and to alter key structural and functional aspects of the plasma membrane that regulate TLR4 signaling. As a result of these actions, marine n-3 PUFAs have been shown to reduce leukocyte chemotaxis, adhesion molecule expression, and production of inflammatory cytokines[37,40]. In experimental studies, EPA and DHA have been shown to counter the effects of saturated fatty acids and bacterial lipopolysaccharides.[37,40]

Observational studies demonstrated an inverse association between intake or status of marine n-3 PUFAs and circulating inflammatory marker concentrations.[41-44] Many, though not all, intervention studies with marine n-3 PUFAs showed lower concentrations of inflammatory markers in the bloodstream and reduced production of inflammatory eicosanoids and cytokines by isolated inflammatory cells.[37,40] Anti-inflammatory actions of plant n-3 PUFA α-linolenic acid appear to require its conversion to the more biologically active EPA.[45,46]

Antioxidant Vitamins

There is a bidirectional interaction between oxidative stress and inflammation. The generation of oxidants is part of the host inflammatory response. Oxidants can damage components of host cells. In turn, oxidants, and oxidized cell components, acting through transcription factors, such as NF-κB, induce inflammation. Thus, one mechanism to control inflammation may be to prevent oxidative stress. This is accomplished through enhancing antioxidant defense mechanisms, including increasing the concentrations of antioxidant vitamins, such as vitamin C (ascorbate), vitamin E (tocopherols and tocotrienols), and carotenoids (β-carotene, lycopene, lutein, astaxanthin).

Antioxidant vitamins act to reduce exposure to oxidants and, thereby, decrease activation of NF-κB and subsequent production of inflammatory cytokines, eicosanoids, and others. Inflammatory cells maintain high intracellular concentrations of vitamin C, although these concentrations become depleted during acute activation. At the site of inflammation, the ratio of oxidized to reduced ascorbate is increased. Observational studies reported inverse relationships between intake or status of antioxidant vitamins (vitamin C, vitamin E, lutein, lycopene) and various circulating markers of inflammation.[47-52] Some studies with vitamin C or vitamin E supplements or with carotenoid rich fruits and vegetables report that they lower CRP and inflammatory cytokines[53,54] but findings are not consistent.[5]

Polyphenols

Polyphenols are secondary metabolites of plants. They include flavanones, flavones, flavanols, and flavonols. In vitro studies suggest that flavonoids have anti-inflammatory activity through several mechanisms, including decreased eicosanoid production through inhibition of phospholipase A_2, COX, and LOX; inhibition of inducible nitric oxide synthase; and inhibition of inflammatory cytokine production. These effects seem to involve inhibition of activation of key proinflammatory transcription factors, such as NF-κB and activator protein-1. Several polyphenols have benefits in animal models of inflammatory processes and disease. Observational data in humans relating intake of polyphenols with inflammatory markers are inconsistent.[5] Human intervention studies investigating the effect of polyphenols on markers of inflammation have often focused on polyphenol-rich foods. For example, the effects of red wine, chocolate, and berries on inflammatory markers are believed to result from their constituent polyphenols.[5] Bioavailability of polyphenols is often poor, and circulating concentrations are low, usually much lower than those used in the in vitro experiments that demonstrated strong anti-inflammatory effects. Very little information on flavonoids exists in relation to inflammatory disorders. Nevertheless, polyphenols may act to regulate inflammation via effects on the gut microbiota, effects that do not require them to be systemically bioavailable.

Gut Microbiota and Inflammation

Gut Microbiota

Up to 10^{14} micro-organisms are contained in the adult human body (several times more microbial cells than human cells), and most of these micro-organisms are in the gastrointestinal tract.[55] More than 1 kg of bacteria is present in the gut, and feces typically comprise 50% bacteria. This means that humans excrete 50 to 100 g of bacteria each day. Micro-organisms are at their lowest numbers ($<10^3$ colony-forming units [CFU]/g) in the stomach because of its low pH and fast transit time, but they reach concentrations as high as 10^{12} CFU/g in the colon, a reflection of the colon's much slower transit time, less acidic pH, and low oxygen. At least 500 different bacterial species have been cultured from human feces (including lactobacilli and bifidobacteria), yet just

40 species account for 99% of those that have been identified. The microbial inhabitants of the intestinal microbiota are still not fully identified because of the limitations of identification techniques and individual variations. Newer molecular typing methods are enabling identification of further gut-associated species.

A major function of the colonic microbiota is the fermentation of dietary substances that have passed undigested through the small intestine, and of mucus produced within the gut. Saccharolytic fermentation produces short-chain fatty acids (SCFAs), such as acetate, propionate, and butyrate, which are essential nutrients for the colonocytes. Interactions also occur between gut microbiota and inflammatory cells present within and beyond the gut epithelium; these interactions can be chemical or may involve direct cell-to-cell contact.[56,57] Such interactions seem to play an important role in defining the inflammatory response in the gut wall and systemically. Gut dysbiosis is described in many inflammatory conditions, an observation that is interpreted to support the role of the microbiota in regulating inflammation. However, reverse causality cannot be ruled out.

Dietary patterns, specific foods, and some dietary components that result in less inflammation often promote a healthy, diverse gut microbiota[58-60] and it is possible that such patterns/foods/components have dual effects acting both via the gut microbiota and systemically.

Probiotics and Prebiotics

Probiotics are "live microorganisms which, when administered in sufficient amounts, confer a health benefit to the host."[61] Lactic acid-producing bacteria, including lactobacilli and bifidobacteria of human intestinal origin, are the most used probiotics, although other bacterial genera and even some yeasts are also used as probiotics. Probiotic bacteria modulate the gastrointestinal microenvironment and release antimicrobial factors such as defensins. Several probiotics successfully preserve epithelial barrier function by induction of mucin secretion and maintenance of tight junction integrity.[62,63] This decreases translocation of agents such as bacterial lipopolysaccharides, which can trigger systemic inflammation, from the gut lumen into the bloodstream. Some probiotic bacteria antagonize NF-κB activation and thereby decrease production of proinflammatory cytokines, while some interact with gut wall-residing dendritic cells to induce their maturation and secretion of IL-10, which is thought to favor the induction of regulatory T cells.[64]

A prebiotic is "a substrate that is selectively utilized by host microorganisms conferring a health benefit."[65] Prebiotics are typically carbohydrates that escape digestion by mammalian enzymes in the small intestine but are hydrolyzed by microbial enzymes in the colon, often to SCFAs. Prebiotics include fructo-oligosaccharides, galacto-oligosaccharides and other oligosaccharides, and even marine n-3 PUFAs and polyphenols are now thought to have probiotic action.[65] Some interventions with prebiotic oligosaccharides in humans have been reported to lower concentrations of inflammatory markers, including CRP.[66]

Dietary Inflammatory Indexes

Chronic low-grade inflammation is linked to diet and is believed to increase the risk of cardiovascular, metabolic, and cognitive diseases. Thus, there has been an interest in classifying diets according to the extent to which they are likely to promote inflammation. Several such classification systems have been proposed. The "dietary inflammatory index" (DII) is a literature-derived, nutrient-based index based on published associations of 45 dietary components (energy, macronutrients, fiber, micronutrients, plant polyphenols, herbs, spices), which are each weighted (positively for increasing inflammation and negatively for decreasing inflammation) in relation to blood concentrations of six inflammatory biomarkers (CRP, TNF, IL-1β, IL-6, IL-4, and IL-10); a higher score indicates a more pro-inflammatory diet.[67,68] The "dietary inflammation score" (DIS) is a literature-derived food and supplement-based score based on biologic plausibility and published associations of intakes of 18 foods or food groups (e.g., leafy greens and cruciferous vegetables, tomatoes, apples and berries, deep yellow or orange vegetables and fruit, legumes, fish, high-fat dairy, low-fat dairy, nuts, red meat, added sugars, refined grains, and starchy vegetables) and micronutrient supplement use, which are each weighted (positively for increasing inflammation and negatively for decreasing inflammation) in relation to blood concentrations of four inflammatory biomarkers (CRP, IL-6, IL-8, and IL-10); a higher score indicates a more pro-inflammatory diet.[69] The "anti-inflammatory dietary index" (AIDI) is a literature-derived, food-based score based on the associations of intakes of 18 foods or food groups (e.g., total fruits and vegetables, whole grain breads, dry fruits, legumes, nuts, unprocessed red meat), which are each weighted (negatively for increasing inflammation and positively for decreasing inflammation) in relation to plasma CRP concentrations; a higher score indicates a more anti-inflammatory diet.[70] The "empirical dietary inflammatory pattern score" (EDIP) is based on regression models predicting the association of intakes of foods and food groups that are each weighted (positively for increasing inflammation and negatively for decreasing inflammation) in relation to blood concentrations of three inflammatory biomarkers (CRP, IL-6, TNF receptor 2); a higher score indicates a more pro-inflammatory diet.[71,72] EDIP, DIS, and AIDI are all based on whole foods, although they are derived differently, while DII is mainly nutrient based. High correlations between DII and DIS have been reported, while there are low-to-moderate correlations between EDIP and DIS and low correlations between EDIP and DII.

Summary

Inflammation is a stereotypical physiologic response to infections and tissue injury. It initiates pathogen killing as well as tissue repair processes and helps to restore homeostasis at infected or damaged sites. Acute inflammatory reactions are usually self-limiting and resolve rapidly. This process involves the activation of negative feedback mechanisms, such as the secretion of immunoregulatory cytokines (e.g., IL-10 and transforming growth factor-β) and SPMs, inhibition of proinflammatory signaling cascades, receptor shedding, and activation of regulatory cells. Inflammatory responses that fail to regulate themselves can become chronic and contribute to the initiation, perpetuation, and progression of disease. This has been well recognized for "high grade" chronic inflammation where the inflammatory response is central to the pathology and the disease is treated with anti-inflammatory medications. However, in the last 2 decades, "low grade" chronic inflammation has been recognized to underpin atherosclerosis, cognitive decline, and loss of muscle mass and to be linked with obesity playing a causal role in its comorbidities. Dietary choices are important in determining the extent of low-grade inflammation and this is likely one reason why diet is a risk factor for these diseases.

A healthy eating pattern characterized by consumption of whole grains, nuts and seeds, fruits and vegetables, and fish is associated with reduced inflammation (**Figure 65.4**), a finding suggesting candidate anti-inflammatory dietary components. In line with this, several dietary indexes of inflammation have been defined based on the known impact of foods or food components on circulating inflammatory markers. These observations indicate that diet plays a role in setting inflammatory tone and in determining risk of at least some diseases that involve inflammation. Among food components, there is good evidence that marine omega-3 PUFAs, antioxidant vitamins, plant polyphenols, and those components that beneficially modify the gut microbiota all reduce inflammation (Figure 65.4). On the other hand, saturated fatty acids, simple sugars, AGEs, and oxidized lipids all increase inflammation (Figure 65.4). Dietary components that enter the circulation act to influence inflammation through a small number of common mechanisms: they may bind to PRRs to trigger inflammation; they promote or mitigate oxidative stress; they act to promote or mitigate inflammatory signaling, for example via NF-κB; or they act as substrates for synthesis of chemicals involved in the inflammatory response. The gut microbiota is important in determining host inflammatory tone, at both the gut and systemic levels. Thus, foods and food components that modify the gut microbiota can influence inflammation. The composition of this microbiota can be modified by intake of various foods as well as prebiotics or probiotics.

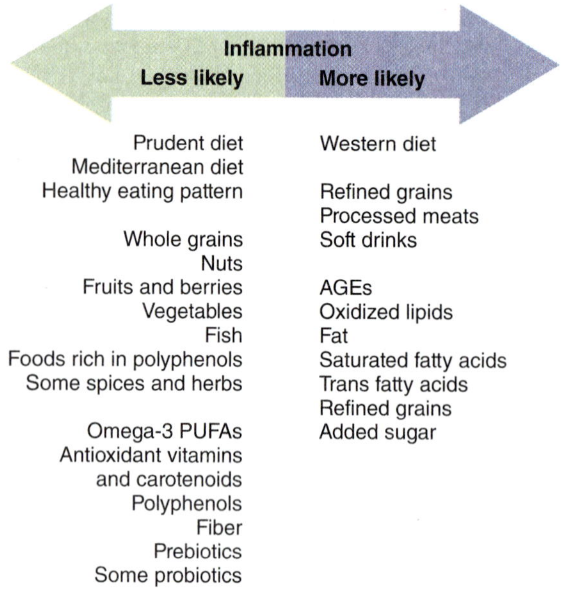

Figure 65.4 The comparative effects of dietary patterns, specific foods, and individual food components on inflammation. The dietary patterns, foods, and food components on the left are associated with lower inflammation while the dietary pattern, foods, and food components on the right are associated with more inflammation. *AGEs*, advanced glycation end products; *PUFAs*, polyunsaturated fatty acids.

References

1. Calder PC, Albers R, Antoine JM, et al. Brit J Nutr 2009;101: S1–45.
2. Calder PC, Ahluwalia N, Albers R, et al. Brit J Nutr 2013;109: S1–S34.
3. Hansson GK. N Engl J Med 2005;352:1685–1695.
4. Tilg H, Moschen AR Nat Rev Immunol 2006;6:772–783.
5. Calder PC, Ahluwalia N, Brouns F, et al. Brit J Nutr 2011;106: S5–78.
6. Kumar A, Takada Y, Boriek AM, et al. J Mol Med 2004;82: 434–448.
7. Blevins HM, Xu Y, Biby S, et al. Front Aging Neuro 2022;14: 879021.
8. Wang S, Dougherty EJ, Danner RL Pharmacol Res 2016;111: 76–85.
9. Barnig C, Bezema T, Calder PC, et al. Front Immunol 2019;10: 1699.
10. Tedner SG, Asarnoj A, Thulin H, et al. J Intern Med 2022;291: 283–302.
11. Iversen R, Sollid LM Ann Rev Pathol 2023;18:47–70.
12. Hwang DH, Kim JA, Lee JY Eur J Pharmacol 2016;785, 24–35.
13. Twarda-Clapa A, Olczak A, Białkowska AM, et al. Cells 2022;11: 1312.
14. Esposito K, Pontillo A, Di Palo C, et al. JAMA 2003;289: 1799–1804.
15. Holloszy JO, Fontana L. Exper Gerontol 2007;42:709–712.
16. Guarente L, Picard F. Cell 2005;120:473–482.
17. Dali-Youcef N, Lagouge M, Froelich S, et al. Ann Med 2007;39, 335–345.
18. Salminen A, Ojala J, Huuskonen J, et al. Cell Mol Life Sci 2008;65: 1049–1058.
19. Kim DH, Kim JY, Yu BP, et al. Biogerontol 2008;9:33–47.
20. Sharman MJ, Volek JS. 2004 Clin Sci 2008;107:365–369.
21. Chrysohoou C, Panagiotakos DB, Pitsavos C, et al. J Am Coll Cardiol 2004;44:152–158.
22. Fung TT, McCullough ML, Newby PK, et al. Am J Clin Nutr 2005;82:163–173.
23. Esposito K, Marfella R, Ciotola M, et al. JAMA 2004;292: 1440–1446.
24. Estruch R, Martínez-González MA, Corella D, et al. Ann Int Med 2006;145:1–11.
25. Dai J, Miller AH, Bremner JD, et al. Circ 2008;117:169–175.
26. Schwingshackl L, Hoffmann G. Nutr Metab Cardiovasc Dis 2014; 24:929–939.
27. Purschwitz K, Rassoul F, Reuter W, et al. [Zeitschrift für Gerontologie und Geriatrie] 2001;34:476–479.
28. Szeto YT, Kwok TC, Benzie IF. Nutr 2004;20:863–866.
29. Ford ES, Mokdad AH, Liu S Eur J Clin Nutr 2005;59:278–283.
30. Fargnoli JL, Fung TT, Olenczuk DM, et al. Am J Clin Nutr 2008;88:1213–1224.
31. Schulze MB, Hoffmann K, Manson JE, et al. Am J Clin Nutr 2005; 82:675–684.
32. Lopez-Garcia E, Schulze MB, Fung TT, et al. Am J Clin Nutr 2004; 80:1029–1035.
33. Calder PC, Bosco N, Bourdet-Sicard R, et al. Age Res Rev 2017; 40:95–119.
34. di Giuseppe R, Di Castelnuovo A, Centritto F, et al. J Nutr 2008; 138:1939–1945.
35. Christie WW, Harwood JL. Essays Biochem 2020;64:401–421.
36. Calder PC. Essays Biochem 2020;64:423–441.
37. Calder PC. [Biochimica et Biophysica Acta] 2015;1851: 469–484.
38. Chiang N, Serhan CN. Essays Biochem 2020;64:443–462.

39. Panigrahy D, Gilligan MM, Serhan CN, et al. Pharmacol Thera 2021;227:107879.
40. Calder PC. Proc Nutr Soc 2020;79:404–416.
41. Yli-Jama P, Seljeflot I, Meyer HE, et al. Atherosclerosis 2002;164: 275–281.
42. Pischon T, Hankinson SE, Hotamisligil GS, et al. Circ 2003; 108:155–160.
43. Lopez-Garcia E, Schulze MB, Manson JE, et al. J Nutr 2004;134: 1806–1811.
44. Ferrucci L, Cherubini A, Bandinelli S, et al. J Clin Endocrin Metabolism 91, 439–446.
45. Baker EJ, Miles EA, Burdge GC, et al. Prog Lipid Res 2016;64: 30–56.
46. Baker EJ, Valenzuela CA, De Souza CO, et al. [Biochimica et Biophysica Acta] 2020;1865:158662.
47. Kritchevsky SB, Bush AJ, Pahor M, et al. Am J Epidemiol 2000;152:1065–1071.
48. Erlinger TP, Guallar E, Miller ER 3rd, et al. Arch Int Med 2001;161:1903–1908.
49. van Herpen-Broekmans WM, Klöpping-Ketelaars IA, Bots ML, et al. Eur J Epidem 2004;19:915–921.
50. Wannamethee SG, Lowe GD, Rumley A, et al. Am J Clin Nutr 2006;83:567–574.
51. Wang L, Gaziano JM, Norkus EP, et al. Am J Clin Nutr 2008;88:747–754.
52. Helmersson J, Arnlöv J, Larsson A, et al. Brit J Nutr 2009;101: 1775–1782.
53. Devaraj S, Jialal I. Free Rad Bio Med 2000;29:790–792.
54. Block G, Jensen C, Dietrich M, et al. J Am Coll Nutr 2004;23: 141–147.
55. Holzapfel WH, Haberer P, Snel J, et al. Int J Food Microbiol 1998;41:85–101.
56. Samuelson DR, Welsh DA, Shellito JE. Front Microbiol 6:1085.
57. Verdu EF, Galipeau HJ, Jabri B. Nat Rev Gastroenterol Hepatol 2015;12:497–506.
58. Wu GD, Chen J, Hoffmann C, et al. Science 2011;334:105–108
59. Gentile CL, Weir TL. Science 2018;362:776–780.
60. Valdes AM, Walter J, Segal E, et al. BMJ 2018;361:k2179.
61. Hill C, Guarer F, Reid G, et al. Nat Rev Gastroenterol Hepatol 2014;11:506–514.
62. Ohland CL, MacNaughton WK. Am J Physiol 2010;298: G807–G819.
63. Ulluwishewa D, Anderson RC, McNabb WC, et al. J Nutr 2011; 141:769–776.
64. Thomas CM, Versalovic J. Gut Microbes 2010;1:148–163.
65. Gibson GR, Hutkins R, Sanders ME, et al. Nat Rev Gastroenterol Hepatol 2017;14:491–502.
66. Dehghan P, Pourghassem Gargari B, Asghari Jafar-badi M. Nutrition 2014;30:418–423.
67. Shivappa N, Steck SE, Hurley TG, et al. Pub Health Nutr 2014; 17:1689–1696.
68. Hébert JR, Shivappa N, Wirth MD, et al. Adv Nutr 2019;10: 185–195.
69. Byrd DA, Judd SE, Flanders WD, et al. J Nutr 2019;149: 2206–2218.
70. Kaluza J, Harris H, Melhus H, et al. Antiox Redox Sig 2018;28: 78–84.
71. Tabung FK, Smith-Warner SA, Chavarro JE, et al. J Nutr 2016; 146:1560–70.
72. Tabung FK, Smith-Warner SA, Chavarro JE, et al. J Nutr 2017; 147:1567–1577.

B. CARDIOVASCULAR DISEASES

CHAPTER 66

Nutrient and Genetic Regulation of Lipoprotein Metabolism

Edward A. Fisher, Raanan Shamir, and Robert A. Hegele

CHAPTER OUTLINE

Abbreviations

ABCA1, ATP-binding cassette protein A1; **ABL**, Abetalipoproteinemia; **ADH**, autosomal dominant hypercholesterolemia; **ARH**, autosomal recessive hypercholesterolemia; **ASCVD**, atherosclerotic cardiovascular disease; **CE**, cholesteryl ester; **CETP**, cholesterol ester transfer protein; **CVD**, cardiovascular disease; **DNA**, Deoxyribonucleic Acid, **FCH**, familial combined hyperlipidemia; **FCS**, familial chylomicronemia syndrome; **FH**, familial hypercholesterolemia; **FHBL**, familial hypobetalipoproteinemia; **GWAS**, genome-wide association studies; **HDL**, high-density lipoprotein; **HL**, hepatic lipase; **IDL**, intermediate-density lipoproteins; **LCAT**, lecithin cholesterol acyltransferase; **LDL**, low-density lipoprotein; **MCS**, multifactorial chylomicronemia; **miRs**, micro RNAs; **MTP**, microsomal triglyceride transfer protein; **RCT**, reverse cholesterol transport; **TG**, triglyceride; **VLDL**, very low-density lipoprotein

Introduction

The relationship among different dietary components and lipoprotein metabolism has been long recognized on both an experimental and observational basis. For example, the INTERHEART study showed that, across many ethnic groups and areas of the world, cardiovascular disease (CVD) risk was inversely related to the consumption of "heart healthy" foods: indeed ~30% of the population attributable risk of CVD was accounted for by unhealthy dietary intake.[1] In classic studies,[2-4] the relationships between dietary cholesterol and specific characteristics of fats (particularly the degree of fatty acid saturation) and plasma concentrations of LDL and HDL cholesterol were established by careful clinical experimentation. Over subsequent years, additional studies were conducted in animals and people to show that the other macronutrients (proteins and carbohydrates), and other dietary components also had effects on plasma lipid and lipoprotein concentrations.[5]

As cellular and molecular biological techniques advanced in the latter quarter of the 20th century, a battery of studies was conducted to approach the mechanistic underpinnings of the clinical observations and intervention results. These studies were greatly expanded in scope by the revolution in the manipulation of the mouse genome, which allowed for the development of models in which candidate genes implicated in the response to nutritional factors could be inserted by transgenesis, or inactivated by homologous recombination, in the context of normal and abnormal backgrounds and conditions (e.g., atherosclerosis). More recently, the sequencing of the human genome, coupled with high-throughput technologies, has led to the next phase of discovery in many areas of physiology and pathophysiology. As of 2010,[6] genome-wide association studies (GWAS) and next-generation DNA sequencing had already established thousands of genetic loci associated with the plasma concentrations of total lipids (cholesterol, triglycerides) and the individual lipoprotein fractions. Some of these loci were already known from previous evidence to be functional players in lipid and lipoprotein metabolism; the regulation of many of these was subject to a nutritional component. Other loci found by modern genetic technologies are completely novel discoveries, with their roles and regulation to be established. In this chapter, we will summarize the major genetic factors that are known to determine or have strong influence on human lipoprotein metabolism. For a detailed summary of the impact of specific nutrients on human plasma lipoprotein concentrations, the reader is referred to Chapter 67 in this textbook.

High Plasma Total and LDL Cholesterol

High blood cholesterol, especially low-density lipoprotein (LDL) cholesterol, is associated with increased risk of premature atherosclerotic cardiovascular disease (ASCVD).[7] Measurement of serum total cholesterol is a reflection of the amount of circulating very low-density lipoproteins (VLDL), LDL, high-density lipoproteins (HDL), and chylomicrons; chylomicrons disappear when measurement is done while fasting. Thus, a fasting lipoprotein profile is needed when hypercholesterolemia is identified. Hypercholesterolemia, both monogenic and multifactorial, affects about 5% of the population, and total blood cholesterol in childhood is associated with increased risk of ASCVD events in adulthood.[8] Compared with individuals with normal lipids, the risk for premature atherosclerotic cardiovascular disease is highest in those with monogenic elevated LDL cholesterol but is also increased in those with polygenic elevated LDL cholesterol.[9]

Monogenic Disorders Causing Elevated LDL Cholesterol

LDL is almost entirely made of cholesteryl esters and apolipoprotein (apo) B 100, and it serves as the main carrier of cholesterol to peripheral tissues, including the liver. Peripheral uptake of cholesterol is dependent on binding of LDL to its receptor. Currently, three monogenic disorders causing autosomal dominant hypercholesterolemia (ADH) have been identified, as well as one autosomal recessive form (**Table 66.1**). Loss of function variants in the LDL receptor gene (*LDLR*) are most common, while rare variants in other genes, i.e., binding defective apo B (*APOB* gene) and gain of function in proprotein convertase subtilisin/kexin 9 (*PCSK9* gene) account for a minority of cases presenting with ADH. In the lipid clinic, when patients are referred with elevated LDL cholesterol and suspected ADH, a pathogenic variant is found in about half of patients, while about one-third have polygenic hypercholesterolemia, a condition that results from the accumulated impact of dozens to hundreds of common DNA susceptibility variants, each of which only modestly raises LDL cholesterol individually.[10]

Familial Hypercholesterolemia (FH) Due to *LDLR* Variants

FH (OMIM # 143890) affects ~ 1:250 - 300 individuals, although the detection rate in the population is low in most countries.[11] FH most commonly results from single copy (heterozygous) pathogenic variants of the *LDLR* gene located on chromosome 19p13.[12] Variants in this gene underlie 80% to 90% of cases of ADH.[10] The transmembrane LDL receptor, present in almost all tissues, controls cholesterol homeostasis by a complex process that includes synthesis of the receptor in the endoplasmic reticulum, migration to the Golgi apparatus and then, to the cell surface, binding of the LDL receptor to LDL via apo B-100, internalization of the receptor-ligand complex, and recycling of the receptor to the cell surface and the LDL to the lysosome.[7]

More than 3,000 variants in *LDLR* have been reported so far,[13] including variants that affect each step in the LDL receptor cellular itinerary. When one allele of the LDL receptor is defective, a 30% increase in LDL production is observed. However, in homozygous (bi-allelic) FH, both complete and virtual absence of receptor function are observed, along with 2- to 3-fold increase in LDL production.[12] Several clinical algorithms permit a diagnosis of heterozygous FH, with a definite diagnosis now made most often by DNA sequencing of the *LDLR* gene.[14] Diagnostic studies of receptor function in fibroblasts are no longer performed.[14] If heterozygous FH is left untreated, myocardial infarction may occur at 30 to 40 years of age, with more than 50% and 15% of affected males and females, respectively, dying before age 60.[12]

In the more severe homozygous form of FH, individuals can develop cutaneous and tendinous xanthomas (a condition characterized by fatty deposits toward the skin surface) in the first decade of life. This condition may result in death from cardiac ischemia and aortic valve involvement as early as the second decade of life, and often before age 30; fortunately, this condition is extremely rare (~1 in 300,000).[12] Dietary management is usually insufficient for treating children with heterozygous FH and statins are recommended, based on elevated LDL cholesterol and the presence of risk factors, usually at the

Table 66.1 Monogenic Disorders Causing Elevated Low-Density Lipoprotein Cholesterol Levels

Disorder	Estimated Incidence	LDL Serum Levels	Clinical Findings	Gene Defect	Treatment
Heterozygous familial hypercholesterolemia (HeFH)	1:500[a]	Usually >200 mg/dL; can be lower in children	Tendon xanthomata (hallmark), xanthelasma, corneal arcus	Autosomal dominant (ADH) mutations in LDL receptor gene	Dietary treatment[b]; drug treatment[c]
Homozygous familial hypercholesterolemia (HoFH)	1 per million	LDL >400 mg/dL (average >600 mg/dL)	Planar, tendon, and tuberous xanthomata by age 6 y; death from coronary disease as early as 10 y age; irreversible aortic valve involvement by age 10 y if untreated	Mutations in LDL receptor gene in both alleles	Dietary treatment; drug treatment when some LDL receptor activity is present; LDL apheresis; liver transplantation
PCSK9 mutations	≤3% of cases with ADH[d]	Like HeFH	Like HeFH	ADH; gain of function mutations	Like HeFH
Familial defective Apo-B	≤7% of cases with ADH[d]	Like HeFH	Like HeFH	ADH; Apo-B gene mutations in LDL receptor binding domain	Like HeFH
Autosomal recessive hypercholesterolemia (ARH)	Few cases	Like HoFH; on average, ~100–150 mg/dL lower than HoFH	Like HoFH, with less aortic valve involvement and slower progression	Mutations in adaptor protein that is essential, in the liver, for clathrin-mediated endocytosis of LDL	Dietary treatment; response to statin therapy

ADH, autosomal dominant hypercholesterolemia; *Apo-B*, apolipoprotein B; *LDL*, low-density lipoprotein.

[a]Can be more frequent than 1:100 in various populations because of a founder effect.
[b]Restriction of dietary saturated fat and dietary cholesterol can reduce LDL serum levels but are insufficient to achieve normal values. They have an added effect with drug therapy.
[c]Statins are the main treatment. Combination of statins with ezetimibe further lowers LDL blood levels. Combination of statins with bile acid resins also has a synergistic effect.
[d]As reported in Rahalkar AR, Hegele RA. Mol Genet Metab 2008;93:282–94.

age of 10 years.[15] Often, a second medication, such as ezetimibe or an inhibitor of PCSK9, is required in patients with heterozygous FH. In homozygous FH patients, current recommendations include diet, multiple medications, LDL aphaeresis and novel biologic therapies, including monoclonal antibody treatments such as evinacumab; liver transplantation is now rarely used.[16]

Familial Defective apo B

Every LDL particle has one molecule of apo B that serves as the ligand for the LDL receptor.[17] Rare pathogenic variants within the *APOB* gene region that encode the LDL receptor binding domain result in reduced affinity for LDL particles and their accumulation in plasma. The resulting phenotype is called familial defective apo B (OMIM # 144010)[18] and is indistinguishable from FH, due to *LDLR* pathogenic variants.[19] By far the most prevalent *APOB* variant causing familial defective apo B is p.Arg3527Gln, which may account for 2% to 7% of all cases of autosomal dominant hypercholesterolemia in Europeans.[12] As with patients who have heterozygous FH from pathogenic variants in *LDLR*, diagnosis is currently most often made by DNA sequencing.[14] Treatment includes diet, statins and, often, a second agent such as ezetimibe or an inhibitor of PCSK9.[14]

Gain-of-Function Variants in PCSK9

PCSK9 is a serine protease that regulates the degradation of the LDL receptor and thus, plays a major role in the control of cholesterol influx into cells.[20] Loss-of-function variants result in increased LDL receptor expression and reduced LDL serum concentration, as well as reduced risk for cardiovascular disease.[21] However, patients with a gain of function variant in *PCSK9* have increased blood LDL cholesterol and similar clinical presentation as heterozygous FH (OMIM # 603776). As with patients who have heterozygous FH from pathogenic variants in *LDLR*, diagnosis of ADH from *PCSK9* gain-of-function is currently most often made by DNA sequencing.[14] As with familial defective apo B, treatment includes diet, statins, and, often, a second agent such as ezetimibe or an inhibitor of PCSK9.[14]

Autosomal Recessive Hypercholesterolemia (ARH)

This extremely rare form of homozygous FH (OMIM# 603813) was first described in Italian patients[22] but has since been reported in other geographic ancestries.[23] This disease results from loss-of-function variants in the LDL receptor adaptor protein 1, encoded by the *LDLRAP1* gene, which is essential for clathrin-mediated endocytosis of LDL in the liver.[23] Clinically, patients with ARH resemble patients with homozygous FH, and ASCVD end points appear to be just as severe and premature.[23] Treatment is identical to the common form of HoFH that results from *LDLR* pathogenic variants.[16]

High Plasma HDL Cholesterol (hyperalphalipoproteinemia)

Plasma HDL cholesterol concentration has heritability estimates of ~50%[8,24]. Among patients referred to specialty clinics with elevated HDL cholesterol, <15% will have a rare pathogenic variant or mutation in a gene, while 30% or more will have a polygenic basis for their high HDL cholesterol phenotype.[25] Extremely high HDL-C is defined as ≥1.4 mmol/L (54.1 mg/dL) in males or ≥1.8 mmol/l (69.6 mg/dL) in females.[25] These thresholds adhere closely to the top and bottom 10th percentiles of HDL-C concentration in a North American population.[26] The salient features of high HDL cholesterol disorders are summarized in **Table 66.2**.

Cholesterol Ester Transfer Protein Deficiency

By far, the best characterized genetic cause of high HDL cholesterol is the deficiency of cholesterol ester transfer protein (CETP). The primary function of CETP is to mediate the one-for-one exchange of a cholesteryl ester (CE) molecule in HDL with a triglyceride molecule in VLDL or LDL. In this way, some HDL-CE, presumably derived from peripheral cells, including foam cells in atherosclerotic plaques, can be directed to the liver indirectly via uptake of the VLDL and LDL by the LDL receptor. The remaining HDL-CE can be directly delivered to the liver through the interaction of HDL with the scavenger receptor SR-B1. The general term for the delivery (direct and indirect) of CE from peripheral cells to the liver is reverse cholesterol transport (RCT), and this property of HDL is thought to be a major contributor to its atheroprotective role in human observational and animal intervention studies.[27]

The relationship of CETP to plasma HDL cholesterol was initially defined in Japan, where CETP deficiency resulting from loss-of-function variants was found to explain >50% of high HDL cholesterol cases, with concentration as high as 120 mg/dL (3 mmol/L) in homozygotes and >70 mg/dL (>1.9 mmol/L) in heterozygotes.[27] In addition to increased HDL, cholesterol, the size of the HDL particles is larger in carriers of variant *CETP*, presumably due to the inability to transfer accumulated CE to VLDL or LDL. In addition to rare variants in *CETP*, common polymorphisms at this gene locus are associated with variation in HDL cholesterol, albeit much smaller in magnitude in GWAS observations.[6]

CETP had been considered an attractive target for the pharmaceutical industry, given that its inhibition can be easily achieved by small molecules, and that this inhibition raises HDL cholesterol. Nonetheless, the failure of several pharmacologic CETP inhibitors to decrease ASCVD risk[28] and mixed evidence for longevity in Japanese homozygotes has suggested that the HDL particles that accumulate in CETP deficiency may be dysfunctional, although several studies in vitro do not support this.[29]

Table 66.2 Genetic Causes of High Plasma Levels of High-Density Lipoprotein Cholesterol

Disorder	Incidence	Lipid Levels	Associated Physical Findings	Gene Defect	Dietary and Other Therapy
CETP deficiency[a]	Most common in Japan, where 2% and 7%, respectively, heterozygotic for mutations that cause complete or partial deficiency	Homozygotes: HDL-C levels usually >120 mg/dL; heterozygotes: HDL-C typically 70–100 mg/dL; in both cases, LDL-C levels may be slightly to modestly decreased	None noted	Complete deficiency most often caused by a splice-site mutation in intron 14; partial deficiency most often with a missense mutation (*D442G*) in exon 15	None at present
Hepatic lipase deficiency[b]	Rare (≤20 known kindreds)	HDL-C can be >70 mg/dL; TG 200–450 mg/dL with TG-rich HDL and β-VLDL particles	None noted	Homozygous defect in the hepatic lipase gene (LIPC)	One report noting that fenofibrate in two patients with complete deficiency substantially improved plasma lipid profile[c]

CETP, cholesterol ester transfer protein; *HDL*, high-density lipoprotein; *HDL-C*, high-density lipoprotein cholesterol; *LDL-C*, low-density lipoprotein cholesterol; *TG*, triglyceride; *VLDL*, very-low-density lipoprotein.

[a]Data from Weissglas-Volkov D, Pajukanta P. J Lipid Res 2010;51:2032–57.

[b]Data from Weissglas-Volkov D, Pajukanta P. J Lipid Res 2010;51:2032–57; Denke MA. Nutrient and genetic regulation of lipoprotein metabolism. In: Shils ME, Shike M, Ross AC et al, eds. Modern Nutrition in Health and Disease. 10th ed. Burlington, MA: Jones & Bartlett Learning, LLC, 2006.

[c]Data from Ruel IL, Lamarche B, Mauger JF et al. Arterioscler Thromb Vasc Biol 2005;25:2600–7.

Only one remaining CETP inhibitor remains under development by the pharmaceutical industry.

Hepatic Lipase Deficiency

Hepatic lipase (HL) has been most studied with triglyceride metabolism. For example, one of its important functions is to remodel VLDL remnants to LDL particles by hydrolyzing triglycerides. Not surprisingly, individuals with deficiency of HL have elevated plasma triglycerides and VLDL remnants (usually called IDL), but they also have elevated HDL triglycerides and HDL cholesterol, the latter of which can be >70 mg/dL (>1.9 mmol/L).[8] Similar to CETP deficiency, the HDL particles are frequently increased in patients with hepatic lipase deficiency. In GWAS studies, common variants of *LIPC* locus encoding hepatic lipase have been identified as associated with slight variations in plasma HDL cholesterol.[6]

Polygenic Predisposition

As noted above, hundreds of genetic loci have been associated with small deviations of plasma HDL cholesterol identified in GWAS. When these loci are combined into a metric called a polygenic score, studies in several cohorts showed that a disproportionate number of individuals with elevated HDL cholesterol have an extremely high polygenic score for this phenotype.[25] This indicates that many cases of elevated HDL cholesterol do not result from a variant in a single gene but rather develop because of the combined influences of multiple variants with individually small effects.[25] In addition to candidate genes that encode proteins, an exciting recent development is that HDL concentration in mice is also controlled by micro RNAs (miRs), most notably by miR-33,[30] that are also conserved in humans. Thus, it is possible that variation in the sequence or expression of miR-33 or other miRs will also be a genetic contributor to elevated HDL cholesterol.

Low Plasma HDL Cholesterol (hypoalphalipoproteinemia)

Monogenic disorders that result in very low plasma HDL cholesterol concentration have a wide range of clinical manifestations and have been variably associated with CVD risk (**Table 66.3**). Extremely low HDL-C is defined as ≤0.8 mmol/L (30.9 mg/dL) in males or ≤1.0 mmol/L (38.7 mg/dL) in females. Polygenic factors are also seen in about one-quarter of patients referred with low HDL cholesterol.[25] The altered gene products are each involved in the formation of HDL particles, including ATP-binding cassette protein A1 (ABCA1), lecithin:cholesterol acyltransferase (LCAT) and apolipoprotein (apo) A-I. Currently available HDL-raising drugs are often not effective for patients with these monogenic disorders.[31] Homozygous or biallelic

Table 66.3 Genetic Causes of Low High-Density Lipoprotein Cholesterol

Disorder	Incidence	Lipid Levels	Associated Physical Findings	Gene Defect	Dietary and Other Therapy
Tangier disease	<100 kindreds	Total cholesterol low to normal; TG normal to moderately elevated; HDL cholesterol <10 mg/dL	Corneal opacities; yellow-orange tonsils; hepatosplenomegaly; cardiovascular disease risk probably increased	Homozygous defect in *ABCA1* gene preventing formation of HDL	LDL cholesterol levels low, but clearance of TG-rich lipoproteins defective; low-fat diets recommended
Familial LCAT deficiency	<40 kindreds	Total cholesterol normal; TG normal to moderately elevated; HDL cholesterol <12 mg/dL	Corneal opacities; normochromic normocytic anemia; proteinuria with increased risk for renal failure by age 40–50 y; corneal opacities; cardiovascular disease risk generally not increased	Homozygous defect in *LCAT* gene leading to absent or markedly reduced LCAT enzyme	Low-fat diet likely prudent; possibly medication to lower TG and LDL cholesterol
Fisheye disease (partial LCAT deficiency)	<15 kindreds	Total cholesterol normal; TG normal to moderately elevated; HDL cholesterol <7 mg/dL	Corneal opacities; cardiovascular disease risk generally not increased	Homozygous defect in *LCAT* leading to normal measurable LCAT levels but ineffective ability to esterify free cholesterol in HDL	Low-fat diet likely prudent; possibly medication to lower TG and LDL cholesterol
Apo-A-I deficiency	<20 kindreds	Total cholesterol and TG normal; HDL cholesterol usually <10 mg/dL	Corneal opacities; diffuse palmar xanthomata in some kindreds; cardiovascular disease risk increased in some kindreds	Multiple distinct homozygous mutations in *APOA1* gene leading to reduced Apo-A-I levels	Low-fat diet likely prudent; possibly medication to lower TG and LDL cholesterol

ABC, adenosine triphosphate–binding cassette; *Apo*, apolipoprotein; *HDL*, high-density lipoprotein; *LCAT*, lecithin:cholesterol acyltransferase; *LDL*, low-density lipoprotein; *TG*, triglyceride.

loss-of-function variants in these genes cause the full-blown syndromes discussed below with extremely low or absent HDL cholesterol, while a single copy (heterozygous) of such a variant causes low but still detectable HDL cholesterol concentration, without systemic features.[14] Thus, a reasonable treatment strategy is to reduce plasma LDL cholesterol using diet and pharmacotherapy, as appropriate.[32]

Tangier Disease

Tangier disease is characterized by a near total absence of HDL particles from the plasma.[33] The molecular lesion in Tangier disease lies in the *ABCA1* gene, which encodes a cell membrane protein that governs cholesterol efflux,[34-36] and is a key component of the RCT pathway.[37] In the absence of functional ABCA1, plasma HDL concentration rarely exceeds 10 mg/dL (0.25 mmol/L) and often approaches zero, as apo A-I fails to bind and remove cholesterol from peripheral cells.[38] Tangier disease classically presents during adolescence or young adulthood with enlarged orange or yellow tonsils and adenoids, hepatosplenomegaly, and peripheral neuropathy that may be either transient or progressive and debilitating.[33,38,39] Accumulation of cholesterol in macrophages results in foam cell formation in the spleen, liver, intestinal epithelium, bone marrow, and other parts of the reticuloendothelial system. Cholesterol also becomes deposited in fibroblasts, neurons, and Schwann cells.[33,38] CVD risk in Tangier disease patients and carrier relatives appears to be increased, although this remains controversial.[40,41]

Deficiencies of LCAT

LCAT deficiency can produce two distinct clinical syndromes, depending on the location of the variant in the *LCAT* gene.[42-44] The first of these, familial LCAT deficiency, is characterized by complete absence of enzyme activity and has a range of clinical features including corneal opacities, dyslipidemia, anemia, mild hypoalbuminuria, and proteinuria, with lipid deposits in renal mesangium, interstitium, and glomeruli. The second of these, fisheye disease, is characterized by a partial deficiency of LCAT activity, and is mainly associated with later-onset corneal opacities arising from deposition of LCAT substrates, such as phospholipids and free cholesterol. In both LCAT deficiency states, HDL is low but detectable and CVD risk is not clearly increased.[45]

Apo A-I Deficiency

Pathogenic variants of the *APOA1* gene that affect the structure of apo A-I are associated with extremely low plasma HDL cholesterol (<10 mg/dL or <0.25 mmol/L). Clinical presentation is variable and includes corneal clouding, tendon or skin-fold xanthomata and, in some kindreds, premature CVD.[44-46]

High Plasma Triglycerides

Lipoprotein triglyceride (TG) comes from two sources: exogenous—from dietary fat—and endogenous—produced by the liver or released from adipose tissue. The most dramatic form of hypertriglyceridemia is chylomicronemia, which refers to pathologic accumulation of intestinally derived chylomicron particles.[47-51] Chylomicronemia syndrome refers to the presence of clinical features in patients with chylomicronemia, including failure to thrive in infancy, nausea and vomiting, abdominal pain, eruptive xanthomas on the trunk and limbs, lipemia retinalis, hepatosplenomegaly and, less commonly, anemia, intestinal bleeding, brain fog and encephalopathy.[47-51] Typically, chylomicrons are present when fasting TG >880 mg/dL (>10 mmol/L),[49] although pancreatitis usually occurs when TG >1760 mg/dL (>20 mmol/L).[50] Plasma appears milky and turbid, or lipemic, due to its high TG content, and other lipoprotein concentrations vary based on etiology.[51] Dietary fat restriction is the mainstay of treatment, although biologic therapies are in development.[48]

Chylomicronemia: Monogenic and Polygenic Forms

Genetic causes of high plasma TG are listed in **Table 66.4**.

The two main clinical presentations of chylomicronemia are familial chylomicronemia syndrome (FCS) and multifactorial chylomicronemia (MCS). FCS (formerly called type 1 hyperlipoproteinemia) accounts for <5% of all cases of severe hypertriglyceridemia and has an estimated prevalence of 1 in 100,000 to 1,000,000.[50-53] FCS develops in childhood, adolescence, or early adulthood, typically in the absence of secondary factors.[50,51] In FCS, lipolytic processing is genetically absent, leading to accumulation of chylomicrons in combination with deficiencies of other lipoproteins, such as VLDL, LDL, and HDL.[50,51] In contrast, MCS (formerly called type 5 hyperlipoproteinemia) is up to 100-times more common than FCS.[53-55] Other lipoproteins, particularly VLDL and remnants, are elevated in MCS,[50,51] which present later in life, often in the presence of secondary factors, such as obesity, poor diet, alcohol use, diabetes, hepatic disease, renal disease, autoimmune disease, and certain medications, such as corticosteroids, immunosuppressants, or antiviral agents.[52-55] The lifetime risk of pancreatitis is higher in FCS than MCS patients, but the latter still run a much greater risk of this complication than the general population.[50,53-55]

FCS is a Mendelian recessive disorder resulting from biallelic rare pathogenic loss-of-function variants in one of five genes, namely: *LPL* (in 60%–80% of cases), *APOC2* encoding apo C-II, which is a cofactor for the activation of LPL,[53] *APOA5* encoding apo A-V, which is thought to stabilize LPL-mediated hydrolysis[53]; and *GPIHBP1* encoding glycosylphosphatidylinositol-anchored HDL-binding protein,[56] which mediates transcytosis of LPL to the capillary surface and *LMF1* encoding lipase maturation factor 1,[57] which is important for the proper folding and assembly of LPL (Table 66.4). By comparison, the much more common MCS is a non-Mendelian condition associated with either: 1) heterozygous pathogenic variants in one of the above genes in 15% to 20% of patients[54,58]; and/or 2) polygenic risk in 30% to 50% of patients.[54,58] The presence or absence of either genetic factor in MCS guarantees neither development of nor protection from severe hypertriglyceridemia, respectively.

In the past, diagnosis of FCS required biochemical assay of lipolytic activity in post-heparin plasma, but today, genomic DNA sequence analysis has become the standard method for diagnosis. DNA sequencing has shown that the huge preponderance of patients with chylomicronemia do not have FCS, but rather MCS.[14,53]

It should also be noted that there is an ultra-rare condition, transient infantile hypertriglyceridemia, caused by biallelic pathogenic variants in *GPD1* (encoding glycerol-3-phosphate

Table 66.4 Monogenetic Causes of High Triglycerides with Fasting Chylomicronemia in Familial Chylomicronemia Syndrome (FCS)

Disorder	Estimated Incidence	Lipid Levels	Associated Findings	Gene Defect	Dietary and Other Therapy
Homozygous familial lipoprotein lipase deficiency	1 in 1 million	TG ≥1,500–4,500 mg/dL	Presents in childhood, often in first year of life; eruptive xanthomata, hepatomegaly, and splenomegaly from TG accumulation in organs; all signs regress with dietary therapy; normal life span if dietary therapy initiated	Homozygous mutations in *LPL* gene causing markedly reduced or absent enzyme activity	Restrict dietary fat to <20 g/d, to provide maintenance of TG levels at <2,000 mg/dL; supplemental vitamin E not required because Apo-B–containing lipoproteins are present; gene therapy using virally mediated transfer of the *LPL* gene in early-phase clinical trials
Apolipoprotein-C-II deficiency	<20 kindreds described	TG ≥1,500–4,500 mg/dL	Presents later in life than homozygous familial lipoprotein lipase deficiency (age 13–60 y); no association with premature atherosclerosis	Homozygous mutations in *APOC2* gene leading to absent or nonfunctional Apo-C-II	Restrict dietary fat to <20 g/d; supplemental vitamin E not required because Apo-B–containing lipoproteins are present; patients can respond to transfusion of plasma from physiologically normal people to supply Apo-C-II and temporarily correct the abnormality
Glycosyl-phosphatidyl-inositol-anchored HDL-binding protein deficiency	<5 kindreds described	TG ≤6,000 mg/dL	Presents later in life	Homozygous mutations in *GPIHBP1* gene	None specifically determined; likely same as for lipoprotein lipase deficiency
Apo-A-V deficiency	<5 kindreds described	TG ≤5,000 mg/dL	Presents later in life	Homozygous mutations in *APOA5* gene	None specifically determined; likely same as for lipoprotein lipase deficiency
Lipase maturation factor-1 deficiency	<5 kindreds described	TG ≤5,000 mg/dL	Presents later in life; neurologic findings in some kindred.	Homozygous mutations in *LMF1* gene	None specifically determined; likely same as for lipoprotein lipase deficiency

APO, apolipoprotein; *HDL*, high-density lipoprotein; *TG*, triglyceride.

dehydrogenase), that manifests at birth with extreme hypertriglyceridemia, and which becomes less severe later in life.[59]

Genetic Causes of Hypertriglyceridemia Without Fasting Chylomicronemia

Several genetic causes of this condition are listed in **Table 66.5**.

Mild-to-moderate hypertriglyceridemia due to primarily elevated VLDL is commonly seen in the clinic; it affects 1 in 20 adults in the population.[53] Mild-to-moderate hypertriglyceridemia is not a single gene disorder like FCS; instead, it is a complex polygenic condition like MCS, but with a lower total genetic burden of common variants.[60] In the past, the term "familial hypertriglyceridemia" was used to define this condition by analogy with "familial hypercholesterolemia," but DNA sequencing and family studies have shown that this was a misnomer, and we suggest that the term should be discarded. The complex genetic architecture in mild-to-moderate hypertriglyceridemia suggests that an individual who carries an excess of specific common and rare susceptibility variants is more likely—in the context of adverse secondary factors, such as obesity, poor diet, alcohol intake, poorly controlled diabetes, and hypothyroidism—to develop common hypertriglyceridemia. Treatment includes management of secondary factors contributing to the trait, together with improved diet.

Familial combined hyperlipidemia (FCH) presents with lipoprotein abnormalities, including increased VLDL, LDL, and apo B, with depressed HDL, associated with an abnormal lipoprotein profile in at least one first-degree relative.[47,61] This is a relatively common phenotype that affects ~1:40 people. FCH patients may sometimes have xanthomas and increased CVD risk. Recent findings suggest that FCH represents a spectrum of disorders for which a range of common and rare genetic variants contribute to susceptibility.[61]

Familial dysbetalipoproteinemia (hyperlipoproteinemia type 3) has a population prevalence of ~1 in 10,000.[47] The main lipoprotein abnormality is an increase in TG-rich lipoprotein remnants, also known as intermediate-density lipoproteins (IDL) or β-VLDL. Affected people often have tuberous or tuberoeruptive xanthomas on the extensor surfaces of their

Table 66.5 Genetic Causes of High Triglycerides Without Fasting Chylomicronemia

Disorder	Estimated Incidence	Lipid Levels	Associated Findings	Gene Defect	Dietary and Other Therapy
Familial combined hyperlipidemia (hyperlipoproteinemia type 2B)	1 in 40	Total cholesterol ≤400 mg/dL; LDL cholesterol ≤320 mg/dL; TG levels ≤800 mg/dL; HDL cholesterol 30–45 mg/dL	Tendon xanthomata, xanthelasmas; early cardiovascular disease a common feature	Complex trait, cumulative susceptibility of small-effect common alleles and rare mutations of multiple genes	Restrictions in dietary fat, reduction of high glycemic index food, cessation of alcohol; higher doses or combinations of drug treatment required to lower both LDL cholesterol and TG
Dysbetalipoproteinemia (hyperlipoproteinemia type 3)	1 in 10,000	Total cholesterol average 450 mg/dL with TG 700 mg/dL; direct measurements of LDL disproportionately low (e.g., 120 mg/dL); this LDL includes IDL particles; most TG-rich lipoproteins are β-VLDL	Palmar xanthomata, tuberous and tuberoeruptive xanthomata; premature atherosclerosis	Cumulative susceptibility of small- effect common alleles of multiple genes as above concurrent with homozygosity for *APOE* E2/E2 isoform; several mutations in the *APOE* gene can produce a dominantly inherited phenotype	Weight loss can cause remission of the overt lipoprotein disorder; high-fat meals worsen dyslipidemia; reduced-fat diet reduces chylomicron production, improving fasting lipids; control of diabetes or hypothyroidism improves the dyslipidemia
Familial hypertriglyceridemia (hyperlipoproteinemia type 4)	1 in 20	TG levels ≤800 mg/dL and HDL cholesterol levels 30–45 mg/dL; some have near normal levels	None	Complex trait, cumulative susceptibility of small-effect common alleles and rare mutations of multiple genes, including *LPL* and *APOA5*, occurring together in the same patient	Not formally studied; restrictions in dietary fat, reduction of high glycemic index food, cessation of alcohol, and avoidance of high-fat splurges expected to reduce TG levels

APO, apolipoprotein; *HDL*, high-density lipoprotein; *IDL*, intermediate-density lipoprotein; *LDL*, low-density lipoprotein; *TG*, triglyceride; *VLDL*, very-low-density lipoprotein.

extremities, plantar or palmar crease xanthomas, and increased CVD risk. People with this disorder typically are homozygous for the binding defective *APOE* E2 isoform. In addition, a range of common and rare genetic variants contribute to susceptibility to this disorder.[47,62] Disease expression often requires accompanying factors such as obesity, type 2 diabetes, or hypothyroidism. Treatment includes low fat, low glycemic index diet, usually together with statins.

Low Plasma Cholesterol or Triglycerides

As with the hypercholesterolemia states, severe rare monogenic conditions are associated with hypolipidemia.[63] ASCVD risk is reduced in individuals who carry these variants.[64] However, in the most severe forms, adverse phenotypes such as fat malabsorption and fat-soluble vitamin deficiencies require therapeutic dietary interventions. Genetic conditions associated with low plasma TG are listed in **Table 66.6**.

Abetalipoproteinemia (ABL)

This ultra-rare disorder is an autosomal recessive disease that results from variants in the *MTP* gene encoding the VLDL assembly factor, microsomal triglyceride transfer protein (MTP). In the homozygous state, very little apo B in either the liver or intestine can be lipidated in the endoplasmic reticulum. The poorly folded protein is directed to the proteasome pathway for degradation.[65] Thus, few apo B-containing lipoproteins can be assembled and secreted, resulting in low plasma chylomicrons, VLDL, and LDL. The patients also have failure to thrive due to fat malabsorption, which results from the failure to form chylomicrons. Because transport of fat-soluble vitamins depends on intact assembly and secretion of apo B-containing lipoproteins, symptomatology follows from prolonged vitamin deficiency. This includes: 1) atypical retinitis pigmentosa and visual impairment from vitamin A deficiency; 2) osteomalacia, rickets, and/or osteoporosis from vitamin D deficiency; 3) sensory loss and ataxia from vitamin E deficiency; and 4) bleeding disorder from vitamin K deficiency. Also, the red blood cells in ABL have a characteristic deformity referred to as "acanthocytosis," which,

Table 66.6 Low Plasma Levels of Cholesterol or Triglycerides

Disorder	Estimated Incidence	Lipid Levels	Associated Findings	Gene Defect	Dietary and Other Therapy
Abetalipoproteinemia (ABL)	<120 cases described	Total cholesterol and TG <50 mg/dL	Fat malabsorption, including vitamin E, deficiency of which leads to ataxia, sensory loss, and retinitis pigmentosa; obligate heterozygotes have no biochemical or clinical phenotype	Autosomal recessive with loss of function mutations in the *MTP* gene, whose product is required for Apo-B lipoprotein assembly in the intestine and liver	Fat-reduced diet or substitute medium-chain TG; vitamin E supplementation
Familial hypobetalipoproteinemia (FHBL)	For the forms related to the *APOB* gene: homozygous form very rare; heterozygous 1:3,000; also rare are three other forms not linked to the APOB gene	Patients homozygous for truncating mutations of Apo-B have lipid levels similar to ABL; for heterozygotes and those with the non–Apo-B forms, milder (~50%–70%) reductions in plasma Apo-B or LDL-C noted	For homozygotes with Apo-B truncations, same as in ABL; some compound heterozygotes may also have less severe fat malabsorption. Hepatic steatosis may also be present in both homozygotes and heterozygotes; Obligate heterozygotes have reduced plasma LDL and Apo-B; Most heterozygotes are generally asymptomatic	For the *APOB* gene–associated forms, truncating protein mutations. For the other forms, there is 1) linkage to chromosome locus 3p21; 2) loss of function mutation of *PCSK9*; or, 3) a familial form that remains unassociated with a gene or chromosomal locus	For the homozygotes with Apo-B truncations, same as for ABL; some heterozygotes also need vitamin E supplementation and dietary fat restriction; no therapies currently recommended for the other forms
Familial combined hypolipidemia[a]	Rare (a few families for each variant)	Heterozygotes: TG <65 mg/dL, LDL-C <75 mg/dL, HDL-C normal; compound heterozygotes: TG <25 mg/dL, LDL-C <35, HDL-C <20 (based on data for patients with ANGPTL3 mutations)	None	Nonsense mutation in *ANGPTL3*	None described

APOB, apolipoprotein-B; HDL-C, high-density lipoprotein cholesterol; LDL-C, low-density lipoprotein cholesterol; TG, triglyceride.

[a]Data from Romeo S, Yin W, Kozlitina J et al. J Clin Invest 2009;119:70–9; and Musunuru K, Pirruccello JP, Do R et al. N Engl J Med 2010;363:2220–7.

Other data from Denke MA. Nutrient and genetic regulation of lipoprotein metabolism. In: Shils ME, Shike M, Ross AC et al, eds. Modern Nutrition in Health and Disease. 10th ed. Burlington: Jones & Bartlett Learning, LLC, 2006; and Schonfeld G, Lin X, Yue P. Cell Mol Life Sci 2005;62:1372–8.

together with low LDL cholesterol, is pathognomonic. Once the diagnosis of ABL is made, the clinical problems can be arrested and ameliorated by administration of water-soluble forms of vitamin E together with high oral doses of other fat-soluble vitamins, which are absorbed through the medium-chain triglyceride pathway into the portal circulation. In the heterozygous state, such as obligate heterozygote parents, plasma apo B-containing lipoproteins are essentially normal and there is complete absence of the spectrum of clinical features seen in homozygotes.[63]

Familial Hypobetalipoproteinemia (FHBL)

This disorder most often results from pathogenic variants in the *APOB* gene encoding apo B and affecting structural integrity of the protein. Heterozygotes have low plasma concentrations (<5th percentile) of both LDL cholesterol or apo B, while homozygotes have a phenotype that is indistinguishable from ABL.[63] The main differentiating feature is that obligate heterozygote parents of a homozygous FHBL child have half-normal LDL cholesterol and apo B concentrations.[63]

FHBL is caused by pathogenic variants in *APOB*, often a nonsense variant resulting in a premature stop codon, although numerous missense variants have been reported more recently. In contrast to the *APOB* variants in the receptor-binding domain that cause familial defective apo B in FH, the FHBL variants in *APOB* result in the production of carboxy-truncated forms of apo B that can vary in length from 2% to 89% of normal of the full length hepatically produced apo B-100. The intestine normally produces a shorter apo B isoform that arises from editing of the *APOB* mRNA, called apo B-48 that is 48% of the hepatic protein. Depending on the position of the truncation variant, there is reduced production of apo B as well as increased clearance from the plasma of lipoproteins containing the truncated species. Interestingly, the defective allele exerts a negative effect on the production of apo B encoded by the normal allele, giving rise to the dominant nature of the defect. The heterozygous state is found in 1:1000 to 3,000 individuals, with the homozygous state being exceedingly rare, probably as rare as ABL. Simple heterozygotes have low plasma total and LDL cholesterol, as well as reduced triglyceride concentration, and are usually asymptomatic although they are protected from ASCVD.[64] Some are reported to have fatty liver.

There are also at least two additional non-*APOB* gene related forms of FHBL (Table 66.6).

PCSK9 Deficiency

In contrast to the very high LDL cholesterol seen with gain-of-function variants in *PCSK9*, heterozygous loss-of-function variants in this gene result in increased concentration of the LDL receptor and enhanced clearance of LDL particles. Heterozygotes for loss-of-function variants in *PCSK9* have markedly depressed LDL cholesterol and apo B, and also have markedly reduced lifetime risk of CAD.[21,64] Only a handful of homozygotes for loss-of-function variants in *PCSK9* variants have been reported, and their main feature is biochemical with very low plasma but not absent LDL cholesterol and apo B, without any of the multisystem manifestations of ABL or homozygous FHBL.

Familial Combined Hypolipidemia

Previous studies had linked rare loss-of-function variants in angiopoietin-like proteins (ANGPTLs; particularly 3 and 4) to low plasma triglycerides. In an exome-sequencing approach in which members of one family with inherited hypobetalipoproteinemia, but no variants in *MTP*, *APOB*, or *PCSK9* were analyzed, it was found that nonsense variants in *ANGPTL3* were associated with low plasma LDL cholesterol and triglycerides in simple heterozygotes, and extremely low plasma LDL cholesterol, HDL cholesterol and triglycerides in compound heterozygotes.[66] There were no other clinical features in these patients. From preclinical studies, a potential mechanism contributing to these changes was proposed to be the loss-of-function of ANGPTL3, an inhibitor of lipoprotein and endothelial lipases, resulting in greater remodeling of apo B-containing and HDL lipoproteins. No specific treatment is needed in familial combined hypolipidemia.

Future Directions

Most areas of human biology and medicine have been affected by the recent progress resulting from the human genome project and related initiatives. Lipoprotein metabolism is no exception—the field has already benefited and will continue to benefit from progress in the genomic and postgenomic research era.

Targeted Therapies

The identification of causal genes for rare extreme human dyslipidemia syndromes has already been translated into targeted drug therapies that are widely available for clinical use. For instance, inhibitors of PCSK9, such as evolocumab, alirocumab, and inclisiran, are now routinely prescribed to reduce elevated LDL cholesterol, particularly in FH patients.[14] Mapping the causal gene for familial combined hypolipidemia has resulted in development of a biological treatment—the anti-ANGPTL3 antibody evinacumab—that is among the most potent agents available for hyperlipidemia.[14,48] Also, drugs that target apo C-III, such as olezarsen and plozasiran, seem to be potentially useful for treatment of severe hypertriglyceridemia.[14,48] Based on these examples, it is likely that other agents will be developed following a similar paradigm.

Genetic Risk Prediction of Dyslipidemia and Atherosclerosis

Early identification of individuals at risk for developing dyslipidemia can provide an opportunity for early lifestyle modification or evidence-based pharmacologic interventions capable of decreasing the prolonged burden of exposure to a suboptimal lipid profile and other risk factors. It is now becoming feasible to integrate all relevant genetic risk variants to determine a patient's overall polygenic score for specific dyslipidemias and/or atherosclerosis risk.[14] Genetic variables may improve risk determinations derived from existing algorithms, such as the Framingham risk score.

Next-Generation DNA Sequencing

Next-generation sequencing of whole exomes (i.e., all coding regions) or whole genomes will generate extensive new information about inter-individual DNA differences. Accounting for both common and rare deleterious or protective alleles will help to more accurately assign risk for development of dyslipidemia or atherosclerosis and may help with identification of subgroups of patients who are more likely to respond to particular drug interventions—an active area of study called "pharmacogenomics." Genetic information is also being included as a covariate in studies of lipoprotein responsiveness to dietary interventions in the evolving area of "nutrigenomics." Tailored dietary advice—in addition to other lifestyle interventions—may one day be offered to dyslipidemic patients, based on their genetic profile.

Challenges and Opportunities Arising from Emerging Genomic Technologies

The need for a mechanistic understanding of how new genes found by next-generation sequencing cause deviations in plasma lipoproteins will challenge our experimental capacity, but it is essential to develop new approaches to rapidly understand gene function ("functional genomics"). Also, there is the potential for unforeseen ethical, legal, and social issues that may arise when complete human genomic information becomes part of a patient's medical record. Taking full advantage of the new opportunities that have arisen through the discoveries from genetic studies will require analogous technologic advances allowing high throughput, reliable and robust functional validation at all stages: in vitro, in vivo, in nonhuman species and humans, and ultimately clinical trials of diet and other therapies.

References

1. Iqbal R, Anand S, Ounpuu S, et al. Circ 2008;118:1929–37.
2. Ahrens EH, Jr., Insull W, Jr., Blomstrand R, et al. Lancet 1957; 272:943–53.
3. Hegsted DM, McGandy RB, Myers ML, et al. Am J Clin Nutr 1965;17:281–95.
4. Anderson JT, Grande F, Keys A. Am J Clin Nutr 1976;29:1184–9.
5. Hegsted DM, Kritchevsky D. Am J Clin Nutr 1997;65:1893–6.
6. Teslovich TM, Musunuru K, Smith AV, et al. Nature 2010; 466:707–13.
7. Ference BA, Ginsberg HN, Graham I, et al. Eur Heart J 2017;38:2459–72.
8. Jacobs DR, Jr., Woo JG, Sinaiko AR, et al. N Engl J Med 2022;386:1877–88.
9. Jacob E, Hegele RA. Curr Atheroscler Rep 2022;24:419–26.
10. Wang J, Dron JS, Ban MR, et al. Arterioscler Thromb Vasc Biol 2016;36:2439–45.
11. Daniels SR, Greer FR. Pediatr 2008;122:198–208.
12. Defesche JC, Gidding SS, Harada-Shiba M, et al. Nat Rev Dis Primers 2017;3:17093.
13. Alves AC, Chora JR, Bourbon M. Curr Opin Lipidol 2019;30: 148–50.
14. Berberich AJ, Hegele RA. Endocr Rev 2022;43:611–53.
15. Khoury M, Bigras JL, Cummings EA, et al. Can J Cardiol 2022;38:1168–79.
16. Cuchel M, Raal FJ, Hegele RA, et al. Eur Heart J 2023:44: 2277–91.
17. Sniderman A, Langlois M, Cobbaert C. Curr Opin Lipidol 2021; 32:226–30.
18. Innerarity TL, Weisgraber KH, Arnold KS, et al. Proc Natl Acad Sci U S A 1987;84:6919–23.
19. Defesche JC, Pricker KL, Hayden MR, et al. Arch Intern Med 1987;153:2349–56.
20. Lambert G. Curr Opin Lipidol 2007;18:304–9.
21. Cohen JC, Boerwinkle E, Mosley TH, Jr., et al. N Engl J Med 2007;354:1264–72.
22. Garcia CK, Wilund K, Arca M, et al. Science 2001;292:1394–8.
23. D'Erasmo L, Di Costanzo A, Arca M. Curr Opin Lipidol 2020;31:56–61.
24. Weissglas-Volkov D, Pajukanta P. J Lipid Res 2010;51:2032–57.
25. Dron JS, Wang J, Low-Kam C, et al. J Lipid Res 2017;58: 2162–70.
26. Rifkind BM, Segal P. JAMA 1983;250:1869–72.
27. Lewis GF, Rader DJ. Circ Res 2005;96:1221–32.
28. Di Bartolo BA, Duong M, Nicholls SJ. Curr Opin Lipidol 2016;27:545–9.
29. Matsuura F, Wang N, Chen W, et al. J Clin Invest 2006;116: 1435–42.
30. Rayner KJ, Suarez Y, Davalos A, et al. Science 2010;328: 1570–3.
31. Franceschini G, Werba JP, D'Acquarica AL, et al. Clin Pharmacol Ther 1995;57:434–40.
32. Pearson GJ, Thanassoulis G, Anderson TJ, et al. Can J Cardiol 2021;37:1129–50.
33. Assmann G, von Eckardstein A, Brewer HB. Scriver CR, Beaudet AL, Sly WS, and Valle D, eds. Familial Analphalipoproteinemia: Tangier Disease. The Metabolic & Molecular Bases of Inherited Disease New York, NY: McGraw-Hill, 2001; 2937–60.
34. Bodzioch M, Orso E, Klucken J, et al. Nat Genet 1999;22: 347–51.
35. Brooks-Wilson A, Marcil M, Clee SM, et al. Nat Genet 1999;22: 336–45.
36. Rust S, Rosier M, Funke H, et al. Nat Genet 1999;22:352–5.
37. Vedhachalam C, Duong PT, Nickel M, et al. J Biol Chem 2007;282:25123–30.
38. Nofer JR, Remaley AT. Cell Mol Life Sci 2005;62:2150–60.
39. Fredrickson DS. J Clin Invest 1964;43:228–36.
40. Serfaty-Lacrosniere C, Civeira F, Lanzberg A, et al. Atherosclerosis 1994;107:85–98.
41. Soumian S, Albrecht C, Davies AH, et al. Vasc Med 2005;10: 109–19.
42. Kuivenhoven JA, Pritchard H, Hill J, et al. J Lipid Res 1997;38: 191–205.
43. Gigante M, Ranieri E, Cerullo G, et al. J Nephrol 2006;19: 375–81.
44. von Eckardstein A. Atherosclerosis 2006;186:231–9.
45. Geller AS, Polisecki EY, Diffenderfer MR, et al. J Lipid Res 2018; 59:2421–35.
46. Zanoni P, von Eckardstein A. Curr Opin Lipidol 2020;31: 62–70.
47. Hegele RA. Nat Rev Genet 2009;10:109–21.
48. Shamsudeen I, Hegele RA. Expert Rev Clin Pharmacol 2022;15: 395–405.
49. Yuan G, Al-Shali KZ, Hegele RA. CMAJ 2007;176:1113–20.
50. Brahm AJ, Hegele RA. Nat Rev Endocrinol 2015;11:352–62.
51. Laufs U, Parhofer KG, Ginsberg HN, et al. Eur Heart J 2020;41: 99c–109c.
52. Baass A, Paquette M, Bernard S, et al. J Intern Med 2020;287: 340–8.
53. Dron JS, Hegele RA. Front Endocrinol (Lausanne) 2020;11:455.
54. Paquette M, Bernard S, Hegele RA, et al. Atherosclerosis 2019; 283:137–42.
55. Hegele RA, Berberich AJ, Ban MR, et al. J Clin Lipidol 2018;12: 920–7 e4.
56. Beigneux AP, Franssen R, Bensadoun A, et al. Arterioscler Thromb Vasc Biol 2009;29:956–62.
57. Peterfy M, Ben-Zeev O, Mao HZ, et al. Nat Genet 2007;39: 1483–7.
58. Dron JS, Wang J, Cao H, et al. J Clin Lipidol 2019;13:80–8.

59. Basel-Vanagaite L, Zevit N, Har Zahav A, et al. Am J Hum Genet 2012;90:49–60.
60. Dron JS, Wang J, McIntyre AD, et al. J Clin Lipidol 2020;14: 28–34 e2.
61. Gill PK, Dron JS, Berberich AJ, et al. J Clin Lipidol 2021;15: 79–87.
62. Johansen CT, Wang J, Lanktree MB, et al. Arterioscler Thromb Vasc Biol 2011;31:1916–26.
63. Gill PK, Hegele RA. Expert Rev Endocrinol Metab 2023;18: 241–53.
64. Dron JS, Patel AP, Zhang Y, et al. JAMA Cardiol 2023;8:258–67.
65. Fisher EA, Ginsberg HN. J Biol Chem 2002;277:17377–80.
66. Musunuru K, Pirruccello JP, Do R, et al. N Engl J Med 2010;363:2220–7.

B. CARDIOVASCULAR DISEASES

CHAPTER 67

Nutrition in the Prevention of Coronary Heart Disease and the Management of Lipoprotein Disorders

Ernst J. Schaefer

CHAPTER OUTLINE

Abbreviations

ACC, American College of Cardiology, **AHA,** American Heart Association, **Apo**, apolipoprotein; **CHD**, coronary heart disease; **CVD**, cardiovascular disease; **DHA**, docosahexaenoic acid; **EPA**, eicosapentaenoic acid; **HDL-C**, high-density lipoprotein cholesterol; **hs-CRP**, C-reactive protein; **Lp(a),** lipoprotein(a), **LDL-C**, low-density lipoprotein cholesterol; **MI**, myocardial infarction; **NT-proBNP**, N-terminal pro-brain natriuretic peptide, **PCSK9**, proprotein convertase subtilisin kexin 9, **sdLDL-C**, small dense LDL-C, **VLDL**, very-low-density lipoprotein, and **WHO,** World Health Organization

Introduction

Cardiovascular disease (CVD) remains the leading cause of death worldwide. Major CVD risk factors include increased age, male sex, hypertension, hypertension treatment, diabetes, smoking, elevated total cholesterol due to increased low density lipoprotein cholesterol (LDL-C) and decreased high-density lipoprotein cholesterol (HDL-C). The American Heart Association (AHA) CVD risk algorithm using these parameters can be accessed at: https://static.heart.org/riskcalc/app/index.html#!/baseline-risk. The cornerstone of CVD prevention is lifestyle modification with diets low in saturated fat (<10% of energy), trans fats, cholesterol, and sugars (<10% of energy), and enriched in plant foods, vegetable oils, and omega-3 fatty acids, salt restriction in those with hypertension, and regular exercise of ≥150 minutes/week. In addition, control of modifiable risk factors with lifestyle modification and, if necessary, with statins for elevated LDL-C, angiotensin-converting enzyme inhibitors for hypertension, and metformin for diabetes, especially in those with established CVD or at high risk for CVD (≥7.5% 10-year risk). Optimal values for CVD prevention are: LDL-C <70 mg/dL, blood pressure <120/80 mmHg, glycosylated hemoglobin <6.5%, and high sensitivity C reactive protein <1.0 mg/L. For additional LDL-C lowering in patients with marked LDL-C elevation in the absence of secondary causes, ezetimibe and proprotein convertase subtilisin kexin 9 (PCSK9) inhibitors can be used along with statin therapy. In this chapter, the evidence for such recommendations is reviewed, as is the diagnosis and management of lipoprotein disorders.

Cardiovascular Disease Pathogenesis and Prediction

The Magnitude of the Problem

According to the World Health Organization (WHO), CVD is the leading cause of death globally.[1] About 18 million people die annually from CVD, approximately 32% of all global deaths. Of these, 85% are due to heart attacks and strokes. Of the 17 million deaths occurring in people under age 70 years due to noncommunicable diseases, about 40% are due to CVD. According to WHO, most CVD can be prevented by addressing behavioral risk factors, including an unhealthy diet, physical inactivity, and cigarette smoking. For CVD prevention, the WHO recommends cessation of smoking, reduction of salt in the diet, eating more fruits and vegetables, regular physical activity, and avoiding harmful use of alcohol. In addition, drug treatment of risk factors is recommended to reduce CVD risk by using essential medications, such as statins, aspirin, beta-blockers, angiotensin-converting enzyme inhibitors, and metformin.[1]

According to the Centers for Disease Control and Prevention (CDC), heart disease is the leading cause of death for men, women, and people of most racial and ethnic groups in the United States.[2] One person dies about every 30 seconds in the United States from CVD, and about 700,000 people in the United States die annually from heart disease—about 1 in every 5 deaths.[2] Annual costs for heart disease in the United States are about $250 billion, including the costs of healthcare services, medicines, and lost productivity due to death. In the United States, someone has a heart attack every 40 seconds. In 2020, approximately 800,000 people in the United States had a heart attack, and of these, 600,000 was a first heart attack, and about 200,000 were recurrent. Moreover, about 1 in 5 heart attacks are silent—the damage is done, but the person is not aware of it.[2] Silent heart attacks are most common in individuals with diabetes. The American Heart Association (AHA) recently released their extensive annual analysis of CVD statistics and risk factors in the United States, documenting almost twofold differences in CVD incidence by state.[3] It should be noted that most patients with diabetes and kidney disease die from CVD. Therefore, if one groups mortality from coronary heart disease, stroke, diabetes, and kidney disease together, this group accounts for almost 50% of all known causes of death.

CVD Pathogenesis

CVD is composed of coronary heart disease (CHD), stroke, and peripheral arterial disease, and is mainly caused by atherosclerosis, a process in which arteries become occluded.[4] The characteristics of this process in the artery wall are the presence of cholesterol-laden macrophages or foam cells, proliferation of smooth muscle cells with excess connective tissue, calcification, and sometimes thrombosis as the terminal event occluding the artery. A heart attack or myocardial infarction (MI) occurs when one or more of the three major coronary arteries becomes blocked. A stroke occurs when one or more of the arteries supplying the brain becomes occluded.[4]

Increased age (≥65 years), high blood pressure (now defined as ≥130/80 mmHg), diabetes (fasting glucose ≥125 mg/dL), and cigarette smoking are all associated with excess damage to the endothelial cell lining of the artery wall and are all independent CVD risk factors.[5,6] LDL, lipoprotein (a) [Lp(a)], and very low-density lipoproteins (VLDL) can all be deposited in the artery wall, especially at sites of damage. High concentration of total serum cholesterol (≥200 mg/dL) due to increased serum concentration of LDL-C (≥130 mg/dL) is a significant independent risk factor for CVD. In addition, serum HDL serves to remove cholesterol from the artery wall. Low concentration of HDL-C (<40 mg/dL in men and <50 mg/dL in women) is also a significant independent CVD risk factor.[4,5] Diets high in animal fats, hydrogenated fats, full-fat dairy products, sugar, and salt have all been associated with excess obesity, elevated blood cholesterol, and high age-adjusted CVD mortality.[4]

All ACC/AHA CVD risk factors and CVD risk enhancers according to the 2018 cholesterol guidelines are shown in **Table 67.1**. CVD risk enhancers include family history, kidney disease, elevated high-sensitivity C reactive protein (hs-CRP), Lp(a), and triglycerides.[5,6] Additional risk enhancers also include elevated serum values of direct LDL-C, small dense LDL-C, homocysteine, troponin-T, and N-terminal pro-brain natriuretic peptide (NT-proBNP).[4-9] In the Atherosclerosis Risk in Communities Study in adults without CVD at baseline, those with elevated concentrations of both NT-proBNP (>160 pg/mL, marker of heart failure), and troponin T [>14.0 ng/L]) marker of heart damage) had 7 times the 10-year risk of CVD mortality, compared with adults who were not in this category.[9]

CVD Risk Prediction

In **Table 67.2** information is provided about the American College of Cardiology (ACC)/AHA pooled cohort equation for CVD risk assessment.[5]

It is important to emphasize that this risk calculator is only valid for adults between 40 and 79 years of age, with total serum cholesterol between 130 and 320 mg/dL, HDL-C between 20 and 100 mg/dL, systolic blood pressure between 90 and 200 mmHg, and diastolic blood pressure between 30 and 140 mmHg.[5] Using this risk calculator, it has been recommended that adults with a 10-year CVD risk of ≥7.5% are candidates for statin therapy, and that for adults with established CVD, intensive

Table 67.1 Cardiovascular Risk Factors[4-9]

1. Age ≥65 years
2. Sex: Being male
3. Race: being African American increases hypertension & stroke risk
4. Serum Total Cholesterol ≥200 mg/dL
5. Serum HDL-C <40 mg/dL in men and <50 mg/dL in women
6. Blood Pressure ≥130/90 mmHg
7. Blood Pressure Treatment
8. Diabetes (Fasting Glucose ≥125 mg/dL)
9. Smoking

Risk Enhancers

1. High sensitivity C-reactive protein (hs-CRP) ≥2.0 mg/L
2. Lp(a) ≥50 mg/dL or ≥125 mmol/L
3. Fasting serum triglycerides ≥175 mg/dL
4. Family history of premature CVD
5. Kidney Disease
6. Direct LDL cholesterol ≥130 mg/dL
7. Small Dense LDL cholesterol >50 mg/dL
8. N-terminal pro-brain natriuretic peptide (NT-ProBNP) >160 pg/mL
9. Troponin T >14.0 ng/L
10. Homocysteine >14.0 µmol/L

Data from Schaefer, Ernst J. "Lipoproteins, nutrition, and heart disease." The American journal of clinical nutrition vol. 75,2 (2002): 191-212. doi:10.1093/ajcn/75.2.191 and Oluleye, Oludamilola W et al. "Troponin T, B-type natriuretic peptide, C-reactive protein, and cause-specific mortality." Annals of epidemiology vol. 23,2 (2013): 66-73. doi:10.1016/j.annepidem.2012.11.004

statin therapy should be used (rosuvastatin 20–40 mg/day or atorvastatin 40–80 mg/day) to obtain a 50% or more reduction in LDL-C values.[5,6] In addition, we have documented that direct LDL-C and small dense LDL-C values provide significantly better CVD risk estimates than calculated LDL-C values.[7,8]

Statin therapy should be given to high-risk adults in addition to lifestyle modification.[6] In the JUPITER (Justification for the Use of Statins in Prevention: an Intervention Trial Evaluating Rosuvastatin) 17,802 apparently healthy men and women with LDL-C <130 mg/dL and hs-CRP ≥2.0 mg/L were randomized to rosuvastatin 20 mg/day or placebo. The combined endpoint of myocardial infarction (MI), stroke, arterial revascularization, hospitalization for unstable angina, and/or CVD death was reduced by 44% over a 1.9-year period in favor of the treatment group.[10,11] The trial was stopped early because of marked differences between treatment groups. LDL-C and hs-CRP concentrations were reduced by 50% and 37%, respectively, vs baseline in the treatment group, with no significant change in the control group.[11] Those adults in the trial achieving LDL-C <70 mg/dL and hs-CRP <1.0 mg/L achieved CVD risk reductions of 79%, compared with the placebo group.[11] A meta-analysis of cholesterol lowering trials indicates that therapy for LDL-C lowering with statins, ezetimibe on top of statin therapy, and proprotein convertase subtilisin kexin 9 (PCSK9) inhibitors on top of statin therapy are all effective in lowering CVD risk and mortality in randomized placebo-controlled trials.[12]

While LDL-C targets were removed from recent guidelines, there is consensus that the goals of therapy are to get LDL-C to <100 mg/dL in adults without CVD and to <70 mg/dL in adults with established CVD (heart disease, stroke, post-angioplasty, or coronary bypass surgery). Not only has statin therapy been shown to be effective in reducing LDL-C concentration by about 50% or more, such therapy also greatly decreases CVD risk and mortality by 30% to 40%.[5,12] Adults on statin therapy should be given 200 mg/day of coenzyme Q10 supplementation because statin therapy has been shown to inhibit coenzyme Q10 formation. Co-enzyme Q10 is important for normal muscle function and supplementation with this agent has been shown to reduce mortality risk in CVD patients.[13] Adults who fail to achieve desired LDL-C can have ezetimibe 10 mg/day and/or proprotein convertase substilisin kexin-9 (PCSK9) inhibitors added to their regimen, both of which have been shown to lower CVD risk.[6,12]

An alternative CVD risk calculator is the Reynolds Risk Score that uses: 1) sex, 2) age, 3) smoking, 4) systolic blood pressure, 5) total cholesterol, 6) HDL-C, 7) hs-CRP, and 8) parental history of CVD prior to age 60 years. This risk calculator can be accessed at: www.reynoldsriskscore.org.[14,15] CVD risk can be further refined by assessing for the actual presence of coronary artery disease using coronary artery calcium score with ultrafast computed tomographic (CT) scanning, which takes about 1 minute. One can then use the risk assessment calculator developed by the Multi-Ethnic Study of Atherosclerosis (MESA) that combined this score with the ACC/AHA standard risk factors and can be accessed at: www.mesa-nhlbi.org.[16,17] Adding the cardiac calcium score to the pooled cohort equation increases the accuracy of CVD risk prediction from about 70% to about 80%. However, cardiac calcium score cannot be used to follow treated patients, as lipid-rich plaques regress and calcify with lifestyle and statin therapy. In this situation, the best method of follow-up is periodic high resolution standardized carotid and femoral ultrasound imaging.[18] There have been significant advances in computed tomographic coronary angiography using intravenous contrast material, which have documented that this noninvasive methodology can be used to image calcified and noncalcified lipid-rich coronary plaques. Moreover, it is the presence of these latter noncalcified, lipid-rich plaques that best predict major adverse coronary events.[19]

Table 67.2 ACC/AHA Pooled Cohort Equation for CVD Risk[5]

1. Ten year and lifetime CVD risk can then be assessed at: https://static.heart.org/riskcalc/app/index.html#!/baseline-risk
2. Enter the following information into the calculator: age, gender, race, total cholesterol, HDL cholesterol, systolic blood pressure, diastolic blood pressure, blood pressure treatment, diabetes, and smoking.
3. Then submit and the algorithm calculates 10-year risk of atherosclerotic CVD.
4. A 10-year risk of ≥7.5% has been defined as high, and such adults are candidates for statin therapy in addition to lifestyle modification.

Data from Goff, David C Jr et al. "2013 ACC/AHA guideline on the assessment of cardiovascular risk: a report of the American College of Cardiology/American Heart Association Task Force on Practice Guidelines." Circulation vol. 129,25 Suppl 2 (2014): S49-73. doi:10.1161/01.cir.0000437741.48606.[98]

Population Studies

Many cross-sectional population studies have examined the interrelationships between diet, blood cholesterol concentrations, and heart disease. The earliest of these was the Seven Countries Study conducted by Ancel Keys.[20] This study clearly documented that the key dietary ingredient linked to prospective CHD in seven different countries and 16 populations (including the Finland, Greece, Italy, Japan, the Netherlands, the United States, and Yugoslavia) was the concentration of saturated fat intake ($r = 0.84$) which, in turn, was strongly linked to blood cholesterol concentrations. Another important early population study was the Ni-Hon-San Study involving men of Japanese descent living in Japan, Hawaii, or California. The investigators documented about three-times higher age-adjusted CHD mortality rates for the men in California than in Japan, with intermediate rates in Hawaii.[20] CHD mortality was strongly associated with increased serum cholesterol concentrations, increased total fat intake (mainly saturated fat) and decreased carbohydrate intake (mainly from rice).[21]

In the 20 Countries Study, Stamler and colleagues reported significant positive correlations between CHD mortality and intake of butter, all dairy products, eggs, meat and poultry, and sugar and syrup, and a significant inverse association with intake of grains, fruit, and starchy and nonstarchy vegetables (**Table 67.3**).[22] The strength of this study and these other earlier studies is that they used 7-day food records, which, if done carefully, provide a more reliable and accurate assessment of actual dietary intake than food frequency questionnaires, which have been used in more recent studies.[23]

In the InterHeart Study, Yusuf and colleagues collected important data on differences between CHD cases and controls Table 67.3).[24] In this study, the ratio of serum apolipoprotein B (apoB, the major protein of LDL) to serum apoA-I (the major protein of HDL) was the most significant predictor of CHD, with smoking, psychosocial stress, diabetes, hypertension, obesity, lack of daily vegetables and fruits, and lack of physical activity all playing significant roles. In aggregate, these factors accounted for about 90% of the attributable risk.[24]

A recent report detailed fasting plasma fatty acid concentrations (Boston Heart Fatty Acid Population Study) for 1,169,621 men and women living in 45 states across the United States (states with <100 adults were excluded).[25] Median fatty acid percentages were 42.5% for n-6 polyunsaturated fatty acids (mainly linoleic acid at 30.9% and arachidonic acid at 8.9%), 31.4% for saturated fatty acids (palmitic acid at 22.9%, stearic acid at 7.7%, and myristic acid at 0.99%), 21.3% for monounsaturated fatty acids (mainly oleic acid at 20.3% and palmitoleic acid at 1.67%), 2.57% for the two major n-3 fatty acids, eicosapentaenoic acid (EPA at 1.32%) and docosahexaenoic acid (DHA at 2.25%), and 0.45% for trans fatty acids (16:1n7 trans fatty acid at 0.26% and 18:1n9 trans fatty acid at 0.19%). Omega-3 fatty acid and saturated fatty acid concentrations were strongly linked to reported statewide CHD mortality rates, with correlation coefficients of −0.57 and +0.45 (both $p<0.01$), respectively.

Women ≥65 years of age, and adults living in northeastern and western states had significantly higher n-3 fatty acid concentrations and significantly lower saturated fatty acid concentrations and CHD mortality rate concentrations than other adults, especially those living in midwestern and southern states.[25] States with the highest CHD mortality (more than twofold higher) and higher saturated fatty acid concentrations were Arkansas, Mississippi, Oklahoma, and Alabama, while the converse was true for states with the lowest CHD mortality rates, namely Massachusetts, Oregon, Alaska, California, and Washington.[25] These latter states also had the highest plasma EPA concentrations. These data indicate that much of the large differences in CHD mortality rates by state relate to variation in serum fatty acids, which are controlled by dietary intake.

Table 67.3 Population Studies—Cardiovascular Risk & Lifestyle

A. 20 Countries Study—Correlations with CVD Death[22]	
1. Sugar and Syrup Intake	0.68
2. All Dairy Products	0.62
3. Eggs	0.59
4. Meat and Poultry	0.56
5. Butter Intake	0.55
6. Vegetable, Fruit, and Grain Intake	−0.63
B. Interheart Study—Relative Risk*[24]	
1. ApoB/ApoA-I Ratio	3.25
2. Smoking	2.87
3. Psychosocial Stress	2.67
4. Diabetes	2.37
5. Hypertension	1.91
6. Obesity	1.62
7. Lack of Daily Vegetables & Fruits	1.43
8. Lack of Physical Activity	1.16

*Based on 15,152 men and women with CHD and 14,820 age- and gender-matched controls in 52 countries. All relative risk values were highly significant ($p<0.0001$). Together, these variables accounted for 90% of the risk in men and 94% of the risk in women.

Controlled Dietary Studies

In studies conducted under controlled circumstances, diets restricted in saturated fat to <10% of energy have been shown to lower LDL-C concentration by 12% to 20%, compared with an average U.S. diet.[4,26-34] Moreover, dietary cholesterol and trans-fatty acids can both raise LDL-C concentration and should be restricted.[29,30,35] Three different groups of investigators published predictive equations based on composite analyses of many controlled diet studies to determine the effects of different constituents of the diet on LDL-C concentration under controlled metabolic ward conditions. The investigators and equations are as follows (**Table 67.4**).

These equations predict that a change in saturated fat intake will have the largest effects on LDL-C, followed by change in polyunsaturated fat intake, followed by changes in monounsaturated fat intake.[32-34] Only Hegsted et al., included dietary cholesterol in their formula. Using the Mensink and Katan formula, if an individual lowers S from 14% to 7% of energy, and raises P from 5% to 12%, with no change in dietary carbohydrate intake, the predicted LDL-C concentration will be reduced by about 12 mg/dL, or by about 10% if the individual has an LDL-C concentration of 120 mg/dL. This equation does not take dietary cholesterol into account. Controlled diet experiments and a recent meta-analysis clearly indicate that increased dietary cholesterol intake raises LDL-C concentration and is associated with increased CVD risk.[35] Even under controlled conditions, marked variability in the LDL-C lowering response to dietary change is noted, in part because of sex, as well as because of differences in APOE and other genotypes.[4,36-38]

Table 67.4 Prediction of Diet Effects on LDL-C

1) **Hegsted Formula**[37]
Change LDL-C (mg/dL) = [1.74 × Change S] − [0.766 × Change P] + [0.0439 × Change C].
2) **Mensink and Katan Formula**[33]**:**
Change in LDL-C (mg/dL) = [1.28 × Change in S] − [0.24 × Change in M] − [0.55 × Change in P].
3) **Yu and Kris-Etherton Formula**[34]**:**
Change in LDL-C (mg/dL) = [1.46 × Change in S] − [0.07 × Change stearic acid] − [0.69 × Change in M] − [0.96 × Change P].

Abbreviations: S = saturated fat intake as a percentage of energy intake when exchanged for carbohydrate, C = change in dietary cholesterol in mg/1,000 kcal, M = monounsaturated fatty acid intake as a percentage of energy (mainly oleic acid or 18:1-n9) when exchanged for carbohydrate, and P = polyunsaturated fatty acid intake as a percentage of energy (mainly as linoleic acid, 18:2-n6; arachidonic acid, 20:4-n6; and a-linolenic acid, 18:3-n3) when exchanged for carbohydrate. In the formula of Yu et al, stearic acid is not included in the saturated fat category, which includes only lauric (12:0), myristic (14:0), and palmitic (16:0) acids.[32-34]

Data from Hegsted, D M et al. "Dietary fat and serum lipids: an evaluation of the experimental data." The American journal of clinical nutrition vol. 57,6 (1993): 875-83. doi:10.1093/ajcn/57.6.875; Mensink, R P, and M B Katan. "Effect of dietary fatty acids on serum lipids and lipoproteins. A meta-analysis of 27 trials." Arteriosclerosis and thrombosis : a journal of vascular biology vol. 12,8 (1992): 911-9. doi:10.1161/01.atv.12.8.911 and Yu, S et al. "Plasma cholesterol-predictive equations demonstrate that stearic acid is neutral and monounsaturated fatty acids are hypocholesterolemic." The American journal of clinical nutrition vol. 61,5 (1995): 1129-39. doi:10.1093/ajcn/61.4.1129

Decreasing the intake of trans-fatty acids is important because these fatty acids, in the form of hydrogenated vegetable oils, are also known to raise LDL-C.[4,30] Almost all soft margarines are now trans-fat free or very low in trans-fatty acids. Therefore, they may be a better option than butter, given that their fatty acid content is usually similar to that of soybean oil. Investigators have also documented that when saturated fatty acids are replaced with polyunsaturated fatty acids in the diet, both LDL Apo-B and HDL Apo-A-I decrease significantly because of enhanced fractional catabolic rate.[4,31] These changes appear to be related to upregulation of the LDL receptor in the liver.[4,31] When one restricts animal fat to lower saturated fat intake, one usually lowers monounsaturated fat intake, because most animal fats have almost as much monounsaturated fat as saturated fat. Therefore, the only logical way to replace saturated fat is with vegetable oil that is rich in both polyunsaturated and monounsaturated fats.

Additional ways to lower LDL-C with lifestyle modification include adding two servings per day of plant stanol or sterol margarine, which will decrease cholesterol absorption,[39] or to increase dietary fiber, including the daily use of psyllium.[40] Another issue is the effect of different types of carbohydrate on lipid concentrations. Dietary fructose is more deleterious in terms of effects on visceral fat, triglyceride concentrations, and HDL-C concentrations than glucose.[41]

Most patients in the high-risk category require medications to lower their LDL-C concentration to established goals.[5,6] Many physicians do not refer patients to a dietitian, in part because one or two visits with a dietitian may have only a small effect on lowering LDL-C concentration. It is now becoming clear that for both LDL-C lowering and weight loss, more intensive group approaches lasting several months are required to achieve any significant type of lifestyle change, as was observed in the Diabetes Prevention Program for weight loss and prevention of diabetes in individuals with prediabetes.[42]

Dietary Intervention Studies

The most compelling data justifying any treatment strategy come from large-scale randomized clinical trials. Surprisingly, the number of large dietary intervention studies to examine CVD risk reduction has been limited, in part because such trials are much more labor intensive and difficult to carry out than placebo-controlled trials with pills. The largest and most significant studies are reviewed below along with recent meta-analyses. In addition, trials using omega-3 fatty acids and vitamin B supplementation are covered, along with some additional meta-analyses.

The first significant studies were the first and second Oslo Diet Heart Studies in men with and without CHD (**Table 67.5**). These studies documented the benefits of replacing animal fats in the diet with vegetable oil (mainly corn oil).[43-47] The second trial also had a smoking cessation component. In the first trial, there was a 44% decrease in mortality from MI after 11 years of follow-up for the intervention group, while in the second trial, there was a 47% reduction in MI over 5 years, and long-term benefits were also seen.[43-47]

The Los Angeles Veterans Administration Study (Table 67.5) was the first diet trial carried out in the United States and enrolled 846 men living in a large domiciliary.[48-51] They were equally randomized to their usual diet or a test diet in which saturated fat was reduced (11% of energy in the test diet vs 18% in the standard diet). The animal fat was replaced by vegetable oil (corn, cottonseed, safflower, and soybean), with 16% of energy as polyunsaturated fat (vs 5% in the usual diet). Both diets contained approximately 40% of energy as total fat.[48-51] Over an average follow-up of 6.5 years, total cholesterol concentration in the treatment group, compared with the control group, was reduced by 13%, and a significant ($P<0.01$) 31% decrease was noted in the endpoints of MI, CHD mortality, and other cardiovascular endpoints including stroke, ruptured aneurysms, and ischemic gangrene.[48-51] There was, however, a higher reported prevalence of gallstones and certain cancers in the treatment group.[50,51]

The Finnish Mental Hospital Study (Table 67.5) was a significantly larger trial in which 5,115 men and women in mental hospital N and 5,497 men and women in mental hospital K were placed on either an experimental diet (hospital N) or the usual Finnish diet (hospital K) for the first 6-year period (1959–1965), whereas during the second period (1965–1971), the hospitals were crossed over, with hospital N adults receiving the usual Finnish diet and hospital K receiving the experimental diet. The goal was to replace the dairy and butter fat in the usual Finnish diet with skimmed milk "filled" with soybean oil instead of full-fat milk and to replace butter with margarine high in soybean oil in the experimental diet.[52-54] Both diets contained about 2,800 energy, with approximately 110 g of fat (35% of energy). The usual Finnish diet contained about 19% of energy as saturated fat and about 4.5% as polyunsaturated fat, with 480 mg of cholesterol per day, while the test diet contained about 9% of energy as saturated fat and 14% of energy as polyunsaturated fat, with 280 mg of cholesterol per day, respectively. In subsets of individuals, the fatty acid content of adipose tissue for linoleic acid and myristic acid were measured and determined to be about 10% and 4.3% of total fatty acids on the usual diet and about 30% and 1.5% of total fatty acids on the experimental diet, respectively.

Mean CHD mortality rates were significantly ($P = 0.002$) lower by 53% on the experimental diet than on the usual diet.

Table 67.5 Selected Dietary Intervention Studies

Oslo Diet Heart Study I[43,44]

1. 412 men with CHD were randomized for 5 years to a standard Norwegian diet or a diet in which animal fat was replaced with vegetable oil (they received dietary counseling).
2. The standard diet was about 18% of energy from saturated fat and was low vegetable oil, while the test diet was about 8% of energy from saturated fat, 21% of energy as polyunsaturated fat, and 10% of energy as monounsaturated fat.
3. The treatment group had a 33% lower MI rate at 5 years and a 44% lower MI mortality rate at 11 years ($p<0.05$).

Oslo Diet Heart Study II[45-47]

1. 1232 men with high serum cholesterol (290–380 mg/dL) without CHD were enrolled with 80% being cigarette smokers.
2. Adults were randomized (5 years) to usual care or advice to replace animal fat with vegetable oil and smoking cessation program.
3. MI (fatal and nonfatal) and sudden death rates were reduced by 47%.

Los Angeles VA Study[48-51]

1. Eight hundred and forty-six men in a large domiciliary were given either their usual high animal fat diet or a test diet high in vegetable oils over 6.5 years.
2. Over 6.5 years, there was a 31% decrease in CVD endpoints.

Finnish Mental Hospital Study[52-54]

1. 5,115 men and women in hospital N were placed on a test diet and 5,497 adults in hospital K were placed on the usual diet for 6 years, and then they were crossed over for the same period.
2. In the test diet, dairy and butter fat was replaced with skimmed milk with added soybean oil and butter was replaced with high soybean oil margarine.
3. CHD mortality rates were reduced by 51% in the test group during the first diet period and by 56% during the second diet period.

Data from Leren P. Acta Med Scand 1966;466:1–92; Leren P. Circulation 1970;42:935–942; Hjermann I, Velve Byre K et al. Lancet 1981;2:1303–1310; Hjermann I, Holme I, Leren P. Am J Med 1986;80:7–11; Holme I, Haaheim LL, Tonstad S, Hjermann I. Nutr Metab Cardiovasc Dis 2006;16:330-338; Dayton S, Pearce ML, Goldman H et al. Lancet 1968;2:1060–1062; Dayton S, Pearce ML, Hashimoto S. Circulation 1969;34(Suppl II):1–63; Pearce ML, Dayton S. Lancet 1971;1:464–467; Sturdevant RA, Pearce ML, Dayton S. N Engl J Med 1973; 288:24–27; Turpeinen O. Circulation 1979;59:1–7.; Turpeinen O, Karvonen MJ, Pekkarinen M et al. Int J Epidemiol 1979;8:99–118; Miettinen M, Turpeinen O, Karvonen MJ et al. Int J Epidemiol 1983;12:7–25.

For hospital K, these rates were 50.6% lower on the experimental diet vs the usual Finnish diet, whereas for hospital K, they were 56.1% lower. Plasma cholesterol concentrations was significantly lower by 12% for hospital K (236 vs 268 mg/dL) and 19% lower for hospital N (216 vs 267 mg/dL) when the experimental diet was compared with the usual diet.[52-54] In women, there was a 34% mean reduction in CHD mortality rates in favor of the experimental diet group, but these differences did not reach statistical significance, in part because of the substantially lower event rates in women.[54]

Another large intervention study, the Minnesota Mental Hospital Study, was an open-label randomized study involving 9,057 men and women of all ages at six mental hospitals and one nursing home.[55] Adults were placed on diets containing about 40% of energy from fat that differed in their polyunsaturated fat content (5% vs 15%), saturated fat (18% vs 9%), and dietary cholesterol (466 vs 166 mg/day).[55] The treatment group had 14% lower serum cholesterol concentration, but no significant difference was noted in CHD morbidity or mortality among the groups.[55] This negative result may have resulted from the relatively normal mean serum cholesterol of the study population (207 mg/dL at baseline), the relatively young age of the study population in which the largest single age group was less than 30 years, and the relatively short duration of the test diet (mean, 384 days).[55] The shorter duration of the study resulted from discharges from the mental hospital in part because of the introduction of the medication chlorpromazine.

The Lyon Diet Heart Trial was a secondary prevention study in 605 men and women who had a prior heart attack. Study adults were randomized to a usual French diet or a more "Mediterranean diet" in which all adults also received two servings per day of a specially prepared margarine high in a-linolenic acid.[56] After a 44-month average follow-up, the diet group had a 76% decrease in cardiac deaths (six deaths in the treatment group vs 19 deaths in the control group; $P<0.01$).[56] The benefit in this trial was related to increases in plasma a-linolenic acid; however, the number of endpoints was very low.[56]

The Women's Health Initiative was the largest dietary intervention trial to ever be conducted in the United States. In contrast to prior studies, no food was provided and the animal or saturated fat was not replaced by vegetable, but by carbohydrate.[57] In this trial, 48,835 postmenopausal women aged 50 to 79 years were randomly allocated to a diet low in fat (40% of total group or 19,541 adults) vs their usual diet (60% of total group or 29,294 adults). All adults in the control group received a copy of *Dietary Guidelines for Americans*, which emphasized the importance of a heart healthy diet. The dietary intervention was implemented by holding group classes and individual interview sessions that included dietary assessments using food frequency questionnaires.[58] The goals of the intervention were to decrease total fat intake to 20% of energy and to increase the intake of vegetables and fruits to five servings per day and grains to six servings per day.[58]

The investigators reported that the adults in the active arm of the study at 6 years of follow-up had: 1) mean total fat intake of 28.8% of energy vs 37.0% in the control group, 2) mean saturated fat intake of 9.5% vs 12.4% in the control group, 3) mean monounsaturated fat intake of 10.8% vs 14.2% in the control group, and 4) mean polyunsaturated fat intake of 6.1% vs 7.5% in the control group.[58] Moreover, adults in the active group had increased their vegetable and fruit intake by 1.1 servings per day and their grain intake by 0.5 servings per day.[58] One of the confounding features of the study was that of those participating in the active dietary arm, 8,052 women also participated in the hormone replacement arm of the Women's Health Initiative, and 5,017 also participated in the calcium and vitamin D arm of this study.[58]

The primary aim of the study was to ascertain whether a low-fat diet would reduce the risk of breast cancer. Over 8.1 years of follow-up, 0.42% women per year developed breast cancer in the diet group vs 0.45% per year in the control group.[58-61] Therefore, adults in the active diet group lowered their risk of developing invasive breast cancer by 9% (hazards ratio, 0.91; confidence interval, 0.83 to 1.01; $P = 0.07$).[58-61] The investigators also assessed the impact of the dietary intervention on CVD.[58]

After 8.1 years of follow-up, the risk of CHD was only reduced by 3% in the intervention group (hazards ratio, 0.97; confidence interval, 0.90 to 1.06), and the risk of stroke was increased by 2% (hazard ratio, 1.02; 95% confidence interval, 0.90 to 1.15).[58] The dietary intervention also had no significant impact on risk of colorectal cancer or the development of diabetes.[57-61] The diet group did have significantly ($P<0.05$) lower LDL-C by 3.6 mg/dL, lower systolic blood pressure by 0.31 mm Hg, and lower factor VIIC by 4.29%, compared with the control group.[58] However, these changes were small. In a subgroup analysis among women who achieved ≤6% of energy as saturated fat, the risk of CHD was reduced by 19% (hazard ratio, 0.81; 95% confidence interval, 0.69 to 0.95; $P<0.01$).[58,59] Such differences were also observed in those adults in the diet group who had the lowest trans-fatty acid intake (hazard ratio, 0.81; 95% confidence interval, 0.69 to 0.95).[58,59]

The final dietary intervention study to be reviewed is the PREDIMED (Prevencion con Dieta Mediterranea) Study conducted at multiple centers in Spain. In this trial, 7,447 adults (55–80 years of age, 57% women) at high CVD risk but without disease, were randomized to one of three diets: 1) a Mediterranean diet supplemented with extra virgin olive oil, 2) a Mediterranean diet supplemented with nuts, and 3) a control diet in which adults received advice to reduce total fat intake. High CVD risk was defined either as having diabetes or having at least three of the following four CVD risk factors: 1) smoking, 2) hypertension, 3) elevated LDL-C, 4) obesity, and/or 5) family history of premature CHD).[62]

For the Mediterranean diet ≥3 servings/day of fresh fruits, ≥2 servings/day of vegetables, ≥3 servings/week of fatty fish or seafood, legumes, and ≥7 glasses of wine/week were recommended. The use of olive and tree nuts and peanuts was encouraged. Participants in the olive oil group were given 1 liter per week of extra virgin olive oil and were instructed to consume 4 tablespoons per day, while those in the nut group were given 30 grams of nuts to consume per day, consisting of 15 grams of walnuts, 7.5 grams of hazelnuts, and 7.5 grams of almonds. Soft drinks, baked goods, sweets, and pastries were discouraged as was red and processed meat and fat spreads. For the low-fat group, ≥3 servings/day of low-fat dairy products as well as breads, potatoes, pasta, and rice were recommended. The fruit, vegetable, and seafood recommendation were the same, except that lean fish was recommended. The low-fat group was discouraged from consuming vegetable oils, nuts, fatty meats, and fat spreads.[62]

Over a median 4.8 years of follow-up, there were 96 events (3.8%) in the olive oil group, 83 events (3.4%) in the nuts group, and 109 events (4.4%) in the low-fat group, resulting in significantly lower event rates in the two Mediterranean diet groups (hazard ratios 0.69 and 0.72, range, 0.53–0.95, $P<0.01$) compared with the low-fat diet group.[62] Subsequently, protocol deviations were identified with lack of proper randomization at two sites out of 11, resulting in the omission of 1,588 participants, leaving a total of 5,859 participants. Despite this adjustment, the results of the reanalysis were similar and consistent with the concept that a Mediterranean diet rich in either olive oil or nuts reduced CVD events compared with a low-fat diet.[63]

In a 2020 Cochrane Systemic Review titled "Reduction in saturated fat for cardiovascular disease," the investigators concluded that: "the findings suggest that reducing saturated fat intake for at least 2 years causes a potentially important reduction in combined cardiovascular events."[64] They also concluded that replacing energy from saturated fat with polyunsaturated fat or carbohydrate both appear to be useful strategies, while effects of replacement with monounsaturated fat are unclear. The reduction in combined cardiovascular events resulting from reducing saturated fat was not altered by study duration, sex, or baseline level of cardiovascular risk, but greater reductions in saturated fat caused greater reductions in cardiovascular events. In this analysis, the investigators included 15 randomized controlled trials with 56,675 participants in their analysis. They did not include the Finnish Mental Hospital Study or the PREDIMED study because of concerns with randomization but did include multiple smaller studies that were mainly assessing coronary angiographic changes.[64] The Finnish Mental Hospital had the best compliance of all these studies, as all of the food was provided to the participants. Although the study was not blinded, the investigators did use a randomized crossover design and, in each 6-year diet period, they documented that replacing saturated fat with polyunsaturated fat resulted in about a 50% reduction in CVD. This finding is in striking contrast to the Women's Health Initiative, which was a negative trial comparing a low-fat diet with a normal American diet.

Dietary Intervention Studies Using Supplements

Omega-3 Fatty Acid Supplementation Trials

The DART Trials (Diet and Atherosclerosis and Reinfarction Trials) conducted in the United Kingdom in more than 2,000 patients with established CHD documented beneficial effects of fish consumption or the use of two fish oil capsules per day in reducing CHD death by 29%.[65] However, this finding was not confirmed in a follow-up study, possibly because of much greater aspirin use in the second study.[66] In the GISSI (Gruppo Italiano per lo Studio della Soppravvivenza nell'Infarto miocardico-Prevenzione) Study of 11,323 Italian patients with a history of MI, the use of 1 g/day of concentrated fish oil (containing 465 mg of eicosapentaenoic acid [EPA] and 375 mg of docasahexaenoic acid [DHA]) was associated with a reduction in overall recurrence of CHD, as well as a 53% reduction in sudden death in the first 4 months after MI in those persons receiving the supplement, vs the control group.[67,68] This product was marked in the United States as LOVAZA and was also shown to be effective as a triglyceride-lowering agent at 4 g/day, with triglyceride lowering of 50% or more in combination with statin therapy in patients with triglyceride concentration >500 mg/dL. The use of vitamin E was also studied in the GISSI trial, but no benefit of this supplementation on CVD outcomes was observed.[68]

In the JELIS (Japan Eicosapentaenoic Acid Lipid Intervention Study) 15,000 men and women without CHD and 3,645 adults with CHD between 40 and 75 years of age, with total cholesterol concentration >250 mg/dL, were all placed on statin therapy and then randomized to receive EPA 1,800 mg/day group or no additional treatment. The primary endpoint was a major CVD event (sudden death, fatal or nonfatal MI, unstable angina, angioplasty, or coronary artery bypass surgery). After 4.6 years of follow-up, the rate of events was 19% lower in the EPA group ($P = 0.011$).[69] No differences were noted in sudden death rates between the groups. In patients with earlier CHD, events were also reduced by 19% by EPA vs no treatment ($P = 0.048$), whereas in patients with a history of earlier MI, the event risk was reduced with EPA by 27% ($P = 0.033$).[69] No effect on stroke risk was noted, except in adults with an earlier stroke, in whom the use of EPA resulted in a 20% relative risk reduction in recurrent stroke ($P<0.05$).[69-74]

In the overall JELIS study, the most benefit was noted in those adults with triglyceride concentration >150 mg/dL and HDL-C concentration <40 mg/dL, in whom the use of EPA reduced CHD events by 53% ($P = 0.043$).[56] The use of EPA also reduced CHD risk by 22% ($P<0.05$) in adults with impaired glucose tolerance (fasting glucose >110 mg/dL).[72] EPA use was not associated with significant effects on key lipid concentrations; however, its use was associated with marked increases in plasma EPA, and study adults with concentrations >150 µg/mL had the lowest risk in the trial.[73]

In the Alpha Omega Trial, 4,837 post-MI patients were randomized to placebo margarine, margarine containing 2 g of a-linolenic acid, margarine containing a total of 400 mg of combined EPA and DHA, or margarine containing the combination of all three of these fatty acids over a 40-month period.[74] No significant differences in CVD endpoints were noted between groups, possibly because the amount of omega-3 fatty acids provided were insufficient.[75]

In REDUCE-IT (Reduction of Cardiovascular Events with Icosapent-Ethyl Intervention Trial) 8,179 patients were enrolled. Of these, 70.9% had established CVD and the remainder had diabetes and other risk factors on statin therapy with fasting triglyceride concentration of 135 to 499 mg/dL and LDL-C of 41 to 100 mg/dL. The patients were randomly assigned to receive 4 grams/day of icosapent ethyl (purified EPA) or mineral oil placebo for a median of 4.9 years. The primary endpoint was a composite of cardiovascular death, nonfatal myocardial infarction, nonfatal stroke, coronary revascularization, or unstable angina. The key secondary endpoint was a composite of cardiovascular death, nonfatal myocardial infarction, or nonfatal stroke.[76]

A primary endpoint event occurred in 17.2% of the patients in the treatment group compared with 22.0% of the patients in the placebo group (hazard ratio, 0.75, $P<0.001$). However, a larger percentage of patients in the icosapent ethyl group than in the placebo group were hospitalized for atrial fibrillation or flutter (3.1% vs 2.1%, $P = 0.004$) and serious bleeding events occurred in 2.7% of the patients in the icosapent ethyl group and in 2.1% in the placebo group ($P = 0.06$). The authors concluded that among patients with elevated triglyceride concentration, despite the use of statins, the risk of ischemic events, including cardiovascular death, was significantly lower among those who received 2 grams of icosapent ethyl twice daily than among those who received placebo.[76]

In the Strength Trial (Study to Assess Statin Residual Risk with Epanova in High Cardiovascular Risk Patients with Hypertriglyceridemia) 13,078 patients with high cardiovascular risk, hypertriglyceridemia, and low HDL-C were randomized to receive either 4 grams/day of omega-3 fatty acids (mixture of EPA and DHA) or corn oil placebo.[77] The primary efficacy measure was a composite of cardiovascular death, nonfatal myocardial infarction, nonfatal stroke, coronary revascularization, or unstable angina requiring hospitalization. When 1,384 patients had experienced a primary endpoint event (of a planned 1,600 events), the trial was prematurely halted based on an interim analysis that indicated a low probability of clinical benefit of omega-3 FA vs the corn oil comparator. Among the 13,078 treated patients, 12,633 (96.6%) completed the trial with ascertainment of primary endpoint status. The primary endpoint occurred in 12.0% of patients treated with omega-3 fatty acids vs 12.2% of patients treated with corn oil (hazard ratio, 0.99, $P = .84$). The investigators concluded that among statin-treated patients at high cardiovascular risk, the addition of omega-3 fatty acids compared with corn oil in patients on usual background therapies resulted in no significant difference in a composite outcome of major adverse cardiovascular events.[77]

Three meta-analyses on the topic were published in 2022. Yokoyama and colleagues reviewed data on 141,009 patients randomized to receive EPA, or EPA and DHA, vs various placebo formulations.[78] They concluded that EPA only reduced CVD risk in comparison with mineral oil, and that combined EPA and DHA reduced risk of CVD death.[78] Markozannes and colleagues reviewed data on 97,709 adults randomized to various trials using omega-3 fatty acids.[79] The authors concluded that, compared with robust evidence that low-dose omega-3 fatty acids reduced CVD risk, higher doses and the use of omega-3 icosapent ethyl ester required further study.[79] Yan and colleagues examined data from 15 randomized controlled trials and concluded that that there is moderate evidence that omega-3 fatty acids reduce CVD risk and mortality. These authors recommended the use of the prescription EPA ethyl ester formulation.[80]

There is strong evidence from JELIS and REDUCE-IT that EPA supplementation at a dose sufficient to increase plasma EPA concentration to >150 µg/mL is associated with significant CVD risk reduction, especially in those with triglyceride concentration >150 mg/dL. In December of 2019, the U.S. Food and Drug Administration approved icosapentyl icosapent ethyl at a dose of 4 grams/day as an adjunct to maximally tolerated statin therapy to reduce the risk of MI, stroke, coronary revascularization, and unstable angina requiring hospitalization in adult patients with elevated fasting triglyceride concentration >150 mg/dL and established CVD or diabetes, with two or more additional risk factors. Similar approvals have occurred in Canada and Europe.

Vitamin Supplementation Trials

Large, randomized, placebo-controlled trials have not shown any significant benefit in terms of CVD risk reduction associated with the use of vitamin E, vitamin C, or a mixture of antioxidant vitamins.[81-84] In a meta-analysis, Clarke and colleagues concluded that such supplementation provides no benefit for CVD risk reduction.[83] However, in a subanalysis of the Heart Outcomes Prevention Evaluation 2 Study, serum homocysteine and the risk of stroke were both reduced by about 25% in 5,522 patients with known CVD randomized to a daily combination of 2.5 mg of folic acid, 50 mg of vitamin B6, and 1 mg of vitamin B_{12}, or matching placebo over a 5-year period.[85] Moreover, in the China Stroke Prevention Study, 20,702 adults with hypertension, but without CVD, were randomized to receive daily folate 0.8 mg and enalapril 10 mg vs enalapril alone. Over a 4.5-year period, adults in the folate group had a significant 21% reduction in stroke risk and a 20% reduction in all CVD events, compared with the control group.[86] In the Framingham Heart Study, stroke was a major risk factor for dementia.[87] A consensus task force concluded that increased plasma total homocysteine reflecting folate, vitamin B_6, and vitamin B_{12} status is a modifiable risk factor for development of cognitive decline, dementia, and Alzheimer's disease in older persons.[88] Spence and Hankey reviewed the vitamin B evidence and concluded that B vitamins can be used to lower serum homocysteine concentration and stroke risk, and that methyl cobalamin or hydroxycobalamin can be used instead of cyanobalamin.[89] Jenkins and colleagues reviewed the evidence on supplemental vitamins and minerals for CVD prevention.[90] They concluded that: 1) folic acid and vitamin B supplementation decreases stroke risk, 2) multivitamin, vitamin D, calcium, or vitamin C supplementation have no effect on CVD risk, and 3) niacin added to statin therapy increases all-cause mortality.[90]

Dietary Guidelines

Summary statements of the most recent World Health Organization dietary guidelines for the prevention of CVD are shown in **Table 67.6**.[1]

Similar lifestyle guidance was provided in the 2013 ACC/AHA guideline on lifestyle management for CVD risk reduction (**Table 67.7**).[91] However, here they recommended saturated fat intake of <7% of energy instead of <10%, in part because these are also guidelines for the treatment of hypercholesteremic adults.[91]

Table 67.6 World Health Organization Guidelines[1]

1. Eat at least 400 grams or five portions of fruit and vegetables/day. Always include vegetables in a meal. Eat a variety of raw fresh fruit and vegetables as snacks.
2. Reduce the amount of fat intake to <30% of total energy to help unhealthy weight gain in the adult population. Reduce saturated fat to <10% of total energy intake. Reduce trans fats to <1% of total energy intake. Replace both saturated and trans fats with unsaturated fat, in particular polyunsaturated fats.
3. Limit the intake of free sugars to <10% of total energy intake.
4. Keep salt intake to <5 grams/day (sodium intake to <2 grams/day).

Data from World Health Organization; www.who.int/health-topics/cardiovascular-diseases

Table 67.7 Acc/AHA 2013 Guideline on Lifestyle Management to Reduce CVD Risk[91]

1. **Overall Dietary Pattern:** Consume a dietary pattern that emphasizes intake of vegetables, fruits, and whole grains, includes low-fat dairy, poultry, fish, legumes, nontropical vegetable oils and nuts, and limited intake of sweets, sugar-sweetened beverage, and red meats.
2. **Advise adults who would benefit from LDL lowering to:** Aim for a dietary pattern that achieves 5% to 6% of energy from saturated fat and reduce the percent of energy from trans fats. Eat at least 400 grams or five portions of fruit and vegetables/day. Always include vegetables in a meal. Eat a variety of raw fresh fruit and vegetables as snacks.
3. **Advise adults who would benefit from BP lowering to:** Consume no more than 2,400 mg of sodium/day, and to further reduce sodium intake to 1,500 mg/day if additional BP reduction is required.
4. **Physical Activity:** Advise adults to engage in aerobic physical activity to reduce LDL-C. non-HDL-C, and BP: three to four sessions per week, lasting on average 30 to 40 minutes per session, and involving moderate-to-vigorous intensity physical activity.

Data from Eckel, Robert H et al. 2013 AHA/ACC guideline on lifestyle management to reduce cardiovascular risk: a report of the American College of Cardiology/American Heart Association Task Force on Practice Guidelines. Circulation vol. 129,25 Suppl 2 (2014): S76-99. doi:10.1161/01.cir.0000437740.48606.d1

Table 67.8 Dietary Guidelines for Americans 2020–2025[92]

1. Follow a healthy dietary pattern at every life stage.
2. Customize and enjoy nutrient-dense food and beverage choices to reflect personal preferences, cultural traditions, and budgetary considerations.
3. Focus on meeting food group needs with nutrient-dense foods and beverages and stay within calorie limits.

The core elements that make up a healthy dietary pattern include:

- **Vegetables of all types**—dark green, red and orange, beans, peas, and lentils, starchy and other vegetables.
- **Fruits**, especially whole fruit.
- **Grains**, at least half of which are whole grain.
- **Dairy**, including fat-free or low-fat milk, yogurt and cheese, and/or lactose-free versions and fortified soy beverages, and soy products.
- **Protein foods**, including lean meats, poultry, and eggs, seafood, beans, peas, and lentils, and nuts, seeds, and soy products.
- **Oils**, including vegetable oils and oils in foods, such as seafood and nuts.

The following foods and beverages should be limited:

- **Limitations:** Foods and beverages higher in added sugars, saturated fat, and sodium, and limit alcoholic beverages.
- **Added sugars:** Less than 10% of energy per day starting at age 2. Avoid foods and beverages with added sugar for those younger than 2 years.
- **Saturated fat:** Less than 10% of energy per day starting at age 2 years.
- **Sodium:** Less than 2,300 mg per day—and even less for children younger than age 14.
- **Alcohol Beverages:** Adults of legal drinking age can choose not to drink or drink in moderation by limiting intake to two drinks or fewer in a day for a man and one drink or fewer in a day for a woman. When alcohol is consumed, drinking less is better for health than drinking more. There are some adults who should not drink alcohol, such as women who are pregnant.

Reproduced from US Dietary Guidelines 2020-2025. www.dietaryguidelines.gov

In **Table 67.8**, the U.S. dietary guidelines for 2020–2025 are presented.[92] Here the focus is on reducing sugar and saturated fat to <10% of energy each. In **Table 67.9** the 2021 AHA guidelines are presented.[93] Unlike the U.S. dietary guidelines, they do not provide specific targets for sugar and saturated fat, but they do emphasize a heart healthy diet.

In **Table 67.10**, 2018 guidelines by Grundy and colleagues focus on the management of cholesterol for CVD risk reduction (6). Their focus is primarily on risk assessment and pharmacologic therapy.

Lipoprotein Disorders

Approach to the Patient

It is important to determine whether the patient has evidence of CVD (history of myocardial infarction, coronary artery bypass, angioplasty, stroke, carotid or peripheral vascular disease

Table 67.9 Summary of 2021 AHA Dietary Guidance to Improve Cardiovascular Health[93]

Feature 1: Energy Intake	Adjust energy intake and expenditure to achieve and maintain a healthy body weight.
Feature 2: Fruits and vegetables	Eat plenty of fruits and vegetables. Choose a wide variety.
Feature 3: Whole grains	Choose foods made mostly from whole grain foods rather than refined grains.
Feature 4: Dietary protein	Choose healthy sources of protein (mostly plants; regular intake of fish and seafood, low-fat and fat-free dairy products, and if meat and poultry is desired, choose lean cuts and avoid processed forms.
Feature 5: Oils and fats	Use liquid plant oils rather than tropical oils (coconut, palm, and palm kernel), animal fats, (butter and lard) and partially hydrogenated fats.
Feature 6: Food processing	Choose minimally processed foods instead of ultra-processed foods.
Feature 7: Sugars	Minimize intake of beverages and foods with added sugar.
Feature 8: Salt	Choose and prepare foods with little or no salt.
Feature 9: Alcohol	If you do not drink alcohol, do not start; if you choose to drink alcohol, limit intake.
Feature 10: Adherence: Adhere to this guidance regardless of wherever food is prepared or consumed.	

Reproduced from Lichtenstein, Alice H et al. 2021 Dietary Guidance to Improve Cardiovascular Health: A Scientific Statement From the American Heart Association. Circulation vol. 144,23 (2021): e472-e487. doi:10.1161/CIR.0000000000001031

Table 67.10 Summary of 2018 AHA/ACC Guideline on Cholesterol Management[6]

Dietary Intake	In all individuals, emphasize a heart-healthy lifestyle across the life course.
Clinical ASCVD	In patients with clinical CVD, reduce LDL-C with high intensity statin therapy (rosuvastatin 20–40 mg/day or atorvastatin 40–80 mg/day) or maximally tolerated statin therapy.
High-Risk ASCVD	In very high-risk ASCVD, use a LDL-C threshold of 70 mg/dL to consider the addition of nonstatins to statin therapy.
Very high LDL-C	In patients with severe primary hypercholesterolemia (LDL-C concentration ≥190 mg/dL) without calculating CVD risk, begin high-intensity statin therapy.
Diabetes mellitus	In patients 40–75 years of age with diabetes mellitus and LDL-C ≥70 mg/dL, start moderate intensity statin therapy.
Primary prevention	In adults 40–75 years of age without diabetes mellitus and with LDL-C concentration ≥70 mg/dL at a 10-year CVD risk of ≥7.5%, start a moderate-intensity statin if a discussion of treatment options favors statin therapy.
Primary prevention—risk-enhancing factors	In adults 40–75 years of age without diabetes mellitus and with LDL-C concentration ≥70 mg/dL at a 10-year CVD risk of 7.5–19.9% (intermediate risk), risk enhancing-factors favor initiation of statin therapy.
Primary prevention—coronary artery calcium score	In adults 40–75 years of age without diabetes mellitus and with LDL-C concentration ≥70-189 mg/dL at a 10-year CVD risk of 7.5-19.9%, if a decision about statin therapy is uncertain, consider measuring coronary artery calcium (CAC) score.
Follow-up	Assess adherence and percentage response to LDL-C lowering medications and lifestyle changes with repeat measurements 4–12 weeks after statin initiation or dose adjustment, repeated every 3–12 months as needed.
Risk-enhancing factors include family history of premature CVD (<65 years), persistently elevated LDL-C ≥160 mg/dL, metabolic syndrome, chronic kidney disease, history of pre-eclampsia or premature menopause (<40 years of age), chronic inflammatory disorders (e.g., rheumatoid arthritis, psoriasis, or chronic HIV), high-risk ethnic groups (south Asian), persistent elevations of serum triglycerides ≥175 mg/dL, serum apoB ≥130 mg/dL, serum hs-CRP ≥2.0 mg/L, ankle brachial index <0.90, and serum Lp(a) >50 mg/dL. A positive coronary artery calcium score favors statin therapy, especially if the score is ≥100 Agatston units.	

Modified from Grundy, Scott M et al. 2018 AHA/ACC/AACVPR/AAPA/ABC/ACPM/ADA/AGS/APhA/ASPC/NLA/PCNA Guideline on the Management of Blood Cholesterol: A Report of the American College of Cardiology/American Heart Association Task Force on Clinical Practice Guidelines. Circulation vol. 139,25 (2019): e1082-e1143. doi:10.1161/CIR.0000000000000625

surgery, or claudication), kidney disease, neurologic disease, eye problems, or a family history of CVD, hypertension, diabetes, or lipid disorders.[94] It is also important to assess status of smoking, hypertension, diabetes, hospitalizations, alcohol intake, and current and past use of medications and supplements. Assessments of blood pressure, body mass index (BMI), waist circumference, the presence of pulses and bruits, xanthomas and arcus, and standard physical examination including ophthalmologic and neurologic status are also important. In high-risk patients, consideration should be given to non-invasive testing to assess for the presence of atherosclerosis using carotid and/or femoral ultrasound imaging, coronary artery calcium scoring, or computed tomographic coronary angiography.[16-19]

Laboratory assessment after an overnight fast is important to establish a diagnosis of lipid disorders. The following measurements should be done: total cholesterol, triglycerides, HDL-C, direct LDL-C, small dense LDL-C, Lp(a), hs-CRP, homocysteine, vitamin B_{12}, folate, thyroid-stimulating hormone, and a comprehensive metabolic profile that includes glucose, creatinine, and liver enzymes. All of these tests, along with serum fatty acids, can now be done using finger-stick testing and dried blood spot cards, if necessary. These tests are important to rule out secondary causes of lipid abnormalities. It is also important to assess laboratory values after an overnight fast because multiple values including glucose, triglycerides, HDL-C, and direct LDL-C can be significantly affected by a recent meal.[95]

Secondary causes of elevated LDL-C (≥60 mg/dL) include excess dietary intake of saturated and trans fats, weight gain, hypothyroidism, anorexia, diuretics, cyclosporin, glucocorticoids, amiodarone, biliary obstruction, nephrotic syndrome, and obesity. One also needs to rule out secondary causes of elevated fasting triglycerides (≥150 mg/dL) and low HDL-C (<40 mg/dL in men and <50 mg/dL in women), which may be due to obesity, diabetes, hypothyroidism, kidney disease, high sugar intake, excessive alcohol intake, oral estrogens, glucocorticoids, protease inhibitors, retinoic acid, anabolic steroids, sirolimus, raloxifene, tamoxifen, beta blockers (except carvedilol), or thiazides.

In patients with hypertension (>130/90 mmHg), it is important to emphasize regular exercise, restriction of alcohol and salt, as well as weight loss, if indicated. The medication of choice is an angiotensin converting enzyme inhibitor, such as enalapril. β-blocker medications (carvedilol and/or metoprolol) are also very effective, especially in those with somewhat elevated heart rates. If necessary, a calcium channel blocker such as amplodipine can be added along with a thiazide-type diuretic such as chlorthalidone or hydrochlorothiazide. In patients on diuretics, serum potassium needs to be monitored and they may need foods rich in potassium or oral potassium supplements. In those with prediabetes, walking 20 minutes per day, eliminating high-fat, high-sugar desserts, and the use of metformin (850 mg po bid) can reduce the risk of developing diabetes by >50%.[42]

Common Polygenic Familial Lipoprotein Disorders Associated with CVD

There are four common polygenic lipid disorders seen in families with premature CVD: 1) familial lipoprotein(a) excess, 2) familial combined hyperlipidemia, 3) familial dyslipidemia, and 4) familial hypoalphalipoproteinemia.[96] Familial Lp(a) excess is observed in about 20% of families with premature CVD.[96] Affected patients have Lp(a) as measured by immunoassay that are >90th percentile of the general population (>80 mg/dL or >192 nmol/L).[96,97] Lp(a) is an apoB-containing particle (mainly LDL) with apo(a) attached to the terminal end of apo-B-100.[94] Lp(a) concentration is in large part determined by the number of apo(a) isoforms, which are inherited.[94] Fewer kringle-4–like repeats results in less intrahepatic degradation of apo(a) and more secretion.[94] Most patients with familial Lp(a) excess have decreased kringle-4 repeats.[94] An Lp(a) value >50 mg/dL or 125 nmol/L has been associated with increased CVD risk.[6,97] The treatment of choice is to optimize all other CVD risk factors, especially LDL-C. PCSK9 inhibitors (alirocumab and evolocumab) will lower Lp(a) about 30% and LDL-C about 60% along with significant CVD risk reduction on top of statin therapy.[6,94] More effective medications are currently being developed for Lp(a) lowering. In addition, a specific polymorphism in the LPA gene, which occurs in about 5% of the population, has been linked to three-times higher median Lp(a) concentration (>100 mg/dL or 250 nmol/L) and twice the CVD risk, which can be normalized with low-dose aspirin therapy (100 mg/day).[98]

Familial combined hyperlipidemia is the most common familial cause of elevated LDL-C (≥160 mg/dL) and is found in about 15% of families with premature CHD (<age 60 years).[96] These patients have been shown to have increased production of both cholesterol and VLDL and LDL apo-B.[99] Cholesterol production can be assessed by measuring plasma lathosterol concentration. Affected family members may have elevated triglycerides (≥150 mg/dL), elevated LDL-C, or both. Moreover, affected family members often have low HDL-C (<40 mg/dL in men and <50 mg/dL in women), and obesity. The ideal therapy for these patients, in addition to lifestyle modification and weight loss, if indicated, is statin treatment.

Familial dyslipidemia is observed in about 15% of families with premature CVD. Affected family members have elevated fasting triglycerides (≥150 mg/dL).[96] These patients also generally have decreased HDL-C concentration (<40 mg/dL in men and <50 mg/dL in women) and elevated sdLDL-C (>50 mg/dL).[96] They often have delayed clearance of VLDL Apo-B-100 and enhanced clearance of HDL Apo-A-I, but some patients may also have overproduction of VLDL Apo-B-100.[4,94] In contrast to patients with familial combined hyperlipidemia, these patients do not have elevated cholesterol production markers, but are often overweight, insulin resistant, or with diabetes. Effective treatment strategies consist of restriction of energy and sugars, along with exercise, optimization of serum glucose concentration, fibrate therapy (micronized fenofibrate 200 mg/day), and omega-3 fatty acid treatment. Statin therapy may also be indicated to optimize LDL-C concentration, especially if they have CVD.[100]

Low HDL-C (<40 mg/dL) is observed in about 50% of patients with premature CHD. Many of these patients are overweight or obese and have elevated insulin concentration.[94,96,101-103] In addition, more than half have familial combined hyperlipidemia or familial dyslipidemia (see earlier). However, about 5% of families with premature CHD have familial hypoalphalipoproteinemia (low HDL-C) with normal LDL-C and triglyceride concentrations and no secondary causes of dyslipidemia.[96] These patients generally have decreased HDL Apo-A-I production, and other than lifestyle modification and weight loss, the most effective therapy is to optimize all other CVD risk factors, including LDL-C concentration.[94]

Monogenic Lipid Disorders Associated with Elevated LDL-C and/or Tendon Xanthomas

Familial monogenic lipid disorders associated with elevated LDL-C, ≥190 mg/dL, include: 1) familial hypercholesterolemia, 2) lysosomal acid lipase deficiency and, in some cases, 3) phytosterolemia.[94] The most common of these disorders is heterozygous familial hypercholesterolemia, observed in about 1 in 250 people in the general population, and in about 1% of families with premature CVD.[94,96] These patients have delayed clearance of LDL associated with defects in the LDLR, APOB, PCSK9, LDLRAP, and/or STAP1 genes.[94] They can develop arcus senilis, tendinous xanthomas in the Achilles tendons and on the hands, as well as xanthelasma, secondary to cholesterol deposition. Heterozygotes with this disorder usually have LDL-C ≥250 mg/dL, whereas homozygotes often have values ≥500 mg/dL.[94,101] Homozygotes are at high risk of developing CHD and aortic stenosis before the age of 20 years unless they are treated, while untreated heterozygotes develop CHD in the 40s or 50s.[94] Optimal therapy in homozygotes includes LDL apheresis as well as trials of ezetimibe, statins, PCSK9 inhibitors, and incliseran.[94] Heterozygotes can usually be effectively treated with lifestyle modification, and the combination of an effective statin, ezetimibe, and, if necessary, a PCSK9 inhibitor.[94,96,101]

Elevated LDL-C may rarely be caused by lysosomal acid lipase deficiency due to defects in the LIPA gene.[94,102] These patients lack lysosomal acid lipase enzymic activity, resulting in lack of intracellular cholesteryl ester hydrolysis and storage of cholesteryl esters in many cells, especially in the liver, resulting over time in liver failure. The liver failure can be effectively prevented and treated with enzyme replacement using sebelipase (Synageva, Lexington, MA).[94,102]

Another cause of elevated LDL-C is phytosterolemia, characterized by elevated plasma β-sitosterol and campesterol (both >10 mg/L) due to defects in the ATP binding cassette (ABC) transporters G5 and G8, and ABCG5 and ABCG8 genes.[94,103] These patients markedly overabsorb both plant sterols and cholesterol and may develop tubo-eruptive xanthomas and premature CVD. They may also have normal LDL-C.[94,103] The most effective treatment is a diet low in plant sterols and cholesterol as well as ezetimibe, which blocks sterol absorption and lowers plant sterol concentrations by 50%.[103] Some of these patients may also require statin therapy along with ezetimibe to normalize LDL-C and reduce their CVD risk. About 5% of patients with serum LDL-C ≥190 mg/dL have evidence of phytosterolemia.[103]

A final rare group of patients that usually have striking tendinous and tuberous xanthoma with normal LDL-C are those with cerebrotendinous xanthomatosis.[104] The diagnosis is made by the finding of elevated plasma cholestanol, usually >10 mg/L. The best laboratory for this measurement is at Oregon Health Sciences University (Dr. Andrea DeBarber). These patients also develop cholestanol deposits in the brain and are at increased risk of developing cataracts and significant neurologic disease.[104] They cannot convert cholesterol to chenodeoxycholate, one of the major bile acids, because of a defect in the sterol-27-hydoxylase gene (CYP27A1) resulting in shunting of the cholesterol into cholestanol production. The treatment of choice is oral chenodeoxycholate (250 mg three times daily), which prevents the development of neurologic disease and normalizes the plasma cholestanol.[104]

Monogenic Lipid Disorders Associated with Hypertriglyceridemia

In the absence of secondary causes of dyslipidemia, patients with marked hypertriglyceridemia (fasting serum triglyceride concentration >500 mg/dL) usually have defects in the enzyme lipoprotein lipase, in apolipoproteins (apoA-V, apoC-II), or the glycosylphosphatidylinositol anchored high-density lipoprotein binding protein 1 (GPIHBP1), which affect the activity or function of the enzyme.[94,100] Functional lipoprotein lipase is necessary for normal hydrolysis of serum triglycerides, and having such defects results in lack of triglyceride catabolism. These patients often have defects in the following genes: LPL, APOC2, and GPIHBP1.[94,100] Plasma cholesterol concentration is usually approximately one-fifth of the triglyceride concentration, with VLDL-C usually being >50 mg/dL.[94,100] These patients may also have fasting chylomicronemia.[100] Direct LDL-C and HDL-C concentrations are often low at <50 mg/dL and <30 mg/dL, respectively.[100] These patients are at significantly increased risk of developing pancreatitis.[100] They can also develop transient eruptive xanthomata and lipemia retinalis (milky plasma that can be visualized in retinal veins). When LPL activity is measured in postheparin plasma, it is usually very low or absent.

The treatment of choice is dietary fat restriction, to less than 15% of energy, and sugar restriction. Some intake of essential fatty acids by using vegetable oil and fish oil capsules is important.[100] Fenofibrate can be effective in these patients in lowering triglycerides. In children, the dose of generic micronized fenofibrate is 67 mg/day, whereas in adults, the standard dose is 200 mg/day. When such patients present in adulthood, they are usually heterozygous for LPL deficiency or have some other genetic defect, and they are often obese and have diabetes. Treatment with a low-energy, low–saturated fat, low–refined carbohydrate diet is indicated in such patients, along with weight loss, if indicated, exercise, tight control of blood glucose in diabetes, and the use of 200 mg/day of generic micronized fenofibrate. If, after treatment with the fibrate, triglycerides are lower than 400 mg/dL and LDL-C is elevated, statin therapy may be required to normalize LDL-C.[100] In addition, fish oil treatment with four or more capsules per day or prescription icosapent ethyl ester can be effective in lowering triglycerides.[100]

Dysbetalipoproteinemia

Patients with dysbetalipoproteinemia have elevations in total plasma cholesterol and triglycerides that are often both in the range of 250 to 500 mg/dL due to increases in VLDL and intermediate density lipoproteins.[94] Their VLDL-C concentration is markedly elevated (often >100 mg/dL), their direct LDL-C is usually low, and their HDL-C is usually relatively normal. These patients may develop tuboeruptive xanthomas and premature CVD, as well as gout and diabetes.[94] They usually have the ApoE2/2 genotype but may rarely have ApoE deficiency (undetectable plasma ApoE) or hepatic lipase deficiency.[94,105,106] In the latter situation, HDL-C may be elevated. The diagnosis is established by next-generation DNA sequencing of the APOE and LIPC genes.[94] Treatment consists of a diet low in cholesterol, saturated fat, and sugar, as well as weight loss, if indicated. These patients are responsive to micronized fenofibrate (200 mg/day) and statin therapy, which can be used in combination.[94]

Monogenic HDL Deficiency Disorders

Marked HDL deficiency (HDL-C <10 mg/dL) is rare. It can be observed in severe hypertriglyceridemia (fasting triglycerides >1,000 mg/dL), liver cirrhosis, marked inflammation, or uncontrolled diabetes.[107-109] In the absence of these conditions, such patients may have: 1) Apo-A-I deficiency, 2) Tangier disease, 3) combined lecithin:cholesterol acyltransferase (LCAT) deficiency, or 4) α-LCAT deficiency, also known as fisheye disease.[94,107-111] Patients who are unable to produce apo-A-I have HDL-C <5 mg/dL, undetectable plasma Apo-A-I, and normal LDL-C and triglyceride concentrations. They may develop planar xanthomas, arcus juvenilis, and severe premature CHD.[107-109] They have defects in the APOAI gene or in the entire APOAI/CIII/AIV gene complex, and the treatment of choice is to optimize their LDL-C with lifestyle modification and statin therapy.[107-109]

Patients with Tangier disease have mild diffuse corneal opacification seen on slit-lamp examination, orange mucosa throughout the gastrointestinal tract, and hepatosplenomegaly. They have defective cellular cholesterol efflux because of defects in adenosine triphosphate–binding cassette protein A1 (ABCA1).[107-109] They have only pre-β HDL in their plasma, low LDL-C (~50% of normal), and moderate hypertriglyceridemia. They often develop CHD in their 50s or 60s. The treatment of choice is to optimize their LDL-C with lifestyle modification and statin therapy.[107-109]

Patients with LCAT deficiency are unable to esterify their plasma cholesterol, and therefore, have elevated plasma, LDL, and HDL free cholesterol concentrations.[107,109,110] They have severe diffuse corneal opacification, abnormal LDL known as Lp-X, marked HDL deficiency, and have little preβ and α-4 HDL present in their plasma.[110] They develop splenomegaly, anemia, and eventually, renal failure.[110] They do not appear to develop premature CVD. The treatment of choice in these patients is a heart healthy diet and control of all other risk factors for kidney disease, especially hypertension if present.

A variant of LCAT disease, known as fisheye disease, is characterized by an inability to esterify the cholesterol only in HDL.[111] These patients also have corneal opacification, but they do not have Lp-X and they do not develop anemia or renal insufficiency. However, they often have elevated LDL-C and triglyceride concentrations and may develop premature CVD. The treatment of choice in these latter patients, in addition to lifestyle modification, is to optimize LDL-C with statin therapy.

Monogenic LDL Deficiency Disorders

There are two disorders associated with marked LDL deficiency (LDL-C <40 mg/dL): 1) abetalipoproteinemia with undetectable apoB concentration, or 2) hypobetalipoproteinemia with detectable but very low apoB concentration (<40 mg/dL). Patients with abetalipoproteinemia have relatively normal HDL-C , but LDL-C that is very low. They are unable to secrete either apoB-48 from the intestines or apoB-100 from the liver because of defective microsomal transfer protein (MTP). MTP allows for the combining of apoB with triglycerides for the secretion of chylomicrons in the intestine and VLDL in the liver.[94] Mean plasma cholesterol and triglyceride values in these patients are about 50 mg/dL and 10 mg/dL, respectively, and the HDL-C is about 50 mg/dL. The diagnosis is established by the finding of undetectable ap-B in the plasma. These patients also have very low concentrations of vitamin A and E, and present in childhood with fat malabsorption, with atypical retinitis pigmentosa by about age 10 years and, if the condition has not been detected at that point, with spinocerebellar ataxia in the third and fourth decades of life. The treatment of choice is supplementation with fat-soluble vitamins (15,000 units of vitamin A per day, 1,000 mg of vitamin E per day, daily use of one tablespoon of vegetable oil as salad dressing on salad, two fish oil capsules per day, and use of vitamin K preoperatively to support adequate clotting or 1 unit of fresh frozen plasma before major surgery).[94]

Patients with hypobetalipoproteinemia usually have total cholesterol and triglycerides of about 80 and 40 mg/dL, respectively, with an HDL-C of about 40 to 50 mg/dL; hence their LDL-C concentration is usually <40 mg/dL.[94] Unlike patients with abetalipoproteinemia, these patients have detectable plasma or serum apoB and usually normal serum concentrations of fat-soluble vitamins. Their defects are due to apoB truncations from APOB gene mutations.[94] No treatment is generally required. However, these patients may have fatty liver disease because of decreased ability to export lipoproteins from the liver.[94]

Conclusions

CVD remains the leading cause of death worldwide, and major risk factors have been defined as increased age, male sex, hypertension, hypertension treatment, diabetes, smoking, elevated total cholesterol, and decreased HDL-C. The ACC/AHA CVD risk calculator should be accessed at: https://static.heart.org/riskcalc/app/index.html#!/baseline-risk. Adults with a 10-year risk of >7.5% are candidates for statin therapy in addition to lifestyle modification. Additional CVD risk factors include family history, kidney disease, and increased serum concentrations of LDL-C, sdLDL-C, Lp(a), hs-CRP, triglycerides, homocysteine, troponin T, and NT-proBNP. The cornerstone of CVD prevention is a diet low in saturated fat, trans fats, cholesterol, and sugars, and enriched in plant foods, vegetable oils, and omega-3 fatty acids, along with salt restriction in those with hypertension, and regular exercise of ≥150 minutes/week. Optimal measures for CVD prevention include LDL-C <70 mg/dL, blood pressure <120/80 mmHg, glycosylated hemoglobin <6.5%, and hs-CRP <1.0 mg/L. For the management of significant lipoprotein disorders, the cornerstone of treatment is lifestyle modification and statin therapy. However, additional LDL-C lowering therapies, including ezetimibe and PCSK9 inhibitors, are now available, as are effective triglyceride-lowering treatments, including omega-3 fatty acids and fenofibrate.

References

1. World Health Organization. www.who.int/health-topics/cardiovascular-diseases#tab=tab_1
2. Centers for Disease Control and Prevention. www.cdc.gov/heartdisease/index.htm
3. Tsao CW, Aday AW, Almarzooq ZI, et al. Circ 2022;145:e153-e639.
4. Schaefer EJ. Am J Clin Nutr. 2002; 75:191–212.
5. Goff DC Jr., Lloyd-Jones DM, Bennett G, et al. Circ 2014;129: S49–S73.
6. Grundy SM, Stone NJ, Bailey AL, et al. Circ 2019;139:e10821143.
7. Ikezaki H, Fisher VA, Lim E, et al. Clin Chem 2019;65: 1102–1114.
8. Ikezaki H, Lim E, Cupples LA, Liu CT, Asztalos BF, Schaefer EJ. J Am Heart Assoc 2021;10:e019140.
9. Oluleye OW, Folsom AR, Nambu V, et al Ann Epidemiol 2013;23:66–73.
10. Ridker PM, Danielson E, Fonseca FA, et al. N Engl J Med 2008;359:2195–2207.
11. Ridker PM, Danielson E, Fonseca FA, et al. Lancet 2009;373: 1175–1182.
12. Silverman MG, Ference BA, Im K, et al. JAMA 2016;316: 1289–1297.
13. Mortensen SA, Rosenfeldt F, Kumar A, et al. JACC 2014;2: 641–649.
14. Ridker PM, Buring JE, Rifai N, et al. JAMA 2007;297:611–619.
15. Ridker PM, Paynter NP, Rifai N, et al. Circ 2008;118:2243–2251.
16. Budoff MJ Shaw LJ, Lou ST, et al. J Am Coll Cardiol 2007;49: 1860–1870.
17. Mesa website. Accessed December 12, 2011. Available at www.mesa-nhlbi.org
18. Willeit P, Tschiderer L, Altura E, et al. Circ 2020;142:621–642.
19. Williams MC, Kwiecinski J, Doris M, et al. Circ 2020;141: 1452–1462.
20. Keys A. Circ 1970;41:1162–1175.
21. Kato H, Tillotson J, Nichaman JZ, et al. Am J Epidemiol 1973;97:372–383.
22. Stamler J. In: Levy RI, Dennis BR, Ernst N, eds. Nutrition, Lipids, and Coronary Heart Disease. New York, NY: Raven Press,1979; 25–88.
23. Schaefer EJ, Augustin JL, Schaefer MM, et al. Am J Clin Nutr 2000;71:746–751.
24. Yusuf S, Hawken S, Ounpuu S, et al. Lancet 2004;364:937–952.
25. Diffenderfer MR, Rajapakse N, Pham E et al. J Clin Lipidol 2022;16:184–197.
26. Schaefer EJ, Lichtenstein AH, Lamon-Fava S, et al. Arterioscler Thromb Vasc Biol 1995;15:1079–1085.
27. Schaefer EJ, Lichtenstein AH, Lamon-Fava, S et al. Am J Clin Nutr 1996;63:234–241.
28. Lichtenstein AH, Ausman LM, Jalbert SM, et al. J Lipid Res 2002;43:264–273.
29. Lichtenstein AH, Ausman LM, Carrasco W, et al. Arterioscler Thromb 1994;14:168–175.
30. Lichtenstein AH, Ausman LM, Jalbert SM, et al. N Engl J Med 1999;340:1933–1940.
31. Ooi EM, Lichtenstein AH, Millar JS, et al. J Lipid Res 2012;53:1958–1967.
32. Hegsted DM, Ausman LM, Johnson JA, et al. Am J Clin Nutr 1993;57:875–883.
33. Mensink RP, Katan MB. Arterioscler Thromb 1992;12:911–919.
34. Yu S, Derr J, Etherton TD, et al. Am J Clin Nutr 1995;61: 1129–39.
35. Zhao B, Gan L, Graubard BI, et al. Circ 2022;145:1506–1520.
36. Schaefer EJ, Lamon-Fava S, Ausman LM, et al. Am J Clin Nutr 1997;65:823–830.
37. Li Z, Otvos JD, Lamon-Fava S, et al. J Nutr 2003;133:3428–3433.
38. Lopez-Miranda J, Ordovas JM, Mata P, et al. J Lipid Res 1994;35:1965–1975.
39. Talati R, Sobieraj DM, Makanji SS, et al. J Am Diet Assoc 2010;110:719–726.
40. Olson BH, Anderson SM, Becker MP, et al. J Nutr 1997;127: 1973–1980.
41. Stanhope Kl, Schwartz JM, Keim NL, et al. J Clin Invest 2009;119:1322–1334.
42. Diabetes Program Prevention Research Group, Knowler WC, Barrett-Connor E, Fowler SE at al. N Engl J Med 2002;346: 393–403.
43. Leren P. Acta Med Scand 1966;466:1–92.
44. Leren P. Circ 1970;42:935–942.
45. Hjermann I, Holme I, Velve Byre K, et al. Lancet 1981;2: 1303–1310.
46. Hjermann I, Holme I, Leren P. Am J Med 1986;80:7–11.
47. Holme I, Haaheim LL, Tonstad S, Hjermann I. Nutr Metab Cardiovasc Dis 2006;16:330–338.
48. Dayton S, Pearce ML, Goldman H, et al. Lancet 1968;2: 1060–1062.
49. Dayton S, Pearce ML, Hashimoto S. Circ 1969;34:1–63.
50. Pearce ML, Dayton S. Lancet 1971;1:464–467.
51. Sturdevant RA, Pearce ML, Dayton S. N Engl J Med 1973;288: 24–27.
52. Turpeinen O. Circ 1979;59:1–7.
53. Turpeinen O, Karvonen MJ, Pekkarinen M, et al. Int J Epidemiol 1979;8:99–118.
54. Miettinen M, Turpeinen O, Karvonen MJ, et al. Int J Epidemiol 1983;12:7–25.
55. Frantz ID Jr, Dawson EA, Ashman PL, et al. Arteriosclerosis 1989;9:129–135.
56. de Lorgeril M, Salen P, Martin JL, et al. Circ 1999;99:779–785.
57. Patterson RE, Kristal AR, Tinker LF, et al. Ann Epidemiol 1999;9:178–187.
58. Howard BV, Van Horn L, Hsia J, et al. JAMA 2006;295: 655–666.
59. Prentice RL, Chlebowski RT, Patterson R, et al. JAMA 2006;295: 629–642.
60. Beresford SM, Johnson KC, Ritterberg C, et al. JAMA 2006;295:643–654.
61. Tinker LF, Bonds DE, Margolis KL, et al. Arch Intern Med 2008;168:1500–1511.
62. Estruch R, Ros E, Salas-Calvado J, et al. N Engl J Med 2013;368:1279–1290.
63. Estruch R, Ros E, Salas-Calvado J, et al. N Engl J Med 2018;378:e34.
64. Hooper L, Martin N, Jimosh O, et al. Cochrane Database Syst Rev 2020;19:CD011737.
65. Burr ML, Gilbert JF, Holliday RM, et al. Lancet 1989;2:757–761.
66. Burr ML. Proc Nutr Soc 2007;66:9–15.
67. GISSI Prevenzione Investigators. Lancet 1999;354:447–455.
68. Marchioli R, Barzi F, Bomba E, et al. Circ 2002;105:1897–1903.
69. Yokoyama M, Origasa H, Matzuzaki M, et al. Lancet 2007;370: 1090–1098.
70. Matsuzaki M, Yokoyama M, Saito Y, et al. Circ J 2009;73: 1283–1290.
71. Tanaka K, Ishikawa Y, Yokoyama M, et al. Stroke 2008;39: 2052–2058.
72. Saito Y, Yokoyama M, Origasa H, et al. Atherosclerosis 2008;200: 135–140.
73. Oikawa S, Yokoyama M, Origasa H, et al. Atherosclerosis 2009;206:535–539.
74. Itakura H, Yokoyama M, Matsuzaki M, et al. J Atheroscler Thromb 2011;18:99–107.

75. Kromhout D, Giltay EJ, Geleijnse JM, et al. N Engl J Med 2010;363:2015–2026.
76. Bhatt D, Steg PG, Miller M, et al. N Engl J Med 2019;380: 11–22.
77. Nicholls SJ, Lincoff AM, Garcia M, et al. JAMA 2020;324: 2268–2280.
78. Yokoyama Y, Kuno T, Morita SX, et al. J Cardiol 2022;80: 416–422.
79. Markozannes G, Ntzani EE, Tsapas A, et al. Clin Nutr 2022;41: 923–930.
80. Yan J, Liu M, Yang D et al. Cardiovasc Drugs Ther. 2022. doi:org/10.1007/s10557-022-07379-z
81. Study of the Effectiveness of Additional Reductions in Cholesterol and Homocystein (SEARCH) Collaborative Group, Armitage JM, Bowman L, Clarke RJ, et al. JAMA 2010;303:2486–2494.
82. Clarke R, Halsey J, Lewington S, et al. Arch Intern Med 2010;170:1622–1631.
83. Clarke R, Bennett D, Parish S, et al. Am J Clin Nutr 2014;100: 657–666.
84. Lonn E, Yusuf S, Arnold MJ, et al. N Engl J Med 2006;354: 1567–1577.
85. Saposnik G, Ray JG, Sheridan P, et al. Stroke 2009;40: 1365–1372.
86. Huo Y, Li J, Qin X, et al. JAMA 2015;313:1325–1335.
87. Li J, Ogrodnik M, Devine S, et al. Alzheimers Dement 2018;14: 35–42.
88. Smith AD, Refsum H, Bottiglieri T, et al. J Alzheimers Dis 2018;62:561–570.
89. Spence DJ, Hankey GJ. Stroke 53:2702–2708.
90. Jenkins DJA, Spence JD, Giovanucci EL, et al. JACC 2021;77: 423–436.
91. Eckel RH, Jakicic JM, Ard JD, et al. Circ 2014;129:S76–99.
92. US Dietary Guidelines 2020-2025. Available at www.dietary guidelines.gov.
93. Lichtenstein AH, Appel LJ, Vadiveloo M, et al. Circ 2021;144: e472–e487.
94. Schaefer EJ, Geller AS, Endress G. Curr Opin Lipidol 2019;30: 56–62.
95. Otokozawa S, Diffenderfer MR, Asztalos BF, et al. Metabolism 2009;58:1536–1542.
96. Genest JJ, Martin-Munley S, McNamara JR, et al. Circ 1992;85: 2025–2033.
97. Lamon-Fava S, Marcovina SM, Albers JJ, et al. J Lipid Res 2011;52:1181–1187.
98. Chasman DI, Shiffman D, Zee RY, et al. Atherosclerosis 2009;203: 371–376.
99. van Himbergen T, Otokozawa S, Matthan NR, et al. Arterioscler Thromb Vasc Biol 2010;30:1113–20.
100. Chyzhyk V, Kozmic S, Brown AS, et al. J Clin Lipidol 2019;13: 89–94.
101. Corral P, Geller AS, Polisecki EY, et al. Atherosclerosis 2018;277: 256–261.
102. Pullinger CR, Stock EO, Movsesyan I, et al. J Clin Lipidol 2015;9:716–726.
103. Brinton EA, Hopkins PN, Hegele RA, et al. J Clin Lipidol 2018;12:152–161.
104. Duell PB, Salen G, Eichler FS, et al. J Clin Lipidol 2018;12: 1169–1178.
105. Mak AC, Pullinger CR, Tang LF, et al. JAMA Neurol 2014;71:1228–1236.
106. Tani M, Horvath KV, Lamarche B, et al. Atherosclerosis 2016;253: 7–14.
107. Schaefer EJ, Santos RD, Asztalos BF. Curr Opin Lipidol 2010;21:289–297.
108. Schaefer EJ, Anthanont P, Diffenderfer MR, at al. Prog Cardiovasc Dis 2016;59:97–106.
109. Geller AS, Polisecki EY, Diffenderfer MR, et al. J Lipid Res 2018;59:2421–2435.
110. Roshan B, Ganda OP, Desilva R, et al. J Clin Lipidol 2011;5: 493–499.
111. Dimick SM, Sallee B, Asztalos BF, et al. J Clin Lipidol 2014:8: 223–230.

B. CARDIOVASCULAR DISEASES

CHAPTER 68

Diet and Blood Pressure

Gabriela Andries, Beth Coplan, and Marc Cohen

CHAPTER OUTLINE

Abbreviations

ACC/AHA, American College of Cardiology/American Heart Association; **BMI**, body mass index; **BP**, blood pressure; **CVD**, cardiovascular disease; **CI**, confidence interval; **DASH**, Dietary Approaches to Stop Hypertension; **FDA**, Food and Drug Administration; **HR**, hazard ratio; **NHANES**, National Health and Nutritional Examination Survey; **RCT**, randomized clinical trials; **HTN**, systemic hypertension; **U.S.**, United States of America

Introduction

Elevated systemic blood pressure (BP), referred to here as systemic hypertension (HTN), is the leading cause of cardiovascular disease (CVD) and chronic kidney disease (CKD). Ironically, it is one of the most preventable risk factors for premature death and disability.[1,2] Worldwide, it is estimated that 1.39 billion adults have HTN, with an even distribution between men and women. Despite a plethora of evidence involving lifestyle modification and medical drug therapies, the global age-standardized prevalence of HTN has **increased** by 5.2%, between years 2000 and 2010. In 2015, elevated BP (≥140 mmHg) was associated with 7.8 million all-cause deaths. Forty percent of those deaths were related to ischemic heart disease, 38% to ischemic stroke, and 43% to hemorrhagic stroke. As the worldwide prevalence of HTN increased, the estimated number of BP-related all-cause and cardiovascular mortalities has significantly increased, especially in the low and middle-income countries.[3]

In the United States, surveys conducted by the National Health and Nutrition Examination Survey (NHANES) showed that 45% of individuals aged 18 and above had HTN, and the prevalence was higher in men (51%) than women (39.7%).[4] The prevalence also increased with age; 75% of adults aged 60 or older had HTN. Non-Hispanic blacks had a higher prevalence of HTN (57.1%) compared with non-Hispanic whites (43.6%) or Hispanic adults (43.7%).[4] The NHANES report provided 2017–2018 HTN prevalence estimates, using the 2017 American College of Cardiology and American Heart Association (ACC/AHA) definition of hypertension and new guidelines. The 2017 guidelines redefined HTN by lowering the previous threshold levels of ≥140/≥90 mmHg down to ≥130/≥80 mmHg. It also reclassified pre-hypertension as elevated blood pressure (120–129/<80 mmHg).[5]

The rationale for this new categorization of HTN was based on 1) an expanding evidence base showing a linear relationship between elevated BP and CVD risk, and 2) evidence that more intensive blood pressure control improves cardiovascular outcomes and survival.[6-9] In the Systolic Blood Pressure Intervention Trial (SPRINT), first published in 2015 with follow-up data published in 2021, 9,361 patients with HTN and baseline elevated cardiovascular risk were randomized to either an intensive systolic BP target (<120 mmHg), or a standard systolic BP target (<140 mmHg).[8,9] At 3.3 years of follow-up, the investigators found that the lower BP target resulted in a significantly lower rate of the composite outcome of myocardial infarction, acute coronary syndrome, stroke, acute decompensated heart failure, or cardiovascular death; 1.77% /year vs 2.4% /year; (hazard ratio [HR] 0.73; 95% confidence interval [CI], 0.63–0.86). Investigators also found lower **all-cause mortality** with intensive systolic BP targets (1.1% per year vs 1.4% per year; HR 0.75; 95% CI 0.61–0.92).[8,9]

Although the cause of HTN is complex and multifactorial, and varies between individuals, evidence appears to consistently highlight an imbalance in dietary habits (high sodium, low potassium, and limited fruit and vegetable intake), physical inactivity, and excess alcohol consumption, as major contributors to the development of HTN across populations.[10-12] Overall, lifestyle changes focused on achieving a healthier diet, increasing physical activity, and moderation in alcohol intake are the cornerstone of initial HTN prevention and treatment.[5,13] Contemporary dietary recommendations have focused on empirical, clinical trial evidence examining foods consumed in combinations, or the overall dietary pattern, as opposed to focusing on individual nutrients. Several dietary regimens have undergone rigorous clinical study and have been endorsed by multiple societal guidelines in relation to HTN: the Mediterranean diet,[14,15] Dietary Approaches to Stop Hypertension (DASH) diet, and the DASH-Sodium diet.[16,17]

Dietary changes that lower BP can prevent normotensive individuals from developing HTN and, thereby, reduce their risk for CVD. In fact, even minor reductions in BP across an entire population could have a tremendous, beneficial public health impact. It has been estimated that just 3 mmHg (average) reduction in systolic BP is associated with an 8% reduction in stroke mortality and a 5% reduction in mortality from Coronary Heart Disease (CHD) (**Figure 68.1**).[18]

Numerous subsequent studies showed similar findings and a meta-analysis of RCTs that studied the effects of the DASH diet (or modified DASH-style dietary patterns) on CVD risk factors demonstrated significant decreases in systolic and diastolic BP, by 5.2 and 2.6 mmHg, respectively. These reductions contributed to a 13% reduction in the 10-year Framingham risk score for CVD.[19]

Although there is ample evidence that dietary changes lower BP, there are fewer data on whether dietary change can reduce the age-related rise in systolic BP, which is approximately 0.6 mmHg per year in adults.[20] Trials of dietary change are often less than 3 years in duration and, therefore, are not of sufficient length to answer this question. Hence, whether the BP reductions observed in these trials have merely shifted the age-related rise in BP curve downward, without a change in slope (**Figure 68.2A**), or have actually reduced its slope (see **Figure 68.2B**) cannot be determined from short-term trials.[21]

Nonetheless, migration studies, ecologic studies, and, most recently, observational analyses of trial data[22] offer some evidence to suggest that dietary factors may indeed reduce the age-associated rise in SBP. This chapter reviews the evidence on the relationships between diet and BP. The summary of evidence and corresponding recommendations largely reflect existing reviews of RCTs and consensus statements, with updates from roughly the past 10 years (2010–2022). **Table 68.1** and **Figure 68.3** provide a summary of this evidence, derived from the Whelton PK, et al., 2017 ACC/AHA/AAPA/ABC/ACPM/AGS/APhA/ASH/ASPC/NMA/PCNA guidelines for the prevention, detection, evaluation, and management of high blood pressure in adults: A report of the American College of Cardiology/American Heart Association Task Force on Clinical Practice Guidelines.[5]

Dietary Factors That Lower Blood Pressure

Weight Loss

Elevated body mass index (BMI) is an independent risk factor for the development of type 2 diabetes, dyslipidemia, CKD, ischemic heart disease, stroke, nonalcoholic fatty liver disease, and obstructive sleep apnea. In addition, there is a high prevalence of HTN in persons with obesity.[23] The clinical significance of this

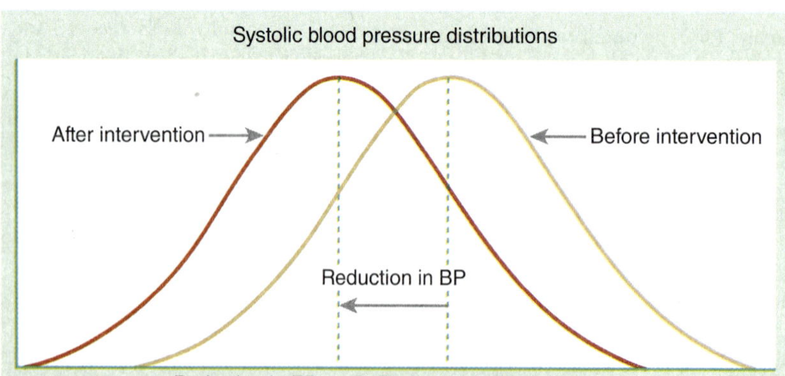

Reduction in BP mmHg	% Reduction in mortality Stroke	CHD	Total
2	–6	–4	–3
3	–8	–5	–4
5	–14	–9	–7

Figure 68.1 Estimated effects of population wide shifts in systolic BP on mortality. CHD, coronary heart disease.

Data from Stamler R. Hypertension 1991;17: I-16–I-20.[18]

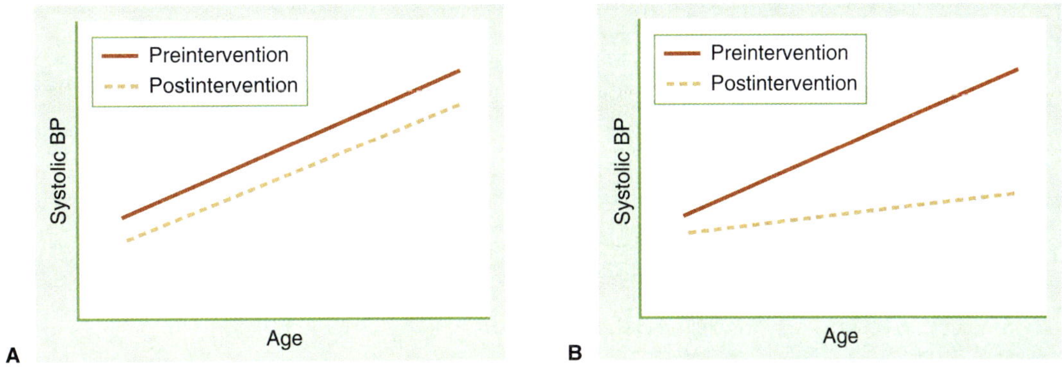

Figure 68.2 A. Model in which a dietary intervention shifts age-BP curve downward without affecting the slope. B. Model in which a dietary intervention shifts age-BP curve downward and reduces the slope.

Reproduced from Appel LJ. Hypertension: A companion to Braunwald's Heart Disease. Philadelphia: Saunders 2007;201–212.[21]

association is magnified by the high and increasing prevalence of obesity worldwide. In 2015, there were 603.7 million obese adults worldwide. Over the past 25 years, the prevalence of obesity in both children and adults has doubled in 73 countries.[24] According to the 2018 National Health and Nutrition Examination Survey, the prevalence of obesity was >40% in the United States.[25] Even more disturbing is the global increase in the prevalence of overweight and obesity in children and adolescents. It was estimated that 41 million children below the age of 5 were overweight (BMI in the 85th–95th percentile for sex and age), or obese (BMI ≥95th percentile) according to the WHO Global Estimate 2014 report.[26]

Blood pressure increases in parallel with increases in body weight. This relationship is multifaceted and closely related to

Table 68.1 Best Proven Nonpharmacologic Interventions for Prevention and Treatment of Hypertension[a]

	Nonpharmacologic Intervention	Dose	Approximate Impact on SBP		
			Hypertension	Normotension	Reference
Weight loss	Weight/body fat	Best goal is ideal body weight, but aim for at least a 1-kg reduction in body weight for most adults who are overweight. Expect about 1 mmHg for every 1-kg reduction in body weight.	~5 mmHg	~2/3 mmHg	S6.2-1
Healthy diet	DASH dietary pattern	Consume a diet rich in fruits, vegetables, whole grains, and low-fat dairy products, with reduced content of saturated and total fat.	~11 mmHg	~3 mmHg	S6.2-6, S6.2-7
Reduced intake of dietary sodium	Dietary sodium	Optimal goal is <1,500 mg/d, but aim for at least a 1,000-mg/d reduction in most adults.	~5-6 mmHg	~2/3 mmHg	S6.2-9, S6.2-10
Enhanced intake of dietary potassium	Dietary potassium	Aim for 3,500-5,000 mg/d, preferably by consumption of a diet rich in potassium.	~4/5 mmHg	~2 mmHg	S6.2-13
Physical activity	Aerobic	90–150 min/wk 65%–75% heart rate reserve	~5/8 mmHg	~2/4 mmHg	S6.2-18, S6.2-22
	Dynamic resistance	90–150 min/wk 50%–80% 1 rep maximum Six exercises, three sets/exercise, 10 repetitions/set	~4 mmHg	~2 mmHg	S6.2-18
	Isometric resistance	4 × 2 min (hand grip), 1 min rest between exercises, 30%–40% maximum voluntary contraction, three sessions/wk 8–10 wk	~5 mmHg	~4 mmHg	S6.2-19, S6.2-31

(continues)

Table 68.1 Best Proven Nonpharmacologic Interventions for Prevention and Treatment of Hypertension[a] *(continued)*

	Nonpharmacologic Intervention	Dose	Approximate Impact on SBP		
			Hypertension	Normotension	Reference
Moderation in alcohol intake	Alcohol consumption	In individuals who drink alcohol, reduce alcohol[b] to: Men: <2 drinks daily Women: <1 drink daily	~4 mmHg	~3 mmHg	S6.2–22, S6.2–24

Your Guide to Lowering Your Blood Pressure With DASH. Available at www.nhlbi.nih.gov/files/docs/public/heart/new_dash.pdf. Accessed Jan 5, 2024.

Top 10 Dash Diet Tips. Available at: https://dashdiet.org/dash-diet-tips.html Accessed Jan 5, 2024.

[a]Type, dose, and expected impact on BP in adults with a normal BP and with hypertension.

[b]In the United States, one "standard" drink contains roughly 14 g of pure alcohol, which is typically found in 12 oz of regular beer (usually about 5% alcohol), 5 oz of wine (usually about 12% alcohol), and 1.5 oz of distilled spirits (usually about 40% alcohol).

DASH, Dietary Approaches to Stop Hypertension; *SBP*, systolic blood pressure.

Whelton PK, Carey RM, Aronow WS, et al. J Am Coll Cardiol 2018;71:e127–248.[5]

Reproduced from Whelton PK, Carey RM, Aronow WS, Casey DE Jr, Collins KJ, Dennison Himmelfarb C, DePalma SM, Gidding S, Jamerson KA, Jones DW, MacLaughlin EJ, Muntner P, Ovbiagele B, Smith SC Jr, Spencer CC, Stafford RS, Taler SJ, Thomas RJ, Williams KA Sr, Williamson JD, Wright JT Jr. 2017 ACC/AHA/AAPA/ABC/ACPM/AGS/APhA/ASH/ASPC/NMA/PCNA guideline for the prevention, detection, evaluation, and management of high blood pressure in adults: a report of the American College of Cardiology/American Heart Association Task Force on Clinical Practice Guidelines. J Am Coll Cardiol 2018;71:e127–248.

Recommendations for nonpharmacological interventions

References that support recommendations are summarized in **Online Data Supplements 9–21.**

COR	LOE	Recommendations
I	A	1. Weight loss is recommended to reduce BP in adults with elevated BP or hypertension who are overweight or obese.[S6.2-1–S6.2-4]
I	A	2. A heart-healthy diet, such as the DASH (Dietary Approaches to Stop Hypertension) diet, that facilitates achieving a desirable weight is recommended for adults with elevated BP or hypertension.[S6.2-5–S6.2-7]
I	A	3. Sodium reduction is recommended for adults with elevated BP or hypertension.[S6.2-8–S6.2-12]
I	A	4. Potassium supplementation, preferably in dietary modification, is recommended for adults with elevated BP or hypertension, unless contraindicated by the presence of CKD or use of drugs that reduce potassium excretion.[S6.2-13–S6.2-17]
I	A	5. Increased physical activity with a structured exercise program is recommended for adults with elevated BP or hypertension.[S6.2-3,S6.2-4,S6.2-12,S6.2-18–S6.2-22]
I	A	6. Adult men and women with elevated BP or hypertension who currently consume alcohol should be advised to drink no more than 2 and 1 standard drinks* per day, respectively.[S6.2-23–S62-28]

*In the United States, 1 "standard" drink contains roughly 14 g of pure alcohol, which is typically found in 12 oz of regular beer (usually about 5% alcohol), 5 oz of wine (usually about 12% alcohol), and 1.5 oz of distilled spirits (usually about 40% alcohol).[S6.2-29]

Figure 68.3 Class of recommendation (COR) and the level of evidence (LOE), for non-pharmacologic interventions.[5]

Reproduced from Whelton PK, Carey RM, Aronow WS, Casey DE Jr, Collins KJ, Dennison Himmelfarb C, DePalma SM, Gidding S, Jamerson KA, Jones DW, MacLaughlin EJ, Muntner P, Ovbiagele B, Smith SC Jr, Spencer CC, Stafford RS, Taler SJ, Thomas RJ, Williams KA Sr, Williamson JD, Wright JT Jr. 2017 ACC/AHA/AAPA/ABC/ACPM/AGS/APhA/ASH/ASPC/NMA/PCNA guideline for the prevention, detection, evaluation, and management of high blood pressure in adults: a report of the American College of Cardiology/American Heart Association Task Force on Clinical Practice Guidelines. J Am Coll Cardiol 2018;71:e127–248.

other comorbidities present in obesity. Multiple studies have shown that the pathophysiologic mechanisms associating obesity with HTN include insulin resistance, inflammation, oxidative stress, adipokines, the sympathetic nervous system, and the renin-angiotensin aldosterone system. Increased visceral adiposity is strongly related with endothelial dysfunction, attributed to elevated inflammatory cytokines, oxidative stress, and reduced nitric oxide availability. In addition to heightened sympathetic nervous system activity, excess adipose tissue is associated with increased angiotensin type 1 and 2 receptor expression, as well as elevated circulating angiotensin II, angiotensin-converting enzyme, and aldosterone concentrations. Elevations in these molecules cause impaired natriuresis, increased sodium reabsorption, and extracellular volume expansion, inciting the development and progression of HTN in obesity.[23,27]

On average, weight loss reduces BP. The reduction in BP occurs before, and even without, attainment of an ideal body weight. In a meta-analysis of 25 randomized controlled trials, an average weight loss of 5.1 kg lowered mean systolic BP by 4.4 mmHg and diastolic BP by 3.6 mmHg. It was estimated that BP would be reduced by ~1 mmHg with every kilogram of weight loss. In subgroup analyses, those who lost more weight (>5 kg) had greater reductions in BP.[28] Clinical trials, dose-response analyses, and observational studies also provide strong evidence that greater weight loss leads to greater BP reduction. The 2017 ACC/AHA Guidelines for Management of Hypertension recommend weight loss as one of the best proven nonpharmacologic interventions to reduce BP in adults with elevated BP or HTN who are overweight or obese (class of recommendation I, level of evidence A).[5]

Lifestyle interventions, such as diet and exercise, however, are limited in their magnitude of effect and **sustainability**. Although patients who undergo lifestyle interventions see initial improvements in BP, behavioral counseling along with diet and exercise often do not result in **persistent** weight loss. As a result, lifestyle interventions often fail to attenuate the long-term adverse cardiovascular outcomes resulting from obesity. In the TOHP II (Trials of Hypertension Prevention phase II), overweight participants with stage 1 HTN were randomized to undergo a 3-year program of group meetings and individual counseling focused on dietary change, physical activity, and social support. The mean weight change in the weight loss intervention group was −4.4 kg at 6 months, −2.0 kg at 18 months, and −0.2 kg at 36 months. The reductions in systolic and diastolic BP were 1.3 and 0.9 mmHg, respectively, at 36 months (**Figure 68.4**).[29]

Unfortunately, the long-term follow-up at 26 years post-trial failed to show any difference in mortality between groups.[30]

Similar results were also seen in the LOOK AHEAD trial (Action for Health in Diabetes).[31] In this trial, 5,145 patients with obesity and type 2 diabetes were randomly assigned to receive intensive lifestyle intervention, vs diabetes support and education. Although the intervention group had larger weight loss initially, there was only a 2.6 kg between-group difference after 10 years. There was no significant difference between the two groups in cardiovascular morbidity and mortality.[31]

Surgical bariatric procedures have been shown to have more profound and long-lasting effects on weight loss. Multiple RCTs and meta-analysis have confirmed the role of bariatric surgery in improving CVD risk factors, including better glycemic, lipid, and BP control.[32-34] In one cohort, the rate of remission from HTN was reported to be as high as 83% with only 13% of patients experiencing resurgence of HTN after 5 years.[35] The impact of bariatric surgery on CVD risk factors was proven to translate into significant reduction in major adverse cardiovascular events. In a large cohort of 190,000 U.S. Medicare patients with morbid obesity and a median follow-up of 4 years, the bariatric surgery group was shown to have a significant 37% risk reduction in **all-cause mortality** (HR: 0.63; 95% CI: 0.60–0.66; $P<0.001$) compared with the control group.[36] In addition, bariatric surgery was associated with a 54% lower risk of new-onset HF hospitalization (HR: 0.46; 95% CI: 0.44–0.49; $P<0.001$), 37% lower risk of MI (HR: 0.63; 95% CI: 0.59–0.68; $P<0.001$), and 29% lower risk of ischemic stroke (HR: 0.71; 95% CI: 0.65–0.79; $P<0.001$). Moreover, the number needed to treat (NNT) with bariatric surgery at 3 years of follow-up to prevent a major adverse cardiovascular event was only 15 patients.[36]

Surgical weight loss does confer risks due to the procedures themselves, including surgical complications, reoperation and reintervention, acute kidney injury, and perioperative mortality.

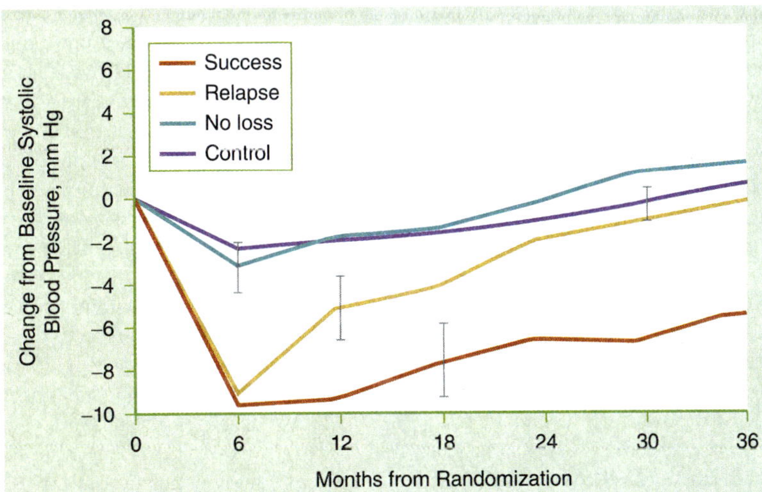

Figure 68.4 Mean systolic BP change in the TOHP2 in four groups of participants: those assigned to weight loss group who successfully maintained weight loss, those assigned to a weight loss group who lost weight experienced relapse, those assigned to weight loss group who never lost weight, and the control group.

Reproduced from Stevens VJ et al. Ann Intern Med 2001;134:1–11.[29]

Nonetheless, these risks have relatively low rates of occurrence with current surgical techniques and are outweighed by the improved morbidity and mortality observed following weight loss surgery.[37,38] Current guidelines recommended consideration of bariatric surgery in patients with BMI ≥40 kg/m^2 or ≥ 35 kg/m^2 with serious obesity-related comorbidities.[39]

Although further study of the effects of weight loss in combating the age-related rise in BP is needed, available evidence, in aggregate, strongly supports weight loss as an effective method for preventing and treating HTN.

Reduced Salt (Sodium Chloride) Intake

The terms salt and sodium are often used interchangeably. Salt is composed of 40% sodium and 60% chloride (1 g sodium = 2.5 g salt). The human body needs a small amount of salt to maintain fluid balance and homeostasis.[40] For 5 million years, the only source of salt for our ancestors was that found naturally in plants and animals. Presumably, they were consuming less than 1 g of salt per day. The addition of salt to food began relatively recently, approximately 5,000 years ago. This was driven by agricultural and population growth and the discovery of salt as a food preserving agent. Today, human consumption of salt is approximately 10 g/day in most countries, representing 10 to 20 times increase, relative to 5,000 years ago. Because the human body had been geared to conserve salt, it is difficult to adjust physiologically to an increase in salt intake, from an evolutionary standpoint.[41] Increased salt intake results in excess salt and water retention and changes in vascular tone, which leads to HTN and organ damage.[42]

Dietary sodium intake has a direct relationship with BP. Evidence of this relationship comes from animal studies, epidemiologic studies, clinical trials, and meta-analyses of more than 50 randomized trials that have been performed to date. In the International Study of Sodium, Potassium, and Blood Pressure (INTERSALT), urinary electrolyte excretion, and BP of 10,079 patients from 52 centers were analyzed. This study found a significant positive relationship between 24-hour urinary sodium excretion and systolic and diastolic BP, after adjustment for age and sex.[43] In a meta-analysis that studied 34 trials with plausible levels of sodium intake, a median reduction in salt intake of 4.4 g/day lowered systolic BP/diastolic BP by 2.42/1.0 mmHg in normotensive individuals, and by 5.39/2.82 mmHg in persons with HTN.[44] A trial of 12 patients with resistant HTN found that reducing sodium consumption by 14.8 g/day to 2.7 g/day for a week lowered SBP/DBP by 22.7/9.1 mmHg.[45]

Rigorously controlled, dose-response studies provide the most compelling evidence of the effects of sodium consumption on BP.[17,46,47] These trials tested different levels of sodium intake. Each trial found statistically significant, direct, progressive, dose-response relationships. The largest of these trials, the DASH-Sodium trial,[17] tested the effects of each of three different sodium intakes on two different diets—the DASH diet[16] and a control diet resembling a typical U.S. diet. As estimated from 24-hour urine collections, the three sodium levels (termed lower, intermediate, and higher) provided 65, 107, and 142 mmol/day (or 1.5, 2.5, and 3.3 g/day), respectively, of sodium. The main results of the DASH-Sodium trial are displayed in **Figure 68.5**.[17] The BP response to lower sodium intake, albeit direct and progressive, was nonlinear. Decreasing sodium consumption by approximately 0.9 g/day (40 mmol/day) caused a greater BP reduction when the starting sodium level was lower than 100 mmol/day than when it was above.

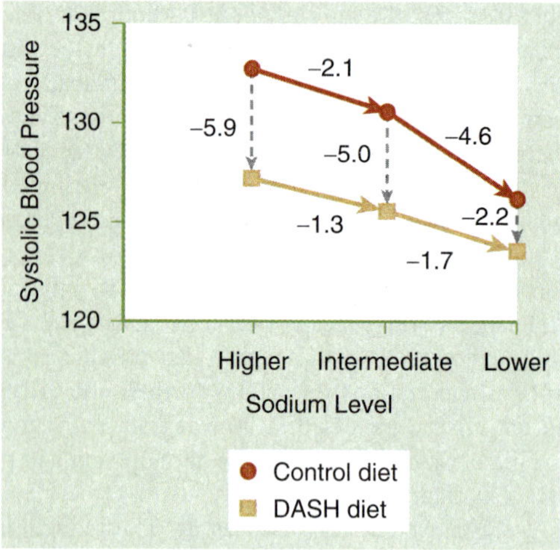

Figure 68.5 Mean systolic BP changes in the DASH-Sodium trial. Solid lines display the effects of sodium reduction in the two diets. Hatched lines display the effects of the DASH diet at each sodium level.

Reproduced from Sacks FM, Svetkey LP, Vollmer WM, et al. N Engl J Med 2001;344:3–10.[17]

In subgroup analyses by race and sex, reduced sodium intake significantly lowered BP in African Americans, non–African Americans, men, and women. Reducing sodium intake significantly lowered BP in nonhypertensive individuals who were following both the DASH and control diets.[48,49] A reduced sodium intake can lower BP in individuals taking antihypertensive medications and can improve HTN control. A randomized double-blind trial in patients with HTN on captopril showed that reducing salt intake by 5.8 g/day over a month resulted in a further lowering of their systolic/diastolic BPs by 13 and 9 mmHg, respectively.[50] In ecologic studies, low sodium intake has also been associated with a reduction in the age-related rise in SBP.[41]

Similar to other interventions, the BP response to change in dietary sodium intake is heterogeneous. Despite attempts to classify individuals in research studies as "salt sensitive" and "salt resistant,"[51] the change in BP from a change in sodium intake is not binary. Instead, change in BP has a continuous distribution, meaning that individuals have a greater or lesser degree of BP reduction. Although the BP response is varied, the magnitude of BP reduction resulting from reduced sodium intake is larger in individuals with HTN, who are Black, and/or elderly, compared with individuals who are normotensive, white, and young. This could partially be attributable to differences in the renin-angiotensin-aldosterone system.[52] Due to their low plasma renin activity, Black patients with HTN are usually considered poor responders to RAAS blockers and better responders to diuretics. However, low-sodium diets seem to mitigate the BP response to RAAS blockers. In an RCT with a crossover design, Black patients with HTN on a lower-sodium diet were found to exhibit a better BP response to lisinopril than to diuretics.[53]

Multiple observational studies and meta-analyses have examined the relationship of sodium intake with CVD outcomes, and their results showed a direct linear relationship between them.[54,55] However, substantial methodologic issues, often related to study population and the accuracy of sodium measurement, have made it methodologically challenging to find direct evidence of a relationship between sodium intake and CVD.[56] For example, the Prospective Urban Rural Epidemiology (PURE) study stirred controversy and confusion, as the investigators reported a J or U-shaped association that both low and high salt intakes were associated with increased mortality risk.[57,58] A major limitation, however, was the inaccurate estimation of salt intake from spot urine using the Kawasaki formula, which has been proven to be unreliable and biased.[59,60] A single measurement of urinary sodium is not accurate enough to reflect the usual salt intake, due to large day-to-day variability in salt consumption.[61] Another limitation is the inclusion of sicker participants with CVD or at high-risk for CVD, which can be a confounder and potentially create a biased conclusion of the association.

Higher mortality and CVD risk were likely explained by these individuals' underlying diseases rather than their reduced salt intake.[40] Despite these challenges, the TOHP Follow-Up study attempted to provide better insight into the association between sodium intake and CVD events. In this study, multiple 24-hour urine collections were used to measure sodium exposure in an attempt to reduce bias from measurement and random error. The investigators found a 32% reduction among those excreting less than 2.3 g/day, and no evidence of a J-shaped association when examining the spline curves.[62]

Long-term randomized trials on the effect of sodium reduction on CVD are scarce, owing to ethical concerns and methodologic challenges, including reduced compliance over many years, cross-contamination, and the need for a large sample size to achieve statistical power. A few moderate-sized RCTs have examined the effects of reduced sodium intake on clinical CVD events in both patients who are normotensive and those with HTN. Pooling of their results showed that reducing salt intake was associated with a 20% reduction in CVD events (random-effects RR=0.81, 95% CI 0.66 to 0.98; I2=0%) (**Figure 68.6**).[63]

Further benefit of sodium reduction is confirmed by natural experiments performed in Finland,[64] and the United Kingdom.[65] Since the late 1970s, various population-wide measures have been implemented to decrease the intake of salt in Finland: including public health campaigns, collaboration with the food industry, and adoption of salt-labeling legislation. This resulted in reduced salt consumption from around 14 g/day in 1972, to about 9 g/day in 2002. This diet change was associated with a 10-mmHg decrease in both systolic and diastolic BP, and an approximately 80% reduction in CV mortality in the middle-aged population, despite an overall increase in both obesity and alcohol consumption at that time.[64] Similar results were seen in the United Kingdom, where a nationwide salt reduction program was conducted in 2003 and resulted in a 15% reduction in population salt intake by 2011 (from 9.5 g/day in 2003 to 8.1 g/day in 2011). The investigators found a 3 mmHg and a 1.4 mmHg drop in systolic and diastolic BP, respectively. This BP reduction was associated with a significant 42% reduction in mortality and a 40% reduction in stroke and ischemic CVD from 2003 to 2011.[65]

Reduced sodium intake could have other health benefits, including reduced risk of subclinical CVD (i.e., left ventricular hypertrophy, ventricular fibrosis, and diastolic dysfunction),[66] kidney disease,[67] gastric cancer,[68] and disordered mineral metabolism (i.e., increased urinary calcium excretion, potentially leading to renal stone[69] and osteoporosis).[70] In particular, LV mass is directly associated with sodium intake in cross-sectional studies, and one small trial in the early 1990s documented that sodium reduction was associated with regression of left ventricular hypertrophy.[71] Reduced sodium intake has been associated

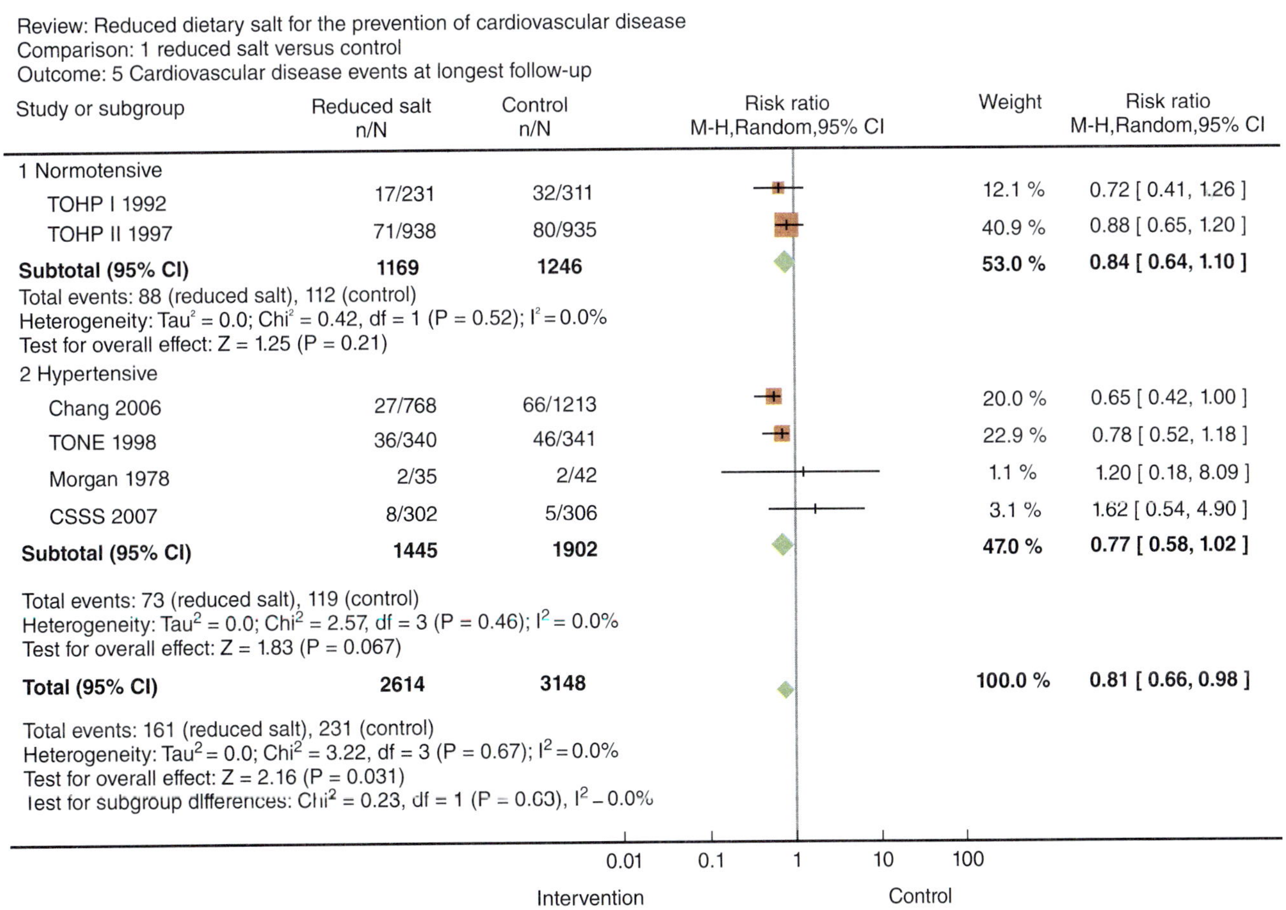

Figure 68.6 Cardiovascular disease events of reduced salt versus control group at longest follow-up.

Reproduced from Adler AJ, Taylor F, Martin N, et al. Cochrane Database Syst Rev. 2014 Dec 18;2014(12):CD009217. doi: 10.1002/14651858.CD009217.pub3. PMID: 25519688; PMCID: PMC6483405.[63]

with reduced risk of heart failure.[72] Despite the strong evidence for benefit in patients with stage A and B heart failure, there is insufficient evidence to recommend sodium reduction (to a target of <1,500 mg/day) for stages C and D heart failure.[73]

In the Study of Dietary Intervention under 100 mmol in Heart Failure (SODIUM-HF) trial, 806 patients with chronic heart failure with New York Heart Association functional class 2–3 were randomized to a low sodium diet group (target of less than 1,500 mg/day) vs the usual care, according to local guidelines.[74] The median difference in sodium intake between groups was only 415 mg/day at 12 months. The investigators found no significant difference between the two groups in the primary outcome composite of CV-related hospitalization, CV-related emergency department visits, and all-cause mortality. The absence of treatment effect for the primary outcome was consistent across prespecified subgroups.[74]

The 2020-2025 Dietary Guidelines for Americans, as well as numerous other organizations, recommend a population-wide reduction in sodium intake including: No more than 1.2 g/day of sodium for ages 1 through 3; 1.5 g/day for ages 4 through 8; 1.8 g/day for ages 9 through 13; and 2.3 g/day for all other age groups.[75] The 2017 ACC/AHA guidelines for HTN set an optimal goal of 1.5 g/day (65 mmol) of sodium as the recommended upper limit of intake for the U.S. population.[5] Reducing sodium intake is arguably one of the most important interventions that will bring substantial benefits in reducing morbidity and mortality. Current average sodium intake of U.S. residents is about 3.4 g/day, which is 50% more than the Chronic Disease Risk Reduction (CDRR) recommended limit of 2.3 g/day.[76,77] More than 90% of people in the United States exceed recommended limits across all age groups.[76] Even the youngest individuals consume excess sodium; more than 95% of children aged 2 to 13 years exceed recommended limits for their age groups, the consequences of which could track into adulthood and influence later health outcomes. See **Figure 68.7**.

As more than 70% of sodium consumption comes from sodium added during food manufacturing and commercial food preparation, any meaningful approach to reducing sodium intake must involve food manufacturers and restaurants.[77] That is why the U.S. FDA announced final guidance for industry on voluntary short-term (2.5 years) sodium reduction targets.[78] This initiative, which was proven to be successful in the United Kingdom, was anticipated to bring a sustained improvement in the population-wide reduction in sodium intake, which will bring enormous benefit on health and well-being of the nation.[78]

Increased Potassium Intake

Potassium, an essential nutrient needed for maintenance of total body fluid volume, acids, electrolyte balance, and normal cell function, has been associated with BP control.[79] Higher dietary intake of potassium may help lower blood pressure by increasing vasodilation and urinary sodium excretion, which reduces plasma volume, especially in salt-sensitive individuals, and prevents kidney stones.[80] The recommendations published in 2005, by the Institute of Medicine, for the United States and

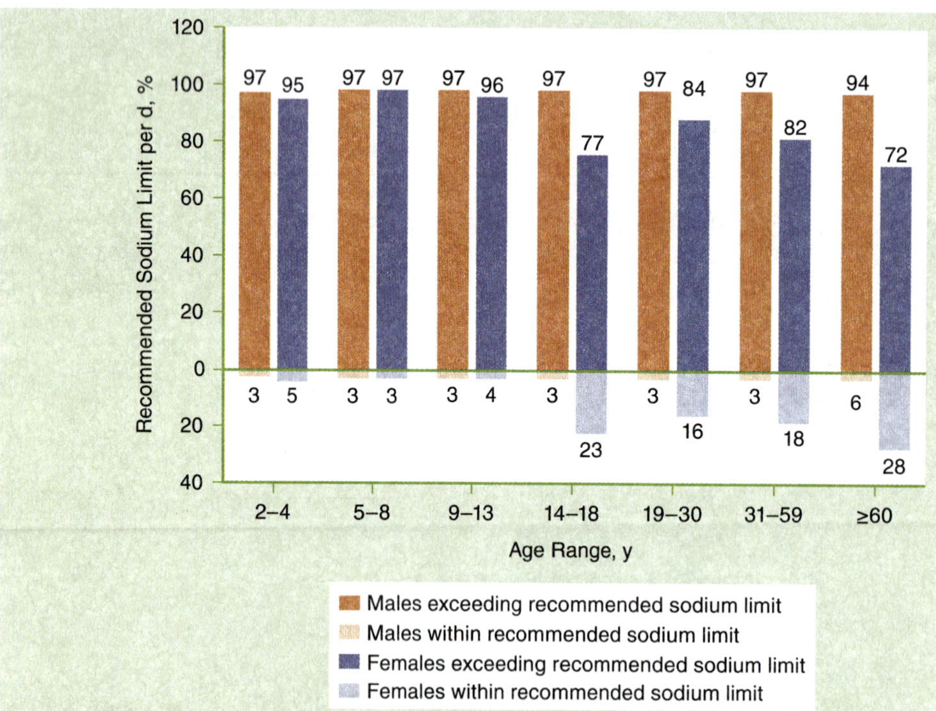

	2–4	5–8	9–13	14–18	19–30	31–59	≥60
Recommended sodium limit, mg/d	1200	1500	1800	2300	2300	2300	2300
Mean sodium intake of males, mg/d	2274	2785	3451	3888	4274	4172	3799
Mean sodium intake of females, mg/d	2061	2525	3030	2875	3142	3062	2802

Figure 68.7 Mean sodium intake in the United States by age, relative to recommended limits.

Reproduced from Mayne ST, McKinnon RA, Woodcock J. JAMA 2021;326:1675–1676,[77] and USDA and DHHS. Dietary Guidelines for Americans, 2020-2025 (9th ed). Available from www.dietaryguidelines.gov/.[75]

Canada for individuals 14 years and older, were for an Adequate Intake (AI) of 4.7 g/day of potassium.[80] Evidence for a relationship has been studied in animal experiments, observational studies, clinical trials, and meta-analyses of the clinical trials. However, the evidence base for a direct relationship between potassium and BP remains inconclusive.[81] Dietary surveys consistently showed that the U.S. population consumes less potassium than is currently recommended. In fact, the 2013–2014 NHANES data showed that the average American adult male consumed 3.0 g/day of potassium, and the average American female consumed 2.3 g/day of potassium.[76] Subsequently, the 2015–2020 Dietary Guidelines for Americans identified potassium as a "nutrient of public health concern."[82] Currently, the National Institute of Health, the U.S. National Academy of Medicine, and the 2020–2025 Dietary Guidelines for Americans recommend 2.6 g/day of potassium for adult females and 3.4 g/day for adult males (AI).[75] Most adults should be able to increase potassium intake through their daily diets to reach the population goals set for potassium intake to benefit from the effects that it has on lowering BP.

Historically, several analyses, dating back 3 decades, have been performed (Figure 68.8). A 1997 meta-analysis concluded

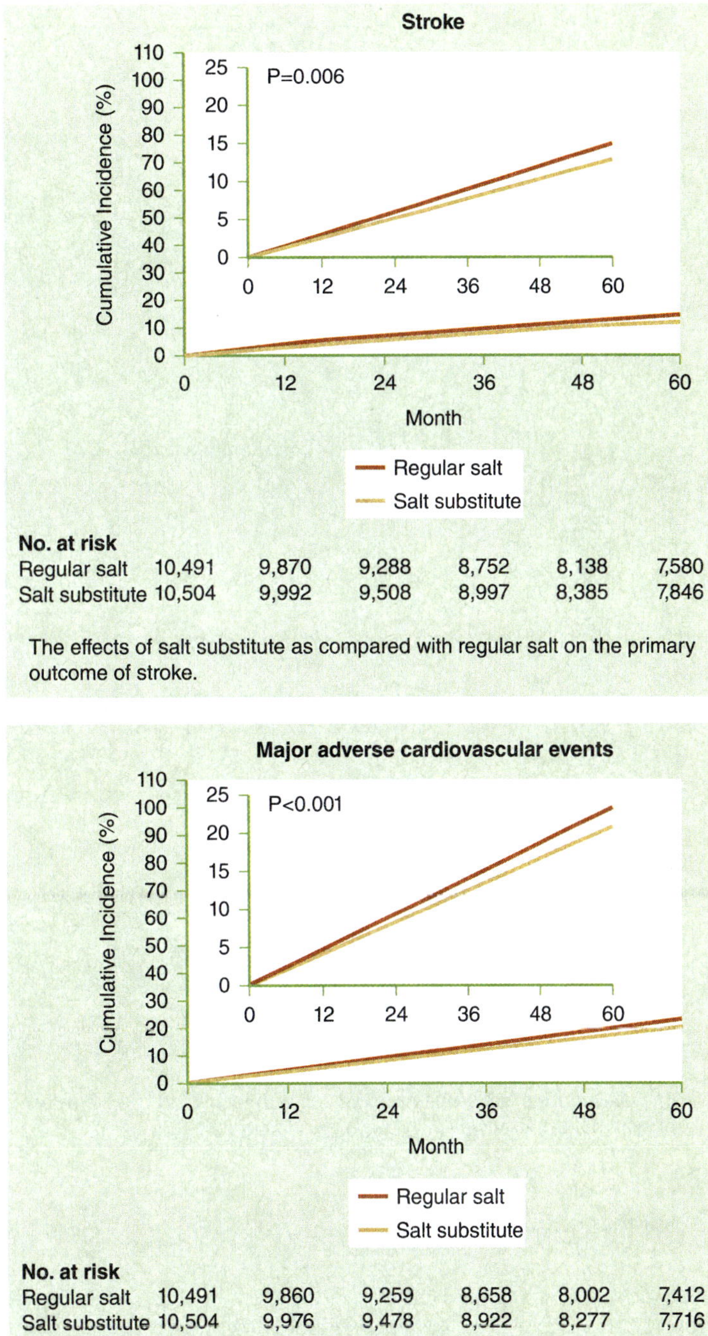

A The effects of salt substitute as compared with regular salt on the primary outcome of stroke.

B The effects of salt substitute as compared with regular salt on the secondary outcomes of major adverse cardiovascular events (a composite of nonfatal stroke, nonfatal acute coronary syndrome or death from vascular causes.

Figure 68.8 Effects of salt substitution on cardiovascular events and death. *(continues)*

Reproduced from Neal B, Wu Y, Feng X, et al. N Engl J Med 2021;385:1067–77. doi:10.1056/NEJMoa2105675.[91]

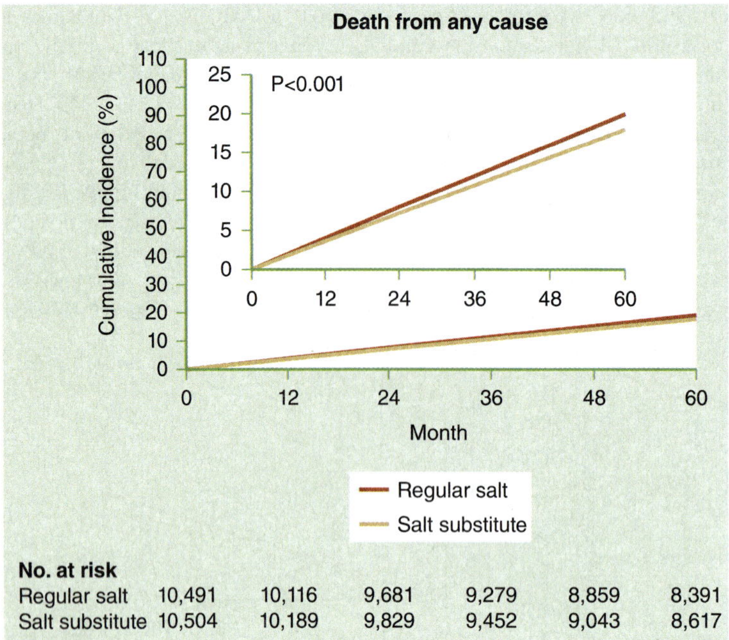

Tho offocto of calt cubctituto ac compared with rogular calt on doath from any cause.

C

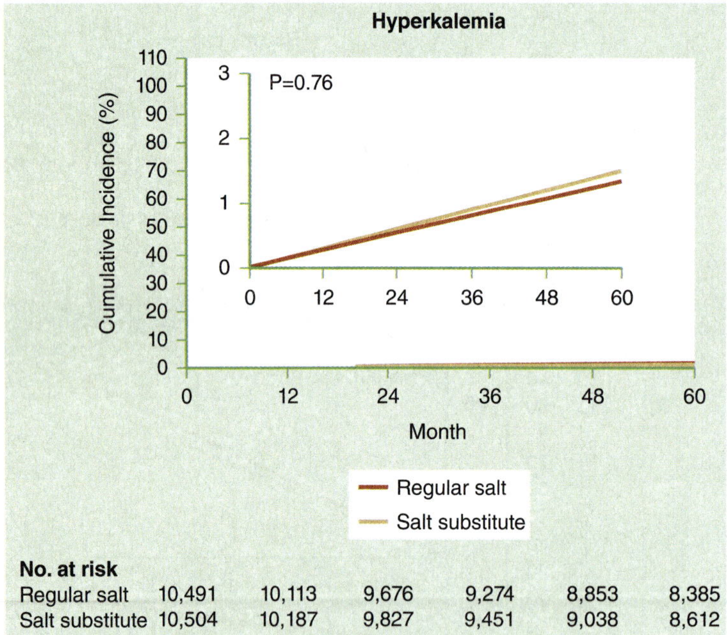

The effects of salt substitute as compared with regular salt on the safety outcome of hyperkalemia.

D

Figure 68.8 *(continued)* Effects of salt substitution on cardiovascular events and death.

Reproduced from Neal B, Wu Y, Feng X, et al. N Engl J Med 2021;385:1067–77. doi:10.1056/NEJMoa2105675.[11]

that potassium supplementation was associated with a significant reduction in mean systolic BP and diastolic BP in people with or without HTN.[83] The authors promoted potassium intake for prevention and treatment of high BP, especially for those who are unable to reduce their risk of sodium sensitivity.[83] In 1999, a systematic review was conducted to provide evidence-based recommendations on dietary consumption and supplementation of potassium in the prevention and treatment of HTN. It was determined that potassium supplementation above the "recommended" daily dietary intake should not be recommended as a treatment for HTN.[84] A 2006 Cochrane meta-analysis, including six randomized controlled trials, reported no effect of potassium supplementation on primary hypertension.[85] Subsequently, a 2013 meta-analysis, including RCTs and cohort studies, reported that increased potassium intake can reduce BP in people with or without HTN, without side effects on blood lipid and catecholamine concentrations, or renal function. It also suggested that increased dietary potassium intake may prevent and control HTN and stroke.[86] Although most studies used potassium supplements to increase daily intake, rather than natural dietary sources of potassium, it may be beneficial to increase potassium intake by consuming foods that are rich in potassium, such as fruits and vegetables.

In the DASH trial, the two "treatment" groups that increased fruit and vegetable consumption and, thereby, increased potassium intake, both lowered BP. The DASH diet provides roughly 4.7 g/day (120 mmol/day) of potassium and lowered systolic BP by an average of 5.5 mm Hg and systolic BP by 3 mmHg.[16] Another trial documented that increased fruit and vegetable intake lowered BP, but it did not specify the amount of potassium that was consumed.[87] In 2017, a meta-analysis focused on the effect of potassium supplementation vs placebo on blood pressure in patients with HTN, to provide recommendations for potassium supplementation as an adjuvant agent. Compared with placebo, potassium supplementation resulted in modest but significant reductions in both systolic BP and diastolic BP and was recommended as an adjuvant agent for patients with HTN.[88] Another 2017 meta-analysis found that participants with HTN were able to lower their systolic BP by 4.5 mmHg and diastolic BP by 2.9 mmHg with potassium supplementation.[89] The most current review, a 2020 meta-analysis, collated the results from 32 trials published between 1982 and 2016.[90] The aim of this large meta-analysis was to try to determine a dose-response relationship between potassium supplementation or higher average daily intake, and BP in RCTs. All included trials were performed over at least a 4-week time period and potassium supplementation ranging from 30 to 120 mmol/day. Even with the varying doses provided, no exact dose-response was determined at this time.[90]

The most current prospective RCT was conducted in China involving 20,995 persons and published in 2022.[91] It was an open-label, cluster-randomized trial involving persons with a history of stroke, or aged 60 years or older with HTN. Six hundred villages were randomly assigned to either use a salt substitute (75% sodium chloride and 25% potassium chloride by mass), or regular salt (100% sodium chloride). After a mean duration of follow-up of 4.7 years, the rate of subsequent stroke was lower with the salt substitute than with regular salt: 29.1 vs 33.7 events per 1,000 person-years; rate ratio, 0.86; 95% confidence interval [CI], 0.77 to 0.96; $P=0.006$). Also improved were the rates of major CV events, (49.1 vs 56.3 events per 1,000 person-years; rate ratio, 0.87; 95% CI, 0.80 to 0.94; $P<0.001$), and all-cause mortality (39.3 s vs 44.6 events per 1,000 person-years; rate ratio, 0.88; 95% CI, 0.82 to 0.95; $P<0.001$). The rate of serious adverse events attributed to hyperkalemia was not significantly higher with the salt substitute than with regular salt (3.35 events vs 3.30).[91]

However, potassium and sodium interact such that the effects of potassium on BP depend on the concurrent intake of sodium and vice versa. Specifically, reduced sodium intake has greater BP-lowering effects when potassium intake is low and has lesser BP-reducing effects when potassium intake is high. In addition, increased potassium intake has greater BP-lowering effects when sodium intake is high and has lesser BP-reducing effects when sodium intake is low. For example, in one trial, high potassium intake (120 mmol/day) diminished the pressor response to increased sodium consumption in non-hypertensive Black men and, to a lesser extent, in non-Blacks (**Figure 68.9**).[92] However, a statistically significant decrease in BP was also found with increased potassium when sodium intake was 2 to 4 gm/day. Therefore, increased potassium can continue to be beneficial in terms of blood pressure, even as individuals decrease their sodium intake.[93,94]

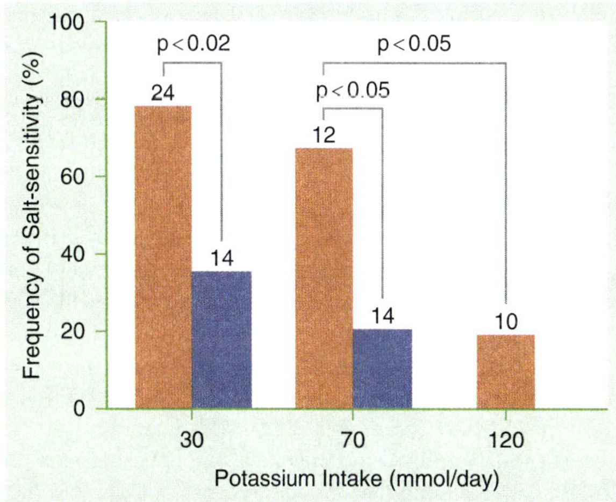

Figure 68.9 Prevalence of sodium sensitivity in normotensive individuals (Blacks, brown bars; Whites, blue bars) at three levels of potassium intake. Sodium sensitivity is defined by a sodium-induced increase in mean arterial pressure of at least 3 mmHg.

Reproduced from Morris RC Jr, Sebastian A, Forman A et al. Hypertension 1999;33:18–23.[92]

Among healthy individuals with normal kidney function, potassium intake from foods greater than 4.7 g/day (120 mmol/day) does not pose a risk, because excess potassium is readily excreted. However, in individuals whose urinary potassium excretion is impaired, such as by drugs or medical conditions, an intake of less than 4.7 g/day (120 mmol/day) is advised because of the risk of adverse cardiac effects (dysrhythmias) from hyperkalemia. Angiotensin-converting enzyme inhibitors, angiotensin receptor blockers, nonsteroidal anti-inflammatory drugs, and potassium-sparing diuretics are drugs that may impair potassium excretion. Impaired renal excretion of potassium is associated with certain medical conditions, such as diabetes, CKD, end-stage renal disease, severe heart failure, and adrenal insufficiency. In addition, elderly individuals are at increased risk of hyperkalemia. Although CKD may impair renal excretion of potassium, available evidence is insufficient to identify the level of kidney function below which hyperkalemia will result from a high dietary intake of potassium. Because of this uncertainty, an expert panel set a wide range of recommended potassium intake (2.0 to 4.0 g/day) for patients with advanced (stage 3 or 4) CKD.[95]

Moderation of Alcohol Consumption

A direct, dose-response relationship between alcohol consumption and BP has been documented through observational and experimental studies, especially in the setting of more than two alcoholic drinks per day.[96] This relationship is independent of potential confounders such as age, obesity, and sodium intake.[97] Epidemiologic evidence suggests that **heavy** alcohol intake is strongly associated with increased risk of HTN. However, the relationship between light to moderate alcohol consumption and incidence of HTN remains uncertain.[98] Current clinical guidelines on the primary prevention of HTN recommend limiting alcohol intake to less than or equal to two drinks/day for men and no more than one drink/day for women. This level of consumption is considered moderate drinking. One drink is defined as 12 oz. of regular beer, 5 oz. of wine (12% alcohol), or 1.5 oz. of 80% proof distilled spirits. Uncertainty remains regarding any benefits or risks attributable to light to moderate alcohol intake on the risk of HTN. Individual studies have yielded conflicting results, ranging from beneficial, to unassociated, to adverse effects.[97,98]

Several systematic reviews and meta–analyses have tried to determine a dose–response relationship between average daily alcohol consumption and the incidence of increasing BP. They found that the association between light to moderate alcohol intake and risk of developing HTN differed by sex. In women, light to moderate alcohol intake modestly lowered HTN risk, while heavy alcohol intake increased HTN. In men, there was no benefit found in light to moderate alcohol consumption and findings showed a trend toward increased risk of HTN. Although the protective effect of light to moderate alcohol consumption in women is modest, it does support light to moderate alcohol consumption as protective.[99,100]

A meta-analysis of 36 trials including 2,865 participants (2,464 men and 401 women), was performed to determine if there is a beneficial effect to reducing alcohol intake on change in BP.[101] Those participants who initially drank no more than two drinks/day and then decreased their intake of alcohol to near abstinence noticed no significant effect on blood pressure. Participants who drank three drinks/day but reduced their alcohol intake to near abstinence sustained a decreased systolic BP (median: 1.18 mmHg) and diastolic BP (median 1.09 mmHg). The third group of participants, who drank at least six drinks/day, experienced the strongest decrease in ystolic BP (median: 5.5mmHg) and diastolic BP (median: 3.97mmHg) by reducing alcohol intake by 50%. The dose response link was apparent in healthy participants, and those with established HTN or other cardiovascular risk factors. These findings support the idea that decreasing alcohol intake lowers BP in a dose-dependent manner, with a possible threshold effect at two drinks/day.[101]

A recent study with more than 17,000 U.S. adults from the third NHANES found that **moderate** intake (7–13 drinks/week) of alcohol increased the risk of HTN.[102] Moderate drinkers were 53% more likely to have stage 1 HTN and twice as likely to have stage 2 HTN, compared with those who never drank. Similarly, compared with those who never drank, heavy drinkers were 69% more likely to have stage 1 HTN and 2.4 times as likely to have stage 2 HTN.[102]

Overall, available evidence remains conflicting as to whether light to moderate alcohol intake is beneficial or harmful for lowering high blood pressure. Studies have shown that lowering daily/weekly alcohol intake will significantly reduce both systolic and diastolic BP, with a possible threshold response at greater than two drinks/day. While there is some evidence that women may benefit from light to moderate alcohol intake, limiting alcohol intake should be advised for both men and women and, specifically, should not exceed two alcoholic drinks/day for men or one alcoholic drink/day for women.

Dietary Patterns

Plant-based Diets

Despite the wide variation in definition, the term "plant-based" diet is generally defined as a dietary pattern that contains lower amounts of animal-sourced foods and higher amounts of plant-sourced foods. Vegetarian diets are a subset of plant-based diets, which exclude intake of some or all animal foods entirely. The vegan diet excludes all animal-derived products, including dairy products, eggs, and honey.[103] Although plant-based diets vary in their specific components and amounts of animal products included, a *healthy* plant-based diet emphasizes high-quality plant foods such as fruits, vegetables, whole grains, legumes, and nuts, and limits saturated fats, sodium, refined grains, and sugar-sweetened foods and beverages, as well as excessive energy intake.[75]

Numerous randomized, controlled trials, and observational studies have studied the effects of plant-based diets on HTN. While the results are variable, a pooled meta-analysis of seven of these clinical trials found that consumption of vegetarian diets was associated with a mean reduction of 4.8 mmHg in systolic BP and 2.2 mmHg in diastolic BP (**Figure 68.10**).[104]

Abundant evidence supports the relationship between healthy plant-based diets and cardiovascular health. Pooling data from five prospective cohort studies, with a mean follow-up of 10.6 years, showed that vegetarians who had followed their diet for more than 5 years had a 24% lower mortality rate from ischemic heart disease than non-vegetarians. Occasional meat eaters and the fish eaters within the nonvegetarian category also had lower mortality from ischemic heart disease than regular meat eaters.[105] A meta-analysis of seven cohort studies on vegetarian diets, with a total of 124,706 participants, found a significant 29% lower ischemic heart disease mortality (RR=0.71, 95% CI 0.56–0.87; $P<0.001$) in vegetarians than in nonvegetarians.[106]

Dietary Approaches to Stop Hypertension Diet (DASH Diet)

In the initial landmark DASH trial, 459 participants were randomized to one of three diets, and the effects of each diet on BP were studied.[16] The most effective diet, now termed the DASH diet, was rich in fruits, vegetables, and low-fat dairy products; it included whole grains, poultry, fish, and nuts; and was reduced in fats, red meat, sweets, and sugar-containing beverages. It was rich in potassium, magnesium, calcium, and fiber, and low in total fat, saturated fat, and cholesterol; it was also slightly increased in protein. Participants following the DASH diet significantly reduced their BP by a mean of 5.5/3.0 mmHg compared with the control group. The reductions in BP resulting from the diets occurred rapidly, taking 2 weeks or fewer to manifest (**Figure 68.11**).[16]

Subgroup analyses of the DASH diet,[16] and the DASH-Sodium trials,[17] showed significantly lowered BP in men, women, Black adults, White adults, persons with HTN, and normotensive individuals.[47,48] Effects of the DASH diet were most striking in the African American participants, with average systolic/diastolic BP reductions of 6.9/3.7 mmHg, significantly larger than those in White participants (3.3/2.4 mmHg). In the subset of persons with HTN, compared with normotensive individuals, the DASH and fruit and vegetable diets were also associated with a dramatic lowering of systolic/diastolic BP by 11.4/5.5, vs 7.2/2.8 mmHg. The beneficial effects of the DASH diet in individuals with HTN have obvious clinical significance, and the corresponding effects in normotensive individuals have major public health implications (see Figure 68.1). The DASH-Sodium trial (described earlier)[17] documented that the DASH diet significantly lowered BP at each of three sodium levels (see Figure 68.3), with the combination of

A

Source	Subgroup within study	Comparison	Statistics for each study: Difference in mean	Lower limit	Upper limit	*P* value
Rouse et al,[6] 1983	Lacto-ovo	Omnivorous	−6.8	−9.6	−4.0	<0.001
Ferdowsian et al,[12] 2010	Vegan	Omnivorous	−5.7	−11.1	−0.3	0.04
Margetts et al,[7] 1986	Lacto-ovo	Omnivorous	−3.5	−6.9	−0.1	0.047
Hakala and Karvetti,[15] 1989	Lacto	Omnivorous	−3.3	−8.3	1.8	0.21
Kestin et al,[8] 1989	Lacto-ovo, men	Omnivorous	−3.0	−9.1	3.1	0.34
Sciarrone et al,[14] 1993	Lacto-ovo, men	Omnivorous	1.5	−15.3	18.3	0.86
Nicholson et al,[13] 1999	Vegan	Omnivorous	8.5	−10.2	27.2	0.37
Total			**−4.8**	**−6.6**	**−3.1**	<0.001

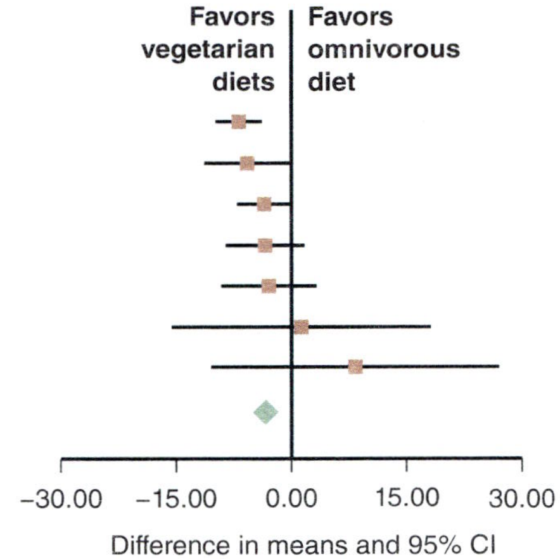

B

Source	Subgroup within study	Comparison	Statistics for each study: Difference in mean	Lower limit	Upper limit	*P* value
Ferdowsian et al,[12] 2010	Vegan	Omnivorous	−5.5	−9.1	−1.9	0.003
Rouse et al,[6] 1983	Lacto-ovo	Omnivorous	−2.7	−4.7	−0.7	0.008
Hakala and Karvetti,[15] 1989	Lacto	Omnivorous	−2.5	−9.2	4.2	0.46
Margetts et al,[7] 1986	Lacto-ovo	Omnivorous	−1.2	−3.1	0.7	0.22
Sciarrone et al,[14] 1993	Lacto-ovo, men	Omnivorous	−1.0	−14.8	12.8	0.89
Kestin et al,[8] 1989	Lacto-ovo, men	Omnivorous	−0.8	−6.0	4.4	0.76
N cholson et al,[13] 1999	Vegan	Omnivorous	4.8	−8.3	17.9	0.47
Total			**−2.2**	**−3.5**	**−1.0**	<0.001

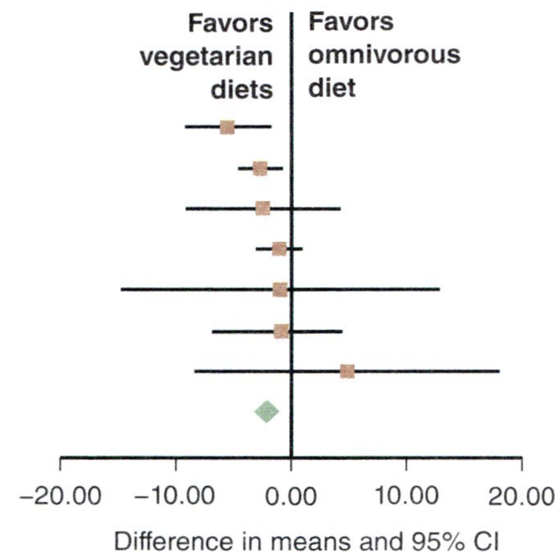

Figure 68.10 Pooled systolic and diastolic blood pressure (BP) responses to vegetarian diets in clinical trials.

Rep oduced from Yokoyama Y, Nishimura K, Barnard ND, et al. JAMA Intern Med. 2014;174:577–87.[104]

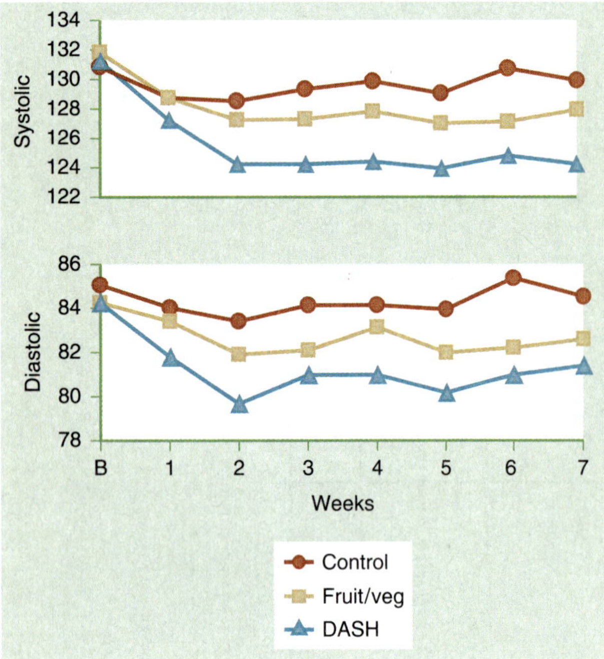

Figure 68.11 Blood pressure by week during the Dietary Approaches to Stop Hypertension (DASH) feeding study in three diets (control diet, fruits and vegetables diet, and the DASH diet.)

Reproduced from Appel LJ, Moore TJ, Obarzanek E et al. N Engl J Med 1997;336:1117.[16]

the DASH diet and the lowest sodium intake yielding the greatest reductions in BP.[17]

There has been a great deal of conjecture about which components of DASH-style diets may be responsible for the BP-lowering effects. The health benefit associated with the DASH diet is probably due to the following. The diet is rich in fruits and vegetables, which provide high potassium, magnesium, and fiber. Because the fruit and vegetable component of the diet accounted for approximately half of the BP-lowering effects of the DASH diet (see Figure 68.11), some other component of the DASH diet must be responsible for further reducing BP. Compared with the fruits and vegetables diet, the DASH diet provided more vegetables, low-fat dairy products, and fish while containing less red meat, sugar, and refined carbohydrates.

The Optimal Macronutrient Intake Trial to Prevent Heart Disease (OmniHeart) examined whether altering macronutrient intake could also improve the DASH diet's BP-lowering effects.[107] Three variants of the DASH diet were studied: 1. A diet rich in carbohydrate (58% of total energy); 2. A diet rich in protein (about half from plant sources); and 3. A diet rich in unsaturated fat (predominantly monounsaturated). Each diet was similar to the original DASH diet in that each was reduced in saturated fat, cholesterol, and sodium, and rich in fruit, vegetables, fiber, and potassium. Although all three OmniHeart diets lowered systolic BP (**Figure 68.12**), substituting some of the carbohydrate (~10% of total kcal) with either protein or unsaturated fat lowered BP more effectively.[108]

After the publication of the DASH and DASH-Sodium trials, numerous subsequent trials have examined the effect of the DASH diet on BP in persons with and without HTN, with different designs, dietary protocols, and patient characteristics. A systematic review and meta-analysis of thirty DASH diet RCTs, with a total of 5,545 participants, was performed.[109] The DASH diet resulted in significant reduction in systolic/diastolic BP (−3.2 mmHg and −2.3 mmHg), and the benefit was consistent across all baseline BP levels.[109] The effect was more pronounced when sodium intake was lower than 2,400 mg/day, in individuals under the age of 50, and in participants with hypertension but without antihypertensive medication. Systematic reviews and meta-analyses of prospective cohort studies have also shown that DASH dietary patterns were associated with decreased incidence of cardiovascular disease (RR 0.80, 95% CI 0.76–0.85), coronary heart disease (RR 0.79, 95% CI 0.71–0.88), stroke (RR 0.81, 95% CI 0.72–0.92), and diabetes (RR 0.82, 95% CI 0.74–0.92).[110]

Due to these consistent and strong observations, the DASH diet has been endorsed by the 2017 ACC/AHA Hypertension Guidelines as an effective diet for controlling BP (class of recommendation 1, level of evidence A).[5] The DASH diet is considered safe and appropriate for the general population. However, the applicability of the DASH diet in patients with

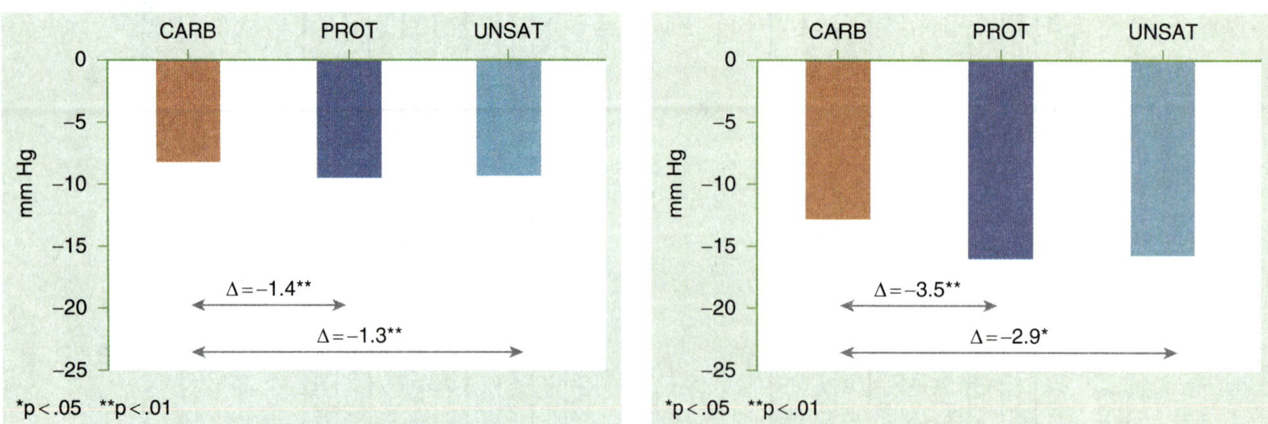

A **B**

Figure 68.12 Effects of three healthy dietary patterns tested in the OmniHeart feeding study on systolic blood pressure (CARB, similar to the DASH diet; PROT, rich in protein, about half from plant sources; and UNSAT, rich in monounsaturated fat) in all participants (A) and in hypertensive participants (B).

Reproduced from Appel LJ, Brands MW, Daniels SR et al. Hypertension 2006;47:296–308.[108]

advanced chronic kidney disease stage 4 and 5 is uncertain, due lack of studies documenting safety and immediate impact on potassium and phosphate balance.[111] The National Kidney Foundation notes that the applicability of the dietary pattern must be determined on an individual basis, especially with regard to serum potassium control and adequacy of protein and energy intake.[112]

Mediterranean Diet

The Mediterranean diet refers to the dietary pattern and culinary techniques practiced by countries and populations living around the Mediterranean Sea. It emphasizes an abundance of plant foods (leafy green vegetables, fruits, whole-grain cereals, legumes, and nuts), and olive oil as a main source of fat. It limits dairy products, but encourages consumption of moderate amounts of fish, poultry, and wine, as well as low amounts of red meat.[113] This dietary pattern was first recognized, in the Seven Countries Study from the 1960s, to be associated with reduced risk of long-term CVD and cardiovascular mortality.[114] Cardiovascular benefit of the Mediterranean diet was further shown in the Lyon Heart Study, a secondary prevention trial, which showed that patients in the Mediterranean diet group had significant reductions in CV mortality and recurrent nonfatal MI, compared to the regular Western diet group (1.24 vs 4.07 events per 100 patient years) after a mean follow-up of 46 months.[115]

To confirm the efficacy, a more recent and larger multicenter RCT, the Prevención con Dieta Mediterannea (PREDIMED) trial was performed.[14] In this trial, 8,713 participants without history of CVD were randomly assigned to one of three dietary groups: a Mediterranean diet supplemented with extra-virgin olive oil; a Mediterranean diet supplemented with nuts; or a control low-fat diet. The median follow-up was 4.8 years. This trial was halted early after a significant reduction in CVD events became overwhelmingly evident in the Mediterranean diet arms of the trial. The investigators found a 30% significant reduction in the composite primary endpoint of myocardial infarction, stroke, and death from cardiovascular causes in patients who received either the Mediterranean diet with extra-virgin olive oil, or the one with nuts, compared with the control diet.[14] These positive results of the PREDIMED study were mainly driven by a significant decrease in the incidence of stroke, an endpoint that is closely related to BP.

Numerous studies have been conducted to evaluate the effect of the Mediterranean diet on BP. In a RCT performed on 1,142 healthy participants, consumption of a Mediterranean diet was associated with a significant −5.5 mmHg reduction in systolic BP in males, but **not** females. Changes in diastolic BP were not statistically significant.[116] A meta-analysis of six trials comparing the Mediterranean to low-fat diets showed a small reduction of 1.7 mmHg in systolic BP and 1.5 mmHg in diastolic BP with the Mediterranean diet.[117] Additionally, in a post-hoc analysis of PREDIMED participants, a statistically significant difference in diastolic BP reduction was seen for both Mediterranean groups compared with control, although the changes were small (−1.5 mmHg for Mediterranean extra-virgin olive oil group and −0.65 mmHg for Mediterranean nuts group).[118]

Due to its positive impact on cardiovascular outcomes, the Mediterranean diet is currently recommended by the Dietary Guidelines for Americans 2020-2025 as one of the healthiest dietary patterns.[75] It was also recommended by the 2019 ACC/AHA Guideline on the Primary Prevention of Cardiovascular Disease to decrease ASCVD risk (class of recommendation 1, level of evidence A).[119] However, for HTN management and prevention, the most recent 2017 ACC/AHA guidelines assessed the level of evidence for the Mediterranean diet as "low" or less persuasive.[5]

Dietary Factors with Limited or Uncertain Effects

Fiber

Fiber consists of the indigestible parts of plant foods. Since hypertension increases the risk for cardiovascular disease mortality, the impact of fiber on blood pressure has been investigated. Observational studies and several trials provide evidence that increased fiber intake, especially soluble (viscous) fiber, may decrease BP.[120,121] In a meta-analysis of 24 trials, Streppel et al., reported that supplemental fiber (average increase of 11.5 g/day) was associated with a net systolic/diastolic BP reduction of 1.1/1.3 mmHg.[121] More recently, a meta-analysis of RCTs was conducted that found an overall lowering effect of viscous soluble fiber on SBP by 1.6 mmHg and DBP by 0.4 mmHg at a median dose of 8.7g/day of fiber. The trial also revealed that systolic BP reductions of 2.4 mmHg were observed only for supplementation using psyllium fiber.[122] Although some studies have demonstrated lowering effects of fiber on BP, it remains unclear if dietary fiber can independently lower BP, as well as the necessary amount and type of dietary fiber.

Calcium and Magnesium

Calcium plays a role in BP regulation, but the importance of supplemental calcium intake for the prevention of HTN is still in question. A 1995 meta-analysis,[123] which examined results from 23 observational studies, documented an inverse association between dietary calcium intake and BP. However, the effect size was small, and there was evidence of publication bias and heterogeneity across studies. Increased dietary calcium intake may have BP-reducing effects, and evidence of this relationship exists in a variety of studies. A more recent meta-analysis,[124,125] including a 2015 Cochrane review of 16 trials with 3,048 participants, was conducted with a median follow-up period of 3.5 months.[125] This review found that higher calcium intake (1,000–2,000 mg/day) significantly reduced both systolic/diastolic BP by 1.4 mmHg and 0.98 mmHg, respectively, in normotensive adults.[125] The reduction was greatest among younger adults (<35 years) and with doses higher than 1,500 mg/day.

Observational epidemiologic studies have found inverse associations between dietary magnesium intake and BP. A meta-analysis consisting of 20 trials concluded that magnesium showed only a small overall reduction in SBP and DBP, of 0.6 mmHg and 0.8 mmHg, respectively, but that a significant reduction was observed in a dose-dependent manner.[126] More recently, a meta-analysis of 12 RCTs, including 545 hypertensive participants, found that magnesium supplementation in doses ranging from 243 to 973 mg/day for 8 to 26 weeks resulted in a small reduction in systolic BP, of 1.3 mmHg, but a more significant decrease in diastolic BP, of 2.2 mmHg.[127] In 2012, another meta-analysis of 22 trials was conducted that showed that the intake of magnesium supplementation exceeding

370 mg daily was more effective in the reduction of BP in persons with HTN than a dose less than 370 mg daily. The mean reduction of systolic/diastolic BP was 3–4/2–3 mmHg after 3 to 24 weeks.[128] Oral magnesium was also shown to enhance the blood pressure-lowering effect of antihypertensive medications in a comprehensive review of 44 human studies.[129] Magnesium supplementation appears to achieve a small reduction in blood pressure, but more evidence is needed through larger, longer duration, and better-quality trials.

Fat Intake

Omega-3 Polyunsaturated Fat, Saturated Fat, Omega-6 Polyunsaturated Fat, and Monounsaturated Fat, Cholesterol

Although early studies focused on the effects of total fat intake on BP, there is a plausible biologic basis to hypothesize that certain types of fat (e.g., omega-3 polyunsaturated fat) may lower BP, whereas other types of fat (e.g., saturated fat) may raise it. Several small trials and meta analyses of these trials[130] have shown that high-dose supplements of omega-3 polyunsaturated fatty acid, commonly termed "*fish oil*," can lower BP in persons with hypertension. In normotensive individuals, BP reductions resulting from fish oil supplements, 3 g/day, tend to be small. A dose-response meta-analysis of randomized controlled trials by Zhang et al., determined that the greatest reduction in BP occurred at moderate doses of omega-3 polyunsaturated fatty acids of 2 to 3 gm daily. These effects were stronger for systolic BP when the average participant age was over 45 years. Omega-3 polyunsaturated fatty acid intakes above the recommended 3 gm/day were not associated with additional benefit, particularly in the normotensive subgroup.[131] Side effects of Omega 3 ingestion included a fishy taste and belching.

The effect of saturated fat on BP in adults has been examined in two prospective observational studies and a few clinical trials.[132] In the Nurses' Health Study and the Health Professionals Follow-up Study, saturated fat intake was not associated with onset of HTN.[133,134] Because most trials tested diets that concurrently reduced saturated fat and increased polyunsaturated fat, the lack of a BP effect also suggests no benefit from polyunsaturated fat. In a large randomized controlled trial of healthy infants, a reduced saturated fat diet led to significantly reduced systolic/diastolic BP, by 1/1 mmHg, over 7 months.[135]

Dietary intake of omega-6 polyunsaturated fat (mostly linoleic acid in Western diets) has been shown to have little or no effect on BP.[132] A Cochrane review in 2015 reported no significant effects on BP in a pooled analysis of two clinical trials that increased omega-6 polyunsaturated fatty acid intake.[136]

Although early trials did not support an association between monounsaturated fat intake and BP, subsequent trials found that diets rich in monounsaturated fats modestly lower BP.[137] The International Study on Macronutrients and Blood Pressure (INTERMAP) supported findings to increase intake of monounsaturated fats from vegetable sources to aid in lowering BP.[138] However, an increase in monounsaturated fat is often linked to a reduction in carbohydrates, potentially with a change in the type of carbohydrate as well.[139] This was also seen in the Omni Heart-Randomized Trial.[107] Therefore, it remains unclear whether the effects of increased monounsaturated fat intake reflect an increase in this nutrient and/or a decrease in carbohydrate intake or change in the type of carbohydrate. The INTERMAP study identified a relationship of dietary cholesterol intake to systolic, but not diastolic, BP for all participants.[138]

Carbohydrate

Both the amount and type of carbohydrates consumed may affect BP. Globally, many populations that consume carbohydrate-rich, low-fat diets have lower BP levels than Western countries.[140] Still, findings of observational studies have been inconsistent.[141] Increasing carbohydrate intake by decreasing total fat typically did not reduce BP in small early trials. In contrast, the OmniHeart feeding trial documented that partial exchange of carbohydrate with monounsaturated fat or protein (roughly half from plant sources) lowered BP (refer back to Figure 68.12). More recent evidence shows a link between increased intakes of added sugars, especially sugar-sweetened beverages, and elevated BP. Studies included animal studies, acute ingestion studies in humans fed high doses of different sugars and, more recently, epidemiologic studies. In cross-sectional studies, drinking more sugar-sweetened beverages was associated with elevated BP in adolescents.[142] In prospective studies, more than one soft drink per day significantly increased the odds of developing high BP.[143] In post hoc analyses of a completed trial, there was a direct relationship between reduction in sugar-sweetened beverage intake and reduction in BP.[144] A systematic review of six cross-sectional and six prospective cohort studies showed that intake of at least 12 fluid ounces daily of sugar-sweetened beverages was associated with increasing risk of HTN by at least 6%, and systolic BP by 1.8 mmHg over 18 months.[145] Reducing intake of all sugars may be of benefit, as another meta-analysis of 12 trials with a follow-up duration of at least 8 weeks showed that diets high in sugar led to a mean increase in BP.[146] In a meta-analysis of trials that iso-calorically substituted fructose for other sugars, a 1.5 mmHg net reduction in diastolic BP and no effect on systolic BP were reported.[147]

Protein

Two major observational studies, the INTERMAP,[148,149] and the Chicago Western Electric Study, noted significant inverse relationships between protein intake and BP.[141] In both these studies, diets higher in protein from plant sources were associated with lower BP, whereas diets higher in protein from animal sources had no significant effect. A systematic review of observational and interventional studies found evidence that dietary protein, especially plant protein, may have a small beneficial effect on BP.[150]

Two trials found that increased protein intake from soy supplements can reduce BP. In one trial of individuals taking antihypertensive agents,[151] supplemental soy protein (25% kcal from protein, half from soy) reduced average 24-hour systolic/diastolic BP by 5.9/2.6 mmHg. In a large trial conducted in China,[152] supplemental soy protein, which increased total protein intake from 12% to 16% kcal, lowered average BP by 4.3/2.7 mmHg, net of a supplemental carbohydrate control group. Overall, clinical trials and observational studies support the hypothesis that increased intake of protein from plant sources can lower BP.

Vitamin C

Laboratory, observational, and depletion-repletion studies suggest that increased vitamin C intake and higher vitamin C concentration are associated with lower BP. A meta-analysis of RCTs showed that vitamin C supplementation of 500 mg/day (median dose) for 8 weeks in persons with HTN showed a significant reduction in systolic and diastolic BP of 4.8 and 1.7 mmHg, respectively.[153] A 2020 systematic review and meta-analysis suggested that 500 mg/day of vitamin C supplementation showed a significant reduction in both systolic and diastolic BP of 4.1 and 2.3 mmHg, respectively.[154]

Patient Subgroups

Children and Adolescents

Systemic hypertension in adulthood is often presaged by HTN in childhood. Therefore, early intervention is important to prevent adverse CV outcomes.[155] The prevalence of HTN in children varies across countries. A recent systematic review reported a pooled worldwide prevalence of HTN of 4% (95% CI, 3.29%–4.78%) in children 19 years and younger.[156] In the NHANES prevalence among children and adolescents, aged 8–12 years, increased from 5.2% in 1999–2002, to 6.2% in 2003–2006, and then decreased to 4.6%.[157] In the same study, a positive correlation was observed between HTN and childhood obesity. It is important to note that about 1% of childhood HTN is secondary to renal disease, endocrine abnormalities, and other conditions.[158]

Diet patterns and quality are important determinants of the development of HTN in childhood. These include high sodium intake, low potassium, and high consumption of sugar-sweetened beverages. Sodium intakes in children and adolescents are well above recommended levels, largely due to processed and prepared foods.[159] Sodium intake is positively associated with HTN among children. This association may be stronger among those who are overweight/obese.[160] The DASH diet can effectively lower BP in adults with HTN,[16,17] and similar effects have been observed in children and adolescents. In a population-based cohort study in 1,068 Dutch children, aged 5 to 7 years, higher adherence to the DASH diet was associated with lower systolic BP and lower odds of having HTN.[161] Other small studies have also demonstrated that DASH diet interventions improved BP, relative to routine care, in adolescents.[162]

High consumption of sugar-sweetened beverages has been associated with HTN, independent of obesity in children.[163] Further health-related correlations were identified by the authors of the, "Added Sugars and Cardiovascular Disease Risk in Children: A Scientific Statement From the American Heart Association."[164] They concluded that "Strong evidence supports the association of added sugars with increased cardiovascular disease risk in children through increased energy intake, increased adiposity, and dyslipidemia.... it is reasonable to recommend that children consume ≤25 g (100 kcal or ≈6 teaspoons) of added sugars per day, and to avoid added sugars for children <2 years of age. Few children achieve such levels, making this an important public health target." In one study, reduction in dietary sweeteners was associated with reduction in BP.[165]

African Americans

African Americans, on average, have higher BP and are at greater risk of BP-related complications, especially stroke, kidney disease, and all-cause and CV mortality than Whites.[166] In an analysis of the MESA cohort, 63% of Black participants reported "a poor diet." In rigorously controlled efficacy trials, African Americans achieved greater BP reductions than Whites with several nonpharmacologic therapies, particularly decreased sodium intake, increased potassium intake, and the DASH diet.[16,17,47,48] Despite awareness, and tools currently available, a recent analysis in patients with clinical heart failure documented high prevalence of HTN, and suboptimal treatment of HTN persists in this group[167] (See **Figure 68.13** and **Table 68.2**).

Rethy et al. showed, in NHANES data from 2017–2018, that the prevalence of systolic BP ≥130 mmHg was 41% (95% CI, 36–47) among heart failure (HF) patients overall and 49%[43-55] among those on treatment. More than 50% of adults

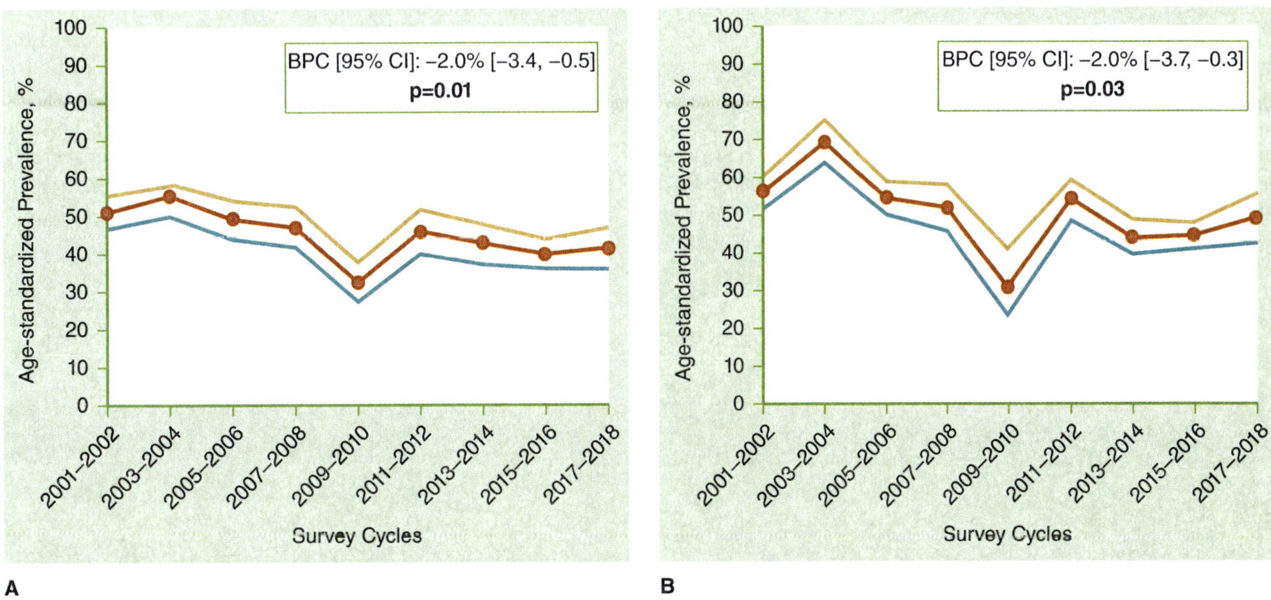

Figure 68.13 Uncontrolled BP was present in 48% (95% ci, 49%–56%) of adults with heart failure (representing 2.3 million people). Prevalence of uncontrolled BP improved over time but uncontrolled risk factors remained high—2017 to 2018: 41% (36%, 47%) and 7% (5%, 12%) had uncontrolled BP and HbA1c, respectively.[167]

Reproduced from Rethy L, et al. Blood pressure and glycemic control among ambulatory US adults with heart failure: National Health and Nutrition Examination Survey 2001 to 2018. (2022) Circulation Heart Failure 15, e009229.[167]

Table 68.2 Adjusted Prevalence Ratios of Uncontrolled BP (SBP ≥130 mmHg) and Uncontrolled HbA1c (HbA1c ≥8%) by Race/Ethnicity Among Individuals With HF Ages ≥20 Years: NHANES 2001–2018

	SBP ≥130 mmHg		HbA1c ≥8%	
	Model 1[a]	Model 2[b]	Model 1[a]	Model 2[b]
Non-Hispanic White adults	1.00 (REF)	1.00 (REF)	1.00 (REF)	1.00 (REF)
Non-Hispanic Black adults	1.18 (1.02–1.37)	1.19 (1.02–1.39)	1.20 (0.79–1.80)	1.03 (0.67–1.58)
Mexican American adults	1.14 (0.90–1.44)	1.08 (0.81–1.43)	1.15 (0.68–1.93)	1.07 (0.61–1.89)

[a]adjusted for age, sex and treatment status.

[b]further adjusted for poverty level, education and health insurance.

Reproduced from: Rethy L, Thanh-Huyen T, Vu NS, et al. Circ Heart Failure;15:e009229.[167]

with HF and HTN had a systolic BP ≥130 mmHg and more than 20% of those on treatment for diabetes had HbA1c ≥8.0%. Prevalence of HTN and diabetes varied by race and ethnicity. Uncontrolled BP among adults with HF was significantly higher among non-Hispanic Black, compared with non-Hispanic White participants. The authors concluded, "We document an unacceptably high prevalence of uncontrolled BP and HbA1c in a nationally representative, ambulatory heart failure sample with significant differences in BP control by race and ethnicity."[167]

In adjusted models, non-Hispanic Black participants were 1.19 (1.02–1.39) times more likely to have uncontrolled BP than non-Hispanic White participants; 53% [48%–58%] compared with 47% [43%–51%], $P < 0.05$).

Conclusions

The U.S. Food and Drug Administration summarized the current challenges that healthcare providers and patients face regarding the relationship between diet and HTN.[77] Poor dietary habits, leading to HTN, CVD, obesity, diabetes, and certain cancers, are among the leading causes of morbidity and mortality in the United States.[76,168] Mayne et al., wrote, "*Good nutrition is a critical foundation for health and wellbeing.*"[77] Reducing excess sodium intake is projected to have substantial benefits in reducing CVD-related morbidity, mortality, and to save billions of dollars in healthcare costs.[75,78] A compelling body of evidence points to additional dietary factors beyond sodium that affect BP. Dietary changes known to lower BP effectively include weight loss, increased potassium intake, moderation of alcohol intake (among those who drink), and DASH-style dietary patterns. Although other dietary elements may also affect BP, existing evidence is inconclusive, and/or their effects are small. Given the direct, progressive relationship of BP with clinical outcomes, strategies to lower BP in individuals with HTN are warranted. Although changes in lifestyle and diet are effective, they are challenging to execute and difficult to maintain.

In contrast to persons with HTN, recent analyses have also clarified the relationship between diet and BP in individuals with normal BP or only mild HTN.[169,170] An updated evidence report and systematic review recently published by the U.S. Preventive Services Task Force noted that, although healthy diet and physical activity interventions were associated with small improvements in BP, LDL-C concentration, and all measures of adiposity in the short term (6 months), further evidence concerning long-term CV morbidity and mortality outcomes was limited.[169] Sustainable, preventative efforts will require systemic, nationwide initiatives, which specifically target food production operations. Such an approach was proven to be technically feasible in the Chinese, Diet, Exercise and Cardiovascular Health (DECIDE-DIET) trial.[170] The CHH diet had reduced energy from fat, increased energy intake from protein, and carbohydrates, and increased dietary fiber and potassium. Importantly, sodium intake was significantly reduced. The researchers identified a significant reduction in systolic and diastolic BP with the CHH "healthy," diet groups. The researchers concluded that it was possible to create a heart healthy, affordable diet menu that aligned with local cuisine culture and *maintained* consumer acceptance, at least during the 4-week duration of this trial.[170]

Acknowledgment

The authors acknowledge Lawrence J Appel, the author of the previous version of this chapter, for his contributions, and also Gouthami Chennu for her contribution to this version.

References

1. Mills KT, Bundy JD, Kelly TN, et al. Circ 2016;134:441–450.
2. Yusuf S, Joseph P, Rangarajan S, et al. Lancet 2020;395:795–808. Erratum in: Lancet 2020;395:784.
3. Mills KT, Stefanescu A, He J. Nat Rev Nephrol 2020;16: 223–237.
4. Fryar CD, Ostchega Y, Hales, et al. Hypertension Prevalence and Control Among Adults: United States, 2015–2016. NCHS Data Brief, no 289. Hyattsville, MD: National Center for Health Statistics, 2017.
5. Whelton PK, Carey RM, Aronow WS, et al. J Am Coll Cardiol 2018;71:e127–248.
6. Ettehad D, Emdin CA, Kiran A, et al. Lancet 2016;387: 957–67.
7. Huang Y, Wang S, Cai X, et al. BMC Med 2013;11:177.

8. Wright JT Jr, Williamson JD, Whelton PK, et al. N Engl J Med 2015;373:2103–16.
9. Lewis CE, Fine LJ, Beddhu S, et al. N Engl J Med 2021;384: 1921–30.
10. Savica V, Bellinghieri G, Kopple JD. Annu Rev Nutr 2010;30: 365–401.
11. Chan Q, Stamler J, Griep LMO, et al. J Atheroscler Thromb 2016;23:276–89.
12. Ozemek C, Laddu DR, Arena R, et al. Curr Opin Cardiol 2018;33:388–393.
13. Jones DW, Whelton PK, Allen N, et al. Hypertens 2021;77: e58–e67.
14. Estruch R, Ros E, Salas-Salvadó J, et al. N Engl J Med 2018; 378:e34.
15. Vincent-Baudry S, Defoort C, Gerber M, et al. Am J Clin Nutr 2005;82:964–71.
16. Appel LJ, Moore TJ, Obarzanek E, et al. N Engl J Med 1997;336:1117–24.
17. Sacks FM, Svetkey LP, Vollmer WM, et al. N Engl J Med 2001;344:3–10.
18. Stamler R. Hypertens 1991;17: I-16–I-20.
19. Siervo M, Lara J, Chowdhury S, et al. Br J Nutr 2015;113:1–15.
20. Appel LJ. J Clin Hypertens 2008;10:7–11.
21. Appel LJ. Hypertension: A Companion to Braunwald's Heart Disease. 1st ed. Philadelphia, PA: Saunders, 2007;202–12.
22. Sacks FM, Campos H. N Engl J Med 2010;362:2102–12.
23. Cohen JB. Curr Cardiol Rep 2017;19:98.
24. GBD 2015 Obesity Collaborators. N Engl J Med 2017;377: 13–27.
25. Hales CM, Carroll MD, Fryar CD, et al. Prevalence of obesity and severe obesity among adults: United States, 2017-2018. NCHS Data Brief. 2020:1–8. PMID: 32487284.
26. Wühl E. Acta Paediatr 2019;108:37–43.
27. DeMarco VG, Aroor AR, Sowers JR. Nat Rev Endocrinol 2014;10: 364–376.
28. Neter JE, Stam BE, Kok FJ, et al. Hypertens 2003;42:878–84.
29. Stevens VJ, Obarzanek E, Cook NR, et al. Ann Intern Med 2001;134:1–11.
30. Cook NR, Appel LJ, Whelton PK. J Clin Hypertens 2018;20: 1666–1673.
31. Look A, Wing RR, Bolin P, et al. N Engl J Med 2013;369:145–154.
32. Ikramuddin S, Korner J, Lee WJ, et al. JAMA 2018;319:266–278.
33. Mingrone G, Panunzi S, De Gaetano A, et al. Lancet 2015;386: 964–73.
34. Schauer PR, Bhatt DL, Kirwan JP, et al. N Engl J Med 2017;376: 641–651.
35. Nudotor RD, Canner JK, Haut ER, et al. Surg Obstet Relat Dis 2021;17:308–318.
36. Mentias A, Aminian A, Youssef D, et al. J Am Coll Cardiol 2022;79:1429–1437.
37. Chang SH, Stoll CR, Song J, et al. JAMA Surg 2014;149: 275–287.
38. Arterburn DE, Telem DA, Kushner RF, et al. JAMA 2020;324: 879–887.
39. Consensus Development Conference Panel. Ann Intern Med 1991;115:956–961.
40. He FJ, Tan M, Ma Y, et al. J Am Coll Cardiol 2020;75:632–647.
41. Roberts WC. Proc (Bayl Univ Med Cent) 2001;14:314–322.
42. Ellison DH, Welling P. N Engl J Med 2021;385:1981–1993.
43. INTERSALT. BMJ 1988;297:319–328.
44. He FJ, Li J, Macgregor GA. BMJ 2013;346:1325.
45. Pimenta E, Gaddam KK, Oparil S. Hypertens 2009;54: 475–81.
46. Johnson AG, Nguyen TV, Davis D. J Hypertens 2001;19:1053–60.
47. MacGregor GA, Markandu ND, Sagnella GA, et al. Lancet 1989;2:1244–7.
48. Vollmer WM, Sacks FM, Ard J, et al. Ann Intern Med 2001;135: 1019–28.
49. Bray GA, Vollmer WM, Sacks FM, et al. Am J Cardiol 2004;94: 222–7.
50. MacGregor GA, Markandu ND, Singer DR, et al. Br Med J (Clin Res Ed) 1987;294:531–34.
51. Elijovich F, Weinberger MH, Anderson CA, et al. Hypertens 2016;68:e7–e46.
52. He FJ, Markandu ND, MacGregor GA. Hypertens 2001;38: 321–5.
53. Falconnet C, Bochud M, Bovet P, et al. J Hypertens 2004;22: 1213–20.
54. Aburto NJ, Ziolkovska A, Hooper L, et al. BMJ 2013;346, f1326.
55. Poggio R, Gutierrez L, Matta MG, et al. Public Health Nutr 2015;18:695–704.
56. Appel LJ. BMJ 2009;339:b4980.
57. Mente A, O'Donnell M, Rangarajan S, et al; PURE, EPIDREAM and ONTARGET/TRANSCEND Investigators. Lancet 2016;388: 465–75.
58. Mente A, O'Donnell M, Rangarajan S, et al. Lancet 2018;39: 496–506.
59. He FJ, Ma Y, Campbell NRC, MacGregor GA, et al. Hypertens 2019;74:572–580.
60. He FJ, Tan M, MacGregor GA. Lancet 2019;393:1293.
61. Birukov A, Rakova N, Lerchl K, et al. Am J Clin Nutr 2016;104: 49–57.
62. Cook NR, Appel LJ, Whelton PK. Circ 2014;129:981–989.
63. Adler AJ, Taylor F, Martin N, et al. Cochrane Database Syst Rev 2014;12:CD009217.
64. Karppanen H, Mervaala E. Prog Cardiovasc Dis 2006;49:59–75.
65. He FJ, Pombo-Rodrigues S, Macgregor GA. BMJ Open 2014;4: e004549.
66. Frohlich ED. Hypertens 2007;50:161–6.
67. Verhave JC, Hillege HL, Burgerhof JG, et al; PREVEND Study Group. J Intern Med 2004;256:324–30.
68. D'Elia L, Rossi G, Ippolito R, et al. Clin Nutr 2012;31:489–98.
69. Cappuccio FP, Kalaitzidis R, Duneclift S, et al. J Nephrol 2000;13:169–77.
70. Devine A, Criddle RA, Dick IM, et al. Am J Clin Nutr 1995;62: 740–5.
71. Jula AM, Karanko HM. Circ 1994;89:1023–1031.
72. He J, Ogden LG, Bazzano LA, et al. Arch Intern Med 2002;162:1619–24.
73. Patel Y, Joseph J. Int J Mol Sci 2020;21:9474.
74. Ezekowitz JA, Colin-Ramirez E, Ross H, et al; SODIUM-HF Investigators. Lancet 2022;399:1391–1400.
75. U.S. Department of Agriculture & U.S. Department of Health and Human Services. (2020). Dietary Guidelines for Americans, 2020–2025, 9th ed. Available from www.dietaryguidelines.gov/
76. U.S. Department of Agriculture, Agricultural Research Service. What We Eat in America, 2013-2014. Available at www.ars.usda .gov/Services/docs.htm?docid=18349, 2017
77. Mayne ST, McKinnon RA, Woodcock J. JAMA 2021;326: 1675–1676.
78. FDA: Voluntary Sodium Reduction Goals: Target mean and upper bound concentrations for sodium in commercially processed, packaged, and prepared foods: Guidance for Industry, 2021. Available at www.fda.gov/sodiumreduction
79. Young DB. Role of Potassium in Preventative Cardiovascular Medicine. Kluwer Academic Publishers, 2001.
80. Food and Nutrition Board, Institute of Medicine. Sodium and chloride. In Dietary Reference Intakes for Water, Potassium, Sodium, Chloride, and Sulfate. Washington, DC: National Academy Press, 2005.
81. Weaver C, Stone M, Lobene A, et al. Nutr Today 53, 184–195.

82. U.S. Department of Health and Human Services, U.S. Department of Agriculture. 2015-2020 Dietary Guidelines for Americans Available at https://health.gov/our-work/food-nutrition/2015-2020-dietary-guidelines, 2015
83. Whelton PK, He J, Cutler JA, et al. JAMA 1997;277:1624–32.
84. Burgess E, Lewanczuk R, Bolli P, G, et al. CMAJ 160:S35–45.
85. Dickinson HO, Nicolson DJ, Campbell F, et al. Cochrane Database Syst Rev 2006;3:Cd004641.
86. Aburto NJ, Hanson S. Gutierrez H, et al. BMJ 2013;346:f1378.
87. John JH, Ziebland S, Yudkin P, et al. Lancet 2002;359: 1969–74.
88. Poorolajal J, Zeraati, F, Soltanian A, et al. PLOS One 2017; 12(4):e0174967. https://doi.org/10.1371/journal.pone.0174967
89. Filippini T, Violi F, D'Amico R, et al. Int J Cardiol 2017;230: 127–35.
90. Filippini T, Naska A, Kasdagli MI, et al. J Am Heart Assoc 2020;9:e015719.
91. Neal B, Wu Y, Feng X, et al. N Engl J Med 2021;385:1067–77.
92. Morris RC Jr, Sebastian A, Forman A, et al. Hypertens 1999;33: 18–23.
93. Kawasaki T, Itoh K, Kawasaki M. Hypertens Res 1998;21: 235–43.
94. Chang HY, Hu YW, Yue CS, et al. Am J Clin Nutr 2006;83: 1289–96.
95. National Kidney Foundation. Am J Kidney Dis 2004;43: S1–S29.
96. Xin X, He J, Frontini MG, et al. Hypertens 2001;38:1112–17.
97. Okubo Y, Miyamoto T, Suwazono Y, et al. Alcohol 2001;23: 149–56.
98. Briasoulis, A, Agarwal, V, Messerli, F. J Clin Hypertens 2012;14: 792–98.
99. Sesso, H, Cook, N, Buring, J, et al. Hypertens 2008;51:1080–87.
100. Husain, K, Ansari, R, Ferder, L. WJC 2014;6:245–52.
101. Roerecke M, Kaczorowski J, Tobe SW, et al. Lancet Public Health 2017;2:e108–20.
102. Aladin, A, Chevli, P, Ahmad, MI, et al. Am J Cardiol 2021;160: 60–66.
103. Satija A, Hu FB. Trends Cardiovasc Med 2018;28:437–41.
104. Yokoyama Y, Nishimura K, Barnard ND, et al. JAMA Intern Med 2014;174:577–87.
105. Key TJ, Fraser GE, Thorogood M, et al. Am J Clin Nutr 1999;70:516S–24S.
106. Huang T, Yang B, Zheng J, et al. Ann Nutr Metab 2012;60: 233–40.
107. Appel LJ, Sacks FM, Carey VJ, et al., OmniHeart Collaborative Research Group. JAMA 2005;294:2455–64.
108. Appel LJ, Brands MW, Daniels SR, et al. Hypertens 2006;47: 296–308.
109. Filippou CD, Tsioufis CP, Thomopoulos CG, et al. Adv Nutr 2020;11:1150–60.
110. Chiavaroli L, Viguiliouk E, Nishi SK, et al. Nutrients 2019; 11:338.
111. Raphael KL. Kidney Int 2019;95:1296–98.
112. Ikizler TA, Burrowes JD, Byham-Gray LD, et al. Am J Kidney Dis 2020;76:S1–S107.
113. Widmer RJ, Flammer AJ, Lerman LO, et al. Am J Med 2015;128:229–38.
114. Kromhout D, Menotti A, Bloemberg B, et al. Prev Med 1995;24: 308–15.
115. de Lorgeril M, Salen P, Martin JL, et al. Circ 1999;99:779–85.
116. Jennings A, Berendsen AM, de Groot LC, et al. Hypertens 2019;73:578–86.
117. Nordmann AJ, Suter-Zimmermann K, Bucher HC, et al. Am J Med 2011;124:841–51.
118. Toledo E, Hu FB, Estruch R, et al. BMC Med 2013;11:207.
119. Arnett DK, Blumenthal RS, Albert MA, et al. Circ 2019;140: e596–e646.
120. Whelton SP, Hyre AD, Pedersen B, et al. J Hypertens 2005;23: 475–81.
121. Streppel Mt, Arends LR, van't Veer P, et al. Arch Intern Med 2005;165:150–56.
122. Khan K, Jovanovski E, Ho HVT, et al. Nutr Metab Cardiovasc Dis 2018;28:3–13.
123. Cappuccio FP, Elliott P, Allender PS, et al. Am J Epidemiol 1995;142:935–45.
124. van Mierlo LAJ, Arends LR, Streppel MT, et al. J Human Hypertension 2006;20:571–80.
125. Cormick G, Ciapponi A, Cafferata ML, et al. Cochrane Data System Rev 2015;6:CD010037.
126. Jee SH, Miller ER 3rd, Guallar E, et al. Am J Hypertens 1999;12:84–92.
127. Dickinson HO, Nicolson D, Campbell F, et al. Cochrane Data System Rev CD004640. doi:10.1002/14651858.CD004640.pub2
128. Kass L, Weekes J, Carpenter L. Eur J Clin Nutr 2012;66: 411–18.
129. Rosanoff A. Magnes Res 2010;23:27–40.
130. Geleijnse JM, Giltay EJ, Grobbee DE, et al. Hypertens 2002;20: 1493–99.
131. Zhang X, Ritonja J, Zhou N, et al. J Am Heart Assoc 2022;11: e025071.
132. Morris MC. J Cardiovasc Risk 1994;1:21–30.
133. Ascherio A, Rimm EB, Giovannucci EL, et al. Circ 1992;86: 1475–84.
134. Ascherio A, Hennekens C, Willett WC, et al. Hypertens 1996;27:1065–72.
135. Niinikoski H, Jula A, Viikari J, et al. Hypertens 2009;53: 918–24.
136. Al-Khudairy L, Hartley L, Clar C, et al. Cochrane Database Syst Rev 2015;16:CD011094.
137. Ferrara LA, Raimondi AS, d'Episcopo L, et al. Intern Med 2000;160:837–42.
138. Miura K, Stamler J, Brown I, et al. J Hypertens 2013;31: 1144–50.
139. Shah M, Adams-Huet B, Garg A. Am J Clin Nutr 2007;85: 1251–56.
140. Sacks FM, Rosner B, Kass EH. Am J Epidemiol 1974;100: 390–98.
141. Stamler J, Liu K, Ruth KJ, et al. Hypertens 2002;39:1000–06.
142. Bremer AA, Auinger P, Byrd RS. J Nutr Metab 2010:196476 [Epub 2009 Sep 6].
143. Dhingra R, Sullivan L, Jacques PF, et al. Circ 2007;116: 480–88.
144. Chen L, Caballero B, Mitchell DC, et al. Circ 2010;121: 2398–406.
145. Malik AH, Akram Y, Shetty S, et al. Am J Cardiol 2014;113: 1574–80.
146. Te Morenga LA, Howatson AJ, Jones RM, et al. Am J Clin Nutr 2014;100:65–79.
147. Ha V, Sievenpiper JL, de Souza RJ, et al. Hypertens 2012;59: 787–95.
148. Sakuri M, Stamler J, Miura K, et al. J Hypertens 2011;29: 222–28.
149. Elliott P, Stamler J, Dyer AR, et al. Arch Intern Med 2006;166: 79–87.
150. Altorf-van der Kuil W, Engberink MF, Brink EJ, et al. PLoS One 2010;5:e12102.
151. Burke V, Hodgson JM, Beilin LJ, et al. Hypertens 2001;38: 821–26.
152. He J, Gu D, Wu X, et al. Ann Intern Med 2005;143:1–9.

153. Juraschek SP, Guallar E, Appel LJ, et al. Am J Clin Nutr 2012;95:1079–88.
154. Guan Y, Dai P, Wang H. Medicine 2020;99:e19274.
155. Chen X, Wang Y. Circ 2008;117:3171–80.
156. Song P, Zhang Y, Yu J, et al. JAMA Pediatr 2019;173:1154–1163.
157. Hardy ST, Sakhuja S, Jaeger BC, et al. JAMA Netw Open 2021;4:e213917.
158. Flynn J, Zhang Y, Solar-Yohay S, et al. Hypertens 2012;60: 1047–54.
159. Cogswell ME, Yuan K, Gunn JP, et al. Morb Mortal Wkly Rep 2014;63:789–97.
160. Yang Q, Zhang Z, Kuklina EV, et al. Pediatr 2012;130:611–9.
161. Zafarmand MH, Spanjer M, Nicolaou M, et al. Hypertens 2020;75:59–70.
162. Couch SC, Saelens BE, Levin L, et al. J Pediatr 2008;152: 494–501.
163. Jalal DI, Smits G, Johnson RJ, et al. J Am Soc Nephrol 2010;21:1543–9.
164. Vos MB, Kaar JL, Welsh JA, et al. Circ 2017;135:e1017–e1034.
165. Chen L, Caballero B, Mitchell DC, et al. Circ 2010;121: 2398–406. Erratum in: Circ 2010;122:e408.
166. Post WS, Watson KE, Hansen S, et al. Circ 2022;146: 229–239.
167. Rethy L, Thanh-Huyen T, Vu NS, et al. Circ Heart 2022; 15, e009229.
168. Kochanek KD, Xu JQ, Arias E. Mortality in the United States. CDC National Center for Health Statistics; 2019. NCHS Data Brief No. 395.
169. Patnode CD, Redmond N, Iacocca MO, et al. JAMA 2022;328: 375–88.
170. Wang Y, Feng L, Zeng G, et al. Circ 2022;146:303–15.

C. PEDIATRIC AND ADOLESCENT DISORDERS

CHAPTER 69

Pediatric Feeding Problems

Richard M. Katz, Jillian Murphy, and Ellen K. Wingert

CHAPTER OUTLINE

Abbreviation

GI, gastrointestinal

Introduction

Feeding children is thought of as natural behavior, a bonding experience between caregiver and child, physiologically necessary and, provided that there is enough nutrition available, a routine process. However, like many disabilities that affect children, routine eating can be disrupted by complex behavior anomalies, physical defects, or by acute and/or chronic disease states. Recognition of these eating deficits in children requires attention to the concerns of caregivers, acknowledgement that aberrant eating and swallowing is a real, not imaginary, issue that is critical for the medical community to recognize.

Recognition of this problem is recent and attempts to diagnose, treat, and classify this spectrum of issues had been incomplete until recently.[1] Current thoughts are that this problem does not necessarily result in undernutrition but can also be seen in children who are gaining weight well and in fact, who are obese. Recent collaboration by providers invested in caring for these children has resulted in a specific diagnostic category of Pediatric Feeding Disorders (PFD)[1] in addition to the previously established diagnostic codes in use. Conventionally, PFD was thought to occur in young children and eating disorders (ED) to more likely occur in older adolescents who may have body image distortion. It is now understood that those age distinctions may be less clear than previously thought. Children having ARFID (Avoidant Restrictive Food Intake Disorder) require different therapy than a patient with an ED.

This is an exciting time on the road to treatment for these complex disorders. This chapter will cover the classic diagnostic processes for this problem as well as describe the newest best practices and standard of care for the evaluation and treatment of PFD.

Pediatric Feeding Disorder is a term used to describe children having trouble in consuming adequate nutrition by mouth (impaired feeding), most of whom are affected by comorbid medical, developmental, and behavioral conditions. The term is often mistaken for eating disorders such as anorexia and bulimia, but it is differentiated in that food avoidance is not motivated for weight loss or body image concerns. Most normally developing children learn to accept and consume a well-balanced, healthful diet to sustain growth and health.[2] They develop the capacity for self-regulation and adapt to various parent and environmental conditions. Satter[3] expanded this concept and outlined the role of the child and the parent during feeding. However, biologic, personal, and social factors can interfere with the principle of self-regulation.

Nearly 25% of infants and children are affected by feeding disorders at some point in their development.[4] The rate is much higher, nearly 80%, among children with developmental disabilities. Further breakdown of the prevalence indicates 52% of toddlers are not consistently hungry at mealtimes, 42% end meals after a brief session, 35% are picky eaters, and 33% show food selectivity.[4] However, severe feeding problems are noted in 3% to 10% of children[5] with greater prevalence in children with physical disabilities (26%–90%) and among those with medical illness and prematurity (10%–49%).[6-8] Children with neurodevelopmental disorders commonly present with oral sensitivity resulting in a limited repertoire of accepted foods based on texture, taste, appearance, or temperature.[9,10]

The consequences of undernutrition on growth and development are well documented[11-13] and cause substantial morbidity and mortality. Feeding disorders affect the whole family, resulting in significant stress and strain in the caregiver–child relationship.[14] Two-thirds of a caregiver's waking time may be spent attending to a child with disordered feeding,[15] and this intense involvement of a primary caregiver takes time away from other family and household duties.

Feeding disorders have multiple etiologies, including medical, nutritional, behavioral, psychological, and environmental factors.[11,12] **Table 69.1** presents examples of common childhood feeding problems. Feeding disorders manifested by children with normal growth parameters may present as restricted food variety, inappropriate food consistency, and disruptive mealtime behaviors. Children who missed the opportunity to experience foods and textures at certain developmental stages, the "critical" or "sensitive" periods, are more resistant to new foods and higher textures. Texture difficulty, in the absence of oral–motor dysfunction, can result from dental problems, developmentally inappropriate food textures, or avoidance because of an aversive experience of gagging and choking. Children with autism spectrum disorder tend to demonstrate difficulty with food textures with or without presenting with delayed oral motor skills or pharyngeal difficulties.[16]

Children with feeding disorders are a heterogeneous group. They range from those without medical problems to those who have gastrointestinal (GI) tract disorders, systemic illness, developmental delay, and physical disabilities. A child's feeding progresses from a strictly biologic need to a process combining maturation and learning in a nurturing social environment. (**Table 69.2**) Thus, feeding disorders should be conceptualized as biopsychosocial problems. Interactions among the three mechanisms pose a challenge to the differential diagnosis and evaluation and treatment. Many, but not all, persistent feeding difficulties in children may have an associated underlying structural, neurologic, or physiologic disorder. Yet, in most children with significant feeding disorders, no clear etiology becomes apparent with even the most thorough evaluation.

The COVID-19 (SARS-CoV-2) pandemic impacted children and adolescents in ways far beyond a severe respiratory disease. There was an astonishing increase in the number of social and mental disturbances in these patients and their families.[17] Social isolation, loss of comradery at school, virtual learning due to quarantines and lack of access to healthcare providers, including behavioral health specialists, all impacted children's ability to cope with new and prolonged stressors. The response to these traumatic events resulted in increased reports of increased selectivity, severe food refusals, and has resulted in many children being admitted to hospitals for malnutrition, weight loss, dehydration, and severe emotional maladaptive behaviors.[18] This unfortunate world-wide pandemic required creative thinking and approaches to diagnosing and treating many disease processes including Pediatric Feeding Disorders.

One meta-analysis of 29 studies has shown a twofold increase of depression and anxiety symptoms in the years 2021 to 2022. These rates were higher in older adolescents and females.[19] Medical admissions for combined medical and restrictive intake disorders are increasingly well-known[20] and can result in delayed diagnosis of medical and psychiatric issues.[21]

Feeding is a complex task requiring a sequential progression of a repertoire of skills to be successful. (**Table 69.3**) Guidance to families based on developmental progressions observed in typical children are inappropriate for children with cerebral palsy, failure to thrive, syndromic disorders, and muscular and neuromuscular problems. Poor coordination of oral structures may interfere with the ability to move food in the mouth, chew, or swallow in a safe and effective manner. Delayed motor skills may interfere with self-feeding.

Successful feeding is often perceived by parents as a measure of parenting competence. Effective feeding is contingent on the ability of both the adult and child to give, read, and interpret others' cues. Neurologic impairment may interfere with the ability to give clear hunger or satiety cues.

Children with developmental disabilities, medical conditions, and severe behavior problems are unlikely to outgrow their feeding problems without intervention. (**Table 69.4**) Therefore, it is important that caregivers and care providers recognize

Table 69.1 **Common Feeding Problems**

Total food refusal
Food refusal by volume
Food refusal by texture
Food refusal by type
Bottle dependency
Lengthy meals
Maladaptive behaviors
Adipsia (Inability or refusal to drink)

Table 69.2 **Optimizing the Feeding Environment**

Quiet room with limited distractions
Developmentally appropriate positioning
Stable and supportive chair
Developmentally appropriate utensils
Child demonstration of readiness cues for feeding
Stable routine and schedules
Caregiver positioned at eye level

Table 69.3 **Basic Pediatric Feeding Guidance**

- Avoid distractions during mealtimes (television, cell phones, etc.)
- Maintain a pleasant neutral attitude throughout meal
- Limit meal duration (20–30 min)
- Four to six meals/snacks a day with only water in between
- Serve age-appropriate foods
- Systematically introduce new foods (up to 8–15 times)
- Encourage self-feeding
- Tolerate age-appropriate mess

Reproduced from Kerzner, B, Milano, K, MacLean, W C, Pediatr 2015; 135(2) 344–353. doi:10.1542/peds.2014-1630

Table 69.4 Medical Conditions Seen in Feeding Problems

Anatomic abnormalities
- Cleft lip or palate

Pierre Robin sequence
CHARGE association
Cardiopulmonary effects
Complex congenital heart disease
Chronic lung disease
COVID-19 (SARS-CoV-2)
Aspiration pneumonia
Neuromuscular disorders
- Cerebral palsy
- Cranial nerve anomalies
- Pseudobulbar palsy
- Intracranial mass lesion

Disorders of esophageal phase of swallowing
- Cricopharyngeal achalasia
- Tracheoesophageal fistula/esophageal atresia repair
- Esophageal mass, stricture, web
- Vascular rings
- Foreign bodies

Motility disorders
Lumen disorders
- Gastroesophageal reflux disease
- Peptic esophagitis/gastritis
- Inflammatory bowel disease

Genetic disorders
- Prader-Willi syndrome
- Trisomy 21
- Velocardiofacial syndrome (22q11.2 deletion syndrome)

Metabolic disorders
Miscellaneous
- Constipation
- Food allergies

CHARGE, coloboma, heart disease, atresia choanae, retarded growth and retarded development and/or CNS anomalies, genital hypoplasia, and ear anomalies and/or deafness.

a child's feeding problem early and have an evaluation to offer early intervention to arrest the downward spiral of the child's feeding problem.

Feeding disorders include various feeding behaviors and characteristics. These behaviors can be categorized into ability (unable to perform) and motivational deficits (unwilling).[17] A child with low energy or fine motor deficits may not self-feed. A child reported as having poor appetite and reduced consumption may have swallow dysfunction, taste aversion, texture sensitivity, dental problems, recurring ear infections, or many other disorders. A child with severe gastroesophageal reflux may actively gag or vomit to relieve the discomfort. Total food refusal is uncommon in normal healthy children, except during illness or transiently when they are emotionally upset. However, a comprehensive assessment is still necessary to rule out physical causes of food refusal.

Assessment

In all presentations of food refusal, it is also important to assess whether the concerns could represent age-typical patterns or true dysfunction. Often, children's behaviors of concern to parents could be developmental variations, such as poor self-feeding or messy eating at 1 year of age. Variability in eating between meals is common in children between the ages of 2 and 5 years. These children are active, easily distracted, and resist staying seated. They demand independence and control and insist on certain utensils and foods. In fact, weight gain slows down and children of this age do not require the same calories as they did as infants. The slope in the World Health Organization's growth chart reflects this decrease in growth velocity.[18] Similarly, picky eating is usually a transient behavior and part of normal development in preschool children.[19] Caregivers who fail to understand developmental variations become anxious and devise novel feeding techniques, often creating a conflict between parent compulsion and child capacity.[20]

A determination as to whether the child's feeding pattern is normal for age and developmental levels should be made by the child's healthcare providers. Beyond typical fussy eating in infancy and toddlerhood, red flags for pediatric feeding disorders include wheezing, choking, gagging with eating, anticipatory gagging at meals, frequent respiratory infections, diarrhea, vomiting, pain with feeding, sudden change in eating habits after recovering from illness or injury, forceful feeding by the caregiver, and faltering growth and nutritional deficiencies.[21] To aid in communication between providers and payors, feeding disorders may be diagnosed by healthcare professionals using ICD criteria for PFD[1] and/or DSM criteria for Avoidant/Restrictive Food Intake Disorder.[22]

Once a feeding problem is identified, it is important to identify medical, anatomical, developmental, psychosocial, and/or behavioral factors that may be either exacerbating the feeding problems or acting as barriers to development of normal feeding behavior. Most often, feeding concerns are related to food selectivity (refusal of age-appropriate nutritious choices), low appetite (eating smaller portions than needed for age), or fear of food (phobic anxiety and avoidance).[21] Low appetite and food selectivity tend to be more chronic concerns. Acute fear of food, however, typically presents as more distressing for children and their caregivers and may have sudden onset and rapid course following a traumatic event.[22] There are many ways to measure severity of feeding concerns, which can aid in diagnosis, communication between providers, and tracking progress over time.[23]

In most cases, a team of experienced professionals from several disciplines, including gastroenterology, nutrition, occupational therapy and/or speech therapy, and psychology, is necessary to establish a differential diagnosis of the presenting symptoms to specify the cause or function. The process of understanding the origin of feeding disorders involves identifying symptoms and behaviors as well as determining the predisposing, precipitating, and perpetuating factors. Contributing factors may be related to environment or caregiver behavior or related to child physical presentation or sensation, temperament, comorbid psychopathology, or trauma history. As medical and developmental issues are responsible for the onset of most feeding problems,[24] child factors should be assessed prior to environmental/caregiver factors.

History

A detailed history helps to formulate the nature of the problem, especially for the conditions that affect nutritional status and feeding care. The focus also should be on the child's developmental level and the caregiver's understanding and knowledge of appropriate feeding regimens, textures, volumes, and methods.

Table 69.5 Medications Causing Interfering Side Effects

Medication	Possible Gastrointestinal Effects
Iron	Nausea, vomiting
Common antibiotics	Nausea, vomiting, pain
Nonsteroidal anti-inflammatories	Mucosal injury
Psychotropics	Lethargy, dysphagia
Stimulants	Depressed appetite

Medication effects on a child's feeding include depressed appetite, nausea, GI irritation, and constipation. Some of the medications and their effects on the GI system and potential for feeding difficulties are presented in **Table 69.5**. In young children, prenatal risk factors and neonatal history should be obtained, especially in premature infants or those with a complicated neonatal course. Surgical history is vital, particularly GI surgery, but any surgery in childhood could result in disordered eating. Finally, a family history could elicit other evidence of feeding and eating disorders.

A profile of the child's feeding history should include the onset, course, frequency, intensity, duration, and variability of feeding behaviors across time, settings, and caregivers. A profile of the family dynamics, caregiver knowledge of child development and feeding practices, family feeding history, and resources also should be gathered. The role of the family, particularly the primary caregiver, is paramount in the management of feeding difficulties. In a typical feeding situation, the parent decides what to serve the child and the child eats to satisfaction.[3] Caregivers' approach to mealtimes can affect feeding interactions and outcomes during meals profoundly.

Medical Evaluation

A thorough history and physical examination should be conducted to rule out organic causes. Any organ or system of the body, especially the GI system, can act as a precipitating factor for feeding problems. Patient observation should focus on child temperament, tactile responses, motor control, oral integrity, coordination, suck/swallow competency, positioning, and signs of pain and discomfort. A physician may observe parts of the feeding anatomy and function that are accessible to clear the patient for safe oral feeding. Speech and occupational therapists may conduct feeding trials to explore oral motor functioning, positioning, sensation, swallow efficiency, muscle strength, suck-swallow-breathing coordination, and self-feeding skills.

Instrumental evaluations can be used along with the clinical assessment. The modified barium swallow study is the most common assessment for evaluating the dynamic phases of swallowing function. This evaluation provides information regarding structural and functioning findings, the risk of aspiration, and the effectiveness of treatment techniques. Fiberoptic endoscopic evaluation of swallowing (FEES) uses a flexible endoscope passed transnasally to allow visualization of the nasal, pharyngeal, and laryngeal structures.[25] Ultrasound as an imaging tool visualizes the relationships between movement patterns of the oral and pharyngeal structures.[25]

Diagnosing medical disorders in children with feeding difficulties is a challenge, especially in preverbal children who are

Table 69.6 Gastrointestinal Disorders Affecting the Process of Nutrition Assimilation

Consumption: appetite, dysphagia, aspiration, craniofacial abnormalities

Retention: emesis, diarrhea, rumination, dysmotility

Digestion: food allergy, FPIES (food protein induced enterocolitis), lactose intolerance, eosinophilic esophagitis (EoE)

Absorption: celiac disease, dumping syndrome

Elimination: constipation, Hirschsprung disease

unable to report on their condition. Table 69.4 lists examples of medical conditions commonly seen in children with complex feeding difficulties. Feeding disorders in children with medical problems are complex because they interact with behavioral and social factors, making the differential diagnosis more challenging. The most common clinical symptoms suggestive of medical conditions are dysphagia, gastroesophageal reflux, diarrhea, and constipation. GI disorders interfere with the process of consumption, retention, digestion, absorption, and elimination (**Table 69.6**), frequently resulting in weight loss, lethargy, illness, and feeding disorders.

Anthropometry

Measurements of height, weight, and weight-to-height ratio are indispensable in determining nutrition and growth status. Much can be learned about the child's growth and development by plotting serial points of these values. A thorough analysis of the serial points can help establish the onset and course as well as the precipitating and perpetuating factors of the feeding problem. Children below the 5th percentile should not be assumed to have a problem. Weight gain should be a concern only if the child's growth velocity falters and falls off the growth curve. However, even well-nourished children often need assistance because they may exhibit problematic feeding behaviors and/or micronutrient deficiencies.

Nutritional assessment of the child's dietary intake is critical. Dietary information can be used to determine calories consumed and nutritional components of the diet. Diet history should include both current and past practices and feeding patterns. Dietary information may be obtained through caregiver recall; however, this may be inaccurate. Children's feeding varies among meals and days. Assessing a single meal or a day's calorie count does not reveal the child's true nutritional status. Therefore, analysis of a 3-day food diary provides the most valid estimate of the child's true intake variations in food types eaten and amounts consumed.[26] Nutritional strategies given to families should consider family preferences, resources, culture, ethnicity, and education.

In some cases, functional disorders may later reveal underlying organic causes of the feeding disorder. For example, delayed gastric emptying without evidence of systemic disease may develop in healthy children.[27] Similar findings of latent physical problems without clinical evidence were identified by Staiano et al.,[27] in their study of upper GI motility in children with progressive muscular dystrophy. It is common for symptoms of a motility disorder, food allergy, and lactose intolerance to emerge once children are taught to eat larger volumes and an increased

Table 69.7 Symptoms of Gastroesophageal Reflux

Vomiting	Coughing
Frequent swallowing	Regurgitation
Dental erosion	Food refusal
Stricture	Barrett esophagus
Anemia	Bleeding
Stridor	Hoarseness
Sandifer Syndrome	Aspiration/extra-esophageal inflammation
Arching	Posturing

variety of foods. Most often a diagnostic evaluation is needed to confirm presumed clinical diagnoses. Food allergies are an increasingly widespread problem in infants and children. Most food allergies do not manifest with classic immunoglobulin E–mediated symptoms such as urticaria or pruritus, but they often manifest with mild-to-moderate GI distress that can lead to food refusal behaviors.[28,29] Presumably, the child was self-regulating and inadvertently avoided detection of medical issues through food refusal. Therefore, most children with food refusal should have allergy evaluations. GI tract disorders beyond gastroesophageal reflux disease can play a role in maladaptive feeding behaviors. Anomalies such as malrotation, esophageal stenosis, vascular anomalies, and motility disorders of the esophagus, stomach, and duodenum all have been associated with food refusal.[30,31]

Emesis is a common presenting symptom in children with GI tract disorders, food allergies, conditioned aversion, rumination, or other underlying disorders. Most common, however, is gastroesophageal reflux. Symptoms of reflux are presented in **Table 69.7**. Children with such symptoms should be referred to a gastroenterologist for diagnosis and therapy.

Dysphagia can occur in one or more phases of swallowing, including the oral phase, initiation of the swallow, pharyngeal phase, and esophageal phase.[25] Causes of dysphagia may include neurologic disorders, anatomic abnormalities, pulmonary disease, and genetic syndromes.[16] Clinical manifestations of dysphagia include oral motor delays, poor suck-swallow-breathe coordination, respiratory distress, food refusals, and food selectivity.[16] **Table 69.8** presents common symptoms of dysphagia in children.

Behavioral Evaluation

The priority of initial behavioral assessment of a child with feeding concerns should be to identify specific appropriate behaviors the child is lacking and/or inappropriate behaviors that are present with feeding.[32] Child and family strengths should also be noted. Feeding expectations are cultural and family-specific, so it is helpful to defer to caregivers on whether a particular feeding issue is interfering with quality of life.

Table 69.8 Symptoms of Dysphagia

Drooling
Chewing difficulty
Food pocketing
Gagging with textures
Coughing during and after swallowing
Wet vocal quality after swallowing

Most parents will have valuable input to allow clinicians to understand their child's feeding problems; however, they may be unable to directly report a cause associated with the feeding concerns, may misidentify a correlating factor as causal, or may perceive a practice as normative that their child experiences as aversive. As such, it is helpful to supplement information gained from parent interviews with standardized forms and direct observational assessment. Standardized forms are helpful in determining clinical severity and tracking progress.[23] Complementary observations such as Functional Behavior Assessment (FBA) or structured parent-child interaction coding are helpful in determining contributing and maintaining factors.[23] Although completing an FBA may be beyond the purview of many generalist clinicians, it is helpful to observe social challenges families may demonstrate in observed meals such as conflict between caregivers, inappropriate expectations, or food insecurity.

Treatment

In an interdisciplinary team, the physician treats underlying medical causes; the dietitian prescribes calories needed, appropriate foods and required nutrients; the occupational or speech therapist develops oral motor and pharyngeal intervention, positioning for feeding, the need for adaptive equipment, and self-feeding skills; and the psychologist works to develop strategies to decrease the child's mealtime anxiety and related food refusal behaviors, increase the motivation to eat and drink, and eliminate disruptive feeding behaviors. For treatment to be effective over the long term, parents and other caregivers must be trained to implement all feeding recommendations in the home, daycare, and school environments. All referral concerns, no matter how simple, should be addressed to avoid more serious problems. Prognosis with early intervention is very favorable for most cases.[33]

Caregiver Education

Education is sometimes as simple as providing positive feedback on growth, medical presentation, developmental skills, and/or behavior. Parents should be reassured if their children are receiving adequate nutrition and are following their growth curve, no matter how small or thin the child may appear. When concerns are minimal, reassurance and routine growth monitoring and follow-up may suffice. The clinician can also provide basic pediatric feeding guidance[21] (see Table 69.3).

For children presenting with PFD, parents should be educated on the complex etiology and treatment options. Parents need to know that many feeding problems are part of a learned response that can be unlearned with support from healthcare providers. Caregivers should be helped to understand that food refusal behaviors are an expression of anxiety or fear, rather than indicative that the child is "bad" or "difficult," or that their fear "is all in their head." Often, problems are worsened unintentionally by caregiver mismanagement.[20] It is important to educate parents and caregivers regarding how the child's feeding problem developed and what they can learn to do to change their child's feeding behavior.[34] In many instances, caregivers experience guilt about their contribution, real or imagined, to the child's feeding problems. They need assurance that a few behavioral changes (e.g., making mealtime pleasant) may improve their child's eating behavior (Table 69.2 and 69.3).[35]

Medical Intervention

Treatment is rarely straightforward, even in situations in which the etiology and functional causes of the feeding disorder have been identified. Gastroenterologists have developed many noninvasive techniques for the evaluation of GI functions.[36] Medical treatment is extensive for organically based feeding difficulties. Occasionally, hospitalization may be needed for objective clinical observation when initial assessment does not provide the answer to the problem or when caregiver report and child status are incongruent. In many cases, management using alternative routes of delivering nutrition may be needed. Enteral feedings should be considered for the following: children who cannot consume sufficient calories by mouth, those who fatigue easily from the effort to chew and swallow, require undue amount of caregiver time, become ill frequently, are medically unsafe to eat or drink. Children with severe malnutrition and failure to thrive and those with severe food allergies also may benefit from enteral feeding. Even when a child is on enteral feedings, nonnutritive and nutritive stimulation should be initiated at the earliest possible opportunity.

Treatment of issues that affect feeding in children include management of chronic constipation, which impairs the desire to eat.[37] This should start during the initial evaluation visit but is most important during any intensive feeding program. Gastrointestinal motility issues are prevalent in these children and can involve the entire GI tract.

Most commonly, however, the upper GI tract is involved with GE reflux and the potential for Sandifer Syndrome[38] Treatment often requires multiple medications including acid suppression and enhancing GI motility. The availability of promotility agents is country dependent, but motilin agonists (erythromycin) are commonly used to stimulate gastric emptying. Gastric accommodation is often enhanced with cyproheptadine and used in some situations as an appetite stimulant.[39,40] Generalized promotility agents such as metoclopramide and cisapride are less commonly used due to their increased incidence of potential CNS or cardiac side effects.

Other medical issues often appear in conjunction with children with PFD either as a causative agent or secondarily to the feeding disorder itself. Children with PFD may have multiple nutrient deficiencies requiring macromolecule and micromolecule supplementation. For example, anemia is often associated with poor nutrient distribution or secondary to food toxicity. This can be seen in young children who consume only untreated milk products, the so-called "milk babies." Such cases will require therapeutic iron supplementation.[41]

Identified food allergens may require avoidance and other treatments (e.g., desensitization) to progress with feeding goals. Eosinophilic esophagitis is a recently noted antecedent to feeding disorders and it too must be recognized and medically treated to enhance successful feeding.

Behavioral Intervention

Behavioral interventions for pediatric feeding problems range in scope and intensity. Children with generalized behavioral dysregulation or generalized anxiety will benefit from broad behavioral treatments. Similarly, in families affected by poverty, trauma, or parental mental illness, children's feeding issues may improve with increased family support. Feeding problems often improve or resolve with treatment of comorbid conditions. For persistent feeding issues, resolution of comorbid concerns may enhance the effectiveness of subsequent feeding interventions.[42,43]

Behavioral feeding interventions should be tailored based on the presenting concern, severity, and family motivation for treatment. For mild presentations of poor appetite or food selectivity, any knowledgeable healthcare provider can provide education and training to caregiver and child regarding appropriate meal schedules and developmentally appropriate expectations for mealtime behavior (see Table 69.3).

For more severe cases and in cases of significant fear and avoidance, a licensed behavioral health provider (psychologist, behavior analyst, or psychotherapist) will be best suited to treat the feeding disorder. Applied Behavior Analysis (ABA) and Cognitive Behavioral Therapy (CBT) have each been successfully applied in treating feeding problems including disruptive mealtime behaviors, food refusal, food selectivity, and mealtime anxiety.[33,44-46] Clinicians may treat children directly or may use a parent-mediated approach throughout, but caregiver training is a component of all successful behavioral treatments for young children.[43] Many behavioral treatment modalities have been shown to be effective in the treatment of feeding disorders, including in-home, hospital-based, school-based, and telehealth interventions.[42,46-48]

Due to the complex nature of most feeding disorders, behavioral health providers should collaborate closely with children's medical providers regarding status and goals. When possible, specialty providers with experience treating Pediatric Feeding Disorder and/or Avoidant Restrictive Food Intake Disorder should be consulted. For complex and severe feeding disorders, specialty intensive multidisciplinary treatment and/or inpatient care is required.[33,42,48]

Oral–Motor/Sensory Intervention

The goal of oral–motor intervention is to improve the quality of feeding skills within the child's functional ability. As part of the therapeutic process, the occupational/speech therapist develops oral motor exercises to strengthen muscles, improve efficiency of chewing, reduce postural effort with proper positioning, and provide adaptive equipment to improve self-feeding skills. In addition, textures may be modified to improve bolus control and swallowing abilities or control the size or flow rate of bolus delivery. Sensory processing problems are common in children with feeding problems.[49-51] Especially for children with neurodevelopmental concerns such as autism spectrum disorder, food refusal may be based on sensory characteristics of food.[16,51] Impairment of oral-sensory function can include excessive gagging, food selectivity, and difficulty transitioning to higher textures with oral hypersensitivity and overstuffing with oral hyposensitivity.[52-54] Treatment techniques should always include caregiver training to ensure consistent carryover at meals. In most cases, there exist effective therapeutic techniques to treat pediatric feeding problems.[55-56] Oral-motor treatment techniques are individualized to the child's feeding difficulties and adapted to the caregiver's ability to follow through with the recommendations.

Summary

Feeding disorders are surprisingly common. The incidence of feeding disorders will continue to rise as medically and developmentally complex children survive because of advances in medical technology. Feeding disorders occur because of medical, sensory, physical, personal, social, and environmental dynamics. These factors rarely operate independently. Prolonged feeding problems have grave consequences for the child's physical, cognitive, and social health, and lead to caregiver stress. The multifactorial etiology of feeding disorders demands a multidisciplinary approach to treat the whole child as well as their caregivers effectively. Caregiver education and training are indispensable to maintenance and generalization. Most children with feeding disorders can be treated effectively by experienced healthcare providers.

Acknowledgments

The authors acknowledge James K. Hyche for his contributions to the previous version of this chapter.

References

1. Goday PS, Huh SY, Silverman A, et al. J Pediatr Gastroenterol Nut 2019;1:124–9.
2. Davis, CM. Can J Med Assoc 1939;41:257–61.
3. Satter E. J Pediatr 1990;117:181–9.
4. Reau NR, Senturia YD, Lebailly SA, et al. J Dev Behav Pediatr 1966;17:149–53.
5. Lindberg L, Bohlin G, Hagekull B. Int J Eat Disord 1991;10: 395–405.
6. Reilley S, Skuse D, Problete X. J Pediatr 1996;129:877–82.
7. Douglas JE, Bryon M. Arch Dis Child 1985;75:304–8.
8. Thommessen M, Heidberg A, Kasse BF, et al. Acta Paediatr 1991;80:527–33.
9. Williams Z. Failla M, Gotham K, Woynaroski T, Cascio C. J Autism Dev Disord 2018;48:4231–49.
10. Schoen S, Miller L, Brett-Green B, Nielsen D. Front Integ Neurosci. 2009;3:1–11.
11. Riordan MM, Iwata BA, Wohl KM, et al. Appl Res Ment Retard 1980;1:95–112.
12. Babbitt RL, Hoch TA, Coe DA, et al. J Dev Behav Pediatr 1994;15: 278–91.
13. Barett DE, Radke-Yarrow M, Klein RE. Dev Psychol 1984;18: 541–56.
14. Greer AJ, Gulotta CS, Laud RBJ. Pediatr Psychol 2008;33: 612–20.
15. Chase HP, Martin H. N Engl J Med 1970;933–9.
16. Lefton-Greif MA. Phys Med Rehab Clin North Am 2008;19: 837–51.
17. Krom H, van Mameren J, Remijn L, et al. J Pediatr Gastroenterol Nut. 2022;75:529–34.
18. Peterson KM, Ibanez VF, Volkert VM. J Appl Behav Anal 2021;54:6–24.
19. Racine N, McArthur BA, Cooke JE, et al. JAMA Pediatr 2021;175:1142–50.
20. Otto AK, Jary JM, Sturza J, et al. Pediatr 2021;148:4e2021052201.
21. Yazdani S, Bloomberg Z, Klauber R et al. Eating and Weight Disorders-Studies on Anorexia, Bulimia and Obesity. https://journals.scholarsportal.info/details/15901262/v27i0007/2943_arfiderismas.xml. Published online.
22. Manikam R, Perman J. J Clin Gastroenterol 2000;11:34–46.
23. WHO Multicentre Growth Reference Study Group. WHO Child Growth Standards: length/height-for-age, weight-for-age, weight-for-length, weight-for-height, and body mass index-for-age: methods and development. Geneva, Switzerland: World Health Organization, 2006.
24. Cardona CS, Tiemeier H, Van Hoeken D, et al. Int J Eat Disord 2015;48:570–9.
25. Powell FC, Farrow CV, Meyer C. Appetite 2011;57:683–92.
26. Milano K, Chatoor I, Kerzner B. Curr Gastroenterol Rep 2019;21: 1–8.
27. American Psychiatric Association. Diagnostic and Statistical Manual of Mental Disorders, 5th ed. (DSM-5). Washington, DC; American Psychiatric Publishing, 2013.
28. Borowitz KC, Borowitz, SM. Pediatr Clin 2018;65:59–72.
29. Volkert VM, Piazza CM. Handbook of Evidence Based Practice in Clinical Psychology. 2013;13:28.
30. Arvedson JC. Dev Disabil Res Rev 2008;14:118–27.
31. Palmer S, Horn S. In: Palmer S, Ekvall S, eds. Feeding Children. Pediatric Nutrition in Developmental Disorders. 6th ed. Springfield, IL: Charles C Thomas, 1978:107–29.
32. Staiano A, Giudice E, Romano A, et al. J Pediatr 1992;121: 720–4.
33. Haas AM. Curr Allerg Asthma Rep 2010;10:258–64.
34. Garcia-Careaga M Jr, Kerner JA Jr. Nutr Clin Pract 2005;5: 526–35.
35. Zangen T, Cairla C, Zangen S, et al. J Pediatr Gastroenterol Nutr 2003;37:225–7.
36. Cucchiara S. Int Semin Paediatr Gastroenterol Nutr 1998;7:2.
37. Piazza, CC, Fisher, WW, Brown, KA, et al. J Appl Behav Anal 2003;36:187–204.
38. Volkert, VM, Piazza CC. In: Sturmey P & Herson M eds. Handbook of Evidence-Based Practice in Clinical Psychology. Hoboken, NJ: Wiley, 2012.
39. Fischer E, Silverman A. Semin Speech Lang 2007;28:223–31.
40. Finnane JM, Jansen J, Mallan KM, Daniels LA. J Nutr Educ Behav 2017;49:11–18.
41. O'Brian F, Azrin NH. J Appl Behav Anal 1972;5:389–99.
42. Krasaclap A, Kovacic K, Goday PS. Nutr Clin Pract 2020; 265–72.
43. Rodriguez L, Diaz J, Nurko S. J Peds 2013;261–7.
44. Kim SY, Yun JM, Lee JW, et al. Clin Thera 2021;1757–72.
45. Bertrand V. Front Pediatr 2021;2296–360.
46. Agostoni C, Turck D. J Pediatr Gastroenterol Nut. 2011;53: 594–600.
47. Sharp WG, Volkert VM, Scahill L, et al. J Pediatr 2017;181: 116–24.
48. Linscheid TR, Budd KS, Rasnake LK. Pediatric Feeding Disorders. In: Roberts MC, ed. Handbook of Pediatric Psychology. New York, NY: Guilford, 1995;501–15.
49. Addison LR, Piazza CC, Patel MR, et al. J Appl Behav Anal 2003;45:455–71.
50. Berth DP, Bachmeyer MH, Kirkwood CA, et al. J App Behav Anal 2019;52:622–41.
51. Davis AM, Bruce A, Cocjin J, et al. Curr Gastroenterol Rep 2010;12:189–94.
52. Yi S-H, Joung Y, Choe Y, Kim E, Kwon J. JPGN 2015;60: 819–24.
53. Seiverling L, Williams K, Hendy H, Adams W, Yusupova S. Kaczor A. Appetite 2019;133:223–30.
54. Farrow C, Coulthard H. Appetite 2012;58(3):842–6.
55. Silverman AH. Ann Nutr Metab 2015;66:33–42.
56. Marshall J, Hill RJ, Ware RS, Ziviani J, Dodrill P. J Pediatr Gast Nutr. 2015;60:680–7.

CHAPTER 70

Protein-Energy Malnutrition

Uzma Rani, Sarah Malik, Aamer Imdad, and Zulfiqar Ahmed Bhutta

CHAPTER OUTLINE

Abbreviations

ART, antiretroviral therapy; **DALYs**, disability-adjusted life years; **HA**, height for age; **HIV**, human immunodeficiency virus; **MUAC**, mid-upper arm circumference; **NG**, nasogastric; **ORS**, oral rehydration solution; **PEM**, protein-energy malnutrition; **ReSoMal**, rehydration solution for malnutrition; **RUSF**, ready-to-use supplementary food; **SAM**, severe acute malnutrition; **UNICEF**, United Nations International Children's Emergency Fund; **WA**, weight for age; **WH**, weight for height; **WHO**, World Health Organization; **ATP**, adenosine triphosphate

Introduction

Malnutrition occurs when an individual's dietary intake is not balanced with nutritional needs, leading to ill effects on health, well-being, and/or productivity.[1] Malnutrition includes undernutrition, overnutrition, and micronutrient deficiencies.[1] Undernutrition can be described as wasting (low weight for height), stunting (low height for age), and underweight (low weight for age).[2] Wasting indicates acute malnutrition, whereas stunting usually results from chronic malnutrition. Wasting can happen in both stunted (acute or chronic malnutrition) and nonstunted children. Protein-energy malnutrition (PEM) describes the condition in which the most salient elements are depletion of body energy stores and tissue proteins, presenting in a wide array of clinical manifestations and usually accompanied by micronutrient deficiencies. Protein-energy malnutrition is used synonymously with severe acute malnutrition or wasting. PEM can directly result from inadequate food intake, known as primary PEM, or be caused by recurrent illness associated with gastrointestinal malabsorption, reduced appetite, and/or increased nutrient needs, known as secondary PEM.[3] Marasmus and Kwashiorkor are both severe manifestations of protein energy malnutrition.[4] Marasmus results from total energy deficiency and presents with loss of appetite, adipose tissues, and low weight for height. Kwashiorkor results from a low protein diet with the hallmark of edema.[5] This chapter will discuss primary PEM, focusing on severe acute malnutrition.

Epidemiology

Early childhood (less than 2 years of age) is the most vulnerable period for malnutrition during the life course because of high nutritional requirements (**Figure 70.1**). The prevalence of wasting and stunting is the highest in the first 2 years of life until reaching a plateau around 24 months. In lower- and middle-income countries, this is further aggravated by recurrent gastrointestinal infections, leading to increased losses and high nutrient demand.[6]

In 2020, an estimated 45.4 million children were wasted, and about 13.6 million were severely wasted.[8] The COVID-19 pandemic is thought to have increased the burden by 15%, due to deterioration in household wealth and disruptions to the availability and affordability of food and essential nutrition services.[8] Globally, approximately 149.2 million children were stunted in 2020, which accounts for about 22% of the global population of children <5 years of age.[8] Compared with 2000, the prevalence of stunting decreased from 33% to 22%; however, wasting increased to 6.7% compared with 5.4% in 2000.[8,9] Approximately 10% of all children under 5 years of age live in high-income countries, while only 1% of wasted and 2% of stunted children do.[9,10] About 72% of all wasted and about 59% of all stunted children live in lower and middle-income countries (see **Figure 70.2**).[11] Acute malnutrition is rare in adolescents and is usually associated with primary illness.

Etiology and Determinants of Malnutrition

The conceptual framework of malnutrition developed by UNICEF in 1990 was recently updated to include the distal determinants of maternal and child nutrition.[12] The framework provides conceptual clarity on the enabling, underlying, and immediate determinants of adequate nutrition and their vertical and horizontal interconnectedness. It describes the outcomes of improved maternal and child nutrition in the form of favorable survival rates and improved growth, development, learning, economic and social outcomes, gender equity, and cohesion in societies (**Figure 70.3**).

Like the socioecological model, the framework includes influencers on nutrition at different societal levels. The framework points to the complexity and intersectionality of the enabling and underlying determinants affecting maternal and child health and nutrition. The immediate determinants include the availability of adequate foods, with good feeding and dietary practices and services. Adequate foods include age-appropriate, nutrient-rich foods, breastmilk and complementary foods for infants, safe drinking water, and household food security for families and children. Age-appropriate feeding, adequate food preparation, food consumption, and hygiene practices are also underlying determinants of nutrition. Services comprise adequate nutrition, health, sanitation, education, and social protection services, with healthy food environments and healthy living environments that prevent disease and promote adequate diets and physical activity. All of the above factors are tied to good governance and political strategy for the availability of resources, with a focus on social, gender, and cultural norms. This framework gives opportunities for improvement in maternal and child nutrition and offers directions for political authorities and public health officials, emphasizing the need for a multi-pronged approach to tackle malnutrition. However, it must be pointed out that globally, the quality of this information is variable, especially around representative diet quality and quantitative information.

Pathophysiology and Adaptation

The underlying mechanism of severe malnutrition has been the focus of many historic studies and is an evolving field of research. The adaptive mechanisms and metabolic changes differ based on the severity and chronicity of the malnutrition (**Figures 70.4a**

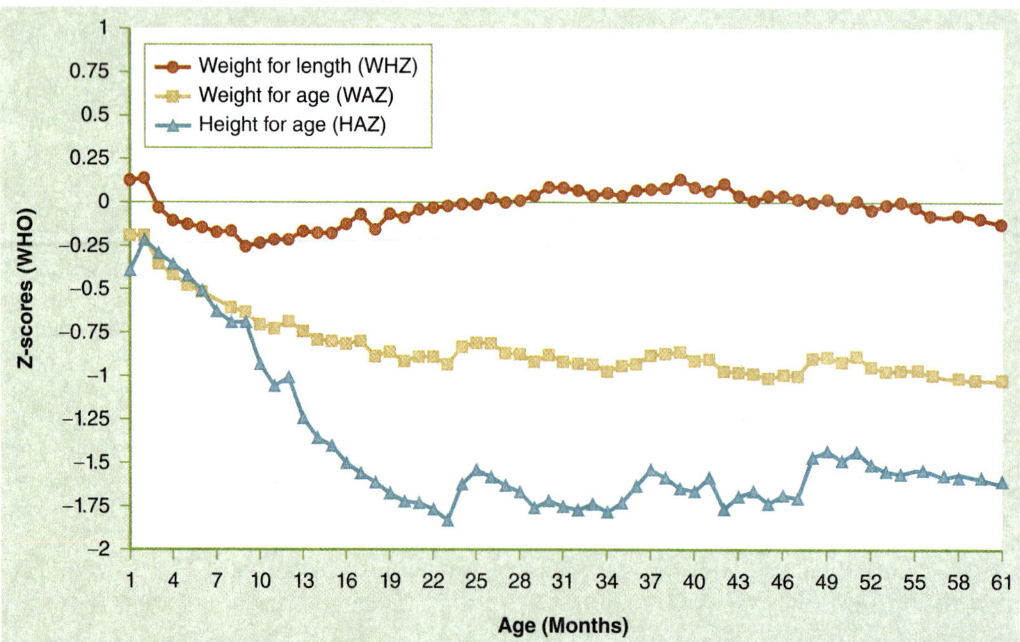

Figure 70.1 Mean anthropometric z scores according to age for all 54 studies, relative to the WHO standard (1 to 59 months).[7]

Reproduced from Victora CG, de Onis M, Hallal PC, et al. Pediatrics 2010;125: e473–80.[7] doi:10.1542/peds.2009-1519

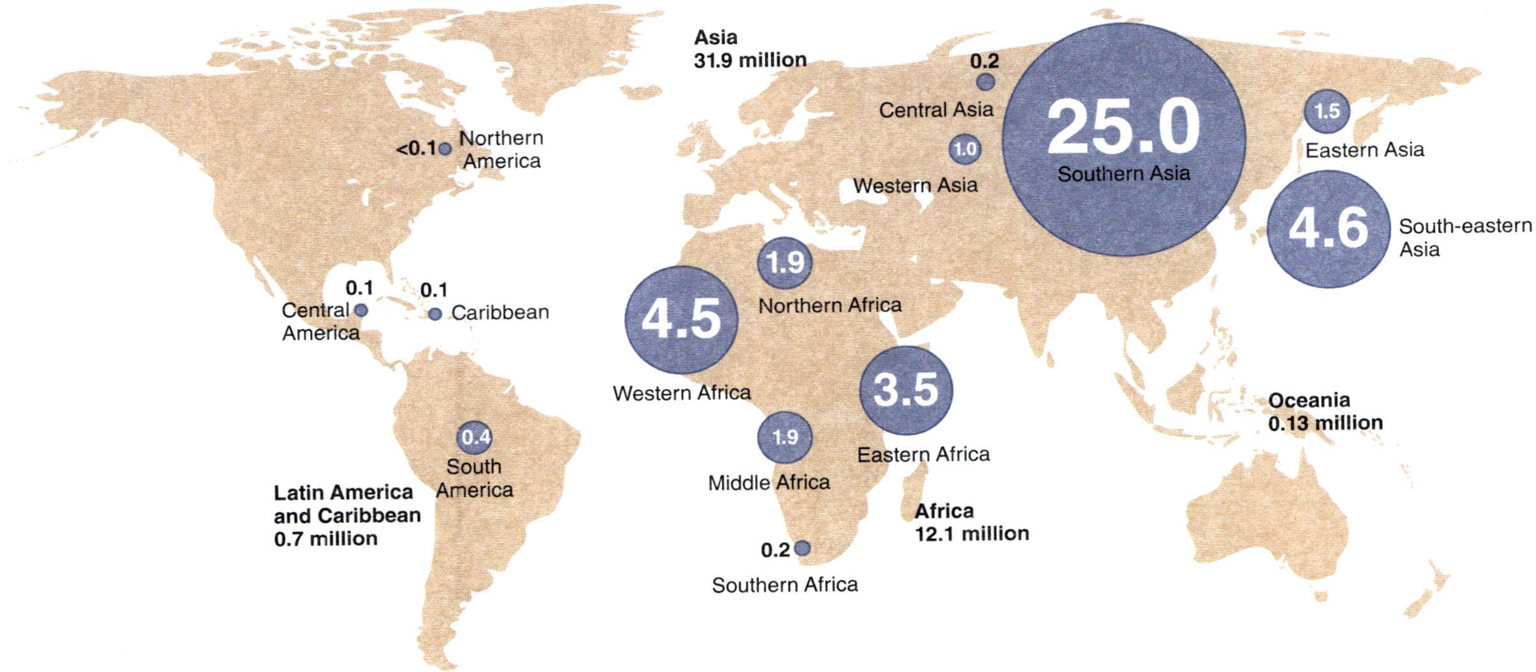

Figure 70.2 Number of children in a million, with wasting in 2020.

The UNICEF/WHO/WB Joint Child Malnutrition Estimates (JME) group released new data for 2021, release May 6, 2021. Available at www.who.int/news/item/06-05-2021-the-unicef-who-wb-joint-child-malnutrition-estimates-group-released-new-data-for-2021. Accessed September 6, 2022.[11]

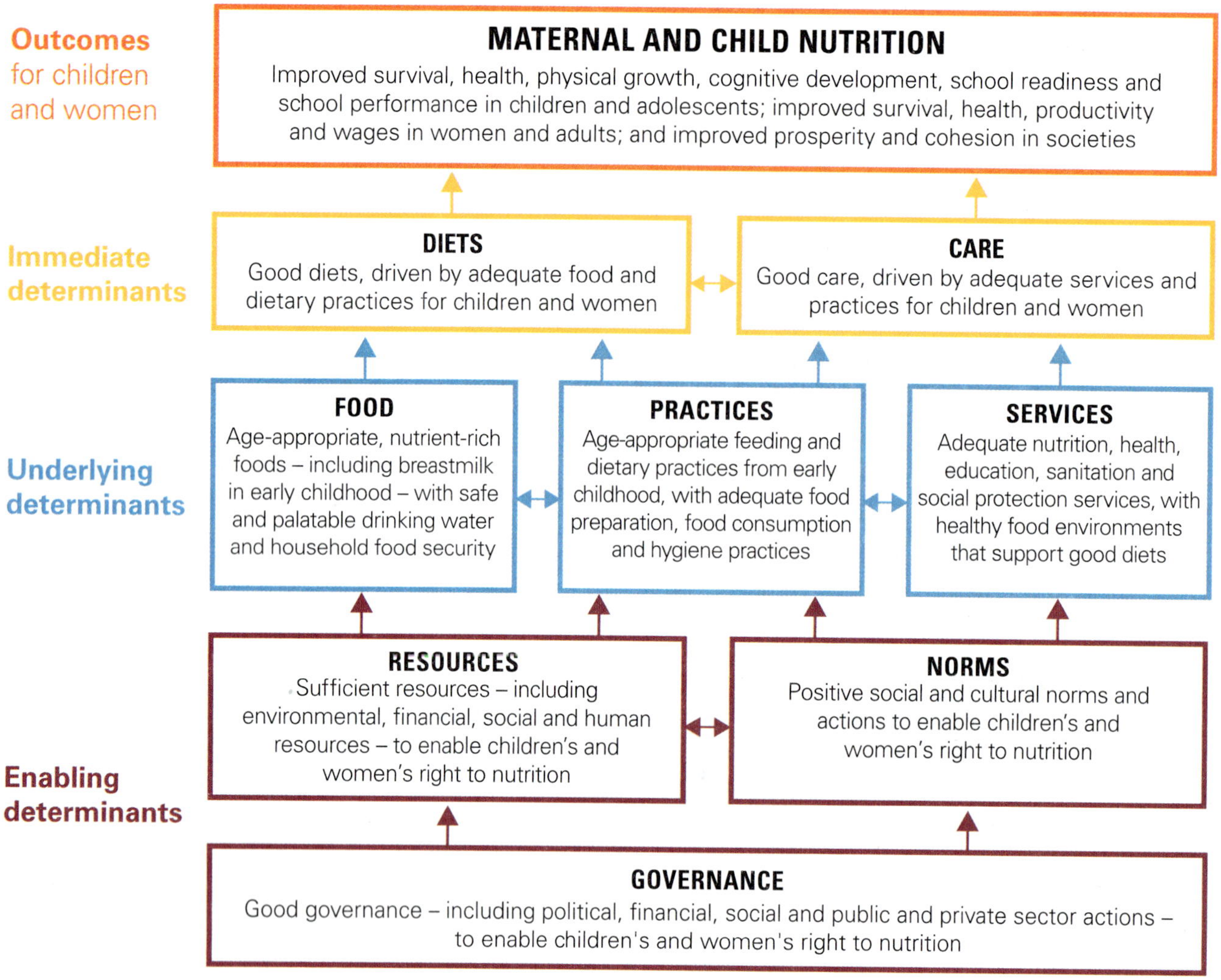

Figure 70.3 UNICEF Conceptual Framework on the Determinants of Maternal and Child Nutrition, 2020.[12] A framework for the prevention of malnutrition in all of its forms.

and **70.4b**).[13] Short-term starvation (up to several days) can lead to the breakdown of myofibrillar protein and adipose tissue to convert them into glucose, a process called gluconeogenesis.[13]

In addition to the breakdown of myofibrillar protein and adipose tissue, the release of many inflammatory substances, such as interleukin-1, interleukin 6, and tumor necrosis factor can impact the body composition and functions, leading to decreased appetite and degradation of myofibrillar protein and adipose tissue.[14,15] Lack of appetite and weakness can further reduce the ability to take food, leading to a cycle of malnutrition. More studies are needed to understand the role of these inflammatory markers and changes in the cellular function of children with severe wasting in developing countries.

Protein intake is essential to cell and organism growth as most enzymes, antibodies, growth factors, and cytokines are made up of proteins. The mechanistic target of rapamycin complex 1 (mTORC1) is a protein complex that senses the availability of protein/energy and regulates protein synthesis.[16] As part of its regulatory pathway, it will repress protein and lipid synthesis when protein is deficient, thus limiting cell and organism growth.[16,17] The pathophysiologic changes in Kwashiorkor are complex and poorly understood, and the exact etiology of why some children develop edema in Kwashiorkor is not well established.[18-20]

Metabolic and Endocrine Changes

The altered endocrine and macronutrient metabolism in Kwashiorkor and Marasmus have been studied. Maladaptive response secondary to a protein-poor diet leading to persistent glycolysis and protein catabolism has been noted.[21] Similarly, reduced essential and nonessential amino acids[22] and altered fatty acid oxidation have been seen in severe wasting.[23] Due to impaired glucose homeostasis, the affected children are at higher risk of developing hypoglycemia.[24] Several studies have also suggested that glucose uptake by tissues, a process regulated by insulin, is impaired in PEM.[25,26] Impaired pancreatic function, although poorly understood, can lead to changes in the glucagon response.[27] Cortisol concentration usually remains normal or high in children with severe malnutrition.[28] Other endocrine changes include reduction in leptin concentration, which can affect appetite regulation and immune function. Thyroid function is usually reduced in severe malnutrition; however, its clinical significance is still unclear.[29]

Oxidative stress is also associated with severe malnutrition. The imbalance between reactive oxygen species and antioxidants can lead to mitochondrial damage and decreased ATP production, leading to impaired cellular function.[30]

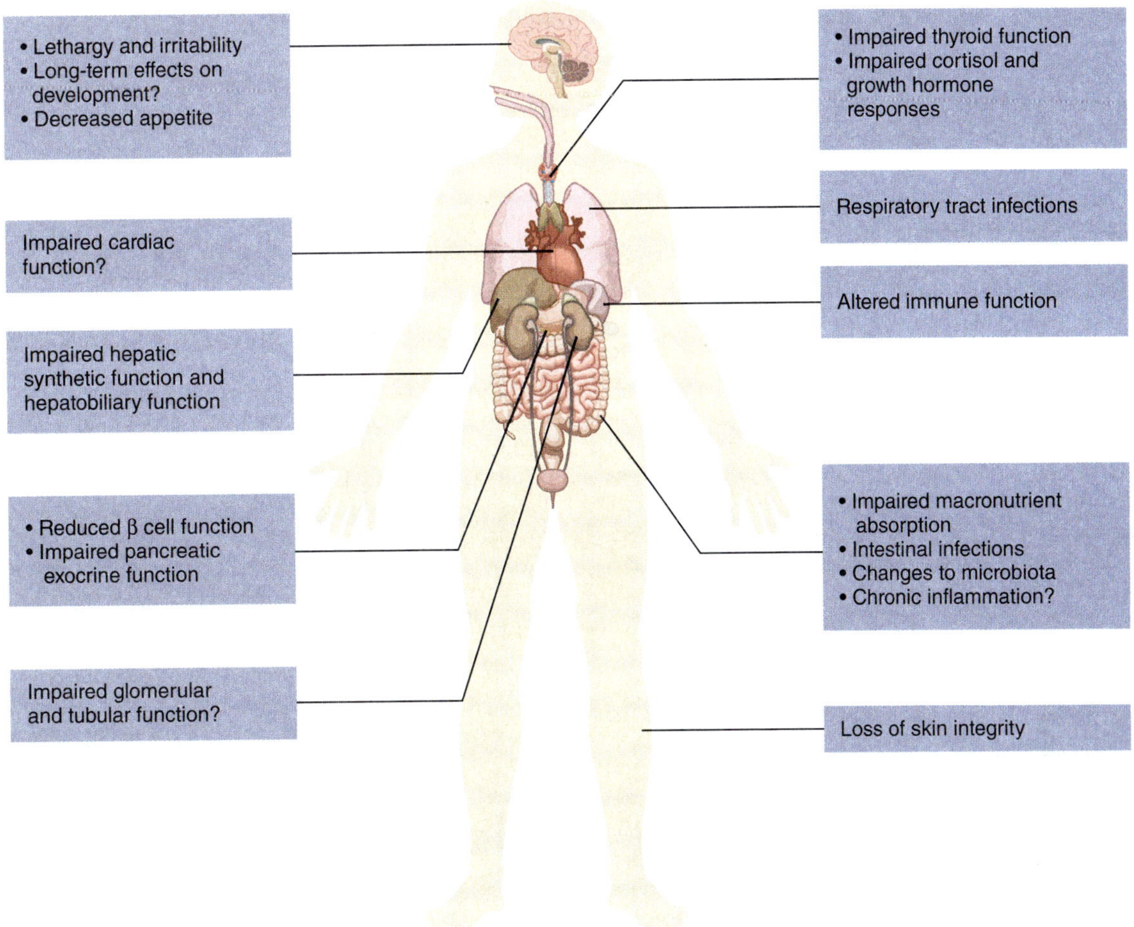

Figure 70.4a Multi-organ involvement in severe malnutrition[5]

Reproduced from Bhutta, Z., Berkley, J., Bandsma, R. et al. Severe childhood malnutrition. Nat Rev Dis Primers 3, 17067 (2017). https://doi.org/10.1038/nrdp.2017.67

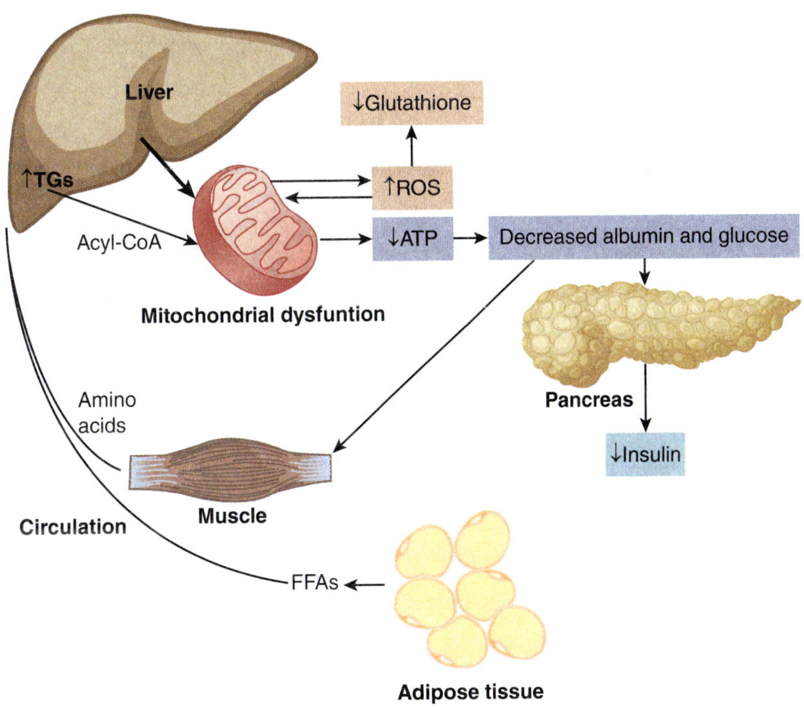

Figure 70.4b Biochemical changes in severe malnutrition.

ROS, reactive oxygen species; *FFAs*, free fatty acids; *TGs*, triglycerides.

Immune System

Impaired metabolic changes, along with decreased protein production and oxidative stress because of a catabolic state during wasting, can lead to secondary immunodeficiency,[31] and increased susceptibility to frequent, life-threatening infections.[32] The immune dysfunction may involve impaired T-cell function, reduced neutrophilic microbicidal response,[33] and altered complement cascade response to infective micro-organisms.[34] Recent studies also suggest that gut microbiome dysfunction (dysbiosis) is associated with malnutrition and can lead to enteric dysfunction with frequent diarrhea, reduced intestinal mucosal barrier integrity, reduced nutrient absorption, and mucosal inflammation.[35-39] Dysbiosis in the form of reduced microbial diversity and increased ratios of pathogenic gram-negative bacteria, specifically Enterobacteriaceae, are associated with malnutrition.[40] These observations have led to the development of microbiota-directed nutritional interventions that showed promising results in reversing dysbiosis-associated effects of malnutrition.[41]

Hemodynamic and Cardiac Function

A global catabolic state results in decreased protein synthesis and increased breakdown of proteins, which can lead to decreased cardiac function and cardiac muscle atrophy in children with severe malnutrition.[42,43] Gruber et al., reported that energy restriction in an animal model led to a proportional decrease in subcellular components of cardiomyocytes.[43,44] Other hemodynamic changes include decreased peripheral circulation, postural hypotension, and reduced heart rate.[23,45,46]

Hepatic Function

Severe malnutrition is associated with hepatic dysfunction that can manifest as impaired hepatic lipid oxidation and synthetic pathways.[23,47] Decreased production of albumin is postulated to be a contributing factor to edema in Kwashiorkor.[19] Hepatic steatosis is observed in children with Kwashiorkor; however, its clinical significance and the underlying pathologic phenomenon are not well established.[47]

Renal and Electrolytes

Many studies have reported low glomerular filtrate rate (GFR) and tubular dysfunction in children with malnutrition; however, the clinical significance of these findings remains unclear.[48] Frequent diarrhea leading to dehydration can contribute to decreased GFR and electrolyte imbalance.[5] The impaired Na/K ATPase pump in the loop of Henle also contributes to the defective fluid and electrolyte balance.[49,50]

Brain and Cognitive Functions

Severe malnutrition is associated with decreased cerebral function, apathy, impaired motor and speech development, and decreased cognitive function.[51-53] The underlying mechanism of these changes is multifactorial and needs further study.

Diagnosis

The diagnosis of severe acute malnutrition is suggested by the history and assessment of risk factors and confirmed by physical examination with a high index of suspicion in resource-poor regions. (**Table 70.1**) History should focus on the complete diet,

Table 70.1 Diagnostic Criteria for Protein-Energy Malnutrition

Age	Moderate Malnutrition	Severe Malnutrition
Infants 0 to 6 months	Criteria are not well established, but Z score[a] for weight for length <−2 SD, wasting, and absence of edema can help make the diagnosis.	■ MUAC <110 mm OR ■ Weight for length age Z-score <−3 SD
Children 6 through 59 months old	■ MUAC between 115 mm and 124 mm OR ■ Weight for height Z-score between −2 to −3 AND ■ Bilateral pitting edema: absent	■ MUAC <115 mm OR ■ Weight for height/weight for length Z- score <−3 SD OR ■ Bilateral pitting edema: present Edema can be graded as follows: 1$^+$=if present in lower body and feet 2$^+$=if present in upper body and arms 3$^+$=if present in face
Children 5 years and older	BMI for age Z-score between −2 SD and −3 SD	BMI for age Z-score <−3 SD

BMI, body mass index; *MUAC*, mid-upper arm circumference; *SD*, standard deviation.[55-57]

[a]Z-score (or standard deviation score) is the deviation of an individual's value from the reference population's median value, divided by the standard deviation for the reference population.

Data from Mwangome, Martha K et al. "Mid-upper arm circumference at age of routine infant vaccination to identify infants at elevated risk of death: a retrospective cohort study in the Gambia." Bulletin of the World Health Organization vol. 90, 12 (2012): 887–94. doi:10.2471/BLT.12.109009; World Health Organization. Pocket book of hospital care for children: Second edition, Guidelines for the management of common childhood illnesses, release Janaury 1, 2013. Available at www.who.int/publications/i/item/978-92-4-154837-3. Accessed May 2, 2022; World Health Organization. Guideline: Updates on the Management of Severe Acute Malnutrition in Infants and Children, release August, 2013. Available at www.who.int/publications/i/item/9789241506328. Accessed May 5, 2022.

development history, and the presence of complications such as diarrhea, blood in stool, vomiting, abdominal distention, facial swelling, cough, seizures, pedal edema, skin rashes, and vision difficulties (including night blindness).[5] The physical examination should include vital signs and triage for life-threatening complications such as shock, central cyanosis, seizures, severe dehydration, and severe pallor. The presence of edema is an essential finding of wasting and should be assessed carefully. Anthropometric measures, including weight, height, and mid-upper arm circumference (MUAC), further help to establish the severity of the PEM.[54]

Treatment

The mainstay of treating severe malnutrition includes the short-term goals of decreasing mortality from life-threatening infections and/or fluid and electrolyte disturbances, and long-term goals of nutritional and neurocognitive rehabilitation. Historically, severe acute malnutrition was managed in inpatient units; however, the recent development of ready-to-use foods (RUTFs) and other supplementary foods, along with a better understanding of the disease process have encouraged community-based outpatient management (**Figure 70.5**). The success of community-based management of severe acute malnutrition is linked to the availability of community resources, backup support of the inpatient centers to manage complicated cases, financial and community support to carry out outpatient follow-ups, and a supply of ready-to-use therapeutic foods.[57]

Approach to the Treatment of Severe Protein-Energy Malnutrition

The treatment of severe acute malnutrition depends on the presence of any medical complication requiring inpatient management. The focus should be placed on triaging the patients

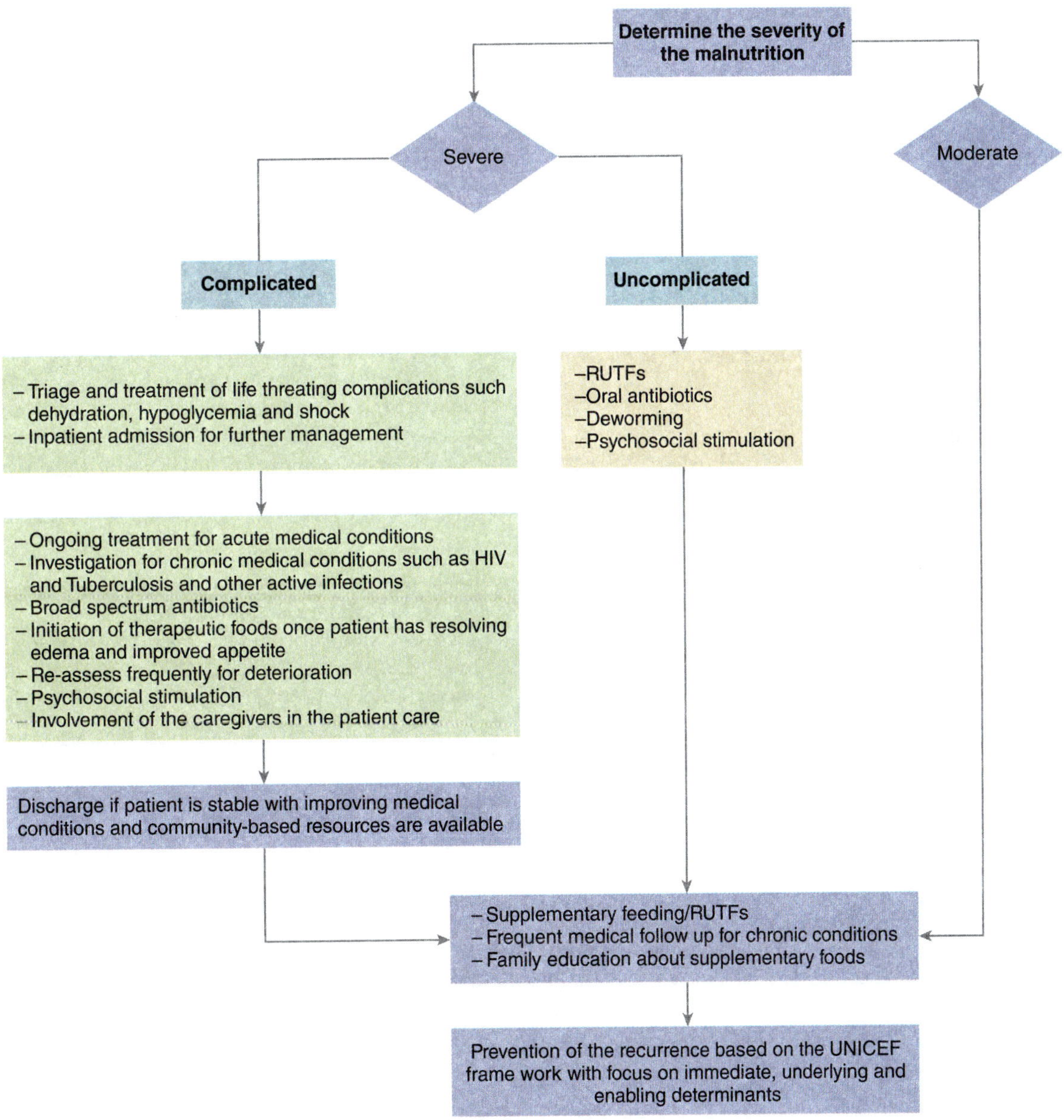

Figure 70.5 Algorithm for management of malnutrition based on severity[5]

Data from Bhutta, Z., Berkley, J., Bandsma, R. et al. Severe childhood malnutrition. Nat Rev Dis Primers 3, 17067 (2017). https://doi.org/10.1038/nrdp.2017.[67]

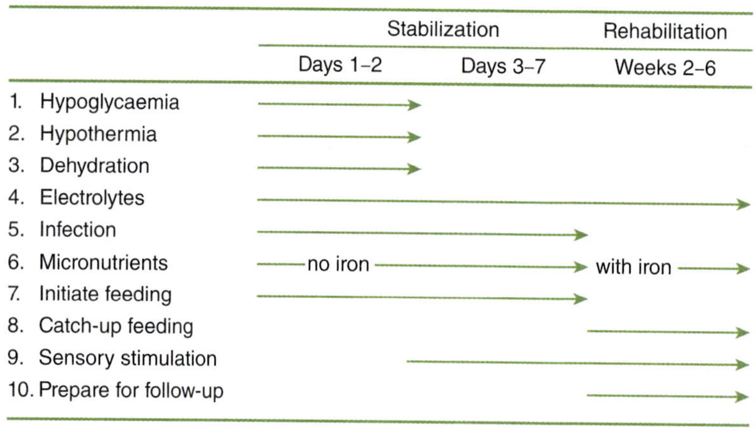

Figure 70.6 Timeframe for managing a child admitted with severe acute malnutrition.

Reproduced from World Health Organization. Pocket book of hospital care for children: Second edition, Guidelines for the management of common childhood illnesses, release Janaury 1, 2013. Available at www.who.int/publications/i/item/978-92-4-154837-3. Accessed May 2, 2022.[54]

and deciding who needs initial inpatient care and who can be managed in community centers. Children who present with severe PEM along with loss of appetite or any medical complication will require inpatient management. Carefully monitored outpatient treatment is appropriate for children who maintain a good appetite and have no medical complications. The goal of inpatient management should be initial stabilization and resolution of complications of severe malnutrition to successfully transition the patient to outpatient care for further management and rehabilitation.

Inpatient Management of Complicated Severe Acute Malnutrition

The care of children with complicated sever acute malnutrition (SAM) can be categorized in phases, including an initial stabilization phase of evaluation and management of hypoglycemia, hypothermia, dehydration, and infection. (**Figure 70.6**) This is followed by a rehabilitation phase in which a carefully designed feeding regimen is initiated with the goal of steady weight gain. Additionally, emotional, and physical stimulation, training of the caregivers, and other preparation for hospital discharge are incorporated into the therapy plan.

Management of complications in the stabilization phase of acute, severe malnutrition is summarized in **Table 70.2**.

Therapeutic Dietary Regimen and Catch-up Growth

Nutritional rehabilitation must progress slowly to gradually reverse the metabolic adjustments in the severely malnourished state. Once stabilized, therapeutic feeding should be cautiously started with milk-based foods. It is best to start with small, frequent feeds of a liquid formula (at 130 mL/kg/day, 100 if severe edema exists) orally or by NG tube.[54] For young infants,

Table 70.2 Management of Complications in Children with Severe Acute Malnutrition Malnutrition[3,56,58-61]

Initial Stabilization	Criteria	Treatment
1. Hypoglycemia	Blood glucose <3 mmol/L (54 mg/dL)	■ If conscious child, administer 50 mL of 10% glucose or sucrose solution orally or by NG tube, followed by feeding as soon as possible ■ If unconscious child, administer 10% glucose or sucrose solution, 5 mL/kg IV or by NG tube; if IV is not possible ■ If oral, IV, or NG administration is not possible, administer one teaspoon of moistened sugar with one to two drops of water sublingually. Repeat every 20 minutes.
2. Hypothermia	Blood temperature <35.5°C (96°F)	■ The child should be wrapped in a heated blanket or put under an incandescent lamp. ■ If these options are not available, the child can be warmed by close contact with the caregiver's body
3. Infection		■ All hospitalized children are given ampicillin (50 mg/kg intramuscularly or IV every 6 hours+gentamicin (7.5 mg/kg intramuscularly or IV once daily) ■ For children who look particularly ill or in areas of antibiotic resistance, ceftriaxone (50 mg/kg intramuscularly or IV once daily) can be used instead.
		■ Administer **metronidazole** (10–12 mg/kg orally every 8 hours)

Initial Stabilization	Criteria	Treatment
	Quick recovery	■ For children who recovered quickly and are discharged before the completion of the parenteral antibiotic course, transition the antibiotic course to oral amoxicillin (40 to 45 mg/kg orally every 12 hours), amoxicillin-clavulanate, or a third-generation cephalosporin to complete at least 1 week of antibiotics ■ Children who remain hospitalized continue broad-spectrum antibiotics for at least 7 to 10 days
	HIV infection	■ Children should be tested for HIV as soon as possible ■ If not already treated with antiretroviral (ARV) drugs, these should be started as soon as possible ■ Doses of ARV should be the same as those without malnutrition
	Malaria infection	■ In areas of endemic malaria, some practitioners recommend antimalarial drugs for all children with malnutrition (artemisinin-combination therapy)
	Anthelminthic	■ Children with malnutrition should be tested for parasites if possible and be treated for intestinal helminth infections
	High Fevers (38.5 to 39°C or higher [>102°F])	■ Antipyretics (acetaminophen/paracetamol, ibuprofen, etc.) can be used to treat fever but should be reserved for high fevers ■ Small risk of renal toxicity (ibuprofen) or hepatic toxicity (acetaminophen) when using these medications, especially in children with Kwashiorkor ■ Aspirin (acetylsalicylic acid) should never be used in malnourished children
	Dehydration	■ Treat with **oral** rehydration ■ The child should drink Rehydration Solution for Malnutrition (ReSoMal) by mouth; an NG tube can also be used (composition in **Table 70.3**). ■ 70–100 ml/kg body weight of ReSoMal solution should be adequate volume to restore hydration ■ Initial bolus: no more than 15 mL/kg over 1 hour. A second bolus can be given if the child is not ready to hydrate by mouth. ■ Caution should be taken with the IV route and monitored for signs of improvement (decrease in pulse and respiration rates).
	Appetite Test	Passed test if children are observed while they consume about 30 g of RUTF without distress or vomiting. ■ Children that do not pass the test are given therapeutic milk formula (F-75) at a daily dose of 80 to 100 kcal/kg broken into small feeding, every 2–3 hours.

IV, intravenous; *NG*, nasogastric; *RUTF*, ready to use therapeutic food; *mL*, milliliter; *kg*, kilogram; *kcal*, kilocalorie.

Data from Protein Energy Malnutrition. In Ramirez-Zea M Caballero B. Modern Nutrition in Health and Disease, 11th ed., Wolters Kluwer Health/Lippincott Williams & Wilkins. 2012:894–905; World Health Organization. Pocket book of hospital care for children: Second edition, Guidelines for the management of common childhood illnesses, release Janaury 1, 2013. Available at www.who.int/publications/i/item/978-92-4-154837-3. Accessed May 2, 2022; Williams PCM, and Berkley JA. (2018) Paediatr Int Child Health 38, S32–S49; World Health Organization. Management of severe malnutrition: a manual for physicians and other senior health workers, release 1999. Available at https://apps.who.int/iris/handle/10665/41999. Accessed September 5, 2022; Zangenberg M, Abdissa A, Johansen OH, et al. (2020) Trop Med Int Health 25, 424–32.

breastfeeding should not be interrupted, and parents should be involved in care and feeding.[58]

The initial inpatient therapeutic dietary regimen consists of two liquid formulas: F-75 (75 kcal and 0.9 g protein/100 mL) during the stabilization phase and F-100 (100 kcal and 2.9 g protein/100 mL) during the rehabilitation phase (**Table 70.4**). In addition to the mineral and vitamin supplement included in the formula F-75, a dose of vitamin A should be given on admission because ocular lesions can develop as a result of increased demands of retinol when adequate protein and energy feeding begins (200,000 IU; 50,000 to 100,000 IU for infants).[62]

In the rehabilitation phase, rapid weight gain is expected (>10 g/kg/day) but should not be a criterion for discharge. Patients should be discharged once complications are successfully managed, and the patient can eat with their mouth, with assurance of long-term follow-up at an outpatient community center. During this rehabilitation phase, children are at higher risk of refeeding syndrome (electrolyte imbalance) and should be closely monitored.[5]

Emotional and Physical Stimulation

The attitude of the person who feeds the patient is important to overcome the patient's lack of appetite. A safe environment and caregiver involvement in feeding can stimulate the admitted children. As soon as the child can move without assistance, they must be encouraged to explore, play, and participate in activities that involve body movements. Physical activity during nutrition

Table 70.3 ReSoMal: Rehydration Solution for Malnutrition

Component	Concentration (mmol/L)
Glucose	125
Sodium	45
Potassium	40
Chloride	70
Citrate	7
Magnesium	3
Zinc	0.3
Copper	0.045
Osmolarity	300

Reproduced from World Health Organization. Management of severe malnutrition: a manual for physicians and other senior health workers, release 1999. Available at https://apps.who.int/iris/handle/10665/41999. Accessed September 5, 2022.

rehabilitation results in faster longitudinal growth and accretion of lean body tissues.[64]

Community Management of Uncomplicated Severe Acute Malnutrition

Community-based management of the uncomplicated severe SAM subtype is usually accomplished by RUTF feeding (175 kcal/kg/day). Patients should attend the community center every week or two for medical evaluation, additional medical treatment if needed, and to receive enough RUTF to last until the next appointment.[65] Health education given to caregivers is essential to ensure the child's recovery.[65]

Ready-to-use therapeutic foods (RUTFs), a dehydrated nutrition supplement mimicking liquid F-100, were developed for use in community settings with minimal risk of contamination.[54] Most RUTFs currently used are oil-based pastes or spreads, which can be made locally using basic technologies.[66] RUTFs have high-energy density (5.5 kcal/g) and have been made from peanuts, milk powder, sugar, oil, and a micronutrient mix. This product can be kept unrefrigerated in simple packaging for several months; it is eaten uncooked, and its low water content prevents bacterial growth.[67]

WHO and UNICEF outline the minimum micronutrient and macronutrient concentrations, safety standards, and quality control for RUTF (**Table 70.5**). While the peanut-based formulation of RUTF is the most common, alternative formulations have been used to manage malnutrition.[67] To improve palpability and reduce cost, some alternatives include substituting locally sourced protein (soy, whey protein) and grain (soya, maize, sorghum).[67] The milk powder component contributes to the high cost of RUTFs; however, clinical trials using the RUTF without milk powder have shown lower efficacy on child growth.[68,69]

The appetite test with RUTFs should be repeated every 1 to 2 days to assess feeding capabilities. If tolerated, feedings may be advanced to RUTF-dominant feedings. For children who fail the appetite test, or where RUTF is not accessible, feedings can proceed with F100 until they pass the appetite test and start on home diets in the hospital.

Children with severe acute malnutrition are susceptible to multiple micronutrient deficiencies, especially iron deficiency, and possess a higher physiologic demand for iron during the catch-up growth phase. The prevalence of anemia is high in SAM children, ranging from 40% to 90%. About half of the anemia in children with SAM is attributed to iron deficiency. Iron deficiency in severely malnourished children could be due to low

Table 70.4 Composition and Preparation of F-75 and F-100 Formula from Basic Foods

Protein Source	Amount (G or Ml)	Cereal Flour (G)	Sugar (G)	Oil (Ml)	Water (Ml)	Mineral Mix (Ml)[a]	Vitamin Mix (Ml)[b]
F-75							
Dried skim milk	25	35	70	27	1,000	20	140
Dried whole milk	35	35	70	17	1,000	20	140
Fluid milk (cow, goat, camel)	300	35	70	17	700	20	140
F-100							
Dried skim milk	80	—	50	60	1,000	20	140
Dried whole milk	110	—	50	30	1,000	20	140
Fluid milk (cow, goat, camel)	880	—	75	20	120	20	140

[a]Mineral mix composition (per liter): potassium chloride, 89.5 g; tripotassium citrate, 32.4 g; magnesium chloride, 30.5 g; zinc acetate, 3.3 g; copper sulfate, 0.56 g; sodium selenate, 10 mg; potassium iodide, 5 mg; and water to make 1000 mL.

[b]Vitamin mix composition (per liter): thiamin 0.7 mg; riboflavin, 2.0 mg; nicotinic acid, 10 mg; pyridoxine, 0.7 mg; cyanocobalamin, 1 μg; folic acid, 0.35 mg; ascorbic acid, 100 mg; pantothenic acid, 3 mg; biotin, 0.1 mg; retinol, 1.5 mg; calciferol, 30 μg; α-tocopherol, 22 mg; vitamin K, 40 μg.

Protein Energy Malnutrition chapter in Modern Nutrition in Health and Disease; 11th Edition, 2012.

Table 70.5 Minimum Macronutrient and Micronutrient Concentrations in RUTFs; Based on the WHO Guidelines.

Energy	52–550 Kcal/100 g
Protein	10%–12% of total energy
Lipids	45%–60% of total energy
Sodium	290mg/100g maximum
Potassium	1110–1400mg/100 g
Calcium	300–600mg/100 g
Phosphorus (excluding phytate)	300–600/100 g
Magnesium	80–140 mg/100 g
Iron	10–14 mg/100 g
Zinc	11–14 mg/100 g
Copper	1.4–1.8 mg/100 g
Selenium	20–40 μg
Iodine	70–140 μg/100 g
Vitamin A	0.8–1.1 mg/100 g
Vitamin B_1	0.5 mg/100 g minimum
Vitamin B_2	1.6 mg/100 g minimum
Vitamin B_6	0.6 mg/100 g minimum
Vitamin B_{12}	1.6 μg /100 g minimum
Vitamin C	50 mg/100 g minimum
Vitamin D	15–20 μg/100 g minimum
Vitamin E	20 mg/100 g minimum
Vitamin K	15–30 μg/100 g minimum
Folic Acid	200 μg/100 g minimum
Niacin	50 mg/100 g minimum
Pantothenic Acid	3 mg/100 g minimum
Biotin	60 μg/100 g minimum
n-6 fatty acids	3%–10% of total energy
n-3 fatty acids	0.3%–2.5% of total energy

Reproduced from Collins S, Sadler K, Dent N, et al. Food Nutr Bull 2006;27: S49-82.

intake, increased losses, higher demand, and/or poor absorption. Iron is essential for adequate catch-up growth and neurologic development in children with severe acute malnutrition. The optimal dose of iron in RUTF for managing iron deficiency is not well established.[69,70] Standard WHO-based RUTF contains about 10 to 14 g/100 g of RUTF. However, recent studies suggest that alternative RUTF formulations may be more effective in correcting anemia and iron deficiency in children with SAM, compared with current standard RUTF formulations.[69-71] A recent meta-analysis showed that a high dose of iron, compared with a standard dose, may lead to better resolution of anemia; however, recovery rates were lower, and mortality was higher in the high iron group compared with the standard dose.[69] The number of included studies was small, and future studies are needed to decide the optimal dose of iron in RUTF.

Prognosis

Children with SAM can have short-term and long-term complications. The case fatality rates range from 4% to 40%, depending on complications at diagnosis and available resources to manage childrens' inpatient and outpatient care.[72-74] Adverse outcome predictors include failed appetite tests, edema, HIV infection, diarrhea, dehydration, and anemia. Children with SAM can have long-term complications. A prospective cohort study from Malawi followed children for 7 years postdischarge after treating SAM. Results showed that children with SAM had poorer growth, physical strength, and school achievement postdischarge from the treatment facility.[75]

Children with SAM typically live in settings of poverty, low maternal education, low food and security, lack of access to clean drinking water, and overstretched health systems. The acute management of SAM in inpatient or outpatient settings does not alter these risk factors, and these children continue to be at substantial risk of recurrence of severe malnutrition and complications. Studies from Malawi, Kenya, Gambia, and Bangladesh have shown that children who recovered from SAM continued to be at high risk of mortality from pneumonia and diarrhea. The potential reasons for this increased risk of morbidity and mortality in post-treatment settings include ongoing exposure to community-level risk factors for malnutrition, an undiagnosed chronic illness, and/or lack of access to secondary and tertiary care hospitals.

Children with severe wasting and stunting are at high risk of neurodevelopmental delays, behavioral problems, and reduced school performance, leading to low earning potential as an adult.[76,77] Studies that have used psychosocial stimulation involving play and giving positive feedback training to mothers, along with prolonged home visiting programs postdischarge, have been shown to improve developmental outcomes in the short term, as well as educational achievement and IQ in adolescence.[78,79] Children with SAM may be deficient in key micronutrients involved in neurodevelopment, such as polyunsaturated fatty acids, thiamine, vitamin D, folate, iron, copper, selenium, iodine, and zinc. Supplementations with these micronutrients are not well studied in children with SAM. Future studies involving supplementation with these micronutrients with long-term follow-up are needed, with outcomes focused on neurodevelopment.[80] Future studies are also needed to assess the optimal duration of treatment with therapeutic foods and to determine the optimal supplementation of macronutrients and micronutrients. Furthermore, the benefits of psychosocial stimulation and physical activity as part of treating severe acute malnutrition need more evaluation, as do the natural history of long-term effects of malnutrition on health risks in adolescence and adulthood.

Prevention

Prevalence of undernutrition has declined in countries that have reduced poverty and have invested in health, nutrition, education, and the social sector. However, much more needs to be done to achieve the sustainable development goal target of reducing the prevalence of child wasting to less than 5% by 2025 and less than 3% by 2030. To reduce the prevalence of wasting, there is an immediate need to accelerate prevention, early detection, and treatment. Interventions to prevent stunting must be focused on pregnancy and the first 2 years of life, including prevention of low birth weight, adoption of appropriate infant feeding practices, and increased child access to essential nutrition, health, water, and sanitation services. A recent publication by UNICEF identifies the three main roadblocks to the prevention of child wasting, including first, a lack of focus to map out the determinants and drivers of wasting, which are context-specific and may vary, depending on the environment in which a child lives.[81] Second is the inability to identify specific vulnerabilities in households and communities where some children may be at high risk of developing wasting and are at increased risk of mortality, especially among those younger than 2 years of age. Finally, the lack of availability of treatment of wasting in the right quantity and quality at the right time and place. The recent COVID pandemic was a major blow for struggling communities and has led to more food insecurity and limited resources in low and middle-income countries.[8] To cross these barriers, the focus should be to prioritize the financial resources available for increasing accessibility, implementing early detection, and receiving treatment. The strategy to achieve these goals should include improving the geospatial allocation of services, integrating essential maternal and child nutrition services and targeted intervention, and financing key nutrition interventions to treat wasting, such as RUTF. The goal should be to scale up prevention, early detection, and treatment of wasting for the most vulnerable children: the youngest of the young, the poorest of the poor, and those left behind by a humanitarian crisis.[81]

Using appropriate prevention and management intervention for SAM can significantly reduce mortality due to wasting. However, in addition to saving a child from SAM, the goal must be to thrive after treatment. To achieve this goal, curative and preventive services must be transformed, and previously neglected populations, such as infants younger than 6 months of age, children with underlying disabilities, and other identified vulnerabilities, must be prioritized. The risk factors for treatment failure, relapse, and readmission must be investigated further. In the long term, strategies for preventing malnutrition must follow a multisectoral approach that involves food production, distribution and availability (food security); preventive medicine; education, social development; and economic improvement. At the national level, effective control and prevention can be achieved only through sustained long-term political commitment and actions to eradicate the underlying causes of undernutrition. The overarching priority should be to make wasting a thing of the past in countries in Africa and Asia, as well as in all countries. It is a global moral imperative to recognize and reduce the burden of morbidity and mortality associated with severe childhood malnutrition. To eradicate PEM, including SAM, political will is required at the global level, along with an investment in reducing conflicts, climate change, and manmade and natural disasters; the issues that should unite nations based on humanity.[5]

Acknowledgments

The authors acknowledge Manuel Ramirez-Zea and Benjamin Caballero, the authors of the previous version of this chapter, for their contributions.

References

1. World Health Organization. Malnutrition. Accessed March 15, 2023. Available at www.who.int/news-room/fact-sheets/detail/malnutrition
2. Maleta K. Malawi Med J 2006;18:189–205.
3. Protein Energy Malnutrition. In: Ramirez-Zea M, Caballero B. Modern Nutrition in Health and Disease, 11th ed. Wolters Kluwer Health/Lippincott Williams & Wilkins. 2012;894–905.
4. Marasmus. Titi-Lartey OA,Gupta V. In StatPearls. Treasure Island, FL: StatPearls Publishing, 2022.
5. Bhutta ZA, Berkley JA, Bandsma RHJ, et al. Nat Rev Dis Primers 2017;3:17067.
6. Rodriguez L, Cervantes E, Ortiz R. Int J Environ Res Public Health 2011;8:1174–205.
7. Victora CG, de Onis M, Hallal PC, et al. Pediatr 2010;125: e473–80.
8. Food and Agriculture Organization of the United Nations, International Fund for Agricultural Development, World Food Programme, World Health Organization, United Nations International Children's Emergency Fund. The State of Food Security and Nutrition in the World 2021. Transforming Food Systems for Food Security, Improved Nutrition and Affordable Healthy Diets for All, Release 2021. Accessed May 5, 2022. Available at https://doi.org/10.4060/cb4474en
9. The United Nations Children's Fund (UNICEF). Child Malnutrition. Accessed September 15, 2022. Available at https://data.unicef.org/topic/nutrition/malnutrition/
10. Isanaka S, Andersen CT, Cousens S, et al. BMJ Glob Health 2021;6.
11. The UNICEF/WHO/WB Joint Child Malnutrition Estimates (JME) group released new data for 2021. Accessed September 6, 2022. Available at www.who.int/news/item/06-05-2021-the-unicef-who-wb-joint-child-malnutrition-estimates-group-released-new-data-for-2021
12. The United Nations International Children's Emergency Fund (UNICEF). Conceptual Framework on Maternal and Child Nutrition. Accessed September 15, 2022. Available at www.unicef.org/documents/conceptual-framework-nutrition. release 2020
13. Cohen S, Nathan JA, Goldberg AL. Nat Rev Drug Discov 2015;14:58–74.
14. Grant RW, Stephens JM. Am J Physiol Endocrinol Metab 2015;309:E205–13.
15. Argiles JM, Busquets S, Lopez-Soriano FJ. Curr Opin Clin Nutr Metab Care 2003;6:401–6.
16. Semba RD, Trehan I, Gonzalez-Freire M, et al. Adv Nutr 2016;7:853–65.
17. Semba RD, Shardell M, Sakr Ashour FA, et al. EBioMedicine 2016;6:246–52.
18. Ciliberto H, Ciliberto M, Briend A, et al. BMJ 2005;330:1109.
19. Golden MH. Paediatr Int Child Health 2015;35:90–109.
20. Smith MI, Yatsunenko T, Manary MJ, et al. Science 2013;339: 548–54.
21. Whitehead RG, Allyene GA. Br Med Bull 1972;28:72–9.

22. Jahoor F, Badaloo A, Reid M, et al. Am J Clin Nutr 2006;84:1400–5.
23. Badaloo AV, Forrester T, Reid M, et al. Am J Clin Nutr 2006; 83:1283–8.
24. Kerpel-Fronius E, Kaiser E. Acta Paediatr Scand 1967;172:19.
25. Baig HA, Edozien JC. Lancet 1965;2:662–5.
26. Spoelstra MN, Mari A, Mendel M, et al. Metab 2012;61:1224–30.
27. Robinson HM, Seakins A. Pediatr Res 1982;16:1011–5.
28. Jaya Rao KS, Srikantia SG, Gopalan C. Arch Dis Child 1968;43: 365–7.
29. Kalk WJ, Hofman KJ, Smit AM, et al. Am J Clin Nutr 1986;43: 406–13.
30. van Zutphen T, Ciapaite J, Bloks VW, et al. J Hepatol 2016; 65:1198–208.
31. Rytter MJ, Kolte L, Briend A, et al. PLoS One 2014;9:e105017.
32. Scrimshaw N. J Nutr 2003;133:316S–21S.
33. Takele Y, Adem E, Getahun M, et al. PLoS One 2016;11:e0157919.
34. Bauer PV, Hamr SC, Duca FA. Cell Mol Life Sci 2016;73:737–55.
35. Blaut M. Proc Nutr Soc 2015;74:227–34.
36. Iddrisu I, Monteagudo-Mera A, Poveda C, et al. Nutrients 2021;13.
37. Kane AV, Dinh DM, Ward HD. Pediatr Res 2015;77:256–62.
38. Yatsunenko T, Rey FE, Manary MJ, et al. Nature 2012;486: 222–7.
39. Million M, Diallo A, Raoult D. Microb Pathog 2017;106:127–38.
40. Chen RY, Mostafa I, Hibberd MC, et al. N Engl J Med 2021; 384:1517–28.
41. Webb JG, Kiess MC, Chan-Yan CC. CMAJ 1986;135:753–8.
42. Silverman JA, Chimalizeni Y, Hawes SE, et al. Arch Dis Child 2016;101:166–71.
43. Gruber C, Nink N, Nikam S, et al. J Anat 2012;220:179–85.
44. Viart P. Am J Clin Nutr 1977;30:334–48.
45. Heymsfield SB, Bethel RA, Ansley JD, et al. Am Heart J 1978;95:584–94.
46. Bandsma RH, Mendel M, Spoelstra MN, et al. Pediatr Res 2010;68:423–8.
47. Klahr S, Alleyne GA. Kidney Int 1973;3:129–41.
48. Kerpel-Fronius Ö. The Pathophysiology of Infantile Malnutrition; Protein-Energy Malnutrition and Failure to Thrive, Akadémiai Kiadó, Hungary, 1983.
49. Patrick J, Golden M. Am J Clin Nutr 1977;30:1478–81.
50. Grantham-McGregor S, Powell C, Walker S, et al. Child Dev 1994;65:428–39.
51. Walker SP, Wachs TD, Gardner JM, et al. Lancet 2007;369: 145–57.
52. Atalabi OM, Lagunju IA, Tongo OO, et al. Int J Neurosci 2010; 120:23–7.
53. Cederholm T, Jensen GL, Correia M, et al. J Cachexia Sarcopenia Muscle 2019;10:207–17.
54. Mwangome M, Ngari M, Fegan G, et al. Am J Clin Nutr 2017; 105:1415–23.
55. Mwangome MK, Fegan G, Fulford T, et al. Bull World Health Organ 2012;90:887–94.
56. World Health Organization. Pocket Book of Hospital Care for Children, 2nd ed. Guidelines for the Management of Common Childhood Illnesses. Accessed May 2, 2022. Available at www.who.int/publications/i/item/978-92-4-154837-3
57. World Health Organization. Guideline: Updates on the Management of Severe Acute Malnutrition in Infants and Children. Accessed May 5, 2022. Available at www.who.int/publications/i/item/9789241506328
58. Williams PCM, Berkley JA. Paediatr Int Child Health 2018;38: S32–S49.
59. World Health Organization. Management of Severe Malnutrition: a Manual for Physicians and Other Senior Health Workers. Accessed September 5, 2022. Available at https://apps.who.int/iris/handle/10665/41999
60. Kabashneh S, Alkassis S, Shanah L, et al. Cureus 2020;12:e8486.
61. Zangenberg M, Abdissa A, Johansen OH, et al. Trop Med Int Health 2020;25:424–32.
62. Community-Based Management of Severe Acute Malnutrition: a Joint Statement by the World Health Organization, the World Food Programme, the United Nations System Standing Committee on Nutrition, and the United Nations Children's Fund. Accessed May 20, 2022. Available at https://apps.who.int/iris/handle/10665/44295
63. Torun B, Viteri FE. Eur J Clin Nutr 1994;48:S186–9.
64. Collins S, Sadler K, Dent N, et al. Food Nutr Bull 2006;27: S49–82.
65. Sandige H, Ndekha MJ, Briend A, et al. J Pediatr Gastroenterol Nutr 2004;39:141–6.
66. Manary MJ. Food Nutr Bull 2006;27:S83–9.
67. Oakley E, Reinking J, Sandige H, et al. J Nutr 2010;140: 2248–52.
68. Bahwere P, Balaluka B, Wells JC, et al. Am J Clin Nutr 2016;103:1145–61.
69. Imdad A, Francois M, Chen FF, et al. BMJ Open 2022;12:e057389.
70. Akomo P, Bahwere P, Murakami H, et al. BMC Public Health 2019;19:806.
71. Kassaw A, Amare D, Birhanu M, et al. BMC Pediatr 2021;21:176.
72. Wagnew F, Dessie G, Takele WW, et al. BMC Public Health 2019;19:1175.
73. Adem F, Edessa D, Bayissa B, et al. Pediatric Health Med Ther 2020;11:235–43.
74. Lelijveld N, Groce N, Patel S, et al. BMJ Glob Health 2020;5.
75. Martorell R. Am J Hum Biol 2017;29.
76. Black RE, Victora CG, Walker SP, et al. Lancet 2013;382: 427–51.
77. Hamadani JD, Huda SN, Khatun F, et al. J Nutr 2006;136: 2645–52.
78. Walker SP, Chang SM, Powell CA, et al. BMJ 2006;333:472.
79. Grantham-McGregor S. J Nutr 1995;125:2233S–8S.
80. The United Nations International Children's Emergency Fund (UNICEF). Early prevention, detection and treatment of child wasting in the most vulnerable countries to the global food and nutrition crisis: UNICEF's acceleration plan 2022–2023. Accessed on October 17, 2022. Available at www.unicef.org/reports/no-time-waste
81. The United Nations Children's Fund (UNICEF). No time to waste: Improving Diets, Services and Practices for the Prevention, Early Detection and Treatment of Wasting in Early Childhood Available at www.unicef.org/documents/no-time-waste New York, NY: 2021

CHAPTER 71

Inherited Metabolic Disease: Amino Acids, Organic Acids, and Galactose

Daniel J. Pomerantz and Gerard T. Berry

CHAPTER OUTLINE

Abbreviations

δ-ALA, δ-aminolevulinic acid; **ARG**, arginine; **ASA**, argininosuccinate lyase deficiency; **ATP**, adenosine triphosphate; **BCAA**, branched-chain amino acid; **BCKA**, branched-chain α-ketoacid; **BCKAD**, branched-chain α-ketoacid dehydrogenase; **BH4**, tetrahydrobiopterin; **CβS**, cystathionine β-synthase; **CH_3-B_{12}**, methylcobalamin; **CIT**, citrulline; **CNS**, central nervous system; **CoA**, coenzyme A; **DHPR**, dihydropteridine reductase; **DNPH**, dinitrophenylhydrazine; **EEG**, electroencephalogram; **ETF**, electron transfer factor; **FAA**, fumarylacetoacetate; **FAH**, fumarylacetoacetic acid hydrolase; **GA-I**, glutaric acidemia type I; **GAL**, galactose; **GAL-1-P**, galactose-1-phosphate; **GALK**, galactokinase; **GALT**, galactose-1-phosphate uridyl transferase; **GCD**, glutaryl-coenzyme A dehydrogenase; **GC/MS**, gas chromatography/mass spectrometry; **GLU**, glucose; **GLU-1-P**, glucose-1-phosphate; **GLY**, glycine; **HC**, head circumference; **HCO_3**, bicarbonate; **HMG**, 3-hydroxy-3-methylglutaryl; **ILE**, isoleucine; **IVA**, isovaleric acid; **IVD**, isovaleryl-coenzyme A dehydrogenase; **IVG**, isovalerylglycine; **l-DOPA**, l-3,4-dihydroxyphenylalanine; **LEU**, leucine; **LNAA**, large neutral amino acid; **LYS**, lysine; **MAT**, methionine-S-adenosyltransferase; **MBG**, 2-methylbutyrylglycinuria; **MET**, methionine; **MMA**, methylmalonic acidemia; **MPKUCS**, Maternal Phenylketonuria

Collaborative Study; **MS/MS**, tandem mass spectrometry; **MSUD**, maple syrup urine disease; **NAD**, nicotinamide adenine dinucleotide; **NADP**, nicotinamide adenine dinucleotide phosphate; **NTBC**, 2-(2-nitro-4-trifluoromethylbenzoyl)-1,3-cyclohexanedione; **OHIVA**, 3-hydroxyisovaleric acid; **OMIM**, Online Mendelian Inheritance in Man; **ORN**, ornithine; **OTC**, ornithine transcarbamylase; **PAH**, phenylalanine hydroxylase; **PHE**, phenylalanine; **PKU**, phenylketonuria; **p-OHPPAD**, propionic acidemia; **RDA**, recommended dietary allowance; **SAM**, S-adenosylmethionine; **THR**, threonine; **TPP**, thiamin pyrophosphate; **TRP**, tryptophan; **TYR**, tyrosine; **UCED**, urea cycle enzyme deficiency; **UDP**, uridine diphosphate; **USDA**, United States Department of Agriculture; **VAL**, valine

Clinical Genetics Perspective

The clinical geneticist's approach to the general subject of nutrition and the specific requirement for nutrients is based on the perspective that the Recommended Dietary Allowance (RDA)[1] for an essential nutrient is not optimum for all individuals. Rather, individuals throughout a population have genetically determined variations in the enzymes involved in nutrient utilization, thus extending nutrient requirements over a wide range throughout populations. This concept arose historically from two older scientific disciplines: human biochemical genetics and nutrition science. The former discipline originated with Sir Archibald Garrod's Croonian lectures of 1908. Garrod defined four "inborn errors of metabolism" (alkaptonuria, cystinuria, albinism and "pentosuria") as clinical disorders with inherited blocks in the normal flow of metabolic processes. Biochemical and clinical expression of these metabolic blocks demonstrated patterns of inheritance consistent with Mendel's predictions for transmission of single genes with a large effect on the phenotype. For example, in alkaptonuria, Garrod observed that the amount of protein ingested was proportional to the darkness of urine and, thus, the amount of alkaptones excreted. While most individuals did not express either the biochemical or clinical phenomena, asymptomatic carriers could be stressed with protein and excrete discernible amounts of alkaptones in the urine. Thus arose the "individualistic" concept that an individual's genes regulated metabolism and that disease states were created by blocks in this metabolic flow that yielded accumulated precursors and deficient products.

Today, we recognize that "inborn errors of metabolism" are discontinuous traits resulting from variations in the amount and function of enzymes and coenzymes.[2-4] The amino acid structure and sequence, as well as the quantity of enzymes, are dictated by both the genes encoding them and epigenetic regulation of said genes. The control of enzyme function is predicated by molecular regulation through gene transcription, post-transcriptional processing of RNA, translation, post-translational modification, cofactor interaction, trafficking, and protein turnover. To date, there are approximately 24,000 genes in the human genome, of which approximately 6,000 have been shown to be associated with monogenic disorders, according to search engines such as Online Mendelian Inheritance in Man (OMIM).[5] The extent of normal variation in genes controlling enzyme activity suggests that approximately 30% of our population is heterozygous for common alleles.[6] Within this continuous diversity, mutations produce discontinuous, relatively rare traits that are expressed as disease under normal environmental conditions.

Mutant gene frequencies vary in populations; for example, mitochondrial branched-chain α-ketoacid dehydrogenase (BCKAD) deficiency (maple syrup urine disease [MSUD]) occurs in one of approximately every 185,000 newborns worldwide, but it occurs in one of 176 in a Mennonite population.[7] The Mennonite mutation is in the *E1α* gene and changes a tyrosine (TYR) at position 194 to asparagine (Y302N). In the homozygous state, it produces extreme toxicity caused by accumulated branched-chain α-ketoacids (BCKAs) if affected newborns are fed the RDA for branched-chain amino acids (BCAAs). Normal development is expected if dietary isoleucine (ILE), leucine (LEU), and valine (VAL) are restricted to 20% to 40% of the RDA during infancy, however.[8] Considerable human variation occurs in the structure and activity of enzymes involved in the catabolism of essential amino acids, but few individuals are so impaired that ingestion of the RDA creates severe disease. Population-based newborn screening and dietary intervention are now applied through public health programs for more than 40 rare inborn errors for which newborn screening predicts genetic susceptibility given a normal diet.[2,3] In contrast to these relatively rare inborn errors, all humans lack the enzyme that converts L-gulono-α-lactone to ascorbic acid, but scurvy does not occur if sufficient vitamin C is ingested and absorbed.[9] Thus, the frequency of genetic susceptibility given a "normal" diet ranges from rare to common and extends to the metabolism of amino acids, nitrogen, carbohydrates, lipids, fatty acids, organic acids, purines, pyrimidines, minerals, and vitamins.

Genetic Disorders Benefited by Dietary Support

More than 400 genetic disorders have been reported in which toxic manifestations relate to accumulation, deficiency, or overproduction of normally occurring substrates and products of metabolic flow.[3] In many of these disorders, modification of the dietary supply alleviates the manifestations. In many others, however, irreversible damage has already occurred by the time symptoms appear. Optimum management of these disorders depends on identifying affected persons while they are presymptomatic, or before irreversible disease has occurred. Because the disorders are heritable, genetic markers are present from the moment of conception, and thus the genetic power of prediction and prevention is applicable. In practice, certain monogenic disorders can be detected in the fetus in the 10th to 16th week of gestation by studies on chorionic villus or amniotic fluid cells. Prenatal diagnosis can be made in the ninth to 12th week of gestation through chorionic villus biopsy.[10] Teratogenic sequelae of an inborn error in a pregnancy, such as birth defects in children of mothers with phenylketonuria (PKU), may be prevented by strict control of blood phenylalanine (PHE) before conception and throughout pregnancy. Other inherited metabolic alterations are detected postnatally in the presymptomatic infant by analysis of blood, urine, erythrocytes, leukocytes, or cultured skin fibroblasts for impaired enzymes, accumulated substrates, or products of alternate metabolic pathways.

A selective search for presymptomatic genetic disease is often undertaken when a family history of inherited disease is present. Selective screening for inherited disease is also initiated for relatively common symptoms, such as poor weight gain, in childhood. Early treatment has proven effective for many diseases such as PKU, galactosemia, isovaleric acidemia, homocystinuria, MSUD, argininosuccinic aciduria, and citrullinemia. Irreversible brain damage occurs if treatment is not initiated in PKU in the early neonatal period. In MSUD, galactosemia, isovaleric acidemia, and disorders of the urea cycle, irreversible damage to the brain may occur within the first week of life, whereas disorders of fatty acid oxidation may remain undetected for weeks until a metabolic stress (e.g., an infection) produces hypoglycemia.

To prevent irreversible damage from inherited metabolic disorders, population-based, newborn screening of dried blood spots from a newborn's heel stick has been performed since the 1960s. The screening technology has progressed from bioassay to tandem mass spectrometry (MS/MS), which has enabled expansion of detection of accumulated analytes from a few amino acids to include fatty acids and organic acids. These compounds are determined from their acylcarnitine profiles, and amino acids are determined from their butylester derivatives. In 2006, the American College of Medical Genetics[11] recommended a uniform screening panel (RUSP) of 29 disorders with primary targets and 25 secondary disorders, guiding the formation of the United States RUSP as guidance for state newborn screening programs. As of 2023, this list includes 36 primary disorders and 25 secondary disorders.[12] Forty-two disorders are screened for by MS/MS of dried blood spots, and all these disorders require rapid retrieval of the screen-positive newborn and urgent confirmation or denial of the screening results with immediate dietary intervention for the confirmed diagnosis. These metabolic conditions were selected through an iterative process among clinicians, laboratory scientists, and nutritionists. Selection was based on sufficient knowledge and availability of screening and confirmatory technology, the possibility for preventive intervention, and the probability that a screen-positive infant for one of these disorders would have a good outcome from early intervention. Examples of the disorders screened for, their MS/MS profiles, and the acute interventions required for a screen-positive infant are outlined in **Table 71.1**.[11-22]

Table 71.1 Core and Secondary Targets of Inborn Errors of Metabolism with Online Mendelian Inheritance in Man Number Recommended for Newborn Screening, Marker Analytes Used with Tandem Mass Spectrometry Screening, and Nutritional Therapy During Diagnosis and Acute Therapy

Inborn Error (Omim No.)	Marker Analyte	Nutritional Therapy During Diagnosis and Acute Illness
Core Targets		
Amino acid disorder		
Argininosuccinic acidemia (No. 207900)	CIT	Rapid intervention is required Blood NH_3 >200 μmol/L Omit dietary protein for 1–2 d only; increase L-ARG and/or L-CIT if not arginase deficient If necessary, IV dextrose and electrolytes at 150 mL/kg/24h to supply 10 mg/kg/min Give oral rehydration solutions or sugar-sweetened, caffeine-free soft drinks with added glucose powder to maintain energy intake at 125%–150% of RDA Initiate oral medical food and complete diet as rapidly as tolerated Sodium benzoate, phenylbutyric acid, or phenylacetic acid is used to help decrease blood NH_3
Citrullinemia (No. 215700)	CIT	Rapid intervention is required Blood NH_3 >200 μmol/L Omit dietary protein for 1–2 days only; increase L-ARG if not arginase deficient If necessary, give IV glucose and electrolytes at 150 mL/kg/24h to supply 10 mg/kg/min Give Pedialyte and sugar-sweetened, caffeine-free soft drinks with added Polycose or Moducal to maintain energy intake at 125%–150% of RDA Initiate oral medical food and complete diet as rapidly as tolerated Sodium benzoate, phenylbutyric acid, or phenylacetic acid is used to help decrease blood NH_3
Homocystinuria (No. 236200)	MET	Maintain adequate hydration Administer 25–100 mg/kg pyridoxine in addition to usual infant diet for 1 month to determine if patient is vitamin B_6 responsive If patient is not vitamin B_6 responsive, restrict MET (20 mg/kg) with medical food, folate, betaine, and complete diet at end of 1 mo

(continues)

Table 71.1 Core and Secondary Targets of Inborn Errors of Metabolism with Online Mendelian Inheritance in Man Number Recommended for Newborn Screening, Marker Analytes Used with Tandem Mass Spectrometry Screening, and Nutritional Therapy During Diagnosis and Acute Therapy *(continued)*

Inborn Error (Omim No.)	Marker Analyte	Nutritional Therapy During Diagnosis and Acute Illness
Core Targets		
Maple syrup urine disease (MSUD) (Nos. 248600, 248611, 248610, 238339)	LEU ± VAL	Rapid intervention is required Omit dietary BCAAs for 1–2 days Correct metabolic acidosis and electrolyte abnormalities Provide adequate energy to suppress endogenous protein catabolism (125%–150% of RDA for age) Monitor hydration status, electrolyte status, and clinical symptoms carefully to prevent neurologic crisis. Add L-ILE to therapy within 1–2 d when plasma ILE concentration reaches ≈105 μmol/L Plasma LEU concentration remains elevated for prolonged period if either ILE or VAL is deficient Suspect concurrent sepsis Initiate oral medical food and complete diet as soon as tolerated
Phenylketonuria (PKU) (No. 261600)	PHE, PHE/TYR	Omit dietary PHE for 1–2 days *only* For infant, offer Pedialyte with added Polycose to maintain electrolyte balance if needed If necessary, give IV glucose, electrolytes at 150 mL/kg/24h to supply a glucose infusion rate of 10 mg/kg/min, and amino acids free of PHE to maintain anabolism Give sugar-sweetened, caffeine-free soft drinks with added Polycose or Moducal to maintain energy intake at 100% of RDA Initiate oral medical food and complete diet as rapidly as tolerated
Tyrosinemia type I (TyrI) (No. 276700)	TYR	Rapid intervention is required Omit dietary PHE, TYR, 1–2 days *only* For infants, offer Pedialyte with added Polycose to maintain electrolyte balance if needed
Guanidinoacetate Methyltransferase Deficiency (GAMT) (No 601240)	GAA	If necessary, give IV glucose and electrolytes at 150 mL/kg/24h to supply a glucose infusion rate of 10 mg/kg/min, and amino acids free of PHE and TYR to maintain anabolism Give sugar-sweetened, caffeine-free soft drinks with added Polycose or Moducal to maintain energy intake at 120%–130% of RDA Initiate oral medical food and complete diet as rapidly as tolerated Initiate supplementation with creatine monohydrate with oral doses of 400–800 mg/kg bw/day in 3–6 divided doses. Initiate supplementation with ornithine with oral doses of 400–800 mg/kg/day in 3–6 divided doses. Initiate dietary restriction of arginine to 15–25 mg/kg/day that corresponds to 0.4–0.7 g/kg/day protein intake. Essential amino acid medical formula should be supplemented (0.5–0.8 g/kg/day) to prevent protein malnutrition.
Fatty acid oxidation disorders		
Carnitine uptake deficiency (CUD) (No. 212140)	C0	Give uncooked cornstarch as needed to help prevent hypoglycemia if ≥6 mo of age Avoid fasting If necessary, give IV glucose at 150 mL/kg to supply a glucose infusion rate of 10 mg/kg/min At home, give frequent feedings of fluids containing 2.5 g carbohydrate per fluid ounce Initiate oral diet as rapidly as possible
Long-chain-hydroxy-acyl-CoA dehydrogenase deficiency (LCHAD) (No. 609016)	C16-OH; C18:1-OH	Give uncooked cornstarch as needed to help prevent hypoglycemia if ≥6 mo of age Avoid fasting If necessary, give IV glucose at 150 mL/kg to supply a glucose infusion rate of 10 mg/kg/min At home, give frequent feedings of fluids containing 2.5 g carbohydrate per fluid ounce Initiate oral diet as rapidly as possible

Inborn Error (Omim No.)	Marker Analyte	Nutritional Therapy During Diagnosis and Acute Illness
Core Targets		
Medium-chain acyl-CoA dehydrogenase deficiency (MCAD) (No. 201450)	C8/C10 ± C6, C10:1, C8	Give uncooked cornstarch as needed to help prevent hypoglycemia if ≥6 mo of age Avoid fasting If necessary, give IV glucose at 150 mL/kg to supply a glucose infusion rate of 10 mg/kg/min At home, give frequent feedings of fluids containing 2.5 g carbohydrate per fluid ounce Initiate oral diet as rapidly as possible Avoid medium-chain triglycerides
Trifunctional protein (TFP) deficiency (No. 609015)	C16-OH, C18:1-OH	Give uncooked cornstarch as needed to help prevent hypoglycemia if ≥6 mo of age Avoid fasting If necessary, give IV glucose at 150 mL/kg to supply a glucose infusion rate of 10 mg/kg/min At home, give frequent feedings of fluids containing 2.5 g carbohydrate per fluid ounce Initiate oral diet as rapidly as possible
Very-long-chain acyl-CoA dehydrogenase deficiency (VLCAD) (No. 201475)	C14:1, C14:1/C12:1	Give uncooked cornstarch as needed to help prevent hypoglycemia if ≥6 mo of age
	± C14, C16, C18:1	Avoid fasting If necessary, give IV glucose at 150 mL/kg to supply a glucose infusion rate of 10 mg/kg/min At home, give frequent feedings of fluids containing 2.5 g carbohydrate per fluid ounce Initiate oral diet as rapidly as possible
Organic acid disorders		
β-Ketothiolase (BKT) deficiency[a] (No. 248600)	C5:1, ± C5OH	Delete dietary ILE 1–2 d only Administer L-carnitine Offer Pedialyte with Polycose to maintain electrolyte balance if needed Give IV glucose and electrolytes at 150 mL/kg/24h to supply a glucose infusion of 10 mg/kg/min, if required Give sugar-sweetened, caffeine-free soft drinks with added Polycose or Moducal to maintain energy intake at 100%–125% of RDA Initiate oral medical food and complete diet as rapidly as possible
β-Methylcrotonyl-CoA carboxylase (3MCC) deficiency[a] (No. 210200)	C5-OH, ± C5:1	Delete dietary LEU 1–2 d only Administer IV L-carnitine Provide vigorous fluid replacement Correct metabolic acidosis and electrolyte abnormalities Provide adequate energy to suppress catabolism (125%–150% of RDA for age) If necessary, give IV glucose, lipid, and L-amino acids free of LEU Initiate oral medical food and complete diet as rapidly as tolerated
Cobalamin A and B (Cbl A, B) defects[a] (Nos. 251100, 251110)	C3, C3/C2	Delete ILE, MET, THR, VAL 1–2 d *only* Initiate oral diet as rapidly as possible and administer pharmacologic doses of folate and IM hydroxycobalamin
Glutaric acidemia type I[a] (GA-I) (No. 231670)	C5-DC	Omit dietary LYS and TRP 1–2 d *only* For infant, offer Pedialyte with added Polycose to maintain electrolyte balance if needed If necessary, give IV glucose and electrolytes at 150 mL/kg/24h to supply a glucose infusion of 10 mg/kg/min, and amino acids free of LYS and TRP Give sugar-sweetened, caffeine-free soft drinks with added Polycose or Moducal to maintain energy intake at 100% Initiate oral medical food and complete diet as rapidly as tolerated

(continues)

Table 71.1 Core and Secondary Targets of Inborn Errors of Metabolism with Online Mendelian Inheritance in Man Number Recommended for Newborn Screening, Marker Analytes Used with Tandem Mass Spectrometry Screening, and Nutritional Therapy During Diagnosis and Acute Therapy *(continued)*

Inborn Error (Omim No.)	Marker Analyte	Nutritional Therapy During Diagnosis and Acute Illness
Core Targets		
HMG-CoA lyase deficiency[a] (No. 246450)	C5-OH, ± C6-DC	Rapid intervention is required Omit dietary LEU 1–2 d only Limit fat intake Administer L-carnitine Vigorous replacement of fluids Correct severe metabolic acidosis and electrolyte abnormalities Provide adequate energy to suppress endogenous protein catabolism (125%–150% of RDA for age) Suspect concurrent sepsis: have a low threshold to treat after obtaining appropriate cultures Initiate oral medical food and complete diet as rapidly as tolerated
Isovaleric acidemia[a] (IVA) (No. 243500)	C5	Omit dietary LEU 1–2 d *only* Administer GLY and L-carnitine For infant, offer Pedialyte with Polycose to maintain electrolyte balance if needed Give sugar-sweetened, caffeine-free soft drinks with added Polycose or Moducal to maintain energy intake at 100%–125% of RDA If necessary, give IV glucose and electrolytes at 150 mL/kg/24h to supply a glucose infusion rate of 10 mg/kg/min, and amino acids free of LEU. Initiate oral medical food and complete diet as rapidly as tolerated
Methylmalonic acidemia[a] (MUT) (No. 251000)	C3, C3/C2	Rapid intervention is required Omit dietary ILE, MET, THR, VAL 1–2 d *only* Administer L-carnitine Provide vigorous replacement of fluids Correct severe metabolic acidosis and electrolyte abnormalities; provide adequate energy to suppress endogenous protein catabolism (125%–150% of RDA for age) Suspect concurrent sepsis: have a low threshold to treat after obtaining appropriate cultures
Multiple carboxylase deficiency[a] (MCD) (No. 253260)	C5-OH, ± C3	Biotin, 10–20 mg/d For infant, offer Pedialyte with added Polycose to maintain electrolyte balance if needed If necessary, give IV glucose and electrolytes at 150 mL/kg/24h to supply a glucose infusion rate of 10 mg/kg/min Give sugar-sweetened, caffeine-free soft drinks with added Polycose or Moducal to maintain energy at 100%–125% of RDA Initiate complete diet as rapidly as tolerated
Propionic acidemia[a] (PPA) (No. 606054)	C3, C3/C2	Rapid intervention is required Omit dietary ILE, MET, THR, VAL 1–2 d Administer L-carnitine Provide vigorous replacement of fluids Correct severe metabolic acidosis and electrolyte abnormalities Provide adequate energy to suppress endogenous protein catabolism (125%–150% of RDA for age) Suspect concurrent sepsis: have a low threshold to treat after obtaining appropriate cultures Initiate complete diet as rapidly as tolerated

Inborn Error (Omim No.)	Marker Analyte	Nutritional Therapy During Diagnosis and Acute Illness
Core Targets		
Other disorders		
Biotinidase (BIOT deficiency (No. 253260)	±C5-OH, C5:1, Biotinidase Activity	Biotin, 10–20 mg/d For infant, offer Pedialyte with added Polycose to maintain electrolyte balance if needed Give sugar-sweetened, caffeine-free soft drinks with added Polycose or Moducal to maintain energy at 100%–125% of RDA If necessary, give IV electrolytes and glucose at 150 mL/kg/24h to supply a glucose infusion rate of 10 mg/kg/min Initiate infant diet as rapidly as tolerated
X-linked Adrenoleukodystrophy (X-ALD) (No. 300371)	C26:0-LPC	Refer to Pediatric Neurology
Galactose-1-phosphate uridyltransferase (GALT) deficiency[b] (No. 606999)	Galactose/ GALT Enzyme Activity	Avoid lactose- and galactose-containing infant formulas. Avoid drugs containing galactose or lactose
Secondary Targets		
Amino acid disorders		
Argininemia (No. 107830)	ARG	Rapid intervention is required Blood NH_3 >200 μmol/L Omit dietary protein 1–2 d *only* Give Pedialyte and sugar-sweetened, caffeine-free soft drinks with added Polycose or Moducal to maintain energy intake at 125%–150% of RDA If necessary, give IV glucose and electrolytes at 150 mL/kg/24h to supply glucose at 10 mg/kg/min Initiate oral medical food and complete diet as rapidly as tolerated Sodium benzoate, phenylbutyric acid, or phenylacetic acid is used to help decrease blood NH_3
Biopterin regeneration (BIOPT REG) deficiency (No. 261630)	PHE, PHE/TYR	
Biopterin synthesis (BS) defect (No. 261630)	PHE, PHE/TYR	
Citrin deficiency (No. 603471)	CIT	High-protein, low-carbohydrate diet
Hypermethioninemia (No. 250850)	MET	Omit dietary MET 1–2 d *only* Initiate oral medical food and complete diet as rapidly as tolerated
Hyperphenylalaninemia (No. 261630)	PHE	Omit dietary PHE 1–2 d *only* For infant, offer Pedialyte with added Polycose to maintain electrolyte balance if needed If necessary, give IV glucose, electrolytes at 150 mL/kg/24h to supply a glucose infusion rate of 10 mg/kg/min, and amino acids free of PHE to maintain anabolism Give sugar-sweetened, caffeine-free soft drinks with added Polycose or Moducal to maintain energy intake at 100% of RDA Initiate oral medical food and complete diet as rapidly as tolerated
Tyrosinemia type II (TyrII) (No. 276600)	TYR	Omit dietary PHE, TYR, 1–2 d *only* For infant, offer Pedialyte with added Polycose to maintain electrolyte balance if needed; give sugar-sweetened, caffeine-free soft drinks with added Polycose or Moducal to maintain energy intake at 120%–130% of RDA

(continues)

Table 71.1 Core and Secondary Targets of Inborn Errors of Metabolism with Online Mendelian Inheritance in Man Number Recommended for Newborn Screening, Marker Analytes Used with Tandem Mass Spectrometry Screening, and Nutritional Therapy During Diagnosis and Acute Therapy *(continued)*

Inborn Error (Omim No.)	Marker Analyte	Nutritional Therapy During Diagnosis and Acute Illness
Secondary Targets		
		If necessary, give IV glucose and electrolytes at 150 mL/kg/24h to supply a glucose infusion rate of 10 mg/kg/min, and amino acids free of PHE and TYR to maintain anabolism Initiate oral medical food and complete diet as rapidly as tolerated
Tyrosinemia type III (TyrIII) (No. 276710)	TYR	Omit dietary PHE, TYR, 1–2 d *only* For infant, offer Pedialyte with added Polycose to maintain electrolyte balance if needed If necessary, give IV glucose and electrolytes at 150 mL/kg/24h to supply a glucose infusion rate of 10 mg/kg/min, and amino acids free of PHE and TYR to maintain anabolism Give sugar-sweetened, caffeine-free soft drinks with added Polycose or Moducal to maintain energy intake at 120%–130% of RDA Initiate oral medical food and complete diet as rapidly as tolerated
Fatty acid oxidation disorders		
Carnitine acylcarnitine transporter (CACT) defect (No. 212138)	C16:1; C18:1	Rapid intervention is required Avoid fasting If necessary, give IV electrolyte and glucose at 150 mL/kg to supply a glucose infusion rate of 10 mg/kg/min. Give uncooked cornstarch as needed to help prevent hypoglycemia if ≥6 mo of age At home, give frequent feedings of fluids containing 2.5 g carbohydrate per fluid ounce Initiate oral diet as rapidly as possible
Carnitine palmitoyl-transferase I (CPT IA) defect (No. 600528)	Carnitine	Give uncooked cornstarch as needed to help prevent hypoglycemia if ≥6 mo of age Avoid fasting If necessary, give IV glucose at 150 mL/kg to supply a glucose infusion rate of 10 mg/kg/min At home, give frequent feedings of fluids containing 2.5 g carbohydrate per fluid ounce Initiate oral diet as rapidly as possible
Carnitine palmitoyl-transferase II (CPT II) defect (No. 255110)	C16:1, C18:1	Give uncooked cornstarch as needed to help prevent hypoglycemia if ≥6 mo of age Avoid fasting If necessary, give IV glucose at 150 mL/kg to supply a glucose infusion rate of 10 mg/kg/min At home, frequent feedings of fluids containing 2.5 g carbohydrate per fluid ounce Initiate oral diet as rapidly as possible
Dienoyl-CoA reductase deficiency (DE RED) (No. 222745)		Give uncooked cornstarch as needed to help prevent hypoglycemia if ≥6 mo of age Avoid fasting If necessary, give IV glucose at 150 mL/kg to supply a glucose infusion rate of 10 mg/kg/min At home, frequent feedings of fluids containing 2.5 g carbohydrate per fluid ounce Initiate oral diet as rapidly as possible
Glutaric acidemia type II[a] (GA-II) (Multiple acyl-CoA dehydrogenase deficiency) (No. 231680)	C4, C5, C5-DC, C6, 8, 12, 14, 16	Omit dietary LYS and TRP 1–2 d *only* Restrict fat Administer L-carnitine and GLY Administer riboflavin Maintain anabolism, electrolyte balance and hydration Initiate oral medical food and complete diet as rapidly as tolerated

Inborn Error (Omim No.)	Marker Analyte	Nutritional Therapy During Diagnosis and Acute Illness
Secondary Targets		
Medium-chain ketoacyl-CoA thiolase deficiency (MCKAT) (No. 602199)	C8, C8/C10, ±C6, C6, C10:1	Give uncooked cornstarch as needed to help prevent hypoglycemia if ≥6 mo of age Avoid fasting If necessary, give IV glucose at 150 mL/kg to supply a glucose infusion rate of 10 mg/kg/min At home, give frequent feedings of fluids containing 2.5 g carbohydrate per fluid ounce Initiate oral diet as rapidly as possible
Medium-/short-chain hydroxyacyl-CoA dehydrogenase deficiency (M/SCHAD) (No. 300256)	C4-OH	Give uncooked cornstarch as needed to help prevent hypoglycemia if ≥6 mo of age Avoid fasting If necessary, give IV glucose at 150 mL/kg to supply a glucose infusion rate of 10 mg/kg/min At home, give frequent feedings of fluids containing 2.5 g carbohydrate per fluid ounce Initiate oral diet as rapidly as possible
Short-chain acyl-CoA dehydrogenase deficiency (SCAD) (No. 201470)	C4	Give uncooked cornstarch as needed to help prevent hypoglycemia if ≥6 mo of age Avoid fasting If necessary, give IV glucose at 150 mL/kg to supply a glucose infusion rate of 10 mg/kg/min At home, give frequent feedings of fluids containing 2.5 g carbohydrate per fluid ounce Initiate oral diet as rapidly as possible
Organic acid disorders		
2-Methyl-3-hydroxybutyric acidemia (2M3HBA)	C5, C5:1, C5-OH	Omit dietary ILE 1–2 d *only* Administer L-carnitine Offer Pedialyte with Polycose to maintain electrolyte balance if needed Give IV glucose and electrolytes at 150 mL/kg/24h to supply a glucose infusion rate of 10 mg/kg/min, if required Give sugar-sweetened, caffeine-free soft drinks with added Polycose or Moducal to maintain energy intake at 100%–125% of RDA Initiate oral medical food and complete diet as rapidly as tolerated
2-Methylbutyryl-CoA dehydrogenase deficiency (2MBG) (No. 600006)	C5	? LEU restriction
3-Methylglutaconyl hydratase deficiency[a] (3MGA) (No. 250950)	C5-OH	Omit dietary LEU 1–2 d *only* Administer IV L-carnitine Provide vigorous fluid replacement Correct metabolic acidosis and electrolyte abnormalities Provide adequate energy to suppress catabolism (125%–150% of RDA for age) If necessary, give IV glucose, lipid, and L-amino acids free of LEU Return to oral medical food and complete diet as rapidly as tolerated
Cobalamin C and D (Cbl C, D) defects[a] (Nos. 277410, 277400)	C3/C2	Omit dietary ILE, MET, THR, VAL 1–2 d *only* Initiate oral diet as rapidly as possible and administer pharmacologic doses of folate and IM hydroxycobalamin

(continues)

Table 71.1 Core and Secondary Targets of Inborn Errors of Metabolism with Online Mendelian Inheritance in Man Number Recommended for Newborn Screening, Marker Analytes Used with Tandem Mass Spectrometry Screening, and Nutritional Therapy During Diagnosis and Acute Therapy *(continued)*

Inborn Error (Omim No.)	Marker Analyte	Nutritional Therapy During Diagnosis and Acute Illness
Secondary Targets		
Isobutyryl-CoA dehydrogenase deficiency (IBG) (No. 611283)	C4	Omit dietary LEU 1–2 d *only* Increase GLY, VAL and L-carnitine For infant, offer Pedialyte with Polycose to maintain electrolyte balance if needed If necessary, give IV glucose and electrolytes at 150 mL/kg/24h to supply a glucose infusion rate of 10 mg/kg/min, and amino acids free of LEU Give sugar-sweetened, caffeine-free soft drinks with added Polycose or Moducal to maintain energy intake at 100%–125% of RDA Initiate oral medical food and complete diet as rapidly as tolerated
Malonic (MAL) acidemia (No. 248360)	C3	Restrict fat Administer L-carnitine and medium-chain triglycerides Avoid fasting Give uncooked cornstarch as needed to help prevent hypoglycemia if ≥6 mo of age If necessary, give IV glucose at 150 mL/kg to supply a glucose infusion rate of 10 mg/kg/min At home, give frequent feedings of fluids containing 2.5 g carbohydrate per fluid ounce Initiate oral diet as rapidly as possible
Other disorders		
Galactokinase (GALK) deficiency[b] (No. 230200)		Same as for normal infant Avoid formula, food, and drugs containing galactose or lactose Initiate oral diet as rapidly as tolerated
Galactose epimerase (GALE) deficiency[b] (No. 230350)		Same as for normal infant Avoid formula, food, and drugs containing galactose or lactose Initiate oral feedings as soon as tolerated

Colon (:) followed by number, double bonds; *ARG*, arginine; *BCAA*, branched-chain amino acid; *C*, acyl group or carbon chain; *CIT*, citrulline; *CoA*, coenzyme A; *DC*, dicarboxyl; *GLY*, glycine; *HMG-CoA*, 3-hydroxy-3-methylglutaryl-coenzyme A; *ILE*, isoleucine; *IM*, intramuscular; *IV*, intravenous; *LEU*, leucine; *LYS*, lysine; *MET*, methionine; *NH3*, ammonia; *O*, oxygen; *OH*, hydroxy; *OMIM*, Online Mendelian Inheritance in Man; *PHE*, phenylalanine; *RDA*, recommended dietary allowance; *THR*, theronine; *TRP*, tryptophan; *TYR*, tyrosine; *VAL*, valine.

[a]One or more amino acids involved in the disorder.

[b]Screened for by measuring blood galactose.

Data from American College of Medical Genetics. Genet Med 2006;8(Suppl 1):1S–252S; HRSA Recommended Uniform Screening Panel (2023, February 21) www.hrsa.gov/advisory-committees/heritable-disorders/rusp Chace DH, Lim T, Hansen CR et al. Clin Chim Acta 2009;402:14–8; Frazier DM. Newborn screening by mass spectrometry. In: Acosta PB, ed. Nutrition Management of Patients with Inherited Metabolic Disorders. Sudbury, MA: Jones and Bartlett, 2010:21–67; Gillingham MB. Nutrition management of patients with inherited disorders of mitochondrial fatty acid oxidation. In: Acosta PB, ed. Nutrition Management of Patients with Inherited Metabolic Disorders. Sudbury, MA: Jones and Bartlett, 2010:369–403; Korman SH. Mol Genet Metab 2006;89:289–99; Matalon KM. Introduction to genetics and genetics of inherited metabolic disorders. In: Acosta PB, ed. Nutrition Management of Patients with Inherited Metabolic Disorders. Sudbury, MA: Jones and Bartlett, 2010:1–19; Miinalainen IJ, Schmitz W, Huotari A et al. PLoS Genet 2009;5:e1000543; Molven A, Matre GE, Duran M et al. Diabetes 2004;53:221–7 and National Center for Biotechnology Information. OMIM: Online Mendelian Inheritance in Man [database]. Available at: www.ncbi.nlm.nih.gov/sites/entrez?db=omim. Accessed July 17, 2023.[2, 12-20]

The impaired enzymes of these are listed in **Table 71.2**.[16,23-38] Because the initiation of nutrition therapy in the newborn changes with diagnostic confirmation and increasing age, some recommended nutritional changes over the age of the child are listed in **Table 71.3**.

Although many patients with inherited disorders of metabolism benefit from specialized nutrition support, each disorder would require a chapter for adequate discussion. Thus, this chapter emphasizes disorders for which population-based screening, retrieval, diagnosis, and dietary therapy are available to prevent irreversible, severe pathologic problems.

General Principles of Genetic Disease Management

Fourteen general approaches to the treatment of inherited metabolic disease are discussed here. Due to the heritable nature of these disorders, all diagnoses of inborn errors of metabolism require genetic counseling of the parents regarding recurrence risks, burden of the disease on the affected child, and alternatives

Table 71.2 Chromosomal Location, Gene Size, Number of Mutations, Tissue Distribution of Genes, and Genotype/Phenotype Correlations

Enzyme	Chromosomal Location	Gene Size (Kb)	Number of Mutations	Tissue Distribution	Genotype/Phenotype Correlation
Enzymes of Amino Acid Metabolism					
Phenylalanine hydroxylase	12q22-q24.1	>90	>500	Liver, kidney	Genotype broadly predicts metabolic and clinical phenotype
Dihydropteridine reductase	4p15.1-p16.1		21	Liver, fibroblasts, erythrocytes, leukocytes, platelets	?
Guanosine triphosphate cyclohydrolase	14q22.1–q22.2	30	42	Liver	None
6-Pyruvoyltetrahydropterin synthase	11q22.3–q23.3	?	>28	Liver, erythrocytes	Genotype associated with phenotype
Pterin-4 α-carbinolamine dehydratase	10q22		7	Lymphocytes, scalp hair root cells	Mild phenotypes
Fumarylacetoacetate hydrolyase	15q23–q25		34	Liver, renal tubules, lymphocytes, erythrocytes	Genotype not clearly associated with phenotype
Maleylacetoacetate isomerase	14q24.3	?	3	Liver, fibroblasts, kidney	?
Tyrosine aminotransferase	16q22.1	10.9	15	Liver, kidney	None
4-Hydroxyphenylpyruvate dioxygenase	12q24–qter	21	?	Liver	?
Cystathionine β-synthase	21q22.3	30	30	Liver, fibroblasts, brain, phytohemagglutinin-stimulated lymphocytes, amniotic fluid cells, chorionic villus cells	Genotype associated with phenotype I278T: vitamin B_6 responsive T353M: African, not vitamin B_6 responsive G307S: Celtic, not vitamin B_6 responsive
Methionine-*S*-adenosyltransferase	10q22	20	17	Liver[22,28]	Genotype not clearly associated with phenotype
Guanidinoacetate Methyltransferase	19p13.3	4.5	57	Brain, Skeletal Muscle, Heart	Genotype not clearly associated with phenotype
Enzymes of Organic Acid Metabolism					
Branched-Chain α-Ketoacid Dehydrogenase Complex					
E1α (decarboxylase)	19q13.3	55	12	Liver, fibroblasts, leukocytes, muscle	Y393W (Mennonite) (classic phenotype)
E1β (stabilizes E1α)	6q1.4	100	4	Liver, fibroblasts, leukocytes, muscle	11bp del → stop
E2 (transacylase)	1p31	68	6	Liver, fibroblasts, leukocytes, muscle	E163X R183P, common in Ashkenazi?
E3 (lipoamide dehydrogenase)	7p22	20	10	Liver, fibroblasts, leukocytes, muscle	Affects pyruvate and α-ketoglutarate dehydrogenases as well

(continues)

Table 71.2 Chromosomal Location, Gene Size, Number of Mutations, Tissue Distribution of Genes, and Genotype/Phenotype Correlations *(continued)*

Enzyme	Chromosomal Location	Gene Size (Kb)	Number of Mutations	Tissue Distribution	Genotype/Phenotype Correlation
Enzymes of Organic Acid Metabolism					
E1α kinase (inactivating)	16p13.12	40	2	Liver, fibroblasts, leukocytes, muscle	Inhibited by tumor necrosis factor α and produces cancer cachexia
E1α phosphatase (activating)	?	?	?	Liver, fibroblasts, leukocytes, muscle	Actives branched-chain-α-ketoacid dehydrogenase complex
Isovaleryl-CoA dehydrogenase	15q14-q15	2.1–4.6	20	Liver, fibroblasts	Genotype not associated with phenotype
3-Methylcrotonyl-CoA carboxylase	?	?	?	Fibroblasts, lymphocytes	Genotype associated with phenotype
3-Methylglutaconyl-CoA-hydratase (type 1)	?	?	?	Fibroblasts, lymphocytes	?
3-Hydroxy-3-methylglutaryl-CoA lyase (HMG-CoA lyase)	1p35.1.36.1	?	?	Liver	Genotype associated with phenotype
2-Methylbutyryl-CoA dehydrogenase (Acyl-CoA dehydrogenase-SBCAD)	10q26.13	20	>12	Fibroblasts[15,21]	None reported
Multiple carboxylase (Holocarboxylase synthetase)	?	?	>30	Liver, fibroblasts, leukocytes[32]	Genotype associated with phenotype
Glutaryl-CoA dehydrogenase	19p13.2	7	>90	Liver, kidney, fibroblasts, leukocytes, amniotic fluid cells, chorionic villus cells	None between genotype and clinical severity Specific mutations correlate with severity of organic aciduria
Propionyl-CoA carboxylase				Heart, kidney, liver cells	None
α-Subunit	13q32	100	?		
β-Subunit	3q13.3q22	?	?		
Methylmalonyl-CoA mutase	6p12–p21.2	?	22	Kidney, liver, placenta cells	Genotype associated with phenotype
Biotinidase	3p25	?	>100	Serum leukocytes, fibroblasts[25]	None reported
β-Ketothiolase (mitochondrial acetoacetyl-CoA thiolase)	11q22.3-q23.1	1.5	>40	Liver[26]	None reported
2-Methyl-3-hydroxybutyryl-CoA dehydrogenase	XP11.2 H517B10 gene[35]	1.3	?	All human tissue Highest in liver and kidney	?
Isobutyrl-CoA dehydrogenase	ACAD8 gene			Fibroblasts	
Malonyl-CoA decarboxylase	MLYCD gene[30]		22	Fibroblasts[29]	None reported
Cobalamin A	4q31.21	?	?	Liver, skeletal muscle[24]	?
Cobalamin B	12q24[24]	1.1	?	Liver, skeletal muscle, fibroblasts[36]	?

Enzyme	Chromosomal Location	Gene Size (Kb)	Number of Mutations	Tissue Distribution	Genotype/Phenotype Correlation
Enzymes of Organic Acid Metabolism					
Cobalamin C	1P34.1 MMACHC protein[34]	?	>42	Fibroblasts	None reported
Cobalamin D	2q23.2 MMADHC protein[23]			Fibroblasts[27]	None reported
Enzymes of Nitrogen Metabolism					
Mitochondrial					
Carbamylphosphate synthetase 1	2q35	122	>32	Liver, intestine, kidney (trace)	Genotype associated with phenotype
N-acetylglutamate synthetase	17q21.31	?	?	Liver, intestine, kidney (trace), spleen	Genotype associated with phenotype
Ornithine transcarbamylase	Xp21.1	73	>230	Liver, intestine, kidney (trace)	Genotype associated with phenotype
Citrin	7q21.3	?	30	Liver, kidney, heart, small intestine[31,33]	?
Cytosol					
Argininosuccinate synthetase	9q34.1 (many pseudogenes) 7cen-q11.2	53	14	Liver, kidney, fibroblasts, brain (trace)	Genotype associated with phenotype
Argininosuccinate lyase	?	?	12	Liver, kidney, brain, fibroblasts	Genotype associated with phenotype
Arginase	6q23	13	2	Liver, erythrocytes, lens, brain (trace)	Genotype associated with phenotype
Enzymes of Galactose Metabolism					
Galactose-4-epimerase	1p36–35	4	9	Erythrocytes, fibroblasts, liver	Unclear
Galactokinase	17p24	7.3	13	Liver	Cataracts only
Galactose-1-phosphate uridyltransferase	9p13	4.3	>150	Erythrocytes, leukocytes, fibroblasts, intestinal mucosa, liver	Genotype associated with phenotype Q188R (white) S135L (black), Δ5 kb (Ashkenazi)

CoA, coenzyme A.

in reproduction. The focus of these therapeutic strategies is on direct nutrition and medical interventions for the diagnosed infant. The choice of therapy depends on the pathophysiologic mechanisms producing disease and the medical approach to returning homeostasis to the whole body. Several therapeutic approaches may be tried sequentially or used simultaneously, depending on the acuteness of the disease process.

1. *Enhancing anabolism and suppressing catabolism*. This combined approach involves the use of high-energy feedings and appropriate amino acid mixtures for aminoacidopathies and organic acidemias. An attempt should be made to prevent the physiologic catabolic state in the 4- to 14-day-old infant, and an anabolic state should be maintained throughout childhood. This therapeutic maneuver should be common to all inborn errors involving catabolic pathways. Caution must be exercised, however, to use high-energy feedings primarily during the acute presentation, but not on a long-term basis, to prevent excessive weight gain and the consequences of obesity-related insulin resistance and other complications.
2. *Correcting the primary imbalance in metabolic relationships*. This correction involves using both dietary restriction to reduce accumulated substrates that are toxic and provision of products that may be deficient. For example, in phenylalanine

Table 71.3 **Approximate Daily Requirements for Selected Nutrients[a] by Infants and Children with Selected Inherited Disorders of Amino Acid and Organic Acid Metabolism**

		Age						
Nutrient	**Unit**	**0 < 6 mo**	**6 < 12 mo**	**1 < 4 y**	**4 < 7 y**	**7 < 11 y**	**11 < 15 y**	**15 < 19 y**
Energy	kcal/kg	145–95	135–80	—	—	—	—	—
	kcal/d	—	—	1,300	1,700	2,400	2,200–2,700	2,100–1,800
	(range)			900–1,800	1,300–2,300	1,650–3,300	1,500–3,700	1,200–3,900
Fluid	mL/kg[b]	160–135	145–120	95	90	75	50–55	50–65
Protein[c]	g/kg	3.5–3.0	3.0–2.5	—	—	—	—	—
	g/d	—	—	30–55	35–65	40–75	50–90	50–95
Fat	g/d	31	30	—	—	—	—	—
Linoleic acid	g/d	4.4	4.6	7.0	10	10–12	16	16
α-Linolenic acid	g/d	0.5	0.5	0.7	0.9	1.0–1.2	1.0–1.2	1.1–1.6
Isoleucine[d]								
MSUD[d]	mg/kg	30–100	25–70	20–60	20–60	20–30	20–30	310–30
PPA/MMA[d]	mg/kg	100–70	90–60	80–50	70–40	60–30	50–25	40–20
Leucine[d]								
MSUD[d]	mg/kg	60–100	35–70	25–55	30–60	30–60	30–50	15–40
Isovaleric acidemia[d]	mg/kg	150–70	130–70	100–60	90–50	80–40	70–30	60–25
Lysine[d]								
GA-I[d]	mg/kg	100–70	90–40	80–30	75–25	65–25	60–20	55–15
Methionine[d]								
HCU[d]	mg/kg	20–35	15–35	10–30	10–20	10–20	10–20	5–10
PPA/MMA[d]	mg/kg	50–30	45–25	40–20	35–15	30–10	25–10	20–10
Phenylalanine[d]								
PKU[d]	mg/day	135–400	135–330	200–320	200–1100	200–1100	200–1100	200–1100
Tyrosinemias[d]	mg/kg	45–95	35–90	30–85	25–80	20–70	20–70	15–65
Threonine[d]								
PPA/MMA[d]	mg/kg	80–50	70–40	60–30	55–25	50–25	45–25	40–25
Tyrosine[d]								
PKU[d]	mg/day	1100–2100	2500–3000	2800–3500	4000–6000	4000–6000	4000–6000	4000–6000
Tyrosinemias	mg/kg	45–95	30–75	30–60	25–55	20–40	15–30	10–30
Tryptophan[d]								
GA-I	mg/kg	40–10	30–10	20–10	15–8	10–6	8–5	8–4
Valine[d]								
MSUD	mg/kg	40–95	30–80	25–70	30–70	25–30	20–30	15–30
PPA/MMA	mg/kg	85–50	80–45	75–45	70–40	60–30	60–30	50–30

Nutrient	Unit	Age						
		0 < 6 mo	6 < 12 mo	1 < 4 y	4 < 7 y	7 < 11 y	11 < 15 y	15 < 19 y

GA-I, glutaric acidemia type I; *HCU*, homocystinuria; *MMA*, methylmalonic acidemia; *MSUD*, maple syrup urine disease; *PKU*, phenylketonuria; *PPA*, propionic acidemia.

[a]All known essential amino acids, essential fatty acids, minerals, and vitamins must be provided in adequate amounts.

[b]At least 1.5 mL of fluid should be offered for each kilocalorie of energy ingested by the infant and 1 mL/kcal by children and adults.

[c]Mean daily protein intake by physiologically normal children by age is as follows: 2 <4 years, 54.9 g; 4 <9 years, 66.1 g; 9 <4 years, 80.9 g; 14 <19 years, 96.5 g.

[d]After 1 to 2 days of deleting appropriate amino acids, introduce those amino acids at the maximum noted for age. Monitor plasma concentrations frequently, and modify the amino acid prescription appropriately.

hydroxylase (PAH) deficiency, PHE is restricted and TYR is supplemented.

3. *Enhancing excretion of accumulated substances that are overproduced.* The kidney can aid as a dialysis organ in removing accumulated toxic precursors. Maintaining diuresis with hydration is a critical component of therapy.
4. *Providing alternative metabolic pathways to decrease accumulated toxic precursors in blocked reaction sequences.* For example, the accumulated ammonia in enzyme defects of the urea cycle is reduced by removing nitrogen through administration of therapeutic amounts of phenylacetic acid (less noxious precursors, such as phenylbutyrate are used) to form phenylacetylglutamine from glutamine with the consequent loss of two nitrogen atoms in urine. Toxic nitrogenous metabolites can also be cleared by benzoic acid, which forces glycine adducts of hippuric acid that result in the per-mole urinary excretion of one nitrogen atom. Similarly, in isovaleric acidemia, innocuous isovalerylglycine (IVG) is formed from accumulating isovaleric acid (IVA) if supplemental glycine (GLY) is provided to drive the ubiquitous glycine-*N*-transacylase. IVG is then excreted in the urine. In cystathionine β-synthase (CβS) deficiency, relatively benign methionine can be formed from methylation of toxic homocysteine by betaine.
5. *Using metabolic inhibitors to lower overproduced products.* For example, allopurinol inhibits xanthine oxidase and decreases overproduction of uric acid in gout; lovastatin and compactin suppress hydroxymethylglutaryl–coenzyme A (CoA) reductase and reduce excess endogenous cholesterol biosynthesis in familial hypercholesterolemia; and 2-(2-nitro-4-trifluoromethyl-benzoyl)-1,3-cyclohexanedione (NTBC) inhibits *p*-hydroxyphenylpyruvic acid dioxygenase (*p*-OHPPAD) and thus the toxic succinylacetone production in type I tyrosinemia.
6. *Supplying products of blocked secondary pathways.* In cystic fibrosis, the exocrine pancreas does not function in a normal manner to produce and secrete digestive enzymes. Administration of these pancreatic enzymes partially corrects this insufficiency and may prevent the sequelae of fat-soluble vitamin deficiencies in the newborn and young child.
7. *Stabilizing altered enzyme proteins.* The rate of biologic synthesis and degradation of holoenzymes depends on their structural conformation. In some holoenzymes, saturation by coenzyme increases their biologic half-life and thus overall enzyme activity at the equilibrium produced by mutant proteins. This therapeutic mechanism is exemplified in PKU, MSUD, and homocystinuria. Tetrahydrobiopterin (BH_4 [Kuvan]) is available as a drug for therapy of patients with non-PKU hyperphenylalaninemia to enhance PAH activity.[37] Pharmacologic intake of vitamin B_1 in MSUD or vitamin B_6 in homocystinuria increases intracellular pyridoxal phosphate or thiamin pyrophosphate (TPP) and increases the specific activity of BCKAD complex or CβS, respectively.[39-42] Another approach is to provide chemical chaperones to stabilize mutant proteins. For example, excess intravenous infusion of D-galactose has increased defective α-galactosidase in the cardiac variant of Fabry disease.[43]
8. *Replacing deficient coenzymes.* Various vitamin-dependent disorders are caused by blocks in coenzyme production and are "cured" by lifetime pharmacologic intake of a specific vitamin precursor. This mechanism presumably involves overcoming a partially impaired enzyme reaction by mass action. If reactions that are required to produce methylcobalamin (CH_3-B_{12}) or adenosylcobalamin are impaired, homocystinuria or methylmalonic aciduria (or both) will result. Daily intake of milligram quantities of vitamin B_{12} may cure both disorders.[44] In biotinidase deficiency, the coenzyme biotin is not released from its covalently bound state. Reviews of "vitamin-dependency syndromes" have been published.[40-42]
9. *Artificially inducing or replacing enzyme production.* If the structural gene or enzyme is intact, but suppressor, enhancer, or promoter elements are not functional, abnormal amounts of enzyme may be produced. It should be possible to "turn on" or "turn off" the structural gene and enable normal enzymatic production to occur. PAH activity has been induced in patients diagnosed with PKU when they received PHE loading for 3 days with intact protein.[45,46] In the acute porphyria of type I tyrosinemia, excessive δ-aminolevulinic acid (δ-ALA) production may be reduced by suppressing transcription of the δ-ALA synthase gene with excess glucose (GLU) and hematin (**Figure 71.1** and **71.2**). In type I tyrosinemia, the overproduction of succinylacetone may be "turned off" by blocking an earlier enzyme *p*-OH-phenylpyruvate oxidase with the drug NBTC.
10. *Replacing enzymes.* Many attempts to replace deficient enzymes by plasma infusion and microencapsulation have been tried, with limited success. Polyethylene glycol-coated phenylalanine ammonia lyase is a recently FDA-approved biologic that converts phenylalanine into ammonia and trans-cinnamic acid to help reduce phenylalanine via enzyme substitution.[47] The use of polyethylene glycol coating of phenylalanine hydroxylase and adenosine deaminase significantly prolonged the biologic half-life of this enzyme in treating PKU and severe combined immunodeficiency, respectively.[47,48] The engineering of β-glucosidase with a high mannan receptor site enables intravenous use of alglucerase (Ceredase) to treat type I Gaucher disease. Human α-galactosidase A replacement therapy in Fabry disease reverses substrate storage in the lysosome.[43] Recombinant human α-glucosidase can prevent progression and improve cardiac and muscle function in Pompe disease.[49]

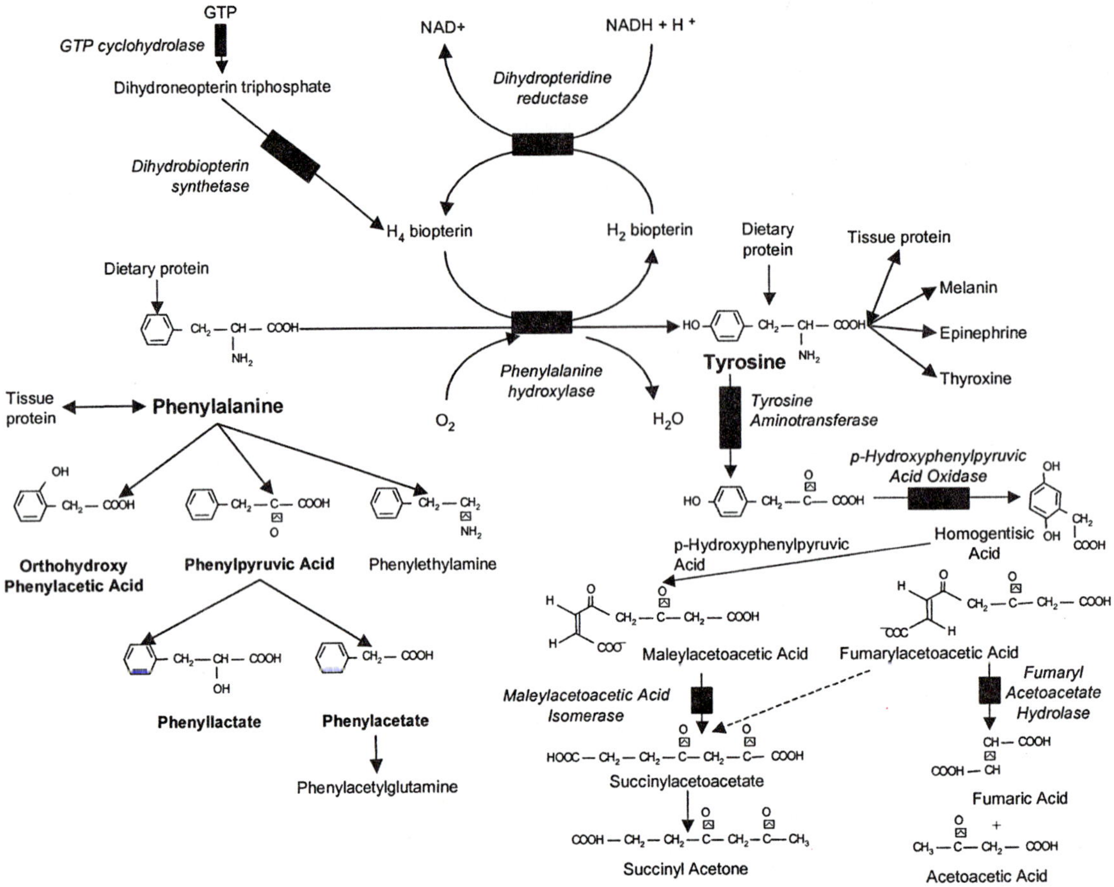

Figure 71.1 Metabolism of aromatic amino acids. The metabolic flow and nutrient interaction in disorders of phenylalanine and tyrosine are diagrammed. The *black bars* represent impaired enzymes in biopterin biosynthesis, phenylketonuria, and tyrosinemia. *GTP*, guanosine triphosphate; *NAD*, nicotinamide adenine dinucleotide (NADH is the reduced form).

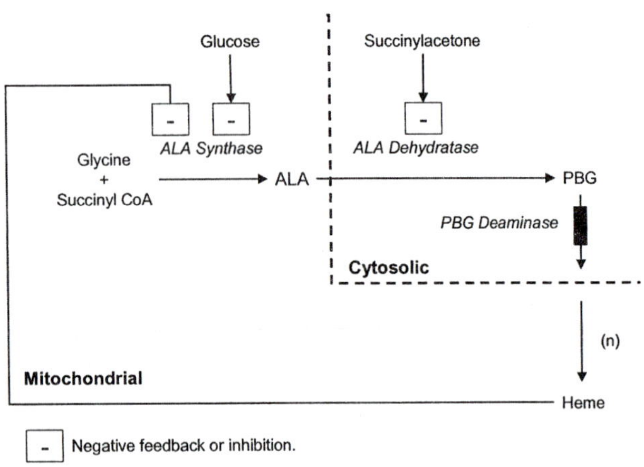

Figure 71.2 Inhibition site in heme biosynthesis of relevance to diagnosis and treatment of tyrosinemia type I. The *black bar* represents the partial block in acute intermittent porphyria with resultant overproduction of δ-aminolevulinic acid (δ-ALA) and porphobilinogen (PBG) with decreased heme biosynthesis. In type I tyrosinemia, succinylacetone is produced and inhibits δ-ALA dehydratase with accumulation of δ-ALA alone, which is neurotoxic. δ-ALA accumulation can be reduced by addition of excess dietary glucose (GLU) and by hematin infusions that negatively control δ-ALA synthase at levels of both enzyme and gene expression. *CoA*, coenzyme A.

11. *Transplanting organs.* Liver transplantation in a host of inherited metabolic disorders benefits systemic metabolism; the return of organ function replaces deficient enzyme activity.[50,51] Bone marrow and renal transplantations may also be of benefit.
12. *Correcting the underlying defect in DNA so the body can manufacture its own functionally normal enzymes.* The DNA for many proteins that are functionally deficient, such as adenosine deaminase, hypoxanthine-guanine phosphoribosyl transferase, and ornithine transcarbamylase (OTC) and the low-density lipoprotein cholesterol receptor, have been cloned, and retroviral constructs containing their cDNA have been transfected into dividing somatic cells from affected individuals. Human gene therapy using CRISPR technology is currently contemplated for these inborn errors, although several technical problems in the toxicity of vectors, gene stability, and gene expression must be solved first. Other molecular approaches, such as homologous recombination to correct mutant sequences or to inhibit RNA for dominant disorders producing antagonism to the normal allele are also future possibilities.[52,53]
13. *Preventing absorption of a nutrient that is toxic in excess.* Large neutral amino acids (LNAAs) free of PHE are available for preventing absorption of PHE from the intestinal tract and from passage across the blood–brain barrier.[54]
14. *Enhancing host metabolism via regulating gut bacterial utilization of metabolites.* Dietary nutrients and macromolecules are

utilized by gut bacteria as an essential step in digestion and metabolism in humans. Antibiotic clearance with metronidazole of propionigenic bacteria has been used to reduce propionate and ammonia production in Propionic Acidemia and Methylmalonic Acidemias. Other inborn errors of metabolism such as PKU, congenital disorders of glycosylation, glycogen storage disorders and other organic acidemias and aminoacidopathies, have been proposed to have dysbiotic microbiomes, which may disrupt metabolic control and promote catabolism. Current research is evaluating the role of antibiotic, prebiotic, probiotic, and postbiotic strategies to address this concern.[55]

Nutritional management remains a principal component in treating all inherited disorders, and some practical considerations for nutrition therapy should be considered. Foremost is the need to maintain normal growth, which cannot be achieved without adequate intake of energy, amino acids, and nitrogen. Energy requirements may be greater than normal when intact protein is restricted and free amino acids supply protein equivalent.[54] Free amino acids administered in one daily dose are oxidized to a much greater extent than when the same dose is divided and administered throughout the day.[56] Nitrogen balance was improved considerably when free amino acids were ingested in several doses throughout the day with intact protein rather than in one dose.[57] With total protein intakes (protein equivalent [g N×6.25 from free amino acids = g protein equivalent] plus intact protein) 25% or more greater than RDA for intact protein for age,[1] and closer to actual intakes by physiologically normal children in the United States,[58] patients with PKU in the United States have grown normally in height.[59] If adequate energy and amino acids cannot be ingested to support normal growth through oral feedings, nasogastric, gastrostomy, or parenteral feedings should be used. Failure to adapt nutrient intake to the individual needs of each patient can result in intellectual disability, metabolic crises, neurologic crises, growth failure and, with some inherited metabolic diseases, death. When specific amino acids or nitrogen require restriction, total elimination of the toxic nutrient for 1 to 3 days in the presence of excess energy intake is the best approach to initiating therapy. Longer elimination or over-restriction may precipitate deficiency of the amino acids or nitrogen. Because the most limiting nutrient in the diet determines growth rate, over-restriction of an amino acid, nitrogen, or energy may be problematic.

Diet restrictions to correct imbalances in metabolic relationships require the use of elemental medical foods. These medical foods must be accompanied by small amounts of intact protein that supply the restricted amino acids. Intact proteins seldom supply more than 50%, and often supply much less, of the protein requirements of patients. Nitrogen-free foods that provide energy are limited in their range of nutrients. Consequently, care must be taken to provide all nutrients required in adequate amounts,[1] particularly because certain minerals are not well absorbed[60] and some vitamins may not be metabolized normally.

Elemental medical foods consist of small molecules that often provide an osmolality that exceeds the physiologic tolerance of the patient. Abdominal cramping, diarrhea, distention, nausea, and vomiting can result from hyperosmolar feedings. Aside from gastrointestinal distress, more serious consequences can occur, such as hypertonic dehydration, hypovolemia, hypernatremia, and death. Osmolalities of selected medical foods intended for inherited diseases of amino acid metabolism have been published.[61,62]

Aromatic Amino Acids

Inborn errors of the aromatic amino acids were historically the first to respond to nutrition support. PKU was discovered in 1933, and the prevention of its resultant intellectual disability by dietary intervention is a classic example of personalized nutrition.

Biochemistry

The essential amino acid phenylalanine (PHE) is used for tissue protein synthesis and hydroxylation to form tyrosine (TYR). The hydroxylation reaction requires Phenylalanine Hydroxylase (PAH), oxygen (O_2), tetrahydrobiopterin (BH_4), dihydropteridine reductase (DHPR), and nicotinamide adenine dinucleotide (NAD), plus hydrogen ion (H+) (Figure 71.1). The normal adult uses only 10% of the RDA for PHE for new protein synthesis, and approximately 90% is hydroxylated to form TYR. The growing child uses 60% of the required PHE for new protein synthesis, and 40% is hydroxylated to form TYR. Mass spectrometry and stable isotope studies of patients with PKU provide information on other pathways available for PHE metabolism. These alternative pathways (Figure 71.1) are minor in the metabolism of PHE at 50 μmol/L concentration in the plasma of physiologically normal individuals. Byproducts become apparent when PHE is not hydroxylated to TYR and accumulates to more than 500 μmol/L, however.[61]

TYR is the normal immediate product of PHE and is essential to five pathways (Figure 71.1), including synthesis of protein, catecholamines, melanin pigment, and thyroid hormones. TYR also provides energy when it is catabolized through *p*-OHPPAD to fumarate and acetoacetate. Enzymes required in this latter degradative pathway include TYR aminotransferase, *p*-OHPPAD, homogentisic acid oxidase, and fumarylacetoacetic acid hydrolase (FAH) (Figure 71.1).

Phenylketonuria

PKU is a group of inherited disorders of PHE metabolism caused by impaired PAH activity. The disease is expressed at 3 to 6 months of age and is characterized by developmental delay, microcephaly, abnormal electroencephalogram (EEG), eczema, musty odor, and hyperactivity. If not treated before 2 weeks of age, the metabolic imbalance produces intellectual disability that is worsened with longer time to treatment. The defect in metabolism in classic PKU is associated with 2% of the activity of normal PAH, and these classic mutations are now defined.[62] The enzyme is expressed primarily in liver.

Molecular Biology

Five of the most frequent mutations in a US clinic include I65T, R408W, Y414C, L348V, and IVS10nt546, which account for more than 50% of mutant PAH alleles. Genotypes with R408W and IVS10nt546 result in more severe PAH impairment, whereas Y414C and I65T have relatively mild phenotypes.[62] Heterozygous parents for classic PKU have 50% enzyme activity, and in the absence of known DNA changes, they can be identified by increased ratios of midday semifasting PHE squared to TYR (P^2/T) in vivo.[63]

The genetic bases for disorders of PAH followed localization of the PAH gene to chromosome 12q22-q24.1 and cloning of the gene, which has 90 kilobases (kb), 13 exons, and 12 introns (Table 71.3). More than 500 different mutations have been identified that cause the "PKU phenotype," and these involve deletions in coding frames, missense mutations, and intron splice site mutations. Ethnic variation occurs in the type and frequency of PAH mutations.[64] Cloning of the PAH gene and identification of different mutations have assisted in genotyping probands, counseling families, and predicting the amount of dietary PHE that will be required.[62] Immediate and lifelong avoidance of excess PHE in the diet continues to be the principal therapy of PKU.

Initial investigations reported responsiveness to oral saptopterin, a tetrabiopterin within 24 to 48 hours at doses of 5 to 10 mg/kg in patients with selected mild mutations in the PAH gene.[65-69] These trials are based on the hypothesis that oral administration of BH_4 makes it possible for the mutant enzymes to suppress their low binding affinity for BH_4, thus enabling patients with this subset of PAH mutations to perform the PHE hydroxylation reaction.[69] Follow-up studies appear to suggest that dosing up to 20 mg/kg of saptopterin is necessary to evaluate for biopterin responsiveness, which is defined as at least 30% reduction in PHE levels on plasma amino acid analysis.[70] These studies have paved the way for FDA approval of saptopterin in the management of PHE levels in patients with PAH deficiency. Although certain genotypes, such as those containing L485, may respond with increased PAH gene expression, other missense mutations may have effects on increased catalysis or increased PAH stability. Many patients with "classic" PKU fail to respond to tetrahydrobiopterin (BH_4).[70]

Other forms of PKU may result from defects in other enzymes involved in the overall reaction. DHPR, an enzyme normally present in many tissues, reduces the quinonoid form of dihydrobiopterin to BH_4 (Figure 71.1). The gene for DHPR is located on chromosome 4p15.1–p16.1. Several other types of PKU result from defects in the synthesis of BH_4 (Figure 71.1; Table 71.3). In addition to functioning as a coenzyme for PAH, BH_4 is also required by TYR hydroxylase and tryptophan (TRP) hydroxylase (Figure 71.1). Because these enzymes produce essential neurotransmitters, defects in biopterin synthesis are associated with progressive neurologic disease unless BH_4, L-3,4-dihydroxyphenylalanine (L-DOPA), and serotonin are replaced.[71]

Although the precise pathogenesis of intellectual disability in classic PKU is not known, accumulation of PHE or its catabolic byproducts, deficiency of TYR or its products, or all four circumstances will produce central nervous system (CNS) damage if PHE accumulates in plasma at greater than normal concentrations during critical periods of brain development. The pathologic consequence varies with the time in brain development at which the chemical insult occurs. Deficient myelination and abnormalities in brain proteolipids or proteins occur in late gestation and during the first 6 to 9 months of life.[72] During this period, oligodendroglia migration may also be impaired, resulting in irreversible brain damage later in childhood. Protein synthesis in the brain is also depressed, probably owing to competitive inhibition by high PHE concentrations on blood–brain barrier transport with consequent imbalance in intraneuronal amino acid concentrations.[73] In the mature brain, neurodegeneration,[74] behavioral difficulties, and prolonged performance times may result from depressed neurotransmitter synthesis. Impairment of these neuropsychologic functions in the mature brain may be reversible when PHE returns to normal concentrations in cells and blood.[75,76]

Screening

The disorders of PHE metabolism require identification, diagnosis, and appropriate therapy before clinical expression of the disease. Nutrition and possibly other therapy should be instituted before the third week of life. Thus, a four-part public health program involving screening, retrieval, diagnosis, and treatment must be coordinated and efficient to prevent intellectual disability.

A screening test of dried blood on filter paper that uses MS/MS[11] detects potential cases in the newborn population. The actions taken in retrieval depend on the concentration of blood PHE, days of age, and protein intake at the time of screening.

Protein ingestion may not be required for a positive PKU screen, but quantitative, normal concentrations during the first 48 hours of life are needed for comparison.[77] Almost all infants with PKU have blood PHE concentration greater than normal during the first day of life, even before the first feeding if they have classic PKU mutations.[71] Neonates with PAH gene mutations resulting in less severely impaired PAH may take longer to develop an elevated blood PHE concentration. Some infants with relatively mild elevation of blood PHE have serious neuropathologic features that are progressive because of a defect in synthesis of BH_4.

Newborn screening in all 50 states, in conjunction with aggressive, rapid approaches to retrieval and diagnosis, has led to early implementation of diet therapy and prevention of intellectual disability.[77] With the present early infant discharge from the newborn nursery and the increase in breastfeeding, lower PHE concentrations, of 121 to 242 μmol/L (2 to 4 mg/dL), are considered positive, and follow-up is initiated.[78] Approximately one in 14,000 non-Hispanic White newborns in the United States is affected with PKU,[79] whereas one in 132,000 Black newborns is affected.[80] For mild non-PKU hyperphenylalaninemia (defined as elevation of Phe between 2 mg/dL and 10 mg/dL) without TYR elevation, the estimated incidence is one in 48,000 for all newborns.[79]

Diagnosis

If the follow-up screening test yields blood PHE higher than 242 μmol/L (4 mg/dL), all plasma amino acids should be quantitated by ion-exchange chromatography with the infant on known PHE intake from intact protein sources. A precise diagnosis is necessary to establish the mode of therapy. A primary classification scheme differentiates PAH deficiency into four categories as follows[81]:

- **Mild hyperphenylalaninemia (MHP).** Affected infants have plasma Phe concentration 120 μmol/L (2 mg/dL) > and <600 μmol/L (10 mg/dL) on a normal diet.
- **Mild PKU.** Affected individuals tolerate 400 to 600 mg of dietary phenylalanine per day.
- **Moderate PKU.** Affected individuals tolerate 350 to 400 mg of dietary phenylalanine per day.
- **Classic PKU.** Affected individuals tolerate <250 to 350 mg of dietary phenylalanine per day to keep plasma concentration of Phe at a safe level of ≤300 μmol/L (5 mg/dL). This most severe form of PKU is secondary to complete or near-complete deficiency of PAH activity. Without dietary treatment, most individuals develop profound, irreversible intellectual disability.

Differential diagnosis requires several laboratory procedures. These include the following: ion-exchange chromatography or MS/MS to determine plasma PHE, TYR, and other amino acid concentrations; urinary organic analysis by gas chromatography/MS (GC/MS) genotyping of parents and proband[61,62,81]; and assays of erythrocyte DHPR and urinary biopterin.[82] For families with an affected child, prenatal diagnosis is available through direct mutational analysis of fetal cell DNA for known PAH genes or indirect restriction fragment length polymorphism analysis of PAH in parents and proband for unknown fetal PAH genes.[61] Because PAH is not expressed in cultured amniotic fluid cells and because PHE concentration does not increase in amniotic fluid until the last trimester, prenatal diagnosis was not possible before molecular techniques became available.

Treatment

Patients with plasma PHE concentration >360 mol with associated low plasma TYR concentration and normal BH_4 and DHPR require prompt treatment with a PHE-restricted diet, with TYR supplementation if levels are consistent below target range. The objective of nutritional therapy of the child with classic PKU is to maintain blood PHE concentration to allow optimum growth and brain development by supplying adequate energy, protein, and other nutrients, while restricting PHE and supplementing TYR intake.[58,83]

Other approaches to therapy include sapropterin supplementation, with dosing up to 20 mg/kg of saptopterin necessary to evaluate for biopterin responsiveness. The majority of individuals with mild or moderate PKU may be responsive to sapropterin, while up to 10% of individuals with classic PKU show a response. Sapropterin responsiveness is defined by at least 30% decrease in PHE on plasma amino acid analysis. Genotyping may be necessary to help predict which individuals will respond to sapropterin.[84] Sapropterin is currently FDA approved exclusively for PAH-related PKU disorders, but not other biopterin responsive PKU syndromes, even as biopterin supplementation for non-DHPR deficiency-related biopterin disorders is supported by the most recent guidelines on tetrahydrobiopterin deficiencies.[85]

Large neutral amino acids (LNAA) may decrease the plasma PHE concentration in affected adolescents and adults; however, this approach should not be used in women of childbearing age. The theoretical mechanism of action of LNAAs is that phenylalanine shares a transporter with other LNAA along the blood-brain barrier. Some individuals exclude excess phenylalanine because they can support high capacity transport across the blood-brain barrier.[86] Reports of LNAA supplementation have demonstrated reduced brain phenylalanine concentration despite consistently high serum concentration of phenylalanine, due to competition along this transporter.[87,88]

Pegvaliase is a pegylated phenylalanine ammonia lyase injection, which metabolizes PHE into transcinnamic acid and ammonia, thereby reducing plasma PHE via a nondietary and nonbiopterin-dependent mechanism. The FDA approved this first-in-class enzyme substitution therapy for adult patients with PKU in 2018, allowing for better PHE control and, in certain cases, dietary liberalization.[47]

Therapy of the child with biopterin-deficient forms of hyperphenylalaninemia requires administration of BH_4 and use of the PHE-restricted, TYR-supplemented diet in combination with L-DOPA and carbidopa.[68,71] Serotonin that is derived from TRP may improve behavior, because TRP hydroxylase is also impaired by diminished BH_4.[70]

Initiation of Nutritional Therapy of Classic Phenylketonuria. Blood PHE concentration at the time of diagnosis may be rapidly lowered by feeding the infant a 20-kcal/oz (67 kcal/dL) PHE-free medical food mixture.[16,83-89] A minimum of 120 kcal/kg/day intake is necessary. Within a mean of 4 days (SD ±3), blood PHE concentration should decrease to treatment range on a PHE-free formula. Treatment should be initiated in hospitalized infants, if possible, to enable adequate parental information transfer, and to monitor blood amino acid concentrations daily. Laboratory results should be available promptly to prevent precipitation of PHE deficiency and to permit rapid replacement of blood PHE and TYR to optimum concentrations.

If the infant or child is not hospitalized for initiation of nutritional therapy, or if only weekly blood PHE concentrations are obtained, 48 hours of a PHE-free medical food mixture followed by a maintenance medical food formula containing adequate PHE should be prescribed. Blood PHE concentration will decrease to treatment range within a mean of 10 days (SD ±5) with this approach.[16,83-89] Blood PHE concentration should be between 1 and 5 mg/dL no later than the third week of life.

Long-term Care. Long-term care of the patient with classic PKU dictates that medical food and intact protein provide all nutrients in required amounts.

Nutrient Requirements. Data in **Table 71.3** outlines the PHE, TYR, protein, energy, and fluid to offer. A prescription must be written that is individualized to the age, specific genotype,[16,81] growth rate, protein intake, and energy needs of each patient. Weekly adjustments in the diet prescription are necessary during the first 6 months of life, based on hunger, growth, development, and laboratory analyses of plasma PHE and TYR concentrations. The prescribed PHE should maintain the 3- to 4-hour postprandial plasma PHE concentration between 120 and 300 μmol.[90] PHE is an essential amino acid[91] and cannot be eliminated from the diet without producing death. Excess restriction produces growth failure, rashes, bone changes, and intellectual disability.[92]

The infant with classic PKU requires 20 to 50 mg PHE per kilogram body weight for growth, and younger infants require a higher amount.[93] The PHE requirement declines rapidly between 3 and 6 months of age as the growth rate declines. Requirements for PHE in the 6- to 12-month-old patient with classic PKU may decrease to 15 mg/kg/day, but they vary considerably (Table 71.3). Frequent monitoring of blood PHE and TYR concentrations, and of intake, is required to prevent excess intake when growth rate decelerates and to prevent inadequate intake when growth rate is at its peak, as in early infancy and during the prepubertal and pubertal growth spurts.

TYR is an essential amino acid for individuals with PKU. For this reason, plasma TYR concentration must be monitored; if low, L-TYR supplements are given. To supply adequate TYR intake to patients with PKU, approximately 8% to 10% of protein prescribed should be as TYR. TYR supplements alone will not prevent intellectual disability in classic PKU.[94]

Protein requirements are increased to more than the 1980 RDA when a free amino acid mix, rather than intact protein, is the primary protein source.[16] Thus, recommendations for protein for nutritional therapy exceed the RDA. A mean protein intake 24% higher than the 1980 RDA for age was associated with greater PHE tolerance and growth in infants with PKU than was found when mean protein intake was near the RDA.[95] Recommendations for energy and fluid intake (**Table 71.4**)[1,83,96-99] are the same as those for physiologically normal individuals.[1]

Adequate intakes of fat and essential fatty acids linoleic and linolenic acids have been suggested.[1] The total fat requirement for children more than 12 months of age has not been determined, but adequate intakes of linoleate and α-linolenate are suggested. Current recommendations are for 31 and 30 g of fat for infants from birth to 6 months of age and from 7 through 12 months of age, respectively (Table 71.3). Minerals and some vitamins often must be considerably greater than the RDA for age[1] to prevent deficiency in patients on elemental diets (Table 71.4).[96,98]

Phenylalanine-Free Medical Foods. Adequate amino acids and nitrogen cannot be obtained from intact protein without ingesting excess PHE (intact proteins contain 2% to 9% PHE by weight).[100] Thus, special elemental medical foods are used to provide protein, minerals, and vitamins. Sources of these products are given in **Table 71.5**.

Table 71.4 Recommended Dietary Intakes for Minerals and Vitamins by Infants and Children

Nutrient	Recommended Intake for Age[a]						
	0–<6 mo	6–<12 mo	1–<4 y	4–<7 y	7–<11 y	11–<15 y	15–<19 y
Minerals							
Calcium[b] (mg)	400	600	800	800	1,300	1,300	1,300
Chloride (mEq)	5.1	16.1	42.3	53.6	53.6	64.9	64.9
Chromium (μg)	0.2	5.5	11	15	20	25	25
Copper (mg)	0.20	0.22	1.0–1.5	1.5–2.0	2.0–2.5	2.0–3.0	2.0–3.0
Iodine (μg)	65	65	65	73	73	73	95
Iron (mg)[c]	10	15	15	10	10	18	18
Magnesium (mg)	50	75	150	200	250	410	410
Manganese (mg)	0.3	0.6	1.2	1.5	1.9	1.9	2.2
Molybdenum (μg)	2.0	3.0	17	22	22	43	45
Phosphorus (mg)	300	500	800	800	1,250	1,250	1,200
Potassium (mEq)	9–24	11–33	76.7	97.1	97.1–115.0	115.0	126.7
Selenium (μg)	15	20	20	30	40	55	55
Sodium (mEq)	5–15	16.1	43.5	52.2	65.2	65.2	65.2
Zinc (mg)	5	5	10	10	10	15	15
Vitamins							
A (μg RE)	400	500	400	500	700	1,000	1,000
D (μg)[b]	5	5	5	5	5	5	5
E (mg α-TE)	6	7	6	7	11	15	15
K (μg)[b]	5	10	30	55	55–60	75	75
Ascorbic acid (mg)	40	50	45	45	45	75	90
Biotin (μg)	35	50	65	85	120	120	120
B_6 (mg)	0.3	0.6	0.9	1.3	1.6	1.8	2.0
B_{12} (μg)	0.5	1.5	2.0	2.5	3.0	3.0	3.0
Choline (mg)	125	150	200	250	375	550	550
Folate (μg)	65	80	150	200	300	400	400
Niacin equiv (mg)[d]	6	8	9	11	16	18	18
Pantothenic acid (mg)	2.0	3.0	3.0	3.0	4.0	5.0	5.0
Riboflavin (mg)	0.4	0.6	0.8	1.0	1.4	1.6	1.7
Thiamin (mg)	0.3	0.5	0.7	0.9	1.2	1.4	1.4

[a]Data from Otten JJ, Hellwig JP, Meyers LD. Washington, DC: National Academies Press, 2006.

[b]For patients deficient in galactose-1-phosphate uridyl transferase, in addition to usual dietary intake of calcium, vitamin D, and vitamin K, add daily 1,000 mg calcium, 10 μg vitamin D, and 1 mg vitamin K.[99]

[c]Iron needs of patients with phenylketonuria may be greater than the recommended dietary allowance. For male patients who are more than 11 years of age, 15 mg/day of iron is recommended.[96,97]

[d]Niacin needs of patients with phenylketonuria are greater than recommended dietary allowances.[98]

Data from Otten JJ, Hellwig JP, Meyers LD. Dietary Reference Intakes: The Essential Guide to Nutrient Requirements. Washington, DC: National Academies Press, 2006.

Table 71.5 Source of Medical Foods for Inherited Metabolic Diseases

Abbott Nutrition: 3300 Stelzer Rd, Columbus, OH 43219; 800-551-5838 http://abbottnutrition.com Medical food for infants, children, and adults.
Applied Nutrition Corp: 10 Saddle Rd, Cedar Knolls, NJ 07927; 800-605-0410 www.medicalfood.com Medical food for children and adults.
Cambrooke Foods: 2 Central St, Framingham, MA 01701; 866-456-9776 www.cambrookefoods.com Medical food for children and adults.
Mead Johnson Nutritionals: 2400 West Lloyd Expressway, Evansville, IN 47721; 800-457-3550 www.mjn.com Medical food for infants, children, and adults.
Nutricia North America: P O Box 117, Gaithersburg, MD 20884; 800-365-7354 www.shsna.com Medical food for infants, children, and adults.
Vitaflo USA LLC: 211 N Union St Suite 100, Alexandria, VA 22314; 888-848-2356 www.vitaflousa.com Medical food for children and adults.

Intact Proteins. Serving lists have long been available to simplify the PHE-restricted diet for families and professional persons guiding them. With the availability of the Internet, however, nutrient composition of foods is now available from the United States Department of Agriculture (USDA).[100]

A diet plan for an infant with PKU may be found in **Table 71.6**. By 4 years of age, patients with some genotypes may require as much as 25 mg/kg/day, and others as little as 15 mg/kg. Of the more than 30 medical foods, most are designed for patients older than 12 months of age, and some contain no minerals and vitamins and lead to deficiencies.

Management Problems. Management problems described for children with PKU may occur in other children with inherited disorders of metabolism. The principles described here apply to children with other disorders as well but are not reiterated in other sections.

Maintenance of an adequate intake of protein and energy is important for the infant and child with PKU, although PHE must be restricted. Excess energy intake may lead to obesity[101] that is difficult to correct, and weight loss results in elevation of plasma PHE concentration. Nutritional therapy must be aggressive and, if intake fails to meet the prescription, a nasogastric or gastrostomy tube should be placed to achieve anabolism. This is important in disorders of BCAAs, nitrogen metabolism, and organic acidemias. Amino acids and nitrogen are obtained from medical foods; therefore, the amount of medical food offered must be varied to provide needs. Nonprotein sources of energy, such as corn syrup, Moducal and Protein-Free Diet Powder (Mead Johnson Nutritionals, Evansville, IN), Polycose Glucose Polymers, Pro-Phree (Abbott Nutrition Products Division, Abbott Laboratories, Columbus, OH), Duocal (Nutricia North America, Gaithersburg, MD), sugar, and pure fats can be added to maintain energy intake and to satisfy the child's hunger without affecting blood PHE concentration.

Factors that may elevate blood PHE concentration include acute infections or trauma with concomitant tissue catabolism, excessive or inadequate PHE intake, and inadequate protein or energy intake. Any infection should be promptly diagnosed and appropriately treated. The best approach to nutritional therapy during short-term infections is to decrease intake of PHE and increase intake of fluids and carbohydrates, through the use of

Table 71.6 Sample Diet: Phenylketonuria (2 Weeks of Age), Weight 3.25 kg

Prescription	Total	Per kg			
Phenylalanine (mg)	179	55			
Tyrosine (mg)	906	278			
Protein (g)	11.4	3.5			
Energy (kcal)	390	120			
Water to make (mL)	600				
Medical Food	**Amount**	**Phenylalanine (mg)**	**Tyrosine[a] (mg)**	**Protein (g)**	**Energy (Kcal)**
Phenex-1[b,c]	49 g	0	735	7.4	235
Similac Advance Infant Formula with Iron Powder[b]	38 g	177	171	4.1	170
Water to make	600 mL				
Total		177	906	11.5	405

[a]Add L-tyrosine to medical food mix only if plasma tyrosine is less than the lower limit of normal. L-Tyrosine is poorly soluble in water. Plasma reference range is 55 to 147 μmol/L.
[b]Abbott Nutrition.
[c]Other medical foods that may be used for infants are Phenyl-Free-1 (Mead Johnson Nutritionals) and Periflex Infant (Nutricia North America).

Pedialyte with added Polycose powder (Abbott Nutrition Products Division, Abbott Laboratories, Columbus, OH), fruit juices, high-carbohydrate and protein-free beverages, and soft drinks that do not contain caffeine.

Excess PHE intake is the most common cause of elevated blood PHE in the older child and adult with PKU. This may be caused by overprescription, misunderstanding of the diet by the caretaker, or dietary noncompliance. Frequent evaluations of blood PHE concentration with accompanying accurate diet diaries for calculation of intake are used to determine the dietary PHE prescription. By monitoring urinary organic acids, increases or decreases in ketone excretion aid in differentiating between ingestion of inadequate energy and excess PHE, respectively.

PHE deficiency associated with inadequate PHE intake has three specific stages of development.[102,103] The first is characterized by decreased blood and urine PHE. Clinically, the child may appear normal, lethargic, or anorectic and may fail to gain length or weight. In the older child, increases in blood alanine and β-hydroxybutyric and acetoacetic acidemia result from muscle alanine production and β-lipolysis. In the second stage, blood PHE concentration increases due to muscle protein degradation, although blood TYR concentration may be low. BCAA concentrations may increase with decreases in other plasma amino acids. Aminoaciduria appears because of renal tubular malabsorption. In this stage, body protein stores are catabolized, energy sources are depleted, and active membrane transport functions are impaired. Eczema is common. In the third stage of PHE deficiency, blood PHE is lower than normal, as are concentrations of other amino acids. Accompanying clinical manifestations include failure to grow, osteopenia, anemia, sparse hair, and, finally, death, if the deficiency is not corrected by supplemental dietary PHE and TYR.

Insufficient protein intake results in an inadequate supply of essential amino acids or nitrogen for growth. When protein synthesis is decreased, PHE is no longer used for growth and accumulates in the blood. When catabolism occurs because of a prolonged lack of nitrogen or amino acid intake, blood PHE concentration increases, because tissue protein contains 5.5% PHE. In instances of protein insufficiency, medical food intake should be increased to supply the required nitrogen or essential amino acids.

When energy is provided as carbohydrate and fat, and if adequate nitrogen is available, nonessential amino acids may be synthesized from their ketoacid precursors. Furthermore, carbohydrate ingestion leads to insulin secretion; and insulin promotes amino acid transport into the cell and consequent protein synthesis.[103] When energy intake is inadequate, tissue catabolism occurs to meet energy needs, leading to an elevated blood PHE concentration. Sufficient energy must be provided through generous but not excess use of nonprotein and low-protein foods to ensure a normal growth rate.

A low blood PHE concentration (<25 μmol/L) may lead to depressed appetite,[104] decreased growth[105] and, if prolonged, intellectual disability.[84] Low blood PHE is often caused by inadequate prescription of PHE. In such cases, the prescription for PHE can be increased by the addition of measured amounts of intact protein.

Assessment of Nutritional Therapy. Along with biweekly assessment of growth through measurement of length, weight, and head circumference (HC) and evaluation of development, adequacy of PHE and TYR intake is determined by twice-weekly quantitation of plasma PHE and TYR during the first 6 months, and weekly thereafter, until the child is 1 year of age. The first year is the period of most rapid growth and greatest vulnerability to nutrition insult. After 1 year, regular, frequent blood tests may suffice for monitoring the diet. If plasma PHE concentration exceeds 300 μmol/L (5 mg/dL), the prescription for PHE is decreased, and frequent blood tests are done until plasma PHE is between 60 and 300 μmol/L.

For blood tests to be of use in adjusting the prescription, laboratory analyses must be both accurate and prompt. Fluorometric, ion-exchange, high-performance liquid chromatography, and MS/MS methods are quantitative and are used to monitor plasma PHE and TYR concentrations. If properly instructed, parents may be given responsibility for obtaining the specimens on filter paper or in microcapillary tubes and mailing them to a central laboratory.

A record of food ingested before blood sampling for PHE and TYR measurement is essential and should be kept by the child's caregiver. The correlation among (a) the child's intake of PHE, TYR, protein, and energy; (b) the child's clinical status; and (c) the plasma PHE and TYR concentrations is considered when the diet is altered.

Results of Therapy. Early diagnosis and treatment (before 2 weeks of age) of infants with PKU with a nutritionally adequate, PHE-restricted, TYR-supplemented diet promotes normal growth and prevents intellectual disability. A national study showed that mean height, weight, HC, and IQ scores of children who were treated early were the same as those of physiologically normal children at 4 years of age.[106,107] Trefz et al.[82] reported that when blood PHE concentration was maintained at <360 μmol/L, no difference was observed in mean IQ of 9-year-old children.

The semisynthetic nature of the PHE-restricted diet has led to questions concerning its adequacy. Mean serum carnitine (total and free) of treated patients was in the reference range when patients were fed medical foods containing carnitine.[108,109] Medical foods containing higher amounts of L-TYR have alleviated the problem of low plasma TYR concentration.[110,111] After an overnight fast, the concentrations of plasma GLY were elevated in patients, one group of whom received a GLY-free medical food (Periflex; Nutricia North America, Gaithersburg, MD).[112] Treated patients with PKU often have lower-than-normal transthyretin when they are fed the RDA for protein.[113] Arnold et al.[114] and Acosta et al.[101] reported a positive correlation between height and plasma transthyretin, with concentration <200 mg/L associated with poor linear growth.

Depressed plasma concentrations of total cholesterol have been reported in treated children and untreated adults with PKU.[115-117] Pregnant women with PKU whose serum cholesterol concentrations fail to increase during early gestation are at high risk of spontaneous abortion,[118] possibly because of an inadequate hormonal response to pregnancy. Castillo et al.[119] reported inhibition of brain and liver 3-hydroxy-3-methylglutaryl-CoA (HMG-CoA) reductase and mevalonate-5-pyrophosphate decarboxylase in experimental hyperphenylalaninemia. According to Artuch et al.,[120] elevated plasma PHE resulted in decreased serum ubiquinone 10 concentration. Hargreaves et al.,[121] however, found no difference in mononuclear cell coenzyme Q_{10} concentration among control, treated, and untreated patients with PKU. Lower-than-normal plasma and erythrocyte docosahexaenoic acid concentrations and higher-than-normal n-6 series fatty acid concentrations have been found in patients undergoing therapy for PKU.[122,123] The significance of these differences is unclear, but they appear to be the result of the low-fat content of the medical food.[124,125]

Iron deficiency has been reported in children undergoing therapy for PKU, despite iron intake greater than the RDA.[126,127] Poor selenium status has been found in children with PKU who were receiving medical foods without added selenium.[128] Patients with PKU and low selenium concentration had elevated concentrations of thyroxine and reverse tri-iodothyronine, which decreased significantly with selenium supplementation.[129] Plasma retinol of children with PKU is often lower than in normal children,[113] but they are within reference ranges with adequate medical food intake.[97]

Bone changes were reported as early as 1956 in treated children with PKU. Preschool children with plasma PHE concentration within treatment range had normal bone mineralization.[130-133] Greeves et al.[134] reported that 25% of patients with PKU, most of whom were more than 8 years of age, had a history of fractures, compared with 18% of physiologically normal siblings. Untreated PKU mice (PAH^{enu-2}) had reduced mean femur weight compared with treated and control mice, and shorter mean femur length than control mice. Femur strength was greater in treated mice compared with control mice.[135] Elevated serum prolactin in patients with poorly controlled PKU resulted in high prevalence of menstrual irregularities in girls.[136] As blood PHE increased in older patients under poor dietary control, values for bone mineral content and bone density were always lower than control values. Amino acid imbalances, inadequate protein intake, the need for phosphorus to buffer organic acids made from excess dietary PHE, and inadequate estrogen because of excess prolactin secretion could have contributed to bone abnormalities.[137] Mussa et al.[138] reported that as plasma PHE concentration increased, osteoclastogenesis increased, leading to bone damage. Mean plasma immunoglobulin A and immunoglobulin M were significantly lower in patients with PKU undergoing therapy,[139] relative to healthy children.

Diet Discontinuation. In the past, certain clinicians suggested that the diet could be discontinued at 4, 6, or 12 years of age with no adverse effects.[140-143] Investigators questioned this possibility because studies showed significant differences in performance and intelligence in children,[107,144] and in neurologic function in adults, who discontinued the diet, and in those who remained on the diet. Severe agoraphobia,[145] reversible by a return to the PHE-restricted diet, was also reported in adults. Vitamin B_{12} deficiency resulting in hematologic changes and neurologic disease occurs in off-diet patients who refuse foods of animal origin, but who fail to ingest PHE-free medical foods.[146,147] Vegan-type methylmalonic aciduria from vitamin B_{12} deficiency is the most likely pathophysiologic mechanism.

In studies using the patient as his or her own control, elevated plasma PHE prolonged performance time on neuropsychologic tests of higher integrative function, reduced the mean power frequency of the EEG, and decreased urinary dopamine excretion and plasma L-DOPA in older treated patients with PKU.[148,149] A correlation was found among high plasma PHE, prolonged performance time on neuropsychologic tests, and decreased urinary dopamine in 10 patients. In a study of eight additional patients, statistically significant decreases were found in the mean power frequency of the EEG and in plasma L-DOPA when plasma PHE increased.[150] EEG slowing occurred in PKU heterozygotes at concentration changes of blood PHE induced by aspartame ingestion (150 μmol/L).[75] These effects were reversible and correlated in the reverse direction when plasma PHE was reduced. Severe neurologic deterioration has occurred in several off-diet patients with PKU.[74,150] Reversal of most of the symptoms occurred in a patient who returned to a PHE-restricted diet that contained medical food.[74]

Maternal Phenylketonuria

Pregnant women with PKU who are untreated at conception and during gestation have children with intrauterine growth retardation, microcephaly, and congenital anomalies, often severe and incompatible with life. Intellectual disability is common in children of mothers whose plasma PHE is higher than the normal range.[151] The pathogenesis of fetal damage is uncertain but is believed to be related to elevated maternal blood PHE concentration,[152] as PHE is actively transported across the placenta to the fetus.[153] Fetal plasma PHE concentration is 1.5 to 2 times those of maternal blood.[154] Such elevated fetal plasma PHE is, again, concentrated two- to fourfold by the fetal blood–brain barrier.[155] Intraneuronal PHE concentration of 600 μmol interferes with brain development by one or more of several mechanisms, including abnormal oligodendroglial migration and myelin and other protein synthesis.[156] Thus, it is important to maintain normal plasma PHE in reproductive-age woman with PKU before conception and throughout gestation. Surviving children of untreated women fail to grow and develop normally.[157] In fact, Kirkman[158] predicted that if the fertility of these women is normal and they are not treated with control of PHE intake, the incidence of PKU-related intellectual disability could return to the prescreening level after only one generation.

In 1984, the Maternal Phenylketonuria Collaborative Study (MPKUCS) was initiated to answer questions related to diet and reproductive outcome in women with PKU.[159] Results supported the premise that a PHE-restricted diet, plasma PHE lower than 360 μmol/L, and the gestational age at which the diet is initiated affect reproductive outcome.[160] Further modifications toward trimester-based guidelines of dietary management have been revised over time, most recently in June of 2022 by the Genetic Metabolic Dieticians International.[83]

Nutritional Therapy of Maternal Phenylketonuria. A PHE-restricted diet should be initiated at least 3 months before a planned pregnancy by women who have PKU, if they have previously discontinued the diet. To obtain adequate protein and fat storage in early pregnancy to support third-trimester fetal growth, careful attention must be paid to diet and nutrition status. Although the plasma PHE concentration most likely to yield the best reproductive outcome is unknown, one group of investigators suggested that these objectives may be achieved by a PHE-restricted diet that maintains plasma PHE between 60 and 180 μmol/L.[83,155] Plasma PHE lower than 60 μmol/L may lead to maternal muscle wasting and poor fetal growth. The recommended PHE intake for initiating therapy is in **Table 71.7**.[1,83,118,124] Other indices of nutrition status should be in the normal range for pregnant women. After initiation of diet with the minimum recommended PHE prescription (Table 71.7), plasma PHE concentration is usually monitored twice weekly to maintain the targeted value.

Even after the plasma PHE is stabilized in the treatment range, frequent changes in the individualized diet prescription are required as pregnancy progresses, based on concentrations of plasma PHE, TYR, and other amino acids, and on weight gain. PHE and TYR requirements of each pregnant woman depend on genotype, age, state of health, protein intake, and trimester of pregnancy.[83,124] At approximately mid-pregnancy, PHE tolerance increases considerably.

Table 71.7 Recommended Phenylalanine, Tyrosine, Protein, Fat, Essential Fatty Acid, and Energy Intakes for Pregnant Women with Phenylketonuria

Trimester	Nutrients							
	Phenylalanine[a] (mg/d)	Tyrosine[a] (mg/d)	Protein (g/d)	Fat (g/d)	Linoleic Acid (g/d)	α-Linolenic Acid (g/d)	Energy[b] (kcal/d)	
							Mean	Range
Trimester 1 (0 <14 wk gestation)	265–770	6,000–7,600	≥70	47–132	13	1.4	2,500	2,100–3,400
Trimester 2 (14 <27 wk gestation)	400–1,650	6,000–7,600	≥70	47–132	13	1.4	2,500	2,100–3,400
Trimester 3 (27 <41 wk gestation)	700–2,275	6,000–7,600	≥70	47–132	13	1.4	2,500	2,100–3,400
Lactation	700–2,275	6,000–7,600	≥70	47–132	13	1.4	2,500	2,100–3,400

[a]Recommended ranges from Genetic Metabolic Dieticians International Guidelines, most recently updated in June of 2022.

[b]Modified from Otten JJ, Hellwig JP, Meyers LD. Dietary Reference Intakes: The Essential Guide to Nutrient Requirements. Washington, DC: National Academies Press, 2006. For some women, energy requirements may be greater than the upper limit of the range given to obtain appropriate weight gain

Data from Otten JJ, Hellwig JP, Meyers LD. Dietary Reference Intakes: The Essential Guide to Nutrient Requirements. Washington, DC: National Academies Press, 2006. For some women, energy requirements may be greater than the upper limit of the range given to obtain appropriate weight gain.

As noted for the child with PKU, the amount of protein prescribed (Table 71.7) exceeds the RDA because of the use of free amino acids as the primary source of protein equivalent. A PHE-free medical food (Table 71.5) is used to provide most of the protein prescribed; and nitrogen-free foods, such as pure sugars and fats, are used to supply the remaining energy needs.

Birth measurements of neonates of women with PKU are negatively correlated with maternal plasma PHE and are positively correlated with maternal energy and protein intakes and weight gain during pregnancy. The 5-year mean head circumference (HC) Z-scores of children of women who attained plasma PHE lower than 360 μmol/L by 10 weeks' gestation and throughout pregnancy were 0.50±1.53 vs 0.30±0.88 at birth, whereas the mean HC Z-scores of children of women with plasma PHE of 360 to 600 μmol/L by 10 weeks' gestation and throughout the remainder of gestation declined from −0.65±0.87 to −0.87±1.97. The 5-year HC of children of women with plasma PHE >600 μmol/L throughout pregnancy decreased from −1.46±1.08 at birth to −2.09±1.57. Over 5 years, catch-up growth did not occur in height of children of women with plasma PHE >360 μmol/L.[161] The plasma PHE concentration of pregnant woman with PKU is negatively correlated with total protein intake,[162] suggesting that total protein intake should minimally be at the amount recommended in Table 71.7 for better control of plasma PHE. Appropriate maternal weight gain is related to height and prepregnancy weight, and is greater for underweight women than for women of normal weight (**Table 71.8**).[163]

Two families of fatty acids, linoleic (C18:2, n-6) and α-linolenic (C18:3, n-3), are essential for humans.[1] Adequate intakes of linoleic and α-linolenic acids are suggested to be 13 and 1.4 g/day, respectively, during pregnancy.[1] Women in the MPKUCS with a good reproductive outcome had greater fat intake throughout pregnancy than women with a poor outcome.[118] Whether the poor outcome was caused by inadequate essential fatty acids is not clear. Because some of the medical foods are devoid of or contain very little fat, the fat used to supply 30% to 40% of energy in the diet should be obtained from cooking and salad oils, margarines, salad dressing, and shortenings with either unhydrogenated canola oil or soybean oil as the first ingredient.[100]

Table 71.8 Recommended Weight Gain During Pregnancy for Women with Phenylketonuria

Weight Status at Conception	Recommended Weight Gain	
	First Trimester (kg)	Total (kg)
Normal weight	1.6	11.4–15.9
Underweight	2.3	12.7–18.2
Overweight	0.9	6.8–11.4
Obese		5.0–9.1

Data from Rasmussen KM, Catalano PM, Yaktine AL. Curr Opin Obstet Gynecol 2009;21:521–6.

PHE-free medical foods that provide the prescribed protein equivalent (nitrogen×6.25) for the pregnant woman with PKU also provide the required amounts of minerals and vitamins. Therefore, a prenatal vitamin capsule containing vitamins A and D should not be prescribed for women ingesting adequate amounts of PHE-free medical food, as this may provide vitamin A at levels approaching those that are teratogenic.[164] Women who do not ingest adequate medical food before and during gestation should be given supplements of folic acid and vitamin B_{12} to help reduce the incidence of congenital heart defects in children.[165]

Monitoring Nutritional Therapy. Ongoing monitoring of women with PKU involves measuring plasma concentrations of PHE and TYR, maternal weight gain, and concentrations of other plasma amino acids, transthyretin, ferritin, and zinc. Because pregnant women with PKU are at risk for premature delivery, they should be treated as high-risk patients, even if their plasma PHE is in the targeted treatment range. Multiple ultrasound studies, beginning at 16 to 20 weeks' gestation, may be requested to monitor fetal head size and intrauterine growth patterns. A level II ultrasound to scan for heart defects and other malformations may also be ordered.

Tyrosinemias

Several known disorders of TYR metabolism (**Table 71.9**) may be amenable to nutritional therapy (Figure 71.1). Precise biochemical diagnosis is important because disorders such as liver disease, scurvy, and prematurity may produce increases in blood TYR concentration that is not caused by permanent specific enzyme defects in TYR metabolism.

Seven clinical forms of hereditary tyrosinemia have been reported (Table 71.9). Type Ia is caused by a primary defect of hepatic FAH with the production of an abnormal metabolite, succinylacetone.[166] The gene for FAH has been localized to chromosome 15q23-25 (Table 71.2).[166] Succinylacetone is formed from the accumulated substrate fumarylacetoacetate (FAA) (Figure 71.1). FAA, at subapoptotic doses, has been reported to induce spindle disturbances and segregational defects in both rodent and human cells, a finding leading to the speculation that FAA functions as a thiol-reacting agent that disturbs the organelle/mitotic spindle.[167] If maleylacetoacetic acid isomerase is functional, succinylacetone is also formed from maleylacetoacetate. Succinylacetone is extremely toxic and is associated with impaired active transport function and disordered hepatic enzymes, including *p*-OHPPAD and δ-ALA dehydratase.[166] Decreased activity of both hepatic and erythrocyte δ-ALA dehydratase has been reported in these patients and is postulated to be the mechanism for development of acute porphyric-like episodes (Figure 71.2).[167] Using the drug NTBC to inhibit *p*-OHPPAD activity has prevented acute porphyric episodes and decreased progression of cirrhosis and Fanconi syndrome.[168]

Tyrosinemia type Ia is characterized by generalized renal tubular impairment with hypophosphatemic rickets, progressive liver failure producing cirrhosis and hepatic cancer, hypertension, episodic behavioral and peripheral nerve deficiencies, and elevated concentrations of blood PHE and TYR, with succinylacetone and δ-ALA excretion in urine.[169] The most common mutant allele is a splice donor site gain in intron 12 (IVS12G A+5). Many other missense and nonsense mutations are known (Table 71.2). Reversion of the IVS12 mutation to normal in noncancerous hepatic nodules is described. FAH is expressed in amniotic and chorionic villus cells, and prenatal diagnosis is available by biochemical or molecular techniques.[166]

Tyrosinemia type Ib, believed to be caused by a deficiency of maleylacetoacetate isomerase, has been reported in one infant.[166] Liver failure, renal tubular disease, and progressive psychomotor retardation occurred before the child's death at 1 year of age. Succinylacetone did not accumulate. If this finding is confirmed, the pathophysiology of tyrosinemia type I will require re-evaluation.

Tyrosinemia type II is characterized by greatly elevated concentrations of blood and urine TYR, and increases in urinary phenolic acids, *N*-acetyltyrosine, and tyramine. A deficiency of hepatic cytosolic TYR aminotransferase has been demonstrated.[166] Characteristic physical findings include stellate corneal erosions and plaques. Persistent keratitis and hyperkeratosis occur on the fingers and palms of the hands and on the soles of the feet.[170] These skin abnormalities respond to restriction of dietary PHE and TYR. Intracellular crystallization of TYR is thought to cause these inflammatory responses. Intellectual disability may occur. The TYR aminotransferase gene is located on human chromosome

Table 71.9 Inherited Disorders Producing Increased Plasma Tyrosine

Designation	Enzyme Defect	Clinical Features
Hepatorenal tyrosinemia (type Ia)	Fumarylacetoacetate hydrolyase	Cirrhosis
		Renal Fanconi syndrome
		Acute porphyria (succinylacetone)
		Hepatocellular carcinoma
Hepatorenal tyrosinemia (type Ib)	Maleylacetoacetate isomerase	Liver failure
		Fanconi syndrome
		Psychomotor retardation (no succinylacetone)
Oculocutaneous tyrosinemia (type II)	Hepatic cytosol tyrosine aminotransferase	Eye and skin disorders with variable intellectual disability
		Hydroxyphenylpyruvic acidemia
Primary *p*-OHPPAD deficiency (type IIIa)	*p*-OHPPAD	Neurologic abnormalities Intellectual disability
Hawkinsinuria (type IIIb)	*p*-OHPPAD	Metabolic acidosis
		Microcephaly
Transient neonatal (type IIIc)	*p*-OHPPAD	Prematurity, possibly benign
Tyrosinosis (Medes)	Possibly type Ia	Myasthenia (possibly acute porphyric attack)

p-OHPPAD, 4-hydroxyphenylpyruvic acid dioxygenase.

16q22.1 (Table 71.2). Missense, deletions, nonsense, and splice site mutations are known.

Three clinical subsets of type III tyrosinemia result from dysfunctions of *p*-OHPPAD (Figure 71.1 and Table 71.9). The most severe is type IIIa with no hepatic *p*-OHPPAD. Neurologic abnormalities, including seizures, ataxia, and intellectual disability have been reported in untreated patients with type IIIa.[171] Hawkinsinuria (type IIIb) is named for the 2-L-cysteinyl 5-1,4-dihydroxycyclohexenyl-acetic acid that presumably is formed from an intermediate of impaired *p*-OHPPAD reaction. Metabolic acidosis, failure to thrive, and an odor resembling a swimming pool are described. PHE and TYR restriction improves the critical condition.

Type IIIc is neonatal tyrosinemia, associated with increased plasma and urinary concentrations of TYR and their metabolites. It occurs in 0.2% to 10% of neonates.[166] Short-term protein restriction to 1.5 to 2.0 g/kg body weight/day has lowered plasma TYR in most patients within 4 weeks of life. Whether added ascorbate will stabilize and increase the activity of *p*-OHPPAD in this disorder is not clear. Persistence of hypertyrosinemia may lead to impaired mental function,[172] and short-term diet and ascorbate therapy are recommended.

Diagnosis

Differential diagnosis, imperative for institution of appropriate therapy, requires quantitation of plasma succinylacetone and plasma amino acids by ion-exchange chromatography, and urinary succinylacetone and urinary organic acids by GC/MS. The more severe tyrosinemia type I may not be detected by newborn screening evaluation of plasma TYR concentration through MS/MS. If blood TYR exceeds 8 mg/dL (440 μmol/L) at 14 days of age, renal tubular and hepatic function are evaluated, as well as urine, by organic acid analysis for the presence of *p*-hydroxyphenyl acids and succinylacetone. However, elevated tyrosine can also be an indicator of liver immaturity and dysfunction in the neonatal period and is not specific for Tyrosinemia Type 1. Rather, plasma succinylacetone has been shown to be a sensitive and specific marker for Tyrosinemia Type 1.[173] Dried blood spot analysis via MS/MS for the presence of succinylacetone, rather than elevated tyrosine, has become the gold standard of newborn screening for this disorder.[173] Prenatal diagnosis of hereditary tyrosinemia type I has been made by measurement of succinylacetone in amniotic fluid,[174] by measurement of FAH activity in cultured amniotic fluid cells, and by molecular analyses. Tyrosinemia type II results in a marked increase in urinary *p*-OH-phenylacids and blood TYR.[174] It increases with age of the infant, whereas type IIIc decreases. Hawkinsinuria is measured by its ninhydrin reaction using ion-exchange chromatography.

Treatment

Therapy of the hereditary tyrosinemias requires a firm diagnosis because the approaches to treatment among the various types are different. The objective of nutritional therapy for the hereditary tyrosinemias is to provide a biochemical environment that allows normal growth and development of intellectual potential. Nutrition management alone prevents pathophysiologic changes only in types II and III, for which the prognosis is excellent. Plasma PHE should be maintained between 40 and 80 μmol/L, and plasma TYR should be between 50 and 150 μmol/L. 2-(2-Nitro-4-trifluoromethylbensylate)-1,3-cyclohexanedione therapy in tyrosinemia type Ia, with concomitant nutrition management to maintain plasma TYR at < 500 μmol/L,[175] has prevented acute porphyria episodes, decreased progression of cirrhosis and Fanconi syndrome, improved the survival of patients, and reduced the need for liver transplantation during early childhood. The homeostatic effects of NTBC on succinylacetone production also decrease, but do not eliminate risk for hepatocarcinoma.

Renal impairment, if present, must also be treated in tyrosinemia type Ia. Generalized renal tubular failure may result in metabolic acidosis, hypophosphatemia, rickets, and hypokalemia unless replacement of bicarbonate, phosphate, 1,25-dihydroxycholecalciferol, and potassium is instituted. Rapid treatment of infection is required to prevent a catastrophic catabolic state with overproduction of succinylacetone.

Many of the porphyric symptoms are caused by overproduction of δ-ALA secondary to the inhibitory effect of succinylacetone on δ-ALA dehydratase, or decreased heme biosynthesis (Figure 71.2). Parenteral nutrition with 20% to 25% dextrose solution may control these acute porphyric attacks.[176] Continued or progressive loss of energy-requiring functions that involve loosely bound heme to heme-protein (plasma membrane transporters, cytochrome P-450) may be caused by rapid turnover and insufficient heme biosynthesis (Figure 71.2). Infusions of hematin have produced transient decreases in δ-ALA and improved acute attacks of intermittent porphyria, but this invasive therapy is not recommended unless NTBC is unavailable.[7,177] Hepatocellular carcinoma, however, is not prevented and requires liver transplantation to prevent metastases.[166] The drug NTBC, which inhibits the activity of *p*-OHPPAD in the treatment of tyrosinemia type I, reduces the need for liver transplantation, and diet therapy is an indicated adjunct to drug therapy.[167] An excellent mouse model exists.[178] Newborn and infantile treatment of tyrosinemia type I with diet and NTBC has been shown to prevent the risk of hepatic failure and need for hepatic transplantation. However, the data on preventing the progression to hepatocarcinoma is more ambiguous.[179] Follow-up of liver α-fetoprotein and enzymes, as well as hepatic sonography, are indicated. Hepatic transplantation may be required.

Nutrient Requirements. A prescription that recommends daily amounts of PHE, TYR, protein, energy, and fluid should be written. The prescription for PHE and TYR is based on blood analyses correlated with intake that indicate the child's requirement or tolerance for each amino acid (Table 71.3).

Because a large portion of PHE is normally hydroxylated to form TYR, PHE must also be restricted in the diet of patients with tyrosinemia. PHE requirements appear to be greater for children with tyrosinemia than for children with PKU. In general, the more distal the block in the catabolic pathway, the more normal the amino acid requirement. The TYR needs of children with tyrosinemia have been inadequately described and vary with metabolic state and the accumulation of succinylacetone. When plasma TYR is inadequately controlled in NTBC-treated patients, symptoms of tyrosinemia type II occur.[175]

Because the primary protein source used for infants is a free amino acid mix, the recommended intake is greater than for physiologically normal infants (Table 71.3). For tyrosinemia type Ia, when NTBC is administered, energy needs are similar to those of physiologically normal infants.[1] Fat, essential fatty acids, and mineral and vitamin recommendations are the same as in the patient with PKU (Table 71.3).

Medical Foods Free of Phenylalanine and Tyrosine. Adequate protein cannot be obtained from intact protein without ingesting excess PHE and TYR (proteins contain, by mass, 1.4% to 5.8% TYR).[100] Thus, special medical foods are used that contain no PHE or TYR. Several medical foods free of PHE and TYR and containing minerals and vitamins are available to provide protein. Sources of medical foods are given in Table 71.5. The composition of is available from the USDA.[94]

Initiation of Nutritional Therapy. The most rapid decline of blood TYR concentration at the time of diagnosis may be obtained by feeding a 20 kcal/oz (67 kcal/dL) PHE- and TYR-free formula with no added sources of PHE and TYR. Total energy intake >120 kcal/kg/day is required to prevent a catabolic phase. Laboratory results of blood PHE and TYR concentrations should be rapidly available or deficiency of PHE and TYR[180] could be precipitated. Catabolism, caused by inadequate intake of protein, energy, PHE, or TYR, is particularly undesirable in treating tyrosinemia type I, because a catabolic phase with overproduction of succinylacetone will worsen the clinical state. Intact protein sources containing 20 to 70 mg PHE and 60 to 80 mg TYR/kg body weight/day are usually required after 1 to 2 days of total restriction in the newborn period with tyrosinemia type II or III. Patients with tyrosinemia type Ia may tolerate more dietary PHE and TYR if they are managed with NTBC.

Assessment of Nutritional Therapy. Frequency of assessment is dictated by the type of tyrosinemia and the clinical course of the patient. In tyrosinemia type I, vital signs, height, weight, HC, neurologic examination, and development are documented weekly for the first 3 months; biweekly for the second 3 months; and monthly between 6 months and 1 year of life. Plasma amino acids are quantitated by ion-exchange chromatography or MS/MS, and succinylacetone and *p*-hydroxyphenyl organic acids are quantitated by GC/MS. Additional laboratory studies include urinary δ-ALA, blood and urine assessment of renal losses (bicarbonate [HCO_3^-], potassium [K^+], and sodium [Na^+]), and liver status (α-fetoprotein and liver function tests). Clinical status, dietary intake, and laboratory data should be monitored and correlated in managing tyrosinemia type Ia at intervals indicated previously. Application for eventual liver transplantation should be initiated within the first year of life for patients with type Ia tyrosinemia.

Outcomes of Nutritional Therapy. Outcomes to date have been variable with tyrosinemia type Ia.[166] Some of this variation is caused by the lack of clear diagnostic criteria in the past to delineate the various types of tyrosinemia. Early detection and diagnosis using GC/MS, PHE and TYR restriction, hematin infusions, and early replacement of renal tubular losses bring success at early ages in patients with treated tyrosinemia type Ia. Although NTBC may be the ultimate treatment, implimenting the treatment immediately after birth may be necessary to prevent hepatocellular carcinomas, and hepatic transplantation will be required if liver nodules progress on hepatic sonography, or if the liver α-fetoprotein concentration rises suddenly.[168] The PHE- and TYR-restricted diet has been successful in several patients with tyrosinemia types II and III, with rapid resolution of clinical signs and symptoms.[166] Neonatal tyrosinemia requires early but transient protein restriction. The efficacy of oral ascorbate at 50 mg/day is unclear. Controlled outcome data are not yet available.

Sulfur-Containing Amino Acids

The biochemistry and nutrition requirements for sulfur-containing amino acids have been largely elucidated in humans by studies of inherited blocks in their metabolic pathways.

Biochemistry

Intact protein contains approximately 0.3% to 5.0% MET.[106] Some dietary MET is used by the body for tissue protein synthesis, but most is used through the trans-sulfuration pathway to form adenosylmethionine, adenosylhomocysteine, homocysteine, cystathionine, α-ketobutyrate, cysteine, and their derivatives (**Figure 71.3**). The first step in the trans-sulfuration pathway is the synthesis of *S*-adenosylmethionine (SAM), a reaction catalyzed by MET-*S*-adenosyltransferase (MAT). Impaired MAT results in hypermethioninemia and variable clinical expression from sulfurous breath odor to intellectual disability. The hepatic isoform of MAT only is deficient.[28] In this reaction, the adenosyl portion of adenosine triphosphate (ATP) is transferred to MET. Biologically important compounds that obtain their methyl group from SAM include creatine, choline, phosphatidylcholine, methylated DNA and RNA, and epinephrine. Decarboxylated SAM is the source of the three carbon moieties of spermidine and spermine. S-adenosylhomocysteine is formed as an intermediary product in this pathway and is hydrolyzed to homocysteine.

Homocysteine has four possible pathways open to it. Homocysteine reacts with serine in the presence of CβS, found in the liver and brain, to form cystathionine (Figure 71.3). CβS requires pyridoxal phosphate as a coenzyme. Homocysteine can also be remethylated to form MET through two different enzymatic reactions. In one reaction, the methyl group is derived from betaine and is catalyzed by betaine-homocysteine methyltransferase. The second reaction requires N^5-methyltetrahydrofolate as a methyl donor to CH_3-B_{12} (Figure 71.3) and is catalyzed by 5-methyltetrahydrofolate-homocysteine methyltransferase. Finkelstein[181] used an in vitro system that approximated in vivo conditions in rat liver to measure the simultaneous product formation by the three enzymes that use homocysteine. In this control system, 5-methyltetrahydrofolate-homocysteine methyltransferase, betaine homocysteine methyltransferase, and CβS accounted for 27%, 27%, and 46% of the homocysteine consumed, respectively. The fourth pathway open to homocysteine is spontaneous oxidation to homocystine (Figure 71.3). This reaction occurs inside cells only when homocysteine is present in abnormal amounts. It is essentially irreversible because the disulfide bond of homocystine is covalent. Homocystine is not further metabolized. CβS metabolizes most homocysteine with high affinity to cystathionine, by using serine as a cosubstrate and pyridoxal phosphate as a stabilizing active coenzyme. Cystathionine is then hydrolyzed to cysteine and α-ketobutyrate. The enzyme cystathioninase, which also uses pyridoxal phosphate as a coenzyme, is required for this reaction (Figure 71.3). A deficiency of cystathioninase results in cystathioninuria, which has no pathologic consequence. α-Ketobutyrate is converted to propionyl-CoA, which is carboxylated to methylmalonyl-CoA and isomerized to succinyl-CoA, a Krebs cycle intermediate. L-Cysteine is catabolized to pyruvate, ammonia (NH_3), and hydrogen sulfide (H_2S).

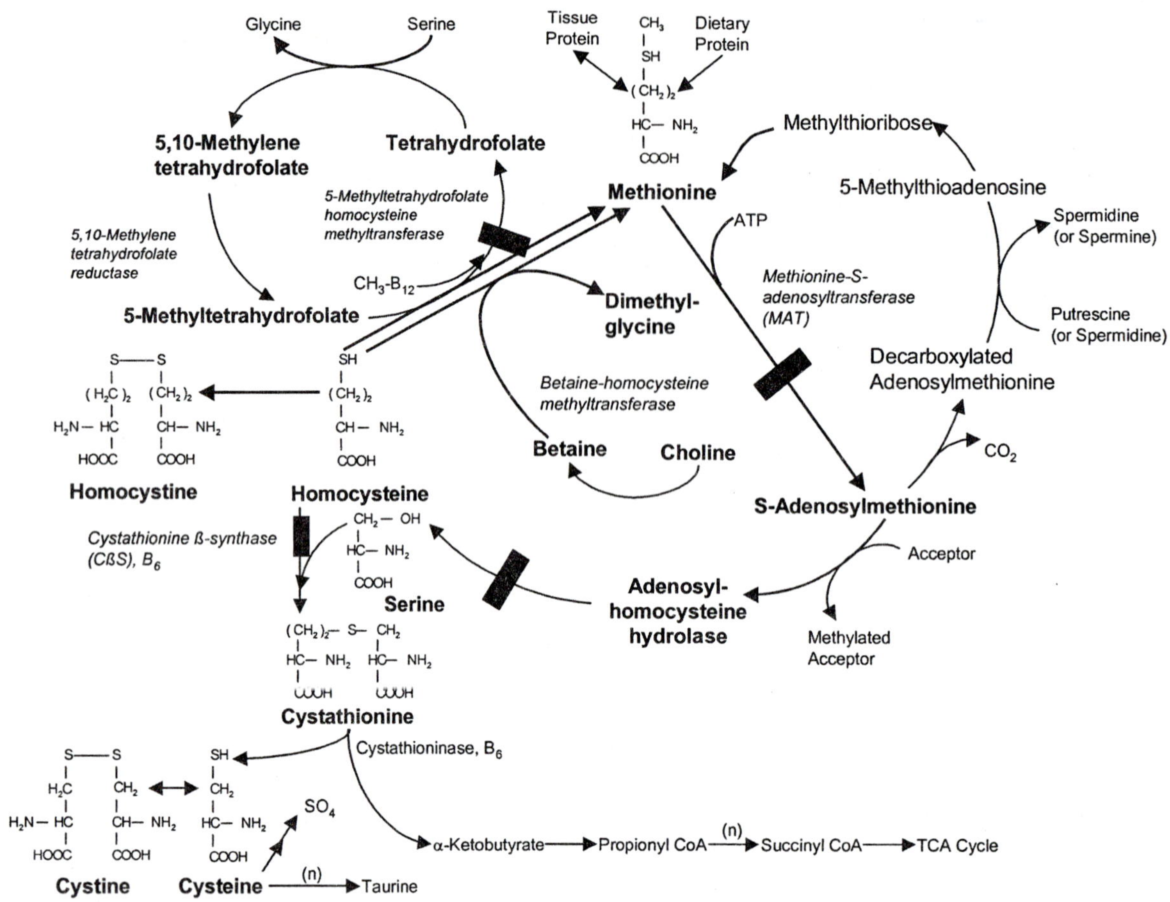

Figure 71.3 Metabolic pathways of sulfur amino acids. The *black bars* represent impaired reactions in three inherited metabolic disorders resulting in hyperhomocyst(e)inemia. *CoA*, coenzyme A; *TCA*, tricarboxylic acid.

Three analytes suggesting inherited disorders of MET metabolism are currently part of newborn screening: MET for homocystinuria and hypermethioninemia and C3 with C3/C2 for cobalamin defects.

Homocystinuria

Defects in the function of CβS or 5-methyltetrahydrofolate-homocysteine methyltransferase result in classic homocystinuria. Impaired activity of the latter enzyme may be caused by failure to synthesize CH_3-B_{12} from vitamin B_{12} or by a deficiency in 5,10-methylenetetrahydrofolate reductase, as well as by mutations in the apoenzyme CβS. Several different defects impair the uptake, transfer, and conversion of dietary vitamin B_{12} to CH_3-B_{12}.[39,182] Both hydroxycobalamin and folate are required to treat these disorders.

The most common form of homocystinuria is caused by a deficiency of CβS. The human CβS locus has been mapped to chromosome 21q-22.3.[183] The gene has been cloned, and more than 90 mutations are characterized in expression systems. Although phenotypes vary for the same genotype, some mutations respond to vitamin B_6 (I278T, P145L, A114V), and others do not (G307S).[176] Severely impaired enzyme function produces accumulation of plasma homocyst(e)ine and MET and decreased cyst(e)ine in cells and physiologic fluids. If this biochemical circumstance is not treated early in life, skeletal changes, dislocated lenses, intravascular thromboses, osteoporosis, malar flushing, and, in some patients, intellectual disability will occur.

The skeletal changes and dislocated lenses are presumably caused by a structural defect in collagen formation produced by α-homocystine interaction with aldose groups on collagen[28] or by irreversible inhibition of lysyl oxidase by homocysteine thiolactone.[184] Intravascular thromboses may occur at any age and have been found in coronary, renal, carotid, and intracranial arteries. The natural history of homocystinuria caused by CβS deficiency was clarified in a large series of patients.[178] However, investigators suggested that in Denmark, most homozygotes for mutation C.833T >C(p.12787) are unaffected or are diagnosed following thromboembolic events occurring in the third decade of life.[185] Heterozygosity for some mutations in CβS (or in tetrahydrofolate reductase) may predispose patients to the development of premature occlusive arterial disease.[186]

It is not known to what degree the intellectual disability seen in homocystinuria is attributable to a metabolic sequela, such as a deficiency of cystathionine in myelin formation, or results from multiple small cerebrovascular thromboses. Mental deficiency may occur in patients with severely impaired CβS as a consequence of multiple cerebral arteriolar obstructions if homocystinemia is not controlled by diet.

Screening

CβS deficiency is inherited as an autosomal recessive disease. Accurate estimates of the incidence for homocystinuria are not available, but newborn screening in 13 countries found one case in 344,000 infants screened.[28] Homocystinuria occurs in many ethnic groups, but it has a higher frequency in persons of Irish descent (one in 58,000) than in other ethnic groups.[28] This finding may be a bias of ascertainment because of the original description of and continued screening for this disorder in the Irish population. Vitamin B_6–responsive mutations in CβS are probably not ascertained by neonatal screening for elevated blood MET concentration.

Selective screening uses the inexpensive urinary nitroprusside reaction. In this reaction, excessive amounts of reduced homocysteine and cysteine form a stable red color with nitroprusside. This selective screening test for sulfur amino acids should be included in the evaluation of any patient with an unknown cause of arterial thrombosis, dislocated lens, marfanoid habitus, or intellectual disability. This test result is also positive in cystinuria, and should be included as a screen for patients with nephrolithiasis.

In a large survey of patients with homocystinuria caused by CβS deficiency, only 13% had a response to vitamin B_6.[28] Most of these patients had "leaky mutations" with residual CβS activity and disease that was expressed in adolescence or young adulthood, rather than in early childhood. Response to vitamin B_6 occurs in several mutations, with some residual enzyme activity. The mechanism involves stabilizing CβS to biologic degradation.[28] The more residual enzyme activity present, the more dramatic the response to vitamin B_6. Hypermethioninemia may not be present in the newborn if CβS activity exceeds 15% of normal. Some patients with CβS deficiency have no activity in the fibroblasts, yet they appear to have a vitamin B_6 response.[187]

Diagnosis

Positive results in a newborn screen must be followed by an assay of plasma amino acids using ion-exchange chromatography or MS/MS because many environmental and genetic variations cause neonatal hypermethioninemia. In patients with a CβS defect, homocystine, cysteine-homocysteine, and MET concentrations are all elevated in plasma and increase with increased protein intake (Figure 71.3). Demonstration of significantly decreased CβS, folate, CH_3-B_{12}, or homocysteine methyltransferases is necessary to confirm the diagnosis and to implement appropriate therapy. MET may be elevated in the absence of homocystinemia in liver disease, and in specific impairment of SAM. In contrast, in defects of homocysteine remethylation to MET, MET is low or normal, whereas homocysteine is elevated. Hyperhomocysteinemia caused by disorders of cobalamin methylation to CH_3-B_{12} or the two homocysteine methyltransferases are not detected by nonselective newborn screening for elevated blood MET but by elevated C3 and the ratio of C3 to C2. Similarly, vitamin B_6-responsive mild CβS deficiency can be missed by newborn screening. Thus, selective screening using urinary nitroprusside reaction or plasma amino acid analysis is indicated for all children or adults with ectopic lens, unexplained vascular occlusions, marfanoid habitus, and mental delay.

Management of CH_3-B_{12} deficiency or impaired methyl transfer with homocystinuria does not include MET-restricted diets. Rather, pharmacologic amounts of vitamin B_{12}, folate, choline, or betaine are added depending on the primary defect. Liver biopsy specimens, transformed lymphoblasts, or cultured skin fibroblasts express CβS and are used to confirm the most common cause of homocystinuria; and molecular screening and sequencing of the CβS gene are useful in predicting management. Prenatal diagnosis can be provided by direct enzyme assay of amniotic fluid cells or by DNA analysis, if the mutations are known.[28,176]

Treatment

If homocystinuria is caused by CβS deficiency expressed in the newborn, the clinical objectives are (a) to prevent the development of skeletal and ocular abnormalities, (b) to prevent intravascular thromboses, and (c) to ensure normal intellectual development.

Pharmacologic doses of pyridoxine should be tried in all patients with hypermethioninemia and homocystinemia.[28,188,189] Some mutations are known to be responsive (Table 71.3). In newborns and early childhood, 25 to 100 mg/day should be tried for 4 weeks before MET restriction. Older children and adults should be given oral pyridoxine (1 g/day). The effect of pyridoxine on plasma MET and homocystine concentrations is monitored weekly on a constant protein intake. Because enzyme stabilization is the most common mechanism of vitamin responsivity, weeks may be required for a biochemical response to occur. If the plasma MET and homocysteine are reduced, pyridoxine should be gradually lowered until the minimum dose required to maintain biochemical normality is reached. Doses of 25 to 750 mg/day have been required for some patients. Excess vitamin B_6 for prolonged periods may cause peripheral neuropathy[189] and liver injury[190]; consequently, if vitamin B_6 is not helpful, it should be discontinued. Betaine supplements (6 g/day) help to maintain postprandial plasma homocysteine concentration in the nearly normal range in vitamin B_6–responsive individuals.[191]

Patients whose condition does not respond completely to pyridoxine require a MET-restricted diet supplemented with L-cysteine. Cysteine becomes an essential amino acid in homocystinuria (Figure 71.3). If plasma folate concentration is lower than normal, owing to excess use in remethylating homocysteine to MET, folate should be added as a supplement.

Betaine is used as adjunctive therapy in patients who are not responsive to vitamin B_6. Doses range from 120 to 150 mg/kg/day and are divided into three doses per day. Total plasma homocyst(e)ine should be followed as the parameter of effective therapy because both free MET and free homocysteine in plasma may decrease to normal, whereas total plasma homocyst(e)ine remains elevated and in concentrations that increase risk of vascular occlusion.[192] Adverse events from betaine treatment are rare but have been reported. In one case,[193] a child who was noncompliant with diet restriction and who had a plasma MET concentration of 3,000 μmol/L developed cerebral edema that resolved when betaine was discontinued. Van Calcar[194] described inherited disorders of sulfur amino acid metabolism, as well as their nutrition management. Liver transplantation has been described to be curative in four reported cases.[195] The role of enzyme replacement therapy and gene therapy as potential curative options for homocystinuria are being evaluated.[196]

Nutrient Requirements. In prescribing and implementing nutrition care plans for infants and children with homocystinuria caused by CβS deficiency, one must consider energy, protein, MET, cysteine, folate, vitamins B_6 and B_{12}, betaine, and fluid needs. Younger infants have a greater MET requirement per kilogram of body weight than older infants. Suggested daily MET intakes range from 35 mg/kg in the young infant to 5 mg/kg in the 15- to 19-year old. Suggested beginning energy, protein, MET, and fluid intakes for infants and children of different ages are given in Table 71.3. If the medical food mixture provides more than 24 kcal/oz, extra fluid should be offered between feedings to meet fluid needs.

Calcium cystinate, a soluble form of L-cystine, should supplement the MET-restricted diet at all ages. The young infant should be offered 300 mg/kg body weight. This amount may be decreased to 200 mg/kg at 6 months of age and 100 mg/kg at 3 years of age and thereafter. The calcium cystinate should be mixed with the MET-free medical food to provide even distribution throughout the day. Older children can sprinkle it in applesauce or other low-protein solids.

Methionine-Free Medical Foods. Several medical foods have been developed as protein sources for patients with homocystinuria. (Table 71.5). MET may be provided for the young infant by the addition of specified amounts of proprietary infant formula to the MET-free medical foods. As growth and development proceed, intact protein-containing foods should be added at the usual ages to supply essential MET. The MET requirement is small, and most foods contain moderate amounts in relation to the requirement.[94] The amount of intact protein that can be ingested; therefore, it is small. See Reference #94 for nutrient composition of foods that may be used to help supply MET and other nutrients. A sample diet for a neonate is given in **Table 71.10**.

Assessment of Nutritional Therapy. After the introduction of the diet and stabilization, plasma MET and cystine concentrations should be monitored twice weekly until 3 months of age, weekly until 6 months of age, and twice monthly thereafter, if blood MET concentrations are stable. Because free MET and homocystine in plasma may revert to normal, whereas total plasma homocyst(e)ine remains elevated, measurement of plasma cystine is recommended when the free MET is normal or the free homocystine is not measurable. After a diet change, plasma MET and cysteine should be measured after 3 days have elapsed. A 3-day diet diary before each blood sample is necessary to evaluate plasma MET and cysteine. Plasma MET should be maintained between 15 and 45 μmol in 2- to 4-hour postfeeding plasma. Little or no homocystine should be present in blood and urine. Total plasma homocyst(e)ine should approach 10 μM/L. Growth and development, as well as clinical evaluation of the pulses, skeletal growth, and ocular lenses, are routinely assessed.

Results of Nutritional Therapy. In a retrospective study of 629 patients with CβS deficiency, MET restriction initiated neonatally prevented intellectual disability, reduced frequency of lens dislocation, and reduced incidence of seizures. Pyridoxine treatment of late-detected vitamin B_6–responsive patients decreased the rate of thromboembolic events.[188] A study in 2020 also confirmed that newborn screen-identified patients had lower frequency of ocular, vascular, central nervous system, and skeletal symptoms, compared with the non-newborn screen group.[197] Hispanic male twins born at 38 weeks' gestation with homocystinuria grew well during the first year of life while on nutrition support. The protein intake of these two patients averaged 3.7 g/kg/day during the first 6 months of life and 2.6 g/kg/day during the second 6 months. Energy intake averaged 131 kcal/kg during the first 6 months and 100 kcal/kg during the second 6 months. Patients with poorly controlled plasma homocysteine may have excessive growth in height. Optimal metabolic control may prevent overgrowth.[198]

Cysteine deficiency was reported in a 3-year-old boy with homocystinuria who received 32 mg/kg/day of L-cysteine, manifested as abnormally low plasma cystine, elevated plasma MET, and weight loss.[199] The Hispanic twins referred to earlier received 58 to 118 mg cystine/kg/day, which resulted in plasma cystine of 19 to 30 μmol/L. Up to 150 mg L-cystine/kg/day may be required to maintain normal plasma cystine.

Elevated plasma copper and ceruloplasmin concentrations were found in 15 patients with homocystinuria, compared with those in age- and sex-matched physiologically normal controls. No relationship with plasma homocysteine could be found.[200] The twins mentioned earlier had elevated serum copper concentrations of 151 and 144 μg/dL at approximately 13 months of age. Low plasma selenium (approximately 15 μmol/L) and erythrocyte glutathione peroxidase activity (approximately 3 U/g hemoglobin)

Table 71.10 Sample Diet: Homocystinuria (2 Weeks of Age), Weight 3.25 kg

Prescription	Total	Per kg			
Methionine (mg)	98	30			
Cystine (mg)	975	300			
Protein (g)	11.4	3.5			
Energy (kcal)	390	120			
Water to make (mL)	600				
Medical Food	**Amount**	**Methionine (mg)**	**Cystine (mg)**	**Protein (g)**	**Energy (kcal)**
XMET Analog[a,b]	52 g	0	183	6.8	247
Enfamil Lipil Powder	47 g	99	61	4.6	221
L-Cystine, 10 mg/mL	73 mL	0	730	0	0
Water to make	600 mL				
Total		99	974	11.4	468

[a]Nutricia North America.

[b]Other medical foods that may be used for infants are Hominex-1 (Abbott Nutrition) and HCY-1 (Mead Johnson Nutritionals).

were found in a child with homocystinuria who was treated with a medical food free of selenium.[201] Administration of 50 μg selenium (in selenium-enriched yeast) every other day was required to maintain normal indices of selenium status. The twins mentioned earlier ingested, on average, 26 μg selenium (as sodium selenite) daily throughout the first year of life. Serum selenium ranged between 60 and 72 μg/L, similar to values reported for physiologically normal babies fed human milk.[202]

Vitamin A absorption tests were carried out in eight untreated patients with homocystinuria, by measuring the elevation in serum after administration of retinol. The explanation proposed for the resulting subnormal serum vitamin A was that retinol was oxidized by −SH groups excreted into the gut.[201] Of the eight plasma retinol values obtained on the twins studied, one was <20 μg/dL and five were between 20 and 30 μg/dL. According to the parental report, vitamin A intake was always more than adequate (1.2 to 5.6 times the RDA for age). Serum transthyretin concentrations were all <20 mg/dL (marginal), and two of the four values obtained were <15 mg/dL (deficient).

Fasting serum folate in eight untreated patients with homocystinuria was abnormally low (4 ng/mL, compared with 8 ng/mL in control children). Two of these patients were treated with 20 mg/day folate, which led to a decrease in urinary homocystine.[203] Severe folate deficiency was found in an untreated infant with homocystinuria who was receiving diluted boiled cow's milk for an episode of gastroenteritis. Excessive use of 5-methyltetrahydrofolate in the remethylation of homocysteine to form MET is proposed as a reason for folate deficiency in untreated patients.[204] The twins noted above had adequate hemoglobin concentration after 4 months of age, and mean corpuscular volume was normal.

Termination of Nutritional Therapy. Most clinicians who treat individuals with homocystinuria believe that patients should be kept on the diet indefinitely. Termination of the diet after growth is achieved may lead to thromboembolisms and ciliary muscle laxity with lens dislocation. Yap et al.,[205] reported that appropriate homocysteine-lowering therapy significantly reduced vascular risk in patients with homocystinuria. When initiation or maintenance of nutritional therapy is not possible, acetylsalicylic acid (1 g/day) and dipyridamole (100 mg/day) increase platelet survival time and decrease thrombotic events.[206] Pharmacologic doses of vitamin E may reduce oxidative stress and platelet activation in patients with homocystinuria.[207] Vitamin B_6 in pharmacologic doses should be continued in vitamin B_6–responsive patients.

Reproductive Performance

Fewer conceptions are reported for both men and women whose condition does not respond to vitamin B_6 than for those who are responsive. Children of male patients do not suffer excess losses and are generally reported to be physiologically normal. A study showed that higher rates of fetal loss occurred in presumptive heterozygous fetuses carried by CβS-deficient mothers than in physiologically normal women.[208] Good reproductive outcome was reported in children of a woman who had tight metabolic control during pregnancy.[209] Whether hypermethioninemia, homocysteinemia, or other metabolic variations in MET metabolism are teratogenic is unclear, but a teratogenic mechanism, as defined for maternal PKU, is possible. In addition, folate-responsive neural tube defects may involve hyperhomocyst(e)inemia as a pathophysiologic mechanism.

Carrier State

Heterozygotes for CβS deficiency may be at risk for premature vascular occlusion. The physician's duty to diagnose, inform, and treat extended family members of probands with CβS deficiency awaits further definition of this risk and outcomes of intervention.[210]

Organic Acids

Several essential amino acids contribute to the synthesis of the acyl groups of organic acids. Among these are the following: the BCAAs, ILE, LEU, and VAL; the sulfur amino acid MET; the hydroxy amino acid threonine (THR); and the dibasic amino acids lysine (LYS) and TRP (Figures 71.3 and **71.4**).

Biochemistry

The BCAAs, ILE, LEU, and VAL are essential nutrients. In the newborn, 75% of those ingested are used for protein synthesis. That present in excess of need for synthetic purposes is degraded to provide energy (Figure 71.4). When acetyl-CoA and succinyl-CoA are not formed from the appropriate BCAA, they are not available to enter the tricarboxylic acid cycle to yield energy because the organic acids are excreted in the urine. The initial step in catabolism is reversible transamination, requiring a specific transaminase and the coenzyme pyridoxal phosphate. The second step is irreversible oxidative decarboxylation, which uses the BCKAD complex. This complex is located in the inner mitochondrial membrane and requires the coenzymes TPP, lipoic acid, CoA, and NAD^+.[211-215] Figure 71.4 diagrams this overall reaction. At least six proteins are involved: E1α, E1β, E2, E3, a kinase, and phosphatase.

Isovaleryl-CoA, synthesized from α-ketoisocaproic acid by BCKAD, is catalyzed to 3-methylcrotonyl-CoA by isovaleryl-CoA dehydrogenase (IVD), a mitochondrial enzyme that requires flavoprotein and uses electron transfer factor (ETF) (Figure 71.4). Mutations in the gene for IVD result in isovaleric acidemia (Table 71.2).[216] According to Xu et al.,[217] approximately 14% of α-ketoisocaproate is catabolized to hydroxymethylbutyrate in homogenized rat and human liver by cytosolic α-ketoisocaproate dehydrogenase. No in vivo data to support this pathway have been published, however.

3-Methylcrotonyl-CoA is carboxylated at the 4-carbon by 3-methylcrotonyl-CoA carboxylase to form 3-methylglutaconyl-CoA (Figure 71.4). This enzyme, associated with the inner mitochondrial membrane, contains covalently bound biotin. Mutations in the gene for 3-methylcrotonyl-CoA carboxylase result in 3-methylcrotonylglycinuria (Table 71.2).

3-Methylglutaconyl-CoA hydratase, presumably located in the mitochondria, hydrates 3-methylglutaconyl-CoA to HMG-CoA. HMG-CoA is cleaved by HMG-CoA lyase to acetoacetic acid and acetyl-CoA (Figure 71.4). Mutations in the gene for 3-methylglutaconyl-CoA hydratase result in 3-methylglutaconic aciduria, whereas mutations in the gene for HMG-CoA lyase result in hydroxymethylglutaric aciduria (Table 71.2).

Complete catabolism of ILE through its major degradative pathway results in the synthesis of acetyl-CoA and succinyl-CoA (Figure 71.4). After synthesis of 2-methylbutyryl-CoA by BCKAD, this organic acid is catalyzed by a dehydrogenase to tiglyl-CoA, which is acted on by a hydratase to form 2-methyl-3-hydroxybutyryl-CoA. This compound is used by a dehydrogenase to form 2-methylacetoacetyl-CoA. Mitochondrial acetoacetyl-CoA thiolase (β-ketothiolase) interconverts

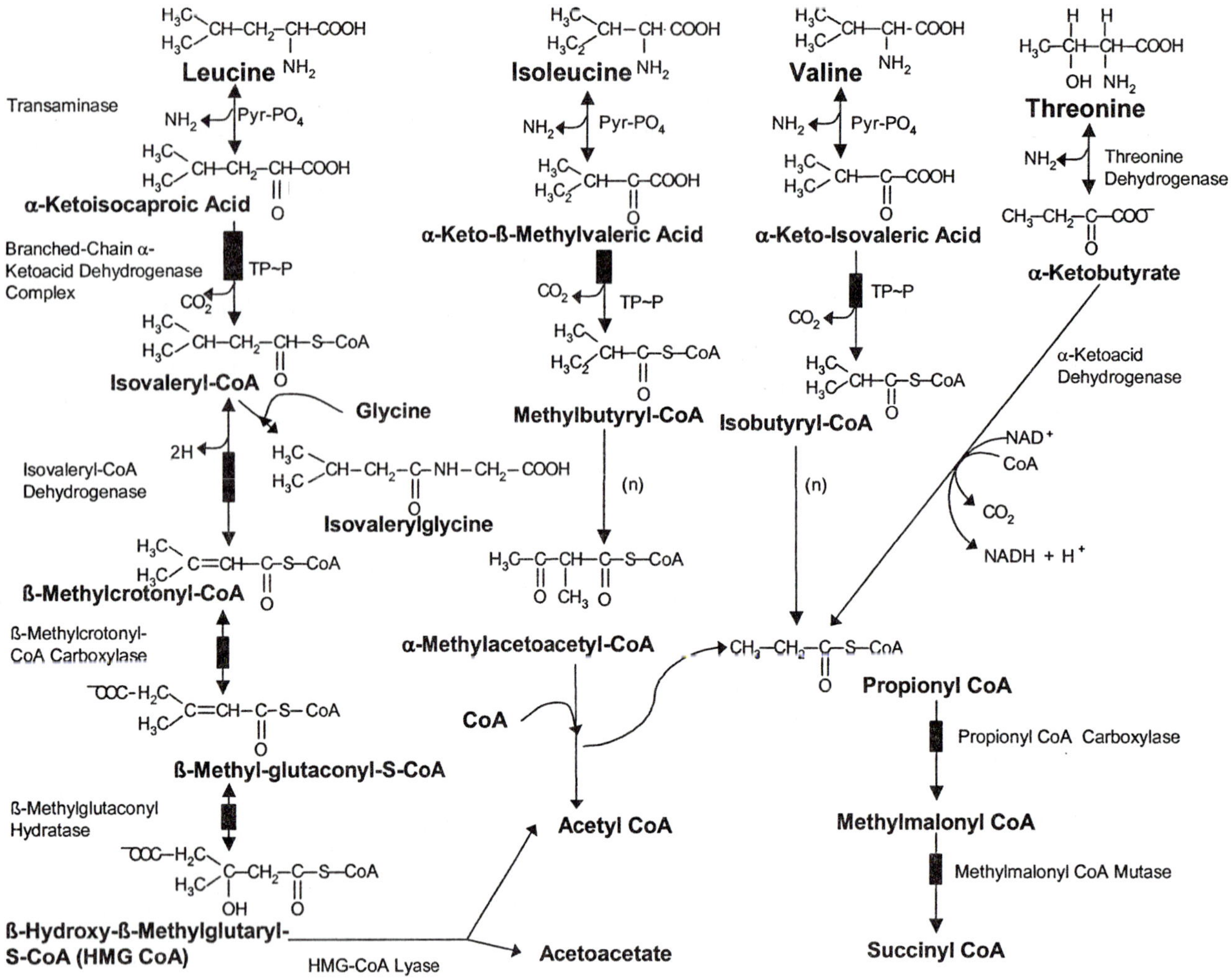

Figure 71.4 Metabolism of branched-chain amino acids and threonine. The *black bars* indicate sites of enzyme defects. *CoA*, coenzyme A; *(n)*, several steps; *NAD*, nicotinamide adenine dinucleotide (NADH is the reduced form).

2-methylacetoacetyl-CoA to acetyl-CoA plus propionyl-CoA. Propionyl-CoA plus HCO_3^-, in the presence of ATP, biotin, and magnesium (Mg^+), is acted on by propionyl-CoA carboxylase to form D-methylmalonyl-CoA. The biotin molecule is responsible for the transfer of the carboxyl group. Isolated deficiency of the propionyl-CoA carboxylase, caused by mutations in the genes encoding its nonidentical α- and β-subunits (Table 71.2), results in propionic acidemia (PPA).[218]

Methylmalonyl-CoA racemase converts D-methylmalonyl-CoA to L-methylmalonyl-CoA, which is isomerized to succinyl-CoA by methylmalonyl-CoA mutase. This dimer contains 1 mol of tightly bound adenosylcobalamin per mole of subunit. Mutations in the gene encoding for methylmalonyl-CoA mutase, as well as those encoding for mitochondrial glutathionylcobalamin reductase and adenosylreductase, result in methylmalonic acidemia (MMA) (Table 71.3 and Figure 71.4). Cytosolic cobalamin reductase/β-ligand transferase deficiency results in homocystinuria and MMA.[218]

After synthesis of isobutyryl-CoA from 2-ketoisovaleric acid by BCKAD (Figure 71.4), it is acted on by isobutyryl-CoA dehydrogenase,[219] followed by a hydratase, a deacylase, and two further dehydrogenases to form propionyl-CoA.[218] A deficiency of isobutyryl-CoA dehydrogenase results in difficulty metabolizing essential VAL.

Two other amino acids, MET and THR, as well as odd-chain fatty acids, thymine, uracil, and the side chain of cholesterol, are catabolized to propionyl-CoA (Figures 71.3 and 71.4). The transamination of MET becomes most prominent when plasma MET exceeds 350 μmol/L and seems to function as a spillover pathway that accounts for only a minor portion of total MET catabolism.[220] Even in the fasting state, with plasma MET in the normal range, some MET is transaminated and decarboxylated to form 3-methylthiopropionate,[221,222] and the α-ketobutyrate formed by the trans-sulfuration pathway is used to synthesize propionyl-CoA (Figure 71.3).

The major pathway of THR catabolism is through oxidation of the hydroxyl group by a specific dehydrogenase to form α-amino-β-ketobutyrate, which subsequently forms acetyl-CoA plus GLY. Another, lesser-used pathway of catabolism is through serine (THR) dehydrogenase and deamination to form α-ketobutyrate,[223] which is acted on by α-ketoacid dehydrogenase to form propionyl-CoA[224] (Figure 71.4). Energy, normally obtained from the oxidation of succinyl-CoA in the tricarboxylic acid cycle, is lost in the urine as propionic acid and methylmalonic acid in PPA and MMA, respectively.

Two essential amino acids contained in food and body protein, LYS and TRP, are precursors of glutaric acid (**Figure 71.5**). The metabolism of TRP does not resemble that of any other metabolite.[223] Its principal degradative pathway occurs in the liver and leads

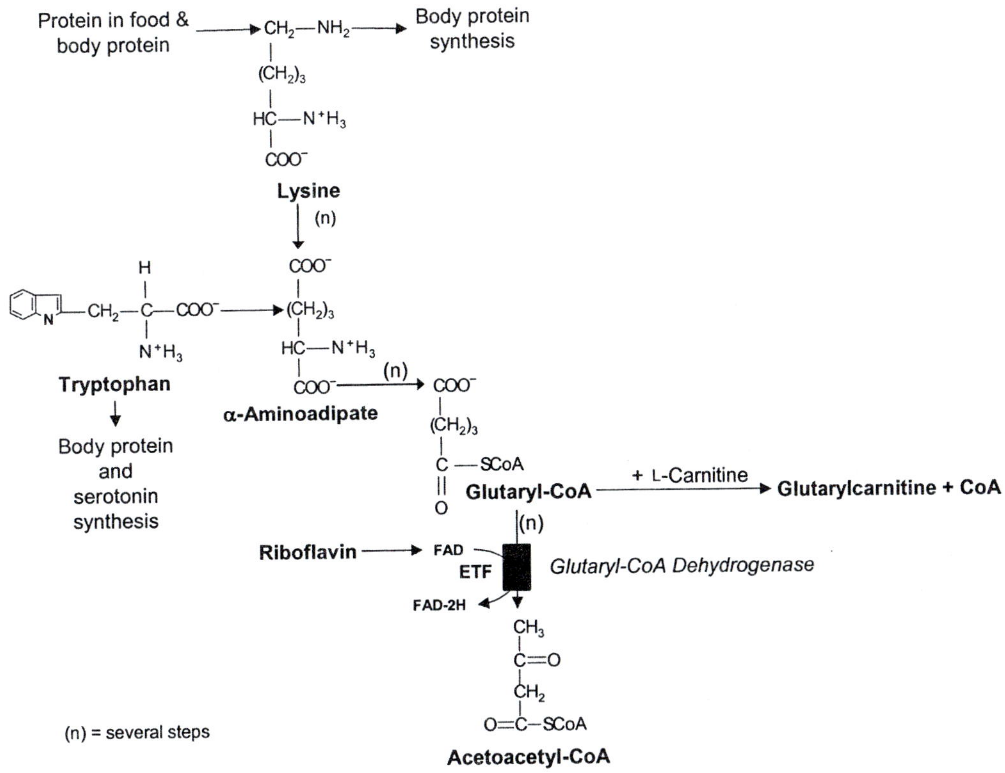

Figure 71.5 Metabolism of lysine and tryptophan. The *black bar* indicates the site of the enzyme defect in glutaric acidemia type I. *L-Carnitine* enhances urinary excretion of glutaric acid. *ETF* is the electron transfer factor that, when impaired, may cause glutaric aciduria and the accumulation of other substrates using ETF. *CoA*, coenzyme A; *FAD*, flavin adenine dinucleotide.

to the formation of nicotinic acid, usually classified as a vitamin, and many. TRP metabolism also leads to serotonin. The initial reaction in the main pathway is oxygenation to form formylkynurenine. The enzyme, TRP oxygenase, contains an iron porphyrin. The formate of N-formylkynurenine is handled as a C_1 fragment by the H_4 folate system. Kynurenine is a branch point; the main pathway continues with a mixed-function oxygenase that includes flavin adenine dinucleotide and uses NAD or NAD phosphate (NADP) as the cosubstrate in the synthesis of 3-hydroxykynurenine. One side branch splits off the bulk of the side chain as alanine through the action of the pyridoxal phosphate enzyme kynureninase, and another side branch removes the α-amino group by transamination (also using pyridoxal phosphate), but the keto group forms a Schiff base with the aromatic amine to form the stable aromatic compound kynurenate. Kynurenate has been found in the brain, where it antagonizes the effects of excitatory amino acids.

3-Hydroxykynurenine is also a branch point. The main pathway now uses the kynureninase that caused a branch earlier to remove alanine but to produce 3-hydroxyanthranilate. The branch is again caused by transamination, which results in a quinoline ring to form xanthurenic acid. A third oxygenase cleaves 3-hydroxyanthranilate to an unstable intermediate, 2-amino-3-carboxymuconic 6-semialdehyde. This unstable intermediate cyclizes to a Schiff base to form quinolinate. An enzyme, picolinic carboxylase, competes with the formation of quinolinate and decarboxylates the intermediate to one that forms picolinate. Most of the decarboxylated material is caught by a dehydrogenase, however, that converts the aldehyde to an acid and leads through α-ketoadipate and glutaryl-CoA to acetoacetyl-CoA (Figure 71.5).

LYS is one of two essential amino acids that have an α-amino group that does not equilibrate with the body pool of amino groups; the other is THR. The amino group of LYS is transferred to other amino acids, but the reverse does not occur. Most degradation of LYS occurs by a unique pathway in which a secondary amine is formed between the ε-amino group and the carbonyl group of α-ketoglutarate. The product, saccharopine, is formed by an enzyme that reduces the hypothetical Schiff base with NADP. Normally, saccharopine does not accumulate but is oxidized by another dehydrogenase that splits the linkage on the other side of the bridge nitrogen. The sum of the reduction-oxidation reactions is effectively a transamination-yielding glutamate and α-aminoadipic semialdehyde. The latter can form a Schiff base, but with the double bond on the side of the nitrogen atom away from the carboxylate. This compound can be oxidized by another dehydrogenase to become α-aminoadipate. A transamination converts this homolog of glutamate to the corresponding α-ketoadipate. In a reaction analogous to the oxidation of α-ketoglutarate to succinyl-CoA, glutaryl-CoA is formed (Figure 71.5). Another oxidation introduces a double bond, forming glutaconyl-CoA, which is decarboxylated to crotonyl-CoA. This unsaturated fatty acyl-CoA is an intermediate in the normal oxidation of fatty acids, and subsequent reactions of the material derived from LYS are those of fatty acid oxidation leading to acetoacetyl-CoA.[223]

Branched-Chain α-Ketoaciduria (Maple Syrup Urine Disease)

Maple Syrup Urine Disease (MSUD) is actually a group of inherited disorders of ILE, LEU, and VAL metabolism. These disorders result from several different gene mutations that impair

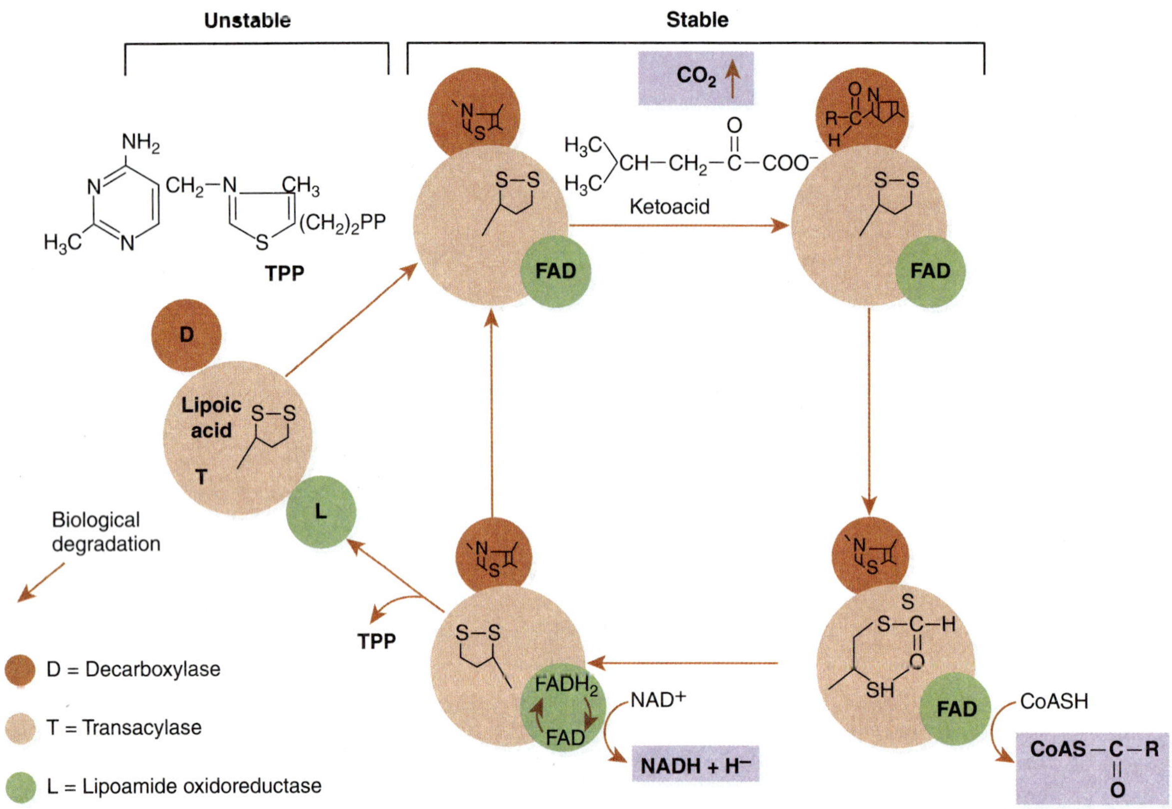

Figure 71.6 Model for stabilization of branched-chain α-ketoacid dehydrogenase by thiamin pyrophosphate (TPP). The multienzyme complex branched-chain α-ketoacid dehydrogenase has a configuration that is more stable to degradation when TPP binding sites on its decarboxylase moiety are occupied. *CoASH*, coenzyme A; *FAD*, flavin adenine dinucleotide (FADH is the reduced form); *NAD*, nicotinamide adenine dinucleotide (NADH is the reduced form).

various components of the multienzyme Branched Chain Alpha-Ketoacid Dehydrogenase Complex (BCKAD) (**Figure 71.6** and Table 71.2). These genes are *E1α*, *E1β*, *E2*, and *E3*.[211] *E1α* is inactivated by a kinase and activated by its phosphatase (Table 71.2). The BCKAD-specific kinase and phosphate have neither been cloned nor do they have chaperonin proteins involved in their mitochondrial assembly process. Almost all mutations in these proteins that produce MSUD are private. The only one common among Mennonites is e in the E1α protein, and it is a substitution of asparagine for TYR at amino acid 393 (Y393N) (Table 71.2). Although most mutant enzymes are immunologically present, one reported patient had absent branched-chain acyltransferase (E2) as a cause of thiamin-resistant MSUD.[213,225,226] An autosomal recessive mode of inheritance was found in all reported cases, a finding supporting a nuclear rather than a mitochondrial origin of these proteins. The cellular mechanisms involved in assembling the products of these nuclear genes into a multienzyme complex in mitochondria are of considerable clinical and fundamental importance and remain unresolved.

Infants with classical MSUD appear normal at birth and are clinically well until after they eat a protein-containing feeding. The most severely impaired enzymes may produce seizures, apnea, and death within 10 days of birth. The disorder is characterized by elevated blood, urine, and cerebrospinal fluid concentrations of the BCKAs, their amino acid precursors, and the pathognomonic alloisoleucine. Progressive neurologic dysfunction and production of fragrant urine with the odor of burnt sugar (caramel) or maple syrup follow. The sweet smell may be evident only in earwax and is easily sensed after otoscopic examination. Neurologic impairment in the newborn is manifested by poor sucking, irregular respiration, rigidity alternating with periods of flaccidity, opisthotonos, progressive loss of the Moro reflex, and seizures.[227]

Several variants, covering a spectrum of impaired mitochondrial BCKAD complexes have been reported. Clinical manifestations are expressed intermittently on protein loading or with febrile illness in patients with partial enzyme activity between 5% and 20% of normal. Patients with 3% to 30% BCKAD complex activity express an intermediate form of the disease. A thiamin-responsive form, with expression similar to the intermediate form, has been described.[211] Whole body LEU-1-^{13}C oxidation to $^{13}CO_2$ may be the best method of ascertaining total body needs, because peripheral cells may not reflect liver and renal BCKAD function.[213,214]

Untreated patients with classic MSUD (<2% BCKAD complex activity) who survive beyond early infancy have retarded physical and mental development.[225,226] Early diagnosis and therapy lead to normal growth and development.[8] If death occurs in the first few days of life, few unique abnormalities are seen in the brain. With prolonged survival, deficient myelination is thought to be caused by enzymes involved in myelin formation, inhibition of amino acid transport, and inhibition by BCKAs of oxidative phosphorylation.[225,226] Jouvet et al.,[228] reported that increased concentrations of BCKAs, particularly α-ketoisocaproic acid, induced apoptosis in glial and neuronal cells in culture and in vivo after intracerebral injection into the developing rat brain.

Screening

Because apnea and death may be the first clinical manifestations of the classic disorder, newborn screening, retrieval, and initiation of therapy are urgent; and all four processes must be completed within the first week of life. Nonselected screening of the newborn population is currently in progress (in some states) using an MS/MS assay for blood LEU concentration.[229] Bedside screening in selected children uses the urinary dinitrophenylhydrazine (DNPH) reaction for branched-chain α-ketoaciduria, although this testing modality is less readily available for patients than in past decades. This reaction can also be used to monitor dietary progress. International newborn screening indicates that the incidence of MSUD is approximately 1 in 185,000.[229] Notably, patients with variant forms, such as intermittent MSUD, can be missed on standard NBS approaches, making it essential for metabolic specialists to consider MSUD on the differential in high suspicion cases with normal NBS results.[230]

Diagnosis

Any infant with a blood LEU concentration > 4mg/dL (305 μmol/L) on the newborn screening test should be evaluated immediately. Most infants with classic disease have >8 mg/dL (610 μmol/L) LEU at 72 hours of age. Diagnosis is confirmed using ion-exchange chromatography to quantitate plasma ILE, LEU, VAL, and alloisoleucine and GC/MS to identify urinary BCKAs. The extent of enzyme impairment should be determined on cultured cells such as dermal fibroblasts to enable future prenatal diagnoses, because prenatal monitoring is available if the cellular phenotype is confirmed in fibroblasts cultured from the patient's skin.[231] Total body LEU oxidation using stable isotopes and the $^{13}CO_2$ breath test is the most reliable diagnostic tool for establishing dietary needs, including thiamine responsivity.[214] However, the most reliable strategy for evaluating MSUD in cases of high suspicion and elevated branched chain amino acids, gene sequencing of the three critical genes associated with MSUD (*BCKDHA, BCKDHB, DBT*), is essential in diagnostics.[227]

Treatment

When TPP saturates its site on E1α, the biologic turnover of BCKAD is decreased (Figure 71.6). Increasing thiamine ingestion increases intracellular TPP, and the TPP binding sites on the decarboxylase (E1) moiety of the BCKAD complex become saturated. When these TPP binding sites are occupied, the multienzyme complex undergoes a conformational change that makes it more resistant to chymotrypsin and heat degradation. The biologic half-life of the enzyme and overall activity are increased when a new equilibrium of enzyme synthesis and degradation is reached. This model has been tested and is supported by clinical, functional, and structural studies (Figure 71.6).[213,215,231,232]

Although hemodialysis with nitrogen-free dialysate or exchange transfusion may be required when diagnosis is delayed, these actions are seldom necessary if screening, retrieval, and diagnosis are completed within 8 to 10 days of life. Because hemodialysis superimposes iatrogenic risk and prolongs a catabolic phase, it is not recommended. BCAA-free orogastric feeding of protein and energy should begin as soon as the diagnosis is made. The objective is to produce anabolism in the infant and thereby prevent accumulation of neurotoxic BCKAs.[233] If orogastric feeding is not acceptable, gastrostomy or a central line for hyperalimentation with dextrose and lipid should be initiated for initial care of classic MSUD during the neonatal period. Except during illness, restricting protein intake to 1.5 g/kg/day may be adequate therapy for patients with 20% or more of enzyme activity.

Long-term therapy for MSUD is dietary. The objective of long-term nutritional therapy in the child with MSUD is to maintain plasma concentrations of BCAAs that will allow maximal development of intellect while supplying adequate energy, protein, and other nutrients for optimal growth. Plasma BCAAs (3 to 4 hours after a meal) should be maintained within the following ranges: ILE, 40 to 90 μmol/L; LEU, 80 to 200 μmol/L; and VAL, and 200 to 425 μmol/L. ILE deficiency results in skin lesions that resemble acrodermatitis enteropathica.[234,235] With deficiency of ILE or VAL, plasma LEU remains elevated. With deletion of the BCAAs, plasma ILE returns to normal first, followed by VAL. When adequate dietary ILE and VAL are then added to maintain normal plasma concentrations, plasma LEU returns to normal in 5 to 10 days.[236]

The objectives of nutritional therapy are met by using a combination of medical foods (Table 71.5 for sources of medical foods and Reference #94 for nutrient composition of other foods) and intact protein. Most patients with MSUD who have detectable BCKAD multienzyme complex by immunoassay respond to oral thiamin administration of 100 to 1,000 mg daily.[213,214,232] Supraphysiologic amounts of oral thiamin should be added for at least a 3-month trial period because they stabilize the enzyme complex (Figure 71.6). Increased residual specific activity of mitochondrial membrane-bound enzymes may require this prolonged period because of the biologic half-life of this subcellular organelle. During this period, decreased sensitivity to dietary BCAAs is usually observed, and more can be added to the diet. Evaluation of total body LEU oxidation before and during thiamin administration gives direct evidence of responsivity.[214] In classic MSUD, thiamine is only an adjunct therapy in certain responsive cases, whereas dietary restriction of LEU with compensatory supplementation of ILE and VAL to reduce blood brain transport of neurotoxic LEU through the transporter LAT1.

Orthoptic liver transplantation resulted in a clear increase in whole-body BCKA oxidation to at least the level of MSUD variants and is globally considered curative. These patients no longer required BCAA restriction, and the risk of metabolic decompensation during catabolism was abolished, although previous neurologic damage is not reversed.[237]

Nutrient Requirements. Data in Table 71.4 outline the suggested amounts of BCAAs, protein, energy, and fluid to offer the infant or child with MSUD. Because the BCAAs are essential, they cannot be deleted from the diet without producing growth failure and death. In planning nutritional therapy of the infant or child with MSUD, a prescription should be written that includes recommended amounts of BCAAs, protein, energy, and fluid for the day. Frequent adjustments in the diet prescription are necessary, and are needed daily during the first few weeks, and biweekly during the first 6 months of life, based on appetite, growth, development, and laboratory analyses of plasma BCAAs and BCKAs. Because LEU residues are more prominent than ILE and VAL in most proteins, supplemental free L-ILE and L-VAL may be necessary in the newborn period and beyond to prevent deficiency of these two essential amino acids. Competition between the free BCAAs at the intestinal cell can cause imbalances in plasma amino acids.[238]

Requirements for ILE, LEU, and VAL vary, depending on age, type, and extent of enzyme defect, protein intake, growth rate, and state of health. Younger infants normally have greater requirements per unit of body weight than older infants.

A rapid decline occurs in requirements for BCAAs between 3 and 6 months of age. Careful monitoring of plasma concentrations and intake of BCAAs is required to prevent excess intake when growth rate declines and to provide adequate intake when growth is accelerated, as in early infancy, during prepuberty and puberty, and during the last half of gestation. Successful pregnancy has been reported in four adults with MSUD.[239-242]

The recommended protein intake for infants with MSUD (Table 71.3) is greater than that for physiologically normal infants and children because the primary protein source consists of free amino acids.[57] The recommended energy intake after the initial acute period is higher than for normal infants and children because the ketoacids of the BCAAs cannot be used for energy synthesis (Table 71.3).[1] During the neonatal acute period, up to 170 kcal/kg/day may be required.[243]

Gropper et al.,[244] reported that 11.0- to 16.6-year-old healthy children ingested 37% of energy as fat, whereas children 7 to 11 years old who had MSUD ingested 29% of energy from fat. Mazer et al.,[245] reported that three patients fed a fat-free medical food ingested 12% to 29% of energy intake from fat, and three patients who ingested medical food with fat received 26% to 46% of energy from fat, with a mean fat intake among the six patients of 28% of energy. The recommended fat intake for persons of this age range is up to 35% of energy.[1] Table 71.3 shows amounts of linoleic and α-linolenic acids to provide adequate intakes.

Because of the organic acidemia that results from poor control of blood BCAA concentrations, poor bone mineralization may result in patients with organic acidemias. Consequently, minerals recommended in Table 71.3 should be consumed. Poor mineral absorption[230] may also cause bone problems.

Medical Foods Free of Branched-Chain Amino Acids. Adequate protein cannot be obtained from ordinary foods without ingesting more BCAAs than are required in classic MSUD. The BCAA content of foods, as a percentage of protein, ranges from approximately 3.5% to 8.5%.[94] Therefore, medical foods are formulated from amino acids free of BCAAs. Sources of these products are given in Table 71.5.

Other Foods. Because BCAAs are essential, intact protein must be prescribed as part of the diet, and other essential nutrients must be added as well. See Reference #100 for the composition of the standard medical foods. For information on estimating the LEU content of foods, see Reference #26. Strauss and Holmes Morton have also advocated for a branched chain free formula supplemented by VAL and ILE to reduce LEU transport into the brain via LAT1, but enriched with other amino acids transported by LAT1 including PHE, TYR, tryptophan, glutamine, alanine, methionine, and threonine, although this is not explicitly discussed by the GMDI in their most recent nutritional guidelines update in 2021.[83,246]

Initiation of Nutritional Therapy. A rapid decline in plasma ILE and VAL can be achieved at the time of diagnosis by feeding medical food free of BCAAs. The plasma LEU will continue to increase over the first 4 days of life, however, even if dietary BCAA restriction is implemented at birth.[236] MSUD is not expressed at birth in most patients, and infants whose screening results are positive are treated at 7 to 14 days of life. In our experience, branched-chain ketoacidosis can be averted by high energy intake with no added BCAAs over a 72-hour period, if this intake is instituted between 8 and 11 days of age. An association exists among the degree of α-ketoisocaproic acid excretion, LEU elevation, and clinical outcome.[247] Laboratory plasma BCAAs should be rapidly available to prevent the predicted deficiency in ILE and VAL. When replacement has begun, these two amino acids may be added as free L-amino acids to increase their ratio to LEU in intact protein. High-energy intake of 140 to 170 kcal/kg of body weight during this period prevents catabolism of body protein. If the osmolality of the medical food mixture permits, protein at 3.0 to 3.5 g/kg should be offered. This regimen lowers the concentrations of BCAAs to nearly normal ranges. If deficiency of either ILE or VAL occurs, plasma LEU will remain elevated as a function of muscle catabolism or decreased protein synthesis.

The LEU required at birth may decrease from 40 to 70 mg/kg/day at 2 years. Medical foods designed for MSUD should be used for therapy. To prevent deficiency of the essential amino acids ILE and VAL, solutions of these two BCAAs may be added back to the medical food mixture. **Table 71.11** for a sample diet for an infant.

Assessment of Nutritional Therapy. Frequency of assessment is dictated by the clinical course and the response of plasma amino acids. Monitoring of therapy should use three combined approaches. Ion-exchange or MS/MS should be used daily to quantitate plasma amino acid concentrations for approximately 3 weeks after birth; these measures help to determine requirements for the individual BCAAs. Urine can be evaluated at bedside for a decline in the DNPH reaction. However, this test has become less accessible for patients in the outpatient setting, leading some to consider Urine Ketones as an alternative screening option. Quantitation of organic acids by GC/MS can detect a decline in BCKAs and the presence of β-lipolysis.

After requirements are established, plasma amino acid concentrations are determined approximately every 2 weeks to ensure that the child has not outgrown the prescription. Samples should be obtained at midday before the noon feeding. Organic acid analysis of urine has been found to be helpful. BCKAs decrease under optimum dietary conditions. When energy or a specific amino acid is overly restricted, evidence of β-lipolysis (acetoacetic acid, β-OH-butyric acid) is found.

After hospital discharge, daily testing of urine with DNPH or ketones by a parent at home screens rapidly for ketoaciduria. As a rule, preventive clinical evaluation of the child for cryptogenic infections before overt ketoacidosis occurs is more effective than attempts to treat the child after a catabolic phase has produced attendant ketoacidosis. If the urine DNPH results are positive, a blood sample should be collected on filter paper for LEU assay, and the urine should be further analyzed by GC/MS to differentiate ketonuria from branched-chain α-ketonuria. With a diet history, a physician's examination, and laboratory analyses, one can usually differentiate among over-restriction, intercurrent infection, or diet under-restriction as a cause for BCKA. Every effort should be made to maintain plasma BCAAs in the normal range. A plasma LEU concentration >600 μmol/L is associated with clinically significant α-ketoacidemia and the appearance of ataxia.[247] One report noted that changes suggesting dysmyelination were found in the white matter, cerebral hemispheres, brainstem, mesencephalon, thalamus, and globus pallidus of patients with chronically elevated BCAAs and BCKAs.[248]

Episodes of infection and trauma can evoke catabolism of tissue protein and increase plasma BCAAs. Clinical improvement is rapid if some BCAAs are administered along with an amino acid mix that provides 150 to 200 kcal/kg/day. Parenteral amino acid solutions free of BCAAs have also caused a rapid

Table 71.11 Sample Diet: Branched-Chain Ketoaciduria (2 Weeks of Age), Weight 3.25 kg

Prescription	Total	Per kg				
Isoleucine (mg)	163	50				
Leucine (mg)	229	70				
Valine (mg)	195	60				
Protein (g)	11.4	3.5				
Energy (kcal)	406	125				
Water to make (mL)	600					
Medical Food	**Amount**	**Isoleucine (mg)**	**Leucine (mg)**	**Valine (mg)**	**Protein (g)**	**Energy (kcal)**
BCAD 1[a,b]	58 g	0	0	0	9.4	290
Enfamil Lipil Powder[a]	20 g	132	229	134	2.0	94
Table sugar	8 g	0	0	0	0.0	30
L-Isoleucine,[c] 10 mg/mL	3.0 mL	30	0	0	0.0	0
L-Valine,[c] 10 mg/mL	6.4 mL	0	0	64	0.0	0
Water to make	600 mL					
Total		162	229	198	11.4	414

[a]Mead Johnson Nutritionals.
[b]Other medical foods that may be used for infants are Ketonex-1 (Abbott Nutrition) and MSUD Analog (Nutricia North America).
[c]L-Isoleucine and L-valine are required to prevent deficiency.

decline in plasma BCAAs during infection, with concomitant clinical improvement.[249] During and after surgery, intravenous glucose administration is important to prevent hypoglycemia.[250] Addition of supplemental alanine and glutamine can be considered in cases of hyperleucinosis with concomitant hypoalaninemia and hypoglutaminemia, as these amino acids may be helpful in reducing LEU transport via LAT1,[240] although this should be discussed with MSUD experts before considering this option.

Outcome of Nutritional Therapy. Patients diagnosed at or before 5 days of age had IQs (97±13 SD) higher than those of physiologically normal siblings or parents.[8] Factors influencing IQ were age at the time of diagnosis, neonatal condition, and long-term metabolic control.

The first attempt to manage an 8-month-old infant with MSUD used pure amino acids totaling approximately 50 g daily, oil and sugar to yield 1,500 kcal/day, and mineral and vitamin mixtures. Plasma BCAAs decreased significantly, and the maple syrup odor disappeared from the urine. During the diet trial, the patient's length and weight increased from the 3rd to 50th percentile.[251]

Nutritional therapy of a neonate with BCKA was subsequently described.[252] Plasma amino acids were measured as a basis for diet changes. Approximate protein intake ranged from 3.5 to 3.0 g/kg/day, and energy intake was approximately 125 kcal/kg/day between 3 and 8 months of age. At birth, the infant was at the 10th percentile for length and weight; during the first year of life, length increased toward the 50th percentile but weight remained at approximately the 10th percentile. Anemia, with a hemoglobin concentration of 85 g/L, was present from 1 to 3 months of age. At 55 weeks, the patient had a developmental quotient of 97, which is in the normal range.

Experience with seven patients with MSUD, three of whom died, was reported.[253] Surviving patients received an amino acid mix free of BCAAs to which corn oil (43% of energy), dextrimaltose (45% of energy), minerals, and vitamins were added. Protein and energy intakes were not reported. Linear growth of one patient by 1 year of age was considerably less than the 10th percentile and three of the four surviving patients had weights lower than the 10th percentile.

Other investigators reported poor growth in treated children with BCKA.[254-259] Few investigators reported protein or energy intake. Intakes of both protein and energy by children with MSUD were lower than those of physiologically normal children of comparable age.[245] Whether failure to thrive is caused by the underlying disease or by iatrogenic dietary effects is unclear. Some investigators reported normal growth when adequate protein and energy were fed.[210,211,250,251] Because of competitive inhibition of mitochondrial oxidative phosphorylation between BCKAs and pyruvate, these patients may need greater energy intake than that recommended by the Institute of Medicine (now the National Academy of Medicine [NAM]).[1]

Selenium deficiency was found in treated patients with MSUD who received selenium-free medical food.[246] Folic acid deficiency was reported in an infant after 4 months on a synthetic diet.[247] Acidosis occurred in an infant treated with an amino acid mix in which several amino acids were provided as the HCl salt.[248]

Termination of Nutritional Therapy. Patients with classic MSUD are unable to terminate the diet, even if they respond to thiamine, except in cases of liver transplantation. Numerous medical complications can arise[252] and the occurrence of death in variants with intermittent MSUD suggests the need for ongoing therapy, even in these relatively stable patients. The BCKAs are acute neurotoxins and probably interfere with oxygen consumption and ATP production in the medullary reticular substance of the brain.[211]

Isovaleric Acidemia

Isovaleric acidemia was first described in 1967, identified by the urinary excretion of IVA.[253] Subsequently, deficiency of IVD was defined in cultured skin fibroblasts.[254] Although a deficiency of electron transfer protein is also reported, mutations in the apoenzyme are specific for isovaleryl-CoA as substrate. IVD deficiency results in a block in the catabolism of LEU at the step after BCKAD complex (Figure 71.4). The IVD gene was assigned to chromosome 15q14-q15[251] (Table 71.2), molecular heterogeneity is defined and several classes are proposed on the basis of the effects of various mutations of the *IVD* gene.[255-257] Class I has normal-sized IVD and missense mutations. Classes II, III, and IV have reduced-size proteins, and class V has no immunologically detectable IVD, whereas the last group may be asymptomatic with mild biochemical abnormalities. IVA, 3-hydroxyisovaleric acid (OHIVA), and the adduct IVG accumulate in body fluids. Gas-liquid chromatography and MS can identify these compounds in body fluids, and the enzyme is quantifiable in cultured dermal fibroblasts.[218,254]

Phenotypic abnormalities result from the toxic accumulation of free IVA. An alternate pathway producing IVG using GLY-*N*-acylase reduces the accumulation of the toxic precursors. Thus, clinical differences in phenotype are caused by the degree of impaired IVD and epigenetic phenomena, such as the degree of detoxification available to this alternative pathway.[218,257,258] Carnityl adducts offer an additional alternate pathway for detoxifying free isovaleric acidemia.[259,260]

Despite molecular advances in understanding the mutations affecting IVD, two disease forms continue to be clinically useful classifications: the acute form and the chronic intermittent form.[218] Patients with the acute form of isovaleric acidemia are generally normal full-term infants. Within the first days of life, poor feeding, tachypnea, vomiting, and a characteristic "sweaty feet" odor (caused by IVA) of the blood and urine are frequently noted. Diarrhea, lethargy, hypotonia, and tremors may also be found. Some patients do not respond to treatment; they may become cyanotic or comatose, and death often results. The exact cause of death is frequently unknown. Severe metabolic acidosis, hyperammonemia, CNS hemorrhage, cardiac arrest, and sepsis are some probable causes. Infants whose conditions are detected early, and who respond to treatment, survive the neonatal period and develop appropriately. If acute neonatal disease is prevented, the condition progresses into the chronic intermittent type of isovaleric acidemia.

In the chronic intermittent form, infants are normal at birth. During late infancy, they may develop episodes of vomiting, acidosis, stupor, and coma. A sweaty-feet odor is usually present, and transient alopecia is occasionally seen. These episodes may begin as early as 2 weeks of age, and the frequency of attacks decrease with age. Urinary tract and upper respiratory infections frequently trigger these episodes, as do excessive intakes of protein and aspirin. Many children affected by the intermittent form prefer fruits and vegetables over meat and milk. The degree of enzyme (IVD) impairment and the capacity of the alternate IVG-producing pathway, as well as intake, probably produce these clinically different presentations.[218]

Several patients with either the acute or the chronic form of isovaleric acidemia have had moderate-to-severe hematologic abnormalities, including leukopenia and thrombocytopenia, with pancytopenia the most common. Isovaleric acidemia inhibits granulopoietic progenitor cell proliferation in bone marrow cultures and may account for the neutropenia often seen in isovaleric acidemia.[261] In one instance, transfusion of packed red cells and platelets prevented further complications. Depressed hemoglobin concentration was also seen in several patients. Transient alopecia is more common with the chronic intermittent form than with the acute form of the disease and may be nutritionally related. Hyperammonemia ($\leq$1200 μmol) has also been reported during neonatal crises.[218]

Screening

Most states now screen for isovaleric acidemia, using MS/MS for elevated C5 in plasma[261] (Figure 71.4). As newborn screening and diagnosis of isovaleric acidemia expands to all states, the incidence of this disorder will become known.

Diagnosis

Because IVG is excreted during both remission and ketotic attacks, measurement of urinary IVG by GC/MS is the best method of diagnosis. Physiologically normal, unaffected 3- to 5-year-old children have no detectable urinary IVG (<2 mg/day). Affected children of the same age excrete 40 to 250 mg/day. During ketotic episodes, urinary 3-OHIVA, 4-OHIVA, and methylsuccinic acid are excreted in large quantities.[218]

Diagnosis has been historically confirmed by measuring the impaired ability of skin fibroblasts cultured from affected individuals to oxidize LEU-2-^{14}C to $^{14}CO_2$.[253,262] A more complicated assay, using mitochondria and 1-^{14}C-IVA, has also been used.[263] High-field proton nuclear magnetic resonance is a promising new technique for the rapid diagnosis of isovaleric acidemia because it can readily detect IVG in a small sample of urine.[264] However, the standard of diagnostic testing lies in gene sequencing the *IVD* gene for biallelic pathogenic variants. Prenatal diagnosis is available by gene sequencing of the *IVD* gene of amniotic fluid and enzyme assay of cultured amniotic fluid cells. A heterozygote for IVD has been detected prenatally.[262]

Treatment

During acute ketotic attacks, parenteral fluid therapy and correction of the metabolic acidosis are indicated as adjuncts to high energy intake and L-carnitine and glycine (GLY) therapy. Serum and urine IVA concentrations are monitored during ketotic attacks. GC/MS analysis is the most accurate means of determining serum and urinary IVA. A special method of GC/MS allows separate quantitation of the two isomers, IVA and 2-methylbutyric acid. Serum IVA ranges from 0.1 to 84 mg/dL,[205] depending on the patient's clinical status. A simple and rapid method of determining 4-OHIVA concentration in the plasma has been devised[263]; however, elevations of this metabolite lag at least 36 hours behind the maximum plasma concentration of IVA, and this delay limits the clinical use of this method. Monitoring urinary IVG provides a good parameter of nutrition therapy. Titration of IVG with free GLY to a stable optimum level is desirable. Excess GLY over free substrate (IVA) may inhibit

IVG production, however. When LEU restriction is optimal and the patient's condition is stable, a dose of approximately 90 mg GLY/kg/day is optimal. During acute disease, a higher intake (300 to 600 mg/kg/day) of GLY may be necessary until infection or dietary LEU indiscretions are removed.[264]

Nutrient Requirements. A low-protein diet of 1.2 to 1.5 g/kg/day in infants improves clinical symptoms, and many patients restrict protein by choice.[205] This amount represents only 60% of the RDA. Protein restriction alone is, therefore, not the best mode of therapy because over-restriction of essential BCAAs (ILE, VAL) is inevitable when LEU is adequately restricted in intact protein and is likely to lead to catabolism,[265] although there is minimal discussion in the literature about supplementing ILE and VAL in the context of BCAA restriction.

LEU restriction and the use of pharmacologic doses of GLY have been reported. In six patients with isovaleric acidemia, GLY therapy resulted in decreased IVA in plasma and urine.[205] Urinary IVG simultaneously increased, often twofold to threefold. Clinical improvement was characterized by increased growth, control of acidosis, and resolution of pancytopenia with GLY supplements and protein restriction over a 2-week period. The loss of 19 isovaleryl and acetyl conjugates in the urine, in conjunction with the use of a protein-restricted and LEU-restricted diet, caused depletion of essential amino acids in patients with IVA and resulted in the need for total protein intake (intact plus protein equivalent given in Table 71.3.[265,266]

GLY used to remove IVA through an alternative pathway is a prototype for nutrition detoxification of accumulated substrates in inborn errors of metabolism.[267,268] The ubiquitous enzyme GLY-*N*-acylase has a broad range of substrates that accumulate in other inborn errors of metabolism and may also be amenable to this approach. The relative amount of GLY required to optimize removal of IVA (or other substrates for the GLY-*N*-acylase reaction) need careful evaluation and will change with the clinical condition of the patient.[267]

Some evidence supports substrate inhibition of the reaction when excess GLY is added under stable conditions. The optimal dose of supplemental GLY was determined for a 9-year-old girl with isovaleric acidemia who was well and who maintained an LEU intake of 54±3.6 mg/kg/day. Supplementation with GLY at less than or more than the range of 50 to 150 mg/kg resulted in a 50% decrease in IVG excretion. Urinary IVA excretion was consistent throughout the study. No β-OHIVA was detected in plasma or urine. These results indicated that (a) the optimal dose of GLY for this patient under these stable clinical and nutrition conditions was 50 to 150 mg/kg; (b) an optimal dose of GLY should be quantitated for specific age, clinical state, degree of enzyme activity, and LEU intake in the treatment of isovaleric acidemia; and (c) GLY supplements >300 mg/kg/day increased plasma and urinary concentrations of GLY, but resulted in decreased IVG excretion, as if this substrate inhibited GLY-*N*-acylase when concentration of its cosubstrate, isovaleryl-CoA, was controlled.[267]

Systemic carnitine deficiency has been demonstrated in several patients with isovaleric acidemia.[269] Although plasma carnitine was low in these patients, the acylcarnitine ester—isovaleryl carnitine—was increased, especially during illness.[269,270] Relative deficiency of muscle carnitine and use of carnitine as an adduct for isovaleric acidemia are two reasons for treating with extra L-carnitine. The relative therapeutic value of L-carnitine was compared with that of GLY in the treatment of isovaleric acidemia in a 4.5-year-old boy.[270] Administration of GLY plus LEU resulted in the excretion of more IVA as IVG than when LEU was administered alone. LEU plus L-carnitine increased isovaleryl carnitine excretion from a pretreatment level of 7 μmol/24 hours to a post-treatment level of 1,470 μmol/24 hours. Large doses of carnitine are needed in the range of 100 to 200 mg/kg/day to accomplish this therapeutic excretion, whereas 100 to 150 mg/kg/day of GLY will suffice. Small doses of carnitine supplements are recommended to prevent deficiency.

A study in 2017 showed that dietary practices in the clinical management of isovaleric acidemia in Europe before the advent of the European E-IMD management guidelines were published was provider-dependent and broad in approach.[271] While the E-IMD guidelines suggest LEU restriction in certain cases, the parameters of how much to restrict were not clearly defined.[272]

Medical Foods Free of Leucine. Four medical foods free of LEU have been designed specifically for the nutritional therapy of patients with isovaleric acidemia and other disorders of LEU catabolism (Table 71.5).

Outcome of Nutritional Therapy. A male infant with isovaleric acidemia, who was treated from the neonatal period with a medical food designed for MSUD with added L-ILE and L-VAL and whole protein to supply essential restricted LEU, had normal growth and development. Height and weight were between the 25th and 50th percentiles. HC was at the 50th percentile. On average, the diet supplied 2.5 to 3.0 g protein/kg/day and 100 mg LEU/kg/day. L-Carnitine and GLY were not a part of the diet regimen.[273]

Growth of a male infant diagnosed prenatally with isovaleric acidemia was reported. At birth, the infant was breastfed ad libitum, and was given 250 mg GLY/kg/day. In spite of the low protein intake from human milk and the GLY supplement, the patient became acidotic and began to vomit and hyperventilate at 3 days of age. Breastfeeding was discontinued, and an LEU-free diet providing 125 kcal/kg, supplemented with 380 mg GLY/kg/day, was begun. His clinical status improved rapidly. Dietary LEU at 45 mg/kg/day with GLY at 250 mg/kg/day and protein at 2.0 g/kg was introduced at 5 days of age. At 2 years of age, the patient was developmentally normal and was higher than the 95th percentile for height and weight. His diet at 2 years of age supplied the following per kilogram: 46 mg LEU, 1.7 g protein, and 72 kcal. Only one hospitalization for vomiting and dehydration was required during the 2-year period.[267]

Outcomes in nine patients with isovaleric acidemia who were managed by protein restriction (1.5 to 2.0 g/kg in infancy and 0.8 to 1.5 g/kg thereafter) and 250 mg GLY/kg/day were reported.[272] Because all had secondary carnitine deficiency (total serum carnitine, 19±3 μmol/L), diets of four of the children were supplemented with 50 mg/kg/day L-carnitine and, in these patients, serum carnitine returned to normal (51±5 μmol/L). Actual height, weight, and HC of patients were not reported, although growth velocities were stated to be normal after diet initiation. Developmental quotients or IQ scores of the five children in whom diet was initiated during the neonatal period ranged from 49 to 115.

Food refusal was reported in a patient with isovaleric acidemia.[274] Both physiologic and behavioral components to feeding problems were reported. The physiologic component involved altered serotonin metabolism. Any factor, such as hyperammonemia or a high-carbohydrate, low-protein diet, that stimulates the transport of TRP, a precursor of serotonin, into the brain could lead to anorexia. A low-TRP diet was suggested as one alternative to the treatment of anorexia.

A study in 2016 demonstrated that the dietary strategies in IVA management are not associated with severe hematologic abnormalities or nutritional insufficiencies. However, there was significant reduction in plasma PUFA, especially in omega-3 and omega-6 fatty acids, along with an elevation in omega-9 fatty acids. These findings were not suggestive of complete essential fatty acid deficiency but, rather, of isolated or combined omega-3 and omega-6 fatty acid depletion.[275]

Of 11 reported French patients with isovaleric acidemia, eight were alive after the neonatal period. LEU restriction and GLY supplementation were used to manage these patients. Of the eight surviving patients, six had normal development.[276]

In 2020, Mutze et al., demonstrated that NBS for isovaleric acidemia reduced mortality in classic I isovaleric acidemia patients, but did not consistently ameliorate the risk of severe neonatal metabolic decompensation; episodes that are strongly correlated with less favorable neurocognitive outcome. However, patients with mild isovaleric acidemia had strong clinical outcomes, regardless of metabolic maintenance therapy, raising questions about the benefits of identifying mild isovaleric acidemia patients via NBS.[261]

Other Organic Acidemias

3-Methylcrotonylglycinuria

3-Methylcrotonylglycinuria, caused by deficiency of 3-methylcrotonyl-CoA carboxylase (Figure 71.4), has a wide spectrum of clinical symptoms. Some neonates may present with severe hypotonia and seizures, whereas other infants may present at a later age with hypotonia and without severe metabolic acidosis. Some infants may be only mildly delayed, with hypoglycemia, failure to thrive, respiratory failure, and hemiparesis during febrile illness. Some patients may not present until adulthood, with fatigue, weakness, myopathy, and fatty liver with elevated liver enzymes. Presymptomatic detection using abnormal acylcarnitine profiles by MS/MS is basedon elevations in C5OH. Some patients are asymptomatic.[218] Treatment with L-carnitine, prevention of fasting, and essential LEU restriction usually results in normal development (Table 71.3). The University of North Carolina developed guidelines for nutritional therapy, based on biochemical phenotypes and experience with previously undiagnosed patients.[26] However, tandem MS newborn screening has revealed an unexpectedly high prevalence of neonates with this condition, the vast majority presenting asymptomatically. Reports of prevalence are as low as 1:40,000 newborns in Australia and Germany and as high as 1:2,400 in the Faroe Islands.[277-279]

3-Methylglutaconic Aciduria

3-methylglutaconic aciduria is known to have at least nine types. Type I, characterized by a deficiency of 3-methylglutaconyl-CoA hydratase (Figure 71.4), has diverse and nonspecific clinical symptoms. The major urinary metabolites are 3-methylglutaconic and 3-hydroxyisovaleric acids. L-Carnitine supplementation and essential LEU restriction may be beneficial (Table 71.4).[218] The other eight types of this disorder do not respond to diet intervention.

3-Hydroxy-3-Methylglutaric Aciduria (3-Hydroxy- 3-Methylglutaryl–Coenzyme A Lyase Deficiency)

One-third of patients with HMG-CoA lyase deficiency (Figure 71.4) present in the neonatal period, and two-thirds are between 3 and 11 months of age, with severe hypoglycemia and metabolic acidosis (but with little or no ketosis), hyperammonemia, vomiting, and hypotonia, which may progress to coma and death. The symptoms resemble those of Reye syndrome. Treatment by restriction of essential LEU and fat, avoidance of fasting, and L-carnitine supplementation generally leads to normal development (Tables 71.3 and 71.4). HMG-CoA lyase deficiency is a disorder of BCAA (LEU) metabolism; its diagnosis is from the major abnormal metabolites in urine, 3-hydroxy-3-methylglutaric, 3-methylglutaconic, and 3-hydroxyisovaleric acids. This disorder is also one of ketone body metabolism.[218] LEU and fat restriction with L-carnitine supplementation following neonatal diagnosis, with ongoing prevention of hypoglycemia and metabolic acidosis, have resulted in normal growth and development.[26]

2-Methylbutyrylglycinuria (MBG)

2-Methylbutyrylglycinuria (MBG) results from a defect in short-branched-chain acyl-CoA dehydrogenase, also called 2-methylbutyryl-CoA dehydrogenase.[21] 2-Methylbutyryl-CoA dehydrogenase transfers electrons from 2-methylbutyryl-CoA formed in ILE metabolism to form tiglyl-CoA.[15] This enzyme deficiency is believed by some investigators to be benign,[30] although it is part of newborn screening in the United States (Table 71.1).

2-Methyl-3-Hydroxybutyryl–Coenzyme A Dehydrogenase Deficiency

2-Methyl-3-hydroxybutyryl-CoA in ILE metabolism and 2-methyl-branched-chain fatty acids are acted on by 2-methyl-3-hydroxybutyryl-CoA dehydrogenase to form 2-methylacetoacetyl-CoA.[280] This X-linked recessive disorder affects the activity of 2-methyl-3-hydroxybutyryl-CoA dehydrogenase (linked to mutations in the *HSD17BD* gene) in 2-methyl-3-hydroxybutyric acidemia (2M3HBA), which may be found on newborn screening (Table 71.1).[281] An ILE-restricted diet has been tried, with some success.[26]

Isobutyryl-CoA Dehydrogenase

Isobutyryl-CoA dehydrogenase deficiency results from a mutation in the *ACAD8* gene and affects the metabolism of VAL.[221] Whether treatment with a VAL-restricted diet is necessary is unknown.[282] However, most cases identified through newborn screening have been asymptomatic, suggesting that, in most cases, it is a biochemical variant not requiring therapy.

Mitochondrial Acetoacetyl-CoA Thiolase Deficiency

Mitochondrial acetoacetyl-CoA thiolase deficiency, also called β-ketothiolase deficiency, is an autosomal recessive disorder of both ILE and ketone body metabolism converting

2-methylacetoacetyl-CoA and acetoacetyl-CoA to acetyl-CoA and acetyl-CoA to acetoacetyl-CoA (Figure 71.4 and Table 71.1). Prevention of fasting, L-carnitine administration, and restriction of ILE result in a favorable outcome.[26]

Enoyl CoA Hydratase (ECHS1) and 3-Hydroxyisobutyryl-CoA Hydrolase (HIBCH)

Enoyl CoA hydratase (ECHS1) and 3-hydroxyisobutyryl-CoA hydrolase (HIBCH) deficiencies are rare disorders of auxillary enzymes in the valine catabolism pathway. To date, there are about 70 known cases in the world. These disorders are associated with a Leigh Syndrome Like phenotype with progressive neurologic sequelae. Recent studies have suggested that valine restriction and dietary fat restriction improve the neurologic outcomes across both of analogous conditions.[283-285]

Other disorders resulting in organic acidemia that are now screened for during the newborn period in some states using MS/MS are described in **Table 71.12**.[29,211,218,220,286-290] This table includes the trivial name of the disorder, the enzyme defect, presenting clinical and laboratory symptoms, diagnostic studies, nutritional therapy during diagnostic workup and illness, long-term therapy, and outcomes. Data in Table 71.3 describe ranges of recommended intakes for major nutrients. Recommendations for mineral and vitamin intakes are in Table 71.4.

Propionic Acidemia

Isolated deficiency of propionyl-CoA carboxylase (Figures 71.3 and 71.4) results in the accumulation of propionate in blood and of 3-hydroxypropionate, methylcitrate, tiglyglycine, and unusual ketone bodies in urine.[220] Two complementation groups, *pccA* and *pccBC*, have been defined among patients deficient in propionyl-CoA carboxylase. These groups correspond to mutations affecting genes coding for the α-subunits and the β-subunits, respectively, of the carboxylase apoprotein. Clinically, the disorder is characterized by severe metabolic ketoacidosis, which often appears in the neonatal period and requires the following: vigorous alkali therapy; restriction of essential ILE, MET, THR, VAL, and odd-chain fatty acids; prevention of fasting and weight loss[291]; L-carnitine supplementation; and adequate energy for growth to cover that lost in the urine as propionylcarnitine (Tables 71.3 and 71.4). Sources of ILE, MET, THR and VAL, as well as other nutrients, may be found in Reference #94. Sources of medical foods available for PPA and MMA are given in Table 71.5. Over-restriction of ILE in the diet may result in acrodermatitis enteropathica–like cutaneous lesions in patients with either PPA or MMA.[292] Sample diets for an infant with PPA or MMA are given in **Table 71.13**. Vitamin D deficiency is common within this disorder and should be monitored accordingly.[291] Oral antibiotic therapy to reduce gut propionate production may also prove useful. Carbaglu can also be useful in helping regulated hyperammonemia associated with excessive propionate production. Screening of almost 1 million newborns for PPA or MMA with MS/MS resulted in finding one of 65,000 infants affected.[293] Orthotopic liver transplant can help ameliorate metabolic decompensations and improve cardiac dysfunction associated with the disorder but does not eliminate the risk of metabolic strokes or metabolic decompensation, raising the need to still continue some form of medical foods.[294]

Foods containing odd-chain fatty acids that must be deleted from the diets of patients with PPA or MMA include the following: some fish oils (menhaden, mullet, tuna); chicken fat; olive oil; lard[295]; the fat of ruminants[296], including butterfat and cream; and any food to which propionate is added to retard spoilage. Dupont and Mathias[297] reported γ-oxidation of ^{14}C-linoleate into methylmalonate to be 20 times greater than from ^{14}C-palmitate. Lower plasma concentration of docosahexaenoic acid (DHA, 22:6n-3) has been reported in treated patients with inherited metabolic disorders, than in controls,[298] but values were not significantly different. The small difference found may result from the lower fat intake by children with the MMA mutation. Table 71.3 for the amounts of fat and essential fatty acids to feed. Minerals and vitamins ingested should meet the recommendations in Table 71.4, because of poor absorption of mineral found with elemental diets[65] and because of loss of bone mineral with metabolic acidosis.

North et al.,[299] reported that no deaths occurred in their cohort of patients with PA. Growth and nutrition status were improved with the use of medical foods and gastrostomy tube feedings. Hypotonia and cognitive delay were still present in all children, however. Pancreatitis, acute or chronic, has been reported as a complication of all organic acidemias.[299] Suppression of granulopoietic progenitor cell proliferation[273,300] and pancytopenia[301,302] have been reported in patients with organic acidemias. A successful pregnancy was reported in a woman with PA.[241]

Thomas et al.,[303] suggested that the energy needs of patients with PA were lower than those of physiologically normal children. Feillet et al.,[304] who used intact protein restriction without the use of medical foods (free amino acids), found resting energy expenditure to be reduced by approximately 20% in patients with MMA or PA. De Koning et al.,[305] found higher resting energy expenditure in patients fed adequate medical food than in those fed none, perhaps because of higher lean body mass. Yannicelli et al.,[306] found normal growth in seven infants and children with PPA or MMA when they were fed energy at 98% of RDA and protein at 115% of the recommendations of the Food and Agriculture Organization, World Health Organization, and United Nations University.

Methylmalonic Acidemia

Neonatal or infantile metabolic acidosis is the clinical hallmark of isolated methylmalonyl-CoA mutase deficiency (Figures 71.3 and 71.4). Cells from some apomutase-deficient children have no functional mutase (designated *mut*^{0}); cells from others contain a structurally altered mutase with reduced affinity for adenosylcobalamin and with reduced stability (*mut*$^-$).[220] Such children have MMA that does not respond to cobalamin supplementation but can be treated with the same diet and medical therapy as PPA. Mild mutase deficiency may be unmasked in infancy by prolonged breastfeeding by a vegan mother.[307] Differential diagnosis may include vitamin B_{12} deficiency caused by many environmental factors (e.g., vegan diets) or transport defects (e.g., intrinsic factor deficiency, ileitis, rare transporter defects). In these situations, MMA and aciduria are accompanied by homocystinemia and homocystinuria caused by CH_3-B_{12} as well as adenosylcobalamin deficiency (Figure 71.3). Progressive renal insufficiency occurs in patients with poorly controlled

Table 71.12 Other Organic Acidemias for Which Newborn Screening Occurs by Tandem Mass Spectrometry in Most States

Disorder (Enzyme Defect Reference)	Presenting Symptoms		Diagnostic Studies	Long-Term Therapy	Nutrition Assessment	Outcome
	Clinical	Laboratory				
3-Methylcrotonyl-CoA (3-methylcrotonyl-CoA carboxylase [MCC])[211]	Variable neonatal onset: irritability drowsiness, difficulty feeding, vomiting, rapid respirations, apnea, seizures, spasms, coma, death, if untreated Later onset: hypotonia; some patients have no clinical symptoms	Metabolic acidosis Hypoglycemia Later onset: mild metabolic acidosis	Identification of blood and urinary 3-methyl-crotonylglycine Evaluation of MCC activity in cells, MCC mutation analysis	Restrict LEU Administer L-carnitine and GLY Prevent fasting Consider not treating if not concerned about symptomatic disease. Table 71.3 for recommended nutrient intakes	Plasma LEU, L-carnitine, GLY, IVG, serum transthyretin, ferritin Bone radiographs of lumbar vertebrae Urinary IVG, 3-hydroxyisovaleric acid Cation/anion gap Dietary intakes of LEU, protein, energy, minerals, and vitamins Growth	Normal growth and development with neonatal diagnosis and excellent therapy with LEU restriction, L-carnitine, and GLY
3-Methylglutaconic aciduria type 1 (3-methylglutaconyl-CoA hydratase)[218]	Coma, psychomotor retardation and/or speech retardation may be present	Metabolic acidosis on fasting may be present Hypoglycemia on fasting may be present	Identification of 3-methylglutaconic acid Assay of hydratase activity in fibroblasts	Restrict LEU Administer L-carnitine and GLY Prevent fasting Table 71.3 for recommended nutrient intakes	Plasma LEU, L-carnitine, GLY, IVG, serum transthyretin, and ferritin Bone radiographs of lumbar vertebrae Urinary IVG and β-hydroxyisovaleric acid Cation/anion gap Dietary intakes of LEU, protein, energy, minerals, and vitamins Growth	Normal growth and development with neonatal diagnosis and excellent therapy with LEU restriction, L-carnitine, and GLY
3-Hydroxy-3-methylglutaric aciduria (3-hydroxy-3-methylglutaryl-CoA lyase [3-HMG-CoA lyase])[218]	Babies appear normal at birth, but symptoms appear during the first week of life, infancy, or childhood Vomiting, hypotonia, lethargy, seizures, and/or coma in 10% of cases, death Symptoms similar to Reye syndrome	Metabolic acidosis with very low blood pH Severe hypoglycemia Hypoketonemia Hypoketonuria Hyperammonemia and/or elevated transaminases may be present	Quantification of urinary 3-hydroxy-3-methylglutaric acid, 3-methylglutaconate, 3-hydroxyisovalerate, and plasma 3-methylglutaryl-carnitine Analysis of enzyme activity in fibroblasts and mutation analysis	Restrict LEU and fat Administer L-carnitine Prevent fasting Table 71.3 for recommended nutrient intakes	Plasma LEU, carnitine (free) Serum transthyretin and ferritin Urinary 3-methylglutaric acid Dietary intake of LEU, protein, fat, energy, minerals, and vitamins. Growth	Normal growth and development with early and lifelong therapy

Disorder (Enzyme Defect Reference)	Presenting Symptoms		Diagnostic Studies	Long-Term Therapy	Nutrition Assessment	Outcome
	Clinical	Laboratory				
Mitochondrial acetoacetyl-CoA thiolase deficiency (β-ketothiolase)[218]	Intermittent acidosis, ketosis, vomiting, diarrhea, hematemesis, and melena Failure to thrive Severe headaches Ataxia Coma	Hyperglycinemia		Prevention of fasting L-Carnitine administration Moderate restriction of ILE L-LEU and L-VAL administration to maintain normal plasma concentrations Energy to maintain anabolism	Plasma ILE, VAL Urine ketoacids by Ketostix Plasma carnitine Serum transthyretin Plasma albumin Plasma ferritin Growth Nutrient intake	Normal growth and development with early and continued therapy
2-Methyl-3-hydroxybutyryl-CoA dehydrogenase deficiency (2-methyl-3-hydroxybutyric aciduria)[218]	Progressive loss of mental and motor skills Spastic dysplegia Hypotonia Cortical blindness Myotonic seizures Retinopathy Ataxia Dystonia Choreoathetosis Cardiomyopathy	Lactic acidemia Urinary excretion of 2-methyl-3-hydroxybutyric acid and tiglyglycine	Elevated blood C5OH and C5:1 Enzyme assay Molecular analysis	Same as above	Plasma ILE, LEU, and VAL Urine ketoacids by Ketostix Plasma carnitine Serum transthyretin Plasma albumin and ferritin Growth Nutrient intake	ILE restriction resulted in decreased abnormal urinary metabolites and stabilization of neurologic symptoms
Isobutyryl-CoA dehydrogenase deficiency[218]	Failure to thrive Dilated cardiomyopathy Muscular hypotonia	C4 increased in blood Increased C4/C2 ratio Carnitine deficiency	Ethylmalonic acid in urine Mutations in *ACAD8* gene Enzyme assay Isobutyryl GLY in urine Elevated urinary C4	? VAL restriction beginning in early infancy L-Carnitine administration L-ILE and L-VAL administration to maintain normal plasma concentrations Adequate energy to support anabolism	Plasma BCAA concentrations if treated Plasma carnitine Growth Serum transthyretin Plasma albumin, ferritin Nutrient intake	May develop normally without therapy

(continues)

Table 71.12 Other Organic Acidemias for Which Newborn Screening Occurs by Tandem Mass Spectrometry in Most States *(continued)*

Disorder (Enzyme Defect Reference)	Presenting Symptoms		Diagnostic Studies	Long-Term Therapy	Nutrition Assessment	Outcome
	Clinical	Laboratory				
Propionic acidemia (propionyl-CoA carboxylase [PCC])[220]	Refusal to feed Vomiting Dehydration Lethargy Hypotonia Seizures Coma Intellectual disability Osteoporosis Hepatomegaly may be present	Severe metabolic acidosis EEG abnormalities Hyperammonemia Hyperglycinemia Neutropenia, anemia, and thrombocytopenia may be present	Measurement of plasma propionate, urinary 3-hydroxypropionate, methylcitrate, and tiglylglycine Analysis of PCC activity and mutations in cultured fibroblasts.	Restrict ILE, MET, THR, VAL, linoleic acid, and odd-chain fatty acids Prevent fasting Administer L-carnitine Table 71.3 for recommended nutrient intakes Avoid breads with calcium or sodium propionate Avoid butter, cream, olive oil, chicken fat, and menhaden oil	Plasma ILE, MET, THR, VAL, and GLY Blood ammonia Cation/anion gap Urinary metabolites of propionate, serum transthyretin, and ferritin Bone radiographs of lumbar vertebrae Dietary intake of ILE, MET, THR, VAL, protein, energy, minerals, and vitamins Growth	Normal growth with early therapy Neurologic outcome improved
Methylmalonic aciduria (methylmalonyl-CoA mutase or methylmalonic acidemia [MMA])[220]	Hypertonia Areflexia, lethargy Failure to thrive Recurrent vomiting Dehydration Severe intellectual disability or death if untreated Coma, hepatomegaly, muscular hypotonia, and respiratory distress may be present	Acidosis Ketosis Ketonuria Hyperammonemia Hyperglycinemia Hypoglycemia Hyperuricemia Neutropenia Thrombocytopenia Pancytopenia	Urinary methylmalonic acid and/or MMA in blood by MS/MS Fibroblast studies of enzymes. Measure total plasma homocysteine, free methionine, and homocystine.	Restrict ILE, MET, THR, VAL, linoleic acid, and odd-chain fatty acids Prevent fasting Administer L-carnitine Table 71.3 for recommended nutrient intakes Energy to enhance anabolism	Plasma ILE, MET, THR, VAL, and GLY Blood ammonia Cation/anion gap Urinary metabolites of propionate and methylmalonate Serum transthyretin and ferritin Bone radiographs of lumbar vertebrae Dietary intake of ILE, MET, THR, VAL, protein, energy, minerals, and vitamins Growth	Normal growth with early therapy Neurologic outcome improved

Disorder (Enzyme Defect Reference)	Presenting Symptoms		Diagnostic Studies	Long-Term Therapy	Nutrition Assessment	Outcome
	Clinical	Laboratory				
Methylmalonic acidemia (cobalamin reductase adenosyltransferase)[220]	Lethargy Failure to thrive Recurrent vomiting Dehydration Respiratory distress Muscular hypotonia, Intellectual disability, hepatomegaly, and coma may be present	Metabolic acidosis Ketonemia/ketonuria Hyperammonemia, leukcpenia, thrombocytopenia, and hypoglycemia may be present	Urinary methylmalonic acid and/or MMA in blood by MS/MS Fibroblast studies of enzymes Measure total plasma homocysteine, free methionine, and homocystine	1–2 mg OH cobalamin daily Modest protein restriction	Plasma ILE, MET, THR, VAL, and GLY Blood ammonia Cation/anion gap Urinary metabolites of propionate and methylmalonate Serum transthyretin and ferritin Bone radiographs of lumbar vertebrae Dietary intake of ILE, MET, THR, VAL, protein, energy, minerals, and vitamins Growth	Normal growth with early therapy Neurologic outcome improved
Cobalamin A, B[185]		Metabolic acidosis Ketonemic/ ketoaciduria Hyperammonemia Hyperglycinuria Leukopenia Anemia Thrombocytopenia	C3, C3/C2 screening Methylmalonyl-CoA mutase activity Methionine synthase activity Plasma MMA	Hydroxycobalamin 1–2 mg IM/day Restriction of ILE, MET, VAL, and THR if not controlled by OHB_{12} Betaine and folate administration	Nutrient intake Growth Plasma B_{12}, BCAAs, and betaine Hemograms	
Cobalamin C, D[185]	Failure to thrive Poor feeding Lethargy Developintellectual disability Seizures Feeding problems Hypotonia Microcephaly Nystagmus Hydrocephalus Heart defects	Hematologic abncrmalities Acidosis ±Acidosis Pigmentary retinopathy Decreased visual acuity	C3, C3/C2 screening Methylmalonyl-CoA mutase activity Methionine synthase activity Plasma MMA	Hydroxycobalamin up to 20 mg/day DO NOT GIVE MEDICAL FOODS for Cobalamin C! Consider restriction of ILE, MET, THR and VAL if not controlled by OHB_{12} Betaine and folate administration	Nutrient intake Growth Plasma B_{12}, BCAAs, betaine Hemograms	

(continues)

Table 71.12 Other Organic Acidemias for Which Newborn Screening Occurs by Tandem Mass Spectrometry in Most States *(continued)*

Disorder (Enzyme Defect Reference)	Presenting Symptoms		Diagnostic Studies	Long-Term Therapy	Nutrition Assessment	Outcome
	Clinical	Laboratory				
Malonyl-CoA decarboxylase (malonyl-CoA decarboxylase deficiency)	Developmental delay Hypotonia Abnormal genitalia Short stature	Acidosis Seizures Hypoglycemia Cardiomyopathy Brain abnormalities[29]	C3 on newborn screening	L-Carnitine administration Fat restriction Essential fatty acids Medium-chain triglycerides Increased carbohydrate[290]	Nutrient intake Growth Erythorocyte fatty acids	Improved cardiomyopathy Improved growth Normal plasma carnitine
Multiple carboxylase deficiency[288]	Feeding and breathing difficulties Hypotonia Seizures Lethargy ±Developmental delay ±Skin rash, alopecia	Metabolic acidosis Organic acidurias Mild hyperammonemia	C5OH, ±C3 on newborn screening	Administer 8–10 mg biotin daily		
Biotinidase deficiency[288]	Seizures Hypotonia Ataxia Developmental delay Vision problems Hearing loss Alopecia Skin rash Candidiasis[279]		C5OH, C51 on newborn screening	5–10 mg oral biotin daily		With newborn screening diagnosis and continuous therapy, patients should remain asymptomatic

BCAA, branched-chain amino acid; *CoA*, coenzyme A; *EEG*, electroencephalogram; *GLY*, glycine; *ILE*, isoleucine; *IM*, intramuscular; *IVG*, isovalerylglycine; *LEU*, leucine; *MET*, methionine; *MS/MS*, tandem mass spectrometry; *THR*, threonine; *VAL*, valine.

Table 71.13 Sample Diet: Propionic and Methylmalonic Acidemia (2 Weeks of Age), Weight 3.25 kg

Prescription	Total	Per kg
Isoleucine (mg)	325	100
Methionine (mg)	162	50
Threonine (mg)	260	80
Valine (mg)	276	85
L-Carnitine (mg)	325	325
Protein (g)	11.4	3.5
Energy (kcal)	455	140
Water to make (mL)	600	

Medical Food	Amount	Isoleucine (mg)	Methionine (mg)	Threonine (mg)	Valine (mg)	L-Carnitine (mg)	Protein (g)	Energy (kcal)
Propimex-1[a,b]	40 g	48	0	40	0	360	6.5	192
Similac Advance Infant Formula with Iron Powder[a]	45 g	259	124	263	288	2	4.9	234
L-Isoleucine,[c] 10 mg/mL	2 mL	20	0	0	0	0	0.0	0
Polycose liquid[a]	12 mL	0	0	0	0	0	0.0	24
Water to make	600 mL							
Total		327	124	303	288	362	11.4	450

[a]Abbott Nutrition.
[b]Other medical foods that may be used for infants are OA 1 (Mead Johnson Nutritionals) and XMTVI Analog (Nutricia North America).
[c]L-Isoleucine is required to prevent deficiency.

conditions and true MMA.[308] Combined liver-kidney transplantation has proven beneficial in such patients with end-stage renal failure.[309]

Pregnancy in a woman with late-onset MMA was reported to have a normal outcome; at 3 years of age, this woman's child was developing normally.[310] Early-onset MMA often results in early death in spite of aggressive therapy with diet, L-carnitine, carbaglu, gastrostomy feeding, and metronidazole.[311] Table 71.3 provides suggested nutrient intakes by age, and Table 71.5 lists sources of medical foods. While dietary composition for patients with isolated methylmalonic acidemia syndromes secondary to deficiencies in Methylmalonic Mutase, Cobalamin A and B are analogous to dietary composition for patients with propionic acidemia[312] patients with cobalamin C deficiency should not be started on medical foods with low protein due to potential toxic complications of depleted methionine, which is associated with decreased head circumference, worse neurodevelopmental outcomes, and shorter stature.[313]

Glutaric Acidemia Type I

Deficiency of glutaryl-CoA dehydrogenase (GCD) (Figure 71.5) causes glutaric acidemia type I (GA-I), a disorder characterized by macrocephaly at birth and dystonia and dyskinesia appearing during the first years of life. Chemically, it is identified by excretion of glutaric and 3-hydroxyglutaric acids in urine and, pathologically, by neuronal degeneration of the caudate and putamen. Computed tomography and magnetic resonance imaging scans often show frontotemporal atrophy or arachnoid cysts before the onset of symptoms. GCD deficiency is inherited as an autosomal recessive trait. More than 60 pathogenic mutations have been identified in the *GCDH* gene (19p13.2); and because no single mutation is prominent outside of consangiuous populations, most GA-I patients are heterozygous for two different mutant alleles.[314] The exact pathophysiology of GA-I is not known, but it may be attributable to toxic effects of glutaric acid, quinolinic acid, or 3-hydroxyglutaric acid, or an abnormality in γ-aminobutyric acid metabolism. The incidence of GA-I is not known, but it may be as high as one in 30,000 live births, with an increased prevalence in an Old Order Amish community. Striatal damage and neurologic phenotype do not develop in all patients, and evidence indicates that the following measures may prevent their development: early supplementation with L-carnitine; vigorous treatment of intercurrent infections with fluids, glucose, and insulin; dietary restriction of essential LYS and TRP; and avoidance of fasting.[315,316] Some genotype–phenotype relationships to outcome exist. In the Ontario mutation (IVS-1+5ntG→T), treatment from birth, after prenatal diagnosis, failed to prevent acute neurologic episodes in early childhood.[317] Table 71.3 describes recommended nutrient intakes, Table 71.5 outlines sources of medical foods, and Reference #94 gives information on food sources of essential LYS, TRP, and other nutrients. TRP must not be overly

Table 71.14 Sample Diet: Glutaric Acidemia Type I (2 Weeks of Age), Weight 3.25 kg

Prescription	Total	Per kg			
Lysine (mg)	228	70			
Tryptophan (mg)	70	10			
Protein (g)	11.4	3.5			
L-Carnitine (mg)	325	100			
Energy (kcal)	413	127			
Water to make (mL)	600				
Medical Food	**Amount**	**Lysine (mg)**	**Tryptophan (mg)**	**Protein (g)**	**Energy (kcal)**
Glutarex-I[a,b,c]	59 g	0	0	8.3	283
Similac advance with Iron Powder	25 g	224	44	2.6	130
L-Tryptophan (10 mg/mL)	2.5 mL	0	25	0	0
Water to make (mL)	600 mL				
Total		224	69	10.9	413

[a]Abbott Nutrition.
[b]Other medical foods that may be used for infants are GAI (Mead Johnson Nutritionals) and XLYS, Trp Analog (Nutricia North America).
[c]Glutarex-1 contains 900 mg L-Carnitine/100 g powder.

restricted because it is an essential amino acid.[318] **Table 71.14** shows a sample diet for an infant with GA-I.

Untreated women with GA-I have given birth. Two infants were physiologically normal, and two others had structural changes in the brain. Mothers had low plasma carnitine and elevated urine glutaric and 3-OH glutaric acid concentrations, with intermittent fatigue.[319,320] Prenatal diagnosis is based on demonstrating increased concentration of glutaric acid in amniotic fluid, deficiency of GCD in cultured amniocytes, or (probably) material obtained by chorionic villus sample, or, in appropriate families, mutation analysis.[312,313]

Ammonia

Nutrition management of disorders involving ammonia fixation and urea production uses traditional rules for treating inborn errors of metabolism. Three essential rules are restricting the toxic precursor, adding deficient product, and encouraging alternative pathways for nitrogen excretion. Furthermore, an anabolic state should be maintained to promote growth and prevent catabolism of lean body mass. Study of the biologic variation imparted on ammonia fixation and the urea cycle by heritable mutations in these biochemical reactions has greatly increased our understanding of the normal physiology, biochemistry, and molecular biology of human nitrogen metabolism.[321]

Biochemistry

Central dogma holds that ammonia is converted to urea in the liver through the Krebs-Henseleit cycle (**Figure 71.7**) and is excreted in the urine. The first three enzymes of the cycle and *N*-acetylglutamate synthetase are mitochondrial. *N*-acetylglutamate synthetase catalyzes the conversion of acetyl-CoA plus glutamate to *N*-acetylglutamate, an essential cofactor for carbamylphosphate synthesis. Carbamylphosphate synthetase I catalyzes the conversion of ammonia, ATP, and bicarbonate to carbamylphosphate. OTC uses carbamylphosphate and ornithine (ORN) as cosubstrates to form citrulline (CIT), which is exported from mitochondria to the cytoplasm, where cytosolic reactions are linked to those three mitochondrial functions. CIT and aspartate form argininosuccinic acid, a reaction catalyzed by argininosuccinic acid synthetase. Fumarate is cleaved from argininosuccinic acid by argininosuccinic acid lyase, yielding arginine (ARG). Urea is then formed by the action of arginase, regenerating cytosolic ORN, which is transported back into the mitochondria to react with OTC.

Citrin is an isoform of the mitochondrial aspartate-glutamate carrier in the inner mitochondrial membrane and is responsible for the exchange of aspartate for cytosolic glutamate and H^+ ion. Patients with a deficiency of citrin (CTNL2) have mutations on chromosome 7q21.3[36] and elevated plasma CIT concentration. In neonates, intrahepatic cholestasis and secondary argininosuccinic acid synthetase deficiency are observed. Coma and death may result from brain edema.[34]

Urea Cycle Enzyme Deficiencies

Disorders of the urea cycle comprise a group of inherited defects in the six enzymes that produce urea (Figure 71.7).[321] With the exception of OTC deficiency, all of these disorders have an autosomal recessive mode of inheritance. OTC deficiency is inherited as an X-linked dominant trait that is usually lethal in male patients. Many of the genes for these enzymes have been localized to the human genome, have been cloned, and have had mutations defined. *OTC*, a 73-kb gene containing 10 exons, is located on Xp21.1. More than 230 mutations have been defined with some genotype–phenotype relationships.[322] For example, the generation of stop codons from R109X and R109Q results

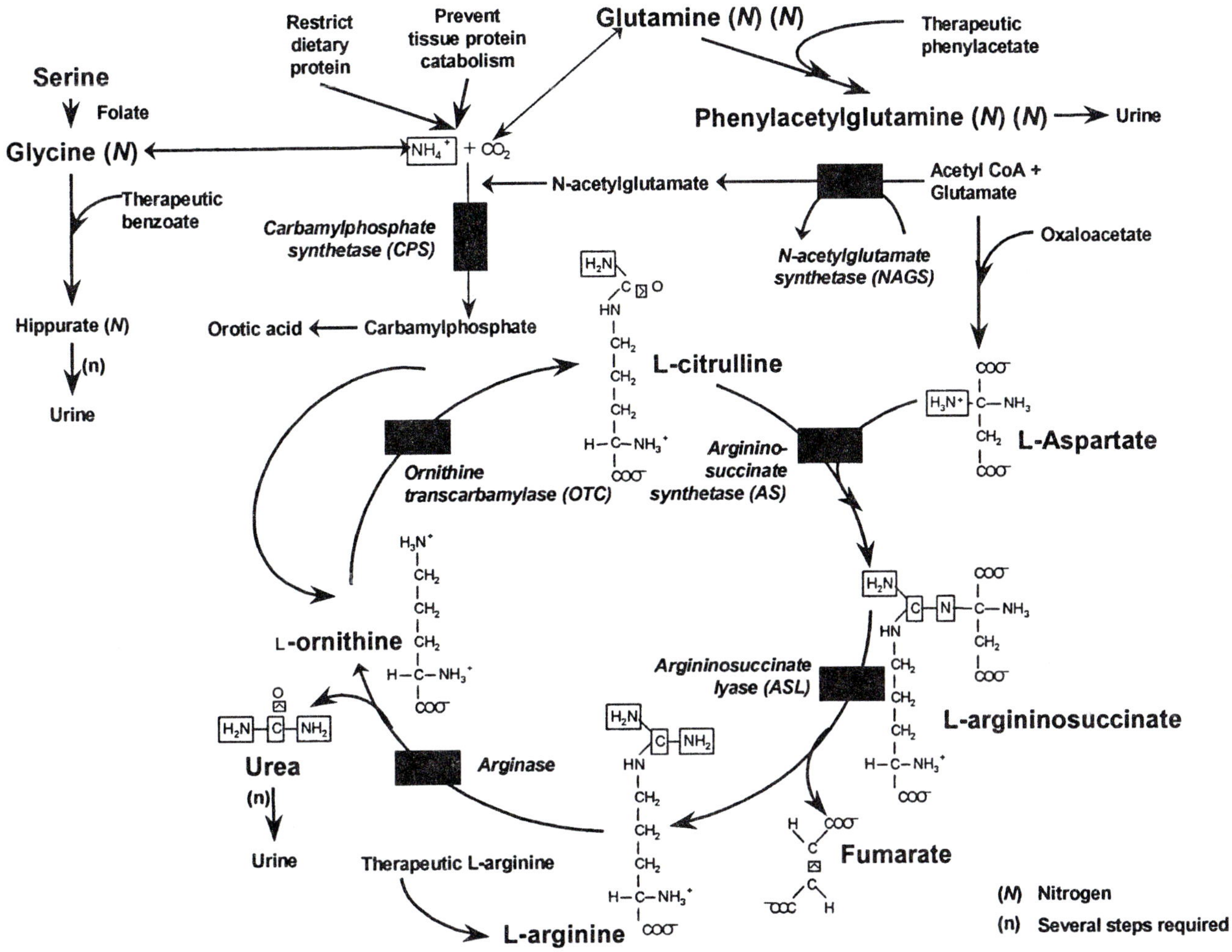

Figure 71.7 Inborn errors in the urea cycle and nutrition approaches to their management. Ammonia fixation and urea production are metabolically cycled, with inherited blocks producing hyperammonemia indicated by *black bars*. Important nitrogen molecules and the biochemical origins are outlined in *boxes*. Mitochondrial enzymes in urea synthesis are carbamylphosphate synthetase, *N*-acetylglutamate synthetase, and ornithine transcarbamylase. Use of benzoate, phenylacetate, and phenylbutyrate is indicated to provide alternate pathways for nitrogen excretion. Dietary arginine is added to provide urea cycle substrate distal to genetically impaired reactions. Restriction of dietary protein and addition of dietary energy to prevent protein catabolism are also indicated. *CoA*, coenzyme A.

in no residual liver enzyme and a severe neonatal presentation. By contrast, the R129-to-histidine mutation, also present in the *spf-ash* mouse model of OTC deficiency, has a milder phenotype and residual liver enzyme activity.[323] Carbamylphosphate synthetase is located on 2q35, argininosuccinic acid synthetase on 9q34, argininosuccinic acid lyase on 7q11, and arginase on 6q23. All are primarily expressed in liver, and all have defined mutations in their respective disorders (Table 71.2).[321,322,324,325] Argininosuccinic acid synthetase has several pseudogenes that confound DNA analysis.[321] In OTC deficiency, defects in the protein include disordered mitochondrial uptake and immunologically absent protein in this organelle.[323] Arginase has two genes with differential expression in liver and erythrocytes. In addition to these defects in ureagenesis, a seventh cause of hyperammonemia is the hyperammonemia, homocitrullinemia, hyperornithinemia syndrome, which is caused by defective mitochondrial uptake of ORN.[326]

Hyperammonemia is a biochemical manifestation of all disorders of the urea cycle. Other biochemical characteristics of each defect follow: carbamylphosphate synthetase I defect causes decreased plasma CIT; OTC deficiency results in orotic aciduria and X-linked patterns of transmission; argininosuccinate synthetase deficiency is associated with increased plasma CIT accompanied by orotic aciduria; argininosuccinate lyase deficiency (ASA) causes increased argininosuccinate in plasma and urine; and arginase deficiency is characterized by increased ARG in plasma and urine. Clinical features in the newborn suggesting urea cycle enzyme deficiency (UCED) occur with protein ingestion. These defects include poor feeding, vomiting, lethargy, hypotonia, stupor, bleeding diatheses, convulsions, coma, shock, and death.[321,327] Intellectual disability occurs in survivors of these disastrous episodes, but successful control of hyperammonemia in the newborn period can prevent this sequela.

Clinical Phenotypes

Hyperammonemia and its clinical sequelae of vomiting, lethargy, and coma relate to excessive protein intake, catabolism, or valproate therapy[328] and are observed in all of the UCEDs.

Biochemical and phenotypic manifestations differ in the individual enzyme deficiencies, however. In ASA, a specific hair abnormality, trichorrhexis nodosa, is evident. This condition is related to ARG deficiency and the relatively high ARG content of normal hair protein. Hair reverts to normal with ARG supplementation. Adult siblings with ASA may have few, if any, clinical manifestations despite identical mutations. In patients with defects in one of the first four enzymes, ARG deficiency is also associated with progressive degeneration of the CNS and a peculiar rash with control of hyperammonemia through protein restriction alone.[329,330]

Each enzyme defect has a spectrum of clinical manifestations ranging from death in the newborn period to cyclic vomiting and migraine in adolescence. For example, the typical male patient with OTC deficiency has <5% normal activity and dies in the neonatal period. A surviving male child with the late-onset form of OTC deficiency has immunologically present OTC with decreased affinity for ORN, a shift of pH optimum, and 25% of normal activity under physiologic conditions.[331] Mutational analyses of the *OTC* gene have differentiated neonatal late-onset phenotypes.[321]

Enzymatic evidence of genetic heterogeneity comes from kinetic studies in fibroblasts of patients with argininosuccinic acid synthetase deficiency. Early biochemical studies showed that enzymes from patients with citrullinemia had decreased binding of CIT or aspartate, but the residual argininosuccinic acid synthetase had a distinct and different curve of activity in each patient.[332] Analyses of RNA in citrullinemic patients showed heterogeneity, and more than 20 different mutations are now defined.[321]

Expression of the Heterozygous State for Ornithine Transcarbamylase Deficiency. The female heterozygote of OTC deficiency may have mild protein intolerance manifested clinically by migraine in adults and by cyclic vomiting with intermittent hyperammonemia in children. When protein or ammonium chloride loads were administered to 15 children with migraine and cyclic vomiting, nine had abnormally high baseline plasma ammonium; the tests produced marked hyperammonemia in eight children, and six developed migraine symptoms. Enzyme assay in seven girls with cyclic vomiting showed three with deficient OTC activity. Heterozygous female patients with OTC deficiency may be asymptomatic, or as severely affected as hemizygous male patients.[333]

Screening

Nonselected screening of all newborns for UCEDs was routinely conducted in Massachusetts using a bacterial auxotroph that required ARG. Among the children tested, nine of 700,000 newborns were found to be homozygous affected or heterozygous for ASA.[334] Problems associated with analysis of ammonia in blood were described by Barsotti.[335] This method can be readily adopted in offices and hospitals for selective screening. Its lack of use in the United States is related to cost and demand. All states now screen for citrullinemia and argininosuccinic aciduria using MS/MS.

The true incidence of UCEDs is not known because population-based screening has not been conducted, and many undiagnosed deaths may be caused by these disorders. The overall incidence is certainly underestimated at one in 30,000 live births.[336] Only three UCEDs are primary screened for by MS/MS (Table 71.1).

Diagnosis

Hyperammonemia, in association with other characteristic biochemical and clinical findings, is diagnostic of specific disorders in the urea cycle.[321] The enzyme defect can be inferred from metabolites (in addition to ammonia) that accumulate in blood and urine: orotic acid in the urine in OTC deficiency; and CIT, argininosuccinic acid, and ARG in the plasma and urine in argininosuccinate synthetase deficiency, ASA, and arginase deficiency, respectively. Carbamylphosphate synthetase, or the rare *N*-acetylglutamate synthetase deficiency, is suggested by exclusion of these four enzymopathies and requires liver biopsy with enzyme analyses for diagnosis.[337] Hyperammonemia can also be caused by acute or chronic liver diseases, galactosemia, neonatal Niemann-Pick disease type IC, tyrosinemia type I, hereditary fructose intolerance, Reye syndrome, asparaginase treatment, PPA, lysinuric protein intolerance, hyperornithinemia, isovaleric acidemia, MMA, long-term use of parenteral amino acids, and many different infectious agents in infancy. Definitive diagnosis depends on clinical acumen, followed by appropriate laboratory studies.[327] A suggested algorithm for differential diagnosis of UCEDs is given in **Figure 71.8**. A suspected UCED is a medical emergency and requires immediate intervention.

Treatment

The treatment of inherited urea cycle enzymopathies can be divided into short-term and long-term therapy, and depends on the specific diagnosis.[325] Valproate therapy should be avoided in patients with UCEDs because the use of this agent may enhance hyperammonemia.[264,315]

Short-Term Therapy. The preferred approach is to begin orogastric perfusion with high energy intake (150 kcal/kg/day) but no protein. Pro-Phree (Abbott Nutrition) or PFD1 (Mead Johnson Nutritionals) is useful for this approach (Table 71.5). L-ARG (350 to 500 mg/kg/day) should be added to this formulation. Sodium benzoate (300 mg/kg/day) can successfully reduce acute hyperammonemia in the neonatal period by conjugation of benzoic acid with GLY to form hippurate. Phenylbutyric acid or phenylacetic acid (500 mg/kg/day) is also given to form phenylacetylglutamine, which is excreted in the urine, thus eliminating from the body two nitrogen atoms per molecule (Figure 71.7 and **Table 71.15**). Urinary potassium loss is enhanced by the excretion of hippurate and phenylacetylglutamine. Consequently, plasma potassium concentrations should be monitored, and supplements should be given, if needed.[338]

Hemodialysis may be useful in the presence of coma in reducing plasma ammonium concentration. Peritoneal dialysis for 7 days in a male neonate with OTC deficiency removed 50 times more ammonia than a single exchange transfusion. Peritoneal dialysis includes risks, such as *Candida* peritonitis and continued catabolism, however. If dialysis is used, parenteral L-ARG HCl and sodium phenylbutyrate should be given as well. Continued intravenous protein-free energy is also recommended.

A priority of newborn therapy is to force the neonate into an anabolic phase with high-energy feedings (**Table 71.16**). Peripheral venous hyperalimentation with 10% to 20% GLU and lipid (2 to 4 g/kg) may be necessary if gavage is not tolerated. As gavage feedings are increased, peripheral alimentation should be decreased. After 1 to 2 days of no-protein, high-energy, L-ARG

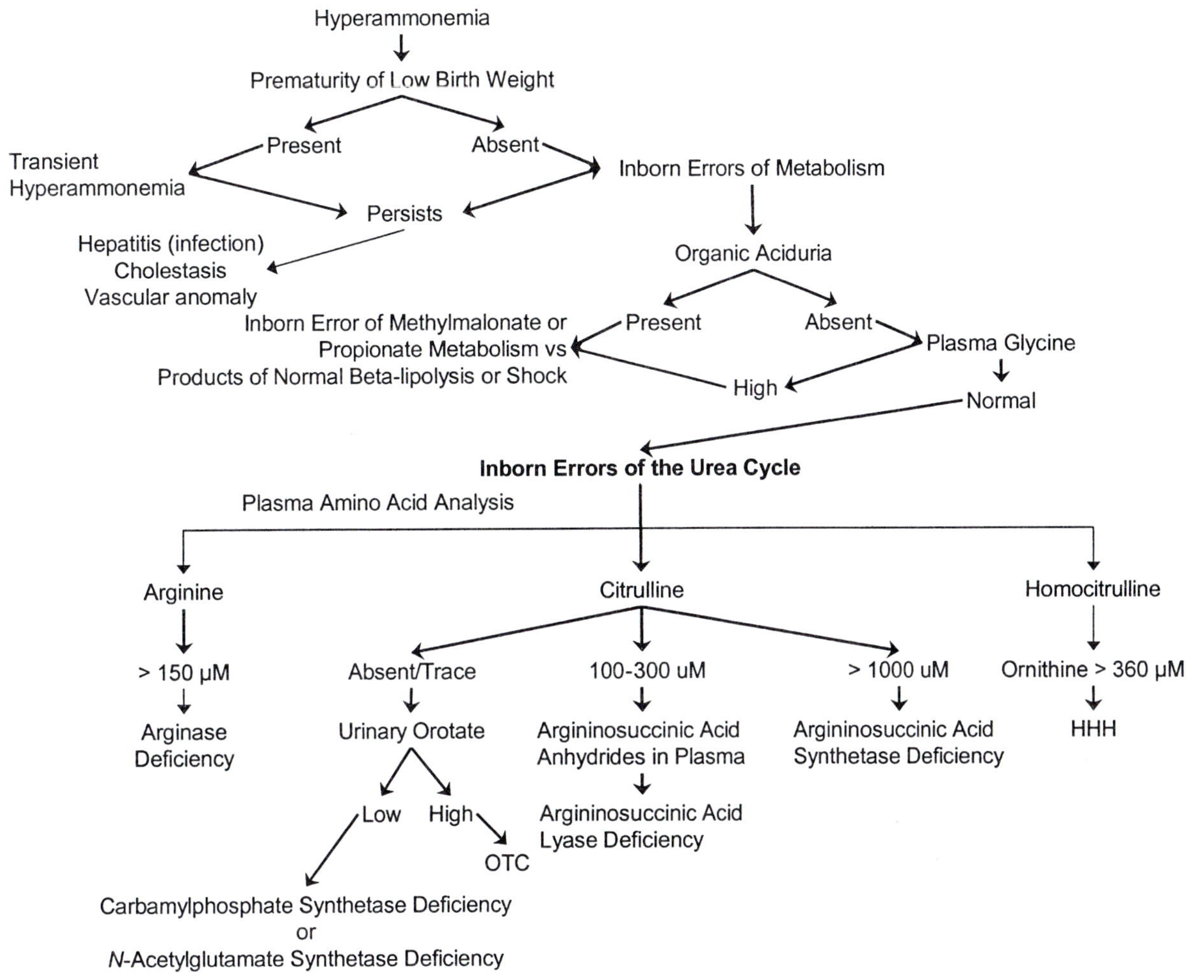

Figure 71.8 Algorithm for differential diagnosis of urea cycle disorders. *HHH*, hyperornithinemia, homocitrullinemia, and hyperammonemia; *OTC*, ornithine transcarbamylase.

Table 71.15 Suppliers of Medications and Nutrition Supplements Required for Treatment of Urea Cycle Disorders

Product	Supplier
Medications	
Sodium benzoate+sodium phenylacetate (Ammonul) (intravenous)	Ucyclyd Pharma 8125 North Hayden Road Scottsdale, AZ 85258-2463 602-667-3914, 888-829-2593 (24 h) http://ucyclyd.com/
Nutrition Supplements	
L-Arginine powder or capsule (free base) L-Citrulline powder[a]	Jo Mar Laboratories 583 Division Street, Suite B Campbell, CA 95008-6915 800-538-4545; FAX: 408-374-5920 www.jomarlabs.com/
L-Arginine HCl[b] (R-Gene-10) (10% pyrogen-free solution)	Pharmacia & Upjohn 100 Route 206 North Peapack, NJ 07977 For emergencies, call 800-821-7000 (direct) Order from pharmaceutical distributor

(continues)

Table 71.15 Suppliers of Medications and Nutrition Supplements Required for Treatment of Urea Cycle Disorders *(continued)*

Product	Supplier
Nutrition Supplements	
L-Citrulline powder[a]	Seybridge Pharmacy 37 New Haven Road Seymour, CT 06483-3469 203-888-0073 www.seybridgepharmacyandgifts.com
L-Isoleucine, L-valine in 50-mg packets	Vitaflo USA LLC 211 N Union St Suite 100, Alexandria, VA 22314 888-848-2356 www.vitaflousa.com/

[a]Amino acid powders have different densities. Consequently, they should be measured on a scale that reads in grams. A 1-week supply may be weighed and placed in a vial. The week's supply may then be mixed to a known volume in boiled water, capped, and stored in a refrigerator. The daily amount required may be measured in a disposable syringe or volumetric flask.
[b]Hyperchloremic acidosis may occur with high-dose L-arginine HCl. Consequently, plasma concentrations of chloride and bicarbonate should be monitored, with bicarbonate administration, if needed.

Table 71.16 Recommended Daily Nutrient Intakes (Ranges) for Infants and Children with Urea Cycle Disorders

	Nutrient		
Age	**L-Arginine[a] (g/kg)**	**Protein[b] (g/kg)**	**Energy[b,c] (kcal/kg)**
Months			
0 < 1	500–100	2.2–1.5	155–120
1 < 2	500–100	2.0–1.5	150–115
2 < 3	500–100	1.9–1.3	145–110
3 < 4	400–100	1.5–1.2	140–105
4 < 5	400–100	1.4–1.0	135–100
5 < 6	400–100	1.3–1.0	130–95
6 < 7	400–100	1.2–0.89	125–90
7 < 8	300–100	1.1–0.8	120–85
8 < 9	300–100	1.1–0.8	115–80
9 < 10	300–100	1.0–0.8	110–80
10 < 11	300–100	1.0–0.8	105–80
11 < 12	300–100	1.0–0.8	105–80
Years			
1 < 2	300–100	1.3–0.8	110–105
2 < 3	300–100	1.2–0.8	105–100
3 < 4	300–100	1.1–0.8	100–95
4 < 7	300–100	1.0–0.7	95–85
7 < 11	300–100	1.0–0.7	85–65
11 < 19	300–100	1.0–0.6	60–40

[a]Not used with arginase deficiency.
[b]Based on the 50th percentile weight for age. Modify as needed to maintain normal linear growth and plasma ammonia concentrations <10 μmol/L.
[c]Supply at least 1.5 mL fluid/kcal to infants and 1.0 mL/kcal to children.

HCl-supplemented and benzoate-supplemented feedings, blood ammonia concentration should revert to nearly normal. Cautious addition of 1.0 to 1.5 g/kg/day of protein is then necessary.

Long-Term Therapy. The objectives of therapy in a child with a UCED are to maintain plasma ammonia as close to normal as possible, and to supply protein and other essential amino acids and nutrients that will allow maximal intellectual development and optimal growth. Four major approaches are used in treating individuals with UCED (Figure 71.7): (a) reducing precursors of ammonia (protein intake), (b) correcting ARG deficiency, (c) enhancing alternate mechanisms of waste nitrogen loss, and (d) accelerating renal excretion of accumulated intermediates.[325]

Methods used to reduce ammonia precursors include protein restriction, prevention of body protein catabolism, and use of essential and semiessential amino acids. Any time that intake of protein or essential amino acids is severely restricted, precursors for synthesis of carnitine (LYS, MET), glutathione (cysteine, glutamate), and taurine (cysteine) may be limiting. Restricted MET intake may result in a decrease in the available pool of labile methyl groups required for synthesis of important metabolic compounds.

L-ARG base supplementation is required in all the UCEDs except arginase deficiency. To maintain normal or slightly elevated plasma ARG concentration, 100 to 500 mg/kg of body weight daily are used.[326] L-ARG can then produce ORN for ammonia fixation and drive the cycle to CIT and argininosuccinate (Figure 71.7). These two amino acids are poorly absorbed by the kidneys and allow nitrogen loss. Acceleration of renal excretion of accumulated intermediates in the impaired cycle is sought. ARG supplementation increases CIT and argininosuccinic acid excretion in argininosuccinic acid synthetase and ASA deficiencies, respectively.

Waste nitrogen urinary loss can be enhanced by the use of sodium benzoate, phenylacetate, or phenylbutyrate (Figure 71.7).[339] GLY conjugates with benzoate through GLY-*N*-acylase, which leads to excretion of a nearly stoichiometric quantity of nitrogen as hippurate (Figure 71.7). Toxicity is low on 200 to 500 mg/kg/day. Folate must be administered to provide a source of one-carbon fragments for synthesis of GLY from serine to prevent GLY depletion. Pyridoxine is necessary for transamination. Pantothenic acid (4 mg/L) in tissue culture media enhanced CoA and hippurate concentrations to a greater extent than smaller or greater amounts.[340] Phenylbutyrate and phenylacetate increase urinary nitrogen excretion as phenylacetylglutamine. The suggested dose is 500 mg/kg of body weight. This efficient alternative pathway removes two molecules of nitrogen per molecule of phenylacetylglutamine and requires monitoring of protein intake to prevent deficiency. Excess use of these nitrogen-binding drugs can lead to nitrogen deficiency and poor growth.

Further, the use of phenylbutyrate leads to excretion of glutamine as phenylacetylglutamine and increased LEU oxidation in normal adults.[341] Plasma BCAA concentrations and protein synthesis are also decreased with the administration of phenylbutyrate at 10 g/day with a protein intake of 0.4 g/kg body weight/day.[342] Oral formulations, such as glycerol phenylbutyrate, which is significantly more palatable than sodium phenylbutyrate but works through the same mechanism, can be useful in controlling ammonia as a long-term management strategy.[343,344] Carbaglu, an analog of carbamyl glutamate, can help replace deficient n-acetylglutamate found in NAGS and CPS-1 deficiency (as well as in Propionic Acidemia and Methylmalonic Acidemia).[344]

Patients with CIT II deficiency require a diet very different from that in other UCEDs; it must be high protein (17%–21% of energy), high fat (40%–47% of energy), and low carbohydrate (33%–40% of energy), with L-ARG supplements (Table 71.3).[34,338]

Catabolism during a febrile illness, or because of poor appetite, may lead to life-threatening elevations in blood ammonia. In addition to prompt diagnosis and treatment of the infection, decreased protein intake (0 g for 1 to 2 days), increased energy intake, and peritoneal dialysis may all be required. Gastrostomy feedings may be required to ensure adequate intake and to prevent inadequate growth or catabolism.

In planning nutritional therapy of the infant or child with a UCED, a formal prescription should be written that includes recommended amounts of protein, energy, fluid, L-ARG, and drugs that enhance nitrogen loss. The prescription for protein should be based on the specific diagnosis, degree of impaired urea cycle, and blood ammonia concentration, and correlated with various parameters of growth, including rates of height and weight increase and hair, nails, teeth, and skin changes.

Protein intake suggested in Table 71.16 are based on amount required to cover obligatory nitrogen losses and growth needs,[345] and are modified based on unpublished data from 10 infants with one of four different enzyme defects diagnosed within the first 10 days of life. Intake may need to be increased if the child does not grow adequately on the recommended intake or if sodium benzoate, phenylacetate, or phenylbutyrate is administered. Over-restriction of protein may lead to catabolism and impaired renal activity to excrete ammonium (NH_4^+). TRP intake should be at the minimal requirement for growth to prevent excess serotonin synthesis that suppresses appetite.[346]

Energy intakes recommended in Table 71.16 are somewhat higher than those for physiologically normal infants and children to provide ketoacid precursors from carbohydrate for synthesis of nonessential amino acids, and to prevent protein degradation. Carbohydrate should not provide more than 50% of the energy to prevent elevated plasma triacylglycerol concentration.

In any situation in which protein-restricted diets are fed, L-carnitine supplements may be necessary. Recommended amounts of supplemental L-carnitine are 50 to 100 mg/kg/day. L-Carnitine supplements are reported to lower blood ammonium concentration.[347,348] Citrate deficiency was reported in patients with ASA, and supplementation was recommended.[349,350]

Medical Foods for Urea Cycle Disorders. Nutritional therapy of UCEDs requires restricting nitrogen intake, which is best accomplished by providing approximately two-thirds of the prescribed protein in the form of essential amino acids only (medical foods). Sources of medical foods for UCEDs are given in Table 71.5. Cyclinex-1 and Cyclinex-2 (Abbott Nutrition) contain 25 mg of L-carnitine per gram of protein, as well as ample BCAAs.

Other Foods. The nutrient content of foods containing intact protein are available from USDA.[94] **Table 71.17** provides a sample diet for an infant with a UCED.

Assessment of Nutritional Therapy. Frequency of assessment is, in part, dictated by the clinical course of the patient. Blood ammonia concentration should be monitored

Table 71.17 Sample Diet: Urea Cycle Enzyme Defect (2 Weeks of Age), Weight 3.25 kg

Prescription	Total	Per kg		
L-Arginine (mg)	975	300		
Protein (g)	7.2	2.2		
Energy (kcal)	470	145		
Water to make (mL)	600			
Medical Food	**Amount**	**L-Arginine (mg)**	**Protein (g)**	**Energy (kcal)**
Cyclinex-1 Powder[a,b]	69 g	0	5.2	352
Similac Advance Infant with Iron Powder[a]	20 g	65	2.2	104
Polycose liquid[a]	7 mL	0	0.0	14
L-Arginine,[c] 10 mg/mL	92 mL	920	0.0	0
Water to make	600 mL			
Total		985	7.2	470

[a]Abbott Nutrition.
[b]Another medical food that may be used for infants is WND (Mead Johnson Nutritionals).
[c]DO NOT use with arginase deficiency.

routinely and maintained at <50 μM. Plasma concentrations of amino acids should be monitored and maintained in the normal range. Plasma albumin and globulin concentrations are indices of protein status and should be evaluated frequently. Plasma transthyretin and retinol-binding protein have shorter half-lives than albumin and can provide information on protein status at an earlier stage in deficiency than albumin. Caretakers should provide diet diaries and records of health status in tandem with blood tests for ammonia and plasma amino acids. Growth and development should be routinely assessed. If evidence of protein deficiency occurs, or if growth is not maintained, increased protein intake will be necessary.

Results of Nutritional Therapy. Results of therapy in infants with complete or nearly complete enzyme deficiencies have been less than optimal, with delayed death and subnormal development. If serious brain swelling and coma are prevented in the neonatal period or if the onset of disease expression is delayed, physical growth and mental development will be more nearly normal with nutrition and pharmacologic support.[321,351-353] In one report, better protein intake improved both growth and protein status without a concomitant increase in plasma ammonia concentration.[354] If the diagnosis is anticipated and treatment is begun during the neonatal period in children with affected siblings with citrullinemia or argininosuccinic acidemia, a relatively normal outcome is observed even in patients with severe enzyme defects.[355] Coma may occur postpartum in the woman with OTC deficiency unless a protein-restricted diet and phenylbutyrate are administered.[356] Female patients with symptomatic OTC deficiency have fewer hyperammonemic episodes and reduced risk of further cognitive decline if they are treated with a protein-restricted diet and drugs that enhance waste nitrogen excretion.[351,357,358] A successful pregnancy outcome was reported in a woman with ASA.[359] Phenylbutyrate and a protein-restricted diet in a female patient with OTC deficiency resulted in a normal infant.[360]

Galactose

Biochemistry

Because lactose from milk is the principal carbohydrate and energy source for infants and young children, galactose (GAL) maintains a central metabolic role in human nutrition. Lactose is hydrolyzed in the intestine by lactase to GLU and GAL, and 0.5 to 1.0 mg GAL/kg per minute is produced endogenously.[361] GAL is converted to uridine diphosphate (UDP) GAL and GLU-1-phosphate (GLU-1-P). UDP-GAL and UDP-GLU are essential building blocks for the post-translational trafficking and function of membrane bound and secreted proteins. GLU-1-P is converted to energy. This evolutionarily conserved pathway occurs primarily in the liver (**Figure 71.9**). First, GAL enters cells by a permease. Then GAL is phosphorylated to GAL-1-phosphate (GAL-1-P) by galactokinase (GALK). Classic galactosemia is caused by impaired GAL-1-P uridyl transferase (GALT). GALT is highly conserved from *Escherichia coli* to humans in its catalytic structure and function. UDP-GLU binds and releases GAL-1-P by dissociation reactions. Then the UMP-GALT complex binds GAL-1-P, releases UDP-GAL, and frees GALT for a subsequent set of bimolecular reactions. UDP-GAL and UDP-GLU are important precursors for glycoproteins and glycolipids; UDP-GAL and UDP-GLU are interconverted by epimerase (Figure 71.9).

Galactosemia

Elevated blood GAL may occur because of deficient functioning of GALK, GALT, or UDP-GAL-4-epimerase[362] (Figure 71.9). Patients with GALK deficiency produce excess galactitol and galactonic acid through alternate pathways, and have only cataracts without hepatocellular dysfunction. GALK deficiency does not produce the acute hepatotoxic manifestations or the accumulation of GAL-1-P seen in GALT deficiency. Several variants with different degrees of function and structure have been described for mutant

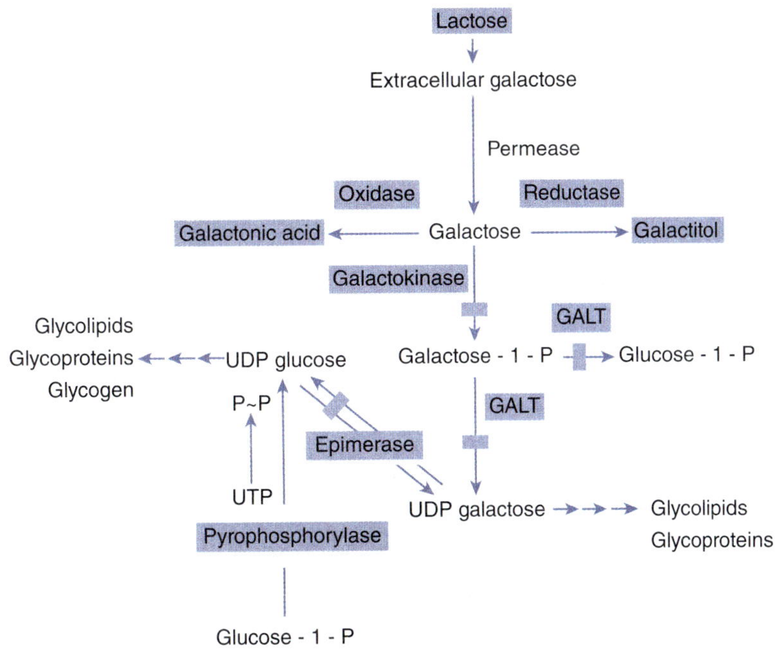

Figure 71.9 Metabolic blocks in galactose metabolism that lead to galactosemia. Classic galactosemia is caused by galactose-1-phosphate uridyl transferase (GALT) deficiency. *ATP*, adenosine triphosphate; *UDP*, uridine diphosphate; *UTP*, uridine triphosphate.

GALT.[348] This gene locus is on chromosome 9p[363] The cDNA and gene for GALT have been accurately sequenced, and many different mutations identified.[364-367] A common missense mutation, Q188R, alters the UMP-GALT complex and prevents the second dissociation reaction.[352]

Patients with GALT deficiency (Q188R/Q188R) appear to synthesize less GAL per kilogram of body weight as infants than do adults.[368,369] The adults synthesize daily five to seven times the amount of GAL than the neonates based on body weight, however.

Galactosemia caused by deficiency of GALT leads to an accumulation of GAL-1-P that is toxic and competes with GLU-1-P for the production of UDP-GLU through competitive inhibition of the UTP-dependent GLU-1-P pyrophosphorylase reaction (Figure 71.9). This situation leads to deficient UDP-GLU and UDP-GAL, and reduced post-translational production of glycoproteins and glycolipids.[370-378] GALT-deficient patients were found to have defective galactosylation of serum transferrin[371-374] and follicle-stimulating hormone.[375] After treatment, the proportion of truncated glycans decreased and the proportion of disialylated biantennary complex type increased and returned almost, but not completely, to normal.[376] In GALT-deficient yeast, gene transfer of the pyrophosphorylase enabled the organisms to grow on GAL, and a similar UDP-hexose deficiency was reported in human galactosemia cells.[377,378] Treatment using GALK inhibitors to reduce endogenous production of GAL-1-P is an active area of current research.[376-379]

Clinical symptoms of the GALT defect appear early in infancy. Acute hepatotoxicity may appear with the start of human milk feeding or feeding of proprietary infant formulas that contain lactose. Prolonged neonatal jaundice at 4 to 10 days of age is common. Hyperbilirubinemia, decreased clotting factor production, and hyperammonemia are consequences of toxic injury to liver cells by GAL-1-P excess. Bleeding diatheses, *E. coli* sepsis, and shock are catastrophic events that may occur during the neonatal period unless GAL is withheld and GAL-1-P concentrations are reduced. Therefore, rapid screening, retrieval, diagnosis, and treatment are essential for population-based newborn screening programs if the clinical sequelae of neonatal galactosemia are to be prevented. Other relatively minor symptoms occur. Approximately 10% of infants with GALT deficiency are born with cataracts. Anemia from various causes is present in approximately 40% of untreated patients. Lethargy, hypotonia, food refusal, vomiting, and diarrhea are common symptoms in infancy. Localized *E. coli* infections may manifest after therapy is initiated. Long-term effects of GALT deficiency include growth delays, dyspraxic speech, ataxia, other neurologic signs, and ovarian insufficiency despite early treatment in some galactosemic patients. Epigenetic, environmental, and modifier genes are actively being studied as causative for these chronic problems in adults with galactosemia.

Retarded mental development and delayed physical growth occur in most of the untreated patients who survive.[380] The pathophysiology of galactosemia remains unclear, but early dietary therapy clearly prevents neonatal sepsis, shock, and bleeding. Some effects of GALT deficiency may occur during embryogenesis.[381] Cataracts occur in approximately 45% of untreated children and are thought to result from formation and accumulation of galactitol in the lens of the eye, which is impermeable to efflux. Galactitol creates an osmotic gradient that allows glutathione to efflux and results in decreased concentration of lens glutathione. When glutathione concentration is decreased, glutathione peroxidase is inactivated, and hydrogen peroxide accumulates to toxic levels. Hydrogen peroxide denatures lens protein and thus produces lenticular cataracts.[362] Hepatomegaly occurs in nearly all patients with GALT deficiency, and cirrhosis develops in untreated patients. Liver damage results in reduced synthesis of liver coagulants and albumin, and decreases in many different liver functions. Because of decreased albumin synthesis and proteinuria, ascites and generalized edema occur in approximately 36% of untreated patients. The albumin synthesized by untreated patients with galactosemia contains large amounts of GAL, whereas albumin of physiologically normal individuals is free of GAL.[382] Untreated patients, or those with poorly controlled galactosemia are susceptible to infection with Gram-negative organisms, probably as a direct result of inhibition by GAL-1-P of post-translational processing of secreted or membrane-bound proteins.[378] GAL-1-P also impairs active renal tubular transport. This impairment results in generalized aminoaciduria, galactosuria, and glucosuria, as well as loss of phosphate, potassium, and bicarbonate. On rare occasions, hypoglycemia occurs. Causes include reduced glycogen stores attributable to depressed UDP-GLU concentration and hyperinsulinemia, that may result from GAL stimulation of pancreatic cells.

Although early diagnosis and removal of GAL during the neonatal period are lifesaving, some patients with classic (G/G, homozygous for GALT alleles) galactosemia have long-term poor outcomes, including infertility in female patients, growth failure, dyspraxic speech, ataxia, and intellectual disability.[380,383] The risk factors for premature ovarian failure in female patients with GALT deficiency have been described as (a) the patient's molecular genotype for GALT, (b) mean GAL-1-P concentrations during dietary treatment, (c) the patient's nutrition environment at diagnosis and during treatment, and (d) the patient's ability to oxidize ^{13}C-galactose to $^{13}CO_2$ in expired air.[384] The enigmatic outcome may be of intrauterine origin and has been associated with severe mutations in GALT. For example, the Q188R mutation in exon 6 is prevalent among non-Hispanic White people. Homozygosity for this mutation is a risk factor for both dyspraxic speech and ovarian failure.[384,385] By contrast, the S135L mutation in exon 5 of the *GALT* gene is associated with good outcomes in patients treated from birth, is prevalent in Black patients, and has differential expression in different organs.[386] Other variant mutations, such as the Duarte variant (N314D), when associated with G alleles such as E203K, may return structural integrity and function to the dimeric GALT protein and produce a good clinical outcome.[363,387,388] More than 150 mutations have been defined to date in patients with galactosemia.[374] A breath test that quantifies total body oxidation of D-GAL to CO_2 and, thus, includes liver function of GALT, may be the best predictor of outcome.[384]

Screening

GALT deficiency of clinical importance to the neonate is found in approximately one in 8,000 births. The most common screening method used is the Beutler fluorescent test for galactosemia.[389] This procedure consists of incubating eluted erythrocytes from dried blood on filter paper with a mixture of UDP-GLU, GLU-6-P dehydrogenase, and NADP. Erythrocytes from physiologically normal persons containing GALT and phosphoglucomutase produce GLU-1-P, GLU-6-P, and fluorescence-reduced NADP through the added (linked) reaction of excess GLU-6-P dehydrogenase. If heat has inactivated this reaction (e.g., in summer), GLU-1-P is added to differentiate between heat inactivation that included endogenous phosphoglucomutase and GALT deficiency alone (true galactosemia).

Positive screening results occur in patients with classic (G/G) GALT deficiency and with variants of GALT that are thermolabile, such as the Duarte/galactosemia compound heterozygote. Confirmation of positive Beutler screening test results requires quantitative enzyme activity, quantitation of erythrocyte GAL-1-P content, and GALT mutational analysis. A combined analysis of the GALT biochemical phenotype and molecular genotype is important for diagnosis, therapy, prognosis, and genetic counseling.[363,385] A test of total body oxidation of GAL to expired CO_2 is the method of choice for prognosis.[384]

Diagnosis

Patients with positive results of the Beutler test should have all lactose removed from their diets immediately while enzyme diagnosis and family workup proceed. Fresh, sterile heparinized blood should be sent to a central laboratory experienced in enzyme, analyte, and molecular analyses. Both the patient and the parents should be evaluated by the center for biochemical phenotype and molecular genotype. Diagnosis of galactosemia is accomplished through measurement of activity of GALT in erythrocytes, erythrocyte GAL-1-P content, and molecular analysis of the *GALT* gene for prevalent mutations. No activity occurs in individuals homozygous for the classic disease (G/G), whereas heterozygotes (G/N) have approximately one-half of normal activity. Because the Q188R allele occurs in 70% of non-Hispanic White people with G/G galactosemia, this allele is important to identify.[367,370,385] The Duarte allele (N314D) has a characteristic isozyme pattern when erythrocytes are studied by isoelectric focusing for the GALT enzyme. The need for therapy in patients with an activity of 25% or less of GALT, as in compound heterozygotes for Duarte/galactosemia alleles, has not been established. GAL should be restricted in any infant with any mutant genotype if erythrocyte GAL-1-P is more than 2 mg/dL, urinary galactitol exceeds 20 mmol/mol creatinine, or hepatotoxicity is present, however.

GALT is expressed in cultured amniotic fluid cells and chorionic villi. Thus, GALT deficiency can be detected prenatally by both direct enzyme assay and DNA analysis if the mutations are known.[390] Amniotic fluid of a fetus with galactosemia was found to have an elevated concentration of galactitol. Assessment of the galactitol content of amniotic fluid by GC/MS provides an ancillary method for prenatal diagnosis and may be related to enigmatic outcome.[391]

Treatment

The objectives of therapy in galactosemia are to ameliorate or prevent acute hepatocellular damage by lowering accumulated intracellular GAL-1-P while simultaneously providing adequate energy and nutrients for normal growth and development.[392] Treatment should begin as early in life as possible and consists of removal of all sources of lactose and GAL from the diet of patients with no GALT enzyme activity. Patients with 5% to 10% of normal GALT activity may tolerate small amounts of GAL. Rates of erythrocyte GAL-1-P decline differ in patients with differing genotypes.[393] Investigators noted that patients receiving the same nutritional therapy whose genotype was Q188R/Q188R had an erythrocyte GAL-1-P concentration at 5 to 8 months of age of 4.9 mg/dL. Patients with Q188R/other at the same age had a concentration of 3.3 mg/dL, and those with other/other had a concentration of 2.5 mg/dL. Patients with nonclassic genotypes had more rapid declines in erythrocyte GAL-1-P concentrations than did patients with classic genotypes.

Nutrient Requirements. Protein, energy, and some other nutrient requirements of infants and children with well-controlled galactosemia may be greater than those for physiologically normal individuals of the same age, sex, and physical activity level.[394] Investigators reported that growth both in weight and in length and height of GALT-deficient patients was less than in physiologically normal children. Whether this phenomenon is caused by a greater need for nutrients or a pathophysiologic effect of abnormally glycosylated peptide growth hormones and receptors or other epigenetic signaling pathways is debated.[379] Table 71.4 shows recommended intakes of protein and energy. Adequate vitamin D_3 and calcium intake must be considered because decreased bone mineral content is found in adults with GALT deficiency. For patients >3 years old, in addition to nutrients present in the diet, the following should be added daily: calcium, 1,000 mg; vitamin D, 10 μg; and vitamin K, 1.0 mg.[104] Following 2 years of supplementation, prepubertal children had increased concentration of carboxylated osteocalcin and a significant increase in the bone mineral content of the lumbar spine. Whether higher than normal energy and protein intakes will prevent the linear growth retardation seen in children with poorly controlled galactosemia is not known. Recommended protein, fat, and essential fatty acid intakes are in Table 71.3. In view of the poor bone mineralization in patients with GALT deficiency,[395,396] calcium recommendations are given in Table 71.4.

Formulas. Human milk contains 6% to 8% lactose, cow's milk contains 3% to 4% lactose, and many proprietary infant formulas contain 7% lactose. These milks must be replaced by a powdered formula free of bioavailable GAL (Similac Sensitive Isomil Soy, Abbott Nutrition; or Enfamil ProSobee powder, Mead Johnson Nutritionals). Both ready-to-feed and concentrated liquid Isomil and Enfamil ProSobee have added carrageenan, which contains 27% GAL; thus, they are not recommended for infants with GALT deficiency.

Powdered formulas containing soy protein isolate have approximately 14 mg GAL/L in the form of raffinose and stachyose, oligosaccharides that contain GAL. As the human intestine has no enzymes to hydrolyze these oligosaccharides, they may be safely used for feeding infants and children with galactosemia. Similac Go & Grow Soy-Based Infant Formula (Abbott Nutrition) and Enfamil Next Step ProSobee Lipil Formula powder (Mead Johnson Nutritionals) may be used by children and adults who require a GAL-free formula. Casein hydrolysates such as Similac Expert Care Alimentum (Abbott Nutrition), Enfamil Nutramigen (Mead Johnson Nutritionals), and Enfamil Pregestimil (Mead Johnson Nutritionals) have been treated to remove lactose, but may still contain small amounts of GAL (<37 mg/L). Liquid Alimentum also contains carrageenan. Alimentum powder contains xanthan gum as a stabilizer, and although the casein hydrolysate contains < 37 mg/L GAL, the xanthan gum provides no bioavailable GAL. Lactose-free infant formulas may contain as much as 75 mg GAL/L and should be avoided for infants with GALT deficiency. The breath test that quantitates [1-^{13}C] GAL to $^{13}CO_2$ provides information on whole body GAL oxidation and may be used to determine the amount of GAL that can be safely fed.[397,398]

Elemental formulas such as EleCare (Abbott Nutrition) and Neocate (Nutricia North America), are lactose and GAL free. Zlatunich and Packman reported than an elemental diet resulted in a rapid decrease in erythrocyte GAL-1-P concentration in an infant who carried the two mutations Q188R and K25N.[399] Urine galactitol also decreased from 368 to 145–227 mol/mmol creatinine.

Free Galactose in the Diet. Milk and milk products are well-known sources of lactose and casein. Less well known is that different forms of casein and cheeses contain GAL in various amounts,[392,400] although aged cheese contains very little. Based on studies by botanists and plant physiologists that began in the early 1950s, it is clear that cereals, fruits, legumes (dried beans and peas), nuts, seeds, tubers, and vegetables contain GAL. Many investigators believe, however, that the GAL in these foods is in linkages resistant to human digestive enzymes. Nonetheless, many fruits and vegetables do contain free (soluble) GAL in amounts ranging from < 0.5 to 35.4 mg/100 g fresh weight.[401] One report noted that 10 of 12 baby foods had detectable amounts of free GAL, with applesauce and squash containing the largest amounts.[402] GAL has also been found in fermented cocoa beans.[403]

Oral loads of raffinose and stachyose were administered to one patient with GALT deficiency.[404] No changes were observed in erythrocyte GAL-1-P, the analyte used to monitor diet compliance. Soybean formula was fed to a 14-year-old patient with GALT deficiency, and an insignificant change in GAL-1-P was found.[405] These observations led clinicians to suggest that legumes could be used freely in the diets of patients with GALT deficiency, although explicit warning of the possible release of absorbable free GAL in the small intestine by bacteria when the patient had diarrhea had been given.[404] The GALT-deficient mouse given a chow diet containing 1.75% of GAL-containing carbohydrates excreted significant amounts of GAL, galactonate, and galactitol in the urine, a finding suggesting the digestion and absorption of bound GAL.[406] Ingestion of legumes over a 5-year period led to a 21% to 100% increase in erythrocyte GAL-1-P in four patients with GALT deficiency, whereas of four patients who did not receive legumes, three had decreases in GAL-1-P of 17%, 19%, and 23%, and one had an increase of 37%.[407] Chickpea flour contains 110 mg free GAL per 100 g flour.[408] Free GAL in six types of cooked legumes ranged from 42 to 444 mg/100 g dry beans.[409]

Free and Bound Galactose in Food and Drugs. Free GAL is present in abundance in the lactulose that is used to treat hyperammonemia. Caution to exclude this medication in neonates is emphasized. Bound GAL is present in lactose, arabinogalactans I and II, feruloylated GAL, galactan, glycolipids, galactinol, galactopinitols, and rhamnogalacturonans I and II.[409] Lactose is found in milk and dairy products, and as an extender in many over-the-counter and prescription drugs.[410] Hash-browned potatoes and some other prepared foods contain added lactose, and some chefs sprinkle meat with lactose before frying for faster and better browning.[411] Calcium lactobionate, the active ingredient in calcium glubionate (Neo-Calglucon), a liquid calcium supplement, is a substrate of β-galactosidases[411] and yields free GAL.

Organ meats such as brain, kidney, liver, pancreas, and spleen contain galactosylcerebrosides, gangliosides, and lactosyl sulfatide. These compounds are constantly turned over in living organisms. In patients with GALT deficiency, the free GAL liberated is metabolized to GAL-1-P or other metabolites such as galactitol.[412] Glycoproteins found in muscle meats contribute significant amounts of GAL to the diet.[413] Commercial meats, prepared for infant consumption, contain microgram amounts of free and bound GAL per 100g.[414] Whether any foods other than milk and all milk products are important in the diet of the adult with galactosemia is unclear, although investigators have suggested that hypoglycosylation could worsen if GAL intake were to increase.[415]

Enzymes that Degrade Compounds Containing Bound Galactose. α-Galactosidases, enzymes that digest carbohydrates with GAL in α-linkages, are found in human tissues and are probably responsible for the degradation of human galactosylcerebrosides, galactosylsulfatides, and gangliosides. α-Galactosidases also appear to be distributed in many plant tissues.[411] α-Galactosidases digest carbohydrates with GAL in α-linkages. Beano (GlaxoSmithKline) is an α-galactosidase isolated from *Aspergillus niger* and is marketed as a food enzyme that reduces gas formation from many vegetables. α-Galactosidases can hydrolyze the terminal GAL attached to digalactosyldiacylglycerol, thus yielding GAL and monogalactosyldiacylglycerol.[411]

β-Galactosidases are found in many foods, including apples and pears, peppers, tomatoes, and cocoa beans.[411] The human intestine also contains β-galactosidases.[416] The β-galactosidases in the human intestine digest lactose, but they also have heterogenous activity on compounds with GAL in β-1,4 linkages.[416] β-Galactosidase preparations from *E. coli* release GAL from ferulic acid and from monogalactosyldiacylglycerols.[394] Galactolipids in foods are hydrolyzed by human pancreatic lipolytic enzymes and duodenal contents.[417]

Results of Nutritional Therapy. Treatment of patients with GALT deficiency, although lifesaving, may not result in complete freedom from the sequelae of the disorder.[392] Infants who are diagnosed and treated early and who maintain excellent dietary control have better intellectual function than those who have poor control, or are diagnosed late.[380] Control is defined on the basis of erythrocyte GAL-1-P concentration and is considered excellent if this is consistently <2 mg/dL.[418] Patients may have difficulty with language, abstract thinking, visual perception, ovarian failure, and cataracts despite early diagnosis and good dietary control.[418-421] These clinical deficits may be related to intrauterine damage from intrauterine accumulation of GAL-1-P or galactitol, or maternal GAL transported into the vulnerable fetus.[381,384] Embryonic membranes are constantly synthesized and degraded and require GAL and UDP-GAL. GAL (i.e., milk) restriction in at-risk pregnant women is generally advised, although little change in outcome has been observed.[380]

The observations that UDP-GLU and UDP-GAL are deficient in cells of patients with galactosemia offer a likely reason for poor outcome, despite adequate lactose restriction.[370-376,384,422] Unfortunately, uridine supplements have not been helpful in therapy.[423,424] More recently, decreased numbers of mature follicles were found in ovarian biopsies of GALT-deficient female patients. It is hoped that long-term stimulation with exogenous follicle-stimulating hormone will assist in follicular maturation and will enable these patients to become pregnant.[425]

Investigators reported that bone mineralization in both female and male patients with galactosemia was less than that of normal age-, sex-, and ethnicity-matched individuals.[395,396] Bone mineralization was positively correlated with calcium intake and, in women, was improved in those patients who received estrogen.[380] Many patients with GALT deficiency fail to ingest adequate calcium to maintain appropriate bone density early in life.[426]

A reported mouse knockout for GALT had no acute hepatotoxicity or apparent ovarian failure. The mouse model differed from that for humans in that it had no aldose reductase and did not overproduce galactitol (Figure 71.9). Thus, a pathologic effect of polyol production and treatment using aldose reductase inhibitors is of considerable research interest for future therapeutic intervention.[427] The mouse has also lost an important signaling protein through evolution for ovarian follicle maturation and cerebellar epithelial cell growth. The *ARHI* gene, when reintroduced into transgenic mice, produced the twirler phenotype, with infertility and growth restriction. Human galactosemic cells overexpress this ARHI when they are stressed by increased intracellular GAL-1-P. Thus, this signaling pathway has become an important area for research.[428]

Assessment of Nutritional Therapy. Evaluations of growth, development, lens development, liver function, erythrocyte GAL-1-P concentration, and urinary galactitol excretion are necessary to determine the adequacy of dietary intervention.[429]

Diet Termination. Although some investigators recommended liberalizing the GAL-restricted diet at 12 to 13 years of age, this approach is not warranted because the potentially damaging effects of accumulated galactitol and GAL-1-P in the lens, liver, kidney, and brain remain. Women with galactosemia must continue treatment with GAL exclusion to reduce possible in utero damage to their future children.[430] Termination of dietary therapy is not recommended.

Acknowledgments

The author(s) acknowledge Phyllis B. Acosta and Louis J. Elsas II, the authors of the previous version of this chapter, for their contributions.

References

1. Otten JJ, Hellwig JP, Meyers LD. Dietary Reference Intakes: The Essential Guide to Nutrient Requirements. Washington, DC: National Academies Press, 2006.
2. American College of Medical Genetics. Genet Med 2006; 8(Suppl 1):1S–252S.
3. Hoffmann GF, Zschocke J, Nyhan WL. Inherited Metabolic Diseases: A Clinical Approach. Berlin, Germany: Springer, 2010.
4. Elsas LJ. Approach to inborn errors of metabolism. In: Goldman L, Ausiello DA, eds. Cecil's Textbook of Medicine. 24th ed. Philadelphia, PA: WB Saunders, 2010.
5. Online Mendelian Inheritance of Man. 2023. Available at www.omim.org/statistics/entry
6. Harris H. The Principles of Human Biochemical Genetics. 3rd ed. Amsterdam, Holland: North Holland Publishing, 1980.
7. Strauss KA, Puffenberger EG, Carson VJ. Maple Syrup Urine Disease. 2006 [Updated 2020 Apr 23]. In: Adam MP, Everman DB, Mirzaa GM, et al., editors. GeneReviews® [Internet]. Seattle (WA): University of Washington, Seattle; 1993-2023. 2023. Available at www.ncbi.nlm.nih.gov/books/NBK1319/
8. Strauss KA, Carson VJ, Soltys K, et al. Mol Genet Metab 2020;129:193–206.
9. Burns JJ. Am J Med 1959;26:740–8.
10. Lipson MH, Kraus J, Rosenberg LE. J Clin Invest 1980;66: 188–93.
11. American College of Medical Genetics. Newborn Screening: Toward a Uniform Screening Panel and System. Bethesda, MD: American College of Medical Genetics, 2006.
12. HRSA Recommended Uniform Screening Panel. 2023. Available at www.hrsa.gov/advisory-committees/heritable-disorders /rusp
13. Chace DH, Lim T, Hansen CR, et al. Clin Chim Acta 2009;402: 14–8.

14. Frazier DM. In: Acosta PB, ed. Newborn screening by mass spectrometry. Nutrition Management of Patients with Inherited Metabolic Disorders. Sudbury, MA: Jones and Bartlett, 2010:21–67.
15. Gillingham MB. In: Acosta PB, ed. Nutrition management of patients with inherited disorders of mitochondrial fatty acid oxidation. Nutrition Management of Patients with Inherited Metabolic Disorders. Sudbury, MA: Jones and Bartlett, 2010: 369–403.
16. Korman SH. Mol Genet Metab 2006;89:289–99.
17. Matalon KM. In: Acosta PB, ed. Introduction to genetics and genetics of inherited metabolic disorders. Nutrition Management of Patients with Inherited Metabolic Disorders. Sudbury, MA: Jones and Bartlett, 2010:1–19.
18. Miinalainen IJ, Schmitz W, Huotari A, et al. PLoS Genet 2009;5:e1000543.
19. Molven A, Matre GE, Duran M, et al. Diabetes 2004;53:221–7.
20. National Center for Biotechnology Information. OMIM: Online Mendelian Inheritance in Man [database]. Available at www.ncbi.nlm.nih.gov/sites/entrez? db=omim Accessed July 17, 2023.
21. Sim KG, Wiley V, Carpenter K, et al. J Inherit Metab Dis 2001;24:51–9.
22. Andresen BS, Christensen E, Corydon TJ, et al. Am J Hum Genet 2000;67:1095–1103.
23. Kemper AR, Brosco J, Comeau AM, et al. Genet Med 2017;19(1):121–126.
24. Coelho D, Suormala T, Stucki M, et al. N Engl J Med 2008;358:1454–64.
25. Dobson CM, Wai T, Leclerc D, et al. Proc Natl Acad Sci U S A 2002;99:15554–9.
26. Hymes J, Stanley CM, Wolf B. Hum Mutat 2001;18:375–81.
27. Marriage B. In: Acosta PB, ed. Nutrition Management of Patients with Inherited Metabolic Disorders of Branched-Chain Amino Acid Metabolism. Nutrition Management of Patients with Inherited Metabolic Disorders. Sudbury, MA: Jones and Bartlett, 2010:175–236.
28. Miousse IR, Watkins D, Coelho D, et al. J Pediatr 2009;154: 551–6.
29. Mudd SH, Levy HL, Kraus JP. In: Scriver CR, Beaudet AL, Sly WS et al, eds. Disorders of Transsulfuration. The Metabolic and Molecular Bases of Inherited Disease. 8th ed. New York, NY: McGraw-Hill, 2001:2007–56.
30. Salomons GS, Jakobs C, Pope LL, et al. J Inherit Metab Dis 2007;30:23–8.
31. Sass JO, Ensenauer R, Roschinger W, et al. Mol Genet Metab 2008;93:30–5.
32. Saheki T, Inoue K, Tushima A, et al. Mol Genet Metab 2010;100:559–64.
33. Suzuki Y, Yang X, Aoki Y, et al. Hum Mutat 2005;26:285–90.
34. Tabata A, Sheng JS, Ushikai M, et al. J Hum Genet 2008;53: 534–45.
35. Weisfeld-Adams JD, Morrissey MA, Kirmse BM, et al. Mol Genet Metab 2010;99:116–23.
36. Yang SY, He XY, Miller D. Mol Genet Metab 2007;92:36–42.
37. Zhang J, Dobson CM, Wu X, et al. Mol Genet Metab 2006;87: 315–22.
38. Blau N, Koch R, Matalon R, et al. Mol Genet Metab 2005; 86(Suppl):S1.
39. Elsas LJ, Danner DJ. Ann N Y Acad Sci 1982;378:404–21.
40. Elsas LJ, McCormick DB. Vitam Horm 1987;43:103–44.
41. Elsas LJ, McCormick D. Vitam Horm 1987;44:145.
42. Fowler B. Eur J Pediatr 1998;157(Suppl 2):S60–6.
43. Desnick RJ, Brady R, Barranger J, et al. Ann Intern Med 2003;138:338–46.
44. Baumgartner ER, Wick H, Linnell JC, et al. Helv Paediatr Acta 1979;34:483–96.
45. Langenbeck U, Burgard P, Wendel U, et al. J Inherit Metab Dis 2009;32:506–13.
46. Gjetting T, Petersen M, Guldberg P, et al. Am J Hum Genet 2001,68:1353–60.
47. Thomas J, Levy H, Amato S, et al. Mol Genet Metab 2018;124: 27–38.
48. Hershfield MS, Buckley RH, Greenberg ML, et al. N Engl J Med 1987;316:589–96.
49. Van den Hout JM, Reuser AJ, de Klerk JB, et al. J Inherit Metab Dis 2001;24:266–74.
50. Burdelski M. Pediatr Transplant 2002;6:361–3.
51. Kayler LK, Merion RM, Lee S, et al. Pediatr Transplant 2002;6:295–300.
52. Elsas LJ. In: Bennett JC, Plum F, eds. Inborn errors of metabolism. Cecil's Textbook of Medicine. 22nd ed. Philadelphia, PA: WB Saunders, 2004.
53. Fraites TJ Jr, Schleissing MR, Shanely RA, et al. Mol Ther 2002;5:571–8.
54. Matalon R, Michals-Matalon K, Bhatia G, et al. J Inherit Metab Dis 2007;30:153–8.
55. Montanari C, Parolisi S, Borghi E, et al. Front in Phys 2021; 12:716520.
56. Pratt EL, Snyderman SE, Cheung MW, et al. J Nutr 1955;56: 231–51.
57. Herrmann ME, Brosicke HG, Keller M, et al. Eur J Pediatr 1994; 153:501–3.
58. Schoeffer A, Herrmann ME, Brosicke HG, et al. J Nutr Med 1994;4:415–8.
59. Fulgoni VL III. Am J Clin Nutr 2008;87:1554S–7S.
60. Acosta PB. In: Acosta PB, ed. Nutrition Management of Patients with Inherited Disorders of Aromatic Amino Acid Metabolism. Nutrition Management of Patients with Inherited Metabolic Disorders. Sudbury, MA: Jones and Bartlett, 2010:119–74.
61. Alexander JW, Clayton BE, Delves HT. Q J Med 1974;169: 80–111.
62. Martin SB, Acosta PB. J Am Diet Assoc 1987;87:48–52.
63. Scriver CR, Kaufman S. In: Scriver CR, Beaudet AL, Sly WS, et al, eds. The Hyperphenylalaninemias. The Metabolic and Molecular Bases of Inherited Disease. 8th ed. New York, NY: McGraw-Hill, 2001:1667–724.
64. Eisensmith RC, Martinez DR, Kuzmin AI, et al. Pediatr 1996;97:512–6.
65. Griffin RF, Elsas LJ. J Pediatr 1975;86:512–7.
66. Scriver CR, Waters PJ, Sarkissian C, et al. *PAHdb*: a locus-specific knowledgebase. Hum Mutat. 2000;15(1):99–104
67. Kure S, Hou DC, Ohura T, et al. J Pediatr 1999;135:375–8.
68. Lassker U, Zschocke J, Blau N, et al. J Inherit Metab Dis 2002;25:65–70.
69. Muntau AC, Roschinger W, Habich M, et al. N Engl J Med 2002;347:2122–32.
70. Trefz FK, Scheible D, Gotz H, et al. J Inherit Metab Dis 2009;32:22–6.
71. Blau N, Thony B, Cotton RGH, et al. In: Scriver CR, Beaudet AL, Sly WS et al, eds. Disorders of Tetrahydrobiopterin and Related Biogenic Amines. The Metabolic and Molecular Bases of Inherited Disease. 8th ed. New York, NY: McGraw-Hill, 2001:1725–76.
72. Dobbing J. In: Davis JA, Dobbing J, eds. The Later Development of the Brain and Its Vulnerability. Scientific Foundations of Paediatrics. London, UK: Heinemann, 1981.
73. Pardridge WM, Choi TB. Fed Proc 1986;45:2073–8.
74. Villasana D, Butler IJ, Williams JC, et al. J Inherit Metab Dis 1989;12:451–7.
75. Krause W, Halminski M, McDonald L, et al. J Clin Invest 1985;75:40–8.
76. Epstein CM, Trotter JF, Averbook A, et al. Electroencephalogr Clin Neurophysiol 1989;72:133–9.

77. Pass K, Levy H. Impact of Early Discharge on Screening for Inborn Errors of Metabolism. Washington, DC: MCH CORN Clearinghouse, 1995.
78. Doherty LB, Rohr FJ, Levy HL. Pediatr 1991;87:240–4.
79. Anonymous. NIH Consens Statement 2000;17:1–33.
80. Fernhoff PM, Fitzmaurice N, Milner J, et al. South Med J 1982;75:529–32.
81. Guldberg P, Rey F, Zschocke J, et al. Am J Hum Gen 1998;63: 71–79.
82. Trefz FK, Burgard P, Konig T, et al. Clin Chim Acta 1993;217: 15–21.
83. Genetic Metabolic Dietitians International. Available at https://gmdi.org/Members/Clinical-Practice-Tools/Nutrition-Guidelines
84. Acosta PB, Wenz E, Williamson M. J Am Diet Assoc 1978;72: 164–9.
85. Opladen T, López-Laso E, Cortès-Saladelafont E, et al. Orphanet J Rare Dis 2020;15:26.
86. Weglage J, Wiedermann D, Denecke J, et al. J Inherit Metab Dis. 2002;25:431–6.
87. Pietz J, Dunckelmann R, Rupp A, et al. Eur J Pediatr 1998;157:824–30.
88. Moats RA, Koch R, Moseley K, et al. J Inherit Metab Dis 2000;23:7–14.
89. Smith I, Beasley MG, Ades AE. Arch Dis Child 1990;65:472–8.
90. Vockley J, Andersson HC, Antshel KM, et al. Genet Med 2014;16:188–200.
91. Rose WC. Nutr Abstr Rev Ser Hum Exp 1957;27:631–47.
92. Hanley WB, Linsao L, Davidson W, et al. Pediatr Res 1970; 4:318–27.
93. Acosta PB. In: Wapnir RA, ed. The Contribution of Therapy of Inherited Amino Acid Disorders to Knowledge of Amino Acid Requirements. Congenital Metabolic Diseases. New York, NY: Marcel Dekker, Inc, 1985.
94. Batshaw ML, Valle D, Bessman SP. J Pediatr 1981;99:159–60.
95. Acosta PB, Yannicelli S. Acta Paediatr Suppl 1994;407:66–7.
96. Acosta PB, Stepnick-Gropper S, Clarke-Sheehan N, et al. JPEN J Parenter Enteral Nutr 1987;11:287–92.
97. Acosta PB, Yannicelli S. Biol Trace Elem Res 1999;67:75–84.
98. Lewis JS, Loskill S, Bunker ML, et al. Fed Proc 1974;33:666A.
99. Panis B, Vermeer C, van Kroonenburgh MJ, et al. Bone 2006;39:1123–9.
100. U.S. Department of Agriculture Agricultural Research Service. Nutrient Data Laboratory. Accessed July 17, 2023. Available at: www.nal.usda.gov/fnic/foodcomp/search
101. Acosta PB, Yannicelli S, Singh R, et al. J Am Diet Assoc 2003;103:1167–73.
102. Umbarger B, Berry HK, Sutherland BS. JAMA 1965;193:128–34.
103. Elsas LJ, MacDonell RC Jr, Rosenberg LE. J Biol Chem 1971;246:6452–9.
104. Nakagawa I, Takahashi T, Suzuki T, et al. J Nutr 1962;77:61–8.
105. Sibinga MS, Friedman CJ, Steisel IM, et al. Dev Med Child Neurol 1971;13:63–70.
106. Holm VA, Kronmal RA, Williamson M, et al. Pediatr 1979; 63:700–7.
107. Azen CG, Koch R, Friedman EG, et al. Am J Dis Child 1991;145:35–9.
108. Schulpis KH, Nounopoulos C, Scarpalezou A, et al. Acta Paediatr Scand 1990;79:930–4.
109. Acosta PB. Phenylalanine Hydroxylase Deficiency Managed by Analog XP. Columbus, OH: Ross Products Division, Abbott Laboratories, 1992.
110. Francois B, Diels M, de la Brassinne M. J Inherit Metab Dis 1989;12(Suppl 2):33–334.
111. Acosta PB, Yannicelli S, Marriage B, et al. J Am Coll Nutr 1999;18:102–7.
112. Buist NR, Prince AP, Huntington KL, et al. Acta Paediatr Suppl 1994;407:75–7.
113. Acosta PB, Greene C, Yannicelli S. Int Pediatr 1993;8:6–16.
114. Arnold GL, Vladutiu CJ, Kirby RS, et al. J Pediatr 2002;141: 243–6.
115. Acosta PB, Alfin-Slater RB, Koch R. J Am Diet Assoc 1973;63: 631–5.
116. Galluzzo CR, Ortisi MT, Castelli L, et al. J Inherit Metab Dis 1985;8(Suppl 2):129.
117. Schulpi KH, Scarpalezou A. Clin Pediatr (Phila) 1989;28:466–9.
118. Acosta PB, Michals-Matalon K, Austin V, et al. In: Platt LD, Koch R, de la Cruz F, eds. Nutrition Findings and Requirements in Pregnant Women with Phenylketonuria. Genetic Disorders and Pregnancy Outcome. New York, NY: Parthenon, 1997: 21–32.
119. Castillo M, Zafra MF, Garcia-Peregrin E. Neurochem Res 1988;13:551–5.
120. Artuch R, Colome C, Vilaseca MA, et al. J Inherit Metab Dis 2001;24:359–66.
121. Hargreaves IP, Lee PJ, Briddon A. Amino Acids 2002;22:109–18.
122. Galli C, Agostoni C, Mosconi C, et al. J Pediatr 1991;119:562–7.
123. Sanjurjo P, Perteagudo L, Rodriguez-Soriano J, et al. J Inherit Metab Dis 1994;17:704–9.
124. Poge AP, Baumann K, Muller E, et al. J Inherit Metab Dis 1998;21:373–81.
125. Acosta PB, Yannicelli S, Singh R, et al. J Pediatr Gastroenterol Nutr 2001;33:253–9.
126. Bodley JL, Austin VJ, Hanley WB, et al. Eur J Pediatr 1993;152:140–3.
127. Gropper SS, Trahms C, Cloud HH, et al. Int Pediatr 1994;9: 237–43.
128. Zachara BA, Wasowicz W, Gromadzinska J, et al. Biomed Biochim Acta 1987;46(Suppl):S209–13.
129. Calomme MR, Vanderpas JB, Francois B, et al. Experientia 1995;51:1208–15.
130. Al Qadreh A, Schulpis KH, Athanasopoulou H, et al. Acta Paediatr 1998;87:1162–6.
131. Carson DJ, Greeves LG, Sweeney LE, et al. Pediatr Radiol 1990;20:598–9.
132. Hillman L, Schlotzhauer C, Lee D, et al. Eur J Pediatr 1996;155(Suppl 1):S148–52.
133. McMurry MP, Chan GM, Leonard CO, et al. Am J Clin Nutr 1992;55:997–1004.
134. Greeves LG, Carson DJ, Magee A, et al. Acta Paediatr 1997;86:242–4.
135. Yannicelli S, Medeiros DM. J Inherit Metab Dis 2002;25: 347–61.
136. Schulpis KH, Papakonstantinou E, Michelakakis H, et al. Clin Endocrinol (Oxf) 1998;48:99–101.
137. Klibanski A, Neer RM, Beitins IZ, et al. N Engl J Med 1980; 303:1511–4.
138. Mussa A, Roato I, Spada M, et al. Mol Genet Metab. 2010;99:227A–8A.
139. Gropper SS, Chaung HC, Bernstein LE, et al. J Am Coll Nutr 1995;14:264–270.
140. Holtzman NA, Welcher DW, Mellits ED. N Engl J Med 1975;293:1121–4.
141. Horner FA, Streamer CW, Alejandrino LL, et al. N Engl J Med 1962;266:79–81.
142. Hudson FP. Arch Dis Child 1967;42:198–200.
143. Holtzman NA, Kronmal RA, van Doorninck W, et al. N Engl J Med 1986;314:593–8.
144. Brumm VL, Grant ML. Mol Genet Metab 2010;99(Suppl 1): S18–21.
145. Waisbren SE, Levy HL. J Inherit Metab Dis 1991;14:755–64.
146. Aung TT, Klied A, McGinn J, et al. J Inherit Metab Dis 1997;20:603–4.
147. Hanley WB, Feigenbaum A, Clarke JT, et al. Lancet 1993;342:997.

148. Krause W, Epstein C, Averbook A, et al. Pediatr Res 1986;20:1112–6.
149. Elsas LJ, Trotter JF. In: Wurtman RJ, Ritter-Walker E, eds. Changes in Physiological Concentrations of Blood Phenylalanine Produce Changes in Sensitive Parameters of Human Brain Function. Dietary Phenylalanine and Brain Function. Boston, MA: Birkhauser, 1988:187–95.
150. Thompson AJ, Smith I, Brenton D, et al. Lancet 1990;336:602–5.
151. Lenke RR, Levy HL. N Engl J Med 1980;303:1202–8.
152. Levy HL. Enzyme 1987;38:312–20.
153. Kudo Y, Boyd CA. J Inherit Metab Dis 1990;13:617–26.
154. Hanley WB, Clarke JTR, Schoonheyt W. Clin Biochem 1987; 20:149–56.
155. Kirby ML, Miyagawa ST. J Inherit Metab Dis 1990;13:634–40.
156. Okano Y, Chow IZ, Isshiki G, et al. J Inherit Metab Dis 1986;9:15–24.
157. Koch R, Trefz F, Waisbren S. Mol Genet Metab 2010; 99(Suppl 1):S68–74.
158. Kirkman HN. Appl Res Ment Retard 1982;3:319–28.
159. Koch R, Friedman EG, Wenz E, et al. J Inherit Metab Dis 1986;9(Suppl 2):159–68.
160. Platt LD, Koch R, Hanley WB, et al. Am J Obstet Gynecol 2000;182:326–33.
161. Michals-Matalon K, Acosta PB, Azen C. J Inherit Metab Dis 2007;30:19A.
162. Michals K, Acosta PB, Austin V, et al. Eur J Pediatr 1996; 155(Suppl 1):S165–8.
163. Rasmussen KM, Catalano PM, Yaktine AL. Curr Opin Obstet Gynecol 2009;21:521–6.
164. Rothman KJ, Moore LL, Singer MR, et al. N Engl J Med 1995;333:1369–73.
165. Matalon KM, Acosta PB, Azen C, et al. Ment Retard Dev Disabil Res Rev 1999;5:122–4.
166. Mitchell GA, Grompe M, Lambert M, et al. In: Scriver CR, Beaudet AL, Sly WS et al, eds. Hypertyrosinemia. The Metabolic and Molecular Bases of Inherited Disease. 8th ed. New York, NY: McGraw-Hill, 2001:1777–805.
167. Jorquera R, Tanguay RM. Hum Mol Genet 2001;10:1741–52.
168. Anonymous. Prescrire Int 2007;16:56–8.
169. Smith I, Glossop J, Beasley M. J Inherit Metab Dis 1990;13: 651–7.
170. Macsai MS, Schwartz TL, Hinkle D, et al. Am J Ophthalmol 2001;132:522–7.
171. Cerone R, Holme E, Schiaffino MC, et al. Acta Paediatr 1997;86:1013–5.
172. Mamunes P, Prince PE, Thornton NH, et al. Pediatr 1976;57: 675–80.
173. Chinsky JM, Singh R, Ficicioglu C. Genet Med 2017;19: 1380–1380.
174. Gagne R, Lescault A, Grenier A, et al. Prenat Diagn 1982;2: 185–8.
175. Holme E, Lindstedt S. Clin Liver Dis 2000;4:805–14.
176. Shemin D. Harvey Lect 1954;50:258–84.
177. Rank JM, Pascual-Leone A, Payne W, et al. J Pediatr 1991;118:136–9.
178. Grompe M, Lindstedt S, al Dhalimy M et al. Nat Genet 1995; 10:453–60.
179. Imseis EM, Bynon JS, Thornhill C. World J Hepatol. 2017;9: 487–90.
180. Cohn RM, Yudkoff M, Yost B, et al. Am J Clin Nutr 1977;30: 209–14.
181. Finkelstein JD. J Nutr Biochem 1990;1:228–37.
182. Rosenblatt DS, Fenton WA. In: Scriver CR, Beaudet AL, Sly WS et al, eds. Inherited Disorders of Folate and Cobalamin Transport and Metabolism. The Metabolic and Molecular Bases of Inherited Disease. 8th ed. New York, NY: McGraw-Hill, 2001:3897–934.
183. Kraus JP. Eur J Pediatr 1998;157(Suppl 2):S50–3.
184. Liu G, Nellaiappan K, Kagan HM. J Biol Chem 1997;272: 32370–7.
185. Skovby F, Gaustadnes M, Mudd SH. Mol Genet Metab 2010;99: 1–3.
186. Boers GH, Smals AG, Trijbels FJ, et al. N Engl J Med 1985;313: 709–15.
187. Kraus JP. J Inherit Metab Dis 1994;17:383–90.
188. Mudd SH, Skovby F, Levy HL, et al. Am J Hum Genet 1985; 37:1–31.
189. Schaumburg H, Kaplan J, Windebank A, et al. N Engl J Med 1983;309:445–8.
190. Yoshida I, Sakaguchi Y, Nakano M, et al. J Inherit Metab Dis 1985;8:91.
191. Matthews A, Johnson TN, Rostami-Hodjegan A, et al. Br J Clin Pharmacol 2002;54:140–6.
192. Singh RH, Kruger WD, Wang L, et al. Genet Med 2004;6:90–5.
193. Braverman NE, Mudd SH, Barker PB, et al. AJNR Am J Neuroradiol 2005;26:2705–6.
194. van Calcar S. In: Acosta PB, ed. Nutrition Management of Patients with Inherited Disorders of Sulfur Amino Acid Metabolism. Nutrition Management of Patients with Inherited Metabolic Disorders. Sudbury, MA: Jones and Bartlett, 2010: 237–281.
195. Yamada K; Yokoyama K; Aoki K, et al. Int J Neonatal Screen 2020;6:60.
196. Kerkvliet SP, Rheault MN, Berry SA. Am J Med Genet Part A 2021;185A:1247–1250.
197. Sacharow SJ, Picker JD, Levy HL. In: Adam MP, Everman DB, Mirzaa GM, et al., eds. Homocystinuria Caused by Cystathionine Beta-Synthase Deficiency. [Updated 2017 May 18]. GeneReviews® [Internet]. Seattle (WA): University of Washington, Seattle; 1993-2023. 2004. Available from: www.ncbi.nlm.nih.gov/books /NBK1524/
198. Topaloglu AK, Sansaricq C, Snyderman SE. Pediatr Res 2001; 49:796–8.
199. Sansaricq C, Garg S, Norton PM, et al. Acta Paediatr Scand 1975; 64:215–8.
200. Dudman NP, Wilcken DE. Clin Chim Acta 1983;127:105–13.
201. Spooner RJ, Fell GS, Halls DJ, et al. Clin Nutr 1986;5:29–32.
202. Smith AM, Picciano MF, Milner JA. Am J Clin Nutr 1982;35: 521–6.
203. Carey MC, Donovan DE, Fitzgerald O, et al. Am J Med 1968;45:7–25.
204. Carey MC, Fennelly JJ, Fitzgerald O. Am J Med 1968;45: 26–31.
205. Wagstaff J, Korson M, Kraus JP, et al. J Pediatr 1991;118:569–72.
206. Yap S, Naughten ER, Wilcken B, et al. Semin Thromb Hemost 2000;26:335–40.
207. Marcus AJ. N Engl J Med 1983;309:1515–7.
208. Davi G, Di Minno G, Coppola A, et al. Circ 2001;104:1124–8.
209. Yap S, Barry-Kinsella C, Naughten ER. Br J Obstet Gynaecol 2001;108:425–8.
210. Kang SS, Wong PW, Malinow MR. Annu Rev Nutr 1992;12: 279–98.
211. Chuang DT, Shih VE. In: Scriver CR, Beaudet AL, Sly WS et al, eds. Maple Syrup Urine Disease (branched-chain ketoaciduria). The Metabolic and Molecular Bases of Inherited Disease. 8th ed. New York, NY: McGraw-Hill, 2001:1971–2005.
212. Danner DJ, Armstrong N, Heffelfinger SC, et al. J Clin Invest 1985;75:858–60.
213. Ellerine NP, Herring WJ, Elsas LJ, et al. Biochem Med Metab Biol 1993;49:363–74.
214. Elsas LJ, Ellerine NP, Klein PD. Pediatr Res 1993;33:445–51.
215. Heffelfinger SC, Sewell ET, Elsas LJ, et al. Am J Hum Genet 1984;36:802–7.

216. Sweetman L, Williams JC. In: Scriver CR, Beaudet AL, Sly WS et al, eds. Branched Chain Organic Acidurias. The Metabolic and Molecular Bases of Inherited Disease. 8th ed. New York, NY: McGraw-Hill, 2001:2125–63.
217. Xu M, Nakai N, Ishigure K, et al. Biochem Biophys Res Commun 2000;276:1080–4.
218. Fenton WA, Gravel RA, Rosenblatt DS. In: Scriver CR, Beaudet AL, Sly WS et al, eds. Disorders of propionate and methylmalonate metabolism. The Metabolic and Molecular Bases of Inherited Disease. 8th ed. New York, NY: McGraw-Hill, 2001:2165–93.
219. Yoo EH, Cho HJ, Ki CS, et al. Clin Chem Lab Med 2007;45: 1495–7.
220. Tangerman A, Wilcken B, Levy HL, et al. Metab 2000;49: 1071–7.
221. Steele RD, Benevenga NJ. J Biol Chem 1978;253:7844–50.
222. Kaji H, Niioka T, Kojima Y, et al. Res Commun Chem Pathol Pharmacol 1987;56:101–9.
223. Coomes MH. In: Devlin TM, ed. Amino Acid Metabolism II: Metabolism of the Individual Amino Acids. Textbook of Biochemistry with Clinical Correlation. New York, NY: Wiley-Liss, 2002:779–824.
224. Baretz BH, Tanaka K. J Biol Chem 1978;253:4203–13.
225. Elsas LJ, Pask BA, Wheeler FB, et al. Metab 1972;21: 929–44.
226. Elsas LJ, Danner D, Lubitz D, et al. In: Walser M, Williamson GR, eds. Metabolic Consequences of Inherited Defects in Branched-Chain Alpha-Ketoacid Dehydrogenase: Mechanisms of Thiamine Action. Metabolism and Clinical Implications of Branched-Chain Amino Acids and Ketoacids. New York, NY: Elsevier, 1981:369–82.
227. Strauss KA, Puffenberger EG, Carson VJ. In: Adam MP, Everman DB, Mirzaa GM, et al., eds. Maple Syrup Urine Disease. [Updated 2020 Apr 23]. GeneReviews® [Internet]. Seattle (WA): University of Washington, Seattle; 1993-2023, 2006. Available at www.ncbi.nlm.nih.gov/books/NBK1319/
228. Jouvet P, Rustin P, Taylor DL, et al. Mol Biol Cell 2000;11: 1919–32.
229. Jones PM, Bennett MJ. Clin Chim Acta 2002;324:121–8.
230. Puckett RL, Lorey F, Rinaldo P, et Al. Mol Gen Metab, 2010; 100(2):136–142.
231. Elsas LJ, Priest JH, Wheeler FB, et al. Metab 1974;23:569–79.
232. Fernhoff PM, Lubitz D, Danner DJ, et al. Pediatr Res 1985;19:1011–6.
233. Thompson GN, Francis DEM, Halliday D. J Pediatr 1991;119: 35–41.
234. Giacoia GP, Berry GT. Am J Dis Child 1993;147:954–6.
235. Koch SE, Packman S, Koch TK, et al. J Am Acad Dermatol 1993;28:289–92.
236. DiGeorge AM, Rezvani I, Garibaldi LR, et al. N Engl J Med 1982;307:1492–5.
237. Strauss KA, Carson VJ, Soltys K, et. Al. Mol Gen Met 2020; 129:193–206.
238. Szmelcman S, Guggenheim K. Biochem J 1966;100:7–11.
239. Zaleski LA, Dancis J, Cox RP, et al. Can Med Assoc J 1973;109:299–300.
240. Lie IE, Haugstad S, Holm H. Hum Nutr Appl Nutr 1985;39: 130–6.
241. van Calcar SC, Harding CO, Davidson SR, et al. Am J Med Genet 1992;44:641–6.
242. Grunewald S, Hinrichs F, Wendel U. J Inherit Metab Dis 1998;21:89–94.
243. Hammersen G, Wille L, Schmidt H, et al. Monogr Hum Genet 1978;9:84–9.
244. Gropper SS, Naglak MC, Nardella M, et al. J Am Coll Nutr 1993;12:108–14.
245. Mazer LM, Yi SH, Singh RH. J Inherit Metab Dis 2010;33:121–7.
246. Strauss KA, Wardley B, Robinson D, et al. Mol Gen Metab 2010;99:333–345.
247. Snyderman SE, Goldstein F, Sansaricq C, et al. Pediatr Res 1984;18:851–3.
248. Treacy E, Clow CL, Reade TR, et al. J Inherit Metab Dis 1992;15:121–35.
249. Berry GT, Heidenreich R, Kaplan P, et al. N Engl J Med 1991;324:175–9.
250. Delaney A, Gal TJ. Anesthesiol 1976;44:83–6.
251. Dent CE, Westall RG. Arch Dis Child 1961;36:259–68.
252. Mazariegos GV, Morton DH, Sindhi R, et Al. J Peds 2012;160: 116–121.
253. Budd MA, Tanaka K, Holmes LB, et al. N Engl J Med 1967;277:321–7.
254. Rhead WJ, Tanaka K. Proc Natl Acad Sci U S A 1980;77:580–3.
255. Kraus JP, Matsubara Y, Barton D, et al. Genomics 1987;1:264–9.
256. Ensenauer R, Vockley J, Willard JM, et al. Am J Hum Genet 2004;75:1136–42.
257. Itoh T, Ito T, Ohba S, et al. Tohoku J Exp Med 1996;179:101–9.
258. Hutchinson RJ, Bunnell K, Thoene JG. J Pediatr 1985;106:62–5.
259. Dubiel B, Dabrowski C, Wetts R, Tanaka K. J Clin Invest 1983;72:1543–52.
260. Ikeda Y, Tanaka K. J Biol Chem 1983;258:1077–85.
261. Muetze U, Henze L, Gleich F, et al. JIMD 2021;44:857–870.
262. Lehnert W, Hunkler D. Eur J Pediatr 1986;145:260–6.
263. Blaskovics ME, Ng WG, Donnell GN. J Inherit Metab Dis 1978;1:9–11.
264. Shigematsu Y, Kikawa Y, Sudo M, et al. Clin Chim Acta 1984;138:333–6.
265. Millington DS, Roe CR, Maltby DA, et al. J Pediatr 1987;110: 56–60.
266. Loots DT, Erasmus E, Mienie LJ. Clin Chem 2005;51:1510–2.
267. Loots DT, Mienie LJ, Erasmus E. Eur J Clin Nutr 2007;61: 1323–7.
268. Naglak M, Salvo R, Madsen K, et al. Pediatr Res 1988;24:9–13.
269. Bartlett K, Gompertz D. Biochem Med 1974;10:15–23.
270. Stanley CA, Hale DE, Coates PM, et al. Pediatr Res 1983;17: 877–84.
271. Pinto A, Daly A, Evans S, et al. Mol Gen Metab Rep 2017;12: 16–22.
272. chrome-extension://efaidnbmnnnibpcajpcglclefindmkaj/www.e-imd.org/files/medias/files/recommendations/IVA%20guideline_Quick%20reference%20guide_Ensenauer_201408.pdf
273. Roe CR, Millington DS, Maltby DA, et al. J Clin Invest 1984; 74:2290–5.
274. Lott IT, Erickson AM, Levy HL. Pediatr 1972;49:616–8.
275. Dercksen M, Kulik W, Mienie LJ, et Al. Euro J Clin Nut 2016;70:1123–1126.
276. Berry GT, Yudkoff M, Segal S. J Pediatr 1988;113:58–64.
277. Schulze A, Lindner M, Kohlmüller D, et al. Pediatr 2003;111: 1399–1406.
278. Wilcken B, Wiley V, Hammond J, Carpenter K. N Engl J Med 2003;348:2304–2312.
279. Wilcken B. JIMD 2016;39:171–172.
280. Hyman SL, Porter CA, Page TJ, et al. J Pediatr 1987;111:558–62.
281. Rousson R, Guibaud P. J Inherit Metab Dis 1984;7(Suppl 1):1012.
282. Zschocke J, Ruiter JP, Brand J, et al. Pediatr Res 2000;48:852–5.
283. Pata S, Flores-Rojas K, Gil A, et al. Orphanet J Rare Dis 2022; 17:340.
284. Sato-Shirai I, Ogawa E, Arisaka A, et al. Brain Dev 2021;43: 308–313.
285. Abdenur JE, Sowa M, Simon M, et al. Mol Gen Met 2020;24.
286. Yang SY, He XY, Olpin SE, et al. Proc Natl Acad Sci U S A 2009;106:14820–4.

287. Sass JO, Sander S, Zschocke J. J Inherit Metab Dis 2004;27: 741–5.
288. Wolf B. In: Scriver CR, Beaudet AL, Sly WS et al, eds. Disorders of Biotin Metabolism. The Metabolic and Molecular Bases of Inherited Disease. 8th ed. New York, NY: McGraw-Hill, 2001: 3935–62.
289. Wolf B. Mol Genet Metab 2010;100:6–13.
290. Yano S, Sweetman L, Thorburn DR, et al. Eur J Pediatr 1997; 156:382–3.
291. Baumgartner MR, Hörster F, Dionisi-Vici C, et al. Orphanet J Rare Dis 2014;9:130.
292. De Raeve L, De Meirleir L, Ramet J, et al. J Pediatr 1994;124: 416–20.
293. Chace DH, DiPerna JC, Kalas TA, et al. Clin Chem 2001;47: 2040–4.
294. Shchelochkov OA, Carrillo N, Venditti C. In: Adam MP, Everman DB, Mirzaa GM, et al., eds. Propionic Acidemia. [Updated 2016 Oct 6]. GeneReviews® [Internet]. Seattle (WA): University of Washington, Seattle; 1993-2023. 2012. Available at www.ncbi.nlm.nih.gov/books/NBK92946/
295. Schlenk H. Fed Proc 1972;3:1431A.
296. Mulder H, Walstra P. The Milk Fat Globule: Emulsion Science Applied to Milk Products and Comparable Foods. Farnham Royal, Bucks, UK: Commonwealth Bureaux, 1974.
297. Dupont J, Mathias MM. Lipids 1969;4:478–83.
298. Vlaardingerbroek H, Hornstra G, de Koning TJ, et al. Mol Genet Metab 2006;88:159–65.
299. North KN, Korson MS, Gopal YR, et al. J Pediatr 1995;126: 916–22.
300. Kahler SG, Sherwood WG, Woolf D, et al. J Pediatr 1994;124: 239–43.
301. Inoue S, Krieger I, Sarnaik A, et al. Pediatr Res 1981;15:95–8.
302. Stork LC, Ambruso DR, Wallner SF, et al. Pediatr Res 1986; 20:783–8.
303. Thomas JA, Bernstein LE, Greene CL, et al. J Am Diet Assoc 2000;100:1074–6.
304. Feillet F, Bodamer OA, Dixon MA, et al. J Pediatr 2000;136: 659–63.
305. de Koning TJ, van Hagen CC, Carbasius-Weber E, et al. J Inherit Metab Dis 2002;25:46.
306. Yannicelli S, Acosta PB, Velazquez A, et al. Mol Genet Metab 2003;80:181–8.
307. Ciani F, Poggi GM, Pasquini E, et al. Clin Nutr 2000;19:137–9.
308. Baumgartner ER, Viardot C. J Inherit Metab Dis 1995;18: 138–42.
309. van't Hoff WG, Dixon M, Taylor J, et al. J Pediatr 1998;132: 1043–4.
310. Diss E, Iams J, Reed N, et al. Am J Obstet Gynecol 1995;172: 1057–9.
311. van der Meer SB, Poggi F, Spada M, et al. J Pediatr 1994;125: 903–8.
312. Manoli I, Sloan JL, Venditti CP. In: Adam MP, Everman DB, Mirzaa GM, et al., eds. Isolated Methylmalonic Acidemia. [Updated 2022 Sep 8]. GeneReviews® [Internet]. Seattle (WA): University of Washington, Seattle; 1993-2023. 2005. Available at www.ncbi.nlm.nih.gov/books/NBK1231/
313. Sloan JL, Carrillo N, Adams D, et al. In: Adam MP, Everman DB, Mirzaa GM, et al., eds. Disorders of Intracellular Cobalamin Metabolism. [Updated 2021 Dec 16]. GeneReviews® [Internet]. Seattle (WA): University of Washington, Seattle; 1993-2023. 2008. Available from: www.ncbi.nlm.nih.gov/books/NBK1328/
314. Goodman SI, Frerman FE. In: Scriver CR, Beaudet AL, Sly WS et al, eds. Organic Acidemias Due to Defects in Lysine Oxidation: 2-Ketoadipic Acidemia and Glutaric Acidemia. The Metabolic and Molecular Bases of Inherited Disease. 8th ed. New York, NY: McGraw-Hill, 2001:2195–204.
315. Baric I, Zschocke J, Christensen E, et al. J Inherit Metab Dis 1998;21:326–40.
316. Hoffmann GF, Zschocke J. J Inherit Metab Dis 1999;22:381–91.
317. Greenberg CR, Reimer D, Singal R, et al. Hum Mol Genet 1995;4:493–5.
318. Yannicelli S. In: Acosta PB, ed. Nutrition Management of Patients with Inherited Disorders of Organic Acid Metabolism. Nutrition Management of Patients with Inherited Metabolic Disorders. Sudbury, MA: Jones and Bartlett, 2010:283–341.
319. Crombez EA, Cederbaum SD, Spector E, et al. Mol Genet Metab 2008;94:132–4.
320. Garcia P, Martins E, Diogo L, et al. Eur J Pediatr 2008;167: 569–73.
321. Brusilow S, Horwich A. In: Scriver CR, Beaudet AL, Sly WS et al, eds. Urea Cycle Enzymes. The Metabolic and Molecular Bases of Inherited Disease. 8th ed. New York, NY: McGraw-Hill, 2001:1909–63.
322. Azevedo L, Stolnaja L, Tietzeova E, et al. Mol Genet Metab 2003;78:152–7.
323. Tuchman M, Berry SA, Thuy LP, et al. Pediatr 1993;91: 664–66.
324. Haberle J, Schmidt E, Pauli S, et al. Hum Mutat 2003;21:444.
325. Haberle J, Schmidt E, Pauli S, et al. Hum Mutat 2003;21:593–7.
326. Valle D, Simell O. In: Scriver CR, Beaudet AL, Sly WS et al, eds. The Hyperornithinemias The Metabolic and Molecular Bases of Inherited Disease. 8th ed. New York, NY: McGraw-Hill, 2001:1857–98.
327. Leonard JV, Morris AA. Semin Neonatol 2002;7:27–35.
328. Oechsner M, Steen C, Sturenburg HJ, et al. J Neurol Neurosurg Psychiatry 1998;64:680–2.
329. Kline JJ, Hug G, Schubert WK, et al. Am J Dis Child 1981;135:437–42.
330. Cederbaum SD, Shaw KN, Valente M. J Pediatr 1977;90: 569–73.
331. Levin B, Abraham JM, Oberholzer VG, et al. Arch Dis Child 1969; 44:152–61.
332. Kennaway NG, Harwood PJ, Ramberg DA, et al. Pediatr Res 1975;9:554–8.
333. Legras A, Labarthe F, Maillot F, et al. Crit Care Med 2002;30: 241–4.
334. Levy HL, Coulombe JT, Shih V. In: Bickel H, Guthrie R, eds. The New England Experience. Neonatal Screening for Inborn Errors of Metabolism. Berlin, Germany: Springer, 1980.
335. Barsotti RJ. J Pediatr 2001;138(Suppl):S11–9.
336. Summar M, Tuchman M. J Pediatr 2001;138(Suppl):S6–10.
337. Steiner RD, Cederbaum SD. J Pediatr 2001;138(Suppl):S21–9.
338. Singh RH. In: Acosta PB, ed. Nutrition Management of Patients with Inherited Disorders of Urea Cycle Enzymes. Nutrition Management of Patients with Inherited Metabolic Disorders. Sudbury, MA: Jones and Bartlett, 2010:405–29.
339. Batshaw ML, MacArthur RB, Tuchman M. J Pediatr 2001; 138(Suppl):S46–54.
340. Palekar A. Pediatr Res 2000;48:357–9.
341. Darmaun D, Welch S, Rini A, et al. Am J Physiol 1998;274:E801–7.
342. Scaglia F. Mol Genet Metab 2010;100(Suppl 1):S72–6.
343. Lichter-Konecki U, Caldovic L, Morizono H, et al. Ornithine Transcarbamylase Deficiency. [Updated 2022 May 26]. In: Adam MP, Everman DB, Mirzaa GM, et al., eds. GeneReviews® [Internet]. Seattle (WA): University of Washington, Seattle; 1993-2023. 2013. Available at www.ncbi.nlm.nih.gov/books/NBK154378/
344. Ah Mew N, Simpson KL, Gropman AL, et al. In: Adam MP, Everman DB, Mirzaa GM, et al., eds. Urea Cycle Disorders Overview. [Updated 2017 Jun 22]. GeneReviews® [Internet]. Seattle (WA): University of Washington, Seattle; 1993-2023. 2003. Available from: www.ncbi.nlm.nih.gov/books/NBK1217/

345. FAO/WHO/UNU Expert Consultation. Protein and Amino Acid Requirements in Human Nutrition. Geneva, Switzerland: World Health Organization, 2007:161–184.
346. Hyman SL, Coyle JT, Parke JC, et al. J Pediatr 1986;108:705–9.
347. Ohtani Y, Ohyanagi K, Yamamoto S, et al. J Pediatr 1988;112:409–14.
348. Ohtsuka Y, Griffith OW. Biochem Pharmacol 1991;41:1957–61.
349. Iafolla AK, Gale DS, Roe CR. J Pediatr 1990;117:102–5.
350. Renner C, Sewell AC, Bervoets K, et al. Eur J Pediatr 1995;154:909–14.
351. Maestri NE, Brusilow SW, Clissold DB, et al. N Engl J Med 1996;335:855–9.
352. Maestri NE, Clissold D, Brusilow SW. J Pediatr 1999;134: 268–72.
353. Nicolaides P, Liebsch D, Dale N, et al. Arch Dis Child 2002;86:54–6.
354. Acosta PB, Yannicelli S, Ryan AS, et al. Mol Genet Metab 2005;86:448–55.
355. Sanjurjo P, Rodriguez-Soriano J, Vallo A, et al. Eur J Pediatr 1991;150:730–1.
356. Arn PH, Hauser ER, Thomas GH, et al. N Engl J Med 1990;322:1652–5.
357. Burlina AB, Ogier H, Korall H, et al. Mol Genet Metab 2001;72:351–5.
358. Scaglia F, Zheng Q, O'Brien WE, et al. Pediatr 2002;109: 150–2.
359. Mardach MR, Roe K, Cederbaum SD. J Inherit Metab Dis 1999;22:102–6.
360. Redonnet-Vernhet I, Rouanet F, Pedespan JM, et al. Neurol 2000;54:1008.
361. Wilson O, Schfert W, Ballatore W, et al. Pediatr Res 1995;37:323A.
362. Holton JB, Walter JH, Tyfield LA. In: Scriver CR, Beaudet AL, Sly WS et al, eds. Galactosemia. The Metabolic and Molecular Bases of Inherited Disease. 8th ed. New York: McGraw-Hill, 2001:1553–87.
363. Elsas LJ, Langley S, Steele E, et al. Am J Hum Genet 1995;56: 630–9.
364. Elsas LJ, Lai K. Genet Med 1998;1:40–8.
365. Flach JE, Reichardt JK, Elsas LJ. Mol Biol Med 1990;7:365–9.
366. Leslie ND, Immerman EB, Flach JE, et al. Genomics 1992; 14:474–80.
367. Lai K, Willis AC, Elsas LJ. J Biol Chem 1999;274:6559–66.
368. Berry GT, Moate PJ, Reynolds RA, et al. Mol Genet Metab 2004;81: 22–30.
369. Schadewaldt P, Kamalanathan L, Hammen HW, et al. Mol Genet Metab 2004;81:31–44.
370. Elsas LJ, Fridovich-Keil J, Leslie ND. Int Pediatr 1993;8:101–9.
371. Kadhom N, Baptista J, Brivet M, et al. Biochem Med Metab Biol 1994;52:140–4.
372. Wolfrom C, Raynaud N, Kadhom N, et al. J Inherit Metab Dis 1993;16:78–90.
373. Charlwood J, Clayton P, Keir G, et al. Glycobiol 1998;8:351–7.
374. Stibler H, von Dobeln U, Kristiansson B, et al. Acta Paediatr 1997;86:1377–8.
375. Prestoz LL, Couto AS, Shin YS, et al. Eur J Pediatr 1997;156: 116–20.
376. Lai KW, Cheng LY, Cheung AL, et al. Cell Tissue Res 2003;311: 417–25.
377. Lai K, Elsas LJ. Biochem Biophys Res Commun 2000;271: 392–400.
378. Lai K, Langley SD, Khwaja FW, et al. Glycobiol 2003;13: 285–94.
379. Lai K, Elsas LJ, Wierenga KJ. IUBMB Life 2009;61:1063–74.
380. Waggoner DD, Buist NR, Donnell GN. J Inherit Metab Dis 1990;13:802–18.
381. Irons M, Levy HL, Pueschel S, et al. J Pediatr 1985;107:261–3.
382. Urbanowski JC, Cohenford MA, Levy HL et al. N Engl J Med 1982;306:84–6.
383. Schweitzer S, Shin Y, Jakobs C, et al. Eur J Pediatr 1993;152: 36–43.
384. Guerrero NV, Singh RH, Manatunga A, et al. J Pediatr 2000;137:833–41.
385. Elsas LJ, Langley S, Paulk EM, et al. Eur J Pediatr 1995;154: S21–7.
386. Lai K, Langley SD, Singh RH, et al. J Pediatr 1996;128:89–95.
387. Elsas LJ, Dembure PP, Langley S, et al. Am J Hum Genet 1994;54:1030–6.
388. Fridovich-Keil JL, Langley SD, Mazur LA, et al. Am J Hum Genet 1995;56:640–6.
389. Beutler E, Baluda MC. J Lab Clin Med 1966;68:137–41.
390. Elsas LJ. Prenat Diagn 2001;21:302–3.
391. Jakobs C, Kleijer WJ, Allen J, et al. Eur J Pediatr 1995; 154(Suppl):S33–6.
392. Acosta PB. In: Acosta PB, ed. Nutrition Management of Patients with Inherited Metabolic Disorders of Galactose Metabolism. Nutrition Management of Patients with Inherited Metabolic Disorders. Sudbury, MA: Jones and Bartlett, 2010:343–67.
393. Singh RH, Kennedy MJ, Jonas CR, et al. J Inherit Metab Dis 2003;26:123A.
394. Panis B, Gerver WJ, Rubio-Gozalbo ME. Eur J Pediatr 2007;166:443–6.
395. Kaufman FR, Loro ML, Azen C, et al. J Pediatr 1993;123: 365–70.
396. Rubio-Gozalbo ME, Hamming S, van Kroonenburgh MJ, et al. Arch Dis Child 2002;87:57–60.
397. Berry GT, Nissim I, Gibson JB, et al. Eur J Pediatr 1997;156 (Suppl 1):S43–9.
398. Berry GT, Singh RH, Mazur AT, et al. Pediatr Res 2000;48: 323–8.
399. Zlatunich CO, Packman S. J Inherit Metab Dis 2005;28:163–8.
400. Portnoi PA, MacDonald A. J Hum Nutr Diet 2009;22:400–8.
401. Gross KC, Acosta PB. J Inherit Metab Dis 1991;14:253–8.
402. Gropper SS, Gross KC, Olds SJ. J Am Diet Assoc 1993;93: 328–30.
403. Cerbulis J. Arch Biochem Biophys 1954;49:442–50.
404. Gitzelmann R, Auricchio S. Pediatr 1965;36:231–5.
405. Koch R, Acosta P, Ragsdale N, et al. J Am Diet Assoc 1963;43: 212–5.
406. Ning C, Reynolds R, Chen J, et al. Pediatr Res 2000;48:211–7.
407. Holton JB. Galactosemia. In: Schaub J, van Hoof F, Vis HL, eds. Inborn Errors of Metabolism. New York, NY: Raven Press, 1991:169–80.
408. Lineback DR, Ke CH. Cereal Chem 1975;52:334–7.
409. Acosta PB, Gross KC. Eur J Pediatr 1995;154(Suppl):S87–92.
410. Kumar A, Weatherly MR, Beaman DC. Pediatr 1991;87:352–60.
411. Harju M. Milchwissenschaft 1990;45:411–5.
412. Segal S. Int Pediatr 1992;7:75–82.
413. Wiesmann UN, Rose-Beutler B, Schluchter R. Eur J Pediatr 1995;154(Suppl):S93–6.
414. Weese SJ, Gosnell K, West P, et al. J Am Diet Assoc 2003;103: 373–5.
415. Rubio-Gozalbo ME, Panis B, Zimmermann LJ, et al. Mol Genet Metab 2006;89:316–22.
416. Asp NG. Biochem J 1971;121:299–308.
417. Andersson L, Bratt C, Arnoldsson KC, et al. J Lipid Res 1995;36:1392–400.
418. Donnell GN, Bergren WR, Perry G, et al. Pediatr 1963;31:802–10.
419. Kaufman FR, Xu YK, Ng WG, et al. J Pediatr 1988;112:754–6.
420. Shield JP, Wadsworth EJ, MacDonald A, et al. Arch Dis Child 2000;83:248–50.
421. Webb AL, Singh RH, Kennedy MJ, et al. Pediatr Res 2003;53: 396–402.

422. Ng WG, Xu YK, Kaufman FR, et al. J Inherit Metab Dis 1989;12:257–66.
423. Gibson JB, Reynolds RA, Palmieri MJ, et al. Metab 1995;44: 597–604.
424. Manis FR, Cohn LB, McBride-Chang C, et al. J Inherit Metab Dis 1997;20:549–55.
425. Rubio-Gozalbo ME, Gubbels CS, Bakker JA, et al. Hum Reprod Update 2010;16:177–88.
426. Rutherford PJ, Davidson DC, Matthai SM. J Hum Nutr Diet 2002;15:39–42.
427. Berry GT. Eur J Pediatr 1995;154(Suppl):S53–64.
428. Lai K, Tang M, Yin X, et al. Biosci Hypotheses 2008;1:263–71.
429. Wehrli SL, Berry GT, Palmieri M, et al. Pediatr Res 1997;42: 855–61.
430. Komrower GA. J Inherit Metab Dis 1982;5(Suppl 2):96–104.

CHAPTER 72

Inherited Lipid Disorders of Beta-Oxidation

Jerry Vockley, Lynne A. Wolfe, and Deborah L. Renaud

CHAPTER OUTLINE

Abbreviations

ABC, adenosine triphosphate binding cassette; **ACAD**, acyl-CoA dehydrogenase; **ACD**, acyl-coenzyme A dehydrogenase; **AFLP**, acute fatty liver of pregnancy; **ATP**, adenosine triphosphate; **CACT**, carnitine-acylcarnitine translocase; **CoA**, coenzyme A; **CPT**, carnitine palmitoyltransferase; **DHA**, docosahexaenoic acid; **EFA**, essential fatty acid; **ETF**, electron transfer flavoprotein; **FAD**, flavin adenine dinucleotide; **GA**, glutaric aciduria; **HELLP**, hemolysis, elevated liver enzymes, low platelet; **HMG**, hydroxymethylglutaryl; **LCAD**, long-chain acyl-coenzyme A dehydrogenase; **LCHAD**, long-chain 3-hydroxyacyl-coenzyme A dehydrogenase; **MADD**, multiple acyl-coenzyme A dehydrogenation disorder; **MCAD**, medium-chain acyl-coenzyme A dehydrogenase; **MCT**, medium-chain triglyceride; **MRI**, magnetic resonance imaging; **PKU**, phenylketonuria; **PTS**, peroxisomal targeting signal; **PEX**, peroxins; **RCDP**, rhizomelic chondrodysplasia punctata; **SCAD**, short-chain acyl-coenzyme A dehydrogenase; **SCOT**, succinyl-coenzyme A:3-ketoacid coenzyme A transferase; **SIDS**, sudden infant death syndrome; **TFP**, trifunctional protein; **VLCAD**, very-long-chain acyl-coenzyme A dehydrogenase; **VLCFA**, very-long-chain fatty acid; **X-ALD**, X-linked adrenoleukodystrophy

Introduction

Fatty acid β-oxidation (FAO), which results in sequential cleavage of two carbon units from fatty acids, occurs in two separate organelles in cells, with different physiologic functions. Mitochondrial FAO is primarily a source of energy for the body during times of fasting and metabolic stress. Free fatty acids released into the blood via catabolism of fat stores or from dietary sources are metabolized in the mitochondria. β-oxidation also functions as a degradative route for complex lipids in a different subcellular compartment, the peroxisomes (sometimes referred to as microbodies). Peroxisomes are subcellular organelles bounded by a single lipid bilayer membrane.[1] Peroxisomal FAO is primarily a pathway for degradation of complex lipids and fatty acids with chain lengths too long for metabolism by mitochondrial FAO.[2] The enzymes involved in each FAO in each organelle are different, and clinical phenotypes related to their dysfunction differ dramatically.[3]

Mitochondria are bounded by two lipid bilayer membranes (the inner and outer mitochondrial membranes).[1] The intermembrane space constitutes a distinct compartment within the mitochondria, while the space bounded by the inner mitochondrial membrane is known as the matrix. Mitochondria are unique among organelles in animals in that they contain their own genetic information and are solely maternally inherited.[4] Most proteins found in the mitochondria, however, are nuclear encoded and thus are inherited in a standard Mendelian fashion. In general, they are synthesized in a larger precursor form containing information in an amino terminal signal peptide necessary for targeting the proteins to the correct location.[5,6] These signal sequences are usually removed after import of the protein into the mitochondrion. More than one targeting signal may be necessary to direct the imported protein to the correct mitochondrial space or membrane.

All peroxisomal proteins are encoded in the nuclear genome, synthesized on free cytoplasmic polyribosomes, and transported to the organelle post-translationally. This process is mediated by specific targeting sequences on proteins and a variety of specific receptor proteins on the peroxisomes.[7] The most common is the presence of a serine-lysine-leucine amino acid motif at the carboxy terminus of the protein. This peroxisomal targeting signal (PTS1), present in more than 95% of proteins destined for the peroxisomal matrix, binds to the cytosolic receptor protein Pex5p. A second mechanism uses an amino terminus targeting signal (PTS2), which binds to the receptor protein Pex7p. The PTS-receptor complexes are stabilized and transported to the peroxisomal membrane where they interact with the docking machinery and are then translocated into the peroxisomal matrix.

To date, 32 proteins involved in these processes, encoded by *PEX* genes and known as peroxins, have been identified and characterized (**Figure 72.1**, **Table 72.1**). These peroxins can be divided into three groups: 1) Pex3, Pex6, and Pex19, which are essential for peroxisome membrane assembly; 2) peroxins responsible for protein import; and 3) Pex11, which interacts with other proteins involved in dynamin-like protein 1 (DLP1, also known as DNM1L), mitochondrial fission factor (Mff) and mitochondrial fission protein 1 (Fis1).[8]

Table 72.1 Enzymes of Peroxisomal β-oxidation

Acyl-CoA synthetases
Acyl-CoA oxidases (straight chain and branched chain)
Acyl-CoA thioesterases
Bifunctional protein
2-enoyl-CoA hydratase
3-hydroxyacyl-CoA dehydrogenase
3-ketoacyl-CoA thiolases
Carnitine acyltransferases
2-methyl-CoA racemase
Enzymes of β-oxidation of unsaturated fats
2,4-dienoyl-CoA reductase
3,2-*Trans*-enoyl-CoA isomerase
2-enoyl-CoA hydratase
2,5-enoyl-CoA reductase
3,5-2,4-dienoyl-CoA isomerase

Peroxisomes and mitochondria arise by division of previously existing organelles and are randomly distributed to daughter cells upon cellular division.[7] Peroxisomes interact with mitochondria at multiple levels. They share common fission factors, which are involved in the division of both peroxisomes and mitochondria. Peroxisomes are metabolically linked to mitochondria

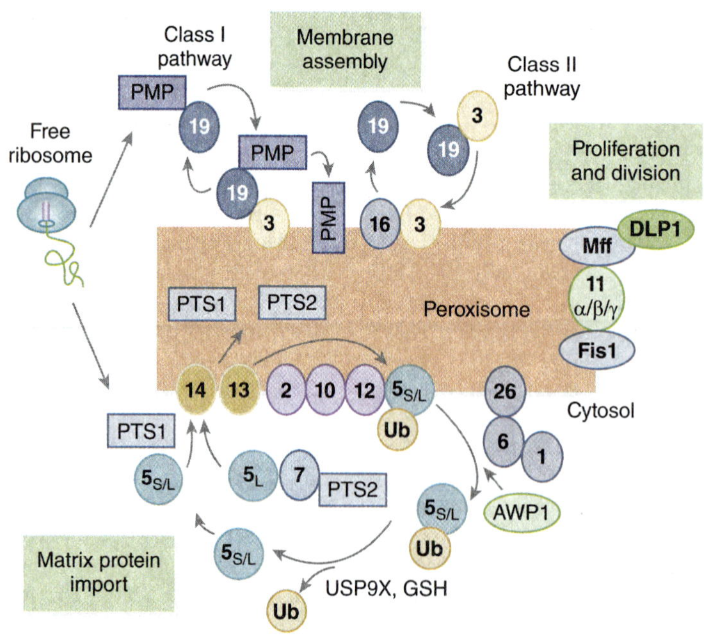

Figure 72.1 Assembly of peroxisomes and import of proteins into the organelle is a complicated process that requires at least 19 *PEX* genes (proxins, numbered from 1–19) and a variety of other steps. Peroxins are divided into three groups by the functions: membrane assembly, matrix protein import, and proliferation and division as described in the text. Mutations in any of these genes impair peroxisomal biogenesis and cause clinical disease.

Reproduced from Okumoto, K., Tamura, S., Honsho, M., Fujiki, Y. (2020). Peroxisome: Metabolic Functions and Biogenesis. In: Lizard, G. (eds) Peroxisome Biology: Experimental Models, Peroxisomal Disorders and Neurological Diseases. Advances in Experimental Medicine and Biology, vol 1299. Springer, Cham. https://doi.org/10.1007/978-3-030-60204-8_1.[8]

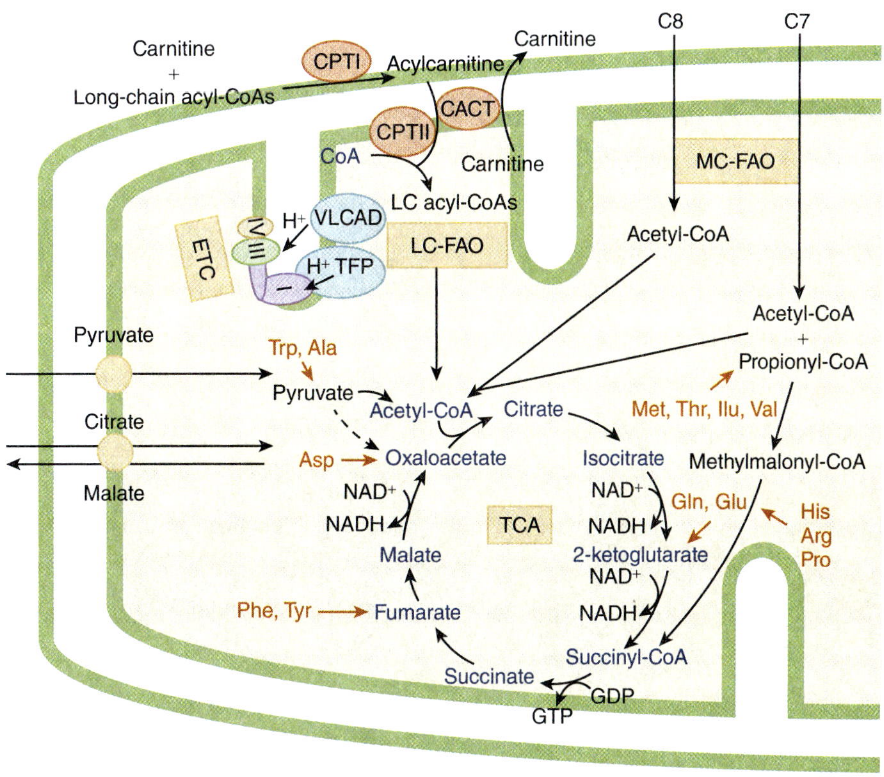

Figure 72.2 Pathway of enzyme and transporter proteins involved in mitochondrial β-oxidation and its interaction with other energy pathways. Abbreviations appear in the text.

Modified from Vockley, J. The changing face of disorders of fatty acid oxidation. Mayo Clinic proceedings vol. 69,3 (1994): 249-57. doi:10.1016/s0025-6196(12)61064-7.[13]

through the beta-oxidation of fatty acids and the metabolism of reactive oxygen species.[9] Secondary mitochondrial changes occur in some peroxisomal biogenesis defects.[10] Approximately 2,000 proteins are involved in mitochondria metabolism and function, at least 25 of which are involved in FAO (**Figure 72.2** and **Table 72.2**).[11] Of these, defects in at least 22 have been shown to cause disease in humans.[12]

The Enzymes of β-Oxidation

Mitochondria

Free fatty acids are transported through the blood after intestinal absorption or mobilization from endogenous stores by the use of albumin as a carrier protein or in the form of triacylglycerols in lipoprotein complexes.[14] Transport of free fatty acids intracellularly and through the cytoplasm is probably accomplished by a specific transport process; however, the mechanism of this step is not well characterized.[15] Before undergoing β-oxidation, free fatty acids must be activated to their corresponding acyl-CoA thioesters (Figure 72.2; Table 72.2). Long chain-specific acyl-CoA synthetases can be found in various subcellular locations but are thought to arise from a single gene product.[16] Short- and medium-chain carboxylic acids enter the mitochondrial matrix directly, where they are activated. In contrast, long-chain fats are activated in the cytoplasm and require active transport into mitochondria. Transport of long-chain acyl-CoAs requires at least two enzymes, a transporter protein, and the use of carnitine as an intermediate carrier molecule, that comprise the carnitine cycle. Carnitine is itself transported intracellularly by a specific transporter protein.[17] Two carnitine transporters have been described, one specific to liver and a second with a more ubiquitous distribution, including kidney, muscle, and fibroblasts. Long-chain acyl-CoAs are conjugated to carnitine by carnitine palmitoyl transferase I (CPT I). This enzyme is located on the inner aspect of the outer mitochondrial membrane. Tissue-specific isoforms of this enzyme exist for muscle, liver, and brain.[17] Long-chain acylcarnitines are then passed to carnitine palmitoyl transferase II (CPT II) in the inner mitochondrial membrane by a translocase.

Once present in the mitochondrial matrix, acyl-CoAs of chain lengths of an even number of carbon atoms on the alpha backbone undergo a series of enzymatic reactions that result in the release of the two-carbon unit acetyl-CoAs and a new acyl-CoA molecule that is two carbons shorter (**Figure 72.3**; Table 72.2). For fats with an odd number of carbons on the alpha chain, the last cycle of FAO generates one propionyl-CoA and one acetyl-CoA. The first step in this cycle is the dehydrogenation of the acyl-CoA to 2-enoyl-CoA. This reaction is catalyzed by a family of related enzymes, the acyl-CoA dehydrogenases (ACAD).[18] Four different members of this family are active in β-oxidation: very long, long, medium, and short chain acyl-CoA dehydrogenases (VLCAD, LCAD, MCAD, and SCAD, respectively), which differ in their chain-length specificity. The role of LCAD in fatty acid β-oxidation is unclear. It is present in much lower concentration than VLCAD in tissues where the two have been separated, and thus would appear to play a minor role in the flux of fatty acids through β-oxidation. LCAD, however, has significant activity toward long, branched chain substrates, unlike VLCAD and, thus, may be more important in their metabolism.[19,20] A final long chain

Table 72.2 Enzymes Involved in Mitochondrial Fatty Acid Oxidation

Enzyme	Proven Clinical Disorder
Fatty acid activation	
Acyl-CoA synthetase	No
Carnitine cycle	
Plasma membrane carnitine transporter	Yes
CPT I	Yes
Carnitine-acylcarnitine translocase	Yes
CPT II	Yes
Mitochondrial β-oxidation spiral	
Very long-chain acyl-CoA dehydrogenase (membrane)	Yes
LCAD (matrix)	No
MCAD	Yes
SCAD	Yes
Trifunctional protein	Yes
Long-chain 2-enoyl CoA hydratase	
Long-chain 3-hydroxyacyl-CoA dehydrogenase	Yes (isolated)
Long-chain 3-ketoacyl-CoA thiolase	
Crotonase (short-chain 2-enoyl-CoA hydratase)	No
SCHAD	Yes
Short-chain 3-ketoacyl-CoA thiolase	Possible
Enzymes of β-oxidation of unsaturated fats	
Long-chain Δ^3,Δ^2-enoyl-CoA isomerase	No
Short-chain Δ^3,Δ^2-enoyl-CoA isomerase	No
2,4-Dienoyl-CoA reductase	Possible
Enzymes of ketone body production	
HMG-CoA synthase	Yes
HMG-CoA lyase	Yes
D-3-Hydroxybutyrate dehydrogenase	

CoA, coenzyme A; *CPT*, carnitine palmitoyltransferase; *HMG*, hydroxymethylglutaryl; *LCAD*, long-chain acyl-coenzyme A dehydrogenase; *MCAD*, medium-chain acyl-coenzyme A dehydrogenase; *SCHAD*, short-chain 3-hydroxyacyl-coenzyme A dehydrogenase; *SCAD*, short-chain acyl-coenzyme A dehydrogenase

ACAD, designated ACAD9, is more active toward unsaturated substrates than saturated ones, but it's primary role in mitochondria appears to be nonenzymatic, acting as a respiratory chain complex I assembly factor.[21]

The ACADs differ from most other dehydrogenases because they use electron transfer flavoprotein (ETF) as a final electron acceptor and, thus, can channel electrons directly into the ubiquinone pool of the electron transport machinery by way of ETF:ubiquinone oxidoreductase (ETF dehydrogenase).[22] The ACDs are homotetramers (except VLCAD and ACAD9, which are homodimers) whose monomers are synthesized in larger precursor form in the cytoplasm from nuclear encoded transcripts, and then transported into mitochondria.[18,23] Once inside the mitochondrial matrix, the leader peptide is removed by one or two specific proteases, and the mature subunits assemble into the active multimer. One molecule of FAD is noncovalently attached to each ACAD subunit. cDNAs for each of these proteins have been cloned and sequence analysis shows that they are approximately 30% to 35% conserved, suggesting evolution from a common primordial gene.[23] Four other members of this gene family are involved in the metabolism of branched chain amino acids, lysine, and tryptophan rather than in β-oxidation.[24]

The 2-enoyl-CoA moieties produced by the ACDs are hydrated to 3-hydroxy-acyl-CoAs. These, in turn, undergo 2,3 dehydrogenation to 2-ketoacyl-CoAs, followed by cleavage of the thioester bond.[25] This releases acetyl-CoA and completes one turn of the recursive β-oxidation cycle. The exact mechanism of these steps varies for substrates of differing chain length. The mitochondrial trifunctional protein (TFP) contains 2-enoyl-CoA hydratase, 3-hydroxy-acyl-CoA dehydrogenase, and 3-ketoacyl-CoA thiolase activities for longer chain acyl-CoA substrates.[26] This complex is a heterotetramer consisting of 2 α and 2 β subunits. LCHAD and 3-enoyl-CoA hydratase activities reside on the α subunit, while 3-ketoacyl-CoA thiolase activity resides on the β subunit. In contrast, individual proteins, which catalyze these reactions for shorter chain substrates, have single activities. These proteins include a short/medium chain 3-hydroxyacyl-CoA dehydrogenase (S/MCHAD), a short chain enoyl-CoA hydratase (also called crotonase), and distinct medium and short chain 3-ketoacyl-CoA thiolases.[26,27] The substrate specificities of many of these enzymes overlap, and additional enzymes with yet different substrate optima possibly exist for some steps of β-oxidation. Enzymes catalyzing several additional sets of reactions are necessary for the complete oxidation of unsaturated fatty acyl-CoA molecules including a 2,4-dienoyl-CoA reductase[28] and a D^3, D^2-enoyl-CoA isomerase.[20] In odd-chain (carbon number) fatty acid oxidation, the final three carbon intermediate propionyl-CoA is metabolized by propionyl-CoA carboxylase. There is evidence that TFP associates with complex I of the respiratory chain, while VLCAD interacts with TFP in a macromolecular protein complex that promotes metabolic channeling of reducing equivalents for oxidative phosphorylation (Figure 72.2).[29,30] ETFDH similarly interacts with complex III of the respiratory chain.[30]

Ketone bodies are produced exclusively in the liver from acetyl-CoA generated by β-oxidation (**Figure 72.4**). Hydroxymethyl glutaryl (HMG)-CoA synthase forms 3-hydroxy-3-methylglutaryl-CoA (HMG-CoA) from acetoacetyl-CoA and acetyl-CoA. Acetyl-CoA and acetoacetate are then produced by cleavage of HMG-CoA by HMG-CoA lyase.[31,32] Finally, acetoacetate is reduced to D-3-hydroxybutyrate by D-3-hydroxybutyrate dehydrogenase within mitochondria.[33]

Several alternative metabolic pathways become important when mitochondrial β-oxidation is impaired. Peroxisomal β-oxidation allows continued metabolism of longer chain fats, while omega oxidation in the cytosol (which proceeds from the opposite end of the fatty acid) results in the production of

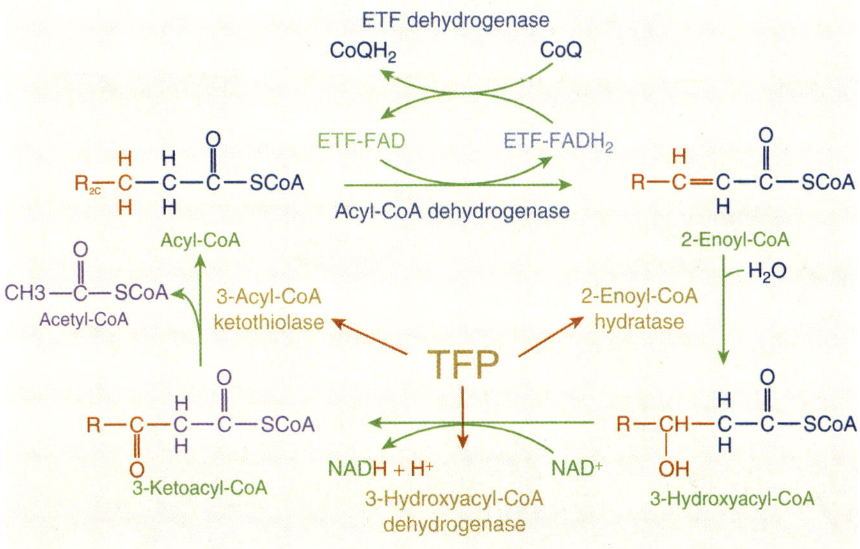

Figure 72.3 The four mitochondrial steps of fatty acid oxidation. The acyl-CoA dehydrogenase used varies, depending on substrate. The next three steps for long chain fat oxidation are contained in the trifunctional protein (TFP). For medium and short chain fats, individual enzymes are active at each step.

Adapted with permission from SIMDNAMA program and the Society for Inherited Metabolic disorders.

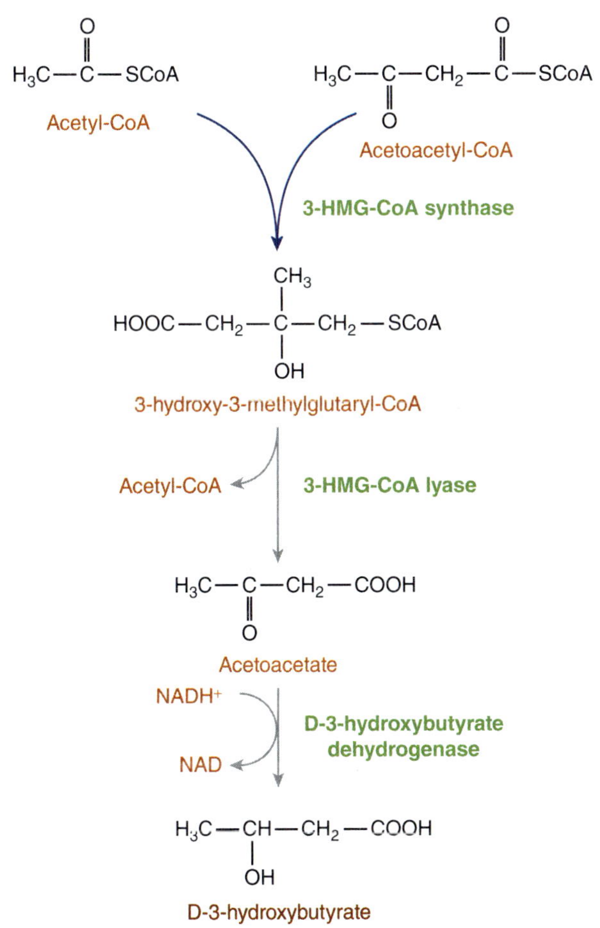

Figure 72.4 Generation of ketone bodies from the products of β-oxidation.

Adapted with permission from SIMDNAMA program and the Society for Inherited Metabolic disorders.

the characteristic dicarboxylic acids, often present in these disorders. In addition, deacylation of acyl-CoA by cytosolic thioesters, and conjugation of acyl-CoAs to glycine and carnitine, become important mechanisms of CoA scavenging and detoxification, respectively.

Peroxisomes

β-Oxidation in peroxisomes and mitochondria similarly involves four steps: dehydrogenation, hydration of the double bond, a second dehydrogenation, and then thiolytic cleavage. However, the β-oxidation cycle in peroxisomes differs from mitochondria in several important ways.[9] Peroxisomes and mitochondria have different substrate specificities catalyzed by different enzymes, encoded by distinct genes (**Figure 72.5**, Table 72.1). Fatty acids are transported across the peroxisomal membrane as acyl-CoAs rather than as acylcarnitines into the mitochondria. The peroxisomal cycle results only in partial chain shortening rather than complete oxidation of fatty acids. As a result, ATP production from peroxisomal substrates is less efficient because electrons produced by peroxisomal oxidases are donated directly to molecular oxygen for hydrogen peroxide production rather than to the respiratory chain. Acetyl-CoA, propionyl-CoA, and medium-chain acyl-CoAs produced by chain-shortening in the peroxisome are shuttled to the mitochondria for full oxidation. Carnitine plays a role in the export of chain-shortened fatty acids from the peroxisome but is not involved in fatty acid uptake.[9] Peroxisomes are assembled and proteins targeted to them by a series of genes including the PEX family (Figure 72.1).[8]

Three peroxisomal half ATP-binding cassette (ABC) transporters have been described in humans, including ABCD1 (ALDP), ABCD2 (ALDRP), and ABCD3(PMP70).[31] The best characterized of these half transporters is ABCD1/ALDP, which is involved in the transport of acyl-CoA esters across the peroxisomal membrane. X-linked adrenoleukodystrophy (X-ALD) is caused by mutations in the *ABCD1* gene coding for ALDP.[9,34] Very long chain fatty acids are synthesized in the endoplasmic reticulum from shorter-chain fatty acids via the fatty acid

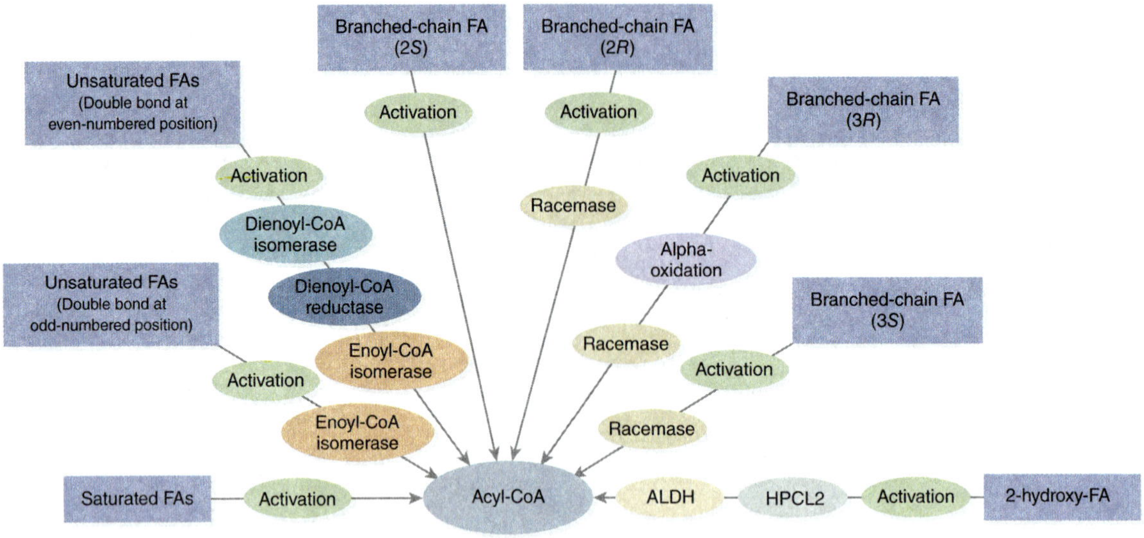

Figure 72.5 β-oxidation of fatty acids and other substrates in the peroxisome. The combination of enzymes used depends on the starting substrate. The end products are typically only partially degraded and are transported to other cellular compartments for additional metabolism. Abbreviations are in the text.

Adapted with permission from SIMDNAMA program and the Society for Inherited Metabolic disorders.

synthase system, and then elongated prior to being shuttled to the peroxisome for β-oxidation.[9] The activation of fatty acids to fatty acyl-CoAs is accomplished by acyl-CoA synthetases in the peroxisomal membrane. The first step of the β-oxidation cycle in the peroxisomal matrix (Figure 72.5) is oxidation by one of three different acyl-CoA oxidases (ACOX1, ACOX2, and ACOX3). The second and third steps of the β-oxidation cycle are conducted by a bifunctional protein complex containing enoyl-CoA hydratase and 3-hydroxyacyl-CoA dehydrogenase activities associated with the inner aspect of the peroxisomal membrane. Peroxisomal specific 3-ketoacyl-CoA thiolases catalyze the final step of the cycle to produce acetyl-CoA and regenerate an acyl-CoA.

Multiple carnitine acyltransferases with different chain-length specificities catalyze the conversion of acetyl-CoA and acyl-CoAs to acetylcarnitine and acylcarnitines, thus facilitating their export from the peroxisome.[9] The chain-shortening step of the synthesis of docosahexaenoic acid (DHA) (C22:6) is achieved by a single cycle of peroxisomal β-oxidation. Additional enzymes involved in the metabolism of unsaturated long-chain fats in peroxisomes include organelle-specific 2,4-dienoyl-CoA reductase, 2 enoyl-CoA isomerase, 2- and methylacyl-CoA racemase.

In addition to β-oxidation, peroxisomes are involved in several important metabolic pathways, including α-oxidation of phytanic acid and other fatty acids, ether-phospholipid biosynthesis (including plasmalogens), detoxification of glyoxylate, pipecolic acid oxidation, biosynthesis of cholesterol and bile acids, and the metabolism of reactive oxygen species.[8]

Defects of Mitochondrial Fatty Acid Metabolism

Defects of the Carnitine Cycle

Defects identified in this pathway include a specific plasma membrane carnitine transporter protein, CPT I, and II, and the carnitine-acylcarnitine translocase (CACT). A single report of two patients with an undefined defect in fatty acid oxidation and free fatty uptake in fibroblasts has been published.[35]

Deficiency of the plasma membrane carnitine transporter represents primary carnitine deficiency.[17,36,37] Carnitine is freely filtered by the kidney and must be reabsorbed from the proximal tubules to preserve plasma concentration. Because the transporter for carnitine is deficient in kidney as well as in muscle and liver, tissues whose carnitine content is highest, this defect results in defective renal reabsorption and reduced tissue storage of carnitine.[38] This then leads to a deficiency of carnitine in end organs, and to impairment of long chain fatty acid metabolism. Patients with carnitine transporter deficiency can present with severe hypoglycemia and dilated cardiomyopathy in infancy or childhood. Alternatively, they can show onset of hypertrophic cardiomyopathy, progressive muscle weakness, and muscle lipid storage with mild elevations of creatine kinase. Prolonged QT syndrome has been reported, leading to sudden unexplained death. Hypertrophic cardiomyopathy has been reported in middle-aged carriers of OCTN2 mutations. Fetal hydrops secondary to this disorder has been reported. Multiple reports of asymptomatic, affected mothers have been identified when newborn screening of their affected or carrier offspring have been positive for severely low free carnitine concentration. Plasma carnitine is extremely low or undetectable in these children but rises dramatically with supplementation with pharmacologic doses of carnitine (100–400 mg/kg /day). Patient symptoms also resolve dramatically with therapy. Outcome is good if diagnosis is prompt and therapy is instituted. Carnitine transporter deficiency can be diagnosed by uptake studies using cultured fibroblast, or by direct molecular analysis of the *OCTN2* gene.

Deficiency of liver CPT I has been reported. Severe disease is seen, but milder variants have been identified in geographically restricted populations. Diagnosis is based on enzymatic and mutation analysis. Severe symptoms include episodic hypoketotic hypoglycemia beginning in infancy and multiorgan system failure.[39-41] Muscle and cardiac symptoms are not present. Organic aciduria is not prominent in this disorder, but hyperammonemia may occur. Plasma carnitine is normal or elevated with a high free fraction. Analysis of samples from patients with CPT I deficiency has revealed normal CPT I in muscle, but low activity in other tissues, including liver.[41] Patients thus far have not responded well to dietary carnitine therapy, but use of the synthetic medium chain triglyceride triheptanoin has been reported. Molecular analysis of CPT1 patients has identified common mutations in the *CPT1A* gene in the Hutterite and Canadian First Nation/Inuit population.[42,43] Identified through newborn screening programs, affected individuals have, for the most part, been well. Patients with isolated muscle or cardiac CPT I deficiency have not been reported.

Deficiency of the CACT was initially reported in newborns, who had a nearly uniform poor outcome, presenting with severe hypoketotic hypoglycemia, hyperammonemia, and cardiac arrhythmias and/or cardiomyopathy.[44] All have had a grossly elevated acylcarnitine to free carnitine ratio, while dicarboxylic aciduria was reported in one. Carnitine supplementation did not appear to improve clinical symptoms. More recently, patients with a milder clinical course have been identified who have responded well to modest carnitine supplementation and dietary therapy.[17] Two affected siblings have been reported in which the younger child was prospectively treated and had not developed any sequelae 2 years later.[45] It appears that these patients have a higher level of residual enzyme activity than the more severely affected patients. Specific diagnosis of this disorder can be made via direct enzyme or molecular analysis. Triheptanoin appears to significantly improve outcome and can normalize blood ammonia concentration.

CPT II deficiency is the most common of this group of disorders. It classically presents in late childhood or early adulthood as episodes of recurrent exercise- or stress-induced myoglobinuria.[17,46] Episodes can be severe enough to lead to acute renal failure due to tubular necrosis. Patients are typically well between episodes. There is no tendency to develop hypoglycemia. Weakness and muscle pain are reported. Cardiomyopathy is rare. The characteristic diagnostic finding in these patients is a low total plasma carnitine with increased acylcarnitine fraction and no dicarboxylic aciduria. Long-chain acylcarnitines may be elevated.[17] More severe CPT II deficiency presents with symptoms similar to those in CACT deficiency. In these patients, presenting symptoms include neonatal hypoglycemia, hyperammonemia, hepatomegaly, and cardiomyopathy. A severe reduction of CPT II activity is found in all tissues tested, including liver, heart, muscle, and fibroblasts. Plasma-free carnitine is usually decreased with an increased acyl fraction. A common mutation leads to an enzyme with partial activity, accounting for the late onset of symptoms, while multiple other reported mutations can cause early onset disease.[47] Carnitine supplementation does not benefit the severe form of CPT II deficiency, but triheptanoin significantly improves outcome.[48-50] Familial phenotypic variation has been reported.[51] Occasional families with apparent autosomal dominant transmission of partial CPT II deficiency have been described, and at least one case appeared to be related to a mutation on one CPT II allele.[47,51,52] It is unknown why these patients exhibit symptoms, although a dominant negative effect on tetramer assembly and/or modifying gene effects have been postulated.[53,54]

Defects of Acyl-CoA Dehydrogenases

VLCAD deficiency presents with three major clinical symptoms that form a spectrum across ages.[55] Hypoglycemia, sometimes accompanied by hyperammonemia, is common in the first 2 to 4 years of life. Hepatic failure is reported, but rare. Muscular symptoms, mostly recurrent rhabdomyolysis, predominate beginning in adolescence, while cardiomyopathy can occur at any age.[56] 3-Hydroxy dicarboxylic acids or saturated dicarboxylic acids are present in the urine during acute episodes but can normalize when well.[57,58] Molecular testing has allowed some correlation of specific genotypes with phenotype, although it is imperfect. Most notably, a recurrent mutation identified through newborn screening typically leads to onset of symptoms in adolescence or adulthood.

ACAD9 deficiency was first reported in a 14-year-old, previously healthy, boy who died of a Reye syndrome-like episode and cerebellar stroke triggered by a mild viral illness and ingestion of aspirin.[59] Subsequently, it was identified as a common cause of respiratory chain deficiency, due to a nonenzymatic function as a complex I assembly factor, now recognized as the most common phenotype.[60-62]

Putative LCAD deficiency has been reported, but all of the patients originally categorized as LCAD deficient subsequently have been proven to have a deficiency of VLCAD instead.[63] Thus, there are no known patients with *bona fide* LCAD deficiency, although it has been proposed to play a role in surfactant deficiency.[64]

Numerous patients with SCAD deficiency have been reported.[65,66] However, the characteristic metabolites of ethylmalonic and methylsuccinic acids of SCAD deficiency are also detected in individuals with normal SCAD activity in fibroblasts.[65] The presence of one of two variants of SCAD, in about one-third of the general population (625 G>A and 511 C>T), also predisposes to excessive ethylmalonic acid production but no apparent disease. Finally, patients with SCAD deficiency identified by newborn screening are generally well.[67,68] Thus, it now appears that this defect leads to a biochemical phenotype but not clinical disease, and that previous reports of varied symptoms in SCAD deficiency were likely coincidental with unrelated other disorders.

MCAD deficiency has emerged as one of the most common inborn errors of metabolism in the United States and Western Europe, and has been extensively reviewed.[12,69,70] The most frequent clinical presentation is one of intermittent hypoketotic hypoglycemia, with peak symptoms in the first 3 years of life.[71] Mild hyperammonemia and coma may or may not be present. These findings often lead to the nonspecific diagnosis of Reye syndrome. The patient is usually well between attacks. Dicarboxylic aciduria is extensive during the attacks but can be undetectable by routine means when the patient is well. Similarly, micro- and macrovesicular hepatic steatosis, muscle weakness, and lipid excess in muscle present during the acute illness resolve between acute episodes. Historically, most patients who died of MCAD deficiency did so after having survived an initial episode. Thus, recurrent Reye syndrome-like episodes especially should trigger suspicion of this disorder. Sudden death in a previously healthy child has been described in numerous cases of MCAD deficiency. This can occur as early as the first day of life and has been seen in a previously healthy adult who was calorie restricted following abdominal surgery. In the appropriate age range, such deaths are often misascribed to "sudden infant death syndrome" (SIDS). Autopsy usually demonstrates the characteristic micro- and macrovesicular steatosis and should suggest the diagnosis. Analysis of the acylcarnitine and acylglycine profile from a bile specimen, as well as enzyme assay in cultured fibroblast (which may be recovered from deep tissues such as the fascia lata of the thigh up to 48 hours postmortem) may be helpful in proving it. Finally, completely asymptomatic individuals have been identified during family studies of patients. The diagnosis of MCAD deficiency in asymptomatic individuals is possible by metabolite analysis of various bodily fluids.[72] This technique has allowed development of newborn screening for MCAD deficiency, with a resultant near elimination of morbidity.[73-75]

Remarkable progress has been made in recent years in our understanding of the molecular mechanisms responsible for MCAD deficiency. Following cloning of the cDNA for MCAD, several groups simultaneously reported that a single common mutant allele was responsible for up to 90% of mutant alleles in patients with this disorder.[76,77] The substitution of a G for an A residue at position 985 (985 A>G) results in the replacement of a lysine by a glutamic acid residue, and production of an unstable protein.[78] Furthermore, screening of newborn blood samples has revealed a high carrier frequency for this disorder in some populations. Allele frequencies for the 985 A>G mutation range from 1/20 in northern European populations to less than 1/100 in Asian and some southern European populations. This mutation leads to misfolding of the protein in mitochondria and subsequent degradation by the mitochondrial protein quality control system. In the United States, the estimated carrier frequency for all mutations in Caucasians is 1/60.[76] This translates to a predicted disease frequency of 1/15,000. MCAD deficiency is much less frequent in African and Asian populations. Based on these studies, the predicted incidence for MCAD is similar to or greater than that for phenylketonuria (PKU).

Deficiency of Other β-oxidation Enzymes

Mitochondrial trifunctional protein (TFP) is composed of the three long chain enzymes following VLCAD in the FAO cycle on two different subunits. Patients can have a defect of one or all of the enzymes.[79-81] One group presents primarily with symptoms of cardiomyopathy, myopathy, and hypoglycemia. Peripheral neuropathy and recurrent myoglobinuria are usually present. These patients are deficient in all three enzymatic activities of TFP. The other group, deficient only in long-chain 3-hydroxyacyl-CoA dehydrogenase deficiency (LCHAD) activity, has hepatocellular disease with hypoglycemia and usually pigmentary retinopathy. Cholestasis and liver fibrosis can also be present.[82] Peripheral neuropathy is later in onset than in TFP deficiency, typically becoming clinically important in late adolescence or young adulthood.[83] Milder cases with adolescent onset of recurrent rhabdomyolysis have been reported.[84] In a large series of patients with isolated LCHAD deficiency, the mean age of clinical presentation was 5.8 months, with seven presenting in the neonatal period.[80] Thirty-nine patients presented with hypoketotic hypoglycemia, while 11 presented with chronic problems, consisting of failure to thrive, feeding difficulties, cholestatic liver disease, and/or hypotonia. Mortality in this series was high, with 38% dying within 3 months of diagnosis. Morbidity in the surviving patients was also high, with recurrent metabolic crises and muscle problems despite therapy. In a series of 21 patients with trifunctional protein deficiency, nine presented with rapidly progressive clinical deterioration. Six of these had hypoketotic hypoglycemia.[81] The remaining 12 patients presented with nonspecific, chronic symptoms, including hypotonia (100%), cardiomyopathy (73%), failure to thrive, or peripheral neuropathy. Ten patients presented in the neonatal period. Mortality was high (76%) and mostly attributable to cardiac disease. Two patients who were diagnosed prenatally died despite treatment. The disorder is now routinely identified by newborn screening in the United States and many industrialized countries, allowing for early treatment and prevention of catastrophic infantile presentations. However, treatment to date does not seem to impact the eye or neurologic symptoms.[83]

A common G to C mutation at nucleotide position 1528 (1528 G>C) of the α-subunit accounts for approximately 60% of mutant alleles identified thus far. Most other molecular defects destabilize the trifunctional protein, resulting in the multiple enzymatic deficiencies seen in some patients.[81,85-87] Heterozygosity for the common LCHAD mutation has been implicated in the development of acute fatty liver of pregnancy (AFLP) or "hemolysis, elevated liver enzymes, low platelet" (HELLP) syndrome (see below). Of note, patients with primary defects in respiratory chain function can have a secondary decrease of LCHAD activity and/or a less specific decrease in fibroblast oxidation of radiolabeled palmitate.[88] Thus, care must be taken to appropriately differentiate these patients from those with primary LCHAD deficiency.

A limited number of patients have been identified with isolated long chain 2,4-dienoyl-CoA reductase or 2-ketacyl-CoA dehydrogenase deficiencies.[89] One patient presented in the newborn period with persistent hypotonia. She was found to have elevated lysine and decreased carnitine in plasma. 2-*trans*,4-*cis*-decadienenoylcarnitine was identified in plasma and urine, and reduced activity of 2,4-dienoyl-CoA reductase was found in liver and muscle. The patient died at 4 months of age of respiratory acidosis. A mouse model of this deficiency leads to severe hypoglycemia.[90,91] A patient with this enzyme deficiency was shown by exome sequencing to have a mutation in the *NADK2* gene involved in NAD biosynthesis.[92] Confirmation of the clinical significance of this defect awaits the identification of additional patients. Patients with isolated short chain 3-hydroxyacyl-CoA dehydrogenase (SCHAD) have been identified based on a clinical phenotype of recurrent hypoglycemia and suggestive metabolite profiles in blood acylcarnitine and urine organic acid profiles. However, the molecular mechanism for this defect differs from other FAO disorders. Rather than having a direct effect on FAO, mutations in this protein impair its interaction with the enzyme glutamate dehydrogenase, an important regulator of insulin secretion, leading to increased sensitivity to insulin and thus hypoglycemia.[93]

Multiple Acyl-CoA Dehydrogenation Disorder (MADD)

Abnormalities of ETF or ETF:ubiquinone oxidoreductase (ETF dehydrogenase) deficiency lead to an *in vivo* deficiency of all of the dehydrogenases that use ETF as an electron acceptor.[94] This includes the ACADs discussed above, as well as isovaleryl-CoA dehydrogenase, 2-methylbutyryl-CoA dehydrogenase, isobutyryl-CoA dehydrogenase, glutaryl-CoA dehydrogenase, dimethylglycine dehydrogenase, and sarcosine dehydrogenase, enzymes involved in the intermediate metabolism of branched chain amino acids, tryptophan, lysine, and choline, respectively. Accumulation of intermediate compounds related to blockages in all of these pathways is seen. Because of the presence of glutaric acid in the urine of some patients, this disorder is frequently referred to as glutaric aciduria type II (GA II) to distinguish it from a primary deficiency of glutaryl-CoA dehydrogenase (GA I). Clinical manifestations of MADD are extremely heterogeneous.[95,96] A neonatal form can be seen with severe hypotonia, dysmorphic features, and cystic kidneys. These infants also exhibit metabolic acidosis and hypoglycemia. Milder variants are common, presenting with nonspecific neurologic signs, lipid storage myopathy, fasting hypoketotic hypoglycemia, and/or intermittent acidosis. In some patients, only fasting hypoketotic hypoglycemia and/or intermittent acidosis is seen and can be of late onset, as can recurrent rhabdomyolysis.[95,97] In these cases, the organic acid profile in times of illness is usually dominated by ethylmalonic and adipic acids, leading to an alternate name of ethylmalonic-adipic aciduria for this disorder. Structural brain abnormalities are common, including agenesis of the cerebellar vermis, hypoplastic temporal lobes, and focal dysplasia of the cerebral cortex.[98] Neuronal migration abnormalities may be present. Neonatal fatal cardiomyopathy has been reported.[99] Some patients with MADD respond dramatically to riboflavin with normalization of clinical symptoms and biochemical markers.[95,100,101]

Genetic and functional analyses of fibroblasts from patients with MADD have identified defects in both protein subunits of ETF and in ETF dehydrogenase.[102] For the most part, riboflavin-responsive patients have been found to have mutations in the ETFDH gene that presumably affect either protein folding or FAD binding. Otherwise, correlation of the mutation identified with severity of clinical symptoms has not been possible. Mutations in other genes in the synthesis of FAD lead to a metabolite pattern like that seen in ETF and ETFDH deficiencies, but with more varied clinical symptoms, including seizures and lactic acidosis. Mutations in genes involved in cellular riboflavin transport and FAD synthesis can lead to a biochemical picture of MADD, but the clinical picture varies.[103,104]

Disorders of Ketone Bodies

3-ketoacyl-CoA thiolase deficiency. Three patients with deficiency of this enzyme were originally described with developmental delay and muscle weakness, one of whom died of a Reye syndrome-like illness with urine metabolite findings suggestive of a defect in mitochondrial 3-ketoacyl-CoA thiolase.[105] Definitive enzyme testing was not performed. Since then, more than 30 patients have been identified with mutations in this gene, also known as short chain 3-ketoacyl-CoA thiolase. Most have presented with recurrent episodes of acidosis exacerbated by intercurrent illness, but with a characteristic persistence of ketone bodies in blood and urine when well.[69] Hypoglycemia has not been common. A single patient with medium chain 3-ketoacyl-CoA thiolase deficiency had metabolic acidosis, liver dysfunction, and rhabdomyolysis associated with vomiting and dehydration.

Deficiency of HMG-CoA lyase, which is also active in the metabolism of leucine and ketone synthesis, has been reported in close to 100 patients.[106] It presents with hypoketotic hypoglycemia with hyperammonemia and acidosis in the first year of life. Seizures and white matter abnormalities have been described in one child and one adult, subsequently diagnosed with HMG-CoA Lyase deficiency.[107,108] Identification of hydroxymethylglutaric acid in the urine is diagnostic. Common mutations exist in Saudi Arabia and Spain/Portugal.[106] Multiple variants have been found in apparently asymptomatic adults. Succinyl-CoA: 3-ketoacid CoA transferase (SCOT) functions in conjunction with mitochondrial acetoacetyl-CoA thiolase to generate ketones in extra-hepatic tissues. SCOT deficiency presents as persistent ketonuria in the first 1 to 2 years of life, while acetoacetyl-CoA thiolase deficiency presents with variable clinical symptoms, including exaggerated ketoacidosis in response to minor physiologic stress.[109] HMG-CoA synthase deficiency has been reported in at least 26 patients who presented with coma, hypoglycemia, and dicarboxylic aciduria with very low ketones.[110] Their acylcarnitine profiles have been reported as normal.

Pregnancy-Induced Symptoms in β-Oxidation Defects

There are now numerous reports of LCHAD deficient patients born following pregnancies with a variety of complications.[111] More than 20 pregnancies complicated by AFLP (Acute Fatty Liver of Pregnancy) or HELLP (Hemolysis, Elevated Liver enzymes, and Low Platelets) syndrome have been reported. The 1528 G>C mutation was present on at least one allele of the TFP α-subunit gene in all 12 affected infants in whom molecular genetic studies were done. Three patients with CPT I deficiency, one patient with CACT deficiency, one patient each with MCAD and SCAD deficiency, and more than one patient with complete trifunctional protein deficiency were born to mothers who developed liver disease during their pregnancies.[112-114] β-oxidation is now recognized as critical to the normal development and function of the placental-fetal unit.[115] During the last trimester of pregnancy, there is an increase in long chain fatty acids considered critical for generating ketones to meet increasing fetal energy demands. There is also a relative deficiency of maternal plasma carnitine. This may contribute to the development of liver-related complications of pregnancy in carriers of β-oxidation disorders. In a retrospective case-controlled study, a greater than 18% increase in pregnancy-related liver disease was seen across the entire β-oxidation spectrum.[116] Low birth weight and premature delivery are also reported in carriers of β-oxidation disorders.[115,116] Thus, pregnancies complicated by AFLP or HELLP should trigger an evaluation for a β-oxidation defect in the baby following delivery. It may also be reasonable to provide known carriers of β-oxidation disorders carnitine supplementation, especially during the last trimester of pregnancy. No clinical studies have identified whether prenatal carnitine supplementation has any impact on preventing liver-related complications or the outcome of the fetus.

Defects of Peroxisomal Fatty Acid Metabolism

Defects of Peroxisomal Biogenesis

Defects of peroxisomal biogenesis with a resulting failure to import all or a subset of matrix enzymes are more common than single enzyme deficiencies. Apparent absence or significant reduction in the number of peroxisomes has been demonstrated in four disorders historically considered to be distinct: Zellweger syndrome, neonatal adrenal leukodystrophy, infantile Refsum disease, and pipecolic acidemia.[117,118] It is now clear that these are related disorders that form a spectrum of clinical severity from the early lethal phenotype seen in Zellweger syndrome to the later onset symptoms of infantile Refsum disease and pipecolic acidemia. Therefore, they are referred to collectively as Zellweger spectrum disorders. Peroxisomal biogenesis defects are now divided into two groups: Zellweger spectrum disorder and rhizomelic chondrodysplasia punctata type 1.

Zellweger syndrome was the first disorder of peroxisomal biogenesis to be identified.[118] Classic findings include characteristic dysmorphic facial features and other malformations accompanied by severe neurologic dysfunction with hypotonia, seizures, and ultimate deterioration. Liver function is abnormal, as is that of the gastrointestinal tract, thus leading to failure to thrive. Neuronal heterotopia and renal cortical cysts are seen pathologically. Proximal limb shortening may be present. Death usually occurs within the first year of life. Although peroxisomes were originally believed to be absent, peroxisomal ghosts have been identified. These consist of peroxisomal membranes containing typical membrane proteins but devoid of their contents, a result of defective import of PTS1 and PTS2 matrix enzymes. Instead, peroxisomal enzymes are mislocalized to the cytosolic compartment and lead to severe impairment of cholesterol synthesis, bile acid metabolism, β-oxidation of very long-chain and branched-chain fats, ether phospholipid synthesis, and phytanic acid and pipecolic acid metabolism.[8,118] Neonatal adrenoleukodystrophy and infantile Refsum disease can manifest in the first 6 months of life similarly to Zellweger syndrome, but with slightly later onset of symptoms.[117,118] These clinical phenotypes, along with pipecolic acidemia, can also appear in early childhood (<3 years of age) with milder symptoms including developmental delay, hypotonia, hearing impairment, retinopathy, and nystagmus.[118] Biochemical findings in patients with neonatal adrenoleukodystrophy, infantile Refsum disease, and pipecolic acidemia are like those of Zellweger syndrome, but they may be more subtle. Classification of Zellweger spectrum disorders by age of onset and presentation is now more common due to molecular and clinical overlap between these historically described phenotypes. Neonatal-infantile, childhood, and adolescent-adult presentations are described and are useful for describing prognosis and counseling of families.[119] Heimler syndrome, a mild Zellweger spectrum disorder associated with variants in *PEX1, PEX6*, and *PEX 26*, is characterized by sensorineural hearing loss, amelogenesis imperfecta, and retinal degeneration without neurologic involvement.[120,121]

Rhizomelic chondrodysplasia punctata (RCDP) type 1, characterized by abnormal facies, developmental delay, cataracts, abnormal calcifications of epiphyses, and severe proximal limb shortening, results from failure to import PTS2-associated enzymes as a result of mutations in the *PEX7* gene.[7] RCDP type 2, dihydroxyacetone-phosphate acyltransferase (DHAPAT) deficiency, and RCDP type 3, alkyl DHAP synthase deficiency, have similar clinical features but are isolated disorders of ether phospholipid biosynthesis, resulting in plasmalogen deficiency with normal phytanic acid.[122]

Thirteen *PEX* genes, encoding proxins, have been associated with Zellweger spectrum disorders. More than 90% of patients with Zellweger syndrome spectrum phenotypes have mutations in PEX1, PEX2, PEX6, PEX10, PEX12, or PEX26, with PEX1 mutations accounting for the majority.[123] Patients with the Zellweger syndrome spectrum phenotypes clinically have also been shown to have single enzyme deficiencies. Peroxisomal assembly is not disrupted in these patients.[7]

Disorders of Single Enzymes

X-ALD, the most common disorder involving peroxisomal β-oxidation, affects all ethnic groups.[7,124] Newborn screening has estimated the incidence of X-ALD as 1 in 10,500, with a range from 1 in 4,845 in Minnesota to 1 in 16,200 in Illinois.[125-130] Initial biochemical studies characterized an impaired ability to activate fatty acids to their acyl-CoA esters and suggested that this was a disorder of an acyl-CoA synthetase. Molecular data, however, identified defects in an ATP-binding transporter located in the peroxisomal membrane as being responsible for this disorder.[34,131]

X-ALD is caused by mutations in the *ABCD1* gene, which encodes ALDP (ABCD1). The most likely function of ALDP, given available data, is the transport of very-long-chain fatty acids (VLCFAs) across the peroxisomal membrane, probably in the acyl-CoA–bound activated form.[34,131] Several clinical phenotypes have been associated with defects in this locus.[132] Classic cerebral childhood adrenoleukodystrophy accounts for 35% of cases. Boys are usually well until the second half of the first decade of life (peak onset, 3 to 10 years of age), when changes in learning and behavior suggestive of attention deficit hyperactivity disorder occur, followed by progressive quadriplegia, blindness, and dementia. Magnetic resonance imaging (MRI) changes usually precede neurologic symptoms. Death generally occurs within 5 years of onset of symptoms in untreated patients.

The cerebral inflammatory form of the disease can also occur in adolescents and adults. In 40% to 45% of male patients, a later onset disorder designated *adrenomyeloneuropathy* is seen. In this disorder, male patients experience progressive spastic paraparesis with peripheral neuropathy beginning in the second decade of life or later. At least 65% of women with X-ALD will show a similar picture of variable severity. Some boys and older males with a defect in this gene will show only primary adrenal insufficiency. Adrenal insufficiency occurs independently of neurologic symptoms. Asymptomatic boys and men also have been described. Great clinical heterogeneity can be noted within families.[132] The lack of genotype–phenotype correlation has been difficult to explain. It is likely that multiple genetic, epigenetic, and environmental factors contribute to a multi-hit hypothesis for the development of individual phenotypes in patients with similar genotypes.

The remaining single enzyme defects in peroxisomal β-oxidation are rare autosomal recessive disorders.[7] Deficiency of acyl-CoA oxidase (ACOX1) is characterized by early-onset seizures, developmental delay with regression of milestones, sensorineural deafness, retinopathy, and hypotonia. Dysmorphic features are not present or are mild. Hepatic peroxisomes are present, and an isolated elevation of VLCFAs is found in blood and urine. MRI studies reveal cerebral or cerebellar white matter abnormalities. The mean age of death is 5 years.[133] Two adult siblings with acyl-CoA oxidase deficiency presented with progressive ataxia, mild cognitive impairment, cataracts, and retinopathy associated with atrophy of the cerebellum and brainstem.[130] Deficiency of peroxisomal acyl-CoA oxidase 2 (ACOX2), an enzyme involved in the shortening of the cholesterol side-chains during bile acid synthesis, results in persistent transaminitis and liver damage, which can be attenuated by treatment with ursodeoxycholic acid.[134] Primary hyperoxaluria type 1, which results from deficiency of hepatic peroxisomal alanine-glyoxylate aminotransferase, leads to kidney stones, nephrocalcinosis, and kidney failure.[135,136]

Peroxisomal D-bifunctional protein deficiency manifests with a Zellweger-like phenotype. Seizures, particularly infantile spasms, are a common feature. An accumulation of VLCFAs, bile acid intermediates, and branched-chain fatty acids is present but with normal plasmalogen synthesis, a finding that distinguishes this disorder from a peroxisomal biogenesis defect.[137,138] Juvenile and adult onset forms of D-bifunctional protein deficiency present with hearing loss, ataxia, and peripheral neuropathy.[139] Patients with isolated deficiency of one of the components of the bifunctional protein have also been described.[140] 2-Methylacyl-CoA racemase deficiency manifests with adult-onset sensory motor neuropathy and pigmentary retinopathy.[141,142] Tremors, seizures, stroke-like episodes, and encephalopathy also have been described.[143]

Classic Refsum disease resulting from deficiency of phytanic acid oxidase is not a defect of peroxisomal β-oxidation itself, but it deserves mention because of the confusing nomenclature. Clinical findings include retinitis pigmentosa, cerebellar ataxia, and peripheral neuropathy, with onset of symptoms varying from childhood to the 5th decade of life. Patients have no cognitive impairment, dysmorphic features, or hepatomegaly. This disorder is characterized by accumulation of phytanic acid in blood and tissues, whereas other peroxisomal enzyme functions are normal.[144,145]

Diagnosis of Defects of Fatty Acid Metabolism

Mitochondria

Diagnosis of defects of mitochondrial β-oxidation requires a high level of suspicion in the appropriate clinical settings (**Table 72.3**).[146] This is crucial because in many instances, the biochemical abnormalities that might suggest a defect of β-oxidation resolve along with clinical symptoms. Thus, individuals with unexplained hypoglycemia, acidosis, hyperammonemia, myopathy with or without recurrent myoglobinuria, or progressive neuropathy should be considered as candidates for one of these disorders. All cases of Reye syndrome, patients with Reye syndrome-like symptoms, "SIDS", and unexplained sudden or near death in childhood should be evaluated for a defect in fat metabolism. It is critical to obtain blood and urine samples acutely for appropriate analysis, as biochemical findings may disappear when the patient is well. These patients should be more extensively evaluated if a diagnosis is not immediately obvious. This is best done in a referral center by a specialist trained in the evaluation of these patients.[72,147,148] Plasma free and total carnitine concentrations can often suggest whether a deficiency is primary or secondary (**Table 72.4**). Laboratory clues to the diagnosis of one of these disorders include dicarboxylic or 3-hydroxy dicarboxylic aciduria, hypoglycemia in the face of hypoketosis, mild lactic acidosis, and mild hyperammonemia. Hyperuricemia is a useful sign if present but is nonspecific. Myoglobin in blood or urine and elevated plasma creatine kinase can be seen when muscle involvement is present. Mass spectrometry can detect minute quantities of intermediates that are specific for certain fatty acid oxidation disorders and forms the basis of newborn screening for these disorders. These intermediates are often not detectable by routine organic acid analysis.[72,147,149] All patients with a suspected defect in fatty acid metabolism should receive an echocardiogram because of the high incidence of cardiomyopathy in many of these disorders. A fasting test with or without a subsequent challenge test with medium or long-chain triglycerides can induce excretion of a diagnostic metabolite but should only be undertaken in a hospital setting by someone with experience in performing such tests. It should be noted that a medium-chain triglyceride loading test must only be performed after specifically excluding the diagnosis of MCAD deficiency, as catastrophic consequences can result in patients with MCAD deficiency. Molecular analysis is readily available for all disorders in this group and is the most appropriate follow-up test in a patient identified by newborn screen or suspected on the basis of clinical symptoms. Specific enzyme analysis can be performed for many of these defects on cultured fibroblasts, lymphocytes, and/or liver or skeletal muscle, but is of limited clinical availability.

Newborn screening for disorders of β-oxidation, using tandem mass spectroscopy of dried blood spots collected prior to discharge from the nursery, has been implemented across the United States and in many European countries. Presymptomatic identification of these patients can, in many instances, prevent catastrophic events, especially sudden death, and is now well recognized to be cost-effective.[150]

Table 72.3 Findings Suggestive of a Defect in Mitochondrial β-Oxidation

Hypoketotic hypoglycemia with fasting or stress
Reye syndrome (especially recurrent)
Hypotonia and/or muscle weakness
Peripheral neuropathy
Chorioretinitis
Coma
Sudden death in infancy or childhood
Cardiomyopathy
Unexplained metabolic acidosis±hyperammonemia
Hyperuricemia
Recurrent myoglobinuria
Elevated serum creatine kinase
Dicarboxylic aciduria
Carnitine deficiency

Table 72.4 Plasma Carnitine Concentrations in Defects of β-Oxidation

Enzyme Defect	Total Carnitine	Free Carnitine	Free/Total Carnitine
Carnitine transporter	Very low	Low	Normal
CPT I	Normal or high	High	High
Translocase and CPT II	Low	Very low	Low
VLCAD, LCAD, MCAD, SCAD, LCHAD, SCHAD, ETF, and ETF dehydrogenase, 2, 4-dienoyl-coenzyme A reductase	Low	Low	Normal or low

CPT, carnitine palmitoyltransferase; *ETF*, electron transfer flavoprotein; *LCAD*, long-chain acyl-coenzyme A dehydrogenase; *LCHAD*, long-chain 3-hydroxyacyl-coenzyme A dehydrogenase; *MCAD*, medium-chain acyl-coenzyme A dehydrogenase; *SCAD*, short-chain acyl-coenzyme A dehydrogenase; *SCHAD*, short-chain 3-hydroxyacyl-coenzyme A dehydrogenase; *VLCAD*, very-long-chain acyl-coenzyme A dehydrogenase

Peroxisomes

Diagnosis of defects of peroxisomal β-oxidation is more straightforward than that of mitochondrial defects because clinical symptoms and biochemical abnormalities are not intermittent. Clinical findings suggestive of a defect of peroxisomal biogenesis or an isolated defect in β-oxidation are summarized in **Table 72.5**. It is important to consider these diagnoses in the setting of unexplained seizures or developmental delay, especially if accompanied by hypotonia.

Biochemical findings in these disorders are outlined in **Table 72.6**. Screening of plasma or urine for VLCFAs, pipecolic acid, phytanic acid, and pristanic acid will identify most patients.[117,122,151] The measurement of plasma VLCFAs should not be used to exclude hemizygosity for X-ALD in female patients. Additional investigations can include quantitation of bile acid and cholesterol synthesis intermediates, red blood cell membrane plasmalogen concentration, and the presence or absence of peroxisomes in cultured skin fibroblasts. Specific enzyme analysis on cultured skin fibroblasts and complementation analysis can be performed to identify the precise nature of the defect in a patient. Molecular mutational analysis is available for most peroxisomal disorders and, therefore, is now preferred to fibroblast studies in most patients.[122]

Table 72.5 Findings Suggestive of a Defect in Peroxisomal β-Oxidation

Neurologic and/or intellectual features
Developmental delay or regression
Seizures
Hypotonia
Peripheral neuropathy
Neuronal migration defects
Demyelination/ leukodystrophy
Dysmorphic features
Large anterior fontanelle
High forehead
Epicanthal folds
Rhizomelic chondrodysplasia
Hepatomegaly and/or hepatic dysfunction
Renal cysts
Adrenal insufficiency
Ophthalmologic abnormalities
Optic atrophy
Cataracts
Retinopathy
Hearing impairment
Failure to thrive

Genetics of Defects of Fatty Acid Metabolism

All defects of mitochondrial fatty acid metabolism identified thus far, and all peroxisomal orders except X-linked adrenoleukodystrophy, are inherited in an autosomal recessive fashion. The recurrence risk for subsequent siblings is, therefore, 25%. X-linked adrenoleukodystrophy is inherited as an X-linked trait. Affected males are hemizygous for one abnormal gene, while carrier women are heterozygous for one normal and one mutant allele. Thus, carrier women face a 50% risk of having an affected male or carrier female with each pregnancy. There can be great clinical heterogeneity within families, and individuals with the severe childhood onset and less-severe myeloneuropathy forms may be present in the same family. To explain this, it has been proposed that one or more additional autosomal genes are responsible for modulating the clinical severity of this disorder.

Sudden Infant Death and Defects of Fatty Acid Metabolism

The role of mitochondrial β-oxidation defects in sudden infant death syndrome (SIDS) bears special mention. As the definition of SIDS includes the presence of normal autopsy findings, many patients with defects of β-oxidation will be eliminated from this population by careful microscopic study of muscle, heart, and liver. Unfortunately, not all deaths reported as "SIDS" receive the same level of scrutiny and, thus, cases of β-oxidation disorders are likely to be under-represented. In addition, some disorders of β-oxidation can clearly present with minimal or no tissue changes, thus making postmortem diagnosis difficult. It appears that approximately 1% to 5% of unexplained sudden deaths in infancy and childhood are related to such disorders.[152] Because of this high frequency, and because of the lack of reliable autopsy findings, all children who die suddenly from unexplained causes (whether or not they are infants) should be evaluated for possible metabolic defects, including disorders of β-oxidation.[152] Postmortem exam should include analysis of blood, urine, or vitreous humor for organic acids, acylglycines, and acylcarnitines. Tissue samples of liver, skeletal muscle, and heart should be rapidly frozen and stored at -80 °C for future enzymatic analysis, and a skin fibroblast culture should be started, when possible. Only with such an intense effort to determine a diagnosis in these cases, can accurate counseling be presented to the family regarding recurrence risks and asymptomatic siblings should be identified.

Treatment of Defects of Fatty Acid Metabolism

Mitochondria

Current practices for mitochondrial FAO disorders include diverse approaches to dietary treatment focused primarily on manipulation of dietary fat, protein, and carbohydrate plus the avoidance of fasting.[55] The avoidance of fasting by prescribing increased frequency of meals is a simple, preventive measure that ensures a constant supply of glucose. Minimizing the need

Table 72.6 Biochemical Features of Peroxisomal Disorders

Enzyme Defect	Vlcfa	Phytanic Acid	Pipecolic Acid	Pristanic Acid	Bile Acid Intermediates	Plasmalogens
Peroxisomal Biogenesis defects	High	High[a]	High	High[a]	High	Low
RCDP type 1	Normal	High	Normal	Normal	Normal	Low
RCDP types 2 and 3	Normal	Normal	Normal	Normal	Normal	Low
X-ALD	High	Normal	Normal	Normal	Normal	Normal
Straight-chain acyl-CoA oxidase deficiency	High	Normal	Normal	Normal	Normal	Normal
D-bifunctional protein deficiency	High	Normal–high[a]	Normal	Normal–high[a]	High	Normal
2-Methylacyl-CoA racemase deficiency	Normal	Normal–mildly elevated	—	Very high	Very high	Normal
Refsum disease	Normal	Very high	Normal	Normal	Normal	Normal

CoA, coenzyme A; *RCDP*, Rhizomelic chondrodysplasia punctata; *VLCFA*, very-long-chain fatty acid; *X-ALD*, X-linked adrenoleukodystrophy
[a]Phytanic and pristanic acids are derived from the diet and may be normal initially in newborns.

to mobilize fatty acids for energy from adipose tissue reduces the accumulation of toxic fatty acid metabolites that, when elevated, can cause vomiting, lethargy, coma, and/or death. The optimal prescribed time between meals is not well established and can vary from one infant to another, children, and adult patients with fatty acid oxidation disorders.[153,154] A regimen of restricted fat and high carbohydrate intake has generally been recommended for long-chain defects.[155] Prescribed fat restrictions are variable, but typical energy from dietary fat is 25% to 30% or less of the daily energy needs for age.[156] Data from LCHAD or TFP-deficient patients suggested use of a diet providing age-appropriate protein, while limiting long chain fatty acid intake to 10% of total energy may be beneficial. A high protein (30% of energy), low carbohydrate (48% of energy) diet over a 6-day period was found to reduce obesity, increase lean muscle mass, and increase metabolic stability in children with long-chain β-oxidation disorders.[157,158] This approach has not been studied long-term. Fat restriction does not appear to be necessary for MCAD deficiency. A useful online resource for information on dietary management is the website of the Genetic Metabolic Dietitians International (GMDI.org).

Use of medium chain triglyceride (MCT) oil as a lipid substrate emphasizes β-oxidation pathways not affected by long chain fat defects, as it is composed of shorter-length fatty acids (C8:0 and C10:0 acids), which are not dependent on carnitine for entry into the mitochondria for subsequent oxidation.[159] Fifteen percent to 18% of total energy from MCT oils is commonly recommended.[154] Recurrent episodes of rhabdomyolysis and exercise-induced rhabdomyolysis were reduced or eliminated when 0.5 grams/kg of MCT oil was given 20 minutes prior to exercise in one study.[160] Triheptanoin (Dojolvi) has been approved by the FDA for treatment of long-chain FAO disorders.[49] Triheptanoin is a triglycerol with C7 side arms that, when metabolized by MCAD and SCAD, leads to production of two molecules of acetyl-CoA and one of propionyl-CoA. The latter is further metabolized to succinate, a TCA cycle intermediate that is depleted in patients with long-chain FAO defects. Compared with an existing supplementation with MCT oil (trioctanoylglycerol), patients treated with triheptanoin showed near elimination of hypoglycemic episodes, improved cardiac function, and fewer episodes of rhabdomyolysis.[49,50,161] Clinical trials were conducted with 25% to 35% of energy from triheptanoin and 10% from natural fats but, in older individuals, this amount of intake can lead to GI symptoms, including bloating and diarrhea. Lower doses are likely to be effective to some extent in this setting. Administration of triheptanoin with food improves tolerability. As with MCT oil, administration 20 to 30 minutes before activity can improve exercise tolerance.

Augmentation of diet with essential fatty acids (EFA) (at 1%–2 % of total energy intake) is often used to reduce risk of EFA deficiency.[162,163] Flaxseed, canola, walnut, or safflower oils are used for this purpose.[154,164] A study of acylcarnitine production in fibroblasts exposed to different fatty acids found that low saturated fats and high polyunsaturated fats reduced the production of toxic acylcarnitines in long chain β-oxidation disorders.[164] A daily multivitamin and mineral supplement including all the fat-soluble vitamins have also been recommended.

Docosahexanoic acid (DHA) deficiency can develop in patients with LCHAD deficiency and has been hypothesized as a cause for retinal degeneration in these patients.[162] One study demonstrated improved vision outcomes in LCHAD patients with higher DHA and lower 3-hydroxyacylcarnitines.[165]

Commercially available medical foods provide formulations composed of modified fats, concentrated proteins, and multivitamins. Formulas with a high percentage of energy from MCT oil have been suggested for patients with long-chain fatty acid oxidation disorders. When prescribed in conjunction with sufficient EFA supply, this approach can meet the essential nutrient needs of patients. Other strategies include a combination of available formulas to produce a diet high in complex carbohydrates, low in fat, and adequate in vitamin and mineral content[154] No clinical studies have been performed to assess the long-term efficacy of these formulas for the unique requirements of patients with β-oxidation disorders.

Increased caloric intake from carbohydrates may be necessary during intercurrent illness, due to increased metabolic demands on the body. This can be met via oral or nasogastric tube administration of an appropriate intravenous formula if oral intake is inadequate. Intravenous infusion of glucose (8–10 mg/kg/minute) can be used when oral intake is interrupted or during acute episodes associated with infection.[153,154]

Carnitine supplementation has long been used in the treatment of disorders of β-oxidation on the rationale that it allows repletion of the intramitochondrial carnitine pool and accelerates the removal of toxic fatty-acid intermediates.[155] However, its use remains controversial and of unproven value except in carnitine transporter deficiency.[153,166] Carnitine supplementation has been reported to normalize plasma carnitine and to increase urinary excretion of acyl-carnitine esters; however, it neither guarantees the prevention of the accumulation of toxic medium chain free fatty acids in plasma, prevents spontaneous episodes of hypoglycemia, nor reduces symptoms of lethargy, hypoglycemia, and vomiting.[167] Conversely, short periods of carnitine supplementation have been suggested to increase ketogenesis and reduce symptoms during periods of fasting hypoglycemia.[168] Concern over the arrhythmogenic potential of long-chain acylcarnitine intermediates remains.[169] Recommended doses range from 30 to 50 mg/kg/day with a maximum dose of 1,000 mg.[154]

Riboflavin is the precursor to flavin adenine dinucleotide (FAD), an essential cofactor for the ACDs, ETF, and ETF dehydrogenase. Several patients with biochemical abnormalities suggestive of β-oxidation defects have responded with clinical improvement to pharmacologic doses of riboflavin (100–200 mg/day). One of these groups has a variant of ETFDH dehydrogenase with a defect in interaction with FAD.[95] Increasing intramitochondrial concentrations of FAD by administration of riboflavin apparently allows enough cofactor binding to restore activity. A second set of patients with a picture of late onset lipid storage myopathy and muscle weakness showing some level of hepatic dysfunction has been described.[170] Again, these patients appear to respond to therapy with riboflavin, but their defect remains undefined. Marked clinical heterogeneity within each group has been documented.

Bezafibrate has been shown to upregulate oxidation rates of long-chain fats with decreased muscle pain and increased physical activity in some patients with long-chain β-oxidation disorders, even those with severe VLCAD missense mutations.[171-174] While an open-label clinical trial of bezafibrate on patients with long-chain fatty acid oxidation disorders reported some positive effects, it failed to show clinical improvement in a double-blind placebo-controlled trial.[175-177]

Peroxisomes

Treatment of patients with defects in peroxisomal β-oxidation has been difficult. Therapy for X-ALD has received the most attention.[178] Adrenal-hormone replacement for adrenal insufficiency is required in most male patients with X-ALD and in 1% to 2% of female carriers.[179] Several inhibitors of VLCFA synthesis have been used to control excess accumulation of VLCFAs in patients. These include oleic acid, glycerol trioleate, and glycerol trierucate, popularly known as "Lorenzo's oil."[180] The first extensive therapeutic trial to address the efficacy of Lorenzo's oil used a diet that provided 10% of energy as fat, with less than 10 to 15 mg of C26:0 fatty acid per day.[165] In addition, 1.7 g/kg body weight/day of glycerol trioleate oil and 0.3 g/kg body weight/day of glycerol trierucate were also given, along with 10 to 15 mL of safflower oil and 2 g fish oil (to avoid a deficiency of EFAs). Using this regimen, VLCFA levels were normalized in patients with either adrenoleukodystrophy or adrenomyeloneuropathy, but little or no clinical improvement could be documented. In view of the lack of definite benefit of Lorenzo's oil and significant safety concerns raised in multiple studies, Lorenzo's oil is no longer recommended for patients with X-ALD. Instead, a heart-healthy diet is recommended. Similarly, multiple small studies have demonstrated a decrease in VLCFA with lovastatin treatment without definite therapeutic benefit. Although elevated VLCFA is a hallmark of X-ALD, VLCFA levels do not appear to correlate with clinical phenotype and; therefore, the role of decreasing VLCFA in the treatment of X-ALD has become less of a focus.

Hematopoietic stem-cell transplantation has shown the most promise for boys with the childhood cerebral form of X-ALD.[181,182] Short-term stabilization of clinical findings and MRI abnormalities have been demonstrated in patients when hematopoietic stem transplantation or cord blood transplantation was performed early in the course. Transplantation of autologous hematopoietic stem cells genetically altered to express ALDP is an available option for boys who do not have a human leukocyte antigen (HLA)–matched donor or for adults with cerebral disease who are at significant risk of mortality with conventional allogeneic hematopoietic stem-cell transplantation.[182] There continues to be no effective treatment for boys with advanced disease. *N*-Acetyl-L-cysteine in conjunction with stem cell transplantation might improve outcomes in boys with advanced cerebral disease.[183] Treatment for adults with adrenomyeloneuropathy is primarily symptomatic. Management of symptoms of myelopathy and peripheral neuropathy includes treatment of gait disturbance associated with spasticity and weakness in the legs, pain management, and treatment of neurogenic bowel and bladder symptoms. Specific therapy for the other single enzyme defects of peroxisomal β-oxidation has not been reported.

Treatment of the disorders of peroxisomal biogenesis has been problematic because of the multisystemic involvement and the numerous metabolic pathways affected in these individuals. Recent treatment guidelines have been published describing the multidisciplinary care of these patients.[118,119] The reduction of VLCFA intake has not been shown to be beneficial but reduction of phytanic acid dietary intake may be indicated for persistently elevated phytanic acid.

Stabilization or improvement in vision and retinal function have been noted with normalization of blood DHA in a number of uncontrolled studies in patients with milder phenotypes, suggesting a role of DHA in the visual pathway.[184] DHA-containing phospholipids bilayers within the retina optimize the kinetics of the metarhodopsin II–G-protein–coupled signaling pathway. Correction of DHA deficiency in patients with peroxisomal biogenesis defects was previously recommended but a benefit has not been consistently demonstrated in controlled studies and; therefore, DHA supplementation is no longer recommended for all patients with peroxisomal biogenesis defects. Supplementation of vitamins A, D, E, and K, as well as carnitine, might be helpful.[118,119] Bile acid supplementation with cholic acid may be beneficial in patients with liver dysfunction by suppressing the synthesis of toxic bile acid intermediates.[185] Patients should be monitored for adrenal insufficiency and treated accordingly. Patients with seizures, when present, should be treated with anticonvulsant medications. Patients with milder phenotypes may benefit from hearing aids, glasses, and developmental services. Detailed prospective studies to evaluate the effectiveness of treatment in these patients are needed. Hematopoietic stem-cell transplantation and gene therapy have not been systematically evaluated in patients with these conditions and would be limited by their multisystem nature.

References

1. Lazarow PB. J Neuropath Exper Neurol 1995;54:720–5.
2. Novikov DK, Vanhove GF, Carchon H, et al. J Biol Chem 1994;269: 27125–35.
3. Kunau WH, Dommes V, Schulz H. Prog Lipid Res 1995;34: 267–342.
4. Wallace DC. Sci 1992;256:628–32.
5. Rosenberg LE, Fenton WA, Horwich AL, et al. Ann NY Acad Sci 1987;488:99–108.
6. Kalousek F, Hendrick JP, Rosenberg LE. Proc Natl Acad Sci U S A 1988;85:7536–40.
7. Braverman NE, D'Agostino MD, Maclean GE. Dev Disabil Res Rev 2013;17:187–96.
8. Okumoto K, Tamura S, Honsho M, et al. Adv Exp Med Biol 2020;1299:3–17.
9. Wanders RJA, Vaz FM, Waterham HR, et al. Adv Exp Med Biol 2020;1299:55–70.
10. Thoms S, Gronborg S, Gartner J. Trends Mol Med 2009;15: 293–302.
11. Calvo S, Jain M, Xie X, et al. Nat Genet 2006;38:576–82.
12. Vockley J, Whiteman DA. NMD 2002;12:235–46.
13. Vockley J. Mayo Clin Proc 1994;69:249–57.
14. McGarry JD, Foster DW. Annu Rev Biochem 1980;49:395–420.
15. Stremmel W, Strohmeyer G, Berk PD. Proc Natl Acad Sci USA 1986;83:3584–8.
16. Abe T, Fujino T, Fukuyama R, et al. J Biochem 1992;111:123–8.
17. Longo N, Amat di San Filippo C, Pasquali M. Am J Med Genet C Semin Med Genet 2006;142C:77–85.
18. Gregersen N, Andresen BS, Pedersen CB, et al. J Inherit Metab Dis 2008;31:643–57.
19. Battaile KP, McBurney M, Van Veldhoven PP, et al. Biochim Biophys Acta 1998;1390:333–8.
20. Lea W, Abbas AS, Sprecher H, et al. BBA Mol Cell Biol L 2000;1485:121–8.
21. Ensenauer R, He M, Willard JM, et al. J Biol Chem 2005;280:32309–16.
22. Beckmann JD, Frerman FE. Biochem 1985;24:3922–5.
23. Tanaka K, Matsubara Y, Indo Y, et al. In: Tanaka K, and Coates PM, eds. The Acyl-CoA Dehydrogenase Family: Homology and Divergence of Primary Sequence of Four Acyl-CoA Dehydrogenases and Consideration of Their Functional Significance. Fatty Acid Oxidation: Clinical, Biochemical and Molecular Aspects. New York, NY: Alan R. Liss, Inc., 1990;577–98.
24. Vockley J, Mohsen al WA, Binzak B, et al. Methods Enzymol 2000;324:241–58.
25. Houten SM, Wanders RJ. J Inherit Metab Dis 2010;33:469–77.
26. Orii KE, Orii KO, Souri M, et al. J Biol Chem 1999;274:8077–84.
27. Filling C, Keller B, Hirschberg D, et al. Biochem Biophys Res Commun 2008;368:6–11.
28. Helander HM, Koivuranta KT, Horelli-Kuitunen N, et al. Genomics 1997;46:112–9.
29. Wang Y, Mohsen AW, Mihalik SJ, et al. J Biol Chem 2010;285: 29834–41.
30. Wang Y, Palmfeldt J, Gregersen N, et al. J Biol Chem 2019;294: 12380–91.
31. Fukao T, Song XQ, Mitchell GA, et al. Pediatr Res 1997;42: 498–502.
32. Mitchell GA, Fukao T. In: Scriver C, Beaudet AL, Sly W, and Valle D, eds. Inborn Errors of Ketone Body Metabolism. The Metabolic and Molecular Basis of Inherited Disease New York, NY: McGraw-Hill, 2001;2327-56..
33. Kassovskabratinova S, Fukao T, Song XQ, et al. Am J Human Genet 1996;59:519–28.
34. Tawbeh A, Gondcaille C, Trompier D, et al. Int J Mol Sci 2021;22.
35. Odaib AA, Shneider BL, Bennett MJ, et al. N Engl J Med 1998;339:1752–7.
36. Longo N. Ann Nutr Metab 2016;68(Suppl 3):5–9.
37. Longo N, Frigeni M, Pasquali M. Biochim Biophys Acta 2016;1863:2422–35.
38. Scaglia F, Wang Y, Longo N. Arch Biochem Biophys 1999;364: 99–106.
39. IJlst L, Mandel H, Oostheim W, et al. J Clin Invest 1998;102: 527–31.
40. Bennett MJ, Boriack RL, Narayan S, et al. Mol Genet Metab 2004;82:59–63.
41. Wieser T, Deschauer M, Olek K, et al. Neurol 2003;60: 1351–3.
42. Prasad C, Johnson JP, Bonnefont JP, et al. Mol Genet Metab 2001;73:55–63.
43. Park JY, Narayan SB, Bennett MJ. Clin Chem Lab Med 2006;44:1090–1.
44. Stanley CA, Hale DE, Berry GT, et al. N Engl J Med 1992;327: 19–23.
45. Pierre G, Macdonald A, Gray G, et al. J Inherit Metab Dis 2007;30:815.
46. Engel AG, Angelini C. Sci 1973;179:899–902.
47. Taggart RT, Smail D, Apolito C, et al. Hum Mutat 1999;13: 210–20.
48. Vockley J, Burton B, Berry GT, et al. Mol Genet Metab 2017; 120:370–7.
49. Vockley J, Burton B, Berry GT, et al. J Inherit Metab Dis 2019; 42:169–77.
50. Vockley J, Burton B, Berry G, et al. J Inherit Metab Dis 2021; 44:253–63.
51. Vladutiu GD, Bennett MJ, Fisher NM, et al. Musc Nerve 2002; 26:492–8.
52. Vladutiu GD, Bennett MJ, Smail D, et al. Mol Genet Metab 2000; 70:134–41.
53. Vladutiu GD. Mol Genet Metab 2001;74:51–63.
54. Vockley J, Rinaldo P, Bennett MJ, et al. Mol Genet Metab 2000; 71:10–8.
55. Vockley J. Am J Manag Care 2020;26:S147–S54.
56. Ogilvie I, Pourfarzam M, Jackson S, et al. Neurol 1994;44: 467–73.
57. Aoyama T, Tynan K, Dietz HC, et al. Hum Mol Genet 1993;2: 2135–40.
58. Bertrand C, Largilliere C, Zabot MT, et al. Biochim Biophys Acta 1993;1180:327–9.
59. He M, Rutledge SL, Kelly DR, et al. Am J Hum Genet 2007;81: 87–103.
60. Repp BM, Mastantuono E, Alston CL, et al. Orphanet J Rare Dis 2018;13:120.
61. Sinsheimer A, Mohsen AW, Bloom K, et al. Mol Genet Metab 2021;134:156–63.
62. Xia C, Lou B, Fu Z, et al. iScience 2021;24:103153.
63. Yamaguchi S, Indo Y, Coates PM, et al. Pediatr Res 1993;34: 111–3.
64. Goetzman ES, Alcorn JF, Bharathi SS, et al. J Biol Chem 2014;289:10668–79.
65. Jethva R, Bennett MJ, Vockley J. Molec Genet Metab 2008;95: 195–200
66. van Maldegem BT, Wanders RJ, Wijburg FA. J Inherit Metab Dis 2010;33:507–11.
67. Waisbren SE, Levy HL, Noble M, et al. Mol Genet Metab 2008; 95:39–45.
68. Gallant NM, Leydiker K, Tang H, et al. Mol Genet Metab 2012; 106:55–61.
69. Bennett MJ, Rinaldo P, Strauss AW. Crit Rev Clin Lab Sci 2000; 37:1–44.
70. Rinaldo P, Matern D, Bennett MJ. Annu Rev Physiol 2002;64: 477–502.

71. Iafolla AK, Thompson RJ, Jr., Roe CR. J Pediatr 1994;124: 409–15.
72. Rinaldo P, Cowan TM, Matern D. Genet Med 2008;10:151–6.
73. Dezateux C. Eur J Pediatr 2003;162(Suppl 1):S25–8.
74. van der Hilst CS, Derks TG, Reijngoud DJ, et al. J Pediatr 2007; 151:115–20, 20 e1–3.
75. Waisbren SE, Landau Y, Wilson J, et al. Dev Disabil Res Rev 2013;17:260–8.
76. Tanaka K, Yokota I, Coates PM, et al. Hum Mutat 1992;1:271–9.
77. Andresen BS, Bross P, Udvari S, et al. Human Molec Genet 1997;6:695–707.
78. Yokota I, Saijo T, Vockley J, et al. J Biol Chem 1992;267: 26004–10.
79. Strauss AW, Bennett MJ, Rinaldo P, et al. Semin Perinatol 1999; 23:100–12.
80. den Boer ME, Wanders RJ, Morris AA, et al. Pediatr 2002;109: 99–104.
81. den Boer ME, Dionisi-Vici C, Chakrapani A, et al. J Pediatr 2003; 142:684–9.
82. Odievre MH, Sevin C, Laurent J, et al. Acta Paediatr 2002;91: 719–22.
83. Grunert SC, Eckenweiler M, Haas D, et al. J Inherit Metab Dis 2021;44:893–902.
84. Miyajima H, Orii KE, Shindo Y, et al. Neurol 1997;49:833–7.
85. Spiekerkoetter U, Bennett MJ, Ben-Zeev B, et al. Muscle Nerve 2004;29:66–72.
86. Angdisen J, Moore VD, Cline JM, et al. Curr Drug Targets Immune Endocr Metabol Disord 2005;5:27–40.
87. Rector RS, Payne RM, Ibdah JA. Adv Drug Deliv Rev 2008;60: 1488–96.
88. Enns GM, Bennett MJ, Hoppel CL, et al. J Pediatr 2000;136: 251–4.
89. Roe CR, Millington DS, Norwood DL, et al. J Clin Invest 1990;85: 1703–7.
90. Miinalainen IJ, Schmitz W, Huotari A, et al. PLoS Genet 2009;5: e1000543.
91. Makela AM, Hohtola E, Minalainen IJ et al. Sci Rep 2019;9:12038.
92. Houten SM, Denis S, Te Brinke H, et al. Hum Mol Genet 2014;23: 5009–16.
93. Velasco K, St-Louis JL, Hovland HN, et al. J Inherit Metab Dis 2021;44:240–52.
94. Frerman FE, Goodman SI. Proc Natl Acad Sci U S A 1985;82: 4517–20.
95. Olsen RKJ, Andresen BS, Christensen E, et al. Hum Mut 2003; 22:12–23.
96. Frerman FE. Biochem Soc Trans 1988;16:416–8.
97. Gregersen N, Bross P, Andresen BS. Euro J Biochem 2004;271: 470–82.
98. Takanashi J, Fujii K, Sugita K, et al. Pediatr Neurol 1999;20: 142–5.
99. Singla M, Guzman G, Griffin AJ, et al. Pediatr Cardiol 2008;29: 446–51.
100. Gregersen M, Rhead W, Christensen E. In: Tanaka K, and Coates PM, eds. Riboflavin Responsive Glutaric Aciduria Type II. Fatty Acid Oxidation: Clinical, Biochemical, and Molecular Aspects. New York, NY: Alan R. Liss, Inc, 1990; 477–94.
101. Wen B, Dai T, Li W, et al. J Neurol Neurosurg Psychiatry 2010;81:231–6.
102. Loehr JP, Goodman SI, Frerman FE. Pediatr Res 1990;27:311–5.
103. Olsen RKJ, Konarikova E, Giancaspero TA, et al. Am J Hum Genet 2016;98:1130–45.
104. Muru K, Reinson K, Kunnapas K, et al. Mol Genet Genomic Med 2019;7:e915.
105. Bennett MJ, Gibson KM, Sherwood WG, et al. J Inherit Metab Dis 1993;16:831–6.
106. Pie J, Lopez-Vinas E, Puisac B, et al. Mol Genet Metab 2007;92:198–209.
107. Yylmaz Y, Ozdemir N, Ekinci G, et al. Pediatr Neurol 2006;35:139–41.
108. Bischof F, Nagele T, Wanders RJ, et al. Ann Neurol 2004;56: 727–30.
109. Grunert SC, Foster W, Schumann A, et al. Biochimie 2021; 183:55–62.
110. Rojnueangnit K, Maneechai P, Thaweekul P, et al. Eur J Med Genet 2020;63:104086.
111. Ibdah JA, Dasouki MJ, Strauss AW. J Inherit Metab Dis 1999; 22:811–4.
112. Walter JH. J Inherit Metab Dis 2000;23:229–36.
113. Matern D, Hart P, Murtha AP, et al. J Pediatr 2001;138:585–8.
114. Matern D, Schehata BM, Shekhawa P, et al. Mol Genet Metab 2001;72:265–8.
115. Shekhawat PS, Matern D, Strauss AW. Pediatr Res 2005;57:78R–86R.
116. Browning MF, Levy HL, Wilkins-Haug LE, et al. Obstet Gynecol 2006;107:115–20.
117. Wanders RJ. Subcell Biochem 2013;69:23–44.
118. Braverman NE, Raymond GV, Rizzo WB, et al. Mol Genet Metab 2016;117:313–21.
119. Klouwer FC, Berendse K, Ferdinandusse S, et al. Orphanet J Rare Dis 2015;10:151.
120. Daich Varela M, Jani P, Zein WM, et al. Am J Med Genet C Semin Med Genet 2020;184:618–30.
121. Mechaussier S, Perrault I, Dollfus H, et al. Adv Exp Med Biol 2020;1299:81–7.
122. Poll-The BT, Gartner J. Biochim Biophys Acta 2012;1822: 1421–9.
123. Cheillan D. Adv Exp Med Biol 2020;1299:71–80.
124. Moser HW, Raymond GV, Lu SE, et al. Arch Neurol 2005;62:1073–80.
125. Wiens K, Berry SA, Choi H, et al. Am J Med Genet A 2019; 179:1205–13.
126. Hall PL, Li H, Hagar AF, et al. Int J Neonatal Screen 2020;6.
127. Lee S, Clinard K, Young SP, et al. JAMA Netw Open 2020;3:e1920356.
128. Baker CV, Cady Keller A, Lutz R, et al. Int J Neonatal Screen 2022;8.
129. Burton BK, Hickey R, Hitchins L, et al. Int J Neonatal Screen 2022;8.
130. Priestley JRC, Adang LA, Drewes Williams S, et al. Int J Neonatal Screen 2022;8.
131. Wanders RJ, Visser WF, van Roermund CW, et al. Pflugers Arch 2007;453:719–34.
132. Moser HW, Mahmood A, Raymond GV. Nat Clin Pract Neurol 2007;3:140–51.
133. Morita A, Enokizono T, Ohto T, et al. Brain Dev 2021;43: 475–81.
134. Alonso-Pena M, Espinosa-Escudero R, Herraez E, et al. Hepatology 2022;76:1259-74.
135. Soliman NA, Mabrouk S. Clin Kidney J 2022;15:i33–i6.
136. Bacchetta J, Lieske JC. Clin Kidney J 2022;15:i17–i22.
137. Landau YE, Heimer G, Barel O, et al. Mol Genet Metab Rep 2020;25:100631.
138. Werner KM, Cox AJ, Qian E, et al. Am J Med Genet A 2022;188:357–63.
139. Matsuda Y, Morino H, Miyamoto R, et al. Neurol Genet 2020;6:e396.
140. Ferdinandusse S, Denis S, Mooyer PA, et al. Ann Neurol 2006;59:92–104.
141. Van Veldhoven PP, Meyhi E, Squires RH, et al. Eur J Clin Invest 2001;31:714–22.

142. Clarke CE, Alger S, Preece MA, et al. Neurol 2004;63:188–9.
143. Tanti MJ, Maguire MJ, Warren DJ, et al. BMJ Case Rep 2022;15.
144. Wierzbicki AS. Biochem Soc Trans 2007;35:881–6.
145. Kruska N, Reiser G. Neurobiol Dis 2011;43:465–72.
146. Saudubray JM, Martin D, de Lonlay P, et al. J Inherit Metab Dis 1999;22:488–502.
147. Smith EH, Matern D. Curr Protoc Hum Genet 2010;17:1–20.
148. Vockley J, Singh RH, Whiteman DA. Curr Opin Clin Nutr Metab Care 2002;5:601–9.
149. Chace D, Barr J, Duncan M, et al. Clin Lab Stand Inst Doc 2007;27:1–79.
150. Wilcken B. J Inherit Metab Dis 2010;33:501–6.
151. McCandless SE, Wright EJ. Birth Defects Res 2020;112:350–66.
152. Bennett MJ, Rinaldo P. Clin Chem 2001;47:1145–6.
153. Catzeflis C, Bachmann C, Hale DE, et al. Eur J Pediatr 1990;149:577–81.
154. Solis JO, Singh RH. J Am Diet Assoc 2002;102:1800–3.
155. Kerner J, Hoppel C. Ann Rev Nutr 1998;18:179–206.
156. Vockley J, Longo N, Madden M, et al. Clin Nutr ESPEN 2021;41:293–8.
157. Gillingham MB, Purnell JQ, Jordan J, et al. Mol Genet Metab 2007;90:64–9.
158. Gillingham MB, Matern D, Harding CO. Top Clin Nutr 2009;24:359–65.
159. Gillingham M, Van Calcar S, Ney D, et al. J Inherit Metab Dis 1999;22:123–31.
160. Gillingham MB, Scott B, Elliott D, et al. Mol Genet Metab 2006;89:58–63.
161. Vockley J, Enns GM, Ramirez AN, et al. Mol Genet Metab 2022;136:152–62.
162. Harding CO, Gillingham MB, van Calcar SC, et al. J Inherit Metab Dis 1999;22:276–80.
163. Uauy R, Treen M, Hoffman DR. Semin Perinatol 1989;13: 118–30.
164. Roe CR, Roe DS, Wallace M, et al. Mol Genet Metab 2007;92: 346–50.
165. Gillingham MB, Weleber RG, Neuringer M, et al. Mol Genet Metab 2005;86:124–33
166. Howat AJ, Bennett MJ, Variend S, et al. Br Med J (Clin Res Ed) 1985;290:1771–3.
167. Rinaldo P, Schmidt-Sommerfeld E, Posca AP, et al. J Pediatr 1993;122:580–4.
168. Nyhan W, Ozand P. Disorders of fatty acid oxidation. Atlas of Metabolic Diseases London, UK: Chapman and Hall Medical, 1998;223–30.
169. Bonnet D, Martin D, Pascale De L, et al. Circ 1999;100:2248–53.
170. Bruno C, Dimauro S. Curr Opin Neurol 2008;21:601–6.
171. Bonnefont JP, Bastin J, Behin A, et al. N Engl J Med 2009;360: 838–40.
172. Bonnefont JP, Bastin J, Laforet P, et al. Clin Pharmacol Ther 2010;88:101–8.
173. Gobin-Limballe S, Djouadi F, Aubey F, et al. Am J Hum Genet 2007;81:1133–43.
174. Gobin-Limballe S, McAndrew RP, Djouadi F, et al. Biochim Biophys Acta 2010;1802:478–84.
175. Shiraishi H, Yamada K, Oki E, et al. Mol Genet Metab Rep 2019;20.
176. Suyama T, Shimura M, Fushimi T, et al. Mol Genet Metab Rep 2020;24.
177. Orngreen MC, Madsen KL, Preisler N, et al. Neurol 2014;82: 607–13.
178. Moser HW. NeuroRx 2006;3:246–53.
179. Kachwala I, Regelmann MO. Int J Neonatal Screen 2022;8.
180. Moser HW, Moser AB, Hollandsworth K, et al. J Mol Neurosci 2007;33:105–13.
181. Cartier N, Aubourg P. Brain Pathol 2010;20:857–62.
182. Eichler F, Duncan C, Musolino PL, et al. N Engl J Med 2017;377:1630–8.
183. Tolar J, Orchard PJ, Bjoraker KJ, et al. Bone Marrow Transpl 2007;39:211–5.
184. Litman BJ, Niu SL, Polozova A, et al. J Mol Neurosci 2001;16:237–42; discussion 79–84.
185. Berendse K, Klouwer FC, Koot BG, et al. J Inherit Metab Dis 2016;39:859–68.

CHAPTER 73

Nutritional Management of Infants and Children with Specific Diseases

Bridget M. Hron

CHAPTER OUTLINE

Abbreviations

AIDS, Acquired Immunodeficiency Syndrome; **ASPEN**, American Society for Parenteral and Enteral Nutrition; **BMI**, body mass index; **BRAT**, bananas, rice, apples, and toast; **CDC**, Centers for Disease Control and Prevention; **DGBI**, disorders of gut-brain interaction; **EFAD**, essential fatty acid deficiency; **ESPGHAN**, European Society for Pediatric Gastroenterology, Hepatology, and Nutrition; **FOD-MAPs**, fermentable oligo-di- mono-saccharides and polyols; **FPIAP**, protein-induced allergic proctocolitis; **FPIES**, food protein-induced enterocolitis syndrome; **GER**, gastroesophageal reflux; **HAART**, highly active antiretroviral therapy; **HIV**, Human immunodeficiency virus; **IBD**, inflammatory bowel diseases; **IFALD**, intestinal failure-associated liver disease; **IgE**, immunoglobin-E or non-IgE mediated; **LCT**, long chain triglycerides; **MCT**, medium chain triglycerides; **NASPGHAN**, North American Society for Pediatric Gastroenterology, Hepatology, and Nutrition

Introduction

The nutritional management of infants and children with specific conditions requires a working knowledge of the underlying pathophysiology, an understanding of how these conditions alter nutritional requirements, and awareness of appropriate nutritional therapies. This chapter reviews the nutritional implications of specific conditions in children that alter dietary requirements and discusses the general approach to nutritional management, with detailed reference both to enteral and parenteral nutritional support.

Specific Diseases and Other Conditions Requiring Nutritional Management

Gastrointestinal Disorders

Gastroesophageal Reflux

Gastroesophageal reflux is the passage of gastric contents into the esophagus, and it occurs as a result of transient relaxation of the lower esophageal sphincter. When symptoms are bothersome, or associated with medical complications, this can be considered gastroesophageal reflux disease. Gastroesophageal reflux disease is common, affecting 25% to 50% of infants and 0% to 38% of older children.[1-4] Additionally, rates are increased with certain pediatric conditions, such as neurologic impairment, Cornelia de Lange syndrome, cystic fibrosis, prematurity, or esophageal atresia.[5,6]

Current recommendations for GER management were published jointly by the North American Society for Pediatric Gastroenterology, Hepatology, and Nutrition (NASPGHAN) and the European Society for Pediatric Gastroenterology, Hepatology, and Nutrition (ESPGHAN) in 2009.[5] After careful history and physical exam to assess warning signs for anatomic and metabolic abnormalities, nutritional strategies are recommended as first steps.

For infants, the first nutritional intervention is thickening diets, which can be done using specialized formulas containing added rice, or adding rice or oatmeal to standard formulas. For older infants, pureed foods or commercial thickeners can be considered. Cereal thickeners are ineffective at thickening breastmilk, as they contain hydrolytic enzymes. While commercial thickeners are effective for thickening breastmilk, they generally cannot be used for premature infants or children under age 1 year (specific age thresholds vary based on thickener type and institution), and require caution with use in older children at risk of impaired gastrointestinal perfusion (e.g., congenital heart disease and/or high risk of necrotizing enterocolitis).[7] The second dietary strategy is use of extensively hydrolyzed or amino acid-based formula (see **Table 73.1**). This is due to the high degree of symptom overlap of food protein-induced allergic proctocolitis or cow's milk intolerance with gastroesophageal reflux.

Table 73.1 Classification of Pediatric Formulas

Category	Protein Source	Carbohydrate Source	Fat Source	Examples	Indications
Infant Formula (age < 1 year)					
Polymeric formula	Milk	Lactose, maltodextrin	LCT (palm olein, coconut, soy and high oleic sunflower, high oleic safflower oils)	Enfamil Infant, Similac Advance or Pro-Advance, Gerber Good Start Gentle Pro	Routine newborn management
Extensively hydrolyzed formula	Milk	Corn maltodextrin, corn syrup solids, cornstarch	LCT (palm olein, coconut, soy and high oleic sunflower, high oleic safflower oils) MCT	Alimentum, Nutramigen	Intolerance to cow's milk, malabsorption
Amino acid based/Elemental formula	Amino acids	Corn syrup solids	LCT (high oleic safflower, high oleic safflower, soy, canola oils) MCT	Elecare Infant, Neocate, Puramino	Intolerance to polymeric or extensively hydrolyzed formula, malabsorption
Pediatric Formulas (age 1–13 years)					
Polymeric formula	Milk, soy, pea	Corn maltodextrin, rice syrup solids, sugar, cornstarch	LCT (canola oil, soybean oil, safflower oil, sunflower oil, flax oil, coconut oil) MCT	Pediasure, Boost Kid Essentials, Kate Farms Pediatric	Routine pediatric management
Extensively hydrolyzed formula	Milk, pea	Corn maltodextrin, sugar, cornstarch	LCT (canola oil, soybean oil, safflower oil, sunflower oil, flax oil, coconut oil, fish oil) MCT	Pediasure Peptide, Kate Farms Pediatric Peptide	Intolerance to cow's milk, malabsorption
Amino acid based/Elemental formula	Amino acids	Corn syrup solids, tapioca starch, potato starch	LCT (safflower oil, soybean oil, coconut oil, palm kernel oil, sunflower oil) MCT	Elecare Jr, Neocate Jr.	Allergy to cow's milk Intolerance to extensively hydrolyzed formula, malabsorption

Category	Protein Source	Carbohydrate Source	Fat Source	Examples	Indications
Pediatric Formulas (age 1–13 years)					
Blenderized tube feeds	Milk, pea, beef, poultry, quinoa, fish, brown rice, hemp, oats, etc.	Fruits, vegetables, rice, grains	LCT (canola oil, olive oil, flax seed oil, fish oil, almond butter, grapeseed oil, beef, poultry, sesame seeds, sunflower seeds, fish, egg, hemp powder MCT	Nourish Real Food Blends Compleat Pediatric Organic Blends Home-made Blend	Parental preference Need for thickened feeds
Selected Formulas Used for Specific Conditions					
Preterm formulas	Milk	Maltodextrin, lactose, corn syrup solids	LCT (high oleic sunflower, high oleic safflower, soy, coconut oils) MCT	Similac Neosure, Enfamil Neuropro Enfacare, Similac Special Care	Prematurity
Soy-based infant formulas	Soy	Corn syrup solids, corn starch, sucrose	LCT (palm olein, coconut, high oleic sunflower, high oleic safflower, coconut, soybean oils)	Enfamil Prosobee, Similac Isomil, Good Start Soy	Galactosemia
Ketogenic formulas	Milk	Corn syrup solids, corn starch	LCT (high oleic sunflower, soy, palm oils) MCT	KetoCal, KetoVie	Intractible epilepsy
Low fat formulas	Milk	Maltodextrin, glucose syrup, corn syrup solids	LCT (walnut, sunflower oils) MCT (refined coconut oil, palm oil)	Nutricia Monogen, Enfaport	Chylous effusions

Abbreviations: *LCT*: long chain triglycerides; *MCT*: medium chain triglycerides.

Data from Klepper CM, Moore J, Gabel ME, et al. Nutr Clin Pract 2-23;38:302–17.[8]

For older children, expert guidelines recommend diet and lifestyle interventions to target a healthy weight, as first-line therapy. Acid suppression therapy may be indicated for both infants and older children, based on response to above therapies or if erosive esophagitis is present. Further diagnostic strategies are also outlined in the expert recommendations.[5]

Allergic Gastrointestinal Disorders

Allergic gastrointestinal diseases are increasing in prevalence. Food allergies can be classified as either immunoglobin-E (IgE) or non-IgE mediated. Classic Ig-E-mediated food allergies can present with anaphylactic symptoms or a range of other cutaneous, ocular, gastrointestinal, respiratory and cardiovascular symptoms, and require strict allergen avoidance[9]. There are several non-IgE mediated allergic conditions that also require nutritional management, as discussed below.

Eosinophilic Esophagitis. Eosinophilic esophagitis is characterized by eosinophilic infiltrates in esophageal mucosa accompanied by clinical symptoms such as dysphagia, food impaction and chest pain, and can result in esophageal strictures.[10] Food is a common trigger for eosinophilic esophagitis. Empiric six-food elimination diets, in which the most common allergens (milk, egg, soy, wheat, peanuts/treenuts, shellfish) are avoided, can induce remission in both adult and pediatric patients, although is difficult to maintain.[11] A study in adults suggests that elimination of milk alone may be as effective as six-food elimination diets.[12] While not effective in adults, targeted testing for specific food allergens may be helpful in pediatrics, although this approach is limited, as blood testing is IgE-based.[9,11,13] Medication options include swallowed steroids, prepared in a slurry formulation to adhere to the esophageal wall and act topically, and proton pump inhibitors. Most recent recommendations were published by the American Gastroenterological Association (AGA) Institute and North American Society of Pediatric Gastroenterology, Hepatology, and Nutrition in 2007.[11]

Food Protein-Induced Allergic Proctocolitis (FPIAP). Food protein-induced allergic proctocolitis, also known as allergic colitis or cow's milk protein intolerance, affects 2% to 3% of infants.[14] This classically presents with hematochezia (blood in stool) in an otherwise well-appearing infant. This most typically occurs in response to ingestion of cow's milk protein, although other dietary antigens (egg, soy, corn, or more than one allergen) may be implicated.[15] While this condition is often termed "cow's milk allergy" or "milk protein allergy," the pathophysiology is distinct from typical IgE/histamine-mediated allergic reactions and likely related to T-cell dysregulation.[16] A colonoscopy is not necessary for the diagnosis; however, if done, one may observe patchy eosinophilic infiltrate in the sigmoid colon.[15] The prognosis for this condition is excellent, with >90% of children achieving tolerance of cow's milk by age 2 years.[17]

Food Protein-Induced Enterocolitis Syndrome (FPIES). Food protein-induced enterocolitis syndrome (FPIES) is a non-IgE mediated food allergy that presents as severe vomiting, diarrhea, and lethargy approximately 1 to 4 hours after allergen consumption, which can progress to dehydration and even hypovolemic shock in severe cases.[18,19] Most cases in infancy are related to cow's milk ingestion, and other foods (such as rice, oats, vegetables, fruits, fish, and meats) may be problematic in older children. As with eosinophilic esophagitis, IgE-based skin prick testing is of limited value. Oral food challenges done under careful supervision by a trained allergist may be necessary if the history is not clear.[19] The majority of children ultimately develop tolerance to food allergens by age 5 to 7 years, although rates vary based on type of allergen and severity of disease.[19,20]

Diarrheal Diseases

Acute Diarrhea. Diarrhea is defined as the passage of loose, watery stools at least three times per day. Acute diarrhea is often self-limited and lasts several days. Acute diarrhea is often caused by viruses (e.g., norovirus and rotavirus), followed by bacteria (e.g., *Salmonella, Shigella, Campylobacter, Escherichia coli* and *Clostridium difficile*) and parasites (e.g., *Cryptosporidium, Entamoeba histolytica* and *Giardia lamblia*). The pathophysiology of most viral diarrheas is both osmotic and secretory.[21] In contrast, the pathophysiology of most toxigenic bacterial diarrheas (e.g., *Salmonella, Shigella,* and *Campylobacter* spp., enteropathogenic *Escherichia coli* serotype O157:H7) is secretory diarrhea resulting from stimulation of the adenylate cyclase system, as occurs in cholera infection.[22]

The major goal of therapy during acute diarrhea is to ensure a normal state of hydration. For patients able to tolerate enteral fluids, an oral rehydration solution containing glucose and electrolytes is the preferred approach. Fluid and electrolyte replacement must keep pace with intestinal losses until the diarrhea abates. Hospitalization and intravenous fluid therapy may be necessary, particularly if fever and/or vomiting accompany the diarrhea.

Continued feeding of a patient with diarrhea is an important way to maintain adequate nutritional status, as well as to provide vital topical nutrition to the recovering gastrointestinal tract.[23] Starvation, or the use of low energy, restrictive diets, such as the so-called BRAT (bananas, rice, apples, and toast) diet, is not recommended. Feeding may result in slightly increased stool output, especially in osmotic or mixed diarrheal conditions; carbohydrate malabsorption in acute diarrhea, while common, is usually transient. It should not preclude breastfeeding or administration of oral rehydration solution. Lactose-free diets are not routinely recommended but may be helpful in some cases of acute diarrhea in young or malnourished children.[24,25] Occasionally, reduced lactase activity persists and the patient develops postinfectious malabsorptive diarrhea. This condition can persist for weeks to months, but can also be managed by a reduced lactose diet.

In a few patients, the acute episode of diarrhea does not resolve in the usual 4 to 5 days. In these patients, nutritional management becomes a more important consideration for prevention of malnutrition. Following an episode of acute diarrhea, children are also at risk for postinfectious gastrointestinal disorders such as postviral gastroparesis and irritable bowel syndrome that can affect nutritional intake.[26]

Persistent Diarrhea. Diarrhea that lasts longer than 14 days is termed persistent diarrhea. Noninfectious causes of persistent diarrhea result from a spectrum of congenital or acquired abnormalities that include changes in villous structure (e.g., celiac disease), anatomic abnormalities (e.g., short bowel syndrome), ultrastructural abnormalities (e.g., microvillous inclusion disease), and abnormalities at the molecular level (e.g., congenital chloride diarrhea). If not managed appropriately, these disorders frequently result in the same abnormalities of mucosal function observed in acute diarrheas. Severe cases, such as seen in congenital enteropathies, can result in severe malnutrition, dehydration, and death without the use of parenteral nutrition support. The nutritional management of illnesses associated with persistent diarrhea, namely celiac disease, inflammatory bowel disease, and short bowel syndrome, are described here. Nutritional management of other chronic diarrheas, in general, must be tailored to the specific cause and pathophysiology of these conditions.

Celiac Disease and Other Gluten Responsive Conditions

Celiac disease is a gluten-induced enteropathy that occurs in a genetically susceptible host. The classical presentation is of a pediatric patient presenting with malnutrition secondary to steatorrhea and malabsorption. However, it is now clear that celiac disease can present with a range of intestinal and extraintestinal symptoms at any age. Untreated celiac disease can present with several associated nutritional complications, including growth failure, malnutrition, deficiencies of iron, folate, vitamin B_{12}, zinc and magnesium, and low bone mineral density.[27] While many of these nutritional issues may resolve with initiation of gluten-free diets, some patients may require ongoing micronutrient supplementation.[27]

Strict gluten-free diets are the standard of care, although they have significant limitations. Many staples of a healthful diet (such as fruits and vegetables) are inherently gluten free. However, gluten-free diets are often rich in highly-processed carbohydrates and saturated fats, low in fiber and multiple micronutrients (such as iron and vitamin D), and may promote weight gain, especially in a susceptible population.[28,29] Gluten-free diets are costly, difficult to follow, especially outside of the home environment, and are often elusive as unintentional ingestion occurs frequently.[30] At present, no pharmacologic strategies exist for celiac disease, although many are in development.

Additional gluten-responsive conditions include 1) wheat allergy and 2) nonceliac gluten sensitivity. Wheat allergy is an allergic reaction related to mast cell activation. Strict gluten avoidance is necessary to prevent life-threatening allergic/anaphylactic reactions. Nonceliac gluten sensitivity refers to a constellation of symptoms (such as diarrhea, fatigue, and headache) that are reported following ingestion of a gluten-containing diet, but without the characteristic intestinal changes seen in celiac disease. Removal of gluten from the diet may be required for symptom management.[31]

Inflammatory Bowel Disease

The pathogenesis of inflammatory bowel diseases (IBD), including ulcerative colitis and Crohn's disease, involves immune dysregulation, environmental triggers, and intestinal luminal microbial antigens in a genetically susceptible host.[32] Mucosal inflammation can lead to malabsorption, particularly with small bowel involvement in Crohn's disease, contributing to malnutrition and growth failure. More than 50% of individuals with IBD demonstrate evidence of micronutrient deficiencies, most commonly iron, zinc, vitamin D, and folic acid.[33] IBD management targeting mucosal healing, in combination with energy and micronutrient supplementation, is critical to achieve optimal growth potential and reverse deficiencies.

Dietary therapies for inflammatory bowel disease are a topic of active research. Exclusive enteral nutrition, or the provision of polymeric or amino acid-based formulas, is recommended as first-line therapy for Crohn's induction treatment strategy by ESPGHAN.[34] Exclusive enteral nutrition promotes mucosal healing with a favorable side-effect profile and a positive effect on growth, compared with steroids. However, exclusive enteral nutrition can be difficult to maintain outside of the induction period. Therefore, other dietary treatment strategies, such as the Crohn's Disease Exclusion Diet and other forms of partial enteral nutrition, the Specific Carbohydrate Diet and the Mediterranean Diet, are currently being studied.

Short Bowel Syndrome

Short bowel syndrome is defined by a marked diminution of the effective absorptive surface area of the bowel secondary to acquired or congenital gastrointestinal disease requiring specialized nutritional therapy.[35] In general, the severity of short bowel syndrome is related inversely to the length of the remaining intestinal segment; however, loss of the ileocecal valve, which acts as a physiologic sphincter to slow transit time and prevent backwash ileitis, increases symptom severity.[36,37]

The early phase of short bowel syndrome immediately after massive resection is usually associated with massive fluid and electrolyte losses, and often requires parenteral nutrition support. As the remaining small bowel gradually adapts, enteral intake usually can be advanced, but this must proceed slowly. In general, continuous feeding through either an indwelling nasogastric tube or a gastrostomy tube is better tolerated during this phase than bolus feeding.[38] Human milk is the preferred diet for infants with short bowel syndrome. When unavailable, amino acid-based formulas (see Table 73.1) are commonly used, due to the reduced bowel surface area and the predilection for food allergies in patients with short bowel syndrome.

Intestinal adaptation is the restructuring of the bowel after significant loss and proceeds by two mechanisms: hypertrophy of the remaining bowel and hyperplasia of the intestinal mucosa, which act synergistically to increase the total intestinal surface area. Teduglutide is a glucagon-like peptide-2 analog that stimulates mucosal growth and has been shown to decrease reliance on parenteral nutrition support.[39] Once maximum adaptation is ultimately achieved, a process that may take several years, more complex proteins and carbohydrates can be introduced. Even during this final phase; however, frequent small feedings may be necessary.

Pharmacologic manipulations (e.g., cholestyramine to chelate bile acids, loperamide to slow transit time, or antibiotics to eradicate small bowel bacterial overgrowth) may provide symptomatic and physiologic improvement. Surgical bowel lengthening operations (e.g., Bianchi procedure, or serial transverse enteroplasty), or, in severe cases, small intestinal transplantation may be required.

Symptoms also result from removal of specific segments of intestine. Because disaccharidase activity is greater in jejunal cells and cholecystokinin and other intestinal hormones are secreted by jejunal sites, removal of the jejunum results in more severe carbohydrate malabsorption, as well as altered motility. Ileal loss results in loss of bile salt uptake and malabsorption of vitamin B12.[35,40-43]

Disorders of Gut-Brain Interaction

Disorders of gut-brain interaction (DGBI), previously known as functional gastrointestinal disorders, are heterogeneous disorders involving motility disturbance, visceral hypersensitivity, altered mucosal and immune function, altered gut microbiota, and altered central nervous system processing.[44] Maintenance of adequate nutrition is of critical importance in many of these conditions.

Irritable bowel syndrome is the most common DGBI, with a reported prevalence of 3% to 16%.[45] It is defined as recurrent abdominal pain that is associated with stooling, and/or associated with change in frequency of stooling. As with other DGBIs, IBS treatment requires a transdisciplinary biopsychosocial approach. Nutritional management has gained popularity. Reducing intake of fermentable oligo- di- mono-saccharides and polyols (FOD-MAPs) found in many fruits and vegetables has been associated with improvement in symptoms.[46,47] This restrictive diet is complex and requires extensive monitoring by a physician and dietitian, although individuals should be carefully screened for eating disorders prior to initiation.[48]

Cyclic vomiting syndrome is increasingly recognized as an important, yet underappreciated, cause of chronic vomiting in childhood.[49] This syndrome is characterized by repetitive, severe episodes of vomiting lasting hours or days that may necessitate hospital admission for intravenous rehydration. Another cause of chronic vomiting in children is rumination syndrome, which is painless and effortless regurgitation.[50] Formerly believed to exist solely in children with developmental delay, it is now recognized to be fairly common in children and adolescents with typical development, and affects 3% of adults.[51] In all such cases, nutritional deficits are caused by inadequate intake and are managed by ensuring adequate provision of protein and energy during periods of quiescence, in conjunction with standard medical therapy.

Neurologic Disorders

Neurologic Impairment and Aspiration

Children with neurologic impairment are at high risk for aspiration, malnutrition, and gastrointestinal symptoms but can achieve adequate growth with adequate provision of energy and management of comorbidities. Therefore, ESPGHAN guidelines recommend close observation of nutritional status, often requiring additional anthropometric criteria (e.g., triceps, skinfold thickness, and tibial length) as weight and height may not be reliable, and bloodwork for micronutrient status.[52] Oropharyngeal dysphagia should be promptly recognized and treated, using sensorimotor interventions, alterations in feeding parameters (e.g., food consistency, posture, and time), and use of enteral tubes in severe cases. Children with neurologic impairment should be monitored for development of gastroesophageal reflux, constipation, and micronutrient deficiencies.

Epilepsy

For children with certain forms of refractory epilepsy (such as infantile spasms, epilepsy with myoclonic-atonic seizures, Rett syndrome, or severe myoclonic epilepsy of infancy), ketogenic diets may be an effective management strategy.[53] The ketogenic diet provides <5% to 10% total energy from carbohydrate and induces hepatic ketone body production to induce a state of ketosis. The mechanism of action is poorly understood but may be due to effects on neurotransmitters, mitochondrial function, inflammatory cytokines, or even the intestinal microbiome.[53] Side effects of the ketogenic diet include constipation, vomiting, diarrhea, hyperlipidemia, hypoglycemia, acidosis, growth problems, and anorexia.[54]

Cardiopulmonary Disorders

The two major cardiopulmonary disorders for which children require nutritional support are congenital heart disease and cystic fibrosis. Other conditions causing cardiac or respiratory failure, such as congenital or acquired cardiomyopathies and interstitial lung diseases, are less common, although a similar approach is generally used for their nutritional management.

Congenital Heart Disease

Growth failure is a common finding in infants and children with congenital heart disease.[55] Children with severe heart disease may have an increased energy requirement, but many cases of associated malnutrition are due to inadequate intake.[56] This may be the result of poor appetite, excessive tiring during feeding, or poor oromotor skills.[55,56] In addition, fluid and sodium intakes are frequently restricted as a part of treatment, and the use of diuretics is common. Either practice, of course, may limit growth even if intake of protein and energy is adequate.

The most common form of nutritional therapy for infants with congenital heart disease is use of a high–nutrient-density commercial formula. Tube feedings through either a nasogastric or a gastrostomy tube are frequently necessary, particularly in infants with excessive tiring or aspiration during feeding. Implementation of standardized protocols to address weight may be effective to identify and treat poor weight gain.[57] In general, if sufficient nutrients are delivered, most such patients will grow at a reasonably normal rate.[58-60] In addition, catch-up growth will occur in most children following corrective surgery for congenital heart disease as long as extra protein and energy are made available.[60]

Cystic Fibrosis

Cystic fibrosis is an autosomal recessive disease caused by mutations in the chloride transporter Cystic Fibrosis Transmembrane conductance Regulator (CFTR) gene. Cystic fibrosis was named after the characteristic findings of cyst formation and fibrosis within the pancreas and results in progressive deterioration in pancreatic and pulmonary function.

Cystic fibrosis can contribute to malnutrition in several key ways.[61] The first is reduced nutrient intake, particularly during acute exacerbations. The second is increased metabolic needs in settings of acute exacerbations.[62] The third is exocrine pancreatic insufficiency, which severely limits the absorption of fat and fat-soluble nutrients and varies in severity based on genotype.[63] Fourth is the presence of gastrointestinal comorbidities, such as gastroesophageal reflux and impaired gastrointestinal motility. Addressing each of these components is critical, as improvement of nutritional status improves muscle strength, long-term growth, and lung function.[64-66] Recommendations have targeted weight-for-length percentile ≥ the 50th percentile on Centers for Disease Control and Prevention (CDC) growth charts.[67]

Pancreatic insufficiency can be addressed with appropriate pancreatic enzyme therapy. For individuals who are fed orally, enteric capsules can be administered with meals. For individuals who receive G-tube feeds, an in-line cartridge containing lipase allows for predigestion of formula and is the preferred method of therapy, although enteric capsules can be used.[68] Attention must be given to adequate supplementation of fat-soluble vitamins, particularly vitamin K, which may be deficient in patients with cystic fibrosis.[69]

The most recent guidelines for nutrition in CF from NAPSGHAN were published in 2002, and by ESPEN/ESPGHAN in 2017.[70,71] However, the landscape of CF treatment is now changing rapidly and positively. With the advent of CFTR modulators (e.g., ivacaftor), adult and pediatric patients with CF have improved lung function, fewer pulmonary exacerbations and hospitalizations, increased body mass index, and higher levels of fecal elastase (indicating less fat malabsorption).[72-75] Future studies will be critical to inform new guidelines for nutritional interventions in this current era.

Critical Illness, Trauma, and Burns

The American Society of Parenteral and Enteral Nutrition published guidelines on optimal nutritional therapy for child with critical illness.[76] In brief, the Society advocates nutritional screening and assessment, determination of energy expenditure ideally by indirect calorimetry, preferential use of the gastrointestinal tract, and longitudinal consultation with a dedicated pediatric nutrition support team. Enteral feedings should be initiated early when medically feasible, advanced according to a stepwise algorithm, continuously monitored, and adjusted when necessary according to a uniform definition and approach to feeding intolerance. Interruptions in enteral feeding resulting from diagnostic procedures, therapeutic interventions, and enteral access malfunction should be minimized. Parenteral nutrition should be reserved for those circumstances when enteral nutrition fails, and not administered within first 24 hours of ICU admission, based on outcomes from the PEPaNIC clinical trial.[77]

Energy intake for the child with critical illness is ideally determined by the use of indirect calorimetry. Standard equations for predicting energy needs, such as the Harris Benedict formula, are frequently inaccurate, and the stress response to illness and injury may not be as pronounced as was previously believed.[78-86] Overfeeding can result in respiratory compromise through excess production and subsequent elimination of carbon dioxide, as well as hyperglycemia, hyperlipidemia, and liver dysfunction.[87] In the absence of indirect calorimetry, experts recommend delivering at least two-thirds of the prescribed daily energy requirement. Protein provision is critically important to prevent catabolism (e.g., breakdown of diaphragmatic, intercostal, and cardiac muscle mass, and its untoward effects on respiratory, cardiac, and immunologic function).[88,89] Protein dose should be 2 to 3 g/kg/d for children 0 to 2 years; 1.5 to 2 g/kg/d for children 2 to 13 years; and 1.5 g/kg/d for children 13 to 18 years.[85]

The protein and energy needs of children with burn injuries have been extensively studied and have, therefore, been definitively established, with recommended protein intakes of 1.5 to 2.0 g/kg/d for adults and 2.5 to 4.0 g/kg/d for children.[90] Energy intake can be estimated from several equations and should account for hypermetabolism associated with burn injury, while avoiding overfeeding.[90] Although administration of greater amounts helps to maintain body weight, it does not appear to result in preservation of lean body mass; as such, protein and energy in excess of these amounts are no longer routinely given.[91,92]

Additional Disorders

Acute Kidney Injury and Chronic Kidney Disease

Adequate provision of energy and protein can prolong life and optimize linear growth in children with renal failure.[93] Children with acute kidney injury or chronic kidney disease can experience decreased appetite.[94] Energy requirements of children with chronic renal failure appear to be no greater than those of normal children, although both acute and chronic renal disease affect nutrient metabolism.[95,96] Multiple aspects of renal disease have deleterious effects on protein metabolism, such as reduced intake, hypercatabolism, metabolic acidosis, reduced anabolism, and others.[96] Current recommendations are to provide the protein dietary reference intake for age, and to recognize that children requiring hemodialysis may need additional protein to account for loss.[97] Excess vitamin A should be avoided, and dietary phosphate should be restricted to avoid development of renal bone disease. Additional supplementation of water-soluble vitamins may be necessary for patients on renal replacement therapy/dialysis. The Pediatric Renal Nutrition Taskforce recently published guidelines for energy and protein intake in children with chronic kidney disease.[97]

HIV/AIDS

Human immunodeficiency virus (HIV) infection and acquired immunodeficiency syndrome (AIDS) can result in malnutrition, growth failure, and abnormalities of body composition from a multitude of factors. Oral, gingival, and esophageal lesions (e.g., ulcers, periodontitis, thrush) can make eating a distasteful and painful experience. Gastritis, nausea, vomiting, and abdominal pain, caused by either infection or irritation from medications, can adversely affect appetite. Pancreatitis, which typically manifests with vomiting, is a known side effect of several highly active antiretroviral therapy (HAART) medications. Chronic diarrhea may also cause ongoing nutrient loss. Finally, encephalopathy, which may be present in advanced HIV disease, can also cause inadequate energy intake.

During the height of the HIV/AIDS epidemic, approximately 90% of all children with HIV/AIDS developed malnutrition at some time during the course of their disease.[98,99] Children with HIV infection may require higher energy intake due to impaired absorption and impaired energy use. Adequate nutrition is not only critical to prevent growth failure but also may improve intestinal absorption and increase CD4 cell numbers.[100] Fortunately, with the widespread use of HAART, substantial improvements in height-for-age, weight-for-age, and weight-for-height values can be achieved.[101] Detailed guidelines on management of the pediatric patient with HIV have been published by ASPEN.[102]

General Approach to Nutritional Therapy

Children with "impaired oral intake that is not age-appropriate, and is associated with medical, nutritional, feeding skill, and/or psychosocial dysfunction," can be diagnosed with a Pediatric Feeding Disorder.[103] This often coexists with other conditions, discussed previously. See Chapter 69 for a detailed review of this condition.

Developing patient-centered nutritional therapy starts with assessment of the patient's underlying nutritional status. This is critical to understand potential complications as well as the urgency and character of interventions. Pediatric undernutrition and overnutrition have been associated with many complications, including increased morbidity and mortality in the hospitalization setting.[104,105] In the pediatric patient, undernutrition can be diagnosed using anthropometric measures, such as age and sex-adjusted low weight for length or BMI (**Table 73.2**), with z-score below −3, consistent with severe undernutrition, or between −3 and −2, consistent with moderate malnutrition.[106] Overnutrition can be diagnosed using body mass index percentiles; BMI in the 85th to 94th percentile is consistent with overweight, 95th percentile or greater, obesity, and ≥120% of the 95th percentile or BMI ≥35 kg/m^2, severe obesity. A nutrition-focused physical exam is critical to interpret anthropometrics, confirm the diagnosis of under- or overnutrition, and assess for clinical evidence of micronutrient deficiencies.

A multidisciplinary team is typically necessary to create a holistic care plan, including a primary care physician; physician specialists, such as a gastroenterologist; and a registered dietitian, registered nurse, social worker, psychologist, and feeding therapist. The role of this group is to support the patient and caregivers to outline and meet goals of care in a culturally and socially sensitive manner.[108]

Determination of Energy Requirement

Resting energy expenditure contributes roughly two-thirds of the overall energy requirement. Additional components include energy required for physical activity, growth, and the thermic effect of food, as well as energy needed to account for losses (e.g., gastrointestinal, or renal).[109] Indirect calorimetry is a highly effective tool to measure resting energy expenditure, but its use is limited by practical considerations, including machine availability. As such, several methods exist to estimate resting energy expenditure from age, sex and weight, and height, in some cases. These include equations published by Harris-Benedict,[110] the World Health Organization,[111] Schofield,[112] and Estimated Energy Requirements from the Institute of Medicine.[113] An activity factor can then be applied to estimate total energy expenditure. These equations are useful tools but may be inaccurate in patients with overweight/obesity.[109] Demonstrating adequate growth on a specific caloric provision remains the gold standard for energy adequacy.

Enteral Nutrition

In individuals fed via the oral route, nutritional support should be attempted using conventional foods, adding commercial oral supplements, if needed. If oral feeds are not tolerated, unsafe or insufficient for metabolic demands, or if alternative nonpalatable nutrition or medicinal interventions are required, placement of an enteral tube may be indicated. Enteral nutrition (i.e., tube feeding) is typically initiated in the inpatient setting and is successfully used in critically ill and acutely ill infants and children. It is preferred to parenteral nutrition because it is both less costly, more physiologic, and less risky.[114-118] Additional details regarding the practice of enteral nutrition support were published by ASPEN.[119]

Selection of Formula

Appropriate selection of formula requires a patient-centered approach with collaboration from multiple disciplines. A wide range of conventional and alternative formulas now exist

Table 73.2 Current Criteria for Diagnosis of Undernutrition

	Mild Malnutrition	Moderate Malnutrition	Severe Acute Malnutrition
World Health Organization (2009)			
Weight for height z-score	N/A	−3≤ to <−2	<−3
Length/height for age z-score	N/A	−3≤ to <−2	N/A
Mid upper arm circumference	N/A	N/A	<115 mm
Edema	N/A	No	Present
American Society for Parenteral and Enteral Nutrition and Academy of Nutrition and Dietetics (2015)			
Single data point available			
Weight for height z-score	−1 to −1.9	−2 to −2.9	−3 or less
BMI for age z-score	−1 to −1.9	−2 to −2.9	−3 or less
Length/height for age z-score	No data	No data	−3 or less
Mid upper arm circumference	−1 to −1.9	−2 to −2.9	−3 or less
Multiple data points available			
Weight gain velocity (<2 years of age)	<75% of expected	<50% of expected	<25% of expected
Weight loss (2–20 years of age)	5% usual body weight	7.5% usual body weight	10% usual body weight
Deceleration in weight for length/height z-score	Decline of 1 z-score	Decline of 2 z-score	Decline of 3 z-score
Inadequate nutrient intake	51%–75% estimated	26%–50% estimated	<25% estimated

Data from Becker P, Carney LN, Corkins MR, et al. Nutr Clin Pract 2015;30:147–61; WHO Child Growth Standards and the Identification of Severe Acute Malnutrition in Infants and Children: A Joint Statement by the World Health Organization and the United Nations Children's Fund, Geneva. 2009.[106,107]

(Table 73.1). A detailed medical history should probe for underlying gastrointestinal, neurologic, metabolic, or other conditions as outlined above that may require specific nutrient composition or alterations in route of delivery (such as need for postpyloric nutrition). A detailed dietary history should include assessment of previous nutritional strategies, with close attention to use of nutrition support and food allergies and intolerances. A detailed social history should include dietary frameworks (such as vegetarian, vegan, Kosher, etc.) that are important to the patient and their caregivers, logistics of meals such as location of feeding (such as school vs home environment), caregivers involved in feeding, regularity of travel, etc. This information will inform the choice of the least restrictive diet that meets nutritional needs.

Route of Administration

Table 73.3 summarizes the indications, contraindications, and adverse events associated with the most common enteral routes. Gastric feedings can be given intermittently (bolus feedings)

Table 73.3 Nutritional Support Indications, Contraindications, and Adverse Events

Nutritional Support	Indications	Contraindications	Adverse Events
Gastric			
Nasogastric	Inability to tolerate oral intake with functional gastrointestinal tract: Anorexia Dysphagia Odynophagia	Facial trauma Basilar skull fracture *Esophageal trauma* *Esophageal obstruction* *Anticoagulation*	Malposition—intracranial, tracheal, pulmonary, esophageal Pneumothorax Discomfort Sinusitis Epistaxis Gastric irritation or ulceration Nasal irritation Nasal ulcers/necrosis

Nutritional Support	Indications	Contraindications	Adverse Events
Gastric			
Gastrostomy tube	Long term need for feeding support and ability to tolerate gastric feeds	Peritonitis Untreated coagulopathy *Gastrointestinal tract obstruction* *Massive ascites/portal hypertension* *Peritoneal metastasis* In addition, for percutaneous endoscopic gastrostomy: Esophageal obstruction Technical inability to approximate stomach to abdominal wall *Large hiatal hernia* *Severe under- or overnutrition* *Previous abdominal surgery including ventriculoperitoneal shunt*	Cellulitis/wound infection Displacement or misplacement Leakage Granulation tissue Peritonitis Necrotizing fasciitis Risks of anesthesia and surgery Tube occlusion Buried bumper syndrome Colocutaneous fistula In addition, for percutaneous endoscopic gastrostomy: Perforation Gastrocolic fistula
Postpyloric			
Nasojejunal	Inability to tolerate gastric feeds: Gastroparesis Gastric outlet obstruction Recurrent aspiration Proximal enteric fistula	Facial trauma Basilar skull fracture Bowel obstruction Bowel perforation *Esophageal trauma* *Esophageal obstruction* *Anticoagulation*	Failure of placement Pneumothorax Displacement or misplacement Clogging of the tube Epistaxis Nasal irritation Nasal ulcers Feeding-related diarrhea Abdominal cramping Metabolic complication, such as hyperglycemia Dumping-like symptoms Otitis media Nasal mucosal pressure sores Esophageal ulcers Sinusitis
Gastro-jejunal tube	Long term need for feeding support and inability to tolerate gastric feeds: Gastroparesis Gastric outlet obstruction Recurrent aspiration Proximal enteric fistula	Bowel obstruction Bowel perforation Peritonitis Untreated coagulopathy	Intussusception Cellulitis Displacement or misplacement Granulation tissue Clogging of the tube Feeding-related diarrhea Abdominal cramping Metabolic complication, such as hyperglycemia Dumping-like symptoms Risks of anesthesia and surgery

(continues)

Table 73.3 Nutritional Support Indications, Contraindications, and Adverse Events *(continued)*

Nutritional Support	Indications	Contraindications	Adverse Events
Postpyloric			
Jejunosotomy tube	Long term need for feeding support and inability to tolerate gastric feeds: Gastroparesis Gastric outlet obstruction Recurrent aspiration Proximal enteric fistula	Ascites Untreated coagulopathy Peritoneal dialysis Peritoneal metastasis	Pain Cellulitis Feeding-related diarrhea Abdominal cramping Clogging of tube Metabolic complications, such as hyperglycemia Displacement Leakage Failure of placement Misplacement Gastric hemorrhage Perforation of internal organs and peritonitis Colocutaneous fistula Risks of intravenous sedation or risks of anesthesia and surgery Hemorrhage at jejunostomy site Pressure sore due to skin disk of jejunostomy Dumping-like symptoms
Parenteral Nutrition	Inability to tolerate gastric or postpyloric feeding	Severe electrolyte derangements Hemodynamic instability Allergy (often to lipid components)	See Table 73.8

Relative contraindications identified in italics.

Data from DeLegge MH. Gastroenterol Clin North Am 2018;47:23–37; Gencosmanoglu R. Neurocrit Care 2004;1:309–17; Niv E, Fireman Z, Vaisman N. World J Gastro 2009;15:1281–8.[124-126]

or continuously (drip feedings), either throughout the day or only during part of the day, depending on the patient's age, condition, and nutritional status, in an individualized approach.[120-123]

For most children unable to feed orally for a short duration, gastric feeding via nasogastric tube (NG) is a reasonable alternative. However, NG tubes can dislodge and cause nasal passage breakdown with prolonged use. Therefore, if the patient requires long-term enteral tube dependence (e.g., static or progressive neurologic impairment), a gastrostomy tube, inserted percutaneously, laparoscopically, or surgically, can be considered.

Jejunal tube feeding can be considered in children in whom gastric feeding is insufficient to meet nutritional needs.[127] Jejunal feeds require continuous formula delivery, due to the lack of distensibility of the small bowel. As with nasogastric tubes, nasojejunal tubes are a short-term solution. These can be placed by nursing at bedside or with fluoroscopic assistance. Gastrojejunal (G-J) tubes can be placed laparoscopically or endoscopically, and consist of a jejunal port for feeding and gastric port for venting. Complications of G-J tubes include clogging and intussusception. A surgically placed jejunostomy may be necessary if G-J tube is contraindicated, but hase high rates of complications such as stoma drainage.[127]

If patients continue to experience feeding intolerance to gastric or jejunal feeds, despite medical and nutritional optimization, parenteral nutrition may be used as a sole source of nutrition or as a supplement to tolerated enteral nutrient combinations.

Parenteral Nutrition

The widespread use of parenteral nutrition is considered to be one of the major contributing factors to the reducing mortality rates of infants born with surgically correctable lesions of the gastrointestinal tract (e.g., omphalocele, gastroschisis, or intestinal atresias) as well as of infants with short bowel syndrome and intractable diarrhea. Parenteral nutrition is a high-risk therapy that requires careful attention to every aspect of the technique, thereby minimizing its hazards and maximizing its benefits. This is best accomplished by consultation with a trained transdisciplinary nutrition support team, which can improve the appropriateness, adequacy and safety of PN use.[108,128] Additional details on the safe administration of parenteral nutrition have been published by ASPEN.[129]

Route of Administration

Parenteral nutrition is most preferably administered via central venous catheter, although it can be administered via peripheral vein. Peripheral administration is limited by the osmolarity of the solution, which is largely determined by glucose and amino acid concentration. Infusion of high osmolarity solutions peripherally can result in phlebitis, thrombosis of vessel and require need for frequent replacement of peripheral lines.[130] Higher provision of energy and protein intakes required by neonates and young children typically require the central venous route. Appropriate locations of the distal tip of the central venous catheter include superior vena cava, inferior vena cava, or right atrium (under certain circumstances).[131] Centrally placed, surgically placed umbilical venous catheters can be used as a reliable short-term option for neonates, with careful monitoring for infection and thrombosis.[130] Choice of surgical catheter is outside of the scope of this chapter and can be found in ASPEN guidelines.

Nutrient Infusate

Parenteral nutrition is a high risk, complex fluid that is best managed by an interdisciplinary team consisting of physicians, dietitians, nurses, and pharmacists. The nutrient infusate is highly personalized for each patient and typically compounded by a specialized pharmacy. The nutrient infusate should include nitrogen source in the form of amino acids and a carbohydrate source as dextrose, electrolytes, minerals, trace elements, and vitamins. Intravenous fat is often infused separately.

Commercially prepared crystalline amino acid mixtures are used as the nitrogen source of parenteral nutrition infusates. These mixtures contain most essential amino acids (exceptions are cysteine and tyrosine, which are either unstable or insoluble in aqueous solution) and varying amounts of nonessential amino acids. Cysteine can be added separately for extremely low birthweight infants (<1000 g). Total protein dose varies based on age, with preterm infants requiring 4 g/kg, infants typically 3 g/kg, with decreasing requirements as children age (**Table 73.4**).

Glucose/dextrose is the preferred nonlipid parenteral energy source. Dextrose provision is typically initiated at 5% to 10% solutions for children and adolescents, or at 4 to 7 mg/kg/min for infants. Dosing can be increased as tolerated by glucose infusion rate of 1 to 2 mg/kg/min, until goal caloric provision (accounting for energy from protein and lipid) has been achieved. Serum glucose should be monitored daily at initiation, at minimum. Some children, particularly during initiation of parenteral nutrition, develop hyperglycemia and osmotic diuresis with concomitant urinary loss of electrolytes when the amount of glucose infused exceeds tolerance. Careful, continuous administration of small doses of insulin may be necessary to use energy.

General electrolyte requirements are presented in Table 73.4. However, this varies considerably between patients and within patients (with changes in clinical status and medication use), and should not be interpreted as absolute requirements. Adjustments are often necessary and should be made on the basis of close monitoring (**Table 73.5**). The amounts of calcium and phosphorus required for optimal skeletal mineralization often cannot be incorporated into the parenteral nutrition infusate because of the chemical incompatibility of calcium and phosphate and solubility profiles of amino acid solutions. This is particularly true for the LBW infant. In this situation, enteral supplementation of calcium and phosphorus may be necessary to achieve goal intakes.

Multivitamins and trace elements should be included daily in the nutrient infusate, although not all trace elements may be

Table 73.4 Estimation of Pediatric Macronutrient and Electrolyte Requirements

	Infants and Toddlers	Children	Adolescents
Calories (kcal/kg/d)	80–120 kcal/kg	60–90 kcal/kg	30–75 kcal/kg
Macronutrients			
Dextrose (mg/kg/min)	7–18	5.5–14	4–14
Amino acids (g/kg/d)	1.5–3.0	1–3	0.8–2.5
Fat (g/kg/d)	0.5–3.0	1.0–2.5	1.0–2.5
Electrolytes			
Sodium (mEq/kg/d)	2–5	2–5	1–2
Potassium (mEq/kg/d)	2–4	2–4	1–2
Calcium (mEq/kg/d)	0.5–4.0*	0.5–4.0	10–20 mEq
Phosphorus (mmol/kg/d)	0.5–2.0*	0.5–2.0	10–40 mmol
Magnesium (mEq/kg/d)	0.3–0.5	0.3–0.5	10–30 mEq
Chloride	As needed to maintain acid-base balance		

*For preterm infants, calcium needs are 2–4 meq/kg, phosphorus 1–2 mmol/kg.

Data from Baker RD, Davis S, Baker S, Davis A. Ped Paren Nutr Chapman & Hall, 1997; Mirtallo J, Canada T, Johnson D, et al. JPEN 2004;28:S39–70.[132,133]

Table 73.5 Suggested Monitoring for Children Requiring Nutrition Support in the Inpatient Setting

Variables to Be Monitored	Suggested Frequency	
	Initial Period	**Follow-Up/Home**
Growth Variables		
Weight	Daily	Daily to monthly
Length	Weekly	Weekly to monthly
Head circumference	Weekly	Weekly to monthly
Body composition	Weekly	Monthly
Metabolic Variables		
Plasma electrolytes	Daily until stable, then weekly	Weekly
Plasma calcium, magnesium, phosphorus	Daily until stable, then weekly	Weekly to monthly
Blood acid–base status	Daily until stable, then weekly	Weekly to monthly
Blood urea nitrogen	Weekly	Monthly
Plasma albumin	Weekly	Monthly
Liver function studies	Weekly	Weekly to monthly
Triglycerides	Daily while increasing lipids	Weekly to monthly
Iron indices	Monthly	Every 3–4 months or as indicated
Hemoglobin	Weekly	Every 3–4 months or as indicated
Urinary glucose	Weekly	As needed
Variables for Detection of Infection		
Clinical observations (activity, temperature, vital signs etc.)	Daily	Daily
White blood cell count	As indicated	As indicated
Cultures	As indicated	As indicated
Catheter site/dressing	6–12 times/d	2–6 times/d

Initial period is the time during which the desired energy intake is being achieved or at a time of metabolic instability.

Data from Baker RD, Davis S, Baker S, Davis A. Ped Paren Nutr Chapman & Hall, 1997.[132]

necessary for short-term use.[134] Parenteral dosing can be found in **Table 73.6**. Routine iron supplementation is not necessary in short-term use. For longer use, iron is ideally administered enterally, although parenteral formulations exist.[135]

Parenteral Lipid Emulsions

Lipid emulsions are an energetically dense source of energy, provide essential fatty acids, and are critically important for brain development. Currently available lipid emulsions are sourced from 1) soybean oil (such as Intralipid, Fresenius Kabi AB, or Nutrilipid, B. Braun); 2) fish oil (Omegaven, Fresenius Kabi); or 3) mixed lipid emulsions (SMOFlipid, Fresenius Kabi, which contains soy, medium chain triglycerides, olive oil, and fish; ClinoLipid, Baxter, which contains 80% olive oil and 20% soybean oil). The nutritional compositions of these oils are presented in **Table 73.7**. Choice of fats depends on clinical condition, duration of therapy, allergy profile, and availability. Use of fish oil lipid emulsions is an effective strategy for treatment of children with intestinal failure-associated liver disease.[42,43]

Lipid emulsion dosing is based on estimated macronutrient needs and choice of emulsion (Tables 73.4 and 73.7). Initial dosing is typically 0.5 to 1 g/kg/d with gradual increases of 0.5 to 1 g/kg/d to achieve goal dosing. For soybean oil, goal dosing is often 2 to 3 g/kg/d. A dose of only 0.5 g/kg/d of soybean oil emulsion is sufficient to prevent classic essential fatty acid deficiency (EFAD), and doses of <1 g/kg/d may be used for patients with intestinal failure-associated liver disease. Goal dosing for mixed lipid emulsions is similar at 2 to 3 g/kg, and these higher doses are necessary to prevent essential fatty acid deficiency.[138] Fish oil lipid emulsions are generally dosed at 1 g/kg/d, and can be increased to 1.5 g/kg for children with poor weight gain or concern for essential fatty acid

Table 73.6 Pediatric Micronutrient and Trace Element Requirements

Nutrient	Preterm Infants	Term Infants (<1 year)	Children and Adolescents (1–18 years)
Micronutrients			
Vitamin A[a]	700–1500 IU/kg/d (227–455 μg/kg/d)	2300 IU/d (150–300 μg/kg/d)	150 μg/d
Vitamin D	200–1000 IU/d (80–400 IU/kg/d)	400 IU/d (40–150 IU/kg/d)	400–600 IU
Vitamin E	2.8–3.5 mg/kg/d (2.8–3.5 IU/kg/d)	2.8–3.5 mg/kg/d (2.8–3.5 IU/kg/d)	11 mg/d (11 IU/d)
Vitamin K	10 μg/kg/d (recommended but not currently possible)	10 μg/kg/d (recommended but not currently possible)	200 μg/d
Vitamin B1 (thiamin)	0.35–0.50 mg/kg/d	0.35–0.50 mg/kg/d	1.2 mg/d
Vitamin B2 (riboflavin)	0.15–0.2 mg/kg/d	0.15–0.2 mg/kg/d	1.4 mg/d
Vitamin B3 (niacin)	4–6.8 mg/kg/d	4–6.8 mg/kg/d	17 mg/d
Vitamin B5 (pantothenate)	2.5 mg/kg/d	2.5 mg/kg/d	5 mg/d
Vitamin B6 (pyridoxine)	0.15–0.2 mg/kg/d	0.15–0.2 mg/kg/d	1.0 mg/d
Vitamin B7 (biotin)	5–8 μg/kg/d	5–8 μg/kg/d	20 μg/d
Vitamin B9 (folate)	56 μg/kg/d	56 μg/kg/d	140 μg/d
Vitamin B12 (cobalamin)	0.3 μg/kg/d	0.3 μg/kg/d	1.0 μg/d
Vitamin C (ascorbic acid)	15–25 mg/kg/d	15–25 mg/kg/d	80 mg/d
Trace Elements[b]			
Zinc	400–500 μg/kg/d	0–3 months: 250 μg/kg/d 3–12 months: 100 μg/kg/d	50 μg/kg/d
Copper	40 μg/kg/d	20 μg/kg/d	20 μg/kg/d
Iodine	1–10 μg/kg/d	1 μg/kg/d	1 μg/kg/d
Selenium	7 μg/kg/d	2–3 μg/kg/d	2–3 μg/kg/d
Manganese	≤1 μg/kg/d	≤1 μg/kg/d	≤1 μg/kg/d
Molybdenum	1 μg/kg/d	0.25 μg/kg/d	0.25 μg/kg/d
Chromium[c]	0.2 μg/kg/d	0.2 μg/kg/d	0.2 μg/kg/d

[a]1 ug RAE (retinol activity equivalent) = 1 μg all-trans retinol = 3.33 IU vitamin A.
[b]Maximum dosing: zinc maximum 5 mg/d; Copper 0.5 mg/d; Selenium μg/d; Manganese 50 μg/d; Molybdenum 5 μg/d; Chromium 5 μg/d.
[c]Dosing from Greene et al,[136] although separate dosing is not typically required, as chromium can be present in adequate amounts via contamination with other trace elements.

Data from Bronsky J, Campoy C, Braegger C, et al. Clin Nutr 2018;37:2366–78; Domellof M, Szitanyi P, Simchowitz V, et al. Clin Nutr 2018;37:2354–9.[134,135]

deficiency. All lipid emulsions should be administered as an infusion, often over 24 hours but can also be administered in a cycled pattern over shorter periods of time (e.g., 12 or 18 hours).

When initiating lipid emulsion therapy, serum triglyceride concentration should be monitored closely to assess tolerance. Children with extreme prematurity, small for gestational age, severe malnutrition, physiologic stress, or those on medications know to alter lipid metabolism (e.g., corticosteroids, tacrolimus) may have difficulty metabolizing lipids.[139,140] In these cases, providing lipid emulsion over a shorter time frame (e.g., 20 hours), or reducing lipid dose, may improve tolerance. Severe hypertriglyceridemia can be associated with pancreatitis, and, as such, serum lipase concentration should be measured.

High dosing of lipids or rapid infusion of lipids can result in fat overload syndrome. This presents clinically with respiratory distress, hepatomegaly, hemorrhage, and fever in

Table 73.7 Comparison of Nutritional Composition of Intravenous Lipid Emulsions

	Soy Lipid Emulsion		Mixed Lipid Emulsions		Fish Oil
Brand Names	**Intralipid**	**Nutrilipid**	**SMOFLipid**	**ClinoLipid**	**Omegaven**
Oil Source					
Soy, %	100	100	30	20	0
Fish Oil, %	0	0	15	0	100
Medium Chain Triglycerides, %	0	0	30	0	0
Olive Oil, %	0	0	25	80	0
ω-3					
Linolenic acid, %	4–11	4–11	1.5–3.5	0.5–4.2	1.1 (mean)
EPA, %	0	0	1–3.5	0	13–26
DHA, %	0	0	1–3.5	0	14–27
ω-6					
Linoleic acid, %	44–62	48–58	14–25	13.8–22	1.5 (mean)
Arachidonic acid, %	0	0	NR	0	0.2–2
Additional Information					
ω-6:ω-3 ratio	7:1	7:1	2.5:1	9:1	1.8:1
Phytosterols, mg/L	422±131	NR	142±15	209±39	0
Energy, kcal/mL	2	2	2	2	1.12
Additives					
Egg phospholipid, g/100 mL	1.2	1.2	1.2	1.2	1.2
Glycerin, g/100 mL	2.25	2.5	2.5	2.25	2.5
α-Tocopherol, mg/100 mL	0	0	16.3–22.5	32	15–30
Sodium oleate, g/100 mL	0	0.03	0.03	0.03	0.03

Data from Koletzko B, Goulet O, Hunt J, et al. J Ped Gastro Nutr2005;41(suppl 2):S1–87.[137]

the context of hypertriglyceridemia. Treatment requires discontinuation of lipid source and is generally supportive in nature. Serum triglyceride concentration should be monitored until normalized.

Complications of Parenteral Nutrition

Parenteral nutrition is a life-saving intervention with significant risks that are summarized in **Table 73.8**. As such, parenteral nutrition is best managed by an interdisciplinary team consisting of physicians, dietitians, nurses, and pharmacists to identify and manage these risks.

Risks of parenteral nutrition are both catheter-related and metabolic. Catheter-related complications can be controlled, but it is difficult to prevent them entirely. Central venous catheter insertion may be complicated by insertion into arterial circulation, pneumothorax, hemothorax, injury to an artery, or hematoma. The most common problem of central vein PN is infection. Careful attention to care of the central catheter, including frequent dressing changes under sterile conditions and regular cap changes, is particularly important for controlling infection. Thrombosis, dislodgment, perforation, infusion leaks (pericardial, pleural, mediastinal), and infections also may occur with use of central vein catheters. The most frequent complications of peripheral vein PN are phlebitis and soft-tissue infiltrates, which require careful and frequent observation of the infusion site.

Metabolic complications result either from the limited metabolic capacity of the patient for the various components of the nutrient infusate or from the infusate itself. In addition to hyperglycemia and electrolyte abnormalities discussed above, hypoglycemia may result with inadequate glucose infusion rate for metabolic needs. Abnormal serum amino acid profiles may be observed, as they reflect the composition of the infusate itself.

Intestinal failure-associated liver disease (IFALD), previously termed "parenteral nutrition associated liver disease" or

Table 73.8 Complications of Short- and Long-Term Parenteral Nutrition Use

Complication	Probable Cause
Disorders Related to Metabolism	
Hyperglycemia	Excessive intake (either excessive concentration or infusion rate), change in metabolic state (e.g., infection, surgical stress)
Hypoglycemia	Sudden cessation of infusion
Azotemia	Excessive nitrogen intake
Electrolyte disorders	Excessive or inadequate intake, change in metabolic state
Mineral disorders	Excessive or inadequate intake
Vitamin disorders	Excessive or inadequate intake
Essential fatty acid deficiency	Failure to provide essential fatty acids
Hyperlipidemia	Excessive intake, change in metabolic state (e.g., stress, sepsis)
Disorders Related to Infusate Components	
Metabolic acidosis	Use of hydrochloride salts of amino acids (e.g., cysteine)
Hyperammonemia	Inadequate arginine intake
Abnormal plasma aminograms	Amino acid pattern of nitrogen source
Hepatic disorders	Unknown; suggested etiologies include prematurity, malnutrition, sepsis, inadequate stimulation of bile flow, toxic effects of amino acids, specific amino acid deficiency, excessive amino acid and/or carbohydrate intake, and nonspecific response to lack of feeding

"TPN cholestasis," is a liver disease marked by steatosis, cholestasis, and sometimes cirrhosis, which occurs in 40% to 60% of children requiring PN use for >1 month.[40] The secondary biliary cirrhosis is a major source of morbidity and mortality for patients with short bowel syndrome who require parenteral nutrition.[41] While the exact mechanism is unclear, several interventions can slow or abate the development of IFALD. These include restoration of intestinal continuity, enteral feeding early in clinical course, administration of parenteral nutrition for <24 hours—a process known as "cycling," avoiding excessive parenteral energy, and limiting the dose of intravenous lipid from a soybean source to 1 g/kg.[35,40] Additional considerations include employing a choleretic agent (i.e., ursodeoxycholic acid), treating intestinal bacterial overgrowth, and preventing sepsis. When IFALD develops, substitution of conventional intravenous lipid emulsions with a parenteral fish oil emulsion can improve liver function.[42,43]

Weaning Children from Parenteral Nutrition

In most infants, administration of parenteral nutrients need not interfere with introducing enteral feedings. As stated previously, trophic or small volume feeds are an important method to reduce risk for intestinal failure-associated liver disease and should be maintained when possible. The volume of enteral feedings can be advanced as tolerated by the infant and the volume of the parenteral nutrition infusate can be carefully decreased. During the period of combined enteral and parenteral nutrition, care should be taken to ensure both that nutrient requirements are met as nearly as possible and that tolerance for both fluids and nutrients is not exceeded.

Home Parenteral Nutrition

Medically stable patients who require parenteral nutrition for prolonged periods of time (e.g., >3 months) may receive this therapy at home. As with inpatient parenteral nutrition, home parenteral nutrition requires careful monitoring by a highly trained, multidisciplinary team consisting of trained medical providers, registered dietitians, registered nurses, and social workers. Small portable infusion pumps are available such that the necessary apparatus can be enclosed in vests, backpacks, and so on, to allow even the patient who requires near constant infusion of parenteral nutrients to lead a reasonably normal life. In many cases, sufficient nutrients can be administered during only a portion of the day, known as "cycling" parenteral nutrition, thus allowing the patient to pursue developmentally appropriate daytime activities.

Discharge on home parenteral nutrition support is high risk, as patients remain at risk for metabolic and catheter-related complications and, therefore, requires that the patient receives a stable parenteral nutrition regimen and that caregivers undergo extensive training. In addition, patients should be able to tolerate at least a few hours off of the infusate without hypoglycemia or dehydration, to reduce the danger of accidental disconnection of the infusion system. Caregivers are given explicit instruction on how to manage catheters (which can be peripherally or centrally inserted central venous access), deliver infusion safely, and assess for catheter and metabolic-related complications.

Standard nutrient infusates are obtained typically from a commercial vendor and stored in a small home refrigerator.

Children have been treated successfully at home for several months to years, often with excellent growth patterns and outcomes. Successful home parenteral nutrition, particularly for the young pediatric patient, requires frequent interval follow-up, including outpatient visits with a multidisciplinary care team. Ongoing assessment of suitability of enteral nutrition intake, as well as long-term sequalae of nutrient deficiencies/excess, is necessary. Metabolic bone disease may be related to inadequate vitamin D (as intravenous formulations are not available in the United States) as well as suboptimal calcium and phosphorus provision (related to solubility of current amino acid solutions).

Summary

The nutritional management of infants and children with specific conditions must take into account the specific nutrient requirements resulting from the presence of these disorders, the reduced nutrient intake that so often accompanies them, and the psychosocial environment in which they exist. The optimal method of delivery of nutrition may vary, but with guiding principle of using the least invasive and least restrictive intervention possible—oral feeding is preferred to enteral feeding which, in turn, is preferred to parenteral feeding. Nutrition support should be delivered by a multidisciplinary team using a patient-centered approach.

Acknowledgment

The author acknowledges Arthur Cooper, Richard L Mones, and William C Heird, the authors of the previous version of this chapter, for their contributions.

References

1. Nelson SP, Chen EH, Syniar GM, et al. Arch Ped Adolesc Med 1997;151:569–72.
2. Martin AJ, Pratt N, Kennedy JD, et al. Pediatr 2002;109:1061–7.
3. Van Howe RS, Storms MR. BMC Pediatr 2010;10:7.
4. Singendonk M, Goudswaard E, Langendam M, et al. J Ped Gastroenterol Nutr 2019;68:811–7.
5. Rosen R, Vandenplas Y, Singendonk M, et al. J Ped Gastroenterol Nutr 2018;66:516–54.
6. Luzzani S, Macchini F, Valade A, et al. Am J Med Genet A 2003;119A:283–7.
7. Duncan DR, Larson K, Rosen RL. Curr Gastroenterol Rep 2019;21:30.
8. Klepper CM, Moore J, Gabel ME, et al. Nutr Clin Pract 2-23;38:302–17.
9. Burks AW, Jones SM, Boyce JA, et al. Pediatr 2011;128:955–65.
10. Lucendo AJ, Molina-Infante J, Arias A, et al. United European Gastroenterol J 2017;5:335–58.
11. Furuta GT, Liacouras CA, Collins MH, et al. Gastroenterol 2007;133:1342–63.
12. Kliewer KL, Gonsalves N, Dellon ES, et al. Lancet Gastroenterol Hepatol 2023;8:408–21.
13. Spergel JM, Brown-Whitehorn TF, Cianferoni A, et al. J Allergy Clin Immunol 2012;130:461–7 e5.
14. Koletzko S, Niggemann B, Arato A, et al. J Ped Gastroenterol Nutr 2012;55:221–9.
15. Lake AM. J Ped Gastroenterol Nutr 2000;30(suppl):S58–60.
16. Mennini M, Fiocchi AG, Cafarotti A, et al. World Allergy Organ J 2020;13:100471.
17. Kaya A, Toyran M, Civelek E, et al. J Ped Gastroenterol Nutr 2015;61:69–73.
18. Sicherer SH, Eigenmann PA, Sampson HA. J Ped 1998;133:214–9.
19. Nowak-Wegrzyn A, Jarocka-Cyrta E, Moschione Castro A. J Investig Allergol Clin Immunol 2017;27:1–18.
20. Caubet JC, Ford LS, Sickles L, et al. J Allergy Clin Immunol 2014;134:382–9.
21. Hamilton JR. Viral Enteritis: A Cause Of Disordered Small Intestinal Epithelial Renewal. Chronic Diarrhea in Children. New York, NY: Raven Press, 1984.
22. Sack RB. (1983) Bacterial and parasitic agents of acute diarrhea. Acute Diarrhea: Its Nutritional Consequences in Infancy. New York, NY: Raven Press, 1983;53–65.
23. Duggan C, Nurko S. J Ped 1997;131:801–8.
24. MacGillivray S, Fahey T, McGuire W. Cochrane data Syst Rev 2013: CD005433.
25. Shane AL, Mody RK, Crump JA, et al. Clin Infect Dis 2017;65: e45–e80.
26. Saps M, Pensabene L, Di Martino L, et al. J Ped 2008;152: 812–6, 6 e1.
27. Caruso R, Pallone F, Stasi E, et al. Ann Med 2013;45:522–31.
28. Vici G, Belli L, Biondi M, et al. Clin Nutr 2016;35:1236–41.
29. Dickey W, Kearney N. Am J Gastro 2006;101:2356–9.
30. Silvester JA, Therrien A, Kelly CP. Am J Gastro 2021;116: 1148–55.
31. Hill ID, Fasano A, Guandalini S, et al. J Pediatric Gastro Nutr 2016;63:156–65.
32. Sartor RB. (2006) Nat Clin Pract Gastroenterol Hepatol 3, 390–407.
33. Ehrlich S, Mark AG, Rinawi F, et al. Nutr Clin Pract 2020;35: 315–22.
34. van Rheenen PF, Aloi M, Assa A, et al. J Crohns Colitis 2020.
35. Duggan CP, Jaksic T. N Engl J Med 2017;377:666–75.
36. Wilmore DW. J Ped 1972;80:88–95.
37. Cooper A, Floyd TF, Ross AJ, 3rd, et al. J Pediatr Surg 1984;19: 711–8.
38. Byrne TA, Wilmore DW, Iyer K, et al. Ann Surg 2005;242: 655–61.
39. Kocoshis SA, Merritt RJ, Hill S, et al. JPEN 2020;44:621–31.
40. Lacaille F, Gupte G, Colomb V, et al. J Ped Gastro Nutr 2015;60: 272–83.
41. Farrell MK, Balistreri WF. Clin Perinatol 1986;13:197–212.
42. Gura KM, Duggan CP, Collier SB, et al. Pediatr 2006;118: e197–201.
43. Gura KM, Parsons SK, Bechard LJ, et al. Clin Nutr 2005;24: 839–47.
44. Drossman DA. Gastroenterol 2016.
45. Devanarayana NM, Rajindrajith S. World J Gastro 2018;24: 2211–35.
46. Varju P, Farkas N, Hegyi P, et al. PloS One 2017;12:e0182942.

47. Schumann D, Klose P, Lauche R, et al. Nutr 2018;45:24–31.
48. Scarlata K, Catsos P, Smith J. Clin Gastroenterol Hepatol 2020;18:543–5.
49. Pareek N, Fleisher DR, Am J GastroAbell T. 2007;102:2832–40.
50. Murray HB, Juarascio AS, Di Lorenzo C, et al. Am J Gastro 2019;114:562–78.
51. Josefsson A, Hreinsson JP, Simren M, et al. Gastroenterol 2022;162:731–42 e9.
52. Romano C, van Wynckel M, Hulst J, et al. J Ped Gastro Nutr 2017;65:242–64.
53. Wells J, Swaminathan A, Paseka J, et al. Nutrients 2020;12.
54. Ruan Y, Chen L, She D, et al. Eur J Clin Nutr 2022;76:1234–44.
55. Mills KI, Kim JH, Fogg K, et al. Pediatr 2022;150.
56. Thommessen M, Heiberg A, Kase BF. Eur J Clin Nutr 1992;46: 457–64.
57. Gongwer RC, Gauvreau K, Huh SY, et al. Cardiol Young 2018;28: 1093–8.
58. Schwarz SM, Gewitz MH, See CC, et al. Pediatr 1990;86: 368–73.
59. Unger R, DeKleermaeker M, Gidding SS, et al. Am J Dis Child 1992; 146:1078–84.
60. Hofner G, Behrens R, Koch A, et al. Pediatr Cardiol 2000;21: 341–6.
61. Mariotti Zani E, Grandinetti R, Cunico D, et al. Nutrients 2023;15.
62. Naon H, Hack S, Shelton MT, et al. Chest 1993;103:1819–25.
63. Augarten A, Ben Tov A, Madgar I, et al. Eur J Gastroenterol Hepatol 2008;20:164–8.
64. Mansell AL, Andersen JC, Muttart CR, et al. J Ped 1984;104: 700–5.
65. McPhail GL, Acton JD, Fenchel MC, et al. J Ped 2008;153:752–7.
66. Farrell PM, Kosorok MR, Rock MJ, et al. Pediatr 2001;107: 1–13.
67. Stallings VA, Stark LJ, Robinson KA, et al. J Am Diet Assoc 2008; 108:832–9.
68. Freswick PN, Reid EK, Mascarenhas MR. Nutrients 2022;14.
69. Wilson DC, Rashid M, Durie PR, et al. J Ped 2001;138:851–5.
70. Borowitz D, Baker RD, Stallings V. J Ped Gastro Nutr 2002;35: 246–59.
71. Turck D, Braegger CP, Colombo C, et al. Clin Nutr 2016;35: 557–77.
72. Volkova N, Moy K, Evans J, et al. J Cyst Fibros 2020;19:68–79.
73. Sawicki GS, McKone EF, Pasta DJ, et al. Am J Respir Crit Care Med 2015;192:836–42.
74. Davies JC, Cunningham S, Harris WT, et al. Lancet Respir Med 2016;4:107–15.
75. Davies JC, Wainwright CE, Sawicki GS, et al. Am J Respir Crit Care Med 2021;203:585–93.
76. Mehta NM, Skillman HE, Irving SY, et al. JPEN 2017;41:706–42.
77. Fivez T, Kerklaan D, Mesotten D, et al. N Engl J Med 2016; 374:1111–22.
78. Chwals WJ, Bistrian BR. Crit Care Med 2000;28:2655–6.
79. Jaksic T, Shew SB, Keshen TH, et al. J Pediatr Surg 2001;36:63–7.
80. Hunter DC, Jaksic T, Lewis D, et al. Br J Surg 1988;75:875–8.
81. Coss-Bu JA, Jefferson LS, Walding D, et al. Am J Clin Nutri- 1998; 67:74–80.
82. Joosten KF, Verhoeven JJ, Hazelzet JA. Nutr 1999;15:444–8.
83. Hardy CM, Dwyer J, Snelling LK, et al. Nutr Clin Pract 2002; 17:182–9.
84. Mehta NM, Bechard LJ, Leavitt K, et al. JPEN 2009;33:336–44.
85. Mehta NM, Compher C, ASPEN JPEN 2009;33:260–76.
86. Coss-Bu JA, Klish WJ, Walding D, et al. Am J Clin Nutr 2001; 74:664–9.
87. Mehta NM, Bechard LJ, Dolan M, et al. Pediatr Crit Care Med 2011; 12:398–405.
88. Duffy B, Pencharz P. Pediatr Res 1986;20:32–5.
89. Keshen TH, Miller RG, Jahoor F, et al. J Pediatr Surg 1997;32: 958–62; discussion 62–3.
90. Clark A, Imran J, Madni T, et al. Burns Trauma 2017;5:11.
91. Patterson BW, Nguyen T, Pierre E, et al. Metab Clin Exper 1997;46:573–8.
92. Hart DW, Wolf SE, Herndon DN, et al. Ann Surg 2002;235: 152–61.
93. Furth SL, Hwang W, Yang C, et al. Pediatr Nephrol 2002;17: 450–5.
94. Rees L, Shaw V. Pediatr Nephrol 2007;22:1689–702.
95. National Kidney Foundation. Am J Kidney Dis 2000;35: S17–S104.
96. Fiaccadori E, Sabatino A, Barazzoni R, et al. Clin nutr 2021;40:1644–68.
97. Shaw V, Polderman N, Renken-Terhaerdt J, et al. Pediatr Nephrol 2020;35:519–31.
98. McKinney RE, Jr., Robertson JW. J Ped 1993;123:579–82.
99. Miller TL, Evans SJ, Orav EJ, et al. Am J Clin Nutr 1993;57: 588–92.
100. Guarino A, Spagnuolo MI, Giacomet V, et al. J Ped Gastro Nutr 2002;34:366–71.
101. Kabue MM, Kekitiinwa A, Maganda A, et al. AIDS Patient Care STDS 2008;22:245–51.
102. Sabery N, Duggan C, American Society for Parenteral and Enteral Nutrition, et al. JPEN 2009;33:588–606.
103. Goday PS, Huh SY, Silverman A, et al. J Ped Gastro Nutr 2019;68:124–9.
104. Mogensen KM, Malone A, Becker P, et al. Nutr Clin Pract 2019;34:657–65.
105. Bechard LJ, Duggan C, Touger-Decker R, et al. Crit Care Med 2016;44:1530–7.
106. Becker P, Carney LN, Corkins MR, et al. Nutr Clin Pract 2015;30:147–61.
107. WHO Child Growth Standards and the Identification of Severe Acute Malnutrition in Infants and Children: A Joint Statement by the World Health Organization and the United Nations Children's Fund, Geneva. 2009.
108. Barrocas A, Schwartz DB, Bistrian BR, et al. Nutr Clin Pract 2023;38:10–26.
109. Bendavid I, Lobo DN, Barazzoni R, et al. Clin Nutr 2021;40: 690–701.
110. Harris JA, Benedict FG. Proc Natl Acad Sci U S A 1918;4:370–3.
111. Energy and protein requirements: report of a joint FAO/WHO/UNU expert consultation. Energy and Protein Requirements: Report of a Joint FAO/WHO/UNU Expert Consultation. 1985; 206.
112. Schofield WN. Hum Nutr Clin Nutr 1985;39(suppl 1):5–41.
113. Dietary reference intakes for energy, carbohydrate, fiber, fat, fatty acids, cholesterol, protein, and amino acids (macronutrients). Washington, DC: National Academy Press, 2005.
114. Chellis MJ, Sanders SV, Webster H, et al. JPEN 1996;20:71–3.
115. Briassoulis GC, Zavras NJ, Hatzis MT. Pediatr Crit Care Med 2001;2:113–21.
116. Briassoulis G, Zavras N, Hatzis T. Nutr 2001;17:548–57.
117. de Lucas C, Moreno M, Lopez-Herce J, et al. J Ped Gastro Nutr 2000;30:175–80.
118. Kawagoe JY, Segre CA, Pereira CR, et al. Am J Infect Control 2001;29:109–14.
119. Bankhead R, Boullata J, Brantley S, et al. JPEN 2009;33:122–67.
120. Martinez EE, Bechard LJ, Brown AM, et al. Clin Nutr 2022;41:2621–7.
121. Brown AM, Irving SY, Pringle C, et al. JPEN 2022;46:1011–21.
122. Horn D, Chaboyer W. Am J Crit Care 2003;12:461–8.
123. Horn D, Chaboyer W, Schluter PJ. Aust Crit Care 2004;17: 98–100, 2–3.

124. DeLegge MH. Gastroenterol Clin North Am 2018;47:23–37.
125. Gencosmanoglu R. Neurocrit Care 2004;1:309–17.
126. Niv E, Fireman Z, Vaisman N. World J Gastro 2009;15:1281–8.
127. Broekaert IJ, Falconer J, Bronsky J, et al. J Ped Gastro Nutr 2019;69:239–58.
128. Kohli-Seth R, Sinha R, Wilson S, et al. Nutr Clin Pract 2009;24:728–32.
129. Ayers P, Adams S, Boullata J, et al. JPEN 2014;38:296–333.
130. Worthington P, Balint J, Bechtold M, et al. JPEN 2017;41: 324–77.
131. Pittiruti M, Hamilton H, Biffi R, et al. Clin Nutr 2009;28: 365–77.
132. Baker RD, Davis S, Baker S, Davis A. Ped Paren Nutr Chapman & Hall, 1997.
133. Mirtallo J, Canada T, Johnson D, et al. JPEN 2004;28:S39–70.
134. Bronsky J, Campoy C, Braegger C, et al. Clin Nutr 2018;37: 2366–78.
135. Domellof M, Szitanyi P, Simchowitz V, et al. Clin Nutr 2018;37:2354–9.
136. Greene HL, Hambidge KM, Schanler R, et al. Am J Clin Nutr 1988;48:1324–42.
137. Koletzko B, Goulet O, Hunt J, et al. J Ped Gastro Nutr 2005;41(suppl 2):S1–87.
138. Carey AN, Rudie C, Mitchell PD, et al. JPEN 2019;43:305–10.
139. Shennan AT, Bryan MH, Angel A. J Ped 1977;91:134–7.
140. Park W, Paust H, Brosicke H, et al. JPEN 1986;10:627–30.

CHAPTER 74

Food Insecurity and Child Development

Rafael Pérez-Escamilla

CHAPTER OUTLINE

Abbreviations

C-SNAP, Children's Sentinel Nutrition Assessment Project; **ECLS**, Early Childhood Longitudinal Study; **ECLS-B**, Early Childhood Longitudinal Study-Birth Cohort; **FAO**, Food and Agricultural Organization; **FI**, food insecurity; **HFI**, Household Food Insecurity; **HFSSM**, Household Food Security Survey Module; **NHANES**, National Health and Nutrition Examination Survey; **USDA**, US Department of Agriculture

Introduction

Decades of research have conclusively shown that child malnutrition has a negative impact on the physical and intellectual development of children.[1,2] Furthermore, the preponderance of the evidence indicates that household food insecurity (HFI) has a strong negative influence on children's dietary quality and child physical health status across the life course.[3-7] Over the past decade, our understanding of the consequences of HFI on child development has substantially improved. For this reason, this chapter examines the influence of HFI on the development of children and youth and discusses potential mediators, specifically maternal depression.

Food and Nutrition Security Definitions

According to the United Nations, food and nutrition security is a fundamental human right that exists "when all people at all times have physical, social, and economic access to food, which is safe and consumed in sufficient quantity and quality to meet their dietary needs and food preferences for a healthy and active life."[8] It is estimated that nearly 2 billion people, or more than one-quarter of the world's population do not have regular access to a nutritious and sufficient food supply.[9]

In the United States, household food security has been defined as "access by all people at all times to enough food for an active, healthy life and includes, at a minimum, (a) the ready availability of nutritionally adequate and safe foods and (b) an assured availability to acquire acceptable foods in socially acceptable ways (e.g., without resorting to emergency food supplies, scavenging, stealing, or other coping strategies)."[10] Thus, HFI exists in situations with "limited or uncertain availability of nutritionally adequate and safe foods or limited or uncertain ability to acquire acceptable foods in socially acceptable ways."[10]

The domestic and global definitions are consistent with and complement each other. Together, they highlight the fact that HFI is a complex construct that includes the following multiple dimensions[8,11,12]: (a) amount of food (energy); (b) nutritional quality of food (dietary quality including macro- and micronutrient content); (c) food safety (free of pathogens or harmful contaminants); (d) cultural acceptability (respecting food preferences); (e) social acceptability (respecting the dignity of people); and (f) equity (access by all people at all times).

HFI is a strong but not the only determinant of nutrition security of individuals, as the latter is also determined by access to healthcare, and other social determinants of health, including housing, transportation, and heating.[5,13,14]

Nutrition security is understood as a condition that occurs when the body tissues are exposed to optimal amounts of nutrients and other essential substances, is the result of household food security, healthcare access security, and access to other basic human needs, including adequate sanitation.[10] Food security and the other determinants of nutrition security interact with each other in multidirectional ways.[15,16] For example, a household with limited resources for purchasing food may decide to not bring a child to the doctor, or not to purchase needed medications. For food security to exist, households need to have access to healthy and nutritious foods. Access to these foods, in turn, depends on having an adequate income and for foods to be available in enough quantities in the country, region, and communities where the households are located. National food availability for human consumption represents the balance between foods locally grown and foods imported, minus foods exported, spoiled, or fed to animals (**Figure 74.1**).

Ultimately, maintenance of an adequate food supply at the global level is paramount for achieving household food security and nutrition security worldwide in the context of environmental sustainability.[17] Global food supplies are strongly influenced by climate change, agricultural commodity price policies, and

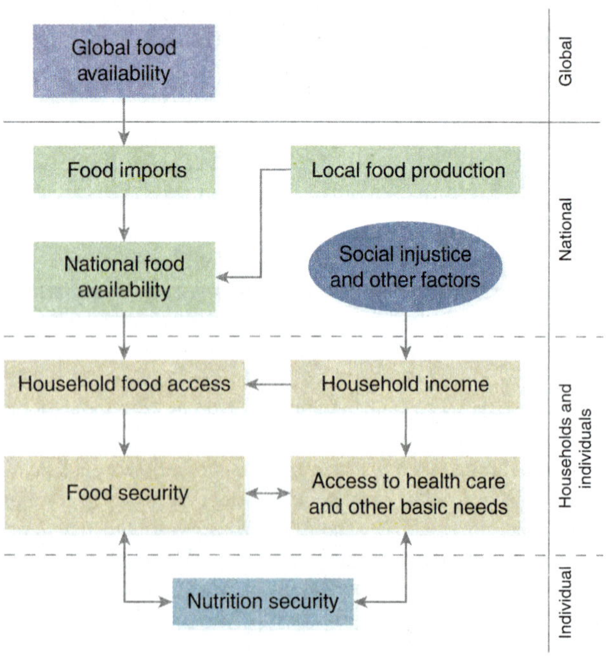

Figure 74.1 The relationships between global food security, household food security, and nutrition security.

Reproduced from Frankenberger TR, Frankel L, Ross S, et al. Arlington, VA: Washington, DC: United States Agency for International Development, 1997.[14]

armed conflicts.[17] Thus, food and nutrition security is underlined by six dimensions[18]: (a) availability: access to nutritious foods by growing, trading, or purchasing them (b) accessibility: social, economic, and physical access to food (c) utilization: food is transformed into meals that are ingested and the nutrients and other food-derived healthy substances observed and used by the person's organism (d) stability: stable access to nutritious and safe foods irrespective of household, civil unrest, or weather conditions (e.g., climate change) (e) agency: people are empowered to choose what they eat and how it is produced with freedom and dignity and (f) sustainability: protection of human and environmental health in the long term. Food insecurity occurs when one or more of these dimensions are compromised.

Food Insecurity Measurement

The five commonly used methods for assessing HFI directly or indirectly are as follows[8]: (a) the Food and Agricultural Organization (FAO) method for estimating average food energy available per capita (b) household income and expenditure surveys (c) dietary intake surveys (d) maternal-child anthropometry and (e) experience-based FI scales. Experience-based scales were originally developed in the United States, and their use has spread globally.[19-22] These scales are strongly grounded on qualitative research conceptualizing HFI as a "managed" household process that moves through a series of stages. Hence, experiences with food insecurity are often measured at the household level, based on the report of a person in charge of food acquisition and preparation for the household, with questions about children directed at the primary caregiver.[8] The process by which a household experiences food insecurity starts with a state of worry or anxiety related to the uncertainty about future access to food and is followed by sacrificing dietary quality, eventually reducing the amount of food consumed, first among adults and then among children.[8]

FI scales include questions that examine each stage or one specific stage of HFI. For each household, a total summative score is estimated based on the number of affirmative responses. These scores can then be converted into FI categories (i.e., food security, mild HFI, moderate HFI, severe FI) based on cutoff points that discriminate among the different stages of HFI experienced. The U.S. Household Food Security Survey Module (HFSSM) was found to have adequate psychometric behavior and validity for use in both large-scale surveys and smaller studies in the United States, and this tool has been adapted and validated in other high-income as well as in low- and middle-income countries.[11,19-22]

Experience-based FI scales directly measure the phenomenon of interest, are easy to apply, and have been well accepted by both target communities and policymakers.[8,23-26] These scales do not capture all of the dimensions of the HFI construct, however. For example, they do not assess the safety of the food supply accessed by the household or measure issues related to water security, although household water security experience scales are now available.[27] In addition, the measure captures FI at the household level but cannot identify individuals within the same household experiencing different degrees of HFI.[8]

This chapter reviews evidence focused on experience-based HFI scales because: (a) they are constructed around the consensus definition of household food security adopted by the international community; (b) they represent a direct measurement of the phenomena of interest; (c) a substantial body of evidence evaluating the impact of HFI on child nutrition and health outcomes, based on these scales, is now available, (d) a critical mass of evidence on HFI (measured with experience-based scales) and early childhood development has continued to emerge.[23]

Global Food Insecurity Trends

The U.S. Census Bureau has reported annual household FI prevalence since 1995, estimated from the 18-item HFSSM applied as the December Current Population Survey Food Security Supplement. The reference period of the HFSSM is the 12 months preceding the survey, and households are classified as having food security, low food security, or very low food security, based on the number of affirmative answers to the HFSSM items.[19] In 2021, 10.2% of U.S. households were food insecure (i.e., with either low or very low food security)—this translates to approximately 14 million households.[28] In 2020, 14.8% were food insecure, up from 13.6% in 2019, interrupting a decade long decline in food insecurity, which was not unexpected given the arrival of the COVID-19 pandemic in the United States.[28] U.S. households are more likely to be food insecure if they are poor, are headed by a single woman, have a head of household who reports being Hispanic or African American, or have children. Indeed, according to the U.S. Department of Agriculture (USDA),[28] in 2020, households with children had almost twice the rate of FI as those without children (21% vs 11.3%, respectively). Thus, the future well-being of a substantial number of low-income children in the United States is at risk because, as shown below, their development is affected by FI through multiple pathways.

Food Insecurity Trends During the COVID-19 Pandemic

Increases in unemployment, poverty, major disruptions in food supply chains, school closures, and the strong deterioration of primary healthcare systems during the COVID-19 pandemic that exploded in early 2020, now combined with the invasion of Russia of Ukraine that started in February of 2022, led to major increases in HFI. This is especially true among vulnerable groups, including women, young children, and ethnic/racial groups that have been historically marginalized by society.[28,29] In Brazil,, for example the prevalence of HFI increased from 10 million in 2019 to 19 million in 2020 and then to 31 million in 2021–2022, particularly affecting households living in poverty, with children and with people of color.[30] In some countries, like the United States, the impact of COVID-19 on HFI was partially mitigated once social protection measures were put into place, COVID-19 vaccines became available, and schools and economies started to reopen.[9,29]

In the past, estimates of HFI based on experience-based scales were difficult to compare across countries because of the use of different scales, time frames, and classification algorithms. However, this situation has changed with the advent of well-standardized scales, such as FAO's Food Insecurity Experience Scale (FIES),[22] which is an adaptation of the adult items of the Latin American and Caribbean Food Security Scale (ELCSA).[31] (**Table 74.1**) This scale is now being used globally to track Sustainable Development Goal 2.2 (ending all forms of malnutrition), with data available from close to 150 countries, covering 90% of the world population.[22]

Table 74.1 Items of the Latin American and Caribbean Food Security Scale (ELCSA)[22] and the Food Insecurity Experience Scale (FIES) Scales[31]

ELCSA Item[1]	FIES Items[2] During the Last 12 Months, was there a Time when[3]:	Meaning of the Item[4]
1. During the last [TIME PERIOD], were you worried that your household would run out of food because of lack of money or other resources to get food?	1. You or others in your household were worried you would not have enough food to eat because of a lack of money or other resources?	Refers to a state of worry or concern by the respondent that the household respondent might run out of food, in light of events that may affect their ability to get food (such as loss of employment). It is not necessary that the household actually ran out of food in order to answer affirmatively to this question.
2. During the last [TIME PERIOD], did your household run out of food because of lack of money or other resources to get food?	2. You or others in your household were unable to eat healthy and nutritious food because of a lack of money or other resources?	Referring to any experiences when there was actually no food in the household because they did not have money or other resources (for example, the household's own production or bartering) to get food.
3. During the last [TIME PERIOD], did your household lack enough money or other resources to get healthy and nutritious food?	3. You or others in your household ate only a few kinds of foods because of a lack of money or other resources?	This question asks the respondent if the household was unable to get foods they considered healthy or those that make a nutritious or balanced diet. This could happen if the household did not have enough money or other resources to get healthy foods. This question refers to the quality of the diet and not the quantity of foods eaten.

(continues)

Table 74.1 Items of the Latin American and Caribbean Food Security Scale (ELCSA)[22] and the Food Insecurity Experience Scale (FIES) Scales[31] *(continued)*

ELCSA Item[1]	FIES Items[2] During the Last 12 Months, was there a Time when[3]:	Meaning of the Item[4]
The Next Five Questions Ask About Adults in the Household, Referring to Persons Over 18 Years of Age[5]		
4. During the last [TIME PERIOD], did you or any adult in your household have to consume a diet based on only a few kinds of foods because of lack of money or other resources to get food?	4. You or others in your household had to skip a meal because there was not enough money or other resources to get food?	Did any adult in the household have a diet that did not have a variety of foods or did any adult have to eat the same foods or just a few kinds of foods every day because the household did not have money or other resources to get food? This question refers to quality of the diet and not the quantity of foods eaten.
5. During the last [TIME PERIOD], did you or any adult in your household not eat breakfast, lunch or dinner [or skip a meal] because of lack of money or other resources to get food?	5. You or others in your household ate less than you thought you should because of a lack of money or other resources?	Did any adult have to skip a major meal (breakfast, lunch, or dinner) because the household did not have money or other resources to get food? This question refers to insufficient quantity of food.
6. During the last [TIME PERIOD], did you or any adult in your household eat less than you thought you should because of a lack of money or other resources to get food?	6. Was there ever no food to eat of any kind in your house because of lack of resources to get food?	Did any adult have to eat less than they considered they should, even if they did not skip a meal, because the household did not have money or other resources to get food? This question does not apply to adults who are eating less to lose weight. This question refers to insufficient quantity of food.
7. During the last [TIME PERIOD], did you or any adult in your household feel hungry but did not eat because of lack of money or other resources to get food?	7. You or others in your household were hungry but did not eat?	Clarifies that the experience of being hungry but not eating was due to lack of money or resources, but not other reasons such as dieting or medical regimens. Did any adult experience the physical sensation of hunger but went without food because there was no way to obtain food—no money or other resources (but not other reasons such as dieting or medical regimens)? This question refers to hunger due to insufficient quantity of food.
8. During the last [TIME PERIOD], did you or any adult in your household eat only one meal in a day or go without eating for a whole day because of lack of money or other resources to get food?	8. You or others ever go a whole day and night without eating anything at all because there was not enough food?	Clarifies that the experience of not eating at all or only one meal in a day was due to lack of money or resources, but not other reasons such as dieting or medical regimens. This question refers to insufficient quantity of food.
In Households with Children Under [xx] Years of Age, Ask the Following Questions to the Same Respondent:		
9. During the last [TIME PERIOD], did any child, aged XX or younger, in your household not eat healthy because of lack of money or other resources to get healthy and nutritious food?		In reference to any child in the household, same meaning as item 3
10. During the last [TIME PERIOD], did any child, aged XX or younger, in your household have to consume a diet based on only a few kinds of foods because of lack of money or other resources to get food?		In reference to any child in the household, same meaning as item 4

ELCSA Item[1]	FIES Items[2] During the Last 12 Months, was there a Time when[3]:	Meaning of the Item[4]
11. During the last [TIME PERIOD], did any child in your household not eat breakfast, lunch, or dinner because of lack of money or other resources to get food?		In reference to any child in the household, same meaning as item 5
12. During the last [TIME PERIOD], did any child, aged XX or younger, in your household eat less than you thought he/she should because of lack of money or other resources to get food?		Refers to whether the adult respondent felt any child had to eat less than the respondent considered they should, even if the child/children did not skip a meal, because the household did not have money or other resources to get food
13. During the last [TIME PERIOD], did you have to serve less food to any child in your household because of lack of money or other resources to get food?		Refers to whether an adult in the household had to serve less food than normal to any child (irrespective of whether any adult was served less or not) because the household did not have money or other resources to get food. "Serving less" could mean smaller portion sizes or fewer meals.
14. During the last [TIME PERIOD], did any child in your household feel hungry but did not eat because of lack of money or other resources to get food?		In reference to any child in the household, same meaning as item 7
15. During the last [TIME PERIOD], did any child in your household eat only one meal in a day or go without eating for a whole day because of lack of money or other resources to get food?		In reference to any child in the household, same meaning as item 8

[1]ELCSA response options: YES, NO, Don't Know, Refused.

[2]FIES response options: YES, NO, Don't Know, Refused.

[3]The period of reference depends on the purpose of the data collection effort. Surveys have typically used a reference period of 3 months or 12 months.

[4]In the case of FIES all items could also be asked just in relationship to the personal experience of the respondent. In that instances items must be interpreted in reference to the experience of the person responding to the survey.

[5]The age at which a society considers a person as an adult or a child may vary across cultures. In the introduction, respondents should be informed that questions 4 to 8 refer to any "adults" and questions 9 to 15 to "children", i.e., persons over or under a certain age (as determined by the investigators within the context of the survey population). Th cut-offs for "children" may also be determined by the objectives of the study, e.g., some studies may focus on children under 5 years of age only.

Data from FAO. The Food Insecurity Experience Scale (FIES). Available at www.fao.org/in-action/voices-of-the-hungry/fies/en/ (Accessed January 16, 2022); and FAO. Escala Latinoamericana Y Caribeña de Seguridad Alimentaria (ELCSA): Manual De Uso Aplicaciones. May 2012. Available at www.fao.org/3/i3065s/i3065s.pdf (Accessed January 16, 2022).

Conceptual Framework

Household food and nutrition security during gestation, infancy, and early childhood are essential for the provision of the nurturing care needed for proper human development.[32,33] HFI can affect a child's physical, mental, social, psychological, and emotional development, through different pathways (**Figure 74.2**).[34] One biological pathway involves the direct link between HFI and poor dietary intake, nutritional status, and overall well-being.[34,35] A second, psychological, pathway involves the worry or anxiety, feeling of deprivation and alienation, distress, and adverse family and social interactions that result when households are exposed to HFI.[36,37] Thus, it is plausible that FI may lead to suboptimal physical and mental development of children.

Indeed, as detailed by Perez-Escamilla and Vianna[34] and Gallegos et al.,[23] HFI has been hypothesized to be linked to poor child development through diverse diet-related mechanisms, involving lack of food availability and suboptimal dietary quality, characterized by excessive consumption of ultra-processed foods and sugar-sweetened beverages, which leads to lower intake of nutrients needed for proper physical and mental development, including cognition[35,36]; and/or higher consumption of energy and nutrients that increase risk of poor child development. The relationship between HFI and child dietary quality is supported by evidence coming from countries across world regions.[37-43]

Another mechanism by which HFI may lead to poor early childhood development (ECD) is stress and anxiety related to uncertainty or worry about where the food will come from, or skipping meals. The latter is expected because food is a key

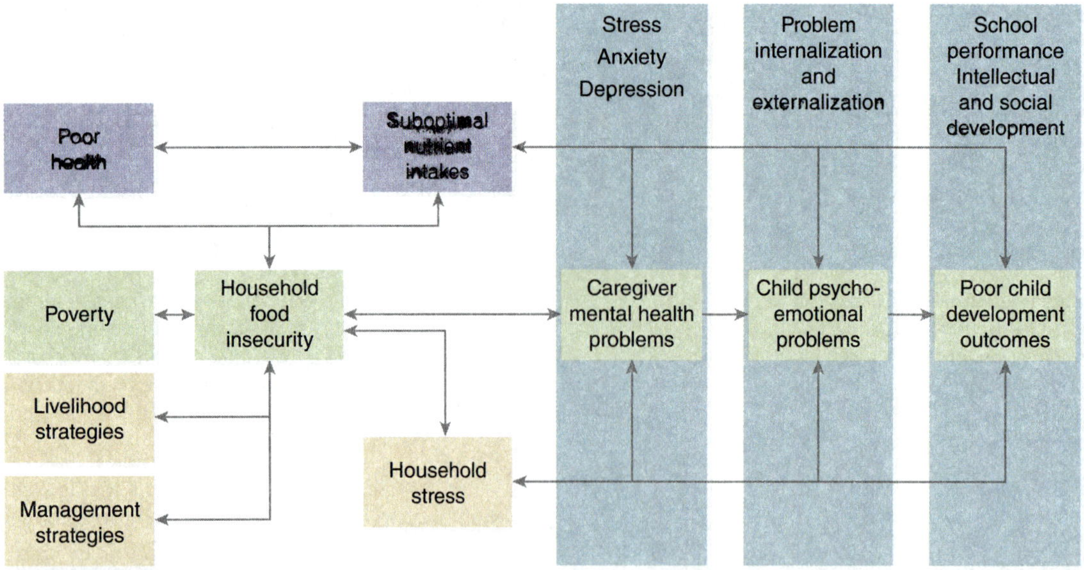

Figure 74.2 Potential pathways mediating the link between household food insecurity and child development, health, and well-being.

Reproduced from Perez-Escamilla F, de Toledo Vianna RP. J Appl Res Child 2012;3(1):9.[34]

social determinant of health (SDOH) that meets a basic human need, and it is well known that the inability of a household to meet its essential needs increases stress, depression, and anxiety among its members,[44-47] activating the stress-related hypothalamic–pituitary axis.[48] Stress and anxiety in the context of high dysfunctionality of the family unit[44,45] can impact parenting skills and, hence, children's psycho-emotional, social, and cognitive development.[23,34]

HFI has been found to negatively affect children's access to health care[6,49] and physical health and well-being, including increased risk of stunting, infectious diseases such as malaria, diarrhea, ear and upper respiratory infections, dental caries, and hospitalizations across countries with diverse socioeconomic, demographic, and cultural characteristics.[3,5-7,45,50,51] Interestingly, although HFI has consistently been shown to negatively impact the quality of children's diets, childhood obesity does not seem to mediate the relationship between HFI and poor child health outcomes.[50,52] Studies are needed to better understand this finding as HFI has been consistently associated with increased risk of obesity among adult women.[53] Poor child health and well-being is another possible mediator of the negative impact of HFI on ECD, as it prevents families from offering nurturing environments with access to healthy food, play, and sleep routines, as well as loving interactions. This, in turn, affects children's learning at home, early childcare and education, school, and other settings.[32,33]

Food Insecurity and Child Development

Seven systematic reviews, published between 2012 and 2022, examined associations between HFI and diverse child development outcomes and potential mediators, including maternal mental health (**Table 74.2**).[23,34,54-58]

Pérez-Escamilla and Vianna[34] conducted a systematic review to examine: (a) the association between HFI and child intellectual, behavioral, and psycho-emotional development, controlling for socioeconomic indicators; (b) the relationship between HFI and maternal mental health, and (c) if the relationship between poverty and poor child development outcomes is mediated by HFI. All studies included in this review were observational and most were cross-sectional and conducted in the United States.

Intellectual, Behavioral, and Psycho-Emotional Development

Cross-sectional, U.S.-based studies have consistently shown independent associations between HFI and diverse child psycho-emotional and academic indicators, although studies have used different HFI experience-based scales. Findings from the Children's Sentinel Nutrition Assistance Program (C-SNAP, now called Children's HealthWatch) study showed that caregivers of infants and children up to 3 years of age were more likely to identify their children as being at risk of poor child development if they lived in food insecure (vs food secure) households, after adjusting for maternal depression and other confounders. Likewise, a cross-sectional study in Arkansas, Louisiana, and Mississippi found that 12- to 17-year olds living in food insecure households had lower psychosocial function. Furthermore, a U.S. multistate study found that food insecure children were more likely to be hyperactive and to be either late or absent from school. Consistently, a study in Pittsburgh found that FI was associated with aggression and anxiety among school-aged children. HFI was also found to be associated with problem internalization among preschool- and school-aged children in Massachusetts, and with anxiety and depression. Multivariable analyses from data collected in low-income households across 18 cities in the United States showed that, among children who were 3 years old on average, there was a dose-response relationship between HFI severity level and child behavioral/mental health problems, including aggressiveness, anxiety, depression, lack of concentration, and/or hyperactivity.[59]

Table 74.2 **Summary of Systematic Reviews on Household Food Insecurity (HFI) and Early Child Development (ECD) Outcomes**

Author(s), Literature Period & Populations	Year of Publication	Primary Objectives	Conclusions
Pérez-Escamilla and Vianna[34]	2012	To examine: 1. The association between HFI and child intellectual, behavioral, and psycho-emotional development, controlling for socioeconomic indicators 2. The relationship between HFI and maternal mental health 3. If the relationship between poverty and poor child development is mediated by HFI	1. HFI is negatively associated with: multiple child development domains outcomes and poor maternal mental health, which is likely to mediate the relationship between HFI and poor child growth and development 2. HFI is a mediator of the relationship between poverty and poor child development
Shankar et al.[56] high-income countries ■ toddlers, preschoolers, school age, and adolescents.	2017	To examine the associations between FI and adverse childhood developmental behavioral outcomes, including early cognitive; development, academic performance; inattention, externalizing behaviors, and depression in 4 groups–infants	From infancy to adolescence, household FI, even at marginal levels, is independently associated with children's: 1. Poor behavioral outcomes 2. Low academic performance 3. Emotional problems
Tamiru & Belachew[57] Studies published between 2011–2015 Ethiopia, Venezuela	2017	Examine the association between food insecurity and school absenteeism	Household food insecurity is associated with students' school absenteeism
De Oliveira et al.[58] Studies published between 2006–20 high-income (HICs, N = 15) & low and middle-income (LMICs, N = 4) countries children <5 years old	2020	Examine the association between HFI and ECD domains and subdomains	HFI was associated with developmental risk and poor math skills in studies conducted in HICs and with poor vocabulary skills in both HICs and LMICs
Gallegos et al.[23] ■ studies published between 2011–21 ■ high-income countries ■ birth-12 years old children	2021	1. Describe the impact of the severity and persistence of food insecurity on child development 2. Use a socio-ecological framework to examine significant proximal and distal factors 3. Outline directions for future research	Transitioning between food security and food insecurity had a significant and lasting negative effect on: 1. Academic and cognitive function and 2. Externalizing behavior
Macassa et al.[54] ■ studies published 2019–2022 ■ South Africa and Swaziland	2022	Examine the relationship between food insecurity and intimate partner violence (IPV) in southern Africa.	HFI associated with intimate partner violence (physical, psychological, and emotional)
Hatcher et al.[55] ■ LMICs	2022	Examine the association between HFI and Violence Against Women and Girls (VAWG)	HFI increased twofold the risk of VAWG

Data from Young SL, Frongillo EA, Jamaluddine Z, et al. Adv Nutr 2021;12(4):1058-1073; Perez-Escamilla F, de Toledo Vianna RP. Journal of Applied Research on Children 2012;3(1):9.; Macassa G, Francisco JD, Militao E et al. Women. 2022;2(4):397-407.; Hatcher AM, Page S, Aletta van Eck L et al. PLOS Glob Public Health 2022;2(9): e0000479.; Shankar P, Chung R, Frank DA. J Dev Behav Pediatr 2017;38(2):135-150.; Tamiru D, Belachew T. Agriculture & Food Security 2017;6(1):1-4.; de Oliveira KHD, de Almeida GM, Gubert MB et al. Matern Child Nutr 2020;16(3):e12967.

Data from the third U.S. National Health and Nutrition Examination Survey (NHANES III) data showed that school-aged children experiencing HFI were more likely to have seen a psychologist and to have more difficulty getting along with their peers.[60] NHANES-III analyses also showed that adolescents living in food insecure households were more likely to have experienced dysthymia, thoughts of death, a desire to die, and suicide attempts.[61]

Academic Performance

Additional NHANES III results showed that 6- to 11-year-old children from food insufficient (vs food sufficient) households had lower arithmetic scores and were more likely to have repeated a grade.[60] Longitudinal studies support the evidence from the cross-sectional studies. Findings from the U.S. Early Childhood Longitudinal Study birth cohort (ECLS-B) showed that HFI at 9 months of age predicted lower maternal attachment and lower mental development at 2 years of age. This association was mediated by maternal depression and poorer parenting practices at 9 months.[62] Another analysis of the ECLS-B found that intermittent HFI (i.e., FI at 2 years but not at 9 months) predicted lower mental development scores among children at 2 years, compared with persistently food secure households.[63]

A longitudinal analysis of the U.S. Early Childhood Longitudinal Study kindergarten cohort (ECLS-K) data[64] documented that HFI in kindergarten predicted lower math scores and social skills in third grade among girls, but not boys. Likewise, girls (but not boys) from persistently food insecure households (i.e., those that were food insecure both at kindergarten and third grade) had lower increases in reading scores, compared with their persistently food secure counterparts. Both girls and boys living in households that were food secure in kindergarten, and then became food insecure by third grade, had lower increases in reading scores, compared with children whose households were persistently food secure. Transitioning from food insecurity to food security during the same period was associated with improved social skills among girls only.[64] A 2-year follow-up study in the United States found that persistent HFI, between 4 and 14 years and 5 and 16 years of age, increased risk of internalizing and externalizing problems.[65]

In summary, the studies included in the review by Perez-Escamilla and Vianna[34] show that HFI, independent of socio-economic status, represents not only a nutritional but also a psycho-emotional and cognitive development challenge to children exposed to it. This, in turn, is likely to translate into poor academic performance and intellectual achievement later in life. All these studies were conducted in the United States, and most included children of racial and ethnic minority groups, highlighting the need to use an anti-racism equity lens when considering solutions to HFI and poor ECD.

Maternal Mental Health

Maternal depression has been identified as a risk factor for sub-optimal child development. Women who are depressed are less likely to stimulate and engage with the development of their children. For example, women experiencing depression may be less likely to play and read to their children or to engage in nurturing interactions during child feeding and the establishment of healthy sleep and hygiene routines.[34,66] Additionally, maternal depression may increase the risk of lower child access to health-care services and food assistance programs[67,68] which, in turn, may negatively affect the development and overall well-being of young children.[69,70]

Epidemiologic studies have consistently found an independent association between HFI and maternal depression. NHANES III analyses found that the prevalence of maternal clinical depression and anxiety symptoms were associated in a dose-response manner with HFI severity.[47] Consistently, there was a dose-response relationship between the severity of HFI and perceived stress, trait anxiety, and depressive symptoms among pregnant women from North Carolina.[70] HFI was also associated with depression among pregnant Latinas living in Connecticut.[71] As previously indicated, longitudinal analyses of the ECLS showed that HFI at 9 months of age was associated with maternal depression which, in turn, mediated the association between HFI and poorer child health, mental development, and obesity outcomes at 2 years of age.[72,73] The C-SNAP study also found that maternal depressive symptoms were associated with HFI,[68] and Whitaker et al.,[59] found that HFI was independently and positively associated, in a dose-response fashion, with maternal clinical depression and anxiety symptoms. Findings from the multistate U.S. longitudinal study "Rural Families Speak," documented a bidirectional relationship between HFI and maternal depression,[74] which is consistent with the conclusions from a recent NHANES analysis[47] and a qualitative study by Leung et al.[46]

A well-known risk factor for maternal depression is intimate partner violence (IPV), which is more likely to be present in households experiencing food insecurity, due to the psycho-emotional stress resulting from HFI and manifesting itself through dysfunctional family dynamics.[44,45] This hypothesis is supported by studies conducted across countries with diverse levels of socio-economic development including Nepal, Ecuador, and the United States.[54] A descriptive review of five studies, four cross-sectional and one longitudinal, tested the association between food insecurity and intimate partner violence in South Africa and Swaziland.[54] They found a consistent association between HFI and different forms of IPV (physical, psychological, and emotional) despite the variability in the way HFI and IPV were measured across studies.[54] These findings are supported by a subsequent systematic review that found that HFI and violence against women and girls (VAWG) were related in low- and middle-income settings (including sub-Saharan Africa and South America).[55]

Poor maternal mental health can exacerbate the negative impact of HFI on human development across the life course. The systematic review by Shankar et al.,[56] sheds additional light on the negative influence of HFI on child development and women's mental health. This review included 23 articles from the United States, the United Kingdom, and Australia, confirming the associations between HFI and poor psycho-social, mental health, and behavioral and cognitive development from infancy through adolescence. Specifically, HFI was negatively associated with: (a) mental development of infants and toddlers (two analyses from the longitudinal ECLS-B; one cross-sectional), (b) behavioral problems among preschoolers (two longitudinal; two cross-sectional); (c) academic performance and behavioral problems among school-aged children (eight longitudinal; nine cross-sectional), (d) psychosocial (including substance abuse and suicidal ideation) and cognitive development among adolescents (one longitudinal, five cross-sectional). The findings from adolescents are concerning as when they become parents themselves they may carry with them the HFI-related mental health problems that may prevent them from practicing nurturing parenting skills with their own children. The study by Pedroso et al. in Brazil[72] showed that, among infants with depressed mothers, those experiencing mild and moderate/severe HFI had higher odds of having early child development delays, compared with infants in food secure households (**Figure 74.3**).[72]

A state-of-the-art review by Gallegos et al.;[23] examined the evidence published between 2011 and 2021 on the

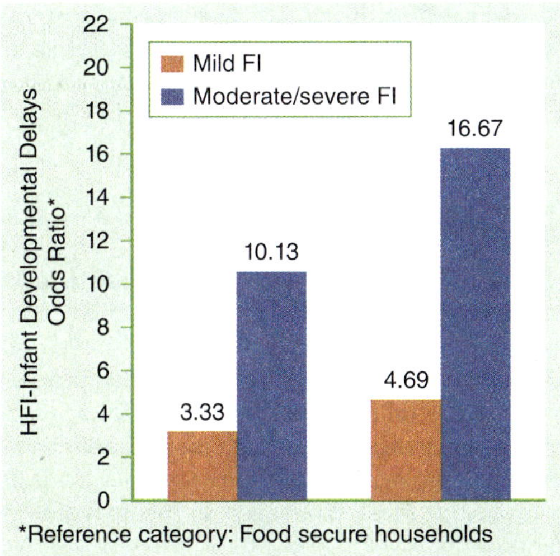

Figure 74.3 Association between HFI and infant developmental delays in Brasilia, Brazil. The left set of bars refers to developmental delays among infants with depressed mothers by mild (OR, 95% CI: 3.33, 1.17-9.46) or moderate/severe FI (10.13, 2.18-47.10) status. The right set of bars refers to infants with mothers with anxiety and depression by mild FI (4.69, 1.41-15.59) or moderate/severe FI (16.67, 2.70-95.66).

Data from Pedroso J, Buccini G, Venancio SI, et al. Matern Child Nutr 2020;16(4):e12997.[72]

relationship between HFI and child development among children ranging in age from newborn to 12 years in high-income countries. Findings from 17 studies documented that transitioning between food security and food insecurity had a negative, significant, and lasting effect on academic and cognitive function and behavior (i.e., externalizing). Hence, this review supports the hypothesis and conclusions from previous reviews[34,56] that there is a relationship between food insecurity and suboptimal child development. It is important to note that 10 of the 17 articles were secondary analyses of data coming from two large cohorts from the United States: The Early Childhood Longitudinal (ECLS) Birth and Kindergarten cohorts and the Fragile Families and Child Wellbeing Study. Fifteen of the 17 articles assessed HFI with the experience-based US-HFSSM. HFI status changes over time were reported in seven articles, but only one study examined HFI status transitions also considering different levels of severity of HFI.[75] In line with the other recent reviews,[34,58] this review found that the scales/items used to assess child development outcomes, and the way they were scored, differed across studies.

School performance and Cognition

Five of the seven studies examining the association between food insecurity and academic or cognitive functioning outcomes were longitudinal. Of these, four found a significant negative association between persistent HFI and child development outcomes, including a poorer approach to learning[76,77] and reading and math scores.[75,78] Consistently, Grineski et al., found that only children in households who transitioned from marginally food secure to food insecure (worsening food insecurity) had lower math and working memory scores.[75] By contrast, Kimbro and Denney[79] found no associations between either persistent or transient food insecurity and academic outcomes (reading, math, or science) across two time points.

School Absenteeism

An important dimension of academic performance is school absenteeism. A meta-analysis of four studies[57] found that students from food secure households were 57% less likely to be absent from school than students from food insecure households. These findings complement the reviews discussed above[23,34] by adding a potential mechanism through which FI may negatively impact academic performance.

Behavioral Development

The association between HFI and child behavior (externalizing, internalizing, self-control, self-regulation, general conduct) was examined by Gallegos et al., in 12 studies (n = 4 cross-sectional and eight longitudinal).[23] Three of the four cross-sectional studies found a positive association between food insecurity and behavioral problems.[80-82]

Among the eight longitudinal studies, two reported that shifting from food security at baseline to HFI at follow-up was associated with increased externalizing behavior[75,79]; with a third study finding this relationship only among boys.[83] Grineski et al.,[75] reported a significant positive association between persistent HFI and externalizing behaviors. Kimbro and Denney reported a positive association between persistent food insecurity on internalizing behaviors[79]; while Huang et al., found such an association among boys only.[83] King[84] found increasing internalizing behaviors in children in households where adults only were food insecure; and increasing externalizing behaviors in children in households where both adults and children were food insecure. Among the remaining longitudinal studies, one[77] found that hyperactivity and conduct problems in kindergarten were associated with food insecurity at an earlier time point, but another did not find such an association.[85,86] Four articles reported negative associations between HFI and self-control.[76-78,86] Interestingly, this association was particularly strong when households fluctuated into and out of HFI.

Overall, findings from this systematic review indicate that persistent and transient HFI may negatively affect children's behavioral development at different ages. Shorter and more transient forms of FI were associated with increased externalizing behaviors, while more persistent FI was associated with internalizing and self-control behavioral issues. Because results were mixed, further research is required to better understand these relationships.

Parental Developmental Concerns

Three cross-sectional studies found that both mild and moderate food insecurity was associated with increased developmental concerns reported by parents, after controlling for the birth weight of the child and feeding and caregiver's age, education, employment, and marital status.[79,87,88]

Psychosocial

In the review conducted by Gallegos et al., four studies assessed the associations between food insecurity and psychosocial outcomes (two cross-sectional and two prospective). One cross-sectional analysis[89] found that declines in pervasive development, but not in the remaining four subdomains of the

Child Behavior Checklist, were associated with FI after adjusting for child, maternal, and household confounders. The other cross-sectional study[90] found that FI, but not marginal FI (vs food security) was associated with the child not having good or excellent health.

One longitudinal study[76] reported that children who transitioned from living in a food insecure household in the first grade to a food secure family in third grade had lower social skills than those living in food secure households across time. Interestingly, these associations were not found when HFI persisted across time. Grineski et al.,[75] found that two HFI transition patterns over time were associated with declines in child interpersonal skills (rated by teachers), after controlling for child and school level confounders. The two patterns were (a) marginally food secure at kindergarten switching to HFI at grade 1; and (b) marginally food secure at both kindergarten and grade 1. Findings from these longitudinal studies suggest that transitioning between food security and food insecurity over time matters for child psychosocial development. Importantly, they also suggest that even marginal food security early on may negatively impact children's interpersonal skills and development later in life, even if they are no longer experiencing HFI.

A recent systematic review and meta-analysis of 19 studies (five in high income countries and four in low-middle income countries (LMICs)[58] showed that HFI was significantly associated with developmental risk and lower vocabulary and math scores (**Figure 74.4**).

HFI was marginally associated with cognitive/school readiness and reading and motor development, and was associated with poor ECD in children under 5 years old. Specifically, HFI was associated with developmental risk and poor math skills in studies conducted in HICs and with poor vocabulary skills in studies conducted in both HICs and LMICs. Prospective studies examining HFI and ECD are needed in LMICs.[58]

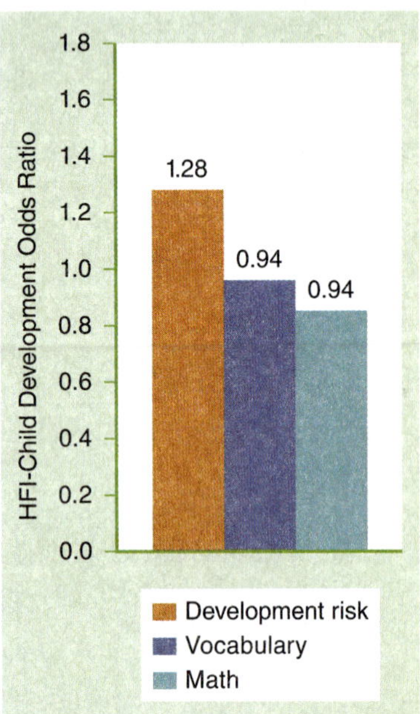

Figure 74.4 Association between HFI and early child development risk in Brasilia, Brazil. The odds ratio's 95% confidence intervals were 1.14-1.45 for memory, 0.90-0.98 for vocabulary, and 0.93-0.96 for math.

Data from de Oliveira KHD, de Almeida GM, Gubert MB, et al. Matern Child Nutr 2020;16(3):e12967.[58]

The findings from this review are consistent with data presented earlier in this chapter. Its meta-analysis makes an important contribution by estimating the effect sizes of the associations of FI with diverse child development domains.

Conclusions

The seven systematic reviews and one meta-analysis informing this chapter examined the associations between HFI and child development over a period of 2 decades.[23,34,54-58] Collectively, the evidence indicates that HFI negatively affects child development, and that this relationship is likely to be mediated by nutritional factors, as HFI has a negative impact on dietary quality and health outcomes, and especially by the strong psycho-emotional stress negatively affecting the dynamics and leading to dysfunction in the family unit.[44-46]

The reviews consistently show that food HFI, even in its mildest form, is associated with increased risk of poor child development outcomes across developmental domains; psycho-emotional, social, behavioral, and cognitive. The fact that these negative impacts of HFI on child development have been detected since infancy and persist throughout childhood and adolescence is of major public health concern. Some of these reviews also document the likelihood that the relationships between HFI and human development are mediated by poor maternal mental health, including anxiety and depression, and overall household dysfunction.[44-47]

Collectively, the reviews consistently note that HFI affects child development even after adjusting for poverty measures, which confirms that experience-based HFI scales capture information that goes beyond socioeconomic status indicators. Importantly, qualitative studies that have documented how children and youth experience FI support the findings from the epidemiologic studies[91-99] in the context of family chaos,[97] a direct outcome of living under poverty conditions that creates toxic stress by limiting access to human basic needs.[98,99]

The systematic reviews strongly support the need to address the mental health of mothers and other caregivers to mitigate the negative impact of HFI on child development and the caregiver's well-being.[34,47,73,74,89] Many studies have found that maternal depression is associated with the "chaotic"[44-46] household's experience of HFI and may partially or fully mediate its negative impact on diverse child development domains. The chronic and cumulative stress on parents unable to provide food security and other basic needs to their households can affect their parenting behaviors.[23,34,89] Indeed, maternal depression has been associated with lower-quality care and maternal-child interactions, less attachment, and child neglect and abuse.[62,72,91,98]

Several limitations of the currently available evidence on HFI and child development are important to recognize. First, most study designs have been cross-sectional and although key evidence has emerged from countries like Brazil,[58,72] South Africa,[54] Ghana,[100] and Bangladesh,[101] there is still not enough representation from middle- and lower-income countries in this area. Second, different scales, cutoff points, and reference time periods have been used to classify households into different food (in) security categories. Specifically, although some studies examined different levels of HFI severity, others have classified households only as food secure or food insecure (i.e., dichotomous). Third, there is lack of standardization of instruments used to assess similar developmental domains across studies.[23,34,58] This limits the comparison of interpretation of findings across studies.

In conclusion, an integration of the evidence reviewed suggests that HFI is indeed a powerful toxic stressor that is likely to have both direct and indirect impacts on the psycho-emotional, social, behavioral, and intellectual development of children. Our review strongly supports the likelihood that the HFI experience affects child development above and beyond the independent effects of poverty and confirms diverse pathways through which FI negatively affects child development including maternal mental health and the lack of access to healthy foods and health care among socio-economically disadvantaged families.

Policy Implications

The findings discussed in this chapter have several policy implications. First, household members who have already been exposed to FI are likely to need family support services to learn to cope constructively with this stressor and its mental health sequelae.[44,46,91,98] Second, all children should be screened for exposure to HFI across sectors (health, nutrition, education, social protection) and referred, as needed, to nutrition, parenting and psychosocial stimulation support services. In the United States, programs such as Head Start and the Supplemental nutrition program for Women, Infants and Children (WIC) may be ideally poised to help address the negative consequence of HFI on the development of children.[102-104]. The WIC program is relevant for this undertaking as it serves low-income pregnant and lactating women and their children under 5 years of age and reaches about half of the births that happen every year in the country.

Third, the impacts of HFI should be taken into consideration in national education and nutrition policy formulation. There is a need to strengthen programs like school feeding, which is particularly important during public health emergencies such as the COVID-19 pandemic.[9,29,105] Fourth, addressing FI should be a top priority for large-scale initiatives based on well-coordinated intersectoral policies and programs that include effective integrated services involving the health, nutrition (e.g., WIC and school meals programs, education (initial education, early stimulation) and social protection sectors (e.g., parenting skills)). This is needed to improve the development and well-being of children at risk of living in families experiencing FI.

Fifth, attention needs to be paid to the modifiable social and physical environment factors; more access to affordable healthy foods; more safe areas to perform leisure time activities, such as in the highly segregated neighborhoods where most FI families live.

Research Recommendations

Further advances in our understanding of how HFI affects the development of children will depend heavily on stronger conceptual frameworks and more robust studies based on prospective research designs with well-standardized measures of child development, based on the lived experience of FI among children and youth, as reported by them[91-96]; with more representation of diverse settings, as high-income countries, as the research has been concentrated in the United States.

Acknowledgments

This publication was supported by the Cooperative Agreement Number 5 U48DP006380-02-00 funded by the Centers for Disease Control and Prevention, Prevention Research Center Program (PI. Rafael Pérez-Escamilla). Its contents are solely the responsibility of the author and do not necessarily represent the official views of the Centers for Disease Control and Prevention or the Department of Health and Human Services. This chapter is dedicated to the memory of Ernesto Pollitt, my mentor, colleague, and friend.

References

1. Pollitt E. J Nutr 2000;130(Suppl):350S–3S.
2. Victora CG, Adair L, Fall C, et al. Lancet 2008;371:340–57.
3. Perez-Escamilla R. Food Insecurity in Children: Impact on Physical, Psychoemotional, and Social Development. Modern Nutrition in Health and Disease 201; Chapter 72:1007–1016
4. Rodríguez LA, Mundo-Rosas V, Méndez-Gómez-Humarán I, et al. Matern Child Nutr 2017;13(4):e12372.
5. Gundersen, Ziliak. Health Aff (Millwood) 2015;34(11):1830–9.
6. Thomas MMC, Miller DP, Morrissey TW. Pediatr 2019;144(4): e20190397.
7. Cook JT, Frank DA, Berkowitz C, et al. J Nutr 2004;134: 1432–8.
8. Perez-Escamilla R, Segall-Corrêa AM. Rev Nutr (Brazil) 2008;21(Suppl):15–26.
9. FAO, IFAD, UNICEF, WFP, and WHO. The State of Food Security and Nutrition in the World 2022. Repurposing Food and Agricultural Policies to Make Healthy Diets More Affordable. Rome, FAO. 2022. Accessed January 6, 2023. Available at www.fao.org/publications/sofi/2022/en/
10. Anderson SA. J Nutr 1990;120:1557–1600.
11. Coates J, Frongillo EA, Rogers BL, et al. J Nutr 2006; 136(Suppl):1438S–48S.
12. Jones AD, Ngure FM, Pelto G, et al. Adv Nutr. 2013;4(5): 481–505.
13. Frank DA, Neault NB, Skalicky A, et al. Pediatr 2006;118(5): e1293–302.
14. Knowles M, Rabinowich J, Ettinger de Cuba, et al. Matern Child Health J 2016;20:25–32.
15. Sharareh N, Wallace AS. Healthcare (Basel). 2022 Feb 17; 10(2):380.
16. Frankenberger TR, Frankel L, Ross S, et al. Household Livelihood Security: A Unifying Conceptual Framework For CARE Programs. Proceedings of the USAID Workshop on Performance Measurement for Food Security, December 11–12, 1995, Arlington, VA. Washington, DC: United States Agency for International Development, 1997.
17. Perez-Escamilla R. Curr Dev Nutr 2017 20;1(7):e000513.
18. Food and Agriculture Organization; High Level Panel of Experts on Food Security and Nutrition (HLPE). Food Security and Nutrition: Building a Global Narrative Towards 2030; Committee

on World Food Security: Rome, Italy, 2020. Accessed January 6, 2023. Available at www.fao.org/3/ca9731en/ca9731en.pdf
19. US Department of Agriculture Food Security Data. Accessed February 11, 2023. Available at USDA ERS - Food Security and Nutrition Assistance
20. Pérez-Escamilla R, Segall-Corrêa AM, Kurdian Maranha L, et al. J Nutr 2004;134:1923–8.
21. Pérez-Escamilla R, Melgar-Quiñonez H, Nord M, et al. Perspect Nutr Hum (Colombia) 2007(Suppl):117–34.
22. FAO. The Food Insecurity Experience Scale (FIES). Accessed January 16, 2022. Available at www.fao.org/in-action/voices-of-the-hungry/fies/en/
23. Gallegos D. Int J Environ Res Public Health 2021;18(17):8990.
24. Pérez-Escamilla R. Glob Food Sec 2012;1(2):120–125.
25. Pérez-Escamilla R, Gubert MB, Rogers B, et al. Global Food Security 2017;14:96–104.
26. Pérez-Escamilla R, Vilar-Compte M, Gaitan-Rossi P. Global Food Security. 2020;26:100459.
27. Young SL, Frongillo EA, Jamaluddine Z, et al. Adv Nutr 2021;12(4):1058–1073.
28. Alisha Coleman-Jensen, Matthew P. Rabbitt, Christian A. Gregory, Anita Singh. Household Food Security in the United States in 2020, ERR-298, U.S. Department of Agriculture, Economic Research Service. 2021. Accessed February 11, 2023. Available at www.ers.usda.gov/webdocs/publications/102076/err-298.pdf?v=1492
29. Pérez-Escamilla R, Cunningham K, Moran VH. Matern Child Nutr 2020;16(3):e13036.
30. Rede PENSSAN. Inquérito Nacional sobre Insegurança Alimentar no Contexto da Pandemia da Covid-19 no Brasil. Accessed February 11, 2023. Available at www12.senado.leg.br/noticias/arquivos/2022/10/14/olheestados-diagramacao-v4-r01-1-14-09-2022.pdf
31. FAO. Escala Latinoamericana Y Caribeña de Seguridad Alimentaria (ELCSA): Manual De Uso Aplicaciones. 2012. Accessed January 16, 2022. Available at www.fao.org/3/i3065s/i3065s.pdf
32. Britto PR, Lye SJ, Proulx K, et al. Lancet 2017;389(10064): 91–102.
33. Black MM, Pérez-Escamilla R, Rao SF. Adv Nutr 2015;6(6): 852–9.
34. Perez-Escamilla F, de Toledo Vianna RP. J Appl Res Child 2012;3(1):9.
35. Prado EL, Dewey KG. Nutr Rev 2014;72267–284.
36. Georgieff MK, Brunette KE, Tran PV. Dev Psychopathol. 2015;27: 411–423.
37. Dave JM, Evans AE, Saunders RP, et al. J Am Diet Assoc 2009;109:697–701.
38. Rosas LG, Harley K, Fernald LC, et al. J Am Diet Assoc 2009;109:2001–9.
39. Oh SY, Hong MJ. Eur J Clin Nutr 2003;57:1598–604.
40. Knueppel D, Demment M, Kaiser L. Public Health Nutr 2010;13:360–7.
41. Pérez-Escamilla R, Segall-Corrêa AM, Kurdian Maranha L, et al. J Nutr 2004;134:1923–8.
42. Melgar-Quiñonez HR, Zubieta AC, MkNelly B, et al. J Nutr 2006;136(Suppl):1431S–7S.
43. Rafiei M, Nord M, Sadeghizadeh A, et al. Nutr J 2009;8:28.
44. Fiese BH, Gundersen C, Koester B, Jones B. Econ Hum Biol 2016;21:147–55.
45. Johnson AD, Markowitz AJ. J Pediatr. 2018;196:275–282.
46. Leung CW, Laraia BA, Feiner C, et al. J Acad Nutr Diet. 2022;122(10):1903–1910.e2.
47. Reeder N, Tolar-Peterson T, Bailey RH, et al. Nutrients 2022;14(15):3081.
48. Stephens MA, Wand G. Alcohol Res 2012;34(4):468–83.
49. Peltz A, Garg A. Pediatr 2019;144(4):e20190347.
50. Drennen CR, Coleman SM, Ettinger de Cuba S. Pediatr 2019;144:20190824.
51. Jackson DB, Testa A. J Public Health Dent 2021;81(2):150–161.
52. Kac G, Schlüssel MM, Pérez-Escamilla R, et al. PLoS One 2012;7(9):e45747.
53. Carvajal-Aldaz D, Cucalon G, Ordonez C. Front Nutr 202;26;9: 1012734.
54. Macassa G, Francisco JD, Militao E, et al. Women 2022;2(4): 397–407.
55. Hatcher AM, Page S, Aletta van Eck L, et al. PLOS Glob Public Health 2022;2(9):e0000479.
56. Shankar P, Chung R, Frank DA. J Dev Behav Pediatr 2017;38(2): 135–150.
57. Tamiru D, Belachew T. Agri Food Secur 2017;6(1):1–4.
58. de Oliveira KHD, de Almeida GM, Gubert MB, et al. Matern Child Nutr 2020;16(3):e12967.
59. Whitaker RC, Phillips SM, Orzol SM. Pediatr 2006;118:e859–68.
60. Alaimo K, Olson CM, Frongillo EA Jr. Pediatr 2001;108:44–53.
61. Alaimo K, Olson CM, Frongillo EA. J Nutr 2002;132(4):719–25.
62. Zaslow M, Bronte-Tinkew J, Capps R, et al. Matern Child Health J 2009;13:66–80.
63. Hernandez DC, Jacknowitz A. J Nutr 2009;139:1517–1524.
64. Jyoti DF, Frongillo EA, Jones SJ. J Nutr 2005;135:2831–9.
65. Slopen N, Fitzmaurice G, Williams DR, et al. J Am Acad Child Adolesc Psychiatr 2010;49:444–452.
66. Tran TD, Luchters S, Fisher J. Child Care Health Dev 2017;43(3):415–426.
67. Casey PH, Szeto KL, Robbins JM, et al. Arch Pediatr Adolesc Med 2005;159:51–6.
68. Casey P, Goolsby S, Berkowitz C, et al. Pediatr 2004;113: 298–304.
69. Cook JT, Frank DA. Ann N Y Acad Sci 2008;1136:193–209.
70. Laraia BA, Siega-Riz AM, Gundersen C, et al. J Nutr 2006;136: 177–82.
71. Hromi-Fiedler A, Bermúdez-Millán A, Segura-Pérez S et al. Matern Child Nutr. 2011;7(4):421–30.
72. Pedroso J, Buccini G, Venancio SI, et al. Matern Child Nutr 2020;16(4):e12997.
73. Bronte-Tinkew J, Zaslow M, Capps R, et al. J Nutr. 2007;137(9): 2160–5.
74. Huddleston-Casas C, Charnigo R, Simmons LA. Public Health Nutr 2009;12:1133–1140.
75. Grineski SE, Morales DX, Collins TW, et al. J Dev Behav Ped 2018;39:715–725.
76. Howard L. Br J Nutr 2011;105:1852–1860.
77. Johnson AD, Markowitz AJ. Child Dev 2018;89:e1–e17.
78. Gee KA. J Fam Issues 2018;39:2437–2460.
79. Kimbro RT, Denney JT. Health Aff 2015;34:1949–1955.
80. Encinger A, Shizu Kutaka T, Chernyavskiy P, et al. Early Educ Dev 2020;31:1264–1280.
81. Ramsey R, Giskes K, Turrell G, et al. J Child Health Care 2011;15:401–416.
82. Huang Y, Potochnick S, Heflin CM. J Fam Iss 2018;39: 1465–1497.
83. Huang J, Vaughn MG. Br J Soc Work 2016;46:993–1008.
84. King C. Hum Biol 2018;28:14–22.
85. Melichor M, Chastang JF, Falissard B, et al. PLoS One 2012;7:e52615.
86. Jackson DB, Newsome J, Vaughn MG, et al. J Crim Justice 2018;56:127–139.
87. Belsky DW, Moffitt TE, Arseneault L, et al. Am J Epid. 2010;172:809–818.
88. Hobbs S, King C. J Nutr Educ Behav 2018;50:687–694.
89. Nagata JM, Gomberg S, Hagan MJ, et al. J Hunger Environ Nutr 2019;14:526–539.

90. Cook JT, Black, M, Chilton M, et al. Adv Nutr 201;4:51–61.
91. Chilton M, Knowles M, Rabinowich J, et al. Public Health Nutr 2015;18(14):2643–53.
92. Bernal J, Frongillo EA, Herrera H, et al. J Nutr 2012;142(7): 1343–9.
93. Bernal J, Frongillo EA, Herrera HA, et al. J Nutr 2014;144(10): 1619–26.
94. Bernal J, Frongillo EA, Rivera JA. Matern Child Nutr 2016;12(3): 546–57.
95. Rosemond TN, Blake CE, Shapiro CJ, et al. J Acad Nutr Diet 2019;119(10):1644–1652.
96. Frongillo EA, Bernal J, Rampalli KK, et al. Ecol Food Nutr 2022;61(1):64–80.
97. Fram MS, Frongillo EA, Jones SJ et al. J Nutr 2011;141(6): 1114.
98. Shonkoff JP, Garner AS, Siegel BS, et al. Pediatr 2012;129: e232–e246.
99. Victora CG, Hartwig FP, Vidaletti LP, et al. Lancet. 2022; 399(10336):1741–1752.
100. Aurino E, Wolf S, Tsinigo E. PLoS One 2020;15(4):e0230965.
101. Saha KK, Tofail F, Frongillo EA, et al. Child Care Health Dev 2010;36(3):309–16.
102. Black MM, Quigg AM, Cook J, et al. Arch Ped Adolesc Med Arch 2012;166:444–451.
103. Metallinos-Katsaras E, Gorman KS, Wilde P, et al. Matern Child Health J 2011;15:627–33.
104. Anderson CE, Martinez CE, Ritchie LD, et al. J Nutr 2022;152(8):1974–1982.
105. Pepetone A, Frongillo EA, Dodd KW, et al. J Nutr 2023;153(4):1231–1243.

D. DISORDERS OF THE ALIMENTARY TRACT

CHAPTER 75

Oral Health and Disease

Teresa Marshall and Riva Touger-Decker

CHAPTER OUTLINE

Abbreviations

ADA, American Dental Association; **AIDS**, acquired immunodeficiency syndrome; **AOP**, anterior occluding pairs; **CCR**, concurrent chemoradiation; **CDC**, Centers for Disease Control and Prevention; **DHHS**, Department of Health and Human Services; **DMFS**, decayed-missing-filled tooth-surfaces; **DMFT**, decayed-missing-filled teeth; **ECC**, early childhood caries; **ERQOL**, eating-related quality of life; **GI**, gastrointestinal; **HIV**, human immunodeficiency virus; **ICU**, intensive care unit; **IOD**, implant-supported dentures; **NCCN**, National Comprehensive Cancer Network; **NHANES**, National Health and Nutrition Examination Survey; **OPC**, oral pharyngeal cancer; **POP**, posterior occluding pairs; **PPM**, parts per million; **SSB**, sugar sweetened beverages; **USDA**, United States Department of Agriculture; **VHNSS**, Vanderbilt Head and Neck Symptom Survey

Introduction

The relationship between oral health and disease and nutrition is multidirectional. As the entry point for the gastrointestinal tract, a healthy oral cavity is essential for eating, drinking, and swallowing. Local and systemic diseases can impact all aspects of oral function and dietary intake and nutrition status. This chapter explores oral health from birth through aging, including local and systemic diseases and their impact on diet and nutrition status.

Cellular and Structural Characteristics of the Oral Soft Tissues and the Role of Nutrition in Craniofacial and Oral Tissue Development and Maintenance

Distinctive characteristics of oral tissues, such as the inability of enamel to remodel, high cellular turnover rate of the oral mucosa, rates of alveolar bone growth, and the production of saliva, make oral tissues a unique indicator of physiologic changes. The oral cavity is a site of symptoms of many chronic diseases, including dental caries, periodontal disease, acquired immunodeficiency syndrome (AIDS), nutritional anemias, herpes, salivary gland disorders, osteoporosis, gastrointestinal diseases, diabetes, and cancers. Congenital anomalies such as cleft lip and palate are birth defects that have a complex genetic and environmental etiology linked to maternal nutrient status.

Associations between oral infection and systemic health are recognized, may be bidirectional, and mechanistic pathways are being explored. Established associations exist between periodontal disease and obesity, obesity-related chronic diseases, and adverse pregnancy outcomes. For example, obesity-induced hyperinflammation and immune dysfunction are associated with an increased risk of periodontal disease,[1] while periodontal disease is associated with an increased prevalence of myocardial infarction, stroke, and adverse pregnancy outcomes.[2,3] The inherent nature of oral infectious diseases requires functioning immune and cellular repair systems, and unequivocal data link nutrient intake to these host defense mechanisms. The relationships between oral and systemic health and nutrition require the careful attention of all health professionals, including physicians, dentists, and registered dietitians.[4]

To fully appreciate these complex relationships, it is vital to understand the structure and function of the craniofacial-oral-dental complex. Teeth are specialized structures necessary for the initial processing of food and are composed of three mineralized tissues: enamel, dentin, and cementum, which encase the highly vascular dental pulp or "nerve." These relationships can be seen in the schematic cross section of a tooth in **Figure 75.1**. The teeth are retained in their bony sockets by means of a fibrous structure termed the *periodontal membrane* or *ligament*. Factors such as bacteria and the inflammatory immune response can affect the integrity of this structure and bone surrounding the socket, resulting in periodontal disease that may progress sufficiently to cause loosening and loss of the teeth.

Each tooth develops from a tooth bud or germ located in the jaws. The bud consists of an epithelial component that arises as an invagination from the surface and produces enamel. The mesenchymal component consists of the dental papilla, which produces the tooth pulp and dentin, and the dental follicle, which produces the cementum and periodontal ligament once the tooth has formed. **Table 75.1** details the chronology of the formation of the human dentition. The primary teeth begin forming at approximately 6 weeks in utero when cells in the primitive oral cavity differentiate to form the dental lamina, which is the site of tooth bud development. The formation of the crown of the tooth begins with the secretion of a dentin matrix containing collagen fibrils. Mineral ions then enter the matrix to form small crystals on or between the collagen fibrils. Enamel formation begins as soon as the first dentin layer has been laid down. This mineralization process constitutes enamel maturation and continues after the matrix is fully formed. The mineralization process begins as early as four months in utero and continues into late adolescence. After the tooth erupts into the oral cavity, it continues to incorporate minerals (including fluoride) from saliva, food, and drinking fluids into its structure.[5]

The life history of a tooth may be divided into three main eras: (a) the period during which its crown is forming and mineralizing in the jaw, (b) the period of maturation when the tooth is erupting into the oral cavity and its root or roots are forming, and (c) the maintenance period while it is functioning in the oral cavity.[5] During the pre-eruptive period, the developing enamel and dentin are subject to nutritional deficiencies or imbalances in the same manner as any other developing tissues. Nutrient deficiencies can affect the secretory and maturation stages of enamel formation. Following eruption into the oral cavity, the enamel is bathed in saliva and exposed to oral micro-organisms and their byproducts as well as food, so nutritional deficiencies or excesses and dietary habits can affect teeth in a local manner.[5]

At least three striking differences exist between the mineralized tissues of teeth and other tissues of the body. First, enamel contains no capillary or lymphatic vessels to act as transport systems; however, the intimate relationships between the organic and the inorganic components of enamel suggest that pathways in the enamel exist for the diffusion of ions and small molecules from saliva and, possibly, from blood. Although the dentin likewise contains no vascular elements, it is more readily permeable to the passage of extracellular fluids from the blood by way of the dentinal tubules that traverse it. The interchange among

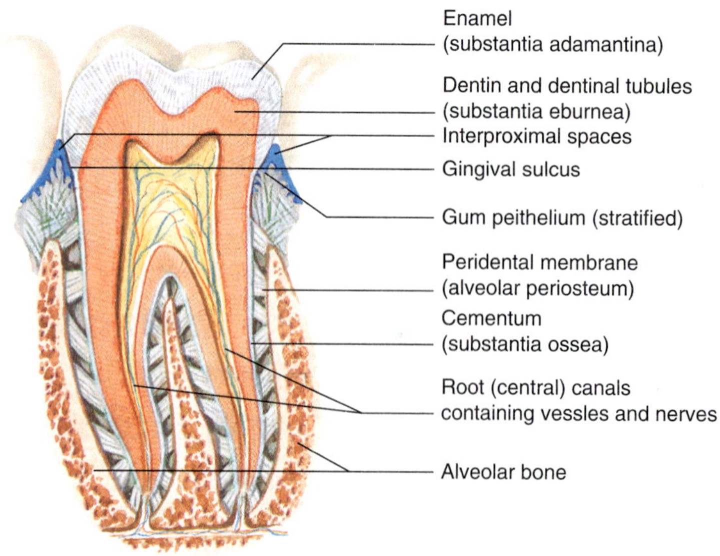

Figure 75.1 Schematic illustration of teeth in contact with the alveolar bone.

Table 75.1 Chronology of Development of the Human Dentition

Tooth	Hard Tissue Formation Begins	Amount of Enamel Formed at Birth	Enamel Completed	Eruption	Root Completed
Primary Dentition					
Maxillary					
Central incisor	4 mo in utero	Five-sixths	1½ mo	7½ mo	1½ y
Lateral incisor	4½ mo in utero	Two-thirds	2½ mo	9 mo	2 y
Cuspid	5 mo in utero	One-third	9 mo	18 mo	3¼ y
First molar	5 mo in utero	Cusps united	6 mo	14 mo	2½ y
Second molar	6 mo in utero	Cusp tips still isolated	11 mo	24 mo	3 y
Mandibular					
Central incisor	4½ mo in utero	Three-fifths	2½ mo	6 mo	1½ y
Lateral incisor	4½ mo in utero	Three-fifths	3 mo	7 mo	1½ y
Cuspid	5 mo in utero	One-third	9 mo	16 mo	3¼ y
First molar	5 mo in utero	Cusps united	5½ mo	12 mo	2¼ y
Second molar	6 mo in utero	Cusp tips still isolated	10 mo	20 mo	3 y
Permanent Dentition					
Maxillary					
Central incisor	3–4 mo	—	4–5 y	7–8 y	10 y
Lateral incisor	10–12 mo	—	4–5 y	8–9 y	11 y
Cuspid	4–5 mo	—	6–7 y	11–12 y	13–15 y
First bicuspid	1½–1¾ y	—	5–6 y	10–11 y	12–13 y
Second bicuspid	2–2¼ y	—	6–7 y	10–12 y	12–14 y
First molar	At birth	Sometimes a trace	2½–3 y	6–7 y	9–10 y
Second molar	2½–3 y	—	7–8 y	12–13 y	14–16 y
Mandibular					
Central incisor	3–4 mo	—	4–5 y	6–7 y	9 y
Lateral incisor	3–4 mo	—	4–5 y	7–8 y	10 y
Cuspid	4–5 mo	—	6–7 y	9–10 y	12–14 y
First bicuspid	1¾–2 y	—	5–6 y	10–12 y	12–13 y
Second bicuspid	2¼–2½ y	—	6–7 y	11–12 y	13–14 y
First molar	At birth	Sometimes a trace	2½ –3 y	6–7 y	9–10 y
Second molar	2½–3 y	—	7–8 y	11–13 y	14–5 y

elements in the enamel occurs through bathing its external surface with saliva. In contrast, the interchange in the dentin occurs by the movement of ions present in the blood supply to the pulp or periodontal membrane.[6] Second, because of the absence of cells, mineralized dental tissues do not have a microscopically or chemically detectable ability to repair improperly formed or mineralized areas, and the tooth does not have the ability to repair itself after a portion has been destroyed by tooth decay or mechanical injury. An exception is the remineralization of slightly demineralized, superficial areas of the enamel where the organic matrix and surface integrity are still intact, commonly referred to as "white spots." Secondary dentin is formed by the odontoblasts, which persist throughout life on the pulpal surface in response to chemical stimuli from an advancing carious lesion, in an effort to mitigate the noxious influence. The lack of ability to repair dental tissues is in direct contrast to bone, with its continual turnover.[6] Third, unlike other tissues, the mineralized tissues of teeth undergo a partial change of environment. When the tooth begins to emerge into the oral cavity, the vascular supply to the enamel organ is severed, and the enamel surface comes in contact with a complex mixture of saliva, micro-organisms, food debris, and epithelial remnants. Thus, instead of a pure systemic environment, the erupted tooth has, in addition, an oral or external environment. Consequently, the enamel and cementum surfaces on which carious lesions are initiated by microbial action are largely outside of the influences of humoral immune systems, so immune relationships with the caries process are primarily limited to those in saliva.[6]

The development and maintenance of the soft tissues and bone that support the teeth are also subject to nutrient deficiencies or excesses. The periodontium, as seen in Figure 75.1, comprises the gingiva; the periodontal ligament (peridental membrane), which joins the root cementum to the alveolar bone; the root cementum, which is a specialized, mineralized tissue like bone that covers the root of the tooth; and the alveolar bone, which forms and supports the sockets of the teeth. The alveolar bone grows in response to dental eruption, is modified by dental changes, and resorbs when teeth are lost. The finite space between the tooth and the gingiva, known as the gingival sulcus, is lined by a nonkeratinized epithelium. In addition, dental plaque, one of the primary agents responsible for initiating both dental caries and gingivitis, contains a high concentration of bacteria which, in the gingival sulcus, are juxtaposed with a "naked" epithelium. Thus, bacteria and their byproducts or antigens can permeate the gingival epithelium and precipitate a classic inflammatory response that denotes periodontal disease. In fact, an intact immune system, which is highly dependent on nutrient status, is vital to maintaining periodontal health. The diversity of hard and soft tissues that comprise the oral structures, and the distinctive nutritional needs of each, contribute to the uniqueness of the mouth as an external reflection of past and present nutritional problems.[4,7]

Nutrients Involved in Oral Tissue and Bone Development

Nutrient deficits can result in defects in tooth and salivary gland development. The most studied conditions and nutrients affecting tooth integrity, enamel solubility, and salivary flow and composition in animal models include protein-energy malnutrition, ascorbic acid, vitamin A, vitamin D, calcium and phosphorus, iron, zinc, and fluoride. Malnutrition—deficiencies of vitamin A, ascorbic acid, vitamin D, and iodine—and fluoride excess have been demonstrated to affect human dentition (**Table 75.2**). The reader is also encouraged to review relevant sections of chapters in this text on each specific nutrient mentioned.

Enamel hypoplastic defects and hypomineralization are the hallmarks of undernutrition and overnutrition during tooth development.[8,9] Vitamin A deficiency is implicated as a critical factor in tooth health because it frequently accompanies protein-energy malnutrition and is known to affect epithelial tissue development, tooth morphogenesis, and odontoblast differentiation.[10] The interference with calcification is manifested clinically by enamel hypoplasia.[11] Additionally, when present during the first trimester of pregnancy, vitamin A excess can result in severe craniofacial and oral clefts and lip defects.[12]

Vitamin D, calcium, and phosphorus deficiencies significantly affect tooth development and decrease resistance to dental caries. If a vitamin D deficiency occurs in utero or in the young infant, there can be delays in tooth eruption, and the quality of the enamel is compromised, thereby increasing caries risk.[13]

Table 75.2 Effects of Nutrient Deficiencies on Tooth Development

Nutrient	Effect on Tissue	Effect on Caries	Human Data
Protein-energy malnutrition	Tooth eruption delayed	Yes	Yes
	Tooth size		
	Enamel solubility decreased		
	Salivary gland dysfunction		
Vitamin A	Decreased epithelial tissue development	Yes	Yes
	Tooth morphogenesis dysfunction		
	Decreased odontoblast differentiation		
	Increased enamel hypoplasia		
Vitamin D/calcium/phosphorus	Lowered plasma calcium	Yes	Yes
	Hypomineralization (hypoplastic defects)		
	Tooth integrity compromised		
	Delayed eruption patterns		
Ascorbic acid	Dental pulpal alterations	No	No
	Odontoblastic degeneration		
	Aberrant dentin		

Nutrient	Effect on Tissue	Effect on Caries	Human Data
Fluoride	Stability of enamel crystal (enamel formation)	Yes	Yes
	Inhibition of demineralization		
	Stimulation of remineralization		
	Mottled enamel (excess)		
	Inhibition of bacterial growth		
Iodine	Delayed tooth eruption	No	Yes
	Altered growth patterns		
	Malocclusion		
Iron	Slow growth	Yes	No
	Tooth integrity		
	Salivary gland dysfunction		

Leaver demonstrated that extreme calcium and phosphorus deficiencies can result in hypomineralization of developing teeth.[14] The deficit, however, must be severe enough to reduce plasma concentrations of calcium and phosphorus. The highly effective homeostatic mechanisms in humans that mobilize calcium from the skeleton to maintain normal plasma calcium concentration make the chances of this occurring rare. Bawden postulated that vitamin D hypovitaminosis is more important in considering hypomineralization resulting from inadequate calcium transport into developing dental tissues.[15] Vitamin D deficiency also increases the risk of enamel hypoplasia, delays eruption patterns of teeth, and increases the risk of early childhood caries.[16-18]

In childhood vitamin D deficiency, teeth are characterized microscopically by a widened layer of predentin, the presence of interglobular dentin, and interference with enamel formation (hypoplastic defects).[19] Eruption of the deciduous teeth is delayed, and the eruption sequence is altered in young children with rickets. The permanent incisors, cuspids, and first molars are usually affected because their development coincides with the age at which rickets is most common. Vitamin D-resistant rickets results in more frequent and severe tooth defects, relative to primary rickets, including large pulps with developmental "exposures."

Vitamin C deficiency also affects tooth development and eruption. Deciduous and permanent teeth of scorbutic infants contain minute pulpal hemorrhages attributable to vitamin C deficiency. In older children with vitamin C deficiencies, the dental pulp undergoes hyperemia, edema, necrosis, and aberrant calcification, whereas the dentin shows odontoblastic degeneration and irregular formation.[20] However, the relationship between vitamin C deficiency and dental caries is poorly defined. Although it is likely that the primary mechanism of vitamin C deficiency–induced tooth, gingival, and bone disease is mediated through the disruption of collagen biosyntheses, no study has clearly demonstrated the relationship between scurvy and dental caries.[21] In areas where goiter is endemic, children born to mothers with severe iodine deficiency are characterized by marked mental and physical growth retardation. Eruption of the primary and secondary teeth is often greatly delayed and precluded. Malocclusion may occur because of altered patterns of craniofacial growth and development.

Nutritional status during development can profoundly affect oral disease when malnutrition is present. Several studies have demonstrated that tooth eruption is delayed, tooth integrity is compromised (especially enamel surface solubility), and the incidence of dental caries is increased in chronically malnourished animals and children.[22,23] In Lima, Peru, Alvarez et al., demonstrated significant delays in tooth eruption and exfoliation in three groups of malnourished children; such delays were associated with and appeared to be the direct cause of a significant temporal delay in caries development in the primary teeth.[24] These data support previous studies on children with malnutrition in India and Guatemala.[25,26]

The development of teeth and salivary glands is intimately associated with the nutrient supply. Teeth subjected to nutritional insult during critical stages of development are at an increased risk of decay. Menaker and Navia reported that impaired salivary function accompanies the morphologic changes in teeth and could increase susceptibility to caries.[27] These data also explain, in part, the positive association between socioeconomic status and prevalence of dental caries in deciduous teeth.[23] Nutritional insults early in life can affect tooth formation, resulting in increased susceptibility to caries. Malnutrition early in life delays tooth development; hence, teeth erupt later, and caries will occur in the older child.[23] As is discussed in the section under dental caries, the incidence of caries is associated with the social determinants of health, which are associated with low-quality diets.

On a broader scale, 3% of babies born in the United States each year have some congenital anomaly evident at birth or later.[28,29] Prominent among these defects are structural, functional, or biochemical abnormalities involving the craniofacial complex. The most common orofacial birth defect is the cleft lip and cleft palate affecting 10.7 infants/10,000 births.[29-31] In addition, select other craniofacial oral-dental disorders such as craniosynostosis, hemifacial microsomia, anodontia, amelogenesis imperfecta, dentinogenesis imperfecta, osteogenesis imperfecta, chondrodystrophies, and juvenile periodontitis represent major challenges to human oral health.[32] Neural tube defects are among the most common birth defects, despite the decline in incidence with folate fortification of grain products[33]; they range in severity and can result in the incomplete formation of cranial bones. Many birth defects have genetic and/or environmental risk factors, and racial and ethnic variations in prevalence exist. Certain nutrients given in excess, especially early in pregnancy (e.g., retinoic acid and other lipophilic molecules such as vitamins K and E), are known to induce craniofacial oral-dental malformations.

The regulatory genes and gene products functioning as transcriptional factors for the bronchial arches that give rise to the midface and lower face are being discovered, and their interaction with nutrients (e.g., retinoic acid via its specific receptors) has been found to be critical to craniofacial oral-dental morphogenesis.[34] Excess exogenous retinoic acid produces significant craniofacial malformations associated with clefting, dental development, hemifacial microsomia, spina bifida, eye defects, and limb morphogenesis.[35] A striking illustration of the need to understand the effects of nutrition on birth defects is the observation that folate supplements provided around the time of conception significantly reduced the occurrence of neural tube defects among high-risk persons in the United Kingdom and elsewhere.[33,36,37] Similar results suggest that folic acid or multivitamins may reduce the risk of congenital craniofacial malformations such as cleft lip and/or cleft palate.[33,37]

Nutrient Deficiencies

The oral mucosa is particularly susceptible to physiologic or anatomic changes resulting from nutritional deficiencies. Because the turnover rate of oral mucosal cells is rapid (gingival sulcular epithelial cells have a turnover rate of 3 to 7 days), sufficient nutrients must be available at appropriate times and in the correct concentration for DNA replication, protein synthesis, and cell and tissue maturation to occur. The oral epithelium acts as an effective barrier against the invasion of toxic substances, particularly antigens derived from oral microbes, into the underlying collagenous connective tissue. Inadequate nutrition can compromise the integrity of the oral epithelia, thus increasing the tissue's susceptibility to infectious disease.

For these reasons, the oral cavity is one of the first regions of the body to exhibit clinical signs of nutrient deficits and malnutrition. Virtually every classic nutrient deficiency or toxicity, including scurvy, beriberi, and pellagra, has signs and symptoms in the oral cavity and surrounding structures. The lips, tongue, oral mucosa, and gingiva may all reflect nutritional aberrations before signs appear elsewhere in the body (**Table 75.3**).

The dorsum of the tongue can undergo changes in size or color and taste due to atrophy or hypertrophy of tongue papillae.

Table 75.3 Nutrition Risk Factors to Consider in Physical Examination

Body Area	Nutrition Risk Symptoms	Nutrition Implications
Hair	Dull, shedding, easily pluckable	Generalized protein energy malnutrition
Face	a. Malar pigmentation (dark skin over cheeks and under eyes) Bitemporal wasting	Niacin, B vitamins, malnutrition
	b. Nasolabial seborrhea	Niacin, riboflavin, vitamin B_6
	c. Edematous	Protein deficiency
	d. Moon face	Corticosteroid impact
	e. Lack of color	Inadequate iron, undernutrition
Eyes	Pale eye membranes	Inadequate iron
	Bitot spots, conjunctival xerosis, keratomalacia, corneal xerosis	Inadequate vitamin A
Lips	Cheilosis (red/swelling)	Inadequate niacin, riboflavin
	Angular fissures	Inadequate niacin, vitamin B_6, riboflavin, iron
Gingiva	Spongy, bleeding, abnormal redness	Inadequate vitamin C
Tongue	a. Glossitis (red, raw, fissured)	Inadequate folate, niacin, riboflavin, iron, vitamins B_6, and B_{12}
	b. Pale, atrophic, smooth/slick (filiform papillary atrophy)	Inadequate iron, vitamin B_{12}, niacin, folate
	c. Magenta color	Inadequate riboflavin
Nails	Spoon shaped, brittle, ridged	Inadequate iron
Back muscles	Bony prominences along shoulder girdle	Malnutrition
	Tendons prominent to palpation	

Reproduced from Touger-Decker R. Dent Clin North Am 2003:47:259–78.

Long-standing nutrient deficiencies can cause atrophy of the papillae and denudation of the dorsum. A bright red, painful tongue and swelling of the oral mucosa can be early symptoms of pernicious anemia from a lack of vitamin B_{12}. Inflammation, a burning sensation, and tenderness of the tongue or palate can be caused by a deficiency of B vitamins, protein, or iron.[38] The mucosa can become pale in iron-, folic acid-, or vitamin B_{12}-induced anemias. Atrophy of the filiform papillae of the tongue (glossitis) is a sign of malnutrition, usually resulting from multiple nutritional deficiencies.

In ascorbic acid deficiency, the classic oral signs of scurvy are first seen in the oral cavity. They include red, swollen interdental papillae that bleed readily and inflamed and swollen marginal and attached gingiva. Low fruit and vegetable intakes are associated with low plasma vitamin C concentration, and smokers generally have lower vitamin C status than nonsmokers.[39,40] Low dietary intake and serum vitamin C concentration are inversely associated with periodontal disease.[41,42] No evidence has been found that consumption of ascorbic acid in excess of the dietary reference intakes is associated with increased oral health.[43]

A clinical sign of malnutrition is of little significance by itself, however, because several etiologic factors usually contribute to a differential diagnosis. For example, inflammation or cracking of the lips can be caused by allergies, licking, or drooling, as well as by nutritional aberrations. Angular cheilosis can result not only from vitamin deficiencies but also when overclosure of the jaw in denture wearers allows the skin folds at the corners of the mouth to provide a moist area for bacterial or fungal infections to develop. **Table 75.4** provides a functional oral assessment tool that can be used as a broad clinical examination guide for nondentist health providers. More detail on nutrition assessment and clinical manifestations of micronutrient deficiencies can be found in chapters on these topics in this text.

Nutrient Excesses

Nutrient excesses can also affect the oral cavity. Vitamin A toxicity can impair the proper development of the oral mucosal epithelium and can result in various oral changes, including delayed wound healing.[44,45]

Table 75.4 Functional Oral Nutrition Risk Evaluation

Structure	Nutrition-Focused Examination	Management
Lips	Observe for dryness, ulcers, cracks, angular fissures, changes in color, texture, moisture, swelling	Alter diet texture and consistency; determine risk for micronutrient deficiencies and order laboratory tests as needed.
Gingiva, labial, and oral mucosa	Observe for bleeding, signs of infection, ulcers, changes in color, moisture, texture, swelling, patient complaints of pain or tenderness, erythema	Alter diet texture, temperature, and consistency; determine risk for oral infections and micronutrient deficiencies and order laboratory tests as needed
	Observe for red or white patches, lesions, or ulceration	Screen for oral cancer, fungal or other infections
Teeth	Ask about tooth pain with biting or chewing; observe for missing teeth Partial or full mandibular or maxillary denture	Adjust diet consistency; evaluate caries risk Determine impact of dentures on food choices and eating ability; ask if dentures are used for eating.
	Observe for partial or complete edentulism	Determine occlusion and impact of edentulism on food choices and eating ability; Adjust diet consistency
Tongue	Observe for swelling, burning, pain, dryness, fissures, lesions, or ulcers, changes in texture and color, erythema, focal pigmentation. Ask about changes in taste	Determine risk for systemic diseases and micronutrient deficiencies; order laboratory tests as needed; alter diet texture
Temporomandibular joint	Observe for limited mandibular opening, pain on palpation, difficulty with or painful mandibular opening; pain or discomfort with biting and chewing	Change diet consistency, food "hardness"; limit "chewy" foods
Saliva	Observe for minimal or excessive saliva, mucosal dryness, drooling, change in consistency; Ask about ability to eat a snack or meal without a drink, altered taste, or dry eyes	Increase fluids; determine risk for systemic disease, evaluate for dysgeusia and dysphagia; limit spices, salty and "hard" foods; review prescription and over the counter medications for risk of hyposalivation
Muscles of mastication	Observe for pain on palpation or with biting or chewing, complaints of difficulty biting or chewing	Change diet consistency, food "hardness"; limit "chewy" foods
Skin	Observe for changes in color, temperature, texture, or moistness, wounds, lesions	Determine risk for micronutrient deficiencies

For all symptoms, consider necessary referrals to physicians, dentists, or dietitians for further evaluation and patient management as appropriate. Ask patients about frequency and duration of symptoms, changes in disease history and changes in medications (prescription and over the counter) and lifestyles that may relate to the symptoms presented.

Fluoride

The decline in coronal caries in industrialized countries over time is associated with the widespread availability of fluoride. Fluoride has a systemic effect before tooth eruption and a topical effect after eruption; the systemic and topical effects are additive. Fluoride is more effective in preventing smooth surface caries than occlusal caries. The primary sources of fluoride include community water supplies, foods, and beverages prepared with fluoridated water, dentifrices, and other dental products. Concentrated topical fluoride products are professionally applied to protect erupted teeth and are not swallowed. The preventive benefits of fluoride depend on its concentration, form of delivery (systematic or topical), and source (water, toothpaste, tablet, drops, rinse, gel).

Mechanisms of Action

Fluoride's caries preventive effects have been extensively investigated, and three mechanisms of action are recognized.[46,47] First, fluoride ions replace some of the hydroxyl groups of the hydroxyapatite of developing teeth to form fluoridated hydroxyapatite. The fluoridated hydroxyapatite is less susceptible to acid demineralization than hydroxyapatite, increasing enamel stability. Second, fluoride present in saliva can decrease the rate of demineralization and enhance remineralization of early carious lesions. Following mild demineralization, fluoride, calcium, and phosphate can diffuse back into the surface layers of enamel to facilitate remineralization. Third, fluoride has an antimicrobial effect on the pathogenic microbiota. In higher concentration, fluoride inhibits the growth of *S. mutans* found in the dental biofilm, while in low concentration, fluoride inhibits bacterial enzymes, thus reducing carbohydrate fermentation and subsequent acid production.

Water Fluoridation

Fluoridation of community water supplies is passive, effective, and cost-effective for preventing tooth decay in the United States.[47] Between 1950 and 1980, clinical studies conducted in 20 countries demonstrated that adding fluoride to community water supplies resulted in 40% to 50% caries reduction in primary teeth and 50% to 60% reduction in permanent teeth.[48] Children with lifelong residence in optimally fluoridated communities had 25% lower DMFS scores than their peers who were not exposed to fluoridated drinking water.[49] Furthermore, cessation of community water fluoridation has been associated with an increased risk of caries, exacerbated by social inequities.[50,51] McLaren et al., has monitored caries and untreated decay in 7-year-old children from Calgary, Canada, which ceased water fluoridation in 2011; and from Edmonton, Canada, which continues to have water fluoridation. Children from Calgary had greater primary tooth decay than their peers from Edmonton within 2 to 3 years following water fluoridation cessation.[50] Over a 7-year period, both caries experience and untreated decay increased in children from Calgary while remaining the same for children from Edmonton.[51] A decline in caries prevalence among children living in communities with non-fluoridated water is attributed to the intake of foods and beverages processed with fluoridated water and the use of fluoride-containing dentifrices and topical fluorides in the dental office.[46,52]

Fluoride Supplementation

Children living in communities with suboptimal water fluoride concentration can be prescribed fluoride supplements to reduce caries risk (**Table 75.5**).[46,52] The level of fluoride supplementation is determined by the child's age and the concentration of fluoride in the water supply. Well water should be tested for fluoride prior to supplementation. In 2010, the American Dental Association's (ADA) Council on Scientific Affairs released evidence-based recommendations for fluoride supplementation (see Table 75.5).[46] The recommendations emphasized that fluoride supplementation be limited to children at high caries risk and exposed to inadequate fluoride in water.[46] Age, caries risk, and other sources of fluoride should be considered before recommending fluoride supplements. Fluoride supplementation is not recommended in areas where the drinking water is at or above 0.6 ppm fluoride. Although evidence exists that "the incidence of caries in both the primary and the permanent teeth of children can be reduced with the use of dietary fluoride supplements, evidence also indicates that the use of dietary fluoride supplements during tooth development increases the potential risk of developing very mild enamel fluorosis."[46]

In general, soy-based formulas have a higher fluoride concentration than milk based formulas. Unlike ready-to-use formulas, which have a fluoride concentration of approximately 0.15 ppm (milk based) and 0.21 ppm (soy based), powdered or liquid-concentrate formulas must be reconstituted with water. If the water contains 1 ppm fluoride, the prepared formulas would have mean concentrations of 1.03 ppm (milk based) and 1.07 ppm (soy based) for powdered versions and 0.64 ppm and 0.75 ppm, respectively, for liquid concentrates.[53] The fluoride concentration of breast milk and cow's milk is low (0.1 ppm). Liquid fluoride supplements (drops) are recommended for infants who are exclusively breastfed and for young children who live in nonfluoridated water areas and are at high risk of dental caries.[46] Health professionals should consult the ADA evidence-based practice guidelines for fluoride supplementation.[46,53]

Dental Fluorosis

Despite fluoride's ability to reduce caries risk, its widespread availability in drinking water, other beverages, and dental products has resulted in an increased risk of dental fluorosis.[46,47,53-56] Fluorosis is characterized by white striations or opaque flecks, white or brown staining, or, in severe cases, pitting of the tooth enamel. Fluorosis occurs during critical periods of tooth

Table 75.5 Supplemental Fluoride Dosage Schedule (Mg/D[a]) According to Fluoride Concentration of Drinking Water

Concentration of Fluoride in Drinking Water (ppm)			
Age	<0.3	0.3–0.6	>0.6
6 mo–2 y	0.25 mg	0	0
3–6 y	0.50 mg	0.25 mg	0
6–16 y	1.00 mg	0.50 mg	0

ppm, parts per million.

[a]2.2 mg sodium fluoride contain 1 mg fluoride.

Data from Rozier RG, Adair S, Graham F et al. J Am Dent Assoc 2010;141:1480–9.

development; the most critical time appears to be the first or second year of life when maturation of upper anterior permanent teeth enamel occurs.[57] Using National Health and Nutrition Examination Survey (NHANES) 1999 to 2004 data, Beltran-Aguilar et al., found an increase in the prevalence (40.7% vs 22.6%) of fluorosis among adolescents aged 12 to 15 years compared with the same age cohort in 1986 to 1987.[56] The prevalence (28.4%, 14.3%, 8.7%) of fluorosis declined with age (20–29, 30–39, 40–49 years, respectively) in the NHANES 1999-2004 cohort.[56] Although the prevalence of fluorosis has increased, severe fluorosis was noted in less than 1% of cases. Current levels of fluorosis are primarily a cosmetic concern and do not increase susceptibility to caries or impact oral health-related quality of life.[47]

The U.S. Department of Health and Human Services (DHHS) recommended an optimal water fluoride concentration of 0.7 mg/L beginning in 2015 to reduce caries risk and limit fluorosis.[47] The lower fluoride concentration acknowledges the impact of fluoride availability from other sources. The fluoride content of foods (hot cereals, instant potatoes) and beverages (concentrated juices, powdered beverages) reconstituted or diluted with fluoridated water is higher than the original products.

The Iowa Fluoride Study, which is the longest longitudinal study of fluoride exposure (dietary and nondietary), fluorosis, and caries conducted in the United States to date, recruited new mothers between 1992 and 1995 and followed the mothers and their infants through age 20 years, periodically assessing dental status and conducting food frequency questionnaires to assess fluoride consumption.[55] Supplemental use of fluoride, along with fluoride from toothpaste, were also recorded. In infants aged 3 to 9 months, most fluoride came from reconstituted infant formulas and other beverages with added fluoridated water. In toddlers (aged 16–36 months), dentifrices contributed a substantial amount of fluoride. Fluoride intake of those with mild fluorosis was higher than those without fluorosis at age 5 years. Almost all of the identified cases of fluorosis in this longitudinal study of 600 children were mild. The authors found no evidence to recommend the avoidance of fluoridated water for reconstituting infant formulas. They suggested that parents concerned about the risk of mild fluorosis who use reconstituted infant formulas consult their dentist or physician for recommendations regarding the use of water with lower fluoride concentration. In addition, parents should supervise the use of fluoridated toothpastes to ensure that no more than a "pea-sized" amount is used, and that the toothpaste is expectorated rather than swallowed.[55]

In early 2011, the ADA Council on Scientific Affairs released evidence-based clinical recommendations regarding fluoride intake and infant formulas.[53] For infants fed exclusively powdered or liquid concentrate formulas that require water reconstituted for use, the clinical recommendations are consistent with the findings of Levy et al.,[55] and the ADA Research Foundation study.[54] The recommendations supported "the continued use of powdered or liquid concentrate infant formulas reconstituted with optimally fluoridated drinking water" provided the individuals are aware of the risk of dental fluorosis.[53] They emphasized the importance of consulting with dentists and physicians about the use of fluoridated water and infant formulas, and advised use of water with low fluoride concentration for reconstituting formulas when there was concern about the risk of fluorosis.[53] It is also prudent to consider other sources of fluoride consumed by toddlers and children, such as reconstituted foods and fluids made with fluoridated water and the use of fluoridated dentifrices.

Diet, Nutrition, and Oral Infectious Diseases

Oral Microbiome

Both dental caries and periodontal disease are infectious diseases; the oral microbiota is essential but not sufficient for either disease to occur. The oral microbiome is a complex community of commensal, symbiotic, and pathogenic micro-organisms. The mouth is initially colonized by bacteria at birth, subsequently colonized by bacteria from the environment, influenced by infant feeding mode, and susceptible to environmental challenges (i.e., diet, tobacco smoke, stress, mechanical disruption, medications). A healthy microbiome is characterized by a predominance of commensals supported by the host to resist colonization by pathogenic microbiota. Disruption of the host-immune response by high added sugar intakes, marginal nutrient intakes, alcohol exposure, tobacco exposure, inflammation, and/or invasion of pathogenic microbiota leads to dysbiosis and an increased risk of oral infectious diseases, including dental caries and periodontal disease.

Dental Caries

Dental caries is "the most common chronic disease in the United States and among the most common diseases in the world" and one of the most frequently occurring preventable infectious diseases of the oral cavity, globally affecting up to 90% of individuals in some countries.[5,58,59] Caries is a major cause of tooth loss in children and adults. In 2000, the U.S. Surgeon General reported that approximately 80% of caries in children and adolescents was present in approximately 25% of the population.[5] According to data from the 2015–2016 NHANES, the prevalence of total dental caries was 21.4%, 50.5%, and 53.8%, respectively, while the prevalence of untreated caries was 8.8%, 15.3%, and 13.4%, respectively, for children aged 2 to 5 years, 6 to 11 years, and 12 to 19 years of age.[60] The prevalence of caries (90% for adults aged 20–64 years and 96% for adults 65 years and older) was substantially higher in adults than in youth, according to 2011–2016 NHANES data.[59] The prevalence of both total and untreated caries tends to be higher among low-income children and adults than among those of higher income strata in the United States and globally.[58,61,62]

Since the mid-1970s, the incidence of dental caries has declined, mainly because of preventive measures such as fluoride and dental sealants.[63-65] Dietary factors increase the risk of dental caries; additionally, preserving the integrity of dentition is essential for overall nutritional status. Untreated tooth decay can lead to pain, eventual tooth loss with subsequent masticatory dysfunction, compromised dietary intake, missed school and work, and lower quality of life.

The rate of tooth loss, or edentulism, has continued to decline as a result of fluoride exposure and better oral healthcare access.[4,58] The prevalence of edentulism for adults aged 20 to 64 years (2%) is less than for adults aged 65 years and older (9%). The risk of edentulism is higher for adults who are poor, less educated, or current smokers.[59] For adults aged 65 years and older, the prevalence of edentulism was 34% among the poor, 35% for those with less than a high school education, and 43% for current smokers.[59]

Caries is a dynamic process that has three phases: (a) demineralization (loss of mineral when plaque pH falls to less than the critical pH of ~5.5), (b) equilibrium, and (c) remineralization (addition of mineral when plaque pH rises above the critical pH) of tooth enamel. During the caries process, pathogenic oral bacteria ferment carbohydrates to produce acid at the tooth surface. The released acid demineralizes the tooth surface and, if unchecked, can proceed to form an overt lesion. During the initial stages of the caries process, incipient or white spot lesions develop. Between periods of microbial fermentation, calcium, phosphorus, and fluoride that have been released from the tooth enamel or present in saliva can be redeposited in the enamel to remineralize the tooth. A clinical cavity (caries) forms in response to progressive demineralization and is the final stage in the disease process.

The extent of caries in a population is measured by decayed-missing-filled teeth (DMFT) or tooth surfaces (DMFS), representing the sum of the number of teeth (out of 20 primary or 32 adult teeth) or permanent tooth surfaces (out of a possible 128 surfaces on 32 teeth) that are decayed, missing, or filled.[66] Caries can occur on the occlusal, chewing, smooth and/or root surfaces of the tooth.

Role of Carbohydrates in Dental Caries

Dental caries is a multifactorial oral infectious disease of dietary origin. Fermentable carbohydrates, along with the oral bacteria and tooth surface, are an essential component of the etiology. Environmental factors, including fluoride exposure, oral hygiene practices, dietary behaviors, the presence and adequacy of saliva, and immune function impact caries risk.

Nutrient intakes, carbohydrate exposures, and dietary behaviors have a primary role in caries risk.[67] Nutrition has a systemic effect, whereas carbohydrate exposures and dietary behaviors have local effects. Nutrition during tooth development can affect the tooth's integrity and salivary glands, subsequently impacting the tooth's susceptibility to a microbial challenge. Systemically, nutrient deficiencies adversely impact the volume, antibacterial properties, and physiochemical properties of saliva. Systemic diseases, medications, and illicit substances (i.e., methamphetamine) affecting the integrity of the oral cavity and salivary flow can also impact oral health and increase caries risk.[67,68]

Fermentable carbohydrates include sugars and starches. Sugars are found naturally within or are added to foods and beverages.[69-71] Examples of added sugars include white or brown sugar, honey, molasses, maple, malt, corn syrup or high-fructose corn syrup, fructose, and dextrose.[70,71] Starches digested by salivary amylase to oligosaccharides are susceptible to fermentation by the oral bacteria. According to Lingstrom et al., gelatinization of starch increases its susceptibility to breakdown by salivary amylase into maltose, maltotriose, and dextrins.[72] Maltodextrins produced during food processing are also susceptible to fermentation.[73]

Oral bacteria are unable to ferment protein or fat. Foods composed primarily of protein and fat are considered noncariogenic and include meats, eggs, fish, some vegetables, cheeses, and nuts. They are also unable to ferment most nutritive or non-nutritive sweeteners, and both are considered low caries risk. Oral bacteria do not ferment most sugar alcohols; however, some, including sorbitol, can be slowly fermented. Xylitol was previously considered caries protective through mechanisms that promote remineralization and decrease microbial growth. However, a recent systematic review suggests otherwise, finding low-quality, insufficient evidence to support the hypothesis that xylitol-containing products protect against caries.[74]

Epidemiologic surveys, animal experiments, and early controlled human studies linked sugars and starches to developing dental caries. The exposure time, that is, the length of time that sugars and starches spend in the oral cavity, is likely the most important factor determining caries risk. However, measurement of exposure time is difficult; both quantity of carbohydrate and frequency of exposure are typically used as proxy measures. The relative validity of quantity vs frequency of intake has been debated. During each eating event, the plaque pH may fall below the critical pH and remain low for up 30 minutes following the last sip or bite, thereby increasing the risk of demineralization. In 2009, Anderson et al., conducted an evidence analysis of sucrose intake and suggested that the frequency of exposure was more important than the quantity of intake.[75] Moynihan and Kelly recently conducted a systematic review on behalf of the World Health Organization to identify the association between the quantity of sugar intake and dental caries.[76] They reported that evidence of moderate quality supported the hypothesis that sugar intake is associated with dental caries and suggested that caries are lower when free sugar intakes is restricted to <10% of energy intake.[76] Furthermore, the quantity and frequency of intake are not independent of each other; the interaction between the frequency and quantity is associated with caries risk.[77]

While sugars and starches are the substrates for oral bacteria, we consume foods and beverages. Palmer et al., reported that the frequency of food exposure was higher in young children with severe early childhood caries compared with their caries-free peers.[78] Marshall et al., similarly reported that higher snack and daily total eating events increased caries risk at ages 1 to 5 years in participants of the Iowa Fluoride Study.[79] In the same study, children with caries at 5 years had higher median sugar-sweetened beverage (SSB) intakes at 2, 3, 4, and 5 years and lower median milk intakes at 2 and 3 years.[80] Hancock et al., conducted a systematic review and reported that higher intakes of processed sugar- and starch-containing foods increased caries risk in children and preadolescents.[81]

The 2001 National Institutes of Health Consensus Development Conference on Caries reported that two studies supported a strong diet–caries relationship, 16 supported a moderate relationship, and 18 supported a weak relationship.[82] The report emphasized that the results of the included studies differed from the results of similar studies published before widespread fluoride availability, concluding that "sugar consumption is likely to be a more powerful indicator for risk of caries infection in persons that don't have regular exposure to fluoride".[68] This relationship has been further supported by other investigators.[75,83,84] Investigation of fluoride's role in reducing caries risk associated with added sugars is complicated by the fact that added sugar-containing foods and beverages are often made with fluoridated water. Marshall et al., recently reported that each additional 8-ounce intake of SSBs throughout childhood increased the risk of expected caries (decayed and filled surfaces) at age 17 years by 42%, while controlling for toothbrushing and total fluoride; the fluoride content of the SSB was excluded from total fluoride for analyses.[84]

Factors Affecting Cariogenicity

The cariogenic potential of a food or beverage is influenced by the clearance time from the oral cavity, nutrient composition, acid content, sequence intake in the meal, and nature of the foodstuff.[85] Clearance time of foods and beverages from the oral cavity impacts caries risk. Clearance time is based on several factors inherent to the individual person and the food or beverage. Individual factors include oral motor skills (the ability to form a cohesive bolus), pocketing behaviors, and beverage swishing. Carbohydrate containing foods lost in the oral cavity, or pocketed, and beverage swishing prolong exposure time. During usual intake, beverages clear rapidly from the oral cavity. The clearance time of foods is influenced by the degree of processing and the presence of sugar. Minimally processed raw and cooked starches (rice, potatoes) are easily cleared and have low retention times. Baked goods composed of minimally processed starches with added sugars (cookies, cakes) and processed foods composed of hydrolyzed starches with (sweetened cereals, commercial pastries) or without added sugars (unsweetened cereals, salty snacks) are more difficult to clear from the mouth, resulting in increased retention times. Retention of the baked sugar-starch combinations within interproximal spaces increases cariogenicity, compared with sugar alone.[67,86,87] Retentiveness is not synonymous with stickiness. Marshmallows and jellies are sticky; however, their retentive properties are low, and they are cleared from the mouth more quickly than the starches within cookies, cakes, and chips. Kashket et al., demonstrated these principles in research that explores the salivary clearance rates of high-starch and high-sugar foods.[88,89]

The nutrient composition of foods and beverages also influences caries risk. Foods containing fats and proteins dilute carbohydrate exposure, reducing the caries risk. Other components within food decrease the enamel solubility (fluoride within tea, phytate, oxalate, milk protein) or stimulate salivary secretion, facilitating enamel remineralization (calcium, protein). Cheese seems to both prevent demineralization and promote remineralization.[90] The protective properties of cheeses are attributed to their texture, which increases salivary secretion, and their protein, calcium, and phosphorus content, which neutralizes plaque acids.[91]

Foods that stimulate chewing neutralize and increase the intraoral pH, thus facilitating remineralization and decreasing caries risk. Chewing gum, particularly that sweetened with xylitol, xylitol/sorbitol mixtures, or non-nutritive sweeteners, can reduce dental plaque and improve oral health.[92,93] Gum chewing stimulates salivary flow, which clears food debris and releases salivary buffers that can neutralize acids produced during carbohydrate fermentation. A systematic review and meta-analyses suggested that chewing sugar-free gum reduces caries risk compared with nonchewing controls, although the feasibility of using sugar-free gum as a public health measure is unclear.[94]

The sequence of eating foods in a meal affects the magnitude of the drop in plaque pH. When foods or beverages containing fermentable carbohydrates are followed by foods high in protein and fat or stimulate saliva, the fall in plaque pH is arrested, and the pH returns to normal. For example, the fall in pH in response to sugared coffee is rapid but will rise if followed by a food having a more neutral (nuts) or higher (cheese) pH. The risk of demineralization is lessened by pairing acidogenic foods and beverages with cariostatic (e.g., nuts) and anticariogenic (e.g., cheese) items.

The presence of adequate saliva in the mouth protects against caries. Unstimulated saliva is present between eating and drinking events, while stimulated saliva is produced during chewing foods, upon exposure to acidic or tart foods and beverages, or in anticipation of preferred foods. Previously healthy individuals who develop xerostomia following head and neck radiation and as a medication side effect often experience caries associated with the decline in saliva. Salivary flow declines during sleep; drinking beverages or consuming foods with sugars before or during sleep increases caries risk.

Saliva protects against caries by one of four mechanisms. First, saliva prevents the aggregation of bacteria at the tooth's surface and speeds the clearance of food particles, starch, and sugars from the mouth. Second, saliva dilutes plaque acids and contains buffering agents (proteins, bicarbonates, and phosphates), which neutralize plaque acids. Third, immunoglobulins present in saliva protect the teeth by depressing bacterial activity. Finally, the presence of calcium, phosphate, and fluoride ions in saliva promotes the remineralization of tooth enamel.

Root Caries

When gingival tissues recede, the root surfaces of teeth are exposed to the oral environment. Because roots lack the protective enamel layer, they are highly susceptible to dental caries. Reports from NHANES 1999 to 2004 suggest that, among adults aged 20 years and older, approximately 18% had root caries, with a prevalence of 31.6% among those age 60 years and older.[95] Older adults at the highest risk of developing root caries include individuals with coronal caries, gingival recession, low salivary flow, low fluoride exposure, frequent intake of fermentable carbohydrates, and poor oral hygiene.

Root caries are a form of caries and susceptible to the same dietary risk factors as coronal caries.[96] In a 2-year longitudinal study of older adults in Boston, MA, individuals in the highest quintile of root caries had significantly higher intakes of sweetened beverages, solid fermentable carbohydrates, and starches than those who were free of root caries.[97] Adults free of root caries ate 50% more cheese and drank 25% more milk than their peers with caries.

Early Childhood Caries (ECC)

Caries in infants and toddlers is a preventable disease associated with inappropriate feeding practices. Historically referred to as nursing caries or baby bottle tooth decay, ECC is defined as the presence of one or more decayed, missing, or filled tooth surfaces in a child less than 6 years of age.[98] Severe ECC is defined as a smooth surface lesion in children less than 3 years or multiple lesions in children 3 to 5 years of age.[98] Children with ECC are at greater risk of developing additional caries than children who are caries free.[99]

In addition to the dietary risk factors previously discussed, risk factors for ECC include allowing access to beverages, particularly SSBs, by bottle or sippy cup throughout the day, night-time feedings, and prolonged on-demand breastfeeding. Inadequate nutrient access in utero increases risk of enamel hypoplasia and caries risk in the primary teeth. Preterm delivery and low socioeconomic status are also associated with increased risk of caries. Social determinants of oral health include health access inequities, food insecurity, pollution, and water insecurity.

Assessment of Diet-Related Caries Risk

Dental caries is a diet-dependent disease; assessment of diet-related caries risk factors should consider the nature, quantity, and frequency of fermentable carbohydrate-containing foods and beverages.[100,101] Not all patients need a full dietary assessment; however, all individuals should be screened to identify those whose food or beverage choices and/or eating behaviors increase their caries risk. Diet-related caries risk assessments should be completed on individuals identified at risk by screening and/or presenting with active caries. The objectives of a diet-related caries risk assessment are to identify alternative dietary practices to reduce caries risk. A diet-related caries risk screen should identify SSB intakes, meal/snack patterns, and meal structure.[100,101] If the individual consumes more SSBs than currently recommended, has more than six meal/snacks per day, or has unstructured meal patterns, a complete diet-related caries risk assessment should be completed.

A complete assessment expands the diet-related screen to identify special dietary choices, recent dietary changes, 24-hour recall, social determinants of oral health that impact ability to access food (transportation, resources), and medication use causing xerostomia and/or impacting appetite. Understanding the patient's dietary preferences and recent changes in diet and/or dietary habits is necessary to make patient-sensitive recommendations. Knowledge gained from the 24-hour recall allows for identification of specific recommendations to reduce the quantity or frequency of fermentable carbohydrate exposures and provide anticipatory guidance considering caffeine and/or energy intake adjustments. For patients with difficulty accessing appropriate food substances, guidance or referral to appropriate resources is necessary to enable appropriate behaviors. Identification of medication side effects might necessitate referral to their dentist, physician, or pharmacist for appropriate compensatory strategies.

Several brief diet-related caries risk assessment tools have been developed that target specific dietary behaviors such as frequent eating/drinking or between meal exposures.[102,103] The simplicity of the tools facilitate implementation, but the lack of depth and specificity of information collected hinder individual dietary recommendations or anticipatory guidance.

Food choices and dietary behaviors promoting oral health and reducing caries risk include:

1. Selecting an age-appropriate diet consistent with the USDA Dietary Guidelines [71] to ensure adequate nutrient intake.
2. Selecting minimally processed foods as the foundation of dietary intake; reserve ultra-processed sweets and snack foods for limited consumption at special occasions.
3. Limiting SSBs to age-appropriate recommendations and consuming within a 30-minute time period. Age-appropriate recommendations for SSB intake are based on ~5% of total energy intake from added sugars and include: 0 oz/day for <3 years; 0 to 4 oz/day for 3 to 5 years; 0 to 6 oz/day for 6 to 12 years; 0 to 8 oz/day for 13+ years.[100]
4. For young children, limiting 100% juice intake to the American Academy of Pediatrics' recommendations and consuming at meal or snack occasions.[104] Provide water between meals and snacks.
5. Consuming all carbohydrate-containing foods and beverages at three structured meals and two to three structured snacks to limit exposure frequency. Meal and snack occasions should not exceed 30 to 40 minutes to limit exposure time.
6. Offering nonfood rewards for preferred behaviors; offering sweets and snacks places inappropriate emphasis on such food items.

Periodontal Disease

The periodontium or periodontal tissue supports the tooth and includes the gingiva, periodontal ligament, and alveolar bone. The classification of periodontal diseases and conditions was revised in 2017 during a World Workshop coordinated by the American and European Periodontal professional organizations.[105] The three contemporary periodontal disease classifications include 1) periodontal health and gingival disease, 2) periodontitis, and 3) other conditions impacting the periodontium. Dental biofilm-induced gingivitis and periodontitis are infectious diseases. Gingivitis is less severe than periodontitis; patients with gingivitis can return to a state of health while patients with periodontitis continue to have disease, although their status might be stable. Periodontal disease is precipitated by a shift from symbiotic to pathogenic microbiota in the gingival crevice; the dysbiosis leads to inflammation (gingivitis). Unchecked and exacerbated by an impaired host immune response, tissue inflammation is exacerbated, bone is resorbed, and tooth loss may occur (periodontitis). Periodontal disease is the primary cause of tooth loss in adults. Approximately 47% of adults over 30 years of age and 70% of adults 65 years and older have some form of periodontal disease.[106] As with dental caries, individuals with lower incomes, less education, and/or currently smoking are at higher risk than their peers with higher incomes, more education, and/or not smoking, respectively.

Periodontitis and Nutrition

Numerous investigations between diet factors, nutrient intakes, and periodontal disease suggest that neither dietary factors nor nutrient deficiencies cause periodontal disease. However, the evidence suggests that individual nutrient deficiencies and overall low diet quality increases risk. Adequate nutrient intakes are necessary for tissue health and to support a functional immune response. Boyd and Madden summarized the nutrient effects on the periodontium in **Table 75.6**.[107] Individual nutrients, including vitamin C and vitamin D, are associated with increased risk of or severity of periodontal disease. For example, Chapple et al., reported that adults in NHANES 1988–1994 in the upper quintile of serum vitamin C concentration had 47% lower odds of having severe periodontitis than their peers in the lower quintile.[42] A recent systematic review with meta-analyses reported that serum vitamin D concentration was significantly lower in patients with chronic periodontitis than their healthy peers.[108] Higher overall diet quality, measured using the Alternative Healthy Eating Index, was associated with lower odds of severe periodontitis in adult Hispanic participants from geographically diverse U.S. metropolitan centers.[109]

Periodontitis and Obesity

Obesity and obesity-related diseases (Type 2 diabetes and metabolic syndrome) are associated with periodontal disease. Obesity-associated chronic, systemic inflammation and/or impaired immune responses could link the two diseases.[110] Reeves et al., reported that both total body weight and waist circumference were associated with periodontal disease in adolescents, aged 17 to 21 years, in NHANES 1988 to 1994; adolescents with periodontitis were 7 kg heavier and had waist circumferences 8 cm greater than their peers without periodontitis.[111] Similarly, NHANES 1988–1994 adults with metabolic syndrome were more likely to have periodontal disease, with

Table 75.6 Nutrient Effects on the Periodontium

Nutrient	Functional Changes Resulting from Inadequate Intake	Groups at Risk of Inadequate Intake	Food Sources of Nutrient
Protein	Compromised antibacterial properties of saliva Impaired acute-phase response to infection ↓Neutrophil function Lag period in the initiation of wound healing ↓Collagen synthesis	Persons with poorly controlled diabetes Patients with advanced cancer Patients with advanced AIDS Patients fasting ≥4 d or those with chronic poor nutrient intakes	Meat Dairy Legumes
Vitamin A	↓Production of γ-interferon ↓Collagen synthesis ↓Epithelialization ↓Incidence of infection	Patients with cystic fibrosis Patients with advanced AIDS Persons with GI conditions with malabsorption Persons taking weight-loss medication: orlistat	Fortified dairy foods Dark green, leafy vegetables Meat and dairy
B-complex vitamins	Inability to produce adequate energy ↓Protein synthesis, including DNA and RNA Breakdown of the mucosal barrier to pathogens	Persons with HIV infection Elderly vegans (vitamin B_{12}) Postgastrectomy patients Patients taking medications: H_2 blockers, phenytoin, methotrexate	Enriched breads and cereals Green, leafy vegetables Meat and dairy
Vitamin C	↓Neutrophil function Breakdown of the mucosal barrier to pathogens ↓Collagen synthesis	Smokers Substance abusers Elderly persons Persons with chronic disease People who avoid fruits and vegetables	Citrus fruits Dark green, leafy vegetables Potatoes Cantaloupe
Vitamin D	Impaired absorption of calcium	Elderly adults in northern latitudes; those with little sun exposure	Fortified milk Eggs Liver
Vitamin E	↓Overall immune response ↓Antibody production	Persons with GI conditions with malabsorption Patients with advanced AIDS	Nuts and seeds Polyunsaturated oils Whole grains
Vitamin K	↓Bone density and possibly bone strength	Persons receiving anticoagulant therapy	Green, leafy vegetables Liver
Boron	Impaired wound healing Possible association with bone calcification		Legumes Fruits, especially dried Vegetables
Calcium	Inadequate formation of peak bone mass Accelerated bone loss postmenopausally Osteoporosis Possible association with tooth loss	Young women Postmenopausal women	Milk and milk products Tofu processed with calcium Legumes Dried fruit
Copper	↓Tensile strength of collagen ↑Bone fragility ↓Proliferation neutrophils	Persons with an increased intake of antacids Persons taking megadoses of iron/zinc Alcohol abusers Patients with cystic fibrosis Patients with short bowel syndrome Postgastric bypass patients Persons taking medications: dexamethasone, penicillamine	Green, leafy vegetables Whole grains Nuts Meats Whole grains Shellfish Chocolate Organ meats

(continues)

Table 75.6 Nutrient Effects on the Periodontium *(continued)*

Nutrient	Functional Changes Resulting from Inadequate Intake	Groups at Risk of Inadequate Intake	Food Sources of Nutrient
Iron	↓Neutrophil phagocytic activity ↓Proliferation lymphocytes	Young children Women of childbearing age	Meat Eggs Legumes Dried fruit
Magnesium	More rapid development of osteopenia	Persons with chronic alcoholism Medications: diuretics Elderly persons Postmenopausal women Persons with diabetes	Green leafy vegetables Whole grains Nuts
Zinc	↑Susceptibility to infection ↓Protein synthesis, including DNA and RNA	Elderly persons? Persons with alcoholism	Meats Whole grains

AIDS, acquired immunodeficiency syndrome; *GI*, gastrointestinal; *HIV*, human immunodeficiency virus.

Modified from Boyd LD, Madden TE. In: Palmer CA, ed. Diet and Nutrition in Oral Health. Englewood Cliffs, NJ: Prentice-Hall 2003:202–12.

abdominal obesity being the only associated criterion.[112] Body mass index and waist-to-hip ratio were associated with periodontal disease progression in male participants of the Veterans Administration Longitudinal Study.[113]

Alveolar Bone Health and Dentition

Alveolar bone remodels in response to occlusal forces associated with chewing. With tooth loss, the alveolar bone is no longer required for tooth support; as a consequence, bone resorption is accelerated and ridge height is diminished. Bone loss is greatest during the first 6 months following tooth extraction. The reduction in residual ridge height is more pronounced in women than in men, and resorption is greater in the mandible than in the maxilla. Positive calcium balance could be especially important to help preserve the integrity of the residual ridges of edentulous postmenopausal women. Low vitamin D has been associated with alveolar bone loss in adult males.[114] Calcium and moderate vitamin D supplementation improve periodontal health and reduce tooth loss; however, their ability to prevent alveolar bone resorption is unclear.[115-117] Severe mandibular resorption makes it difficult to construct a mandibular denture with good stability and retention, or to place an implant. Individuals electing implant placement following tooth loss for either single tooth replacement or to support a denture may undergo alveolar ridge preservation at the time of extraction to preserve bone or later bone grafting to replace lost bone to ensure adequate implant support.

Periodontitis and Osteoporosis

Periodontal disease and osteoporosis are both inflammatory diseases characterized by bone loss. Alveolar bone loss, alveolar bone density, and clinical attachment loss are each correlated with systemic bone loss.[118] The alveolar process (crest of the maxilla and mandible) is composed primarily of trabecular bone. Histologically, it is the same type of bone found in the distal radius, neck of the femur, and vertebrae. When negative calcium balance occurs in the body, calcium is more easily mobilized from skeletal sites consisting of trabecular rather than cortical bone. Thus, the alveolar bone provides a potential labile source of calcium available to meet other tissue needs. Because the alveolar process is thought to undergo resorption before other bones, change detected in the alveolar process may be used for early diagnosis of osteoporosis. Jabbar et al.,[119] in a cohort of postmenopausal women, found that periodontal disease was more frequent in women with osteoporosis than in a matched cohort without osteoporosis. In women, aveolar bone mass is highly correlated with total bone mass; women with low bone density have fewer teeth. Women with severe residual ridge resorption have osteopenia on the iliac crest, and those with severe postmenopausal osteoporosis are three times more likely to be edentulous than physiologically normal control individuals.[120] Dietrich found that higher serum 25(OH)D_3 concentration was associated with decreased loss of tooth attachment in adults more than 50 years old.[121] Payne et al., over a 24-month period, noted that women with osteoporosis lost significantly more alveolar bone than those without the condition.[122]

Diet assessment for periodontal disease evaluates the adequacy of nutrient and energy intake. A diet consistent with the USDA Dietary Guidelines food group recommendations, providing appropriate energy balance to support tissue health is recommended.[71]

Tooth Wear: Enamel Erosion

Mechanical and chemical forces at the tooth surface can result in tooth wear and subsequent enamel loss. Abrasion is a type of mechanical wear that occurs when the tooth surface is in repeated contact with a nontooth material. For example, enamel is lost during aggressive tooth brushing, eating particularly fibrous foods, or habitual chewing on hard objects. Abfraction is a type of mechanical wear associated with uneven biting forces. For example, if teeth are not aligned appropriately, repeated biting or chewing will result in a cervical lesion at the juncture of coronal and root surfaces in response to the biting force. Dental erosion is defined as the loss of surface enamel or dentin in response to repeated acid exposure. Acids might be of internal or external origin but are not associated with bacterial fermentation of carbohydrates.

Internal acids are of gastrointestinal origin, and exposure generally occurs during repeated vomiting, rumination, or gastrointestinal reflux.[123] Episodes of vomiting during bacterial or viral illness are generally of short duration and unlikely to cause erosion. However, vomiting associated with eating disorders, including anorexia nervosa, bulimia nervosa, and binge eating disorder, is a risk factor for erosion. Similarly, the presence of rumination syndrome and untreated gastroesophageal reflux increases risk of erosion. Silent or asymptomatic reflux is of particular concern, as the patient might not recognize the reflux symptoms. Medical conditions including pregnancy, obesity, and scoliosis, as well as some food and beverage items, increase risk of or exacerbate existing reflux disorders.

The most obvious clinical symptom of erosion associated with intrinsic acids is loss of tooth enamel (perimolysis) and dentin on the lingual and incisal surfaces of anterior teeth and occlusal surfaces of posterior teeth. Perioral stomatitis, along with swollen salivary, parotid, and submandibular glands, can occur as a result of hypersalivation. The degree of gland swelling is typically relative to the frequency of vomiting. Redness of the throat and calluses on the knuckles may also be additional signs of self-induced vomiting. Following exposure to intrinsic acids, rinsing with water or bicarbonate is recommended to cleanse the mouth of acid. Toothbrushing immediately following exposure to intrinsic acids can increase risk of mechanical abrasion of soft enamel and is not recommended for at least 20 minutes following the vomiting episode.

Extrinsic acids contributing to dental erosion are typically of dietary origin, although environmental contaminants can increase the risk of erosion.[123] Acidic fruits, fruit juices, vegetables, carbonated beverages, sports drinks, tart or sour candies, and vitamin C supplements can contribute to erosion. However, few observational investigations have documented erosion in response to acidic food or beverage consumption. Generally, an induced salivary response to the acidic food or beverage clears the item, buffers the acid, and facilitates remineralization. The lack of supporting evidence linking acidic foods and beverages to erosion should not be interpreted as evidence of no causation. Indeed, erosion has been noted in individuals with xerostomia or behaviors that increase the length of exposure to the acidic food or beverage. Behaviors noted to increase exposure time, including frequent intake, prolonged eating episodes, repeated swishing or holding acidic beverages, and pocketing of acidic foods such as lemons.

Identification of the etiology of erosion is necessary to prevent further tooth damage as well as maintain systemic health. Because eating disorders, rumination, and gastroesophageal reflux impact systemic health, morbidity, and mortality, all patients with erosion should be screened for systemic conditions and referred, as appropriate. If the patient does not have a medical condition causing the erosion, dietary food and beverage choices and eating behaviors should be investigated. Eating behaviors that prolong exposure to the acidic food or beverage require modification to reduce exposure time. In addition, consuming acidic foods and beverages as part of a mixed diet will reduce the exposure.

Diabetes and Oral Health

The relationship between oral health and disease and diabetes is multifaceted; blood glucose control impacts the mouth, and oral infectious diseases (e.g., caries, periodontal disease, fungal infections) may likewise impact blood glucose concentration, ultimately affecting diabetes control. Oral complications of diabetes include increased risk and incidence of soft tissue and fungal infections, caries, periodontal disease, delayed wound healing, hyposalivation, xerostomia, alterations in taste, and burning mouth or tongue.[124-126]

Caries risk, in this case, may be due to salivary hyperglycemia, hyposalivation, and xerostomia. Candidiasis is a fungal infection associated with hyperglycemia, alterations in immune response, and diminished salivary flow.[125] Periodontal disease is sometimes referred to as the "sixth complication of diabetes"[127] because individuals with diabetes are at greater risk than those without diabetes, and disease severity is associated with glucose control.[128] Tasting alterations may occur secondary to xerostomia, hyposalivation, altered saliva, burning mouth, and fungal infections. All potential causes merit exploration to determine if their cause is due to a comorbid condition(s) beyond diabetes.

Glycemic control is essential to reduce the incidence and severity of oral complications of diabetes. It should be evaluated in persons with prediabetes, type 1 or type 2 diabetes who complain of xerostomia, burning tongue, dysgeusia, or present with fungal infections. Appropriate intervention includes lifestyle management (diet, hygiene, physical activity). Adherence to an individualized diet plan, conscientious and effective oral hygiene, topical fluoride, and smoking cessation, when indicated, can maintain oral health integrity throughout life.[126]

Oral Surgery

The type, extent, and location of the oral surgery, the patient's nutritional well-being pre-operatively, and any other comorbid conditions contribute to the degree of oral impairment following surgery and nutrition care needs. In general, except for maxillofacial or orthognathic surgeries, oral intake is compromised for a relatively short period, and the risk of nutrient deficiencies is low in individuals without pre-existing malnutrition or other nutritional disorders.

Maxillofacial Trauma Surgery

For the patient undergoing maxillofacial trauma surgery, the ability to consume a solid food diet may be impaired for several weeks to months. Energy needs may also be increased, depending on the trauma or condition that led to the surgery. For individuals with severe trauma who have a fractured jaw, the subsequent surgery and fixation of the jaw increase energy and macro- and micronutrient needs considerably because of the trauma and the need for nutrients for wound healing. As a result, weight loss is common; limitations in eating ability, reduced appetite due to pain and swelling, and the need for a liquid diet contribute to weight loss in these patients. If patients cannot meet their energy and nutrient needs orally, enteral nutrition support with a tube feeding is needed.[129] Early intervention with nutrition is essential to minimize weight loss, maximize wound healing and recovery, and reduce the risk of infection.

There is a paucity of research exploring nutrition status in patients undergoing maxillofacial and other oral surgeries. Christensen et al., examined weight loss in patients aged 15 years and older following surgery for mandibular fractures.[130] Study participants in the trauma intensive care unit (ICU) lost, on average, 8.8% of their body weight, while those who were not in the ICU lost a mean of 4.6% of their weight, with the greatest weight loss at 7 weeks postsurgery. Limitations of this and other studies include their retrospective nature and the fact that patients undergoing maxillofacial trauma surgery may experience other trauma and have other comorbid conditions. In a second prospective study, the same authors[131] explored weight change and markers of nutrition status, including prealbumin and albumin, in patients with mandibular fractures. Similar to their prior findings, weight loss was, on average, 4.1% of the presurgery weight. There were no significant changes in serum albumin concentration; prealbumin concentration was normal but low at baseline, likely due to inflammation from the trauma; they rose significantly over the 9 weeks of the study.

There are no published guidelines for nutrition support or supplementation for patients undergoing maxillofacial or other oral surgery. Following surgery for mandibular fractures, a liquid diet consumed with a straw is needed, so the jaw remains immobilized, and healing can occur. The diet must provide sufficient energy, nutrients, and fluids for wound healing, and all foods must be blenderized, strained, and able to go through a straw. High-protein, high-energy meal replacement and modular formulas can help provide the additional energy and protein needed for wound healing. Dietary fiber needs may be met by blenderizing, liquifying, and straining vegetables, fruits, grains, and cereals. Fluid needs should be calculated, and some free water provided to help ensure an adequate fluid intake. Liquid forms of vitamin-mineral supplements can be used to meet micronutrient needs unless the patient receives an adequate amount through their oral diet and liquid meal replacement formulas.

Dental Implant Surgery

The number and location of implants will dictate the extent of oral impairment following surgery. Implants placed in the anterior (or front) teeth will affect the mandibular opening and biting ability for a short time. In contrast, implants placed in the posterior (or back) teeth affect the mandibular opening and chewing ability. Important principles to remember are avoiding foods and fluids that can irritate the mucosa (citrus, spices) during the immediate postoperative healing phase and limiting foods such as popcorn, nuts, or seeds that may get caught in stitches. The impact is typically limited to the first 7 to 10 days following surgery. There is no published research on specific nutrient needs for individuals undergoing implant surgery. One scoping review[132] explored whether dietary supplements and micronutrients affected implant osseointegration. The authors concluded that other than vitamin D, there were a lack of data to support the need for any specific micronutrients or dietary supplements. In adults with a vitamin D deficiency, supplemental vitamin D enhanced the likelihood of successful implant osseointegration.[132]

Tooth Extraction Surgery

Like with implants, the extent of impairment in eating ability will depend on the number and location of teeth extracted. In those with third molars (or wisdom teeth) extracted, the ability to eat is impaired by pain and swelling. Ibikunie et al.,[133] evaluated oral health-related quality of life of 186 individuals aged 18 to 51 years who underwent third molar surgery and did or did not receive prednisolone. Seventy-five percent or more of all patients had difficulty with mandibular opening, chewing, and swallowing immediately postoperatively; by day three, more than 50% had difficulty chewing and swallowing. Those receiving prednisolone returned to chewing and swallowing ability sooner. Although the immediate impact on eating ability is significant following third-molar surgery, the duration is usually 1 week or less, so supplemental nutrients are not typically necessary.

In those who have teeth extracted in preparation for dentures, the duration of impairment in eating depends on how long it will take for the new dentures to be fabricated and whether a temporary denture will be provided. While tissues are healing in the immediate postoperative phase (3 to 5 days), spicy and citrus foods and fluids should be avoided because they may cause discomfort and irritate the oral mucosa. During this same time, a graduated diet should be used, starting with liquids like soups, ice creams/sorbets, smoothies, yogurts, and puddings, and progressing to soft cheeses, mashed potatoes, and other semisolid foods tolerated. Gradually, soft, nonirritating foods such as chopped and cooked fruits and vegetables, along with pasta, other grains, and meats/fish/poultry with gravies and sauces, can be introduced. The progression in types of foods depends on the individual's tolerance and ability to bite and chew without risk of choking. Modifications in both cooking and food preparation can be used to compensate for difficulties with biting and chewing.

A similar diet regimen can be used for patients with teeth extracted and immediate denture placement. It is important to remember that healing is individualized, and diets should be as well; the primary principles to keep in mind are avoiding irritation to the mucosa (citrus, spices, coarse breads) during healing and limiting foods such as nuts or seeds that may get caught in stitches.

Oral and Pharyngeal Cancers

The role of diet and nutrition in preventing and treating cancers of the head and neck depends on whether one is discussing prevention, treatment, or survivorship phases.

Prevention

Processed meats are associated with an increased risk of oral and pharyngeal cancers (OPC), while diets rich in fruit and vegetables, particularly foods rich in vitamins C and E and folate, are inversely associated with OPC risk.[134-139] Antioxidant nutrients and phytochemicals, including vitamins C and E, β-carotene, and flavonoids in some fruits and vegetables, have been suggested as protective dietary components against oral cavity cancers globally.[134,139,140] Although epidemiology studies have demonstrated protective effects of fruits and vegetables for OPC, the nutrient impact is based primarily on food, not supplemental sources. The primary dietary sources of folate include leafy greens, legumes, fruits, and fortified grains. Using data from the International Head and Neck Cancer Epidemiology (INHANCE) Consortium, Galeone, et al., showed an inverse relationship between overall OPC risk and total folate intake from foods, including fortified foods and dietary supplements. Clinical trials and longitudinal studies are needed to support the protective effects found in these epidemiologic studies.[134,136,139]

Treatment

OPC treatment typically involves surgery, chemotherapy, and radiation therapy.[141-143] These are associated with varying degrees of compromise to the integrity of the oral cavity and nutritional well-being. Oral toxicities are frequent, occurring independently and interdependently; they can occur in the acute, intermediate, and late stages of recovery. Nutrition status is compromised by the oral toxicities and symptom burden experienced by the patient, and the emotional and social challenges accompanying a cancer diagnosis and treatment. Anorexia may also occur. An interprofessional approach is critical for the optimal management of individuals with OPC.[144-146] A comprehensive nutrition assessment by a dietitian is critical to determine the optimal route(s) of perioperative nutrition support to help minimize weight loss and eating difficulties before, during, and following cancer treatment therapies.[144-146] At diagnosis, patients may have already experienced weight loss and have varying levels of malnutrition. Because of the location of OPCs (in and around the oral cavity and surrounding tissues, muscles, and joints), individuals undergoing treatment for an OPC are at risk for eating and drinking difficulties and subsequent malnutrition throughout all stages of the cancer care continuum.[147] It is essential to consider oral toxicities, difficulty eating or swallowing, that occur throughout treatment, and those that may be permanent[143] when prescribing medical nutrition therapy.

The National Comprehensive Cancer Network (NCCN) guidelines[145] recommend that all individuals with OPC undergo nutrition assessment and continued monitoring along with evaluation of speech, swallowing, and oral health. Any individual with 5% weight loss over the preceding month or 10% over the preceding 6 months is at risk of malnutrition. The Patient-Generated Subjective Global Assessment of PG-SGA[147] is a valid and reliable tool to assess and monitor nutrition status in patients with OPC.

Surgeries for OPC can cause changes to physical appearance and functional ability to consume foods and fluids, along with dysphagia, which may also occur due to cancer treatments. The effects of surgery on functional and sensory ability to consume foods and fluids vary with the degree and location of resection and impact on function and salivary glands. They may also be modulated by loss of neurologic function and sensations in certain areas and tissue fibrosis.

Radiation therapy can adversely affect nutrition and oral health status, causing changes in both sensory and functional abilities. Oral mucositis and its attendant pain, candidiasis, tissue fibrosis, caries, periodontal disease, xerostomia, dysgeusia, ageusia, and osteoradionecrosis[148,149] can often occur during and after treatment and be long term. If not treated before irradiation, chronic pulpal and periodontal infections can become acute because of radiation-associated changes and can lead to osteoradionecrosis. Oral mucosal pain occurs during therapy and can continue in long-term survivors. Fibrosis postradiation can limit mandibular opening and tongue movement and increase difficulty with eating and swallowing.[148] Oral toxicities may vary, depending on the location of the oropharyngeal area undergoing radiation, and are the direct consequence of the effect of treatment on otherwise normal tissues and structures, including the oral mucosa, bones, joints, and salivary glands and musculature. The *Vanderbilt Head and Neck Symptom Survey* (VHNSS) is a validated tool to assess patient symptoms and outcomes.[150] In clinical practice, it can provide the clinician insight into the nature and extent of symptoms impacting individual patients' eating abilities and other activities of daily living.

Chemotherapy, whether completed as concurrent chemoradiation (CCR) or individually, has similar associated toxicities affecting the oral mucosa, with acute and chronic impacts. Toxicities include mucositis, oral infections, altered taste, and xerostomia. Oral mucositis, one of the most common conditions, makes the tissues more susceptible to physical trauma from sharp tooth edges and hard pieces of food. Ulceration, secondary infection, and painful stomatitis may result. Generally, the more intense the cytotoxic therapy, the more common the oral complications. Through proper prophylactic therapy, including medications, bacteriostatic oral rinses, and a team-based approach to nutritional, medical, and dental care, many side effects can be reduced, resulting in fewer problems with eating and drinking.[142] Diet and nutrition support approaches may change throughout the treatment course due to the nutrition status of the patient and the presence and extent of oral toxicities impacting the individual's ability to eat.[142,143] More information on nutrition and diet and cancer and the survivorship phases of cancer is in the chapter on nutrition support of the patient with cancer.

Oral Lesions and Related Systemic Conditions

Oral infections such as herpes simplex can result in painful oral lesions that impair the desire and ability to eat for a short time. Usually, palliative oral care and appropriate food choices (non-spicy, temperate, easily masticated) can help maintain nutritional status. However, in individuals with immune system disorders, such as human immunodeficiency virus (HIV) infection and AIDS, and those with autoimmune diseases, oral lesions may be recurrent and impact their ability to eat for longer durations, thus compromising nutritional status.[144]

People living with HIV infection may have oral lesions manifesting from the disease process (Kaposi sarcoma, recurrent aphthous stomatitis, major aphthous ulcers), pathogens causing infections (candidiasis or herpes simplex viruses), oral infectious diseases (dental caries or periodontal disease), or due to side effects of select medications used to treat HIV.[144,151] The frequency of occurrence of many of these lesions can be reduced in incidence and severity by prophylactic regimens. However, when they are present, appetite and intake may be compromised due to pain and increased risks of dehydration and weight loss. Ensuring a diet balanced in nutrients and modified to meet individual needs for consistency, flavor, texture, and spice, along with appropriate between-meal (and snacks) oral hygiene, can improve oral health.

Ulcerations in the mouth due to systemic GI diseases may be evident and may precede GI symptoms.[152,153] Mays et al., reported buccal mucosal edema, mucogingivitis, deep linear ulcerations, and lip swelling associated with Crohn's disease.[152] These lesions can affect the ability to eat and appetite, increasing risks for dehydration, weight loss, and compromised nutritional well-being. Careful attention to the nutrition needs of people with Crohn's disease is essential to reduce the risk of further nutrition compromise resulting from oral lesions. A team-based approach to care is needed; dentists, physicians, dietitians, and other health professionals can coordinate treatment modalities to maximize eating ability, reduce pain and GI sequelae, and attain and maintain nutritional well-being.

Celiac disease, another autoimmune disorder, also has gastrointestinal manifestations that can begin in the mouth. Sometimes referred to as gluten-sensitive enteropathy, it manifests as an immune response to gluten consumption and can be well-controlled with a gluten-free diet. Oral manifestations of celiac disease include dental enamel hypoplasia and aphthous ulcers. Dietary control of the disease may prevent aphthous ulcers.[153]

dos Santos et al., conducted a living systematic review in patients with COVID-19. They found that although taste disorders and xerostomia are known side effects, oral lesions (aphthous-like ulcers, herpes-like lesions, candidiasis, glossitis) are another common symptom in individuals with COVID-19.[154] The most common locations for the oral lesions identified were the tongue, followed by the lips and palate.[154] Whether these oral manifestations are specific to COVID-19 or its impact on the immune system is a limitation of the study, further research is needed.

Effects of Saliva on Oral Health and Nutrition

Saliva is a primary factor for both the function and maintenance of the oral cavity. In addition to saliva's importance in speech and deglutition, antimicrobial and nonantimicrobial systems protect both the hard and soft tissues. Cessation or a severe decrease of salivary flow from such causes as surgical removal of salivary glands, radiation therapy to the salivary glands, uncontrolled diabetes, medications, autoimmune diseases including Sjogren's syndrome, or other systemic diseases like amyloidosis, can lead to microbial infection of the oral cavity, erosion, caries, and dysgeusia, as well as an inability to lubricate, masticate, and swallow food.[155-158]

Hyposalivation is a reduction in saliva secondary to altered function of the salivary glands,[159] whereas xerostomia is the subjective feeling of dry mouth.[155] Both can be acute and short-term, as seen with individuals with uncontrolled diabetes or medications causing xerostomia. They are long-term in individuals with Sjogren's syndrome and those who undergo radiation therapy to the salivary glands. All these conditions profoundly affect the selection of and ingestion of foods and, thus, nutritional status and systemic health. Additionally, comfort in wearing dentures depends on saliva's lubrication of the soft tissues. Patients with either hyposalivation or xerostomia have poor retention of their dentures. They may develop ulcerations at denture borders, thus making biting and chewing difficult and painful, possibly affecting dietary intake and nutrition status.

Historically, there was a belief that the aging process brought a "natural" reduction of salivary flow, but salivary flow in healthy persons does not always decrease with age.[160] De Martins Souza Pina[159] et al.'s systematic review explored the prevalence of hyposalivation in older adults who were non-institutionalized. After excluding studies with diseases known to cause hyposalivation or xerostomia and those addressing cancer or cancer therapies, they found an overall prevalence of 33.7%.[159] Limitations included the inability to control for polypharmacy, which can contribute to both hyposalivation and xerostomia. Many classes of medications can cause xerostomia to varying degrees, including anticholinergics (e.g., antihistamines, tricyclic antidepressants, and antipsychotics), as well as α-adrenergic receptor blockers (narcotics, antianxiety and hypnotic agents, antiemetics, antihypertensives, diuretics, beta, and calcium channel blockers, antivirals, antihistamines, and selective serotonin reuptake inhibitor antidepressants).[155,156]

Although salivary flow may be decreased because of medications, it generally does not interfere with the ability to eat or impact the integrity of the oral cavity. Individuals taking medications with xerostomia as a side effect should be advised to use sugar-free gums containing relatively nonfermentable sugar alcohols such as sorbitol, mannitol, or xylitol. To prevent caries, aggressive oral hygiene, topical fluoride, and the use of sugar-free gums, along with frequent water consumption, can help to reduce caries risk and risk of erosion.[156] The use of artificial saliva, although not a panacea, may be beneficial to some of these patients.

Tooth Loss

The prevalence of full (or complete loss of teeth) and partial edentulism (loss of some teeth) has declined in the United States and Globally.[5,161] Parker et al.,[161] examined three measures of tooth loss in adults aged 50 years and older during two NHANES periods, from 1999 to 2004 (6,283 adults) and 2011 to 2016 (7,443 adults). Approximately 11% (10.8%) had complete edentulism, almost one-third (31.9%) lacked functional dentition (<20 teeth), and 16.7% had severe tooth loss (eight or fewer teeth).[161] Adults with compromised general health or select comorbid conditions had a greater prevalence of severe tooth loss and lack of functional dentition than those who were otherwise healthy. The Centers for Disease Control and Prevention (CDC) reported similar statistics for the period from 2017 to 2018; 12.9% of adults 65 years and older were completely edentulous.[162] The prevalence of tooth loss increased with age (grouped at 65–69 years, 70–74 years, and 75 years and older), going from 8.9% to 10.6% and 17.8%, respectively.

Edentulism has been and continues to be more prevalent among those from the lowest income groups.[163] Other disparities exist for different racial and ethnic groups. Non-Hispanic Black men and women have a greater tooth loss prevalence than non-Hispanic white and Hispanic men.[162] Dementia risk may also be increased in individuals with tooth loss. In a meta-analysis, Chen et al.,[164] found that the risk of dementia was increased 1.34-fold by tooth loss. These findings, while limited, provide insight for further research into the causes of this association.

The impact of tooth loss on diet and nutrition status can be profound. The majority of the research on nutrition, diet and tooth loss, and replacement has been done with older adults. As tooth loss increases, occlusion is increasingly compromised, and difficulty with mastication increases. Prior research has demonstrated that loss of natural teeth adversely affects food variety and diet quality, with compromises in dietary fiber, fruits and vegetables, and micronutrients.[165-168] Tooth loss, particularly lack of functional dentition, negatively influences eating habits, diet quality, and quality of life.[163,167,169,170] With the loss of anterior teeth (canine to canine), the functional ability to bite food is reduced. With the loss of posterior teeth (premolars and molars), the ability to chew food properly is compromised. Combined, both can increase the risk of dysphagia, weight loss, and malnutrition.

Zelig et al., found that older adults with tooth loss experienced positive adaptive and maladaptive behaviors.[169] Adaptive behaviors included choosing softer foods, cooking foods until they were soft enough to puree or blenderize, and chopping, shredding, and grinding foods. Individuals took longer to eat and were more likely to eat before going out rather than eat with others. Maladaptive behaviors focused on avoiding some fruits, many vegetables, nuts and seeds, whole grains, and meats. Eating-related quality of life (ERQOL) is also negatively impacted, as individuals with tooth loss may be embarrassed by their appearance and reduce the frequency of smiling, eating with others, talking, and eating in public environments.[169]

Systematic reviews[171] and cross-sectional studies[172-174] have demonstrated a greater risk and presence of malnutrition in older adults with tooth loss than in older adults without tooth loss. Honeywell et al,[172] conducted a cross-sectional study of patient data for 305 older adults who came for care at a U.S. dental school clinic. Using several measures of dentition status (functional dentition and totals for the number of teeth, posterior occluding pairs of teeth, and anterior occluding pairs of teeth), they found that all were related to risk for malnutrition (with the Self-Mini Nutritional Assessment). The odds of malnutrition decreased by 7% for every additional increase in the number of remaining teeth. Consistent with this finding, the risk for or presence of malnutrition was more likely in those without functional dentition than those with functional dentition. Similarly, those with more posterior pairs of occluding teeth (POP) and those with more anterior pairs (AOP) were more likely to have normal nutrition status than those with less POP or AOP. In a sample of hospitalized adults in Italy, Tonaizzo[175] similarly found that patients with malnutrition had fewer teeth than those without malnutrition. The findings of these studies[172,173,175] are supported by systematic reviews and meta-analyses on this topic.[171,176]

Dietary guidance for adults with tooth loss must be tailored to the individual's nutrient needs, other comorbid conditions, and lifestyle. Counseling patients to eat "soft foods" is insufficient, as not all soft foods are easy to eat. Soft rolls or white bread can be difficult to chew and swallow. The focus should be on modifications in food textures in raw or cooked foods and cooking techniques. This can be achieved by chopping, mincing, shredding, mashing, or even grating foods like raw fruits and vegetables. Cooking foods until soft and using gravies and sauces can make tough-to-chew foods easier to masticate. Fruit or vegetable smoothies provide a way to consume fresh fruits and vegetables. Clinicians can address tooth loss and approaches to eating during patient visits; asking patients for a diet recall can provide insights into adaptive and maladaptive behavior so that dietary guidance can be tailored to individual needs to achieve a nutritionally adequate diet. **Table 75.7** offers general guidelines for individuals with tooth loss that can be readily adapted.[177]

Table 75.7 Guidance for Eating Healthier with Tooth Loss[177]

Tips to Improve Fruit and Vegetable Intake:

- Choose softer fruits and vegetables like bananas or avocados
- Peel or remove hard-to-chew skin
- Cut or chop into bite-sized pieces
- Cook to a softer consistency
- Blend or puree into a smoothie or a cold soup
- Buy canned (without added salt or sugar) or frozen, as needed

Tips To Improve Whole Grain Intake:

- Choose cooled whole grains like oatmeal, brown rice, quinoa, or couscous
- Cut bagels and rolls into smaller pieces
- Cook pasta and rice until soft
- Allow cereals to soften in milk or other liquid
- Toast lightly and add a spread like butter or cream cheese
- Have sips of fluids while eating to add moisture

Tips To Improve Protein Intake:

- Choose easy-to-eat and drink, protein rich foods like ground meat or poultry, fish or seafood, eggs, cooked beans, tofu, or dairy products like cottage cheese, cow or other dairy, or nondairy alternative milks, yogurts, and cheese
- Remove hard-to-chew skin
- Cook foods that will shred or flake easily, such as fish, shredded or pulled beef, chicken, or pork, or vegetable protein alternatives
- Cook to a softer consistency or until tender; try a slow cooker
- Add sauces and gravies to moisten
- Use a knife and fork or mini-food chopper to cut foods into bite-sized pieces

Tips to become more comfortable eating around others:

- Share eating challenges with family and friends so you can enjoy eating with them
- Plan meals with family and friends where everyone can eat the same thing
- When going to a party or event, eat first or bring something you can eat just in case there are no foods you can tolerate
- Leave yourself enough time to eat
- Go to restaurants where the menu includes foods you can eat and don't be afraid to ask how foods are prepared and specify how you would like them cooked

Reproduced from Zelig, R., Honeywell, S., Touger-Decker, R. Nutrition and Oral Health. Oral Health and Aging. Springer, Cham. 2022. Available at https://doi.org/10.1007/978-3-030-85993-0_3[177]

Dentures

The impact of new or replacement dentures may improve certain characteristics of the eating experience but not necessarily nutrient intake or nutritional status. Al-Sultani and colleagues[178] found that patients who received new (replacement) complete dentures experienced improvements in their eating experience 1 month postinsertion. Although some patients reported reduced feelings of embarrassment and interruptions in meals, and improved confidence and ability to eat, tough or hard foods continued to cause problems. In a systematic review, Bezerra et al., looked at the impact of implant-supported dentures (IOD) compared with edentulism on nutrition status and nutrient intake.[179] They found that nutrient bioavailability of foods was not significantly affected when comparing individuals with IODs to those who are edentulous. In contrast, those with IODs were more likely to be well-nourished and less likely to have malnutrition[179] than those who were edentulous. Oral health-related quality of life and functional ability to eat may be improved when a mandibular implant-supported denture and complete removable maxillary denture are provided compared with maxillary and mandibular complete removable dentures, but the impact on nutrient intake and nutrition status remains equivocal.[180-184] Potential explanations for the lack of changes in nutrient intake and nutrition status when a combination of conventional or -implant supported dentures are provided may be multifactorial. Both formation of dietary behaviors and tooth loss occurs over time. Hence, once patterns are formed and maladaptive behaviors like avoidance of fresh fruits and vegetables are adopted, provision of artificial teeth in any form without education and counseling may not change the patterns. Gaps exist globally on research exploring associations between different types of dentures (complete and partial with and without implant support), removal of dentures for eating and diet quality, nutrient intake, and nutrition status.

There are limited data on guidelines for eating with dentures.[178,184] Kanazawa et al., randomized adults getting new dentures[184] to either a control group that received new dentures and advice on their new dentures, or an experimental group that received new dentures along with diet advice using a diet education pamphlet. The researchers found that those who received the diet education pamphlet had a significantly higher nutrient intake 3 months post-treatment than those who did not. However, at 6 months posttreatment, intake of some nutrients declined. This is an area for continued research.

When adults receive new dentures, dietary guidance (**Table 75.8**) should be provided in a graduated approach.[177,185] Understanding the maladaptive and adaptive behaviors a patient has when they get the dentures can help the clinician provide targeted guidance to help the patient achieve a nutritionally adequate diet and use their dentures for eating.[177,185] Food texture modifications can help during the adaptive phases of learning to eat with new dentures.[177,185] All health professionals should ask patients with dentures if they use them for eating and, if not, why. Those with difficulty eating with their dentures should be referred to a dentist for further evaluation and, as needed, to a dietitian for medical nutrition therapy.

Table 75.8 Guidance for Eating Following Denture Provision[177]

Tips for Diet Progression Following Provision of New Dentures:
Day #1: May need to eat chopped, minced, pureed, or blenderized foods; smoothies, soups, puddings
Day #2–3: Choose soft or cut up foods as tolerated ■ Animal and vegetable protein sources: cook until soft,* cut into small pieces, moisten with gravy or sauces ■ Dairy and nondairy alternatives: milk, yogurt, spreads, cheeses, hot/cold cereal with milk, smoothies ■ Fruit and vegetables: soups, smoothies, peel and cook until mashable or fork tender, as toleratedGrains: mashed potatoes, oatmeal, cream of wheat, pastas
Day #4–5: Progress to raw cut-up foods as tolerated ■ Peel and cut or chop raw fruits and vegetablesUse a knife and fork to cut food into bite-sized pieces
Foods to Avoid (at Least Initially):
■ Dry foods that fall apart in the mouth like rice, quinoa, grains, biscuits, crackers, and muffins ■ Hard, crunchy, tough-to-chew foods like crusty breads, stringy meats, nuts, and seeds ■ Soft, sticky foods like white bread and soft rolls, nut butters ■ Foods that may get stuck like seeds and pulp, or other foods that break into small pieces ■ Chewing gum and hard candies
Other Tips for Eating Well with Dentures:**
■ Cut food using a knife and fork and take small bites ■ Chew foods at corners of the mouth to take the pressure off the front teeth ■ Balance the biteforce on both sides of the mouth ■ Fluids should also be encouraged to help moisten food ■ Choose softer* foods as tolerated ■ Cook foods to a softer* consistency ■ Modify the way the food is prepared and consumed ■ Practice good oral hygiene after meals and snacks
Troubleshooting Pain or Discomfort, and Denture Instability:
■ Use fixatives as needed under the guidance of an oral healthcare professional ■ Follow up with a dentist for denture adjustment, as needed

*The narrative describes how to interpret the word "soft."
**See Table 75.2 above for further discussion of appropriate food modifications for patients with tooth loss and replacement.

Reproduced from Zelig, R., Honeywell, S., Touger-Decker, R. (2022). Nutrition and Oral Health. In: Hogue, CM., Ruiz, J.G. (eds) Oral Health and Aging. Springer.[177]

Older Adults and Oral Health

The aging process involves physiologic changes that can affect and be affected by oral health. According to the U.S. National Council on Aging, 80% of those 65 years and older have at least one chronic comorbid condition, and 68% have more than one.[5,186] Sixty-five percent of older adults take two or more medications, and more than one-third take more than five.[187] The most common medication categories were for pain, followed by those for the central nervous system, cardiovascular system, and anticholinergics. The potential for oral manifestations of chronic comorbid conditions and side effects of polypharmacy on the oral cavity can individually and together impact oral health, diet quality, and nutritional status.

Dysgeusia, xerostomia, and hyposalivation (see prior sections of this chapter) may be due to comorbid conditions and medications with these known side effects, rather than the aging process itself. Tooth loss, pain, joint dysfunction, partial and full dental prostheses, caries, periodontal disease, and xerostomia can all impair appetite, ability to eat and drink and nutrition status, and eating-related quality of life.[5] Nutrition assessment should include exploring adaptive and maladaptive eating behaviors and individualized guidance on a diet that addresses any limitations in sensory or mechanical ability to eat.

Summary

Diet and nutritional status can affect individuals' oral health, and the integrity of the oral cavity likewise can impact eating and drinking ability, systemic health, eating-related quality of life, and nutrition status. Local oral and systemic diseases can mediate the relationship between nutrition and oral health. An interprofessional team-based approach is needed for patient care. Collaboration and coordination of patient care between physicians, dentists, dietitians, and other healthcare team members is essential for patient-centered care.

Acknowledgments

The authors acknowledge Diane Rigassio Radler and Dominick P DePaola for their contributions to the previous versions of this chapter.

References

1. Ganesan SM, Vazana S, Stuhr S. Periodontol 2021;87:299–314.
2. Sanz M, Eel Castillo AM, Jepsen S, et al. J Clin Peridontol 2020;47:268–88.
3. Bobetsis YA, Graziani F, Gürsoy M, et al. Periodontol 2020;83: 154–74.
4. Institute of Medicine of the National Academies. Advancing Oral Health in America. Washington, DC: National Academies Press, 2011.
5. U.S. Department of Health and Human Services, National Institute of Dental and Craniofacial Research, National Institutes of Health. Oral Health in America: A Report of the Surgeon General. Rockville, MD: US Department of Health and Human Services, 2000:23–60.
6. Shaw JH, Sweeney EA. In: Shils ME, Young VR, eds. Modern Nutrition in Health and Disease. 7th ed. Philadelphia, PA: Lea & Febiger, 1988:1070–1.
7. Moynihan P, Thomason M, Walls A, et al. J Dent 2009;37: 237–49.
8. Sweeney EA, Saffir AJ, deLeon R. Am J Clin Nutr 1971;4:29–31.
9. Sawyer DR, Kwoku AL. J Dent Child 1985;54:141–5.
10. Punysingh JT, Hoffman S, Harris SS, et al. J Oral Pathol 1984;13:40–57.
11. Boyle PE. J Dent Res 1933;13:139–50.
12. Slavkin H. J Am Dent Assoc 1996;127:681–2.
13. Misra M, Pacaud D, Petryk A, et al. Pediatr 2008;122:398–417.
14. Leaver AG. Clin Orthop 1971;8:90–107.
15. Bawden JW. Anat Rec 1989;24:226–33.
16. Fulton A, Amlani M, Parekh S. Br Dent J 2020;228:515–8.
17. Mellanby M. Br Dent J 1923;44:1031–41.
18. Singleton R, Day G, Thomas T, et al. J Dent Res 2019;98: 549–55.
19. Wolf JJ. Am J Dis Child 1935;49:905–11.
20. Boyle PE. J Dent Res 1934;14:172.
21. Schiltz JR, Rosenbloom J, Levinson GE. J Embryol Exp Morphol 1977;37:49–57.
22. Alvarez JO, Eguren JC, Caleda J, et al. J Dent Res 1990;69: 1564–6.
23. Alvarez JO, Navia JM. Am J Clin Nutr 1989;49:417–26.
24. Alvarez JO, Lewis CA, Saman C, et al. Am J Clin Nutr 1988;48:368–72.
25. Rami-Reddy V, Vijayalakshmi PB, Chandrasekhar-Reddy BK. Odont Pediatr 1986;7:1–5.
26. Delgado H, Habicht JP, Yarbrough C, et al. Am J Clin Nutr 1975;38:216–24.
27. Menaker L, Navia JM. J Dent Res 1973;52:688–91.
28. Centers for Disease Control and Prevention. Data & Statistics on Birth Defects. Accessed August, 7, 2022. Available at www.cdc.gov/ncbddd/birthdefects/data.html
29. Mai CT, Isenburg JL, Canfield MA, et al. Birth Defects Res 2019;111:1420–35.
30. National Institute of Dental Research. Mineralized Tissues, Craniofacial Development, Dentofacial Malrelations and Trauma. Broadening the Scope: Long-Range Research Plan for the Nineties. Washington, DC: US Government Printing Office, 1990. NIH publication 90–1188.
31. National Institute of Dental and Craniofacial Research. Prevalence of Cleft Lip & Palate. Accessed August 8, 2022. Available at: www.nidcr.nih.gov/research/data-statistics/craniofacial-birth-defects/prevalence
32. Slavkin HC. Cleft Palate J 1990;27:101–9.
33. Rader JI, Schneeman BO. Pediatr 2006;1174:1394–99.
34. Akam M. Cell 1989;57:347–9.
35. Lammer EJ, Chen DT, Hoar RM, et al. N Engl J Med 1985;313: 837–41.
36. MRC Vitamin Study Research Group. Lancet 1991;338: 131–7.
37. Taparia S, Gelineau-van Waes J, Rosenquist TH. Clin Chem Lab Med 2007;45:1717–27.
38. Moynihan PJ, Lingstrom P. Oral consequences of nutritional well-being. Touger-Decker R, Sirois DA, Mobley CC, eds. Nutrition and Oral Medicine. Totowa, NJ: Humana Press, 2005:107–125.
39. McGrath AJ, Hamill LL, Cardwell CR, et al. Eur J Nut 2016;55: 1377–88.
40. Schleicher RL, Carroll MD, Ford ES, et al. Am J Clin Nutr 2009;90:1252–63.
41. Chapple ILC, Bouchard P, Cagetti MG. J Clin Periodontol 2017;44:S39–51.
42. Chapple IL, Milward MR, Dietrich T. J Nutr 2007;137:657–64.
43. Food and Nutrition Board, Institute of Medicine. Dietary Reference Intakes for Vitamin C, Vitamin E, Selenium and Carotenoids. Washington, DC: National Academies Press, 2000.
44. Hathcock J, Hattan DC, Jenkins M, et al. Am J Clin Nutr 1990;52:183–202.
45. Food and Nutrition Board, Institute of Medicine. Dietary Reference Intakes for Vitamin A, Vitamin K, Arsenic, Boron, Chromium, Copper, Iodine, Iron, Manganese, Molybdenum, Nickel, Silicon, Vanadium and Zinc. Washington, DC: National Academies Press, 2001.

46. Rozier RG, Adair S, Graham F, et al. J Am Dent Assoc 2010;141: 1480–89.
47. U.S. Department of Health and Human Services Federal Panel on Community Water Fluoridation. Public Health Rep 2015;130:318–331.
48. Murray JJ, Rugg-Gunn AJ. Fluorides and Dental Caries. 2nd ed. Bristol, UK: John Wright & Sons, 1982.
49. Brunelle JA, Carlos JP. J Dent Res 1990;69:723–7.
50. McLaren L, Patterson S, Thawer S, et al. Community Dent Oral Epidemiol 2016;44:274–82.
51. McLaren L, Patterson SK, Faris P, et al. Can J Public Health 2022. Available at https://doi.org/10.17269/s41997-022-00654-4
52. Ismail AI, Hasson H. J Am Dent Assoc 2008;139:1457–68.
53. Berg J, Gerweck C, Hujoel PP, et al. J Am Dent Assoc 2011;142: 79–87.
54. Siew C, Strock S, Ristic H, et al. J Am Dent Assoc 2009;140: 1228–36.
55. Levy S, Broffit B, Marshall TA, et al. J Am Dent Assoc 2010;141:1190–1201.
56. Beltran-Aguilar ED, Barker L, Dye BA. Prevalence and Severity of Enamel Fluorosis in the United States, 1986–2004. NCHS data brief no 53. Hyattsville, MD: National Center for Health Statistics. 2010. Accessed August 8, 2022. Available at www.cdc.gov/nchs/products/databriefs/db53.htm
57. Ismail Al, Messer JG. J Public Health Dent 1996;56:22–7.
58. National Institute of Dental and Craniofacial Research. Oral Health in America. Accessed August 8, 2022. Available at www.nidcr.nih.gov/research/oralhealthinamerica
59. Centers for Disease Control and Prevention. Dental Caries Among Adults and Older Adults: Oral Health Surveillance Report. Accessed August 8, 2022. Available at: www.cdc.gov/oralhealth/publications/OHSR-2019-dental-carries-adults.html
60. Fleming E, Afful J. Prevalence of Total and Untreated Dental Caries Among Youth: United States, 2015-16. NCHS Data Brief, no 307. Accessed August 8, 2022. Available at www.cdc.gov/nchs/products/databriefs/db307.htm
61. Peres MA, Macpherson LMD, Weyant RJ, et al. Lancet 2019; 394:249–60.
62. Bashir NZ. J Am Dent Assoc 2022;154:300–8.
63. Petersen PE. The World Oral Health Report. Geneva: World Health Organization, 2003.
64. U.S. Department of Health and Human Services, Centers for Disease Control and Prevention. Trends in Oral Health Status: United States, 1988–1994 and 1999–2004. Series 11, no. 248, 2007.
65. National Institutes of Health Consensus Statement. Diagnosis and Management of Dental Caries throughout Life, 2001. Accessed August 8, 2022. Available at https://consensus.nih.gov/2001/2001DentalCaries115PDF.pdf
66. WHO Oral Health Programme. Global Data on Dental Caries Prevalence (DMFT) in Children Aged 12 Years. Accessed August 8, 2022. Available at https://apps.who.int/iris/handle/10665/66520
67. Moynihan P. Bulletin of the World Health Organization. 2005;83:694–9.
68. Touger-Decker R, Mobley C. J Am Diet Assoc 2007;107:1418–28.
69. Johnson RK, Frary C. J Nutr 2001;131:2766S–71S.
70. Marshall TA. J Am Dent Assoc. 2015;146:61–4.
71. U.S. Department of Agriculture. MyPlate. Accessed August 8, 2022. Accessed August 8, 2022. Available at www.myplate.gov/
72. Lingstrom P, van Houte J, Kashket S. Crit Rev Oral Biol Med 2000;11:366–80.
73. Al-Khatib GR, Duggal M, Toumba KJ. J Dent 2001;29:409–14.
74. Riley P, Moore D, Ahmed F, et al. Cochrane Data System Rev 2015;3:CDO10743. Accessed August 8, 2022. Available at www.cochranelibrary.com/cdsr/doi/10.1002/14651858.CD010743.pub2/full
75. Anderson CA, Curzon ME, Van Loveren C, et al. Obe Rev 2009;10(suppl 1):41–54.
76. Moynihan PJ, Kelly SAM. J Dent Res. 2014;93:8–18.
77. Al-Zahrani A, Al-Qahtani M, Al-Barti M. Sci World J. 2022; ID5447723.
78. Palmer CA, Kent R, Loo CY, et al. J Dent Res 2010;89: 1224–9.
79. Marshall TA, Broffitt B, Eichenberger-Gilmore, et al. J Public Health Dent 2005;65:166–73.
80. Marshall TA, Levy SM, Broffitt B, et al. Pediatr 2003;112: e184–91.
81. Hancock S, Zinn C, Schofield G. Eur J Oral Sci 2020;128: 467–75.
82. Burt BA, Pai S. J Dent Educ 2001;65:1017–24.
83. Bagramian RA, Garcia-Godoy F, Volpe A. Am J Dent 2009;22: 3–8.
84. Marshall TA, Curtis AM, Cavanaugh JE, et al. J Acad Nutr Diet 2021;121:253–60.
85. Lussi A, Jaeggi T, Zero D. Caries Res 2004 38(suppl 1):34–44.
86. Mundorff SA, Featherstone JDB, Bibby BC, et al. Caries Res 1990;24:344–55.
87. Pollard MA, Imfeld T, Higham SM, et al. Caries Res 1996;30: 132–7.
88. Kashket S, van Houte J, Lopez LR, et al. J Dent Res 1991;70: 1314–9.
89. Kashket S, Zhang J, van Houte J. J Dent Res 1996;75: 1885–91.
90. Silva MF, Jenkins GN, Burgess RC, et al. Caries Res 1986;20: 263–9.
91. DePaola DP, Kashket S. Nutr Rev 2002;60:97–103.
92. Hildebrandt GH, Sparks BS. J Am Diet Assoc 2000;131: 909–16.
93. Ly KA, Milgrom P, Rothen M. J Am Dent Assoc 2008;139: 553–63.
94. Newton JT, Awojobi O, Nasseripour M, et al. JDR Clin Trans Res 2020;5:214–23.
95. U.S. Department of Health and Human Services. Centers for Disease Control and Prevention. MMWR Surveillance Summaries. Surveillance for Dental Caries, Dental Sealants, Tooth Retention, Edentulism and Enamel Fluorosis United States, 1988–1994 and 1999–2002. 2005;54:1–44.
96. Ritter AV, Shugars DA, Bader JD. Community Dent Oral Epidemiol 2010;38:383–97.
97. Papas AS, Palmer CA, Rounds MC, et al. Ann N Y Acad Sci 1989;561:124–42.
98. Council on Clinical Affairs. American Academy of Pediatric Dentistry. Definition of Early Childhood Caries (ECC). Accessed August 8, 2022. Available at www.aapd.org/assets/1/7/d_ecc.pdf
99. O'Sullivan DV, Tinanoff N. J Dent Res 1993;72:1577–8.
100. Marshall TA. Dent Clin North Am 2019;63:595–605.
101. Marshall TA. J Am Dent Assoc 2009;140:670–4.
102. Council on Clinical Affairs. Am Acad Pediatr Dent 2002; 32:101–8. Accessed June 3, 2011. Available at: www.aapd.org/media/Policies_Guidelines/G_CariesRiskAssessment.pdf
103. Featherstone JDB, Adair SM, Anderson MH, et al. J Calif Dent Assoc 2003;31:257–69.
104. Heyman MB, Abrams SA, Section on Gastroenterology, Hepatology and Nutrition, et al. Pediatr 2017;139:e20170967.
105. Caton JG, Armitage G, Beglundh T, et al. J Periodontol 2018;89:S1–8.
106. Eke PI, Dye BA, Wei L, et al. J Dent Res 2012;91:914–20.
107. Boyd LD, Madden TE. In: Palmer CA, ed. Nutrition and the Periodontium. Diet and Nutrition in Oral Health. Englewood Cliffs, NJ: Prentice-Hall, 2003:202–12.
108. Machado V, Lobo S, Proenca L, et al. Nutrients 2020;12:2177. doi 10.3390/nu1082177

109. Salazar CR, Laniado N, Mossavar-Rahmani Y, et al. J Clin Periodontal 2018;45:780–790.
110. Pink C, Kocher T, Meisel P, et al. J Clin Periodontol 2015;42: 988–97.
111. Reeves AF, Rees JM, Schiff M, et al. Arch Pediatr Adolesc Med. 2006;160:894–9.
112. Andriankaja OM, Sreenivasa S, Dunford R, et al. Aust Dent J 2010;55:252–9.
113. Gorman A, Kaye EK, Apovian C, et al. J Clin Periodontol 2012;39:107–14.
114. Alshouibi EN, Kaye EK, Cabral HJ, et al. J Dent Res. 2013;92;689–93.
115. Intini G, Katsuragi Y, Kirkwood KL, et al. Adv Dent Res 2014;26:38–46.
116. Botelho J, Machado V, Proenca L, et al. Nutrients 2020;12:1471. doi: 10.3390/nu12051471
117. Miley DD, Garcia MN, Hildebolt CF, et al. J Periodontol 2009;80:1433–9.
118. Yu B, Wang CY. Periodontol 2000 2022;89:99–113.
119. Jabbar S, Drury J, Fordham J, et al. J Periodont Res 2011;46: 97–104.
120. Jeffcoat MK, Chesnut C. J Am Dent Assoc 1993;124:49–56.
121. Dietrich T. Am J Clin Nutr 2004;80:108–13.
122. Payne JB, Reinhardt RA, Nummikoski PV, et al. Osteoporosis Int 1999;10:34–40.
123. Marshall TA. J Am Dent Assoc 2018;149:148–52.
124. Mauri-Obradors E, Estruga-Devesa A, Jane-Salas E, et al. Med Oral Patol Oral Cir Bucal 2017;22:e586–94.
125. Kudiyirickal MG, Pappachan JM. Endocrine 2015;49:27–34.
126. Lalla E, Hsu WC, Lamster IB. In: Genco RJ, Williams RC, eds. Dental and Medical Comanagement of Patients with Diabetes. Periodontal Disease and Overall Health: A Clinician's Guide. Yardley, PA: Professional Audience Communications, 2010: 216–34.
127. Loe H. Diabet Care 1993;16:329–34.
128. Nguyen ATM, Akhter R, Garde S, et al. Diabet Res Clin Pract 2020;165;2–9.
129. McNeish, J., Shapiro, DS. In: Ferneini, E., Bennett, J, eds. ICU Management of the Maxillofacial Surgery Patient. Perioperative Assessment of the Maxillofacial Surgery Patient. Cham: Springer, Cham, 2018:71–80.
130. Christensen BJ, Chapple AG, King BJ. J Oral Maxillofac Surg 2019;77:777–782.
131. Christensen BJ, Chapple AG, King BJ. J Oral Maxillofac Surg 2019;77:1227.e1–1227.e6.
132. Nastri L, Moretti A, Migliaccio S, et al. Nutrients 2020;12: 268–284.
133. Ibikunle, AA, Wasiu LA, Akinola LL. Oral Maxillofac Surg 2017;20:343–352.
134. De Vito R, Lee YCA, Parpinel M, et al. Epidemiol 2019;30: 93–102.
135. World Cancer Research Fund/American Institute for Cancer Research. Food, Nutrition, Physical Activity, and the Prevention of Cancer: A Global Perspective, 2018. Accessed July 21, 2022. Available at www.wcrf.org/wp-content/uploads/2021/02/Summary-of-Third-Expert-Report-2018.pdf
136. Bravi F, Lee YCA, Hashibe M, et al. Oral Dis 2021:27:73–93.
137. Xu J, Xin-xin Y, Yun-Gang W, et al. PLoS One 2014;9: 495048.
138. Chuang SC, Jenab M, Heck J. Cancer Causes Control 2012;23: 69–88.
139. Galeone C, Edefonti V, Parpinel M, et al. Int J Cancer 2015:136(4):904–914.
140. Petridou E, Zavras A, Lefatzis D, et al. Cancer 2002;94:2981–8.
141. Parmar A, Macluskey M, Goldrick N, et al. Cochrane Data System Rev 2021;: CD006386. doi: 10.1002/14651858.CD006386
142. Ganzer H, Epstein JB, Touger-Decker R. Nutrition Management of the Cancer Patient. Nutrition and Oral Medicine, 2nd ed. New York, NY: Humana Press, 2014:235–253.
143. Ganzer H, Rothpletz-Puglia P, Byham-Gray L, et al. Supp Care Cancer 2015;23:3257–3268.
144. Sroussi HY, Kaste LM, Epstein J, et al. In: Touger-Decker R, Mobley C, Epstein J, eds. Human Immunodeficiency Virus/HIV. Nutrition and Oral Medicine. 2nd ed. New York, NY: Humana Press, 2014:255–276.
145. Phister DG, Spencer S, Adkins D, et al. NCCN Guidelines Version 2.2022 Head and Neck Cancers. 2022.
146. Gill A, Givi B, Moore MG. Head Neck 2019;41:577–583.
147. Jager-Wittenaar HJ, de Bats HF, Welink-Lamberts BJ. Nutr Clin Pract 2020;35:353–362.
148. Sroussi HY, Epstein JB, Bensadoun RJ, et al. Cancer Med 2017;6(12)2918–2931.
149. Anderson G, Ebadi M, Vo K, et al. Cancers 2021;13:4912.
150. Murphy BA, Dietrich MS, Wells N, et al. Head Neck 2010;32(1):26–37.
151. Moosazadeh M, Shafaroudi AM, Gorji NE, et al. EBD 2021. Available at https://doi.org/10.1038/s41432-021-0209-8
152. Mays JW, Sarmadi M, Moutsopoulos NM. J Evid Base Dent Pract 2012:51:265–282.
153. Guandalini S, Assiri A. JAMA Pediatr 2014:168(3)272–278.
154. dos Santos JA, Normando AGC, Carvalho da Silva RL, et al. J Dent Res 2021;100(12):1321–1329.
155. Wolff A, Joshi RK, Ekstrom J, et al. Drugs R D 2017;17:1–28.
156. Plemons JM, Al-Hashimi I, Marel CL. JADA 2014;145: 867–873.
157. Negrini S, Emmi G, Greco M, et al. Clin Exper Med 2022;22: 9–25.
158. Proctor GB, Shaalan AM. J Dent Res 2021;100(11): 1201–1209.
159. De Martins Souza Pina G, Mota Carvalho R, de Freitas Silva BS, et al. Gerodontology 2020;37:317–331.
160. Heft MW, Baum BJ. J Dent Res 1984;63:1182–5.
161. Parker ML, Thonton-Evans G, Wei L, et al. MMWR 2020;69.
162. Fleming E, Afful J, Griffin SO. Prevalence of tooth loss among older adults: United States, 2015–2018. NCHS Data Brief, no 368. Hyattsville, MD: National Center for Health Statistics. 2020. Available at file:///Users/decker/Downloads/cdc_89421_DS1%20(1).pdf
163. Dye BA, Wetherspoon DJ, Mitnik GL. J Am Dent Assoc 2019;150(1):9–23.
164. Chen J, Ren CJ, Wu L, et al. Front Aging Neurosci. 2018;10(415):1–9.
165. Zhu Y, Hollis JH. Tooth loss and its association with dietary intake and diet quality in American adults. J Dent 2014;42: 1428–1435.
166. Ervin RB, Dye BA. Gerodontol 2012;29:e693–e702. PMID: 21923863
167. Tada A, Miura H. Arch Gerontol Geriatr 2014;59:497–505.
168. Watson S, McGowan L, McCrum LA, et al. Intl J Behav Nutr Phys Activ 2019;16:43.
169. Zelig R, Jones VM, Touger-Decker R, et al. J Dent Res Clin Translat Res 2019. E-published ahead of print 2/28/2019. Available at https://doi.org/10.1177/2380084419827532 PMID:30931718
170. Sahyoun NR, Lin CL, Krall E. J Am Diet Assoc 2003;103: 61–6.
171. Tonaizzo MP, Amorim PS, Muniz FW, Weidlich P. Clin Nutr 2018;37(3):824–30.
172. Honeywell S, Samavat H, Touger-Decker R, et al. J Dent Res Clin Translat Res doi: https://doi.org/10.1177/23800844211063859 [E-pub January 10, 2022].
173. Kikutani T, Yoshida M, Enoki H, et al. Geriatr Gerontol Intl. 2013;13(1):50–54.

174. Zelig R, Byham-Gray L, Singer SR, et al. J Aging Res Clin Pract 2018;7:107–114.
175. Toniazzo MP, Amorim PS, Rost JF, et al. Nutr Clin Pract. 2022; 1–10.
176. Zelig R, Goldstein S, Touger-Decker R, et al. JDR Clin Trans Res. 2020:2380084420981016.
177. Zelig R, Honeywell S, Touger-Decker R. In: Hogue CM, Ruiz JG, eds. Nutrition and Oral Health. Oral Health and Aging. Cham: Springer. 2022:29–45.
178. Al-Sultani HF, Field JC, Thomason JM, et al. J Dent Res Clin Translat Res 2019;4(1):29–40.
179. Bezerra AP, Gama LT, Pereira LJ. Clin Nutr 2021;40:3235–3249.
180. Awad MA, Morais JA, Wollin S, et al. J Dent Res 2012;91:39–46.
181. Kroll P, Hou L, Radaideh H, et al. J Oral Implant 2018;44: 313–324.
182. Boven GC, Raghoebar GM, Vissink A, et al. J Oral Rehab 2015;42:220–233.
183. Moynihan P, Varghese R. JDR Clin Translat Res 2021; 23800844211026608.
184. Kanazawa M, Suzuki H, Komogamine Y, et al. Clinical Oral Investigations 2019;23:2245–2252.
185. Mobley C, Dounis G. JADA 2013;144(2):e11–e15.
186. National Council on Aging. The Top 10 Most Common Chronic Conditions in Older Adults. 2022. Available at www.ncoa.org/article/the-top-10-most-common-chronic-conditions-in-older-adults
187. Young EH, Pan S, Yap AG, Reveles KR, Bhakta K. PLoS One 2021;16:e0255642.

CHAPTER 76

Esophagus and Stomach

Vikram Rangan and Andrew Ukleja

CHAPTER OUTLINE

Abbreviations

CagA, cytotoxin-associated gene; **CCK**, cholecystokinin; **CVS**, cyclic vomiting syndrome; **ECL**, enterochromaffin-like; **EGD**, esophagogastroduodenoscopy; **EMR**, endoscopic mucosal resection; **EndoFLIP**, endoscopic functional lumen imaging probe; **EoE**, Eosinophilic esophagitis; **eos/hpf**, eosinophils per high power field; **ESD**, endoscopic submucosal dissection; **FDA**, Food and Drug Administration; **GERD**, Gastroesophageal Reflux Disorder; **GES**, Gastric electrical stimulation; **GLP-1**, glucagon like peptide-1; **GP**, Gastroparesis; **GU**, Gastric ulcer; **ICCs**, interstitial cells of Cajal; **IEM**, ineffective esophageal motility; **IF**, intrinsic factor; **IL**, interleukin; **IM**, intestinal metaplasia; **LES**, lower esophageal sphincter; **MMCs**, myoelectrical complexes; **NSAID**, nonsteroidal anti-inflammatory drug; **PN**, parenteral nutrition; **POEM**, Per Oral Endoscopic Myotomy; **PPI**, proton pump inhibitor; **PUD**, peptic ulcer disease; **TIF**, transoral incisionless fundoplication; **TRPV1**, transient receptor potential cation channel subfamily V member 1; **UES**, upper esophageal sphincter; **VacA**, vacuolating cytotoxin A

Introduction

The esophagus and stomach are critical structures involved in oral intake and digestion. Absorption of nutrients in these organs is minimal. Without a properly functioning esophagus and stomach, the ability to eat and initially digest food can be significantly impaired. Additionally, partial or total surgical resection of the esophagus or stomach for symptomatic disease states can also markedly affect a person's ability to eat or drink. Physiologic and anatomic aspects of the esophagus and stomach are covered in detail in the chapter on the nutritional physiology of the alimentary tract.

Esophagus

Anatomy

The esophagus is a tubular structure approximately 30 cm in length (**Figure 76.1**).[1] The organ is artificially divided into the proximal, mid, and distal esophagus. This structure is extremely muscular, which correlates to its primary function: propulsion. The funnel-shaped pharynx joins the mouth to the esophagus. There is a ring of tissue known as the upper esophageal sphincter (UES), where the pharynx connects to the esophagus. The sphincter opens to allow food or fluids to pass during swallowing. The upper esophagus consists of striated muscle. There is an inner circular layer and an outer longitudinal layer of muscle. At the level of the aortic arch, the esophageal striated muscle transforms into smooth muscle. A ring of thick, smooth muscle called the lower esophageal sphincter (LES), sits at the bottom of the esophagus and is approximately 40 cm from the incisors (teeth).

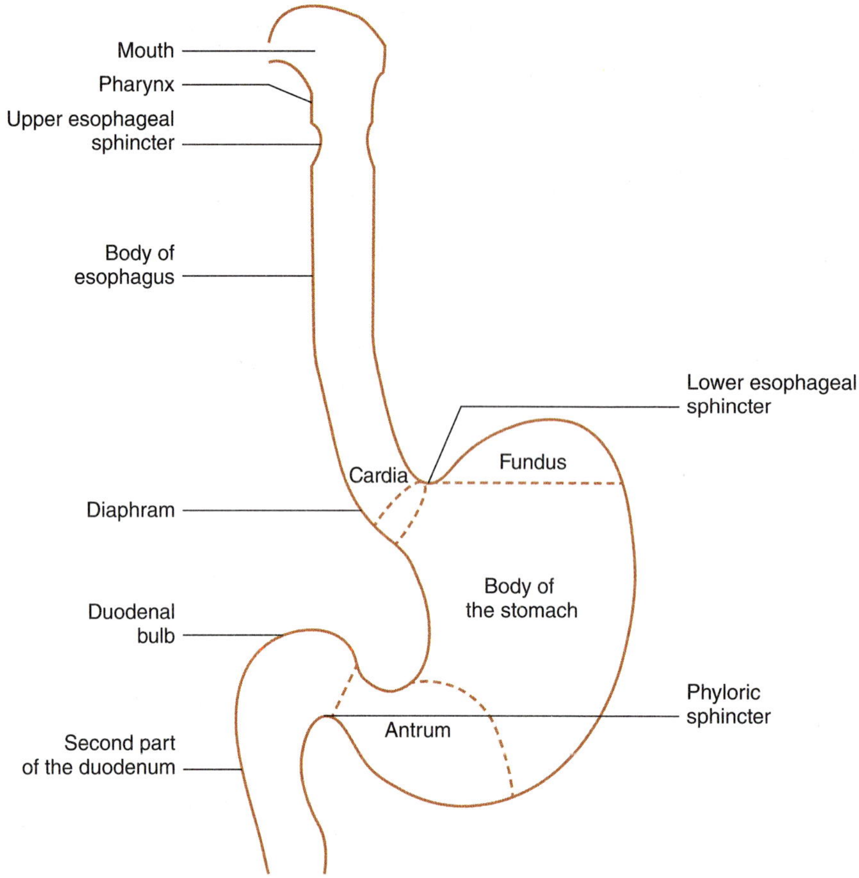

Figure 76.1 Anatomy of the human upper gastrointestinal tract.

The LES serves to allow food and fluids to flow from the esophagus to the stomach when it is relaxed and to prevent regurgitation of gastric materials back into the esophagus when it is in a contracted state. The muscle of the esophagus receives its innervation from cranial nerve X.[2] This nerve originates in the dorsal motor nucleus of the vagus nerve and synapses on the myenteric plexus (esophageal nerve system).

Disease

When disease of the esophagus exists, it generally involves a process that affects the esophageal lining (mucosa) or the muscular component. Diseases of the esophagus can give symptoms that include trouble swallowing (dysphagia), noncardiac chest pain, heartburn, regurgitation, and weight loss.

Muscular (Motor)

Achalasia

The classic muscle disorder of the esophagus is known as achalasia.[3] The primary features are a lack of propulsive esophageal muscle contractions (peristalsis) in combination with the failure to relax the LES, resulting from loss of inhibitory ganglion cells of the distal esophagus. This disrupts the flow of food and liquid from the mouth through the esophagus and into the stomach, resulting in classic symptoms of dysphagia (to both solids and liquids) and regurgitation. Atypical chest pain and weight loss are less common symptoms.

Three different subtypes of achalasia have been described and are differentiated based on the behavior of the esophageal body. In type I, there is no meaningful muscle activity within the esophageal body. In type 2, there is panpressurization of the esophageal body. In type 3, there is spastic contraction of the esophageal body. These three subtypes of achalasia are distinguished based on high-resolution esophageal manometry, which is the definitive test for evaluating this condition.[4] The test entails transnasal placement of an esophageal probe that measures pressures within both the esophageal body and LES with liquid swallows.

While not diagnostic, an upper endoscopy is required in all patients with suspected achalasia to rule out malignancy or other structural abnormality of the distal esophagus that can present with achalasia-type symptoms (frequently labeled "pseudoachalasia"). A barium esophagram is a helpful adjunct in achalasia diagnosis and classically shows a dilated esophagus, delayed clearance of barium, or other contrast material, lack of organized peristalsis, and a tight LES that does not adequately relax, with the resulting appearance of "bird's beak" esophagus. However, a barium esophagram is not a fully sensitive test for achalasia, as many individuals (particularly those in the early stage of the disease) may not have this classic "bird's beak" appearance. Upper endoscopy with an endoscopic functional lumen imaging probe (EndoFLIP) can provide additional diagnostic information.[5] This newer diagnostic modality entails using a fluid-filled balloon to measure distensibility (i.e., the ratio of cross-sectional area to balloon pressure) of the lower esophageal sphincter and secondary peristalsis within the esophageal body.[6] This diagnostic modality is particularly helpful in cases where manometry and barium swallow are equivocal.

There are three definitive treatments for achalasia. The classic treatment is Heller Myotomy, which entails surgical myotomy of the lower esophageal sphincter, usually in combination with an anti-reflux intervention (like a Toupet or Dor fundoplication). This option is preferred for those with concomitant hiatal hernia or for whom there is concern over postmyotomy reflux. Pneumatic balloon dilation (with the goal of disrupting the lower esophageal sphincter) is another potential intervention for achalasia. Although it has the advantage of being quicker and less invasive, it is associated with an estimated risk of perforation of 1.9%,[7] with higher rates noted in older studies. It also frequently requires multiple attempts (with a progressive increase in the size of dilating balloon) before effective intervention is achieved.[8] This treatment is generally reserved for those with achalasia type 2 and is not recommended for type 3.[7] The newest treatment is Per Oral Endoscopic Myotomy (POEM), which entails endoscopic myotomy of the longitudinal muscle fibers of the distal esophagus. POEM is considered the most effective treatment for achalasia, as it allows for a longer myotomy than Heller Myotomy. It was shown in a randomized, controlled trial to be superior to pneumatic dilation in treatment-naïve achalasia[9] and has been shown to be superior to Heller myotomy as well in both type 1 and type 3 achalasia,[10,11] with success rates over 90% for all three achalasia subtypes. It is the preferred treatment for those with Achalasia type 3.[7,10,11] However, it is also associated with an increased likelihood of gastroesophageal reflux disorder (GERD) following the procedure,[12] with more than 50% of individuals undergoing this procedure developing objective evidence of pathologic reflux after 1 year.[13]

Regardless of treatment modality, patients with achalasia will usually continue to have poor esophageal propulsion, even after intervention on the LES. These patients generally remain dependent on gravity and must stay upright after eating to allow materials to pass from the mouth to the stomach. Patients with achalasia are at increased lifetime risk of developing esophageal cancer, even after treatment, although the overall risk is low (noted to be 1.3% in one study).[14] They are particularly predisposed to the development of squamous cell cancer of the esophagus, with the rate increasing with increased length of follow-up.[15] As a result, periodic endoscopy for cancer surveillance is often done in patients with achalasia, although cancer surveillance on a universal basis is not currently recommended.[7] In those who are not candidates for definitive intervention (generally those patients with a large number of comorbidities thought to be poor surgical candidates), other interventions can include a trial of botox injection to the lower esophageal sphincter, which has been proven to be effective, but with a high rate of symptom relapse, especially 6 to 12 months or more after treatment.[16] Medications with smooth muscle relaxant properties (such as nitrates or calcium channel blockers) can also be used, although they are generally less effective than the definitive interventions discussed above.

Other Esophageal Motor Disorders

EGJ outflow obstruction is another condition that is frequently noted on esophageal manometry and is characterized by a lack of adequate deglutitive relaxation of the LES. Unlike achalasia; however, this condition is notable for normal peristalsis within the body of the esophagus. EGJ outflow obstruction can be incidentally noted in asymptomatic individuals, and its clinical relevance is uncertain. In many cases, this condition may be caused by opioid medications, which are associated with impaired LES relaxation,[17,18] and spastic peristalsis of the esophageal body.[17] However, symptomatic individuals with EGJ outflow obstruction can present with dysphagia, retrosternal pain, and heartburn/regurgitation. Once a mechanical cause has been ruled out, treatment options are like those for achalasia in those with persistent dysphagia symptoms, including POEM, Heller myotomy, pneumatic dilation, and a trial of botox injection to the lower esophagus. However, it has been estimated that more than half of patients with EGJ outflow obstruction may have spontaneous resolution of symptoms over 6 months to 2 years, and thus only a minority require an intervention.[19]

Patients may also have spastic activity of the body of the esophagus in the setting of normal relaxation of the lower esophageal sphincter, a condition known as distal esophageal spasm, where inadequate inhibitory nerve signaling to the distal esophagus results in premature contraction after swallowing.[20] Diagnosis is made based on high-resolution esophageal manometry, and barium esophagram will sometimes show a characteristic corkscrew appearance of the esophagus. Treatment is aimed at underlying symptoms, which generally include dysphagia, chest pain, and GERD, although clinical presentation varies significantly, and many patients are asymptomatic. Smooth muscle relaxants, like nitrates, phosphodiesterase-5 inhibitors, and calcium channel blockers, can be used, although large-scale studies of their efficacy for this condition are lacking. Tricyclic antidepressants can also be used in patients with pain predominant symptoms.[21] A trial of antireflux medication is also reasonable for patients with concomitant reflux symptoms.[20] Botox injection and surgical myotomy can be considered in rare cases where patients have persistent symptoms refractory to medications.[20]

Hypomotility (decreased muscle contractions) of the esophagus can occur with systemic disease in which the muscle and nerves of the esophagus are affected, resulting in ineffective esophageal motility (IEM) or absent contractility.[22] The latter occurs commonly in scleroderma and other connective tissue diseases. There are no known effective treatments for hypomotility disorder effects on the esophagus.[23] Some data suggest that medications such as Buspirone, Prucalopride, Bethanechol, and Pyridostigmine can augment esophageal contractility. However, research is limited, and none of these medications have been shown to improve symptoms of dysphagia.[24-29] There is a known correlation between GERD and ineffective esophageal motility (although clear causation has not been established), and it is recommended that underlying acid reflux be adequately controlled in all patients with this condition.[23]

Mucosal Disease

Gastroesophageal Reflux Disorder (GERD)

GERD was previously broadly defined as reflux of stomach contents causing troublesome symptoms and complications.[30] Classic symptoms of GERD include heartburn and dysphagia. Less common (but still frequently noted) symptoms include chest pain, odynophagia, and nausea/vomiting.[31] It is also possible for GERD to cause extraesophageal symptoms like cough and hoarseness. GERD is a frequently encountered condition. Although estimates of its prevalence vary widely, one recent meta-analysis estimated a pooled worldwide prevalence of 13.3% of at least weekly GERD symptoms, with age >50 years, smoking, nonsteroidal anti-inflammatory drug (NSAID) use, and obesity all serving as risk factors.[32]

GERD is caused by a combination of the esophagogastric pressure gradient (often seen in patients with central obesity and increased intra-abdominal pressure), as well as impaired barrier function of the esophagogastric junction (frequently seen in those with hiatal hernia).[33] Approximately 18% of the U.S. population report GERD symptoms weekly.[34] GERD can be investigated by esophagogastroduodenoscopy (EGD), which assesses for inflammation and mucosal damage of the distal esophagus. However, the majority of patients with GERD have nonerosive reflux disease,[35] where pathologic reflux of stomach contents causes bothersome symptoms but no objective evidence of mucosal damage to the distal esophagus. Ambulatory pH monitoring can provide further diagnostic information by objectively measuring the amount of acid exposure in the distal esophagus. This can be accomplished with a transnasal probe or a wireless capsule attached to the distal esophagus. The former has the advantage of collecting impedance data, which provide insight into the type of reflux (liquid, gas, or mixed), the proximal extent of reflux in the esophagus, and the frequency of nonacidic reflux events. The latter has the advantage of generally being more comfortable for the patient, allowing for a period of data collection of up to 96 hours (testing using a transnasal probe generally provides only 24 hours of data). The more recent Lyon consensus incorporates these objective measures of reflux, defining GERD as the presence of advanced grade erosive esophagitis, long segment Barrett's esophagus, peptic stricture, or distal esophageal acid exposure time of over 6% on ambulatory pH testing.[36]

Patients with GERD may have impaired quality of life because of intermittent pain and nausea. Anorexia may develop, because chronic discomfort and nausea from reflux can affect appetite. In addition, if the GERD symptoms are worse with eating, this will also result in reduced oral intake. More severe manifestations of GERD are esophageal ulceration and stricture formation.[37] GERD-based strictures are more commonly seen in the distal esophagus. These strictures may require balloon dilation with the use of an endoscope. The "obstruction" caused by these strictures can lead to a change in diet or a markedly reduced oral intake, resulting in malnutrition and weight loss. Effective medical treatment strategies for GERD exist, such as proton pump inhibitors (PPI) and H2 receptor antagonists. However, many patients with GERD are recalcitrant to medications, particularly those with nonerosive reflux disease.[35] Corrective surgical therapy may be required for these patients. The most common surgery is a Nissen fundoplication.[38] Newer antireflux procedures, including transoral incisionless fundoplication (TIF), magnetic sphincter augmentation, and radiofrequency ablation of the LES (with the goal of remodeling to prevent GERD), have similar outcomes to traditional Nissen fundoplication.[39]

Barrett's Esophagus

Patients with longstanding GERD are at risk of developing a condition called Barrett's esophagus, where the normal stratified squamous epithelium of the distal esophagus is replaced with metaplastic intestinal columnar epithelial cells. This condition is associated with an increased eventual risk of developing adenocarcinoma of the esophagus. There is thought to be a progression from GERD-related inflammatory changes of the distal esophagus to metaplasia, dysplasia, and eventually to cancer. The risk of cancer progression has been estimated to be 0.2% to 0.5% per year in those with Barrett's esophagus without dysplasia and 7% per year in those with high-grade dysplasia.[40] Increased length of Barrett's esophagus is also associated with an increased risk of developing esophageal cancer. In addition to chronic GERD, risk factors for Barrett's esophagus include obesity, family history of Barrett's esophagus or esophageal cancer, smoking, male gender, and Caucasian race.[40] Barrett's esophagus itself does not cause any symptoms, and an endoscopy with biopsy is needed for a definitive diagnosis.

Esophageal Cancer

Mucosal-based disease can also affect a person's ability to eat. The best example of this is cancer of the esophagus. Adenocarcinoma is now the most common cause of esophageal cancer in the United States, while worldwide, squamous cell carcinoma is the most common form. Esophageal cancer is most frequently caused by tobacco and alcohol. Less common associations include achalasia, lack of trace metals (especially selenium), lye ingestion, ionizing radiation, and human papillomavirus. Deficiencies of vitamins A and C are associated with the development of squamous cell cancer of the esophagus. Overall, cancer of the esophagus is more common in men by a 3:1 ratio.[41]

In patients with esophageal cancer, oral intake can be compromised because of esophageal obstruction. The tumor itself can be shrunk temporarily with the use of endoscopic therapies, including tumor tissue ablation. The placement of an esophageal stent across the tumor can also temporarily open the esophagus.[42] Frequently, the patient cannot eat or drink enough to maintain adequate nutritional status, and placement of a gastrostomy feeding tube is needed. Some surgeons prefer a jejunostomy feeding tube to a gastrostomy feeding tube for patients who will receive surgery, to avoid a hole in the stomach requiring repair before being pulled up into the thoracic cavity after esophagectomy.[43]

Esophagectomy is the primary treatment for esophageal cancer and is generally reserved for patients with an optimal chance of complete cure.[44] Ivor-Lewis esophagectomy entails the removal of most of the esophagus and a pull-up of the stomach into the thoracic cavity and the attachment of the stomach to the very upper portion of the esophagus just below the UES. This surgery carries a 5% to 10% mortality rate. Morbidity from this surgery can include anastomotic leakage, pulmonary problems, and coronary events. Patients can develop anastomotic strictures, gastroparesis, and regurgitation after surgery. This problem can become significant enough to require a jejunal feeding tube. Treatment outcomes in esophageal cancer are poor if curative surgery is not an option. Radiation is rarely used as monotherapy for esophageal cancer. Chemotherapy can be used as monotherapy but is more commonly used as systemic therapy for patients with metastatic disease. Combined radiotherapy and chemotherapy are used for patients with regional metastasis.

Eosinophilic Esophagitis

Eosinophilic esophagitis (EoE) is a chronic immune/antigen-mediated disease characterized by eosinophil-predominate esophageal inflammation and dysfunction. In adults, classic symptoms are dysphagia and food impaction. Other manifestations include abdominal pain, chest pain, heartburn, and vomiting. Diagnosis of EoE is based on symptoms of esophageal dysfunction and esophageal biopsies showing high eosinophils per high power field (eos/hpf) and persistence of eosinophilic infiltration after a trial of proton pump inhibitors in the absence of secondary causes of eosinophilia.[45] In EoE, the esophageal inflammation can progress over time to esophageal fibrotic remodeling, resulting in narrow caliber esophagus and esophageal stricture. Patients may present with inflammatory or fibrostenotic phenotypes. There is an interacting relationship between GERD and EoE, which is complex. It has been suggested that GERD may contribute to the pathogenesis of EoE by causing esophageal mucosal integrity changes, promoting trans-epithelial allergen permeation, and subsequent allergic immune activation.[46] Both GERD and EoE are more common in males.

Treatment approaches include elimination diets, pharmacotherapy, and esophageal dilation in case of esophageal stricture.[47] After a trial of PPIs, topical corticosteroids, typically asthma steroid preparations swallowed, not inhaled, are used as a mainstay of therapy.[48] Oral budesonide has been used as maintenance therapy in refractory cases of EoE. PPI therapy leads to clinical and histologic (<15 eos/hpf) remission in half of the patients with suspected EoE. Therefore, PPIs can be considered a first-line treatment due to their safety profile, easy administration, and high response rates. The choice of therapy is often made after the patient is informed about the pros and cons of alternative therapies, such as diet and topical steroids. In case of unresponsiveness to PPI therapy, a choice between topical steroid or dietary therapy should be made based on the severity of the disease (severe symptoms should be treated with topical steroid therapy) or the patient's lifestyle and preferences. Dupilumab, a VelocImmune-derived human monoclonal antibody against the interleukin (IL) 4 receptor that inhibits IL4 and IL13 signaling, has been recently approved by the U.S. Food and Drug Administration (FDA) for the treatment of EoE.[49] Dupilumab can reduce dysphagia, eosinophilic infiltration, and abnormal endoscopic features and is generally well tolerated. Diet modifications have been used successfully, especially in the pediatric population. The concept is to remove food allergy triggers of EoE from the diet. The elimination diet eliminates the six foods most associated with food allergy (cow's milk protein, wheat, egg, soy, peanut/tree nuts, and seafood).[50] The most common trigger is cow's milk, followed by wheat, eggs, and legumes. Often, allergy skin testing is performed to identify EoE food triggers. Once the food responsible for EoE has been identified, long-term avoidance has been recommended to maintain disease remission. Diet elimination has been effective in up to half of adult patients and even more in pediatric populations.

Stomach

Anatomy

The stomach is a sac-like organ, a J-shaped dilation of the digestive tract between the esophagus and the duodenum. It acts as a reservoir that stores ingested food, grinds it, and releases digested food gradually into the duodenum in a controlled fashion. The stomach volume ranges from 1.5 to 2 L in adults. It can be divided into four regions defined by anatomic or histologic landmarks.[51] The cardia is a small area of the stomach immediately adjacent to its junction with the esophagus. The fundus is a dome-shaped area of the superior portion of the stomach. The body or corpus of the stomach, which is the largest portion of the stomach, is located below the fundus. The last region is the antrum, which extends from the body to the junction of the pylorus with the duodenal bulb. The pylorus is a tubular structure between the duodenum and the stomach that contains a circular muscle called the pyloric sphincter (see **Figure 76.2**). There are variations in position and contour of the stomach, depending on body habitus. For example, atonic or hypotonic types are associated with a more longitudinal axis of the stomach, while hypertonic types with a more transverse axis. The anatomic landmarks also correspond roughly with the mucosal differences observed in histology. The gastric wall consists of four layers: mucosa, submucosa, muscularis propria, and serosa.[52] Mucosa, the inner layer, covers the inside of the gastric lumen and includes surface epithelium, lamina propria, and muscularis mucosae. The underlying submucosa contains connective tissue (elastin fibers), immune cells, arterioles, venules, lymphatics, and the submucosal (Meissner's) plexus. The third layer, the muscularis propria, contains three muscle layers, including inner oblique, middle circular, and external longitudinal and myenteric (Auerbach's) plexus. The external layer of the stomach, serosa, is a continuation of the visceral peritoneum, extending beyond the

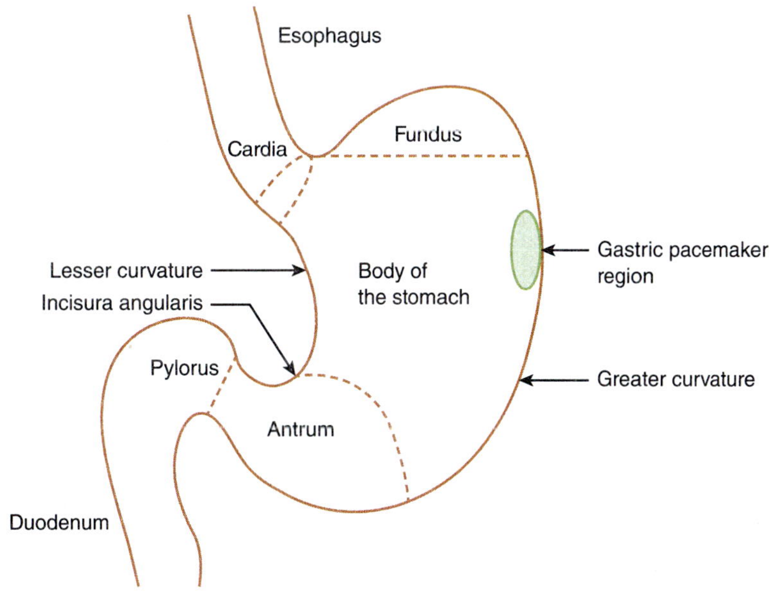

Figure 76.2 Parts of the stomach.

stomach as a double layer from lesser curvature as lesser omentum, and from the greater curvature as greater omentum. The gastric mucosal surface is composed primarily of a single layer of columnar epithelial cells (tunica propria). The mucosa contains mucous cells and gastric glands. The surface mucous cells secrete mucus, which acts as a mechanical barrier, a lubricant for food, and antibacterial protection. Therefore, mucus plays an important cytoprotective role from acid, pepsin, ingested substances, and pathogens.

The gastric glands in the mucosa layer form tubular structures that are present throughout the stomach and contain different types of specialized epithelial cells, depending on the region. Three types of gastric glands are recognized: cardiac, fundic (oxyntic), and pyloric. The cardiac glands are populated mainly by mucous-producing cells and undifferentiated cells. The fundus and body, the acid-secreting segment of the stomach, contain the oxyntic or parietal glands. They are the most common gastric glands responsible for the secretion of acid, gastric enzymes (lipase, pepsinogens), and intrinsic factor (IF). Parietal cells, the main cell type in the oxyntic glands, produce hydrochloric acid and contain H^+-K^+-ATPase in the apical microvillus membrane. Acid secretion begins in response to hormonal, chemical, or neurologic stimuli. Parietal cells also secrete IF. Deeper layers of the oxyntic glands contain chief (zymogenic) cells, which are responsible for the synthesis and secretion of pepsinogens I and II. Both pepsinogens are stored in zymogen granules and are released into the gastric lumen, where they are converted to pepsin in the contact with gastric acid. Endocrine or enteroendocrine cells, somatostatin-containing D cells, and histamine-secreting enterochromaffin-like (ECL) cells are scattered throughout the oxyntic glands. The antrum and pylorus contain the pyloric (antral) glands, composed of mucous and endocrine cells, including gastrin-producing G cells. The G cells play a vital role in the control of gastric acid secretion.[53] After meal ingestion, gastrin is released rapidly into the bloodstream in response to multiple stimuli such as gastric distention, amino acids and peptides, and vagal stimulation. Gastrin is also secreted directly into the stomach lumen, where it acts as gastric growth and differentiation factor. In proximity to G cells are antral D cells, which secrete somatostatin.[54] Somatostatin acts as a potent inhibitor of gastrin and acid secretion. The effects of somatostatin involve complex mechanisms such as the paracrine effect (indirect action on G cells and direct action on ECL cells) or endocrine effect (distant action on parietal cells).[55]

The arterial blood supply to the stomach is derived from branches of the celiac artery, including common hepatic, left gastric, and splenic arteries. The stomach has diversified collateral pathways of arterial blood supply. The venous system of the stomach drains into the portal vein, and the splenic and superior mesenteric veins. Lymph from the gastric wall drains via lymphatics (dense subperitoneal plexus) into regional lymph nodes around the stomach. The innervation of the stomach is complex. The stomach's autonomic innervation (sympathetic and parasympathetic) comes through the nerves along the visceral arteries.[56] The sympathetic supply is derived from preganglionic fibers arising from anterior spinal nerves located in the thoracic segments 6th to 9th, which synapse in the celiac plexus and ganglia. Postganglionic fibers course through the celiac plexus alongside the branches of the celiac and superior mesenteric arteries to the stomach. The parasympathetic supply arises in the dorsal vagal nucleus on the floor of the fourth ventricle and comes from the right and left vagus nerves, esophageal plexus, and anterior and posterior vagal trunks at the level of the gastric cardia. The vagal trunks continue as the anterior and posterior nerves of Latarjet form gastric and pyloric branches to the stomach. Preganglionic fibers synapse with the ganglion cells in the submucosal and myenteric plexuses, and postganglionic fibers supply gastric glands and smooth muscles of the stomach wall.

Gastric Neuromuscular Function

Gastric neuromuscular function is a complex and highly regulated process that allows gastric emptying in a controlled fashion for further optimal digestion and absorption of the nutrients in the small bowel (see **Figure 76.3**).[57] The major gastric activities include receptive relaxation of the fundus to receive the food, peristaltic waves of the body and antrum to grind the food into small particles, and antral contractions toward the pylorus allowing for chyme emptying gradually into the duodenum. The neuromuscular activity of the stomach is regulated by the central nervous system, parasympathetic and sympathetic nervous system, and the enteric nervous system. The interstitial cells of Cajal (ICCs) form "the pacemaker region" on the greater curvature of the stomach between the fundus and the proximal body, and they play the key role in the generation of action potentials, regulation of the frequency of smooth muscle contractions, and initiation of peristaltic activity.[58] Slow waves or "gastric myoelectrical activity" generated by the ICCs regulate, control, and pace gastric smooth muscle contractions. In the normal stomach, the slow waves occur at approximately three cycles per minute and propagate distally toward the pylorus. After food is swallowed, vagal nerve stimuli induce relaxation of the stomach fundus, and the fasting migrating motor complex pattern (discussed later) is switched to the fed pattern, which lasts from 2 to 8 hours, depending on meal composition. The ingested food must be reduced to liquid or semiliquid consistency before emptying into the duodenum. The rate of gastric emptying is regulated by gastrointestinal hormones secreted by the stomach and duodenum, as well as input from the autonomic nervous system. Gastric emptying is inhibited by the chyme's physiochemical properties, such as large volume and large particles, high acidity, high fat content, hypertonicity, and high-nutrient concentration. Neurotransmitters ultimately regulate the contractions and relaxation of gastric smooth muscles. Antro-duodenal contractions are controlled by GI hormones and input from the nervous system.[59] For example, cholecystokinin (CCK), a gastric braking hormone, is released from neuroendocrine cells in the duodenum in response to exposure to gastric acid, amino acids, and fatty acids. CCK leads to gastric inhibition by slowing gastric emptying and by stimulating glucagon like peptide-1 (GLP-1) secretion, which also has a potent inhibitory effect. During fasting, electrical and contractile activities of the stomach occur in a regular pattern called migrating myoelectrical complexes (MMCs), including three phases, which repeat approximately every 90 to 120 minutes.[60] In phase 1, there is little or no contractile activity. Random and irregular contractions occur during phase 2, while phase 3 is characterized by a burst of regular, short-lasting (5–10 minutes), high-amplitude, phasic contractions. The MMCs travel through the small intestine before terminating in the distal ileum. In gastroparesis, a case of neuromuscular dysfunction of the stomach, dysregulation of the gastric emptying for liquids and solids is observed.[61]

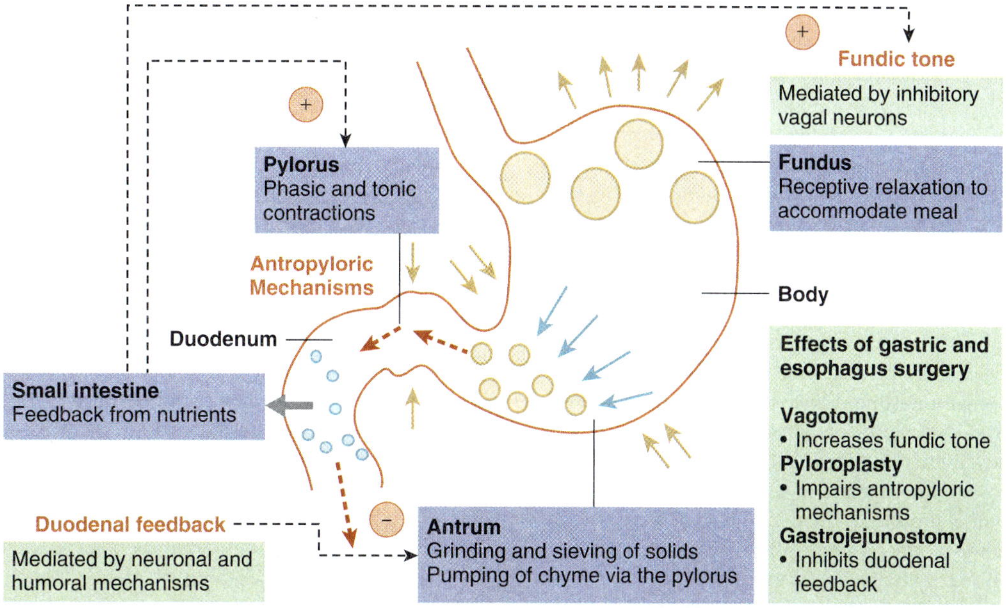

Figure 76.3 Mechanisms involved in a meal passage through the stomach in health. Fundus and proximal body of stomach relaxes to accommodate a large volume of ingested meal without change in intragastric pressure. The food is slowly transferred in to the body and antrum where peristaltic waves mix the food with gastric acid and grinding of the solids takes place into small particles. Antropyloric mechanisms allow 1 to 2 mm particles empty into the duodenum. Duodenal feedback in response to the load of nutrients is associated with increase in pyloric tone, inhibition in antral contractions, and modification in fundic tone; therefore, allows for slowing down emptying of gastric content. Examples of gastric surgery effects on responses involved in gastric emptying of ingested meal.

Gastric Digestion

Gastric digestion plays an important role in energy and nutrient absorption from ingested food. This complex process involves both physical and chemical breakdown of the food. Relevant digestion of carbohydrates, proteins, and fats takes place in the stomach. Normal gastric secretion and motility are needed for proper food digestion. Gastric acid allows softening of food particles. Digestive enzymes pepsin and gastric lipase initiate the hydrolysis of proteins and fats, while peristaltic antral contractions grind food particles into a smaller size to allow gastric emptying. In addition, food structure impacts the breakdown of nutrients and their release from ingested food. Following oral mastication, the rest of the food breakdown occurs in the stomach. The rate of food breakdown in the stomach impacts gastric emptying and further nutrient absorption in the intestine. The food grinding and mixing is under a complex feedback system. Gastric acid is necessary to soften food and break it down at a very low gastric pH of ~2. The stomach produces approximately 2 to 3 L of fluids daily. Gastric fluid is composed of mucus, acid, enzymes, and intrinsic factor.

The major digestive activity of the stomach is proteolysis. Protein digestion begins in the stomach by the action of an endopeptidase, pepsin.[62] In response to a meal and low gastric pH, the enzyme precursor pepsinogen is secreted by the chief cells, which is activated to pepsin by hydrochloric acid present in the lumen of the stomach. Optimal activity of pepsin is at a pH between 1 and 3. Pepsin is neutralized on contact with alkaline pancreatic juices in the proximal small bowel when the pH exceeds 5. Fat digestion also starts in the stomach.[63] Lingual lipase swallowed with food remains active in the stomach initially. Gastric lipase produced by the body and fundus cells contributes to the enzymatic digestion of triglycerides. The optimal lipolytic activity of gastric lipase is within a gastric pH between 3 and 6. Gastric lipase breaks down triacylglycerols into diglycerides and fatty acids. The diglycerides derived in this process serve as further emulsifiers. Except for short-chain fatty acids, no fat absorption occurs in the stomach. In adults, gastric lipase plays only a minor role in fat digestion. However, the contribution of gastric lipase can be significant in pancreatic lipase deficiency or lipase inactivation (Zollinger-Ellison syndrome). Starch digestion begins in the mouth with the action of salivary amylase. Both salivary and pancreatic amylases have optimum activities close to a neutral pH of 7. Although salivary amylase is destroyed by gastric acid, some enzymatic activity still occurs within a bolus of food before mixing with gastric acid. Most of the stomach's digestive processes depend on adequate acid production.

Gastric acid production is a complex process that occurs in response to stimulation by gastrin produced by antral cells, histamine from mast cells and enterochromaffin-like cells, and acetylcholine release from vagal nerve stimulation.[54,59] A significant increase in gastric acid secretion is observed with oral intake compared with a baseline secretory output during fasting. Three phases of gastric acid secretion have been recognized: cephalic, gastric, and intestinal.[64] The cephalic phase results from the stimulation of gastric acid secretion in response to the smell, taste, or thoughts of food, and it is mediated via the vagal nerve. In case of inadvertent vagal nerve injury (e.g., Nissen fundoplication), no increase in gastric acid production is seen during the cephalic phase. Sham feeding can confirm this, when food is not swallowed but only smelled, chewed, and spat. In the gastric phase, the direct presence of food in the stomach stimulates acid production in response to gastric distention via stretch receptors and peptide stimulation via gastric chemoreceptors. During the intestinal phase, nutrients entering the small intestine cause intestinal distention, peptides stimulate intestinal chemoreceptors, and hydrochloric acid stimulates the small intestine by activating negative feedback mechanisms resulting in the inhibition of gastric acid secretion. The chyme in the proximal intestine stimulates the release of secretin and cholecystokinin and, in combination with cholinergic nerve, stimuli cause the pancreas to secrete bicarbonate and digestive enzymes into the intestinal lumen, both acting as negative feedback by slowing gastric emptying. Gastric acid also facilitates the absorption of iron, vitamin B_{12}, calcium, and medications (e.g., thyroxin) and potentially prevents enteric infections and bacterial overgrowth.[65] A low gastric pH is required to convert the ferric to a ferrous form of iron, which allows efficient iron absorption. Gastric acid imbalance linked to high pH in the small intestine may impact the gut microbiome and result in dysbiosis. Reduced gastric acid production seen in pathologic conditions or altered gastric anatomy will impact nutrient digestion and absorption in the intestine. After a total gastrectomy, no IF is produced. Therefore, enteral absorption of vitamin B_{12} from the GI tract is impossible. Parenteral vitamin B_{12} supplementation (intravenous, IM, or intranasal) is necessary for this setting. Similarly, the use of high doses of PPIs, which results in low acid production, has been linked to impaired absorption of vitamin B_{12} and calcium.

Gastritis

Gastritis is a common acute or chronic inflammatory condition of the gastric mucosal layer. The extent and distribution of gastritis are linked to the underlying cause and host responses. It can be classified based on the etiology, topographic distribution, and morphology.[66] Acute gastritis may result from infection, alcohol or drug-related ischemia, radiation, or exposure to corrosive agents. The typical symptoms include severe epigastric pain, nausea, vomiting, or anorexia. The illness typically develops suddenly and resolves rapidly. The treatment of acute gastritis includes temporarily avoiding solid food followed by a slow diet advancement, antispasmodics, antiemetics, sedatives, and fluids. In severe cases, intravenous hydration and medications may be needed. Common causes of chronic gastritis include the use of aspirin or NSAIDs, *Helicobacter pylori* infection, or pernicious anemia (autoimmune).[67] The symptoms often include epigastric discomfort or burning, fullness, upper abdominal pain, and poor appetite. The treatment of chronic gastritis depends on its cause. Antacids are the mainstay of therapy, including histamine-2 blockers or PPIs. They are effective in promoting healing and resolving symptoms. Drugs that cause gastric mucosal injury should be stopped and avoided. An antibiotic regimen is necessary to treat chronic gastritis caused by *H. pylori* infection.[68] *Helicobacter pylori* infection is the most common chronic bacterial infection in humans and is associated with the development and recurrence of peptic ulcer, atrophic gastritis, or gastric cancer.[69,70] Fortunately, a decline in *H. pylori* infection has been observed in the United States over the last few decades. *H. pylori* infection can be diagnosed by invasive methods (endoscopy with biopsy for histology or culture, or rapid urease test) or by noninvasive tests (serology, urea breath test, or stool antigen test).[71] When *H. pylori* infection is confirmed, eradication is recommended to prevent further complications, such as peptic ulcer, chronic active gastritis, mucosa-associated lymphoma, and gastric cancer.[72]

Treatment typically consists of triple or quadruple therapy containing two antibiotics and an adequate dose of PPIs with or without bismuth.[73,74] Antibiotics should be selected carefully, depending on the history of drug allergies, local antibiotic resistance, prior antibiotic exposure, and safety profile.[74] Therapy failures are relatively common due to noncompliance, drug intolerance and their adverse effects, previous antibiotic exposure, short therapy course, and high drug resistance rates. High virulence factors for *H. pylori* include the presence of a cytotoxin-associated gene (CagA) and vacuolating cytotoxin A (VacA). In many countries, the standard triple therapy has been replaced by quadruple therapy due to high resistance rates.[75] New therapies are emerging, including rifabutin-based triple regimen, which appears to achieve much higher *H. pylori* eradication rates.[76,77] While rifabutin is currently considered a rescue therapy, it may become a first line of treatment in the era of antibiotic resistance. The eradication of *H. Pylori* should be confirmed, especially in cases of peptic ulcer or family history of gastric cancer.

Other infectious gastritis includes viral (CMV, HSV) or bacterial (Syphilis, tuberculosis) infections.[66] Viral gastritis predominantly affects immunocompromised patients. Syphilitic gastritis has been reported in conjunction with hepatitis, proctitis, and HIV infection.[78] Acute gastritis of early secondary syphilis may cause diffuse thickening of gastric folds, ulcerations, or mid-stomach stricture. A variety of granulomatous diseases can affect the stomach, with Crohn's disease and sarcoidosis as the most common. Bacterial infections with spirochetes and mycobacteria, fungi, parasites, foreign bodies, Langerhans cell histiocytosis, and eosinophilic granulomatosis can result in granuloma formation. Lymphocytic gastritis is a rare disorder characterized by dense lymphocytic infiltration of the gastric epithelium.[79] In adults, lymphocytic gastritis is typically associated with *H. pylori* infection or celiac disease. Therefore, *H. Pylori* eradication often causes significant improvement in symptoms and resolution of inflammation and gastric intraepithelial lymphocytosis. Eosinophilic gastritis is a rare condition, typically part of eosinophilic gastroenteritis, which is characterized by diffuse eosinophilic infiltration of the GI tract, peripheral blood eosinophilia, and upper GI symptoms in the absence of known causes for eosinophilia (parasitic infection, cow's milk protein allergy, or IBD).[80] Chronic atrophic gastritis related to pernicious anemia is treated with vitamin B_{12} supplementation. It is important to recognize that the low acid state observed in severe chronic atrophic gastritis may result in malabsorption of essential vitamins such as vitamin B_{12}, micronutrients (iron, calcium, magnesium, zinc), macronutrients, and medications.

Gastric Ulcer

Gastric ulcer (GU) results from a defect in the gastric mucosa that extends through the muscularis mucosa into the deeper layers of the gastric wall.[81] Gastric ulcers may be asymptomatic or present with constant epigastric pain, food-provoked epigastric discomfort, nausea, or early satiety. A GU should be suspected in patients with dyspeptic symptoms in the setting of NSAID use or a history of *H. pylori* infection.[82] Patients may present initially with an ulcer-related complication such as hemorrhage, gastric outlet obstruction, or perforation. Diagnosis of GU is based on direct visualization by endoscopy or indirectly by radiologic imaging (upper GI study). All patients with GU disease should be tested for *H. pylori* infection and malignancy, especially in those with advanced age and with no typical risk factors for peptic ulcer disease (PUD). *H. pylori* eradication should be confirmed in all cases of peptic ulcers.[73,83] Genetic factors and variations in proinflammatory cytokines and tumor necrosis factor are important in predisposing to *H. pylori* infection.[84] Proton pump inhibitors are the mainstay of therapy with high rates of ulcer cure after 6 to 8 weeks of treatment.[85] An antibiotic regimen should be used for *H. pylori* eradication. NSAID use should be discontinued if appropriate. Diet modification (low fiber, small meals) should be considered if obstructive symptoms are present. There is no evidence to support that a bland diet or dietary restrictions may prevent peptic ulcers. In addition to *H. pylori* infection and NSAID use, smoking, alcohol, genetic, and potentially, dietary factors may contribute to the development of gastric ulcers.[86,87] Smoking is an independent risk factor for PUD. In addition, ulcers in smokers are more refractory to therapy and associated with higher recurrence.[88] Alcohol causes direct injury to the gastric mucosal barrier.

Dietary factors have been proposed to impact the frequency of ulcer disease. Certain foods may have protective effects, while others contain potential toxins. High consumption of fruits and vegetables, dietary fiber, and vitamin A has been linked to lower risk of ulcer disease.[89,90] Coffee consumption is not a risk factor for PUD.[91] There is also evidence that psychosocial factors may play a role in ulcer disease.[92] Most ulcers result from superimposed processes such as *H. pylori* infection, and NSAIDs, which inhibit prostaglandin synthesis and reduce the production of gastric acid and bicarbonate, impairing glutathione release, the integrity of the mucosal barrier, and mucosal blood flow. The primary malfunction of secretory, defense, or repair mechanisms is a rare cause of ulcers. Most patients with *H. pylori* infection have pangastritis associated with reduced gastric acid secretion, while only 15% have predominant antral gastritis linked to low somatostatin and high production of gastric acid.[93] Increased acid secretion after successfully eradicating *H. pylori* may contribute to ulcer recurrence.

Gastric Cancer

Gastric cancer is the third leading cause of cancer mortality worldwide.[94] The incidence of gastric cancer in the United States has decreased dramatically over the past decades. This is thought to be related to the increased use of refrigeration for food storage, eating fewer salted and smoked foods, and the decline in the number of individuals with *H. pylori* infection (a major cause of gastric cancer). Approximately 26,000 new cases of stomach cancer have been diagnosed each year, and almost 11,000 deaths per year have been reported in the United States.[95] Gastric cancer affects mainly older individuals, with a median age of diagnosis of 68 years.[95] It is more common in males than females, and Black people (double) than White people in the United States. A higher risk of gastric cancer has also been found in Native Americans and Hispanics than in White people. Most patients with gastric cancer have advanced, often incurable disease at the initial presentation. Overall, the 5-year survival rate in the United States from gastric cancer is approximately 27%.[96]

Gastric cancers can be divided into proximal tumors (esophagogastric junction and gastric cardia) and distal type (fundus, body, and antrum). Weight loss and abdominal pain are the most frequent manifestations. Weight loss results mainly from reduced oral intake related to anorexia, abdominal pain, early satiety, nausea, or dysphagia (cancers in the proximal stomach or at the esophagogastric junction). Vomiting may be the initial presentation due to distal obstruction by the tumor (gastric outlet obstruction). In addition to malnutrition, iron deficiency anemia is common, while overt GI bleeding is a rare presentation. The diagnosis of gastric cancer requires histologic examination of tumor tissue obtained from gastric biopsies during upper endoscopy.[97] The extent of tumor involvement is defined by computed tomography. There are numerous dietary, environmental, and genetic risk factors for the development of gastric carcinoma.[98,99] However, the predominant risk factor is *H. pylori* infection leading to chronic active inflammation of the stomach, which is necessary for progressing from atrophy to gastric cancer.[69,100] The risk of development of gastric cancer due to *H. pylori* infection depends on multiple factors, such as genetics, duration of infection, the strain of bacteria, and environmental factors (diet, smoking).[100,101]

Numerous dietary factors have been proposed as potential risk factors for gastric cancer. Recognizable protective factors include higher intake of fresh fruits and vegetables and lower intake of pickled or salted foods (lower content of nitrates, salt, and polycyclic aromatic amines).[102,103] Risk factors for cancer include high intake of fried food, fats, red meat, and aflatoxins.[104,105] However, currently available data are insufficient to make definitive conclusions regarding the impact of diet on the development of gastric cancer.

Two premalignant conditions, atrophic gastritis and intestinal metaplasia of the stomach, are linked to gastric cancer. Chronic atrophic gastritis is the loss of specialized glandular tissue in its appropriate gastric region.[106] The *H. pylori* infection is associated with an approximately 10-fold increased risk of atrophic gastritis.[107] The incidence of progression to gastric cancer is approximately 0.1% to 1.0% per year in patients with extensive mucosal atrophic changes.[108] Autoimmune metaplastic atrophic gastritis associated with anti-parietal cell and intrinsic factor antibodies is limited to the body and fundus. Autoimmune gastritis associated with pernicious anemia carries the risk of gastric cancer, but at a much lower rate than observed with *H. Pylori* infection.[107,109] Proposed mechanisms contributing to the development of gastric cancer in atrophic gastritis include low acid output, increased rate of bacterial overgrowth, formation of nitroso compounds, diminished secretion of ascorbate, and high gastrin from the low acid output.[107] The other premalignant condition is intestinal metaplasia (IM) of the stomach, which can be divided into three types.[110,111] Type I (complete) IM is not a risk factor for gastric cancer. Type II (incomplete) IM contains columnar "intermediate" cells in various stages of differentiation and goblet cells secreting sialomucins. Type III (incomplete) IM is characterized by the presence of less differentiated cells than type II. Type II and III are associated with a nearly 20-fold increased risk of gastric cancer, with early gastric cancer reported in almost 42% of patients with type III IM within 5 years of diagnosis.[112,113]

Long-term endoscopic surveillance is required in patients at higher risk of cancer, including incomplete IM, family history of gastric cancer, and people who are Asian or Black.[106] Surgical resection is the primary curative treatment for gastric cancer. However, surgically curable early gastric cancers are usually asymptomatic and rarely detected.[114] Advanced endoscopic techniques, endoscopic mucosal resection (EMR), and endoscopic submucosal dissection (ESD) may be applied as curative therapies for highly selected early gastric cancers.[115,116] The criteria for EMR include cancer limited to mucosa without lymph node involvement, size of the tumor less than 2 cm, and no presence of multiple gastric cancers.[116] Approximately 75% of patients with gastric cancer have spread of disease to the perigastric lymph nodes or distant metastases at the time of diagnosis, and they are excluded from surgery for a cure. The TNM classification is used to stratify the disease and to predict prognosis in patients treated with gastrectomy.[114] For example, large tumor size (>5 cm) is associated with worse survival independently of nodal status or tumor stage.[117] Survival after surgery alone remains poor, between

20% to 50% at 5 years.[118] Patients with resectable cancer are often treated with surgery in combination with perioperative chemotherapy. A few meta-analyses suggested a 15% to 20% reduction in mortality with neoadjuvant chemotherapy.[119,120] Docetaxel containing regimen FLOT (fluorouracil/leucovorin oxaliplatin, docetaxel) is currently a preferred regimen in the perioperative period in resectable carcinomas.[119]

As a general rule, total gastrectomy is performed for proximal gastric tumors and diffuse gastric cancer, while partial gastrectomy is reserved for tumors in the distal stomach. Surgical resection allows for effective palliation of symptoms in those patients with obstruction. However, the role of gastric resection is questionable in cases of extensive involvement by cancer, bulky metastatic disease, retroperitoneal invasion, or peritoneal carcinomatosis. To improve outcomes for this group of patients, using perioperative chemotherapy or chemoradiotherapy is imperative. Malnutrition is common in patients with gastric cancer. Nutritional interventions may include oral nutritional supplements, perioperative enteral feeding (via nasoenteric tubes, percutaneous gastrostomy, or jejunostomy tubes), or parenteral nutrition (PN) in patients with GI tract dysfunction. Post gastrectomy syndromes are often associated with weight loss, hypoalbuminemia, and nutritional deficiencies (iron, calcium, vitamin B_{12}). Postoperatively, patients may suffer from issues related to gastric motility, such as gastric stasis, gastroparesis, and dumping syndrome requiring lifelong dietary modifications and medical therapies.

Gastroparesis

Gastroparesis (GP) is defined by the presence of upper GI symptoms and an objective finding of delayed gastric emptying without a mechanical obstruction.[121] Gastroparesis is a chronic disease divided into diabetic, idiopathic, or postsurgical types based on presumed etiology. Idiopathic GP is the most common type. In addition, infiltrative and neurologic disorders such as amyloidosis and Parkinson's disease may be associated with gastroparesis. The exact prevalence of GP is unknown, as it is underdiagnosed. It is believed that GP affects approximately 1% to 2% of the adult US population.[122] Gastroparesis is more common in women (four times) than in men. Typical manifestations include nausea, vomiting, abdominal pain, early satiety, postprandial fullness, and bloating. In severe cases, weight loss may be a prominent symptom, due to poor oral intake. Patients with idiopathic GP report more early satiety, postprandial fullness, and abdominal pain, while patients with diabetic GP suffer more often from retching and vomiting. Diagnosis of GP is established by documenting delayed gastric emptying of solids by standard gastric scintigraphy.[123] Gastric retention above 60% at 2 hours and retention >10% at 4 hours is consistent with delayed gastric emptying. In patients who cannot tolerate a solid meal or have an allergy to eggs, a liquid nutrient meal with an isotope may be used to measure gastric emptying. Other methods used to assess gastric emptying include a C13 breath test or wireless capsule, both reliable in assessing gastric transit time. Management of GP can be challenging and often requires a multistep approach. Patients with refractory vomiting and abdominal pain require frequent emergency center visits and hospitalizations for rehydration, electrolyte replacement, and vomiting control. Providing oral nutrition may be difficult in this setting. Short- or long-term postpyloric feeding may be needed for patients at nutritional risk. Initial management of gastroparesis consists of dietary modification, optimization of glycemic control, hydration, and pharmacologic therapy with prokinetics and antiemetics.[124] Opiates should be avoided, or tapered if appropriate, as they inhibit gastric emptying.

Diet modification is considered first-line therapy in patients with GP. Patient compliance may be a problem. Consultation with a dietitian should be arranged. In clinical practice, only modest symptom improvement was found with dietary adjustments. Several dietary modifications are beneficial (**Table 76.1**). Frequent and small meals are preferred. A high fiber (raw fruits and vegetables) diet and high-fat foods should be avoided because of the risk of provoking GP symptoms. Insoluble fiber is difficult to digest without effective gastric motility, and a high-fat meal inhibits gastric emptying. The diet should be high in protein to preserve muscle mass. For patients who cannot tolerate solids, blenderizing foods into liquids is beneficial, as liquid emptying is often preserved. Refractory vomiting, nausea, and chronic abdominal pain significantly impair oral intake, leading to hypokalemia, metabolic alkalosis, dehydration, and malnutrition. In patients with refractory symptoms, micronutrient and vitamin deficiencies have been observed.[125] Therefore, multivitamin supplementation should be provided to those at nutritional risk. Nutritional intervention and enteral or parenteral nutrition may be needed in patients with refractory symptoms at high nutritional risk, or with overt malnutrition. In diabetic gastroparesis, optimization of glycemic control is critical. Acute hyperglycemia with glucose concentration >270 mg/dL slows the gastric emptying rate by impairing the function of the gastric pacemaker resulting in gastric dysrhythmias and inhibition of antral contractions.[126] In addition, patients

Table 76.1 Dietary Interventions in Patients with Gastroparesis

Dietary Recommendations	Preferred Food (Improving or Nonprovoking Symptoms in Gastroparesis)[128]
■ Eat 5–6 meals a day	■ Saltine and Graham crackers
■ Eat small meals (low volume)	■ Jello
■ Avoid large and late meals (after 6 pm)	■ Popsicles
■ Chew food very well	■ Applesauce
■ Liquid food preferred over solids	■ Ginger ale
■ Pureed or blenderized food preferred	■ Tea
■ Avoid high fiber food (raw fruit and vegetables)	■ Clear soup
■ Avoid high-fat food	■ Potatoes, sweet potatoes
■ Eat protein-rich foods	■ Pretzels
■ Caloric supplements	■ Gluten-free foods
■ Sit up for 1 hour after eating a meal	■ White rice
■ Consider a short walk after meals	■ White fish and salmon

with diabetes should avoid GLP-1 analogs or agonists (liraglutide, exenatide) and incretin-based drugs (pramlintide; amylin analog), as they delay gastric emptying and may cause nausea.[127]

Pharmacologic therapy is necessary for patients who fail dietary modification (**Table 76.2**). Prokinetic agents, D_2 dopamine receptor antagonists, metoclopramide, and domperidone have been shown to accelerate the rate of gastric emptying. Prokinetics should be administered 15 to 30 minutes before meals and bedtime and used short term, if possible. A liquid formulation of metoclopramide is preferred by allowing easy dose titration and being more effective with less gastric retention and more predictable absorption. Metoclopramide is the only prokinetic approved for GP by the FDA. However, neurologic adverse effects, including a black box warning for tardive dyskinesia, limit its use in GP.[129] Lack of response or intolerance is common with metoclopramide. Domperidone is not available for routine use in the United States.[130] For acute exacerbation of gastroparesis, erythromycin (motilin agonist) IV can be administered at a low dose (3 mg/kg) once daily. Intravenous erythromycin can improve acute gastric stasis and symptoms in up to 40% of patients.[131] Erythromycin IV is even more effective in combination with metoclopramide IV (10 mg three times daily). Oral erythromycin is not as effective as the IV form and is associated with tachyphylaxis (loss of effectiveness over time). New prokinetics have been effective in GP in clinical trials, but they must get approved for this indication.[132] The novel drug relamorelin (ghrelin agonist) has been shown to improve symptoms and gastric emptying in patients with diabetes and GP and has an appetite-stimulating effect.[133,134] Tegaserod and prucalopride, 5-HT4 agonists, have also shown promise in GP by improving symptoms and gastric emptying rate.[135,136] Antiemetics have been used successfully in gastroparesis to control nausea and vomiting in outpatient or hospital settings.[137] 5-HT3 antagonists available in oral and IV form (Ondansetron 4–8 mg orally tid; granisetron

Table 76.2 Pharmacologic, Endoscopic, and Surgical Management Options in Severe Refractory Gastroparesis.[139,144]

Drug Therapy	Endoscopic and Surgical Interventions
Prokinetics ■ Metoclopromide 5–20 mg IV or PO; qid ■ Domperidone* 10–20 mg PO qid ■ Erythromycin 70–250 mg IV qd; 250 mg PO tid ■ Prucalopride** 2 mg PO q d	**Endoscopic** ■ Intrapyloric botox injection (100–200 Units)
Antiemetics ■ Ondansetron 4 mg IV q 8 hr; 4–8 mg PO tid ■ Granisetron 1 mg IV; q 12 hr; 1 mg PO bid ■ Prochlorperazine 10 mg IV; q 6–8 hr, 5–10 mg PO qid ■ Trimethobenzamide 200 mg IV q 8 hr; 300 mg PO tid ■ Chlorpromazine 10–25 mg PO tid ■ Meclizine 12.5–25 mg PO tid ■ Scopolamine 1.5 mg patch every 3 days ■ Aprepitant*** 40–80 mg IV or PO qd ■ Ginger 1g PO bid	**Endoscopic Tube Placement** ■ PEG (Gastrostomy tube placement) for decompression ■ PEJ (Jejunal tube placement; direct or indirect) for feeding
Other Medications (Nausea and Vomiting) ■ Ativan 1mg IV q 8–12 hr ■ Haloperidol 1-2 mg IV q 8–12 hr	**Endoscopic** ■ Gastric per-oral endoscopic myotomy ■ (G-POEM)
Neuromodulators (Visceral Abdominal Pain) **Antidepressants** Tricyclic ■ Amitriptyline 25–100 mg/d ■ Nortriptyline 25–100 mg/d ■ Desipramine 25–75 mg/d ■ Imipramine 25–100 mg/d Serotonin/norepinephrine reuptake inhibitors ■ Duloxetine 60–120 mg /d Others ■ Mirtazapine 7.5–30 mg/d **Anticonvulsants** ■ Gabapentin 900–1,200 mg/day ■ Pregabalin 100–300 mg/day	**Surgical** ■ Gastric electrical stimulation**** (GES) ■ Pyloromyotomy (laparoscopic) ■ Pyloromyotomy+GES (laparoscopic) ■ Gastric resection

*Not FDA approved.
**Available under FDA investigational drug protocol.
***Not approved for gastroparesis.
**** Compassionate use for nausea and vomiting only, approved under a Humanitarian Device Exemption.

Data from Lacy BE, Tack J, Gyawali CP. (2022) Clin Gastroenterol Hepatol 20:491-500; Camilleri, M, Parkman HP, Shafi MA, et al. (2013) Am J Gastroenterol 108:18–37.

2 mg PO bid) are the most effective antiemetic drugs. Cardiac monitoring is recommended in patients taking 5-HT3 antagonists with a history of hypokalemia, hypomagnesemia, heart failure, arrhythmias, or concomitant use of medications that prolong the QT interval. Prochlorperazine 5 to 10 mg orally or rectally can also be used successfully. A new drug, aprepitant (neurokinin-1 receptor antagonist; oral or IV), currently approved only for postoperative and chemotherapy-induced emesis, has been very helpful for nausea control; however, it has not been well studied in gastroparesis.[138] Compliance with dietary modification and pharmacotherapy should be monitored. In patients with refractory symptoms, despite diet modification, prokinetics, and antiemetics, consideration should be given to further interventions such as intrapyloric botox injection, perioral endoscopic pyloromyotomy (G-POEM), gastric electrical stimulation, or surgical intervention.[139] The use of intrapyloric Botox injections in GP has been controversial, as two small, placebo-controlled, randomized trials showed no improvement in gastric emptying.[140,141] However, Botox injections are often performed in clinical practice due to very limited therapeutic options. In retrospective studies, more than 50% of patients had clinical improvement in symptoms with intrapyloric Botox injections.[142,143]

Patients with refractory symptoms and malnutrition (>10% of body weight loss over 6 months) may require placement of a jejunostomy tube for enteral nutrition and a venting gastrostomy tube for gastric decompression to control vomiting and abdominal pain (see Table 76.2). Before placement of a percutaneous jejunostomy tube, a trial of nasojejunal feeding should be performed to ensure that enteral feeding is tolerated at a goal rate. Patients should receive jejunal feeds as continuous cyclic feeding, typically overnight. Parenteral nutrition is rarely indicated and restricted only to patients who failed enteral nutrition or who need nutrition support perioperatively. Long-term parenteral nutrition is required in patients with combined gastric and intestinal dysmotility, such as visceral myopathy or systemic sclerosis. Generally, gastric peroral endoscopic myotomy (G-POEM) or gastric electrical stimulation is attempted before considering a long-term tube placement for feeding and venting. Endoscopic intervention, G-POEM, has been recently incorporated into the armamentarium for refractory gastroparesis. This minimally invasive technique allows cutting of the pyloric muscle through the submucosal channel created during endoscopy.[145] In several studies and meta-analyses, G-POEM has been associated with significant improvement in gastric emptying rate and GP symptom reduction.[146-148] The endoscopic procedure appears to be safe and effective, with promising outcomes regarding clinical response and normalization of gastric emptying.[146] However, further trials are needed to predict which patients respond the best and assess the procedure's long-term efficacy. Gastric electrical stimulation (GES) can only be considered in patients with refractory nausea and vomiting. This is a laparoscopic procedure involving the placement of two leads on the external surface of the gastric wall and the insertion of a gastric pacemaker device in the abdominal wall.[149] The results of GES in GP have been mixed.[150,151] However, a large meta-analysis found a reduction in parenteral nutrition requirement, improvement in symptoms, and quality of life.[152] Complications of GES include infection, gastric leads erosions and migration, dislodgement, and gastric wall perforation. GES placement can be combined with laparoscopic pyloroplasty in refractory cases.[153] Gastric surgery is rarely indicated in patients with gastroparesis. Indications for surgery include complete or subtotal gastrectomy, pyloroplasty, and gastrojejunostomy to relieve refractory nausea and vomiting.[154] Patients with morbid obesity and gastroparesis may benefit from Roux-en-Y gastric bypass as a primary intervention.[155]

Table 76.3 summarizes all interventions for gastroparesis. When patients present with nausea and vomiting and normal gastric emptying, consideration must be given to other conditions, including functional dyspepsia or cyclic vomiting syndrome, potential psychiatric diseases such as depression, eating disorders, psychogenic vomiting, side effects of psychotropic medications, and rumination syndrome. An eating disorder should be suspected, particularly in young women with weight loss and altered body perception. Rumination syndrome is a behavioral disorder more commonly seen among adolescents and adults, particularly high achievers or those with high stress levels.[156] The behavior is characterized by effortless regurgitation of undigested food within minutes of starting or finishing meal ingestion without nausea or vomiting. Symptoms of GP often overlap with those of functional dyspepsia, with predominant symptoms of epigastric pain, early satiety, and postprandial fullness. A slight delay in gastric emptying can be seen in 30% of patients with functional dyspepsia.[157] Cyclic vomiting syndrome (CVS) is recurrent episodes of intense nausea and vomiting lasting for hours to a few days, separated by symptom-free periods of variable intervals.[158] Cannabinoid hyperemesis syndrome from long-term marijuana use has been recognized as a

Table 76.3 Nutritional Support in Refractory Gastroparesis[125]

Indications for Initiation of Nutritional Support	Type of Nutritional Support/Intervention
▪ Severe malnutrition (weight loss >5%–10% of usual body weight over 6 months) ▪ Frequent emergency room visits or hospitalizations requiring intravenous hydration, electrolyte replacement, and prokinetics (erythromycin/metoclopramide) ▪ Overall poor quality of life	▪ Enteral nutrition • Short-term Nasoenteric tubes • Long-term Percutaneous jejunal tubes Gastrostomy tube for venting ▪ Parenteral nutrition • Short-term Via Peripheral line or PICC line • Long-term (rare) Via Central venous access

Data from Bharadwaj, S., Meka, K., Tandon, P., Rathur, A., Rivas, J. M., Vallabh, H., Jevenn, A., Guirguis, J., Sunesara, I., Nischnick, A., & Ukleja, A. (2016). Management of gastroparesis-associated malnutrition. Journal of digestive diseases, 17(5), 285–294. https://doi.org/10.1111/1751-2980.12344

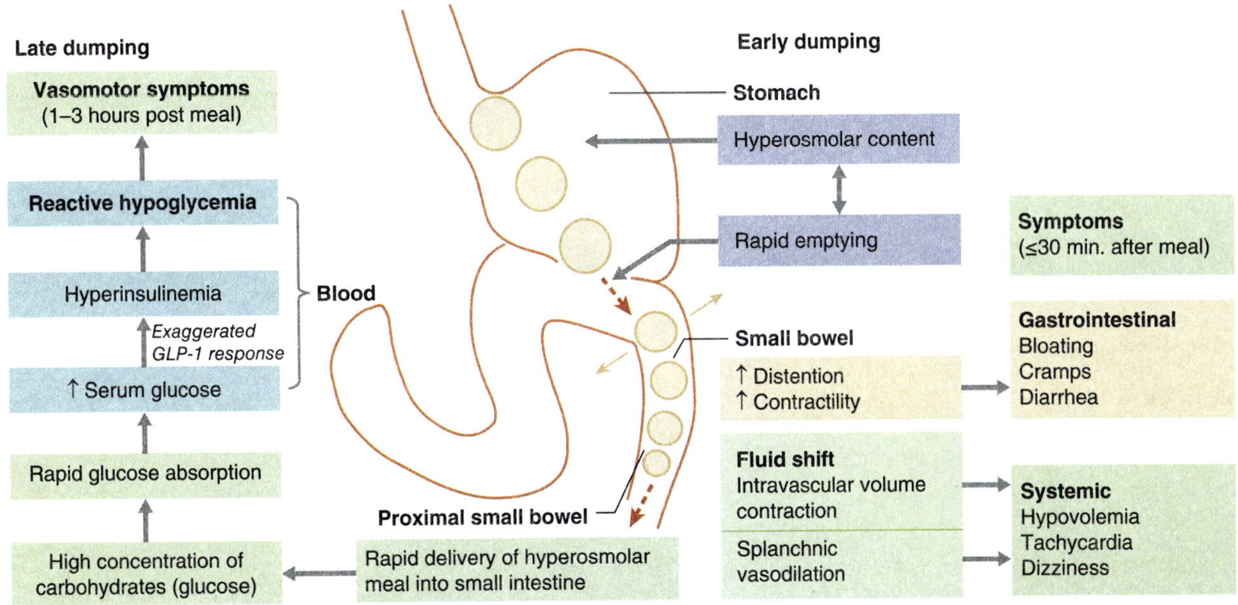

Figure 76.4 Mechanisms of early and late dumping syndrome and related symptoms after gastric resection (Billroth II surgery). Simple sugar and high meal osmolarity, and accelerated emptying of a meal lead to increased small bowel distensibility and peristalsis. There is a rapid fluid shift after hyperosmolar chyme reaches the proximal small bowel. In response to fluid shifts, there is an activation of serotonin and kallikrein-kinin system associated with increased levels of gut hormones and an activation of rennin-aldosterone axis resulting in hypovolemia contributing to systemic symptoms in early dumping. Late dumping is a result of rapid glucose absorption from small bowel and rapid increase in insulin secretion leading to reactive hypoglycemia. Exaggerated GLP-1 response is a main contributing factor leading to hyperinsulinemia (Incretin effect). Hyperinsulinemic hypoglycemia is responsible for neurologic and vasomotor symptoms seen in late dumping.

separate condition, but clinically mimics CVS.[159] This condition has been reported more often in males and can be associated with prodromal anxiety symptoms. Symptoms may partially improve with hot water showers and topical capsaicin cream, which works by activating transient receptor potential cation channel subfamily V member 1 (TRPV1) receptors as in hot water bathing.[160]

Gastric Stasis

Impaired gastric emptying after gastric surgery may be a result of postsurgical atony, vagal denervation, anastomotic stricture, or the presence of a small gastric remnant. Symptoms are like those of gastroparesis. The diagnostic work-up should include an upper GI series with small bowel series to assess postsurgical anatomy and exclude mechanical obstruction; and an upper endoscopy to exclude anastomotic stricture or marginal ulcer. Delayed gastric emptying is diagnosed by standard solid meal gastric scintigraphy. Dietary modifications, including small, frequent meals and a low-fiber diet, are the mainstay of therapy. Prokinetics, metoclopramide, and erythromycin can be potentially helpful in gastric atony. Gastrectomy or revision surgery may be required in cases of a failure of diet and medical therapies.

Dumping Syndrome

Various types of gastric resections and reconstructions for benign and malignant conditions of the stomach and esophagus can result in long-term postoperative complications. Dumping syndrome is a result of changes in gastric anatomy, innervation, and function. Those alterations lead to accelerated gastric emptying of undigested food into the small intestine, rapid fluid, and glucose shifts, and modified hormonal response. Clinically significant dumping symptoms have been reported in about 20% of patients after pyloroplasty or distal gastrectomy, and refractory dumping is present in 5% of cases.[161,162] Dumping symptoms after R-Y gastric bypass typically resolve within 1 to 2 years after surgery. Dumping syndrome presents with a constellation of manifestations divided into early and late symptoms.[163] Early dumping includes both gastrointestinal (epigastric discomfort and cramping, bloating, nausea, diarrhea, borborygmi) as well as vasomotor symptoms (diaphoresis, palpitations, flushing, desire to lie down within 60 minutes after a meal) (see **Figure 76.4**). Complex mechanisms responsible for early dumping include osmotic effects of the duodenal content leading to fluid shift, duodenal distention, GI hormone release, and autonomic neuronal responses. Early dumping is more common and observed in up to 75% of patients.[164] Late dumping, typically observed within 1 to 3 hours postmeal, is characterized by mainly systemic symptoms related to reactive hypoglycemia following a postprandial insulin peak. An exaggerated GLP-1 response is the key process responsible for hyperinsulinemia and hypoglycemic effects in late dumping. Symptoms of late dumping can be related to neuroglucopenia, which include fatigue, confusion, weakness, syncope, and those of a systemic type, which are palpitations, tremors, irritability, and sweating from vagal and sympathetic activation.[165] Isolated late dumping may affect up to 25% of patients.[164] The diagnosis of dumping syndrome is made primarily on clinical presentation, prior history of gastric surgery, and tests to document rapid GI transit such as upper gastrointestinal series, gastric emptying study, and modified oral glucose tolerance test to assess for hypoglycemia and systemic symptoms.[164] Most patients with dumping syndrome are treated conservatively with dietary modifications to slow down gastric and small bowel transit

Table 76.4 Dietary Modifications and Drug Therapy for Dumping Syndrome

Dietary Interventions	Pharmacotherapy
■ Frequent 5–6 meals a day ■ Small (low volume) meals ■ Separate liquids from solid meal by 30 minutes ■ Avoid high sugar content and calorie-dense food ■ Avoid simple sugars (candies, sweets, milk shakes) ■ Preferred high fiber diet and high complex carbohydrates (raw vegetables) ■ Increase protein intake (chicken, fish) ■ Chew food very well ■ Higher fat intake, if tolerated	Early dumping Octreotide ■ 50–200 μg SC tid 15–30 minutes before meals ■ LAR 10–20 mg IM every month Pasireotide[172] ■ 50–200 μg SC tid before meals ■ 10–20 mg IM every month
Dietary supplements ■ Pectin, guar gum, glucomannan	Late dumping ■ Octreotide (as above dosing) ■ Pasireotide (as above dosing) ■ Acarbose 50–100 mg tid ■ Diazoxide 50–100 mg bid/tid
Lying supine for 30 minutes after meals (if severe vasomotor symptoms)	

SC, subcutaneous; *IM*, intramuscular

and to modify meal composition and size. Frequent and small (low volume) meals, separation of liquids from solids, high fiber, protein-rich foods, and a diet low in simple carbohydrates are recommended. Dietary supplements that increase the viscosity of food (pectin, guar gum, glucomannan) are beneficial to control symptoms of dumping as second-line therapy after diet.[166] In addition, patients with early dumping should be advised to lie down for 30 minutes after meals to slow down gastric emptying and reduce symptoms of hypovolemia.

Pharmacologic therapy is reserved for patients who fail dietary modification. A somatostatin analog, octreotide, has been associated with significant symptom improvement in both early and late dumping.[167] The effects of octreotide are multidirectional and include delayed accelerated gastric and small bowel transit, inhibition of enteral hormones and insulin secretion, prevention of postprandial splanchnic vasodilation, and increased intestinal absorption of water and sodium. Therefore, octreotide is a preferred therapy for dumping syndrome. Short-acting octreotide is administered initially subcutaneously three times daily within 30 minutes before meals. The need for repeat and often painful injections is a major limiting factor to using it. A long-acting formulation, Octreotide LAR 20 mg, given as a monthly intramuscular injection, has been used with good results, including a significant reduction in symptoms and improved quality of life.[168] Small studies have shown the benefits of acarbose, an alpha-glycosidase inhibitor, in late dumping. Acarbose delays monoglyceride release from complex carbohydrates, slowing down carbohydrate absorption and preventing reactive hypoglycemia by delaying the release of insulin.[169] Acarbose is given orally with meals at 50 to 100 mg (tid). Another drug to consider in late dumping is diazoxide, a potassium channel activator that inhibits calcium-induced insulin secretion from beta cells with subsequent prevention of reactive hypoglycemia[170] (see **Table 76.4**).

For patients with severe malnutrition, gastric or jejunal feeding via G- or J tube can be helpful to reverse malnutrition. Nonresponders to diet and pharmacotherapy associated with intractable dumping symptoms may require surgical reintervention. As patients may experience symptom improvement over time, surgical intervention should be delayed for at least 1 year. The choice of surgical procedure will depend on the type of initial intervention with a concept to rectify the underlying problem. For example, conversion from a loop gastrojejunostomy to a Roux-en-Y reconstruction is preferred in patients with a distal gastrectomy. There are no head-to-head studies comparing outcomes of different surgeries in dumping syndrome. Patients with severe reactive hypoglycemia after Roux-en-Y gastric bypass and pancreatic nesidioblastosis (Islet cells hyperplasia related to excessive GLP-1 secretion) are unique, and they may benefit from surgical reintervention. A meta-analysis of 14 studies revealed resolution of reactive hypoglycemia in 82% of patients after gastric pouch restriction, 76% after gastric bypass reversal, and 67% after pancreatic resection.[171]

Functional Dyspepsia

The diagnosis of functional "nonulcer" dyspepsia requires the exclusion of other organic causes of dyspeptic symptoms.[55] Dyspepsia is defined by the presence of one or more of the following: postprandial fullness, early satiation, epigastric pain, burning, and bloating without evidence of structural disease. The pathophysiology of functional dyspepsia is not well understood. Several potential mechanisms have been suggested, such as impaired gastric accommodation, mild delay in gastric emptying, rapid gastric emptying, antral hypomotility, gastric dysrhythmias, vagal dysfunction, visceral hypersensitivity, and psychosocial factors (generalized anxiety disorder, somatization, major depression). The management of patients with functional dyspepsia is controversial. Patients should be tested and treated for H. pylori as an initial approach. The eradication of *H. pylori* may improve dyspeptic symptoms by alteration in acid secretion or potentially by modification of gut microflora. Antisecretory therapy with PPIs is a first-line therapy and can be moderately effective.[173] A once-daily PPI for several weeks may sufficiently relieve dyspeptic symptoms. Subsequently, those patients with inadequate response to PPIs can be considered for neuromodulation therapy such as tricyclic antidepressants (amitriptyline, desipramine) or an azapirone (buspirone) to improve postprandial symptoms.[174,175] The anti-depressant mirtazapine is beneficial by reducing global symptoms, early satiety, GI-specific anxiety, and stimulating appetite and weight gain.

Prokinetics (metoclopramide 5–10 mg tid.) should be reserved for patients who failed the above therapies and should be used only in the short term. If postprandial nausea is a predominant symptom, a trial of antiemetics is reasonable, although data from clinical trials are lacking. New drug therapies for dyspepsia are in the pipeline, including prokinetic serotonergic drugs (velusetrag, prucalopride) and acotiamide, which enhances gastric accommodation.[176] Functional dyspepsia is a chronic disease with symptoms that vary in severity over time. Retreatment may be necessary.[175]

Acknowledgments

The authors acknowledge Mark H Delegge, the author of the previous version of this chapter, for his contributions.

References

1. Patti MG, Gantert W, Way LW. Surg Clin North Am 1997;77: 959–70.
2. Hamdy S, Aziz Q, Rothwell JC, et al. Am J Physiol 1997;272: G802–808.
3. Vaezi MF, Richter JE. J Clin Gastroenterol 1998;27:21–35.
4. Yadlapati R, Kahrilas PJ, Fox MR, et al. Neurogastroenterol Motil Off J Eur Gastrointest Motil Soc 2021;33:e14058.
5. Desprez C, Roman S, Leroi AM, Gourcerol G. Neurogastroenterol Motil Off J Eur Gastrointest Motil Soc 2020;32:e13980.
6. Hirano I, Pandolfino JE, Boeckxstaens GE. Clin Gastroenterol Hepatol Off Clin Pract J Am Gastroenterol Assoc 2017;15: 325–34.
7. Vaezi MF, Pandolfino JE, Yadlapati RH, et al. Am J Gastroenterol 2020;115:1393–411.
8. Felix VN. Ann N Y Acad Sci 2018;143:124–31.
9. Ponds FA, Fockens P, Lei A, et al. JAMA 2019;322:134–44.
10. Andolfi C, Fisichella PM. Br J Surg 2019;106:332–41.
11. Kumbhari V, Tieu AH, Onimaru M, et al. Endosc Int Open 2015;3:E195–201.
12. Schlottmann F, Luckett DJ, Fine J, et al. Ann Surg 2018;267: 451–60.
13. Kumbhari V, Familiari P, Bjerregaard NC, et al. Endosc 2017;49: 634–42.
14. Markar SR, Wiggins T, MacKenzie H, et al. Dis Esophagus Off J Int Soc Dis Esophagus 2019;32:doy106.
15. Leeuwenburgh I, Scholten P, Alderliesten J, et al. Am J Gastroenterol 2010;105:2144–9.
16. Pasricha PJ, Rai R, Ravich WJ, et al. Gastroenterol 1996;110: 1410–5.
17. Snyder DL, Vela MF. Curr Opin Gastroenterol 2020;36:344–50.
18. Ratuapli SK, Crowell MD, DiBaise JK, et al. Am J Gastroenterol 2015;110:979–84.
19. Patcharatrakul T, Alkaddour A, Pitisuttithum P, et al. Ann N Y Acad Sci 2020;1481:210–23.
20. Roman S, Kahrilas PJ. Dysphagia 2012;27:115–23.
21. Khalaf M, Chowdhary S, Elias PS, Castell D. Am J Med 2018;131:1034–40.
22. Richter JE. Lancet Lond Engl 2001;358:823–8.
23. Gyawali CP, Sifrim D, Carlson DA, et al. Neurogastroenterol Motil Off J Eur Gastrointest Motil Soc 2019;31:e13584.
24. Di Stefano M, Papathanasopoulos A, Blondeau K, et al. Dis Esophagus Off J Int Soc Dis Esophagus 2012;25:470–6.
25. Agrawal A, Hila A, Tutuian R, et al. J Clin Gastroenterol 2007;41:366–70.
26. Lei WY, Hung JS, Liu TT, et al. J Gastroenterol Hepatol 2018;33:650–5.
27. Chen CL, Yi CH, Liu TT, Orr WC. Scand J Gastroenterol 2013;48:1363–70.
28. Dhar SI, Nativ-Zeltzer N, Mehdizadeh OB, et al. Dysphagia 2022;37:4–10.
29. Aggarwal N, Thota PN, Lopez R, Gabbard S. Neurogastroenterol Motil Off J Eur Gastrointest Motil Soc 2018;30(2).
30. Vakil N, van Zanten SV, Kahrilas P, et al. Am J Gastroenterol 2006;101:1900–20.
31. Richter JE, Rubenstein JH. Gastroenterol 2018;154:267–76.
32. Eusebi LH, Ratnakumaran R, Yuan Y, et al. Gut 2018;67:430–40.
33. Lottrup C, Khan A, Rangan V, Clarke JO. Ann N Y Acad Sci 2020;1481:182–97.
34. Friedenberg FK, Hanlon A, Vanar V, et al. Dig Dis Sci 2010;55:1911–7.
35. Fass R. J Clin Gastroenterol 2007;41:131–7.
36. Gyawali CP, Kahrilas PJ, Savarino E, et al. Gut 2018;67:1351–62.
37. Nayyar AK, Royston C, Bardhan KD. Dig Liver Dis Off J Ital Soc Gastroenterol Ital Assoc Study Liver 2003;35:143–50.
38. Broeders JA, Draaisma WA, Bredenoord AJ, et al. Br J Surg 2010;97:845–52.
39. Zimmermann CJ, Lidor A. Gastroenterol Clin North Am 2021;50: 809–23.
40. Shaheen NJ, Falk GW, Iyer PG, et al. Am J Gastroenterol 2016;111:30–50.
41. Brown LM, Swanson CA, Gridley G, et al. J Natl Cancer Inst 1995;87:104–9.
42. Bower M, Jones W, Vessels B, et al. Ann Surg Oncol 2009;16: 3161–8.
43. Siddiqui AA, Glynn C, Loren D, Kowalski T. Dis Esophagus Off J Int Soc Dis Esophagus 2009;22:216–22.
44. Ng T, Vezeridis MP. J Surg Oncol 2010;101:725–9.
45. Dellon ES, Liacouras CA, Molina-Infante J, et al. Gastroenterol 2018;155:1022–1033.e10.
46. Lucendo AJ, Molina-Infante J, Arias Á, et al. United European Gastroenterol J 2017;5:335–358.
47. Molina-Infante J, Lucendo AJ. Gastroenterol Hepatol 2015;38: 388–397.
48. Moawad FJ, Schoepfer AM, Safroneeva E, et al. Aliment Pharmacol Ther 2014;39:603–608.
49. Hirano I, Dellon ES, Hamilton JD, et al. Gastroenterol 2020;158:111–122.e10.
50. Arias A, Gonzalez-Cervera J, Tenias JM, et al. Gastroenterol 2014;146:1639–1648.
51. Soybel DI. Surg Clin North Am 2005;85:875–94.
52. Wood LD, Montgomery EA. Structure and Innervation of Hollow Viscera. Gastrointestinal Anatomy and Physiology. Oxford, United Kingdom: John Wiley & Sons, Ltd, 2014;1–14,.
53. Ramsay PT, Carr A. Surg Clin North Am 2011;91:977–82.
54. Feldman M. Am J Gastroenterol 2013;108:347–352.
55. Dockray, GJ. Best Pract Res Clin Endocrinol Metab 2004;18: 555–568.
56. Goyal RK, Hirano I. N Engl J Med 1996;334:1106–1115.
57. Goyal RK, Guo Y, Mashimo H. Neurogastroenterol Motil 2019;31:e13546.

58. Sarna SK. Am J Physiol Gastrointest Liver Physiol 2008;294: G372–G390.
59. Camilleri M. Curr Opin Endocrinol Diabetes Obes 2019;26: 3–10.
60. Deloose E, Janssen P, Depoortere I, et al. Nat Rev Gastroenterol Hepatol 2012;9:271—285.
61. Hunt RH, Camilleri M, Crowe SE, et al. Gut 2015;64:1650–68.
62. Richter C, Tanaka T, Yada RY. Biochem J 1998;335:481–490.
63. Armand M. Curr Opin Clin Nutr Metab Care 2008;10:156–164.
64. Richardson CT, Walsh JH, Cooper KA, et. al. J Clin Invest 1977;60:435–441.
65. Carabotti M, Annibale B, Lahner E. Nutrients 2021;13:208.
66. Pennelli G, Grillo F, Galuppini F, et al. Pathologica 2020;112: 153–165.
67. Sipponen P, Kekki M, Siurala M. J Gastroenterol Hepatol 1991;6:244–51.
68. Guevara B, Cogdill AG. Dig Dis Sci 2020;65:1917–1931.
69. Correa P, Piazuelo MB. J Dig Dis 2012;13:2–9.
70. El-Serag HB, Kao JY, Kanwal F, et al. Clin Gastroenterol Hepatol 2018;16:992–1002.
71. Pohl D, Keller PM, Bordier V, Wagner K. World J Gastroenterol 2019;25:4629–4660.
72. Ford AC, Forman D, Hunt RH, et al. BMJ 2014;348:g3174.
73. Malfertheiner P, Megraud F, O'Morain CA, et al. Gut 2017;66: 6–30.
74. Kim SY, Chung JW. Antibiot (Basel) 2020;9:436.
75. Thung I, Aramin H, Vavinskaya V, et al. Aliment Pharmacol Ther 2016;43:514–533.
76. Shah SC, Iyer PG, Moss SF. Gastroenterol 2021;160:1831–1841.
77. Ribaldone DG, Fagoonee S, Astegiano M, et al. J Clin Med 2019;8:199.
78. Yu HJ, Kim SJ, Oh HH, et al. Medicine (Baltimore) 2021;100: e28212.
79. Feeley KM, Heneghan MA, Stevens FM, McCarthy CF. J Clin Pathol 1998;51:207–10.
80. Zhang M, Li Y. J Gastroenterol Hepatol 2017;32:64–72.
81. Kamada T, Satoh K, Itoh T, et al. J Gastroenterol 2021;56: 303–322.
82. Huang JQ, Sridhar S, Hunt RH. Lancet 2002;359:14.
83. Ford AC, Gurusamy KS, Delaney B, et al. Cochrane Database Syst Rev 2016;4:CD003840. doi: 10.1002/14651858
84. Miftahussurur M, Yamaoka Y. Expert Rev Gastroenterol Hepatol 2015;9:1535.
85. Scally B, Emberson JR, Spata E, et al. Lancet Gastroenterol Hepatol 2018;3:231–241.
86. Martin DF, Montgomery E, Dobek AS, et al. Am J Gastroenterol 1989;84:1268–72.
87. Lee SP, Sung IK, Kim JH, et al. Clin Endosc 2017;50:578–584.
88. Anda RF, Williamson DF, Escobedo LG, et al. Arch Intern Med 1990;150:1437–41.
89. Aldoori WH, Giovannucci EL, Stampfer MJ, et al. Am J Epidemiol 1997;145:42–50.
90. Rosenstock SJ, Bonnevie T, Bonnevie O, et al. Gut 2003;52: 186–93.
91. Shimamoto T, Yamamichi N, Kodashima S, et al. PLoS One 2013;8:e65996.
92. Levenstein S, Rosenstock S, Jacobsen RK, Jorgensen T. Clin Gastroenterol Hepatol 2015;13:498-506.e.
93. Moss SF, Legon S, Bishop AE, et al. Lancet 1992;340:930–2
94. Thrift AP, El-Serag HB. Clin Gastroenterol Hepatol 2020;18: 534–542.
95. Howlader N, Noone AM, Krapcho M, et al, eds. SEER Cancer Statistics Review, 1975-2014, National Cancer Institute. Bethesda, MD, 2017. Available at https://seer.cancer.gov/csr/1975_2014/
96. American Cancer Society. Cancer Facts & Figures 2022. (2022) Atlanta, Ga: American Cancer Society [Internet]. Accessed June 2022.
97. Smyth EC, Verheij M, Allum W, et al. Ann Oncol 2016;27:38.
98. Yusefi AR, Bagheri Lankarani K, et al. Asian Pac J Cancer Prev 2018;19:591–603.
99. Thrift AP, Nguyen TH. Gastrointest Endosc Clin N Am 2021;31:425–439.
100. Khatoon J, Rai RP, Prasad KN. World J Gastrointest Oncol 2016;8:147–58.
101. Ladeiras-Lopes R, Pereira AK, Nogueira A, et al. (2008) Cancer Causes Control 19:689–701.
102. Tsugane S, Sasazuki S. Gastric Cancer 2007;10:75–83.
103. Tay SW, Li JW, Fock KM. Curr Opin Gastroenterol 2021;37: 158–163.
104. Kim SR, Kim K, Lee SA, et al. Nutrients 2019;11:826.
105. Rock CL, Thomson C, Gansler T, et al. CA Cancer J Clin 2020;70:245–271.
106. Shah SC, Piazuelo MB, Kuipers EJ, Li D. Gastroenterol 2021;161:1325–1332.
107. Kuipers EJ. Clin Gastroenterol Hepatol 2015;13:2290–2.
108. Rugge M, Meggio A, Pravadelli C, et al. Gut 2019;68:11–17.
109. Annibale B, Esposito G, Lahner E. Expert Rev Gastroenterol Hepatol 2020;14:93–102.
110. Jass JR, Filipe MI. Histochem J 1981;13:931–9.
111. Gawron AJ, Shah SC, Altayar O, et al. Gastroenterol 2020;158: 705–30.e5.
112. Rokkas T, Filipe MI, Sladen GE. Gut 1991;32:1110–1113.
113. Shao L, Li P, Ye J, et al. Int J Cancer 2018;143:1671–7.
114. Amin MB, Greene FL, Edge SB, et al. CA Cancer J Clin 2017;67:93–99.
115. Park CH, Yang DH, Kim JW, et al. Clin Endosc 2020;53: 142–166.
116. Ono H, Yao K, Fujishiro M, et al. Dig Endosc 2021;33:4–20.
117. Zu H, Wang F, Ma Y, Xue Y. PLoS One 2013;8:e54502.
118. Sexton RE, Al Hallak MN, Diab M, Azmi AS. Cancer Metastasis Rev 2020;39:1179–1203.
119. National Comprehensive Cancer Network (NCCN). NCCN Clinical Practice Guidelines in Oncology. Accessed on June 21, 2022. Available at www.nccn.org/professionals/physician_gls)
120. Wagner AD, Syn NL, Moehler M, et al. Cochrane Database Syst Rev 2017;8:CD004064.
121. Pasricha PJ, Parkman HP. Gastroenterol Clin North Am 2015;44:1–7.
122. Rey E, Choung RS, Schleck CD, et al. J Neurogastroenterol Motil 2012;18:34–42.
123. Abell TL, Camilleri M, Donohoe K, et al. J Nucl Med Technol 2008;36:44–54.
124. Hasler WL. Gastroenterol Clin North Am 2014;44:113–26.
125. Bharadwaj S, Meka K, Tandon P, et al. J Dig Dis 2016;17: 285–94.
126. Jebbink RJA, Samsom M, Bruijs PPM, et al. Gastroenterol 1994;107:1390–397.
127. Camilleri, M. N Engl J Med 2007;356:820–29.
128. Wytiaz V, Homko C, Duffy F, et al. Dig Dis Sci 2015;60: 1052–58.
129. Acosta A, Camilleri M. Gastroenterol Clin North Am 2015;44: 97–111.
130. Schey R, Saadi M, Midani D, et al. Dig Dis Sci 2016;61: 3545–551.
131. Richards RD, Davenport K, McCallum RW. Am J Gastroenterol 1993;88:203–7.
132. Sanger GJ, Pasricha PJ. Expert Opin Investig Drugs 2017;26: 331–42.

133. Lembo A, Camilleri M, McCallum R, et al. Gastroenterol 2016;151:87–96.e6.
134. Nelson AD, Camilleri M, Acosta A, et al. Neurogastroenterol Motil 2016;28:1705–13.
135. Banh HL, MacLean C, Topp T, et al. Clin Pharmacol Ther 2005;77:583–6.
136. Carbone F, Van den Houte K, Clevers E, et al. Am J Gastroenterol 2019;114:1265–74.
137. Hasler WL. Gastroenterol Clin North Am 2015;44:113–26.
138. Fahler J, Wall GC, Leman BI. Ann Pharmacother 2012; 46:e38.
139. Lacy BE, Tack J, Gyawali CP. Clin Gastroenterol Hepatol 2022;20:491–500.
140. Friedenberg FK, Palit A, Parkman HP, et al. Am J Gastroenterol 2008;103:416–23.
141. Arts J, Holvoet L, Caenepeel P, et al. Aliment Pharmacol Ther 2007;26:1251–258.
142. Ukleja A, Tandon K, Shah K, et al. World J Gastrointest Endosc 2015;7:790–98.
143. Coleski R, Anderson MA, Hasler WL. Dig Dis Sci 2009;54: 2634–2642.
144. Camilleri, M, Parkman HP, Shafi MA, et al. Am J Gastroenterol 2013;108:18–37.
145. Khashab MA, Ngamruengphong S, Carr-Locke D, et al. Gastrointest Endosc 2017;85:123–28.
146. Aghaie Meybodi M, Qumseya BJ, Shakoor D, et al. Endosc Int Open 2019;7:E322–E329.
147. Kamal F, Khan MA, Lee-Smith W, et al. Aliment Pharmacol Ther 2022;55:168–177.
148. Vosoughi K, Ichkhanian Y, Benias P, et al. Gut 2022;71:25–33.
149. Abell T, Mccallum R, Hocking M, et al. Gastroenterol 2003;125:421–28.
150. Heckert J, Sankineni A, Hughes WB, et al. Dig Dis Sci 2016;61: 168–75.
151. Lal N, Livemore S, Dunne D, Khan L. Gastroenterol Res Pract 2015;2015:762972.
152. Levinthal DJ, Bielefeldt K. Autonomic Neurosci 2017;202: 45–55.
153. Davis BR, Sarosiek I, Bashashati M, et al. J Gastrointest Surg 2017;21:222–27.
154. Jones, MP, Maganti K. Am J Gastroenterol 2003;98:2122–129.
155. Papasavas PK, Ng JS, Stone AM, et al. Surg Obes Relat Dis 2014;10:795–9.
156. Murray HB, Juarascio AS, Di Lorenzo C, et al. Am J Gastroenterol 2019;114:562–578.
157. Wauters L, Dickman R, Drug V, et al. United Euro Gastroenterol J 2021;9:307–331.
158. Venkatesan T, Levinthal DJ, Tarbell SE, et al. Neurogastroenterol Motil 2019;31(Suppl 2):e13604.
159. Perisetti A, Gajendran M, Dasari CS, et al. Ann Gastroenterol 2020;33:571–578.
160. Stumpf JL, Williams LD. J Pharm Pract 2021;34:786–793.
161. Tack J, Arts J, Caenepeel P, et al. Nat Rev Gastroenterol Hepatol 2009;6:583–590.
162. van Beek AP, Emous M, Laville M, Tack J. Obes Rev 2017;18: 68–85.
163. Vecht J, Masclee AA, Lamers CB. Scand J Gastroenterol Suppl 1997;223:21–27.
164. Scarpellini E, Arts J, Karamanolis G, et al. Nature Rev Endocrinol 2020;16:448–466.
165. Mine S, Sano T, Tsutsumi K, et al. J Am Coll Surg 2010;211: 628–36.
166. Ukleja A. Nutr Clin Pract 2005;20:517–25.
167. Didden P, Penning C, Masclee AA. Aliment Pharmacol Ther 2006;24:1367–75.
168. Penning C, Vecht J, Masclee AA. Aliment Pharmacol Ther 2005;22:963–9.
169. Hasegawa T, Yoneda M, Nakamura K, et al. J Gastroenterol Hepatol 1998;13:1201–6.
170. Thondam SK, Nair S, Wile D, Gll GV. QJM 2013;106:855–8.
171. Mala T. Surg Obes Relat Dis 2014;10:1220–5.
172. Tack J, Aberle J, Arts J, et al. Aliment Pharmacol Ther 2018;47: 1661–1672.
173. Pinto-Sanchez MI, Yuan Y, Hassan A, et al. Cochrane Database Syst Rev 2017;11:CD011194. doi: 10.1002/14651858
174. Lacy BE, Talley NJ, Locke GR 3rd, et al. Aliment Pharmacol Ther 2012;36:3–15.
175. Ford AC, Mahadeva S, Carbone MF, et al. Lancet 2020;396: 1689–1702.
176. Tack J, Camilleri M. Curr Opin Pharmacol 2018;43:111–117.

D. DISORDERS OF THE ALIMENTARY TRACT

CHAPTER 77

Assessment of Malabsorption

John Dibaise

CHAPTER OUTLINE

Abbreviations

7αHCO, 7α-hydroxy-4-cholesten-3-one; **A1AT**, α1-antitrypsin; **BAM**, bile acid malabsorption; **CT**, computed tomography; **ePFT**, endoscopic pancreatic function test; **ERCP**, endoscopic retrograde cholangiopancreatography; **MRCP**, magnetic resonance cholangiopancreatography; **MRI**, magnetic resonance imaging; **PABA**, para-aminobenzoic acid; **PLE**, protein-losing enteropathy; **^{75}SeHCAT**, selenium-75-homocholic acid taurine; **SIBO**, small intestinal bacterial overgrowth

Introduction

Disorders of malabsorption represent important challenges to clinicians, both in diagnosis and treatment. Malabsorptive conditions remain challenging diagnoses, in part because their clinical features may be identical, while their underlying pathophysiologies and, therefore, treatment may be different. Digestion and absorption involve numerous active components, including mechanical mixing and intestinal motility, digestive enzyme and bile acid production, mucosal function, blood supply, and the presence of the commensal gut microbiota. These components work together to allow the normal processing of nutrients that can be summarized into three major steps; (a) luminal and brush border digestion, primarily by pancreatic and biliary secretions (digestion phase), (b) absorption into the intestinal mucosa to be processed and packaged (mucosal phase), and (c) transport into the vascular or lymphatic circulation (postabsorptive or distributive phase). (Details on the physiology of intestinal food and nutrient digestion and absorption are covered elsewhere in this textbook.) Importantly, malabsorption can result from defects in one or more of the three phases described. Although maldigestion and malabsorption differ pathogenetically because the digestion and absorption of nutrients are tightly interconnected processes, consideration of maldigestive and malabsorptive pathologies separately may be overly simplistic. Therefore, the term "malabsorption" often is used clinically to denote derangements in both processes and is used in this manner throughout this chapter. The optimal approach to the diagnosis of malabsorption continues to evolve as new tests are developed. In the sections that follow, specific diagnostic tests are discussed and practical suggestions for the assessment of malabsorption are presented.

When to Suspect Malabsorption

The clinical presentation of malabsorption may vary tremendously, from severe steatorrhea (bulky, foul-smelling, greasy stools) and weight loss, to gas-bloat and chronic diarrhea, to signs of micronutrient deficiencies, to an incidental asymptomatic abnormality on laboratory testing (**Table 77.1**). This wide spectrum exists because the clinical features of malabsorption depend on the underlying cause and severity of the malabsorptive process. It is important to recognize that the classical presentation of steatorrhea and weight loss despite an adequate dietary intake is uncommon, at least in industrialized countries. In contrast, most patients with malabsorption present with mild, nonspecific symptoms or mainly extraintestinal manifestations, such as anemia or bone loss, which may lead to erroneous or delayed diagnosis.

Gastrointestinal symptoms are the most common indicators seen in malabsorptive conditions; however, occasionally they may be absent or minimal. Although chronic diarrhea is the most common symptom prompting an evaluation of malabsorption, chronic diarrhea is not, in most cases, due to a primary malabsorptive condition. In contrast, steatorrhea is a hallmark of malabsorption and is characterized by a porridge-like appearance with a very foul smell. Steatorrhea results from fat malabsorption, whereas watery diarrhea that occurs because of malabsorption may result from the osmotic effects of malabsorbed carbohydrates or the secretory effects of malabsorbed bile acids. Increased gaseousness with excessive belching or flatulence and abdominal bloating and/or distention occurs commonly, resulting from the fermentation of malabsorbed carbohydrates by colonic microbiota.

Table 77.1 Symptoms and Laboratory Evaluation of Specific Nutrient Malabsorption

Malabsorbed Nutrient	Clinical Manifestations	Laboratory Tests
Fat	Pale, bulk greasy malodorous stool; diarrhea without distention or gas	Stool Sudan III staining, acid steatocrit, fecal fat
Protein	Edema, muscle atrophy	Fecal nitrogen, serum albumin, fecal α1-antitrypsin clearance
Carbohydrate	Watery diarrhea, flatus, borborygmi, bloating, abdominal distension	Specific breath tests
Vitamin B_{12}	Macrocytic anemia, fatigue, loss of vibratory or position sense, glossitis	Hemoglobin, MCV, serum vitamin B_{12}, methylmalonic acid
Folic acid	Macrocytic anemia, glossitis	Hemoglobin, MCV, serum, and red blood cell folate
B-complex vitamins	Angular cheilosis, painless glossitis, acrodermatitis, dermatitis, ataxia, symmetric motor or sensory dysfunction	Serum vitamin levels
Iron	Microcytic anemia, fatigue, painful glossitis, koilonychia	Hemoglobin, MCV, serum ferritin, serum iron, and iron-binding capacity
Zinc	Altered taste (dysgeusia), dermatitis, hair loss	Serum zinc
Magnesium	Paresthesias, tetany	Serum magnesium, 24-h urinary magnesium
Selenium	Muscle pain, cardiomyopathy	Serum selenium level
Copper	Weakness, paresthesias, gait abnormalities	Serum copper and ceruloplasmin levels
Calcium and vitamin D	Paresthesias, tetany, bone pain or fractures, positive Chvostek/Trousseau signs, muscle cramps	Serum calcium, phosphorus, 25-hydroxyvitamin D, alkaline phosphatase, and parathyroid hormone
Vitamin C	Scurvy, perifollicular hemorrhage, corkscrew body hair, easy bruising	Serum vitamin C (ascorbic acid)
Vitamin A	Night blindness, follicular hyperkeratosis	Serum retinol, retinal esters
Vitamin E	Decreased deep tendon reflexes	Serum tocopherol
Vitamin K	Easy bruising, hemorrhage	Prothrombin time/international normalized ratio
Fluid, electrolytes	Tachycardia, carpal pedal spasm, dry mouth, paresthesias	Serum electrolyte panel, creatinine, urea nitrogen
Bile salts	Watery diarrhea	Serum 7α-hydroxy-4-cholesten-3-one, fecal bile acids

MCV, mean corpuscular volume.

Although weight loss is prevalent in individuals with severe forms of malabsorption, those with less severe forms may not present with weight loss. The pattern of weight loss also may vary, with some individuals losing weight early in the process with subsequent stabilization, and others with progressive weight loss. Abdominal pain, aside from cramping associated with defecation, is not a common complaint among those with malabsorption. Major exceptions include individuals with chronic pancreatitis or Crohn disease who have undergone multiple bowel resections. Nausea, emesis, borborygmi, and anorexia (or hyperphagia) also may be seen occasionally. In addition to the classic symptoms of steatorrhea noted, other characteristics of stool output to note include typical volume, number of stools per day, characteristics (e.g., watery, semiformed), presence of visible food, incontinence, and temporal characteristics (e.g., stool outputs in relation to meals).

Other nonspecific signs and symptoms that should prompt a consideration of the presence of a malabsorptive condition include easy bruising or bleeding, edema, hyporeflexia or hyper-reflexia, bone pain, unexpected fractures, poor wound healing, paresthesias, tetany, and dysgeusia.

Classification and Diagnostic Approach to Malabsorption

Malabsorptive disorders may be classified based on the deranged step in the digestive and/or absorptive process (i.e., luminal, mucosal, or postabsorptive), the macronutrient (i.e., carbohydrate, fat, or protein) or micronutrient (i.e., vitamin or trace element) affected, whether multiple nutrients or isolated nutrients are malabsorbed (i.e., global, partial, or selective), or based on clinical manifestations (i.e., overt, subclinical, or asymptomatic). Although an intimate understanding of the normal digestive and absorptive processes is not necessary to allow a clinician to diagnose a malabsorptive syndrome, this knowledge allows a clinician to achieve a broader differential diagnosis and to investigate suspected malabsorption in a more expeditious and, presumably, cost-effective manner (**Table 77.2**).

The diagnostic approach to malabsorption is twofold: (a) to confirm that malabsorption is present and (b) to determine its cause. Because the presence of malabsorption is not always readily apparent, the determination of its cause is often the major issue. The etiology of malabsorption may be determined from a detailed history and physical examination. Importantly, the history should include information regarding timing of symptom onset; bowel habits and stool characteristics; presence of growth failure, delayed sexual maturation, and weight loss or gain (e.g., fluid retention); associated gastrointestinal and other systemic symptoms; presence of concomitant chronic systemic, gastrointestinal, pancreaticobiliary, or liver disorders; prior surgery to the gastrointestinal tract; history of radiation exposure to the gut; travel history; diet; prescription, alcohol, and illicit drug use; high-risk sexual behavior; and family history. Pertinent aspects on examination include muscle wasting, skin rash/lesions, oral lesions, edema, abdominal distention, tenderness, and organomegaly and other potential signs of micronutrient deficiencies. Together with the history and examination, initial "routine" blood tests such as a complete blood count, chemistry panel, prothrombin time, magnesium, ferritin, folate, and vitamin B_{12} may provide evidence in support of the presence of malabsorption and further focus the investigative testing required

Table 77.2 Causes of Malabsorption

Small intestinal mucosal disease
Abetalipoproteinemia
Celiac disease and related syndromes
Chronic mesenteric ischemia
Crohn's disease
Common variable immunodeficiency
Disaccharidase deficiency
Eosinophilic gastroenteritis
Graft-versus-host disease
Lymphoma
Lymphangiectasia
Radiation enteritis
Terminal ileal disease
Infectious disorders
AIDS (HIV) enteropathy
Mycobacterial diseases
Parasitic diseases
Tropical sprue
Whipple disease
Luminal digestion disorders
Atrophic gastritis
Bile acid malabsorption
Chronic cholestasis
Hypersecretory states
Pancreatic insufficiency
Small intestinal bacterial overgrowth
Postsurgical disorders
Intestinal bypass
Postgastrectomy syndromes
Short bowel syndrome
Terminal ileum resection
Systemic disorders
Addison disease
Amyloidosis
Carcinoid syndrome
Connective tissue diseases

(continues)

Table 77.2 Causes of Malabsorption *(continued)*

Cronkhite-Canada syndrome
Diabetic gastroenteropathy
Hyperthyroidism and hypothyroidism
Hypoparathyroidism
Mastocytosis
Neuroendocrine tumors
Neurofibromatosis type 1
Vasculitis
Protein-losing gastroenteropathies
Erosive gastrointestinal diseases
Nonerosive gastrointestinal diseases
Increased interstitial pressure
Medication-induced conditions
Acarbose
Cholestyramine
Colchicine
Neomycin
Orlistat
Somatostatin analogues

AIDS, acquired immunodeficiency syndrome; *HIV*, human immunodeficiency virus.

to identify the specific malabsorptive process involved. Celiac serology also should be considered as a first-line diagnostic test in patients suspected of having malabsorption (see chapter 81). Stool testing for occult blood, calprotectin, elastase, and chronic infectious etiologies also should be considered at this stage.

Various tests, both invasive and noninvasive, are available to determine the specific cause of malabsorption. When the history is suggestive of a particular cause, testing can be used to confirm the diagnosis; however, further testing may not be necessary when the presence of malabsorption and its cause are obvious. Clearly, both the sequence of testing and the specific test chosen should be based on the clinical circumstances.

Functional Tests of Malabsorption

Fat Malabsorption Studies

Fat malabsorption is most determined by demonstrating fat in the stool; however, a serum test is available to screen for fat malabsorption. Diminished serum carotene concentration suggests deficiency in the absorption of fat-soluble vitamins, but also could be related to dietary deficiency. Serum carotene has not been studied extensively in patients with malabsorptive conditions, but concentrations typically are low in patients with steatorrhea.[1] The sensitivity and specificity of this test, compared with the gold standard fecal fat collection, remains to be determined.

The determination of fecal fat can be made both qualitatively and quantitatively. The advantages of qualitative approaches that make them more acceptable in clinical practice include their simplicity, lower cost, and less cumbersome collection methods. Disadvantages include their lower sensitivity, reproducibility, and reliability, particularly for mild-to-moderate steatorrhea, making them useful only when positive.[2] The most used qualitative approach involves the application of a fat-soluble stain such as Sudan III to a microscope slide smeared with stool mixed with glacial acetic acid, which is then examined under a microscope for stained fat globules. This technique can provide a semiquantitative assessment of fat excretion by determining both the number and size of the fat globules. Other qualitative techniques include the acid steatocrit,[3] an inexpensive, semiquantitative, gravimetric assay that uses an aliquot of centrifuged, homogenized stool to determine the percentage of fat; near-infrared reflectance analysis, which allows determination not only of fat excretion but also nitrogen and carbohydrate in a single sample[4]; and the ^{14}C- (or ^{13}C)-triolein breath test,[5] which involves the measurement of breath CO_2 following the ingestion of radiolabeled triolein, a triglyceride. Despite their simplicity and reportedly good test characteristics, none of these tests are widely used in the United States.

In contrast, the quantitative approach (Van de Kamer method) is more reliable and remains the gold standard for diagnosing fat malabsorption but is more complicated and expensive to perform and, therefore, often reserved for use in situations in which other tests have provided conflicting results. A 72-hour stool collection period is recommended to reduce variability and allow a better estimation of daily stool weight. A shorter period may be reasonable when moderate-to-severe diarrhea is present. *Ingestion of a diet with a known daily fat content (100 g/day often is used for convenience) is necessary for accurate stool fat determination.* This requires patient instruction on the proper diet or recording of all dietary intake during the stool collection, from which the daily fat intake can be estimated. Furthermore, because the ability to process and analyze (e.g., nuclear magnetic resonance spectroscopy, titrimetric, radioisotopic, spectrophotometric) the specimen is often not available locally, these samples typically need to be sent to an outside reference laboratory. A normal fat excretion per 100 g/day of dietary fat is less than 7 g/day; however, because fat excretion increases as stool weight increases, values of up to 14 g/day may result solely from large stool weight. Unfortunately, because of considerable overlap in values, quantitative fecal fat measurement is generally unable to discriminate among causes of steatorrhea. Ultimately, the quantitative fecal fat test often is not needed because other methods are available to diagnose the most relevant diseases of the pancreas, liver, and small bowel. Despite its limitations, the quantitative assessment of fecal fat may be useful in certain clinical scenarios, such as to establish the presence of malabsorption when there are no overt features, to monitor treatment in patients with established fat malabsorption, to estimate fecal energy loss in patients with severe malabsorption and malnutrition, and to differentiate fat from bile acid malabsorption in patients with an ileal resection.

Carbohydrate Malabsorption Studies

Carbohydrate malabsorption generally occurs either as a part of a global malabsorptive process or as a selective defect, usually localized to the epithelium. As mentioned, near-infrared reflectance analysis of the stool has been shown to be a reliable test of carbohydrate malabsorption.[6] However, because the direct measurement of carbohydrate excretion in the stool does not accurately measure small intestinal carbohydrate malabsorption, indirect tests are used. Indirect tests take advantage of the ability of colonic bacteria to metabolize malabsorbed carbohydrate, which can result in a decrease in stool pH, an increase in stool osmolality, and/or an increase in breath hydrogen or CO_2 excretion following a carbohydrate load. A stool pH less than 5.5 and a stool osmolar gap [stool (sodium+potassium)×2–280 mOsm/L] greater than 50 are characteristic of carbohydrate malabsorption, but although suggestive, are insufficiently sensitive or specific to confirm its presence.

Because carbohydrate malabsorption in the presence of a more global malabsorptive process is of lesser importance diagnostically, testing for carbohydrate malabsorption generally focuses on the detection of specific disaccharidase deficiencies. This can be accomplished by breath testing and oral tolerance testing. Although gene testing for lactase persistence/deficiency has been described in both family and case-control studies, until its clinical role and cost-effectiveness become clearer, its use is currently limited to research purposes in the United States.[7]

Breath testing involves the oral ingestion of a specific carbohydrate solution (e.g., lactose, fructose, sucrose) followed by the serial collection and measurement of hydrogen in expired breath samples over a set period. These substrates may also be labeled using radioactive (^{14}C) or stable (^{13}C) isotopes with subsequent measurement of breath CO_2. Malabsorbed carbohydrate results in a rise in hydrogen of more than 20 ppm.[8] Breath hydrogen testing may be affected by the presence of hydrogen-scavenging methanogenic microorganisms, resulting in a false-negative study. The concomitant assessment of breath methane may improve the accuracy of this test.[8] Because these indirect tests rely on bacterial fermentation of malabsorbed carbohydrate, the concurrent use of antibiotics may affect the results. The impact of potent antisecretory agents (e.g., proton pump inhibitors) and probiotics remains less certain.[9] Oral tolerance tests, which have lower sensitivity than breath tests, have largely been replaced by breath testing. In this procedure, the test sugar (most commonly lactose) is ingested, and blood glucose concentrations are measured over time. An increase in serum glucose of less than 20 mg/dL plus the development of symptoms is positive for malabsorption. However, the results of either breath or oral tolerance tests for carbohydrate malabsorption do not prove causality in symptom generation. Proof requires either simultaneous assessment of symptoms along with breath or serum samples, or clinical response to carbohydrate restriction (e.g., lactose or fructose) or enzyme (e.g., lactase) supplementation.

As noted, breath testing or oral tolerance testing are used mainly when evaluating specific disaccharidase deficiency syndromes and do not necessarily reflect global carbohydrate malabsorption. In contrast, the D-xylose test measures the functional absorptive capacity of the proximal small bowel and is considered to be the initial screening test of choice when malabsorption is suspected.[10] D-xylose is absorbed by passive diffusion; about 50% is absorbed and excreted in the urine. Conventionally, D-xylose in the serum and urine is measured following the oral administration of 25 g of D-xylose. Because of high false-positive rates, mostly related to errors with the urine collection, the use of only serum testing at 1 hour and 3 hours after D-xylose ingestion has been suggested.[11] A D-xylose concentration of less than 20 mg/dL at 1 hour, or less than 22.5 mg/dL at 3 hours, is considered abnormal and, given the high sensitivity and specificity demonstrated, suggests the presence of an intestinal mucosal (i.e., rather than a luminal digestive) malabsorptive process. Therefore, D-xylose absorption is normal in individuals with pancreatic exocrine insufficiency. However, an abnormal result does not indicate the cause of the malabsorption. Although this test is well tolerated and relatively inexpensive, it is not widely used in contemporary clinical practice because it does not identify the cause of malabsorption, given the availability of other tests that can (e.g., celiac disease antibodies and intestinal biopsy) and the fact that it generally does not alter the diagnostic algorithm for many clinicians when evaluating suspected malabsorption.

Protein Malabsorption and Protein-Losing Enteropathy Studies

Intestinal protein loss occurs more commonly because of loss of protein through the gut epithelium, rather than from the malabsorption of ingested protein. In healthy individuals, loss of protein through the gut epithelium plays only a minor role in total protein metabolism; daily enteric loss of serum proteins accounts for approximately 1% to 2% of the entire serum protein pool and enteric loss of albumin accounts for less than 10% of total albumin.[12] In contrast, gastrointestinal protein loss in protein-losing enteropathy (PLE) has been reported to involve up to 60% of the total albumin pool. The serum proteins most affected by this process are those with limited ability to rapidly respond to such losses and generally have longer half-lives, such as albumin, most immunoglobulins, and ceruloplasmin.

The detection of protein malabsorption is difficult. It requires the performance of nutrient balance studies and special analytic testing and is generally reserved for the research setting. Like carbohydrate malabsorption, detection of protein malabsorption is of lesser importance because protein malabsorption usually occurs as part of a global malabsorptive process. In contrast, the detection of protein loss via the gut (i.e., PLE) is of more clinical importance.[13] **Table 77.3** lists a selection of potential causes of PLE, divided by primary pathogenic mechanism.

The initial step in the evaluation of the patient with hypoproteinemia and/or hypoalbuminemia is to exclude other, more common, etiologies such as malnutrition, liver, and renal diseases. At present, when no other cause has been identified and concern about possible PLE exists, the most used and reliable method to determine enteric protein loss is to determine the clearance of α_1-antitrypsin (A1AT) from plasma.[13] A1AT is a protein synthesized in the liver that is neither actively secreted nor absorbed, has a molecular weight like albumin, and does not undergo proteolysis or degradation in the gut, thereby allowing for its intact elimination and detection in feces. The measurement of A1AT clearance requires both a blood sample to determine plasma A1AT and a 24-hour stool collection to determine stool volume and stool A1AT. Methods for determination of A1AT include nephelometry and radial immunodiffusion. The analysis of a random fecal A1AT concentration is not a reliable substitute for measurement of A1AT clearance, due to its insufficient sensitivity. However, when elevated, it may be a more convenient surveillance method to monitor treatment response or remission.[14] Normal A1AT clearance is ≤27 mL/24 hours. As diarrhea from any cause can increase the clearance of A1AT, in the setting of diarrhea, the normal A1AT clearance increases to

Table 77.3 Selected Causes of Protein Losing Enteropathy

Erosive Gastrointestinal Disease
Graft-versus-host disease
Inflammatory bowel disease
NSAID-enteropathy
Sarcoidosis
Ulcerative jejunoileitis
Nonerosive Gastrointestinal Disease
Amyloidosis
Bacterial infections (small intestinal bacterial overgrowth, strongyloides, tuberculosis, *Helicobacter pylori*, Whipple's disease)
Celiac disease
Collagenous colitis
Eosinophilic gastroenteritis
Gastric polyposis
Ménétrier disease
Lymphatic Congestion/Obstruction
Portal hypertension
Primary lymphangiectasia
Secondary lymphangiectasia (heart failure, mesenteric panniculitis, retroperitoneal fibrosis, tuberculous infiltration, lymphoma, neoplastic lymphoid invasion)
Thoracic duct obstruction
Cardiac Disease
Congenital heart disease
Congestive heart failure (particularly right sided)
Constrictive pericarditis
Fontan operation for univentricular heart

≤56 mL/24 hours. Because A1AT is degraded in environments with a pH of less than 3.5, if a gastric source of protein loss is suspected or an acid hypersecretory state is known to be present, it is recommended that the test be performed while the individual is receiving acid suppressive therapy.[15]

Bile Acid Malabsorption Studies

Bile acid malabsorption (BAM) is an important, often unrecognized, cause of chronic diarrhea.[16] Three main types of BAM have been described: Type 1 in those with distal ileal disease or resection (classically when <100 cm distal ileum resected), Type 2 in those with no overt disease (also referred to as idiopathic or primary bile acid diarrhea), and Type 3 in those with other disorders or medications responsible.[16] Intestinal absorption of bile acids occurs through a combination of passive diffusion throughout the small intestine and, predominantly, active absorption in the distal ileum, resulting in <5% of bile acids entering the colon. The diarrhea is thought to occur mainly via the effects of di-α-hydroxy bile acids, such as chenodeoxycholic and deoxycholic acid, on the induction of colonic secretion and, possibly, propulsive colonic contractions. In general, testing for BAM/diarrhea is reserved for those with diarrhea but without overt ileal disease or steatorrhea.

Although the direct measurement of fecal bile acids or fecal excretion of radiolabeled bile acids is a sensitive and reliable test for BAM, because of its unpleasant and labor-intensive nature, it was, until recently, generally reserved for research purposes. A 48-hour stool collection while on 100 gram of fat/day diet for measurement of bile acids is now available at some commercial laboratories and can be done simultaneously with a quantitative fecal fat study. The ^{14}C-taurocholate bile acid absorption test has been described but is limited by the need for radioactivity and a 72-hour stool collection following ingestion of the radiolabeled bile acid.[17] The test most widely used clinically worldwide to both detect BAM and monitor response to a bile acid sequestrant (e.g., cholestyramine, colestipol, colesevelam) is the measurement of selenium-75-homocholic acid taurine (^{75}SeHCAT); however, it is not available in the United States. In this test, a radiolabeled homolog of taurocholate, a marker of active absorption of bile acids from the ileum that is resistant to bacterial deconjugation, is administered orally and total body retention is measured over several days. Retention of less than 50% after 3 days or fewer than 5% after 7 days is abnormal. Disadvantages of ^{75}SeHCAT include its long half-life (180 days) and problems with interpretation in the setting of liver disease.

Recently, the measurement of serum 7α-hydroxy-4-cholesten-3-one (7α-HCO or 7αC4 or C4) has been introduced as a validated nonradioactive method for monitoring the enzymatic activity of hepatic cholesterol 7α-hydroxylase, the rate-limiting step and major regulatory enzyme in the synthesis of bile acids that is closely related to the fecal loss of bile acids.[18] Therefore, production of 7α-HCO reflects bile acid synthesis, and an increase is expected in individuals with significant bile acid loss. In a direct comparison with the ^{75}SeHCAT test, increased serum 7α-HCO has a sensitivity of 90% and specificity of 79% in diagnosing BAM.[19] The usefulness of this test, available in some commercial laboratories, in more substantial distal small intestinal resections is unclear. At present, this test is not widely used, perhaps because of clinicians' common use of empiric therapeutic trial of bile acid sequestrants in patients with suspected BAM. The measurement of serum fibroblast growth factor 19 (FGF19), which is released by ileal enterocytes in response to bile acid uptake and provides feedback inhibition of bile acid synthesis, has also been evaluated as a screening test for BAM but was found to have lower sensitivity and specificity than 7α-HCO and is not recommended as a screening tool.[20] Lastly, the combination of a random stool sample for total bile acids and serum 7α-HCO is currently under investigation as a diagnostic test.

Due to the limitations of available tests for BAM, treatment response to bile acid sequestrants has been commonly used as an indirect "test" in clinical practice. However, problems with the approach of a "therapeutic trial" include poor palatability and side effects of bile acid sequestrants causing a false negative result, and lack of specificity of bile acid sequestrants, which may bind to and inactivate other factors (e.g., *Clostridioides difficile* toxin), causing a false positive result. A recent clinical guideline on the management of bile acid diarrhea from the Canadian Association of Gastroenterology recommended testing rather than a therapeutic trial with bile acid sequestrant, using SeHCAT (where available) or fasting plasma 7α-HCO.[21]

Indirect Tests for Small Intestinal Bacterial Overgrowth

Small intestinal bacterial overgrowth (SIBO) is a common cause of malabsorption, albeit generally mild, and should be considered in any individual with consistent symptoms, regardless of the presence of overt malabsorption, particularly when a predisposing anatomic or functional condition exists (**Table 77.4**).[22] Potential nutritional consequences of SIBO include malabsorption of fat and carbohydrate, micronutrient deficiencies, and weight loss. The accurate diagnosis of SIBO remains problematic.[23] Due to limitations in SIBO diagnosis by small bowel fluid aspiration and culture (see below), breath testing for hydrogen and methane has become the most widely used method to diagnose SIBO in clinical practice. Breath testing is low risk, relatively inexpensive, portable, and easy to perform.[8,22] Breath testing uses an orally ingested carbohydrate (e.g., glucose, lactulose, xylose) as a substrate, which is metabolized in the presence of excessive bacteria in the small bowel, releasing hydrogen (and methane and sulfide when present), which is subsequently absorbed and then released into the expired air. A rise in hydrogen of >20 ppm within 90 minutes of ingestion in the breath sample is consistent with SIBO.[8] As noted previously, the concomitant assessment of breath methane may improve the accuracy of this test. High fasting hydrogen (>20 ppm) is also common in SIBO but lacks sensitivity and specificity. Numerous factors may influence the results of this test, including diet, exercise, tobacco smoking, recent use of antibiotics, rapid orocecal transit, and the diagnostic criteria used.[8] In general, breath tests have demonstrated wide variations in sensitivity and specificity and disappointing abilities to predict the results of small bowel culture.[24]

Other noninvasive alternatives are based on the detection of bacterial metabolites of either endogenous or exogenous substrates. The ^{14}C (and ^{13}C)–D-xylose breath test measures pulmonary excretion of radiolabeled CO_2 produced from bacterial fermentation of the labeled substrate. Initial reports using this technique suggested considerably better performance than the hydrogen breath test. However, more recent reports suggest widely ranging sensitivities and specificities.[25,26] Disappointing results have also been seen with the measurement of products of luminal bacterial metabolism in urine (e.g., elevated indicans and choloyl-para-aminobenzoic acid [PABA]) or blood (e.g., elevated folate, D-lactate, short-chain fatty acids, unconjugated bile acids), and the ^{14}C-glycocholate breath test.[22] Because of the limitations of diagnostic tests for SIBO, it is common clinical practice to provide empiric antibiotic treatment (i.e., "therapeutic trial") for individuals suspected of having SIBO. Unfortunately, like the empiric bile acid sequestrant trial for suspected bile acid malabsorption, this empiric approach in SIBO is fraught with problems and, unless a high index of suspicion is present, is generally not recommended.[22] Notably, it is important to identify and treat the underlying cause of SIBO and to identify and correct nutritional deficiencies, when present.

Table 77.4 Risk Factors for Small Intestinal Bacterial Overgrowth

Gastric acid hyposecretion
Small bowel dysmotility
Altered small bowel anatomy
Immunodeficiency
Multifactorial

Direct and Indirect Tests for Pancreatic Exocrine Insufficiency

Various tests are available for the detection of pancreatic exocrine insufficiency. The quantitative pancreatic stimulation tests are considered the gold standard. These tests require gastrointestinal intubation followed by stimulation of pancreatic secretion using intravenous secretin, cholecystokinin, or a liquid test meal (Lundh test), with subsequent collection of duodenal fluid, which is analyzed for pancreatic secretions including enzymes and/or bicarbonate, depending on the secretagogue used.[27] Because these tests require radiation exposure to place the tube and are invasive, time-consuming, expensive, and performed in only a few academic centers, they are seldom performed in clinical practice. The endoscopic pancreatic function test (ePFT) has been developed to facilitate the performance of this quantitative pancreatic stimulation test. Initial studies comparing results among conventional Dreiling tube collection and the ePFT in both healthy individuals and patients with chronic pancreatitis have demonstrated promising results.[28,29] Although the 1-hour endoscopic collection period is likely not feasible for wide-spread implementation in clinical practice, a simplified procedure requiring endoscopic collection of duodenal fluid at only two time points, 30 and 45 minutes after secretin administration, has been shown to produce 92% accuracy compared with the standard 1-hour protocol.[30] Recently, the combination of EUS and secretin-ePFT results were able to predict the development of classic chronic pancreatitis structural changes over time.[31]

Indirect pancreatic function tests are more widely available; however, they are less sensitive than direct tests because they become abnormal only after about 90% of pancreatic enzyme output has declined. Given their limited sensitivities, these tests may be most useful in conjunction with other tests to support the diagnosis of pancreatic insufficiency. The most readily available indirect tests involve random stool collection for chymotrypsin or elastase. Decreased concentrations are found in the setting of pancreatic insufficiency. Conflicting information regarding the sensitivity and specificity of both tests for mild-to-moderate exocrine dysfunction exists, leading some authorities to suggest that neither test is suitable for screening of chronic pancreatitis.[32,33] Nevertheless, there appears to be consensus that fecal elastase testing is superior to fecal chymotrypsin testing in overall sensitivity, and the results are not affected by oral pancreatic enzyme use. The bentiromide and pancreolauryl tests are based on the effect of pancreatic enzymes on ingested oral substrates, with subsequent detection of metabolites in urine, plasma, or breath. Like the stool tests, these tests are of limited value in the detection of mild pancreatic insufficiency. Falsely abnormal results are also problematic. Furthermore, the bentiromide test is no longer available because of allergic concerns related to the use of PABA and the pancreolauryl test is not widely available in the United States. The dual-labeled Schilling test also can be used to diagnose pancreatic insufficiency as a cause of vitamin B_{12} deficiency, based on the differential absorption of orally administered radiolabeled cobalamin bound to intrinsic factor and R protein.[34] However, like the other indirect tests, its ability to detect mild-to-moderate pancreatic insufficiency is limited. In general, the Schilling test for the diagnosis of causes of vitamin B_{12} deficiency (e.g., inadequate dietary intake, inadequate production of intrinsic factor, or presence of SIBO, pancreatic insufficiency or terminal ileal disease/resection) is mainly of historical interest given its complexity, the availability of simpler means of diagnosing causes of vitamin B_{12} deficiency, and the present lack of availability of radiolabeled cobalamin. Because of the limitations of the available tests of pancreatic insufficiency, an empiric trial of oral pancreatic enzymes is frequently employed in clinical practice.

Morphologic Tests of Malabsorption

Although functional tests are useful to establish the presence of malabsorption, a specific diagnosis is provided infrequently. Unless the underlying cause is readily apparent from the history, physical examination and initial blood and stool testing, additional evaluation usually is needed. Morphologic tests can be divided into those that provide anatomic information (i.e., radiologic and endoscopic imaging) and those that provide histologic information (i.e., tissue biopsy).

Endoscopy

Direct visualization of the gastrointestinal tract, although seldom permitting a definitive diagnosis, may allow demonstration of a small bowel mucosal process and enables the collection of small bowel aspirates and/or mucosal (or deeper) biopsies (**Figure 77.1**). The endoscopic evaluation has been facilitated by the development of wireless capsule endoscopy, which allows examination of the entire small bowel, but not the ability to obtain fluid or tissue samples; and high resolution magnified endoscopic systems that allow total or near total enteroscopy (e.g., double-balloon and single-balloon enteroscopes) while also permitting both diagnostic and therapeutic procedures through the endoscope.[35,36] Wireless capsule endoscopy appears to be superior to small bowel radiography to detect subtle mucosal changes[37] but should be avoided in individuals with known or suspected strictures. At present, limited data exist on the diagnostic value of wireless capsule endoscopy and either conventional push enteroscopy or balloon-assisted enteroscopy for the evaluation of small bowel malabsorptive or diarrheal disorders. A high index of suspicion is needed when considering these enteroscopy approaches, and their routine use in the evaluation of chronic diarrhea and/or suspected malabsorption is not recommended.

While jejunal biopsies are sometimes diagnostic when duodenal biopsies are not,[38] most small intestinal mucosal malabsorptive processes are diffuse and affect the duodenum, allowing inspection for mucosal abnormalities and collection of biopsy specimens using a conventional upper endoscope. Importantly, a normal-appearing small bowel fold pattern is not uncommon in small bowel malabsorptive conditions and should not deter the endoscopist from taking biopsies. Although inflammatory changes (e.g., ulcers, erosions) and strictures may be seen, the most common mucosal abnormalities suggesting a malabsorptive process are features of atrophy, including a reduction in the number or absence of valvulae conniventes, a mosaic appearance to the mucosa, and the presence of scalloped folds, with a reported sensitivity of 94% and specificity of 92% in celiac disease, the most common diffuse small bowel malabsorptive condition (**Figure 77.2**).[39] Given that some malabsorptive conditions may be patchy, the use of magnification endoscopy combined with either chromoendoscopy[40] or water immersion[41] has been suggested to help both identify patients with partial atrophy and direct biopsies. These methods require additional testing to prove their diagnostic utility in a cost- and time-efficient manner.[42]

Although a colonic source is generally not considered when evaluating suspected malabsorption, colonoscopy with retrograde ileoscopy may be useful in the setting

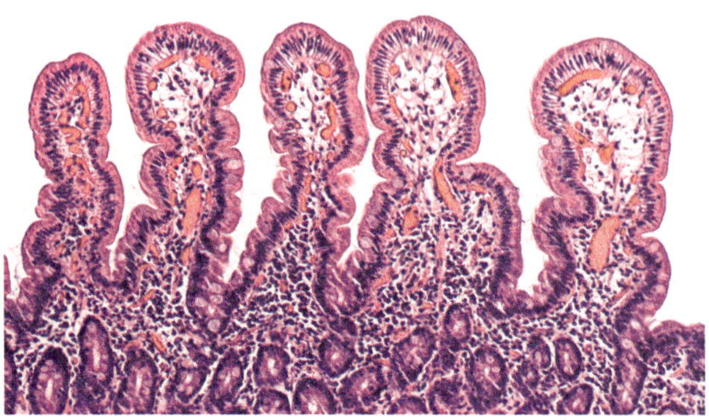

Figure 77.1 Normal small bowel biopsy. This shows normal villi and crypts of the jejunum (hematoxylin and eosin).

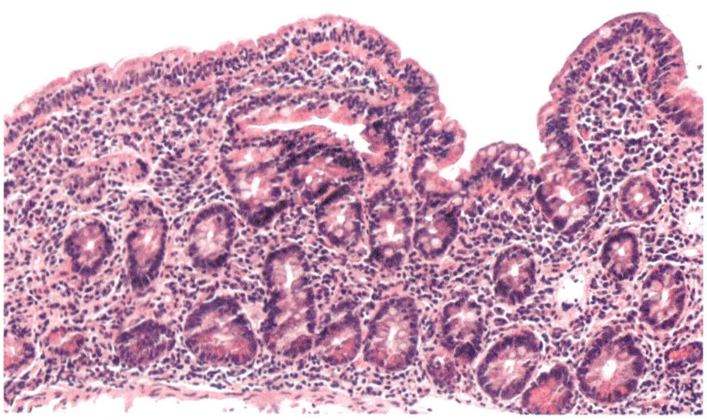

Figure 77.2 Celiac disease, small bowel biopsy. The total villous atrophy, crypt hyperplasia, and relatively squat, cuboidal surface epithelium are all features indicative of mucosal injury. The prominent intraepithelial lymphocytosis and the similar inflammatory infiltrate in the lamina propria are suggestive of celiac disease (hematoxylin and eosin).

of chronic diarrhea. Terminal ileal inspection and biopsy appear to be of the most value when Crohn disease or infection is suspected, abnormal findings are noted on radiologic imaging of the terminal ileum, or abnormal-appearing terminal ileal mucosa is identified on colonoscopy.[43] Routine terminal ileal biopsies of normal-appearing mucosa appear to be of low yield.

Endoscopic imaging of the pancreas, as with endoscopic retrograde cholangiopancreatography (ERCP) and endoscopic ultrasound, may be helpful when pancreatic exocrine insufficiency is being considered. Both ductal and parenchymal abnormalities have been described that are suggestive of the presence of chronic pancreatitis[44]; however, a normal examination does not exclude the presence of pancreatic exocrine insufficiency, and additional pancreatic function testing may be required. Both diagnostic and therapeutic procedures also are possible using these techniques.

Endoscopic Sampling

Although inspection of the upper gut mucosa during endoscopy will generally not permit a definitive diagnosis, endoscopy does allow collection of small bowel fluid and mucosal samples that may permit a definitive diagnosis. These biopsies may be diagnostic or strongly suggestive of specific intestinal malabsorptive conditions (**Table 77.5**). It is important to recognize, however, that biopsies may miss the diagnosis, given the patchy nature and the nonspecific histologic features of some diseases.[45] Although not always pathognomonic, intestinal biopsies may be helpful in identifying disease and directing the diagnostic evaluation and treatment plan, and should be performed even when the endoscopic appearance is normal. Standard forceps biopsy via the endoscope is usually sufficient, and the use of suction biopsy without endoscopic control is no longer necessary. It has been recommended that at least four mucosal biopsies from different duodenal sites be obtained to optimize the likelihood of securing a diagnosis.[46] Reports suggest that biopsies from the duodenal bulb may aid in the diagnosis of celiac disease, although this remains controversial, given the potential for confusion related to acid-peptic duodenitis and compression from Brunner glands.[46] Although conventional endoscopic biopsies usually sample only the mucosa, deeper biopsies of the mucosa and submucosa may be obtained endoscopically using relatively simple techniques that may allow the diagnosis of conditions that previously required surgery for definitive diagnosis (e.g., Ménétrier disease).

Table 77.5 Small Bowel Histology to Diagnose Malabsorptive Conditions

Diagnostic or supportive
Abetalipoproteinemia
Amyloidosis
Eosinophilic enteritis
Giardiasis
Lymphangiectasia
Lymphoma
Mastocytosis
Mycobacterial disease
Parasitic diseases
Whipple disease
Suggestive but not diagnostic
Acquired immune deficiency syndrome enteropathy
Autoimmune enteropathy
Celiac disease
Crohn disease
Collagenous sprue
Common variable immunodeficiency
Radiation enteritis
Tropical sprue
Viral gastroenteritis
Generally normal
Liver disease
Pancreatic disease
Primary disaccharidase deficiency
Small intestinal bacterial overgrowth

SIBO is characterized by the presence of an excessive concentration of either overall bacteria or specific types of bacteria in the small bowel. The culture of aspirated small bowel fluid (classically collected from the jejunum but usually collected from the duodenum in clinical practice) has traditionally been considered the gold standard in the diagnosis of SIBO and, despite significant limitations, is still considered by most experts to be the preferred diagnostic test.[47] A small bowel aspirate can be obtained easily during endoscopy by passing a sterile aspiration catheter through the working channel of the endoscope. Limitations include its invasiveness, expense, potential for contamination, potential for not detecting SIBO occurring in the more distal small intestine, and difficulty in culturing the enteric micro-organisms. Because of these limitations, the reliability of the technique has been questioned, and indirect methods of detecting SIBO have been developed as potential alternatives (see previous section). It will be interesting to see whether recent improvements in the culture-independent, molecular microbial fingerprinting methods currently reserved for research on the gut luminal microbiome, can be applied clinically in the diagnosis of SIBO and improve the utility of the small bowel aspirate. Additionally, the small bowel aspirate can also be examined microscopically for the presence of *Giardia lamblia*.

Radiology

Radiologic imaging is often used in the evaluation of suspected malabsorption. Barium-contrast studies of the small bowel (e.g., small bowel follow-through, enteroclysis) and cross-sectional abdominal imaging (e.g., computed tomography

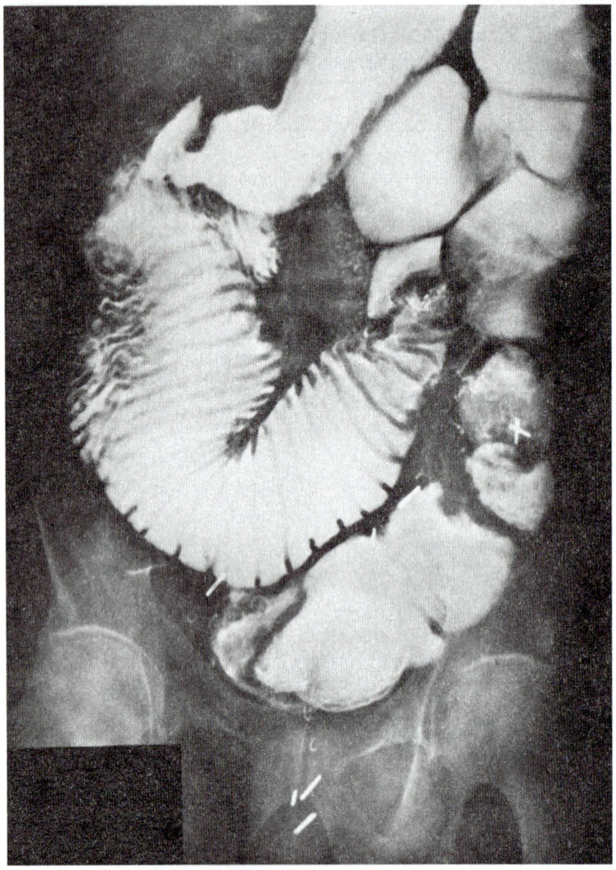

Figure 77.3 Short bowel syndrome. A small bowel follow-through x-ray study demonstrates the approximate residual length of this very short bowel, as well as marked dilation of the residual duodenum and jejunum. The jejunum is anastomosed to the distal residual colon.

[CT] and magnetic resonance imaging [MRI]) are most used (**Figure 77.3**). Although barium studies can help define the diffuse or segmental nature of a condition and presence of anatomic abnormalities (e.g., stricture, dilation, diverticula, fistula, ulceration, tumor, and stagnant loops of bowel), and provide a crude assessment of transit, classical findings of malabsorption (e.g., segmentation, flocculation, dilation of bowel loops) lack both sensitivity and specificity to diagnose specific conditions. Cross-sectional imaging techniques, particularly when using an enterography protocol, have the added ability to evaluate the full thickness of the bowel wall, extraintestinal organs (e.g., pancreas, liver, lymph nodes, mesentery) and the mesenteric vasculature.[48] CT or MR enteroclysis, where a nasoenteral tube is placed into the duodenum or jejunum for direct instillation of contrast, is also available but is infrequently used, likely due to the need for nasal intubation and the good results from use of the enterography protocol alone. Magnetic resonance cholangiopancreatography (MRCP), much like the endoscopic variety, can provide an assessment of pancreatic parenchymal and ductal abnormalities. Although MRCP does not allow for collection of specimens or therapeutic procedures, given its noninvasive nature and improvements in the quality of the images, MRCP has generally replaced ERCP as the diagnostic test of choice for suspected pancreaticobiliary disease.[44] Secretin-enhanced MRCP has been developed to assess exocrine pancreatic function; however, the results correlate better with severe disease, and this technique has been suggested to be too insensitive to diagnose mild disease.[49] Similarly, another report questioned the ability of MRCP to characterize early chronic pancreatitis accurately compared with the ePFT.[50] MRI also has been studied as a diagnostic test for PLE, primarily in children with lymphangiectasia. Imaging was able to detect abnormalities in both mesenteric and intestinal swelling clearly and to document the presence of dilated and prominent lymphatics. Investigators suggested that these findings, in the setting of an appropriate clinical presentation, could be used to support the diagnosis of PLE without having to proceed to invasive testing such as biopsy or endoscopy.[51] Lymphangiography may also be helpful in select cases. The use of technetium-99m-labeled albumin to scintigraphically document protein loss in the gut has been shown to be useful both in the diagnosis of PLE and in monitoring response to treatment. However, at present, this test is mainly reserved for research use because the cost and overall accuracy has yet to be determined.[52] Finally, if a neuroendocrine tumor (e.g., gastrinoma or somatostatinoma) is suspected as the cause of malabsorption, an ^{111}indium octreotide scan, positron emission tomography scan, or EUS of the pancreas, may be helpful to establish the diagnosis and extent of the disease.

Practical Strategy When Evaluating Suspected Malabsorption

A primary goal in the evaluation of patients with suspected malabsorption is to establish a definitive diagnosis quickly and inexpensively. The optimal strategy for the diagnosis of a malabsorptive process is still evolving and, in general, the order of tests must be individualized. A detailed history and physical examination is the first step in this process. In most cases, this helps determine whether malabsorption is present, focuses the subsequent evaluation, and occasionally provides the diagnosis. If the history suggests a specific cause, testing can be directed to confirm the diagnosis. Next, an evaluation of routine blood and stool tests may provide additional hints to the diagnosis and further focus additional testing, given that symptoms may be absent or may mimic other diseases. Serologic testing for celiac disease should be considered at this stage given the myriad clinical manifestations of this disorder (see chapter 81). If the presence of steatorrhea remains unclear, fecal fat assessment is recommended. Similarly, if PLE is under consideration, fecal A1AT clearance is recommended. Breath tests for specific carbohydrate malabsorption could also be considered, depending on clinical suspicion. Next, abdominal imaging should be considered. Depending on the availability of tests, CT enterography may be the single most informative imaging study. Upper and lower gastrointestinal endoscopy with collection of a small bowel aspirate and biopsies of the proximal and distal small bowel and colon are often necessary to obtain a definitive diagnosis. Pancreatic function testing or a therapeutic trial of pancreatic enzymes may be considered, depending on the clinical scenario. Additional testing may be considered on an individual basis, depending on the preceding findings.

General Treatment Strategies for Malabsorption

Briefly, optimal treatment of the variety of conditions that may cause or worsen malabsorption requires accurate diagnosis and treatment of the underlying disorder. Adjustment of dietary intake to maximize intake of a well-balanced diet that does not increase net stool output and worsen malabsorption may require the assistance of a registered dietitian to determine details of habitual nutrient intake, and to work with the physician and patient on appropriate food and nutrient intake strategies. This may require specific oral nutrient supplements to correct or prevent macronutrient or micronutrient depletion (e.g., complete liquid or solid nutritional supplements; specific energy or protein supplements; specific or complete vitamin, mineral, and trace element supplements). In severe cases, a specialized nutrition support service may be needed to help initiate and manage use of complete parenteral nutrition formulations and/or intravenous hydration and micronutrient/electrolyte administration (see chapters 83 and 84). In all cases, serial evaluation of body weight and physical examination changes indicative of nutrient depletion/repletion, serial laboratory tests of specific micronutrient concentrations in blood, and stool analysis, as outlined above, are important to monitor the efficacy of therapy in patient management.

References

1. Naveh Y, Ken-Dror A, Zinder O, et al. J Pediatr Gastroenterol Nutr 1996;5:210–13.
2. Simko V. Am J Gastroenterol 1981;75:204–8.
3. Amann ST, Josephson SA, Toskes PP. Am J Gastroenterol 1997;92:2280–4.
4. Stein J, Purschian B, Bieniek U, et al. Eur J Gastroenterol Hepatol 1994;6:889–94.
5. Mylvaganam K, Hudson PR, Ross A, et al. Gut 1986;27:1347–52.
6. Stein J, Purschian B, Zeuzem S, et al. Clin Chem 1996;42: 309–12.
7. Larvela IE. Ann Med 2005;37:179–85.
8. Rezaie A, Buresi M, Lembo A, et al. Am J Gastroenterol 2017; 112:775–84.
9. Romagnuolo J, Schiller D, Bailey RH. Am J Gastroenterol 2002; 97:1113–26.
10. Craig RM, Ehrenpreis ED. J Clin Gastroenterol 1999;29:143–50.
11. Ehrenpreis ED, Salvino M, Craig RM. J Clin Gastroenterol 2001; 33:36–40.
12. Schmidt PN, Blirup-Jensen S, Svendsen PJ, et al. Scand J Clin Lab Invest 1995;55:35–45.
13. Copland AP, DiBaise JK. Pract Gastroenterol 2017;41:22–37.
14. Strygler B, Nicor MJ, Santangelo WC, et al. Gastroenterol 1990;99:1380–7.
15. Takeda H, Nishise S, Furukawa M, et al. Dig Dis Sci 1999;44: 2313–8.
16. DiBaise JK. Pract Gastroenterol 2020;10–24.
17. Schiller LR, Hogan RB, Morawski SG, et al. Gastroenterol 1987;92:151–60.
18. Brydon WG, Nyhlin H, Eastwood MA, et al. Eur J Gastroenterol Hepatol 1996;8:117–23.
19. Eusufzai S, Axelson M, Angelin B, Einarsson K. Gut 1993;34: 698–701.
20. Vijayvargiya P, Camilleri M, Carlson P, et al. Aliment Pharmacol Ther 2017;46:581–8.
21. Sadowski DC, Camilleri M, Chey WD, et al. Clin Gastroenterol Hepatol 2020;18:24–41.
22. Adike A, DiBaise JK. Gastroenterol Clin N Am 2018;47:193–208.
23. Khoshini R, Dai SC, Lezcano S, et al. Dig Dis Sci 2008;53: 1443–54.
24. Corazza GR, Menozzi MG, Strocchi A, et al. Gastroenterol 1990; 98:302–9.
25. King CE, Toskes PP, King CE, et al. Gastroenterol 1986;91: 1447–51.
26. Riordan SM, McIver CJ, Duncombe VM, et al. Am J Gastroenterol 1995;90:1455–60.
27. Domínguez Muñoz JE. Best Pract Res Clin Gastroenterol 2010;24:233–41.
28. Stevens T, Conwell DL, Zuccaro Jr G, et al. Am J Gastroenterol 2006;101:351–5.
29. Stevens T, Conwell DL, Zuccaro Jr G, et al. Gastrointest Endosc 2008;67:458–66.
30. Stevens T, Conwell DL, Zuccaro Jr G, et al. Am J Gastroenterol 2007;102:297–301.
31. Monachese M, Lee PJ, Harris K, et al. Endosc Ultrasound 2021;10:116–23.
32. Luth S, Teyssen S, Forssmann K, et al. Scand J Gastroenterol 2001;36:1092–9.
33. Gredal C, Madsen LG, Larsen S. Pancreatol 2003;3:389–94.
34. Chen WL, Morishita R, Eguchi T, et al. Gastroenterol 1989;96:1337–45.
35. Maiden L, Elliott T, McLaughlin SD, et al. Dig Dis Sci 2009;54:1280–3.
36. May A, Färber M, Aschmoneit I, et al. Am J Gastroenterol 2010;105:575–81.
37. Marmo R, Rotondano G, Piscopo R, et al. Aliment Pharmacol Ther 2005;22:595–604.
38. Fry LC, Bellutti M, Neumann H, et al. Dig Dis 2008;26:134–9.
39. Maurino E, Capizzano H, Niveloni S, et al. Dig Dis Sci 1993;38:2028–33.
40. Siegel LM, Stevens PD, Lightdale CJ, et al. Gastrointest Endosc 1997;46:226–30.
41. Cammarota G, Cesaro P, Cazzato A, et al. J Clin Gastroenterol 2009;43:244–8.
42. Rubio-Tapia A, Murray JA. Gastrointest Endosc 2007;66:382–6.
43. McHugh JB, Appelman HD, McKenna BJ. Am J Gastroenterol 2007;102:1084–9.
44. Choueiri NE, Balci NC, Alkaade S, et al. Curr Gastroenterol Rep 2010;12:114–20.
45. Owens SR, Greenson JK. Histopathol 2007;50:64–82.
46. Hopper AD, Cross SS, Sanders DS. Endoscopy 2008;40:219–24.
47. Riordan SM, McIver CJ, Walker BM, et al. Am J Gastroenterol 1996;91:1795–803.
48. Dave-Verma H, Moore S, Singh A, et al. Curr Probl Diagn Radiol 2008;37:279–87.
49. Sanyal R, Stevens T, Novak E, et al. AJR Am J Roentgenol 2012;198:124–32.
50. Zuccaro P, Stevens T, Repas K, et al. Pancreatol 2009;9:764–9.
51. Liu NF, Lu Q, Wang CG, Zhou J-G. Lymphol 2008;41: 111–115.
52. Seok JW, Kim S, Lee SH, et al. Clin Nucl Med 2002;27:431–3.

CHAPTER 78

Diet and Intestinal Disaccharidases

Steve Hertzler, Monica Ramakrishnan, Yeonsoo Kim, and Dennis Savaiano

CHAPTER OUTLINE

Abbreviations

LNP, lactase nonpersistence; **LP**, lactase persistence; **LPH**, lactase-phlorizin hydrolase; **PBS**, phosphate-buffered saline; **PNG**, pyridoxine-5′-β-D-glucoside; **RER**, rough endoplasmic reticulum; **SI**, sucrase-isomaltase; **SNP**, single nucleotide polymorphism; **BCM-7**, beta-casomorphin-7

Introduction

Disaccharides are a significant source of carbohydrates in the diet. The major disaccharides include sucrose (*O*-α-D-glucopyranosyl-[1→2]-β-fructofuranoside), lactose (*O*-β-D-galactopyranosyl-[1→4]-β-glucopyranose), maltose (*O*-α-D-glucopyranosyl-[1→4]-α-glucopyranose), and trehalose (*O*-α-D-glucopyranosyl-[1→1]-α-glucopyranose). Lactose is the primary carbohydrate in human milk, which contains approximately 7% lactose by weight,[1] among the highest of all mammalian milks (**Table 78.1**). Sucrose, lactose, and maltose comprise approximately 50% of the total carbohydrate in the diet consumed by a healthy adult in United States.[2] Most of the maltose present in the intestine is derived from starch digestion, with only small amounts being contributed from grains and fermented beverages. Mushroom, honey, and shellfish are some dietary sources of trehalose.[3]

Because the small intestine is normally impermeable to disaccharides, intestinal disaccharidase activity is required for absorption of their component monosaccharides.[6] In humans and other mammals, four known enzymes or enzyme complexes exist for disaccharide digestion: sucrase-isomaltase (SI), lactase-phlorizin hydrolase (LPH; lactase), maltase-glucoamylase, and trehalase.[7] In contrast to the other enzymes that hydrolyze alpha-glucosidic bonds, lactase hydrolyzes beta-glucosidic bonds.[8] Low concentrations of any of these enzymes in the intestinal mucosa result in carbohydrate malabsorption that also may be associated with clinical symptoms such as diarrhea, abdominal pain, and flatulence.[9]

Development of Brush-Border Digestive Enzymes

The carbohydrate-digesting enzymes of the small intestine are anchored in the brush border. The disaccharidases present include sucrase, lactase, glucoamylase, isomaltase, and trehalase. The substrates and products of each disaccharidase are shown in **Table 78.2**.[10] SI activity at 34 weeks of gestation reaches 70% of the adult level, rising to the adult level at birth.[11] Glucoamylase

Table 78.1 Lactose Content of Selected Dairy Foods

Food	Typical Serving Size	Lactose Content Per Serving (G)	References
Whole milk	249 g (1 cup)	12	FDC, USDA[4]
2% reduced-fat milk	245 g (1 cup)	12	FDC, USDA[4]
1% reduced-fat milk	246 g (1 cup)	12.2	FDC, USDA[4]
Nonfat milk	246 g (1 cup)	12.4	FDC, USDA[4]
Lactose-reduced milk			
70% lactose reduced	245 g (1 cup)	3–4	Moore BJ[5]
100% lactose reduced	245 g (1 cup)	0–1	Moore BJ[5]
Low-fat Greek yogurt	245 g (1 cup)	6.4	FDC, USDA[4]
Cheese			
Parmesan	56.7 g (2 oz)	0	FDC, USDA[4]
American	56.7 g (2 oz)	2.05	FDC, USDA[4]
Blue, parmesan	56.7 g (2 oz)	1–2	Moore BJ[5]
Camembert	56.7 g (2 oz)	0–1	Moore BJ[5]
Cheddar	56.7 g (2 oz)	0.09	FDC, USDA[4]
Cottage cheese	220 g (1 cup)	8.76	Moore BJ[5]
Ice cream—10% fat	133 g (1 cup)	9–10	Moore BJ[5]
Ice cream—16% fat	148 g (1 cup)	9–10	Moore BJ[5]
Ice milk	132 g (1 cup)	9–10	Moore BJ[5]

Data from FoodData Central, USDA, Nutrient Database (https://fdc.nal.usda.gov/) accessed on March 7th, 2022 and adapted some information with permission from Moore BJ. Dairy foods: are they politically correct? Nutr Today 2003;38:82–90.

and trehalase activities are detected at 13 weeks of gestation.[12] Lactase activity develops later in gestation. The lactase activity is only 30% that of a full-term infant at 34 weeks of gestation and is still just 70% of the full-term activity by 35 to 38 weeks.[11]

The activity of disaccharidases in the brush border is recognized as the rate-limiting step in disaccharide digestion.[13] Thus, congenital or acquired enzyme deficiencies cause poor absorption of disaccharides. In addition, loss of disaccharidase activity can

Table 78.2 Role of Brush-Border Enzymes in Digestion of Disaccharides and Starch

Enzyme	Enzymatic Activity	Substrate	Products
Lactase	β-(1-4) Galactosidase	Lactose	Glucose, galactose
Sucrase	α-(1-4) Glucosidase Hydrolysis of the α-1, β-2 glucose-fructose bond in sucrose	Sucrose, maltose, maltotriose, α-limit dextrins with terminal α 1-4 links	Glucose, fructose, malto-oligosaccharides with terminal α 1-6 linkages
Glucoamylase	α-(1-4) Glucosidase	Maltose, maltotriose, Malto-oligosaccharide	Glucose, malto-oligosaccharide with terminal α 1-6 linkage
Isomaltase	α-(1-6) Glucosidase	Maltose, isomaltose, α-limit dextrins (malto-oligosaccharide with terminal α 1-6 links)	Glucose, malto-oligosaccharides
Trehalase	α- and β-Glucosidase (tested on renal trehalase)	Trehalose	Glucose

Reproduced from Treem WR. J Ped Gastroenterol Nutr 1995;21:1–14. doi:10.1097/00005176-199507000-00001

occur secondary to damage to the intestinal mucosa because of certain diseases (e.g., alcoholism, celiac disease), infections, medications, surgery, or radiation exposure.[14]

Lactase-Phlorizin Hydrolase

Location and Functions

The highest activity of LPH in humans is found in the jejunum, approximately 50 to 200 cm distal to the ligament of Treitz. Its activity is 25% lower at Treitz's ligament and is minimal in the ileum.[15-17] The gene for LPH is located on chromosome 2 and it directs the synthesis of a pre-pro form of LPH in the enterocytes. The pre-pro LPH is processed intracellularly (and possibly by pancreatic proteases) into the mature form that is anchored in the cell membrane at the brush border. The human enzyme has two catalytic sites, both on the luminal side of the enterocyte cell membrane. These active sites, β-galactosidase (EC 3.2.1.23) and phlorizin hydrolase (EC 3.2.1.62), comprise Glu1749 in domain IV and Glu1273 in domain III, respectively.[18] The β-galactosidase portion can hydrolyze lactose, cellobiose, *o*-nitro-phenyl-β-glucopyranoside, and *o*-nitro-phenyl-β-galactopyranoside.[17,18] Phlorizin hydrolase can hydrolyze phlorizin, β-glycopyranosyl-ceramides, and *m*-nitro-phenyl-β-glucopyranoside.[17,18]

Beyond its well-recognized role in lactose digestion, evidence indicates that LPH could be involved in the hydrolysis of other nutritionally important β-glucosides. For example, although the glycosylated forms of isoflavones and flavonoids occur in nature, only the aglycone forms can be absorbed from the intestine. It was previously assumed that the colonic microflora was mainly responsible for this deconjugation. However, two studies[19,20] demonstrated that the lactase catalytic site of LPH can hydrolyze glycosylated isoflavones and flavonoids, making them available for absorption in the small intestine. Similarly, the hydrolysis of a β-glucosidic linkage is necessary to release pyridoxine from pyridoxine-5′-β-D-glucoside (PNG), an important step in increasing the bioavailability of this form of vitamin B_6 that provides roughly 15% of the total vitamin B_6 in a mixed diet.[21] Mackey et al.,[21] reported that LPH purified from rat small intestinal mucosa possesses the capability to hydrolyze PNG in vitro.

Types of Hypolactasia and Lactase Nonpersistence

Full-term infants, regardless of racial or ethnic background, generally possess high levels of lactase activity. Congenital lactase deficiency is the rare condition in which lactase is absent at birth. Even in Finland, where the condition is most common, only 42 cases were reported from 1966 to 1998.[22] In these infants, lactase activity in jejunal biopsy specimens is reduced to 0 to 10 IU/g protein, and severe diarrhea results from unabsorbed lactose.[22] Treatment with a lactose-free formula eliminates the diarrhea and promotes normal growth and development.

Congenital lactase deficiency has an incidence of one in 60,000 newborns.[23] Mutations in the LPH gene could cause congenital lactase deficiency and studies have reported a total of seven mutations in the Finnish population.[23,24] Congenital lactase deficiency is a separate clinical entity from congenital lactose intolerance. Congenital lactose intolerance is a rare and serious disease with vomiting, failure to thrive, dehydration, disacchariduria including lactosuria, renal tubular acidosis, aminoaciduria, liver damage, and cataracts as possible clinical sequelae.[25-28] The cause of this disorder is not lactase deficiency but rather the gastric absorption of intact lactose.[26] Although this condition can be fatal in early infancy if not recognized, a milk-free diet leads to rapid recovery; and often, patients may be able to tolerate a normal diet (with milk) after 6 months of age.[25]

The loss of intestinal lactase activity (hypolactasia) is either congenital or acquired, and lactase is the only digestive enzyme for which greatly reduced activity in adulthood is common. Acquired hypolactasia is subdefined as primary or secondary. Primary hypolactasia (also referred to as lactase nonpersistence [LNP]) is a genetically programmed, irreversible loss of the majority (90%–95%) of intestinal lactase activity that occurs sometime after weaning, usually between 3 and 5 years of age.[29,30] LNP affects approximately 75% of the world's population (**Table 78.3**). Interestingly, most Northern Europeans and a few pastoral tribes in African and the Middle East are lactase persistent, maintaining high lactase concentrations throughout life.[31] Because the loss of lactase is the normal pattern in mammalian physiology (humans are the only known mammalian species to have a subpopulation that retains lactase activity) and is not pathogenic, the use of the term "lactase deficiency" to describe the primary loss of lactase is incorrect. Last, the terms lactase nonpersistence and lactose intolerance should not be used interchangeably. The former term simply describes the loss of lactase activity, whereas the latter pertains to the development of clinical symptoms resulting from lactose maldigestion. Two major hypotheses exist to explain the pattern of LNP distribution in the world population. The first—the geographic hypothesis—was proposed by Simoons in 1978.[32] According to this hypothesis, mutations for lactose persistence occurred

Table 78.3 Prevalence of Lactase Nonpersistence in Various Ethnic Groups

Group	Prevalence (%)
Northern European	2–7
White (United States)	6–22
Central European	9–23
Indian (Indian subcontinent)	
Northern	20–30
Southern	60–70
Hispanics	50–80
Ashkenazi Jews	60–80
African-American	60–80
Black African	70–95
Native American	80–100
Asian	85–100

Data from Srinivasan R, Minocha A. Postgrad Med 1998;104:109–23. Also contains information from Sahi T. Scand J Gastroenterol 1994;29:7–20.

Table 78.4 Prevalence of Lactose Malabsorption in Various Geographic Regions

Region	Prevalence (95% CI)
Asia (except Middle East)	64% (54–74)
Eastern Europe, Russia, and former Soviet Republics	47% (33–61)
Latin America	38% (18–57)
Middle East	70% (57–83)
Northern Africa	66% (45–88)
Northern America	42% (13–71)
Oceania	45% (19–71)
Sub-Saharan Africa	63% (54–72)
Northern, Southern and Western Europe	28% (19–37)

Data from Storhaug CL, Fosse SK, Fadnes LT. Lancet Gastroenterol Hepatol 2017;2:738–46.

several thousand years ago, during the origin of dairying. In those geographic regions where dairy farming was practiced, individuals with the mutation coding for lactose digestion had improved tolerance to milk and gained a selective advantage over their counterparts, especially when living under marginal nutritional conditions.

Although there are no recent studies reporting the prevalence of lactase nonpersistence, Storhaug et al.,[33] reported a meta-analysis of the prevalence of lactose malabsorption, a common condition caused by lactase nonpersistence.[34] This meta-analysis included 450 study populations from 89 countries. Lactose malabsorption was assessed by genotyping in 231 study populations, hydrogen breath tests in 83, lactose tolerance tests in 101, and 35 used alternative procedures. Overall, the estimate of global prevalence of lactose malabsorption was 68% (**Table 78.4**) and the estimate was 74% with the genotyping data. The prevalence was highest in Asia, with the lowest at 58% in Pakistan and the highest being 100% in South Korea.

The malaria hypothesis was proposed by Anderson and Vullo.[35] The authors suggest that malaria selected for LNP. Noting that LNP is common in geographic areas of the world that are endemic for malaria, the authors posit that the genetic tendency for LNP would cause lactose maldigestion and intolerance symptoms leading to a corresponding decline in milk intake in affected individuals. Because milk products are excellent sources of riboflavin, it is further proposed that many of these individuals may have had marginal riboflavin deficiency. A state of marginal riboflavin deficiency that could be tolerated by the person and yet still lead to localized flavin deficiency in the erythrocytes, is theorized to inhibit the multiplication of malaria parasites and, thus, reduce mortality from malaria. Although this hypothesis is interesting, a study conducted in Northern Sardinia showed no differences in the prevalence of LNP in villages with a history of low, intermediate, or high malaria morbidity and mortality.[36] Similar results were reported in a study conducted in the Faluni population, an African ethnic group, who are mostly resistant to malaria.[37] The presence of malarial parasite *Plasmodium falciparum* in the asymptomatic Faluni population was more common in individuals with LNP genotype; however, there was no statistical significance.[37] A further commentary on this study[38] notes that, in contrast to the data on LNP, the frequencies of glucose-6-phosphate dehydrogenase deficiency and β-thalassemia trait (two disorders that are known to be selected by malaria) were strikingly higher in the areas with a high past malarial endemicity vs the area with low malarial endemicity. Thus, the limited evidence presented so far does not support the malaria hypothesis.

In LNP individuals, lactase activity in the jejunal enterocytes is found in a mosaic-type pattern, meaning that some jejunal enterocytes produce high levels of lactase, whereas others, even those sharing the same villus, do not produce lactase.[39,40] Thus, rather than a uniform reduction in lactase production among all enterocytes, LNP individuals may have a patchy distribution of lactase-producing enterocytes that are low in number relative to the non–lactase-producing enterocytes. However, in lactase-persistent individuals, all enterocytes may produce lactase.

The molecular basis for LNP has received much attention. LNP is an autosomal recessive trait and the gene for human LPH is located on chromosome 2q21.[41] Initial studies suggested that alterations in the post-translational modifications of LPH were responsible for the low lactase activity in hypolactasia.[42,43] Rossi et al.,[39] found that intestinal biopsies of hypolactasic individuals in Southern Italy had substantial levels of lactase mRNA. Thus, it was thought that hypolactasic persons do synthesize the lactase protein, but that post-translational modifications cause it to be either malfolded and enzymatically inactive or result in its intracellular degradation.[44] Sebastio et al.,[45] studied individuals with the hypolactasic phenotype and the lactase-persistent phenotype. There was no clear difference in the lactase mRNA concentration in the intestinal biopsies of individuals with either phenotype. The authors concluded that expression of lactase is controlled at the post-transcriptional level.

Despite this evidence, the current opinion is that lactase regulation is primarily at the level of transcription. Numerous studies[46-48] have demonstrated the importance of the presence of adequate lactase mRNA to have expression of LPH activity. Krasinski et al.,[49] found that LPH mRNA in rats was abundant before weaning but declined twofold to fourfold during weaning. The LPH activity observed in the rats corresponded with the protein and mRNA concentrations at the different life stages. Thus, transcriptional mechanisms were thought to be responsible for regulation of lactase biosynthesis. Escher et al.,[50] studied lactase-specific activity and lactase mRNA in Asian, Black, and White patients. They observed that lactase activity always corresponded with the lactase mRNA concentration, thereby suggesting that transcriptional regulation is responsible for variable lactase activity. Furthermore, studies of the porcine LPH gene identified a sequence CE-LPH1 in the promoter region, which binds a *trans*-acting nuclear factor NF-LPH1. High NF-LPH1 was found in enterocytes of newborn pigs that had high lactase activity, whereas it was lower in adult pigs that had low lactase activity. It was suggested that nuclear factor NF-LPH1 could be implicated in the lowering of lactase activity at weaning and could provide an explanation for the molecular regulation of hypolactasia.[51] Subsequent studies have shown that other nuclear factors also may interact with the CE-LPH1 promoter region.[52]

The newer finding in the genetics of lactose intolerance is the discovery of single nucleotide polymorphisms (SNP) that appear to define individuals who will maintain or lose lactase activity after weaning. The first of these to be discovered is C/T-13910, in European populations.[53] A one base-pair polymorphism 13.9 kb upstream from the lactase gene, specifically found in intron 13 of the mini chromosome maintenance complex component 6 (*MCM6*) gene, on chromosome 2, appears to be responsible for determining lactose digestion status in many European populations.[53] The location of the SNP appears to be the binding site for the transcription factor Oct 1. The expression of the lactase gene is several times higher for the T-13910 allele. Both T/T-13910 and C/T-13910 allow enough transcription binding that nonpersistence is seen only with the homozygous C/C-13910 allele. This is a plausible molecular explanation for the dominance of tolerance long observed in lactose genetics. A second essential finding is that SNPs vary by racial group around the world. Initial thought was that maybe three different SNPs existed for Europeans, Middle Eastern populations, and Africans. However, to date, at least eight unique SNPs have been identified.[54] It is possible that much of the variation around age at onset and degree of intolerance could be related to the specific SNPs.

Most recent findings indicate a genetic-epigenetic interaction that leads to LNP or LP.[55] It has been reported that LNP could be caused by the age-related buildup of methylation, which is reliant on DNA variation.[55] The C-13910 allele in the LNP haplotype accumulates modified cytosines that silence regulatory elements in the LCT and MCM6 gene. However, the T-13910 allele in the LP haplotype exhibit modifications, depending on age, to maintain the activity of LCT.[56] It has been discovered that DNA methylation could compete with transcription factor binding.[57] Therefore, the increased affinity of T-13910 allele for transcription factor binding could avoid the accumulation of modified cytosines at LCT–MCM6 regulatory regions.[56]

Secondary acquired hypolactasia occurs because of enterocyte damage resulting from disease, medication, surgery, or radiation.[58] **Table 78.5** lists some of the causes of secondary hypolactasia. In one study of malnourished patients, lactase was reduced to a greater degree than other disaccharidases and was the last of the disaccharidases to recover.[59] A possible explanation is that the activity of lactase is only about 50% as high as the other disaccharidases, even in lactase persistent individuals.[60] A key issue in the management of secondary hypolactasia is lactose restriction in the diet. Although removing lactose-containing dairy foods may improve clinical tolerance, it also may deprive a malnourished patient of the nutritional value of these foods. Secondary hypolactasia can be reversed once the underlying problem has been resolved, but the process is slow and can take 6 months or longer.[58]

Table 78.5 Potential Causes of Secondary Hypolactasia

Diseases		
Small Bowel	**Multisystem**	**Iatrogenic**
HIV enteropathy	Carcinoid syndrome	Chemotherapy
Bacterial or viral enteritis (e.g., rotavirus)	Cystic fibrosis	Radiation enteritis
Sprue (celiac and tropical)	Diabetic gastropathy	Surgical resection of intestine
Whipple diseases (intestinal lipodystrophy)	Protein energy malnutrition	Medications
Ascaris lumbricoides infection	Zollinger-Ellison syndrome	Colchicine (antigout)
Blind loop syndrome	Alcoholism	Neomycin (antibiotic)
Parasitic disease (e.g., giardiasis, cryptosporidiosis)	Iron deficiency	Kanamycin (antibiotic)
Infectious diarrhea		Aminosalicylic acid (antibiotic)
Short gut		Polymycin
Inflammatory bowel disease (Crohn's disease, ulcerative colitis)		Tetracycline
		Post-surgical conditions (e.g., stagnant loop syndrome or short bowel syndrome)

Modified from Savaiano D, Hertzler S, Jackson KA et al. In: Coulston AM, Rock CL, Monsen ER, eds. Nutrition in the Prevention and Treatment of Disease. San Diego: Academic Press, 2001:563–75.

Clinical Assessment of Lactase Activity

Lactase activity is assessed by either direct or indirect methods. The direct measurement of lactase activity obtained by a small intestinal mucosal biopsy or intestinal perfusion is the most accurate, but these methods are invasive and carry risk for complications, such as intestinal bleeding.[61] Thus, direct methods are rarely performed clinically. Genetic tests could be used to identify lactase nonpersistence and lactase persistence or C/T-13910 polymorphism.[62]

Indirect methods for assessing the digestion of a dose of lactose include breath tests (hydrogen, $^{13}CO_2$, $^{14}CO_2$), blood tests (glucose and galactose), urine tests (galactose, lactose/lactulose ratio), fecal tests (pH, reducing substances), and intolerance symptoms. Of these methods, the breath hydrogen test has been widely used. The test is based on the principle that lactose, which escapes digestion in the small intestine, is fermented by the colonic bacteria, resulting in the generation of hydrogen gas (the only known source of molecular hydrogen in the body). A portion of this hydrogen gas diffuses into the blood and is excreted via the lungs. The method has excellent sensitivity and specificity, but careful attention to the test protocol is required.[60] A recent study assessed lactase activity by measuring galactonate and galactitol in serum samples and urine samples.[63] Galactose and two of its metabolites, galactitol and galactonate, have been shown to have a bimodal (high or low) postprandial response in serum and urine after milk consumption. These reactions matched the lactase persistence genotype at rs4988235. Therefore, measuring galactonate and galactitol, the metabolites of galactose, following a lactose overload could be a noninvasive strategy to assess lactase activity (LA).[63]

Lactose Maldigestion and Symptoms of Lactose Intolerance

The well-documented high prevalence of LNP in much of the world's population has unfortunately misled many to believe that lactose intolerance is equally common. However, a large body of evidence now exists demonstrating that symptoms of lactose intolerance in response to physiologic amounts of lactose (8 to 16 fl oz of milk) affect only a small fraction of lactose maldigesters. An example is a study by Carroccio et al.,[64] where 323 Sicilian individuals (72 children aged 5 to 16 years, 141 adults aged 17 to 64 years, and 110 elderly adults aged 65 to 85 years) underwent breath hydrogen testing with a 25-g lactose dose (1 g/kg for children) to determine lactose digestion status and were queried for the presence of lactose intolerance in the ensuing 24-hour period. Of the 323 individuals, 117 (36%) were classified as lactose maldigesters. Just 13 of the lactose maldigesters experienced symptoms of lactose intolerance, which was 4% of the total study group and 11% of the lactose maldigesters.

Another concern is that many individuals self-diagnose lactose intolerance when they may not be lactose maldigesters. A study conducted by Jianqin et al.,[65] in a Chinese population showed that 23 individuals were identified with lactose maldigestion after a urinary galactose test among the 45 individuals who self-reported lactose intolerance. Similarly, a study in the U.S. population found that only 25 adults were both lactose intolerant and lactose maldigesters among the 94 adults who had self-perceived lactose intolerance.[66] Furthermore, two studies by Suarez et al., and another by Johnson et al.,[67-69] demonstrated that 30% to 33% of individuals who claim to have lactose intolerance were, in fact, lactose digesters. Finally, of the 49 individuals in the Carroccio et al., study[64] who had self-reported milk intolerance at entry into the study, only five were both lactose maldigesters and lactose intolerant. It is likely that some individuals who self-diagnose with lactose intolerance may have an underlying bowel condition. The findings indicate that diagnosing lactose intolerance solely on the basis of reported symptoms after milk ingestion is unreliable. Objective tests of lactose maldigestion, or the evaluation of symptoms in a double-masked, placebo-controlled study were necessary.[70] Recently, NIH proposed to conduct a double-blinded placebo study to accurately determine lactose intolerance in human individuals. The results indicated that of the 148 individuals who were self-identified as lactose intolerant, 81 were confirmed to be lactose maldigesters after a hydrogen breath test. Of the 81 lactose maldigesters, 60 had symptoms after consumption of dairy products and 45 of the 60 did not have symptoms after the placebo treatment. Therefore, 45 individuals were identified as lactose intolerant.[71] Symptoms of lactose intolerance include flatulence, cramping, abdominal pain, nausea, distention, bloating, and diarrhea[72] and appear to be related to the ability of the colonic microflora to process undigested lactose.[73] Differences in the types of bacteria and/or their metabolic activities affect how lactose is fermented. It has been proposed that a dominance of lactic acid bacteria in the colon would improve the fermentation of lactose into short-chain fatty acids and other products that are absorbed easily from the colon.[74] Thus, the potential for osmotic diarrhea caused by unfermented lactose would diminish. Furthermore, lactic acid bacteria may also reduce intestinal gas production either directly or indirectly. Lactic acid bacteria are able to ferment lactose without producing hydrogen[75] and the pH-lowering effect of lactic acid may inhibit bacteria that are major hydrogen producers (e.g., clostridia, *Escherichia coli*).[76,77] Given that hydrogen gas can account for 50% or more of total colonic gas during active fermentation,[78] reducing hydrogen production should appreciably lower the volume of gas produced and, therefore, reduce flatulence. However, some studies suggest that an individual's symptoms may result from increased sensitivity to gas, not an increase in the absolute volume of gas.[79,80]

It is common for many lactose-maldigesting individuals to believe that any amount of lactose will cause intolerance symptoms. However, the relationship between lactose maldigestion and the development of lactose intolerance is complex. The symptom response is influenced by several physiologic and psychosocial factors, including dose, transit, residual lactase, and colonic capacity.

Dietary Approaches for Overcoming Lactose Intolerance

Dose of Lactose

Historically, a 50-g lactose challenge (equivalent to 1 L of milk) has been used in lactose tolerance testing. Between 80% and 100% of lactose maldigesters experience intolerance symptoms when such an unphysiologic amount of lactose is fed on an empty stomach.[29] However, the symptom responses to a typical serving of milk (e.g., 8 fl oz of milk [240 mL] containing 12 g lactose) are typically much lower and are often absent in lactose maldigesters. Residual lactase activity may help to explain tolerance to smaller lactose loads. Bond and Levitt,[81] using an intestinal intubation technique, demonstrated that lactose maldigesters may absorb anywhere from 42% to 75% of a 12.5-g lactose dose. The combination of the relatively small amount of lactose, coupled with residual lactase activity, usually results in negligible symptom responses at lactose doses of 12 g or less. Some studies[82-84] have identified a small number of individuals who may be sensitive to as little as 3 to 5 g lactose. However, one study[82] did not have appropriate masking of treatment identities, whereas another double-blind study,[83] found that only 3/59 lactose maldigesters had a positive symptom response to 3 g lactose, a number that was not different from the 0-g lactose placebo. Furthermore, a double-blind study by Suarez et al.,[67] found that even individuals who claimed to be severely lactose intolerant did not report more symptoms when 8 fl oz per day of regular vs 100% lactose-hydrolyzed milk was fed for 7 days. Hertzler et al.,[85] using breath hydrogen analysis, determined that a 2-g lactose dose was completely absorbed, whereas there was some evidence of lactose maldigestion (without symptoms) at a 6-g dose. Good tolerance to doses of lactose up to 7 g has been confirmed.[86]

A lactose dose of 12 to 15 g ingested in a single meal is a safe limit to exhibit no symptoms.[62] Larger but still physiologic loads of lactose (e.g., 15 to 25 g) cause symptoms in approximately 50% of lactose maldigesters.[87] In general, 12 g or greater of lactose can cause abdominal pain in some lactose maldigesters, whereas significant increases in flatulence symptoms may not be present until the lactose dose reaches 20 g.[85,88] However, if a total of approximately 25 g lactose is consumed as two separate doses of 12 g each at breakfast and dinner, symptoms are minimal.[68] Lactose maldigesters may be able to tolerate even a large amount of lactose ($\geq$34 g) daily, if the total dose is subdivided into smaller doses (i.e., $\leq$12 g) throughout the day.[89,90]

Gastrointestinal Transit Effects

Gastrointestinal transit time is negatively correlated with peak breath hydrogen and intolerance symptoms in lactose maldigesters.[91,92] Delaying gastrointestinal transit may enhance lactose digestion by increasing the time in which residual lactase is able to hydrolyze lactose. Greater digestion of lactose in the small intestine decreases the amount of lactose reaching the colon, thus decreasing the osmotic load as well as limiting hydrogen production. In addition, slower fermentation of lactose in the colon could allow for more efficient disposal of fermentation gases, thus reducing the potential for symptoms.

The most successful approach for slowing the transit of lactose to the colon is for milk to be consumed with meals. Consuming lactose-containing foods as part of a meal has been shown to increase the time to peak breath hydrogen, decrease overall hydrogen production, and decrease symptoms of intolerance in lactose maldigesters.[93-96] Other gastrointestinal transit-related factors that have been studied, including altering the energy content, viscosity, temperature, and fat content of milk. Increasing the energy content or viscosity of milk does not improve lactose digestion or tolerance, although increasing the energy content may slow gastric emptying somewhat.[97,98] A study of lactose solutions (50 g) at 2° to 3°C (cold), 20° to 21°C (room temperature), and 55° to 58°C (hot) detected no effect of temperature on lactose digestion or overall gastrointestinal symptoms, although the cold solution was associated with somewhat greater abdominal pain and less flatulence than the other solutions.[99] An initial study by Leichter[100] reported improved lactose tolerance to whole milk vs skim milk. However, there was no mention of randomization in this study and no statistical evaluation of intolerance symptoms. Subsequent studies have found no significant effects of whole vs skim milk on lactose digestion[96] or tolerance.[101,102]

Cocoa has been studied for its ability to increase tolerance to lactose. One mechanism by which cocoa may exert its effect is by slowing the rate of gastric emptying. Chocolate milk has been shown in two studies to decrease breath hydrogen and symptoms of intolerance, compared with plain milk,[96,103] but masking treatment identities is difficult in these types of studies and differences in taste and appearance cannot be excluded as possible confounding variables. Milk chocolate also was studied for its ability to increase tolerance to lactose in self-described lactose intolerants.[104] No differences in gastrointestinal symptoms or stool frequency and consistency were found when individuals consumed 100 g chocolate samples containing either 12 g of lactose or 2 g of lactose. The results of this study provide additional support for the theory that chocolate may improve lactose digestion; however, objective evidence (i.e., breath hydrogen) was not obtained.

Microbe-Containing Dairy Foods

The lactose in yogurt with live cultures is digested better than lactose in milk and is well tolerated by those who are lactose intolerant.[105-111] During manufacturing, most commercial yogurts are enriched to 6% lactose before fermentation because of the addition of milk solids. However, as the number of lactic acid bacteria (*Lactobacillus delbrueckii* subsp. *bulgaricus* and *Streptococcus thermophilus*) increase to approximately 100 million cells/mL, 20% to 30% of the lactose is used, resulting in a lactose concentration near 4% in the final product,[111,112] a concentration similar to milk. During fermentation, the β-galactosidase activity of yogurt increases significantly.[113] Because of the buffering of gastric acid by yogurt's calcium and phosphorus, substantial numbers of live lactic acid bacteria may enter the duodenum.[114] On entry into the duodenum, the intact bacterial cells interact with bile acids, causing disruption of the cell membrane and access of lactose to the β-galactosidase. This process is termed "autodigestion" and is limited to the amount of lactose normally present in yogurt.[107,115] Although the inherent β-galactosidase activity of yogurt is an important determinant of lactose digestion, it is not the only consideration. Yogurts with widely differing β-galactosidase activities,[109] and even pasteurized yogurt, all tend to improve tolerance, even if lactose maldigestion increases somewhat.[108,110,116] Ingesting yogurt containing L casei Shirota and B. breve for 4 weeks indicated fewer symptoms and less maldigestion. In addition, the positive effects appeared to endure for at least 3 months after the probiotic was discontinued.[117] Thus, factors such as physical form, gelling, or the energy density of yogurt may be important. Most commercial "frozen yogurts," however, are an exception to this rule. Because these products are typically pasteurized after culturing, β-galactosidase activity is reduced to zero, and the breath hydrogen and symptom responses are similar to those of ice milk or ice cream.[118]

Kefir is a fermented dairy beverage that, compared with yogurt, contains a more diverse array of micro-organisms, typically including a yeast, in its starter culture.[119] One study demonstrated that kefir possessed β-galactosidase activity equal to or higher than yogurt, and that it improved both lactose digestion and lactose intolerance symptoms.[120] Thus, kefir represents a potential alternative to yogurt for lactose-intolerant individuals.

Unlike fermented dairy beverages, unfermented acidophilus milks are made by adding *Lactobacillus acidophilus* cells to cold milk, with no multiplication of the organism unless the storage temperature exceeds 40°F (5°C).[121] Most studies that have evaluated the effect of acidophilus milk on lactose digestion and tolerance have reported no improvement.[106,121-125] One likely explanation for these findings is that many commercial products do not have enough *L. acidophilus*. It is notable that studies reporting a positive effect of acidophilus milk on lactose digestion used concentrations that were much higher than what is available commercially.[126,127] In addition, the strains of *L. acidophilus* used may not be sensitive to bile acids in the intestine.[121] Thus, bile acids do not disrupt the bacterial cell membrane in the intestinal lumen, thereby preventing exposure of lactose to bacterial β-galactosidase. Feeding milk containing sonicated *L. acidophilus* cells improved lactose digestion compared with the same milk that had unsonicated cells.[122] Subsequent research has focused on developing unfermented milks with different types of bacteria that have high β-galactosidase activity[128] or strains of *L. acidophilus* having different degrees of bile acid sensitivity or lactose uptake.[129]

Lactose-Hydrolyzed Milk and Lactase Enzyme Supplements

The level of intact lactose in milk can be reduced greatly or eliminated by the incubation of regular milk with lactases derived from yeasts (e.g., *Kluyveromyces fragilis, Kluyveromyces lactis*) or other fungi (e.g., *Aspergillus oryzae, Aspergillus niger*).[93,130-132] In addition, tablets or capsules containing lactase can be taken with lactose-containing meals. Use of enzymes and milks treated with them has been affirmed as Generally Recognized as Safe by the U.S. Food and Drug Administration.[133] The most commonly available products are milks treated to achieve 70% or 100% lactose hydrolysis[134,135] and caplets containing different concentrations of lactase.[135] New products are appearing regularly.

Lactose digestion and lactose intolerance symptoms are improved by lactose hydrolyzed milk (50% to 100% lactose hydrolysis).[136-147] Generally, the same is true for lactase supplements,[148] especially those containing doses of lactase in the range of 3,000 to 6,000 FCC units (Up to 9,000 FCC units may be required for doses of lactose exceeding 20 g).[149-154] However, some studies have shown that the commercial lactase supplements available currently were not successful in reducing symptoms of lactose intolerance, suggesting the need for an increase in the dose of lactase supplements for complete hydrolysis of lactose.[155,156] Purified lactase from the strain *Kluyveromyces marxianus* has been reported to be more efficient in digesting lactose by being more active in the small intestine compared with commercial lactase supplements.[131,156] Hydrolysis of 80% of the lactose in a commercial meal replacement supplement significantly reduced breath hydrogen excretion and tended to decrease lactose intolerance symptoms.[157] Although lactose-hydrolyzed milks and lactase supplements generally are effective for improving lactose digestion, these products are rarely necessary for lactose maldigesters, unless large amounts of lactose are going to be consumed, or the lactose will be consumed in the absence of other foods, as could be the case with a lactose-containing meal replacement. Another consideration is the increased cost of lactase supplements or lactose-hydrolyzed milk vs regular milk.[67] Finally, the added sweetness caused by hydrolysis of lactose to glucose and galactose could either increase[158] or decrease[147] its acceptability.

Colonic Adaptation to Malabsorbed Lactose

The reduction of lactase in LNP is permanent. Several studies have demonstrated that the ingestion of 50 g lactose or more per day for periods of 1 to 14 months had no impact on jejunal lactase activity, as measured by intestinal biopsies.[139-161] Despite this fact, milk-feeding programs to school children in Ethiopia, India, and China (each having a high prevalence of lactose maldigesters) have shown that the tolerance to lactose improves greatly within weeks of the initiation of milk.[161-163] In addition, Johnson et al.[164] reported that 77% of lactose intolerant African American individuals could tolerate ≥12 g lactose if lactose was increased gradually and fed daily over a period of 6 to 12 weeks. The postulated mechanism for this improved tolerance is colonic bacterial adaptation to malabsorbed lactose.

Good evidence exists for colonic bacterial adaptation to lactose in humans. Hertzler et al.,[90] fed 20 lactose maldigesters either lactose or dextrose for 10 days in a randomized crossover study. The dose of each carbohydrate was gradually increased from 0.6 to 1.0 g/kg/day over the course of the feeding period. The lactose feeding period dramatically decreased the breath hydrogen response to a 0.35 g/kg lactose challenge. In addition, fecal β-galactosidase activity increased threefold and flatulence symptoms in response to the lactose challenge were decreased by 50%. A similar decrease in breath hydrogen excretion was reported in a study of African American adolescent girls who were fed a dairy-rich diet containing approximately 33 g lactose per day for 21 days.[165] Briet et al.[166] reported colonic adaptation to a dose of 34 g lactose per day for 13 days and a corresponding drop in lactose intolerance symptoms in response to a 50-g lactose challenge. However, a control group of individuals fed sucrose for the same length of time also showed a decrease in the symptom response to the lactose challenge, even though there was no evidence of metabolic adaptation in that group. Finally, a recent study[167] of the feeding of gradually increasing lactose doses (6 to 24 g/d over 12 wk) to LNP individuals demonstrated similar metabolic adaptations (e.g., increased fecal beta-galactosidase, decreased breath hydrogen) as those observed by Hertzler and Savaiano[90]. In addition, there was a near doubling of total fecal bifidobacteria populations after prolonged lactose feeding and a nonsignificant trend toward improvement of gastrointestinal symptoms.

Prebiotics are primarily maldigested carbohydrates like oligosaccharides or galaoligosaccharides that are selectively metabolized by particular lactic acid-generating bacteria (bifidobacteria and lactobacilli) in the colon.[168] One of the most promising novel treatments for LI, is colonic bacterial adaptability in response to prebiotic therapy.[169] When a patient with LM develops microbial adaptation in the human intestinal tract, the changed population of anaerobic and microaerophilic bacteria boosts intraluminal beta-galactosidase activity, improving digestion and lowering fermentation product synthesis.[170] Briet et al.,[166] proposed increasing lactose exposure to the colon bacteria to promote colon bacteria adaptability to improve lactose digestion and tolerance. A randomized double-blinded clinical trial demonstrated the improved lactose intolerance and maldigestion due to RP-G28, a novel galacto-ologosaccharide.[171]

Milk Containing A2 Beta-Casein

Cow's milk contains two beta-caseins: A1 beta-casein and A2 beta-casein. Milk containing only A2 beta-casein or A2 milk has recently acquired popularity because of its health benefits. A2 beta-casein is the original beta-casein and all the other beta-caseins were formed due to mutations.[172] The presence of proline at position 67 in A2 beta-casein prevents the release of a proinflammatory peptide, beta-casomorphin-7 (BCM-7). However, histidine at position 67 in A1 beta-casein, allows the release of BCM-7.[173] In 2007, Cieślińska and colleagues reported that A1 beta-casein produced four times more BCM-7 than A2.[174] In 2012, the same group demonstrated very low BCM-7 in raw milk, somewhat higher BCM-7 in pepsin-digested milk, and 20 to 30 times greater BCM-7 in milk digested with pepsin, trypsin, and elastase.[175] According to Jinsmaa and Yoshikawa, elastase is a critical pancreatin enzyme that cleaves BCM-7 from longer betacasomorphin peptides.[176]

Milk containing A1 beta-casein has been associated with increasing symptoms of lactose intolerance. A double-blinded randomized crossover trial conducted in Chinese adults, aged between 20 and 50 years, demonstrated an increase in gastrointestinal symptoms **after** consumption of milk containing a mixture of A1 and A2 beta-casein (58:42 ratio), compared with milk containing 100% A2 beta-casein,[177] **consistent with results of several other studies**.[65,66,178,179] Additionally, milk with 100% A2 beta-casein exhibited **less maldigestion** in lactose maldigesters.[66] Furthermore, studies reported an increase in proinflammatory markers and glutathione with A1-containing milk.[65,180] BCM-7 can reduce the uptake of cysteine, which is a rate-limiting precursor in the synthesis of glutathione.[180] It is hypothesized that the digestive byproduct of A1 beta-casein (BCM-7) can elevate symptoms of lactose intolerance by increasing inflammation. However, there is no evidence yet on BCM-7 independently increasing symptoms of lactose intolerance.

Gene Therapy for Lactose Intolerance

During et al.,[181] explored the possibility of using gene therapy as a potential cure for LNP. An adeno-associated virus vector containing the lactase gene (*AAVlac*) was administered via an orogastric tube to hypolactasic rats. An *AAV* with the luciferase (*AAVluc*) gene and phosphate-buffered saline (PBS) served as controls. The adeno-associated virus vector was chosen because it is a defective, helper-dependent virus and because it is nonpathogenic in humans and other species. Following a single administration of *AAVlac*, four of four rats tested were positive for *lacZ* mRNA within 3 days, whereas none of the *AAVluc* or PBS rats were positive. An acute lactose challenge was given on day 7 after administration, and an elevation of plasma glucose from 114±4 mg/dL to 130±3 mg/dL in 30 minutes was observed in the *AAVlac* rats. The control rats had a flat glucose curve. The positive blood glucose response to a lactose challenge and the positive *lacZ* mRNA persisted at 4 and 6 months, respectively, after the vector administration. There have been no human studies to date, but the concept is intriguing.

Genetic Circuit for Lactose Intolerance

The optimal pH of β-galactosidase for hydrolysis of lactose is 6.5. A drop in the pH of the intestine can affect the activity of β-galactosidase. Therefore, Cheng et al.,[182] engineered a tri-stable-switch circuit in the plasmid *pETDuet1-1* to respond to intestinal lactose and pH signals. This genetic switch was derived from *lambda* phage.[183,184] Two mutants were cloned into *pETDuet1-1* to sense signals: lactose-inducible promoter *placm* and pH-responsive promoter *patp2*. β-galactosidase and L-lactate dehydrogenase were the enzymes included in the system. The final engineered strain was BL21: *pETDuet1-1* and it was formed by transforming the circuit into *E. Coli* BL21. BL21: *pETDuet1-1* strain was used to test the functionality of circuit to switch between β-galactosidase and L-lactate dehydrogenase with respect to lactose and pH in vitro (bacterial cell culture) and in vivo (mice model). The circuit performed well in vitro at different pH levels, and it prevented a pH drop in intestine and decrease in β-galactosidase activity in mice, which were administered with lactose. It was also noted that the gut microbiota of the mice was restored from effects of excessive lactose intake. Therefore, the authors suggest that this technique could treat lactose intolerance.

Sucrase-Isomaltase and Maltase-Glucoamylase

SI is an integral enzyme in the small intestinal brush border that contains a sucrase (EC 3.2.1.48) domain for the hydrolysis of the α-1,2 glucose to fructose bond in sucrose and an isomaltase (EC 3.2.1.10) domain for hydrolyzing maltose and α-1,6–linked limit dextrins.[10] The activity of SI is distributed along the entire length of the small intestine. The highest activity is found in the jejunum, with 20% to 30% less activity found either proximal to the ligament of Treitz or in the distal ileum.[15] The function of SI c overlaps considerably with the maltase-glucoamylase complex (EC 3.2.1.3 and 3.2.1.20) that primarily hydrolyzes maltose. In fact, approximately 80% of the maltase activity is accounted for by SI and only 20% by maltase-glucoamylase.[10] The activity of maltase-glucoamylase progressively increases to its maximal level in the distal ileum.[15]

The gene encoding human SI has been localized to the long arm of chromosome 3.[185] In the rough endoplasmic reticulum (RER), SI is synthesized as a long polypeptide chain carrying two similar but not identical active sites (pro-SI).[10] The pro-SI is inserted into the RER via its N-terminal region. In the RER, the peptide elongates and is glycosylated with mannose residues at asparagine sites. The glycoprotein then migrates to the Golgi complex, where mannose residues are cleaved and glycosylation with N-acetyl galactosamine and sialic acid occurs. After this glycosylation, the pro-SI is inserted into the enterocyte membrane, with the sucrase domain protruding farthest out into the lumen. Pro-SI is then rapidly processed by trypsin, yielding the two subunits of sucrase and isomaltase.

The regulation of SI expression involves factors at the level of transcription, translation, glycosylation, and processing by luminal proteases. Several descriptions of various derangements of the intracellular processing and transport of pro-SI can lead to deficient enzyme activity.[186-190] In addition, some evidence has been presented that sucrase, more than other intestinal disaccharidases, may be inducible to some degree by dietary sucrose in animals and in healthy adults with normal sucrase activity.[10,191,192] However, neither sucrose nor fructose increases sucrase activity in patients with SI deficiency.[10]

Much less is known about SI deficiency compared with hypolactasia, but the research base is growing.[193] Although reports of acquired or adult-onset SI deficiency exist,[194-196] congenital SI deficiency is most common. Recently, the mutation *SI* c.273_274delAG was found to be most common in an Inuit population with congenital SI deficiency.[197] More than 25 mutations in the human sucrose gene associated with congenital SI deficiency phenotype have been discovered.[198] In congenital SI deficiency, a nearly complete absence of sucrase activity exists, whereas isomaltase activity can vary from traces of activity to nearly normal levels.[10] Congenital SI deficiency is inherited as an autosomal recessive and occurs in only a few population groups. Estimates of the prevalence of congenital SI deficiency[10,199] are as follows: Greenland Eskimos (2% to 10%), Native Alaskans (3%), Canadian native peoples (3.6% to 7.1%), Danes (<0.1%), North Americans (≤0.2%), and Americans of European descent (2%–9%). It has been hypothesized that SI deficiency is high only in Arctic regions because of a historic animal-based diet. Thus, individuals with sucrose malabsorption who depended on animal foods would not have intolerance symptoms or a selective disadvantage.[200]

Dietary therapy for SI deficiency consists of sucrose restriction. It is not usually necessary to make the diet starch free except in infants or in older children who do not respond to a sucrose-free diet.[10] In glucoamylase deficiency, however, elimination of dietary starch appears to be more important and has successfully resolved carbohydrate malabsorption symptoms in children diagnosed with this condition.[201] In some reported cases of congenital glucoamylase deficiency in infants and children, there have been coexisting lactase and sucrase deficiencies as well, necessitating more comprehensive diet therapy.[202,203]

An additional dietary management strategy for SI deficiency involves the use of enzyme supplement therapy.[204] Sacrosidase is a liquid product prepared from the yeast *Saccharomyces cerevisiae* that contains approximately 9,000 IU of sucrase activity/mL. Although sacrosidase promotes sucrose hydrolysis, it is unlike the human enzyme in that it does not have isomaltase activity. However, this product reduces the breath hydrogen response to a sucrose challenge and decreases sucrose intolerance symptoms such as diarrhea, flatulence, abdominal pain, and bloating.[205,206]

Trehalase

Trehalase (trehalose 1-glucohydrolase; EC 3.2.1.28) is a disaccharidase that hydrolyses trehalose to two molecules of glucose.[207] The enzyme is present in the brush-border membrane of the small intestine and proximal convoluted tubules of the kidney.[208] The highest activity of this enzyme is observed in the proximal part of the jejunum.[209] As with SI deficiency, trehalase deficiency is rare outside of Arctic populations. Gudmand-Høyer et al., found that trehalose activity was low in 14 of the 29 Greenlandic individuals.[207] Trehalose maldigestion resulting from trehalase deficiency causes symptoms like that of lactose maldigestion.[210] However, young mushrooms are the only significant contributors of trehalose in the present-day diet, perhaps contributing as much as 6 g trehalose in a 150-g serving.[210] Thus, trehalose ingestion from foods is not likely to exceed 20 g/day and, as per a study conducted by Oku and Nakamura, this amount is insufficient to cause diarrhea or abdominal symptoms.[211]

Trehalase is the only disaccharidase present in human plasma.[208] The significance of plasma trehalase is yet undefined, but trehalase activity in plasma may be associated with glucose metabolism. In one study, high plasma trehalase activity in an individual was associated with a greater likelihood of developing diabetes.[208] Serum trehalase activity has been found to be low in patients with rheumatoid arthritis, suggesting that processes leading to inflammation could affect enzyme activity.[212] Intestinal trehalase deficiency is rare, and because dietary trehalose intake is also low, very little clinical or nutritional significance of this disorder exists.

Summary

Unexplained gastrointestinal symptoms such as flatulence, abdominal pain, and diarrhea can often be traced to low levels of intestinal disaccharidase activity. However, hypolactasia is the only disaccharidase depletion that affects significant numbers of people worldwide. Fortunately, many dietary approaches exist for the management of disaccharidase depletions (**Table 78.6**).

Table 78.6 Dietary Strategies for Lactose Intolerance

Factors Affecting Lactose Digestion	Dietary Strategy	References
Dose of lactose	Consume a cup of milk or less at a time, containing up to 12 g lactose.	Suarez et al., (1995)[67] Hertzler et al., (1996)[85] Suarez et al., (1997)[68]
Intestinal transit	Consume milk with other foods, rather than alone, to slow the intestinal transit of lactose.	Solomons et al. (1985)[93] Martini and Savaiano (1988)[94] Dehkordi et al., (1995)[95]
Yogurts	Consume yogurts containing active bacterial cultures. A serving, or even two, should be well tolerated. Lactose in yogurts is better digested than the lactose in milks. Pasteurized yogurts do not improve lactose digestion; however, these products, when consumed, produce little to no symptoms.	Kolars et al., (1984)[106] Gilliland and Kim (1984)[110] Savaiano et al., (1984)[107] Shermak et al. (1995)[116]
Digestive aids	Over-the-counter lactase supplements (pills, capsules, and drops) may be used when large doses of lactose (>12 g) are consumed at once. Lactose-hydrolyzed milks are also well tolerated.	Moskovitz et al. (1987)[154] Lin et al., (1993)[151] Ramirez et al., (1994)[149] Nielsen et al., (1984)[158] Biller et al. (1987)[137] Rosado et al., (1989)[145] Brand and Holt (1991)[141]
Colon adaptation	Consume lactose-containing foods daily or prebiotics to increase the colon bacteria's ability to metabolize undigested lactose.	Perman et al., (1981)[76] Johnson et al., (1993)[69] Hertzler et al., (1996)[90] Pribila et al., (2000)[165] JanssenDuijghuijsen, et al., (2024)[167] Savaiano and Ritter (2013)[171]
A2 milk	Milk containing 100% A2 beta-casein are well tolerated by lactose maldigesters.	Jianqin et al., (2016)[65] He et al., (2017)[177] Ho et al., (2014)[178] Sheng et al., (2019)[179] Ramakrishnan et al., (2020)[66]

Modified from Savaiano D, Hertzler S, Jackson KA et al. In: Coulston AM, Rock CL, Monsen ER, eds. Nutrition in the Prevention and Treatment of Disease. San Diego: Academic Press, 2001:563–75.

Acknowledgments

The authors acknowledge Rubina Khan and Michelle Asp for their contributions to the previous version of this chapter.

References

1. Berger PK, Plows JF, Demerath EW, et al. Curr Opin Clin Nutr Metab Care 2020;23:277–81.
2. Johnson LR. Sugar Absorption. In: Barrett KE, Merhant JL, Ghishan FK, Said HM, Wood JD, eds. Physiology of the Gastrointestinal Tract. 2006; 1653, Elsevier Academic Press, USA.
3. Collins J, Danhof H, Britton RA. Gut Microbes 2019;10: 204–9.
4. Food Data Central, USDA, Nutrient Database. Last accessed March 7, 2022. Available at https://fdc.nal.usda.gov/
5. Moore BJ. Nutr Today 2003;38:82–90.

6. Kluch M, Socha-Banasiak A, Pacześ K, et al. Polski Merkuriusz Lekarski: Organ Polskiego Towarzystwa Lekarskiego 2020;49: 275–8.
7. Elferink H, Bruekers J, Veeneman GH, et al. Cell Molec Life Sci 2020;77:4799–4826.
8. Forsgård RA. Amer J Clin Nutr 2019;110:273–9.
9. Born P. World J Gastroenterol 2007;13:5687–91.
10. Treem WR. J Ped Gastroenterol Nutr 1995;21:1–14.
11. Kien CL, Heitlinger LA, Li BU, et al. Semin Perinatol 1989;13: 78–87.
12. Tso P, Crissinger K. In: Stipanuk MH, ed. Overview of Digestion and Absorption. Biochemical and Physiological Aspects of Human Nutrition. Philadelphia, PA: Saunders, 2000; 75–90.
13. Bayless TM, Christopher NL. Am J Clin Nutr 1969;22:181–90.
14. Srinivasan R, Minocha A. Postgrad Med 1998;104:109–23.
15. Skovbjerg H. Clin Chim Acta 1981;112:205–12.
16. Montgomery RK, Krasinski, SD, Hirschhorn JN, et al. Ped Gastroenterol Nutr 2007;45:S131–37.
17. Ugidos-Rodríguez S, Matallana-González MC, Sánchez-Mata MC. Food Function 2018;9:4056–68.
18. Zecca L, Mesonero JE, Stutz A, et al. FEBS Letters 1998;435: 225–8.
19. Day AJ, Cañada FJ, Diaz JC, et al. FEBS Letters 2000;468: 166–70.
20. Wilkinson AP, Gee JM, Dupont S, et al. Xenobiotica 2003;33: 255–64.
21. Mackey AD, Henderson GN, Gregory JF. Biol Chem 2002;277: 26858–64.
22. Järvelä I, Enattah NS, Kokkonen J, et al. Am J Human Genet 1998;63:1078–85.
23. Kuokkanen M, Kokkonen J, Enattah NS, et al. Am J Human Genet 2006;78:339–44.
24. Torniainen S, Freddara R, Routi T, et al. BMC Gastroenterol 2009;9:8.
25. Hoskova A, Sabacky J, Mrskos A, et al. Arch Dis Child 1980;55:304–16.
26. Berg NO, Dahlqvist A, Lindberg T, et al. Acta Paediatr Scand 1969;58:525–7.
27. Hirashima Y, Shinozuka S, Ieiri T, et al. Eur J Pediatr 1979; 130:41–5.
28. Russo G, Mollica F, Mazzone D, et al. Acta Paediatr Scand 1974; 63:457–60.
29. Newcomer AD, McGill DB. Clin Nutr 1984;3:53–8.
30. Gilat T, Russo S, Gelman-Malachi E, et al. Gastroenterol 1972; 62:1125–7.
31. Sahi T. Scand J Gastroenterol 1994;29:7–20.
32. Simoons FJ. Dig Dis 1978;23:963–80.
33. Storhaug CL, Fosse SK, Fadnes LT. Lancet Gastroenterol Hepatol 2017;2:738–46.
34. Misselwitz B, Pohl D, Frühauf H, et al. United Euro Gastroenterol J 2013;1:151–9.
35. Anderson B, Vullo C. Gut 1994;35:1487–9.
36. Meloni T, Colombo C, Ruggiu G, et al. Ital J Gastroenterol Hepatol 1998;30:490–3.
37. Lokki AI, Järvelä I, Israelsson E, et al. Malar J 2011;10:9.
38. Auricchio S. Ital J Gastroenterol Hepatol 1998;30:494–5.
39. Rossi M, Maiuri L, Fusco MI, et al. Gastroenterol 1997;112: 1506–14.
40. Maiuri L, Rossi M, Raia V, et al. Gastroenterol 1994;107: 54–60.
41. Grand RJ, Montgomery RK, Chitkara DK, et al. Gut 2003;52: 617–19.
42. Witte J, Lloyd M, Lorenzsonn V, et al. J Clin Invest 1990;86: 1338–42.
43. Lorenzsonn V, Lloyd M, Olen WA. Gastroenterol 1993: 105:51–9.
44. Naim HY. Histol Histopathol 2001;16:553–61.
45. Sebastio G, Villa M, Sartorio R, et al. Am J Hum Genet 1989; 45:489–97.
46. Montgomery RK, Büller HA, Rings EH, et al. FASEB J 1991;5: 2824–32.
47. Lloyd M, Mevissen G, Fischer M, et al. J Clin Invest 1992;89: 524–9.
48. Fajardo O, Naim HY, Lacey SW. Gastroenterol 1994;106: 1233–41.
49. Krasinski SD, Estrada G, Yeh KY, et al. Am J Physiol 1994;267: G584–G94.
50. Escher JC, de Koning ND, van Engen CG, et al. J Clin Invest 1992;89:480–3.
51. Troelsen JT, Olsen J, Norén O, et al. J Biol Chem 1992;267: 20407–11.
52. Troelsen JT, Mitchelmore C, Spodsberg N, et al. Biochem J 1997;322:833–8.
53. Enattah NS, Sahi T, Savilahti, et al. Nat Genet 2002;30:233–7.
54. Tishkoff SA, Reed FA, Ranciaro A, et al. Nat Genet 2007; 39:31–40.
55. Kuchay R. Drug Discov Ther 2020;14:1–7.
56. Labrie V, Buske OJ, Oh E, et al. Nat Struct Mol Biol 2016;23: 566–73.
57. Domcke S, Bardet AF, Adrian Ginno P, et al. Nature 2015;528: 575–9.
58. Scrimshaw NS, Murray EB. Am J Clin Nutr 1988;48:1083–159.
59. Mehra R, Khambadkone MR, Jain MK, et al. Ind Pediatr 1994;31: 1351–5.
60. Gray G. Gastroenterol 1993;105:931.
61. Arola H, Scand J. Gastroenterol 1994;29:26–35.
62. Robles L, Priefer R. Diagnostics 2020;10:412.
63. Vionnet N, Münger LH, Freiburghaus C, et al. Am J Clin Nutr 2019;109:470–7.
64. Carroccio A, Montalto G, Cavera G, et al. J Am Coll Nutr 1998;17:631–6.
65. Jianqin S, Leiming X, Lu X, et al. Nutr J 2016;15:35.
66. Ramakrishnan M, Eaton TK, Sermet OM, et al. Nutrients 2020;12:3855.
67. Suarez FL, Savaiano DA, Levitt MD. N Engl J Med 1995;333: 1–4.
68. Suarez FL, Savaiano DA, Arbisi P, et al. Am J Clin Nutr 1997; 65:1502–6.
69. Johnson AO, Semenya JG, Buchowski MS, et al. Am J Clin Nutr 1993;57:399–401.
70. Suarez F, Levitt M. Gut 1997;41:715–6.
71. Rocco A, Compare D, Sgamato C, et al. Nutrients 2021;13: 1653.
72. Srinivasan R, Minocha A. Postgrad Med 1998;104:109–23.
73. Vonk RJ, Priebe MG, Koetse HA, et al. Eur J Clin Invest 2003;33:70–5.
74. Hill MJ. In: Delmont J, ed. Bacterial Adaptation to Lactase Deficiency. Milk Intolerances and Rejection. New York, NY: Karger, 1983:22–6.
75. Ballongue J. In: Selminen S, von Wright A, eds. Bifidobacteria and Probiotic Action. New York, NY: Marcel Dekker, 1993: 369–71.
76. Perman J, Modler S, Olson AC. J Clin Invest 1981;67:643–50.
77. Vogelsang H, Ferenci P, Frotz S, et al. Gut 1988;29:21–6.
78. Tomlin J, Lowis C, Read NW. Gut 1991;32:665–9.
79. Hammer HF, Petritsch W, Pristautz H, et al. Wien Klin Wochenschr 1996;108:175–9.
80. Levitt MD, Furne J, Olsson S. Ann Intern Med 1996;124:422–4.
81. Bond JH, Levitt MD. Gastroenterol 1976;70:1058–62.
82. Bedine MS, Bayless TM. Gastroenterol 1973;65:735–43.
83. Newcomer AD, McGill DB, Thomas PJ, et al. Gastroenterol 1978;74:44–6.

84. Gudmand–Høyer E, Simony K. Dig Dis 1977;22:177–81.
85. Hertzler SR, Huynh B-C, Savaiano DA. J Am Diet Assoc 1996;96:243–6.
86. Vesa TH, Korpela RA, Sahi T. Am J Clin Nutr 1996;64:197–201.
87. Savaiano DA, Levitt MD. J Dairy Sci 1987;70:397–406.
88. Gremse DA, Greer AS, Vacik J, et al. Clin Pediatr 2003;42: 341–5.
89. Suarez FL, Adshead J, Furne JK, et al. Am J Clin Nutr 1998;68:1118–22.
90. Hertzler SR, Savaiano DA. Am J Clin Nutr 1996;64:232–6.
91. Ladas S, Papanikos J, Arapakis G. Gut 1982;23:968–73.
92. Labayen I, Forga L, Gonzalez A, et al. Aliment Pharmacol Ther 2001;15:543–9.
93. Solomons NW, Guerrero AM, Torun B. Am J Clin Nutr 1985;41:199–208.
94. Martini MC, Savaiano DA. Am J Clin Nutr 1988;7:57–60.
95. Dehkordi N, Rao DR, Warren AP, et al. J Am Diet Assoc 1995;95:484–6.
96. Silanikove N, Leitner G, Merin U. Nutrients 2015;7:7312–31.
97. Vesa TH, Marteau PR, Briet FB, et al. J Nutr 1997;127:2316–20.
98. Vesa TH, Marteau PR, Briet FB, et al. Am J Clin Nutr 1997;66:123–6.
99. Peuhkuri K, Vapaatalo H, Nevala R, et al. Scand J Clin Invest 2000;60:75–80.
100. Leichter JL. Am J Clin Nutr 1973;26:393–6.
101. Cavalli-Sforza LT, Strata A. Hum Nutr Clin Nutr 1986;40C: 19–30.
102. Vesa TH, Lember M, Korpela R. Eur J Clin Nutr 1997;51:633–6.
103. Lee CM, Hardy CM. Am J Clin Nutr 1989;49:840–4.
104. Jarvinen RM, Loukaskorpi M, Uusitupa MI. (2003) Eur J Clin Nutr 1989;57:701–5.
105. Gallagher CR, Molleson AL, Caldwell JH. J Am Diet Assoc 1974;65:418–9.
106. Kolars JC, Levitt MD, Aouji M, et al. N Engl J Med 1984;310:1–3.
107. Savaiano DA, Abou El Anouar A, Smith DE, et al. Am J Clin Nutr 1984;40:1219–23.
108. Lerebours E, Ndam CND, Lavoine A, et al. Am J Clin Nutr 1989; 49:823–7.
109. Martini MC, Lerebours EC, Lin WJ, et al. Am J Clin Nutr 1991;54:1041–6.
110. Gilliland SE, Kim HS. J Dairy Sci 1984;67:1–6.
111. Verruck S, Balthazar CF, Rocha RS, et al. Adv Food Nutr Res 2019;89:95–164.
112. Rasic JL, Kurmann JA. The Nutritional-Physiological Value of Yoghurt. Yogurt: Scientific Grounds, Technology, Manufacture, and Preparations. Copenhagen, Denmark: Technical Dairy Publishing, 1978:99–137.
113. Pimenta FS, Luaces-Regueira M, Ton AM, et al. Cell Physiol Biochem 2018;48:1901–14.
114. Martini MC, Bollweg GL, Levitt MD, et al. Am J Clin Nutr 1987;45:432–6.
115. Martini MC, Kukielka D, Savaiano DA. Am J Clin Nutr 1991;53:1253–8.
116. Shermak MA, Saavedra JM, Jackson TL, et al. Am J Clin Nutr 1995;62:1003–6.
117. Almeida CC, Lorena SL, Pavan CR, et al. Nutr Clin Pract 2012;27:247–51.
118. Martini MC, Smith DE, Savaiano DA. Am J Clin Nutr 1987;46:36–40.
119. Guzel-Seydim ZB, Kok-Tas T, Greene AK, et al. Crit Rev Food Sci Nutr 2011;51:261–8.
120. Hertzler SR, Clancy SM. J Am Diet Assoc 2003;103:582–7.
121. Newcomer AD, Park HS, O'Brien PC. Am J Clin Nutr 1983;38:257–63.
122. Mc Donough FE, Hitchins AD, Wong NP. Am J Clin Nutr 1987;45:570–4.
123. Lin MY, Savaiano DA, Harlander S. J Dairy Sci 1991;74:87–95.
124. Onwulata CI, Rao DR, Vankineni P. Am J Clin Nutr 1989;49: 1233–7.
125. Saltzman JR, Russell RM, Golner B, et al. Am J Clin Nutr 1999;69:140–6.
126. Montes RG, Bayless TM, Saavedra JM, et al. J Dairy Sci 1995; 78:1657–64.
127. Kim HS, Gilliland SE. J Dairy Sci 1983;66:959–66.
128. Jiang T, Mustapha A, Savaiano DA. J Dairy Sci 1996;79:750–7.
129. Mustapha A, Jiang T, Savaiano DA. J Dairy Sci 1997;80: 1537–45.
130. Rao D. Food Sci Technol Intl 1997;3:87–92.
131. Ibrahim SA, Gyawali R. In: Park YW, and Haenlein, GFW, eds. Lactose Intolerance. Milk and Dairy Products in Human Nutrition: Production, Composition and Health. New York, NY: John Wiley and Sons Limited, 2013: 246–60.
132. Ibrahim SA, Gyawali R, Awaisheh SS, et al. J Dairy Res 2021;88:357–65.
133. Geel TM, McLaughlin PM, de Leij LF, et al. Mol Genet Metab 2007;92:299–307.
134. Land O Lakes. Dairy Ease Nutritional Information. Accessed July 12, 2012. Available at www.dairyease.com/benefits/nutritional_whole.html
135. Lactaid. Lactaid products. Accessed July 12, 2012. Available at www.lactaid.com
136. Rosado JL, Morales M, Pasquetti A, et al. La Rev Invest Clin 1988;40:141–7.
137. Biller JA, King S, Rosenthal A, et al. J Pediatr 1987;111:91–4.
138. Payne-Bose D, Welsh JD, Gearhart HL, et al. Am J Clin Nutr 1977;30:695–7.
139. Payne DL, Welsh JD, Manion CV. Am J Clin Nutr 1981;34: 2711–5.
140. Pedersen ER, Jensen BH, Jensen HJ. Scand J Gastroenterol 1982;17:861–4.
141. Brand JC, Holt S. Am J Clin Nutr 1991;54:148–51.
142. Turner SJ, Daly T, Hourigan JA, et al. Am J Clin Nutr 1976;29:739–44.
143. Paige DM, Bayless TM, Mellits ED, et al. J Agric Food Chem 1979;27:677–80.
144. Cheng AHR, Brunser O, Espinoza J, et al. Am J Clin Nutr 1979;32:1989–93.
145. Rosado JL, Morales M, Pasquetti A. JPEN J Parenter Enteral Nutr 1989;13:157–61.
146. Nagy L, Mozsik G, Garamszegi M, et al. Acta Med Hung 1983;40:239–45.
147. Reasoner J, Maculan TP, Rand AG, et al. Am J Clin Nutr 1981;34:54–60.
148. Baijal R, Tandon RK. JGH Open 2020;5:143–48.
149. Ramirez FC, Lee K, Graham DY. Am J Gastroenterol 1994;89:566–70.
150. Gao KP, Mitsui T, Fujiki K, et al. Nagoya J Med Sci 2002;65: 21–8.
151. Lin MY, DiPalma JA, Martini MC, et al. Dig Dis Sci 1993;38: 2022–7.
152. Sanders SW, Tolman KG, Reitburg DP. Clin Pharmacol 1992; 11:533–8.
153. DiPalma JA, Collins MS. J Clin Gastroenterol 1989;11:290–3.
154. Moskovitz M, Curtis C, Gavaler J. Am J Gastroenterol 1987;82:632–5.
155. O'Connell S, and Walsh G. Applied Biochem Biotechnol 2006; 134:179–191.
156. O'Connell S, Walsh G. Appl Biochem Biotechnol 2006;141: 1–14.
157. Suarez FL, Zumarraga LM, Furne JK, et al. J Am Diet Assoc 2001;101:1447–52.

158. Nielsen OH, Schiotz PO, Rasmussen SN, et al. J Pediatr Gastroenterol Nutr 1984;3:219–23.
159. Gilat F, Russo S, Gelman-Malachi E, et al. Gastroenterol 1972;62:1125–7.
160. Keusch GT, Troncale FJ, Thavaramara B, et al. Am J Clin Nutr 1969;22:638–41.
161. Reddy V, Pershad J. Am J Clin Nutr 1972;25:114–9.
162. Habte D, Sterky G, Hjalmarsson B. Acta Paediatr Scand 1973;62:649–54.
163. Greenfield H. Nutr Today 2003;38:77–81.
164. Johnson AO, Semenya JG, Buchowski MS, et al. Am J Clin Nutr 1993;58:879–81.
165. Pribila BA, Hertzler SR, Martin BR, et al. J Am Diet Assoc 2000;100:524–28.
166. Briet F, Pochart P, Marteau P, et al. Gut 1997;41:632–5.
167. JanssenDuijghuijsen L, Looijsteijn E, van den Belt M, et al. Am J Clin Nutr 2024;119:702–10.
168. Gibson PR, Shepherd SJ. J Gastroenterol Hepatol 2010;25: 252–8.
169. Briet F, Flourie B, Achour L, et al. Gastroenterol 1995;109: 1446–53.
170. Hertzler SR, Savaiano DA, Levitt MD. Dig Dis Sci 1997;42: 348–53.
171. Savaiano DA, Ritter AJ, Klaenhammer TR, et al. Nutr J 2013; 12:160.
172. Formaggioni P, Summer A, Malacarne M, et al. Ann Fac Med Vet Univ Parma 1999;19:127–65.
173. Sodhi M, Mukesh M, Kataria RS, et al. Indian J Endocrinol Metab 2012;16:856.
174. Cieslinska A, Kaminski S, Kostyra E, et al. Milchwissenschaft 2007;62:125–7.
175. Cieslinska A, Kostyra E, Kostyra H, et al. Int J Food Sci Nutr 2012;63:426–30.
176. Jinsmaa Y, Yoshikawa M. Peptides 2012;20:957–62.
177. He M, Sun J, Jiang ZQ, et al. Nutr J 2017;16:72.
178. Ho S, Woodford K, Kukuljan S, et al. Eur J Clin Nutr 2014; 68:994–1000.
179. Sheng X, Li Z, Ni J, et al. J Pediatr Gastroenterol Nutr 2019;69:375–82.
180. Deth R, Clarke A, Ni J, et al. Nutr J 2016;15:82.
181. During MJ, Xu R, Young D, et al. Nat Med 1998;4:1131–5.
182. Cheng M, Cheng Z, Yu Y, et al. BMC Biol 2021;19:137.
183. Ptashne M. Control Circuits – Setting the Switch. A Genetic Switch: Gene Control and Phage Lambda. Cell Press & Blackwell Scientific Publications, 1986: 49–67.
184. Casjens SR, Hendrix RW. Virol 2015;479–480:310–30.
185. West LF, Davis MB, Green FR, et al. Ann Hum Genet 1988;52: 57–61.
186. Green F, Edwards Y, Hauri HP, et al. Gene 1987;57:101–10.
187. Lloyd ML, Olsen WA. N Engl J Med 1987;316:438–42.
188. Jacob R, Zimmer KP, Schmitz J, et al. J Clin Invest 2000;106: 281–7.
189. Ouwendijk J, Moolenaar CEC, Peters WJ, et al. J Clin Invest 1996;97:633–41.
190. Moolenaar CEC, Ouwendijk J, Wittpoth M, et al. J Cell Sci 1997;110:557–67.
191. Rosensweig NS, Herman RH. J Clin Invest 1968;47:2253–62.
192. Schmitz J, Odievre M, Rey J. Gastroenterol 1972;62:389–92.
193. Cohen SA. Mol Cell Pediatr 2016;3:5.
194. Muldoon C, Maguire P, Gleeson F. Am J Gastroenterol 1999; 94:2298–9.
195. Ringrose RE, Preiser H, Welsh JD. Dig Dis Sci 1980;25:384–7.
196. Cooper BT, Scott J, Hopkins J, et al. Dig Dis Sci 1983;28:473–7.
197. Marcadier JL, Boland M, Scott CR, et al. CMAJ 2015;187:102–7.
198. Sander P, Alfalah M, Keiser M, et al. Hum Mutat 2006;27:119.
199. DeJonge AM, Norris KS, Hernandez K, et al. Gastroenterol 2014;146:S705.
200. Gudmand-Høyer E, Fenger HJ, Kern-Hansen P, et al. Scand J Gastroenterol 1987;22:24–8.
201. Lebenthal E, Khin-Maung U, Zheng BY, et al. J Pediatr 1994;124:541–6.
202. Nichols BL, Avery SE, Karnsakul W, et al. J Pediatr Gastroenterol Nutr 2002;35:573–9.
203. Karsakul W, Luginbuehl U, Hahn D, et al. J Pediatr Gastroenterol Nutr 2002;35:551–6.
204. Puntis JW, Zamvar V. Arch Dis Child 2015;100:869–71.
205. Treem WR, McAdams L, Stanford L, et al. J Pediatric Gastroenterol Nutr 1999;28:137–42.
206. Treem WR, Ahsan N, Sullivan B, et al. Gastroenterol 1993;105:1061–8.
207. Gudmand-Høyer E, Fenger HJ, Skovbjerg H, et al. Scand J Gastroenterol 1988;23:775–8.
208. Eze LC. Biochem Genet 1989;27:487–95.
209. Gudmand-Hoyer E, Skovberg H. Scand J Gastroenterol 1996;31:111–21.
210. Arola H, Koivula T, Karvonen AL, et al. Scand J Gastroenterol 1999;34:898–903.
211. Oku T, Nakamura S. Eur J Clin Nutr 2000;54:783–8.
212. Yoshida K, Mizukawa H, Haruki E. Clin Chim Acta 1993; 215:123–4.

D. DISORDERS OF THE ALIMENTARY TRACT

CHAPTER 79

Short Bowel Syndrome

Alexa N. Sasson, Johane P. Allard, and Khursheed N. Jeejeebhoy

CHAPTER OUTLINE

Abbreviations

CI, confidence interval; **CRBS**, catheter related blood stream; **DFD**, defined formula diet; **EGF**, epidermal growth factor; **GH**, growth hormone; **GLP**, glucagon-like peptide; **HEN**, home enteral nutrition; **HGH**, human growth hormone; **HPN**, home parenteral nutrition; **IGF-I**, insulinlike growth factor-I; **IT**, intestinal transplantation; **Na**, sodium; **NPO**, *nil per os*; **ORS**, oral rehydration solution; **PN**, parenteral nutrition; **PYY**, peptide YY; **SBS**, short bowel syndrome; **SCFA**, short-chain fatty acid; **TGF-a**, transforming growth factor-a

Introduction

Short bowel syndrome is a complex malabsorptive condition. The management of this disorder requires a multidisciplinary approach to patient care. There continues to be advancement in the medical and surgical management of patients living with this condition. The goal of this chapter is to advance the knowledge of healthcare providers and provide an educational background for the diagnosis and treatment of this disorder.

Definition

A consensus paper[1] provided the following definitions: Short-bowel syndrome (SBS) results from surgical resection(s) and is characterized by the inability to maintain protein-energy, fluid, electrolyte, or micronutrient balances when on a conventionally accepted, normal diet. Short bowel syndrome is the most frequent cause of chronic intestinal failure,[2] which can also be induced by intestinal obstruction, dysmotility, surgical complications, congenital defect, or disease-associated loss of absorption. Intestinal failure is also defined as the inability to maintain protein-energy, fluid, electrolyte, or micronutrient balance. The major difference between intestinal failure and SBS is that intestinal failure is the result of a variety of conditions, whereas the definition of SBS implies a length of functional intestinal surface under 200 cm.[2,3] The different types of SBS are outlined in **Table 79.1**.

Etiology

The principal causes of SBS are outlined in **Table 79.2**. The two major causes of surgical short bowel are inflammatory bowel disease and vascular disease. The risk factors for vascular disease leading to resection of the intestine are the same as those for other vascular diseases: increasing age, smoking, cardiac disease leading to low output or predisposing to embolization, hypercoagulable states, diabetes, and vasculitis.

Table 79.1 Types of Short Bowel Syndrome

Terminal jejunostomy	Complete ileal and colonic resection with jejunum preserved
Jejunocolic anastamosis	Complete ileal resection and part of colon preserved
Jejunoileal anastamosis	Jejunal resection, with at least 10 cm of terminal ileum with ileocecal valve maintained and colon retained

Table 79.2 Causes of Short Bowel Syndrome*

Mesenteric infarction
Radiation enteritis
Crohn's disease
Neoplasms
Abdominal wall trauma
Surgical complications

*Listed above are the most common causes in older children and adults. Neonatal conditions include necrotizing enterocolitis, gut volvulus, intestinal atresias, gastroschisis, and congenital enteropathy.

Pathophysiologic Considerations

To understand and treat SBS, it is necessary to understand normal function and how it is altered by surgical resection.

Gastric Emptying

The rate at which a meal enters the intestine is regulated by the rate of gastric emptying. Gastric emptying of liquids depends on their osmolarity. For digestible solids, the emptying is regulated by the particle size. However, intestinal contents entering the distal small intestine prolong gastric emptying[4] through regulation by neuroendocrine peptides of the gut. More specifically, release of peptide YY (PYY) and glucagon-like peptide 1 (GLP-1) from ileal L cells in the distal small bowel work to slow gastric emptying as well as slow intestinal motility.[5] These peptides are hypothesized to mediate the "ileal brake." Therefore, significant resection of the ileum and right colon will accelerate gastric emptying and intestinal transit because of depressed circulation of these regulatory hormones. In the case of massive small bowel resection, gastric hypersecretion occurs due to disruptions of the circulating hormones of PYY and enteroglucagon-1 and enteroglucagon-2,[6] which can contribute to rapid gastric emptying in addition to reducing nutrient absorption by inactivating pancreatic enzymes.

Small Bowel

Small bowel motility is three times slower in the ileum than in the jejunum.[7] In addition, the ileocecal valve may slow transit, especially when the ileum has been resected.[8]

The adult small bowel receives approximately 5 to 6 L of endogenous secretions and 2 to 3 L of exogenous fluids per day. It reabsorbs most of this volume in the small bowel. The amount reabsorbed in the small intestine depends on the nature of the meal.[9] With a meat- and salad–based meal, most of the fluid is absorbed in the jejunum, whereas with a milk and baked goods meal, less is absorbed proximally and more distally. This is due to the composition of the diet whereby ingestion of simple carbohydrates such as glucose, fructose, and lactose have a greater hyperosmotic effect compared with complex carbohydrates.[10] In addition, the absorptive processes differ in the jejunum and the ileum. These differences depend partly on the nature of the electrolyte transport processes and partly on the permeability of the intercellular junctions. In general, water absorption is a passive process resulting from the active transport of nutrients and electrolytes. The transport of sodium creates an electrochemical gradient and drives the uptake of carbohydrates and amino acids across the intestinal mucosa. In addition, the ileum has neutral sodium chloride absorption. However, the net absorption not only depends on these processes but also on the extent of back diffusion of the transported material into the intestinal lumen through leaky intercellular junctions. In the jejunum, these junctions are leaky and, thus, jejunal contents are always isotonic.

Fluid absorption in this region of the bowel is inefficient compared with the ileum. It has been estimated that the efficiency of water absorption is 44% and 70% of the ingested load in the jejunum and ileum, respectively. For sodium, the corresponding estimates are 13% and 72%.[9] Hence, the ileum is important in the conservation of fluid and electrolytes.

Unique Functions of the Ileum

The terminal ileum uniquely absorbs vitamin B_{12} and bile salts. Bile salts are essential for the efficient absorption of fats and fat-soluble vitamins. Normally, the demand for bile salts imposed by fat absorption cannot be met by synthesis alone. This full need is met only by ileal resorption of bile salts, which are then recycled into the intestine. With ileal resection, the loss of bile salts increases and is not met by an increase in synthesis. The bile salt pool is depleted, and fat absorption is reduced. In addition, loss of bile salts into the colon affects the colonocytes and reduces the ability of the colon to reabsorb salt and water. The result is increased diarrhea. In the colon, bile salts are also dehydroxylated to deoxy bile salts, which induce colonic water secretion.

Colon

The colon has the slowest transit, varying between 24 and 150 hours. The intercellular junctions are the tightest in this part of the bowel, and the efficiency of water and salt absorption in the colon exceeds 90%.[11] In addition, carbohydrate is fermented in the colon to short-chain fatty acids (SCFAs) having two important actions. First, SCFAs enhance salt and water absorption.[12] Second, the energy content of malabsorbed carbohydrates is salvaged by being absorbed as SCFAs. Data suggest that in patients with SBS, this salvage may be greater than in physiologically normal persons.[13] Thus, the colon becomes an important organ for fluid and electrolyte conservation and for the salvage of malabsorbed energy substrates in patients with a short bowel.

Effects of Intestinal Resection

Motility

Gastric motility is enhanced following small bowel resection.[14] Whereas proximal small bowel resection does not increase the rate of intestinal transit, ileal resection significantly accelerates intestinal transit.[14,15] In this situation, the colon aids in slowing intestinal transit so that in patients with a short bowel without a colon, a marker fed by mouth was completely excreted in a few hours.[16]

Absorption of Fluid and Electrolytes

The effect of intestinal resection depends on the extent and site of resection. Proximal resection results in no bowel disturbance because the ileum and colon absorb the increased fluid and electrolyte load efficiently. The remaining ileum continues to absorb bile salts, and thus, there is little reaching the colon to impede salt and water resorption. In contrast, when the ileum is resected, the colon receives a much larger load of fluid and electrolytes and receives bile salts that reduce its ability to absorb salt and water, with resulting diarrhea. In addition, if the colon is resected, the ability to maintain fluid and electrolyte homeostasis is severely impaired.[17]

Absorption of Nutrients

Absorption of nutrients occurs throughout the small bowel, and the removal of the jejunum alone causes the ileum to take over most of the lost function. In this situation, no significant

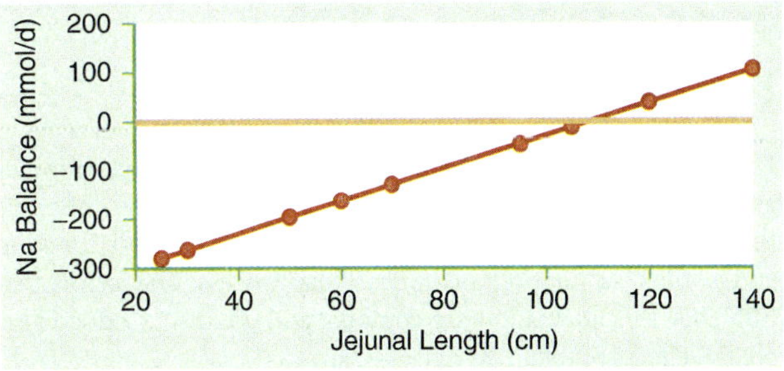

Figure 79.1 Jejunal efflux or absorption of sodium (Na) in relation to length of remaining small bowel. The *dark solid line* is the demarcation at which sodium secretion changes to absorption.

Data from Nightingale JM, Lennard-Jones JE, Walker ER et al. Lancet 1990;336:765–8.

malabsorption occurs.[18] In contrast, even a loss of a 100 cm of ileum causes steatorrhea.[19] The degree of malabsorption increases with the length of small bowel resection, and the variety of nutrients malabsorbed increases.[20,21] Balance studies of energy absorption showed that the absorption of fat and carbohydrate was equally reduced to between 50% and 75% of intake.[22] However, nitrogen absorption was reduced to a lesser extent, namely, to 81% of intake.[22] In the study of Ladefoged et al.,[21] the degrees of calcium, magnesium, zinc, and phosphorus absorption were reduced, but did not correlate with the remaining length of bowel; and it was recommended that parenteral nutrition (PN) be mandatory in these patients. Other studies showed similar reduction in absorption, but only half of the patients required parenteral replacement.

Data recalculated from Nightingale et al.,[23] in patients with a jejunostomy indicate that fluid balance can be maintained by the oral route if the remaining small bowel exceeds 110 cm (**Figure 79.1**), but nutrient balance can be maintained even if the remaining bowel is as short as 60 cm (**Figure 79.2**). The data suggest that it is easier to meet needs for energy and nitrogen by increasing oral intake than the needs for electrolytes and divalent ions. A review of the literature before the availability of PN showed that small bowel resections of up to 33% result in no malnutrition, and resections of up to 50% of normal length can be tolerated without special aids; however, patients with small bowel resection more than 75% of normal length require nutritional support to avoid severe malnutrition.[24-34]

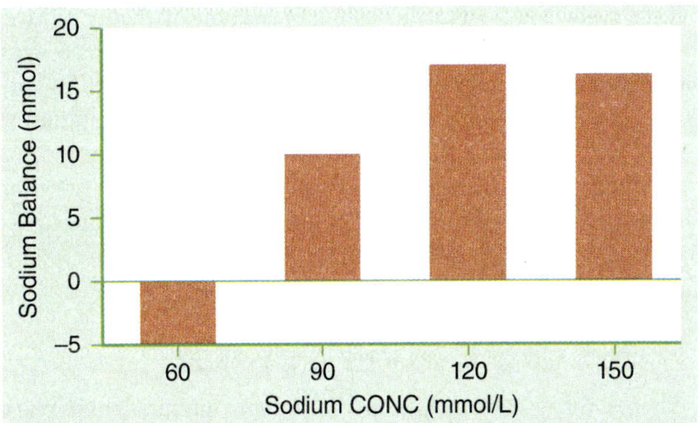

Figure 79.2 Effect of increasing the sodium content of oral rehydration solution on fluid absorption. *CONC*, concentration.

Reproduced from Lennard-Jones JE. Clin Ther 1990;12(Suppl A):129–38.

Adaptation of the Intestine

Following resection, the remaining small bowel hypertrophies and increases in absorptive function.[35-38] This process enhances the ability of the remaining bowel to recover the lost function and is, thus, an important compensatory process. The factors that influence this adaptation are complex and are discussed later, as are the effects of total PN.

Eating exposes the gastrointestinal tract to a unique set of stimuli that do not occur when the bowel is kept constantly empty, a process called bowel rest. The advent of PN resulted in the ability to rest the bowel for short or long periods of time without causing malnutrition, a situation that had not been possible previously. This process nourished the body but excluded the gut from nutrient and hormonal stimuli that occur during the ingestion of an oral diet. The advent of defined formula diets (DFDs) without residue, and diets composed of monomers such as glucose instead of polymeric starch, modified the stimuli received by the gut when exposed to a normal diet. Because nutrients are absorbed progressively along the length of the bowel, the jejunum is exposed to a higher concentration of nutrients than the ileum. Resection of the proximal bowel causes the ileum to receive more nutrients. Resection of the ileum, conversely, does not alter the jejunal nutrient load but may reduce stimuli from hormones released by the ileum.

Hormonal Response of the Ileum and Colon

A major advance in the understanding of intestinal adaptation has evolved with studies of the role of intestinotrophic hormones. These include growth hormone (GH),[39] insulin-like growth factor-I (IGF-I),[40] epidermal growth factor (EGF),[41] transforming growth factor-α (TGF-α),[42] and glucagon-like peptide-2 (GLP-2).[43] In mice, Drucker et al.,[44] showed that GLP-2 was a potent intestinotrophic hormone and native GLP-2 was rapidly degraded by dipeptidyl peptidase, making externally injected hormone effects transitory. The hormone was modified to prevent this degradation and became the most potent intestinotrophic factor. Jeppesen et al.,[45] showed that, in normal individuals, GLP-2 in blood rose with a meal. In contrast, there was an absence of such a response in patients with combined ileal and colonic resection. In contrast, patients who had an ileal resection but retained the colon had elevated fasting and meal-induced GLP-2 concentration.[46] These studies show that ileocolonic resection markedly reduces the likelihood of intestinal adaptation and causes patients with a jejunostomy to continue to have severe malabsorption. In contrast, preservation of the colon allows the remaining jejunum to adapt and helps to explain why patients with a remaining colon can often avoid permanent PN. Finally, isolated jejunal resection leaves the hormonal machinery of the ileum and colon intact.

Effect of Excluding Food from the Bowel Lumen

When food is excluded from the lumen, intestinal hypoplasia occurs in experimental animals. At the same time, body composition can be simultaneously maintained using PN. These facts have been extensively documented, and the interested reader is referred to a review by Tappenden.[47]

In growing or neonatal animals, PN and bowel rest maintained normal body growth but resulted in reduced bowel length and gastric and pancreatic hypoplasia.[48-51] Despite the occurrence of mucosal hypoplasia, the development of disaccharidase enzymes and glucose transport was accelerated, and mucosal concentrations of these enzymes increased in neonatal animals receiving PN.[49,51] Hypoplasia occurred mainly in the proximal small bowel and was less evident distally.[50] In adult animals, the effect of PN and bowel rest diminished mucosal mass but stimulated glucose absorption per milligram of mucosal protein.[52] In addition, PN and bowel rest increased intestinal permeability[53] and altered the response to endotoxin.[54]

Does the Nature of Enteral Nutrition Cause Hypoplasia?

It is not only the lack of food but also the nature of the diet that influences mucosal bulk. In neonatal studies, mother's milk was no better than formula.[50] However, refined intragastric liquid feeds caused relative hypoplasia, compared with a solid diet.[48,55]

Factors Influencing Bowel Atrophy

In general, it appears that the decreased digestive and absorptive activities of the mucosa during bowel rest are the major reasons for hypoplasia. This concept is supported by the finding that simply increasing the tonicity of the bowel contents results in an increase of the mucosal mass.[56] Absorption of amino acids leads to a nonspecific increase of mucosal function and mass.[57] Finally, disaccharide hydrolysis followed by absorption stimulates mucosal growth to a greater extent than equivalent monosaccharide absorption.[58]

Another factor affecting the mucosa appears to be biliary-pancreatic secretion. Transplantation of the ampulla causes mucosal hypoplasia, whereas infusion of cholecystokinin and secretin stimulates mucosal growth.[59,60] SCFAs were shown to prevent or reduce mucosal atrophy in animals receiving PN and bowel rest, even when SCFAs were given parenterally.[61-63] Dietary fiber is the main source of colonic fermentable substrate for SCFA production. Therefore, fiber in the diet aids the maintenance of mucosal mass, and DFD are not as good as a solid diet in this regard. Glutamine is a nutrient for the bowel mucosa, and the supplementation of PN with glutamine preserves gastric and colonic mass in solely PN-fed rats but does not preserve small bowel mucosal height.[64]

Does Bowel Rest Induce Gut Atrophy in Humans?

In the rat, bowel rest with PN caused atrophy in days,[65] but in humans, even after 21 days of bowel rest with PN, there was no change in gut hormone production after a meal[66] or any histologic atrophy.[67,68] In children, bowel rest caused atrophy only when it was prolonged beyond 9 months.[68] However, investigators noted a reduction in the size of the microvilli and a fall in brush-border enzyme activity.[67]

In summary, animal data suggest that when the bowel is not used, it atrophies. Mucosal atrophy results from a combination of the lack of functional stimulation and the absence of hormonal, biliary, and pancreatic secretion. In addition to food in the lumen in general, the only convincing specific nutrient trophic factors are SCFAs and perhaps glutamine. Finally, the dramatic mucosal atrophy seen in animals on bowel rest while receiving PN does not occur in humans, even after a few weeks of bowel rest. Thus, few data suggest that patients receiving PN for short periods need to be fed enteral diets before introduction of a normal diet to avoid malabsorption from bowel mucosal atrophy.

Specific Aspects of Management

Control of Gastric Hypersecretion and Motility

Reducing acid secretion improves absorption in patients with a short bowel.[69] Furthermore, hypersecretion can cause nausea, reflux, and hemorrhage from severe esophageal ulceration; these effects are prevented by proton pump inhibitors.[70]

Jejunal Resection with Intact Ileum and Colon

Patients in this category can be fed orally immediately and rarely have any problems.

Ileal Resection of Less than 100 cm with Colon Largely Intact

Patients in this category have so-called bile salt-induced diarrhea, and they are best helped by the administration of 4 g of cholestyramine one to three times a day to bind bile salts. Bile salts are absorbed by the ileum and are recycled. After ileal resection, bile salts remain unabsorbed and enter the colon, where they cause water secretion and diarrhea. Vitamin B_{12} absorption is also impaired in some patients.

Ileal Resection Between 100 and 200 cm with Colon Largely Intact

These patients have little difficulty in maintaining nutrition with an oral diet, but they have almost complete malabsorption of bile salts. In consequence, they have a deficiency of bile salts in the intestinal lumen because bile salt synthesis alone without recycling of bile salts through the ileum is unable to maintain an adequate concentration in the intestinal lumen. These patients have colonic water secretion resulting from entry of bile salt into the colon, as well as malabsorption of fatty acid caused by a low concentration of bile salts in the intestinal lumen. Malabsorbed fatty acids entering the colon increase water secretion. For such patients, a low-fat diet may be considered to reduce diarrhea. In patients with a larger resection, the bile salt pool is depleted, and cholestyramine alone is no longer able to prevent diarrhea. Patients often require antimotility agents such as loperamide, phenoxylate, or codeine. Parenteral vitamin B_{12} replacement is also needed.

Resection in Excess of 200 cm of Small Bowel and Partial Colectomy

Patients in this group require the graduated adaptation program indicated later, in the discussion of general management considerations.

Resection Leaving Less than 60 cm Small Bowel or Only Duodenum: Massive Bowel Resection

Patients in this category need home PN (HPN) indefinitely. However, many patients in this category show a surprising degree of adaptation. They may require less PN over time and may benefit from orally absorbed nutrients. PN can be reduced when weight gain is excessive, and cautious reduction of HPN does not cause electrolyte imbalance and dehydration.

General Aspects of Management

Control of Diarrhea

Diarrhea results from a combination of increased secretions, increased motility, and osmotic stimulation of water secretion secondary to malabsorption of luminal contents. Initially, after massive bowel resection, diarrhea is controlled by keeping the patient *nil per os* (NPO) to reduce any osmotic component. Gastric hypersecretion can be controlled by the continuous infusion of appropriate doses of proton pump inhibitors. In addition, loperamide can be used to slow gastric and intestinal transit. If loperamide does not work, then codeine or phenoxylate may be tried. Doses can be escalated as tolerated because many patients appear to have a threshold below which these agents are ineffective. Oral antidiarrheal medications must be taken 20 to 30 minutes before meals to be maximally effective.

Intravenous Fluids

In the immediate postoperative period, all patients with SBS require intravenous fluids and electrolytes to replace losses. Sodium and potassium chloride, as well as magnesium, are the most important ions to be replaced; and plasma electrolyte concentrations should be monitored frequently. Fluid is infused according to measured losses and to maintain adequate urine output. The infusion is tapered as oral intake is increased.

General Oral Nutrition Considerations

Strategies to Maintain Fluid and Electrolyte Balance by the Oral Route

The next consideration is to determine the nature of oral feedings. In patients who have more than 100 cm of residual jejunum, refeeding should be progressive, with a view ultimately to

Table 79.3 Composition of a Typical Oral Rehydration Solution

Glucose	100 mmol/L
Sodium chloride	60 mmol/L
Sodium citrate	60 mmol/L
Magnesium (as gluconate salt)	30 mmol/d
Sodium chloride tablets with food	

Table 79.4 Major Issues Regarding Oral Intake in Patients With a Short Bowel

Bowel adaptation requires constant adjustment and individualization of the oral feeding regimen.
Avoid high-fiber, nutrient-poor foods.
Add salt to diet to make the diet "isotonic."
Restrict both hypotonic and hypertonic fluids.
Sip liquids throughout the day.
Avoid specific foods that seem to worsen diarrhea.

feeding a normal oral diet. In patients with less than 100 cm of jejunum as the only small bowel remaining, dietary intake and fluids cause increased fluid loss[23] (see Figure 79.1). By contrast, in patients who have little small bowel left, the initial target should be small-volume isotonic feedings containing a glucose-electrolyte content as in oral rehydration solution (ORS) (**Table 79.3**). Fluid absorption has been shown to improve with increasing sodium concentration (see Figure 79.2). In addition, to provide sufficient sodium to absorb dietary carbohydrate, it is necessary to ingest 10 to 15 g of sodium chloride as tablets with meals. Such a regimen prevents osmotic stimulation of secretion and yet stimulates the bowel to absorb nutrients.[71]

Maintenance of Energy Balance

The absorption of energy is better preserved than that of fluid and electrolytes and becomes limiting when the bowel is very short (**Figure 79.3**). Comparison of Figures 79.1 and 79.3 show that fluid and electrolyte losses become limiting when 100 cm of jejunum remains, whereas energy absorption becomes limiting when jejunal length is less than 60 to 70 cm. Hyperphagia is the key to meeting energy requirements. In a group of patients with short bowel, the absorption of carbohydrate and fat energy was about 60% of intake, whereas protein absorption was about 80%.[22] Unpublished data in patients receiving HPN suggest that body weight equilibrates at an absorbed energy intake at about 32 kcal/kg/day. In a 60-kg physiologically normal person, about 1,800 kcal/day will be sufficient to maintain weight. If absorption is about 60% of intake, it will be necessary to increase oral energy intake to 3,000 to 4,000 kcal/day to absorb about 1,800 kcal/day.

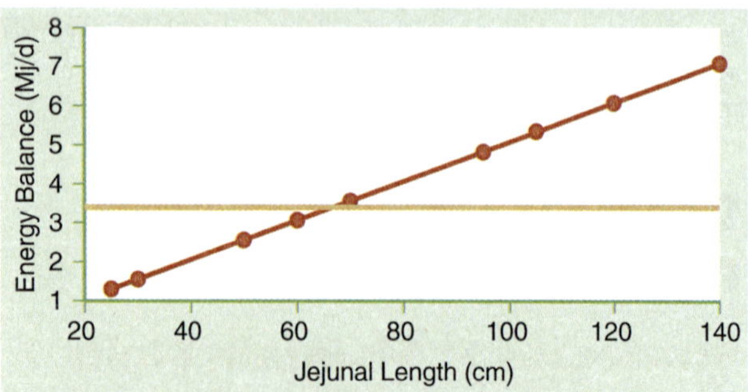

Figure 79.3 The relationship of energy absorption to length of remaining small bowel. The *dark solid line* shows the level at which energy absorption will meet the requirements of an average adult.

Data from Nightingale JM, Lennard-Jones JE, Walker ER et al. Lancet 1990;336:765–8.

Progressive feeding should be attempted, as shown in **Tables 79.4** and **79.5**. The diet should generally be lactose free because lactase is reduced in such patients.[72]

Carbohydrate vs Fat Feeding

The ability of the colon to salvage malabsorbed carbohydrate can be used to improve nutrition in patients with a short bowel who have a colon. In these patients, hyperphagia with complex carbohydrates can result in the salvage of as much as 1,000 additional kilocalories.[73] In contrast, in patients with a jejunostomy and no colon, a high-fat diet was as well absorbed as carbohydrate,[74] and it did not result in additional loss of divalent ions. Because of its palatability and high-energy density, a high-fat diet is more readily consumed.

Fat restriction has been advocated, especially in patients with a remaining colon, because malabsorbed long-chain fatty acids can cause colonic water secretion and may bind divalent ions such as magnesium and calcium.[75] However, in two crossover studies of patients with short bowel with variable lengths of remaining colon, a high-fat diet was comparable to a high-carbohydrate diet with regard to total fluid, energy, nitrogen, sodium, potassium, and divalent ion absorption.[16,22] Therefore, a low-lactose diet containing high energy from both fat and complex carbohydrate and high protein intake is recommended in most patients with a short bowel. The objective is to promote hyperphagia by making the diet palatable and acceptable. In adults who require about 30 kcal/kg/day, the aim is to increase intake gradually to about 60 kcal/kg/day to provide sufficient absorbed energy despite malabsorption. The rationale for this approach is discussed by Woolf et al..[22]

Table 79.5 Energy Intake and the Effects of Dietary Fat and Carbohydrate

Encourage "hyperphagia" to compensate for malabsorption.
Avoid soluble carbohydrates: sugar and lactose.
Increase both fat and complex carbohydrate intake.
Fat slows intestinal transit time.
Fat allows intake of more energy for the same volume.
Carbohydrate is absorbed by the colon as SCFA after microbial metabolism.
Fat may cause water secretion in colon.
Unabsorbed carbohydrate and excess SCFA can cause osmotic diarrhea.

SCFA, short-chain fatty acid.

Micronutrient Supplements

A summary of micronutrients that commonly require supplementation in SBS is given in **Table 79.6**. Vitamin B_{12} concentration should be measured, and if subnormal, injections of 200 to 1,000 μg/month should be instituted (all patients without terminal ileum require vitamin B_{12} supplementation for life). Supplements of potassium, magnesium, and zinc are given as needed to normalized blood concentrations with serial monitoring. Potassium, as gluconate, may be added to a concentration of 12 mmol/L in the ORS. In addition, magnesium heptogluconate is useful as a supplement to correct hypomagnesemia without causing diarrhea. It is possible to add 30 mmol of magnesium/L of ORS, to be sipped over the day.

Enteral Nutrition

Studies by McIntyre et al.,[74] showed that enteral feedings are not absorbed better than a solid diet in patients with a jejunostomy. However, a more recent study by Joly et al.[76] showed that even in patients who were studied 3 months after intestinal resection, a tube-fed polymeric diet was better absorbed than an oral diet and, when combined with an oral diet, allowed seven of nine patients studied to absorb enough energy and protein to become independent of PN. The difference between an oral diet and a tube-fed diet was better absorption of fat and protein, but not carbohydrate, which was well absorbed from an oral diet. Hence, tube feeding or home enteral nutrition (HEN) can potentially be used in many patients to meet energy and protein intake, but it often fails to allow fluid and electrolyte balance. The fluid and electrolyte requirements make the parenteral route essential in many patients. In a large study from northern Alberta,[77] only nine of 797 patients received HEN for a short bowel. Furthermore, of the patients receiving HEN because of gastrointestinal disorders, 82% of the 89% who survived either went back to an oral diet (77%) or were changed to PN (4.6%). Hence, tube feeding is rarely a long-term option for patients with SBS.

Table 79.6 Micronutrient Supplementation in Patients With Short Bowel Syndrome

When zinc, copper, and selenium losses are high:	
Zinc gluconate	100 mg/d
Copper	1,000–3,000 μg/d
Selenium	60–100 μg/d
Malabsorption of fat-soluble vitamins:	
Vitamin A	10,000 IU/d
Vitamin D	1,25-OH vitamin D 0.25–0.5 μg/d or ergocalciferol 50,000 IU up to several times weekly orally
Vitamin E	1,200 IU/d
Calcium gluconate	1,500 mg/d
Severe osteoporosis:	
Alendronate 70 mg orally weekly; consider zoledronate 5 mg intravenously yearly	

Parenteral Nutrition

In patients with less than 100 cm of remaining jejunum and in those with a combined small bowel and colon resection, PN is lifesaving. PN is started in such patients within a few days of the resection, and initially, approximately 32 kcal/kg of a mixed energy substrate and 1 g/kg amino acids are infused with sodium 150 to 200 mM, potassium 60 to 100 mM, calcium 9 to 11 mM, magnesium 7 to 15 mM, and zinc 70 to 100 μmol/day.

Macronutrients, provide energy as a combination of dextrose, protein, and lipid. Several types of lipid emulsions are available, which vary by oil content. A lipid emulsion can contain one type of oil, such as 100% soybean oil or a combination of two oils such as 80% olive oil with 20% soybean oil. Mixed oil lipid emulsion are being used more frequently. These consist of various combinations of soybean oil, medium-chain triglycerides, olive oil, and fish oil. It has a more diverse lipid profile in comparison to soybean oil emulsions, which contain higher amounts of polyunsaturated fatty acids, principally omega-6, as well as small amounts of phytosterols[78] shown to either be more proinflammatory, immunosuppressive, or contributing to cholestasis. At least in the short term, mixed oil emulsion has been shown to decrease cholestasis and risk of intestinal failure related liver disease.[79] However, long-term effects need to be further investigated, as it was recently reported that it may increase risk of hospitalization in HPN patients, when compared with soybean oil emulsion.[80]

Among trace elements, zinc is the most important because large losses have been noted in patients with a high endogenous output of intestinal fluids. Oral feedings are simultaneously started, and attempts are made to reduce PN as oral feedings are increased. It will become apparent whether the patient requires PN on a long-term basis. If that is the case, the patient should be started on a program of HPN. As the bowel adapts over months and even years, the patient requires less PN, and ultimately about 30% of patients who initially required HPN can be weaned off HPN by using up to 2 L of ORS; high-energy diet; individualized dietary modification; and supplements of potassium, magnesium, calcium, fat-soluble vitamins, and zinc. These individuals are monitored regularly until their body weight is stable, urine output is adequate, and blood electrolytes are in balance.

Hypomagnesemia is a particularly serious problem in these patients. Ingestion of magnesium salts orally may increase diarrhea and; therefore, it often becomes difficult to use magnesium supplements orally. Magnesium heptogluconate has been used successfully for this purpose. This preparation is available as a palatable liquid that is added to the ORS supplement in quantities of 30 mM/day. If this approach is not successful, magnesium sulfate can be injected intramuscularly in doses of 12 mM one to three times a week or given intravenously to supplement the oral intake.

Vitamin supplementation needs comment. These patients can absorb water-soluble vitamins but have difficulty absorbing fat-soluble vitamins. They may require large doses of vitamin A, D, and E to maintain normal concentrations. In addition, pills often pass through whole in these patients; hence liquid preparations must be used. Measurement of these vitamin status and supplementation with aqueous preparations of vitamin A and E (Aqasol A and E) and 1,25-dihydroxy-vitamin D in doses that normalize the plasma concentrations are recommended. Normalization may not be possible with oral vitamins in some individuals, especially vitamin E. In other patients, an oral diet with intravenous fluid and electrolytes becomes necessary, and in the remainder, full PN may be required for 3 or more days weekly.

In general, patients with SBS who require PN are best managed with the assistance of an experienced multidisciplinary nutrition support team.

Hormone Therapy

Somatostatin Analog

Long-acting somatostatin analog has become available and can be given subcutaneously. All studies with somatostatin analog showed a reduction in the volume of stool output and an increase in sodium or chloride absorption.[81-83] However, the reduction was not sufficient to avoid PN in patients who required it.[82]

Human Growth Hormone

Byrne et al.,[84,85] in an observational study and subsequently in a controlled trial, found that a combination of human GH (HGH) and glutamine allowed patients with a short bowel to reduce PN. A systematic review[86] of five controlled clinical trials showed that HGH, with or without glutamine, increased body weight by 1.66 kg (confidence interval [CI], 0.69 to 2.63; $P=0.0008$), lean body mass by 1.93 kg (CI, 0.97 to 2.90; $P=0.0001$), energy absorption by 4.42 kcal (CI, 0.26 to 8.58; $P=0.04$), and nitrogen absorption by 44 g (95% CI, 0.20 to 9.49; $P=0.04$). These minor benefits were associated with a 77% incidence of edema and a 32% incidence of carpel tunnel syndrome. Although all studies showed significant weight gain, only one showed a significant increase in absorption. While glutamine supplementation appears to be of little benefit in individuals with SBS, HGH seemed to benefit patients mainly by its well-known somatotrophic effect instead of by significantly improving absorption. Further, significant side effects were associated with the use of HGH.

Glucagon-like Peptide-2 Analog

GLP-2 analog is an intestinotrophic hormone that has been shown to be an overall safe and effective therapeutic option that improves nutrient absorption, increases intestinal absorptive capacity and modestly, but significantly, decreases the need for PN in patients with SBS who previously required stable PN.[87,88] Its mechanism of action includes stimulating hypertrophy of bowel mucosa increasing villus height and depth of crypts, increasing intestinal and portal blood flow, and delaying gastric emptying. Multicenter, double-blind randomized controlled trials have been completed on the efficacy of daily subcutaneous injections of an analog of GLP-2, teduglutide, which is resistant to enzymatic degradation by enzyme dipeptidyl peptidase IV.[87,88] A recent meta-analysis on the available evidence of teduglutide, including ten studies on GLP-2 use, revealed a response rate of 64% at 6 months, 77% at 1 year and, 82% at ≥2 years. The weaning rate was estimated to be 11% at 6 months, 17% at 1 year, and 21% at ≥2 years.[89] The efficacy of a novel long-acting GLP-2 analogue, glepaglutide, is currently under study.[90] When initiating GLP-2 analog therapy, it is important that individual nutrition and hydration status be closely monitored and that both oral and parental support volume be adjusted as needed. This includes monitoring for signs and symptoms of fluid imbalance and fluid overload, monitoring fluid input and output along with daily weights. Lab work monitoring including electrolytes, trace elements, vitamins and minerals is important. For every percentage decrease of PN achieved, enteral energy intake must increase to meet the demands and work toward a goal of achieving oral autonomy.

Complications

Gastric Hypersecretion and Peptic Ulceration

Gastric hypersecretion occurs immediately after intestinal resection and tends to be transient. However, peptic ulceration may occur in some patients. Treatment with histamine (H_2)-blockers and proton pump inhibitors have been found to be successful.[70,91]

Cholelithiasis

After ileal resection, the enterohepatic cycle of bile salts is interrupted. In consequence, bile salt loss occurs beyond the ability of the liver to increase synthesis, and the bile salt concentration in bile falls. The reduction of the concentration of chenodeoxycholate in the bile increases cholesterol secretion.[92] This combination makes the bile lithogenic.[93] Clinically, in this situation, an increased incidence of gallstones has been observed. A study in experimental animals has shown an increased incidence of pigment stones.[94]

Renal Stones

Hyperoxaluria occurs in patients with SBS when the small intestinal remnant is reanastomosed to the colon resulting in an increased absorption of oxalate by the colon, due to bile salts in the colon.[95,96] Hyperoxaluria is associated with renal stone formation, and the propensity to form stones is increased by reduced citrate in the urine.[97] Treatment involves a low-oxalate diet, cholestyramine to bind bile salts, and use of citrate (e.g., as calcium citrate) to prevent stone formation.

D-Lactic Acidosis

In some patients with SBS, when the small intestinal remnant is reanastomosed to the colon, a syndrome of slurred speech, ataxia, and altered affect may occur episodically.[98] Superficially, the patient appears "drunk." The cause of this syndrome is microbial fermentation of malabsorbed carbohydrate in the colon to D-lactate with absorption of this metabolite.[99] The treatment of this condition involves the use of a low-carbohydrate diet.[100]

Metabolic Bone Disease

Individuals with SBS are at increased risk of abnormal bone health. The metabolic changes and loss of functional absorptive capacity can deplete micronutrients such as calcium, magnesium, and vitamin D, which may lead to bone demineralization and development of osteoporosis, osteomalacia, or both conditions.[101] Routine bone mineral density monitoring is essential.

Intestinal Failure Associated Liver Disease

Liver disease can occur in this patient population and is mainly seen in those on long-term administration of PN.[102] It is characterized by progressive intrahepatic cholestasis and steatohepatitis. The pathogenesis of IFALD is multifactorial and recent studies have focused on dose modification of lipid to less than 1 g/kg/day, use of emulsions not 100% soybean oil based, along with cycling PN infusion to decrease risk.[103,104] In a deeply jaundiced patient on HPN, infusion of a pure omega-3 fatty acid emulsion (Omegaven) reversed clinical and liver biopsy changes of cholestasis.[105]

Small Intestinal Bacterial Overgrowth

Bacterial overgrowth is a risk factor in SBS, given the disruption of gastrointestinal anatomy, motility, and bacteriostatic actions of the gut.[106] The colonization of excessive microbes can lead to both clinical and nutritional consequences and antimicrobial regimens are the main stay of therapy.

Parenteral Complications

Although lifesaving, PN associated complications can be life threatening. Catheter related blood stream (CRBS) infection risk is the most frequent complication, often requiring hospitalization due to sepsis. Other catheter-related complications include venous thrombosis, catheter dislodgement, or breakage. Recurrent CRBS infections can also lead to significant reduction in venous access, leading to consideration for intestinal transplantation.

Impact on Quality of Life and Survival

The complexities surrounding a diagnosis of SBS is multifactorial and is met with great psychosocial burden.[107,108] Managing physical symptoms and emotional concerns significantly impacts ones quality of life. In addition, requiring PN support compounds the above concerns with the technologic burden and risks of being on long-term PN. Comprehensive patient education and involvement of a multidisciplinary team to provide patient care and support is essential. Teduglutide can help decrease patients' PN requirements and offer more PN independence. A study assessing the impact of teduglutide on QoL vs placebo showed evidence that the most pronounced improvement in QoL were in those individuals who initially had the highest PN volume requirements, as well as subgroup of patients with the diagnosis of inflammatory bowel disease.[109]

Individuals with SBS experience reduced long-term survival when compared with the general population. Five-year survival on HPN depends on the primary diagnosis. Previously published data show that the rates of survival at 3 years and 5 years in PN-dependent patients were approximately 70% and 63%, respectively; death is principally caused by sepsis and, in the case of small intestinal transplantation, rejection, and lymphoma.[110] More recent data evaluating survival in individuals with nonmalignant SBS estimated a 5-year survival of 81.9%.[110] Subgroup analysis did not identify a significant association between patients' survival and known intestinal anatomy, age, or sex.

Surgical Management: Role of Intestinal Transplantation

SBS and HPN are associated with complications that include progressive steatohepatitis resulting in cirrhosis and liver failure,[111,112] catheter-related complications, and repeated sepsis leading to poor venous access, uncontrollable intestinal output, and inability to cope with the HPN regimen.[111-114] These complications can result in failure of HPN and progressive malnutrition. Under these circumstances, if intestinal rehabilitation with hormonotherapy is not possible or failed, the only alternative is intestinal transplantation (IT). In theory, IT is the ideal solution for the treatment of intestinal failure. The patient who has had an IT can eat and enjoy normal food, does not need complicated machinery to deliver intravenous nutrition, and will avoid the previously mentioned complications of HPN and have an improved quality of life.[115] In practice, however, there are associated risks and challenges. Referral to a specialized intestinal transplant center is required for evaluation, surgery, and follow-up care. The success of IT depends on pretransplant status, treatment center size, immunosuppressive regimen, and transplant type (isolated intestine, intestine-liver, multi-visceral).[115-120] Patients who undergo IT do well initially, but have survival in the range of 50% at 5-years[115,120] and 43% at 10 years. Considering that graft survival is even lower, HPN remains the better option, with better long-term outcome[115] and the option of intestinal rehabilitation with hormonotherapy. Therefore, IT should be reserved for those who have significant complications.

At the present time, based on recommendations by Medicare and Medicaid, the initial therapy for intestinal failure is HPN; and IT is recommended when HPN fails, as defined by the following conditions[120]:

1. Impending or overt liver failure secondary to liver injury from PN
2. Thrombosis of two or more central veins
3. Two or more episodes per year of catheter-related systemic sepsis requiring hospitalization
4. A single episode of line-related fungemia, septic shock, or acute respiratory distress syndrome
5. Frequent episodes of severe dehydration despite intravenous fluid in addition to HPN

Other factors to be considered, according to the American Society of Transplantation, include the following:

1. High risk of death attributable to the underlying disease
2. Ultrashort bowel syndrome (gastrostomy, duodenostomy, residual small bowel <10 cm in infants and <20 cm in adults)
3. Intestinal failure with frequent hospitalization, narcotic dependency, or pseudo-obstruction
4. The patient's unwillingness to accept long-term HPN

The contraindications for IT are the same as those in patients eligible for solid organ transplantation.

In summary, SBS is a complex and variable condition that can range from mild, as seen following terminal ileal resection, to a very debilitating condition, following total ileal and colonic resection with an end-jejunostomy. Management varies with the extent and site of resection and the adaptation of the remaining bowel. Complex patients require management by a multidisciplinary team that should include expertise in specialized enteral nutrition, PN support, and intestinal rehabilitation that may require corrective surgery or hormonotherapy.

References

1. O'Keefe SJ, Buchman AL, Fishbein TM. Clin Gastroenterol Hepatol 2006;4:6–10.
2. Pironi L, Arends J, Bozzetti F, ESPEN guidelines on chronic intestinal failure in adults. Clin Nutr 2016;35(2):247–307.
3. Cuerda C, Pironi L, Arends J, ESPEN practical guideline: Clin Nutr. 2021;40(9):5196–5220.
4. Malagelada JR. Gastric, pancreatic and biliary response to a meal. In: Johnson JR, ed. Physiology of the Gastrointestinal Tract. New York, NY: Raven Press, 1981.
5. Tappenden KA. JPEN 2014;38(1 Suppl):14S–22S.
6. Massironi S, Cavalcoli F, Rausa E, et al. Dig Liver Dis 2020;52(3):253–261.
7. Summers RW, Kent TH, Osborne JW. Gastroenterol 1970;59: 731–9.
8. Ricotta J, Zuidema GD, Gadacz TR et al. Surg Gynecol Obstet 1981;152:310–4.
9. Fordtran JS, Locklear TW. Am J Dig Dis 1966;11:503–21.
10. Overduin J, Tylee TS, Frayo RS, et al. Am J Physiol Gastrointest Liver Physiol 2014;306(12):G1108–16.
11. Powell DW. Physiology of the Gastrointestinal Tract. 2nd ed. New York, NY: Raven Press, 1987.
12. Binder HJ, Mehta P. Gastroenterol 1989;96:989–96.
13. Royall D, Wolever TMS, Jeejeebhoy KN. Am J Gastroenterol 1992;87:751–6.
14. Nylander G. Acta Chir Scand 1967;133:131–8.
15. Reynell PC, Spray GH. Gastroenterol 1956;31:361–8.
16. Woolf GM, Miller C, Kurian R, et al. Gastroenterol 1983;84: 823–8.
17. Cummings JH, James WPT, Wiggins HS. Lancet 1973;1:344–7.
18. Booth CC, Aldis D, Read AE. Gut 1961;2:168–74.
19. Hoffman AF, Poley JR. Gastroenterol 1972;62:918–34.
20. Hylander E, Ladefoged K, Jarnum S. Scand J Gastroenterol 1980;15:853–8.
21. Ladefoged K, Nicolaidou P, Jarnum S. Am J Clin Nutr 1980;33:2137–44.
22. Woolf GM, Miller C, Kurian R, et al. Dig Dis Sci 1987;32:8–15.
23. Nightingale JM, Lennard-Jones JE, Walker ER, et al. Lancet 1990;336:765–8.
24. Haymond HE. Surg Gynecol Obstet 1953;61:693–705.
25. McClenahan JE, Fisher B. Am J Surg 1950;79:684–8.
26. Trafford HS. Br J Surg 1956;44:10–13.
27. West ES, Montague JR, Judy FR. Am J Dig Dis 1938;5:690–2.
28. Pilling GP, Cresson SL. Pediatr 1957;19:940–8.
29. Martin JR, Patee CJ, Gardner C, et al. Can Med Assoc J 1953;69: 429–33.
30. Kinney JM, Goldwyn RM, Barr JS, et al. JAMA 1962;179:529–32.
31. Walker-Smith J. Med J Aust 1967;1:857–60.
32. Clayton BE, Cotton DA. Gut 1961;2:18–22.
33. Anderson CM. Br Med J 1965;5432:419–22.
34. Meyer HW. Surgery 1962;51:755–9.
35. Flint JM. Johns Hopkins Med J 1912;23:127–44.
36. Porus RL. Gastroenterol 1965;48:753–7.
37. Booth CC, Evans KT, Menzies T, et al. Br J Surg 1959;46:403–10.
38. Althausen TL, Doig RK, Uyeyama K, et al. Gastroenterol 1950; 16:126–39.
39. Yeh KY, Moog F. Dev Biol 1975;472:156–72.
40. Winesett DE, Ulshen MH, Hoyt EC, et al. Am J Physiol 1995;268:G631–40.
41. Ulshen MH, Raasch RH. Clin Sci (Lond) 1996;90:427–31.
42. Potten CS, Owen G, Hewitt D, et al. Gut 1995;36:864–73.
43. Drucker DJ, Ehrlich P, Asa SL, et al. Proc Natl Acad Sci U S A 1996;93:7911–16.
44. Drucker DJ, DeForest L, Brubaker PL. Am J Physiol 1997;273: G1252–62.
45. Jeppesen PB, Hartmann B, Hansen BS, et al. Gut 1999;45: 559–63.
46. Jeppesen PB, Hartmann B, Thulesen J, et al. Gut 2000;47:370–6.
47. Tappenden KA. Gastroenterol 2006;130(suppl 1):S93–9.
48. Goldstein RM, Hebiguchi T, Luk G, et al. J Pediatr Surg 1985;20: 785–91.
49. Shulman RJ. Gastroenterol 1988;95:85–92.
50. Morgan W, Yardley J, Luk G, et al. Pediatr Surg 1987;22:541–5.
51. Gall DG, Chung M, O'Laughlin EV, et al. Biol Neonate 1987;51: 286–96.
52. Kolter DP, Levine GM, Shiau YF. Am J Physiol 1981;240:432–36.
53. Purandare S, Offenbartl K, Westrom B, et al. Scand J Gastroenterol 1989;24:678–82.
54. Fong YM, Marano MA, Barber A, et al. Ann Surg 1989;210: 449–56.
55. Hosoda N, Nishi M, Nakagawa M, et al. J Surg Res 1989;47: 129–33.
56. Weser E, Babbitt J, Vandeventer A. Dig Dis Sci 1985;30:675–81.
57. Levine GM. Gastroenterology 1986;91:49–55.
58. Weser E, Babbit J, Vandeventer A. Gastroenterol 1986;91: 521–7.
59. Hughes CA, Bates T, Dowling RH. Gastroenterol 1978;75: 34–41.
60. Weser E, Bell D, Tawil T. Dig Dis Sci 26:1981;409–16.
61. Koruda MJ, Rolandelli RH, Bliss DZ, et al. Am J Clin Nutr 1990;51:685–9.
62. Koruda MJ, Rolandelli RH, Settle RG, et al. Gastroenterol 1988;95:715–20.
63. Kripke SA, Fox AD, Berman JM. J Surg Res 1988;44:436–44.
64. Grant JP, Snyder PJ. J Surg Res 1988;44:506–13.
65. Hughes CA, Prince A, Dowling RH. Clin Sci 1980;59:329–36.
66. Greenberg GR, Wolman SL, Cristofides ND, et al. Gastroenterol 1981;80:988–93.
67. Guedon C, Schmitz J, Lerebours E, et al. Gastroenterol 1986; 90:373–78.
68. Rossi TM, Lee PC, Young C, et al. Dig Dis Sci 1993;38:1608–13.
69. Cortot A, Fleming CR, Malagelada JR. N Engl J Med 1979;300: 79–80.
70. Tang SJ, Jose M. Nieto JM, et al. J Clin Gastroenterol 2002;34: 62–3.
71. Griffin GE, Fagan EF, Hodgson AJ. Dig Dis Sci 1982;27:902–8.
72. Richards AJ, Condon JR, Mallinson CN. Br J Surg 1971;58:493–4.
73. Nordgaard I, Hansen BS, Mortensen PB. Lancet 1994;343:373–6.
74. McIntyre PB, Fitchew M, Lennard-Jones JE. Gastroenterol 1986; 91:25–33.
75. Ovesen L, Chu R, Howard L. Am J Clin Nutr 1983;38:823–8.
76. Joly F, Dray X, Corcos O, et al. Gastroenterol 2009;136: 824–31.
77. Cawsey SI, Soo J, Gramlich LM. Nutr Clin Pract 2010;25: 296–300.
78. Diamond IR, Grant RC, Pencharz PB, et al. JPEN 2017;41(5): 866–877.
79. Daniel S, Svoboda L, Chen J. J Pediatr Pharmacol Ther 2021; 26(3):258–264.
80. Clermont-Dejean NM, Schwenger KJP, Salazar E, et al. Clin Nutr. 2021;40(7):4616–4623.
81. Rodrigues CA, Lennard-Jones JE, Thompson DG, et al. Aliment Pharmacol Ther 1989;3:159–69.
82. Ladefoged K, Christensen KC, Hegnhoj J, et al. Gut 1989;30: 943–9.
83. Dharmsathaphorn K, Gorelick FS, Sherwin RS, et al. J Clin Gastroenterol 1982;4:521–4.
84. Byrne TA, Persinger RL, Young LS, et al. Ann Surg 1995;222: 243–55.

85. Byrne TA, Wilmore DW, Iyer K, et al. Ann Surg 2005;242: 655–61.
86. Wales PW, Nasr A, de Silva N, et al. Cochrane Database Syst Rev 2010;(6):CD006321.
87. Jeppesen PB, Gilroy R, Pertkiewicz M, et al. Gut 2011;60: 902–14.
88. Vipperla K, O'Keefe SJ. Expert Rev Gastroenterol Hepatol 2011; 5:665–78.
89. Bioletto F, D'Eusebio C, Merlo FD, et al. Nutrients. 2022; 14(4):796.
90. Naimi RM, Hvistendahl M, Enevoldsen LH, et al. Lancet GH 2019;4(5):354–363.
91. Murphy JP Jr, King DR, Dubois A. N Engl J Med 1979;300:80–1.
92. Jeejeebhoy KN. Trop Gastroenterol 2010;31:244–8.
93. Roslyn JJ, Pitt HA, Mann LL. Gastroenterol 1983;84:148–54.
94. Pitt HA, Lewinski MA, Muller EL, et al. Surgery 1984;96: 154–62.
95. Dobbins JW, Binder HJ. N Engl J Med 1977;296:298–301.
96. Chadwick VS, Gaginella TS, Carlson GL, et al. J Lab Clin Med 1979;94:661–74.
97. Pak CYC, Peterson R, Sakhaee K, et al. Am J Med 1985;79: 284–8.
98. Traube M, Bock JL, Boyer JL. Ann Intern Med 1983;98:171–3.
99. Satoh T, Narisawa K, Konno T, et al. Eur J Pediatr 1982;138: 324–6.
100. Ramakrishnan T, Stokes P. JPEN J Parenter Enteral Nutr 1985;9:361–3.
101. Johnson E, Vu L, Matarese LE. Nutr Clin Pract. 2018 Aug;33(4):454–466.
102. Lee WS, Chew KS, Ng RT, et al. Hepatol Int 2020;14(3): 305–316.
103. Bielawska B, Allard JP. Nutrients 2017;9(5):466.
104. Wanten G, Calder PC, Forbes A. BMJ 2011;342:d1447.
105. Jurewitsch B, Gardiner G, Naccarato M, et al. JPRN 2011;35(3):386–90.
106. Dibaise JK, Young RJ, Vanderhoof JA. Clin Gastroenterol Hepatol 2006;4(1):11–20.
107. MacRitchie KJ. Can Psychiatr Assoc J 1978;23:373–9.
108. Detsky AS, McLaughlin JR, Abrams HB, et al. J Gen Intern Med 1986;1(1):26–33.
109. Chen K, Mu F, Xie J, et al. JPEN 2020;44(1):119–128.
110. Noelting J, Gramlich L, Whittaker S, et al. JPEN 2021;45(5): 1083–1088.
111. Bowyer BA, Fleming CR, Ludwig J, et al. JPEN 1985;9:11–17.
112. Cavicchi M, Beau P, Crenn P, et al. Ann Intern Med 2000;132: 525–32.
113. Howard L. Gastroenterology 2006;130(suppl 1):S52–9.
114. Messing B, Lemann M, Landais P, et al. Gastroenterol 1995; 108:1005–10.
115. O'Keefe SJ, Emerling M, Koritsky D, et al. Am J Gastroenterol 2007;102:1093–100.
116. Freeman RB, Steffick DE, Guidinger MK, et al. Am J Transplant 2008;8:958–76.
117. Jeejeebhoy K, Allard J, Gramlich L. In: Bozetti F, Staun M, Van Gossum A, eds. Home parenteral nutrition in Canada. Home Parenteral Nutrition. Cambridge, MA: CAB International, 2006:36–42.
118. Grant DW, Shah SA. In: Langnas AN, Goulet O, Quigley EMM, Tappenden KA, eds. Results of intestinal transplantation. Intestinal Failure. Malden, MA: Blackwell, 2008:349–56.
119. DeLegge M, Alsolaiman MM, Barbour E, et al. Dig Dis Sci 2007;52:876–92.
120. Fishbein TM. N Engl J Med 2009;361:998–1008.

CHAPTER 80

Nutrition in Inflammatory Bowel Diseases

Berkeley N. Limketkai, Alyssa M. Parian, and Gerard E. Mullin

CHAPTER OUTLINE

Abbreviations

CD, Crohn's Disease; **CDAI**, Crohn's disease activity index; **EN**, enteral nutrition; **IBD**, inflammatory bowel disease; **IOIBD**, International Organization for the Study of Inflammatory Bowel Diseases; **ESPEN**, European Society for Clinical Nutrition and Metabolism; **FOS**, fructo-oligosaccharides; **GBF**, Germinated barley foodstuff; **GOS**, galacto-oligosaccharides; **LGG**, Lactobacillus rhamnosus GG; **PEM**, protein energy malnutrition; **PN**, parenteral nutrition; **PUFA**, polyunsaturated fatty acids; **RCT**, randomized controlled trial; **SCFA**, short chain fatty acids; **UC**, ulcerative colitis

Introduction

Inflammatory bowel disease (IBD) is a condition that leads to chronic inflammation of the gastrointestinal tract.[1] The two predominant types of IBD include Crohn's disease (CD) and ulcerative colitis (UC). Inflammation-related disruptions in the integrity and function of the gastrointestinal tract often result in symptoms of abdominal pain, increased frequency of bowel movements, urgency, loose stools, and rectal bleeding. A difference between CD and UC is the location of the inflammation. CD can affect any segment of the gastrointestinal tract from the mouth to the anus, while UC is limited to the colon. CD has additional manifestations not typically found in UC, including strictures and fistulas.

Maintenance of human health requires continuous ingestion of nutrients and appropriate digestion and assimilation, both of which require an appropriately functioning digestive tract. However, when these processes are compromised, nutritional deficiencies can arise. Conversely, foods consumed can exert positive and negative effects on disease activity in IBD. This chapter discusses the three roles nutrition plays in IBD: instigator, victim, and healer.

Role of Nutrition in the Etiology of Inflammatory Bowel Disease

The incidence and prevalence of IBD have been rising globally, although the reasons for these increases are unclear.[2] The prevailing hypothesis suggests that IBD stems from complex interactions among an individual's genetic susceptibility, host immune response, gut microbiome, and environmental factors. More than 200 genetic loci have been associated with IBD, although studies involving identical twins—born with identical sets of genes—have revealed only partial concordance between twins in the risk of developing IBD.[3-5] On the other hand, the gut microbiome has a more direct relationship with the health of the intestinal mucosa. In healthy intestines, bacteria in the Firmicutes and Bacteroidetes phyla promote production of epithelial substrates. The microbiome in patients with CD tends to have less Firmicutes and Bacteroidetes and more of the pathogenic enterobacteria, while in UC, there is more *Escherichia coli*.[6] Germ-free animals susceptible to IBD remain inflammation free until exposed to bacteria. However, humans must coexist with their gut flora for many reasons, such as food digestion, vitamin K production, and pathogen protection. Given diet's strong influence on the gut microbiome and diet's direct and indirect effects on inflammatory pathways, nutrition has been investigated as an environmental trigger of IBD.

Dietary habits typified by the Western diet (low in fiber and high in saturated fats, refined carbohydrates, and other processed foods) have been implicated in the rise in IBD. A systematic review found that diets high in fat and ω-6 polyunsaturated fatty acids (PUFAs) are associated with increased risk of IBD, while diets high in fruits and fiber appeared protective.[7] Japan has similarly witnessed a significant rise in cases of IBD over recent decades, accompanied by large-scale, rapid dietary changes undertaken by the Japanese population.[8,9] In the prospective Nurses' Health Study in the United States, with 269 incident cases of CD and 338 incident cases of UC, there was no significant association between long-term fatty acid consumption and risk of IBD.[10] Long-term dietary fiber intake was, nonetheless, associated with lower risk of CD, but not UC.[11] One putative mechanism for the proinflammatory effect of fatty acid consumption relates to ω-6 PUFAs induction of arachidonic acid metabolism and synthesis of proinflammatory leukotrienes.[12] The effect of PUFAs on inflammation, however, differs depending on their type, as ω-3 PUFAs may, instead, have anti-inflammatory properties. The protective role of fiber may be mediated by short-chain fatty acids (SCFAs), which are metabolic products of fermentation by anaerobic bacteria in the colon.[13] SCFAs have notable anti-inflammatory effects and help fortify the intestinal epithelial barrier through the production of mucin and other defense peptides.[14]

An earlier proposed explanation that attempts to link food choices with the increasing incidence of IBD is the cold-chain hypothesis.[15] This hypothesis notes that post-World War II, economic prosperity made the refrigerator widely available. Refrigeration quickly transformed food consumption from a pattern of daily use of perishables to dependence on long-term refrigerated storage. Support for this theory rests on the identification of organisms that thrive at near-freezing temperatures, such as *Yersinia* and *Legionella*, which have been associated with IBD-like infections of the gut. These data demonstrate that food choices likely play a role in IBD susceptibility, but the elimination of individual food items will not likely alter disease course.

Impact of Inflammatory Bowel Disease on Nutritional Status

Protein-energy malnutrition (PEM) and other nutritional deficiencies are common complications of IBD. In a retrospective study of 182 patients with newly diagnosed IBD, 36% met the criteria for malnutrition, while 78% had at least one micronutrient deficiency.[16] The effect of IBD on an individual's nutritional status occurs through at least three different mechanisms: (i) altered dietary intake, (ii) intestinal malabsorption, and (iii) increased energy consumption. Active inflammatory disease can lead to increased abdominal pain, frequency of bowel movements, urgency, and loose stools. The presence of small bowel strictures in CD can additionally lead to nausea, vomiting, and bloating. All of these symptoms can be further aggravated by the consumption of foods, thus leading to a habituated aversion to eating and a heightened risk of avoidance/restrictive food intake disorder.[17] There is also a common effort by patients with IBD to pursue exclusion diets as a form of treatment, similarly exposing themselves to the risk of nutritional inadequacy.[18,19] In addition to abnormal food intake, mucosal disease-related nutrient malabsorption/loss and systemic cytokine-driven increases in metabolic requirements can lead to deficiencies in macronutrients, vitamins, and minerals (**Table 80.1**). Although iron deficiency is more commonly found in UC (up to 81%) than in CD, this form of anemia is common in both disease states, affecting at least two-thirds of patients. Anemia from folate or vitamin B_{12} deficiencies is more likely in CD.[20] Medications used to treat IBD also play a significant role in exacerbating nutrient deficiencies through a variety of mechanisms (**Table 80.2**). In addition to macronutrient therapy, targeted micronutrient supplementation is often necessary to replete specific deficiencies acquired in the course of IBD.[21,22]

Table 80.1 Micronutrients Commonly Affected by Inflammatory Bowel Disease

Micronutrient	Frequency of Deficiency		
		Adult	
	Pediatric	CD	UC
Water Soluble			
Iron	17%	39%	81%
Zinc	No data	50%	No data
Folate	0%–2%	67%	30%–40%
Vitamin B_{12}	0%	48%	5%
Fat Soluble			
Vitamin A	14%	11%	No data
Vitamin D	16%–35%	75%	35%
Vitamin E	6%	No data	No data
Vitamin K	No data	No data	No data

CD, Crohn's disease; *UC*, ulcerative colitis.

Data from Mallon DP, Suskind DL. Nutr Clin Pract 2010;25:335–9; Vagianos K, Bector S, McConnell J et al. JPEN J Parenter Enteral Nutr 2007;31:311–9.

Nutrition-Based Therapy for Inflammatory Bowel Disease

IBD therapy optimally induces rapid disease remission, maintains that remission, and improves quality of life.[23,24] Given nutrition's influence on disease activity in IBD, nutrition has

Table 80.2 Medication-Associated Nutrient Effects in Inflammatory Bowel Disease

Drug	Nutrient Affected	Mechanism
Sulfasalazine	Folate	Competitive inhibition of jejunal folate conjugation enzyme
Corticosteroids	Calcium	Steroids suppress calcium absorption by small bowel, increase urinary excretion
Cholestyramine	Fat, calcium, fat-soluble vitamins	Impair absorption of fat by bile sequestration
Azathioprine 6-Mercaptopurine	General energy intake	Can induce nausea, vomiting, dyspepsia

been investigated as primary and adjunctive interventions for IBD. Historically, parenteral nutrition (PN) and enteral nutrition (EN) were studied and employed as forms of nutritional therapy for IBD. However, the concept of nutritional therapy has more recently broadened to include other tools for an IBD treatment plan, including diets and nutritional supplements that alter intestinal epithelial function, enhance enteric flora, or reduce intestinal epithelial inflammation. These options provide clinicians with multiple options for providing nutritional prescriptions to patients with IBD.

Parenteral Nutrition

PN, in conjunction with "bowel rest" (absence of oral food intake), had historically been a form of treatment for severe cases of CD,[25,26] although there is no clear benefit for UC.[27] The original premise behind this strategy is that the elimination of antigenic stimuli from intact food molecules would allow healing of the inflamed tissue. Early studies were, however, plagued by inconsistent disease activity measurements and significant confounding. Currently, evidence-based guidelines do not recommend the use of PN as a primary mode of medical nutrition therapy for IBD.[28] The role of PN in patients with IBD would instead be limited to cases of EN intolerance, intestinal obstruction, moderate-to-severe PEM refractory to enteral routes of nutrition, short bowel syndrome, and preoperative nutrition optimization. PN otherwise provides no advantage over EN for reduction of intestinal inflammation while increasing the risk of complications (e.g., infections, PN-associated liver disease), inconvenience, and cost. Nutritional interventions for intestinal inflammation in IBD, particularly CD, now focus on EN and dietary approaches.

Enteral Nutrition

There is good evidence demonstrating the efficacy of exclusive EN for the induction of remission in CD. Exclusive EN replaces a patient's entire energy and nutrient intake with a processed liquid nutritional supplement administered by mouth or feeding tube. Exclusive EN theoretically prevents small bowel mucosal villous atrophy from long-term bowel rest, maintains epithelial integrity, and reduces intestinal immune system activation. The first randomized, controlled clinical trial demonstrated that exclusive EN could achieve remission as frequently as corticosteroids.[29] A systematic review and meta-analysis by the Cochrane Collaboration found exclusive EN to be more effective than corticosteroids for inducing remission in children with CD but less effective than corticosteroids for inducing remission in adults with CD.[30] The investigators nonetheless noted the quality of evidence to be very low, suggesting that more research is needed. Exclusive EN has been recommended as a first-line treatment for active CD in children.[28,31] The roles of exclusive EN for helping maintain remission in CD or for treatment of UC are still unclear.[32,33]

Compliance rates vary greatly among adult patients with CD from country to country. Although patients in the United States rarely accept long-term exclusive EN, this approach is an important first-line therapy for both inducing and maintaining CD remission in Japan.[34,35] Similarly, European guidelines promote exclusive EN for adults with active CD complicated by effects of steroids, as supplemental exclusive EN for undernourished children with CD, or as first-line therapy for children with active CD to induce remission.[28]

When examined in aggregate as a primary therapy, exclusive EN induces remission almost as effectively as corticosteroid therapy in CD, without steroid-related side effects. Exclusive EN can be administered via tube or mouth in polymeric, semi-elemental, or elemental forms.[30] In children, EN improves growth velocity hampered by corticosteroid therapy.[36] However, cultural preferences, palatability, feeding-tube aversion, and cost hinder the widespread use of exclusive EN in the United States.

Given the challenges of consuming exclusive EN, the possibility of using partial EN for the treatment of IBD has been investigated. Partial EN includes the use of a combination of EN and a solid food diet. The solid food component generally includes a "healthy" diet. In a randomized controlled trial of 74 children with mild-to-moderate CD who either received exclusive EN or partial EN (50% of energy from EN and 50% of energy from an exclusion diet) for 6 weeks, both interventions had equivalent efficacy for achieving corticosteroid-free remission.[37] Partial EN was, however, reported to be significantly more tolerable than exclusive EN; there was otherwise no difference in compliance. These findings suggest that an alternative and more palatable option may exist for induction of remission in CD.

Diet

Unlike exclusive EN, where medical and nutrition society guidelines recommend its use as a treatment for IBD, there is less clarity on which solid food diets might be effective for IBD.[31] A systematic review and meta-analysis in 2019 by the Cochrane Collaboration of RCTs, which evaluated dietary interventions for IBD, did not find adequate evidence to support any particular diet for induction or maintenance of remission.[38] These findings were consistent with clinical guidelines by the European Society for Clinical Nutrition and Metabolism (ESPEN) that could not recommend an "IBD diet."[28] In an effort to provide some dietary guidance for patients with IBD, the International Organization for the Study of Inflammatory Bowel Diseases (IOIBD) synthesized recommendations on food groups based on the existing research literature.[39] The group acknowledged that the level of evidence was generally low or very low but provided some recommendations. For CD, they recommended increasing the consumption of fruits and vegetables, reducing consumption of saturated fats, maltodextrin-containing foods, artificial sweeteners, emulsifiers, carrageenan, and nanoparticles, and avoiding trans-fat and unpasteurized milk. For UC, they recommended increased consumption of foods rich in ω-3 PUFAs (e.g., marine fish); reduction of red or processed meat, myristic acid (palm oil, coconut oil, dairy fats), maltodextrin-containing foods, artificial sweeteners, emulsifiers, carrageenan, and nanoparticles; and avoidance of trans fat and unpasteurized dairy products. A more recent RCT with 194 participants with CD and mild-to-moderate symptoms found the Mediterranean diet equivalent to the Specific Carbohydrate Diet for achieving symptomatic remission at week 6 (43.5% vs 46.5%; $P=0.77$).[40] Among the few participants who had their fecal calprotectin (stool test for inflammation) measured, there was no difference in the proportion of those experiencing meaningful reduction. This study highlighted a similarity between two popular diets for IBD, although they were not compared with a control diet. More research is needed to better understand the role of solid food diets in the treatment of IBD.

Nutrition Supplements for Inflammatory Bowel Disease

Probiotics

Probiotics are live micro-organisms that confer a health benefit on the host when administered in adequate amounts. To be labeled as a probiotic, several criteria must be satisfied.[41] Each probiotic is strictly specified at the genus, species, and strain levels. Probiotics contain live cultures when administered and should be shown to be viable before consumption, during intestinal transit, and be able to colonize the intestines successfully. There should be human-controlled trials evaluating the health benefits of probiotics. Probiotics should be safe for their intended use.[42] There have been longstanding challenges with the oversight, regulation, quality control, safety concerns, and conflicting outcome studies with probiotics.[43]

VSL #3

VSL #3 is a commercial product invented by Dr. Claudio De Simone, which consists of three genera of bacteria and a total of eight different species: Lactobacilli (*L. casei*, *L. plantarum*, *L. acidophilus*, *L. delbrueckii*), Bifidobacteria (*B. longum*, *B. breve*, *B. infantis*), and Streptococcus (*S. salivarius*). VSL#3 has been studied for CD, UC, and pouchitis, with the most compelling evidence for pouchitis. In CD, several meta-analyses have not shown a benefit of probiotics use.[44,45] Several RCTs demonstrated the benefit of using VSL#3 in treating active UC vs placebo.[46-50] A systematic review with meta-analysis concluded that VSL#3, at a dose of 3.6×10^{12} CFU/day, is superior in achieving response and remission in mild to moderate UC. VSL#3 is an adjuvant therapy that is efficacious as monotherapy for active UC. There is some evidence for the use of VSL#3 favoring the maintenance of remission for UC.[51,52] Clinical trials report that the most common side effect is symptomatic bloating without serious adverse events.[53] VSL#3 has substantial evidence to support its use in UC patients undergoing ileal pouch-anal anastomosis with pouchitis. The exact cause is still unknown, but mucosal ischemia, genetic factors, and gut and bacteria dysbiosis are thought to contribute to disease pathogenesis. Antibiotics temporarily alleviate the symptoms, but they quickly return after cessation of treatment.[41] Clinical guidelines suggest using VSL#3 for the induction of remission and to prevent recurrent pouchitis, although they do not yet have enough evidence to recommend the treatment of acute pouchitis.[54]

In 2016, Dr. De Simone terminated commercial agreements with VSL Pharmaceuticals, which changed the formulation, retained the VSL#3 brand name, and marketed the product as having the original formulation. Extensive litigation disclosed the alterations in the VSL#3 product, and judgments favored Dr. Desimone, whose original formulation is now labeled as Visbiome and sold as Vivomixx®. There has been a paucity of original controlled clinical human studies using the altered VSL#3 formulation since the inception of litigation. One such study showed a higher readmission rate in perioperative patients receiving VSL#3 compared with placebo.[55] Panetta et al., performed a meta-analysis of safety data from double-blind, randomized, placebo-controlled trials. Despite 64.5% of patients consuming VSL#3 experiencing an adverse event, the authors concluded that VSL#3 was a safe and well-tolerated agent.[56] Given the lead author's conflict of interest and possible bias, the publication was subsequently retracted by the journal's editor.[57]

The American Gastroenterological Association (AGA) updated their guidelines regarding probiotics in digestive diseases in 2020. For IBD, the AGA only recommended the use of probiotics in the context of a clinical trial,[58] except for pouchitis, for which the AGA only recommends the original formulation of the eight-strain combination (*L paracasei* subsp *paracasei*, *L plantarum*, *L acidophilus*, *L delbrueckii* subsp *bulgaricus*, *B longum* subsp *longum*, *B breve*, *B longum* subsp *infantis*, and *S salivarius* subsp *thermophilus*) (presently labeled as Visbiome). The recommendations have considerably softened since the negative outcome data were published by the Weitzman Institute, reporting the lack of colonic colonization by probiotics and their apparent interference with postantibiotic ecological restoration.[59]

Lactobacillus GG

Lactobacillus rhamnosus GG (LGG) was first isolated in 1983 and is known to have a strong avidity for human intestinal cells, and it can survive acid and bile environments.[60] It is thought to be one of the most studied probiotics, with research showing it could alleviate rotavirus diarrhea in children,[61,62] prevent atopic dermatitis,[63] protect against urinary tract infections,[64,65] and improve symptoms of IBD.[66-68] LGG appears safe and effective in prolonging remission in UC patients.[69-72] LGG was also studied in CD but was found to be no better than placebo.[73,74] A negative large clinical trial of LGG in pediatric viral gastroenteritis has cast doubt on the efficacy of LGG in rotavirus enteritis.[75]

Saccharomyces boulardii

Saccharomyces boulardii is a tropical, nonpathogenic strain of yeast, which has been successfully used as a probiotic for alleviating traveler's diarrhea,[76] HIV-induced diarrhea,[77] antibiotic-associated diarrhea,[78] and even in the prevention of *Clostridioides difficile* relapses.[79] Animal models have shown hope for the use of *S. boulardii* in the treatment of IBD.[80-83] A pilot study demonstrated it in mild-to-moderate UC.[84] A case series of six patients with left-sided UC who were intolerant to mesalamine responded to

S. boulardii 500 mg plus rifaximin 400 mg.[85] In CD, *S. boulardii* decreased the frequency of bowel movements.[86,87] Its addition to mesalamine helped maintain remission of CD, but not as sole therapy.[88] Interestingly, *S. boulardii* improves intestinal permeability in CD.[89] The largest prospective, randomized, placebo-controlled trial of 165 Crohn's patients in remission followed for 1 year found that *S. boulardii* was no better than placebo at preventing relapse.[87] In a large cohort study, Oh and colleagues demonstrated that *S. boulardii* improved the Crohn's disease activity index (CDAI) and serum hemoglobin in CD patients with quiescent disease.[90]

Prebiotics

Prebiotics are defined as nondigestible fibrous products, which selectively promote the growth or activity of beneficial bacteria and improve the well-being of the host. Prebiotics, generally, are not digested and absorbed into the intestines. Instead, they act as nutrition for the beneficial bacteria (Lactobacillus and Bifidobacterium) in the distal small bowel and colon. Some examples of prebiotics include bran, psyllium husk, inulin, fructo-oligosaccharides (FOS), and galacto-oligosaccharides (GOS). Prebiotics are commonly used to regulate the bowels in constipation, where they increase fecal water content, and diarrhea, where they add bulk to the stool. Plantago ovata, also known as psyllium, has been studied to treat UC. One small study showed that a 4-month treatment improved symptoms in patients with UC in remission.[91] An open-label trial was performed on 105 patients with quiescent UC who were randomized to one of three groups: 10 g twice daily of Plantago ovata, 500 mg three times daily of mesalamine, or Plantago ovata plus mesalamine had similar rates of continued remission.[92] Inulin (12 g daily) has decreased fecal calprotectin in active UC patients and was well tolerated.[93] In a mouse model of UC, inulin reduced the amount of nitric oxide in the colon and the severity of the colitis, suggesting that inulin may help treat active UC.[94] Inulin also has shown promise in the treatment of pouchitis.[95] An open-label trial of FOS 15 g daily for 3 weeks was shown to improve symptoms in 10 CD patients with active ileocolonic disease.[96] In contrast, a controlled trial of 103 patients found no difference in clinical response, but patients treated with FOS had higher drop-out rates. Several rodent studies demonstrated decreases in intestinal inflammation and improvement in colonic bacterial composition after FOS treatment,[97-99] therefore, further studies are needed to investigate the potential efficacy of FOS in IBD. Germinated barley foodstuff (GBF) was initially discovered to attenuate colitis by increasing luminal short-chain fatty acid production in animal models.[100,101] A pilot study of 10 patients with active UC, who received 30 g/day of GBF for 4 weeks, was found to improve clinical activity scores, endoscopic scores, and increased stool butyrate concentration.[102] In a study by Kanauchi et al., 20 to 30 grams daily of GBF improved clinical scores and increased Bifidobacterium in patients with UC.[103] In a follow-up, open-label study with 21 patients with UC, GBF improved symptomatic scores, decreased rectal bleeding, and lowered rates of nocturnal diarrhea.[104] To investigate the efficacy of GBF in maintaining remission in UC, 59 patients were given either GBF (20 g daily) plus standard therapy or placebo plus standard therapy for 12 months. Improved clinical activity index scores were seen in the GBF group at each time point, with lower recurrence rates.[105] Faghfoori et al., found that GBF reduced serum tumor necrosis factor (TNF)-alpha, interleukin (IL)-6, and IL-8. The most recent clinical study of GBF found a reduction in C-reactive protein (CRP), abdominal pain, and cramping, compared with placebo.[106] A rodent model has also shown that GBF may prevent the development of colitis-associated neoplasia.[107] However, this decreased risk of neoplasia may be due to the decrease in inflammation and not necessarily a chemopreventative effect. Despite the potential benefit of prebiotic fibers on the course of IBD, pediatric RCTs are lacking.[108] Prebiotics must be used with caution in CD patients with stricturing disease, as high fiber loads can lead to obstruction. Additionally, prebiotics cause gas and bloating when commenced, and starting at lower doses may alleviate these gastrointestinal effects.

Synbiotics

Synbiotics are a combination of prebiotics and probiotics that inhibit the growth of pathogenic bacteria while fostering the growth of beneficial organisms.[109,110] Synbiotic therapy (*B. longum* and psyllium) has been reported to be superior in improving IBD questionnaire scores and CRP concentration.[111] The synbiotic Synergy 1 (FOS/inulin mix and *B. longum*) improved active UC; sigmoidoscopy scores and inflammatory markers improved.[112] The synbiotic containing 109 CFU of the *B. breve* strain and 5.5 g of GOS improved colonoscopic findings and fecal concentration of Bacteroidaceae.[113] In a small study of 10 patients with active CD refractory to mesalamine and steroids, participants were given a synbiotic containing 75 billion CFU of either Bifidobacterium or Lactobacillus and 9.9 g of psyllium. Participants were allowed to adjust the prebiotic and probiotic as needed. Six were able to decrease or stop other therapy, including prednisone they were taking for CD. Despite the clinical improvement, serum inflammatory markers did not improve.[114] A randomized, double-blind, placebo-controlled trial of 35 CD patients found the synbiotic Synergy 1 improved CDAI scores and histologic scores.[115] To investigate the role of synbiotics in preventing postoperative recurrence of CD, 30 patients were randomized to Synbiotic 2000 or placebo in a 2:1 ratio. Synbiotic 2000 consists of four lactic acid bacteria and four fermentable fibers. After 24 months of postoperative follow-up, there was no effect on postoperative recurrence based on symptomatic, endoscopic, laboratory, and histologic criteria.[116] Experimental colitis has improved after administering a symbiotic mixture.[117] Speculation about the efficacy of synbiotics in IBD continues to appear in the literature. However, more randomized clinical trials are required to change guidelines and clinical practice.[118,119]

Omega-3 Fatty Acids

There are two main essential fatty acids that the body cannot produce: linoleic acid (LA) and α-linolenic acid (ALA). LA is a precursor to ω-6 PUFAs, while ALA is a precursor to ω-3 PUFAs. Optimal consumption of LA to ALA occurs at a 4:1 ratio.[90] However, the heavy plant oil consumption found in the Western diet has altered this ratio to 15 to 16:1, likely spurring a worldwide increase in inflammatory disorders, such as CD and UC.[120] Saturated ω-6 FA precursors of arachidonic acid (AA) accumulate in mammalian cell membranes. Although AA is an important signaling molecule playing a crucial role in regulating inflammation, excessive AA can trigger uncontrolled inflammation. Downstream products of AA metabolism include the proinflammatory molecules leukotriene B_4 (LTB_4) and prostaglandin E_2 (PGE_2), which are implicated in the activation of inflammatory cells as well as their subsequent vascular adhesion, migration, and chemotaxis.[121] By contrast, anti-inflammatory properties of ω-3 PUFAs have been attributed to their downstream metabolites: resolvins and protectins.[122] These metabolites are exclusively derived from the ω-3

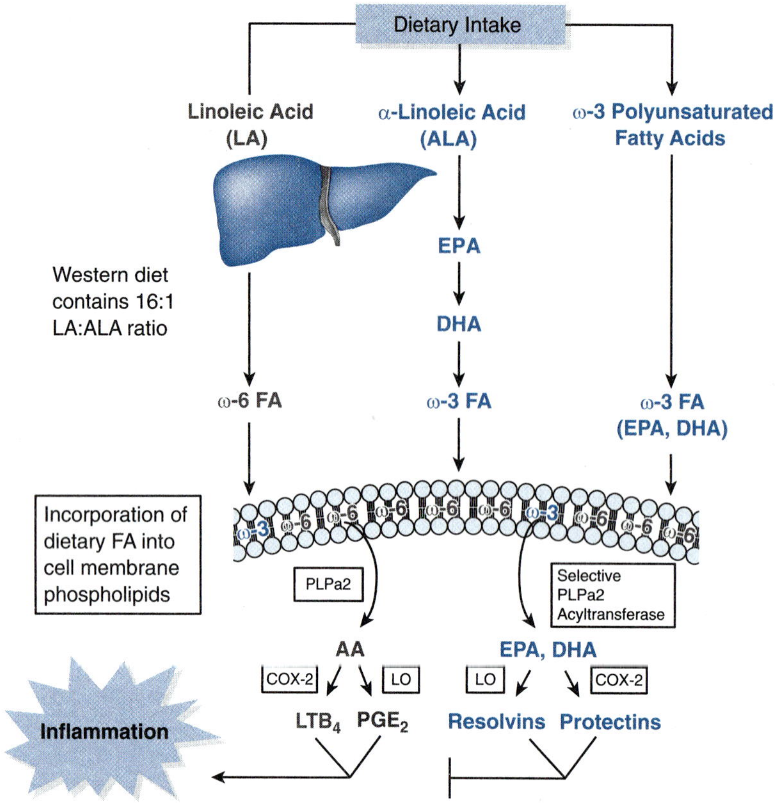

Figure 80.1 Dietary intake of essential fatty acids and their impact on inflammation. *AA*, arachidonic acid; *COX-2*, cyclo-oxygenase-2; *DHA*, docosahexaenoic acid; *EPA*, eicosapentaenoic acid; *FA*, fatty acid; *LO*, lipoxygenase; *LTB*$_4$, leukotriene B$_4$; *PGE*$_2$, prostaglandin E$_2$; *PLP*, phospholipase.

PUFAs eicosapentaenoic acid (EPA) and docosahexaenoic acid (DHA) (**Figure 80.1**).

Omega-3 PUFAs are found most commonly in fatty fishes, such as wild salmon, mackerel, striped bass, herring, sardines, and anchovies. Fish oil, a supplement rich in EPA and DHA, has been studied for the treatment of IBD. The Epanova Program in Crohn's study (EPIC) 1 and 2 were RCTs that found no benefit in the use of fish oil supplements in the maintenance of remission in CD.[123] A subsequent Cochrane review evaluating the role of fish oil for CD concluded that there might be a possible benefit; however, there was a high risk of bias and heterogeneity among the studies.[124] The benefits of fish oil supplements in the treatment of active UC is more promising.[125-127] Studies have found that fish oil reduced proinflammatory LTB$_4$,[126] improved clinical disease activity scores,[125] promoted healthy weight gain,[126] and allowed patients to reduce their prescription therapy of UC.[125] A recent study found that fish oil added to standard therapy decreased fecal calprotectin and maintained remission compared with placebo.[128] Larger RCTs are required to better study this as a therapeutic agent.

Curcumin

Curcumin is a polyphenol isolated from the plant *Curcuma longa* and has been used in ayurvedic medicine for many years due to its anti-inflammatory and antioxidant properties. Curcumin has been shown to have immunosuppressive as well as anticarcinogenic properties.[129,130] Its mechanism of action appears to be related to the downregulation of nuclear factor-kappa B, TNF-α, interferon gamma, IL-1β, IL-6, IL-12, and IL-17.[131-133] Mucosal barrier function is improved through the upregulation of tight junction proteins.[134] It has poor systemic absorption and acts locally within the intestines.[130] There has been one randomized, placebo-controlled trial of oral curcumin in CD patients that found that 360 mg/day of curcumin led to a significant improvement in clinical disease activity, clinical remission, and endoscopic disease activity at 12 weeks without any adverse events.[135] Multiple RCTs in UC have shown efficacy of curcumin in ameliorating clinical symptoms and serologic markers of inflammation.[136-138] Curcumin, when taken concomitantly with mesalamine, was found to improve clinical symptoms scores.[137] In patients with mild-to-moderate endoscopically active UC who received 450 mg of oral curcumin daily compared with placebo had improvement in disease activity scores and mucosal healing.[138] A systematic review of placebo-controlled RCTs found curcumin to be effective for maintaining remission when administered with standard IBD therapies.[139] There was a wide range of doses used in the studies, leaving some uncertainty on optimal dosing. Overall, curcumin appears to have a role in ameliorating inflammation in the colon and is a promising adjuvant therapy with a good safety profile.

Summary

Nutrition has an important bidirectional relationship with IBD. On one hand, disease activity and complications of IBD can compromise nutritional status. On the other hand, nutrition has been implicated as a modifiable factor in the pathogenesis of IBD and as a potential treatment. The role of PN is currently limited to the optimization of nutritional status, whereas exclusive EN has been recommended by European medical and nutritional societies as a first-line corticosteroid-sparing treatment for active CD in children. Due to the limitations of poor palatability with exclusive EN, partial EN has been investigated and found to be a viable alternative. Solid food diets have also been investigated with limited success thus far, although more studies are admittedly still needed. Dietary supplements are alternative options for nutritional therapy, although the evidence is still sparse. As summarized by Alastair et al., "Nutrition therapy in adults with IBD is probably both undervalued and underused, but the evidence base needs to be strengthened to confirm its efficacy, determine better those patients most likely to benefit, and optimize the regimens to be employed."[140]

Acknowledgments

The authors acknowledge Gerald W Dryden and Douglas L Seidner, the authors of the previous version of this chapter, for their contributions.

References

1. Abraham C, Cho JH. N Engl J Med 2009;361:2066–78.
2. Molodecky NA, Soon IS, Rabi DM, et al. Gastroenterol 2012;142: 46–54 e42; quiz e30.
3. Jostins L, Ripke S, Weersma RK, et al. Nature 2012;491:119–24.
4. Tysk C, Lindberg E, Jarnerot G, et al. Gut 1988;29:990–6.
5. Halfvarson J, Bresso F, D'Amato M, et al. Digest Liver Dis 2005;37:768–72.
6. Martinez C, Antolin M, Santos J, et al. Am J Gastroenterol 2008;103:643–8.
7. Hou JK, Abraham B, El-Serag H. Am J Gastroenterol 2011;106:563–73.
8. Sakamoto N, Kono S, Wakai K, et al. Inflamm Bowel Dis 2005;11:154–63.
9. Shoda R, Matsueda K, Yamato S, et al. Am J Clin Nutr 1996;63:741–5.
10. Ananthakrishnan AN, Khalili H, Konijeti GG, et al. Gut 2014;63:776–84.
11. Ananthakrishnan AN, Khalili H, Konijeti GG, et al. Gastroenterology 2013;145:970–7.
12. Jeffery NM, Newsholme EA, Calder PC. Prostaglandins Leukot Essent Fatty Acids 1997;57:149–60.
13. Wong JM, de Souza R, Kendall CW, et al. J Clin Gastroenterol 2006;40:235–43.
14. Willemsen LE, Koetsier MA, van Deventer SJ, et al. Gut 2003;52:1442–7.
15. Hugot JP, Alberti C, Berrebi D, et al. Lancet 2003;362:2012–5.
16. Gold SL, Rabinowitz LG, Manning L, et al. Inflamm Bowel Dis. 2022;29:423–429.
17. Yelencich E, Truong E, Widaman AM, et al. Clin Gastroenterol Hepatol 2022;20:1282–9 e1.
18. Braly K, Williamson N, Shaffer ML, et al. J Pediatr Gastroenterol Nutr 2017;65:533–8.
19. Gibson PR. J Gastroenterol Hepatol 2017;32(suppl 1):40–2.
20. Tighe MP, Cummings JR, Afzal NA. Curr Opin Clin Nutr Metab Care 2011;14:491–6.
21. Mallon DP, Suskind DL. Nutr Clin Pract 2010;25:335–9.
22. Vagianos K, Bector S, McConnell J, et al. J Parenter Enteral Nutr 2007;31:311–9.
23. Lichtenstein GR, Loftus EV, Isaacs KL, et al. Am J Gastroenterol 2018;113:481–517.
24. Rubin DT, Ananthakrishnan AN, Siegel CA, et al. Am J Gastroenterol 2019;114:384–413.
25. Greenberg GR, Fleming CR, Jeejeebhoy KN, et al. Gut 1988;29:1309–15.
26. Jones VA. Dig Dis Sci 1987;32:100S–7S.
27. Dickinson RJ, Ashton MG, Axon AT, et al. Gastroenterol 1980;79:1199–204.
28. Bischoff SC, Escher J, Hebuterne X, et al. Clin Nutr 2020;39: 632–53.
29. O'Morain C, Segal AW, Levi AJ. Br Med J (Clin Res Ed) 1984;288:1859–62.
30. Narula N, Dhillon A, Zhang D, et al. Cochrane Database Syst Rev 2018;4:CD000542.
31. Limketkai BN, Gordon M, Mutlu EA, et al. Inflamm Bowel Dis 2020;26:510–4.
32. Akobeng AK, Zhang D, Gordon M, et al. Cochrane Database Syst Rev 2018;8:CD005984.
33. Sahu P, Kedia S, Vuyyuru SK, et al. Aliment Pharmacol Ther 2021;53:568–76.
34. Matsui T, Sakurai T, Yao T. J Gastroenterol 2005;40(suppl 16): 25–31.
35. Watanabe O, Ando T, Ishiguro K, et al. J Gastroenterol Hepatol 2010;25(suppl 1):S134–7.
36. Sanderson IR, Udeen S, Davies PS, et al. Arch Dis Child 1987;62:123–7.
37. Levine A, Wine E, Assa A, et al. Gastroenterol 2019;157: 440–50 e8.
38. Limketkai BN, Iheozor-Ejiofor Z, Gjuladin-Hellon T, et al. Cochrane Database Syst Rev 2019;2:CD012839.
39. Levine A, Rhodes JM, Lindsay JO, et al. Clin Gastroenterol Hepatol 2020;18:1381–92.
40. Lewis JD, Sandler RS, Brotherton C, et al. Gastroenterol 2021;161:837–52 e9.
41. Orel R, Kamhi Trop T. World J Gastroenterol 2014;20:11505–24.
42. Hill C, Guarner F, Reid G, et al. Nat reviews. Gastroenterol & hepatol 11, 506–14.
43. Freedman SB, Schnadower D, Tarr PI. JAMA 2020;323:823–4.
44. Limketkai BN, Akobeng AK, Gordon M, et al. Cochrane Database Syst Rev 2020;7:CD006634.
45. Rolfe VE, Fortun PJ, Hawkey CJ, et al. Cochrane Data System Rev 2006;CD004826.
46. Tursi A, Brandimarte G, Giorgetti GM, et al. Medical Sci Monit 2004;10:PI126–31.
47. Tursi A, Brandimarte G, Papa A, et al. Am J Gastroenterol 2010;105:2218–27.
48. Bibiloni R, Fedorak RN, Tannock GW, et al. Am J Gastroenterol 2005;100:1539–46.
49. Sood A, Midha V, Makharia GK, et al. Clin Gastroenterol Hepatol 2009;7:1202–9, 9 e1.
50. Ng SC, Plamondon S, Kamm MA, et al. Inflam Bowel Dis 2010;16:1286–98.
51. Miele E, Pascarella F, Giannetti E, et al. Am J Gastroenterol 2009;104:437–43.
52. Venturi A, Gionchetti P, Rizzello F, et al. Aliment Pharmacol Thera 1999;13:1103–8.
53. Mardini HE, Grigorian AY. Inflam Bowel Dis 2014;20:1562–7.
54. Pardi DS, D'Haens G, Shen B, et al. Inflam Bowel Dis 2009;15: 1424–31.

55. Franko J, Raman S, Krishnan N, et al. J Am Coll Surg 2019;229:533–40 e1.
56. Panetta V, Bacchieri A, Papetti S, et al. Eur Rev Med Pharmacol Sci 2020;24:963–73.
57. Panetta V, Bacchieri A, Papetti S, et al. Eur Rev Med Pharmacol Sci 2021;25:6443.
58. Su GL, Ko CW, Bercik P, et al. Gastroenterol 2020;159: 697–705.
59. Suez J, Zmora N, Zilberman-Schapira G, et al. Cell 2018;174: 1406–23 e16.
60. Capurso L. J Clin Gastroenterol 2019;53(suppl 1):S1–S41.
61. Pant N, Marcotte H, Brussow H, et al. BMC Microbiol 2007; 7:86.
62. Li YT, Xu H, Ye JZ, et al. World J Gastroenterol 2019;25: 4999–5016.
63. Sawada J, Morita H, Tanaka A, et al. Clin Exp Allergy 2007;37: 296–303.
64. Karlsson M, Scherbak N, Reid G, et al. BMC Microbiol 2012; 12:15.
65. Marelli G, Papaleo E, Ferrari A. Eur Rev Med Pharmacol Sci 2004;8:87–95.
66. Horvath A, Dziechciarz P, Szajewska H. Aliment Pharmacol Ther 2011;33:1302–10.
67. Francavilla R, Miniello V, Magista AM, et al. Pediatr 2010;126: e1445–52.
68. Gawronska A, Dziechciarz P, Horvath A, et al. Aliment Pharmacol Ther 2007;25:177–84.
69. Zocco MA, dal Verme LZ, Cremonini F, et al. Aliment Pharmacol Ther 2006;23:1567–74.
70. Malin M, Suomalainen H, Saxelin M, et al. Ann Nutr Metab 1996;40:137–45.
71. Gupta P, Andrew H, Kirschner BS, et al. J Pediatr Gastroenterol Nutr 2000;31:453–7.
72. Guandalini S. Dig Liver Dis 2002;34(suppl 2):S63–5.
73. Prantera C, Scribano ML, Falasco G, et al. Gut 2002;51:405–9.
74. Bousvaros A, Guandalini S, Baldassano RN, et al. Inflam Bowel Dis 2005;11:833–9.
75. Freedman SB, Williamson-Urquhart S, Farion KJ, et al. N Engl J Med 2018;379:2015–26.
76. Kirchhelle A, Fruhwein N, Toburen D. Fortschr Med 1996;114: 136–40.
77. James JS. AIDS Treat News 1995;1–4.
78. Micklefield G. MMW Fortschr Med 2014;156(suppl 1):18–22.
79. McFarland LV, Surawicz CM, Greenberg RN, et al. JAMA 1994;271:1913–8.
80. Soyturk M, Saygili SM, Baskin H, et al. World J Gastroenterol 2012;18:6452–60; discussion p 9.
81. Thomas S, Metzke D, Schmitz J, et al. Am J Physiol Gastrointest Liver Physiol 2011;301:G1083–92.
82. Lee SK, Kim HJ, Chi SG. Korean J Gastroenterol 2010;55:33–45.
83. Lee SK, Kim YW, Chi SG, et al. Digest Dis Sci 2009;54:255–63.
84. Guslandi M, Giollo P, Testoni PA. Eur J Gastroenterol Hepatol 2003;15:697–8.
85. Guslandi M. J Clin Gastroenterol 2010;44:385.
86. Plein K, Hotz J. Z Gastroenterol 1993;31:129–34.
87. Bourreille A, Cadiot G, Le Dreau G, et al. Clin Gastroenterol Hepatol 2013;11:982–7.
88. Guslandi M, Mezzi G, Sorghi M, et al. Dig Dis Sci 2000;45: 1462–4.
89. Garcia Vilela E, De Lourdes De Abreu Ferrari M, Oswaldo Da Gama Torres H, et al. Scand J Gastroenterol 2008;43:842–8.
90. Oh GM, Moon W, Seo KI, et al. Korean J Gastroenterol 2020;76: 314–21.
91. Hallert C, Kaldma M, Petersson BG. Scand J Gastroenterol 1991;26: 747–50.
92. Fernandez-Banares F, Hinojosa J, Sanchez-Lombrana JL, et al. Am J Gastroenterol 1999;94:427–33.
93. Casellas F, Borruel N, Torrejon A, et al. Aliment Pharmacol Ther 2007;25:1061–7.
94. Abdelouhab K, Rafa H, Toumi R, et al. Immunopharmacol Immunotoxicol 2012;34:590–7.
95. Welters CF, Heineman E, Thunnissen FB, et al. Dis Colon Rectum 2002;45:621–7.
96. Lindsay JO, Whelan K, Stagg AJ, et al. Gut 2006;55:348–55.
97. Johnson-Henry KC, Pinnell LJ, Waskow AM, et al. J Nutr 2014;144:1725–33.
98. Winkler J, Butler R, Symonds E. Dig Dis Sci 2007;52:52–8.
99. Koleva PT, Valcheva RS, Sun X, et al. Br J Nutr 2012;108: 1633–43.
100. Araki Y, Fujiyama Y, Andoh A, et al. Scand J Gastroenterol 2000;35:1060–7.
101. Araki Y, Andoh A, Koyama S, et al. Biosci Biotechnol Biochem 2000;64:1794–800.
102. Mitsuyama K, Saiki T, Kanauchi O, et al. Aliment Pharmacol Ther 1998;12:1225–30.
103. Kanauchi O, Suga T, Tochihara M, et al. J Gastroenterol 2002;37(suppl 14):67–72.
104. Kanauchi O, Mitsuyama K, Homma T, et al. Int J Mol Med 2003;12:701–4.
105. Hanai H, Kanauchi O, Mitsuyama K, et al. Int J Mol Med 2004;13:643–7.
106. Faghfoori Z, Shakerhosseini R, Navai L, et al. Health Promot Perspect 2014;4:116–21.
107. Koh SJ, Kim JS. J Gastroenterol Hepatol 2011;26:1219–20.
108. Healey GR, Celiberto LS, Lee SM, et al. Nutrients 2020; 12:3204.
109. Arai S, Morinaga Y, Yoshikawa T, et al. Biosci Biotechnol Biochem 2002;66:2017–29.
110. Bomba A, Nemcova R, Gancarcikova S, et al. Br J Nutr 2002;88(suppl 1):S95–9.
111. Fujimori S, Gudis K, Mitsui K, et al. Nutr 2009;25:520–5.
112. Furrie E, Macfarlane S, Kennedy A, et al. Gut 2005;54:242–9.
113. Ishikawa H, Matsumoto S, Ohashi Y, et al. Digestion 2011;84:128–33.
114. Fujimori S, Tatsuguchi A, Gudis K, et al. J Gastroenterol Hepatol 2007;22:1199–204.
115. Steed H, Macfarlane GT, Blackett KL, et al. Aliment Pharmacol Ther 2010;32:872–83.
116. Chermesh I, Tamir A, Reshef R, et al. Dig Dis Sci 2007;52: 385–9.
117. Varnalidis I, Ioannidis O, Papadopoulou A, et al. Cir Cir 2020;88:410–9.
118. Martyniak A, Medynska-Przeczek A, Wedrychowicz A, et al. Biomolecules 2021;11:1903.
119. Akutko K, Stawarski A. J Clin Med 2021;10:2466.
120. Simopoulos AP. World Rev Nutr Diet 2003;92:1–22.
121. Chen M, Lam BK, Kanaoka Y, et al. J Exp Med 2006;203: 837–42.
122. Serhan CN, Chiang N, Van Dyke TE. Nat Rev Immunol 2008;8:349–61.
123. Feagan BG, Sandborn WJ, Mittmann U, et al. JAMA 2008;299: 1690–7.
124. Lev-Tzion R, Griffiths AM, Leder O, et al. Cochrane Data System Rev 2014;CD006320.
125. Aslan A, Triadafilopoulos G. Am J Gastroenterol 1992;87: 432–7.
126. Stenson WF, Cort D, Rodgers J, et al. Ann Intern Med 1992;116:609–14.
127. Barbosa DS, Cecchini R, El Kadri MZ, et al. Nutr 2003;19: 837–42.

128. Scaioli E, Sartini A, Bellanova M, et al. Clin Gastroenterol Hepatol 2018;16:1268–75 e2.
129. Jobin C, Bradham CA, Russo MP, et al. J Immunol 1999;163: 3474–83.
130. Midura-Kiela MT, Radhakrishnan VM, Larmonier CB, et al. Am J Physiol 2012;302:G85–96.
131. Lanjekar AB, Bhowate RR, Bakhle S, et al. J Contemp Dent Pract 2020;21:83–90.
132. Asadi A, Yaghobi Nezhad D, Rafie Javazm A, et al. Iran J Allergy Asthma Immunol 2020;19:84–93.
133. Stohs SJ, Chen O, Ray SD, et al. Molecules 2020;25:1397.
134. Wang N, Wang G, Hao J, et al. Dig Dis Sci 2012;57:1792–801.
135. Sugimoto K, Ikeya K, Bamba S, et al. J Crohn's Colitis 2020;14: 1693–701.
136. Sadeghi N, Mansoori A, Shayesteh A, et al. Phytother Res 2020;34:1123–33.
137. Masoodi M, Mahdiabadi MA, Mokhtare M, et al. J Cell Biochem 2018;119:9552–9.
138. Kedia S, Bhatia V, Thareja S, et al. World J Gastro Pharmacol Thera 2017;8:147–54.
139. Coelho MR, Romi MD, Ferreira D, et al. Nutrients 2020;12: 2296.
140. Alastair F, Emma G, Emma P. J Parenter Enteral Nutr 2011;35: 571–80.

CHAPTER 81

Celiac Disease

David Hakimian and Carol E. Semrad

CHAPTER OUTLINE

Abbreviations

AGA, antigliadin antibody; **DGP**, deamidated gliadin peptide; **EMA**, endomysial antibody; **HLA**, human leukocyte antigen; **Ig**, immunoglobulin; **IL**, interleukin; **tTG**, tissue transglutaminase; **RCT**, randomized control trial; **CD**, celiac disease; **IBS**, irritable bowel syndrome; **SIBO**, small intestine bowel overgrowth

Introduction

Coeliac affection was first described as a diarrheal illness related to diet by Gee in 1888.[1] In 1950, Dicke[2,3] reported that wheat was the cause. He observed that, during the grain shortage of World War II, children with celiac disease miraculously improved in weight and stature on a potato diet, only to relapse when bread was reintroduced after the war. In the intervening years, many advances have been made to characterize the component of wheat responsible for triggering celiac disease and the intestinal immune response. Based on this work, celiac disease is defined as a T-cell–mediated inflammatory disease predominantly of the small bowel, triggered by gluten in the diet from wheat, rye, and barley in persons who are genetically susceptible. The consequences of chronic small bowel inflammation are mucosal atrophy and malabsorption of macronutrients, vitamins, and minerals. This chapter focuses on the clinical presentation and pathogenesis of celiac disease, its nutritional consequences, and the gluten-free diet.

Pathogenesis

Epidemiology

The prevalence of celiac disease in the United States is 1% and has been increasing over time, perhaps related to environmental factors.[4] A higher prevalence has been reported in Sweden, Punjab India, and the West African Sahara.[5] Celiac disease remains underdiagnosed due to unrecognized variability of its clinical presentation and asymptomatic disease. Celiac disease most commonly manifests at age 1 to 2 years, when gluten is first introduced into the diet, and in young adult life, but it can become evident at any age.[6] It is more common in females than males, with a ratio to 2:1.[7]

Pathobiology

The development of celiac disease requires the ingestion of gluten from wheat (includes spelt, triticale, semolina, and kamut), rye (including malt), or barley (**Figure 81.1**) and a susceptible

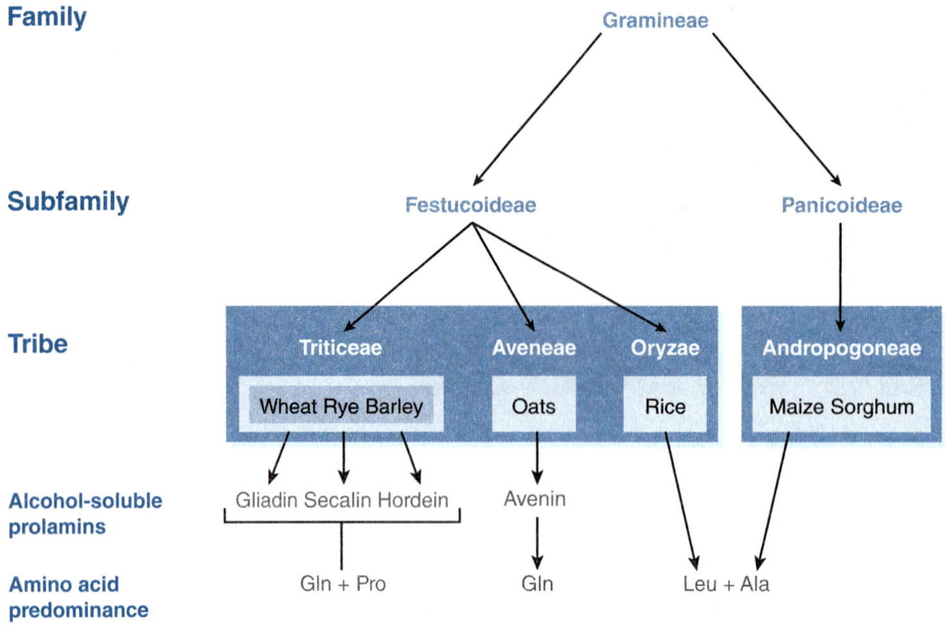

Figure 81.1 Family of grasses. *Ala*, alanine; *Gln*, glutamine; *Leu*, leucine; *Pro*, proline.

genetic background.[5,8] Other environmental risk factors include reovirus[9] and other infections, the amount of gluten ingested, and the microbiome.[5] Retinoic acid may act as a coadjuvant to IL15 to break tolerance to gliadin peptides in individuals genetically predisposed to celiac disease.[10]

Gluten is the protein in wheat that provides the elastic quality highly desirable for baked goods. Repeat stretches of the amino acids proline and glutamine are characteristic of the gluten protein structure. The alcohol-soluble fraction of wheat gluten, gliadin, and its related prolamins from rye and barley are toxic in celiac disease.[11,12] Many gliadin peptides (also referred to as gluten peptides) have been reported to trigger inflammation in celiac disease by in vivo and in vitro studies. One such peptide, a 33-mer peptide from α_2-gliadin[13] contains three previously reported toxic epitopes and cannot be digested by human pancreatic or intestinal proteases, thus rendering it available to trigger an immune response.

A genetic predisposition for celiac disease is suggested by the high concordance of celiac disease in identical twins[14] and the increased risk in first-degree relatives (10%–15%) compared with the general population.[15] The human leukocyte antigen HLA-DQ2 (A1*05 or *02/B1*02) and HLA-DQ8 (A1*03:01/B1*03:02) alleles have the strongest genetic association with celiac disease.[16] However, not all individuals with these risk alleles develop the disease. Non-HLA genes have been identified in genome-wide studies but individually these have small effects.[17] Genetic associations have also been reported between celiac disease and diabetes type 1[18] and Crohn's disease.[19]

Gliadin triggers both adaptive and innate immune responses in celiac intestinal mucosa (**Figure 81.2**). How gluten peptides cross the intestinal epithelium is unknown. Once in the lamina propria, gluten peptides are rendered more negatively charged by conversion of glutamine to glutamic acid by tissue transglutaminase 2 (tTG orTG2) and bind to positively charged pockets of HLA-DQ2 and DQ8 molecules expressed on the outer membrane of antigen-presenting cells (Figure 81.2).[6,20] These bound antigens activate specific T cells that result in release of interferon-γ, IL12, and tumor necrosis factor (TNF). Gliadin is also capable of activating the innate immune system by reprogramming intestinal epithelial cytotoxic T lymphocytes into natural killer–like cells.[21] Release of the cytokine interleukin-15 (IL-15) from the epithelium participates in intestinal inflammation.

The main pathophysiologic consequence of chronic intestinal inflammation is villous atrophy with decreased surface area for nutrient absorption. Chronic inflammation may also downregulate nutrient transport proteins in the intestinal epithelium.[22] Pancreatic insufficiency caused by decreased release of secretin and cholecystokinin from atrophied duodenal epithelium and bacterial overgrowth may contribute to nutrient malabsorption.

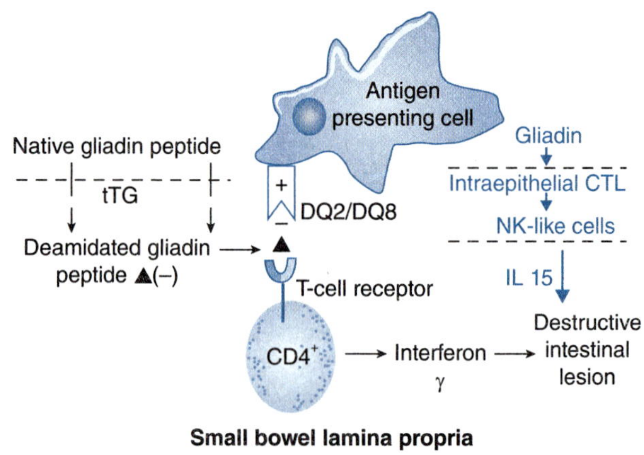

Figure 81.2 Pathogenesis of celiac disease. Poorly digestible gluten peptides cross the intestinal epithelium and reach the lamina propria. Native peptides are deamidated by TG2 and bind to the positively charged human leukocyte antigen (HLA)-DQ2 or DQ8 molecules expressed on the outer membrane of antigen-presenting cells and are recognized by $CD4^+$ T cells. T-cell activation results in release of interferon-γ and destruction of the epithelium. Gliadin also acts directly on the intestinal epithelium to stimulate release of interleukin-15 (IL-15) that can synergize the immune response in the lamina propria. *CTL*, cytotoxic lymphocyte; *NK*, natural killer.

Reproduced from Green PH, Jabri B. Lancet 2003;362:383–91. doi:10.1016/S0140-6736(03)14027-5

Clinical Presentation

The classic clinical presentation of celiac disease is diarrhea, gas and bloating, and weight loss.[23] Since the 1980s, there has been a shift to a nonclassical clinical presentation of celiac disease.[24] Most patients now present with nonspecific or minor digestive complaints, constipation, iron deficiency anemia, or bone disease.[5] Other clinical presentations include fatigue, dyspepsia, unexplained hepatitis, neuropathy, ataxia, dental hypoplasia, and infertility. Individuals with celiac disease may be obese on diagnosis.[25] Virtually all individuals with the burning, itching rash of dermatitis herpetiformis have celiac disease. Approximately 20% of those diagnosed with celiac disease are asymptomatic and come to detection by screening as a first-degree relative.[24]

High-risk groups for the development of celiac disease include first-degree relatives, type 1 diabetes, autoimmune thyroid disease, and Down, William, and Turner syndromes.[26] Other autoimmune diseases associated with celiac disease include autoimmune hepatitis, primary biliary cholangitis, Sjogren syndrome, and systemic lupus erythematosus.[5] Microscopic colitis is prevalent in celiac disease with a pooled event rate of 6.7% (95% CI 4.4–10.0%; $P<0.001$).[27]

Diagnosis

Antibody Testing

Over the years, antibodies have been identified that are associated with celiac disease, most are of the immunoglobulin A (IgA) class.[4] The tissue transglutaminase (tTG) IgA antibody has high sensitivity and specificity (>95%) for active celiac disease in adults, children, and infants eating gluten. It is the best first-line test for screening and diagnosis of active celiac disease when combined with a total IgA level.[5] The endomysial antibody (EMA) recognizes tTG as its autoantigen and has high specificity for celiac disease but is operator dependent and less suited as a screening test.[4] Deamidated gliadin peptide (DGP) IgA and IgG antibodies recognize recombinant toxic gluten peptides deamidated by tTG but have lower sensitivity and specificity compared with the tTG/EMA antibodies. The DGP and/or TTG IgG antibody test is useful in the diagnosis of celiac disease in those with complete IgA deficiency.[28] A false-positive TTG IgA can be of low titer in other inflammatory conditions, such as duodenal Crohn's disease and primary biliary cirrhosis.

Small Intestinal Biopsy

Intestinal inflammation related to celiac disease can vary from mild inflammation (normal villous architecture with an increase in intraepithelial lymphocytes at the villous tips) to varying degrees of villous atrophy with crypt hyperplasia, increased intraepithelial lymphocytes, and increased lymphocytes and plasma cells in the lamina propria found in most patients.[24] Multiple duodenal biopsies obtained when on a gluten-containing diet are recommended to ensure adequate tissue for diagnosis.[5]

These inflammatory changes are not specific for celiac disease and may be found in other conditions, such as tropical sprue, drug-induced enteropathy (particularly olmesartan), CVID, and infections.[24] With more sensitive and specific antibody testing, there has been a paradigm change in the need for duodenal biopsy to diagnose celiac disease. TTG IgA antibody levels that are >10 times the upper limit of normal have been shown to have a high correlation with villous atrophy on duodenal biopsy in children. This allows no biopsy approach for the diagnosis of celiac disease when the TTG IgA antibody is >10 times the upper limit of normal and confirmed on a separate day with a positive EMA test.[29] This approach appears to apply to adults as well. In a recent study in adult low- and high-risk cohorts, TTG IgA levels ≥10 times the upper limit of normal (ULN) had a high specificity and positive predictive value (approaching 100%) in identifying villous atrophy and celiac disease.[30] There were no missed diagnoses of other diseases using a no-biopsy approach with TTG IgA levels >10 ULN.[31] Duodenal biopsy is required for the diagnosis of celiac disease in all other patients (tTG IgA <10 ULN, positive DGP IgA or IgG antibody, IgA deficiency, or red flags in clinical presentation).

Genetic Testing

Virtually all individuals with celiac disease carry the HLA-DQ2 or DQ8 risk alleles,[32] but 40% of the normal population carry these risk alleles and never get the disease and, therefore, it is not a diagnostic test. Genetic testing is useful to exclude the diagnosis of celiac disease in individuals already following a gluten-free diet, to assess risk in first-degree relatives, and to help with diagnosis when duodenal biopsy findings are equivocal.[33]

Nutritional Consequences

Celiac disease always affects the duodenum or proximal jejunum, but it can involve the entire small intestine.[5] The nutrients malabsorbed and the clinical consequences depend on the site and extent of villous atrophy. Disease limited to the duodenum and proximal jejunum results in iron, folate, and calcium malabsorption because high-affinity transport mechanisms for their uptake are expressed only at these sites.[34] The clinical consequences are iron deficiency or macrocytic anemia and bone mass loss.[35] When celiac disease involves the entire small intestine, all macronutrients (carbohydrate, fat, protein) and vitamins and minerals are malabsorbed. The clinical consequences are diarrhea, gas and bloating caused by carbohydrate malabsorption, edema caused by hypoproteinemia, skeletal muscle and body weight loss, and manifestations of vitamin and mineral deficiencies (**Table 81.1**). Vitamin B_{12} deficiency occurs in 10% to 40% of individuals with celiac disease.[35-37] Chronic diarrhea may cause zinc deficiency that slows mucosal healing. Patients with severe malabsorption may develop copper deficiency and neurologic abnormalities.[38]

The most common findings in celiac disease are iron deficiency anemia and metabolic bone loss. In premenopausal women, iron deficiency anemia may be misattributed to menstrual losses or childbirth, thus delaying diagnosis. Celiac disease should be suspected when anemia in women or men does not correct with oral iron therapy or is associated with gastrointestinal symptoms.[39]

Low bone mass (osteopenia, osteoporosis) is found in up to 70% of adults with celiac disease[40-42] and in 16% of children.[43] Osteomalacia is less common. The loss of bone mass is due to chronic inflammation that promotes bone breakdown and secondary hyperparathyroidism due to low calcium and vitamin D absorption. A case report of osteoporosis in a patient with celiac disease due to neutralizing autoantibodies against osteoprotegerin showed bone breakdown.[44] Measurement of the radius forearm in men with celiac disease correlates with the degree of villous

Table 81.1 Nutritional Problems Related to Celiac Disease

Nutrient Deficiencies at Diagnosis	Clinical Manifestations	Potential Complications Associated with a Strict Gluten-Free Diet
Common		Obesity
Iron	Anemia	Hyperlipidemia
Folate	Cheilitis, anemia	Low fiber intake
Calcium Vitamin D	Tetany, bone loss Bone loss	Inadequate intake of folate, thiamin, niacin, riboflavin
Severe disease		
Vitamin B_{12}	Anemia, neuropathy, glossitis, ataxia	
Zinc	Dermatitis, dysgeusia	
Potassium	Weakness, arrhythmia	
Magnesium	Tetany, paresthesias	
Copper	Ataxia, paresthesias	
Vitamin A	Night blindness	
Vitamin E	Ataxia, neuropathy	
Vitamin K	Ecchymoses	
Protein	Edema, muscle wasting	
Carbohydrate	Gas/bloating	

atrophy and suggests a higher prevalence of osteoporosis in men than previously reported.[45]

Adults with celiac disease have an approximately twofold higher risk of bone fracture than the general population. Most fractures occurred prior to diagnosis or in patients nonadherent to a GFD.[46] High impact sports must be avoided until fracture risk has been assessed.

Nutritional Evaluation

Initial evaluation for patients with celiac disease includes the following: anthropometric measurements, including height, weight, muscle mass, and volume status; blood tests, including a complete blood count, electrolyte determinations, liver tests, albumin, and calcium concentrations; and a bone mineral density study in adults. Additional testing is guided by the severity of disease or suspicion of particular vitamin or mineral deficiencies (Table 81.1). Total iron-binding capacity, and ferritin are obtained in patients with microcytic anemia; folate and vitamin B_{12} are measured in patients with normocytic or macrocytic anemia. In patients with evidence of bone loss, tests of 25-OH vitamin D, phosphorus, magnesium, and parathyroid hormone are warranted.[5,24] In patients with severe diarrhea and weight loss, testing for all vitamins and minerals should be performed to include fat-soluble vitamins (D, A, K, E), folate, vitamin B_{12}, zinc, and copper.[47]

Management

The only current treatment for celiac disease is a strict gluten-free diet. A secure diagnosis of celiac disease must be made before the diet is started. A gluten-free diet is complex and costly and it may worsen quality of life, particularly in asymptomatic individuals.[48] Once a diagnosis of celiac disease is made, six key elements (CELIAC) for management are recommended: consultation with a knowledgeable dietitian, education on the disease, a lifelong gluten-free diet, identification of nutritional deficiencies, access to a reputable support group, and continuous follow-up.[49]

Gluten-Free Diet

A gluten-free diet excludes the grains wheat, barley, and rye (see Figure 81.1 and **Table 81.2**). Rice and corn grains are not toxic unless they are contaminated by wheat.[50] Wheat starch (wheat carbohydrate that can bind to gluten) when tested at 20 ppm has been reported to be safe in individuals with celiac disease.[51,52]

Uncontaminated oats are tolerated by most patients with celiac disease and provide food choices along with soluble fiber, polyunsaturated oil, B vitamins, iron, and thiamine.[50-52] There are rare case reports of celiac patients with a specific T-cell response to avenin in oats[53] and intestinal inflammation on challenge with uncontaminated oats.[54] Commercially available oats have variable contamination with wheat.[55] Only gluten-free oats are recommended. Introduction of oats in a gluten-free diet may cause

Table 81.2 Gluten-Free Diet

Grains/Starches Allowed	Grains Not Allowed
Rice (brown, white, wild)	Wheat (including bran germ)
Corn	Spelt
Quinoa	Semolina
Amaranth	Durum
Legumes	Rye
Millet	Barley
Sorghum	Triticale
Arrowroot	Kamut
Buckwheat/kasha	Bulgur
Indian rice grass	Couscous
Tef/teff	Einkorn wheat
Potato	Emmer wheat
Legumes	Farro
Dried beans	Farina

Table 81.3 Dose-Dependent Effect of Gliadin on the Small Bowel

Gliadin Dose (mg)	Symptoms	Intestinal Permeability	Intestinal Biopsy Findings
10[a]	No	ND	Normal
25	Yes	ND	Minimal changes
100	Yes	Normal	Minimal changes
500	Yes	Increased	More pronounced changes

ND, not done.

[a]Equivalent to less than ⅛ teaspoon flour.

Data from Pietzak MM. Gastroenterol 2005;128:S135–S41. doi:10.1053/j.gastro.2005.02.025; Fabiani E, Catassi C, Villari A, et al. Acta Paediatr 1996;85:65–7. doi:10.1111/j.1651-2227.1996.tb14256.x and Olsson C, Hornell A, Ivarsson A, et al. J Human Nutr Diet 2008;21:359–67. doi:10.1111/j.1365-277x.2008.00867.x.[61-63]

abdominal gas and bloating because of increased fiber in the diet that may be confused with activation of celiac disease. Therefore, oats should be gradually introduced into the GFD and patients monitored for symptoms.[56]

The dose of gliadin capable of triggering intestinal inflammation in human studies has been reported to be between 10 and 25 mg (20 to 50 mg gluten) (**Table 81.3**).[57-59] Rarely, patients with celiac disease develop symptoms with 10 mg of gluten in the diet.[58] A dose of 20 mg of gluten correlates with less than one-eighth of a teaspoon of flour; and it is easy to exceed the toxic limit. Foods labeled *gluten free* have been federally regulated since 2014. The Codex Alimentarius Commission has adopted gluten-free standards of fewer than 20 ppm gluten in food.[60] Education on the gluten-free diet requires dietary evaluation and teaching by a knowledgeable registered dietitian, as well as referral to reliable Internet sites and a support group. Individuals must learn to read label ingredients, identify sources of hidden gluten, and avoid cross contamination when preparing gluten-free foods and eating out. Adherence to a gluten-free diet is highest in those diagnosed as children[61] and is most challenging in teenagers.[62,63]

A gluten-free diet poses a risk for weight gain due to improved intestinal absorption and the ingestion of carbohydrate-dense, gluten-free processed foods (see Table 81.1). Smaller portion size is recommended, as is increased exercise. Alternative grains (see Table 81.2), as well as fruits and vegetables, are ideal sources of fiber and micronutrients. Hyperlipidemia may result from increased absorption of cholesterol, but may be beneficial.[64]

Gluten Contamination

A gluten-free diet is difficult to follow because gluten is ubiquitous in the Western diet and may be present in medications, over-the-counter drugs, and other items such as communion wafers.[65] Manufacturers are not required to list nonactive ingredients such as wheat, rye, or barley on labels of drugs, supplements or hygiene products. Personal products such as toothpaste, mouthwash, and lipstick were examined for gluten and most contained <20 ppm.[66]

Cross contamination may occur accidently during gluten-free meal preparation. Traditionally, a dedicated gluten-free environment (separate toasters, colanders, cutting boards, etc.) or separate GF kitchens were used to reduce the risk of gluten contamination from shared equipment. Recent studies suggest that gluten cross contamination does not occur with shared knives, wooden spoons, toasters, and fryers.[67,68] Shared jars of sandwich spreads such as mayonnaise and peanut butter were found to be a source of gluten cross contamination with more than 20 mg/kg of gluten detected in each sample from a shared jar. Cooking gluten-free pasta with shared water showed 100% contamination with gluten. Contamination was eliminated when rinsing the gluten-free pasta with cold water after cooking in shared water and/or washing the cooking pot with soap and water and then cooking gluten-free pasta in water.[68]

Another substantial risk for gluten cross contamination is eating outside of the home. Measurement of gluten contamination in restaurants showed that 32% of foods listed as GF tested positive for gluten (>20 ppm), particularly pasta and pizza served during dinner.[69]

Quality of life

Patients with gastrointestinal symptoms report an improved quality of life on a gluten-free diet[70] but there is a significant burden in social living. Surveys of patients with celiac disease show that 81% avoid eating in restaurants, 38% avoid travel, and 94% bring GF food from home when traveling.[71,72] Eating disorders are increased in those with celiac disease.[73] A recent meta-analysis reported that quality of life is decreased in adults with celiac disease with or without adherence to a GFD with the need to assess psychosocial functioning.[74]

Gluten-free products are relatively expensive, with an average cost 183% higher than gluten-containing products.[75] This is a burden to persons of lower socioeconomic status and may result in worse related health and higher healthcare costs.[76]

Incomplete Response to GFD

Up to 30% of those with celiac disease on a GFD have incomplete resolution of symptoms and/or persistent villous atrophy, or positive antibodies.[77,78] The most common cause of failure to improve on a gluten-free diet is ongoing occult or intentional gluten ingestion.

Some patients are exquisitely sensitive to gluten, with reports of as little as 0.015 mg/day to cause clinical effect.[79] This low threshold for toxicity is difficult to achieve without a naturally gluten-free diet, which includes only fresh, unprocessed foods. Food must be prepared under gluten-free conditions, without additives other than plain salt, olive oil, and fresh herbs. Such patients should follow this diet for at least 3 to 4 weeks and, if gastrointestinal symptoms resolve, reintroduction of GF products should be started under the guidance of a registered dietitian, with monitoring for symptoms and possible nutrient deficiencies.

When gastrointestinal symptoms persist despite a stricter GF diet with negative serology and normal biopsy, other conditions should be considered. If gas/bloating, consider constipation, lactose or fructose intolerance, IBS or small intestinal bacterial overgrowth. If diarrhea, consider microscopic colitis or IBD.

Monitoring

Most patients with celiac disease improve on a strict gluten-free diet within the first 2 weeks. Vitamin and mineral deficiencies, including iron deficiency and bone mass, often improve on a

Table 81.4 Long-Term Monitoring of the Patient with Celiac Disease

Physician	Dietitian
Serial body weight (growth parameters in children)	Serial body weight
Specific blood micronutrient tests	Healthy diet instruction
Antibody tests (tTG IgA antibody; DGP IgG antibody if complete IgA deficiency)	Adherence to gluten-free diet
Blood lipid panel	Appropriate exercise
Serial bone density in adults	
Repeat duodenal biopsy if clinically indicated	

DGP, deamidated gliadin peptide; *Ig*, immunoglobulin A; *tTG*, tissue transglutaminase.

gluten-free diet alone.[42,80] More than 80% will have improvement in bone mineral density within 1 year on a GFD.[46,81] Children may reach peak bone mass if celiac disease is detected and treated early in life before puberty.[42] Treatment with medications may be required in patients with severe osteoporosis or older patients with celiac disease. Caution should be used with bisphosphonate treatment early in disease or in patients with persistent inflammation because intestinal calcium absorption may not be adequate to maintain normal blood calcium concentration.[82]

In patients with malabsorptive diarrhea, poor adherence to a gluten-free diet, or an unbalanced diet, a multiple vitamin with trace elements and minerals, and a separate calcium supplement with vitamin D, are recommended. The calcium with vitamin D supplement is also recommended in patients with lactose intolerance. Women of childbearing age require supplementation of folic acid (800 µg) daily to limit the risk for neural tube defects.

Individuals with celiac disease are monitored by a physician and dietitian for adherence to the gluten-free diet, nutritional status, and complications (**Table 81.4**). Recommendations are for periodic follow-up (3–6 months) in the first year and then every 1 to 2 years to monitor for adherence and adequacy of the GFD and resolution of symptoms and signs. Certain vitamin deficiencies may develop because gluten-free products are particularly low in iron, folate, and calcium.[83] Improvement in symptoms and conversion of antibodies to negative is correlated with improvement in histologic features. However, in one study, complete resolution of intestinal inflammation was identified in only 65% of adult patients with celiac disease at 2 years and 85% at 5 years on a gluten-free diet.[84] A uniform dietary survey to assess adherence may be useful for patient follow-up.[85] Commercial kits are now available to detect gluten in food[86] and gluten peptides in stool and urine.[87,88] However, gluten detection devices for monitoring are not recommended at present due to the high rate of false-positive and false-negative results.[86,89]

Complications

Celiac disease is associated with other autoimmune diseases such as type 1 diabetes and thyroiditis.[90] There is an increased risk for bone fractures and lymphoproliferative malignancy in those with persistent villous atrophy on a GFD. Individuals with celiac disease have a small increased mortality risk compared with the general population[91] and a small increase in cancer risk, including non-Hodgkins Lymphoma (especially T-cell), small bowel adenocarcinoma,[92] and hepatobiliary and pancreas cancers. The cancer risk is highest after age 40 years and within the first year of diagnosis.[93]

Refractory celiac disease is defined as persistent symptoms of malabsorption with ongoing villous atrophy despite a strict gluten-free diet for 12 months or after initially responding to a gluten-free diet.[94] Severe malabsorptive diarrhea and weight loss are characteristic features. Autoimmune enteropathy, combined variable immunodeficiency, and secondary hypogammaglobulinemia have similar clinical presentations and may be misdiagnosed as refractory celiac disease. Treatment often requires nutrition support and steroids. Special marker studies may help identify a subset of patients with a T-cell monoclonal population or receptor gamma gene rearrangements that portend a poor prognosis. These patients require referral to a tertiary hospital, as they are often unresponsive to steroids, are at high risk for enteropathy-associated T-cell lymphoma, and often require parenteral nutrition for survival.

Future Therapies

Prevention of celiac disease may be possible by identifying and intervening on environmental triggers such as infection and the microbiome in those at high risk. The GFD is highly effective in the treatment of celiac disease but difficult to follow due to the ubiquitous use of wheat in the diet of western cultures. Reducing immunogenic gluten by creating new lines of wheat using gene-editing technologies may improve quality of life as an alternative to GF products. Oral enzyme therapy using prolyl endopeptidases from micro-organisms[6,95] or germinating barley are capable of cleaving the gliadin 33-mer to nontoxic fragments to limit gluten contamination. Monoclonal antibodies to IL 15, TG 2 inhibitors, or the restoration of oral tolerance to gluten are under investigation to block the inflammatory reaction triggered by gluten. Development of novel therapies may be helpful to limit the effects of gluten contamination, treat poorly responsive or refractory celiac disease, or promote tolerance to allow the ingestion of gluten in the future.

References

1. Gee SJ. St Bartholomew's Hospital Report 1888;24:17–20.
2. Dicke WK. Thesis (in Dutch). The Netherlands: University of Utrecht. 1950.
3. Dicke WK, Weijers H, Kamer JvD. Acta Paediatr 1953;42:34–42.
4. Lebwohl B, Rubio-Tapia A. Gastroenterol 2021;160:63–75.
5. Catassi C, Verdu EF, Bai JC, Lionetti E. Lancet 2022;399:2413-26.
6. Schuppan D, Junker Y, Barisani D. Gastroenterol 2009;137: 1912–33.
7. Jansson-Knodell CL, Hujoel IA, West CP, et al. Clin Gastroenterol Hepatol 2019;17:1954-68. e13

8. Rubio-Tapia A, Kyle RA, Kaplan EL, et al. Gastroenterol 2009;137:88–93.
9. Bouziat R, Hinterleitner R, Brown JJ, et al. Science 2017;356: 44–50.
10. DePaolo R, Abadie V, Tang F, et al. Nature 2011;471:220–4.
11. Vader LW, Stepniak DT, Bunnik EM, et al. Gastroenterol 2003;125:1105–13.
12. Wieser H. Acta Paediatr 1996;85:3–9.
13. Shan L, Molberg Ø, Parrot I, et al. Science 2002;297:2275–9.
14. Nisticò L, Fagnani C, Coto I, et al. Gut 2006;55:803–8.
15. Singh P, Arora S, Lal S, et al. J Am Coll Gastroenterol 2015;110:1539–48.
16. Tollefsen S, Arentz-Hansen H, Fleckenstein B, et al. J Clin Invest 2006;116:2226–36.
17. Ricaño-Ponce I, Gutierrez-Achury J, Costa AF, et al. Eur J Human Genet 2020;28:313–23.
18. Smyth DJ, Plagnol V, Walker NM, et al. N Engl J Med 2008;359:2767–77.
19. Pinto-Sanchez MI, Seiler CL, Santesso N, et al. Gastroenterol 2020;159:884–903.
20. Kim C-Y, Quarsten H, Bergseng E, et al. Proc Natl Acad Sci 2004;101:4175–9.
21. Meresse B, Curran SA, Ciszewski C, et al. J Exp Med 2006;203:1343–55.
22. Yoo D, Lo W, Goodman S, et al. Am J Physiol Gastro Liver Physiol 2000;279:G1323–G32.
23. Trier JS, Falchuk ZM, Carey MC et al. Gastroenterol 1978;75: 307–16.
24. Caio G, Volta U, Sapone A, et al. BMC Med 2019;17:1–20.
25. Drosdak A, Satyavada S, Ismail M, et al. Intl J Obes 2022;46: 441–3.
26. Rostom A, Murray JA, Kagnoff MF. Gastroenterol 2006;131: 1981–2002.
27. Nimri FM, Muhanna A, Almomani Z, et al. Ann Gastroenterol 2022;35:281.
28. Di Tola M, Bizzaro N, Gaudio M, et al. Digest Dis Sci 2021;66:3234–41.
29. Husby S, Koletzko S, Korponay-Szabó I, et al. J Ped Gastro Nutr 2020;70:141–56.
30. Penny HA, Raju SA, Lau MS, et al. Gut 2021;70:876–83.
31. Hoyle A, Gillett P, Gillett HR, et al. Frontline Gastroenterol 2023;14:97-102.
32. Sollid LM, Thorsby E. Tiss Antigens 1990;36:136–7.
33. Brown NK, Guandalini S, Semrad C, et al. J Am Coll Gastroenterol 2019;114:1587–92.
34. Kreutz JM, Adriaanse MP, van der Ploeg EM, et al. Nutrients 2020;12:500.
35. Bode S, Gudmand-Høyer E. Scand J Gastroenterol 1996;31: 54–60.
36. Dahele A, Ghosh S. Am J Gastroenterol 2001;96:745–50.
37. Dickey W. Eur J Gastroenterol Hepatol 2002;14:425–7.
38. Halfdanarson TR, Kumar N, Hogan WJ, et al. J Clin Gastroenterol 2009;43:162–4.
39. Montoro-Huguet MA, Santolaria-Piedrafita S, Cañamares-Orbis P, et al. Nutrients 2021;13:3437.
40. Corazza GR, Di Sario A, Cecchetti L, et al. Gastroenterol 1995;109:122–8.
41. McFarlane X, Bhalla A, Reeves D, et al. Gut 1995;36:710–4.
42. Capriles VD, Martini LA, Arêas JA. Nutr Rev 2009;67:599–606.
43. Turner J, Pellerin G, Mager D. J Pediatr Gastroenterol Nutri 2009;49:589–93.
44. Riches PL, McRorie E, Fraser WD, et al. N Engl J Med 2009;361:1459–65.
45. Walker MD, Williams J, Lewis SK, et al. Clin Gastroenterol Hepatol 2020;18:99–106.
46. Micic D, Rao VL, Semrad CE. J Clin Densitom 2020;23:190–9.
47. Al-Toma A, Volta U, Auricchio R, et al. United Euro Gastroenterol J 2019;7:583–613.
48. Demirkesen I, Ozkaya B. Crit Rev Food Sci Nutr 2022;62: 571 97.
49. James SP. Gastroenterol 2005;128:6.
50. Collin P, Thorell L, Kaukinen K, et al. Alimen Pharmacol Ther 2004;19:1277–83.
51. Rubio-Tapia A, Rahim MW, See JA, et al. Am J Gastroenterol 2010;105:1412.
52. Kupper C. Gastroenterol 2005;128:S121–S7.
53. Arentz-Hansen H, Fleckenstein B, Molberg Ø, et al. PLoS Med 2004;1:e1.
54. Lundin K, Nilsen E, Scott H, et al. Gut 2003;52:1649–52.
55. Thompson T. N Engl J Med 2004;351:2021–2.
56. Fritz RD, Chen Y. Nutr Res 2018;60:54–67.
57. Ciclitira P, Evans D, Fagg N, et al. Clin Sci (London, England: 1979) 1984;66:357–64.
58. Catassi C, Fabiani E, Iacono G, et al. Am J Clin Nutr 2007;85: 160–6.
59. Catassi C, Rossini M, Rätsch I, et al. Gut 1993;34:1515–9.
60. World Health Organization (WHO). Revised in 2008. Available at www.fao.org/fao-who-codexalimentarius/sh-proxy/en/?lnk=1&url=https%253A%252F%252Fworkspace.fao.org%252Fsites%252Fcodex%252FStandards%252FCXS%2B118-1979%252FCXS_118e_2015.pdf
61. Pietzak MM. Gastroenterology 2005;128:S135–S41.
62. Fabiani E, Catassi C, Villari A, et al. Acta Paediatr 1996;85: 65–7.
63. Olsson C, Hörnell A, Ivarsson A, et al. J Human Nutr Diet 2008;21:359–67.
64. Brar P, Kwon GY, Holleran S, et al. Am J Med 2006;119:786–90.
65. Wieser H, Segura V, Ruiz-Carnicer Á, et al. Nutrients 2021;13: 2244.
66. Verma AK, Lionetti E, Gatti S, et al. J Ped Gastroenterol 2019;68: 26–9.
67. Studerus D, Hampe EI, Fahrer D, et al. J Food Protect 2018;81: 1679–84.
68. Weisbrod VM, Silvester JA, Raber C, et al. Gastroenterol 2020;158:273–5.
69. Lerner BA, Vo LP, Yates S, et al. Am J Gastroenterol 2019;114:792.
70. Ukkola A, Mäki M, Kurppa K, et al. Clin Gastroenterol Hepatol 2011;9:118–23.
71. Cranney A, Zarkadas M, Graham ID, et al. Digest Dis Sci 2007;52:1087–95.
72. Lee A, Ng D, Diamond B, et al. J Human Nutr Diet 2012;25: 233–8.
73. Nikniaz Z, Beheshti S, Abbasalizad Farhangi M, et al. Intl J Eat Disord 2021;54:1563–74.
74. Möller SP, Hayes B, Wilding H, et al. J Psychosom Res 2021;147:110537.
75. Lee AR, Wolf RL, Lebwohl B, et al. Nutrients 2019;11:399.
76. Oza SS, Akbari M, Kelly CP, et al. J Clin Gastroenterol 2016;50:307–12.
77. Leffler DA, Dennis M, Hyett B, et al. Clin Gastroenterol Hepatol 2007;5:445–50.
78. Leonard MM, Weir DC, DeGroote M, et al. J Ped Gastroenterol Nutr 2017;64:286.
79. Bascuñán KA, Vespa MC, Araya M. Eur J Nutr 2017;56:449–59.
80. Annibale B, Severi C, Chistolini A, et al. Am J Gastroenterol 2001;96:132–7.
81. Valdimarsson T, Löfman O, Toss G, et al. Gut 1996;38:322–7.
82. Kumar M, Rastogi A, Bhadada SK, et al. Indian J Med Res 2013;138:882.
83. Thompson T. J Acad Nutr Diet 2000;100:1389.
84. Wahab PJ, Meijer JW, Mulder CJ. Am J Clin Pathol 2002;118: 459–63.

85. Leffler DA, Dennis M, George JBE, et al. Clin Gastroenterol Hepatol 2009;7:530–6.
86. Wolf R, Vipperman-Cohen A, Green P, et al. J Human Nutr Diet 2020;33:876–80.
87. Comino I, Real A, Vivas S, et al. Am J Clin Nutr 2012;95: 670–7.
88. de Lourdes Moreno M, Cebolla Á, Muñoz-Suano A, et al. Gut 2017;66:250–7.
89. Monachesi C, Verma AK, Catassi GN, et al. Clin Translat Gastroenterol 2021;12.
90. Ventura A, Magazzù G, Greco L. Gastroenterol 1999;117:297–303.
91. Lebwohl B, Green PH, Söderling J, et al. JAMA 2020;323: 1277–85.
92. Emilsson L, Semrad C, Lebwohl B, et al. Gastroenterol 2020;159:1686–94.
93. Lebwohl B, Green PH, Emilsson L, et al. Clin Gastroenterol Hepatol 2022;20:e111–e31.
94. Green PH, Paski S, Ko CW, Rubio-Tapia A. Gastroenterol 2022;163:1461-9.
95. Helmerhorst EJ, Zamakhchari M, Schuppan D, et al. PloS One 2010;5:e13264.

D. DISORDERS OF THE ALIMENTARY TRACT

CHAPTER 82

Irritable Bowel Syndrome and Diverticular Disease

David Hakimian and Carol E. Semrad

CHAPTER OUTLINE

Abbreviations

CI, confidence interval; **COLAP**, colonoscopic allergen provocation; **FOD-MAPs**, fermentable oligosaccharides, disaccharides, monosaccharides, and polyols; **HLA**, human leukocyte antigen; **IEL**, intraepithelial lymphocytes; **IBS**, irritable bowel syndrome; **IBS-C**, IBS with constipation predominance; **IBS-D**, IBS with diarrhea predominance; number needed to treat NNT; **IBS-M**, IBS with mixed constipation and diarrhea; **Ig**, immunoglobulin; **RAST**, radioallergosorbent test; **RCT**, randomized control trial; **RR**, relative risk

Irritable Bowel Syndrome

Irritable bowel syndrome (IBS) is a chronic gastrointestinal disorder characterized by a combination of abdominal pain or discomfort and altered bowel habits over a period of at least 3 months that is not explained by structural, histologic, or biochemical abnormalities.[1] IBS is a common condition, affecting 4.4% to 4.8% of individuals in western countries, and greatly affects healthcare use.[1] It is considered a functional bowel disorder.[1] Individuals are classified into one of three subtypes: IBS with constipation predominance (IBS-C), IBS with diarrhea predominance (IBS-D), and IBS with mixed constipation and diarrhea (IBS-M). The pathophysiology of IBS is uncertain. The current disease model focuses on the brain–gut axis, a bidirectional pathway of communication between the central and enteric nervous systems. This pathway is influenced by various physiologic and psychosocial factors that contribute to heightened visceral sensitivity and disordered defecation. Dietary factors, changes in the intestinal microbiome, inflammation, altered motility and sensation, anxiety, and depression are among the variables believed to impinge on this pathway and exacerbate IBS symptoms.[2]

Diet and Irritable Bowel Syndrome Symptoms

Up to two-thirds of patients who suffer from IBS consider their gastrointestinal symptoms to be food related and modify their diets to avoid symptom triggers. Among 955 patients with IBS in Sweden, 13.2% reported severe food avoidance and restriction, and most consumed inadequate or unbalanced diets.[3-5] Commonly identified culprit foods include milk products, raw vegetables (especially onions, cabbage, and beans), fatty foods, spicy foods, coffee, and alcohol. These items have been linked to excessive gas-bloat symptoms and abdominal pain, followed by dyspepsia and loose stools.[4] In most instances, reported food intolerances are not substantiated by formal testing for food allergies, malabsorption, or celiac disease.[5] Perception of food intolerance is higher in the IBS population (60%–70%) than in the general

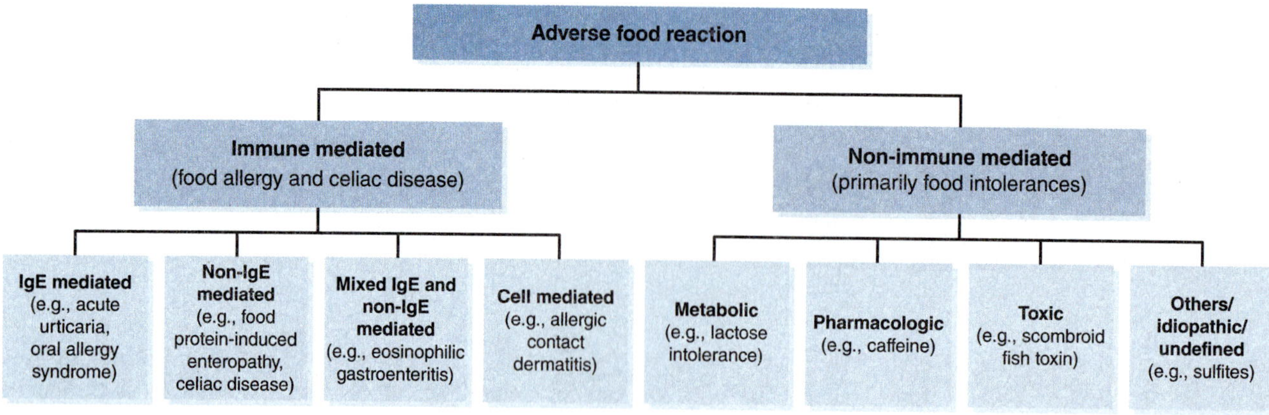

Figure 82.1 Categories of adverse reactions to food. *IgE*, immunoglobulin E.

population (20%–25%).[5-7] This difference is not explained by differences in food consumption. Individuals with functional gastrointestinal (GI) disorders consume similar proportions of foods containing wheat, lactose, caffeine, fructose, alcohol, and bioactive substances (e.g., serotonin, tryptophan) as case controls.[8]

Diet and Irritable Bowel Syndrome Pathophysiology

Adverse reactions to foods are classified as either immune mediated, such as food allergy, or nonimmune mediated, such as a specific food intolerance. A general schema of adverse food reactions and associated mechanisms is depicted in **Figure 82.1**.[9] These varied mechanisms cause gastrointestinal symptoms compatible with IBS (**Figure 82.2**).

Immune-Mediated Conditions

Food Allergy

The role of food allergy in the pathogenesis of IBS is controversial. Although data from dietary elimination and food challenge studies support the role of dietary triggers in a subgroup of IBS

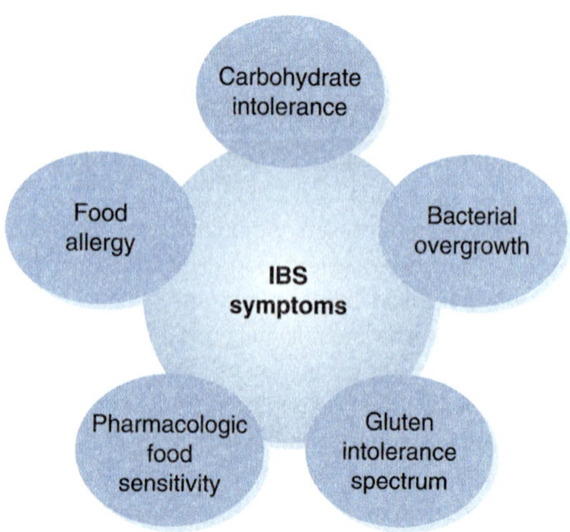

Figure 82.2 Symptom overlap among irritable bowel syndrome (IBS), adverse food reactions, and bacterial overgrowth.

patients, correlation with allergy testing has been inconsistent. Only one of three food elimination-challenge studies performing skin-prick or radioallergosorbent test (RAST) detected a positive correlation between diagnostics and dietary challenge.[10-13] Studies of oral cromolyn sodium, an inhibitor of mast cell degranulation, demonstrated efficacy in modulating IBS symptoms. In a large multicenter study of patients with diarrhea-predominant IBS, cromolyn sodium was as effective as an elimination diet in improving symptoms (67% vs 60%).[14] Treatment response was greater in patients with a positive skin-prick test to food allergens, a finding suggesting that a subset of patients with IBS had allergy-related symptoms.

Diagnostic tests may help clarify the role of food allergy in IBS. Cecal biopsies from patients with IBSD and atopic disease had significantly more mucosal mast cells, increased tryptase released from mast cells, greater colonic permeability and more severe diarrhea compared with those without atopy. These data support a role for allergy in the severity of IBS-D.[15]

Increased fecal immunoglobulin E (IgE) concentration has been reported in patients with food hypersensitivity, based on history, skin-prick test, and RAST test, but not in healthy controls.[16] Of the patients with IBS included in the study, 68% had detectable fecal IgE fragments. The colonoscopic allergen provocation (COLAP) test also demonstrated hypersensitivity reactions at the level of the intestine that are not detectable by skin tests or serum IgE. Injection of food extracts into the colonic submucosa produced weal and flare responses in 74% of patients with IBS and in no healthy controls. Biopsies of these sites confirmed mast cell and eosinophil activation and elimination of suspected food agents resulted in symptom improvement in most COLAP-positive patients.[10,17] An in vitro assay that quantifies basophil activation by food antigens based on CD63 expression identified food hypersensitivity in patients with IBS, with high sensitivity, specificity, and accuracy.[18] Endoscopic instillation of various foods in the small bowel while performing confocal laser endoscopy (CLE) demonstrated that different foods caused mucosal barrier dysfunction in IBS patients. Those who reacted to food antigens had increased intraepithelial lymphocytes and increased claudin, a protein involved in epithelial barrier permeability. They also had four times more atopy history than the healthy group. This suggests that IBS patients may suffer from atypical food allergy, without evidence of positive conventional allergy testing.[19]

IgG-mediated hypersensitivity has been proposed as an alternative mechanism of food allergy in IBS.[20] Although IgG antibodies are commonly considered a physiologic response to food antigen exposure, higher IgG4 in response to common food items has been detected in IBS patients.[21] Greater reduction in IBS symptom score was observed following an elimination diet based on IgG antibodies, compared with scores following a sham diet.[22] Current IBS guidelines do not recommend testing for food allergies and sensitivities unless there are symptoms of food allergy or anaphylaxis.[1,23]

Celiac Disease

Approximately 75% of patients with celiac disease present with gastrointestinal symptoms that overlap with IBS, including abdominal pain, bloating, and altered bowel habits.[24] In the United States, the prevalence of celiac disease in the general population approaches 1%, but it may be as high as 4% in the IBS population.[25] In a systematic review and meta-analysis, a fourfold higher risk of histologically confirmed celiac disease was seen among patients with IBS compared with controls.[26] The exact relationship between celiac disease and IBS is uncertain. IBS possibly represents a misdiagnosis of celiac disease, or celiac disease may coexist with IBS. Other conditions, such as inflammatory bowel disease, gastroenteritis, and small intestinal bacterial overgrowth, have been associated with IBS, suggesting that IBS is a downstream effect of various forms of mucosal inflammation or infection.[27] The highest prevalence of celiac disease is in IBS patients with diarrhea. Recommendations are for serologic screening for celiac disease in this subset of IBS patients.[1]

More controversial is the link between gluten sensitivity and IBS. Gluten sensitivity is a yet ill-defined condition characterized by gastrointestinal symptoms exacerbated by gluten ingestion and alleviated by gluten exclusion, in the absence of serologic and histologic evidence of celiac disease. In a study of patients with IBS-D and negative celiac serologic test results, 35% were human leukocyte antigen (HLA)-DQ2 positive, 23% had increased intraepithelial lymphocytes, and 30% had celiac-related antibodies in duodenal aspirates.[28] A gluten-free diet in individuals with IBS-D who were HLA-DQ2 positive or with positive intestinal aspirate antibodies resulted in significant improvement in diarrheal symptoms, compared with those without these features.[28,29] However, this may have been related to decreased fiber in the diet. A randomized controlled trial (RCT) examining the effect of a gluten-free diet vs a regular gluten diet in IBS-D, demonstrated that patients on a gluten-containing diet had more bowel movements and a defective mucosal barrier compared with those on a gluten-free diet, particularly in HLA DQ 2/8 positive patients.[30] In transgenic mice expressing HLA-DQ8, T-cell proliferation is enhanced without generating villous atrophy and acetylcholine release in the myenteric plexus is increased on gluten exposure.[31,32] These findings offer a potential inflammatory and neuromotor basis for IBS-type symptoms.

Food Intolerance

Food intolerance refers to adverse food reactions resulting from various nonimmune mechanisms. Such mechanisms include direct effects of toxins, pharmacologic agents in foods (e.g., caffeine, tyramine), malabsorption caused by host enzyme or transport deficiency (e.g., lactase, fructose), and idiosyncratic reactions.

Carbohydrate Malabsorption

Malabsorption of lactose, fructose, and sugar alcohols (e.g., sorbitol, xylitol) has been implicated as an underlying cause or trigger for IBS symptoms. The reported prevalence of lactose intolerance in IBS is variable, ranging from 17% to 82%, with most studies showing 25% to 35% in U.S. and European populations.[33-39] Patients with IBS have higher symptom scores from carbohydrate malabsorption, compared with controls who do not have IBS, perhaps related to underlying visceral hyperalgesia.[33] Patients with IBS-D and lactose intolerance had lower rectal distention threshold than healthy patients after ingestion of 20 grams of lactose.[40] Lactose breath testing is recommended in patients with IBS with diarrhea and bloating who lack a clear diagnosis based on food diary data alone.[2] A positive breath test without symptoms suggests lactose maldigestion of uncertain significance, whereas a positive test result accompanied by symptoms suggests true lactose intolerance.

Fructose and sorbitol malabsorption also have been demonstrated in patients with IBS. Fructose is a naturally occurring monosaccharide abundant in fruit and honey and the preferred sweetener in sodas and juices. Sorbitol is a sugar substitute used in many dietetic foods. Rates of fructose and combined fructose–sorbitol malabsorption are similar between healthy volunteers and patients with IBS (30%–40%), ranging from 30% to 40% and 40% to 92%, respectively.[34,37,41,42] Symptom intensity, however, is more pronounced in patients with IBSS. Proposed mechanisms include enhanced colonic microflora fermentative capacity (greater H_2 per unit carbohydrate fermented),[33,43] small intestinal bacterial overgrowth,[44] or intestinal dysmotility and visceral hypersensitivity.[45,46]

Nutritional Consequences of Irritable Bowel Syndrome

Malnutrition is an uncommon finding among patients with IBS. The body weight of people with IBS varies from normal to overweight, with different degrees of obesity. In rare instances, patients with IBS who perceive food intolerance may overly restrict their diet, leading to inadequate intake and malnutrition.[4] This outcome, however, is the exception rather than the rule. In patients presenting with weight loss or significant micronutrient deficiencies, alternative diagnoses must be considered. Further workup should include laboratory, imaging, or endoscopic studies to exclude conditions such as celiac disease, intestinal infection, inflammatory bowel disease, or gastrointestinal neoplasm.

Dietary Management of Irritable Bowel Syndrome

Dietary therapy is a helpful adjunct to pharmacologic treatment for IBS. Pharmacologic treatments are based on the individual symptom pattern and include antidiarrheal agents, stool softeners, microbial manipulations (probiotics or antibiotics), neuromodulators, and antispasmodic agents.[2] Dietary assessment can identify potential food triggers and dispel preconceived notions of food intolerances that may result in unnecessary restrictions. In general, patients with IBS should eat a balanced diet with few restrictions. Dietary modification should be based on the dominant gastrointestinal symptom. Reduction in specific food or beverage items known to exacerbate symptoms should be advised, such as decreased caffeine consumption in patients with IBS-D[47]

Increased intake of food items with therapeutic potential should be considered, such as soluble fiber intake for constipation. Exclusion of a specific food item should only be considered when objective evidence of allergy or intolerance exists.[2]

Exclusion Diets

An exclusion (elimination) diet followed by a food challenge has been used to identify food items that induce IBS symptoms. A systematic review of exclusion diets in patients with IBS demonstrated response rates of 12% to 67%.[48] Double-blind, placebo-controlled challenges identified food intolerance in 6% to 58% of patients with IBS, the majority of whom had sensitivities to milk, wheat, and eggs.[11,12,49] The wide range of responses observed, the lack of a standardized protocol for food or dietary restriction, and major limitations in trial design to date limit interpretation of these studies. On the basis of these data and the potential for macronutrient and micronutrient deficiencies resulting from this approach, generalized exclusion diets are not recommended for routine use in patients with IBS.[2,48,50]

FODMAP Diet

Although generalized exclusion diets are not advised, diets that specifically avoid unabsorbed, fermentable, short-chain carbohydrates, collectively termed fermentable oligosaccharides, disaccharides, monosaccharides, and polyols (FODMAP), may be of benefit to patients with IBS. FODMAPs include fructose and lactose, when malabsorbed, poorly absorbed polyols (sorbitol, xylitol), and fructo-oligosaccharides (fructans) and galacto-oligosaccharides (raffinose) not cleaved by human hydrolases and, thus, poorly absorbed.[51] When not digested or poorly absorbed in the small bowel, these carbohydrates stimulate colonic fluid secretion and increase bacterial fermentation with short chain fatty acid and gas production that can lead to luminal distension and symptoms in patients with IBS.[51]

Support for the use of FODMAP-restricted diets was initially derived from uncontrolled studies examining restriction of lactose[39] or of fructose with or without sorbitol.[37,41] A double-blind, randomized, quadruple-arm, placebo-controlled rechallenge trial demonstrated that 70% or more of patients with IBS who received fructose, fructans, or a combination of both, experienced inadequate symptom control compared with the 14% who received glucose.[52] These findings suggest that the effect of the low FODMAP diet resulted from sugar restriction and not from a placebo effect.

In one systematic review and meta-analysis that included six RCTs and 16 nonrandomized trials, IBS symptoms improved on a low FODMAP diet in both the RCTs (OR 0.44, 95% CI 0.25–0.76; $I2=35.5$, $P=0.001$) and nonrandomized trials (OR 0.03, 95% CI 0.01–0.2; $I2=69.1$, $P=0.02$).[53] In a more recent systematic review and meta-analysis of seven RCTs that compared a low FODMAP diet with various control diets in 397 participants, the low FODMAP diet significantly reduced global IBS symptoms compared with control diets, particularly abdominal pain and bloating.[54]

In a double-blind, placebo-controlled, randomized, three-way crossover study of 103 IBS patients on a low FODMAP gluten-free diet, reinstitution of FODMAPs into the diet caused significantly higher IBS symptom scores, compared with gluten or placebo. This study supports FODMAPs and not gluten as the causative food component and supports recommendations against a gluten-free diet for the treatment of IBS.[55] A recent meta-analysis showed that a low FODMAP diet improved IBS symptoms such as bloating, bowel distention and bowel habits, compared with other diets.[56] Based on the current evidence, the ACG guidelines suggest a limited trial of a low FODMAP diet to improve global IBS symptoms.[1]

While no adverse events have been associated with a low FODMAP diet in IBS studies, the complexity of the diet and long-term use may lead to micronutrient deficiencies without proper dietary counseling and monitoring. Education on the diet should be done by a knowledgeable registered dietitian or high-quality instruction materials. Responses to a low FODMAP diet can be identified within 2 to 6 weeks. In those responding to diet, individual FODMAPs should be reintroduced to assess tolerance and liberalize the diet. The FODMAP diet should be individualized according to tolerance for long-term use.[47,51]

Dietary Fiber

Fiber is defined as carbohydrate with a degree of polymerization of ten or more monomers that are not digested or absorbed. It is divided into insoluble and soluble fibers. The latter dissolve easily in water, have high fermentability, contain gum and pectin, and are found predominantly in vegetables and fruits. Insoluble fibers have low fermentability, contain cellulose, and are found predominantly in green plants.[57]

Fiber supplementation is prescribed as dietary treatment for IBS. Its use depends on the patient's dominant symptom. Fiber supplementation is most often used for patients with constipation and, to a lesser extent, for those with diarrhea. Fiber is believed to alleviate constipation by adding bulk to the stool, which accelerates oral–anal transit and decreases intracolonic pressure.[58] For patients with diarrhea, soluble fiber creates a more viscous stool and can improve stool output. When pain, gas, and bloating predominate, use of fiber may actually be counterproductive because fermentation of fiber generates increased gas production.

Evidence supporting the use of fiber in IBS is weak. Several systematic reviews and meta-analyses have been performed to formulate more definitive conclusions.[2,59-64] A meta-analysis of the effects of fiber (typically insoluble wheat bran or soluble ispaghula husk), vs placebo, in patients with IBS reported that ispaghula husk, but not wheat bran, improved global IBS symptoms over a 6- to 12-week period. However, when only higher quality studies were included, the beneficial effect of ispaghula husk was no longer significant.[64] In a randomized study of soluble fiber (psyllium), insoluble fiber (bran), or placebo in patients with IBS over a 12-week period, those treated with soluble fiber had significantly higher relief of abdominal pain compared with the placebo or bran supplementation groups.[65] A systematic review and meta-analysis of 14 RCTs involving 906 IBS patients treated with fiber, compared with placebo or control therapy, reported a significant benefit with soluble fiber. No benefit was found with insoluble fiber.[66] These studies suggest that soluble, but not insoluble, fiber offers symptom relief in a subset of patients with IBS.[57]

Based on the available data, soluble fiber is a recommended first-line approach to the treatment of IBS.[1,23] In patients with IBS-C, fiber intake should be titrated gradually to minimize the common adverse effects of bloating, abdominal distention, and flatulence.

Peppermint Oil

Methanol, the active ingredient of peppermint oil, induces smooth muscle relaxation in the human colon by inhibiting calcium channels. Other possible mechanisms of action include modulation of voltage-regulated channels that control visceral sensation, direct antimicrobial and anti-inflammatory effects, and psychosocial behavioral effects.[67]

In a recent international systematic meta-analysis comparing peppermint oil with placebo, which included 12 randomized trials with 835 patients, peppermint oil significantly improved global symptoms and abdominal pain in IBS patients, compared with placebo.[68] In Weerts et al.'s randomized, double-blind trial on the efficacy of peppermint oil, both the small intestinal and ileocolonic release forms failed to significantly reduce abdominal pain or global symptoms in IBS with the U.S. Food and Drug Administration/European Medicines Agency more robust responder response primary endpoints. The small intestinal release form of peppermint oil did significantly reduce abdominal pain and overall symptom relief using less rigorous responder secondary endpoints.[69] There were adverse events with peppermint oil, most commonly reflux, belching with or without a minty taste, and headache. Peppermint oil is commonly used in the treatment of IBS, but recent studies question its effectiveness. It may be of benefit to those with abdominal pain.

Probiotics and Prebiotics

The intestinal microbiome may have a pathogenic role in the development of IBS. Manipulation of the gut microflora using prebiotics and probiotics has potential therapeutic value. Probiotics, commonly live *Bifidobacteria* and *Lactobacillus* spp, are available in capsule form or in food items such as yogurts. Prebiotics, commonly nondigestible fructo-oligosaccharides (oligofructose and inulin) and galacto-oligosaccharides in food, stimulate the growth and activity of select intestinal bacteria.

A large meta-analysis assessing the efficacy of probiotics vs placebo on global IBS symptoms reported statistically significant reduction in IBS symptoms across all probiotics used (*Lactobacillus, Bifidobacterium, Streptococcus*, and combinations), compared with placebo.[70] These data, however, displayed significant heterogeneity and likely overestimated the treatment effect. In trials that reported IBS symptoms as a continuous variable, probiotic use resulted in a statistically significant improvement, compared with placebo.[70,71] The optimal probiotic regimen could not be delineated from these trials.

The prebiotic, *trans*-galacto-oligosaccharide has been reported to change stool consistency and improve flatulence and bloating significantly in patients with IBS, compared with placebo. Concurrent with these effects was an increase in fecal bifidobacterial concentration, a finding suggesting a potential benefit from the altered microbiota.[72]

Diverticular Disease

Colonic diverticulosis refers to the development of saclike outpouchings in the colonic wall resulting from herniation of mucosa and submucosa through defects in the muscle layer. Although most patients with diverticulosis remain asymptomatic, up to 10% to 25% develop clinically significant disease. The spectrum of diverticular disease includes a complex of abdominal pain and altered bowel habits like IBS; diverticulitis with or without complications, diverticular hemorrhage, and diverticular colitis (**Figure 82.3**).

Pathophysiology

Fiber deficiency is believed to be central to the development of diverticulosis and diverticular disease. The fiber hypothesis was first proposed by physicians Dennis Burkitt and NS Painter in 1971 while on a medical mission in rural Africa. The team noted that diverticular disease, although prevalent in England, was virtually nonexistent in the Ugandan population. This difference was attributed to high-fiber consumption among Ugandans than among the English.[73,74]

The emergence of diverticular disease appeared to parallel a decrease in dietary fiber consumption. Autopsy studies in the early 20th century detected diverticulosis in 2% to 10% of a Western population, whereas an autopsy series in the 1960s revealed prevalence of 10% to 66%, depending on the age group studied.[74,75] Over time, diverticular disease increased in developing countries that adopted the Western diet.[76] A similar shift was also observed in Japan. In a retrospective study of 1,289 barium enemas performed between 1960 and 1970, the prevalence of diverticular disease was 2.6% among Japanese patients who were more than 40 years old, whereas prevalence rates

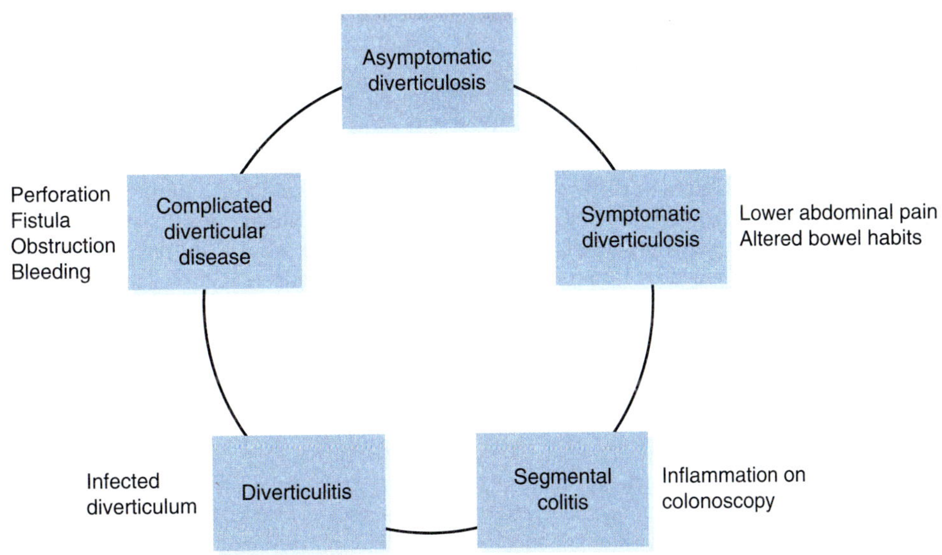

Figure 82.3 Spectrum of diverticular disease.

reached 7% to 12% between 1980 and 1990. This rise in diverticular disease followed a reduction in mean daily fiber intake in 1972 to approximately 70% of the 1952 level.[73,77-79]

A large, prospective study confirmed the inverse relationship between diverticular disease and fiber consumption.[80] Unlike epidemiologic studies, this analysis focused solely on patients with symptomatic diverticular disease. Among an estimated 48,000 men participating in the Health Professionals Follow-Up Study, 385 men with dietary data developed symptomatic diverticular disease over a 4-year period. Patients in the highest quintile of fiber intake had a significantly reduced risk of developing diverticular disease, compared with patients in the lowest quintile (relative risk [RR], 0.58; 95% confidence interval [CI], 0.41, 0.83; $P=0.01$). Conversely, the RR for men on a high-total-fat, low-fiber diet was 2.35 (95% CI, 1.38, 3.98) compared with those on a low-total-fat, high-fiber diet.[80] This association was primarily attributable to consumption of fruit and vegetable fiber rather than cereal fiber.[81] A recent systematic review and analysis of prospective trials that included 865,000 participants reported an inverse risk between high fiber consumption and diverticular disease. There was a significant reduction of diverticular disease of 41% when consuming 30 g/day of fiber, compared with a low-fiber diet of 7.5 g/day. When examining the type of fiber, 10 g/day of fiber from fruit provided the greatest benefit (RR 0.56, 95% CI 0.37–0.84), followed by cereals (RR 0.74, 95% CI 0.67–0.81) and vegetables (RR 0.80, 95% CI 0.45–1.44).[82]

Fiber deficiency is believed to promote diverticula formation by generating low stool mass and reduced luminal size. In this setting, peristaltic forces are transmitted to the colon wall rather than to luminal contents, thus causing mucosa and submucosa to herniate through the weakest points and create sacculations. Less well understood are the variables influencing progression from asymptomatic to symptomatic diverticular disease. One theory is that fiber deficiency alters the gut flora and leads to immunologic changes that stimulate inflammation and diverticulitis.[73] A high-fiber diet may have vitamins and minerals that alter the gut flora and, thereby, decrease the risk for diverticulitis.[83] Further studies are needed to understand the relationship between the diet–microbiome and symptomatic diverticular disease.

Diet in the Management of Diverticular Disease

Prevention of Symptomatic Diverticular Disease

Historically, physicians have advised patients with diverticulosis to avoid consumption of nuts, seeds, popcorn, and other high-fiber foods to prevent symptoms or complications of diverticular disease. This is no longer a valid practice. As part of the Health Professionals Follow-up Study, a cohort of 47,228 men completed periodic medical and dietary questionnaires over an 18-year period.[84] During this interval, 801 incident cases of diverticulitis and 383 incident cases of diverticular bleeding were reported. Review of available dietary records failed to establish an association between corn consumption and diverticulitis or between nut, corn, or popcorn consumption and diverticular hemorrhage. In fact, an inverse relationship was found between nut and popcorn consumption and diverticulitis risk, suggesting a protective effect.[84]

Dietary fiber supplementation has been advocated as a means to prevent formation of diverticula and symptomatic diverticular disease.[85] At least six studies have examined the role of dietary fiber in the management of diverticular disease, including two randomized controlled trials. The first trial randomized 18 patients with diverticulosis to either wheat or bran crisp bread over a 3-month period. Patients receiving the high-fiber bran product experienced greater improvements in pain and total symptom scores, compared with the low-fiber group.[86] In contrast, a second randomized trial showed no difference in pain, lower bowel, or total symptom scores in response to a high-fiber diet, although constipation and stool consistency improved.[87]

Additional data supporting fiber supplementation in diverticular disease are derived from four retrospective, uncontrolled trials, only one of which included patients with complicated diverticular disease (i.e., diverticulitis).[85,88] In this trial, 60 of 100 patients admitted to the hospital with symptomatic diverticular disease had diverticulitis. Seventy-five patients were managed medically whereas 25 required surgery. Following treatment, the medical and surgical groups were started on a high-fiber diet, and 91% remained symptom free after 5 to 7 years of follow up.[88] Overall, the evidence is limited and conflicting for fiber intake and the treatment of symptomatic diverticular disease or prevention of acute diverticulitis.[57]

Diet and Acute Complicated Diverticular Disease

During an acute attack of diverticulitis or diverticular hemorrhage, patients are placed on bowel rest, and their diet is advanced as clinical parameters improve.[89] Although data on dietary advancement are lacking, patients are generally started on a low-fiber diet (10 g/day). As the patient recovers, daily fiber intake can be advanced by 5 g each week until a target intake of 25 to 35 g is achieved.[85] As noted, avoidance of nuts and seeds is not necessary.

Summary

IBS is a common clinical disorder with varied presentations, including predominant constipation, predominant diarrhea, or a mixed picture. Significant numbers of patients with IBS consider their gastrointestinal symptoms to be food related and modify their diets. In most instances, reported food intolerances are not substantiated by formal testing for food allergies, malabsorption, or celiac disease. Individualized dietary counseling is a helpful adjunct to pharmacologic treatments for IBS.[47] Patients with IBS should eat a balanced diet with few restrictions. Dietary modifications should be based on the dominant gastrointestinal symptoms in individual patients and may include the FODMAP diet in those with abdominal pain and bloating, or the use of increased dietary fiber in those with constipation.

The development of diverticular disease is intimately tied to dietary fiber consumption. Less well established is the role of fiber deficiency in the progression or recurrence of symptomatic diverticular disease A high-fiber diet is recommended to prevent diverticular disease and in the management of diverticular disease with constipation or a history of diverticulitis. It is no longer recommended to avoid seeds, nuts, and popcorn in the diet.[89,90]

Acknowledgments

The authors acknowledge Lauren Schwartz for her contributions to the previous version of this chapter.

References

1. Lacy BE, Pimentel M, Brenner DM, et al. J Am Coll Gastroenterol 2021;116:17–44.
2. Camilleri M. JAMA 2021;325:865–77.
3. Melchior C, Algera J, Colomier E, et al. Clin Gastroenterol Hepatol 2022;20:1290–8. e4
4. Simrén M, Månsson A, Langkilde AM, et al. Digestion 2001;63: 108–15.
5. Monsbakken K, Vandvik P, Farup P. Eur J Clin Nutr 2006;60: 667–72.
6. Locke III GR, Zinsmeister AR, Talley NJ, et al. Am J Gastroenterol 2000;95:157–65.
7. Young E, Stoneham MD, Petruckevitch A, et al. Lancet 1994;343: 1127–30.
8. YA S. Am J Gastroenterol 2005;100:2743–8.
9. Boyce JA, Assa'ad A, Burks AW, et al. J Am Acad Dermatol 2011;64:175–92.
10. Zar S, Kumar D, Benson M. Alim Pharmacol Ther 2001;15: 439–49.
11. Bentley S, Pearson D, Rix KB. Lancet 1983;322:295–7.
12. Farah D, Calder I, Benson L, et al. Gut 1985;26:164–8.
13. Petitpierre M, Gumowski P, Girard J. Ann Allergy 1985;54: 538–40.
14. Stefanini G, Saggioro A, Alvisi V, et al. Scand J Gastroenterol 1995;30:535–41.
15. Vivinus-Nébot M, Dainese R, Anty R, et al. J Am Coll Gastroenterol 2012;107:75–81.
16. André F, André C, Colin L, et al. Allergy 1995;50:328–33.
17. Bischoff SC, Mayer J, Meier PN, et al. Intl Arch Allergy Immunol 1997;113:348–51.
18. Carroccio A, Brusca I, Mansueto P, et al. Clin Gastroenterol Hepatol 2010;8:254–60.
19. Fritscher-Ravens A, Pflaum T, Mösinger M, et al. Gastroenterol 2019;157:109–18. e5
20. Ostrowska L, Wasiluk D, Lieners CF, et al. J Clin Med 2021; 10:4317.
21. Zar S, Benson MJ, Kumar D. J Am Coll Gastroenterol 2005;100:1550–7.
22. Atkinson W, Sheldon T, Shaath N, et al. Gut 2004;53:1459–64.
23. Kindt S, Louis H, De Schepper H, et al. Acta gastro-enterologica Belgica 2022;85.
24. Zipser RD, Patel S, Yahya KZ, et al. Digest Dis Sci 2003;48: 761–4.
25. Rubio-Tapia A, Ludvigsson JF, Brantner TL, et al. J Am Coll Gastroenterol 2012;107:1538–44.
26. Ford AC, Chey WD, Talley NJ, et al. Arch Int Med 2009;169: 651–8.
27. O'leary C, Quigley EM. Am J Gastroenterol 2003;98:720.
28. Wahnschaffe U, Ullrich R, Riecken E, et al. Gastroenterol 2001;121:1329–38.
29. Wahnschaffe U, Schulzke JD, Zeitz M, et al. Clin Gastroenterol Hepatol 2007;5:844–50.
30. Vazquez–Roque MI, Camilleri M, Smyrk T, et al. Gastroenterol 2013;144:903–11. e3
31. Black KE, Murray JA, David CS. J Immunol 2002;169:5595–600.
32. Verdu EF, Huang X, Natividad J, et al. Am J Physiol Gastro Liver Physiol 2008;294:G217–G25.
33. Fernandez-Banares F, Esteve-Pardo M, De Leon R, et al. Am J Gastroenterol (Springer Nature) 1993;88.
34. Corlew-Roath M, Di Palma JA. South Med J 2009;102:1010–2.
35. Böhmer CJ, Tuynman HA. Eur J Gastroenterol Hepatol 2001;13: 941–4.
36. Böhmer C, Tuynman H. Eur J Gastroenterol Hepatol 1996;8: 1013–6.
37. Goldstein R, Braverman D, Stankiewicz H. Israel Med Assoc J 2000;2:583–7.
38. Parker TJ, Woolner JT, Prevost AT, et al. Eur J Gastroenterol Hepatol 2001;13:219–25.
39. Vernia P, Di Camillo M, Marinaro Dig Liver Dis 2001;33: 234–9.
40. Yang J, Fox M, Cong Y, et al. Alim Pharmacol Ther 2014;39: 302–11.
41. Fernández-Bañares F, Rosinach M, Esteve M, et al. Clin Nutr 2006;25:824–31.
42. Nelis GF, Vermeeren MA, Jansen W. Gastroenterol 1990;99: 1016–20.
43. King T, Elia M, Hunter J. Lancet 1998;352:1187–9.
44. Nucera G, Gabrielli M, Lupascu A, et al. Alim Pharmacol Ther 2005;21:1391–5.
45. Evans P, Piesse C, Bak Y-T, et al. Scand J Gastroenterol 1998;33:1158–63.
46. Lasser RB, Bond JH, Levitt MD. N Engl J Med 1975;293:524–6.
47. Chey WD, Hashash JG, Manning L, et al. Gastroenterol 2022; 162:1737–45.
48. Park MI, Camilleri M. Neurogastroenterol Motil 2006;18: 595–607.
49. Jones VA, Shorthouse M, McLaughlan P, et al. Lancet 1982;320: 1115–7.
50. Cabré E. Curr Opin Clin Nutr Metab Care 2010;13:581–7.
51. Staudacher HM, Whelan K. Gut 2017;66:1517–27.
52. Shepherd SJ, Parker FC, Muir JG, et al. Clin Gastroenterol Hepatol 2008;6:765–71.
53. Marsh A, Eslick EM, Eslick GD. Eur J Nutr 2016;55:897–906.
54. Dionne J, Ford AC, Yuan Y, et al. J A Coll Gastroenterol 2018;113:1290–300.
55. Nordin E, Brunius C, Landberg R, et al. Am J Clin Nutr 2022;115:344–52.
56. Black CJ, Staudacher HM, Ford AC. Gut 2022;71:1117–26.
57. Gill SK, Rossi M, Bajka B, et al. Nat Rev Gastroenterol Hepatol 2021;18:101–16.
58. Zuckerman MJ. J Clin Gastroenterol 2006;40:104–8.
59. Akehurst R, Kaltenthaler E. Gut 2001;48:272–82.
60. Bijkerk C, Muris J, Knottnerus J, et al. Alim Pharmacol Ther 2004;19:245–51.
61. Jailwala J, Imperiale TF, Kroenke K. Ann Intern Med 2000;133: 136–47.
62. Lesbros-Pantoflickova D, Michetti P, Fried M, et al. Alim Pharmacol Ther 2004;20:1253–69.
63. AO Q. Cochrane Data System Rev 2004;2.
64. Ac F. BMJ 2008;337:a2313.
65. Bijkerk C, De Wit N, Muris J, et al. BMJ 2009;339:b3154.
66. Moayyedi P, Quigley EM, Lacy BE, et al. J Am Coll Gastroenterol 2014;109:1367–74.
67. Chumpitazi BP, Kearns G, Shulman RJ. Alim Pharmacol Ther 2018;47:738–52.
68. Alammar N, Wang L, Saberi B, et al. BMC Comp Alt Med 2019;19:1–10.

69. Weerts ZZR, Masclee AA, Witteman BJ, et al. Gastroenterol 2020; 158:123–36.
70. Moayyedi P, Ford AC, Talley NJ, et al. Gut 2010;59:325–32.
71. Didari T, Mozaffari S, Nikfar S, et al. World J Gastroenterol 2015;21:3072.
72. Silk D, Davis A, Vulevic J, et al. Alim Pharmacol Ther 2009;29: 508–18.
73. Korzenik JR. J Clin Gastroenterol 2006;40:S112–S6.
74. Painter NS, Burkitt DP. BMJ 1971;2:450.
75. Parks T. Clin Gastroenterol 1975;4:53–69.
76. Segal I, Solomon A, Hunt J. Gastroenterol 1977;72:215–9.
77. Nakaji S, Sugawara K, Saito D, et al. Eur J Nutr 2002;41:222–7.
78. Miura S, Kodaira S, Shatari T, et al. Dis colon rectum 2000;43:1383–9.
79. Nakada I, Ubukata H, Goto Y, et al. Dis colon rectum 1995;38:755–9.
80. Aldoori WH, Giovannucci EL, Rimm EB, et al. Am J Clin Nutr 1994;60:757–64.
81. Aldoori WH, Giovannucci EL, Rockett HRH, et al. J Nutr 1998;128:714–9.
82. Aune D, Sen A, Norat T, et al. Eur J Nutr 2020;59:421–32.
83. Wegermann K, Roper J. Gastroenterol 2020;158:1167–8.
84. Strate LL, Liu YL, Syngal S, et al. JAMA 2008;300:907–14.
85. Tarleton S, DiBaise JK. Nutr Clin Pract 2011;26:137–42.
86. Brodribb A. Lancet 1977;309:664–6.
87. Ornstein M, Littlewood E, Baird I, et al. Br Med J (Clin Res Ed) 1981;282:1353–6.
88. Hyland J, Taylor I. Brit J Surg 1980;67:77–9.
89. Peery AF, Shaukat A, Strate LL. Gastroenterol 2021;160: 906–11. e1
90. Hall J, Hardiman K, Lee S, et al. Dis Colon Rectum 2020;63: 728–47.

D. DISORDERS OF THE ALIMENTARY TRACT

CHAPTER 83

Nutrition in Pancreatic Diseases

Stephen J.D. O'Keefe

CHAPTER OUTLINE

Abbreviations

AP, acute pancreatitis; **ARDS**, acute respiratory distress syndrome; **CCK**, cholecystokinin; **CP**, chronic pancreatitis; **CRP**, C-reactive protein; **EN**, enteral nutrition; **GI**, gastrointestinal; **GLP-1**, glucagon-like peptide-1; **ICU**, intensive care unit; **IL**, interleukin; **MAP**, mitogen-activated protein; **MCP-1**, monocyte chemotactic protein-1; **MOFS**, multiorgan failure syndrome; **NG**, nasogastric; **NGJ**, nasogastrojejunal; **PEG**, percutaneous endoscopic gastrostomy; **PN**, parenteral nutrition; **PYY**, peptide YY; **RCT**, randomized controlled trial; **SAP**, severe acute pancreatitis; **SIRS**, systemic inflammatory response; **TNF-α**, tumor necrosis factor-α; **TPN**, total parenteral nutrition

Introduction

The pancreas, a retroperitoneal glandular organ, serves both endocrine and exocrine functions. The exocrine pancreas secretes at least 10 digestive enzymes, which are essential for the digestion and absorption of nutrients. The endocrine pancreas also secretes several hormones that play a key role in maintaining the metabolic homeostasis of the body. Three major pancreatic disorders discussed in this chapter are acute pancreatitis (AP), chronic pancreatitis (CP), and pancreatic cancer. These diseases, by altering the functioning of the pancreas, can all lead to major derangements in nutritional and metabolic homeostasis, although the underlying pathophysiologic mechanisms are different. This chapter at the outset briefly outlines the role that pancreatic enzymes play in the absorption of nutrients, and then focuses on pathophysiology, clinical presentation, assessment of nutritional status, and the principles of nutritional therapy in these three major pancreatic diseases.

Physiology of Pancreatic Secretion

Knowledge of the mechanisms of pancreatic secretion is fundamental to the management of patients with pancreatic disease, especially AP. Pancreatic secretion is precisely orchestrated by

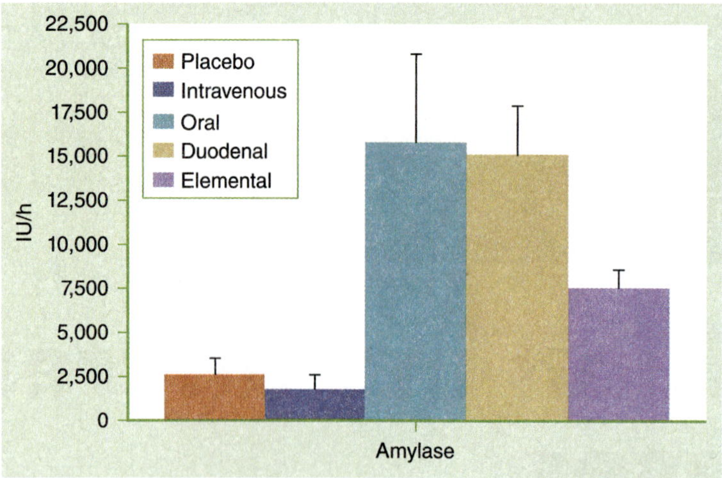

Figure 83.1 Amylase secretion in response to enteral and parenteral nutrition. Relative amylase secretory responses to enteral and parenteral feeding, illustrating no difference between oral and duodenal feeding of a complex diet, an intermediate response to duodenal elemental diet feeding, and no stimulatory effect of intravenous feeding compared with placebo saline.

the release of peptide hormones and neurotransmitters from the gastrointestinal (GI) tract after contact with ingested food. Traditionally, pancreatic stimulation is divided into three phases: the cephalic, gastric, and intestinal phases. The thought of food primes the pancreas to commence the process of zymogen aggregation and secretion. Next, the ingestion and swallowing of food, followed by expansion of the stomach wall, induce vagally mediated pancreatic secretion. Finally, the entry of food into the duodenum induces the most powerful stimulation, associated with acetylcholine and cholecystokinin (CCK) release by the mucosa and propelled by migrating motor complexes. In physiologic studies in healthy volunteers, a liquid formula diet was delivered to different regions of the upper GI tract, and the secretory response was greatest if a polymeric diet was infused into the duodenum (**Figure 83.1**).[1] Furthermore, the secretory response could be reduced significantly if the composition was changed to low-fat elemental. Additional studies showed that total parenteral nutrition (TPN) had no stimulatory effect, and that pancreatic rest could be maintained if the enteral feeding was delivered more than 40 cm past the ligament of Treitz.[2] Finally, ileal delivery stimulated the ileal brake peptides peptide YY (PYY) and glucagon-like peptide-1 (GLP-1), resulting in inhibition of secretion.

Digestion of Nutrients

Without the pancreas, humans cannot survive because the gland is essential for the digestion of food. The normal pancreas secretes digestive enzymes, together with water and electrolytes, predominantly bicarbonate, which enhances luminal enzyme function by neutralizing gastric acid. The most active enzymes are lipase, amylase, and trypsin. Amylase (α-amylase) hydrolyzes dietary starch into disaccharides and trisaccharides, which are then broken down by enzymes on the brush border to absorbable forms, as glucose and maltose. Pancreatic lipase hydrolyzes fat molecules. Bile salts secreted by the liver aid the digestive action of lipase by coating and emulsifying large fat droplets into smaller droplets, thus increasing the overall surface area for lipase to work. Fat hydrolysis results in formation of monomers (two free fatty acids and one 2-monoacylglycerol), which are then absorbed downstream into the lymphatic system by the lacteals. The main proteolytic enzyme, trypsin, is synthesized in the pancreas in an inactive form, such as trypsinogen. Following a meal, when the pancreas is stimulated by CCK and cholinergic reflexes, trypsinogen is released from zymogen stores in the acinar cells and is secreted into the duodenum. Once in the small intestine, the intestinal enzyme enteropeptidase activates it into trypsin by proteolytic cleavage. Trypsin, then, by the process of autocatalysis, activates more trypsinogen molecules. Once activated, the trypsin breaks down food proteins and peptides (proteins broken down to peptides in the stomach by pepsin) to amino acids, which are then absorbed by active transport systems.

Acute Pancreatitis

Demographic and Clinical Presentation

AP is an acute inflammatory process of the pancreas that may involve the peripancreatic tissue and even remote organs. In the United States, approximately 75% to 80% cases of AP are attributed to alcohol abuse or gallstones.[3-5] Other factors associated with AP include medications, trauma, infections, and metabolic causes.[6] Biliary pancreatitis AP is more common in female patients, and alcoholic AP is more common in male patients.[7] Clinical presentation typically consists of severe upper abdominal pain, nausea, and vomiting. Laboratory testing reveals elevated lipase and amylase in the bloodstream.

About 75% of cases of AP in patients admitted to hospitals are mild (edematous and interstitial pancreatitis) and follow a benign, self-limited course, with discharge home by day 4.[8] The remaining 25% cases, called severe AP (SAP), progress with the development of a profound systemic inflammatory response (SIRS), commonly associated with pancreatic gland necrosis, acute peripancreatic fluid collections, and multiorgan failure syndrome (MOFS). All of the mortality ($\leq$50%) from the condition is associated with these complications. The inflamed and swollen pancreatic gland may, itself or by development of acute fluid collections, compress the stomach and duodenum. The result is obstruction to the outflow from the stomach, and patients present clinically with nausea and vomiting. The SIRS is usually associated with ileus and increased mucosal permeability. These critically ill patients often spend weeks in intensive care units (ICUs) and frequently need surgery for pancreatic necrosis and infection. However, early surgery is associated with elevated mortality rates because it is extremely difficult, and every effort should be made to manage patients conservatively for more than 4 weeks with enteral nutrition (EN) until the area of pancreatic necrosis or fluid collection becomes walled off, thus allowing a more definitive approach.

Pathophysiology

Basic understanding of the pathophysiology of AP is essential to comprehend the principles of nutritional therapy in these patients. **Figure 83.2** helps illustrate some of what is known to happen in the evolution of the severe disease. AP is initiated by premature activation of trypsinogen within the acinar cells. Once trypsinogen, which normally is stored in zymogens within the acinar cells in an inactive form, is stimulated within the acinar cells, it autoactivates other trypsinogen molecules and autodigestion of the cell.[9,10] Intracellular injury results in generation of a cascade of proinflammatory cytokines, such as interleukin-1β (IL-1β), tumor necrosis factor-α (TNF-α), IL-8, IL-15, monocyte chemotactic protein-1 (MCP-1), and IL-18,[11-17] via activation of periacinar myofibrocytic nuclear factor κ-light-chain-enhancer of activated B cells (NF-κB) and mitogen-activated protein (MAP) kinase.[18] The intense inflammatory response results in arterial constriction with resultant apoptosis which, in an extreme situation, may lead to pancreatic necrosis. If the inflammation is contained within the pancreatic bed, the disease process is far less serious. Unfortunately, the cytokines are released into the circulation, and a secondary response commencing approximately 48 hours later leads to the generation of prostaglandin-2, thromboxane, leukotriene B_4, and oxygen-derived free radicals within the bronchial and intestinal mucosa that produce cytotoxic lung injury.[19,20]

Substantial evidence indicates that the gut plays a pivotal role in the systemic response. Increased mucosal permeability is characteristic of AP, and the combination of mucosal injury secondary to ischemia and antegrade colonization of the stagnant bowel (ileus) both amplifies the proinflammatory cytokine response and liberates toxic bacterial products into the bloodstream.[21,22] Later in the disease, the leaky mucosa predisposes to bacterial translocation, thus accounting for the observation that most infective complications are caused by gut-derived organisms.[21] The situation is compounded further by the release of proteolytic enzymes, such as trypsin, elastase, phospholipase, and caspase-l, into the circulation, which leads to amplification of cell injury within the lung (second hit) and acute respiratory distress syndrome (ARDS). The systemic injury is accompanied by massive outpouring of vascular fluids, which exacerbate the pulmonary dysfunction and lead to edema and prerenal failure. These complications of ARDS, acute renal failure and intestinal failure, account for nearly all of the mortality associated with the disease, which can approach 30% to 50%, in patients with severe necrotizing pancreatitis.

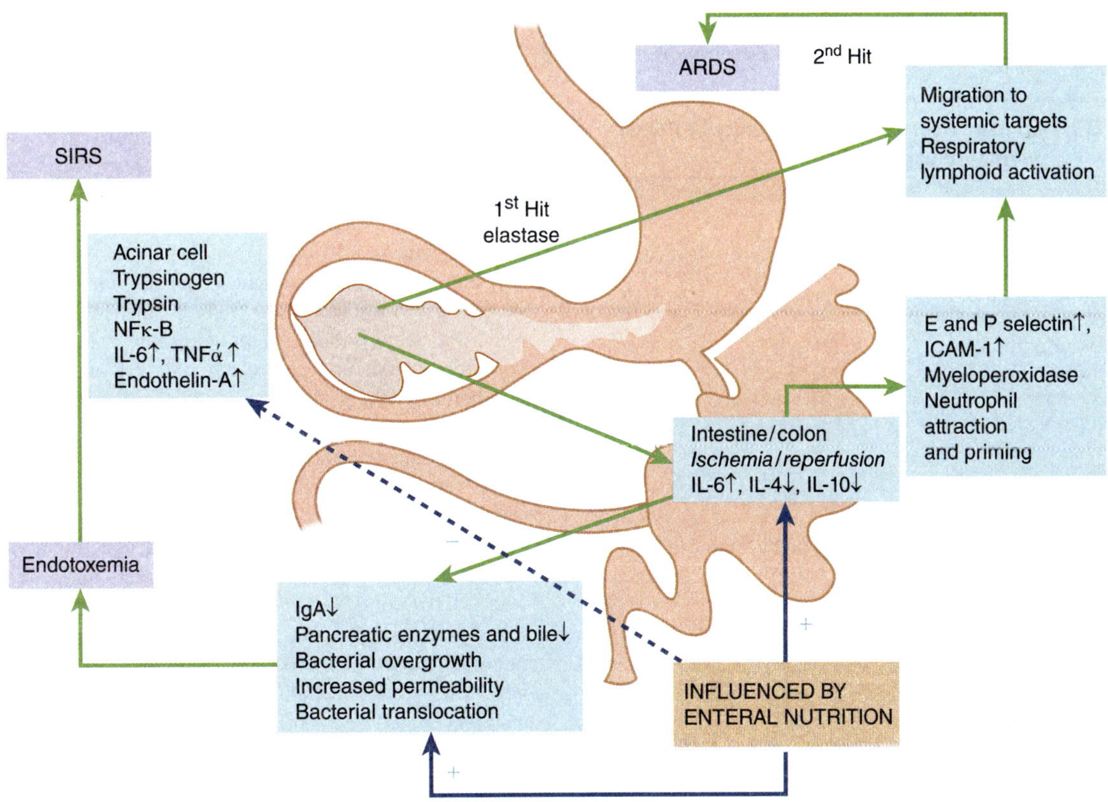

Figure 83.2 Generation of the systemic inflammatory response (SIRS) and acute respiratory distress syndrome (ARDS) by acute pancreatitis and sites where enteral feeding may favorably influence these responses. *ICAM-1*, intercellular adhesion molecule 1; *Ig*, immunoglobulin; *IL*, interleukin; *NF-κB*, nuclear factor κ-light-chain-enhancer of activated B cells; *TNF-α*, tumor necrosis factor-α.

Understanding the pathologic response to AP makes it clear that, in the early stages, pancreatic stimulation should be avoided because it can lead to production and activation of more proteolytic enzymes and can thereby exacerbate the pancreatic inflammation. Once established, AP complicated by MOFS represents one of the most catabolic diseases seen in ICUs, and protein deficiency can occur within the first week unless nutritional support is commenced. A study showed that early EN (i.e., ≤5 days of the onset of symptoms) can prevent some of the progression of these events and thus improve outcome.[23] Patients with SAP often spend weeks in the ICU on ventilators and can suffer severe protein energy malnutrition in the absence of timely nutritional therapy.

Predicting Severity of Pancreatitis and Nutritional Implications

Some patients with SAP present with MOFS and profound SIRS at the outset. Others present with mild AP and then go on to exhibit features of SAP over the next few days. Predicting which of these patients will develop a severe course should help determine the urgency for intervention with specialized nutritional support. Although clinical judgment remains the most used tool, it often underpredicts the severity of pancreatitis.[24] Among laboratory studies, lipase and amylase are most widely used. Although elevated amylase and lipase are used to diagnose AP, there is no correlation between the severity of disease and elevation of these enzymes. Daily monitoring of amylase and lipase does not help in determining the prognosis or the course of the disease.[25]

Various scoring systems (e.g., Balthazar Score, Acute Physiology, and Chronic Health Evaluation [APACHE II] score, Ranson score [**Table 83.1** and **Table 83.2**]) have been employed over the years to predict severity of pancreatitis.[26-31] Some scoring tests can be performed on the day of admission, and some need 48 to 72 hours to complete. None of these scoring systems is consistently accurate, but they are somewhat better than clinical judgment.[32] This situation has led to the search for a simpler measurement.

Table 83.1 Ranson Score (includes equally [1 Point] weighted variables)[a]

0 h	Age	>55 y
	White blood cell count	>16,000 cells/mm^3
	Blood glucose	>200 mg/dL (11.1 mmol/L)
	Serum lactate dehydrogenase	>350 U/L
	Serum aspartate aminotransferase (AST)	>250 U/L
48 h	Hematocrit	Fall by ≥10%
	Blood urea nitrogen	Increase by ≥5 mg/dL (1.8 mmol/L) despite fluids
	Serum calcium	<8 mg/dL (2 mmol/L)
	PO_2	<60 mm Hg
	Base deficit	>4 mEq/L
	Fluid sequestration	>6,000 mL

PO_2, partial oxygen pressure.

[a]Some variables are assessed at presentation and others at 48 hours. Mortality is 0% to 3% with a Ranson score of less than 3, 11% to 15% a with score of 3 or more, and 40% with a score higher than 6, respectively.[28]

Table 83.2 Computed Tomography Severity Score (Balthazar Score)[a]

Prognostic Indicator	CT Grade Points
Pancreatic inflammation	
Normal CT findings	0
Focal or diffuse enlargement of the pancreas	1
Pancreatic gland abnormalities and peripancreatic inflammation	2
Fluid collection in a single location or phlegmon	3
Two or more fluid collections and/or gas bubbles in or adjacent to pancreas	4
Pancreatic necrosis	
No necrosis	0
<30%	2
30%–50%	4
≥50	6

CT, computed tomography.

[a]Score is based on the degree of necrosis, inflammation, and the presence of fluid collections. In a study involving 268 patients with acute pancreatitis, patients with a score higher than 5 were eight times more likely to die, 17 times more likely to have a prolonged hospital course, and 10 times more likely to undergo pancreatic necrosis resection than were patients with scores lower than 5.[33]

Elevated serum creatinine concentration (>1.8 g/dL) was suggested by Muddana et al.,[34] to be predictive of development of pancreatic necrosis, but a subsequent study by Lankish et al.,[35] did not find a similar association. Wu et al.,[36] found that the admission hematocrit (e.g., hematocrit >44 during the first 48 hours) and blood urea nitrogen (BUN) concentrations were predictive of progression to severe complicated disease. Other laboratory blood tests, such as C-reactive protein (CRP), polymorphonuclear elastase, human pancreas-specific protein/procarboxypeptidase B, α_2-macroglobulin, and serum macrophage migration inhibitory factor, have been researched as markers for development of necrosis, but only CRP is clinically available at present.

Among the imaging studies, bolus-contrasted (pancreatitis protocol) computed tomography (CT) of the pancreas can accurately differentiate interstitial from necrotizing pancreatitis. This imaging test, however, is not possible to obtain in patients who develop renal insufficiency as a complication of pancreatitis. Studies have shown that the persistence of the SIRS (with two or more of the following features: temperature >38°C or <36°C; heart rate >90 beats/minute; respiratory rate >20 breaths/minute; (4) white blood cell count <4,000 cells/mm^3 or >12,000/mm^3; partial pressure of carbon dioxide <32 mm Hg); or organ failure after initial resuscitation, is the strongest predictor of mortality.[29,30]

Nutritional Therapy

Pancreatic rest, as mentioned earlier, is a critical step in the initial management of all patients with AP. The need for nutritional therapy in these patients is determined by the severity of their disease and their nutritional status.

Mild Acute Pancreatitis

In the mild form of AP, management is mainly supportive. Management strategy includes pain control, aggressive hydration with intravenous fluids, and nothing by mouth for pancreatic rest for 48 hours. In most cases, the pain improves, nausea and vomiting resolve, serum lipase and amylase concentrations drop, the patient feels better, and oral intake can be retested. Traditionally, patients have been started on a clear liquid diet and then advanced to a low-fat soft diet. More recent randomized controlled trials (RCTs) have, however, reported that starting these patients on a low-fat soft diet is as safe as a clear liquid diet, with varying effect on length of hospital stay.[37-39] These findings suggest that after 2 to 3 days of pancreatic rest, patients can be tried on a low-fat, soft diet with close monitoring for abdominal pain, nausea, vomiting, or any other complication. Once patients exhibit tolerance to this diet, they can be advanced to a regular diet over the next 3 to 4 days.

Severe Acute Pancreatitis

As discussed earlier, patients with SAP represent a very sick group of patients who often go on to develop complications such as acute fluid collections, pseudocyst collections, pancreatic necrosis, and infection of fluid collection or MOSF. Mortality rates vary between 5% and 50%, depending on access to modern ICU management. Aggressive management in the ICU, especially aggressive hydration with several liters of intravenous fluids to correct hypotension and maintain urine output, is crucial in initial management to prevent progression to organ failure. Patients commonly have nausea and vomiting and eating exacerbates the pain. Consequently, they can be effectively fed only by the intravenous route or by jejunal feeding combined with gastric decompression.

Studies based on critically ill patients[8] and metanalyses have clearly shown that early EN is safer, more effective, and less costly than parenteral nutrition (PN).[40,41] What is fascinating is that EN appears to suppress SIRSs, and given that SIRS is predictive of survival, one could expect EN to reduce mortality.[42,43] Few studies have been powered to show mortality differences, but the most recent study comparing EN and PN, from Russia, showed a significantly lower mortality rate with EN in a study limited to 69 randomized subjects.[44] A subsequent meta-analysis of five RCTs concluded that EN results in clinically relevant and statistically significant risk reduction for infectious complications, pancreatic infections, and mortality in patients with predicted SAP.[45]

EN in patients with SAP, as mentioned earlier, raises the fear of stimulating the pancreas and making the pancreatic inflammation worse. In an early study, Ragins et al.,[46] first reported that in a canine model, jejunal feeding did not stimulate pancreatic secretion, as opposed to intragastric or intraduodenal feeding. However, there are marked differences in the control of pancreatic secretion among mammalian species and, in humans, investigators have shown that the semi elemental diet must be delivered more than 40 cm past the ligament of Treitz before pancreatic stimulation is avoided.[2,47] Consequently, conventional jejunal feeding is still stimulatory because it is generally given into the duodenum or proximal jejunum (**Figure 83.3**). Although TPN does maintain pancreatic rest, its serious septic and metabolic complications outweigh potential benefits.[48,49]

The publication of two RCTs of nasogastric (NG) vs nasojejunal feeding, which showed no difference in outcome, was suggested by some investigators to refute the importance of pancreatic rest.[50,51] However, both feeding arms of the trial were stimulatory because feeding, despite the titles, was generally given into the duodenum or, at best, the proximal jejunum, which is stimulatory.[50,51] Furthermore, there is concern that the overall mortality rates were unacceptably high (25%–35%), a finding

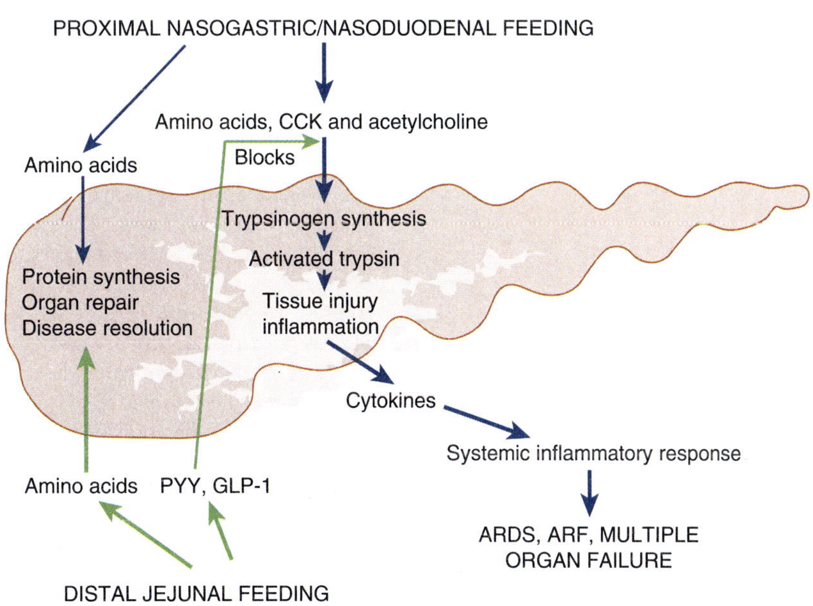

Figure 83.3 The relative effects of proximal vs distal jejunal feeding on the injured pancreas. *ARDS*, acute respiratory distress syndrome; *ARF*, acute renal failure; *CCK*, cholecystokinin; *GLP-1*, glucagon-like peptide-1; *PYY*, peptide YY.

suggesting that neither form of feeding improved outcome. Concern was heightened when it was found, by isotope labeling techniques, that trypsin continued to be synthesized in patients with necrotizing pancreatitis and that the reduced luminal secretion was primarily the result of loss of secretions into the inflammatory mass and circulation.[52]

Careful review of the literature suggests that placement of feeding tubes well below the ligament of Treitz may be beneficial, because the two studies in which this was documented, namely Kalfarentzos et al.,[53] and Petrov et al.,[44] were associated with much lower mortality rates, of 5% to 6%. Despite these concerns, many European centers consider pancreatic rest a minor issue and feed patients with NG feeding tubes. The advantage of this approach is that it facilitates early EN and may, therefore, suppress gut inflammation before it becomes established, and there is no doubt that the earlier EN commences, the easier it is to maintain gut function. Our own practice, as interventional nutritional gastroenterologists, is to place jejunal feeding tubes via transnasal endoscopy as far down the intestine as possible (usually >40 cm past the ligament of Treitz) as soon as the patient is admitted to the hospital. An often forgotten but equally important maneuver is to simultaneously decompress the stomach, because gastric secretions may lead to vomiting and possible aspiration. To ensure this, a double-lumen NG decompression-jejunal feeding tube system is used.[54] However, there is clearly a need for a sufficiently powered prospective RCT of proximal vs distal feeding with analysis of final outcomes to determine whether pancreatic rest allows pancreatitis to resolve more effectively. To answer this challenge, a multicenter study (NIH NCT#00580749) was organized in the United States to compare NG feeding with distal jejunal DJ feeding. The study proved difficult to recruit for because it employed stringent entry criteria in an attempt to keep the study group as homogenous as possible. After 5 years, 28 patients had successfully completed the study. The results showed clearly that patients with severe complicated acute pancreatitis cannot tolerate NG feeding, whereas DJ feeding with BG decompression was highly successful. Feeding failure occurred in 0/14 DJ patients and in 6/11 NG patients ($P = 0.0026$). NG failures were crossed over to DJ feeding with good tolerance and eventual recovery. As a result, the quantity of feed delivered was significantly higher in DJ patients ($P<0.05$). We thus conclude that, because of upper GI compression and pancreatic rest, a jejunal feeding tube should be placed beyond the ligament of Treitz with radiologic or endoscopic confirmation in all patients with this severity of disease.[55] In our practice, TPN is rarely needed, thus minimizing the risk of metabolic and septic complications.

Feeding Techniques

Distal Jejunal Feeding Tube Placement with Gastric Decompression

A nasogastrojejunal (NGJ) feeding tube system (16F/9F-Kendall Dobhoff) with simultaneous gastric decompression (for nausea, vomiting) and jejunal feeding is the optimal approach to feeding malnourished patients with either severe complicated pancreatitis or in survivors left with distortion of the upper GI tract due to necrosis and cyst formation.[43,54] In patients without symptoms of gastric outlet obstruction, a soft nasoenteric tube is used to feed well down the jejunum. Endoscopists often

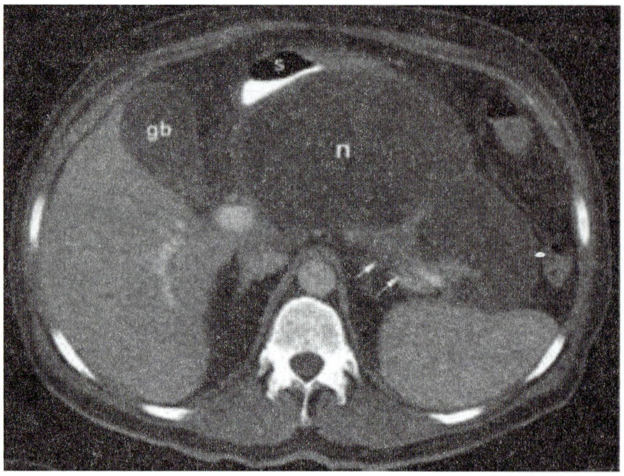

Figure 83.4 Computed tomography scan of patient with necrolysis of the pancreas *(n)* forming a cystic mass *(arrows)* compressing the stomach *(s)*. *gb*, gallbladder.

experience considerable difficulty placing jejunal feeding tubes in patients with AP, because the inflammatory mass distorts the normal contour of the stomach and compresses the duodenum, resulting in functional gastric outlet obstruction (**Figure 83.4**). This problem can be overcome by using transnasal endoscopy, with an ultraslim 5.8-mm video endoscope. This enables cannulation of the nose, stomach, and obstructed duodenum with successful deployment of a guidewire into the jejunum to permit over-the-wire passage of a double-lumen NG decompression jejunal feeding tube (**Figure 83.5**). Our original study showed successful placement of an NGJ tube system in 46 of the initial 51 attempts in ICU patients, who were originally referred for approval of TPN because of clinically presumed ileus.[54] Once the tube is placed, its position is confirmed endoscopically or by imaging. Another major advance is the use of a nasal bridle to prevent accidental dislodgment.[56]

For more detailed illustration open this link: (https://youtu.be/YlkiY6YTnYk, and https://youtu.be/UjVjDkdDLW0.

When Should Interventional Feeding Be Commenced?

The overuse of interventional feeding has recently been questioned by several authorities. For example, the use of "early" full enteral feeding has been questioned by **two** large ICU studies. A recently reported multicenter Dutch study, which included 208 SAP patients from 19 hospitals, showed that more feeding was tolerated by early tube feeding, but that there were no significant subsequent reductions in complications or mortality.[57] Furthermore, the conduct of two large, well-powered randomized clinical trials has provided compelling evidence that early aggressive enteral feeding may actually worsen outcome in ICU patients.[58,59] Finally, Compher et al., recently published the results of their analysis of differences in outcome from the ICU in patients given variable quantities and formulas of nutrition support. Their database was large: 2,320 citations were identified, and data were abstracted from 36 trials, which included 20,578 participants. Remarkably, the outcome was not significantly different if patients were prescribed higher or lower protein or energy doses.[60]

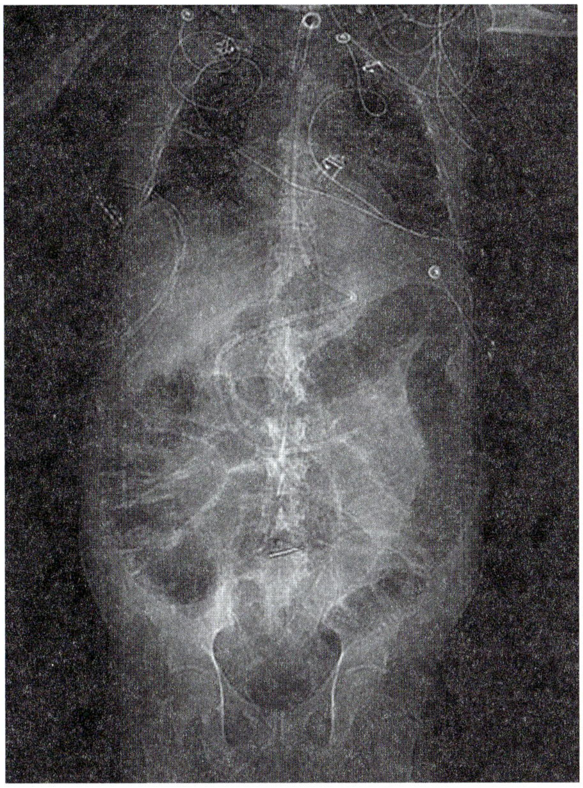

Figure 83.5 Radiographic view of a double-lumen tube (the opaque tube tip is visible at midline) that was successfully placed by transnasal endoscopy in a patient with vomiting, abdominal pain, acute respiratory distress syndrome, and ileus resulting from severe acute pancreatitis, which had also caused duodenal compression.

We have argued that most acute pancreatitis patients admitted to the ICU today are overweight and have sufficient body stores to survive without interventional feeding for 2 weeks. Supportive IV fluids, electrolytes, and micronutrients should be sufficient, unless they were previously malnourished with BMI below 18 kg/m^2.[61]

Feeding

1. A high-protein, low-fat, semielemental feeding formula is recommended.
2. Tube feeding should be started slowly at 20 to 25 mL/hour for 24 hours. If it is well tolerated, rates should be increased with 25-mL increments to the goal, which is estimated to be 25 kcal/kg ideal body weight/day.
3. Ileus is not a contraindication to EN; in fact, EN is the treatment for ileus, but the feeding must be advanced slowly.
4. Every effort should be made to initiate EN as early as possible following admission, because early feeding, compared with feedings after 72 hours, is associated with less gut permeability and diminished activation and release of inflammatory cytokines.[40]
5. Most of these sick patients are in and out of the ICU for procedures and imaging studies and often do not receive 100% of goal for energy by the enteral route. Every effort should be made to maximize the delivery of EN because providing energy and protein closer to goal has been shown to improve outcome.[62,63]
6. Patients should be watched for signs of intolerance, including gastric residual volumes greater than 400 mL, vomiting, and diarrhea (>5 watery stools or >500 mL/24 hours).
7. Patients who have hemodynamic instability, low blood pressures, and are receiving pressor medications to support blood pressure, require closer supervision for signs of intolerance (abdomen distention, hypoactive bowel sounds), because these patients may be at risk of developing gut dysfunction and bowel ischemia.

Aspiration Prevention, Surveillance, and Maintenance of the Tube System

One of the most important ways to prevent aspiration is to elevate the head of the bed 30 to 45 degrees. The second most important action is to prevent overfilling of the stomach with fluid, whether it is feeding or secretions. However, more recent studies have shown that gastric residual volumes are often higher than normal in critically ill patients because of delayed gastric emptying and are well tolerated up to 500 mL.[64] This concern is avoided if NGJ feeding is used, because low-pressure (50 mm Hg) intermittent suction is attached to the gastric port. However, in NG feeding, it is critical to check gastric residual volumes every 4 hours and to hold NG feeding if gastric residual volumes are greater than 400 mL. Feeding tubes should be flushed with 10 to 20 mL tap water every 4 hours. Although NG feeding tubes can be used for medications (crushed if necessary), jejunal tubes must be reserved for feeding alone, unless no alternative route is available and provided the medication comes in liquid form. The tube must be immediately flushed with 10 mL tap water. High gastric residual volumes can be treated by antisecretory agents (e.g., proton pump inhibitors), with consideration of adding the opioid antagonist naloxone to the feeding if narcotics are being used.

Severe Diarrhea

Severe diarrhea, defined as more than five liquid stools in a 24-hour period, more than 500 mL/volume[43] or, in patients with rectal tubes, volumes greater than 500 mL/24 hours, is rarely related to feeding in critically ill patients. In one report, approximately 25% of patients with necrotizing pancreatitis developed severe diarrhea during tube feeding.[65] In a subgroup of these patients, colonic microbiota and fermentation were measured, and both were found to be dramatically suppressed, probably because of the effects of the prolonged use of semielemental or elemental diets deficient in fiber, together with the widespread use of prophylactic antibiotics and proton pump inhibitors. From this finding, the chief cause of diarrhea is suspected to be disturbance of the microbiota and dysbiosis that forms a permissive environment for *Clostridium difficile* infection.[66] Consequently, commercially available soluble fiber (e.g., 10 to 20 g/day) should be added to the tube feedings after the first week of feeding, together with avoidance of prophylactic antibiotics and proton pump inhibitor drugs, although this practice is not supported by controlled trials. Another cause of diarrhea is sorbitol-containing medications.

Severe diarrhea is managed by first checking medications and making the appropriate changes (e.g., stop prophylactic antibiotics, discontinuing sorbitol-containing medications), ruling out *C. difficile* infection, and then using antidiarrheal agents, such as liquid loperamide, added to the tube feeding. If diarrhea continues after 3 days of the foregoing, the feeding can be held for 24 hours. If the diarrhea resolves, feeding can be restarted at 25 mL/hour and advanced with tolerance. If severe diarrhea returns, feeding should be continued as a slower rate (e.g., 20 to 40 mL/hour) with fiber supplementation, and the balance of nutrition should be given by PN until the diarrhea abates.

Home Care

In a 6-month practice review, our service placed 63 distal-jejunal feeding tubes in patients with either AP ($n = 28$), CP (pseudocysts with duodenal compression [$n = 14$], or recurrent AP [$n = 21$]). Fifteen patients with stable pancreatic fluid collections or pseudocysts were discharged home with feeding continuing from 2 weeks to 3 months until the collection either resolved spontaneously or was drained surgically, with surprisingly good tolerance in most.[67] By using this strategy, PN was avoided in 97% of patients initially referred for PN. However, patients and caregivers must be trained, as with patients receiving home TPN, to detect early complications and avoid tube blockages with frequent flushes.

Chronic Pancreatitis

Demographic and Clinical Presentation

CP is a persistent inflammatory state of the pancreas that is characterized by progressive, irreversible damage to the pancreas and that leads to extensive fibrosis and progressive exocrine and endocrine insufficiency. The annual incidence ranges from 3.5 to 10 cases per 10,000.[68] Among the Western population, alcohol consumption accounts for 70% to 90% of cases. Clinical presentations of these patients include abdominal pain, diarrhea, and weight loss. Patients often experience intermittent flare-ups of pancreatitis and often develop a state of chronic abdominal pain. Malabsorption develops because of exocrine gland destruction, and this can lead to steatorrhea, weight loss, and deficiency of fat-soluble vitamins. Ongoing smoking and alcohol intake expedite the loss of function. Over years, these patients can develop diabetes, which complicates nutritional management further; finally, CP is a risk factor for development of pancreatic cancer.[69,70]

Pathophysiology

Protein-energy malnutrition is common in patients with CP. Most patients with CP have alcoholism and poor nutritional status even in the absence of pancreatitis.[71] In the presence of pancreatitis, additional factors contribute toward the development of protein-energy malnutrition. Chronic inflammation in the pancreas can lead to abdominal pain, aversion to food (sitophobia), poor oral intake, and a hypermetabolic state.[72] Progressive glandular destruction leads to both exocrine and endocrine glandular destruction. Exocrine insufficiency is clinically apparent after 90% of the gland atrophies and manifests as malabsorption of carbohydrates, fats, and proteins.[73,74] Endocrine insufficiency usually manifests 7 to 15 years after diagnosis of CP as glucose intolerance (diabetes). Gastric dysmotility is yet another associated finding and has been shown to be present in as many as 44% patients with small duct CP.[75,76]

Basal metabolic rate has been shown to be higher in patients with CP compared with physiologically normal people. This is particularly true for underweight alcoholic patients with CP.[77] Hence, 35 kcal/kg/day of energy intake and 1.0 to 1.5 g/kg/day of protein are generally recommended for these patients.[78] A diet rich in carbohydrate and proteins is recommended, although concurrent diabetes may limit intake of carbohydrates. Patients with CP are at risk for developing deficiencies of vitamins and trace elements. Fat malabsorption puts these patients at risk for malabsorption of fat-soluble vitamins (A, D, E, and K).[79-82] Case reports of vitamin K deficiency with bleeding have been reported. About 20% of patients with CP have clinical vitamin D deficiency with secondary hyperparathyroidism.[83] Vitamin B_{12} deficiency can be seen in as many as 30% of patients with exocrine pancreatic insufficiency.[84] This is explained by the lack of proteolytic pancreatic enzymes in the duodenum that are normally responsible for cleavage of the cobalamin–R-protein complex and allow cobalamin to bind intrinsic factor in the small bowel for absorption in terminal ileum.

Nutritional Assessment

Proper assessment of patients' nutritional status should be done before initiating nutritional support. Patients should ideally be seen in a multidisciplinary clinic containing a dietitian, a pain management specialist, and a pancreatologist. History taking should focus on dietary habits, postprandial abdominal pain, diarrhea, weight loss or failure to gain weight, and diabetes. Questions should be tailored to discern the presence of pancreatic exocrine insufficiency (pale, bulky stools, oil droplets). If malabsorption is suspected, fat absorption should be measured by collection of 72-hour stools while the patient consumes a 100-g fat diet.[85] Alternatively, absorption can be measured by an oral carbon-13–labeled triolein breath test.[85] Pancreatic enzyme secretion can be assessed from the content of enzymes in stimulated duodenal juice at the time of endoscopy[86] or by fecal elastase. Because alcohol is the most common cause in these patients, close attention should be paid to its ongoing use. A dietary diary may be helpful.

Physical examination begins with weight and height measurement. Body mass index (kg/m^2) should be calculated, and the patient assigned as: underweight (<18.5 kg/m^2), normal weight (18.5 to 24.9 kg/m^2), overweight (25.0 to 29.9 kg/m^2), or obese (>30.0 kg/m^2). The presence of loss of subcutaneous fat stores (e.g., triceps skinfold thicknesses), poor dentition, generalized muscle loss, loss of hair, generalized pitting edema, and cutaneous bruising (vitamin K deficiency) should be noted and documented.

Blood laboratory studies including complete blood count, serum albumin concentration, and serum electrolyte concentrations should be obtained. At the initial visit, serum vitamin B_{12} and fat-soluble vitamins, such as retinol, prothrombin time (index of vitamin K status), and 25-hydroxy-cholecalciferol (index of vitamin D status), should be obtained as well. Bone mineral density should also be measured by dual-emission x-ray absorptiometry (DEXA) scan. Total energy requirements can be calculated using the Harris-Benedict formula (**Table 83.3**).

Nutritional Therapy

A key step in management of these patients is complete abstinence from alcohol. Some studies have shown that continued alcohol ingestion can lead to progression of disease, but this

Table 83.3 Harris-Benedict Equation for Calculating the Basal Metabolic Rate[a]

Males	BMR = 66.5 + (13.8 × body weight in kg) + (5.0 × height in cm) − (6.8 × age in y)
Females	BMR = 655 + (9.6 × body weight in kg) + (1.8 × height in cm) − (4.7 × age in y)

BMR, basal metabolic rate.

[a]Once BMR is calculated, caloric needs are calculated by multiplying the BMR by the activity factors as following: sedentary or little or no exercise, multiply BMR by 1.2; lightly active, light exercise, or sports 1 to 3 days per week, multiply BMR by 1.37; moderately active, moderate exercise, or sports 3 to 5 days per week, multiply BMR by 1.55; heavy exercise, hard exercise, or sports 6 to 7 days per week, multiply BMR by 1.72.

finding has not been seen in all studies.[87-89] Smoking has been shown to be a critical factor in the progression of AP to CP and in the progression of disease.[90] Another important component is management of abdominal pain. Patients with CP often develop postprandial abdominal pain that lead to sitophobia. The origin of abdominal pain in CP is complex; some patients do not have pain, whereas others are moribund with uncontrollable pain. Unfortunately, many patients are receiving narcotics, and narcotics reduce motility, suppress gastric emptying, and thus aggravate anorexia. As the disease progresses, pain may also progress, but sometimes it diminishes. By this time, however, digestion becomes impaired by exocrine insufficiency, and the assimilation of absorbed nutrients becomes defective secondary to endocrine failure. Thus, the nutritional management changes with the phase of disease.

Early Chronic Pancreatitis

Key recommendations in this phase include complete abstinence from alcohol and smoking and limitation of narcotic medications. Patients should be referred to a dietitian to supplement the diet appropriately. Dietary modification, small and frequent nutritionally balanced meals, and avoidance of fatty foods are key. Multivitamin-mineral supplements may be required.

Later Disease: Pancreatic Insufficiency

The reserve of the pancreas is considerable: steatorrhea secondary to fat maldigestion occurs only when more than 90% of lipase secretion is lost.[76] The cornerstone of therapy is exogenous pancreatic enzyme supplementation. Conventional supplements are of animal origin, but progress is being made with genetic manipulation of algae. Even though the enzymes are active in experiments outside of the body, their efficacy in reversing steatorrhea is never complete: they reduce fat malabsorption, but rarely reverse it,[85] possibly because they cannot mimic exactly pancreatic physiologic secretion of enzymes in a highly controlled manner to match the quantity of food entering the duodenum. For example, in patients with severe malabsorption, two capsules of an enteric preparation containing 10,000 U lipase are given before and two after meals; two more capsules are given with snacks, which adds up to 40,000 U per meals, like what the pancreas would secrete. Despite this regimen, fat absorption often remains abnormal. This response is difficult to understand because only 10% of normal exocrine secretion is believed to be required to maintain fat digestion and absorption. Some preparations are not enteric coated and risk being degraded by gastric acid in patients with normal acid secretion unless patients are concomitantly treated with acid suppressants.[85] Patients with CP often have small bowel bacterial overgrowth secondary to dysmotility, and this may also interfere with the efficacy of supplements.

Approximately 10% to 15% of patients are unable to sustain adequate nutritional intake with a low-fat diet and pancreatic enzyme supplementation. These patients can be managed by feeding into the jejunum using a nasoenteric feeding tube, percutaneous endoscopic gastrostomy (PEG), or a PEG with a jejunal feeding tube extension. Investigators showed that long-term feeding with PEG-J tubes not only improved nutrition but also reduced analgesia consumption.[86] In rare instances, TPN is required for nutritional therapy, but this should be limited to short-term use to replenish nutrient stores.

Late Disease: "Brittle" Diabetes

The progression from chronic to calcific end-stage pancreatitis is an extraordinarily difficult nutritional management problem. In essence, the pancreas, with the liver, may be considered the "heart of nutrition and metabolism." Pancreatic endocrine failure usually occurs some years after exocrine failure. This results in loss not only of insulin secretion but also of glucagon secretion, which leads to "brittle" diabetes and life-threatening hypoglycemia when insulin is administered, and the patient misses a meal.[91] Glucagon triggers glycogenolysis in hypoglycemia and thus counteracts the actions of insulin. The improvements of digestion and absorption with potent pancreatic enzyme supplements increases glucose absorption and provoke hyperglycemia in diabetes patients. Although insulin therapy suppresses hyperglycemia, its overuse results in profound hypoglycemia because no glucagon is secreted.

Regular food intake and tailored insulin therapy with close blood glucose monitoring are essential to maintaining nutritional status and to preventing neurologic injury from neuroglycopenia. For this reason, gastroenterologists must work closely with endocrinologists in this group of patients. In patients who have access to a center, pancreas transplantation may be the best answer to both endocrine and exocrine dilemmas.

Pancreatic Cancer

Pancreatic carcinoma is the fourth leading cause of cancer-related deaths in the United States. Most (85%) of the tumors are ductal in origin.[92] Approximately 42,000 people are diagnosed with pancreatic cancer every year, and most of them die within 1 year.[92] Risk factors for developing pancreatic cancer include smoking, heavy alcohol use, a family history of pancreatic or other cancers, a history of diabetes, and CP.[93-96] Regarding dietary factors, an increased risk is associated with high consumption of grilled meat (heterocyclic amines), cholesterol, fried foods, and other foods containing nitrosamines.[97]

Clinical presentations include abdominal pain, weight loss, and jaundice. Abdominal pain often radiates to the back, can be persistent or intermittent, and is sometimes precipitated by eating. Weight loss can be profound and is often associated with anorexia and early satiety. Diabetes, diarrhea, and steatorrhea may occur. Prognosis is poor, and surgical resection (pancreaticoduodenectomy) is the only potential curative treatment. Most of these patients present late, and the option of pancreaticoduodenectomy with complete tumor removal is available to only 15% to 20% of patients. The 5-year survival following pancreaticoduodenectomy is only about 25% to 30% for lymph node–negative and 10% for lymph node–positive tumors.[98,99]

From a nutritional standpoint, these patients suffer from abdominal pain, nausea, emesis, poor oral intake, change in taste sensations, early satiety, fatigue, gastric outlet obstruction, malabsorption, and maldigestion. The constellation of these symptoms is often referred to as the anorexia-cachexia syndrome and is associated with poor response to chemotherapy and overall prognosis. Because patients with pancreatic cancer are diagnosed late, as many as 80% have the anorexia-cachexia syndrome at presentation.[100,101] Poor appetite is often secondary to abdominal pain, pancreatic cancer–related duodenal obstruction, exocrine pancreatic insufficiency, depression, and side effects of treatment such as radiation therapy or chemotherapy. These patients often report changes in smell and taste and resultant poor oral intake. Tumor-generated cytokines (e.g., IL-1, IL-6, TNF-α, interferon-γ) also seem to contribute to the catabolic state by altering energy metabolism.

Patients who have lost significant (>10% of their baseline) weight or who have reduced oral energy intake over a 2- to 24-week period are at increased risk for both micronutrient and macronutrient deficiencies. Nutritional therapy can help stabilize these patients and improve their functional status, response to therapy, and prognosis. Nutritional evaluation includes generic evaluation parameters such as body mass index, weight change, midarm muscle circumference, triceps skinfold thickness and laboratory measurements such as prealbumin and albumin, although the last two values better reflect disease activity and prognosis.

Nutritional therapy for pancreatic cancer involves pain and nausea control. Megestrol acetate has shown promise in treating these patients by stimulation of appetite and antagonism of the metabolic effects of the catabolic cytokines produced by cancer and immune cells.[102] Fish oil can reduce production of proinflammatory cytokines in patients with cancer and may be helpful in reversing the weight loss by modulating the metabolic responses to feeding.[103,104] Clinical studies have yielded mixed findings, however. They probably are helpful, but the optimum dose to maximize weight gain and minimize side effects has not been determined.[103-106]

Gordon et al.,[107] studied use of thalidomide (anti-TNF agent) in patients with advanced pancreatic cancer–related cachexia. Fifty patients with advanced pancreatic cancer who had lost at least 10% of their body weight were randomized to receive thalidomide (200 mg daily) or placebo for 24 weeks in a single-center, double-blind RCT. At 8 weeks, patients taking thalidomide had significantly less weight loss and significantly more arm muscle mass compared with placebo.[107]

Another key component in the management of these patients is to recognize and treat pancreatic exocrine insufficiency. Because most of pancreatic cancers are in the head of the pancreas, the tumor can lead to exocrine glandular insufficiency by blocking the pancreatic duct.[33] These patients and those who have undergone pancreaticoduodenectomy benefit from optimum pancreatic enzyme supplementation. Preoperative nutritional support is important to correct fluid, electrolyte, and micronutrient deficiencies, but time should not be wasted trying to correct body mass when the cancer may be at a resectable stage. Postoperative feeding is important because many patients take time to tolerate normal food after a Whipple procedure. Use of an NGJ feeding system is recommended immediately after surgery because gastric outlet dysfunction nearly always occurs for a variable period.

References

1. O'Keefe SJ, Lee RB, Anderson FP, et al. Am J Physiol 2003;284: G27–G36.
2. Kaushik N, Pietraszewski M, Holst JJ, et al. Pancreas 2005;31: 353–9.
3. Lankisch PG, Lowenfels AB, Maisonneuve P. Pancreas 2002;25: 411–2.
4. Spanier BW, Dijkgraaf MG, Bruno MJ. Best Pract Res 2008;22: 45–63.
5. Venneman NG, Buskens E, Besselink MG, et al. Am J Gastroenterol 2005;100:2540–50.
6. Walker WA, Goulet O, Kleinman RE, et al, eds. Pediatric Gastrointestinal Disease: Pathophysiology, Diagnosis, Management. 4th ed. Hamilton, Ontario, Canada: BC Decker, 2004:1584–97.
7. National Institutes of Health. Drinking in the United States: Main Findings from the 1992 National Longitudinal Alcohol Epidemiologic Survey (NLAES), vol. 6. Bethesda, MD: National Institutes of Health, 1998. NIH publication 99-3519.
8. Abou-Assi S, Craig K, O'Keefe SJ. Am J Gastroenterol 2002;97: 2255–62.
9. Gorelick FS, Modlin IM, Leach SD, et al. Yale J Biol Med 1992;65:407–20.
10. Leach SD, Modlin IM, Scheele GA, et al. J Clin Invest 1991;87: 362–6.
11. Aoun E, Chen J, Reighard D, et al. Pancreatol 2009;9:777–85.
12. Pezzilli R, Billi P, Miniero R, et al. Dig Dis Sci 1995;40:2341–8.
13. Stoelben E, Nagel M, Ockert D, et al. Chirurg 1996;67:1231–6.
14. Martin MA, Saracibar E, Santamaria A, et al. Rev Esp Enferm Dig 2008;100:768–73 [Spanish].
15. Ueda T, Takeyama Y, Yasuda T, et al. Surgery 2007;142: 319–26.
16. Andoh A, Takaya H, Saotome T, et al. Gastroenterol 2000;119: 211–9.
17. Osman MO, Gesser B, Mortensen JT, et al. Cytokine 2002; 17:53–9.
18. Shi C, Zhao X, Wang X, et al. Scand J Gastroenterol 2005;40: 103–8.
19. Makhija R, Kingsnorth AN. J Hepatobiliary Pancreat Surg 2002;9: 401–10.
20. Bhatia M, Wong FL, Cao Y, et al. Pancreatol 2005;5:132–44.
21. Rahman SH, Ammori BJ, Holmfield J, et al. J Gastrointest Surg 2003;7:26–35.
22. Fong YM, Marano MA, Moldawer LL, et al. J Clin Invest 1990;85:1896–04.
23. Hegazi R, Raina A, Graham T, et al. JPEN J Parenter Enteral Nutr 2011;35:91–6.
24. Shah U, Shenoy-Bhangle AS. N Engl J Med 2011;365:1528–1536.
25. Yadav D, Agarwal N, Pitchumoni CS. Am J Gastroenterol 2002;97:1309–18.
26. Wu BU, Johannes RS, Sun X et al. Gut 2008;57:1698–703.
27. Ranson JH, Rifkind KM, Roses DF, et al. Surg Gynecol Obstet 1974;139:69–81.
28. Banks PA, Freeman ML. Am J Gastroenterol 2006;101:2379–400.

29. Mofidi R, Duff MD, Wigmore SJ, et al. Br J Surg 2006;93: 738–44.
30. Buter A, Imrie CW, Carter CR, et al. Br J Surg 2002;89: 298–302.
31. Balthazar EJ, Robinson DL, Megibow AJ, et al. Radiol 1990;174: 331–6.
32. Robert JH, Frossard JL, Mermillod B, et al. World J Surg 2002;26: 612–9.
33. Simchuk EJ, Traverso LW, Nukui Y, et al. Am J Surg 2000;179: 352–5.
34. Muddana V, Whitcomb DC, Khalid A, et al. Am J Gastroenterol 2009;104:164–70.
35. Lankisch PG, Weber-Dany B, Maisonneuve P, et al. Am J Gastroenterol 2010;105:1196–200.
36. Wu BU, Johannes RS, Sun X, et al. Gastroenterol 2009;137: 129–35.
37. Teich N, Aghdassi A, Fischer J, et al. Pancreas 2010;39: 1088–92.
38. Jacobson BC, Vander Vliet MB, Hughes MD, et al. Clin Gastroenterol Hepatol 2007;5:946–51.
39. Sathiaraj E, Murthy S, Mansard MJ, et al. Aliment Pharmacol Ther 2008;28:777–81.
40. Heyland DK, Dhaliwal R, Drover JW, et al. JPEN J Parenter Enteral Nutr 2003;27:355–73.
41. Marik PE, Zaloga GP. Crit Care Med 2001;29:2264–70.
42. Hegazi RA, O'Keefe SJ. Curr Gastroenterol Rep 2007;9:99–106.
43. O'Keefe SJ. Nat Rev Gastroenterol Hepatol 2009;6:207–15.
44. Petrov MS, Kukosh MV, Emelyanov NV. Dig Surg 2006;23: 336–44.
45. Petrov MS, van Santvoort HC, Besselink MG, et al. Arch Surg 2008;143:1111–7.
46. Ragins H, Levenson SM, Signer R, et al. Am J Surg 1973;126: 606–14.
47. Vu MK, van der Veek PP, Frolich M, et al. Eur J Clin Invest 1999;29:1053–9.
48. O'Keefe SJ. Nat Clin Prac 2007;4:488–9.
49. O'Keefe SJ. Curr Opin Clin Nutr Metab Care 2006;9:622–8.
50. Kumar A, Singh N, Prakash S, et al. J Clin Gastroenterol 2006;40:431–4.
51. Eatock FC, Chong P, Menezes N, et al. Am J Gastroenterol 2005;100:432–9.
52. O'Keefe SJ, Lee RB, Li J, et al. Am J Physiol 2005;289:G181–7.
53. Kalfarentzos F, Kehagias J, Mead N, et al. Br J Surg 1997;84: 1665–9.
54 O'Keefe SJ, Foody W, Gill S. JPEN J Parenter Enteral Nutr 2003;27:349–54.
55. O'Keefe SJ, Cariem AK, Levy M. J Clin Gastroenterol 2001;32: 319–23.
56. Seder CW, Stockdale W, Hale L, et al. Crit Care Med 2010;38: 797–801.
57. Bakker OJ, van Brunschot S, van Santvoort HC, et al. N Engl J Med 2014;371:1983–93.
58. Casaer MP, Mesotten D, Hermans G, et al. N Engl J Med 2011;365:506–17.
59. Rice TW, The National Heart, Lung, and Blood Institute Acute Respiratory Distress Syndrome (ARDS) Clinical Trials Network. JAMA 2012;307:795–803.
60. Compher C, Bingham AL, McCall M, et al. JPEN J Parenter Enteral Nutr 2022;46:12–41.
61. O'Keefe SJ. The Principles and Practice of Nutritional Support. New York, NY: Springer. 2015; 1:1–241.
62. Artinian V, Krayem H, DiGiovine B. Chest 2006;129:960–7.
63. Barr J, Hecht M, Flavin KE, et al. Chest 2004;125:1446–57.
64. Martindale RG, McClave SA, Vanek VW, et al. Crit Care Med 2009;37:1757–61.
65. Hegazi R, Raina A, Graham T, et al. JPEN J Parenter Enteral Nutr 2011;35:91–6.
66. O'Keefe SJ, Ou J, Delany JP, et al. World J Gastrointest Pathophysiol 2011;2:138–45.
67. Rolniak S RA, Hegazi R, Centa-Wagner P, et al. Gastroenterol 2009;136:A-76.
68. Etemad B, Whitcomb DC. Gastroenterol 2001;120:682–707.
69. Lankisch PG, Lohr-Happe A, Otto J, et al. Digestion 1993;54: 148–55.
70. Wakasugi H, Funakoshi A, Iguchi H. J Gastroenterol 1998;33: 254–9.
71. Manari AP, Preedy VR, Peters TJ. Addict Biol 2003;8:201–10.
72. Duggan S, O'Sullivan M, Feehan S, et al. Nutr Clin Pract 2010;25:362–70.
73. Petersen JM, Forsmark CE. Semin Gastrointest Dis 2002;13: 191–199.
74. DiMagno EP, Go VL, Summerskill WH. N Engl J Med 1973;288: 813–5.
75. Regan PT, Malagelada JR, Dimagno EP, et al. Gut 1979;20: 249–54.
76. Chowdhury RS, Forsmark CE, Davis RH, et al. Pancreas 2003;26: 235–8.
77. Hebuterne X, Hastier P, Peroux JL, et al. Dig Dis Sci 1996;41: 533–9.
78. Giger U, Stanga Z, DeLegge MH. Nutr Clin Pract 2004;19: 37–49.
79. Kalvaria I, Labadarios D, Shephard GS, et al. Int J Pancreatol 1986;1:119–28.
80. Marotta F, Labadarios D, Frazer L, et al. Dig Dis Sci 1994;39: 993–8.
81. Nakamura T, Takebe K, Imamura K, et al. Acta Gastroenterol Belg 1996;59:10–4.
82. Yokota T, Tsuchiya K, Furukawa T, et al. J Neurol 1990;237: 103–6.
83. Mann ST, Stracke H, Lange U, et al. Metab 2003;52:579–85.
84. Glasbrenner B, Malfertheiner P, Buchler M, et al. Klin Wochenschr 1991;69:168–72.
85. Marotta F, O'Keefe SJ, Marks IN, et al. Dig Dis Sci 1989;34: 456–61.
86. Stanga Z, Giger U, Marx A, et al. JPEN J Parenter Enteral Nutr 2005;29:12–20.
87. Nordback I, Pelli H, Lappalainen-Lehto R, et al. Gastroenterol 2009;136:848–55.
88. Takeyama Y. Clin Gastroenterol Hepatol 2009;7(suppl):S15–7.
89. Pelli H, Sand J, Laippala P, et al. Scand J Gastroenterol 2000;35:552–5.
90. Yadav D, Whitcomb DC. Nat Rev Gastroenterol Hepatol 2010;7:131–45.
91. O'Keefe SJ, Cariem AK, Levy M. J Clin Gastroenterol 2001;32: 319–23.
92. Jemal A, Siegel R, Ward E, et al. CA Cancer J Clin 2009;59: 225–49.
93. Lynch SM, Vrieling A, Lubin JH, et al. Am J Epidemiol 2009;170:403–13.
94. Nothlings U, Wilkens LR, Murphy SP, et al. J Natl Cancer Inst 2005;97:1458–65.
95. Ghadirian P, Boyle P, Simard A, et al. Int J Pancreatol 1991;10: 183–96.
96. Stevens RJ, Roddam AW, Beral V. Br J Cancer 2007;96:507–9.
97. Zheng W, Lee SA. Nutr Cancer 2009;61:437–46.
98. Trede M, Schwall G, Saeger HD. Ann Surg 1990;211:447–58.
99. Yeo CJ, Cameron JL, Lillemoe KD, et al. Ann Surg 1995;221: 721–31.
100. Ryan DP, Grossbard ML. Oncologist 1998;3:178–88.
101. Splinter TA. Ann Oncol 1992;3(suppl 3):25–7.

102. Berenstein EG, Ortiz Z. Cochrane Database System Rev 2005;(2): CD004310.
103. Fearon KC, Von Meyenfeldt MF, Moses AG, et al. Gut 2003;52: 1479–86.
104. Brown TT, Zelnik DL, Dobs AS. Int J Gastrointest Cancer 2003;34:143–50.
105. Wigmore SJ, Ross JA, Falconer JS, et al. Nutr 1996;12(suppl): S27–30.
106. Harle L, Brown T, Laheru D, et al. J Altern Complement Med 2005;11:1039–46.
107. Gordon JN, Trebble TM, Ellis RD, et al. Gut 2005;54:540–5.

D. DISORDERS OF THE ALIMENTARY TRACT

CHAPTER 84

Alcohol Use Disorder and Associated Liver Disease

Craig J. McClain, Loretta L. Jophlin, Endashaw M. Omer, and Irina Kirpich

CHAPTER OUTLINE

Abbreviations

AC, alcohol-associated cirrhosis; **ADH**, alcohol dehydrogenase; **AH**, alcohol associated hepatitis; **ALD**, alcohol-associated liver disease; **ALDH**, aldehyde dehydrogenase; **AUD**, alcohol use disorder; **BA**, bile acids; **BCAA**, branched-chain amino acid; **CNS**, central nervous system; **EN**, enteral nutrition; **FMT**, fecal microbiota transplant; **GSH**, glutathione; **HE**, hepatic encephalopathy; **LES**, late evening snack; **LGG**, Lactobacillus rhamnosus Gorbach-Goldin; **LPS**, lipopolysaccharide; **NAD$^+$**, oxidized nicotinamide adenine dinucleotide; **NADH**, reduced nicotinamide adenine dinucleotide; **NADP$^+$**, oxidized nicotinamide adenine dinucleotide phosphate; **NAFLD**, nonalcoholic fatty liver disease; **NASH**, nonalcoholic steatohepatitis; **PAMPs**, pathogen-associated molecular patterns; **PEM**, protein-energy malnutrition; **RNS**, reactive nitrogen species; **ROS**, reactive oxygen species; **RQ**, respiratory quotient; **SAM**, *S*-adenosylmethionine; **SCFAs**, short chain fatty acids; **SEC**, sinusoidal endothelial cell; **TLR-4**, Toll-like receptor-4; **TNF**, tumor necrosis factor; **VA**, Veterans Affairs; **WKS**, Wernicke–Korsakoff syndrome

Overview of the Liver and Alcohol-Associated Liver Disease (ALD)

The liver is the largest organ in the body, and it has a unique dual blood supply, perfused by both the portal vein (directly exposed to absorbed nutrients) and the hepatic artery. The liver is composed of multiple cell types with differing functions. Hepatocytes make up more than 80% of total liver mass and play a critical role in the metabolism of alcohol, amino acids and ammonia, lipids, carbohydrates, vitamins, minerals, hormones, and detoxification of a variety of drugs and xenobiotics. Hepatic stellate cells comprise the major storehouse for vitamin A in the body, and they play a critical role in collagen formation during liver injury and fibrosis. Sinusoidal endothelial cells (SECs) make up approximately half of the nonparenchymal cells of the liver and play an important role in controlling the exchange of materials (including nutrients) between the bloodstream and the liver parenchyma. The SECs express scavenger receptors act as antigen-presenting cells, to name only a few of the important immune functions of these cells. Hepatic Kupffer cells comprise the largest reservoir of fixed macrophages in the body. They play a protective role against gut-derived toxins that have escaped into the portal circulation, and they are a major producer of cytokines, which can markedly influence nutritional status. Bile duct epithelial cells play a major role in transport (e.g., water, bile), express a variety of transporters, and have important immune functions. All of these cell types interact in a coordinated fashion to protect against gut-derived toxins and autoimmune reactions (tolerance), and they modulate hormonal and nutritional status. These cells also play a critical role in the wound-healing response (fibrosis). Liver cells both metabolize alcohol and respond to direct or indirect toxins generated by alcohol metabolism. Importantly, misuse of alcohol is the most common etiology of cirrhosis in the developed world. A recent study from the United States found that between 1999 and 2016, annual deaths from cirrhosis increased by 65%, and rates of hepatocellular carcinoma doubled.[1] Alcohol was a major contributing factor to these deaths.[2] The COVID-19 pandemic further increased alcohol misue and rates of hospitalization for ALD.[3] Thus, liver diseases, including ALD, are a highly relevant and increasing problem.

This chapter reviews alcohol consumption and metabolism, and mechanisms of alcohol-induced liver injury, with a focus on the gut-liver axis, general malnutrition in ALD, deficiencies of specific nutrients, and nutrition support in ALD. It also highlights how many principles of altered nutrition and nutrition support in ALD apply to other advanced liver diseases (e.g., hepatitis C, nonalcoholic steatohepatitis [NASH]).

Alcohol Consumption and Metabolism

Alcohol consumption is common worldwide and in the United States.[1,2] The amounts and patterns of alcohol intake are very important. The 2020–2025 Dietary Guidelines for Americans (U.S. Department of Health and Human Services and U.S. Department of Agriculture, 2020) highlight the concept of the standard drink and the fact that if alcohol is consumed, it should be in moderation (i.e., up to one drink/day for women and two drinks/day for men in adults of legal drinking age) (**Figure 84.1**). Importantly, a standard drink contains five ounces or 14 grams of alcohol. Moderate drinking, as defined here, is unlikely to lead to liver injury and, in otherwise healthy individuals, may actually be protective against some health issues, such as cardiovascular disease.[4] On the other hand, long-term heavy alcohol abuse can cause organ injury, which may, at least in part, result from alcohol-nutrient interactions and alcohol-related nutrient deficiencies. People who misuse alcohol frequently consume large amounts of alcohol, which may contribute to the displacement of needed nutrients. Indeed, a recent analysis of nutritional status and alcohol consumption in individuals with alcohol use disorder (AUD) who were admitted to a treatment program demonstrated that the participants generally had a normal body mass index (BMI) and were not overtly malnourished. However, they consumed about 14 drinks per day, equivalent to about 2,000 kcal per day of "empty calories".[5] Indeed, while they were not overtly malnourished, many had low levels of individual nutrients, such as zinc or magnesium. Moreover, those with individual nutrient deficiencies had early alcohol-induced liver injury, as assessed by liver enzyme abnormalities.[6,7] This suggests a possible role for malnutrition in the development/progression of ALD, which is discussed in more detail subsequently.

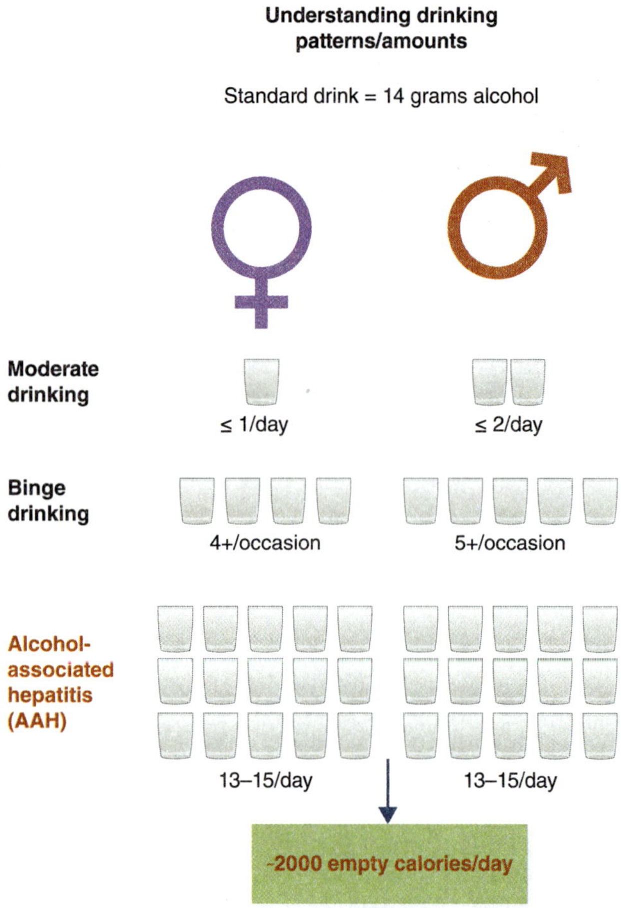

Figure 84.1 Drinking levels and their consequences. In the United States, drinking levels are expressed in terms of standard drinks consumed—that is, the number of alcoholic beverages drunk, each containing about 0.6 fluid ounce or 14 grams of pure alcohol. The Dietary Guidelines for Americans 2020–2025 defined moderate drinking as consuming up to two drinks/day for men and up to one drink/day for women. The Substance Abuse and Mental Health Services Administration defines binge drinking as consuming five or more (for men) or four or more (for women) alcoholic drinks on the same occasion on at least 1 day in the past 30 days. Patients hospitalized for severe AH in the NIAAA-funded DASH study were drinking approximately 13 to 15 drinks/day.

Data from NIH website on drinking: www.rethinkingdrinking.niaaa.nih.gov.

The liver is the main organ for ethanol metabolism. Ethanol is primarily metabolized (~80%) by alcohol dehydrogenase (ADH), oxidizing ethanol to acetaldehyde. However, the cytochrome P-450 systems (mainly CYP2E1) and catalase also contribute to the production of acetaldehyde in ethanol metabolism. Whereas ADH transfers electrons from ethanol to the reducing equivalent oxidized nicotinamide adenine dinucleotide phosphate ($NADP^1$), the cytochrome P-450 system transfers the electrons to molecular oxygen (O_2), and catalase reduces hydrogen peroxide (H_2O_2) to water (**Figure 84.2**). Ethanol metabolism is considered zero order at physiologically relevant doses, meaning that the oxidation of ethanol is saturated at blood alcohol concentrations that cause significant central nervous system (CNS) effects. Whereas the oxidation of ethanol to acetaldehyde

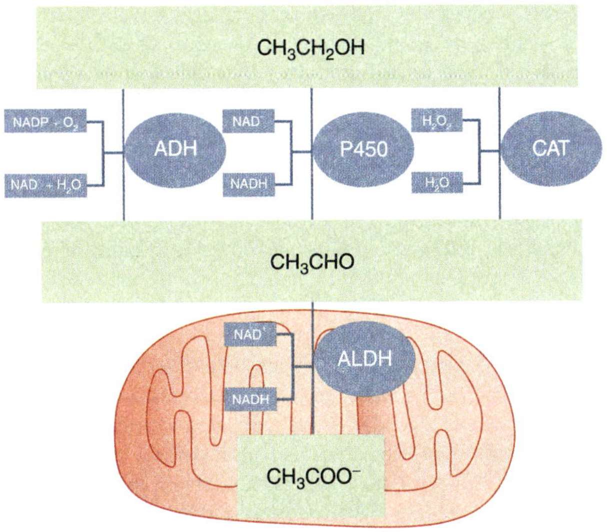

Figure 84.2 Oxidative metabolism of alcohol by the liver. Alcohol (CH_3CH_2OH) is oxidized to acetaldehyde (CH_3CHO) by three enzyme systems: the microsomal ethanol oxidizing system (MEOS), alcohol dehydrogenase (ADH), and catalase (CAT). Acetaldehyde, in turn, is metabolized to acetate (CH_3COO^-) by aldehyde dehydrogenase (ALDH) in the mitochondria. The metabolic and biochemical effects of alcohol metabolism may contribute to ADH. *NAD*, nicotinamide adenine dinucleotide; *NADH*, reduced nicotinamide adenine dinucleotide; *NADP*, nicotinamide adenine dinucleotide phosphate.

is mediated by three distinct enzyme systems, only one enzyme, aldehyde dehydrogenase (ALDH), oxidizes acetaldehyde to acetate (see Figure 84.2). Analogous to ADH, ALDH uses oxidized nicotinamide adenine dinucleotide (NAD^+) as the electron acceptor for this reaction; however, ALDH is located in the mitochondria of the cell (see Figure 84.2). Metabolism of alcohol can alter the redox state to cause hepatotoxicity. Alcohol metabolites (e.g., acetaldehyde) can directly cause hepatotoxicity, and alcohol metabolism can adversely affect distant organs (e.g., gut) to indirectly cause hepatotoxicity, as described below.

Mechanisms of Liver Injury with a Focus on the Gut-Liver Axis

The mechanisms for alcohol-induced liver injury are multiple, highly complex, and often interactive. Almost everyone who drinks heavily for at least 2 to 3 weeks develops fatty liver. Continued alcohol misuse, along with subsequent (often multiple) insults or "hits," can cause some patients to progress from fatty liver to steatohepatitis and/or cirrhosis (**Figure 84.3**). This usually occurs over years of heavy drinking. Here, we provide an overview of selected mechanisms, with a focus on nutrition and the gut:liver axis.

Oxidative stress is one of the best-studied mechanisms of ALD. Oxidative stress is an imbalance between pro-oxidants and antioxidants[4] Reactive oxygen species (ROS) and reactive nitrogen species (RNS) are products of normal metabolism and can be beneficial to the host (e.g., by contributing to bacterial killing),[4] but overproduction or inadequate antioxidant status or both can lead to liver injury. Direct metabolic products of alcohol (acetaldehyde), or toxic lipids generated as a result of alcohol metabolism (often associated with mitochondrial dysfunction), are important causes of hepatotoxicity.[8,9] Acrolein is one such lipid metabolite that induces gut barrier dysfunction and hepatotoxicity, in part, through induction of the endoplasmic reticulum stress response (ER stress).[10] Both genetic and epigenetic factors are important for disease pathogenesis and progression in ALD. Single nucleotide polymorphisms (SNPs) in Patatin-like phospholipase domain-containing protein 3 ($PNPLA_3$), which is involved in lipid metabolism and epigenetic effects related to microRNAs, DNA methylation and histone modifications, play an important role in ALD.[4] Altered immune function and increased production of proinflammatory cytokines, such as Interleukin-1 (IL-1) and Tumor Necrosis Factor-alpha (TNF-α), from multiple cell types, including Kupffer cells and macrophages, play a role in liver injury and systemic inflammation.[4] Bile acids can act as both signaling molecules and hepatotoxins. Concentrations are elevated in AH and correlate with disease severity.[11] Lastly, exposure to environmental toxicants can cause fatty liver and progression

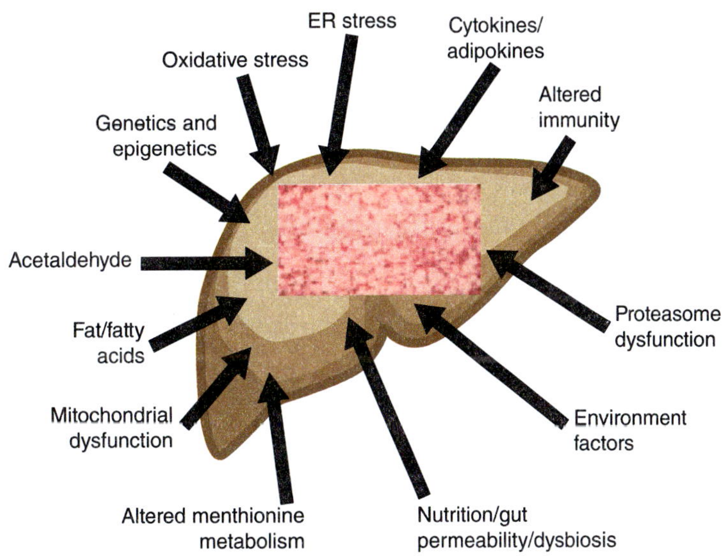

Figure 84.3 Mechanisms for liver disease "2nd hit."

Reproduced from Barve, A., Marsano, L.S., Parajuli, D., et al. (2017). Alcoholic Liver Disease. In: Saeian, K., Shaker, R. (eds) Liver Disorders. Springer, Cham. https://doi.org/10.1007/978-3-319-30103-7_13.

to cirrhosis (termed TASH) similar to excess alcohol consumption.[12] Environmental toxicants can also act synergistically with high fat diet or alcohol to cause liver injury.[13]

We focus our review of mechanisms for ALD on the gut microbiome and barrier dysfunction, which can be markedly impacted by both alcohol and nutrition. Alcohol consumption can result in gradual damage to gut barrier components leading to increased intestinal permeability and intestinal dysbiosis, including pathogenic microbial overgrowth and translocation of bacterial and fungi pathogen-associated molecular patterns (PAMPs) or endotoxins into the systemic circulation and to the liver (**Figure 84.4**). There are several PAMPs associated with ALD, e.g., well-known endotoxin lipopolysaccharide (LPS) from gram-negative bacteria, recently discovered bacterial endotoxin cytolysin secreted by *Enterococcus faecalis (E. faecalis)*,[14] and fungal exotoxin candidalysin produced by *Candida albicans*.[15] PAMPs enter the portal circulation and stimulate hepatic production of tumor necrosis factor (TNF) and other proinflammatory cytokines through Toll-like receptor-4 (TLR-4) signaling, resulting in overwhelming inflammation[16] and may also stimulate profibrotic changes[17] which, along with impaired liver capacity for pathogen clearance and alterations in hepatic metabolism, play a critical role in the development and progression of ALD. Endo toxemia, (elevated LPS) is a common feature of patients with ALD. Our recent study demonstrated that among long-term alcohol consumers with similar levels of alcohol consumption (~15 drinks/day for ~15 years of heavy drinking), those with liver injury (based on ALT levels >40 U/L) had elevated blood LPS, compared with heavy drinking individuals without liver injury.[18] Elevated levels of cytolysin and candidalysin, which may cause direct hepatocyte cell damage and death, were found in patients with more advanced stages of ALD, such as AH.[14,15] Fecal positivity for the cytolysin producing *E. faecalis* and the candidalysin coding gene ECE1 are associated with more severe disease and mortality in AH patients.[14,15]

Alcohol-induced intestinal permeability has been postulated as the primary cause of elevated endotoxemia in ALD,[19,20] and gut dysbiosis is a critical factor contributing to both alcohol-induced "leaky gut" and endotoxemia.[19-21] An early study demonstrated that patients with ALD had significantly increased gut permeability (as assessed by urinary sugar probes) compared with healthy controls and heavy alcohol drinkers without ALD.[19] Another study showed that gut permeability (assessed by recovery of high molecular weight polyethylene glycols) was increased in patients with ALD, and this was correlated with elevated LPS concentration.[20] Increased intestinal permeability (assessed by chromium EDTA) associated with gut microbiota alterations was demonstrated in a subset of alcohol-dependent individuals who were admitted for an alcohol detoxification and rehabilitation program.[22] Those with higher gut permeability and endotoxemia also showed more severe psychological disturbances (e.g., depression, anxiety, and craving), which only partially improved with abstinence and attenuated gut permeability and endotoxemia by the end of the 3-week treatment period.[22,23] It has also been shown that, in heavy drinking individuals with ALD, increased intestinal permeability and elevated markers of microbial translocation were associated with fecal microbiota dysbiosis.[21] Actively drinking cirrhotic patients had higher endotoxemia associated with significant duodenal, ileal, and colonic mucosal and fecal dysbiosis.[24]

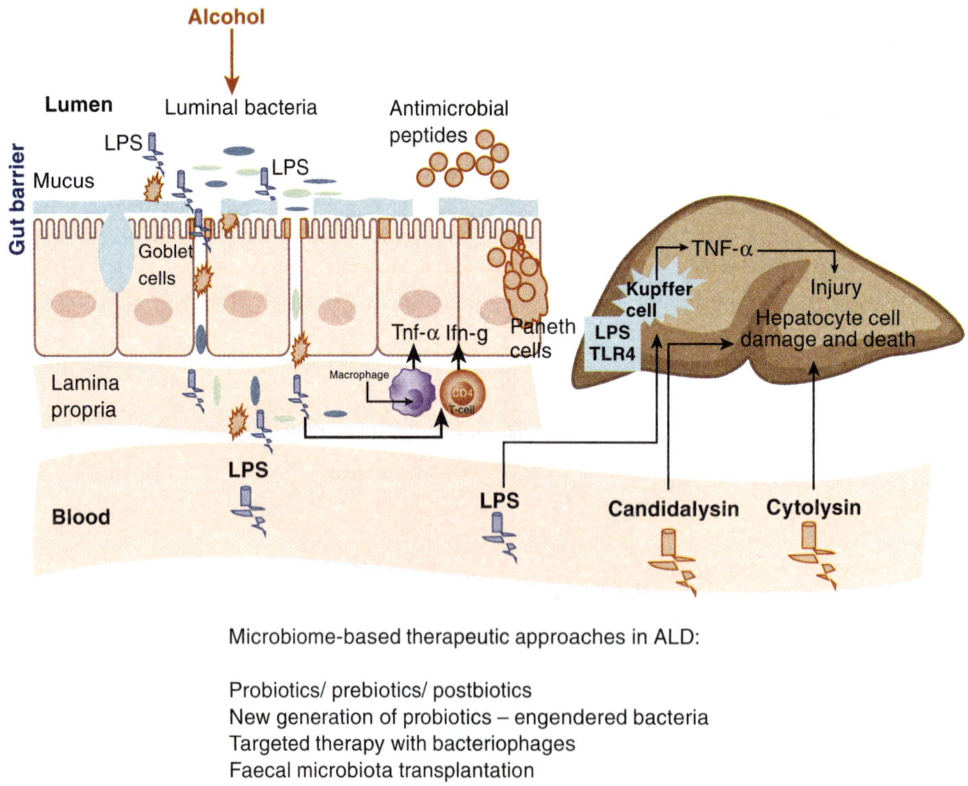

Figure 84.4 The Gut-Liver Axis and ALD. Alcohol consumption may cause dysbiosis and bacterial overgrowth. Intestinal bacteria alterations result in disruption of gut barrier integrity with the subsequent increase in gut permeability to microbial toxins, including LPS, cytolysin, and candidalysin. In the liver, LPS causes hepatocellular inflammation by stimulating hepatic Kupffer cells to release proinflammatory cytokines via TLR-4-mediated mechanisms leading to liver injury. Elevated levels of cytolysin and candidalysin, which may cause a direct hepatocyte cell damage and death. *LPS*, lipopolysaccharides; *TLR-4*, Toll-like receptor-4; *IFN-γ*, Interferon gamma.

Both acute and chronic alcohol consumption are also associated with alterations in the gut microbiota composition and function, contributing to the abnormal gut-liver axis, and thereby exacerbating alcohol-induced liver injury. A study from France[25] reported distinct gut microbiota communities in patients with distinct degrees of severity of alcohol-induced liver injury. Thus, patients with severe AH (sAH) had elevated abundance of *Bifidobacteria* and *Streptococci*, and less *Atopobium* compared with patients with no AH (noAH). *Streptococci* and *Enterobacteria* were positively correlated with the AH severity. *Atopobium* and *Clostridium leptum*, known for anti-inflammatory effects in the intestine, were negatively correlated with blood bilirubin and with fibrosis score, respectively. This study[25] also elegantly demonstrated that intestinal microbiota contribute to individual susceptibility to ALD, that the degree of susceptibility is transmissible from patients to mice, and that this susceptibility could be reversed with normalization of the microbiota. Mice that received microbiota from a patient with sAH developed more severe alcohol-induced liver injury, inflammation, greater intestinal permeability, and higher translocation of bacteria than mice that received microbiota from a patient having AUD without AH (noAH). Less liver injury in noAH mice was associated with *Faecalibacterium* genus, a known anti-inflammatory bacterium, and elevations of the primary bile acid, chenodeoxycholic acid (CDCA), and the secondary bile acid, ursodeoxycholic acid (UDCA), with known hepatoprotective properties. The authors suggested that the specific transferred bacteria associated with the sAH phenotype, the metabolites produced, or a combination of these factors might be transmissible factors contributing to ALD susceptibility.

Various other products of microbial function and metabolism, such as short-chain fatty acids (SCFAs), bile acids (BAs), derivatives from various amino acids (e.g., tryptophan) are significant communicators between the intestine and the liver. These molecules may play a critical role in the modulation of the gut microbiota and directly affect the host. Gut microbiota-associated metabolites were found to be different in alcohol drinking individuals compared with nondrinking healthy controls. Specifically, the SCFAs, propionate and isobutyrate, involved in intestinal epithelial cell homeostasis and gut barrier integrity, were less abundant in the feces of alcohol drinking individuals.[26] Actively drinking patients with cirrhosis had a significantly higher total fecal BA, with an increased secondary/primary BA ratio,[24] and this was accompanied by lower amounts of stool metabolites involved in bioenergetics (citrate, malate, phosphate), amino acids (threonine, ornithine, serine), and pyrimidine intermediates (ribosine, orotic acid, hexanoate). Ethanol-associated dysbiosis correlates with alterations in bacterial tryptophan metabolism; lower fecal indole-3-acetic acid (IAA), and indole-3-lactic acid were found in patients with AH, compared with healthy nondrinking controls. Moreover, restoration of IAA, which is an aryl hydrocarbon receptor (AhR) agonist, improved alcohol-induced liver injury in mice,[27] suggesting that AhR ligand-producing bacteria may be a promising new therapeutic approach in ALD. Furthermore, microbial metabolites along with changes in the gut microbiota composition may also serve as prognostic or diagnostic factors in liver disease. Using integrative and comprehensive analysis of the gut microbiome and metabolome, investigators found multiple correlations between metabolic and microbial signatures, which predicted progressive liver disease in ALD patients.[28] Elevated serum glutamic acid was selected as the most important variable for the prediction.

Several microbiome-based approaches have been developed for the treatment of ALD, including various probiotic/prebiotic/postbiotic supplementation and fecal microbiota transplantation (FMT) to alter the gut microbiota, as well as targeted therapy with phages and specific engineered bacteria to modulate specific bacterial enzymes and metabolic pathways. To date, multiple animal studies have demonstrated that probiotics (nonpathogenic live micro-organisms, e.g., Lactobacillus rhamnosus Gorbach-Goldin (LGG)),[29] or LGG supernatant[30] and LGG-derived nanoparticles,[31,32] ameliorated alcohol-induced liver injury, inflammation, and gut leakiness in mice. However, there are still few clinical studies demonstrating favorable effects of probiotics in human ALD. Among those, one study showed that short-term supplementation with *Bifidobacterium bifidum* and *Lactobacillus plantarum 8PA3* improved concentrations of liver enzymes in patients with mild ALD,[33] and several studies showed improvement of liver pathology in patients with alcohol-associated liver cirrhosis (AC) using different probiotics, e.g., VSL#3, a mix of multiple bacteria,[34] *Lactobacillus casei Shirota*,[35] and Escherichia coli Nissle.[36] Notably, a clinical trial is currently ongoing to investigate the effects of LGG supplementation in patients with AUD and ALD (NCT05178069), aiming to determine LGG effects on both alcohol consumption and liver injury.

Small pilot clinical trials on FMT from healthy donor(s) to ALD patients, specifically patients with AH, have demonstrated safety and beneficial effects on gut microbiome and liver function.[37] In a double-blind, randomized clinical trial, FMT was associated with short-term reduction in alcohol craving and consumption, with favorable microbial changes vs placebo in patients with AC with alcohol misuse.[38] Retrospective analysis of long-term outcomes of FMT and the gut microbiota changes in patients with severe AH demonstrated that FMT improves 3-year survival in patients with AH (65.7% vs 38.5% in the standard-of-care group) associated with reduced incidences of ascites, HE, and infection episodes. Furthermore, alcohol relapse was lower (28.6% vs 53.8%), and the time to relapse was longer in the FMT than in the standard-of-care group.[39]

Several recent studies have demonstrated that targeting specific microbes could be a promising novel therapeutic approach for patients with ALD. For example, precise targeting of *Enterococcus faecalis*, the microbial species that secretes endotoxin cytolysin, attenuated ALD in a mouse model.[14] In a humanized mouse model of ALD, where mice were colonized with bacteria from the feces of AH patients, bacteriophage treatment decreased cytolysin in the liver and abolished ethanol-induced liver injury and steatosis. Moreover, the presence of cytolytic *E. faecalis* correlated with both the severity of liver disease and with mortality in AH patients; most cytolysin-positive patients with AH (89%) died within 180 days of admission.[14] Lastly, synthetic bacteria that are genetically modified to produce specific factors might also be a promising approach in ALD.[27]

In summary, the gut-liver axis plays an important role in development and progression of ALD. Growing evidence suggests that modulation of the gut microbiota composition and function is a promising therapeutic option for patients with ALD to combat liver pathology as well as alleviate physiologic factors underlying alcohol craving and consumption; however, large randomized, controlled clinical trials are needed to further investigate these promising areas of research.

Malnutrition and Nutrition Therapy in ALD

Global Malnutrition in ALD

The most detailed information concerning malnutrition in liver disease comes from two large Veterans Affairs (VA) Cooperative Studies Program studies in patients with AH.[40-43] The first of these (Study 119) demonstrated that virtually every patient with AH had some degree of malnutrition.[40] Patients (284 with complete nutritional assessments) were divided into groups with mild, moderate, or severe AH, based on clinical and biochemical parameters. Almost 50% of their energy intake was derived from alcohol. Although energy intake was frequently not inadequate, there was often deficient intakes of protein and critical micronutrients. Indeed, the mechanisms for malnutrition for these patients are multifactorial, and leading causes of malnutrition in AH and many types of advanced liver disease are noted in **Table 84.1**. The severity of liver disease generally correlated with the severity of malnutrition, and worsening protein-energy malnutrition (PEM) correlated with worse clinical outcomes (**Figure 84.5**).[42] Similar data were generated in a follow-up VA study on AH (Study 275).[44] Voluntary oral food intake during hospitalization inversely correlated in a stepwise fashion with 6-month mortality data. Thus, patients who voluntarily consumed more than 3,000 kcal/day had virtually no mortality, whereas those consuming fewer than 1,000 kcal/day had more than 80% 6-month mortality rate.[42] The degree of malnutrition correlated with the development of serious complications such as encephalopathy, ascites, and hepatorenal syndrome.[42] Recent studies using a national registry of more than 20,000 hospitalized patients in the United States confirmed that malnutrition is an independent risk factor for infectious complications and mortality.[45]

When evaluating the prevalence of malnutrition in patients with liver disease, including ALD, it is important to use tests that accurately define nutritional status. Unfortunately, almost all of the most widely used tests can be influenced either by underlying liver disease or by factors that may be causing the liver disease, such as chronic alcohol consumption (possibly with superimposed hepatitis C infection) (**Table 84.2**). Visceral proteins are frequently, but erroneously, used for evaluating nutritional status, especially protein malnutrition. Visceral proteins such as albumin, prealbumin, and retinol-binding protein are all produced in the liver and correlate better with severity of underlying liver disease than with malnutrition status.[46]

Table 84.1 Causes of Malnutrition

- Anorexia
- Altered taste or smell
- Nausea or vomiting
- Diarrhea or malabsorption
- Poor food availability or quality
- Metabolic disturbances (e.g., hypermetabolism, catabolism)
- Cytokine effects
- Complications of liver disease (portal systemic encephalopathy, ascites, gastrointestinal bleeding)
- Unpalatable diets (low sodium, protein)
- Fasting for procedures

Data from Mendenhall CL, Roselle GA, Gartside P, et al. Alcohol Clin Exp Res 1995;19:635–41.

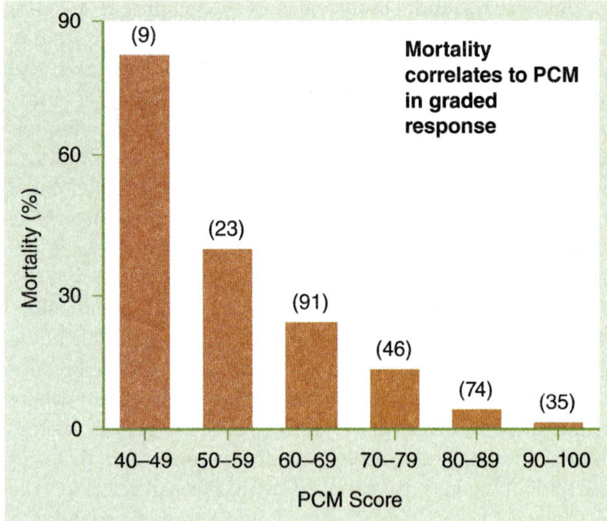

Figure 84.5 A protein/calorie malnutrition score demonstrates that protein/calorie malnutrition relates in a graded response with mortality (a perfect score is 100).

Modified from Mendenhall CL, Roselle GA, Gartside P, et al. Alcohol Clin Exp Res 1995;19:635–41.

The technology and availability of bioelectric impedance has substantially improved. It can also assess fluid status and can be used in patients with ascites. Handgrip dynomometry is inexpensive and evaluates functional status. Metabolic biomarkers have great promise in the diagnosis of malnutrition.[47] Because there is no ideal single indicator of malnutrition in liver disease, often the subjective global assessment in conjunction with a combination of tests is most appropriate for the individual patient.[48-50] The subjective global assessment evaluates PEM based on clinical findings such as muscle wasting, edema, loss of subcutaneous fat, and glossitis or cheilosis.

General Nutritional Supplementation in ALD

Initial interest in nutrition therapy for ALD was stimulated in 1948 by Patek et al.,[51] who demonstrated that a "nutritious diet" improved the 5-year outcome of patients with AC compared with historic controls. Subsequently, nutritional supplementation through a feeding tube was shown to improve liver function significantly in inpatients with ALD, as assessed by serum bilirubin concentration and antipyrine clearance, compared with inpatients who ate a hospital diet.[52] Probably the

Table 84.2 Tests of Nutritional Status or Malnutrition in Alcohol-Associated Liver Disease

- Anthropometry (e.g., triceps skinfold thickness)
- 24-hour urinary creatinine height index
- Assessment of muscle strength
- Bioelectrical impedance
- Dual-energy x-ray absorptiometry
- Biologic parameters (e.g., visceral proteins)
- Subjective global assessment
- Energy balance
- Metabolomics

most important data supporting nutrition therapy come from a multicenter study by Cabré et al.,[53] who randomized patients with severe AH to receive either prednisone, 40 mg daily, or a liver-specific formula containing 2,000 kcal/day through a feeding tube. This polymeric enteral solution was enriched in BCAAs and was energy dense (1.3 kcal/mL) and low in fat and sodium. The 1-month mortality was the same in both groups, but the 1-year mortality was significantly lower in the enteral nutrition group compared with the glucocorticoid group, mainly because of fewer infectious complications. This study clearly documents the importance of enteral nutrition in severe AH. In a recent multicenter trial, patients with severe AH were treated with either intensive enteral nutrition (EN) plus methylprednisolone or conventional nutrition plus methylprednisolone. EN was given via feeding tube for 14 days. The 6-month mortality (primary endpoint) was numerically but not statistically lower in the EN group (44.4%) compared with the control group (52.1%). Importantly, patients from either group receiving <21.5 kcal/kg/day had significantly lower survival rates, as did those receiving <77.6 gm of protein per day. Thus, it is clear that meeting energy, especially protein, requirements is important for optimal outcomes.

Oral or enteral nutrition is desired over parenteral nutrition because of cost, risk of sepsis from the parenteral nutrition line, preservation of the integrity of the gut mucosa, and prevention of bacterial translocation and multiple organ failure. Moreover, total parenteral nutrition can, in some instances, cause liver disease as one of its complications.

Upon hospital admission, it is important to immediately evaluate for electrolyte disturbances and for nutritional status. EN via tube feeding should be considered within 24 to 48 hours following hospitalization in patients unable to maintain oral nutritional intake, with the aim of providing >80% of estimated or calculated goal energy and protein within 48 to 72 hrs. Protein intake is usually recommended at 1.2 to 1.5 g/kg body weight per day (g/kgBW/d). A target of 35 to 40 kcal/kg body weight per day (cal/kgBW/d) is recommended in hospitalized, nonobese patients, especially in the ICU.[54] Obese patients with AH/cirrhosis represent an increasing nutrition challenge. Current guidelines recommend hypocaloric, high protein nutrition therapy in an attempt to preserve lean body mass and to minimize overfeeding complications.[55] Multiple recent society guidelines are available that further detail feeding recommendations.[56-60]

In the outpatient setting, EN supplements have also been shown to improve nutritional status and immune function in patients with AC as well as other etiologies and to reduce hospitalization.[61,62] The concept of a late evening snack (LES) before bedtime was established after studies demonstrated altered energy metabolism in people with liver cirrhosis. Depleted glycogen stores force dependence on fat and protein stores, thus leading to catabolism in times of fasting. Yamanaka et al.[63] established that patients with cirrhosis have a lower basal respiratory quotient (RQ) following overnight fasting compared with controls, a finding indicating reliance on fat stores as an energy substrate. The low RQ resolved following a meal.[63] An earlier study demonstrated that patients with stable cirrhosis maintained positive nitrogen balance when receiving four to six meals a day that included an LES as opposed to three equicaloric and equinitrogenous meals without an LES.[64] An important randomized, controlled trial tested the hypothesis that provision of an LES over a 12-month period would improve body protein stores in patients with cirrhosis. Total body protein was measured by neutron activation analysis at baseline, 3, 6, and 12 months. Provision of an LES to patients with cirrhosis resulted in body protein accrual equivalent to about 2 kg of lean tissue sustained over 12 months, and this benefit was not observed with daytime snacks. Thus, LESs are valuable nutritional interventions in outpatients with cirrhosis.[65] A study from Japan showed that using a multidisciplinary nutrition support team with recommendations that included LES improved 5-year mortality in patients with cirrhosis.[66] Our nutritional recommendations for patients having a spectrum of liver disease (from steatosis to HCC) now includes the incorporation of coffee into the diet. There are multiple studies documenting the efficacy of coffee in multiple types of liver disease, with beneficial effects ranging from decreasing fat and fibrosis to decreasing liver cancer risk.[67]

Selected Mineral Deficiency, Vitamin, or Amino Acid Deficiencies

Zinc

Zinc is an essential trace element required for normal cell growth, development, and differentiation. It is involved in DNA synthesis, RNA transcription, and cell division and activation. It is a critical component in many zinc proteins and enzymes, including critical zinc transcription factors. Zinc acts as a cofactor for two of the urea cycle enzymes and glutamine synthetase, which contributes to ammonia removal in muscle and liver.[68] Zinc deficiency or altered metabolism is observed in many types of liver disease, including ALD and viral liver disease. Some of the mechanisms for zinc deficiency or altered metabolism include decreased dietary intake, increased urinary excretion, activation of certain zinc transporters, and induction of hepatic metallothionein. Zinc deficiency is common in cirrhotic patients, with a prevalence of 84% to 96% in recent studies.[69] Zinc deficiency may manifest in many ways in liver disease, ranging from skin lesions to poor wound healing or liver regeneration, to altered mental status, to altered immune function. Zinc supplementation has been documented to block or attenuate experimental ALD through multiple processes, including stabilizing gut-barrier function, decreasing endotoxemia, decreasing proinflammatory cytokine production, decreasing oxidative stress, and attenuating apoptotic hepatocyte death.[69] Clinical trials in human liver disease are limited in size and quality; however, it is clear that zinc supplementation reverses clinical signs of zinc deficiency in patients with liver disease.[69] According to a meta-analysis including four trials with 247 cirrhotic patients, zinc supplementation combined with lactulose can improve number connection tests in patients with mild HE compared with lactulose monotherapy.[70] This agrees with a previous meta-analysis of four trials that concluded that oral zinc improved this psychometric test but did not decrease PSE.[71] Therefore, zinc supplementation at a dose of 50 mg elemental zinc orally every day taken with a meal to decrease the potential side effect of nausea can be considered in cirrhotic patients.[72]

Magnesium

Magnesium deficiency is common among persons with AUD; mechanisms include decreased intake or accumulation of saturated fatty acids on cell membranes.[73] Moreover, urinary excretion of magnesium and calcium was reported to be increased in a dose-dependent manner following even moderate amounts of alcohol intake, and this can contribute to magnesium deficiency in persons with AUD.[74] Magnesium deficiency is frequently associated with peripheral insulin resistance, which is commonly noted in ALD. Indeed, patients without diabetes but with insulin resistance have been treated with magnesium, with an improvement of peripheral insulin resistance.[75] Magnesium deficiency also was reported to decrease aspartate aminotransaminase significantly in some studies.[76] Finally, magnesium is often associated with muscle cramps, and magnesium supplementation improved muscle cramps in some clinical studies.[77]

Vitamin E

Vitamin E deficiency has been well documented in ALD.[78] Vitamin E has experimentally proven antioxidant activity with hepatoprotective capabilities, including membrane stabilization, reduced nuclear factor-κB (NF-κB) activation and TNF production, and inhibition of HSC activation.[78-81] An early study examining vitamin E supplementation in decompensated ambulatory patients with AC failed to show benefit at 1-year follow-up.[82] In a study of patients with mild-to-moderate AH, 1,000 IU/day of vitamin E improved serum hyaluronic acid but had no beneficial effects on tests of liver function or mortality at the 3-month follow-up, compared with placebo.[83] However, a large multicenter National Institutes of Health-sponsored trial of vitamin E (800 IU/day) in patients with NASH without diabetes showed improvement in liver enzymes, as well as in steatosis and inflammation on liver biopsy.[84] That is the therapy we currently recommend in these patients.

Vitamin B_1 (Thiamin)

Thiamin deficiency is common in many forms of cirrhosis, including AC. In a comparison study of patients with AC, patients with hepatitis C virus-induced cirrhosis, and patients with hepatitis C without cirrhosis, thiamin deficiency incidence was similar regardless of cause.[85] Thiamin deficiency can be caused by inadequate intake, decreased hepatic storage, and both acute and chronic impairment of intestinal absorption by ethanol.[86]

Wernicke-Korsakoff syndrome (WKS) is a severe mental disturbance caused by thiamin deficiency and is often associated with alcoholism. WKS can be described in two distinct stages. The first stage is Wernicke encephalopathy, resulting from acute thiamin deficiency, which is generally reversible by treatment with large doses of thiamin. If this treatment is not given, long-term damage to brain tissue may occur, resulting in Korsakoff psychosis. Supplementation with thiamin will not reverse the effects of Korsakoff psychosis because of permanent brain damage. Signs and symptoms of WKS include acute confusion, nystagmus, ophthalmoplegia (paralysis of the ocular muscles), ataxia, short-term memory loss, and even death, although not all of these symptoms need to be present to diagnose WKS.

Vitamin A (Retinol)

The liver is the major storehouse for vitamin A, with high levels in HSCs. As quiescent stellate cells become activated, they lose their vitamin A stores and can then produce collagen and subsequent fibrosis. Vitamin A deficiency (low serum concentration or abnormal dark adaptation) is present in approximately 50% of patients with AC,[87] and patients with alcoholism have been shown to have very low concentration of hepatic vitamin A at all stages of their disease.[88] Because vitamin A can also cause hepatotoxicity, care must be taken not to provide excess supplementation.

Vitamin D

Vitamin D deficiency in ALD may be related to several causes, including impaired hepatic hydroxylation of vitamin D, dietary insufficiency, malabsorption, reduced hepatic production of vitamin D-binding protein, and impaired cutaneous production secondary to reduced exposure to sunlight.[89] Decreased vitamin D stores in ALD may contribute to lowered bone density and mass as well as increased susceptibility to osteonecrosis and bone fractures.[90] In one study, 85% of patients with AC had compromised vitamin D status.[88] Patients with ALD had more vitamin D deficiency than patients with primary biliary cirrhosis (47%). In a study of 392 patients with CLD, Vidot et al., showed that there was significant correlation between 25-OHD deficiency and presence of HE ($P<0.0001$). In fact, there was a higher incidence of overt HE in patients with low levels of 25-OHD at similar levels of disease severity.[91] When the results were stratified by Child-Pugh class, an association was observed between vitamin D deficiency and severity of liver disease. Vitamin D deficiency in ALD may have several causes, including impaired hepatic hydroxylation of vitamin D, dietary insufficiency, malabsorption, reduced hepatic production of vitamin D-binding protein, and impaired cutaneous production secondary to reduced exposure to sunlight.[88]

Carnitine

Carnitine is a vitamin-like compound that regulates lipid and energy metabolism. Patients with cirrhosis are at high risk of developing carnitine deficiency, which has a detrimental effect on patient outcomes. Conversely, L-carnitine supplementation has shown some efficacy in improving hyperammonemia, HE, sarcopenia, and frailty, and further trials are needed.[92-94]

Branched-Chain Amino Acids

Branched-chain amino acid (BCAA) supplementation is a long-studied therapy for liver disease, primarily used to improve disease-related cachexia and hepatic encephalopathy (HE). More recent studies have shown that supplementation may delay the progression of cirrhosis.

Three large-scale studies have shown favorable responses. A study from Italy found that 12 months of 14.4 g/day of oral BCAA supplementation in patients with cirrhosis yielded significantly reduced rates of death, deterioration, hospital admissions, and duration of hospitalization, compared with equinitrogenous lactoalbumin supplementation. Significant improvements in health-related quality of life were also seen.[95] A study from Japan involving 646 patients with decompensated cirrhosis showed that a group receiving 12 g/day BCAA supplementation had significantly better rates of event-free survival. These patients also had improvement in health-related quality of life and serum albumin over the 2-year course of the study.[96] Another study from Japan showed that BCAA supplementation preserved liver function in patients with Child Class A cirrhosis.[97] These studies provide some evidence that BCAA supplementation can reduce the rate of decompensation in patients with cirrhosis and progression to higher healthcare utilization and decreased quality of life.

Limitations include the bitter taste and subsequent poor palatability of the product (which has been reduced by the creation of granules), as well as high cost. Currently, there are no recommendations regarding BCAA supplementation in the United States, and BCAA supplementation is generally not widely used because of cost and lack of conviction concerning efficacy.

Specific Conditions

Hepatic Encephalopathy

Factors involved in the pathophysiology of HE continue to be elucidated, but the well-known components include increased gut-derived toxins such as ammonia, resulting from decreased detoxification/increased production, increased ratio of aromatic amino acids to BCAAs, and gut:brain axis/neuroinflammation. The finding that these factors can be altered by nutritional intervention has led to new nutrition research in this area.

The foremost topic in nutrition support and HE has been the long-standing "dogma" of protein restriction. Given the evidence that patients with cirrhosis become catabolic and develop a negative nitrogen balance following a short overnight fast, this concept has since been challenged.[63,64] A study of 30 cirrhotic patients hospitalized for HE showed no difference in the course of the disease between those receiving a normal amount of protein (1.2 g/kg/day) and those initially receiving a low-protein diet.[98] Gheorghe et al.,[99] fed cirrhotic patients with HE a 30 kcal/kg/day and 1.2 g protein/kg/day diet divided into four meals, including an LES. Of 153 individuals, 122 improved; notably, those with higher grade HE had the most improvement. Subsequently, the European Society for Parenteral and Enteral Nutrition (ESPEN) recommended avoiding protein restriction "at all cost" and for as short a time as possible in cases of severe protein intolerance. For those patients, it is recommended that BCAAs be given at 0.25 g/kg/day until the patient is able to resume normal protein intake.[54,100]

Current recommendations for the clinical management of HE for lowering ammonia concentration in patients with cirrhosis and HE include lactulose (prebiotic) as the first-line treatment and rifaximin (nonabsorbable antibiotic) as an add-on therapy. Unfortunately, side effects of lactulose, including nausea and diarrhea, make it intolerable for some patients, with subsequent decreased adherence to therapy. Several emerging treatments that target ammonia production such as supplementation with probiotics, prebiotics, and synbiotics (the combination of probiotics and prebiotics) have emerged as an alternative therapy because of the potential to alter gut flora and subsequently the pH of the gut lumen. These changes can decrease ammonia production/absorption. Many studies involve mild HE as revealed by psychometric testing, as opposed to overt HE. One study evaluated the use of a probiotic-containing yogurt for patients with non-alcohol-associated cirrhosis and mild HE and found reversal of symptoms in 12 of 17 study patients and an 88% adherence to the regimen.[101] Liu et al.,[102] studied synbiotics (probiotics plus a fermentable fiber prebiotic) in patients with various forms of liver disease and mild HE (determined by psychometric testing and brainstem auditory evoked potential). The groups receiving either synbiotics or the prebiotic were found to have decreased fecal pH, venous ammonia, and serum endotoxin, compared with placebo. Fifty percent of the individuals in the study groups reversed their mild HE significantly more than with placebo.[102] Finally, a meta-analysis reviewed nine studies involving 349 patients that evaluated synbiotics, probiotics, and lactulose. All studies demonstrated improvement in mild HE compared with placebo; however, limitations noted included short duration of the studies and small numbers of patients.[103] Multiple other potential agents may decrease gut-derived toxins, ranging from zinc to sodium benzoate.[104] FMT has also shown promise in HE and may reduce drinking, as discussed earlier.[105]

Evidence is mixed regarding the direct impact of BCAAs on HE.[106] A 2003 meta-analysis of 11 randomized trials (including 556 patients) demonstrated an overall significant improvement in HE following BCAA supplementation, but without an impact on survival or on the length of time to effect. However, because of "significant statistical heterogeneity" among these studies, the authors of the meta-analysis concluded that there was no "convincing evidence" of effect.[107] As BCAA supplementation continues to be researched, with more recent favorable studies on its impact on disease progression in liver cirrhosis, more consistent data may emerge (see also the previous section on BCAAs).

In summary, hyperammonemia appears to act in combination with systemic- and neuro-inflammation, oxidative and ER stress, and gut dysbiosis in the pathogenesis of HE. A key nutritional concept for patients with HE is avoiding protein restriction, except in patients who are severely intolerant of protein supplementation. Lactulose and rifaximin are the cornerstones of treatment. Probiotics, prebiotics, and synbiotics appear to have a role in mild HE and can be delivered in something as simple as a yogurt supplement. The impact of other nutritional supplementation in HE, including BCAAs, zinc, L-carnitine, and others, is still being evaluated.

Pre- and Post-Transplant

Malnutrition in the preliver transplant (LT) period is common, with approximately 30% to 80% of LT recipients showing some evidence of malnutrition.[108,109] Sarcopenia is a state of reduced muscle mass and function, and it is the phenotype of malnutrition in cirrhosis. It is a major modulator of frailty in patients with cirrhosis. Sarcopenia occurs in approximately 10% to 70% of patients with cirrhosis, depending on severity of liver disease and sex.[110] Somewhat counterintuitively, it is often present in obese patients (sarcopenic obesity) but is more difficult to detect. Indeed, overnutrition and obesity are often problems for patients undergoing LT. Methods for evaluating malnutrition and sarcopenia in the pretransplant patient are the same as those described previously for cirrhosis The mechanisms for sarcopenia are multifactorial and include immobility, endocrine factors, cytokines/inflammation, aging, etc.

Malnutrition has a major impact on morbidity and mortality in the LT patient. LT recipients who are malnourished required more blood products during surgery, had a prolonged postoperative hospital stay, and were at higher risk for postoperative infections.[111,112] It has recently been reported that malnutrition, as assessed by a preoperative prognostic nutritional index, may be a strong predictor of hepatocellular carcinoma recurrence following LT.[113] Moreover, half of all liver transplant patients have sarcopenia, an independent risk factor postoperative sepsis, neurologic complications, longer ICU stay, and need for ventilatory support.[114]

A "Nutrition Prescription" has been proposed for nonobese and obese individuals in the pretransplant setting. For nonobese persons, the goal is 35 kcal/kg/day in a high-energy, high-protein diet and possible BCAA-rich protein supplementation. Goals for this population are increased meal frequency and a late-evening snack.[115] For obese individuals, a tailored, moderately hypocaloric diet of 500 to 800 kcal/day vs a BMI stratified intake of 25 to 35 kcal/kg if between 30–40kg/m^2 and 20–25 kcal/kg if >40 kg/m^2 have been proposed.[115] For most pretransplant individuals, an appropriate protein intake target is approximately 1.2 to 1.5 g/kg/day. Optimally, patients should participate in low-resistance aerobic exercise 3 to 4 times per week, and lift light weights (e.g., 2 lb. hand weights), 2 to 3 times per week. Unfortunately, exercise is often limited by frailty.

Following LT, nutrition status is expected to improve because many metabolic alterations causing malnutrition are corrected by a normal, functioning liver. In the early post-transplant period, the nutrition regimen should be like previously described protocols/guidelines for hospitalized patients with cirrhosis in the ICU. However, due to the high postoperative catabolic state, a protein goal of 1.5 to 2 g/kg/day for the first weeks after transplant has been proposed.[116] Nutritional supplements should be continued until the patient can maintain adequate oral intake.

Restoration of body cell mass does occur but is often incomplete[117,118] and, unfortunately, malnutrition often persists after LT, usually with progressive weight gain and continued skeletal muscle deficiency. Changes in body composition of transplant recipients are characterized by an early and inappropriate gain in fat mass. In a study of 53 patients, 66% had sarcopenia before transplant, 6% had reversal of sarcopenia, and 70% of those who were not sarcopenic before LT developed sarcopenia after LT.[119] While the etiology of persistent sarcopenia is multifactorial, several studies have shown that patients greater than 6 months post-transplant have a daily basal energy expenditure of 100 to 400 kcal lower than prior to transplant[120] creating a physiologic setup for sarcopenic obesity and poor outcomes. Analysis of 18,172 transplanted patients showed that primary nonfunction and 1- and 2-year mortality were significantly higher in morbidly obese patients (BMI >40Kg/m^2), and long-term mortality in obese patients was increased, largely due to cardiovascular events.[121] The 15-year mortality post liver transplant is nearly double in individuals who are obese post-transplant, compared with normal weight and overweight persons.[122] Unfortunately, about half of patients transplanted for NASH have allograft NAFLD 10 years post-transplant[123] and a small fraction of these individuals progress to allograft cirrhosis. Moreover, for newly abstinent individuals transplanted for alcohol liver disease, "addiction transfer" may occur where alcohol reward replaces food reward.[124] A multifactorial approach involving nutrition and addiction experts is needed for such individuals. Finally, the metabolic impacts of steroid and certain anti-rejection drugs should be recognized with modification of regimens to the least obesogenic in post-transplant patients at highest risk for sarcopenia and obesity.[125] In summary, a multidisciplinary approach involving all stakeholders is necessary to ensure that patients have adequate nutritional support before and after LT.

Conclusions

Malnutrition, both PEM and deficiencies in individual nutrients, is a frequent complication of ALD. Severity of malnutrition correlates with severity of ALD. Malnutrition also occurs in patients with cirrhosis from causes other than alcohol. The mechanisms for malnutrition are multifactorial, and malnutrition frequently worsens in the hospital because of fasting for procedures and metabolic complications of liver disease, such as HE. Aggressive nutritional support is indicated in inpatients with ALD, and they often need to be fed through an enteral feeding tube to achieve protein and energy goals. Enteral nutritional support improves nutrition status and may improve clinical outcome. Moreover, LESs in outpatients with cirrhosis improve nutritional status and lean body mass. With no U.S. Food and Drug Administration-approved therapy for ALD, aggressive nutritional intervention should be considered as frontline therapy, along with discontinuation of alcohol consumption. Many of the observations and therapeutic strategies concerning malnutrition and ALD apply to cirrhosis of other causes, especially hepatitis C.

Acknowledgments

The authors acknowledge Juliane I Beier, Sarah Landes, and Mohamad Mohammad, for their contributions to the previous version of this chapter.

References

1. Tapper EB, Parikh ND. BMJ 2018;362:k2817.
2. Kim D, Li AA, Gadiparthi C, et al. Gastroenterol 2018;155: 1154–63 e3.
3. Bittermann T, Mahmud N, Abt P. JAMA Netw Open 2021;4: e2118713.
4. Barve S, Chen SY, Kirpich I, et al. Alcohol Res 2017;38: 289–302.
5. Vatsalya V, Song M, Schwandt ML, et al. Alcohol Clin Exp Res 2016;40:2085–93.
6. Vatsalya V, Kong M, Cave MC, et al. J Nutr Biochem 2018;59: 49–55.
7. Vatsalya V, Gala KS, Mishra M, et al. Alcohol Alcohol 2020;55: 164–70.
8. Warner DR, Liu H, Miller ME, et al. Am J Pathol 2017;187: 2232–45.
9. Vatsalya V, Kong M, Gobejishvili L, et al. Am J Physiol Gastrointest Liver Physiol 2019;316:G115–G22.
10. Moghe A, Ghare S, Lamoreau B, et al. Toxicol Sci 2015;143: 242–55.
11. He L, Vatsalya V, Ma X, et al. Hepatol Commun 2021;5: 798–811.
12. Cave M, Falkner KC, Ray M, et al. Hepatol 2010;51:474–81.
13. Joshi-Barve S, Kirpich I, Cave MC, et al. Cell Mol Gastroenterol Hepatol 2015;1:356–67.
14. Duan Y, Llorente C, Lang S, et al. Nature 2019;575:505–11.
15. Chu H, Duan Y, Lang S, et al. J Hepatol 2020;72:391–400.
16. McClain CJ, Cohen DA. Hepatol 1989;9:349–51.
17. Seki E, Schnabl B. J Physiol 2012;590:447–58.
18. Kirpich IA, McClain CJ, Vatsalya V, et al. Alcohol Clin Exp Res 2017;41:747–57.
19. Keshavarzian A, Holmes EW, Patel M, et al. Am J Gastroenterol 1999;94:200–7.
20. Parlesak A, Schafer C, Schutz T, et al. J Hepatol 2000;32: 742–7.

21. Maccioni L, Gao B, Leclercq S, et al. Gut Microbes 2020;12: 1782157.
22. Leclercq S, Matamoros S, Cani PD, et al. Proc Natl Acad Sci U S A 2014;111:E4485–93.
23. Leclercq S, Cani PD, Neyrinck AM, et al. Brain Behav Immun 2012;26:911–8.
24. Bajaj JS, Kakiyama G, Zhao D, et al. Alcohol Clin Exp Res 2017;41:1857–65.
25. Llopis M, Cassard AM, Wrzosek L, et al. Gut 2016;65:830–9.
26. Couch RD, Dailey A, Zaidi F, et al. PLoS One 2015;10:e0119362.
27. Hendrikx T, Duan Y, Wang Y, et al. Gut 2019;68:1504–15.
28. Gao B, Zhu Y, Gao N, et al. Metabolites 2021;11:766.
29. Wang Y, Liu Y, Kirpich I, et al. J Nutr Biochem 2013;24: 1609–15.
30. Zhao H, Zhao C, Dong Y, et al. Toxicol Lett 2015;234:194–200.
31. Gu Z, Li F, Liu Y, et al. Hepatol Commun 2021;5:846–64.
32. Jiang M, Li F, Liu Y, et al. Hepatol 2022;77:1164–80.
33. Kirpich IA, Solovieva NV, Leikhter SN, et al. Alcohol 2008;42: 675–82.
34. Dhiman RK, Rana B, Agrawal S, et al. Gastroenterol 2014;147: 1327–37 e3.
35. Stadlbauer V, Mookerjee RP, Hodges S, et al. J Hepatol 2008;48: 945–51.
36. Lata J, Novotny I, Pribramska V, et al. Eur J Gastroenterol Hepatol 2007;19:1111–3.
37. Philips CA, Phadke N, Ganesan K, et al. Indian J Gastroenterol 2018;37:215–25.
38. Bajaj JS, Gavis EA, Fagan A, et al. Hepatol 2021;73:1688–700.
39. Philips CA, Ahamed R, Rajesh S, et al. J Clin Exp Hepatol 2022;12:1124–32.
40. Mendenhall CL, Anderson S, Weesner RE, et al. Am J Med 1984;76:211–22.
41. Mendenhall CL, Moritz TE, Roselle GA, et al. JPEN J Parenter Enteral Nutr 1995;19:258–65.
42. Mendenhall C, Roselle GA, Gartside P, et al. Alcohol Clin Exp Res 1995;19:635–41.
43. Mendenhall CL, Tosch T, Weesner RE, et al. Am J Clin Nutr 1986;43:213–8.
44. Mendenhall CL, Moritz TE, Roselle GA, et al. Hepatol 1993;17: 564–76.
45. Lee DU, Fan GH, Hastie DJ, et al. J Clin Gastroenterol 2022;56:349–59.
46. Merli M, Romiti A, Riggio O, et al. JPEN J Parenter Enteral Nutr 1987;11:130S–4S.
47. Mogensen KM, Lasky-Su J, Rogers AJ, et al. JPEN J Parenter Enteral Nutr 2017;41:188–97.
48. Baker JP, Detsky AS, Wesson DE, et al. N Engl J Med 1982;306: 969–72.
49. Camillo B. In: Preedy VR, Lakshman R, Srirajaskanthan R, and Watson RR, eds. Assessment of Nutritional Status and Diagnosis of Malnutrition in Patients with Liver Disease. Nutrition, Diet Therapy, and the Liver. Boca Raton, FL: CRC Press, 2010;22–46.
50. Detsky AS, McLaughlin JR, Baker JP, et al. JPEN J Parenter Enteral Nutr 1987;11:8–13.
51. Patek AJ, Jr., Post J, Ratnoff OD, Mankin H, Hillman RW. J Am Med Assoc 1948;138:543–9.
52. Kearns PJ, Young H, Garcia G, et al. Gastroenterol 1992;102: 200–5.
53. Cabré E, Rodríguez-Iglesias P, Caballería J, et al. Hepatol 2000;32: 36–42.
54. Plauth M, Cabre E, Riggio O, et al. Clin Nutr 2006;25:285–94.
55. McClave SA, Taylor BE, Martindale RG, et al. JPEN J Parenter Enteral Nutr 2016;40:159–211.
56. Lai JC, Tandon P, Bernal W, et al. Hepatol 2021;74:1611–44.
57. Singal AK, Bataller R, Ahn J, et al. Am J Gastroenterol 2018;113:175–94.
58. Bischoff SC, Bernal W, Dasarathy S, et al. Nutr Hosp 2022;39: 434–72.
59. Bischoff SC, Bernal W, Dasarathy S, et al. Clin Nutr 2020;39: 3533–62.
60. Carey EJ, Lai JC, Sonnenday C, et al. Hepatol 2019;70:1816–29.
61. Hirsch S, Bunout D, de la Maza P, et al. JPEN J Parenter Enteral Nutr 1993;17:119–24.
62. Hirsch S, de la Maza MP, Gattás V, et al. J Am Coll Nutr 1999;18:434–41.
63. Yamanaka H, Genjida K, Yokota K, et al. Nutrition 1999;15: 749–54.
64. Swart GR, Zillikens MC, van Vuure JK, et al. BMJ 1989;299: 1202–3.
65. Plank LD, Gane EJ, Peng S, et al. Hepatol 2008;48:557–66.
66. Iwasa M, Iwata K, Hara N, et al. Nutrition 2013;29:1418–21.
67. Contaldo F, Santarpia L, Pasanisi F. Curr Opin Clin Nutr Metab Care 2019;22:389–92.
68. Amodio P, Bemeur C, Butterworth R, et al. Hepatol 2013;58: 325–36.
69. Mohammad MK, Zhou Z, Cave M, et al. Nutr Clin Pract 2012;27:8–20.
70. Shen YC, Chang YH, Fang CJ, et al. Nutr J 2019;18:34.
71. Chavez-Tapia NC, Cesar-Arce A, Barrientos-Gutiérrez T, et al. Nutr J 2013;12:74.
72. Marchesini G, Fabbri A, Bianchi G, et al. Hepatol 1996;23: 1084–92.
73. Wells IC. Can J Physiol Pharmacol 2008;86:16–24.
74. Rylander R, Megevand Y, Lasserre B, et al. Scand J Clin Lab Invest 61:401–5.
75. Guerrero-Romero F, Tamez-Perez HE, Gonzalez-Gonzalez G, et al. Diabetes Metab 2004;30:253–8.
76. Poikolainen K, Alho H. Subst Abuse Treat Prev Policy 2008;3:1.
77. Dahle LO, Berg G, Hammar M, et al. Am J Obstet Gynecol 1995;173:175–80.
78. Arteel G, Marsano L, Mendez C, et al. Best Pract Res Clin Gastroenterol 2003;17:625–47.
79. Evstigneeva RP, Volkov IM, Chudinova VV. Membr Cell Biol 1998;12:151–72.
80. Hill DB, Devalaraja R, Joshi-Barve S, et al. Clin Biochem 1999;32:563–70.
81. Lee KS, Buck M, Houglum K, et al. J Clin Invest 1995;96: 2461–8.
82. de la Maza MP, Petermann M, Bunout D, et al. J Am Coll Nutr 1995;14:192–6.
83. Mezey E, Potter JJ, Rennie-Tankersley L, et al. J Hepatol 2004;40: 40–6.
84. Sanyal AJ, Chalasani N, Kowdley KV, et al. N Engl J Med 2010;362:1675–85.
85. Levy S, Herve C, Delacoux E, et al. Dig Dis Sci 2002;47:543–8.
86. Hoyumpa AM, Jr. Am J Clin Nutr 1980;33:2750–61.
87. Russell RM. Am J Clin Nutr 1980;33:2741–9.
88. Malham M, Jorgensen SP, Ott P, et al. World J Gastroenterol 2011;17:922–5.
89. Malham M, Jørgensen SP, Ott P, et al. World J Gastroenterol 2011;17:922–5.
90. Leevy CM, Moroianu SA. Clin Liver Dis 2005;9:67–81.
91. Vidot H, Potter A, Cheng R, et al. World J Hepatol 2017;9: 510–8.
92. Shiraki M, Shimizu M, Moriwaki H, et al. Hepatol Res 2017;47:321–7.
93. Ohara M, Ogawa K, Suda G, et al. Hepatol Commun 2018;2: 906–18.
94. Hiramatsu A, Aikata H, Uchikawa S, et al. Hepatol Commun 2019;3:348–55.
95. Marchesini G, Bianchi G, Merli M, et al. Gastroenterol 2003;124:1792–801.

96. Muto Y, Sato S, Watanabe A, et al. Clin Gastroenterol Hepatol 2005;3:705–13.
97. Kawamura E, Habu D, Morikawa H, et al. Liver Transpl 2009;15:790–7.
98. Cordoba J, Lopez-Hellin J, Planas M, et al. J Hepatol 2004;41:38–43.
99. Gheorghe L, Iacob R, Vadan R, et al. Rom J Gastroenterol 2005;14:231–8.
100. Plauth M, Merli M, Kondrup J, et al. Clin Nutr 1997;16:43–55.
101. Bajaj JS, Saeian K, Christensen KM, et al. Am J Gastroenterol 2008;103:1707–15.
102. Liu Q, Duan ZP, Ha DK, et al. Hepatol 2004;39:1441–9.
103. Shukla S, Shukla A, Mehboob S, et al. Aliment Pharmacol Ther 2011;33:662–71.
104. Said VJ, Garcia-Trujillo E. Gastroenterol Nurs 2019;42:277–85.
105. Bajaj JS, Kassam Z, Fagan A, et al. Hepatol 2017;66:1727–38.
106. Holecek M. Nutr 2010;26:482–90.
107. Als-Nielsen B, Koretz RL, Kjaergard LL, et al. Cochrane Database Syst Rev 2003;Cd001939.
108. Lim HS, Kim HC, Park YH, et al. Clin Nutr Res 2015;4:242–9.
109. Merli M, Giusto M, Gentili F, et al. Liver Int 2010;30:208–14.
110. Tandon P, Ney M, Irwin I, et al. Liver Transpl 2012;18:1209–16.
111. Harrison J, McKiernan J, Neuberger JM. Transpl Int 1997;10:369–71.
112. Stephenson GR, Moretti EW, El-Moalem H, et al. Transplantation 2001;72:666–70.
113. Kornberg A, Kaschny L, Kornberg J, et al. J Hepatocell Carcinoma 2022;9:649–60.
114. Kumar V, Benjamin J, Shasthry V, et al. J Clin Exp Hepatol 2020;10:467–76.
115. Tandon P, Montano-Loza AJ, Lai JC, et al. J Hepatol 2021;75(suppl 1):S147–S62.
116. Hughes MJ, McNally S, Wigmore SJ. HPB (Oxford) 2014;16:699–706.
117. Schutz T, Hudjetz H, Roske AE, et al. Nutr 2012;28:378–83.
118. Ferreira LG, Santos LF, Anastacio LR, et al. Transplantation 2013;96:579–85.
119. Tsien C, Garber A, Narayanan A, et al. J Gastroenterol Hepatol 2014;29:1250–7.
120. Santos BC, Correia M, Anastacio LR. JPEN J Parenter Enteral Nutr 2021;45:456–64.
121. Nair S, Verma S, Thuluvath PJ. Hepatol 2002;35:105–9.
122. van Son J, Stam SP, Gomes-Neto AW, et al. Metab 2020;106:154204.
123. Narayanan P, Mara K, Izzy M, et al. Transplantation 2019;103:e14–e21.
124. Brunault P, Salame E, Jaafari N, et al. Med Hypotheses 2015;85:68–75.
125. Bhat M, Usmani SE, Azhie A, et al. Endocr Rev 2021;42:171–97.

D. DISORDERS OF THE ALIMENTARY TRACT

CHAPTER 85

Enteral Feeding

Laura E. Matarese and Beth Taylor

CHAPTER OUTLINE

Abbreviations

AAA, aromatic amino acid; **ARDS**, acute respiratory distress syndrome; **BCAA**, branched-chain amino acid; **COPD**, chronic obstructive pulmonary disease; **DM**, diabetes mellitus; **EN**, enteral nutrition; **GI**, gastrointestinal; **HE**, hepatic encephalopathy; **ICU**, intensive care unit; **MUFA**, monounsaturated fatty acid; **PEG**, percutaneous endoscopic gastrostomy; **PN**, parenteral nutrition; **RRT**, renal replacement therapy; **SBS**, short bowel syndrome; **VCO2**, carbon dioxide production

Introduction

For various reasons, many patients are unable to meet their nutrition requirements via an oral diet alone. Current clinical practice guidelines encompassing a plethora of conditions strongly recommend enteral nutrition (EN) support for patients who cannot meet their nutrient needs through voluntary oral intake (for purposes of this chapter, EN refers to tube-feeding methods).[1-3] The timing to start EN in the absence of adequate oral intake varies based on a patient's nutrition status and metabolic demands. For example, in trauma, surgery, and critically ill patients, inflammatory processes may lead to a hypercatabolic state, causing a loss of protein stores and disease-related malnutrition (refer to chapter 65). Similar to patients with preexisting malnutrition, serious consequences may occur, including increased risk of infection, poor wound healing, and immune system dysfunction.[4,5] Traditionally, nutrition was regarded as adjunctive care designed to stabilize the patient during repair. More recently, nutrition support has evolved as a cornerstone of care and, in some high-risk patients, an emergent medical intervention designed to help attenuate the catabolic response to stress, prevent oxidative injury, protect the gastrointestinal (GI) mucosa, modulate the immune response, and promote wound healing (refer to chapters 44, 45).[6]

This chapter provides guidelines for the use, implementation time, access device, substrate characteristics, disease specific formulations, and various administration tips that are important to consider during EN support.

Route of Feeding: Enteral vs Parenteral Nutrition

Besides the GI tract's role in digestion, absorption, and secretion, it is now recognized that the gut is a metabolically active organ that performs an important function in nutrient transport as well as immune defense.[7,8] The deliberation on whether to use EN vs parenteral nutrition (PN) for feeding is largely academic because of the physiologic advantages associated with using normal digestive and absorptive pathways. In practice, if the GI tract is functional, accessible, and safe to use, EN should be given primary consideration with the mindset that PN-supplied nutrients can be used as adjunctive support. Justification for directing nutrition support to the GI tract includes the fact that enteral nutrients experience first-pass metabolism in the liver, thus maximizing utilization (refer to chapter 41). Furthermore, direct exposure of the small intestine to nutrition stimuli supports the functional integrity of the gut, enhances blood flow, and induces the release of endogenous trophic agents (e.g., cholecystokinin, gastrin, bombesin, and bile salts), which are protective against organ failure in critically ill patients (refer to chapter 41).[9] Luminal nutrients and prebiotics delivered via EN help maintain normal intestinal pH and gut microbiota. Altered microbiota or dysbiosis may increase a patient's susceptibility to infections and organ dysfunction (refer to chapter 37).[10] From a practical standpoint, enteral formulas mimic oral intake and supply intact nutrients such as fiber, whole proteins, dipeptides, and specialized fatty acids that cannot be delivered parenterally.

The positive effects of EN, alone or given in combination with supplementary PN, when compared with PN alone are, in general, related to maintaining the integrity of the GI tract and cost (**Table 85.1**).[11,12] In patients who would not require intravenous access for another reason, EN carries a decreased risk of infectious complications compared to PN, a difference in mortality has not been clearly demonstrated.[13,14] In critically ill patients with shock, no difference in infectious complications were noted in a large, randomized trial between early EN and PN fed patients.[15] EN has been associated with reductions in hyperglycemia.[16-18] Historically, EN was associated with reduction in length of stay; however, more recent evidence demonstrates no difference, as technologies and practices to control complications of PN have improved.[11,19]

Table 85.1 Potential Benefits of Using the Enteral Route for Nutrition Support

Physiologic
Maintains GI mucosal integrity
Preserves gut barrier function
First-pass metabolism in the liver
Stimulates release of cholecystokinin
Promotes digestive and absorptive capabilities of the GI tract
Augments cellular antioxidant systems
Decreases incidence of hyperglycemia compared with PN
Certain nutrients are not available in PN form (ie., fiber)
Immunologic
Decreased bacterial translocation
Support of gut-associated lymphoid tissue (GALT) and mucosa-associated lymphoid tissue (MALT) for preservation of mucosal immunologic functions
Lowered risk of infection
Improved wound healing
Cost benefit
Simplified procedures and equipment
Less expensive than PN

Indications and Contraindications

EN should be the method of choice for all individuals with adequate digestive and absorptive capacity of the GI tract presenting with clinical conditions in which oral intake is impossible, inadequate, or unsafe.[1,20] Determining which patients should receive EN requires consideration of several factors, including the patient's clinical status, diagnosis, presence of malnutrition, projected time of insufficient intake, prognosis, risk-benefit ratio, discharge plans, quality of life, ethical considerations, and the patient/family wishes.[1,21-24] Specific indications for EN are listed in **Table 85.2**.

Contraindications to enteral feeding relate primarily to the presence and degree of malnutrition, the patient's ability to consume adequate nutrition by mouth, and GI integrity and functional capacity. Relative contraindications to enteral intervention include an expectation of early resumption of oral intake in a previously well-nourished patient, mechanical intestinal obstruction, intractable severe diarrhea, severe short bowel syndrome

Table 85.2 Indications and Contraindications for Enteral Nutrition Support

Indications	Contraindications
Severe dysphagia from obstruction or dysfunction of the oropharynx or esophagus	Inability to gain access to the GI tract
Neurologic impairment, coma, or delirious state	Nonoperative mechanical GI obstruction that cannot be bypassed with a feeding tube
Persistent anorexia from nausea, poor appetite or taste changes related to disease or treatments	Intractable vomiting or severe GI malabsorption
Gastric ileus	Adynamic ileus
Low-output enterocutaneous fistulas	Distal high-output fistulas that cannot be bypassed with a feeding tube
Increased nutritional requirements such as burns or trauma	Severe GI bleeding
Organ system failure	Aggressive nutrition intervention is not consistent with prognosis or patient wishes

(SBS; < 100 cm of small intestine), or severe GI bleeding (see Table 85.2).[25-27] Nevertheless, some of the potential barriers to EN can be circumvented with careful selection of the enteral access device, formula and route of administration.

Special Considerations

Critically Ill Patients

EN and PN guidelines for critically ill patients have been published by the European Society of Parenteral and Enteral Nutrition (ESPEN) and the American Society for Parenteral and Enteral Nutrition (ASPEN) jointly with the Society of Critical Care Medicine (SCCM).[1,28] These guidelines make well-supported recommendations to initiate normal food intake or enteral feeding early, and that EN is the preferred route of nutrition support over PN.

Low Flow States

Certain clinical conditions frequently observed in critically ill patients. such as hypovolemia, hypotension, or hemorrhagic or septic shock, pose a risk for low splanchnic blood flow that can lead to GI dysmotility, increased mucosal permeability, endotoxemia, and multiple system organ failure. Reperfusion injury and low-flow states in which hypoperfusion of the GI tract is suspected must be considered when initiating enteral feeding. Disproportionate vasoconstriction can occur in response to an insult sustained during critical illness, raising concern that initiation of EN may be poorly tolerated by an under perfused intestine and lead to intestinal ischemia, a potential but rare complication.[29-32]

Despite these concerns, evidence indicates that with appropriate patient selection, careful initiation and close monitoring, EN can be used successfully in critically ill patients.[6,33] EN may be provided guardedly to patients who are receiving inotropes or pressor agents, while observing for signs of intolerance or gut ischemia.[15,34] Guidelines suggest that a patient in the early acute phase of critical illness should start with a conservative dose of EN, with stepwise progression once the patient is fully resuscitated and stable.[35] For patients requiring significant hemodynamic support (including high-dose vasopressor use, alone or in combination with large volume fluid or blood product resuscitation to maintain cellular perfusion), EN should be withheld or delivered at a very low rate of infusion until the patient is fully resuscitated and stable.

Early Enteral Nutrition

The importance of early EN in critically ill patients, especially those with a longer stay, has been widely acknowledged.[1,6,22,35] Benefits seen with early EN, initiated within 24–48 hours of admission, have been attributed in part to secretory IgA, gut-associated lymphoid tissue (GALT), and mucosa-associated lymphoid tissue (MALT), which are stimulated by enteral feedings and help fight infection locally in the gut, and at distant sites as well.[7,36-38] Meta-analyses of randomized controlled trials show that early, compared to late, EN was associated with reduction in infectious complications although, given many methodology limitations and heterogeneity differences in the individual trials, the evidence continues to be of low quality.[1,39-41] In spite of this, guidelines and expert consensus recommend starting low-dose early EN.[1,6,22]

Gastrointestinal Dysmotility

Gastrointestinal (GI) dysmotility often causes enteral feeding intolerance in critically ill patients.[42] Although "enteral feeding intolerance" is not consistently defined in the literature, per a systematic review, three categories are most often used: 1) gastric residual volume (GRV) and/or GI symptoms; 2) ability to achieve EN prescribed volume and 3) some combination of 1 and 2 (more discussion in complications).[43] GI motility is the process of smooth muscle contraction and relaxation within the GI tract, which propels the luminal contents forward with the coordination of peristaltic and non-peristaltic flow. In critically ill patients, this process may be impaired in the stomach (delayed gastric emptying), small bowel (slow transit time, ileus) and large bowel (functional obstruction).[42] This dysmotility may be due to endogenous factors (e.g., pain, inflammation, hypokalemia, etc.) or exogenous factors (e.g., opiate analgesia, vasopressors, etc.) and will be compounded by preexisting conditions (e.g., diabetic gastroparesis) and treatments (e.g., surgery).[44] Prokinetic agents have been shown to accelerate gastric emptying and improve EN tolerance in critically ill patients with dysmotility, although use is "off label" and has not been proven in large trials to reduce mortality and morbidity.[45,46]

Altered Intestinal Anatomy or Function

Patients who have had GI anatomy or function altered either by disease, surgical resection, reconstruction, or replacement with intestinal transplantation can be fed enterally. The choice of enteral feeding will largely depend on the GI anatomy and the health of the mucosa. An intact and functioning stomach, pancreas, and liver will enhance digestion and absorption.

Short Bowel Syndrome (SBS), Anatomical

For the patient with SBS (defined as less than 200 cm of remaining small bowel in continuity), several factors influence tolerance to EN (refer to chapter 79).[47-49] In general, adults with more than 100 cm of small bowel ending in stoma or more than 60 cm of small bowel anastomosed to the colon can tolerate EN (via oral food or tube feedings) and achieve adequate nutrition with close monitoring and individualization of the diet. Patients with less than this amount may require long-term PN.[48,49] Pharmacologic treatments (antisecretory, antidiarrheal, antibiotics, and growth factors) may allow some patients time off PN with oral or EN supplementation to meet their nutritional needs.[49,50] The use of oral food or EN must be balanced with an individual patient's GI output and ability to maintain their hydration and electrolyte status. In patients with only a partial dependence on PN, EN with or without oral diet may result in increased intestinal macronutrient absorption compared to oral diet alone, allowing patients to wean from PN. For some patients, managing large volumes of PN or increased GI output associated with EN lessens their quality of life and, thus, should be part of the decision-making process.[51,52]

If EN is used, a standard polymeric formula with complex nutrients (intact proteins, typically casein or whey, as well as glucose polymers and a mixture of long- and medium-chain triglycerides) can be used for most patients with compromised intestinal function (refer to chapter 77).[49] Additional soluble fiber (e.g., pectin) is not currently recommended.[49] If EN or an oral diet is the sole source of nutrition, the use of a high sodium oral rehydration solution may be needed to replace losses in GI output.[49] These patients are at high risk of vitamin and mineral deficiencies and should be monitored. Digestion and absorption of most nutrients occurs in the duodenum and proximal jejunum, whereas the distal 100 cm of ileum is responsible for the absorption of vitamin B_{12} and bile salts. Malabsorption of bile salts can cause fat malabsorption, steatorrhea, and loss of fat-soluble vitamins.[53] As the terminal ileum is also the site of vitamin B_{12} absorption, the patient will require supplemental parenteral or nasal vitamin B_{12} for life if the terminal ileum is resected.[54]

Short Bowel Syndrome, Functional

Although the length of the remaining intestine is critical for successful EN in SBS, the health of the intestinal mucosa is also an important consideration. Certain diseases and conditions (e.g., Crohn's disease, radiation enteritis) lead to diseased bowel mucosa, impairing absorption (refer to chapter 80). Crohn's disease (CD) typically affects the distal part of the GI tract; however, it can impact the upper GI tract. Steroid treatment is preferred over exclusive EN for inducing clinical remission in adults with active CD, although the evidentiary support is of low quality.[2] Frequent use of low to moderate dose steroids may lead to serious side effects (e.g., osteoporosis, hyperglycemia, infections, etc.).[55] In addition, corticosteroids, unlike EN, do not promote mucosal healing. Therefore, some clinicians and patients prefer a trial of exclusive EN prior to starting medication. A standard polymeric formula may be used orally or per tube.[2] Most patients will prefer to drink the EN, as opposed to having a feeding tube placed. Decreased palatability is often given as the reason for poor oral compliance, however, this may be improved with dietitian support to tailor the regimen to the patient's preferences.[56] Some limited evidence suggests that the use of exclusive EN pre-operatively in patients with CD may reduce inflammation and post-operative infections.[57]

Gastrointestinal Cancers

Patients with gastrointestinal cancers are at increased risk of malnutrition due to the disease and the impact of anticancer therapies (e.g., chemotherapy, radiation, surgery). These patients may suffer from dry mouth, taste changes, nausea, vomiting, anorexia, constipation, diarrhea, and increased fatigue, all impairing their ability to consume adequate oral nutrition (refer to chapter 103). EN should be considered to provide adequate energy and protein. If surgery impacts the ability to adequately consume an oral diet, early EN is recommended in the post-operative period. EN enriched with nutrients to aid in stimulating the immune response (arginine, glutamine, omega-3 polyunsaturated fatty acids, nucleotides, and antioxidants), known as immunonutrition (see **Table 85.3**), has been shown to reduce postoperative infectious complications and hospital length of stay and improve wound healing.[58,59]

Enteral Access, Equipment, and Feeding Methods

Enteral Access

Enteral access devices may be placed at the patient's bedside, in the radiology department, the endoscopy suite, or the operating room.[60,61] The type of enteral access chosen depends on the anticipated length of time that enteral feeding will be required, the patient's anatomy and the potential risk of aspiration. Enteral access devices vary in the type of construction material, length, inner diameter (French size; one French unit = 0.33 mm), presence of a stylet or guidewire to stiffen the tube during placement, weighting of the tube, and number of lumens (see Table 85.3).[60-62] In general, all tubes designed for enteral feeding are either completely radiopaque or have a radiopaque strip to facilitate radiographic confirmation of placement. New guidelines from the Global Enterprise Device Supplies Association (GEDSA) strongly encourage all manufacturers and patients to transition to a new ENFit connector for all enteral feeding tubes, delivery systems and syringes by 2022 to improve patient safety by minimizing medical tubing misconnections.[62,63] Generally, EN for a period of less than 4 to 6 weeks can be provided using nasal tubes, whereas feedings for longer periods require more permanent tubes, such as those placed percutaneously or during surgery.

Table 85.3 Enteral Access Device Characteristics[60-62]

Tube Type	Material	Length	French (Fr) Size	Stylet or Guidewire
Short-Term				
Nasogastric	Polyvinyl chloride or silicone	36 to 48 inches	12 to 20 Fr	No
Salem sump	Polyurethane, silicone or mixture	36 to 60 inches	8 to 14 Fr	Yes/No
Nasoenteric				
Feeding tube				
Long-Term				
Percutaneous endoscopic gastrostomy (PEG)	Silicone or polyurethane	Cut to fit	14 to 24 Fr	Yes/No
Gastrostomy	Silicone or polyurethane	4 to 6 inches	12 to 28 Fr	No
Jejunostomy	Silicone or polyurethane	6 to 10 inches	8 to 14 Fr	No
Jejunostomy via PEG	Silicone or polyurethane	35 to 45 inches	8 to 10 Fr	Yes
Gastrojejunostomy	Silicone or polyurethane	Jejunal length 35 to 45 inches	Gport: 22 to 24 Fr Jport: 8-12 Fr	Yes

Table 85.4 Gastric vs Small Bowel Feeding[64,65]

Gastric Feeding	Small Bowel Feeding
Majority of patients able to protect their airway (for example) ■ Cerebrovascular accident ■ Head and neck surgery ■ Short bowel syndrome (to maximize surface area fed)	Patients with increased aspiration risk or failure of gastric feeding (for example) ■ Gastroparesis/severe GERD ■ Severe acute pancreatitis ■ Heavy sedation ■ Severe respiratory dysfunction

The tip of the feeding tube may be positioned either in the stomach (gastric or pre-pyloric) or the duodenum or proximal jejunum (small bowel or postpyloric). Most patients can tolerate gastric feeding. However, in some critically ill patients, a lower risk of pneumonia and higher delivery of nutrition has been reported with small bowel feeding.[64,65] Some indications for feeding tube tip placement are listed in **Table 85.4**.[64,65]

Despite several technologies to assist with blind placement of nasogastric or nasoenteric feeding tubes at bedside,[66-68] radiological position confirmation continues to be recommended before starting tube feeding. After this initial confirmation, other methods (such as pH, length of tube extending from nares, color of aspirate) may be used to determine potential migration of tube out of position. If there is concern of migration, repeat confirmation via radiograph is warranted. Auscultation should not be used to confirm tube placement.[69] A nasal bridle, a feeding tube retaining device, may be beneficial in some patients that are at high risk for pulling out the feeding tube (e.g., dementia/confusion, inability to secure with tape).[70,71]

Enteral Feeding Method and Equipment

EN is not without potential complications, and requires proper administration and monitoring.[62] The choice of administration method is dictated by the type and site of access and patient tolerance. Enteral feedings can be administered via bolus, intermittent, or continuous methods.[72] Programmable feeding pumps may be used for any method. Gravity drip feeding bags may also be used for any method, however, obtaining a specific drip rate is more difficult. For the bolus feeding method, 60 mL (milliliter) syringes are used, which should be ENFit compatible. EN formulas are offered in an "open system" (individual 240 mL containers that must be open and added to a syringe or a gravity or feeding pump bag) or a "closed system" (one-liter bags that are spiked by an enteral feeding set and set for gravity drip or delivered via a feeding pump). Formula from individual containers should not hang longer than 8 hours, whereas a "closed system" bag may hang for 24 to 48 hours.

Bolus feedings are administered over a shorter time period (5–10 minutes). This administration method should only be used for gastric delivery because the stomach can act as a reservoir to handle larger volumes of formula over a short time. Generally, the patient is fed a volume of 250 to 500 mL of feeding four to six times daily. Feedings provided by this method may result in adverse GI effects because of the sudden delivery of a large, hyperosmolar formula. Intermittent feedings are administered over 20 to 30 minutes and are primarily used for gastric feeds, and may lessen risk of GI intolerance associated with bolus feeds.[62] Continuous feedings are delivered slowly over 12 to 24 hours, usually with an enteral feeding pump. The continuous method is required for tubes placed within the small bowel due to the lack of a reservoir for large volumes. Infusion rates over 120 to 150 mL per hour into the small bowel may lead to cramping or diarrhea. Continuous feedings may be cycled over a portion of the day (generally 8–12 hours). Cycled feedings may be used for patients transitioning from tube to oral feeding, in an attempt to stimulate appetite, or for those receiving home EN, to allow bowel rest and time off the pump. Often the feedings are administered at night and discontinued during the day to afford the patient greater mobility and, if allowed, an opportunity to eat.

Enteral Nutrition Complications, Prevention, Monitoring, and Treatment

The use of an EN protocol is recommended for the timely initiation and advancement of feedings, regardless of method or patient type to help prevent complications.[1,72-74] Factors included in an EN protocol are presented in **Table 85.5**. A standardized protocol may outline an initial EN volume based on a patient's condition. For example, EN should be started at a lower rate (10–20 mL/ hour or 100–150 mL every 4–6 hours) and advanced slowly (increase 10–20 mL/ hour or 50mL, every 4–12 hours) in patients with conditions placing them at risk for refeeding syndrome (e.g., severe malnutrition), hyperglycemia (e.g., diabetes), GI intolerance or malabsorption (e.g., pancreatitis, bowel surgery), or bowel ischemia (e.g., shock).[62,72] To prevent a clogged feeding tube, a minimum flush of 30 mL of water every 4 hours is recommended. Sterile or purified water is recommended to decrease the risk of bacterial contamination.[62] Volume-based protocols may accompany a standard EN protocol. These protocols titrate feeding rate to a volume goal for 24 hours or allow "catch-up" rates when EN is held for any reason other than intolerance (e.g., tests, procedures).[75]

Another preventive measure is maintaining a 45-degree elevation of the head of the bed (HOB) or using the reverse

Table 85.5 EN Protocol Components

1. Criteria and contraindications for EN initiation
2. Route of infusion (gastric vs small bowel), criteria
3. Type of EN formula, criteria
4. Target amount of EN formula
5. Initial bolus or flow rate and rate of advancement
6. Minimum flush volume and frequency
7. Volume-based "catch-up" to account for missed feedings (optional)
8. Evaluation of tolerance (e.g., abdominal distention, nausea, diarrhea)
9. Response to complications (e.g., reduce rate, change formula, start antidiarrheals)

Trendelenburg position during EN infusion to lessen the risk of tracheobronchial aspiration.[62] A nasal tube passing through the upper and lower esophageal sphincters, which normally prevent tracheobronchial aspiration in recumbent positions, may increase regurgitation and aspiration risk.[76] Patients receiving gastric bolus feedings should not recline for at least 2 hours after feeding. Sometimes a preventive maneuver carries a risk that requires additional monitoring; for example, prolonged HOB elevation with continuous EN may increase the risk of sacral pressure ulcer development.

Monitoring for and treatment of complications associated with EN intolerance should also be done in a standardized fashion (**Table 85.6**).[1,6,22,43,46,77-80] Although the complications outlined in Table 85.6 may be initially attributed to the EN, in reality,

Table 85.6 Common EN Complications

Complication	Monitoring	Potential Cause	Treatments
Diarrhea (more than 4–5 moderate loose to soft-formed stools daily)	If recent use of antibiotics—check for *Clostridioides difficile (c. diff)* Steatorrhea Serum electrolytes Na, K, Mg If >7 days, consider micronutrient losses, zinc, water-soluble vitamins Daily weight Daily input/output	Medications (per tube): liquid with sorbitol, magnesium, laxatives, antibiotics, lactulose, etc. Fecal impaction *c. diff* Pancreatic insufficiency or malabsorption Recent bowel surgery Altered gut flora Formula contamination	Pharmacy review of medications Disimpaction, bowel regimen If *c. diff* ruled out antidiarrheals Pancreatic enzymes Increased fluid and electrolyte replacement Change in EN formula ■ Isotonic ■ Peptide-based ■ Soluble fiber
Constipation (no bowel movement in >3 days after EN advanced to goal, or hard stools)	Rectal exam to determine if impacted, Signs of dehydrations—increased serum Na, Cl, BUN, dark and/or decreased UOP Daily input/output	Medications: narcotics, calcium channel blockers, etc. Dehydration Decreased physical activity Inadequate fiber intake	Pharmacy review of medications Maintain adequate hydration Stool softeners, laxatives, enemas, bowel regimen Change EN formula ■ Less concentrated ■ Fiber mixture soluble/insoluble
Nausea/vomiting or abdominal distention	Abdominal exam, emesis amount and appearance GRV Serum electrolytes Na, K, Cl, BUN Daily input/output	Medications: opiates, anticholinergics Gastroparesis Upper GI surgery Impacted stool Obstruction	Pharmacy review of medications, consider prokinetic, antinausea Temporarily hold or reduce EN infusion rate or volume If gastric, change to small bowel feeds Change EN formula ■ Low fat, low fiber ■ Ensure room temperature
Hyperglycemia or Hypoglycemia	Blood glucose checks per hospital protocol Daily inputs/outputs Urine glucose	Medications: Glucocorticoids, insulin Acute infection Diabetes/DKA	Pharmacy review of medications Hospital glycemic control protocol Insulin to carb ratio adjustment Change EN formula ■ Avoid overfeeding, decrease rate ■ If known diabetic with poor control, diabetes formula
Refeeding	Serum electrolytes: K, Mg and PO4 before feeding if patient at risk, daily until stable Vital signs Daily weights Daily input/output	ICU patient Malnutrition present No intake for 5–7 days	Limit dextrose in intravenous fluids Replete low serum electrolytes according to hospital protocol Initiate EN feeding slowly; 10–20 kcal/kg for 1st 24 hours, followed by slow advancement as electrolytes stabilize Give thiamin 100 mg per tube daily for 5–7 days Give multivitamin for 10 days

the cause is likely multifactorial. Thus, the interventions may be included in non-EN protocols.

Whenever a patient receiving EN is experiencing complications, a review of the medication list is prudent. For example, the most commonly reported complication is diarrhea, however, patients receiving EN often receive sorbitol or magnesium containing liquid medications through the feeding tube, which may be causing the diarrhea, not the EN formula. If possible, a crushable form of the medication should be considered, with caution taken not to clog the feeding tube (see below). Gastric residual volumes (GRV) are no longer a recommended daily monitoring technique for EN tolerance.[1,22,43,77] Evidence suggests that there is poor correlation between GRV and risk of aspiration.[81] However, in patients with abdominal distention, nausea or vomiting; GRV volume and appearance (formula versus gastric secretions) may help determine delayed gastric emptying, versus increased gastric secretions. Delayed emptying may be treated with a prokinetic or transition to small bowel feeding, whereas a proton pump inhibitor may be prescribed for a hypersecretory state.[1,22,62]

To prevent clogged feeding tubes, appropriate steps should be taken to maintain tube patency. Medications should never be added directly to the enteral formula, but instead given separately via a y-port or while feedings are held.[62] The feeding tube should be flushed after administration of medications and prior to restarting the enteral feedings. When feasible, medications should be provided in liquid form, following directions for dilution if delivered into the small bowel or if medication has a high viscosity.[62] In general, medications delivered into the small bowel should be diluted with a minimum of 30 mL fluid. If nonliquid medications are used, a pill-crushing device should be used to grind the pills into a fine powder which is then dissolved in a diluent (water or saline) for administration. Capsules may need to be opened and contents dissolved in the diluent. Some medications are unsuitable for crushing (e.g., enteric-coated tablets).[62] If a feeding tube does become clogged, hospital policy should be followed to unclog the tube. Most policies start with an attempt to withdraw any enteral solution remaining in the tube, to then apply the warm water method (using 30 mL syringe to flush using a gentle back and forth motion) and then, if the tube remains obstructed, to use the alkalinized enzyme method (sodium bicarbonate + non-enteric coated pancreatic enzyme + water).[62,82]

Enteral Nutrition Formulas

The selection of an enteral formula is based on matching the composition of a formula to the patient's clinical status, GI function, and nutritional requirements. A large number of commercial enteral formulas with different characteristics are available, and new ones are frequently introduced. Some patients may also prefer to use home blended formulas. Although no standard terminology exists, these formulas can be classified into four general categories: standard, elemental/semi-elemental, disease-specific, and immune-modulating. Other definitions used to classify EN formulas include low residue/fiber supplemented, energy density (products range from 1.0 to 2.0 kilocalories (kcal) per mL, protein content (10%–35% of total kcal), and osmolality.

Whole proteins and complex polysaccharides contribute relatively few milliosmoles (mOsm) to a formula. In contrast, free amino acids and monosaccharides contribute many more for the same number of kcals. As a result, the osmolalities of elemental/semi-elemental diets tend to be higher than standard formulas with intact proteins and complex carbohydrates. Osmolality is reported as mOsm/kg. Isotonic formulas have an osmolality of about 300 mOsmol/kg, which is similar to that of blood. Hypertonic formulas have an osmolality >400 mOsmol/kg and in some instances may lead to EN intolerance (e.g., nausea, abdominal distention, or diarrhea).

Standard Formulas

Standard formulas, also known as polymeric formulas, are the most common, lowest cost group of enteral products, and are typically used for hospitalized, long-term care, and ambulatory patients. These products contain nutrient profiles that mimic a healthy diet so that a daily intake of 1,500 to 2,000 kcal (or 1 to 2L of total formula volume) provides the dietary reference intake for most macro- and micronutrients. In the non-specialty products (discussed later), protein constitutes approximately 10% to 25% of total energy and is supplied as intact protein from eggs, milk, pureed meat, or protein isolates from casein, whey, lactalbumin, soy protein, or egg white. Carbohydrate provides approximately 50% of total energy in the form of corn syrup solids, hydrolyzed cornstarch, or maltodextrin. Fat compromises 25% to 40%, with approximately 80% coming from long-chain triglycerides (LCT) and 20% from medium-chain triglycerides (MCT) using various oils, including borage, canola, corn, fish, safflower, soybean, and sunflower oil. Standard formulas range from 1.0 Kcal/mL (isotonic), to high-energy-density mixtures 1.2 to ~2 Kcal/mL (hypertonic). Concentrated formulas are useful for patients requiring fluid restriction, intolerant to high volume feeds, or for optimizing nutrition provided in a reduced period of time (e.g., cycled feedings).

Because of the fixed nutrient and synthetic nature of these formulas, they may lack some phytochemicals, micronutrients, and unidentified growth factors that are present in food. Therefore, for those individuals on long-term EN, the use of homemade and commercially blenderized tube feeding derived from whole foods has been suggested. Currently, commercial products that are lactose-free, gluten free with either animal or protein-based protein options are available. These formulas tend to be thicker or more viscous and will flow easier through a feeding tube with a 10 Fr or higher diameter.[83,84] If used as the only source of nutrition, blenderized tube feeding (especially if homemade) may be nutritionally inadequate for macro- and micronutrients, which could lead to weight loss or nutrient deficiencies.[85]

Hydrolyzed or Elemental/ Semi-elemental Formulas

Hydrolyzed or elemental/semi-elemental formulas, also known as peptide-based, monomeric, oligomeric, predigested, or chemically defined formulas, are more expensive than standard formulas. Protein sources include crystalline amino acids or hydrolyzed protein from casein, whey, or lactalbumin to provide short-chain peptides and free amino acids, and supply 15% to 35% of total energy. Sources of carbohydrate include hydrolyzed cornstarch, dextrin, and fructose, providing approximately 15% to 35% of total energy. The fat content in these formulas is comprised of some combination of long- and medium-chain triglycerides, with a higher proportion of MCT (40%–50%) than standard formulas. Hydrolyzed formulas were designed for patients with malabsorptive conditions (e.g., pancreatitis, short bowel syndrome) or with demonstrated intolerance to standard formulas. These formulas are theoretically easier to digest due to hydrolyzed protein into small chain peptides, which may facilitate gastric emptying and increased MCT, requiring less digestion and improving tolerance, specifically decreased rate of diarrhea.[86-89] While evidence does not support a benefit when compared to patients tolerating a standard formula,[90] feeding with a semi-elemental formula performed as well or better and was less costly, than an elemental formula in terms of tolerance and digestion.[86] Current recommendations suggest a trial of a standard formula with frequent monitoring first in malabsorptive conditions (e.g., pancreatitis), changing to a hydrolyzed formula with signs of GI intolerance.[86,91]

Fiber and Prebiotic-Supplemented Formulas

Fiber is any non-digestible carbohydrate present in plant foods that are resistant to human enzymatic digestion and enter the colon.[92] Functionally, fiber is classified by its solubility in water. Soluble fibers such as pectin and hydrolyzed guar gum are water soluble, have low viscosity, and can easily be added to enteral formulas. They prolong gastric emptying, delay the absorption of dietary glucose, and are rapidly fermented by colonic bacteria to short-chain fatty acids (mainly acetate, propionate, and butyrate) that serve as energy sources for the colonocytes. Insoluble fibers such as soy polysaccharide and cellulose pass largely unchanged. They increase fecal weight, soften the stool, and shorten transit time. Insoluble fiber is frequently used in long-term enterally fed patients as a means to prevent constipation.[92]

A prebiotic will beneficially alter the colonic microflora of the host conferring a health benefit (refer to chapter 37).[92] The colonic microflora contributes to the barrier that prevents pathogenic bacteria from invading the GI tract.[93] Most prebiotics are fiber, however there are nonfiber prebiotics, including polyphenols and lactulose. Many polymeric formulas have been supplemented with purified fiber to promote bowel regularity, to control diarrhea, or prevent constipation.[94] The fiber content of enteral formulas varies considerably, both in amount and type. The research evaluating fiber-containing enteral formulas in the management of diarrhea has not demonstrated consistent results. A meta-analysis of nineteen trials comparing enteral formulas with and without fiber (soluble, insoluble or mixed) in critically ill patients showed less diarrhea in patients receiving fiber (pooled results of four randomized controlled trials) and no difference in the incidence of various GI complications (distention, constipation, etc.) between the groups (pooled results of five randomized controlled trials).[95] No increased risk of adverse events were noted with the use of fiber-containing formulas.

Disease Specific Formulations

A variety of commercial formulas have been developed for use in disease-specific conditions.[96] The macronutrient portion and, in some cases, the micronutrients have been altered to promote improved tolerance and optimize conditions for recovery. Although guidelines do not recommend routine use of these formulas, some subsets of patients may benefit from their use.[1,22]

Diabetes and Formulas for Glucose Control

Formulas developed to optimize glucose control in patients with diabetes are lower in carbohydrate, higher in fat and have added insoluble fiber. Carbohydrate sources include complex forms, such as oligosaccharides, fructose, and soy polysaccharides, to prevent rapid glucose absorption and rise in blood glucose concentration. The fat content is higher in monounsaturated fatty acids (MUFAs) and lower in polyunsaturated and saturated fatty acids. Lipids provide approximately 45% of the energy. Inconsistent study results and heterogeneity of studies have led to insufficient evidence to make a definitive statement on the potential benefit of the products.[97,98] Potential use should be determined on a case-by-case basis.

Renal Disease Formulas

Renal products have been developed to provide optimal nutrition to patients with a reduced capacity for clearance of various metabolites.[99,100] Compared to standard formulas, these are typically lower in total protein, potassium, magnesium and phosphorus, but enhanced with essential amino acids and histidine to minimize uremic symptoms. They are also energy dense for fluid management. Restricted protein and electrolytes are often not warranted in hospital patients receiving renal replacement therapy (RRT).[96] In the absence of elevated potassium, magnesium, and phosphorus, patients undergoing RRT should receive a standard or high-protein formula. However, renal formulas may be useful in those circumstances where RRT is delayed or must be avoided all together.

Hepatic Disease Formulas

Patients with chronic liver disease present unique challenges, in that they are often malnourished and suffer from micronutrient deficiencies (e.g., thiamin, folate, vitamin C, zinc). Protein restriction will worsen the degree of malnutrition and subsequent clinical outcomes in these patients.[101] Historically, there has been a hesitancy to provide adequate protein because concentrations of branched-chain amino acids (BCAAs) are decreased, and aromatic amino acids (AAAs) are increased, resulting in an imbalance contributing to hepatic encephalopathy (HE). Given the high cost and lack of noted benefit with these formulas, their use is not recommended.[102,103] HE should be medically managed. Only in those patients who are refractory to routine medical therapy for HE and are unable to tolerate standard protein formulas without precipitation of HE, is the use of BCAA formulas warranted.[103]

Pulmonary Disease Formulas

A unit of carbohydrate, when metabolized, produces more carbon dioxide than a unit of fat. In theory, rapid administration of a large carbohydrate load to a patient with severe pulmonary disease can adversely affect respiratory function. Specialized pulmonary formulas replace carbohydrates with increased amounts of fat to reduce the added burden on compromised lungs. However, the usefulness of these formulas compared to standard products has never been proven, and they are rarely used in clinical practice.[99] Because the amount of carbon dioxide produced is more a function of the number of kcal consumed than of the macronutrient source of the energy, it is more important simply to avoid overfeeding in patients with pulmonary compromise. The use of immunonutrition in acute respiratory distress syndrome (ARDS) has been heavily studied and will be discussed in the following section.

Immune-Enhancing Formulas

So-called immune-enhancing enteral nutrition (IMEN) refers to EN products containing varying amounts of specific immune-modulating nutrients (also known as pharmaconutrients) such as omega-3 fatty acids, glutamine, arginine, probiotics, and/or antioxidants believed to exert beneficial effects on the immune, inflammatory, and metabolic processes (**Table 85.7**). Agreement is still lacking about the value of individual substrates or combinations for different specific patient populations.[58,59,100,104] Commercial formulas vary considerably in their composition and dosage of individual nutrients. Numerous studies have been conducted on heterogeneous patient populations within the perioperative space (trauma, cancer surgery, cardiac surgery, GI surgery, etc.) and nonsurgical (ARDS, medicine, cardiology, etc.) populations, using enteral feeding regimens containing either a single pharmaconutrient or a combination measuring a variety of outcome measures.

Although benefit (reduced infections and LOS) has been noted in some patient populations (trauma, surgery, burns, and surgical ICU patients), these benefits have not been realized in other populations (non-surgical medical, cardiac, or medical ICU patients). Even in the surgical populations where benefit has been shown, it is not consistent across studies. For example, in preoperative oncology patients (regardless of the type of cancer surgery), a large meta-analysis of 69 studies included a subset of 31 studies (18 RCTs) with an immunonutrition or pharmaconutrition intervention.[100] The specific interventions varied, however most included arginine, omega-3 fatty acids or dietary nucleotides alone, or in combination. Twelve of the RCTs demonstrated no benefit, whereas the other six showed decreased infectious complications and LOS. No safety concerns were documented.[100] A large meta-analysis of 74 trials, including more than 7,500 surgical patients, demonstrated a decrease in infections and risk of anastomotic leak, but no difference in LOS.[105]

Most nonsurgical data come from medical ICUs, which does not support any recommendation for the use of IMEN to impact infectious complications or length of stay.[106-109] This holds true for specific patient populations in a severe proinflammatory state such as sepsis or ARDS.[110,111] ARDS, which is characterized by hypoxemia, results from a cascade of events involving free radical proliferation and proinflammatory eicosanoid production. A meta-analysis of 10 RCTs with 1,015 patients compared standard formulas to an enteral formula with additional immune-modulating nutrients. The authors found that the intervention produced little or no difference in all-cause mortality between the groups, and only very low-quality evidence suggesting a potential benefit in ventilator days or ICU length of stay.[111]

Despite extensive preclinical and patient intervention research, a straightforward application of evidence to practice is not presently possible, although some studies identified specific patients in whom immune formulas offer benefits. The use of IMEN should be considered in patients experiencing major elective GI surgery, trauma (blunt, penetrating, head injury), head and neck cancer surgery, and severe burns.[1,104]

Table 85.7 Some Characteristics of Components of Immune-Enhancing Tube Feedings

Arginine
May become conditionally essential in catabolic states
Necessary for normal T- and B-lymphocyte and macrophage functions
Role in synthesis of nitric oxide
Glutamine
May become conditionally essential in catabolic states
Primary fuel in rapidly proliferating cells such as enterocytes and immune cells
Protects gut mucosa from injury in animal models
Improves immune and barrier function of gastrointestinal tract, hence reducing bacterial translocation and sepsis in animal models
Omega-3 fatty acids
Attenuates inflammatory injury via anti-inflammatory cytokine: ecosanoid network
Modulates immunity via lymphocyte proliferation and phagocytic activity
Improves gut barrier function in some animal models
Probiotics and prebiotics
May prevent or reverse microbiota dysbiosis
May modulate intestinal immune system
May enhance gut barrier function
Antioxidants (vitamin A/β-carotene, Vitamin C, Vitamin E, selenium, glutamine)
Antioxidants may have cytoprotective effects and decrease organ dysfunction.
May negate toxic effects of oxygen free radicals produced during critical illness

Summary

EN is a safe and efficacious way of feeding patients who are unable to eat. It has become not only a food replacement alternative but also a therapy to optimize metabolic support and aid in preventing complications in patients. It is the preferred technique for nourishing patients when a functional intestinal tract can be safely accessed. A systematic approach should be taken to the initiation, monitoring, and progression of EN therapy, which considers timing, feeding site, length of treatment, disease state, risk of complications, and potential benefits.

Acknowledgments

The authors acknowledge Michele M Gottschlich for her contributions to the previous version of this chapter.

References

1. Taylor BE, McClave SA, Martindale RG, et al. Crit Care Med 2016;44:390–438.
2. Narula N, Dhillon A, Zhang D, et al. Cochrane Database Syst Rev 2018;4:CD000542.
3. Liu M, Gao C. Ann Palliat Med 2021;10:10779–88.
4. Uhl S, Siddique SM, McKeever L, et al. Malnutrition in Hospitalized Adults: A Systematic Review. Rockville, MD: 2021.
5. Williams DGA, Molinger J, Wischmeyer PE. Curr Opin Anaesthesiol 2019;32:405–11.
6. Preiser JC, Arabi YM, Berger MM, et al. Crit Care 2021;25:424.
7. Otani S, Coopersmith CM. J Intens Care 2019;7:17.
8. Xin F, Mzee SAS, Botwe G, et al. World J Surg Oncol 2019; 17:114.
9. Patel JJ, Shukla A, Heyland DK. JPEN J Parenter Enteral Nutr 2021;45:74–8.
10. Wischmeyer PE, McDonald D, Knight R. Curr Opin Crit Care 2016;22:347–53.
11. Cadena AJ, Habib S, Rincon F, et al. J Intensive Care Med 2020;35:615–26.
12. Braga M, Gianotti L, Gentilini O, et al. Crit Care Med 2001;29: 242–8.
13. Dissanaike S, Shelton M, Warner K, et al. Crit Care 2007; 11:R114.
14. Lewis SR, Schofield-Robinson OJ, Alderson P, et al. Cochrane Database Syst Rev 2018;6:CD012276.
15. Reignier J, Boisrame-Helms J, Brisard L, et al. Lancet 2018;391:133–43.
16. Laesser CI, Cumming P, Reber E, et al. J Clin Med 2019;8.
17. Gramlich L, Kichian K, Pinilla J, et al. Nutr 2004;20:843–8.
18. Li K, Wang D, Zhang X, et al. Nutr Diet 2022;79:129–39.
19. Evans RA, McAuley H, Harrison EM, et al. Lancet Respir Med 2021;9:1275–87.
20. Doley J. Nutrients 2022;14.
21. Schwartz DB. Nutr Clin Pract 2018;33:377–87.
22. Singer P, Blaser AR, Berger MM, et al. Clin Nutr 2019;38: 48–79.
23. Blake C, Edwards A, Treleaven E, et al. Nutr Diet 2021.
24. Song J, Zhong Y, Lu X, et al. Medicine (Baltimore) 2018;97: e11871.
25. Pironi L, Corcos O, Forbes A, et al. Clin Nutr 2018;37: 1798–809.
26. Bischoff SC, Escher J, Hebuterne X, et al. Clin Nutr 2020;39: 632–53.
27. Uhl S, Siddique SM, Bloschichak A, et al. J Hosp Med 2022.
28. Compher C, Bingham AL, McCall M, et al. JPEN J Parenter Enteral Nutr 2022;4612–41.
29. Patel JJ, Rice T, Heyland DK. JPEN J Parenter Enteral Nutr 2020;44:779–84.
30. Kurita D, Fujita T, Horikiri Y, et al. Surg Case Rep 2019;5:36.
31. Amenu E, Karim S, Da Silva RC. Cureus 2022;14:e24313.
32. Albrecht HC, Trawa M, Gretschel S. J Int Med Res 2020;48:300060520929128.
33. Haines K, Parker V, Ohnuma T, et al. Crit Care Explor 2022;4:e0683.
34. Wischmeyer PE. Crit Care Med 2020;48:122–5.
35. Koekkoek K, van Zanten ARH. Curr Opin Anaesthesiol 2018;31:136–43.
36. Alverdy JC. Surg Infect (Larchmt) 2018;19:163–7.
37. Alverdy JC, Chang EB. J Leukoc Biol 2008;83:461–6.
38. Miller WD, Keskey R, Alverdy JC. J Infect Dis 2021;223:S264–S9.
39. Tian F, Heighes PT, Allingstrup MJ, et al. Crit Care Med 2018;46:1049–56.
40. Reintam Blaser A, Starkopf J, Alhazzani W, et al. Intensive Care Med 2017;43:380–98.
41. Fuentes Padilla P, Martinez G, Vernooij RW, et al. Cochrane Database Syst Rev 2019;2019.
42. Deane AM, Chapman MJ, Reintam Blaser A, et al. Nutr Clin Pract 2019;34:23–36.
43. Jenkins B, Calder PC, Marino LV. Clin Nutr ESPEN 2022;49:92–102.
44. McClave SA, Gualdoni J, Nagengast A, et al. Curr Gastroenterol Rep 2020;22:1.
45. Gungabissoon U, Hacquoil K, Bains C, et al. JPEN J Parenter Enteral Nutr 2015;39:441–8.
46. Lewis K, Alqahtani Z, McIntyre L, et al. Crit Care 2016;20:259.
47. Jeppesen PB. JPEN J Parenter Enteral Nutr 2014;38:8S–13S.
48. Jeppesen PB, Fuglsang KA. Gastroenterol Clin North Am 2018;47:61–75.
49. Cuerda C, Pironi L, Arends J, et al. Clin Nutr 2021;40:5196–220.
50. Vorre MM, Johansson KS, Fuglsang KA, et al. Dan Med J 2022;69.
51. Jeppesen PB, Chen K, Murphy R, et al. JPEN J Parenter Enteral Nutr 2022;46:905–14.
52. Stanner H, Zelig R, Rigassio Radler D. JPEN J Parenter Enteral Nutr 2022;46:757–70.
53. Skouras T, Dodd S, Prasad Y, et al. Int J Colorectal Dis 2019;34:185–8.
54. Duerksen DR, Fallows G, Bernstein CN. Nutr 2006;22:1210–3.
55. Yasir M, Goyal A, Sonthalia S. Corticosteroid Adverse Effects. Treasure Island, FL: StatPearls, 2022.
56. Purcell L, Mutsekwa R, Angus R, et al. J Hum Nutr Diet 2022;35:435–43.
57. Gordon-Dixon A, Gore-Rodney J, Hampal R, et al. Clin Nutr ESPEN 2021;46:99–105.
58. Osland E, Hossain MB, Khan S, et al. JPEN J Parenter Enteral Nutr 2014;38:53–69.
59. Osland E, Memon B, Memon MA. Transl Gastroenterol Hepatol 2018;3:52.
60. Lord LM. Nutr Clin Pract 2018;33:16–38.
61. Tariq N, Ali A, Chen C. Surg Clin North Am 2020;100: 1091–113.
62. Boullata JI, Carrera AL, Harvey L, et al. JPEN J Parenter Enteral Nutr 2017;41:15–103.
63. Guenter P, Lyman B. Nutr Clin Pract 2016;31:769–72.
64. Alkhawaja S, Martin C, Butler RJ, et al. Cochrane Database Syst Rev 2015;CD008875.
65. Liu Y, Wang Y, Zhang B, et al. Eur J Clin Nutr 2021;75:1337–48.
66. Powers J, Luebbehusen M, Aguirre L, et al. Nutr Clin Pract 2018;33:268–73.
67. Slingerland-Boot R, Bouw-Ruiter M, van Manen C, et al. Clin Nutr 2021;40:5000–7.
68. Jacobson LE, Olayan M, Williams JM, et al. Trauma Surg Acute Care Open 2019;4:e000330.
69. Bourgault AM, Powers J, Aguirre L, et al. Dimens Crit Care Nurs 2020;39:329–38.
70. Chauhan D, Varma S, Dani M, et al. Curr Gerontol Geriatr Res 2021;2021:6650675.
71. Inayat F, Ur Rahman A, Almas T, et al. Cureus 2020;12:e8325.
72. Mueller CME. The ASPEN Nutrition Support Core Curriculum. Silver Spring, MD: American Society of Parenteral and Enteral Nutrition, 2017.

73. Yeh DD, Ortiz LA, Lee JM, et al. JPEN J Parenter Enteral Nutr 2020;44:197–204.
74. Jiang L, Huang X, Wu C, et al. J Crit Care 2020;56:249–56.
75. Sachdev G, Backes K, Thomas BW, et al. JPEN J Parenter Enteral Nutr 2020;44:874–9.
76. Tatsumi H. J Intens Care 2019;7:30.
77. Murthy TA, Chapple LS, Lange K, et al. Am J Clin Nutr 2022;116:589–598.
78. Seifi N, Jafarzadeh Esfahani A, Sedaghat A, et al. Syst Rev 2021;10:95.
79. Heyland DK, Ortiz A, Stoppe C, et al. Crit Care Med 2021;49: 49–59.
80. da Silva JSV, Seres DS, Sabino K, et al. Nutr Clin Pract 2020;35:178–95.
81. Reignier J, Mercier E, Le Gouge A, et al. JAMA 2013;309: 249–56.
82. McClave SA, DiBaise JK, Mullin GE, et al. Am J Gastroenterol 2016;111:315–34; quiz 35.
83. Weston SC, Crespo A, Harwin C. J Pediatr Gastroenterol Nutr 2022;75:110–2.
84. Spurlock AY, Johnson TW, Pritchett A, et al. Nutr Clin Pract 2022;37:615–24.
85. Brown T, Zelig R, Radler DR. Nutr Clin Pract 2020;35:442–53.
86. Alexander DD, Bylsma LC, Elkayam L, et al. World J Gastrointest Pharmacol Ther 2016;7:306–19.
87. Tiengou LE, Gloro R, Pouzoulet J, et al. JPEN J Parenter Enteral Nutr 2006;30:1–5.
88. Cortot A, Phillips SF, Malagelada JR. Gastroenterol 1981;80: 922–7.
89. Heimburger DC, Geels VJ, Bilbrey J, et al. JPEN J Parenter Enteral Nutr 1997;21:162–7.
90. Carteron L, Samain E, Winiszewski H, et al. Crit Care 2021; 25:31.
91. Roberts KM, Nahikian-Nelms M, Ukleja A, et al. Gastroenterol Clin North Am 47, 77–94.
92. Slavin J. Nutrients 2013;5:1417–35.
93. Klurfeld DM, Davis CD, Karp RW, et al. Am J Physiol Endocrinol Metab 2018;315:E1087–E97.
94. Koecher KJ, Thomas W, and Slavin JL. JPEN J Parenter Enteral Nutr 2015;39:337–43.
95. Cara KC, Beauchesne AR, Wallace TC, et al. JPEN J Parenter Enteral Nutr 2021;45:882–906.
96. Fiaccadori E, Sabatino A, Barazzoni R, et al. Clin Nutr 2021;40:1644–68.
97. McMahon MM, Nystrom E, Braunschweig C, et al. JPEN J Parenter Enteral Nutr 2013;37:23–36.
98. Burslem R, Rigassio Radler D, Parker A, et al. Nutr Clin Pract 2022;37:68–80.
99. Cohen DA, Byham-Gray L, Denmark RM. J Hum Nutr Diet 2013;26:286–93.
100. Brajcich BC, Stigall K, Walsh DS, et al. J Am Coll Surg 2022;234:384–94.
101. Ghufran A. Nutr Clin Pract 2020;35:211–7.
102. Plauth M, Bernal W, Dasarathy S, et al. Clin Nutr 2019;38: 485–521.
103. Nanchal R, Subramanian R, Karvellas CJ, et al. Crit Care Med 2020;48:e173–e91.
104. McCarthy MS, Martindale RG. Nutr Clin Pract 2018;33: 348–58.
105. Mazaki T, Ishii Y, Murai I. Ann Surg 2015;261:662–9.
106. Heyland D, Muscedere J, Wischmeyer PE, et al. N Engl J Med 2013;368:1489–97.
107. van Zanten AR, Sztark F, Kaisers UX, et al. JAMA 2014;312: 514–24.
108. Lopez-Delgado JC, Grau-Carmona T, Trujillano-Cabello J, et al. Nutrients 2022;14.
109. Grau-Carmona T, Moran-Garcia V, Garcia-de-Lorenzo A, et al. Clin Nutr 2011;30:578–84.
110. Rice TW, Wheeler AP, Thompson BT, et al. JAMA 2011;306: 1574–81.
111. Dushianthan A, Cusack R, Burgess VA, et al. Respir Care 2020;65:99–110.

D. DISORDERS OF THE ALIMENTARY TRACT

CHAPTER 86

Parenteral Nutrition

Vivian M. Zhao and Thomas R. Ziegler

CHAPTER OUTLINE

Abbreviations

AMA, American Medical Association; **ASPEN**, American Society for Parenteral and Enteral Nutrition; **BG**, blood glucose; **BSI**, bloodstream infections; **CDC**, Centers for Disease Control and Prevention; **Cr^{3+}**, chromium; **EFA**, essential fatty acid; **EN**, enteral nutrition; **ESPEN**, European Society for Parenteral and Enteral Nutrition; **FDA**, Food and Drug Administration; **FFA**, free fatty acid; **GI**, gastrointestinal; **GSHPx**, glutathione peroxidase; **HPN**, home parenteral nutrition; **ICU**, intensive care unit; **IV**, intravenous; **LBW**, low birth weight; **LCT**, long-chain triglyceride; **MCT**, medium-chain triglyceride; **IVFE**, intravenous fat emulsion; **MBD**, metabolic bone disease; **Mn^{2+}**, manganese; **MVI**, multivitamin infusion; **MTE**, mixed trace elements; **PICC**, peripherally inserted central venous catheter; **PN**, parenteral nutrition; **CPN**, central vein parenteral nutrition; **PPN**, peripheral vein parenteral nutrition; **PVC**, polyvinyl chloride; **RCT**, randomized controlled trial; **RES**, reticuloendothelial system; **TNA**, total nutrient admixture

Introduction

Malnutrition (e.g., loss of significant lean body mass and depletion or frank deficiency of specific essential vitamins, minerals, and trace elements) is common in hospitalized patients and patients unable to maintain adequate nutrition and hydration by the enteral route. Various factors contribute to protein-energy malnutrition and loss of micronutrients in these settings, including catabolic hormonal and cytokine signals, resistance to anabolic hormones, lack of adequate oral food intake during illness, abnormal nutrient losses (e.g., via drains, renal replacement therapies, wounds, emesis, polyuria), *nil per os* (NPO) status resulting from diagnostic or therapeutic tests and procedures, and increased macronutrient and micronutrient needs during specific illnesses.

Current recommended energy goals for patients who are stable and critically ill are between 20 to 30 and 12 to 25 kcal/kg/day, respectively. Protein (amino acids) provision between 1.2 and 1.5 g/kg/day is appropriate for most adult patients in the hospital and at home. Most patients can meet these goals with conventional PN methodologies. Adequate vitamins, minerals, electrolytes, essential amino acids, and fats must be provided based on recommended allowances for healthy individuals. Conventional PN provides all of these substrates and fluids. Metabolic, infectious, and mechanical complications can occur with PN but can be prevented or diminished with careful monitoring and adherence to current standards of practice.

Historical Aspects

Vinnars, Wilmore, and Bistrian summarized significant historical aspects of PN therapy.[1,2] Beidle and Krauts first described intravenous (IV) administration of glucose in 1896, and Elman described IV administration of protein as fibrinogen hydrolysate in 1937. The first IV fat emulsion (IVFE) in the United States, Lipomul, was developed in 1960.[1] Various clinicians used peripheral vein PN (PPN) using 5% or 10% glucose, protein hydrolysates, IVFE, electrolytes, and multivitamins from 1955 to 1965. Serious side effects led to the withdrawal of IV Lipomul from the U.S. market in the early 1960s. This created a serious problem requiring that glucose be given either in large, relatively isotonic volumes for peripheral vein infusion or in a hyperosmolar form requiring infusion into a major vein. Although central catheters threaded into veins had been used as early as 1944, they were uncommon. A safe and effective IVFE, Intralipid, was developed by Wretlind in 1961[2] and approved for clinical use in most European countries by 1963. It was not approved for use in the United States Until 1972. The availability of Intralipid in Europe led to increased PN use through peripheral veins.[1,2]

Widespread interest and increased PN use occurred after the publication of reports by Dudrick et al., from the University of Pennsylvania in 1968.[3] Using central percutaneous catheters to deliver nutrient solutions with glucose as the source of nonprotein energy, plus micronutrients (vitamins, minerals, and mixed trace elements [MTE]), these investigators demonstrated convincingly that PN, as the sole source of nutrients, resulted in good growth in infants with malnutrition, positive nitrogen balance, and nutritional and clinical improvement in adults with malnutrition over periods of many weeks.[3]

Clinical Practice Guidelines

Since its formation in 1978, the American Society for Parenteral and Enteral Nutrition (ASPEN) continues to promote the safe and efficacious use of specialized nutrition support (PN and EN) through publishing new clinical guidelines and revisions of old guidelines for specialized nutrition support in patients with critical illness and other conditions.[4-9] All ASPEN guidelines can be accessed through www.nutritioncare.org. Similarly, the European Society for Parenteral and Enteral Nutrition (ESPEN: www.espen.org) and the Canadian Clinical Practice Guidelines Committee has published clinical practice guidelines on PN uses in several conditions[10-13] and critical illnesses (www.criticalcarenutrition.com),[14] respectively.

Indications

PN is primarily used to maintain or improve the nutritional and metabolic status of patients who cannot be adequately nourished by oral or enteral feeding (EN) for a short term or long term over a lifetime (e.g., ≥90 days). The uses of PN in managing patients with specific clinical problems are also discussed in relevant chapters in this book and the ASPEN, ESPEN, and Canadian Clinical Practice Guidelines (**Table 86.1**).[15-18]

Table 86.1 Common Indications for the Use of Parenteral Nutrition

- Following massive small bowel resection with or without colonic resection (short bowel syndrome)
- Severe intestinal motility disorders (e.g., intestinal pseudo-obstruction)
- Immature gastrointestinal tract, gastrointestinal atresia (pediatric patients)
- Proximal high-output fistulas or perforated small bowel
- Severe diarrhea or emesis, partial or complete small bowel obstruction, severe ileus
- Severe gastrointestinal bleeding or severe hemodynamic instability precluding adequate enteral feeding

Prescribing PN requires consideration of various factors, such as the patient's diagnosis and prognosis. It is not a defensible substitute for oral or EN when adequate provision by either of these methods is feasible. Most (>85%–90%) patients in the hospital setting, including those in the intensive care unit (ICU), can be fed enterally via oral diet, oral nutrient supplements, or gastric or intestinal tubes. However, EN is contraindicated in approximately 10% to 15% of patients in the hospital deemed to require PN via PPN or, commonly in ICU patients, via a central vein (CPN).[7-10]

The common PN indications listed in Table 86.1 are generally accepted worldwide. Therapeutic effects of PN occur via the combined provision of energy (such as amino acids, dextrose, and essential fatty acids (EFAs) from lipids), vitamins, minerals, and trace elements.[5,15] PN supports vital organ and cell functions, immunity, protein synthesis, skeletal and respiratory muscle capacity, wound healing, and tissue repair.

PN is an adjunctive treatment in most cases. A meta-analysis of PN in patients with critical illness demonstrated that PN did not affect mortality, but decreased complication rates, especially in patients with malnutrition.[16] Some of the data that reflect negatively on PN may be because of overfeeding and PN-induced hyperglycemia, rather than harm caused by using PN. Rigorous studies in critically ill patients have not resolved the controversies concerning the appropriate timing (early vs late) and composition of PN. The clinical benefits in terms of morbidities and mortality remain unclear.[8,17-19]

Despite the routine use of PN in many clinical settings globally (e.g., in ICU and non-ICU hospital settings, in-home patients with Short Bowel Syndrome (SBS) or other forms of intestinal failure), many areas of uncertainty concerning its optimal use remain, given the current lack of rigorous comparative effectiveness research in large, well-powered patient populations (**Table 86.2**).[20,21] Problems with earlier PN studies included small sample sizes, uncertain or heterogeneous patient characteristics (including varying definitions of illness severity), and inappropriate or inadequate blinding.[6-22]

Generally accepted, although not evidence-based, contraindications for PN are as follows: when the GI tract is functional and accessible for enteral feeding; when PN is unlikely to be required for more than 5 to 7 days with GI function expected to normalize; if the patient cannot tolerate the intravenous fluid (IVF) load required for PN; when the patient has severe hyperglycemia or marked electrolyte abnormalities at the time initiation

Table 86.2 Current Areas of Uncertainty Regarding Use of Parenteral Nutrition

- Clinical impact of various durations on minimal or no feeding in hospital patients
- Optimal timing for the initiation of PN, alone and in relation to enteral nutrition
- Efficacy of various doses of energy, amino acids, fat, and micronutrients (vitamins, minerals, trace elements) in various clinical settings
- Impact of altered amounts or overall composition of essential and nonessential amino acids in PN formulas
- Efficacy of alternative lipid emulsions in PN (e.g., olive oil, fish oil, structured lipids, medium-chain triglycerides, alone and in combination)
- Efficacy of longer-term PN in the post-ICU hospital or home phase
- Clinical efficacy of parenteral administration of specific nutrients (e.g., specific micronutrients or amino acids) to supplement enteral feedings

ICU, intensive care units; *PN*, parenteral nutrition.

Data from Ziegler TR. N Engl J Med 2009;361:1088–97; Ziegler TR. N Engl J Med 2011;365: 562–4.[20,21]

Table 86.3 Composition of Typical Parenteral Nutrition Formulations

Component	Peripheral Vein Parenteral Nutrition	Central Vein Parenteral Nutrition
Volume (L/d)	2–3	1.0–1.5
Dextrose (%)	5	10–25
Amino acids (%)	2.5–3.0	3–8
Lipid (%)	3.5–5.0	2.5–5.0
Sodium (mEq/L)	50–150	50–150
Potassium (mEq/L)	20–35	30–50
Phosphorus (mmol/L)	5–10	10–30
Magnesium (mEq/L)	8–10	10–20
Calcium (mEq/L)	2.5–5.0	2.5–5.0
Vitamins		
Trace elements[a]		

[a]Conventional commercial preparations containing mixtures of essential vitamins and trace elements are used in both peripheral and central vein parenteral nutrition (PN). Typical ranges are shown above, but amounts of specific components outside the ranges shown may be used in some cases. See text for additional details on PN composition.

of PN is planned; and if the new placement of a central venous catheter solely for PN is deemed to pose undue risks.[5,20]

Compared with PN, EN is less expensive, may better maintain intestinal mucosal structure and gut absorptive and barrier functions (as shown in animal studies), and is associated with fewer mechanical, metabolic, and infectious complications.[20] Therefore, after starting PN, it is standard practice for EN to be initiated with the return of adequate GI function and advanced as tolerated, while the amount of administered PN is correspondingly reduced.

Comparison of Peripheral Vein and Central Vein Parenteral Nutrition

PN may be administered via a peripheral or central vein. CPN is typically infused through the subclavian or internal jugular vein. Limited rigorous clinical comparative effectiveness trials are available on PPN efficacy. PPN may be helpful for a patient who needs short-term nutritional support and has good peripheral veins.[22,23] PPN does not require the insertion and maintenance of a central venous catheter. To minimize peripheral vein damage, 10% or 20% isotonic IVFE as the primary energy source is substituted for the hypertonic glucose solutions typically used in CPN. PPN and CPN provide all essential and nonessential amino acids, with higher amino acid concentrations provided via CPN. A comparison of the typical composition of PPN and CPN is shown in **Table 86.3**.

Electrolytes in PN are adjusted, as indicated, by GI losses, renal function changes, and other clinical parameters, coupled with serial monitoring of blood concentrations, to maintain serum or plasma concentrations within the normal range.[5,8] In the presence of elevated blood concentrations, lower doses (or elimination) of specific electrolytes in PN may be indicated until blood concentrations normalize. Higher dextrose concentrations in CPN typically increase potassium, magnesium, and phosphorus requirements; hence, higher doses of these nutrients might be required in CPN compared with PPN. In both PPN and CPN, the concentrations of sodium and potassium salts as chloride can be increased to correct metabolic alkalosis, and the concentration of salts as acetate can be increased to correct metabolic acidosis.[5,15] Regular insulin can be added to PN, as needed, to maintain blood glucose (BG) concentration within the goal range of 140 to 180 mg/dL in the hospital setting.[24]

Historically, trace elements added daily to PN were mixtures of chromium, copper, manganese, selenium, and zinc. In September of 2020, the U.S. Food and Drug Administration (FDA) reformulated and approved a new MTE without chromium. These minerals remain available as individual additives, except for manganese. Daily MVI added to PN are mixtures of vitamins A, B_1 (thiamin), B_2 (riboflavin), B_3 (niacinamide), B_6 (pyridoxine), B_9 (folate), B_{12}, C, D, E, biotin, and pantothenic acid. Vitamin K may be present in some multivitamin preparations. Specific vitamins (e.g., B_1, B_6, B_9, B_{12}, C, and K) can also be supplemented separately in PN.[5,15] The specific parenteral recommendations for trace elements and vitamins for adult and pediatric patients have been published in expert consensus papers.[25,26]

A typical 2.5-L PPN preparation composed of 3% amino acids (4 kcal/g), 4% lipid (as a 20% or 30% IVFEs, 10 kcal/g), and 5% dextrose (3.4 kcal/g) provides 75 g of protein and 1,725 kilocalories (kcal), with an osmolality of 600 mOsm/L (osmolality should be ≤900 mOsm/L). A typical 1.5-L CPN may be composed of 5% to 7% amino acids, 3% to 4% lipid, and 15% to 20% dextrose, with a higher osmolality of approximately 1,700 mOsm/L. Such high-osmolality CPN solutions cannot be infused through peripheral veins because of the development of phlebitis. PPN requires frequent (e.g., every 3 to 4 days) peripheral vein catheter changes to prevent thrombophlebitis. Despite limited evidence, some institutions add low-dose heparin infusion (e.g., 1,000 units/liter) and/or corticosteroids (e.g., 5 mg hydrocortisone/bag) in PPN to prevent phlebitis.

PPN usage necessitates significant fluid administration (e.g., 2 to 3 L/day) to provide adequate energy and amino acids in most patients. Given the low dextrose and amino acid concentrations in PPN, most energy is provided by the lipid emulsion. However, stable hospital patients with body weight ≤70 kg can often be given amino acid and energy doses approaching estimated needs with as little as 2 L PPN/day. For example, a 60-kg patient is estimated to require 25 kcal/kg/day and 1.2 g amino acid/day to meet energy and amino acid needs, respectively. The energy goal would be 1,500 kcal/day (25 kcal/kg×60 kg), and the protein (amino acid) goal is 72 g/day (1.2 g×60 kg). A two-liter PPN solution typically consists of 3% amino acids and provides 240 kcal/day (1.0 g/kg/day; 60 g×4 kcal/g); the typical dextrose concentration of 5% in PPN provides 340 kcal (100 g dextrose/2-L×3.4 kcal/g), and the typical PPN lipid concentration of 4% provides 800 kcal (80 g lipid/2-L×10 kcal/g). Thus, such a 2-liter PPN formulation provides 1,380 kcal/day (or 23 kcal/kg/day).

Similar calculations can be used to determine proportions and percentages of amino acids, dextrose, and lipid to meet estimated goals when CPN is used. Thus, for a 60-kg patient, an energy goal of 1,500 kcal/day can be met with a 1.5-L/day CPN. CPN provides 6% amino acids, which provides 360 kcal (1.5 g/kg/day; 72 g/day × 4 kcal/g), 15% dextrose provides 765 kcal (67% nonprotein energy; 225 g dextrose/1.5 L×3.4 kcal/g), and 2.5% lipid provides 375 kcal (37.5 g lipid/1.5 L×10 kcal/g). As shown earlier, providing sufficient energy by the central route in CPN without providing a large percentage as lipid necessitates infusing hypertonic glucose solutions (typically 10% to 20% dextrose; see Table 86.3).

Catheters for Parenteral Nutrition Administration

PN is given through different vascular access sites, such as the peripheral line, midline, and, most preferably, a central line through the subclavian, jugular, or femoral veins. The catheter tip must be in a vessel with high blood flow causing rapid dilution; this minimizes the occurrence of phlebitis and thrombosis.

Peripherally inserted central venous catheters (PICC) are used for short-term PN (e.g., <30 days). If PICCs are used for PN, the catheter tip must be positioned in a central vein, such as the superior or inferior vena cava. Central catheters placed via the subclavian vein are reported to have lower bacterial colonization compared with internal jugular, or femoral vein approaches.[27]

Several reviews and guidelines on preventing catheter complications have been published.[28-30] Central catheters reduce infection incidence and can remain in place and functional for a long time. These are surgically placed within the subclavian or jugular vein, with the tip in the superior vena cava. The extravascular portion of the catheter is tunneled before being brought out through the skin. The catheter is typically anchored at the skin exit with a Dacron cuff, which eliminates the need for sutures and acts as a barrier to bacteria.[28-30] Another type of central catheter, called a port, is implanted subcutaneously with silicone chambers or other elastomers. A catheter connects the chamber and is usually placed into the subclavian vein with its tip in the superior vena cava.[31] The PN solution is infused into the chamber via special needles inserted through the skin.

Insertion and utilization of an indwelling central venous catheter poses various risks to the patient, including pneumothorax, hemothorax, thrombosis, infection, vascular or nerve injury, hypersensitivity reactions, and microbial contamination. Reported complication rates range from 0.3% to 19.3% and vary according to the definition of complications, physician expertise, the preparation used, the frequency of manipulation of the catheter, and other factors.[30] Thrombogenicity varies with the catheter material; the earlier and stiffer polyvinyl and polyethylene catheters were associated with more thrombus formation than silicone or polyurethane catheters. Multi-lumen catheters have increased use by providing additional access for infusing medications and blood, and for blood sampling without interfering with PN administration. There are conflicting reports about the frequency of catheter-related sepsis and placement site-related infection rates (e.g., femoral and jugular sites tend to have higher rates than the subclavian site).[30,32] Tunneled catheters (e.g., jugular) reduce the incidence of line sepsis[33] and other problems, such as dislodgment.

Several approaches have been adopted to prevent catheter-related bloodstream infections (CRBSIs) in patients receiving short- and long-term CPN.[34-39] These include the use of strict catheter insertion and maintenance protocols, appropriate hand hygiene and aseptic technique with catheter insertion and manipulation, use of tunneled, cuffed catheters for chronic PN infusions (e.g., in-home patients requiring CPN), avoidance of femoral vein catheters, and use of various types of catheters impregnated with antimicrobial agents or chlorhexidine.[34-39] Other interventions for CRBSI prevention include using different catheter site dressings with antibiotic or antiseptic ointments and catheter locks containing heparin, vancomycin, sodium citrate, taurolidine, or ethanol.[34-37,39] In 2014, the U.S. Centers for Disease Control and Prevention (CDC) updated its 2011 comprehensive guidelines for preventing CRBSI.[40]

Delivery Systems

PNs are compounded in unique EVA (ethylene-vinyl acetate) PN bags, which are made with a copolymer of ethylene and vinyl acetate, characterized by high elasticity, resistance to wear and tear, antimold and antibacterial. CPNs are generally delivered via various electronic peristaltic pumps, which have become increasingly sophisticated, automated, and expensive. They ensure even flow rates, overcome the increased resistance of filters of small porosity (especially with continued use), minimize the likelihood of clotting at the catheter tip, and reduce the need for frequent nursing surveillance. Most have an air-in-line alarm system that prevents the occurrence of air embolism.

Dual-chambered plastic bags allow admixture of macronutrients immediately before PN infusions are available. These bags are manufactured either empty or with macronutrients (e.g., dextrose in one chamber and amino acids in the other chamber). They are convenient for home PN use, especially for patients who receive IVFEs regularly. When lipid is used, the dextrose, amino acids, and electrolytes are added to the bottom chamber of the empty bag, and the desired IVFEs dose to the upper chamber. Before administration, the plastic divider is removed, and the admixture with lipid is prepared. This increases stability, because the PN admixture is not ready until just before infusion.

In-line IV filters continue to be recommended when administrating PN.[41] Filters remove or reduce the infusion of particulate matter, air, and micro-organisms into the patient. Particulates have been found to clog pulmonary capillaries and cause pulmonary embolus when they exceed 5 μm. They could potentially deposit in other soft tissues such as the brain, spleen, renal medulla, and lung. For PPN, in-line filters have been reported to decrease the incidence of phlebitis.[5,15,41] Historically, two commonly used filters during PN infusions were 0.22- and 1.2-micron filters, but they led to associated confusion and errors related to using two filters. The current guideline recommends using a 1.2-micron filter for all PNs.[41] The 1.2-micron filter is also effective in removing organisms such as *Candida albicans*, as they are large particles in the range of 3 to 6 μm.

Insulin adsorption to the catheter varies appreciably, depending on the binding characteristics of the nutrients present, the type of plastic in the delivery system, filters, and the insulin concentration added.[24] Insulin adsorption is minimal in TNAs. When insulin is added to PN formulations for patients with diabetes, the dosage must be closely monitored until properly adjusted.[24]

Components and Requirements

Fluid

Fluid requirements for most adults are approximately 30 to 40 mL/kg/day.[15] Fluid in PN plus other IVFs, with or without enteral fluid intake, must meet individual requirements determined by evaluating the clinical and laboratory data (e.g., physical examination findings relevant to fluid status and plasma sodium and urea concentrations). Consideration of the close interrelationships of water, electrolytes, hormonal factors, and organ function is critical when prescribing PN. In addition to clinical factors that could cause excessive retention or loss, consideration must be given to fluid intake with medications, "keep-vein-open" infusions, and changes in insensible water loss. Meticulous recording of fluid intake and output is necessary. Assessment of volume status by hemodynamic monitoring may be required in some critically ill patients.

Standard PN admixtures can be administered to patients with increased fluid needs, especially when extra renal losses are involved, with supplemental IVFs to meet needs in the acute care setting. Additional fluid can be added to PN or given separately. For patients with fluid overload, PN should be made as concentrated as possible. Expansion of extracellular fluid is common in hospitalized patients with malnutrition. It increases body weight and decreases serum concentrations of albumin, prealbumin, and other proteins, independent of nutritional status.

Energy and Macronutrient Requirements

Current guidelines estimate energy goals in clinically stable adult patients at approximately 20 to 30 kcal/kg/day. For ICU patients, precise energy needs for improved clinical outcomes remain unclear, but hypocaloric feeding (e.g., 12–25 kcal/kg/day) is recommended by the European and American/Canadian clinical practice guidelines.[8,10,20,42] The recommended energy goal for patients with obesity is <14 kcal/kg actual weight.[7] The energy goals for infants and children requiring PN are beyond the scope of this chapter, but they have been reviewed in the ASPEN guidelines.[4,43] Although precise energy, protein, fat, and micronutrient needs in various patients requiring PN are not well defined, based on decades of experience, PN appears to be generally safe and effective for most patients (**Table 86.4**).[4,5,8,10,11,14,20,21,44,45]

Table 86.4 General Guidelines for Macronutrient Dosing in Adult Hospital Patients

Substrate	Energy Dose
Energy Dose (g/kg/day)[a,b,c]	
Clinically stable patient	REE×1–1.3 (or 20–30 kcal/kg/day)
ICU patient	REE (or 12–25 kcal/kg/day)
Obese patient	50%–70% REE (or <14 kcal/kg/day)
Amino Acid Dose (Essential+Nonessential) (g/kg/day)[d,e]	
Adult without organ failure	1.2–1.5
ICU patient	1.2–2
Acute encephalopathy	0.6
Renal failure, not on renal replacement therapy (based on renal function)	0.6–1.2
Renal failure, on renal replacement therapy	
Peritoneal dialysis (PD)	1.2–1.4
Hemodialysis (HD)	1.2–1.5
Continuous renal replacement therapy (CRRT)	1.8–2.5

ICU, intensive care unit; *PN*, parenteral nutrition; *REE*, resting energy expenditure.
Initial PN order nonprotein energy: 60%–70% as dextrose and 30%–40% as lipid.
[a]PN amino acids provide 4 kcal/g, dextrose 3.4 kcal/g, and lipid emulsion 10 kcal/g.
[b]Energy needs can be estimated by indirect calorimetry; these measurements can be inaccurate in mechanically ventilated patients receiving high levels of inspired oxygen or as a result of air leaks or other technical issues with the ventilator.
[c]The Harris-Benedict equation can be used to estimate REE:
Males (kcal/day): 66.5+(13.75×weight in kg)+(5.003×height in cm)–(6.75×age)
Females (kcal/day): 655+(9.563×weight in kg)+(1.85×height in cm)–(4.676×age)
[d]In obese individuals (actual body weight (ABW) is ≥25% of ideal body weight (IBW)), adjusted body weight should be used to calculate REE:
Adjusted body weight=IBW (use standard equations)+0.25×(ABW–IBW)
[e]Some clinical guidelines recommend amino acid doses ≥2.0 g/kg/day in specific subgroups, such as burn injury or renal replacement therapy.

Data from Ziegler TR. N Engl J Med 2009;361:1088–97; Fiaccadori E, Sabatino A, Barazzoni R, et al. Clin Nutr 2021;40:1644–68; Ikizler TA, Burrowes JD, Byham-Gray LD, et al. Am J Kidney Dis 2020;76:S1–S107.[20,46,47]

Amino Acids

Conventional PNs provide essential and nonessential amino acids. Intravenous amino acid solutions have evolved from the original hydrolysates of casein or blood fibrin to crystalline L-amino acids of different compositions and varying concentrations, based partly on the amino acid composition of high-quality dietary proteins. Formulations of crystalline L-amino acids have been developed for specific clinical problems, with varying claims for superiority over more general formulas for use in infants with renal and hepatic failure or trauma, and for growth.[4-8] Commercial formulations differ among and within manufacturers in amino acid composition and concentrations, and some with added electrolytes (**Table 86.5**). Many pharmacies that use automated compounders will stock one strength

Table 86.5 Composition of Some Commercial Amino Acid Products for Parenteral Nutrition

Product	Prosol 20%	Clinisol 15%	Aminosyn II 15%	Trophamine 10%[a]	Freamine 10%
Manufacturer	Baxter	Baxter	ICU Medical	B. Braun	B. Braun
Essential Amino Acids (g/100 mL)					
Valine	1.44	0.96	0.75	0.78	0.66
Lysine	1.35	1.18	1.575	0.82	0.73
Histidine	1.18	0.894	0.45	0.48	0.28
Isoleucine	1.08	0.749	0.99	0.82	0.69
Leucine	1.08	1.04	1.5	1.4	0.91
Phenylalanine	1	1.04	0.447	0.48	0.56
Threonine	0.98	0.749	0.6	0.42	0.4
Methionine	0.76	0.749	0.258	0.34	0.53
Tryptophan	0.32	0.25	0.3	0.2	0.15
Nonessential Amino Acids (g/100 mL)					
Alanine	2.76	2.17	1.49	0.54	0.71
Glycine	2.06	1.04	0.75	0.36	1.4
Arginine	1.96	1.47	1.527	1.2	0.95
Proline	1.34	0.894	1.083	0.68	1.12
Glutamic acid	1.02	0	1.107	0.5	0
Serine	1.02	0.592	0.795	0.38	0.59
Aspartic acid	0.6	0	1.05	0.32	0
Tyrosine[b]	0.05	0.039	0.405	0.24	0
Taurine	0	0	0	0.02	0
Cysteine[b]	0	0	0	< 0.016	< 0.016
Phosphoric acid	0	0	0	0	0.12
Electrolytes (Anion profiles per liter)					
Acetate	140 mEq	127 mEq	107.6 mEq	96.2 mEq	89 mEq
Chloride	0	0	0	5 mEq	< 3 mEq
Phosphate	0	0	0	0	10 mMol
Sodium	0	0	50 mEq	< 3 mEq	10 mEq
Other					
pH (range)	6 (5.5–6.5)	6 (5–7)	5.8 (5–6.5)	5.5 (5–6)	6.5 (6–7)
Osmolarity, mOsm/L	1835	1357	1270	875	950
Total Amino Acid, g/L	200	150	150	100	97
Nitrogen, g/L	32.1	23.7	23	15.5	15.3

[a]Designed for infants and young children (including those of low birth weight).
[b]Conditionally essential.
All information was obtained from package inserts.

(usually 15% or 20%) of standard amino acids to make all PN formulations.[4,5,48]

The typical recommended dose of amino acids for adults is 1.2 to 1.5 g kg/day. Under special circumstances, such as renal failure (with or without dialysis) or burn injury (see Chapter 95), higher amino acid doses (approaching 2.5 g kg/day) have been recommended.[6,8,49,50] Higher PN amino acid doses are also required for infants and growing children.[4] The amino acid concentration in PN is adjusted downward or upward as a function of the amino acid dosing goal and the degree of renal dysfunction.[4,5,8,15,20] Some guidelines recommend the routine addition of glutamine as a conditionally essential amino acid in ICU patients (see the chapter on glutamine).[10,50] Although it is not questioned that essential and sufficient nonessential amino acids should be provided in PN in amounts needed to sustain adequate protein synthesis and intermediary metabolism, surprisingly limited data from rigorous, adequately powered RCTs to define optimal doses of total or individual amino acids in PN are available.[50,51]

Carbohydrates

Glucose (dextrose) is the carbohydrate used for energy contribution in PN. It is usually the primary energy source, typically ordered as 60% to 70% of the PN nonprotein energy (see Table 86.4). Parenteral glucose is in the form of a monohydrate; one gram provides 3.4 kcal. It is available in various concentrations, relatively inexpensive, and is rapidly metabolized by most patients (**Table 86.6**).

Glucose Metabolism and Hormonal Changes

Glucose infused into humans increases insulin secretion, thus leading to increased serum insulin concentration. In patients who are stable, this adaptive response is often adequate for maintaining normal serum glucose. Abrupt PN cessation can result in rebound hypoglycemia in some patients because the insulin secretion is not blunted immediately with the withdrawal of the PN infusion. Therefore, clinical practice dictates that IV dextrose 5% or 10% be administered after PN withdrawal to prevent hypoglycemia, unless the patient eats some carbohydrate-containing food or is being tube fed.[32] Because such patients are not uncommon, glucose tolerance must be checked before large amounts are infused cyclically. ASPEN clinical guidelines on preventing hyperglycemia and hypoglycemia in neonates and adults receiving PN have been published.[24,53] Sudden increases or decreases in glucose infusion can be averted by using infusion pumps that can automatically increase the infusion of the admixture and taper it without changing the pump settings.

Glucose metabolism in patients with trauma, injury, burns, sepsis, or advanced cancer that induces weight loss differs markedly from physiologically normal persons. Patients who are stable can efficiently oxidize infused glucose to carbon dioxide (CO_2) up to approximately 14 mg/kg/minute. In contrast, patients with critical illnesses have only about half that capacity, 5 mg/kg/minute in patients with burns and 6 to 7 mg/kg/minute in patients postsurgery. Infusion above the limiting rate results in the conversion of glucose to fat, with an increase in energy expenditure and respiratory quotient above one. The conversion of excess glucose to fat is energy-dependent. After oxidation of the resultant fat, the derived energy (as adenosine triphosphate sources) is 30% of that theoretically obtained by direct oxidation of the converted glucose.

Table 86.6 Osmolalities and Energy Values of Intravenous Dextrose Preparations[a]

Dextrose (%)	Osmolality (mOsm/kg H_2O)	kcal/L
5	278	170
10	523	340
15	896	510
20	1,010	680
25	1,410	850
50	2,523	1,700
70[b]	3530	2,380

[a]Dextrose: 3.4 kcal/g.
[b]Product commonly used for compounding PN.

Control of Blood Glucose Concentrations

Control of hyperglycemia is an essential component of managing patients receiving PN. PN administration with a substantial amount of dextrose contributes to hyperglycemia; however, this dextrose-induced hyperglycemia is often complicated by an acute disease state (trauma, burns, and sepsis), a chronic disease condition (diabetes) or concomitant pharmacotherapy (corticosteroids, protease inhibitors) that often results in moderate-to-severe hyperglycemia. Patients who are critically ill are particularly at risk for stress-induced hyperglycemia, which is well known to predispose them to infection, possibly by impairing immune functions.[5,54,55]

A landmark trial addressing hyperglycemia in patients with critical illness was published by Van den Berghe et al., in 2001, in which 1,548 ICU patients were randomized to intensive insulin therapy (BG range: 80–110 mg/dL or 4.4–6.1 mmol/L) or standard insulin treatment (BG range: 180–200 mg/dL or 10–11.1 mmol/L).[56] This study demonstrated decreased mortality in the group receiving intensive insulin therapy.[56] Other morbidity improvements were observed in patients who resided in the ICU for >5 days and received intensive insulin therapy, including decreased ICU length of stay, reduced time receiving ventilator support, decreased prevalence of acute renal failure, and decreased bacteremia.[56] Subsequent studies showed improvements in mortality and clinical outcomes resulting from the control of hyperglycemia without the administration of insulin.[57,58]

In a large multicenter, multinational study of 6,104 patients in the ICU by the NICE-SUGAR investigators, patients were randomized within 24 hours after admission to undergo either intensive glucose control, with target BGs of 81 to 108 mg/dL (4.5–6.0 mmol/L), or conventional glucose control, with target BGs of ≤180 mg/dL (≤10.0 mmol/L).[59] Mortality was slightly but significantly higher in the intensive glucose control group at 90 days (intensive control 27.5% vs conventional control 24.9%; $P=0.02$)[60] than in the conventional glucose control group. There were no other significant clinical outcome differences between the two treatment groups.

Based on available data, the standard of care in patients in ICU is to maintain BG between 140 and 180 mg/dL (7.8 and 10.0 mmol/L) via insulin infusion, and to provide very close clinical monitoring to prevent hypoglycemia.[61] The most recent clinical practice guidelines by the Endocrine Society recommend a BG target of 100 to 180 mg/dL (5.6–10.0 mmol/L) for hospitalized patients with noncritical illness (with or without diabetes).[62] However, no data are available to evaluate tight BG control parameters, specifically in patients receiving PN.

Lipid Emulsions

The clinical use and metabolic or oxidative, immune, and clinical effects of the variety of IVFEs are now available. They have been comprehensively reviewed.[63-67] IVFEs, a source of essential fatty acids (EFAs) and energy dense nonprotein kcal, are an integral part of PN. Current guidelines suggest that IVFEs are indicated in PN during the first week of ICU stay.[8,68] Conventional IVFEs provide essential linoleic and α-linolenic fatty acids. Generally, more than 3% of total energy provided as EFAs are required to prevent EFA deficiency (see the section in this chapter).

Composition

IVFEs consist of tiny droplets (≤0.5 μm) with triglyceride as the core and cholesterol derived from egg yolk phosphatides surrounded by a solubilizing and stabilizing surface layer of the emulsifying phospholipids.[69] In the United States, soybean oil-based or soybean/safflower oil-based emulsions were the only IVFE available for PN use historically, primarily providing omega-6 as the long-chain fatty acid source. Other IVFEs available in Europe and Canada since the 1990s were not approved for use in the United States until 2013. These new FDA-approved IVFEs include olive oil/soybean oil, approved for use in adults in October 2013; soybean oil/medium-chain triglyceride (MCT)/olive oil/fish oil, approved for adult uses in July 2016 and pediatric use in March 2022; and, fish oil, approved for pediatric patients with PN-associated cholestasis in August of 2018. They are made from different lipid sources; allergic reactions to any ingredients must be considered, including soy, olive oil, coconut, fish, shellfish, eggs, and peanuts (**Table 86.7**).

Table 86.7 Composition of Commercial Intravenous Fat Emulsion Products for Parenteral Nutrition

	Intralipid 10%, 20%, 30% Nutralipid 20%	Clinolipid 20%	Smoflipid* 20%	Omegaven* 10%
Manufacturer	Fresenius Kabi B. Braun	Baxter	Fresenius Kabi	Fresenius Kabi
Lipid Source	100% soybean oil	20% soybean oil 80% olive oil	30% soybean oil 25% olive oil 30% MCT 15% fish oil	100% fish oil
Omega-6: Omega 3 ratio	7: 1	9: 1	2.5: 1	1: 8
α-Linolenic acid (Omega-3)	4–11	2–2.3	2.5	1.8
DHA (Omega-3)	0	0	1–3.5	14–27
EPA (Omega-3)	0	0	1–3.5	13–26
Linoleic acid (Omega-6)	44–62	18.5	21.4	4.4
Oleic acid (Omega-9)	24.5	44.3–79.5	23–45	4–11
α-tocopherol (mg/L)	35–38	32	160–230	150–296
Phytosterols (mg/mL)	348±33	327±8	179±10	0
Manufacturer's dosing guidelines (g/kg/day)	1–1.5 (Max: 2.5)	1–1.5 (Max: 2)	1–2 (Max: 2.5)	Not approved for adult use in the United States
Minimum dose to prevent EFAD in adults (mL/1,000 kcals)	38	113	113	No data

DHA, docosahexaenoic acid; *EFAD*, essential fatty acid deficiency; *EPA*, eicosapentaenoic acid; *MCT*, medium-chain triglyceride.
*Approved for Pediatric use.
α-tocopherol: antioxidant properties; prevents lipid peroxidation.[70]
Phytosterols: associated with PN-associated liver dysfunction.[71]
Information obtained from product package inserts.

Data from Biesalski HK. (2009) Gastroenterology 137, S92–104; Zaloga GP. (2015) JPEN J Parenter Enteral Nutr 39, 39s–60s.

CPN typically provides 30% to 40% of non–amino acid energy as fat emulsions (Table 86.4). The recommended dose of IVFEs is 1 g/kg/day and does not exceed 2 to 2.5 g/kg/day (Table 86.7). When PN compounders are used, 20% or 30% lipid emulsions may be mixed with dextrose, amino acids, and micronutrients in the same infusion bag as TNA ("all-in-one" or "3-in-1") solutions and are typically given over 24 hours.[4,5,8] IVFEs can be given as separate infusions as a 20% lipid emulsion over 10 to 12 hours/day via peripheral or central venous access.

Different IVFEs consist of varying percentages and ratios of fatty acid profiles, and additives may have different therapeutic and inflammatory effects. The inflammatory properties of eicosanoids produced by fatty acid metabolism differ between omega-3, omega-6, and omega-9 fatty acids.[67,72] Published studies on clinical outcome differences among IVFEs remain limited.

Systematic reviews and meta-analyses have been published to evaluate the role of omega-3 fatty acid-supplemented PN in patients who are critically ill.[73-75] Two reviews showed no improvement in mortality, infectious complications, or ICU length of stay with the PN solutions containing omega-3 fatty acids.[73,74] Fish oil-based IVFEs demonstrated a trend toward reducing ventilation days and mortality,[74] while the other reviews showed no difference in mortality.[73] The most recent meta-analysis included 49 RCTs (n=3,641 patients); omega-3 fatty acids were associated with a 40% reduction in infections, 56% reduction in sepsis, and both ICU and hospital length of stay were shortened by approximately 2 days.[75] This meta-analysis also suggested that omega-3 fatty acids may have hepatoprotective effects, with significant benefits in liver enzymes and higher α-tocopherol concentration, and lower inflammatory markers.[75] To benefit from omega-3 fatty acids, the doses of fish oil should be between 0.1 and 0.2 g/kg/day in PN.[68]

In a double-blind, parallel-group RCT of 100 medical or surgical ICU patients requiring PN, patients receiving olive oil-based IVFE in PN demonstrated similar clinical outcomes (infections, organ dysfunction, mortality) and plasma inflammatory or oxidative stress indexes, and neutrophil functions, compared with patients receiving PN with soybean oil-based IVFE.[76]

Stability and Safety Factors

The relevant properties of the phosphatide emulsifiers and various factors influencing the stability of IVFEs in the presence of various additives in the admixture have been reviewed.[77,78] However, the available stability analyses of IVFEs varied with combinations of PN components and employed different analytical techniques. IVFEs can be destabilized by physical properties, including temperature (room temperature vs refrigeration), time, and viscosity.[77] Several studies examined the effects of many additives, including amino acids, lipids, dextrose, and monovalent, divalent, and trivalent ions. Iron dextran was the most disruptive component of PN when IVFE was present and led to PN destabilization. Hence, iron dextran should not be added to IVFE-containing PN.[78] The ASPEN and ESPEN guidelines address specific PN stability concerning the administration of IVFEs.[77,78]

Potential Adverse Effects of Lipid Infusions

The IVFE infusion rate should not exceed 0.08 to 0.13 g/kg/hour to avoid fat overload syndrome, which results from impaired triglyceride elimination.[65] Triglyceride concentration should be monitored at baseline and weekly, particularly in patients with known lipid disorders, pancreatitis, or liver or renal disease, to assess triglyceride clearance.[20] Fish oil-based and soy-based or olive oil-based IVFEs should be reduced or discontinued when triglyceride concentration is high.[65,68] This lipid emulsion dose in PN can be decreased from daily to two to three times weekly (or can be temporarily discontinued) if plasma triglycerides rise to more than 400 mg/dL.[15,20]

The ability to metabolize these emulsions in infants is directly related to infant maturity. The risk of lipid accumulation in blood and its sequelae is most significant in the premature infant, the small for gestational age low birth weight (LBW) infant, and the nutritionally depleted older child. Lipid accumulation in the hepatic reticuloendothelial system (RES) with likelihood of depressed immune responses and competition with bilirubin and other substances for albumin binding has been described. Cases have been reported, primarily in young children, of bleeding dyscrasia associated with high plasma lipid concentrations and platelets engorged with lipids.[15] Reports of altered pulmonary function during acute hyperlipidemia have varied. Some have noted decreased pulmonary diffusion capacity, while other investigators found no change in lung dynamics, but decreased arterial oxygenation.[15]

Patients with acute respiratory distress syndrome (ARDS) who received IVFE for more than 6 to 8 hours (maximum rate suggested in package inserts) demonstrated significant decreases in arterial partial pressure of oxygen (PaO_2), fraction of inspired oxygen (FiO_2), and mean pulmonary arterial pressure; and substantial increases in pulmonary vascular resistance and pulmonary venous admixture.[79,80] It is recommended that IVFEs be administered cautiously over at least 12 hours to patients with ARD and infused over 24 hours as an all-in-one PN.[68] Considering conventional energy goals in those in the ICU (12 to 25 kcal/kg/day maximum), 250 mL of a 20% IVFE provides 50 g lipid (500 kcal), which is a reasonable daily dose.

Lipids as Pharmacologic Vehicles

Some IV medications are lipid-based, which influences the overall energy provision. One example is propofol, initially marketed for anesthesia induction but used commonly for sedation in the ICU setting. It is manufactured in 10% fat emulsion (1.1 kcal/mL), and some patients who require large doses of propofol receive substantial energy per day (e.g., 20 mL/hour of propofol provides 528 kcal/day).[81] Clevidipine is an intravenous calcium channel blocker, sometimes used to control postoperative hypertension. It is formulated in a 20% lipid emulsion (2 kcal/mL). Thus, the lipid energy provided by these agents must be considered in PN orders to avoid complications associated with overfeeding (see other sections in this chapter). In addition, several lipid formulations of amphotericin B have been available for over 50 years. However, the energy contribution of lipid formulations of amphotericin B is negligible.

Minerals

All essential electrolytes, trace elements, and vitamins are supplied in conventional PN. However, optimal intakes of specific micronutrients to meet individual needs in clinical settings are essentially unknown, and more data are needed. Clinically, a reasonable approach is to maintain specific micronutrients, when measured, within the normal range. A deficiency of an essential nutrient may lead to imbalances. For example, a lack of potassium, sodium, phosphate, or nitrogen impairs or abolishes the retention of other elements, and zinc depletion can cause a negative nitrogen balance.[82,83]

Sodium, potassium, calcium, magnesium, phosphate, and chloride are essential nutrients provided in conventional PN individualized to the patient (see Table 86.3). Detailed information on electrolyte and mineral depletion and repletion is given in the chapter on electrolytes. Because many patients on PN have GI malabsorption, impaired renal reabsorption, or both, often associated with large fluid ionic losses, a continuing concern in the care of such patients is the adequacy of fluid and electrolyte balance.[8,15,84] Generally, patients with critical illness require daily monitoring of electrolytes and dosing adjustments and close, serial monitoring. Excessive intestine or kidney losses or abnormal retention require appropriate changes with monitoring. Mineral needs for infants and children have been summarized in the literature.[26]

Infants need relatively large amounts of calcium and phosphorus (inorganic phosphate), so solubility in the PN solution can be problematic. It has been recommended that glycerophosphate or glucose phosphate be given together with calcium gluconate, as more soluble forms of these nutrients. Solubility curves have been developed to estimate calcium and phosphate compatibility in PN solutions. Recommended pediatric calcium, phosphorus, and magnesium concentrations for PN solutions have been published.[5,15,85]

Hypophosphatemia has multiple causes (see the chapter on phosphorus); however, in patients receiving PN, it is often associated with a sudden infusion of glucose, the metabolism of which stimulates the transfer of phosphate from plasma to cells (see the section on refeeding syndrome in this chapter). When initiating PN, 0.3 to 0.6 mMol/kg/day or 10 to 15 mMol/1000 kcal of phosphorus should be provided.[86,87]

Magnesium and magnesium balance are appreciated more, especially in the critical care setting, where serum magnesium concentrations are being assessed with increased frequency. Risk factors for magnesium depletion are well recognized (see the chapter on magnesium). Although serum magnesium does not accurately reflect magnesium status, a low concentration usually indicates magnesium deficiency. Management of moderate to severe magnesium deficiency in various clinical situations for different age groups is given in the chapter on magnesium and elsewhere, including the ASPEN guidelines for daily electrolyte and mineral requirements for adults and children.[15,25,26] Hypermagnesemia may occur with fluctuating or progressively deteriorating renal function or decreasing intestinal or renal magnesium loss; periodic monitoring of magnesium concentration and adjustment in PN are necessary.

Trace Elements

In 1979, commercial intravenous solutions of zinc, copper, Mn^{2+}, and Cr^{3+} became available after an expert committee of the Nutrition Advisory Group of the American Medical Association (AMA) recommended them to the FDA.[88] Trace minerals such as iron, iodide, zinc, copper, Cr^{3+}, and selenium are essential human nutrients.[15] The biochemical and physiologic roles of these trace elements, and the effects of their depletion in humans and other species, are reviewed in relevant chapters. The trace element needs of infants, children, and adults are summarized in recent reviews and chapters on these elements.[25,26] MTE products containing six trace elements (molybdenum, zinc, copper, Mn^{2+}, Cr^{3+}, and selenium) or seven trace elements (iodine plus the six mentioned earlier) were commercially available until the new FDA-approved four trace elements mixture (copper, zinc, manganese, and selenium) became available in July of 2020. **Table 86.8** summarizes daily trace element recommendations for PN and provides the composition of the current MTE product. Individual trace elements (copper, zinc, selenium, chromium, iron) can be added to PN when patients' requirements are above commercial MTE.

Table 86.8 Daily Trace Element Supplementation Recommendations for Adult Parenteral Nutrition Formulations

Trace Element	Standard Daily Intake Dosage Recommendation	Dose Provided in Conventional Commercial Product[a] (Per 1 mL)
Chromium (μg)	10–15	0
Copper (mg)	0.3–0.5	0.3
Iron (mg)	1–1.2	0[c]
Manganese (μg)	55–100[b]	55.0
Selenium (μg)	20–100	60
Zinc (mg)	2.5–6.5	3

[a]Tralement® (American Regent, Inc., Shirley, NY).
[b]Manganese contamination in various components of parenteral nutrition (PN) can contribute to total intake. Serial blood monitoring of manganese concentrations is indicated with long-term PN use.
[c]Iron dextran can be added to PN formulations that do not contain lipid emulsion.

In their salt forms, the cationic trace elements (iron, zinc, copper, Cr^{3+}, and Mn^{2+}) are highly regulated and tend to be absorbed in small amounts from food by the normal intestine. When in excess in the body, all of these elements can be toxic. Intravenous supplementation increases the risk of excessive retention because intestinal controls are bypassed. Iron is particularly poorly excreted in the urine after an IV infusion. Copper, Mn^{2+}, and, to a smaller extent, molybdate are excreted through the bile into the intestinal tract; hence continued administration of the usual amounts of copper and Mn^{2+} in the presence of excretory liver dysfunction imposes a risk of toxicity (see later). In contrast, the anionic forms of trace elements (iodide, selenite, or molybdate) are well absorbed and excreted in the urine. Many trace elements are present in variable amounts as contaminants in PN components.

The IV iron requirement for the term infant is estimated to be about 50 to 100 μg/kg/day, while the premature infant may need more than double the dose.[26] Older children need 1 to 2 mg/day.[15] Nonmenstruating women and men whose condition is stable need about 1 mg, and menstruating women need double that amount per day. Iron loss through frequent venipuncture for various tests may be estimated based on 1 mg of iron lost for every 1 mL of packed red cells removed (see the chapter on iron).

Patients on chronic PN without iron, and who eat very little, invariably become iron deficient over time. Patients with a duodenum and proximal jejunum who eat a regular diet while on chronic PN usually absorb enough iron to prevent deficiency.

Routine measurement of hemoglobin, iron studies, and serum ferritin concentration help assess iron status. During acute stress, such as infections, measuring serum iron and ferritin may not help diagnose iron deficiency because serum iron concentration decreases, whereas serum ferritin concentration increases. For a confirmed iron deficiency, IV iron may be given as iron dextran in varying amounts after ensuring that the patient has no hypersensitivity to a test dose. Iron dextran (100 to 200 mg/day) can be added to non–lipid-containing PN to prevent or treat iron deficiency. Other IV iron products include iron sucrose and ferric gluconate, which have not been tested for their stabilities and solubilities in PN. Hence, they should be administered separately.

The original AMA recommendations for zinc were reasonable for patients who are stable and those who are hypermetabolic. Still, much remains to be learned about specific micronutrient needs for optimal biochemical concentrations and clinical outcomes in patients in the hospital.[20,89,90] As in adults, severe diarrhea secondary to infectious disease and SBS in children are associated with increased zinc losses and need.[89] The 2004 ASPEN clinical practice guidelines for PN help estimate intestinal losses of zinc, which may approach 12 to 15 mg/L stool.[15] Balance studies have evaluated the requirements of zinc to be 3 mg/day in patients without GI losses and 12 mg/day in patients with diarrhea and fistula losses.[91] Zinc contamination of PN additives is variable, depending on the specific sources. Periodic checks of serum zinc concentration are essential to proper dosing to meet patients' needs.

Unlike zinc, increased stool volume is not associated with significant increases in copper excretion, and urinary losses tend to be low.[92] Thus, copper will accumulate in the body when the dose exceeds maintenance needs. For stable patients, the daily dosage recommendation for copper is 0.3 to 0.5 mg/day (see Table 86.8). Caution in copper administration in obstructive jaundice is emphasized because the significant excretory route is through bile. Copper deficiency has also been observed in patients receiving jejunal feeding (which bypasses the duodenum and proximal jejunum, where most copper is absorbed), in various malabsorptive conditions, with certain medications, after partial or complete gastric resection, and following roux-en-Y gastric bypass surgery for obesity.[93-96]

Mn^{2+} is essential for all experimental species studied, but clear evidence of Mn^{2+} deficiency in humans is lacking. Conversely, Mn^{2+} toxicity in patients receiving PN is well known. Mn^{2+} was a common contaminant in adult, pediatric, and neonatal PN (25.18, 9.92, and 1.37 μg, respectively).[97] Excessive Mn^{2+} can lead to hypermanganesemia and reversible Parkinson-like symptoms. Exposure to Mn^{2+} doses at five times the current daily requirement, together with the PN contamination, can lead to neurotoxicity. Whole-blood Mn^{2+} concentration is accurate for monitoring, and correlates well with Mn^{2+} accumulation in the brain by magnetic resonance imaging.[98] Current FDA-approved intravenous trace element mixtures contain 55 μg Mn^{2+}, almost a 10X reduction from previous products (500 μg). There have been numerous reports of Mn^{2+} toxicities in pediatrics and adults on long-term PN (>14 days).[98] Routine monitoring of whole-blood Mn^{2+} is recommended, and reducing or omitting supplementary Mn^{2+} might be necessary for some patients, especially those with liver diseases.

The original AMA qualitative suggestions for Cr^{3+} were based on estimates from balance data on healthy persons.[88] Newer quantitative data on Cr^{3+} requirements are still lacking. The lack of information could be because of the difficulty of measuring plasma Cr^{3+} concentration (normally very low), lack of information on tissue concentrations of this ion, and lack of controlled studies on very low Cr^{3+} intake. Cr^{3+} functions as a regulator of insulin action.[89,99,100] Few well-documented symptomatic Cr^{3+} depletion cases have occurred in adult patients receiving long-term PN with little or no supplementary Cr^{3+}.[89] Cr^{3+} depletion was associated with sudden glucose intolerance, glycosuria, weight loss, and neurologic symptoms, especially peripheral neuropathy. Development of symptoms appeared to be related to prolonged glucose infusions and intestinal fluid losses, both of which increase Cr^{3+} needs. The patients responded well to Cr^{3+}, often 250 μg Cr^{3+} infused daily for weeks.[99] Cr^{3+} toxicity has not been observed, even with doses greater than 250 μg/day. Although some patients with malabsorption may become deficient, little information on the appropriate dosage for using Cr^{3+} in PN is available. Chromium is another contaminant commonly present in adult, pediatric, and neonatal PN (4.85, 1.5, and 0.28 μg, respectively).[97] Cr^{3+} is no longer in the current FDA-approved trace element mixture (Table 86.8).

Selenium recommendations were not made in the 1979 AMA report. In the same year, there were reports in English relating selenium deficiency to Keshan disease in China (see the chapter on selenium), and in a patient receiving PN.[101] Considerable clinical and biochemical information has accrued since then, including selenium deficiency in patients receiving PN, with some deaths associated with cardiomyopathy and reports of muscle tenderness and weakness. The normal range of serum selenium for adults is 110 to 165 mcg/L. Selenious acid (as selenite salts) has been available for intravenous use and was approved by the FDA in July of 2019. ASPEN recommends adding 20 to 100 μg/day in PN for most patients. Still, higher doses may be needed for patients requiring long-term PN or home parenteral nutrition (HPN) to maintain normal serum selenium concentration (see Table 86.8).

Selenoprotein P concentration in plasma (measured by radioimmunoassay) correlates well with extracellular selenium-dependent glutathione peroxidase (GSHPx) and can be a marker for selenium status in deficient patients receiving long-term PN. The benefits of high-dose (400 to 4,000 μg/day) selenium in sepsis or critical illness were reviewed in a meta-analysis that included 19 RCTs.[102] RCTs suggested that high-dose selenium supplementation decreased total mortality and shortened hospital length of stay. However, it may not reduce 28-day mortality, incidence of new infections, or length of stay in patients in the ICU or on mechanical ventilation.

The aluminum toxicity issue in PN is decades old and is still unresolved because it bypasses the GI tract's normal barriers. Aluminum toxicity was well delineated in patients with renal disease who were treated with aluminum-containing, phosphate-binding antacids and received aluminum-contaminated water in hemodialysis. Manifestations of aluminum toxicity include neurologic changes (apraxic motor abnormalities involving speech, myoclonus, and seizures), Alzheimer's disease, metabolic bone disease, dyslipidemia, and genotoxic activity.[103]

Widely varying aluminum concentrations have been found in PN ingredients. Many PN ingredients contain aluminum, including calcium gluconate, inorganic phosphates, dextrose solutions, IVFEs, amino acid solutions, potassium, magnesium, and cysteine hydrochloride.[103-105] A study of infants given PN showed mean aluminum contamination of 14.0 μg/kg/day, 3X the FDA-advised maximum exposure (5 μg/kg/day).[104] Another study showed that the total aluminum concentration in the final PN was at least 40% higher than the total aluminum content of the ingredients added, with a minimum amount of 233 μg/kg/day of aluminum.[105]

Vitamins

The chapters in this book on specific vitamins give detailed information on metabolism, factors associated with depletion, and specific requirements of vitamins in infants, children, and adults. The original parenteral multivitamin formulations in the United States were based on those proposed in 1975 by the Nutrition Advisory Group of the AMA.[106] The adult formulation was approved by the FDA in 1979, and the pediatric formula was approved in 1984. In 1985, an FDA/AMA-sponsored workshop proposed several changes for parenteral formulations of vitamins. The new guidelines were published in the Federal Register in 2000.[15] The significant difference was to include vitamin K (150 μg) in the adult formulation, which closely mimics what the normal patient would require via the GI tract in health. The adequate daily intake of vitamin K via the GI tract in health is 90 μg for women and 120 μg for men; however, bacteria in the GI tract synthesize vitamin K. **Table 86.9** summarizes current daily vitamin recommendations for PN formulations and provides the composition of a conventional commercial mixed vitamin product.[25,26] Individual vitamins can also be added to the PN solution for repletion.

Adsorption and destruction of individual vitamins may occur through contact with plastic containers and tubes, filters, exposure to light and heat, and interactions with other substances present in solutions. Factors affecting the solubility and stability of vitamins in various pharmaceutical preparations have been reviewed.[78] ASPEN recommends that complete photoprotection of PN should be maintained throughout the compounding process.[107] Appreciable amounts of retinol appear to be lost from the solution by a combination of adsorption and photodegradation when the flow through the tubing is slow.[15,107] This is particularly true with the increased light intensity used in neonatal nurseries. With bright sunlight exposure, 100% of PN solutions can be lost in 3 hours. The use of polyolefin tubing rather than polyvinyl tubing reduces retinol loss. Little loss of vitamin D occurs in plastic delivery systems. Riboflavin and pyridoxine are unstable when exposed to direct sunlight for hours, and 50% of vitamin K can be lost from PN solutions in 3 hours. Thiamin, folate, riboflavin, and pyridoxine are stable under fluorescent light.[15,107]

Adding IV multivitamins to PN presents the possibility of some loss of specific vitamins. Thiamin is split and loses biologic activity in the presence of sulfite compounds that are components of amino acid solutions in the United States; hence, IV multivitamins should not be added directly to undiluted amino acid solutions, but rather to the ultimate solution shortly before infusion into the patient. Ascorbic acid is progressively lost in the presence of copper and oxygen.

ASPEN clinical guidelines and expert opinions have been followed to define doses of vitamins used in PN (see Table 86.9).[8,25,26,108] Varying amounts of vitamins have been administered to patients, with differing intervals between the times of vitamin infusion and blood samplings. Few investigations have completely surveyed all 13 vitamins.

Vitamin D is essential for bone health by improving the efficiency of the absorption of calcium and phosphate by the small intestine. Calcitriol, the active form of vitamin D, plays an intimate role with parathyroid hormone in bone turnover (see the chapter on vitamin D). A high incidence of vitamin D deficiency is observed in patients receiving PN, which leads to an increased risk of bone disease.[109-111] Studies reported that 30% to 87%

Table 86.9 Daily Vitamin Supplementation Recommendations for Adult Parenteral Nutrition Formulations

Vitamin	Standard Daily Intake Dosage Recommendation[a]	Dose Provided in Conventional Commercial Product[b]
Thiamin (vitamin B_1) (mg)	3–6	6
Riboflavin (vitamin B_2) (mg)	3.6–5	3.6
Niacin (vitamin B_3) (mg)	40–47	40
Biotin (μg)	60	60
Cyanocobalamin (vitamin B_{12}) (μg)	5–6	5
Folic acid (μg)	400–600	600
Pantothenic acid (mg)	15–17	15
Pyridoxine (vitamin B_6) (mg)	3–6	6
Vitamin A (IU)	3,300–3500	3,300 (as palmitate)
Vitamin C (ascorbic acid) (mg)	110–200	200
Vitamin D (IU)	200	200 (as cholecalciferol)
Vitamin E (mg)	10	10 IU (as α-tocopherol)
Vitamin K	Individual assessment	150 μg (as vitamin K_1)

IU: International units.
[a]Minimal dosages recommended by the FDA for an effective parenteral vitamin product and recent consensus recommendations.[25]
[b]Infuvite Adult Pharmacy Bulk Package Multiple Vitamins for Infusion (Sandoz Canada, Inc., Boucherville, QC, Canada).

Data from Mirtallo J, Canada T, Johnson D, et al. JPEN J Parenter Enteral Nutr 2004;28:S39–S70.[15]

of adults and 20% to 35% of children on PN were vitamin D deficient.[109,110,112] Vitamin D deficiency has been associated with malabsorption, limited sun exposure, and inadequate supplementation. Whether the current parenteral dosage recommendations are adequate for these functions remains to be demonstrated.

There is evidence that serum choline concentration is depressed in patients receiving long-term PN.[113,114] Impairment of the trans-sulfuration pathway of methionine to choline appears to be a possible mechanism for this disorder. In a small pilot study, choline supplementation for patients receiving HPN demonstrated significant improvement in a test measuring verbal and visual memory, compared with patients receiving HPN who did not receive choline.[115] Compher et al.,[116] suggested an association between vitamin B_{12} and choline deficiencies in patients receiving long-term HPN with severe malabsorption syndrome.

There continues to be reports of severe thiamin deficiency in adult patients receiving PN. This occurs despite (a) the availability and safety of IV multivitamins or individual forms of vitamins, (b) widespread knowledge of its increased need with the use of high-glucose formulas, (c) its known instability in contact with the sulfite in amino acids that necessitates minimum storage at room temperature before use, and (d) the reasonably rapid onset (1 to 2 months) of life-threatening metabolic changes characteristic of its deficiency. Unfortunately, the problem was complicated by periodic shortages of intravenous multivitamin preparations. Such issues continue to be reported when vitamins are not added to PN.[117] Losing thiamin and other water-soluble vitamins occurs with diuretic and renal replacement therapy.[118,119] Thiamin requirements also increase as a function of PN dextrose dosing, and refeeding-induced thiamin depletion can easily result from using this vitamin in carbohydrate metabolism pathways.[15,20,87]

Monitoring of Parenteral Nutrition Therapy

Comprehensive Nutritional Monitoring

PN monitoring in hospitals requires daily assessment of multiple factors to determine the continued need for PN (vs EN or a combination of EN and PN) and PN compositional changes that may be required, given clinical conditions (**Table 86.10**). Blood concentrations of albumin and prealbumin in hospitalized patients, especially in the ICU setting, are markedly decreased by non-nutritional factors (inflammation, infection, decreased hepatic synthesis, and increased clearance from blood) and may also be increased (volume depletion) or decreased (volume overload) based on fluid status. Because of the long-circulating half-life of albumin (18 to 21 days), blood concentration may remain low for weeks despite adequate feeding, and is slow to respond to nutritional repletion. Serial blood prealbumin concentration (3-day half-life) may be helpful as an index of general protein status in clinically stable, euvolemic patients in the outpatient setting, to monitor response to PN with or without EN.

Integration of the factors in Table 86.10 allows a subjective determination of whether patients' nutritional regimens need modification. For hospitalized patients receiving PN, BG, blood electrolytes, and renal function tests should be determined daily, or at least several times weekly. Liver function tests should be measured at least a few times a week. When arterial blood gas pH measurements are available, patients on ventilators should be monitored daily. Routine monitoring of copper, selenium, zinc, thiamin, vitamin B_6, vitamin C, and 25-hydroxyvitamin D concentrations is indicated for patients with various comorbidities and long-term PN.[20,25]

Table 86.10 Comprehensive Nutritional Monitoring of Hospitalized Patients

- Review past medical and surgical history, the tempo of current illness, expected hospital course, and upcoming operations or procedures
- Determine dietary intake pattern and previous use of specialized nutrition support
- Obtain daily body weight (the tempo of weight loss or regain)[a]
- Perform physical examination with attention to organ function, fluid status, and evidence of protein-energy malnutrition and lesions consistent with vitamin and mineral deficiencies
- Evaluate gastrointestinal tract function for estimated ability to tolerate enteral feeding
- Determine ambulatory capacity, mental status
- Measure and evaluate standard blood tests (organ function indices, electrolytes, pH, triglycerides, selected vitamins, trace elements, and minerals if at risk for depletion)
- Re-assess calorie and protein requirements (see Table 86.4)
- Estimate micronutrient needs in light of clinical conditions and blood tests (see Tables 86.8 and 86.9)
- Evaluate enteral and parenteral access for nutrient delivery via EN and/or PN

EN, enteral nutrition; *PN*, parenteral nutrition.

[a]In patients with involuntary body weight loss or weight gain, the weight change can reflect volume status and/or the need to adjust energy in PN.

Complications

Complications related to vascular access, catheter-related sepsis, PN stabilities, and underlying diseases were discussed earlier. They can be prevented or minimized by having responsibility vested in experienced physicians, nurses, dietitians, and pharmacists working as a team to closely supervise inpatients and outpatients requiring PN,[4,5,8,15,20,120] and by attention to the factors discussed earlier and outlined in Table 86.10. Disorders related directly to PN, including liver dysfunction, gallstones, and decreased bone mass, are addressed in this section.

Metabolic and Clinical Complications of the Overfeeding and Refeeding Syndromes

Table 86.11 shows common metabolic and clinical complications associated with overfeeding and refeeding syndrome (RS), which remain unfortunately common.[20,86,87,121,122] Adverse metabolic effects are less common in patients receiving PPN than

Table 86.11 Clinical and Metabolic Complications of Overfeeding and Refeeding Syndromes in Patients Receiving Parenteral Nutrition

- Hyperglycemia from excess dextrose provision with or without inadequate insulin administration
- Intracellular shift of magnesium, phosphorus, and/or potassium from rapid or excess dextrose infusion, with consequent refeeding hyperinsulinemia
- Immune cell dysfunction and infection (from effects of hyperglycemia)
- Cardiac failure or arrhythmias (excess fluid, sodium, and other electrolytes; refeeding-induced electrolyte shifts)
- Neuromuscular dysfunction (refeeding-induced electrolyte shifts; thiamin depletion)
- Azotemia (excess amino acid, inadequate energy provision in relation to amino acid dose)
- Fluid retention (refeeding hyperinsulinemia; excess PN fluid and/or sodium)
- Elevated liver function tests and/or hepatic steatosis (excess dextrose, fat emulsion, or total energy)
- Increased blood ammonia concentrations (excess amino acids, especially in patients with liver dysfunction and with inadequate nonamino acid energy provision)
- Hypercapnia (excess total kilocalorie)
- Respiratory insufficiency (refeeding-induced hypophosphatemia; excess fluid, kilocalorie, carbohydrate, or fat emulsion)
- Hypertriglyceridemia (excess carbohydrate or fat emulsion administration in PN)

PN, parenteral nutrition.

those receiving CPN, given the lower doses of dextrose. However, these effects can still occur and must be monitored. High energy, dextrose, amino acid, and fat loads are readily administered via a central vein. Risk factors for PN-induced hyperglycemia include hypertonic dextrose use in patients with obesity, diabetes, and sepsis; patients with poorly controlled BG at PN initiation; initial use of high dextrose loads (e.g., >150 g/day); insufficient insulin administration and adequate monitoring of BG; and concomitant administration of corticosteroids and pressor agents.

Overfeeding can induce severe metabolic complications, affecting several organ systems.[20] A large study found that PN use, overfeeding, and sepsis were the major risk factors for liver dysfunction in critically ill patients.[123] Thus, PN should be advanced carefully to meet nutrient intake goals. The composition should be adjusted based on the daily results of close metabolic and clinical monitoring (see Table 86.10).[20,124] The energy provided by IVFs (e.g., 5% dextrose), lipid-based medications, and EN must be considered in the PN prescription to avoid overfeeding.

Refeeding syndrome (RS) is common in at-risk patients, including those with pre-existing malnutrition, electrolyte depletion, prolonged periods of starvation, or IVF alone.[86,87,121,122] RS in these settings is primarily mediated by initially excessive administration of intravenous dextrose (e.g., >150g dextrose/day). This markedly stimulates insulin secretion, which may rapidly lower blood potassium, magnesium, and especially phosphorus concentrations because of intracellular shifts and use in metabolic pathways. Accelerated carbohydrate metabolism increases thiamin use and can precipitate symptoms and signs of thiamin deficiency.[20,87]

Prevention of RS requires identification of at-risk patients, low PN dextrose (e.g., 100–150 g dextrose) until electrolyte concentrations are stable; provision of higher PN doses of potassium, magnesium, and phosphorus, based on blood concentrations and renal function; and supplemental PN thiamin (e.g., 100 mg/day for 3 to 5 days after PN initiation).[87]

Hepatic Dysfunction

Steatosis, intrahepatic cholestasis, and portal inflammation can occur with overfeeding, excessive carbohydrates, specific nutrient deficiencies, and long-term PN, particularly in children and adults.[123,125-127] It can progress to portal tract fibrosis and infiltration, liver failure, and death. Extensive and continuing literature has confirmed PN as a contributing risk factor for hepatobiliary dysfunction of varying degrees and incidence.[123,125-127] Reviews of the biochemical, clinical, and histopathologic changes in adults and children have emphasized the multifactorial nature of the problem.[123,125] Increases in serum transaminase, alkaline phosphatase, γ-glutamyl-transferase and, less frequently, bilirubin are indicators of hepatic dysfunction.

In children, the degree of prematurity, infection, inability to consume food orally, intestinal dysfunction, number of surgical procedures, duration of PN, and excessive long-term energy administration are risk factors for the development of PN-associated liver disease (PNLD).[125-127] Immaturity of the hepatic excretory function and the enterohepatic circulation, particularly in the neonate, is one reason for the development of cholestasis. Small studies markedly reduced this complication with the administration of fish oil-based PN substituted for soybean oil-based PN.[128-131] Simply reducing the soybean oil-based lipid may be responsible for some or most of these benefits.[131,132] The most current ESPEN guidelines have addressed the short-term and long-term usage of fish oil-based IVFE.[133] Specific IVFE components (e.g., phytosterol) may also play a role in liver dysfunction.[71,134] In adults, pre-existing liver and other diseases, sepsis, pre-existing malnutrition, the extent of bowel resection or damage, excess nonprotein energy, little or no oral intake, and PN duration are also associated risk factors for PNLD.[123,126,127]

Various agents have been tested on patients receiving PN who have developed evidence of significant hepatic dysfunction and require continued PN.[108,125] Giving a mixture of MCT and LCT to PN-treated patients as the IVFEs did not cause a change in liver size or grayscale value, whereas LCT infusion increased both.[108] Because metronidazole could depress the formation by the intestinal bacteria of potentially damaging bile acids, this drug has been tested in patients receiving PN.

Gallstones

Decreased gallbladder contractility has been observed repeatedly as a PN- and bowel rest-associated risk factor; this situation can progress to cholelithiasis as the duration of PN increases.[15] During fasting, there is decreased secretion of cholecystokinin, a peptide hormone secreted in the duodenum in response to meals that induce gallbladder contractility. Patients who receive long-term PN because of resection or terminal ileum disease usually have bile salt malabsorption. Thus, the bile salt pool decreases, and lower concentrations of bile salts are present in the gallbladder. This situation, in turn, increases the tendency of cholesterol to precipitate in the bile, thus forming the nidus of gallstones.[127]

In one study, cholelithiasis developed in 6.2%, 21.2%, and 37.8% at 6, 12, and 24 months of the 119 patients receiving HPN, respectively.[135] Another study reported that 36.9% of 292 patients on long-term HPN had cholelithiasis, of which 14.4% required cholecystectomy.[136] Although 5.5% of patients receiving HPN had no signs of cholecystitis, prophylactic cholecystectomy was performed.

Cholelithiasis can potentially be managed as follows: nutrition should be provided enterally whenever possible to decrease biliary stasis, liver function studies should be checked periodically, ultrasound should be used liberally when a cholestatic picture is evident if gallstones are detected, and if laparotomy is to be done for any reason, cholecystectomy should be considered at that time.

Metabolic Bone Disease

Metabolic bone disease (MBD) is usually a long-term complication of PN administration. MBD, including osteomalacia, osteopenia, and osteoporosis, was first reported in the 1970s.[137] The prevalence of MBD in patients on long-term PN has been reported as 45% or 84% among pediatric and adult patients, respectively.[137,138] MBD is more prevalent in patients with specific comorbidities, including SBS, inflammatory bowel diseases, and cancer.[108,138,139] In a study of 165 adult patients receiving long-term PN, 84% had T-scores via dual-energy x-ray (DXA) absorptiometry that were decreased by more than one standard deviation from normal, which meets the criteria for osteopenia. Further, 41% had T-scores reduced by >2.5 standard deviations from normal, meeting the criteria for osteoporosis.[140]

The exact cause of MBD in patients on long-term PN is unknown, but it is considered a multifactorial syndrome. Several factors may contribute to altered bone metabolism, including the inadequate provision of calcium and phosphorus, cyclic PN infusion, excessive protein and vitamin D, and aluminum toxicity from aluminum contamination in PN components. Rickets has been described in infants receiving PN. The causative factor appeared to be a need for more calcium and phosphate in the small fluid volume required by the neonate, rather than more vitamin D.[137] Negative calcium balance related to hypercalciuria may occur in adults receiving PN, especially during cyclic PN infusion. Supplementation with either sodium or potassium acetate (replacing NaCl or KCl) resulted in significant decreases in urinary calcium in patients receiving 24-hour and cyclic PN, primarily because of increased renal tubular reabsorption with reduced excretion to near-infusion levels.[141] High doses of protein, especially when given via cyclic PN, increase urinary excretion of calcium and potentially contribute to the development of MBD. Aluminum toxicity causes osteomalacia and aluminum contamination from PN components has been discussed.

Administration of corticosteroids has also been associated with MBD because these drugs increase bone resorption, impair osteoblastic activity, and inhibit the GI tract's vitamin D and calcium absorption.[138,142] Phosphorus, a major component of bone formation, likely induces this favorable response by increasing renal tubular calcium absorption.[143] Amenorrhea and smoking are also factors.

Routine assessment and monitoring for MBD should be incorporated as the standard of care for patients receiving long-term PN. Routine blood studies of calcium, phosphorus, and magnesium can help to detect MBD and evaluate the adequacy of minerals provided in PN. DXA should be used to assess bone mineral density, if indicated.

Drug Compatibility

The frequency of drug interventions for coexistent illnesses or complications of PN requires ensuring that administering a drug as part of the PN solution or in conjunction with that solution will not produce incompatibility or an adverse reaction. Regular insulin is the most commonly added drug in PN, and it effectively controls BG in most patients, with or without a separate intravenous insulin infusion.[5,8,15,20] PN drug compatibility was recently reviewed regarding which medicines can be added in PN, which needed to be coinfused (Y-site) and which should be given separately from PN.[78] PN and drug compatibility are not addressed here because the varied mixtures provide the potential for numerous chemical and physicochemical interactions, which can compromise the safety and efficacy of the medicine. Minimal combinations of drugs and different PN solutions have been studied. In addition, some drugs are compatible with traditional dextrose and amino acid PN solutions but not in TNAs (e.g., iron dextran).

Home Parenteral Nutrition

Since the first patients were discharged from the hospital to home on PN in 1969 and the early 1970s in the U.S. and Canada,[1,2] HPN remains vital for outpatient nutritional support. An HPN registry for the United States and Canada was established at the New York Academy of Medicine from 1978 to 1983 to collect and compile longitudinal data on patients receiving HPN. In 1984, this registry became a joint effort of the Oley Foundation and ASPEN, originally known as the OASIS Registry. Since 1985, it has been known as the North American Home Parenteral and Enteral Nutrition Patient Registry (HPEN Registry), produced by the Oley Foundation. The last HPEN registry report was published in 1991; it is challenging to determine the prevalence of patients receiving HPN. Based on the data from Medicare for 2013, insurance providers, and home infusion providers, there were 4,129 pediatric patients and 20,883 adult patients receiving HPN.[144] There are no recent data available.

HPN has various complications, including short-term and long-term metabolic disorders. Short-term metabolic complications include fluid imbalance, electrolyte disturbances, and glucose abnormalities, while long-term complications include hepatobiliary disorders, metabolic bone disease, iron deficiency anemia, and manganese toxicity.

HPN presents various stresses to the patient and family members,[145] including the sudden need to cope with the technical aspects, time demands, and safety issues of HPN after hospital discharge, management of handicaps resulting from primary and secondary illness and their treatments, concerns about meeting costs, patient dependency, and excessive dependence on others. A smooth transition to home care requires adequate predischarge assessment and training of the patient and family in HPN management, and close support by the health-care team by telephone contact and follow-up at home or in the physician's office to ensure that the patient's condition remains satisfactory. Questionnaires on various quality-of-life measures in HPN-treated pediatric and adult patients have been published.[146,147] A survey of 466 adults receiving HPN showed that the quality of life for patients on long-term HPN depends on the frequency of HPN infusions per week and the underlying disease.[147]

The cost of HPN for adults varies significantly, from $19,000 to $253,000 per patient-year, while the PN cost in pediatrics is reported as $83,000 per year.[148] Many factors impact the HPN cost, including the number of days of PN/week required, PN-associated complications and monitoring, and other common items (e.g., HPN bags, personnel, consumables, follow-up, etc.). HPN is a lifesaving treatment for many patients with conditions associated with intestinal failure.[149] HPN has also been much more cost-effective than PN used in hospitals.[148,150]

Future Perspectives

PN is a common component of clinical care worldwide. Yet many essential areas of uncertainty remain concerning the optimal use of complete PN and parenteral macronutrients and micronutrients in combination with EN. The 2022 ASPEN guidelines recommend an energy goal of 12 to 25 kcal/day for the first 7 to 10 days of ICU stays.[8] Recent studies suggest using newly innovated indirect calorimetry (IC) technology to accurately determine the nutrition needs of ICU patients and deliver individualized nutrition therapy.[151,152] Indirect calorimetry is considered the "gold standard" to assess energy needs and is recommended by all societal guidelines. However, some suggested that routine utilization of IC might not change clinical outcomes in the ICU.[153] The optimal energy needs for ICU patients remain controversial, with a most recent study demonstrating that hypocaloric (<70% of 25 kcal/kg) feeding was associated with increased 28-day mortality risk.[154] New methods, such as high-resolution plasma metabolomics, will be essential to assess nutritional status, nutrition-related pathophysiology, and the response of adults and children receiving parenteral nutrition support.[155,156]

Acknowledgments

The authors acknowledge Rex O. Brown and Gayle Minard for their contributions to the previous version of this chapter.

References

1. Vinnars E, Wilmore D. JPEN J Parenter Enteral Nutr 2003;27:225–31.
2. Bistrian BR. Nestle Nutr Workshop Ser Clin Perform Programme 2009;12:127–36.
3. Dudrick SJ, Wilmore DW, Vars HM, et al. Surgery 1968;64:134–42.
4. Druyan ME, Compher C, Boullata JI, et al. JPEN J Parenter Enteral Nutr 2012;36:77–80.
5. Boullata JI, Gilbert K, Sacks G, et al. JPEN J Parenter Enteral Nutr 2014;38:334–77.
6. Brown RO, Compher C. JPEN J Parenter Enteral Nutr 2010;34:366–77.
7. Choban P, Dickerson R, Malone A, et al. JPEN J Parenter Enteral Nutr 2013;37:714–44.
8. Compher C, Bingham AL, McCall M, et al. JPEN J Parenter Enteral Nutr 2022;46:12–41.
9. Kumpf VJ, de Aguilar-Nascimento JE, Diaz-Pizarro Graf JI, et al. JPEN J Parenter Enteral Nutr 2017;41:104–12.
10. Preiser JC, Schneider SM. Clin Nutr 2011;30:549–52.
11. Cuerda C, Pironi L, Arends J, et al. Clin Nutr 2021;40:5196–220.
12. Muscaritoli M, Arends J, Bachmann P, et al. Clin Nutr 2021;40:2898–913.
13. Weimann A, Braga M, Carli F, et al. Clin Nutr 2021;40:4745–61.
14. Cahill NE, Narasimhan S, Dhaliwal R, et al. JPEN J Parenter Enteral Nutr 2010;34:685–96.
15. Mirtallo J, Canada T, Johnson D, et al. JPEN J Parenter Enteral Nutr 2004;28:S39–S70.
16. Heyland DK, MacDonald S, Keefe L, et al. JAMA 1998;280:2013–9.
17. Worthington P, Balint J, Bechtold M, et al. JPEN J Parenter Enteral Nutr 2017;41:324–77.
18. Lewis SR, Schofield-Robinson OJ, Alderson P, et al. Cochrane Database Syst Rev 2018;6:Cd012276.
19. Patkova A, Joskova V, Havel E, et al. Adv Nutr 2017;8:624–34.
20. Ziegler TR. N Engl J Med 2009;361:1088–97.
21. Ziegler TR. N Engl J Med 2011;365:562–4.
22. Gura KM. Nutr Clin Pract 2009;24:709–17.
23. Culebras JM, Martin-Peña G, Garcia-de-Lorenzo A, et al. Curr Opin Clin Nutr Metab Care 2004;7:303–7.
24. McMahon MM, Nystrom E, Braunschweig C, et al. JPEN J Parenter Enteral Nutr 2013;37:23–36.
25. Blaauw R, Osland E, Sriram K, et al. JPEN J Parenter Enteral Nutr 2019;43(suppl 1):S5–S23.
26. Hardy G, Wong T, Morrissey H, et al. JPEN J Parenter Enteral Nutr 2020;44(suppl 2):S5–S23.
27. Norwood S, Wilkins HE, 3rd, Vallina VL, et al. Crit Care Med 2000;28:1376–82.
28. Pittiruti M, Hamilton H, Biffi R, et al. Clin Nutr 2009;28:365–77.
29. Kovacevich DS, Corrigan M, Ross VM, et al. JPEN J Parenter Enteral Nutr 2019;43:15–31.
30. Patel AR, Patel AR, Singh S, et al. Cureus 2019;11:e4717.
31. Fonkalsrud EW, Berquist W, Burke M, et al. Am J Surg 1982;143:209–11.
32. Marik PE, Flemmer M, Harrison W. Crit Care Med 2012;40:2479–85.
33. Timsit JF, Sebille V, Farkas JC, et al. JAMA 1996;276:1416–20.
34. Timsit JF, Schwebel C, Bouadma L, et al. JAMA 2009;301:1231–41.
35. Camargo LF, Marra AR, Büchele GL, et al. J Hosp Infect 2009;72:227–33.
36. Oliveira C, Nasr A, Brindle M, et al. Pediatr 2012;129:318–29.
37. Gundogan K, Dave NJ, Griffith DP, et al. JPEN J Parenter Enteral Nutr 2020;44:661–7.
38. Bell T, O'Grady NP. Infect Dis Clin North Am 2017;31:551–9.
39. Daoud DC, Wanten G, Joly F. Nutrients 2020;12:439.
40. Marschall J, Mermel LA, Fakih M, et al. Infect Control Hosp Epidemiol 2014;35:753–71.
41. Worthington P, Gura KM, Kraft MD, et al. Nutr Clin Pract 2021;36:29–39.
42. Casaer MP, Mesotten D, Hermans G, et al. N Engl J Med 2011;365:506–17.
43. Mehta NM, Skillman HE, Irving SY, et al. J Parenter Enter Nutr 2017;41:706–42.
44. Kutsogiannis J, Alberda C, Gramlich L, et al. Crit Care Med 2011;39:2691–9.
45. Thibault R, Pichard C. Crit Care Clin 2010;26:467–80, viii.
46. Fiaccadori E, Sabatino A, Barazzoni R, et al. Clin Nutr 2021;40:1644–68.

47. Ikizler TA, Burrowes JD, Byham-Gray LD, et al. Am J Kidney Dis 2020;76:S1–S107.
48. Directors ABo. JPEN J Parenter Enteral Nutr 2009;33:255–9.
49. Rodriguez NA, Jeschke MG, Williams FN, et al. JPEN J Parenter Enteral Nutr 2011;35:704–14.
50. Iacone R, Scanzano C, Santarpia L, et al. Nutrients 2020; 12:772.
51. Yarandi SS, Zhao VM, Hebbar G, et al. Curr Opin Clin Nutr Metab Care 2011;14:75–82.
52. Kauffmann RM, Hayes RM, Jenkins JM, et al. JPEN J Parenter Enteral Nutr 2011;35:686–94.
53. Arsenault D, Brenn M, Kim S, et al. JPEN J Parenter Enteral Nutr 2012;36:81–95.
54. McCowen KC, Malhotra A, Bistrian BR. Crit Care Clin 2001;17: 107–24.
55. Mifsud S, Schembri EL, Gruppetta M. Br J Hosp Med (Lond) 2018;79:634–9.
56. van den Berghe G, Wouters P, Weekers F, et al. N Engl J Med 2001;345:1359–67.
57. Van den Berghe G, Wouters PJ, Bouillon R, et al. Crit Care Med 2003;31:359–66.
58. Finney SJ, Zekveld C, Elia A, et al. JAMA 2003;290:2041–7.
59. Investigators N-SS. N Engl J Med 2009;360:1283–97.
60. Finfer S, Chittock DR, Su SY, et al. N Engl J Med 2009;360: 1283–97.
61. Kavanagh BP, McCowen KC. N Engl J Med 2010;363:2540–6.
62. Korytkowski MT, Muniyappa R, Antinori-Lent K, et al. J Clin Endocrinol Metab 2022;107:2101–28.
63. Mundi MS, Martindale RG, Hurt RT. JPEN J Parenter Enteral Nutr 2017;41:3S–13S.
64. Mirtallo JM, Ayers P, Boullata J, et al. Nutr Clin Pract 2020;35: 769–82.
65. Raman M, Almutairdi A, Mulesa L, et al. Nutrients 2017;9:388.
66. Biesboer AN, Stoehr NA. Nutr Clin Pract 2016;31:610–8.
67. Anez-Bustillos L, Dao DT, Baker MA, et al. Nutr Clin Pract 2016;31:596–609.
68. Mayer K, Klek S, García-de-Lorenzo A, et al. JPEN J Parenter Enteral Nutr 2020;44(suppl 1):S28–S38.
69. Mirtallo JM, Dasta JF, Kleinschmidt KC, et al. Ann Pharmacother 2010;44:688–700.
70. Biesalski HK. Gastroenterol 2009;137:S92–S104.
71. Zaloga GP. JPEN J Parenter Enteral Nutr 2015;39:39S–60S.
72. Calder PC, Waitzberg DL, Klek S, et al. JPEN J Parenter Enteral Nutr 2020;44(suppl 1):S21–S7.
73. Palmer AJ, Ho CK, Ajibola O, et al. Crit Care Med 2013;41: 307–16.
74. Manzanares W, Dhaliwal R, Jurewitsch B, et al. JPEN J Parenter Enteral Nutr 2014;38:20–8.
75. Pradelli L, Mayer K, Klek S, et al. JPEN J Parenter Enteral Nutr 2020;44:44–57.
76. Umpierrez GE, Spiegelman R, Zhao V, et al. Crit Care Med 2012;40:1792–8.
77. Otero-Millán L, Lago Rivero N, Blanco Rodicio A, et al. Clin Nutr ESPEN 2021;45:19–25.
78. Boullata JI, Mirtallo JM, Sacks GS, et al. JPEN J Parenter Enteral Nutr 2022;46:273–99.
79. Suchner U, Katz DP, Fürst P, et al. Crit Care Med 2001;29: 1569–74.
80. Mayer K, Seeger W. Crit Care 2010;14:128.
81. Cohen HJ, Brown MR, Hamilton D, et al. Am J Clin Nutr 1989;49:132–9.
82. Rudman D, Millikan WJ, Richardson TJ, et al. J Clin Invest 1975;55:94–104.
83. Starker PM, LaSala PA, Forse RA, et al. JPEN J Parenter Enteral Nutr 1985;9:300–2.
84. Maroulis J, Kalfarentzos F. Clin Nutr 2000;19:295–304.
85. Schmidt GL. Fluids and Electrolytes. Silver Spring, MD: 2015.
86. Zhou YQ, Zhang XM, Chen ZQ, et al. Supp Care Cancer 2022;30:6995–7003.
87. da Silva JSV, Seres DS, Sabino K, et al. Nutr Clin Pract 2020;35:178–95.
88. Shils ME, Burke AW, Greene HL, et al. JAMA 1979;241: 2051–4.
89. Btaiche IF, Carver PL, Welch KB. JPEN J Parenter Enteral Nutr 2011;35:736–47.
90. Livingstone C. Nutr Clin Pract 2015;30:371–82.
91. Jeejeebhoy K. Gastroenterol 2009;137:S7–12.
92. Shike M, Roulet M, Kurian R, et al. Gastroenterol 1981;81: 290–7.
93. Spiegel JE, Willenbucher RF. J Parenter Enteral Nutr 1999;23: 169–72.
94. Griffith DP, Liff DA, Ziegler TR, et al. Obesity (Silver Spring) 2009;17:827–31.
95. Gletsu-Miller N, Broderius M, Frediani JK, et al. Int J Obes (Lond) 2012;36:328–35.
96. Sherf Dagan S, Goldenshluger A, Globus I, et al. Adv Nutr 2017;8:382–94.
97. Olson LM, Wieruszewski PM, Jannetto PJ, et al. J Parenter Enteral Nutr 2019;43:970–6.
98. Hardy G. Gastroenterol 2009;137:S29–35.
99. Verhage AH, Cheong WK, Jeejeebhoy KN. JPEN J Parenter Enteral Nutr 1996;20:123–7.
100. Moukarzel A. Gastroenterol 2009;137:S18–28.
101. Van Rij AM, Thompson CD, McKenzie JM, Robinson MF. Am J Clin Nutr 1979;43:2076–85.
102. Zhao Y, Yang M, Mao Z, et al. Medicine 2019;98:e15473.
103. Hernández-Sánchez A, Tejada-González P, Arteta-Jiménez M. Eur J Clin Nutr 2013;67:230–8.
104. Hall AR, Arnold CJ, Miller GG, et al. JPEN J Parenter Enteral Nutr 2017;41:1228–33.
105. Aşut E, Köksal N, Dorum BA, et al. Turk J Pediatr 2018;60: 385–91.
106. Vanamee P SM, Burke AW. JPEN J Parenter Enteral Nutr 1979;3:258–62.
107. Robinson DT, Ayers P, Fleming B, et al. Nutr Clin Pract 2021;36:927–41.
108. Wanten G, Calder PC, Forbes A. BMJ 2011;342:d1447.
109. Braga CB, Vannucchi H, Freire CM, et al. JPEN J Parenter Enteral Nutr 2011;35:493–8.
110. Bharadwaj S, Gohel TD, Deen OJ, et al. Nutr Clin Pract 2014;29:681–5.
111. Thomson P, Duerksen DR. JPEN J Parenter Enteral Nutr 2011;35:499–504.
112. Ubesie AC, Kocoshis SA, Mezoff AG, et al. J Pediatr 2013;163: 1692–6.
113. Buchman AL. Gastroenterol 2009;137:S119–28.
114. Glade MJ. Nutrition 1999;15:92–6.
115. Buchman AL, Sohel M, Brown M, et al. JPEN J Parenter Enteral Nutr 2001;25:30–5.
116. Compher CW, Kinosian BP, Stoner NE, et al. JPEN J Parenter Enteral Nutr 2002;26:57–62.
117. Ferrie S. Nutr Clin Pract 2012;27:65–8.
118. Gundogan K, Akbudak IH, Bulut K, et al. Nutr Clin Pract 2019;34:565–71.
119. Kamel AY, Dave NJ, Zhao VM, et al. Nutr Clin Pract 2018;33: 439–46.
120. Directors ABo. JPEN J Parenter Enteral Nutr 2002;26: 1sa–138sa.
121. Cioffi I, Ponzo V, Pellegrini M, et al. Clin Nutr 2021;40: 3688–701.
122. Ponzo V, Pellegrini M, Cioffi I, et al. Intern Emerg Med 2021;16:49–60.

123. Grau T, Bonet A, Rubio M, et al. Crit Care 2007;11:R10.
124. Zhao VM, Ziegler TR. Crit Care Nurs Clin North Am 2010;22:369–80.
125. Rangel SJ, Calkins CM, Cowles RA, et al. J Pediatr Surg 2012;47:225–40.
126. Buchman A. JPEN J Parenter Enteral Nutr 2002;26:S43–8.
127. Cavicchi M, Beau P, Crenn P, et al. Ann Intern Med 2000;132: 525–32.
128. de Meijer VE, Gura KM, Meisel JA, et al. Arch Surg 2010;145: 547–51.
129. Deshpande G, Simmer K. Curr Opin Clin Nutr Metab Care 2011;14:145–50.
130. Park KT, Nespor C, Kerner J, Jr. J Perinatol 2011;31(suppl 1): S57–60.
131. Rochling FA. Nutrients 2021;13:895.
132. Cober MP, Killu G, Brattain A, et al. J Pediatr 2012;160: 421–7.
133. Lapillonne A, Fidler Mis N, Goulet O, et al. Clin Nutr 2018;37:2324–36.
134. Calkins KL, DeBarber A, Steiner RD, et al. JPEN J Parenter Enteral Nutr 2018;42:633–41.
135. Dray X, Joly F, Reijasse D, et al. J Am Coll Surg 2007;204: 13–21.
136. Lawiński M, Jachnis A, Ukleja A, et al. Pol Przegl Chir 2014;86:111–5.
137. Gatti S, Quattrini S, Palpacelli A, et al. Nutrients 2022;14:995.
138. Acca M, Ragno A, Francucci CM, et al. J Endocrinol Invest 2007;30:54–9.
139. Diamanti A, Bizzarri C, Basso MS, et al. J Bone Miner Metab 2010;28:351–8.
140. Pironi L, Labate AM, Pertkiewicz M, et al. Clin Nutr 2002;21:289–96.
141. Berkelhammer CH, Wood RJ, Sitrin MD. Am J Clin Nutr 1988;48:1482–9.
142. Buchman AL, Moukarzel A. Clin Nutr 2000;19:217–31.
143. Berkelhammer C, Wood RJ, Sitrin MD. JPEN J Parenter Enteral Nutr 1998;22:142–6.
144. Mundi MS, Pattinson A, McMahon MT, et al. Nutr Clin Pract 2017;32:799–805.
145. Srp F, Steiger E, Gulledge AD, et al. Nurs Clin North Am 1989;24:447–59.
146. Nagelkerke SCJ, van Oers HA, Haverman L, et al. J Pediatr Gastroenterol Nutr 2022;74:116–22.
147. Burden ST, Jones DJ, Gittins M, et al. Clin Nutr 2019;38: 1433–8.
148. Arhip L, Serrano-Moreno C, Romero I, et al. Clin Nutr 2021;40: 339–49.
149. Rhoda KM, Suryadevara S, Steiger E. Surg Clin North Am 2011;91:913–32, ix–x.
150. Li W, Guo H, Li L, et al. Front Oncol 2022;12:858712.
151. De Waele E, Jonckheer J, Wischmeyer PE. Curr Opin Crit Care 2021;27:334–43.
152. Wischmeyer PE, Molinger J, Haines K. Nutr Clin Pract 2021;36:275–81.
153. McClave SA, Omer E. Nutr Clin Pract 2021;36:268–74.
154. Lv C, Jiang X, Long Y, et al. Front Nutr 2022;9:902986.
155. Jones DP, Park Y, Ziegler TR. Annu Rev Nutr 2012;32: 183–202.
156. Bellissimo MP, Jones DP, Martin GS, et al. JPEN J Parenter Enteral Nutr 2021;45:1635–44.

CHAPTER 87

Bone Biology in Health and Disease

Kate A. Ward

CHAPTER OUTLINE

Abbreviations

Al, aluminum; **BALP**, bone-specific alkaline phosphatase; **BMC**, bone mineral content; **BMD**, bone mineral density; **BMP**, bone morphogenic proteins; **[Ca^{2+}]**, concentration of calcium ions; **CT**, calcitonin; **CTx-Ia**, C-telopeptide of collagen cross-links; **DMP-1**, dentin matric protein; **DXA**, dual-energy x-ray absorptiometry; **ECF**, extracellular fluid; **FGF-23**, fibroblast growth factor; **FRAX**, Fracture Risk Assessment Tool; **HIV**, human immunodeficiency virus; **ICTP**, **IGF-1**, insulin like growth factor-I; **NTx**, N-telopeptide of collagen cross-links; **MSC**, mesenchymal stem cell; **OC**, osteocalcin; **OI**, osteogenesis imperfecta; **PICP**, carboxy-terminal propetide of type I collagen; **PINP**, amino-terminal propetide of type I collagen; **PTH**, parathyroid hormone; **PTHrP**, parathyroid hormone receptor protein; **RANKL**, rank ligand; **TALP**, total alkaline phosphatase; **TGF**, transforming growth factor beta; **TRAP-5b**, tartrate-resistant acid phosphatase 5b

Bone Composition and Structure

Bone is a tissue in which cells make up only 2% to 5% of the volume, and nonliving material makes up 95% to 98%. It is the nonliving material that gives the bone its basic mechanical properties of hardness, stiffness, and resilience. This nonliving material consists of a mineral-encrusted protein matrix (also called osteoid), with the mineral comprising about half the volume and the organic matrix the other half. Unlike other connective tissues, virtually no free water is present in the bony material itself. Embedded in this solid material are cells, called osteocytes, residing in lacunae in the matrix and communicating with one another through an extensive network of long cellular processes lying in channels called canaliculi, which ramify throughout the bone. Because of this arrangement, virtually no volume of normal bone is more than a few micrometers from a living cell. Furthermore, even in the dense bone of the shafts of long bones, an extensive network of vascular channels exists, so the most remote osteocyte is typically no more than 90 μm away from a capillary.

Bone Mineral

The mineral of bone is a carbonate-rich, imperfect hydroxyapatite with variable stoichiometry. Calcium comprises 37% to 40%, phosphate 50% to 58%, and carbonate 2% to 8% of this mineral. These values vary somewhat from species to species, and the carbonate component is particularly sensitive to systemic acid–base status (decreasing in acidosis and increasing in alkalosis). In addition, bone mineral contains small amounts of sodium, potassium, magnesium, citrate, and other ions present in the extracellular fluid (ECF) at the time when the mineral was deposited, adsorbed onto the crystal surfaces, and trapped there, as the water in the recently deposited matrix is displaced by the growing mineral crystals.

Protein Matrix

The protein matrix of bone, as for tendons, ligaments, and dermis, consists predominantly of collagen, which comprises approximately 90% of the organic matrix. For bone, the collagen is type I. Collagen is a long, fibrous protein, coiled as a triple helix. For the molecules of the protein to coil tightly, no side chains can project from the peptide backbone on the side facing inward. Hence, every third amino acid in the body of the collagen molecule is glycine, which has no side chain. However, projecting outward are the side chains of various other amino acids, such as lysine, which allow the posttranslational formation of tight, covalent bonds between collagen fibers. This cross-linking helps to prevent fibers from sliding along one another when bone is stressed along the axis of the fibers.

Noncollagenous Matrix Proteins

Noncollagenous proteins make up approximately 10% to 15% of the organic matrix of bone and consist of glycolsylated proteins (e.g., alkaline phosphatase), proteoglycans, which may have a role in osteoblast function and a family of proteins in which glutamic acid residues are carboxylated in the γ position, called gla-proteins.[1] The best studied of gla-proteins is osteocalcin (or bone gla-protein), which comprises approximately 1.5% of the matrix proteins. Other gla-proteins include osteonectin, fibronectin, matrix gla-protein, osteopontin, and bone sialoprotein. The functions of these many constituents are not entirely clear. Some doubtless serve as chemoattractants for osteoclasts or as points of osteoclast attachment, whereas others stimulate osteoblasts to lay down new bone. Because of these properties of the matrix proteins, bone seems to contain some of the chemical signals for its own remodeling.

The shape and three-dimensional structure of bone are determined by its protein matrix. A bone that has been completely demineralized in the laboratory (by soaking in ethylenediaminetetra-acetic or other acid) looks entirely normal; and when sectioned, stained, and examined under a microscope, it reveals all of the fine structure of bone. In fact, prior demineralization has been the traditional first step in studying bone histologically (because mineralized bone tends to damage the microtome knives used by histologists to make their sections).

Bone Cells and Their Functions

The five principal bone cells are osteoprogenitor cells, lining cells, osteoblasts, osteoclasts, and osteocytes. They are responsible both for maintaining the mechanical properties and mediating the calcium homeostatic function of bone and are involved in the auto and paracrine actions of bone.

Osteoprogenitor cells are mesenchymal stem cells (MSCs) found in the periosteum and bone marrow. Various growth factors, cytokines, and hormones [including transforming growth factor-β_1 (TGF-β_1), fibroblast growth factors (FGFs), several bone morphogenetic proteins (BMPs), the active form of vitamin D, 1,25-dihydroxyvitamin D, and parathyroid hormone (PTH)] are responsible for controlling the proliferation and differentiation of these MSC's into preosteoblasts, osteoblasts, and osteocytes.

Lining cells are flat, fibrocyte-like cells covering free surfaces of bone. They are most probably derived from, or closely related to, the osteoblast cell line. They form a membrane that completely covers free bone surfaces and insulates bone from the cells and hormones in the general circulation. They demarcate a virtual compartment between the lining cells on one side and mature bone on the other. This compartment is continuous with the space in canaliculi surrounding osteocyte processes and may have different ionic composition from that of the ECF located outside, that is, between the lining cells and the capillaries of bone. It is possible that lining cells, by adjusting ion fluxes between the ECF and the bone compartment, contribute to the maintenance of calcium ion concentrations in the ECF.

Osteoblasts are derived from mesenchymal stem cells, the cells that lay down bone, first by synthesizing, depositing, and orienting the fibrous proteins of the matrix, and then by initiating changes that render the matrix capable of mineralization. They express a wide panel of genes for structural, enzymatic, and regulatory proteins including collagen type I, alkaline phosphatase, bone sialoprotein, osteopontin, osteonectin, and osteocalcin; some of which are used as markers of bone formation. As well as being controlled by several factors, osteoblasts themselves secrete autocrine, paracrine, and hormonal factors and, in addition to involvement in bone modeling unit formation, they also link the skeleton to the function of other organ systems.

Osteoblasts deposit this matrix between and beneath themselves on a pre-existing bone surface, thereby pushing themselves backward as they add new bone.[2] Bone matrix, when freshly deposited, consists of about half protein and half water and is not immediately mineralizable, just as the similar collagen-based structures, tendon and ligament, do not normally calcify. Therefore, the osteoblast still has more work to do after forming and depositing the matrix. The details of the process are not completely clear, but they involve secretion of proteins by the osteoblast into the matrix that it had just previously laid down. These help to create a three-dimensional configuration that allows calcium and phosphate ions in the ECF to arrange themselves in the apatite crystal habitus. Osteoblasts also secrete an enzyme called alkaline phosphatase that hydrolyzes various phosphate compounds in the local environment, thereby increasing phosphate ion concentration at the mineralizing site and at the same time removing natural crystal inhibitors (e.g., pyrophosphate). Finally, as mineral is deposited, it displaces the water of the original matrix. The apatite crystals that form are spindle shaped and oriented parallel to and between the collagen fibers.

Osteoclasts are derived from the monocyte–macrophage line of cells, are usually multinucleated, and resorb bone. They do this first by attaching firmly to a microscopic bony surface and then walling off a small region of that surface. The attachment involves linkage between proteins called integrins in the cell membrane of the osteoclast with proteins in the bone matrix, such as osteopontin, that exhibit a particular amino acid sequence (RGD, i.e., arginine-glycine-asparagine). Once firmly attached, osteoclasts secrete acid and proteolytic and lysosomal enzymes (such as collagenase, tartrate-resistant acid phosphatase and cathepsin K) and hydrogen ions into this confined space. These dissolve the mineral and digest the matrix. The osteoclasts then release the breakdown products into the ECF around the resorption site, where they are carried away by the circulating blood. After working for a short period of time (measured in days), the osteoclasts undergo programmed cell death (apoptosis), leaving their excavation to be refilled by osteoblasts. The calcium they dissolve from bone mineral seems to trigger or augment this apoptotic process, as osteoclasts blocked from producing acid accumulate on bone surfaces and have longer life spans. When examined histologically in sections of bone, these processes necessarily appear localized. However, the activity is more typically a "project," moving along a bony surface, with osteoclastic work preceding and the osteoblastic formation filling in behind.

The calcium and phosphorus released into the bloodstream at a resorption site will usually be used to mineralize remodeling sites elsewhere in the skeleton, currently in their formation phase. However, the protein fragments are metabolized or excreted. Some of the amino acids released in collagen degradation re-enter the body's amino acid pool and can be reused for protein synthesis elsewhere. However, those that have undergone post-translational modification (e.g., proline to hydroxyproline and the amino acids involved in collagen cross-linking) cannot be reused. Some of those products, circulating in plasma, or excreted in urine, are used for assessment of bone turnover. For this reason, bone remodeling requires a continuing supply and intake of dietary protein.

Osteocytes are osteoblasts that have stopped matrix synthesis and have become embedded in bone as the other bone-forming cells around them continue to add new layers of matrix.[3] Osteocytes make up 95% of bone cells in the adult skeleton and control remodeling in the skeleton. They have an important role in mineralization, rather than control of embryonic development and postnatal growth.[4] Osteocytes have dendritic processes, which are embedded in matrix, connecting with other bone cells and the marrow; this interconnecting network is crucial for their role of mechanosensing through the WNT/ β-catenin signaling pathway, and in bone remodeling. They control renal handling of phosphate and osteoblast formation. Their autocrine and paracrine action is through secretion of many factors, including sclerostin, RANKL, BMPs, DMP-1, and FGF-23. Osteocytes monitor the amount of strain (bending) that occurs in their domains when bone is mechanically loaded and for reporting that information to lining cells on nearby anatomic bone surfaces, which may then initiate local bone remodeling projects; Sclerostin is the main factor involved in the response to load; it reduces osteoblast function. Newer therapies for treatment of osteoporosis target two of the pathways in which osteocytes have a key role, antibodies for RANK-L and sclerostin.

Table 87.1 Humoral Factors Acting on Bone Cells

Osteoblasts	Osteoclasts
Parathyroid hormone	Calcitonin
1,25(OH)$_2$ vitamin D	
Glucocorticoids	Interleukin-1, 6, TNFα
Insulin-like growth factors 1(IGFs)	Colony-stimulating factor-1 (CSF-1)
Transforming growth beta	TGF-α factor-β (TGF-β)
Interleukin-6	TGF-β RANK ligand
Parathyroid hormone–related (PTHrP)	
Osteoprotegerin Bone morphogenetic protein (BMP) Sclerostin Thyroid hormone Estrogen Prostaglandin E2	PTH, PTHrP 1,25(OH)D$_3$ BMP 2 Osteoprotegerin Estrogen IGF-1 Prostaglandin E2

The activity of these bone cells is influenced by many systemic and local hormonal agents. Additionally, the cells influence the activity of one another. **Table 87.1** lists a few of the many agents influencing osteoblasts and osteoclasts. This is a rapidly developing field of investigation, and much is still to be learned. The osteoblasts, or cells of the osteoblast lineage, occupy a central position, not only in forming bone but also in processing systemic signals to the bone remodeling apparatus (see the later section on revision of bony material). Thus, although parathyroid hormone (PTH) is responsible for stimulating bone resorption, no PTH receptors are present on the osteoclast. Rather, they are found on osteoblasts (and related cells) which, in response to PTH binding, release or express on their surface agents (e.g., RANK ligand [RANKL] OPG pathway) that stimulate osteoclast activity. By contrast, osteoclasts do possess calcitonin (CT) receptors and are, thus, able to respond rapidly to the antiresorptive signal provided by secreted CT.

Bone Architecture

Bone consists of a dense outer shell, or cortex, and an internal, chambered system of interconnected plates, rods, and spicules called cancellous or trabecular bone (**Figure 87.1**). In the shafts of the long bones, the cortical component predominates, creating a hollow tube, whereas closer to the joints, the cortex thins and the interior is made up of an extensive latticework of cancellous bone. Bones such as the vertebrae, pelvis, sternum, and shoulder

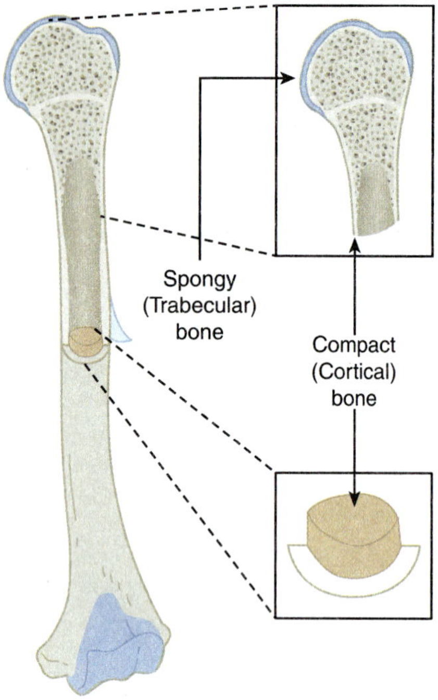

Figure 87.1 Gross and microarchitecture of a typical long bone.

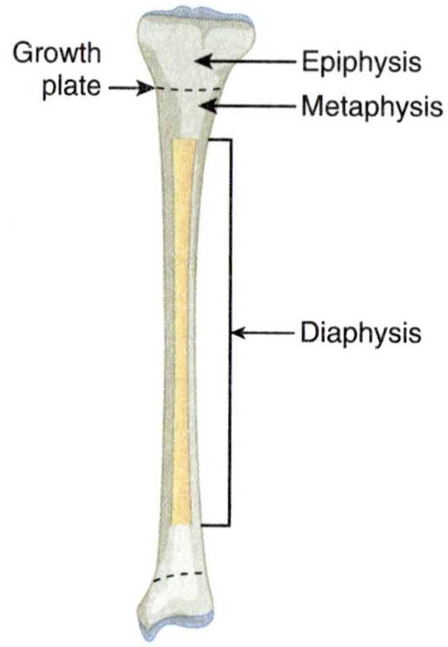

Figure 87.2 Principal regions of a growing long bone.

blades possess a thin outer rind of cortex and an even distribution of cancellous bone on the inside. The internal, three-dimensional architecture of cancellous bone is arranged along the lines of force (compressive, bending, tension) that a particular bone experiences and, hence, provides maximum structural strength with minimum material. Cortical bone sites also adapt to those forces, by increasing their cross section and/or mineral content.

The proportions of mineral and matrix and of calcium and phosphorus are essentially identical in cortical and cancellous bone. Sometimes, the issue has been confused in the literature because of the difficulty of removing adherent marrow elements from cancellous bone samples before chemical analysis. Fundamentally, however, bone is bone. On the other hand, cancellous bone turns over (remodels) much more rapidly than cortical bone. This is partly because of the greater surface area of cancellous bone. (Remodeling always starts from a microanatomic bone surface and burrows into the bony material.) It is also partly the result of the generally greater contact with hemopoietic marrow in cancellous bone. In fact, the partition of remodeling between bone with and without red marrow is probably more important than the partition between cortical and trabecular bone.

The end segments of bones are called epiphyses (**Figure 87.2**). The shafts of long bones are called diaphyses, and the flared portion of the shaft merging with the region of the growth plate is called a metaphysis. The lining cells on the outside of the bone form a tough sheet or membrane, called the periosteum, whereas the cells on the inside surfaces of both cortical and trabecular bone are called the endosteum.

The spaces between the trabecular plates and spicules are filled with bone marrow. Early in life, much of that marrow is hemopoietic, but later, the blood-producing marrow is confined to the bones of the trunk, and the peripheral skeletal marrow spaces are filled mostly with fat. In dense cortical bone, remodeling over the years produces a series of internal structures called osteons or Haversian systems, in which concentric cylindric layers of bone are laid down along the course of a capillary. These are produced by the usual remodeling process (see later, in the section on Bone Remodeling). First, a tubular hole is created in the bone by osteoclastic resorption; then, it is filled in by successive waves of osteoblasts moving from the outside inward.

At their ends, where bones meet one another in a joint, the bony surface is covered with a layer of cartilage rather than with periosteum. In health, this cartilage, is highly hydrated and is lubricated by synovial fluid held there by a tough connective tissue sac called the joint capsule. This arrangement ensures that the bones move on one another smoothly.

Bone Development

In utero, most bones are formed first as cartilage models, which are gradually replaced by bone. In this process, blood vessels invade the cartilage, calcification ensues, and the calcified cartilage is then removed by osteoclasts and replaced by bone laid down by osteoblasts. In infancy and childhood, bone growth and development follow a similar pattern. However, to accommodate growth, most bones possess one or more plates of cartilage perpendicular to the main axis of growth, separating, for example, the bone at the ends of a long bone from the bone of the shaft. This structure, called a growth plate (Figure 87.2), consists of rapidly proliferating cartilage cells that push the ends of the bones away from the shafts as they multiply. Blood vessels invade these columns of proliferating cartilage cells from the shaft side and initiate calcification of the cartilage and bony replacement in a process like what occurs in utero. Meanwhile, the ends of the bone are being pushed still further away. Thus, there continues to be an anatomic and temporal sequence consisting of growth plate, proliferating cartilage, calcifying cartilage, and bony replacement, progressing down the metaphysis from the epiphyses toward the diaphysis.

Growth stops when the ossification process catches up with the formation of new cartilage at the growth plate and bony bridges develop across the growth plate, thus firmly anchoring the ends to the shaft of the bone. This closure process is initiated by the estrogen produced at puberty in both girls and boys. When growth in length is occurring, the process of modeling (see later) shapes the outsides of the bones to keep them in proper proportion with the growth in length with a subsequent period of addition of mineral.[5,6] At midshaft, this typically involves periosteal new bone formation and endosteal resorption, whereas at the metaphysis, it is the other way about, with periosteal resorption and endosteal formation.

Bone Remodeling

Although the intercellular bony material, which makes up 95% to 98% of the volume of bone, is nonliving, nevertheless, it is capable of being revised and replaced, just as are soft tissues, whose cellular constituents are turning over constantly and invisibly. In bone, since the revision process occurs at discrete locations and is readily visualizable microscopically, it ensures repair and replacement of old fatigued bone. The process normally follows a stereotyped sequence of activation first, then resorption, then reversal, and finally, formation. This sequence is shown diagrammatically in **Figure 87.3**.

In the activation step, lining cells on a bone surface retract, exposing the bony substance directly to the circulating blood. Initiation of the process is by osteocytes. Mineralized bone serves as a chemoattractant for osteoclast precursors, which migrate to the exposed site, coalesce, attach to bone as osteoclasts, and begin to erode into the bone. After removing a suitable volume of bone, osteoclasts undergo apoptosis and disappear from the site. (However, as noted earlier, the process drifts along a bony surface, and new osteoclasts are recruited to keep the process going.) A reversal phase then follows. Osteoblasts move in and begin to replace the bone removed from the cavity. They orient the collagen fibers in parallel arrays, layer by layer, often alternating direction of the fibers every few micrometers. In this way, they build what is termed *lamellar bone*, which is similar in a sense to plywood (in which the grain of the wood runs in different directions in different layers). In the adult, osteoblasts advance at a rate of approximately 0.5 μm/day, and mineralization lags approximately 10 days behind the advancing matrix deposition. Some of the osteoblasts, as already noted, stay behind in cavities within the bone and become osteocytes.

When this revision process is finished at a particular site, the remaining surface osteoblasts become quiescent, flatten out, and turn into lining cells, effectively sealing the new bone surface until, at some later time, a new remodeling project is initiated locally. In healthy adults, this sequence takes approximately 3 months

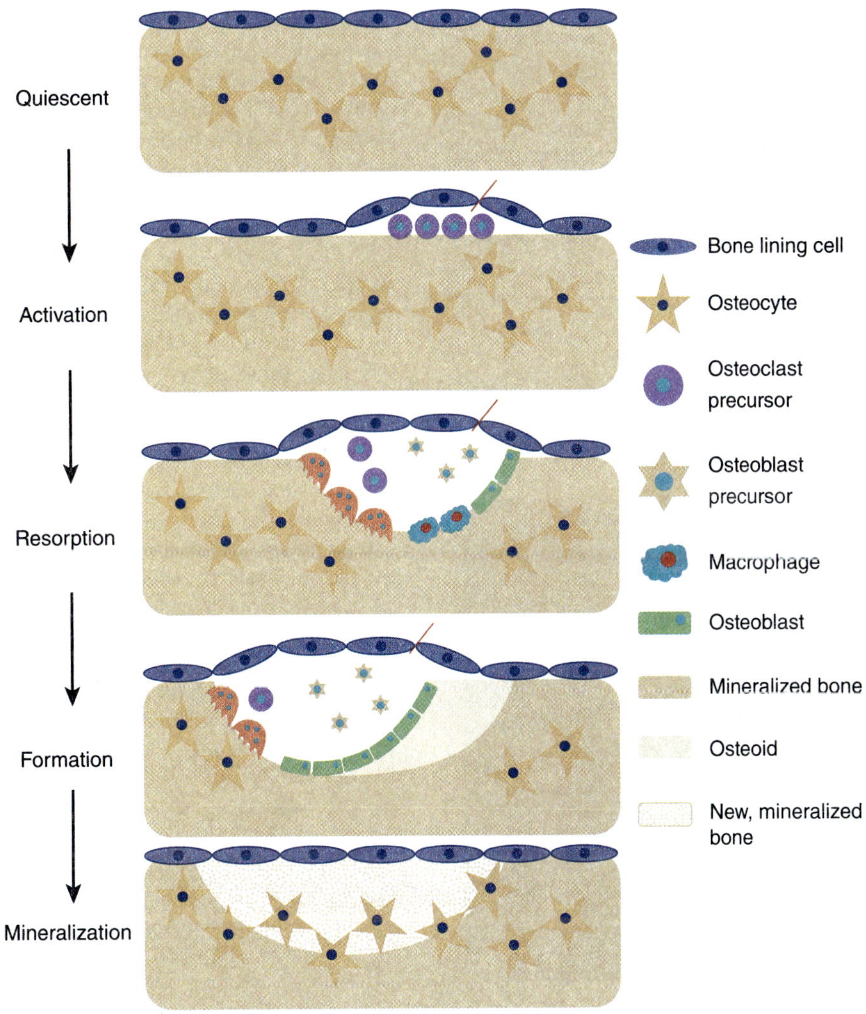

Figure 87.3 Schematic diagram of some of the major bone cells, indicating their role in the bone remodeling sequence. In the resting state, the bone surface is covered by lining cells and osteocytes lie embedded in the matrix. These retract at activated sites. Multinucleated osteoclasts move in and excavate a cavity. Then, after reversal, columnar osteoblasts deposit new bone matrix. Finally, the resting state is re-established.

Reproduced from Owen R, Reilly GC. In vitro Models of Bone Remodelling and Associated Disorders. Front. Bioeng. Biotechnol. 2018;6:134. doi: 10.3389/fbioe.2018.00134.

at any given site from start to finish. The resorptive phase takes approximately 2 to 3 weeks, and formation takes 2 to 3 months. The process is faster during periods of rapid growth, infancy, and adolescence, and is typically slower in elderly persons.

The coordination of this remodeling activity is mediated by osteocytes that initiate cell-to-cell signals and interactions that both stimulate and inhibit the forming and resorbing processes. Briefly, osteocytes secrete two substances: RANKL and osteoprotegerin. The former initiates osteoclast cell formation and differentiation and is necessary for the continued function of osteoclastic resorption. Osteoprotegerin complexes with RANKL and prevents its docking with RANK on the osteoclast surface. Thus, osteoblasts and narrow stromal cells orchestrate bone resorption. At the same time, they secrete FGF-23 and BMPs to stimulate osteoblast function, osteoblasts in turn secrete RANK, which stimulates osteoclast action. Finally, osteocytes also secrete sclerostin, a glycoprotein that blocks the stimulation of osteoblastic new bone formation.

This process is technically called *remodeling*. One of its principal purposes is to replace damaged volumes of bone with fresh new material. During growth; however, and indeed at any time when the shape of the bone is being revised, the resorptive and formative processes occur, not at the same site, but in different parts of the bone. For example, the small shafts of the long bones of a child are enlarged to adult size by osteoblastic formation on the periosteal surface and by corresponding osteoclastic resorption on the endosteal surface. This process is technically called *modeling* and is like remodeling in most respects, except the new bone is laid down at a location different from the site of resorption.

One of the ways in which this revision process can be readily visualized is by the administration of carefully timed doses of a fluorescent compound such as one of the tetracycline antibiotics. The molecules of these markers bind to mineralizing bone sites and are trapped as the site is buried in layers of new bone. The markers become readily visible, appearing like tree rings, when undecalcified bone biopsy specimens are viewed microscopically under ultraviolet illumination (**Figure 87.4**).

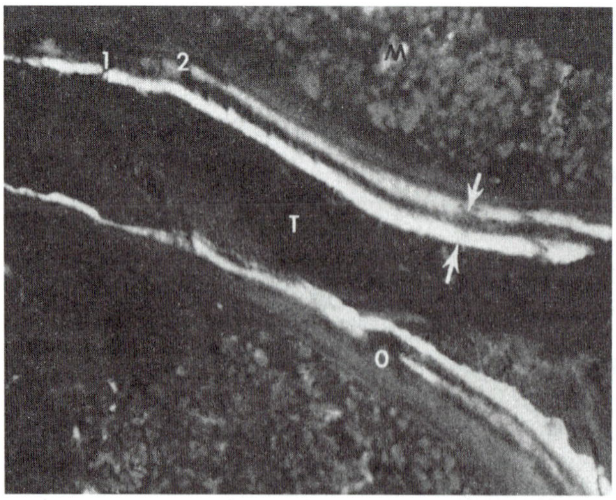

Figure 87.4 Section through trabecular bone obtained from the iliac crest. The distance between the *arrows* represents the amount of bone deposited during the time between administration of the two tetracycline labels. *M*, marrow; *T*, trabecular bone; *1* and *2*, first and second tetracycline labels.

Bone Functions

Bone serves two distinct functions: the provision of mechanical rigidity and stiffness to our bodies; and the provision of a homeostatic buffer, particularly to help our bodies maintain a constant concentration of calcium in the circulating body fluids, and to provide a reserve supply of phosphorus. The mechanical function is necessary so that we can resist gravity and move about on dry land. (Strictly speaking, a rigid skeleton would not be necessary in a fully buoyant medium such as water. For example, some originally bony fish, such as the sturgeon, have lost nearly all their skeletons over the millennia of evolution. Still their mechanical function is maintained.) The homeostatic function of bone is the older of the two, from the standpoint of evolution, and is the more fundamental, because the body will sacrifice the structural function before it risks losing the homeostatic one. In other words, the body will weaken the bone structurally to maintain the calcium concentrations of the blood and ECF.

Mechanical Functions

In the mechanical function of bone, nature strikes a balance between a skeleton so massive that it would resist most forces, but be too heavy to carry around, and one so flimsy that, although adequate to meet calcium homeostatic needs, it would be too fragile to sustain the mechanical forces of exertion or of minor injuries. Bone finds the middle ground by adjusting its mass (through change in mineral content, bone shape, distribution, or size) using a classic negative feedback loop (the Mechanostat) adapting to strains (changes in length) under routine use by approximately 0.05% to 0.10% in compression or tension and by 0.10% to 0.20% in shear loading (**Figure 87.5**).[6,7] This strain set point is a major determinant of bone size, shape and density during growth and bone density during adult remodeling. In the operation of this control loop, the osteocytes detect the amount of bending (or strain) in a local region of bone and send signals to nearby cells of the remodeling apparatus, which mediate the balance between resorption and formation. Such changes will induce adaptation in size, shape, and mineral through modeling and/or remodeling. Any change to the magnitude of strain, for example through immobilization, which reduces strain (after a stroke or spinal cord injury) or through under-mineralization (e.g., in osteomalacia), which reduces bone stiffness, will change the strain set point and trigger bone loss or gain. Increased strain, for example in racquet sports, causes differences in the loaded limb, predominantly through an increase in bone size, but also in bone density (**Figure 87.6**).

Homeostatic Functions

The bone remodeling process is also a major mediator of the homeostatic function of bone involving all bone cells, osteoclasts, osteoblasts, osteocytes, and bone lining cells. This is a complex function and involves autocrine and paracrine factors, including cytokines, growth factors, and hormones. Osteocytes play a key role, regulating both osteoblast and osteoclast function.[4] In contrast, the pattern of strain under loading is the major determinant of *where* bone remodeling occurs; PTH is the principal determinant of *how much* remodeling is occurring and how readily bone cells in any given locus respond to local stimuli to initiate a remodeling project. PTH secretion is directly responsive to the body's need for calcium; however, calcium is never simply removed from bone. Instead, whole volumes of

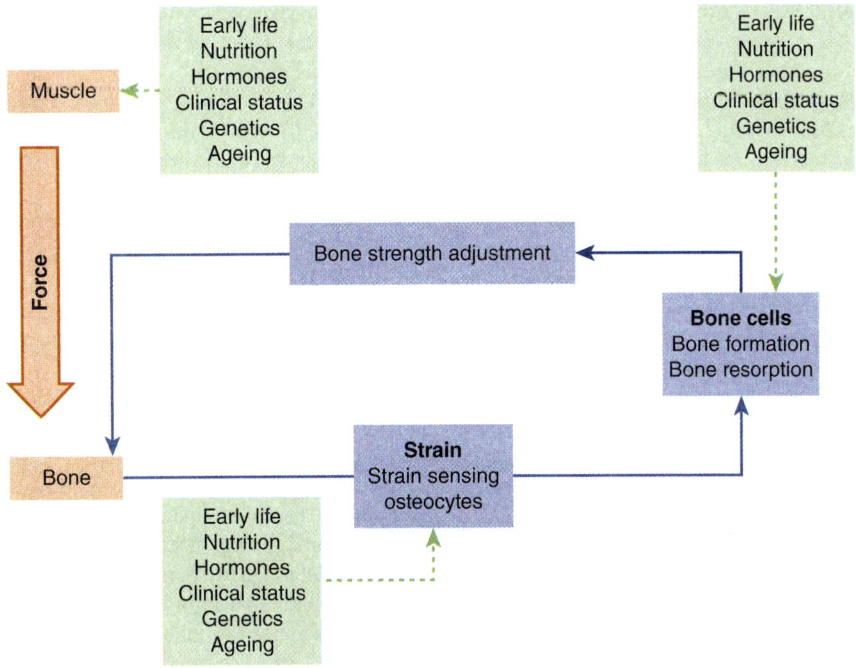

Figure 87.5 Schematic diagram of mechanostat negative feedback loop, showing adaptation to changes in strain, induced when the bone undergoes increased or decreased loading. Also highlighted are the points at which life stage, environment, and genetics may change response to loading.

Reproduced from Ward, K. Proc Nutr Soc 2012;71:27–37. doi:10.1017/S0029665111003375.

bone are removed, and their calcium is then scavenged to meet the body's needs.

As noted, resorption precedes formation. This produces local asynchrony of mineral movement. Remodeling at any given site first makes its calcium and phosphorus available to the body (as mineral is removed during resorption). The same site later creates a demand for calcium and phosphorus as the mineralizing site extracts mineral from the blood flowing past it. Averaged over the whole skeleton, these processes are about equal at any given time. New remodeling sites release as much calcium as the older sites deposit. However, this local asynchrony also means that,

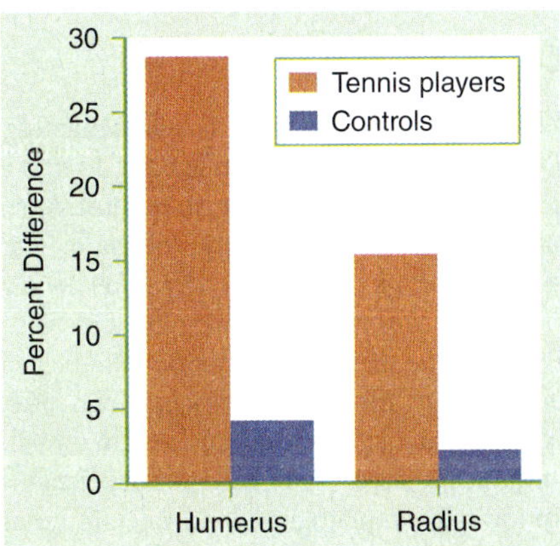

Figure 87.6 Difference in humeral and radial bone mass between dominant and nondominant arms of world-class tennis players and nonathletic controls. Even in the controls, the greater use of the dominant arm results in more bone on that side. This difference is greatly amplified in the tennis players.

Data from Kannus P, Haapasalo H, Sievanen H, et al. Bone 1994;15:279–84.

when remodeling activity is increased, resorption changes first, and more calcium is released from bone than previously initiated remodeling sites need. This makes an additional supply of calcium temporarily available to the body. Conversely, when remodeling is acutely suppressed, resorption drops immediately while previously initiated formation continues, and thus, the skeleton can soak up a temporary excess of calcium from the blood. Modulation of remodeling in this way by the hormones PTH and CT is a major part of the basis of the regulation of blood calcium concentration by the skeleton.

There is a second, fallback mechanism by which bone supports ECF [Ca^{++}], i.e., by the physical-chemical equilibrium between calcium crystals on bone surfaces and calcium ions in body fluids.[8] Because of the low solubility of hydroxyapatite, this mechanism will support a concentration with ECF only 50% to 60% of normal (i.e., serum calcium concentrations of 5–6 mg/dL), which is why patients with hypoparathyroidism have low serum calcium concentration. However, if the pH in the space under the lining cells on bone surfaces (or in the spaces around osteocytes) falls, even slightly toward a more acidic pH, bone mineral solubility increases and there is a selective inhibition of alkaline phosphatase.[9] The ultimate importance of this second mechanism in health and disease remains poorly elucidated, with direct evidence of the impact of changes in acid-base through dietary intervention giving mixed results.

Two examples with nutritional relevance illustrate how the better understood, cell-based system operates. During the absorption of large quantities of calcium from milk in infant feeding, CT is secreted to suppress osteoclastic resorptive activity, thereby reducing release of calcium from bone. This allows the demands of rapidly mineralizing bone to be met by absorbed calcium from the milk, and at the same time, it prevents the absorbed calcium from causing a dangerous increase in the concentration of calcium ions ([Ca^{2+}]) in the blood. Later, during the postabsorptive phase, CT levels drop and PTH rises to stimulate resorption again. This action sustains the blood [Ca^{2+}] calcium in the presence of steady mineral demands of bone formation and wide swings in calcium input from the intestine.

Another example is found during antler formation in deer. Because the skeletal remodeling cycle lasts several weeks at any given site, the asynchrony between calcium availability in resorption and calcium demand in formation is capable of prolonged support of unbalanced calcium needs. Each spring, the demands of the growing antler exceed the calcium available from the late winter foliage that makes up the deer's diet. Therefore, a burst of PTH-mediated remodeling occurs throughout the skeleton at the time antler mineralization begins. This creates a temporary surplus of calcium from the resorptive phase of skeletal remodeling, which is used to mineralize the bone of the rapidly forming antler. Later, when the new bony remodeling sites throughout the skeleton enter their own formation phase and now have a net demand for calcium, they are filled in with calcium made available from the then calcium-rich summer grasses and foliage.

Nutrients Important to Skeletal Integrity

Total nutritional status influences bone cell function just as it does the function of other tissues. However, cellular malnutrition affects bone currently being remodeled, whereas the strength of bony structures at any given time is dependent, not so much on current cell function, as on the mass, shape, and distribution of bony material accumulated by bone cellular activity over many years.[6] For that reason, acute nutritional stresses or deficiencies rarely produce overt skeletal symptoms in adults, even when they severely compromise bone cell function. Effects manifest more promptly in the growing skeleton, particularly during periods of rapid growth (infancy and adolescence), both because they have less skeletal capital in their bone banks and because they are forming and revising it much more rapidly. Nevertheless, a few nutrients, when deficient, are more likely than others to produce skeletal manifestations. These include calcium, phosphorus, vitamin D, and certain other trace nutrients.

Calcium and Phosphorus

In addition to buffering absorptive oscillations in blood calcium concentration, bone serves as a nutrient reserve for both calcium and phosphorus. This reserve is to the calcium (and phosphorus) functions of the body as body fat is to energy metabolism. Unlike most nutrient reserves, however, bone has acquired a distinct function, that is, mechanical and structural support. In other words, we walk around on our calcium reserve. It follows that any influence, nutritional or otherwise, that alters the size of the calcium reserve will alter bone strength.

Bone is a rich source of calcium. Total skeletal calcium averages 1,100 to 1,500 g, and each cubic centimeter of bone contains more calcium than the entire circulating blood volume in an adult. Thus, in comparison with other nutrients, the calcium reserve is huge. Although low-calcium diets relative to population habitual calcium intakes usually deplete the bony reserves, they do so slowly. Thus, whereas the *population-level* risk of fracture rises immediately, it will take many years for bone strength to be sufficiently reduced to lead to a perceptible increase in an *individual person's* risk of fracture. The slow expression of the effect of calcium deficiency led many nutritional scientists in the past to the erroneous (and seemingly nonsensical) conclusion that calcium was not important for adult bone strength. In many populations that are habituated to low calcium intakes, data are rare on fracture prevalence and incidence.[10] Nevertheless, a nutritional deficiency that develops over 30 years is just as truly a deficiency as one that develops over 30 days.

Low intakes of calcium and phosphorus can both limit bone acquisition during growth and cause bone loss after maturity. Calcium intake operates most directly through modulation of remodeling via PTH, as described earlier. In the antler example, if summer foliage were not calcium rich, each cycle of antler formation would deplete the skeleton, bony replacement would fail to occur, and bone mass would fall progressively over the animal's adult life. Because human calcium requirements rise during periods of rapid growth and with age,[3] and because calcium intakes tend to fall in elderly persons, precisely, such depletion occurs in most human populations as they age.

Inadequate phosphorus availability also affects bone, but in a different way. The osteoblast environment is one of continuous mineralization, with the matrix extracting phosphate (as well as calcium) from the fluid bathing the bone-forming cells. Although calcium makes up approximately 40% of the bone mineral, phosphate (PO_4^{3-}) accounts for nearly 60%. Thus, phosphorus is fully as important for bone building as is calcium. Rapid growth is not possible without a high blood phosphate concentration, a fact that explains the substantially higher blood phosphate values in children. When phosphate concentration in the blood entering bone is low, mineralization extracts as much phosphate from the blood as it can but, in so doing, creates a local environment severely depleted of phosphate. Osteoblasts, like all cells, need phosphate for their own metabolism. The result is serious interference with osteoblast function: matrix deposition is slowed, and osteoblast initiation of mineralization is reduced even more. Osteocytes, play an important part in the homeostasis of phosphate via Fibroblast growth factor-23 (FGF-23) and 1,25 (OH)D pathways. Abnormalities in the mineralization of matrix, through inadequate supply of calcium, phosphate or vitamin D produce the typical histologic pattern of rickets and osteomalacia in bone.

Vitamin D

Vitamin D has many bone benefits, such as facilitating the development of osteoclast precursors at an activated remodeling locus and augmenting osteoclast response to resorptive stimuli. The hormone also stimulates synthesis and release of osteocalcin by osteoblasts. However, its major importance for bone is the facilitation of intestinal absorption of calcium (and to some extent phosphorus) from the diet. Severe vitamin D deficiency, with or without low calcium intake, causes rickets and osteomalacia.[11] Milder shortages of the vitamin reduce calcium availability to the body and produce a situation of calcium deficiency, driving up bone resorption rates to maintain blood calcium homeostasis via PTH and, with inadequate capacity to increased bone formation, results in osteoporosis. Rickets (disorder of the growth plate usually in infancy and early childhood) and osteomalacia (defects in cortical bone mineralization occurring in adolescents and older adults), are the clinical manifestations of vitamin D deficiency and are qualitative defects of bone (inadequate mineralization) rather than quantitative (osteoporosis). There is much debate regarding the definitions of vitamin D deficiency, with subclinical concentration, referred to as *insufficient* concentration, associated with clinical manifestation.

Vitamin K

Vitamin K functions in the γ-carboxylation of the glutamic acid residues of several bone gla-proteins, in the context of bone, the most important being osteocalcin and matrix Gla protein. Vitamin K deficiency results both in undercarboxylation of osteocalcin and in reduced osteocalcin synthesis by osteoblasts, which impacts synthesis and mineralization of the bone matrix. Matrix Gla protein is involved in calcium binding of hydroxyapatite. Recent data suggest an interaction between vitamin K status and the microbiome, although data in human studies are limited to date (see later discussion on single micronutrients vs dietary patterns/quality). However, low vitamin K status is associated in epidemiologic studies with low bone mass, increased hip fracture risk, and increased cardiovascular mortality.[4]

Micronutrients

Vitamin C and certain trace minerals (notably copper, zinc, and manganese) are important cofactors for the synthesis or cross-linking of matrix proteins. Copper is the cofactor for lysyl oxidase, the enzyme responsible for cross-linking collagen fibrils. Interference with cross-linking results in structurally weak bone. Ascorbic acid is a required cofactor for the cross-linking of collagen fibrils and, in its absence, bone strength is impaired. In the presence of deficiencies of these micronutrients during growth, severe bone abnormalities can result, including stunting of growth, deformity of bones, and epiphyseal dysplasia. Whether adults can develop sufficient deficiencies of these nutrients to interfere significantly with bone integrity remains unknown.

Skeletal Disorders and their Nutritional Correlates

Osteoporosis

Osteoporosis is a multifactorial condition of the skeleton in which skeletal strength is reduced sufficiently so that fractures occur on minor trauma. There are three main types of primary osteoporosis, postmenopausal, age-related, or juvenile. Osteoporosis through secondary causes such as comorbidities or their treatments have multifactorial etiology. One in two women and one in five men over the age of 50 years will suffer an osteoporotic fracture in their lifetime. Generally, osteoporosis exhibits reduced bone mass (i.e., both matrix and mineral) as well as various microstructural disturbances of bony architecture.[5] By current World Health Organization standards, bone mass is estimated by x-ray absorptiometry as an areal bone mineral density (g/cm^2), (BMD, sometimes aBMD). Osteopenia is characterized by a BMD value at hip or spine between 1 and 2.5 standard deviations below the young normal adult mean. Osteoporosis is defined as a BMD value more than 2.5 standard deviations below young adult normal, whether or not a fracture is present. Areal BMD is, unfortunately, a poor way to represent bone structural strength, because it explicitly eliminates the important influence of bone size and its depth as the measurement is a two-dimensional projection of a three-dimensional shape. In addition, the definition of osteoporosis was derived in postmenopausal, white women, and so may not be applicable in men, premenopausal women or in other racial and ethnic groups.[12]

As well as the assessment of bone density, several fracture risk assessment tools have been developed; Fracture Risk Assessment Tool (FRAX), Garvan, and Q-Fracture are the most widely applied. The methods can be used alone or in conjunction with DXA femoral neck BMD. These are particularly attractive as they don't rely on specialist technology for the assessment of an individual's fracture risk. FRAX is now available in over 66 countries globally and covers 80% of the world's population. By taking a combination of the most common risk factors for osteoporosis, FRAX provides an individual with predicted 10-year risk of any osteoporotic fracture or a hip fracture.[13] Risk factors included in FRAX are: age, sex, height, weight, fracture history, parental hip fracture history, smoking, glucocorticoids, rheumatoid arthritis, secondary osteoporosis, alcohol >3 units per day, and femoral neck BMD (optional).

A common feature of most cases of osteoporosis is elevated bone remodeling, particularly in postmenopausal women.[6] Remodeling activity, although designed to repair bone, makes it temporarily weaker during the remodeling process; and when remodeling is in excess of mechanical need, it causes only weakness and can lead to fragility fractures without obvious trauma, such as those of the vertebrae. Estrogen deficiency, low calcium intake, and vitamin D deficiency all contribute to a harmful postmenopausal rise in bone remodeling. In men, reduction in testosterone and estrogen plus lifestyle factors such as excessive alcohol intake, are also risk factors.

Lack of weight bearing exercise, smoking and high intake of alcohol are important risk factors to the development of osteoporosis. Extremes of body weight and both low and high BMI are also risk factors for fracture. While the pattern of fractures differs somewhat in obese and overweight individuals, it is now clear that fracture risk is elevated, despite previously being considered as protective to bone. Lifestyle choices can impact both peak bone mass accrual and later disease risk.[6]

In addition to the single micronutrients considered in this chapter, recent interest has shifted to understanding better how the diet and understanding how diet quality and patterns impact fracture risk. Data from multiple epidemiologic studies show the importance of a nutrient-rich dietary pattern for skeletal health and prevention of fractures. From a public health perspective, interpreting what is meant by a healthy diet, rather than focusing on a single micronutrient for a specific condition, presents an opportunity to target multiple systems at the same time. The microbiome, and its interaction/ interdependence with micronutrient absorption and protein intake is also of increasing importance to bone health and ageing.[14] Intervention opportunities include changes in diet quality/patterns to improve nutritional content, altering the microbiome for improvements in inflammation and nutrient absorption, and behavioral intervention.

Rickets and Osteomalacia

Rickets is a disorder of the growth apparatus of bone in which the growth cartilage fails to mature and mineralize normally.[7] Rickets usually manifests in infants and early childhood. Growth is stunted, and various deformities about the growth plates occur, commonly including widening of the growth plate causing bulges at the wrist or ankle. Osteomalacia is the corresponding disorder in adolescents and adults, in whom newly deposited bone matrix fails to mineralize adequately. New matrix formation is slowed in both conditions, but mineralization is retarded even more; thus, unmineralized matrix accumulates on microscopic bone surfaces. For this reason, the proportion of mineral to matrix drops. In severe cases, unmineralized bone may constitute so large a proportion of the skeleton that individual bones lose their stiffness and become severely misshapen, further altering their biomechanical capacity (bowed legs, knock-knees, and misshapen pelvis, which presents problems during childbirth). A common occurrence with vitamin D deficiency is a weakening in muscles, causing proximal muscle weakness and reduced loading to bone.

The most well-known forms of nutritional rickets and osteomalacia are those associated with vitamin D deficiency with the presence or absence of low dietary calcium intake. The principal pathogenesis of these common forms follows from insufficient intestinal absorption mainly of calcium (and to some extent phosphorus) from the diet. In attempting to keep blood calcium concentration close to normal, the body raises PTH secretion. Because one of the effects of PTH is to increase renal clearance of phosphate, this adaptive response drives the already-reduced blood phosphate concentration down such that severe phosphate deficiency develops, first locally in the vicinity of the osteoblasts and chondrocytes and then in other tissues as well (producing, e.g., muscle weakness, tenderness, and pain).

Rickets and osteomalacia also develop for reasons other than vitamin D deficiency, including extreme calcium deficiency, dietary phosphate deficiency, fluoride excess, aluminum toxicity, and cadmium poisoning, as well as in association with certain rare vascular malignant diseases. Toxins or products of tumor metabolism, such as fibroblast growth factor-23 (FGF-23), interfere with normal osteoblast function (tumor-induced osteomalacia) or, alternatively, lower the renal phosphorus threshold. Typical of the latter mechanism is a group of heritable abnormalities of renal phosphate transport, the most common of which is X-linked hypophosphatemia.[8] These conditions have in common the inability to maintain the blood phosphate concentration required for growth. Such forms of rickets, just like that owing to vitamin D deficiency, produce their bony effects primarily because of severe hypophosphatemia. Therapy is directed at elevating serum phosphate using active vitamin D and/ phosphate supplementation or using a monoclonal antibody therapy to downregulate the action of FGF-23.

Paget's Disease of Bone

Paget's disease is a local but often multifocal disorder of the bone remodeling process of uncertain etiology. Resorption proceeds erratically, with formation filling in with new bone behind it. Bone architecture and even external bone shape are disordered. During the early resorptive phase, the bone is excessively fragile and may fracture readily. The high level of bone remodeling is usually accompanied by high concentrations of remodeling markers, particularly serum alkaline phosphatase. When the process involves the skull, bony growths may constrict the cranial nerve passages and may lead to deafness, for example. No nutritional correlates of this disorder are known.

Parathyroid Dysfunction

Because PTH is the principal determinant of the amount of remodeling activity in the skeleton, one could expect significant skeletal manifestations of parathyroid functional disorders. However, the reality is complicated. Patients with hypoparathyroidism have reduced remodeling, slightly greater than average bone mass, and probably fewer fractures. By contrast, patients with severe, long-standing hyperparathyroidism may have reduced bone mass; subperiosteal bone loss; widespread, very active bone remodeling; and even cysts in bone filled with osteoclast-like cells. Cases of this extreme type are rare today, and generally patients with primary hyperparathyroidism have no abnormalities of the skeleton detectable by ordinary x-ray studies. However, untreated mild primary hyperparathyroidism is now known to result in increased fracture risk, not for reasons of decreased bone mass but because of increased PTH-mediated bone remodeling. When PTH hypersecretion is pulsatile, and the peaks are short lived, PTH is trophic for bone and can produce large increases in spine bone density. For that reason, PTH 1-34 is now an approved therapy for osteoporosis.

Osteogenesis Imperfecta

Osteogenesis imperfecta (OI) is a group of heritable disorders in which one of several mutations may occur in the gene encoding for the collagen type 1 alpha molecules that comprise the bulk of bone matrix.[15] Mutations in 18 other genes can lead to an OI phenotype. Patients with OI have short stature, bony deformities and fragile skeletons with reduced bone mass. In a common form of OI, long bones typically have narrow shafts as well as reduced mass. Patients with OI suffer many fractures throughout life, in the most severe cases often starting in utero. Epidemiologic data show increased fracture risk through adulthood. When some other amino acid replaces glycine in the collagen molecule, the triple helix will not coil properly, collagen synthesis is reduced, and the matrix is abnormal. The severity of the defect depends on the position of the substitution in the chain of the collagen molecule. Fractures heal normally. Other collagen-based connective tissues are also affected in certain forms of the disease, including dentin, ligaments and tendons, and sclerae. Despite the manifest abnormality of the bone matrix, the mechanisms for the reduction in bone mass are still being elucidated, although recent data show that some of the affected genes may play a role in osteoblast function and Wnt signaling.

Bony Manifestations of Diseases of Nonskeletal Systems

Many clinical risk factors for osteoporosis should trigger fracture risk assessment, ranging from lung disease through neurological/psychiatric diseases and HIV. These include patients with chronic liver disease, but especially with biliary cirrhosis, commonly have a bone disease that is basically osteoporosis.[16] Patients who are to undergo liver transplantation often have severe osteoporosis, attributable to a combination of the underlying disease, the immobilization that inevitably accompanies the severe disability of these very sick patients, and the treatments they have received.

Patients with end-stage renal disease often have a varying mixture of osteosclerosis, osteomalacia, and hyperparathyroid bone disease.[17] Exact expression of these varied abnormalities depends on the medical regimens the patients receive, specifically the way in which these regimens manage calcium, phosphorus, and vitamin D metabolism for the patient.

Patients with a variety of disorders of the small intestine, but especially those with gluten-sensitive enteropathy, malabsorb fat-soluble vitamins and hypersecrete calcium and magnesium into the digestive juices. As a result, these patients are commonly deficient in vitamin D, calcium, and magnesium. They often have severe osteoporosis and may have osteomalacia as well. Patients with diabetes (Type 1 and Type 2) also have increased fracture risk, independent of BMD.

Patients with muscle disease such as myopathies or sarcopenia are at higher risk of falling, but also reduction in loading to bone will also reduce bone mass and increase fracture risk.

Patients who have had organ transplants commonly have osteoporosis,[18] in part because they present for organ transplantation with already reduced bone mass and in part because the immunosuppressive therapy used to sustain the transplant itself causes bone loss.

Skeletal Manifestations of Systemic Nutrient Deficiencies

Protein-Calorie Malnutrition

As noted, the cells of bone are as dependent on total nutrition as are other cells, and bone suffers in starvation just as do other tissues. However, bone strength is not immediately affected in acute malnutrition, especially in adults. The bony effects of protein-calorie malnutrition are most obvious during growth, when both growth rates and bone mass accumulation are retarded by malnutrition; and in the repair of fractures, especially in elderly persons. Protein-calorie malnutrition is common among elderly persons, and when they break a bone, such as the hip, serious complications and even death may ensue. Protein supplementation has been shown to reduce these complications substantially, and it is an important and necessary component of the treatment of most patients with hip fracture.[19,20] The reason for the trophic effect of protein on bone is partly that dietary protein helps to sustain normal insulin like growth factor-I (IGF-1) concentration, needed for bone growth and repair,[19] and partly that, as discussed, bone formation requires fresh dietary protein.

Magnesium Deficiency

Magnesium deficiency occurs in severe intestinal malabsorption (e.g., gluten-sensitive enteropathy, fistulas, or ileal resection, especially with high-fat diets) or with urinary losses from renal tubular defects. Initially, magnesium deficiency impairs bony responsiveness to PTH and thus leads to hypocalcemia despite a rising PTH. As deficiency progresses, parathyroid response falters, and PTH secretion falls. The hypocalcemia of magnesium deficiency is thus the result of impairment of the calcium regulatory system and is unresponsive to calcium supplementation.[19,21] Less severe degrees of magnesium deficiency in these same syndromes are associated with reduced bone mass, also unresponsive to calcium supplementation. In addition to other needed treatments (e.g., calcium), magnesium supplements are necessary in these patients. Finally, silent magnesium deficiency often accompanies low vitamin D status. The mechanism is uncertain. The deficiency manifests itself as a failure to elevate PTH secretion in response to the poor calcium absorption of vitamin D deficiency.

Investigation of Nutrient Effects on the Skeleton

As noted, the effects of nutrient deficiencies on the skeleton express themselves slowly in adults. For this same reason, nutrient effects on the skeleton of any kind are difficult to detect and easy to misinterpret.

Change in Bone Mass

Bone is a composite of mineral and matrix, and bone mass refers to the quantity of such bone present in the whole organism (or in a particular body region). Technically, changes in bone mass itself cannot be measured in vivo because no way of detecting the organic component of the composite is known. However, in conditions of health, and in most bone diseases, the proportion of mineral and matrix is about the same (50:50, by volume). In addition, good methods are available for measuring bone mineral content either for the whole skeleton or for various regions of interest by x-ray absorptiometric methods.

Dual-energy x-ray absorptiometry (DXA) is a method where tightly collimated beams of x-rays of two-energies are passed back and forth across the body (or one of its regions), and absorption (known as attenuation) of its photons is measured by a detector on the side opposite the x-ray source. Absorption is a function of the amount of mineral, and body composition (fat and fat-free mass) present in the path of the beam. This measures, for example, femoral neck bone mineral density in less than 3 minutes, lateral spine scan in less than 5 minutes and spine BMD in less than 2 minutes. The technique has a reproducibility on the order of 1% to 2%. While not the focus of this chapter, DXA also measures body composition, providing measures of lean mass (as a surrogate marker of muscle mass) and fat mass, and the resolution is now sufficient to allow assessment of visceral fat.

Because total body calcium in an adult is in the range of 900 to 1,500 g, and because change in bone mass (positive or negative calcium balance) is rarely more than approximately 100 mg/day (and usually much less), it follows that closely spaced, repeat measurements by DXA will produce results within the reproducibility error of the method. For that reason, measurements in persons are usually recommended to be separated by 12 to 24 months (a minimum of 6 months).

Most commonly, DXA is used to assess *change* in bone mineral density. Assessment of femoral neck BMD is now the recommended method of assessment for clinical diagnosis of osteoporosis. In addition to BMD assessment, newer DXA technologies also provide high-resolution images of the lateral spine, allowing semiquantitative assessment of vertebral fractures. Vertebral fracture assessment by DXA is now indicated in many patient groups, allowing a low-radiation, convenient assessment of the spine. Methodologies to assess bone size, distribution, and to estimate strength are available, and produce three-dimensional images of the bone, and allow assessment of the separate bone compartments.[13] The most common of those methodologies is quantitative computed tomography, with standalone peripheral methods predominating.[13] Single-slice pQCT uses rotate-translate technology; more recently, high resolution pQCT has been available, providing a "virtual bone biopsy" and having a resolution that allows direct measurement of trabecular bone. While these haven't yet replaced DXA as a clinical tool to predict individual fracture risk, in research, they have led to increased understanding of the causes of bone fragility in both children and in adults.

The classic nutritional approach to nutrient status is the measurement of metabolic balance, in this case, calcium (or phosphorus) balance. Because more than 99% of body calcium is in the skeleton, total body calcium balance reflects predominantly bone balance. Moreover, because calcium is essentially never removed from or added to preformed bone tissue (rather, units of tissue itself are removed or added), it follows that body calcium balance is a direct measure of bone tissue balance. The balance method is theoretically the ideal way of assessing change in bone mass, but it is expensive, and balance studies are difficult to perform accurately. The principal source of this difficulty, for poorly absorbed nutrients such as calcium, is that most of the ingested calcium ends up in the feces. Accurate timing of fecal excretion is nearly impossible. Moreover, the lag time between ingestion and fecal excretion averages several days in healthy adults, and failure to take that lag time into account leads to serious misinterpretation of calcium balance results.[16] Colored dye markers demarcating treatment periods are not adequate safeguards. Rather, continuous intake markers (e.g., polyethylene glycol [PEG 3350]) are required, and fecal output must be adjusted both for its polyethylene glycol content and the time lag.

The Remodeling Transient

Any intervention, nutritional or otherwise, that alters bone remodeling activity will produce a transient change[18] in calcium balance (or bone mass), which results from the asynchrony of the remodeling cycle. Because of the temporal separation of resorption and formation at each remodeling site, suppression of remodeling will produce a prompt but temporary increase in bone mass (**Figure 87.7**). If this follows, for example, the administration of supplemental calcium, phosphorus, or vitamin D, the retention of bone mineral should not be interpreted to mean that the patient has a pre-existing deficiency. (Such a deficiency may be present, of course, but positive balance will occur whether deficiency is present or not, simply because initially, the resorptive component of remodeling is reduced more than the formative component.) Because the remodeling cycle lasts at least 3 months in healthy young adults, bone mass and calcium balance continue to change under the influence of this asynchronous remodeling for at least that long. The process may take a year or more in elderly persons for formation and resorption once again to come into equilibrium. Response to nutritional interventions can be interpreted only *after* the transient is complete. This is evidenced well in studies of calcium supplementation in children and adolescents, which predominantly show that effects are not sustained if a follow-up period, post supplementation is included.[6] Trials with lasting effects tend to be where a protein-based calcium supplement is administered, or in populations with habitually low calcium or low calcium intakes relative to their population norm. If balance is more positive at that time (or bone mass by DXA is still increasing), only then can one safely conclude that the individuals needed more of the nutrient than they had previously been receiving. Because these constraints are rarely followed, much of the calcium literature is contradictory and confusing.

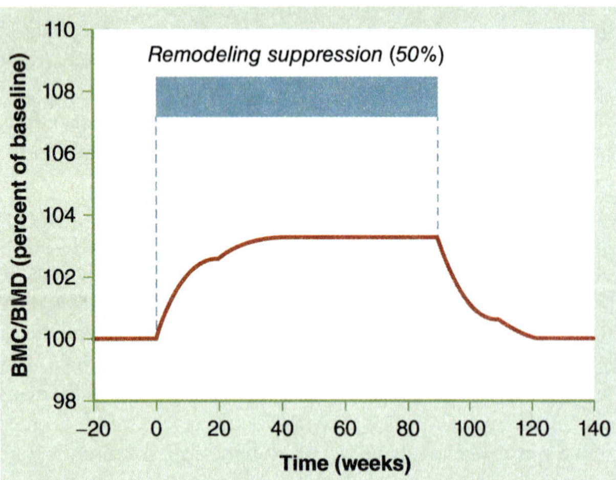

Figure 87.7 Positive bone remodeling transient in a healthy person who is not calcium deficient, in response first to a large increase in calcium intake (sufficient to suppress remodeling by 50%) and then to its later withdrawal. The vertical axis is bone mass (i.e., either bone mineral content [BMC] or bone mineral density [BMD]), expressed as a percentage of the baseline value. The initial rise in bone mass does not continue past one remodeling cycle (40 weeks in this illustration), and the bone gained by remodeling suppression is lost once again when remodeling returns to its previous level.

Bone Histomorphometry

The term *histomorphometry* means the measurement of shapes on histologic specimens of bone.[19] As noted, many substances attach to bone crystals as they are forming and then become trapped as more bone is laid down on top of them. Some of those substances, like the tetracycline antibiotics, fluoresce brilliantly when they are illuminated by ultraviolet light. Histomorphometry takes advantage of that property by giving patients paired, timed doses of tetracycline several days before obtaining a bone biopsy (typically from the iliac crest). Specimens are sectioned on special microtomes without first removing the mineral and are then examined with an ultraviolet microscope. Figure 87.4 presents a typical photomicrograph from such a labeled biopsy. Because the distance between fluorescent lines can be measured with a calibrated eyepiece or by shape-sensing software, and because the times of administration are known, one can derive a direct and reasonably precise estimate of the rate at which the remodeling cells are working and how active the remodeling process may be. Among other histologic features, measurements are not only made of the distance between labels but also of the extent of bone surface covered with a label. This method is very useful for studying bone biology and disease, but it has limited applicability to the study of nutritional problems affecting bone.

Biochemical Markers of Bone Remodeling

Bone turnover markers present a noninvasive assessment of bone metabolism. Over the past 3 decades, improvements in the methodology and sensitivity of the assays have been made.[22] In the synthesis of bone collagen, the ends of the collagen molecules are clipped off as the triple helix is assembled, the proline molecules in the peptide chain are converted to hydroxyproline, and crosslinks are developed between the side chains of adjacent collagen fibrils, involving particularly lysine and hydroxylysine protruding from the backbone of the peptide chain. Additionally, both alkaline phosphatase and noncollagenous proteins are secreted into the matrix; in this process, some of these substances leak into the bloodstream, where they can be measured. Later, when bone is broken down, hydroxyproline residues and cross-links, because they cannot be recycled, are metabolized or excreted. All these activities leave residues or produce effects that can be measured in serum or urine. Collectively, these circulating or excreted substances are called biochemical biomarkers of bone remodeling.[22] They reflect the level of bone remodeling activity. Each one reflects different stages of bone turnover, which can impact interpretation and selection of when to use. **Table 87.2** summarizes the principal markers currently in use together with the component of remodeling they are thought to reflect most directly. In this connection, when remodeling activity changes, ultimately, both resorption and formation generally change, almost always in the same direction, and often to very nearly the same extent. Under steady-state conditions, therefore, a marker for either formation or resorption may be used as an index of remodeling activity.

Although measurements of bone biomarkers can be a relatively inexpensive way of assessing bone remodeling activity under differing nutritional conditions, they are at most only semiquantitative and lack sensitivity and specificity. Bone turnover markers can be used to assess bone turnover clinically and to monitor treatment adherence to oral bisphosphonates but cannot directly diagnose osteoporosis. In other words, a 50% drop in a resorption marker does not mean a 50% reduction in the amount of bone resorbed. In addition, the markers exhibit important discrepancies among themselves potentially due to the different stages of activity they represent. For example, serum alkaline phosphatase is high in nutritional rickets despite a generally low level of new bone apposition, and 1,25(OH)$_2$ vitamin D elevates serum osteocalcin apparently without altering actual bone-forming activity.[23]

Table 87.2 Biochemical Markers of Bone Remodeling[22]

Formation
■ Serum/plasma
• Total alkaline phosphatase (TALP)
• Bone-specific alkaline phosphatase (BALP)
• Osteocalcin (OC)
• Carboxy-terminal propeptide of type I collagen (PICP)
• Amino-terminal propeptide of type I collagen (PINP)
Resorption
■ Serum/plasma
• C-telopeptide of collagen cross-links (CTx-Ia)
• Cross-linked C-telopeptide of type I collagen (ICTP)
• Tartrate-resistant acid phosphatase 5b (TRAP-5b)
• N-telopeptide of collagen cross-links (NTx)
■ Urine
• N-telopeptide of collagen cross-links (NTx)
• C-telopeptide of collagen cross-links (CTx-Ia)

Acknowledgments

I gratefully acknowledge the author of the "Bone Biology in Health and Disease" chapter in the previous editions, Dr. Robert Heaney, for substantial contributions to the present chapter.

References

1. Gokhale JA, Boskey AL. In: Marcus R FKJ, ed. The Biochemistry of Bone. Osteoporosis. 2nd ed. San Diego, CA: Academic Press, 107–88;2001.
2. Donsante S, Palmisano B, Serafini M, et al. Int J Mol Sci 2021;22.
3. Bonewald LF, Johnson ML. Bone 2008;42:606–15.
4. Dallas SL, Prideaux M, Bonewald LF. Endocr Rev 2013;34:658–90.
5. Rauch F, Bailey DA, Baxter-Jones A, et al. Bone 2004;34:771–5.
6. Ward K. Proc Nutr Soc 2012;71:27–37.
7. Frost H. Anat Rec 1987;219:1–9.
8. Blair HC, Larrouture QC, Tourkova IL, et al. Am J Physiol Cell Physiol 2018;315:C587–C97.
9. Arnett TR. J Nutr 2008;138:415S–8S.
10. Balk EM, Adam GP, Langberg VN, et al. Osteoporos Int 2017;28:3315–24.
11. Uday S, Hogler W. J Steroid Biochem Mol Biol 2019;188:141–6.
12. Lewiecki EM, Wright NC, Singer AJ. Osteoporos Int 2020;31:2069–71.
13. Fuggle NR, Curtis EM, Ward KA, et al. Nat Rev Endocrinol 2019;15:535–47.
14. Cronin O, Lanham-New SA, Corfe BM, et al. Calcif Tissue Int 2022;110:273–84.
15. Tauer JT, Robinson ME, Rauch F. JBMR Plus 2019;3:e10174.
16. Herlong H, Recker R, Maddrey W. Gastroenterol 1982;83:103–8.
17. Goodman WG, Coburn JW, Slatopolsky E. Renal Osteodystrophy in Adults and Children. Primer on the Metabolic Bone Diseases and Disorders of Mineral Metabolism 5th ed. Washington, DC: American Society for Bone and Mineral Research, 2003:430–47.
18. Epstein S. J Bone Miner Res 1996;11:1–7.
19. Rizzoli R, Biver E, Brennan-Speranza TC. Lancet Diabet Endocrinol 2021;9:606–21.
20. Delmi M, Rapin C, Bengoa Jea. Lancet 1990;335:1013–6.
21. Rude RK. Magnesium Depletion and Hypermagnesemia. In: Compston JE, Lian JB, eds. Primer on the Metabolic Bone Diseases and Disorders of Mineral Metabolism 5th ed. Washington, DC: American Society for Bone and Mineral Research, 2003;292–5.
22. Vasikaran S, Eastell R, Bruyere O, et al. Osteoporos Int 2011;22:391–420.
23. Pettifor JM, Prentice A, Ward KA, Manghan A. Chapter 12 Skeletal System. Nutrition and Metabolism. 3rd ed. The Nutrition Society. In: Roche HM, Macdonald IA, Annemie MW, Schols J, Lanham-New SA. © 2023 The Nutrition Society. Published 2023 by John Wiley & Sons Ltd. Available at www.wiley.com/go/roche/nutritionandmetabolism

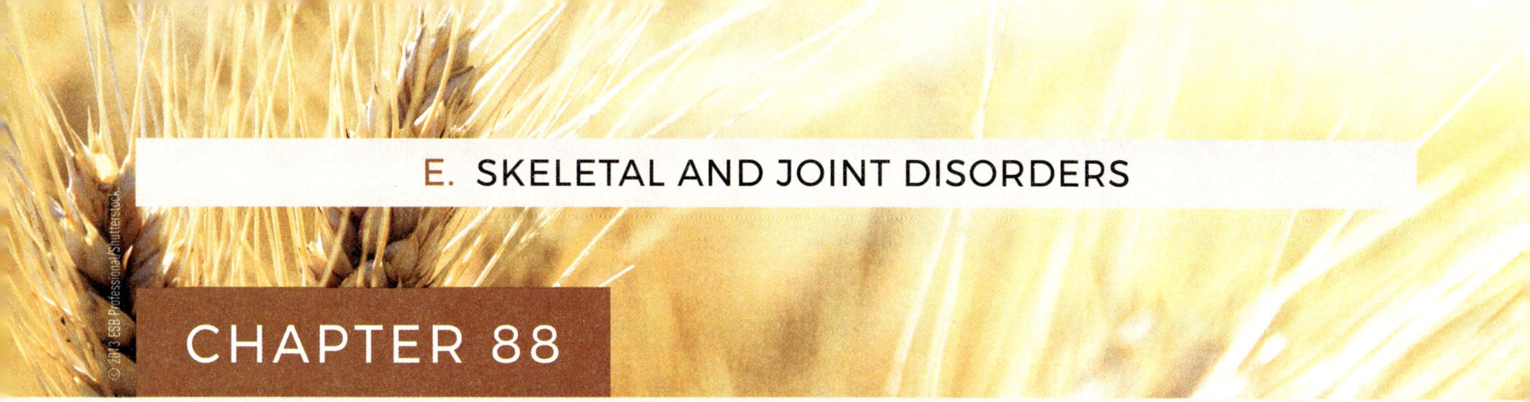

E. SKELETAL AND JOINT DISORDERS

CHAPTER 88

Prevention and Management of Osteoporosis

Katherine L. Tucker and Bess Dawson-Hughes

CHAPTER OUTLINE

Abbreviations

BMC, bone mineral content; **BMD**, bone mineral density; **BMC**, bone mineral content; **BMI**, body mass index; **DASH**, Dietary Approaches to Stop Hypertension; **DXA**, dual energy x-ray absorptiometry; **EPA**, eicosapentaenoic acid; **ESR1**, Estrogen Receptor Alpha; **FOS**, Framingham Osteoporosis Study; **FRAX**, Fracture Risk Assessment Tool; **IGF-I**, insulinlike growth factor-I; **MRI**, magnetic resonance imaging; **NAM**, National Academy of Medicine; **NHANES**, National Health and Nutrition Examination Survey; **NHS**, Nurses' Health Study; **OPG**, osteoprotegerin; **PTH**, parathyroid hormone; **RANKL**, receptor activator of nuclear factor B ligand; **RDA**, recommended dietary allowance; **TBS**, trabecular bone score; **UVB**, ultraviolet B rays; **WHI**, Women's Health Initiative; **WHO**, World Health Organization

Strategies for the Prevention and Management of Osteoporosis

Osteoporosis is a progressive deterioration in bone microarchitecture associated with loss of bone mineral density (BMD), leading to increasing risk of fracture with time. The prevalence of this condition in the United States exceeds 12 million adults 50 years old or older, with more than 40 million additional older adults at higher risk of developing osteoporosis because of low BMD. Total incident fractures for the U.S. population 50 years old or older were estimated at more than 2 million in 2005, with 71% of these in women.[1] Extrapolating to 2025, approximately four of every 10 women more than 50 years old in the United States are expected to experience a fracture. Black adults have a lower prevalence of osteoporosis and fracture than non-Hispanic White adults.[2] It had been assumed that BMD and fracture risk in Hispanics was between that of non-Hispanic White and Black adults. However, data from the National Health and Nutrition Examination Survey (NHANES) III (1988 to 1994) and the NHANES 2005 to 2008 showed that prevalence of low BMD was decreasing in non-Hispanic White people, but increasing in Hispanic people.[3,4] Most recently, data from the NHANES 2005–2010, showed that the highest age-adjusted prevalence of osteoporosis among adults, aged 50 years and older, was among older Mexican American women (16%), while women in the Boston Puerto Rican Health Study had prevalence of 10.7%, compared with 10.1% of NHANES non-Hispanic White, and 3.8% of non-Hispanic Black women. Surprisingly, Puerto Rican men had much higher prevalence of osteoporosis (8.6%) than other racial ethnic groups; 3.9% in Mexican American, 2.3% in non-Hispanic White, and 1.3% in non-Hispanic Black men.[5] More research is needed to understand these apparent differences in risk.

In the United States, the burden of osteoporosis resulting in hip fractures was estimated at 17 to 20 billion dollars annually, including acute and rehabilitative care, with costs projected to increase by 50% by 2025.[6] Medical costs aside, the effect of a hip fracture can be devastating for the individual. Large percentages of older hip fracture patients do not regain the ability to walk unassisted, approximately one-third are admitted into long-term care, and excess mortality ranges from 10% to 20% during the following year.[7]

Bone Mineral Density and Osteoporosis

Osteoporosis is characterized by low BMD and compromised bone strength, leading to increasing risk of fracture. Osteoporotic bone tissue shows deterioration of microarchitecture, with thinner trabeculae, reduced mineralization, and thinning of cortical surfaces associated with increased cortical porosity.[8] Total BMD is the result of a delicate balance between bone resorption by osteoclasts and bone formation by osteoblasts during continuous remodeling. During childhood, bone growth requires a balance in favor of bone acquisition, whereas in young adults who have achieved their peak bone mass, BMD tends to be relatively stable until around age 50 years. Thereafter, osteoclast activity begins to exceed that of osteoblasts and bone loss occurs.[9] After the onset of menopause in women, bone loss accelerates to two to six times premenopausal rates, and then gradually slows to about 1% annually by 10 years after menopause.[10,11] In contrast, longitudinal studies in older men suggest consistent albeit slow bone loss (i.e., ~1% per year).[11]

Individually, changes in bone mass also reflect numerous exposures that affect the remodeling balance. Therefore, osteoporosis prevention depends on optimizing peak bone mass, minimizing exposures that lead to bone loss, and optimizing nutritional exposures for bone maintenance throughout life. For more discussion of bone biology and composition, see the chapter on bone and joint biology. An overview of associated factors, discussed in more detail later, can be found in **Table 88.1**.

Measuring Bone Mineral Density

In the last 4 decades, measurement of BMD has improved significantly. The most widely used method is dual energy x-ray absorptiometry (DXA), which captures the energy absorbed as x-rays pass through the bone from an energy source on one side to a detector on the other.[12] A newer addition to DXA is the trabecular bone score (TBS) of the spine, which correlates with bone microarchitecture and provides additional information above BMD.[13] Single x-ray absorptiometry also has been widely used but is appropriate only for areas without much overlying tissue, such as the wrist or heel. Newer techniques include quantitative computed tomography to measure metabolically active trabecular bone,[14] quantitative ultrasound, which captures the modulation of sound waves as they pass through the tissue,[15] and magnetic resonance imaging (MRI) of bone microarchitecture, which also provides information on cortical porosity and marrow fat.[16]

DXA, which continues to be the most used measure, provides measures of BMD at specific locations of the hip and spine, in g/cm^2. These are compared with population standards to provide a T-score, used by physicians in defining osteoporosis or osteopenia. Scores indicate the extent to which an individual is above or below the mean optimal density using standard deviation units. A score of +1 to −1 is normal, −1 to −2.5 is osteopenia (low bone mass), and −2.5 and below is osteoporosis. The World Health Organization (WHO) international reference standard for the description of osteoporosis in postmenopausal women and in men 50 years old or older, uses DXA measures of the femoral neck from non-Hispanic White women, 20 to 29 years old, in the NHANES III.[17] Z-scores, which relate bone density to healthy individuals of the same age and sex, are often used in younger individuals. Apart from the NHANES, few studies have been conducted in racial and ethnic minorities and, although practitioners discuss whether different racial and ethnic groups should have different reference standards, the WHO currently recommends use of a single standard for optimal comparison across groups. An analysis by race/ethnicity in the Women's Health Initiative, however, showed that the T-score underestimated risk of major osteoporotic fracture over 19 years of follow-up for all women, but particularly among non-Hispanic Black and Hispanic older women.[18]

Bone Mineral Density and Risk of Fracture

Measuring BMD to define osteoporosis is important because an inverse relationship exists with risk of fracture in older adults. A meta-analysis of 12 cohorts in diverse populations showed that DXA femoral neck BMD was a strong predictor of subsequent fracture risk for both men and women.[19] Although hip fracture is the most serious, other fractures may also have important effects on health and independence. Vertebral compression factures, which shift vertebrae into a wedge shape leading to kyphosis, or curved spine, can cause chronic pain and disability and are more common in women than men.[20]

Risk of fracture increases with age because of changes in bone quality, declining bone density, and falls, which increase with aging because of declining muscle strength, loss of balance, gait difficulties, arthritis, poor vision, and use of medications.[21] In the Framingham Osteoporosis Study (FOS) of older adults,

Table 88.1 Key Messages for Bone Health

Risk statistics:
- More than 12 million adults, 50 years and older, have osteoporosis in the United States.
- An estimated 4 of 10 women more than 50 years old may experience a fracture.
- A large percentage of older patients with hip fracture do not regain the ability to walk unassisted and are admitted into long-term care.
- Excess mortality ranges from 10% to 20% during the year after a hip fracture.

General risk factors:
- Aging increases risk because of declining muscle strength, loss of balance, gait difficulties, arthritis, poor vision, and use of medications.
- The WHO FRAX (fracture risk assessment tool) includes older age, female sex, low BMI, prior fracture, parental hip fracture, current smoking, long-term use of glucocorticoids, rheumatoid arthritis, conditions leading to secondary osteoporosis (type 1 diabetes, osteogenesis imperfecta, untreated hyperthyroidism, hypogonadism or premature menopause, chronic malnutrition or malabsorption, chronic liver disease), and consumption of three or more alcoholic drinks per day in calculating fracture risk.

Nutrients and bone health:
- Clearly protective nutrients include calcium, magnesium, potassium, vitamin D, and vitamin K.
- Likely protective nutrients include silicon, strontium, vitamin C, vitamin E, vitamin B_{12}, vitamin B_6, folate, carotenoids, and protein.
- Possible negative effects may be seen with high intakes of sodium, phosphorus, iron, fluoride, and vitamin A.

Caffeine and alcohol:
- Excessive caffeine is a risk factor but may be offset with calcium intake.
- Moderate alcohol intake appears protective, but heavy alcohol intake poses a significant risk.

Body weight and composition:
- Low body mass index and weight loss are risk factors for low bone mineral density and fracture.
- For any given weight, however, abdominal fat mass may contribute to risk of fracture.

Physical activity:
- Weight-bearing physical activity and resistance exercises are protective of bone mineral density.
- Strength and balance exercises improve muscle function and reduce falls.
- Aerobic exercise is important during weight reduction to protect against bone loss with weight loss.

Genetics:
- Family history of fracture and identification of bone-sensitive gene polymorphisms show that genetic risk factors are important.
- However, this risk may be mitigated with optimal diet, physical activity, restriction to moderate alcohol consumption, and avoidance of smoking.

important risk factors associated with bone loss over time in women included age, lower weight, and weight loss, whereas estrogen use was protective; in men, bone loss was associated with smoking. Surprisingly, no association was seen in either men or women between bone loss and physical activity, caffeine intake, calcium intake, or serum 25-OH vitamin D concentration.[22] In the Rotterdam Study of older adults, bone loss was associated with lower weight and smoking in both men and women, whereas calcium intake was protective in men but not in women.[23] A large study of 9,516 older U.S. women found that risk of fracture was associated with previous fracture, greater height, fair or poor self-rated health, hyperthyroidism, treatment with benzodiazepines or anticonvulsants, greater caffeine intake, and spending up to 4 hours/day standing.[24] More recently, an assessment of genetic and clinical risk factors for hip fracture, using Mendelian randomization, identified five genetic signals, one associated with falls and four with bone biology. They noted a strong causal relationship between risk of hip fracture and low femoral neck BMD, with moderate causal effects with Alzheimer's disease, and with past or current smoking.[25] Further work with gene-environment interactions may help to clarify risk and individuals who may benefit from specific interventions.

To better assess risk, the WHO developed the Fracture Risk Assessment Tool (FRAX).[26] This tool calculates the 10-year risk of fracture with weighted scores for age, sex, BMI, prior fragility fracture, parental hip fracture, current smoking, long-term use of glucocorticoids, rheumatoid arthritis, conditions leading to secondary osteoporosis (type 1 diabetes, osteogenesis imperfecta, untreated hyperthyroidism, hypogonadism, premature menopause [<45 years], chronic malnutrition or malabsorption, or chronic liver disease), consumption of three or more alcoholic drinks per day and, BMD if available.[27] Although helpful and widely used, this tool is constantly being updated and adapted as more information becomes available.[28] A recent review noted that, while the use of FRAX has been heterogeneous across countries, it has proved useful in predicting fracture.[29] The FRAX tool does not consider nutritional determinants of risk. However, it provides a useful base from which to investigate the effects of nutritional variables, after other contributors to risk are considered.

Nutritional Determinants of Bone Density and Fracture Risk

As living tissue, with constant resorption and rebuilding, bone appears to be responsive to a wide range of nutrients. Some of these have only recently been understood and others continue to be actively investigated. Food sources of the minerals and vitamins most clearly associated with bone status are provided in **Table 88.2**.

Table 88.2 Good Dietary Sources of Key Nutrients for Bone Health

	Daily Value[a]	Foods
Calcium	1,000 mg	Milk, yogurt, and cheese
		Small or canned fish with edible bones (sardines, salmon)
		Calcium set tofu
		Fortified soy milk
Magnesium	400 mg	Whole grains and whole grain cereals (wheat bran, wheat germ, brown rice, quinoa, oatmeal, raisin bran, shredded wheat)
		Nuts (almonds, cashews, peanuts, peanut butter)
		Mature beans and peas (soybeans, pinto beans, kidney beans, black-eyed peas, lentils)
		Dark green leafy vegetables (spinach, collards, kale, Swiss chard)
		Fish (halibut, pollock, tuna, haddock)
		Dark chocolate, cocoa
Potassium	3,500 mg	Baked potato, sweet potato
		Tomato paste, tomato sauce
		Mature beans (kidney beans, white beans, soy beans, lima beans, lentils)
		Yogurt, milk
		Fish (halibut, rockfish, cod, trout)
		Winter squash
		Orange juice
		Banana
Vitamin D	400 IU (10 μg)	Fatty fish (herring, salmon, sardines, swordfish)
		Fortified milk and yogurt
		Fortified breakfast cereals
Vitamin K	80 μg	Dark green leafy vegetables (kale, Swiss chard, collard greens, spinach)
		Dark salad greens (leaf lettuce, watercress, raw spinach)
		Cruciferous vegetables (broccoli, brussels sprouts)
		Vegetable oils (soybean oil, canola oil)

[a]The daily value is the suggested intake for a 2,000-kcal (8.374 MJ) diet and is the amount used on food labels. Individual requirements vary.

Minerals

Although calcium and vitamin D have long been known to be important for long-term fracture risk, more recent research has shown that bone mass is, in fact, sensitive to a wide variety of nutritional exposures. Dietary intake is a centrally important modifiable factor in the development of peak bone mass and in the protection against bone loss with aging. Because bone undergoes continuous remodeling, an adequate supply of nutrients is required to support bone formation and retention. The bone matrix is composed of calcium, phosphorus, protein, and trace minerals, including magnesium, and these are of primary importance. However, additional dietary components affect the remodeling balance through effects on calcium regulation, inflammation, DNA methylation, and other regulatory processes that stimulate bone resorption or formation. Understanding these relationships is important, because it points to dietary quality as a critical factor in bone status, as opposed to an earlier focus mainly on calcium supplementation for prevention of bone loss.[30,31]

Calcium

Calcium is the major mineral component of bone mass, and nearly 99% of the calcium in the adult human body is contained in bones in the form of hydroxyapatite. Children need relatively large amounts of calcium to lay down new bone with rapid growth. The 1997 Food and Nutrition Board set adequate intakes at 1,300 mg (32.5 mmol) per day for children 9 to 18 years old, for maximizing peak bone mass to protect against osteoporosis later in life.[32] However, supplementation studies in children have shown mixed results. A 2000 review of calcium supplementation and bone concluded that calcium contributed to higher BMD primarily at cortical bone sites, was most effective in populations with low baseline calcium intake, and was more effective in pubertal than prepubertal children.[33] A review of 19 calcium supplementation trials in 2,859 children found that calcium supplementation had a small effect on upper limb BMD but no effect on femoral neck or lumbar spine.[34]

A broader review of calcium trials, including adults, found that all but two of 52 trials showed better bone balance with calcium intervention, greater bone gain during growth, reduced bone loss with aging, or reduced fracture.[35] In contrast, a 2017 meta-analysis of 33 clinical trials with 51,145 participants showed no significant association of calcium with hip or total fracture, vs placebo.[36] However, few studies have considered baseline status. It seems likely that those with inadequate calcium intake would benefit more from supplements than those with adequate intakes.

Dietary sources of calcium may be more effective than calcium supplements. An NHANES III follow-up analysis found that low recalled milk intake during childhood and adolescence was associated with significantly lower hip BMD and a doubling of fracture risk among women 50 years of age and older.[37] A large 5-year study of British adults found that fracture risk was 75% higher among women with baseline calcium intakes less than 525 vs 1,200 mg or more per day, and the association was stronger for women less than 50 years of age than for older women.[38] However, in the Nurses' Health Study (NHS), women who reported drinking two or more glasses of milk per day, vs up to one per week, did not differ significantly in hip fracture incidence.[39]

Intervention studies with calcium-rich foods have shown beneficial effects on bone. In one, spinal bone loss was significantly lower in premenopausal women who used dairy foods to raise calcium intake from 900 to 1,500 mg (22.5 to 37.5 mmol) per day, relative to controls.[40] In another, three additional servings of yogurt per day led to significant reduction in urinary excretion of bone resorption markers in older women.[41] Calcium in foods like milk and yogurt may be used more effectively than supplements because it comes packaged with other important nutrients that work together, including vitamin D, protein, potassium, and magnesium. A recent review noted that dairy products, particularly fermented products, like yogurt, which provide probiotics in addition to calcium and protein, appear to be associated with lower risk of hip fracture.[42]

Phosphorus

Phosphorus is essential for bone, but too much phosphorus in combination with low calcium intake can lead to reduced calcium bioavailability and potential bone loss. Although uncommon, phosphorus deficiency can lead to reduced mineralization and bone resorption. Deficiency has been seen in older adults with malnutrition, intestinal malabsorption, or long-term use of medications that bind phosphorus, including antacids.[43] In the general population, excess phosphorus is more of a concern than deficiency. The U.S. diet tends to be high in phosphorus relative to calcium. Mean intakes of phosphorus in the NHANES 2007 to 2008 study were 1,123 and 1,550 mg/day, for women and men, relative to a recommended dietary allowance (RDA) of 700 mg, whereas mean intakes of calcium were 833 and 1,038 mg for women and men relative to an RDA of 1,000 to 1,200 mg.[44,45]

Excess phosphate forms complexes with calcium that interfere with calcium absorption, which may, in turn, lower serum calcium and lead to secretion of parathyroid hormone (PTH), lower 1,25(OH)$_2$D production, lower intestinal calcium absorption and, consequently, bone resorption to release calcium from bone.[46] Short-term metabolic studies have documented some of these mechanisms.[47,48] A more recent review noted that higher dietary phosphorus, relative to calcium, is associated with higher parathyroid hormone and FGF23, with a negative effect on bone health.[49]

One major source of excess phosphorus in the U.S. diet is phosphoric acid from cola drinks. Two studies in teenage girls found that cola consumption significantly increased the odds of fracture.[50] In the FOS, women consuming cola daily had significantly lower hip BMD than those who consumed cola less than once per week.[51] In contrast, a short-term metabolic study showed negligible effects of phosphoric acid–containing beverages on urinary calcium excretion, and concluded that the effects seen in observational studies may be caused by milk displacement.[52] However, milk displacement did not explain the significant negative effects seen for cola in the FOS, and there was no effect of other soft drinks. It is likely that regular exposure to phosphoric acid may cause small BMD losses, which accumulate to measurable losses over time. A 2017 review of phosphorus intake and bone health concluded that increasing intake from additives is associated with negative effects on bone in both humans and animals.[53]

Magnesium

Magnesium is important to the formation of hydroxyapatite and may enhance bone strength through its role in crystallization.[54] It also is known to regulate active intestinal calcium transport. In animal studies, experimental magnesium deficiency decreased bone volume and trabecular thickness, bone mass, PTH, 1,25(OH)$_2$ vitamin D concentration, and osteoprotegerin (OPG), whereas it increased receptor activator of nuclear factor B ligand (RANKL) and osteoclastogenesis.[55-58]

Circulating magnesium concentration was significantly lower in women with osteoporosis than in those with normal bone mass.[59] In observational studies, magnesium intake was significantly positively associated with BMD and protective against bone loss.[60-62] This is important because magnesium intakes tend to be low; daily median intakes from NHANES data ranged from 177 mg among African American women to 326 mg among non-Hispanic White men, relative to recommendations of 320 and 420 mg for women and men, respectively.[63] Some intervention studies with magnesium have shown benefit on bone mass in adolescent girls,[64] in suppressing bone turnover markers in young men[65] and protecting against bone loss in osteoporotic women,[66] findings suggesting that this important mineral may be underappreciated for its important role in bone health. A review of 28 studies, nine with blood Mg, 12 with Mg intake, and seven with Mg supplementation, showed that low serum Mg concentration was associated with osteoporosis; that low Mg intake was associated with lower BMD and higher fracture risk, and that Mg supplementation did result in improved BMD and lower fracture risk.[67] Overall, Mg appears to be an important nutrient for bone health, particularly as it is a shortfall nutrient in terms of intake in the U.S. population.

Potassium

Potassium promotes renal calcium retention and potassium bicarbonate is important in neutralizing the acid load of most diets, which may protect against calcium loss from the bones. Potassium administration increased the excretion of the bone formation marker serum osteocalcin, and decreased the resorption marker urinary hydroxyproline.[68] Several population-based studies have demonstrated protective associations between potassium intake and bone status. In premenopausal women, a difference of 8% in femoral neck BMD between the highest and lowest quartiles of potassium intake was seen.[69] In perimenopausal and early postmenopausal women, dietary potassium was associated with lower bone resorption and greater BMD.[68] In older adults in the FOS, potassium showed a protective association with BMD in men and women at baseline, and lower BMD loss over time in men.[62] In another study of elderly women, higher urinary potassium excretion at baseline was associated with 4% greater total body BMD and 11% greater trabecular BMD at 5 years.[70] An analysis of the Korean NHANES found that in adults over 50 years of age, there was a significant positive association between potassium intake and BMD, as well as a lower risk of osteoporosis in the women, but not the men.[71] The sodium/potassium intake ratio appears to be important for bone. One analysis noted lower BMD with a higher ratio in middle aged Korean women.[72] Another noted that, relative to preagricultural humans, the modern human diet is deficient in potassium (2,500 mg vs 7,000 mg/day) and contains excess sodium (~4000 mg vs 600 mg/day).[73] This combination may have particularly negative effects on bone.

Sodium

Sodium intake in the United States is considerably higher than recommended. NHANES 2007 to 2008 data showed that, in contrast to recommendations of approximately 1,500 mg, mean daily sodium intakes in U.S. adults were 4,043 mg for men and 2,884 mg for women.[45,74] This likely contributes to renal calcium excretion. Studies have shown that each 1,000 mg of additional sodium was associated with a 20-mg increase in urinary calcium loss—the amount likely to be absorbed from 80 mg of dietary calcium,[75] and consequently with lower BMD. The optimal intake balance for protecting bone was approximately 1,000 mg calcium and less than 2,000 mg sodium per day.

As noted above, the effect of sodium may depend on potassium exposure. A metabolic study found that postmenopausal women given 5,175 mg of sodium per day had increases in urinary calcium and the resorption marker N-telopeptide, whereas those given sodium plus potassium citrate had decreases in urinary calcium and no increase in N-telopeptide.[76] In the Dietary Approaches to Stop Hypertension (DASH) trial, a diet high in fruit, vegetables, low-fat dairy, and thus high in potassium, was randomly assigned, vs a control diet, for 30 days. The DASH diet significantly reduced serum markers of bone turnover. DASH plus sodium reduction lowered PTH in the control group and lowered urinary calcium in both.[77] In another study of postmenopausal women who reduced sodium intake to less than 2,000 mg/day for 6 months, urinary calcium excretion and bone turnover markers were reduced.[78] However, other studies have been less clear about the effects of sodium on bone[79,80]; and one study showed no adverse effect on BMD of 3,000 mg of sodium, compared with 1,500 mg/day, when participants were supplemented to ensure adequate calcium and vitamin D intakes.[81]

Fluoride

In the correct amount, fluoride has long been known to prevent tooth decay and has been added to most water supplies in the United States. Fluoride substitutes for the hydroxyl group in hydroxyapatite, forming fluorapatite. Fluoride has been shown to result in bone with larger crystals and higher BMD, but lower elasticity.[82] The effect of fluoride on fracture risk has been controversial, with both protective effects[83,84] and increased risk[85-88] being reported, likely depending on the total exposure. In the largest randomized, placebo-controlled trial of sodium fluoride in postmenopausal women with osteoporosis, spine BMD increased, but so did vertebral fracture risk.[86] A meta-analysis of 25 studies showed that fluoride treatment increased spine and hip BMD, but with no effect on fracture risk, although the protective effect was seen with low doses (<20 mg/day of fluoride equivalents).[89] A comparison of bone tissue from individuals in municipalities with and without fluoridated water, showed no differences in the physical characteristics of bone.[90] In contrast, a study in 4,306 Swedish women found that although baseline BMD was higher, risk of fracture was also higher with greater fluoride exposure.[91] It is important to note that while fluoride appears protective in areas with low fluoride, excessive fluoride, which does occur in some water supplies, will cause skeletal and tooth fluorosis, making bones brittle and teeth mottled.[92] Fluoride supplementation, either in short- or long-acting forms, is not approved by the U.S. Food and Drug Administration for the prevention or treatment of osteoporosis.

Iron

Iron is an important cofactor for hydroxylases in collagen formation. Both low iron intake and iron overload have been negatively associated with bone. Iron overload has been associated with low BMD in patients with genetic hemochromatosis and with African hemosiderosis.[93,94] However, low iron is more of a concern in the general population. Rats fed iron-deficient diets showed compromised bone morphology, strength and density, and decreased bone formation, assessed by serum osteocalcin.[95,96] Studies in postmenopausal women have reported that higher iron intake was associated with greater baseline BMD[97] and, prospectively, with lower loss of BMD in a subset of women using hormone replacement therapy and taking 800 mg calcium per day.[98] In contrast, however, another study showed no association between iron status and BMD in women.[84] A review on this topic noted that both deficiency and excess are associated with low BMD and greater fracture risk.[99]

Silicon

Silicon is important for collagen and glycosaminoglycan formation in bone and cartilage, influencing the formation of the organic matrix. Silicon is also a major ion of osteogenic cells. Orthosilicic acid, the form of silicon absorbed in the diet, appears to be associated with bone formation through increased synthesis of collagen type I and stimulation of osteoblasts.[87,88] Chicks fed a silicon-deficient diet developed abnormally shaped bones,[100] whereas the addition of silicon to the diet of depleted rats resulted in fewer osteoclasts, increased bone formation, decreased bone turnover, and increased BMD.[101,102]

Few studies have been conducted in humans, but those studies have shown protective effects. In the FOS, dietary silicon was positively associated with BMD at hip sites for men and premenopausal women but not postmenopausal women.[103] French patients with osteoporosis showed significant improvements in trabecular bone volume with silicon treatment,[104] and femoral BMD increased in female osteoporosis patients given intramuscular silicon, compared with others given fluoride, oral magnesium, or controls.[105] These results suggest that higher silicon intake may be protective of BMD. A review including eight studies, four with dietary intake and four with Si supplementation, concluded that an adequate intake of Si for beneficial effects on bone may be around 25 mg/day. However, they note that more trials are needed.[106]

Other Minerals

Copper is a cofactor for lysyl oxidase, which catalyzes cross-linking of lysine and hydroxyproline in collagen. Animals with induced copper deficiency have reduced bone strength[107] and greater bone loss with aging.[108] In women, plasma copper concentration was correlated with lumbar spine BMD,[109] and lower copper status has been noted in elderly patients with fracture, relative to matched controls.[110] A controlled feeding study in men showed increased activity of bone resorption markers when moved from a high (6 mg/day) to low (0.7 mg) copper diet, and this was reversed by returning to the high copper diet.[111] However, another study did not replicate this effect.[112]

Zinc may affect bone through its role in nucleic acid and protein metabolism.[113] Lower serum and bone zinc and higher urinary zinc have been noted in patients with osteoporosis.[114] In animals, zinc increased alkaline phosphatase and DNA synthesis, which may stimulate bone formation.[115] Supplementation with zinc gluconate has been shown to increase alkaline phosphatase activity.[116] In one study, postmenopausal women were randomized to treatment with calcium plus copper and zinc vus calcium plus corn starch. After 2 years, women with usual daily zinc intakes less than 8.0 mg benefited from the copper and zinc supplements, but women consuming adequate amounts of dietary zinc lost more BMD than control women.[117]

Boron intake may protect bone by decreasing urinary calcium, phosphorus, and magnesium losses and increasing serum estradiol.[118] In rats, boron deprivation altered trabecular bone and reduced the force needed to break the femur, confirming the importance of boron to cortical bone strength and trabecular bone microarchitecture.[119] The systemic administration of boric acid reduced alveolar bone loss in periodontal disease in rats.[120] However, no randomized trials of boron supplementation to prevent bone loss or prevent fractures exist.

Manganese also may contribute to bone status, although it has rarely been examined independently from other trace minerals. In rats, supplementation with manganese led to significantly increased BMD of the lumbar vertebrae and increased serum osteocalcin, suggesting that manganese contributes to bone formation.[121] One study randomized older postmenopausal women to daily supplementation with trace minerals (15 mg zinc, 5 mg manganese, 2.5 mg copper), placebo, 1,000 mg elemental calcium, or calcium plus trace minerals for 2 years. Relative to other groups, women receiving calcium plus trace minerals had less bone loss and had increases in spinal BMD.[122]

Vitamins

Vitamin D

Vitamin D promotes positive calcium balance and stimulates bone formation. Hence, it is often considered a protective vitamin for bone. In the skeleton, activated vitamin D [1,25$(OH)_2$D (calcitriol)] stimulates bone resorption and enhances mineralization and bone formation.[44] Vitamin D is obtained from the diet mainly as cholecalciferol (vitamin D_3) from animal sources but also as ergocalciferol (vitamin D_2) from plant sources (notably mushrooms that have been exposed to ultraviolet light). Uniquely, vitamin D_3 is also synthesized cutaneously from 7-dehydrocholesterol with sunlight (ultraviolet B rays, UVB) exposure (see chapter on vitamin D). Because dietary sources are limited, sunlight exposure is important to vitamin D status. Vitamin D insufficiency may be more common than previously appreciated. A review showed that 25(OH)D concentration less than 30 ng/mL (75 nmol/L) was prevalent in every region of the world studied, and concentration less than 10 ng/mL (25 nmol/L) was most common in South Asia and the Middle East. Other factors associated with low vitamin D included obesity, older age, female sex, higher latitude, winter season, darker skin pigmentation, less sunlight exposure, and low dietary intake of vitamin D.[123]

A systematic review[124] concluded that there was good evidence that vitamin D–fortified foods increase serum 25(OH)D in young and older adults and adolescents, that serum 25(OH)D concentration is associated with BMD in older children, and that it is inversely associated with serum PTH. There is fair evidence that serum 25(OH)D is associated with BMD or change in BMD in adolescents and older adults, but inconsistent evidence in relation to fracture. The 2011 Institute of Medicine (IOM, now the National Academy of Medicine (NAM)) report on calcium and vitamin D noted that more recent observational studies are generally consistent with these findings, but clinical trials show mixed results.[44] A 2021 randomized controlled trial, with 379 adults aged ≥70 years in England, found no difference in change in BMD over 1 year, between 3 doses (between 12,000 to 48,000 IU, given once/month) of vitamin D, suggesting no effect of higher doses.[125]

Despite these findings, several studies show that serum vitamin D status is associated with bone health. A case-cohort study found that elderly women with concentration of the active metabolite of vitamin D 1,25(OH_2)D ≤ 23 pg/mL (60 pmol/L) were twice as likely to have a hip fracture over approximately 4 years.[126] In elderly Dutch adults, serum 25(OH)D ≤12 ng/mL (30 nmol/L), vs higher, was associated with three times the risk of fracture over 6 years.[127] Similarly, in the NHANES III follow-up, serum 25(OH)D concentration greater than 24 ng/mL (60 nmol/L), vs lower, was associated with 36% lower risk of hip fracture in elderly adults.[128] In contrast; however, no significant association between serum 25(OH)D and 11-year risk of fracture was seen in postmenopausal women in the Os des Femmes de Lyon study.[129]

Dietary intake of vitamin D may be protective. One large prospective study of 76,507 postmenopausal women found that vitamin D intake (diet plus supplements) greater than 600 IU (15 μg), vs less than 200 IU (5 μg), was associated with 27% lower odds of osteoporosis.[130] An 18-year prospective analysis of 72,337 postmenopausal women found that intake of 500 IU (12.5 μg) or more per day vs less than 140 IU (3.5 μg) of vitamin D was associated with 37% lower risk of hip fracture.[39]

In contrast, 400 IU (10 μg) per day vitamin D_3 supplementation had no effect on fracture in Dutch elders,[131] and other low-dose vitamin D trials also have had negative results.[132,133] Several large clinical trials showed no benefit of vitamin D supplementation even at higher doses (800 to 1,100 IU [20 to 27.5 μg]) on fracture, BMD, bone turnover markers, or intestinal calcium absorption.[134-138]

Meta-analyses of 12 trials on the efficacy of oral supplemental vitamin D in elderly adults for nonvertebral fracture and eight trials for hip fracture concluded that doses greater than 400 IU (10 μg) per day resulted in 18% to 20% reduction in hip and nonvertebral fracture, respectively, independent of additional calcium supplementation.[139] A meta-analysis of six studies in children found a trend toward higher BMD with vitamin D supplementation (132 to 400 IU [0.33 to 10 μg]) per day vs placebo, but this was significant only for those with low baseline vitamin D status.[140] Together, these results suggest that vitamin D supplementation may be effective in individuals with low vitamin D status, but that doses greater than 400 IU (10 μg) may be needed for adults. A threshold concentration of serum 25OHD or dose of vitamin D supplementation that prevents fractures in older adults has not been firmly established, although the IOM (NAM) concluded that, in adults of more than 70 years of age, 1,200 mg of calcium and 800 IU of vitamin D may prevent fractures. However, high doses of vitamin D administered once yearly are not recommended. A randomized, placebo-controlled trial in elderly women at high risk for fracture demonstrated that 500,000 units of oral cholecalciferol administered annually was associated with a 25% greater risk of fracture and a 15% greater likelihood of falling.[141]

Vitamin D Plus Calcium

Because calcium and vitamin D work together, several intervention studies have included both. A study in Greece showed that inclusion of dairy foods fortified with 1,200 mg calcium and 22.5 μg vitamin D per day resulted in significant improvement in BMD, relative to controls.[142] Combined supplementation with vitamin D_3 and calcium significantly reduced 3-year nonvertebral fracture in U.S. elders,[143] and similar results were found in elderly French women,[144] the United Kingdom,[145] and Finland.[146] However, the WHI saw no reduction in fracture risk over 7 years in all postmenopausal women randomized to 1,000 mg of calcium plus 400 IU of vitamin D_3, vs placebo, although, in a subset of women more than 60 years old, calcium and vitamin D were protective against hip fractures.[147] A follow-up after completion of a large, 3-year, placebo-controlled trial of calcium and vitamin D supplementation in older men and women showed that most of the BMD benefits accrued during the trial were lost 2 years after supplementation ended.[148] Most recently, an umbrella review concluded that the combination of calcium and vitamin D reduced the risk of hip fracture in eight of 12 systematic reviews or meta analyses, with most response among institutionalized individuals, while no reduction in fracture risk was seen in trials exclusively with community dwelling individuals or with vitamin D alone.[149] Together, these studies suggest that the benefits of calcium and vitamin D against fracture risk may be dependent on other situational factors.

Vitamin K

Vitamin K may be important for bone health as a coenzyme for glutamate carboxylase, which mediates the conversion of glutamate to γ-carboxyglutamate (Gla).[150] This γ-carboxylation is essential for proteins to attract Ca^{2+} for incorporation into hydroxyapatite crystals. A high concentration of undercarboxylated osteocalcin is an indicator of inadequate vitamin K status.[151]

In food, vitamin K is found as phylloquinone (vitamin K_1) in green leafy vegetables and other plants, and menaquinone (vitamin K_2) in liver, meats, and fermented foods, such as cheese. In several observational studies, low vitamin K intake was associated with low BMD and/or greater fracture.[152,153] Higher concentration of undercarboxylated osteocalcin also has been correlated with lower BMD and greater risk of hip fracture.[154,155] Despite evidence that vitamin K has improved bone turnover profiles and decreased undercarboxylated osteocalcin,[150] many vitamin K supplementation trials have shown no effect on BMD[156,157]; although one, with a menaquinone-4 dose of 45 mg/day, showed reduced bone loss in postmenopausal women.[158] However, a review focusing on vitamin K_2 noted that this form of the vitamin stimulates bone formation by promoting differentiation in osteoblasts and reducing differentiation in osteoblasts[159]; and a recent meta-analysis of K_2 supplementation showed positive effects on BMD and prevention of fracture among postmenopausal women.[160]

Vitamin C

Vitamin C is an essential cofactor for hydroxylating lysine and proline residues in procollagen.[161] Deficiency in animals has been associated with defective tissue collagen synthesis and low BMD.[162] Indeed, ascorbic acid is an essential nutrient for in vitro osteogenesis. Epidemiologic studies of vitamin C in relation to BMD and fracture have shown mixed and sometimes complex results, which may be related to other coexposures. Older adults in the FOS within the highest tertile of total vitamin C intake (median, 313 mg/day) had significantly lower risk of fracture over 15 to 17 years of follow-up than those in the lowest tertile (median, 94 mg/day).[163] The large Swedish Mammography Cohort reported that hip fracture among current smokers with low vitamin C intake was three times greater than for nonsmokers with adequate vitamin C intake; with almost five times greater risk in current smokers with low intakes of both vitamin E and C.[164] The Utah Study of Nutrition and Bone Health showed that up to 488 mg/day of vitamin C intake was protective in older adult former or current smokers, after which the relationship plateaued.[165] The Boston Puerto Rican Health Study found that low plasma vitamin C, but not intake, was associated with lower hip BMD.[166] However, a meta-analysis concluded that higher dietary intake of vitamin C was associated with lower risk of osteoporosis or hip fracture.[167] Together, the evidence suggests that vitamin C may be protective of BMD and against fracture, but these relationships are complex and may be dependent on other factors, including adequacy of other nutrients (calcium and vitamin E), smoking status, and estrogen.

Vitamin E

Vitamin E is a powerful antioxidant and, like vitamin C, may protect against the negative effects of oxidative stress on bone resorption. Bone loss with aging has been associated with prostaglandins, cytokines, and growth factors in the bone microenvironment.[168] Evidence exists that vitamin E can reduce macrophage prostaglandin E_2 production and cyclo-oxygenase activity, which may, in turn, protect bone.[169] Vitamin E has been shown to enhance bone quality in old mice, with greater bone weight, protein and mRNA transcripts for osteocalcin, type I collagen, and insulin-like growth factor-I (IGF-I),[170] and to maintain BMD in ovariectomized rats.[171]

In humans, vitamin E supplements have been associated with lower serum C-telopeptide, a marker of bone resorption.[172] No association was seen between vitamin E intake and BMD in the WHI.[173] Older adult smokers in Utah in the highest vs the lowest quintile of vitamin E intake had an approximately 70% lower likelihood risk of hip fracture, but no association was seen in nonsmokers. Similarly, in Sweden, vitamin E appeared to protect against hip fracture in smokers and as noted for vitamin C, the effect was enhanced when intake of both vitamins was high.[164] A 2014 review noted that supplementation with a-tocopherol may be a "double edged sword," in that plasma status or dietary intake were related with greater BMD, and lower fracture risk, while high-dose supplementation may be detrimental by interfering with vitamin K and blocking other isomers of vitamin E.[174]

B Vitamins

Folate, vitamin B_{12}, and vitamin B_6 play important roles in the one-carbon metabolism pathway, which is critical for DNA synthesis, methylation, and repair, and, therefore, may affect bone formation.[175] Early case reports identified vitamin B_{12} deficiency in patients with osteoporosis, greater fracture risk in patients with pernicious anemia, and improvement with vitamin B_{12} therapy.[176] Vitamin B_{12} affects alkaline phosphatase and osteoblastic activity[177,178]; also, some evidence shows that low vitamin B_{12} and B_6 may stimulate increased osteoclast activity.[179] A recent analysis of femoral heads from patients undergoing hip arthroplasty showed significantly lower trabecular thickness and area associated with less than 7.6 ng/mL (17.3 nmol/L), vs higher serum folate concentration, and significantly lower trabecular number with less than 5.3 ng/mL (21.5 nmol/L), vs higher serum vitamin B_6 concentration.

Several studies have identified relationships between folate status and BMD,[180,181] but others have not.[182] In Denmark, associations were seen at baseline, but not longitudinally.[180] Similarly, some, but not all, studies have shown associations with fracture risk. In Norway,[183] women with folate concentration less than 1.3 vs ≥2.9 ng/mL (2.9 vs 6.6 nmol/L) were 2.4 times more likely to have a hip fracture over 12 years. In Italy, adults with serum folate intake ≤4.1 ng/mL (9.3 nmol/L), vs higher, had double the fracture risk over 4 years.[184] In contrast, another study with almost 1,000 elderly women did not show an association between folate and fracture risk.[185]

Vitamin B_{12} has also been associated with bone status in epidemiologic studies. NHANES III participants with, vs without, osteoporosis had significantly lower serum vitamin B_{12} and higher methylmalonic acid levels.[186] In the FOS, plasma B_{12} less than 200 pg/mL (148 pmol/L), vs higher, was significantly associated with lower BMD at the hip in men and at the spine in women.[187] In Dutch elders, the prevalence of osteoporosis was almost seven times higher in women with vitamin B_{12} less than 285 pg/mL (210 pmol/L) vs greater than 434 pg/mL (320 pmol/L).[188] Longitudinally, elderly U.S. women with vitamin B_{12} concentration ≤280 pg/mL (206 pmol/L), vs higher, had greater decline in total hip BMD over 42 months.[189] In contrast, vitamin B_{12} was not associated with longitudinal change in BMD in perimenopausal women, or with fracture risk in elderly men and women in two European studies.[185]

Fewer studies have been conducted with vitamin B_6. Rotterdam Study participants in the highest quartile of vitamin B_6 intake, vs lower, had higher femoral neck BMD and 23% to 45% lower risk of fracture.[190] In the FOS, 4-year bone loss was inversely associated with vitamin B_6 and hip fracture risk, but these associations were attenuated after control for baseline BMD, serum vitamin D, and homocysteine.[191] In another study, patients with hip fracture had lower vitamin B_6 concentrations and fewer enzymatic cross-links than controls (6.5 vs 12.3 ng/mL [26.4 vs 49.8 nmol/L]), suggesting that vitamin B_6 may operate through different mechanisms than folate and vitamin B_{12}.[192]

All three of these vitamins are required to maintain low concentration of homocysteine, and homocysteine has been associated with impaired collagen cross-linking, reducing bone strength, and increasing susceptibility to fracture.[193] In Finland, women in the highest quartile vs lowest quartile, for serum homocysteine had 2.4 times the likelihood of hip fracture over 12.6 years,[183,194] whereas in another cohort of elderly women, hip bone loss was 2.6 times greater in the highest tertile of homocysteine, vs the lowest, but there was no association with fracture].[195] In contrast, homocysteine was associated with fracture risk but not with 4-year BMD loss in the FOS.[191] A case-control subset of the WHI also found a significant association between homocysteine and hip fracture risk, although it was attenuated with adjustment for kidney function.[196] However, there was no association between homocysteine and fracture in a French cohort of older women.[197]

Despite strong evidence of association, supplementation studies with B vitamins have not been supportive. Two studies, with combination folate, vitamin B_{12}, and vitamin B_6 supplements for 1 and 2 years, respectively, showed no beneficial effect on bone turnover markers.[198,199] A large trial with 5,522 older adults found that supplementation with all these B vitamins showed no association with hip fracture over 5 years.[200] A more recent trial, the Women's Antioxidant and Folic Acid Cardiovascular Study, with 4,810 women, similarly found no association with fracture risk or bone turnover markers.[201] A 2015 meta-analysis, including four randomized controlled trials, also noted no associations with bone turnover or fracture.[202]

Vitamin A

Although vitamin A deficiency is relatively common worldwide, it is retinol excess that is of most concern to bone. Preformed vitamin A is obtained from the diet as retinol and retinyl esters from animal foods and derived from metabolism mainly in the intestine of provitamin A carotenoids (see chapters on vitamin A and carotenoids).

Retinol has been shown to accelerate bone resorption and inhibit bone formation in several animal studies.[177,203] Bone has nuclear receptors for vitamins A and D. The active forms of vitamins A and D may compete at the level of availability of the retinol receptors used for gene regulation by both retinoic acid and vitamin D.[204]

In humans, administration of retinyl palmitate decreased the plasma calcium increase in response to $1,25(OH)_2D$.[205] In observational studies, higher retinol intake was associated with greater bone loss and risk of hip fracture. In Swedish women with vitamin A intakes greater than 1,500 μg/day vs less than 500 μg/day, BMD was 10% to 14% lower at the femoral neck and lumbar spine, respectively, and the risk of hip fracture was doubled.[206] Swedish men with serum retinol greater than 75.6 vs 62.2 to 67.6 μg/dL (>2.7 vs 2.2 to 2.4 μmol/L) were almost 2.5 times as likely to have a hip fracture over 30 years.[207] In the NHS, women with intakes greater than 2,000 vs less than

500 μg/day of retinol were 1.9 times more likely to have a hip fracture over 18 years, whereas carotenoid intake did not contribute to fracture risk.[208]

Two studies showed U-shaped relationships between vitamin A and bone status; one showed lower baseline BMD and greater 4-year BMD loss at both low and high vitamin A intakes,[209] and another showed twice the risk of hip fracture in the lowest or highest quintile of serum retinol vs the middle quintile categories.[210] However, several studies showed no association between vitamin A and bone status, including no association between indicators of vitamin A status and bone health.[211-213] Incident analyses with more than 75,000 postmenopausal women in the WHI did not find overall associations between vitamin A intake and fracture, but women with vitamin D intake of 11 μg/day or less who were in the highest quintile of vitamin A intake did show a significant 20% greater risk of fracture.[214]

A recent review concluded that the negative effect of retinol supplementation on bone was greater among those with obesity or vitamin D deficiency.[215] However, they concluded that adequate vitamin A intake is needed to maintain healthy bones, and that provitamin A carotenoids appear protective. Together, the evidence suggests a possible increased risk of fracture with excess vitamin A. Variations in effects may result from differences in baseline vitamin A status in the study populations, measurement error, lack of power, or limited follow-up. The findings of risk are sufficient to generate concern about overexposure to retinol, which may occur primarily from supplement use.

Provitamin A Carotenoids

Carotenoid-rich foods do not contribute to vitamin A toxicity; rather, carotenoids may have a positive effect on bone through their antioxidant activity. B-Cryptoxanthin has been shown to have an anabolic effect on bone calcification by increasing alkaline phosphatase activity and calcium content in rat femoral tissue and directly stimulating bone formation and inhibiting bone resorption.[216,217]

Several observational studies have reported protective effects of carotenoids. Significant positive associations have been reported for serum β-cryptoxanthin and β-carotene with BMD in Japanese postmenopausal women[218]; dietary β-carotene intake with lumbar spine BMD in Australian postmenopausal women[219]; and lycopene intake with lower 4-year loss in lumbar spine BMD in women; and intakes of total carotenoids with lower 4-year loss in trochanter BMD in men in the FOS.[220] Framingham participants in the highest, vs lowest, tertile of total carotenoid intake had almost 50% lower risk of hip fracture during a 17-year follow-up.[221] A protective association between serum lycopene and oxidative stress and bone resorption markers also was shown in another study of postmenopausal women.[222] In contrast, the WHI reported no protective associations between serum carotenoids and BMD in their large sample of postmenopausal women.[173] A 2023 meta-analysis found that b-carotene was associated with greater BMD and less osteoporosis among Asians, but not in Western groups.[223]

In controlled trials, intake of β-cryptoxanthin–fortified juice led to a significant increase in β-carboxylated osteocalcin concentration and decrease in the bone resorption markers, serum bone tartrate–resistant acid phosphatase activity and N-telopeptide of type I collagen,[224] whereas supplementation with lycopene decreased oxidative stress and N-telopeptide of type I collagen.[225] A 2021 systematic review and meta-analysis concluded that higher intake of β-cryptoxanthin was associated with lower risk of osteoporosis and hip fracture, and that significance remained in subgroup analysis by sex and region of origin.[226] Together with observed protective effects of fruit and vegetable intake on bone (see later), the evidence suggests that carotenoids appear to be protective of bone status.

Protein

Metabolic studies have demonstrated that high protein leads to calcium loss, and it has been assumed that calcium is drawn from the skeleton to maintain serum calcium concentration in the face of an acid load.[227] Conversely, protein has been shown to increase circulating IGF-I, which is associated with bone growth; and this may counteract the effect of the acid load on bone resorption. Furthermore, it has since been shown that protein intake promotes calcium absorption, thereby explaining the increase in calciuria without a negative effect on bone.[228] Studies relating protein intake to bone turnover markers have shown mixed effects: one showed a decrease in deoxypyridinoline, a marker of bone resorption,[229] and another reported no change in bone turnover markers.[230]

Observational studies have found that higher protein intakes were associated with greater BMD and lower risk of fracture. In the FOS, 4-year bone loss was lowest in older adults with the highest protein intakes.[231] Furthermore, those in the highest three quartiles of protein intake had significantly lower risk of hip fracture compared with those with the lowest intake.[232] Similarly, an inverse relation was observed across quartiles of protein intake and risk of hip fracture in Iowan women.[233] In patients with recent hip fracture, protein supplementation for 6 months significantly lowered bone loss from the hip over 1 year.[234] In contrast, the NHS showed no association between protein intake and risk of hip fracture over 12 years.[235] However, a more recent systematic review and meta-analysis of 13 prospective cohort studies, found that high, vs low, protein intakes were significantly associated with 16% lower risk of hip fracture, and noted that protein intake above the current RDA may be beneficial for bone health.[236]

In the context of usual diets, current evidence suggests that higher protein is protective rather than harmful to bone. In French postmenopausal women, increased risk of fracture with higher protein intake was seen only among those with low calcium intake.[237] In a U.S. study, dietary protein appeared most protective against bone loss only among those using calcium supplements.[148] The overall acid–base balance and calcium content of the diet appear to affect the response skeletal to protein, with better results in the presence of alkaline-forming foods and adequate calcium intake.

Fatty Acids

Fatty acids are important to numerous aspects of metabolism. The n-3 and n-6 polyunsaturated fatty acids may influence bone health through several complex mechanisms, including opposing effects on inflammatory cytokines,[238,239] modulation of prostaglandin E2 production,[240,241] and enhancement of calcium transport and retention.[242,243] Essential n-3 and n-6 fatty acids and their derivatives also act as ligands for peroxisome proliferator-activator receptors,[244] which are involved in the differentiation of mesenchymal stem cells to adipocytes or osteoblasts.[245-247] Lipoxins are synthesized from arachidonic acid, and resolvins are synthesized from eicosapentaenoic acid (EPA) and docosahexaenoic acid (DHA).[248,249] Both have anti-inflammatory effects and have been associated with lower bone loss induced by periodontitis in animal models.[250-252]

Several animal studies support a protective association between n-3 fatty acids or higher n-3 to n-6 ratio and bone health. Few studies have been conducted in humans. Increasing ratios of n-6 to n-3 fatty acid intakes were associated with lower hip BMD in the Rancho Bernardo Study,[253] and intake of fish containing n-3 fatty acids was associated with less BMD loss in astronauts.[254] In the FOS, three or more, vs fewer, servings of fatty fish per week was associated with protection against 4-year loss of femoral neck BMD, although interactions between n-3 fatty acids with arachidonic acid were seen. Women with relatively high EPA+DHA intakes had higher baseline BMD when arachidonic acid intakes were also high; whereas men with the lowest EPA+DHA intakes in combination with high arachidonic acid intakes lost more BMD over time than those with low arachidonic acid, suggesting that both types of fatty acids must be adequate for optimal protection of bone.[255] It may be that in the presence of high n-3 intakes, production of prostaglandin E2 is suppressed, thus allowing other positive effects of arachidonic acid. Arachidonic acid can be synthesized from linoleic acid, which itself has been hypothesized to have negative effects on bone in high amounts, because of activation of NF-κβ.[256,257]

Studies of fatty acid intakes and fracture have reported contradictory findings.[258-261] The Age, Gene/Environment Susceptibility Study, with 1,438 participants, reported that the highest tertiles of PUFA, n=3 and eicosatetraenoic acid were associated with lower fracture risk in men, while daily fish oil use was associated with lower fracture in both men and women.[262] Protective effects with supplementation with fish oil or n-3 fatty acids on BMD have been reported in postmenopausal women,[263,264] but another study saw no effect on BMD or bone turnover markers in women given fish oil and calcium vs calcium alone for 12 months.[265] Adults randomly assigned to a diet high in α-linolenic acid vs an average U.S. diet had significantly lower serum N-telopeptide concentration.[266] One review concluded that high saturated fat intake may significantly impair bone regeneration, while n-3 fatty acids were associated with improved bone regeneration, microarchitecture, and strength,[267] and another[268] showed a negative effect of saturated fat on risk of hip fracture.

Other Food Constituents and Dietary Patterns

Caffeine

Some evidence shows that caffeine has a negative effect on bone, although study results vary. BMD was significantly lower in growing rats supplemented with 0.2% caffeine for 20 weeks, compared with controls, and osteoclastogenesis of bone marrow cells isolated from caffeine-treated rats was enhanced.[269] The viability of bone marrow–derived mesenchymal stem cells from rats decreased in a concentration-dependent manner with higher caffeine, with negative effects on osteoblastogenesis.[270] In human osteoblast cells, increasing caffeine incrementally decreased $1,25(OH)_2D_3$–induced VDR expression and alkaline phosphatase activity, affecting osteoblastic function.[271] A 2022 review concluded that although caffeine may affect bone through antagonism of adenosine receptors, the evidence of caffeine on bone is mixed, with some evidence of a modest adverse effect.[272]

In humans, caffeine is known to increase short-term urinary calcium excretion.[273] However, a 2002 review concluded that caffeine had no effect on total 24-hour urinary calcium excretion, and the negative effect of caffeine on calcium absorption may be offset by "as little as 1 to 2 tablespoons of milk".[274] Consistently, observational studies of postmenopausal women found that consumption of two or more cups of coffee per day was associated with lower BMD only in those who did not regularly drink milk,[275] and that bone loss was higher with two or more cups of coffee per day only when calcium intakes were less than 800 mg/day.[273] Similarly, in a cohort of 31,527 Swedish women, intake of four or more cups of coffee daily was associated modestly with increased fracture risk, but mainly in women with low calcium intake.[276] In Framingham, caffeine intake was associated with risk of hip fracture in elderly women but not in men.[277] In contrast, however, elderly Swedish men, but not women, consuming four or more cups of coffee per day had significantly lower hip BMD. This association was modified by genotypes for cytochrome P-450 1A2 (CYP1A2), associated with metabolism of caffeine, but not with calcium intake.[278]

Alcohol

Osteoporosis is commonly seen in chronic alcoholism.[279] Heavy use of alcohol is associated with multiple nutritional deficiencies which, as described throughout this chapter, are likely to have their own negative effects on bone. In addition, ethanol itself appears to have direct effects on bone remodeling, affecting both BMD and bone strength. Long-term administration (3 months) of alcohol at a dose roughly equivalent to 1 L of wine per day in male adult rats showed a 10% reduction in bone density and a 12% reduction of mechanical strength of the femur.[280] A review in 2012 concluded that the decrease in bone mass and strength seen with heavy alcohol consumption is due to a decrease in bone formation, with evidence of osteocyte apoptosis, oxidative stress, and Wnt signaling pathway modulation.[281]

In contrast, moderate alcohol use has been associated with higher BMD in several studies of older adults[282,283] and with lower bone loss over time.[284] In the FOS, hip BMD was 3% to 5% greater in men consuming one to two drinks per day of total alcohol or beer, and hip and spine BMD were 5% to 8% greater in postmenopausal women consuming more than two drinks per day of total alcohol or wine. However, in men, more than two drinks per day of distilled liquor were associated with significantly lower BMD. The tendency toward stronger associations between BMD and beer or wine, relative to liquor, suggests that constituents other than ethanol may contribute to bone health.[285] Together, the evidence is strong to support beneficial effects of moderate alcohol consumption but harmful effects of heavy alcohol consumption on bone status. Mechanisms for protective effects could include the estrogenic effects of ethanol, which stimulates the conversion of androstenedione to estrone. Additional compounds, such as silicon or polyphenols specifically in beer and wine, may offer additional benefit.

Although beneficial for BMD, less is known about moderate alcohol use and fracture. A meta-analysis of three cohort studies from Canada, Australia, and the Netherlands found that alcohol intake of more than two drinks per day vs less was associated with almost 70% greater hip fracture risk.[286] Another study found a U-shaped association, in which hip fracture was lowest among drinkers consuming up to two drinks per day relative to either abstainers or heavier drinkers.[287] A recent review of 11 studies on alcohol and BMD, and eight on fracture, concluded that consumption of up to two drinks/day was correlated with higher BMD, while higher risk of hip fracture was seen with three or more drinks/day.[288]

Dietary Patterns

It is increasingly clear that bone is sensitive to a wide variety of nutrients and exposures. Although studies usually have examined nutrients one at a time to understand associations and mechanisms, the final effect depends on many together. Therefore, it is useful to look at the total dietary pattern to see if combinations of these nutrients in the diet have an effect that may be greater than that of any single nutrient. Because so many food constituents are likely to affect bone, including an array of phytonutrients not yet fully explored, it may be expected that healthy diets may provide better overall support for bone growth, repair, and maintenance than any single nutrient or even food product.

In the FOS, individuals were divided into dietary pattern groups using cluster analysis.[289] Men consuming a diet high in fruits, vegetables, and breakfast cereals had significantly greater hip BMD than all other diet pattern groups. Women consuming the candy pattern had the lowest BMD, whereas an alcohol pattern (moderate intake), in addition to the fruit and vegetable pattern, had the highest BMD. Dairy and meat groups tended to have intermediate BMD. Consistent with the evidence for individual nutrients, the fruit and vegetable group had the highest intakes of magnesium, potassium, vitamin C, and vitamin K. Similarly, in Scottish women 50 to 59 years old, a "healthy" pattern was associated with lower bone resorption, whereas a pattern characterized by snack foods was associated with lower femoral neck BMD.[290] A U.K. study of postmenopausal twin pairs also showed a positive association between hip BMD and a pattern containing wine but a negative association with a traditional 20th-century English diet (fried fish, fried potatoes, baked beans, red and processed meats, savory pies, and cruciferous vegetables).[291] Another U.K. study found significant associations between a pattern high in fruit intake and spine bone mineral content (BMC) in adolescents and older women and femoral neck BMC in adolescent boys.[292] In Australian women 18 to 65 years old, a dietary pattern consisting of refined cereals, soft drinks, fried potatoes, sausages and processed meat, vegetable oils, beer, and takeout foods was significantly associated with lower total body BMC relative to other patterns. In contrast, a pattern consisting of legumes, seafood, seeds, nuts, wine, rice and rice dishes, other vegetables, and vegetable dishes was positively associated with BMD at the hip and spine and with total body BMC.[293] An analysis of the NHANES 1999 to 2002 data found no association between the 2005 Healthy Eating Index score and bone turnover markers. However, milk intake was significantly negatively associated with bone resorption, assessed as the urinary N-telopeptide-to-creatinine ratio, and women with the highest added sugars intake had the highest serum bone–specific alkaline phosphatase, associated with higher bone turnover, where resorption exceeds formation.[294]

A Canadian cohort study showed that a baseline dietary pattern characterized by fruits, vegetables, and whole grains was associated with 14% lower risk of fracture in women, with a similar pattern that did not reach significance in men.[295] Plant-based dietary patterns have been increasingly recommended for health and ecological sustainability. However, a recent NHANES analysis comparing three plant-based dietary indices and bone loss[296] found that both the overall and healthy plant-based diets were associated with greater osteopenia, and that the unhealthy plant-based diet was associated with greater risk of osteoporosis, concluding that a plant-based diet may have negative effects on bone. This is consistent with the conclusions of a 2014 review of vegetarian diets and bone, suggesting that vegetarians, and particularly vegans, may be at greater risk of lower BMD and fracture.[297]

Dietary pattern intervention also has been shown to be effective. The DASH diet (high in fruit, vegetables, and low-fat dairy products) was shown to significantly reduce bone turnover. The addition of low sodium intake reduced calcium excretion in both the DASH and usual diet groups and further reduced serum osteocalcin in the DASH group.[77] Another study compared the effects on bone turnover with 14 weeks of a low-sodium DASH-type diet vs a high-carbohydrate low-fat diet (both with >800 mg of calcium per day). The DASH group showed significant decreases in urinary sodium and calcium, an increase in urinary potassium, and lower bone turnover markers relative to the high-carbohydrate, low-fat diet group.[298]

Together, the evidence supports the role of an overall healthy diet, beyond adequate calcium intake, for optimal bone health. Pattern analysis is consistent with nutrient analyses showing that diets rich in fruits, vegetables, and low-fat dairy products; moderate in alcohol; and low in sodium appear to be important.

Other Risk Factors for Osteoporotic Fracture

Body Weight and Body Composition

Numerous studies show that higher total body weight is directly associated with greater BMD and lower risk of fracture. Furthermore, weight loss is associated with loss in BMD. One review suggested that each 10% weight loss may lead to 1% to 2% bone loss.[299] However, total weight consists of fat-free mass and fat tissue, and it is likely that these components may have differing effects on bone. With the recognition of fat as an endocrine organ, the effect of fat mass on bone may extend beyond its mechanical load on the skeleton.

Abdominal fat mass is more metabolically active, and produces hormones, chemokines, and cytokines that may contribute to inflammation[300] and insulin resistance,[301] both of which may negatively affect bone and increase risk of fracture.[302,303] A few studies[304,305] have shown that when the mechanical loading effect of body weight was statistically removed, fat mass was negatively associated with bone. In Dutch men and women, most initial relationships between measures of fat distribution and BMD were positive but became negative after BMI adjustment. However, insulin or adiponectin concentrations did not explain either the positive or negative associations.[306] Other studies noted that visceral, but not subcutaneous, fat had negative associations with femur strength in young women, hip and spine BMD in obese adolescent girls, and BMD in Korean men and women.[307-310] In Puerto Rican adults in Boston, the likelihood of osteopenia or osteoporosis at the hip increased by 10% to 16% for every 100 g increase in body weight–adjusted abdominal fat mass.[311] Although total body weight improves BMD, central obesity may have negative effects on bone status.

Importantly, weight loss may be associated with bone loss unless balanced with physical activity. A 2015 systematic review concluded that there is modest hip bone loss with diet-induced weight loss.[312] Most recently, the effects of weight loss on bone due to bariatric surgery have been examined. One study found that adolescents with sleeve gastrectomy had significant 2 y loss in bone density and strength, relative to a control group.[313]

An editorial emphasizes the importance of this issue, given the increasing frequency of bariatric surgery among youth.[314]

Physical Activity

In addition to the gravitational effect of body weight, a clear association exists between weight-bearing exercise and BMD.[315] Improvements in bone mass also have been shown with resistance exercise.[316] Strength and balance exercises improve muscle function and reduce falls, further reducing the risk of fracture.[317] Increased aerobic exercise during weight loss is important to mitigate the bone loss otherwise seen with weight reduction.[318] A genetically controlled study with twins showed that long-term physical activity during adulthood led to thicker cortices in long bone shafts, higher bending strength, and higher trabecular density and compressive strength in distal bones.[319] One review concluded that high-force (weightlifting) or high-impact (jumping) exercises have the greatest effect on bone, whereas non–weight-bearing exercises, such as swimming, do not affect bone.[315]

Genetics

Although bone responds to many environmental exposures, investigators have estimated that approximately 80% of the variance in peak bone mass, and a smaller proportion of the variance in bone loss with aging, may result from heritability.[320] The identification of important genetic variants and of gene–environment interactions is an active area of research. A family history of hip fracture may be predictive of a twofold increased risk of fracture.[321] Studies of candidate gene polymorphisms and genome-wide association studies have identified genes associated with BMD or fragility fracture, including genes coding for the low-density lipoprotein receptor–related protein 5 (LRP5), estrogen receptor alpha (ESR1), and OPG.[321] A large quantitative trait analysis resulted in confirmation for 74 SNPs at 32 loci in replication sets of Icelandic, Danish, and Australian adults, with three regions near genes previously identified: the *RANKL* gene, OPG, and ESR1, and two additional regions. Several loci were also associated with osteoporotic fractures, including loci close to the *RANK* gene.[322] A review noted that most of the known genes involved in osteoporosis encode components of pathways involved in bone synthesis or resorption, but that only a small proportion of the total genetic variation involved in osteoporosis has been identified.[323]

One of the earliest genetic markers examined in relation to a dietary interaction was that for the vitamin D receptor, which has been associated with BMD in several populations.[324] Women with the low BMD vitamin D receptor genotype had more rapid bone loss and failed to increase calcium absorption in response to lower calcium intake.[325,326] In another example, Greek postmenopausal carriers of the A allele vs the GG homozygotes of the rs4988321 polymorphism in the LRP5 had lower spine BMD only when calcium intake was less than 680 mg/day, a finding suggesting that adequate calcium intake may eliminate the negative effect of this polymorphism on BMD.[327] Other reported gene–diet interactions affecting bone include observations that a high-fat diet may be detrimental or beneficial to bone mass, depending on the presence of specific allelic variants in the *PPARG* gene, as shown in both mice and humans[328]; evidence that *Alox5*$^{-/-}$ mice placed on a high-fat diet gain more fat mass and lose more bone mass, relative to wild-type controls[329]; and evidence that high saturated fat intake in women with the APOE –219T/T and +113C/C genotypes is detrimental to bone status.[330]

Strategies for the Prevention and Management of Osteoporosis

An extensive body of research has identified numerous modifiable risk factors for osteoporosis and fracture. It is now clear that many aspects of diet and nutrition are important for bone health, including not only adequate intake of calcium and vitamin D but also of magnesium, potassium, other trace minerals, vitamin K, B vitamins, carotenoids, vitamins C and E, protein, and essential fatty acids. A dietary pattern that promotes acid-base balance is also important for bone and muscle. This pattern includes a diet rich in fruits and vegetables, as seen in the Mediterranean Diet and the Healthy Eating Index. At the same time, it is important to avoid excess intakes of phosphorus, vitamin A, and sodium, and to maintain moderate alcohol intake. Although osteoporosis is one area in which BMI has been shown to be protective, we now understand that abdominal adiposity can have negative effects on bone through the release of certain cytokines. Weight-bearing physical activity is protective and is particularly important during weight loss, which may otherwise lead to bone loss as well. Muscle-strengthening exercises also can help to protect against weight loss–associated bone loss and help to strengthen muscles to reduce the risk of falling. Increased muscle mass also may contribute to greater bone strength among individuals across all age groups. Finally, we are in the early stages of understanding gene–nutrition interactions and expect that in the future, more personalized advice may be given based on genetic profiles for osteoporosis prevention.

With respect to management of osteoporosis, particularly for individuals who continue to be at high risk based on family history, age, previous fractures, or low bone mass, dietary recommendations for calcium and vitamin D remain an important cornerstone of the treatment approach, which often includes pharmacologic agents such as bisphosphonates. In the pivotal (registration) trials that established the anti-fracture efficacy of osteoporosis therapies, including bisphosphonates,[331-334] raloxifene,[333] denosumab,[336] romozoumab,[337] parathyroid peptides,[338,339] and hormone replacement therapy, supplemental calcium in amounts of 500 to 1,500 mg and vitamin D_3 in amounts of 250 to 1,200 IU per day were part of the base therapy (given to both intervention and control groups). The anti-fracture efficacy of these drugs in patients with inadequate intakes of vitamin D and/or calcium has not been established. Most societies and advocacy groups recommend a minimum of 1,200 mg of calcium per day either from the diet or with supplements, and at least 600 to 800 IU of vitamin D per day, both of which are in line with the IOM recommendations for the general population.[2,45,340-342]

Dietary calcium is the preferred form for obtaining required calcium, with supplements used only as needed to achieve the target intake of 1,200 mg per day. Data from the WHI indicate that adding 1,200 mg per day of supplemental calcium in women with an average calcium intake of 1,100 mg/d resulted in a 16% increase in risk of nephrolithiasis.[343] Serum 25(OH)D concentrations in patients with osteoporosis should be maintained at or above 20 ng/mL (50 nmol/L). This usually can be achieved by adequate sunlight exposure and careful attention to diet, but frequently requires addition of a supplement containing 200 to 400 IU of vitamin D_3. In patients with malabsorption, gluten enteropathy, liver disease, gastric or intestinal bypass, or patients taking long-term antiseizure medications, 1,000 to 2,000 IU/day or more of vitamin D may be required to maintain serum 25(OH)D concentration higher than 20 ng/mL. In summary, calcium and vitamin D are essential components of any treatment regimen for osteoporosis. However, research suggests that many aspects of the diet and numerous nutrients contribute to bone health, emphasizing the importance of focusing on overall dietary quality in addition to medical approaches. Combining supplementation of vitamin D and calcium, as needed, with lifestyle changes and in some cases, drugs, to prevent further bone loss can reduce the risk of devastating osteoporotic fractures.

Acknowledgments

The authors would like to acknowledge Clifford Rosen for his contributions to the previous version of this chapter.

References

1. Burge R, Dawson-Hughes B, Solomon DH, et al. J Bone Miner Res 2007;22:465–75.
2. *Bone Health and Osteoporosis: A Report of the Surgeon General*, Rockville, MD: 2004.
3. Looker AC, Melton LJ, 3rd, Borrud LG, et al. Osteoporos Int 2012;23:771–80.
4. Looker AC, Melton LJ, 3rd, Borrud LG, et al. Osteoporos Int 2012;23:1351–60.
5. Noel SE, Mangano KM, Griffith JL, et al. J Bone Miner Res 2018;33:396–403.
6. Becker DJ, Kilgore ML, Morrisey MA. Curr Rheumatol Rep 2010;12:186–91.
7. Dempster DW. Am J Manag Care 2011;17(suppl 6):S164–9.
8. Dempster DW, Shane E, Horbert W, et al. J Bone Miner Res 1986;1:15–21.
9. Heaney RP, Abrams S, Dawson-Hughes B, et al. Osteoporos Int 2000;11:985–1009.
10. Ensrud KE, Palermo L, Black DM, et al. J Bone Miner Res 1995;10:1778–87.
11. Jones G, Nguyen T, Sambrook P, et al. BMJ 1994;309:691–5.
12. Jain RK, Vokes T. J Clin Densitom 2017;20:291–303.
13. Palomo T, Muszkat P, Weiler FG, et al. Arch Endocrinol Metab 2022;66:694–706.
14. Keaveny TM, Clarke BL, Cosman F, et al. Osteoporos Int 2020;31:1025–48.
15. Hans D, Baim S. J Clin Densitom 2017;20:322–33.
16. Martel D, Monga A, Chang G. Radiol Clin North Am 2022;60: 537–45.
17. Kanis JA, McCloskey EV, Johansson H, et al. Bone 2008;42: 467–75.
18. Wu Q, Xiao X, Xu Y. J Clin Med 2020;9.
19. Johnell O, Kanis JA, Oden A, et al. J Bone Miner Res 2005;20: 1185–94.
20. Cooper C, Atkinson EJ, O'Fallon WM, et al. J Bone Miner Res 1992;7:221–7.
21. Grisso JA, Kelsey JL, Strom BL, et al. N Engl J Med 1991;324: 1326–31.
22. Hannan MT, Felson DT, Dawson-Hughes B, et al. J Bone Miner Res 2000;15:710–20.
23. Burger H, de Laet CE, van Daele PL, et al. Am J Epidemiol 1998;147:871–9.
24. Cummings SR, Nevitt MC, Browner WS, et al. N Engl J Med 1995;332:767–73.
25. Nethander M, Coward E, Reimann E, et al. Cell Rep Med 2022;3:100776.
26. World Health Organization. Fracture Risk Assessment Tool (FRAX). World Health Organization Collaborating Centre for Metabolic Bone Diseases. United Kingdom: University of Sheffield. Accessed July 13, 2023. Available at www.shef.ac.uk /FRAX/
27. Kanis JA, McCloskey EV, Johansson H, et al. Osteoporos Int 2010;21(suppl 2):S407–13.
28. Kanis JA, Hans D, Cooper C, et al. Osteoporos Int 2011;22: 2395–411.
29. Kanis JA, Harvey NC, Johansson H, et al. Aging Clin Exp Res 2020;32:187–96.
30. Tucker KL, Bhupathiraju SN. In: Cho KH, Michel JP, and Buldau J et al, eds. Micronutrients and Bone. Chapter 18. Textbook of Geriatric Medicine. Seoul, Korea: Argos, 2010.
31. Tucker KL. Curr Osteoporos Rep 2009;7:111–7.
32. Dietary Reference Intakes for Calcium, Phosphorus, Magnesium, Vitamin D, and Fluoride. Washington, DC: 1997.
33. Wosje KS, Specker BL. Nutr Rev 2000;58:253–68.
34. Winzenberg T, Shaw K, Fryer J, et al. BMJ 2006;333:775.
35. Heaney RP. J Am Coll Nutr 2000;19:83S–99S.
36. Zhao JG, Zeng XT, Wang J, et al. JAMA 2017;318:2466–82.
37. Kalkwarf HJ, Khoury JC, Lanphear BP. Am J Clin Nutr 2003;77: 257–65.
38. Key TJ, Appleby PN, Spencer EA, et al. Public Health Nutr 2007; 10:1314–20.
39. Feskanich D, Willett WC, Colditz GA. Am J Clin Nutr 2003;77: 504–11.
40. Baran D, Sorensen A, Grimes J, et al. J Clin Endocrinol Metab 1990;70:264–70.
41. Heaney RP, Rafferty K, Dowell MS. J Am Diet Assoc 2002;102: 1672–4.
42. Rizzoli R. Aging Clin Exp Res 2022;34:9–24.
43. Lotz M, Zisman E, Bartter FC. N Engl J Med 1968;278:409–15.
44. Ross AC, Taylor CL, Yaktine AL, et al eds. Dietary Reference Intakes for Calcium and Vitamin D. Washington, DC: 2011.
45. U.S. Department of Agriculture. What We Eat in America, NHANES 2007–2008. Nutrient Intakes from Food: Mean Amounts Consumed per Individual by Race/Ethnicity and Age, in the United States, 2007–2008. 2010. Accessed July 13, 2023. Available at: www.ars.usda.gov/ba/bhnrc/fsrg
46. Clark I. Am J Physiol 1969;217:865–70.
47. Kemi VE, Karkkainen MU, Rita HJ, et al. Br J Nutr 2010;103: 561–8
48. Kemi VE, Karkkainen MU, Karp HJ, et al. Br J Nutr 2008;99: 832–9.
49. Takeda E, Yamamoto H, Yamanaka-Okumura H, et al. Adv Nutr 2014;5:92–7.
50. Wyshak G, Frisch RE. J Adolesc Health 1994;15:210–5.
51. Tucker KL, Morita K, Qiao N, et al. Am J Clin Nutr 2006;84: 936–42.
52. Heaney RP, Rafferty K. Am J Clin Nutr 2001;74:343–7.
53. Vorland CJ, Stremke ER, Moorthi RN, et al. Curr Osteoporos Rep 2017;15:473–82.
54. Li M, Hasegawa T, Masuki H, et al. J Oral Biosci 2010;52:94–9.
55. Rude RK, Gruber HE, Norton HJ, et al. Bone 2005;37:211–9.

56. Rude RK, Gruber HE, Wei LY, et al. Nutr Metab (Lond) 2005; 2:24.
57. Rude RK, Kirchen ME, Gruber HE, et al. Miner Electrolyte Metab 1998;24:314–20.
58. Creedon A, Flynn A, Cashman K. Br J Nutr 1999;82:63–71.
59. Mutlu M, Argun M, Kilic E, et al. J Int Med Res 2007;35: 692–5.
60. New SA, Robins SP, Campbell MK, et al. Am J Clin Nutr 2000;71: 142–51.
61. Ryder KM, Shorr RI, Bush AJ, et al. J Am Geriatr Soc 2005;53: 1875–80.
62. Tucker KL, Hannan MT, Chen H, et al. Am J Clin Nutr 1999;69: 727–36.
63. Ford ES, Mokdad AH. J Nutr 2003;133:2879–82.
64. Carpenter TO, DeLucia MC, Zhang JH, et al. J Clin Endocrinol Metab 2006;91:4866–72.
65. Dimai HP, Porta S, Wirnsberger G, et al. J Clin Endocrinol Metab 1998;83:2742–8.
66. Stendig-Lindberg G, Tepper R, Leichter I. Magnes Res 1993;6: 155–63.
67. Rondanelli M, Faliva MA, Tartara A, et al. Biometals 2021;34: 715–36.
68. Sebastian A, Harris ST, Ottaway JH, et al. N Engl J Med 1994;330:1776–81.
69. Macdonald HM, New SA, Fraser WD, et al. Am J Clin Nutr 2005;81:923–33.
70. Zhu K, Devine A, Prince RL. Osteoporos Int 2009;20:335–40.
71. Ha J, Kim SA, Lim K, et al. Nutr Res Pract 2020;14:55–61.
72. Choi SS, Kim YA, Kim HJ, et al. Korean J Fam Med 2023;44: 21–7.
73. Lanham-New SA. J Nutr 2008;138:172S–7S.
74. In: Henney JE, Taylor CL, and Boon CS, eds. Strategies to Reduce Sodium Intake in the United States. Washington, DC: 2010.
75. Devine A, Criddle RA, Dick IM, et al. Am J Clin Nutr 1995;62: 740–5.
76. Sellmeyer DE, Schloetter M, Sebastian A. J Clin Endocrinol Metab 2002;87:2008–12.
77. Lin PH, Ginty F, Appel LJ, et al. J Nutr 2003;133:3130–6.
78. Carbone LD, Barrow KD, Bush AJ, et al. J Bone Miner Metab 2005;23:506–13.
79. Carbone LD, Bush AJ, Barrow KD, et al. J Bone Miner Metab 2003;21:415–20.
80. Greendale GA, Barrett-Connor E, Edelstein S, et al. J Am Geriatr Soc 1994;42:1050–5.
81. Ilich JZ, Brownbill RA, Coster DC. Eur J Appl Physiol 2010;109:745–55.
82. Grynpas MD, Chachra D, Limeback H. In: Henderson JE, and Goltzman D, eds. The Action of Fluoride on Bone. Chapter 23. The Osteoporosis Primer New York, NY: Cambridge University Press, 2000.
83. Pak CY, Sakhaee K, Adams-Huet B, et al. Ann Intern Med 1995;123:401–8.
84. Unfer TC, Muller EI, de Moraes Flores EM, et al. Clin Chim Acta 2007;384:113–7.
85. Kleerekoper M, Peterson EL, Nelson DA, et al. Osteoporos Int 1991;1:155–61.
86. Riggs BL, Hodgson SF, O'Fallon WM, et al. N Engl J Med 1990;322:802–9.
87. Reffitt DM, Ogston N, Jugdaohsingh R, et al. Bone 2003;32: 127–35.
88. Jugdaohsingh R. J Nutr Health Aging 2007;11:99–110.
89. Vestergaard P, Jorgensen NR, Schwarz P, et al. Osteoporos Int 2008;19:257–68.
90. Chachra D, Limeback H, Willett TL, et al. J Dent Res 2010;89:1219–23.
91. Helte E, Donat Vargas C, Kippler M, et al. Environ Health Perspect 2021;129:47005.
92. Everett ET. J Dent Res 2011;90:552–60.
93. Schnitzler CM, Macphail AP, Shires R, et al. J Bone Miner Res 1994;9:1865–73.
94. Guggenbuhl P, Deugnier Y, Boisdet JF, et al. Osteoporos Int 2005;16:1809–14.
95. Katsumata S, Tsuboi R, Uehara M, et al. Biosci Biotechnol Biochem 2006;70:2547–50.
96. Medeiros DM, Plattner A, Jennings D, et al. J Nutr 2002;132: 3135–41.
97. Harris MM, Houtkooper LB, Stanford VA, et al. J Nutr 2003;133: 3598–602.
98. Maurer J, Harris MM, Stanford VA, et al. J Nutr 2005;135: 863–9.
99. Ledesma-Colunga MG, Weidner H, Vujic Spasic M, et al. Semin Hematol 2021;58:188–200.
100. Carlisle EM. Ciba Found Symp 1986;121:123–39.
101. Rico H, Gallego-Lago JL, Hernandez ER, et al. Calcif Tissue Int 2000;66:53–5.
102. Hott M, de Pollak C, Modrowski D, et al. Calcif Tissue Int 1993;53:174–9.
103. Jugdaohsingh R, Tucker KL, Qiao N, et al. J Bone Miner Res 2004;19:297–307.
104. Schiano A, Eisinger F, Detolle P, et al. Rev Rhum Mal Osteoartic 1979;46:483–6.
105. Eisinger J, Clairet D. Magnes Res 1993;6:247–9.
106. Rondanelli M, Faliva MA, Peroni G, et al. Exp Biol Med (Maywood) 2021;246:1500–11.
107. Jonas J, Burns J, Abel EW, et al. Ann Nutr Metab 1993;37: 245–52.
108. Rico H, Roca-Botran C, Hernandez ER, et al. Menopause 2000;7:413–6.
109. Chaudhri MA, Kemmler W, Harsch I, et al. Biol Trace Elem Res 2009;129:94–8.
110. Conlan D, Korula R, Tallentire D. Age Ageing 1990;19:212–4.
111. Baker A, Harvey L, Majask-Newman G, et al. Eur J Clin Nutr 1999;53:408–12.
112. Cashman KD, Baker A, Ginty F, et al. Eur J Clin Nutr 2001;55:525–31.
113. Beattie JH, Avenell A. Nutr Res Rev 1992;5:167–88.
114. Atik OS. J Am Geriatr Soc 1983;31:790–1.
115. Yamaguchi M, Yamaguchi R. Biochem Pharmacol 1986;35: 773–7.
116. Peretz A, Papadopoulos T, Willems D, et al. J Trace Elem Med Biol 2001;15:175–8.
117. Nielsen FH, Lukaski HC, Johnson LK, et al. Br J Nutr 2011;106: 1872–9.
118. Nielsen FH, Hunt CD, Mullen LM, et al. FASEB J 1987;1: 394–7.
119. Nielsen FH, Stoecker BJ. J Trace Elem Med Biol 2009;23: 195–203.
120. Demirer S, Kara MI, Erciyas K, et al. Arch Oral Biol 2012;57: 60–5.
121. Bae YJ, Kim MH. Biol Trace Elem Res 2008;124:28–34.
122. Strause L, Saltman P, Smith KT, et al. J Nutr 1994;124:1060–4.
123. Mithal A, Wahl DA, Bonjour JP, et al. Osteoporos Int 2009;20: 1807–20.
124. Cranney A, Horsley T, O'Donnell S, et al. Evid Rep Technol Assess (Full Rep) 2007;1–235.
125. Aspray TJ, Chadwick T, Francis RM, et al. Am J Clin Nutr 2019;109:207–17.
126. Cummings SR, Browner WS, Bauer D, et al. N Engl J Med 1998;339:733–8.
127. van Schoor NM, Visser M, Pluijm SM, et al. Bone 2008;42: 260–6.

128. Looker AC, Mussolino ME. J Bone Miner Res 2008;23:143–50.
129. Garnero P, Munoz F, Sornay-Rendu E, et al. Bone 2007;40: 716–22.
130. Nieves JW, Barrett-Connor E, Siris ES, et al. Osteoporos Int 2008;19:673–9.
131. Lips P, Graafmans WC, Ooms ME, et al. Ann Intern Med 1996;124:400–6.
132. Komulainen MH, Kroger H, Tuppurainen MT, et al. Maturitas 1998;31:45–54.
133. Meyer HE, Smedshaug GB, Kvaavik E, et al. J Bone Miner Res 2002;17:709–15.
134. Zhu K, Bruce D, Austin N, et al. J Bone Miner Res 2008;23: 1343–8.
135. Law M, Withers H, Morris J, et al. Age Ageing 2006;35:482–6.
136. Lyons RA, Johansen A, Brophy S, et al. Osteoporos Int 2007;18: 811–8.
137. Porthouse J, Cockayne S, King C, et al. BMJ 2005;330:1003.
138. Grant AM, Avenell A, Campbell MK, et al. Lancet 2005;365: 1621–8.
139. Bischoff-Ferrari HA, Willett WC, Wong JB, et al. Arch Intern Med 2009;169:551–61.
140. Winzenberg T, Powell S, Shaw KA, et al. BMJ 2011;342:c7254.
141. Sanders KM, Stuart AL, Williamson EJ, et al. JAMA 2010;303: 1815–22.
142. Moschonis G, Katsaroli I, Lyritis GP, et al. Br J Nutr 2010;104: 100–7.
143. Dawson-Hughes B, Harris SS, Krall EA, et al. N Engl J Med 1997;337:670–6.
144. Chapuy MC, Pamphile R, Paris E, et al. Osteoporos Int 2002;13:257–64.
145. Trivedi DP, Doll R, Khaw KT. BMJ 2003;326:469.
146. Heikinheimo RJ, Inkovaara JA, Harju EJ, et al. Calcif Tissue Int 1992;51:105–10.
147. Jackson RD, LaCroix AZ, Gass M, et al. N Engl J Med 2006;354:669–83.
148. Dawson-Hughes B, Harris SS, Krall EA, et al. Am J Clin Nutr 2000;72:745–50.
149. Chakhtoura M, Bacha DS, Gharios C, et al. J Clin Endocrinol Metab 2022;107:882–98.
150. Bugel S. Vitam Horm 2008;78:393–416.
151. Sokoll LJ, Sadowski JA. Am J Clin Nutr 1996;63:566–73.
152. Feskanich D, Weber P, Willett WC, et al. Am J Clin Nutr 1999;69:74–9.
153. Booth SL, Broe KE, Gagnon DR, et al. Am J Clin Nutr 2003;77:512–6.
154. Luukinen H, Kakonen SM, Pettersson K, et al. J Bone Miner Res 2000;15:2473–8.
155. Szulc P, Chapuy MC, Meunier PJ, et al. Bone 1996;18:487–8.
156. Emaus N, Gjesdal CG, Almas B, et al. Osteoporos Int 2010;21: 1731–40.
157. Cashman KD, O'Connor E. Nutr Rev 2008;66:532–8.
158. Shiraki M, Shiraki Y, Aoki C, et al. J Bone Miner Res 2000;15: 515–21.
159. Villa JKD, Diaz MAN, Pizziolo VR, et al. Crit Rev Food Sci Nutr 2017;57:3959–70.
160. Ma ML, Ma ZJ, He YL, et al. Front Public Health 2022;10: 979649.
161. Peterkofsky B. Am J Clin Nutr 1991;54:1135S–40S.
162. Kipp DE, McElvain M, Kimmel DB, et al. Bone 1996;18: 281–8.
163. Sahni S, Hannan MT, Gagnon D, et al. Osteoporos Int 2009;20:1853–61.
164. Melhus H, Michaelsson K, Holmberg L, et al. J Bone Miner Res 1999;14:129–35.
165. Zhang J, Munger RG, West NA, et al. Am J Epidemiol 2006;163:9–17.
166. Mangano KM, Noel SE, Dawson-Hughes B, et al. J Nutr 2021;151:3764–72.
167. Malmir H, Shab-Bidar S, Djafarian K. Br J Nutr 2018;119: 847–58.
168. Raisz LG. J Bone Miner Res 1993;8(suppl 2):S457–S65.
169. Wu D, Mura C, Beharka AA, et al. Am J Physiol 1998;275: C661–8.
170. Arjmandi B, Juma S, Beharka A, et al. J Nutr Biochem 2002; 13:543.
171. Norazlina M, Ima-Nirwana S, Gapor MT, et al. Exp Clin Endocrinol Diabetes 2000;108:305–10.
172. Pasco JA, Henry MJ, Wilkinson LK, et al. J Womens Health (Larchmt) 2006;15:295–300.
173. Wolf RL, Cauley JA, Pettinger M, et al. Am J Clin Nutr 2005;82:581–8.
174. Chin KY, Ima-Nirwana S. Nutrients 2014;6:1424–41.
175. Lucock M. Mol Genet Metab 2000;71:121–38.
176. Eastell R, Vieira NE, Yergey AL, et al. Clin Sci (Lond) 1992;82:681–5.
177. Kim GS, Kim CH, Park JY, et al. Metab 1996;45:1443–6.
178. Carmel R. Arch Intern Med 1988;148:1712–4.
179. Herrmann M, Schmidt J, Umanskaya N, et al. Bone 2007;41: 584–91.
180. Rejnmark L, Vestergaard P, Hermann AP, et al. Calcif Tissue Int 2008;82:1–11.
181. Cagnacci A, Baldassari F, Rivolta G, et al. Bone 2003;33: 956–9.
182. Bozkurt N, Erdem M, Yilmaz E, et al. Arch Gynecol Obstet 2009;280:381–7.
183. Gjesdal CG, Vollset SE, Ueland PM, et al. J Bone Miner Res 2007;22:747–56.
184. Ravaglia G, Forti P, Maioli F, et al. J Gerontol A Biol Sci Med Sci 2005;60:1458–62.
185. Yazdanpanah N, Zillikens MC, Rivadeneira F, et al. Bone 2007;41:987–94.
186. Morris MS, Jacques PF, Selhub J. Bone 2005;37:234–42.
187. Tucker KL, Hannan MT, Qiao N, et al. J Bone Miner Res 2005;20:152–8.
188. Dhonukshe-Rutten RA, Lips M, de Jong N, et al. J Nutr 2003;133:801–7.
189. Stone KL, Bauer DC, Sellmeyer D, et al. J Clin Endocrinol Metab 2004;89:1217–21.
190. Macdonald HM, McGuigan FE, Fraser WD, et al. Bone 2004;35:957–64.
191. McLean RR, Jacques PF, Selhub J, et al. J Clin Endocrinol Metab 2008;93:2206–12.
192. Saito M, Fujii K, Marumo K. Calcif Tissue Int 2006;79:160–8.
193. Lubec B, Fang-Kircher S, Lubec T, et al. Biochim Biophys Acta 1996;1315:159–62.
194. Elshorbagy AK, Gjesdal CG, Nurk E, et al. Bone 2009;44: 954–8.
195. Zhu K, Beilby J, Dick IM, et al. Osteoporos Int 2009;20: 1183–91.
196. Leboff MS, Narweker R, LaCroix A, et al. J Clin Endocrinol Metab 2009;94:1207–13.
197. Perier MA, Gineyts E, Munoz F, et al. Osteoporos Int 2007;18:1329–36.
198. Green TJ, McMahon JA, Skeaff CM, et al. Am J Clin Nutr 2007;85:460–4.
199. Herrmann M, Umanskaya N, Traber L, et al. Clin Chem Lab Med 2007;45:1785–92.
200. Sawka AM, Ray JG, Yi Q, et al. Arch Intern Med 2007;167: 2136–9.
201. Stone KL, Lui LY, Christen WG, et al. J Bone Miner Res 2017;32:2331–8.
202. Ruan J, Gong X, Kong J, et al. Med Sci Monit 2015;21:875–81.

203. Binkley N, Krueger D. Nutr Rev 2000;58:138–44.
204. Thompson PD, Jurutka PW, Haussler CA, et al. J Biol Chem 1998;273:8483–91.
205. Johansson S, Melhus H. J Bone Miner Res 2001;16:1899–905.
206. Melhus H, Michaelsson K, Kindmark A, et al. Ann Intern Med 1998;129:770–8.
207. Michaelsson K, Lithell H, Vessby B, et al. N Engl J Med 2003;348:287–94.
208. Feskanich D, Singh V, Willett WC, et al. JAMA 2002;287: 47–54.
209. Promislow JH, Goodman-Gruen D, Slymen DJ, et al. J Bone Miner Res 2002;17:1349–58.
210. Opotowsky AR, Bilezikian JP, Study NIf-u. Am J Med 2004;117:169–74.
211. Barker ME, McCloskey E, Saha S, et al. J Bone Miner Res 2005;20:913–20.
212. Sowers MF, Wallace RB. J Clin Epidemiol 1990;43:693–9.
213. Ballew C, Galuska D, Gillespie C. J Bone Miner Res 2001;16:2306–12.
214. Caire-Juvera G, Ritenbaugh C, Wactawski-Wende J, et al. Am J Clin Nutr 2009;89:323–30.
215. Yee MMF, Chin KY, Ima-Nirwana S, et al. Molecules 2021;26.
216. Uchiyama S, Sumida T, Yamaguchi M. Biol Pharm Bull 2004;27:232–5.
217. Yamaguchi M, Uchiyama S. Biol Pharm Bull 2003;26:1188–91.
218. Sugiura M, Nakamura M, Ogawa K, et al. Osteoporos Int 2008;19:211–9.
219. Wattanapenpaiboon N, Lukito W, Wahlqvist ML, et al. Asia Pac J Clin Nutr 2003;12:467–73.
220. Sahni S, Hannan MT, Blumberg J, et al. Am J Clin Nutr 2009;89:416–24.
221. Sahni S, Hannan MT, Blumberg J, et al. J Bone Miner Res 2009;24:1086–94.
222. Rao LG, Mackinnon ES, Josse RG, et al. Osteoporos Int 2007;18:109–15.
223. Gao SS, Zhao Y. Osteoporos Int 2023;34:627–39.
224. Yamaguchi M, Igarashi A, Uchiyama S, et al. J Health Sci 2004;50:619–24.
225. Mackinnon ES, Rao AV, Josse RG, et al. Osteoporos Int 2011;22:1091–101.
226. Kim SJ, Anh NH, Diem NC, et al. Foods 2021;10.
227. Barzel US. J Bone Miner Res 1995;10:1431–6.
228. Kerstetter JE, O'Brien KO, Caseria DM, et al. J Clin Endocrinol Metab 2005;90:26–31.
229. Arjmandi BH, Khalil DA, Smith BJ, et al. J Clin Endocrinol Metab 2003;88:1048–54.
230. Roughead ZK, Johnson LK, Lykken GI, et al. J Nutr 2003;133: 1020–6.
231. Hannan MT, Tucker KL, Dawson-Hughes B, et al. J Bone Miner Res 2000;15:2504–12.
232. Misra D, Berry SD, Broe KE, et al. Osteoporos Int 2011;22: 345–9.
233. Munger RG, Cerhan JR, Chiu BC. Am J Clin Nutr 1999;69: 147–52.
234. Schurch MA, Rizzoli R, Slosman D, et al. Ann Intern Med 1998;128:801–9.
235. Feskanich D, Willett WC, Stampfer MJ, et al. Am J Epidemiol 1996;143:472–9.
236. Wallace TC, Frankenfeld CL. J Am Coll Nutr 2017;36: 481–96.
237. Dargent-Molina P, Sabia S, Touvier M, et al. J Bone Miner Res 2008;23:1915–22.
238. Albertazzi P, Coupland K. Maturitas 2002;42:13–22.
239. Kettler DB. Altern Med Rev 2001;6:61–77.
240. Watkins BA, Li Y, Lippman HE, et al. Prostaglandins Leukot Essent Fatty Acids 2003;68:387–98.
241. Watkins BA, Lippman HE, Le Bouteiller L, et al. Prog Lipid Res 2001;40:125–48.
242. Baggio B, Budakovic A, Nassuato MA, et al. Kidney Int 2000;58:1278–84.
243. Coetzer H, Claassen N, van Papendorp DH, et al. Prostaglandins Leukot Essent Fatty Acids 1994;50:257–66.
244. Krey G, Braissant O, L'Horset F, et al. Mol Endocrinol 1997;11: 779–91.
245. Rosen ED, Sarraf P, Troy AE, et al. Mol Cell 1999;4:611–7.
246. Barak Y, Nelson MC, Ong ES, et al. Mol Cell 1999;4:585–95.
247. Lecka-Czernik B. Curr Osteoporos Rep 2010;8:84–90.
248. Hong S, Gronert K, Devchand PR, et al. J Biol Chem 2003;278:14677–87.
249. Serhan CN, Hong S, Gronert K, et al. J Exp Med 2002;196: 1025–37.
250. Herrera BS, Ohira T, Gao L, et al. Br J Pharmacol 2008;155: 1214–23.
251. Hasturk H, Kantarci A, Ohira T, et al. FASEB J 2006;20:401–3.
252. Serhan CN, Jain A, Marleau S, et al. J Immunol 2003;171: 6856–65.
253. Weiss LA, Barrett-Connor E, von Muhlen D. Am J Clin Nutr 2005;81:934–8.
254. Zwart SR, Pierson D, Mehta S, et al. J Bone Miner Res 2010;25:1049–57.
255. Farina EK, Kiel DP, Roubenoff R, et al. Am J Clin Nutr 2011;93:1142–51.
256. Park HJ, Lee YW, Hennig B, et al. Nutr Cancer 2001;41:126–34.
257. Dichtl W, Ares MP, Jonson AN, et al. Metab 2002;51:327–33.
258. Farina EK, Kiel DP, Roubenoff R, et al. J Nutr 2011;141: 1146–53.
259. Virtanen JK, Mozaffarian D, Cauley JA, et al. J Bone Miner Res 2010;25:1972–9.
260. Orchard TS, Cauley JA, Frank GC, et al. Am J Clin Nutr 2010;92:1452–60.
261. Martinez-Ramirez MJ, Palma S, Martinez-Gonzalez MA, et al. Eur J Clin Nutr 2007;61:1114–20.
262. Harris TB, Song X, Reinders I, et al. Am J Clin Nutr 2015;101:947–55.
263. Terano T. World Rev Nutr Diet 2001;88:141–7.
264. Kruger MC, Coetzer H, de Winter R, et al. Aging (Milano) 1998;10:385–94.
265. Bassey EJ, Littlewood JJ, Rothwell MC, et al. Br J Nutr 2000;83:629–35.
266. Griel AE, Kris-Etherton PM, Hilpert KF, et al. Nutr J 2007;6:2.
267. Martyniak K, Wei F, Ballesteros A, et al. Bone 2021;143: 115736.
268. Mozaffari H, Djafarian K, Mofrad MD, et al. Osteoporos Int 2018;29:1949–61.
269. Liu SH, Chen C, Yang RS, et al. J Orthop Res 2011;29: 954–60.
270. Zhou Y, Guan XX, Zhu ZL, et al. Br J Pharmacol 2010;161: 1542–52.
271. Rapuri PB, Gallagher JC, Nawaz Z. J Steroid Biochem Mol Biol 2007;103:368–71.
272. Berman NK, Honig S, Cronstein BN, et al. Osteoporos Int 2022;33:1235–41.
273. Harris SS, Dawson-Hughes B. Am J Clin Nutr 1994;60:573–8.
274. Heaney RP. Food Chem Toxicol 2002;40:1263–70.
275. Barrett-Connor E, Chang JC, Edelstein SL. JAMA 1994;271: 280–3.
276. Hallstrom H, Wolk A, Glynn A, et al. Osteoporos Int 2006;17: 1055–64.
277. Kiel DP, Felson DT, Hannan MT, et al. Am J Epidemiol 1990;132:675–84.
278. Hallstrom H, Melhus H, Glynn A, et al. Nutr Metab (Lond) 2010;7:12.

279. Kelepouris N, Harper KD, Gannon F, et al. Ann Intern Med 1995;123:452–60.
280. Broulik PD, Vondrova J, Ruzicka P, et al. Physiol Res 2010;59:599–604.
281. Maurel DB, Boisseau N, Benhamou CL, et al. Osteoporos Int 2012;23:1–16.
282. Holbrook TL, Barrett-Connor E. BMJ 1993;306:1506–9.
283. Felson DT, Zhang Y, Hannan MT, et al. Am J Epidemiol 1995;142:485–92.
284. Hansen MA, Overgaard K, Riis BJ, et al. Osteoporos Int 1991;1:95–102.
285. Tucker KL, Jugdaohsingh R, Powell JJ, et al. Am J Clin Nutr 2009;89:1188–96.
286. Kanis JA, Johansson H, Johnell O, et al. Osteoporos Int 2005;16:737–42.
287. Mukamal KJ, Robbins JA, Cauley JA, et al. Osteoporos Int 2007;18:593–602.
288. Godos J, Giampieri F, Chisari E, et al. Int J Environ Res Public Health 2022;19.
289. Tucker KL, Chen H, Hannan MT, et al. Am J Clin Nutr 2002;76:245–52.
290. Hardcastle AC, Aucott L, Fraser WD, et al. Eur J Clin Nutr 2011;65:378–85.
291. Fairweather-Tait SJ, Skinner J, Guile GR, et al. Am J Clin Nutr 2011;94:1371–5.
292. Prynne CJ, Mishra GD, O'Connell MA, et al. Am J Clin Nutr 2006;83:1420–8.
293. McNaughton SA, Wattanapenpaiboon N, Wark JD, et al. J Nutr 2011;141:1516–23.
294. Hamidi M, Tarasuk V, Corey P, et al. Am J Clin Nutr 2011;94:199–208.
295. Langsetmo L, Hanley DA, Prior JC, et al. Am J Clin Nutr 2011;93:192–9.
296. Zheng Y, Wang J, Wang Y, et al. Nutrients 2023;15.
297. Tucker KL. Am J Clin Nutr 2014;100(suppl 1):329S–35S.
298. Nowson CA, Patchett A, Wattanapenpaiboon N. Br J Nutr 2009;102:1161–70.
299. Shapses SA, Riedt CS. J Nutr 2006;136:1453–6.
300. Perry CD, Alekel DL, Ritland LM, et al. Menopause 2008;15:619–27.
301. Girard J, Lafontan M. Diabetes Metab 2008;34:439–45.
302. Melton LJ, 3rd, Leibson CL, Achenbach SJ, et al. J Bone Miner Res 2008;23:1334–42.
303. Faulhaber GA, Premaor MO, Moser Filho HL, et al. Bone Marrow Transpl 2009;43:953–7.
304. Zhao LJ, Liu YJ, Liu PY, et al. J Clin Endocrinol Metab 2007;92:1640–6.
305. Hsu YH, Venners SA, Terwedow HA, et al. Am J Clin Nutr 2006;83:146–54.
306. Zillikens MC, Uitterlinden AG, van Leeuwen JP, et al. Calcif Tissue Int 2010;86:116–25.
307. Kim CJ, Oh KW, Rhee EJ, et al. Clin Endocrinol (Oxf) 2009;71:18–26.
308. Seo HJ, Kim SG, Kim CS. BMC Public Health 2008;8:253.
309. Russell M, Mendes N, Miller KK, et al. J Clin Endocrinol Metab 2010;95:1247–55.
310. Gilsanz V, Chalfant J, Mo AO, et al. J Clin Endocrinol Metab 2009;94:3387–93.
311. Bhupathiraju SN, Dawson-Hughes B, Hannan MT, et al. Am J Clin Nutr 2011;94:1063–70.
312. Zibellini J, Seimon RV, Lee CM, et al. J Bone Miner Res 2015;30:2168–78.
313. Huber FA, Singhal V, Tuli S, et al. Radiol 2023;307:e223256.
314. Hornick MA, Van Name M. J Bone Miner Res 2023;38:931–2.
315. Guadalupe-Grau A, Fuentes T, Guerra B, et al. Sports Med 2009;39:439–68.
316. Layne JE, Nelson ME. Med Sci Sports Exerc 1999;31:25–30.
317. Campbell AJ, Robertson MC, Gardner MM, et al. BMJ 1997;315:1065–9.
318. Ryan AS, Nicklas BJ, Dennis KE. J Appl Physiol (1985) 1998;84:1305–10.
319. Ma H, Leskinen T, Alen M, et al. J Bone Miner Res 2009;24:1427–33.
320. Eisman JA. Endocr Rev 1999;20:788–804.
321. Ferrari S. Best Pract Res Clin Endocrinol Metab 2008;22:723–35.
322. Styrkarsdottir U, Halldorsson BV, Gretarsdottir S, et al. N Engl J Med 2008;358:2355–65.
323. Duncan EL, Brown MA. J Clin Endocrinol Metab 2010;95:2576–87.
324. Riggs BL, Nguyen TV, Melton LJ, 3rd, et al. J Bone Miner Res 1995;10:991–6.
325. Dawson-Hughes B, Harris SS, Finneran S. J Clin Endocrinol Metab 1995;80:3657–61.
326. Krall EA, Parry P, Lichter JB, et al. J Bone Miner Res 1995;10:978–84.
327. Stathopoulou MG, Dedoussis GV, Trovas G, et al. J Am Diet Assoc 2010;110:1078–83.
328. Ackert-Bicknell CL, Demissie S, Marin de Evsikova C, et al. J Bone Miner Res 2008;23:1398–408.
329. Le P, Kawai M, Bornstein S, et al. Endocrinol 2012;153:6–16.
330. Tolonen S, Mikkila V, Laaksonen M, et al. Bone 2011;48:1058–65.
331. Lyles KW, Colon-Emeric CS, Magaziner JS, et al. N Engl J Med 2007;357:1799–809.
332. Chesnut CH, 3rd, Skag A, Christiansen C, et al. J Bone Miner Res 2004;19:1241–9.
333. McClung MR, Geusens P, Miller PD, et al. N Engl J Med 2001;344:333–40.
334. Reginster J, Minne HW, Sorensen OH, et al. Osteoporos Int 2000;11:83–91.
335. Ettinger B, Black DM, Mitlak BH, et al. JAMA 1999;282:637–45.
336. Cummings SR, San Martin J, McClung MR, et al. N Engl J Med 2009;361:756–65.
337. Cosman F, Crittenden DB, Adachi JD, et al. N Engl J Med 2016;375:1532–43.
338. Miller PD, Hattersley G, Riis BJ, et al. JAMA 2016;316:722–33.
339. Neer RM, Arnaud CD, Zanchetta JR, et al. N Engl J Med 2001;344:1434–41.
340. Dawson-Hughes B, Mithal A, Bonjour JP, et al. Osteoporos Int 2010;21:1151–4.
341. Hanley DA, Cranney A, Jones G, et al. CMAJ 2010;182:E610–8.
342. Norman AW, Bouillon R, Whiting SJ, et al. J Steroid Biochem Mol Biol 2007;103:204–5.
343. Wallace RB, Wactawski-Wende J, O'Sullivan MJ, et al. Am J Clin Nutr 2011;94:270–7.

E. SKELETAL AND JOINT DISORDERS

CHAPTER 89

Nutrition and Diet in Arthritis

Amanda S. Alexander, Joseph E. Baggott, and Sarah L. Morgan

CHAPTER OUTLINE

Abbreviations

AA, arachadonic acid; **ACR**, American College of Rheumatology; **AICA**, aminoimidazolecarboxamide; **AICAR**, AICA ribotide; **ADA**, adenosine deaminate; **BIA**, bioelectrical impedance analysis; **BID**, twice daily by mouth; **BMI**, body mass index; **BOKS**, Boston Osteoarthritis of the Knee Study; **DHA**, docosahexaenoic acid; **DHFR**, dihydrofolate reductase; **DMARD**, disease-modifying antirheumatic drug; **DXA**, dual-energy x-ray absorptiometry; **EPA**, eicosapentaenoic acid; **GAIT**, Arthritis Intervention Trial; **HAQ**, Health Assessment Questionnaire; **hs-CRP**, high sensitivity C-reactive protein; **Ig**, immunoglobulin; **IDEA**, Intensive Diet and Exercise for Arthritis; **IL**, Interleukin; **ILI**, intensive lifestyle intervention; **kg**, kilogram; **LED**, low energy diet; **MNA**, mini nutritional assessment; **MRI**, magnetic resonance imaging; **MTX**, methotrexate; **NHANES III**, Third National Health and Nutrition Examination Survey; **NHIS**, National Health Interview Survey; **OA**, osteoarthritis; **RA**, rheumatoid arthritis; **RDA**, recommended dietary allowance; **RDI**, recommended dietary intake; **REE**, resting energy expenditure; **SGA**, subjective global assessment; **T2D**, type 2 diabetes; **TNF**, tumor necrosis factor; **VLED**, very low energy diet; **WOMAC**, Western Ontario and McMaster University Osteoarthritis Index

Introduction/Definition of Rheumatic Disease/Arthritis

Definition and Prevalence

"Rheumatic diseases are characterized by inflammation (redness and/or heat, swelling, and pain) and loss of function of one or more connecting or supporting structures of the body. Rheumatic diseases typically affect joints, tendons, ligaments, bones, and muscles, producing symptoms such as pain, swelling, and stiffness and may involve internal organs".[1] Some examples of rheumatic diseases are infectious arthritis, osteoarthritis (OA), psoriatic arthritis, rheumatoid arthritis (RA), fibromyalgia, systemic lupus erythematosus, gout, polymyositis, bursitis, and tendonitis. Arthritis refers to inflammation of a joint.[2] The Arthritis Foundation lists over 100 types of arthritis and related diseases.[3,4] The National Arthritis Data Workgroup, using data from the National Health Interview Survey (NHIS), estimated that approximately 46.4 million Americans have self-reported or physician-diagnosed arthritis.[5]

Table 89.1 Arthritis Drug-Nutrient Interactions[10,21-23]

Drug	Affected Nutrient Status
Aspirin	↓folate, ↓iron, ↓vitamin C
Salicylate	↓folate
Nonsteroidal Anti-inflammatory Drugs	↓Iron, ↓folate
Sulfasalazine	↓folate
Methotrexate	↓folate
Corticosteroids	↓calcium, ↓vitamin D, ↓potassium, ↓zinc, ↓vitamin C, ↓magnesium, ↓folate, ↓selenium
Tetracycline	↓calcium, ↓magnesium, ↓iron
Colchicine	↓vitamin B_{12}, ↓sodium, ↓potassium
D-penicillamine	↓vitamin B_6, ↓magnesium, ↓zinc, ↓copper

Arthritis Effects on Nutritional Status

There are several mechanisms by which arthritis may affect nutritional status. Food intake may be impaired because it is difficult to chew, and meal preparation may be difficult because of arthritis-related joint pain or swelling. In addition, inflammation from arthritis may increase energy expenditure in performing activities of daily living, and there may be increased resting energy expenditure, and protein catabolism.[6-9] An additional nutritional correlate includes the effects of obesity on arthritis in OA and gout.

There are some general strategies to help individuals with arthritis in the preparation of food.[10] A wide variety of items, including extended drinking straws, over-the-sink cutting boards, jar lid openers, rolling trivets, nonskid pads, easy grip utensils, pan holders, and handles for milk cartons are available to make food preparation easier.[11]

Drug Nutrient Interactions

Drugs used to treat arthritis may interact with and affect the need for a wide variety of nutrients.[12,13] Arthritis drug therapies may cause side effects, such as nausea and vomiting, which may affect nutrient intake.[14] In addition, drug therapies affect nutritional status (see folic acid-methotrexate (MTX) interaction, below).[15-20] An example of a common drug-nutrient effect is iron deficiency from gastrointestinal bleeding from chronic aspirin or nonsteroidal anti-inflammatory drug use. **Table 89.1** outlines possible drug-nutrient interactions.

Methotrexate (MTX) is viewed as the gold standard and anchor therapy in the treatment of RA.[23-25] Evidence suggests that low dose MTX blocks purine nucleotide biosynthesis at a folate-dependent step catalyzed by AICAR transformylase, and that this produces immunosuppression via adenosine accumulation (**Figure 89.1**).[26,27]

Because MTX is an antifolate, the potential for the drug to produce a functional deficiency of folate exists. Morgan et al., found that functional folate deficiency did exist in MTX-treated patients[17] and that certain MTX toxicities, such as cytopenias, resembled those found in uncomplicated folate deficiency, while others, such as liver toxicity, are probably a toxic effect of MTX.[28,29] As MTX toxicity is a major reason for discontinuing

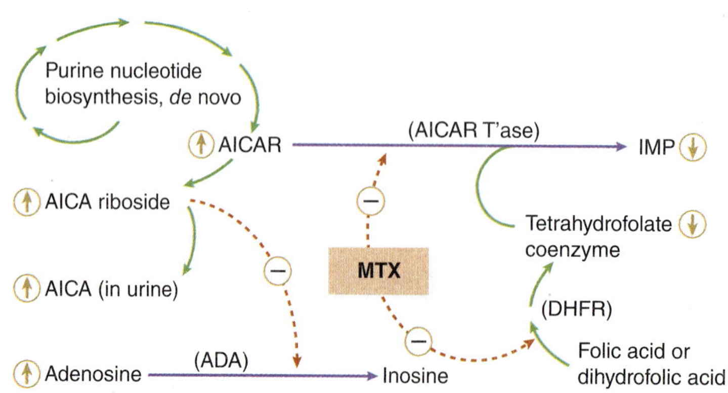

Figure 89.1 Proposed mechanism of action of low-dose mtx therapy for rheumatoid arthritis. *AICA*, aminoimidazolecarboxamide; *AICAR*, AICA ribotide; *AICAR T'ase*, AICAR transformylase; *ADA*, adenosine deaminase; *DHFR*, dihydrofolate reductase; *MTX*, methotrexate. (Dotted lines with negative signs in circles indicate enzyme inhibition; circles with arrows indicate increased or decreased in vivo levels.

the drug, there are compelling reasons to ameliorate these side effects.[14]

Several studies have demonstrated that folic acid supplementation of MTX-treated RA patients lowers drug toxicity while preserving efficacy.[30-34] The exact mechanism is not known; however, it is possible that folic acid replenishes folate coenzyme concentrations in tissues susceptible to MTX-induced folate deficiency (e.g., enterocytes, liver) while it has little effect on MTX cytotoxicity in immunologic cells. However, supplementation with the carbon-substituted, fully reduced folate, folinic acid (leucovorin factor) can negate efficacy and cause a flare in arthritis.[20,35] It is now routinely recommended that patients with RA receiving MTX also be supplemented with folic acid.[36-40]

Nutrient Effects on Arthritis

Diet and nutrients are implicated in the development of arthritis, and dietary and nutrient interventions are often used as therapeutic strategies.[41] This chapter will focus on the nutritional correlates of three types of arthritis: gout, OA, and RA.

Effects of Study Design on Nutritional Outcomes

In studying the effects of nutrients, foods, and dietary patterns on arthritis, it is first important to consider the effect of study design on nutritional intervention in chronic disease processes.[42-46] The use of evidence-based medicine and randomized, controlled trials is useful for drug trials where there is randomization, blinding, and other measures of quality. Randomized controlled trials are less useful for evaluating nutrient interventions for the following reasons and these concepts should be kept in mind in the evaluation of the trials reported in the remainder of the chapter:

1. Nutrient effects are related to inadequate intakes and insufficient nutritional status, whereas drug trials evaluate the effect on a disease that is not caused by the absence of a drug. This means that drug effects tend to be large and have specific outcomes that lend themselves to evidence-based medicine clinical trials. Nutrient interventions affect many tissues and the outcomes are often "within the 'noise' of biological variability" and therefore, nutrient effects may be lost in clinical trial outcomes.[42]
2. Nutrient effects follow a sigmoid response curve while drugs usually have a dose response curve (**Figure 89.2**). This means that disease response depends upon baseline nutrient status and that an adequate dose must be given to see an effect. If a population is already sufficient in a nutrient, no effect of supplementation on a disease process is expected. In a deficient population, larger doses may be required to see a response than may be used in a clinical trial. In a drug study, a drug can be tested against a placebo group that does not receive the drug. In a nutritional study, the placebo group would need to have zero nutrient intake or be documented to be deficient, which is generally unethical.[42,45,46]
3. Nutrient effects generally manifest themselves as small biologic differences over a long period of time; therefore, a short-term, randomized, controlled trial may not see a measurable outcome.
4. Nutrients work together (i.e., calcium and vitamin D) so it is often difficult to evaluate the effect of a single nutrient on an outcome.[44]

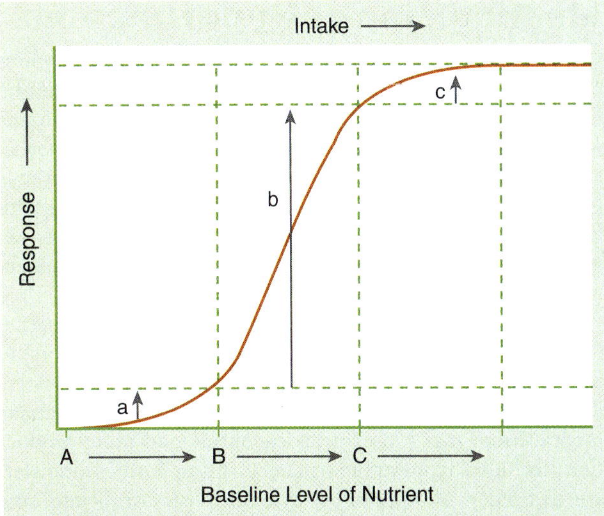

Figure 89.2 The sigmoid response curve for a typical nutrient.

Reproduced from Heaney RP. NEJM 2012;367:77–8. doi:10.1056/NEJMe1206858

5. Nutrient effects tend to be polyvalent with nutrients affecting many tissues and different tissues may have different response curves.[47]
6. Nutritional effects may take years or decades to manifest, while drug studies have relatively shorter drug outcome vistas. This means that the effect of a nutrient on an outcome like arthritis could have a long latency and require a long study to determine usefulness.[43,44]

Meta-analyses often include clinical trials where the baseline nutritional status of a population was not known, the intervention dose may not have been adequate, and the length of the nutritional intervention was short; these facts should be taken into account in the evaluation of the outcome of meta-analyses.[45]

A hierarchy of nutritional interventions has been proposed, starting with a single nutrient or phytochemical, progressing to a food, food groups, and dietary patterns.[48,49] The concept of food synergy suggests that interactions among constituents in food are important, meaning that trials of food patterns are reasonable, as opposed to clinical trials of a single nutrient or food.

Gout

Definition

Gout is a crystalline-type arthritis that results from deposition of uric acid in joints when uric acid reaches its limit of solubility.[50] The uric acid crystals cause inflammation, swelling, and pain in the affected joint, especially the first metatarsal phalangeal joint (podagra).[51-54] The clinical spectrum of gout can range from acute arthritis to accumulation of uric acid crystalline deposits called tophi, or uric acid urolithiasis and, less frequently, renal failure (gouty nephropathy).[55]

Epidemiology

Using data from the National Health and Nutrition Survey (NHANES 2015–2016), the prevalence of gout in adults was 3.9% (9.2 million) in the United States and was more prevalent in men (5.2%) than in women (2.7%).[56]

Mechanisms of Hyperuricemia

Uric acid is the product of purine catabolism in humans. Xanthine oxidase catalyzes the oxidation of hypoxanthine to xanthine and ultimately to uric acid. Hyperuricemia is defined as a serum uric acid concentration >7.0 mg/dL in men or > 6.0 mg/dL in women and is a common feature in all types of gout. While hyperuricemia is a risk factor for gouty arthritis, the majority of individuals with hyperuricemia will not develop gout.[52,57] Overt manifestations of gout are more likely when serum urate is >9.0 mg/dL and urinary uric acid excretion is >800 mg/day.[58]

The basic mechanisms of hyperuricemia are overproduction (~10% of patients) and under excretion (~90% of patients) of uric acid. Overproduction could result from myeloproliferative disorders, malignant diseases, or hemolytic anemias. Metabolic inborn errors that cause increased purine and urate production include hypoxanthine-guanine phosphoribosyltransferase deficiency, phosphoribosylpyrophosphate synthetase over activity, and glucose-6-phosphatase deficiency. Inhibitors of xanthine oxidase are frequently used when individuals overproduce uric acid. Impaired urate clearance may result from renal failure, dehydration, diabetic ketoacidosis, ethanol intake, diuretics, or use of specific medications (see below).[59] Uricosuric agents may be considered for the treatment of impaired urate excretion.

Gout may be classified as primary or secondary.[52] Primary gout is related to an inborn error or an acquired metabolic defect. Examples of secondary gout include the use of drugs that increase serum urate concentration (diuretics, tacrolimus, methoxyflurane, cyclosporine, ethambutol, pyrazinamide, cytotoxic chemotherapies, ethanol, low-dose salicylates, levodopa, ribavirin, interferon, or teriparatide).[52] Other drugs and nutrients are known to lower serum urate concentration (ascorbic acid, benzbromarone, calcitonin, citrate, estrogen, fenofibrate, losartan, probenecid, high-dose salicylates, and sulfinpyrazone).[60]

Metabolic and Lifestyle Profiles of Individuals with Hyperuricemia/Gout

The classic profile of an individual with gout is an obese, hypertensive, middle-aged man who eats rich foods and drinks alcohol.[55] However, a study of serum metabolites identified 7 to 13 that could differentiate gout from hyperuricemia and from normouricemia, suggesting that changes in complex metabolic pathways operate synergistically to predispose individuals to gout.[61]

Metabolic Syndrome and Gout

There are strong correlations between obesity, insulin-resistance, and serum uric acid concentration, and between renal uric acid clearance and degree of insulin resistance.[62] Choi et al.,[63] determined the relationship between metabolic syndrome (defined as ≥3 of the following criteria: abdominal obesity, hypertriglyceridemia, low high-density lipoprotein cholesterol, hypertension, or high fasting glucose[64] and gout prevalence. The prevalence of metabolic syndrome was 62.8% in patients with gout and 25.4% in patients without gout. In addition, a history of gout was shown to be an independent risk factor for the development of type 2 diabetes.[65]

Dietary Correlates of Hyperuricemia and Gout

Dietary purines can contribute up to one-third of the uric acid in the body.[66] However, individuals with gout receiving a purine-free diet had little change in plasma uric acid, compared with those receiving a regular diet.[67] A variety of foods have been reported to increase serum uric acid and to increase gouty attacks, including meats, seafood, yeast, yeast extracts, peas, beans, lentils, asparagus, spinach, mushrooms, and beer and alcoholic drinks.[68] Data from the NHANES III survey showed that, after adjusting for age, serum uric acid concentration in adults differed between extreme quintiles by 0.48 mg/dL for total meat intake (95% CI 0.34–0.61, *P* for trend <0.001), 0.16 mg/dL for seafood (95% CI 0.06–0.27, *P* for trend 0.005), and −0.21 mg/dL for total dairy food intake (95% CI −0.37–−0.04, *P* for trend=0.02), indicating that higher meat and seafood consumption, but not total protein, may be related to hyperuricemia.[69,70] In the Health Professionals Follow-up Study, men in the highest quintile of meat and seafood intakes had 41% and 51% increased risk of gout, respectively, while the consumption of purine-rich vegetables was not associated with risk.[69,71] In Taiwan, a study of vegans, lacto-ovo vegetarians, and non-vegetarians found that differences in mean serum uric acid were only 0.13 to 0.27 mg/dL lower in the vegetarians, but the risk of gout was significantly lower (RR 0.61).[72]

Individuals consuming a serving of milk one or more times a day had lower serum uric acid than those consuming no milk.[70] Yogurt consumption at least every other day was also associated with lower serum uric acid compared with individuals not consuming yogurt.[70] Data from the NHANES III also showed that the adjusted odds ratio for hyperuricemia in individuals consuming ≥6 cups of coffee per day, compared with no coffee consumption, was 0.57 (*P* for trend 0.001).[73] Tea consumption was not associated with hyperuricemia, and the authors suggested that the inverse association with coffee intake may be mediated by components other than caffeine. Another study using NHANES III data found that consumption of soft drinks containing fructose, but not artificially sweetened soft drinks, was associated with higher serum uric acid concentration.[74] A meta-analysis concluded that substantial fructose consumption is significantly associated with an increased risk of gout (RR ≥1.62).[75] The metabolism of fructose has been shown to generate uric acid more than other sugars.[76]

Data from the Health Professionals Follow-up Study showed that greater vitamin C intake was associated with lower uric acid.[77] At intake higher than 400 to 500 mg/day of vitamin C, no change serum uric acid concentration occurred.[68,78] A review of 15 small (mostly in <100 individuals) vitamin C intervention studies (most ≤500 mg/day) showed some decreases in serum uric acid.[79]

Risk of gout was prospectively evaluated over a 23-year period in 47,150 men who had no history of gout at baseline.[71] Increased risk was positively associated with higher consumption of meat and inversely associated with higher intake of dairy products. There were no relationships with intakes of total protein or purine-rich vegetables. As meat has highly bioavailable iron, this might partially explain the significant positive association of serum ferritin with serum uric acid.[80]

Alcohol and Hyperuricemia and Gout

Various mechanisms linking hyperuricemia and gout to alcohol consumption have been postulated. These include: 1) production of temporary lactic acidosis; 2) stimulation of purine production by accelerated degradation of ATP to AMP by conversion of acetate to acetyl CoA during alcohol metabolism; 3) high purine content in beer, particularly guanosine; and 4) consumption of lead-contaminated alcohol, causing a reduction in renal urate excretion and subsequent hyperuricemia.[68]

Data from the Health Professionals Follow-up Study suggested a positive dose response relation between total alcohol intake and gout.[81] The relative risk for hyperuricemia with beer or liquor consumption was greater than for wine intake. Analyses from the Nutrition and Health Survey in Taiwan found that increasing beer intake was independently associated with hyperuricemia.[78]

Dietary Therapy of Gout

Dietary therapies may have additive benefits to pharmacologic therapy, although they rarely lower serum urate concentration by more than 1 mg/dL, even with severe purine restriction.[55,67,82,83] Dietary management of gout is particularly useful during a gouty flare.[84-86] **Figure 89.3** displays the impact of dietary components on gout risk. The American Dietetic Association Nutrition Care Manual[87] recommends that during an acute gouty attack, the patient should: 1) consume 8 to 16 cups of fluid/day, at least half as water; 2) abstain from alcohol; 3) limit animal foods; 4) eat a moderate amount of protein with recommended sources as low-fat or nonfat dairy, tofu, eggs, and nut butters; and 5) limit meat, fish, and poultry to 4 to 6 ounces/day. During remission from a gout flare, the recommendations are to 1) consume 8 to 16 cups of fluid/day, at least half as water; 2) abstain from alcohol; 3) follow a well-balanced eating plan following the Dietary Guidelines for Americans and, as tolerated, consume animal foods and continue to eat a moderate amount of protein; and 4) maintain a desirable body weight and avoid fasting or high-protein diets for weight loss.[87]

Several studies have demonstrated that the Dietary Approaches to Stop Hypertension (DASH) diet and the Mediterranean diet, with subsequent weight loss, lowers serum uric acid and reduces the prevalence of gout and flares, as expected.[88-93]

Choi et al.,[85] recommend the following guidelines for patients with gout: 1) exercise daily and reduce weight; 2) limit red meat intake; 3) tailor seafood intake to individual risk for cardiovascular disease and consider omega-3 fatty acid supplements; 4) drink skim milk or consume other low-fat dairy products daily up to two servings per day; 5) consume vegetable protein, nuts, legumes, and purine-rich vegetables; 6) reduce alcoholic beverages to less than one to two drinks/day for men or one drink for women; 7) limit sugar-sweetened soft drinks and other beverages containing high-fructose corn syrup; 8) allow coffee drinking if already drinking coffee; and 9) consider taking vitamin C supplements.

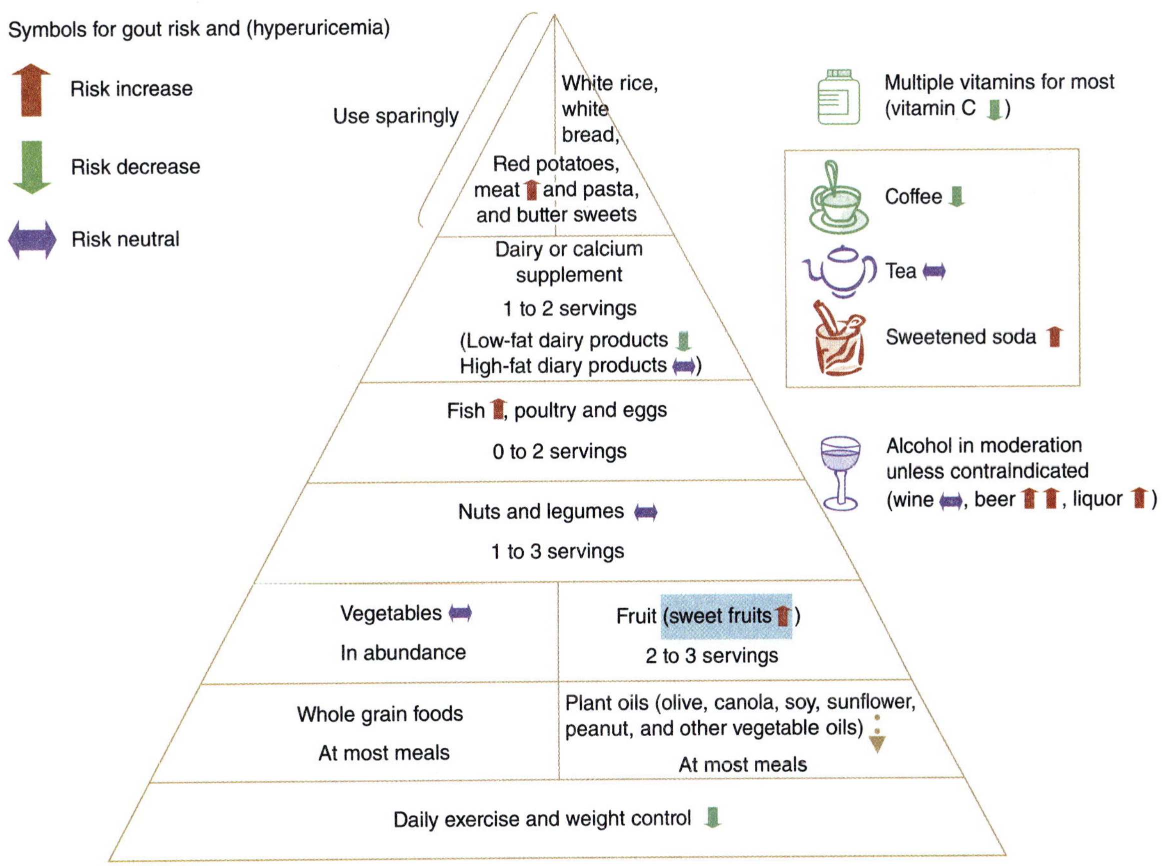

Figure 89.3 Dietary gout risk and the healthy eating pyramid.[85]

A wide variety of natural products also have been reported to be anti-inflammatory, to inhibit xanthine oxidase, or to be efficacious in the hyperuricemic mouse model.[94] Perhaps the most significant human study is the consumption of cherries or its extracts.[95-99] Most, but not all,[97] show a decrease in serum urate or gout risk or flares. However, most trials are small, in number and duration, so larger studies are indicated.[95,99]

Summary

Unfortunately the actual dietary practices of individuals with gout appear to be inconsistent with current recommended diet therapy recommendations.[100] New goals for gout management include a major emphasis on diet therapy to reduce insulin resistance and produce weight loss in overweight patients.

Osteoarthritis

Definition

Osteoarthritis (OA) is a progressive disorder of the joints and surrounding tissues with damage and gradual loss of articular cartilage, resulting in the development of bony spurs, remodeling of subarticular bone, laxity of ligaments, and weakening of periarticular muscles and synovial inflammation.[101-103] OA can be defined radiologically, subjectively by symptoms, or by clinical examination findings.[103]

Epidemiology and Demographic Factors

OA is the most common arthritis worldwide.[104] The estimate of OA prevalence depends on whether the subjective, radiologic, or clinical definition is used.[105] Data from the NHANES III and the NHIS estimate that approximately 27 million adults in the United States have OA.[106] In NHANES II, weight-bearing radiographs were used and joint space narrowing was found in 37.4% of adults ≥69 years.[107] The most affected joints are knees, hips, and hands. OA prevalence and incidence increases with age, with individuals ≥50 years being more affected.[106] Prior to age 50, men are more affected by OA, but after age 50, women are more affected.[108-110] OA of the knee has doubled in prevalence since the mid-20th century; increased longevity and elevated BMI are implicated as likely causes.[111]

Nutrient Intakes of Individuals with OA

Twelve patients with OA were evaluated for dietary intake and it was found that >50% consumed diets with <67% of the Recommended Dietary Allowances (RDA) for iron, zinc, vitamin E, folate, and vitamin B_6.[112] Seventy-nine percent of 77 patients participating in a multidisciplinary program for the management of OA were obese, and the degree of obesity was related to the amount of arthritis pain. Dietary intakes of vitamin D, folate, pyridoxine, and zinc averaged <80% of the RDA.[113]

Comee et al., evaluated adherence to the Mediterranean diet and diet quality using the Healthy Eating Index-2015 (based on adherence to the 2015 Dietary Guidelines) in a cross section of individuals from the National Health and Nutrition Examination (NHANES) 2007-2014[114] and arthritis prevalence was evaluated. Adherence to a Mediterranean diet pattern was worse among individuals with arthritis, compared with those without arthritis. Individuals without arthritis had higher scores for total fruit, whole fruit, total vegetable, greens and beans, whole grains, seafood, and plant protein intake, compared with individuals with arthritis. Individuals with arthritis consumed more total sugar, total fat, and less dietary fiber and beta-carotene than individuals without arthritis.

Consumption of a Western vs a prudent diet (higher in fruits, vegetables, and whole grains) was associated with progression of knee OA, gauged by symptoms and radiographic questions in a longitudinal cohort from the Osteoarthritis Initiative database (https://oai.ucsf.edu/oai-osteoarthritis-initiative and https://nda.nih.gov/oai).[115] The authors suggest that eating a diet rich in vegetables, fruits, whole grains, and legumes might be related to decreased OA disease progression.

Nutritional Correlates and OA

Local risk factors like joint injury, obesity, mechanical factors (i.e., joint alignment), systemic risk factors like diet and bone metabolism, and nonmodifiable risk factors like age and gender are related to OA development.[41] Diseases such as obesity, diabetes, dyslipidemia, and gut dysbiosis, which have strong nutritional correlates, are also related to OA.[41,104,116,117]

Obesity and OA

Obesity is a risk factor for OA.[109,110,116,117] In a study comparing individuals with OA vs RA, OA patients were, on average, 15 pounds (6.8 kg) overweight, while patients with RA averaged 10 pounds (4.5 kg) underweight.[118] Reyes et al., observed 1,764,061 Spanish primary care individuals for a median of 4.45 years and evaluated incidence rates of knee and hand OA.[119] Compared with individuals of normal weight, the risk of knee, hip, and hand OA were increased by factors of 2, 3.1, and 4.7, respectively, for overweight (BMI >25 and <30) and grades 1 (BMI 30 to <35) and grade 2 (BMI 35 to 40) obesity.

Increased weight, with abnormal loading of cartilage, is likely a cause of OA.[116] However, as obesity increases, the risk of OA in non–weight-bearing joints, such as the hands, metabolic factors including inflammatory factors from the adipose tissue are also likely involved.[116,120]

Type 2 Diabetes Mellitus (T2D) and OA

The prevalence of OA is higher in individuals with type 2 diabetes (T2D) than in those without diabetes.[121] T2D was a predictor of joint space narrowing in men with established OA.[122] Individuals with T2D are frequently obese, which may be one of the causes of the relationship. However, chronic hyperglycemia causing oxidative stress, insulin resistance, and a low-grade inflammatory state, also likely have causative effects.[123,124]

Dyslipidemia and OA

Obese patients with T2D have higher risks of having dyslipidemia and there is a twofold higher prevalence of dyslipidemia in patients with OA compared with patients without OA.[125] It is hypothesized that hypercholesterolemia may cause atherosclerosis and reduce blood flow to subchondral bone and also enhance the inflammatory response in OA.[104]

Microbiome and OA

The gut microbiome is also postulated to play a role in OA development and progression and there are data emerging from animal and human studies.[126-129] Lei et al., gave the probiotic *Lactobacillus casei Shirota* to 537 individuals with knee OA in a double-blind, placebo-controlled, 24-week trial.[130] There was significant reduction in high sensitivity C-reactive protein (hs-CRP) in the probiotic group, compared with the placebo group. Pain scores were significantly improved in the probiotic group, and it was suggested that improvements were mediated through significant reduction in hs-CRP. However, gut microbiome DNA profiles were not completed in this study.

Nutritional Management of OA and Effects of Individual Nutrients on OA

This section will review clinical trials with single nutrients, foods, and dietary patterns on OA.

Vitamin C

Vitamin C, L-ascorbic acid, is a water-soluble vitamin found in fruits and vegetables. Vitamin C deficiency is called scurvy.[131] Vitamin C is involved in collagen biosynthesis and acts as an antioxidant. The effect of vitamin C on OA incidence and OA progression is not clear and studies have given conflicting results.[132] There have been no double-blind, placebo-controlled trials of vitamin C related to OA. Given the fact that most individuals consuming a reasonable diet are not vitamin C deficient, it is not surprising that OA outcomes may be variable.

Vitamin D

Vitamin D (ergocalciferol and cholecalciferol) is a fat-soluble vitamin necessary for calcium absorption, bone mineralization, and collagen metabolism. Vitamin D is synthesized in the skin after sun exposure and the preformed vitamin is found in fatty fish and fish liver oils. Egg yolks, liver, and cheese have smaller amounts of vitamin D.[133]

Because of the importance of vitamin D in collagen and bone metabolism, there has been an interest in the effect of vitamin D status on OA incidence and progression. Participants in the Framingham Heart Study with low dietary intakes of vitamin D and low 25-hydroxyvitamin D (<30 ng/mL) were more likely to have progression of established OA. However, these factors did not affect the risk for development of OA.[134] The authors hypothesized that low vitamin D status may hinder the repair response of bone and cartilage to damage.[134]

Data from the Study for Osteoporotic Fractures Research Group found that elderly women in the lowest tertile of serum 25-hydroxyvitamin D were three times more likely to develop hip joint space narrowing than women in the highest tertile.[135] Data from both the Framingham Osteoarthritis Study and the Boston Osteoarthritis of the Knee Study (BOKS) found no association between 25-OH-vitamin D and radiographic worsening of arthritis.[136] Additionally, in the BOKS trial, there was no relationship of 25-OH vitamin D to cartilage loss, as measured by magnetic resonance imaging (MRI). In 880 individuals from the Tasmanian Older Adult Cohort Study, with a mean age of 61 years, sunlight exposure and serum 25-OH vitamin D concentration were associated with medial and lateral tibial cartilage loss in women and in individuals with knee pain, but not in men or individuals without OA on knee radiographs or knee pain.[137]

Clinical trials have evaluated the use of vitamin D supplements for OA therapy. McAlindon et al., conducted a 2-year, placebo-controlled, double-blind, clinical trial of 146 individuals with knee OA who were randomized to receive placebo or cholecalciferol 2,000 IU (50μg) daily to elevate the 25-OH vitamin D to >36 ng/mL.[138] The primary outcomes were knee pain and cartilage volume loss measured by magnetic resonance imaging. The change in knee pain was not significantly different between the placebo and vitamin D-treated groups and cartilage volume decreased to a similar extent in both groups over time. The authors did acknowledge that a possible cause of the null result was that individuals were already replete in vitamin D, that the intervention dose was insufficient, or that the placebo group took supplements. However, they also indicated that vitamin D concentration did not increase in the placebo group during the trial.

Arden et al., conducted the VIDEO study, a 3-year, randomized, double-blind, placebo-controlled trial of 474 individuals over the age of 50 with knee OA. Participants were given 800 (20 μg) IU of cholecalciferol vs placebo. Approximately 50% of both treatment groups had a 25-OH vitamin D level <20 μg/L, therefore the placebo group was not entirely a deficient group. The rate of medial knee joint space narrowing was not different between the two groups. There was no difference in joint pain between the two groups. A post-hoc analysis of the VIDEO trial suggested that participants who maintained sufficient 25-OH vitamin D status over 2 years had less loss of tibial cartilage volume and less increase in knee effusion volume compared with those who did not maintain sufficient vitamin D concentration.[139] However, another analysis of VIDEO trial participants found that vitamin D supplementation had no effect on knee synovitis or on subchondral bone marrow lesions determined by magnetic resonance imaging.[140] Therefore, the results of vitamin D supplementation for OA are inconclusive.

Vitamin E

Vitamin E includes a group of fat-soluble vitamins found in vegetable oils, nuts, and seeds, with antioxidant properties.[141] It has been hypothesized that the cartilage degeneration in OA is related to oxidative stress.[142] Observational studies have had conflicting results related to vitamin E intake and OA progression,[142] and there have been no properly conducted human trials.

Overall, there are variable results related to the effect of single nutrients (vitamins C, D, and E) on OA incidence and progression, and no firm conclusions can be drawn about the use of these vitamins for OA therapy.

Dietary Components

Fiber

Dietary fiber is the remains of plant cells that are resistant to digestion in the small intestine. Some fiber is absorbed or digested in the colon.[143] Dietary fiber is postulated to affect OA by effects on body weight and inflammation.

Dai et al., evaluated the effect of fiber on knee pain in the Osteoarthritis Initiative (https://oai.ucsf.edu/oai-osteoarthritis-initiative and https://nda.nih.gov/oai), which was a cohort of

4,769 men and women aged 45 to 70 y, who were at risk of or had knee OA, followed for 8 years.[144] Dietary fiber intake was evaluated at baseline with a food frequency questionnaire and Western Ontario and McMaster Universities Osteoarthritis Index (WOMAC) pain scores were evaluated over time. Total dietary fiber intake was inversely related to having moderate or severe pain. Compared with individuals in the lowest quartile of fiber intake, individuals in the highest quartile had lower risk of having moderate pain. The authors concluded that individual's taking in ≥25 g of fiber per day had lower risk of developing moderate or severe knee pain over time.

Another study evaluated the effect of dietary fiber on symptomatic knee OA and radiographic knee OA in the Framingham Offspring OA study and the Osteoarthritis Initiative (https://oai.ucsf.edu/oai-osteoarthritis-initiative and https://nda.nih.gov/oai).[145] In both cohorts, there was significantly lower symptomatic knee arthritis in the highest vs lower quartile of fiber intake. There was no relationship of fiber intake to radiographic OA. These studies are interesting from a nutritional trial design perspective because they compare groups that are relatively deficient to more sufficient groups.

Lipids/Omega 3 Fatty Acids

Saturated fatty acid intake is pro-inflammatory, while omega 3 fatty acids have anti-inflammatory effects.[146] In a case-controlled study in Saudi Arabia, the radiographic severity of knee OA was related to dietary antioxidant intake by a food frequency questionnaire.[147] Individuals with radiographic knee OA consumed fewer antioxidants and more saturated and mono unsaturated fatty acids as a percent of total energy intake.

Akbar et al., reviewed four studies using omega-3 fatty acids in OA treatment from 1992–2006.[148] Because of issues with clinical trial design and implementation, they concluded that the use of fish oil supplements (higher omega-3 fatty acid content) in OA is inconclusive. No substantial adverse effects related to fish oil supplements were reported in any of the trials.

The Multicenter Osteoarthritis Study[149] evaluated the risk of development of knee OA[150] and evaluated serum fatty acid profiles, knee x-rays and magnetic resonance imaging results, and knee pain measures, during 60 months of follow-up. Associations of omega 3- and omega-6 fatty acid concentrations, adjusted for age, sex, BMI, education, race, baseline pain, and depressive symptoms, with development of symptomatic and radiographic OA were evaluated. Serum fatty acids had no association with risk of later knee or hand OA. They concluded that there is no role for fatty supplements in knee OA. Therefore, additional studies seem warranted related to fatty acid supplementation and OA.

Foods

Milk/Soy Milk

It has been postulated that, as dairy foods are a good source of calcium, vitamin D, and protein, that they may be beneficial for OA.[151] A cross sectional study of 655 Turkish men and women, aged ≥50 years, suggested that milk consumption was protective against symptomatic knee arthritis.[152] In another cross-sectional study of 5,764 individuals, soy milk intake was assessed by food frequency questionnaire and the presence of joint space narrowing and osteophytes were assessed radiographically.[153] Soy milk intake was negatively associated with the presence of osteophytes, but there was no relationship to joint space narrowing. A study of 2,148 participants in the Osteoarthritis Initiative (www.oai.ucsf.edu and https://nda.hih.gov/oai) followed individuals with OA documented by radiographs for 48 months.[154] Baseline dietary intake data from a food frequency questionnaire was available. There was a dose-response between baseline milk intake and a decrease in joint space width narrowing in women, when baseline disease severity, BMI, dietary factors, and other potential confounders were adjusted. In men, there was no significant association between decreases in joint space width and milk consumption.

Dietary Patterns

Mediterranean Diet

A Mediterranean diet, which includes plentiful intakes of fruits, vegetables, legumes, grains, fish/seafood, extra virgin olive oil, and a moderate intake of red wine, has been found to be beneficial for a variety of diseases.[155] Veronese et al., evaluated 2,527 community-dwelling females with a mean age of 61.2 years from the Osteoarthritis Initiative (www.oai.ucsf.edu and https://nda.hih.gov/oai).[156] Knee OA was diagnosed clinically and radiologically, and a food frequency questionnaire was completed at the baseline visit. A Mediterranean diet score was calculated.[157] Participants in the highest quartile of adherence to the Mediterranean diet had significantly lower prevalence of knee OA than those in the lowest quartile.

In an extension of the Osteoarthritis Initiative (www.oai.ucsf.edu and https://nda.hih.gov/oai), 4,330 individuals with a mean age of 61.1 years were followed for 4 years.[158] Knee OA outcomes were radiographic findings, onset of a combination of a painful knee and radiographic osteoarthritis, and knee pain worsening. A Mediterranean diet adherence measure was again calculated.[157] Those individuals who were more adherent to a Mediterranean diet had lower risk of knee pain worsening. In a total of 2,994 individuals free from symptomatic knee patient at baseline, better adherence to a Mediterranean pattern had a 9% lower risk of symptomatic knee OA during follow-up. There were no reported associations between radiographic knee OA and Mediterranean diet.

Dyer et al., evaluated 99 individuals (aged 31–90 years) with a clinical diagnosis of OA who were randomized to a Mediterranean vs normal diet for 16 weeks.[159] An Arthritis Impact Measurement scale was completed before, after, and in the middle of the dietary intervention. There was no difference in any component of the measurement scale between the dietary groups, However, IL-1α (a proinflammatory cytokine) was significantly lower in the Mediterranean diet group compared with the control group. There was also a significant improvement in hip rotation range of motion and knee flexion in the Mediterranean diet group.

Weight Reduction and OA

Strategies to reduce weight have been shown to both decrease the onset and progression of OA in weight-bearing joints. In the Framingham cohort, Felson et al., found that weight loss of 5.1 kg over 10 years decreased the odds of incident knee OA by 50%.[160]

In 142 obese patients with radiographic evidence of knee OA and reported disability related to the arthritis,[161] each pound of weight loss was associated with an approximate four-unit reduction of knee joint forces including compressive force, resultant force, abduction moment, and medial rotation moment.

The 18-month Arthritis, Diet and Activity Promotion Trial of 316 community-dwelling individuals ≥60 years of age[162] demonstrated that a diet plus exercise intervention improved WOMAC scores significantly, relative to an exercise only, diet only, or control group. Both the diet only and the diet plus exercise groups had significant weight loss, but only the diet plus exercise group had significant reduction in self-reported pain ($P \leq 0.05$) with a 30.3% decrease in pain over the 18-month intervention. There was no difference in radiographic progression across intervention groups. The same study found that serum biomarkers (cartilage oligomeric protein, hyaluronan, antigenic keratin sulfate, and transforming growth factor β1 (TGF-βB1)) were relatively stable over 18 months; however, higher serum hyaluronan was correlated with worse radiographic classification, lending support for the idea that OA is a metabolically active disease.[163]

Another clinical trial of obese OA patients, aged >50 years, compared a very low energy diet (VLED) (415 kcal/day) to a low energy diet (LED) (810 kcal/day) for 16 weeks.[164] After 8 weeks, both groups were shifted to a 1,200 kcal/day diet of normal foods and meal replacements. There was no significant difference between diets in pain response, and both groups had significant improvement in symptoms.

The Action for Health in Diabetes Trial prospectively evaluated 2,203 obese individuals with baseline knee pain, randomized to an intensive lifestyle intervention program (ILI) (with support for behavioral change in diet and physical activity) vs an education group (general education about healthy eating and physical activity).[165] The ILI group had greater weight loss (−9.02 vs −0.78 kg, $P<0.001$) more improvement in WOMAC score ($P<0.001$), and in physical function, relative to the education group.

Messier et al., conducted the Intensive Diet and Exercise for Arthritis (IDEA) trial, which was a single-blinded, randomized, controlled trial of diet plus exercise, diet, or exercise alone between 2006 and 2011.[166] At 18 months, knee compression forces were lower in the diet alone group and interleukin 6 was lower in the diet and diet plus exercise groups, compared with other groups.[167] Weight loss from diet had caused the most significant load reductions at the hip, knee, and ankle.[168] There appeared to be a dose response to weight loss.[168,169] Long term weight loss of 10% to 19.9% of baseline body weight was more optimal than lesser amounts of weight loss. An additional 10% of loss caused reductions in pain and improved function.[168,169]

Bariatric surgical procedures are increasingly being used to cause weight loss.[170] Hooper et al., followed 47 obese women and one man (mean age 44±9 years, mean BMI=51±8 kg/m^2) who had an open or laparoscopic roux-en-Y gastric bypass 12 months previously.[171] The mean weight loss in the women was 41±15 kg. They reported a 51% reduction in WOMAC pain score, a 64% decrease in stiffness, and a 74% improvement in function. Another study followed 53 patients who underwent laparoscopic gastric banding for 2 years and reported significant weight loss and substantial improvement in OA.[172]

A cohort study from the Health Improvement Network of individuals with OA found that the rate of analgesic prescription was lower in the bariatric surgery group compared with the non-bariatric surgery group. The number of analgesic prescriptions was also lower during a follow-up of 4.3 years.[173]

Given the findings above, various consensus panels have endorsed weight loss as part of a comprehensive treatment program for OA. The American College of Rheumatology's hand, hip, and knee guidelines strongly recommend weight loss for overweight or obese individuals.[174] The European Alliance of Associations for Rheumatology suggested specific guidelines for individuals with hip and knee OA, including exercise, self-monitoring, meal plans, and nutritional and psychological evaluation and education.[175,176] The Osteoarthritis Research Society International has also recommended dietary weight management as part of OA therapy.[177,178]

Nutritional Supplements

Glucosamine/Chondroitin Sulfate

Glucosamine and chondroitin sulfate are components of cartilage and results have been mixed related to their use for treatment of both knee and hip OA.[179]

Clegg et al.[180] conducted a multicenter, double-blind, placebo and celecoxib-controlled Arthritis Intervention Trial (GAIT). The 1,583 patients who had symptomatic knee OA were randomly assigned to 1,500 mg of glucosamine, 1,200 mg of chondroitin sulfate, both, 200 mg of celecoxib (a nonsteroidal anti-inflammatory medication) daily, or placebo for 24 weeks. The outcome measure was a 20% decrease in knee pain. Chondroitin sulfate or glucosamine alone or in combination did not reduce knee pain. The authors noted a high placebo response rate and noted that most patients had relatively low knee pain scores at baseline. Subanalysis suggested that glucosamine may have benefit in individuals with moderate to severe knee pain.

Fransen et al.[181] conducted a double-blind, placebo-controlled, randomized 2-year trial of 605 participants of 1,500 mg of glucosamine sulfate, 800 mg of chondroitin sulfate, both, or placebo. The combination showed a significant reduction in joint space narrowing ($P=0.046$) compared with placebo. There were no significant differences in knee pain between the treatment groups compared with placebo.

Meta-analyses related to glucosamine/chondroitin[182-184] have shown that glucosamine, chondroitin, or the combination are generally less effective than celecoxib for pain relief and improvement of physical function. Glucosamine supplementation was associated with improvements related to stiffness. Simental-Mendia[182] found that glucosamine, chondroitin, or the combination did not affect the WOMAC index but suggested that both glucosamine and chondroitin sulfate alone improved knee pain. McCarty et al.,[185] suggested that the dose of glucosamine of 1.5 g daily is likely not adequate to demonstrate efficacy. They suggest that dose-ranging studies with glucosamine are important for future evaluation.

Curcumin

Many natural supplements have been suggested for use in OA. Turmeric (*curcuma longa*) is a plant in the ginger family found in Southeast Asia that has been suggested as a treatment for OA.[186] Curcumin is unstable and has low bioavailability,[186] and trials have used various formulations, which makes the interpretation of curcumin outcomes in OA difficult. Clinical trials have generally been low in quality and meta-analyses have suggested that the low numbers of patients and the poor quality of trials have not allowed for definitive conclusions related to efficacy.[187,188] However, a recent meta-analysis suggested that curcumin and its extracts may be a safe and effective supplement for use in OA.[189] However additional studies are required to draw a conclusion about the usefulness of curcumin.

Summary

Many of the studies of nutrient supplementation related to OA have the disadvantage of not being studied in a deficient population or may have had inadequate dosing. Therefore, outcomes related to nutritional supplements in OA are difficult to interpret. The strongest evidence exists for weight loss and adherence to a Mediterranean diet as dietary interventions for OA. The efficacy of nutritional supplements such as glucosamine and chondroitin sulfate and curcumin is controversial and requires further study.

Rheumatoid Arthritis (RA)

Definition

RA is an inflammatory disease of the synovial lining of the joint and results in pain, stiffness, swelling, joint damage, and loss of function. Inflammation most often affects joints of the hands and feet and tends to be symmetrical. This symmetry helps distinguish RA from other forms of the disease.[190] Approximately 1% of the U.S. population, (approximately 2.1 million people) has RA.[191] Patients with RA may have excess death, radiographic deterioration of joints, and reduced functional status, despite disease-modifying antirheumatic drug (DMARD) treatment.[192-194]

Mechanisms Affecting Nutritional Status in RA

RA can affect nutritional status through several mechanisms.[195-197] Joint swelling and tenderness can impair the ability to prepare food, involvement of the temporomandibular joint can impair chewing, and general joint involvement can impair self-feeding. In addition, patients may have xerostomia that can impair food intake. Many of the pharmacologic therapies used to treat RA also have side effects that further impair food intake, such as nausea with MTX or proteolysis with corticosteroid therapy.[198,199] The catabolic and inflammatory aspects of the disease also affect nutritional status (see below). **Table 89.2** displays mechanisms by which RA may affect nutritional status.

Nutrient Intakes and Vitamin Levels in Patients with RA

In 1943, 31 patients with RA were evaluated to determine their dietary history for the year prior to onset.[200] Although their dietary intakes were not substantially different from those of a cross-section of families in the North Atlantic States, more than two-thirds of patients had diets that were low in calcium, thiamine, and riboflavin, and approximately 50% had low intake of vitamin C. In 1996, Kremer et al.,[201] found that patients with RA had diets that were high in total fat, low in fiber, and deficient in micronutrients. Among 41 patients with active RA, diets were low in pyridoxine, zinc, magnesium, folate, and copper.

Morgan et al.[197,202] evaluated the nutrient intake and status of 32 MTX treated adults with RA.[197] Intakes of pyridoxine, calcium, magnesium, and zinc were less than 33% of the 1989 RDA. Thirty percent had deficient plasma folate and 23% had deficient erythrocyte folate prior to starting MTX therapy. In a subsequent study, 79 patients (mean age 53 y and disease duration 9 y) were followed for one year after starting MTX therapy.[202] Based on the average of five 24-hour dietary recalls, the patients consumed <67% of the Recommended Dietary Allowance (RDA) for folate, vitamin B_{12}, vitamin E, calcium, iron, magnesium, copper, and zinc. Prior to the start of MTX therapy, 47% of patients had deficient plasma folate and 11% had deficient erythrocyte folate concentrations.

Nutritional Status of Patients with RA and Juvenile RA and Body Composition

Comorbidities of RA include loss of lean body mass, termed rheumatoid cachexia, which is the result of the hypermetabolism associated with this disease.[203-207] Sir James Paget first described

Table 89.2 Adverse Effects of Rheumatoid Arthritis on Nutritional Status

Increased nutrient requirements
Increased metabolism and nitrogen losses increase protein requirements.
Inflammation increases micronutrient requirements (i.e., antioxidant vitamins).
Diminished intake
Articular disease diminishes the ability to self-feed, shop, and prepare foods.
Morning stiffness may decrease morning appetite.
Temporomandibular joint disease may impair the ability to chew.
Depression from chronic disease may impair food intake.
Sjogren's syndrome and xerostomia may impair food intake.
Diminished absorption
Abnormalities of the small bowel, liver and pancreas may decrease absorption of nutrients.
Drug therapy such as salicylazosulfpyridine may inhibit nutrient absorption.
Inadequate utilization
Drug therapy such as MTX, salicylazosulfapyridine, and nonsteroidal anti-inflammatory drugs may inhibit enzymes of intermediary metabolism.
Vitamin B_{12} deficiency can reduce folate uptake and retention by the cells.
Increased excretion
Urinary excretion of nutrients may be increased in the catabolic state of active disease and by some drug treatments such as prednisone.
Chronic blood loss from nonsteroidal anti-inflammatory agents may increase the need for hematopoietic precursors (e.g., iron).

rheumatoid cachexia in 1873.[208] In patients with rheumatoid cachexia, there may be an increase in fat mass with coexisting loss of lean body mass called cachectic obesity.[207] Rheumatoid cachexia may also be influenced by disease activity and physical function of the individual.[209]

Hallowell et al.[210] evaluated 50 patients with RA and 50 normal controls. Upper arm muscle circumference was low in 14% of patients and most of the biochemical measures (serum albumin, transferrin, retinol-binding protein, thyroxine-binding prealbumin, zinc, and folic acid) were low in the patients. Thirteen of 50 patients (26%) met the criteria for malnutrition (low anthropometry plus two or more biochemical measures), vs none of the controls.[210] In another study,[211] a likelihood of malnutrition index was calculated for 38 hospitalized patients with RA that included serum folate and vitamin C, triceps skin fold or BMI, arm muscle area corrected for bone, total lymphocyte count, serum albumin and hematocrit. Twenty-seven of 38 patients (71%) had a high likelihood of malnutrition, based on this index.

Mody et al.,[212] assessed the nutritional status of 220 patients with RA using triceps skinfold, upper arm muscle circumference, BMI, percentage ideal body weight, and serum albumin. Forty-five of 220 patients (20%) had a low value in one or more anthropometric measures and six (3%) had low albumin. Obesity, defined as a BMI >30, was found in 10% of this population. In contrast, Kalla et al.,[213] compared 65 patients with RA with 71 matched controls using anthropometric measurements and serum protein concentrations. Lean body mass was similar in the controls and patients, and corticosteroid therapy had no effect on anthropometric measurements.

In another study,[214] 24% of 75 RA outpatients had anthropometric measurements below the 10th percentile, and 14% below the 5th percentile. Patients with poorer functional class, more severe radiographic disease, or with extra articular disease, had worse nutritional status and loss of lean body mass than patients with less severe disease. Similarly, more than half of RA patients in another study had values in the lowest 10th percentile for upper arm muscle area; reflecting a loss in somatic muscle stores and the catabolic nature of the disease.[215] In female patients, there was a correlation between low fat mass and elevated erythrocyte sedimentation rates and C-reactive protein. The wasted patients were more limited in their ability to perform activities of daily living, as assessed by the Health Assessment Questionnaire (HAQ).

Roubenoff et al.[203] found that in 24 patients, 67% were cachectic and their lean body mass was inversely associated with numbers of swollen joints. This suggests that protein catabolism and gluconeogenesis may be required to maintain disease activity. Chronic energy deficits were apparently not a cause of lower lean body mass. Elevated tumor necrosis factor (TNF)-α was found in three of five patients with flaring joints. TNF-α was not elevated in patients with less active disease. A later study showed that lean body mass was 13% lower in RA patients considered to be in control (without change in medication over the past year) than in matched controls, indicating that there was a loss of approximately one-third of mobilizable muscle cell mass.[152] Cytokine production from peripheral blood mononuclear cells was also increased in these patients and higher interleukin-1β and TNF-α production was associated with increased resting energy expenditure.

Morgan et al.[202] measured BMI in 79 patients starting MTX for RA. Sixty percent had BMI <25, 10% were moderately underweight (BMI 18–20.9), and 4% were severely underweight (BMI <18); but 30% were overweight (BMI 25–29.9), and 10% obese (BMI ≥30). Roubenoff et al.[216] found that low physical activity by women with RA results in lower total energy expenditure. While patients with RA are at risk for cachexia, poor food choices and inactivity may also put them at risk for obesity.[217,218]

Another study[219] evaluated body composition by dual-energy x-ray absorptiometry (DXA) in 60 patients hospitalized for RA. The mean BMIs for women and men, respectively, were 24.4 and 26.9. However, by calculation of the fat-free mass index (fat-free mass/kg/m^2), 52% of the women and 30% of the men were malnourished. They also evaluated the Subjective Global Assessment (SGA)[220-222] and Mini Nutritional Assessment (MNA)[223,224] malnutrition screening tools. The MNA had poor specificity and good sensitivity, while the SGA had good specificity and poor sensitivity to detect malnutrition. The authors suggested that the MNA be used as a screening tool, followed by body composition analysis with DXA. In another study of 80 outpatients with RA, the same investigators[225] completed DXA, bioelectrical impedance (BIA), BMI, MNA, and waist circumference. Twenty-six percent of women and 21% of men had low fat-free mass index and, in this cohort, the MNA and waist circumference were poor predictors of rheumatoid cachexia. DXA and BIA had good general agreement, but as limits of agreement were wide, they concluded that BIA may be of limited usefulness in clinical practice.

Other investigations have generally confirmed the relationship between chronic inflammation and rheumatoid cachexia/cachectic obesity.[207] A cross-sectional study of 14 patients with RA and 14 patients with noninflammatory arthritis,[226] using joint counts for tenderness and swelling and the modified Stanford Health Assessment questionnaire (mHAQ),[227,228] found that REE, adjusted for fat-free mass, was significantly higher in the patients with RA compared with patients with noninflammatory arthritis (1498±162 kcal vs 1330±206 kcal, $P<0.031$). In addition, REE and IL-6 concentrations were higher and fat-free mass lower in patients with active arthritis, suggesting that hypermetabolism and loss of fat-free mass are associated with active inflammatory disease.

Giles et al., matched 189 men and women with RA to non-RA controls and used DXA to determine body composition.[229] Women, and older individuals with RA, had higher likelihood of presarcopenia and sarcopenia,[230] obesity, and sarcopenic obesity than those without RA. Abnormal body composition was associated with higher self-reported disability scores (m-HAQ), higher C-reactive protein, greater joint deformity, rheumatoid factor seropositivity, and lack of current treatment with a DMARD.[229] In another study, higher levels of malnutrition were associated with activity limitations, frailty, dementia, and lower quality of life scores.[231] Additionally, being overweight or obese has been associated with higher tender and swollen joint counts in individuals with RA.[232] In another investigation, in 197 individuals,[233] disability score by m-HAQ was 0.52 units higher for individuals in the highest vs the lowest quartile of fat mass ($P<0.001$) and 0.81 units higher for individuals in the lowest vs highest quartile of lean body mass ($P<0.001$). Various inflammatory mediators have been associated with disability and cachexia, including low IFG-1/GFBP-1 ratio[234] (insulin-like growth factor IGF-1 and its regulating binding protein IGFBP-1 insulin-like growth factor -1 binding protein).

Toms et al., evaluated 400 patients with RA for metabolic syndrome.[235] Older age and high HAQ score were associated with higher prevalence, while MTX therapy was associated with reduced likelihood of metabolic syndrome. In contrast, Stavropolulos-Kalinoglou correlated physical activity, dietary intake from a 3-day food diary, and inflammatory status (IL-6, IL-1, TNF-α levels) with BMI and body fat determined by bioelectrical impedance.[236] They found no relationship between inflammatory markers and BMI and body fat. Instead, higher levels of activity had the highest association with lower body fat and low BMI.

Several strategies have been used to try to alter the detrimental effects of chronic stress of RA on body composition.[237,238] Progressive resistance exercise has been shown to increase strength but is not accompanied by changes in body composition.[237,238] Rall et al.,[239] showed that treatment with MTX normalized leucine kinetics, probably as a result of a reduction in protein catabolism. In a study of obese patients with RA, a weight reduction program that emphasized caloric restriction, protein supplements, and physical exercise for 12 weeks was demonstrated to maintain lean body cell mass and maintain physical functioning.[240]

One intervention study, using beta-hydroxy-butyric acid, glutamine, and arginine vs a placebo mixture of alanine, glutamic acid, glycine, and serine for 12 weeks to treat the hypermetabolism in RA.[241] The beta-hydroxy-butyric acid, glutamine, and arginine were not more effective than the placebo mixture in reversing cachexia in patients with RA.

More recently, a study looking at the effects of tocilizumab, a humanized anti-interleukin 6 (IL-6) receptor antibody, on body composition and metabolic profile in RA patients found that RA patients gained weight via muscle mass and modification in fat distribution, but not an increase in fat, during treatment with tocilizumab, suggesting that blocking IL-6 might help treat sarcopenia in RA patients.[242]

Diet Therapy for RA

Food Allergy

That allergic responses to food may be related to RA disease activity is a relatively old idea. In the early 1900s, many reports of arthritis caused by specific foods appeared.[243-251] Van de Laar et al.,[252] reported circumstantial evidence for a relationship between RA and food allergy. The human gut contains immunologically active cells in the Peyer's patches.[252-256] Mechanisms that may alter the handling of food antigens include atrophic gastritis, increased intestinal permeability and alteration in gut flora.

Thirty-five patients who reported food-related gastrointestinal symptoms were evaluated to determine if there was a relationship between joint swelling and arthralgias and the presence of circulating immune complexes.[257] The number of individuals with circulating immune complexes was higher in patients with arthralgias than in patients without arthralgias, and there were higher levels of complexes in those with arthralgias. This supports the idea that food-related gastrointestinal symptoms might involve immunologic/allergic mechanisms. However, the success of elemental diets, which do not contain antigenic proteins, has been variable.[258,259]

In a study of 704 patients relating food intakes to arthritis symptoms, 28% felt that there was an association between a specific food and their disease activity.[260] The foods that were thought to have an unfavorable effect included: preservatives, beef, pork, food additives, milk, and sugar. Patients with a history of allergy, drug allergy, or a family history of allergy were more likely to report an association between foods and disease activity.

Panush et al.[261] described a patient with inflammatory arthritis whose joint symptoms worsened with milk intake. The patient was fasted for several days and then given an elemental diet for 33 days followed by blinded food challenges. Placebo foods such as lettuce and carrots caused no exacerbations in joint symptoms. Four blinded milk challenges caused increases in morning stiffness and swollen joints. Immune studies suggested delayed and cutaneous hyper-reactivity to milk. An additional case study documented a patient whose RA was worsened by intake of milk and cheese.[262] In contrast, a diet recall study by Kong et al., found no notable correlation between dairy consumption and patients' self-reported disease activity by RAPID3 questionnaire.[263]

The mucosal production of nitric oxide and release of myeloperoxidase and eosinophil cationic protein were measured after rectal challenges with cow's milk and gluten in 27 individuals with RA and 18 healthy controls,[264] but no relationship between self-reported food intolerance symptoms and mucosal reactivity was seen. In a study of 60 children with juvenile RA, one of 60 patients had milk intolerance, which was associated with knee swelling.[265]

In conclusion, while food allergy may affect a small percentage of patients with RA, it is not likely a major factor in most. If patients indicate that they have an arthritis flare associated with a food, a prudent course of action would be to avoid the offending food. Referral to a registered/licensed dietitian can be beneficial, to make sure that the diet is nutritionally adequate and that food groups are not completely excluded.

Omega-3 Fatty Acids as Therapy for RA

The anti-inflammatory effects of omega 3 fatty acids in RA are due to its replacement of omega-6 fatty acids (primarily arachidonic acid) at enzymes that metabolize omega-6 to proinflammatory mediators.[266] This, in effect, inhibits omega-6 metabolism. Omega 3 fatty acids may also inhibit the production of cytokines and reactive oxygen species.[267] The typical dose is 2 to 5 grams per day of eicosapentaenoic acid (EPA) and 1 to 3 grams per day of docosahexaenoic acid (DHA), usually oral, although intravenous infusion as an emulsion was also effective in the treatment of adjuvant arthritis.[268] A diet relatively low in AA along with omega-3 fatty acid supplements was shown to produce better clinical results, along with lower amounts of leukotriene, thromboxanes, and prostaglandin metabolism compared with a Western diet.[269]

In a double-blind, placebo-controlled crossover study, patients with RA received dairy products with or without added omega-3 supplements.[270] Biomarkers for DNA damage and oxidative stress were not changed. However, excretion of hydroxypyridinium crosslinks was reduced by this treatment indicating protection of bone and cartilage collagen.

Flaxseed, which contains α-linolenic acid, a short-chain ω-3 fatty acid, as part of an anti-inflammatory diet may be useful as an adjuvant therapy in treatment of RA, with one study showing decreases in pain severity, morning stiffness, and improvement in quality of life compared with a control group.[271]

The addition of olive oil (contains oleic acid an ω-9 monounsaturated fatty acid) to omega- 3 supplements was reported to produce better clinical changes in patients with RA[272] and a study from Japan showed that eating fish more than two times/week was negatively correlated with frailty in RA.[273] The omega-9 series of fatty acid may be metabolized to compounds that inhibit the synthesis of proinflammatory mediators from the omega-6 series of fatty acids. A report from Sweden suggests that moderate intake of oily fish may prevent RA.[274] This agrees with the low prevalence of RA in the fish-eating Greenland Eskimos.

A prospective, double-blind randomized-controlled trial with 60 female patients with RA in Serbia found that a 12-week supplementation with fish oil (omega-3) or fish oil with primrose evening oil resulted in a decrease in the DAS 28 score compared with a control group with no supplementation.[275] Most recently, a systematic review from 2021 found that administration of omega-3 polyunsaturated fatty acids at high doses (≥2.7 g/d) may reduce RA disease activity.[276]

The Effect of Specific Foods on Rheumatic Symptoms

In a study of 742 patients with RA, juvenile RA, ankylosing spondylitis, psoriatic arthropathy, primary fibromyalgia, and OA, one-third of RA patients reported a worsening of symptoms with specific foods.[277] Forty-three percent of patients with juvenile RA reported an exacerbation of symptoms with specific foods. The foods that were most frequently cited in worsening joint symptoms were meat, wine, alcohol, coffee, sweets, sugar, chocolate, apples, and citrus fruits.[180]

Popular Diets as Treatment for RA

Panush et al.,[278] conducted a double-blind, placebo-controlled trial of diet therapy for patients with active RA who met American College of Rheumatology criteria. The experimental diet was modeled after the "Dong Diet"[279,280] and was free of additives, preservatives, fruit, red meat, herbs, and dairy products. The placebo diet also excluded items from some major food groups so that it somewhat mimicked the experimental dietary intervention. Clinical evaluations occurred at entry and termination of the 10-week trial showed no differences between the experimental vs the placebo diet in patients' global assessment of disease, examiner's assessment of disease, duration of morning stiffness, 50-foot walking time, grip strength, numbers of tender or swollen joints, and erythrocyte sedimentation rates. However, two patients improved markedly on the experimental diet and elected to continue it.

Fifty patients with RA were followed in a 24-week, double-blind, randomized trial of a well-balanced diet vs a diet low in saturated fat, high in polyunsaturated fat, and with "hypoallergenic" foods.[281] The caloric content of the experimental diet was set to facilitate weight loss and achieve the patient's ideal body weight. There was a significant reduction in the number of tender joints and erythrocyte sedimentation rate between the experimental diet and the balanced diet. No other measures of disease activity differed. It was speculated that weight loss during the protocol may have had a beneficial effect on disease activity.

A recent trial[282] to examine whether a weight loss intervention program may improve rheumatoid arthritis disease activity may add further evidence to weight loss' beneficial effect on disease activity. Obese RA patients (BMI≥30) with DAS28≥3.2 were randomized to a diet intervention of a hypocaloric diet of 1,000 to 1,500 kcal/day and high protein meal replacements or control diet. The diet intervention group lost an average of 9.5 kg vs control group (0.5 kg) ($P<0.001$) along with improvements in RAPID3, decrease in serum leptin, and increase in serum adiponectin after 12 weeks.[282]

Fasting and Vegetarian Diets

The use of fasting and vegetarian diets has been extensively reviewed.[243,245] The preliminary reports of improvements in disease activity with fasting have led to speculation on the relationship of RA to dietary antigens, as fasting would lower the antigen stimulus to the gut mucosa.

In 1979, Sköldstam et al.[283] randomized RA patients to a control diet or to a 10-day fast, followed by a lactovegetarian diet. Fasting individuals received 800 kcal/day from fruit and vegetable juices, were purged with castor oil and received five water enemas on the first day of the fast. After fasting, the individuals began a lactovegetarian diet for 1 week and then were discharged with instructions to follow the diet for an additional 9 weeks. Yogurt was allowed, but other dairy products were discouraged. Ten patients served as controls and consumed a more typical diet. When compared with controls, five of 15 patients had a 10% decrease in erythrocyte sedimentation rate and a five-point or more decrease in the Ritchie articular index of joint tenderness. However, only one of 14 in the diet group had persistent improvement after 9 weeks. Fasting patients lost a mean of 3.5 kg during the fasting period, which could itself have altered disease activity.

In an additional study by the same group,[284] 20 inpatients were followed for several months on an ordinary diet and then, after fasting for 10 days, began a vegan diet. The vegan diet did not include refined sugar, corn flour, salt, strong spices, preservatives, alcoholic beverages, tea, or coffee. Patients improved in their pain perception, but there were no significant changes in grip strength, tender joint count, or an index of joint tenderness, compared with the control period. Because improvements in disease activity in trials have frequently been accompanied by weight loss, the authors completed a follow-up study using data from clinical trials that evaluated the effects of vegan, lactovegetarian, or Mediterranean diets.[285] Loss of body weight did not significantly contribute to disease improvement in this study. However, in two randomized clinical trials in Mexico, women with RA who followed a Mediterranean diet and participated in a dynamic exercise program had decreases in weight and waist circumference and an increase in hand grip strength[286] and improvement in quality of life scores.[287]

Kjeldsen-Kragh et al.,[288] completed a randomized, single-blinded, controlled trial of patients who stayed at a health farm for 4 weeks. After an initial 10-day fast, patients were changed to a gluten-free vegan diet for 3.5 months and then a lactovegetarian diet for the remainder. A control group stayed at a convalescent home and ate a general diet. Compared with the control group, at 1 month, there was a significant improvement in number of tender joints, Ritchie's articular index, and the number of swollen joints, the pain score, and the duration of morning stiffness, grip strength, erythrocyte sedimentation rate, C-reactive protein, and HAQ score. The individuals in the convalescent home had a decrease only in the pain score. At 13 months, there was a persistent improvement in the diet group compared with the control group in most of the indices. The radiographic score deteriorated slightly in both groups at the end of the trial. Although food allergy was suspected, the authors stated that it probably did not explain the results. They speculated that changes in fatty acids in the diet or weight loss could have affected disease activity. The BMI and triceps skinfold significantly decreased in the diet group vs control group; however, the upper arm muscle area was not different.[289] There were no significant differences in serum albumin, hemoglobin, ferritin, zinc, or copper concentrations between the two groups, suggesting that the fasting and vegetarian diet had only a minor effect on nutritional status. Platelet counts, white blood cell counts, total immunoglobulin (Ig)G, IgM rheumatoid factor, and complement factors C3 and C4 were significantly lower in the diet group after 1 month.[290] At 1 year, only white blood cell count,

IgM rheumatoid factor, and C3 and C4 were significantly different. The diet group was further classified as diet responders and diet nonresponders, based on subjective improvement in joint disease activity. There were greater decreases in platelet count, C3, C4, in the responders vs nonresponders. Fecal flora changed markedly when the patients changed from the omnivorous diet to the vegan diet and differed markedly among patients with high and low response.[291] The authors speculated that changes in gut flora might be beneficial to disease activity by reducing intestinal inflammation and reducing absorption of bacterial and dietary antigens.

Differences in psychological factors among the individuals who volunteered for the trial vs other patients without RA not in the trial, and diet responders vs nonresponders were evaluated.[292] The study patients had a stronger belief in their ability to influence their health status, lower belief that chance affects their health, and greater belief in the efficacy of alternative forms of treatment. Responders had a significantly lower belief in the effectiveness of traditional medical care than nonresponders. This suggests that patients with specific psychological characteristics were selected for the clinical trial and that the expectations concerning the efficacy of alternative therapies could have contributed to the effect of the vegetarian diet. A 2-year follow-up of the patients found that all of the diet responders, and half of the diet nonresponders, were still following a modified diet.[292] No food item was identified that worsened joint disease. There was also a significant difference in the pain score, duration of morning stiffness, number of tender joints, Ritchie's articular index, and the number of swollen joints between diet responders vs the control patients. This suggests that some subgroup of RA patients can benefit from a vegetarian diet and that the effects of the diet can be sustained in a subgroup of patients.

Nenonen et al.,[293] randomized 43 patients to a control diet or to an uncooked vegan diet, rich in lactobacilli. One-half of the dietary group experienced nausea or diarrhea during the dietary intervention and stopped the diet prematurely. The HAQ score, duration of morning stiffness and pain at rest and with movement, the Ritchie articular index, and C-reactive protein were not different from the control and diet groups at 2 months. However, a composite Disease Activity Score that measured subjective changes in arthritis was significantly improved in the dietary intervention group, compared with the control group. There was a significant difference in fecal flora between patients with improvement vs little improvement after 1 month of the diet.[294]

Hafstrom et al., randomized 66 patients with RA to a gluten-free vegan diet or a nonvegan diet for 1 year (22 in the gluten-free vegan group and 25 in the nonvegan diet group completed 9 months of therapy).[295] Also, 40.5% of the vegan group had an ACR20 (a measure of joint improvement) improvement in symptoms compared with 4% in the nonvegan group. While antibodies against gliadin and β-lactoglobulin decreased in the gluten-free vegan group, there was no change in radiologic progression in either group.

McDougall et al.,[296] reported on a 4-week, very low fat (10% of total energy) vegan diet intervention in free-living RA patients. After 4 weeks, there was a significant loss in body weight, but erythrocyte sedimentation rate, C-reactive protein, and rheumatoid factor did not change. The joint tenderness score, joint swelling score, and severity of morning stiffness decreased significantly, but there was no significant change in duration of morning stiffness. Patients with the largest degree of improvement had the most active disease at the beginning of the study; individuals with long-standing disease showed little improvement. The authors concluded that a vegan diet could improve symptoms and speculated that changes in intestinal permeability to antigens and levels of food and bacterial antigens could be mechanisms.

A study in Italy recruited 40 patients with long-standing RA with stable symptoms treated with conventional and biologic DMARDS and evaluated the effect of a 3-month diet, avoiding meat, gluten, and lactose, compared with a controlled, balanced diet including meat, gluten, and lactose and found that the elimination diet group had an improvement in the Short Form Health survey (SF-36), but not Disease Activity Score of 28 joints (DAS28). Circulating leukocytes, neutrophils, and hs-C-reactive protein was decreased after the 3-month diet intervention.[297]

A Cochrane review on dietary interventions for RA concluded that it was difficult to pool data from separate studies, but noted that two randomized controlled trials related to fasting and Mediterranean diet improved pain.[298] They, and another review[219] concluded that the effects of vegan and elimination diets were uncertain and that many of the diets evaluated had a high drop-out rate and weight loss as unexpected side effects. A more recent review found that fasting resulted in significant but transient subjective improvements, the Mediterranean diet resulted in improvements in some RA disease activity measures, and vegetarian, elimination, peptide, and elemental diets resulted in varied responses.[276]

In conclusion, fasting and a vegetarian type of diet may have beneficial effects in a subset of patients. Food allergy may be related to disease activity in a minority of patients. Fasting is not a long-term therapeutic approach to the management of RA.

Complementary/Alternative Nutritional Therapies for RA

The use of complementary/alternative diet approaches for treatment of RA is common. In a sample of 296 patients, 60.5% admitted to the use of some sort of alternative therapy such as herbs/algae or cartilage components.[299] Various compounds have been studied for their efficacy in RA.[300] While many have preliminary data, stronger human studies are necessary for most. The effect of probiotic supplementation on disease activity in RA has been mixed. Some studies have shown *Lactobacillus casei* supplementation and the combination of *L. casei* and *Bifidobacterium bificum* improve disease activity and inflammatory status.[301-303] In another study, probiotic supplements were not found to improve symptoms in RA.[304]

Ginger as Therapy for RA

A double-blind, placebo-controlled trial by Aryaeian et al.,[305] looked at the effect of ginger supplementation on gene expression of genes involved in immune mediation, NF-Kβ, PPAR-γ, FOXP3, T-bet, GATA-3, and RORγt in 70 active RA patients in Iran. Study participants were divided into two groups, either ginger (1,500 mg daily as two capsules) or placebo. After the 12-week intervention, reduction of disease activity was statistically significant and gene expression of FoxP3 increased significantly and T-bet and RORγt gene expression decreased significantly between the two groups.[305]

Summary

Reduced capacity for food preparation and poor intake results in poor nutritional status for some patients with arthritis. Medications used in treatment may also affect nutritional status. In recent decades, both overweight and underweight patients with arthritis have appeared. For the overweight patient, it is especially important to lose weight and normalize insulin sensitivity, which may alleviate some symptoms of gout, OA, and possibly RA.

The hypermetabolic state of RA may lead to loss of lean body mass. The removal of specific foods from the diet and fasting followed by a vegetarian diet may help a small subset of patients. Specific diets and glucosamine and chondroitin sulfate supplements remain unproven therapies for the treatment of RA or OA. On the other hand, omega-3 fatty acid supplements have some benefit in the treatment of RA and quality of life of RA patients. An example of a preventable adverse drug-nutrient interaction is the use of folic acid supplements to prevent MTX toxicity in RA patients.

References

1. The Free Dictionary. Available at http://medical-dictionary.thefreedictionary.com/Rheumatic+diseases.
2. The Free Dictionary. Rheumatic disease. Accessed March 11, 2024. Available at http://medical-dictionary.thefreedictionary.com/Rheumatic+diseases
3. MedicineNet. Arthritis (joint inflammation). Accessed March 11, 2024. Available at www.medicinenet.com/arthritis/article.htm
4. Arthritis Foundation. What is arthritis. Accessed March 11, 2024. Available at www.arthritis.org/health-wellness/about-arthritis/understanding-arthritis/what-is-arthritis
5. Helmick CG, Felson DT, Lawrence RC, et al. Arthritis Rheum 2008;58:15–25.
6. Rall LC, Rosen CJ, Dolnikowski G, et al. Arthritis Rheum 1996;39:1115–24.
7. Rall LC, Roubenoff R. Rheumatol (Oxford, England) 2004;43:1219–23.
8. Roubenoff R, Roubenoff RA, Cannon JG., et al. J Clin Invest 1994;93:2379–86.
9. Roubenoff R, Roubenoff RA, Ward LJ, et al. Am J Clin Nutr 1990;52:1113–7.
10. Arthritis Center and Department of Nutrition Sciences at The University of Alabama at Birmingham. The Essential Arthritis Cookbook. Mankato, MN: Appletree Press, Inc., 1999.
11. Arthritis Center and The Department of Nutrition Sciences TUoAaB. The Essential Arthritis Cookbook Kitchen Basics for People with Arthritis, Fibromyalgia, and Other Chronic Pain and Fatigue. Mankato, MN: Appletree Press, Inc.
12. Alonso-Aperte E, Varela-Moreiras G. Eur Journal Clin Nutr 2004;54:S69–74.
13. Mason P. Proc Nutr Soc 2010. Epub ahead of print.
14. Alarcon G, Tracy I, Blackburn W. Arthritis Rheum 1989;32:671–6.
15. Kremer JM, Galivan J, Streckfuss A, et al. Arthritis Rheum 1986;29:832–5.
16. Leeb BF, Witzmann G, Ogris E, et al. Clin Exp Rheumatol 1995;13:459–63.
17. Morgan SL, Baggott JE, Altz-Smith M. Arthritis Rheum 1987;30:1348–56.
18. Morgan SL, Baggott JE, Lee JY, et al. J Rheumatol 1998;25:441–6.
19. van Ede AE, Laan RFJM, Blom HJ, et al. Sem Arthrit Rheum 1998;27:277–92.
20. Tishler M, Caspi D, Fishel B, et al. Arthrit Rheum 1988;31:906–8.
21. Roe D. Diet and Drug Interactions. Ithaca, NY: Van Nostrand Reinhold, 1989.
22. Roe D. Drug-induced Nutritional Deficiencies. Ithaca, NY: AVI Publishing Company, 1985.
23. Pelton R, LaValle JB, Hawkins EB, et al. Drug-induced Nutrient Depletion Handbook. Ohio: Leexsi-Comp, Inc., 2000:416–26.
24. Bijlsma JW, Jacobs JW. Joint, Bone, Spine : Revue du Rhumatisme 2009;76:452–4.
25. Pincus T, Yazici Y, Sokka T, et al. Clin Exper Rheum 2003;21:S179–85.
26. Baggott JE, Morgan SL, Ha TS, et al. Clin Exp Rheum 1993;11:101–5.
27. Baggott JE, Morgan SL, Sams WM, et al. Arch Dermatol 1999;135:813–7.
28. Morgan SL, Baggott JE, Vaughn WH, et al. Arthritis Rheum 1990;33:9–18.
29. Morgan SL, Baggott JE, Vaughn WH, et al. Ann Intern Med 1994;121:833–41.
30. Hornung N, Ellingsen T, Stengaard-Pedersen K, et al. J Rheumatol 2004;31:2374–81.
31. Hunt PG, Rose CD, McIlvain-Simpson G, et al. J Rheumatol 1997;24:2230–2.
32. Morgan SL, Baggott JE, Vaughn WH, et al. Ann Intern Med 1994;121:833–41.
33. Morgan SL, Baggott JE, Vaughn WH, et al. Arthrit Reum 1990;33:9–18.
34. Stewart KA, MacKenzie AH, Clough JD, et al. Sem Arthrit Rheum 1991;20:332–38.
35. Shiroky JB, Neville C, Esdaile JM, et al. Arthritis Rheum 1993;36:795–803.
36. Morgan SL, Baggott JE, Alarcon G. BioDrugs 1997;8:164–75.
37. Diaz-Borjon A. Drugs Aging 2009;26:273–93.
38. Niehues T, Lankisch P. Pediatr Drugs 2006;8:347–56.
39. Pavy S, Constantin A, Pham T, et al. Joint, Bone, Spine: Revue du Rhumatisme 2006;73:388–95.
40. Visser K, Katchamart W, Loza E, et al. Ann Rheum Dis 2009;68:1086–93.
41. Abramoff B, Caldera FE. Med Clin North Am 2020;104:293–311.
42. Blumberg J, Heaney RP, Huncharek M, et al. Nutr Rev 2010;68:478–84.
43. Heaney RP. Am J Clin Nutr 2003;78:912–9.
44. Heaney RP. J Nutr 2008;138:1591–5.
45. Heaney RP. N Engl J Med 2012;367:77–8.
46. Lappe JM, Heaney RP. Dermato-Endocrinol 2012;4:95–100.
47. Heaney RP. Curr Opin Clin Nutr Metab Care 2011;14:440–4.
48. Jacobs DR, Jr., Gross MD, Tapsell LC. Am J Clin Nutr 2009;89:1543S–8S.
49. Jacobs DR, Jr, Steffen LM. Am J Clin Nutr 2003;78:508S–13S.
50. Luk AJ, Simkin PA. Am J Manag Care 2005;11:S436–S42.
51. National Institute of Arthritis and Musculoskeletal and Skin Diseases. NIAMS Health Topics. Available at www.niams.nih.gov/hi/topics/arthritis/artrheu.htm. Bethesda, MD: National Institute of Arthritis and Musculoskeletal and Skin Diseases, 2002.
52. Richette P, Bardin T. Lancet 2010;375:318–28.
53. National Institute of Arthritis and Musculoskeletal and Skin Diseases. Gout. Accessed March 11, 2024. Available at www.niams.nih.gov/health-topics/gout
54. Pascual E, Sivera F. Curr Opin Rheum 2007;19:122–7.
55. Boulware DW, Becker MA, Edwards NL. In: Koopman WJ, Boulware DW, Heudebert GR, eds. Gout and Crystal Induced Synovitis. Clinical Primer of Rheumatology. Philadelphia, PA: Lippincott Williams & Wilkins, 2003:262–77.
56. Chen-Xu M, Yokose C, Rai SK, et al. Arthrit Rheum (Hoboken, NJ) 2019;71:991–9.

57. Eggebeen AT. Am Acad Family Phys 2007;76:801–8.
58. Gutman A. Postgrad Med 1972;51, 61–6.
59. Becker MA, Roessier BJ. In: Scriver CR, Beaudet AL, Sly WS, and Valle D, eds. Chapter 49, Hyperuricemia and Gout. The Metabolic and Molecular Bases of Inherited Disease. 1995;2:1655–77.
60. Terkeltaub R, Zelmlan D, Scavulli J, et al. Joint, Bone, Spine: Revue du Rhumatisme 2009;76:444–46.
61. Shen X, Wang C, Liang N, et al. Arthrit Rheumatol (Hoboken, NJ) 2021;73:1738–48.
62. Lee SJ, Terkeltaub RA, Kavanaugh A. Curr Opin Rheumatol 2006;18:193–98.
63. Choi HK, Ford ES, Li C, et al. Arthritis Rheum 2007;57:109–15.
64. Third Report of the National Cholesterol Education Program (NCEP) Expert Panel on Detection, Evaluation, and Treatment of High Blood Cholesterol in Adults (Adult Treatment Panel III) final report. Circ 2002;106:3143–421.
65. Choi HK, De Vera MA, Krishnan E. Rheumatol 2008;47:1567–70.
66. Fam AG. J Rheumatol 2002;29:1350–54.
67. Porcelli B, Vannoni D, Leoncini R, et al. In: Sahota A, and Taylor M, eds. Purine and Pyrimidine Metabolism in Man VIII. New York, NY: Plenum Press, 1995:47–52.
68. Fam AG. J Rheum 2002;29:1350–5.
69. Choi HK, Curhan G. Curr Opin Rheum 2005;17:341–5.
70. Choi HK, Liu S, Curhan G. Arthritis Rheum 2005;52:283–89
71. Choi HK, Atkinson K, Karlson EW, et al. N Engl J Med 2004;350:1093–103.
72. Chiu THT, Liu CH, Chang CC, et al. Clin Nutr (Edinburgh, Scotland) 2020;39:837–44.
73. Choi HK, Curhan G. Arthritis Care Res 2007;57:816–21.
74. Choi JWJ, Ford ES, Gao X, et al. Arthritis Care Res 2008;59: 109–16.
75. Jamnik J, Rehman S, Blanco Mejia S, et al. BMJ Open 2016;6: e013191.
76. Pillinger MH, Keenan RT. Bull NYU Hospital Joint Dis 2008;66:231–9.
77. Gao X, Curhan G, Forman JP, et al. J Rheumatol 2008;35:1853–8.
78. Yu K-H, Se L-C, Huang Y-C, et al. Semin Arthritis Rheum 2008;37:243–50.
79. Brzezińska O, Styrzyński F, Makowska J, et al. Nutrients 2021;13.
80. Fatima T, McKinney C, Major TJ, et al. Arthrit Res Ther 2018;20:179.
81. Choi HK, Atkinson K, Karlson EW, et al. Lancet 2004;363: 1277–81.
82. Emerson BT. N Engl J Med 1996;334:445–51.
83. Yu T-F. Am J Med 1974;56:676–85.
84. Saag KG, Choi H. Arthrit Res Ther 2006;8(suppl 1):S2–S8.
85. Choi HK. Curr Opin Rheumatol 2010:22:165–72.
86. Zhang W, Dohert M, Bardin T. Ann Rheum Dis 2006;65:1312–24.
87. Slupper A. Nutrition Care Manual, Chicago, IL: American Dietetic Association, 2008.
88. Danve A, Sehra ST, Neogi T. Best Pract Res Clin Rheum 2021;35:101723.
89. Kakutani-Hatayama M, Kadoya M, Okazaki H, et al. Am J Lifestyle Med 2017;11:321–9.
90. McCormick N, Rai SK, Lu N, et al. JAMA Netw Open 2020;3:e2027421.
91. Rai SK, Fung TT, Lu N, et al. BMJ 2017;357:j1794.
92. Tang O, Miller ER, 3rd, Gelber AC, et al. Clin Rheumatol 2017;36:1413–7.
93. Yokose C, McCormick N, Choi HK. Curr Rheum Rep 2021; 23:51.
94. El-Tantawy WH. Arch Physiol Biochem 2021;127:61–72.
95. Chen PE, Liu CY, Chien WH, et al. Evid Based Complement Alternat Med 2019;2019:9896757.
96. Singh JA, Green C, Morgan S, et al. (2020) J Clin Rheum 2019;26: 147–56.
97. Stamp LK, Chapman P, Frampton C, et al. Rheumatol (Oxford, England) 2020;59:2374–80.
98. Zhang Y, Neogi T, Chen C, et al. Arthrit Rheum 2012;64: 4004–11.
99. Collins MW, Saag KG, STher Adv Musculoskel Dis 2019;11: 1759720x19847018.
100. Shulten P, Thomas J, Miller M, et al. J Hum Nutri Diet 2009; 22:3–11.
101. The Free Dictionary. Available at http://medical-dictionary.thefreedictionary.com/osteoarthritis
102. Hutton CW. Ann Rheum Dis 1989;48:958–61.
103. Litwic A, Edwards MH, Dennison EM, et al. Brit Med Bull 2013;105:185–99.
104. Bortoluzzi A, Furini F, Scirè CA. Autoimmun Rev 2018;17: 1097–104.
105. Vina ER, Kwoh CK. Curr Opin Rheumatol 2018;30:160–7.
106. Lawrence RC, Felson DT, Helmick CG, et al. Arthritis Rheum 2008;58:26–35.
107. Dillon CF, Rasch EK, Gu Q, et al. J Rheumatol 2006;33: 2271–9.
108. Felson DT. Epidemiol Rev 1988;10:1–28.
109. Felson DT, Anderson JJ, Naimark A, et al. Ann Intern Med 1988;109:18–24.
110. Felson DT, Zhang Y. Arthritis Rheum 1998;41:1343–55.
111. Wallace IJ, Worthington S, Felson DT, et al. Proc Natl Acad Sci Un S A 2017;114:9332–6.
112. Kowsari B, Finnie S, Carter R, et al. J Am Dietetic Assoc 1983;82:657–59.
113. White-O-Connor B, Sobal J. Clin Ther 1986;9:30–43.
114. Comee L, Taylor CA, Nahikian-Nelms M, et al. Nutrition (Burbank, Los Angeles County, Calif.) 2019;67–68:110533.
115. Xu C, Marchand NE, Driban JB, et al. Am J Clin Nutr 2020;111:667–76.
116. Berenbaum F, Wallace IJ, Lieberman DE, et al. Nat Rev Rheumatol 2018;14:674–81.
117. Thomas S, Browne H, Mobasheri A, et al. Rheumatol (Oxford, England) 2018;57:iv61–iv74.
118. Eising L. J Bone Joint Surg 1963;45:69–81.
119. Reyes C, Leyland KM, Peat G, et al. Arthrit Rheumatol (Hoboken, NJ) 2016;68:1869–75.
120. Yusuf E, Nelissen RG, Ioan-Facsinay A, et al. Ann Rheumat Dis 2010;69:761–5.
121. Louati K, Vidal C, Berenbaum F, et al. RMD Open 2015;1:e000077.
122. Eymard F, Parsons C, Edwards MH, et al. Osteoarthrit Cart 2015;23:851–9.
123. Courties A, Sellam J. Diabetes Research Clin Pract 2016;122: 198–206.
124. Veronese N, Cooper C, Reginster JY, et al. Sem Arthritis Rheum 2019;49:9–19.
125. Baudart P, Louati K, Marcelli C, et al. RMD Open 2017;3: e000442.
126. Binvignat M, Sokol H, Mariotti-Ferrandiz E, et al. Joint, Bone, Spine: Revue du Rhumatisme 2021;88:105203.
127. Biver E, Berenbaum F, Valdes AM, et al. Ageing Res Rev 2019;55:100946.
128. de Sire A, de Sire R, Petito V, et al. Nutrients 2020;12.
129. Tan TC, Chong TKY, Low AHL, et al. Intl J Rheum Dis 2021;24:984–1003.
130. Lei M, Guo C, Wang D, et al. Benef Microbes 2017;8: 697–703.
131. Nathional Institutes of Health. Vitamin C. Accessed March 11, 2024. Available at https://ods.od.nih.gov/factsheets/VitaminC-HealthProfessional/.
132. Dunlap B, Patterson GT, Kumar S, et al. Ther Adv Chronic Dis 2021;12:20406223211047026.

133. National Institutes of Health. Accessed March 11, 2024. Available at https://ods.od.nih.gov/factsheets/Vitaminc-Consumer/
134. McAlindon TE, Felson DT, Zhang Y, et al. Ann Intern Med 1996;125:353–9.
135. Lane N, Gore R, Cummings S, et al. Arthritis Rheum 1999;42:854–60.
136. Felson DT, Niu J, Clancy M, et al. Arthritis Rheum 2007;56: 129–36.
137. Ding C, Cicuttini F, Parameswaran V, et al. Arthritis Rheum 2009;60:1381–89.
138. McAlindon T, LaValley M, Schneider E, et al. JAMA 2013;309:155–62.
139. Zheng S, Jin X, Cicuttini F, et al. JAMA 2017;130:1211–8.
140. Perry TA, Parkes MJ, Hodgson R, et al. BMC Musculoskel Disord 2019;20:76.
141. National Institutes of Health. Vitamin E. Accessed March 11, 2024. Available at https://ods.od.nih.gov/factsheets/VitaminE-Consumer/
142. Chin KY, Ima-Nirwana S. Front Pharmacol 2018;9:946.
143. Institute of Medicine Panel on the Definition of Dietary F, and the Standing Committee on the Scientific Evaluation of Dietary Reference I. Dietary Reference Intakes Proposed Definition of Dietary Fiber. Washington, DC: National Academies Press (US), 2001.
144. Dai Z, Lu N, Niu J, et al. Arthrit Care Res 2017;69:1331–9.
145. Dai Z, Niu J, Zhang Y, et al. Ann Rheum Dis 2017;76:1411–9.
146. Mustonen AM, Nieminen P. Curr Rheumatol Rep 2021;23:41.
147. Alissa EM, Alzughaibi LS, Marzouki ZM. J Hum Nutr Diet 2020;33:431–8.
148. Akbar U, Yang M, Kurian D, et al. J Clin Rheumatol 2017;23: 330–9.
149. Segal NA, Nevitt MC, Gross KD, et al. PM&R 2013;5:647–54.
150. Felson DT, Misra D, LaValley M, et al. Osteoarthrit Cart 2021;29:973–8.
151. Sahni S, McLean RR. Arthrit Care Res 2014;66:795–6.
152. Kaçar C, Gilgil E, Tuncer T, et al. Clin Exper Rheumatol 2004;22:473–6.
153. Li H, Zeng C, Wei J, et al. Rheumatol Intl 2016;36:1215–22.
154. Lu B, Driban JB, Duryea J, et al. Arthrit Care Res 2014;66: 802–9.
155. Schwingshackl L, Morze J, Hoffmann G. Brit J Pharmacol 2020;177:1241–57.
156. Veronese N, Stubbs B, Noale M, et al. Clin Nutr (Edinburgh, Scotland) 2017;36:1609–14.
157. Panagiotakos DB, Pitsavos C, Stefanadis C. Nutr Metab Cardiovasc Dis 2006;16:559–68.
158. Veronese N, Koyanagi A, Stubbs B, et al. Clin Nutr 2019;38: 2735–9.
159. Dyer J, Davison G, Marcora SM, et al. J Nutr Health Aging 2017;21:562–6.
160. Felson DT. Bull Rheum Dis 1992;4:6–7.
161. Messier SP, Gutekunst DJ, Davis C, et al. Arthrit Rheum 2005;52:2026–32.
162. Messier SP, Loeser RF, Miller GD, et al. Arthrit Rheum 2004;50:1501–10.
163. Chua SD, Jr., Messier SP, Legault C, et al. Osteoarthrit Cart 2008;16:1047–53.
164. Riecke BF, Christensen R, Christensen P, et al. Osteoarthrit Cart 2010;18:746–54.
165. Foy CG, Lewis CE, Hairston KG, et al. Obes 2011;19:83–93.
166. Messier SP, Legault C, Mihalko S, et al. BMC Musculoskel Disord 2009;10:93.
167. Messier SP, Mihalko SL, Legault C, et al. JAMA 2013;310: 1263–73.
168. Messier SP, Beavers DP, Mihalko SL, et al. J Biomechan 2020;98:109477.
169. Messier SP, Resnik AE, Beavers DP, et al. Arthrit Care Res 2018;70:1569–75.
170. Bult MJF, van Dalen T, Muller AF. Eur J Endocrin 2008;158: 135–45.
171. Hooper MM, Stellato TA, Hallowell PT, et al. Intl J Obes 2007;31:114–20.
172. Sultan S, Parikh M, Youn H, et al. Surg Endosc 2009;23: 1569–73.
173. Zeng C, Lane NE, Li X, et al. Osteoarthrit Cart 2021;29: 1412–7.
174. Kolasinski SL, Neogi T, Hochberg MC, et al. Arthrit Care Res 2020;72:149–62.
175. Fernandes L, Hagen KB, Bijlsma JW, et al. Ann Rheum Dis 2013;72:1125–35.
176. Jordan KM, Arden NK, Doherty M, et al. Ann Rheum Dis 2003;62:1145–55.
177. Bannuru RR, Osani MC, Vaysbrot EE, et al. Osteoarthrit Cart 2003;27:1578–89.
178. McAlindon TE, Bannuru R, Sullivan MC, et al. Osteoarthrit Cart 2014;22:363–88.
179. National Center for Complementary and Integrative Health. Glucosamine and chondroitin for osteoarthritis: what yoy need to know. Accessed March 11, 2024. Available at www.nccih.nih.gov/health/glucosamine-and-chondroitin-for-osteoarthritis-what-you-need-to-know
180. Clegg DO, Reda DJ, Harris CL, et al. New Eng J Med 2006;354:795–808.
181. Fransen M, Agaliotis M, Nairn L, et al. Ann Rheum Dis 2015;74:851–8.
182. Simental-Mendía M, Sánchez-García A, Vilchez-Cavazos F, et al. Rheum Int 2018;38:1413–28.
183. Zhu X, Sang L, Wu D, et al. J Orthopaed Res 2018;13:170.
184. Zhu X, Wu D, Sang L, et al. Clin Exp Rheumatol 2018;36: 595–602.
185. McCarty MF, O'Keefe JH, DiNicolantonio JJ. J Diet Suppl 2019;16:179–92.
186. Hsiao A-F, Lien Y-C, Tzeng I-S, et al. Complement Ther Med 2021;63:102775.
187. Bannuru RR, Osani MC, Al-Eid F, et al. Sem Arthrit Rheum 2018;48:416–29.
188. Daily JW, Yang M, Park S. J Med Food 2016;19:717–29.
189. Zeng L, Yu G, Hao W, et al. Biosci Rep 2021;41:BSR20210817.
190. Arthritis Foundation. Rheumatoid arthritis patient education & resources. Accessed March 11, 2024. Available at www.arthritis.org/rheumatoid-arthritis-patient-education
191. Helmick CG, Felson DT, Lawrence RC, et al. Arthrit Rheum 2008;58:15–25.
192. Myasoedova E, Davis JM, 3rd, Crowson CS, et al. Curr Rheum Rep 2010;12:379–85.
193. Sokka T, Abelson B, Pincus T. Clin Exper Rheum 2009;26: S35–61.
194. Sokka T, Kautiainen H, Pincus T, et al. Arthrit Res Ther 2010;12:R42.
195. Martin R. Proc Nutr Soc 1998;57:231–4.
196. Touger-Decker R. J Am Diet Assoc 1988;88:327–31.
197. Morgan SL, Hine RJ, Vaughn WH, et al. Arthritis Care Res 1993;6:4–10.
198. Heimburger DC, Weinsier RL. In: Shanahan J, ed. Chapter 11, Therapeutic Diets. Handbook of Clinical Nutrition. St. Louis, MO: Mosby, 1997;235–66.
199. Roubenoff R, Roubenoff RA, Ward LM, et al. Am J Clin Nutr 1990;52:1113–7.
200. Bayles TB, Richardson H, Hall FC. N Engl J Med 1943;229: 319–24.
201. Kremer J. Lipids 1996;31:S243–7.

202. Morgan SL, Anderson AM, Hood SM, et al. Arthritis Care Res 1997;10:9–17.
203. Roubenoff R, Roubenoff RA, Cannon JG, et al. J Clin Invest 1994;93:2379–86.
204. Walsmith J, Roubenoff R. Int J Cardiol 2002;85:89–99.
205. Rall L, Walsmith J, Snydman L, et al. Arthritis Rheum 2002;46:2574–7.
206. Zoico E, Roubenoff R. Nutr Rev 2002;60:39–51.
207. Summers GD, Deighton CM, Rennie MJ, et al. Rheumatol 2008; 47:1124–31.
208. Paget J. Lancet 1973;2:727–9.
209. Santo RC, Silva JM, Lora PS, et al. Clin Rheumatol 2020;39: 3603–13.
210. Helliwell M, Coombes E, Moody B, et al. Ann Rheum Dis 1984;43:386–90.
211. Collins R, Dunn T, Harrell W, et al. Clin Rheumatol 1987;6: 391–8.
212. Mody GM, Brown GM, Meyers OL, et al. S Afr Med J 1989;76: 255–7.
213. Kalla AA, Brown GM, and Meyers OL. (1992) S Afr Med J 82, 411–4.
214. Hernandez-Beriain J, Segura-Garcia C, Rodriguez-Lozano B, et al. Scand J Rheumatol 1996;25:383–7.
215. Munro R, Capell H. Ann Rheum Dis 1997;56:326–9
216. Roubenoff R, Walsmith J, Lundgren N, et al. Am J Clin Nutr 2002;76:774–9.
217. Ahluwalia IB, Mack KA, Murphy W, et al. MMWR CDC Surveill Summ 2003;52:1–80.
218. Haffner S, Taegtmeyer H. Circ 2003;108:1541–5.
219. Elkan A-C, Engvall I-L, Tegstrand B, et al. Eur J Clin Nutr 2008;62:1239–47.
220. Detsky AS, McLaughlin JR, Baker JP, et al. JPEN 1987;11: 8–13.
221. Detsky AS, McLaughlin JR, Baker JP, et al. Nutricion Hospitalaria 2008;23:400–7.
222. Persson MD, Brismar KE, Katzarski KS, et al. J Am Geriatr Soc 2002;50:1996–2002.
223. Guigoz Y, Vellas B, Garry PJ. Nutr Rev 1996;54:S59–65.
224. Vellas B, Guigoz Y, Garry PJ, et al. Nutr (Burbank, Los Angeles County, Calif.) 1999;15:116–22.
225. Elkan A-C, Engvall I-L, Cederhol T, et al. Eur J Nutr 2009;48:315–22.
226. Arshad A, Rashid R, Benjamin K. Modern Rheumatol Japan Rheum Assoc 2007;17:470–75.
227. Bruce B, Fries J. Clin Exper Rheumatol 2005;23:S14–8.
228. Pincus T, Summey JA, Soraci SA, Jr., et al. Arthritis Rheum 1983;26:1346–53.
229. Giles JT, Ling SM, Ferrucci L, et al. Arthritis Care Res 2008;59:807–15
230. Tekgoz E, Colak S, Ozalp Ates FS, et al. Int J Rheum Dis 2020;23:1685–91.
231. Tański W, Wójciga J, Jankowska-Polańska B. Nutrients 2021;13: 1259.
232. Alvarez-Nemegyei J, Pacheco-Pantoja E, González-Salazar M, et al. Rheumatol Clin (Engl Ed) 2020;16:462–7.
233. Giles JT, Bartlett SJ, Andersen RE, et al. Arthritis Rheum 2008;59:1407–15.
234. Engvall I-L, Elkan A-C, Tengstrand B, et al. Scan J Rheumatol 2008;37:321–28.
235. Toms TE, Panoulas VF, John H, et al. Arthritis Res Ther 2009;11:R110.
236. Stavropoulos-Kalinoglou A, Metsios GS, Smith JP, et al. Int J Obes 2010;34:295–301.
237. Rall LC, Roubenoff R. Arthritis Care Res 1996;9:151–6.
238. Rall L, Roubenoff R, Cannon JG, et al. Med Sci Sports Exerc 1996;28:1356–65.
239. Rall LC, Rosen CJ, Dolnikowski G, et al. Arthritis Rheum 1996;39:1115–24.
240. Engelhart M, Kondrup J, Hoie LH, et al. Clin Exp Rheum 1996;14:289–93.
241. Marcora S, Lemmey A, Maddison P. Clin Nutr (Edinburgh, Scotland) 2005;24;442–54.
242. Tournadre A, Pereira B, Dutheil F, et al. J Cachexia Sarcopenia Muscle 2017;8:639–46.
243. Henderson C, Panush R. Rheum Dis Clin North Am 1999;25:937–65.
244. Panush R. Rheum Dis Clin North Am 1991;17:259–72.
245. Panush R. In Koopman WJ, ed. Arthritis and Allied Conditions, A Textbook of Rheumatology Vol. 1 pp., Philadelphia, PA: Lippincott Williams & Wilkins, 2001;965–86.
246. Lewin P, Taub SJ. JAMA 1936;106:2144.
247. Hench BS, Bauer W, Dawson MH, et al. Ann Intern Med 1940;13:1838–90.
248. Hench BS, Bauer W, Boland E, et al. Ann Intern Med 1941;15:1002–108.
249. Marquardt JL, Snyderman R, Oppenheim JJ. Cellular Immunol 1973;9:263–72.
250. Epstein S. Ann Allergy 1969;26:343–439.
251. Pottenger RT. Ann Intern Med 1928;12;323–33.
252. van de Laar M, van der Korst J. Sem Arthritis Rheum 1991;21: 12–23.
253. Cunningham-Rundles C. Rheum Dis Clin North Am 1991;17;287–307.
254. Inman RD. Rheum Dis Clin North Am 1991;17:309–21.
255. Sköldstam L, Magnusson KE. Rheum Dis Clin North Am 1991;17:363–71.
256. Delafuente J. Rheum Dis Clin North Am 1991;17:203–12.
257. Bengtsson U, Hanson L, Ahlstedt S. Clin Exp Allergy 1996;26: 1387–94.
258. Kavanagh R, Workman E, Nash P, et al. British J Rheumatol 1995;34:270–3.
259. Holst-Jensen SE, Pfeiffer-Jensen M, Monsrud M, et al. Scand J Rheumatol 1998;27:329–36.
260. Tanner SB, Callahan LF, Panush R, et al. Arthritis Care Res 1990;3:189–95.
261. Panush RS, Stroud RM, Webster EM. Arthritis Rheum 1986;29:220–6.
262. Parke AL, Hughes G. British Med J 1981;282:2027–9.
263. Kong SS, Robinson M, Hosterman T, et al. Cureus 2021;13:e15010.
264. Liden M, Kristjansson G, Valtysdottir S, et al. Scand J Rheumatol 2010;39:292–8.
265. Schrander JJP, Marcelis C, De Vries MP, et al. Br J Rheumatol 1997;36:905–8.
266. Lee S, Gura KM, Kim S, et al. Nutr Clin Prac 2006;21:323–41.
267. Calder PC. Am J Clin Nutr 2006;83:1505S–19S.
268. Bahadori B, Uitz E, Thonhofer R, et al. J Parenter Enteral Nutr 2010;34:151–5.
269. Adam O. Eur J Clin Nutr 1995;49:703–17.
270. Dawczynski C, Schubert R, Hein G, et al. Br J Nutr 2009;101:1517–26.
271. Ghaseminasab-Parizi M, Nazarinia MA, Akhlaghi M. Eur J Nutr 2022;61:1377–89.
272. Berbert AA, Konda CRM, Almendra CL, et al. Nutr 2005;21: 131–6.
273. Minamino H, Katsushima M, Torii M, et al. Sci Rep 2021;11: 5104.
274. Rosell M, Wesley AM, Rydin K, et al. Epidemiol (Cambridge, Mass.) 2009;20:896–901.
275. Veselinovic M, Vasiljevic D, Vucic V, et al. Nutrients 2017;9.
276. Philippou E, Petersson SD, Rodomar C, et al. Nutr Rev 2012;79:410–28.

277. Haugen M, Kjeldsen-Kragh J, Nordvåg B, et al. Clin Rheumatol 1991;10:401–7.
278. Panush R, Carter R, Katz P, et al. Arthritis Rheum 1983;26: 462–71.
279. Dong CH, Banks J. The Arthritic's Cookbook. New York, NY: Thomas Y. Crowell Company, 1973.
280. Dong CH, Banks J. New Hope for the Arthritic. New York, NY: Thomas Y. Crowell Company, 1975.
281. Sarzi-Puttini P, Comi D, Boccassini L, et al. Scand J Rheumatol 2000;29:302–7.
282. Ranganath VK, La Cava A, Vangala S, et al. Rheumatol (Oxford, England) 2023;62:565–74.
283. Sköldstam L, Larsson L, Lindström F. Scand J Rheumatol 1979;8:249–55.
284. Sköldstam L. Scand J Rheumatol 1986;15:219–23.
285. Skoldstam L, Brudin L, Hagfors L, et al. Nutr J 2005;4:15.
286. Pineda-Juárez JA, Lozada-Mellado M, Hinojosa-Azaola A, et al. Physiother Theory Pract 2022;38:504–12.
287. García-Morales JM, Lozada-Mellado M, Hinojosa-Azaola A, et al. J Clin Rheum 2020;26:S116–S22.
288. Kjeldsen-Kragh J, Borchgrevink C, Laerum E, et al. Lancet 1991;338:899–902.
289. Haugen M, Kjeldsen-Kragh J, Skakkebek N, et al. Clin Rheumatol 1993;12:62–9.
290. Kjeldsen-Kragh J, Mellbye O, Haugen M, et al. Scand J Rheumatol 1995;24:85–93.
291. Peltonen R, Kjeldsen-Kragh J, Haugen M, et al. Br J Rheumatol 1994;33:638–43.
292. Kjeldsen-Kragh J, Haugen M, Forre O, et al. Br J Rheumatol 1994;33:569–75.
293. Nenonen M, Helve T, Rauma A-L, et al. Br J Rheumatol 1998;37:274–81.
294. Peltonen R, Nenonen M, Helve T, et al. British J Rheumatol 1997;36.
295. Hafstrom I, Ringertz B, Spanberg A, et al. Rheumatol 2001;40: 1175–9.
296. McDougall J, Bruce B, Spiller G, et al. J Alternative Complementary Med 2002;8:71–5.
297. Guagnano MT, D'Angelo C, Caniglia D, et al. Nutrients 2021;13.
298. Hagen KB, Byfuglien MG, Falzon L, et al. Cochrane Data System Rev 2009;CD006400.
299. Ikuyama S, Imamura-Takase E, Tokunaga S, et al. Mod Rheumatol 2009;19:253–9.
300. Efthimiou P, Kukar M. Rheumatol Int 2010;30:571–86.
301. Alipour B, Homayouni-Rad A, Vaghef-Mehrabany E, et al. Int J Rheum Dis 2014;17:519–27.
302. Vaghef-Mehrabany E, Alipour B, Homayouni-Rad A, et al. Nutr (Burbank, Los Angeles County, Calif.) 2014;30:430–5.
303. Zamani B, Golkar HR, Farshbaf S, et al. Int J Rheum Dis 2016;19:869–79.
304. Lomax AR, Calder PC. Curr Pharm Design 2009;15: 1428–518.
305. Aryaeian N, Shahram F, Mahmoudi M, et al. Gene 2019;698: 179–85.

F. MALNUTRITION IN SURGERY, TRAUMA, AND DISEASE

CHAPTER 90

Malnutrition in Disease and Inflammatory States

Gordon Jensen and Leah Gramlich

CHAPTER OUTLINE

Abbreviations

AAIM, Academy-ASPEN Indicators for Malnutrition; **ASPEN**, American Society for Parenteral and Enteral Nutrition; **BMI**, body mass index; **CT**, computed tomography; **DEXA**, dual-energy x-ray absorptiometry; **ESPEN**, European Society for Parenteral and Enteral Nutrition; **GLIM**, Global Leadership Initiative on Malnutrition; **HCUP**, Healthcare Cost and Utilization Project; **MCC**, Malnutrition Clinical Characteristics; **MRI**, magnetic resonance imaging; **REE**, resting energy expenditure; **SGA**, Subjective Global Assessment; **TNF**, tumor necrosis factor-a

Preface

The disease-related inflammation and malnutrition construct has continued to gain acceptance across the globe.[1-10] There were more than 22,880 PubMed citations from 2012 to 2022 with the keywords malnutrition and disease and 4,000 of these citations also included the keyword inflammation. Malnutrition in the setting of disease is common and is associated with adverse medical treatment outcomes. Recognition of the contributions of disease and inflammation to malnutrition are of fundamental importance in conducting risk assessment for developing malnutrition, diagnosis of malnutrition, selection of appropriate interventions, ongoing monitoring, and guiding outcome expectations. A historic perspective is imperative to appreciate the advances that have been made, as well as the key research priorities that remain. We will use malnutrition to refer to protein-energy undernutrition throughout this chapter. Cachexia is discussed in detail in Chapter 88. Cachexia is a malnutrition overlap syndrome that is an example of disease and inflammation-related malnutrition.[11] Sarcopenia is another malnutrition overlap syndrome that is characterized by decline in muscle function and mass.[11] First described in older persons, malnutrition and inflammation may be contributing factors. Further information is provided in Chapter 56.

A New Understanding

The stage for the consideration and articulation of the relationship between disease, inflammation, and malnutrition was set in the 1970s, with recognition of the high prevalence of malnutrition in hospital settings. Among the most highly cited papers in the clinical nutrition field are two, by Bruce Bistrian and George

Blackburn, that placed the prevalence of protein-energy malnutrition in general surgical (1974) and medical patients (1976) in the 44% to 50% range.[12,13] Importantly, many such patients were not being identified as malnourished by health practitioners and were, therefore, not receiving appropriate nutrition intervention. A call to action was shared by Charles Butterworth in his 1974 paper "The Skeleton in the Hospital Closet".[14] Efforts were then undertaken to develop and validate approaches to malnutrition diagnosis that could be applied in healthcare settings. A Subjective Global Assessment tool was developed by Khursheed Jeejeebhoy and colleagues in the 1980s.[15,16] This widely used approach to diagnosis incorporates nutrient intake, weight change, oral/gastrointestinal symptoms, functional capacity, metabolic requirement, and physical examination. This was one of the first approaches to incorporate a metabolic stress of disease component and it has long been considered the reference standard for comparison for newer approaches.

As we entered the 21st century, there was growing awareness in the medical community of research findings that implicated inflammatory response as an etiologic factor in the development of a host of medical conditions and their associated complications. Breakthroughs in medical treatment targeting inflammation have since evolved that have changed treatment paradigms and altered the course of disease. Gordon Jensen first synthesized the interactions of inflammation with disease-associated malnutrition in a series of papers that began with a 2006 paper titled, "Inflammation as the key interface of the medical and nutrition universes: a provocative examination of the future of clinical nutrition and medicine."[1] Leading approaches to malnutrition diagnosis now include an appreciation for the contributions of disease and inflammation.

Malnutrition Has Continued to Evolve

Famine and starvation have been prevalent causes of malnutrition throughout history and remain so today. Natural disasters, global warming, drought, poverty, and food insecurity present continued challenges to our food supply and to food access. Conflict, war, and genocide also compromise availability of food and deprivation of food is sometimes cruelly wielded as a weapon. In 2020, the UN Refugee Agency estimated that 82.4 million people were forcibly displaced.[17] Despite improvements in agriculture, education, public health, health care, and living standards that have supported increased longevity, disease-related malnutrition, including disorders of weight excess, are now common in health care and other settings across the globe.[6,18] Disease-related malnutrition is associated with poor outcomes that include increased mortality, complications, length of hospital stay, and healthcare cost, as well as poor quality of life. Recently, the International Working Group for Patients' Right to Nutrition Care has articulated that nutrition care is a human right intrinsically linked to the right to food and the right to health.[19]

The COVID-19 pandemic illustrates the complex interplay of a global infectious disease and malnutrition. Longevity actually decreased across all 50 states over the course of the pandemic.[20] Feeding America estimated that 17 million Americans would become food insecure during the pandemic.[21] Contributing factors include social distancing, lockdowns, quarantine, supply-chain breakdown, loss of employment, and increased cost of food. Food insecurity and malnutrition are associated with chronic diseases that are, in turn, associated with higher risk for more severe complications of COVID-19.[22] Hospitalized COVID-19 patients have a high prevalence of malnutrition that is associated with increased risk of mortality.[23]

The growing prevalence of obesity has also been projected to decrease life expectancy.[24] The United Kingdom Overseas Development Institute reported that the number of adults who are obese or overweight in the developing world more than tripled between 1980 and 2008. The global prevalence of obesity has increased substantially over the past 40 years, from less than 1% in 1975, to 6% to 8% among girls and boys in 2016 and from 3% to 11% among men, and from 6% to 15% among women over the same period.[25] Developing countries now face a dual burden of prevalent classic malnutrition (undernutrition) as well as growing obesity.[26] Both malnutrition and obesity are now associated with excess morbidity, mortality, and cost across the globe. It must be noted that undernutrition and obesity often occur simultaneously in obese persons who consume poor-quality diets and among obese persons with disease, surgery, or injury.[27-29] It is also common to find sarcopenic obesity among obese older persons in such undernourished states.[30-34]

Diagnosis of Disease-Associated Malnutrition

Historic approaches to diagnosis of malnutrition syndromes have made use of criteria that are currently understood to lack full validity, including poor specificity, sensitivity, and inter-observer reliability.[2] Misdiagnosis is common because there are multiple syndrome definitions in the literature and healthcare practitioners receive little training in the recognition and diagnosis of malnutrition. One consequence is that, despite high prevalence, malnourished patients in the hospital are often not identified and; therefore, do not receive appropriate intervention. Corkins and colleagues[35] investigated ICD-9 malnutrition diagnosis codes from the 2010 Healthcare Cost and Utilization Project (HCUP) that contained discharge data from 1,051 hospitals in 45 states. They found that only 3.2% of hospital discharges had a malnutrition diagnosis. Prevalence rates in the 30% to 50% range[12,13,36,37] are typically reported for malnutrition in hospital settings, so many malnourished patients in HCUP were not coded appropriately. From such a dataset, it is not possible to discern the number of missed diagnoses. Those HCUP patients who had a malnutrition diagnosis were older, had longer lengths of stay, and incurred greater healthcare costs. They were also more likely to have comorbidity, be discharged to home care, and suffer mortality. Only 13.4% of the patients who received a malnutrition diagnosis received either parenteral or enteral nutrition.

Progress Toward Building Consensus

As mentioned previously, the Subjective Global Assessment[15,16] was developed in the 1980s and includes a metabolic stress of disease component that may be considered a proxy indicator of inflammatory status. In 2010, an International Consensus Guideline Committee[3,4] incorporated an appreciation for the role of inflammatory response into their proposed etiologic nomenclature for malnutrition diagnosis in adults in clinical practice settings as follows: *starvation-associated malnutrition*, when there is chronic starvation without inflammation; *chronic disease-associated malnutrition*, when inflammation is chronic and of mild to moderate degree; and *acute disease or injury-associated malnutrition*, when inflammation is acute and of severe degree. This paper highlighted two key observations:

- "The pathophysiology of malnutrition that is associated with disease or injury invariably consists of a combination of varying degrees of under- or over-nutrition and acute or chronic inflammation, leading to altered body composition and diminished biological function".[3,4,38]
- "The point at which the severity or persistence of inflammation results in a decrease in lean body mass associated with functional impairment would be considered disease-related malnutrition."[3,4]

In 2012, the Academy of Nutrition and Dietetics and the American Society for Parenteral and Enteral Nutrition (ASPEN) extended this approach with suggested Academy-ASPEN Indicators for Malnutrition (AAIM) to support diagnosis that includes the presence of illness or injury, poor food intake, weight loss, and physical findings of fat loss, muscle loss, edema, or reduced grip strength.[39] These indicators have also been referred to as the Malnutrition Clinical Characteristics (MCC). In 2017, the European Society for Parenteral and Enteral Nutrition (ESPEN) formally adopted a disease/inflammation-based guideline similar to these earlier approaches.[18] In 2016, the Global Leadership Initiative on Malnutrition (GLIM), a collaborative effort of ASPEN, ESPEN, the Latin American Federation of Parenteral and Enteral Nutrition, the Parenteral and Enteral Nutrition Society of Asia, and other nutrition societies, embarked on an effort to build global consensus around commonly used evidence-based criteria for diagnosis of malnutrition in adults in clinical practice settings. Weight loss, low BMI, and reduced muscle mass were selected as phenotypic criteria, while reduced food intake and disease burden/inflammation were selected as etiologic criteria. One phenotypic and one etiologic criterion were deemed requirements for the preliminary diagnosis of malnutrition. Where available, this diagnosis is intended to trigger comprehensive nutrition assessment by a skilled nutrition professional. However, the primary objective is to offer a simple approach that can be readily used in global settings with limited clinical nutrition resources. The GLIM diagnostic construct and guidance for testing and validation were published in 2019 and 2020, respectively.[40-42] The most current iteration of the diagnostic categories of malnutrition, by underlying etiology, include chronic disease with inflammation, chronic disease with minimal or no perceived inflammation, acute disease or injury with severe inflammation, and starvation related to hunger/food shortage associated with socioeconomic or environmental factors.[10,39-41] Additional information regarding the criteria used for diagnosing malnutrition and the impact of inflammation is found in **Tables 90.1** and **90.2**.

While regional and institutional preferences have historically driven selection of approaches to diagnosis of malnutrition, recent studies suggest that these newer approaches to diagnosis

Table 90.1 Criteria for Diagnosing Malnutrition

	Phenotype or Etiology	MCC	SGA	GLIM
Disease Burden/Inflammation	E	X	X	X
Unintentional Weight Loss	P	X	X	X
Reduced Food Intake	E	X	X	X
Muscle Loss	P	X	X	X
Reduced Functional Capacity	P	X	X	
Fat Loss	P	X	X	
GI Symptoms (>2 weeks)	E		X	
Edema*	P	X		
Low BMI	P			X

- First four characteristics are consistent for all three criteria
- Need at least two characteristics to diagnose moderate or severe malnutrition

Questions and clinical observations to consider when evaluating for malnutrition:

- Have you lost weight? Over what period of time?
- Have you been eating less than usual?
- Do you have nausea, vomiting, or diarrhea persisting longer than 2 weeks?
- Do you have reduced functional capacity or low energy level?
- Does the patient have a loss of subcutaneous fat? (triceps, face, ribs)
- Does the patient have muscle wasting (temporal, pectoralis, deltoid, quadriceps, gastrocnemius)

E = etiology of malnutrition, P = phenotypic presentation of malnutrition, MCC = Malnutrition Clinical Characteristics, SGA = Subjective Global Assessment, GLIM = Global Leadership Initiative on Malnutrition

*Edema that cannot be explained by disease process such as heart failure, cirrhosis, or kidney disease.

Reproduced from Moulder CL, Stotts MJ. Nutr Iss Gastroenterol 2022;Series #221:28–43.[43]

Table 90.2 Malnutrition with and without Inflammation

Marker	Malnutrition with Inflammation	Malnutrition without Inflammation
Catecholamines	↑	↑
Cytokines	↑	↔
REE*	↑	↓
Positive Acute Phase Proteins	↑	↔
Negative Acute Phase Proteins	↓	↔
Albumin	↓	↓↑
Edema	+	–
Body Cell Mass	↓	↓
Amino Acids for Fuel	↑	↓
Catabolism	↑	↓

*REE = resting energy expenditure.

Reproduced from Moulder CL, Stotts MJ. Nutr Iss Gastroenterol 2022;Series #221:28–43.[43]

Table 90.3 Assessment of Disease-Related Malnutrition and Inflammation

- A number of conditions or diseases are characterized by severe acute inflammatory response.
  - Critical illness, major infection/sepsis, adult respiratory distress syndrome, systemic inflammatory response syndrome, severe burns, major abdominal surgery, multi-trauma, and closed head injury.
- Many conditions or diseases are more typically associated with mild to moderate chronic inflammatory response.
  - Cardiovascular disease, congestive heart failure, cystic fibrosis, inflammatory bowel disease, celiac disease, chronic pancreatitis, rheumatoid arthritis, solid tumors, hematologic malignancies, sarcopenic obesity, diabetes mellitus, metabolic syndrome, cerebrovascular accident, neuromuscular disease, dementia, organ failure/transplant—kidney, liver, heart, lung, or gut, periodontal disease, pressure wounds, and chronic obstructive pulmonary disease. Examples of starvation-associated conditions that generally have little or no discernable inflammatory component include anorexia nervosa or compromised intake in the setting of major depression.
- Look for conditions or disease that can place one at nutritional risk secondary to increased requirements, or compromised intake or assimilation.
  - Critical illness, severe burns, major abdominal surgery, multitrauma, closed head injury, previous gastrointestinal surgery, severe gastrointestinal hemorrhage, enterocutaneous fistula, gastrointestinal obstruction, mesenteric ischemia, severe acute pancreatitis, chronic pancreatitis, inflammatory bowel disease, celiac disease, bacterial overgrowth, solid or hematologic malignancy, bone marrow transplant, acquired immune deficiency syndrome, and organ failure/transplant—kidney, liver, heart, lung, or gut.

Reproduced from Jensen GL, Hsiao PY, Wheeler D. JPEN J Parenter Enteral Nutr 2012;36:267–274.[5]

of malnutrition have similar use in identifying malnourished individuals and in predicting adverse outcomes.[44-55] This is anticipated, as they share a number of common criteria, including a disease component that is a proxy indicator of inflammation and associated feeding difficulties. Irrespective of the approach that is chosen, assessment of patients with disease-related malnutrition can be facilitated using the indicators of malnutrition and inflammation described below.

Assessment of Disease-Related Malnutrition and Inflammation

This section will focus specifically on those aspects of assessment directly relevant to disease-related malnutrition and inflammation in adults, as highlighted in **Table 90.3**. A comprehensive review of clinical nutrition assessment is found in Chapter 58.

Medical History

Knowledge of a patient's medical history is especially useful in discerning the likelihood of developing malnutrition and the potential impacts of disease and inflammation. Nonvolitional weight loss is a well-validated indicator of malnutrition and is often also associated with underlying disease or inflammation. The clinical significance of weight loss is determined by its degree and duration. A 10% loss of body weight over 6 months is notable, while a 30% loss of body weight over the same duration is severe and life threatening. To obtain a reliable weight history, it is often necessary to query the patient as well as the medical records, family, and caregivers. Some diseases and conditions are characterized by severe, acute inflammatory response whereas others are usually associated with a chronic inflammatory response that is mild to moderate in degree and may be relapsing and remitting (Table 90.3). Medical/surgical conditions or diseases that place the patient at increased risk to become malnourished due to increased nutritional requirements, or compromised intake or assimilation, should not be overlooked (Table 90.3). Assessment should also include a review of medications, with attention to undesirable side effects including anorexia, xerostomia, nausea, diarrhea, and constipation. Potential drug/nutrient interactions should also be identified (see Chapter 106).

It is also common for acute inflammatory insults to be superimposed on those with chronic conditions. As might be anticipated, a synergistic association has been described between inflammation, malnutrition, and mortality in patients with diabetes.[56] The inflammatory milieu, especially when severe, may modify nutrient requirements by elevating resting energy expenditure and increasing muscle catabolism and nitrogen losses. Inflammation also promotes anorexia, decreasing food intake and further compromising nutritional status. Derangements of micronutrient status including vitamin D, iron, and zinc are also well described. A deteriorating course may result because the presence of inflammation may reduce the benefit of nutritional interventions and the associated malnutrition may, in turn, diminish the effectiveness of medical therapies. Nutritional supplementation alone is ineffective in preventing muscle protein loss in the presence of significant inflammation.[57] A recent report[58] examined the association of baseline inflammation as assessed by admission C-reactive protein (CRP) with the effectiveness of early nutritional support among 1,950 hospitalized patients. Compared with the control group, patients receiving nutritional support showed a significant reduction in 30-day mortality regardless of CRP concentration. However, the subgroup of patients with high inflammation (CRP >10 mg/dL) had no beneficial effect of nutritional support on 30-day mortality.

Clinical Signs and Physical Examination

Fever, hypothermia, and tachycardia are among the nonspecific clinical signs of inflammatory response. The nutrition-focused physical examination should identify disease and inflammation-associated edema, as well as signs of weight gain/loss and specific nutrient deficiencies. Physical findings of weight loss associated with decreased muscle and subcutaneous fat mass should not be missed, although these changes may not be readily appreciated when significant edema is manifest.

Anthropometric Data

Because nonvolitional weight loss is often associated with an underlying disease or inflammatory condition, measurements of body weight are recommended with each clinic visit or hospitalization. Body weight was historically standardized for height to obtain an ideal weight for comparison, but available reference tables require subjective assessment of frame size and offer limited reference data for many relevant populations. A simple measure of body size and an indirect measure of body fatness is provided by body mass index (BMI), defined as weight (kg)/height (m^2), but low BMI is of limited use for assessment of malnutrition in countries with high rates of overweight and obesity. It must also be noted that being overweight or obese does not mean that one cannot be severely malnourished.[26] Such individuals may also suffer impairment of nutrition intake or assimilation and, therefore, underweight status is not required for the diagnosis of malnutrition. Diagnostic criteria for sarcopenia in individuals with obesity were recently reviewed.[34] Consensus on plausible approaches to define the prevalence of the condition, its clinical and functional relevance as well as most effective prevention and treatment strategies were identified as priorities to advance this work.

Inflammation promotes erosion of muscle mass, so that assessment of muscle mass is especially pertinent in disease and inflammatory states. Classical anthropometric measurements including skin-folds and circumferences can be helpful, but their application in routine clinical settings has been limited because practitioner training is required to achieve acceptable skill and reliability. Body composition assessment technologies in current use include bioelectrical impedance analysis (BIA), dual-energy x-ray absorptiometry (DEXA), computed tomography (CT), and magnetic resonance imaging (MRI).[59] The imaging modalities have become state of the art for precise measurements of muscle mass, but portability, access, and cost remain significant limitations to widespread adoption for evaluation of muscle mass. It is, however, possible to take advantage of CT or MRI studies that are being done for other clinical purposes to evaluate musculature. The importance of muscle in the assessment of malnutrition in disease and inflammatory conditions is illustrated by an innovative use of CT assessment from Vickie Baracos at University of Alberta and her collaborators at M.D. Anderson Cancer Center.[60] Lumbar skeletal muscle index (skeletal muscle area in cm^2 at L3/height in m^2) imaged by CT was predictive of survival in 104 patients with advanced cancer. Of particular note, 63% of these patients had BMI ≥25 kg/m^2 and reduced muscle was associated with decreased survival irrespective of body size.

Laboratory Indicators

Reduced serum albumin and prealbumin concentrations have historically been promoted as indicators of protein malnutrition but are now widely recognized as lacking validity for this indication, due to poor sensitivity and specificity in clinical settings.[61] These proteins are negative acute phase reactants that are readily reduced by systemic response to disease, injury, or inflammatory condition. To help determine whether active inflammation is present, the positive acute phase reactant, C-reactive protein, may be measured. If C-reactive protein is elevated, and albumin or prealbumin is decreased, inflammation is likely to be a contributing factor. Variable cut-off values have been proposed for CRP as a marker of inflammation.[56,62] Using the GLIM approach to diagnosis of malnutrition, with CRP as a marker of inflammation, Cederholm and colleagues,[63] assessed the relationship between CRP and low food intake in acutely ill, older hospitalized patients. Their findings identified a serum CRP of 3.0 mg/dL as a reasonable threshold for acute inflammation leading to reduced food intake. However, C-reactive protein suffers limitations as a point in time measure, and trends over the clinical course may best be monitored. Leukocytosis and hyperglycemia are nonspecific laboratory indicators that are often associated with inflammatory response. Some investigators propose the use of neutrophil/lymphocyte ratio to determine the presence of inflammation.[64] Additional tests that may be considered to further support the presence of inflammatory response include 24-h urine urea nitrogen and energy expenditure measurements using indirect calorimetry. If severe acute systemic inflammatory response is manifest, negative nitrogen balance and elevated resting energy expenditure are anticipated.

Dietary Assessment

In the context of disease-related inflammation and malnutrition, food intake may be limited due to anorexia, gastrointestinal dysfunction, procedures, surgeries, and treatment or medication side effects. While dietary assessment in clinical settings can be challenging, 24-hour recall and diet history approaches may be conducted when feasible. Depending on the situation, it is often necessary to go beyond the patient to obtain relevant diet history information by accessing medical records, family, and/or caregivers. Thorough review of dietary practices and supplements for potential inadequacies and toxicities is indicated. Consultation with a registered dietitian/nutritionist to conduct diet assessment is highly recommended. As many patients present to healthcare providers with acute medical events superimposed upon chronic health conditions, it is common for patients to have had decreased food intake and malnutrition for weeks or months prior to evaluation. Recognition of poor dietary intake is essential so that appropriate intervention may be undertaken.

Functional Outcomes

Strength and physical performance measures can be used to confirm impaired muscle function that may accompany advanced malnutrition in disease and inflammatory states. Increasingly, functional status assessment is considered to be central to nutritional assessment.[65] The most commonly used practical clinical assessment is hand-grip strength measured with a simple hand-grip dynamometer. Physical performance tests, like timed gait, chair stands, and stair steps, are used in the comprehensive assessment of integrated functions in frail older persons. The decline in overall functional status observed in advanced malnutrition is associated with nutrient deficiencies and impairment of organ system functions. Poor wound healing and immune compromise are relevant examples of such impairments. Improved wound healing parameters and responsiveness to recall antigens by delayed hypersensitivity testing may be measured to demonstrate improvements with nutritional repletion, although it must be recognized that these are both multivariable outcomes for which improved nutritional status is but one variable.

Future Directions in Inflammation

In the face of disease-related inflammation and malnutrition, a favorable outcome will generally require treatment of the underlying disease or condition and nutrition intervention. It is important to recognize that while anti-inflammatory interventions offer benefits, they may also have undesirable side effects. For example, anti-inflammatory interventions may also blunt desirable aspects of inflammation. Raising the white blood cell count, mounting a fever, synthesizing needed acute reactants for immune response, wound healing, and breaking down muscle to provide substrate are all potentially adaptive, acute, physiologic responses, but sustained severe inflammation is clearly associated with poor outcomes. The ability to recognize risk for development of severe persistent inflammation would be a significant advance and could provide opportunities for early pre-emptive intervention. Type of disease or injury, acuity, severity, duration, degree of associated malnutrition, and underlying genotype and phenotype, are likely to be contributing risk factors. Potential biomarkers for recognition of inflammation include C-reactive protein, IL-6 and other cytokines, and procalcitonin. Decline in albumin or prealbumin and associated expansion of the extravascular, extracellular fluid space are also observed with inflammatory response. Other potential indicators include measures of immune dysfunction, oxidative stress, micronutrient status, mitochondrial function, organ injury, and microbiota composition.

Some individuals may be predisposed to more severe inflammatory response. Gene polymorphisms have been identified that appear to be associated with more robust inflammatory response[66] and weight loss in cancer cachexia.[67] Distinctive gene expression arrays have also been associated with systemic inflammatory response.[68-70] Systems biology approaches using metabolic phenotyping have seen growing application.[68,71-75] Metabolomics have been used to predict early onset of systemic inflammatory response syndrome in patients with major trauma, according to variation in levels of carbohydrate, amino acids, glucose, lactate, glutamine signals, and fatty acyl chains and lipids.[71] Large-scale discovery proteomics using high-resolution liquid chromatography-mass-spectrometry and multiplex cytokine analysis identified a variety of candidate biomarker proteins that were associated with survival in severe burn patients.[72] Metabolic profiling revealed multiple metabolites and lipid species that were associated with inflammatory markers in coronary artery disease.[75]

Opportunities for Intervention

Anti-inflammatory interventions include a variety of nutrition-related modulators (**Table 90.4**). While a detailed discussion of these topics is beyond the scope of this chapter, please see Chapter 65, for further information. The appreciation of an inflammatory component in our understanding of malnutrition establishes a rationale for the use of anti-inflammatory nutrition interventions and provides a compelling basis for nutrition practitioners to be integral members of the medical team. The stage is set for precision nutrition management of inflammatory states through dietary and other manipulations.[76]

Conclusions

Malnutrition in the setting of disease and inflammation is fundamentally different. Inflammation modifies nutrient requirements by elevating resting energy expenditure and increasing muscle catabolism and nitrogen losses. Inflammation also promotes anorexia, decreasing food intake and further compromising nutritional status. Adverse outcomes may result because the presence of inflammation may reduce the benefit of nutritional interventions and the associated malnutrition may, in turn, diminish the effectiveness of medical therapies. Favorable outcomes

Table 90.4 Anti-Inflammatory Interventions

Anti-inflammatory medications

- Non-steroidal anti-inflammatory drugs
- Steroids
- Cox-2 inhibitors

Anti-cytokines

- Anti-TNF
- Blocking IL-1 and IL-6

Glycemic control

- Insulin

Nutrition modulators

- n3 fatty acids
- Antioxidants
- Bioactive food components and modified diets
- Caloric restriction
- Enteral preferred to parenteral feeding

Microbiota

- Prebiotics
- Probiotics
- Transplant

Physical activity/weight loss

Data from Jensen GL. JPEN J Parenter Enteral Nutr 2006;30:453–463.[1]

Table 90.5 Research Priorities

Health systems research

- Can we fully integrate screening, assessment, intervention, and monitoring to promote improved outcomes for patients with disease-related inflammation and malnutrition?
- Can we develop standardized approaches to diagnosis and data reporting?
- Can we better train practitioners? There is need for a validated approach to diagnosis that incorporates key core variables that are amenable to use by practitioners with limited nutrition expertise or resources.
- Is there opportunity for a streamlined hybrid approach that integrates screening and assessment components for a preliminary malnutrition diagnosis? In settings where feasible, this diagnosis should trigger comprehensive assessment.
- Appropriate and useful outcome measures are needed. Mortality, length of stay, complications, and readmission are complex multivariable outcomes that are impacted by many other variables besides nutritional status.

Clinical/applied research

- Tailored population and disease-specific interventions warrant further testing. There is a pressing need to identify those types of patients who are most likely to accrue benefits from such nutrition support, so that precision nutrition intervention will be possible.
- Technology development is needed to provide portable, valid, reliable, and inexpensive assessment of muscle mass and function that is applicable in diverse settings.

Basic research

- To support precision nutrition interventions, it will be necessary to identify and develop relevant biomarker panels, metabolic profiles, and genomic analyses. These advances will help to facilitate risk assessment, diagnosis, and monitoring of undernutrition and inflammation.
- Identify and test anti-inflammatory approaches, including dietary modifications and nutrient supplementation.

Data from National Institutes of Health "Malnutrition in Clinical Settings: Research Gaps and Opportunities Workshop", September 12, 2022.

require both treatment of underlying disease and appropriate nutrition interventions. Disease-related malnutrition is now common across the globe in health care and other settings. Poor outcomes associated with disease-related malnutrition include increased mortality, complications, length of hospital stay, and healthcare cost, as well as poor quality of life. Recognition of the contributions of disease and inflammation to malnutrition will aid in conducting risk assessment for developing malnutrition, diagnosis of malnutrition, selection of appropriate interventions, ongoing monitoring, and guiding outcome expectations. While we are beginning to appreciate the complexities of disease-related inflammation and malnutrition, our research agenda must address key remaining deficiencies in diagnostics, biomarkers, and integration, as summarized in **Table 90.5**.

Acknowledgments

Some of the material presented in this chapter is adapted from both an earlier review[6] and a book chapter[7] authored by Gordon Jensen.

References

1. Jensen GL. JPEN J Parenter Enteral Nutr 2006;30:453–463.
2. Jensen GL, Bistrian B, Roubenoff R, et al. JPEN J Parenter Enteral Nutr 2009;33:710–716.
3. Jensen GL, Mirtallo J, Compher C, et al. JPEN J Parenter Enteral Nutr 2010;34:156–9.
4. Jensen GL, Mirtallo J, Compher C, et al. Clin Nutr 2010;29: 151–153.
5. Jensen GL, Hsiao PY, Wheeler D. JPEN J Parenter Enteral Nutr 2012;36:267–274.
6. Jensen GL. J Parenter Enteral Nutr 2015;39:56–62.
7. Jensen GL. In: Loscalzo J, Fauci A, Kasper D, Hauser S, Longo D, Jameson J, eds. Malnutrition and Nutritional Assessment. Harrison's Principles of Internal Medicine. 21st ed. McGraw Hill, 2022. Accessed September 24, 2022. Available at https://accesspharmacy.mhmedical.com/content.aspx?bookid=3095&sectionid=264532265
8. Cederholm T, Barazzoni R, Austin, P, et al. Clin Nutr 2017;36: 49–64.
9. Jensen GL, Cederholm T, Correia MITD, et al. JPEN J Parenter Enteral Nutr 2019;43:32–40.
10. Cederholm T, Jensen GL, Correia MITD, et al. Clin Nutr 2019;38:1–9.
11. Jensen GL, Cederholm T. Am J Clin Nutr 2018;108:1157–1158.
12. Bistrian BR, Blackburn GL, Hallowell E, at al. JAMA 1974;230: 858–860.
13. Bistrian BR, Blackburn GL, Vitale J, et al. JAMA 1976;235: 1567–1570.
14. Butterworth CE. Nutr Today 1974;9:4–8.
15. Baker JP, Detsky AS, Wesson DE, et al. N Engl J Med 1982;306: 969–972.
16. Detsky AS, McLaughlin JR, Baker JP, et al. JPEN J Parenter Enteral Nutr 1987;11:8–13.
17. United Nations. Refugees, Asylum-Seekers, and Internally Displaced Persons. Accessed August 29, 2022. Available at www.un.org/en/fight-racism/vulnerable-groups/refugees-asylum-seekers-internally-displaced
18. Popkin BM, Corvalan C, Grummer-Strawn LM. Lancet 2020;395: 65–74.
19. Cardenas D, Correia MITD, Ochoa JB, et al. Clin Nutr 2021;40:4029–4036.

20. Arias E, Tejada-Vera B, Kochanek KD, et al. Provisional Life Expectancy Estimates for 2021. Vital Statistics Rapid Release. Report No. 23, pp.1–6, August 2022. Accessed September 14, 2022. Available at www.cdc.gov/nchs/data/vsrr/vsrr023.pdf
21. Feeding America. March 2021. Accessed September 14, 2022. Available at www.feedingamerica.org/sites/default/files/2021-03/National%20Projections%20Brief_3.9.2021_0.pdf
22. Rodriguez-Leyva D, Pierce GN. Nutrients 2021;13:1752.
23. Mechanick JI, Carbone S, Dickerson RN, et al. JPEN J Parenter Enteral Nutr. 2021;45:13–31.
24. Preston SH, Vierboom YC, Stokes A. PNAS 2018;115:957–961.
25. Jaacks LM, Vandevijvere S, Pan A, et al. Lancet Diabetes Endocrinol 2019;7:231–240.
26. Barazzoni R, Gortan Cappellari G. Rev Endocr Metab Disord 2020;21:307–313.
27. Ledikwe JH, Smiciklas-Wright H, Mitchell DC, et al. J Am Geriatr Soc 2004;52:589–595.
28. Correia Horvath JD, Dias de Castro ML, Kops N, et al. Nutr Hosp 2014;29:292–299.
29. Leibovitz E, Giryes S, Makhline R, et al. Eur J Clin Nut 2013;67:620–4.
30. Roubenoff R. Obes Res 2004;12:887–8.
31. Stenholm S, Harris TB, Rantanen T, Visser M, Kritchevsky SB, Ferrucci L. Curr Opin Clin Nutr Metab Car 2008;11:693–700.
32. Jensen GL, Hsiao PY. (2010). Curr Opin Clin Nutr Metab Care 2008;13:46–51.
33. Kohara K. Endocr 2014;45:15–25.
34. Donini LM, Busetto L, Bischoff SC, et al. Clin Nutr 2022;41: 990–1000.
35. Corkins MR, Guenter P, Dimaria-Ghalili RA, et al. JPEN J Parenter Enteral Nutr 2014;38:186–95.
36. Barker LA, Gout BS, Crowe TC. Int J Environ Res Public Health 2011;8:514–27.
37. Allard J, Keller H, Jeejeebhoy K, et al. JPEN J Parenter Enteral Nutr 2016;40:487–497.
38. Soeters PB, Reijven PLM, van der Schueren MAE, et al. Clin Nutr 2008;27:706–16.
39. White JV, Guenter P, Jensen G, et al. JPEN J Parenter Enteral Nutr 2012;36:275–283.
40. Jensen GL, Cederholm T, Correia MITD, et al. JPEN J Parenter Enteral Nutr 2019;43:32–40.
41. Cederholm T, Jensen GL, Correia MITD, et al. Clin Nutr 2019;38:1–9.
42. Keller H, de van der Schueren MAE, GLIM Consortium, et al. JPEN J Parenter Enteral Nutr 2020;44:992–1003.
43. Moulder CL, Stotts MJ. Nutr Iss Gastroenterol. 2022;221:28–43.
44. Guerra RS, Fonseca I, Pichel F, et al. J Acad Nutr Diet 2015;115:927–38.
45. Contreras-Bolívar V, Sánchez-Torralvo FJ, Ruiz-Vico M, et al. Nutrients 2019;11:2043.
46. Ceniccola GD, Okamura AB, Sepúlveda Neta JDS, et al. JPEN J Parenter Enteral Nutr 2020;44:1347–1354.
47. Hipskind P, Rath M, JeVenn A, et al. J Am Coll Nutr 2020;39: 518–527.
48. Henrique JR, Pereira RG, Ferreira RS, et al. Nutr 2020; Nov–Dec:79–80.
49. Allard JP, Keller H, Gramlich L, Jeejeebhoy KN, Laporte M, Duerksen DR. Clin Nutr 2020;39:2771–2777.
50. Brito JE, Burgel CF, Lima J, et al. Clin Nutr 2021;40:4366–72.
51. Burgel CF, Eckert IDC, Brito JE, Rodrigues FW, Silva FM. J Hum Nutr Diet 2021;34:935–944.
52. Vavruk AM, Martins C, Mazza do Nascimento M. Nutr Clin Pract 2021;36:993–1002.
53. Shahbazi S, Hajimohammadebrahim-Ketabforoush M, Vahdat Shariatpanahi M, Shahbazi E, et al. Clin Nutr 2021;43:377–382.
54. de Araújo BE, Kowalski V, Leites GM, da Silva Fink J, Silva FM. Eur J Clin Nutr 2022;76:685–692.
55. Correia MITD, Tappenden KA, Malone A, et al. Clin Nutr 2022;41:687–697.
56. Wang J, Chen L, Huang Z, Lu J, et al. Front Nutr 2022;9: 872512.
57. Puthucheary ZA, Rawal J, McPhail M, et al. JAMA 2013;310: 1591–1600.
58. Merker M, Felder M, Gueissaz L, et al. JAMA Network Open 2020;3:e200663.
59. Sheean P, Gonzalez MC, Prado CM, et al. JPEN J Parenter Enteral Nutr 2020;44:12–43.
60. Parsons HA, Baracos VE, Dhillon N, Hong DS, Kurzrock R. PLoS One. 2012;7:e29330.
61. Evans DC, Corkins MR, Malone A, et al. Nutr Clin Pract 2021;36:22–28.
62. Xin W, Ming Y, Yizhong G, et al. Front Oncol 2021;11:766398.
63. Pourhassan M, Cederholm T, Trampisch U, et al. Eur J Clin Nutr 2022;76:397–400.
64. Y Lin, Song C, Cui J, et al. Front Nutr 2022;9. Available at www.frontiersin.org/articles/10.3389/fnut.2022.860285 doi:10.3389/fnut.2022.860285
65. Russell MK. Nutr Clin Pract 2015;30:211–8.
66. Paoloni-Giacobino A, Grimble R, Pichard C. Clin Nutr 2003;22: 507–514.
67. Deans DA, Tan BH, Ross JA, et al. Am J Clin Nutr 2009;89: 1164–1167.
68. LaRosa SP, Opal SM. Crit Care Clin 2011;27:407–419.
69. Xiao W, Mindrinos MN, Seok J, et al. J Exp Med 2011;208: 2581–2590.
70. Wong HR. Crit Care Med 2012;40:1618–1626.
71. Mao H, Wang H, Wang B, et al. J Proteome Res 2009;8: 5423–5430.
72. Finnerty CC, Jeschke MG, Qian WJ, et al. Crit Care Med 2013;41:1421–1434.
73. Nicholson JK, Holmes E, Kinross JM, Darzi AW, Takats Z, Lindon JC. Nature 2012;491:384–92.
74. Mirnezami R, Jiménez B, Li JV, et al. Ann Surg 2014;259: 1138–49.
75. Zhu Q, Wu Y, Mai J, et al. Front Immunol 2022;13:829425.
76. Demetrowitsch TJ, Schlicht K, Knappe C, et al. Front Immunol 2020;11:587895.

F. MALNUTRITION IN SURGERY, TRAUMA, AND DISEASE

CHAPTER 91

Nutrition Support for the Patient with Surgery, Trauma, or Sepsis

Nilesh M. Mehta

CHAPTER OUTLINE

Abbreviations

ARDS, acute respiratory distress syndrome; **ASPEN**, American Society for Parenteral and Enteral Nutrition; **ASPEN/SCCM**, American Society for Parenteral and Enteral Nutrition and the Society of Critical Care Medicine; **BEE**, basal energy expenditure; **CRP**, C-reactive protein; **EAT-ICU**, Early Goal-Directed Nutrition in ICU Patients; **EN**, enteral nutrition; **EPaNIC**, Early Parenteral Nutrition Completing Enteral Nutrition in Adult Critically Ill Patients; **ESPEN**, European Society for Parenteral and Enteral Nutrition; **GI**, gastrointestinal; **GRV**, gastric residual volume; **HAT**, hydrocortisone+ascorbic acid or vitamin C+thiamine; **IC**, indirect calorimetry; **ICU**, intensive care unit; **ILE**, Intravenous lipid emulsions; **IVLE**, IV lipid emulsion; injectable emulsions; **LBM**, lean body mass; **MODS,** multiple organ dysfunction syndrome; **PICU**, pediatric intensive care unit; **PN**, parenteral nutrition; **RQ**, respiratory quotient; **SCCM**, Society of Critical Care Medicine; **SGA**, subjective global assessment; **SPN**, supplemental parenteral nutrition; **TGC**, tight glucose control

Introduction

Specialized nutrition support plays an integral role in the perioperative management of patients undergoing major surgical procedures and in patients hospitalized for severe injuries and sepsis. Baseline nutritional status has been associated with outcomes from surgery, trauma, and sepsis; hence, nutritional history may be an important variable in determining risk for adverse outcomes. Volitional oral intake is compromised in these groups, making them dependent on the delivery of nutrients via enteral feeding devices and/or the parenteral route. Optimal nutrient delivery has been associated with improved wound healing, reduction in septic complications, and other morbidities, especially in patients with trauma and severe injuries. There continues to be significant interest in the role of nutrition in immune modulation and outcomes from sepsis. However, the benefits of *immunonutrition,* particularly its role in improving outcomes, has yet to be definitively proven. Furthermore, specialized nutritional therapies are not without side effects and are resource intensive. Nutritional therapies often compete with other therapeutic interventions during acute illness. Finally, the evidence-base for many nutritional therapies related to the type, amount, route, and timing of nutrient delivery is scant, and recommendations are often based on low-level evidence of benefit. Fortunately, the risks from nutrition therapy are minimized, and benefits are increased when experienced professionals deliver this complex technical therapy to appropriate patient populations. With the application of evidence-based strategies and careful selection of patients, the benefits of specialized nutrition may be realized. This chapter will review current guidelines or best practice recommendations related to nutrition therapy in adults with critical illness, based on available evidence.

General Guidelines/Best Practice—Key Concepts

Nutrition therapy during the acute phase of critical illness must be aimed at correcting micronutrient deficiencies and optimizing protein and nonprotein energy, with the goal of attenuating muscle loss and promoting recovery with improved outcomes. In recent years, there has been a steady increase in the number of studies examining optimal nutrition therapies in adults with critical illness. However, several questions remain unanswered due to the paucity of well-designed trials; therefore, some areas of nutrition delivery are guided by low-level evidence or expert opinion.[1,2] The following concepts have been incorporated into these best practice recommendations. Specialized nutrient delivery starts with the assessment of patients on admission to the intensive care unit (ICU). Nutrition screening and assessment using validated tools may help detect patients who are most likely to benefit from specialized nutrition. The next step is to determine the optimal macro (energy and protein requirements) and micronutrient requirements and the prescription of the nutrient delivery goals for individual patients. Individualized prescriptions vs a *one size, fits all* strategy is preferred.[3] Where available, indirect calorimetry must be used to guide the energy prescription.[4,5] Following the prescription of nutrition therapy goals, it is essential to determine the optimal timing and route of feeding for energy and protein delivery.

Many centers have implemented stepwise algorithms that help select the timing for initiation and advancement of enteral nutrition (EN) and assist with monitoring for tolerance. Implementation of these guidelines have been associated with improved nutrient delivery in children with trauma, surgery, and sepsis admitted to the pediatric intensive care unit (PICU).[6] Recent studies have highlighted the optimal strategy for parenteral nutrition (PN) use in children and adults who are critically ill and unable to meet nutrient delivery goals via EN.[7-13] The role of specific nutrients, micronutrients, and immunonutrition cocktails remains under investigation and may be clarified in the future.[14,15] Until then, a pragmatic approach to nutrition therapy that provides timely intervention by a nutrition team, avoids under- and overfeeding, helps monitor tolerance to nutrient intake, and documents relevant outcomes is desirable. Recovery from critical illness, with minimal loss of lean body mass and preserved long-term functional status, is the goal.[16] We will explore some of these concepts in subsequent sections of this chapter

Identification of the at-Risk Patient

The objective of nutrition screening is to identify patients with malnutrition or at risk of malnutrition who may benefit from specialized nutrition support. Screening is recommended within the first 1 to 2 days after hospital admission to identify patients who should be referred to a nutrition specialist for further assessment. Several screening tools have been developed to evaluate nutrition status in adults (see Chapter 58).[17] These tools include parameters such as recent weight loss, poor intake, lack of appetite, body weight or height, and provide a numerical score to categorize risk of malnutrition. **Table 91.1** describes the characteristics of commonly used screening tools.

Patients screened as high risk should be evaluated using a validated nutrition assessment tool (see Chapter 58). Anthropometry, commonly used to assess nutritional status, has several disadvantages in patients who are critically ill. Weight and other measurements require expertise, may be inaccurate in patients who are critically ill with fluid shifts, and suffer from interobserver variability. The Subjective Global Assessment (SGA) incorporates changes in organ function and body composition, gastrointestinal (GI) symptoms, and recent nutrient intake to predict nutrition status.[28] With adequate training in obtaining medical history and physical examination, the SGA can be applied in various settings with good intra- and inter-rater reliability.[29-31]

Despite initial interest in admission concentrations of serum albumin and prealbumin in patients in the hospital setting, their use cannot be recommended as surrogates for malnutrition in critical illness.[32] Protein-energy malnutrition decreases albumin synthesis, but a decrease in protein degradation can maintain serum concentration. This occurs with marasmus when protein and energy intakes are severely restricted. Lower availability of constitutive transport proteins such as albumin ($t_{1/2}$ = 21 days), transferrin (TFN; $t_{1/2}$ = 8 days), or thyroxin-binding prealbumin ($t_{1/2}$ = 2–3 days) may reflect the degree of malnutrition.[33] However, inflammatory conditions (e.g., trauma, sepsis, peritonitis) increase serum interleukin-6 (IL-6), which stimulates the acute phase protein response to increase C-reactive protein (CRP) and α-1-acid glycoprotein (AAG), and reduce constitutive protein production.[34] Serum albumin trends are inversely correlated with markers of inflammation and are not correlated with the adequacy of nutrient delivery.[35] Low constitutive proteins with low CRP are more likely to indicate pre-existing malnutrition. Elevated CRP with depressed constitutive proteins may reflect inflammation, protein-energy malnutrition, or both. Systemic markers of inflammation are associated with organ injury and have been proposed as important biomarker candidates for nutritional interventions.[36,37] Some scores have incorporated inflammatory markers (such as IL-6) to identify patients most likely to benefit from nutritional therapies.[26]

The Physiologic Response to Surgery and Injury

A sound understanding of the metabolic response to illness is essential when determining nutritional requirements. The goal of nutritional therapy is to offset the catabolic characteristics of this stress response to illness to preserve muscle mass and function. The metabolic, physiologic, inflammatory, and cytokine responses to surgery and injury have been well described.[38,39] **Figure 91.1** provides a simplified illustration of key pathways in the stress response to critical illness.

Table 91.1 Nutritional Screening Tools for Critically Ill Patients

Tool	Parameters Included	Score and Risk of Malnutrition	Validity
Malnutrition Screening Tool (MST)[18]	Recent weight loss (score 0-1) and poor dietary intake (score 0-4)	2=at risk of malnutrition	In patients with cancer (N=106) undergoing radiotherapy, MST was compared with the subjective global assessment (SGA) of nutritional status.[19] The MST had a sensitivity of 100% and a specificity of 81%. The positive predictive value was 0.4 and the negative predictive value was 1.0.
Mini Nutritional Assessment–Short Form (MNA-SF)[20]	Recent dietary intake, recent weight loss, mobility, recent acute disease or psychological stress, neuropsychological problems, and BMI Score (0-3) for each variable	<11=at risk This score prompts moving to the full MNA assessment	During preoperative evaluation of ambulatory patients over 60 years of age, the MNA-SF predicted absence of overt malnutrition revealed by the MNA, with 100% sensitivity and negative predictive value (NPV).[21] MNA-SF may be used as the first step of a two-step nutritional evaluation.
Nutrition Risk Screening (NRS-2002)[22]	Recent weight loss (%); Recent poor intake (%); BMI Disease severity Elderly Score 0–3 for each parameter	3 = indication to initiate specialized nutrition support	Tool predicts likelihood of reduced length of stay in patients at risk by the NRS who were provided nutrition support.
Malnutrition Universal Screening Tool (MUST)[23]	BMI Weight loss (%) acute disease effect score Score 0–3 for each parameter	>2=high risk 1=medium risk 0=low risk	Good agreement with other tools.[24] Predicts mortality risk and length of stay in acutely ill adults.[25]
Nutrition risk in intensive care (NUTRIC) score[26]	Age; APACHE II, SOFA; comorbidities; days from hospital to ICU admission. IL-6 (this parameter is excluded in the modified NUTRIC score)[27]	0–5 Low score 6–10 high score associated with worse outcomes and likely to benefit from specialized nutrition in therapy	Developed and validated specifically for patients in ICUs. May be useful in identifying critically ill patients most likely to benefit from optimal amounts of macronutrients when considering mortality as an outcome.

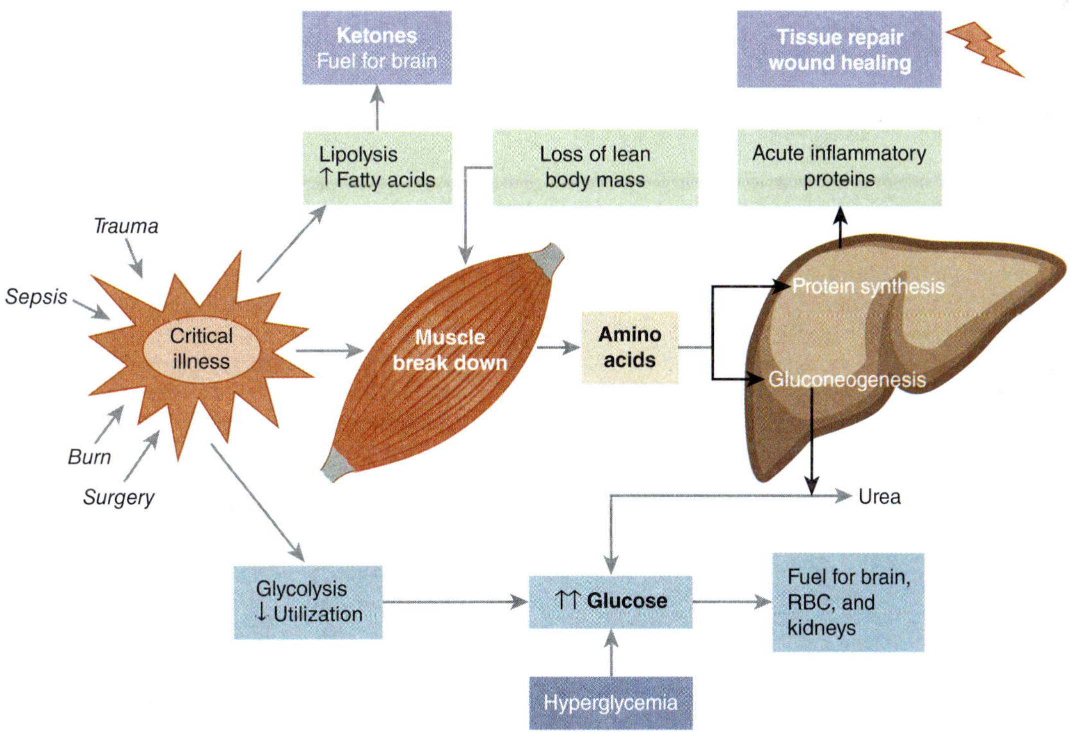

Figure 91.1 Stress response to critical illness.

Mehta NM, Jaksic T. In: Duggan CP, Walker W, eds. The Critically ill Child. Nutrition in Pediatrics. Hamilton, ON: B. C. Decker Inc, 2008; 663–73.[18]

The metabolic response to stress is an evolutionary adaptive response to critical illness that was essential for survival in prehistoric humans. The prehistoric humans (hunters-gatherers in the preagricultural age) depended on this response to overcome periods of illness or injury when extrinsic nutrients were not available. The response is driven by a complex neurohormonal cascade that helps release alternative energy substrates from indigenous sources. Serum insulin, glucagon, cortisol, catecholamines, and proinflammatory cytokines increase and result in the breakdown of endogenous stores of protein, carbohydrate, and fat. Key characteristics of the stress response are intense catabolism with muscle breakdown to provide free amino acids. The free amino acids can be channeled through the liver, where the carbon skeletons are used for gluconeogenesis to produce glucose for the brain, erythrocytes, and renal medulla. The amino acids are also reprioritized to help tissue repair and produce acute-phase reactant and inflammatory proteins. However, prolonged protein breakdown that is not offset by the ability to synthesize protein can result in a net negative balance, with loss of lean body mass. The goal of optimal protein delivery during the acute phase of illness is to offset the catabolic loss and preserve lean body mass. Changes in energy expenditure, hyperglycemia, and resistance to anabolic signals, including insulin, are other metabolic consequences of the stress response that may impact nutritional requirements.

Energy expenditure may be influenced by temperature, agitation, shivering, and other systemic effects of illness. Although previously thought to be universally increased during stress, the actual energy expenditure is unpredictable and may often be decreased (hypometabolism). Sedation, assisted ventilation, and muscle relaxation in the ICU, may further decrease energy expenditure. Furthermore, there is some endogenous energy production from substrate breakdown during acute critical illness. Hence, patients may be at risk of unintentional overfeeding due to overestimation of their energy requirement. The use of an indirect calorimeter (metabolic cart) allows accurate assessment of energy expenditure and may be used to assess individual requirements and energy prescription in patients who are critically ill.[4,5,40,41]

Nutritional Requirements

Energy Requirements and Prescription

The nutrient prescription should meet the metabolic demands of the patient. Overfeeding increases oxygen consumption, generates hepatic lipogenesis, produces immunosuppression (secondary to hyperglycemia or lipid deposition), and increases CO_2 production, and should be avoided. There are many equations derived from healthy populations that have been used to estimate energy expenditure in patients with critical illness. The Harris-Benedict formula is one such method to estimate basal energy expenditure (BEE):

$$\text{In men: BEE} = 66.5 + (13.8 \times \text{body weight in kg}) + (5.0 \times \text{height in cm}) - (6.8 \times \text{age in years})$$

$$\text{In women: } 665 + (9.6 \times \text{body weight in kg}) + (1.8 \times \text{height in cm}) - (4.7 \times \text{age in years})$$

In earlier decades, these estimated values were multiplied by stress and activity factors, but studies using indirect calorimetry have shown that the use of predictive equations, particularly with stress and activity factors added, often results in overfeeding in patients with critical illness.[42-45]

Indirect calorimetry using a metabolic cart has been employed for decades and is now gaining renewed interest as a tool to determine accurate energy prescription in patients who are critically ill.[40] Indirect calorimetry measures CO_2 production and O_2 consumption via expired gas to determine the overall resting energy expenditure via the Weir equation: Energy expenditure (kcal/day) = $3.941 \times VO_2$ (L/min) + $1.11 \times VCO_2$ (L/min) $\times$ 1,440. Measurements under controlled conditions approximate the Harris-Benedict equation within 5% to 10%, demonstrating that large stress factors are unnecessary. The respiratory quotient (RQ) ratio analyzes the substrate used by the patient because each fuel has a characteristic RQ during metabolism (carbohydrate RQ = 1.0; protein RQ = 0.8; fat RQ = 0.7). Lipogenesis has an RQ value of approximately 8; a calculated RQ greater than 1 is diagnostic of overfeeding. The technique is labor intensive and requires defined protocols. Unfortunately, errors in gas exchange measurements because of air leaks or administration of a high O_2 concentration can produce significant measurement error.[46] Institutional capabilities for IC are variable and may be limited by availability of the device or expertise. In the absence of indirect calorimetry (IC), predictive equations or a simplistic weight-based equation (25–30 kcal/kg/d) may be used to determine energy requirements with vigilance for early detection of unintended underfeeding or overfeeding. Recent studies of measured energy expenditure, VCO_2-based energy expenditure estimations, and technological advances in IC may help overcome some of these limitations.[47-49] Modern calorimeters need to be re-examined for their role in critical illness.

The role of IC has not been proven to improve outcomes in critical illness, and recent trials have shown divergent results.[50] In the Early Goal-Directed Nutrition in ICU Patients (EAT-ICU) trial, adults who were critically ill and mechanically ventilated were randomized to *early goal-directed nutrition* (aimed at delivering 100% of energy requirements based on IC and protein requirements based on 24-h urinary urea, from day 1 by EN and PN), or *standard nutrition* (aimed at delivering 25 kcal/kg/day by EN).[51] There was no significant difference in the primary (physical quality of life at 6 months) and other important outcomes between the two groups. These trials suffer from heterogeneity in relation to protein intake, route, and timing of feeding, and outcomes assessed, which may account for the results. Despite the lack of objective data, there is an increasing emphasis on measuring energy expenditure in all guidelines for patients who are critically ill. With the advent of newer validated devices,[47] there is an urgent need for well-designed trials comparing isocaloric and hypocaloric diets guided by indirect calorimetry.[52]

Until objective data are available, a pragmatic approach to energy delivery is recommended. Energy requirements are met by administering fat (10 kcal/g IV or 9.1 kcal/g enteral), carbohydrate (4.0 kcal/g enteral and 3.4 kcal/g hydrated glucose), and protein (4 kcal/g). In the absence of IC, the 2016 American Society for Parenteral and Enteral Nutrition and the Society of Critical Care Medicine (ASPEN/SCCM) guidelines recommended the use of predictive equations or a weight-based equation (25–30 kcal/kg/day) to estimate energy requirement.[2] The 2019 European Society for Clinical Nutrition and Metabolism (ESPEN) guidelines recommend measured energy expenditure by calorimetry or, in its absence, VO_2 (from the pulmonary arterial catheter) or VCO_2 (from ventilator module) to guide energy prescriptions.[53] Trials in adult critical care populations have challenged the benefits of overzealous energy administration during the acute phase of illness. Trials comparing high energy intakes with low energy intakes have failed to show any significant differences in clinical outcomes. Based on a more recent analysis of trials that have shown no difference in clinical outcomes between patients with higher vs lower energy delivery, the 2022 ASPEN/SCCM guidelines recommend feeding 12 to 25 kcal/kg in the first 7 to 10 days of ICU stay (moderate evidence grade and weak strength).[1]

Glucose Requirements

Hyperglycemia is a known response to the stress of critical illness and may be associated with worse outcomes. Hepatic gluconeogenesis produces hyperglycemia as glucose production increases from 2 to 2.5 mg/kg/min normally to 4 to 5 mg/kg/min during stress.[54] The maximal rate of glucose oxidation is 5 mg/kg/min (7.2 g/kg/day), which is easily exceeded. Traditional recommendations have been to maintain blood glucose values lower than 200 mg/dL. However, results of a 2001 trial suggested that tighter glucose control (80 to 120 mg/dL) with insulin improves clinical outcomes.[55] In this landmark trial, tight glucose control (TGC) (80 to 110 mg/dL) with intensive insulin therapy was associated with lower incidence of sepsis, ICU length of stay, and hospital mortality, compared with conventional insulin therapy (blood glucose <200 mg/dL). It was unclear whether insulin had the primary effect on the cardiac response or through other effects because patients who underwent general surgery in that study (including trauma, vascular, and other intra-abdominal procedures) showed no significant improvement with the aggressive insulin treatment. The single-center unblinded trial was remarkable for high mortality in both arms, and patients received high doses (200–300 g) of intravenous dextrose soon after surgery. The results were deemed controversial, especially when multiple subsequent trials failed to replicate the benefits of TGC, with increased incidence of hypoglycemia.[56-59] Based on available evidence, a target blood glucose of 140 or 150 to 180 mg/dL is recommended for the general adult ICU patient.[2]

Protein Requirements and Prescription

The metabolic response to stress from critical illness, trauma, or surgery, is characterized by protein catabolism and resultant loss of lean body mass (Figure 91.1). This is the theoretical foundation for the perception that protein requirements may be higher in patients who are critically ill. Administering enough protein to offset catabolic losses during acute critical illness may be beneficial as it offsets catabolic losses, supports anabolism and, thereby, prevents loss of lean body mass in critical illness.

The 2016 ASPEN/SCCM guideline[2] recommended a protein dose of 1.2 to 2 g/kg/day for adults with critical illness. Higher amounts may be required for patients with burns, obesity, or trauma. However, these recommendations for optimal protein dose during critical illness are based on low-level evidence. Large prospective observational studies demonstrated a significant association between higher protein intake and improved clinical outcomes.[60-63] However, in the absence of any effect shown in small RCTs and a paucity of well-designed trials of protein dosing with equivalent energy intake, the 2022 ASPEN/SCCM guidelines suggest an individualized protein prescription based on clinical assessment.[1] There are ongoing trials in adult and pediatric ICU populations that are attempting to answer this question.

Fat Requirements

Glucose should provide approximately 50% to 60% of total energy (approximately 70% to 80% of nonprotein energy). The balance of nonprotein energy should be given as 1 to 1.5 g/kg/day of fat with triglyceride concentration less than 300 mg/dL.[64] Hyperlipidemia with triglycerides greater than 500 mg/dL mandates withholding IV lipid emulsion (IVLE) in PN until triglycerides decrease to a safer range. The maximum recommended dose of IV lipid is 2.5 g/kg/day in the adult, but this should be used rarely.[65] Fat energy can be increased to 50% of requirement in select patients with severe hyperglycemia or high CO_2 production, but with risks of hyperlipidemia, cholestasis, immunosuppression, and increased infection. Suspected overfeeding with increased CO_2 should be treated by a reduction in total energy.[64-66]

Intravenous lipid emulsions (ILE) are safe and provide parenteral energy by decreasing the reliance on excessive glucose provision. An adequate ILE dose is also required to prevent an essential fatty acid deficiency in patients who are critically ill and reliant on PN. The recent availability of mixed oil and fish oil-based lipid formulations as an alternative to the soy-based 18-carbon omega-6 fatty acid preparation in the United States was met with enthusiasm. These products provide the potential for reducing the adverse effects of the traditional soy-based PN and the perceived benefits of olive oil or fish oil in lipid formulations. Although large observational studies have demonstrated these benefits in patients with trauma and other critical illness, there is a paucity of positive RCTs that have confirmed their impact on clinical outcomes.[2] The 2022 ASPEN/SCCM guidelines recommend either mixed oil or soybean oil ILE; and fish oil or nonfish oil ILE for patients with critical illness who are deemed candidates for PN.[1]

Micronutrient Requirements in Critical Illness

A delicate balance between an inflammatory state, oxidative stress, and the immune response is crucial for the recovery from critical illness. The antioxidant capacity may be compromised, and serum micronutrients and vitamins are often significantly decreased in this population. This observation is the basis of the theoretical concept that repletion or pharmacologic supplementation of antioxidants and other micronutrients may improve clinical outcomes in patients who are critically ill.[15,67-69] However, micronutrient provision in this population is widely variable. Low micronutrient concentrations in plasma or serum may not reflect true deficiency, especially in the absence of recorded losses in patients with critical illness. Trials examining the role of micronutrient supplementation have shown mixed, often negative, and controversial results.[70] Recent clinical trials examining the role of high doses of single micronutrients such as selenium, thiamine, vitamin C, and vitamin D have failed to demonstrate clinical benefits in sepsis and are not recommended. Based on these trial results, the 2016 ASPEN/SCCM guidelines made no recommendations regarding selenium, zinc, and antioxidant supplementation in sepsis.[2] High-dose copper, selenium, zinc, vitamin E, and vitamin C were not recommended in the absence of confirmed deficiency by the 2018 ESPEN guidelines.[53]

Immunonutrition

The role of nutrition as an agent to support host defense mechanisms and prevent infectious complications in patients who are critically ill has been explored in several trials. Specialized formulas containing a combination of trace elements, antioxidants, amino acids (glutamine, arginine), or potentially less inflammatory (ω-3) fatty acids have been tested for their ability to alter or attenuate the immune and inflammatory response seen in critical illness. Production of glutamine, the most abundant free amino acid in the body, increases during stress and sepsis. However, serum and intracellular levels drop as this amino acid becomes "conditionally essential" during times of high metabolic rate.[71] It provides a substrate for enterocytes, rapidly proliferating cells, and T lymphocytes. Both enteral and parenteral feedings supplemented with glutamine have been shown to be effective.[71-73] Unfortunately, glutamine is difficult to solubilize and degrades into ammonia and pyroglutamate, two toxic products during heat sterilization that have limited its use with IV feeding. Arginine is a precursor of nitric oxide, nitrites, and nitrates and promotes the proliferation of stimulated T cells. It plays a role in cellular immunity and fibroblast proliferation and serves as a secretagogue for growth hormone, prolactin, glucagon, and insulin production.[74]

The 2016 ASPEN/SCCM guidelines recommend immune-modulating formulas for select populations, such as those with traumatic brain injury and patients following surgery.[2] The guidelines do not recommend arginine supplementation in sepsis due to a lack of consistent evidence of benefit. The 2018 ESPEN guidelines recommend that additional enteral glutamine and high-dose ω-3–enriched enteral formulas should not be given on a routine basis to patients who are critically ill.[53] Glutamine may be used in patients with trauma and severe burn injury.[53] However, trials have shown conflicting results in these populations. In a recently concluded double-blind, randomized, placebo-controlled trial of enteral glutamine supplementation in deep second- or third-degree burn patients, 1,200 patients were randomized to 0.5g/kg glutamine or placebo within 72 hours of admission.[75] Supplemental glutamine did not reduce the time to discharge from the hospital.

Vitamins

The vitamin content of the original formula for patients with critical illness was based on the known nutritional needs of healthy individuals to prevent nutrient deficiency. Clinicians suggested that requirements may be greater in patients who are seriously ill, requiring specialized nutrition.[76] Higher dosages of vitamin C or vitamin E may play a critical role in antioxidant defense. Patients following trauma or emergency surgery were randomized to vitamins C and E (α-tocopherol 1,000 IU every 8 hours orally and ascorbic acid 1,000 mg every 8 hours IV during an ICU stay up to 28 days) or nothing.[77] No difference in pneumonia or acute respiratory distress syndrome (ARDS) was detected, but multiple organ dysfunction syndrome (MODS) was significantly lower with vitamin treatment, although the incidence of MODS occurring in both groups was low.

The effect of the acute-phase response on fluid distribution and measured vitamin concentrations has been studied after orthopedic surgical procedures.[78] Plasma vitamin A, vitamin E, and pyridoxal-5′-phosphate drop in association with an increase in CRP over a 7-day period. Vitamin concentrations returned to baseline as the acute-phase response resolved. Reductions in vitamin concentrations should be interpreted with care during the acute-phase response. Supplemental vitamins may be indicated when low vitamin concentrations persist during the absence of an acute-phase response and especially in the setting of recorded losses.

In 2017, a highly publicized study reported dramatic improvements in organ injury, time-to-shock reversal, and mortality in adults with sepsis who received a combination HAT therapy (hydrocortisone+ascorbic acid or vitamin C+ thiamine), compared with historic controls.[79] Several subsequent trials designed to study either a single vitamin or a combination regimen with hydrocortisone have been published, having enrolled more than 1,300 adults with sepsis or septic shock-associated organ dysfunction. **Table 91.2** describes the key features of these trials. The results are mixed, with an overall lack of demonstrable benefit for patient-centered outcomes. The heterogeneity of these trials does not allow for a meta-analysis of the combined data. Until more data are available, combination therapies may not be routinely applied in adults with critical illness. The sample sizes of trials were small and, therefore, do not allow interpretation of their benefits to select populations.

Route of Nutrition Support

Exclusive Enteral vs Parenteral Nutrition

In general, the enteral route is preferable to parenteral feeding if access to and function of the GI tract allows enteral feeding.[2,89] Although gastroparesis occurs during critical illness or after injury, access directly into the small intestine allows adequate nutrition to meet nutrient needs. Experimentally, parenteral feeding decreases brush border enzyme activity, probably increases mucosal permeability, reduces villous height, and generates inhibitory effects on gut mucosal immunity.[90] Each factor is benefited by enteral feeding. Parenteral nutrition bypasses the gut, and; therefore, the beneficial effects of EN may not be seen with the potential additional cost and infectious burden.

Substantial clinical data support the use of EN. Many studies were performed in patients with trauma,[73,91,92] in whom small bowel cannulation during celiotomy allowed randomization to early enteral feeding, parenteral feeding, or no specialized nutrition. Almost uniformly, studies showed significant reductions in pneumonia and intra-abdominal abscesses (if at risk of intra-abdominal abscess) with enteral feeding. In particular, patients most at risk for infectious complications benefited the most.[91] Gastroparesis usually resolves within 3 to 4 hours after an isolated closed head injury unless the patient is in a barbiturate coma; enteral feeding should be instituted as gastroparesis resolves.

Similar results have been shown with patients following general surgery. In several trials, early EN improved postoperative infections, immunologic parameters, and GI tract function.[93-95] Early postoperative nutrition precludes further loss in lean body mass (LBM) owing to starvation, and should be delivered enterally rather than parenterally when possible. Enteral and parenteral feeding has been studied in inflammatory bowel disease,[24,25] following transplantation,[27,88] and in patients with mild-to-moderate pancreatitis.[80] Uniformly, reports show beneficial effects with enteral feeding when clinically tolerated.

In several past RCTs, including critically ill patient populations such as those with trauma, burns, head injury, major surgery, and acute pancreatitis, EN was associated with reductions in infectious morbidity (generally, pneumonia and central line infections in most patient populations; specifically, abdominal abscess in trauma patients) and ICU length of stay. In a meta-analysis of 18 RCTs comparing PN with EN in 3,347 patients, the use

Table 91.2 Summary of Studies Examining Effects of Vitamins Alone, or in Combination with Thiamine and Hydrocortisone in Critical Illness

Study (Year)	Design	Patient Population	Intervention	Comparator Group	Primary Outcome(s)	Comments
RCT (Completed)						
CITRIS-ALI (2019)[80]	Randomized, double-blind, placebo-controlled, multicenter (seven sites) trial	Adults (n = 167) with sepsis and ARDS	IV Vitamin C (50 mg/kg in dextrose 5% in water, every 6 hours for 96 hours. (n = 84)	Placebo (dextrose 5% in water only) every 6 hours for 96 hours (n = 83)	Change in modified SOFA score at 96 hours; 2) C-reactive protein; 3) thrombomodulin. No differences among groups.	Mortality was lower in intervention group. No adverse events.
Vitamins (2020)[81]	Open label, randomized clinical trial Multicenter (10 ICUs in Australia, New Zealand, and Brazil)	Adults (n=216) with septic shock Enrolled within 24 hours of diagnosis with septic shock	HAT therapy IV Vitamin C (1.5 g every 6 hours), Hydrocortisone (50mg every 6 hours), and Thiamine (200mg every 12 hours), until shock resolution or up to 10 d. (n=109)	Hydrocortisone (50mg every 6 hours) alone until shock resolution or up to 10 days. (n = 107)	Duration of time alive and free of vasopressor up to day 7. No significant difference among groups.	No serious adverse events.
ORANGES (2020)[82]	Randomized, double-blind, placebo-controlled trial, in two sites (community hospitals)	Adults with sepsis or septic shock: therapy initiated in the ED Average time to study drug was 10 h	HAT therapy Hydrocortisone 50 mg q6h, Vitamin C 1.5g q 6h, and Thiamine 200 mg q 12 hours, for a maximum of 4 days. (n = 68)	Matching saline placebo for a maximum of 4 days. (n = 69)	1) Resolution of shock (vasopressor discontinuation)—significantly quicker in HAT therapy (27 vs 53 h, $P < 0.001$). 2) Change in SOFA score on Day 4. No difference among groups.	No significant diff in rates of RRT or oxaluria among the groups.
ACTS (2020)[83]	Randomized, blinded trial, multicenter (n=14)	Adults (n = 205) with septic shock Average time from vasopressor initiation to study drug—13.5 h	HAT therapy hydrocortisone (50 mg), vitamin C (1,500 mg), and Thiamine (100 mg) every 6 hours for 4 days (n = 103)	Placebo Volume matched (n = 102)	Change in SOFA scores at 72 hours. No significant difference.	No difference in major adverse events. Open-label corticosteroids by the clinicians was allowed.

Table 91.2 Summary of Studies Examining Effects of Vitamins Alone, or in Combination with Thiamine and Hydrocortisone in Critical Illness

Study (Year)	Design	Patient Population	Intervention	Comparator Group	Primary Outcome(s)	Comments
RCT (Completed)						
ATESS (2020)[84]	Randomized, controlled trial, double-blind, multicenter (four sites, ED)	Adults (n = 111) with septic shock Median time to study drug was 9 h	IV Vitamin C (50 mg/kg) and Thiamine (200 mg) q 12 h for 48 h (n=53)	Placebo (identical volume of 0.9% saline) q 12 h for 48 h (n=58)	Change in SOFA score at 72 hrs. No significant difference.	Open-label hydrocortisone in more than 50% of the patients (balanced between the two groups).
VICTAS (2021)[85]	Randomized, double-blind, adaptive, placebo-controlled, multicenter (N = 43 sites), trial Planned enrollment—2,000 patients	Adults with sepsis-induced acute respiratory and/ or cardiovascular dysfunction. Median time to study drugs was 14.7 hours	Hydrocortisone (50 mg), IV vitamin C (1.5 g), and thiamine (100 mg), within 4 h of randomization and then every 6 h up to 96 h. Open-label corticosteroids by the clinicians was allowed.	Placebo (matched volume) every 6 hours up to 96 hours.	Ventilator- and vasopressor-free days (VFDs) in the first 30 days following randomization. Underpowered to detect difference due to early termination.	After 501 participants were enrolled, additional funding for the trial was withheld due to a change in the funder's priorities.
RCT (Ongoing)						
RESPOND[86]	Open label, randomized controlled, multicenter (ANZICS—PSG, Brazil, Korea), 3-arm trial	Children (8 days–17 yrs.) admitted to PICU with septic shock treated with inotropes for at least 2 hours	Standard therapy, plus 1) Hydrocortisone + vitamin C 2) Hydrocortisone alone 1:1:1 allocation to the two interventions and standard therapy arms.	Standard therapy	Time alive and free of vasopressors, censored at 7 days.	Long-term outcomes: quality of life, developmental scale and functional score.
VITaCCA	Randomized placebo-controlled, double-blind, 3-arm with 1:1:1 allocation, multi-center (seven sites, Netherlands) trial	Adults with cardiac arrest admitted to the ED, and with ROSC (n=270)	IV Vitamin C at 1) supplemental dose (3g/day) and 2) pharmacologic dose (10d/day).	Placebo	Change in R-SOFA score at 96 h.	
VICToRY	Pilot study, randomized, blinded, multicenter (20 Burn Units worldwide)	Adults (n = 180) with severe burn injury	IV Vitamin C high dose (200mg/kg/day × 96 hours) + Standard burn care	Placebo + Standard burn care	28-day composite outcome of Persistent Organ Dysfunction (POD) and all-cause mortality.	Pilot feasibility trial prior to larger RCT.

Other Studies						
Marik (2017)[79]	Pre-post retrospective, propensity-matched study, single center	Adults with sepsis or septic shock (n = 94)	Pre-cohort (n = 47)–HAT Hydrocortisone (50 mg q6h), Vitamin C (1.5 g q6h) and Thiamine (200 mg q 12h) Post-cohort: Historical (n = 47)	NA	Hospital mortality—significantly lower in the treatment group (8.5% vs 40.4%), $P < 0.001$. The propensity adjusted odds of mortality in vitamin C group was 0.13 (95% CI, 0.04–0.48; $P = 0.002$).	SOFA score decreased vasopressors weaned faster in vitamin C group.
Coloretti (2020)[87]	Pre-post retrospective, propensity-matched study, single center	Adults with sepsis (n = 112) matched 1:1 using a propensity score model	Pre-cohort: hydrocortisone alone (n = 56) Post-cohort: HAT therapy (n = 56)	NA	Length of mechanical ventilation significantly lower in Post-cohort. Hospital mortality No significant difference	

Reproduced from Mehta NM. Pediatr Crit Care Med 2022;23:385–9.[88] doi:10.1097/PCC.0000000000002943.

of EN, compared with PN, had no effect on overall mortality but decreased infectious complications and ICU length of stay.[96] The reduction in infectious complications with EN was not observed when the energy administered by the two routes was similar. Furthermore, appropriate use of PN in select patients with attention to aseptic precautions, such as diligent central line care and avoidance of overfeeding, has allowed for safe and effective use of PN in recent times.[97] In two large multicenter trials comparing early PN with EN, 4,798 patients who were critically ill and in medical ICUs were randomized to PN or EN within 36 hours of admission[98] or 24 hours after intubation.[99] The energy and protein delivery were very similar in these trials. There were no significant differences in clinical outcomes in these studies. However, enteral feeding intolerance was more common in the EN treatment group, and EN was associated with a fourfold increase in ischemic bowel and colonic pseudo-obstruction, compared with early PN.[99] The low cost of providing EN continues to drive recommendations for its use in patients with critical illness where it is not contraindicated and is tolerated. The 2022 ASPEN/SCCM Guidelines recommend that nutrition support therapy in the form of early EN be initiated within 24 to 48 hours in patients with critical illness who are unable to maintain volitional intake.

Enteral Nutrition Practical Considerations

EN is preferred in patients who are critically ill but is often impeded by a variety of challenges in the ICU setting.[100,101] Based on large cohort studies, early initiation of EN (within 24–48 hours of ICU admission) with the goal of achieving up to two-thirds of the nutrient delivery goal in the first week of critical illness has been associated with improved clinical outcomes.[89] Delivery of EN into the stomach (gastric feeding) via feeding tubes is preferred. However, in some patients, gastric feeding is either contraindicated (e.g., high risk or history of aspiration) or has not been tolerated. Gastroparesis complicates the course of some patients who are critically ill, which may preclude early intragastric feeding. When gastroparesis occurs, or gastric EN is not tolerated, enteral access beyond the ligament of Treitz allows enteral feeding. Successful feeding into the small intestine (postpyloric EN) has been reported by several centers. In a meta-analysis of multiple RCTs, postpyloric tube feeding was associated with the delivery of higher proportions of energy targets and reduction in the gastric residual volume (GRV).[102] However, postpyloric feeding was not associated with any benefits to mortality, new-onset pneumonia, or aspiration. In centers with resources and expertise, the use of postpyloric feeding may be indicated when gastric feeding is not feasible or not tolerated. Prokinetic agents and gastric air insufflation are techniques used to increase the success rate of postpyloric placement. Enteral nutrition may be delivered intermittently as boluses or as a continuous infusion. While continuous feeding is used in many centers, intermittent or bolus feeding is thought to be more physiologic.[103] There are currently no trials demonstrating the superiority of any one form of EN.[104] **Table 91.3** summarizes some key evidence-based guidelines for optimal EN delivery in patients with critical illness.

The use of stepwise EN algorithms helps with the incorporation of evidence-based strategies and has been shown to improve nutrient delivery. These algorithms or guidelines provide recommendations for initiating EN, stepwise advancement at regular points in time, with guidance on detecting and managing EN intolerance. These guidelines help minimize or avoid interruptions to EN and provide indications for supplemental PN and its timing. **Figure 91.2** describes one such stepwise algorithm developed in a PICU and associated with improved EN delivery.[105]

Table 91.3 Practical Recommendations for Enteral Nutrition Delivery in Critical Illness[2,53,89]

	Recommendation
1	EN may be initiated early (within 24–48 hrs. of admission) in most patients without contraindications.
2	In the setting of hemodynamic compromise or instability, EN should be withheld until the patient is fully resuscitated and/or stable.
3	Presence of bowel sounds and evidence of bowel function (i.e., passing flatus or stool) are not required for initiation of EN.
4	In most critically ill patients, intragastric feeding is the preferred initial mode of EN. Postpyloric feeding may be used in patients with high risk of aspiration or gastric EN intolerance.
5	Inappropriate interruptions to EN must be avoided. Elevated gastric residual volumes may not be used as an isolated marker of EN intolerance.
6	Enteral feeding algorithms or stepwise protocols may improve EN delivery in the ICU patient.
7	For all intubated ICU patients receiving EN, the head of the bed should be elevated 30°–45° and use of chlorhexidine mouthwash twice a day should be considered.
8	The use of prokinetic agents, postpyloric feeding, or transition to continuous feeding may be considered in individualized patients to optimize EN delivery or minimize aspiration risk.

Supplemental Parenteral Nutrition

PN is used as an alternative or supplement (SPN) to insufficient EN during critical illness in patients who cannot tolerate EN, or when EN is contraindicated. There is significant controversy regarding the optimal timing of initiating supplemental PN in such cases. The costs and infectious burden of PN needs to be balanced against its use to meet energy and protein delivery goals. With modern PN practices, there has been a reduction in the risk of bacteremia and hyperglycemia, which were commonly seen in the early years of PN administration. The optimal timing (early vs late) for SPN has been examined in adult and pediatric ICU trials.[8,12] In the Impact of Early Parenteral Nutrition Completing Enteral Nutrition in Adult Critically Ill Patients (EPaNIC) trial, 4,500 adult patients with critical illness were randomized to either early PN (within 48 h) or late PN (after day 7 in the ICU) to supplement insufficient EN.[106] The late PN group was more likely to be discharged earlier from the ICU and hospital and had fewer ICU infections, lower

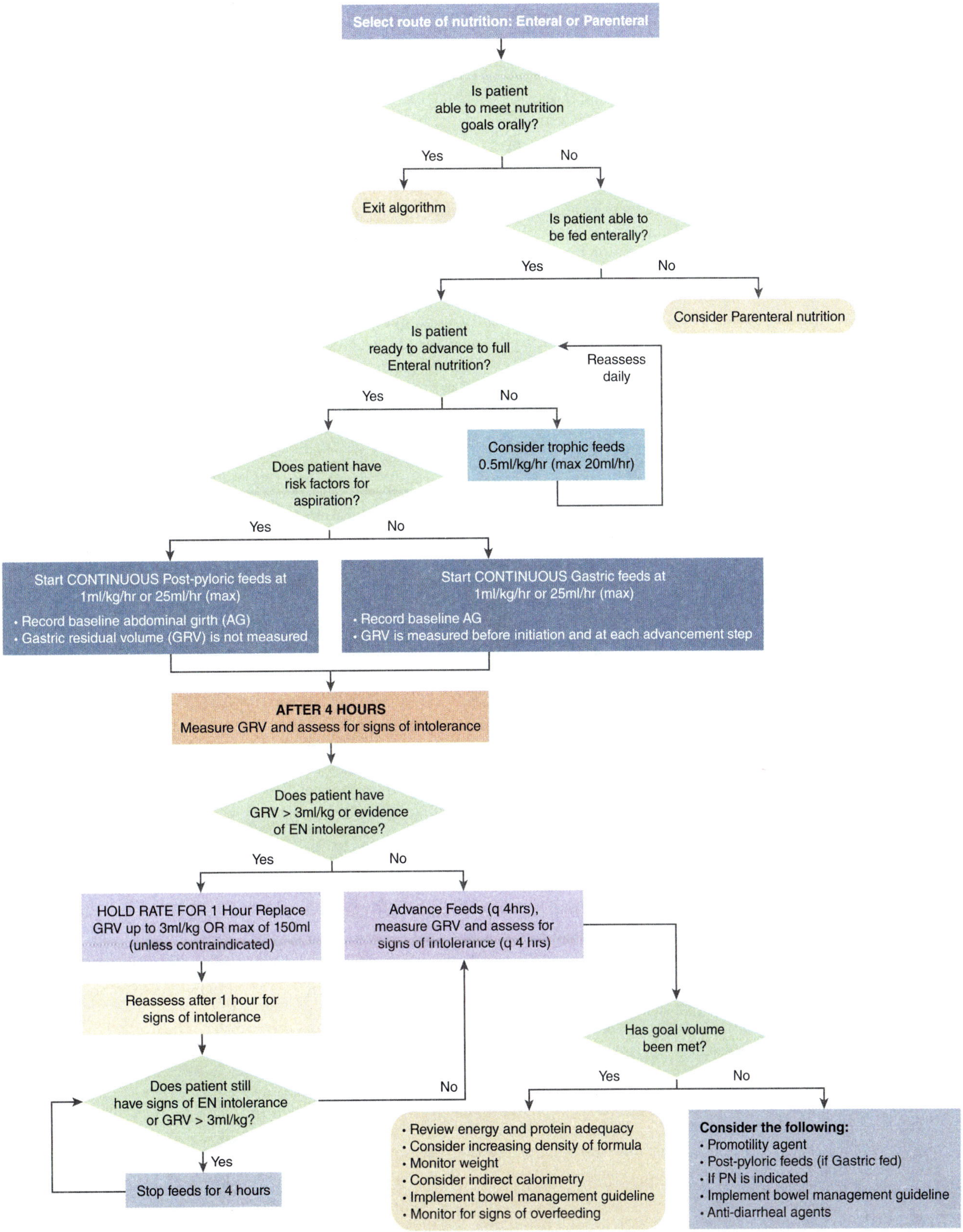

Figure 91.2 Stepwise enteral nutrition algorithm for the pediatric intensive care unit.

Reproduced from Hamilton S, McAleer DM, Ariagno K, et al. Pediatr Crit Care Med 2014;15:583–9.[185] doi:10.1097/PCC.0000000000000179.

incidence of cholestasis, and shorter duration of mechanical ventilation and renal replacement therapy. Patients in both groups received similar intravenous vitamins, minerals, and trace elements. There was no difference in mortality. In another RCT in adults with critical illness, the use of supplemental PN to optimize energy provision was associated with reduction in nosocomial infections in patients who failed to achieve energy delivery goals with EN alone.[107]

The 2016 ASPEN/SCCM guidelines reviewed results from four other RCTs that examined early vs late SPN. Based on findings of no significant differences in infectious complications or mortality in the groups, the guidelines recommend not initiating SPN prior to day 7 of ICU admission and after implementing all strategies to maximize EN. The 2018 ESPEN guidelines recommended the initiation of PN within 3 to 7 days in patients with contraindications to EN.[53] Based on the results of a single RCT,[12] the ASPEN/SCCM pediatric guidelines suggest that SPN should be delayed until 1 week after PICU admission for patients with a normal baseline nutrition state and low risk of nutrition deterioration. For children with severe malnutrition or those at risk of nutrition deterioration, PN may be supplemented in the first week if they are unable to advance past low volumes of EN. Earlier PN supplementation, between 4 and 7 days of ICU admission, may benefit some patients (e.g., those who do not tolerate full EN and after the early acute phase of critical illness) by achieving prudent nutrient delivery goals.[108] Future studies aimed at these subpopulations, with attention to the timing of achieving nutrient delivery goals and prevention of overfeeding, may help answer this question.

Summary/Conclusions

Nutrition therapy impacts the disease course and outcomes in patients who are critically ill. The delivery of optimal nutrition is an essential goal of critical care. Recent studies have helped clarify many aspects of optimal critical care nutrition. However, the amounts of macronutrients and micronutrients required to achieve improved outcomes remains unclear. Furthermore, questions about the right time for nutrient delivery, the appropriate route, and the role of specialized formulations need further clarification. **Tables 91.4** and **91.5** summarize key recommendations from recent nutrition guidelines for critically ill adults and children, respectively. The field

Table 91.4 Recommendations for Nutrition Delivery in Critically Ill Adult Patients

	Clinical Question	Recommendation	Evidence Grade	Strength of Recommendation
1	Does provision of higher vs lower energy intake impact clinical outcomes?	No significant difference in clinical outcomes was found between patients with higher vs lower levels of energy intake. We suggest feeding between 12 and 25 kcal/kg (i.e., the range of mean energy intakes examined) in the first 7–10 days of ICU stay.	Moderate	Weak
2	Does provision of higher vs lower protein intake impact clinical outcomes?	There was no difference in clinical outcomes in the relatively limited data. Because of a paucity of trials with high-quality evidence, we cannot make a new recommendation at this time beyond the 2016 guideline suggestion for 1.2–2.0 g/kg/day.	Low	Weak
3	In adult critically ill patients who are candidates for EN, does similar energy intake by PN vs EN as the primary feeding modality in the first week of critical illness impact clinical outcomes?	There was no significant difference in clinical outcomes. Because similar energy intake provided as PN is not superior to EN and no differences in harm were identified, we recommend that either PN or EN is acceptable.	High	Strong
4	In adult critically ill patients receiving EN, does provision of SPN, compared with no SPN during the first week of critical illness, impact clinical outcomes?	There was no significant difference in clinical outcomes. Based on findings of no clinically important benefit in providing SPN early in the ICU admission, we recommend not initiating SPN prior to day 7 of ICU admission.	High	Strong
5	In adult critically ill patients receiving PN, does provision of mixed-oil ILEs (i.e., medium-chain triglycerides, olive oil, FO, mixtures of oils), compared with 100% soybean-oil ILE, impact clinical outcomes?	Because of limited statistically or clinically significant differences in key outcomes, we suggest that either mixed-oil ILE or 100% soybean-oil ILE be provided.	Low	Weak
6	In adult critically ill patients receiving PN, does provision of FO-containing ILE, compared with non–FO-containing ILE, impact clinical outcomes?	We suggest that either FO- or non–FO-containing ILE be provided to critically ill patients who are appropriate candidates for initiation of PN.	Low	Weak

Reproduced from Compher C, Bingham AL, McCall M, et al. JPEN J Parenter Enteral Nutr 2022;46:12–41.[1]

Table 91.5 Recommendations for Nutrition Delivery in Critically Ill Pediatric Patients

	Clinical Question	Recommendation	Evidence Grade	Strength of Recommendation
1	What is the recommended energy requirement for critically ill children?	Based on observational cohort studies, we suggest that measured energy expenditure by IC be used to determine energy requirements and guide prescription of the daily energy goal. If IC measurement of resting energy expenditure is not feasible, we suggest that the Schofield or Food Agriculture Organization/World Health Organization/United Nations University equations may be used without the addition of stress factors to estimate energy expenditure.	Low	Weak
2	What is the target energy intake in critically ill children?	Based on observational cohort studies, we suggest achieving delivery of at least two-thirds of the prescribed daily energy requirement by the end of the first week in the PICU. Cumulative energy deficits during the first week of critical illness may be associated with poor clinical and nutrition outcomes.	Low	Weak
3	What is the minimum recommended protein requirement for critically ill children?	Based on evidence from RCTs and as supported by observational cohort studies, we recommend a minimum protein intake of 1.5 g/kg/d. Protein intake higher than this threshold has been shown to prevent cumulative negative protein balance in RCTs.	Moderate	Strong
4	What is the optimum method for advancing EN in the PICU population?	Based on observational studies, we suggest the use of a stepwise algorithmic approach to advance EN in children admitted to the PICU. The stepwise algorithm must include bedside support to guide the detection and management of EN intolerance and the optimal rate of increase in EN delivery.	Low	Weak
5	What is the best site for EN delivery: gastric or small bowel?	Existing data are insufficient to make universal recommendations regarding the optimal site to deliver EN to critically ill children. Based on observational studies, we suggest that the gastric route be the preferred site for EN in patients in the PICU. The postpyloric or small intestinal site for EN may be used in patients unable to tolerate gastric feeding or those at high risk for aspiration. Existing data are insufficient to make recommendations regarding the use of continuous vs intermittent gastric feeding.	Low	Weak
6	What is the indication for and optimal timing of PN in critically ill children?	Based on a single RCT, we do not recommend the initiation of PN within 24 hours of PICU admission.	Moderate	Strong
7.	What is the role of PN as a supplement to inadequate EN?	The threshold for and timing of PN initiation should be individualized. Based on a single RCT, supplemental PN should be delayed until 1 week after PICU admission for patients with normal baseline nutrition state and low risk of nutrition deterioration. Based on expert consensus, we suggest PN supplementation for children who are unable to receive any EN during the first week in the PICU. For patients who are severely malnourished or at risk of nutrition deterioration, PN may be supplemented in the first week if they are unable to advance past low volumes of EN.	Low	Weak

Reproduced from Mehta NM, Skillman HE, Irving SY, et al. Pediatr Crit Care Med 2017;18:675–715.[89] doi:10.1177/0148607117711387.

of critical care nutrition is in urgent need of homogenous, well-designed trials.

Based on the currently available evidence, a prudent strategy for nutrient delivery to patients with critical illness is illustrated in **Figure 91.3**. Such a strategy may include the following principles: (1) Early screening and assessment of patients on admission to the ICU to identify high-risk patients who might benefit from nutritional interventions; (2) Individualized prescription of macronutrient (energy and protein) and micronutrient delivery goals based on underlying nutritional status and disease characteristics. Avoidance

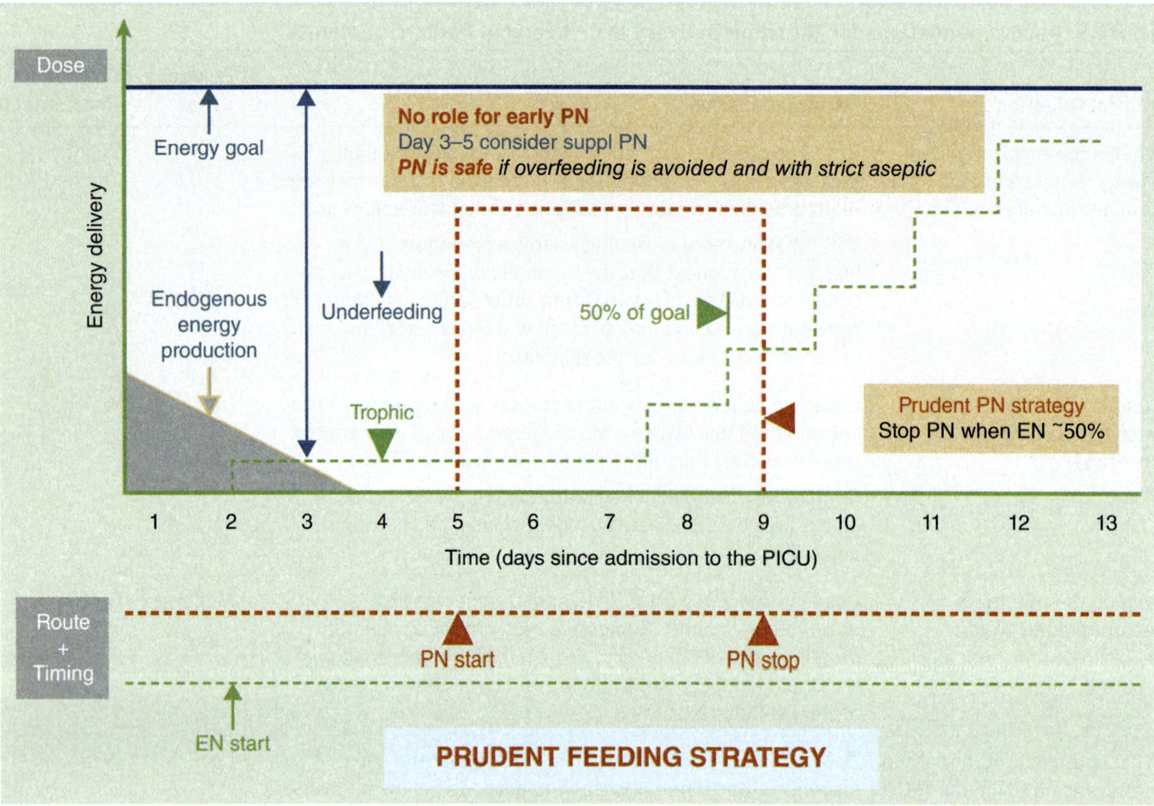

Figure 91.3 Prudent nutrition strategy for critically ill patients.

of overzealous macronutrient intake during the acute phase of critical illness; (3) A general preference for the enteral route for feeding, initiated early (within 24–48 hours) and advanced using a stepwise algorithm to achieve prudent goals (e.g., >60% of target) in the first week; (4) Judicious use of supplemental PN (preferably after day 4) in select patients with contraindication to EN or when EN is inadequate in the first week; and (5) Monitoring for signs of intolerance to EN or PN, and refeeding syndrome in high-risk patients. The use of immune-enhancing enteral formulas and supplementation with large doses of micronutrients and antioxidants should not be provided on a routine basis. Patients with trauma, surgery, or sepsis are characterized by heterogeneity of illness severity and physiologic variations across different phases of critical illness. Hence, a "one size fits all" approach to nutrition is flawed, and nutrition therapy (prescription, delivery, and monitoring) must be individualized.

References

1. Compher C, Bingham AL, McCall M, et al. JPEN J Parenter Enteral Nutr 2022;46:12–41.
2. McClave SA, Taylor BE, Martindale RG, et al. JPEN J Parenter Enteral Nutr 2016;40:159–211.
3. Wischmeyer PE. Crit Care 2017;21:316.
4. Fraipont V, Preiser JC. JPEN J Parenter Enteral Nutr 2013;37:705–13.
5. Mehta NM, Bechard LJ, Leavitt K, et al. JPEN J Parenter Enteral Nutr 2009;33:336–44.
6. Martinez EE, Bechard LJ, Mehta NM. Nutr Clin Pract 2014;29:360–7.
7. Berger MM, Pantet O, Jacquelin-Ravel N, et al. Clin Nutr 2019;38:2408–16.
8. Casaer MP, Mesotten D, Hermans G, et al. N Engl J Med 2011;365:506–17.
9. Jimenez L, Mehta NM, Duggan CP. Curr Opin Clin Nutr Metab Care 2017;20:227–31.
10. Ridley EJ, Davies AR, Parke R, et al. Crit Care 2018;22:12.
11. Ariagno K, Bechard L, Quinn N, et al. JPEN J Parenter Enteral Nutr 2022;46:190–6.
12. Fivez T, Kerklaan D, Mesotten D, et al. N Engl J Med 2016;374:1111–22.
13. Jacobs A, Verlinden I, Vanhorebeek I, Van den Berghe G. J Clin Med 2019;8:830.
14. Manzanares W, Heyland DK. Nutr 2013;29:359–60.
15. Dresen E, Pimiento JM, Patel JJ, Heyland DK, Rice TW, Stoppe C. JPEN J Parenter Enteral Nutr 2023;47:S38–S49.
16. Wischmeyer PE. Curr Opin Crit Care 2016;22:279–84.
17. Anthony PS. Nutr Clin Pract 2008;23:373–82.
18. Ferguson M, Capra S, Bauer J, et al. Nutr 1999;15:458–64.
19. Ferguson ML, Bauer J, Gallagher B, et al. Australas Radiol 1999;43:325–7.
20. Rubenstein LZ, Harker JO, Salva A, et al. J Gerontol A Biol Sci Med Sci 2001;56:M366–M72.
21. Cohendy R, Rubenstein LZ, Eledjam JJ. Aging (Milano) 2001;13:293–7.
22. Kondrup J, Rasmussen HH, Hamberg O, et al. Clin Nutr 2003;22:321–36.
23. Weekes CE, Elia M, Emery PW. Clin Nutr 2004;23:1104–12.

24. Stratton RJ, Hackston A, Longmore D, et al. Br J Nutr 2004;92:799–808.
25. Stratton RJ, King CL, Stroud MA, et al. Br J Nutr 2006;95:325–30.
26. Heyland DK, Dhaliwal R, Jiang X, et al. Crit Care 2011;15:R268.
27. Rahman A, Hasan RM, Agarwala R, et al. Clin Nutr 2016;35:158–62.
28. Detsky AS, McLaughlin JR, Baker JP, et al. JPEN J Parenter Enteral Nutr 1987;11:8–13.
29. Bauer J, Capra S, Ferguson M. Eur J Clin Nutr 2002;56:779–85.
30. Norman K, Schutz T, Kemps M, et al. Clin Nutr 2005;24:143–50.
31. Sezer S, Ozdemir FN, Afsar B, et al. Transplant Proc 2006;38:517–20.
32. Bretschera C, Boesiger F, Kaegi-Braun N, et al. EClinicalMedicine 2022;45:101301.
33. Jacobs DG, Jacobs DO, Kudsk KA, et al. J Trauma 2004;57:660–78; discussion 79.
34. Kudsk KA, Minard G, Wojtysiak SL, et al. Surg 1994;116:516–23.
35. Yeh DD, Johnson E, Harrison T, et al. Nutr Clin Pract 2018;33:419–25.
36. Arabi YM, Casaer MP, Chapman M, et al. Intensive Care Med 2017;43:1239–56.
37. Stoppe C, Wendt S, Mehta NM, et al. Crit Care 2020;24:499.
38. Mehta NM, Jaksic T. In: Duggan CP, Walker W, eds. The Critically ill Child. Nutrition in Pediatrics. Hamilton, ON: B. C. Decker Inc, 2008; 663–73.
39. Preiser JC, Ichai C, Orban JC, et al. Br J Anaesth 2014;113:945–54.
40. Oshima T, Berger MM, De Waele E, et al. Clin Nutr 2017;36:651–62.
41. Singer P, Pichard C, De Waele E. Clin Nutr Exp 2020;33:18–23.
42. Chwals WJ. New Horiz 1994;2:147–55.
43. Kerklaan D, Hulst JM, Verhoeven JJ, et al. J Pediatr Gastroenterol Nutr 2016;63:445–50.
44. Klein CJ, Stanek GS, Wiles CE, 3rd. J Am Diet Assoc 1998;98:795–806.
45. Mehta NM, Bechard LJ, Dolan M, et al. Pediatr Crit Care Med 2011;12:398–405.
46. Campbell SM, Kudsk KA. JPEN J Parenter Enteral Nutr 1988;12:610–2.
47. Oshima T, Dupertuis YM, Delsoglio M, et al. Clin Nutr ESPEN 2019;32:50–5.
48. Mehta NM, Smallwood CD, Joosten KF, et al. Clin Nutr 2015;34:151–5.
49. Smallwood CD, Martinez EE, Mehta NM. Respir Care 2016;61:354–8.
50. Singer P, Pichard C, Rattanachaiwong S. Curr Opin Clin Nutr Metab Care 2020;23:91–5.
51. Allingstrup MJ, Kondrup J, Wiis J, et al. Intensive Care Med 2017;43:1637–47.
52. Wischmeyer PE, Molinger J, Haines K. Nutr Clin Pract 2021;36:275–81.
53. Singer P, Blaser AR, Berger MM, et al. Clin Nutr 2019;38:48–79.
54. Kavanagh BP, McCowen KC. N Engl J Med 2010;363:2540–6.
55. van den Berghe G, Wouters P, Weekers F, et al. N Engl J Med 2001;345:1359--67.
56. Australian N-SSIft, New Zealand Intensive Care Society Clinical Trials G, the Canadian Critical Care Trials G, et al. Intensive Care Med 2015;41:1037–47.
57. Azevedo JR, Lima ER, Cossetti RJ, et al. Arq Neuropsiquiatr 2007;65:733–8.
58. Bilotta F, Caramia R, Paoloni FP, et al. Anesthesiol 2009;110:611–9.
59. Investigators CS, Annane D, Cariou A, et al. JAMA 2010;303:341–8.
60. Allingstrup MJ, Esmailzadeh N, Wilkens Knudsen A, et al. Clin Nutr 2012;31:462–8.
61. Bechard LJ, Parrott JS, Mehta NM. J Pediatr 2012;161:333–9 e1.
62. Mehta NM, Bechard LJ, Zurakowski D, et al. Am J Clin Nutr 2015;102:199–206.
63. Weijs PJ, Sauerwein HP, Kondrup J. Clin Nutr 2012;31:774–5.
64. McClave SA, Martindale RG, Vanek VW, et al. JPEN J Parenter Enteral Nutr 2009;33:277–316.
65. Singer P, Berger MM, Van den Berghe G, et al. Clin Nutr 2009;28:387–400.
66. Ziegler TR. N Engl J Med 2009;361:1088–97.
67. Berger MM, Chiolero RL. Nestle Nutr Workshop Ser Clin Perform Programme 2003;8:99–111; discussion 7.
68. Berger MM, Shenkin A. Nutr 2006;22:952–5.
69. Heyland DK, Dhaliwal R, Suchner U, et al. Intensive Care Med 2005;31:327–37.
70. Berger MM, Manzanares W. Curr Opin Clin Nutr Metab Care 2021;24:165–75.
71. Yarandi SS, Zhao VM, Hebbar G, et al. Curr Opin Clin Nutr Metab Care 2011;14:75–82.
72. Griffiths RD, Jones C, Palmer TE. Nutr 1997;13:295–302.
73. Kudsk KA, Minard G, Croce MA, et al. Ann Surg 1996;224:531–40; discussion 40–3.
74. Kirk SJ, Barbul A. JPEN J Parenter Enteral Nutr 1990;14:226S–9S.
75. Heyland DK, Wibbenmeyer L, Pollack JA, et al. N Engl J Med 2022;387:1001–10.
76. Luo M, Fernandez-Estivariz C, Jones DP, et al. Nutr 2008;24:37–44.
77. Nathens AB, Neff MJ, Jurkovich GJ, et al. Ann Surg 2002;236:814–22.
78. Louw JA, Werbeck A, Louw ME, et al. Crit Care Med 1992;20:934–41.
79. Marik PE, Khangoora V, Rivera R, et al. Chest 2017;151:1229–38.
80. Fowler AA, 3rd, Truwit JD, Hite RD, et al. JAMA 2019;322:1261–70.
81. Fujii T, Luethi N, Young PJ, et al. JAMA 2020;323:423–31.
82. Iglesias J, Vassallo AV, Patel VV, et al. Chest 2020;158:164–73.
83. Moskowitz A, Huang DT, Hou PC, et al. JAMA 2020;324:642–50.
84. Hwang SY, Ryoo SM, Park JE, et al. Intensive Care Med 2020;46:2015–25.
85. Sevransky JE, Rothman RE, Hager DN, et al. JAMA 2021;325:742–50.
86. Schlapbach LJ, Gibbons K, Ridolfi R, et al. Front Pediatr 2021;9:663435.
87. Coloretti I, Biagioni E, Venturelli S, et al. J Crit Care 2020;59:37–41.
88. Mehta NM. Pediatr Crit Care Med 2022;23:385–9.
89. Mehta NM, Skillman HE, Irving SY, et al. Pediatr Crit Care Med 2017;18:675–715.
90. Hegazi RA, Wischmeyer PE. Crit Care 2011;15:234.
91. Kudsk KA, Croce MA, Fabian TC, et al. Ann Surg 1992;215:503–11; discussion 11–3.
92. Moore FA, Feliciano DV, Andrassy RJ, et al. Ann Surg 1992;216:172–83.
93. Daly JM, Lieberman MD, Goldfine J, et al. Surg 1992;112:56–67.
94. Daly JM, Weintraub FN, Shou J, et al. Ann Surg 1995;221:327–38.
95. Fan ST, Lo CM, Lai EC, et al. N Engl J Med 1994;331:1547–52.
96. Elke G, van Zanten AR, Lemieux M, et al. Crit Care 2016;20:117.
97. Guenter P, Worthington P, Ayers P, et al. Nutr Clin Pract 2018;33:295–304.

98. Harvey SE, Parrott F, Harrison DA, et al. Health Technol Assess 2016;20:1–144.
99. Reignier J, Boisrame-Helms J, Brisard L, et al. Lancet 2018;391:133–43.
100. Mehta NM, McAleer D, Hamilton S, et al. JPEN J Parenter Enteral Nutr 2010;34:38–45.
101. Wischmeyer PE. Curr Opin Crit Care 2021;27:169–76.
102. Zhang Z, Xu X, Ding J, et al. Nutr Clin Pract 2013;28:371–80.
103. Ichimaru S. Nutr Clin Pract 2018;33:790–5.
104. Brown AM, Irving SY, Pringle C, et al. JPEN J Parenter Enteral Nutr 2021.
105. Hamilton S, McAleer DM, Ariagno K, et al. Pediatr Crit Care Med 2014;15:583–9.
106. Marik PE, Zaloga GP. Intensive Care Med 2008;34:1980–90.
107. Heidegger CP, Berger MM, Graf S, et al. Lancet 2013;381: 385–93.
108. Bechard LJ, Staffa SJ, Zurakowski D, et al. Am J Clin Nutr 2021;114:1859–67.

CHAPTER 92

Malnutrition in Burn Injury

Dalia Barayan and Marc G. Jeschke

CHAPTER OUTLINE

Abbreviations

ASPEN, American Society for Parenteral and enteral nutrition; **BCAA**, branched-chain amino acid; **EN**, enteral nutrition; **ESPEN**, European Society for Parenteral and Enteral Nutrition; **FA**, fatty acid; **FFA**, free fatty acids; **IC**, indirect calorimetry; **IU**, international unit; **IV**, intravenous; **LBM**, lean body mass; **LOS**, length of stay; **MPB**, muscle protein breakdown; **MPS**, muscle protein synthesis; **PN**, parenteral nutrition; **REE**, resting energy expenditure; **RQ**, respiratory quotient; **TBSA**, total body surface area

Introduction

Severe burns represent one of the most debilitating forms of trauma, affecting nearly 11 million people each year worldwide.[1] According to the World Health Organization, burn injuries account for approximately 180,000 deaths per year, with more than 1.1 million cases requiring serious medical attention in the United States annually.[2] Each year in Canada, there are an estimated 43,000 emergency visits and over 2,000 hospitalizations attributed to flame burns alone.[3] Over the last 20 years, patient outcomes have improved due to the establishment of specialized burn centers and improvements in resuscitation, dedicated burn-specific protocols, improved wound coverage and infection control, and improved management of inhalation injury.[4] Nonetheless, patients with burns still experience unacceptably high morbidity and mortality rates after injury.[1-4]

Severe burns, or burns encompassing more than 20% of the total body surface area (TBSA), trigger a hypermetabolic response that far surpasses that induced by any other injury or disease state in magnitude and persistence.[4,5] Generally, postburn hypermetabolism is characterized by increased resting energy expenditure (REE) with accompanying hyperglycemia, adipose wasting and lipolysis, and protein catabolism. This stress response is initiated to provide sufficient energy for maintaining organ function and whole-body homeostasis under demanding trauma conditions.[5,6] While initially ubiquitous and essential, the metabolic need of a burn patient is the highest of any medical state, approaching 140% of that predicted.[7] Accumulating evidence suggests that patients with severe burns can remain hypermetabolic for up to 3 years postinjury.[4,7-9] Significant nutritional support to help meet these prolonged and heightened energy demands is imperative for survival of patients with severe burns.[10-12] Without intervention, burn patients would universally experience undernutrition, ultimately causing cumulative energy deficits that can lead to vast cachexia, multiorgan failure, and eventually death.[10,11] Unfortunately, our knowledge about the complicated physiology of nutrition is currently lacking, and clinical regimens differ widely among burn centers.[12,13] Many critical questions regarding the optimal timing, route, amount, and diet composition in the burn population remain unanswered.[13] An improved understanding of postburn hypermetabolism, accompanying changes, and nutritional and medical management is paramount to optimizing care and improving outcomes for patients with severe burns.

This chapter provides an overview of the complex hypermetabolic response to burn injury and discusses key mediators and sequelae, focusing on alterations in glucose, fat, and amino acid metabolism. We then review the current state of knowledge of nutrition in burn patients gained from clinical trials, retrospective analyses, and studies in other patient populations, especially trauma and critical care. Areas where further investigation is still needed with regards to optimal nutritional support and the ideal management of postburn hypermetabolism are noted.

Complexity of the Systemic Response to Burn Injury

The systemic response to burn injury is not dissimilar to other traumas and follows the biphasic pattern of metabolic regulation initially described by Sir David Cuthbertson in 1942 (**Figure 92.1**).[14] The first phase, or "ebb phase," usually occurs within 48 hours postburn and is characterized by decreases in cardiac output, oxygen consumption, and metabolic rate as well as impaired glucose tolerance associated with its hyperglycemic state (hypometabolism characteristic of early shock).[15,16] However, within the first 5 days postinjury, these metabolic variables gradually increase to a plateau phase.[14,17] This second phase, or "flow phase," is associated with hyperdynamic circulation and a profound hypermetabolic state, in which body temperature, oxygen, and glucose consumption, CO_2 production, and futile substrate cycling are all substantially increased.[15-17] Typically, patients are considered hypermetabolic when their REE, expressed as percent predicted, is ≥10% above normal.[18]

The magnitude and duration of this stress response are directly proportional to the degree of burn injury, quantified based on TBSA with second- and third-degree burns.[19,20] For patients with >40% TBSA third-degree burns, resting metabolic rate at thermal neutral temperature (30°C) has been shown to surpass 140% of normal at admission and is reduced to 130% once wounds are fully healed, to 120% at 6 months, and then to 110% at 12 months postinjury.[20] In addition to burn size, factors including age,[21] body composition,[22] genetics,[23] and the presence of smoke inhalation injury or another insult like infection,[24] can further accentuate and prolong the increased metabolic rate after injury. If left unmitigated, prolonged hypermetabolism after burn trauma is a major concern, as it is associated with a profound catabolic state in almost every organ system of the body.[4,25] To date, the cellular and molecular mechanisms underlying burn-induced hypermetabolism have not been fully identified. Despite improved clinical care, the detrimental sequela of this complex response is a significant contributor to postburn morbidity and mortality.[26]

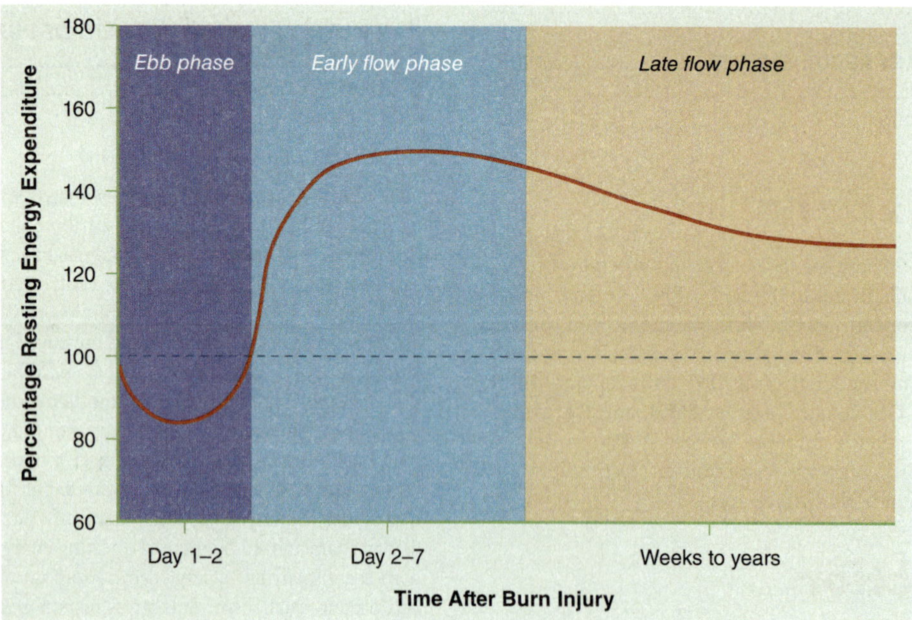

Figure 92.1 Overview of the systemic response to burn injury. Severe burn induces a unique and remarkably complex response that follows a biphasic pattern of metabolic regulation proposed by Cuthbertson et al. A short "ebb" phase characterized by hypometabolism occurs immediately after injury and is characterized by a decrease in metabolic rate, oxygen consumption, body temperature, and enzymatic activity. The ebb phase is followed by a longer hypermetabolic "flow" phase marked by an increased catabolism, with a high oxygen consumption and an elevated REE rate.

Created with BioRender.com.

Mediators and Consequences of Postburn Hypermetabolism

Accumulating evidence suggests that marked and sustained increases in catecholamine,[27] glucocorticoid,[28] glucagon,[29] and dopamine secretion[30] are involved in initiating the acute hypermetabolic response and its ensuing catabolic state after injury. The rise in these catabolic hormones is accompanied by a decrease in the normal endogenous activity of anabolic agents, primarily human growth hormone[31,32] and testosterone,[33] which combine to result in a large net protein loss.[34] Cytokines such as interleukin 6 (IL-6) and tumor necrosis factor (TNF),[35] endotoxin,[36] neutrophil-adherence complexes,[37] reactive oxygen species,[38] nitric oxide,[39] and coagulation[40] as well as complement cascades[41] have also been implicated in mediating and maintaining the postburn hypermetabolic state. Once these cascades of events are initiated, their mediators and by-products further drive the persistent and increased metabolic rate accompanied by alterations in glucose, lipid, and amino acid metabolism.[25] Failure to address these metabolic changes can have catastrophic consequences (**Figure 92.2**).[4,15,26]

Glucose Metabolism

In healthy individuals, glucose homeostasis is tightly regulated. Under normal circumstances, postprandial elevations in circulating glucose concentration stimulate pancreatic β-cells to release insulin, a potent anabolic hormone.[42] Insulin release promotes peripheral glucose uptake into skeletal muscle and adipose tissue and suppresses glucose production (i.e., gluconeogenesis) in the liver, thereby restoring blood glucose concentration.[42,43] Following thermal injury; however, metabolic alterations cause significant shifts in energy substrate metabolism to provide glucose, a primary fuel source, to vital organs.[25] Glucose production in the liver is approximately twofold greater in patients with burns than in healthy controls.[44] This increased glucose production is excessive for tissue use and leads to the "diabetes of stress."[45]

Moreover, unlike healthy individuals, exogenous glucose infusion does not fully attenuate hepatic gluconeogenesis in patients with burns (only about 50%).[44] While the exact cause is not entirely defined, it appears that these patients undergo a "double hit" where both central and peripheral insulin sensitivity is impaired after injury, resulting in poor glucose control.[45] Specifically, insulin exerts a diminished ability to suppress hepatic glucose output (central insulin resistance) and a reduced ability to stimulate glucose disposal into skeletal muscle (peripheral insulin resistance).[42-45] Therefore, although glucose remains a preferred substitute for tissue energy postburn, there is a well-defined limit to its use, given the insulin resistance.

Of major importance is recent evidence suggesting that impaired glucose metabolism postburn is detrimental and is associated with poor survival after injury.[46] These findings were further supported by prospective randomized trials demonstrating that postburn treatment with anti-diabetes drugs, like insulin[47] or metformin,[48] was beneficial for restoring postburn

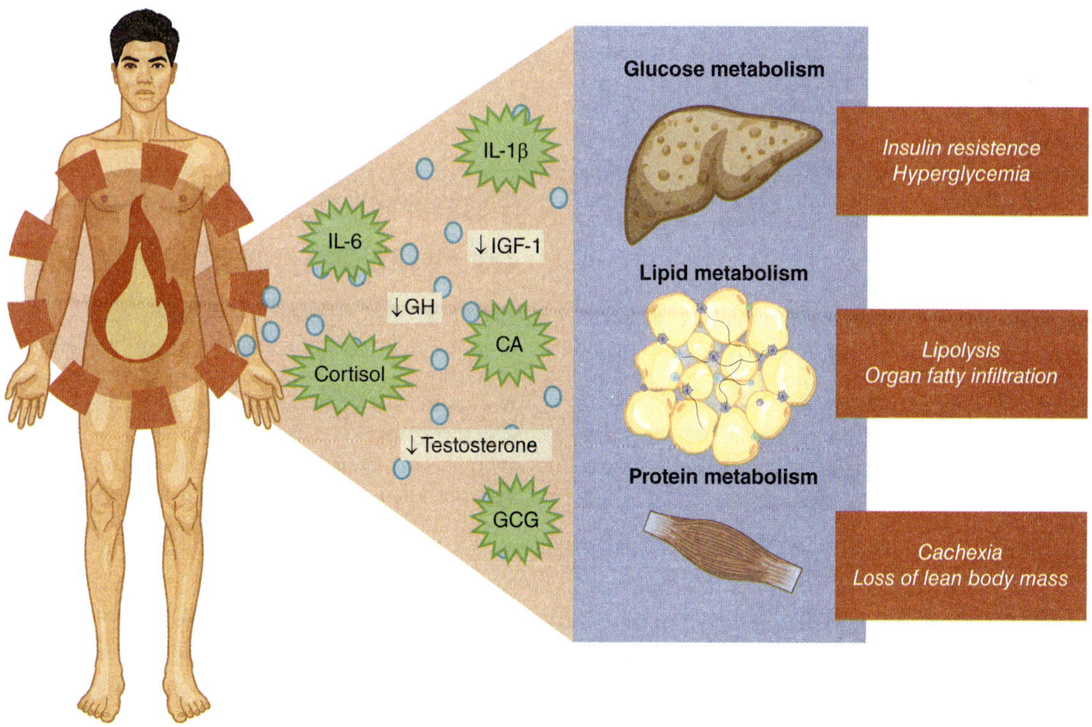

Figure 92.2 Mediators and metabolic consequences of burn-induced hypermetabolism. Marked and sustained increases in stress hormones and pro-inflammatory mediators, combined with a decrease in anabolic agents, are involved in initiating the hypermetabolic response to burn injury and its ensuing catabolic state. Once released, these mediators and by-products further drive the persistent and increased metabolic rate accompanied by alterations in glucose, lipid, and amino acid metabolism. If left unmitigated, this protective stress response becomes auto destructive, driving an entire spectrum of metabolic abnormalities that lead to vast cachexia, multi-organ failure, and mortality.

Created with BioRender.com.

glucose homeostasis and organ function and morbidity outcomes. Retrospective cohort studies also confirmed survival benefits of tight glycemic control in patients with severe burns.[49,50] Together, these studies highlight the significant clinical challenges associated with postburn insulin resistance and hyperglycemia and the importance of glucose control and treatment after injury.

Lipid Metabolism

Fat metabolism is another critical metabolic pathway that is significantly altered during the postburn hypermetabolic period. As with other forms of trauma, the rise in postburn catecholamines leads to activation of adipose tissue lipolysis, mainly via β2 and β3 adrenergic stimulation.[51,52] Lipolysis refers to the enzymatic breakdown (hydrolysis) of triglycerides into free fatty acids (FFAs) and glycerol molecules. These substrates are then released into the circulation to be used as fuel by other tissues in the body through beta-oxidation, gluconeogenesis, and glycolysis.[25,53] Similar to glucose; however, the amount of exogenous fat that can be used as an energy substitute after major burns is limited; only 30% of FFAs, at most, is oxidized for fuel, while the remaining 70% is recycled.[53,54] This is a marked difference from starvation, during which 90% of lipolysis-derived FFA are used for fuel.[25,54] Accordingly, burn-induced lipolysis and FFA mobilization have been linked to several detrimental complications.[4] For instance, excessive FFA release may overwhelm glucose transport activity after injury, which further impairs insulin-mediated glucose uptake, contributing to postburn insulin resistance.[54,55] This has been supported by research on type 2 diabetes, demonstrating that elevated FFA are predictive of the incidence and severity of disease.[56,57]

A clinical analysis by Kraft et al., demonstrated a strong correlation between liberated FFA after burn injury and increased fatty infiltration into vital organs, especially the liver.[58] Liver fatty infiltration and dysfunction, or hepatic steatosis, remains a leading cause of morbidity and mortality in burn patients.[4] These authors also found that, in children with severe burns, elevated triglycerides were associated with worsened organ function and glucose metabolism, and increased incidence of infection, sepsis, and poor outcomes.[58] Interestingly, a more recent study showed that pharmacologic inhibition of burn-induced adipose lipolysis can effectively ameliorate ectopic fat deposition in the livers of postburn mice, thereby improving organ function following injury.[59] Thus, while adipose tissue was previously ignored, it is becoming increasingly evident that this organ plays a central role in meditating postburn hypermetabolic responses.

Protein Metabolism

The main source of fuel in the patient with burns is protein/amino acids from skeletal muscle.[60] Chronic catabolism of skeletal muscle is pathognomonic of severe burn trauma, which leads to marked wasting of muscle protein and, consequently, lean body mass (LBM) within days after injury.[61] The prolonged erosion of LBM and resultant wasting leaves burn survivors cachectic and deconditioned.[9] This metabolic abnormality, in turn, significantly delays wound healing and contributes to long-term morbidity after injury.[45] Surprisingly, studies have shown that lean body mass is reduced for up to 2 to 3 years postburn.[4,7,9] In patients with severe burns (>30% TBSA), this cachectic state can be observed for several years after injury.[25,61] The clinical consequences of persistent protein catabolism can be severe, depending on the degree of muscle protein loss. A 10% to 15% loss in LBM has been shown to decrease wound healing,[62] while a further increase in LBM loss of ~30% is associated with significant increases in infections, profound weakness, skin breakdown (pressure sores), and diminished wound healing.[63] A LBM loss of 40% or greater usually becomes fatal.[63] Mortality is further augmented by severe muscle weakness, which can prolong mechanical ventilator requirement, inhibit sufficient cough reflexes, and delay mobilization in patients with protein malnutrition.[64] Decreases in muscle mass may also be responsible for the delay in growth commonly observed in pediatric patients with burns, even at 2 years postburn.[65] As skeletal muscle is responsible for 70% to 80% of whole-body insulin-stimulated glucose uptake, profound muscle wasting may further augment persistent insulin resistance after injury.[66]

While the specific mechanism is unclear, likely mediators are stress hormones, cytokines, and oxidants.[4] From a mechanistic view, burn injury induces a concurrent increase in skeletal muscle protein synthesis (MPS) and breakdown (MPB) rates. However, MPB significantly surpass MPS, resulting in a significant net loss of total body protein.[67] Moreover, increasing protein intake does not further increase MPS, indicating that the catabolic drive cannot be overcome with more substrate synthesis.[68] The exception would be adding an anabolic agent, such as a growth hormone[69] or testosterone,[33] which would be more likely to overcome the net amino acid effect. Modulation of this hypermetabolic, hypercatabolic response after injury using several different therapeutic approaches, including nutritional support in conjunction with pharmacologic agents is, thus, paramount for the restoration of normal metabolic structure and function in patients with severe burns.[68,70] Non-nutritional management of burn-induced hypermetabolism will be discussed further later in this chapter.

Timing and Route of Nutritional Therapy

Metabolic rates of patients with burns can surpass twice normal, and failure to fulfill these energy requirements can lead to impaired wound healing, organ dysfunction, and increased susceptibility to infection and death.[70] While there is a lack of widespread appreciation among clinicians in favor of nutritional support over pharmacologic agents to resolve burn-induced hypermetabolism, there exists a consensus that adequate assessment and provision of dietary needs is imperative for patient outcome, particularly during the pivotal period of muscle wasting and wound repair.[13,71,72] So, an important question arises: *how*, *when*, and *what* should a patient with severe burns be fed?

Enteral vs Parenteral Nutrition

In the 1960s and 1970s, parenteral nutrition (PN) was routinely used for patients with burns, but this quickly changed in the 1980s, when PN became associated with adverse complications and infections.[73] Since then, enteral nutrition (EN) has become the preferred route of nutrition in burns.[74] PN is generally defined as providing nourishment via central and peripheral veins, whereas EN includes nourishment into the gut and may include diet, oral supplementation, and tube feeding.[71,74] The preference for EN over PN is unanimously accepted because of its compelling advantages of being safer and cost-effective, with lower risk of complications.[74] Most of the benefits of EN are associated with food actually being present in the gut, as this has an essential role in preserving normal physiologic immune and inflammatory functions.[75] For instance, proponents of EN use will point out that by nourishing the bowel mucosa, EN improves gut-associated immune and inflammatory status.[74,76]

Moreover, EN maintains the integrity of the gastrointestinal (GI) tract, thereby preventing the translocation of bacteremia and decreasing the risk of developing infections.[72] This notion is supported by several trials demonstrating that providing EN in adult patients with burns is associated with shorter ICU stays and reduced incidence of wound infections.[72] In contrast, parenteral formulas are more expensive and carry more significant risks, including complications with gaining central venous access, line infections, and poorer glycemic control.[72,73] Several studies have also found that PN is associated with overfeeding, liver and immune dysfunction, and a threefold increase in mortality.[72,74] Thus, a parenteral route should only be used in patients with burns with contraindication to enteral feeding.

Early vs Delayed Treatment

A major determinant of patient outcome after severe burns is time to treatment, including time to nutritional support. Following injury, substantial intestinal mucosal damage and increased bacterial translocation occur during the acute phase, resulting in decreased nutrient absorption.[75] With this in mind, early EN is beneficial for decreasing circulating catecholamines, cortisol, and glucagon, and maintaining intestinal mucosal integrity, motility, and blood flow.[72,77-80] Hence, providing even low rates of EN early after admission has a theoretical advantage. While current evidence remains inconclusive about the benefits of "early" (as defined by feeding in the first 24 hrs) vs "late" (as defined by more than 24 hrs postburn) EN with regard to its effects on length of stay (LOS) and mortality, accumulating evidence suggests that initiation of support within the first 12 hours postinjury is optimal and effectively attenuates the hypermetabolic response as well as reduces infectious complications.[81,82] The greatest support for early vs delayed EN in patients with burns comes from a meta-analysis encompassing 753 patients posttrauma.[83] This study showed that patients who received EN within 36 hours had significantly decreased propensity for infection and complications.[83]

Besides clinical data, animal studies show that early initiation of EN can effectively mitigate the adverse postburn hypermetabolic response discussed above.[84] Mochizuki et al., demonstrated that guinea pigs who received EN 2 hours after burn had a significantly lower metabolic rate at 2 weeks postinjury, compared with those who started receiving nutritional support at 3 days after burn.[84] However, it is important to emphasize that these are only a few studies, and the benefits of early feeding in hypermetabolic burn patients have not been studied adequately.[71] The main reason is that the burn population's heterogeneity makes research designs problematic and inherently flawed for conducting large-scale studies that will unravel the relationship between early nutrition and improved patient outcomes.[71] Furthermore, although several studies have documented the benefits of early nutrition on several outcome measures, there is a body of literature that contradicts these findings.[71] Notably, recent clinical trials have found no significant advantage to early enteral nutrition in patients with burns.[85,86] Notwithstanding these limitations, most burn clinicians agree that, until such studies are conducted, early EN feeding initiated within the first 48 hours postinjury is most favorable.

Estimating Energy Requirements

The primary goal of nutritional support after burn injury is to provide adequate energy and nutrients necessary to maintain vital organ and immune function, minimize further catabolism and tissue loss, and promote successful healing and recovery.[10-12,70] Early EN (within 12 hours) ameliorates postburn catabolism and improves outcomes, but it is important to note that this is a delicate balance.[13,72] Undernutrition reduces immunocompetence, delays healing, and increases mechanical ventilation dependency and risk of infection,[87-89] whereas overfeeding, in the form of excess energy or protein, is associated with hyperglycemia, carbon dioxide retention, fatty infiltration of organs, and azotemia.[90,91] Therefore, adequate nutrition is essential to alleviate hypermetabolism, but too much feeding is detrimental and should be avoided.[92,93] Nutritional support must be individualized, monitored, and adjusted throughout recovery.

Predictive Equations

Currently, the appropriate energy amount is still under discussion, as is the composition of nutritional support, and existing feeding regimens rely on REE to calculate energy requirements. With this line of thought, numerous formulas to estimate the energy needs of burn survivors have been developed and used throughout the years (**Table 92.1**).[94] For example, the Curreri formula was first created in 1972 by retrospectively studying nine patients and reverse calculating the approximate kilocalories that would have been needed to compensate for their weight loss.[95] Interestingly, the Curreri formula and many other older formulas were found to overestimate the present metabolic requirements of patients with burns and other, more sophisticated formulas with different variables have since been introduced, including the Toronto, Schofield, American Society for Parenteral and Enteral Nutrition (ASPEN), and Ireton-Jones formulas.[94] Still, another study compared 46 different formulas for predicting dietary requirements and showed that none correlated well with measured energy expenditure in 24 patients with burns after injury.[96] Because energy expenditure fluctuates after burn, using fixed formulas often result in underfeeding during periods of highest energy use and overfeeding later in the clinical course. A comparison of supplementation values for a standardized patient (30 years old, 72 kg, 170 cm, 40% TBSA) using the different formulas demonstrates a wide range, from 2,099 Kcal/day (Ireton-Jones) to 5,229 Kcal/day (Modified Schofield).[97]

Indirect Calorimetry

According to the European Society for Parenteral and Enteral Nutrition (ESPEN), the gold standard for the measurement of energy expenditure is indirect calorimetry (IC), which measures

Table 92.1 REE Formulas and Patient Specifications

Kcals: Energy intake in past 24 hours
Harris Benedict: Basal requirements in Kcals using the Harris Benedict formula with no stress factors or activity factors
T: Body temperature in degree Celsius
Y: Age in years
S: Male = 1/Female = 0
TR: Trauma present = 1/absent = 0
O: Obesity present = 1/absent = 0
B: Burn present = 1/absent = 0

Formula Name	Specifications	Equation and Factors
Harris-Benedict		BMR × Activity factor × Injury factor
	Male BMR	66 + (13.7 × weight in kg) + (5 × height in cm) − (6.8×age)
	Female BMR	665 + (9.6 × weight in kg) + (1.8 × height in cm) − (4.7 x age)
		Activity factor: Confined to bed: 1.2; Minimal ambulation: 1.3
		Injury factor: < 20% TBSA: 1.5; 20%–40% TBSA: 1.6; > 40% TBSA: 1.7
Curreri	All patients	(25 × weight in kg) + (40 × %TBSA)
Pennisi	*Adults*	
	Kilocalories	(20 × weight in kg) + (70 × % TBSA)
	Protein	(1 g × weight in kg)+(3g × %TBSA)
	Children	
	Kilocalories	(60 × weight in kg) + (35 × %TBSA)
	Protein	(3g × weight in kg) + (1g × %TBSA)
Toronto Formula	All patients	(− 4343 + (10.5 × % TBSA) + (0.23 × Kcal intake) + (0.84 × Harris Benedict) + (114 × T [°C]) − (4.5 × days postburn) × Activity Factor
		Activity factor
		Ventilated 1.2
		Nonventilated:
		Confined to bed: 1.2
		Minimal ambulation: 1.3
		Moderate activity: 1.4
Modified Schofield	Male	10–18 Y = (0.074 × weight in kg) + 2.754
		18–30 Y = (0.063 × weight in kg) + 2.896
		30–60 Y = (0.048 × weight in kg) + 3.653
		> 60 Y = (0.049 × weight in kg) + 2.459
	Female	10–18 Y = (0.056 × weight in kg) + 2.898
		18–30 Y = (0.062 × weight in kg) + 2.036
		30–60 Y = (0.034 × weight in kg) + 3.538
		> 60 Y = (0.038 × weight in kg) + 2.755
		Injury factors
		< 10% TBSA = 1.2
		11–20% TBSA = 1.3
		21–30% TBSA = 1.5
		31–50% TBSA = 1.8
		> 50% TBSA = 2.0
ASPEN	All patients	25-35 Kcal/kg/day
Ireton-Jones Formula	Nonventilated	629 − (11 × Y) + (25 × weight in kg) − (609 × O)
	Ventilated	1784 − (11 × Y) + (25 × weight in kg) + (244 × S) + (239 × TR) + (804 × B)

Modified from Mendonca Machado N, Gragnani A, Masako Ferreira L. Nutr Hosp 2011;26:692–700.[97]

the volume of expired gas as well as the inhaled and exhaled concentrations of oxygen (VO_2) and carbon dioxide (VCO_2) via tight-fitting face masks or ventilators and, therefore, metabolic rate.[98,99] To calculate energy requirements in thermally injured patients, IC should be performed in the fed state, at rest, with a factor of 1.2 to 1.4 applied to account for activity, the stress of wound care, and other nursing tasks.[100] IC can also be used to confirm adequate nutritional support (underfeeding/overfeeding) by calculating the ratio of VCO_2/VO_2, also known as the respiratory quotient (RQ).[92,98-101] RQ is affected by the body's metabolism of specific substrates; for instance, fat is used as a major energy source during unstressed starvation, producing an RQ of <0.7.[100,101] In contrast, the normal metabolism of mixed substrates yields an RQ of around 0.75 to 0.90.[100,101] Finally, overfeeding is typified by the synthesis of fat from carbohydrates, resulting in an RQ of >1.0.[100,101] This explains one feared complication of overfeeding: difficulty in weaning from ventilatory support.[90] Despite this concern, one study found that high-carbohydrate diets in pediatric patients with burns led to decreased muscle wasting and did not result in RQs above 1.05 or any respiratory complications.[102] It is also important to note that IC is not practical to perform daily. Under such circumstances, the Toronto equation, which considers anthropometrics, age, energy intake, temperature, TBSA, and days postburn, is validated and recommended as an alternative.[98,99] Regardless of whether energy expenditure is measured using IC or estimated by fixed formulas, it is typically reassessed more than once a week so that strategies to optimize energy and protein intake can be implemented.[72]

Composition of Postburn Feeds: Macronutrients

Carbohydrates, lipids, and proteins are the main biological building blocks used to fuel complex metabolic processes and generate energy. Each of these three macronutrients must be supplemented in the patient with a burn injury in specific ratios once an energy goal has been determined.

Carbohydrates

Carbohydrate supplementation in patients with burns is critical, as it serves as a vital nutrient source and spares amino acid and protein stores in the skeletal muscle relative to high-fat diets. While the number of evidence-based studies on exact carbohydrate requirements is scarce, it is estimated that patients with severe burns have energy requirements that far exceed the body's ability to oxidize glucose sufficiently.[98,103,104] When providing carbohydrates (primarily via the parenteral route), it is essential to not exceed the maximum rate at which glucose can be assimilated, which has been reported to be 5 mg/kg/min or approximately 7 g/kg/day (2,240 kcal for an 80 kg man). Excessive amounts can have deleterious effects including hyperglycemia, conversion of glucose to fat, polyuria resulting in poor fluid balance, glucosuria, and excessive carbon dioxide production, making ventilation and weaning from ventilator support more difficult.[104] At the same time, failure to provide adequate carbohydrates can lead to greater rates of catabolism, loss of LBM, impaired wound healing, sepsis, infection, and death.[98, 103] Thus, providing the appropriate amount of carbohydrates after an injury is a fine balancing act. ESPEN recommends close monitoring of blood glucose, especially when providing carbohydrates via the parenteral route, with continuous intravenous insulin infusion, as appropriate, to target upper limit glucose concentration in the range of 4.5–8.0mmol/L.[98]

Lipids

Although evidence concerning fat intake in patients with burns is limited, lipid supplementation is usually recommended to minimize fatty acid deficiency, but this should be no more than 15% of total energy.[26,98] Like glucose, the amount of exogenous fat that can be used as an energy substitute after burn is limited; only about 30% of FFAs are oxidized for fuel, while the remaining 70% are merely recycled.[54,55] Additionally, adverse outcomes have been reported in patients with burns supplied with lipid intakes reaching 35% of total energy requirement, compared with 15%.[105] Indeed, excess fat supplementation can worsen immune function and result in hepatic steatosis.[98,105] Fat should comprise <25% of total nonprotein energy in these patients.[73,98] In addition to quantity, the fatty acid subtype must be considered. Omega-6-containing formulas can generate a proinflammatory response, while formulas with a greater proportion of omega-3 are associated with improved immune response, glycemic regulation, and patient outcomes.[106] Some enteral formulas have an omega-6:omega-3 ratio between 2.5:1 and 6:1, while "immune-enhancing" formulas have a more balanced composition of 1:1.[107] It is important to note that, although studies suggest that relatively low omega-6, but sufficient lipid supplementation is beneficial, there is still debate regarding enteral formula composition and lipid quantity in patients with burns.[108]

Protein

Protein is also crucial in patients with burns due to enhanced muscle catabolism in response to injury, necessitating supplementation to maintain adequate wound healing and immune function and to reduce LBM loss.[72] Administering supraphysiologic doses improves protein synthesis and nitrogen balance, although it does not eliminate catabolism.[109] It is important to note that protein (1.5–2.0 g/kg/day for adults, 2.5–4.0 g/kg/day for children) should be given with sufficient carbohydrates and fat to minimize its use as an energy source rather than as nutrients to maintain muscle mass and promote wound healing.[73,98] While the current body of literature concerning amino acid supplementation is controversial, several studies have highlighted the immune-enhancing benefits of glutamine and arginine, which are markedly depleted following burn injury.[67,69] In particular, glutamine provides energy for lymphocytes and enterocytes, and serves as a precursor for the antioxidant glutathione.[110] Clinically, supplementation of glutamine improves patient outcomes and minimizes LOS.[111] Conversely, while arginine supplementation improves burn wound healing and immune responsiveness, it is potentially harmful overall and is not recommended.[73,98]

Composition of Postburn Feeds: Micronutrients

In addition to macronutrients, supplementation of "micronutrients" (vitamins and trace elements) is also recommended after burn injury, as it plays an important role in oxidative stress, immunity, and wound healing but is lost in large quantities with exudative burn wound and fluid losses.[72,98,112]

Vitamins

Patients with severe burns incur significant oxidative stress due to the injury and fluid resuscitation early in the treatment process.[38] To protect organs from oxidative damage, antioxidants that scavenge free radicals or inhibit their formation must be provided to the patient with burns.[113] To that effect, the provision of vitamins A, C, and E as part of the postburn nutrition prescription is important in decreasing oxidative stress after injury and can prevent unnecessary tissue damage.[114,115] Both vitamins A and C are involved in wound healing.[114,115] Vitamin A has been shown to promote epithelialization, fibroplasia, angiogenesis, and collagen synthesis; however, there is conflicting evidence regarding whether supplementation improves postburn healing outcomes.[113] Likewise, vitamin C plays an active role in collagen synthesis and cross-linking, and concentration can decrease by up to 50% in patients with severe burns.[115] This is further exacerbated by the fact that a disproportionately high number of burn victims smoke or use alcohol in excess and are already at risk of baseline vitamin C deficiency.[113] As such, vitamin C supplementation of 500 mg twice daily via the enteral route is recommended.[98,113-115] Vitamin E helps improve immune function, but no specific dosing guidelines exist.[113] Nonetheless, several studies have shown that supplementation of vitamin E and C, along with the trace element zinc, exerts several beneficial effects on the pediatric burn population.[98,115] Supplementation with vitamin E alone after injury effectively decreased lipid peroxidation and time associated with wound healing, compared with children who did not receive an antioxidant supplementation containing vitamin E, vitamin C, and zinc.[98,115]

Thiamin and folate are both B vitamins with beneficial effects in patients with burns and can be adequately supplemented using a standard oral/enteral multivitamin.[113] Thiamin has decreased lactate and pyruvate concentrations after injury, while folate promotes protein and DNA synthesis and, thus, postburn wound healing.[112,113] Vitamin D deficiency can have immediate and long-term effects in patients with severe burns.[116] At first, vitamin D depletion can contribute to an increased risk for infection and poor glucose control.[116] In the long term, patients with burns continue to be at risk for vitamin D deficiency and its associated complications, as burn scar tissue has been shown to produce less vitamin D than healthy tissue.[117] Although it is well established that standard multivitamins contain inadequate amounts of vitamin D for repletion, no specific dosing guidelines have been established to date.[98,113]

Copper

The trace element copper (Cu) plays an integral role in wound healing, with adequate concentration required for initiating collagen and elastin cross-linking through the extracellular enzyme lysyl oxidase.[118] Within the first week following a burn injury; however, wound exudate and urinary losses often lead to a 20% to 40% decrease in the body's normal Cu stores.[119] Deficiencies in Cu have been linked to altered immunity, fatal arrhythmias, and poor outcomes.[118,119] While supplementation of trace elements like copper is recommended in patients with burns, exact dosing guidelines are currently unavailable.[98,118]

Selenium

Selenium (Se) is another essential trace element, due to its powerful antioxidant effects mediated by activating glutathione peroxidase, along with being a constituent part of other seleneoproteins.[120] Se is involved in cell-mediated immunity, and dietary Se influences leukocyte functions such as adherence, migration, and cytokine secretion.[118,120] Like copper, patients with burns lose significant amounts of Se through wound exudate.[119] Selenium deficiency is linked with immune compromise, which is of paramount importance given that burn patients are prone to nosocomial infection.[120] Thus, ensuring that Se is supplemented during their hospital stay after injury is vital. Patients with burns who receive selenium have been found to have a significantly shorter LOS in the intensive care unit.[98,113,118]

Zinc

Zinc (Zn) has a wide range of functions within the body including, but not limited to, regulation of immune cell responses, DNA synthesis, protein, and collagen synthesis, cellular proliferation, and stimulating bone formation/mineralization, as well as wound healing.[118] Following a burn injury; however, the combined deficits of Zn through urine and wound exudate account for approximately 5% to 10% of the normal body stores.[121] Interestingly, this depletion of Zn content and subsequent decrease in serum concentration appears to be unrelated to the size of the burn injury.[122] Yet, within rat models, Zn depletion elicits greater burn-induced insulin resistance with associated protein catabolism and hyperglycemia than rats with normal zinc concentration.[121] Furthermore, studies examining the critical-care population have found an inverse relationship between serum zinc and Sequential Organ Failure Assessment scores.[122] A recent Canadian study showed that enteral supplementation of 50 mg of Zn daily sufficiently normalized serum concentration by discharge in 82% of patients with burn sizes ranging from 10% to 93% TBSA.[98,123] It was also found that this amount of enteral zinc supplementation is safe and does not interfere with copper concentration, when administered with a multiple vitamin and micronutrient supplement.[123] Notably, Berger proposed the use of an IV antioxidant cocktail containing Zn (30mg), Cu (4mg), and Se (500μg), which is now becoming increasingly popular in burn care.[115] A summary of recommendations is in **Table 92.2**.

Non-Nutritional Management of Postburn Hypermetabolism

Besides enteral nutritional support, non-nutritional strategies are often used to help alleviate the hypermetabolic, hypercatabolic state in patients with severe burns.[5,68,98] Due to the life-threatening nature of major burns, patient prognosis largely depends on the excision of burned skin and subsequent restoration of the skin barrier.[4,68] Early excision and grafting promote wound closure while reducing fluid loss, inflammation, risk of infection, and overall metabolic rate.[4,68] Although there is a lack of consensus, and clinical practices vary globally, several anabolic agents have also shown promise in counteracting postburn morbidities like muscle catabolism and insulin resistance (**Table 92.3**).[70]

Table 92.2 Summary of Nutritional Recommendations in Burn Patients

Topic	Nutritional Recommendations
Route	EN preferred; PN should only be used in burn patients with contraindication
Timing	Early; within first 48 hours of injury
REE	
Indirect calorimetry	Gold standard; performed in fed state, at rest, with a factor of 1.2–1.4
Predictive equation	In absence of IC, the Toronto equation is recommended as an alternative
Macronutrients	Both EN and PN
Carbohydrate	55%–60% of total energy; glucose infusion rate <5 mg/kg/min (7g/kg/day)
Fat	<15%–25% of total energy
Protein	1.5–2.5 g/kg/day (*adults*); 2.5–4.0 g/kg/day (*children*)
Micronutrients	EN/Oral (*unless specified*)
Vitamin A	No consensus; consider 10,000-15,000 IU/day for major burns
Vitamin C	500 mg twice daily
Vitamin D	No consensus; consider 600–1,000 IU/day
Vitamin E	No consensus; consider 800–3,000 IU/day for major burns
Thiamin	No consensus; consider 5–10 mg/day
Folate	No consensus; consider 1–2 mg/day
Zinc	No consensus; consider 25–50 mg/day
Copper	No consensus; consider up to 4 mg/day
Selenium	No consensus; consider 400–1,000 μg/day
Antioxidant Cocktail	IV containing 30 mg zinc, 4 mg copper, 500 μg selenium for 2–3 weeks postburn
Immune-Modulating Nutrients	
Arginine	Not recommended
BCAAs	Not recommended
ω-3FAs	Not recommended
Dietary Nucleotides	Not recommended
Glutamine	No consensus; consider 0.3-0.57 g/kg/day of ornithine-α-ketoglutarate

BCAA, branched-chain amino acid; *EN*, enteral nutrition; *FA*, fatty acid; *IC*, indirect calorimetry; *IU*, international unit; *IV*, intravenous; *PN*, parenteral nutrition; *REE*, resting energy expenditure.
Modified from Hayhoe R, Maniatis K, Shahrokhi S, Jeschke, M. Nutrition support in burn injury. In: Hickson M, Smith S, Whelan K, eds. Advanced Nutrition and Dietetics in Nutrition Support Available at https://doi.org/10.1002/9781118993880.ch5.14.[124]

For instance, oxandrolone, an anabolic androgenic steroid, effectively promotes protein synthesis, LBM maintenance, and wound healing, and decreases hospital LOS and mortality in patients with burns.[125] Similarly, treatment with propranolol, a nonspecific β-blockader, impedes catecholamine surges, improves wound healing, and decreases heart rate, cardiac index, cytokine release, and adipose thermogenesis postinjury.[52] Other non-nutritional strategies to manage postburn hypermetabolism include thermoregulation of ambient temperature, pain control, and early rehabilitation.[68,72,126]

Summary

Nutritional support is uniquely important but extremely challenging for patients with burn injuries. Unlike other forms of trauma, severe burns trigger a profound and persistent hypermetabolic stress response that affects nearly every organ system of the body, causing substantial morbidity and mortality. Early assessment and provision of the right balance of carbohydrates, lipids, and protein, along with additional micronutrients, is essential to ameliorate and manage burn-induced hypermetabolism and its ensuing catabolic state following injury. However, there is no general agreement on the gold standard timing, route, amount, and composition of nutritional support for burn patients. Most clinicians advocate for early EN (within 48 hours) with high-carbohydrate formulas. Nevertheless, accurate nutritional endpoints and goals still need to be further investigated and validated before an optimal, universal feeding regimen for patients with burns can be ascertained. More importantly, given that the pathologic metabolic alterations are not homogenous, treatment of these patients must be individualized and should not only exclusively use nutritional support. Ideally, a more holistic approach to attenuate the catastrophic metabolic response in severely injured burn patients should be adopted by combining various effective therapeutic strategies, such as using anabolic pharmacologic agents and nutritional support.

Table 92.3 Common Anabolic Agents in Burn Care

Examples of widely used and accepted anabolic agents used in conjugation with nutritional therapy to mitigate the hypermetabolic response in burn patients.

Anabolic Agent	Target Tissue	Effect
Insulin	Liver	↓ Proteolysis
	Muscle tissue	↑ Lean body mass
	Fat tissue	↓ Fat oxidation
Insulin-like growth-factor-1	Muscle tissue	↓ Proteolysis
		↑ Lean body mass
Oxandrolone	Muscle tissue	↓ Proteolysis
		↑ Lean body mass
		↓ Fat oxidation
Propanolol	Fat tissue	↑ Glucose homeostasis
Metformin	Liver	↑ Glucose homeostasis
	Muscle tissue	↓ Proteolysis
	Fat tissue	↓ Fat oxidation

Modified from Abdullahi A, Jeschke MG. Nutr Clin Pract 2014;29(5):621–30. doi: 10.1177/0884533614533129 PMID: 25606644.[70]

Acknowledgments

The authors acknowledge Celeste C Finnerty, Rachael A Harrison, and David N Herndon for their contributions to the previous version of this chapter.

References

1. Mehta K, Arega H, Smith NL, et al. Am J Surg 2022;223:157–163.
2. Hebron C, Mehta K, Stewart B, et al. Ann Glob Health 2022;88: 34.
3. Banfield J, Rehou S, Gomez M, et al. J Burn Care Res 2015;36:213–217.
4. Jeschke MG, van Baar ME, Choudhry MA, et al. Nat Rev Dis Primers 2020;6:11.
5. Jeschke MG. J Burn Care Res 2016;37:86–96.
6. Coss-Bu JA, Jefferson LS, Walding D, et al. Nutr 1998;14:649–652.
7. Jeschke MG, Gauglitz GG, Kulp GA, et al. PLoS One 2011;6:e21245.
8. Gauglitz GG, Herndon DN, Kulp GA, et al. J Clin Endocrinol Metab 2009;94:1656–1664.
9. Knuth CM, Auger C, Chi L, et al. Shock 2021;56:744–754.
10. Herndon DN, Tompkins RG. Lancet 2004;363:1895–1902.
11. Wilmore DW, Curreri PW, Spitzer KW, et al. Surg Gynecol Obstet 1971;132:881–886.
12. Cartwright MM. Crit Care Nurs Clin North Am 2004;16:467–487.
13. Williams FN, Branski LK, Jeschke MG, et al. Surg Clin North Am 2011;91:609–629.
14. Cuthbertson DP, Angeles Valero Zanuy MA, León Sanz ML. Nutr Hosp 2001;16:176–182.
15. Jeschke MG, Chinkes DL, Finnerty CC, et al. Ann Surg 2008;248:387–401.
16. Wilmore DW, Aulick LH. Surg Clin of North Am 1978;58:1173–1187.
17. Monk DN, Plank LD, Franch-Arcas G, et al. Ann Surg 1996;223:395–405.
18. Long CL, Schaffel N, Geiger JW, et al. JPEN J Parenter Enteral Nutr 1979;3:452–456.
19. Smith A, Barclay C, Quaba A, et al. Burns 1997;23:291–294.
20. Jeschke MG, Mlcak RP, Finnerty CC, et al. Crit Care 2007;11:R90.
21. Mlcak RP, Jeschke MG, Barrow RE, et al. Ann Surg 2006;244:121–130.
22. Cambiaso-Daniel J, Malagaris I, Rivas E, et al. Pediatr Crit Care Med 2017;18:e598–e605.
23. Schwacha MG, Holland LT, Chaudry IH, et al. Shock 2005;23:123–128.
24. The American Burn Association Consensus Conference on Burn Sepsis and Infection Group. J Burn Care Res 2007;28:776–790.
25. Wolfe RR, Herndon DN, Jahoor F, et al. N Engl J Med 1987;317:403–408.
26. Demling RH, Seigne P. World J Surg 2000;24:673–680.
27. Goodall M, Stone C, Haynes BW. Ann Surg 1957;145:479–487.
28. Coombes EJ, Batstone GF. Burns Incl Therm Inj 1982;8:333–337.
29. Jeschke MG, Klein D, Herndon DN. Ann Surg 2004;239:553–560.
30. Hamberger B, Farnebo L-O, Liljedahl S-O. Burns 1980;7:20–24.
31. Jeevanandam M, Ramias L, Shamos RF, et al. Surg 1992;111:495–502.

32. Jeffries MK, and Vance ML. J Burn Care Rehabil 1992;13: 391–395.
33. Ferrando AA, Sheffield Moore M, Wolf SE, et al. Criti Care Med 2001;29:1936–1942.
34. Dolecek R. Keio J Med 1989;38:262–276.
35. Hur J, Yang HT, Chun W, et al. Ann Lab Med 2015;35:105–110.
36. Winchurch RA, Thupari JN, Munster AM. Surg 1987;102: 808–812.
37. Mulder PPG, Vlig M, Boekema BKHL, et al. Front Immunol 2021;11.
38. Parihar A, Parihar MS, Milner S, et al. Burns 2008;34:6–17.
39. Rawlingson A. Burns 2003;29:631–640.
40. Lavrentieva A, Depetris N, Kaimakamis E, et al. Ann Burns Fire Disasters 2016;29:172–177.
41. Mannes M, Schmidt CQ, Nilsson B, et al. Semin Immunopathol 2021;43:773–788.
42. Gearhart MM, Parbhoo SK. AACN Clin Issues 2006;17:50–55.
43. Robinson LE, van Soeren MH. AACN Clin Issues 2004;15: 45–62.
44. Wolfe RR, Jahoor F, Herndon DN, et al. Surg 1991;110:54–67.
45. Porter C, Tompkins RG, Finnerty CC, et al. Lancet 2016;388: 1417–1426.
46. Gore DC, Chinkes D, Heggers J, et al. J Trauma Acute Care Surg 2001;51:540–544.
47. Cochran A, Davis L, Morris SE, et al. J Burn Care Res 2008;29: 187–191.
48. Jeschke MG, Abdullahi A, Burnett M, et al. Ann Surg 2016;264: 518–527.
49. Pham TN, Warren AJ, Phan HH, et al. J Trauma 2005;59: 1148–1154.
50. Hemmila MR, Taddonio MA, Arbabi S, et al. Surg 2008;144: 629–637.
51. Herndon DN, Nguyen TT, Wolfe RR, et al. Arch Surg 1994;129: 1301–1304.
52. Herndon DN, Hart DW, Wolf SE, et al. (2001) N Engl J Med 345, 1223–1229.
53. Galster AD, Bier DM, Cryer PE, et al. J Trauma 1984;24: 938–945.
54. Cree MG, Wolfe RR. Am J Physiol Endocrinol Metab 2008;294: E1–E9.
55. Cree MG, Aarsland A, Herndon DN, et al. Crit Care Med 2007;35: S476–483.
56. Shah P, Vella A, Basu A, et al. Diabetes 2002;51:301–310.
57. Boden G, Chen X, Ruiz J, et al. J Clin Endocrinol Metab 1994;78: 657–663.
58. Kraft R, Herndon DN, Finnerty CC, et al. J Clin Endocrinol Metab 2013;98:314–321.
59. Barayan D, Vinaik R, Auger C, et al. Shock 2020;53:137–145.
60. Kinney JM, Long CL, Gump FE, et al. Ann Surg 1968;168: 459–474.
61. Porter C, Herndon DN, Børsheim E, et al. J Burn Care Res 2016; 37:53–63.
62. Pollack SV. J Dermatol Surg Oncol 1979;5:615–619.
63. Windsor JA, Hill GL. Ann Surg 1988;207:290–296.
64. Arora NS, Rochester DF. Am Rev Respir Dis 1982;126:5–8.
65. DeFronzo RA, Jacot E, Jequier E, et al. Diabetes 1981;30: 1000–1007.
66. Diaz EC, Herndon DN, Lee J, et al. J Trauma Acute Care Surg 2015;78:816–822.
67. Jahoor F, Desai M, Herndon DN, et al. Metab 1988;37: 330–337.
68. Bakhtyar N, Sivayoganathan T, Jeschke MG. Intens Crit Care 2015;1–6
69. Herndon DN, Ramzy PI, DebRoy MA, et al. Ann Surg 1999;229: 713–720.
70. Abdullahi A, Jeschke MG. Nutr Clin Pract 2014;29:621–630.
71. Abdullahi A, Jeschke MG. In: Rajendram R, Preedy VR, and Patel VB, eds. Enteral Nutrition Support in Burns. Diet and Nutrition in Critical Care. New York, NY: Springer New York, 2014:1–13.
72. Rodriguez NA, Jeschke MG, Williams FN, et al. JPEN J Parenter Enteral Nutr 2011;35:704–714.
73. Jeschke MG, Kamolz LP, Sjöberg F, Wolf SE eds. Handbook of Burns Vienna, Austria: Springer Vienna, 2012.
74. Seres DS, Valcarcel M, Guillaume A. Therap Adv Gastroenterol 2013;6:157–167.
75. Magnotti LJ, Deitch EA. J Burn Care Rehabil 2005;26:383–391.
76. Alverdy J, Chi HS, Sheldon GF. Ann Surg 1985;202:681–684.
77. Vicic VK, Radman M, Kovacic V. Asia Pac J Clin Nutr 2013;22:543–547.
78. Gottschlich MM, Jenkins ME, Mayes T, et al. J Burn Care Rehabil 2002;23:401–415.
79. Andel D, Kamolz L-P, Donner A, et al. Burns 2005;31:302–305.
80. Hansbrough WB, Hansbrough JF. J Burn Care Rehabil 1993;14: 512–516.
81. Moore EE, Jones TN. J Trauma 1986;26:874–881.
82. Todd SR, Kozar RA, Moore FA. Nutr Clin Pract 2006;21: 421–429.
83. Marik PE, Zaloga GP. Crit Care Med 2001;29:2264–2270.
84. Mochizuki H, Trocki O, Dominioni L, et al. Ann Surg 1984;200: 297–310.
85. Wasiak J, Cleland H, Jeffery R. Cochrane Database Syst Rev 2006;CD005489.
86. Peck MD, Kessler M, Cairns BA, et al. J Trauma 2004;57: 1143–1148.
87. Berger MM, Chioléro RL. Curr Opin Crit Care 2007;13: 180–186.
88. Dickerson RN, Boschert KJ, Kudsk KA, et al. Nutr 2002; 18:241–246.
89. McCowen KC, Friel C, Sternberg J, et al. Crit Care Med 2000; 28:3606–3611.
90. Graf S, Pichard C, Genton L, et al. Clin Nutr 2017;36:224–228.
91. Hart DW, Wolf SE, Herndon DN, et al. Ann Surg 2002;235: 152–161.
92. Clark A, Imran J, Madni T, et al. Burns Trauma 2017;5:11.
93. Saffle JR, Graves C. Nutritional Support of the Burned Patient. Total Burn Care. Elsevier, 2007: 398–419.
94. Ireton-Jones CS, Turner WW, Liepa GU, et al. J Burn Care Rehabil 1992;13:330–333.
95. Curreri PW. J Trauma 1990;30:S20–23.
96. Dickerson RN, Gervasio JM, Riley ML, et al. JPEN J Parenter Enteral Nutr 2002;26:17–29.
97. Mendonça Machado N, Gragnani A, Masako Ferreira L. Nutr Hosp 2011;26:692–700.
98. Rousseau A-F, Losser M-R, Ichai C, et al. Clin Nutr 2013;32: 497–502.
99. Suman OE, Mlcak RP, Chinkes DL, et al. Burns 2006;32: 335–342.
100. McClave SA, Snider HL. Nutr Clin Pract 1992;7:207–221.
101. Saffle JR, Medina E, Raymond J, et al. J Trauma 1985;25: 32–39.
102. Hart DW, Wolf SE, Zhang XJ, et al. Crit Care Med 2001;29: 1318–1324.
103. Sheridan RL, Yu YM, Prelack K, et al. JPEN J Parenter Enteral Nutr 1998;22:212–216.
104. Burke JF, Wolfe RR, Mullany CJ, et al. Ann Surg 1979;190: 274–285.
105. Garrel DR, Razi M, Larivière F, et al. JPEN J Parenter Enteral Nutr 1995;19:482–491.
106. Alexander JW, Saito H, Trocki O, et al. Ann Surg 1986;204:1–8.
107. Alexander JW, Gottschlich MM. Crit Care Med 1990;18: S149–153.
108. Heyland DK, Samis A. Intensive Care Med 2003;29:669–671.

109. Patterson BW, Nguyen T, Pierre E, et al. Metab 1997;46: 573–578.
110. Gore DC, Jahoor F. Arch Surg 1994;129:1318–1323.
111. Windle EM. J Burn Care Res 2006;27:764–772.
112. Gamliel Z, DeBiasse MA, Demling RH. J Burn Care Rehabil 1996;17:264–272.
113. Nordlund MJ, Pham TN, Gibran NS. J Burn Care Res 2014;35: 121–133.
114. Rock CL, Dechert RE, Khilnani R, et al. J Burn Care Rehabil 1997; 18:269–278.
115. Berger MM. Nutr Clin Pract 2006;21:438–449.
116. Klein GL. Adv Nutr 2011;2:457–462.
117. Klein GL, Chen TC, Holick MF, et al. Lancet 2004;363:291–292.
118. Berger MM, Shenkin A. J Trace Elem Med Biol 2007;21(suppl 1(44–48).
119. Sampson B, Constantinescu MA, Chandarana I, et al. Ann Clin Biochem 1996;33(pt 5):462–464.
120. Huang Z, Rose AH, Hoffmann PR. Antioxid Redox Signal 2012;16:705–743.
121. Claeyssen R, Andriollo-Sanchez M, Arnaud J, et al. Biol Trace Elem Res 2009;127:132–142.
122. Cander B, Dundar ZD, Gul M, et al. J Crit Care 2011;26:42–46.
123. Caldis-Coutris N, Gawaziuk JP, Logsetty S. J Burn Care Res 2012;33:678–682.
124. Hayhoe R, Maniatis K, Shahrokhi S, Jeschke, M. Nutrition support in burn injury. In: Hickson M, Smith S, Whelan K, eds. Advanced Nutrition and Dietetics in Nutrition Support Available at https://doi.org/10.1002/9781118993880.ch5.14.
125. Hart DW, Wolf SE, Ramzy PI, et al. Ann Surg 2001;233: 556–564.
126. Williams FN, Jeschke MG, Chinkes DL, et al. J Am Coll Surg 2009;208:489–502.

G. BEHAVIORAL, PSYCHIATRIC, AND NEUROLOGIC DISORDERS

CHAPTER 93

Nutrition and Neurologic Diseases

Liang Sun and Xiang Gao

CHAPTER OUTLINE

Abbreviations

AD, Alzheimer's disease; **AHA**, American Heart Association; **aHEI**, alternate healthy eating index; **AIBL**, Australian Imaging, Biomarkers, and Lifestyle; **aMED**, alternate Mediterranean diet; **CVD**, cardiovascular disease; **DALY**, disability-adjusted life years; **DASH**, dietary approaches to stop hypertension; **DHA**, docosahexenoic acid; **FA**, fatty acids; **FI**, food insecurity; **FINGER**, Finnish Geriatric Intervention Study to Prevent Cognitive Impairment and Disability; **GBD**, global burden of disease; **HPA**, hypothalamic-pituitary-adrenal; **HPFS**, health professionals follow-up study; **iPDGC**, International Parkinson Disease Genomics Consortium; **LCPUFA**, long-chain polyunsaturated fatty acid; **MCI**, mild cognitive impairment; **MIND**, Mediterranean-DASH intervention for neurological delay; **MMSE**, Mini-Mental State Examination; **MUFA**, monounsaturated fatty acid; **NHS**, Nurses health study; **PCA**, principal components analysis; **PD**, Parkinson's disease; **PUFA**, polyunsaturated fatty acid; **QOL**, quality of life; **RBD**, REM sleep behavior disorder; **RCT**, randomized controlled trial; **RLS**, restless leg syndrome; **SD**, standard deviation; **SNP**, single-nucleotide polymorphism; **SWS**, slow wave sleep; **WHICAP**, Washington Heights–Inwood Columbia Aging Project

Introduction

In parallel with the extended life expectancy, the number of the world's older adult population (≥65 years) has dramatically increased; it is estimated to reach 1 billion (12% of the total global population) by 2030 and 1.6 billion (16.7%) by 2050.[1] This demographic change presents a public health challenge, with increased prevalence of age-related diseases, including diabetes, cardiovascular diseases (CVD), and neurologic diseases.

Neurologic diseases, such as stroke and neurodegenerative diseases (e.g., dementias and Parkinson's disease (PD)),[2] were already ranked as the leading global causes of disability, and the second leading cause of death, in 2016.[3] From 2010 to 2019, the total number of deaths and disability-adjusted life years (DALYs) from neurologic diseases increased by 15% and 7%, respectively.[2] Mechanistic studies have suggested several common pathophysiologic or etiologic pathways shared by diverse neurologic diseases, involving brain energy metabolism, oxidative stress, neuroinflammation, epigenetic modification (DNA methylation, histone modification, noncoding RNA), mitochondrial dysfunction, protein clearance, and the gut-brain axis.[4-9] However, global efforts to develop effective treatments still face enormous challenges.

A growing body of evidence has highlighted the role of nutrition as an important modifiable factor for preserving cognition, delaying neurologic diseases, and preventing cardiometabolic diseases linked to dementia risk,[10] thus reducing the attributable death and burden of these diseases. Many nutrients (e.g., B vitamins, vitamins C and E, carotenoids, n-3 fatty acids), food ingredients (e.g., polyphenols), and uric acid have been suggested as being beneficial for neurologic function or promoting healthy aging via antioxidative and anti-inflammatory properties.[1,10-13] However, the evidence from clinical studies is less conclusive. Due to the complexity of diet and synergistic interactions among nutrients, pinpointing the specific effective nutrients may be difficult.[14,15] Therefore, current studies have paid more attention to dietary patterns, which target multiple physiologic pathways and risk factors simultaneously,[16] better reflect daily eating behaviors,[1] and are more applicable and understandable for dietary recommendations.[17] The most extensively studied dietary patterns are the Mediterranean, the dietary approaches to stop hypertension (DASH), and the newly developed Mediterranean-DASH diet intervention for neurologic delay (MIND) diets.

This chapter focuses on the associations of diet, including dietary patterns and individual nutrients/food components, with the prevention and management of major neurologic diseases such as Alzheimer's disease (AD), PD, and stroke. Given the current attention on circadian rhythm misalignment, and as sleep disturbance is usually considered prodromal or a risk factor for neurologic diseases,[18-21] the bidirectional relationship between nutrition and sleep health is addressed in the last part of this chapter.

Alzheimer's Disease

Epidemiology of Dementia and Alzheimer's Disease

The prevalence of dementia, which severely impacts one's daily activities (memory, language, thinking skills) and quality of life (QOL), has escalated rapidly over the past several decades. It is estimated that more than 55 million people have dementia worldwide, and that it will reach 78 million by 2030 and 139 million by 2050.[22] According to the Global Burden of Disease (GBD) study in 2019, dementia has contributed to 17.3% of neurologic deaths and 9.3% of neurologic DALYs. The hallmarks of aging, such as alterations in brain structure and metabolism, are in parallel with the development and progression of dementia.[23-25] However, no curative treatment is available for dementias or age-related cognitive decline.[1,5,14-16,24] Moreover, the GBD 2019 suggested that 32.8% of dementia DALYs are attributable to risk factors such as smoking (15.1%), obesity (12.5%), and high fasting plasma glucose (9.7%).[2] According to the 2020 report of the Lancet Commission on dementia prevention, intervention, and care, 12 modifiable risk factors, together, have accounted for approximately 40% of dementia cases worldwide,[26] highlighting the importance of shifting the research focus toward prevention.

AD is the most common form of dementia, accounting for an estimated 60% to 80% of dementia cases.[27] It is clinically characterized by progressive memory loss and cognitive deterioration.[18] The hallmark pathologies are an accumulation of extracellular amyloid plaques (protein β-amyloid, Aβ) and intracellular neurofibrillary tangles (protein tau) in the brain,[1] accompanied by the death of neurons and progressive cerebral atrophy.[28] Other potential mechanisms, including oxidative stress, neuroinflammation, and impaired glucose metabolism, may also play vital roles in AD pathogenesis.[29] The strongest nonmodifiable risk factors for AD are advanced age (≥65 years), female gender (especially when age ≥80), and APOE ε4 allele. APOE ε4 allele has been shown to have the strongest disease association among the ~40 AD-associated genetic risk loci, and 60% to 80% of risk of AD is thought to depend on heritable factors.[30] Adhering to dietary patterns like the Mediterranean, the DASH, and the MIND diets, with higher consumption of nutrients or food components rich in antioxidant and anti-inflammatory properties (e.g., B vitamins, n-3 fatty acids, and polyphenols) are thought to be associated with lower risks of AD and cognitive decline.

Nutrition and Risk of Alzheimer's Disease/Cognitive Decline

Dietary Patterns

Plant-based dietary patterns, such as the Mediterranean and the DASH diets, are known to have favorable associations with cardiometabolic diseases and have been studied in association with cognitive function and AD risk.[31-33] The Mediterranean diet is characterized by extra-virgin olive oil as the major source of fat, along with high intakes of vegetables, fruits, legumes, nuts/seeds, unrefined cereal products, marine fish, moderate intake of dairy products and alcohol, and limited consumption of meat and sweets.[31,32] The DASH diet emphasizes plant-based foods and low-fat dairy and limits the intake of fat, sweets, and sodium.[32] The hybrid MIND diet is modified toward neuroprotection, emphasizing plant-based foods (especially green leafy vegetables and berries) and restricting animal foods and saturated fat.[32]

The Mediterranean diet is one of the most extensively studied dietary patterns. In a recent systematic review, which included seven randomized controlled trials (RCTs) and 38 longitudinal studies, most studies showed that adherence to the Mediterranean diet was associated with improved cognitive health, especially for global cognitive function.[34] However, mild cognitive impairment (MCI) and AD results were mixed. Interestingly, in the Boston Puerto Rican Health Study (n=913), the association between higher adherence to the Mediterranean diet and greater 2-year change in global cognitive function was only seen in participants with type 2 diabetes ($\beta \pm SE = 0.027 \pm 0.01$; $P=0.02$), but not in those without type 2 diabetes, suggesting effect modification of managing type 2 diabetes on this association.[35] In the German DZNE-Longitudinal Cognitive Impairment and Dementia Study, the favorable associations of Mediterranean diet adherence with mediotemporal gray matter volume and memory were correlated with decreased amyloid and pTau181 pathology,[36] which merits replication in future research. Higher adherence to the DASH and MIND diets have also been associated with slower cognitive decline and lower risk of AD, with stronger associations observed for the MIND diet, compared with the Mediterranean and the DASH diets.[32] Other healthy dietary patterns, generated either from a priori (e.g., Healthy Eating Index) or a posteriori (e.g., principal components analysis (PCA), reduced rank regression) approaches, were also associated with lower risks of cognitive decline and dementia.[14] In a meta-analysis of 12 prospective cohort studies (66,930 participants), higher adherence to a healthy dietary pattern was associated with 18% reduced risk of overall dementia (12 studies), and 39% lower likelihood of developing AD (six studies), when comparing extreme categories.[37]

Animal models have suggested that energy restriction may modulate the nutrient-sensing pathways in the brain and have beneficial effects on synaptic plasticity, neurogenesis, and neuroprotection.[25,38,39] A few clinical trials also observed improvement in cognitive performance in nonobese or elderly adults after reducing energy intake (30%–50%).[39] The brain is vulnerable to stress, and exposure to stress during early life is associated with lifelong changes in brain structure and cognitive function.[9] The Boston Puerto Rican Health Study, which enrolled 1,358 participants, assessed food insecurity (FI) with the US Household Food Security Scale.[40] Cognitive performance was measured with a battery of seven tests. FI was inversely associated with global cognitive performance and with the Mini-Mental State Examination (MMSE) score at baseline. In the same cohort, 597 participants with baseline MMSE scores ≥24 were selected to be followed for 2 years.[41] Individuals in the very low food security group (≥6 positive responses of the 10 questions) were found to have faster 2-year cognitive decline [mean difference: −0.26 (95% CI: −0.41, −0.10)] compared with those in the food-secure group. A systematic review including eight additional studies suggested that FI in early or later life was associated with global cognitive decline, and that later-life FI was also related to the deterioration of executive function and memory.[42]

Individual Nutrients/Food Components

Fat is necessary for the structure of neuronal membranes.[25] Among fatty acids (FA), monounsaturated (MUFA) and polyunsaturated (PUFA) are usually considered "good fats," and n-3 long-chain PUFAs (LCPUFA) are believed to have neuroprotective effects.[17,43,44] In the Women's Health Study, dietary intakes of saturated FA (SFA), MUFA, PUFA, and trans-FA were assessed in association with 4-year cognitive trajectory.[45] Participants in the top quintile of SFA intake had higher risk of 4-year decline in global cognition (64%) and verbal memory (65%) than those in the bottom quintile. In contrast, MUFA intake (but not PUFA) was associated with 44% to 48% lower risk of the worst decline in global cognition and verbal memory, defined as the lowest 10% of the distributions of the change scores. In a dose-response meta-analysis of 21 prospective studies (4,438 cases), each 8 g/d higher intake in PUFA was associated with 29% reduced risk of mild cognitive impairment (MCI).[46] Each 0.1 g/d higher intake of dietary docosahexenoic acid (DHA) was associated with 14% to 27% lower risk of dementia and AD. However, the evidence was less conclusive in clinical studies. A meta-analysis of 38 RCTs that explored the effect of ≥6 months of n-3 LCPUFAs on neurocognitive outcomes or cognitive decline found limited to no beneficial effect.[47]

Among micronutrients, B vitamins have been a focus, based on their roles as cofactors in all the core pathologies of AD involving one-carbon metabolism, mitochondrial function, lipid and phospholipid metabolism, neurotransmitter synthesis and synaptogenesis, DNA synthesis, and repair, as well as amyloidosis and Tau protein phosphorylation,[23,24] providing mechanistic clues for their protective roles against dementia. Although observational studies have shown inverse associations of folate, vitamin B_6 and B_{12} with cognitive decline, results from RCTs were inconsistent.[23] Notably, some evidence suggested that the effect of B vitamin supplementation on improving cognitive function was more pronounced when baseline homocysteine concentration was >11.3 μmol/L or serum folate was <12 nmol/L,[23,48] which remains to be further tested. Vitamin D was also theorized as a protective factor for cognitive health.[23] Cross-sectional studies found lower 25(OH)D concentration in individuals with lower cognition, and several, but not all, longitudinal studies have reported an inverse association between 25(OH)D concentration and cognitive decline.[23,49] The heterogeneity in study design, follow-up duration, and outcome assessments (cognitive tests) may partially explain the mixed results.

Emerging evidence has also addressed the role of polyphenols (e.g., flavonoids) as procognition compounds, due to their antioxidant and anti-inflammatory properties and interactions with the gut-brain axis.[23,50,51] Polyphenol-rich foods (fruit juices, berries, cherries, cocoa, green tea, coffee, and nuts), common features of plant-based dietary patterns such as the Mediterranean, DASH, and MIND diets, were found to be associated with improved cognitive performance.[51] However, RCTs using polyphenol supplements generated controversial results regarding their effects on global cognitive functions and distinct cognitive domains, possibly influenced by the varied evaluation scores across trials.[52]

The association between alcohol consumption and cognitive impairment is controversial. A dose-response meta-analysis including 16 prospective studies found a U-shape curve association between alcohol consumption and risk of cognitive deficit ($P_{non\text{-}linearity}$ <0.05).[53] Only low consumption of alcohol (<11g/day), but not heavier drinking, was associated with lower incidence of cognitive deficits and dementia, compared with nondrinking. In contrast, in the Chinese Longitudinal Healthy Longevity Survey (CLHLS), which enrolled 15,758 participants aged 65 years or older, compared with current drinkers, alcohol cessation for ≥5 years was associated with an 18% to 21% reduced risk of cognitive impairment.[54]

Nutrition and Progression of Alzheimer's Disease/Cognitive Decline

There is limited published research on the effects of diet quality on slowing disease progression in those who already have symptoms of cognitive decline or have been diagnosed with AD. In the Washington Heights–Inwood Columbia Aging Project (WHICAP), 482 participants with mild cognitive impairment (MCI), an intermediate stage between normal aging and early dementia, were followed for 4.3 years.[55] Higher adherence to the Mediterranean diet was associated with 48% lower AD risk, when comparing the highest with the lowest tertiles of adherence score (adjusted HR=0.52; 95% CI, 0.30–0.91; P=0.02). In the "Nutrition, the unrecognized determinant in Alzheimer disease" (NUDAD) study, which included 551 patients (219 with subjective cognitive decline, 135 with MCI, and 197 with AD), lower diet quality (less adherence to the Dutch guidelines for a Healthy Diet) was associated with 22% higher risk of clinical progression (diet quality score <53 vs ≥53) during a mean follow-up of 2.2 years.[56] The Australian Imaging, Biomarkers, and Lifestyle (AIBL) Study of Aging evaluated the association between Mediterranean diet adherence and cerebral Aβ accumulation in 77 individuals identified as "Aβ accumulators" by Pittsburgh Compound B positron emission tomography.[57] Higher adherence to the Mediterranean diet and intake of fruits among the individual components were associated with less Aβ accumulation during a 36-month follow-up.

Deterioration of brain energy metabolism is considered a presymptom of AD, which specifically influences glucose metabolism but does not affect that of ketone bodies.[15,16,58] Therefore, the ketogenic diet has been suggested as a way to rescue brain energy. The effects of ketogenic therapies on delaying AD progression and alleviating cognitive function were evaluated in clinical trials. A systematic review of 10 RCTs conducted in patients with AD/MCI suggested improved general cognition (assessed via the Alzheimer's Disease Assessment Scale-Cognitive) and improved episodic and secondary memory as long-term effects.[59] Interestingly, the increase in ketone concentration and improvement in cognitive performance was more pronounced in those with *APOE* ε4(−) than *APOE* ε4(+), suggesting effect modification by *APOE* status. In a recent randomized crossover trial in patients with a clinically confirmed diagnosis of AD, those following a ketogenic diet showed improved daily function (AD Cooperative Study - Activities of Daily Living inventory score) and QOL (Quality of Life in AD score), compared with usual diet plus low-fat, healthy-eating guidelines.[60]

Parkinson Disease

Epidemiology of Parkinson's Disease

PD is the second most common neurodegenerative disease worldwide and will affect more than 1 million Americans by 2030.[61] PD is characterized by movement symptoms of bradykinesia, tremor, rigidity, and postural instability.[62,63] The hallmark of PD pathology is the presence of Lewy bodies aggregated by various proteins (primarily α-synuclein) in the substantia nigra pars compacta (SNc) region of the midbrain, which results in progressive degeneration of dopaminergic neurons, and depletion of dopamine in the nigrostriatal pathways.[63-66] Hypopigmentation in the SNc is also observed.[67] Notably, PD has a long prodromal phase characterized by atypical symptoms, including constipation, hyposmia, and sleep disturbance, which could appear ~20 years before the clinically recognized motor symptoms.[13] In nonmotor manifestations of PD, cognitive decline, and dementia are also common,[18,68] but with less prominent deficits in language and memory.[69]

In contrast to AD, males are more prone to develop PD, which may be due to the lack of protection of estrogen on dopaminergic neurons.[5] Although 90% of PD cases have no clearly identified genetic cause,[70] some have argued that PD might have a genetic component. More than 40 genetic loci have been discovered to be linked to PD and parkinsonism, among which the most common PD-associated mutations were found in *GBA* and *LRRK2* genes.[66] Among nongenetic risk factors, exposure to pesticides and histories of melanoma or traumatic brain injury were associated with increased PD risk. In contrast, smoking (nicotine), physical activity, and using medications such as ibuprofen were associated with reduced PD risk.[13] The associations of red hair and *MC1R* variants with PD risk have provided a novel pathological clue, namely the Pigmentation hypothesis,[67] which may inspire therapeutic innovation for PD. Nutritionally, adherence to healthy dietary patterns, dietary flavonoids, caffeine, nicotine, and blood urate concentrations were inversely associated. In contrast, consumption of dairy products was positively associated with PD risk or its prodromal features.

Nutrition and Parkinson Disease

Dietary Patterns

Although not as extensively studied as AD, the association between diet quality and PD risk has also been assessed by a priori and a posteriori approaches. In the Nurses' Health Study (NHS) and the Health Professionals Follow-up Study (HPFS), which enrolled 81,676 women and 49,692 men, respectively, free of PD at baseline, the alternate Health Eating Index (AHEI) and the alternate Mediterranean Diet Score (aMed) were predefined.[71] Two dietary patterns, namely the prudent dietary patterns (characterized by high plant-based foods and fish) and the Western dietary pattern, were generated by PCA. After a 16-year follow-up with 508 new PD cases identified, the prudent diet score and the AHEI and aMed scores were inversely associated with PD risk (RRs 0.70–0.78 comparing extreme quintiles), suggesting protective effects of healthy dietary patterns. Similarly, in the Women's Lifestyle and Health study of more than 47,000 Swedish women, high adherence (score 6–9) to the Mediterranean dietary pattern in middle age compared with low adherence (score 0–3) was associated with a 46% lower risk of PD in later life, after ~20 years of follow-up.[72]

A Canadian case-control study (167 cases and 119 controls) compared the association of PD onset age with adherence to the MIND diet, the original Mediterranean diet (OMeDi), and the Greek Mediterranean diet (GMeDi).[73] In women, higher adherence to the MIND and GMeDi diets were associated with later onset age. In contrast, in men, only adherence to the GMeDi was inversely associated with the onset age of PD. In the Pennsylvania Geisinger Rural Aging Study longitudinal cohort, diet quality was assessed with a Dietary Screening Tool, consisting of 25 food- and behavior-specific questions.[74] Among older adults, mean age 81.5 years, individuals in the top tertile of diet quality score had a 61% lower risk of PD incidence than those in the bottom tertile. The authors conducted a meta-analysis including four additional studies and a total of 140,617 participants. They reported a pooled RR of 0.64 (95% CI: 0.49, 0.83) when comparing high with low diet quality/pattern scores.

Individual Nutrients/Food Components

In human studies, high homocysteine concentration and genetic risk alleles in folate metabolism were associated with high PD risk. Animal studies reported elevated vulnerability of dopaminergic cells via increased oxidative stress and altered methylation during B-vitamin deficiency.[64] Therefore, B vitamins in one-carbon metabolism (folate, vitamin B6, and vitamin B_{12}) may have beneficial effects by reducing homocysteine. Nevertheless, evidence from epidemiologic studies was inconclusive. For instance, in the NHS and the HPFS cohorts (415 incident cases), dietary intake of folate, vitamin B_6, or vitamin B_{12} were not associated with PD incidence during 14 to 18 years of follow-up.[75] Similar results for dietary folate and vitamin B_{12} were observed in the Rotterdam Study, which included 5,289 participants aged 55 years and older followed for 9.7 years.[76] However, the authors documented that per standard deviation (SD), a higher intake of vitamin B_6 was associated with 31% lower risk of incident PD, but only in smokers, in stratified analysis.

Nutrients with antioxidative properties have shown limited results. In the NHS and HPFS cohorts, with over 1,000 incident

PD cases, dietary intakes of vitamin E and carotenoids were not associated with PD risk.[77] The inverse association between vitamin C intake and PD risk became nonsignificant after excluding the first 4 years of follow-up, suggesting potential reverse causality.

Notably, associations between specific food components and PD risk have shown heterogeneity between men and women, and the underlying mechanisms remain unclear. In the NHS and HPFS cohorts, the association between the intake of flavonoids and PD risk was evaluated based on the consumption of tea, berry fruits, apples, red wine, and orange/orange juice.[78] Interestingly, men in the highest quintile of total flavonoid intake had 40% reduced risk of incident PD than those in the lowest quintile. However, no association was observed in women. Similarly, higher intake of flavonoids and flavonoid-rich foods was associated with 47% lower mortality risk in male patients with PD when comparing extreme quartiles, but not in female patients with PD ($P_{heterogeneity}=0.01$).[79] For caffeine, a neuroprotective factor targeting adenosine 2A receptors in the brain,[80] the association with PD risk was also stronger among men than women. In the Cancer Prevention Study II Nutrition Cohort, with 112,122 participants (317 cases), the inverse association between caffeine consumption and PD risk was significant in men ($P_{trend}<0.002$) but marginally in women ($P_{trend}=0.05$).[81] It was suggested that hormone replacement therapy might modify the association in women, with a stronger association in nonusers than users. However, the $P_{interaction}$ did not reach statistical significance.

Another interesting example is urate, which is also considered an antioxidant. In the HPFS cohort (47,406 men; 248 PD cases), higher dietary urate index was associated with 53% lower risk of PD incidence, when comparing extreme quintiles,[82] while a Mendelian Randomization analysis suggested a protective effect of high urate concentration on PD progression.[83] In a nested case-control study within the NHS, HPFS, and the Cancer Prevention Study II Nutrition, plasma urate was measured in 388 incident PD cases, and 1,267 controls.[84] The inverse association between plasma urate and PD risk was more pronounced in men than in women ($P_{heterogeneity}=0.001$). This was further confirmed in a meta-analysis including three additional cohorts. In contrast, the association between PD risk and dietary nicotine, believed to regulate striatal function via stimulation of nicotinic acetylcholine receptors,[85] was found to be stronger in women than in men.[86] In the never-smoker participants from the NHS and HPFS cohorts (601 cases), dietary nicotine intake was calculated from the intakes of peppers, tomatoes, processed tomatoes, potatoes, and tea.[86] During 26 years of follow-up, higher dietary nicotine intake was associated with lower PD risk in women, but not men. It remains to be elucidated whether the sex-specific association reflects an actual biologic difference, other risk factors, or is due to chance.

Dairy products were associated with high PD risk,[13] which may be attributed to their urate-lowering effects.[87] In the NHS and HPFS cohorts (1,036 cases), individuals with high intake of low-fat dairy foods (3+servings/day) had 34% increased PD risk, relative to nonconsumers.[88] When they pooled the results with three additional studies, higher intake of total milk and dairy were associated with 56% and 27% elevated risk of PD incidence, respectively. In a two-sample Mendelian randomization, a single-nucleotide polymorphism (SNP) in the lactase gene (rs4988235) was used as an instrumental variable for estimating the causal relationship between dairy intake and PD risk.[89] Each one-serving increase of genetically predicted dairy intake was associated with 70% higher PD risk in the Courage-PD consortium (23 studies, 9,823 cases, and 8,376 controls). The positive association was further confirmed with data from the International Parkinson Disease Genomics Consortium (iPDGC).

Nutrition and Prodromal Parkinson Disease

There is limited but promising data on associations between nutrition and prodromal PD symptoms regarding healthy dietary patterns. In the NHS and HPFS (47,679 participants), adherence to dietary patterns was calculated using the alternate Mediterranean diet (aMED) and AHEI scores.[90] Prodromal PD features were assessed by questions regarding probable REM sleep behavior disorder (RBD) and constipation, hyposmia, impaired color vision, excessive daytime sleepiness, body pain, and depressive symptoms. Individuals in the top quintile of aMED and AHEI had fewer prodromal features (≥3 vs 0, RR 0.72–0.82 for baseline and 0.66–0.67 for long-term adherence) and lower risk of individual features (constipation, excessive daytime sleepiness, and depression) compared with those in the bottom quintile, suggesting the protective role of healthy dietary patterns in early prevention of PD. In the Kailuan study (71,640 Chinese adults), better diet quality was also associated with lower odds of having prodromal PD (such as RBD, insomnia, excessive daytime sleepiness, constipation, and depressive symptom).[91]

In the cohort of Hellenic Longitudinal Investigation of Aging and Diet (HELIAD) that included 1,731 older Greek participants (≥65 years), the probability of prodromal PD (pPD) was evaluated according to the criteria of the International Parkinson and Movement Disorder Society, including five risk markers and seven prodromal markers.[92] Higher adherence to a Mediterranean diet was associated with a 21% lower probability of pPD when comparing extreme quartiles of adherence score, which may be attributed to daytime somnolence, constipation, urinary dysfunction, and depression. In the same cohort (HELIAD), 1,030 participants with pPD probability <30% were further included in the prospective analysis of the Dietary Inflammatory Index (DII) and pPD incidence (pPD probability ≥30%) during a mean follow-up of 3.1 years.[93] The authors documented a 20.3% higher risk of pPD per unit increase of DII score, with 1.6 times greater risk when comparing the highest with the lowest tertiles of DII score, suggesting that a more proinflammatory diet was associated with higher pPD risk.

Nutrition and Progression and Mortality of Parkinson Disease

Studies evaluating the association between nutrition and PD progression or mortality are limited. Among 1,251 individuals with PD from the NHS and HPFS, a healthy dietary pattern assessed by AHEI was associated with 31% to 43% lower risk of all-cause mortality (prediagnosis analyses HR 0.69, 95% CI=0.56–0.85, and postdiagnosis analyses HR 0.57, 95% CI=0.42–0.78), which was even stronger when combined with an active lifestyle.[94] In a cross-sectional analysis including 1,053 individuals with idiopathic PD, a patient-reported outcome in PD (PRO-PD) score adjusted for years since diagnosis was used to define PD progression.[95] Higher intakes of Mediterranean diet ingredients such as fresh vegetables and fruit, nuts, fish, olive oil, wine, and fish oil supplementation were associated with slower progression of PD. In contrast, higher consumption of canned vegetables, fruits, soda, fried foods, beef, ice cream, yogurt, and cheese, and iron supplementation were associated with more rapid progression. In the Rush Memory and

Aging Project, the parkinsonism progression of 689 participants aged 59 to 97 years was assessed by change in a global parkinsonian score that included four signs.[96] The authors found a significant inverse association between the MIND diet score and the annual progression rate of parkinsonism ($P=0.04$). In contrast, the Mediterranean diet score was only marginally associated ($P=0.06$), and the DASH score was nonsignificant.

A meta-analysis including four prospective studies from European and U.S. populations assessed the association between coffee/caffeine intake and PD progression (transition to Hoehn and Yahr stage 3, initiation of L-dopa treatment, motor fluctuation, and dyskinesia).[97] Overall, higher caffeine consumption was associated with 17% slower progression of PD (HR=0.83, 95% CI=0.71–0.98) among the individuals at an early stage of PD.

Stroke

Epidemiology of Stroke

The global burden of stroke increased substantially from 1990 to 2019, with the number of incident cases, prevalent cases, and death rising by 70%, 85%, and 43%, respectively.[98] Notably, there were 6.55 million stroke deaths and 143 million stroke-related DALYs, making stroke the second-leading cause of death and the third-leading cause of death and disability combined. Stroke has also accounted for 69.8% of neurologic deaths and 52.3% of neurologic DALYs; the combination of stroke and dementia is the leading cause of neurologic disability and death globally.[2]

Surprisingly, 87.0% of total stroke DALYs (and similar in its subtypes) were attributable to 19 risk factors, among which the five leading risk factors were high systolic blood pressure, high BMI, high fasting plasma glucose, air pollution, and smoking.[98] Although slightly decreased from 1990 (32.6%), the combination of six dietary factors (diets high in sodium and red meat, and low in fruits, vegetables, whole grains, and alcohol consumption) still contributed to 30.6% of the population attributable fraction (PAF) of stroke-related DALYs, second only to hypertension (55.5%). Interestingly, stroke and dementia have several shared modifiable risk factors, providing an opportunity for delaying both diseases.

Although evidence linking improved diet quality and recurrent stroke is less conclusive, optimal nutrition could be an important strategy for retarding the onset and development of the first stroke, such as salt reduction and adherence to the Mediterranean diet.[44,99] In addition to a shared risk factor like hypertension, stroke subtypes may differ in physiologic mechanisms. For example, lowering plasma LDL-C to within the optimal range (<100 mg/dL) and antiplatelet therapy could be beneficial for ischemic stroke (IS), but harmful for intracerebral hemorrhage (ICH).[100-102]

Nutrition and Stroke

Dietary Patterns

A growing body of evidence suggests that adherence to plant-based diets, such as the Mediterranean, DASH, prudent, or Nordic diet, is associated with lower stroke risk.[43,44,103,104] In contrast, the Western diet, characterized by high intakes of red and processed meats and refined carbohydrates was associated with high stroke risk.[103] In the prospective INTERSTROKE study recruiting participants from 32 countries (13,447 cases and 13,472 controls), higher adherence to a healthy diet was associated with 40% lower risk of acute first stroke (comparing extreme tertiles of the modified AHEI).[105] Moreover, individuals in the lowest two tertiles of modified AHEI contributed to 23.2% of population attributable risk (PAR) for stroke, compared with those in the highest tertile, which was stronger in participants aged ≥55 years ($P_{interaction}$ <0.01). Surprisingly, the diet-stroke association was inverse in South Asia compared with other regions, raising concerns about the influence of food preparation methods and whether the diet scores developed in Caucasian populations apply to populations in other areas of the world. In the Women's Health Initiative Dietary Modification Trial (48,835 postmenopausal women, aged 50 to 79 years), reducing total fat intake to 20% of energy together with increasing intakes of vegetables, fruits, and grains for 8.3 years failed to show any difference in incident stroke compared with usual diet.[106] In the Prevención con Dieta Mediterránea (PREDIMED) study, participants allocated to an energy-unrestricted Mediterranean diet (supplemented with extra-virgin olive oil or mixed nuts) had a 42% lower risk of stroke compared with those assigned to the low-fat diet,[107] suggesting that adopting a whole dietary pattern is more effective than manipulating a single macronutrient. RCTs have also suggested that plant-based diets (e.g., the DASH, the Mediterranean, or the Nordic diets) can reduce blood pressure modestly,[104] through improved vasodilation, antioxidant components, modified baroreceptors, renin-angiotensin, the sympathetic nervous system, and increased insulin sensitivity,[108] and may further reduce stroke risk.

Overnutrition is known to increase stroke risk by aggravating obesity, dyslipidemia, diabetes, and hypertension, whereas undernutrition in utero and early life may also increase one's lifelong stroke risk.[43] For instance, in the Kailuan Study among 97,399 participants (724 cases), participants with in-utero exposure to the Great Chinese Famine (1959 to 1961) had a 99% increased risk of ICH in midlife compared with those without exposure to famine.[109] Therefore, beyond healthy dietary patterns, adequate intake throughout life (but not exceeding energy requirements) is also crucial to optimize diet quality and promote health.

Individual Nutrients/Food Components

Consistent with healthy dietary patterns, foods rich in fiber were inversely associated with incident stroke risk in observational studies; in contrast, intakes of red and processed meat and sugar-sweetened beverages (SSB) were positively associated.[44,103,104] A comprehensive meta-analysis of 123 cohort studies examined associations between 12 major food groups (whole grains, refined grains, vegetables, fruits, nuts, legumes, eggs, dairy, fish, red meat, processed meat, and SSB) and risk of first stroke.[110] They reported that the highest intakes of vegetables (16 studies, 12,442 cases), fruits (17 studies, 30,523 cases), and fish (20 studies, 14,360 cases) were associated with 5% to 17% reduced risk of first stroke, compared with the lowest intakes. The dose-response analysis found a nonlinear association between stroke and dietary vegetable intake, which was less steep when intake exceeded 200 g/day ($P_{non\text{-}linearity} = 0.03$). In contrast, individuals with the highest consumption of red meat (seven studies, 10,541 cases), processed meat (six studies, 9,492 cases), and SSBs (seven studies, 11,187 cases) had 9% to 16% increased risk, compared with those with the lowest consumption. Animal food intake (such as red meat) may facilitate the gut microbiota to produce toxic metabolites like trimethylamine N-oxide (TMAO), which may increase stroke risk.[111]

More consistent results were found for dietary fiber, n-3 LCPUFAs, and potassium (inversely), as well as sodium and refined sugars (positively),[43,44,104] consistent with the foods mentioned in the previous paragraph. A meta-analysis of prospective cohort studies suggested an inverse association between dietary intake of n-3 PUFAs and stroke risk.[112] In a pooled analysis, the circulating concentration of n-6 PUFA linoleic acid was inversely associated with IS risk.[113] Interestingly, in the prospective China Kadoorie Biobank, with 8,159 participants, there were large variations in erythrocyte PUFA across 10 geographic areas.[114] The researchers found that elevated erythrocyte linoleic acid was associated with 31% lower risk of IS but with 1.33 times higher risk of ICH, when comparing extreme quintiles. Moreover, they documented an inverted-U shape curve association between erythrocyte eicosapentaenoic acid (EPA) concentration and IS, with a threshold of 0.70%, suggesting that the protective effect of n-3 PUFAs might only be found at a high level. However, in a meta-analysis of RCTs, supplementation of n-3 PUFA (ranging from 0.5 to >5 g/day) had no beneficial effects on stroke risk (RR 1.02, 95% CI 0.94 to 1.12, 31 RCTs).[115] In the 2×2 factorial VITAL trial (vitamin D and Omega-3 Trial), supplementation of 1 g/d n-3 PUFAs for 5.3 years did not reduce the risk of total stroke, IS, ICH, or fatal stroke.[116] In the Reduction of Cardiovascular Events With Icosapent Ethyl-Intervention Trial (REDUCE-IT), which tested the effects of high-dose purified EPA ethyl ester (icosapent ethyl, 4 g/d) in 8,179 statin users with elevated triglycerides, a 28% reduction in total stroke was found after 4.9-years.[117] Notably, 70.7% of the REDUCE-IT participants had prior CVD, and the rest had diabetes and other cardiovascular risk factors, suggesting that the beneficial effects of n-3 PUFAs on stroke may be restricted to certain diseases, which merits further study.

High sodium intake is a well-known risk factor for hypertension. Therefore, salt reduction or replacing sodium with potassium could be an effective strategy for stroke prevention by lowering systolic and diastolic blood pressure.[43,118] Interestingly, in the Kailuan study, researchers generated five distinct trajectories by repeated salt intake.[119] Individuals with a moderate-decreasing salt intake trajectory had a 24% lower ICH risk (adjusted HR 0.76; 95% CI, 0.63–0.92) than those with the moderate-stable salt intake trajectory. For other micronutrients, such as antioxidant vitamins, B vitamins, vitamin D, and calcium, results have been less conclusive.[43,44] Of note, the participants' nutritional status should be considered before any intervention. For instance, in the VITAL trial, supplementation of 2,000 IU/d vitamin D3 for 5.3 years did not reduce stroke risk (total stroke, IS, and ICH).[116] However, most VITAL participants (>87%) had serum total 25(OH)D concentration above 20 ng/mL at baseline, which already reached the level of cardiovascular benefit suggested by observational studies.[120] Another important example is folic acid supplementation, which has been suggested to reduce stroke risk only in areas without folate fortification.[44] Meta-analyses of clinical trials suggested that folic acid supplementation alone or in combination with a low dose (≤0.05 mg/day) of vitamin B_{12} was effective in reducing stroke risk, especially in regions that lack folic acid fortification.[121,122] In line with this, a meta-analysis of genetic studies also found a stronger association between the MTHFR 677C→T variant and increased homocysteine concentration in populations with low folate intake (such as Asia), which had a higher risk of stroke (1.68, 95% CI 1.44–1.97) than other regions.[123]

Nutrition and Stroke Progression and Mortality

Although the association between nutrition and the risk of recurrent stroke is less conclusive and evidence from RCTs is limited, higher adherence to a Mediterranean diet, reducing salt intake, and using folic acid supplementation in regions of low folate intake may lower risk of secondary stroke, according to the GRADE (Grading of Recommendations Assessment, Development, and Evaluation) approach.[44] A similar association between diet quality and stroke mortality has also been observed, although the outcome was usually analyzed using combined CVD mortality. For instance, in the NHS cohort, with 74,886 women followed for 20 years (1984–2004), higher adherence to the Mediterranean diet (Alternate Mediterranean Diet Score) was associated with 13% lower risk of incident stroke and 39% reduced CVD mortality, when comparing top and bottom quintiles.[124] The associations of 10 dietary factors (fruits, vegetables, nuts/seeds, whole grains, unprocessed red meats, processed meats, SSBs, polyunsaturated fats, seafood n-3 fats, and sodium) with mortality due to cardiometabolic diseases were evaluated using nationwide data of the United States.[125] The total number of stroke-related deaths attributed to an overall suboptimal diet was 66,547 per year (51.9%). For IS and ICH, 7,170 per year (44.5%) and 19,863 (60.9%), respectively. Among the dietary factors, the largest numbers of death due to total stroke were associated with low consumption of vegetables (28,039 [21.9%]) and fruits (28,741 [22.4%]) and high sodium intake (13,787 [10.7%]).

Sleep Disturbances

Sleep Health and Neurologic Diseases

Sleep disorder is another common neurologic disease. During the aging process, sleep disturbance and other neurologic diseases share common mechanisms and pose risks to each other. Patients with neurodegenerative diseases usually reported having sleep disorders, such as circadian rhythm misalignment, insomnia, hypersomnolence, sleep-disordered breathing, motor disturbances of sleep (e.g., restless leg syndrome, RLS), and parasomnias (e.g., RBD),[18] which escalate with the progression of the disease.[65,126] On the other hand, healthy sleep, defined as onset within 30 minutes and consolidated for 7 to 9 hours, plays a vital role in maintaining brain metabolic homeostasis via its cellular and metabolic restorative function,[65] such as regulating waste clearance (e.g., β-amyloid in AD and α-synuclein in PD) through the glymphatic system.[127,128] It can also regulate neural activity to optimize synaptic plasticity and promote memory consolidation and integration.[129]

Interestingly, better sleep consolidation (less sleep fragmentation as measured by actigraphy) could attenuate the effects of the *APOE* ε4 allele on incident AD, cognitive decline, and development of neurofibrillary tangle density, observed in 698 dementia-free older adults followed for 6 years[130]. However, with sleep disturbance, the perturbed brain homeostasis involves neuroinflammation, accumulated neurotoxic waste, substantia nigra neuronal loss, changes in neurogenesis and synaptic plasticity, blood-brain barrier function, and cerebral network function, as well as intermittent hypoxemia induced by obstructive sleep apnea,[65] which are further associated with deterioration in psychomotor responses, attention, memory, and cognitive function.[131] Poor sleep quantity and quality are known as prodromal features or early clinical features of cognitive decline or neurodegeneration and could exacerbate disease processes.[18-21] For instance, RBD, characterized by loss of muscle atonia and abnormal behaviors during REM sleep, is a prodromal symptom of synucleinopathies, including PD.[66] Therefore, healthy sleep should also be considered an important component of multidomain approaches targeting multifactorial neurologic diseases.

Notably, an unhealthy diet has been suggested to be associated with poor sleep quantity and quality. Disrupted sleep and circadian misalignment are also believed to modify eating behaviors and promote excessive energy intake.[132] Therefore, diet and sleep have interactive effects on various health outcomes, which last for a lifetime.[133] Understanding the bidirectional relationship between nutrition and sleep will help achieve optimal diet and sleep and facilitate the prevention of related cardiometabolic and neurologic diseases.

Nutrition and Sleep Disturbance

Emerging evidence suggests that diet quality and specific nutrients may play roles in the regulatory hormonal pathways and further alter sleep quantity and quality.[128] This is the case for caffeine, an antagonist to adenosine, which regulates sleep-wake cycles.[133] The western diet was suggested as one that disturbs the regulation of the hypothalamic-pituitary-adrenal (HPA) axis and increases the secretion of glucocorticoids like cortisol, which further increases the craving for low-quality diets, and leads to disrupted sleep and the circadian clock.[128,134] In contrast, adhering to healthy dietary patterns (American Heart Association recommendations or the Mediterranean diet) was associated with lower cortisol secretion.[135,136] Some clinical trials using phospholipid phosphatidylserine, n-3 PUFAs, magnesium, potassium, and flax seed cultivars have also shown reduced release of cortisol[128] and, thus, may improve sleep quality. Other mechanisms involve downstream metabolites that can be bioactive in sleep regulation (e.g., prostaglandin D2 for PUFAs) and the dietary-related, anti-inflammation, and anti-oxidation properties closely related to sleep health.[137,138] The association between urate concentration, a natural antioxidant, and possible RBD (pRBD) risk was evaluated in the Kailuan study, which included 12,923 Chinese adults free of PD and dementia.[139] The cumulative average urate concentration in plasma was measured in 2006, 2008, and 2010, and the pRBD risk was assessed by validated RBD questionnaire–Hong Kong. Individuals in the top urate quintile had 57% lower odds of having pRBD in 2012 than those in the bottom quintile. However, no significant association was found with new-onset pRBD (symptom onset within 1 year), possibly due to residual confounding.

Effects of dietary patterns and specific foods on sleep architecture and quality were evaluated in short-term clinical trials.[140] In observational studies, adherence to the Mediterranean diet was associated with lower risk of insomnia.[141] A low carbohydrate/high-fat diet was associated with reduced REM sleep and sleep efficiency (SE) but increased slow wave sleep (SWS) and arousals.[140,142] In contrast, a high carbohydrate/low-fat diet was related to reduced sleep-onset latency (SOL) and SWS, and increased REM sleep, which might be attributed to altered substrate oxidation and hormone secretion due to the manipulation of the carbohydrate and fat content of the diet. Moreover, dietary fat and carbohydrate quality may also impact sleep duration and quality.[133,138] Higher intake of SFAs and sugar was associated with less restorative and lighter sleep. In contrast, higher PUFAs (especially n-3 LCPUFAs) and fiber intakes were associated with more restorative and deeper sleep. A few clinical trials have also suggested improved sleep quality, quantity, and SOL when supplementing with vitamin D, melatonin, magnesium, or zinc.[133] Observational studies found that deficiency of other vitamins (e.g., B vitamins, and vitamins A, C, E) and minerals (e.g., iron and selenium) were also associated with shorter sleep duration or reduced sleep quality.[133] In the Kailuan Study, individuals with anemia (defined as hemoglobin <12.0 g/dL) had 32% higher risk of having insomnia, compared with their nonanemia counterparts, which was further confirmed by a meta-analysis including three additional cross-sectional studies with a pooled OR of 1.39 (95% CI 1.22–1.57).[143] Select foods, including dairy products, fatty fish, fruit (tart cherries and kiwifruit), and vegetables, may have a beneficial effect on sleep quality and duration, possibly due to their rich content of vitamins, n-3 PUFAs, or melatonin.[140,144]

Alcohol, considered a sedative with a sleep-promoting effect, was found to decrease sleep latency and disrupt the architecture of electrophysiologic sleep, possibly through its regulation of serotonin and norepinephrine secretions.[133,145] It was also associated with insomnia, circadian abnormalities, and breathing-related sleep problems in observational studies.[145] In the Kailuan Study, including 11,905 participants free of stroke, cancer, PD, dementia, and head injury, moderate and heavy drinkers had a 29% to 53% higher likelihood of having pRBD than nondrinkers.[146] In the same cohort, heavy drinkers were also associated with 31% to 47% higher risks of short sleep duration, snoring, and obstructive sleep apnea than nondrinkers.[147] However, only hard liquor, but not beer or wine, was positively associated with poor sleep quality. Interestingly, in a meta-analysis of 21 studies, higher alcohol consumption was associated with 25% increased risk of sleep apnea (pooled RR 1.25, 95% CI 1.13–1.38), which was restricted to low/middle-income countries in subgroup analysis.[148] In the NHS and HPFS cohorts, heavy drinkers (>30g/day) had a 25% lower risk of developing RLS compared with nondrinkers (95% CI 0.57–0.99, P_{trend}=0.05).[149] As most previous studies were conducted in case-control and cross-sectional designs with varied drinking habits (amount and type), more longitudinal studies and repeated measurements are needed to explore the association between alcohol and sleep quality.

Chrono-nutrition is an emerging research field linking the timing of eating with health outcomes. Skipping breakfast, but eating late at night were associated with higher risk of obesity and cardiometabolic diseases.[150] Among 7,771 participants of the Kailuan Study, those having a night-eating frequency 6+ times per week had faster progress of arterial stiffness (14.1 (95% CI, 0.6–27.5) cm/s per year) compared with those who never or rarely ate at night.[151] A study of 3,129 Japanese female workers suggested that unhealthy eating habits such as skipping breakfast and eating irregularly were associated with poor sleep quality.[152] In the nationally representative American Time Use Survey, 24-hour activities were recorded in 124,239 participants, among which eating or drinking was identified as <1, <2, and <3 hours before bedtime.[153] Compared with those who did not eat/drink before bed, individuals who ate/drank <1 hour before bedtime were twice as likely (OR 2.26, 95% CI: 1.93, 2.64) to report waking after sleep onset >30 minutes and longer sleep duration (30 minutes, 95% CI 95% CI 27, 33) which might be compensatory increases for inadequate sleep after late-night eating.

Sleep and Eating Behaviors

Biologic factors regulate eating behaviors under circadian rhythm (or circadian clock) and, in turn, affect nutritional status.[133] Circadian rhythm creates a synchronized network of biologic processes (e.g., metabolic and energetic processes), which runs on a 24-h rhythm.[134] Therefore, the feed-fast cycle, which regulates eating behaviors via hormone regulation, is coordinated with the sleep-wake cycle. However, due to work or social activities, millions worldwide are estimated to have insufficient or disrupted sleep, or sleep at biologically inappropriate times.[132,138] Sleep restriction and disturbance, as well as circadian misalignment, will disturb the secretion of appetitive and satiety hormones such as leptin and ghrelin, which promote excessive energy intake and positive energy balance,[132,133,138] thus increasing the risk of weight gain and related cardiometabolic and neurologic outcomes. Besides increased energy intake, individuals with deprived sleep or circadian misalignment were also prone to energy-dense and low-quality food choices. Sleep deprivation and circadian misalignment may also contribute to metabolic dysregulation through activation of the HPA axis, leading to perturbed downstream glucose, insulin, and adiponectin, which aggregates the vicious cycle between unhealthy diet and sleep disorders and contributes to an increased risk of CVD.[133]

In a meta-analysis of 11 clinical trials, the 24-hour energy intake of individuals following partial sleep deprivation was significantly increased by 385 kcal (95% CI 252, 517) compared with control individuals, which was accompanied by higher fat and lower protein intakes.[154] In 41 adolescents aged 14 to 16 years in a crossover design RCT, sleep restriction to 6.5 hours was associated with dietary choices of higher glycemic index and glycemic load, particularly more desserts and sweets, with no differences in fat or protein intake, compared with the healthy sleep duration (10 hours).[155] In the HPFS cohort, which enrolled 15,273 men free of cancer and cardiometabolic diseases, probable insomnia was defined as having nonrestorative sleep plus ≥one of the three nocturnal insomnia symptoms.[156] Participants with probable insomnia were found to have higher total energy intake (35.8 kcal/d, 95% CI: 17.4, 54.1 kcal/d) and worse scores in AHEI components of trans fat (higher intake), sodium (higher intake), and vegetables (lower intake) compared with their noninsomnia counterparts. In the American Heart Association (AHA) Go Red for Women Strategically Focused Research Network cohort, the chronotype was accessed by the Morningness-Eveningness Questionnaire.[157] Compared with those having morning/intermediate chronotype, individuals with the evening chronotype had poorer diet quality, including higher dietary energy density and lower consumption of plant protein, fiber, and fruits and vegetables. Interestingly, in a trial with 25 healthy participants, short sleep duration was associated with increased 2-oleoylglycerol (2-OG, an endocannabinoid system compound), enhanced functional MRI response to food odor in the piriform cortex, and higher energy density of food choices.[158] The association between changed 2-OG and food choices may be mediated by the odor-evoked connection between the piriform and insula cortex, providing a mechanistic pathway linking sleep deprivation with food preference. Another potential link connecting sleep quality and eating behaviors could be taste dysfunction, which was suggested in a cross-sectional analysis from the Kailuan study.[159] Among 11,030 participants free of neurodegenerative diseases, those with overall low sleep quality showed higher likelihood of having altered taste perception (adjusted OR 2.03, 95% CI 1.42–2.91) compared with those with high sleep quality. Individual symptoms like insomnia, daytime sleepiness, and short sleep duration were also significantly related to altered taste perception, which may further impact food choice; however, the results of such research remain to be seen.

Conclusions and Perspectives for Future Study

According to existing evidence, nutrition is one of the key modifiable factors in delaying and preventing neurologic diseases. Although suggested to be beneficial for neurologic health via several potential mechanisms, the effects of the individual micronutrients may be too small or may interact with each other in the daily diet,[6,7] resulting in controversial findings. For macronutrients such as fats and carbohydrates, quality and quantity are both important.[17,43,44,138,140] The balance of macronutrients may be more critical to optimize brain health, as it is impossible to manipulate the intake of a single macronutrient without influencing the other two.[25] Believed to be more effective than their components, the effects of dietary patterns have been evaluated more in observational studies than in RCTs. However, the latter can bring more reliable evidence for causal inference than the former.[14,43,44] Some have argued that dietary patterns such as the Mediterranean and the DASH diets were designed for overall health but not particularly for neurologic health.[17,160] Therefore, the observed associations could be due to the direct effect on brain function or through improvement in cardiometabolic risk factors.[14,16]

Future studies should pay close attention to methodologic challenges when assessing the relationships between nutrition and neurologic health.[9,10,14,15,17] For observational studies, the measurement error of dietary intakes, the method of preparing foods (e.g., fried and salted fish), and the heterogeneity in generating dietary adherence scores may influence the accuracy of exposure assessment. For RCTs, the sample size, the duration, optimal time or window for intervention, the dosage, and type of supplementation, as well as the nutritional status of participants, may partially explain the controversial results.[17,23] For instance, the effect of B vitamins in RCTs could be influenced by the baseline status of folic acid, vitamin B_{12}, and homocysteine.[1,12,15,17,44,48,111] Meanwhile, the protective effect of certain nutrients might be greater in individuals with either nutrient deficiencies (e.g., B vitamins), a genetic predisposition to cognitive impairment (e.g., *APOE* ε4), or at an early stage or prodromal phase of the diseases.[1,10,13,16,24] It is also noteworthy that using varied outcome assessments (e.g., different cognitive scores, brain imaging, serum markers) or composite outcomes (major CVD or total stroke) may produce mixed results.[10,16,43] Therefore, more precise and concordant methods or tools to evaluate nutritional status and neurologic outcomes are needed.

Emerging research areas, including fluid intake, hydration status, and eating behaviors (chrono-nutrition), may also provide interesting hints linking nutrition and neurologic health.[10,15] However, it should be noted that optimal diet is not an isolated factor but a component of a healthy lifestyle, which also includes physical activity, healthy sleep patterns, appropriate stress management, social interaction, and other multidomain approaches,[4,5,9,10,15,30,161] to pursue healthy aging and delay the onset of neurologic diseases. The synergistic association effect of these lifestyle behaviors has not been extensively evaluated. A proof-of-concept RCT, the Finnish Geriatric Intervention Study to Prevent Cognitive Impairment and Disability (FINGER), was the first such attempt and improved cognitive function in high-risk older adults.[162]

Gut microbiota, which likely play a key role in linking diet and neurologic diseases, through the gut-brain axis, has attracted growing research interest.[5,16,39,163,164] Their chronologic changes may modulate aging-related changes in cognitive function.[12] Dysbiosis of gut microbiota may also be a characteristic of AD.[39] It is associated with the risk of stroke and cognitive impairment, bringing novel targets for treatment.[165] The diversity in gut microbiota, together with differences in genetic and epigenetic variations, environmental exposure, and individual metabolic status, may explain the heterogeneity of personal susceptibility/response to a given intervention approach, which highlights the necessity of precision nutrition strategies in the future.[9,15,38]

Optimum nutrition is a critical constituent of multidomain approaches targeting multifactorial neurologic diseases. More well-designed large RCTs are needed to determine the appropriate nutritional strategies for various stages of disease progression, with methodologic issues and other novel dietary factors being considered.

References

1. Power R, Prado-Cabrero A, Mulcahy R, et al. Ann Rev Food Sci Tech 2019;10:619–39.
2. Avan A, Hachinski V. Alzheimer Dementia 2021;17:1072–6.
3. Global Burden of Disease 2016 Neurology Collaborators. Lancet Neurology 2019;18:459–80.
4. Ramirez-Salazar SA, Herren C, McCartney J, et al. Curr Nerol Neurosci Rep 2021;21:55.
5. Mao XY, Yin XX, Guan QW, et al. Pharmacol Ther 2021;226:107861.
6. Athanasopoulos D, Karagiannis G, Tsolaki M. Adv Nutr 2016;7:917–27.
7. Ashleigh T, Swerdlow RH, Beal MF. Alzheimer Dementia 2022;333–342.
8. Irwin MR, Vitiello MV. Lancet Neurol 2019;18:296–306.
9. Miquel S, Champ C, Day J, et al. Age Res Rev 2018;42:40–55.
10. Scarmeas N, Anastasiou CA, Yannakoulia M. Lancet. Neurol 2018;17:1006–15.
11. Lalkovicova M, Danielisova V. Neural Regen Res 2016;11:865–74.
12. Businaro R, Vauzour D, Sarris J, et al. Front Nutr 2021;8:669846.
13. Ascherio A, Schwarzschild MA. Lancet Neurol 2016;15:1257–72.
14. van de Rest O, Berendsen AA, Haveman-Nies A, et al. Adv Nutr 2015;6:154–68.
15. Flanagan E, Lamport D, Brennan L, et al. Age Res Rev 2020;62:101079.
16. Jennings A, Cunnane SC, Minihane AM. BMJ 2020;369:m2269.
17. Morris MC. Ann NY Acad Sci 2016;1367:31–7.
18. Porter VR, Buxton WG, Avidan AY. Curr Psychiatr Rep 2015;17:97.
19. Dauvilliers Y, Schenck CH, Postuma RB, et al. Nature Rev Dis Primers 2018;4:19.
20. Alonso-Navarro H, Garcia-Martin E, Agundez JAG, et al. Neurol 2019;92:948–64.
21. Trenkwalder C, Allen R, Hogl B, et al. Lancet Neurol 2018;17:994–1005.
22. World Health Organization. Global Status Report on the Public Health Response to Dementia. Geneva, Switzerland, 2021.
23. Moore K, Hughes CF, Ward M, et al. Proc Nutr Soc 2018;77:152–63.
24. Fenech M. Adv Nutr 2017;8:958–70.
25. Wahl D, Cogger VC, Solon-Biet SM, et al. Age Res Rev 2016;31:80–92.
26. Livingston G, Huntley J, Sommerlad A, et al. Lancet 2020;396:413–46.
27. Alzheimer's Association. Alzheimers Dement 2022;18:700–89.
28. Daulatzai MA. Neurochem Res 2012;37:2627–58.
29. Teipel S, Gustafson D, Ossenkoppele R, et al. J Nucl Med 2022;63:981–5.
30. Scheltens P, De Strooper B, Kivipelto M, et al. Lancet 2021;397:1577–90.
31. Petersson SD, Philippou E. 2021 Adv Nutr 2021;7:889–904.
32. van den Brink AC, Brouwer-Brolsma EM, Berendsen AAM, et al. Adv Nutr 2019;10:1040–65.
33. Lichtenstein AH, Appel LJ, Vadiveloo M, et al. Circ 2021;144:e472–e87.
34. Limongi F, Siviero P, Bozanic A, et al. J Am Med Dir Assoc 2020;21:1402–9.
35. Mattei J, Bigornia SJ, Sotos-Prieto M, et al. Diabet Care 2019;42:1372–9.
36. Ballarini T, Melo van Lent D, Brunner J, et al. Neurol 2021;96:e2920–e32.
37. Liu YH, Gao X, Na M, et al. J Alzheimer Dis 2020;78:151–68.
38. Milosevic M, Arsic A, Cvetkovic Z, et al. Front Nutr 2021;8:688086.
39. Chu CQ, Yu LL, Qi GY, et al. Neurosci Biobehav Rev 2022;135:104556.
40. Gao X, Scott T, Falcon LM, et al. Am J Clin Nutr 2009;89:1197–203.
41. Wong JC, Scott T, Wilde P, et al. J Nutr 2016;146:1740–5.
42. Na M, Dou N, Ji N, et al. Adv Nutr 2020;11:667–76.
43. Hankey GJ. Lancet Neurol 2012;11:66–81.
44. English C, MacDonald-Wicks L, Patterson A, et al. Lancet Neurol 2021;20:150–60.
45. Okereke OI, Rosner BA, Kim DH, et al. Ann Neurol 2012;72:124–34.
46. Zhang Y, Chen J, Qiu J, et al. Am J Clin Nutr 2016;103:330–40.
47. Brainard JS, Jimoh OF, Deane KHO, et al. J Am Med Dir Assoc 2020;21:1439–50 e21.
48. Barnes JL, Tian M, Edens NK, et al. Nutr Rev 2014;72:707–19.
49. Palacios N, Scott T, Sahasrabudhe N, et al. Nutr Neurosci 2020;23:688–95.
50. Grinan-Ferre C, Bellver-Sanchis A, Izquierdo V, et al. Age Res Rev 2021;67:101271.
51. Rajaram S, Jones J, Lee GJ. Adv Nutr 2019;10:S422–S36.
52. Poti F, Santi D, Spaggiari G, et al. Intl J Molec Sci 2019;20:351.
53. Ran LS, Liu WH, Fang YY, et al. Epidemiol Psychiatr Sci 2021;30:e13.
54. Zhang XC, Gao X, Lyu YB, et al. Biomed Environ Sci 2021;34:509–19.
55. Scarmeas N, Stern Y, Mayeux R, et al. Arch Neurol 2009;66:216–25.
56. Doorduijn AS, de van der Schueren MAE, van de Rest O, et al. J Am Med Dir Assoc 2020;S1525–S8610.
57. Rainey-Smith SR, Gu Y, Gardener SL, et al. Transl Psychiatry 2018;8:238.
58. Lilamand M, Mouton-Liger F, Di Valentin E, et al. Front Nutr 2021;8:807970.
59. Grammatikopoulou MG, Goulis DG, Gkiouras K, et al. Adv Nutr 2020;11:1583–602.
60. Phillips MCL, Deprez LM, Mortimer GMN, et al. Alzheimers Res Ther 2021;13:51.
61. Marras C, Beck JC, Bower JH, et al. NPJ Parkinsons Dis 2018;4:21.

62. Yemula N, Dietrich C, Dostal V, et al. J Parkinson Dis 2021; 11:1491–505.
63. Olanow CW, Tatton WG. Ann Rev Neurosci 1999;22:123–44.
64. Murray LK, Jadavji NM. Nutr Res Rev 2019;32.218–30.
65. Lajoie AC, Lafontaine AL, Kaminska M. Chest 2021;159:818–27.
66. Gan-Or Z, Alcalay RN, Rouleau GA, et al. Sleep Med Rev 2018;41:101–12.
67. Gao X, Simon KC, Han J, et al. Ann Neurol 2009;65:76–82.
68. Zuzuarregui JRP, During EH. Neurotherapeutics 2020;17: 1480–94.
69. Gauthier S, Rosa-Neto P, Morais J, et al. World Alzheimer Report 2021: Journey Through the Diagnosis of Dementia. London, England: Alzheimer Disease International, 2021.
70. Klein C, Westenberger A. Cold Spring Harbor Perspect Med 2012;2:a008888.
71. Gao X, Chen H, Fung TT, et al. Am J Clin Nutr 2007;86:1486–94.
72. Yin W, Lof M, Pedersen NL, et al. Movement Disorders 2021; 36:255–60.
73. Metcalfe-Roach A, Yu AC, Golz E, et al. Movement Disorders 2021;36:977–84.
74. Liu YH, Jensen GL, Na M, et al. J Parkinson Dis 2021;11:337–47.
75. Chen H, Zhang SM, Schwarzschild MA, et al. Am J Epidemiol 2004;160:368–75.
76. de Lau LM, Koudstaal PJ, Witteman JC, et al. Neurol 2006; 67:315–8.
77. Hughes KC, Gao X, Kim IY, et al. Movement Disorders 2016;31: 1909–14.
78. Gao X, Cassidy A, Schwarzschild MA, et al. Neurol 2012;78: 1138–45.
79. Zhang X, Molsberry SA, Yeh TS, et al. Neurol 2022;98: e1064–e76.
80. Xu K, Bastia E, Schwarzschild M. Pharmacol Ther 2005;105: 267–310.
81. Palacios N, Gao X, McCullough ML, et al. Movement Disorders 2012;27:1276–82.
82. Gao X, Chen H, Choi HK, et al. Am J Epidemiol 2008;167:831–8.
83. Simon KC, Eberly S, Gao X, et al. Ann Neurol 2014;76:862–8.
84. Gao X, O'Reilly EJ, Schwarzschild MA, et al. Neurol 2016; 86:520–6.
85. Ma C, Liu Y, Neumann S, et al. Translat Neurodegen 2017;6:18.
86. Ma C, Molsberry S, Li Y, et al. Am J Clin Nutr 2020;112:1080–7.
87. Dalbeth N, Wong S, Gamble GD, et al. Ann Rheum Dis 2010; 69:1677–82.
88. Hughes KC, Gao X, Kim IY, et al. Neurol 2017;89:46–52.
89. Domenighetti C, Sugier PE, Ashok Kumar Sreelatha A, et al. Movement Disorders 2022;37:857–64.
90. Molsberry S, Bjornevik K, Hughes KC, et al. Neurol 2020;95: e2095–e108.
91. Zhang X, Xu J, Liu Y, et al. Movement Disorders 2022;37: 2367–75.
92. Maraki MI, Yannakoulia M, Stamelou M, et al. Movement Disorders 2019;34:48–57.
93. Balomenos V, Bounou L, Charisis S, et al. J Nutr Biochem 2022;105:108994.
94. Zhang X, Molsberry SA, Schwarzschild MA, et al. JAMA Network Open 2022;5:e2227738.
95. Mischley LK, Lau RC, Bennett RD. Oxid Med Cell Longev 2017;2017:6405278.
96. Agarwal P, Wang Y, Buchman AS, et al. J Nutr Health Aging 2018;22;1211–5.
97. Hong CT, Chan L, Bai CH. Nutrients 2020;12:1860.
98. Collaborators GBDS. Lancet Neurol 20:795–820.
99. Pandian JD, Gall SL, Kate MP, et al. Lancet 2018;392:1269–78.
100. Wu Z, Huang Z, Lichtenstein AH, et al. BMC Med 2021;19:142.
101. Diener HC, Hankey GJ. J Am Coll Cardiol 2020;75:1804–18.
102. Sun L, Clarke R, Bennett D, et al. Nat Med 2019;25:569–74.
103. Sherzai A, Heim LT, Boothby C, et al. Nutr Rev 2012;70:423–35.
104. Hankey GJ. Stroke 2017;48:3168–74.
105. O'Donnell MJ, Chin SL, Rangarajan S, et al. Lancet 2016;388:761–75.
106. Howard BV, Van Horn L, Hsia J, et al. JAMA 2006;295:655–66.
107. Estruch R, Ros E, Salas-Salvado J, et al. N Engl J Med 2018;378:e34.
108. Alexander S, Ostfeld RJ, Allen K, et al. J Geriatr Cardiol 2017;14:327–30.
109. Li Y, Li Y, Gurol ME, et al. Neurol 2020;94:e1996–e2004.
110. Bechthold A, Boeing H, Schwedhelm C, et al. Crit Rev Food Sci Nutr 2019;59:1071–90.
111. Spence JD. Nutrients 2019;11:647.
112. Cheng P, Huang W, Bai S, et al. Sci Rep 2015;5:14161.
113. Marklund M, Wu JHY, Imamura F, et al. Circ 2019;139: 2422–36.
114. Sun L, Du H, Zong G, et al. Eur J Nutr 2022;61:3235–46.
115. Abdelhamid AS, Brown TJ, Brainard JS, et al. Cochrane Database Syst Rev 2020;3:CD003177.
116. Manson JE, Bassuk SS, Cook NR, et al. Circ Res 2020;126: 112–28.
117. Bhatt DL, Steg PG, Miller M, et al. N Engl J Med 2019;380: 11–22.
118. Aburto NJ, Ziolkovska A, Hooper L, et al. BMJ 2013;346: f1326.
119. Li Y, Huang Z, Jin C, et al. Stroke 2018;49:1332–9.
120. Zhang R, Li B, Gao X, et al. Am J Clin Nutr 2017;105:810–9.
121. Zhao M, Wu G, Li Y, et al. Neurol 2017;88:1830–8.
122. Hsu CY, Chiu SW, Hong KS, et al. J Stroke 2018;20:99–109.
123. Holmes MV, Newcombe P, Hubacek JA, et al. Lancet 2011;378:584–94.
124. Fung TT, Rexrode KM, Mantzoros CS, et al. Circ 2009;119: 1093–100.
125. Micha R, Penalvo JL, Cudhea F, et al. JAMA 2017;317:912–24.
126. Brzecka A, Leszek J, Ashraf GM, et al. Front Neurosci 2018;12:330.
127. Xie L, Kang H, Xu Q, et al. Science 2013;342:373–7.
128. Pistollato F, Sumalla Cano S, Elio I, et al. Adv Nutr 2016;7: 679–89.
129. Zahed H, Zuzuarregui JRP, Gilron R, et al. Movement Disorders 2021;36:1526–42.
130. Lim AS, Yu L, Kowgier M, et al. JAMA Neurol 2013;70:1544–51.
131. Veatch OJ, Keenan BT, Gehrman PR, et al. Lancet Neurol 2017; 16:158–70.
132. McHill AW, Wright KP, Jr. Obes Rev 2017;18(suppl 1):15–24.
133. Frank S, Gonzalez K, Lee-Ang L, et al. Front Neurol 2017;8:393.
134. Pratt CA, Brown AGM, Dixit S, et al. Adv Nutr 2022;13: 1402–14.
135. Garcia Prieto MD, Tebar FJ, Nicolas F, et al. Clin Endocrinol (Oxf) 2007;66:185–91.
136. Mattei J, Bhupathiraju S, Tucker KL. J Nutr 2013;143:1753–9.
137. Zhao M, Tuo H, Wang S, et al. Mediators Inflamm 2020;2020:3142874.
138. Muscogiuri G, Barrea L, Annunziata G, et al. Crit Rev Food Sci Nutr 2019;59:2158–65.
139. Shen Y, Li J, Schwarzschild M, et al. Ann Clin Transl Neurol 2019;6:2368–76.
140. St-Onge MP, Mikic A, Pietrolungo CE. Adv Nutr 2016;7:938–49.
141. Jaussent I, Dauvilliers Y, Ancelin ML, et al. Am J Geriatr Psychiatry 2011;19:88–97.
142. Afaghi A, O'Connor H, Chow CM. Nutr Neurosci 2008;11: 146–54.
143. Neumann SN, Li JJ, Yuan XD, et al. Chin Med J (Engl) 2020;134:675–81.
144. Komada Y, Okajima I, Kuwata T. Int J Environ Res Public Health 2020;17:9440.

145. He S, Hasler BP, Chakravorty S. Curr Opin Psychol 2019;30: 117–22.
146. Ma C, Pavlova M, Li J, et al. Ann Clin Translat Neurol 2018;5:1176–83.
147. Zheng D, Yuan X, Ma C, et al. Public Health Nutr 2021;24: 4851–8.
148. Simou E, Britton J, Leonardi-Bee J. Sleep Med 2018;42:38–46.
149. Batool-Anwar S, Li Y, De Vito K, et al. J Clin Sleep Med 2016;12:187–94.
150. Pot GK. Proc Nutr Soc 2018;77:189–98.
151. Zhang X, Wu Y, Na M, et al. J Am Heart Assoc 2020;9:e016455.
152. Katagiri R, Asakura K, Kobayashi S, et al. J Occup Health 2014;56:359–68.
153. Iao SI, Jansen E, Shedden K, et al. Br J Nutr 2021;127:1–10.
154. Al Khatib HK, Harding SV, Darzi J, et al. Eur J Clin Nutr 2017; 71:614–24.
155. Beebe DW, Simon S, Summer S, et al. Sleep 2013;36:827–34.
156. Cheng FW, Li Y, Winkelman JW, et al. Am J Clin Nutr 2016;104:462–9.
157. Zuraikat FM, St-Onge MP, Makarem N, et al. J Nutr 2021;151:1150–8.
158. Bhutani S, Howard JD, Reynolds R, et al. Elife 2019;8:e49053.
159. Gao Y, Li J, Cheng FW, et al. J Acad Nutr Dietet 2021;121: 435–45.
160. Cichon N, Dziedzic A, Gorniak L, et al. Intl J Molec Sci 2021;22:10707.
161. Yannakoulia M, Kontogianni M, Scarmeas N. Age Res Rev 2015;20:74–8.
162. Ngandu T, Lehtisalo J, Solomon A, et al. Lancet 2015;385: 2255–63.
163. Tengeler AC, Kozicz T, Kiliaan AJ. Nutr Rev 2018;76:603–17.
164. Gentile F, Doneddu PE, Riva N, et al. Intl J Molec Sci 2020; 21:7471.
165. Koszewicz M, Jaroch J, Brzecka A, et al. Pharmacol Res 2021;164:105277.

G. BEHAVIORAL, PSYCHIATRIC, AND NEUROLOGIC DISORDERS

CHAPTER 94

Nutrition and Intellectual and Developmental Disabilities

Susan L. Hyman, Linda G. Bandini, Aviva Must, and Carol Curtin

CHAPTER OUTLINE

Abbreviations

ARFID, avoidant restrictive food intake disorder; **ASD**, autism spectrum disorder; **BMI**, body mass index; **ESPGHAN**, European Society for Paediatric Gastroenterology Hepatology and Nutrition; **FFM**, fat-free mass; **I/DD**, intellectual and developmental disabilities; **NCCIH**, National Center for Complementary and Integrative Health; **PWS**, Prader-Willi Syndrome; **RD**, registered dietitian; **RMR**, resting metabolic rate; **SGA**, second-generation antipsychotics; **UDL**, universal design for learning; **WHO**, World Health Organization

Introduction

Practitioners and clinical researchers must understand the nutritional needs of individuals with intellectual and developmental disabilities (I/DD) to prevent comorbidities and promote wellness. I/DD is an umbrella term encompassing a range of developmental disabilities that present before the age of 18 years. The limitations experienced by individuals with I/DD range from mild to severe; they can involve intellectual or cognitive development and functioning, sensory domains (i.e., vision, hearing), physical development and mobility, communication, social development, and behavior.[1]

Not all people with a developmental disability have an intellectual disability (ID); the latter is characterized by significant limitations in both intellectual functioning and adaptive behavior that arise before the age of 22 years. Intellectual functioning includes mental or cognitive skills such as learning, reasoning, and problem-solving. Adaptive behavior refers to daily living skills that people perform in everyday life and includes social skills, self-care skills, decision-making, occupational skills, and the ability to function in the community.[2]

The causes of I/DD are multifactorial, including genetics, maternal prenatal health (e.g., infections or maternal

Table 94.1 Prevalence Estimates of Intellectual and Developmental Disabilities (I/DD)[4]

Diagnosis	Prevalence
Developmental disability overall	1 in 6
Autism Spectrum Disorder (ASD)	1 in 44
Cerebral Palsy	1–4 per 1,000
Down syndrome	1 in 700
Intellectual disability (ID)	1 to 3 per 100
Prader-Willi syndrome	6 per 100,000
Spina bifida	1 in 2,758

exposures), prematurity, complications during birth or early infant infections, untreated conditions such as jaundice in newborns, and/or exposure to environmental toxins (e.g., lead). Although some causes are identifiable, such as genetic and chromosomal disorders (e.g., Down syndrome and Fragile X syndrome) and toxic exposures during pregnancy (e.g., Fetal Alcohol Syndrome), many causes are unknown or are considered idiopathic.[3] **Table 94.1** shows the prevalence of the disabilities covered in this chapter.

The intellectual, behavioral, social, sensory, and motor impairments often associated with I/DD and the social exclusion often experienced by this population can limit access to educational, recreational, medical, and community resources. Healthcare professionals may also lack the knowledge necessary to accommodate their needs. The health disparities experienced by this population are well documented.[5-7] Although individuals with I/DD are generally healthy, i.e., their developmental condition should not be considered a "disease," this population is at risk for poor diet, low levels of physical activity,[8-10] and increased sedentary behavior.[11,12] In addition, individuals with I/DD often have medical complexities, with significant chronic health problems that affect multiple organ systems, often resulting in functional limitations and increased use of health resources.[13]

Physiologic, genetic, social, medication management, behavioral, and other risk factors often give rise to complex medical conditions, including increased risk for obesity,[14-23] metabolic syndrome,[9] type 2 diabetes, hypertension, and hyperlipidemia.[24] By adulthood, individuals with I/DD are at extremely high risk of multiple co-occurring conditions that may shorten life expectancy and negatively impact quality of life.[25] Medical complexity calls for coordination of care across specialties and disciplines and the incorporation of shared decision making with families. Proactive dietary guidance is critical to health promotion and the treatment of co-occurring conditions. Clinicians must consider the context of individuals' lives, which may include residential care, interdisciplinary care teams, and the inclusion and support of caregivers in their care.

In this chapter, we highlight several conditions that fall under the I/DD umbrella as illustrative examples of particular nutritional issues that are common to individuals with I/DD: autism spectrum disorder (ASD) as a biobehavioral disorder with severe food selectivity attributed to both sensory and behavioral factors; Down syndrome as a genetically determined cause of I/DD and hypotonia; Prader-Willi syndrome as an example of a genetic disorder associated with I/DD and obesity with a characteristic behavioral phenotype of high insatiability; and spina bifida and cerebral palsy as developmental disorders with motor implications for growth and nutrition.

Assessment of Eating and Feeding Concerns

The dietary history is an important component of health promotion to ensure adequate nutrient intake and in the clinical management of over- or underweight status. Well child and adult health maintenance visits with a primary care provider should routinely include a dietary history. Understanding how an individual eats and drinks (e.g., is fed by a caregiver, self-feeds with hands, or uses utensils), when they eat (mealtimes, snacks throughout the day), and the variety of foods they eat are all important pieces of information in the history. This information is used to determine the individual's nutritional sufficiency and what, if anything, should be done to address behavior, setting, or mechanics related to eating. Whereas medical clinicians might obtain a narrative history in the initial assessment, registered dietitians (RDs) use a standard approach to assess intake, such as a 24-hour recall or a food record (See Chapter 49). People with I/DD are at increased risk for nutritional compromise, so attention to healthy eating should start in early childhood and continue throughout the life course.

Medical and Social History

When there is concern about over- or undernutrition or problems with eating, the history should include the timing and circumstances of initial caregiver concerns. Changes in appetite or the mechanics of eating may occur with intercurrent medical events, in response to medications or changes in the individual's social situation. Many families of children with I/DD have lower incomes and less social and family support; thus, it is important to consider food insecurity and other environmental and social stressors with any change in growth, weight, or eating behavior. It is especially important to include adverse life events, including transitions in caregivers, in the history.[26]

Growth Parameters

It is common to plot height and weight to assess growth in infants, children, and adolescents using World Health Organization (WHO) growth charts[27] for infants and children up to the age of 2 years, and the Center for Disease Control and Promotion (CDC) growth charts thereafter.[28] The trajectory of weight gain may slow prior to delays in linear growth when energy intake is low. A change in growth trajectory in childhood or weight after adult height is reached may help identify factors temporally related to new onset eating or nutritional concerns.

Some populations with genetic and/or neurologic etiologies for their disability, such as Down syndrome,[29] Prader-Willi syndrome,[30] or cerebral palsy, have associated differences in body habitus that may support the use of disorder-specific growth charts, as discussed in the condition-specific sections that follow. The etiologies of the specific developmental disability need to be considered in assessing growth history. For example, larger-than-average head circumference and height are seen with Fragile X syndrome (an X-linked intellectual disability),[31] and children with autism spectrum disorder may have larger than typical head size in early childhood.[32]

Physical and Laboratory Examinations

Both physical and neurologic examinations are used to evaluate deficits or changes in nutritional intake. The medical, social, and dietary history may guide the examination. Routine laboratory monitoring of individuals with obesity should follow guidelines for the general population.[33,34] For individuals with certain types of developmental disorders, such as spina bifida or cerebral palsy, obtaining accurate height may be difficult due to an inability to stand, scoliosis, kyphosis, or contractures. The European Society for Paediatric Gastroenterology Hepatology and Nutrition (ESPGHAN) guidelines recommend using segmental measurements when accurate measures of height cannot be obtained due to joint contractures or scoliosis.[35]

Neurologic Assessment

From birth through early childhood, there is a neurologic progression of oral motor skills, moving from suckling to munching to being able to move a bolus of food with the tongue to be chewed by molars and safely swallowed. Oral motor problems such as those found in cerebral palsy, or genetic disorders associated with hypotonia, may result in delays in this developmental progression. These oral motor difficulties can lead to choking or gagging if individuals are presented with foods they cannot manage motorically. This may also lead to inadequate intake, aspiration, and/or food refusal. Caregivers may also be overly cautious and delay the advancement of food complexity; this can result in a child being food selective due to limited eating experience. Assessment includes oral motor evaluation of tongue movement and coordination of chewing and swallowing as well as the seated position that best protects the airway. This evaluation is often done in the context of an observational feeding assessment by an occupational or speech therapist in collaboration with a physician or advanced practice provider. A cine swallowing study may be necessary to determine if there is aspiration or another reason to alter the thickness or texture of foods or beverages.

Aerodigestive and Gastrointestinal Assessment

Assessment of the coordination of breathing and swallowing may be necessary to determine the safety of oral feeding in individuals with motor impairments. Discomfort from gastroesophageal reflux or eosinophilic esophagitis can be responsible for food aversion in individuals with I/DD. The latter can cause severe food selectivity but continued acceptance of liquids. Medications used for seizures or psychopharmacologic agents used as a component of a behavioral management plan can worsen constipation and result in gastrointestinal discomfort, with secondary implications for eating behavior.

Behavioral and Sensory Assessment

Food selectivity, or what parents may term "picky eating," is characterized by food refusal and a diet with a limited variety of foods. Resistance or refusal based on texture, taste, smell, color, or other characteristics can be mild to severe. Although common in individuals with I/DD, it is most prominent in autistic individuals.[36,37] A detailed dietary history and behavioral observation of the feeding interaction are important components of the assessment of food selectivity. The narrative history of sensory symptoms can be supported by questionnaires completed by the caregiver or patient that capture sensory aversions, sensory seeking, and the sensory experience.

Assessment of Energy Needs

There has been limited research on the energy needs of children and adolescents with I/DD. Several researchers have provided estimates of energy needs based on height for youth with Down syndrome,[38] Spina Bifida,[39] cerebral palsy (CP),[40] and Prader-Willi syndrome.[41] However, these estimates were based on studies with very small sample sizes that measured energy intake, not energy expenditure. Limited work has been done for adults with these conditions.

There are no evidence-based guidelines to determine energy needs for individuals with I/DD.[42] Because of alterations in body composition, variability in ambulatory status and physical activity level, and the presence of other clinical factors (e.g., medications), the predictive equations for resting metabolic rate (RMR) or the estimated energy requirements (EER) provided by the Dietary Reference Intakes (DRI)[43] are likely to yield an overestimation of energy needs. Consequently, adjustments to either the calculated RMR or EER should be based on the considerations indicated above, informed by a nutritional assessment and clinical judgment. Regular monitoring of changes in weight is needed to determine whether the estimates are appropriate or require further adjustment.

The Approach to Treatment of Nutritional Disorders in People with I/DD

Access and Communication

Access to quality health care remains challenging for individuals with I/DD. They face many barriers, including inadequate training of healthcare professionals, poor communication between patients and providers, and a complex healthcare system that is difficult to navigate.[44]

Practitioners must educate themselves about the patient's needs and take time to communicate in ways that facilitate positive rapport and the patient's engagement in the clinical encounter. Some individuals with I/DD may not speak verbally and may use augmentative and alternative communication. Aided approaches may include communication boards that use symbols, objects, pictures, or words that the individuals can point at to convey their thoughts and preferences. To support the cognitive and literacy challenges that some individuals with I/DD experience, using materials that have been created using universal design for learning (UDL) principles can be helpful. UDL is an evidence-based framework[45-47] that aims to maximize individuals' understanding, expression, and application of learning by providing multiple ways to acquire information and demonstrate their comprehension.

Virtual or telehealth delivery of services and care is growing in practice and may be preferred by some individuals with I/DD and their caregivers, as it addresses the logistic, financial, and other burdens often associated with scheduling and traveling for in-person visits. Inclusive telehealth should be considered if the individual prefers to be seen remotely and a physical exam is not required for the visit.

Clinicians should also be aware that the use of language and labels to describe people with disabilities often shift; trends have been changing in the use of person-first vs identity-first language, which reflects the desires and preferences of different constituent groups regarding disability terminology. For example, many in the autistic self-advocacy movement have moved away from person-first language (i.e., "a person with autism") to identity-first language (i.e., "autistic person") as a way to assert that being autistic is integral to one's identity and not a disorder to be "cured."[48,49] Some have argued that person-first language may accentuate, rather than diminish, stigma.[50] In this chapter, both person-first and identity-first terminology are used in an effort to acknowledge these preferences. Clinicians should use terminology and language consistent with the desires of the individuals and populations with whom they are working, and should seek the input of their patients to guide the use of acceptable, nonoffensive terminology.

Obesity

Guidelines for the neurotypical population, such as the 2007 American Academy of Pediatrics recommendations[33] and the U.S. Preventive Services Task Force's recommendations to promote weight loss for the prevention of obesity-related morbidity and mortality in adults,[34] are applicable for people with I/DD with modifications for developmental level, motor, and behavior-related factors. For example, adaptations might entail using picture schedules for meals and snacks or a behaviorally supported exercise program.[51]

Attention to portion control and minimizing the use of food rewards as reinforcement for behavioral compliance need to be emphasized beginning in early childhood. As with all populations, obesity prevention and management include limiting sugar-sweetened beverages and providing healthy snacks and foods high in fiber. As many children and adults with I/DD depend on caregivers to set a routine and to provide food and beverages, collaboration between the home, school, and/or day setting is essential.

Several weight loss studies of both youth and adults with I/DD have demonstrated that this population can be engaged in individual or group-based multicomponent interventions that focus on making changes in diet and physical activity.[52-59] Successful interventions with this population must consider the context of individuals' lives, focus on multicomponent interventions, include and support caregivers' involvement, provide nutrition education and counseling in understandable formats (e.g., using plain language and visual supports), promote physical activity, and use positive behavioral techniques to facilitate the adoption and maintenance of healthy lifestyle behaviors.[60]

Many individuals with I/DD, including those with ASD, are prescribed medications such as second-generation antipsychotics (SGAs) for behavioral challenges. These medications, in general, are associated with weight gain and metabolic syndrome.[61] Weight gain with SGAs occurs for unclear reasons that include the impact on neurotransmitters that regulate appetite and satiety, and is reported to be more common in individuals with ASD.[62] While dietary management and exercise is the recommended initial intervention, clinical trials have examined the potential benefit of adding metformin to address weight gain in this setting.[62-64] Metformin is used in the general population to help address weight gain with SGAs because of an overall positive effect of metformin on lowering BMI in the literature. Further research is needed to determine how best to combine dietary and medical management to prevent weight gain with prescription of these medications.[63]

Underweight

Energy intake and expenditure imbalance may be responsible for poor weight gain in individuals with oral motor dysfunction, food selectivity, and/or significant hyperactivity. Increasing energy intake using preferred foods may require the RD to examine products and modes of food preparation with caregivers to provide maximal nutritional benefit through diet, and to consider oral motor status and feeding behaviors.[65]

Nutritional Deficiencies

Food selectivity places individuals with I/DD at risk for inadequate macro- and micronutrient intake. Severe food selectivity seen in ASD has been associated with rickets, scurvy, vitamin A deficiency, vitamin K deficiency, and others.[66,67] Thus, analysis of dietary intake is an important part of nutritional assessment in this population. More commonly, however, the deficits seen in the general population are also found in the I/DD population, including insufficient intakes of calcium and vitamin D.[68] Inborn errors of metabolism that are etiologically responsible for I/DD may require restricted diets for treatment (e.g., phenylketonuria) and require ongoing counseling by an RD knowledgeable in managing patients with these disorders. (See Chapter 71)

Constipation

Constipation is a common co-occurring condition in multiple disorders associated with I/DD. In addition to motoric reasons for poor gastrointestinal motility (e.g., hypotonia, spasticity), other factors may include sedentary behavior, diets low in fiber, inadequate fluid intake, and the potential for altered gut microbiome.[69] Addressing these factors through dietary and behavioral changes is the focus of prevention and ongoing management. The use of bulk agents or laxatives for the initial treatment of constipation is common. Discomfort with passing a hard or large stool may cause pain, leading to continued stool withholding behavior. Interprofessional care might include specialists (e.g., gastroenterologists, psychologists) who can consult on or develop supportive behavioral approaches to ameliorate these difficulties.

Constipation has been reported as the primary gastrointestinal comorbidity with ASD and may be a greater concern for individuals with greater severity of autistic symptoms and less language.[69] Constipation is also associated with behavioral challenges and sleep problems in this population.[70] Gastrointestinal symptoms, including constipation, have been attributed to bacterial dysbiosis, altered stress response, intestinal inflammatory response, and impaired disaccharidase activity in autistic individuals.[69] Further research is needed to determine which individuals have these and other factors as the cause of their GI symptoms.

Motor and Positioning Interventions

As discussed above, neurologic motor impairment in individuals with hypotonia or hypertonia may impact their ability to eat safely. Aspiration of thin liquids may require the addition of a thickener. Safety and the need for water to manage constipation, for example, are among the parameters monitored. In addition to the neuromotor concerns about aspiration, individuals with I/DD are at high risk for poor oral hygiene, dental caries, periodontal disease, and/or tooth loss. They may also overpack their mouths with food and/or may exhibit impulsivity that precludes careful chewing.

Eating vs Feeding Behavioral Disorders

Eating disorders in people with I/DD typically refer to behavioral refusal of foods, as in Avoidant Restrictive Food Intake Disorder (ARFID).[71] There may be an increased risk for food intake disorders like anorexia nervosa (see Chapter 95), but typically, the food selectivity and overeating seen in people with I/DD are not associated with altered body image perception.[72]

Sleep Disorders

Sleep problems are both a cause of and consequence of obesity; inadequate sleep increases insulin resistance and is associated with sedentary lifestyle and late night snacking. (See Chapter 60). Sleep onset and sleep maintenance problems are common in ASD (>70%)[73]; evaluation for sleep apnea is recommended for all individuals with Down syndrome.[74] Medications, such as stimulants used for attention deficit/hyperactivity disorder symptoms, may result in delayed sleep onset.

Complementary and Integrative Practices

Evidence-based dietary counseling by an RD is standard clinical practice. However, the use of unproven therapies by families and caregivers to manage or ameliorate developmental and behavioral symptoms of developmental disorders is common. The use of therapies without an evidence base may derive from the hope for a cure or symptom improvement espoused by the providers of these therapies that some individuals or caregivers are seeking. Additionally, the use of these therapies may be based on the belief that "natural" interventions like diet and dietary supplement use have fewer side effects than the therapies/medications recommended by allopathic practitioners, the self-empowerment related to individual choice, and belief in the mechanisms for these therapies that are promoted as fact by the proponents of as yet unproven practices. The categories of complementary and integrative therapies include mind-body therapies, energy therapies, and nutritional/"biomedical" interventions. Complementary and integrative interventions include restrictive diets and/or dietary supplements. Nowhere in clinical practice are dietary interventions that have not yet been proven more controversial than for symptoms of ASD, where more than half of families report using dietary supplements.[75]

Popular dietary supplements change frequently in an attempt to mirror emerging basic and translational research that has not yet been adequately investigated. Some dietary supplements, like melatonin, have been studied with evidence to recommend its use for sleep onset.[73] Others, like omega 3 fatty acids[76] and probiotics,[77] have not been studied sufficiently to recommend their use. Similarly, there is insufficient evidence to support the use of gluten- or casein-restricted diets for autistic children to promote language or improve behavior.[76] Of note, if restrictive diets are elected, alternative sources of necessary nutrients need to be identified. The safety of supplements also needs to be addressed in dietary consultation. RDs should consult with caregivers regarding the safety and potential medication-supplement interactions that could be harmful. The National Center for Complementary and Integrative Health (NCCIH) maintains web-based information on novel interventions,[78] specific herbs, and dietary supplements.[79] The Natural Medicines Database also provides comprehensive, evidence-based information on foods, herbs, supplements, complementary and alternative treatment topics, safety, effectiveness, and interactions.[80]

Complementary and integrative interventions are also used for symptoms of other developmental disorders such as Down syndrome, with almost half of families reporting supplement use[81] to attempt to improve developmental outcomes, alter oxidative stress, or decrease inflammation, among other symptoms. Although many different products are used, often in combination,[81] there is insufficient evidence to date to support the use of dietary supplements for specific symptoms of Down syndrome.

It is the responsibility of health professionals to engage in shared decision making with caregivers and people with I/DD. A thorough dietary history, including supplements and over-the-counter proprietary products, will permit the dietitian to assess dietary excess and deficits that families who elect unproven therapies need to be aware of and address. **Table 94.2** summarizes the key assessment and management of nutritional concerns in individuals with I/DD.

Specific Conditions

Evidence-based nutritional standards of care for disorders that include I/DD are important goals for the field. This section describes factors the clinician must consider in providing care for this population. The Academy of Nutrition and Dietetics provides guidelines for RDs working with individuals with I/DD.[82] The following section highlights the nutrition-related concerns for the five aforementioned I/DD conditions.

Autism Spectrum Disorder

Autism spectrum disorder (ASD) is a heterogeneous developmental condition characterized by difficulties with communication and social interaction, restricted interests, and repetitive behaviors.[71] The symptoms associated with ASD impact an individual's ability to function in school, work, and other life domains. ASD is regarded as a "spectrum" disorder because of the wide variation in the manifestation and severity of symptoms. Not all autistic individuals have an intellectual disability; approximately

Table 94.2 Summary of Assessment and Management of Nutritional Concerns in Individuals with I/DD

Parameter	Assessment	Management
Growth	■ Use of WHO and CDC growth charts and in specific conditions consider additional use of specialized growth charts. ■ In individuals with scoliosis, contractures, and/or an inability to stand, consider other methods to assess height (e.g., segmental length, arm span, knee height). ■ Use of anthropometric measures (e.g., skinfold thickness, arm circumference) can be used in addition to height and weight to assess fat and lean tissue.	Longitudinal monitoring of weight and length (or height)
Oral Motor Function	■ Neurologic assessment for chewing and swallowing ■ Promoting good oral health	■ Oral motor therapies ■ Positioning ■ Texture modifications ■ Regular oral health visits
Gastrointestinal function	■ Reflux ■ Constipation ■ Pain and motility issues	Dietary modification and medical management
Bone Health	Bone density and biomarkers (biochemical tests)	Dietary modification, medical management, and weight bearing
Behavior	■ Interview parents and caretakers about mealtime behaviors, including food aversions, selective eating, anxiety, sensory sensitivities, eating disorders, disruptive behavior ■ Assess nutrient adequacy ■ Diet history and 24-hour dietary intake (food recall or food diary)	■ Dietary intervention to address nutrient inadequacy ■ OT/SLP intervention for management of sensory issues related to food. ■ Behavioral intervention for disruptive mealtime behavior anxiety around food and eating disorders
Weight status	■ Use of CDC BMI charts to assess weight status for children over 2 years in most contexts ■ Assessment of fat and lean tissue, as described for growth ■ Assess medication use and its relation to weight status	Dietary and behavioral interventions to promote healthy weight
Food allergies and intolerances	Check for food allergies, celiac disease, and other intolerances and whether dietary restrictions impact on food intake	Dietary intervention if dietary restrictions impact on nutrient adequacy and diet quality
Alternative and integrative therapies	Interview parent or caretaker to determine whether alternative and integrative therapies are being used	Discuss safety concerns regarding products with parent/caretaker

OT/SLP: occupational therapy/speech-language pathology.

two-thirds of the population has intellectual functioning in the average or above range.

Feeding Problems

Feeding problems are five times more common among autistic youth[37] than in neurotypical youth, ranging from 30% to 84%.[83] The most common feeding problem is food selectivity, described earlier in this chapter. Food selectivity may place individuals at risk for inadequate nutrient intake.[36,66,84-89] Although picky eating is common in toddlers with typical development, toddlers diagnosed with ASD may have very limited diets before 18 months of age. It may be reported before initial concerns about social communication and other restricted and repetitive behaviors that characterize ASD are raised.[90] Food selectivity is also associated with mealtime behavior problems.[91] Autistic children may refuse foods based on the texture, smell, and the appearance of the food.[83,92] These food aversions may be based on true differences in sensory perception relative to taste or texture,[66,93] perseveration or repetitive behaviors, anxiety, fear of change, fear of new foods, or a response to discomfort (e.g., pain with reflux). Aversions to sensory stimuli, demands, or changes in routine may lead to disruptive mealtime behaviors, which may impact food intake and family mealtimes.

Nutrient inadequacy is directly related to the severity of food selectivity. Case studies of children with severe food selectivity have reported several nutritional deficiencies[94-96] including vitamin C deficiency resulting in scurvy.[97] A 2021 meta-analysis suggested that sensory sensitivity is highly correlated with food selectivity,[98] which tends to persist from early childhood through adolescence.[99] Given the high prevalence of feeding problems among autistic youth and the risk for inadequate nutrient intake, it is recommended that these children be screened for food selectivity.[32,67]

Obesity

In the context of the prevalence of feeding problems among autistic youth, the prevalence of obesity has consistently been shown to be significantly higher in these youth than their peers without ASD.[19,61,100] Although the reason for the higher prevalence of obesity in autistic youth is unclear, it has been hypothesized that preference for high energy foods and low intake of vegetables, in particular, may be contributing factors.[101] Longitudinal studies to confirm these associations are needed.

Autistic youth are also less likely to meet recommended physical activity levels than their typically developing peers. Parents report numerous barriers to physical activity participation in this population, including a lack of opportunities, staff, and coaches familiar with working with autistic youth.[102] Screen time also appears to be higher in this population.[12] In addition to the weight loss interventions mentioned in the section on Obesity above, the Healthy Weight Research Network for Autism and Developmental Disabilities (https://HWRN.org) published expert recommendations for primary care practitioners on the management of overweight and obesity in autistic children.[51]

Bone Health

Individuals with ASD have been reported to have increased fractures in adulthood[103] and, like other children and youth in the United States, have insufficient vitamin D and calcium intake.[68] Vitamin D and calcium intake may be further diminished in this population by commonly used but unproven dietary interventions that eliminate casein.[66]

Down Syndrome

Down syndrome, also referred to as Trisomy 21, is a genetic disorder caused by an extra copy of chromosome 21. If the trisomy is present in some but not all cells, it results in mosaic trisomy 21, which has a more variable phenotype. Individuals with Down syndrome usually have an intellectual disability in the mild-to-moderate range. They have several physical attributes, including short stature, a small head and neck, macroglossia, epicanthal eyelid folds, small hands and feet, and poor muscle tone. Down syndrome is often diagnosed prenatally through genetic testing or is identified at birth.[74]

Feeding Problems

Feeding problems are common in infancy and associated with abnormalities of oropharyngeal function[104-106] and hypotonia,[74] which increase the risk for aspiration.[105] Oral sensory sensitivity is common in children with Down syndrome and may lead to food selectivity. About half of babies with Down syndrome have congenital heart defects, and 1% to 5% are born with duodenal atresia. These medical problems require dietary consultation.[74] Screening for feeding problems early is essential to ensure appropriate interventions. An interprofessional team is important to help families navigate the complex problems associated with feeding disorders.

Growth

Individuals with Down syndrome are shorter than their typically developing peers.[29,107,108] Body composition studies in youth with Down syndrome have shown less lean tissue and a greater percentage of body fat in children[109] and adolescents than in those without Down syndrome.[110] Down syndrome-specific growth charts have been developed in the United States[79] and the United Kingdom.[111] In the United States, current clinical guidelines[74] advise using Down syndrome-specific growth charts to monitor height, weight, and head circumference for children up to 10 years of age. After the age of 10 years, the median (50th percentile) on the Down syndrome-specific Body Mass Index (BMI) growth curves corresponds to the 85th percentile BMI on the CDC charts.[28] Therefore, for youth with Down syndrome over the age of 10 years, the CDC BMI charts will have better specificity, as the Down syndrome-specific charts will miss many children with obesity.

Energy Expenditure

The few studies that have measured resting metabolic rate (RMR) in children with Down syndrome have been small but suggest that RMR is reduced.[109,112-114] One study[109] followed 28 prepubertal children with Down syndrome over 3 years and did not find that the lower RMR predicted gain in fat mass over time. In adults with Down syndrome, findings of lower RMR have been equivocal.[108,115,116] It has been suggested that the lower RMR observed in children with Down syndrome may reflect their hypotonicity.[112]

Obesity

Obesity is common in children[107,117-119] and adults[120-122] with Down syndrome, with onset early in childhood.[107] Alterations in body composition and lower energy needs in this population increase the risk for obesity. Children with Down syndrome and obesity are at increased risk for obstructive sleep apnea.[107,123] Promoting healthy eating habits and engagement in physical activity is important to prevent the development of obesity in young children with Down syndrome. As mentioned earlier, a few weight loss programs have been developed and tested for this population with promising results.[52,53,55,56,124]

Celiac Disease

Children with Down syndrome are at an increased risk for celiac disease and other autoimmune disorders such as Type 1 diabetes.[125] A 2018 meta-analysis of individuals with Down syndrome that included 31 studies, mainly from Europe and the United States with over 4,000 individuals, found the prevalence of celiac disease to be 5.8%.[126] The 2022 clinical guidance from the American Academy of Pediatrics suggests clinical monitoring for a broad range of symptoms associated with celiac disease and screening when clinically indicated.[74]

Prader-Willi Syndrome

Prader-Willi syndrome (PWS) is a rare, complex neurodevelopmental disorder caused by hypothalamic impairment that impacts many physiologic functions,[127] including growth and appetite regulation. PWS is associated with I/DD, hyperphagia, hypogonadism, hypothyroidism, sleep difficulties, behavioral problems, and hormonal changes.[128] The diagnosis of PWS is made through genetic testing and is characterized by deletion or no expression of the paternal genes in the region 11-Q13 on chromosome 15.[129,130] PWS is frequently considered in the differential diagnosis of children and adults with I/DD and obesity.[131]

Growth

Growth hormone deficiency is common in individuals with PWS.[132] Individuals with PWS have short stature[131] and low fat-free mass.[133,134] However, growth hormone treatment is now common in individuals with PWS and is associated with reduced BMI and increased fat-free mass.[132] Appetite and weight gain occur in five distinct phases, some with subphases.[129] Phase 0 occurs *in utero* and is associated with decreased fetal movement and reductions in weight and length. Phase 1 occurs in infancy and has two sub-phases: in phase 1a, the infant has difficulty feeding, which can lead to failure to thrive; and sub-phase 1b, where feeding occurs without difficulties and the infant begins to grow at a normal rate. In Phase 2a in early childhood, from 2.1 to 4.5 years, weight increases without an increase in appetite; in Phase 2b, between ages 4.5 to 8 years, both weight gain and appetite increase. Phase 3, after age 8 years, is characterized by decreased satiety and hyperphagia. Lastly, in Phase 4, in adulthood, individuals who were previously hyperphagic and food seeking can feel full. Although not all individuals will exhibit all phases, knowledge of these phases allows practitioners to provide anticipatory guidance to families at an early age to prevent the development of obesity.

Satiety

Appetite dysregulation in PWS is thought to result from a complex interplay of neuronal regulation of satiety in the hypothalamic and limbic systems, hormonal signals that regulate energy use and adiposity, and peripheral regulation of satiety by hormones and gut peptides. Clinically, children and adults with PWS present with persistent and insatiable appetites with hyperphagia. They engage in food seeking and food stealing behaviors and may eat nonfood items.[132]

Energy Expenditure and Energy Needs

Decreased metabolic rates observed in individuals with PWS have been attributed to decreases in FFM.[133-135] Total energy expenditure (which includes both the energy costs of resting metabolic rate, the thermic effect of food, and the energy spent on physical activity) also appears to be lower in these individuals, due to both decreases in resting metabolic rate and decreases in physical activity.[133,134,136,137] Other studies in individuals with PWS measuring physical activity have found lower levels than recommended.[138,139]

Dietary Management

Because of the low resting metabolic rate and physical activity levels observed in youth with PWS, a low-energy diet is recommended to prevent weight gain. The decreased satiety and very strong drive to eat in this population requires a structured environment around food to prevent overeating. Food stealing, sneaking, and hoarding are commonly observed in individuals with PWS. The need for close supervision required places a significant burden on caregivers.[113,116] Early guidance by an RD for parents is imperative to promote healthy eating patterns and limit opportunities of overeating. In addition, attention is needed to ensure adequate nutrient intake for children and adults receiving low-energy diets.

Spina Bifida

Spina bifida, also called myelomeningocele, is a neural tube defect resulting from the failure of the neural tube to close properly during the first month of gestation.[140] Spina bifida may range from mild to severe; the severity depends on what level of the spinal cord the defect occurred and the extent of damage to neural tissue. Individuals with Spina bifida due to lesions in the lumbar region may need orthotics to walk, and those with lesions in the thoracic region are likely to ambulate via wheelchair. Both bowel and bladder function are typically affected. Diagnosis may occur during pregnancy or after birth. A minority of individuals with Spina bifida have an intellectual disability, but learning disabilities are common.[140] Infants with Spina bifida may also present with a Chiari II malformation in the brain that may affect the ability to suck and swallow and may lead to dysphagia.[141]

Growth

Linear growth is decreased in children with Spina bifida,[142,143] and by adulthood, individuals with Spina bifida present with short stature.[144] Body composition studies report decreases in fat-free mass, compared with children without Spina bifida,[145-147] which becomes apparent after the age of 4 years.[146]

Energy Needs

Although evidence is sparse, the daily energy needs of youth with Spina bifida appear to be lower than those of their typically developing peers.[145,148] Some of these lower needs are attributable to their decreased resting metabolic rate due to decreases in FFM.[145-147] Low levels of physical activity[149] and impairments in mobility[145,148] also contribute to lower energy needs.

Obesity

Overweight and obesity are common problems in children[142,143,150,151] and adults with Spina bifida.[152] Fat distribution also appears to be altered, with greater deposition in the legs of children.[153] Assessment of obesity is complicated by these alterations in body composition; the use of BMI to assess obesity may underestimate body fat in these youth.[148,154] Nonetheless, the Spina Bifida Association recommends using BMI to screen for obesity and suggests using segmental lengths or arm span when height cannot be obtained for calculating BMI.[155]

Bone Health

Risk factors for reduced bone density in individuals with Spina bifida include being nonambulatory, inadequate calcium and vitamin D intakes, and lack of weight-bearing and physical activity. Assessment of factors that affect bone health is important to consider in managing these individuals.

Cerebral Palsy

Cerebral palsy (CP) is a group of disorders in which muscle tone and neuromotor control impair movement and posture. CP is also called "static encephalopathy" due to an insult to the developing brain that is not progressive. The diagnosis of CP is based on a clinical neurologic examination, which may be supported by neuroimaging.[156] Persons with CP may exhibit exaggerated reflexes, low tone or spasticity of the limbs and trunk, involuntary movements, limited range of motion, unsteady walking, or a combination of these. Difficulties with swallowing, as well as eye muscle imbalance, are often seen. The impact of CP varies greatly, from mild to severe. Some individuals with CP can ambulate relatively easily, while others may need assistance, cannot walk, and/or use a wheelchair. The degree of motor impairment is functionally graded using the Gross Motor Function Classification System.[156] Neurologic function and orthopedic contractures may change with growth and development. Some individuals with CP may also be blind, deaf, have ID, and/or epilepsy. Due to oral motor dysfunction, individuals with CP may have difficulties with speaking, saliva control, sucking, chewing, and/or swallowing.

Feeding Problems

Feeding problems in children with CP are common.[157-159] Oropharyngeal dysphagia is a common cause of feeding problems.[157,160] Feeding problems have been associated with poor nutritional status.[157] Other factors, such as motor impairment that limits self-feeding, poor trunk control, and an inability to communicate hunger and thirst, impact food intake. A significant percentage of children with CP have underweight and/or malnutrition, which is directly related to feeding ability.[161] Inadequate intake of micronutrients is common[162] and reflects oral motor problems and/or a decreased volume of food intake. When children with CP cannot meet their nutritional needs orally, enteral feeds may be needed. ESPGHAN provides guidelines on when and how to provide enteral nutrition.[35] Effective engagement of parents and caregivers is also a critical component of enteral nutrition.[35,163]

Growth

CP is associated with short stature, although the relative contributions of neurologic causes, mobility limitations, and undernutrition are not fully understood.[164,165] Where length assessment may be difficult to obtain because of contractures and scoliosis,[166] segmented measures can provide more accurate measures of length.[167] Specialty growth charts are not recommended to assess growth because they are based on small samples that included children with over- and undernutrition.[168]

Body Composition

Assessment of nutritional status in individuals with CP is complex, given the variability in the severity of motor impairment seen in this population. Body composition measures such as skinfolds or bioelectric impedance analysis are recommended to assess nutritional status,[35,166] as measures of weight alone cannot distinguish between muscle and fat mass[164,169] and validated tools to assess under-, or overnutrition or body composition are unavailable.[170] Equations to calculate body adiposity based on triceps skinfolds for children with CP are available[171] and are recommended by ESPGHAN.[35] However, concerns about the accuracy and precision of these estimates have been raised.[172] Because of decreases in FFM observed in youth with CP,[173] BMI may underestimate the true prevalence of obesity.[174] Although consensus on the best tools to monitor nutritional status is lacking, longitudinal measures of weight, height, and body composition can be used to monitor growth and changes in nutritional status in children with CP and to ensure that they are growing on their own growth curve.[175]

Energy Expenditure and Energy Needs

Although the evidence base for energy expenditure in children and adults with CP is limited, it appears that energy needs in this population are lower than in the general population.[145,173,176-179] Energy needs vary considerably in individuals with CP and are associated with the degree of ambulation and alterations in body composition. General guidelines for determining energy requirements must be adapted to take these factors into account. It should be noted that for children with CP fed via gastrostomy, overestimation of energy needs results in overfeeding.[180] Monitoring growth and weight changes are essential to ensure that individuals with CP receive adequate nutrition to prevent over- or undernutrition.[165]

Bone Health

Bone health is of particular concern in individuals with CP who, like many people in the general population, may have inadequate intakes of calcium and vitamin D. Bone health is also compromised by altered bone density due to diminished weight bearing with resultant increased risk for fractures.[161] Dual-energy X-ray absorptiometry, a scan used to determine bone density, is recommended for children with CP.[35]

Constipation

Oral motor problems may result in diets with lower fiber and fluid intake. These diets, in combination with decreased physical activity and intestinal motility, are often associated with constipation. Increasing intake of fiber and fluids, along with medical management, should be employed to address problems with constipation.

Conclusions

Individuals with I/DD may present to RDs, primary care practitioners, and medical specialists with problems related to over- or undernutrition and a broad range of medical and behavioral problems that have nutritional implications. Clinicians need to be familiar with the common causes of I/DD (where known), the biologic impact of the underlying diagnoses and co-occurring conditions, and the biobehavioral risk factors that need to be considered in the nutritional management of these individuals. Despite the heterogeneity of the population of individuals with I/DD, they often share a common set of deficits in communication, cognition, motor, sensory, and/or adaptive domains. Successful dietary intervention requires the clinician to understand how to communicate with people who may have communication challenges, some of whom may use alternative or augmentative communication modes. Additionally, it is paramount to understand the living or home environment of individuals with I/DD, how they spend their days, and the role of caregivers in intervention and health promotion efforts. An interprofessional team is often needed to address the multifaceted needs faced by people with I/DD. The life course approach requires attention to healthy eating and physical activity patterns that begin in childhood and advocacy for the needs of people with disabilities in community settings.

Acknowledgment

The writing of this chapter was supported in part by the Health Resources and Services Administration (HRSA) of the U.S. Department of Health and Human Services (HHS) under cooperative agreement UT4MC41898, Healthy Weight Research Network (HWRN). The information, content, and/or conclusions are those of the authors and should not be construed as the official position or policy of, nor should any endorsements be inferred by, HRSA, HHS, or the U.S. Government.

References

1. Durkin M, Rubenstein E. In: Abbeduto L, McIntyre LL, Tasse MJ, eds. Epidemiology of intellectual and developmental disabilities. APA handbook for intellectual and developmental disabilities: foundations. American Psychological Association, 2021:73–87.
2. American Association on Intellectual and Developmental Disabilities. Accessed October 6, 2022. Available at www.aaidd.org/intellectual-disability/definition
3. Centers for Disease Control and Prevention. Developmental disabilities: causes and risk factors. Accessed October 6, 2022. Available at www.cdc.gov/ncbddd/developmentaldisabilities/causes-and-risk-factors.html
4. Centers for Disease Control and Prevention. Developmental disabilities: selected conditions in developmental disabilities. Accessed October 13, 2022. Available at www.cdc.gov/ncbddd/developmentaldisabilities/conditions.html
5. Krahn GL, Hammond L, Turner A. Mental Retard Develop Dis Res Rev 2006;12:70–82.
6. Krahn GL, Fox MH. J Appl Res Intell Dis 2014;27:431–446.
7. Krahn GL, Walker DK, Correa-De-Araujo R. Am J Public Health 2015;105(suppl 2):S198–S206.
8. Corvey K, Menear KS, Preskitt J, et al. Mat Child Health J 2015;20:466–476.
9. Kim Y, Conners RT, Hart PD, et al. Dis Health J 2013;6:253–259.
10. Segal M, Eliasziw M, Phillips S, et al. Dis Health J 2016;9:392–398.
11. Bandini L, Gleason J, Curtin C, et al. Autism 2013;17:44–54.
12. Must A, Phillips SM, Curtin C, et al. Autism 2014;18:376–384.
13. Kuo DZ, Houtrow AJ, Norwood KW, et al. Pediatr 2016;138:e20163021.
14. Broder-Fingert S, Brazauskas K, Lindgren K, et al. Acad Ped 2014;14:408–414.
15. Dreyer Gillette ML, Borner KB, et al. J Dev Behav Ped 2015;36:489–496.
16. Levy SE, Pinto-Martin JA, Bradley CB, et al. J Ped 2019;205:202–209.
17. Bandini L, Curtin C, Hamad CD, et al. J Ped 2005;146:738–743.
18. Curtin C, Anderson SE, Must A, Bandini L. BMC Ped 2010;10:11.
19. Must A, Eliasziw M, Phillips SM, et al. Child Obes 2017;13:25–35.
20. Phillips KL, Schieve LA, Visser S, et al. Mater Child Health J 2014;18:1964–1975.
21. Kahathuduwa CN, West BD, Blume J, et al. Obes Rev 2019;20:1667–1679.
22. Li YJ, Xie XN, Lei X, et al. Obes Rev 2020;21:e13123.
23. Maïano C. Obes Rev 2011;12:189–197.
24. Shedlock K, Susi A, Gorman GH, et al. J Ped 2016;178:183–187.e181.
25. Croen LA, Zerbo O, Qian Y, et al. Autism 2015;19:814–823.
26. Morgart K, Harrison JN, Hoon Jr, AH, et al. Develop Med Child Neurol 2021;63:1149–1154.
27. World Health Organization. Accessed November 17, 2022. Available at www.who.int/tools/child-growth-standards/standards
28. Centers for Disease Control and Prevention. Reviewed June 16, 2017. Accessed November 17, 2022. Available at www.cdc.gov/growthcharts/clinical_charts.htm
29. Zemel BS, Pipan M, Stallings VA, et al. Pediatr 2015;136:e1204–e1211.
30. Butler MG, Lee J, Manzardo AM, et al. Pediatr 2015;135:e126–e135.
31. Hersh JH, Saul RA, Committee on Genetics. Pediatr 2011;127:994–1006.
32. Hyman SL, Levy SE, Myers SM, et al. Pediatr 2020;145:e20193447.
33. Barlow SE. Pediatrics 2007;120:S164–S92.
34. U.S. Preventive Services Task Force. 2018. Accessed November 17, 2022. Available at www.uspreventiveservicestaskforce.org/uspstf/recommendation/obesity-in-adults-interventions
35. Romano C, Van Wynckel M, Hulst J, et al. J Ped Gastroenterol Nutr 2017;65:242–264.
36. Bandini L, Anderson S, Curtin C, et al. J Ped 2010;157:259–264.
37. Sharp WG, Berry RC, McCracken C, et al. J Autism Dev Dis 2013;43:2159–73.
38. Culley WJ, Goyal K, Jolly DH, et al. J Ped 1965;66:772–775.
39. Grogan CB, Ekvall SM. J Am Coll Nutr 1999;18:316–323.
40. Culley WJ, Middleton TO. J Ped 1969;75:380–384.
41. Hoffman CJ, Aultman D, Pipes P. J Am Diet Assoc 1992;92:823–830, 833.
42. Academy of Nutrition and Dietetics Chicago. Pocket Guide to Children with Special Health Care and Nutritional Needs, 2nd ed. Chicago, IL: Academy of Nutrition and Dietetics, 2021.
43. Institute of Medicine Dietary reference intakes for energy, carbohydrate, fiber, fat, fatty acids, cholesterol, protein, and amino acids. Institute of Medicine, Food and Nutrition Board. Washington, DC: The National Academies Press, 2005.
44. Ervin DA, Hennen B, Merrick J, et al. Front Public Health 2014;2:1–8.
45. UDL. 2024. Accessed October 6, 2022. Available at http://udlguidelines.cast.org
46. Ok MW, Rao M, Bryant BR, et al. Exceptionality 2016;25:116–138.
47. Rao K, Smith SJ, Lowrey KA. Intell Dev Dis 2017;55:37–47.
48. Botha M, Hanlon J, Williams GL. J Autism Dev Dis 2021;1–9.
49. Taboas A, Doepke K, Zimmerman C. Autism 2022;13623613221130845.
50. Gernsbacher MA. J Child Psychol Psychiatr 2017;58:859–861.
51. Curtin C, Hyman SL, Boas DD, et al. Pediatr 2020;145:S126–S39.
52. Bandini LG, Eliasziw M, Dittrich GA, et al. Ped Obes 2021;16:e12816.
53. Curtin C, Bandini LG, Must A, et al. J Ped 2013;163:1402–1408:e1401.
54. Killian HJ, Pallotto IK, Sweeney BR, et al. J Autism Dev Dis 2022;52:791–799.
55. Pona AA, Dreyer Gillette ML, Odar Stough C, et al. 2022 Child Obes 2022;13:455–461.
56. Ptomey LT, Saunders RR, Saunders M, et al. J App Res Intell Dis 2018;31:82–96.
57. Ptomey LT, Washburn RA, Lee J, et al. Contemp Clin Trial 2019;84:105817.
58. Maïano C, Normand CL, Aimé A, et al. Res Dev Dis 2014;35:1914–1926.
59. Spanos D, Melville CA, Hankey CR. Nutr J 2013;12:132.

60. Curtin C, Bowling AB, Boutelle KN, et al. Intl Rev Res Dev Dis 2021;61:223–261.
61. Curtin C, Jojic M, Bandini LG. Harvard Rev Psychiatr 2014;22: 93–103.
62. Goltz J, Ivanov I, Rice TR. Intl J Dev Dis 2021;67:159–167.
63. Fitzgerald I, O'Connell J, Keating D, et al. Evid Bas Mental Health 2022;25:15–22.
64. Handen BL, Anagnostou E, Aman MG, et al. J Am Acad Child Adolesc Psychiatr 2017;56:849–856:e846.
65. Speyer R, Cordier R, Kim JH, et al. Dev Med Child Neurol 2019;61:1249–1258.
66. Hyman SL, Stewart PA, Schmidt B, et al. Pediatr 2012;130 (suppl 2):145.
67. Yule S, Wanik J, Holm EM, et al. J Acad Nutr Diet 2021;121: 467–492.
68. Stewart PA, Hyman SL, Schmidt BL, et al. J Acad Nutr Diet 2015;115:1237–48.
69. Madra M, Ringel R, Margolis KG. Child Adolesc Psychiatr Clin North Am 2020;29:501–513.
70. Mazurek MO, Vasa RA, Kalb LG, et al. J Abnorm Child Psychol 2013;41:165–176.
71. American Psychiatric Association. Diagnostic and Statistical Manual of Mental Disorders - 5th edition, Washington, DC: American Psychiatric Association, 2013.
72. Baraskewich J, von Ranson KM, McCrimmon A, et al. Autism 2021;25:1505–1519.
73. Williams Buckley A, Hirtz D, Oskoui M, et al. Neurol 2020;94: 392–404.
74. Bull MJ, Trotter T, Santoro SL, et al. Pediatr 2022;149.
75. Höfer J, Hoffmann F, Bachmann C. Autism 2017;21:387–402.
76. Sathe N, Andrews JC, McPheeters ML, et al. Medicina 2019; 55:129.
77. Ng QX, Loke W, Venkatanarayanan N, et al. Medicina 2019; 55:129.
78. National Center for Complementary and Integrative Health. 2021. Accessed September 22, 2022. Available at www.nccih.nih.gov/health/providers/digest/autism-spectrum-disorder-and-complementary-health-approaches
79. National Center for Complementary and Integrative Health. 2024. Accessed September 22, 2022. Available at www.nccih.nih.gov/health/atoz
80. NatMed. Accessed November 17, 2022. Available at https://naturalmedicines.therapeuticresearch.com/
81. Lewanda AF, Gallegos MF, Summar M. J Ped 2018;201:100–105: e130.
82. Conway C, Lemons S, Terrazas L. J Acad Nutr Diet 2020;120: 2061–2075:e2057.
83. Mayes SD, Zickgraf H. Res Autism Spectrum Dis 2019;64: 76–83.
84. Marí-Bauset S, Llopis-González A, Zazpe I, et al. Autism 2017;21: 310–322.
85. Sharp WG, Postorino V, McCracken CE, et al. J Acad Nutr Diet 2018;118:1943–1950.
86. Zimmer MH, Hart LC, Manning-Courtney P, et al. J Autism Dev Dis 2012;42:549–556.
87. Lockner DW, Crowe TK, Skipper BJ. J Am Diet Assoc 2008;108:1360–1363.
88. Herndon AC, DiGuiseppi C, Johnson SL, et al. J Autism Dev Dis 2009;39:212–222.
89. Graf-Myles J, Farmer C, Thurm A, et al. J Dev Behav Ped 2013;34:449–459.
90. Emond A, Emmett P, Steer C, et al. Pediatr 2010;126:337.
91. Curtin C, Hubbard K, Anderson S, et al. J Autism Dev Dis 2015;45:3308–3315.
92. Hubbard KL, Anderson SE, Curtin C, et al. J Acad Nutr Diet 2014;114:1981–1987.
93. Chistol LT, Bandini LG, Must A, et al. J Autism Dev Dis 2017;48:583–591.
94. Duignan E, Kenna P, Watson R, et al. Case Rep Ophthalmol 2015;6:24–29.
95. Pineles SL, Avery RA, Liu GT. Pediatr 2010;126:e967–70.
96. Zaenglein A, Martin A, Carlson L, et al. Ped Dermatol 2020;37: 698–700.
97. Ma N, Thompson C, Weston S. J Autism Dev Dis 2016;46: 1464–1470.
98. Page SD, Souders MC, Kral TV, et al. J Autism Dev Dis 2021; 1–20.
99. Bandini LG, Curtin C, Phillips S, et al. J Autism Dev Dis 2017;47:439–446.
100. Sammels O, Karjalainen L, Dahlgren J, Wentz E. Obes Facts 2022;15:305–320.
101. Evans EW, Must A, Anderson SE, et al. Res Autism Spect Dis 2012;6:399–405.
102. Must A, Phillips S, Curtin C, et al. J Phys Act Health 2015;12: 529–534.
103. Neumeyer AM, O'Rourke JA, Massa A, et al. J Autism Dev Dis 2015;45:881–887.
104. Ravel A, Mircher C, Rebillat AS, et al. Archives de Pédiatrie 2020;27:53–60.
105. Jackson A, Maybee J, Moran MK, et al. Dysphagia 2016;31: 663–671.
106. Stanley MA, Shepherd N, Duvall N, et al. Am J Med Genet Part A 2019;179:177–182.
107. Basil JS, Santoro SL, Martin LJ, et al. J Ped 2016;173:143–148.
108. Fernhall B, Figueroa A, Collier S, et al. Mental Retard 2005; 43:391–400.
109. Hill DL, Parks EP, Zemel BS, et al. Euro J Clin Nutr 2013;67: 1087–1091.
110. González-Agüero A, Ara I, Moreno LA, et al. Res Dev Dis 2011; 32:1685–1693.
111. Styles ME, Cole TJ, Dennis J, et al. Arch Dis Child 2002;87: 104–108.
112. Luke A, Roizen NJ, Sutton M, et al. J Ped 1994;125:829–838.
113. Chad K, Jobling A, Frail H. Am J Mental Retard 1990;95: 228–235.
114. Bauer J, Teufel U, Doege C, et al. J Ped 2003;143:264–266.
115. Allison DB, Gomez JE, Heshka S, et al. Intl J Obes Rel Metabol Dis 1995;19:858–861.
116. Schapiro MB, Rapoport SI. J Intell Dis Res 1989;33:211–219.
117. Hsieh K, Rimmer JH, Heller T. J Intell Dis Res 2014;58:851–863.
118. Hatch-Stein JA, Zemel BS, Prasad D, et al. Pediatr 2016;138.
119. Bertapelli F, Pitetti K, Agiovlasitis S, et al. Res Dev Dis 2016;57: 181–192.
120. Bell AJ, Bhate MS. J Intell Dis Res 1992;36:359–364.
121. Prasher VP. J Intell Dis Res 1995;39:437–441.
122. Rubin SS, Rimmer JH, Chicoine B, et al Mental Retard 1998;36: 175–181.
123. Dyken ME, Lin-Dyken DC, Poulton S, et al. Arch Ped Adolesc Med 2003;157:655–660.
124. Ptomey LT, Sullivan DK, Lee J, et al. J Acad Nutr Diet 2015;115: 112–118.
125. Bergholdt R, Eising S, Nerup J, Pociot F. Diabetologia 2006;49: 1179–1182.
126. Du Y, Shan L-F, Cao Z-Z, et al. Oncotarget 2018;9:5387.
127. Tauber M, Hoybye C. Lancet Diabet Endocrinol 2021;9: 235–46.
128. Fussell JR. Itasca, IL: American Academy of Pediatrics, 2018.
129. Miller JL, Lynn CH, Driscoll DC, et al. Am J Med Genet Part A 2011;155:1040–1049.
130. Mackay J, McCallum Z, Ambler GR, et al. J Paed Child Health 2019;55:1029–1037.
131. Butler M, Forster JL. Curr Ped Rev 2019;15:207–244.

132. Muscogiuri G, Barrea L, Faggiano F, et al. J Endocrinol Invest 2021;44:2057–2070.
133. Schoeller DA, Levitsky LL, Bandini LG, et al. Metab 1988;37: 115–120.
134. van Mil EG, Westerterp KR, Kester AD, et al. Intl J Obes 2000;24:429–434.
135. Goldstone AP, Brynes AE, Thomas EL, et al. Am J Clin Nutr 2002;75:468–475.
136. Butler MG, Theodoro MF, Bittel DC, et al. Am J Med Genet Part A 2007;143:449–459.
137. Davies PS, Joughin C. Am J Mental Retard 1993;98:349–353.
138. Castner DM, Tucker JM, Wilson KS, et al. Res Dev Dis 2014;35: 3081–3088.
139. Nordstrøm M, Hansen BH, Paus B, et al. Res Dev Dis 2013;34: 4395–4403.
140. Iskandar BJ, Finnell RH. N Engl J Med 2022;387:444–450.
141. Stevenson KL. Neurosurgical Focus 2004;16:E5.
142. Hayes-Allen MC, Tring FC. Br J Prevent Soc Med 1973;27:192.
143. Hayes-Allen MC. Dev Med Child Neurol 1972;14;59–64.
144. Rotenstein D, Adams M, Reigel DH. Eur J Ped 1995;154: 398–402.
145. Bandini LG, Schoeller DA, Fukagawa NK, et al. Ped Res 1991;29:70–77.
146. Shepherd K, Roberts D, Golding S, et al. Am J Clin Nutr 1991;53:1–6.
147. Amanda Liusuwan R, Widman LM, Ted Abresch R, et al. J Spinal Cord Med 2007;30:S105–11.
148. Polfuss M, Sawin KJ, Papanek PE, et al. Dis Health J 2018;11: 442–446.
149. van den Berg-Emons HJ, Bussmann JB, Brobbel AS., et al. J Ped 2001;139:880–886.
150. Fiore P, Picco P, Castagnola E, et al. Eur J Ped Surg 1998;8: 34–36.
151. Mita K, Akataki K, Itoh K, et al. Dev Med Child Neurol 1993; 35:305–311.
152. Dosa NP, Foley JT, Eckrich M, et al. Dis Rehab 2009;31: 914–920.
153. Mueske NM, Ryan DD, Van Speybroeck AL, et al. Dev Med Child Neurol 2015;57:273–278.
154. Dopier Nelson M, Widman LM, Ted Abresch R, et al. J Spinal Cord Med 2007;30:S127–S139.
155. McPherson AC, Chen L, O'Neil J, Vanderbom KA. Accessed November, 17, 2022. Available at www.spinabifidaassociation.org/wp-content/uploads/Nutrition-Metabolic-Syndrome-and-Obesity.pdf
156. Sheu J, Cohen D, Sousa T, et al. Ped Rev 2022;43:572–581.
157. Fung EB, Samson-Fang L, Stallings VA, et al. J Am Diet Assoc 2002;102:361–373.
158. Reilly S, Skuse D, Poblete X. J Ped 1996;129:877–882.
159. Sullivan PB, Lambert B, Rose M, et al. Dev Med Child Neurol 2000;42:674–680.
160. Benfer KA, Weir KA, Bell KL, et al. Pediatr 2013;131: e1553–62.
161. Jesus AO, Stevenson RD. Phys Med Rehab 2020;31:25–37.
162. Marchand V, Motil KJ. J Ped Gastroenterol Nutr 2006;43: 123–135.
163. Mahant S, Cohen E, Nelson KE, et al. Paed Child Health 2018;23:209–213.
164. Samson-Fang L, Bell KL. Eur J Clin Nutr 2013;67(suppl 2): S5–8.
165. Kuperminc MN, Gottrand F, Samson-Fang L, et al. Eur J Clin Nutr 2013;67(suppl 2):S21–3.
166. Romano C, Dipasquale V, Gottrand F, et al. Dev Med Child Neur 2018;60:892–896.
167. Stevenson RD. Arch Ped Adolesc Med 1995;149:658–662.
168. Samson-Fang L, Fung E, Stallings VA, et al. J Ped 2002;141: 637–643.
169. Must A, Anderson SE. Clin Obes Adults Child 2005;2:215–230.
170. Sørensen SJ, Brekke G, Kok K, et al. Dev Med Child Neur 2021;63:1374–1381.
171. Gurka MJ, Kuperminc MN, Busby MG, et al. Dev Med Child Neur 2010;52:e35–e41.
172. Snik DA, de Roos NM. Ann Phys Rehab Med 2021;64:101271.
173. Stallings VA, Zemel BS, Davies JC, et al. Am J Clin Nutr 1996;64:627–634.
174. Whitney DG, Miller F, Pohlig RT, et al. Int J Obes 2019;43: 82–90.
175. Kuperminc MN, Stevenson RD. Dev Dis Res Rev 2008;14: 137–146.
176. Van den Berg-Emons H. J, Saris W. H, De Barbanson DC, et al. J Ped 1995;127:578–584.
177. Azcue MP, Zello GA, Levy LD, et al. J Ped 1996;129:870–876.
178. Bell KL, Davies P. S. Am J Clin Nutr 2010;92:313–319.
179. Walker JL, Bell KL, Boyd RN, et al. Am J Clin Nutr 2012;96: 1309–1315.
180. Sullivan PB, Alder N, Bachlet AM, et al. Dev Med Child Neurol 2006;48:877–882.

CHAPTER 95

Eating Disorders

Sarah C. Collica, Kimberly R. Smith, and Angela S. Guarda

CHAPTER OUTLINE

Abbreviations

ABA, activity-based anorexia; **AN**, anorexia nervosa; **ARFID**, avoidant/restrictive food intake disorder; **BED**, binge-eating disorder; **BMI**, body mass index; **BN**, bulimia nervosa; **CBT**, cognitive behavioral therapy; **CCK**, cholecystokinin; **FBT**, family behavioral therapy; **GLP-1**, glucagon-like peptide 1; **GWAS**, genome-wide association studies; **IPT**, interpersonal therapy; **NGT**, nasogastric tube; **OSFED**, other specified feeding or eating disorders; **PYY**, peptide tyrosine tyrosine; **RDN**, registered dietitian nutritionist; **TPN**, total parenteral nutrition

Introduction

Eating disorders are driven behavioral disorders resulting in significant functional impairment and, in extreme cases, death. The fifth edition of the *Diagnostic and Statistical Manual of Mental Disorders*[1] distinguishes several major categories of eating disorders, including anorexia nervosa (AN), bulimia nervosa (BN), binge-eating disorders (BED), avoidant/restrictive food intake disorder (ARFID), and other specified feeding or eating disorders (OSFED). AN and BN are best thought of as "dieting disorders"[2] as both these conditions are characterized by an overvalued fear of fatness that drives a set of disturbed eating and weight control behaviors. Besides restricting food intake, eating and weight control behaviors can include binge eating; excessive exercise; self-induced vomiting; or misuse of laxatives, diuretics, and diet pills. Engagement in these behaviors, coupled with the physiologic consequences of starvation or the binge-purge-restrict cycle, sustains and heightens food preoccupation and body image disturbance. BED is characterized by regular binge eating but lacks the compensatory behaviors to prevent weight gain found in BN and is frequently associated with obesity. ARFID is not driven by a disturbance in how one's body weight or shape is experienced. It is, however, a behavioral disturbance in food intake that results in inadequate nutritional intake and can have serious physical and mental health impacts. OSFED is a heterogenous catch-all category that includes partial syndrome AN or BN cases. This chapter reviews the diagnosis, epidemiology, etiology, complications, and treatment of AN, BN, BED, ARFID, and OSFED.

Overview of Eating Disorders

Anorexia Nervosa

AN is a syndrome of self-starvation characterized by weight loss to an underweight level in the context of age, sex, developmental trajectory, and physical health. Weight loss is accompanied by an overvalued fear of fatness and often by amenorrhea or the absence of three or more consecutive menstrual cycles. AN is further subdivided into a restricting (AN-R) or a binge-eating/purging (AN-P) subtype. Individuals with AN-R restrict food intake and frequently exercise excessively or fidget in the service of weight loss but do not binge or engage in purging behaviors. By contrast, individuals with AN-P additionally regularly binge eat or purge by self-induced vomiting or abuse of laxatives, diuretics, and enemas. Persons with AN restrict their food repertoire and the amount of food consumed and typically favor low-energy density foods.

Bulimia Nervosa

BN is a dieting disorder characterized by restrictive eating and episodic (weekly or more) binge eating followed by compensatory behaviors to prevent weight gain. Binge eating is defined as the consumption of an amount of food larger than most people would eat in a similar period under similar circumstances. It is associated with a sense of loss of control over eating. Typical binge foods are high-fat, high-energy, "forbidden" foods, and amounts consumed are often 2,000 kilocalories or more per binge.[3] Between binges, individuals with bulimia tend to limit the quantity and type of food intake, consuming "safe," low-energy, low-fat foods. Other compensatory behaviors following a binge, besides restricting, and skipping meals or fasting, can include excessive exercise, purging, vomiting, or abuse of laxatives and diuretics. As in AN, dieting and preoccupation with thinness develop into a consuming passion that is difficult to interrupt and impairs psychologic and social function. The distinction between AN-P and BN is primarily one of weight. Individuals who binge and purge and are underweight (body mass index (BMI) of <18.5 in an adult female) have AN-P, whereas persons who are normal weight or overweight have BN.

Binge-Eating Disorder

BED is defined as regular binge eating (weekly or more often) associated with a subjective sense of loss of control over eating and associated feelings of shame and guilt but lacking the compensatory behaviors to prevent weight gain typical of BN. Patients with BED overeat more consistently throughout the day than those with BN and are more likely to be overweight or obese.[4]

Avoidant/Restrictive Food Intake Disorder

ARFID is an eating disorder characterized by selective or inadequate food intake with heterogeneous causes. Three preliminary subtypes have been described based on the primary motive for avoiding eating. These include food restriction (type, quantity, or both) due to low appetitive drive, fear of aversive consequences of eating (vomiting, choking, or abdominal pain), or fears related to certain sensory characteristics of food (textural food selectivity as often seen in autistic children). Most cases of ARFID, however, are mixed and do not fall neatly into only one of these subtypes. ARFID can lead to significant weight loss or nutritional deficiencies. In extreme cases, individuals with ARFID may become dependent on enteral feeding or oral supplements and may experience marked interference with psychosocial functioning.[5]

Other Specified Feeding or Eating Disorders

OSFED is a heterogeneous diagnostic category.[6] It includes partial-syndrome cases of AN, BN, and BED that do not meet full criteria for one of these diagnoses. Patients with "atypical AN" experience strong fear of fatness and have lost a significant percentage of body weight yet are still normal or above normal weight. Medical complications in atypical AN are like those seen in AN and are often more closely related to speed and percent of weight lost than to absolute BMI at presentation. Purging disorder describes individuals trying to influence weight or shape via vomiting or the use of laxatives in the absence of binge eating, and who are not underweight and, therefore, do not meet full criteria for AN-purging type or BN.

Epidemiology

Eating disorders are relatively common, with a global estimated lifetime prevalence of more than 8% in women and 2% in men.[7] Each year, an estimated 3.3 million healthy person-years are lost to disability because of the protracted nature of eating disorder pathology.[8] In the general adult population of the United States, the 12-month and lifetime prevalence of AN are estimated to be 0.05% and 0.8%, respectively.[9] The highest incidence of anorexia nervosa is around 15 years of age,[10] and it more commonly affects women.[9]

The 12-month prevalence of BN is 0.14%, and the lifetime prevalence is 0.28%.[9] The median age of onset of bulimia nervosa is 18 to 20 years.[9] BN more commonly affects women[9] and may be preceded by a brief period of AN in up to 40% of patients. In these individuals, however, the challenges of maintaining a low body weight through restrictive behaviors lead to binge eating, weight regain and adoption of purging, and other compensatory behaviors.[11]

The 12-month and lifetime prevalence rates of BED in the United States are 0.44% and 0.85%.[9] The prevalence of BED is highest in individuals with obesity. It is tenfold higher in individuals with class III obesity (BMI >40) than in those with class I obesity (BMI=30–34.9).[12] BED is two to three times more likely to affect women than men.[9,12]

ARFID has been relatively recently defined and has a heterogenous presentation; thus, knowledge regarding its prevalence, comorbidity, and course is limited.[13] A recent scoping review shows wide variation in prevalence estimates among clinical ED populations, ranging from 1.5% to 64%.[14] Unlike AN, BN, and BED, ARFID is more prevalent in males, and the average age of onset is earlier.[14]

OSFED is a phenomenologically heterogenous group, and epidemiologic information is limited. A recent study estimated the number of OSFED global cases as nearly 25 million in 2019,[15] making it the most commonly diagnosed eating disorder.[7] It is important to realize that eating disorders can occur in people of all genders, ages, races, or ethnicities, body weights, and socioeconomic groups.

Etiology: Risk and Maintaining Factors

Knowledge of the pathogenesis of eating disorders remains limited; however, their etiology is multifactorial and includes the interaction of both genetic and environmental risk factors. Biologic factors and learning are increasingly implicated in the escalating and driven nature of these behavioral disorders. One way to conceptualize vulnerability and factors leading to an eating disorder is in terms of predisposing, precipitating, and perpetuating factors. Predisposing factors include genetic vulnerability. Precipitating factors include developmental influences (such as puberty-induced hormonal changes) and environmental influences (e.g., trauma, peer, or sociocultural influences). Perpetuating factors often include learned behavior and physiologic consequences of engaging in eating and weight control behaviors that feed forward to sustain the disorder as it becomes chronic.

Predisposing Factors

Family, twin, and molecular studies support genetic vulnerability to eating disorders. Early family studies demonstrated cross-transmission within families of AN, BN, and other eating disorders, suggesting shared familial liability. The strongest heritability was observed for AN, with a prevalence of eating disorders in first-degree relatives of probands with AN approximately 7 to 12 times that of controls.[16] Subsequent twin studies found significantly higher concordance rates in monozygotic than in dizygotic twins, with heritability estimates of 0.28–0.74 for AN, 0.55–0.62 for BN, and 0.39–0.45 for BED.[17] As ARFID is a relatively new diagnosis, little is known about its familial or genetic risk. Early genetic research examining candidate genes presumed to be potential biologic markers for AN and BN focused on polymorphisms in neurotransmitter genes (serotonin and dopamine related genes) and on genes involved in appetite and satiety signaling and weight regulation (leptin- and estrogen-related genes). Results, however, have proved inconsistent.[17] More recent genome-wide association studies (GWAS) for AN have identified several significant loci and genetic correlations with physical activity, other psychiatric disorders, and metabolic, personality, and anthropomorphic traits, suggesting both psychiatric and metabolic contributions to this disorder.[18-20] Heritable personality traits frequently associated with eating disorders include elevated harm avoidance, neuroticism, and cognitive rigidity.[21] Perfectionism, conscientiousness, persistence, and obsessive qualities are often discriminating features of restricting type AN, whereas elevated impulsivity, novelty seeking, negative emotionality, and stress reactivity are more commonly associated with conditions marked by binge purge behaviors.[22] Comorbid anxiety, mood, and substance use disorders are elevated in individuals who develop an eating disorder and often precede its onset, complicating treatment, and prognosis.[23] Substance use has been associated with higher mortality risk in persons with eating disorders.[24]

Precipitating Factors

Age of onset for AN and BN rarely precedes the start of pubertal changes, and most cases occur in females, suggesting a role for ovarian hormones and sexual development in the activation of disordered eating. Dieting behavior, body dissatisfaction, and internalization of the thin ideal have all been associated with an increased risk of an eating disorder.[25] In older adolescents, the transition to college, eating outside of the home, performance and achievement demands, role and identity changes, and peer influences are additional factors that can contribute to disordered eating.[26] Stress, physical illness, or trauma may affect appetite and initiate the process of weight loss or the onset of purging behavior. Engaging in new forms of exercise, for instance, joining the track team often precedes dieting in AN. Stress and changes in work and school routines during the recent COVID-19 pandemic have been associated with a significant increase in eating disorders, especially among adolescents.[27]

Perpetuating Factors

Over time, as eating and weight control behaviors escalate, secondary physiologic, psychological, and social consequences come into play, which contribute to sustaining an eating disorder. Physiologic consequences of starvation, for example, contribute to gastrointestinal slowing which, in turn, promotes early satiety and preoccupation with gastrointestinal function (constipation, abdominal pain, bloating). Similarly, frequent purging by vomiting can worsen gastroesophageal reflux, leading to regurgitation and rumination after meals, heartburn, and increased ease of vomiting. Starvation or binge eating induces neuroendocrine and neurotransmitter changes that may contribute to the driven nature of an eating disorder. Neurocognitive and affective consequences of starvation include difficulty with attention, indecisiveness, concentration, and labile mood. These may contribute to the driven nature of the disorder in the face of increasingly negative consequences on health and function. Over time, prolonged illness may exhaust family members and social supports and interfere with school or work performance, isolating the affected individual and increasing feelings of loneliness, shame, or low self-esteem.[28] Finally, eating disorders are increasingly recognized as learning disorders, with both stimulus-response and action-outcome learning increasingly implicated in maintaining and reinforcing thoughts and behaviors associated with the disorder.[29]

Appetitive Signaling and Neurobiology of Eating Disorders

Animal models allow scientists to use experimental methods to examine physiologic factors contributing to the onset and maintenance of human behavioral conditions. The most widely studied animal model of AN is the activity-based anorexia (ABA) rodent model, which has provided critical insight into the neurobiology of this disorder. Although animal models cannot mimic cognitive aspects of AN, such as fear of fatness or body dissatisfaction, the ABA model demonstrates how access to exercise in the setting of food restriction can become rewarding and result in escalating voluntary food restriction and resultant starvation. In the ABA paradigm, when food access is limited to 1.5 hours a day and animals are provided unlimited access to a running wheel, they paradoxically run more and eat less than control animals. Individual vulnerability to ABA varies; however, adolescent female rodents that are the most active at baseline are more susceptible than adult, male, or sedentary animals, consistent with the observation that the onset of AN is more common in young females who exercise, with exercise often preceding the onset of dieting behavior. Furthermore, ABA animals demonstrate disturbances in both the homeostatic and hedonic controls of food intake, some of which may play a sustaining role in the driven behavioral phenomenology that is common to both ABA and AN.

Homeostatic mechanisms controlling food intake include gut peptides and hormones that influence hunger (orexigenic) or satiety (anorexigenic) signaling. Ghrelin is the sole identified peripheral orexigenic hormone to date, while anorexigenic compounds include cholecystokinin (CCK), glucagon-like peptide 1 (GLP-1), and peptide tyrosine tyrosine (PYY). In ABA animals, serum ghrelin is elevated relative to control animals, and ghrelin concentration is positively correlated with food anticipatory activity or hyperactivity exhibited before the onset of scheduled food availability.[30] These data from the ABA model suggest that increased ghrelin, a compensatory response to starvation that would normally induce increased food intake may, instead, be contributing primarily to the "hyperactivity" phenomenology in AN.

As in ABA, human studies indicate that physiologic consequences of eating disorder behaviors include dysregulated hunger and satiety signaling, which may impact hypothalamic feeding regulatory pathways to perpetuate disordered eating and weight control behaviors. In AN, for example, ghrelin is elevated in both the fasted and fed state, whereas anorexigenic peptides appear largely elevated relative to that observed in healthy controls.[31] Early satiety induced by elevated anorexigenic signals may result in insufficient energy intake for weight gain, thus perpetuating AN. Although data are mixed, in both BN and BED, ghrelin appears dysregulated, which may maintain hunger and increase vulnerability to binge eating behaviors.[32] In BN, studies show reduced postprandial satiety responses; this insufficient satiety signal may be inadequate to reduce food intake, thereby facilitating continued binge eating behaviors.[31]

In both ABA and human neuroimaging studies of individuals with eating disorders, hedonic circuits controlling feeding behavior also appear dysregulated. Recent data demonstrate that excitation of the mesolimbic reward pathway via designer drugs can prevent the development of ABA.[33] In humans, structural reductions in the dorsal striatum, as measured by MRI, have been associated with reduced sensitivity to taste reward in AN and BN, relative to healthy controls.[34] In AN, increased fronto-accumbal structural connectivity has been associated with more severe eating disorder symptomatology.[35] Functional MRI studies show that individuals with AN demonstrate greater activation of reward circuits in response to food cues (smell or sight of food) but reduced activation in response to food consumption (taste) relative to healthy controls. In BN and BED, the inverse appears true.[36,37]

Consequences and Complications

Physical Complications and Signs

Physical complications of eating disorders arise because of starvation or binge/purge behaviors. Therefore, the diagnostic group at highest risk includes those individuals who are severely underweight and who also purge. Rapidity and amount of weight lost, as well as the frequency of purging, also impact risk. Mortality is significantly elevated for both AN and BN.[10] Besides complications of the eating disorder itself, treatment and refeeding are associated with additional potential medical risks.

Starvation-Related Complications

Malnutrition and starvation in AN are associated with numerous physical signs and symptoms. Patients at very low BMI will appear cachectic, with muscle wasting and weakness on examination. Physiologic responses to self-starvation are aimed at conserving energy and affect multiple organ systems. Cardiovascular complications include bradycardia, low blood pressure, and orthostatic hypotension. The latter can result from fluid restriction, vagal instability, or decreased venous return attributable to atrophy of peripheral muscles.[38] Estrogen, follicle-stimulating hormone, and luteinizing hormone revert to prepubertal concentrations due to disturbances in gonadotropin-releasing hormone pulsatility, leading to amenorrhea and infertility. In prepubertal patients, normal secondary sexual characteristics, such as breast development and height, may be stunted by malnutrition.[39] In the gastrointestinal tract, starvation results in delayed gastric emptying and delayed gastrointestinal transit times,[40] resulting in early satiety, abdominal pain, bloating, and constipation.[41] Anemia is common, and pancytopenia and bone marrow suppression can occur in patients with severe malnutrition.[42] Osteoporosis is a largely irreversible consequence of AN, occurring relatively early in the course of the disorder[43] and resulting in elevated fracture risk. Patients with chronic AN may incur debilitating hip and spinal compression fractures. Hypoglycemia is common in starvation and, when severe, can result in syncope, seizure, brain damage, or even coma and death.[44] Chronic hypoglycemia may underlie some of the neuroendocrine disruptions observed in eating disorders. Disturbances of regulatory hormones seen in AN include alterations in growth hormone, cortisol, and catecholamines. These changes may, in turn, contribute to maintaining anorexic behaviors and cognition.[45] Starvation and rapid weight loss can result in electrolyte imbalances, and patients presenting in an acutely starved state may have hypomagnesemia, hypocalcemia, or hyponatremia. The latter may be due to sodium depletion, diuretic abuse, or water intoxication, with severe hyponatremia resulting in seizures as a potential presenting symptom.[46] Excessive water ingestion is often motivated by a desire to increase satiety, make weight, or "eliminate toxins."

Purging-Related Complications

Patients who vomit may have visible bilateral parotid and salivary gland hypertrophy and Russell's sign on exam. Russell's sign, the presence of calluses on the dorsum of the hand, results from reflexive biting during manual stimulation of the gag reflex to induce vomiting. Dental caries and enamel erosion of the lingual surface of the teeth by gastric acid are also common. Recurrent vomiting can lead to reflux, esophagitis, hematemesis, and Mallory Weiss tears. Additionally, some patients may ruminate or chew and spit food as part of their disorder.

Frequent ingestion of large quantities of laxatives results in colonic dysmotility and laxative dependence. Rebound constipation can occur when laxatives are discontinued. Other gastrointestinal complications arising from excessive laxative misuse include hemorrhoids and rectal prolapse.

Individuals who engage in frequent vomiting and laxative or diuretic use can present with acute dehydration and presyncopal or syncopal symptoms. Chronic laxative and diuretic abuse can lead to renal damage, nephrocalcinosis, and electrolyte and acid–base imbalances.[47] Hypokalemia is the most common, serious electrolyte imbalance, which increases the risk of potentially lethal cardiac arrhythmias.

Refeeding-Related Complications

Refeeding syndrome is the term used to describe the constellation of severe metabolic abnormalities that may arise in AN as a result of weight restoration.[48] Severe hypophosphatemia, a serious potential complication of refeeding and a hallmark of the refeeding syndrome, results from the intracellular shift of serum phosphate needed for the regeneration of adenosine triphosphate, 2,3-diphosphoglycerate and glycerol-3-phosphate involved in cellular anabolic processes. Hypophosphatemia is associated with cardiac and neuromuscular dysfunction and hematologic red and white blood cell dysfunction. Intracellular shifts in potassium and magnesium, leading to low serum concentrations of these two electrolytes, contribute to an increased risk of cardiac arrhythmias and gastrointestinal and neuromuscular complications. Rapid extracellular fluid expansion, resulting in peripheral edema, is common in malnourished patients during early refeeding. In extreme cases, congestive heart failure is a risk. Abrupt cessation of laxative or diuretic abuse and vomiting behavior can further compound fluid retention. Severely underweight patients are at risk of Wernicke's encephalopathy related to thiamine deficiency. Acute gastric dilatation is a potentially life-threatening complication of refeeding that can result in infarction by compression of the mesenteric artery or in rupture of the stomach. Both pancreatitis and transaminitis are potential consequences of starvation and may transiently worsen with refeeding before improving.

Psychological Complications

Individuals with eating disorders describe a consuming and constant preoccupation with food, weight, or the consequences of eating that occupies much of their waking time. Furthermore, starvation can result in a syndrome indistinguishable from major depression, characterized by low mood, apathy, anhedonia, and decreased concentration and energy that responds to nutritional rehabilitation.[49] Mood, anxiety, substance use, and personality disorders are also often comorbid or may predate the onset of an eating disorder.[50,51] Finally, demoralization and loss of self-esteem frequently accompany patients' unsuccessful attempts to control their behaviors as increasingly negative consequences of the disorder accrue over time and impair an individual's function and quality of life.

Social and Developmental Complications

Eating is a highly social activity, and interpersonal relationships often become impaired by an eating disorder. Affected individuals become increasingly isolated in attempting to hide or avoid confrontation over their food choices or amounts eaten. They spend time engaged in solitary eating rituals and exercise routines that take precedence over age-appropriate social engagements. Formation of intimate relationships and sexual function are also frequently impaired by starvation's effect on libido or by heightened body image concerns. Because the age of onset of eating disorders is often in youth, they can interrupt normal developmental tasks, including separation-individuation from parents, identity formation, and the development of meaningful peer relationships.

Treatment

Initial treatment goals for all eating disorders include medical stabilization, normalizing eating patterns and weight control behaviors, and restoring weight for underweight individuals. The primary modalities of treatment are behavioral psychotherapeutic interventions and nutritional guidance, although there is a limited role for medications in the binge spectrum eating disorders. Psychotherapy aimed at elucidating underlying individual vulnerabilities to these disorders can provide a meaningful narrative to patients to help them understand the development of their disorder, but this alone is unlikely to bring about behavioral change. Primarily insight-focused therapies are best reserved for later stages of treatment once patients have been medically stabilized and there has been progress in normalizing eating and weight control behaviors. Treatment of co-occurring psychiatric disorders (mood, anxiety, or substance use disorders) or medical conditions (diabetes or celiac disease) is also important as these may complicate treatment response or alter nutritional guidance.

Some ambivalence toward behavior change is common in most eating disorders and can undermine engagement in treatment. This ambivalence can result from the driven nature of the disorder or anxiety about the consequences of behavior change. For example, in AN and BN, patients who value restrictive eating and exercise may be reluctant to stop these behaviors and to incorporate more variety and energy-dense foods in their meals. In ARFID, and often in AN, patients may be fearful that they cannot tolerate most foods and that eating them will result in gastrointestinal discomfort. Successful treatment, therefore, requires a cognitive shift from viewing eating and weight-related disordered behaviors as a coping strategy to appreciating that they, in fact, represent the primary impairment to healthy functioning. The motivated nature of eating disorders and ambivalence about change may lead to patients minimizing the severity of their problems, rationalizing their behavior, or denying they have an eating disorder. Role induction, the formation of a therapeutic alliance, and maximizing motivation to change behavior are, thus, important foci of any treatment approach.

Evidence-Based Practice

Several comprehensive practice guidelines for the treatment of eating disorders have been published. Examples include the American Psychiatric Association Practice Guidelines,[52] the United Kingdom National Institute for Health and Care Excellence guidelines,[53] and the Royal Australian and New Zealand College of Psychiatrists clinical practice guidelines.[54] Professionals familiar with the psychiatric, behavioral, and medical factors associated with eating disorders should make decisions about the appropriate setting for treatment and the introduction of pharmacotherapy.

Medical Stabilization and Hospitalization

The first goal of treatment is the accurate assessment and determination of medical stability.[55] Dangerously low BMI, rapid and profound recent weight loss, metabolic abnormalities (hypokalemia, hypophosphatemia, or hypoglycemia), and unstable vital signs (severe bradycardia or recent syncope) may warrant hospitalization. Interventions may include cardiac monitoring, rehydration, electrolyte replacement, and the initiation of nutritional rehabilitation.[56] Patients with very low BMI, or those who have rapidly lost large amounts of weight by fasting, are at risk for Wernicke's encephalopathy and should receive thiamine supplementation intramuscularly or intravenously before the administration of food or intravenous dextrose solutions.[57] Co-occurring severe psychiatric disorders, suicidality, self-injury, pregnancy, or type 1 diabetes, are other reasons an initial inpatient hospitalization might be appropriate. Failure to respond to a trial of outpatient treatment may be an additional indication for inpatient or residential behavioral treatment. Behavioral specialty units for the treatment of eating disorders or dedicated adolescent medicine eating disorders services are preferable, as specialty care is generally more effective than care on a general psychiatric or medical service. Most patients admitted to inpatient specialty units for eating disorders have AN, atypical AN, or low weight ARFID. Uncomplicated BN, BED, and nonunderweight ARFID are more commonly treated in the outpatient setting or at intermediate levels of care, as they are less likely to be medically unstable. Higher levels of care in behavioral specialty programs for eating disorders may include inpatient, residential, partial hospitalization, or intensive outpatient treatment. Patients commonly transition between several levels of care during a course of treatment, depending on illness severity and clinical presentation.

Nutritional Rehabilitation and Weight Restoration

Nutritional rehabilitation and weight restoration are necessary, if not sufficient, for recovery from AN. Achievement of a healthy BMI is the strongest positive predictor of treatment outcome, although relapse risk remains elevated at 30% for the initial year following weight restoration.[58] Wider variety and fat content of foods consumed at discharge from intensive treatment in weight-restored patients with AN have also been associated with better treatment outcomes at 1 year follow-up.[59] The preferred and safest mode of feeding is meal-based oral refeeding aimed at broadening food choices, extinguishing fears of consuming energy-dense foods, normalizing eating and weight control behaviors, and restoring weight. Specialty inpatient and residential treatment programs for eating disorders provide meal supervision, nutritional guidance, and group therapies delivered by a multidisciplinary treatment team in the context of a structured behavioral treatment protocol. Most such programs achieve average weight gains of 2 to 4 lbs. per week.

Some medical and psychiatric specialty programs additionally use supplemental nasogastric tube (NGT) feeding to boost lower weight gain rates. In emergent settings, or when access to a specialty behavioral protocol is limited, the use of NGT feeding may be lifesaving. However, prolonged NGT use can be associated with complications such as nasal irritation, epistaxis, electrolyte disturbance, patient distress, and patient-initiated tube removal.[60] Additionally, as underweight patients with either AN or ARFID are often nervous about consuming a balanced weight-gain diet, NGT feeds do not address anxiety about food choices or target the restricted eating repertoire and avoidance of high-energy foods that are hallmarks of eating disorders. Total parenteral nutrition (TPN) should rarely, if ever, be used for patients with eating disorders, as severely cachectic patients are immunocompromised and at elevated risk of sepsis and opportunistic infections, such as candidiasis.[61] Additionally, both TPN and NGT feeding incur a greater risk of refeeding syndrome than meal-based refeeding. They are best considered temporary measures to achieve a stable weight before transferring to a specialized meal-based behavioral treatment program or to outpatient specialty care. For outpatients, reasonable weight gain targets are lower, and expectations of 1 to 2 lbs. per week gain are consistent with effective progress. For adolescent patients, the involvement of family in monitoring and supervising meals and providing meal support is an important aspect of treatment.

Historically, starting energy intake prescriptions for weight restoration were low, and energy intake advancement protocols were slow, for fear of precipitating the refeeding syndrome in patients with severe malnutrition. Recent data, however, support the safety of faster refeeding energy intake advancement prescriptions.[62,63] Faster rates of refeeding and early weight gain have also been associated with shorter hospital stays and a higher likelihood of achieving weight restoration.[64-67] Recommended starting energy levels are 1,200 to 2,000 kilocalories per day, advancing to a weight gain diet of 3,500 to 4,000 kilocalories per day by day 7 to 10. Hypophosphatemia, a cardinal element of the refeeding syndrome, is more closely linked to BMI <14 than to the energy advancement rate. Serum phosphate should, however, be closely monitored and repleted orally as needed during the first week of renourishment or until stable, especially in individuals with low weight or who have lost a large amount of weight rapidly.[66,68] Early in the refeeding process, patients should also be monitored closely for elevations in transaminases, postprandial hypoglycemia, and peripheral edema, which are common complications. Constipation and abdominal discomfort may be treated with nonstimulant osmotic laxatives and antacids.

As patients who are underweight progress in treatment, the determination of target weight is important. Relevant data in determining an individualized target weight include family and personal weight history and developmental trajectory, such as weight at menarche.[69] In patients with AN and secondary amenorrhea, resumption of menses occurs on average at about 4 to 5 lbs. above weight at menarche. BMI percentile is often used to set a target weight for an adolescent if growth charts are not available, with BMI-for-age percentile of 14 to 39 associated with resumption of menses in 50% of patients at 1-year follow-up in one study.[70] For adults with AN, a target BMI of 19 to 20 is generally associated with better outcomes than lower BMI. However, a slightly higher target may be more appropriate for patients with a premorbid history of obesity or being overweight. Once weight gain is achieved, continued treatment to maintain appropriate energy intake, dietary variety, and fat intake is key in decreasing the risk of relapse.[71]

Psychotherapy

Psychotherapy should focus on normalizing eating, weight-control behaviors, body image, and eating-related cognitions rather than on searching for underlying root causes of the eating disorder, as understanding alone is unlikely to bring about behavior change. Psychotherapy can be helpful at all stages of treatment; it can help motivate and engage patients in changing their behavior, decrease food- and body-image related anxiety and dysphoric feelings, identify, and manage triggers and cognitive distortions, and develop healthy coping skills.

Randomized clinical trials provide strong empirical evidence in support of cognitive behavioral therapy (CBT) for the treatment of eating disorders, particularly for BN and BED.[72] Self-help manualized CBT interventions for BN and BED are also often effective. CBT involves several components: normalization of eating behavior, self-monitoring and maintaining food records, correcting cognitive distortions that sustain disordered eating habits, and relapse-prevention techniques. Interpersonal psychotherapy (IPT), a treatment that focuses on maladaptive interpersonal patterns and relationships, is an effective alternative to CBT for the treatment of both BN and BED; however, IPT is typically of longer duration and has fewer studies supporting its efficacy in comparison with CBT.[73]

CBT for AN has more modest efficacy, and no specific psychological treatment for adult outpatients with AN is superior.[74] For adolescents with AN, outpatient family therapy, where parents take control of their child's food intake, has proven to be effective for adolescents with short illness duration.[75] Data on the effectiveness of psychotherapeutic treatments for OSFED and ARFID are scant, but there is preliminary evidence to support family behavioral therapy (FBT) and CBT as promising.[76]

Pharmacology

There is limited support for pharmacologic treatment of eating disorders, although antidepressants and antianxiety medications can help treat co-occurring anxiety and depressive disorders. For BN and BED, high-dose fluoxetine has the strongest evidence. It is effective in decreasing binge eating and vomiting frequency at doses of 40 to 60 mg per day, regardless of comorbid depression or anxiety.[77] However, abstinence rates from binge eating and compensatory behaviors are lower than for CBT.[78] Fluoxetine can be prescribed alone, in conjunction with psychotherapy, or as an adjunctive treatment if psychotherapy alone is not effective. There is also some support, mainly from small studies, for the efficacy of a range of antidepressants, including most serotonergic reuptake inhibitors and some tricyclic antidepressants.[78]

For BED, lisdexamfetamine was found to reduce binge eating days and increase weight in three short-term, randomized control trials; however, discontinuation rates due to side effects were significant, and studies have been criticized for bias that may limit the generalizability of results and overall safety and tolerability.[79] The use of antidepressants in individuals with BED has also been shown to improve the efficacy of treatment, independent of the presence of co-occurring depression or anxiety.[80]

For AN, there is no strong evidence for pharmacotherapy. The best evidence is for olanzapine, which demonstrated statistically significant but very modest increases in weight in a multisite randomized control trial. No serious side effects, including metabolic syndrome, akathisia, or extrapyramidal effects, were noted in this study.[81]

The Multidisciplinary Team Approach to Treatment

Eating disorders are complex illnesses with medical, psychological, and behavioral aspects that are best managed by different specialists in close collaboration. To start, these patients may not overtly volunteer their eating disorder history. They may present to treatment requesting help with physical or psychological complications of their illness (such as abdominal pain, depression, or tooth decay). The initial presentation may be to a primary care doctor, a registered dietitian nutritionist (RDN), a therapist, or a dentist.[82] It is, therefore, important that clinicians be aware of warning signs that may warrant screening for an eating disorder. These include frequent symptoms seen in these patients as a result of fasting, binge eating, vomiting, or excessive exercise, along with other signs, including rapid weight changes, adoption of self-imposed diets (gluten-free or vegan), or restrictive fad diets, new onset of gastrointestinal symptoms, electrolyte disturbances (especially hypokalemia), and participation in sports that have weight criteria (e.g. gymnastics or crew).[83] Once individuals with eating disorders enter care, their treatment team may include psychiatrists, psychologists, or other mental health practitioners; pediatricians; general practitioners; RDNs; and occupational therapists.[84]

In addition to addressing the eating disorder, psychiatrists can diagnose and treat comorbid psychiatric conditions and assess for safety and the need for hospitalization. Pediatricians, internists, and general practitioners may be involved in managing medical issues.[85] Depending on the complexity of the complications, cardiologists, endocrinologists, or other specialists might also be part of the treatment team. Specific psychological therapies used in treating patients with eating disorders, such as CBT or FBT, are best implemented by trained therapists.[86]

Nutritional counseling is an essential component of treatment for an eating disorder, and RDNs are crucial members of the treatment team.[87] RDNs who choose to work with this patient population should be aware of the evidence-based psychological models used in eating-disorder treatment[88] and seek a professional clinical supervisory relationship with expert RDNs in ED to further their knowledge and skills.[89] An excellent resource regarding general principles and clinical practice standards for RDNs treating patients with EDs is outlined by the Australia & New Zealand Academy for Eating Disorders.[83] Occupational therapists can play an important role in providing eating-related interventions, including practice with eating socially and cooking skills to support sustained cognitive and behavioral changes.[90]

The goals of the team members may overlap, especially when it comes to nutritional and behavioral counseling. The various clinical specialists of the multidisciplinary team should be aware of some general principles to ensure a consistent treatment approach and to maintain good communication with one another. Interventions should be designed to normalize eating patterns by scheduling three meals a day, eating normal portion sizes, and minimizing exclusion of "fear" foods, including those that are nutrient poor (like ice cream or French fries).[83] Patients should be discouraged from consuming fat-free or sugar-free diet products. Although most patients with BN, and many with BED, are dissatisfied with their body shape and want to lose weight, it is important to stress that weight maintenance and interrupting binge eating and purging are the initial goals of treatment, even in patients with overweight. Restricting intake and very-low-energy diets will likely exacerbate binge eating behavior unless bulimia is in remission. In the case of BED, data are mixed; however, many clinicians believe that behavioral weight loss management strategies can exacerbate binge eating unless they are introduced after a normal eating pattern is established. Underweight individuals with AN or ARFID should be discouraged from exercising during weight restoration, or they are unlikely to achieve significant weight gain. Compulsive daily weighing should be discouraged. Ideally, weight should be monitored weekly by a clinician with whom the patient can discuss and process their reactions to changes in weight.

As multiple clinicians are likely to contribute to this multidisciplinary team approach and their roles overlap, collaboration and uniform recommendations are key. Doctors, therapists, and RDNs should be prepared to develop and communicate a unified and consistent treatment plan and assess progress as a team. Family members, especially in the case of adolescents, should also be included in this team approach. Streamlining communication pathways is crucial in treating these complex disorders effectively.[84]

Acknowledgment

The authors acknowledge Janelle W. Coughlin and Margaret Seide for their contributions to the previous version of this chapter.

References

1. Diagnostic and Statistical Manual of Mental Disorders: DSM-5. Washington, D.C: American Psychiatric Association, 2013.
2. Beumont PJ, Touyz SW. Eur Child Adolesc Psychiatry 2003;12(suppl 1):I20–4.
3. Fairburn CG, Harrison PJ. Lancet 2003;361:407–16.
4. Walsh BT, Boudreau G. Int J Eat Disord 2003;34(suppl):S30–8.
5. Eddy KT, Harshman SG, Becker KR, et al. Int J Eat Disord 2019;52:361–6.
6. Jenkins ZM, Mancuso SG, Phillipou A, et al. BJPsych Open 2021;7:e147.
7. Galmiche M, Dechelotte P, Lambert G, et al. Am J Clin Nutr 2019;109:1402–13.
8. van Hoeken D, Hoek HW. Curr Opin Psychiatry 2020;33:521–7.
9. Udo T, Grilo CM. Biol Psychiatry 2018;84:345–54.
10. van Eeden AE, van Hoeken D, Hoek HW. Curr Opin Psychiatry 2021;34:515–24.
11. Eckert ED, Halmi KA, Marchi P, et al. Psychol Med 1995;25: 143–56.
12. Duncan AE, Ziobrowski HN, Nicol G. Eur Eat Disord Rev 2017;25:165–71.
13. Brigham KS, Manzo LD, Eddy KT, et al. Curr Pediatr Rep 2018;6:107–13.
14. Bourne L, Bryant-Waugh R, Cook J, et al. Psychiatry Res 2020;288:112961.
15. Santomauro DF, Melen S, Mitchison D, et al. Lancet Psychiatry 2021;8:320–8.
16. Strober M, Freeman R, Lampert C, et al. Am J Psychiatry 2000;157:393–401.
17. Yilmaz Z, Hardaway JA, Bulik CM. Adv Genomics Genet 2015;5:131–50.
18. Bulik CM, Blake L, Austin J. Psychiatr Clin North Am 2019;42: 59–73.
19. Duncan L, Yilmaz Z, Gaspar H, et al. Am J Psychiatry 2017;174:850–8.
20. Watson HJ, Yilmaz Z, Thornton LM, et al. Nat Genet 2019;51:1207–14.
21. Culbert KM, Racine SE, Klump KL. J Child Psychol Psychiatry 2015;56:1141–64.
22. Cassin SE, von Ranson KM. Clin Psychol Rev 2005;25:895–916.
23. Hudson JI, Hiripi E, Pope HG, Jr., et al. Biol Psychiatry 2007;61:348–58.
24. Mellentin AI, Mejldal A, Guala MM, et al. Am J Psychiatry 2022;179:46–57.
25. Stice E, Gau JM, Rohde P, et al. J Abnorm Psychol 2017;126: 38–51.
26. Herpertz-Dahlmann B, Seitz J, and Konrad K. Eur Arch Psychiatry Clin Neurosci 2011;261 Suppl 2:S177–81.
27. Taquet M, Geddes JR, Luciano S, et al. Br J Psychiatry 2021;1–3.
28. Treasure J, Stein D, Maguire S. Early Interv Psychiatry 2015;9: 173–84.
29. Walsh BT. Am J Psychiatry 2013;170:477–84.
30. Pardo M, Roca-Rivada A, Al-Massadi O, et al. Peptides 2010;31: 1912–9.
31. Smith KR, Moran TH. Physiol Behav 2021;238:113456.
32. Keel PK, Eckel LA, Hildebrandt BA, et al. Int J Eat Disord 2018;51:53–61.
33. Foldi CJ, Milton LK, Oldfield BJ. Neuropsychopharmacol 2017; 42:2292–300.
34. Frank GK, Shott ME, Hagman JO, et al. Am J Psychiatry 2013;170:1152–60.
35. Cha J, Ide JS, Bowman FD, et al. Hum Brain Mapp 2016;37: 3835–46.
36. Simon JJ, Skunde M, Walther S, et al. Soc Cogn Affect Neurosci 2016;11:1393–401.
37. Wierenga CE, Ely A, Bischoff-Grethe A, et al. Front Behav Neurosci 2014;8:410.
38. Mitchell JE, Crow S. Curr Opin Psychiatry 2006;19:438–43.
39. Russell GF. J Psychiatr Res 1985;19:363–9.
40. Hadley SJ, Walsh BT. Curr Drug Targets CNS Neurol Disord 2003;2:1–9.
41. Cooper M, Collison AO, Collica SC, et al. Int J Eat Disord 2021;54:1055–62.
42. De Filippo E, Marra M, Alfinito F, et al. Eur J Clin Nutr 2016;70:1305–8.
43. Mehler PS, Cleary BS, Gaudiani JL. Eat Disord 2011;19:194–202.
44. Gaudiani JL, Sabel AL, Mehler PS. Int J Eat Disord 2014;47: 148–56.
45. Mattingly D, Bhanji S. J R Soc Med 1995;88:191–5.
46. Winston AP. Ann Clin Biochem 2012;49:132–43.
47. Stheneur C, Bergeron S, Lapeyraque AL. Eat Weight Disord 2014; 19:455–60.
48. Runde J, Sentongo T. Pediatr Ann 2019;48:e448–e54.
49. Brockmeyer T, Holtforth MG, Bents H, et al. Compr Psychiatry 2012;53:496–501.
50. Cooney M, Lieberman M, Guimond T, et al. J Eat Disord 2018;6:7.
51. Marucci S, Ragione LD, De Iaco G, et al. Endocr Metab Immune Disord Drug Targets 2018;18:316–24.
52. American Psychological Association (APA). Am J Psychiatry 2000;157:1–39.
53. Eating Disorders: Recognition and Treatment. London, United Kingdom: UK NICE, 2017.
54. Hay P, Chinn D, Forbes D, et al. Aust N Z J Psychiatry 2014; 48:977–1008.
55. Society for Adolescent Health and Medicine, Golden NH, et al. J Adolesc Health 2015;56:121–5.
56. Strandjord SE, Sieke EH, Richmond M, et al. Eat Weight Disord 2016;21:403–10.
57. Oudman E, Wijnia JW, Oey MJ, et al. Psychiatry Clin Neurosci 2018;72:774–9.
58. Berends T, Boonstra N, van Elburg A. Curr Opin Psychiatry 2018;31:445–55.
59. Schebendach JE, Mayer LE, Devlin MJ, et al. J Am Diet Assoc 2011;111:732–6.
60. Hindley K, Fenton C, McIntosh J. J Eat Disord 2021;9:90.
61. Michihata N, Matsui H, Fushimi K, et al. Eat Weight Disord 2014;19:473–8.
62. Garber AK, Sawyer SM, Golden NH, et al. Int J Eat Disord 2016;49:293–310.
63. Golden NH, Cheng J, Kapphahn CJ, et al. Pediatr 2021; 147:e2020037135.
64. Wade TD, Allen K, Crosby RD, et al. Eur Eat Disord Rev 2021;29:472–81.
65. Garber AK, Mauldin K, Michihata N, et al. J Adolesc Health 2013; 53:579–84.
66. Redgrave GW, Coughlin JW, Schreyer CC, et al. Int J Eat Disord 2015;48:866–73.
67. Lund BC, Hernandez ER, Yates WR, et al. Int J Eat Disord 2009;42:301–5.
68. Kells M, Gregas M, Wolfe BE, et al. Nutr Clin Pract 2022;37: 470–8.

69. Marion M, Lacroix S, Caquard M, et al. J Eat Disord 2020;8:42.
70. Golden NH, Jacobson MS, Sterling WM, et al. Int J Eat Disord 2008;41:301–6.
71. Mayer LE, Schebendach J, Bodell LP, et al. Int J Eat Disord 2012;45:290–3.
72. Eating Disorders: Core Interventions in the Treatment and Management of Anorexia Nervosa, Bulimia Nervosa and Related Eating Disorders. Leicester, United Kingdom: UK National Collaborating Centre for Mental Health, 2004.
73. Murphy R, Straebler S, Basden S, et al. Clin Psychol Psychother 2012;19:150–8.
74. Solmi M, Wade TD, Byrne S, et al. Lancet Psychiatry 2021;8: 215–24.
75. Lock J, Le Grange D, Agras WS, et al. Arch Gen Psychiatry 2010;67:1025–32.
76. Thomas JJ, Becker KR, Breithaupt L, et al. J Behav Cogn Ther 2021;31:47–55.
77. Bello NT, Yeomans BL. Expert Opin Drug Saf 2018;17:17–23.
78. Svaldi J, Schmitz F, Baur J, et al. Psychol Med 2019;49:898–910.
79. Fornaro M, Solmi M, Perna G, et al. Neuropsychiatr Dis Treat 2016;12:1827–36.
80. McElroy SL. J Clin Psychiatry 2017;78(suppl 1):14–9.
81. Attia E, Steinglass JE, Walsh BT, et al. Am J Psychiatry 2019;176:449–56.
82. Ranalli DN, Studen-Pavlovich D. Dent Clin North Am 2021; 65:689–703.
83. Heruc G, Hart S, Stiles G, et al. J Eat Disord 2020;8:77.
84. McDevitt S, Passi V. Ir J Psychol Med 2018;35:289–99.
85. Klein DA, Sylvester JE, Schvey NA. Am Fam Physician 2021;103:22–32.
86. Hay P. Intern Med J 2020;50:24–9.
87. Ozier AD, Henry BW, American Dietetic Association. J Am Diet Assoc 2011;111:1236–41.
88. Hilbert A, Hoek HW, Schmidt R. Curr Opin Psychiatry 2017;30:423–37.
89. Hackert AN, Kniskern MA, Beasley TM. J Acad Nutr Diet 2020;120:1902–19:e54.
90. Biddiscombe RJ, Scanlan JN, Ross J, et al. Aust Occup Ther J 2018;65:98–106.

H. OTHER SYSTEMIC DISORDERS

CHAPTER 96

Nutrition, Diet, and the Kidney

Anuja Shah, Connie M. Rhee, and Joel D. Kopple

CHAPTER OUTLINE

Abbreviations

aBW, actual body weight; **ACE**, angiotensin-converting enzyme; **ACEI**, angiotensin-converting enzyme inhibitor; **AKI**, acute kidney injury; **Apo**, apolipoprotein; **ARB**, angiotensin receptor blocker; **BMI**, body mass index; **BW**, body weight; **CAPD**, continuous ambulatory peritoneal dialysis; **CKD**, chronic kidney disease; **CPD**, chronic peritoneal dialysis; **CRF**, chronic renal failure; **CRP**, C-reactive protein; **CVVH**, continuous venovenous hemofiltration; **CVVHD**, continuous venovenous hemofiltration with concurrent hemodialysis; **DKD**, diabetic kidney disease; **DM**, diabetes mellitus; **DRI**, dietary reference intake; **ESA**, erythropoiesis-stimulating agents; **ESKD**, end-stage kidney disease; **ESRD**, end stage renal disease; **GFR**, glomerular filtration rate; **GH**, growth hormone; **HD**, hemodialysis; **HDL**, high-density lipoprotein; **ICU**, intensive care unit; **IDL**, intermediate-density lipoprotein; **IGF-I**, insulin-like growth factor-I; **IL**, interleukin; **ISRNM**, international society of renal nutrition and metabolism; **KDIGO**, kidney disease improving global outcomes; **KDOQI**, Kidney Disease Outcome Quality Initiative; **LDL**, low-density lipoprotein; **Lp(a)**, lipoprotein (a); **LPD**, low-protein diet; **MD**, maintenance dialysis; **MDRD**, Modification of Diet in Renal Disease; **MHD**, maintenance hemodialysis; **MUFA**, monounsaturated fatty acid; **NDD-CKD**, non-dialysis dependent chronic kidney disease; **NKF**, National Kidney Foundation; **nPNA**, protein equivalent of total nitrogen appearance; **ONTARGET**, Ongoing Telmisartan Alone and in Combination with Ramipril Global Endpoint Trial; **PD**, peritoneal dialysis; **PEM**, protein-energy malnutrition; **PEW**, protein-energy wasting; **PN**, parenteral nutrition; **PTH**, parathyroid hormone; **PUFA**, polyunsaturated fatty acid; **RDA**, recommended dietary allowance; **SUN**, serum urea nitrogen; **TGF-β**, transforming growth factor-β; **TLC**, Therapeutic Lifestyle Changes; **TNF-α**, tumor necrosis factor-α; **UNA**, urea nitrogen appearance; **VLDL**, very-low-density lipoprotein

Kidney Function

The kidney has three primary functions: excretory, endocrine, and metabolic. All three functions may be impaired in renal disease and affect the patient's nutritional status and management. When injury, necrosis, and scarring of the renal parenchyma cause a loss of renal function, the quantity of the substances that are filtered by the kidney fails. However, many aspects of renal function undergo adaptive changes that preserve homeostasis and minimize the derangements in plasma and tissue concentrations of substances that normally are excreted by the kidney. Prominent among these adaptions are nephron hypertrophy and an increase in blood flow and glomerular filtration rate (GFR) in those nephrons that are still functional. Chronic kidney disease (CKD) has been classified into five stages, as shown in **Table 96.1**.[1] The presence of proteinuria is associated with an increase in the likelihood of developing CKD and accelerated progression of CKD to chronic renal failure (CRF).

Water and many organic compounds and minerals accumulate in renal failure.[3] Low-protein diets (LPDs), various minerals, and other compounds reduce the accumulation of many of these substances. Eventually, renal failure may become so severe that the adaptive mechanisms are no longer adequate to maintain homeostasis, even with special diet therapy restricting fluid, minerals, and protein intake. The accumulation of these compounds and minerals, the endocrine and metabolic disturbances, and the clinical signs and symptoms that result from renal failure are referred to as uremia. If this condition is not treated by maintenance hemodialysis (MHD), chronic peritoneal dialysis (CPD), or renal transplantation, clinical deterioration and death will eventually supervene. Diet therapy with a low protein diet and mineral restricted intake may ameliorate uremic symptoms in people with CKD who have some residual renal function.

Excretion and regulation of body water, minerals, and organic compounds are the most important functions of the kidney. Without renal excretory function, patients rarely live longer than 4 to 5 weeks, and often less than 10 days, particularly if they are hypercatabolic. In contrast, anephric patients can be kept alive for years with intermittent MHD or CPD, even though many endocrine and metabolic disorders that occur with a failing kidney are not completely corrected.

The kidney produces hormones with diverse metabolic effects, including 1,25-dihydroxycholecalciferol, erythropoietin, and renin. These have been reviewed elsewhere.[4-6] Vitamin D_3 (cholecalciferol) is hydroxylated in the liver to form 25-hydroxycholecalciferol. This compound is then converted in the kidney to 1,25-dihydroxycholecalciferol (also called 1,25-dihydroxy vitamin D or calcitriol), the most potent natural form of vitamin D (see Chapter 18). In renal failure, impaired synthesis of 1,25-dihydroxy vitamin D contributes to a vitamin D–deficient state associated with impaired intestinal calcium absorption, hyperparathyroidism, resistance to the actions of parathyroid hormone (PTH) on bone, and the development of renal osteodystrophy. Epidemiologic data suggest that 25-hydroxyvitamin D and 1,25-dihydroxy vitamin D may have other beneficial effects, including reducing cancer risk, cardiovascular disease (CVD), and mortality.[7-9] These findings need confirmation by controlled prospective clinical trials.[10,11] Synthesis and secretion of a hormone from bone, fibroblast growth factor-3 (FGF-23), increases early during CRF in response to hyperphosphatemia. Elevated serum FGF-23 suppresses the renal synthesis of 1,25-dihydroxycholecalciferol.[12]

Erythropoietin stimulates erythropoiesis in bone marrow,[5] and the anemia of CRF is primarily caused by impaired erythropoiesis resulting from reduced erythropoietin production by the diseased kidneys. In renal failure, compounds that accumulate may also suppress erythropoiesis, and mild hemolysis often contributes to the anemia. Recombinant DNA-synthesized human erythropoietin is commonly used to increase blood hemoglobin of patients with advanced CKD and those undergoing maintenance dialysis (MD).[13] Other erythropoiesis-stimulating agents (ESA) have recently been developed to promote erythropoiesis and reduce anemia in patients with CRF receiving MD.[14]

Renin stimulates the conversion of angiotensinogen to angiotensin I which, in turn, is converted to angiotensin II by the angiotensin-converting enzyme (ACE). Angiotensin II is a potent

Table 96.1 Stages of Chronic Kidney Disease[a]

Stage	Description	GFR (mL/min/1.73 m²)	Treatment
1	Kidney damage with normal or increased GFR	≥90	Medicinal and Dietary Therapy
2	Kidney damage with mildly decreased GFR	60–89	Medicinal and Dietary Therapy
3a	Moderately decreased GFR	45–59	Medicinal and Dietary Therapy
3b	Moderately decreased GFR	30–44	Medicinal and Dietary Therapy
4	Severely decreased GFR	15–29	Medicinal and Dietary Therapy
5	Kidney failure	<15 (or receiving dialysis therapy)	5D if receiving chronic hemodialysis or chronic peritoneal dialysis, 5T if kidney transplant recipient

[a]Chronic kidney disease is defined as kidney damage or a glomerular filtration rate (GFR) <60 mL/min/1.72 m² for at least 3 months. Kidney damage is defined as pathologic abnormalities or markers of damage, including abnormalities in blood or urine tests or imaging studies.

Reproduced from American Journal of Kidney Diseases from the National Kidney Foundation KDOQI US Commentary on the 2012 KDIGO Clinical Practice Guideline for the Evaluation and Management of CKD. Am J Kidney Dis 2014;63:P713–35.[2]

vasoconstrictive agent that raises blood pressure and may stimulate collagen formation and cell proliferation in the kidney and other tissues. Renal renin secretion is stimulated by renal ischemia (e.g., in renal artery stenosis) and sometimes other renal diseases; increased plasma renin can cause hypertension. Renal disease, particularly renal failure, may engender hypertension by other mechanisms, including sodium chloride and water retention.

Interrelationships between Nutrients and Kidney Function

Kidney function regulates and is influenced by the body's pools and concentrations of water, minerals, and many other nutrients and metabolites.

Effects of Malnutrition on the Kidney

Malnutrition can have important, but usually reversible, effects on renal function in individuals with healthy kidneys.[15] In humans, malnutrition decreases the GFR and the capacity to concentrate and acidify urine.[15,16] If protein intake increases, these functions may normalize. GFR falls reversibly in individuals with obesity treated with weight reduction diets or bariatric surgery.[17]

Low protein intake reduces renal blood flow and GFR.[15,16] A reduction in insulin-like growth factor-I (IGF-I) may contribute to these changes with low-protein diets.[18,19] Reduction in extracellular body water and circulating blood volume may also decrease renal blood flow and GFR. Increased sodium chloride and water intake often reverse this condition.

Individuals with malnutrition frequently have lower specific gravity in random urine specimens and, therefore, increased daily urine volumes. Impaired concentrating ability, which occurs with malnutrition, probably contributes to the nocturia that may occur. The inability to concentrate urine normally in patients with malnutrition appears to result from the low protein intake and consequent low rate of urea synthesis.[20] Urea is critical for normal urinary concentrating ability. Some of the urea filtered by the glomerulus is reabsorbed in the renal tubule. The urea accumulates in the interstitium of the renal medulla where, when the collecting duct aquaporin receptors are expressed, urea and other chemicals attract water from the lumen of the distal tubule and collecting duct by osmotic pressure. The loss of water from the lumen of the distal tubule and collecting duct by osmotic pressure increases the urine concentration. When less urea accumulates in the renal medullary interstitium, the ability to attract water from the collecting duct is reduced. Hence, the urinary concentrating ability is diminished. The capacity to dilute urine is normal in malnutrition.

Those with malnutrition are more likely to develop acidosis after an acid load.[16] Urinary phosphate and ammonia are primary carriers of acid (protons) in the urine. Hydrogen ion secretion into the distal nephron's lumen lowers the tubular fluid's pH, converts HPO4–to H2PO4–and stimulates ammonia production and conversion of ammonia to NH_4^+. In individuals with low phosphorus intake, the phosphate filtered in the kidney is largely reabsorbed; less phosphorus is excreted in the urine, and this reduces the capacity of the kidney to excrete acid. Phosphate infusion increases the urinary excretion of titratable acid in patients with malnutrition.[16] Renal production and excretion of ammonium are also reduced in malnutrition, both under basal conditions and after an acid load.[16]

During prolonged starvation, the kidney may account for up to 45% of endogenous glucose production. However, part of the rise in the renal contribution to total glucose synthesis results from a fall in total body glucose production.[21] In extended starvation, net renal extraction of lactate, pyruvate, amino acids, and glycerol also occurs.[19] The carbon skeleton in these compounds is virtually completely converted into glucose. During prolonged starvation, the kidney extracts free fatty acids and β-hydroxybutyrate, and acetoacetate is released.[21]

Ketoacidosis can develop when energy intake, especially carbohydrate intake, is insufficient to match needs or during abnormal fuel utilization states, including diabetic ketoacidosis, carbohydrate restriction, and SGLT 2 inhibitor therapy. Ketoacidosis is caused, in large part, by reduced insulin. Insulin suppresses the release of fatty acids from adipose tissue, and ketoacids and hydroxy acids are metabolites of fatty acid metabolism. With deficient serum insulin, more fatty acids are released from adipose tissue, and the production of ketoacids and hydroxy acids increases. As glycogen stores are depleted, gluconeogenesis from amino acids increases, leading to more acid production and renal ammonium generation.

Acute starvation and other conditions associated with increased catabolism of nucleic acids, purines, and amino acids, such as with chemotherapy for leukemia and certain other tumors, can cause a marked increase in uric acid production. Hyperuricemia can lead to the deposition of uric acid sludge in the kidney and lower urinary tract and may cause acute kidney injury (AKI). Treatment consists of agents to decrease hyperuricemia for maintenance of good hydration and urine flow.[22]

Effects of Protein and Amino Acid Intake on Renal Function

Protein intake appears to cause an immediate and more long-term increase in renal blood flow and GFR in humans. Normally, a transient increase in renal blood flow and GFR of about 20% to 28% occurs about 2 hours after ingestion of a protein or amino acid load, and generally lasts about 1 hour.[23,24] Renal blood flow and GFR increase transiently and more quickly following an intravenous infusion of a mixture of essential and nonessential amino acids[25] or a 30-minute infusion of arginine hydrochloride.[26] The increase in GFR and renal plasma flow occur in a dose-dependent manner.[27] Cooking converts creatine in meat to creatinine and, thereby, changes creatinine concentration. Therefore, using serum or urine creatinine to estimate the acute changes in GFR after a meal with cooked meat rather than measuring true GFR can give falsely low or high GFR values.[28]

High Protein and Ketogenic Diets and the Risk of CKD

High protein, low carbohydrate diets have been studied as one way to promote weight loss and to improve blood glucose management in people with type 2 diabetes. There is a concern that high protein intake may accelerate the progression of renal failure in patients with CKD and increase the risk of CKD in people with normal kidneys.[29] Low protein diets for people with CKD are discussed elsewhere in this chapter. Ketogenic diets are low in carbohydrate and high in protein, with protein intakes up to about 2 grams/kilogram/day. The goal of ketogenic diets is to shift to ketone metabolism and increase the use of fat stores for energy generation. Energy intake is often lower while on these diets. The potential clinical benefits of these diets include weight loss, improved insulin sensitivity, reduced HbA1C in people with diabetes, decreased serum triglycerides and visceral adiposity, and potentially beneficial changes to the gut microbiome.[30] The renal hemodynamic effects of protein intake are well established. Increased glomerular blood flow and filtration rate occur after high-protein meals in animals and humans. The hyperfiltration is mediated by increased intraglomerular pressure. These hemodynamics can further damage diseased kidneys and injure even normal ones over time.[31,32]

Several studies have examined whether high protein intake can damage kidneys; the findings are mixed. Observational studies like the Korean Genome and Epidemiology Study of 9,226 people in the general population showed a higher risk of decline of renal function in those in the highest quartile of protein intake.[33] An Iranian cohort study of 1,800 people ingesting low-carbohydrate, high-protein diets showed the highest risk for incident CKD in those consuming the highest tertile of protein intake.[34] However, these findings were not confirmed in the setting of controlled studies. Low carbohydrate diets were found to be similar to low-fat and Mediterranean diets regarding changes in renal function in patients with moderate obesity in the Dietary Intervention Randomized Control Trial.[35] Similarly, a high-protein, low-carbohydrate diet in healthy people with obesity did not increase the incidence of albuminuria or impaired renal function, compared with individuals eating a low-fat diet.[36] Very low ketogenic diets (i.e., diets containing a very low carbohydrate content of <20 g/day), 1–1.5 g protein/kg ideal body weight (estimated based on gender and height), 15–30 g fat/day, and about 500 to 800 kilocalories/day did not affect kidney function in one study of patients with mild kidney dysfunction, or those with normal kidney function who were studied for 3 months.[37]

Protein-induced hyperfiltration and the use of GFR as the measure of kidney function may confound these results. High-protein diets, by causing hyperfiltration and thereby increasing GFR, may obscure injury or further damage to the kidney for short or even extended periods. The beneficial effects of weight (fat) loss on blood pressure and insulin sensitivity may counterbalance the deleterious effects of hyperfiltration on the kidney. Overall, current data do not preclude using a high-protein, low-carbohydrate diet for people with normal renal function. Care should be taken to evaluate and monitor renal function, urine sediment, and overall nutrient intake in people with healthy kidneys who consume these diets for long periods. Randomized, prospective, controlled, long-term intervention trials are needed to more definitively study these diets' potential to harm the kidney. People with advanced CKD may become symptomatically uremic if they ingest diets providing 1.0 to 2.0 g protein/kg/day.

Effect of Nutritional Intake on the Rate of Progression of Renal Failure

Mechanisms of Progression

Individuals with CKD who have sustained substantial loss of GFR often continue to lose renal function inexorably until they develop terminal renal failure.[38-40] Although the rate of progression of renal failure varies greatly among patients, in many individuals, the decline in kidney function is approximately linear.[38,41] The percentage of patients with renal insufficiency who will progress to renal failure is unknown. Still, most patients who sustain a loss of GFR of 60% or greater will show the continued progression of kidney failure. Renal failure may progress because of continued activity of the underlying renal disease or the superimposition of other diseases that may contribute to renal injury, such as hypertension, adverse effects of nephrotoxic medicines (e.g., antibiotics or radiocontrast material), urinary tract obstruction, kidney infection, hypercalcemia, or hyperuricemia. There are maladaptive responses that include inflammation and renal fibrosis that continue even after the initial cause of the renal disease seems to have disappeared, and when superimposed illnesses are not present.[42-44] Other risk factors have been identified, including proteinuria, socioeconomic factors, and hypertension; therapies explicitly directed at these causes are important but are often not successful in stopping progression.[45] Progression of renal failure may continue in patients who have relief of urinary tract obstruction, cessation of nephrotoxic medications, or partial recovery from acute renal failure.

Studies of animal and in vitro models of CKD or renal failure have led to the following observations. There is a common set of physiologic and biochemical responses to chronic renal function loss, largely independent of the underlying type of kidney disease. When the loss of functioning nephrons becomes sufficient to cause renal insufficiency, the remaining individual functioning nephrons generally undergo an increase in the glomerular plasma flow and GFR, and an enlargement in size of the glomeruli and the tubules (i.e., nephron hypertrophy).[46-48] The capillary blood pressure and the blood pressure gradient across the capillary wall in the remaining functional glomeruli increase.[46,49] In addition, the chemical, electrical, and pore-size barriers to the movement of plasma proteins across the glomerulus and into the renal tubule lumen are impaired. This can lead to proteinuria (leakage of protein into the urine) and deposition of proteins in renal tissue. This, in turn, can stimulate the migration of leukocytes and monocytes, collagen deposition, cellular proliferation, and other inflammatory and scarifying changes.[50,51] These processes may contribute to progressive renal damage. Endothelin-1, TGF-beta, monocyte chemoattractant protein-1, RANTES, and interleukin 8 are some of the inflammatory factors that mediate this process.[52] Many of the preceding changes, which could be considered adaptive physiologic responses, are believed to promote further renal injury and fibrosis. These factors, as well as continuing activity of the underlying renal disease and associated illnesses like hypertension, diabetes, and obesity, can lead to progressive renal failure.

Transforming growth factor-β (TGF-β) plays a central role in scarring of the diseased kidney.[53] TGF-β acts on many other mediators that promote renal fibrosis and protein matrix accumulation.[54] Potential causes of progressive renal failure are summarized in **Table 96.2**. Most of these processes have been investigated only in animal models, and it is inferred that they play a role in human progressive renal disease. It is pertinent that many of these mechanisms appear to be susceptible to amelioration or reversal by nutritional therapy (see Table 96.2). For example, protein-restricted diets have been shown to reduce the adaptive increase in glomerular blood flow and GFR and proteinuria in humans with renal disease.[26] Albuminuria is a risk factor for more rapid progression of CKD, and the magnitude of albuminuria is correlated with the rate of progression of a given type of CKD. Sodium restriction may decrease blood pressure and subsequent renal injury. Obesity appears to be associated with more rapid progression of kidney failure, and diabetes, which is often modifiable by good nutritional practice, may accelerate the progression of kidney failure.

Table 96.2 Potential Causes and Mechanisms of Progression of Renal Failure[a]

Continued activity of the underlying renal disease(s)
Systemic hypertension[b]
High-protein diet[b]
High-phosphorus diet[b]
High-total-fat or high-cholesterol diet[b]
High-calcium-phosphorus product in sera[b]
Excessive Vitamin D intake (causing hypercalcemia)[b]
High serum oxalate concentration (can be enhanced by high ascorbic acid intake)[b]
Proteinuria[b]
Angiotensin II[b]
Aldosterone[b]
Hyperuricemia
Acidemia
Inflammatory response in the kidney with release of cytokines and monokines
Platelet aggregation in kidney
Increased mesangial matrix production
Deposition of other proteins in glomerulus
Progressive interstitial fibrosis
Lipoprotein and lipid deposition in glomerulus
Release of growth factors in the kidney
Glomerular and tubular hypertrophy
Intraglomerular capillary pressure and capillary blood flow
Intraglomerular transcapillary hydraulic pressure
Increased generation of reactive oxygen metabolites in remaining functional nephrons
Calcium phosphate or calcium oxalate deposition in the kidney
Nephrotoxic medicines (e.g., radiocontrast material, aminoglycoside antibiotics)
Enhanced renal tubular generation of ammonia leading to complement activation
Intrarenal release of endothelin-1
Lead, cadmium toxicity

[a]For many of these factors, evidence that they may cause progressive renal failure is derived from animal models or in vitro systems.

[b]These causes of progressive kidney failure may act through more than one of the mechanisms listed in the lower half of this table.

Experimental Evidence for Effects of Nutritional Intake and Status on Progression of Chronic Kidney Failure

Proteins and Amino Acids

In experimental animals with kidney disease, a high-protein diet stimulates an increase in GFR, glomerular capillary blood flow, blood pressure gradients across the glomerular capillary wall, and enlargement of individual nephrons, whereas a LPD blunts or prevents these responses.[46,55] Normal rats with a renal injury who are fed a high-protein diet develop renal failure. When such animals are fed a LPD, the progression of renal failure is retarded or arrested, even if the underlying kidney disease is not due to a nutritional cause.[56,57] It has been hypothesized that high protein intake increases glomerular capillary blood flow and transcapillary glomerular hydraulic pressure, which causes progressive renal injury to the basement membrane (filtering wall) of the glomerulus.[29,31,58,59] Observational studies in humans have shown an association between higher protein intake and kidney decline in those with pre-existing CKD.[33,60,61] However, clinical trials in humans have not demonstrated an increase in proteinuria as evidence of glomerular basement membrane damage in those with baseline normal kidney function, possibly due partly to small sample sizes and the short duration of the studies.

High-protein diets may also promote renal insufficiency by other mechanisms. These include the following: induction of nephron hypertrophy with activation of growth factors that stimulate cell hypertrophy, proliferation, and scarring in the glomerulus; enhanced oxidation rates in the nephron, with increased generation of reactive oxygen species[62]; an acid load that stimulates both renal ammonium production with activation of complement, endothelin-1, angiotensin II, and aldosterone pathways.[63,64] A LPD may reduce these processes.[65-67]

The source of dietary protein may also be important. The Atherosclerosis Risk in Communities study and the Singapore Chinese Health Study showed a higher risk of CKD with higher red meat consumption.[68,69] Diets providing soy protein, a vegetable protein, compared with casein, an animal protein, may more effectively retard the progression of kidney failure in rats with remnant kidneys[70] and possibly in humans with CKD. This may reflect the alkalinizing effects of high vegetable and fruit diets.

Metabolites of tryptophan, particularly indoxyl sulfate, may cause renal fibrosis and more rapid progression of kidney disease. A medication containing activated charcoal that binds these metabolites in the gastrointestinal tract has been shown to reduce fibrosis in the kidneys of rats with CKD, and to retard CKD progression in both rats and humans. A large clinical trial in the United States to examine the effects of this charcoal binder, which binds both indoxyl sulfate and P-cresyl sulfate, failed to show an impact on the progression of CKD, the primary outcome of the trial.[71] However, in post hoc analyses, compliance with medicinal intake and variability in the causes of CKD may have confounded the results.[72]

Acidosis may enhance the rate of progression of CKD. One mechanism seems to involve the increased renal production of ammonia to excrete a large acid load. Ammonium ion stimulates complement production in rats and promotes progressive renal injury.[73] Acidosis also stimulates the release of the profibrotic cytokine, endothelin-1, in the kidney.[74] Several randomized prospective clinical trials (RCTs) in people with CKD and reduced GFR indicate that the daily administration of alkali, either sodium bicarbonate or citrate, slows the rate of progression of chronic kidney failure (CKF).[75-77] Goraya et al., showed that diets very high in vegetables and fruits, which provide an alkaline load to the patient, slow the rate of progression of CKF.[78]

Glucose

Rats with diabetes and moderate hyperglycemia develop renal hypertrophy and increased hemodynamics.[79] Similar abnormalities occur in the intact kidneys of humans with diabetes. Early during the course of diabetes, patients develop increased renal blood flow, increased GFR, and large kidneys.[80] Ultimately, in many of these individuals, glomerulosclerosis occurs; about 40% of people with type 1 and type 2 diabetes in the United States develop CKD.[81] Approximately 35% of people with CKD in the United States have underlying diabetes.[82] In the early stages of diabetes, strict glucose control may prevent or reverse these phenomena.[80] SGLT-2 inhibitors may reverse or retard the hemodynamic changes associated with hyperglycemia, and decrease the progression of GFR loss and proteinuria in people with diabetes.[83]

Phosphorus and Calcium

A low phosphorus intake independent of protein intake appears to retard the progression of renal failure.[84,85] One theory on the mechanism of action of a low phosphorus intake is that it decreases the deposition of calcium phosphate in kidney tissue, which may cause further renal damage.[85-88] In renal tissue obtained by biopsy or autopsy, there is a direct correlation between the calcium content and serum creatinine concentration, a measurement inversely related to the GFR.[86] In general, the calcium concentration of renal tissue is increased more in diseased kidneys with more severe renal histopathologic changes. Medications that bind phosphate in the intestinal tract and decrease intestinal phosphorus absorption are a valuable adjunct to dietary phosphorus restriction, because it is difficult to lower dietary phosphorus intake to the levels that, by itself, are necessary to maintain normal serum phosphorus values without making diets highly restrictive, unpalatable, and difficult for adherence.

Obesity and Excess Energy Intake

Obesity is a common contributing factor to the development of CKD. Individuals who are obese frequently have increased renal plasma flow, GFR, and filtration fraction, presumably due to their higher dietary intake of protein and possibly other nutrients.[89] Albuminuria, glomerulomegaly, and glomerulosclerosis are more likely to be present in people with obesity. Obesity increases the risk of CKD by predisposing individuals to diabetic nephropathy, hypertensive nephrosclerosis, and focal and segmental glomerular sclerosis, even when diabetes is not present.[89] People with obesity are also at increased risk for calcium oxalate and urate stones, and renal cell carcinoma. In people with established CKD, obesity is more likely to be associated with a greater albuminuria and more rapid progression of kidney failure.

Lipids and Lipoproteins

Human studies have shown a classic pattern of lipids in patients with CKD; high triglycerides, low and dysfunctional high-density lipoproteins, and low-density lipoprotein concentration that does not correlate with the high CVD burden. Lipid-lowering therapy does not improve outcomes in patients with end-stage kidney disease (ESKD), and there is much debate about their use in patients with CKD who do not need dialysis.[90] Lipid metabolism is altered in proteinuric states and may affect podocyte function.[91] After intra-arterial procedures, cholesterol crystal deposition can cause acute kidney injury.[92] Although altered lipid metabolism is well-described in CKD, optimal medical and nutritional interventions to ameliorate adverse effects of lipids remain an area for study.

Water Intake

There is a common notion that high water intake should protect the kidneys. Actual interventional data on the effects of different amounts of water intake are limited. For certain health conditions, such as nephrolithiasis, the data are clear that higher water intake helps prevent the supersaturation of minerals and precipitation of more stones. The National Kidney Foundation suggests a total daily water intake of 3 L for men and 2.2 for women with CKD. However, in the REIN cohort study, both the lowest and highest water intakes were associated with faster CKD progression.[93] In the MDRD study, individuals with CKD in the highest quartile for water intake had the fastest rate of GFR decline. For normal individuals, the data are also dependent on nonrandomized trials, but there is some suggestion that higher water intake may be beneficial.[94] Overall, the data do not support an indiscriminately higher water intake to decrease the rate of progression in those with established CKD.

Medications

Although studies in animals indicate an important role for dietary restriction of protein and phosphorus and reduction or increase of certain dietary fats to control progressive renal failure, there is evidence that certain medications may substitute for, or possibly add to, the benefits of nutrient restriction.[59] Angiotensin II causes vasoconstriction, alters glomerular permeability to serum proteins, and may stimulate mesangial cell proliferation and aldosterone secretion.[95,96] ACE inhibitors (ACEIs) [medications that decrease blood pressure by inhibiting the enzyme that catalyzes the conversion of angiotensin I to angiotensin II] and angiotensin receptor blockers (ARBs) also lower glomerular capillary blood flow and blood pressure gradients across the glomerular capillary wall in rats with renal insufficiency.[96,97] These agents decrease high blood pressure and retard progressive renal failure in animals and humans with and without diabetes.[98-102] ACEIs and ARBs reduce urinary protein excretion in people with CKD and reduce or abolish microalbuminuria in those with and without diabetes.[103,104] The mineral corticosteroid, aldosterone, increases the production of cytokines that promote renal fibrosis and collagen matrix formation in the kidney.[105]

Medications that block the binding of aldosterone to its receptors reduce proteinuria in rats and in people with CKD, and slow the progression of CKD in rats.[106,107] Whether medicines that block receptors for aldosterone retard the progression of CKD in humans is not well established. Patients receiving aldosterone receptor blockers must be monitored for hyperkalemia.[106,107] Finerenone, a non-steroidal aldosterone receptor blocker, has been shown to lower the risk of CKD progression in people with CKD and type 2 diabetes.[108] Antihypertensive medicines also slow the progression of CKD by reducing blood pressure.

Human Studies on the Effect of Dietary Therapy on Progression of Chronic Kidney Failure

From the mid-1970s to the present, many, but not all, dietary studies in humans with renal insufficiency have indicated that low intakes of protein and phosphorus will retard the rate of progression of renal failure.[67,109-119] Some evidence indicates that low protein and phosphorus intakes may act separately to slow progressive renal failure.[120] Diet studies have evaluated diets that provided about 0.50 to 0.60 g protein/kg body weight/day or a very LPD containing about 0.28 g/kg/day (e.g., ~16 to 25 g protein/day). The very LPD diet was supplemented with 10 to 20 g/day of the nine essential amino acids, or mixtures of several essential amino acids and ketoacid or hydroxyacid analogs of the other five essential amino acids.[109,111-114,116,119] These diets were compared with either a more liberal diet containing approximately 1.0 g or more of protein/kg/day and more phosphorus, or an ad libitum diet.

The ketoacid or hydroxyacid analogs are structurally identical to their corresponding essential amino acids, except that the amino (NH_2) group attached to the second α-carbon of the amino acid is replaced with a keto group or a hydroxy group. These ketoacid and hydroxyacid analogs can be transaminated in the body to the respective amino acids, although a proportion of the analogs are degraded rather than transaminated. Because the ketoacids and hydroxyacids lack the nitrogen-containing amino group on the α– carbon, they provide the patient with a lesser nitrogen load. As they are degraded in the body mainly to carbon dioxide and water, they should engender fewer waste products that normally accumulate in renal failure. Ketoacid analogs of the branched-chain amino acids, especially leucine, may be more likely to promote protein accrual, possibly by decreasing protein degradation.[121,122]

The largest and most intensive examination of whether low-protein, low-phosphorus diets retard the rate of progression of renal disease was the National Institutes of Health–funded Modification of Diet in Renal Disease (MDRD) Study.[95,119] Using an intention-to-treat analysis, investigators studied the effects of three levels of dietary protein and phosphorus intake and two blood pressure management goals on the progression of CKD. A total of 840 adults with renal disease, excluding people with diabetes who were taking insulin, were divided into two study groups according to their GFR.

In Study A, 585 patients with GFR, measured by iodine-125 iothalamate clearances, of 25 to 55 mL/1.73m²/min were randomly assigned to either a usual-protein, usual-phosphorus diet (1.3 g/kg standard body weight/day of protein and 16 to 20 mg/kg/day of phosphorus) or a low-protein, low-phosphorus diet (LPD, 0.58 g/kg/day of protein and 5 to 10 mg/kg/day of phosphorus). Participants were also randomized to be treated with either a moderate or a strict blood pressure goal: mean arterial blood pressure 107 mm Hg (113 mm Hg for those ≥61 years of age) or 92 mm Hg (98 mm Hg for those ≥61 years of age). Study B included 255 patients with baseline GFR of 13 to 24 mL/1.73 m²/min. Patients were randomly assigned to the same low-protein, low-phosphorus diet or to the very-low-protein, very-low-phosphorus diet (VLPD, 0.28 g/kg/day of protein and 4 to 9 mg/kg/day of phosphorus) with a ketoacid-amino acid supplement (0.28 g/kg/day). They were randomly assigned to the same moderate or strict blood pressure control groups, as in Study A. Adherence to the dietary prescriptions was good.[119]

Participants in Study A prescribed the LPD had significantly faster declines in GFR during the first 4 months than those assigned to the usual-protein diet. Thereafter, the rate of decline of the GFR in the LPD group was significantly slower than in the group fed the usual-protein, usual-phosphorus diet. Over the entire treatment period, there was no difference in the overall rate of progression of renal failure. However, it is likely that the initial greater fall in GFR in the patients prescribed the LPD reflected a hemodynamic response to the reduction in protein intake rather than a greater rate of progression of the parenchymal renal disease. This response could be beneficial, reflecting a decrease of intrarenal hyperfiltration and intrarenal hypertension. In Study B, the VLPD group had a slightly slower decline of GFR than the LPD group, but the average rate of decline did not differ significantly between the two groups.

In the MDRD Study, the very-low-protein, ketoacid-amino acid supplemented–diet (VLPD) was not compared with the usual protein intake.[95,119] It is possible that the lack of significant effect of the LPD on the progression of renal failure could reflect the short mean duration of treatment in the MDRD study (2.2 years). If the trend toward the slower progression of renal failure in the LPD groups present at the termination of the MDRD study had persisted during a longer follow-up period, a statistically significantly slower progression may have been observed with the 0.58 g/kg protein diet (LPD) in Study A, and the VLPD in Study B. Several other characteristics of the patient population and study design of the MDRD trial may have led to the lack of statistically significant difference in the progression of renal failure between the diet groups.[123]

Vegetarian LPDs providing soy protein may retard the progression of CRF more effectively than diets with similar protein content containing animal protein.[70,124-127] The mechanisms of such an effect are unknown, but they may be related to the total content and different composition of fats in the vegetarian diet. The vegetarian diet is reported to improve the serum lipid profile in patients with CKD and nephrotic syndrome.[125,128] It is also possible that a soy protein diet provides less of an acid load (see above).

Several published meta-analyses have evaluated clinical trials of the effects of LPDs on the rate of progression of kidney failure. In general, the LPDs were also low in phosphorus. These meta-analyses evaluated different series of clinical trials, only some of which included the MDRD study.[67,129-132] Three of the meta-analyses used the onset of end-stage renal disease (ESRD), defined as when the CRF patient started treatment with MHD or CPD or received a kidney transplant,[129,130,132] as the key outcome. These meta-analyses reported statistically significant reductions of 0.54, 0.67, and 0.61, respectively, in the relative risk of a patient with CRF assigned to the LPDs, compared with higher protein intakes, on reaching this ESRD endpoint. The patients ingesting LPDs were reported to be started on MHD or CPD at lower GFRs than individuals with higher protein intakes. In contrast, one meta-analysis used the rate of decrease in GFR as the key outcome.[131] This last study described a slowing in the progression of renal failure of only 6% which, although statistically significant, was considered of questionable clinical significance.

The discrepancy between these two sets of findings may be partly explained by the fact that ingestion of LPDs leads to a reduction in the generation of protein and amino acid metabolism products, some of which may be toxic. Hence, LPDs may ameliorate uremic symptoms and delay the need to inaugurate dialysis treatment or renal transplantation at a GFR at which higher protein intakes will make patients sick from uremia, thereby inducing the physician to inaugurate renal replacement therapy (RRT, i.e., dialysis therapy or kidney transplantation). Starting RRT at a lower GFR can be considered an advantage to the LPDs, even if there is little or no delay in the rate of progression of CKD. In another study, elderly patients with GFRs of 5 to 7 mL/min/1.73 m² were randomly assigned to start treatment with MHD or to begin therapy with a very low-protein ketoacid-amino acid diet. The latter patients were treated with the diet for 1.0 to 58.1 months (median, 10.7).[133] Those assigned to the ketoacid diet appeared to fare as well, clinically and nutritionally, as did those assigned initially to MHD.

These meta-analyses also examined clinical trials that used an intention-to-treat analysis, whereby the data from individuals assigned to a given dietary intake were included in the results, whether they adhered to the dietary prescription or even were available for follow-up testing. Thus, these studies do not rule out the possibility that people who adhere closely to these LPDs may have a more significant delay in the rate of progression of their CKD.

Another meta-analysis analyzed the results of five prospective clinical trials of the effects of LPDs on the progression of diabetic CKD in patients with insulin-dependent diabetes.[130] This study indicated that LPDs retard progression in these individuals. However, smaller numbers of patients were analyzed; two of the trials did not have a randomized concurrent control group, and the key endpoints were less definitive.

In a secondary analysis of Study B, in which the decrease in GFR was correlated with the actual quantity of protein ingested, there was no statistically significant effect of consuming the LPD vs the very LPD supplemented with ketoacids and amino acids on the progression of renal failure.[95] However, in Study A, when the data from the two groups were combined and then analyzed, there was a significant inverse relationship between the protein intake actually ingested, as determined from the urea nitrogen appearance (UNA) and the rate of decline in GFR.[95] The actual dietary protein intake associated with the lowest rate of decline in GFR was about 0.62 g/kg/day. More recently, at least one other small-scale randomized controlled study compared treatment with a very-low-protein, ketoacid-amino acid–supplemented diet to a 0.6 g protein/kg/day diet in patients with CKD. The results indicated that the ketoacid diet was more effective at slowing the progression of CKD.[134]

The mechanism of action by which ketoacid diets retard the progression of renal failure is unclear. Studies suggest that alkalinization of the urine can retard the progression of renal failure.[75,77,135] Because ketoacids in the dietary supplements are present as alkaline salts, perhaps it is the alkali in these supplements that retard the progression of renal failure.

A 12-year follow-up analysis was conducted in Study A MDRD patients. The hazard ratio was adjusted for baseline characteristics of developing ESRD or a combination of either ESRD or all-cause mortality.[95] ESRD was defined as commencing MD therapy or receiving a kidney transplant. This study indicated that, during the first 6 years after the onset of the dietary protein prescription, there was a statistically significant adjusted lower hazard ratio of incurring ESRD, or the combination of either ESRD or mortality in those assigned to the 0.60 g protein/kg/day diet vs those assigned to the 1.3 g protein/kg/day diet.[95] This difference tended to reverse itself in the second 6-year period of follow-up.

In Study B patients, those assigned to the ketoacid-supplemented diet had a significantly greater hazard ratio, adjusted for baseline factors, for death after they developed ESRD during the 12 years after assignment to their diet prescription.[136] These data are intriguing because patients were treated in the study, on average, only for the first 2.2 years of follow-up and were then referred to their usual physicians. With few exceptions, ketoacid mixtures were unavailable to patients in the United States after the MDRD Study ended. Thus, the Study B patients could not have taken ketoacids during most of this 12-year observation period. It is difficult to know whether the greater 12-year mortality in the ketoacid-assigned group was a coincidence. A large, retrospective study from France did not confirm any difference in long-term mortality rate in patients with CKD who were prescribed ketoacid diets.[137] However, this latter study compared the survival of patients fed ketoacid-feds with survival data from the French dialysis registry, and with patients who underwent transplant procedures in Bordeaux.

The Nurses' Health Study compared protein intakes of individuals with different levels of GFR, determined from their serum creatinine concentration.[60] In this study, 1,624 women, aged 42 to 68 years, had protein intake measured in 1990 and again in 1994 using a semiquantitative food frequency questionnaire. Women with baseline estimated GFR >55 mL/min/1.73 m^2 but <80 mL/min/1.73 m^2 showed a fall in GFR of −1.69 mL/min/1.73 m^2 per 10-g increase in protein intake. However, after adjustment for measurement error, the change in estimated GFR was −7.72 mL/min/1.73 m^2 per 10-g increase in protein intake. This association was of borderline statistical significance. High intake of nondairy animal protein in people with mild renal insufficiency was associated with a significantly greater fall in estimated GFR (−1.21 mL/min/1.73 m^2 per 10-g increase in nondairy animal protein intake). A retrospective study in renal transplant recipients indicated that those recipients who ingested higher protein diets experienced greater losses of GFR.[138]

Taken together, the post hoc analysis of the MDRD Study data, the results of the approximately 12 years of follow-up to the MDRD Study, the Nurses' Health Study, the study in renal transplant recipients, the four meta-analyses, and the newer randomized ketoacid studies all point to the probability that LPDs will retard the rate of progression of renal failure in individuals with CKD. As these LPDs often engender sufficiently lower uremic toxicity for a given level of reduced renal function, patients fed these diets may be able to avoid MHD, CPD, or renal transplantation at GFR levels that would require patients with CKD who consume higher protein intakes to commence such therapy.

Another unresolved issue is whether LPDs will retard progression of renal failure in patients receiving ACEIs and/or ARBs. Because dietary protein restriction exerts many of the same hemodynamic and other physiologic effects on the kidney as ACEIs and ARBs,[31,139,140] it is possible that the renal-protective effects of LPDs, when combined with ACEIs and ARBs, are replicative, rather than additive. In a small study, 82 patients with type 1 diabetes were randomly assigned to a LPD (0.60 g protein/kg/day) or a usual-protein diet[141]; most patients were receiving ACEIs. In this 4-year trial, the low-protein group experienced a 10% incidence of death or ESRD, compared with a 27% incidence in the patients eating the usual protein diet (P <0.042). However, there was no difference in the rate of decline in GFR in the two groups.[141]

Another potential advantage of a LPD is that it can reduce proteinuria, which is associated with greater progression of renal failure and increased risk of CVD. Although ACEIs and ARBs, singly or in combination, and lowering of high blood pressure reduce proteinuria, these treatments do not eradicate it. Thus, there may also be an antiproteinuric role for LPDs, at least in patients with persistent proteinuria. Similarly, low sodium intake may reduce the magnitude of proteinuria.

Nutritional Alterations in the Nephrotic Syndrome

The nephrotic syndrome is a kidney disorder characterized by losses of large quantities of protein in the urine (≥3.0 g/day), low serum albumin, high serum cholesterol and other fats, and accumulation of excess body water to form edema.[142] This condition is caused by diseases that affect the glomerulus and increase glomerular permeability to protein. Because people with nephrotic syndrome have large urinary protein losses and their appetite is frequently poor, they often develop protein wasting and debility.

Certain vitamins and most trace elements are protein bound in plasma, and patients are also at risk for developing deficiencies of these nutrients. Excessive urinary iron, copper, and vitamin D losses and vitamin D deficiency have been reported in people with nephrotic syndrome.[142,143] Malnutrition may occur in these patients, even when they do not have advanced kidney failure. For a given type of renal disease, heavy proteinuria is associated with a more rapid progression of renal failure, possibly because of the incorporation of proteins into the glomerular mesangium that may cause sclerosis or inflammatory responses.[144] Many growth factors and other bioactive substances are also bound to proteins that are filtered by the leaky glomerulus in patients with nephrotic syndrome. It is postulated that some of these bioactive compounds, when filtered and exposed to the renal tubular lumen or renal interstitium, may promote progressive renal damage.[145] Protein-restricted diets, ACEIs, ARBs, and aldosterone receptor antagonists, blood pressure control independently of these medicines, and reduced urinary sodium excretion may each reduce renal losses of protein.[146-148] Use of these medications in patients with nephrotic syndrome to reduce proteinuria, while maintaining a sufficiently high protein intake to increase both albumin and total body protein mass may be desirable.

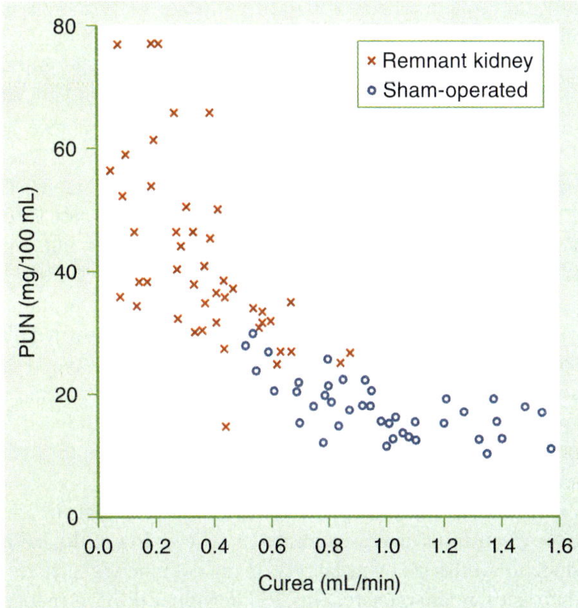

Figure 96.1 Relationship between plasma urea nitrogen (PUN) and the glomerular filtration rate as indicated by urea clearance in Sprague-Dawley rats with chronic renal insufficiency and sham-operated controls. Chronic renal failure was produced by ligation of two-thirds to three-fourths of the arterial supply to the left kidney and contralateral nephrectomy.

Reproduced from Kopple JD. In: Alfin-Slater BB, Kritchevsky D, eds. Nutrition and the Kidney. Human Nutrition: A Comprehensive Treatise. New York, NY: Plenum Publishing, 1979;4:409–57.

Nutritional and Metabolic Consequences of Chronic Renal Failure

Clinical, Nutritional, and Metabolic Disorders

CRF causes pervasive nutritional and metabolic disorders that may affect virtually every organ system. Advanced CRF is a complex disorder caused by a marked reduction in the excretory, endocrine, and metabolic functions of the kidney. Patients with CRF eventually develop uremia, the accumulation of nitrogenous and other metabolites in the blood, in combination with the clinical signs and symptoms of advanced renal failure. Most of these compounds are products of amino acid and protein metabolism. Quantitatively, the most prominent are urea, creatinine, other guanidine compounds, and uric acid (**Figure 96.1**). It is generally believed that some of these compounds are toxic in high concentrations.

The signs and symptoms of uremia include weakness, insomnia, fatigue, loss of appetite, nausea, vomiting, weight loss, diarrhea, itching, muscle cramps, hiccups, twitching or jerking of the extremities, fasciculations, tremors, emotional irritability, and decreased mental concentration and comprehension. A characteristic fetid breath caused, at least partly, by the exhalation of methylamines, is often present.

Altered serum concentrations of other electrolytes and acidemia (excessive accumulation of hydrogen ions in the blood) can occur. They can have profound and life-threatening effects on the physiologic processes and metabolism of the body (see Figure 96.1). Abnormalities in water and electrolyte balance and acidemia are caused by impaired ability of the failing kidney to regulate, by excretion, the content of water, salts, and acids in the body. The sodium and water disturbances associated with renal failure can lead to edema in the extremities and lungs, congestive heart failure, and hypertension or, if excessive sodium depletion occurs, reduction in extracellular fluid volume and a fall in blood pressure. When renal failure is not in the end stage, most of these clinical and metabolic disorders can be ameliorated or prevented with diet and medications. Severe, untreated uremia can lead to lethargy, loss of consciousness, coma, convulsions, and death.

Advanced CRF causes pervasive alterations in the absorption, excretion, or metabolism of many nutrients. These disorders include the following: accumulation of chemical products of protein metabolism,[3] decreased ability of the kidney to excrete a large sodium load or to conserve sodium rigorously when dietary sodium is restricted[149]; impaired renal ability to excrete water, potassium, calcium, magnesium, phosphorus, trace elements, acids, and other compounds; a tendency to retain phosphorus[12,150-154]; decreased intestinal absorption of calcium[154] and possibly iron[153]; and high risk for developing vitamin deficiencies, particularly vitamins B_6, C, and folic acid, and vitamin D.[152,155] The patient with CRF is also likely to accumulate potentially toxic chemicals, such as aluminum, that are normally ingested in small amounts and excreted in the urine.[153]

Uremia is also a polyendocrinopathy, and many metabolic and clinical manifestations of uremia are caused by endocrine disorders. Many hormone concentrations are elevated in renal failure, particularly peptide hormones, because of the impaired ability of the kidney to degrade peptides. These include PTH, leptin, glucagon, insulin, growth hormone (GH), prolactin, luteinizing hormone, often follicle-stimulating hormone (FSH), and gastrin.[154,156-162] Increased secretion of some hormones, particularly PTH and, to some degree, insulin, may contribute to elevated plasma concentrations. Patients with CRF have altered thyroid hormone concentration, similar to the euthyroid sick syndrome, but mild hypothyroidism may not be uncommon.[163] Of the hormones elaborated by the kidney, plasma erythropoietin and 1,25-dihydroxycholecalciferol are reduced[7,8,96,152]; and plasma renin activity may be increased, normal, or decreased.[6]

Serum GH is elevated, and serum IGF-I is usually normal in renal failure, but there is resistance to the activity of both GH and IGF-I.[164,165] The impaired sensitivity to glucagon is reversed by hemodialysis, although hyperglucagonemia persists.[157] There is resistance to the peripheral actions of insulin.[166] These effects on insulin and glucagon contribute to the mild glucose intolerance usually present in patients without diabetes but with CRF.[167] Obesity, common in patients with CKD, may contribute to glucose intolerance. Impaired actions of hormones in uremia may result from circulating inhibitors in serum, downregulation of receptor numbers, or postreceptor defects in the signal transduction system. Cytosolic calcium participates in certain cell signaling systems. Elevated basal cytosolic calcium, induced by hyperparathyroidism, appears to be one of the postreceptor signal transduction disorders caused by CRF.[168]

Most of the products of metabolism that accumulate in renal failure result from decreased excretion. The ability of the failing kidney to synthesize or metabolize many compounds, including amino acids, is also impaired. In CRF, the kidney displays reduced glutamine catabolism, impaired alanine synthesis, and decreased glycine conversion to serine and citrulline to urea.[169,170] Serum and tissue taurine concentrations are often low.

Quantitatively, the most important end product of nitrogen metabolism is urea.[171] In a clinically stable patient with CRF who eats at least 40 g of protein/day, the net quantity of urea produced daily contains an amount of nitrogen equal to about 80% to 90% of the daily nitrogen intake. Guanidines are the next most abundant product of nitrogen metabolism. Guanidine compounds include creatinine, creatine, guanidinosuccinic acid, and methylguanidine. Many polypeptides and small proteins also accumulate in CRF.[172] Prime suspects for uremic toxins include urea, guanidine compounds, phenolic acids, proinflammatory cytokines, carbonyl compounds, oxidants, and some of the hormones elevated in uremic plasma, especially PTH and possibly glucagon.[3,157,168,173,174] Urea is probably only a mild toxin and increased or reduced concentrations of many compounds contribute to the uremic syndrome.

Disordered gastrointestinal function may affect nitrogen metabolism in patients with CRF. The gastrointestinal tract metabolizes urea, uric acid, creatinine, and choline. It synthesizes or releases from larger molecules dimethylamine, trimethylamine, trimethylamine oxide, ammonium, sarcosine, methylamine, methylguanidine, p-cresol sulfate, p-cresyl-sulfate, p-cresyl-glucuronide, indoxyl sulfate, and indoxyl-3-aldehyde.[172] The gut metabolism or synthesis of many of these compounds is increased in CRF, possibly because of an altered gut microbiome, which includes a rise in the quantity of intestinal bacterial flora.[175,176]

Some of the metabolic alterations in uremia are adaptive homeostatic responses that offer both benefits and disadvantages to the patient.[173] Hyperparathyroidism is an example. As the kidneys fail, impaired excretion of phosphorus leads to phosphorus retention. Concomitantly, the diseased and scarred renal parenchyma is less able to convert 25-hydroxycholecalciferol to the most potent metabolite of vitamin D, 1,25-dihydroxycholecalciferol. Low plasma 1,25-dihydroxycholecalciferol leads to an increase in PTH secretion.[154]

Deficiency of 1,25-dihydroxycholecalciferol impairs intestinal calcium absorption and causes resistance to the actions of PTH in bone.[152,154] These alterations, as well as increased serum phosphorus, promote hypocalcemia and contribute to the development of hyperparathyroidism. Elevated serum PTH reduces renal tubular reabsorption of phosphorus (enhancing urine phosphorus excretion), lowers serum phosphorus, promotes renal synthesis of 1,25-dihydroxycholecalciferol, mobilizes calcium from bone, and increases intestinal calcium absorption. However, intestinal calcium absorption usually remains low or, in mild renal insufficiency, normal. The benefits derived from these homeostatic actions are that more normal plasma phosphorus and calcium concentrations are maintained in patients with mild to more advanced renal insufficiency. The tradeoff is the development of hyperparathyroidism.[173,174] PTH has been implicated as a pervasive uremic toxin that adversely affects many organs and tissues and contributes to the uremic syndrome.[174] Fibroblast growth factor-23 (FGF-23) is a bone-derived hormone that also regulates phosphorus homeostasis by reducing renal tubular phosphorus reabsorption and promoting urinary phosphorus excretion. FGF-23, which is αKlotho-dependent, also suppresses 1,25-dihydroxycholecalciferol synthesis in the kidney, stimulates calcium and sodium reabsorption in the distal renal tubules, and plays an important role in bone mineralization.[154]

Anemia, which usually is primarily, but not exclusively, the result of impaired erythropoiesis caused by deficiency of erythropoietin, can be treated effectively with this hormone.[8] To reduce risk of adverse events, including the need for blood transfusions, and the high costs of therapy, sufficient erythropoietin analogs are usually given to raise hemoglobin concentration, but only to maintain hemoglobin to no greater than 11.5 g/dL.[2,8,177,178] Large doses of iron are usually given intravenously to patients undergoing MD, and orally or intravenously to patients with CKD who are not receiving dialysis, to attain a higher serum total iron binding saturation. Erythropoietin and other erythropoiesis-stimulating agents are expensive and, in large doses, possibly hazardous. Higher serum iron will often decrease the amount of these medications needed to maintain target hemoglobin concentration.[179]

When kidney failure is a complication of an underlying systemic disease, such as diabetes, hypertension, or lupus erythematosus, other manifestations of these underlying diseases may also adversely affect the patient and be progressive. These problems do not affect every patient; many people with CRF or who are undergoing dialysis lead full and productive lives.

With the institution of a special diet or treatment with MHD or CPD, blood concentrations of many metabolic products that accumulate in uremic plasma decrease, and the patient may experience clinical improvement. MHD or CPD enables people without renal function to live for many years. Despite such improvement, however, many clinical and metabolic disorders may persist or even progress. These include oxidative and carbonyl stress, an inflammatory state; type IV hyperlipidemia and other disorders of lipid metabolism[180,181]; cardiovascular, cerebrovascular, and peripheral vascular diseases[182]; osteodystrophy with disordered bone architecture, osteoporosis, or osteomalacia (aluminum toxicity often contributes to the osteomalacia)[154]; anemia[7,8]; impaired immune function and decreased resistance to infection; mildly impaired peripheral and central nervous system function; muscle weakness and atrophy; frequent viral hepatitis[183]; sexual impotence and infertility; generalized protein-energy wasting (PEW)[184-191]; depression; and poor rehabilitation.[192] Most of these complications are aggravated by a nutritionally poor diet and improved with good nutrition.

The intestinal absorption, excretion, and/or metabolism of many nutrients may be altered in CRF. In addition, decreased food intake and excessive intake of certain minerals, such as phosphorus, sodium, or potassium, may alter clinical and nutritional status. Medicinal therapy may adversely affect nutrient metabolism in renal failure. Anticonvulsant medications may cause deficiencies of vitamin D and folic acid; hydralazine, isoniazid, and other medications may cause vitamin B_6 deficiency.[193] Part of the challenge of diet therapy is to provide for the altered requirements and tolerance of many nutrients that occur in CRF.

Protein-Energy Wasting (PEW)

Patients with CRF, particularly those undergoing MHD or CPD, frequently show evidence of wasting, particularly PEW (**Table 96.3**). The term PEW is used because not all causes of PEW result from inadequate nutrient intake.[184-191,194-196] Evidence for PEW includes decreased relative body weight, BMI, skinfold thickness, arm muscle mass, total body nitrogen, and potassium, subjective global nutritional assessment (SGA), low growth rate in children, decreased serum concentrations of proteins including albumin, transthyretin (prealbumin), and transferrin, and low muscle alkali-soluble protein.[197] The plasma amino acid pattern, which is pathognomonic for renal failure, is similar to that seen with malnutrition.[198] PEW is sometimes observed in patients with CRF who are not receiving dialysis, but symptoms are more prevalent in patients undergoing MHD or CPD.[199] Most surveys of patients undergoing MD indicate that they have increased prevalence of mild to moderate PEW,[184-191,194,195,200-202] compared with patients not receiving dialysis. Epidemiologic data show that the prevalence of PEW increases as kidney function declines, spanning from <2% in CKD stages 1–2 to 11%–54% in CKD stages 3 to 5.[200] In addition to PEW, patients with CRF are at higher risk for deficiencies of iron, zinc, and vitamins B_6, C, folic acid, and 1,25-dihydroxycholecalciferol, and possibly, carnitine and other nutrients.[203-207]

There are many causes of wasting in CRF (**Table 96.4**).[194,208] Dietary intake is often inadequate to meet increased energy needs.[184,188,209,210] Low dietary intake is primarily the result of

Table 96.3 Evidence for Protein-Energy Wasting in Patients with Advanced Chronic Kidney Failure[a]

- Decreased
  - Body weight
  - Height (children)
  - Growth (children)
  - Body fat (skinfold thickness)
  - Fat-free solids
  - Intracellular water
  - Muscle mass (midarm muscle circumference)
  - Total body potassium (nondialyzed patients)
  - Total body nitrogen (patients receiving chronic peritoneal dialysis)
  - Total albumin mass, synthesis, and catabolism
  - Valine pools (nondialyzed patients)
  - Serum
    - Total protein
    - Albumin
    - Transferrin
    - Prealbumin (transthyretin)
    - C3
    - C3 activator
    - Cholinesterase
  - Plasma
    - Leucine
    - Isoleucine
    - Total tryptophan
    - Valine
    - Tyrosine
    - Valine/glycine ratio
    - Essential/nonessential amino acid ratio
  - Muscle
    - Alkali-soluble protein
    - RNA/DNA ratio
    - Valine
    - Tyrosine
- Normal to increased
  - Plasma
    - Glycine

[a]People with CRF may have normal values for these parameters, but statistical comparisons indicate that the concentrations are often abnormal.

Table 96.4 Causes of Protein-Energy Wasting (PEW)

1. Decrease protein and energy intake
   a. Anorexia
      i. Retained products of metabolic processes
         a. Especially products of nitrogen metabolism
      ii. Reduced naturally occurring orexigens or increased naturally occurring anorexigens
      iii. Increased proinflammatory cytokines
   b. Restriction of food groups that reduce essential nutrient intake
   c. Emotional depression or severe anxiety
   d. Diseased or impaired organ function that participates in nutrient intake
   e. Intercurrent illnesses
   f. Physical or financial inability to obtain or prepare food
2. Hypermetabolism
   a. Increased energy expenditure due to
      i. Inflammation
      ii. Intercurrent illnesses
      iii. Altered adiponectin and resistin metabolism
3. Impaired activity of anabolic hormones
   i. Insulin resistance
   ii. Growth hormone and IGF-I resistance
   iii. Hypothyroidism—often subclinical
   iv. Testosterone deficiency
4. Enhanced activity of catabolic hormones
   i. Hyperparathyroidism
   ii. Hyperglucagonemia
   iii. Increased sensitivity to the catabolic actions of glucagon
   iv. Increased sympathetic nervous system (adrenalin) activity
5. Metabolic acidosis with acidemia
6. Decreased physical activity
7. Other frequent intercurrent or comorbid illnesses
   i. Comorbidities (diabetes mellitus, heart failure, pneumonia, obstructive sleep apnea, peripheral vascular disease)
8. Dialysis procedure
   a. Nutrient losses into dialysate
   b. Dialysis-induced inflammation
9. Dialysis-related hypermetabolism

(continues)

Table 96.4 Causes of Protein-Energy Wasting (PEW) *(continued)*

10. Decrease protein and energy intake
    a. Anorexia
        i. Retained products of metabolic processes
            a. Especially products of nitrogen metabolism
        ii. Reduced naturally occurring orexigens or increased naturally occurring anorexigens
        iii. Increased proinflammatory cytokines
    b. Restriction of food groups that reduce essential nutrient intake
    c. Emotional depression or severe anxiety
    d. Diseased or impaired organ function that participates in nutrient intake
    e. Intercurrent illnesses
    f. Physical or financial inability to obtain or prepare food
11. Hypermetabolism
    a. Increased energy expenditure due to
        i. Inflammation
        ii. Intercurrent illnesses
        iii. Altered adiponectin and resistin metabolism
12. Impaired activity of anabolic hormones
        i. Insulin resistance
        ii. Growth hormone and IGF-I resistance
        iii. Hypothyroidism—often subclinical
        iv. Testosterone deficiency
13. Enhanced activity of catabolic hormones
        i. Hyperparathyroidism
        ii. Hyperglucagonemia
        iii. Increased sensitivity to the catabolic actions of glucagon
        iv. Increased sympathetic nervous system (adrenalin) activity
14. Metabolic acidosis with acidemia
15. Decreased physical activity
16. Other frequent intercurrent or comorbid illnesses
        i. Comorbidities (diabetes mellitus, heart failure, pneumonia, obstructive sleep apnea, peripheral vascular disease)
17. Dialysis procedure
    a. Nutrient losses into dialysate
    b. Dialysis-induced inflammation
18. Dialysis-related hypermetabolism

anorexia, most often caused by uremic toxicity, a CRF-related inflammatory state, the anorexigenic effects of acute or chronic superimposed illnesses, and depression.

The superimposed illnesses that people with CRF frequently sustain often induce a catabolic state.[200-202,211-213] Dialysis itself may induce wasting. Hemodialysis and peritoneal dialysis remove free amino acids, peptides, or bound amino acids,[214-217] water-soluble vitamins,[155] proteins (with peritoneal dialysis and, to a lesser extent, with hemodialysis),[215,218] glucose (during hemodialysis with glucose-free dialysate),[219] and other bioactive compounds. Hemodialysis also increases net protein breakdown, mainly by activating the complement cascade system and inducing the release of catabolic cytokines.[220,221] Excessive acid accumulation in the blood (acidemia) engenders protein catabolism.[222] Blood losses occur due to frequent blood drawing for laboratory testing, common occult gastrointestinal bleeding, and the sequestration of blood in the hemodialyzer and blood tubing after a hemodialysis treatment.[223] Because blood is a rich source of protein, these losses may contribute to protein depletion.

Other possible but not well established causes of wasting include altered endocrine activity, particularly resistance to insulin,[166] GH, and IGF-I,[164-166] hyperglucagonemia,[157] hyperparathyroidism,[154,168,173,174] deficiency of 1,25-dihydroxycholecalciferol,[152] endogenous and exogenous uremic toxins, such as aluminum; and loss of metabolic functions of the kidney. Because the kidney is a metabolic organ that synthesizes or degrades many biologically valuable compounds, including amino acids,[169,170] the loss of these activities in kidney failure could disrupt the body's metabolism and promote wasting.

Inflammation and Oxidant and Carbonyl Stress in Chronic Renal Failure and Patients Undergoing Maintenance Dialysis

People with CRF and those undergoing MD frequently show evidence of inflammation. This evidence includes increased serum concentrations of acute-phase proteins, including C-reactive protein (CRP), serum amyloid A, and ceruloplasmin. Serum concentrations of negative acute-phase proteins, including albumin, transferrin, transthyretin (prealbumin), and cholesterol-carrying lipoproteins, may decrease as a result of PEW and as a direct result of inflammation.[210,224-230] Most surveys suggest that serum CRP is increased in about 30% to 50% of U.S. and European patients, and a lower proportion of Asian patients, treated with MHD.[207,225] Serum proinflammatory cytokines, including tumor necrosis factor-α (TNF-α), interleukin-1 (IL-1), and IL-6, are commonly elevated in people with advanced CRF.[207,225] In patients with advanced CRF and those receiving MD, there is also an accumulation in the serum of compounds that cause oxidant or carbonyl stress.[225,231,232] Oxidant stress refers to cellular injury caused by exposure of the cell to compounds that oxidize chemicals in the cell.[232] Carbonyl stress refers to cellular injury caused by carbon-containing compounds that react with compounds in the cell.[231] Homocysteine is such a carbonyl-reactive compound that is increased in the plasma of people with CRF or receiving MD treatment and which exerts several adverse effects on the vascular endothelium.[233-237]

There are several causes of inflammation in individuals with CRF. They include comorbid illnesses, CRF (which itself leads to an increase in serum concentrations of several oxidants, reactive carbonyl compounds, and proinflammatory cytokines), oxidant and carbonyl stress, chronic low-grade infections (e.g., by chlamydia), and reaction to the vascular access prostheses that are necessary to perform hemodialysis, the hemodialyzer itself, the peritoneal dialysis catheter (for CPD-treated patients) and, for patients treated with MHD or CPD, dialyzer tubing or impure dialysate.[224,228-230]

Protein-Energy Wasting and Inflammation

PEW and inflammation are associated with increased risk of morbidity and mortality in people undergoing MD.[9,200-202,227,232,238] Markers of inflammation have been linked to atherosclerosis and cardiovascular morbidity and mortality.[225,226,230,237] Laboratory research indicates that certain acute-phase proteins, oxidants, reactive carbonyl compounds, and proinflammatory cytokines can be toxic to the endothelium. These compounds may cause inflammation, cellular proliferation, and increased matrix deposition in the endothelium, with the formation of inflamed atherosclerotic plaques that are likely to rupture and increase the likelihood of myocardial infarction or stroke. The increased risk of morbidity and mortality in patients undergoing MD who have PEW and inflammation is of importance because the adjusted mortality rate for patients treated with MD in the United States is 21% to 22% annually.[213]

There is an overlap between the manifestations of protein-energy malnutrition (PEM) and inflammation. PEM and inflammation display reduced serum concentrations of negative acute-phase proteins such as albumin, transferrin, transthyretin, and cholesterol-carrying lipoproteins.[210,227-230,239] The lowest serum concentrations of these proteins are generally found with inflammation rather than PEM. Inflammation may cause PEW by inducing anorexia (e.g., TNF-α and IL-6 are anorexigenic) and engendering a hypercatabolic state.[230,239-241] The relative contributions of PEW and inflammation to the high morbidity and mortality of patients with CRF are controversial, especially because PEW shares many clinical manifestations with inflammation. These considerations have led some investigators to question whether PEW is hazardous or is only an important risk factor for morbidity and mortality when it is associated with inflammation.[242] Intuitively, it seems that nutrient intake inadequate to maintain healthy body protein mass must eventually place the patient at risk for increased morbidity and mortality. This is a question that demands further investigation. PEM and inflammation occur together so commonly in patients with ESRD that investigators have described them as components of a single syndrome referred to as the malnutrition-inflammation complex syndrome (MICS).[243] Several studies indicate that nutritional support improves PEW in patients undergoing MD and those with CRF not receiving dialysis.[244] Data from well-designed retrospective studies suggest that nutritional treatment of PEW may improve survival in patients receiving MHD with PEW.[244]

Numerous researchers have pointed out that relationships between traditional risk factors and mortality are markedly altered or even reversed in patients undergoing MD.[245,246] These altered risk factor patterns have been observed for BMI, predialysis serum total cholesterol, LDL cholesterol, predialysis creatinine, serum urea nitrogen (SUN), UNA (net urea production), blood pressure, magnitude of acidosis, and possibly serum homocysteine and PTH.[227,228,235,238,247-254] Although several explanations have been advanced to explain these phenomena, the most likely mechanism is related to malnutrition-inflammation complex syndrome.[247,249,252,253] Patients undergoing MD with nutritionally inadequate diets, who have major comorbid conditions or evidence of systemic inflammation, are more likely to have low BMI and be at greater risk for mortality. Healthier individuals are more likely to have a greater appetite and have higher body weight, greater protein intake, and hence higher UNA and serum urea, and greater metabolic acid production and metabolic acidosis.

Survival bias may also account for some paradoxical risk factors (i.e., those individuals with abnormal levels of these risk factors may be more likely to die earlier before they start MD therapy). These explanations cannot completely account for the altered risk factor patterns in patients treated with MD. As an example, individuals with obesity have lower mortality rates than those patients with overweight.[245,246,250,255] Further research is needed in this area.

Dietary Management of Chronic Kidney Disease

This section describes approaches to the dietary management of patients who are non-dialysis dependent (NDD) and those who are dialysis-dependent with chronic kidney disease (CKD). A recommended plan for nutrient intake for patients with NDD-CKD receiving maintenance dialysis is provided in **Table 96.5**.

General Principles of Dietary Management

The impaired capacity of the kidneys to excrete minerals and metabolic products of many ingested nutrients, and the high prevalence of metabolic disorders, PEW, obesity, diabetes, and hypertension, along with the greater risk of cardiovascular complications, and evidence that dietary intake may ameliorate the progression of CKD, indicate that nutritional management is a critical aspect of the treatment of patients with CKD and those receiving chronic dialysis. The five goals of dietary therapy are to 1) prevent or treat obesity and diabetes, 2) prevent or minimize uremic toxicity and the metabolic derangements of renal failure, 3) reduce risk of adverse cardiac and vascular events, 4) prevent or correct PEW, 5) slow/delay the rate of CKD progression and need for renal replacement therapy.[256-258]

Adherence to prescribed diets may be challenging for patients with CKD and their care partners. Patients may undergo major changes in their lifestyle with negative consequences to their health-related quality of life (HRQOL). For example, they may need to severely limit or eliminate favorite foods to avoid metabolic derangements (e.g., hyperphosphatemia, excess sodium intake with subsequent edema and worsened hypertension, acidemia). They may need to procure difficult-to-obtain and more expensive foods, prepare special recipes, and consume foods they do not like. Food preparations may make demands on the time, effort, and emotional support system of their care partners. Therefore, it is of paramount importance that clinicians and interprofessional care team members use a patient-centered approach to the diet and nutritional management of patients with CKD that 1) prioritizes shared decision making, 2) does not impose radical changes in dietary intake without a clear indication that such modifications may be beneficial, and 3) incorporates the expertise of specialty-trained renal registered dietitians (RDs). To ensure effective and safe implementation of nutritional management, patients and their care partners should receive counseling and training on the principles of dietary therapy and the design and preparation of diets, as well as continuous encouragement regarding dietary adherence by an RD. Careful monitoring of diet and intake of specific nutrients is essential to avoid declines in energy and either excessive or deficient nutrient intakes, which can set the stage for malnutrition and metabolic and clinical disorders that may be encountered in CKD.

A team-based approach to diet management is essential to the effective and safe implementation of prescribed diets. The team should include patients and their care partners, nephrologists, specialty-trained renal RDs, nurses and, where available, educators (i.e., for management of diabetes and dialysis treatment), and social workers. Diet plans should be tailored to the individual patient, using a personalized approach. At routine visits, clinicians and interprofessional team members should monitor dietary intake and discuss results with the patients and their

Table 96.5 Recommended Dietary Intake for Patients with CKD Who Are Non-Dialysis Dependent or Receiving Maintenance Dialysis

	Non-Dialysis-Dependent Chronic Kidney Disease	Maintenance Hemodialysis or Chronic Peritoneal Dialysis
Protein	Stages 3-5 CKD without DM: Low protein diet 0.55–0.60 g/kg/day, OR Very low-protein diet 0.28–0.43 g/kg/day+additional keto acid/amino acid analogs to meet protein requirements (0.55–0.60 g/kg/day) Stage 3–5 CKD with DM: Low protein diet 0.6–0.8 g/kg/day	Hemodialysis and Peritoneal Dialysis Without DM: 1.0–1.2 g/kg/day Hemodialysis and Peritoneal Dialysis With DM: 1.0–1.2 g/kg/day For patients with DM at risk of hyper- and/or hypoglycemia, higher levels of dietary protein intake may need to be considered to maintain glycemic control.
Energy	25–35 kcal/kg/day based on age, sex, level of physical activity, body composition, weight status goals, CKD stage, and concurrent illness or presence of inflammation	
Fat (% of total energy intake)*	30%	30%
Polyunsaturated/ saturated fatty acid ratio	1.0:1.0	1.0:1.0
Carbohydrates	Rest of nonprotein energy	
Total fiber intake	20–30 g/day	20–30 g/day
Minerals	Range of intake	
Sodium	<100 mmol/day (or <2.3 g/day)	<100 mmol/day (or <2.3 g/day)
Potassium	Adjust dietary potassium intake to maintain serum potassium within the normal range; usually ≤70–100 mEq K/day depending on the stage of CKD	Adjust dietary potassium intake to maintain serum potassium within the normal range; usually ≤70–80 mEq K/day
Phosphorus	Adjust dietary phosphorus intake to maintain serum phosphorus concentrations in the normal range; usually ≤800 mg P/day†	Adjust dietary phosphorus intake to maintain serum phosphorus concentrations in the normal range; usually ≤1,000 mg P/day†
Calcium	800–1,000 mg/day (including dietary calcium, calcium supplementation, and calcium-based phosphate binders)	800–1,000 mg/day (including dietary calcium, calcium supplementation, and calcium-based phosphate binders)
Magnesium	200 mg/day	200–250 mg/day
Water	≤3,000 mL/day usually, as tolerated	750–1,500 mL/day
Vitamins	**NOTE: The following is from the KDOQI Guidelines. The authors' recommendations and justification for their recommendations are given in the text.**	
Vitamin B_{12}	Routine supplementation not necessary, unless needed to correct folate of B_{12} deficiency/insufficiency	Routine supplementation not necessary, unless needed to correct folate of B_{12} deficiency/ insufficiency
Vitamin C	If at risk of vitamin C deficiency, 75 mg/day in women and 90 mg/day in men	If at risk of vitamin C deficiency, 75 mg/day in women and 90 mg/day in men
Folic acid	Routine supplementation not necessary, unless needed to correct folate of B_{12} deficiency/insufficiency	Routine supplementation not necessary, unless needed to correct folate of B_{12} deficiency/ insufficiency
Vitamin A	No routine supplementation	No routine supplementation
Vitamin D	See text	See text
Vitamin E	No routine supplementation	No routine supplementation
Vitamin K	No routine supplementation	No routine supplementation

	Non-Dialysis-Dependent Chronic Kidney Disease	Maintenance Hemodialysis or Chronic Peritoneal Dialysis

*If patients have hypertriglyceridemia, fat intake should be <15% of total energy; fish oils can be used to replace some long-chain triglycerides.

†To maintain serum phosphorus within the normal range, intestinal phosphate binders or inhibitors of intestinal phosphate absorption, in addition to dietary phosphorus restriction, are often necessary in non-dialysis dependent patients with advanced CKD and are almost always necessary for dialysis-dependent patients.

CKD, chronic kidney disease; *DM*, diabetes mellitus.

Note: Some researchers in the field of renal nutrition take a decidedly different position from the 2020 KDOQI recommendations on vitamin supplementation. These researchers recommend that advanced CKD patients and MD patients should be routinely prescribed supplements of the water soluble vitamins as well as vitamin D compounds. Also, increased quantities of vitamin B_6 (pyridoxine HCl) should be taken at 5 mg per day for NDD-CKD patients and chronic peritoneal dialysis patients and 10 mg per day for MHD patients. The rationale for these recommendations is as follows. The prescribed dialysis diet for many NDD-CKD patients and chronic dialysis patients is low in many water-soluble vitamins, in part because of the need to restrict intake of high potassium foods (e.g., fruits and vegetables). In addition, advanced CKD and MD patients often undergo periods of time where they become anorexic and eat poorly, due to intercurrent illnesses or emotional distress. Moreover, these patients are often prescribed medicines that antagonize or interfere with vitamins. The increased need for vitamin B_6 intake is based on several studies describing vitamin B_6 deficiency if these patients do not take vitamin B_6 supplements.[204] The dose of vitamin B_6 intake recommended is based on a dose-finding study.[204] The rationale for the routine intake of vitamin D compounds is given below.

Modified from Ikizler TA, Burrowes JD, Byham-Gray LD, et al. Am J Kidney Dis 2020;76:S1–S107.[260]

care partners. Interprofessional team members should support the renal RDs' efforts to train and counsel patients and monitor dietary adherence. Clinicians and interprofessional team members should assume an energetic, positive, and sympathetic approach to promote adherence to prescribed diets. Evidence has underscored that the preceding techniques can enable many patients to attain acceptable levels of dietary adherence.[259]

Patients with advanced CKD (i.e., stages 4–5 CKD) are at high risk for inadequate energy intake. Because prescribed diets are sometimes marginally low in certain nutrients (e.g., protein) and high in others (e.g., fat and select carbohydrates), and PEW is common, it is important to routinely evaluate the nutritional adequacy of the diet and the patient's nutritional status. Such evaluation should include an assessment of protein-energy status, mineral and bone metabolism parameters, risk factors for CVD, and urine albumin excretion, including the presence of microalbuminuria and the RD's comprehensive nutrition assessment. The updated 2020 National Kidney Foundation (NKF) Kidney Disease Outcome Quality Initiative (KDOQI) Clinical Practice Guidelines for Nutrition in CKD recommendations for the evaluation of protein-energy status in patients undergoing chronic dialysis are shown in **Table 96.6**.[260,261] To maintain

Table 96.6 Recommended Measures for Monitoring Nutritional Status of Maintenance Dialysis Patients

Category	Measure	Minimum Frequency of Measurement
I. Measurements that should be routinely performed in all patients as usual care	Body weight and body mass index	Monthly
	Body composition	Monthly
	Specialty dietitian administered comprehensive nutrition assessment including appetite, dietary intake, body weight/body mass index, biochemical data, anthropometry, nutrition-focused findings	Within first 90-days of dialysis, annually, or when indicated by nutrition screening or provider referral
II. Technical devices and anthropometry measurements to assess body composition	Skinfold thickness and waist circumference	As needed
	Creatinine kinetics	As needed
	Conicity index	As needed
	Dual-energy x-ray absorptiometry	As needed
	Bioelectric impedance	As needed (hemodialysis only)
III. Laboratory measurements	Serum albumin	Monthly or as needed
	Normalized protein nitrogen appearance (normalized protein catabolic rate)	Every 3 months or as needed
	Serum prealbumin	As needed
	Serum urea nitrogen (SUN)	Monthly or as needed
	Serum creatinine	Monthly or as needed
	Serum potassium	Monthly or as needed
	Serum phosphorus	Monthly or as needed

(continues)

Table 96.6 Recommended Measures for Monitoring Nutritional Status of Maintenance Dialysis Patients *(continued)*

Category	Measure	Minimum Frequency of Measurement
IV. Handgrip strength	Handgrip strength	As needed
V. Energy requirements	Indirect calorimetry	As needed
	Disease-specific predictive energy equations	As needed
VI. Composite nutritional indices	7-Point Subjective Global Assessment	As needed
	Malnutrition Inflammation Score	As needed
VII. Protein and energy intake assessment methods	3-Day Food Record	As needed
	24-Hour Food Recall	As needed
	Food Frequency Questionnaire	As needed

Data from Ikizler TA, Burrowes JD, Byham-Gray LD, et al. Am J Kidney Dis 2020;76:S1–S107.[260]

good dietary compliance, monitor fluid and electrolyte disorders, and adequate assessment of clinical and nutritional status, it is often preferable for clinicians and RDs to see patients with stages 3 to 5 CKD frequently, often monthly. Patients with slowly progressive mild or moderate stages 2, 3, or even 4 CKD may be able to see the physician and RD less frequently. Cross-sectional studies indicate that dietary protein and energy intake begins to decline, and protein-energy status begins to deteriorate when the glomerular filtration rate (GFR) decreases to roughly 30 to 35 mL/min/1.73m^2).[262-264] The decline in nutritional status is gradual and usually mild to moderate until the GFR falls severely, to less than ~10 mL/min/1.73m^2, and when patients with advanced CKD transition to dialysis therapy.[209,262,265,266] Patients with CKD with greater decreases in body weight have increased risk of mortality.[263,264] Although nutritional status may improve during the first few months of dialysis therapy,[267] adequate protein and energy intake and nutritional status at the onset of maintenance dialysis treatment appears to be a reliable predictor of both nutritional status 2 to 3 years later and of longevity.[265,268] Hence, dedicated attention and concerted efforts should be made to prevent malnutrition when patients with advanced CKD approach the time of dialysis transition and during the first few weeks of maintenance dialysis therapy. Such efforts should be directed toward maintaining good nutritional intake, rapidly instituting therapy for supervening illnesses, and maintaining adequate nutritional intake during such illnesses.

Urea Nitrogen Appearance and the Ratio of Serum Urea Nitrogen to Serum Creatinine

The control of dietary protein intake is pivotal to nutritional management of patients with NDD and dialysis-dependent CKD, as well as those with acute kidney injury. Hence, one must accurately monitor nitrogen intake. Those who are in nitrogen balance should have total nitrogen output (appearance) equal to nitrogen intake minus about 0.5 g nitrogen per day for unmeasured losses from growth of skin, hair, and nails and from sweat, respiration, flatus, and blood drawing.[269] For clinical purposes, a slightly positive or negative balance does not substantially alter the use of the nitrogen output to estimate intake. If patients are in very positive or negative balance, such as from pregnancy or severe infection, nitrogen output may not reflect intake.

The measurement of total nitrogen output or appearance is too laborious, inconvenient, and expensive to be broadly applied in the clinical setting. However, because urea is the major nitrogenous product of protein and amino acid degradation, urea nitrogen appearance (UNA) can be used to estimate total nitrogen output and, hence, nitrogen intake.[270-272] UNA refers to the amount of urea that appears or accumulates in body fluids and all outputs, such as urine, dialysate, and fistula drainage. The term UNA is used rather than urea production or generation because some urea is degraded in the gastrointestinal tract; the ammonia released from urea is largely transported to the liver and converted back to urea.[273,274] Thus, the enterohepatic urea cycle has little effect on urea or total nitrogen economy, and this cycle can be ignored without compromising the ability of the UNA to estimate total nitrogen output or intake accurately. Moreover, the recycling of urea cannot be measured without costly and time-consuming isotope studies. UNA is calculated as follows:

Equation 1:

$$\begin{aligned}\text{UNA (g per day)} &= \text{urinary urea nitrogen (grams per day)}\\ &\quad + \text{dialysate urea nitrogen (grams per day)}\\ &\quad + \text{change in body urea nitrogen (grams per day)}\end{aligned}$$

Equation 2:

$$\begin{aligned}&\text{Change in body urea nitrogen (grams per day)} =\\ &\quad (\text{SUN}_f - \text{SUN}_i\text{, g/L per day}) \times \text{BW}_i\,(\text{kg}) \times (0.60\ \text{L/kg})\\ &\quad + (\text{BW}_f - \text{BW}_i\text{, kg per day}) \times \text{SUN}_f\,(\text{g/L}) \times (1.0\ \text{L/kg})\end{aligned}$$

where i and f are the initial and final values for the period of measurement, SUN is serum urea nitrogen (grams per liter), BW is body weight (kilograms), 0.60 is an estimate of the fraction of body weight that is water, and 1.0 is the fractional distribution of urea in the weight that is gained or lost (i.e., 100%).

The estimated proportion of body weight that is water may be increased in patients who are edematous or lean, and decreased in individuals who are obese or in children. Changes in body weight during the 1 to 3 day measurement of UNA are assumed to be due to changes in body water. In patients undergoing hemodialysis, the urea concentration in the dialysate is low and difficult to measure accurately, and UNA can be calculated during the interdialytic interval and then normalized to 24 hours. Because many patients undergoing dialysis have little or no urinary excretion, the equation for calculating their UNA during the interdialytic interval can often be simplified to Equation 2.

In the authors' metabolic studies, the relationship between UNA and total nitrogen appearance (output) in chronically uremic patients not undergoing dialysis is as follows[270]:

Equation 3:

$$\text{Total nitrogen appearance (g per day)} = 1.19\ \text{UNA (g per day)} + 1.27\ \text{(g per day)}$$

If the individual is in neutral nitrogen balance, the UNA also will correlate closely with nitrogen intake. Equation 4 describes observed relationships between UNA and dietary nitrogen intake in patients not receiving dialysis who are clinically stable, primarily with stage 5 CKD, and who are in neutral protein balance.

Equation 4:

$$\text{Dietary nitrogen intake (g per day)} = 1.20\ \text{UNA (g per day)} + 1.74$$

Multiplication of Equation 3 by 6.25 will convert total nitrogen output to net protein degradation (grams per day), that is, the difference between the absolute rates of body protein degradation and protein synthesis. Multiplication of Equation 4 by 6.25 will convert dietary nitrogen to dietary protein intake (grams per day). When both nitrogen intake and UNA are known, nitrogen balance can be estimated from the difference between nitrogen intake and total nitrogen appearance estimated from the UNA. If the patient is markedly anabolic, such as during pregnancy, particularly in its later stages, Equation 4 will underestimate nitrogen intake. For patients with large protein losses, such as from nephrotic syndrome or peritoneal dialysis, or who are acidemic and have sufficient kidney function to excrete large quantities of ammonium, Equations 3 and 4 will underestimate both nitrogen appearance and nitrogen intake. Under most circumstances; however, the UNA provides a powerful tool for monitoring nitrogen output and intake or estimating balance. Maroni et al. and other researchers described similar techniques for monitoring these parameters.[269,271]

The relationships among the UNA, total nitrogen appearance, and dietary nitrogen intake in patients undergoing continuous ambulatory peritoneal dialysis (CAPD) are shown in Equations 5 and 6. Other researchers have described similar equations.[275] Because protein losses in peritoneal dialysate are variable, some equations include an independent term for daily nitrogen losses from protein in the peritoneal dialysate. Multiplication of these terms by 6.25 can convert the equations to net protein output (grams per day) or, in clinically stable patients who are in approximately neutral protein balance, to dietary protein intake (grams per day).

Equation 5 for clinically stable CAPD patients:

$$\text{Total nitrogen appearance (g per day)} = 0.94\ \text{UNA} + 5.54$$

Equation 6 for clinically stable CAPD patients:

$$\text{Dietary nitrogen intake (g per day)} = 0.97\ \text{UNA} + 6.80$$

The UNA (also called Gu) can be estimated in patients receiving hemodialysis by urea kinetic modeling.[272,276] This technique involves the predialysis and post-dialysis SUN and body weight, the urea clearance characteristics of the hemodialyzer, and the blood flow, dialysate flow, and duration of the dialysis therapy.

The relationships among UNA, net protein degradation, and dietary protein intake in patients receiving MHD have been described in other studies.[272,276] Net protein degradation in patients undergoing dialysis, normalized to body weight, is usually referred to as the nPNA (normalized protein equivalent of total nitrogen appearance) or nPCR (normalized protein catabolic rate).[277] A critique of the precision and reproducibility of these calculations is presented elsewhere.[270,276] The nPNA or nPCR, indicating net total body protein degradation, commonly underestimates total protein intake in MHD and CPD patients (compare Equation 4 with Equation 6). This occurs because more nitrogen is excreted in the breath at higher protein intakes.

The SUN-to-serum creatinine ratio also correlates closely with dietary protein or amino acid intake in patients with CKD who are NDD. This relationship can be used to estimate the recent daily protein intake of these patients. Although this ratio is not as precise as the UNA and can be influenced by certain clinical factors,[278] it is easy and inexpensive to measure.

Dietary Prescription

Body Weight Assessment in Nutritional Management of CKD

For purposes of nutritional prescription, the body weights in this chapter refer to median body weights of average Americans for a given height, age, sex, and frame size, as reported by National Health and Nutrition Examination Survey (NHANES) data (**Table 96.7**).[65,260] For patients with obesity or underweight (i.e., >115% or <95% of ideal or standard body weight, respectively), the 2020 KDOQI Clinical Practice Guidelines for Nutrition in CKD advise the use of adjusted body weight (ABW).[260] The use of ABW is based on the rationale that about 25% of excess body weight (i.e., adipose tissue) in patients with obesity is metabolically active tissue. However, ABW has not been validated for use in CKD and may overestimate or underestimate dietary protein and energy requirements.

Dietary Protein Intake

Non-Dialysis Dependent Chronic Kidney Disease. A large body of evidence generated from observational studies, clinical trials, and meta-analyses shows the effect of low-protein diets on reduced risk of ESKD, need for renal replacement therapy, mortality, and CKD-related metabolic complications, as well as improved HRQOL.[67,279] In addition, metabolic balance studies in adults with and without CKD have shown that, in the context of adequate energy intake (≥30 kcal/kg/day), (A) dietary protein intake can be safely reduced to 0.55 to 0.60 kg/day,[224,280,281] and (B) further reduction of dietary protein intake to 0.3 to 0.4 grams/kg/day with the addition of some essential amino acids (EAAs) and ketoacid or hydroxy acid analogs (KAAs) of the other EAAs can also ensure a sufficient intake of protein and amino acids to meet protein requirements.[282] Different compositions of KAAs and EAAs have been tested in the setting of CKD, with most containing four ketoacid analogs (of the EAAs isoleucine, leucine, phenylalanine, and valine), one hydroxyacid analog (of the EAA methionine), and four amino acids considered essential in CKD

Table 96.7 Body Weight Metrics in Chronic Kidney Disease

Metric	Definition/Formula	Strengths and Limitations
Ideal body weight	▪ Body weight associated with the lowest mortality for a given height, age, sex, and frame size and is based on the Metropolitan Life Insurance Height and Weight Tables for normal people; Hamwi method can be used to estimate desirable body weight. Women: 100 lb (45.36 kg) for first 5′0″ (127 cm) and add 5 lb (2.27 kg) for each additional inch (25.4 cm) >5′0″ ▪ Men: 106 lb (48.08 kg) for first 5′0″ (127 cm) and add 6 lb (2.72 kg) for each additional inch (25.4 cm) >5′0″	▪ Not generalizable to the CKD population and data-gathering methods were not standardized. ▪ None of these methods has been validated for CKD or chronic dialysis patients
Standard body weight	▪ Average 50th percentile weights for normal men and women by age, height, and frame size in the United States (based on NHANES II data)	▪ Used in the original KDOQI nutrition guideline ▪ Although data are validated and standardized and use a large database of ethnically diverse groups, data are provided only on what individuals weigh, not what they should weigh to reduce morbidity and mortality. Also, the data are derived from people without CKD, not from people with CKD.
Desirable body weight	▪ Based on body mass index	▪ In studies of dialysis patients showing that higher body mass index is associated with lower mortality, researchers may not have statistically adjusted for all confounders related to comorbid conditions, and it is unclear how it may relate to patients with CKD not receiving dialysis
Adjusted body weight	Adjusted BW = ideal BW + [(actual BW − ideal BW) × 0.25] ▪ It is recommended that body weight should be adjusted for calculation of nutrient recommendation if patient's weight is <95% or >115% of ideal/standard BW: adjusted BW = edema-free BW + [(standard BW − edema-free BW) × 0.25]	▪ Based on the theory that only 25% of the excess body weight (adipose tissue) in obese patients is metabolically active tissue. This has not been validated for use in CKD and may either overestimate or underestimate energy and protein requirements
Edema-free body weight	▪ Analogous to estimated dry weight in the patient being treated by renal replacement therapies	
% of usual body weight	▪ Percent usual BW = {(usual BW − current BW)/usual BW} × 100	

CKD, chronic kidney disease; *NHANES*, National Health and Nutrition Examination Survey; *BW*, body weight.

Reproduced from Ikizler TA, Burrowes JD, Byham-Gray LD, et al. Am J Kidney Dis 2020;76:S1–S107.[260]

(tryptophan, threonine, histidine, and tyrosine).[260,282] Collectively, these supplements are referred to as ketoacids.

Among adults with stages 3 to 5 NDD-CKD without underlying diabetes who are metabolically stable, the 2020 KDOQI Clinical Practice Guidelines for Nutrition in CKD recommend a low or very low protein diet supplemented with KAAs (**Table 96.8**) to reduce the risk of ESKD and death based on Grade 1A evidence (i.e., strong recommendation; very confident that the true effect lies close to that of the estimate of the effect) and also to improve HRQOL based on Grade 2C evidence (i.e., weak recommendation; moderately confident in the effect estimate: the true effect is likely to be close to the estimate of the effect, but there is a possibility that it is substantially different).[260] Specifically, they recommend (1) a low protein diet providing 0.55 to 0.60 g/kg of body weight/day or (2) a very low-protein diet providing 0.28 to 0.43 dietary protein intake/kg/day with additional keto acids/amino acid analogs. Other clinical practice guidelines from the International Society of Renal Nutrition and Metabolism[283] and Kidney Disease Improving Global Outcomes (KDIGO)[284] provide slightly varying recommendations for dietary protein intake in NDD-CKD (Table 96.8).

Among adults with stages 3 to 5 NDD-CKD with underlying diabetes, the updated 2020 KDOQI nutrition guidelines also advise that it is reasonable to prescribe dietary protein intake of 0.6 to 0.8 g/kg/day under close supervision. With the advent of strong evidence showing the cardiorenal benefit of sodium-glucose cotransporter-2 (SGLT2) inhibitors in patients with CKD with and without diabetes[285-288] and nonsteroidal mineralocorticoid antagonists (ns-MRAs) in patients with CKD and diabetes,[289,290] further research is needed to determine the combined effects of these pharmacotherapies with low protein diets in CKD.

Table 96.8 Recommendations for Dietary Protein Intake and Other Major Macronutrients in Adult Chronic Kidney Disease Patients Across Clinical Practice Guidelines

	NKF KDOQI[65]	KDIGO[284]	ISRNM[283]
Protein intake for NDD-CKD patients (g/kg/day)	Stages 3–5 CKD without DM: Low protein diet 0.55–0.60 g/kg/day, OR Very low-protein diet 0.28–0.43 g/kg/day+ additional keto acid/amino acid analogs to meet protein requirements (0.55–0.60 g/kg/day) Stage 3–5 CKD with DM: Low protein diet 0.6–0.8 g/kg/day	Stages 4–5 CKD with or without DM: Low-protein diet 0.8 g/kg/day CKD patients at risk of progression Avoid high-protein intake (>1.3 g/kg/day)	Low protein diet 0.6–0.8 g/kg/day
Protein intake for chronic dialysis patients (g/kg/day)	Hemodialysis and Peritoneal Dialysis Without DM: 1.0–1.2 g/kg/day Hemodialysis and Peritoneal Dialysis With DM: 1.0–1.2 g/kg/day For patients with DM at risk of hyper- and/or hypoglycemia, higher levels of dietary protein intake may need to be considered to maintain glycemic control	No recommendation	>1.2 g/kg/day
Energy intake (kcal/kg/day)	25–35 kcal/kg body weight per day based on age, sex, level of physical activity, body composition, weight status goals, CKD stage, and concurrent illness or presence of inflammation	No recommendation	30–35 kcal/kg/day
Sodium intake (mmol/day)	Stages 3–5 CKD, dialysis, or post-transplantation: <100 mmol/day (or <2.3 g/day)	<90 mmol/day (<2 g/day) of sodium (corresponding to 5 g/day of sodium chloride)	80–100 mmol/day
Potassium intake (mmol/kg/day)	Stages 3–5 CKD, dialysis, or posttransplantation: Adjust dietary potassium intake to maintain serum potassium within the normal range	Recommend expert dietary advice and information in the context of an education program, tailored to severity of CKD and the need to intervene on potassium where indicated	<1 mmol/kg/day if increased in NDD-CKD and HD patients; usually not an issue in PD patients
Phosphorus intake (mg/day)	Stages 3–5 CKD, dialysis, or posttransplantation: Adjust dietary phosphorus intake to maintain serum phosphate concentrations in the normal range	Recommend expert dietary advice and information in the context of an education program, tailored to severity of CKD and the need to intervene on phosphorus, where indicated	800–1,000 mg/day in addition to phosphate binders if serum phosphorus is increased

NKF KDOQI, National Kidney Foundation Kidney Disease Outcomes Quality Initiative; *KDIGO*, Kidney Disease Improving Global Outcomes; *ISRNM*, International Society of Renal Nutrition and Metabolism; *NDD-CKD*, non-dialysis dependent CKD; *HD*, hemodialysis; *PD*, peritoneal dialysis.

Reproduced from Ikizler, T Alp et al. Am J Kidney Dis 2020;76(3 suppl 1):S1–S10; Ikizler TA, Cano NJ, Franch H, et al. Kidney Int 2013;84:1096–107; Kidney Disease: Improving Global Outcomes (KDIGO) CKD Work Group. Kidney Int 2013;3:1–150.

While the 2020 KDOQI nutrition guidelines do not distinguish among recommendations for dietary protein intake across varying estimated GFR (eGFR) levels within the stages of 3 to 5 NDD-CKD, the potential advantages of using a low-protein diet become even more compelling at more advanced levels of kidney dysfunction (i.e., eGFR <25 mL/min/1.73m^2) for several reasons. First, at this degree of kidney dysfunction, potentially toxic products of nitrogen metabolism begin to accumulate in larger quantities. Second, because low-protein diets generally contain less phosphorus, phosphorus intake can be reduced more readily with these diets (see later sections on recommended phosphorus intake). As a result of these first two factors, the patient may be less uremic for a given low level of GFR and may be able to delay the onset of dialysis treatments using a conservative management approach.[67,257,258,291] These diets may also retard the progression of CKD. Third, some patients with CKD eat too little protein rather than too much; hence, specific training and encouragement to follow a prescribed diet may increase the likelihood that the patient will not ingest too little protein.

Low Protein Diets in the Conservative Management of Patients with Advanced CKD. When the eGFR declines to less than 5 mL/min/1.73m^2, there has been uncertainty as to whether patients fare as well with low protein diets as they will do with inaugurating routine dialysis treatment and commencing higher dietary protein intakes. Because patients with these low eGFR levels may be at high risk for PEW and long-term sequelae of uremic toxicity,[209,262,265,266] many nephrologists recommend commencing renal replacement therapy at this time. However, there is growing evidence that supports the use of dietary intervention and conservative non-dialytic management as an alternative approach for people with advanced CKD. One study showed that older adults with GFRs of 5 to 7 mL/min/1.73m^2 fared as well clinically when they were fed very low protein diets supplemented with ketoacid-amino acid mixes, as they did when they were started on chronic dialysis therapy and prescribed higher dietary protein intakes.[133] Although there may be concern that older adults with CKD are at heightened risk for PEW, growing research has shown that low-protein diets may be effective in reducing CKD progression in both older and younger adults.[292] In a study of 352 patients with NDD-CKD who received moderate, low, and very low protein diets, after a median follow-up 4.2 years, the lower protein intakes were associated with significantly slower decline in eGFR in the overall cohort and in the older (>65 years) patients.[293] In another study of 56 patients with NDD-CKD aged >70 years who were followed for a median period of 26.5 months, those who received a ketoacid/EAA supplemented very low protein diet did well clinically with a delay in the commencement of chronic dialysis by an average of >10 months.[133]

If patients cannot maintain sufficiently high energy intake to maintain their body weight and there is no other identifiable cause for weight loss, transition to dialysis therapy should be considered.[133] This is important because PEW at the commencement of maintenance dialysis therapy is a predictor for higher mortality.[244,294,295] Patients receiving MHD may require nutritional supplementation (i.e., oral nutritional supplements, intradialytic parenteral nutrition).[296] The distinct dietary protein needs of patients treated with various modalities of chronic dialysis, including incremental hemodialysis, are discussed next.[291,297,298]

Maintenance Dialysis: Hemodialysis, Peritoneal Dialysis, and Incremental Dialysis. Based on earlier[299,300] and more contemporary studies of dietary protein requirements of patients receiving MHD,[301,302] it has been hypothesized that this population has greater protein needs because of the removal of amino acids and peptides by dialysis procedures,[214-216] catabolic and inflammatory stimuli engendered by the dialysis procedure (e.g., complement activation) and, acidemia.[221,303] Prior research studies have shown that most clinically stable patients receiving MHD will maintain protein balance with a diet providing adequate energy intake and 1.0 g protein/kg/day, but a safe intake that will maintain balance in most patients is probably about 1.15 g protein/kg/day. Patients undergoing chronic peritoneal dialysis have daily losses into peritoneal dialysate of about 9 g of protein, a small amount of bound amino acids (e.g., peptides), and about 2.5 to 4.0 g/day of free amino acids. They are also subjected to inflammatory and other catabolic stimuli.[216,218]

The updated 2020 KDOQI nutrition guidelines advise that, in adult patients with stage 5 CKD receiving hemodialysis or peritoneal dialysis with or without diabetes who are metabolically stable, dietary protein intake of 1.0 to 1.2 g/kg/day should be prescribed to maintain stable body protein, based on opinion-level evidence.[260] Practice guidelines from the International Society for Renal Nutrition and Metabolism[283] and KDIGO recommend slightly higher dietary protein targets (Table 96.8).[284] KDOQI guidelines also advise that for patients at high risk of hyper- and hypoglycemia, higher dietary protein intakes may be helpful to maintain glycemic control.[260]

There has been growing interest in incremental hemodialysis as a personalized, patient-centered treatment approach.[304] The concept of incremental dialysis was first described for people undergoing CPD. The KDOQI Clinical Practice Guidelines for Peritoneal Dialysis Adequacy recommended that, for patients with significant residual kidney function, consideration may be given to starting CPD therapy incrementally.[305-307] Similarly, for incremental hemodialysis, residual kidney function can be used as a guide to establish and adjust the hemodialysis prescription. The 24-hour urine urea clearance (Kru) may be the threshold for initiating treatment with twice-weekly hemodialysis (i.e., at a Kru ≤3.0 ml/min/1.73m^2); twice-weekly hemodialysis treatment can be maintained as long as the Kru is >2.0 ml/min/1.73m^2.[291,297,298] Given that more frequent hemodialysis may potentially accelerate loss of residual kidney function due to ischemic insults (i.e., from intradialytic hypotension, postdialysis hypovolemia), release of nephrotoxic mediators during hemodialysis (i.e., due to exposure to dialysis tubing and impurities), decreased circulating urea leading to reduction in osmotic diuresis, and deactivation of remaining nephrons,[291,297,298] the main rationale for a less frequent hemodialysis regimen is to preserve residual kidney function. Preserved residual kidney function may, in turn, lead to better nutritional status. The improved nutritional status would be due to a combination of increased removal of uremic toxins and, therefore, decreased anorexia, earlier removal of middle molecules (due to earlier start of chronic dialysis), better water volume and blood pressure control, reduced left ventricular hypertrophy, preservation of healthy endocrine function, and improved HR-QOL.[308-311] Further research examining the effects of incremental hemodialysis on these endpoints is needed.

Caria et al., examined the effects of integrating nutritional therapy with incremental dialysis on residual kidney function. In a multicenter, prospective study of 68 patients with advanced CKD and eGFR of 5 to 10 mL/min/1.73m^2, 38 patients joined a Combined Diet Dialysis Program (CDDP) in which they underwent weekly hemodialysis paired with a dietary protein restriction of 0.6 g/kg/day. These patients were compared with 30 patients with advanced CKD who transitioned to standard thrice-weekly hemodialysis.[312] Over a 24-month follow-up period, patients in the CDDP group demonstrated greater maintenance of residual kidney function and urine output, lower erythropoietin resistance index, decreased serum β2 microglobulin concentration, reduced need for cinacalcet and phosphate binders, and stable parameters of nutritional status. Further studies are needed to determine the optimal diet prescription for patients with ESRD undergoing various frequencies of hemodialysis treatment.

Dietary Protein Source: Plant-Based Protein Sources. In addition to the quantity of dietary protein ingested, there is growing interest in dietary protein sources for patients with CKD. In patients with NDD-CKD, vegetables and fruits may have salutary effects on CKD outcomes by reducing the dietary acid load and ameliorating the adverse effects of acidosis, including slowing the rate of CKD progression.[313] Animal-based proteins increase dietary acid load, whereas many vegetables and fruits contain proteins that are more acid-neutral or provide an alkali load. In contrast, grain-based foods appear to be mildly acid-producing. Nuts, wild rice, lentils, corn, and coconuts contain phytate, which binds to phosphate in foods and decreases intestinal absorption of phosphorus.[313,314] Greater emphasis on vegetable-, fruit-, and other plant-based proteins may also have favorable effects on other adverse CKD-related disorders, including reducing hyperglycemia, hypertension, elevated body mass index, and nephrolithiasis.[126,127,315,316]

Although the 2020 KDOQI nutrition guidelines indicate that there is insufficient evidence to recommend a particular source of dietary protein (plant vs animal) in patients with stage 1 to 5 CKD, including those receiving dialysis,[260] there is a growing body of largely epidemiologic evidence suggesting that diets containing large amounts of vegetables and fruits[32,126,313] may be associated with improved kidney and metabolic parameters in patients with NDD-CKD. For example, in patients with nephrotic syndrome, earlier studies have shown that a vegetarian, soy-based LPD reduced proteinuria and dyslipidemia.[124,125,128] In a small clinical trial of 41 patients with diabetic kidney disease (DKD) randomized to a soy protein diet containing 35% animal, 35% soy, and 30% vegetable proteins, or a control diet providing 70% animal and 30% vegetable proteins, and followed for 4 years, those randomized to the soy protein group showed significant reduction in serum glucose, lipids, CRP, urinary creatinine, and proteinuria.[317] A subsequent observational study of 420 patients with type 2 diabetes from the Dutch DIALECT-1 cohort showed that among those who consumed dietary protein intake of about 0.8 g protein/kg/day, greater consumption of vegetable protein was associated with lower prevalence of CKD, as defined by a eGFR < 60 mL/min/1.73^2.[318]

The Ongoing Telmisartan Alone and in Combination with Ramipril Global Endpoint Trial (ONTARGET) quantified the effect of lifestyle and dietary modifications on kidney disease by estimating population-attributable characteristics. A secondary analysis of 6,916 adults without diabetes indicated that greater dietary vegetable intake was associated with lower risk of CKD and mortality.[319] Most recently, in a nationally representative sample of 1,994 adults with NDD-CKD selected from the continuous NHANES, both the amount and source of dietary protein were associated with mortality.[279] Compared with participants with dietary protein intake of 0.6 to $<$1.0 g/kg/day, those with incrementally greater protein intake had increasingly worse survival. Additionally, participants in the highest tertile of high biologic value proteins (i.e., proteins typically from animal sources) had higher mortality risk, compared with those with intakes in the lowest tertile.

A note of caution must be made because LPDs for patients with CKD that contain less than 0.8 g protein/kg/day provided exclusively or almost entirely from plant proteins may be deficient in one or more essential amino acids and other essential nutrients.[320,321] Hence, if the dietary protein is mainly derived or exclusively from plant protein, it is important that a dietitian help patients design diets to meet all of their nutritional needs.

Dietary Energy Intake

Studies in patients with NDD-CKD and MHD indicate that energy expenditure is normal or nearly normal when patients are lying in bed, sitting, following ingestion of a standard meal, and during defined exercise.[281,322,323] Nitrogen balance studies in patients with stages 4 to 5 NDD-CKD who were ingesting diets providing 0.55 to 0.60 g of dietary protein/kg/day and 15, 25, 35, or 45 kcal/kg/day indicate that the energy intake necessary to ensure neutral or positive nitrogen balance is approximately 35 kcal/kg/day.[281] Similar findings were obtained in nitrogen balance studies of patients receiving MHD who were ingesting 1.1 g protein/kg/day and 25, 35, or 45 kcal/kg/day.[324] However, virtually every survey of energy intake in patients with stages 4 to 5 NDD-CKD and MHD indicates that, on average, the dietary intake is lower than this level, usually substantially lower than 30 kcal/kg/day.[325-328] In patients with NDD-CKD or undergoing MHD, decreased body fat is one of the more prominent alterations in nutritional status, supporting the contention that these patients require more energy than they usually ingest.[325,326,328] Conversely, patients who receive chronic peritoneal dialysis (CPD) often gain fat due to the additional energy they receive from glucose absorbed from the dialysate in the peritoneal cavity plus the subsequent increase in circulating insulin.

The updated 2020 KDOQI Guidelines recommend that patients with stages 1 to 5 CKD, including those receiving dialysis, and patients who are postkidney transplant who are metabolically stable, should have an energy intake of 25 to 35 kcal/kg body weight/day based on age, sex, physical activity level, body composition, body weight goals and stage of CKD to maintain normal nutritional status.[260] Concurrent illnesses, including hyperparathyroidism, infection, and other causes of inflammation, may increase energy expenditure and requirements. These factors should be considered when designing the dietary energy prescription for an individual patient. Patients should be monitored routinely to assess whether energy requirements are being met.

Dietary Carbohydrate Intake

Clinical practice guidelines provide limited guidance regarding the optimal amount and type of carbohydrate intake for patients with NDD-CKD and those on chronic dialysis, due to limited evidence in this area.[260] However, based on principles developed for the general population, it is suggested that patients with CKD should be encouraged to eat complex rather than refined carbohydrates to reduce triglyceride synthesis and to improve glucose tolerance.

Dietary Fiber Intake. Dietary fiber is a nondigestible form of carbohydrate found in plant foods, including fruits, vegetables, whole grain products, beans, nuts, and seeds.[126,127,329,330] Fiber may increase satiety and bulk in the intestinal contents, which may improve gastrointestinal motility and reduce constipation. Select dietary fibers may also reduce cholesterol and glucose absorption and trap bile acids, subsequently improving serum glucose and lipid profiles. Fiber can alter the intestinal microbiota composition and function by increasing gut microbiome-induced production of short-chain acids, leading to lower pH in the colonic lumen, increased bioavailability of some minerals, and modulation of the immune system.[126,127,329-333] Studies in the physiologically normal population also suggest that high dietary fiber intake may reduce the incidence of constipation, irritable bowel syndrome, diverticulitis, and colonic neoplasia.[329-333] One observational study of a nationally representative cohort of 2,677 adults with CKD and 34,478 adults without CKD showed that people who were in the lowest tertile of dietary fiber intake had higher risk of all-cause death than those with fiber intake in the highest tertile.[329]

Soluble fiber, which is soluble in the intestinal lumen but not absorbed, includes pectins, certain gums, and psyllium.[126,127,329-333]

Supplemental soluble dietary fiber may also reduce plasma total and LDL cholesterol and triglyceride concentrations. A high dietary fiber intake may also reduce colonic bacterial ammonia generation and enhance fecal nitrogen excretion. Foods high in fiber are often high in potassium, phosphorus, and low-quality protein. Thus, one must pay attention to the daily mineral intake when prescribing high-fiber diets to patients with CKD. More rapid satiety associated with increased fiber intake may suppress energy intake, which may not be desirable for patients receiving chronic dialysis who have inadequate energy intake.

Although the updated 2020 KDOQI nutrition guidelines do not directly suggest an optimal amount of dietary fiber intake for patients with CKD,[260] other clinical practice guidelines provide mixed recommendations due to limited evidence in this area.[334] The KDOQI Guidelines on Dyslipidemia in CKD recommend a daily intake of 20 to 30 g of fiber/day.[335] The KDIGO Clinical Practice Guidelines for Diabetes and CKD advise that for patients with diabetic kidney disease (DKD), 40% to 50% of energy from carbohydrates should come from whole grains. The European Society for Clinical Nutrition and Metabolism recommends the inclusion of fiber-rich foods in the diet,[336] whereas the Kidney Health Australia-Caring for Australians with Renal Impairment (CARI) group recommends a balanced diet rich in fruits and vegetables.[337]

Dietary Fat Intake

Various practice guidelines provide recommendations for dietary fat intake for patients with CKD.[334] The KDOQI guidelines for diabetes and CKD recommend that, for patients with DKD, 30% of energy intake should come from dietary fat, with 5%, 5%, 10%, and 10% of fat derived from saturated fat, omega-6, omega-3, and omega-9 fatty acids, respectively.[338] In contrast, the KDIGO guidelines advise that <15% of total energy intake come from dietary fat in patients with CKD and hypertriglyceridemia, and, that fish oils should replace some of the long-chain triglycerides.[339] The European Society for Clinical Nutrition and Metabolism Guidelines on Enteral Nutrition advises that there should be a distribution between lipids and carbohydrates (i.e., 30% and 55%–60%, respectively), with an emphasis on unsaturated fatty acids; for patients with elevated serum cholesterol, they recommend that ≤30% of energy should be derived from dietary fat, with <10% saturated fatty acids, >10% monounsaturated fatty acids (MUFAs), and <300 mg/day of cholesterol.[336] The Australian and New Zealand Renal Task Force recommends that 1) dietary fat intake should be <30% and saturated fat intake <10% of total energy intake in patients with stage 3 to 4 CKD, and 2) saturated fats be <7%, polyunsaturated fatty acids (PUFAs) be ≤10%, and MUFAs be ≤20% of total energy intake.[340] In contrast, the updated 2020 KDOQI Nutrition guidelines specifically focus on recommendations for PUFAs in CKD.[260] They advise that adults with stages 3 to 5 CKD, including those receiving chronic dialysis or who are postkidney transplant, should not be routinely prescribed long-chain (LC) n-3 PUFAs, including those derived from fish or flaxseed and other oils. These recommendations are based on grade 2B-2C evidence for patients with NDD-CKD, MHD, and those who are post-transplantation, and opinion-level evidence for patients with CPD. However, for patients with CKD with abnormal lipid profiles, they advise 1) prescribing ~2 g/day of LC n-3 PUFAs for patients with stage 3 to 5 NDD-CKD to improve hypertriglyceridemia (based on grade 2C evidence), and 2) prescribing 1.3 to 1.4 g/day of LC n-3 PUFAs to improve hypertriglyceridemia for patients receiving MHD (based on grade 2C–2D evidence) and CPD (opinion-level evidence). Further research is needed to determine the optimal amounts and types of dietary fat intake for patients with NDD-CKD receiving chronic dialysis.

Dietary Potassium

The kidneys play a major role in potassium homeostasis, with dietary potassium intake matched by regulated potassium excretion by the kidneys, and a smaller amount excreted in the stool.[341-343] In the context of normal kidney function, 90% of dietary potassium intake is excreted by the kidneys in a circadian pattern, with excretion highest at noon and lowest at midnight; the gastrointestinal tract excretes the remainder. However, as kidney function declines, the proportion of potassium excreted by the gastrointestinal tract increases (i.e., from about 7–10 mEq/day to about 17 mEq/day in patients with advanced CKD and ESKD).[280]

Hyperkalemia is uncommon when the eGFR is greater than 60 mL/min/1.73m^2 due to the kidneys' remarkable capacity to increase potassium excretion with dietary potassium excess.[341-343] However, hyperkalemia may ensue when there is a reduction in GFR, tubular flow rate, sodium delivery to the distal nephron, or decreased expression of potassium and sodium channels in the apical cell membrane or the sodium/potassium pump in the basolateral membrane, as well as with altered urinary pH.[344] Renin, angiotensin, and aldosterone inhibitors, particularly when taken in combination, may cause hyperkalemia, particularly in patients with CKD and those receiving chronic dialysis.

There is an ongoing debate as to whether common dietary intake of potassium may cause hyperkalemia, especially when the intake is within the broad range of normal.[330,345] In this regard, the following factors should be considered: first, there is no precise relationship between dietary potassium intake and serum potassium concentration in people with NDD-CKD and ESKD; observational studies show only weak correlations.[345] Second, population-based studies indicate that dietary potassium intake is generally low in patients with NDD-CKD and ESKD, which may be due to guidelines recommending dietary potassium restriction in these populations.[330,346] Third, in healthy adults without underlying CKD, the Institute of Medicine recommended a higher dietary potassium intake of 4.7 g/day,[347] and heart-healthy diets, such as the DASH diet, which is rich in fruits and vegetables, also have a higher dietary potassium content of roughly 4.5 g/day.[348] Fourth, the advent of the new potassium binders for hyperkalemia may allow for liberalization of dietary potassium intake, greater consumption of heart-healthy diets, more prevalent use of reno-protective renin-angiotensin-aldosterone system inhibitors, and avoidance or delay for the need of dialysis.[330,349] Finally, there may be varying bioavailability of dietary potassium, depending on the types of concomitant nutrients ingested.[341-343,345] For example, dietary potassium intake from fruits and vegetables is less likely to engender hyperkalemia than dietary potassium from meat sources, due to the effect of the latter foods on increasing intracellular potassium entry. In contrast, with high fruit and vegetable diets, fecal potassium excretion is often increased due to the greater fecal volume from the higher fiber present in these potassium-rich foods. Conversely, "hidden sources" of dietary potassium may substantially contribute to the daily potassium intake. These include potassium preservatives added to prepared meat and other foods (i.e., which may add 300–575 mg of potassium per 100 g of meat intake) and the use of potassium chloride as a salt substitute.

There is a paucity of evidence supporting the longstanding paradigm of dietary potassium restriction for patients with NDD-CKD and ESKD. Epidemiologic data suggest that lower dietary potassium intake is associated with worse outcomes in these populations. In a multicenter cohort of 415 patients receiving MHD who completed food frequency questionnaires, those with dietary potassium intake in the lowest tertile had higher mortality risk, compared with those in the highest tertile of intake of potassium, independent of case-mix, laboratory, and nutritional covariates.[330] In a retrospective analysis of nationally representative adults from the continuous NHANES cohort, among 3,172 and 34,721 participants with impaired and normal kidney function, respectively, the lowest tertile of dietary potassium scaled to energy intake was associated with higher mortality compared with the highest tertile of intake.[346] Compared with high potassium intake from plant-dominant sources, participants with low potassium intake from animal-dominant sources had higher mortality irrespective of kidney function. Nonetheless, very high potassium intakes (e.g., ≥~120–130 mmol/day) from high potassium foods, such as packages of dried fruits or large volumes of smoothies, may cause severe, life-threatening hyperkalemia in patients with advanced NDD-CKD and those receiving chronic dialysis. The 2020 KDOQI nutrition guidelines recommend that dietary potassium intake be adjusted to maintain serum potassium concentration within the normal range, based on opinion-level evidence and that, for those with hyper- or hypokalemia, dietary or supplemental potassium intake should be based on patients' individual needs and clinician judgment.[260] Further research is needed to determine the effect of dietary potassium on the health and well-being of patients with NDD-CKD and ESKD to better inform nutritional guidelines for this population. It is important to remember that sudden death is a not uncommon cause of mortality in patients receiving MHD; whether altered serum potassium plays a role in these deaths is unknown.[350]

Dietary Phosphorus

Dietary phosphorus is typically categorized as being from organic or inorganic sources.[351,352] Organic sources largely include animal-based and plant-based foods. Phosphate in animal-based and plant-based foods leads to different degrees of net intestinal phosphate absorption, roughly 46% to 60% with animal foods and 20% to 40% with plant-based foods. The lower bioavailability of plant-based phosphorus is due to the presence of phytate in many plants. Phytate forms an unusually strong bond with phosphate. The absence of the degrading enzyme, phytase, in the intestinal tract of humans reduces the ability of the intestinal tract to sever the phytate-phosphate bond. In contrast, inorganic sources of phosphorus largely originate from food additives. They have a much higher bioavailability, contributing a proportion of the dietary phosphorus absorbed from the intestinal tract in U.S. diets.

In patients with CKD and those receiving chronic dialysis, high dietary phosphorus intake can lead to elevated plasma phosphorus.[353,354] Hyperphosphatemia, by lowering serum calcium, provides a strong stimulus to development of secondary hyperparathyroidism which, by increasing calcium release from bone and enhancing intestinal absorption of calcium, will promote a rise in serum calcium. The combination of elevated serum phosphorus and hyperparathyroidism can lead to an increased serum calcium-phosphorus product with a greater risk of calcium phosphate deposition in the arteries and other soft tissues, as well as the highly morbid and fatal complication of calciphylaxis. Experimental data indicate that hyperphosphatemia can directly lead to vascular calcification, with downstream effects on vascular stiffness and left ventricular hypertrophy. Epidemiologic studies in people with NDD-CKD have shown that higher serum phosphorus concentration is associated with increasing mortality[355]; a U-shaped association has been observed in patients receiving chronic dialysis (i.e., both lower and higher serum phosphorus is associated with mortality).[356]

Given the established association of high serum phosphorus with worse survival among patients with NDD-CKD or ESKD,[355-357] clinical practice recommendations often promote dietary phosphorus restriction to mitigate hyperphosphatemia. The International Society of Renal Nutrition and Metabolism has recommended a daily dietary phosphorus intake of 800 to 1,000 mg/day in patients with NDD-CKD and those receiving chronic dialysis.[283] However, the KDIGO guidelines state that there is low-quality evidence in support of this practice, making it a weak recommendation.[358] The 2020 KDOQI nutrition clinical practice guidelines advised that in patients with stages 3 to 5 CKD, including those receiving dialysis, dietary phosphorus intake should be adjusted to maintain serum phosphate in the normal range, based on grade 1B evidence.[260] They also advise that among adults with stages 1 to 5 CKD, including those receiving chronic dialysis, it is reasonable to consider the bioavailability of phosphorus sources (i.e., animal, vegetable, additives) when making decisions regarding phosphorus restriction based on opinion-level evidence. To maintain normal serum phosphorus in patients with stage 5 CKD and for those receiving chronic dialysis, individuals will usually require a combination of dietary phosphorus restriction and intestinal binders of phosphate or inhibitors of intestinal phosphate absorption.

Studies of dietary phosphate restriction in ESKD have shown mixed findings. In a secondary analysis of the HEMO study, prescribed low dietary phosphorus intake was not associated with a survival benefit among patients receiving MHD.[359] Shinaberger et al., studied more than 30,000 patients receiving MHD from a large U.S. dialysis provider organization. Compared with patients with a 6-month increase in serum phosphorus and dietary protein intake, those patients with increased serum phosphorus and decreased protein intake had an 11% higher mortality; those with a concomitant decrease in both serum phosphorus and protein intake had 6% higher mortality, and those with decreased serum phosphorus but increased protein intake had 10% greater survival.[360] A multicenter prospective cohort of patients receiving chronic hemodialysis who completed food frequency questionnaires[301] found that those with the lowest tertile of phosphorus intake or dietary phosphorus-to-protein ratio (mg/g) had greater mortality than those with dietary phosphorus consumption in the highest tertiles. In contrast, an earlier 5-year prospective cohort of 224 patients receiving MHD suggested that higher intake of dietary phosphorus and higher dietary phosphorus-to-protein ratios were both associated with higher-mortality risk.[361]

By limiting dietary phosphorus intake, patients may concomitantly reduce consumption of other critical nutrients such as protein, fat, complex carbohydrates, fiber, and various vitamins and minerals. Observational dietary studies of patients receiving MHD have suggested that dietary restrictions are associated with decreased consumption of fruits and vegetables[362] and overall fiber.[363] Use of pharmacotherapies such as phosphate binders, may potentially optimize dietary intake while also maintaining more optimal CKD-mineral bone disease (MBD) parameters. For example, in a small clinical trial of 110 adults receiving thrice-weekly hemodialysis with low serum albumin, <4.0 g/dL, patients were randomized to high-protein (50–55 g) meals during dialysis, providing 400 to 500 mg phosphorus, combined with a potent phosphorus binder (lanthanum carbonate) or a low-protein (<1 g) meal during dialysis, providing <20 mg phosphorus. The effects of these meals on the composite outcome of a rise in serum albumin of ≥0.2 g/dL, with serum phosphorus maintained between 3.5 and 5.5 mg/dL, were compared.[302] After 8 weeks, those in the high-protein and phosphorus meal group who were also given a potent phosphorus binder experienced more frequent attainment of the primary outcome (27% vs 12%, respectively). Further study is needed to define the optimal amount and source of dietary phosphorus intake. Until more definitive information is available, the authors recommend that patients with advanced NDD-CKD and those receiving CPD and MHD thrice-weekly should be limited to 800 to 1,000 mg/day of phosphorus.

Phosphorus Binders. In patients with advanced CKD, particularly those receiving maintenance hemodialysis, the combination of dialysis and dietary phosphorus restriction often does not maintain serum phosphorus within the normal range.[364,365] Hence, phosphate binders, which bind phosphate in the intestinal tract and render it less available for absorption, are important tools in the CKD-MBD management of patients with advanced CKD.[301,351,352,366,367] The 2017 KDIGO Clinical Practice Guideline Update for the Diagnosis, Evaluation, Prevention, and Treatment of CKD-MBD advises against the use of calcium-based binders due to concern regarding excessive accumulation of calcium in soft tissues while also recommending that serum phosphate targets for patients with stages 3 to 5 CKD, including those receiving chronic dialysis, be maintained within the normal range.[368] Aluminum-based binders are discouraged because of the potential risk that aluminum phosphate binders can induce osteomalacia, anemia, and dementia. Medicines that interfere with intestinal phosphate transport may be available in the near future.[369]

Prioritizing Dietary Goals

The number and magnitude of dietary modifications for patients with CKD are so great that if they are all presented to the patient at one time, demoralization and noncompliance may occur. A patient-centered approach that emphasizes shared decision making and prioritizes dietary treatment goals tailored to the individual is essential. Control of protein, energy, sodium, potassium, and phosphorus intake is emphasized. Calcium needs and magnesium restriction are usually addressed in patients prescribed protein-restricted diets. Trace element and vitamin needs can be addressed with dietary supplements, although further research concerning the efficacy and safety of vitamins[370-372] and other dietary supplements, including iron,[373] and pre- and probiotics[374] on the nutritional health and clinical outcomes of patients with CKD and ESRD are needed. Acidosis can be prevented and treated with alkalizing compounds or, if the patient is highly motivated, with a high vegetable and fruit diet. Ongoing and further research is needed to better define the optimal amount and types of dietary carbohydrate, fiber, and fat intake for patients with CKD.

Obesity, Chronic Kidney Disease, and Hypertension

Obesity predisposes patients to CKD and hypertension. Either diabetes or hypertension is the cause of ESKD in approximately 75% of patients.[375] Many cases of hypertension are nutrition related due to either obesity or excessive sodium intake.[65,89,375,376]

Obesity has many effects on the kidney.[65,89] It predisposes people to diabetes and diabetic nephropathy and can cause hypertension and hypertensive nephrosclerosis. Obesity is associated with increased incidence of CKD, nondiabetic focal and segmental glomerulosclerosis, more rapid progression of kidney failure, an increased incidence of proteinuria and a greater magnitude of proteinuria for a given type of CKD, urinary tract stones, and renal cell carcinoma. Individuals with obesity and advanced kidney failure, especially those undergoing MD, actually have greater survival.[254,377-380] Several hypotheses have been proposed to explain this altered risk-factor pattern, but the mechanisms responsible for this phenomenon have not been established.[210,239,381,382] Blood pressure has been directly correlated with fat cell size and/or number.[383,384] Furthermore, waist-to-hip circumference ratio correlates directly with blood pressure.[376,385]

Nutritional Therapy in Acute Kidney Injury

AKI is characterized by a sudden reduction or cessation in GFR. The most common causes of AKI include shock, severe infection, trauma, medicines, obstruction, vasculitis, and certain types of glomerulonephritis. The histopathologic renal lesions associated with AKI vary according to the underlying causes(s). In many instances, a patient who survives the underlying diseases will recover from the AKI. When patients sustain AKI, they are likely to develop fluid and electrolyte disorders, uremic toxicity, and PEW. These disorders are prone to develop when the patient is both oliguric and hypercatabolic, which are common complications of AKI. The most current definitions and staging of AKI come from the *Kidney Disease: Improving Global Outcomes* (KDIGO) clinical practice guidelines on AKI.[386] There are three stages of AKI based on the serum creatinine and urine output; Stage 1: Serum creatinine of 1.5 to 1.9 times baseline or an increase of creatinine of ≥0.3 mg/dL or a urine output <0.5 mL/kg/h for 6 to 12 hours; Stage 2: Serum 2.0 to 2.9 times baseline [or urine output <0.5 mL/kg/h for ≥12 hours]; Stage 3: ≥3.0 times baseline or increase in serum creatinine to 4.0 mg/d or initiation of renal replacement therapy or patients <18 years, a decrease in eGFR to <35 mL/min per 1.73 m² or urine output <0.3 mL/kg/h for 24 hours or anuria for ≥12 hours.[386] Complications and mortality are related to the severity of AKI; staging should be considered when making management decisions for patients.

Patients with AKI, and particularly those with underlying catabolic illnesses, frequently undergo metabolic changes that enhance net degradation of protein and amino acids and consumption of fuel substrates.[387,388] Energy expenditure is often increased.[385] In vitro studies of rat muscle indicate that protein degradation is enhanced and protein synthesis is reduced.[379,389] In addition, hepatic gluconeogenesis is increased. If the liver from these animals is perfused or incubated with amino acids, the elevated hepatic glucose and urea production is further enhanced.[380]

As a result of these metabolic derangements, these patients often cannot use protein, amino acids, and energy substrates efficiently. Hence, it is difficult to maintain and improve the nutritional status by enteral or parenteral nutrition.[382,384,390]

It is important to recognize that AKI can occur in people who do not appear sick and are not hypercatabolic. AKI may be caused by many factors, including intake of new medicines or a transient fall in blood pressure due to adjustment of antihypertensive medications. Sometimes there is no apparent cause. Under these circumstances, patients are usually not very catabolic and almost always recover their renal function, often without the need for dialysis therapy. These cases are usually Stage 1 or Stage 2 AKI.

General Principles Regarding the Nutritional and Metabolic Management of AKI

Fluid and mineral balance should be carefully monitored in patients with AKI to prevent overhydration or electrolyte disorders. Water intake, in general, should equal output from urine and all other measured sources (e.g., nasogastric aspirate, fistula drainage) plus 400 mL/day. This regimen accounts for the contributions of endogenous water production from metabolism and the insensible water losses (from respiration and skin losses) to water balance. In general, if the patient is catabolic, weight should be allowed to decrease by 0.2 to 0.5 kg/day to avoid excessive accumulation of fluid. Sodium, potassium, phosphorus, and magnesium intake should be restricted to prevent the accumulation of these minerals. Energy and, if feasible, protein intake, should be at a level to meet the patient's nutritional requirements. By controlling the water and electrolyte intake and lowering the UNA, one may be able to reduce the need for dialysis.

Energy and macro- and micronutrient needs will depend on the patient's nutritional status, catabolic rate, residual GFR, and the clinical indications for initiating dialysis therapy. A patient with AKI who has a high residual GFR may receive larger quantities of nutrients, because there is less risk of developing fluid and electrolyte disorders or accumulating potentially toxic metabolites. For a patient who has little or no urine flow and who is not very catabolic or uremic, the intake of small quantities of water, minerals, and amino acids may reduce the need for dialysis; this approach may be particularly beneficial if it is anticipated that the patient will not tolerate dialysis well. Similarly, a patient who is starting to recover from AKI may be given this latter treatment to avoid dialysis for a few days until renal function becomes adequate. In these patients, high-energy diets providing small amounts of essential amino acids or ketoacids with little or no protein may be used for short periods. Whenever feasible, patients with AKI should receive nutrition orally. If the patient will not eat adequately, the use of oral nutritional supplements, elemental diets, and enteral tube-feeding should be considered (see **Table 96.9**).

Specific Nutrients

Protein and Amino Acids. Different quantities of nitrogen and various compositions of amino acid formulations have been proposed for enteral or parenteral administration to patients

Table 96.9 Typical Composition of Solutions for Parenteral Nutrition in Patients with Acute Kidney Injury[a]

	Daily Quantity or Concentration to Be Infused	
Volume	Liters	1.0
Essential and nonessential free crystalline amino acids (4.25%–5.0%)[b] or	g/L	42.5–50
Essential amino acids (5%)[b]	g/L	12.5–25
Dextrose (D-glucose)[c]	g/L	350
Lipid emulsion[c]	10% or 20%	50 or 100 g/500 mL
Energy (approx.)[c]	kcal/L	1,140
Electrolytes[d]		
Sodium[e]	mmol/L	40–50
Chloride[e]	mmol/L	25–35
Potassium	mmol/d	≥35
Acetate	mmol/d	35–40
Calcium	mmol/d	5
Phosphorus	mmol/d	8
Magnesium	mmol/d	4
Iron	mmol/d	2
Trace elements		See text

(continues)

Table 96.9 Typical Composition of Solutions for Parenteral Nutrition in Patients with Acute Kidney Injury[a] *(continued)*

	Daily Quantity or Concentration to Be Infused	
Vitamins		
Vitamin A[f]		See text
Vitamin D		See text
Vitamin K[g]	mg/wk	7.5
Vitamin E[h]	IU/d	10
Niacin	mg/d	20
Thiamin HCl (B_1)	mg/d	2
Riboflavin (B_2)	mg/d	2
Pantothenic acid (B_3)	mg/d	10
Pyridoxine HCl (B_6)	mg/d	10
Ascorbic acid (C)	mg/d	70
Biotin	μg/d	100
Folic acid	mg/d	2
Vitamin B_{12}	μg/d	3

[a]These nutrients are present in each bottle containing 500 mL of 8.5% to 10% crystalline amino acids or 250 to 500 mL of 5% essential amino acids and 500 mL of 70% dextrose. The vitamins and trace elements are an exception because they are added to only one bottle per day. The patient's fluid status and serum electrolytes and glucose must be monitored closely. The composition and volume of the intravenous infusate may need to be changed if the patient is very uremic, acidemic, or volume overloaded; if the serum electrolyte concentrations are not normal or if they are changing; or if dialysis therapy is not readily available or is particularly hazardous to the patient (see text).

[b]For patients who are more catabolic (e.g., urea nitrogen appearance ≥5 g/day), who are or will be undergoing regular dialysis treatments (particularly for ≥2 weeks), or who have PEW, essential and nonessential amino acids are infused: about 1.0 to 1.2 g/kg/day for hemodialysis-treated patients and 1.2 to 1.3 g/kg/day for peritoneal dialysis-treated patients (see text). For patients who are less catabolic, for whom one is trying to avoid instituting dialysis treatment, and who will not be receiving PN for more than 2 or 3 weeks, 0.30 to 0.40 g/kg/day of the nine essential amino acids (preferably with arginine) may be infused. Patients undergoing continuous venovenous hemofiltration or more than thrice-weekly hemodialysis may be given 1.5 to 2.5 g/kg/day of essential and nonessential amino acids, depending on their clinical and metabolic status. See text for discussion of the formulations of amino acids.

[c]To attain an energy intake of 30 to 35 kcal/kg/day, 70% dextrose is added as necessary (see text). Lower energy intakes may be used for patients with obesity or when larger amounts of amino acids are given (see text). For the higher levels of energy intake (i.e., 35 kcal/kg/day), an additional 70% dextrose may be added to the solutions. To balance the sources of energy and to prevent essential fatty acid deficiency, lipid emulsions are used. For patients who are septic or at high risk for sepsis, about 10% to 20% of energy or less may be given as lipids. For more stable patients, 20% to 30% of energy may be given as lipids. The lipid emulsions probably should be infused over at least 12 hours, if not 24 hours, to reduce the hyperlipidemia that may occur with intravenous infusion of lipid emulsions (see text). The lipid emulsions may be infused in a separate line or mixed with the amino acid and dextrose solutions and infused soon after mixing (see text). A 20% lipid emulsion may be used to reduce the water load. The approximate energy values are dextrose monohydrate, 3.4 kcal/g; amino acids, 3.5 kcal/g; lipid emulsions 10%, 1.1 kcal/mL; and 20%, 2.0 kcal/mL.

[d]When electrolytes are being added, the amounts intrinsically present in the amino acid solution should be considered.

[e]Refers to the final concentrations of electrolytes after any additional 70% dextrose or other solutions have been added.

[f]Vitamin A probably should be avoided unless PN is continued for more than several days (see text).

[g]Should be given orally or parenterally and not in the PN solution because of antagonisms.

[h]May need to be increased with use of lipid emulsions.

with AKI. Observations regarding the optimal type and amount of nutritional support for patients with AKI are conflicting.[391] This is probably the result of several factors; the clinical and metabolic status and the clinical course of patients with AKI is so complex and variable that it would be necessary to study large numbers of patients to show statistically significant benefits of any type of nutritional therapy. In addition, many of these studies were retrospective or not randomly controlled, and study designs may have led to unintentional biases. Patients who are catabolic with AKI may need both good nutrition and metabolic intervention to suppress catabolic processes and to promote anabolism; providing nutrients without metabolic intervention may not have a beneficial effect on nutritional status or clinical outcome, particularly in the first days after the onset of AKI.

The current policy for amino acid or protein intake in patients with AKI at the authors' institution is as follows (Table 96.9): Patients may be prescribed a low enteral or intravenous nitrogen intake if there is a low UNA (i.e., <4 to 5 g N per day), if they have no evidence of moderate or severe protein malnutrition, if one anticipates that the patient will recover renal function within the next 1 or 2 weeks, and if there is a clinical justification for avoiding dialysis therapy.[390] Under these conditions, 0.3 to 0.5 g/kg/day of primarily high-quality protein or essential amino acids, preferably with arginine, may be prescribed. We do not give more than 0.4 g/kg/day of essential amino acids as the sole nitrogen source because larger quantities of the nine essential amino acids without nonessential amino acids may cause serious amino acid imbalances.[390,392]

Diets providing 0.10 to 0.30 g/kg/day of miscellaneous protein and 10 to 20 g/day of essential amino acids or ketoacids may also be used in patients who can ingest orally. These regimens should minimize the rate of accumulation of nitrogenous metabolites. Unless patients are severely catabolic, they will usually maintain a neutral or only mildly negative nitrogen balance. Hence, the need for dialysis may be minimized or avoided. Because intravenous solutions containing only the nine essential amino acids are not readily available, this therapeutic approach is not commonly tried. If the patient has substantial residual renal function (e.g., GFR of 5 to 10 mL/min) or earlier stage AKI and is not very catabolic, we may treat the individual like a patient with CRF who is not receiving dialysis. The individual would receive 0.55 to 0.60 g protein or amino acids/kg desirable body weight per day.

For patients who are more catabolic and have higher UNA (>5 g N per day), have severe PEW, or are undergoing thrice-weekly hemodialysis therapy and, with higher stages of AKI who either already have AKI or are anticipated to have AKI for more than 2 weeks, a higher protein or amino acid intake may be prescribed, up to 1.0 to 1.2 g/kg body weight/day. If tolerated, 1.2 g protein or amino acids/kg/day is preferable. Compared with small quantities of essential amino acids, these larger nitrogen intakes may improve nitrogen balance, particularly after the first 1 or 2 weeks of dialysis treatments. However, the UNA almost invariably rises, and the increased azotemia and—in those patients receiving PN—the larger volumes of fluid necessary to provide this amount of amino acids may increase the need for dialysis therapy.

Continuous Venovenous Hemofiltration with or without Concurrent Hemodialysis

Continuous venovenous hemofiltration (CVVH) or CVVH with concurrent hemodialysis using low dialysate flow rates (CVVHD) is commonly used for the management of patients who are critically ill with AKI or have other conditions with fluid or nitrogen intolerance (e.g., severe liver or congestive heart failure). With CVVH or CVVHD, catheters are placed into a large flow vein, such as the femoral vein.[393] The blood flows through a small filtering apparatus, where some of the plasma water is filtered; the remaining blood is returned to the vein.

The following are among the advantages of CVVH and CVVHD: Unlike hemodialysis treatments, patients receiving CVVH or CVVHD are usually given this treatment 24 hours per day. Hence, large quantities of water, electrolytes, and metabolic products may be removed each day, even with smaller rates of removal of these chemicals. Because the rate of removal of water and electrolytes is slow, CVVH or CVVHD is less likely to cause or worsen hypotension or induce other adverse physiologic changes (e.g., cardiac arrhythmias). The high quantities of water and small molecules removed over a 24-hour period with CVVHD, including metabolic waste products, allow one to more safely administer large amounts of amino acids and other nutrients to the patient. Indeed, with CVVHD, such patients, even when hypercatabolic with severe AKI, may be nutritionally treated as one would treat hypercatabolic patients who do not have reduced renal function. For patients receiving CVVHD, we often prescribe 1.5 to 2.5 g/kg/day of mixtures of essential and nonessential amino acids intravenously or similar amounts of protein given enterally. Amino acid losses with CVVH or CVVHD are generally about 4 to 7 g/day and are slightly higher when patients are receiving amino acid infusions compared with when they are not.[394,395]

Patients with AKI who receive nutritional support may also receive hemodialysis treatment for several hours daily instead of CVVHD.[387,388] The hemodialysis may be administered as often as every day rather than three times weekly, which is the usual treatment for clinically stable patients receiving MHD. If AKI persists for more than 2 to 3 weeks, patients undergoing regular dialysis treatment are often treated as are patients who are undergoing MHD, with about 1.0 to 1.2 g/kg/day of protein or amino acids for those being treated with hemodialysis or 1.2 to 1.5 g/kg/day for those receiving peritoneal dialysis.

Other Nutritional Techniques to Reduce Hypercatabolism

Some investigators have proposed adding amino acids and additional glucose to the dialysate of patients undergoing CPD or MHD.[396,397] The nutrients diffuse into the body during dialysis. Currently, these techniques may provide supplemental nutrition but cannot be used for total nutritional support.

Because the metabolic status of patients with AKI often enhances the net catabolism of protein, amino acids, and other energy substrates,[379,380,382,384,389,390,398] there may be advantages to administering agents that promote anabolic processes or reduce catabolic pathways. The nitrogen intake appears to be used more efficiently if a greater proportion of the administered amino acids is essential.[382,390,399] However, this hypothesis has not been tested clinically. In addition, studies in patients who are catabolic without renal failure suggest that intravenous infusions in which a large proportion of the amino acids are composed of branched-chain amino acids (i.e., isoleucine, leucine, and valine) may have a specific anabolic effect.[400,401] Research is heterogeneous in this regard. Ketoacid analogs of the branched-chain amino acids have also been shown to promote anabolism when studied by in vitro preparations and when given to patients without uremia who are not hypercatabolic.[121,122]

The intravenous infusion of the salt complex of α-ketoglutarate and ornithine in patients who are postoperative and receiving PN is reported to reduce UNA and to increase nitrogen balance.[402] Patients with severe stress without renal failure display a rapid fall in intracellular muscle glutamine.[403] Administration of glutamine has been reported to improve protein balance in these patients.[403,404] Arginine also may increase nitrogen balance.[405] Whether these individual amino acid treatments change morbidity or mortality has not been established.

Energy. Patients with AKI appear to have the same energy expenditure and needs as individuals with similar types and severities of coexisting illnesses but who do not have renal failure. The exception to this is the patient with AKI who is receiving small amounts of amino acids or protein to decrease the generation of uremic toxins and, thereby, reduce or avoid the need for dialysis treatment (as described in the previous section). Studies suggest, but have not definitively demonstrated, that when small amounts of amino acids or protein are taken, higher energy intake may enhance the efficiency of nutrient use and reduce protein wasting.[382,384,385,398]

Higher energy intake (i.e., about 35 kcal/kg/day) may be used for patients treated with small amounts of amino acids or protein to improve the efficiency of use of these nutrients or to meet the higher energy needs of a patient. The small amounts of amino acids or protein are given to reduce the need for dialysis treatment. Patient energy needs may be estimated by multiplying the Harris-Benedict equation[406] or the World Health Organization equations[407] for calculating the daily energy requirements of normal individuals by a stress factor to adjust for the patient's illness[408] and then by 1.25. The 1.25 is included to ensure sufficient energy is provided to promote anabolism or diminish the rate of catabolism; the value of this term has not been demonstrated. Energy expenditure, measured by indirect calorimetry, can also be multiplied by 1.25 to estimate the daily energy requirement. Some authorities recommend lower energy intakes for AKI patients, closer to 20 to 30 kcal/kg per day,[391] although large clinical trials have not tested the value of such lower intakes.

With the almost routine use of CVVHD or higher dialysis frequency exceeding three times weekly to treat patients who are hypercatabolic with AKI who, for example, have a systemic inflammatory response syndrome (SIRS), amino acid or protein intakes are commonly increased to about 1.5 to 2.5 g/kg/day. At this level of amino acid and protein intake, the need for increased energy intake (e.g., determined by multiplying calculated energy needs by 1.25) may not be necessary. Prospective studies to test this hypothesis have not been published.

Larger energy intakes are not used because there appears to be little nutritional advantage to administering more energy to patients who are catabolic. Because high energy intakes generate more carbon dioxide from the infused carbohydrate and fat, they can promote hypercapnia if pulmonary function is impaired.[409] Carbon dioxide retention is particularly likely to occur with very high carbohydrate loads. In addition, high-energy intakes may cause obesity and fatty liver,[410] increasing the patient's water load.

Because most patients with AKI do not tolerate large water volumes, glucose is usually administered in a 70% solution. The common exception to this is the patient receiving CVVH or CVVHD or frequent (e.g., daily) hemodialysis or peritoneal dialysis. The glucose and amino acid solutions are mixed, providing the amino acids and energy simultaneously (see Table 96.9). Patients receiving PN for more than 5 days should receive about 25 g/day of an intravenous lipid solution so they do not become deficient in essential fatty acids. For patients who are septic or at high risk of severe sepsis, likely no more than 10% to 20% of total energy should be provided from fat. Those without sepsis and not at high risk of infection may be given about 20% to 30% of energy as lipid emulsions. Some investigators have recommended giving up 30% to 40% of energy as lipids to provide sufficient fatty acids to organs that normally use lipids as their main energy source, and to approximate the normal U.S. dietary intake more closely. However, some researchers have reported that infusions of large amounts of fat emulsions, such as 50 g over 8 to 12 hours, may impair the function of the reticuloendothelial system[411]; they have questioned whether infusion of lipid emulsions could lower host resistance. A prudent approach may be to infuse lipid emulsions over at least 12 to 24 hours, to prevent marked increases in plasma lipids. With careful attention to aseptic control, the lipid emulsions may be mixed with glucose and amino acids.[412]

Minerals. A mineral prescription for PN for patients with AKI is shown in Table 96.9. Any mineral recommendations are tentative and must be adjusted according to the patient's clinical status. If the serum concentration of an electrolyte is increased, it may be advisable to reduce the quantity infused or not administer it at the onset of parenteral nutrition. The patient must be monitored closely because the hormonal and metabolic changes that often occur with the initiation of PN may cause the serum electrolytes to fall rapidly, particularly serum potassium and phosphorus. Metabolic changes and impaired GFR can lead to a rapid rise in serum concentrations during repletion.

Trace elements are probably not necessary in PN solutions given to patients who are catabolic with AKI unless this is the sole source of nutritional support for at least 2 to 3 weeks, or the patients are known to have depletion of certain trace elements. The nutritional requirements for trace elements have yet to be established for patients with AKI or CRF receiving PN.

Vitamins. Vitamin requirements have yet to be well defined for patients with AKI. Tentative recommendations for vitamins for patients receiving PN are shown in Table 96.9. Much of the recommended intake is based on information obtained from studies in patients with CRF, physiologically normal individuals, or patients who are acutely ill without CRF. Vitamin A is probably best avoided for the first several days of nutritional support because serum vitamin A concentration is elevated in CRF, and small doses of vitamin A have been reported to cause toxicity in patients undergoing MD.[413] After the first several days of nutritional therapy, a dose of vitamin A between 50% and 100% of the RDA for vitamin A[414] for normal individuals may be given daily.

Vitamin D is fat-soluble, and vitamin stores should not become depleted during the few days to weeks that most patients with AKI receive PN. However, many individuals without CKD or AKI have been shown to have insufficient (serum concentrations, 15 to <30 nM) or deficient (<15 nM) 25-hydroxycholecalciferol. It may be helpful to check serum concentration for this hormone in patients with AKI and give cholecalciferol or ergocalciferol supplements, if low. The turnover of the active metabolite, 1,25-dihydroxycholecalciferol (calcitriol), is much faster than 25-hydroxycholecalciferol. Hence, calcitriol also may be needed in patients with AKI.[415]

Although vitamin K is fat-soluble, vitamin K deficiency has been reported in patients without CRF who are not eating and are receiving antibiotics.[416] Vitamin K, therefore, should be given routinely to patients receiving PN (see Table 96.9). Ten milligrams per day of pyridoxine hydrochloride (8.2 mg/day of pyridoxine) is recommended because studies in clinically stable or sick patients undergoing MHD indicate that this quantity may be necessary to prevent or correct a vitamin B_6 deficiency.[204] Patients should probably not receive more than about 60 mg of ascorbic acid per day because of the risk of increased oxalate production.[417,418] The nutrient intake of patients with AKI must be carefully reevaluated each day and sometimes more frequently. This reevaluation is particularly important for patients with AKI because they may undergo rapid changes in their clinical and metabolic condition.

Peripheral Parenteral Nutrition

Peripheral parenteral nutrition may be advantageous for patients with AKI who cannot meet their daily nutritional requirements via diet and/or tube feeding. Peripheral infusions may enable these patients to receive adequate nutrition without resorting to PN through a large flow vein. In these patients, it is often most practical to infuse an 8.5% to 10% amino acid solution or a 20% lipid emulsion into a peripheral vein and to administer as much as possible of the other essential nutrients, including carbohydrates, enterally.

The peripheral vascular access used for hemodialysis can also be used for PN. Because there is a high blood flow through the vascular access used for hemodialysis, hypertonic solutions can be used, and the water load to the patient can be reduced. This technique increases the risk of infection or thrombosis in the vascular access. It should not be used in patients needing hemodialysis access for extended periods.

Supplemental or Intradialytic Parenteral Nutrition

Infusion of amino acids, glucose, and/or lipids may be given as a nutritional supplement to patients with AKI or CRF who cannot meet their energy and nutrient needs orally. Supplemental amino acids, glucose, and/or lipids can be infused conveniently during the hemodialysis procedure. Because most patients in need of oral nutritional supplements have decreased intakes of both amino acids and energy, the recommended approach is to infuse 40 to 42 g of essential and nonessential amino acids and 200 g of D-glucose (150 g of D-glucose if the hemodialysate contains glucose). This preparation is infused throughout the hemodialysis procedure at a constant rate into the blood leaving the dialyzer. Such a technique minimizes the normal fall in amino acid and glucose pools resulting from dialysis losses of these nutrients. Most of the infused glucose and amino acids are retained; the amino acid losses into dialysate increase by only about 4 to 5 g.[214] Lipid infusions can be substituted for some infused glucose, but this may pose a risk of reducing host resistance to infection.[411] Patients with low serum phosphorus or potassium concentrations at the onset of dialysis treatment may require supplements of these electrolytes during the amino acid and glucose supplementation. To prevent reactive hypoglycemia, the infusion should not be stopped until the end of hemodialysis. Ideally, the patient should eat a carbohydrate source 20 to 30 minutes before the end of the infusion.

There has been debate about whether intravenous supplements with amino acids, glucose, and/or lipids thrice weekly for about 3 to 4 hours during hemodialysis benefit patients undergoing MHD who eat poorly.[244,419] Retrospective analyses suggest that intradialytic PN or oral nutritional supplements may reduce the mortality rate of patients undergoing MHD who have PEW.[244,420,421] One study indicated that this benefit was observed only when the serum albumin was 3.3 g/dL or lower.[420] Intradialytic PN should be used only in patients who cannot increase their intake of foods or take oral or enteral supplements. The intravenous PN should be continued only if nutritional or clinical assessment indicates that this benefits the patient. This area requires further investigation.

References

1. National Kidney Foundation. Am J Kidney Dis 2003;42:S1–201.
2. Inker LA, Astor BC, Fox CH, et al. Am J Kidney Dis 2014;63:713–35.
3. Glassock R. In: Kopple JD, Kalantar-Zadeh K, Fouque D, eds. Uremic Toxins: An Integrated Overiew of Classification and Pathobiology. Nutritional Management of Nutritional Disease New York, NY: Academic Press, 2022.
4. Nigwekar SU. In: Rhee C K-ZK, Brent G, eds. Vitamin D and Parathyroid Hormone in Kidney Disease. Endocrine Disorders in Kidney Disease Springer, Cham, 2019:223–9.
5. Singh AK. In: Singh AK WG, ed. Erythropoiesis: The Roles of Erythropoietin and Iron. ITextbook of Nephro-Endocrinology. Academic Press, 2018:207–15.
6. Carey RM Padia SH. In: Singh AK WG, ed. Physiology and Regulation of the Renin–Angiotensin–Aldosterone System. Textbook of Nephro-Endocrinology. Academic Press, 2018:1–25.
7. Jean G, Terrat JC, Vanel T, et al. Nephrol Dial Transplant 2008;23:3670–6.
8. Wolf M, Betancourt J, Chang Y, et al. J Am Soc Nephrol 2008;19:1379–88.
9. St Peter WL, Li S, Liu J, et al. Pharmacother 2009;29:154–64.
10. Palmer SC, McGregor DO, Craig JC, et al. Cochrane Database Syst Rev 2009;CD008175.
11. Palmer SC, McGregor DO, Craig JC, et al. Cochrane Database Syst Rev 2009;CD005633.
12. Bansal A, Chonchol M. Phosphorus metabolism and fibroblast growth factor 23 in chronic kidney disease. Nutritional Management of Renal Disease. Elsevier, 2022:383–96.
13. Eschbach JW, Kelly MR, Haley NR, et al. N Engl J Med 1989;321:158–63.
14. Foley RN. Nat Rev Nephrol 2010;6:218–23.
15. Narasaki Y Rhee CM. In: Kopple JD MS, Kalantar-Zadeh K, Fouque D, eds. Effect of Nutritional Status and Changes in Nutrient Intake on Renal Function. Nutritional Mnanagement of Renal Disease. Academic Press, 2022:225–45.
16. Benabe JE, Martinez-Maldonado M. Miner Electrolyte Metab 1998;24:20–6.
17. Ibrahim HN, Weber ML. Curr Opin Nephrol Hypertens 2010;19:534–8.
18. Hirschberg R, Kopple JD, Blantz RC, et al. J Clin Invest 1991; 87:1200–6.
19. Hirschberg R, Kopple JD. J Am Soc Nephrol 1991;1:1034–40.
20. Klahr S, Tripathy K, Garcia FT, et al. Am J Med 1967;43: 84–96.
21. Owen OE, Felig P, Morgan AP, et al. J Clin Invest 1969;48: 574–83.
22. Alakel N, Middeke JM, Schetelig J, et al. Onco Targets Ther 2017;10:597–605.
23. Bosch JP, Lew S, Glabman S, et al. Am J Med 1986;81:809–15.
24. Smoyer WE, Brouhard BH, Rassin DK, et al. J Lab Clin Med 1991;118:166–75.
25. Castellino P, Hunt W, DeFronzo RA. Kidney Int Suppl 1987;22:S15–20.
26. Hirschberg R, Kopple JD. Kidney Int 1987;32:382–7.
27. Giordano M, Castellino P, McConnell EL, et al. Am J Physiol 1994;267:F703–8.
28. Preiss DJ, Godber IM, Lamb EJ, et al. Ann Clin Biochem 2007;44:35–42.
29. Ko GJ, Rhee CM, Kalantar-Zadeh K, et al. J Am Soc Nephrol 2020;31:1667–79.
30. Dowis K, Banga S. Nutrients 2021;13.
31. Brenner BM, Meyer TW, Hostetter TH. N Engl J Med 1982;307:652–9.
32. Ko GJ, Kalantar-Zadeh K, Goldstein-Fuchs J, et al. Nutrients 2017;9.
33. Jhee JH, Kee YK, Park S, et al. Nephrol Dial Transplant 2020;35:98–106.
34. Farhadnejad H, Asghari G, Emamat H, et al. J Ren Nutr 2019;29:343–9.
35. Tirosh A, Golan R, Harman-Boehm I, et al. Diabet Care 2013;36:2225–32.
36. Friedman AN, Ogden LG, Foster GD, et al. Clin J Am Soc Nephrol 2012;7:1103–11.
37. Bruci A, Tuccinardi D, Tozzi R, et al. Nutrients 2020;12.
38. Mitch WE, Walser M, Buffington GA, et al. Lancet 1976;2: 1326–8.
39. Barsotti G, Guiducci A, Ciardella F, et al. Nephron 1981;27: 113–7.

40. Sato Y, Takahashi M, Yanagita M. Semin Nephrol 2020;40:206–15; erratum:328.
41. Collister D, Ferguson T, Komenda P, et al. Semin Nephrol 2016;36:273–82.
42. McCormack LJ, Beland JE, Schneckloth RE, et al. Am J Pathol 1958;34:1011–21.
43. Singh M, Karakala N, Shah SV. J Ren Nutr 2017;27:462–4.
44. Bazzi C, Rizza V, Casellato D, et al. J Nephrol 2012;25:810–8.
45. Hannan M, Ansari S, Meza N, et al. Clin J Am Soc Nephrol 2021;16:648–59.
46. Hostetter TH, Olson JL, Rennke HG, et al. Am J Physiol 1981;241:F85–93.
47. Fu Y, Tang C, Cai J, et al. Am J Physiol Renal Physiol 2018;315:F1098–F106.
48. Kanzaki G, Puelles VG, Cullen-McEwen LA, et al. JCI Insight 2017;2.
49. Helal I, Fick-Brosnahan GM, Reed-Gitomer B, et al. Nat Rev Nephrol 2012;8:293–300.
50. Liu Y. Nat Rev Nephrol 2011;7:684–96.
51. Olson JL, Hostetter TH, Rennke HG, et al. Kidney Int 1982;22:112–26.
52. Zoja C, Abbate M, Remuzzi G. Curr Opin Nephrol Hypertens 2006;15:250–7.
53. Wang S, Hirschberg R. J Biol Chem 2004;279:23200–6.
54. Wang S, Wilkes MC, Leof EB, et al. FASEB J 2005;19:1–11.
55. Dunger A, Berg S, Kloting I, et al. Exp Clin Endocrinol Diabetes 1997;105(suppl 2):48–50.
56. Farr LE, Smadel JE. J Exp Med 1939;70:615–27.
57. Kusano K, Segawa H, Ohnishi R, et al. J Nutr Sci Vitaminol (Tokyo) 2008;54:237–43.
58. Hostetter TH, Troy JL, Brenner BM. Kidney Int 1981;19:410–5.
59. Kalantar-Zadeh K, Fouque D. N Engl J Med 2017;377:1765–76.
60. Knight EL, Stampfer MJ, Hankinson SE, et al. Ann Intern Med 2003;138:460–7.
61. Kamper AL, Strandgaard S. Annu Rev Nutr 2017;37:347–69.
62. Schrier RW, Harris DC, Chan L, et al. Am J Kidney Dis 1988;12:243–9.
63. Paller MS, Hostetter TH. Am J Physiol 1986;251:F34–9.
64. Ravikumar NPG, Pao AC, Raphael KL. Adv Chronic Kidney Dis 2022;29:406–15.
65. KDOQI Work Group. Clinical practice guidelines for nutrition in chronic renal failure. K/DOQI, National Kidney Foundation. Am J Kidney Dis 2000;35:S1–S140.
66. Mocanu CA, Cuiban E, Paul R, et al. Am J Clin Nutr 2022;116: 836–7.
67. Rhee CM, Ahmadi SF, Kovesdy CP, et al. J Cachexia Sarcopenia Muscle 2018;9:235–45.
68. Haring B, Selvin E, Liang M, et al. J Ren Nutr 2017;27:233–42.
69. Lew QJ, Jafar TH, Koh HW, et al. J Am Soc Nephrol 2017;28: 304–12.
70. Walls J, Williams AJ. Contrib Nephrol 1988;60:179–87.
71. Niwa T. Panminerva Med 2017;59:139–48.
72. Asai M, Kumakura S, Kikuchi M. Ren Fail 2019;41:47–56.
73. Nath KA, Hostetter MK, Hostetter TH. J Clin Invest 1985;76: 667–75.
74. Wesson DE, Buysse JM, Bushinsky DA. J Am Soc Nephrol 2020;31:469–82.
75. Phisitkul S, Khanna A, Simoni J, et al. Kidney Int 2010;77: 617–23.
76. Di Iorio BR, Bellasi A, Raphael KL, et al. J Nephrol 2019;32: 989–1001.
77. de Brito-Ashurst I, Varagunam M, Raftery MJ, et al. J Am Soc Nephrol 2009;20:2075–84.
78. Goraya N, Simoni J, Jo CH, et al. Kidney Int 2014;86:1031–8.
79. Anderson S, Vora JP. J Diabetes Complications 1995;9:304–7.
80. Stackhouse S, Miller PL, Park SK, et al. Diabetes 1990;39:989–95.
81. Alicic RZ, Rooney MT, Tuttle KR. Clin J Am Soc Nephrol 2017;12:2032–45.
82. United States Renal Data System. 2022 USRDS Annual Data Report: epidemiology of kidney disease in the United States. National Institutes of Health, National Institute of Diabetes, and Digestive and Kidney Diseases, Bethesda, MD.
83. Piperidou A, Loutradis C, Sarafidis P. J Hum Hypertens 2021;35:12–25.
84. Ibels LS, Alfrey AC, Haut L, et al. N Engl J Med 1978;298: 122–6.
85. Lumlertgul D, Burke TJ, Gillum DM, et al. Kidney Int 1986;29:658–66.
86. Gimenez LF, Solez K, Walker WG. Kidney Int 1987;31:93–9.
87. Harris DC, Hammond WS, Burke TJ, et al. Kidney Int 1987;31:41–6.
88. Chang AR, Anderson C. Annu Rev Nutr 2017;37:321–46.
89. Kopple JD, Feroze U. J Ren Nutr 2011;21:66–71.
90. Kochan Z, Szupryczynska N, Malgorzewicz S, et al. Nutrients 2021;13.
91. Kim JJ, Wilbon SS, Fornoni A. Kidney360 2021;2:755–62.
92. Li X, Bayliss G, Zhuang S. Int J Mol Sci 2017;18.
93. Wagner S, Merkling T, Metzger M, et al. Nephrol Dial Transplant 2022;37:730–9.
94. Sontrop JM, Dixon SN, Garg AX, et al. Am J Nephrol 2013;37:434–42.
95. Levey AS, Greene T, Sarnak MJ, et al. Am J Kidney Dis 2006;48:879–88.
96. Don BR SM, Lo JC Endocrine hypertension: effects of hormones on renal function. Basic and Clinical Endocrinology. New York, NY: McGraw-Hill, 2004:414–38.
97. Anderson S, Meyer TW, Rennke HG, et al. J Clin Invest 1985;76:612–9.
98. Tolins JP, Raij L. Hypertens 1990;16:452–61.
99. Ruggenenti P, Perna A, Benini R, et al. J Am Soc Nephrol 1999;10:997–1006,
100. Brenner BM, Cooper ME, de Zeeuw D, et al. N Engl J Med 2001;345:861–9.
101. Lewis EJ, Hunsicker LG, Clarke WR, et al. N Engl J Med 2001;345:851–60.
102. Ruggenenti P, Perna A, Gherardi G, et al. Am J Kidney Dis 2000;35:1155–65.
103. Lewis EJ, Hunsicker LG, Bain RP, et al. N Engl J Med 1993;329:1456–62.
104. Viberti G, Mogensen CE, Groop LC, et al. JAMA 1994;271: 275–9.
105. Blasi ER, Rocha R, Rudolph AE, et al. Kidney Int 2003;63: 1791–800.
106. Campese VM, Park J. J Hypertens 2006;24:2157–9.
107. Ku E, Campese VM. Pediatr Nephrol 2009;24:2301–7.
108. Bakris GL, Agarwal R, Anker SD, et al. N Engl J Med 2020;383:2219–29.
109. Walser M. Clin Nephrol 1975;3:180–6.
110. Maschio G, Oldrizzi L, Tessitore N, et al. Kidney Int 1982;22:371–6.
111. Alvestrand A, Ahlberg M, Bergstrom J. Kidney Int Suppl 1983;16:S268–72.
112. Barsotti G, Morelli E, Giannoni A, et al. Kidney Int Suppl 1983;16:S278–84.
113. Gretz N, Korb E, Strauch M. Kidney Int Suppl 1983;16:S263–7.
114. Mitch WE, Walser M, Steinman TI, et al. N Engl J Med 1984;311:623–9.
115. Rosman JB, ter Wee PM, Meijer S, et al. Lancet 1984;2:1291–6.
116. Walser M, LaFrance ND, Ward L, et al. Kidney Int 1987;32: 123–8.
117. Ihle BU, Becker GJ, Whitworth JA, et al. N Engl J Med 1989;321:1773–7.

118. Zeller K, Whittaker E, Sullivan L, et al. N Engl J Med 1991;324:78–84.
119. Klahr S, Levey AS, Beck GJ, et al. N Engl J Med 1994;330. 877–84.
120. Barsotti G, Giannoni A, Morelli E, et al. Clin Nephrol 1984;21:54–9.
121. Mitch WE, Walser M, Sapir DG. J Clin Invest 1981;67:553–62.
122. Tischler ME, Desautels M, Goldberg AL. J Biol Chem 1982;257:1613–21.
123. Kopple J. Nutritional Management of Nondialyzed Patients with Chronic Renal Failure. Philadelphia, PA: Lippincott Williams and Wilkins, 2004.
124. Williams AJ, Baker F, Walls J. Nephron 1987;46:83–90.
125. D'Amico G, Gentile MG, Manna G, et al. Lancet 1992;339: 1131–4.
126. Kalantar-Zadeh K, Joshi S, Schlueter R, et al. Nutrients 2020;12.
127. Kalantar-Zadeh K, Rhee CM, Joshi S, et al. Curr Opin Nephrol Hypertens 2022;31:26–35.
128. D'Amico G, Remuzzi G, Maschio G, et al. Clin Nephrol 1991;35:237–42.
129. Fouque D, Laville M, Boissel JP, et al. BMJ 1992;304:216–20.
130. Pedrini MT, Levey AS, Lau J, et al. Ann Intern Med 1996;124: 627–32.
131. Kasiske BL, Lakatua JD, Ma JZ, et al. Am J Kidney Dis 1998;31:954–61.
132. Fouque D, Wang P, Laville M, et al. Nephrol Dial Transplant 2000;15:1986–92.
133. Brunori G, Viola BF, Parrinello G, et al. Am J Kidney Dis 2007;49:569–80.
134. Mircescu G, Garneata L, Stancu SH, et al. J Ren Nutr 2007;17:179–88.
135. Shah SN, Abramowitz M, Hostetter TH, et al. Am J Kidney Dis 2009;54:270–7.
136. Menon V, Kopple JD, Wang X, et al. Am J Kidney Dis 2009;53:208–17.
137. Chauveau P, Couzi L, Vendrely B, et al. Am J Clin Nutr 2009;90:969–74.
138. Bernardi A, Biasia F, Pati T, et al. Am J Kidney Dis 2003;41: S146–52.
139. Meyer TW, Lawrence WE, Brenner BM. Kidney Int Suppl 1983;16:S243–7.
140. Hostetter TH, Rosenberg ME. J Am Soc Nephrol 1990;1:S55–8.
141. Hansen HP, Tauber-Lassen E, Jensen BR, et al. Kidney Int 2002;62:220–8.
142. Schnaper HW. In: Schrier RW, ed. Nephrotic Syndrome: Minimal Change Disease, Focal Glomerulosclerosis, and Related Disorders. Diseases of the Kidney. Philadelphia, PA: Lippincott Williams & Wilkins, 2001;2:1773–831.
143. GA K. Nutritional and non-nutrtional management of the nephrotic syndrome. New York, NY: Academic Press, 2013.
144. Ibels LS, Gyory AZ. Medicine (Baltimore) 1994;73:79–102.
145. Hirschberg R, Wang S. Curr Opin Nephrol Hypertens 2005;14:43–52.
146. Taguma Y, Kitamoto Y, Futaki G, et al. N Engl J Med 1985; 313:1617–20.
147. Kaysen GA, Gambertoglio J, Jimenez I, et al. Kidney Int 1986;29:572–7.
148. Kaysen GA, Davies RW. J Am Soc Nephrol 1990;1:S75–9.
149. Gonick HC, Maxwell MH, Rubini ME, et al. Nephron 1966; 3:137–52.
150. Burnier Sr M. Nutritional Management of Sodium, Chloride, and Water in Kidney Disease and Kidney Failure. Nutritional Management of Renal Disease, Elsevier, 2022;313–28.
151. St-Jules DE, Shah A. Management of Potassium in Chronic Kidney Disease and Acute Kidney Injury. Nutritional Management of Renal Disease. Elsevier, 2022:329–43.
152. Scialla J. Vitamin D in kidney disease. Nutritional Management of Renal Disease. Elsevier, 2022:397–411.
153. Swaminathan S, Singh M. Trace elements, toxic metals, and metalloids in kidney disease. Nutritional Management of Renal Disease Elsevier, 2022:467–88.
154. Bellorin-Font E, Voinescu A, Martin KJ. Nutr Manag Renal Dis 2022;353–81.
155. Chazot C, Steiber AL, Kopple JD. Vitamin metabolism and requirements in chronic kidney disease and kidney failure. Nutritional Management of Renal Disease. Elsevier, 2022: 413–65.
156. Rabkin R, Simon NM, Steiner S, et al. N Eng J Med 1970;282: 182–7.
157. Sherwin RS, Bastl C, Finkelstein FO, et al. J Clin Invest 1976; 57:722–31.
158. Cuttelod S, Lemarchand-Beraud T, Magnenat P, et al. Metab 1974;23:101–13.
159. Davidson WD, Moore TC, Shippey W, et al. Gastroenterol 1974; 66:522–5.
160. Samaan NA, Freeman RM. Metab 1970;19:102–13.
161. Lim VS, Fang VS. Am J Med 1975;58:655–62.
162. Tourkantonis A, Spiliopoulos A, Pharmakiotis A, et al. Nephron 1981;27:271–2.
163. Hershman JM, Krugman LG, Kopple JD, et al. Metab 1978; 27:755–9.
164. Fouque D, Peng SC, Kopple JD. Kidney Int 1995;47:876–83.
165. Ding H, Gao XL, Hirschberg R, et al. J Clin Invest 1996;97: 1064–75.
166. McCaleb ML, Wish JB, Lockwood DH. Endocr Res 1985;11: 113–25.
167. Vajda FJ, Martin TJ, Melick RA. Endocrinol 1969;84:162–4.
168. Fadda GZ, Hajjar SM, Perna AF, et al. J Clin Invest 1991;87: 255–61.
169. Kopple JD, Fukuda S. Am J Clin Nutr 1980;33:1363–72.
170. Tizianello A, De Ferrari G, Garibotto G, et al. J Clin Invest 1980;65:1162–73.
171. Weiner ID, Mitch WE, Sands JM. Clin J Am Soc Nephrol 2015;10:1444–58.
172. Kopple JD. In: Massry SG, Glassock RJ, eds. Products of Nitrogen Metabolism and Their Toxicity. Massry & Glassock's Textbook of Nephrology. Philadelphia, PA: Lippincott Williams & Wilkins, 2001:1262–78.
173. Bricker NS. N Engl J Med 1972;286:1093–9.
174. Massry SG, Smogorzewski M. Semin Nephrol 1994;14:219–31.
175. Simenhoff ML, Saukkonen JJ, Burke JF, et al. Lancet 1976;2: 818–21.
176. Noce A, Marchetti M, Marrone G, et al. Eur Rev Med Pharmacol Sci 2022;26:2057–74.
177. Locatelli F, Aljama P, Canaud B, et al. Nephrol Dial Transplant 2010;25:2846–50.
178. Clement FM, Klarenbach S, Tonelli M, et al. Arch Intern Med 2009;169:1104–12.
179. Besarab A, Coyne DW. Nat Rev Nephrol 2010;6:699–710.
180. Krol E, Rutkowski B, Wroblewska M, et al. Miner Electrolyte Metab 1996;22:13–5.
181. Attman PO, Alaupovic P. Nephron 1991;57:401–10.
182. Lindner A, Charra B, Sherrard DJ, et al. N Engl J Med 1974;290:697–701.
183. Briggs WA, Lazarus JM, Birtch AG, et al. Arch Intern Med 1973;132:21–8.
184. Kopple JD. Miner Electrolyte Metab 1992;18:269–75.
185. Churchill DN, Taylor DW, Keshaviah PR, et al. J Am Soc Nephrol 1996;7:198–207.
186. Woodrow G, Oldroyd B, Turney JH, et al. Nephrol Dial Transplant 1996;11:1613–8.
187. Palop L, Martinez JA. Am J Clin Nutr 1997;66:498S–503S.

188. Dwyer JT, Cunniff PJ, Maroni BJ, et al. J Ren Nutr 1998;8:11–20.
189. Aparicio M, Cano N, Chauveau P, et al. Nephrol Dial Transplant 1999;14:1679–86.
190. Chung SH, Na MH, Lee SH, et al. Perit Dial Int 1999; 19(suppl 2):S517–22.
191. Williams AJ, McArley A. J Ren Nutr 1999;9:157–62.
192. Carlson DM, Duncan DA, Naessens JM, et al. Mayo Clin Proc 1984;59:769–75.
193. Hirschberg R. Drug–Nutrient Interactions in Renal Failure. Nutritional Management of Renal Disease. Elsevier, 2022: 921–33.
194. Cianciaruso B, Brunori G, Kopple JD, et al. Am J Kidney Dis 1995;26:475–86.
195. Lowrie EG, Lew NL. Am J Kidney Dis 1990;15:458–82.
196. Fouque D, Kalantar-Zadeh K, Kopple J, et al. Kidney Int 2008;73:391–8.
197. Gracia-Iguacel C, Gonzalez-Parra E, Mahillo I, et al. Br J Nutr 2019;121:1271–8.
198. Laidlaw SA, Berg RL, Kopple JD, et al. Am J Kidney Dis 1994;23:504–13.
199. MacLaughlin HL, Friedman AN, Ikizler TA. Am J Kidney Dis 2022;79:437–49.
200. Koppe L, Fouque D, Kalantar-Zadeh K. J Cachexia Sarcopenia Muscle 2019;10:479–84.
201. Hanna RM, Ghobry L, Wassef O, et al. Blood Purif 2020;49: 202–11.
202. Workeneh BT, Kalantar-Zadeh K, Moore LW. J Ren Nutr 2021;31:335–9.
203. Mahajan SK, Prasad AS, Lambujon J, et al. Am J Clin Nutr 1980;33:1517–21.
204. Kopple JD, Mercurio K, Blumenkrantz MJ, et al. Kidney Int 1981;19:694–704.
205. Sprenger KB, Bundschu D, Lewis K, et al. Kidney Int Suppl 1983;16:S315–8.
206. Bellinghieri G, Savica V, Mallamace A, et al. Am J Clin Nutr 1983;38:523–31.
207. Batchelor EK, Kapitsinou P, Pergola PE, et al. J Am Soc Nephrol 2020;31:456–68.
208. Obi Y, Qader H, Kovesdy CP, et al. Curr Opin Clin Nutr Metab Care 2015;18:254–62.
209. Kopple JD, Berg R, Houser H, et al. Kidney Int Suppl 1989;27:S184–94.
210. Sumida K, Kovesdy CP. Causes and Treatment of Protein-Energy Wasting in Kidney Disease. Nutritional Management of Renal Disease. Elsevier, 2022, 191–206.
211. Grodstein GP, Blumenkrantz MJ, Kopple JD. Am J Clin Nutr 1980;33:1411–6.
212. Keane WF, Collins AJ. 1994 Am J Kidney Dis 1994;24: 1010–8.
213. Collins AJ, Kasiske B, Herzog C, et al. Am J Kidney Dis 2005;45:A5–7, S1–280.
214. Wolfson M, Jones MR, Kopple JD. Kidney Int 1982;21:500–6.
215. Ikizler TA, Flakoll PJ, Parker RA, et al. Kidney Int 1994;46: 830–7.
216. Kopple JD, Blumenkrantz MJ, Jones MR, et al. Am J Clin Nutr 1982;36:395–402.
217. Chazot C, Shahmir E, Matias B, et al. Kidney Int 1997;52: 1663–70.
218. Blumenkrantz MJ, Gahl GM, Kopple JD, et al. Kidney Int 1981;19:593–602.
219. Wathen RL, Keshaviah P, Hommeyer P, et al. Am J Clin Nutr 1978;31:1870–5.
220. Gutierrez A, Alvestrand A, Wahren J, et al. Kidney Int 1990;38:487–94
221. Gutierrez A, Bergstrom J, Alvestrand A. Clin Nephrol 1992;38:20–9.
222. Mehrotra R, Kopple JD, Wolfson M. Kidney Int Suppl 2003; S13–25.
223. Linton AL, Clark WF, Driedger AA, et al. Nephron 1977;19: 95–8.
224. Kopple JD, Coburn JW. Medicine (Baltimore) 1973;52:597–607.
225. Kalantar-Zadeh K, Stenvinkel P, Pillon L, et al. Adv Ren Replace Ther 2003;10:155–69.
226. Kalantar-Zadeh K, Kopple JD, Kamranpour N, et al. Kidney Int 2007;72:1149–56.
227. Rambod M, Kovesdy CP, Bross R, et al. Am J Clin Nutr 2008;88:1485–94.
228. Rambod M, Bross R, Zitterkoph J, et al. Am J Kidney Dis 2009;53:298–309.
229. Dukkipati R, Spatola L, Kalantar-Zadeh K, et al. Nutritional Management of Maintenance Hemodialysis Patients. Nutritional Management of Renal Disease. Elsevier, 2022:557–91.
230. Carrero JJ, Kistler B, Stenvinkel P. Inflammation in Chronic Kidney Disease. Nutritional Management of Renal Disease. Elsevier, 2022, 91–105.
231. Miyata T, Kurokawa K. Carbonyl Stress in Uremia. Nutritional Management of Renal Disease. Elsevier, 2022:121–6.
232. Liakopoulos V, Roumeliotis S, Gorny X, et al. Oxidat Med Cell Longev 2017:3081856.
233. Bachmann J, Tepel M, Raidt H, et al. J Am Soc Nephrol 1995;6:121–5.
234. Robinson K, Gupta A, Dennis V, et al. Circ 1996;94:2743–8.
235. Kalantar-Zadeh K, Block G, Humphreys MH, et al. J Am Soc Nephrol 2004;15:442–53.
236. Welch GN, Loscalzo J. N Engl J Med 1998;338:1042–50.
237. Kopple JD. ASAIO J 1997;43:246–50.
238. Noori N, Kovesdy CP, Dukkipati R, et al. Am J Clin Nutr 2010;92:1060–70.
239. Bologa RM, Levine DM, Parker TS, et al. Am J Kidney Dis 1998;32:107–14.
240. Garcia-Martinez C, Llovera M, Agell N, et al. Biochem Biophys Res Commun 1994;201:682–6.
241. Sarraf P, Frederich RC, Turner EM, et al. J Exp Med 1997;185:171–5.
242. Stenvinkel P, Heimburger O, Lindholm B, et al. Nephrol Dial Transplant 2000;15:953–60.
243. Kalantar-Zadeh K, Kopple JD. Am J Kidney Dis 2001;38: 1343–50.
244. Lodebo BT, Shah A, Kopple JD. J Ren Nutr 2018;28:369–79.
245. Rhee CM, Ahmadi SF, Kalantar-Zadeh K. Curr Opin Nephrol Hypertens 2016;25:208–16.
246. Kalantar-Zadeh K, Rhee CM, Chou J, et al. Kidney Int Rep 2017;2:271–81.
247. Kopple JD, Zhu X, Lew NL, et al. Kidney Int 1999;56: 1136–48.
248. Port FK, Ashby VB, Dhingra RK, et al. J Am Soc Nephrol 2002;13:1061–6.
249. Kalantar-Zadeh K, Block G, Humphreys MH, et al. Kidney Int 2003;63:793–808.
250. Liu Y, Coresh J, Eustace JA, et al. JAMA 2004;291:451–9.
251. Kalantar-Zadeh K, Kopple JD, Humphreys MH, et al. Nephrol Dial Transplant 2004;19:1507–19.
252. Johansen KL, Young B, Kaysen GA, et al. Am J Clin Nutr 2004;80:324–32.
253. Leavey SF, McCullough K, Hecking E, et al. Nephrol Dial Transplant 2001;16:2386–94.
254. Kalantar-Zadeh K, Kopple JD, Kilpatrick RD, et al. Am J Kidney Dis 2005;46:489–500.
255. Kopple JD. Am J Clin Nutr 2005;81:1257–66.
256. Bolasco P, Kalantar-Zadeh K, Rhee C. Panminerva Med 59, 115.
257. Rhee CM, Nguyen DV, Nyamathi A, et al. (2020) Curr Opin Nephrol Hypertens 29, 92–102.

258. Zarantonello D, Rhee CM, Kalantar-Zadeh K, et al. Curr Opin Nephrol Hypertens 2021;30:97–107.
259. Martino S, Chwastiak L, Finkelstein F. In: Kopple JD, Massry SG, Kalantar-Zadeh K, eds. Motivating the Kidney Disease Patient to Dietary Adherence and Other Healthy Lifestyle Activities. Nutritional Management of Renal Disease. 3rd ed. New York, NY: Academic Press.
260. Ikizler TA, Burrowes JD, Byham-Gray LD, et al. Am J Kidney Dis 2020;76:S1–S107.
261. Kistler BM, Moore LW, Benner D, et al. J Ren Nutr 2021;31: 116–20:el.
262. Kopple JD, Greene T, Chumlea WC, et al. Kidney Int 2000;57:1688–703.
263. Ku E, Kopple JD, Johansen KL, et al. Am J Kidney Dis 2018;71:657–65.
264. Ku E, Kopple JD, McCulloch CE, et al. Am J Kidney Dis 2018;71:648–56.
265. Kopple JD. In: Robinson RR, ed. Nutrition in Renal Failure: Causes of Catabolism and Wasting in Acute or Chronic Renal Failure. Nephrology. Proceedings of the IXth International Congress of Nephrology. New York, NY: Springer, 1984;2: 1498–515.
266. Ikizler TA, Greene JH, Wingard RL, et al. J Am Soc Nephrol 1995;6:1386–91.
267. Mehrotra R, Berman N, Alistwani A, et al. Am J Kidney Dis 2002;40:133–42.
268. Hsiung JT, Kleine CE, Naderi N, et al. J Ren Nutr 2019;29: 310–21.
269. Calloway DH, Odell AC, Margen S. J Nutr 1971;101:775–86.
270. Kopple JD, Gao XL, Qing DP. Kidney Int 52:486–94.
271. Maroni BJ, Steinman TI, Mitch WE. Kidney Int 1985;27:58–65.
272. Sargent JA, Gotch FA. J Am Diet Assoc 1979;75:547–51.
273. Varcoe R, Halliday D, Carson ER, et al. Clin Sci Mol Med 1975;48:379–90.
274. Walser M. J Clin Invest 1974;53:1385–92.
275. Bergstrom J, Furst P, Alvestrand A, et al. Kidney Int 1993;44:1048–57.
276. Blake P, Daugirdas J. In: Jacobs C, Kjellstrand CM, Koch KM et al. eds. Quantification and Prescription: General Principles. Replacement of Renal Function by Dialysis. 4th rev ed. Dordrecht, Kluwer Academic Publishers, 1996:619–56.
277. Kopple JD, Jones MR, Keshaviah PR, et al. Am J Kidney Dis 1995;26:963–81.
278. Kopple JD, Coburn JW. JAMA 1974;227:41–4.
279. Narasaki Y, Okuda Y, Moore LW, et al. Am J Clin Nutr 2021;114:303–13.
280. Kopple JD, Coburn JW. Medicine (Baltimore) 1973;52:583–95.
281. Kopple JD, Monteon FJ, Shaib JK. Kidney Int 1986;29:734–42.
282. Shah AP, Kalantar-Zadeh K, Kopple JD. Am J Kidney Dis 2015;65:659–73.
283. Ikizler TA, Cano NJ, Franch H, et al. Kidney Int 2013;84: 1096–107.
284. Kidney Disease: Improving Global Outcomes (KDIGO) CKD Work Group. Kidney Int 2013;3:1–150.
285. Heerspink HJL, Stefansson BV, Correa-Rotter R, et al. N Engl J Med 2020;383:1436–46.
286. McMurray JJV, Solomon SD, Inzucchi SE, et al. N Engl J Med 2019;381:1995–2008.
287. Neumiller JJ, Rhee CM, Kalantar-Zadeh K. Am J Nephrol 2017;46:459–61.
288. Perkovic V, Jardine MJ, Neal B, et al. N Engl J Med 2019;380:2295–306.
289. Bakris GL, Agarwal R, Filippatos G, et al. N Engl J Med 2021;384:e42.
290. Pitt B, Filippatos G, Agarwal R, et al. N Engl J Med 2021; 385:2252–63.
291. Rhee CM, Ghahremani-Ghajar M, Obi Y, et al. Panminerva Med 2017;59:188–96.
292. Narasaki Y, Rhee CM, Kramer H, et al. Curr Opin Clin Nutr Metab Care 2021;24:10–7.
293. Watanabe D, Machida S, Matsumoto N, et al. Nutrients 2018;10.
294. Kalantar-Zadeh K, Rhee C, Sim JJ, et al. J Cachexia Sarcopenia Muscle 2013;4:89–94.
295. Kim JC, Kalantar-Zadeh K, Kopple JD. J Am Soc Nephrol 2013;24:337–51.
296. Kittiskulnam P, Banjongjit A, Metta K, et al. Sci Rep 2022;12: 4529.
297. Kalantar-Zadeh K, Crowley ST, Beddhu S, et al. Semin Dial 2017;30:251–61.
298. Rhee CM, Unruh M, Chen J, et al. Semin Dial 2013;26:720–7.
299. Borah MF, Schoenfeld PY, Gotch FA, et al. Kidney Int 1978;14:491–500.
300. Kopple JD, Shinaberger JH, Coburn JW, et al. Trans Am Soc Artif Intern Organs 1969;15:302–8.
301. Brown-Tortorici AR, Narasaki Y, You AS, et al. Nutrients 2022;14.
302. Rhee CM, You AS, Koontz Parsons T, et al. Nephrol Dial Transplant 2017;32:1233–43.
303. Kaysen GA, al Bander H. Am J Nephrol 1990;10(suppl 1): 36–42.
304. Rhee CM, Obi Y, Mathew AT, et al. Semin Nephrol 2018;38: 325–35.
305. Peritoneal Dialysis Adequacy Work Group. Am J Kidney Dis 2006;48(suppl 1):S98–129.
306. Peritoneal Dialysis Adequacy Work Group. Am J Kidney Dis 2006;48(suppl 1):S91–7.
307. National Kidney Foundation. Am J Kidney Dis 2006; 48(suppl 1):S130–58.
308. Lee YJ, Rhee CM, Kalantar-Zadeh K. Ann Transl Med 2018;6:317.
309. Obi Y, Rhee CM, Mathew AT, et al. J Am Soc Nephrol 2016;27:3758–68.
310. Wang M, Obi Y, Streja E, et al. Nephrol Dial Transplant 2018;33:1823–31.
311. You AS, Kalantar-Zadeh K, Obi Y, et al. Kidney Int Rep 2020;5:643–53.
312. Caria S, Cupisti A, Sau G, et al. BMC Nephrol 2014;15:172.
313. Wesson DE. Clin J Am Soc Nephrol 2021;16:1292–9.
314. Moe SM, Zidehsarai MP, Chambers MA, et al. Clin J Am Soc Nephrol 2011;6:257–64.
315. Chauveau P, Koppe L, Combe C, et al. Nephrol Dial Transplant 2019;34:199–207.
316. Joshi S, McMacken M, Kalantar-Zadeh K. Am J Kidney Dis 2021;77:287–96.
317. Azadbakht L, Atabak S, Esmaillzadeh A. Diabet Care 2008;31:648–54.
318. Oosterwijk MM, Soedamah-Muthu SS, Geleijnse JM, et al. Kidney Int Rep 2019;4:710–9.
319. Dunkler D, Kohl M, Teo KK, et al. Am J Kidney Dis 2016;68: 29–40.
320. Tallman DA, Khor B-H, Karupaiah T, Khosla P, Chan M, Kopple JD. J Renal Nutr. In Press.
321. Khor BH, Tallman DA, Karupaiah T, et al. Nutrients 2021;13.
322. Monteon FJ, Laidlaw SA, Shaib JK, et al. Kidney Int 1986;30: 741–7.
323. Schneeweiss B, Graninger W, Stockenhuber F, et al. Am J Clin Nutr 1990;52:596–601.
324. Slomowitz LA, Monteon FJ, Grosvenor M, et al. Kidney Int 1989;35:704–11.
325. Kluthe R, Luttgen FM, Capetianu T, et al. Am J Clin Nutr 1978;31:1812–20.
326. Kopple JD. Kidney Int 1978;14:340–8.
327. Marckmann P. Clin Nephrol 1988;29:75–8.

328. Wolfson M, Strong CJ, Minturn D, et al. Am J Clin Nutr 1984;39:547–55.
329. Narasaki Y, Kalantar-Zadeh K, Ivaturi K, et al. Dietary Fiber and Mortality Across Kidney Function in a Nationally Representative Cohort. Abstract presented at the National Kidney Foundation Spring Clinical Meeting. March 2020.
330. Narasaki Y, Okuda Y, Kalantar SS, et al. J Ren Nutr 2021;31: 411–20.
331. Anderson JW, Baird P, Davis RH, Jr., et al. Nutr Rev 2009;67: 188–205.
332. Aune D, Chan DS, Lau R, et al. BMJ 2011;343:d6617.
333. Barber TM, Kabisch S, Pfeiffer AFH, et al. Nutrients 2020;12.
334. Sabatino A, Regolisti G, Gandolfini I, et al. J Nephrol 2017;30:743–54.
335. National Kidney Foundation. Am J Kidney Dis 2003;41:S1–92.
336. Toigo G, Aparicio M, Attman P-O et al. Clin Nutr 2000;19: 197–207.
337. Johnson DW, Atai E, Chan M, et al. Nephrol (Carlton) 2013;18:340–50.
338. National Kidney Foundation. Am J Kidney Dis 2017;49: S1–S180.
339. Kidney Disease: Improving Global Outcomes (KDIGO). Kidney Int Suppl 2013;3:259–305.
340. Ash S, Campbell K, MacLaughlin H, et al. Nutr Diet 2006;63: S35–S45.
341. Palmer BF, Carrero JJ, Clegg DJ, et al. Mayo Clin Proc 2021;96:744–62.
342. Palmer BF, Clegg DJ. Adv Physiol Educ 2016;40:480–90.
343. Palmer BF, Clegg DJ. Am J Kidney Dis Am J Kidney Dis 2019;74:682–95.
344. Hunter RW, Bailey MA. Nephrol Dial Transplant 2019;34: iii2–iii11.
345. Clase CM, Carrero JJ, Ellison DH, et al. Kidney Int 2020;97: 42–61.
346. Narasaki Y, You AS, Malik S, et al. Am J Clin Nutr 2022;116: 1123–1134.
347. Kovesdy CP, Appel LJ, Grams ME, et al. J Am Soc Hypertens 2017;11:783–800.
348. Staruschenko A. Hypertens 2018;71:1015–22.
349. Sussman EJ, Singh B, Clegg D, et al. J Ren Nutr 2020;30: 475–83.
350. Makar MS, Pun PH. Am J Kidney Dis 2017;69:684–95.
351. Kalantar-Zadeh K, Gutekunst L, Mehrotra R, et al. Clin J Am Soc Nephrol 2010;5:519–30.
352. Kalantar-Zadeh K. Patient Prefer Adherence 2013;7:379–90.
353. Houston J, Isakova T, Wolf F. In: Kopple JD, Massry SG, Kalantar-Zadeh K, eds. Phosphorus and FGF23 in Chronic Kidney Disease. Nutritional Management of Renal Disease. 3rd ed. New York, NY: Academic Press, 2013.
354. Martin K. In: Kopple JD, Massry SG, Kalantar-Zadeh K, eds. Calcium, Parathyroid Hormone and Bone Disease in Kidney Disease and Chronic Kidney Failure. Nutritional Management of Renal Disease. 3rd ed. New York, NY: Academic Press, 2013.
355. Kovesdy CP, Anderson JE, Kalantar-Zadeh K. Clin Nephrol 2010;73:268–75.
356. Kalantar-Zadeh K, Kuwae N, Regidor DL, et al. Kidney Int 2006;70:771–80.
357. Block GA, Klassen PS, Lazarus JM, et al. J Am Soc Nephrol 2004;15:2208–18.
358. Kidney Disease: Improving Global Outcomes CKDMBDWG. Kidney Int Suppl, 2009;S1–130.
359. Lynch KE, Lynch R, Curhan GC, et al. Clin J Am Soc Nephro 2011;6:620–9.
360. Shinaberger CS, Greenland S, Kopple JD, et al. Am J Clin Nutr 2008;88:1511–8.
361. Noori N, Kalantar-Zadeh K, Kovesdy CP, et al. Am J Kidney Dis 2010;55:A85–A.
362. Kalantar-Zadeh K, Kopple JD, Deepak S, et al. J Ren Nutr 2002;12:17–31.
363. Khoueiry G, Waked A, Goldman M, et al. J Ren Nutr 2011;21:438–47.
364. Narasaki Y, Rhee CM. Clin J Am Soc Nephrol 2020;16:9–11.
365. Waheed AA, Pedraza F, Lenz O, et al. Nephrol Dial Transplant 2013;28:2961–8.
366. Rhee CM, Zhou M, Woznick R, et al. Int Urol Nephrol 2023;55:377–387.
367. Gutekunst L. J Ren Nutr 2016;26:209–18.
368. Kidney Disease: Improving Global Outcomes CKDMBDUWG. Kidney Int Suppl 2011;7:1–59.
369. Block GA, Rosenbaum DP, Yan A, et al. J Am Soc Nephrol 2019;30:641–52.
370. Roumeliotis S, Duni A, Vaios V, et al. Nutrients 2022;14.
371. Roumeliotis S, Liakopoulos V, Schurgers LJ. Curr Vasc Pharmacol 2022;20:121–6.
372. Chazot C, Steiber A, Kopple JD. J Ren Nutr. 2022.
373. Ribeiro M, Fonseca L, Anjos JS, et al. Nutr Clin Pract 2022;37: 81–93.
374. Tian N, Li L, Ng JK, et al. Nutrients 2022;14.
375. Burrows NR, Koyama A, Pavkov ME. MMWR Morb Mortal Wkly Rep 2022;71:412–5.
376. Savica V, Bellinghieri G, Kopple JD. Annu Rev Nutr 2010;30: 365–401.
377. Kalantar-Zadeh K, Kopple JD, Kilpatrick RD, et al. Am J Kidney Dis 2005;46:489–500.
378. Kalantar-Zadeh K, Kuwae N, Wu DY, et al. Am J Clin Nutr 2006;83:202–10.
379. Clark AS, Mitch WE. J Clin Invest 1983;72:836–45.
380. Frohlich J, Scholmerich J, Hoppe-Seyler G, et al. Eur J Clin Invest 1974;4:453–8.
381. Kovesdy CP, Kalantar-Zadeh K. Semin Nephrol 2009;29:3–14.
382. Feinstein EI, Blumenkrantz MJ, Healy M, et al. Medicine (Baltimore) 1981;60:124–37.
383. Dornfeld LP, Maxwell MH, Waks A, et al. Kidney Int Suppl 1987;22:S254–8.
384. Feinstein EI, Kopple JD, Silberman H, et al. Kidney Int Suppl 1983;16:S319–23.
385. Mault JR, Bartlett RH, Dechert RE, et al. Trans Am Soc Artif Intern Organs 1983;29:390–5.
386. Palevsky PM, Liu KD, Brophy PD, et al. Am J Kidney Dis 2013;61:649–72.
387. Fiaccadori E, Regolisti G, Cabassi A. NDT Plus 2010;3:1–7.
388. Fiaccadori E, Sabatino A, Barazzoni R, et al. Clin Nutr 2021;40:1644–68.
389. Flugel-Link RM, Salusky IB, Jones MR, et al. Am J Physiol 1983;244:E615–23.
390. Kopple JD. JPEN J Parenter Enteral Nutr 1996;20:3–12.
391. Druml W. Metabolic Management and Nutritional Support in Acute Kidney Injury. Nutritional Management of Renal Disease. Elsevier, 2022, 659–84.
392. Nakasaki H, Katayama T, Yokoyama S, et al. JPEN J Parenter Enteral Nutr 1993;17:86–90.
393. Mehta RL. Semin Nephrol 1994;14:64–82.
394. Davenport A, Roberts NB. Crit Care Med 1989;17:1010–4.
395. Davies SP, Reaveley DA, Brown EA, et al. Crit Care Med 1991;19:1510–5.
396. Kopple JD, Bernard D, Messana J, et al. Kidney Int 1995;47: 1148–57.
397. Tjiong HL, van den Berg JW, Wattimena JL, et al. J Am Soc Nephrol 2005;16:1486–93.
398. Leonard CD, Luke RG, Siegel RR. Urology 1975;6:154–7.
399. Kopple JD, Swendseid ME. Am J Clin Nutr 1974;27:806–12.

400. Cerra FB, Upson D, Angelico R, et al. Surgery 1982;92:192–9.
401. Daly JM, Mihranian MH, Kehoe JE, et al. Surg 1983;94: 151–8.
402. Leander U, Furst P, Vesterberg K, et al. Clin Nutr 1985;4:43–51.
403. Hammarqvist F, Wernerman J, Ali R, et al. Ann Surg 1989;209:455–61.
404. Stehle P, Zander J, Mertes N, et al. Lancet 1989;1:231–3.
405. Daly JM, Reynolds J, Thom A, et al. Ann Surg 1988;208:512–23.
406. Harris JA, Benedict FG. A Biometric Study of Basal Metabolism in Man. Carnegie Institution of Washington, 1919.
407. Garrel DR, Jobin N, de Jonge LH. Nutr Clin Pract 1996;11: 99–103.
408. Wilmore DW. The Metabolic Management of the Critically Ill. Springer Science & Business Media, 2012.
409. Askanazi J, Elwyn DH, Silverberg PA, et al. Surg 1980;87: 596–8.
410. Jeejeebhoy KN, Langer B, Tsallas G, et al. Gastroenterol 1976; 71:943–53.
411. Seidner DL, Mascioli EA, Istfan NW, et al. JPEN J Parenter Enteral Nutr 1989;13:614–9.
412. Driscoll DF, Baptista RJ, Bistrian BR, et al. Am J Hosp Pharm 1986;43:416–9.
413. Smith FR, Goodman DS. J Clin Invest 1971;50:2426–36.
414. Dietary Reference Intakes for Vitamin A, Vitamin K, Arsenic, Boron, Chromium, Copper, Iodine, Iron, Manganese, Molybdenum, Nickel, Silicon, Vanadium, and Zinc Washington, DC: 2001.
415. Reichel H, Koeffler HP, Norman AW. N Engl J Med 1989;320: 980–91.
416. Udall JA. JAMA 1965;194:127–9.
417. Balcke P, Schmidt P, Zazgornik J, et al. Ann Intern Med 1984;101:344–5.
418. Pru C, Eaton J, Kjellstrand C. Nephron 1985;39:112–6.
419. Dukkipati R, Kalantar-Zadeh K, Kopple JD. Am J Kidney Dis 2010;55:352–64.
420. Chertow GM, Ling J, Lew NL, et al. Am J Kidney Dis 1994;24:912–20.
421. Capelli JP, Kushner H, Camiscioli TC, et al. Am J Kidney Dis 1994;23:808–16.

CHAPTER 97

Hematologic Aspects of Common Nutritional Anemias

Aśok C. Antony

CHAPTER OUTLINE

Abbreviations

DMT1, divalent metal transporter 1; **DNA**, deoxyribonucleic acid; **MCV**, mean corpuscular volume; **NSAID**, nonsteroidal anti-inflammatory drug; **NTD**, neural-tube defect; **RBC**, red blood cell; **RDA**, recommended dietary allowance; **RNA**, ribonucleic acid; **sTfR**, soluble transferrin receptor; **ZPP**, zinc protoporphyrin

Introduction

The formation of blood cells (hematopoiesis) occurs in the bone marrow medullary cavity of virtually all bones in the newborn, but in adults, active blood formation is confined to the central skeleton (skull, vertebral column, ribs, and pelvis) and upper ends of the humerus and femur. All hematopoietic cells arise from a very small population of self-renewing stem cells that, under the influence of a range of growth factors, give rise to RBCs, white blood cells (polymorphonuclear neutrophils, eosinophils, basophils, monocytes, lymphocytes), and platelets. Because the turnover of blood cells is normally greater than that of other tissues in the body, deficiency of any of the three important nutrients—iron, vitamin-B_{12} (cobalamin), and folate (folic acid, pteroylglutamic acid)—can become limiting for hematopoiesis.

Anemia is defined as a low concentration of hemoglobin in the blood. Optimum hemoglobin values at different ages are presented in **Table 97.1**.

From the public health point of view, the main hematopoietic nutrients causing nutritional anemias are iron, folate, and vitamin B_{12}. Although general nutritional aspects related to these nutrients are discussed in separate chapters, hematologic principles and related issues are discussed in this chapter. Whereas other trace metals and vitamins contribute in small part to normal hematopoiesis, their clinical role independent of the three major nutrients is of lesser public health importance. Nevertheless, when patients continue to have anemia despite repletion with iron, folate, and vitamin B_{12}, and other medical causes for anemia have been ruled out, the potential role of other trace metals and vitamins may be sought.

Table 97.1 Hemoglobin Cut-offs Used to Define Anemia at Sea Level[a]

Age or Sex Group	Hemoglobin Lower Than the Values (G/Dl) Defines Anemia
Men	13.0
Nonpregnant women	12.0
Pregnant women	11.0
Children (6 mo–5 y)	11.0
Children (5–11 y)	11.5
Children (12–13 y)	12.0

[a]The optimal hemoglobin value should be higher than these values for various ages and sex. The hematocrit is usually three times the hemoglobin value. Individuals living at an altitude of 4,000 ft or more have higher normal blood values; the hemoglobin should be approximately 1 g/dL *higher* for people residing at approximately 6,500 to 7,000 ft altitude.

Data from Gleason G, Scrimshaw N. In: K Kraeme MZ, ed. Nutritional Anemia. Basel, Switzerland: Sight and Life Press, 2007:45–58; and Biesalki HK, Erhardt J. In: Kraemer K, Zimmerman MB, eds. Diagnosis of Nutritional Anemia: Laboratory Assessment of Iron Status. Nutritional Anemia. Basel, Switzerland: Sight and Life Press, 2007:37–43.

Common Nutritional Anemias—General Considerations

Iron, folate, and vitamin B_{12} are the three key micronutrients that are critically required for the optimization of erythropoiesis[1-7]; therefore, a deficiency of either iron, or folate, or vitamin B_{12} can independently lead to anemia.[5,6] However, and particularly in resource-limited countries, all three micronutrient deficiencies are commonly found in the same individual; this usually arises from chronic consumption of a relatively monotonous diet that is not sufficiently balanced with an optimum combination of grains, vegetables, lentils and beans, fruits, and animal-source foods (such as meats, seafood, and dairy products). Most populations living in resource-limited countries who indicate they are "nonvegetarians" actually consume a "near-vegetarian" diet,[4,7] which contains relatively small quantities of meat only once or twice a week. Such a diet usually cannot sustain an optimum iron and vitamin B_{12} balance. Moreover, the folate content of such diets is also low, resulting from a meager consumption of fresh green salads, vegetables, and fruit, and common cooking practices that destroy folate.

Physiologic factors, including growth spurts in childhood and adolescence, and in women during menstruation, pregnancy, and lactation, can lead to an increased requirement of these three micronutrients, predisposing to deficiency. Nutritional anemia in pregnancy increases the risk for bleeding at delivery[8] and shock, following even mild hemorrhage, can be fatal without immediate blood replacement. Maternal anemia may also increase the risk of postpartum hemorrhage. Anemia with hemoglobin less than 7 g/dL places pregnant women at serious risk for maternal mortality (cardiac failure during labor from anemia, poor tolerance of hemorrhagic blood loss during labor, slower wound-healing time, and increased risk for infection). Hence, it is important to proactively find and treat all young adolescent women for anemia, and to identify all pregnant women with anemia at the earliest time possible.

There are also pathologic causes that trigger increased requirements. These include chronic diseases like malaria, which lead to increased hemolysis and anemia, where the increased demand to support erythropoiesis warrants prophylactic therapy with folate, and chronic hookworm infestation of the small intestine, or schistosomiasis in the urinary bladder, that leads to daily blood loss over months to years.

When individuals with several serious chronic infectious or other illnesses suffer from anorexia, this leads to reduced food intake and aggravates these micronutrient deficiencies. When their daily diet is imbalanced to begin with, such individuals need to rely on supplements of iron, folate, and vitamin B_{12}, which are best delivered via fortified foods. However, in resource-limited countries where food insecurity can be prevalent in a sizeable fraction of the population, the options of increased food intake, available or affordable dietary supplements, or equitably distributed fortified-food—are often not possible. As a result, these three deficiencies become ensconced in such individuals.

When nutritional deficiency of either one, two, or all three key micronutrients (iron, folate, or vitamin B_{12}) is the cause of anemia,[5,6] it is an axiom that all three micronutrients must be given in sufficient amounts to restore normal hemoglobin concentration.[5,6,9] Therefore, "full therapeutic replacement doses" of each depleted or deficient micronutrient must be administered as a first step to ensure that micronutrient stores in the body are rapidly replenished.[5,6,9] Only after this is accomplished should "lower supplemental maintenance doses" be administered to compensate for ongoing dietary insufficiency.

If this critical two-step intervention is not instituted, there is a serious risk for propagation of *intergenerational* iron-, folate-, and vitamin B_{12}-deficiency in adolescent girls and women, who are at high risk for vertically propagating these deficiencies to the next generation, in a perpetual cycle. In this scenario, pregnant women pass down their borderline-to-deficient micronutrient stores (akin to passing down a depleted "bank balance") to their newborn baby girls.[10] These girls grow up consuming the same unbalanced diet as their mothers at home, and then go through childhood and adolescence carrying their burden of low micronutrient stores with them. When each of these young women now enters pregnancy, they bestow a seriously depleted legacy of micronutrients to their own babies.[10] This is the reason why reversing the cycle of chronic deficiency of each of these three micronutrients should ideally begin among adolescent girls, as the most efficient way to permanently halt such vertical transmission of intergenerational nutritional anemias.

During pregnancy, the micronutrient needs of the developing baby induce further depletion of already meager-to-deficient maternal micronutrient stores. This invariably triggers a serious health crisis with nutritional anemia and risks to the pregnant woman as well as her baby *in utero*. Apart from the fact that deficiency of any or all three micronutrients can independently lead to anemia, they can also trigger pregnancy complications and negatively influence pregnancy outcomes. This is because deficiency of either iron, or folate, or vitamin-B_{12} in a pregnant woman can adversely affect *in utero* neurodevelopment of her baby, leading to postnatal neurologic, or psychologic, or psychiatric consequences in childhood. So, when all three micronutrients are insufficient during pregnancy—as is the case in the majority of women in resource-limited countries—one can predict poor outcomes in their progeny. It is now also well established that folate insufficiency before a woman becomes pregnant places her at risk for having a baby with devastating neural-tube defects (NTDs), as well as psychological-psychiatric manifestations in childhood.[6] Because nutritional vitamin B_{12} deficiency, which is recognized as a worldwide problem, is critically required for the intracellular function of folate, it is plausible that vitamin B_{12} deficiency poses a similar risk to the developing embryo/fetus as documented for folate deficiency.

When these three micronutrient deficiencies extend into childhood, there is poor growth,[11] suboptimal intellectual development and reduced gross motor and problem-solving skills.[12] In high population density resource-limited countries (like India), combined low iron-, folate-, and vitamin B_{12}-status in women is the norm[13]; and because these three micronutrient deficiencies remain incompletely treated during pregnancy, it is plausible that many among the ~26-million babies born in India each year could exhibit such problems. In addition, younger-to-middle-aged adults and the elderly invariably continue to fail to perform at maximum potential due to anemia. Although neurologic disease due to vitamin B_{12} deficiency can develop in any age group, the combined effects of both vitamin B_{12} and folate deficiency lead to chronic elevation of the metabolite homocysteine in the blood, which has myriad effects on cellular function. Parenthetically, it has been long understood that inherited metabolic diseases that lead to hyperhomocysteinemia in children consistently give rise to severe neurologic manifestations. It is now well established that chronic hyperhomocysteinemia places middle aged and elderly individuals at risk for occlusive small-vessel cerebrovascular disease–related strokes[14]; more recently, chronic hyperhomocysteinemia has also been identified as a risk factor for dementia and Alzheimer's disease.[15-20]

Despite correct diagnosis and therapy of a specific nutritional anemia, it is possible that some patients may not normalize their hemoglobin. This can occur when an unrecognized hemoglobinopathy, such as sickle cell trait and/or α- or β-thalassemia minor[21] in either mother or infant with normalized iron, vitamin B_{12}, and folate stores, precludes complete normalization of hemoglobin in patients with anemia.[21] These hemoglobinopathies—with persistent suboptimal hemoglobin and low red cell indices despite iron replenishment—are common in sub-Saharan Africa and India, and protect against malaria.[22] In Uganda, the incidence (with substantial variation) is 13.3%.[23] A significant number of individuals in resource-limited countries could have the sickle cell trait and/or thalassemia trait. Most of these hemoglobinopathies can be detected by a complete blood count and peripheral smear, and a hemoglobin electrophoresis.[21]

General Concepts with Iron Deficiency Anemia

Iron deficiency looms largest among the factors that contribute to the global burden of anemia, which affects one-third of the world's population and has serious consequences for maternal and perinatal health and child development.[24] In many resource-limited countries, more than half of specific groups of individuals are affected. Even in resource-rich countries, the problem of iron deficiency is significant (**Table 97.2**). The World Health Organization quotation (**Box 97.1**) encapsulates the fundamental problem that this chapter aims to highlight. Coincident with the diagnosis of iron deficiency is the imperative for identification of all underlying root causes, which must be addressed to prevent a recurrence of deficiency after iron replacement.

Iron Homeostasis

Iron is primarily absorbed from duodenal mucosal cells (and, to a lesser extent, from upper jejunal mucosal cells) via the divalent metal transporter 1 (DMT1).[25,26] Once iron is in enterocytes, it is transported out of the basolateral surface of cells into the blood by ferroportin. In the blood, iron is bound by transferrin, which delivers it to all cells via cell surface membrane transferrin receptors. The homeostasis of iron is regulated at the level of both intestinal absorption and macrophage release of iron by a key hormone primarily synthesized in the liver, hepcidin. Upon its release under conditions of iron sufficiency (or inflammation), hepcidin binds to ferroportin on enterocytes and degrades its normal function to mediate the transfer of absorbed iron into the plasma. At the same time, the ferroportin on macrophages is bound and degraded by hepcidin, which leads to the retention of macrophage iron that is stored as ferritin. Conversely, low stores of iron lead to suppression of hepcidin synthesis, which allows for an increase in enterocyte absorption of iron and delivery via ferroportin into plasma, as well as allowing for ferroportin-mediated transfer of iron out of macrophages into plasma.

Table 97.2 Prevalence of Iron Deficiency from the United States National Health and Nutrition Examination Surveys, 1988 to 1994 and 1999 to 2000[a]

Sex and Age Groups (y)	1988–1994		1999–2000	
	%	(95% CI)[a]	%	(95% CI)
Both sexes				
1–2	9	(6–11)	7	(3–11)
3–5	3	(2–4)	5	(2–7)
6–11	2	(1–3)	4	(1–7)
Males				
12–15	1	(0.1–2.0)	5	(2–8)
16–69	1	(0.6–1.0)	2	(1–3)
>70	4	(2–3)	3	(2–7)
Females (nonpregnant)				
12–15	9	(6–12)	9	(5–12)
16–19	11	(7–14)	16	(10–22)
20–49	11	(10–13)	12	(10–16)
White, non-Hispanic	8	(7–9)	10	(7–13)
African American, non-Hispanic	15	(13–17)	19	(14 –24)
Mexican American	19	(17–21)	22	(17–27)
50–69	5	(4–7)	9	(5–12)
>70	7	(5–8)	6	(4–9)

CI, confidence interval.

[a]All racial and ethnic groups, except where noted.

Centers for Disease Control and Prevention. MMWR Morb Mortal Wkly Rep 2002;51:897–9.

Box 97.1 Iron Deficiency

Iron deficiency is the most common and widespread nutritional disorder in the world. As well as affecting a large number of children and women in developing countries, it is the only nutrient deficiency that is also significantly prevalent in industrialized countries. The numbers are staggering: 2 billion people—over 30% of the world's population—are anemic, many due to iron deficiency, and in resource-poor areas, this is frequently exacerbated by infectious diseases. Malaria, human immunodeficiency virus/acquired immunodeficiency syndrome, hookworm infestation, schistosomiasis, and other infections such as tuberculosis are particularly important factors contributing to the high prevalence of anemia in some areas.

Iron deficiency affects more people than any other condition, constituting a public health condition of epidemic proportions. More subtle in its manifestations than, for example, protein-energy malnutrition, iron deficiency exacts its heaviest overall toll in terms of ill health, premature death, and lost earnings.

Iron deficiency and anemia reduce the work capacity of individuals and entire populations, bringing serious economic consequences and obstacles to national development. Overall, it is the most vulnerable, the poorest and the least educated who are disproportionately affected by iron deficiency, and it is they who stand to gain the most by its reduction.

Reproduced from World Health Organization Report on Iron deficiency Anemia[27]

Iron and Hematopoiesis

Of a total of 4 to 5 g of iron in the body, about 3 g of iron is in hemoglobin—which is the oxygen carrier in erythrocytes (red blood cells [RBCs]). Whereas some iron is in myoglobin, the remainder is in various heme and nonheme enzymes, including cytochromes that are necessary for oxidative reactions producing energy. Iron is also critical for cell proliferation because ribonucleotide reductase, an iron-dependent enzyme, is required for DNA synthesis. Because the major requirement for iron in the body is for hemoglobin in RBCs, an early manifestation of iron deficiency is anemia.

The bone marrow, with its prodigious production of RBCs (2×10^{11} RBCs/day/adult), can further augment erythropoiesis in response to blood loss or hemolysis sixfold by using stored iron from ferritin and hemosiderin in the liver, spleen, and bone marrow. Iron is incorporated into hemoglobin-synthesizing immature erythroid cells in the marrow via transferrin receptor-mediated endocytosis of transferrin-bound iron. Transferrin receptor expression is greatest in the early stages of erythroid development when hemoglobin synthesis occurs and is lost as these cells mature into RBCs that are released to the circulation.[28] This high density of transferrin receptors likely accounts for the preferential delivery of iron to erythroid precursors for hemoglobin synthesis when iron is in short supply. As a consequence, there can be an [iron deficiency-related] dysfunction of other [less privileged] iron-dependent enzymes/proteins in non-hematologic tissues—such as skeletal muscle, heart, and brain—which can manifest in muscle fatigue, heart failure, cognitive dysfunction, and behavioral changes *even before* iron deficiency anemia manifests clinically.[29-31]

Iron Deficiency and Its Clinical Consequences

Physiologic Adaptation to Iron Deficiency

Impairment of normal RBC production results in a fall in the hemoglobin concentration (anemia), RBC count, and packed red cell volume (hematocrit) below the levels shown in Table 97.1. The decrease in hemoglobin leads to a reduction in the oxygen-carrying capacity of the blood and impairment of oxygen delivery to tissues. Three separate physiologic responses are called into play to compensate for the development of anemia: First, there is a change in the binding of hemoglobin to oxygen so that oxygen is more efficiently unloaded from RBCs to tissues. Second, the cardiac output of blood increases so that more blood (and oxygen) is delivered to the rest of the body.

Third, an increase in the level of the erythropoiesis-stimulating hormone, erythropoietin, stimulates RBC production in the bone marrow.[32]

There are specific mechanisms for the sensing of low iron content in the blood, which trigger homeostatic mechanisms designed to increase iron absorption, upregulate transferrin receptors, and reduce hepcidin to facilitate iron absorption and its release from macrophages into blood.[25,26,33]

When anemia develops over a prolonged duration, these compensatory mechanisms allow a patient to continue normal activity with minimal or no symptoms. Indeed, the severity of the anemia is often disproportionate to the symptoms. Most patients, however, become symptomatic when the hemoglobin drops to less than 7 to 8 g/dL. Other factors such as a patient's age, physiologic conditioning, and the presence of other medical problems will often determine whether symptoms develop at a particular hemoglobin concentration. For example, patients with underlying cardiac dysfunction may develop symptoms related to anemia when the hemoglobin falls to less than 10 g/dL.

Causes of Iron Deficiency

As in most micronutrient-mineral deficiencies, the major causes of iron deficiency involve inadequate dietary intake, increased requirement, increased loss, malabsorption, (or, rarely, defects in iron use). More often, several causes may be found in the same individual.

Chief among the factors contributing to widespread iron deficiency in resource-limited countries is the consumption of a diet that is intrinsically low in bioavailable iron. When additional life events—such as pregnancy and growth, menorrhagia in women of childbearing age, and gastrointestinal blood loss—are superimposed on such individuals with minimal iron stores, the increased demand for iron cannot be met by dietary intake, leading to iron deficiency. Malabsorption of iron may occur following resection of the stomach (total or partial gastrectomy) or surgical bypass of the duodenum, autoimmune gastritis with gastric achlorhydria or hypochlorhydria, and with celiac (nontropical) sprue.

Common causes of gastrointestinal tract blood loss include erosions from chronic use of nonsteroidal anti-inflammatory drugs (NSAIDs, including aspirin) or steroids and *Helicobacter pylori* infection, gastric gastroesophageal hiatal hernia, colonic diverticulosis, polyps, gastrointestinal mucosal angiodysplasia, inflammatory bowel disease, malignant diseases of the gastrointestinal tract, and chronic parasitism. Urinary iron losses can result from lesions within the genitourinary tract (including stones) and intravascular RBC hemolysis (especially from malaria).

Polyparasitism and Parasite-Related Blood Loss

In many resource-limited countries, gastrointestinal blood loss from hookworm infestation (*Ancylostoma duodenale* or *Necator americanus*) is a major cause of iron deficiency anemia. Approximately 740 million people worldwide are infected with hookworm. Each worm attaches itself to the duodenum and jejunum of the gastrointestinal tract and sucks blood and serum protein, with resulting iron deficiency and malnutrition. With light infestation by hookworm, the blood loss is 2 mL/day, but with heavy infestation it can be up to 100 mL/day. *N. americanus* consumes 0.03 mL blood/worm/day, whereas *A. duodenale* consumes much more, at 0.15 to 0.23 mL blood/worm/day. Hemoglobin values of less than 11 g/dL can be seen with infestation by 40 to 160 worms. The number of hookworm eggs in the stool can serve as an indirect measure of hookworm burden. With approximately 1,000 eggs/g of feces, the patient may have a blood loss equivalent to 1 mg iron lost/day. *Trichuris trichiura* (whipworm) roundworm infestation can lead to a loss of 0.005 mL blood/worm/day. Schistosomiasis (*Schistosoma hematobium*) of the bladder can also lead to significant blood loss in the urine, causing iron deficiency.[34]

Infection and infestation with multiple parasites are the norm in the tropics and subtropics and are likely to have a major impact generally on nutrition and more particularly on anemia.[5] More than half of children studied in various geographic regions harbor high worm burdens of *Ascaris, Trichuris*, and hookworm,[35,36] as well as amebiasis (which can induce gastrointestinal bleeding) and, in endemic areas, schistosomiasis. These can persist into young adulthood; thus, more than 44 million pregnant women are infected with hookworm worldwide, with 7.5 million in sub-Saharan Africa alone.[37]

A successful large-scale universal program that was built into an existing public health structure addressed the problem of anemia among 52,000 Vietnamese women of reproductive age.[38,39] The key was improving the hemoglobin and iron status of these women before they became pregnant by giving them weekly iron and folic acid supplementation combined with a regular deworming program. This type of innovative approach is applicable to thousands of women of reproductive age in Southeast Asia and Africa, nearly half of whom have iron deficiency that is frequently accompanied by folate and vitamin B_{12} deficiency. Such programs must be expanded and sustained to benefit these populations over the long run; this remains for the most part an elusive challenge and goal.

Clinical Presentations of Iron Deficiency

In the early stages of iron deficiency, individuals can be entirely asymptomatic, and the incidental discovery of iron deficiency anemia is usually made on a routine blood count. In early iron depletion, women may experience symptoms of fatigue, irritability, an inability to concentrate on a task, and even note hair loss. Eventually, patients with iron deficiency will develop symptoms ascribed to anemia. An initial complaint of most patients with significant anemia is that they feel tired or tire easily with exertion, have a headache, complain of restless legs, and admit to pica.[40,41] Pica—an almost obsessive craving for and eating of materials such as ice (pagophagia), clay and soil (geophagia), or paper—may be seen particularly in children. Pagophagia is believed to be specific to iron deficiency and disappears within 1 to 2 weeks of iron treatment; it was found in up to 8% of African American women with iron deficiency in pregnancy.[40,41] In concordance with the high prevalence of iron deficiency in resource-limited countries, pica is common in childhood, and geophagia much more so in pregnancy. In Nairobi, 793 of 1,071 pregnant women admitted to daily pica.[42] Indeed, all supermarkets in Nairobi have a separate section on the shelf displaying various forms of clay that are routinely consumed during pregnancy.[13] Geophagia in Kenyan women included soft stones ("*odowa*") (in 89.8%); soil (61.2%); and others (9.6%). Because pica is largely due to chronic iron deficiency, "resolution of pica" may be used as a clinical measure of successful replenishment of iron.

Patients with increasing iron deficiency may also complain of a sensation of pins and needles (paresthesias) in the hands

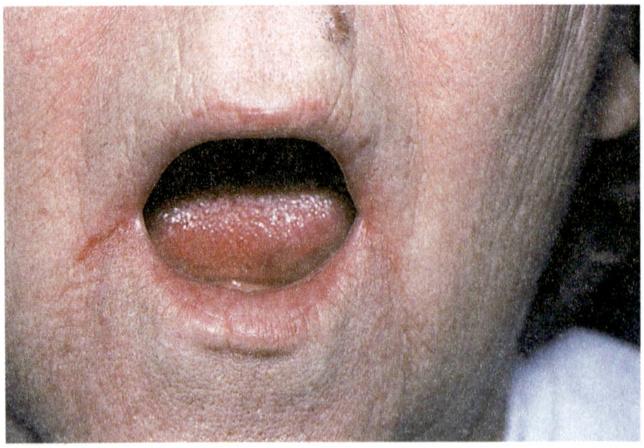

Figure 97.1 Angular stomatitis (fissures at the angles of the mouth) in iron deficiency.

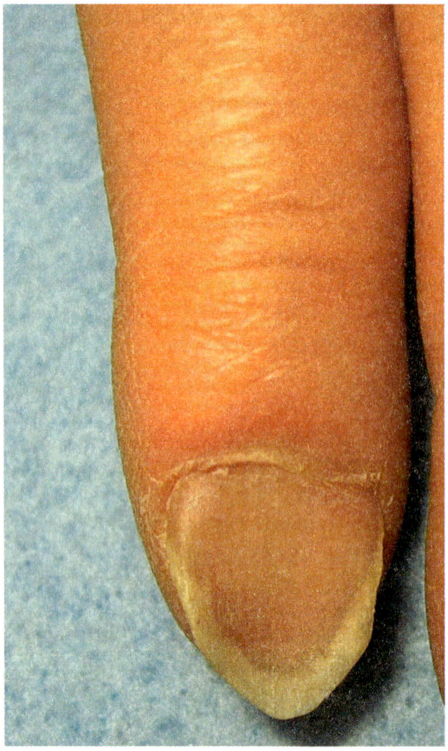

Figure 97.3 Koilonychia.

and feet. Palpitations (awareness of one's own heartbeat), tinnitus (ringing or whistling noise in the ears or head), and even more headache, irritability, dizziness, and generalized weakness may be present. Anemia may produce shortness of breath on exertion accompanied by a rapid heartbeat. In older individuals with severe anemia, cardiac pain (angina) and cardiac failure may develop. Additional symptoms related to iron deficiency include cracks at the angles of the mouth (angular stomatitis) (**Figure 97.1**), difficulty in swallowing, and a sensation of a lump in the throat (dysphagia). Some patients complain of soreness of the mouth and tongue (from impaired squamous-cell regeneration from iron deficiency) that may be aggravated by hot drinks or spicy foods (**Figure 97.2**).

Physical examination of individuals with iron deficiency anemia reveals pallor of the conjunctiva and skin, pale blue sclera, and misshapen nails that break easily and may be spoon shaped (koilonychia) (**Figure 97.3**). The heart and pulse rate may be rapid, and a systolic heart murmur may be detected on cardiac auscultation. Heart failure, characterized by pulmonary edema, congested liver, and edema of the lower extremities may develop. In patients with long-standing severe anemia, the spleen may be enlarged, with the spleen tip palpable under the left costal margin. Radiographic evaluation of patients for dysphagia may reveal the cause as a post-cricoid web in the esophagus (**Figure 97.4**); these patients are at increased risk for developing esophageal cancer.

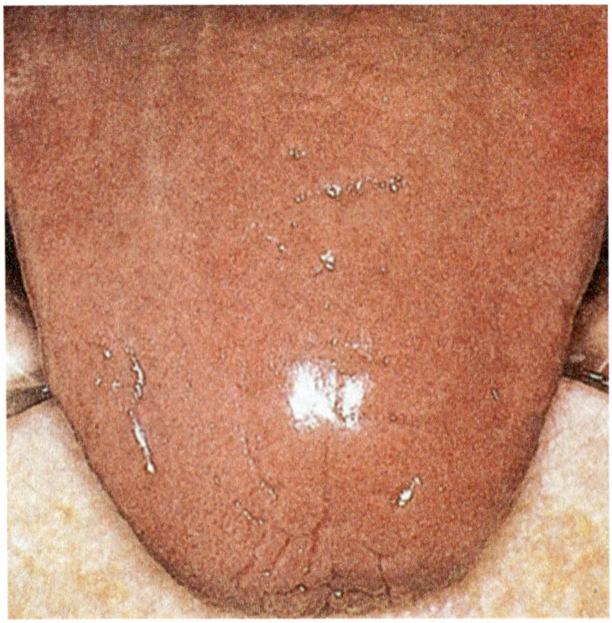

Figure 97.2 Glossitis.

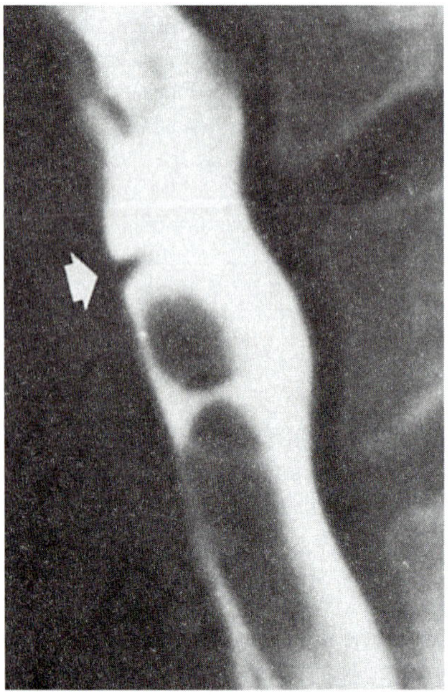

Figure 97.4 Esophageal web in iron deficiency.

Patients with iron deficiency anemia may have additional clinical complaints and findings that are not the result of anemia itself but are related to the underlying cause of iron deficiency. For example, tarry-colored or bright red stools suggest iron loss from gastrointestinal bleeding, whereas bloody or *Coca-Cola*-colored urine suggests urinary iron loss from genitourinary bleeding or intravascular hemolysis, respectively.

Signs and Symptoms of Iron Deficiency among Special Groups

Children can present with cognitive impairment, decreased physical capacity, and reduced immunity. The motor and social development of children (and young adults) with iron deficiency is generally poor. These children are less responsive, show less activity on the playground, have increased fatigue (and other nonspecific symptoms related to increased worm burden), and anorexia.

There is increased susceptibility to infection among children with iron deficiency.[44] Such children are more frequently absent (because of frequent infections) and are less likely to attend school; and when they do attend school, their overall poor school performance may be poor. Improving school children's performance by iron supplements led to substantial improvement in 3 months[45]; anemic Thai children had lower scores in Thai language, mathematics, and other subjects compared with iron-replete children. Another feature of severe iron deficiency among children is their inability to maintain body temperature in cold climates (they need protective clothing).

Iron deficiency leads to defective information processing and impaired cognitive development in adult women, and there is improvement in verbal learning and memory with administration of iron.[31] Young women should be targeted for optimization of their iron status; such interventions can improve their physical endurance, mood, and ability to concentrate, even when they do not have frank anemia.

Among adults working in the field (rubber tappers, workers in tea plantations, and rice fields) in Indonesia and Sri Lanka, iron replenishment led to an overall improvement in productivity that led to increased wages.[44] Even when a financial incentive was removed, the overall performance was better among those adults treated with iron.

Nonhematologic Effects of Iron Deficiency

Work Performance

Both human and animal studies have determined that tissue iron deficiency produces a decrease in work performance even when the hemoglobin is within the normal range, and that work performance improves following iron supplementation.[44] Indeed, in women with ferritin concentration in the iron-deficient range without anemia, parenteral iron improved symptoms of fatigue,[46] and oral iron supplementation improved physical performance[47]—reflecting the importance of iron for optimizing skeletal muscle function. Athletes with chronic iron deficiency without overt anemia may experience symptoms of decreased endurance. In addition, iron deficiency can precipitate heart failure, which reflects the importance of iron for cardiac muscle.[48]

Neurologic Effects

Iron deficiency can also cause cognitive dysfunction and trigger restless leg syndrome, even when there is no anemia. This is because iron is present in large quantities in certain parts of the brain and is required for the activity of enzymes that are components of neurotransmitter metabolism, which is critical for normal cognition. The iron content of parts of the brain is comparable to that of liver and continues to increase until the third decade of life. In children, impaired mental development was found when hemoglobin concentration was less than 10 g/dL at 5 years.[49] Iron-deficient children showed reduced attention and poor learning performance. This was manifested in lower scores in mental and motor function that returned to normal when tests were repeated after iron therapy.

Immunity

Several factors predispose those with iron deficiency to increased susceptibility to infection. Cell-mediated (T-lymphocyte) immunity and neutrophil killing of phagocytosed bacteria may be impaired in iron-deficient patients.[35] Neutrophils ingest and kill organisms by the production of reactive oxygen species, a process referred to as the oxidative burst, which depends on the activity of iron-containing enzymes. Conversely, because micro-organisms also require iron for their growth, parenteral iron treatment can precipitate latent infection.[50] Several studies have shown an increase in malarial attacks in children following treatment of iron deficiency when simultaneous antimalarial prophylaxis was not used.[51]

Life Stage–Specific Issues in Iron Deficiency

Anemia in Pregnancy

Anemia during pregnancy is defined as a hemoglobin concentration (Hb) less than 11 g/dL in the first trimester, less than 10.5 g/dL after 12 weeks (i.e., second and third trimesters), and less than 10 g/dL immediately postpartum.

Iron use is significantly increased during pregnancy to support fetal growth and development,[52] as well as for increased maternal erythropoiesis and other tissue needs. Thus, women need 0.8 mg/day in the first trimester of pregnancy, 4 to 5 mg/day in the second trimester, and more than 6 mg/day in the third trimester. Therefore, women require an extra 1,000 mg iron to meet the needs of a normal pregnancy (220 mg iron during pregnancy going to term), to expand their RBC mass (500 mg), and to provide iron for the placenta and developing fetus. The full-term fetus has approximately 290 mg iron, and the placenta has approximately 25 mg iron. During lactation, women need 2.4 mg/day.[53] All this must be set against the total body iron of 2,500 mg present in an adult woman with normal iron absorption of 1 to 2 mg/day. This amount of iron is more than can be provided by a good diet, so serum iron and ferritin concentration both fall steadily throughout pregnancy. Thus, iron deficiency in pregnancy results from increased demands for iron that cannot be met from stores and diet.

The recommended dietary allowance (RDA) for iron in nonpregnant women (aged 19–50 years) is 18 mg/day, but in pregnancy, it is 27 mg/day.[53,54] Because every pregnancy requires an additional 1,000 mg of iron, this warrants supplemental iron.

For each 325 mg ferrous sulfate tablet containing 65 mg of elemental iron, only approximately 25 mg can be absorbed and used.

Iron deficiency anemia, which affects more than half of all women living in resource-limited countries, can lead to serious consequences for both the mother and her baby. Under ideal conditions, this warrants administration of iron replacement and prophylaxis well before menarche.[27] Severe iron deficiency anemia has been associated with an increase in mortality for both mother and baby via premature labor, placental abruption, and low-birthweight infants.[55-57] There is also an association between severe antenatal anemia, stillbirth, and perinatal death, as well as maternal postpartum hemorrhage.[58-61]

Most resource-limited countries lack an adequate blood-banking infrastructure to support pregnancies. As a result, there is a perpetual paucity of blood transfusions for obstetrical emergencies. Moreover, a culture of regular voluntary blood donation is practically nonexistent. As a result, in 2018, anemia contributed to ~25,000 maternal deaths in India,[62] and "pregnancy-related complications" were the leading cause of death in Indian females between 15 and 19 years of age.[62] In some countries, transfusion is especially risky within the context of an 8% or higher prevalence of human immunodeficiency virus in the general population, and an even higher incidence among "professional" blood donors.

Both animal and human studies have demonstrated that maternal iron deficiency late in pregnancy is associated with neurodevelopmental impairment of the fetus.[63] Not unexpectedly, in humans, maternal iron deficiency at delivery is associated with lower serum ferritin in cord blood of neonates.[64] This has several important potential implications for the future neurodevelopment of the infant. The late fetal and early postnatal period have long been recognized as a critical time of rapid brain development and high nutritional requirement, which is also when there is a high degree of neural plasticity.[65,66]

A double-blind, cluster-randomized controlled trial from rural Western China demonstrated longer gestation and reduced in risk of early neonatal mortality by 54% in women who took iron and folic acid, compared with those taking folic acid alone.[55] Similar findings in the United States among women who were not anemic demonstrated lower incidence of low birth weight and preterm deliveries in women taking iron supplements.[56,57]

Although low hemoglobin, mean cell volume (MCV), mean cell hemoglobin, and mean cell hemoglobin concentration, can suggest iron deficiency, these may be entirely absent during pregnancy because of the known physiologic increase of MCV in pregnancy—of approximately six femtoliters.[67] In addition, and especially in resource-limited countries, microcytic, hypochromic indices may also occur in hemoglobinopathies. The best way to diagnose iron deficiency is with serum ferritin <30 ng/mL, which is diagnostic of iron deficiency in pregnancy.[68] However, a normal ferritin concentration does not exclude iron deficiency, as pregnancy is associated with a physiologic rise in acute-phase proteins,[69] which influence serum ferritin concentration. Neither the use of transferrin saturation nor of soluble transferrin receptor (sTfR) concentration or reticulocyte hemoglobin content is recommended.[68] Finally, when anemia of pregnancy is due to combined iron and folate deficiency, dimorphic anemia is the more frequent clinical presentation.

Neonates and Infants

A normal newborn has approximately 75 mg iron/kg body weight, of which two-thirds is in hemoglobin. Of significance, approximately 80% of iron is transported from mother to fetus in the 3rd trimester; therefore, if a mother is iron deficient during the 3rd trimester, there is a high likelihood that net fetal transfer of iron will be significantly compromised. Infants born of women with iron deficiency anemia are at increased risk for neonatal death from premature birth and low birth weight. The fact that maternal to fetal transfer occurs in the 3rd trimester points to several inter-related issues:

a. Although iron deficiency is reversible in early infancy, when left untreated, it can lead to irreversible neurocognitive problems by school age,[31] manifested by defective intellectual, cognitive, and social development.[70,71] Therefore, attempts to optimize the newborn's iron stores must be a high priority; this has clinical implications for both iron-replete and iron-deficient mothers.
b. Maternal iron stores of an iron-replete mother become exhausted after exclusively breastfeeding their full-term healthy infants for the first 4 months. After this time, these infants are now at risk for developing iron deficiency. (However, if the mother is anemic, the child will also be depleted of iron much earlier). For this reason, the American Academy of Pediatrics recommended that <u>all</u> exclusively breastfed babies be supplemented with oral iron after 4 months of exclusive breastfeeding, until they start consuming age-appropriate iron-containing foods.[72]
c. Prematurely born infants of iron-replete mothers are endowed with relatively lower iron at birth; so, depending on the extent of prematurity, guidelines exist for administration of either enteral or intravenous iron.[73]
d. Whenever maternal iron deficiency extends into the 3rd-trimester, the full-term newborn is at high risk for iron deficiency, which warrants iron replacement as soon as possible. However, without affordable and widely available iron preparations, millions of children of iron-deficient mothers are destined to develop adverse neurologic sequelae from inadequately treated iron deficiency.[44,70,71]
e. Maternal iron-, folate-, and vitamin-B_{12} deficiency is commonly present in the 3rd trimester of mothers living in resource-limited countries. This leads to premature birth, which places further restrictions on maternal-to-fetal transfer of these micronutrients. Therefore, the iron-, folate-, and vitamin-B_{12} status of these infants must be rapidly normalized to prevent adverse neurologic sequelae that will otherwise develop from a deficiency of all three micronutrients.[44,70,71]
f. Delayed cord clamping for a full 120 seconds, measured from the time of delivery of the infant's shoulders, has now been shown to have a beneficial effect on the infant's iron stores, from between 27 to 47 mg when measured at 6 months.[74] This procedure had the impact of a higher RBC, mean corpuscular volume (MCV), ferritin, and total body iron. Delayed cord clamping also showed raised mean venous hematocrit immediately after birth.[23] This is a simple, efficient way to increase the net iron stores for the baby at birth.[75]
g. Human breast milk contains about 0.5 mg iron/L. The bioavailability is high, with 50% of the iron absorbed. This contrasts with cow's milk formulas or unfortified cow's milk, in which only 10% to 20% of the iron is available for absorption. Hence, cow's milk formulas are usually fortified with iron to supply 6 to 12 mg/L. Breast milk supplies sufficient iron to full-term infants to meet their needs for the first 4 to 5 months of life. Because weaning foods often consist of cereals with iron of low bioavailability, many of these cereals contain added iron and ascorbate. Such foods are a major source of iron in the first 1 or 2 years of life.

Toddlers and Children

During infancy, between 6 and 18 months is the period of highest risk, when nutritional iron deficiency can manifest in serious and potentially permanent intellectual impairment with cognitive defects and poor social development.[70] Even infants in the United States with borderline iron deficiency who did not have frank anemia responded to supplemental iron with a significant increase in mental development index scores (+21.6 points) as measured by the Bayley Scales of Mental Development.[76]

The question of whether iron supplements can increase the risk of infection has been answered. A systematic review of 22 randomized controlled trials, involving more than 7,000 children, concluded that iron supplementation (via oral or parenteral iron) or fortified formula milk or cereals has *no* apparent harmful effect on the overall incidence of infectious illnesses in children, except for a small risk of diarrhea, which was deemed unlikely to have an impact on public health.[77] In addition, all children found to have iron deficiency anemia in areas of high malaria prevalence should be given supplementary iron as a component of a proactive intervention that *also* includes provision of treated bed nets and antimalarial drugs.[44] The growth spurt that occurs in adolescence and at menarche marks another phase when iron requirement peaks. This is when a young woman is at risk for additional complications if she becomes pregnant, a not uncommon occurrence, especially in resource-limited settings. A cross-sectional study among adolescents in Sri Lanka revealed that 50% of boys and nearly 60% of girls were anemic, predominantly from iron deficiency.[78]

Elderly Persons

Whereas an adult man has approximately 500 to 1,000 mg of storage iron, and older men have even more, the iron stores in women seldom reach 500 mg. In elderly persons, anemia resulting in a reduction of oxygen-carrying capacity of the blood leads to fatigue, cardiovascular complications (including heart failure), and poor physical capacity arising from reduced muscle strength,[79] increased disability with decreased performance, and increased all-cause mortality among nursing home residents.[80] Persons with iron deficiency and mild or moderate heart failure with ventricular systolic dysfunction were able to increase their distance in the 6-minute walk test by more than 30 m within 4 weeks of initiation of intravenous iron replacement.[48]

Diagnosis of Iron Deficiency

Laboratory Tests

The development of iron deficiency anemia with its characteristic morphologic changes in RBCs is preceded by stages of iron depletion and iron-deficient erythropoiesis. In the early stages of *iron depletion*, a decrease in iron stores in the bone marrow reticuloendothelial cells occurs, with a decrease in plasma ferritin to approximately 20 μg/mL. A marked change in serum iron concentration is generally not seen at this early stage. However, a decrease in the percentage of transferrin saturated with iron (transferrin saturation) and an increase in the iron-binding capacity of transferrin are noted. An increase in iron absorption also occurs.

With progression to *iron-deficient erythropoiesis*, iron stores in reticuloendothelial cells are depleted, and ferritin and plasma iron concentrations decrease to below normal values. (The normal range for ferritin is 30 to 400 ng/mL for men and 13 to 150 ng/mL for women. The normal range for iron is 65 to 175 μg/dL for men and 50 to 170 μg/dL for women). In addition, transferrin saturation decreases to less than 15% (normal is 16% to 50%), and RBC zinc protoporphyrin (ZPP) is increased to more than 30 μg/dL. At this stage, a mild decrease in hemoglobin is seen, but the RBCs appear normal in size and appearance (referred to as normocytic and normochromic cells).

With the appearance of frank *iron deficiency anemia*, there is a further decrease in hemoglobin with a decrease in plasma ferritin to less than 10 μg/mL and MCV to less than 80 femtoliters/cell. These smaller RBCs (referred to as microcytes) contain less hemoglobin than normal RBCs, so the mean corpuscular hemoglobin concentration is also reduced. A further decrease in serum iron and transferrin saturation to less than 10%, with an increase in transferrin iron-binding capacity and increased RBC ZPP, is also seen at this stage.

Although the development of microcytic and hypochromic RBCs is characteristic of iron deficiency anemia, the decrease in hemoglobin generally precedes the appearance of microcytosis. Hence, in early iron deficiency anemia, the MCV is normal, and the population of smaller RBCs in the circulation may be fully hemoglobinized and not hypochromic. However, at a later stage, the peripheral blood smear shows RBCs with reduced hemoglobin content, which appear under the microscope as an increased area of central pallor (hypochromia) in the RBC (Normochromic RBCs have central pallor that occupies one-third of the cell). In severe iron deficiency anemia, RBC morphology may become significantly altered, with RBCs appearing as thin rings with bizarre shapes (**Figure 97.5**).

When iron deficiency anemia coexists with either vitamin B_{12} or folate deficiency, neither the classic features of hypochromic microcytic anemia or megaloblastic anemia are seen. Instead, the peripheral blood smear shows an increase in nuclear segmentation in neutrophils (hypersegmentation) (**Figure 97.6**)—a hallmark of megaloblastic anemia. In contrast to pure iron deficiency anemia, which produces RBC microcytosis, an associated folate or vitamin-B_{12} deficiency can be associated with some cells having an

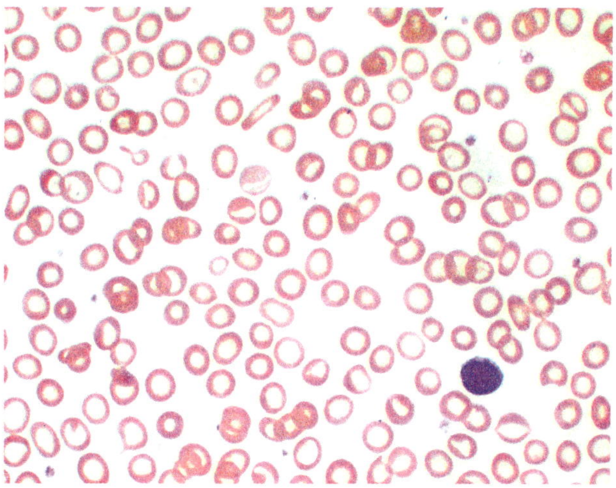

Figure 97.5 Iron deficiency anemia: red blood cells.

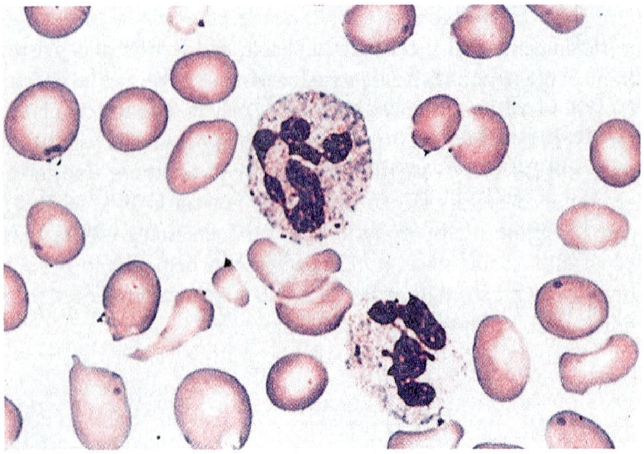

Figure 97.6 Blood smear in vitamin-B_{12} deficiency showing hypersegmentation (more than five lobes in the nuclei) of the neutrophils.

increase in MCV (macro-ovalocytes); so now, both types of cells may be observed on the peripheral smear (dimorphic anemia). Hence, with combined iron and vitamin-B_{12} or folate deficiency, the average MCV of microcytes and macrocytes—when measured by an automated counter—is usually in the normal range (masked megaloblastosis).[6] Iron deficiency anemia is not associated with changes in white blood cell number; however, an increase in the platelet count is often seen.

Although the diagnosis of iron deficiency as the cause of anemia can usually be established by measurements of serum iron, transferrin, and serum iron-binding capacity and a review of the peripheral blood smear, under certain conditions, these measurements may not be reliable indicators of iron deficiency. For example, in chronic inflammatory conditions (e.g., tuberculosis, arthritis, hepatitis, enteritis), there may be a mild microcytic anemia suggestive of iron deficiency. However, in these situations, the plasma ferritin concentration may be elevated as an acute phase reactant and is, therefore, not a reliable marker of iron status. An analysis of the bone marrow to demonstrate the *absence* of stainable iron in the marrow has long been considered the "gold standard" for establishing iron deficiency. **Figure 97.7** shows normal staining of iron in the marrow.

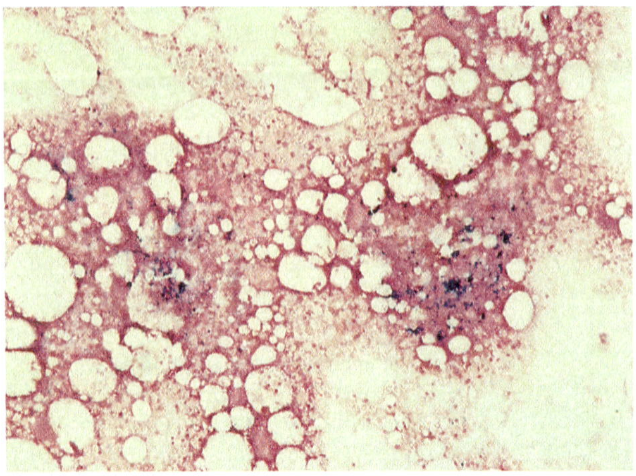

Figure 97.7 Prussian blue staining of iron in the bone marrow.

Although not widely used in clinical medicine, some other tests to detect early iron deficiency have been shown to be of value, particularly in nutrition-based research. ZPP is normally present in trace amounts in RBCs during hemoglobin synthesis. With failure of iron delivery for hemoglobin synthesis, the level of ZPP accumulates in RBCs, thus resulting in an increase in the RBC ZPP/heme ratio. Studies in pediatric populations have shown that an increased ZPP/heme ratio is a reliable indicator of iron-deficient erythropoiesis.[81-83] However, concerns about the use of RBC ZPP as a marker of iron deficiency exist, as the test is sensitive to interference by drug and plasma components, and a significant number of RBCs being analyzed must be new RBCs synthesized in the iron-deficient state.[84]

Another marker, the soluble transferrin receptor (sTfR), has been shown to be present in increased concentration in the circulation in iron deficiency. Measurement of serum sTfR and calculation of the ratio of sTfR/log ferritin (i.e., logarithmic transformation of serum ferritin) is of value in estimating body iron stores, especially when serum iron and transferrin concentration are not reliable as a result of the clinical conditions described earlier, and analysis of the marrow is not feasible. The sTfR/log ferritin ratio is increased in iron deficiency. An obstacle to the routine use of sTfR measurement for determining iron deficiency is a lack of standardization among available assays[85] and its significant cost. This test is also falsely elevated during hemolysis.

Clinical Considerations in Diagnosis and Treatment

Individuals with pre-existing marginal iron stores can be drawn into a state of frank iron deficiency when there is a further increase in demand for iron. This can occur under several clinical circumstances: when individuals contract malaria or have blood losses from any cause; when women are in the last trimester of pregnancy, are lactating, or are weaning their infants; and in adolescents during their growth spurt.

As noted above, in resource-limited settings, multiple micronutrient deficiencies frequently coexist in a single individual, so it is usually impossible to distinguish the etiopathological factors that led to anemia. It is estimated that iron deficiency—resulting from inadequate iron intake that often coexists with infection-related hemolysis of RBCs (malaria) and excess blood loss from parasitic infestation of the intestine (hookworm, whipworm) or bladder (schistosomiasis)—is primarily responsible for one-half of the cases of anemia in these settings.[86]

Deficiency of other micronutrients (e.g., vitamin-B_{12}, folate, riboflavin, and vitamin A, individually or in combination) or infectious diseases (e.g., malaria, human immunodeficiency virus infection, or tuberculosis, individually or in combination), and hemoglobinopathies that protect against malaria (sickle cell disease, thalassemia, glucose-6-phosphate dehydrogenase deficiency) are the cause of the remainder. Vitamin A and zinc deficiency result in increased susceptibility to respiratory and diarrheal illness[87,88] that further contributes a component of chronic disease anemia. A good example of the plethora of etiologic factors contributing to anemia in children was reported from Malawi.[89] Therefore, treatment must be individualized, with all coexisting factors that perpetuate the anemia addressed together with supplementation of the deficient micronutrients.

Differential Diagnosis of Iron Deficiency

Whereas microcytic hypochromic anemia is the hallmark finding of iron deficiency anemia, similar RBC morphology can be seen in a few other conditions that must be differentiated from iron deficiency to avoid inappropriate diagnostic testing and treatment. For example, the decrease in globin synthesis that occurs in thalassemic syndromes also produces mild to moderate microcytic hypochromic anemia in individuals with a thalassemia trait. The diagnosis of thalassemia can be excluded by hemoglobin electrophoresis and more specialized tests.

Anemia of chronic disease may resemble iron deficiency because both have small RBCs and low serum iron. However, the serum iron-binding capacity is normal or low in the anemia of chronic disorders, whereas in iron deficiency, it is raised. Because ferritin is an acute-phase reactant, its concentration may be elevated in the anemia secondary to chronic disease. The sTfR concentration is not elevated in chronic disease (in contrast to ferritin), and therefore, measurement of the sTfR/log ferritin ratio has been suggested to discriminate between these conditions. Distinguishing iron deficiency from anemia of chronic disease may still require direct examination of the marrow for iron stores for confirmation. Measurement of markers of inflammation such as C-reactive protein may also be useful in establishing the presence of inflammation as a cause of anemia.[85]

Optimization of Dietary Iron Intake

Dietary Modulation to Optimize Iron Intake

Humans depend on a steady source of dietary iron for absorption, which is dependent on the type of iron in food; reduced net iron intake is a significant cause of iron deficiency. Nonheme iron, which is found in an inorganic form in plant foods, is also present in cereals, pulses, fruits, and vegetables. Among strict vegetarians, nonheme iron is the sole source of iron. In a less-restricted mixed near-vegetarian diet, nonheme iron can provide 85% to 90% of dietary iron, but only 2% to 10% is absorbed, depending on the presence of extrinsic enhancers and inhibitors of absorption. This is far less than the 20% to 30% absorption of heme iron found in traditional nonvegetarian diets.

The absorption of plant food-derived nonheme iron can be inhibited by phytates, calcium, and polyphenolics, which are found in tea, coffee, herb teas, and cocoa.

Therefore, the best option is to employ methods known to reduce the phytate content of food and thereby increase iron bioavailability. These approaches include soaking and sprouting beans, grains, and seeds, and using yeast to leaven dough to make sourdough breads.[7] Alternative fermented foods can include miso, tempeh, and pickled tofu (consumed across Southeast and East Asia). The addition of organic acids from citrus and sour fruits and vegetables (lime, lemon, orange, sweet lime, mango, tamarind, and tomatoes), or condiments (consisting of acetic acid- or vinegar-preserved lime, mango, or other mixed vegetables in spices) are commonly used throughout Asia and Africa. Foods such as spinach and leaves of the drumstick tree (*Moringa oleifera*) are rich in vitamin C and iron[90] as well as use of vinegar-preserved vegetables, soy sauce, and even ethanol can enhance nonheme iron absorption.[7] Finally, the use of cast-iron cooking pots/woks[91] or even small iron ingots (such as the Lucky Iron Fish) left in the cooking vessel are among the many ways to improve iron intake; this innovation remains a work in progress. For those consuming a near-vegetarian diet, the addition of a small serving of meat or fish usually improves overall iron absorption.

Iron Supplementation and Fortification of Foods

Food fortification and dietary supplementation are the traditional methods for replenishing specific vulnerable groups who are deficient in one or more micronutrients. Although iron supplements (syrups or tablets or an iron-containing snack) can provide sufficient amounts of this micronutrient, the number of individuals who can be reached is limited. By contrast, food fortification, which covers large populations at a far lower cost than dietary supplements, is superior; the key requirement is that the vehicle used for fortification must reach the entire target population.[92] Various traditionally eaten staple foods that have been used as vehicles: these include soy and fish sauce, wheat and corn (maize) flour and cereals, rice, salt, curry powder, and powdered milk. Targeted fortification of specific groups, such as infants, with multinutrient premixes (Sprinkles) for addition to food at home (also known as home fortification) falls between the two traditional forms of replenishment.[93]

Food fortification with iron is among the most cost-effective of interventions to prevent iron deficiency. However, success in reducing morbidity and mortality of vulnerable women and children requires additional programs that protect them. This requires the integration of services that involve the following: clean water; improved sanitation; control of infectious diseases; education on nutrition in schools; for adolescents, stressing the importance of optimizing nutrition; education on the value of breastfeeding; and improving food security. Other programs to reduce poverty, including micro-financing and small animal farming, are integral to such initiatives.

Because iron-rich, non-vegetarian foods are generally more expensive, most low-income families in developing countries consume an iron-poor, monotonous, vegetarian, or semi-vegetarian diet out of poverty.[4] In these populations, education of adolescent girls on the concept of adequate preparation for pregnancy should include information on breastfeeding, the importance of consuming a diverse nutritious diet, and how using a few simple strategies[94] can improve the bioavailability of iron in their diets, as discussed above. This can also include teaching about cultivation of home kitchen–based fruit and vegetable gardens. School gardening linked with a school feeding program can produce fruits and vegetables rich in iron and vitamin C. If culturally acceptable, such education should also include specific instruction on the rearing and consumption of small animals (e.g., goats, poultry, rabbits, fish, guinea pigs) through community-based fish ponds and animal husbandry. Finally, social marketing is critically important to publicize the value of such large-scale interventions.[38]

Treatment of Iron Deficiency with Specific Iron Compounds

Oral Iron

Iron deficiency anemia is treated with iron salts: ferrous sulfate (325-mg tablet containing 65 mg of elemental iron), ferrous gluconate (325-mg tablet containing 36 mg of elemental iron), ferrous fumarate (325-mg tablet containing 108 mg of elemental iron), and iron polysaccharide capsules that are usually dosed by iron content (one preparation contains 150 mg of elemental iron). Ferrous sulfate is the least expensive and most frequently prescribed.

The major limitation to the use of oral iron over the long term is the frequent persistent side effects of nausea, vomiting, diarrhea or constipation, abdominal cramping, and a metallic taste in more than 70% of women—particularly when single daily or multiple daily doses of iron are administered. This has led to poor compliance in most affected women. The lack of a more palatable iron preparation has contributed to the high number of women with resource-limited iron deficiency.

Now there may be a simple solution. Recent studies demonstrate that in mild iron deficiency, it is acceptable to use a single 65-mg tablet containing elemental iron (40–80 mg) when given on an empty stomach, especially with an organic acid like vitamin C, every other day as therapy—as opposed to the longstanding conventional recommendations of daily dosing or multiple daily doses.[94,95] One plausible explanation for the greater efficacy of alternate-day dosing appears to be as follows: Each single tablet of oral iron raises hepcidin concentration, which then reduces the absorption of additional oral iron doses (in multiple daily dosing strategies). Because this oral iron-induced rise in hepcidin actually takes up to 48 hours to fully return to normal, it is felt that dosing patients every 48 hours keeps hepcidin concentration as low as possible, and thereby ensures optimum iron absorption and recovery of hemoglobin.[96] Importantly, this less frequent dosing approach is more tolerable as an optimum strategy to improve compliance and net iron absorption over the long run. Iron taken with meals or close to meals is absorbed less than iron taken on an empty stomach, but it is better tolerated and hence more likely to be acceptable to patients. The key is to continue therapy for at least 6 months after normalization of hemoglobin to fill up iron stores and normalize serum ferritin. The optimum methods for treatment of more severe iron deficiency, using the newer low-dose alternate day regimen vs traditional methods (multiple daily oral dosing vs intravenous iron) are awaited. In the past, using multiple-day dosing, a response to iron therapy has been characterized by a rise in hemoglobin of up to 1 gm/dL/week.

Helicobacter pylori infection has the capacity to induce bleeding from gastric micro-erosions. Because more than one-half of the entire world's population likely is infected with *H. pylori*, eradication of this organism may be indicated for recalcitrant iron deficiency despite therapy. For those with signs of decompensation due to severe iron deficiency anemia, the immediate therapy should include transfusion with packed red cells. While correcting hypoxia, each unit provides between 200 and 300 mg of iron (1 mL of PBC contains 1 mg of iron). The recommended daily dose for children is 3 mg per kilogram of body weight of a liquid preparation.

Parenteral Iron

There is a plethora of safe parenteral iron preparations today.[97] Indeed, because current intravenous iron preparations have been shown to be safe (the risk of anaphylaxis is estimated at 1:200,000) many hematologists have a low threshold to use of intravenous iron whenever this option is easily available and feasible. The indications for use of parenteral iron therapy include documented failure of oral therapy or intolerance due to side effects, or frank noncompliance, inability to absorb iron due to associated food-vitamin-B_{12} malabsorption, atrophic gastritis from pernicious anemia, gastrectomy, or bariatric surgery-induced duodenal bypass, and celiac disease. This is also used when bleeding exceeds the capacity of oral iron to keep up with blood loss, as can occur with inflammatory bowel disease or arteriovenous malformations, and it is also useful for those patients with renal failure undergoing dialysis who cannot achieve a positive iron balance with oral therapy alone.

In developing countries, where blood transfusion facilities are limited, or when the local testing of blood is inadequate to rule out infectious diseases such as human immunodeficiency virus infection, hepatitis, or malaria in donor blood, the presence of severe iron deficiency is yet another indication for parenteral iron therapy. In the United States and abroad, Jehovah's Witnesses who will not accept blood transfusions because of their religious belief are similarly given parenteral iron with erythropoietin. Finally, parenteral iron is used for patients with chronic kidney disease (or cancer) who have a functional iron deficiency and need iron to support the use of erythropoiesis-stimulating agents (usually ferritin needs to be over 100 ng/mL). In addition, intravenous iron can be administered during the second or third trimester when there is need for rapid replenishment of the mother to prevent anemia and deliver optimum amount to the fetus. (Intravenous iron must be avoided in the first trimester of pregnancy, as there is insufficient information on the safety for the developing embryo-fetus).

The role of intravenous iron in heart failure is now established[98]; Four randomized, controlled trials have demonstrated symptomatic benefit and modest reduction in hospitalizations after patients with iron deficiency and heart failure with reduced ejection fraction were treated with intravenous iron. Ongoing studies are designed to assess if intravenous iron clearly improves survival, reduces nonfatal heart failure events, improves myocardial function, or alters the natural history of this form of heart failure.

Acceptable parenteral iron compounds[97] include a low-molecular-weight iron-dextran complex (InFeD), ferric sodium gluconate (Ferrlecit), iron sucrose (Venofer), and ferumoxytol (Feraheme), ferric carboxymaltose (Injectafer), and iron isomaltoside.[99] Among these, low-molecular-weight iron-dextran is the cheapest and easy to administer; following a test dose, 1 gram of iron-dextran can be delivered in an hour. When individuals have an allergic diathesis or reaction to two or more drugs, premedication with steroids is given. Some institutions administer H_2-receptor antagonists like famotidine, which have fewer side effects of hypotension (than diphenhydramine, which can itself mimic an allergic reaction). Other preparations, like Venofer or Ferraheme, require several visits to complete a course of intravenous iron replacement, which can be a disadvantage. On the other hand, those preparations that can be completed in less than an hour cost more. Whereas ferric carboxymaltose can produce hypophosphatemia in up to one-half of individuals, the presence of benzyl alcohol in ferric sodium gluconate (Ferrlecit) argues against its administration during pregnancy (to protect the fetus).

The total amount of iron (in milligrams) needed to correct the hemoglobin deficit can be determined by a calculator that is available online at www.globalrph.com/irondextran.htm. Alternatively, the following formula to restore hemoglobin to normal and to replenish body iron stores can be used:

Body weight (kg) × 2.3 × (15 – patient's hemoglobin in grams/deciliter) + 500 mg

Iron as a Potential Two-Edged Sword

There are no prospective data to support any contention that intravenous iron exacerbates infections; indeed, several studies suggest just the opposite.[100-102] A systematic review and meta-analysis of 103 randomized trials involving iron infusions (1965–2013) included 10,390 patients treated with intravenous iron.[103] The conclusion was that intravenous iron therapy is not associated with an increased risk of serious adverse events or infections. Moreover, daily oral iron supplementation of pregnant Kenyan women resulted in no significant differences in overall maternal *Plasmodium* infection risk, and led to increased birth weight.[104] In addition, every international United Nations service agency, World Health Organization, and the World Food Program all recognize the importance of reversing even mild iron deficiency anemia, particularly during infancy, because of the potential for permanent intellectual impairment. A meta-analysis identified no evidence of adverse effects of iron treatment in children from malaria-endemic areas.[105] Another important study also demonstrated the significant benefit of antenatal iron and folic acid supplements combined with appropriate intermittent preventive treatment of malaria during pregnancy to reduce neonatal mortality in malaria-endemic regions.[106] Finally, a scientifically validated solution for easily replenishing children and adolescents with iron before they reach sexual maturity has remained elusive. Fortified flour, rice and salt[107] are under consideration for fortification with all three micronutrients.

Megaloblastic Anemias—General Considerations

The term megaloblastic anemia describes a group of disorders characterized by a distinct morphologic pattern in hematopoietic cells, which arises from (usually an ongoing) perturbation of deoxyribonucleic acid (DNA) synthesis. Both vitamin B_{12} and folate contribute to DNA synthesis, so a deficiency of either vitamin can lead to megaloblastic anemia. Megaloblastic hematopoiesis commonly manifests as anemia, but this feature is only a manifestation of a more global defect in DNA synthesis that affects all proliferating cells. When DNA synthesis in a cell is perturbed, cell division is impaired; however, the relatively unaffected ribonucleic acid (RNA) in these cells continues to synthesize proteins. This results in a state of unbalanced cell growth

Most megaloblastic cells tend to continually make several attempts (in vain) to double their DNA, but this leads to frequent arrest usually in the S phase of the cell cycle. This eventually results in such cells having a net higher DNA (that is nevertheless insufficient to allow the cell to divide into two daughter cells). Such increased DNA content is visualized under the microscope as a cell having a larger-than-normal nucleus with finely particulate chromatin, indicating that nuclear maturation is arrested; (this is referred to as an "immature" nucleus). In folate- or vitamin-B_{12}-deficiency, the DNA in this immature nucleus is also fragmented in many places and cannot repair itself. By contrast, the relatively unimpaired RNA and protein synthesis results in cells containing a greater amount of "mature" looking cytoplasm and cell volume. The resulting nuclear-cytoplasmic dissociation leads to the condition of "megaloblastosis" that is best observed during the development of RBC (erythroid) precursors as they mature. Normally, the nucleus of erythroid precursors gradually becomes smaller as it undergoes pyknosis (i.e., where there is an irreversible condensation of nuclear chromatin, leading eventually to complete loss of the nucleus in mature RBCs). By contrast, megaloblastic erythroid precursors continue to retain their immature nucleus cells as the cytoplasm continues to fill up with hemoglobin. The net result are megaloblastic erythroid precursor cells (seen in a stained bone marrow smear) that have a large, immature, sieve-like nucleus but with a mature cytoplasm containing progressively more reddish-colored hemoglobin.

Megaloblastic changes in the bone marrow are reflected in the peripheral blood by large RBCs that are filled with hemoglobin; these so called macro-ovalocytes (up to 14 μm) have an increase in mean corpuscular volume (MCV) and are much larger-than-the-normal 7 μm-sized RBCs. They are filled with hemoglobin so that the central pallor, which normally occupies one-third of the cell, is now reduced. The same megaloblastic changes are found in immature white blood cell precursors, which can become 20 to 30 μm giant metamyelocyte and band forms; these are pathognomonic for megaloblastosis in the bone marrow. Normally, the nucleus of more mature leukocytes (the infection-fighting white cells) in the blood, called polymorphonuclear neutrophilic cells, develop between three and five segmented lobes when mature. However, in megaloblastic neutrophils, there is an exaggeration of the nuclear segmentation to a hypersegmented polymorphonuclear neutrophil containing five or more lobes. Normally, fewer than 5% of neutrophils have more than five lobes, and no cells have more than six lobes in the peripheral blood. However, when greater than 5% neutrophils have more than five lobes or 1% of them have more than six lobes, this suggests that megaloblastosis is present. Thus, the combined finding of macro-ovalocytes and hypersegmented polymorphonuclear leukocytes in the peripheral smear reliably points to a megaloblastic process that is likely operative in the bone marrow.

Among megakaryocytes (which eventually give rise to platelets) perturbed DNA synthesis also leads to complex hypersegmentation (i.e., pseudohyperdiploidy) of these cells with liberation of large fragments of cytoplasm from these cells, as well as giant platelets into the circulation. When there is thalassemia or iron deficiency, all the classic expectations of megaloblastosis are masked (masked megaloblastosis) because of the associated perturbed hemoglobin synthesis; however, megaloblastic leukopoiesis with hypersegmented polymorphonuclear leukocytes is still observed.

Most megaloblastic precursor cells do not enter the circulation. It is estimated that most megaloblastic erythroid bone marrow cells (80%–90%) die within the bone marrow, where marrow macrophages effectively scavenge dead or partially disintegrated megaloblasts. This is the basis for finding signs of an underlying hemolytic process (reflecting intramedullary hemolysis) in the peripheral blood on laboratory test results. This is reflected by a lowered absolute reticulocyte count, increased bilirubin (up to 2 mg/dL), decreased haptoglobin, and increased lactate dehydrogenase often above 1,000 units/mL.

The next challenge is to determine from the history, physical examination, and the judicious use of laboratory tests—(using serum folate and vitamin-B_{12} concentrations, and the two ancillary tests on metabolites, serum homocysteine and serum methylmalonic acid)—as to whether there is an underlying folate or vitamin-B_{12} deficiency or both; these issues are covered in other chapters within this textbook dedicated to nutritional issues related to folate and vitamin B_{12}. From the clinical standpoint, the pre-existing stores of these vitamins in the body will dictate the speed with which overt deficiency develops. This principle is particularly relevant in resource-limited countries among pregnant women and children with pre-existing borderline stores of folate and vitamin B_{12}.

It should be noted that megaloblastic anemia is accompanied by an ineffective use of iron; this results in an increased percentage of saturation of transferrin and increased iron stores. But if there is an associated iron deficiency (common in resource-limited settings), the MCV may be normal, and only iron therapy will unmask megaloblastic manifestations in the peripheral blood.

Therapy of a patient with decompensated heart failure, who is also suspected of severe megaloblastic anemia, is usually initiated with slow transfusion of one unit of packed RBCs under the cover of diuretic therapy, as well as the initiation of empiric therapy with full doses of daily parenteral vitamin B_{12} (1 mg/day) and (either parenteral or oral) folate (folic acid, 1 to 5 mg). This regimen, which is usually reserved for the severely ill patient[6]—even before knowledge of the specific type of vitamin deficiency is confirmed—can be continued for up to 10 days. If there is no response by way of a brisk reticulocytosis between the 5th to 8th day, neither vitamin is likely to be etiologically significant in the patient's clinical presentation.

Folate Deficiency Anemia

Nutritional Folate Deficiency

The RDA for folic acid is 400 μg/day.[54] Nutritional folate deficiency is prevalent in resource-limited countries at all ages, especially among women of childbearing age.[3,6,13,108,109] This primarily stems from reduced intake of folate-rich fresh green leafy vegetables and animal-source foods. Although folates can tolerate cooking and baking for 10 to 15 minutes, prolonged cooking of green leaves, vegetables, lentils, or beans beyond 30 minutes destroys the folate in these foods.[6] Curiously, refrigerated green leafy food exposed to artificial fluorescent light in supermarkets have higher folate content, increasing by 84% to 100% by the 10th day.[110]

Body stores of folate are adequate for only about 4 months[111] although those with higher folate stores could take longer to become frankly deficient.[112] Individuals who are chronically in negative folate balance may only require a brief "nudge"—from superimposition of increased requirements or of an associated illness that leads to hemolysis (malaria), anorexia, or folate malabsorption—to "tip" them into folate deficiency. In resource-rich countries, food faddism, alcoholism, or unbalanced slimming diets lead to decreased folate intake in young to middle-aged individuals.[111] Edentulous or infirm persons, neglected older adults who are too ill to prepare their meals, and patients with psychiatric problems are at risk for nutritional folate deficiency.[111]

Before the advent of routine folate supplementation during pregnancy, the incidence of megaloblastic marrows in the United States, Canada, and the United Kingdom during late pregnancy was about 25%, while in South India, it was about 55%.[111] Folate deficiency varies among countries and within regions of a country and is influenced by economic status and net folate content of the ethnic diet consumed by the population under study. In African countries, the prevalence of folate deficiency ranged from a low of 20% in Benin, to 30% in Zimbabwe, to nearly 60% in Sudan[113]; (Note: these differences are likely due to the variable masking of folate deficiency by associated vitamin-B_{12} deficiency and/or hemolysis, as will be discussed below).[109] In Sri Lanka, one-half of schoolchildren had low-folate status, and in India, 60% to 80% of women of childbearing age had nutritional folate deficiency.[109,111]

Folate fortification of foods in the West has led to widespread elimination of folate deficiency and related anemia,[114,115] leading to questions of whether testing for folate deficiency is even justified.[116,117] However, vigilance must be exercised, particularly among women on low carbohydrate diets (because they do not consume sufficient folate-fortified flour), and among the elderly who are at risk for both nutritional folate and vitamin-B_{12} deficiency.[118,119]

Clinical Features of Folate Deficiency

Folate deficiency is often recognized during clinical presentations that predispose to negative folate balance and subsequent deficiency. It is, therefore, best to conceptualize cellular folate deficiency as arising from etiologic categories of decreased supply (i.e., reduced intake, absorption, transport, or use) or increased requirement (i.e., metabolic consumption, destruction, or excretion). However, in the same patient, more than one mechanism may result in net folate deficiency. The precise contribution of one mechanism over the other is often not obvious, and specific tests to define each mechanism are not routinely available for clinical use. Therefore, the clinical context is especially important.

Megaloblastic manifestations of folate deficiency are similar to those of vitamin-B_{12} deficiency but can be differentiated based on the clinical context[6]; the history and physical examination, and laboratory data can usually point to folate deficiency in a patient whose disease onset can be identified within the past 6 months. Cases of neuropathy in adults attributed to folate deficiency are rarely encountered. When they are, the possibility of alcoholism with thiamine deficiency must be considered. Every patient with neuropathy, myelopathy, or psychiatric manifestations, either with or without being associated with megaloblastosis, must be investigated to rule out vitamin-B_{12} deficiency.

Folate Requirements During Pregnancy and Infancy

Folate is critically required during pregnancy for growth of maternal and fetal tissues and after birth throughout all life stages. Thus, pregnancy and lactation are associated with significantly higher folate requirements (more than 400 μg/day) for growth of the fetus, placenta, breast, and other maternal tissues.[120] Folate requirement increases throughout pregnancy and is maximal near term. There is also a threefold increased urinary loss of folate during pregnancy because of a lower renal threshold.

Low-folate status associated with short interpregnancy intervals also predisposes to preterm births.[121] Folate deficiency is eight times as high in twin pregnancies. Multiparity and hyperemesis gravidarum commonly lead to folate deficiency.

When anemia of pregnancy is due to combined iron and folate deficiency, dimorphic anemia is the more frequent clinical presentation. Increased use of folates by the newborn leads to a drop in serum folate concentration by about 6 weeks of age. This drop is exaggerated in premature infants (who have feeding difficulties, infection, or hemolytic disease leading to pure folate deficiency); hence supplementation is routine for them.[111]

Despite the increased demand for folate, which must somehow be met by increased folate intake, most (over 90%) of pregnant women in resource-poor countries consume less than the estimated average requirement of folate[122]; in addition, a substantial number also consume less than optimum amounts of other micronutrients, including iron and vitamin B_{12}. For example, studies on women with low socioeconomic status from North India[108,123] have estimated that the daily intake of folic acid ranged between 75 μg and 167 μg, far lower than the 400 μg/day required to prevent birth defects.[124] Not surprisingly, pregnancy with poor folate [and vitamin-B_{12} intake] are the common cause of megaloblastic anemia in resource-limited countries.

As with adaptations for improving iron homeostasis during iron deficiency, there are simultaneous mechanisms for the sensing of folate deficiency within cells, which triggers the translational upregulation of folate receptors on proliferating hematopoietic precursors and other cells.[6,125] The placenta has a large number of folate receptors,[124] which facilitate binding and transport of folates to all tissues in the developing fetus, including the developing neural tube and neural crest cells.[126-128] The rapidly proliferating tissues in children also have an absolute requirement for exogenously supplied folate. Although human milk of folate-replete mothers can maintain folate balance in breastfed infants, the breast milk content of folate is low when the mother's folate status is poor.

Poor "preparation for pregnancy", with an imbalanced diet and pre-existing multifactorial nutritional anemia (involving both folate- and vitamin-B_{12} deficiency) that remains unaddressed, is a major factor accounting for serious pregnancy complications and adverse birth outcomes, particularly in resource-limited settings. Therefore, additional folate supplements during pregnancy are required to prevent both *pregnancy complications* (pre-eclampsia and spontaneous pregnancy loss,[129-132] placental abruption,[130,131] recurrent pregnancy loss),[130,131,133] and *poor pregnancy outcomes* (fetal growth restriction,[131,134] preterm birth,[135] and stillbirth).[134] Additional adverse outcomes for the baby include NTDs and congenital malformations.[131,134] Normalization of maternal homocysteine with vitamin B_{12} and folate and improvement of these pregnancy outcomes for some complications[131,133] indicate that some of these risks can be reduced with good nutrition.

An Important Caveat in Laboratory Testing for Folate Deficiency

The various methods to test for clinical folate deficiency and to distinguish this from vitamin-B_{12} deficiency are detailed in other dedicated chapters in this textbook to folate and vitamin-B_{12} nutrition. However, one important point warrants special mention:

Recently, the genesis of a serious erroneous and dangerous myth in the contemporary medical literature—which maintained that folate deficiency is uncommon in African countries and across India—was traced to its origin and thereby unmasked.[109] This error originally developed following a <u>misinterpretation</u> of the significance of a "normal serum folate level" in malnourished patients with folate deficiency who also had hemolysis of RBCs and/or vitamin-B_{12} deficiency. Both hemolysis of RBCs and vitamin-B_{12} deficiency can independently <u>elevate</u> the serum folate concentration and give an erroneous impression that the individual's folate status is normal—even when there is severe bona fide tissue folate deficiency.[109] Some background can help provide a better understanding of this important issue:

i. Ever since tests for RBC folate and serum folate concentrations were developed in the 1960s, it was known that RBCs have a 30-fold higher folate content than that present in serum. Therefore, hemolysis of RBC from any cause—such as hemoglobinopathies, or malaria, or even during blood collection/storage prior to testing—can falsely raise the serum folate into the normal range, <u>even</u> in individuals with severe nutritional folate deficiency.[109] This had not posed a diagnostic dilemma because hematologists had long recognized that all patients with hemolysis also required prophylactic daily folic acid supplements to support the ensuing compensatory erythropoiesis in the bone marrow.[6] However, when some investigators erroneously used the results of "a normal serum folate" in severely undernourished individuals with malarial hemolysis to <u>reject</u> the diagnosis of folate deficiency (and thereby withhold critically needed folate supplements), a grave iatrogenic diagnostic-and-therapeutic error was perpetrated.[6,109]
ii. Vitamin-B_{12} deficiency leads to the inactivation of a key (vitamin-B_{12} dependent) intracellular enzyme, methionine synthase[6]; this enzyme normally processes serum 5-methy-tetrahydrofolate, which enters cells in a form suitable for cellular retention and participation in many folate-dependent enzyme reactions that lead to DNA synthesis and methylation, among others.[6] Therefore, during pure vitamin-B_{12} deficiency (pernicious anemia), the 5-methyl-tetrahydrofolate that cannot be used in cells leaks back into serum, thereby resulting in a higher-than-baseline serum folate concentration.[6,109] However, upon vitamin-B_{12} replacement, resumption of intracellular metabolism can result in a dramatic drop in the patient's serum folate within 24 hours to the true baseline value, which can be as low as <u>one-third</u> of the pretreatment level. This has also been confirmed among Indian children with combined nutritional folate and vitamin-B_{12} deficiency.[109] Because nutritional folate <u>and</u> vitamin-B_{12} deficiency is so common across Africa and India,[6,13,109] tests for serum folate often give falsely normal values—even among individuals with genuine tissue folate deficiency.

As a result of this grave clinical error, physicians have withheld administration of folate (and vitamin B_{12}) supplements to hundreds of millions of women and children with nutritional folate deficiency in the clinical setting of hemolysis—(from malaria, hemoglobinopathies like thalassemia and sickle cell diseases, or glucose-6-phosphate dehydrogenase deficiency)—and/or vitamin-B_{12} deficiency (over several decades) in Africa and India.[6,109] This profound blunder cautions against placing an inappropriate over-reliance on the results of serum folate concentration over a careful dietary history and physical examination.[109]

There are additional clinical ramifications from understanding this relationship. Because vitamin-B_{12} deficiency raises serum folate, when individuals are found to have a low serum folate level in the setting of dietary vitamin-B_{12} deficiency, it means that these individuals are actually very severely deficient. Indeed, it has been clinically observed that when such individuals are given vitamin B_{12}, their serum folate concentration can drop further. Thus, it is important to note that nutritional folate deficiency remains very common in most resource-limited countries.[109]

Folates and Perturbed Neurodevelopment in Utero

All inborn errors of folate metabolism, which result in reduced folate availability to the developing human brain, give rise to mental retardation and related mental health problems. Therefore, reduced maternal delivery of folate to the highly proliferative developing embryonic and fetal brain *in utero* can be predicted to have serious anomalies in development. Maternal folate deficiency leads to a spectrum of neurologic abnormalities in the embryo and fetus, ranging from full-blown neural-tube defects (NTDs)[136] to subtle changes that manifest in childhood as behavioral abnormalities.[126,137,138]

(i), *Folate-Responsive Neural-Tube Defects and Neurocristopathies:* NTDs are the most common major congenital malformation of the central nervous system. For the most part, NTDs arise from insufficient delivery of folate to embryonic neural tube cells via folate receptors; this leads to disturbances in neurulation that result in incomplete closure of neural tissues, leading to major midline defects.[126-128,136,139] Such defects originate in the first month of pregnancy (before many women know they are pregnant). Therefore, it is critical for a woman to have enough folate in her body before conception to ensure sufficient availability for the embryo. Anencephaly and spina bifida,[136] the commonest NTDs, are important factors in fetal mortality. Fortunately, landmark studies have established the preventive role of periconceptional folates in both the *recurrence* of NTD (using folic acid 4,000 μg/day) and the *first occurrence* of NTD (using a minimum of folic acid 400 μg/day).[140-142] Of significance, the greatest protection by folates occurs in those regions with the highest rates of NTDs.[143] Based on both preclinical studies[126-128] and clinical trials,[141,142] the relationship between folate replacement and prevention of NTDs fulfills Koch's postulates on the causation of disease.[144]

The spectrum of presentations of NTDs range from the most severe anencephaly with complete rachischisis, anencephaly alone, encephalocele, spina bifida with neural plaque, lipomyelomeningocele, myelomeningocele, diastematomyelia, meningocele, to tethered cord, and spina bifida occulta (See examples in Figures 1–4 of Reference[136]). Depending on the severity and level of the lesion(s), there can be varying degrees of neural deficits with resultant hydrocephalus, motor and sensory loss in lower limbs, loss of bladder and/or bowel control, and lower limb deformities.

(ii), *Folate Depletion in Utero and Psychological-Psychiatric Manifestations in Childhood:* Experimental studies in mice[137] indicate that folate deficiency in dams significantly compromises early pregnancy outcomes (rates of pregnancy, rate of implantation, and effects on the number of live births). In addition, increased apoptotic fetal brain loss (by 20%) and profound abnormalities in the white matter—reflecting perturbed neuronal development in the progeny of folate-deficient dams[137]—has been documented. Surprisingly, despite postnatal folate replenishment, these [progeny] mice exhibited an anxiety phenotype in young adulthood; in another study, rodents exhibited learning problems.[145,146] In concordance with these animal studies, human studies have consistently revealed that lower maternal folate status in early pregnancy is associated with several psychological-neuropsychiatric disorders in childhood. These manifestations in children include hyperactivity/inattention and peer problems,[147] hearing loss,[148] learning deficits, with anxiety or depression, and aggressive behavior.[63,149-154] Conversely, women who took folic acid during the early stages of pregnancy appear to have provided their babies with an *in utero* environment that impacts positively on the child's intellect, emotional intelligence, cognition, language, and academic performance.[63,151,155] These findings are consistent with the Barker Hypothesis on the developmental origins of disease.

Folate Fortification of Food

In January of 1998, fortification of foods (i.e., flour, pasta, macaroni, breads, with folic acid at 140 μg/100 g of food) became part of American law. Subsequent evaluation has demonstrated that flour fortification with folic acid has had multiple salutatory effects during human development and that major congenital abnormalities are prevented.[120] However, there remains concern about the effectiveness of the folic acid fortification program for women in the 15- to 35-year-old age group in preventing NTDs. This stems from their incomplete knowledge base[156] and their tendency to consume low-carbohydrate foods—the very foods that are fortified with folic acid. This is the basis for recommendations to continue to educate women of childbearing age to take folic-acid supplements at 400 μg/day (beyond what they receive through folate-fortified flour).

Some concern has been expressed that folic-acid fortification may raise the risk of colon cancer. However, folic-acid supplements protect against the development of colorectal neoplasia in high-risk patients with ulcerative colitis.[157] Three large, prospective studies[158-160] suggest that long-term folate intake *decreases the risk* for initiation or early development of colorectal cancer. In addition, there appears to be a diminished influence on (precancerous) adenomas. A meta-analysis of 13 randomized trials that compared the incidence of cancer of all types in ~50,000 individuals led to the conclusion that folic acid supplementation does not significantly increase or decrease the risk of cancer over a 5-year treatment period.[161] Collectively, these papers provide reassurance that the fortification of food to prevent NTDs in women of childbearing age has not led to harm among the remaining "nontargeted" population of adults.

However, despite implementation of folate food fortification of flour in more than 80 countries, more than 90% of pregnant women worldwide still do not receive sufficient folate,[122] and more than one-half of the world's population still does not have easy access to folate-fortified foods. Not surprisingly, two population-based studies conducted a decade apart in the most populous state in India identified that the incidence of NTDs was up to 8.21 and 7.48 per 1,000 live births, respectively, which is among the highest worldwide.[162,163]

The recent identification of the efficacy of a quintessentially Indian beverage (a daily cup of tea that is fortified with folate and vitamin B_{12}) to rapidly eliminate both folate and vitamin-B_{12} deficiency among Indian women in a few months is under active study.[13,164] This beverage is consumed by most Indians every day and can be centrally fortified in tea plantations in the highlands of primarily four states in India. Importantly, the fact that a cup of tea is acceptable to all Indian women of every geographic-, social-, economic-, cultural-, and ethnic- distinction—whose diet is dictated by these very differences—provides much hope. It is estimated that vitamin-fortified tea has the potential to dramatically improve the folate- and vitamin-B_{12} status of more than 500 million Indian women, optimize vitamin-dependent embryonic-fetal neurodevelopment *in utero* in 26-million pregnancies annually, reduce complications of pregnancy and delivery, allow more than 100,000 babies to develop normally (without NTDs) each year, and positively impact the mental and physical health of children and middle-aged and elderly Indians.[13,136,164] As tea is the second most common beverage consumed worldwide after water and is commonly consumed every day among people living in other resource-limited countries where these nutritional deficiencies are common, vitamin-fortified tea has the potential to eliminate folate and vitamin-B_{12} deficiency altogether.[13,136,164]

Recent studies[165,166] show that even for the minority of Indian women with a normal vitamin-B_{12} status, those with plasma folate concentration within the normal range—(i.e., between the lower limit of normal and 11.3 ng/mL)—still remain at significant risk for having babies with folate-responsive NTDs. Indeed, for Indian women who have combined low-folate and low vitamin-B_{12} status,[13,167,168] those with an even (falsely)-higher basal plasma folate value[109] will likely continue to remain at high risk for folate-responsive NTD. Vitamin-fortified tea may have major benefits in this population. In the meantime, a prudent approach is to recommend that all Indian women (and those living in resource-limited countries that have not uniformly implemented folate fortification of food) take vitamin B_{12} and folate supplements to improve their chances for having a normal baby.[136]

Therapy of Folate Deficiency

An appropriate regimen for moderate to severe folate deficiency is 2 mg folic acid daily for about 3 to 4 months to rapidly fill up stores. However, with mild folate deficiency, using 1 mg daily oral folic acid is sufficient. Later, if the deficiency is solely due to dietary insufficiency, a lower daily dose of 400 μg is needed indefinitely.

In addition to hematologic diseases leading to increased folate requirement (e.g., autoimmune hemolytic anemia, β-thalassemia, sickle cell disease, etc.), folic acid supplements reduce the hepatotoxicity and gastrointestinal intolerance of methotrexate in psoriasis[157,169] and rheumatoid arthritis[170] without impairing the efficacy of methotrexate.

There appear to be no inherent risks to administering folate supplements in malarial regions in order to support compensatory hematopoiesis in malaria. A recent meta-analysis of 19 studies showed no effect of iron-folic acid supplements and malarial progression[106] and a Cochrane analysis identified _no_ evidence that this is a problem when combined with antimalarial drugs and insecticide-treated bed nets.[171] The safety of low-dose folic acid (1 mg/day) in 467 pregnant women with malaria has also been affirmed[172]; however, high-dose folic acid (5 mg/day) can allow for *Plasmodium* resistance, and have adverse outcomes in children.[173-175]

Therefore, given the propensity to have multiple deficiencies that continue into pregnancy, all nonpregnant women who require antimalarial treatment[176] should be treated for both malaria as well as iron, folate, and vitamin B_{12} to enable optimal hematopoiesis. Likewise, all pregnant women living in resource-limited malarial areas should routinely be given intermittent preventive treatment according to the latest guidelines (currently it is sulfadoxine-pyrimethamine)[177] together with insecticide-treated bed nets, plus 1 mg folic acid, prophylactic doses of 10 to 25 μg vitamin B_{12}, and replacement doses of iron (65 mg elemental iron) orally either daily or every other day once their iron stores are filled.[178]

Vitamin-B_{12} (Cobalamin) Deficiency Anemia

Nutritional Aspects of Vitamin-B_{12} Deficiency

Vitamin B_{12} is essential for the proper intracellular metabolism of folate, i.e., for DNA synthesis, methylation, and other important enzyme reactions. Therefore, vitamin-B_{12} deficiency results in a functional folate deficiency. Vitamin-B_{12} deficiency has a significant adverse impact on the bone marrow (by inducing megaloblastic hematopoiesis) and also on the nervous system of children, adults, and the elderly[6] through mechanisms that are not fully understood.

Humans primarily receive vitamin B_{12} from animal-source foods[179,180] and although the RDA for vitamin-B_{12} is 2.4 μg/day, approximately 6 to 7 μg/day is believed to be the optimal amount.[181,182] Because vegetarian diets provide only up to 0.5 μg/day of vitamin B_{12}, vegetarians are at risk for vitamin-B_{12} deficiency and must receive daily supplements.[4]

Among common vegetarian foods, there is no unfortified plant food—including fermented soy products, tempeh/tempe, seaweed, nori (red algae), chlorella (green algae), spirulina (blue-green algae), or other organic produce—that can consistently provide a sufficient amount of active vitamin B_{12} to support daily requirements.[7] Some vegetarians and herbivores who consume legumes and root-related vegetables can avail of small amounts of bacteria-derived sources of vitamin B_{12}; however, the vitamin B_{12} from cow manure (which is widely used as an organic fertilizer and contaminates green leafy vegetables) does not contribute to meaningful dietary intake of vitamin B_{12} in humans.[7] Therefore, vegetarians should be viewed as being in various stages of progressive vitamin-B_{12} depletion moving toward vitamin-B_{12} deficiency,[4,7,183] and, therefore, should receive supplemental vitamin B_{12}. Of related significance, most residents of resource-limited countries who indicate they are "nonvegetarians" actually consume a "near-vegetarian" diet,[4,7] which contains relatively small quantities of meat only once or twice a week. Such a nonvegetarian diet cannot sustain optimum vitamin-B_{12} balance.

Nutritional vitamin-B_{12} deficiency is a worldwide problem.[6,135,179] Indeed, it has been widely reported among 45% of Northern Chinese women,[184] and in up to 85% of Indian adolescents[185] and 65% of newborns.[186] In Pakistan,[187] among those with megaloblastic anemia (hemoglobin less than 8 g/dL), nearly 80% had vitamin-B_{12} deficiency. Surprisingly, even those consuming (apparently healthy) "Mediterranean diets" rich in fruits and vegetables, but low in animal-source protein, are at risk for vitamin-B_{12} deficiency; among 180 pregnant women, vitamin-B_{12} deficiency was found in 72% of mothers and 41% of babies.[188] Although more than 75% of women in India have documented vitamin-B_{12} depletion or deficiency,[13] there are _no_ robust programs designed to improve the vitamin-B_{12} status of children, adolescent girls, or women of childbearing age in these regions. Hence, the importance of food fortification initiatives involving this vitamin.

Clinical Features of Vitamin-B_{12} Deficiency

Because megaloblastic anemia caused by either folate or vitamin-B_{12} deficiency leads to a functional folate deficiency, the hematological manifestations from either of these two deficiencies are understandably indistinguishable. However, only vitamin-B_{12} deficiency results in a patchy demyelination process, which is expressed clinically as cerebral abnormalities and subacute combined degeneration of the spinal cord. Vitamin-B_{12} deficiency in adults[189] can mimic several neurologic diseases and it is common for nutritional vitamin-B_{12} deficiency to be first diagnosed by neurologists.

Pure vitamin-B_{12} deficiency can first present with megaloblastic anemia followed by development of overt neurologic findings when progressive demyelination of the sheaths of nerves continues over time. However, nutritional vitamin-B_{12} deficiency among vegetarians and near-vegetarians in resource-limited countries is often accompanied by iron deficiency, and malnourished individuals will also have folate deficiency. Individuals with nutritional vitamin-B_{12} deficiency may present in the second and third decades with pancytopenia, mild hepatosplenomegaly, fever, and even occasionally with bleeding (from thrombocytopenia). Alternatively, a neurologic and psychiatric syndrome may develop simultaneously with or even independent of anemia; indeed, this is not uncommonly observed in resource-rich countries where the higher the hematocrit (less anemia), the more severe the neurologic presentation.

Those individuals with vitamin-B_{12} deficiency can present with various symptoms of anxiety, apathy, aggression, irritability, hallucinations, delusions, and features of disinhibition. Paresthesia or ataxia are common first symptoms and diminished vibratory sensation (to a 256 cycles per second tuning fork) and proprioception in the lower extremities (i.e., loss of position sense in the index toe even before great toe involvement), which is elicited by passive movement, are common early objective signs. The full spectrum of signs can include loss of fine or coarse touch, decreased or increased deep tendon reflexes with spasticity or muscle weakness, urinary or fecal incontinence, orthostatic hypotension, amaurosis, global cognitive impairment, dementia, psychosis, or mood disturbances. Although neurologic deficits tend to be mild in most cases, the severity is greater the longer the symptoms are present before diagnosis. Those with the shorter duration of symptoms respond most to vitamin-B_{12} replacement.

Vitamin-B_{12} Nutrition in Pregnancy and Infancy

The fetus is dependent on the mother's vitamin-B_{12} stores for a sufficient quota of vitamin-B_{12} at birth. There is a close correlation between low vitamin-B_{12} concentration in both maternal serum and breast milk and low vitamin B_{12} in the infant.[190,191] Therefore, when mothers do not consume sufficient animal-source foods, they themselves are at risk for nutritional vitamin-B_{12} deficiency, and their infants have smaller stores of the vitamin at birth.[111,192] When these infants subsequently receive a low vitamin-B_{12} content from breast-milk,[181] their reasoning, abstract thinking, and learning can be impaired.[193] And when they are weaned to a diet that is similar to their mothers, which is intrinsically low in vitamin B_{12},[3,4,194] this chronic low-vitamin-B_{12} status is perpetuated in young girls through childhood into adulthood, and into their own reproductive cycles. Therefore, once again, as noted with chronically poor iron and folate nutrition, this vertical intergenerational cycle of vitamin-B_{12} deficiency tends to continue in perpetuity.

A large percentage of mothers, particularly in resource-limited countries, have evidence of vitamin-B_{12} deficiency.[4,195,196] Three-quarters of a cross section of 366 pregnant urban women in South India had vitamin-B_{12} deficiency.[197] Important studies have shown that treatment of the mother during pregnancy[198] or even the infant shortly after delivery will promptly reverse pre-existing vitamin-B_{12} deficiency.[199]

Even among women from resource-rich countries who presumably consume a balanced nonvegetarian diet, longitudinal studies confirm that pregnancy places an additional stress on the mother's vitamin-B_{12} stores, which can lead to metabolic evidence of vitamin-B_{12} deficiency[181,200,201] and negatively affect their breastfed infants' vitamin-B_{12} status at 6 weeks. More than two-thirds of Norwegian infants of otherwise healthy mothers had a metabolic profile consistent with vitamin-B_{12} deficiency, which reverted to normal after vitamin-B_{12} replenishment.[181,201] Thus, it is likely that many more breastfed infants need vitamin-B_{12} supplements early in life, as do their mothers in preparation for pregnancy.[182]

Clinical Features of Vitamin-B_{12} Deficiency

Vitamin-B_{12}-deficient infants present with a spectrum of clinical findings, ranging from feeding difficulties and refusal of both breast milk and complementary food by regurgitation—(which results in further failure to thrive)—to motor and social retardation, reflecting a developmental delay.[4] Among infants with only minor developmental delays and feeding difficulties and biochemical evidence of vitamin-B_{12} deficiency, those treated with vitamin-B_{12} responded with significant clinical benefit, a fact that underscores the importance of this vitamin in postnatal neurodevelopment.[194]

Evidence of megaloblastic anemia may be masked with superimposed iron deficiency. If left untreated, there is growth retardation with reduced height and weight and reduced head circumference with cranial magnetic resonance imaging showing delayed myelination and frontoparietal cortical atrophy in affected infants.[202] These can also be reversed within 3 months of vitamin-B_{12} replacement. Treatment results in a dramatic increase in alertness and responsiveness of the child. Previous abnormal movements (tremors, chorea, or myoclonus) may regress but transiently return within a few days to affect the face or tongue; however, these will resolve in 2 to 3 months.[203] Vitamin-B_{12}- (and folate-) deficiency often resurfaces during wartime, which invariably leaves women and their infants malnourished.[204]

Infants in the West fed a macrobiotic diet (vegan-like with occasional servings of fish) must be rapidly replenished with full doses of vitamin B_{12} before switching to a vitamin-B_{12}-rich diet. Otherwise, up to 20% continue to have low vitamin-B_{12} status, which can lead to impaired psychomotor functioning well into youth and later adolescence with compromise in faculties related to reasoning, abstract thinking, and learning ability.[205]

Therapy of Vitamin-B_{12} Deficiency

An appropriate regimen for severe vitamin-B_{12} replenishment is 1 mg of intramuscular or subcutaneous vitamin-B_{12} (cyanocobalamin) per day (week 1), 1 mg twice weekly (week 2), 1 mg/week for 4 weeks, and then 1 mg per month for life; (about 15%, or 150 μg is retained 48 hours after each 1 mg vitamin-B_{12} injection). In ambulatory patients with mild deficiency, an alternative approach is to use daily oral therapy, initially beginning with 2 mg daily oral vitamin B_{12} for 4 months to rapidly fill stores.[206] Later maintenance therapy with vitamin B_{12} is dictated by the likely etiology. If the deficiency is due to malabsorption of food vitamin B_{12}, daily oral vitamin B_{12} is indicated in full doses of 1 mg daily indefinitely (or monthly injections of 1 mg vitamin B_{12} subcutaneously or intramuscularly). Alternatively, if the deficiency is solely due to dietary insufficiency, lower doses (25 to 50 μg or even less) may be used indefinitely.[6]

For schoolchildren in resource-limited countries, simple community-level interventions, such as micronutrient fortification (using a premix added to school lunches), that build upon the infrastructure of an existing program have proven to be acceptable and efficacious in improving folate and vitamin B_{12}

(and vitamin A and iron) status in Himalayan villages.[207] A broadly acceptable food vehicle (such as tea, flour, rice) for fortification with vitamin B_{12} (in addition to folate) is critically required for most resource-limited countries.[136]

Restricted Diets and Nutritional Anemia

Alcoholism invariably leads to an imbalanced and monotonous diet that is lacking in folate, zinc, or vitamin C; scurvy can lead to gastrointestinal bleeding and iron deficiency from intestinal bleeding.[208] Among those with psychiatric disease, the sole consumption of potato chips (one among many examples) can also lead to scurvy; sole consumption of bok choy can lead to severe clinical iodine deficiency and hypothyroidism-related anemia[209]; and folate deficiency among the elderly or infirm who solely consume tea-and-toast diets is well recognized.[6] Infants who are fed goat's milk in preference to human or cow's milk can suffer from goat's milk anemia, due to the low folate content of goat's milk. Similarly, many premenopausal American women who avoid red meat are at risk for iron and zinc deficiencies.[104] Zinc deficiency can also develop in patients on long-term total parenteral nutrition, where zinc supplements were omitted.

Less Common Nutritional Anemias

Combined or pure nutritional deficiencies of minor micronutrients like vitamin B_6, vitamin B_1, vitamin A, zinc, or copper comprise less than 5% of nutritional anemias. In practice, only after failing to normalize hemoglobin values following complete replacement of iron, folate, and vitamin B_{12}, and after ruling out other chronic medical conditions common in the region (such as malaria, alcoholism, hypothyroidism, cirrhosis, and hemoglobinopathies, other chronic diseases, etc.) are these other deficiencies pursued. When obtaining tests for iron, folate, and vitamin B_{12} deficiency is difficult due to poverty or nonavailability, or when results of folate are likely to be spuriously elevated in hemolytic states or vitamin-B_{12} deficiency, a response to empiric replacement with the three major micronutrients will, for the most part, suffice to rule out minor micronutrient deficiencies. The following section deals with the clinical context of hematologic presentations of less common mineral-vitamin deficiencies. More details on the nutritional aspect of each of these minerals and vitamins are found in other chapters in this textbook.

Zinc

Zinc is required to catalyze enzymes. It forms an integral part of proteins and nucleic acids, is a regulator of gene transcription, and is required for cell proliferation, differentiation, and cellular metabolism. Zinc deficiency can profoundly affect fetal growth and development of tissues (including the brain), and by impairing the integrity of the immune system, it results in reduced cell-mediated and humoral immunity.[56]

An estimated 2 billion people worldwide are affected by dietary zinc deficiency. Low intake of meat and concomitant high dietary intakes of phytates and polyphenols can predispose to both dietary iron and zinc deficiencies.[57] These deficiencies often coexist, more so among vegetarians- compared with nonvegetarian-diets.[58] Marginal zinc deficiency is common (up to 80%) among mother-baby dyads in developing countries and contributes to health problems, especially in children.[87,88,210,211] Zinc deficiency is associated with increased risk for diarrhea and respiratory diseases, and zinc supplementation reduces the incidence of diarrhea and pneumonia, the two leading causes of death in children.[41] Low zinc also predisposes to frequent infections that can lead to the anemia of chronic disease. A majority of South Asian populations suffer from concurrent iron and zinc deficiency.[212] Even in the United States, many premenopausal women have low iron and zinc intakes accompanied by low serum ferritin; the avoidance of red meat intake increases the risk of iron and zinc deficiencies.[60] There appears to be a poorly understood mutually exclusive relationship between the absorption of zinc and iron presumed to result from competition for absorption; likewise, zinc administration can inhibit copper absorption. Thus, there can be unintended consequences of administering oral iron and marginal zinc deficiency could be aggravated among women given iron supplements; it remains to be seen if alternative day iron therapy can allow more zinc intake on the alternate days that iron is not given. Because the daily RDA for zinc is 12 mg for pregnant women, all such women living in resource-limited settings at risk for zinc deficiency should be supplemented with higher doses (the tolerable upper limit of zinc is 40 mg/day).

Zinc deficiency can manifest with acrodermatitis, delayed wound healing, and hematologic abnormalities that mimic iron deficiency anemia. Zinc deficiency can also manifest as multifactorial anemia thought to result from a combination of hypoproliferation (arising from a reduction in erythroid precursors associated with a reduction in erythropoietin), an arrest of differentiation, and increased apoptosis with increased hemolysis arising from decreased RBC span from oxidative damage. However, the direct effect of zinc on erythropoiesis independent of iron is still unclear.

In the absence of associated chronic disease, the presence of zinc deficiency with iron deficiency can complicate diagnosis of iron deficiency, in which a low total iron-binding capacity is associated with low iron and ferritin concentration. Hence, a patient who is not fully responsive to iron may respond to administered zinc with a larger increase in hemoglobin and reticulocytosis, compared with response to either mineral alone.[61]

Copper

Because of the rarity of copper deficiency in the population, the public health importance of copper deficiency is limited to special groups of malnourished individuals.

Acquired copper deficiency occurs primarily in *preterm infants with very low birth weight* as the result of low copper stores conferred by the mother. This deficiency can also be found in infants exclusively fed cow's milk, which has intrinsically lower copper content than human milk. Such infants and children can present with prolonged and/or recurrent diarrhea for which no pathogenic organisms are found. Continued diarrhea invariably aggravates copper deficiency and leads to inadequate availability of copper to support the growth spurt in the latter half of infancy. In the event that the infant is given supplements of zinc and iron, which can further decrease copper absorption, these supplements can also predispose to copper deficiency.

Copper deficiency can also develop when patients are given *long-term enteral or total parenteral alimentation* that is not supplemented with copper. These patients can present years later with microcytic hypochromic anemia associated with mild neutropenia and thrombocytopenia with ringed sideroblasts in the bone marrow.[213,214] The anemia, which can be severe with hemoglobin of approximately 6.5 g/dL and can lead to transfusion dependence, responds well to dietary supplementation with copper. In another instance, prolonged zinc therapy has induced a copper deficiency.[215] In addition, children and adults can develop copper deficiency when they have severe malabsorption syndromes (tropical and nontropical [celiac] sprue), cystic fibrosis, or postsurgical resection of the bowel that interferes with copper absorption.

When *malnourished children* are given nutritional supplements, there is a prompt growth spurt that is estimated to be at least 10 times greater than normal for the age of the child. During this recovery period, an acute deficiency of copper can be induced. The child can then present with copper-responsive anemia with either normocytes or macrocytes or, rarely, even some hypochromic microcytes, and the reticulocyte count is low. The iron concentration is low (but there is no response to iron alone). Neutropenia is also found in most infants. Examination of the bone marrow can reveal megaloblastic changes together with significant vacuolization of the erythroid and myeloid progenitors associated with maturation arrest of myeloid precursors, and Perl's Prussian blue stains can reveal ringed sideroblasts. These changes reverse with copper supplementation.[216]

Riboflavin (Vitamin-B_2)

Riboflavin participates in a diversity of redox reactions that are central to human metabolism (through the cofactors flavin adenine dinucleotide and flavin mononucleotide, which act as electron carriers). Riboflavin is abundant in diets that contain milk or milk products, meat, fish, vegetables, or fruit. Riboflavin deficiency, if severe and prolonged from dietary inadequacy, can produce normocytic normochromic anemia, possibly by interfering with iron absorption or iron mobilization from existing stores. Indeed, correcting riboflavin deficiency in pregnant or lactating women, adult men, and school-age children improved the overall hematologic response to iron supplements.[217] Because one-third of Kenyan children have low RBC riboflavin,[218] and riboflavin supplementation given concurrently with iron improves the response to iron supplementation in adult men and schoolchildren,[219] replacing riboflavin is clearly important in resource-limited countries.

Iodine, Hypothyroidism, and Iron

Foods such as soybeans, cruciferous vegetables, and sweet potatoes contain natural goitrogens. An imbalance in diet through overconsumption of these foods, or the combination of a monotonous consumption of these foods among residents in goitrous belts where dietary iodine is low, can lead to hypothyroidism; this, in turn, can lead to metromenorrhagia, with resulting iron deficiency anemia among affected women.[220]

A randomized controlled study showed that, for those patients with iron deficiency who do not respond to iron replacement, searching for and treating subclinical hypothyroidism will significantly improve their therapeutic outcome within 12 weeks.[221] Indeed, the combination of levothyroxine and ferrous sulfate led to an impressive rise of hemoglobin approximately 1.9 g/dL with combined therapy, compared with a rise of hemoglobin approximately 0.4 g/dL in those individuals given iron alone.

Pyridoxine (Vitamin-B_6)

Pyridoxine has a role as a cofactor for a rate-limiting enzyme in heme biosynthesis (δ-aminolevulinic acid synthase), and pyridoxine deficiency can lead to sideroblastic anemia (with hemoglobin concentration ranging from 4 to 10 g/dL) in the setting of malnutrition. However, the more dominant presentation consists of peripheral neuropathy and dermatitis. The blood smear may reveal basophilic stippling, hypochromia, and microcytosis, but cells may be normocytic. In those with homocysteine elevation that cannot be explained by folate or vitamin-B_{12} or renal dysfunction, pyridoxine deficiency should be sought and treated, if found.

Vitamin C (Ascorbic Acid)

Vitamin C enhances the absorption of dietary iron. Therefore, in scurvy, there can be an independent effect of vitamin C deficiency in reducing iron absorption and another mechanism related to chronic bleeding from the friable mucosa of the gastrointestinal tract.[208] Importantly, the cause of iron deficiency should not be investigated by colonoscopy in a patient with vitamin C deficiency unless the vitamin C is fully replaced, as there is a real danger of perforation of the friable bowel wall as a result of vitamin C deficiency.[208]

Vitamin A

Vitamin A is important for innate and acquired immune response. With deficiency, chronic inflammation-induced anemia can occur. A well-controlled Moroccan study in children suggested that vitamin A supplementation can modestly improve hemoglobin (approximately 0.2 to 1 g/dL) and thereby reduce anemia.[222] Moreover, although the increase was small (approximately 0.5 g/dL), treating Indonesian pregnant women with both iron plus vitamin A for 8 weeks almost doubled the increase in hemoglobin, compared with giving iron alone[223]; this finding highlights the importance of vitamin A in the prevention of anemia, although the mechanism is still unclear.[224]

Vitamin E

Vitamin E deficiency in the premature infant is associated with hemolytic anemia that responds to specific replacement by a rise in the hemoglobin and a fall in the reticulocyte count. Hence, it has become routine to include vitamin E supplements for premature infants.[225] Therapeutic doses of iron (without vitamin E) can increase RBC hemolysis during vitamin E deficiency in infancy[226] in vegetarians who live in resource-limited regions.

Poverty and Nutritional Anemia Is a Common Problem Worldwide

Poverty is a major determinant of food choices and the dietary basis for anemia is closely tied to the economic well-being of the population under study. This is most common in resource-limited countries where a sizeable population at any time has chronic food insecurity. Indeed, almost one-half of the world's population—more than 3 billion people—live on less than $2.50 a day, and therefore subsist on restricted diets imposed by poverty.

This invariably leads to an imbalanced daily consumption of the cheapest locally available staple foods (such as wheat, rice, millet, sorghum, or corn) with very small servings of cooked lentils and/or seasonal vegetables and a small serving of fermented or pickled condiment. These diets are deficient in several micronutrients (iron, folate, vitamin B_{12}, vitamin C) as well as dietary fat, which is crucial for the absorption of fat-soluble vitamins (vitamins A/D/E/K) and is also needed for energy and protein. Such individuals will predictably have multiple causes for nutritional anemia, all of which must be addressed to optimize hemoglobin values. With the availability of more sensitive diagnostic biomarkers in the blood, and a growing understanding of the clinical context, there is increasing recognition of the widespread deficiency of iron, vitamin B_{12}, and folate worldwide.[3,6,7,179,180]

Acknowledgments

The authors would like to acknowledge Christopher R Chitambar for his contributions to the previous version of this chapter.

References

1. Global Nutrition Report. The 2020 Global Nutrition Report-India. Accessed August 20, 2022. Available at https://globalnutritionreport.org/resources/nutrition-profiles/asia/southern-asia/india/
2. Charu-Bahri/IndiaSpend. Business Standard 2017. Accessed August 20, 2022. Available at www.business-standard.com/article/current-affairs/50-indian-women-are-anaemic-why-it-persists-after-70-yrs-of-freedom-117112100219_1.html
3. Antony AC. Am J Clin Nutr 2001;74:157–9.
4. Antony AC. Am J Clin Nutr 2003;78:3–6.
5. Chitambar CR, Antony AC. In: Shils ME, Shike M, Ross AC, et al., eds. Nutritional Aspects of Hematologic Diseases. Modern Nutrition in Health and Disease, 10th ed. Baltimore, Philadelphia: Lippicott Williams & Wilkins. 2006:1436–61.
6. Antony AC. In: Hoffman R, Benz (Jr) EJ, Silberstein LE, et al., eds. Megaloblastic Anemias. Hematology: Basic Principles and Practice. 7th ed. Philadelphia: Elsevier; 2018:514–45.
7. Antony AC. In: Robert T. Means J, ed. Vegetarianism and Other Restricted Diets. Nutritional Anemia—Scientific Principles, Clinical Practice, and Public Health: United Kingdom: Cambridge University Press, 2019:153–74.
8. Kavle JA, Stoltzfus RJ, Witter F, et al. J Health Popul Nutr 2008;26:232–40.
9. Antony AC. In: Goldman L, Schafer AI, eds. Megaloblastic Anemias. Goldman-Cecil Medicine, 25th Ed. New York, NY: Elsevier Saunders, 2015:1104–14.
10. Delisle HF. Ann N Y Acad Sci 2008;1136:172–84.
11. Strand TA, Taneja S, Kumar T, et al. Pediatr 2015;135:e918–26.
12. Kvestad I, Taneja S, Kumar T, et al. PLoS One 2015;10:e0129915.
13. Vora RM, Alappattu MJ, Zarkar AD, et al. BMJ Nutr Prev Health 2021;4:293–306.
14. Khan U, Crossley C, Kalra L, et al. Stroke 2008;39:2943–9.
15. Seshadri S, Beiser A, Selhub J, et al. N Engl J Med 2002;346:476–83.
16. Smith AD, Smith SM, de Jager CA, et al. PLoS One 2010;5:e12244.
17. de Jager CA, Oulhaj A, Jacoby R, et al. Int J Geriatr Psychiatry 2012;27:592–600.
18. Durga J, van Boxtel MP, Schouten EG, et al. Lancet 2007;369:208–16.
19. Walker JG, Batterham PJ, Mackinnon AJ, et al. Am J Clin Nutr 2012;95:194–203.
20. Smith AD, Refsum H, Bottiglieri T, et al. J Alzheimers Dis 2018;62:561–70.
21. Heidemann D, Baker-Genaw K, Joseph N, et al. Consultant 2014;57:837–40. Accessed August 10, 2022. Available at www.consultant360.com/articles/increasing-cost-sensitivity-diagnostic-evaluation-microcytic-anemia
22. Veenemans J, Andang'o PE, Mbugi EV, et al. J Infect Dis 2008;198:401–8.
23. Ndeezi G, Kiyaga C, Hernandez AG, et al. Lancet Glob Health 2016;4:e195–200.
24. World Health Organization. The World Health Report 2002 - Reducing Risks, Promoting Healthy Life. World Health Organization. Accessed August 10, 2022. Available at https://apps.who.int/iris/handle/10665/67454
25. Pasricha SR, Tye-Din J, Muckenthaler MU et al. Lancet 2021;397:233–48.
26. Camaschella C. N Engl J Med 2015;372:1832–43.
27. Zimmermann MB, Hurrell RF. Lancet 2007;370:511–20.
28. Horton MA. Exp Cell Res 1983;144:361–6.
29. Rao R, Georgieff MK. Acta Paediatr Suppl 2002;91:124–9.
30. Lozoff B, Jimenez E, Smith JB. Arch Pediatr Adolesc Med 2006;160:1108–13.
31. Bruner AB, Joffe A, Duggan AK, et al. Lancet 1996;348:992–6.
32. Haase VH. Blood Rev 2013;27:41–53.
33. Muckenthaler MU, Galy B, Hentze MW. Annu Rev Nutr 2008;28:197–213.
34. Hotez PJ, Brooker S, Bethony JM, et al. N Engl J Med 2004;351:799–807.
35. Crompton DW, Nesheim MC. Annu Rev Nutr 2002;22:35–59.
36. Kightlinger LK, Seed JR, Kightlinger MB. J Parasitol 1995;81:159–69.
37. Bundy DA, Chan MS, Savioli L. Trans R Soc Trop Med Hyg 1995;89:521–2.
38. Casey GJ, Phuc TQ, Macgregor L, et al. BMC Public Health 2009;9:261.
39. Phuc TQ, Mihrshahi S, Casey GJ, et al. BMC Public Health 2009;9:266.
40. Lumish RA, Young SL, Lee S, et al. J Nutr 2014;144:1533–9.
41. Edwards CH, Johnson AA, Knight EM, et al. J Nutr 1994;124:954S–962S.
42. Omuga D. J Coll of Medicine (Nairobi, Kenya) 2009;14:24–32. Accessed August 10, 2022. Available at https://profiles.uonbi.ac.ke/osogo/publications/pica-practices-pregnant-women-nairobi-kenya
43. Karama B. Chewing up stones in Kenya - VPRO Metropolis. 2009. Accessed on August 2, 2017. Available at www.youtube.com/watch?v=MDbLpw39UGo
44. Gleason G, Scrimshaw N. In: K Kraeme MZ, ed. Nutritional Anemia. Basel, Switzerland: Sight and Life Press, 2007:45–58.
45. Pollitt E, Hathirat P, Kotchabhakdi NJ, et al. Am J Clin Nutr 1989;50:687–96; discussion 696–7.
46. Pratt JJ, Khan KS. Eur J Haematol 2016;96:618–28.
47. Pasricha SR, Low M, Thompson J, et al. J Nutr 2014;144:906–14.
48. Anker SD, Comin Colet J, Filippatos G, et al. N Engl J Med 2009;361:2436–48.
49. Lozoff B, Jimenez E, Wolf AW. N Engl J Med 1991;325:687–94.
50. Brock J. In: Brock JH, Halliday JW, Pippard MJ et al., eds. Iron in Infection, Immunity, Inflammation, and Neoplasia. Iron

Metabolism in Health and Disease. Philadephia, PA: WB Saunders, 1994:353–89.
51. Prentice AM. J Nutr 2008;138:2537–41.
52. Scholl TO. Am J Clin Nutr 2005;81:1218S–1222S.
53. Bothwell TH. Am J Clin Nutr 2000;72:257S–264S.
54. Food and Nutrition Board. Dietary Reference Intakes for Thiamin, Riboflavin, Niacin, Vitamin B_6, Folate, Vitamin B_{12}, Pantothenic Acid, Biotin, and Choline. Washington, DC: National Academies Press; 1998. Accessed August 10, 2022. Available at www.ncbi.nlm.nih.gov/books/NBK114310/
55. Zeng L, Dibley MJ, Cheng Y, et al. BMJ 2008;337:a2001.
56. Cogswell ME, Parvanta I, Ickes L, et al. Am J Clin Nutr 2003;78:773–81.
57. Siega-Riz AM, Hartzema AG, Turnbull C, et al. Am J Obstet Gynecol 2006;194:512–9.
58. Lieberman E, Ryan KJ, Monson RR, et al. Amer J Obstet Gynecol 1988;159:107–14.
59. Nair M, Choudhury MK, Choudhury SS, et al. BMJ Glob Health 2016;1:e000026.
60. Nair M, Churchill D, Robinson S, et al. Br J Haematol 2017;179:829–37.
61. Rahman MM, Abe SK, Rahman MS, et al. Am J Clin Nutr 2016;103:495–504.
62. UNICEF. Maternal Health 2019. Accessed on August 10, 2022. Available at www.unicef.org/india/what-we-do/maternal-health
63. Christian P, Murray-Kolb LE, Khatry SK, et al. JAMA 2010;304:2716–23.
64. Mireku MO, Davidson LL, Boivin MJ, et al. Pediatr 2016;138: e20161319.
65. Gluckman P, Hanson M. Discov Med 2004;4:401–7.
66. Georgieff MK, Brunette KE, Tran PV. Dev Psychopathol 2015;27:411–23.
67. Chanarin I, McFadyen IR, Kyle R. Br J Obstet Gynaecol 1977;84:504–8.
68. Pavord S, Daru J, Prasannan N, et al. Br J Haematol 2020;188: 819–830.
69. Kaestel P, Aaby P, Ritz C, Friis H. Br J Nutr 2015;114:1072–9.
70. Lozoff B, Jimenez E, Hagen J, et al. Pediatr 2000;105:E51.
71. Chang S, Wang L, Wang Y, et al. Pediatr 2011;127:e927–33.
72. Baker RD, Greer FR. Pediatr 2010;126:1040–50.
73. Rao R, Georgieff MK. Clin Perinatol 2009;36:27–42.
74. Chaparro CM, Neufeld LM, Tena Alavez G, et al. Lancet 2006;367:1997–2004.
75. Liyanage SK, Ninan K, McDonald SD. Pediatr 2020;146: e20201429.
76. Oski FA, Honig AS, Helu B, et al. Pediatr 1983;71:877–80.
77. Gera T, Sachdev HP. BMJ 2002;325:1142.
78. Hettiarachchi M, Liyanage C, Wickremasinghe R, et al. Asia Pac J Clin Nutr 2006;15:56–63.
79. Cesari M, Penninx BW, Lauretani F, et al. J Gerontol A Biol Sci Med Sci 2004;59:249–54.
80. Lipschitz D. J Am Geriatr Soc 2003;51:S10–3.
81. Labbe RF, Dewanji A, McLaughlin K. Clin Chem 1999;45:146–8.
82. Rettmer RL, Carlson TH, Origenes ML, et al. Pediatr 1999; 104:e37.
83. Griffin IJ, Reid MM, McCormick KP, et al. Arch Dis Child Fetal Neonatal Ed 2002;87:F49–51.
84. Goodnough LT, Skikne B, Brugnara C. Blood 2000;96:823–33.
85. Cook JD, Flowers CH, Skikne BS. Blood 2003;101:3359–64.
86. Zimmermann MB, Hurrell RF. Lancet 2007;370:511–20.
87. Bhutta ZA, Black RE, Brown KH, et al. J Pediatr 1999;135: 689–97.
88. Bhandari N, Bahl R, Taneja S, et al. BMJ 2002;324:1358.
89. Calis JC, Phiri KS, Faragher EB, et al. N Engl J Med 2008;358: 888–99.
90. Nnam N. Int J Food Safety, Nutr Public Hlth 2009;2:158–64.
91. Kroger-Ohlsen M, Trugvason T, Skibsted L, et al. J Food Sci 2002;67:3301–3303.
92. Dary O. In: K Kraeme MZ, ed. The importance and Limitations of Food Fortification for the Management of Nutritional Anemias. Nutritional Anemia. Basel, Switzerland: Sight and Life Press, 2007:315–36.
93. Zlotkin S, Tondeur M. In: K Kraeme MZ, ed. Successful Approaches: Sprinkles. Nutritional Anemia. Basel, Switzerland: Sight and Life Press, 2007:269–83.
94. Sandberg AS, Brune M, Carlsson NG, et al. Am J Clin Nutr 1999;70:240–6.
95. Moretti D, Goede JS, Zeder C, et al. Blood 2015;126:1981–9.
96. Stoffel NU, Cercamondi CI, Brittenham G, et al. Lancet Haematol 2017;4:e524–e533.
97. Blumenstein I, Shanbhag S, Langguth P, et al. Expert Opin Drug Saf 2021;20:757–69.
98. Sawicki KT, Ardehali H. Circ 2021;144:253–5.
99. Holm C, Thomsen LL, Langhoff-Roos J. J Matern Fetal Neonatal Med 2019;32:2797–2804.
100. Hoen B, Paul-Dauphin A, Kessler M. Clin Nephrol 2002;57: 457–61.
101. Feldman HI, Joffe M, Robinson B, et al. J Am Soc Nephrol 2004;15:1623–32.
102. Brookhart MA, Freburger JK, Ellis AR, et al. J Am Soc Nephrol 2013;24:1151–8.
103. Avni T, Bieber A, Grossman A, et al. Mayo Clin Proc 2015;90: 12–23.
104. Mwangi MN, Roth JM, Smit MR, et al. JAMA 2015;314: 1009–20.
105. Ojukwu JU, Okebe JU, Yahav D, et al. Cochrane Database Syst Rev 2009:CD006589.
106. Titaley CR, Dibley MJ, Roberts CL, et al. Am J Clin Nutr 2010;92:235–43.
107. Global Fortification Data Exchange. Accessed July 11, 2022. Available at www.fortificationdata.org
108. Misra A, Vikram NK, Pandey RM, et al. Eur J Nutr 2002;41: 68–77.
109. Antony AC. Nutr Rev 2017;75:600–15.
110. Lester GE, Makus DJ, Hodges DM. J Agric Food Chem 2010;58:2980–7.
111. Antony AC. In: Hoffman R, Benz (Jr.) EJ, Shattil SJ, et al., eds. Megaloblastic Anemias. Hematology: Basic Principles and Practice. 1st ed. New York, NY: Churchill-Livingstone, 1991:392–422.
112. Golding PH. Springerplus 2014;3:442.
113. Abdelrahim, II, Adam GK, Mohmmed AA, et al. Trans R Soc Trop Med Hyg 2009;103:493–6.
114. Odewole OA, Williamson RS, Zakai NA, et al. Am J Clin Nutr 2013;98:1042–7.
115. McNamee T, Hyland T, Harrington J, et al. PLoS One 2013;8:e77743.
116. Gudgeon P, Cavalcanti R. Am J Med 2015;128:56–9.
117. Gilfix BM. Clin Biochem 2014;47:533–8.
118. Bach V, Schruckmayer G, Sam I, et al. Clin Interv Aging 2014;9:1187–96.
119. Meziere A, Audureau E, Vairelles S, et al. J Gerontol A Biol Sci Med Sci 2014;69:1576–85.
120. Tamura T, Picciano MF. Am J Clin Nutr 2006;83:993–1016.
121. van Eijsden M, Smits LJ, van der Wal MF, et al. Am J Clin Nutr 2008;88:147–53.
122. Torheim LE, Ferguson EL, Penrose K, et al. J Nutr 2010;140:2051S–8S.
123. Chakravarty I, Sinha RK. Nutr Rev 2002;60:S53–8.
124. Antony AC. Am J Clin Nutr 2007;85:598S–603S.
125. Tang YS, Khan RA, Zhang Y, et al. J Biol Chem 2011;286: 39100–15.
126. Antony AC, Hansen DK. Teratol 2000;62:42–50.

127. Hansen DK, Streck RD, Antony AC. Birth Defects Res A Clin Mol Teratol 2003;67:475–87.
128. Piedrahita JA, Oetama B, Bennett GD, et al. Nat Genet 1999;23:228–32.
129. George L, Mills JL, Johansson AL, et al. JAMA 2002;288: 1867–73.
130. Ray JG, Laskin CA. Placenta 1999;20:519–29.
131. de la Calle M, Usandizaga R, Sancha M, et al. Eur J Obstet Gynecol Reprod Biol 2003;107:125–34.
132. Bennett M. J Reprod Med 2001;46:209–12.
133. Quere I, Mercier E, Bellet H, et al. Fertil Steril 2001;75:823–5.
134. Vollset SE, Refsum H, Irgens LM, et al. Am J Clin Nutr 2000;71:962–8.
135. Bondevik GT, Schneede J, Refsum H, et al. Eur J Clin Nutr 2001;55:856–64.
136. Antony AC, Vora RM, Karmarkar SJ. Lancet Reg Health Southeast Asia 2022;6:100071.
137. Xiao S, Hansen DK, Horsley ET, et al. Birth Defects Res A Clin Mol Teratol 2005;73:6–28.
138. Craciunescu C, Brown E, Mar M, et al. J Nutr 2004;134:162–6.
139. Iskandar BJ, Finnell RH. N Engl J Med 2022;387(5):444–50.
140. Wald N, Sneddon J. Lancet 1991;338:131–7.
141. Czeizel AE, Dudas I. N Engl J Med 1992;327:1832–5.
142. Berry RJ, Li Z, Erickson JD, et al. N Engl J Med 1999;341:1485–90.
143. Crider KS, Bailey LB, Berry RJ. Nutrients 2011;3:370–84.
144. Evans AS. Am J Epidemiol 1978;108:249–58.
145. Ferguson SA, Berry KJ, Hansen DK, et al. Birth Defects Res A Clin Mol Teratol 2005;73:249–52.
146. Berrocal-Zaragoza MI, Sequeira JM, Murphy MM, et al. Br J Nutr 2014;112:1323–32.
147. Schlotz W, Jones A, Phillips DI, et al. J Child Psychol Psychiatry 2010;51:594–602.
148. Martinez-Vega R, Garrido F, Partearroyo T, et al. FASEB J 2015;29:418–32.
149. Roza SJ, van Batenburg-Eddes T, Steegers EA, et al. Br J Nutr 2010;103:445–52.
150. Suren P, Roth C, Bresnahan M, et al. JAMA 2013;309:570–7.
151. Roth C, Magnus P, Schjolberg S, et al. JAMA 2011;306:1566–73.
152. Schmidt RJ, Tancredi DJ, Ozonoff S, et al. Am J Clin Nutr 2012;96:80–9.
153. Veena SR, Krishnaveni GV, Srinivasan K, et al. J Nutr 2010;140:1014–22.
154. Steenweg-de Graaff J, Roza SJ, Steegers EA, et al. Am J Clin Nutr 2012;95:1413–21.
155. Nilsson TK, Yngve A, Bottiger AK, et al. Pediatr 2011;128: e358–65.
156. Feldkamp M, Friedrichs M, Marti K. Am J Med Genet 2002;107:67–9.
157. Antony AC. In: Hoffman R, Benz (Jr.) EJ, Shattil SJ, et al., eds. Megaloblastic Anemias. Hematology. Basic Principles and Practice. 4th ed. Philadelphia, PA: Elsevier Churchill Livingstone, 2005:519–56.
158. Lee JE, Chan AT. Gastroenterol 2011;141:16–20.
159. Lee JE, Willett WC, Fuchs CS, et al. Am J Clin Nutr 2011;93: 817–25.
160. Stevens VL, McCullough ML, Sun J, et al. Gastroenterol 2011; 141:98–105.
161. Vollset SE, Clarke R, Lewington S, et al. Lancet 2013;381: 1029–36.
162. Cherian A, Seena S, Bullock RK, et al. Lancet 2005;366:930–1.
163. Rai SK, Singh R, Pandey S, et al. Asian J Neurosurg 2016;11: 352–55.
164. Vora RM, Antony AC. J Indian Assoc Pediatr Surg 2022;27:1–8.
165. Crider KS, Devine O, Hao L, et al. BMJ 2014;349:g4554.
166. Chen MY, Rose CE, Qi YP, et al. Am J Clin Nutr 2019;109: 1452–61.
167. Finkelstein JL, Fothergill A, Johnson CB, et al. Curr Dev Nutr 2021;5:nzab069.
168. Singh S, Geddam J, Reddy G, et al. BMC Nutr 2017;3:50:1–9.
169. Strober BE, Menon K. J Am Acad Dermatol 2005;53:652–9.
170. Whittle SL, Hughes RA. Rheumatol (Oxford) 2004;43:267–71.
171. Okebe JU, Yahav D, Shbita R, et al. Cochrane Database Syst Rev 2011:CD006589.
172. van Eijk AM, Ouma PO, Williamson J, et al. J Infect Dis 2008;198:1550–3.
173. Metz J. Food Nutr Bull 2007;28:S540–9.
174. van Hensbroek MB, Morris-Jones S, Meisner S, et al. Trans R Soc Trop Med Hyg 1995;89:672–6.
175. Carter JY, Loolpapit MP, Lema OE, et al. Am J Trop Med Hyg 2005;73:166–70.
176. Centers Disease Control and Prevention. Treatment of Malaria: Guidelines for Clinicians (United States). Accessed August 10, 2022. Available at www.cdc.gov/malaria/diagnosis_treatment/clinicians1.html
177. Centers Disease Control and Prevention. Intermittent Preventive Treatment of Malaria for Pregnant Women (IPTp) Accessed August 10, 2022. Available at www.cdc.gov/malaria/malaria_worldwide/reduction/iptp.html
178. Vanderjagt DJ, Ujah IA, Patel A, et al. J Obstet Gynaecol 2009;29:288–95.
179. Stabler SP, Allen RH. Annu Rev Nutr 2004;24:299–326.
180. Stabler SP. N Engl J Med 2013;368:149–60.
181. Bjorke-Monsen AL, Torsvik I, Saetran H, et al. Pediatr 2008;122: 83–91.
182. Bor MV, von Castel-Roberts KM, Kauwell GP, et al. Am J Clin Nutr 2010;91:571–7.
183. Pawlak R, Parrott SJ, Raj S, et al. Nutr Rev 2013;71:110–7.
184. Dang S, Yan H, Zeng L, et al. PLoS One 2014;9:e112586.
185. Kapil U, Bhadoria AS. J Family Med Prim Care 2014;3:247–9.
186. Sukla KK, Tiwari PK, Kumar A, et al. PLoS One 2013;8:e71587.
187. Iqbal SP, Kakepoto GN, Iqbal SP. J Ayub Med Coll Abbottabad 2009;21:92–4.
188. Koc A, Kocyigit A, Soran M, et al. Eur J Nutr 2006;45:291–7.
189. Kalita J, Misra UK. J Neurol 2008;255:353–9.
190. Allen LH, Miller JW, de Groot L, et al. J Nutr 2018;148:1995S–2027S.
191. Bjorke Monsen AL, Ueland PM, Vollset SE, et al. Pediatr 2001; 108:624–30.
192. Hay G, Clausen T, Whitelaw A, et al. J Nutr 2010;140:557–64.
193. Murphy SP, Gewa C, Liang LJ, et al. J Nutr 2003;133:3950S–3956S.
194. Torsvik I, Ueland PM, Markestad T, et al. Am J Clin Nutr 2013;98:1233–40.
195. Taneja S, Bhandari N, Strand TA, et al. Am J Clin Nutr 2007;86:1302–9.
196. Jones KM, Ramirez-Zea M, Zuleta C, et al. J Nutr 2007;137: 1307–13.
197. Samuel TM, Duggan C, Thomas T, et al. Ann Nutr Metab 2013;62:113–22.
198. Shahab-Ferdows S, Anaya-Loyola MA, Vergara-Castaneda H, et al. J Nutr 2012;142:1881–7.
199. Duggan C, Srinivasan K, Thomas T, et al. J Nutr 2014;144: 758–64.
200. Murphy MM, Molloy AM, Ueland PM, et al. J Nutr 2007;137:1863–7.
201. Bjorke-Monsen AL, Ueland PM. J Inherit Metab Dis 2010;34:111–9.
202. Avci Z, Turul T, Aysun S, et al. Pediatr 2003;112:684–6.
203. Yavuz H. Seizure 2008;17(8):748–9.
204. Amin NM, Zeki JM. Indian J Pediatr 2005;72:839–42.
205. Louwman MW, van Dusseldorp M, van de Vijver FJ, et al. Am J Clin Nutr 2000;72:762–9.
206. Wang H, Li L, Qin LL, et al. Cochrane Database Syst Rev 2018;3:CD004655.

207. Osei AK, Rosenberg IH, Houser RF, et al. J Nutr 2010;140: 1146–54.
208. Berkram P, Bedano PM, Kahi CJ, et al. Gastrointest Endosc 2007;66:1065–6.
209. Chu M, Seltzer TF. N Engl J Med 2010;362:1945–6.
210. Prasad AS. BMJ 2003;326:409–10.
211. Shah D, Sachdev HP. Nutr Rev 2006;64:15–30.
212. Akhtar S. J Health Popul Nutr 2013;31:139–49.
213. Fuhrman MP, Herrmann V, Masidonski P, et al. J Parenter Enteral Nutr 2000;24:361–6.
214. Miyoshi I, Saito T, Iwahara Y. Br J Haematol 2004;125:106.
215. Todd LM, Godber IM, Gunn IR. Ann Clin Biochem 2004;41: 414–6.
216. Cordano A. Am J Clin Nutr 1998;67:1012S–1016S.
217. Powers HJ. Am J Clin Nutr 2003;77:1352–60.
218. Siekmann JH, Allen LH, Bwibo NO, et al. J Nutr 2003;133: 3972S–3980S.
219. Ajayi OA, Okike OC, Yusuf Y. Eur J Haematol 1990;44: 209–12.
220. Messina M, Redmond G. Thyroid 2006;16:249–58.
221. Cinemre H, Bilir C, Gokosmanoglu F, et al. J Clin Endocrinol Metab 2009;94:151–6.
222. Zimmermann MB, Biebinger R, Rohner F, et al. Am J Clin Nutr 2006;84:580–6.
223. Suharno D, Muhilal, Karyadi D, et al. Lancet 1993;342: 1325–8.
224. Semba RD, Bloem MW. Eur J Clin Nutr 2002;56:271–81.
225. Oski FA, Barness LA. Am J Clin Nutr 1968;21:45–50.
226. Prache S. Vet Res 1994;25:497–520.

CHAPTER 98

Nutrition in Respiratory Diseases

Molly G. Mailes, Alejandra Yu Lee Mateus, and Margaret M. Johnson

CHAPTER OUTLINE

Abbreviations

ARDS, acute respiratory distress syndrome; **BMI**, body mass index; **CF**, cystic fibrosis; **CO**, cardiac output; **CO_2**, carbon dioxide; **COPD**, chronic obstructive pulmonary disease; **FEV_1**, Forced expiratory volume in 1 second; **FVC**, forced vital capacity; **ICU**, intensive care unit; **MEP**, maximum expiratory pressure; **MIP**, maximum inspiratory pressure **O_2**, oxygen; **PFT**, pulmonary function test; **REE**, resting energy expenditure; **RQ**, respiratory quotient; **SV**, stroke volume; **VA**, alveolar minute ventilation; **VD**, dead-space ventilation; **VE**, total minute ventilation; **VT**, tidal volume

Introduction

Cellular respiration is essential for the functioning of all tissues. Food substrate is converted to usable energy by the formation of high-energy phosphate bonds. Oxygen (O_2) fuels this process and carbon dioxide (CO_2) is produced as a byproduct. The respiratory system supplies the necessary O_2 and eliminates the CO_2 produced. Adequate nutrition is essential for development, growth, and function of the respiratory system. This chapter summarizes the normal structure and function of the respiratory system, the structural and functional changes encountered with common diseases, and the impact of nutritional status on the epidemiology and pathophysiology of pulmonary disease.

Structure and Function of the Respiratory System

Control of Breathing

Breathing is an automatic, rhythmic process that is controlled by afferent signals arising from the pontomedullary portion of the brainstem. Voluntary and involuntary input from higher cerebral centers, and changes in pH and partial pressure of arterial O_2 (PaO_2) and CO_2 ($PaCO_2$), can override these rhythmic impulses to alter respiratory patterns in response to the changing metabolic demands of the organism.

Respiratory Muscles

Inspiration is achieved when a negative intrathoracic pressure is generated by the active contraction of the inspiratory muscles, creating a pressure gradient between the mouth, and the distal air spaces, the alveoli. Moving down this pressure gradient, air fills the lungs until the alveoli and atmospheric pressure equilibrate. Relaxation of the inspiratory muscles returns the thoracic cage to its resting position, thus reversing the pressure gradient and leading to exhalation.

The diaphragm, the primary muscle of respiration, is innervated by the phrenic nerve and is dome shaped at rest in healthy individuals. With contraction, it flattens and descends, increasing both the vertical and anterior-posterior dimensions of the thoracic cage. The diaphragm does not have intrinsic automaticity properties like cardiac muscle and, thus, can fatigue when demand exceeds supply. Other skeletal muscles, including the external intercostal muscles, can also help to facilitate inspiration. Exhalation is typically a passive process, which occurs due to the elastic recoil of the lung. Contraction of the internal intercostal and abdominal muscles participate in active exhalation. Muscle fatigue,[1] a reversible inability of a muscle to generate a prior attainable force, and muscle weakness, the chronic inability to attain adequate force, may both compromise respiration. Inadequate ventilation can occur from muscle fatigue or weakness, aberrations in the function of the nerves innervating the muscles of respiration, or dysfunction of the neuromuscular junction.

Lung Parenchyma

The lungs are composed of the conducting airways, the alveoli, and the capillary beds, which form the gas-exchanging units, the supporting interstitial structures, the pulmonary vasculature, and immune effector cells.

Tracheobronchial Tree (Conducting Airways)

The conducting airways are a series of progressively, dichotomously branching tubular structures extending from the trachea. The trachea and proximal main airways, the bronchi, offer structural support to the airways but do not participate in gas exchange. Gas exchange occurs at the level of the distal respiratory bronchioles and the alveoli. The tracheobronchial tree is lined with ciliated columnar bronchial epithelial cells and submucosal glands that humidify, warm, and filter the inspired air and contribute to the bronchial mucus layer.[2]

Smooth muscle, innervated by the parasympathetic and the nonadrenergic, noncholinergic nervous pathways, lines the tracheobronchial tree. Muscle contraction imparts rigidity to the airways and reduces the caliber of the airway lumen, resulting in increased resistance to bulk gas flow. Additionally, airway edema, inflammation, and excessive mucus may also narrow the airway[2] and diminish gas flow.

Terminal Respiratory Units

The terminal respiratory unit, consisting of respiratory bronchioles, alveolar ducts, and alveoli, is the gas-exchanging unit. Gas exchange occurs at the alveolar–capillary membrane, which consists of the alveolar epithelium and capillary endothelium, their basement membranes, and the contiguous interstitial space.[2] Surfactant, a complex phospholipid and protein mixture that lines the alveolus, reduces its surface tension, decreasing its tendency to collapse at low lung volumes.

Pulmonary Physiology

The primary function of the respiratory system is to transfer O_2 from the inspired air to the bloodstream and CO_2 from the bloodstream to the exhaled air. Coordinated function of the cardiac and respiratory systems continuously supply oxygenated blood to the peripheral tissues in response to metabolic demands. After circulating to the periphery where O_2 is extracted, blood is returned to the right side of the heart and is pumped through the pulmonary arteries to the pulmonary capillaries. At the alveolar-capillary interface, O_2 diffuses down a concentration gradient from the O_2-rich alveolar gas to the pulmonary capillary blood. Most of the transferred O_2 binds to hemoglobin in the red blood cells; a small percentage is dissolved in the plasma. Simultaneously, CO_2 diffuses down a concentration gradient from the capillary blood to the alveolus and is subsequently exhaled.

O_2-enriched gas is continuously replenished at the alveolar level through inspiration. Approximately 30% of each inspired breath remains in the conducting airways and thus does not participate in gas exchange; this is anatomic dead space. A small fraction of each inspired breath reaches alveoli that are not perfused, and thus, do not allow gas transfer. This volume of gas is physiologic dead space. Effective alveolar minute ventilation (V_A) is the difference between the total minute ventilation (V_E) and the sum of anatomic and physiologic dead space ventilation (**Table 98.1**).

Efficient gas exchange is contingent on optimal matching of alveolar ventilation and perfusion. Inadequate gas flow causes perfused alveoli not to be ventilated. The complete absence of ventilation to perfused alveoli is called shunt, which is one mechanism by which hypoxemia may occur. Supplying inspired air to nonperfused alveoli increases physiologic dead space, thereby decreasing the effective tidal volume (V_T). Increased physiologic dead space is commonly seen in emphysema due to obliteration of the capillary bed by proteolytic enzymes.

Increased V_E can initially compensate for mismatched gas and blood flow. Ultimately, however, if metabolic demands exceed these compensatory mechanisms, gas exchange abnormalities will develop.

Additional Respiratory System Functions

In addition to gas exchange functions, the lung acts as a "filter" for the blood and has extensive metabolic functions. The lung synthesizes surfactant and other substances, including histamine and arachidonic acid, which are key mediators in airway inflammation.

Table 98.1 Definition of Respiratory Physiology Terms and Abbreviations

Term	Definition
Tidal volume (V_T)	Volume of gas moved during a single respiration
Minute ventilation (V_E)	Amount of air moved in and out of the lungs in 1 minute; $V_E = V_T \times$ respiratory rate (RR) per minute
Dead-space ventilation (V_D)	Amount of inspired gas that does not participate in gas exchange; ventilation of nonperfused alveoli
V_D/V_T	Fraction of each tidal volume that is dead space
Alveolar minute ventilation (V_A)	Amount of inspired air able to participate in gas exchange; alveolar ventilation is the difference between total minute ventilation and dead-space ventilation
Forced vital capacity (FVC)	Volume of gas that can be forcibly exhaled after a maximal inhalation
Forced expiratory volume in 1 second (FEV_1)	Volume of gas expired in the first second of a forced expiration
Stroke volume (SV)	Amount of blood pumped by the heart in a single beat
Cardiac output (CO)	Volume of blood pumped by the heart in 1 minute (heart rate [HR] × SV)
PaO_2	Partial pressure of oxygen in the arterial blood
$PaCO_2$	Partial pressure of carbon dioxide in the arterial blood
$\dot{V}O_2$	Oxygen consumption (mL/min)
$\dot{V}CO_2$	Carbon dioxide production (mL/min)
Compliance	Change in volume per unit change of pressure
Respiratory quotient (RQ)	Molecules of oxygen used/molecule of carbon dioxide produced
Mixed venous blood	Deoxygenated blood returned to the heart; samples for measurements are obtained from a catheter in the pulmonary artery

Common Pulmonary Pathophysiology

Various diseases can affect the respiratory system and ultimately compromise gas exchange. Bulk gas flow can be diminished by airway obstruction or restriction. In obstructive disease, the airway lumen is narrowed, creating increased resistance to flow. Prototypical obstructive diseases include asthma, emphysema, and chronic bronchitis. Decreased compliance of the respiratory system leads to restrictive disease. The loss of compliance can result from abnormalities in the lung parenchyma, respiratory muscles, or chest wall. Pulmonary fibrosis, the prototypical restrictive disease, is characterized by loss of compliance of the lung parenchyma as a result of interstitial fibrosis and impaired gas exchange, due to alterations in the alveolar capillary interface. Other diseases, such as pneumonia and pulmonary edema, decrease the lung compliance and impair gas exchange by filling the alveolar spaces. Decreased pulmonary compliance increases the work required to maintain adequate bulk gas flow.

Pulmonary function tests (PFTs) measure exhaled gas volumes and flow rates. Characteristic findings on PFTs can distinguish obstructive from restrictive processes. Abnormalities in respiratory muscle function decrease the maximum pressures attainable with inspiration and expiration. Arterial blood gas determinations, which measure the arterial pH, $PaCO_2$, and PaO_2, can evaluate the efficiency of gas exchange. Other tests, including timed walk tests or cardiopulmonary exercise tests, can more comprehensively assess the interaction of the respiratory and cardiac systems with the peripheral muscles. These tests allow one to measure the impact of various insults or interventions, including nutrition, on lung function.

Effects of Malnutrition on Development, Structure, and Function of the Respiratory System

Structural Development

Inadequate nutrition during fetal development is associated with impaired lung development.[3] In animal models, fetal malnutrition results in pulmonary hypoplasia.[4] Inadequate protein during development diminishes collagen and elastin synthesis, causing pathologic changes similar to emphysema.[5] Inadequate prenatal nutrition most commonly manifests as intrauterine growth retardation due to placental insufficiency. Placental insufficiency typically occurs in the latter half of pregnancy during development of the acinar units and alveoli. Thus, prenatal malnutrition most severely affects the distal lung.[6,7] In humans, a direct correlation exists between low birth weight and subsequent decreases in pulmonary function.[8] Inadequate nutrition in the neonatal period is associated with an increased incidence of bronchopulmonary dysplasia.[9]

In animal models, undernutrition leads to diaphragm atrophy and changes in the muscle's contractile strength and fatiguability.[10] The diaphragm is composed of both type I and type II fibers, and malnutrition affects these fibers differently. In grossly underfed and protein-malnourished animals, the cross-sectional area of both types of fibers is greatly reduced, but the fast-twitch fibers (type II) are quantitatively more affected.[11,12] These observations suggest that malnutrition diminishes peak pressure generation but has a more limited impact on endurance. Goldberg, et al., showed that poorly nourished patients have diminished maximal respiratory muscle strength, as measured by maximum inspiratory pressure (MIP) and maximum expiratory pressure (MEP).[13] Notably, the extent of muscle strength loss exceeds the loss of muscle mass, a finding suggesting coexistent myopathy of the remaining muscle.[13]

Ventilatory Drive

The impact of nutrition on respiratory drive is incompletely understood. In healthy individuals, energy and nutrient restrictions decrease the hypoxic respiratory drive.[14,15] Severe anorexia nervosa decreases V_E, which may reverse on refeeding.[16]

Host Defenses

Malnutrition increases general susceptibility to infection, but it specifically alters pulmonary defense mechanisms. Animal models of severe malnourishment have demonstrated decreased alveolar macrophage counts,[17] phagocytosis, and microbial killing.[18] In patients with tracheostomies, nutritional status inversely correlates with lower respiratory tract bacterial colonization.[19] Malnourished patients are also predisposed to pulmonary infection because of inadequate clearance of respiratory secretions, resulting from ineffective cough due to muscle weakness, and a greater propensity for alveolar collapse.

Prototypical Lung Diseases: Relation to Nutritional Status

Nutrition in Critical Illness and Acute Respiratory Distress Syndrome

Nutritional support in critically ill patients is provided with the goal of improving the patient's clinical outcome. Defining optimal nutritional support in this population is challenging due to differences in nutritional status prior to illness, temporal changes in nutritional needs over the course of illness, and impacts of therapeutic interventions, such as surgery, physical therapy, or medications on nutritional needs.

However, although critically ill patients are not a homogenous group, they share many nutritional challenges, including estimation of energy needs, ideal composition of nutritional support, and optimal route and timing of feeding. Investigations of nutritional needs and interventions in critical illness often include patients with multisystem organ failure, including respiratory failure. Respiratory failure is commonly the result of acute respiratory distress syndrome (ARDS), a syndrome defined by hypoxemic respiratory failure in the setting of a severe systemic critical illness or isolated pulmonary disease. This section reviews nutritional supplementation in the critically ill patient, with specific attention to ARDS.

The nutritional milieu of critical illness is characterized by hypermetabolism, protein catabolism, and insulin resistance leading to impaired glucose use and hyperglycemia. Because of the inherent complications of both underfeeding and overfeeding in this population, proper estimation of energy requirements is an essential but challenging task. Energy requirements can be estimated using standard population-based regression formulas, such as the Harris-Benedict equation.[20] However, predictive formulas were derived from physiologically normal individuals at rest and do not accurately predict the energy needs imparted by the stress and hypercatabolism of critical illness. "Correction stress factors," ranging from 1.2 to 1.5 times the calculated resting energy expenditure (REE) are suggested, but their correlation with indirect calorimetry measurements is often suboptimal.[21,22]

O_2 consumption ($\dot{V}O_2$), which can be used as an estimate of energy needs, can be calculated by the Fick equation:

$$\dot{V}O_2 = CO \div (Cao_2 - Cvo_2)$$

where CO is the cardiac output and Cao_2 and Cvo_2 are the O_2 content of arterial and mixed venous blood, respectively. This approach requires invasive monitoring with a pulmonary artery catheter, and a relatively stable patient.

Alternatively, $\dot{V}O_2$ can be assessed with indirect calorimetry using a metabolic cart that measures exhaled gases directly. This technique is not universally available, and it requires dedicated equipment, technical expertise, and a stable fraction of inspired O_2. Despite these limitations, this technique offers the advantage of continuous measurements rather than intermittent snapshots of one's energy needs. Indirect calorimetry is strongly recommended as the preferred method of estimating energy expenditure in critically ill patients on mechanical ventilation.[23]

The $\dot{V}O_2$ (mL/minute) obtained by either the Fick equation or by the gas exchange method is converted to kcal/day using the energy value of O_2 (4.69 to 5.05 kcal/L of O_2 consumed) or the modified Weir equation if Vco_2 (CO_2 production) is also known.[24]

Substrate Supplementation: Implications for Ventilatory Requirements

Enhanced protein degradation is common in critical illness[25] and is believed to contribute to muscle wasting. Protein degradation ranging from 12 to 16 grams of nitrogen/day has been described in early stages of critical illness but may reach 30 grams of nitrogen/day.[26]

Nutritional strategies to mitigate proteolysis have been investigated, but the ideal approach is not clear and does not remain constant over the course of illness.[25] Intravenous fat emulsions, if administered with a minimum of 500 kcal/day of carbohydrate and exogenous protein supplementation, can also limit proteolysis.[27] Protein intake at 1.5 g/kg/day, especially during the early stage of critical illness, may limit proteolysis.[28] The American Society of Parenteral and Enteral Nutrition currently recommends protein intake of 1.2 to 2.0 g/kg/day and 12 and 25 kcal/kg in the first 7 to 10 days of ICU stay for critically ill adult patients.[29]

The appropriate mix of carbohydrate, fat, and protein must be individualized, recognizing that the metabolism of these substrates produce different amounts of CO_2. Carbohydrates produce more CO_2 during oxidation than fat or protein. For every molecule of glucose completely oxidized, six molecules of CO_2 are produced, giving a respiratory quotient of 1 whereas the oxidation of fat and protein produces less CO_2, with a respiratory quotient of 0.7 and 0.8, respectively. V_A must be increased when CO_2 production increases to maintain a normal partial pressure of arterial $PaCO_2$. In the presence of underlying lung disease, the ability to increase V_A may be limited, although mechanical ventilation settings can be adjusted to accommodate increased CO_2 production. However, the use of a low carbohydrate nutrition formula failed to improve weaning from mechanical ventilation and is not widely recommended.[30]

Route of Nutritional Support

Malnutrition at the onset of critical illness is associated with poor outcomes, and improved clinical outcomes are associated with nutritional support.[31] Recently, it has been recognized that many patients who survive an ICU stay are left with long term multisystem complications, termed the post-ICU syndrome.[32] Thus, the efficacy of ICU interventions is now viewed through the wider lens of impact on PICS. Nutritional support is suggested to mitigate the development and severity of PICS.[33] However, the optimal composition and timing of the initiation of feedings remains uncertain.[34]

Enteral feeding is usually accomplished through a nasoenteric tube terminating in the stomach or post pyloric. Potential mechanical risks are associated with enteric feeding tubes, including misplacement in the tracheobronchial tree or pleural space; thus, radiographic confirmation of proper placement is mandatory before initiation of feeding. Placement of nasoenteric tubes postpyloric is often challenging and the benefits are uncertain.[35] Postpyloric feedings should be considered in those with significant gastroesophageal reflux disease, high risk for aspiration, on high doses of sedatives/paralytics, or with intolerance to gastric feeding. Maintaining patients in a semirecumbent position, rather than supine, decreases the risk of aspiration.[36]

Various formulations of enteral nutrition with differing characteristics and composition are commercially available. Formulas using peptides rather than intact proteins, addition of fiber, and coadministration of pre- or probiotics have been investigated but not routinely recommended.[37-39]

Parenteral nutrition is delivered through a central or peripheral vein. Central vein infusions allow for the delivery of more concentrated solutions; thus, it minimizes obligate fluid requirements. In patients with ARDS, limited fluid intake shortens the duration of mechanical ventilation.[40] The addition of heparin,[41] sterile line placement, and restriction of catheter use exclusively to alimentation[42] may limit catheter-associated complications such as thrombosis and infection. Infusion of lipid emulsions decrease diffusing capacity and oxygen saturation by causing ventilation and perfusion mismatch; thus, its use is typically avoided if possible.

Acute Respiratory Distress Syndrome

Despite numerous investigations, optimal nutritional support in ARDS is not yet defined. Patients with ARDS have lower intakes of dietary antioxidants, including vitamin E, vitamin C, retinol, and β-carotene, than healthy controls,[43] lower plasma concentration of α-tocopherol and elevated lipoperoxides, indicative of oxidative damage.[44] These observations lead to speculation that antioxidant supplementation may be beneficial. Although a prospective randomized trial examining the efficacy of supplementation with α-tocopherol and vitamin C did not decrease pulmonary mortality or the development of ARDS, the intervention group did have significantly lower incidence of multisystem organ failure and shorter duration of mechanical ventilation and intensive care unit (ICU) stay.[45]

A systematic review of 10 randomized controlled trials with 1015 participants analyzed the effect of immune-enhanced nutritional formula including supplemental omega-3 fatty acids, eicosapentaenoic acid (EPA), docosahexaenoic acid (DHA), gamma-linolenic acid (GLA), and antioxidants. Evidence from these investigations did not convincingly demonstrate benefit[46] and immune-enhanced nutritional support is not recommended for routine use.[47]

Because of difficulty in achieving full feedings and suggested benefit of hypoenergy feeding, coupled with an understanding that trophic feeding may have befit in preserving the integrity of the gut mucosa, trophic vs full strength feeding was compared in the EDEN study. In this randomized, multicenter trial of 1,000 patients with ARDS, lower volume trophic enteral feeds were compared with full enteral feeding. Trophic feeding did not improve ventilator-free days, 60-day mortality, or infectious complications, but was associated with less gastrointestinal intolerance.[48]

Chronic Lung Diseases

As described above, chronic lung disease is generally classified as obstructive or restrictive, based on the primary physiologic abnormality. Obstructive lung diseases lead to impaired expiratory airflow and, often, associated chronic airway inflammation and recurrent infections. Obstructive lung diseases include asthma, chronic bronchitis, emphysema, cystic fibrosis (CF), and bronchiectasis. Emphysema and chronic bronchitis are most commonly the result of tobacco abuse and are collectively labeled chronic obstructive pulmonary disease (COPD).

Restrictive diseases include infiltrative or fibrotic diseases of the lung parenchyma as well as extrapulmonary processes such as muscular weakness, thoracic cage abnormalities, and neurologic diseases that result in similar physiologic impairments. Fibrosis of the lung parenchyma is often idiopathic but may be due to autoimmune diseases or result from various occupational or environmental exposures.

Chronic Obstructive Pulmonary Disease

COPD is a tremendous global health challenge, affecting 16 million people in the United States[49] and 300 million worldwide.[50] In 2020, 150,000 deaths in the United States were attributable to chronic obstructive lung disease.[51] Tobacco use is the most common cause of COPD, although occupational and environmental exposures contribute substantially to the development of COPD worldwide.

Impact of Nutrition on Development of COPD

COPD collectively includes emphysema and chronic bronchitis. Interestingly, most chronic smokers do not develop clinically relevant COPD,[52] and the amount or intensity of tobacco exposure does not fully account for disease development. Therefore, it is hypothesized that other factors, including one's diet, may serve either a predisposing or protective role in COPD development. Supporting this hypothesis is the observation of pathologic changes like emphysema in patients with anorexia nervosa.[53]

An imbalance of proteases and antiproteases resulting in destruction of the elastin and collagen lung matrix causes emphysema. Oxidants inhaled from tobacco smoke and released from activated inflammatory cells recruited into the airways impair endogenous antiproteases. Several naturally occurring antioxidants are present in the lower respiratory tract to counter inhaled oxidants. Thus, the relationship between dietary antioxidants and the development of COPD have been investigated. Serum β-carotene and retinol correlate with the maintenance of ventilatory function.[54,55] Antioxidant ingestion has been shown in some investigations to be inversely associated with the development of COPD[56,57] but this association has not been universally confirmed.[58] Concerns exist about deleterious effects of β-carotene supplementation (see the section on lung cancer).

Because the type of ingested dietary fat may affect systemic inflammation, the Atherosclerosis Risk in Communities study investigated the relationship between dietary intake of n-3 fatty acids and the subsequent development of COPD in 8,960 current or former smokers. There was a quantity-dependent inverse relationship between the two after controlling for the intensity of tobacco use.[59] Sharp, et al., and Schwartz, et al., demonstrated similar results in 6,346 and 2,526 individuals, respectively.[60,61] The National Health and Nutrition Examination Survey demonstrated an independent association of increased intake of cured meats with an increased risk of newly diagnosed COPD in both men and women in prospective U.S. cohorts, independent of Western dietary or other associated dietary intakes.[62] The role of numerous other dietary factors including coffee, alcohol, wholegrains, minerals, and vitamin D in the development of COPD have been investigated but are limited by the methodologic challenges of dietary studies.[63]

Malnutrition in COPD: Prevalence and Interventions

Malnutrition is common in patients with COPD,[64] and associated with poor outcomes[65,66], including increased exacerbation rate,[67] decreased functional status, mortality, and healthcare costs.[68] Sarcopenia may occur in the presence or absence of low BMI and is strongly associated with functional status. In a cohort of 270 stable patients with COPD, body composition phenotype was associated with physical function, with sarcopenic-obese patients demonstrating the greatest functional impairment.[69] Furthermore, weight loss may be more predictive of mortality than low BMI.[70] A recent systematic review provides a narrative overview of the prevalence and effects of malnutrition in COPD.[68]

Numerous factors contribute to the development of malnutrition in COPD. The increased work of breathing and greater energy expenditure for respiratory muscle activity account for the bulk of the increase in REE.[71] Additionally, diet-induced thermogenesis and the O_2 cost associated with eating limit the ability to consume adequate energy to affect weight gain.

COPD is characterized by an enhanced inflammatory state, primarily resulting from chronic airway inflammation, and circulating inflammatory mediators that may contribute to malnutrition.[72] Skeletal muscle loss has been correlated with circulating inflammatory mediators in patients with stable COPD, a finding suggesting that weight loss and loss of fat-free mass are associated with the host's amplified inflammatory state.[73] Additionally, tumor necrosis factor-α, a cytokine that can induce cachexia, is elevated in weight-losing patients with COPD even in the absence of an acute infection.[74]

Other factors that may contribute to malnutrition in COPD, including early satiety due to lung hyperexpansion leading to compression of the stomach, alteration in taste from the use of inhaled therapies, frailty limiting access to food sources and preparation, and depression. Depression is increasingly recognized as a frequent comorbidity in patients with COPD, which is strongly associated with poor outcomes.[75] However, the relationship between nutrition and depression is this population is inadequately understood.

Due to the high prevalence of malnutrition and strong association with deleterious outcomes in COPD, numerous nutritional interventions have been extensively investigated. These investigations have mostly focused on assuring adequacy of energy intake, nutritional supplements, and the use of growth hormone, anabolic steroids, and vitamin D.

Numerous investigators have prospectively studied the effects of nutritional supplementation on patients with COPD. A review of available data suggested some measurable benefits in patients with stable COPD, but persistent meaningful benefit is uncertain.[76] Consensus guidelines suggest that nutritional supplementation may improve respiratory muscle strength and overall status in malnourished patients (Evidence B)[77] and attempts to increase energy and nutritional intake in undernourished patients with COPD, with or without specific nutritional supplements, is recommended. No specific nutritional supplement has been identified to be superior. As described previously, the oxidation of carbohydrates produces more CO_2 for each mole of O_2 consumed than fats. Thus, it had been hypothesized that a high-fat diet would be superior to a high-carbohydrate diet in patients with limited ventilatory reserve, but confirmatory data that this diet approach confers a clinical benefit is lacking.

Recognizing individual variability and tailoring dietary interventions based on specific needs are likely to result in the most appropriate interventions. Slinde et al., demonstrated significant anthropometric and functional improvements with a 1-year individualized dietary intervention in patients undergoing rehabilitation.[78] Advanced age, anorexia, and an elevated systemic inflammatory response characterize patients who are unable to gain weight.[79]

Administration of growth hormone and anabolic steroids has been studied as adjuvant therapy to improve nutritional status in patients with COPD. Various investigators have shown improvements in nitrogen balance, body weight, increased lean body mass, and decreased adipose tissue mass, but sustained clinical benefit is lacking. Thus, routine use of either growth hormone or anabolic steroids is not currently recommended.[80-85]

Limited data suggest a beneficial effect of the appetite stimulant megestrol acetate. Herrejon and Weisberg demonstrated weight gain and increased appetite in patients with COPD but no improvement in respiratory muscle function.[86,87]

Total vitamin D intake has been inversely associated with COPD. However, in a prospective trial of 155 patients with low serum Vitamin D concentration, supplementation did not decrease exacerbations.[88]

Electrolyte deficiencies, including hypokalemia, hypocalcemia, and hypophosphatemia, can adversely affect respiratory muscle function (see elsewhere in this volume). Hypophosphatemia is especially common during acute exacerbation and is associated with the need for mechanical ventilation and mortality.[89] Monitoring and appropriate supplementation of phosphorous during acute exacerbations is prudent.

Clear nutritional guidance in patients with COPD is lacking. Because of the strong association of malnutrition and weight loss with poor outcomes, nutritional assessment is recommended as part of comprehensive COPD care. Optimal disease management, including control of inflammation and incorporation of physical activity as well as individualized interventions to restore or maintain a normal BMI, are all likely of benefit. There is insufficient evidence to support the routine use of growth hormone, anabolic steroids, or vitamin D. Further investigations into the use of ghrelin and anamorelin hold promise.

Asthma

Asthma is an obstructive lung disease characterized by intermittent wheezing, cough, or chest tightness in the setting of reversible airflow obstruction. It affects more than 260 million people worldwide and is responsible for nearly half a million deaths.[90] Prostanoids released from inflammatory cells contribute greatly to the pathophysiologic changes of asthma, and dietary fat composition may alter prostanoid production. Epidemiologic studies from the 1960s and 1970s demonstrated a low incidence of asthma in populations whose diets were rich in fish oil, a robust source of the n-3 fatty acid eicosapentaenoic acid.[91,92] Clinical investigations of fish oil supplementation show changes in mediators released by inflammatory cells, but variable effects on clinical markers of disease.[93-95] A subsequent review of randomized trials failed to substantiate a therapeutic role for either dietary or supplemental eicosapentaenoic acid.[96]

Both inhaled and intravenous magnesium supplements have been suggested to augment standard β-agonist therapy to reverse bronchoconstriction in acute asthma. Although the efficacy of magnesium therapy has been more fully established in pediatric populations than in adults,[97] a very small study of adults with asthma showed that oral supplementation with magnesium improved objective measures of bronchial reactivity to methacholine, peak expiratory flow measurements, and subjective measures of asthma control and quality of life.[98]

Antioxidant supplementation has been suggested to modulate effects of airway injury in patients with asthma who are exposed to ozone and other air pollutants. Children with asthma in Mexico City were randomized to receive either a combined vitamin E and C supplement or placebo. Supplementation attenuated the loss of small airway function in those children with moderate-to-severe disease.[99] Subsequent investigations have yielded mixed results.[100,101]

Epidemiologic studies strongly link asthma and obesity[102] although our understanding of this relationship is far from complete. Adipose tissue releases proinflammatory mediators, including IL-6, leptin, and tumor necrosis factor-alpha, which promote systemic inflammation[103], thought to worsen airway obstruction. Elevated circulating IL-6 is associated with poor asthma control.[104] Efforts to help patients achieve and maintain an appropriate body weight is a critical component of asthma management.

Cystic Fibrosis

Cystic fibrosis is an autosomal recessive disorder characterized by impaired chloride transport, which affects various organs. Common manifestations include bronchiectasis, recurrent respiratory infections, sinus disease, and pancreatic dysfunction resulting in fat-soluble vitamin deficiency.

Historically, patients with CF, especially those with pancreatic insufficiency, struggled to achieve adequate nutrition and body weight. In addition to pancreatic dysfunction, which is present in 85% newborns with CF, an increased REE and abnormal cellular mitochondrial function due to CF gene dysfunction contribute to malnutrition.[105] Like patients with COPD, patients with CF have been shown to have 25% to 85% increase in REE, 25% to 80% attributed to the increased work of breathing.[106] Additionally, chronic infection, circulating inflammatory cytokines, catecholamines, and cortisol are elevated in CF, with concomitant increases in REE.[107] Antimicrobial treatment of pulmonary infections can reduce REE and promote weight gain,[106] but the chronicity and recurrence of infection historically made it difficult to achieve sustained improvements.

Because nutritional parameters, including BMI, correlate with lung function and outcomes,[108] assuring nutritional adequacy is paramount. Nutritional status should be assessed at the time of diagnosis and annually, as well as during times of clinical instability. Historically, it had been recommended that a patient with CF consume 120% to 150% of the recommended energy intake of age- and sex-matched controls[109] but those recommendations are changing as therapies advance. For those failing to achieve adequate weight gain and growth, nutritional supplementation, including the use of enteral feedings, may aid in weight maintenance and improve lung function.[110]

Pancreatic insufficiency decreases the absorption of fat-soluble nutrients such as β-carotene and vitamins K, A, D, and E. β-Carotene deficiency is common in patients with CF, and supplementation with pancreatic enzymes taken with meals and snacks is recommended.[111] Vitamin K deficiency typically occurs in conjunction with antibiotic use, thus, supplementation is recommended at those times.[112] However, with the introduction of modulator therapies, pancreatic restoration has been identified.[113]

Up to 50% of patients have vitamin A deficiency, leading to visual defects in up to 18%.[114] Low vitamin A concentration correlates with poor lung function and may have prognostic significance.[115] Serum measurements of vitamin A or retinol-binding protein should be obtained when the patient does not have an active infection state, because concentrations vary widely in the presence of inflammation.

Osteopenia and osteoporosis associated with vitamin D deficiency affect 24% of adults with CF.[116] Malabsorption, inadequate exposure to sunlight, and poor hepatic function also contribute to this deficiency. Periodic assessment of bone density and appropriate supplementation of both calcium and vitamin D is recommended to ensure bone health.

Vitamin E is highly fat soluble and correlates well with fat malabsorption. Severely deficient states are characterized by hemolytic anemia, neuromuscular degeneration, and cognitive deficits. Annual assessment of vitamin E concentration and supplementation, when deficient, are recommended.

Iron and zinc deficiency are common in CF. Iron supplements are indicated in the presence of iron deficiency anemia. Zinc supplementation is suggested in children who have failed to meet developmental landmarks.[109] Supplementation with n-3 fatty acids may be beneficial, but confirmatory research is necessary.[117]

The clinical outcomes for patients with CF have changed dramatically over the last decade, due to numerous advancements in therapies. Based on 2019 Cystic Fibrosis Foundation Registry data, the life expectancy of people with CF born between 2015 and 2019 is predicted to be 46 years compared with a life expectancy of 32 years for those born between 1995 and 1999.[118] The development of CF transmembrane conductance regulator (CFTR) modulators, which are engineered to correct the defective protein caused by the CFTR mutation,[119] have transformed the landscape of CF care but their full impact has not yet been realized. Mutations in the CFTR gene may result in a defective protein, insufficient protein formation, or no protein formation at all. Different modulators target specific CFTR mutations. The two classes of available modulators are potentiators, which help open the chloride channel to allow ion transfer to the cell surface, and correctors, which shape the protein to allow it to remain at the cell surface longer.[119] Ivacaftor, introduced in 2012, was the first modulator but it was indicated for a very select number of patients, based on the frequency of mutation it targeted, and it was modestly effective. Highly effective modulator triple therapy, elexacaftor-tezacaftor-ivacaftor (ETI), became commercially available in 2019. Triple therapy is profoundly effective and greatly expanded the number of patients for whom modulator therapy is effective. Today, 90% of the CF population has access to modulator therapy.[120] In addition to improvement in lung function, the two pivotal trials of ETI have shown increases in BMI 0·60 kg/m^2.[121,122] Previously described increases in REE in patients with CF appear to be mitigated by modulator therapy.[120]

With changes in CF therapies, especially with the addition of modulator therapies, the focus on assuring adequate weight gain has now shifted. Excess weight and obesity are becoming more prevalent with the CF Registry reporting an increase from 7% to 16% in overweight CF patients between 1998 and 2017.[120] Changing the dietary habits of patients who had been counseled since childhood to consume a high-fat, high-energy diet is challenging for CF teams and patients. Although some patients with CF still struggle with inadequate weight, dietary recommendations need to be individualized.[120]

Bronchiectasis

Bronchiectasis, an increasingly recognized cause of respiratory impairment, typically manifests with cough, fatigue, and respiratory infection and is characterized by structural changes to the airways. In a cohort of 496 patients enrolled in the United States Bronchiectasis Research Registry, 12.3% had BMI <18.5 kg/m^2.[123] In a prospective trial of 128 patients with bronchiectasis, mean total energy intakes were below estimated requirements and BMI correlated with lung function.[124] Compared with normal weight, underweight in bronchiectasis was associated with increased all-cause mortality (hazard ratio [HR] = 2.60; 95% confidence interval [CI] = 1.92-3.54) in 2,769 individuals with bronchiectasis.[125] Infections with nontuberculous mycobacteria are common in patients with bronchiectasis and low BMI is described more commonly in those with NTM.[126]

Pulmonary Fibrosis

Fibrotic lung disease is a common and, typically, progressive restrictive lung disease, which may result from a variety of identified insults or may be idiopathic. An understanding of the nutritional challenges of patients with fibrotic lung disease is limited by a paucity of investigations and the heterogeneity of disease. Evidence of malnutrition was seen in 30% of a single-center trial of patients with IPF,[127] and weight loss is associated with mortality.[128] Like patients with COPD, patients with IPF have ongoing inflammation, increased work of breathing, deconditioning and, often, hypoxemia that likely contributing to suboptimal nutritional status. Complicating the optimization of nutrition in patients with IPF is the gastrointestinal side effects commonly seen with the approved antifibrotic medications, nintedanib and pirfenidone, which may impair nutrition. In a series of 80 patients on antifibrotic therapy with one of these medications and at least 1 year of follow-up, 35 (44%) lost more than 5% of their baseline body weight, and 11 (19%) lost more than 10%.[129] It is important to highlight that this association does not imply causality, as progressive weight loss is not uncommon in patients with IPF.

In separate animal models, both impaired zinc transport[130] and a high-fat diet[131] contribute to pulmonary fibrosis, suggesting potential dietary strategies to mitigate disease development or progression.

Lung Cancer

In the United States, lung cancer is the leading cause of cancer deaths in both genders. Approximately 80% to 85% of cases in the United States are attributable to tobacco abuse, with a direct relationship between exposure intensity and cancer incidence.[132] However, only approximately 10% to 20% of heavy smokers develop lung cancer, and worldwide, nearly half of patients with lung cancer have never smoked.[133] Together, these findings suggest that factors other than tobacco abuse, including genetic, environmental, and dietary habits, likely contribute to cancer development.

Retinol, Carotenoids, and α-Tocopherol

Epidemiologic studies suggested that low intake and serum carotenoids or retinoids correlated with increased incidence of lung cancer. Initial studies found a greater disease-free interval and fewer new malignancies related to tobacco with the use of retinoids in the prevention of secondary tumors in patients with lung cancer, with no improvement in overall survival.[134] However, a subsequent randomized National Cancer Institute trial comparing isotretinoin (30 mg/day) with placebo for 3 years in 1,166 patients with lung cancer did not improve rates of recurrence or mortality.[135] In smokers, retinoids have been found to increase incidence of lung cancer, suggesting that retinoids could cause specific activation of cigarette carcinogen.[136,137]

Subsequently, the Alpha-Tocopherol, Beta-Carotene Cancer Prevention Group randomized 29,133 male smokers to α-tocopherol, β-carotene, both agents, or placebo.[138] Supplementation with α-tocopherol or β-carotene did not decrease the incidence of lung cancer. However, total mortality was 8% higher in those participants who received β-carotene than in those who did not, primarily from an increase in mortality from lung cancer and ischemic heart disease. In the Beta-Carotene and Retinol Efficacy Trial, former smokers, current smokers, and asbestos workers were randomized to receive β-carotene and vitamin A or placebo. The intervention group had a 1.28 greater relative risk for lung cancer development.[139] These troubling results raised still unresolved concerns regarding pro-oncogenic effects of β-carotene and prompted early termination of the trial. Finally, 1,024 asbestos workers were studied in South Africa in a non-blinded fashion, with a comparison made between supplemental β-carotene (30 mg) and retinol (25,000 IU). The incidence of malignant mesothelioma was significantly lower in the retinol group compared with the β-carotene group, but overall lung cancer incidence was the same.[140] A meta-analysis concluded that there is insufficient evidence to support the use of retinoids for either prevention or treatment of lung cancer.[141]

Alternatively, the Shanghai Women's Health prospective cohort study of 44,942 women reported that greater total dietary tocopherol was inversely associated with lung cancer risk, with the strongest protective effect at 14 mg/day or more of dietary tocopherol intake. In contrast, the use of vitamin E supplements was significantly correlated with an increased lung cancer risk.[142] Currently, β-carotene, α-tocopherol supplementation, or other studied supplements such as flavonoids, *N*-acetylcysteine, or isotretinoin cannot be recommended for lung cancer chemoprevention.

Fruits and Vegetables

Diets rich in fruits and vegetables may decrease the incidence of lung cancer. More than 10 studies have demonstrated the inverse relationship between high consumption of fruits, vegetables, breakfast cereals, and dietary fiber and the incidence of lung malignancy irrespective of tobacco history.[143-148] Special attention has been paid to intake of carrots and yogurt. Consumption of five or more carrots a week in women was associated with a decreased risk of lung cancer.[149,150] Dietary intake of fiber and yogurt was inversely associated to lung cancer, and additionally, high concurrent intake of both fiber and yogurt could potentially decrease the risk of lung cancer by 33%.[151]

Dietary Fat

Epidemiologic data in the 1980s serendipitously uncovered a potential relationship between dietary intake of saturated fat and lung cancer.[152] Subsequent prospective cohort studies demonstrated a positive association between total and saturated fat intake and lung cancer incidence[153-156] and attributable mortality. In a multi-country study including more than 12,000 participants, lung cancer mortality correlated with saturated fat consumption.[157] Conversely, replacement of saturated fat with polyunsaturated or monosaturated fat may reduce lung cancer risk.[156] Finally, specific polyunsaturated fatty acids potentially improve nutritional status and regulate inflammation in patients with lung cancer undergoing radiotherapy and chemotherapy.[158]

Malnutrition and loss of weight and fat-free mass is common in patients with lung cancer undergoing chemotherapy or radiation therapy, and malnutrition correlates with diminished quality of life.[159] However, the value of specific interventions to improve nutritional status in these patients remains uncertain.[160]

Lung Transplantation

Lung transplantation is a therapeutic option for patients with end-stage lung disease resulting from a variety of illnesses. Malnutrition, cachexia, protein wasting, and obesity are common before and after lung transplantation, and may directly impact a patient's clinical course.[161] Compared with normal-weight individuals, mortality was found to be 15% and 22% higher in overweight and obese patients.[161] However, 1 year mortality was not elevated in a multivariate analysis of 9,000 transplant recipients with a BMI of 30 to 34.9 kg/m^2. In that analysis, BMI was only 26% sensitive for total body fat defined obesity by whole body dual x-ray absorptiometry (DXA) scan, highlighting the limitations of use of BMI to define obesity.[162] Acknowledging the limitations of BMI, other tools, including lean psoas area, creatinine-height index, and leptin, in addition to whole body DXA, are being investigated to improve the nutritional risk assessment pretransplant.[163] The Nutritional Risk Index (NRI) pretransplant is a validated tool to predict post-transplant morbidity and mortality.[164]

Cognizant that BMI and nutritional parameters directly impact transplant outcomes and listing eligibility, attempts to optimize those factors is routinely incorporated into pretransplant care. It is unclear whether improving malnutrition perioperatively improves outcomes[165,166] and delaying lung transplantation to improve nutritional status may adversely affect the outcome.[167] In overweight or obese patients, weight reduction alone may improve function sufficiently to delay or obviate the need for transplantation.[168] Current international guidelines for transplant recipient selection identify a BMI $\geq$35 kg/m^2 or $<$16 kg/m^2 in transplant candidates as a risk factor with a high or substantially increased risk; transplant for these patients should only be considered in specialized centers.[169]

Diminished bone density is more common in patients listed for lung transplant than matched controls[170] and, when severe, may affect listing eligibility. Nutritional supplementation, in addition to pharmacologic therapy and exercise, is indicated pretransplant to improve bone density.

Lung transplantation may reverse malnutrition and greater weight gain following transplantation has been associated with improved survival.[171,172] However, elevated BMI or obesity occurs in approximately 30% of patients following lung transplant and increases risk of obesity-associated diseases.[173] Thus, measures to optimize body composition and weight post-transplant are critical. The impact of weight and the nutritional status of lung donors is insufficiently understood. However, preliminary data suggest that lungs obtained from obese donors may be associated with suboptimal outcomes.[174]

Other Clinical Considerations

Compromised nutritional status may indirectly compromise respiratory function. For example, protein malnutrition causing hypoalbuminemia alters the threshold for transudation of fluid into the lung parenchyma and pleural space and results in the development of pulmonary edema and pleural effusions, which impairs gas exchange and increases the work of breathing.

Patients with malignant disease metastatic to the pleura may drain large amounts of protein into the pleural space. Repeated thoracentesis to drain this fluid results in severe protein wasting. Patients with disruption of the thoracic duct develop a chylothorax, and may lose massive amounts of protein, fat, and electrolytes into the pleural space. Parenteral alimentation or oral medium-chain triglycerides are often beneficial in replacing the lost nutrients.

Summary

Nutritional factors, weight, and body composition may all impact the structure and function of the respiratory system as well as the risk and outcomes of various respiratory illnesses. Much of our understanding of these relationships is derived from observational studies demonstrating associations. These studies are often limited by incomplete dietary recall, use of BMI to define underweight or obese status, or focus on a selected dietary component. Despite these limitations, there are clear associations between weight, nutrition, and lung health. Future controlled explorations of the protective or therapeutic role of nutritional interventions offer exciting opportunities to optimize respiratory health. Maturation of our understanding of individual differences, guided by developments in personalized medicine coupled with enhanced comprehension of the microbiome, will likely be integral to these advances.

Acknowledgments

The authors acknowledge Neal M. Patel, for his contribution to the previous version of this chapter.

References

1. Roussos C, Macklem PT. N Engl J Med 1982;307:786–97.
2. Murray JF. The Normal Lung: The Basis for Diagnosis and Treatment of Pulmonary Disease. 2nd ed. Philadelphia, PA: WB Saunders, 1986.
3. Arigliani M, Spinelli AM, Liguoro I, Cogo P. 2018;10:919.
4. Lechner AJ, Winston DC, Bauman JE. J Appl Physiol 1986;60:1610–4.
5. Kalenga M, Shaheen S, Barker DJ. Thorax 1994;49:533–6.
6. Briana DD, Malamitsi-Puchner. Paediatr Respir Rev 2013;14: 256–262.
7. Pike K, Pillow JJ, Lucas JS. Semin Fetal Neonatal Med 2012:17: 92–98.
8. Shaheen S, Barker DJ. Thorax 1994;49:533–6.
9. Ehrenkranz RA, Dusick AM, Vohr BR, et al. Pediatr 2006;117: 1253–1261.
10. Lewis MI, Belman MJ. Clin Chest Med 1988;9(2):337–48.
11. Lewis MI, Sieck GC, Fournier M, et al. J Appl Physiol 1986;60:596–603.
12. Oldfors A, Mair WG, Sourander P. J Neurol Sci 1983;59:291–302.
13. Goldberg AL, Odessey R. Am J Physiol 1972;223:1384–91.
14. Doekel RC Jr, Zwillich CW, Scoggin CH, et al. N Engl J Med 1976;295:358–61.
15. Bair H, Somani P. Chest 1984;85:222–5.
16. Ryan CF, Whittaker JS, Road JD. Chest 1992;102:1286–8.
17. Moriguchi S, Sone S, Kishino Y. J Nutr 1983;113:40–6.
18. Shennib H, Chiu RC, Mulder DS, et al. Surg Gynecol Obstet 1984;158:535–40.
19. Niederman MS, Merrill WW, Ferranti RD, et al. Ann Intern Med 1984;100:795–800.
20. Harris JA, Benedict FG. Standard Basal Metabolism Constants for Physiologists and Clinicians: A Biometric Study of Basal Metabolism in Man. Philadelphia, PA: JB Lippincott, 1919:223.
21. Pirat A, Tucker AM, Taylor KA, et al. Respir Care 2009;54: 487–94.
22. Liggett SB, Renfro AD. Chest 1990;98:682–6.
23. Singer P, Blaser AR, Berger MM, et al. Clin Nutr. 2019;38:48–79.
24. Damask MC, Schwarz Y, Weissman C. Crit Care Clin 1987;3: 71–96.
25. Liebau F, Deane AM, Rooyackers O. Curr Opin Clin Nutr Metab Care. 2021;24:71–78.
26. Puthucheary ZA, Rawal J, McPhail M, et al. JAMA 2013;310: 1591–1600.
27. Edens NK, Gil KM, Elwyn DH. Clin Chest Med 1986;7:3–17.
28. Singer P, Hiesmayr M, Biolo G, et al. Clin Nutr 2014;33:246–51.
29. Compher C, Bingham AL, McCall M, et al. JPEN J Parenter Enteral Nutr 2022;46:12.
30. van den Berg B, Bogaard JM, Hop WC. Intensive Care Med. 1994;20:470.
31. Alberda C, Gramlich L, Jones N, et al. Int Care Med 2009;35: 1728–37.
32. Needham DM, Davidson J, Cohen H, et al. Crit Care Med 2012;40:502–509.
33. van Zanten AR, De Waele E, Wischmeyer PE. Crit Care 2019;23:368.
34. Heighes PT, Doig GS, Simpson F, et al. Anaesth Int Care 2010;38:167–74.
35. Alkhawaja S, Martin C, Butler RJ, Gwadry-Sridhar F. Cochrane Database Syst Rev 2015;8:CD008875.
36. Williams TA, Leslie GD. Intensive Crit Care Nurs 2005;21:5–15.
37. Ford EG, Hull SF, Jennings LM, Andrassy RJ. J Am Coll Nutr 1992;11:11.
38. Hart GK, Dobb GJ. JPEN J Parenter Enteral Nutr 1988;12:465.
39. Barraud D, Bollaert PE, Gibot S. Chest 2013;143:646.
40. Wiedemann HP, Wheeler AP. N Engl J Med 2006;354:2564–75.
41. Imperial J, Bistrian BR, Bothe A Jr, et al. J Am Coll Nutr 1983;2: 63–73.
42. Kruse JA, Shah NJ. Nutr Clin Pract 1993;8:163–70.
43. Metnitz PG, Bartens C, Fischer M, et al. Int Care Med 1999;25:180–5.
44. Richard C, Lemonnier F, Thibault M, et al. Crit Care Med 1990;18:4–9.
45. Nathens AB, Neff MJ, Jurkovich GJ, et al. Ann Surg 2002;236: 814–22.
46. Dushianthan A, Cusack R, Burgess VA, Grocott MP, Calder PC. Cochrane Database Syst Rev 2019;1.
47. Aqeel M. Shahryar Ahmad AM, Pate JJ, Rice TW. Curr Pulmonol Rep 2017;6:113–123.
48. Rice TW, Wheeler AP, Thompson BT, et al. JAMA 2012;;307: 795–803.
49. Wheaton AG, Cunningham TJ, Ford ES, Croft JB; Centers for Disease Control and Prevention (CDC). MMWR Morb Mortal Wkly Rep. 2015;64:289–95.
50. Ruvuna L, Sood A. Clin Chest Med 2020;41:315–327.
51. Ahmad FB, Anderson RN. JAMA. 2021;325:1829–1830.
52. Løkke A, Lange P, Scharling H, Fabricius P, Vestbo J. Thorax 2006;61:935–9.
53. Coxson HO, Chan IH, Mayo JR, et al. Am J Respir Crit Care Med. 2004;170:748–52.
54. Chuwers P, Barnhart S, Blanc P, et al. Am J Respir Crit Care Med 1997;155:1066–71.
55. Morabia A, Menkes MJ, Comstock GW, et al. Am J Epidemiol 1990;132:77–82.
56. Kaluza J, Larsson SC, Orsini N, Linden A, Wolk A. Thorax 2017;72:500–509.
57. Kaluza J, Harris HR, Linden A, Wolk A. Int J Epidemiol. 2018:47:1897–1909.
58. Shahar E, Folsom AR, Melnick SL, et al. Am J Respir Crit Care Med 1994;150:978–82.
59. Shahar E, Folsom AR, Melnick SL, et al. N Engl J Med 1994;331:228–33.
60. Sharp DS, Rodriguez BL, Shahar E, et al. Am J Respir Crit Care Med 1994;150:983–7.
61. Schwartz J, Weiss ST. Am J Clin Nutr 1994;59:110–4.
62. Jiang R, Paik DC, Hankinson JL, Barr RG. Am J Respir Crit Care Med 2007;175:798–804.
63. Scoditti E, Massaro M, Garbarino S, Toraldo DM. Nutrients. 2019;11:1357.
64. Collins PF, Elia M, Stratton RJ. Respirol 2013;18:616.
65. Chailleux E, Laaban JP, Veale D. Chest 2003;123:1460–6.
66. Thomas DR. Clin Geriatr Med 2002;18:835–9.
67. Hallin R, Janson C. Respir Med 2006;100:561–7.
68. Keogh E, Williams ME. Respir Med. 2021;176:106248.
69. Machado FV, Schneider LP, Fonseca J, et al. Eur J Clin Nutr 2019;73:1512–1519.
70. McDonald MN, Wouters EFM, Rutten E, et al. Respir Res 2019;20:100.
71. Donahoe M, Rogers RM, Wilson DO, et al. Am Rev Respir Dis 1989;140:385–91.

72. Vernooy JH, Kucukaycan M, Jacobs JA, et al. Am J Respir Crit Care Med 2002;166:1218–24.
73. Eid AA, Ionescu AA, Nixon LS, et al. Am J Respir Crit Care Med 2001;164:1414–8.
74. Di Francia M, Barbier D, Mege JL, et al. Am J Respir Crit Care Med 1994;150:1453–5.
75. Blakemore A, Dickens C, Chew-Graham CA, et al. Int J Chron Obstruct Pulmon Dis 2019;14:1343–1353.
76. King DA, Cordova F, Scharf SM. Proc Am Thorac Soc 2008;5:519–23.
77. 2022 GOLD Reports - Global Initiative for Chronic Obstructive Lung Disease website. Accessed August 29, 2022. Available at https://goldcopd.org/2022-gold-reports-2/
78. Slinde F, Gronberg AM, Engstrom CR, et al. Respir Med 2002;96:330–6.
79. Creutzberg EC, Schols AM, Weling-Scheepers CA, et al. Am J Respir Crit Care Med 2000;161:745–52.
80. Suchner U, Rothkopf MM, Stanislaus G, et al. Arch Intern Med 1990;150:1225–30.
81. Casaburi R, Porszasz J, Burns MR, et al. Am J Respir Crit Care Med 1997;155:1541–51.
82. Burdet L, de Muralt B, Schutz Y, et al. Am J Respir Crit Care Med 1997;156:1800–6.
83. Pape GS, Friedman M, Underwood LE, et al. Chest 1991;99: 1495–500.
84. Rudman D, Fellor AG, Angraj HS. N Engl J Med 1991;323:1–6.
85. Schols AM, Soeters PB, Mostert R, et al. Am J Respir Crit Care Med 1995;152:1268–74.
86. Weisberg J, Wanger J, Olson J, et al. Chest 2002;121:1070–8.
87. Herrejon A, Palop J, Inchaurraga I, et al. Med Clin (Barc). 2011;137:193–8.
88. Rafiq R, Aleva FE, Schrumpf JA, et al. Am J Clin Nutr. 2022;116:491–499.
89. Farah R, Khamisy-Farah R, Arraf Z, Jacobson L, Makhoul N. Clin Respir J. 2013;7:407–15.
90. World Health Organization. Asthma. Accessed August 28, 2022. Available at www.who.int/news-room/fact-sheets/detail /asthma
91. Kromann N, Green A. Acta Med Scand 1980;208:401–406.
92. Schwartz J, Weiss ST. Eur Respir J 1994;7:1821–4.
93. Arm JP, Horton CE, Spur BW, et al. Am Rev Respir Dis 1989;139:1395–400.
94. Thien FC, Mencia-Huerta JM, Lee TH. Am Rev Respir Dis 1993;147:1138–43.
95. Nagakura T, Matsuda S, Shichijyo K, et al. Eur Respir J 2000;; 16:861–5.
96. Woods RK, Thien FC, Abramson MJ. Cochrane Database Syst Rev 2002:CD001283.
97. Ciarallo L, Sauer AH, Shannon MW. J Pediatr 1996;129:809–14.
98. Kazaks AG, Uriu-Adams JY, et al. J Asthma 2010;47:83–92.
99. Romieu I, Sienra-Monge JJ, Ramirez-Aguilar M et al. Am J Respir Crit Care Med 2002;166:703–9.
100. Pearson PJ, Lewis SA, Britton J, Fogarty A. Thorax 2004;59:652.
101. Trenga CA, Koenig JQ, Williams PV. Arch Environ Health 2001;56:242.
102. Dixon AE, Poynter ME. Am J Respir Cell Mol Biol 2016;54: 601–608.
103. Tilg H, Moschen AR. Nat Rev Immunol 2006;6:772–783.
104. Peters MC, McGrath KW, Hawkins GA, et al. Lancet Respir Med 2016;4:574.
105. Feigal RJ, Shapiro BL. Nature 1979;278:276-7.
106. Bell SC, Saunders MJ, Elborn JS, et al. Thorax 1996;51:126–31.
107. Elborn JS, Cordon SM, Western PJ, et al. Clin Sci (Lond) 1993;85:563–8.
108. Corey M, McLaughlin FJ, Williams M, et al. J Clin Epidemiol 1988;41:583–91.
109. Borowitz D, Baker RD, Stallings V. J Pediatr Gastroenterol Nutr 2002;35:246–59.
110. Steinkamp G, von der Hardt H. J Pediatr 1994;124:244–7.
111. Renner S, Rath R, Rust P, et al. Thorax 2001;56:48–52.
112. Beker LT, Ahrens RA, Fink RJ, et al. J Pediatr Gastroenterol Nutr 1997;24:512–7.
113. Shteinberg M, Haq IJ, Polineni D, Davies JC. Lancet 2021;397: 2195–2211.
114. Rayner RJ, Tyrrell JC, Hiller EJ, et al. Arch Dis Child 1989;64:1151–6.
115. Duggan C, Colin AA, Agil A, et al. Am J Clin Nutr 1996;64: 635–9.
116. Cystic Fibrosis Foundation. Cystic Fibrosis Foundation Patient Registry: 2020 annual data report. Bethesda, MD: Cystic Fibrosis Foundation; 2021. Accessed August 27, 2022. Available at www .cff.org/sites/default/files/2021-11/Patient-Registry-Annual -Data-Report.pdf
117. Innis SM, Davidson AG. Annu Rev Nutr 2008;28:55–72.
118. Cystic Fibrosis Foundation. United States. [Web Archive] Retrieved from the Cystic Fibrosis Foundation. Accessed August 27, 2022. Available at www.cff.org/managing-cf /understanding-changes-life-expectancy
119. Leonard A, Schindler T, Padula L, et al. Cystic Fibrosis Nutrition 101: Getting Started. CF Nutrition 101. Accessed August 18, 2022. Available at https://my.cff.org/cfx-docrep/document/detail /CF%20Nutrition%20101.pdf
120. Bailey J, Krick S, Fontaine KR. Nutrients 2022;14:1216.
121. Heijerman HG, McKone EF, Downey DG, et al. Lancet 2019; 394:1940–1948.
122. Middleton PG, Mall MA, Drevinek P, et al. N Engl J Med 2019;381:1809–1819.
123. Despotes KA. Chron Obstruct Pulmon Dis 2020;7:390–403.
124. King L, White H, Clifton I, et al. Clin Nutr 2021;40: 5162–5168.
125. Lee JM, Lee SA, Han CH, et al. Respir Med 2021;180:106370.
126. Aksamit TR, O'Donnell AE, Barker A, et al. Chest 2017;151: 982–992.
127. Jouneau S, Kerjouan M, Rousseau C, et al. Nutr 2019;62: 115–121.
128. Nakatsuka Y, Handa T, Kokosi M, et al. Respir 2018;96: 338–347.
129. Perelas A, Glennie J, van Kerkhove K, et al. Pulm Pharmacol Ther 2019:59:101839.
130. Liang J, Huang G, Liu X, et al. J Clin Invest 2022;132:e157338.
131. Chu SG, Villalba JA, Liang X, et al. Am J Respir Cell Mol Biol 2019;61:737–746.
132. Mattson ME, Pollack ES, Cullen JW. Am J Public Health 1987;77:425.
133. Sung H, Ferlay J, Siegel RL, et al. CA Cancer J Clin 2021;71: 209–249.
134. Pastorino U, Infante M, Maioli M, et al. J Clin Oncol 1993;11: 1216–22.
135. Lippman SM, Lee JJ, Karp DD, et al. J Natl Cancer Inst 2001; 93:605–18.
136. De Vries N, Van Zandwijk N, Pastorino U. Am J Otolaryngol 1993;14:62–6.
137. Bushue N, Wan YJ. Adv Drug Deliv Rev 2010;62:1285–98.
138. Alpha-Tocopherol, Beta-Carotene Cancer Prevention Study Group. N Engl J Med 1994;330:1029–35.
139. Omenn GS, Goodman GE, Thornquist MD, et al. N Engl J Med 1996;334:1150–5.
140. de Klerk NH, Musk AW, Ambrosini GL, et al. Int J Cancer 1998; 75:362–7.
141. Fritz H, Kennedy D, Fergusson D, et al. PLoS ONE 2011;6: e21107.
142. Wu QJ, Xiang YB, Yang G, et al. Int J Cancer 2015;136:610–7.

143. Brennan P, Fortes C, Butler J, et al. Cancer Causes Control 2000;11:49–58.
144. Jansen MC, Bueno-de-Mesquita HB, Rasanen L, et al. Int J Cancer 2001;92:913–8.
145. Voorrips LE, Goldbohm RA, Verhoeven DT, et al. Cancer Causes Control 2000;11:101–5.
146. Wei X, Zhu C, Ji M, et al. J Clin Nutr 2021:nqab298.
147. Tu H, Heymach JV, Wen CP, et al. Sci Rep 2016;6:26760.
148. De Stefani E, Deneo-Pellegrini H, Ronco AL. Asian Pac J Cancer Prev. 2014;15:2765–9.
149. Speizer FE, Colditz GA, Hunter DJ, et al. Cancer Causes Control 1999;10:475–82.
150. Rachtan J. Acta Oncol 2002;41:389–94.
151. Yang JJ, Yu D, Xiang YB, et al. JAMA Oncol 2020;6:e194107.
152. Byers TE, Graham S, Haughey BP, et al. Am J Epidemiol 1987;125:351–63.
153. Hinds MW, Kolonel LN, Lee J, et al. Am J Clin Nutr 1983;37: 192–3.
154. Jain M, Burch JD, Howe GR et al. Int J Cancer 1990;45:287–93.
155. Stefani ED, Boffetta P, Deneo-Pellegrini H, et al. Nutr Cancer 1999;34:100–10.
156. Yang JJ, Yu D, Takata Y, et al. Clin Oncol. 2017;35:3055–3064.
157. Mulder I, Jansen MC, Smit HA, et al. Int J Cancer 2000;88: 665–71.
158. Tao X, Zhou Q, Rao Z. Int J Clin Pract 2022;2022:6564466.
159. Polanski J, Jankowska-Polanska B, Mazur G. Cancer Manag Res 2021;13:1407–1416.
160. Kiss N, Curtis A. Adv Nutr (Bethesda). 2022:nmac070.
161. Lederer DJ, Wilt JS, D'Ovidio F, et al. Amer Journal of Resp Crit Care Med 2009. 180:887–95.
162. Singer JP, Peterson ER, Snyder ME, et al. Am J Respir Crit Care Med. 2014 Nov;190:1012–21.
163. Weber Gulling M, Schaefer M, Bishop-Simo L, Keller BC. Nutrients 2019;11:2884.
164. Bigelow B, Toci G, Etchill E, et al. Ann Thorac Surg 2021;112: 214–220.
165. Forli L, Bjortuft O, Vatn M, et al. Ann Nutr Metab 2001;45: 159–68.
166. Forli L, Pedersen JI, Bjortuft O, et al. Respir 2001;68: 51–7.
167. Snell GI, Bennetts K, Bartolo J, et al. J Heart Lung Transplant 1998;17:1097–103.
168. Forsythe J, Cooley K, Greaver B. Prog Transplant 2000;10: 234–8.
169. Leard LE, Holm AM, Valapour M, et al. J Heart Lung Transplant 2021;40:1349–1379.
170. Lakey WC, Spratt S, Vinson EN, et al. Clin Transplant 2011;25:426–35.
171. Kyle UG, Nicod L, Romand JA, et al. Transplantation 2003;75: 821–8.
172. Singer LG, Brazelton TR, Doyle RL, et al. J Heart Lung Transplant 2003;22:894–902.
173. Bossenbroek L, den Ouden ME, de Greef MH, et al. Respir 2011; 82:28–35.
174. Okamoto T, Ayyat KS, Sakanoue I, et al. J Heart Lung Transplant 2022;41:818–828.

CHAPTER 99

Nutrition and Infectious Diseases

Alice M. Tang and Ellen Smit

CHAPTER OUTLINE

Abbreviations

AIDS, acquired immunodeficiency syndrome; **ARI**, acute respiratory infection; **ART**, antiretroviral therapy; **DALY**, disability-adjusted life year; **HAART**, highly active antiretroviral therapy; **HIV**, human immunodeficiency virus; **IFN**, interferon; **IL**, interleukin; **MHC**, major histocompatibility complex; **NF-κB**, nuclear factor-κB; **PWH**, people living with HIV; **TB**, tuberculosis; **TNF**, tumor necrosis factor; **UNICEF**, United Nations Children's Fund; **VDR**, vitamin D receptor; **WHO**, World Health Organization

Historical Overview

The relationship between nutrition and infectious diseases was largely characterized in the 20th century. In the early part of the century, the idea of certain "substances," "accessory food factors," or "vitamins" essential for health emerged[1-3]; and knowledge of vitamins and deficiency diseases grew rapidly in the period following 1912.[4] The study of nutrition and infectious diseases was facilitated by developments in immunology, such as the description of humoral antibodies, serologic tests for the diagnosis of infectious diseases, and assays for measurements of immunologic protection.[5] By the late 1930s and early 1940s, it became generally accepted that some dietary deficiencies could increase the risk of infectious diseases.[6-8] With fortification of foods, improvement in diet, and a general increase in the standard of living, micronutrient deficiencies declined and became less of a public health problem in developed countries.

In 1968, a World Health Organization (WHO) expert committee reviewed the interactions between nutrition and infection[9] and concluded the following:

> "Infections are likely to have more serious consequences among persons with clinical or subclinical malnutrition, and infectious diseases have the capacity to turn borderline nutritional deficiencies into severe malnutrition. In this way, malnutrition and infection can be mutually aggravating and produce more serious consequences for the patient than would be expected from a summation of the independent effects of the two."

This work provided the foundation for research on nutrition and infection that has followed to the present, much of which has taken place in low- and middle-income countries, where micronutrient deficiencies are more prevalent.

General Principles

Infection with a pathogen usually triggers a series of responses; these vary depending on the infectious agent. Response to bacterial infection commences with the release of microbial products such as lipopolysaccharides and peptidoglycans from bacterial cell walls, bacterial DNA, and exotoxins; viral infections start with the release of viral double-stranded RNA and viral glycoproteins. Depending on the type of pathogen and its ability to evade the immune response, the location of the infection, host immune status, and other factors, a localized response with influx of polymorphonuclear leukocytes, macrophages, and natural killer cells may occur. Polymorphonuclear cells release inflammatory mediators from granules, including reactive oxygen intermediates such as hydroxyl radicals, hydrogen peroxide, reactive nitrogen, and superoxide anion, and antimicrobial enzymes such as lysozyme, proteases, collagenases, and phospholipases. Macrophages may phagocytize antigens and express cytokines such as interleukin (IL)-1 that attract T lymphocytes to the site of inflammation. The complement system may be activated. If the pathogen is not contained, these localized events may amplify and lead to a larger systemic condition—the acute phase response—which is characterized by fever, somnolence, anorexia, and cachexia and is accompanied by metabolic alterations such as loss of muscle and negative nitrogen balance, fat catabolism, and impaired gluconeogenesis. The acute phase response can also be triggered by other factors, including trauma, burns, surgery, tissue infarction, exposure to chemicals or radiation, and advanced cancer.

During the acute-phase response, changes occur in many plasma proteins, or acute-phase proteins, which have been defined as proteins whose plasma concentrations increase (positive acute-phase proteins) or decrease (negative acute-phase proteins) by at least one-fourth during inflammatory disorders.[10,11] The liver is the main site of synthesis of most acute phase proteins. C-reactive protein,[10] α-1-acid glycoprotein,[12] and serum amyloid A[13] are among the best characterized positive acute-phase proteins. Other positive acute-phase proteins include ferritin, haptoglobin, and ceruloplasmin, and negative acute phase proteins include albumin and transthyretin.

Cytokines play an important role in the acute phase response as inducers of acute phase proteins and hormones, as modulators of inflammation, and as activators and inhibitors of central nervous system functions related to appetite and metabolism.[14] Cytokines that have been implicated in the anorexia that occurs during infection include IL-1β, IL-6, and tumor necrosis factor-α (TNF-α).[14] IL-1β and TNF-α are produced primarily by macrophages and can be triggered by stimuli such as bacterial wall products, lipopolysaccharide, viruses, parasites, and microbial superantigens. IL-6 is produced by monocytes, T lymphocytes, endothelial cells, and fibroblasts and is a major stimulator of the production of acute phase proteins in the liver. These cytokines may have a direct effect on hypothalamic neurons involved in appetite and other neural mediators, such as serotonin, corticotropin-releasing factor, and β-melanocyte–stimulating hormone, and decreases in dopamine and neuropeptide Y. There may be increased cytokines in the circulation as well as the brain that modulate appetite during the acute phase response.[15]

Malnutrition and Specific Infectious Diseases

The nutritional consequences of infection, no matter which micro-organism is causing it, tend to be predictable. Any infection, whether symptomatic or asymptomatic, is accompanied by losses of some nutrients from the body and redistribution of other nutrients. The magnitude of these changes depends on the severity and duration of the infection. Metabolic and nutritional responses that are specific to certain organisms occur when the infection becomes localized within a single organ system. For example, diarrheal infections cause sizeable losses of fluid and electrolytes, whereas paralytic forms of infection result in wasting of bone and muscle. If the infection can be cured or eliminated naturally by the host immune system, lost body nutrients can then be replenished over a period of weeks to months. If the infectious process is not eliminated and becomes chronic, however, body composition can become markedly altered, and a new equilibrium of body nutrient balances is reached at a cachectic, or extremely wasted, level.

The impact of malnutrition on the severity of infection has been investigated most extensively in children with measles, diarrheal disease, respiratory infections, and malaria, and in children and adults with tuberculosis (TB) and human immunodeficiency virus (HIV) infection.

Measles

Measles is a highly contagious disease caused by a virus. An estimated 142,000 measles-related deaths were reported during 2018, mostly children younger than 5 years of age.[16] The number of deaths in 2018 reflects a decrease of 73% since 2000, after the launch of the Measles Initiative, a vaccination campaign in high-risk countries jointly led by the American Red Cross, United Nations Foundation, U.S. Centers for Disease Control and Prevention, United Nations Children's Fund (UNICEF), and WHO. Most measles deaths continue to occur in parts of Africa and Asia.[16] Deaths from measles are largely the result of an increased susceptibility to secondary bacterial and viral infections, and the underlying mechanism includes immune suppression related to malnutrition, especially vitamin A deficiency.[17] Complications may occur in 10% to 30% of cases and include pneumonia, diarrhea, malnutrition, otitis media, mouth ulcers, corneal epithelial keratitis, corneal ulceration, and blindness. During the COVID-19 pandemic, many countries put off the measles vaccination campaigns for 2020 and 2021. This may result in an increase of measles outbreaks globally.[18]

Measles virus is an enveloped RNA virus in the genus *Morbillivirus* of the family paramyxoviridae. One known serotype and eight clades (A to H) of measles exists. Measles is spread when a susceptible person inhales aerosolized droplets containing measles virus. Viral replication occurs initially in macrophages in the lymphoid tissue of the nasopharynx, respiratory mucosa, and lungs.[19] The measles virus enters human cells through a cell surface receptor known as signaling lymphocyte activation molecule.[20] Viremia allows measles virus to spread to multiple organs, including the skin, liver, and conjunctiva, and a prodrome of fever, cough, and conjunctivitis occurs approximately 14 days after infection.

The immune response to measles is thought to be consistent with T-helper type 2–like immune responses in which antibody responses predominate and are driven by IL-4, IL-6, and IL-10. Antibody responses against measles virus proteins are detectable at the onset of the rash. Immune suppression often accompanies measles infection and increases the susceptibility to secondary infections. Delayed-type hypersensitivity skin test responses and in vitro proliferation of lymphocytes to viral antigens are often minimal or absent in measles infection.[19] Infants are protected against measles virus infection by passively acquired maternal antibody to measles.

Although most people recover from measles, those with malnutrition or coinfections are at increased risk of complications.[21-24] In the classic early investigation of a measles outbreak in the Faroe Islands by Panum and Manicus in 1846, the most severe diarrheal disease and highest mortality were described among those patients with greatest poverty and poor diet.[25] Malnourished children have more severe disease and higher mortality.[26,27] More persistent measles infection and viral shedding also have[23] been reported in malnourished children.[21] A close synergism exists between measles and vitamin A deficiency. One study in Zaire demonstrated that severely vitamin A-deficient children were three times more likely to die of measles than children with better vitamin A status before infection.[28] Children with measles who are vitamin A deficient also have a much higher risk of xerophthalmia, corneal ulceration, keratomalacia, and subsequent blindness.[29]

Vitamin A supplementation has consistently been shown to reduce the morbidity and mortality of measles among preschool children.[30-34] Vitamin A supplementation appears to reduce the infectious complications associated with measles immune suppression, such as pneumonia and diarrheal disease, and these effects have been associated with modulation of immune responses by vitamin A.[35,36] In a randomized controlled trial in South Africa, children with severe measles who received vitamin A supplementation recovered more quickly from complications and were hospitalized for significantly fewer days. Their risk of death or a major complication was half that of their untreated counterparts.[37] A meta-analysis of studies in the United Kingdom, South Africa, and Tanzania showed that vitamin A therapy for measles reduced mortality by 67%.[38] A meta-analysis of five randomized controlled clinical trials found that at least two doses of 200,000 IU of vitamin A on consecutive days were needed for significant reduction in measles mortality. Although one dose did not have a significant effect, two doses resulted in a 62% reduction in mortality.[39] Furthermore, updated meta-analyses continue to show the benefits of vitamin A supplementation on measles mortality reduction in children under 5 years of age.[30] High-dose vitamin A therapy continues to be accepted as part of standard recommended treatment of measles infection in low- and middle-income countries.[16] Improving vitamin A status and immunization coverage are also cornerstones of current measles prevention efforts.

Malaria

Malaria is caused by a parasitic infection with protozoan organisms of the genus *Plasmodium*. In 2020, there were an estimated 241 million cases of malaria and approximately 627,000 deaths from malaria worldwide. More than 95% of malaria cases occur in Africa and 80% of all malaria deaths were among children under 5 years old.[40] Of the four *Plasmodium* species that infect humans (*P. falciparum, P. vivax, P. malariae*, and *P. ovale*), the most serious morbidity and mortality are caused by *P. falciparum*.[41] Malarial infection begins when a female *Anopheles* mosquito bites a human and releases sporozoites from the salivary gland into the circulating blood of the host. Sporozoites then enter hepatocytes and, in the pre-erythrocytic cycle, sporozoites grow and develop in the liver into thousands of merozoites. The hepatocytes then release merozoites, which invade erythrocytes. In the asexual erythrocytic cycle, merozoites develop into schizonts, and erythrocytes rupture, releasing more merozoites that can invade erythrocytes. In the sexual stage, some merozoites develop into gametocytes that are taken up in a blood meal by a mosquito, and within the mosquito, these gametocytes develop into microgametes and macrogametes.

In addition to the type of parasite, the extent of transmission and illness depends on the species of the *Anopheles* vector, human immunity, and climate. Specifically, vector species that have a long lifespan and species that prefer to bite humans over animals increase the chances of human transmission. Adults can develop some, but not complete immunity to malaria after years of exposure, and transmission is highest in higher temperature and humid conditions.[42] Transmission can be significantly reduced by insecticide-treated mosquito nets and indoor spraying with insecticides. Vaccination of all children living in high risk regions has been recommended by the WHO since 2021.[40]

Malaria symptoms include fever, chills, and flu-like illness and, if left untreated, can progress to severe illness and death. While treatment generally includes artemisinin-based combination therapy, resistance to antimalarial drugs has become more evident over the past 10 years.[40] Natural history studies suggest that the severity of malaria may be related to nutritional status. Some studies in the older literature have reported that malnutrition, and specifically protein energy malnutrition, is protective against malaria. In particular, refeeding after famine appeared to increase the risk of malaria, especially among persons who are carriers of the disease.[43,44] Nevertheless, some reviews on protein energy malnutrition show that persons with better nutritional status have less severe malaria and lower risk of death.[41,45-47]

Specific nutrients have been shown to affect malaria morbidity. Vitamin A, in particular, influences the incidence as well as the morbidity related to malaria.[48-51] An early randomized, placebo-controlled clinical trial was conducted in Papua New Guinea to examine the effects of vitamin A supplementation (60 mg retinol equivalents every 3 months) on malarial morbidity in preschool children.[52] Children between 6 and 60 months of age were randomly allocated to receive vitamin A or placebo every 3 months. A weekly morbidity surveillance and clinic-based surveillance were established for monitoring acute malaria, and children were followed for 1 year. Vitamin A significantly reduced the incidence of malaria attacks by about 20% to 50% for all except those with extremely high levels of parasitemia. Similarly, vitamin A supplementation reduced clinic-based malaria attacks, which consisted of self-solicited visits to the clinic by mothers who thought that their children should be seen because of fever. Vitamin A supplementation had little impact in children less than 12 months of age and the greatest effect among children aged 13 to 36 months. Recent studies have found that children with severe malaria had lower vitamin A concentration,[51] and vitamin A supplementation reduced the development of clinical malaria in women living with HIV.[53]

Studies of the benefits of zinc on malaria incidence and morbidity have shown conflicting results. Several studies have shown that zinc supplementation reduces the morbidity of malaria.[54-58] In the Gambia, 110 children were randomized to receive 70 mg zinc or placebo twice weekly for 1.24 years. Zinc supplementation was associated with a 30% reduction in the incidence of clinical malaria episodes.[54] In Papua New Guinea, 274 preschool children, aged 6 to 60 months, received 10 mg zinc or placebo/day for 46 weeks. Zinc supplementation resulted in a 38% reduction in malaria morbidity.[58] Other randomized controlled trials of zinc supplementation in children showed a 50% reduction in malaria attacks (*P. vivax*) and a reduction in malaria mortality.[56,57] In contrast, a clinical trial from Burkina Faso in which children were supplemented with 12.5 mg zinc or placebo/day for 6 months did not show an effect of zinc supplementation on clinical malaria episodes.[59] Zinc supplementation also did not appear to reduce the morbidity of malaria when it was given as adjunct therapy with chloroquine to children with malaria, aged 6 months to 5 years.[60,61] Methodologic differences in assessment of malaria may explain the conflicting results, and more research is needed to determine the benefits of zinc supplementation on malaria.

Iron-deficiency anemia is widespread among malaria endemic regions, and iron supplementation is often a key component of relief efforts. A meta-analysis of iron supplementation in malaria endemic regions, however, showed a tendency toward higher parasite counts in individuals who received iron supplements.[62] In contrast, a recent review of studies published between 2012 and 2016 reported that there was no increased risk of malaria with iron supplementation.[63]

Diarrheal Disease

Diarrheal disease, the second leading cause of death in children younger than 5 years of age, causes an estimated 525,000 deaths in children every year.[64] The main causes of diarrheal diseases among children in developing countries are rotavirus, *Escherichia coli*, *Shigella*, *Vibrio cholerae*, *Salmonella*, and *Entamoeba histolytica*.[65] The epidemiology, clinical features, immunology, and pathogenesis of diarrhea differ according to the characteristics of the pathogen in terms of production of toxins, tissue invasion, fluid and electrolyte loss, and location of infection.

Diarrhea is generally defined as the condition in which three or more liquid stools are passed in a 24-hour period.[66] In developing countries, community-based studies suggest that children less than 5 years of age have a median incidence of three diarrheal episodes per year.[66] In general, host defenses in the gut include gastric acidity, the presence of normal microflora, gut motility, mucus production, integrity of microvilli, local secretion of antibody, and cell-mediated immunity; micronutrient malnutrition may impair some of these host defenses. Malnourished infants and children are at higher risk for diarrhea which, in turn, can exacerbate malnutrition and result in more severe disease and higher mortality from diarrhea.[64,67,68]

Diarrhea can lead to dehydration, electrolyte imbalance, and malabsorption of both macronutrients and micronutrients. Vitamins A, D, B_{12}, copper, folate, iron, magnesium, zinc, and selenium deficiencies have all been described during diarrheal disease.[69] Frequent episodes of diarrheal disease can also cause significant damage to the lining of the gut and disrupt normal gut flora colonization, changes that further exacerbate poor nutrition status and immune function. Exclusive breastfeeding for the first 4 to 6 months of life decreases both the severity and the incidence of diarrhea among infants.[70] In particular, breastfeeding reduces sanitation concerns with other forms of infant feeding and eliminates the need for clean sources of drinking water.[71]

Supplementation of specific micronutrients may play an important therapeutic role in managing diarrheal disease. Several studies have shown zinc supplementation to be beneficial in acute diarrhea.[72] In a community-based study in India, children who received micronutrient supplementation with zinc had a 23% faster recovery following a diarrheal episode and a 39% reduction of likelihood that the episode lasted longer than 7 days, compared with children who received micronutrient supplementation without zinc.[73] Zinc was also associated with a reduction in the severity of diarrheal disease. In an analysis of four clinical trials on children with persistent diarrhea, zinc was associated with 29% faster recovery and 49% lower rate of treatment failure or death.[74] Since 2004, the WHO recommends that treatment for all children with diarrheal disease should include clean water, sugar, and salt, along with 10 to 14 days of supplemental zinc (20 mg) to reduce duration and to promote better outcomes.[64]

Vitamin A deficiency is associated with an increased incidence of diarrheal disease, and diarrheal disease is associated with an increased incidence of vitamin A deficiency in children.[30,75-79] Imdad and colleagues regularly update their original 2010 meta-analysis of randomized clinical trials on vitamin A supplementation and the most recent update continues to show a reduction in incidence and mortality related to diarrhea with supplementation in children ages 5 years and younger.[30,80] The benefit of vitamin A appears to be limited to children who are vitamin A deficient or malnourished or to children with shigellosis,[66,81] and the current WHO recommendation does not include vitamin A as part of treatment of diarrhea.

While there is growing evidence on the benefits of probiotics and synbiotics[82-84] as well as benefits of other nutritional interventions on diarrhea,[85] further studies are needed to confirm their role in the treatment of diarrhea.

Acute Respiratory Infections

Acute respiratory infections (ARIs) are major causes of morbidity and mortality among infants and children in developing countries.[86] The main causes of acute lower respiratory infections in children are respiratory syncytial virus, adenovirus, parainfluenza virus, influenza virus, *Streptococcus pneumoniae*, and *Haemophilus influenzae*.[69] The nonspecific immune defenses of the respiratory tract consist of cilia in the tracheobronchial tree, mucus secretion by goblet cells, and phagocytic activity by alveolar macrophages.[69]

Several nutritional factors may play a role in the incidence of ARIs, with the strongest protective evidence for zinc. Controlled clinical trials show that zinc supplementation reduced the incidence of ARIs and pneumonia among children.[87-93] A pooled analysis of four randomized trials showed that zinc supplementation decreased the incidence of pneumonia by 41%.[94] A review of clinical trials reported that zinc supplementation in zinc-deficient populations could prevent 25% of ARIs.[95] Other factors that have been shown to be protective include prevention of low birth weight and reduction of malnutrition.[69] In contrast, vitamin A supplementation has shown little to no effect on reducing the incidence of ARIs among preschool children.[95,96]

Nutrition also plays a role in the treatment and reduction of morbidity and mortality associated with ARIs. Among infants and young children, zinc supplementation reduced the morbidity of severe acute lower respiratory infection and severe pneumonia.[97,98] Although some trials showed that zinc lozenges taken within 24 hours of onset of a cold can reduce the duration and severity of the symptoms, reviews of the effect of zinc were inconsistent.[99-103] Little evidence for the benefit of vitamin A in the treatment of ARIs exists unless measles infection is concurrent.[81]

There has been renewed interest in nutrition and acute respiratory infections since the start of the coronavirus disease 2019 (COVID-19) pandemic. Most of this research has focused on vitamin D and there is some evidence to suggest a protective effect of vitamin D supplementation on the severity of COVID among people with low serum vitamin D status.[104,105] However, micronutrient supplementation with vitamin C, zinc, and vitamin D does not appear to reduce COVID mortality.[106] Further studies are needed to determine the role of nutrition on COVID incidence, severity, and mortality.

Hookworm Infection

The CDC estimates that hookworm infection affects 576 to 740 million people worldwide.[107] Hookworm infection is a major cause of anemia and protein malnutrition. Two species of hookworm, *Ancylostoma duodenale* and *Necator americanus*, account for significant human morbidity and mortality. Hookworm is usually spread from person to person through contamination of soil and vegetation with feces that contain hookworm eggs. In the soil, the eggs develop into larvae that can penetrate human skin on contact. The infective larvae may leave a pruritic, papulovesicular eruption on the skin ("ground itch"). The hookworm larvae migrate through the skin to the alveoli of the lungs, where they may be coughed up and ingested. The hookworms attach to the intestinal mucosa and cause chronic blood loss and depletion of iron stores.[108] Hookworms elicit antibody and cellular immune responses in humans, but limited evidence suggests that these immune responses give any protection against infection.[109] Hookworm infection is associated with growth retardation among infants and children, iron deficiency and impaired mental development in school children, and fatigue and decreased work capacity among adults.[110-112] Poor hygiene is the major risk factor for hookworm infection, and hookworm infection, in turn, increases the risk of iron and other micronutrient deficiencies.

Pioneering studies conducted in the early part of the twentieth century in the United States showed that deworming of heavily infected school children improved hemoglobin concentration and increased growth.[113-115] Hookworm infection was associated with poor mental development,[115,116] and treatment for hookworm in infected children improved performance on mental development tests compared with infected children who were not treated.[117] These studies were conducted at a time when hookworm infection was common in the southern United States.[118] Studies in developing countries showed that treatment of school children for hookworm infection was associated with improved iron status[119] and better growth,[120] and a combination of hookworm treatment and iron supplementation was associated with improved cognitive performance[121] and growth.[122] Recent systematic reviews on the effect of nutritional supplements failed to find significant reductions on reinfection rates of helminths in children.[123,124]

Tuberculosis (TB)

TB is one of the most widespread infectious diseases globally. Nearly 2 billion people, or close to a quarter of the world's population, are infected with *Mycobacterium tuberculosis* or related bacteria; and an estimated 9.9 million new cases of TB occurred worldwide in 2020 alone.[125] The COVID-19 pandemic resulted in a sharp decline in newly diagnosed and reported cases of TB in 2020, with the South-East Asia and the Western Pacific regions accounting for 84% of the global reduction in reports of newly diagnosed cases.[126] In 2020, deaths from TB increased for the first time since 2005, with an estimated 1.3 million deaths occurring among HIV-negative people and 214,000 deaths among people living with HIV.[126] Poverty, overcrowding, malnutrition, tobacco and alcohol use, and underlying HIV infection are among the strongest risk factors for contracting TB.[127-129]

The spectrum of TB ranges from asymptomatic latent infection to active disease. *M. tuberculosis* remains latent in most people who are infected. Once infected, a person has an approximate 10% lifetime risk of developing active disease. Active TB, the form of disease that disrupts normal host physiology to produce symptoms, is generally classified as pulmonary or extrapulmonary disease. Pulmonary disease accounts for approximately 80% of cases, and clinical manifestations include chronic cough and sometimes hemoptysis, dyspnea, and chest pain. Other constitutional symptoms include fever, night sweats, and weight loss. Extrapulmonary disease accounts for approximately 20% of cases of active TB. The most common extrapulmonary sites are the cervical lymph nodes, pleura, kidneys, meninges, and bones or joints. Active TB is curable, and treatment consists of daily doses of multiple antibiotics for 6 to 8 months. Interruption or poor adherence to treatment can lead to multidrug-resistant TB, which is currently a widespread problem in at least 117 countries.[130]

M. tuberculosis is transmitted through aerosolized particles from infectious cases to susceptible contacts. When infected individuals with active disease cough, they produce an aerosol of tiny droplet particles that contain live bacilli. Susceptible contacts become infected by inhaling minimal amounts of these bacilli, which become lodged deep in the passageways of the lungs. The initial interaction between the host immune system and *M. tuberculosis* occurs in the alveoli, where pulmonary macrophages engulf, process, and present mycobacterial antigens in conjunction with major histocompatibility complex (MHC) class II molecules.

When the complex of mycobacterial antigen and MHC molecule is recognized by antigen-specific CD4 lymphocytes, the CD4 lymphocytes release interferon-γ (IFN-γ) and IL-2. The IFN-γ serves to activate macrophages and enhance their ability to contain the mycobacteria. However, the mycobacteria have their own defense mechanisms that prevent the normal destruction of the bacilli by the macrophages.[131,132] The activated macrophages release certain cytokines, including TNF-α,[133] IL-1,[134] and IL-6,[135] which are responsible for the recruitment of cells to the site of infection and formation of a granuloma to wall off spread of the infection. The infection can then remain latent for many years until a state of weakened immune function (e.g., caused by HIV or malnutrition) leads to activation of the disease.

The link between TB and malnutrition is well documented in the literature. Several comprehensive reviews of this topic are available.[128,136-140] Briefly, in the absence of effective chemotherapy, TB is characterized by wasting and high mortality. The association of poor nutrition with TB is evident in older terms for TB such as the Greek term *phthisis* (to waste away), and *consumption*. In the preantibiotic era, studies on treatments for TB focused largely on nutritional therapies.[141-144] Cod liver oil (rich in vitamins A and D) was a widely accepted therapy for TB, and high-protein diets rich in egg and milk were often recommended.[137] However, with the advent of cheap and effective drug therapies for managing active TB, the attention largely shifted to the delivery and uptake of medications.

Both generalized malnutrition and deficiencies of specific micronutrients can lead to impaired immune response to TB infection and increased risk of developing active disease. Active TB has been associated with deficiencies of specific micronutrients, such as vitamins A, B complex, C, D, E, and selenium.[145,146] Interest remains in the role of vitamin D in TB.[147,148] Immune cells, such as macrophages, produce an enzyme that converts circulating 25-OH vitamin D to the more active form of 1,25$(OH)_2$ vitamin D. This 1,25$(OH)_2$ form of vitamin D, in turn, upregulates the vitamin D receptor (VDR), which induces the production of an antimycobacterial peptide, cathelicidin.[131,149,150] Investigators have hypothesized that the increased risk of TB associated with vitamin D deficiency is associated with a lack of appropriate VDR-dependent antimycobacterial activity.

Although the association between malnutrition and TB is well known, evidence from clinical trials investigating whether improved nutrition will reduce the risk of developing active disease or improve the clinical outcome of TB has been inconsistent. A Cochrane review of randomized controlled trials was conducted comparing any oral nutritional supplement with no nutritional intervention, placebo, or dietary advice only among people being treated for active TB. The results showed effects of high energy supplements or combinations of multiple micronutrients (including zinc and vitamin A) on weight gain, but little evidence of effect on other clinical outcomes of TB.[151] Some evidence has shown that supplementation with multiple micronutrients, excluding zinc, reduced the number with a sputum-culture positive test between 1 and 8 months among HIV-positive participants who became culture negative after 1 month of treatment.[152] In addition, multiple micronutrients, including zinc, may reduce mortality in people with HIV and TB coinfection.[153] These results need to be replicated in other populations.

In summary, no clear evidence exists for the role of nutritional support in TB. Although rates of TB transmission are highest among impoverished and malnourished populations, whether correction of malnutrition leads to improved outcomes in the course of TB is uncertain.

Human Immunodeficiency Virus Infection

In 2021, an estimated 38.4 million people were living with HIV worldwide, with 1.5 million new infections and 650,000 AIDS-related deaths in that year alone.[154] HIV infection causes a progressive decline in immunity that can lead to the acquired immunodeficiency syndrome (AIDS) and, if left untreated, death. HIV is spread from person to person by three major routes: sexual contact, mother-to-child transmission, and blood products such as shared needles and syringes. Wasting syndrome, characterized by unintentional loss of more than 10% body weight, was common among HIV-infected adults in industrialized countries before the advent of combination antiretroviral therapy (ART) in the mid-1990s. With appropriate care and treatment, the prevalence of wasting syndrome has drastically diminished but is still documented among people living with HIV (PWH) in areas where access to HIV treatment remains limited, and among certain subgroups of PWH, such as children in low resource settings.[155] Similar to the general population, the prevalence and incidence of weight gain and obesity in PWH is currently on the rise, largely due to improved survival on HIV treatment and partly due to the introduction of new agents in the treatment of HIV.[156-159]

Poor nutrition has been shown to play an important role in disease progression in HIV-infected persons.[160] In untreated HIV infection, nutritional intake may be affected by anorexia, central nervous system disease, dysphagia, and odynophagia (painful swallowing).[161] Decreased food intake may occur even in asymptomatic HIV-infected adults and has been associated with significant weight loss.[162] In the early 1990s, many HIV-infected individuals did not consume at least the recommended dietary allowance for some B-complex vitamins, vitamin E, and zinc,[163,164] and investigators suggested that the recommended dietary allowances should be higher for persons with HIV infection.[165] If HIV infection is left untreated, diarrhea and malabsorption of fats, carbohydrates, and vitamin B_{12} can occur in all stages of infection. Cryptosporidia, microsporidia, *Cytomegalovirus*, and *Mycobacterium avium-intracellulare* are major causes of diarrhea in untreated patients with AIDS and can lead to severe weight loss and death.[166] With continued care and treatment, many of the more severe nutritional problems associated with HIV infection have declined in prevalence.

Energy Requirements

In May of 2003, the WHO, after technical consultations with experts in the field, issued recommendations on nutrient requirements for people living with HIV.[167] Based on a review of the published data on resting energy expenditure among asymptomatic HIV-infected adults, the committee of experts recommended that energy requirements should be increased by approximately 10% to maintain body weight and physical activity in asymptomatic adults and growth in asymptomatic children. In the symptomatic stages of HIV and AIDS, the WHO recommended that energy requirements increase by approximately 20% to 30% to maintain adult body weight; and for children experiencing weight loss, the recommendation was to increase energy intake by 50% to 100% over normal requirements. Since these recommendations came out, a few other studies demonstrated that resting energy expenditure is higher in PWH compared with healthy controls.[168-170] However, these findings need to be re-evaluated in light of the ongoing changes in treatment regimens and increase in obesity observed among PWH.

Micronutrient Requirements

Before the era of HAART, micronutrient deficiencies were reported in many HIV-infected risk groups.[160] Vitamin A deficiency was common among HIV-infected pregnant women[171,172] and children[173] in developing countries. A few studies suggested that serum vitamin C concentration was lower among HIV-infected adults compared with healthy controls.[174] Many HIV-infected adults appeared to have low vitamin B_6,[175,176] vitamin B_{12},[177] and folate.[178] Iron deficiency was common among HIV-infected pregnant women,[179] female injection drug users,[180] and children.[181,182] Low blood zinc concentration and low serum or plasma selenium, consistent with deficiency, were described in HIV-infected adults.[175,183,184]

Nutrients can be involved in the pathogenesis of HIV infection through their roles as antioxidants and in immune function. Antioxidant nutrients such as the carotenoids, tocopherols, vitamin C, and selenium have been implicated in the pathogenesis of HIV infection through their interactions with reactive oxygen intermediates. Nuclear factor-κB (NF-κB) is a pro-inflammatory transcription factor involved in oxidative stress-induced inflammatory responses present in many conditions, including HIV. Antioxidants have been hypothesized as a potential treatment to modulate the effects of reactive oxygen species on the signaling pathway of NF-κB, with subsequent downregulation of HIV expression.[185] Vitamin A plays a central role in the growth and function of T and B cells, antibody responses, and maintenance of mucosal epithelium, including that of the respiratory, gastrointestinal, and genitourinary tracts.[186] Zinc plays an important role in the growth, development, and function of neutrophils, macrophages, natural killer cells, and T and B lymphocytes.[187] Vitamin D is also an important regulator of immune function through its role in regulating cell proliferation, differentiation, and apoptosis.[188]

Early observational studies suggested that certain micronutrient deficiencies were associated with increased morbidity and mortality during HIV infection. Low plasma or serum vitamin A concentrations were associated with accelerated HIV progression,[189] increased mortality,[190] and, in children, growth failure.[191] High serum vitamin E was associated with lower risk of progression to AIDS.[192] The risk of progression to AIDS was higher in those persons with vitamin B_{12} deficiency.[176] Low serum zinc levels were associated with reduced secretory function of the thymus[193] and HIV disease progression.[189,194] Low serum or plasma selenium concentrations were associated with an increased risk of progression to AIDS and higher mortality.[195-197]

In 2003, the report issued by the WHO also included recommended micronutrient requirements for people living with HIV infection.[167] At the time, little scientific evidence existed in the form of randomized controlled clinical trials to recommend micronutrient intakes higher than the recommended dietary allowance. Since then, data from several clinical trials of multiple micronutrient supplementation in HIV-infected adults have been published.[198-205] The earliest and largest of these, a trial of HIV-infected pregnant women in Tanzania, showed that multivitamin supplementation compared with placebo had several beneficial outcomes in HIV-infected pregnant women, including greater increases in T-cell counts during and after pregnancy, better birth outcomes (reduced fetal deaths and low birth weight),[206] and improved weight gain during pregnancy.[207] Further follow-up of the same women over several years demonstrated continued benefits of multiple micronutrients on T-cell counts, lower viral loads, slower HIV progression, and improved overall survival.[198] Additional analyses of these trial participants showed a beneficial effect of micronutrients on maternal wasting,[208] maternal and child hemoglobin status,[209] as well as maternal depression and quality of life.[210]

Since 2003, other clinical trials have been conducted in different geographic regions (Southeast Asia, Africa, United States) with differing population characteristics (including differing levels of micronutrient deficiency at baseline) and using different mixes and doses of multiple micronutrients. Overall, studies so far have not shown any consistent effects of micronutrient supplementation in PWH beyond what is recommended for the general population.[211-213]

Summary

The causes of malnutrition are complex and involve food, society, health, and caring practices. The theoretic framework for the causes of malnutrition used by UNICEF classifies the causes as immediate (diet and care), underlying (available food, practices, and services), and enabling (political, financial, social, cultural, and environmental conditions).[214] A malnourished child has reduced immunity to infection and can have more severe and frequent episodes of illness. Illnesses such as diarrheal disease can reduce appetite, increase malabsorption of nutrients, hasten losses of nutrients and, thus, can further perpetuate a cycle of malnutrition and infection. On the level of the household, problems may exist with food security (i.e., access to safe food of sufficient quality and quantity to ensure adequate health for all family members). The quality and quantity of food should meet the requirements for protein, energy, and micronutrients. Access to food can depend on financial, social, and physical factors. Given the close relationship between malnutrition and infection, factors that affect hygiene, such as clean water and sanitation, affect malnutrition. Inadequate maternal and childcare practices are also underlying household causes of malnutrition, including lack, or early cessation, of breastfeeding; lack of safe, high-quality complementary foods; withholding of food and liquids to a child with diarrhea; inadequate practices of food sharing at the table; poor knowledge of personal hygiene; and lack of immunizations. On the societal level, low status of women, inadequate health services, lack of jobs, and poverty are causes that underlie malnutrition. The theoretic framework of UNICEF for the causes of malnutrition suggests multiple levels and strategies for interventions to reduce morbidity and mortality from infectious diseases, ranging from improvements in sanitation, infrastructure, maternal education, and immunizations to food fortification and provision of micronutrient supplementation.

Acknowledgment

The authors thank Ms. Victoria Zhang for conducting the literature search, summarizing the studies, and providing constructive feedback.

References

1. Grijns G. Geneeskundig Tijdschrift Nederl Indië 1901;41:3.
2. Hopkins FG. Analyst 1906;31:385–404.
3. Funk C. J State Med 1912;20:341–68.
4. Carpenter KJ. J Nutr 2003;133:3023–32.
5. Silverstein AM. A History of Immunology. San Diego, CA: Academic Press, 1989.
6. Heilbron IM, Jones WE, Bacharach AL. Vitam Horm 1944;2: 155–213.
7. Robertson EC. Medicine 1934;13:123–206.
8. Clausen SW. Physiol Rev 1934;14:309–50.
9. Scrimshaw NS, Taylor CE, Gordon JE. Interactions of Nutrition and Infection. Geneva, Switzerland: World Health Organization, 1968.
10. Gabay C, Kushner I. N Engl J Med 1999;340:448–54.
11. Gitlin JD, Colten HR. In: Pick E, Landy M, eds. Molecular Biology of Scute Phase Plasma Proteins. Lymphokines. vol 14 San Diego, CA: Academic Press, 1987:123–53.
12. Fournier T, Medjoubi N, Porquet D. Biochim Biophys Acta 2000;1482:157–71.
13. Uhlar CM, Whitehead AS. Eur J Biochem 1999;265:501–23.
14. Langhans W. Nutr 2000;16:996–1005.
15. Plata-Salaman CR. Intl J Obes Rel Metab Dis 2001;25:S48–S52.
16. World Health Organization. 2019. Accessed September 14, 2022. Available at www.who.int/news-room/fact-sheets/detail/measles
17. Perry RT, Halsey NA. J Infect Dis 2004;189:S4–S16.
18. Centers for Disease Control and Prevention (CDC). 2022. Accessed September 14, 2022. Available at www.cdc.gov/globalhealth/measles/data/global-measles-outbreaks.html
19. Griffin DE. Curr Topics Microbiol Immunol 1995;191:117–34.
20. Tatsuo H, Ono N, Tanaka K, et al. Nature 2000;406:893–7.
21. Dossetor J, Whittle HC, Greenwood BM. Br Med J 1977;1:1633–5.
22. Chen LC, Rahman M, Sarder AM. Int J Epidemiol 1980;9: 25–33.
23. Koster FT, Curlin GC, Aziz KM, et al. Bull World Health Organ 1981;59:901–8.
24. World Health Organization. Accessed March 18, 2024. Available at https://iris.who.int/bitstream/handle/10665/70535/WHO_IVB_2010_eng.pdf
25. Manicus A. Ugeskrift Laeger 1847;6:189–210.
26. Smedman L, Lindeberg A, Jeppsson O, et al. Ann Trop Paediatr 1983;3:169–76.
27. Alwar AJ. East Afr Med J 1992;69:415–8.
28. Markowitz LE, Nzilambi N, Driskell WJ, et al. J Trop Pediatr 1989;35:109–12.
29. Semba RD, Bloem MW. Surv Ophthalmol 2004;49:243–55.
30. Imdad A, Mayo-Wilson E, Haykal MR, et al. Cochrane Database Syst Rev 2022;3:Cd008524.
31. Ellison JB. Br Med J 1932;2:708–11.
32. Barclay AJ, Foster A, Sommer A. Br Med J (Clin Res Ed) 1987;294:294–6.

33. Coutsoudis A, Broughton M, Coovadia HM. Am J Clin Nutr 1991; 54:890–5.
34. Ogaro FO, Orinda VA, Onyango FE, et al. Trop Geogr Med 1993;45:283–6.
35. Coutsoudis A, Kiepiela P, Coovadia HM, et al. Pediatr Infect Dis J 1992;11:203–9.
36. Benn CS, Balde A, George E, et al. Lancet 2002;359:1313–4.
37. Hussey GD, Klein M. N Engl J Med 1990;323:160–4.
38. Glasziou PP, Mackerras DE. BMJ 1993;306:366–70.
39. Sudfeld CR, Navar AM, Halsey NA. Int J Epidemiol 2010;39 (suppl 1):i48–i55.
40. World Health Organization. Malaria. 2022. Accessed July 15, 2022. Available at www.who.int/news-room/fact-sheets/detail /malaria
41. Shankar AH. Malaria. In: Semba RD, Bloem MW, eds. Nutrition and Health in Developing Countries. Totowa, NJ: Humana Press, 2001;177–207.
42. World Health Organization. Accessed March 18, 2024. Avilable at www.who.int/publications/i/item/9789241564106
43. Murray MJ, Murray AB, Murray NJ, et al. Am J Clin Nutr 1978;31:57–61.
44. Murray J, Murray A. Perspect Biol Med 1977;20:471–83.
45. Ferreira E, Alexandre MA, Salinas JL, et al. Malar J 2015;14:346.
46. Caulfield LE, Richard SA, Black RE. Am J Trop Med Hyg 2004;71:55–63.
47. Kojom Foko LP, Nolla NP, Nyabeyeu Nyabeyeu H, et al. Biomed Res Int 2021;2021:5553344.
48. Stürchler D, Tanner M, Hanck A, et al. Acta Trop 1987;44: 213–27.
49. Galan P, Samba C, Luzeau R, et al. Int J Vitam Nutr Res 1990; 60:224–8.
50. Friis H, Mwaniki D, Omondi B, et al. Am J Clin Nutr 1997; 66:665–71.
51. Lawal OA, Adegoke SA, Oseni SB, et al. J Infect Dev Ctries 2018; 12:365–72.
52. Shankar AH, Genton B, Semba RD, et al. Lancet 1999;354: 203–9.
53. Owusu-Agyei S, Newton S, Mahama E, et al. Nutr J 2013;12:131.
54. Bates CJ, Evans PH, Dardenne M, et al. Br J Nutr 1993;69: 243–55.
55. Shankar AH. J Infect Dis 2000;182(suppl 1):S37–S53.
56. Sazawal S, Black RE, Ramsan M, et al. Lancet 2007;369:927–34.
57. Richard SA, Zavaleta N, Caulfield LE, et al. Am J Trop Med Hyg 2006;75:126–32.
58. Shankar AH, Genton B, Baisor M, et al. Am J Trop Med Hyg 2000;62:663–9.
59. Müller O, Becher H, van Zweeden AB, et al. BMJ 2001;322:1567.
60. Group ZAPS. Am J Clin Nutr 2002;76:805–12.
61. Somé JW, Abbeddou S, Yakes Jimenez E, et al. BMJ Open 2015;5:e007828.
62. Group INAC. Safety of Iron Supplementation Programs in Malaria-Endemic Regions. Washington, DC: International Life Science Institute, 2000.
63. Moya-Alvarez V, Bodeau-Livinec F, Cot M. Nutr Rev 2016;74: 612–23.
64. World Health Organization. Diarrhoeal Disease. 2017. Accessed: September 14, 2022. Available at www.who.int/news-room/fact -sheets/detail/diarrhoeal-disease
65. Lanata CF, Fischer-Walker CL, Olascoaga AC, et al. PLoS One 2013;8:e72788.
66. Lanata CF, Black RE. Semba RD, and Bloem MW, eds. Diarrheal Diseases. Nutrition and Health in Developing Countries. Totowa, NJ: Humana Press, 2008:139–78.
67. Black RE, Brown KH, Becker S. Am J Clin Nutr 1984;39:87–94.
68. Chen LC, Huq E, Huffman SL. Am J Epidemiol 1981;114: 284–92.
69. Lanata CF, Black RE. Acute Lower-Respiratory Infections. Nutrition and Health in Developing Countries. Totowa, NJ: Humana Press, 2008:179–214.
70. Gartner LM, Morton J, Lawrence RA, et al. Pediatr 2005;115: 496–506.
71. Ahmed KY, Page A, Arora A, et al. PLoS One 2020;15:e0230978.
72. Lazzerini M. Curr Opin Clin Nutr Metab Care 2016;19: 239–43.
73. Sazawal S, Black RE, Bhan MK, et al. N Engl J Med 1995;333: 839–44.
74. Bhutta ZA, Bird SM, Black RE, et al. Am J Clin Nutr 2000;72:1516–22.
75. McLaren DS, Shirajian E, Tchalian M, et al. Am J Clin Nutr 1965;17:117–30.
76. Cohen N, Rahman H, Sprague J, et al. World Health Stat Q 1985;38:317–30.
77. Khatry SK, West KP, Jr., Katz J, et al. Arch Ophthalmol 1995;113:425–9.
78. Schaumberg DA, O'Connor J, Semba RD. Eur J Clin Nutr 1996;50:761–4.
79. Sommer A. Nutritional Blindness: Xerophthalmia and Keratomalacia. New York, NY: Oxford University Press, 1982.
80. Imdad A, Herzer K, Mayo-Wilson E, et al. Cochrane Database Syst Rev 2010;Cd008524.
81. World Health Organization. Accessed March 18, 2024. Available at www.who.int/publications/i/item/978924156410.6
82. Kambale RM, Nancy FI, Ngaboyeka GA, et al. Clin Nutr 2021;40:3158–69.
83. Applegate JA, Fischer Walker CL, Ambikapathi R, et al. BMC Public Health 2013;13(suppl 3):S16.
84. Li Z, Zhu G, Li C, et al. Nutrients 2021;13:4319.
85. Florez ID, Veroniki AA, Al Khalifah R, et al. PLoS One 2018;13:e0207701.
86. Cherian T. Geneva, Switzerland. WHO Report, 2009.
87. Brown KH, Peerson JM, Baker SK, et al. Food Nutr Bull 2009;30:S12–S40.
88. Ninh NX, Thissen JP, Collette L, et al. Am J Clin Nutr 1996;63:514–9.
89. Sazawal S, Black RE, Bhan MK, et al. Am J Clin Nutr 1997;66:413–8.
90. Sazawal S, Black RE, Jalla S, et al. Pediatr 1998;102:1–5.
91. Meeks GJ, Witter MM, Ramdath DD. Eur J Clin Nutr 1998;52:34–9.
92. Penny ME, Peerson JM, Marin RM, et al. J Pediatr 1999;135: 208–17.
93. Bhandari N, Bahl R, Taneja S, et al. BMJ 2002;324:1358–60.
94. Bhutta ZA, Black RE, Brown KH, et al. J Pediatr 1999;135:689–97.
95. Roth DE, Caulfield LE, Ezzati M, et al. Bull World Health Organ 2008;86:356–64.
96. Vitamin A and Pneumonia Working Group. Bull World Health Organ 1995;73:609–19.
97. Mahalanabis D, Lahiri M, Paul D, et al. Am J Clin Nutr 2004;79: 430–6.
98. Brooks WA, Yunus M, Santosham M, et al. Lancet 2004;363: 1683–8.
99. Marshall I. Cochrane. Database Syst Rev 2000;CD001364.
100. Hulisz D. J Am Pharm Assoc 2004;44:594–603.
101. Basnet S, Mathisen M, Strand TA. J Trace Elem Med Biol 2015;31:163–6.
102. Moshtagh M, Amiri R. Tanaffos 2020;19:1–9.
103. Haider BA, Lassi ZS, Ahmed A, et al. Cochrane Database Syst Rev 2011;2011:CD007368.
104. Kaya MO, Pamukçu E, Yakar B. Epidemiol Health 2021;43:e2021074.
105. Pereira M, Dantas Damascena A, Galvão Azevedo LM, et al. Crit Rev Food Sci Nutr 2022;62:1308–16.

106. Beran A, Mhanna M, Srour O, et al. Clin Nutr ESPEN 2022;48:167–77.
107. Centers for Disease Control and Prevention. Hookworm. 2022. Accessed September 9, 2022. Available at www.cdc.gov/parasites/hookworm/
108. Pawlowski ZS, Schad GA, Stott GJ. Hookworm Infection and Anemia: Approaches to Prevention and Control. Last accessed September 30, 1991. Available at https://iris.who.int/bitstream/handle/10665/40857/9241544155.pdf?sequence=1&isAllowed=y
109. Loukas A, Prociv P. Clin Microbiol Rev 2001;14:689–703.
110. Hotez PJ, Brooker S, Bethony JM, et al. N Engl J Med 2004;351:799–807.
111. Bharti B, Bharti S, Khurana S. Indian J Pediatr 2018;85: 1017–24.
112. Kuong K, Fiorentino M, Perignon M, et al. Am J Trop Med Hyg 2016;95:856–63.
113. Smillie WG, Augustine DL. Am J Dis Child 1926;31:151–68.
114. Smillie WG, Augustine DL. South Med J 1926;19:19–28.
115. Smillie WG, Spencer CR. J Educ Psychol 1926;17:314–21.
116. Waite JH, Neilson IL. JAMA 1919;73:1877–9.
117. Strong EK. Effects of Hookworm Disease on the Mental and Physical Development of Children. 1916. Last accessed February 29, 2024.
118. Semba RD. In: Ward JW, Warren C, eds. Nutrition: Epidemiology and Public Health Overview. Safer and Healthier America: The Advancement of Public Health in the 20th Century. New York, NY: Oxford University Press, 2005.
119. Stoltzfus RJ, Albonico M, Chwaya HM, et al. Am J Clin Nutr 1998;68:179–86.
120. Stephenson LS, Latham MC, Kurz KM, et al. Am J Trop Med Hyg 1989;41:78–87.
121. Boivin MJ, Giordani B. J Pediatr Psychol 1993;18:249–64.
122. Dossa RA, Ategbo EA, de Koning FL, et al. Eur J Clin Nutr 2001;55:223–8.
123. Isah AUJ, Ekwunife OI, Ejie IL, et al. PLoS One 2020;15: e0237112.
124. Yap P, Utzinger J, Hattendorf J, et al. Parasit Vectors 2014;7:229.
125. World Health Organization. Accessed March 18, 2024. Available at www.cdc.gov/globalhivtb/index.html. Last accessed Mar 18, 2024
126. World Health Organization. Global Tuberculosis Report 2021. Accessed September 26, 2022. Available at www.who.int/publications/i/item/9789240037021
127. Atun R, Raviglione M, Marais B, et al. Lancet 2010;376:940–1.
128. USAID. USAID Africa's Health in 2010 project. Nutrition and tuberculosis: a review of the literature and considerations for TB control programs. 2008. Accessed December 9, 2010. Available at https://digitalcommons.calpoly.edu/cgi/viewcontent.cgi?article=1009&context=fsn_fac
129. Duarte R, Lonnroth K, Carvalho C, et al. Pulmonol 2018; 24:115–9.
130. World Health Organization. Tuberculosis: Multidrug-Resistant Tuberculosis (MDR-TB). 2018. Accessed September 26, 2022. Available at www.who.int/news-room/questions-and-answers/item/tuberculosis-multidrug-resistant-tuberculosis-(mdr-tb)
131. Chocano-Bedoya P, Ronnenberg AG. Nutr Rev 2009;67: 289–93.
132. Houben EN, Nguyen L, Pieters J. Curr Opin Microbiol 2006;9: 76–85.
133. Takashima T, Ueta C, Tsuyuguchi I, et al. Infect Immun 1990;58:3286–92.
134. Wallis RS, Vjecha M, mir-Tahmasseb M, et al. J Infect Dis 1993;167:43–8.
135. Ogawa T, Uchida H, Kusumoto Y, et al. Infect Immun 1991;59: 3021–5.
136. Macallan DC. Diagn Microbiol Infect Dis 1999;34:153–7.
137. van Lettow M, Whalen C. Semba RD, and Bloem MW, eds. Tuberculosis. Nutrition and Health in Developing Countries Totawa, NJ: Humana Press, 2008:275–306.
138. Gupta KB, Gupta R, Atreja A, et al. Lung India 2009;26:9–16.
139. van Lettow M, Fawzi WW, Semba RD. Nutr Rev 2003;61:81–90.
140. Sinha P, Davis J, Saag L, et al. J Infect Dis 2019;219:1356–63.
141. Munro WT, Leitch I. Proc Nutr Soc 1945;3:155–64.
142. Getz HR, Long ER, Henderson HJ. Am Rev Tuberc 1951;64: 381–93.
143. Thorn PA, Brookes VS, Waterhouse JA. Br Med J 1956;1: 603–8.
144. Downes J. Milbank Mem Fund Q 1950;28:127–59.
145. Cegielski JP, McMurray DN. Int J Tuberc Lung Dis 2004;8: 286–98.
146. Haiqing C, Chen L, Yin C, et al. Tuberculosis (Edinb) 2020;125: 101994.
147. Facchini L, Venturini E, Galli L, et al. J Chemother 2015;27: 128–38.
148. Martineau AR. Proc Nutr Soc 2012;71:84–9.
149. Blumenthal A, Isovski F, Rhee KY. Transl Res 2009;154:7–14.
150. Liu PT, Stenger S, Li H, et al. Science 2006;311:1770–3.
151. Grobler L, Nagpal S, Sudarsanam TD, et al. Cochrane Database Syst Rev 2016;2016:Cd006086.
152. Villamor E, Mugusi F, Urassa W, et al. J Infect Dis 2008;197: 1499–505.
153. Range N, Andersen AB, Magnussen P, et al. Trop Med Int Health 2005;10:826–32.
154. IN DANGER: UNAIDS Global AIDS Update 2022. Accessed September 26, 2022. Available at www.unaids.org/sites/default/files/media_asset/2022-global-aids-update_en.pdf
155. Abate BB, Aragie TG, Tesfaw G. PLoS One 2020;15:e0238403.
156. Bailin SS, Gabriel CL, Wanjalla CN, et al. Curr HIV/AIDS Rep 2020;17:138–50.
157. Kanters S, Renaud F, Rangaraj A, et al. EClinicalMedicine 2022; 48:101412.
158. Sax PE, Erlandson KM, Lake JE, et al. Clin Infect Dis 2020;71: 1379–89.
159. Wood BR, Huhn GD. Open Forum Infect Dis 2021;8:ofab542.
160. Semba RD, Tang AM. Br J Nutr 1999;81:181–9.
161. Castetbon K, Kadio A, Bondurand A, et al. Eur J Clin Nutr 1997;51:81–6.
162. McCorkindale C, Dybevik K, Coulston AM, et al. J Am Diet Assoc 1990;90:1236–41.
163. Tang AM, Graham NMH, Kirby AJ, et al. Am J Epidemiol 1993;138:937–51.
164. Baum MK, Cassetti L, Bonvehi P, et al. Nutr 1994;10: 16–20.
165. Baum MK, Shor-Posner G, Bonvehi P, et al. Ann NY Acad Sci 1992;669:165–74.
166. Sharpstone D, Gazzard B. Lancet 1996;348:379–83.
167. World Health Organization. Nutrient Requirements for People Living with HIV/AIDS: Report of a Technical Consultation, 13–15 May 2003. 2004. Accessed September 26, 2022. Available at www.who.int/publications/i/item/9241591196
168. Crenn P, Rakotoanbinina B, Raynaud JJ, et al. J Nutr 2004; 134:2301–6.
169. Ware LJ, Jackson AG, Wootton SA, et al. Br J Nutr 2009;102: 1038–46.
170. Mittelsteadt AL, Hileman CO, Harris SR, et al. J Acad Nutr Diet 2013;113:1037–43.
171. Semba RD, Miotti PG, Chiphangwi JD, et al. Lancet 1994;343: 1593–7.
172. Phuapradit W, Chaturachinda K, Taneepanichskul S, et al. Obstet Gynecol 1996;87:564–7.
173. Semba RD, Ndugwa C, Perry RT, et al. Nutr 2005;21: 25–31.

174. Tang AM, Smit E. Vitamins C and E, and HIV infection. In: Friis H, ed. Micronutrients and HIV Infection. Boca Raton, FL: CRC Press, 2001:111–34.
175. Beach RS, Mantero-Atienza E, Shor-Posner G, et al. AIDS 1992;6:701–8.
176. Tang AM, Graham NMH, Chandra RK, et al. J Nutr 1997;127:345–51.
177. Paltiel O, Falutz J, Veilleux M, et al. Am J Hematol 1995;49:318–22.
178. Boudes P, Zittoun J, Sobel A. Lancet 1990;335:1401–2.
179. Semba RD, Kumwenda N, Hoover DR, et al. Eur J Clin Nutr 2000;54:872–7.
180. Semba RD, Shah N, Strathdee SA, et al. J Acquir Immune Defic Syndr 2002;29:142–4.
181. Mueller BU, Tannenbaum S, Pizzo PA. J Pediatr Hematol Oncol 1996;18:266–71.
182. Totin D, Ndugwa C, Mmiro F, et al. J Nutr 2002;132:423–9.
183. Koch J, Neal EA, Schlott MJ, et al. Nutr 1996;12:515–8.
184. Mantero-Atienza E, Sotomayor MG, Shor-Posner G, et al. Nutr Res 1991;11:1237–50.
185. Wang CR, Zhou R, Ng TB, et al. Environ Toxicol Pharmacol 2014;37:626–37.
186. Semba RD. Nutr Rev 1998;56:S38–S48.
187. Shankar AH, Prasad AS. Am J Clin Nutr 1998;68:447S–63S.
188. National Institutes of Health. Vitamin D: Fact Sheet for Health Professionals. 2022. Accessed September 20, 2022. Available at https://ods.od.nih.gov/factsheets/VitaminD-HealthProfessional/
189. Baum MK, Shor-Posner G, Lu Y, et al. AIDS 1995;9:1051–6.
190. Semba RD, Graham NMH, Caiaffa WT, et al. Arch Intern Med 1993;153:2149–54.
191. Semba RD, Miotti P, Chiphangwi JD, et al. J Acquir Immune Defic Syndr Hum Retrovirol 1997;14:219–22.
192. Tang AM, Graham NMH, Semba RD, et al. AIDS 1997;11:613–20.
193. Falutz J, Tsoukas C, Gold P. J Am Med Assoc 1988;259:2850–1.
194. Graham NMH, Sorenson D, Odaka N, et al. J Acquir Immune Defic Syndr 1991;4:976–80.
195. Baum MK, Shor-Posner G, Lai S, et al. J Acquir Immune Defic Syndr 1997;15:370–4.
196. Campa A, Shor-Posner G, Indacochea F, et al. J Acquir Immune Defic Syndr Hum Retrovirol 1999;20:508–13.
197. Kupka R, Msamanga GI, Spiegelman D, et al. J Nutr 2004;134:2556–60.
198. Fawzi WW, Msamanga GI, Spiegelman D, et al. N Engl J Med 2004;351:23–32.
199. Jiamton S, Pepin J, Suttent R, et al. AIDS 2003;17:2461–9.
200. McClelland RS, Baeten JM, Overbaugh J, et al. J Acquir Immune Defic Syndr 2004;37:1657–63.
201. Range N, Changalucha J, Krarup H, et al. Br J Nutr 2006;95:762–70.
202. Semba RD, Kumwenda J, Zijlstra E, et al. Int J Tuberc Lung Dis 2007;11:854–9.
203. Kelly P, Katubulushi M, Todd J, et al. Am J Clin Nutr 2008;88:1010–7.
204. Kawai K, Kupka R, Mugusi F, et al. Am J Clin Nutr 2010;91:391–7.
205. Kaiser JD, Campa AM, Ondercin JP, et al. J Acquir Immune Defic Syndr 2006;42:523–8.
206. Fawzi WW, Msamanga GI, Spiegelman D, et al. Lancet 1998;351:1477–82.
207. Villamor E, Msamanga G, Spiegelman D, et al. Am J Clin Nutr 2002;76:1082–90.
208. Villamor E, Saathoff E, Manji K, et al. Am J Clin Nutr 2005;82:857–65.
209. Fawzi WW, Msamanga GI, Kupka R, et al. Am J Clin Nutr 2007;85:1335–43.
210. Smith Fawzi MC, Kaaya SF, Mbwambo J, et al. HIV Med 2007;8:203–12.
211. Visser ME, Durao S, Sinclair D, et al. Cochrane Database Syst Rev 2017;5:Cd003650.
212. Wobeser WL, McBane JE, Balfour L, et al. PLoS One 2022;17:e0270590.
213. Muzembo BA, Ngatu NR, Januka K, et al. Clin Nutr ESPEN 2019;34:1–7.
214. United Nations Children's Fund (UNICEF). UNICEF Conceptual Framework on Maternal and Child Nutrition. 2021. Accessed September 26, 2022. Available at www.unicef.org/documents/conceptual-framework-nutrition

CHAPTER 100

Nutrition and Age-Related Eye Diseases

Uday Kumar Chekkilla, Elizabeth Whitcomb, Jasper Weinberg, and Allen Taylor

CHAPTER OUTLINE

Abbreviations

AGE, advanced glycation end product; **AMD**, age-related macular degeneration; **ARC**, age-related cataracts; **ARED**, age-related eye disease; **ER**, endoplasmic reticulum; **GI**, glycemic index; **GIP**, gastric inhibitory peptide; **GLP-1**, glucagon-like peptide-1; **HIF**, hypoxia-induced factor; **NIH**, National Institutes of Health; **PA**, physical activity; **RPE**, retinal pigment epithelium; **SCFA**, short chain fatty acids; **sHSP**, small heat shock proteins; **VEGF**, vascular endothelial growth factor

Introduction

Among the five senses of humans, vision can be considered the primary source of acquiring knowledge. The 2023 WHO fact sheet on blindness and visual impairment indicates that 2.2 billion people have impaired vision, with an estimated financial burden of $411 billion, of which $1 billion could have been prevented. The majority of those with impaired vision are in the over-50 age group. Contemporary knowledge suggests that it may be possible to alleviate the burdens imposed by eye conditions including congenital visual impairment, age-related macular degeneration (AMD), age-related cataracts (ARC), diabetic retinopathy, glaucoma, uncorrected refractive errors, and strabismus, by improved nutrition.

Eye Diseases

Malfunctioning of any part of the eye, including the conjunctiva, cornea, sclera, iris, ciliary body, choroid, lens, retina, optic nerve, or vitreous can lead to loss of sight. Eye disease may affect the eyeball as a whole, lids, lachrymal apparatus, eye muscles, or orbit. However, the major sources of dysfunction are the lens (due to cataract development) and retina (due to loss of function).

According to the National Institute of Health (NIH), lost or reduced function of the retina can be referred to as a retinal disease and includes colour blindness, AMD (**Figure 100.1**), diabetic retinopathy, retinitis pigmentosa, retinoblastoma, retinal detachment, and inherited retinal diseases (IRDs), tumors, chorioretinal inflammatory and toxic disorders, paraneoplastic retinopathies and retinal involvement in central nervous system diseases such as autism, phacomatoses and multiple sclerosis.[1]

A clouding or opacification of the eye lens (Figure 100.1), which impacts the passage of light is called a cataract. Types of

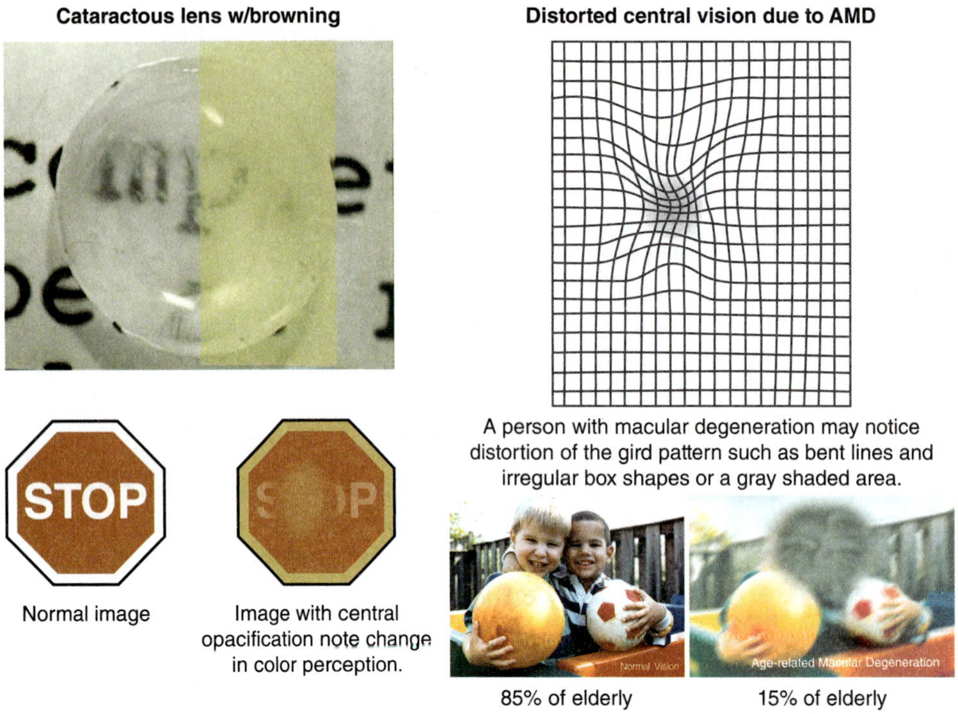

Figure 100.1 Impaired vision in ocular diseases. An example of a cataract demonstrates that kynurenines and browning lead to opacification of the lens (top left) and visual impairment (bottom left). Examples of how AMD results in central visual impairment due to macular degeneration (right).

cataracts include congenital cataracts, age-related cataracts, diabetic cataracts, drug-induced cataracts (including alcohol and smoking disorder-associated cataracts), and cataracts associated with traumatic injury, systemic diseases, endocrine diseases, or cataracts associated with nutritional insufficiency.[2]

Risk Factors Associated with Age-Related Eye Complications

Aging itself is a major risk factor for the development of eye complications, however, other conditions like obesity,[3-5] diabetes,[6, 7] malnutrition,[8-11] reduced physical activity,[12-13] drugs,[14-16] and alcohol consumption,[17] may also accelerate the development of eye complications. Aging is also associated with the development of metabolic diseases.[18] Studies have identified a key role for dietary glycaemia and alterations in the gut microbiome in the development of ocular diseases.[9, 19-22]

Role of Dietary Carbohydrates in Developing Eye Complications

Although the lens and retina are metabolically and structurally different—the lens is made of a single type of cell that expresses a restricted number of gene products, whereas the retina is composed of dozens of different types of cells, each of which expresses a myriad of different gene products—both are adversely affected by high glycemic index (GI) carbohydrate intake.[20] It is well established that carbohydrates play a major role in the development of diabetic complications, including diabetic retinopathy and diabetic cataract,[23-25] and several studies have shown that these complications can be delayed by adjusting the circulating glucose concentration via dietary modulations.[26]

The glycemic index measures the rate at which sugar enters the blood stream after consuming a specific food. It is indexed to 50g of sugar or a slice of white bread. Epidemiological studies have established that there is a strong association between GI and the risk of developing cardiovascular diseases, type-2 diabetes, and AMD.[19] Several studies have shown that consuming foods and diets with a high GI also increases the risk of developing AMD[8, 27-29] and cataracts.[9,30-32] (**Figure 100.2.**)

Dietary patterns, such as the Mediterranean and Prudent diet patterns with higher consumption of fruits and vegetables and lower high GI carbohydrates, were also found to be protective against AMD, and it seems probable that benefits from consuming lower GI diets are complementary to gains from consuming more fruits and vegetables.[9,33] Russet potatoes, white bread, white rice, and short grain rice are generally considered to be higher GI foods, while low GI foods include non-starchy vegetables, fruits, and whole wheat.[34] It should be noted, however, that foods like whole potatoes provide multiple nutrients, including vitamins.

Additional studies exploring the role of hyperglycemia in the development of eye complications indicate that relatively small increases in blood glucose concentration may also lead to eye complications like retinal degeneration and cataract in obese animals.[3,35-36]

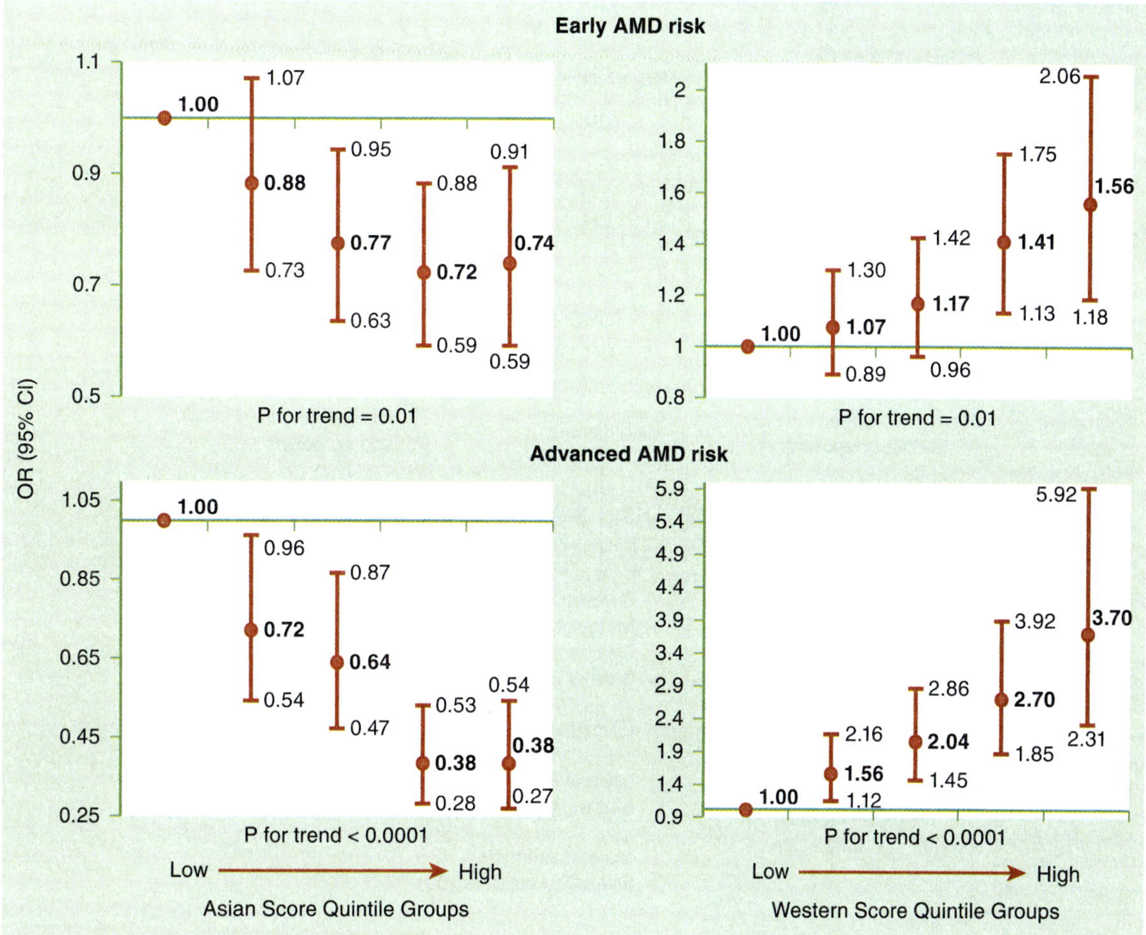

Figure 100.2 Risk of AMD based on dietary patterns in Americans. Western dietary pattern was shown to increase the risk of developing AMD. The image shows that the odds ratio for developing either early AMD (top) or advanced AMD (bottom) is decreased with increased adherence to an Asian diet (left) and increased with increased adherence to a Western dietary pattern (right).

Mechanisms Associated with Dietary Carbohydrates in Eye Complications

Glycaemic Stress and Glycation

Animal studies suggest that the link between higher GI diets and enhanced risk for progression of age-related eye diseases (AREDs) is causal. Researchers proposed that the production of intestinal hormones like GIP (Gastric inhibitory peptide) and GLP-1 (Glucagon-like peptide-1) is correlated with the dietary glycaemic index. High GI foods may result in the development of metabolic syndrome, which comprises several negative metabolic alterations like insulin resistance, fatty liver, obesity, cardiovascular diseases, etc.[37]

It was also anticipated that the hypoxia-induced factor (HIF) pathway, through hyperglycemia, plays a key role in the process. HIF transcription factors are known to play a significant role in glucose metabolism regulation in response to the depletion of cellular oxygen by increasing the expression of downstream genes, such as vascular endothelial growth factor (VEGF), leading to angiogenesis.[19]

But, there are also more "direct" effects of high GI diets and the elevated glucose they deliver rapidly. The Maillard reaction is a non-enzymatic reaction between the free amino acids of biomolecules and reducing sugars, which results in the formation of Amadori products. The Amadori products undergo further rearrangements resulting in the formation of advanced glycation end products (AGEs)[8,21,38] (**Figure 100.3a**). Rowan et al., showed accumulation of multiple AGEs in tissues from mice fed high GI diets.[39] Many of these are considered cytotoxic. A second source of cytotoxicity associated with glycative and oxidative stress is inhibition or insufficiency of proteolytic capacities and accumulation of protein aggregates[21,40,41] (**Figure 100.3b**). The ubiquitin proteasome system is a major protein degradation pathway and, in coordination with small heat shock proteins (sHSP) and autophagic and lysosomal proteolytic pathways, plays a key role in the process of protein triage.[42] Failure of these systems will result in accumulation of protein aggregates, which cause cataractous opacities in the lens and cytotoxicity in both the lens and retina.[43]

Glycation may also interfere with additional important cellular functions including: the SCAP/SREBP lipogenic pathway, inflammatory responses, sphingolipid metabolism, mitochondrial function, and oxidative stress responses.[44] The accumulation of AGEs is also a key factor in glaucoma, diabetic retinopathy, and cataracts. AGEs can mediate angiogenesis and damage the inner blood-retinal barrier, abnormally crosslink extracellular matrix proteins, and impair endothelial cell function.[45-47]

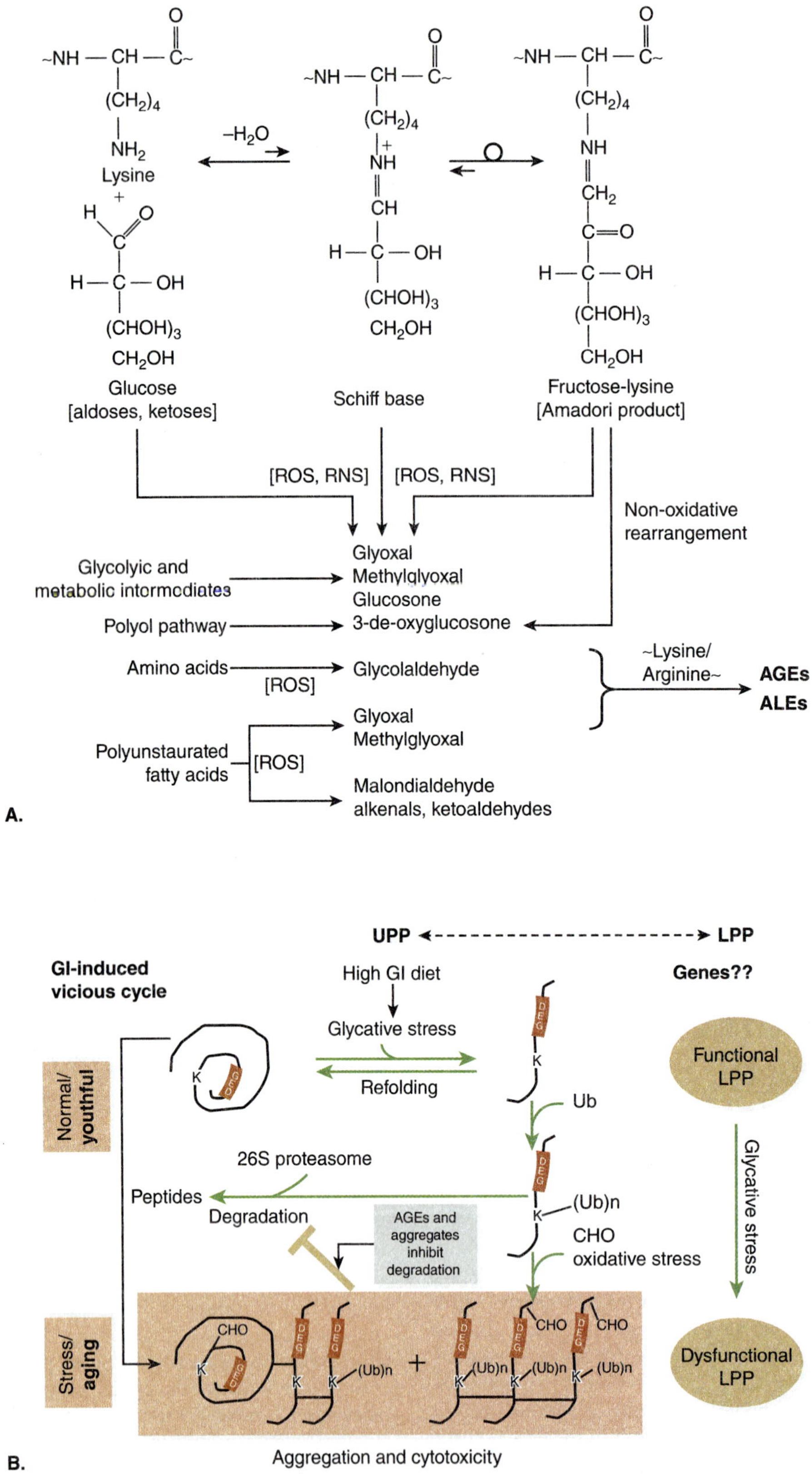

Figure 100.3 Glycative stress-AGE formation and Proteopoise. A. Guan Y, Dai P, Wang H. Medicine (Baltimore) 2020;99:e19274. **B.** Dietary intake of high GI food will increase the protein damage through glycation and oxidation, the toxicity will be neutralized in presence of limited damaged protein through lysosomal/autophagic proteolytic degradation. The figure illustrates how chronic glycative stress leads to proteasome functional insufficiency due to accumulation of AGEs, conjugates, and high molecular aggregates, and results in damage to the proteasome machinery and cellular toxicity. Accumulation of AGEs can be seen in age, obesity, and diabetes-related complications or diseases.

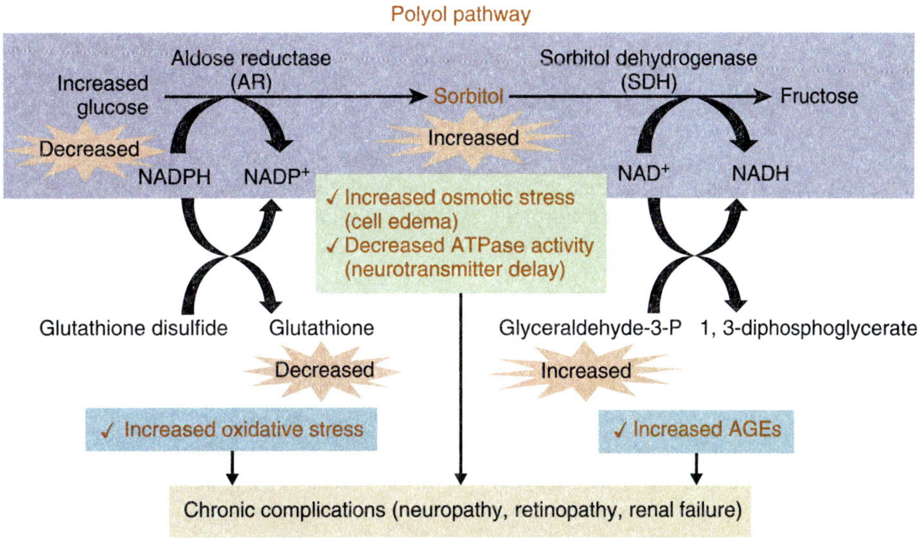

Figure 100.4 Role of the polyol pathway in developing diabetic complications. The figure illustrates primary causes like oxidative stress, osmotic stress, and AGE formation in developing vascular complications which are in turn induced by hyperglycemia in diabetic condition.

http://pubs.rsc.org

Many receptors can bind AGEs, including FEEL-1/FEEL-2, AGE-R2/80K-H/p90, AGE-R1/OST-48/p60, SR-A (I/II), SR-B/CD36, SR-B/CD36AGE-R3/ galectin-3, SR-A (I/II), SR-B/CD36, SR-BI, and SR-E/LOX-1.[48-52] However, RAGE is the most well-studied AGE receptor. AGEs are known to activate numerous signalling mechanisms through RAGE receptor engagement including Rac1/Cdc42, Ras/MAPL/NF-Kb, and JAK/STAT.[53-55]

It has been well established that hyperglycemia in diabetes activates the polyol pathway, which increases the formation of AGEs in eye diseases[48,56,57] (**Figure 100.4**). Some studies have reported that even prediabetes and obese conditions without diabetes may trigger the activation of the polyol pathway.[3,58,59] A long-term high GI diet can mimic prediabetes, as the blood glucose concentration will be at the higher side of normal. However, the role of dietary carbohydrates with a high GI in the activation of the polyol pathway needs to be investigated further.

Oxidative Stress

Many of the insults and responses to insult indicated above involve oxidation. Therefore, it is not surprising that oxidative stress is a widely accepted mechanism in the pathogenesis of age and AREDs.[60] A failure of the biological detoxification machinery to scavenge reactive oxygen species will lead to oxidative stress and the accumulation of free radical molecules like H_2O_2 (hydrogen peroxide), NO (nitric oxide), and $O2^{\bullet-}$ (superoxide). Free radicals react with and alter the function of cellular macromolecules, including proteins, lipids, carbohydrates, and nucleic acids, resulting in accumulation of damaged macromolecules, reduced function and, eventually, degeneration of tissues and organs.[61,62] Photoreceptors and retinal ganglion cells are prone to oxidative stress, due to their high rates of metabolism, as they sit at the nexus of interaction between light, chemical and electrical energy. The retinal pigment epithelium (RPE) also has a high rate of metabolism, additional photo-oxidative stress resulting from its role in absorbing light via melanosomes, and functions to digest a large burden of daily shed rod outer segment tips. Accordingly, it is not surprising that the etiology of AMD involves oxidative stress and damage, particularly to these two layers of retinal cells.[41,63]

Endoplasmic Reticulum Stress

In addition to damaging bulk proteins and proteolytic and protein editing capacities, hyperglycaemia can induce endoplasmic reticulum (ER) stress through increased glycation of proteins in conditions like diabetes and obesity, with deleterious results.[64-68] recent studies have demonstrated a role for ER stress in age-related cataracts and aging retinas.[69-70] Further elucidating the role of glycemic stress in developing eye complications, including trigging the unfolded protein response, could inform about new therapies for AREDs.

Modification of the Gut Microbiome

The microorganisms that inhabit the human intestines evolve in a host-specific way, and are susceptible to modification during the lifespan of the host.[71] Besides protecting against pathogenic microorganisms, the gut microbiota help in maintaining intestinal barrier integrity, in nutrient absorption, and produce important nutrients including short-chain fatty acids (SCFA).[72-74] Gut microbiota change in response to diet.[39] Alterations in the gut microbiome have been associated with diabetic retinopathy and age-related macular degeneration.[39,75,76] Diets with low GI and rich in dietary fiber are known to help in maintaining normal circulatory glucose in diabetes, obesity, and aging,[77] Dietary fiber plays an important role in regulating metabolism in humans by increasing production of SCFA, altering the Prevotella/Bacteroides ratio, and modulating the gut microbiome.[74] Dietary fiber, probiotics, and gut microbiome fermentation were proven to help in glycemic control in diabetes,[74,78,79] which, in turn, may help to protect against AREDs.

Preventive Strategies for Eye Complications

The eye resembles an advanced camera which transforms the light signal into a neuronal signal and transmits it to the brain to produce an image. Like most advanced instruments, it requires careful maintenance to avoid stress and dysfunction. Stresses associated with accelerated aging include smoking, alcohol, poor sleep, pollution, and unhealthy food habits. These factors are also impacted by socio-demographic, environmental, and biopsychosocial status.[80] hyperopia, lens opacities, female sex, Caucasian ancestry, high body mass index, illiteracy, increased thyroid hormone, and antacid usage are known to be associated with AREDs.[81] Aging also affects the cornea (dry eye syndrome, Fuch's dystrophy), and the optic nerve (glaucoma).[82] As most AREDs are irreversible and treatment options are limited, it is crucial to prevent or delay their progression to maintain healthful visual function.

Low Glycemic Index Foods

Food is a major lifestyle factor that impacts our present and future health, and diet-related conditions such as hyperglycaemia play an important role in the development of eye complications. A wealth of information indicates that consuming lower GI diets is associated with prolonged retina and lens function, but this should be corroborated in intervention trials (Figure 100.2).[20]

Physical Activity

Regular physical activity (PA) helps maintain overall health, and numerous research studies have confirmed that worse visual outcomes, including AMD, glaucoma, and retinopathy, are associated with lower PA.[83] One meta-analysis showed reduced risk of age-related cataract development was correlated with increased physical activity.[12] The link between PA and eye complications is bidirectional, because reduced vision will also lead to reduced PA. Retinal diseases like glaucoma and AMD are known to cause a decrease in PA.[83] On the other hand, PA is known to have beneficial effects on retinal ganglion cells, as well as neurodegenerative diseases like Alzheimer's and vascular dementia in aging.[84-86] The development of AMD is linked to oxidative stress, increased glycemia, and glycation, which can potentially be reduced by PA, as it helps with the maintenance of healthy blood glucose concentration and normal lipid profile.[87-89]

Vitamins and Minerals

Vitamins are critical for biological functions, and deficiencies may lead to disease. Roles for numerous vitamins in eye health have been identified.[90,91] Perhaps the most robust indication of crucial roles for sufficient levels of vitamins and zinc come from the large age related eye disease studies (AREDS), which were randomized clinical trials of older adults with cataract, AMD, or both, who were followed for five to 10 years. AREDS1 randomly assigned participants to one of four groups: antioxidants (vitamins C, E, b-carotene), antioxidants + zinc/copper, zinc/copper alone, or placebo. They found that the group with both the vitamins and minerals had a 25% reduced risk of progressing to advanced AMD, but saw no effect on cataracts. Importantly, those who were former smokers and took formulas with b-carotene had a higher incidence of lung cancer. Therefore, AREDS2 randomized participants to add either omega-3 fatty acids or lutein+zeaxanthin, or both, or placebo to the AREDS1 formula without the b-carotene and with less zinc. They found that adding omega-3 fatty acids or lutein+zeathanthin to the AREDS1 formula, or reducing the zinc, had no additional effect on risk of advanced AMD. However, those who took AREDS1 vitamins without b-carotene but with luitein+zeaxanthin had more benefit. Therefore, the investigators now recommend a new formula with vitamins C and E, lutein+zeaxanthin, zinc and copper to help delay progression of intermediate AMD to advanced AMD.[92-93] This AREDS2 formulation is commercially available.

Analyses of epidemiologic data for associations between intake of many vitamins or minerals and risk for AMD have not been conclusive. These are summarized briefly below.

The best-studied vitamin in eye function is vitamin A because it is crucial for the visual cycle. Vitamin A derivatives provide the chemical elements that trigger photoreceptor responses. It is essential for the synthesis of rhodopsin in rod cells, which triggers phototransduction and assists with color and night vision. An early indicator of vitamin A deficiency is night blindness. Vitamin A is fat-soluble, and AMD and ARC were found to be reduced with increased vitamin A intake.[94-96] A meta-analysis suggested a role for vitamin A and carotenoids in preventing age-related cataracts (ARC).[97] Recent evidence shows that vitamin A supplementation plays a key role in preventing RPE atrophic changes in AMD.[98,99]

Ascorbic acid is a reducing agent and is effective in neutralizing free radicals. Importantly, vitamin C is included in the AREDs vitamins that have been shown to decrease risk for progression of intermediate AMD to advanced disease.[92,93] Research findings suggest that AREDs such as AMD and ARC are correlated with deficiency in vitamin C, and that vitamin C may be involved in age-related eye disease development and/or progression.[8,100,101] Some studies suggest that dietary supplementation of vitamin C, which is relatively nontoxic even at high doses, can delay or prevent the development of cataracts, but the data are less clear for AMD.[102]

The prevalence of deficiency of the fat-soluble vitamin D is high throughout the world.[103,104] Vitamin D deficiency is linked with many diseases in humans and, in recent years, researchers have focused on the association between eye complications and vitamin D deficiency.[105-111] Foods like milk and fish are the main dietary sources of vitamin D and have been correlated with decreased risk for the development of AMD.[110,111] However, the role of vitamin D in aging-associated eye diseases needs to be investigated further.

The fat-soluble vitamin E is best known for its potential anti-oxidant and anti-inflammatory properties.[112] Research has revealed a role for vitamin E in preventing or reducing the risk of developing AMD, and vitamin E deficiency may alter photoreceptor function in the retina.[113-116] Importantly, vitamin E is among the AREDs supplement that have been shown to delay progression of AMD, noted above.[92,93] Current research has not demonstrated a beneficial effect for Vitamin E supplementation alone on AMD, and further clinical research is needed.[117,118]

Thiamine plays a key role in energy metabolism and rich dietary sources of thiamine include whole grain cereals, beans, lentils, sunflower seeds, fish, and pork. Thiamine deficiency is not prevalent in developed countries, but people who have undergone gastric bypass surgery are more prone to develop thiamine deficiency.[119,120] Thiamine deficiency leads to serious complications, which include eye problems.[121] Thiamine deficiency, in association with the consumption of alcohol may cause eye muscle paralysis, double vision, dancing eyes or nystagmus, vision loss, and optic nerve head swelling.[122] Thiamine supplementation was reported to prevent damage in diabetes and can potentially inhibit cataract and retinopathy.[123-126] Nevertheless, the role of thiamine supplementation in preventing or delaying the development of AREDs needs to be investigated, and the important role of thiamine in

63. Datta S, Cano M, Ebrahimi K, et al. Prog Retin Eye Res 2017;60:201–18.
64. Kumar CU, Reddy SS, Reddy GB. Proc Nat Acad Sci, India Section B: Biological Sciences 2022;92:523–32.
65. Yuan M, Gong M, Zhang Z, et al. Oxid Med Cell Longev 2020: 6569728.
66. Kong DQ, Li L, Liu Y, et al. Int J Ophthalmol 2018;11:1704–10.
67. Mulhern ML, Madson CJ, Danford A, et al. Invest Ophthalmol Vis Sci 2006;47:3951–9.
68. Taylor A, Bejarano E. Neural Regen Res 2022;17:320–2.
69. Periyasamy P, Shinohara T. Prog Retin Eye Res 2017;60:1–19.
70. Alavi MV, Chiang W–C, Kroeger H, et al. Invest Ophthal Vis Sci 2014;55:4565.
71. Sekirov I, Russell SL, Antunes LC, et al. Physiol Rev 2010;90: 859–904.
72. Lurz E, Horne RG, Maattanen P, et al. Front Nutr 2020;7:83.
73. Degnan PH, Taga ME, Goodman AL. Cell Metab 2014;20: 769–78.
74. Myhrstad MC, Tunsjø H, Charnock C, et al. Nutrients 2020;12:859.
75. Zheng Y, Xiang S, Zhang H, et al. J Agri Food Chem 2021; 69:2204–12.
76. Satyanarayana A, Balakrishna N, Pitla S, et al. PloS One 2011; 6:e26747.
77. Feinman RD, Pogozelski WK, Astrup A, et al. Nutr 2015;31: 1–13.
78. Liu Y, Wang Y, Ni Y, et al. Cell Metab 2020;31:77–91.e5.
79. Bordalo Tonucci L, Dos Santos KM, De Luces Fortes Ferreira CL, et al. Crit Rev Food Sci Nutr 2017;57:2296–309.
80. Hahad O, Frenis K, Kuntic M, et al. Int J Mol Sci 2021:22.
81. Anand R, Bressler SB, Davis, MD et al. Age-Related Eye Disease Study Research Group, Ophthalmol 2000;107:2224–32.
82. Ando R, Szabo B, Laszlo GM, et al. Clin Nutr ESPEN 2016;1 1:e72.
83. Ong SR, Crowston JG, Loprinzi PD, et al. Eye (London, England) 2018;32:1296–303.
84. Hillman CH, Erickson KI, Kramer AF. Neurosci 2008;9:58–65.
85. Heijl A, Leske MC, Bengtsson B, et al. Arch Ophthal 2002; 120:1268–79.
86. Vasudevan SK, Gupta V, Crowston JG. Indian J Ophthal 2011;59:S102–13.
87. Simioni C, Zauli G, Martelli AM, et al. Oncotarget 2018;9: 17181–98.
88. Drenth H, Zuidema SU, Krijnen WP, et al. J Gerontol A Biol Sci Med Sci 2018;73:1545-51.
89. Mainous AG, 3rd, Tanner RJ Anton SD, et al. Am J Preven Med. 2017;53:42–7.
90. Chiu CJ, Taylor A. Exp Eye Res 2007;84:229–45.
91. Chiu CJ, Milton RC, Klein R, et al. Am J Clin Nutr 2007;86: 1210–8.
92. Age-Related Eye Disease Study Research Group. Arch Ophthalmol 2001;119:1417–36.
93. Age-Related Eye Disease Study 2 Research Group. JAMA 2013; 309:2005–15.
94. Vanden Langenberg GM, Mares–Perlman JA, Klein R, et al. Am J Epidemiol 1998;148:204–14.
95. Vingerling JR, Dielemans I, Bots ML, et al. Am J Epidem 1995;142:404–9.
96. Dong W, Gao J, Zhou Z, et al. PLoS One 2018;13:e0209890.
97. Jiang H, Yin Y, Wu CR, et al. Am J Clin Nutr 2019;109:43–54.
98. Zhang D, Robinson K, Washington I. Transl Vis Sci Technol 2021;10:8.
99. Sparrow JR. Proc Natl Acad Sci USA 2016;113:4564–9.
100. Fan X, Sell DR, Hao C, et al. Aging Cell 2020;19: e13176.
101. Lim JC, Caballero Arredondo M, Braakhuis AJ, et al. Nutrients 2020:12.
102. Dosedĕl M, Jirkovský E, Macáková K, et al. Nutrients 2021:13.
103. Amrein K, Scherkl M, Hoffmann M, et al. Eur J Clin Nutr 2020;74:1498–513.
104. Siddiqee MH, Bhattacharjee B, Siddiqi UR, et al. BMC Public Health 2021;21:1823.
105. Parekh N, Chappell RJ, Millen AE, et al. Arch Ophthalmol 2007;125:661–9.
106. Millen AE, Voland R, Sondel SA, et al. Arch Ophthalmol 2011;129:481–9.
107. Skowron K, Pawlicka I, Gil K. Folia Medica Cracoviensia 2018;58:103–18.
108. Reins RY, McDermott AM. Exp Eye Res 2015;134:101–10.
109. Chan HN, Zhang XJ, Ling XT, et al. Int J Mol Sci 2022:23.
110. Zhu W, Wu Y, Meng YF, et al. Nutrients 2016:8.
111. Gopinath B, Flood VM, Louie JCY, et al. Br J Nutr 2014;111: 1673–9.
112. Singh U, Devaraj S, Jialal I. Annu Rev Nutr 2005;25:151–74.
113. Robison WG, Jr., Kuwabara T, Bieri JG. Invest Ophthalmol Vis Sci 1979;18:683–90.
114. Michikawa T, Ishida S, Nishiwaki Y, et al. Asia Pac J Clin Nutr 2009;18:1–7.
115. Simonelli F, Zarrilli F, Mazzeo S, et al. Clinica Chimica Acta; Int J Clinical Chem 2002;320:111–5.
116. Belda JI, Romá J, Vilela C, et al. Mech Age Devel 1999;107: 159–64. Available at www.sciencedirect.com/journal/mechanisms-of-ageing-and-development
117. Tan JSL, Wang JJ, Flood V, et al. Ophthalmol 2008;115: 334–41.
118. Klein BE, Knudtson MD, Lee KE, et al. Ophthalmol 2008;115: 1203–8.
119. Nath A, Tran T, Shope TR, et al. Nutr Res 2017;37:29–36.
120. Albaugh VL, Williams DB, Aher CV, et al. Surg Obes Rel Dis 2021;17:653–8.
121. Whitfield KC, Bourassa MW, Adamolekun B, et al. Ann NY Acad Sci 2018;1430:3–43.
122. Gratton SM, Lam BL. Clin Ophthalmol 2014;8:1021–4.
123. Obrenovich ME, Monnier VM. Sci Ag Knowl Envir 2003:pe6.
124. Hammes HP, Du X, Edelstein D, et al. Nat Med 2003;9: 294–9.
125. Berrone E, Beltramo E, Solimine C, et al. J Biol Chem 2006;281: 9307–13.
126. Bakker SJL. Ophthalmol 2001;108:1167.
127. Tan AG, Mitchell P, Rochtchina E, et al. Invest Ophthalmol Vis Sci 2014;56:216–20.
128. Sen SK, Pukazhvanthen P, Abraham R. Indian J Clin Biochem 2008;23:255–7.
129. Sivaprasad M, Shalini T, Balakrishna N, et al. Ann NutrMetab 2016;68:94–102.
130. Zampatti S, Ricci F, Cusumano A, et al. Nutr Res 2014;34: 95–105.
131. King JC. Am J Clin Nutr 2011;94:679S–84S.
132. Mares–Perlman JA, Klein R, Klein BE, et al. Arch Ophthal (Chicago, Ill.: 1960). 1996;114: 991–7.
133. Kuzniarz M, Mitchell P, Flood VM, et al. Ophthal Epidemiol 2002;9:283–95.
134. Arch Ophthalmol 2001;119:1439–52.
135. Fujita T. J Bone Miner Metab 2000;18:234–6.
136. Duncan G, Jacob TJ. Ciba Found Symp 1984;106:132–52.
137. Nemet AY, Hanhart J, Kaiserman I, et al. Clin Ophthal (Auckland, N.Z.) 2013:7: 2079–84.
138. Roy D, Chiesa R, Spector A. Biochem Biophys Res Commun 1983;116:204–9.
139. Truscott RJ, Marcantonio JM, Tomlinson J, et al. Invest Ophthalmol Vis Sci 1990;31:2405–11.
140. Wride MA. Differentiation 1996;61:77–93.
141. Hightower KR, Hind D. Curr Eye Res 1982;2:239–46.

142. Gupta PD, Johar K, Vasavada A. Acta Pharmacol Sinica 2004;25:1250–6.
143. Tisdale AK, Agrón E, Sunshine SB, et al. JAMA Ophthal 2019;137:543–50.
144. Liu K, Lyu L, Chin D, et al. Proc Natl Acad Sci USA 2015;112:1071–6.
145. Van Leeuwen R, Boekhoorn S, Vingerling JR, et al. JAMA 2005;294:3101–7.
146. Slavin JL, Lloyd B. Adv Nutr 2012;3:506–16.
147. Liu RH. Am J Clin Nutr 2003;78:517S–20S.
148. Boeing H, Bechthold A, Bub A, et al. Eur J Nutr 2012;51: 637–63.
149. Mukherjee PK, Harwansh RK, Bahadur S, et al. J Ethnopharmacol 2017;197:10–24.
150. Rahman S, Jahan N, Makbul SA, et al. J Ethnopharmacol 2020;258:112880.
151. Renoux H. J Facul Homeopathy 2021;110:222–6.
152. Maxion–Bergemann S, Bornhöft G, Sonderegger E, et al. Forschende Komplementarmedizin 2006;13:30–41.
153. Atanasov AG, Waltenberger B, Pferschy–Wenzig EM, et al. Biotech Adv 2015;33:1582–614.
154. Zhao Y, Wu Y, Wang M. Handbook of Food Chemistry. Cheung PCK, Mehta BM, eds. Springer Berlin Heidelberg. 2015; 967—1008.
155. Arulselvan P, Fard MT, Tan WS, et al. Oxid Med Cell Longev 2016;5276130.
156. Kumar PA, Kumar MS, Reddy GB. Biochem J 2007;408:251–8.
157. Muthenna P, Akileshwari C, Reddy GB. Biochem J 2012;442: 221–30.
158. Muthenna P, Raghu G, Akileshwari C, et al. IUBMB Life 2013; 65:941–50.
159. Raghu G, Akileshwari C, Reddy VS, et al. J Food Sci Tech 2017; 54: 2411–21.
160. Kumar PA, Reddy PY, Srinivas PN, et al. J Nutr Biochem 2009;20: 553–62.
161. Suryanarayana P, Krishnaswamy K, Reddy GB. Mol Vis 2003;9:223–30.
162. Saraswat M, Suryanarayana P, Reddy PY, et al. Mol Vis 2010;16:1525–37.
163. Saraswat M, Muthenna P, Suryanarayana P, et al. Asia Pacific J Clin Nutr 2008;17:558–65.
164. Akileshwari C, Muthenna P, Nastasijević B, et al. Exp Diabetes Res 2012:147965.
165. Muthenna P, Suryanarayana P, Gunda SK, et al. FEBS Lett 2009;583:3637–42.
166. Puppala M, Ponder J, Suryanarayana P, et al. PloS One 2012;7:e31399.
167. Suryanarayana P, Kumar PA, Saraswat M, et al. Mol Vis 2004;10:148–54.
168. Saraswat M, Reddy PY, Muthenna P, et al. Br J Nutr 2009;101: 1714–21.

Index

Page numbers followed by *f* denote figures; those followed by *t* denote tables.

A

C

D

E

F

G

H

J

K

L

M

N

P

Q

R

S

T

U

V

W

X

Y

Z